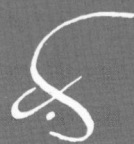
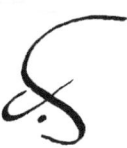

International Directory of Arts

34th Edition 2010

Volume 3

De Gruyter • Saur

Redaktion:
Walter de Gruyter GmbH & Co. KG
Redaktion International Directory of Arts
Mies-van-der-Rohe-Str. 1
80807 München
Tel.: +49 (0)89/76902-316
Fax: +49 (0)89/76902-333
E-mail: artaddress@degruyter.com

Advertising Representatives:

Worldwide:
Media-Agentur Schaefer
Lange Straße 14
D-04103 Leipzig
Tel.: +49 (0) 341 3010620
Fax: +49 (0) 341 3010621
E-Mail: marlen.kuenitz@media-schaefer.de
www.media-schaefer.de

Bibliographic information published by the Deutsche Nationalbibliothek
The Deutsche Nationalbibliothek lists this publication in the Deutsche Nationalbibliografie;
detailed bibliographic data are available in the internet at http://dnb.d-nb.de.

Printed on acid-free paper

Typesetting: M. Wündisch, P. Tanovski, K. Eichfeld, Leipzig

Printed and bound by Strauss GmbH, Mörlenbach

ISBN: 978-3-598-23120-9

Contents

page

Volume 1

Abbrevations VI
Preface VII
Service Providers XI
Museums and Public Galleries 1
Universities, Academies, Colleges 1059
Associations 1125
List of Advertisers 1169

Volume 2

Abbrevations IV
Preface V
Art and Antique Dealers, Numismatics 1
Art and Antiques Fairs 417
Galleries 437
Auctioneers 753
Restorers 797
Art Publishers 911
Art Periodicals 947
Antiquarians and Art Booksellers 969
List of Advertisers 1051

Volume 3

Index of Persons (Directors, curators, presidents and scientific staff of museums) 1
Index of Institutions and Companies 125

Inhalt

Seite

Band 1

Abkürzungen VIII
Vorwort IX
Dienstleisterverzeichnis XI
Museen und Öffentliche Galerien 1
Universitäten, Akademien, Schulen 1059
Vereinigungen 1125
Verzeichnis der Inserenten 1169

Band 2

Abkürzungen VI
Vorwort VII
Kunst- und Antiquitätenhandel, Numismatik 1
Kunst- und Antiquitätenmessen 417
Galerien 437
Auktionatoren 753
Restauratoren 797
Kunstverleger 911
Kunstzeitschriften 947
Antiquariate und Kunstbuchhandlungen 969
Verzeichnis der Inserenten 1051

Band 3

Register der Personen (Direktoren, Kuratoren, Präsidenten und
wissenschaftliche Mitarbeiter der Museen) 1
Register der Institutionen und Firmen 125

Index of Persons
Register der Personen

Directors, curators, presidents and scientific staff of museums
Direktoren, Kuratoren, Vorsitzende und wissenschaftliche Mitarbeiter der Museen

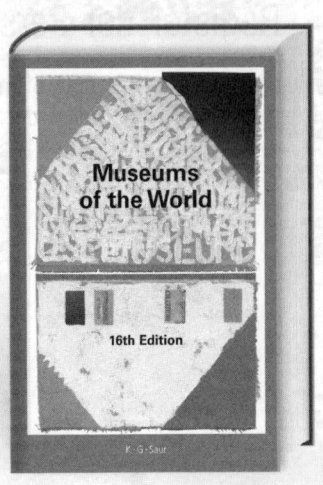

■ Museums of the World

16th Edition 2009
2 volumes. Cplt. xxii, 1,449 pages
Hardcover € 479.00 [D] / *US$ 671.00
Standing Order / Update price € 409.00 [D] / *US$ 573.00
ISBN 978-3-598-20696-2
eBookPLUS € 536.00 [D] / *US$ 671.00
Standing Order / Update price € 459.00 [D] / *US$ 573.00
ISBN 978-3-598-44100-4**

Museums of the World covers in its 16th edition approx. 55,000 museums in 202 countries, listed hierarchically by country and place, and within places, alphabetically by name. A separate chapter records about 500 museum organizations in 132 countries with addresses. A typical entry contains the following details: name of the museum, address, telephone number, fax, eMail address and URL, museum type, year of foundation, name of the director and museum staff, special collections and equipment, number of the entry.

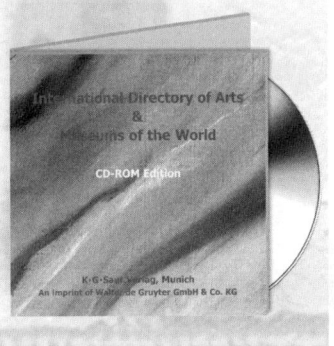

■ International Directory of Arts & Museums of the World

20th Edition 2009
CD-ROM. 2 editions per year
€ 449.00 [D] / *US$ 629.00
Standing Order / Update price € 409.00 [D] / *US$ 573.00
ISBN 978-3-598-40982-0
Also available for network use (LAN)

This CD-ROM provides fast and easy access to over 143,000 addresses in the international world of arts and museums. Users will find information on public and private institutions, such as museums, galleries, universities and academies, art publishers, art journals, auction houses, restorers, art and antiques dealers, antiquarian and art booksellers as well as art fairs and their organizers.

Museums are recorded along with their main areas of collection, directors, academic staff and special fields. Specific indexes facilitate access to the special fields of art and antiques dealers, private galleries, restorers, antiquarian and art booksellers as well as art publishers. From the URLs provided in the documents displayed, users have direct access to the homepages of institutions.

*for orders placed in North America
**currently only available for Libraries / Institutions
Prices are recommended retail prices
Prices are subject to change
Prices do not include postage and handling

DE GRUYTER
SAUR

Aalund, Dorothy 050986
Aamodt, David 049670
Aaraas, Olav 033850, 033851
Aarhus, Aasne 033541
Aarsen, John W. 048787
Aarsrud, Christian 041131
Aarstrand, Kjetil 033753
Aarts, H.J. 033128
Aartsen, Judith 032181
Aas, Scientist Christian K. 033847
Aasberg, Dagny 033616
Aaserud, Anne 034027
Aashagen, Kjersti Haugen 033892
Aasland, Odd 005662
Aasted Schanz, Elsebeth 010006
Aav, Marianne 010815
Abad, Dr. Andres 010436
Abad, Dr. Vicent 039177
Abadesa, R.M. 039167
Abadese, M. 039737
Abadie, Myriam 014386
Abadom, Monica 033429
Abadzic, Sabaheta 004283
Abadzieva, Sonia 030630
Abakumov, Vladimir 037310
Abal de Russo, Prof. Clara 000453
Abaño, Isidro C. 034549
Abaraonye, Dr. Felicia Ihuoma . 033418
Abate, Richard d' 052281, 052286
Abba, Sylvie 012364
Abbas, Muhammad 034136
Abbassi, Q. 024694
Abbate, Dr. Vincenzo 027176
Abbott, Christine 033222
Abbott, Diane 047133
Abbott, Philip 044784
Abbott, Phyliss 046830
Abbott, Sonia 000884
Abbott, Will 049502
Abcouwer, Dr. N.F. 032592
Abderhalden, Michele 041696
Abdo Mahgoub, Prof. Dr.
    Mohamed 010520
Abdolalipour, Mohammad 024744,
    024747, 024751, 024753, 024755,
    024758, 024762, 024767, 024770,
    024783
Abdoolrahaman, S. 030787, 030790,
    030791, 030794, 030795, 030796
Abdulla, K. 042640
Abdullah, Mohamed Yusuf 030699
Abdullin, Rif Mudarisovič 037977
Abdullina, Svetlana
    Aleksandrovna 038076
Abdunov, Georgi 005102
Abe, Hiroshi 028702
Abegg, Dr. Angelika 021663
Abejide, Ola 033443
Abel, Marianne 017781
Abel, Dr. Susanne 020481
Abel, Tiina 010668
Abelanet, Jean 015921
Abele, S. 022067
Abele, Susan 051553
Abeler, Jürgen 022826
Abell, Carl T. 050762
Abell, Sherry S. 050762
Abella, Carmencita T. 034547
Abella, Jordi 039391
Abels, Peggy 049909
Abenov, Kalkabaj Altaevič 029825
Abensour, Dominique 015017
Åberg, Anders 040872
Aberka, Astrid 022666
Abgottspon, Elisabeth 041630
Abgrall 011886
Abildgaard, Hanne 010166
Abita, Salvatore 026086
Abitz, Gerald 049948
Ablinger, Andreas 002047
Abner, Steve 047781
Abraham-Roelants, Josiane 003435
Abraham-Silver, Linda 047815
Abraham, Debbie 000867

Abraham, Julian 043527
Abrahams-Willis, Gabeba 038685
Abrahams, Katy 038718
Abrahams, Ruth K. 050700
Abrahamson, Dana 049076
Abram, Ruth J. 051418
Abramek, Bogusław 035392
Abramović-Miletić, Ljiljana 038168
Abrams, Linda C. 050389
Abramsson, Görel 040739
Abrão, Flávio Jorge 004388
Abravanel, Claude 025096
Abreu Nunes, Dr. Maria Luisa . 035781
Abreu, Jorge M. de 036061
Abreu, Juana Ines 031198
Abric, Christian 012523
Abril, Joana Mari 040396
Abrojan, Élvira Vartanovna 037001
Abruzzi, Sarah K. 051876
Absi, Nuha 029776
Abston, Fleet 050785
Abu Hadba, Abdul Aziz H. 034155
Abubakar, Ali Mohamed 029865
Abueg, Dr. Nelle F. 034573
Abylkasymov, Mengiš Maulenovič 029839
Acalija, Sanja 008926
Acampora, Nick 052252
Accardi, Dr. Giuseppina 025852
Acevedo Basurto, Sara 034363
Aceves Romero, David 031167
Achaah Ndumbi, Patrick 005291
Achatz, Erika 001949
Achenbach, Dr. Nora von 019011
Achenbach, Dr. Sigrid-Ute 017209
Acher, Jean-Pierre 015170
Achi, Y. 025113
Achleitner, Gunda 003037
Achuri, Julio César 008700, 008701
Acidini, Dr. Cristina 026278
Ackerman, Andrew S. 051360
Ackermann, Cornelia 041643
Ackermann, Dr. Marion 022045
Ackermann, Paul R. 054322
Ackerson, Rex D. 053767
Ackley, Clifford S. 047123
Ackroyd, Paul 045057
Ackroyd, Sally 044583
Acorn, Eleanor 006710
Acosta Guerrero, Elena 039902
Acosta Mohalem, José de Jesús 008743
Acquavina, Dr. Rosario 025660
Acquisti, Jon 053013
Actman, Phyllis 052051
Acton, David 054548
Adachi, Hogara 028803
Adachi, Takanori 029706
Ådahl, Dr. Karin 041019
Adair, Mary J. 050185
Adam, Patricia 003511
Adam, Thomas 017635
Adamczyk-Pogorzelska, Bożena 034986
Adami, Valerio 026755
Adamo, Giuseppe 027391
Adams Muranaka, Dr. Therese . 052981
Adams, Anne 049585
Adams, Dr. Beverly 052120
Adams, Celeste M. 049191
Adams, Cheryl 048168
Adams, Cindy 050069
Adams, Connie 049726
Adams, David 044390
Adams, Diana 001307
Adams, E. Charles 053823
Adams, Eloise 051147
Adams, F. 005586
Adams, G. Rollie 052622
Adams, Jay 046668
Adams, Jeremy 051369
Adams, Kathy 052233
Adams, Dr. Kenneth 048804
Adams, Kenneth 006612
Adams, Larry 047089
Adams, Lorraine Margaret 054852
Adams, Mazie M. 050103

Adams, Mel 052231
Adams, Nancy 047121
Adams, Patricia 052780
Adams, Patsy 051285
Adams, Roxana 050619
Adams, Stephen 043408
Adams, Virgil 050575
Adams, Will 044876
Adamsom, Noel 001420
Adamson, Alberta 054366
Adamson, J. 044354
Adamson, John 053569
Adamson, Tony 043400
Adamyan, M.S. 000747
Adán García, Angel Manuel 040318
Adar, Yossi 025013
Addenbrooke, Rosie 044019
Adderley, D.E. 007319
Addison, David 044283
Addison, K.A. 022924
Addison, Laura 053113
Addoms, Andrew 049488
Adelantar, Jonas O. 034597
Adelsberger, Paul 018233
Aderman, Ella 053200
Adesina, Bode 033425
Adey, Dr. Elizabeth 045168
Adey, Elizabeth 045166
Adiletta, Dawn C. 049385
Adisumarto, Dr. Soenartono ... 024578
Adkin, Tammy 006213
Adlbauer, Dr. Karl 001984
Adleman, Debra 051042
Adler, Frances Beatty 051379
Adler, Peter 001330
Adler, Tracy L. 051405
Adlı, Nadir 042885
Adolfsson, Stig 040665
Adolphs, Dr. Volker 017495
Adorno, Prof. Francesco 026291
Adriaens Pannier, Anne 003460
Adriaensens, Annemie 003319
Adriaenssens, Werner 003492
Adrian, Dolab Jorge 000178
Adriani, Prof. Dr. Götz 022220
Adriansen, Inge 010323
Adriázola, Teresa 004222
Adrimi-Sismani, Vassiliki 023314
Adsett, Virginia 043925
Adunka, Roland 001707
Adžemlerski, Svetoslav 005140
Aebersold, Marianne 042192
Aebi, Richard E. 041981
Aerts, Dirk 003281
Aeschlimann, Trudi 041387
Aescht, Dr. Erna 002316
Afonso, Elsa 035488
Afuhs, Eva 042205, 042215, 042217
Afzal, Dr. Muhammad 034119
Agar, Will 050933
Agarwal, R.C. 024318
Ageeva, Marina Vladimirovna . 037566
Ageeva, Zinaida Michajlovna .. 037381
Agejkina, Marija Vladimirovna . 029830
Agenjo, Xavier 040112
Agliati Ruggia, Dr. Mariangela . 041857
Agnelli Caracciolo, Marella ... 028321
Agnelli, Marella 028145
Agner, Raul Dumas 034600
Agnesi, Prof. V. 027188
Agostini, Giuseppe 027244
Agostini, Dr. Grazia 026220
Agostini, Dr. Silvano 027174
Ágota, Kasper 024060
Ågotnes, Dr. Anne 033530, 033532
Agrech, F. 015706
Agrenius, Helen 040719
Agresti, Giannicola 025326
Agten, Beate 020825
Agudo Cadarso, Christina 040334
Ague, Jay 051265
Aguero Moras, Alfonso 034237
Aguiar Branco, Dr. Fernando .. 035947
Aguilar Piedra, Raúl 008847

Aguilar, Dr. Hugo 008856
Aguilar, Dr. Pedro 036046
Aguilera, Kristin 051434
Aguillaume, Cécile 041687
Aguiló Ribas, Catalina 039881
Aguiló Victory, Magdalena 039886
Aguiló, Anna 038938
Aguinaldo, Rosalinda 034489
Aguirre, Gustavo 008642
Aguirre, Joel O. 030864
Aguirre, Pierre 012574
Agúndez García, José Antonio . 039712
Agusti, Jorge 040028
Agustín, Sandy 050939
Ahern, Maureen 049913
Ahlberg, Leif 040764
Ahlbrecht, Helmut 021826
Ahlefeldt Laurvig Bille, Count
    Michael 010213
Ahlers, Dr. Jens 019607
Ahlers, Silvia 004564
Ahlfort, Barbro 040652
Ahlqvist, Ingemar 040924
Ahlsten, Eva 041120
Ahlstrand, J.T. 040814
Ahmad, Mushtaq 034129
Ahmad, Dr. Razi 024432
Ahmady, B. 024781
Ahmaogak, George N. 046467
Ahmas, Kristina 011019
Ahmedzade, T.Z. 003156
Ahola, Tuija 011502
Ahonon, Leonard 004181
Ahrens, Carsten 017603
Ahrens, Claus 020991
Ahrens, Regi 049881
Ahrndt, Prof. Dr. Wiebke 017601
Ahronson, Meir 025175
Åhsberger, Douglas 040618
Ahtela, Eero 011122
Ahtola-Moorhouse, Leena 010813
Ahumada Valencia, Jacob 007557
Aichner, Dr. Ernst 019407
Aida, Susumu 029211
Aidara, Abdoul Hadir 038121
Aide, Inese 030251
Aigneaux, Jean d' 015626
Aigner, Carl 002697
Aigner, Ottilie 002260
Aigner, Uli 020545
Aigremont, Robert d' 013604
Aiken, Dr. Edward A. 053634
Aikens, C. Melvin 048622
Aikey, Michael 053158
Aillagon, Jean-Jacques 028321
Ailonen, Riitta . 010814, 010861, 010870,
    010879, 011008, 011343
Aimé About, Assamoa 008871
Ainhirn, Karola 001734
Airey, Laura 046375
Airlie, Louise 045532
Aistars, John 047470
Aitchison, Julie 001066
Aitken, Catherine 041243
Aitken, Terry 054557
Aizpurua, Inaki 038897
Ajmetov, Ramis Kijamovič 037117
Ajzenberg, Prof. Dr. Elza 004878
Akam, Prof. Michael 043850
Akamati-Lilimbaki, Dr. Maria .. 023218
Akanbiemu, Martin Ogunayo .. 033430
Akasaka, Norio 028465
Akasaki, Yoko 029609
Akasheh, Anahid 051433
Akazijeva, Tetjana Abašťajevna . 043138
Akbarnia , Ladan 047253
Akbulut, Nazım 043020
Akers, Charlene 053374
Akers, John 053707
Akers, Lloyd 052228
Akersten, Dr. William A. 052222
Åkesson, Bernt-A. 040946
Akgül, Nebi 042920
Akgün, Ünal 042989

Akimbek, Azat	029794	Albinson, Cassandra	051266	Ali, Mohamed	042629	Almosara, Emelita V.	034540
Akın, Gülşen	042898	Albiol Sampietro, Vicente F.	039789	Ali, Samirah Mohd	030697	Almquist, Katarina	040931
Akin, Joseph	052762	Albisetti, Guiseppe	041198	Ali, Sanaa	010511	Almquist, Kathleen	053890
Akinade, Dr. Olalekan	033432	Alborino, Dr. V.	018095	Alibaeva, Sara Kuandykovna	029817	Alms, Barbara	017852
Akinpelu, David Ano	033430	Albrecht-Weinberger, Dr. Karl	003011,	Alibhai, Dr. Abdulssatar Mussá	031803	Alonso Martín, Juan José	039689
Akins, Ken	048151		003049	Alicante, Apolinario	034551	Alonso, Alejandro G.	009204
Akins, Rebecca	050810	Albrecht, Donald	051368	Aliev, Kazi Magomet		Alonso, Alicia	039614
Akira, Nagoya	029511	Albrecht, Hansjörg	020739	Ibragimovich	036947	Alonso, Jésus	040428
Akira, Tatehata	029259	Albrecht, Marvin	054360	Alig, Joyce L.	047494	Alòs Trepat, Carme	039043
Akishige, Kunikazu	029412	Albrecht, Silke	021179	Alimonti, Keith	006071	Alowan, Zoe	051976
Akiyama, Yoshi	050631	Albrechtsen, Bo	023342	Alin, Margareta	040809	Alper, Mara	051155
Akkaya, Mustafa	042923, 042928	Albright, Dr. Barry	048740	Alisch, Joachim	017681	Alperin-Hochman, Marlene	047255
Akker, C.	032867	Albright, Liz	051782	Aliyasak, Munasor	030670	Alpers, Rojann R.	053698
Akmamedova, Maja	043099	Albright, Shirley	053799	Alkema, J.	032027	Alphons, Alois	002087
Akomolafe, Tunde	033439	Albright, Terrence	051599	Allaire, Kay	051368	Alpözen, T. O.	042849
Akpan, Aniefiok Udo	033438	Albufera, E. d'	014262	Allan, Laurie	046611	Alram, Dr. Michael	003026
Akramov, N.	054648	Alcolea Blanch, Santiago	039066	Allan, Randy	046970	Alsam, Ali	045057
Akre, Lorne	007001	Aldeguer Gordiola, Daniel	038941	Allan, Shona	045466	Alsobrook, Dr. David E.	047872
Aksajskij, Ivan Grigorjevič	037936	Aldema, Yehuda Levy	025108	Allanfranchini, Patrice	041356	Alston, Joseph	049173
Aksënova, Alisa Ivanovna	038028	Alder, Barbara	041678	Allansson, Jón	024061	Alsvik, Elling	034038
Aksoy, Sule	042963	Alder, M.	044496	Allard, Jeanine	052774	Alswang, Hope	053300
Aksu, Selahattin Eyüp	042835	Alderdice, Dr. David	001547	Allard, Sarah	044082	Altamont, Earl of	025001
Al-Arifi, Dr. Nassir S.	038109	Alderfer, Joel D.	049359	Allart, Dr. B.	032957	Alteen, Glenn	007285
Al-Bubu, Sana'	029784	Alderton, Michéle	006761	Allas, Anu	014040	Alteneder, Prof. Susanne	003041
Al-Dahes, Mohammed	030067	Aldhizer, H.T.	048858	Allason-Jones, Lindsay	045351, 045352	Altenkirch, Gunter	018668
Al-Diban, Martina	020642	Aldin, Shiraz	000005	Alleaume	012306	Alterio, Italo	008825
Al-Esawi, Talaat	010574	Aldini, Tobia	026331	Allegrini, Prof. Piero	026744	Altermatt, Franz	041805
Al-Farsi, Said Ali Said	034100	Aldrich, William M.	051655	Allemand, Eolo	025449	Althaus, Marceline	041800
Al-Hameed, Hazim A.	024809	Aldridge, Carolyn	045918	Allemand, Evelyne-Dorothée	016006	Altherr, Hans Werner	017639
Al-Maamari, Moussa	030271	Aldridge, G.E.	043785	Allemand, Jacqueline	015661	Althöfer, Marie	018044
Al-Makhawi, Shatha R.	043345	Aldridge, Martha	047489	Allemann, Remo	042015	Althoff, Bruno	022676
Al-Qaisi, Rabie	024804	Aldunate del Solar, Carlos	007536	Allen Fleming, Kathy	052749	Althoff, Ralf	018057
Al-Robaae, Dr. Khalaf	024807	Alebo, Lena	040947	Allen, Anne	046532	Altmann, Ferdinand	002391
Al-Saud, Abdullah S.	038111	Alegria, Dr. Ricardo E.	036161	Allen, Babara	053538	Altmann, Katja	021255
Al-Shibli, Alae	024802	Aleksandrova, Marija	005270	Allen, Barbara	053824	Altorfer, Dr. Andreas	041322
Al-Sid, Siddig M. Gasm	040497	Aleksandrova, Nataša	005144	Allen, Brian	054419	Altringer, Lothar	017499
Al-Wohaibi, Dr. Fahed	030068	Aleksandrovič, Natalja Jurjevna	037161	Allen, Brian T.	046479	Altripp, Prof. Dr. Dr. Michael	018809
Alam, Shamim	034120	Aleksandrova, Marija . 005269, 005271	Allen, Bruce	053320	Alvarado-Dupuy, Dr. André	003026
Alam, Dr. Shamsul	003223, 003229,	Alekseev, Nikolaj Makarovič	037699	Allen, Catherine	047870	Alvarado Gonzalo, Manuel de	039183
	003231, 003233	Alekseev, Vladimir Petrovič	036917	Allen, Christopher	044978	Álvares Sánchez, Roberto	039023
Alam, Syamsir	024626	Alekseeva, Aida Nikolaevna	037056	Allen, Dale	048220	Alvárez Arévalo, Miguel	023374
Alaminos López, Eduardo	039667	Alekseeva, Galina Ivanovna	036891	Allen, Dick	053932	Alvárez Masso, Montserrat	039084
Alarcó, Paloma	039686	Aleksiuk, Galina A.	038043	Allen, Donald T.	050580	Alvárez Polo, Giesela	009256
Alarcón Lotero, José Félix	008756	Aleksoska-Baceva, Zaharinka	030630	Allen, Kay	050421	Alvarez Rojas, Antonio	039190
Alarcón Osorio, Gonzalo	031165	Alemany, Uiso	040404	Allen, Kim	006335	Alvarez, Aliciz Lidia	000357
Alarcón, Enrique	039075	Alemany, Véronique	013966	Allen, Louise	044432	Alvarez, Helena	041018
Alario, Prof. Leonardo	025805	Alencar, Ana Maria M. de	004743	Allen, Lucy	049756	Alvarez, José Carlos	035812
Alasalmi, Jikka	011521	Alencar, Vera de . 004679, 004687	Allen, Dr. Marti L.	052406	Alvarez, Josefina	000560
Alaurin, Alice A.	034474	Alessi, Vincent	000912	Allen, Nancy L.	051620	Alvarez, Juan S.	048441
Alausy, Abdalla	029871	Alessio, Alberto	028142	Allen, Richard	048124, 049597	Alvarez, Sergio Antonio	000146
Alba Alvarez, Dionisio	040439	Alex, Erdmute	017868	Allen, Susan	053146	Alves Costa, Catarina	035808
Alba, Dominique	014796	Alex, Dr. Reinhard	017868	Allen, Timothy S.	053812	Alves de Sousa, Dr. Pio G.	035576
Albacete, M.J.	047411	Alexander, Beatrix	019693	Allen, Trudy J.	049135	Alves Gomes, José António	004551
Alban, Marc . 013612, 014139	Alexander, Darvan	047071	Allen, Virginia S.	047092	Alves Pinto, José F.	035728
Albani, Julia	041287	Alexander, G.	046146	Allen, William M.	052598	Alves, C. Douglass	053377
Albani, Lino	026041	Alexander, Helene	044966	Allen, Zack	051070	Alves, Hélio Osvaldo	035728
Albany, James	046412	Alexander, Irene O. . 047899, 047900	Allerman, Cheryl	049772	Alves, Manuel	035893
Albarian, Kimberly	050038	Alexander, Jake F.	052906	Allevi, Dr. Gabriele	025495	Alves, Dr. Sónia	035925
Albendea Solis, Juan Manuel	040164	Alexander, Jane . 001208, 045793	Allgermissen, Hilge	019011	Alvina, Corazon S.	034535
Albendea Solis, Juan Maurel	040233	Alexander, Jean	053763	Allgood, Louise	001577	Alvina, Corazón S.	034541
Alber, Merike	010671	Alexander, Kimberly	052880	Allie, Stephen J.	048831	Alybekova, Aina	030075
Alberdi, Juan B.	000670	Alexander, Kristi	047852	Allikvee, Anu	010668	Alymer, John	043700
Alberg, Dr. Werner	018044	Alexander, Nancy	052505	Allington, Peter	045684	Alzamarova, Raušan Ergazievna	029822
Alberigi, Merry	053062	Alexander, Prof. Richard D.	046493	Allione, Patricia	013371	Amack, Rex	051206
Alberrán, Fernando	031699	Alexander, Robert	001011	Allison Rogers, Traude	054857	Amadei, Daniel	006594
Albers, Birgit	001793	Alexander, Terry Pink	046936	Allison, Brian	001273	Amakasu, Ken	029171
Albers, Dr. M. . 032429, 032430, 032887	Alexandre, Claude	016176	Allison, C.	046133	Amalla, Joesilo C.	034444
Albers, Mary	006749	Alexandru, I.	036650	Allison, David	033147	Amamra	000041
Albers, Thomas	018832	Alexis Grace, Trudie	047865	Allison, Glenn	007362	Aman, Karen	054560
Albert, Karen T.	049437	Alexis, Gerald	023392	Allison, N.	033246	Amanberdìyev, Annamyrat	043098
Albert, Peggy	015393	Alfageme Sánchez, Vitaliano	040437	Allman, William G.	054177	Amann, Bernd	002098
Albert, Shea	038716	Alfaro Castillo, Andrés José	008610	Alloisi, Dr. Sivigliano	027567	Amano, Yukichi	029022
Albertelli, Prof. Giancarlo	026414	Alfaro, Ana	040308	Allsop, Julie	045722	Amaral, Maria Conceição Alves	036032
Alberti, Lluís	040079	Alfen, Peter van	051328	Allwood, Rosamond	044813	Amarelli, Pina	027678
Alberti, Dr. Walter	026805	Alfoldy, Andy	005789	Alm, Göran	040992	Amaro das Neves, António	035728
Albertin, Anne-Marie	012999	Alfoldy, Elaine	005789	Almarri, Ali	043343	Amato, Ann	052304
Albertini, L.	041233	Alfred, Darrin	048258	Almeda, Ermelo	034557	Amato, Prof. Domenico	052588
Albertis, Serge de	014496	Alfter, Dr. Dieter	016843	Almeev, Robert V.	054629	Amato, Pietro . 054682, 054688, 054694
Alberto, Dr. Giordano	026361	Algeri, Dr. Giuliana	026394	Almeida Campos, Diogenes de	004724	Ambroise, Guillaume	014808
Alberts, Rod	048191	Alguacil Ruiz, Pedro	038900	Almeida-Matos, Lúcia	035934	Ambrose, David	054065
Albertsson, Kristinn J.	024065	Alhainen, Kari M.	010972	Almeida, Joel	035779	Ambrose, Pamela E.	052822
Albertus, Dana	020000	Alho, Anna-Liisa	010866	Almeida, José Maria	007504	Ambrosetti, Dr. Giancarlo	027500
Albertus, Jürgen	018611	Ali, H.-A.	024803	Almeras	013771	Ambrosini, Lynne	047749
Alberty, Beth	047251	Ali, Mazhar	034154	Almirante, Julián	040308	Ambrosino, Guy	053105

Index of Persons

Ambrosio, Prof. Augusto C.
– Aranburu Lasa, Juan Jose

Ambrosio, Prof. Augusto C. 027406
Ambrosio, Piero 028270
Amdreoli, Prof. Annararie 026372
Amelin, Olov 041027
Amell, Alexander 048402
Amemiya, Tadashi 029493
Amend, Brad 048498
Amend, Debbie 048498
Amendolagine, Dr. Beatrice 026747
Amer, Prof. Mordechai 025171
Ames, Gerhard 021557
Ametrano, Dr. Silvia 000413
Amias, Elizabeth 043895
Amic, S. 014298
Amick, Alison 051734
Amick, Henry 048300
Amidon, Catherine 052216
Amidon, Eleanor 049347
Aminova, Svetlana Gašimovna ... 037210
Amir, H. 030792
Amir, Mohammad 024546
Amiranišvili, S.J. 016420
Amiri, M. 024735
Amlow, Lothar 018085
Ammann, Christopher 041610
Ammann, Prof. Dr. Gert 002182
Ammann, Hugo 041247, 041249
Ammann, Sandra 041243
Amnehäll, Jan 040663
Amo, Dr. Toshino 029483
Amon, Prof. Dr. Karl 001734
Amonson, John 054040
Amor, Kebbour 000073
Amoretti, Guido 028138
Amoriñ, Prof. Dr. José L. 000232
Amoroso, Dr. Domenico 025700
Amoroso, Richard L. 051801
Amouyel, Prof. 013806
Amprimoz, M.F.X. 015976
Ampuero, Gonzalo 007525
Amsallem, Christiane 015850
Amselgruber, Prof. Dr. 022070
Amsellem, Patrick 047253
Amsler, Cory 048367
Amstad, Hans 042224
Amstutz, Damian 041715
An, Tran Huy 054803
An, Yingnan 008469
An, Zemin 008505
Anand, Dr. Y.P. 024229
Anati, Prof. Emmanuel 025751
Anaversa, Inginera 035759
Anawalt, Patricia 050422
Ancheotti, Dr. Angelo 026208
Ančiporov, A.I. 037665
Ancuţa, Maria 036448
Anda, Raquel de 053000
Anda, Tone 033486
Andami, Parisa 024771
Anday, Anton 053525
Ander, Dr. Angelika 003053
Anderegg, Fred 041733
Andergassen, Dr. Leo 025639
Anderjack, George M. 053091
Anderl, Dick 054550
Anderle, Charlotte 002244
Anderle, Peter 002060
Anderlik, Heidemarie 017299
Anders, Ines 018735, 018736
Anders, Win 007460
Andersen, Britta 010017
Andersen, H.C. 010092
Andersen, Håkon 034034
Andersen, Jeffrey W. 051749
Andersen, Jens 010097
Andersen, Jesper Munk 010145
Andersen, John 010074
Andersen, Kjell 033612
Andersen, Lise 010095, 010096
Andersen, Michael 010179
Andersen, Nikki 054544
Andersén, Peter 011179
Andersen, Randi 033754, 033755
Andersen, Thyge 010245

Andersen, Verginia 004610
Anderson-Densmore, Mari C. ... 050025
Anderson, Annette 047833
Anderson, Betsy 048279, 050365
Anderson, Carol 046188, 046496
Anderson, Christine 053353
Anderson, Claudia 053706
Anderson, Craig 047682
Anderson, Dale W. 049392
Anderson, David 045123, 051741
Anderson, Donald B. 052689
Anderson, Douglas 047217
Anderson, Dr. Duane 053116
Anderson, Duane 053114
Anderson, Elizabeth 005577
Anderson, Ellen 007170
Anderson, Eve 053735
Anderson, Fiona 045322
Anderson, Garry 005663
Anderson, Harold 001120
Anderson, Hilary 050256
Anderson, J. 032990
Anderson, Jaime A. 046601
Anderson, Joe 050195
Anderson, Dr. John P. 054493
Anderson, Judy 050665, 050667
Anderson, Kathi 050288
Anderson, Kris 050100
Anderson, Linda 052469
Anderson, Lynn 053651
Anderson, Michael ... 049706, 049707
Anderson, Neil 043443
Anderson, Owen 001575
Anderson, Pam 054403
Anderson, Peter 053255
Anderson, Philip J. 047700
Anderson, Raymond L. 053755
Anderson, Richard R. 053372
Anderson, Rolf 054201
Anderson, Steve 052516
Anderson, Steven R. 046545
Anderson, Susan K. 046790
Anderson, Susan M. 047052
Anderson, W.M. 053306
Andersson Flygare, Irene 041118
Andersson Møller, Vibeke 010208
Andersson, Beritne 010779
Andersson, Karin 040903
Andersson, Lars 040731, 040763
Andersson, Solveig 040669
Andes, Brad 054485
Andjal, Gabriela Vasylivna 043314
Andjelkovič Dmitrijević,
  Branislava 038170
Ando, György 023441
Andorfer, Gregory Paul 046760
Andrade, Rosemary 040508
Andral, Jean-Louis ... 011686, 011689
Andrási, Gábor 023532, 023533
André-Salvini, Béatrice 014749
André-Schellner, Prof. Brigitte .. 003041
André, Frédéric 003890
André, Dr. Georges 003529
André, Mervyn 001065
Andre, Simone 043751
Andreadaki-Vlasaki, Maria 023038,
  023242
Andreas, Düspohl 016437
Andreasen, Vagn 010226
Andreassen, Dag 033854
Andreassi, Dr. Giuseppe 025454, 028055
Andreasson, Sven-Glov 040560
Andreescu, Laurentiu 036571
Andreeva, Antonina Borisovna .. 038017
Andrefsky, William 052415
Andrejev, Dr. Aleksandr 037754
Andrell, Pauline 033282
Andren, Martin 040809
Andres, Louis M. 050520
Andresen, Thomas Bjorkan ... 033953
Andreu Chazarra, Antonio 039800
Andreu-Lanoe, Guillemette ... 014749
Andreu, Dr. Jaime Sancho 040307
Andrew, Marcia 054003

Andrews, Bernice 005548
Andrews , Gail 047005
Andrews, Janine 005767
Andrews, Dr. Peter 043654
Andrews, Rebecca 000762
Andrews, Richard 053235
Andrews, Thomas F. 050450
Andrews, Toni 053630
Andriamanantena, Ranaivoarivelo 030647
Andrich, Dr. Uwe ... 018820, 018822
Andrien, Joseph 003891
Andries, Pool 003287
Andrieu, Esther 000493
Andrishak, Steve 006715
Andrist, Jean-François 041405
Andronic, Tasia 036485
Androutsou, Elina 023023
Andrušaite, Dzintra 030197
Andújar Persinal, Carlos 010395
Anelli, Dr. Giuseppe 027945
Anene, Francis 033414
Anfang, Werner 002047
Angama, Henri 008866
Angarita Jiménez, Edgar 008736
Angel Aranaz, Miguel 039433
Angel de Marco, Dr. Miguel ... 000220
Angelett, Prof. Luciana Rita ... 027616
Angelis, Gudrun 017489
Angell, Callie 051491
Angelo, Therese 033166
Angeloni, Araldo 027770, 027771
Angelov, Anastas 005011
Angelov, Emil 005239
Angelucci, Sara 007174
Anger, Maksymilian 034946
Anger, Virginia 006680
Angerbauer, Dr. W. 020931
Angerer, Dr. Birgit 020633
Angerer, Ferdl 002741
Angerer, Dr. Martin .. 021322, 021324,
  021329, 021330
Angert, Rina 025146
Anghel, Elena 036527
Anglin Burgard, Timothy 053009, 053033
Anguiano Lozano, Juan Manuel . 031230
Angus, Barbara J. 048509
Angus, E. 045979
Anh, Tran Hong 054813
Anheuser, Kilian 041530
Anibas, Martin 003143
Anić, Hanja 008993
Anilir, Burhanettin 042907
Aniol, Roland 021663
Anisi, Ali Reza 024775
Aniz Montes, Mercè 039384
Anken, Thomas 041478
Ankert, Kerstin 040620
Ankudinova, Elena Andreevna . 037061
Anlagan, Çetin 042955
Anna, Dr. Susanne .. 018043, 018044
Annear, Judy 001535
Annemans, Jan 001094
Annibaldi, Dr. Cesare . 027537, 028321
Annicchiarico, Silvana 026791
Annison, F. 033236
Anosovskaja, Alla Vasiljevna .. 037063
Anraku, Yoshio 029349
Ansel, Dr. Valerie 015880
Ansell, Andy 043645
Ansell, Erynne 049599
Anselm, Gudrun 040888
Anselm, Paul S. 051700
Ansenk, Emily 032897
Ansett, Tony 001094
Ansorge, Gabriele 020171
Anspach, Isabelle 003461
Ansseau, Myriam 003789
Anthenaise, Claude d' . 012425, 014700
Anthoine, Robert 051350
Anthony, Dr. David 051779
Anthony, Ginger 053881
Anthony, James 044573
Anthony, Vincent 046612
Antier, Caroline 015369

Antihi, Adriana 036709, 036710
Antillano Armas, Sergio 054726
Antipas, Archimandrit 023214
Antl, Prof. Gerhard 002794
Antognini, Marco 041696
Antoime, Janeen 052991
Antoine, J.-L. 003896
Antoli Candela, Fernando 040404
Antonakos, Mary 047518
Antonelli, Dr. Angelo 027480
Antonen, Rusty 050083
Antoniadou, Sofia 023027
Antonić, Dr. Dragomir 038129
Antonietti, Thomas ... 041615, 041993
Antonino, Dr. Biancastella ... 025534,
  025567
Antoniuk, Aleksander 034640
Antonoglou, Manolis 023249
Antonov, Anton 005021
Antonov, Teodosi 005094
Antonova, Irina A. 037395
Antonova, Ljubov Vladimirovna . 038012
Antonsen, Lasse B. 051227
Antonuccio, Domenico 027549
Antti-Poika, Jaakko 010869
Anttila, Ritva 005359, 010925
Antunes, José R. ... 035584, 035585
Antweiler, Dr. Wolfgang 019246
Anyanwu, Dr. Starling E.N. ... 033417
Anzai, Ikuro 028960
Anzola, Sonia 054726
Anzuoni, Robert 048812
Ao, Bingquan 008093
Aoki, Yoshiharu 029039
Aoyagi, Masanori 029536
Apalkov, Jurij Aleksandrovič .. 036909
Apap Bologna, Daniela 030757
Aparin, Dr. Boris F. 037723
Apel, Gefion 017877
Apel, Dr. Michael 020555
Apel, Rolf 022149
Apelbaum, Laura C. 054130
Apetrei, Maria 036613
Apfel, Prof. Kurt 002958
Aplan, Peg 048844
Aponte Galvis, Alonso 008811
Aponte, Dr. Nilda E. 036135
Apostolou, Eva 023023
Apostolova, Anna 004986
Apostolova, Lili 005036
Apovnik, Dorothea 018446
Appel, G. 032637
Appel, H.H. 010199
Appel, Dr. Michaela 020579
Appelboom, Prof. Thiérry ... 003465
Appelhof, Ruth 048432
Appenzeller, Hans-Ingo 021880
Appignanesi, Paolo 025986
Appleby, Roger 044944
Appleton, Maggie 045168
April, Jeniffer P. 048890
Aprill, Gregg 047421
Aprison, Dr. Barry 047681
Apschnikat, Kenneth 049498
Apsega, Alejo 000511
Apsemetov, Azhibay 029850
Aquilera Cerni, Vicente 040404
Aquiloni, Laura 026966
Aquilué i Abadies, Xavier ... 039825
Aquilué, Dr. Xavier 039379
Aquin, Stéphane 006391
Aquino, Amelia M. 034513
Arab, Mona 000059
Arai, Toshizo 028524
Arakčeev, Boris Serafimovič .. 037741,
  037743
Aralica, Miodrag 038257
Aramă, Ivona Elena 036510
Aramaki, Teiichi 028524
Arambaşa, Dumitru 036238
Aramova, Lidija Grigorjevna .. 038081,
  038082, 038083
Aranburu Lasa, Juan Jose 040460

Aranburu, Xabier 039130
Aranda-Alvarado, Rocio 049819
Aranega, Prof. Dr. César I. . . . 000319
Arango Trujillo, Ignacio 008653
Arango, Jaime 008731
Arango, Placido 039676
Aranne, Heikki 010949
Arapoyanni, Xeni 022969
Araújo, Emmanoel 004910
Araújo, Dr. José António de . . 036054
Araujo, Marcelo 004905
Araújo, María Helena 035711
Araújo, Ricardo 035709
Araujo, Samuel M. 004686
Aravantinos, Dr. Vassilis 023282
Arbart, Kazimierz 034719
Arbelo de Mazzaro, Prof. Aurora
   Catalina 000347
Arbogast, Pierre 011636
Årbu, Grete 033826
Arbuckle, John B. 050229
Arbucó, Susana B. 000268
Arbulla, Dr. Deborah 028203
Arbury, Steve 052442
Arcand, Jessica 052153
Arcangel, Norberto S. . 034516, 034517
Arcangeli, Lino 025388
Arce Torres, Susana 034288
Arces, Prof. Antonio 026478
Arch, Nigel 045090
Archabal, Nina M. 052846
Archambault, Alan H. 048838
Archangelskij, Venjamin 037562
Archer, James Ronald 033734
Archer, Kent 006974
Archibald, Dr. Joanna 043926
Archibald, Dr. Robert R. 052810
Archimanaritis, Alexios 023169
Archino, Sarah 051405
Archontidou-Argyri, Aglaia . . . 023044,
   023182, 023186
Arciniegas de Giraldo, Julia . . . 008807
Arco-Zinneberg, Ulrich Graf . . . 002603
Arcuri, Arthur 004521
Arcuşi, Marin 036430
Ardila, Marsha T. 047916
Ardizzon, Dr. Paolo 025971
Arduini, Dr. Franca 026240
Arefyeva, Galyna V. 043305
Arell, Berndt 010836
Arena, Albert A. 054053
Arena, Dr. Maria Stella 027596
Arenas Abello, Rosaema 008617
Arenas i Sampera, Josep A. . . 039079
Arenas Morales, Lía Esther . . . 008699
Arends, Dr. T. 032570
Arens, A. 018679
Arenz, Lic. Susana 000264
Ares, Felix 039351
Arévalo Pabón, Dr. Emildo . . . 008683
Arevshatyan, Ruben 000732
Arfelli, Guido Arch 028080
Arfmann, Peter 022076
Argun, A.A. 016394
Argungu, Emir of 033415
Arhaldo, Liboni 027030
Arhip, Cristian 036507
Ariano, Terry 053380
Arickx-George, M. 003897
Arih, Prof. Aleš 038594
Aringer, Dr. Klement 020583
Aristizábal Giraldo, Tulio 008659
Arizzoli-Clémentel, Pierre . . . 016180,
   016183
Arkadjevna, A.A. 038000
Arkio-Laine, Leena . . . 010820, 010826,
   010835, 010838, 010842, 010856,
   010859, 010860, 010877, 010878,
   010883
Arkley, J. 045658
Arko-Pijevac, Milvana 008979
Arlauskienė, Laimutė 030327
Arman, Kıvanç 042933
Armand-Delille, Lionel 013969

Armater, Raymond J. 052339
Armella Maza, Virginia 031210
Armenta Camacho, Juan 031399
Armin, Karlen 042063
Arminio, Roberta Y. 051835
Armirail, Sandrine 013778
Armitage, Helen 043900
Armitage, Dr. Philip L. 043756
Armour, Erica 048692
Armstrong-Gillis, Kathy 052943, 052952
Armstrong Touchine, Maxine . . 047733
Armstrong, Alison C. 044667
Armstrong, Charlotte 051430
Armstrong, Chris 000950
Armstrong, Elizabeth 051544
Armstrong, Fredrick 047564
Armstrong, Joan T. 050696
Armstrong, Richard 052157
Armstrong, Steve 052659
Arna, Dr. Angela 022811
Arnaldo, Javier 039686
Arnaud, Jean-Yves 013069
Arnauld, Claude 006132
Arnauld, Marie-Paul 014779
Arnault, Jean-Charles 012550
Arnăutu, Nicoleta 036235
Arndt, Elke 022418
Arnemann, Heinrich 016796
Arneodo, Sergio 027806
Arnesson, Ingemar 041009
Arnesson, Kerstin 040617
Arnheim, Avigail 025212
Arnhold, Dr. Hermann 020601
Arning, Bill 047364
Arnold-Becker, Dr. Alice 018518
Arnold, Beate C. 022772
Arnold, Dr. Charles D. 007496
Arnold, Dorothea 051424
Arnold, Francis 049934
Arnold, Gotthard 021879
Arnold, J. 045323
Arnold, Karl Heinz 018437
Arnold, Lukas 041990
Arnold, Mary Bea 049157
Arnold, Peter S. 048417
Arnold, Dr. Sven 017159
Arnold, Ute U. 022870
Arnoldi, Clelia 028118
Arnolli, G. 032570
Arnon, Carmella 025126
Arnot, Sally 048627
Arnott, Alastair 045829, 045830
Arnould, Marie-Agnès 012522
Arnoult, Hubert 012345
Arnoux, François 012444
Arnst, Dr. Caris-Beatrice 017065
Arobba, Daniele 026237
Arocha Rovira, José A. 009365
Arodaky, Badr-Eddine 014724
Aron, Kathy 049892
Arons, Dr. Kārlis-Ēriks 030213
Aronson, Julie 047739
Aronsson, Annika 040691
Arpin, Odette 012231
Arpin, Pierre 007318
Arpin, Roland 006735
Arraras, Maria Teresa 036166
Arrasmith, Anne 047009
Arrate-Hernandez, Danilo M. . . 009317
Arrecis, April 048224
Arredondo, Joe R. 050516
Arriaga, Ana 040463
Arriaga, Brenda 047587
Arrieta, Katrin 021494
Arriola, Juan Carlos Juárez . . . 039568
Arriola, Julia 053835
Arroues, Heidi 049349
Arroyo Martín, Dr. José Víctor . 039147
Arroyo, Marta 040305
Arsak, Mychajlo Mychajlovyč . . 043186
Arsenault, Linda 006687
Arseneau, Steve 048360

Arseneault, Gérald . . . 003692, 003706
Arslan, Dr. Ermanno A. 026782, 026785
Art, Sue Backhouse 001150
Arteaga, Agustín 000164, 031216
Arteaga, Agustín 036141
Artëmov, Evgenij 037742
Arterburn, Lesa 046844
Artero-Pamplona, Rafael 039837
Arthur, John E. 049643
Arthur, Lloyd 005926
Articus, Dr. Rüdiger 018995
Artman jr, Paul C. 049262
Artmann, Dr. 002242
Artykov, B.R. 054638
Aru, Krista 010689
Arunga Sumba, Owen 029861
Aruz, Joan 051424
Arvi, Barbara 050469
Arvidsson, Elisabeth 040774
Arvola, Silja 033698
Arx, Christine von 041667
Ary, Max L. 051730
Arzadeh, Kamal 024780
Arzur, Anna-Vari 014913
Asakura, Tetsuo 028874
Ašanin, Snezana 038188
Asbury, Michael J. 045501
Ascencio, Michaelle 054724
Aščeulova, Tatjana Stanislavovna 037574
Asch, Georges 014920
Aschauer, Manfred 002298
Aschengreen Piacenti, Dr. Kirsten 026297
Aschinger 017268
Aschkenes, Anna M. 052147
Asensi Catalán, Maria Dolores . . 039690
Asercion, Allisen 052531
Ash Ezersky, Lorinda 051381
Ash, Jeanelle 052224
Ash, Nancy 052081
Ashdjian, Peter H. 009408
Ashe, James S. 050186
Ashe, Janet 047808
Asher, Dave 033295
Asher, Dorothy J. 048563
Asher, Toni 050029
Ashfaq, Prof. Dr. Muhammad . . 034110
Ashgar, Amanollah Mohammad . 000008
Ashley, Brian 045419
Ashley, Catherine 048370
Ashley, Raymond E. 052975
Ashooh Lazos, Marylou 050628
Ashraf, Dr. Mohammad 034152
Ashton, Alan 045180, 045182
Ashton, Joanne G. 051229
Ashton, Kate 001269
Ashton, Lynn 050492
Ashton, Lynne 045712
Ashton, Rand 054302
Ashton, Richard 044192
Ashton, Dr. S. 043842
Ashton, Vicki 052247
Ashworth, Jim 001152
Ashworth, Meg 045722
Ašimbaeva, N. 037750
Askali, Nurhusih 034483
Aslamatzidou, Z. 023139, 023192
Aslani, Q. 024716
Asmaramah, Rasanubari 030682
Asmus, Cathy 053519
Asmuss, Dr. Burkhard 017299
Asngari, Ir. Moch. 024579
Aso, Keiko 029639
Asolati, Dr. Michele 027146
Asomaddin, Abdugayyumi 042619
Asparuchov, Milko 005119
Asparuzov, Dr. Milko 005122
Aspe, Virginia Armella de 031269
Aspenwall, Jim 051097
Aspes, Dr. Alessandra 028347
Asplund, Lars 053555
Asprov, Ognian 005078, 005111
Ass, Karl-J. 021221
Assaàd, Khaled 042263
Assadulina, Maýa 043095

Assaf, Amnon 025140
Assamoa Aime, About 008874
Assamoi, Aimé About 008872
Assatouroff, Corinne ter 003480
Assel, Dr. Marina von 016995
Assémat, Jean 015654
Aßkamp, Dr. Rudolf 018973
Assmann, Dr. Peter . . . 002316, 002323,
   002330
Assmann, Mag. Dr. Peter 002328
Assoghba, Romain-Philippe . . . 004188
Assouline, Hadassah 025073
Assunção, Dr. Ana Paula 035829, 035985
Assunto, Richard 049944
Ast, Hiltraud 002031
Astaa, Mikko 010822
Astachova, Irina Valerjevna . . . 038040
Aster, Robert 051230
Astika, Tjokorda Bagus 024669
Astini, Piero 026653
Astley Cooper, Patricia 045126
Astore, Silvio 027378, 027379
Astorino, Jean 046646
Astramskas, Arunas 030441
Astro, C. 014489
Åström, Kjell 040974
Åströmad, Kenneth 040609
Astrua, Dr. Paola 028116
Astruc, Franck 016336
Astruc, Jean-Yves 014964
Astrup Bull, Knut 034034
Astrup, Yngve 033824
Astui, Aingeru 039131
Asunción Mateo, María 039981
Asyrov, Yagsymyrad 043093
Asztalos, István 023409
Atac, Robert 053591
Atanasova Tsoneva, Neli 005104
Atanga, Peter S. 005289
Atasever, Yaşar 043013
Ateş, Sevim 042865
Athanasiadis, Tasos 023025
Athanasiadis, Yanni 000987
Athani, Shankar S. 024202
Athayde, Sylvia 004793
Athens, Joe 047733
Atherton, Lois 052214
Atik, Dr. Şeniz 042941
Atiti, Abel 029862
Atiyeh, Walid 045017
Atkins, David 046123
Atkinson, Elmer F. 052054
Atkinson, J. 001548
Atkinson, Lesley 045525
Atkinson, Scott 052978
Atlan, Eva 018430
Atsuo, Shioiri 029524
Attack, Paul H. 006317
Attale, Véronique 016147
Attali, Zohara 025018
Atter, Kerri 048268
Attewell, Alex 044970
Attila, Dimény 036787
Attum, Elaine 048468
Attuquayefio, Dr. Dan K. 022939
Attwood, Maryon 054546
Atvur, Orhan 043057
Atwater, William F. 046301
Atwell, Richard 043830
Atwood, Leland 047868
Atwood, Valdine C. 050554
Aubagnac, Gilles 012834
Aubert, Ben 048495
Aubert, Fabrice 014598
Aubert, Glenda 005440
Aubert, L. 041533
Aubin, Maride 014108
Aubrecht, Dr. Gerhard . 002316, 002316,
   002328
Aubrun, Max 012534
Aubry, Françoise 003485
Aucella, Frank J. 054178
Auchterlonie, Fiona 044227
Audinet, Gérard 014697

Audisio, Aldo 028137
Audley, Paul 047201
Auel, Lisa B. 049002
Auer, Dr. Alfred 002136, 003026
Auer, Dr. Anita 022334, 022335
Auer, Helmut 001687
Auer, Wilfried 021676
Auerbach, Dr. Konrad . 021827, 021828
Aues, Dr. Anita 022339
Auffant, Luisa 010386
Auffret, M.J. 013626
Aufgebauer, Peter 017525
Aufseß, Eckhart Baron von 016645
Augé, Jean-Louis ... 012364, 012366
Auger, Michele 050513
Augier, Jean-Pierre 013779
Augustat, Dr. Claudia 003046
Augusti, Dr. Adriana 028297
Augustin, Josef 001758
Augustin, Stephan 019228
Aujard-Catot, Eliane 011840
Aulagner, Renaud 014946
Aulan, Comte d' 011789
Aulich, C. 022583
Ault, Lee 043761
Aumasson, Pascal 012822
Aumüller, Prof. Dr. Gerhard .. 020271,
                                020275
Aunela, Jorma A. 011274
Aupetitallot, Yves 013177
Auping, Michael 048906
Aurandt, David 006860
Aurél, Szakál 023749
Aurell, Pirkko 011508
Auserve, Philippe . 014387, 014388
Ausfeld, Margaret Lynne ... 051025
Ausiàs, Vera 012942
Ausich, Prof. Bill 047942
Auska, Adolf 001957
Ausman, Mary Beth 047577
Austad, Dan 049098, 053580
Austermayer, H. Josef . 019224, 019225
Austigard, Bjørn 033783
Austin, Fran 049067
Austin, Pat 050013
Auth, Dr. Susan 051499
Autio, Tuula 041110, 041118
Autry, Brick 052132
Autzen, Henning 022057
Auvergnat, Dr. Arielle 015989
Auwärter, Konrad 019867
Auwera, Joost Vande 003459
Auxerre, Alain 012831
Avalos Valenzuela, Ana 007556
Avant, Joanne 052935
Avdeeva, M.V. 037575
Avdeeva, Taisija Dmitrievna . 037897
Avendaño, Clarissa L. 034549
Averjanova, Élvira Valentinovna .. 037143,
                                037653
Averoff-Ioannou, Tatiana 022979, 023171
Avery, Dr. D.M. 038686
Avery, Dr. G.A. 038686
Avery, Joseph T. 051395
Avery, Kim 005937
Avery, Laura 053151
Avery, Louise 006127
Avery, Susan 054525
Avesani, Ricciarda ... 006972, 026146
Avest, H.P. ter 032413
Avić, Mirsad 004283
Avila Bardají, José R. 039634
Avila de Monroy, María 008711
Avila Martínez, Jairo Hernando . 008713
Avramenko, Mychajlo Pavlovyč . 043296
Avrandt, David 006566
Avril, Ellen 049739
Awald, John C. 047244
Awouters, Maria 003474
Awwal, Dr. Iftikhar ul . 003215, 003218
Ax, Goeran 040934
Ax, Michael 010306
Axel, Hungerbeuhler 053842
Axsom, Richard H. 049191

Ayalon, Dr. E. 025204, 025205
Aydın, M. Sezai 042851
Ayerbe, Santiago 008765
Ayers, Betty 005572
Ayers, Marilyn L. 046910
Ayerve, José M. 010423
Ayesha, Abdulla 043333
Aykaş, Satı 042860
Ayling, Emma 046141
Aymon, Marie-F. 041722
Ayob, Dato Muslim 030696
Ayrault, Jacqueline ... 011628, 011629
Ayres, William 053548
Ayscue, Gene 054443
Aytar, Sefer 043081
Aytiyev, Satar Gaparovich 030076
Ayvaz, Melahat 043068
Ayvilera, Fernando Gómez ... 040212
Azarcon, Romualdo A. 034560
Azarmehr, G. 024701
Azcue, Leticia 039676
Azeredo, Paula 035952
Azevedo Abreu, Arienei 004650
Azevedo de Silva, Dr. Nearch .. 004719
Azevedo, José Henrique 035735
Azimi, Esmail 024773
Ažimova, Irina 038042
Aziz, Kapt Nazri 030696
Azizbekova, P.A. 003153
Aznar-Alfonso, Dr. Lydia ... 034467
Azón i Masoliver, Marisa ... 039095
Azuar, R. 034944
Azuma, Hisao 029711
Baad Pedersen, Margit 010098
Baagøe, Jette 010130
Baalbergen, W. 032700
Baartman, H. 038720
Baartvedt, Randi 034053
Baatz, Henning 016521
Bab, Bettina 017489
Babajceva, Nadežda Borisovna . 037927
Babanazarova, Marinika 054637
Babans'ka, Natalija 043198
Babâtă, Ioan 036816
Babboni, Dr. Marina 025783
Babcock, Catherine 050137
Babcock, Elizabeth 052568
Babcock, Sidney H. 051428
Babel, Jerzy 034926
Babelon, Jean-Pierre 013002
Babenko, Valerij Volodymyrovyč . 043171
Baber, Charlotte 045828
Babias, Marius 017263
Babič, Prof. Franka 008908
Babić, Mirko 004259
Babin, Anikca 008926
Babini, Tino 027693
Bacallado Aránega, Juan José . 040099
Bacalles, Lorraine 051344
Baccetti, Dr. Nicola 027137
Baccheschi, Dr. Edi 026406
Bacchetta, Véronique 041519
Baccini, Dr. Stefano 026903
Baccrabere, G. 015986
Bačeva, Irina 005178
Bach-Nielsen, Hanne 010317
Bach Østergaard, Bjarne 010195
Bach, Beverly 051870
Bach, Dana 021010
Bach, Kathy 053762
Bach, Dr. Thomas 019444
Bachčevanov, Ivan 005207
Bachelder, Laura 052640
Bacher, Fabian 016545
Bacher, Wilhelmine 001885
Bachhiesl, MMag. DDr. Christian 001975
Bachl-Hofmann, Mag. Christina . 002954
Bachman, Dona R. 047425
Bachmann-Tonelli, Karen 007059
Bachmann, Alois 041624
Bachmann, Birgit 002765
Bachmann, Prof. Lutz 033847
Bachmisov, Éduard

Konstantinovič 036930, 036931
Bachner, Franziska 001991, 001992,
                                017751
Bachorz, Hanna 035437
Bacigalupi, Don 053760
Bacin, Mark S. 051872
Bâciu, Dr. Florea 036403, 036404
Backer, Julie P. 047796
Backer, Kathleen L. 054079
Backes, Joan 051224
Backes, Peter 022349
Backhaus, Fritz 018430, 018442
Backhouse, Andrew 045917
Backhurst, Timothy D. 051084
Backlund, Connie 048744
Backström, Ragnar 011032
Bacon, Weldon 005711
Bacourou, Emilia 023185
Bacquet, Emmanuel 014374
Bačvarov, Ivan 005035, 005163
Bączyk, Jarosław 035123
Badaró Nadal, Enrique 046251
Badas, Ubaldo 028394
Badault, Albert 012426
Badder, Susan 054098
Bade, Mary 049972, 050603
Badel, Doris 019638
Bader, Joerg 041518
Bader, Judith 022176
Badham, Mark 006106
Badoüin, Gerhard 021298
Badovinac, Zdenka .. 038545, 038549
Badre, Dr. Leila 030265
Badura-Triska, Dr. Eva 003050
Bähr, Gustav-Adolf 020762
Baehrend, John 047624
Baeke, Jan 031939
Bækgaard, Bjarne 010005
Bækkelund, Björn 033586
Baele, Annemie 004167
Baena Alcántara, María Dolores . 039305
Baensch, Antje 018773
Baensch, Norbert 018749
Bär, Dr. Frank P. 020853
Baer, Jan Hodges 049446
Baer, Dr. Rudolf 041229
Bärnighausen, Hendrik 020478
Baerwaldt, Wayne 007195
Baeten, Herman 003947
Baets, August de 003863
Baetz, Tracy L. 049920
Báez Plaza, Anastasio 039115
Baganz, Bruce P. 054164
Bagdanavičienė, Asta 030408
Bagdonienė, Rūta 030482
Baggaley, Sean 045435
Bagge, Christian 018965
Baggenstoss, Josef 041543
Bagheri-Zenouz, Dr. E. 024704
Bagle, Eyvind 033854
Bagnall, Dinah Ryle 045669
Bagnatori, Paola 053013
Bagnell, Warren 006225
Bagnoli, Dr. Alessandro ... 027961
Bahadar Khan, Mohammad 034111
Bahadur Shah, Prof. Karan .. 031875
Baharin Bin Buyong, Dr. Kamarul 030691,
                         030719, 030732
Bahe, Michelle 046670
Bahena García, Jaime Arturo . 031197
Bahm, Linda 046372
Bahn, Erich 003109
Bahn, Dr. Peter 017622
Bahnmaier, Paul M. 050210
Bahrami, M. 024778
Baier, Berthold 019358
Baijot, Joseph 003671
Baik, Jo Gang 029882
Bail, Anne le 016363
Bailer, Dr. Brigitte 002980
Bailey Lewis, Pamela 052458
Bailey, Angela 000892
Bailey, Ann S. 051511

Bailey, Billie K. 054204, 054205, 054206,
                                054207
Bailey, Chris H. 047211
Bailey, Colin B. 051390
Bailey, Courtney 052444
Bailey, David 049182
Bailey, Elizabeth 047419
Bailey, Erin 046621
Bailey, Jack C. 053489
Bailey, Jann L.M. 006056
Bailey, Jeniffer A. 053969
Bailey, Margaret 051596
Bailey, Michael 043461, 049294
Bailey, Patricia 005465
Bailey, Patricia E. 049207
Bailey, Peter 045356
Bailey, Prof. Reeve M. 046493
Bailey, Tom 046558
Bailleux-Delbecq, Martine ... 012931
Bailly-Maitre, Marie-Christine . 013537
Bain, Heather 050503
Bainbridge, Carol 051563
Bainbridge, David S. .. 053392, 053393,
                                053396
Bainbridge, H.M. 045600
Bainbridge, R.A. 045600
Baiocco, Simone 028118
Baioni, Marco 026376
Baird, Vaughan L. 007436
Bairlein, Prof. Dr. Franz .. 022659
Bairrad Ruiro, Marina 035757
Bairsto, Rachel 044921
Baitchiba-Afalna, Kloumtouin .. 007511
Baj, Janina 034839
Bajada, John 030764
Bajanik, Dr. Stanislav 038312
Bajcurová, Dr. Katarína 038328
Bajic, Svetlana 004283
Bajko, Daria 051486
Bajpayee, Aparna 024168
Bajramović, Muhamed 004293
Bajuk, Laura 050478
Bak, Aase 009999
Bak, Jolán 023635
Bak, Dr. Saskia 032570
Bakajutova, Ljudmila M. 037724
Bakal, Jeniffer 051399
Bakalarz, Krystyna 035041
Bakalov, Michail 005206
Bakanauskas, Alvydas 030447
Bakanova, Irina Viktorovna ... 037365
Bakärdžieva, Cenka 005066
Bakärdžijev, Dimitär 005047
Bakarić, Lidija 009025
Bakarich, Carolyn C. 053977
Bakarich, Michael N. 053977
Bakay, Prof. Dr. Kornél 023764
Bakels, Dr. P. 032087
Baker, Alexander 052078
Baker, Dr. Allan 007200
Baker, Alyson 050385
Baker, Barbara J. 046385
Baker, Billy P. 049111
Baker, Bob 050003
Baker, Charlotte 046419
Baker, Chris 044068
Baker, Christopher 047925
Baker, Chuck 047925
Baker, David 047411
Baker, Diane 046605
Baker, Gary E. 051573
Baker, James . 052023, 052873, 052874
Baker, James W. 048413
Baker, Dr. Janet 052120
Baker, Janet 046196, 053086
Baker, Jeanne 054263
Baker, John 045340, 045875
Baker, Kendall 047347
Baker, Larry L. 047035
Baker, Linda 005427
Baker, Paul 045412
Baker, Peggy M. 052211
Baker, Rachel 047671
Baker, Raymond 054263

Baker, Dr. Robert 050517
Baker, Rusty 049373
Baker, Scott 054122
Baker, Sue 048060
Baker, Susan K. 054233
Baker, Warren L. 046385
Baki, Péter 023728
Bakić, Ljiljana 038257
Bakke, Prof. Tor A. 033847
Bakker, C. 032510
Bakker, G. 031931
Bakker, Dr. Lothar 016666
Bakker, Robert J. Roozemond
den 032194
Bakker, W. 032055
Bakkevig, Sverre 033972
Bakouti, Moncef 042770
Bakr, Nemat 010509
Bakreeva, Ljubov Anatoljevna .. 037227
Baksa-Soós, Dr. Veruschka .. 017097
Bákula Budge, Cecilia 034325
Bal, Danielle 013181
Balabanov, Dimitar 005156
Balagtas, Alex L. .. 034512, 034514
Balakina, Natalja Michajlovna .. 037784
Bălan, Filofteia 036449
Balan, Sandrine 014231
Bălan, Zamfir 036260, 036261
Balanza, Ferran .. 039729, 039730
Balao Gonzáles, Angel 039689
Balard, Prof. Michel 015889
Balas, Mary 051722
Balas, Shawna 007470
Balaštíková, Eva 009872
Balatti, Ambrogio 025959
Balázs, Dr. György 023530
Balbaugh, Joe 049862
Balčaitienė, Rasa 030314
Balciunas, Vytautas 030549
Bald, H. 020109
Baldaia, Peter J. 049635
Baldarrago Umpire, Prof. Miguel 034212
Baldellou, Vicente 039499
Baldemair, Vinzenz 002373
Balden, Friedesine 018830
Baldeón, Dr. Amelia 040426
Balderi, Felice 026640
Balderstone, Michael 001360
Baldeschi, Nino 028237
Baldewijns, Jeannine 003640
Baldini, Marino 008968
Baldino, Renzo 041426
Baldit, Alain 015034
Baldoni, Lorenzo 026869
Baldry, Anita 045741
Baldwin Rehbein, Valerie B. .. 052525
Baldwin, Deirdre 049076
Baldwin, Graham 045960
Baldwin, Jane 046741
Baldwin, Nicholas 044984
Baldwin, Phyllis 046718
Bale, N. 001634
Balen, Dubravka 009025
Balen, Jacqueline 009025
Balen, Tim 050371
Balestra, Nino 027665
Balevičius, Mindaugas 030513
Baley, Susan G. 051586
Balher, Gisela Carina 000484
Balic, Haris 003059
Balie, Dr. Isaac 038717
Baligand, Françoise 012821
Balint, Lăcrămioara 036504
Bálint, Magdolna 023562
Bálint, Tamás 023478
Balint, Valerie 049602
Balistrieri, Bruce 051945
Balkanska, Veneta 005092
Ball, Cobie 053407
Ball, Daniela 042248
Ball, Dr. Daniela 041975
Ball, David 053201
Ball, Elizabeth 005579
Ball, Felicity 045940

Ball, Glenys 001612
Ball, JoAnn 052188
Ball, Rex 052939
Ball, Sabine 019115
Ballantine, Dr. David 036135
Ballantine, Thomas 005942
Ballantyne, Dorothy 049494
Ballantyne, Weyman 033291
Ballard, Mark 049765
Ballbé, Xavier 039201
Ballerini, Rosangela 026640
Ballerstedt, Mareike 017585
Ballesta Gómiz, Pedro 039799
Ballesta, J. 040381
Ballesteros, Prof. Miguel ... 000547
Ballestra 014129
Balletto, Prof. Emilio 028131
Ballinger, J.C. 050695
Ballinger, James K. 052120
Ballot, Prof. Dr. G.M. 038813
Balmer, John 054563
Balog, Jelena 009036
Balog, Daniela 002207
Balogh, Mihály 023538
Balogh, Sándor 023573
Balsamo, Prof. Renato 026080
Balshaw, Dr. Maria 045219
Baltasar Coll Tomas, M.I.S.D. .. 039888
Baltensperger, Marianne 042025, 042026
Baltes, Peter 017792
Băltuţă, Steliana 036256
Baltz, Gerald W. 052803
Baltzegar, Mack 049276
Baltzer, Doralies 016815
Balza, Raimundas 030434, 030461,
030462, 030471, 030474, 030560
Balzardi, Elena 041323
Balzekas, Robert 047640
Balzekas, Stanley 047640
Bamala, Louis 008841
Bambara, Parfait Z. 005278
Bamberger, Gerald 017372
Bambery, Anneke 044082
Bamesa, Prof. Tshungu 008838
Bamforth, Janice 046902
Ban, Gheorghe 036698, 036703, 036704
Bana, Giovanni 025374
Banach, Wiesław 035192
Banach, Witold 035085
Banai, Merav 025055
Bańbura, Prof. Dr. Jerzy ... 034974
Banchieri, Daria 028276
Bancroft, Evelyn 001054
Bandaranaike, Sunethra ... 040474
Bandeira, Dr. Dione da Rocha .. 004513
Bandes, Dr. Susan J. 048440
Bandrowski, Margaret 053469
Bandt, J. 001306
Bandurraga, Peter L. 052509
Bandy, Mary Lea 051427
Bandyopadhyay, Dr. Banya .. 024145,
024513
Banerji, Dr. Jayasri 024377
Banfi, Enrico 026799
Banfield, Audray M. 000782
Bang Andersen, Sveinung ... 033972
Bang Larsen, Bent 010299
Bang, Mogens 010099
Bangcas, Salamon L. 034477
Bangcaya, Regino Z. 034585
Bange, Maja 033839
Bangert, Dr. Susanne 010145
Banghard, Karl 020963
Banghart, Bob 049848
Banker, Amy 048252
Banker, Maureen 052447
Bankert, John 047412
Banks, Peter 005740
Banks, R. 001144
Banks, Robert 053769
Bannicke, Elke 017088
Bannister, R. 043659
Bannon, Anthony 052617

Banova, Vera 005262, 005263
Banović, Miroslav 038203
Bantel, Prof. Dietrich 020902
Bantle, Tom 049195
Banu Syed, Rehana 031874
Banz, Richard 054587, 054588
Banzato, Dr. Davide .. 025378, 027142,
027144, 027160
Bao, Yingjian 007830
Baptista Pereira, Fernando
António 036028
Baptista, Dr. Teresa 035635
Baquedano, Enrique 038913
Baquero Velásquez, Ramón .. 008634
Baquero, Dr. Angeles Ramos .. 034180
Bar-Gera, Dov 025019
Bar-Gera, Kenda 025019
Bar-Or, Galia 025035
Bar Semech, Vered 025030
Barabaş, Dr. Neculai .. 036213, 036216,
036639
Barahona, Rolando 008855
Baral, Jody 050439
Barale, Joseph 013164
Baran, Iwona 035263
Barandun, Plasch 042103
Barangulova, Farida F. 037980
Baranova, M.N. 037833
Baranowska, Janina 045078
Bárány, Anders 041027
Barbagallo, Dr. Alfredo I. .. 000241
Barbalić, Željko 008978
Barbanera, Prof. Marcello .. 027597
Barbara, Sibille 014534, 014535
Barbare, Inese 030188
Barbaret, Hervé 014749
Barbeau, Bruce 007173
Barber, Charles D. 052048
Barber, Dan 051112
Barber, Diane 049576
Barber, John 001671
Barber, R.G. 045738
Barber, T. 044901
Barberet-Girardin, E. 016090
Barberini, Maria Giulia 027629
Barbier-Ludwig, Georges ... 012696
Barbier-Mueller, Jean-Gabriel .. 041529
Barbier, Jean 036186, 036188
Barbier, Marjolijn 003289
Barbier, Michel 003913
Barbieri, Frances T. 053271
Barbieux, José 016004
Barbina, Prof. Alfredo 027557
Barbizet, Laure 014741
Barblan, Otto 042115
Barbosa Godefroid, Leonardo .. 004603
Barbosa, Artemio 034494
Barbour Strickland, C. 049263
Barbour, Mark 047441
Barbry jr, Earl J. 050698
Barbu, Cristian 036320
Barca, Ana 036310
Barcal, Alexandra 042196
Barcel, Ellen 053424
Barcellos, Lauro 004774
Barclay, Craig 044190
Barclay, Don 048331
Barclay, Mitzi 052004
Barcley, Craig 044191
Barcroft, Arnold 043463
Barcus, Leisha 048283
Bardelli, Alvaro 025361
Bardiaux, David 015844
Bardiés-Frarty, Isabelle ... 014780
Bardies-Fronty, Isabelle ... 014152
Bardiová, Dr. Mariana 038278
Bardo, Antoine 013563
Bardwell, Sarah 044993
Bareinz, Reinhard 017711
Barendsen, J.E. 032420
Barens, H. 051717
Barents, Els 031947
Barger, Judith M. 050097
Bargero, Vicente 034277

Barghini, Sandra 043512
Bargholz, Christina 019006
Bargiel, Réjane 014713
Barilleaux, Rene Paul 049751
Barina, Miloslav 009719
Barisch, G. 017374
Barisch, M. 017374
Barison, Mara 025431
Baritault du Carpia, M. de 014102,
014103
Barjou, François 013773
Barkefeld, Anna 021917
Barkemeyer, Dr. Werner ... 018373
Barker, Alex 048116
Barker, Deborah 051843
Barker, G. 001543
Barker, Garry 051052
Barker, Jane 046791
Barker, Dr. Nicolas 044027
Barker, Ray 046219
Barker, Rod 048400
Barker, Sally L. 052383
Barker, Stuart 043776
Barker, Wendy 048606
Barketov, Vladimir Aleksandrovič 037847
Barkhofen, Dr. Eva-Maria .. 017085
Barkley, Terry 047206
Barkoun, Gennady 003245
Barksdale, Daryl 050677
Barkwill, Ken 047087
Bârlea, Gheorghe Mihai 036707
Barlek, Josip 009028
Barlier, Gabriel 013054
Barlow, Erin 051381
Barlow, Geraldine 000971
Barlow, Howard 054298
Barlow, Keith 045387
Barlow, Martin 044879
Barlow, Patricia S. 047894
Barlow, Victoria 044982
Barmore, Jim 047444
Barndollar, Lue 047857, 047858
Barnea, Dr. Jackob 002396
Barnes, Althemese 053663
Barnes, James H. 048858
Barnes, Judy 050757
Barnes, Lolly 046991
Barnes, Lucinda 046933
Barnes, Marion 043540
Barnes, Richard 049607
Barnet, Peter 051364, 051424
Barnett, Eric B. 048488
Barnett, George ... 044799
Barnett, James F. ... 051195
Barnett, June 053303
Barnett, Lawrence .. 054008
Barnett, Patricia ... 051390
Barnett, Steve 049677
Barney, Edres 052133
Barnfather, Margaret .. 043443
Barnham, George ... 044645
Barnhart, Audrey ... 054425
Barnhill-West, Barbara .. 051786
Barnhill, Michael ... 046856
Barnosky, Antony ... 046939
Barns, Laurie 048301
Barnum, Lorie 052623
Barnwell, Andrea D. .. 046636
Barocelli, Prof. Francesco .. 027221
Baroff, Deborah ... 050195
Barollo, G. 026465
Baron, David 006777
Baron, Donna 054356
Baron, Evelyne ... 015282
Baron, Dr. Jacques .. 014380
Barón, Javier 039676
Baron, Richard ... 048503
Baroncini, Dr. Claudia .. 026507
Barone, Constance B. .. 052719
Baroni, Fabio 025800
Baroni, Laura 026143
Baronne, Geriool .. 051300
Barons, Richard I. .. 053421
Barovier, Jacopo ... 028301

Barr, Caroline 044397, 044419
Barr, Cheryl B. 046934
Barr, Edward 053774
Barr, Mike 043377
Barr, Norma 005699
Barr, Dr. Peter J. 046321
Barr, Tom 048841
Barracane, Gaetano 025459
Barrachina, Jaime 039939
Barrack, Pamela 046842
Barraclough, Bruc 051389
Barral Iglesias, Alejandro ... 040126
Barrand, Janine 001271, 001281
Barreiros, Mariluz 040291
Barrera Bassols, Marco 031186
Barrera, Alberto 034188
Barrere, Eliza 014483
Barrere, Sylvie 012108
Barreto, Ricardo D. 047136
Barrett-Lennard, Amy 001471
Barrett, Candace 050415
Barrett, Joanna 044708
Barrett, Maurine 053442
Barrett, Michael ... 000787, 049900
Barrett, Wendy 052903
Barreyre, J.Y. 014629
Barrie, Brooke 049323
Barrientos, J. Oscar 023373
Barriga, Prof. Fernando J.A.S. .. 035805
Barringer, Jane S. 053083
Barrington, Robin 006285
Barrois, Alain 014032
Barron, G.D. 033198
Barron, Stephanie 050444
Barroso Panatieri, Angélica ... 004633
Barrowclough, Dr. George F. .. 051327
Barry Byrne, Brenda 005792
Barry, Alexis 053614
Barryte, Bernard 053488
Barsalou, Victoria 006618
Barsch, Dr. Barbara ... 017175, 017177
Barsch, Wolfgang 016531
Barskaja, Marina Iljinična 037502
Barson, Jerry 053105
Barsotti, Gianfranco 026607
Barstow, Norman 049307
Bärsu, Mariana 036375
Barsuglia, Dr. Bruno 026640
Barszcz, Janusz ... 035175, 035302
Barszczewska, Krystyna 035440
Bart, Barbara M. 049631
Barta-Fliedl, Dr. Ilsebill 003012
Barta, Andrew 054425
Barta, Dr. Ilsebill 003008
Bárta, Karel 009954
Bartelink, Drs. J.T.N. 032949
Bartels, Dr. Christoph 017433
Bartenev, Aleksandr Igorevič .. 037767
Bartenhagen, Lynn 051132
Barth, Dr. Angelika 022135
Barth, Frank 018991
Bárth, Dr. János 023724, 023726,
                    023729, 023732
Barth, Linda J. 050470
Barth, Dr. Robert 041988
Barth, Dr. Robert L. 052312
Barth, Dr. Stephanie 020994
Bartha, Dr. István 023943
Barthel, J. 020629
Barthelmess, Dr. Stephan N. .. 019285
Barthelson, Karen 051145
Bartholomew, Ed 046226
Bartik, Dr. Juraj 038299
Bartków-Domagala, Jadwiga .. 035427
Bartkowski, Gregor 017638
Bartkus, Stephen J. 054212
Bartl, Anna 041268
Bartl, Johanna 017864
Bartlett, Prof. Christopher 053788
Bartlett, Diane 050393
Bartlett, Margaret 000970
Bartley, James A. 054478
Bartlit, Nancy 050402
Bartolf, M. 006607

Bartolini Bussi, Prof. Maria
   Giuseppina 026847
Bartolomé Arraiza, Alberto ... 039672
Barton Billman, Hilary 049917
Barton Thurber, T. 049353
Barton, Anita S. 050673
Barton, C. Michael 053701
Barton, Elizabeth 049518
Barton, William R. 051907
Bartoś, Magdalena 035019
Bartos, Paul I. 049146
Bartoš, Václav 009757, 009857
Bartosz, Adam 035267
Bartosz, Anna 035268
Bartram, Rita 048985
Bartsch, Joel A. 049581
Bartuś, Marta 034645
Bartz, Rolf-Peter 022434
Barwell, Victoria 043523
Barwick, Kent 051431
Bary, Bertrand de 015180
Barzakova, Maria 005000
Barzen, Jeb 046786
Bas, Michael le 043654
Basamakov, Christo 004970
Basamakov, Ilko 005132
Basanieri, Giovanni 026085
Basante, Enar 008748
Basart 004880
Basbous, Alfred 030279
Bascha, Nadya 016432
Baschnagel, H. 018789
Bascom, Mansfield 050616
Bascou, Marc 014749
Başer, İsmail 043034
Basile, Corrado 027969
Basile, Kenneth A. 052901, 052902
Basir, Abdul 024679
Baškirova, Marija Michajlovna .. 036997
Bašnjak, Jurij Michajlovič 036897
Basque, Maurice 006351
Bass, Anneta Jakovlevna 037715
Bass, Carol 054500
Bassegoda Nonell, Juan 039072
Bassett, Donald E. 047876
Bassett, Fannie 047626
Bassett, Dr. M.G. 043863
Bassett, Nancy W. 047876
Bassett, Ralph 051812
Bassi, George 050180
Bassi, Raffaella 026625
Bassi, Dr. Roberto 028305
Bassin, Aleksander ... 038540, 038547
Bassinger, Marilyn 051709
Bassler, Roseli Fischer 044427
Basso, Dr. Laura 026802
Bassús, María M. 000179
Bast, Douglas G. 047091
Bastard, Laurent ... 016018, 016019
Basterretxea-Moreno, Dr. Amaia 039143
Bastia, France 003453
Bastidas, Hugo Xavier 049818
Bastin, Donald 052533
Basu Ray Chaudhuri, S. 024424
Bata, Philippe 012886, 014919
Bata, Sonja 007159
Bataillard, Marianne 041453
Batchelor, Elisabeth 049888
Batchelor, Trevor 006611
Bateman, Prof. Richard 045061
Bateman, Suellen 049175
Baten, F.O. Abdul 003213
Batenin, Nikita Sergeevič 037014
Bater, John 048925
Bates, Brenda 048075
Bates, Charles 050279
Bates, Craig D. 054596
Bates, F. 045796
Bates, Geoffrey 050353
Bates, Michael L. 051328
Bates, Tom 050899
Bateva, Mimi 005083
Bath, Marquess of 046055
Bathe, I. 006563

Batista, Alvaro 035474
Batkin, Jonathan 053125
Batkin, Norton 046497
Batori, Dr. Armicla 027602
Batova, Anna Nikolaevna 037631
Batrd, Nicola 045740
Battaglini, Dr. Giuseppe 027423
Battali, Kathleen 048741
Battcher, Ray 047215
Battel, Angelo 027802
Batterson, Lorna 050317
Battinou, Zanet 022987
Battis, Nicholas 051461
Battle, Dr. Thomas C. 054123
Batty, Theresa 004175
Batty, Tim 044115, 044120
Baty, Catherine 054342
Bauch, Herbert 019892
Bauch, Hermann 002246
Bauche, Dean 006503
Bauchiller, Dr. Rafael 039632
Baucum-Fields, Penny 051886
Baud-Berthier, Gilles 012135
Baud, Arlette 041241
Baud, Christiane 041294
Baudement, Marie-Laure 011824
Bauder, Charles C. W. 049069
Baudet, Jean-Charles 016099
Baudet, Jean-Michel 014883
Baudin-Reneau, Katia . 015868, 019705
Baudisson, C. 013490
Bauduin, Jean-Loup .. 013480, 013481,
                              013484
Bauer, Christian 002234
Bauer, Christoph 021877
Bauer, Douglas F. 048764
Bauer, Fritz 022916
Bauer, Gudrun 019189
Bauer, Hans 017331
Bauer, Hans-Rudolf 041828
Bauer, Heidrun ... 022094, 022095
Bauer, Dr. Helmut 020551
Bauer, Herbert 016558
Bauer, Prof. Dr. Ingolf 020912
Bauer, Dr. Markus 018737
Bauer, Martina 020877
Bauer, Peter 020088
Bauer, Richard 053707
Bauer, Sigrud 017331
Bauer, Stéphane 017206
Bauer, Susan 006300
Bauermeister, Janneke 020608
Bauernfeind, M.A. Günther ... 020714
Baughan, Clark 053124
Baulard, Hervé 012453
Baule, John A. 054566
Baum, Prof. Dr. Eckhard 022713
Baum, Kristina ... 018248, 018251
Baum, Rainer 041268
Baum, Wolfgang 021849
Bauman Taylor, Paula 047630
Bauman, Victor-Henrich 036798
Baumane, Inta 030137
Baumann, Anton 002498
Baumann, Daniel 041318
Baumann, Evelies 017762
Baumann, Klaus 018803
Baumann, Dr. Maria .. 021314, 021317
Baumann, Rudolf 041638, 041639
Baumbach, Susan M. 045514
Baume, Nicholas 047116
Baumeister, Dr. Ralf 016741
Baumgärtel, Dr. Bettina 018046
Baumgärtner, Iris 021281, 021283,
                              021285
Baumgardner, George 047444
Baumgardner, Stan 047564
Baumgarten, Heiko 020375
Baumgarten, Dr. K. 016840
Baumgarten, Ute 019348, 019349
Baumgartner, Mag. Lothar ... 001828
Baumstark, Dr. Brigitte 019520

Baumstark, Prof. Dr. Reinhold .. 016958,
              020503, 020508, 020560, 020569,
                              020570
Baur, Andreas 018322
Baus, Dr. Christine 018322
Bausart, Lucie 003291
Bause, George S. 051925
Bausier, Karine 003326
Bauske, Clay R. 049666
Baussan, Magali 016116
Baußmann, Dr. Edda 020608
Bautrènas, Justinas 030529
Bauziene, Morta 030510
Bavnshøj, Peter 010016
Bavoni, Dr. Umberto 028425
Bawden, Dr. Garth L. 046364
Baxter, Alan 043812
Baxter, April 050038
Baxter, Clare 043401, 043402
Baxter, David 006594
Baxter, Ellen 052157
Baxter, Marilynn 052268
Bayard, Jean 007279
Bayard, Pierre 005293
Bayarri, Carlos 040053
Baybut, John 052318
Bayden, Ian 054039
Bayer-Niemeier, Dr. Eva 019820
Bayer, Dagmar 022385
Bayet 004086
Baylis Ashby, Lisa 048748
Bayliss, Joan 043492
Bayliss, Dr. Paul 038739
Baymuradov, N. 042623
Bayona, Wilhelm A. 034611
Bayou, Charlie 015042
Bayram, Metin 042914
Baysal, H. Hüseyin 042877
Bayuelo Castellar, Soraya 008673
Baz, Samir 030277
Bazančuk, Galina 037430
Bazanov, Valerij Anatoljevič .. 036970
Bazely, Sue 006102
Bazennerye, Dr. Jean 016227
Bazin de Jessey, Hervé 012484
Bazin, J. 014183
Bažkenova, Nesipbiken
   Orazchanovna 029824, 029838
Bazley, Shirley 001286
Bazzoni, Maurizio 027484
Bazzotti, Prof. Ugo 026697
Beach, James 050186
Beadle, Kristine 048197
Beady, Charles H. 052143
Beairsto, Murray 052858
Beal, Graham W.J. 048301
Beal, Carolyn 048067
Beall, William 050008
Beam, Michael J. 051102
Beam, Michael N. 047611
Beamis, Frances D. 047294
Beams, Daniel J. 053198
Bean, Jeanet 051689
Bean, Dr. Susan 052880
Beanchard, Laurent 013204
Bearcroft, Don 043369
Beard-Simpkins, Doreen 053642
Beard, David 052057
Beard, Dr. James S. 050727
Beardsley, Robert 048535
Beardsley, Theodore S. 051404
Beardsmore, R.J. 033266
Beasley, James M. 053468
Beasley, Pam 053367
Beaton, Stuart 005441
Beattie, Keith 043480
Beattie, Natalie 001401
Beatty, Brett 049644
Beatty, Thomas P. 053994
Beatyy, John T. 052653
Beaud, Marie-Claude 030588
Beaudin Saeedpour, Dr. Vera .. 047261
Beaufort-Spontin, Dr. Christian .. 003026
Beaufort, Dr. Christian 003007

Beaumont, Helen 024898
Beaumont, Robert N. 054041
Beausoleil, Raymond R. 007092
Beauvau, Princesse de 013230
Beavers, Randell A. 049593
Beavin, H.A. 044588
Beavis, Dr. Ian 046003
Bebout, Terry L. 051185
Beccari, Giuseppe 028419
Beccaria, Marcella 027537
Beccart, Christine 013145
Becchi, Casimiro 027278
Bećević, Azra 004283
Bech, Marianne 010288
Bechelet, Florence 045709
Becher, Inge 018645
Bechers, H.H 032931
Bechet, Georges 030594
Bechhoefer, Melissa 050043
Bechly, Dr. Günter 022056
Bechtelheimer, Daryl 052718
Bechtold, Dr. Ekkehard 001826
Bečić, Berina 004283
Beck-Lipsi, Gabrielle 015157
Beck, Dr. Andreas 020513
Beck, Bettina 017362
Beck, Elisha 049413
Beck, Gary 052983
Beck, Jane C. 050859
Beck, Jerry 050512
Beck, Lane 053823
Beck, Mike 049626
Beck, Noldi 030310
Beck, Rolf 022615
Beck, Stefan 041839
Beck, Tom 046726
Beck, Walter 030301
Beckemeyer, Klemens 020391
Becken, Ellen 010025
Becker-Jákli, Dr. Barbara 019707
Becker, Dr. Annette 018411
Becker, Christine 020021
Becker, Dr. Christoph 042207
Becker, Edwin 031996
Becker, Gail R. 050495
Becker, Dr. H. 018588, 018589
Becker, Horst R. 018424
Becker, Dr. Ingeborg 017093
Becker, Dr. Jack 051174
Becker, Jürgen 018286
Becker, Julia 049217
Becker, Lawrence 054548
Becker, Lisa 050514, 050542
Becker, Dr. Michael 002004
Becker, Dr. Peter-René 017601
Becker, Dr. Rayda 038749
Becker, Sara 046396
Becker, Scott R. 054186
Becker, Dr. Ulrich 001964
Beckers, Marion 017313
Beckerson, Dr. John . 044536, 046084
Beckert, Harriet 051558
Beckett, D. 044284
Beckett, Michael D. 050193
Beckford, Chip 054288
Beckley, Frank J. 045632
Becklund, Stacey 050017
Beckmann, Isrun 017137
Beckmann, Michael 001167
Beckmann, Dr. Ralf 018351
Beckmann, Dr. Uwe 018917
Beckmann, Werner J. 018053
Becks, Jürgen . 022592, 022593, 022594
Becks, Dr. Leonie 019677
Beckstrøm, Alf 033978
Becucci, Dr. Sandra . 025841, 027997
Bedal, Albrecht 021754
Bedal, Prof. Dr. Konrad 016912, 016913
Bedaux, Dr. R. 032585
Bedbur, Dr. Eckart 019602
Bedell, Lisa 051430
Bedenham, M.D. 043918
Bedenig, Dr. K. 042236
Bedford, Emma 038687

Bedia García, Juana 039497
Bednarek, Monika 034891
Bedouin, Regan 048406
Bedoya de Rendón, Martha Elena 008758
Bedoya, José 004222
Bedula, Jane 051064
Bee, Dr. Andreas 018438
Beebe, John 048599
Beebe, Mary L. 050041
Beech, Mike 044367
Beeden, Keith 045776
Beedenbender, Britt 047500
Beeks, Bonny 049150
Beer, Dr. Manuela 019706
Beerli, Klaus 041772
Beernaert, Veerle 003708
Beerts, Harry 004003
Beesch, Ruth 051412
Beesley, Bruce A. . . . 053989, 053991
Begay, Clarenda 054469
Beggs, Heather 049521
Beggs, Jo 045219
Begonja, Zlatko 009023
Begouin, Bernard 012284
Begović, Akad. Miroslav 009033
Begue, Christelle 015955
Béguin, Gilles 014689
Béguin, Pierre Henri 041357
Beha, Wolfram 019637
Behler, Mark 054281
Behlmann, Cletus 052944
Behm, Stefan 002020
Behne, Dr. Axel 021042
Behnke, Arleen 052004
Behnke, Martin 018851
Behr, Lothar 022297
Behrends, Carey F. 051710
Behrendt, Martina 017293
Behrens, Friedrich 021634
Behrens, Sarah 051921
Behrens, Todd 050097
Behrman, Sharon 049413
Behrndt, Helle 010180
Behroozy, N. 025207
Beier-de Haan, Prof. Dr.
Rosmarie 017299
Beier, Aasmund 033879, 033880
Beier, Dr. H.-J. 022570
Beier, Stefan 019652
Beiersdörfer, Dr. Kurt 021061
Beiersdorf, Maciej 034888
Beil, Judith 053686
Beil, Dr. Ralf 017823
Beilmann-Schöner, Dr. Mechthild 021396
Beimel, Alice 052830
Bein, Peter von 022590
Beiranvand, Asadollah 024710
Beirer, Ingrid 017098
Beirne, T. 044527
Beisel, P. 020652
Beisken, Angela 019099
Béistegui de Robles, Dolores . . 031159
Beitchman, Marsha 051324
Beitl, Matthias 003057
Beitz, Dr. Uwe 018090
Bejbosynov, Kuzembaj
Bajbosynovič 029851
Bejenaru, Ionel 036444, 036538, 036686
Bekmuradov, I. 054642
Bekouan, Hortense Zagbayou . . 008869
Bektureeva, Roza Abutalipovna . 029803
Bela, Dr. Zbigniew 034885
Belai, George 023458
Belaja, Z.A. 037900
Belart, Curdin 041222
Belbin, J. 001347
Belčeski, Rubinčo 030623
Belčeva, Tonja 005022, 005023, 005024
Belcher, Helen 049637
Belder, Marie de 004035
Belenis 023141
Beleznay, Andor 023454
Belfield, John 001343
Belfry, Craig 006694

Belgado Camblor, Victoria 039223
Belger, Kathleen 019973
Belgin, Dr. Tayfun . . 002234, 018915,
018916, 018919
Belie, Liesbeth de 003459
Belier, Corinne 014779
Belinskis, Andrejs 030218
Belinskyj, Valeriy 043128
Belisario, Alessandro 026299
Beljaev, A.A. 037908
Beljaeva, Nadešda V. 037644
Beljaeva, Svetlana A. 037914
Belk, Brad 049841, 049843
Bell, Allan 000886
Bell, Brendan 038802
Bell, Chee Chee 049655, 049656
Bell, Christine 043483
Bell, Elizabeth 047448
Bell, Eunice 051699
Bell, Dr. Graham A.C. 006415
Bell, Greg 053646
Bell, Gregory 049983
Bell, Lynn 046688
Bell, Margaret 033257
Bell, Michael 006584
Bell, Michael S. 052740
Bell, Nicholas 047774
Bell, P.J. 010211
Bell, Rae 046406
Bell, Robert 000937
Bell, Theresa H. 050733
Bell, Valerie 048656, 050342
Bella, Miguel Angel di 000544
Bellah, Suzanne 051871
Bellaigue, Dominique de 012095
Bellais, Leslie 050601
Bellamore, Stan 033279
Bellamy, Gene 049220
Bellamy, Dr. Martin 045729
Bellando, Francesco 025449
Bellanger, Françoise 014616
Bellani, Giancarlo 026491
Bellanová, S. 038486
Bellar, Gail 049363
Bellas, Gerald M. 047748
Bellaton, Jean-Louis 015331
Bellavia, Kim 050000
Bellazzi, Dr. Pietro 028382
Bellec, Henri 005625
Bellehumeur, Alain 005420
Bellenger, Sylvain 047808
Beller, Janet 050507
Bellerby, Greg 007280
Bellettini, Dr. Pierangelo 025530, 025537,
025540
Belli, Dr. Gabriella . . 027681, 027684,
028171
Belli, Larry 049968
Bellier, C. 004091
Bellifemine, Graziano 025521
Bellin, Ines 017086
Bellinger, Donald F. 049865
Bellini, Lucette 012314
Bello Dièguez, José M. 039318
Bello Morales, Marinelly 054714
Bellod, Anne 014692
Belloni, Luigi Mario 027125
Bellucci, Alberto Guillermo . . . 000259
Bellucci, Dr. Bruno 027786
Belluscio, Lynne J. 050197
Bellwald, Tony 041615
Belmondo, Dr. Rosalba 027880
Belmonte, Patty 051757
Bělochová, Olga 009614
Belonovič, Éleonora Nikolaevna . 037844
Belonovich, Galina Ivanovna . . 037145
Belotti, Enrico 025273
Belotti, Dr. Ione 025951
Beloubek-Hammer, Dr. Anita . . 017209
Belousova, Natalj Aleksandrovna 037126
Belov, Vladimir Ivanovič 038050
Belova, N.H. 037802
Belser, Eduard J. 041275
Belsham, J. W. 045845

Belsley, Kathy 051988
Belt, Virginia 046664
Belting, Dr. Isabella 020551
Beltrán Figueredo, Martín 008726
Beltrán Lloris, Miguel . 040448, 040449
Beltrán, Carlos 031014
Beltran, Josel 034560
Beltrani, Francesco . . . 026729, 026730
Beltz, Dr, Johannes 042221
Belušić, Andreja 009036
Belver, Manuel H. 039679
Belvin, Robert 051243
Bem, Marek 035404
Beman, Lynn S. 046455
Bemis, Todd 053298
Bemis, William 046452
Bemporad, Dr. Dora Liscia 026290
Ben-Barak, I. 025133
Ben-Chaôbane, Abderrazzak . . . 031781
Ben-Gal, Dr. Ely 025123
Ben-Nuftali, Aya 025217
Ben, Dr. Arie 025201
Benachetti, Sandra 016103
Benalioua, Selim 013888
Benaron, Su 053823
Benavides Becquis, Maria del
Carmen 008601, 008776
Benayahu, Dr. Y. 025214
Benbow, Colin 004190
Benda, Susanne 053773
Bender-Bardenhagen, Claudia . . 018044
Bender, Joseph 053476
Bender, Patricia L. 048557
Bender, Phil 048272
Bendersky, Lydia 054090
Bendig, William C. 049743
Bendiksen, Katriina 033847
Bendixen, Maren 019336
Bendixen, Uwe 019336
Bendyukova, Natalia
Oleksandrivna 043291
Benedetti, Prof. Laudomia 026738
Benedetti, Sergio 024900
Benedetti, Susan 053171
Benedict, Andreas 018064
Benedict, Hope 052908
Benedik, Dr. A. Christian 002931
Benedik, Yvonne 003053
Beneš, Dr. Luděk 009699
Benesch, Dr. Evelyn 002944
Benetti, Attilio 028288
Benezra, Neal 053025
Bengston, Rod 046333
Bengtson, Keith 049477
Bengtson, Prof. S.A. 040815
Bengtsson, Bengt 040984
Bengtsson, Eva-Lena 041011
Benington, Jon 043528
Benitez-Johannot, Purissima . 034504,
034505
Benítez, Antonio Maria 008797
Benitez, Moya 034523
Benito Goerlich, Dr. Daniel 040313
Benito, Fernando 040308
Benjadol, Piyaluk 042650
Benjamin, Bart 052586
Benjamin, Brent 052818
Benjamin, Jeremy 051372
Benjamin, Dr. Tritobia H. 054122
Benke, Arnold 016861
Benke, Dr. Harald 022009
Benke, Uwe 017104
Benn, Carl 007187
Benner, Sue 000887
Bennerhag, Carina 040804
Bennet, Alex 053230
Benneth, Dr. Solbritt 041046
Bennett, Alan 044216
Bennett, Betsy 049171
Bennett, Dr. Betsy 052452
Bennett, Charles 053116, 053118
Bennett, Christobel 001526
Bennett, Dorthy E. 047373
Bennett, Fernanda 052685

Index of Persons
Bennett, Graham
– Beuing, Dr. Raphael

Bennett, Graham 046026
Bennett, Harold D. 047373
Bennett, Isabel 024827
Bennett, James 049104
Bennett, James A. 045455
Bennett, Kate 052620
Bennett, Linda 045184
Bennett, Michael 047808
Bennett, Michael W. 050355
Bennett, Dr. Nicholas 044829
Bennett, Sally B. 051069
Bennett, Sandra 046637
Bennett, Sharon 047543
Bennett, Shelley 053053
Bennett, Swannee 050335
Bennett, Tom 046469
Bennett, Tony 003199
Bennett, Valerie 046699
Bennett, William P. 049461
Benninghoff, Wolfgang 041716
Bennington, Jeanette 049529
Bennison, B. 045267
Bennyman, James C. 053865
Benoist, Guillaume 013445
Benoit, Jean-Louis 016338
Benova, Katarina 038328
Beňovský, Jozef 038395
Benscheidt, Dr. Anja .. 017607, 017609
Bensen, Pat 053154
Benson, Barbara E. 054435
Benson, Benjamin Foley 053139
Benson, Claire 046595
Benson, Dave 005604, 006470
Benson, Joanne 045176
Benson, Lisa 033852
Benson, Mike 044621
Benson, Patricia 048520
Benson, Tammy 048823
Bentabet, Mohamed 000041
Bente, Prof. Dr. Klaus 020001
Bentheim, Oskar Prinz zu 016719
Bentini, Dr. Jadranka 026175
Bentley, Alan 044547
Bentley, Jane 052979
Bentley, Rachel 045461
Bentz, Abigail L. 048260
Bentz, Sylviane 015128
Benvenuti, Prof. Carlo 027353
Benvie, Rachel 045293, 045294
Benyahia, Boualem 000069
Benyovsky, Lucija 009036
Benz, Edward 047097
Benz, Paul 021591
Benz, Robert G. 054532
Benzel, S. 019620
Benzner, Sonja 016440
Beran, Antonia 018639
Beran, Helmut 016999
Beran, Pavel 009888
Beránek, Petr 009740
Beránková, Dr. Helena 009468
Beránková, Dr. Jana .. 009557, 009662
Bérard-Azzouz, Odile 011537
Béraud, Isabelle 013048
Berbion, Phillipe 012108
Berchoux, Pierre 012256
Berchtold, Adriana 042052
Berchtold, Marianne 041315
Berdjajeva, Tatjana Jurjevna ... 037883
Bereczky, Dr. Loránd 023518
Berendsen, Th. J. 032248
Berens, Ingo 016981, 017000
Berents, Dr. Catharina 018718
Beresford, P. 046151
Bereska, Norbert 020641
Bereza, Vasyl Panasovyč 043172
Berežanskij, Aleksandr
Samuilovič 037198
Berg Lofnes, Solveig .. 033940, 033942
Berg, Bettina 017587
Berg, Dan 047000
Berg, Hans-Jürgen 021944
Berg, Kaare 033834
Berg, Karin 033861
Berg, Kristian 041005, 041043
Berg, Manfred 017034
Berg, Peter 016854
Berg, Ralph B. 053034
Berg, Dr. Richard B. 047336
Berg, Roger 049896
Berg, Prof. Dr. Stephan 017495
Bergamaschi, Cirillo 026496
Bergamini, Dr. Giuseppe 028225
Bergbauer, Bertrand 012855
Bergdahl-Nörby, Kerstin 040969
Bergé, Pierre 014108, 031781
Bergen, Terry . 046513, 047491, 049904
Bergenthal, Dr. T. 019430
Bergeon, Annick 012112
Berger-Klingler, Martina 002630, 002631
Berger, Aaron 046338
Berger, Andrea 001781
Berger, Dr. Andreas 019386
Berger, Claudio 041745
Berger, Erwin 016779
Berger, Dr. Eva 021015
Berger, Dr. Frank 018428
Berger, Georg 018121
Berger, Ingrid 033919
Berger, Jerry A. 053468
Berger, Karin 016779
Berger, Mag. Karl 002142
Berger, Michael 016578
Berger, Moshe Tzvi 025092
Berger, Paul 041343
Berger, Shelly 054390
Berger, Teresa J. 050855
Berger, Theodor 020138
Berger, Tobias 033140
Berger, Torge 016779
Berger, Dr. Ursel 017148
Berger, Vicki L. 053699
Bergeret, Anne-Marie . 013268, 013269,
013270, 013271, 013272
Bergeret, Jean 013829, 015337
Bergero, Susana 000340
Bergeron, André 006390, 006531
Bergès, Catherine 011537
Berges, Sylvie 012045
Berggård, Ingeborg 034050
Bergggruen, Oliver 018446
Berggren, Jan 041054
Bergh, Susan 047808
Bergheger, Brian F. 048562
Bergholm, Synnöve 011275
Berghout, Dr. P.-J. 032279
Bergiers, Marcel 004128
Bergil, Catharina 040663
Bergin, Christer 040620
Bergman, Ingela 040536
Bergman, Kaj 040639
Bergmann, Eva-Maria 020811
Bergmann, Jóhanna 024073
Bergmann, Prof. Marianne 018757
Bergmann, Sabine 017338
Bergmann, Sven 018032, 018033
Bergroth, Tom 011449
Bergs, Hans 022678
Bergs, Lilita 052652
Bergstedt, Dr. Clemens 022878
Bergsten, Wayne 050929
Bergström, Anna-Lena 040888
Bergström, Anne 011232, 011233,
011234
Bergström, Dr. Carin .. 040960, 040976,
040981, 041003, 041012, 041022,
041034, 041037, 041040
Bergström, Eva-Lena 041024
Bergström, Lars 040631
Bergström, Matti 010891
Bergstrøm, Ture 010185
Bergundthal, Hermann 041587
Beringer, Emile 012974
Berke, Debra 054167
Berkel, R.R. 032932
Berkes, Peter 041277
Berkey, Vernon 053383
Berkhahn, Bernhart 019961
Berkmen, Çetin 042857
Berkovitz, Sam 052175
Berlichingen, Hans Reinhard Freiherr
von 019441
Berliet, Paul 013932
Berlin, K.W. 001190
Berliner, Nancy 052880
Bērliņš, Guntis 030139
Berlmont, Karleen 053946
Berman-Hardman, Richard 001280
Berman-Miller, Cynthia 049513
Berman, Jerôme 046972
Bern, José Manuel 034176
Berná Garcia, María T. 039008
Bernabò Brea, Dr. Luigi 028054
Bernabò Brea, Dr. Maria 027211
Bernard, A. 003471
Bernard, Charles 012989
Bernard, Christian 041527
Bernard, Goetz 014234
Bernard, Jean 012761
Bernard, Loic 006192
Bernard, Marie-Thérèse 015616
Bernard, Michel 003604, 012938
Bernard, Raymond 014395
Bernardi, Dr. Vito 028071
Bernardini, Dr. Carla 025569
Bernardini, Dr. Maria Grazia ... 026849
Bernardini, Paolo 027838
Bernardinis, Domenico de 041560
Bernasconi, John G. 044717
Bernat, Clark 006491
Bernaudeau, Chloé 013826
Bernauer-Keller, Christine 017808
Berndes, Claske 054075
Berndt, Harald 021201
Berndt, Randall 050596
Berndt, Siegfried 022552
Berner, Dr. Hermann 020447
Bernesconi, César 046295
Bernhard, Friedolin 018708
Bernhardt, Harold O. 047458
Bernhardt, Marcia A. 047458
Bernhausen, Sara 017102
Bernier, Ronald R. 054407
Bernier, Suzie 005644, 006494
Berninger, Gudrun 019647
Bernini, C. 042236
Bernklau, Maria 003053
Bernnett, James 000762
Berns, Christiane 030593
Berns, Maria C. 050422
Bernsmeier, Dr. Uta 017585
Bernstein, Bruce 051448
Berntsen, Gunnar 033925
Beron, Prof. Petar 005198
Beroza, Barbara L. 054596
Berraute, Luis Bernardo 000582
Berres, Marge 051664
Berreth, David S. 048951
Berrettoni, Giulio 027123
Berridge, Peter 043988
Berrigan, Douglas 006153
Berrin, Kathleen 053009, 053033
Berry, Barbara 054238
Berry, Bridget 001365
Berry, D.L. 005458
Berry, Diane 048467, 048469
Berry, Ian 044038, 053161
Berry, James M. 049787
Berry, Susan 053338
Berry, Warren 044095
Berryman, Jill 052512
Berryman, Val Roy 048441
Bersirov, A.M. 037217
Berswordt-Wallrabe, Prof. Dr. Kornelia
von 020144, 021796
Bert, James 053966
Bert, Putter 046889
Bertalan, Karin 020957
Bertani, Licia 026242
Bertaud, Henri 011616
Bertazzolo, Prof. Luigi 027667
Berthaud, Gérard 014106
Bertheaud, Michael 054192
Berthelier, Jean 014008
Berthelier, Nadine 012471
Berthelsen, David 001092
Berthod, Dr. Bernard 013935
Berthold-Hilpert, Monika 018548, 021690
Bertholi, Jean 015563
Berthramieux, André 014252
Berti, Dr. Fausto 026947
Bertie, Dr. David M. .. 044379, 045279,
045492
Bertini, Dr. Pierluigi 026676
Bertling, Dr. Markus 020597
Bertola, Carinne 041806
Bertoletti, Marina 027581
Bertolini, Madalen 053496
Bertolucci, Serena 028168
Bertout, Ernest 014009
Bertram-Neunzig, Dr. Evelyn ... 019710
Bertram, Dr. Marion 017245
Bertrand-Sabiani, Julie 014585
Bertrand, Gérard 014426
Bertrant, Albert 004070
Bertsch, Patricia 051108
Bertschi, Victor 041544
Bertucio, Barbara 050304
Bertuleit, Dr. Sigrid 021783
Bertus, Jean-Paul 014014
Beruete Calleja, Francisco 039387
Berup, Christina 040739
Berwick, Hildegarde 045270
Beryt, Andrzej 035004
Besch, Prof. Eckart 019126
Bescós, Alberto 040043
Beširov, Ilija 005266
Beskrovnaja, Nina Sergeevna .. 037141
Besom, Bob 053447
Bessborough, Madeleine 044213
Besse, Nadine 015302
Besse, Pamela 049292
Besseiche, Daniel 016202
Bessette, Yvonne 006077
Bessir, Mark H. C. 050232
Besson, Christine 011655
Bessone, Dr. Silvana 035814
Bessor, Joyce M. 054615
Best, Sherry L. 053769
Bešťáková, Dr. Kamila 009728
Besteher, Axel 017159
Besten, G.J. den 032698
Beštok, Chabas Korneevič 037533
Bestrom, Beth 052424
Betancaert Nuñez, Dr. Claudia
Gabriele 019011
Betancourt, Julián 008606
Betancur, Víctor Raúl 008571
Betanio, Nemesio V. 034555
Bétemps, Alexis 027701
Bethell, Chris 045038
Bethell, Philip 046212
Bethge, Dr. Ulrich 020013
Bethmann-Hollweg, C. von 016537
Betrim 031839
Betschart, Madeleine 041341
Betscher, Hermann ... 020051, 020052
Betsky, Aaron 032816
Betta, Dr. Carlo 025883
Bettendorf, Josef 020421
Bettens, Doug 001412
Bettison, Dr. Cynthia Ann 053339
Betton, Richard A. 049256
Betton, Thomas 028438
Betts, Jonathan 044337
Betts, Shirley 006587
Betts, Wendy 005841
Betz, Dr. Astrid 020847
Betz, Eden 051717
Betz, Julia P. 051888
Betz, Martin 050609
Beu, Liliana Marinela 036646
Beuchat, Dr. Paul 041782
Beuf, Jean-Claude ... 015509, 015510
Beugeling, Dr. Niels 031977
Beuing, Dr. Raphael 003078

Beukers, Prof. H.	032579
Beukes, Engela	038843
Beumer, A.C.	033100
Beumers, C.E.L.	032830
Beurskens, N.	032372
Beuster, Stephanie	021575
Beutelspacher, Prof. Dr. Albrecht	018693
Beutelspacher, Martin	020418
Beutl, Richard	002300
Beuze, Lyne-Rose	030779
Bevan, Dr. Alex	001403
Bevan, Alison	045482
Bevan, Jennifer	001000
Beverly, Roberta	051716
Bevers, Dr. Holm	017209
Bevilacqua, Jean-Paul	012392
Bevilecqua, Emanuele	027656
Bex, Florent	003299
Beyeler, Christophe	013013
Beyeler, Ernst	041879
Beyer, Patrick	006540
Beyer, René	042239
Beyer, Dr. Roland	017950
Beyer, Teresa	050146
Beynon, Jon	045957
Bezbach, Illja Vasil'ovyč	043320
Bezemek, Dr. Ernst	002102
Bezerra Rolim, Valéria Laena	004463
Bezin, C.	016084, 016085
Bezlaj Krevel, Ljudmila	038562
Bezzola, Dr. Tobia	042207
Bhan, Kuldeep K.	024509
Bhandari, Dr. J.S.	024218
Bhargava, P.C.	024179
Bhat, Prof. H. Krishna	024505
Bhatt, Vikas	024170
Bhattacharya-Stettler, Dr. Therese	041318
Bhaumik, P.K.	024234
Bhave, Dr. S.M.	024443
Bhering Cardoso, Marcio	004696
Bhikul, Tassanee	042686
Bhreathnach-Lynch, Dr. Sighle	024900
Bhullar, Akbar Ali	034136
Biagini, Lorena	027777
Biah, C. Calixte	004181
Bialas, Sebastian	018063
Bian, Xisuo	008034
Bianchi, Loredana	041301
Bianchini, Dr. Gregorio	028182
Bianchino, Dena	050845
Bianco, Dr. Anna Lo	027566
Bianco, Prof. Vito Donato	027720
Biancolini, Daniela	025246, 028143
Bianconi, Teresa	027624
Biasini, Cathérine	021931
Biass-Fabiani, Sophie	011836, 011842
Biba, Prof. Dr. Otto	002997
Bibas, Dr. David	050412
Bibby, Miriam	044758
Biberin, Vladimir Nikolaevič	037974
Bibler, Helen Ann	049322
Bibra, G. von	001234
Bicher, Kristina	052717
Bichot, Denise	012481
Bickel, John	043442
Bickel, Susanne	041496
Bicker, Roxane	020578
Bickford, Christpher P.	049987
Bickhoff, Dr. N.	022034
Bicklmeier, Lori	005684
Bicknell, Terry	033350
Bidar, A.	014491
Biddlecombe, Julie	044627
Biderman, Ellen	053121
Bidner, Stefan	002128
Bidstrup, Wendy	050691
Bidwell, John	051428
Bidwell, Paul	045819
Bieber, Susanneh	048979
Biebl, Klaus	020871
Biedenharn, Murray	050996
Biefel, H.	018226
Biefel, I.	018226

Biegel, Prof. Dr. h.c. Gerd	022559
Biegel, Dr. Gerd	022730
Biegel, Susanne	016677
Bieger, Dr. Alfons	041219
Biehl, Martin	033840
Bielamowicz, Zbigniew	035145
Bielański, Janusz	034894
Bielby, Jane	045808
Bielejec, Stephanie	046652
Biemans, Johan	032064
Biemans, Prof. Dr. Jos A.A.M.	031984
Bienedell, Arturo Alberto	000576
Biener, Herbert	001956
Bieńkowski, Joachim	035309
Bier, F.	012866
Bierens de Haan, J.C.	032560, 032960
Bieri Thomson, Helen	041848
Bieri, Susanne	041323
Biesboer, Dr. P.	032392
Biesen, Alden	003365
Bieske, Dr. Dorothee	018372
Biezaitis, M.	000896
Biffen, Lyndia	033290
Biffiger, Steffan	042029
Biganti, Dr. Tiziana	027256
Bigay, Victorino	034426
Bigda, Daisy R.	047560
Bigeck, Heather	052129
Biggart, Norman	047321
Bigger, Mary Beth	046692
Biggs-Craft, Katherine	006894
Bigham, Julia	046181
Bignasca, Dr. Andrea	041261
Bihari, József	023499
Bihler, A.	018126
Bijelić, Borislav	008896
Bijelić, Dubravka	038168
Bijleveld van Lexmond, Caspar	041612
Bijleveld, W.	031972
Biktaševa, Natalija Nikolaevna	037698
Bilanoha, Ewald	003089
Bilashini Devi, Dr. Haobam	024188
Bilek, Alena	002938, 003037
Bilenduke, Florence	006622
Biles, LeAnna S.	048878
Bilgin, Mehmet	043047
Biličenko, Valentina Andreevna	037761
Bilich, Ivanka	008926
Bilici, Mehmet	042890
Billa, Lawrenece	038739
Billard, Marie Claude	011833
Billerbeck, Barb	051510
Billich, Dorie	005895
Billier, Philippe	015718
Billig, Pamela	047226
Billing, Jerome	052817
Billings, Doug	054386
Billings, Kathy	046327
Billings, Loren	047680
Billings, Vernon N.	050133
Billingsley, Pam	052688
Billisics-Rosenits, G.	001861
Billops, Camille	051403
Bills, Danny	054398
Billups, Di	045638, 045640
Biloparlović, Ivo	008979
Biloslav, Toni	038528, 038586
Bilyk, Zorjana	043242
Bin Ismail, Halal	030690
Binaud, Daniel	012067
Binder-Krieglstein, Dr. Monika	001985
Binder, Christian	018510
Binder, Doug	044500
Binder, Fred	054569
Binder, Gerhard	022848
Binder, Hanspeter	041833
Binder, Stefi	041254
Binder, M.A. Tanja	020249
Bindija, Mosar	038193
Bindley, Kate	043747
Bines, Dr. Joan P.	054349
Bing, Jens	010066
Bingham, David S.	048860
Binikowski	022446

Binkert, Almut	022541
Binkley, David	054142
Binnendijk, G.A.	032949
Binon, Erik	004004
Bins, Coleen	048489
Binti Sirin, Dayang Hajah Zasia	004963
Bintz, Carol	053760
Biran, Prof. Avraham	025109
Biraud, Guy	011649
Birchall, Heather	045219
Bircher, Prof. Dr. Martin	041421
Birchwood-Harper, Barbara	044208
Bird, David M.	048534
Bird, Gill	044005
Bird, Maggie	045036
Bird, Richard	047508
Bird, William	048573
Birdick, Birdy	051774
Birdseye, William S.	054107
Bireley, Ellen K.	049691
Birk, Sherry C.	054155
Birkebæk, Frank A.	010289, 010291, 010292
Birkenstock, Eva	020179
Birker, Susanne	019029
Birklid, Richard	048862
Birkmaier, Leslie E.	047201
Birkner, T.	017155
Birley, Robin	043493
Birman de Bessudo, Raquel	031204
Birnbaum, Dr. Daniel	018444
Birnie, Ian	050444
Bíró, György	023520
Birol, Çiğdem	042937
Birthälmer, Dr. Antje	022825
Birtwhistle, Marjorie	054463
Bis-Worch, Christiane	030593
Bisanz-Prakken, Dr. Marian	002931
Bisbee, Fay	049762
Bischler, Gaston	018947
Bischof, Amy	049454
Bischof, Klaus	022054
Bischof, Petra	020540
Bischofberger, Emil	041216
Bischoff, Dieter	021919
Bischoff, Dr. Franz	017202
Bischoff, Dr. Harald	018238
Bischoff, Dr. Michael	020031
Bischoff, Dr. Ulrich	017982
Bischoff, Walter	022854
Bisconti, Chino	027412
Bisetto, Christine	049723
Bishof, John	053261
Bishop, Adrian	043883
Bishop, Art	052972
Bishop, Dr. C.	043491
Bishop, Eliza H.	048068
Bishop, Gale A.	052472
Bishop, Henry	005679
Bishop, Janet C.	053025
Bishop, Kenny	048924
Bishop, Knox	048117
Bishop, Dr. Michael	045365, 045366
Bishop, Norman	053089
Bishop, Ron	048851
Bishop, Russell J.	053311
Bishop, Tommy	053892
Bisig, Xaver	042126
Biskupić, Margareta	009016
Bismarck, Prof. Dr. Beatrice von	019985
Bisol, David	053091
Bissell, James	047809
Bissett, George Ann	048395
Bissière, Caroline	014159
Bisson, Gäétan	006132
Bissonette, France	007268
Bissonnette, Anne	049933
Bissonnette, Daniel	007309
Biström, Prof. Olof	010818
Bitar, Dr. T.K.	024514
Biswas, Dr. T.K.	024514
Bitler, Dorothea	052451
Bitomsky, Michael	017120
Bitsch, Dr. Helmut	021312
Bitter, Adriana	050383

Bitter, Edwin	050383
Bitterli, Konrad	041923
Bittner, Jan	001338
Bivens, Peg	050002
Bivingou-Nzeingui	008845
Bizjak, Vjekoslav	008961
Bizzarri, Laura	025449
Bjarnason, Páll V.	024092
Bjarnesen, Lynne	048147
Bjelić, Dobrila	004261
Bjerg, Ida	010057
Bjerkestrand, Steinar	033633, 033634
Bjerregaard, Jørgen O.	010018
Bjerregaard, Kirsten	010036
Bjerring-Hansen, Jacob	010166
Bjerring Jensen, Knud	010300
Bjerrkjar, Anne	010143
Bjørgen, Asle B.	033748, 033749
Björk, Barbro	041066
Björklund, Anders	040998, 041039
Björklund, Prof. Kjell R.	033847
Björkman, Anna	040620
Björkman, Sten	010771
Björkroth, Maria	040620
Björn, Eva	041113
Björnemalm, Barbro	041118
Bjork, Alan	052696
Bjork, Tim	052130
Blaas, Karl	042130
Blaauw, J.P.	032572
Blacharska, Wiktoria	034739
Black Petersen, S.	010102
Black, Barbara	053383
Black, Bettye	050132
Black, Bill	001447
Black, Prof. C.M.	045051
Black, George B.	051164
Black, Jack	048712
Black, Jane A.	048171
Black, Jay	054022
Black, Joe	049180
Black, Kenneth N.	052643
Black, L.	001215
Black, Peter	044401
Black, Robin	046305
Black, William N.	048872
Blackaby, Anita D.	049372
Blackburn, Dr. Bob	046428, 046799, 048321, 048401, 048588, 048811, 049298, 049299, 049426, 049427, 049975, 051737, 051951, 052907
Blackburn, M.A.S.	043842
Blackburn, Rachael	049884
Blackett, Diane	044345
Blackman, Barbara W.	052960
Blackman, Joni	049786
Blackman, Leona	048757
Blackwelder, Martha	052949
Blackwood, Frank E.	005923
Blackwood, R. Duke	053346
Bladen, Martha	053972
Blades, John M.	051892
Blades, Julian	046187
Blades, Kent	005555
Bladt, Inger K.	009998
Blæsild, Benno	010039
Blättler, Alexandra	042147
Blaettler, Roland	041528
Blagg, Margaret	046350
Blagg, Tim	049245
Blagov, Jurij Alekseevič	037121
Blahuta, Renee	048663
Blair, Allan	006472
Blair, David	050422
Blair, Kay	054383
Blair, Robert	001455
Blair, William	043478
Blaise, Nicolet	041724
Blajan, Ion	036320
Blakar, Antony	016031
Blake Roberts, Gaye	045886
Blake, Leo D.	047381
Blake, Paul	046213
Blake, Ruby	052708

Blakeley, Erik 044826
Blakey, Ellen Sue 053730
Blanc, Jean 013498
Blanc, Micheline 015940
Blanc, Monique 014733
Blanchard, Cathérine 014601
Blanchard, Edith 047490
Blanchard, Marian 033284
Blanchard, Raoul 041572
Blanchard, Vincent 011956
Blanchebarbe, Prof. Dr. Ursula . 021854
Blanchegorge, Eric . . . 012643, 012644,
. . . . 012719
Blanchette, Normand 005720
Blanchette, Odile 045713
Blanchy, Béatrice 031721
Blanck, Denis 016348
Blanco Sansa, Helena 039098
Blanco, Bobby 034607
Blanco, Cristina 039647
Blanco, Hipólito S. del 000356
Bland, Bart 053505
Bland, Bruce F. 048421
Bland, Gaye K. 052662
Bland, Julia W. 051297
Bland, Sharon 050202
Blank, Dr. Claudia 020518
Blank, Maxine 049990
Blanke, Dr. H. 018944
Blankenfeld, Ruth 021663
Blankenship, Blanton 049294
Blankenship jr, James H. 052016
Blankenship, Julie 002016
Blankenship, Ken 047607
Blanquez, Juan 040073
Blanshard, Catherine 044778
Blanshard, Emily 045892
Bláquez, Jimena 040360
Blas Pascual, Javier 040061
Blas, Javier 039605
Blaschke, Inge 042093
Blaschke, Ursula Ruth 018950
Blaschke, Wolfgang 016583
Blasco, Santiago Mateu 039392
Blass, Hillary 051491
Błaszczyk, Bożena 035167
Blatchford, Ian 045123
Blatt, Martin 047105
Blattner, Dr. Evamaria 022223
Blatz, Hermana 006703
Blauert, Elke 017202
Blaugrund, Dr. Annette 051444
Blavia, Francisco 054705
Błażewicz, Sławomir 035342
Blazewiczea, Stanisław 034717
Blazwick, Iwona 045134
Blazy, Guy . . . 013938, 013944, 015483
Blazy, Simone 013946
Bleakley, Bruce 052970
Bleathman, William 001150
Bled, Max 015842
Bleecker Blades, Margaret 048890
Bleiberg, Edward 047253
Bleiberg, Elizabeth W. 050564
Bleibler, Jürgen 018528
Bleich, Gerhard 017685
Bleile, Dr. Ralf 021663
Bleistein, Joachim 020866
Blendea, Ion 036492
Bles, H. de 032207
Blesa de la Parra, Miguel 039691
Bleser, Maleen de 003289
Blesser, Carol 046898
Blessing, Jennifer 051470
Blethen, H. Tyler 048087
Bley, Karl 020812
Bleymehl-Eiler, Dr. Martina 016873
Blicher-Moritz, Tine 010188
Blicke, Linda 047295
Blickle, K.H. 019121
Blickle, Ursula 019778
Bliemeister, Jeffrey 054256
Blinderman, Barry 051584
Blinov, Viktor Filippovič 036886

Bliss, Shannon 052568
Blitz, Amy 051393
Bljudova, Ljudmila Grigorjevna . 036973
Bloch Ravn, Thomas 010006
Bloch, Dorete 010742
Blocher, Miriam 041288
Block, Christian 012112
Block, Diana 048249
Block, Isabel 050845
Block, Merle 049167
Block, Phillip 051406
Blöchlinger, Peter 042090
Bloem, H. 032306
Bloem, M. 031986
Bloemsma van Nes, Lauren . . . 053696
Blohm Pultz, Janet 048208
Blok, R.S. 014635
Blok, Dr. W. 033017
Blom, J. 032088
Blom, P. 032978
Blomberg, Dr. Katja 017158
Blomberg, Maj-Gun . . . 040622, 040623
Blomberg, Nancy 048258
Blome, Jens 019218
Blome, Prof. Peter 041261
Blommaert, André O. 003709
Blommaert, Georges 003701
Blommel, Henry H. 047964
Blomquist, Fredrik 040888
Blomstedt, Severi 010866
Blond, Anne 010343
Blondel, Madeleine . . . 012783, 012784
Blondin, Linda 007257
Bloodgood, Craig 048414
Bloomfield, Sara J. 054168
Bloor, Paul M. 054276
Blosser, Jack 051799
Blount, Robert 053670
Blue Stanfield, Linda 051081
Blue, John 007085
Blübaum, Dr. Dirk . . . 018528, 021792
Blühm, Dr. Andreas 019715
Blümel, Karin 042199
Bluhm, Hans-Georg 019576
Blum, Dilys 052081
Blum, Jürgen 019357
Blume, Dr. Eugen 017154
Blume, Peter F. 051113
Blume, Sharon 053484
Blumenthal, Arthur R. 054498
Blumenthal, Prof. Dr. W. Michael 017183
Blumentritt, R. 021639
Blundstone, Judith 001073
Blustain, Malinda S. 046481
Bluthardt, Robert 052924
Blyth-Hill, Victoria 050444
Blyth, Mary 045785
Boa, Valerie N.S. 044482
Boal Lee, Mathilde 047063
Boal, Robert 051687
Boano, Giovanni 025773
Boardman, Penelope 047662
Boarman, Don 049438
Boarowy, Innocenta 035298
Boas, Heinz 021512
Boas, Natasha 053144
Boase, Colin 001615
Boast, Mair 045539
Boatright, Neva 048281
Bobar, Maja 004282
Bober, Anna 034659
Bober, Jonathan 046693
Boberg, Scott 050145
Bobinger, Wanda L. 050347
Bobkina, Tatjana Gennadjevna . 038011
Bobkov, Konstantin Vasilevič . . 037222
Bobovec, Ana 008943
Bobzieri, Geraldine 001353
Bočarov, Lev Nikolaevič 038036
Bocca, Massimo 025941
Boccanera, Prof. Giacomo 025715
Boccardo, Dr. Piero 026426
Bocchieri, Prof. Franco 028202
Boccolari, Prof. Giorgio 026863

Bochmann, Deb 046415
Bochnakov, Nikola 005135
Bock, Claudia 022349
Bock, M.A. Martin 020744
Bock, Michael 022049
Bock, Dr. Robert 018406
Bockelman, James 053278
Bockius, Dr. Ronald . . 020231, 020233
Bockrath, Dr. Cornelia 002192
Bockstefl, Mag. Gregor-Anatol . . 002267,
. . . . 002269
Bockwoldt, Michael 048190
Bocquier 012151
Bocvarova-Plavevska, Marika . . 030630
Bod, Roy 047695
Bode-Hempton, Sharon 050152, 050154
Bode, Monika 018965
Boden, Karen 050650
Bodenmüller, Carola 022043
Bodenstein, Anton 002906
Bodenstein, John Gilbert 020817
Bodilsen, Inge 010072
Bodine jr., William B. . 052156, 052164
Bodingbauer, Prof. Adolf Karl . . 002785
Bodini, Prof. Arturo 026795
Bodini, Prof. Floriano 025785
Bodini, Mauro 027983
Bodish, Glenn 047334
Bodner, Connie 051112
Bodner, Helen 006475
Bodó, Dr. Sándor . . . 023466, 023501
Bodoky, Dr. Tamás 023472
Bodorová, Olga 038435
Bodrova, Zoja Evgenjevna 037846
Bodsch, Dr. Ingrid . . 017488, 017502
Böcher, Friederike 016803
Böcher, Hans-Georg 019133
Böck, Elfriede 019033
Boeck, Lorraine 053490
Böckle, Karlheinz 021879
Bödecker, Ehrhardt 022832
Bödewadt, Uwe 041009
Bøe, Kaare 033517
Bøge Henriksen, Kirsten 010281
Böge, Rainer 016513
Boeggild Johannsen, Dr. Kristine 010198
Boeglin, Yves 015927
Böhle, Hans-Theo 017874
Böhle, Rainer 016851
Böhler, Dr. Bernhard 002981
Böhm 021417
Böhm, Dr. András 038484
Boehm, Barbara 051364
Böhm, Helmut 002985
Böhm, Klaus-Dieter 016888
Böhme-Netzel, Susanne 022779
Böhme, Dr. Ernst 018758
Boehme, Dr. Sarah E. 047855
Böhme, Prof. Dr. Wolfgang 017504
Böhmer, Christiane 016926
Böhmer, Otmar 010832
Böhmer, Sylvia 016445
Böhnert, Marlies M. 018410
Böhnke, Bärbel 018164
Böhnlein, E. 020813
Boelen, Jan 003684
Boelens, Christ 032875
Böller, Susanne 020580
Bömerová, Jana 009804
Böndel, Prof. Dr. Dirk 017105
Bönig, Dr. Jürgen . . . 019006, 019022
Böning, Heinrich 021241
Bönnen, Dr. Gerold 022763
Bönninghausen, D.E.J. von 031964
Boer, A. den 032267
Boer, Dr. Anne-Marie 032181
Boer, F.G. de 032723
Boer, P. den 032239
Börjesson, Klas 040728
Boero, Prof. Ferdinando 027415
Boers, T. 031962
Börschlein, Eva-Maria 040024
Börste, Dr. Norbert . . 021058, 021062
Böse, Hans-Joachim 022699

Böser, Wolfgang 018697, 021990
Boesmans, Annick 003618
Böth, Dr. Gitta 018917
Böthig, Dr. Peter 021400, 021402
Boëthius, Lena 040652
Böttcher, Andreas 017564
Boettcher, Chris 053018
Boettcher, Claudia 004642
Boettcher, Graham C. 047005
Böttcher, Dr. Ronald 022056
Böttner, Waltraud 019292
Böttrich, Prof. Dr. Christfried . . . 018807
Boffa, Prof. Giovanni 041198
Bofill, Teresa Blanch 040024
Bogan, Alan T. 048849
Bogan, Dr. Michael A. 046365
Bogardus, Lorraine B. 051068
Bogatyrev, E. 037295, 037337
Bogdanov, Pavel 037250
Bogdanova, Lidija Aleksandrovna 037972
Bogdanova, Marija 004987
Bogdanovna, Vera Ivanovna . . . 037890
Bogdanowicz, Prof. Wiesław . . . 035339
Boggio, Stefano 028271
Boggs, Glen 054524
Boggs, Mac 053428
Boggs, Rex 048792
Bogman, Suzanne 031996
Bognanni, Filippo 027306
Bogner, Peter 003019
Bogoljubskaja, A.V. 038071
Bogurat, W. 035275
Bogus, Vasile 036712
Bogusky, Alfred M. 006125
Boguszewicz, Ireneusz 034672
Bogutz, Ruth 049303
Boguwolski, Ryszard . . 034774, 034775,
. . . . 034776
Boheme, Prof. Dr. Enrico 027651
Bohlmann, Dieter-Theodor 019470
Bohlmann, Dr. Heinz 021772
Bohnenkamp-Renken, Prof. Dr.
Anne 018421
Bohnert, G. 019853, 019855
Bohórquez, Jorge Alfonso 008704
Bohr, G. Ted 051762
Boie, Charles 050920
Boileau, Nanette E. . . . 050484, 052812
Boillot, M. 014146
Boire, Janine 047074
Bois, M. de 032035
Boissard, Jerome de 013089
Boisselle, Lynn 006631
Boissevain, C.F.C.G. . . 032217, 032218
Boissière, Yves 013058
Boitani, Dr. F. 027633
Boits, Paul 003308
Bojani, Prof. Gian Carlo 027264
Bojarczuk, Helena 034779
Bojarskij, Dr. Viktor I. 037817
Boje, Corinna E. 022057
Bojko, Elena Michajlovna 037893
Bojko, Nina Vladimirovna 037172
Bojkova, Irina Petrovna 038063
Bojkova, Margarita Alekseevna . 036982
Bojović, Nevenka 038188, 038188
Bojović, Radivoje 038188
Bok, Franciszek 035241
Bok, Klaus 019338
Bokbot, Dr. Youssef 031785
Bokchun, Youngja 029984
Bokum, Anne-Marie ten 003319
Bol, Marsha C. 053113
Boland, Raymond 047389
Bolandrini, Santina 041847
Bolas, Gerald D. 042185
Bolderman, J.C.L. 032026
Boldt, Erwin 016876, 016878
Bolduan, Anka 017601
Boley, Mark 050897
Boley, Shawn 047001
Bolge, George S. 044847
Bolger, Doreen 046733
Bolger, Dr. Francis 005596

Bolin, Jerry 050562
Boling, Ron 052929
Bolipata, Alfonso Corpus 034588
Bolívar, Julián Gil 008581
Bolla, Dr. Margherita . . 028345, 028350,
028351
Bolla, Stefano 041821
Bolland, Lionel 045537
Bolle, Sandrine 003741
Bollendorf, Jeanne M. 049465
Bolliger Schreyer, Sabine 041315
Bolliger, Dr. Thomas 042234
Bolling, Richard G. 049256
Bologa, Adela 036400
Bologna, Alberto 026587
Bolognani, Betty 052488
Bolohiş, Ioan 036528
Bolongaita, Prof. Johny P. 034446
Bolotov, Valentin A. 037385
Bolotova, G.P. 037729
Bolster, Jane 051323
Bolt, Ed L. 052126
Boltar, Barbara 038596
Bolten Rempt, Dr. H. 032586
Boltesilverman, Diane 048672
Bolton, Bruce D. . . 006360, 006411
Bolton, Chris 053868
Bolton, David 044312
Bolufer Marqués, Joaquim 040430
Boly-Lecocq, Anny 003957
Bolz, Dr. Peter 017112
Bombard, LuAnn 054295
Bomberger, Bruce 050123
Bommerlyn, Loren 048058
Bon, Dr. Mauro 028312
Bona Dea, Artemis 046471
Bonaccorsi, R. 013490
Bonačić Mandinić, Maja 008993
Bonafede, Irja 051025
Bonagura, Pasquale 028321
Bonami, Francesco 028114
Bonamini, Eros 030302
Bonanno, Dr. Carmela 026153
Bonansinga, Kate . . 048510, 048516
Bond-Fahlberg, Margareta 041027
Bond, Anthony 001535
Bond, Debra 051984
Bond, Hallie 047059
Bond, Karen 049059
Bond, Kaye 051995
Bond, Wanda L. 049278
Bondar, Dr. Vitalij Vjačeslavovič . 037159
Bondareva, Valentina Nikolaevna 037008
Bondaz, Gilles 015948
Bondil, Nathalie 006391
Bondjers, Rune 040620
Bone, Laurie 048749
Bone, Oliver 045967
Bonelius, Elina 011409, 011410
Bonemaison, Michel 013933
Bonet Armengol, Jordi 039107
Bonet Rosado, Helena 040310
Bonev, Ivan 005219
Boney, Roger 045432
Bongaarts, J. 032460
Bongaerts, Ursula 027556
Bongard, Rolf B. 017355
Bonham, Rebecca 053395
Bonias, Zisis 023123
Boniface, Didier 015747
Bonifay, Suzanne 015281
Bonilla, Irene 040305
Bonilla, Mary Yamilet 054765
Bonin, Flavio 038587
Bonisławski, Wojciech 034731
Bonn, Greg 050084, 052893
Bonnard Yersin, Pascale 042109
Bonnard, Jacques 014655
Bonnefille, J. Lovis 015255
Bonnefoi, Geneviève 013110
Bonnema, Hollie 049401
Bonner, Deanna 046667
Bonner, Ian 044619
Bonnet Borel, Francoise 042092

Bonnet, P. 032145
Bonnett, Dr. Richard B. 049625
Bonnevialle, Gilles 015270
Bonnier, Bernadette 003903
Bonnin, Mirta 000321
Bono, Sam 046366
Bonofiglio, Marta Maria 000538
Bonoli, Prof. Fabrizio 025546
Bonomi, Prof. Alfredo 027252
Bonomi, Dr. S. 025244
Bonstecht, Andreas 017216
Bonvalet, Louis-Georges 014100
Bonzel, Helen 018563
Boo, Drs. M. de 032949
Booknight, Alice 047912
Boomstra Van Raak, M.A.C. . . . 032569
Boonstra, Dr. J.R. 031932
Boos, Dr. Andreas 021322
Boos, Kurt 017552
Boot, C. 032929
Boot, M. 031986
Booth, Amanda 045369
Booth, Beverley 033174
Booth, C.M. 045631
Booth, Helen 006054
Booth, William H. 053758
Booty, Peter 047498
Bora, Dr. D.K. 024316
Borbolla, Dr. Sol Rubin de la . . 031243
Borchardt, Dr. Albert 019132
Borchardt, Ernstine 003588
Borchardt, Dr. Stefan 019111
Borchardt, Susan A. 050736
Borchers, Karl-Hans 022303
Borchert, hanns 019921
Borchert, Till Holger . . 003403, 003411,
003412
Borčić, Goran 049036
Bordallo Huidrobo, José Antonio 039689
Bordaz, Odile 016301
Bordeon, Tamika 051312
Bordinskich, Gennadij
Aleksandrovič 037879
Bordon, Bob 054297
Bore, Ove Magnus . . 033974, 033975,
033984
Boreham, Peter 045628
Borel-Menut 013426
Borel, Dr. Gilles 041654
Borellini, Alessia 041394
Borén, Ingjerd 040660
Borg, Prof. Dr. Vince 030740
Borgegård, Eva 041138
Borgen, Prof. Liv 033847
Borges Rosales, Prof. J.C. 054763
Borges, Grethe 033623
Borges, Richard 046920
Borges, Wolfgang 019889
Borggrefe, Dr. Heiner 020031
Borghi, Gian Paolo 026210
Borgmann, Jan 018866
Borick, Carl P. 047543
Borisenko, Irina Gavrilovna . . . 037483
Borisenko, Viktor Ivanovič 037963
Borisov, Boris 005142
Borisova, Diana 005016, 005017, 005018
Borisova, Tanja 005137
Borisova, Valentina Vladimirovna 036856
Borja-Santiago, Monina Katherina 034537
Borja-Villel, Manuel J. 039073
Börjesson, Sven 041081
Bork, Prof. Dr. Hans Rudolf . . . 021346
Borman, R.Th.A. 032027
Born, J.G.M. 032436
Bornacelli Guerrero, Humberto . 008791
Bornecrantz, Mona 041066
Bornfriend, Carl 048986
Bornhorst, Dr. Theodore J. 049568
Bornman, A.R. 038649
Bornscheuer, Dr. Marion 018064
Borobia, Mar 039686
Borodavkin, Andrej Viktorovič . . 037591
Borodina, Valentina Ivanovna . . 036957

Boroghină, Emil 036422
Borošak Marjanović, Jelena . . . 009036
Borovich, Yael 025212
Borovinskich, Nadežda Petrovna 037952
Borowski, Batya 025070
Borowski, Wiesław 035318
Borowsky, Gwen 052074
Borpuzari, D. 024278
Borrachero, Germán M. 039722
Borremans, Dr. R. 003667
Borriello, Dr. Mariarosaria 026992
Borromeo, Dr. Georgina 052387
Borrowdale-Cox, Deborah 050252
Borsanyi, Jackie 050770
Borsboom, Leo 033057
Borščenko, Lidija Mychajlivna . . 043229
Borsch, Prof. Thomas 017090
Borsdorf, Prof. Dr. Ulrich 018307, 018311
Borselli, Lucilla 027901
Borsig, Josef 009436
Bortolan, Gino 028314
Bortolotti, Dr. Marco 025549
Borusiewicz, Mirosław 034976, 034980
Borut, Križ 038585
Bory, Jean-René 041427
Borys, Stephen 051691
Borysova, Galyna Stepanivna . . 043324
Borzan, Maria 036652
Bos, Agnès 012855
Bos, W. 032882
Bosano Ansaldo, Dr. Carlos de . 000116
Bosch i Cornellà, Salvador . . . 040088
Boschi, Prof. Enzo 026156
Boschiero, Dr. Gemma 025404, 025406,
025407
Bose, Krishna 024365
Bose, Prof. Sugata 024365
Boser, Dr. Elisabeth . . . 017796, 017798
Boserup, Dr. Ivan 010169
Boshoff, Jaco 038685
Boshouwers, F. 032374
Bosi, Erio 026640
Bosioković, Sava 038187
Boškova, Tatjana 030633
Bošković, Dora 009036
Bosma, M. 032944
Bosmans, Wim 004392
Bosnar Salihagić, Željka 009052
Bošniak, Jurij Michailovič 037975
Boss, George 044193
Boss, Dr. M. 018258
Bossa Herazzo, Donaldo 008658
Bossaert 003822
Bossan, Marie-Josèphe 015139
Bossart, Hans Peter 042207
Bossé, Laurence 014697
Bosshard-Frischknecht, Hulda . 042085
Bosshard, Daniel 041836
Bosshard, Reymond 041766
Bossine, Lesley 044686
Boßler, Sascha 022363
Bostanci, Gülten 033471
Bostander, Nellie 038875
Bostic, Connie 046552
Bostick, Christine 052301
Bostin, Franky 004165
Boston, David 046098
Bostrem, Lidija Andreevna 036865
Boswell Minnerly, Vet 048446
Boswell, Mary Rose 050055
Boswell, Peter 050836
Bot Riera, Issa 040397
Bot, Nadia 041523
Botea, Alin 036602
Botelho, Joce M. 052386
Botelle Ortega, Daniel 039599
Botero, Dr. Clara Isabel 008616
Botha, R. 038654
Bothe, Günter 016781
Bothe, Heinrich 017444
Bothmer, Dietrich von 051424
Botineau, Pierre 021210
Botirov, I.T. 054649
Botsaris, Markos S. 022976

Botsen, Jean-Pierre 014860
Botte, Prof. Virgilio 027011
Bottengård, Else 034030
Bottini, Dr. Angelo 026267
Bottlaender, Michael 019029
Bottrel Tostes, Vera Lúcia 004751
Botwinick, Michael 054577
Bouaïche, Claire-Lise . 041854, 041855
Bouchard-Acodas, Louise 005702
Bouchard, Guy 006813
Bouchard, Marie 014686
Bouchard, Yann 015717
Boucher, Celine 005565
Boucher, George 048437
Boucher, Michel . . . 012187, 012188
Boucher, Sid 001413
Bouchet, Denis 013750
Bouchet, J.-C. 011613
Bouchon, Chantal 014733
Bouchon, Juan Santiago 000478
Bouck, Jill 048472
Bouckaert, Odile-Marie 003417
Boudet, Sylvie 014224
Boudier, Pierre 012472
Boudou, Dominique 014343
Boudreau, Luce Marie 005610
Boudrot, Donna 005349
Bouffault, Colette 014565
Boughton, Jim 052715
Boughton, Peter 043933
Bouglé, Frédéric 015933
Bouillon, Philippe 003766
Boujamid, Azzouz 031773
Boukli, Aïcha 000044
Boulanger, Patrick 014048
Boulanger, Sylvie 012521
Boulbes, M. 012907
Bould, Gillian 045767
Boulet, Jean 016105
Boulliard, Jean-Claude 014634
Boulton Elliott, Helen 007346
Bounous, Jewell P. 053924
Bourassa, Marie 006658
Bourbon Busset, Jean-Louis de . 011877
Bourbon, Prof. Giulio 025790
Bourbonnais, Alain 005397
Bourbonnais, Caroline 012765
Bourcier, Paul 050601
Bourdais, Jany 012720
Bourdon, Don 005382
Bourdon, Kate 006797
Bourdon, Michel 012553
Boureima, Konfé 005276
Bourel, Yves 015482
Bouret, Brigitte 015110
Bourgade, Véronique 011802
Bourgat, Robert 014844
Bourgeau, Jef 052237
Bourgeois, Jean 004132
Bourgeois, Valérie 007243
Bourgoin, Colette 005545
Bourgon, Jacques 014222
Bourgougnon, Eric . . 012037, 014272
Bourgoz, Michel 041240
Bourke, Marie 024900
Bourlet, Michel 012764
Bournazel, Comtesse de 014994
Bourne, Susan 043794
Bouše, Petr 009691
Bouska, Marianne 047728
Bousmanne, Bernard 003435
Bousquet, Nicolas 011600
Boussard 014565
Bousska, Prof. Dr. Vladimíra . . 002971
Boustany, Bernadette 013503
Bouteiller, Hubert 012738
Bouthat, Chantal 006372
Boutry, Pierre 038087
Bouttemy, Marie-Françoise . . . 013807
Bouvette, Henry d'Otreppe de . . 004111
Bouvier, Prof. Dr. Beatrix 022200
Bouwman, J.W.M. 031883
Bouxin, Marc 015056
Bouyack, Kay 052319

Bouzón Gallego, Avelino 040279
Bouzy, Olivier 014584
Bovewizer, Austin 024981
Bovey, Patricia E. 007476
Bowden, Cindy 046631
Bowdery, J. 043988
Bowdion Gil, Carol 047757
Bowditch, John 046486
Bowdrey, Jeremy 043991
Bowen, Allyson 051380
Bowen, Duncan 044461
Bowen, Graham 043602
Bowen, Jane 043415
Bowen, Linnell R. 046498
Bower, Alan 045945
Bower, Barbara A. 048809
Bower, Corinna 046069
Bower, Finn 007017
Bower, Mary 048645
Bower, Thomas 048640
Bowers, David 045277
Bowes-Crick, C.P. 045093
Bowes, Murray 001625
Bowie, Dr. Gavin 043798
Bowle, Jean 044317
Bowles, Evelyn 048487
Bowles, G. 043918, 043919
Bowman 044766
Bowman, Dana 007292
Bowman, Jack 048382
Bowman, Leslie Green 054502
Bowman, Midge 053234
Bowman, Travis 047802
Bownan, JoAnn 051067
Bowron, Julian 001291
Box, David 001168
Boxall, Michael 043591
Boxberg, Hendrik von 018306
Boxer, David 028443
Boxer, Graham 044861
Boxhall, Patricia 000914
Boxley, Robert F. 054466
Boyce, Connie 047192
Boyce, David Y. 050215
Boychev, Petr 005233, 005234
Boyd, Beatrice 053970
Boyd, Sir John 044923
Boyd, L. 006033
Boyd, Peter 045798
Boyd, Robert 046914
Boyd, Willard L. 049713
Boydell, Owen 044214
Boyden, Dr. P. 045056
Boyer, Anne Marie 013007
Boyer, Cindy 052615, 052621
Boyer, Jean-Marc 015794
Boyer, John Vincent 049386
Boyer, Sylvain 011835
Boyette, Todd 046556
Boylan, James 053547
Boylan, Olive S. 051119
Boyle, Dorothy 052898
Boyle, Gail 043747
Boyle, William J.S. 007228
Boynton, Hugh 044538
Boysen, Jesper 010138
Boysen, Renate 019912
Boysson, Bernadette de 012114
Bozan, Maria 036700
Božič, Bojan 038530
Božko, Boris Stepanovyč 043286
Bozkurt, Batuhan 043015
Bozok, Erdoğan 042942
Braack, Craig R. 046900
Braam, Dr. P. 032280
Braarud, Reni 033812
Braat-Voorhuis, T.M.A. 033101
Braaten, Alicia L. 047764
Braaten, Ivar Gunnar 033486
Braaten, Trond 033518
Brabant, Diane 006793
Brabec, Jiří 009858
Bracegirdle, Hilary 046001
Bracegirdle, Robert 045153

Brachaczek, Zofia 035240
Bracht, Dr. Christian 020268
Brack, Dr. P. 042195
Bracke-Logeman, P. 032066
Bracke, W. 003435
Brackenreg, Philip 000805
Bracker de Hugo, M. 015491
Brăďăţan, Aura 036733, 036806
Bradbury, Christopher 046149
Bradbury, David 044988
Bradbury, J. Robert 006469
Bradbury, Jade 050478
Bradbury, Karen 044854
Braddick, Lyndon 044266
Braddock, Pam 044450, 045540
Brade, Dr. Johanna 018737
Brademann, Margret 017657
Braden, Gerald 052497
Bradfield, Gerald 050078
Bradford, Anna 005968, 005971
Bradford, Dr. Bruce C. 048184
Bradford, Jan 051292, 051293
Bradford, Jolene 049449
Bradford, W.J. 005732
Bradley, Allen 051510
Bradley, Anne 045510
Bradley, Betsy 049751
Bradley, Bob 045226
Bradley, Dean 006780
Bradley, Fiona 044241
Bradley, Graeme 033249
Bradley, J. 045438
Bradley, J.M. 044094
Bradley, Larry 053962
Bradley, Laurel 051630
Bradley, Richard 051139
Bradley, Dr. Robert 050517
Bradley, Ron 007332
Bradman, Susan 044030
Bradner, Anne 052418
Bradshaw, Graham 045346
Bradt, Rachelle 051493
Brady, Carolyn 052506
Brady, Christine 044712
Brady, Edith 049466
Brady, Jim 052278
Brady, Tess 053490
Braekeleer, Catherine de 003763
Brändle, Christian .. 042205, 042217,
                              042218
Bräuer, Jörg 016589, 016591
Bräutigam, Karl 018261
Brafford, C.J. 051041
Bragagnolo, Dr. Pietro 026229
Bragg, Cheryl H. 046504
Bragg, Gary O. 051989
Braham, George O. 007240
Brahim, Natalia Majluf 034330
Brahm-Lindberg, Lisa 054551
Brahma Chary, P. 024341
Braica, Silvio 008994
Braid, Dennis 051858
Braillard, Marie-Christine 015662
Braipson, Roger 004072
Braithwaite, Bevan 044102
Brakke, Jaap 019062
Bralić, Zdenko 008998
Bramah, Edward 044919
Bramberger, Alfred 001927
Brancaccio, Jim 050247
Brancaleoni, V. 004044
Brancati, Prof. Antonio 027265
Brancato, Prof. Francesco ... 027181
Brancel, Esther 054339
Brancelj, Andreja 038576
Brancucci, Dr. Michel 041282
Brand-Claussen, Dr. Bettina ... 019144
Brand, M.A. Elke 017709
Brand, Dr. Joachim 017202
Brand, Margrith 041900
Brand, Michael 021915
Brand, Dr. Michael 052565
Brand, Dr. Ruth 041370
Brânda, Nicolae 036239

Brandão Aguirre, Maria Beatriz . 004863
Brandão, Daniela 035893
Brandão, Fátima 035893
Brandaó, Maria Luiza 004478
Brandeller, Paul 003962
Brandén, Peter 041029
Brandenburg, Dr. Hajo 020823
Brandenburg, Dr. Winfried ... 021599
Brandenburger-Eisele, Dr.
  Gerlinde 020978
Brandenstein, Ameli von 021675
Brander, Prof. A. .. 010780, 010782
Brander, Alli 010921
Brandl, Andrea 021781
Brandl, Franz 001785
Brandmeyer, Dennis 053877
Brandoli, Susan 007315
Brandon, Cath 000964
Brandon, G. 005634
Brandson, Lorraine 005618
Brandt, Prof. Dr. Angelika 019025
Brandt, Dr. Jochen 018995
Brandt, Prof. Dr. Michael 019247
Brandt, Monique 005727
Brandt, Dr. Reinhard 022540
Brandt, Thomas 020745
Brandt, Tova 048208
Brandt, Wayne 048664
Branscome, Curtis 053545
Bransom, Jennifer 048121
Brantley, R. Brit. 047591
Braquilanges, Guy de 012382
Brascamp, M.H. 032545
Brasel, Ronald 047523
Brasie, Jeff 047738
Brasil do Amaral, Paulo Cesar . 004643
Braslavsky, Emma 017118
Braslavsky, Noam 017118
Braß, Dr. M. 022067
Brassel, Dirk 017534
Brassine, Edouard 004135
Bratčenko, Tetjana Georgiïvna . 043139
Bråthen, Magne 033705
Bratschi-Tschirren, Max 041470
Brattig, Dr. Patricia 019700
Brattmyhr, John 041001
Bratuž, Ciril 038587
Brauer, Gerald J. 048224
Braun, Dr. Andreas ... 002900, 020511
Braun, Annette 013349
Braun, Axel 017823
Braun, Dr. Claudia 020255
Braun, Dr. Emanuel 018119
Braun, Hannes 002643
Braun, Prof. Dr. Harald 018507
Braun, Dr. J.K. 051588
Braun, Joseph 013348
Braun, Karen 052441
Braun, Märta 041167
Braun, Mary E. 052836
Braun, Matthew 049737
Braun, Monika 019518
Braun, Peter 002627
Braun, Thomas 016894
Braun, Volker 017066
Braun, Wilhelm 020462
Braunegger, Reinhold 002504
Braunlein, John 049047, 051308
Braunová, Helena 009615
Brauns, Dmitri G. 053015
Braunstein, Susan L. 051412
Braunsteiner, Dr. Michael 001685
Braut, Else 033742
Bravo, John 053216
Braxton, Gail G. 048955, 048958,
                     048960, 048961
Bray, Alida 053038
Bray, Bill 048761
Bray, Rick 048135
Bray, Tamara 048313
Bray, Xavier 045057
Brayer, Hermione 013743
Brayner, Vânia 004667
Braz Teixeira, Dr. Madalena ... 035813

Brazauskas, Pijus 030479
Brázdová, Marie 009971
Brearley, Anna 044462
Brebeck, Wulff E. 017673
Brecciaroli, Dr. Luisa 028124
Brecht, Fatima 051432
Bredenkamp, Dalene 038669
Bredereck, Wolfgang 000990
Bredow, Florentine C. 019288
Bredsdorff, Eva B. 044892, 045377,
                              046083
Breede, Claus 005889
Breen, Evelyn 024838
Breen, John D. 053175
Breese, Dr. Kathryn 049755
Bregar, Marjeta 038585
Bregeaut, Pierre 012291
Bregman, B. 033004
Bregovac Pisk, Marina 009036
Brehm, Georgia 050522
Brehm, Dr. Thomas 020853
Brehme, Sylvia 017072
Breidbach, Prof. Dr. Dr. O. ... 019444
Breiler, Ronald 021511
Breinegaard, Jens 010106
Breinl, Karl 016535
Breiter, Fritz 041636
Breithaupt, Brent H. 050142
Breithaupt, Julia 018032
Breitkopf-Bruckschen , Elisabeth 019693
Breitkreuz, Petra 018451
Breitmeier, Uwe 017819
Brekelmans-Bijveld 033054
Breker, Uwe H. 019703
Brekke, Borghild 033814
Brembeck, Josef 020424
Bremer, D. Neil 048561
Bremer, Stanley 032830
Bremner, R.W. 044143
Brendel-Pandich, Susanne ... 053686
Brendow, Dr. Volker .. 016904, 016905
Brendstrup, Dagmar 010040
Brener, Lidija Davydivna 043246
Brengle, Anne B. 051225
Brenke, Johannes 016857
Brennan, Anita 051067
Brennan, Anne G. 054446
Brennan, Edward 051463
Brennan, Nancy 052212
Brennan, Robert 024916
Brenneman, Dan 053823
Brenneman, Jan 048908
Brenner Wilczek, Dr. Sabine ... 018551
Brenner, Itzhak 025182
Brenner, Dr. Klaus-Peter 018753
Brenner, Stefan 001699
Brent, Marty 006524
Bréon, Emmanuel 012136
Bresc, Geneviéve 014749
Breščak, Mateja 038552
Breslauer, David 051064
Bresler, Zinaida Ivanovna 036853
Bresnahan, Bob 050806
Bresse, Robert 014718
Bresser, G.J. de 032477
Bressi, Dr. Nicola 028203
Bressor, Carol 049172
Breth, Renata 053143
Breton, Gérard 013647
Bretón, Prof. Raúl 031164
Bretscher, Peter 041490
Bretscher, Renzo 041546
Bretschneider, Jürgen 019082
Brett, Ruth 005782
Brettner, Friedrich 001941
Breuel, Margitta 017410
Breuer, Dieter 022635
Breuer, Heiko 018965
Breuker, P. 032328
Breunig, Robert G. 048740
Breuze, Marc 014031
Brevik, Bernt 033969
Breville, Tristan 030793
Brewer, Billy 046434

Brewer, Dr. Douglas J.	053909	Broadey, R.	043680	Browes, Judi	044019	Brownell, Carlton C.	050328
Brewer, Esther Vivian	048310	Broady, Duncan	045203	Brown Hirst, Kaye	052905	Browning, Dawn C.	050745
Brewer, R.	043863	Brobbel, Joni	045628	Brown, A.	001150	Browning, Mark	050887
Brewster, Carrie	051049	Brobst, William A.	052334	Brown, Alastair	033570	Brownlee, Reuben	054026
Brewster, Christine	043601	Brocas, Jon	033237	Brown, Alex	045975	Brownlow, Mary	044389, 044736
Brewster, Pete	005380	Brock, Mary	052134	Brown, Ann	049177	Brownson, Ron	033141
Brewster, Rae	001524	Brocke, Dr. Edna	018287	Brown, Anne	052520	Broy, Erich	019970
Brewster, Suzanne G.	046309	Brockhaus, Prof. Dr. Christoph	018064	Brown, Ashley	046593	Brubaker, Bob	054252
Brewster, William F.	050603	Brockhoff, Joan	007088	Brown, Bern Will	005638	Brubaker, Sandra	051663
Brey, Karen	052499	Brockman, Prof. Christian	033847	Brown, Bruce	052645	Bruce, Donald	051312
Brez, Wendy	048767	Brocksch, Dr. Dieter	020901	Brown, Charles B.	049925	Bruce, Elspeth	045488
Breza, Bogusław	035387	Brodanec, Vlado	009028	Brown, Charles T.	050892	Bruce, Michael	049298, 049299
Breza, Michael	051829	Brodbeck, Andrea	022057	Brown, Dr. Charlotte V.	052453	Bruch, Claudia	018327
Briand, Henri	015445	Brodeur, Bill	006625	Brown, Dr. Christopher	045448	Bruchhäuser, A.	017353
Briant, Julius	045123	Brodie, Ian J.	033332	Brown, Dr. Claudia	052120	Bruck, Anton	003067
Brice, Alison	043583	Brodl, Michaela	003053	Brown, Curtis E.	052066	Bruckner, Alfred	020871
Brichet, Michel	003464	Brodneva, Adelja Vladimirovna	037167	Brown, Dan	052902	Brueck, Siegfried	021909
Brichtová, Dobromila	009695	Brody, Aaron	046931	Brown, Daniel	052567	Brückner, Frauke	017499
Brick, Elizabeth M.	051618	Broege, Wallace W.	052590	Brown, Dr. Dann	052272	Brückner, Sigrid	022101
Bricka, Olena Mykolaïvna	043142	Broeker, Dr. Holger	022744	Brown, David	028443, 047457	Brückner, Uwe	022542
Brickman, Penny	053189	Broekgaarden, A.	031950	Brown, David C.	051535	Brüderlin, Prof. Dr. Markus	022744
Brickwood, Kevin	005511	Brogiolo, Prof. G. Pietro	027713	Brown, David J.	054495	Brüggerhoff, Dr. Stefan	017433
Bridenstine, James A.	049859	Broglie, Charles Edouard Prince		Brown, Don	049204	Brühlmann, Fritz	041251
Bridger, Tozzy	044333	de	012422	Brown, Dorothy	007274	Brülisauer, Annemarie	041842
Bridges, Dr. E.M.	044320	Brohel, Edward R.	052199	Brown, Doug	049226	Bruer, Dr. Stephanie-Gerrit	021993,
Bridges, Edwin C.	051023	Broidl, Erich	002799	Brown, Elizabeth A.	053235		021995
Bridges, Mary	046421	Brok, Dr. Barbara den	041678	Brown, Eric	043634	Bruesehoff, Alison	049118
Bridges, Tim	046195	Broker, David	001078	Brown, Forrest	051631	Brugge, Dr. J.P. ter	032991
Bridgford, Todd C.	049344	Brolsma, Allen	048070	Brown, Geneva	048310	Bruggeman, Jean	016259
Bridgman, Norma	038722	Brom, Jodi	054482, 054483	Brown, Geoffrey I.	054469	Brugger, Dr. Ingried	002944
Bridwell, Shawn	050315	Bromberg, Dr. Anne	048115	Brown, Georgina	005712	Brugnoli Bailoni, Francisco	007538
Briedenhann, M.M.	038658, 038659	Brombierstäudl, Andreas	019420	Brown, Gordon	045584, 045585	Brugnoli, Renzo	027685
Briedis, Agris	030180	Bromley, Fran	049367	Brown, Hannah	053859	Brugo, Carlo	027663
Briend, J.	014699	Bromovsky, Fabia	046027	Brown, Harold	006698, 049496	Bruguière, Philippe	014707
Briere, Florence	013620	Brondeau, Armelle de	011583	Brown, Harold L.	049368	Bruhin, Felix	041494
Briese, Jodie G.	054516	Broo, Dr. Hanno	017822	Brown, Helen	043925	Bruhn, Thomas P.	053555
Briest, Francis	016189	Brook, Griff W.T.	001136	Brown, Ian	045758	Bruijn, J.A.	032215
Brigger, Lilli	041558	Brooke, Janet M.	006095	Brown, James A.	049400	Bruin, Dr. R.E. de	032944
Briggs-Jude, Wendy	007388	Brooke, Patrick	045982	Brown, Jane	047555	Bruinjes, Wilhelm	018074
Briggs Murray, James	051466	Brookes-Sullivan, Rebecca	043664	Brown, Janene	049872	Bruintjes, Jaap C.N.	032252
Briggs, Kate Halle	053425	Brookman, Philip	054098	Brown, Jean	044767	Bruist, Gwen	045962
Briggs, Dr. Peter	050517	Brooks Joyner, J.	051765	Brown, Jocelyn	033277	Brukwicki, Jerzy	035319
Brigham, Ann	048388	Brooks-Myers, Inez	051681	Brown, Johanna	054491	Brumbauch, Mary	050531
Brigham, David	054548	Brooks, Anthony T.P.	054406	Brown, Karon	053716	Brumberg, Esther	051439
Brigham, David R.	046408	Brooks, April	045699	Brown, Kate	046069	Brumder, Robert B.	054239
Bright, Bill	048927	Brooks, Carol	054610	Brown, Kevin	044899, 049338	Brumelis, Andrew	007185
Bright, Paul	054487	Brooks, Carolyn	050489	Brown, Laura K.	046597	Brumgardt, Dr. John R.	047543
Brighton, Terence	044463	Brooks, Daniel F.	047003	Brown, Leanne	054023	Brummer, Hans Henrik	041033
Brill, Rainer	019542	Brooks, Eric	050240	Brown, Lee N.	014505	Brun-Frandji, Hélène	014062
Brillonet, Jean-Marc	015553	Brooks, Dr. Jim	051193	Brown, Leslie	047130	Brunacci, Prof. Dr. Aldo	025398
Brimm, Steve	053434	Brooks, Leonard L.	051257	Brown, M.	005366	Brundage, Barbara	053937
Brincard, Marie Therese	052418	Brooks, Lindsay	045720	Brown, Margaret	043596, 043597,	Brundin, Martha	048161
Brind, Rosemary	043550	Brooks, Nick	045647		043599	Brune, Thomas	019153, 022047, 022385
Brindeau, Irène	013629	Brooks, Peggy	049075	Brown, Marlene	053915	Brunecker, Frank	017356
Brindley, Lynne	044922	Brooks, Philip R.B.	046209	Brown, Marlin	005826	Brunel, Georges	014690
Brindley, Ronan	043806	Brooks, Ruth	054328	Brown, Michael K.	049588	Brunet, P.	014331
Bringager, Frithjof	033845	Brooks jr., William F.	050856	Brown, Nancy	048964	Brunette, T.	007294
Brink, Bob	054564	Brooslin, Michael	053462	Brown, Neil	007109	Brunetti, Armando	025961
Brink, Siegfried	022601	Brose, Barbara H.	048107	Brown, Nicolas	052383	Brungard, Charlene K.	046879
Brinklow, R.	044164	Brose, Dr. David S.	047040	Brown, Patricia	047804	Bruni, Giovanni	025878
Brinkmaier, Ulrich	049390	Brose, Lawrence F.	047304	Brown, Peter B.	046223	Bruni, Luca	041821
Brinkman, L.D.	049943	Brosens, Annemarie	033116	Brown, R.J.	052772	Brunin, Ph.	004087
Brinkmann, Hermann	022740	Brosi, Dr. Sibylle	019508, 019517	Brown, Rich	052155	Brunius, Staffan	040998
Brinkmann, Prof. Dr. Vinzenz	018449	Brosnahan, Seán	033195	Brown, Rick	050028, 050031	Brunn-Lundgren, Maria	040648
Brinkmann, Dr. W.	042227	Brosnan, Susan	051263	Brown, Robert	043543	Brunnemann, Eric	023359
Brisch, Brigitte	018401	Brost, Jeff	052525	Brown, Robert M.	050959	Brunner, Alois	021080
Brissard, Michel	016208	Brost, Leif	040700	Brown, Ron	051951	Brunner, Dieter	019160, 019164
Brissette, Serge	012982	Brosthaus, Dr. Karl-Heinz	020304	Brown, Rondalyn	006820	Brunner, Dorla	046988
Brisson	014816	Brothers, Ann	001278	Brown, Dr. S.	043619	Brunner, H.	021743
Brisson, Steven C.	050556	Brothers, Leslie A.	054600	Brown, Scott	046232	Brunner, Dr. Max	021085
Brito Joao, Benedito	031812	Brotherton, Barbara	053248	Brown, Shirley	050012	Brunner, Dr. Michael	022242, 022243
Brito Moreira, Alvaro de	036008	Broucke, Remi	003522	Brown, Tamara	049250	Brunner, Peter	041611
Britschgi, Dr. Jorrit	042221	Broughton, Jamie	043439	Brown, Tammi	053619	Bruno, Agnès	013340, 013846, 014425,
Britschgi, Markus	041466	Brouillet, Johanne	007022	Brown, Terrence	051435		015279, 016032
Britton, Marion	050359	Broun, Dr. Elizabeth	054160	Brown, Tim	047482	Bruno, Prof. Gianfranco	026406
Britz, Günther	017007	Brousseau, Francine	005870, 005871	Brown, Vinita	048771	Brunovský, Ferdinand	038471
Brivio, Dr. Ernesto	026806	Brouwer, Dr. M.	032687	Brown, Vivian A.	034575	Bruns, James	046608
Brix, Thorsten	014330, 014332	Brouwer, M.	032591	Brown, W.	045391	Brunscheen, Scott	048286
Břízová, Zuzana	009549	Brouwer, Norman	051472	Brown, William	045390, 049321	Brunschwiler, Willi	041485
Brizzi, Prof. Gian Caolo	025564	Brouwer, T. de	032703	Brown, Dr. William	049531	Brunschwyler, Greta	050170
Brkić, Dragana	004282	Brovčenkova, Valentina Ivanovna	036933	Brownawell, Christopher J.	048454	Brunson Frisch, Marianne	052206
Brkić, Staniša	038562	Brovinsky, Boris	038562	Browne, Charles C.	052777	Brunson, Jeana	053662
Broach, Barbara K.	048752, 048755	Brower, Ken	054316	Browne, Mary-Louise	033149	Brunson, Neal E.	049816
Broad, Sandy	007053	Brower, Paul	046894	Browne, Patrick	048415	Brunson, Theodore	049816

Index of Persons

Brunt, Kevin
– Busching, Dr. Dr. rer. nat. habil. Wolf-Dieter

Brunt, Kevin 005831
Brupbacher, Adolf 041733
Brusca, Richard C. 053822
Bruschetti, Dr. Paolo 026084
Bruscia, Maria Giuseppina ... 027581
Bruse, Ralf 020683
Bruseker, Annemarie 045121
Brusila, Johannes 011442
Brussa, Carlos A. 046282
Brussee, G.J. 032741
Brust, Alexander 041276
Bruton, Prof. David L. 033847
Bruun Beyer, Mette 010178
Brüvere, Solvita 030172
Bruwier, Marie-Cécile 003890
Bruyère, Edgard 003884
Bruyn, Dr. Jean-Pierre de ... 003628
Bruyn, M.A. de 032851
Bruyn, Vera de 034036
Bryan Hood, Mary 051854
Bryan, Anne 052448
Bryan, Dr. Betsy 046728, 046754
Bryan, Dr. Charles F. 052564
Bryan, Tracey 052564
Bryant, Arthur H. 051793
Bryant, James 052596
Bryant, Jim 052270, 052274
Bryck, Jack 046721
Bryde, John A. 033900
Bryer, Stephen 046119
Bryman, Donna 005697
Bryson, Mary 053808
Bryzensky, Nicole 001518
Brzeziński, Prof. Tadeusz 035250
Brzeziński, Dr. Wojciech 035377
Brzon, Narveen 052517
Brzozowska, Anna 018028
Brzozowski, Jerzy 035234
Brzozowski, Leszek 034770
Bsteh, Prof. Dr. Andreas 002400
Bu, He 008376
Bu, Yangwu 007866
Buberl, Dr. Brigitte 017962
Buccellato, Prof. Laura 000183
Buccheri, Prof. G. 027188
Bucci, Jonathan 054176
Bucciferro, Rose 050352
Buccino Grimaldi, Marchese
  Luigi 027994
Buccleuch, Duke of ... 045755, 045757
Buch, Carl E. 033643, 033644
Buchan, Andrew 044820
Buchanan, Dennis 043786
Buchanan, John 001127
Buchanan jr., John E. 052302
Buchanan, Karen 050534
Buchel, M. 031974
Bucher-Häberli, Elisabeth ... 041576
Bucher, Friedrich 001724
Bucher, Prof. Dr. H. 042227
Bucher, Hans 018515
Bucher, Regina 041748
Buchholtz, Annegret 019705
Buchholz-Todoroska, Małgorzata 035210
Buchholz, Sabine 019543
Buchi, Danielle 012180
Buchmann, Anna 041862
Buchmann, Vagn 010166
Buchmayr, Dr. Friedrich 002708
Buchner, Dr. Gabriele 021150
Buchner, U. Ernest 005795
Buchroithner, Dr. Sonia 002141
Buchs, Denis 041382
Buchtmann, Hans-Georg 022302
Bučinskaja, Valentina
  Stanislavovna 029801
Buck, Angela 046969
Buck, Dennis 046671
Buck, Prof. Dr. Elmar 019714
Buck, Kurt 021071
Buck, Lisa 047742
Buckaloo, Terence 053520
Buckellew, Shari Spaniol 047041
Buckie, Valerie 005333

Buckley, Annette 043534
Buckley, Kerry W. 051625
Buckley, Laurence 051560
Buckley, Laurene 048769, 051240
Buckley, Tracy 047754
Bucknill, Greg 045491
Buckridge, Martin 045523
Buckson, Deborah 051708
Bucur, Corneliu-Ioan .. 036697, 036699,
                    036700, 036705
Buczkowski, Hans 002523
Buczynski, Bodo 017089
Budarov, Viktor Viktorovič 038064,
                           038066
Buday, Kathleen 053791
Budde, Bernhard 017583
Buddle, Allan 049703
Buderer, Dr. Hans-Jürgen ... 020255
Budiastra, Putu 024555
Budiati, Dr. Tinia 024572
Budintis 030545
Budka, Mikołaj 034646
Budko, Prof. Dr. Anatolij A. ... 037825
Budney, Jen 006588
Budriunas, Algimantas 030433
Budriūnas, Algimantas 030522
Budrys, Milda 047670
Budrys, Romualdas 030524
Budzynski, Jan 047977
Büche, Wolfgang 018969
Büchler-Mattmann, Dr. Helene . 041333
Büchler, Dr. Hans 041671
Bücken, Véronique 003459
Buecker, Thomas R. 048052
Bückling, Dr. Maraike 018449
Bühl, Gudrun 054108
Bühl, Herbert 041189
Buehler, J. Marshall 054509
Bühler, Dr. Kathleen 041412
Buehler, Dr. Kathleen 041318
Buehler, Ken 048385
Bühlmann, Isolde 041462
Bühner, Dr. Jan-A. 022026
Buelens, Paul 003545
Buell, Pamela 006704
Buendía García, Mercedes ... 040348
Buenrostro Alba, Manuel 031212
Büren, Daniela von 041565
Büren, Guido von 019472
Bürger, Werner 016595
Bürgin, Christoph 041717
Bürgin, Dr. Toni 041926
Buerlein, Robert A. 052542
Buermeyer, Chris 049301
Buesing, Kay 050377
Bütikofer, Prof. Alfred 042161
Büttner, Dr. Philippe 041879
Bützer, Michael 018041
Büyükçapar, Ahmet 042979
Bufano, Ralph A. 053239
Buff, Urs 041755
Buffa, Gérard 014981
Buford, Elizabeth F. 052451
Bugaeva, Galina Alekseevna ... 037905
Bugajov, Oleksandr Sergijovyč . 043267
Bugallo, Bernardino 039964
Bugalová, Dr. Edita 038311
Buganu, Melitina 036449
Bugár-Mészáros, Károly 023512
Bugarski, Astrida 004283
Bugg, Mathias 041937
Buggeland, Tord 033742
Buggle, Claudia 022520
Buhl Andersen, Christine 010203
Buhl, Henry 052159
Buhler, Leslie L. 054165
Buhlmann, Dr. Britta E. 019480
Buhr, Johannes 022564
Buhrmann, Connie 050720
Buhrs, Michael 020558
Buijs, H.W.J. 032975
Buijs, J. 014635
Buisson, Fernand 014936
Buisson, René 006408

Bujakowska, Beata 034780
Bujić, Marija 038168
Bujok, Dr. Elke 020579
Bujold, Michel 006664
Bukač, Václav M. 009568
Bukačová, Dr. Irena .. 009689, 009690
Bukowska, Halina Lubomiła ... 035029
Bulaty, Dr. Milan 017312
Bulau, Cheryl 050326
Bulei, Gheorghe 036410, 036477,
          036636, 036746, 036747, 036749,
          036750, 036751, 036753, 036754,
                                036829
Bulfon, Christiane 001885
Bulgarelli, Grazia Maria 027634
Bulger, Ralph 006552
Bulgin, Lloyd 005734
Buljević, Zrinka 008993
Bull, Milan 048668
Bullard, E. John 051302
Buller-Focht, Brenda 052596
Buller, Osmo 011389
Bullock, Margaret 052302
Bulova, Gretchen M. 046389
Bultereys, Veerle 003621
Bulthuis, P. 033086
Bultmann, Andreas 002626
Bumgardner, Tim 049740
Bumiller, Manfred 016952
Bunch, Lonnie G. 047650
Bunch, Nancy B. 047909
Bunch, Rosemary 054537
Bund, Dr. Konrad 018804
Bundesen, Laura 048238
Bung, Maria-Hildegard 016925
Bunger, James 053002
Bungert, Petra 003438
Bunke, Jim 053254
Bunker-Lohrenz, Jackie 052714
Bunkichi, Ito 029175
Bunyan, Vicki 001254
Buol-Wischenau, Mag. Karin .. 001985
Buonanno, Prof. Roberto 027577
Burak, Marek 035424
Buran, Dr. Dusan 038328
Buranelli, Dr. Francesco 054678, 054682
Burbank, Jane 052568
Burbano, Ignacio 008753
Burch, Danielle 049429
Burch, Ivan 006402
Burch, Wanda 049830
Burchard, Helen 046868
Burchenal, Margaret 047117
Burchfield, Stanley G. . 052879, 052882
Burckhardt, Dr. Daniel 041282
Burckhardt, Nicole 041231
Burdak, Elisabeth 042247
Burdick, Funi 048651
Burdick, Richard L. 052677
Burdina, Lidija Nikolaevna 037140
Burdychová, Milena 009529
Burel, Bernard 014953
Burford, Beverley 044985
Burg, Dr. Tobias 018306
Burganov, Aleksandr 037400
Burge, Roisin 044575
Burge, Sarah 043693
Burger, Richard L. 051265
Burger, W. 019390
Burgère, André 012687
Burges-Smith, Karen 048653
Burgess, Brian 001174
Burgess, Cynthia 054012
Burgess, Jerry G. 048835
Burgess, Jo 047051
Burgess, Robert 051317
Burgess, Susan M. 044286
Burghardt, Glenn 051671
Burgin, William 049790
Burgiss, Ken 033202
Burgos Cantor, Roberto 008596
Burgraff, Joseph 003644
Buri, Dr. phil. Andreas 041705
Burik, Tatjana Borisovna 037895

Burkard, Adolf 041374
Burkard, W. 017228
Burkarth, Dr. Axel 022047
Burke, Bonnie 005487
Burke, Daniel 052060
Burke, Darrell 006649
Burke, Gregory 033256
Burke, Jo 048225
Burke, Joan T. 047600
Burke, Kathleen M. 049491
Burke, Marcus B. 051404
Burke, Susan M. 006124
Burkel, Don 046807
Burkert, Richard A. .. 049832, 049833
Burkhalter, Françoise 041754
Burkhardt-Aebi, Rosmarie 041602
Burkhardt, Elisabeth 041998
Burkhardt, Guido 017038
Burkhardt, Pia 022418
Burkhart, Linda 047605
Burki, Erika 041386
Burleson, D. 005959
Burlingham, Cynthia 050426
Burluaux-Blanchard, Odile ... 014697
Burman, Rickie 045010
Burmann, Peter 040946
Burmeister, Artur 019896
Burmeister, Bernd 017299
Burmeister, Christina 021289
Burmeister, Helmut 019292
Burmeister, Dr. Ralf 017085
Burn-Murdoch, R.I. 045695
Burn, Dr. L. 043842
Burnet, Marie-Jeanne 041907
Burnett, Ed 045865
Burnett, Kay Taylor 051194
Burnett, La Jean 049495
Burnett, Richard 044456
Burnett, Roger K. 052691
Burnett, Wayne 052773
Burnette, David 051099
Burnham, Laura 048280, 049810
Burnichioiu, Sevastița 036350
Burnie, John 043675
Burnod, Véronique 012303
Burns, Courtney 053158
Burns, Diane 000988
Burns, Emma-Kate 043718
Burns, Jane 048533
Burns, Kelly S. 049643
Burns, Laura 046768
Burns, McAndrew 046582, 046585,
                             046586
Burns, Nancy 051976
Burns, Norman O. 051173
Burns, Pamela 006429
Burns, Richard 043806
Burnside, Madeleine H. 049957
Burny, Jeannine 003488
Buron, Gildas 011908
Buron, René 015496, 015497
Burpee, Jane 005726
Burrell, Jane 045341
Burresi, Dr. Mariagiulia 027358
Burri, Katri 042147
Burroughs, Stuart 043520
Bursey, M.L. 038709
Burt, Ann 051814
Burt, Steve 044784
Burtch, Michael 006987
Burton, John 048741, 050095
Burton, Mary 054299
Burton, Val 000899
Burwick, Georg 052591
Busack, Dr. Stephen 052452
Busceme, Greg 046852
Busch, Anna 041027
Busch, Fritz B. 022727
Busch, Dr. Ina 017822
Busch, Ingo 019059
Busch, Jean 053778
Busch, Wolf-Rüdiger 018601
Buschhoff, Dr. Anne 017592
Busching, Dr. Dr. rer. nat. habil.

Wolf-Dieter 019748	Bykovceva, L.P. 037372, 037411	Calbiac, Christoph de 013850	Cameron, Joyce M. 053442
Buschmann, Harald 018230	Bylicki, Tomasz 035315	Calboutin, A. 038878	Cameron, Rosanna 001151
Bush Tomio, Kimberley 053875	Byman, Bertyl 011145	Caldeira, Ernesto 054536	Cameron, Ross 000844
Bush, Dianne 006275	Bynum Robertson, Marci 050336	Caldeira, Prof. Dr. Helena 035624	Camfield, Paul 048950
Bush, Kate 044912	Byong, Ha Min 030048	Caldeira, Vasco 004879	Camhi, Prof. Jeffrey 025097
Bush, Michelle 006902	Byrczek, Jakub 034821	Calder, Jacqueline 051036	Camilo, Eva 010391
Bush, Walter 052132	Byrd, Eugene 053492	Calderan, Gianni 027645	Camin, Philippe 014208, 015667
Bushman, David 051441	Byrd, Gloria 050196	Calderer Serra, Joaquim 040192	Camm, Simon 045598
Busine, Laurent . . . 003710, 003711	Byrn, Brian 048533	Caldwell, Brenda B. 053866	Cammarata, Paolo 027707
Busjan, Béatrice 022693	Byrne, Joseph 049391	Caldwell, Desiree 047951	Cammarota, Dr. Giampiero 025556
Busk Laursen, Bodil 010172	Byrne, Lisa 000878, 001250	Caldwell, Frank 046087	Cammuso, Philomena M. 053271
Buskirk, Dr. William H. 052534	Byrne, Mary Jo 050709	Caldwell, Dr. J.P. 051588	Camnitzer, Luis 051379
Buskov, Poul 010362	Byrne, Sue 044432	Caldwell, James 047374	Camoys, Lord 044567
Busquet, Andoni 039853	Byrtusová, Jana 009513	Caldwell, Janet 046870	Camp, Kimberly 050807
Buss, Leo W. 051265	Byström, Lars 041021	Caldwell, Joan G. 051306	Campagna, Damon A. 054078
Bussac, Monique 011666	Bystrou, Daniela 017086	Caldwell, Judy 050774	Campagna, Dr. Francesca 026774
Busse, Dr. Klaus 017504	Bytkovskij, Oleg 037610	Caldwell, K. 044059	Campagnolle, Delphine 011994
Busse, Prof. Rido 021902	Byvanck, Dr. F.V. 032651	Caldwell, Karen 044324	Campagnolo, Matteo 041530
Busse, Valerie 050717	Byzewski, Leah 049175	Caldwell, Roy 046939	Campanelli, Dr. Adele . 025966, 025967
Bussel, Gerard W. van 003046	Caballero Barnard, Prof. José M. 031595	Calhoun, Angela 048939	Campanelli, Elide 026167
Bussers, Dr. Helena . . 003457, 003459	Cabanes, Francisco 040322	Calhoun, Ralph 050996	Campanero, Amy 047194
Bussinger, Julia 046915	Cabanis, Prof. E.A. 014693	Caliaa, Prof. Roberto 025281	Campbell Coyle, Heather 054433
Busslinger, Ernst 041738	Cabano-Dempsey, Annali 038648	Caligiuri, Olimpia 041797	Campbell-Ferguson, Gail 047189
Bussy, Diana de 000906	Cabaret, Michel 015064	Calisi, Dr. Marinella 027577	Campbell McMurray, H. 045558
Bustard, Wendy 051150	Cabel, Oswalda A. 034484	Calkins, Jack 005768	Campbell-Shoaf, Heidi 048945
Bušueva, Margarita Viktorovna . 037606	Cabello Carro, Paz 039689	Call, Linda 054573	Campbell, Bruce 045361
Busuladžić, Adnan 004283	Cabello Carro, Dr. Paz 039635	Callaghan, Dr. Amanda 045588	Campbell, C.A. 045869
But, Aurel 036376	Cabezas, Hervé 013406, 015525	Callaghan, Madelein 006997	Campbell, Dr. Caroline 044951
Buta, Sanda Maria 036270	Cable, Christopher B. 046472	Callaghan, Richard . 044217, 044219	Campbell, Chris 051184
Butala, Daniel 008925	Cabot, Ian 045708	Callahan, Ann 051171	Campbell, Colin 054411
Butave, Denis 015184	Cabot, Paul 007214	Callahan, Christine 051538	Campbell, Connie 006489
Butchko, Tom 048525	Cabral, Dr. Elisabeth . 035761, 035781	Callahan, Colin J. 047957	Campbell, D. 044549
Butkevičius, Gintaras 030325	Cabrera Cachón, Teresa 031621	Callahan, Colleen 052561	Campbell, Daniel T. 053382
Butko, Trina 000779	Cabrera, E. 048685	Callahan, David 051558	Campbell, Deborah 046485
Butler, Amena 051731	Cabrera, Geles 031222	Callahan, Elizabeth 048444	Campbell, Donald 001363
Butler, Charles Thomas 047923	Caccia, Antonietta 027899	Callahan, Martin L. 048867	Campbell, Francis D. 051328
Butler, David 045292	Cachola Schmal, Peter 018411	Callan, Josi Irene 053647	Campbell, George 006080
Butler, Gay 047587	Cad, Gabriel 034473	Callanan, Mary 024930	Campbell, Ina 047034
Butler, John 043592, 046015	Cadavid López, Susana 008800	Callataÿ, François de 003437	Campbell, Kathleen 052503
Butler, Joyce 052286	Cadbury, Tom 044312	Callatÿ, François de 003435	Campbell, Kay 001508
Butler, Maria V. 051983	Cadby-Sorensen, Robin 053740	Callebert, Fernand 003404	Campbell, Loraine 053810
Butler, Mary 046825	Cadenet, Anne 012108	Calleen, Dr. Justinus Maria 022702	Campbell, Lorne 045057
Butler, R. 038793	Cadenne 012462	Callegher, Dr. Bruno 027146	Campbell, Mabel 006621
Butler, Roger 000937	Cadenne, Ivan . 012463, 012608, 015831	Calleja Ansótequi, Zoilo 040424	Campbell, Mei 050517
Butler, S.R. 044818	Cadier, Gérard 014991	Callejo, Marcilyn 051781	Campbell, Mognon 001565
Butler, Samuel 051456	Cady, Don 046951, 050902	Callen, Bonnie 005802	Campbell, Mungo 044401
Butler, T. 044646	Caes, Ward 003283	Callewaert, Ferdy 003975	Campbell, Nancy 054255
Butler, Tony 045365, 045895	Caffarelli, Prof. Roberto Vergara 027354	Callies, Prof. Dr. H. 021943	Campbell, Nathan 046859
Butnaru, Ioan 036593	Caffrey, James 047086, 047088	Callis, Charles 033357	Campbell, Neil 006232
Butt, Raymond 045538	Cagigós i Soró, Antonio 040163	Callis, Eleanor 051687	Campbell, Richard 050944
Buttacavoli, Eva 046685	Cagli, Prof. Bruno 027269	Callis, Marion M. 054422	Campbell, Roy G. 052452
Buttar, Lyn 033359	Çağman, Dr. Filiz 042961	Callison, Chris 053910	Campbell, Sara 051935
Buttarazzi, Silvestro 025794	Cahen-Delhaye, Anne . 003466, 003479,	Calloway, Jim 051764	Campbell, Shearon 047731
Buttel, Larry 048970	003492, 003497, 003498	Calloway, William 047918	Campbell, Tacie N. 048376
Butter, Jelena 017246	Cahen, J. 031946	Calomiris, Ellen 050375	Campbell, Tom 043502
Butterfield, Tom 047029	Cahen, Joël J. 031948	Calonne, Sylvain 016260	Campbell, Wallace 051956
Butterweck, Dagmar 003057	Cahill, Ann 045425	Caltik, Heidi 003033	Camphausen, Ute 019992
Buttini, Eric 030594	Cahill, Kevin M. 051326	Calvarin, Margaret . . . 012249, 012694	Campins, Lluís 040082, 040083
Buttlar-Elberberg, Dr. Gertrud . 003107	Cai, Changkui 008324	Calvesi, Prof. Maurizio 026000	Campione, Prof. Dr. F. Paolo . . 041394
Button, Kerri 007390	Cai, Hetian 008309	Calvi di Bergolo, Ippolito 027844	Campo, Marta 039061
Butts, H. Daniel 050656	Cai, Mingkang 007950	Calvin, Prudence K. 054218	Campodonico, Dr. Piero 026420
Butz, Dr. Herbert 017239	Cai, Xiaofang 007660	Calvo, B. 039665	Campofiorito, Italo 004590
Buurma, Hilbrand 032886	Cai, Zhengquan 007845	Calvo, Ernesto 008855	Campos, Prof. Antonio Celso de
Buurman, H. 032171	Caiado, José Pedro 035808	Calvo, Luis María 000615	Arruda 004571
Buus, Henriette 010085	Caicedo de Gómez, Hilda María 008646	Calvocoressi, Richard . 044236, 044273	Campos, Maria Lucia Rocha . . . 004886
Buvelot, Q. 032180	Caicedo Lourido, María Isabel . . 008655	Calzada, S. 039070	Campos, Nelson 036059
Buxton, Judith 005807	Caicedo, Iván 008670	Camacho Cadaval, Fernando . . 040178	Campredon, Elisabeth 012747
Buyandelger, U. 031743	Caille, R.T. 011891, 011893	Camacho, Dr. João Carlos 035604	Camps Extremera, Antonio 039295
Buyko, Malgorzata 034861	Cain, Glena 052123	Camacho, Dr. Paula Cristina . . . 035793	Campuzano Ruiz, Enrique 040138
Buys, C. 032585	Cain, Prof. Dr. Hans-Ulrich 019974	Camara, Dr. Abdoulaye 038120	Camus, Marie-France 013553
Buyssens, Danielle . 041514, 041537	Caines, M. 044185	Camara, Calvin 046510	Canales, Rocio Gonzalez de . . . 031290
Buzenschi, Florentina 036612	Caisley, Sherry 051074	Camara, Prof. Edeltrudes 004346	Cañamás, Maïta 040305
Buzzi, Alessandro 027407	Caivano, Felice 049391	Camartin, Prof. Iso 041989	Canaves, Margalida 039954
Buzzoni, Dr. Andrea . . 026206, 026215,	Caji, Jaroslav 038125	Camasse, Dr. Michele M. 026570	Canbey, Maşuk 043079
026216, 026219	Čajka, Vladimír Michailovič 038033,	Čambálová, Dr. Daniela 038473	Canca Lara, Juan Enrique 040376
Byambasüren, P. 031730	038044, 038046	Camber, Diane W. 050846	Cândea, Ionel . 036260, 036261, 036262,
Byck, Dr. David B. 053173	Čajkovskij, Bogdan M. 043241	Camenzind, Cornelia 042076	036263, 036264, 036265
Byčkova, Marina Nikolaevna . . . 037994	Çakmak, Hacı 042838	Cameron of Lochiel, Sir Donald . 045847	Candeloro, Ilaria 027687
Byčkova, Olga Nikolaevna 037386	Caladio, Prof. Cosimo 026571	Cameron, Dan 051452	Cândido, Maria João 036028
Bye, Joellen 051221	Calahan, Deborah 047011	Cameron, E. 033142	Candon, Anthony 045439
Byers, Bill 047191	Calame, Caroline 041685	Cameron, Elizabeth 044160	Candrowicz, Krzysztof 034967
Byh, Per-Olov 040743	Calandra, Eliana 027187	Cameron, Gary L. 050718	Candy, Ron 007313
Bykov, Dr. Aleksandr 038056	Calatayud Cases, José Vte 039999	Cameron, Jane 045870	Caneer, Ann 052235
Bykova, Jekaterina Vladimirovna 037552	Calbet, Jean-Marie 013713	Cameron, John 043651	Cangialosi, Patricia 000173

Canini, Alberto 027599
Canizales Montoya, Luis
  Fernando 008759
Cankova, Tatjana 005030
Cannan, Gwynedd 051485
Cannata, Dr. Roberto 027569
Cannatella, David C. 046701
Cannegieter, Dr. D.A.S. 032310
Canning-Lacroix, Johane 006885
Cannistraro, Dr. Philip V. 051410
Cannon-Brookes, Peter 044743
Cannon, Annetta 051648
Cannon, Janet 005927
Cano de Martín, Marta 000324
Cano Rivero, Ignacio 040173
Cano Salas, Gilda 031205
Cano, Luis Martín 031268
Cano, Rose Marie 050465
Canogar, Rafael 040404
Canonne, Xavier 003512
Canovai, Dr. Roberto 026640
Canright, Steve 053023
Canseliet, Ovide 003719
Cant, Dr. John G.H. 036151
Cantinho, Dr. Manuela 035798
Cantor, Helen 048936
Canty, Scott 050446, 052461
Canulli, Richard G. 053986
Canuto, Marcello A. 051265
Cao, Chuansong 008042
Cao, Sanzhuang 008554
Cao, Zhongkui 008391
Caouette, Muriel 049050
Capa Eiriz, Dr. José 039332
Capa, José 039890
Capaloi, Franco 014948
Capanna, Prof. Ernesto 027603
Capano, Dr. Antonio 028329
Capazza, Gérard 014398
Capdevila, Dr. Ricardo 000202
Capella Molas, Anna 039399
Capenberghs, Joris . . . 003413, 003414
Capers, Pierre 004138
Capin, Djordje 031755, 031756
Capin, Jean 015974
Capitani, Silvano 026205
Capitanio, Fabrizio 025496
Čapková, Dr. Jana 009572
Čaplinska, Olga Mychajlivna . . . 043259
Capomolla, Soccorso 026956
Capon, Barry 043767
Capon, Edmund 001535, 001536
Capon, S.H. 046096
Capone, Dr. Patricia 047370
Capot, Stéphane 013817
Capovilla, Prof. Sergio Carlos . . 000570
Cappelletti, Prof. Vincenzo 027350
Cappola, Dr. Gianpietro 025300
Capponi, Dr. Carlo 026823
Capps, Gene 054030
Capps, Jennifer 049697
Capps, Larry 049636
Capralos, Souli 022943
Caproni di Taliedo Guasti,
  Firmina 027574
Carabo, Lucille 046918, 046919
Caradec, Marie-Anne 012739
Caramazza, O.G. 026740
Caramelle, Dr. Franz 002121
Caramelli, Fernando 027329
Caramore, B. 028387
Carazzetti, Prof. Riccardo . . 041683,
                          041684, 041688
Carbajal, Londi 053121
Carbajal, Sergio 031218
Carballo, José Rodriguez 040127
Carbert, R.W. 006316
Carbery, David 024922
Carbone, Teresa 047253
Card, Nan 048976
Cardarelli, Andrea . . . 026853, 026854
Cardelli, Alessandra 027634
Cardelli, Patrice 013955
Carden, Willie F. 047750

Cárdenas Marcaida, María
  Caridad 009106
Cárdenas Tovar, Jairo Simón . . . 008821
Cardenas Troncoso, Gloria 007516
Cardenas, Dr. Lourdes B. 034471
Cardenas, Mary R. 051958
Carder, Julia 044852
Cardilli, Dr. Luisa 027589
Cardim Ribeiro, Dr. José 036038
Cardinaels, Marie-Jeanne 003399
Cardini, Giuliano 028025
Cardon, Véronique 003460
Cardona, Hermana Martha 008795
Cardone, Veronique . . . 012009, 013609
Cardoner, F. 039054
Cardoni, Edmund 047305
Cardoso Botelho, Americo G. . . . 035542
Cardoso, Prof. Dr. António 035524
Cardot, Annie 042096
Carduner, Michel 015423
Cardwell, E. 045438
Carengna, Dr. Daniela 027697
Carevic Rivera, Alvaro 007523
Carew, John E. 007203
Carey, Betsy 053678
Carfano, M. Susan 053900
Caridad, Jorge 010396, 010399
Cariello, Dr. Rubina 027994
Carignani, Dr. Mario 026996
Carion, P. 003463
Cariou, André 015022
Carl, Christine S. 049033
Carlano, Annie 053116
Carles, Gilbert 016013
Carlestam, Håkan 040675
Carleton, Dr. Don 046688, 053918
Carli, Dr. Laura 028038
Carlier, Sylvie 016250
Carlier, Yves 013013
Carlile, Janet 005352
Carlosama Mora, Rosa Cecilia . . 008702
Carlsen, Dr. Charles 051846
Carlsen, Margaret . . . 049981, 052370
Carlson-Reynolds, Alexis 048636
Carlson, Anne 047082
Carlson, Erling 021231
Carlson, Jim 053338
Carlson, Joseph 048532
Carlson, Linda 048797
Carlson, Norman P. 049786
Carlson, Robert 047084
Carlson, Dr. Shawn 054190
Carlson, Shirley 047458
Carlsson, Ann-Britt 041098
Carlsson, Eva 040620
Carlström, Dr. Ann-Kristin 040570
Carlstrom, Terry 049365, 053288,
                054114, 054115, 054134, 054156
Carlton, Jamon 054621
Carlton, Rosemary 053361
Carlton, Stewart 053209
Carlweitz, Pontus 041096
Carman, Andrea 051025
Carmen Lema Pensado, María
  del 046277
Carmer, Craig 050399
Carmichael, J. 045003
Carmona Lancheros, Carlos
  Alberto 008643
Carnap-Bornheim, Prof. Dr. Claus
  von 017704, 021663, 021667
Carnazzi, Umberto 025511
Carneiro, Dr. Deolinda Maria
  Veloso 035965
Carneiro, Paula Dias 035952
Carnerini, Carla 027221
Carnes, Griffiths 049942
Carnes, John 050279
Carney, Barbara 053721
Carney, Dr. Margaret 046399
Carney, Terry J. 044468
Carniello, Peter 005822
Carolan, Janet 044106
Carollo, Albert H. 054010

Carollo, Eugene R. 054010
Caron, Euclid 051828
Caron, Jacqueline 006052
Carowick, Sandra L. 049612
Carpenter Correa, Laura 047237
Carpenter Troccoli, Joan 048258
Carpenter, Dean 048462
Carpenter, Elizabeth 050948
Carpenter, Frank 044111
Carpenter, Joan 054603
Carpenter, Dr. M.A. 043848
Carpenter, Matthew 046513
Carpenter, T.F. 043914
Carpentier, Régine 003434
Carpio, Agustín 000487
Carr-Griffin, Paul 044884
Carr-Whitworth, Caroline 044108
Carr, Carolyn 054150
Carr, Christian 053626
Carr, Dr. Dawson 045057
Carr, Josh 051393
Carr, Lucia C. 050574
Carr, Shelvie 048177
Carr, Ted 044176
Carradice, Prof. Ian A. 045678
Carrara, Dr. Giancarlo 025965
Carraro, Dr. Francine 049764
Carré, Dr. Enrique González . . . 034361
Carreau, Alain 015738
Carreau, Raphaële . . . 012528, 012529
Carreck, Marjorie W. 044565
Carregha, Luz 031463
Carreira, Dr. Isabel 035635
Carrel-Lautelme, Suzanne 013846
Carrel, Michel 015530
Carreño Díaz, Prof. Esboardo . . 030947
Carreño, Irving 054711
Carreño, Maria 053000
Carrera Hontana, Enrique de . . . 039653
Carretero Rebés, Salvador 040113
Carretero, Andrés 039661
Carrier, Alan 051819
Carrier, Nicole M. 049740
Carrlee, Ellen 049849, 049850
Carrlee, Scott 049848
Carrocera Fernández, Elias 039814
Carrol, Brian 047279
Carroll, Daine L. 053797
Carroll, Jon 048979
Carroll, M. Lyda 049276
Carroll, Rowan 033288
Carroll, Sue 052398
Carroll, William B. 044598
Carron, Christian G. 049195
Carruthers, Karen 043657
Carruthers, Ray 007127
Carruzzo, S. 042106
Carson, Heather 045049, 045131
Carson, John 052475
Carson, Michael 047699
Carson, Sully 033192
Carstensen, Dr. Jan 017877
Carswell, A.L. 044257
Cartacci, Marcello 027652
Cartagena, María Fernanda 010481,
                                      010484
Carter, Anne 050336
Carter, Beth L. 053508
Carter, Dr. Curtis L. 050923
Carter, Earl N. 049285
Carter, Hazel 050688
Carter, John S. 052057
Carter, Jon . . . 043450, 045690, 045694,
                          045701, 045704
Carter, Jonathan 045691, 045692
Carter, Kathy 050036
Carter, Martha C. 053443
Carter, Michael 051364
Carter, Penelope 044255
Carter, R. 000894
Carter, Susan 053668
Carter, Warrick L. 047677
Cartwright, Colin M. 043582

Cartwright, Derrick R. . 049353, 052978
Cartwright, Thomas 048940
Carty, Paul 024884
Caruana, Sebastian 030772
Carucci, Arturo 027711
Carus, Benno 017167
Carusi, Corina 047662
Caruso, Dr. Ida 026019
Caruso, Romy 054262
Caruso, Sarah 052845
Carustrom, Terry 046539
Carvalho, Adelaide 035952
Carvalho, Luís Manuel Mendonça 035553
Carvalho, Maria Judite de 035913
Carvallo, Henri 016217
Carver, Kate 044602
Carver, Pat 051132
Carvey-Stewart, Elizabeth 052628
Cary, PeeWee 048392
Carzou, Jean-Marie 013992
Čas, Alenka 038501
Casabán, Consuelo Císcar 040305
Casadei, Dr. Sauro 026178
Casado Soto, José Luis 040115
Casagrande, Louis B. 047103
Casagrande, Rodrigo Fernando . 004936
Casagrande, Tiziana . 026192, 026193
Casal Tatlock, Alvaro 046259
Casal, Gabriel 034521
Casaldi, Silvano 027031
Casale, Tommaso 028050
Casalini, Dr. Antonio 027210
Casalino, Vittorio 026415
Casanelles, Eusebi 040222
Casanovas, María Antonia 039076
Casaro, Massimo 026822
Casarola, Albino 026967
Casartelli, Marisa Helena Cestari 004820
Cascales Ayala, Manuel 038976
Casciu, Dr. Stefano . . . 026260, 026274
Caselli Wijsbek, H.E.J. 032965
Casetti, Christoph 041415
Casey, Laura 053293
Casey, Lynn R. 049044
Casey, Michael 024975
Casey, Prof. Michael T. 024965
Cash, Laurisa 049717
Cash, Sarah 054098
Cashbaugh, Todd M. 046489
Cashman, William 046863
Casini, Dr. Stefania 025493
Caskey, Phyllis 050430
Caskey, Tom 048360
Câşlaru, Constantin 036224
Casler, Jennifer R. 048904
Caslin, Jean 049579
Caspary, Dr. Gundula 021850
Caspers, Martha 018428
Caspersen, Kim 044235
Cass-Hassol, Nicholas 051451
Cass, Naomi 001071
Cass, Wilfred 044445
Cassan, Dagmar de 002301
Cassan, Ferdinand de 002301
Cassanelli, Dr. Roberto 026973
Cassata, Giovanni 027190
Cassell, Valerie 049574
Cassels, Patrick 043514
Casseminho Meira, Mariana . . . 004650
Casser, Anja 019504
Cassiano, Dr. Antonio 026581
Cassidy, Christina 053135
Cassidy, Gary N. 052351
Cassin, Michael 054419
Cassin, Virginia 051666
Cassway, Nick 052076
Castagnoli, Prof. Pier Giovanni . 028115
Căstăian, Mihai 036599
Castañeda, Lochin 034267
Castaño Cruz, Hermana Cecilia . 008568
Castañón Albo, Luis 039634
Castanyer Bachs, Narcis 040368
Castel-Branco Pereira, Dr. João . 035763
Castel, Marie-Chantal 014867

Castele, Theodore J. 047816
Castell, Edmon 038936
Castellà Reial, Josep 039971
Castellana, Dr. Gruiseppe .. 025252
Castellano i Tresserra, Anna .. 039101
Castellano Zapater, Eustaquio .. 038909
Castelletti, Dr. Lanfredo 026047, 026050,
026051, 026052
Castello, Rosa María 038946
Castells, Ruben 040447, 040450
Castelnuovo, Sheri 050598
Casteret, Norbert 013005
Castex, Jessica 014697
Castex, Nicole 012904
Castier, Annie 013807
Castilla, Francisca 040308
Castillo Reyes, Samuel 034226
Castillo Tovar, Carlos Arturo ... 008685
Castillo, Anna 053101
Castillos, Prof. Juan José ... 046263
Castle, Charles E. 050040
Castle, Lynn P. 046851
Castle, Myra 049275
Caston, Dr. Ellie 054019
Castor González, Jesús 031193, 031214
Castor, A. 014843
Castrillón Urrera, Carlos Alberto 008580
Castro Benítez, Daniel . 008583, 008585
Castro Mejía, María Eugenia ... 008579
Castro Meza, Raúl 031487
Castro Otero, Gumersindo 040134
Castro, Prof. Dr. Aníbal Pinto de 036095
Castro, Dr. Inés de 019250
Castro, Maria da Graça Salema
de 035992
Castro, Milagros 036172
Castronovo, Simona 028118
Català Gorgues, Angel 040301
Cataldi, Dr. Maria 028057
Catchpole, Julie 033275, 033276
Cates, David L. 003201
Catford, B.L. 001387
Cathcart, Rosemary 044298
Cathrin, Marcel 015467
Cathy, Salter-Jansens 038829
Catleugh, Jon 044956
Catlin-Legutko, Chinnamon ... 048053
Catlin, Erica 047262
Cato, John S. 053600
Cato, Dr. Paisley 052980
Catolin, Eufemia B. 034618
Catron, Joanna D. 048951
Catroppa, Frank 046625
Catrysse, Bernard 029240
Catta, Javier 010463
Cattaneo, Claudia 042150
Cattaneo, Pascal 041744
Cattelain, P. 004090, 004091
Catton, Jonathan 044468
Cau, Ettore 028422, 028423
Cauchie, S. 003950
Caudoux, Michel 016134
Cauk, Françoise 011865
Caulkins, Bethel I. 052529
Caurie, Philippa 046198
Čauševa, Dora 005049
Caussinus, Dany 011644
Cauwe, Nicolas 003492
Cavagnis Sotgiu, Dr. Maria Carla 027585
Cavaillès, Maria 014804
Cavalcante, Raquel 004369
Cavalchini, Pieranna 047117
Cavalier, Odile 011835
Cavalier, Sebastien 013952
Cavalieri, Corlis 051500
Cavalli-Björkman, Åsa 040995
Cavalli, Attilio 026809
Cavalli, Germano 028392
Cavanaugh, Joe 048949
Cavatrunci, Claudio 027634
Cavelti, Dr. Urs J. 041555
Cavendish, Kim L. 048828
Cavens, Travis 049916

Cavish, Jackie 051872
Cavnor, Julie Ann 046757
Cavour Aramayo, Ernesto 004213
Cavruc, Valeriu 036693
Cay Henry, Mary 048027
Cayer, Richard 046413
Çaylak, İsmail 042868
Cayphas, Jean-Paul 003732
Cazacu, Silvia 036320
Cazakoff, Ingrid 007010
Cazan, Mag. Ildikó 003046
Cazé, Sophie 013299
Cazenare, Sabine 011633
Cazenave, Sabine 011632
Cazes, Daniel 015998
Ceballos Carrero, Rosa Maria ... 031039
Ceballos, Orlando Toro 008781
Cebucoprea, Dr. Alexandru ... 036291
Ceccarelli, Giampiero 028007
Cecchetto, Dr. Giacinto 025824
Ceccini de Dallo, Ana María .. 000622
Cech, Christine 003089
Cecil, William A.V. 046551
Čecov, Prof. Dr. V.P. 037354
Cederholm, Katherine 050556
Cederlund, Dr. Johan . 040854, 041122
Cedersand, Kerstin 040998
Cederstrøm, Elisabeth 010195
Cedro, Dr. Bernard 035249
Ceesay, B.A. 016384
Ceesay, Hassoum 016384
Cega, Fani Celio 009009
Čekerinac, Bora 038245
Čeko, Franko 008987
Celant, Germano 051470
Celayir, İsmail Hakkı ... 042934, 042969
Celestian, Susan 052113
Céleyran, Tapié de 014435
Celi, Maria 051372
Celio, Franco 041545
Celis, Augustin 040404
Celli, Jeanann 053499
Celma, Cécile 030775
Celuzza, Dr. Mariagrazia 026473, 026476
Cembellín, Juan Manuel
Gonzalez 039146
Cembrano, Rita 034442, 034445
Cembrola, Bob 051537
Cencic, Jasmine 001736
Cendo, Nicolas 014044
Ceni, Giuseppe Riccardo 028349
Ceniceros Herreros, Javier ... 039813
Cenni, Mario 026640
Centellas Bodas, Aurelio 040433
Cepec, Dr. Martin 017602
Čepič, Taja 038563
Cepon, Miloska 038562
Čeran, Branko 009043
Cerar, Estera 038562
Čerba, Larisa Valerjanovna ... 037970
Cercega, Petru 036370
Cerepak, Mag. Susanne 002953
Cerezo Aranda, Dr. José Antonio 039782
Cérino, Christophe 013872
Ceriotti, Guido 025666
Cerişer, Nicolae 036497
Čerkalin, Sergej Dmitrievič .. 037700
Čerkasova, Olga Anatoljevna ... 037051
Cerméle, Dominic M. 052096
Cermesoni, Barbara 026155
Černá, Margit 009700
Černavin, Lev Davidovyč 037782
Černín, Dr. Svatomír 009897
Černjak, Éduard Isaakevič ... 037953
Černobai, Prof. Dr. Juri 043257
Černova, Vita Vasylivna 043178
Cerny, Zbynek 009515
Cerón Mireles, Prof. José Luis . 031189
Ceron, Clodualdo N. 034558
Cerón, Pablo 031191
Ceroni, Nadia 027485
Cerovac, Branko 008977
Cerović, Dr. Ljubivoje 038218

Cerretelli, Claudio 027450, 027451
Ceruti, Mary 050384
Cerutti, Dr. Antonio 027470
Cerutti, Lilia P. de 000684
Cervellino, Miguel 007522
Cervetti, Dr. Valerio . 027203, 027209
Cervini, Dr. Fulvio 028111
Cesarino, José 054726
Cesbron, Pierre 015405
Češková, Stanislava 009855
Česnova, Ljudmila Alekseevna . 036990
Cession Louppe, Jacqueline ... 003890
Cetinić, Ljiljana 038160
Ceulemans, Louisa 003694
Ceunynck, Christoph de 012722, 012723
Ceut 013107
Cevrim, Dr. Mustafa 017383
Ceynowa, Dr. Klaus 020507
Ceynowa, Dr. Tatjana 016479
Chabard, Dominique 011812
Chabert, Noëlle 014791
Chabot, Jacques 013993
Chačikjan, Svetlana Melkonovna 037695
Chadbourne, Janice 047106
Chadli, Mohammed 031774
Chadwick-Case, Wendy 049786
Chadwick, Adam 045035
Chadwick, Joseph 047179
Chadwick, P.J. 005642
Chadwick, Phyllis 051825
Chadžiev, N. 005189
Chafe, Dr. K.S. 033431
Chaguaceda Toledano, Ana ... 040037
Chai, Paris K. 049537
Chaik, Pak In 029880
Chaimerez, M. G. 025150
Chainé, Françoise 007249
Chait, Karen 051393
Chakarova, Veselina 005080
Chakhmouridian, Dr. Anton ... 007458
Chakraborti, R.M. 024408
Chakravarti, Rasmohan 003214
Chakravarti, Dr. S.K. 024361
Chald, Elisabeth 045681
Chalfant, Rhonda 053259
Chalker, Mary 001563
Challen, John 043974, 043976
Challender, Nancy 049342
Challener, Elisbeth 053153
Challis, Liz 001023
Challis, Paul 001023
Chalmers, Kim 047160
Chalmers, Peggy 045058
Chalupa, Josef 009720
Chalupar, Erwin 002650
Chalverat, Joseph 041845
Chamay, Jaques 041259
Chambareau, René 011785
Chamberlain, Amelia E. 051524
Chamberlain, Lori 006938
Chamberlin, Edward M. 049041
Chambers, Carol 045953
Chambers, Coleen B. 049673
Chambers, Cullen 053872
Chambers, Elizabeth 052302
Chambers, Emma 006487
Chambers, Keith 000982
Chambers, Kevin 051701
Chambers, Michael A. 006579
Chambers, Natalie 043484
Chambon, Cati 013623
Chambon, Fabrice 014314
Chambon, Marion S. 051286
Chamil, Iskhakov 009459
Champanil, Vira 042653
Chan, Elise 054215
Chan, Peng Sam 008071
Chance, Delores 047487
Chancel-Bardelot, Béatrice de . 012161,
012164, 012166
Chandès, Hervé 014649
Chandžieva, Evelina 005015
Chandžieva, Radka 004998
Chanel, Sylvie 014126

Chang, In-Kyung 029922
Chang, Zhenming 008081
Chanler, James 051517
Channing, Susan R 047824
Channon, Ann 043914
Chanson, Robert 041942
Chantrain, Jean-Pascal 003517
Chanzit, Gwen 048258
Chapalain, Jean-Philippe 013849
Chapin, Angela 006810
Chapin, Jean 054084
Chapin, Mona 047739
Chapman Banks, Barbara 050293
Chapman, A.T. 005462
Chapman, Alan 043437
Chapman, Allison 050608
Chapman, Ashley 053598
Chapman, Dennis 054555
Chapman, Gordon 001617
Chapman, J.R. 045610
Chapman, Dr. Jefferson 050011
Chapman, Jim 045355
Chapman, Peter 049228
Chapman, Prof. Stanley 045645
Chapman, Dr. Sydney T. 045478
Chapnick, Harold 048767
Chapoose, Betsy 048807
Chapp, Belena S. 051494
Chappa, Kristen 053008
Chappaz, Jean-Luc 041530
Chappey, Frédéric 013832
Chappuis Waeber, Dominique . 041502
Chapron, Philippe 011921
Chaput, Leo W. 050082
Charauel, P. 013942
Charbonneau, Diane 006391
Charbonneau, Reine 006352
Charest, Marcie 050622
Charette, Joanne 006594
Charette, Luc A. 046346
Chariot, C. 004115
Chariot, Constantin 003793
Charitantos, Katerina 023179
Charitonov, Christo . 005255, 005257
Charland, Stuart 049793, 049794
Charles, Dorothée 014733
Charles, Mary 006641
Charles, Steve 053246
Charlet, Jean-Michel 042095
Charlier, Paul 003908
Charlot, Claude 014712
Charlot, Colette 014289
Charlton, Dr. Christopher ... 044048,
044049
Charlton, K. 033322
Charmayne Young, J. 052765
Charmelo, Julie 047685
Charpentier, Corinne 012751
Charpentier, Dominique 011903
Charrière, Edmond 041406
Chartier, Catherine 015415
Chartrain, Michèle 013152
Charzynska, Jadwiga 034723
Chasansky, Matthew 046670
Chase, Ed 046563
Chase, Harriet 052465
Chase, Jim 047496
Chaserant, Françoise . 013653, 013655,
013656, 013657
Chassaing, Jean-François 013317
Chassé, Sonia 006200
Chasseval, Geneviève de 013370
Château, Georges 015420
Chateau, Micheline 013175
Chateigne, Yann 012108
Chatelain, A. 042106
Chatelain, Prof. E. 014012
Chatelain, Paul 012395
Chatenet, Georges 015993
Chatfield, Linda 048349
Chatland, Gordon 048731
Chatman, Patrina 048295
Chatterjee, Dr. H.J. 044983
Chatterjee, Dr. Sankar 050517

Chatterjee, Sunil Kumar 024471
Chaturvedi, Naresh 024407
Chaucer, S. 045363
Chaudhry, Jamil Ahmed 034116
Chauffeteau, Claude 012113
Chaumet, Danièle 014885
Chaussee, Gérard 014536
Chauvel, Marie Claire 013375
Chauvin, Dr. Robert S. 048184
Chauviniére, Comte de la 015905
Chavanne, Blandine 014416
Chavannes, Virgile Alexandre . 000584
Chavarria, Antonio 053114
Chaves, Dr. Julio César 034208
Chaves, Manuel Wenceslao ... 034202
Chávez Chávez, Dr. José Antonio
Rodolfo 034218
Chavez Posada, Zoraida 008829
Chavez, Anja 054266
Chavez, Michael 049147
Chavez, Thomas 053118
Chavez, Thomas E. 046367
Chazal, Gilles 014800
Cheatham, Jim 048319
Chebanova, N. 054667
Chebitwey, Jane 029860
Checefsky, Bruce 047821
Chefdebien, Anne de 014774
Cheikh, Aboubakar 008831
Chek, Dr. Deborah 053960
Chelbi, Fethi 042738
Chelius, Margot 020674
Chellah, Mwimanji Ndota 054834
Chemera, Lyn 048423
Chen-Lee, Margaret 042518
Chen, Bilin 008270
Chen, Changgen 008501
Chen, Changjin 007655
Chen, Chu-Ming 042372
Chen, Chunfa 007680
Chen, Chung-Tsang 042304
Chen, Chuyuan 007765
Chen, Dayuan 008055
Chen, Demao 008384
Chen, Deshi 008281
Chen, Dongquan 007729
Chen, Jian Ming 007674
Chen, Jianbiao 008283
Chen, Jiang 008112
Chen, Jianhong 008196
Chen, Jieshi 008185
Chen, Jinsheng 008548
Chen, Jui-Wen 042366
Chen, Julie 051479
Chen, Jun-Ling 042394
Chen, Junqing 008014
Chen, K.C. 042552
Chen, Kevin B. 053005
Chen, Lucia 051479
Chen, Pao-Chung 042497
Chen, Po 007826
Chen, Quanxi 007753
Chen, Rongqi 008050
Chen, S.S.H. 032176
Chen, Shaoming 008502
Chen, Shenghua 007906
Chen, Shih-Bey 042542
Chen, Shilong 007819
Chen, Stone 042277
Chen, Wei 008492
Chen, Xianchun 007761
Chen, Xinming 007748
Chen, Xixin 008326
Chen, Xue'ai 008292
Chen, Zhengyu 008150
Chen, Zhiliang 007770
Chen, Zhimin 008512
Chen, Zhixin 008205
Chen, Zhuxuan 008493
Chen, Zu-E 008300
Chenault, Amanda 047849
Cheney, Dorothy 049023
Cheney, Rebecca 043913
Cheng, Jian 007910

Cheng, Sibao 008451
Chenoufi, Brahim 000083
Chenoweth, Bob 053426
Chenoweth, Diana 052951
Chenoweth, Richard 053129
Chepp, Mark 053471
Chera, Constantin 036189, 036394,
036398
Cherchi Usai, Paolo 052617
Cherek, Dr. Janina ... 034688, 034690
Cherif, Housine Med 000079
Cherix, Christophe 041516
Cherix, Dr. Daniel L. 041655
Chermak, Glenda L. 050058
Cherrier, Anne-Marie 014828
Cherry, Stella 024857
Chesebrough, David 047302
Chesi, Prof. Gert 002740
Chesneau-Dupin, Laurence . 012611,
012991
Chester, Timothy J. 049195
Chesters, Dr. Robin 044870
Cheval, François ... 012405, 012406
Chevalier, Alain 015930, 016317
Chevalier, Bernard ... 013284, 015190
Chevallier, Denis 014054
Chevillot, Catherine 014743
Chew Matsuda, Fay 051438
Chew, Nigel 045733
Chew, Ron 053252
Chez Checor, José 010389
Chia, Khoo Boo 030723
Chiacchierini, Prof. Ernesto ... 027611
Chiapperino, Dr. Paola . 025659, 028295,
028310
Chiappini, Dr. Marco 025375
Chiara Bettini, Dr. Maria 025774
Chiaramonte, Dr. Giuseppe . 027187
Chiarchiaro, Charles 051855
Chiarelli, Dr. Caterina 026257
Chiarucci, Dr. Giuseppe 025266, 026556
Chiba, Mizuo 029635
Chicharro Chamorro, José Luis . 039524
Chidavaenzi, R.P. 054847
Chiego, William J. 052940
Chiesi, Dr. Giuseppe .. 041297, 041300
Chikhaoui-Trachene, Prof. Henia 031775
Chikkamare Gowda, B.M. 024150
Chilcote, Doug 052785, 052786
Chilcote, Gary 052785, 052786
Childress, Ann 048097
Childrey, Brooke 049732
Childs, Victoria W. 052789
Chilton, D.G. 045726
Chilton, Meredith 007176
Chimay, Prince de 003520
Chimello, Sylvain 015937
Chimier, Brigitte 016078
Chiminelli, Baldassare 025825
Chimisliu, Cornelia 036419
Chin Foo, Brenda 016378
Chin, Song-gi 029957
Chin, Wanda 048662
Chinea Brito, Carmen Dolores .. 039541
Chinen, Satoru 028652
Chinery, Derek 044601
Chintãuan, Ioan 036539, 036677
Chinwe Anyaegbuna, Felicia ... 033420
Chiocchetti, Dr. Fabio 027434
Chiong, Dr. Chew Tuan 038275
Chipperfield, Donna 051526
Chiriac, Aurel . 036596, 036597, 036598,
036673
Chiriac, Mihai 036289
Chirilã, Dorina 036767
Chiritã, Lenuta 036519
Chisholm, Rachel 045376
Chişiu, Ambrozia 036641
Chitajlenko, Nikolaj Nikolaevič . 037474
Chitaley, Dr. Shya 047809
Chitoshi, Matsuo 028750
Chiu, Lei-Mei Julia 029115
Chiu, Melissa 051340

Chivers, Jim 033135
Chižnjak, Olga Romanivna 043160
Chlebarova, Marija 004992
Chlebecek, Friedrich 002167
Chlebinskas, Antanas 030556
Chlobystin, Andrej 037797
Chłodnicki, Dr. Marek 035121
Chlopinska, Ljudmyla Dmytrivna 043266
Chmara, Gitta von 019337
Chmelinova, Dr. Katarina 038328
Chmelnickaja, Ljudmila 003255
Chmelyk, Roman 043256
Chmielewska, Wiesława 035181
Chmielewski-Hagius, Dr. Anita . 019437
Chmurzyński, Wojciech 035341
Cho, Byoung Soon 030047
Choate, J.R. 049416, 049416
Chocholáč, Ivo 009718
Chodasiewicz, Jerzy 035242
Chodžaev, Muchtar Bachadirovič 029842
Choe, Sung Eun 029993
Choetf, Betsy 049672
Choi, Prof. Duck-Keun 030040
Choi, Eric 008066
Choi, Il Sung 029910
Choi, Kim Sang 029875
Choi, Man-Li 029939
Choinière, France 006370
Choiselat, Roland 006498
Choisne, Franck 015715
Cholette, Yvon 007122
Cholewa, Anita 050940
Cholewa, Kurt 022622
Cholodinskienė, Adele 030420
Chomenko, Tamara Jevgenivna . 043204
Chomič, Svetlana Viktorovna .. 037147
Chong, Alan 047117
Chongkol, Chira 042670
Choremi, Dr. Alkestis 022994
Choremis, Alketis 023024
Chorunženko, V.A. 037010
Chou, Ju-Hsi 047808
Choudhury, B. 024270
Choudhury, Dr. R.D. 024272
Choulet, Jean-Marie 013071
Chow, Christina 051459
Chow, Kim-Fung 007878
Chowdhuri, Dr. M.K. 024358
Chowdhury, Prof. Dr. Saifuddin . 003232
Chrest, Gwen 047836
Chrétien, Isabelle 016014
Chrisanov, Petr Aleksandrovič . 037878
Chrisman, D. Kent 052603
Christ, Hans D. 022068
Christ, Magdalene 017053
Christ, Nelda M. 047046
Christen, Benoît 012012
Christen, François 041800
Christen, Gian Franco 041348
Christen, Judith 041704
Christen, Thomas 041380
Christensen, Barbara 051132
Christensen, Christa 047567
Christensen, Henry 033241
Christensen, J.O. 010199
Christensen, Lars K. 010208
Christensen, Lisbeth 010092
Christensen, Margaret 006036
Christensen, Peter M. M. ... 010138
Christensen, Ruth 030302
Christensen, Svend 010138
Christenson, Elroy 053242
Christenson, Goran 040829
Christiaens, Daniel 003290
Christian Pedersen, Sune 010191
Christian, Jay 048702
Christiansen, Calvin 054210
Christiansen, Jette 010188
Christiansen, Sherie 053290
Christie, David 051976
Christie, Gordon A. 044352
Christie, Nancy 051976
Christman, David C. 049437
Christmann, John 007205

Christmeier, Martina 020847
Christodoulakos, Yannis 023038
Christoffel, Udo 017317
Christoffer, Ed 005827
Christoffersen, Erna 010118
Christoforou, Dionysia 045219
Christophersen, Axel 034041
Christou, Dr. Demos 009434
Christov-Bakargiev, Carolyn ... 027537
Christov, Christo 005025
Christova, Emilia 005052
Christy, Strum 053119
Chrzanowski, Jerzy 035457
Chu, Dr. Christina 007871
Chu, Doris 047135
Chu, Guiqing 007778
Chu, Shibin 007996
Chu, Xinjia 008260
Chubbuck Weinstein, Elizabeth . 046814
Chubrich, Michael 052316
Chudjakova, Lidija Aleksandrovna 036984
Chudy, Tomasz 035314
Chughtai, Arif Rahman 034127
Chuma-Ibe, Gloria Chianu ... 033433
Chunchu, Ai 007601
Chung, Estella 050455
Chung, Prof. Hsiung Yih 042450
Chung, Prof. Jing-Won 029894
Chung, Yang-Ho 030029
Chuoc, Cu Van 054796
Chupin, Olivier 011662, 012290
Church, Dorian 045774
Church, Rena J. 046672
Church, Susan 049662
Churcher, Andrea May 001313
Chushupal, E.N. 030084
Chusnutdinova, Rušanija
Salichovna 037111
Chutiwongs, Dr. N. 032585
Chvanov, Michail Andreevič .. 037978
Chwałek, Roman 035187
Chwaliński, Andrzej 035055
Chwalisz, Zofia 035409
Chylak, Ana 043429
Ciaffoni, Mirko 026877, 026878
Ciamos, Christian 011605
Ciancio, Dr. Angela 026443
Cianciulli, John 048937
Ciannella, Joeen 049493
Ciapponi Landi, Bruno 026670
Ciaran, Dr. Fiona 033311
Ciardi, Dr. Roberto Paolo ... 027787
Ciarla, Dr. Roberto 027627
Ciavolino, Prof. Giuseppe ... 028150
Cibulová, Dr. Vlasta 009833
Cicala Campagna, Dr. Francesca 027967
Cicala, Francesco 027196, 027197
Cicali, Francesco 027198
Çiçeksız, İbrahim 042921
Cicelyn, Eduardo 026990
Cichecka, Barbara 035383
Ciciero, Carla 046940
Cicierski, Ernest 007470
Čičimov, Aleksandar 030640
Cicimurri, David 047797
Cicognani, Giorgio 025643
Cidenova, Rita Badmaevna ... 037986
Ciecior, Gabriele 019055
Ciepiela, Elżbieta 035443
Cierna, Katarína 038328
Cieślak, Bogdan 035412
Cieślicki, Dorothy 048458
Cieslik, Gérard 011648
Cieślikowska-Ryży, Julita ... 034794
Cifelli, Dr. R.L. 051588
Cifuentes, Joaquín 031229
Ciganok, Nina 043145
Cigarroa, Melissa 050147
Cigliano, Flavia 047126
Cihan Özsayner, Zübeyde ... 042939
Čikin, Vladimir Iljič 043289
Čilašvili, L.A. 016422
Cilona, Filippo Ferdinando ... 026191
Cimdiņa, Ruta 030184

Činč-Juhant, Dr. Breda 038556
Cinc-Mars, François 007139
Cindi, F. 038770
Cini, Amato 028240
Cini, Stephen 030760
Činovec, Igor 009880
Cinquepalmi, Angela 026189
Cintulova, Erika 038360, 038362
Ciobanu, Adrian 036660
Cioce, Dr. Pasquale 025522
Cioffari, Gerardo 025455
Cioppi, Dr. Elisabeth 026281
Ciorăscu, Nicolae 036486
Cipoletti, Christopher 047889
Cipollone, Dr. Rosa 027104
Ciprian, Dr. Pavel 009472, 009477,
                         009484
Cirafini, Sandrina 041335
Cirani, Simona 027727
Cirbian, Susanna 040365
Cîrciu, Gratiela Ionelia 036405
Cirill, T. 023829
Cirillo, Lena 001370
Cirjakovic, Ivana 038188
Čirkova, Larisa I. 037757
Ciro Galotta, Brás 004904
Cîrule, Astrīda 030140
Ciseri, Dr. Ilaria 026902
Cisewski, Gene 049639
Cisilino, Dr. Daniela 025417
Cisneros, Jose 049809
Cisneros, Mario Toledo ... 031438
Cisneros, Rosa María 033385
Cist, O. 002312
Ciubotă, Dr. Viorel 036250, 036363,
   036607, 036682, 036683, 036684,
                    036810, 036811
Ciubotaru, Neculai 036831
Ciucă, Răzvan 036722
Ciugudean, Horia Ion 036199
Civai, Mauro 027954
Čiž, Marian 038459
Claassen, Garth 047345
Clabby, Paul 051262
Claes, Gabrielle 003478
Claessens Peré, Nanette ... 003319
Claesson, Anna Maria . 040719, 040952
Clair, Cynthia 051249
Clalüna, Alfons 041918
Clam-Martinic, Dr. M.C.D. ... 002199
Clancey, Erin 050468
Clancy, John L. 053979
Clap, Sylvestre 011837
Clapperton, Raymond 033201
Clarck, Trinkett 046451
Clare, Roy 045059, 045097
Clarés, María Jesús 040308
Claret de Gouvêia, Antônio M. ... 004604
Claret, Sawador 040185
Clarin jr., Oligario L. 034497
Clark-Bedard, Sharon 006507
Clark-Langager, Sarah 046894
Clark, Alexia 044574
Clark, Andrew 001503
Clark, Becky 051688
Clark, Bill 048338
Clark, Carroll 051555
Clark, Charles T. 047790
Clark, Elaine R. 049239
Clark, Eleanor 045083
Clark, Gayle 052571
Clark, H. Nicholas B. 046449
Clark, Jackie 051214
Clark, Jane E. 048934
Clark, Jeanne 050104
Clark, Joann 050882
Clark, Kathy M. 051798
Clark, L.R. 001215
Clark, Martin 045496, 045700
Clark, Michael J. 053922, 053923
Clark, Nancy 046521
Clark, Nora 047400
Clark, Patrick S. 053185
Clark, Peter 048964

Clark, Peter P. 047975
Clark, Phillip 050389
Clark, Richard H. 054322
Clark, Robert 044066
Clark, Robert A. 044032
Clark, Robin 050040
Clark, Roger 043698
Clark, Roger W. 048740
Clark, Scott 052784
Clark, Dr. Thomas D. 050246
Clark, Tom 053948
Clark, Vicky A. 052171
Clarke Di Vico, Carol 046794
Clarke, A.H. 000904
Clarke, Alicia 053081
Clarke, B. 043842
Clarke, Cliff 005778
Clarke, David 033134
Clarke, Deb 045884
Clarke, Ignatius 024961
Clarke, Jeremy 045628
Clarke, Jo 045722
Clarke, Judith 045478
Clarke, Lance 049210
Clarke, Michael 044255
Clarke, Robert 043375
Clarke, Ron 044019, 044515
Clarke, Sandi 000925
Clarke, T. Henry 054245
Clary, Owen 046323
Clason, Anders 040654
Clatworthy, J.C. 043621
Claude, Yves 030572
Claus, Elisabeth 016620
Clausen, Claus 010123
Clausen, H. Erhardt 010015
Clausen, Jochen 019888
Clausen, Kim . 010110, 010311, 010313,
                 010314, 010315
Clausen, Kirstin 005483
Clausen, Willy 041473
Clausnitzer, Regina 018838
Claussen, Hans-Joachim ... 022275
Clavedetscher, Silvia 041859
Clavel, Sylvie 014701
Claveria, Josef 003021
Clavier, Michel 003871
Clavijo Colom, Moraima ... 009201
Clavijo, Martha Isabel 008804
Clawson, Rita 046844
Clay Chace, Helen 051390
Clayden, Stephen 006893
Clayton, George C. 048658
Clayton, Rachelle 005316
Cleal, Rosamund 043452
Clear, Celia 045115, 045116
Clearwater, Bonnie 050839
Cleary, Chris 001450
Cleasby, I.W.J. 043424
Clegg, Dr. W.H. 045983
Cleland, Yvonne 043533
Clemens, Darlene 052244
Clemens, M. 014023
Clemens, William 046939
Clement, Constance 051266
Clément, Lise 005614
Clement, M. 014276
Clement, Morag 044675
Clement, T.A. 046180
Clemente, Milagros 034436
Clements, Diane 045043
Clements, Elaine 046480
Clements, Gayle 053848
Clements, Louis 052521
Clements, Lynn B. 054000
Clendaniel, Dianne 046784
Clephas, Egon 020606
Clergue, Yolande 011728
Clerke, Rachel 045677
Clette, Michele 003473
Cleve, Thomas 019906
Cleveland, James C. 052520
Cleveland, Richard L. 053234
Cleven, E. 032098

Clevenger, Martha R. 050390
Clews, Stephen 043525
Cleyet-Merle, J.J. 013749
Cliffe, L. 045797
Clifford, Bridget 045088
Clifford, Lord Thomas 043959
Clifton-Harvey, Melanie ... 053528
Clifton, Dan 048425
Clifton, Dr. James 049595
Clifton, James 049588
Clifton, Julia 053114
Clin, Marie-Véronique . 014584, 014742
Cline, Garry 046461
Cline, Raymond 051660
Clinefelter, Pamela 050108
Clinger, Melinda 052606
Clinkingbeard, Brion 050493
Clitheroe, Gordon 045502
Clizbe, William C. 049289
Clodfelter, Lee 047236
Cloître, M.C. 014905
Clos Llombart, Jordi 039093
Cloşcă, Gabriela 036264
Close, Elisabeth 001020
Close, Freddy 003565
Close, Helen 045168
Close, Oliver 000903
Close, Timothy 047073
Cloudman, Ruth 050502
Clouter, Carl 005857
Clow, Wayne 053586
Cluff, Raydene 053727
Clute, Richard 046420
Cluver, Dr. M.A. 038691
Clyve, Lynda 046778
Cnops, Pieter 003444
Coakley, Stephanie 053837
Coaña deValicourt, Yago Pico de . 039689
Coates, G. 005922
Coates, Ross 052416
Cobarg, Dr. Merete 020682
Cobb, Elvan 049483
Cobb, Michael 049341
Cobb, Myrna 005578
Cobban, Janet 005991
Cobbe, Toy L. 050726
Cobble, Kelly 052435
Cobbold, A.G.B. 043811
Cobert, William 051326
Cobo, Maria 047262
Cobos Colmenares, Carlos
   Eduardo 008771
Cobos Wilkins, Juan 039756
Cobridges, Robert 051057
Coburn, Dr. Carol 049895
Coburn, Oakley H. 053430
Cobzaru, Dumitru 036463
Cocartă, Adrian 036457, 036458, 036459,
                 036460, 036461
Cocea, Sergio Eduardo ... 000687
Cochat, L. 015946
Cochener, Margaret M. ... 052602
Cochet, Vincent 013013
Cochran, Dorothy 051484
Cochran, Glenn 053753
Cochrane, G. 044138
Cochrane, Dr- Timothy ... 049187
Cockrill, Pauline 000786
Cockrum, Terry 051739
Cocordas, Eleni 051411
Cocuz, Prof. Ioan ... 036551, 036715,
   036733, 036735, 036737, 036806
Codagnone, Brian 047134
Coday, Jean C. 050655
Codding, Mitchell A. 051404
Codega, Samuele 027456
Codicé Pinelli, Dr. Ferruccio . 025653
Codomi, Shelley 048481
Codreanu, Elena 036575
Coe, Erin M. 049123
Coe, John G. 047631
Coekelbergs, Prof. R. 004152
Coelewij, L. 031986
Coelho, Conceição 035971

Coen, Dr. Vittoria 025531
Coenen-Staß, Prof. Dr. Dieter . 020183
Coenen, Luc 003816, 003817
Coënen, Michel 016352
Coers, Frits 032945
Coetzee, F.P. 038850
Coetzee, Marc 050848
Cofax, Debbie 052572
Coffelt, Jon 047001
Coffey, Jeanne 049487
Coffey, John W. 052450
Coffin, Anne 051407
Coffrini, Ermes 025625
Cogan, Stanley ... 048766, 048768
Cogar, William 051146
Cogdell, Sabine 051673
Cogeval, Guy 006391
Cohalan, Mary Lou 048438
Cohen-Straytner, Barbara . 051456
Cohen, A. 025177
Cohen, Christopher 000825
Cohen, David 051457
Cohen, Françoise . 011497, 014501
Cohen, Jack 052080
Cohen, Janie 047328
Cohen, Judith 025183
Cohen, Marvin H. 050866
Cohen, Nancy 053599
Cohen, Patricia 000262
Cohen, Ronald A. 050884
Cohen, Sandor 046696
Cohn, Arthur B. 053958
Cohn, Marjorie B. 047362
Cohn, Terri 046933
Cohon, Dr. Robert 049888
Cohrs, Karin 019368
Cohrs, Peter 019368
Coibanu, Doina 036337
Cojocaru, Adriana 036741
Cokendolpher, James ... 050517
Coker, Gylbert 053738
Colaianni, Elynn 052353
Čolakov, Dimităr 005224
Colantuono, Patrizio ... 025537
Colardelle, Michel 014778
Colardelle, Renée 013179
Colato, Juan 050474
Colazo, Susana 000527
Colbert, Brigitte de 014294
Colburn, Bolton 050066
Colby, James 049785
Coldren, James 048552
Cole-Will, Rebecca 046781
Cole, Felicity 045372
Cole, Garrett 046777
Cole, Geoffrey 043466
Cole, Glory 049499
Cole, Jane 006449
Cole, K. 038709
Cole, Mark 047936
Cole, Natalie 046179
Cole, Polly 043743
Cole, Richard 045054
Cole, Vicki 053051
Coleman, C. Allen 052126
Coleman, Don 048762
Coleman, Evelyn 024946
Coleman, Mary Ruth ... 046384
Coleman, Rhonda 052936
Coleman, Sally 045883
Coleman, Winifred E. .. 054304
Coles, Alec ... 045347, 045349, 045819,
                 045926
Coles, Kerrie 001353
Coley, Dwaine C. 049463
Colin, N. 003605
Colina, Gay 054198
Colitti, Raffaele 025729
Coll Conesa, Dr. Jaume . 040321
Coll Gormley, Jean 049424
Coll, Tom 054591
Colla, Francis 003596
Collado, Elisa 039861
Collas, Josiane 012808

Colle, Jean-Pierre 013110
Colle, Marcella 003698
Collens, David R. 051109
Collett, Keith L. 000897
Collett, Morine 049717
Collette, Dr. Alfred T. . . 053634, 053638
Collette, Alice 049577
Collette, Louis Jeffery 050828
Collier, H. 045602
Collier, Ric 052916
Collin, Fernand 003599
Collin, Dr. Gérard 012023
Collin, Ludo 003638
Colling, Guy 030594
Collings, Donna 006354
Collins Blackmon, Mary 046689
Collins Coleman, Clara 052807
Collins, Anne 046081
Collins, Craig 047095
Collins, D. Cheryl 050636, 050638,
050641, 050642, 050643
Collins, David 051253
Collins, Dr. Des 007200
Collins, George 049241
Collins, J. Thomas 049455
Collins, James H. . . . 048332, 050987
Collins, Lin 045338
Collins, Louanne 045187
Collins, Lynn 001607
Collins, Mary 052415
Collins, Michael 050523
Collins, Paul W. 053093
Collins, Philip 045252
Collins, Phillip 048108
Collins, Toni I. 048591
Collins, Vickie 052073
Collinson, Howard 049712
Collison, Nika 007045
Collu, Cristiana 027074
Colman, Devin A. 052843
Colom, Connie 050538
Colombara, Franco 025993
Colombi Ferretti, Dr. Anna 026176
Colombié, Ervé 015478
Colombino, Carlos . . . 034193, 034200
Colombo Dougoud, R. 041533
Colombo, Dr. Silvano 025858
Colomer, Laia 038927
Colon, Anibal 052764
Colon, Constance B. 053546
Colon, Evangelina 036167
Colon, Helen 045715
Colonna Ceccaldi, Michel 016179
Colonna, Cécile 015165
Colonna, Vittoria 025719
Colossi, David 051399
Colpaert, Hilde 003734, 003735
Colton, Chris 007233
Colton, Fiona . 044544, 044665, 045759
Columbo, Colleen 005890
Colvia, Dr. Anna 027622
Colvin, Richard D. 050607
Colwell, Bruce 051633
Comanducci, Dr. Francesco 027812
Comas Estatella, Dr. Antonio . . . 040386
Comba, Steve 047758
Combe, Anette 041829, 041830
Combe, Jean 042097
Combe, M. 015082
Combs, Jim 049940
Combs, Robert K. 048836
Comeau, Cy 005396
Comeault, Maurice 006923
Comedown, Alberta 054058
Comer, Karen 046614
Comer, Michael 051282
Cometti, Jorge 000182
Cometti, Marta 041394
Comiller, Carla 003089
Comingore, B. 053264
Comlan Godonou, Alain 004187
Commandeur, Beatrix 017052
Commock, Tracy 028443, 028450
Communeau, Matthieu 012750

Compagno-Roeleveld, M.A. 038686
Compagno, Dr. L.J.V. 038686
Compaore, Benilde 005278
Comparin, Guy 013172
Compston, Nici 000762
Compton, Richard 045620
Comte, Philippe 014806
Comte, Samantha 001072
Comtois, George S. 050255
Cona, Dr. Cherubino 027837
Conable, Mary 054073
Conaway, Mary Ellen . 049051, 049073
Conceição Correia, Maria da . . . 035666
Conceição Duarte Rosa, Maria
da 004598
Conconi, Maria Pia 041886
Conde, Victor 039655
Condeescu, Dan Alexandru 036279,
036280, 036281, 036282, 036283,
036308, 036816
Conder, Marjorie D. 052915
Conder, Tony 044925
Condos, Lena 001013, 001014
Condurache, Val 036517, 036521,
036668
Cone Busch, Ellen 051874
Cone Skelton, Annette 046627
Cones-Browne, Rodney 001355
Coneva, Tanja 005173
Confer, Deborah 047144
Conforti, Michael 054419
Cong, Pham Van 054806
Conger, Ivan A. 051856
Conger, Ken 053297
Conihout, Isabelle de 014622
Conijn, G.H. 032264
Conijn, N.W. 032013
Conkle Davis, Gloria 046590
Conklin, David W. 048293
Conklin, Jo-Ann 052381
Conkright, Lance 054294
Conlan, Lynne 043966
Conley, Alston 047625
Conley, Maria 046316
Conlon, Beth 047785
Conlon, Helen 045712
Connaughton, Vincent 024960
Conneau, Jean-Claude 012654
Connell, Andrew 044277
Connell, Dr. D.P. 044711
Connell, John 006332
Connemann, Elfi 017722
Connemann, Hans-Joachim 017722
Conner, Michael 047519
Conner, Roberta 051970
Connolly, Angela 048284
Connolly, Thomas 048622
Connor, Brian 033193
Connor, James T. H. 054146
Connor, Kathy 052617
Connor, Lorraine 050587
Connors, Dennis 053637
Connors, Dr. Thomas 047475
Conocido, Domicilio 030947
Conover, Margaret 053274
Conrad, Dr. Christofer 022057
Conrad, Ernesto 041846
Conrad, Geoffrey 047050
Conrad, Dr. J. 018695
Conrads, Margaret 049888
Conradt, Dirk-Michael 022737
Conrath-Scholl, Gabriele 019676
Conrot, Valérie 030588
Conroy, Christopher J. 046940
Consagre, Franesca 052818
Consey, Kevin E. 046933
Considine, Brian 050436
Conso, Prof. Giovanni 027662
Console, Dr. Ester Maria 054682
Constable, Linda 043682
Constable, S. 045391
Constant Noanti, Mahoun 004183
Constant, Carol 049519
Constantin, Alupei 036251

Constantinescu, Corina 036392
Constantino De La Penna, Carol 040191
Constantopoulos, Dr. Jim 052271
Consuelo Dominguez Cuanalo, Delia
del 031401
Contarino, Dr. Giuseppe 025234
Contassot, Cécile 013173
Conte, Cynthia 047582
Conte, Dr. Robert S. 054375
Conte, Roxana Diaz 000520
Contessa, Gregorio 026690
Conti, Prof. Dr. Luigi Francesco . 026380
Conti, Richard C. 051579
Conti, Roberto 026972
Conticelli, Dr. Valentina 026256
Conticello, Prof. Dr. Baldassarre . 025612,
028000
Conticello, Dr. Baldo 026843
Contiguglia, Georgianna 048255, 053804,
053805, 053807
Contini, Dr. Roberto 017147
Contreras Martínez, José
Eduardo 031615
Contreras, Mamerto 034572
Contreras, Pilar 031195
Contro, Antonia 047672
Conway, Allen C. 051220
Conway, Cari 051518
Conway, James 048300
Conway, Rosemary 054006
Conway, Steve 004198
Conway, Steven 004195
Conwell, Idella M. 050659
Conyers, Wayne 050576
Conyngham, Frank E.P. 054406
Conzelmann, Peter 017453
Conzen, Friedrich G. 018026
Conzen, Dr. Ina 022057
Cook, Darrell J. 048076
Cook, Deborah 045814
Cook, Emma 046073, 046087
Cook, Janel 050540
Cook, Jennifer 049147
Cook, Joyce A. 048768
Cook, Mary Jo 051117
Cook, Prof. Nigel 033847
Cook, Patrick 046138
Cook, Robert 051919, 054296
Cook, Silas B. 052291
Cook, Steven 005716
Cook, Thomas 047905
Cook, Wendy 046193
Cooke, Bill 050245
Cooke, Carolyn 049699
Cooke, Glenn 001503
Cooke, Lynne 046841, 051378
Cooke, Michael 028440, 028444, 028446,
028456
Cooke, Pat 024895, 024905
Cooke, Patrick 044114
Cooke, Prof. Robin 001142
Cooke, Dr. Steven 001509
Cooke, Susan 045151
Cooke, Vina 045337
Cookendorfer, Betty 049376
Cool, F. 003974
Coolidge, Hope 048918
Coolidge, Jennifer 048182
Coolidge, John 050124
Coombe, JoAnne 048386
Coombs, Gary B. 049153
Coombs, Margery C. 046452
Coombs, Robert 054380
Coomer, Robert 048556, 053950
Cooney Frelinghuysen, Alice . . . 051424
Cooper Case, Carolyn 052539
Cooper, Andrea 047875
Cooper, Avis 033200
Cooper, Bernard F. 053350
Cooper, Charles E. 044118
Cooper, Grant 033178
Cooper, Harry 047362
Cooper, Helen 051267
Cooper, Dr. James H. 051485

Cooper, Jo-Ann 001272
Cooper, John 044860
Cooper, Julie 047502
Cooper, Katie 049221
Cooper, Linda G. 054595
Cooper, Norman 006238
Cooper, Rhonda 053550
Cooper, Robert L.D. 044243
Cooper, Ronna 052028
Cooper, Rosemary 045364
Cooper, S. 044614
Cooper, Timothy O. 047292
Cooper, Valerie 005515
Cooper, Virginia 007166, 051132
Coopmans, H. 032022
Coote, Sylvia 045908
Cope, Derek 046109
Cope, Michael W. 044632
Copeland-Burns, Carla 050260
Copley, Dr. David C. 052967
Coppens, Marguerite 003492
Coppernoll, Lee 050923
Coppieters, Pierre 003884
Coppock, C. 045473
Coppola, Gianfranco 027736
Coppola, Regina 046454
Coppus, G.M. 032904
Copson, R. 044284
Coquery, Emmanuel . . 016051, 016052,
016053, 016054
Coral, Laurent de 015376
Corazza, Dr. Carla 026207
Corbat, Marc 041841
Corbett-Cyr, Susan 006623
Corbett, Agnes 047383
Corbier, René 012313
Corbin, Dr. James E. 051149
Corbin, Kendall W. 050940
Corbishley, Grant 033281
Corblet de Fallerans, J. 016000
Corbu, Emilia 036723
Corcoran, M. 024902
Corcos, Sidney 025090
Corcuera, Catalina 031146
Cordaro, Dr. Michele 027561
Cordell, Prof. Linda S. 047144
Corder, Russ 052204
Cordero, Carlos 050371
Cordero, Dr. Mario 026116
Cordes, David 048289
Cordewener, Raphaël 003958
Córdoba, Elsie D. 034168
Cordonnier, Aude 012844
Cordova-Gonzalez, Julia 007515
Cordova, Rudy 054046
Coreth, Peter 002893
Corey, Mark 049239
Coria, Rodolfo 000495
Corio, Amedeo 028335
Corkern, Wilton C. 046314
Corkum, Kate 051226
Cormack, Malcolm 052565
Cormack, Peter 045135
Cormier, Angèle 006348
Cormier, Bernard J. 006890
Cormier, George Ann 052254
Cornamusaz, François 041563
Cornejo Gutiérrez, Miguel Angel 034265
Cornejo, Stephanie 052729
Cornelis, J.P. 003937
Cornelissen, Leo M. 031989
Cornelius, Betty L. 051927
Cornelius, Robert E. 049728
Cornelius, Dr. St. 038790
Cornelius, Steffi 017349
Corner, Rita 007150
Cornet, Josep 039074
Cornetto, Elisabeth 011905
Cornez, G. 003979
Corniani, Maurizio 026698
Cornick, N. 003953
Cornini, Dr. Guido 054682
Cornu, Bernard 013983
Cornwell, John 001187

Coroama, Radu	036314	Couderc, Sylvie	011632, 011633	Crampton, Julian	043742	Crocker, Piers	033979
Coroiu, Dr. Ioan	036385	Coudray, Robert	013843	Crampton, Lewis S.W.	052635	Croft, Brenda L.	000937
Corominas de Basso, Raquel	000420	Coudrot, Jean-Louis	012520	Crampton, Sharon	038661, 038663	Croft, Evelyn	052741
Coronel, Maria Delia	034463	Coudurier, Yvonne	014399	Crandall, Daniel T.	052301	Crofts, Nicholas	041530
Coronilla Castro, Manuel	039302	Coughlan, M.	001347	Crane, Louise	053735	Croke, Fionnuala	024900
Corrain, Prof. C.	026465, 028011,	Coughran-Rayden, Sharon	053956	Crane, Prof. Peter	044687	Croker, R.C.	033156
	028233, 028387	Coulibaly, Seydou	008867	Crane, Susan	053212	Crokett, James	046745
Corre, Caroline	016164	Coulombe, Marie-Christine	007265	Cranson, Diane	000781	Croll, Helen	000934
Correa Sutil, Dr. Sofía	007545	Coulombier, Jean-Luc	015385	Cranston, Peter S.	048163	Cromie, Cheryl-Anne	038740
Correia, Dr. Alberto	036114	Coulon, François	015067	Craughwell-Varda, Kathy	053189	Cronin, Michael	051454
Correia, Dr. Margarita Rebelo	035952	Coulon, Gérard	013459, 015196	Craven, Jane	049771	Cronin, Thomas E.	054039
Correia, Dr. Maria João Pinto	036115	Counot, Bernard	013794	Craven, K.	045722	Cronlund, Carl-Johan	041031
Correia, Dr. Virgilio H.	035639	Count, Mary Ellen	051001	Craven, Tim	045828	Crook, Rosemary	044499
Corrigan, David J.	049388	Cour Dragsbo, Peter la	010323	Crawford-Gore, Betsy	050150	Crook, Rosie	044503
Corrin, Lisa	053248, 054420	Courson, Bruce	053075	Crawford, Bob	052975	Cropera, Dr. Carla	026222
Corriveau, Linda	051976	Courtelarre, Henry	012300	Crawford, C.	006789	Cropp, Aubrey	033165
Corry, Corrine	006796	Courtial, M.C.	013911	Crawford, Henry B.	050517	Cros, Caroline	014697
Corry, L.	045438	Courtiour, Jon	045990	Crawford, James	047395	Crosby, Philippa	044480
Corser, Elizabeth	005635	Courtney, Al	053272	Crawford, Joan	047194	Crosman, Christopher B.	052644
Corsini, Principessa	026254	Courtney, Vernon	054402	Crawford, Louise J.	050984	Cross, Christine	046512
Corsini, Deborah	053044	Courtois, Jean-Pierre	003645	Crawford, Madelyn	050405	Cross, Dave	007492
Corsolini, Prof. Rodolfo P.	000145	Couser, Carol	050336	Crawford, Nina	005683	Cross, Louise	049119
Cort, Cynthia	049210	Cousin, M.	001347	Crawford, R.W.K.	045003	Cross, Sharon	043835
Cort, Louise	054091, 054116	Cousino, Beate	011926	Crawford, Dr. Richard	049178	Cross, Tony	043406, 043407, 045494
Cortenova, Prof. Giorgio	028343	Cousinou, Olivier	014044	Crawford, Robert	045206	Crossland, Leonard B.	022938
Cortés Gomez, Carlos	031149	Coutagne, Denis	011563	Crawford, Shirley	046376	Crossland, Maynard	050220
Cortes, Prof. Erlinda S.	034610	Coutheillas, Didier	012920	Crawford, Vivian	028443	Crosson, David	053038
Corteville, Julie	014267	Couto, Dr. Matilde Tomáz do	035588	Crawley, Tracey	045859	Crosson, Diana	047871
Cortez, Miguel	047692	Coutts, Herbert	044279	Crawshaw, Clare	044677	Crothers, George M.	050254
Cortina, Leonor	031168	Coutts, Howard	043495	Creal, Carolyn	053429	Croton, G.	033215
Cory, Jeff	047644	Couturas, Francis	014829	Creamer, Winifred	048223	Crotti, Pierre	041653
Corzo, Miguel-Angel	052092	Couture, André	006725	Crean, Kristen	052284	Crouch	000998
Cosandier, Juliane	041651	Couture, Llyr E.	048353	Creary, Marilou	051646	Crouch, Ned	047776
Coşereanu, Valentin	036522	Covaci, Heruvima	036566	Creasey, Michael	052396	Crouch, Philip	044681, 046196
Cosgrove, Patricia	046655	Covato, Prof. Carmela	027644	Creddes, Jean	006017	Crous, Dr. D.	038680
Coski, Dr. John M.	052557	Coventry, Susan	045014	Credland, Arthur G.	044710, 044716,	Crouse, George	049877
Cosma, Aurelia	036310	Covey, Dustin	005756		044718	Crouter, Philip	043708
Cosmetatos, Helen	022967	Cowan, Cornelius	049699	Credle, Jamie	053174	Crouy-Chanel, Bernard de	013430
Cossa, Egidio	027634	Cowan, Joan	005784	Creech, C.L.	050005	Crow, Pamela B.	054016
Cossart, Graham	000909	Cowan, Lina	049786	Creecy, Kerry	001123	Crowdy, Hannat	043933
Cossignani, Dr. Tiziano	026120	Cowart, Jack	054098	Creemers, Guido	004074	Crowe, K.L.	045836
Cosson, Maurice	012480	Cowburn, Ian	013574	Crégut, Dr. Evelyne	011841	Crowe, Lilah J.	049197
Costa, Diane	053077	Cowden, Chris	046705	Creighton, Martha	054047	Crowell, Claudia	049183
Costa, Elisabeth	035781	Cowell, Lorraine	054031	Creighton, Dr. Philip D.	048774	Crowell, Darryl	005525
Costa, Dr. Gian Paolo	026172	Cowell, R.	044869	Crelot, Stefan	003869	Crowell, Stephen	052424
Costa, Prof. Giovanni	025871	Cowell, Tom	007410	Cremer, Kurt	016446	Crowell, Stewart A.	051308
Costa, John	052522	Cowie, Kate	000940	Cremers, Filip	004098	Crowell, Todd	047730
Costa jr., Luís da	031809	Cowie, Robert G.	051914	Cremoncini, Roberta	044965	Crowley, Noel	024920
Costa, Dr. Susana Faro da	035589	Cowley, Paul	044157	Crenzien, Helle	010146	Crowley, Peter D.	046446
Costamagna, Dr. Alba	027622	Cowlin, Wendy	046100	Crépin-Leblond, Thierry	012076, 012078,	Crowley, Terry	001108
Costamagna, Liliana	028005	Cowling, A.	045391		012081, 012855	Croy, David E.	048029
Costantini, Dr. Lorenzo	027627	Cowling, Dr. Mary	044281	Crépy, Edvard de	016195	Crozat, Oliver	014754
Costar, John	033306	Cowling, Sue	051790	Crespo-Burgos, Dr. Antonio	010492	Crozemarie, Gilles	012567
Costard, Matthew P.	045682	Cox, Arrington	047910, 047913	Crespo Guliérrez, Victoria	039681	Crozzolin, Umberto	028188
Costas Loyola, José E.	036140	Cox, Dan	050675	Crespo, Michael	046822	Crué, Emmanuel	012587
Coste, Pierre	011894	Cox, Dr. Douglas C.	052404	Cressey, Dr. Pamela J.	046381	Crüzer, Cristina	042022
Costello, Lisa	047520	Cox, Janet	047803	Cresswell, J.	043680	Cruikshanks, Randal	053050
Costello, Noel	001177	Cox, Janson L.	047017	Creutzburg, Gerlinde	016481	Cruise, Sandra	045636
Costelloe, Jean	024979, 024980	Cox, Katherine	049626	Crevecœur, Claude F.	004143	Crul, Yves	003363
Costely, Nigel	045985	Cox, Marilyn	051040	Crew, Keith	052105	Crum McCabe, Elsie	050381
Costersian, John	049109, 054480	Cox, Nigel R.	044433	Crews, Gerald	005898	Crum, Katherine	053420
Costescu, Cornelia	036666	Cox, Dr. Patrick	047083, 053918	Cribb, J.	044923	Crum, Rebecca	054187
Costeur, Dr. Loïc	041282	Cox, Richard	043865, 047259	Crick, Micheal	044684	Crump, Caroline	038753
Costie, Gord	005538	Coxe, Trudy	051539	Crick, Monique	041524	Crusan, Ronald L.	051277
Costoulas, Costas	023090	Coxson, Michael	046215	Crippa, Donna	026374	Crusoe, Lewis D.	050557
Coté, Michel	013954	Coy, John	050782	Crippen, Donna	048500	Crutchfield, Margo	047818
Cote, Richard	054102, 054103	Cozzolino, Giorgio	027486	Crişan, Aurelia	036771	Crutchfield, Sharon	046640
Cotie, Donna	006932	Crabtree, Andrew	043677	Crişan, Ciprian Petru	036700	Cruvinel, Márcio Borges	004939
Cott, Peter	044470	Crăciun, Georgeta	036516	Crişan, Eva	036378	Cruz, Amparo Magdaleno de la	040332
Cotter, Patrick	024859	Cracket, Alan	045878	Crişan, Veronica	036203	Cruz, Fe M. de la	034536
Cotter, Suzanne	045453	Craft, John J.	052037	Crispini, Prof. Giuseppe	027756	Cruz, Martin	040356
Cotterill, F.D.P.	054847	Crago, Jody A.	054363	Crissman, Erin	046341	Cruz, Prudenciana C.	034525, 034546
Cottet, Jean-Pierre	014339	Craig, Currell	053842	Cristache, Florin	036548	Cruz, Sumra I. de la	034487
Cottino, Prof. Alberto	028127	Craig, Joanna	047385	Criste, Colleen	051406	Csaplár, Dr. Ferenc	023494
Cottino, Peter	025316	Craig, Len	001441	Cristofoli, Raoul	028120	Csar, Mary	047065
Cottle, Gerry W.	046189	Craig, Martin	054311	Cristofoli, Vincent	013431, 014518,	Cséfalvay, Pál	023631
Cotton, Areca	054620	Craig, Neil	045873		014520	Cserei, Dr. Ladislav	038438
Cotton, Carol	049103	Craighead, Linda	051903	Critchlow, William H.	006288	Csergö, Tibor-András	036475
Cotton, Gordon A.	053975	Crailsheim, Freiherr Ortholf von	016575	Crites, Dr. Gary	050011	Cseri, Dr. Miklós	023960
Cottrell, Robert	053674	Crain, Sandra	049694	Crittin, Jean-François	041400	Cserményi, Vajk	023941
Cotty, Céline	013525	Cramer-Petersen, Lars	010027	Crivelli, Paolo	041390	Cservenyák, László	023772
Couailhac, C. le	014291	Cramer, Carole	052377	Crnković, Prof. Vladimir	009034	Csillag, Péter	023527
Couch, Janet	050484	Cramer, Jeffrey S.	050288	Crociata, Mariano	026718	Csobádi, Jozef	038437
		Cramer, Wendy	051271	Crocker, Margaret S.	046997	Csütörtöky, Jozef	038364

Čtvrtník, Dr. Pavel 009836, 009967
Cuadrado, Emeterio 039793
Čubareva, Roza Petrovna 037607
Cubberley, Peddyr 044037
Cubells, Raymond 013714
Cubillo, Ángeles Vera 039033
Čubrík, Liljana 038174
Cucchi, Franco 027947
Cuccuini, Dr. Piero 026280
Čučković, Lazo 008925
Cuda, Dr. Maria Teresa 025937
Cudahy, Michael J. 050912
Cudbill, Carolyn 043721
Čudnova, Lidija Aleksandrovna . 037887
Cudworth, Keith R. 050325
Cuello, Juan Pablo 046283
Cuenca, Carmen 052964
Cuervo de Jaramillo, Elvira 008630
Cuevas y Lara, Mónica 031249
Cuevas, Elena 034544
Čuhelová, Blanka 009901
Cui, Baiting 008252
Cui, Junfeng 008522
Cui, Youliang 007712
Cui, Zhenkun 007997
Cukić, Milica 038168
Čuković, Petar 031753
Cukrov, Tončika 009048
Culbertoon, Thomas J. 048976
Čulig, Igor 008925
Cullen, Deborah 051432
Cullen, M.L. 046021
Cullen, Maggie 044082
Culligan, Jenine 049626
Cullinan, Deborah 053005
Cullingford, Neville A. 044222
Cullman, Lewis B. 053420
Cullmann, Dorothy 053420
Culot, Maurice 003443, 003458
Culp, David A. 046423
Culpepper, Marilyn 050976
Culver, Michael 051720
Culver, Ph.D. Michael 051163
Cumbey, Susan G. 046386
Cumming, Glen 007424
Cummings, Bonnie 047809
Cummings, Dorothy A. 038263
Cummings, Ernst E. 048686
Cummings, Fern P. 047320
Cummings, Mary Lou 051263
Cummins McNitt, R. 046425
Cummins, Alissandra 003240
Cummins, Heather 045807
Cummins, Joan 047253
Cunard, Gail C. 049166
Cunia, Claudia 003104
Cunliffe-Lister, E.S. 043801
Cunningham Lavdar, Colleen .. 048458
Cunningham, Deirdre 052617
Cunningham, Denyse M. 048825
Cunningham, Eleanor W. 049248
Cunningham, Jewel .. 006899, 006913
Cunningham, Jo 045926
Cunningham, Kim 049779
Cunningham, Noreen 045371
Cunningham, Patrick J. 047301
Cunningham, Scott 049779
Cunz, Dr. Reiner 019063
Cupp, Franzas 053627
Curbastro, Riccardo Ricci 025762
Curbelo, Dr. Jorge 046288, 046289
Curd, Rolin 047509
Curda, Dr. Andreas 002989
Curdy, Philippe 041992
Curiger, Bice 042207
Curnow, Jack 001062
Curran, Donald 049461
Curran, Emily 047127
Curran, John 051962
Currat, Joseph 042097
Currell, M. 000824
Currey, David 051188
Currie, Gloria 006963
Currin, Robert 049024

Curry, Larrie Spier 049381
Curry, Michael 050363
Curry, Michael P. 049340
Curtet, Jean-Christophe 041515
Curtin, Donna D. 052213
Curtis, Brian 050813
Curtis, Dr. J. 044923
Curtis, John 046404
Curtis, Linda 005310
Curtis, Nan ... 047874, 049229, 052292
Curtis, Neil 043359
Curtis, Nellie K. 051698
Curtis, Penelope 044780
Curtis, Ron 000876
Curtis, Sue 044060
Curtis, Vaughn 045609
Curtiss, Courtney P. 047066
Curtius, Dr. Andreas 020852
Curto i Homedes, Albert 040270
Curto, Argeo 008982
Curto, Dr. Carles 039077
Curto, Domenico 025772
Curtz, Arley 047147
Ćus Rukonić, Jasminka 008895, 008952
Čušenkov, Valerij Gennadjevič . 037530
Cush, Dr. Jane 001116
Cushing, Dr. Paula E. 048261
Cushing, Raymond L. 047814
Cushing, Stanley E. 047102
Cushman, Brad 050338
Cushman, Paula N. 050265
Cusiac, Dragoș 036640
Cusick, Bonnie A. 048381
Cusming, J.R. 044321
Čusova, Elena 037279
Cussner, Alice 054567
Cutler, Scott 048504
Cutler, Walter L. 054135
Čvančara, Alois 009667
Cvetkova, Natalija 005027
Cvetković, Branislav 038195
Cvetković, Marija 038139
Cvjetićanin, Tatjana .. 038134, 038176
Ćvorović, Božena 004283
Cybulski, Antoni 035406
Cygański, Janusz 035069
Cyncar, Nadia 005764
Cypher, Paul 046734
Cypress, Billy L. 046975
Cyr, Jean-Claude 005436
Cytacki, Joe 048828
Cywiński, Dr. Piotr M.A. 035088
Czajka, Jacek 035216
Czajkowski, Józef 035336
Czankus, Ruth 046901
Czaplewski, Dr. N.J. 051588
Czapska, Zofia 035194
Czarny, Eugenusz 034786
Czchodzińska-Przybysławska,
Anna 035177
Czerannowski, Dr. Barbara ... 017867
Czére, Dr. Andrea 023553
Czerniak, Robert Michał 035160
Czernin, Dr. Martin 003047
Czerw, Ewa 035395
Czerwinski, Bogdan 035225
Czichos, Raymond L. 054461
Czoma, Dr. László 023738
Czopek, Sylwester 035185
Czub, Zofia 035190
Czubaczyński, Ryszard 034968
Czuma, Stanislaw 047808
Czymmek, Dr. Götz 019715
D' Agnencia, Jill 050435
D'Ambrosio, Paul 047973
D'Angelo, Salvatore 051747
Da Pra Galanti, Dr. Lorenzo ... 027538
Daalder, Dr. R. 031972
Dabell, Simon 043643
Dabić, Ljubica 038185
Dabney, Mary K. 052053
Dabović, Mirjana 031753
Dąbrowska, Anna 034693
Dąbrowski, Wojciech 034716

Daby, Rebecca 001672
Daccó, Dr. Gian Luigi . 026584, 026585,
.... 026586
Dacey, Timothy J. 046770
Dachnenko, Lidija Oleksandrivna 043326
Dae, Sung Kwon 030024
Daele, Durandy van den 013566
Daelemans, Danny 003582
Daelemans, Frank 003435
Daems, Gérard 015807
Daenens, Lieven 003622
Daenzer, Denise 042223
D'Afflitto, Dr. Chiara 027362
Dagan, Zvi 025083
Dagdar, L. 024752
Dage, Carol J. 049664
Daggett, John E. 048855
Dagnicourt, Eric 014119
Dagon, Yoav 025208
Dagorne, Richard .. 014221, 014223
Daguerre de Hureaux, Alain .. 015992
Dahan, Arie . .025157, 025165, 025215
Dahesh, Dr. A.S.M. .. 030264, 030267
Dahl, Erling 033875
Dahl, Jonas 040663
Dahl, Torveig 033692
Dahlbring, Magnus 040663
Dahle, Sigrio 007433
Dahlgren, Dorothy 047856
Dahlitz, Ray 000821
Dahlström, Björn 030588
Dahlström, Per 040660
Dahm, Klaus 020507
Dahmali, Ech-cherki 031789
Dahmani, Dr. Saïd 000050
Daho, Keltoum Kitouni 000057
Dai, Linyan 008374
Dai, Qingming 007854
Dai, Zengqun 008483
Daijoji, Yoshifumi 029745
Dailey, John R. 054138
Dailledouze 015564
Daily, Charles 053110
Daily, Steven L. 050917
Daim, Prof. Dr. Falko 020233
Dajani, Virginia 051323
Daker, Emma 046179
Dakin Hastings, Lynne 053787
Dalborg, Björn 041042
Dalby, Åge 033743
Dale Zemlansky, Denise 049326
Dale, Alex P. 043367
Dale, Charles 052751
Dale, Michal 053467
Dalen, Mary G. 006128
Daley, Jean 048524
Daley, Teddie 049313
Dalhousie, Earl of 044163
Dalichow, Dr. Bärbel 021182
Dalla Vecchia, Fabio M. 026889
Dall'Agnolo, Daniel 041552
Dallago, Antonio 028369
Dallaj, Dr. Arnalda 026785
Dallas, John 043467
Dallas, Linda 052448
Dalleo, Bruce 051250
Dallesandro, Robert 047429
Dallett, Timothy 006981
Dalli, Alberto 027834, 027836
Dallinger, Dr. Petra-Maria 002313
Dallman, Dr. John E. 050600
Dallmeier, Dr. Martin 021319
Dalous, Pierre 015999
Daloze, M. 003391
Dalsgaard Larsen, Keld 010300
Dalton, Bryan 047325
Dalton, Carolyn 044109
Dalton, Jake 048579
Dalton, Ron 050854
Dalton, Shirley B. 052267
Daly, Beth 053794
Daly, James 051429
Daly, Tess 033309

Daly, Thomas M. 053934, 054253
Dalyell, Kathleen 044838
Dalzell, John 052792
Dam, H. ten 032221
Dama, Prof. Salvatore Cirillo ... 007532
Damamme, Jeanne 014917
Damaschin, Violeta 036253
Damaskinos 023136
D'Amato, Dr. Alfonso 025557
D'Amato, Aurelia 027695
Dambaev, Munko Žartam 036841
Dambergs, Guntis 030253
D'Ambrosio, Anna T. 053916
D'Ambrosio, Dr. Antonio 027393
D'Ambrosio, Paul 047972
Damdinsuren, Tsedmaa 031734
Damdoumi, Vasiliki 023309
Dame, Susan 048334
Damen, L. 032826
Damgaard Sørensen, Tinna ... 010724
Damgaard, Ellen 010218, 010219
Damian, Dr. Paul 036314
Damiani, Dr. Giovanna 027208
Damianov, Nikola 005171
D'Amico, Alessandro 027557
D'Amico, Prof. Claudio 025555
Damito, Dr. Ronaldo 004752
Damjanova, Katerina 004993
Dammel, Werner 020643
D'Amore, Dr. Paola 027627
Dampman, Allan 049640
Damsgaard, Nina 010354
Damy, Ken 025635
Dana, Rae 054278
Dana, Richard 047353
Danailov, Boris 005177, 005200
Dance, Pat 007469
Dancey, Jenny 045246
Danchev, Dancho 005059
Dăncuș, Mihai 036708
Dandaura, M. 033430
Dander, Dr. Marilena 026534
Dandwate, P.P. 024442
Danelzik-Brüggemann, Dr.
Christoph 018046
Daneo, Angelica 048258
Daners, Peter 022057
Danesi, Claudio 026128
Dangel, Gerhard 018478
D'Angelo, Starlyn 046346
Danh, Prof. Dr. Trinh 054794
Daniel, Alain 011786
Daniel, Françoise 012207
Daniel, James 054346
Daniel, Mary 048227
Daniel, Rochette 003456
D'Aniello, Dr. Antonia 027785
Daniels, Christopher 049263
Daniels, John P. 051980
Daniels, R. 044526
Danielson, Jeanette 041126
Danielsson, Ing-Mari 041066
Danielsson, Ralf 040574, 040576
Danielsson, Dr. Roy 040815
Danilow, Susan 051398
Danini, Gianluca 026513
Daniylopoulou, Olga 023067
Danjo, Prof. Tatsuo 028503
Dankbar, Johannes 022190
Danker-Carstensen, Dr. Peter . 021495
Dankl, Dr. Günther 002141
Dann, Alan O. 050701
Dann, Gaston 013792
Danna, Gina 049805
Danna, Tracy 046852
Danne, Rainer 019428
Danneel, M. 003463
Danner, Elisabeth 041710
Dannhorn, Beate 018412
Dannhus, H. 020028
Dano, Dr. Ján 038377
Dansako, Tetsuya 029695
Danson, Stuart 001635
Danšova, Olga Nikolajevna ... 037590

Dante Coronel, Hugo 000123
Dantinne, Claude 013711
Danzer, Dr. Gudrun 001985
Danzl, Barbara 002743
Dapkyunas, Jeanne 003249
Dapsens, J.-C. 003846
Darakčieva, Uljana 005095
Darbé, Louis 003577
Darby, Daniel 007494
Darde, Dominique 014500
Dardenne, Geneviève 003656
Darding, Nicole 003793
Darga, Dr. Robert 021857
Darling-Finan, N. 045783
Darling, Denise L. 048173
Darling, L.A. 007232
Darling, Sharon 053193
Darr, Alan P. 048301
Darre, Pierre 015148
Darrow, Bernard 052410
Darst, Elisabeth 049034
Dart, John 043870
Dartevelle, Patrice 003906
Darwish, Ebrahim 010513
Dary, Anne . . . 011705, 012800, 013868,
                                      015650
Das, Chandranath 024210
Dasbach, Theo P. 047772
Dăscălescu, Gheorghe 036460
Date, Tatumi 029551
Dattilio, Dan 049971
Daubendiek, James 049797
Daubert, Debra 051829
Dauda, J.Y. 033435
Daudelin, Robert 006369
Daudry, Damiano 025338
Dauge, Alexandra 013404
Daugelat, Dr. Friederike 017596
Daugelis, Osvaldas 030371
Daughdrill, Kay 048149
Daughterman-Maguire, Dr.
  Eunice 046754
Daum, Patrick 015067
Daum, Sigrid 019828
Dauphin, Jean 003776
Daure, Regine 012349
Dauschek, Dr. Anja . . . 022036, 022058,
                                      022059
Dautermann Ricciardi, Dana . . 048919
Dautermann, Dr. Christoph . . . 019787
Dauth, Louise 000941
Dautović, Andrea 004283
Dauwe, Marc 003780
Dauzenroth, Willi 020267
Dávalos de Camacho, Maya . . 031240
Davaras, Dr. C. 022949
Davenport, Kimberly 049592
Davenport, Leonard 047196
Davey, Arnold 045562
Davey, Ranson 024900
David-Jones, Carmina 039708
David-Weill, Hélène . . 014705, 014713,
                        014733, 014785
David, Clifford C. 048365
David, Johan 003658
David, Josiane 014156
David, Karen 048529
David, Mihai 036259
David, Norma Jean 049260
David, Romeo G. 034448
David, Wendy 014449
David, Dr. Wolfgang 020242
Davidov, Dr. Dinko 038139
Davidović, Jasna 038245
Davidović, Jelena 038204
Davidow, Joan 048112
Davidse, I. 032734
Davidson, Anne 051905
Davidson, Dan 048049
Davidson, Kathryn 046705
Davidson, Paul 033162
Davidson, Scott 005921
Davies, Andrew 044747
Davies, Brian 045524

Davies, Gerald 046150
Davies, Hugh M. 050040
Davies, Dr. Hugh M. 052967
Davies, Jane . . 005815, 006802, 045568
Davies, Jenny 045173
Davies, Kate 048362
Davies, Lucy 044897
Davies, Marjorie 044161
Davies, Nada 051950
Davies, Pascal 011787
Davies, Richard 045843
Davies, Rick 050867
Davies, Rita 007168
Davies, S. 001543
Davies, Simon 045429, 045430
Davies, Steve 045173
Davies, Su 044683, 044685
Davies, Suzanne 001282
Davies, Dr. W.V. 044923
Davignon, Étienne 003496
Dávila, Alejandro 000335
Davila, Julio V. 034356
Davis Anderson, Brooke 051325, 051383
Davis Gardner, Katie 047884
Davis MacNaughton, Mary 047760
Davis, Dr. Ann 005531
Davis, Ann 046602
Davis, Audrey P. 046382
Davis, Beryl 007117
Davis, Beth 047892
Davis, Betty J. 051763
Davis, Carol A. 048679
Davis, Cheryy 046378
Davis, Darwin R. 052781
Davis, Dustin P. 048990
Davis, Elizabeth 053230
Davis, Elliot 047123
Davis, Gordon 046670
Davis, Israel 049194
Davis, Jacqueline B. 048867
Davis, Janet 007379
Davis, Jaqueline Z. 051456
Davis, Joanne 054247
Davis, Kate 006770
Davis, Katherine 051461
Davis, Kay 047853
Davis, Kieth 033225
Davis, Dr. Leroy 051795
Davis, Lesia 005541
Davis, Dr. Leslie B. 047166
Davis, M. 025058
Davis, Marsh 049030, 049039
Davis, Michael 053546
Davis, Nancy 046764
Davis, P.A.C. 033294
Davis, Phillip V. 053455
Davis, Priscilla 053366
Davis, Robert O. 052915
Davis, Roger 051897
Davis, Ron 006477
Davis, Ruth 046375
Davis, Sarah 051355
Davis, Susan 051897
Davis, Susan S. 049973
Davis, Suzanne 046490
Davis, Todd A. 007333
Davis, Walter R. 047414
Davis, William J. 051574
Davis, Zina 054301
Davison, Fiona 044990
Davison, Liane 007102
Davison, Dr. Patricia 038685
Davison, Walter 049992
Davletov, B. 054673
Davson, Victor 051496
Davy, Geukens 003760
Davy, Mark 051879
Davydova, Natalja Alekseevna . . 037674
Dawans, Francine 003802
Dawe, Dr. Chris 050957
Daws, Russell S. 053664
Dawson Penniman, H. 052061
Dawson, Amy 053018

Dawson, Barbara 024887
Dawson, Beverly 049125
Dawson, Chris 047812
Dawson, David 043722
Dawson, E.A. 043902
Dawson, Frank C. 048442
Dawson, Gay 053141
Dawson, J. 043842, 044002
Dawson, Dr. Mary R. 052158
Dawson, P. 024940
Dawson, Valerie 044742
Day Vincent, Kris 048982
Day, Brenda 046722
Day, Dr. David 053282
Day, Edward 052587
Day, Ginette 013266
Day, Jacqueline 053548
Day, Dr. James 050251
Day, John A. 053961
Day, Karen 048555
Day, Dr. Lucille L. 046935
Day, Michael 044210, 044995
Day, Patricia 049256
Day, Richard 043923, 053990
Day, Sally 043386
Day, Teresa 054389
Dayan, Prof. T. 025214
Dayrit, Marina 034528
Dayson, Diane H. 051474
De Angeli, Cristina Frua 027914
De Angelis, Prof. Giuseppe 027530
De Baere, Bart 003299
De Block, Ann 003257
De Boer, A. Spil. 032641
De Coster, Annie 004154
De Cuyper, Chris 003369
De Cuyper, Meve Jo 003258
De Filippis, Dr. Elena 028274
De Franceschi, Antonio Fernando 004695
De Gelder, Dr. F.A. 032951
De Goede, J. 032807
De Herdt, René 003635
De Jong, H. 032822
De Jonge, Jurgen 003573
De Jonghe, Marcel 003370
De Keerle, Monique 003558
De Kelver, Jan 003560
De Kleijn, G. 031907
de Linos Escofet, Mónica 039633
De Lorentiis, Decio 026673
De Luca, Dr. Bianca 025968
De Luca, Dr. Francesca 026256
De Luca, Maurizio 054682
De Luca, Prof. Paolo 027009
De Luca, Renzo 029085
De Lucia Brolli, Maria Anna . . . 026015
De Marchi, Dr. Andrea G. 027565
De Marco, Dr. Marco 026226
De Martin, Prof. Giancandido . . . 024833,
                        027321, 027323, 027324
De Mier Riaño, Elisa 008621
De Paepe, Pantyika 012631
De Palma, Prof. Luigi Michele . . 026874
De Palma, Dr. Maria Camilla . . 026407
De Paolis, Rosario 026579
De Pasquale, Vicenzo 026750
De Polo Saibanti, Claudio 026294
De Pompeis, Claudio 027274
De Poorter, Alexandra 003492
De Reymaeker, Michel 003876, 003883
De Roeck, Harald 003420
De Ruyter, Rob 003997
de Rycke, Jean-Pierre 004085
De Santi, Claudio . . . 026884, 026885,
                                      026886
De Siena, Antonio 025499
De Simone, Dr. Emilio 025307
De Smet, Urbain 004141
De Strobel, Dr. Anna Maria 054682
De Strycker, Pierre 003314
De Vincentiis, Prof. Italo 027660
De Vito, Natalie 007189
De Waal, Dr. Lydia M. . . 038840, 038841
De Wilde, Karel 003681

De Wilde, Karin 004063
De Wit, Jean-Louis 003388
De Witte, Hubert 003405, 003406,
                        003407, 003408, 003410
De Wolf, Walter 004148
Deacy-Quinn, Christa 053909
Deakin, G.T. 001230
Dealy, Anne F. 049068
Dean Krute, Carol 049390
Dean Stock, Michele 052877
Dean, Deborah 045409
Dean, Jim 051055
Dean, Kevin 053151
Dean, M. 044098
Deaner, Beth J. 051274
Dearing, Vicki 047288
Dearinger, Dr. David 051444
Dearinger, David 047102
Deaton, Linda 053664
Debain, Yvette 005866
Debarge, René 013229
Deblanc Magnée, Marie-Paule . . 004106,
                                      004107
Debo, Thomas A. 052224
Debočicki, Valentin . . . 005062, 005063,
                        005064, 005065, 005067
DeBow, Arthur 052294
Dębowska, Alina 034647
Debrabandère, Béatrice 014484
Debrah, Dr. Issac N. 022925
DeBrari, Frank 054559
Debray, Cécile 014697
DeBree, Gretchen C. 047830
Debrincat, Saviour 030761
DeBruyn, David L. 049195
DeBruyn, Raphaël 003783
Debry, Jaques 012507
Debryn, Johan 004024
Debus-Steinberg, Astrid 022062
DeBuse Potter, Gail 047511
Deca, Eugen . 036222, 036231, 036327,
              036409, 036494, 036643, 036645,
                                      036646
Decames, Jean-Claude 014082
DeCamp, Alta J. 051211
Decán Gambús, Ivanova 054744
Decavele, Dr. Johan 003630
Dechambre, Jeanne 012389
Dechaux, Carine 003971
Deck, Linda T. 052222
Decker, Carla 051512
Decker, E. 015696, 015697
Decker, Friedrich 002281
Decker, Raymond 016178
Decker, Scott E. 049673
Deckers, Ann 003287
Deckers, Yolande 003289
Deckert, Bernd 018025
Deckert, Inge 018025
Deckert, Michael 022418
Declercq, Joost 003543
DeCocker, Dean 053855
Decosterd, Denis 041913
Decoud, Dr. Marcelo 000510
Decroix, Philippe 015219
Decron, Benoît 013765
Decrouez, Danielle 041539
Decter, Avi Y. 046753
Dectot, Xavier 014780
Děd, Stanislav 009521
Dedenroth-Schou, Poul 010206
Dedíková, Soňa 009492, 009493
Deedes, C.M.J. 044110
Deegan, Dr. Ann 047724, 052457,
                        052497, 054608
Deer, Naomi 049718
Deerpalsingh, Salni 030788
Deeva, Valentina Kirillovna 037043
DeFalla, Josie E. 053818
Defauwes, Georges 003530
Defeo, Prof. Rúben D.F. 034580
Defferrard, Jacky 013730
DeFloren, Rude 046317
Deford, T. M. 053227

Degel, Hermann 022057
Degel, Kirsten 010146
Degen, Kurt 021614
Degen, Dr. Thomas 018960
Deggelsegger, M. 002686
DeGrazia, Diane 049692
Degreif, Dr. Uwe 017356
DeGroft, Aaron 053149
Degueurce, C. 013975
Deguglielmo, M. 011880
Deguillaume, Marie-Pierre 015894
Deharde, Ewald 017705
DeHart, Rob 051188
Dehollander, Stef 003748
Dehrkoop, Prestene 048513
Deiber-Kumm, Michele 047605
Deicher, Patrick 041902
Deigendesch, Dr. Roland 020587, 020588,
020589
Deiker, Dr. Tom 047604
Deimel, Dr. Claus 018001, 019228,
019994
Deisenberger, Nicole 022385
Deiser, Leopoldine 003127
Deitrich, Darcy 006260
Deitsch, M.J. 051345
Deitz, Philippe 003808
Dejardin, Dr. Fiona M. 051779
Dejardin, Ian 044961
Dekiert, Dr. Markus 020508
Dekiss, Jean-Paul 011631
Dekker, Annet 031971
Dekker, Anthony P. 032537
Dekker, H. 032418
Dekker, R. 032582
Deknop, A. 003436, 003448
Del Bagno, Rober 051455
Del Falco, Vittorio 027648
Del Frate, Dr. Gabriella 027194
Del-Prete, Sandro 041608
Del Testa, Luise 022873, 022874
Dela, Feliks 035027
Delacoste, Raymond 041750
Delacote, Goery 052999
Delaender, Georges 004123
Delage, Dominique 015838
Delahant, John M. 053146
Delahaye, Marie-Claude 011817
Delaine, Joël 014376, 014379
Delaire 011966
Delaney, Chris 043877
Delangle, J.F. 012038
Delannoy, Agnès 015341
Delany, Max 000971
Delaousie, Renée 014560
Delapierre, Emmanuelle 016096
Delaquis, Christian 007437
Delarge, Alexandre 013056
Delattre, Jean-Luc 003914
Delaval, Eric 011685
Delay, Vincent 041691
Delbaere, Sebastian 032025
Delcourt, Dr. Marianne 004082
Delecouls, Sonia 013586
Delee, Kathryn 046338
Delehanty, Suzanne 050836
Delépine, Francine 003454
Del'Furia, Lucienne 014071
Delgado Restrepo, Gloria 008647
Delgado, Avella 039523, 039525
Delgado, Eric 011784
Delgado, James P. 007302
Delgado, Jane 051393
Delgado, Lino Torres 030807
Delgado, Sally 046599
Delhaise, Christophe 003393
Delić Gozze, Vesna 008904
Delić, Nermin 004257
Delisle, André J. 006401
DeLisle, Godfrey 005632
Delivorrias, Prof. Dr. Angelos . . 023002
Dell, Roger 048737
Dell, Sharon 033335
Della Casa, Bettina 041695

Della Fina, Giuseppe M. 027119
Della Toffola, Leomberto 028308
Dellacroce, Raymond 048679
Dell'Antonia, Dario 016271
Dellantonio, Elio 027454
Dellbeck, Johan 041118
Dellebuur, Amy 007136
Dellinger, Mary H. 048953
Dell'Oca Fiordi, Dr. Angela 027986
Dellsperger, Dr. Yvonne 017362
Delmas, Jean 015129
Delmas, Luc 016232
Delmont, Patrick 016322
Delobel, Jean 011875
Deloffre, Véronique 011917
Deloing, Evelyne 011736
Deloncle, Jacques-Gaspard 014839
DeLong, Dr. Marilyn 052839
Delormas, Jerôme 014028
DeLorme, Harry 053182
Delorme, Yves 015109
Delperdange, Roland 003342
Delporte, Luc 004096
Delroy, Ann 001403
Deltedesco, Franco 026602
Deltour-Levie, Claudine 003492
Delucchi, Gina 053540
DeLucia, Francis 052463
Delye, Emmanuel 003800
DeMaio, Bonnie 052695
Demard, Jean-Christophe 012439
Demaret, Nicole 004081
Demarez, Leonce 003333, 003372
DeMars, Louise L. 047212
DeMarsche, Michael 047883
Demarteau, Jo 032130
DeMersman, Jim 049419, 049420
Demetis, Ioannis 023322
Demetriou, Maro 009410
Demetz, Dr. Stefan 025582
Demeurie, Dirk 003552
Demidenko, Olga Moiseeva 037528
Deminey, E. 038719, 038760
Deming, Diane 052510
Demirci, Süheyla 043063
Demke, Elisabeth 047021
Demo, Dr. Željko 009025
Dempsey, Rey 024904
Dempsey, Terrénce E. . . 052813, 052814
Demski, Tadeusz 035201
Demuth, Renate 021394
DeMuzio, David 052081
Denaro, Dolores 041337
Denelli, Gjergj 000029
Denford, Dr. G.T. 046156
Denford, Geoffrey 046149
Deng, Binquan 007801
Deng, Chaoyuan 007852
Deng, Liming 008085
Deng, Murong 007679
Deng, Qingyun 008390
Deng, Yongfa 008213
Dengler, Thomas 018339
Denhaen, G. 003435
Denis, Paul 007200
Denizli, Hikmet 042813
Denman, Ron 005616
Dennerly, Peter 033185
Dennett, Anthea 044750
Denning, Kylee 053171
Dennis, Linda 052638
Dennis, Sue M. 051484
Dennison, Foster K. 006723
Dennison, Lisa 051470

Dennler, William V.A. 053759
Denny, Colin 001145
Dennys, Vincent 011656
DeNoon, Betty 047349
Densapa, Tashi 024258
Denscher, Dr. Bernhard 003037
Densen, Paul M. 053068
Dent, Dr. D.L. 033020
Dent, G.J.H. 045500
Dentressangle, Franck 012941
Denys, Lama 011756
Denysenko, Jevgen Ivanovyč . . . 043153
Denzel, H. 001709
Deon, Luca 041703
DeOrsay, Paul B. 047866
Deparpe, Patrice 013704
Depasquale, Suzannah 030753
Depierre, Marie-Colette 015122, 015124
Depoid, Henry 012648
Deppe, Dipl.Ing. Frauke 018279
Deppeler-Lang, Walter 042048
Depreter, R. 003463
Depuox, Christian 042231
Dequidt, Luk 003952
Derbic, Rupert 001810
Derbier, Josselin 015117
Derby, Newton 053563
Derbyshire, Janine . . . 044082, 044084
D'Ercole, Dr. Vincenzo 025967
Dercon, Chris 020529
Dercourt, J. 013248
Derda, Dr. Hans-Jürgen 017553
Derefaka, Dr. Abi Alabo 033445
Derenthal, Dr. Ludger . 017202, 017241
Deriks, M. 032905
Dering, Dr. Florian 020551
Derjabin, Vladimir Ignatevič 037720
Derjužina, Alla Vasylivna 043140
Derksen, J. 032242
Derksen, J.J.V.M. 032943
Dermitzakis, Prof. Dr. M.D. 023026
Derr, Rex 051942
Derriks, Claire 003890
Dertazzo, Ludovico M. 027145
Des Rochers, Jacques 006391
Desai, Vishakha N. 051340
Desaint, Jeanne 014101
DeSalvo, Donna 051491
Desaulniers, Guy 005363
Desbenoit, Martine 012154
Desborough, Jeff 054398
Desbrosse, René 013339
Descamps, Patrick 013245
Descatoire, Christine 014780
Deschamps, Stephanie 011744
Descheemaeker, Dominique . . . 003551
Desel, Jochen 016795
Deseyne, Alex . 003925, 003926, 003929
DeSeyter, Craig 049506
Desfossés Le Page, Carmen . . . 007245
DeShaw, Larry 053572
DeShon, Richard N. 052787
Deshpande, Dr. P.N. 024249
Désirat, Georges 015660
Desjardins-Ménégalli,
Marie-Hélène . . . 012959, 012960
Desjardins, Robert Y. 007011
Deslarzes, Bertrand 041397
Desmarais, Charles 047253
Desmarais, Joan 047959
Desmet, Viviane 003493
Desmett, Don 049859
Desmond, Tim 045417
Desmules, Olivier 014013
Desnier, Jean-Luc 014706
Desormeau, Monique 048748
Despotidis, Emmanouil 022956
DesRoches, C. 006276
Desse, Daniel 016194
Dessemme, Christine 012632
Dessens, H.J.A. 031972
Desserud, Thore 033581
Dessornes, Maria E. 053739
Dessy, Marco 033373

Destefanis, Joseph 031719
DeStefano, Giacomo R. 051945
Desvigne, M. 015686
Desy, Dr. Elizabeth A. 050712
Detch, Rosalie S. 050228
Dethlefs, Dr. Gerd 020601
Dethmann, Donald 047835
Detrani, Geoffrey 051489
Detrée, Jean-François 015579
Detrich, Kalman 051442
Dettmann, Julia 018918
Dettmer, Dr. Frauke 021372
Dettmer, Hans-Georg 018771
Deuchar, Stephen 045115
Deuchert, Dr. Norbert 017698
Deudon, G. 003597
Deuker, Horst 014140
Deutmann, Karl Heinrich 017960, 017962
Deutsch, Franz 003129
Deutschbauer, Franz 002736
Deutschle, Knut 021440
Dev, Anand 024236
Devanik, Dave 006423
Devé, Jean 013109
Deveau, Bertha 053068
Deveau, Carole 006345
Devendorf, Meredith 050883
Devereux, Dr. B.J. 043840
Devereux, David 044730, 044731
Deveselleer, Jacques 004034
DeVille, Sue 051787
Devin, Pierre 012824
Devine-Reed, Pat 047642
Devine, Jackie 048289
Devlin, Dr. Felicity 024897
Devoghel, Philippe 013195
DeVonyar-Zansky, Jill 052206
Devory, Marian 052004
Devos, Danny 003306
Devos, E. 003984, 003985, 003986
Devries, Fred 050852
DeVries, Karl 053181, 053183
Devroye-Stilz, Anne 014480
Devynck, Danièle 011585
Dewald, Ann L. 051890
Dewald, Gabi 021833
Dewald, Rolf 042229
DeWalt, Billie R. 052158
Deweerdt, Jean 012845
Dewey, David 051818
Dewey, Judy 046036
Dewey, Stuart 046036
Dewhurst, Dr. C. Kurt 048441
Dewilde, Annick 003758
Dewilde, Jan . 003724, 003725, 003726,
003728, 003729
Dewing, David 044975
Dewitz, Dr. Bodo von 019705
Dewsbery, Sara 045453
Dewsbury, R. 045414
Dexter, Susan 048736
Deyle, Betty 046917
Deysenroth, Paul 054345
Dezellus, Eric 015182
Dhaene, Sylvie 003625
Dhaliwal, Theresa 045023
Dhaulakhandi, V.N. 024436
Dholakia, P.V. 024333
Di Belgiojoso, Dr. Giuseppe B. . . 027835
Di Benedetto, Gino 028157
Di Carluccio, Dr. Antonietta 025289
Di Castro, Dr. Daniela 027619
Di Fabio, Clario 026403, 026412, 026423
Di Felice, Dr. Paola 028070
Di Flumeri, Dr. Gabriella 027627
Di Girolami, Dr. Paola . 025859, 026055,
026479, 026907, 026951, 027531,
027680, 027722
Di Grazia Costa, Ivani 004879
Di Ianni, Angelo 027287
Di Lorenzo, Maria Teresa 027793
Di Malfetta, Dr. Felice 026874
Di Marco, Antoinette 013919
Di Mario, Dr. Mario 028081

Di Martino, Maria 027337
Di Nepi, Simona 045057
Di Niro, Dr. Angela 025730
Di Paolo, Luciana 027329
Di Pascale, Paolo 025449
Di Pietrantonio, Dr. Giacinto . . 025492
Di Pietro, Gaetano 027276
Di Stefano, Dr. Carmela Angela . 027177
Di Tommaso, Francis 051489
Diaconescu, Nicolae 036322
Diaconescu, Petru 036747
Diaconu, Ion 036832
Diakite, Sinkoun 023384
Diallo, Ibrahima 052977
Diamantakis, Stavros 002492
Diamantopoulos, Aella 052078
Diamond, Dr. Judy 050302
Dianto, Herkus 024585
Dianzinga, Dias-Théodore 008845
Diaper, Dr. Hilary 044789
Diaper, Robin 044713
Dias Duarte, Luiz Fernando . . . 004754
Dias Pereira, Prof. Carlos 035632
Dias, Dr. Maduro 035530
Dias, Robério 004770
Diaw, Amadou Tahirou 038116
Díaz Arriola, Luisa 034345
Díaz de Mazzoldi, Olga 008747
Díaz Díaz, José Antonio 039560
Díaz García, Luís 039817
Díaz-Guerra Álvarez, Luis 039659
Díaz Pardo, Isaac 038989
Díaz Rodríguez, Anselmo 034352
Díaz-Tusham, Rafael 051323
Díaz Ungría, Dr. Adelaida de . . 054765
Díaz, Dr. Jimena Perera Diaz . . 046279
Díaz, Joaquín 040287
Díaz, Luis H. 008667
Díaz, Miguel 036171
Dibbets, W.A.G. 032040
Dibert, James 053191
Dičeva, Vera 005194
DiChristina Hager, Joan 047729
Dick, Alvin 051105
Dick, James 052693
Dick, Jutta 018935
Dicke, James F. 051237
Dickens, Denise 052445
Dickenson, Joanne 052269
Dickenson, Victoria 006406
Dickerman, Dr. Robert W. 046365
Dickermann, Dr. Fred 002191
Dickerson Lind, Anne 053170
Dickerson, Patricia 051740
Dickey, Bobby 048884
Dickey, Michael 046546, 047090
Dickhaut, Monique F.A. 032624
Dickie, Kerry 047391, 047392
Dickinson, Cindy 046448
Dickinson, Gary 051316
Dickinson, Pat 007160
Dickinson, Robert 045941
Dickinson, Roger 051252
Dickson III, Ephriam D. 048806
Dickson, Carl A. 049370
DiCosimo, Joanne 005869, 006578
Didier, Alain 015016
Didier, Prof. Arturo 028064
Didier, Hebert 013812
Didrichsen, Maria 010816
Didrichsen, Peter 010816
Diebel, C. 033142
Diederen, Roger 051372
Diederen, Dr. Roger 020541
Diederichs, Dr. Urs 021362
Diederichsen, Detlef 017159
Diedrich, Dr. Stephan 019705
Diegel, Julie 050303
Diehl, Jörg 042114
Diehl, Reinhard 020118
Diehl, Dr. Richard 053856, 053858
Diekmann, Rolf 022072
Dieleman, Peter A. 032693
Diemer, Dr. Kurt 016871

Diene, Baidy 038117
Dienst, Robert 002279
Dierick, Charles 003439
Dierking, Gary R. 048864
Dierks-Staiger, Regine 022057
Dießner, Dr. Petra 020017
Dietel, Volker 020701
Dieter Jr., Raymond A. 047666
Diethorn, Karie 052056
Dietl, Dr. Gerhard 022056
Dietrich, Bev 005932, 005933
Dietrich, Charlotte 021289
Dietrich, Frank 022057
Dietrich, Dr. Gerhard 019700
Dietrich, Liselotte 041729
Dietrich, Nikola 018444
Dietrich, Ute 022713
Dietz, Karen 052829
Dietz, Karl-Heinz 018062
Dietz, Stefanie 022459
Dietz, Steffen 016471
Dietz, Ulysses G. 051499
Dietze, Falk 020534
Dietzfelbinger, Dr. Eckardt 020847
Diez Canseco, Victoria Mujica . . 034369
Diez Valdez, Oscar 034235
Díez, José Luis 039676
Díez, Sofía 039708
DiFerdinando, Charles A. 048399
Diffily, James P. 048902
Difuccia, Mike A. 050991
Digby, Linda 005736
Digel, M.A. Brigitte 022747
Digger, Jo 046041
Dignef, Leo 003352
Dijkstra, J.G. 032682
Dijkstra, Kol.J.A. 031953
Dijkstra, Rob 032378
Dijols, Pierre 013527
Điković, Mario 038143
Dikowitsch, Hermann 002711
Dikun, Z.M. 037149
Dilas, Milica 008977
Dill, Chris 049222, 049223
Dill, Gregor 041292
Dillard Rech, Lori 052033
Diller, Douglas 053091
Diller, Dr. Stephan 021213
Dilley, Cherry 038829
Dilley, Mary Lou 050703
Dillmann, Claudia 018412
Dillon, Alice 053599
Dillon, Douglas 010417
Dillon, Idelle 051978
Dillon, John 053018
Dilly, Georges 011987
Dimaandal, Carmelita 034530
Dimaki, Sofia 022974
DiMaria, Tom 051676
DiMeo Carlozzi, Annette 046693
Dimitrijević, Bora N. 038258
Dimitrov, Prof. Dr. Bojidar 005191
Dimitrov, Ivan 005143
Dimitrova, Stanka 005241
Dimmick, Walter 050186
Dimova, Ljubov Ivanovna 038060
Dimt, Dr. Heidelinde 001870
Dimuchametova, Svetlana
  Aleksandrovna 037645
Dinard, N. Jean 013346
Dinçer, Hüseyin 042912
Diner, Mónica 031233
Ding, Ke 007763
Ding, Rong 007796
Ding, Yucai 008406
Dingemans-Bakels, Drs. F.N. . . 032626
Dingertz, Stig 040831
Dinges, Christiane 019672
Dingwall, Fiona 044361
Dinh, Le Quang 054807
Dinhobl, Dr. Günter 002421
Dinkelspiel, Florence 050378
Dinkhauser, Lienhard 002917
Dinkins, Dr. Leon 052150

Dinsmoie, Elizabeth 005781
Dintaman, Jeniffer 049311
Dinu, Cristina 036800
Dinwerth, Shaun 052535
Diogo Ribeiro, José 035639
Diogo, Joanne 047379
Dion, Jean-Noël 006876
DiPaolo, Michael 050223
Dippold Vilar, Dalmo 004898
Dippold, Günter 016998
Dirmeikis, Edvardas 030396
Dirnbacher, Karl 001940
Dirnbacher, Sonja 001726, 001727
Dirnberger, Rotraut . . . 002402, 002405
Dirnböck, Mag. Robert 002007
Dirosa, Hervé 015786
Dirschel, Angelika 019137
Dirun, Moh Lativ 030693
Dishman, Jay 050819
Dismukes, Sandra B. 047611
Disney, Betty 048102
Disney, Jenny 052898
Diss, Gabriel 014036, 016197
Disston, Deborah 047950
Distel, Matt 047742
Dite, Dr. Tibor 038476
Dittberner, Stephen 052348
Dittemore, Diane 053823
Dittertová, Dr. Eva . . . 009514, 009515
Dittler, Ingeborg 017973
Dittmann, Britta 020155
Dittmar, Monika 022315
Dittmar, Volker 020302
Dittrich, Dr. Erika 018524, 018525,
                                        018526
Dittrich, Rainer 019029
Divelbiss, Maggie 052411
Divincenzo, Dr. Concetta 025466
Divola, John 052597
Dix-Williams, Traci 045954
Dix, Andrea 018583
Dixon, Barbara 046973
Dixon, Bonnie 049045
Dixon, Carol 050645
Dixon, Charlie 049753
Dixon, Christine 000937
Dixon, Claudine 050426
Dixon, David J. 046996
Dixon, Debbie 053148
Dixon, Elaine 052420
Dixon, Gerald 049160
Dixon, Hugh 045256
Dixon, Jenny 050379
Dixon, Kate 050780, 050781
Dixon, Margaret 043994
Dixon, Dr. Michael 045061
Dixon, Wendy M.C. 001207
Diz Ardid, Emilio 039844
Djabbar, Abd. 024666
Djačenko, N.D. 043301
Djačkov, Ivan Ivanovič 029841
Djadaibaev, Amir Jalinovich . . . 029801
Djalal, Dr. Zaidir 024577
Djamba K. Shungu, Joseph . . . 008838
Djanguenane, Nayondjoua 042730
Djokić, Danica 038226
Djordjević, Nenad 038204
Djuraboyev, Abdullajon 054625
Djurdjekanović, Sladjana 038187
Djurić, Gordana 038194
Djuve, Kjetil 033934
Dlin, Elliott 048114
Dlugač, Vladimir Vladimirovič . . 036866
Długoszewska-Nadratowska,
  Hanna 034698
Dłużwska-Sobczak, Anna 034937
Dmitriev, Prof. I.S. 037765
Dmitrovic, Katarina 038188
Dmytrykiw, Danylo 049345
Doane, Erin 053159
Dobbelstein, H. 032299
Dobbie, Dave 006431
Dobbratz, Alferna 049558
Dobbs, Jerry 052907

Dobigny, Sylvaine 011632
Dobigny, Sylviane 011633
Dobke, Dr. Dirk 018982
Dobkowski, Mieczysław 034675
Doblin, Bruce 047689
Dobosi, Dr. Viola 024038
Dobras, Dr. Wolfgang 020230
Dobrea, George 047823
Dobrev, Marin 005210
Dobrianowa-Bauer, Dr. Snejanka
  D. 018429
Dobrik, Dr. István 023782, 023791,
                                        023793, 023794
Dobrileit, Margitta 018047
Dobrogosz, Glenn 049256
Docavo Alberti, Ignacio 040329
Docenko, Alina Vasylivna 043137
Dočev, Spas 005084
Dočeva-Peeva, Elka 005191
Dočeva, Penka 005068
Docksey, Graham 000822
Dodd, Dr. Donald B. 047008
Dodd, Hanut 001095
Dodd, Julie 045611
Dodd, Nick 045778
Dodd, Roger 043995
Dodds, E. 044646
Dodge, Laura 048579
Dodier, Virginia 047433
Dodina, Tatjana Alekseevna . . . 038052,
                                        038054
Dodson, Gin 054235
Dodson, Howard 051466
Dodsworth, Roger C. 044147
Dodwell, A. 045024
Dodwell, P. 045024
Döllinger, Theodor 021076
Dönmez, Asuman 042941
Dörfer, Dr. Anja 020830
Dörhöfer, Friedrich 017659
Döring, Dr. Jürgen 019011
Döring, Martel 020886
Döring, Dr. Thomas 017558
Dörnenburg, Manuela 017395
Dörr, Dr. Cornelia 018136, 019552
Dörr, Cornelia 018135
Doerr, Marjorie A. 048569
Doetsch, Dr. Rainer 019665
Doetzer-Berweger, Oliver 019220,
                                        019221
Doezema, Marianne 053405
Doffey, Marie-Christine 041323
Dogan, Salih 042968
Dogariu, Octavian . . . 036782, 036783,
                                        036784, 036786
Doherty, Peggy M. 047690
Doherty, Roisin 024968, 044090, 044091
Doherty, Trafford 049333
Dohertyld, Erin 033163
Dohm, Katharina 018446
Dohnal, Karel 009583
Doiron, Sue 006870
Doix, Jean Michel 016038
Dojčev, Nejčo 005105, 005108
Dokken Urdal, Haldis 033611
Dolan, Douglas C. . . . 048364, 048367
Dolan, Joseph J. 048426
Dolan, Susan 053837
Dolata, Jan 035290
Dolby, Joe 038687
Dolby, Malcolm J. 046199
Dolce, Dr. Sergio 028192, 028203
Dolci, Alberto 025959
Dolczewski, Zygmunt 035135
Dolenz, Univ.-Doz. Dr. Heimo . . 002508
Dolenz, Dr. Heimo 002192
Doletsky-Rackham, Nina 048323
Doleva, John 053462
Dolgopolova, Zinaida Stepanovna 037885
Dolkart, Judith 047253
Doll, Nancy 049257
Dollins-Colton, Sonya 048457
Dolor, Danny 034495
Doltenkov, V.V. 036890

Dolženko, Fëdor Nikolaevič 036972
Domaas, Ruwe 041137
Domaine, Elmo 026194
Domanig, Andrea 002995
Domański, Edward 035356
Dombrowski, Dieter 022328
Domenech de Celles, Yann 013185
Domènech, Glòria 039057
Domenz, Silke 022249
Dominder, Christer 040880
Domingo Monteverde, Coia ... 039996
Domingos, Tadeu 000107
Domínguez Cabrera, Gonzalo .. 009257
Domínguez Cadeño, Nelson ... 009144
Domínguez, Antonio Franco ... 039037
Domínguez, German 007543
Dominjak, Aleksandr Vladimirovič 037640
Domitruk, Daryl 006427
Dommisse, E. 038791
Dompierre, Louise 005966
Dompnier, Pierre 015380
Domröse, Ulrich 017085
Don Angelo, Maria 026481
Don Che, Kyu 029900
Donabauer, Michael 020430
Donadio, Emmie 050858
Donahue, Rachel L. 048901
Donaldson, E. 000833
Donaldson, John 043845
Donaldson, Susan 053444
Donath, David A. 054532
Donath, Wolfram 017273
Donati, Prof. Bruno 041547
Donati, Dario 042241
Donches, Stephen G. 046963
Donchez Mowers, Charlene ... 046958,
            046959, 046960, 046962
Dondero, Veronica 049421
Donders van den Aardweg, M.G. 032029
Dondo-Tardiff, Michelle 006574
Donegá, Dr. Giuseppe 028321
Donelli, Ivo 008993
Donelly, P. 044752
Donelson, Patty J. 048028
Donevski, Peti . 005220, 005221, 005222,
                            005223
Dong, Daochang 008473
Dong, Fuping 008147
Dong, Pengyun 008357
Dong, Sen 008179
Dong, Xiaowen 007563
Dong, Ziru 007916
Dong, Zso 042309
Donkersloot, B.J. 032300
Donlon, Dr. D. 001540
Donneau, Olivien 003345
Donnellon, Maureen 000859
Donnelly, George J. 050049
Donnelly, John 053948
Donnelly, Peter 044695, 044751
Donner, Drs. E.M. 031952
Donner, Josef 002580, 001834
Donnermeyer, Reinhold 020392
Donoghue, Gail 053908
Donoghue, Michael 051265
Donohue, Deidre 051406
Donohue, Lorena 050340
Donoso, Ricardo 041107
Dony, Thierry 003993
Donzé, Frédéric 041908
Donzelli, Rodolfo 027662
Donzelli, Silvia 017118
Dooley, Bill 053861
Dooley, Cassandra 050614
Dooley, Max 054439
Doosry, Dr. Yasmin 020853
Dopffer, Anna 012071
Doppelfeld, Dr. Basilius 021769
Doppelfeld, Peter 042235
Doppelstein, Jürgen 022447
Doppler, Prof. Dr. Alfred 002624
Doppler, Elke 003095
Doppler, Margarete 001792
Dor, Ehud 025134

Doračić, Damir 009025
Dorado Ortega, Oscar Javier ... 008651
Doran, John 045813
Doran, Patricia A. 050025
Dorbolo, Cheryl 053472
Dordević, Radisav 038231
Dore, Dr. Anna 025538
Dore, Giovanni 028048
Dorée, Yves 015669
Doren, Jo-an' Van 047842
Dorgerloh, Prof. Dr. Hartmut .. 017079,
        017181, 017222, 017264, 017283,
        017284, 017286, 017288, 019597,
        019718, 021002, 021175, 021176,
        021178, 021180, 021181, 021183,
        021185, 021186, 021189, 021194,
        021195, 021196, 021197, 021199,
        021200, 021201, 021202, 021203,
                021204, 021404, 021806
Dorhs, Michael 019292
Doriot Galofaro, Sylvie 041665
Dorling, M. 043848
Dornan, Duncan 044206
Dornan, Vorin 051820
Dorner, Ernst 002248
Dorner, Friederike-Andrea 021011
Dorner, Markus 016804
Dornert, Nadine 012047
Dornieden, Horst 022118
Dornisch, Richard 052830
Dorociak, Heather 049218
Dorosiev, J. 005073
Dorra, Srarfi 042752
Dorsaz, Pierre 042101
Dorsch, Dr. Klaus J. .. 021292, 021294
Dorsett, A.G. 043877
Dorsett, Ann 044881
Dorsey, Robert W. 047737
Dorsky, David A. 051309
Dorst, A. 032848
Dortch, Steve 048588
Dose, Dr. Hanna 016837, 016838
Doseva, Iva 005191
Dosik, Jeffrey 051474
Dossi, Dr. Barbara 002931
Dossi, Dr. Cristina 028300
Doth, Diane 007450
Doublard du Vigneau, Simone .. 013206
Doucet, C. 016090
Doucet, Donna 005908
Doucet, Jean 012054
Doucet, M. 016089
Doudiet, Norman 047462
Dougherty, Erin 051065
Dougherty, J. Terry 048840
Dougherty, Molly 052062
Dougherty, Sandra 049671
Dougherty, Thomas A. 051412
Doughty, P.S. 043564
Douglas, Annabel 044582
Douglas, Bill 051504
Douglas, Deborah 047368
Douglass, Amy A. 053707
Douglass, Ardith 050766
Douglass, Larry 047208
Douglass, Michael 052346
Doulcet, Philippe 012548
Doum, Freddy 003958
Douma, Marian 032998
Douroux, Xavier 012777, 012779
Dourthe, Pacal 014360
Doval Sànchez, Gonzalo 039633
Dovgyj, Mykola Mychajlovyč .. 043176
Dovre, Terry 047272
Dow, George F. 051568
Doway, Stella 038262
Dowdeswell, Prof. J.A. 043847
Dowie, Peggy 045837
Dowlan, Neil 044781
Dowler, H. 044752
Dowling, Valerie 024897
Downend, Allan 045657
Downer, Dr. 024834
Downes, Elizabeth 001221

Downes, Marcia W. 051507
Downey, Martha J. 047015
Downey, Dr. Robert H. 047176
Downham, John 045568
Downie Banks, Dr. Margaret .. 053960
Downs, Alex 046311, 046313
Downs, Linda 048154
Downs, Stuart 049378
Doyle, Louise 000924
Doyle, Maura 048432
Doyle, Paul 024845
Doyle, Dr. Richard 050955
Doyon, Jean-Pierre 006359
Dräger, Dr. Olaf 047499
Dräger, Ulf 018969
Drăgan, Tereza 036190
Draganić, Danica 009017
Draganov, Dr. Dimitar . 005041, 005042
Dragićević, Mate 004267
Drăgoi, Dr. Livia 036382
Dragomir, Petre 036524
Dragoni, Prof. G. 025554
Dragotă, Gheorghe 036456
Dragsbo, Peter 010026
Dragt, G.I.W. 032237, 032655
Draguet, Michel 003459, 003460
Drahoňovský, Jan 009679
Drahošova, Viera 038445
Drake, Ellen Mary 006174
Drake, Pia 019060
Drake, Robert M. 049196
Drake, Tommi 049206
Dran, Jean-Luc 013124
Draper, Amanda 045565
Draper, Jill 044989
Draper, Jo 045174
Draper, Steven C. 048814
Drapkin, Adrienn 049711
Drapkin, Adrienne 049708
Draps, M. Willem 003484
Drašković, Dragan 038205
Dratwa, Daniel 003487
Draughn, Raymond 052658
Draževa, Conja G. 004995
Dražić, Jovanka 038245
Dražin Trbuljak, Lada 009048
Drda, Dr. Miloš 009886, 009887, 009902,
                            009948
Drechsel, Kerstin 017733
Drechsel, Manfred 021669
Drechsler, Dr. Wolfgang 003050
Dreckow, T. 000974
Drees, Dr. Jan 021664
Dreezen, J. 003696
Dreher, Derick 052091
Dreher, Nancy 051072
Dreier, Bruno 019556
Dreishpoon, Douglas 047298
Drejholt, Nils 041015
Drempetić-Hrcic, Nikola 008915
Drenker-Nagels, Dr. Klara ... 017482
Drennan, Emily 047477
Drenzek, Dennis 048820
Dresch, Dr. Jutta 019505
Drescher, Georg 021785
Dresco, Jean-Pierre 041831
Drese, Friedrich 020238
Dressel, Barry 046656, 049688
Dressel, Ina-Maria 016609
Dressler, Iris 022068
Dreuzy, Jehan de 015318
Drew, Alejandrina 048515
Drew, Jane 047449
Drew, Stephen E. 052724
Drewes, C. 038810
Drewitz, Britta 018832
Drewnowski, Benedykt 042004
Drews, Gayle 046883
Drews, Ute 021663
Drexel, Hermann 002574
Drexler, Ray 049918
Drexler, Toni 021709
Dreycus, Dominique 013863

Dreydoppel, Susan M. . 051204, 051205
Dreyer, Dr. Andreas 033635
Dreyer, Chris 051662
Dreyer, Prof. L. 038855
Dreyer, Dr. Wolfgang 019612
Dreyfus, Renee 053009, 053033
Drezgić, Olivera 038245
Drias, Lakhdar 000045
Driesbach, Janice 050300
Driesel, Hans 019637
Drieselmann, Jörg 017296
Driesens, F.J.J. 032780
Driessen, Hendrik 032930
Driessens, Mevrouw Colette .. 003586
Driewer, Ginny L. 054585
Drinjaković, Milica 038188
Drinkard, Joel F. 050491
Drinnenberg, Julia 019292
Driscoll, David 050601
Driskell, Julia 047054
Driver, Rod 001029
Drobney, Dr. Jeffrey A. 049923
Drögemüller, Inka 018446
Dröscher, Elke 019016
Drößler, Stefan 020524, 020551
Droguet, Vincent 013013
Dromgoole, Dominic 045105
Dromgoole, P.S.B.F. 044388
Dronkers, Ben 031942
Dronkers, S.L. 031897
Dror, Mira 025093
Drosinou, Paraskevi 023038
Droste-Vischering-Galen,
    Hubertus 001960
Drozd, Henryk 035193
Dru Burns, Mary 046401
Druček, A.A. 037417
Drucha-Graber, Barbara 002753
Drucker, R. David 051651
Druesedow, Jean 049933
Drumm, Mel 046486
Drummond, H. Maury 046815
Drutt, Matthew 049585
Družak, Antonija 008924, 008925
Dryden, Jackie 045726
Dryfhout, John 047996
Dsuve, Kjetil 033993
Du Bau, Giorgio 028206
du Fayet de la Tour, Alain ... 012997
Du, Gencheng 008348
Du, Jiuming 007574
Duan, Bilin 007749
Duan, Shuangyin 008054
Duan, Xinshu 008285
Duan, Xinya 008166
Duan, Xuelin 008430
Duarte Burró, Ramón M. 034203, 034204
Duarte Ferreira, Selma 004929
Duarte Rodríguez, Rafael Enrique 008623
Duarte, Berta 035622
Duarte, Carlos F. 054721
Dubach, Margaretha 041877
Dubal Bon, Vicente 039755
Dubbi, Franz-Josef 022426
Dubé, Edouard 046408
Dubedat, Paul 015565
Dubick, David 046572, 048470
Dubois, Arlette 015167
Dubois, Bernard 003387
Dubois, Danielle 006409
Dubois, Danuta 014685
Dubois, Oliver 015039
Dubreuil, Nicole 014065, 014066
Dubrisay, Pascal 013845
Dubrović, Ervin 008976
Dubrovin, Sergej Michajlovič .. 037580
Dubrovin, V.M. 036854
Dubuc, André 013785
Dubuisson, Laurente 003327
Ducasse, Sonia 014224
Ducastelle, Jean-Pierre 003326, 003328,
                            003387
Ducey, Dr. Peter K. 048013
Duchamp, Luc 014034

Duchemin-Philippe, Catherine . . 012948
Duchemin, Michael 050455
Duchesne, Prof. Jean-Patrick . . 003792,
003810
Duckers, Peter 045799
Duckworth, Elisabeth 006057
Duclos, Jean-Claude . . 013180, 013182
Duco, D.H. 031979
Ducor, J. 041533
Ducourau, Vincent . . . 011923, 011925
Ducreux, Anne-Claire 013995
Duda, Eugeniusz 034882
Duda, Igor Mykytovyč 043308
Dudant, Didier . 003275, 003276, 003277
Dudar, Dr. Wiesław 035098
Dudeck, Dr. Volker 022889
Dudenhöffer, Franz 021932
Dūdiņš, Andis 030252
Dudley, Andrew 001503
Dudley, Clarissa 049549
Dudnik, I.E. 043218
Dudź, Jerzy 035256
Dudzik, Przemysław 034791
Dueck, Lois 005407
Dückershoff, Michael 017962
Dühr, Dr. Elisabeth 022205
Duell, Charles H.P. 047557
Duell, Marshall 053087
Duenkel, Bob 048826
Dünnebier, Dr. Michael 018015
Dünser, Hans 001826
Duensing, Darrell 052347
Duerr, Mark 050599
Dürüce, Servet 043085
Duesberg, Baron François 003885
Düspohl, Martin 017197
Düsterwald, Dr. Brigitte 019982
Dütschler, H.R. 042058
Dufek, Mike R. 049063
Duff, Ana 047237
Duff, Doris 054264
Duff, James H. 047506
Duffin, Marion 046070
Duffy, Prof. A. 038864
Duffy, Henry J. 047996
Dufilho, Diane 053320
Dufour, Albertine 053539
Dufour, Christophe 041798
Dufour, Gary 001401
Dufour, Pierre 007266
Dufrasne, F. 004044
Dugajeva, Tetjana Igorivna 043124
Dugall, Berndt 018455
Duggan, Ervin S. 051893
Duggen Below, Ina 021132
Dugmore, I.M. 038714
Dugot, Joël 014707
Dugué, Prof. J. 014772
Dugulin, Dr. Adriano . . 028191, 028193,
028194, 028195, 028196, 028197,
028199, 028200, 028204
Duits, Th. te 032826
Duivesteijn, John 032201
Dujardin, Paul 003496
Dujmović-Kosovac, Ljubica 008977
Dukat, Zdenka 009025
Duke, Jacqueline 053115
Duke, Judy 047969
Dulaban, Alta 047840
Dulaney, W. Marvin 047541
Dulau, Robert 014779
Dulda, Mustafa 042790
Dulgheru, Lia Maria . . 036276, 036331,
036621, 036622, 036623, 036624,
036626, 036628, 036755, 036808,
036813, 036814, 036815
Dulibić, Ljerka 009050
Dulieu, René 003956
Dulin, Étienne 016236
Dumalag, Filemon R. 034427
Dumanoir, Thierry 013444, 015057
Dumaret, Isabelle 041391
Dumas, Cyril . . 013739, 013740, 013742,
013743

Dumas, Sylvia 020656
Dumbravă, Ion 036503
Dumbravă, Lucía 036605
Dumitrescu, Corina 036786
Dumitrescu, Horia 036350, 036454,
036465, 036466, 036467
Dumitrescu, Ilinca 036317
Dumitroaia, Gheorghe . 036243, 036609,
036610, 036614, 036615, 036660,
036766, 036767, 036768, 036769
Dumitru, Țeicu 036655
Dummer, S. 044646
Dumon, Noël 003415
Dumont Saint-Priest, F. 011944
DuMont, Bruce 047675
Dumont, Elizabeth R. 046452
Dumont, F. 003805
Dumont, Fabienne 003440
Dumont, Marcel 042067
Dumortier, Christian 015357
Dumortier, Claire 003492
Dumortier, Philippe 015357
Dumoulin, J. 004088
Dumville, John P. . . . 048675, 049600,
052699, 053474, 053561, 054473
Dunand, Michele 013475
Duncan, Jan 052927
Duncan, Jenepher 001401
Duncan, Larry 053598
Duncan, Oliver 054571
Dundass, Kay 033251
Dundon, Margo 049775
Dundore, Mary Margaret 052289
Dungan, Erica Howe 047165
Dunham, Elizabeth L. 053671
Dunina, Elena Michajlovna 037181
Dunkelman, Arthur 050850
Dunkerton, Jill 045057
Dunkley, Diane L. 054101
Dunlap, Joanne 052384
Dunlap, Margo 051683
Dunlap, Melissa L. 050354
Dunlap, Paula 049508
Dunlap, Susan 052660
Dunlop, Harry S. . . . 044414, 044418
Dunlop, J.M. 044297
Dunn, Cathey 050390
Dunn, David W. 053564
Dunn, Deborah F. 049535
Dunn, Forrest 053319
Dunn, Hedy 050402
Dunn, Jayne 044976
Dunn, Lawren . 048706, 049029, 052225
Dunn, Lindsley A. 047766
Dunn, Madeline 048350
Dunn, Nancy 046547
Dunn, Richard 043913, 044386
Dunn, Sharon 048981
Dunne, John 024963
Dunne, Maxine 005805
Dunning, Louise 044082
Dunning, Lynn 043498
Dunstall, David 045791
Dunton, Anna Mary 047394
Duparc, F.J. 032180, 032197
Dupe, Nicole 050203
Duperray, Eve . 012104, 013004, 013006
Dupeux, Cécile 015877
Duplančić, Arsen 008993
Dupont-Bauverie, Caroline 012042
DuPont, Diana 053092
Dupont, George 050090
Dupont, Jean-Pierre 016162
Dupont, Joel 015559
Dupont, Paule 003714
Dupont, Roland 016300
Dupré, Hans Jörg 017809
Duprat, Andrés 000133, 000136
Duprat, Jean-Marie 015838
Dupraz, Cathérine 012496
Dupre la Tour, Yves 013932
Dupree, Sherry 013199
Dupret, Sandra 048297, 048304
Dupriez, Françoise 004137

Dupuis-Sabron, Geneviève 012112
Duquenne, Raymond 004065
Duquesnay, Jean Claude 015569
Duquesnoye, Jeannot 003673
Dura, Ulrike 020021
Duraj, Paweł 035107
Durajková, Dr. Vladimíra 009989, 009990,
009991, 009992
Duran Alcalá, Francisco 039974
Durán Gómez, Eduardo 008637
Durán Nájera, Juan José 031135
Durán Solís, Leonel 031248
Durand, Agnes 014045, 014055
Durand, Alain 012909
Durand, André 012238
Durand, Micheline 011822, 011823
Durand, Régis 014665, 014666
Durandy van den Daele, Baroness
C. 013566
Durant, Graham 000942
Durant, John 047368
Durante, Aldo 026918
Durante, Dr. Antonio 025696
Duranthon, Francis 015999
Đuraš, Prof. Mišo 008910
Durben, Silvan 051849
Durbin, Brian H. 047288
Durden, Chris J. 046701
Durden, Robert 051887
Durdyèeva, Tuvakbibi
Kurbanovna 043090
Duret, Evelyne 015542
Duret, Michel 012954
Durham, A. Scott 051111
Durhône, Marise 011940
Durić, Vladimir 038128
Durkin, Brigid 048136
Durkin, D. 045004
Durko, Prof. Janusz 035338
Durlić, Paun Es 038209
Durman, Prof. Mick 043614
Durmaz, Ş. Nihal 043039
Durnford, Louise 007134
Durnoff, M. 013034
Durocher, Kristina 048737
Duron, Maurice 011897, 012347, 012348
Durracq, Roger-Pierre 013118
Durrani, Prof. Farzaud 034141
Durrleman, Sophie . . . 014705, 014713,
014733, 014785
Durst, Duane R. 049351
Dursum, Brian A. 047986
Durville 014611
Durye, Joseph 016273
Dusar, Mark 004030
Duscher, Michael 002211
Dušejko, Stanislav Dmitrievič . . 037388
Dusheshe, Tanny 023359
Dutarte, Albert 013199
Duțescu, Mihai 036414, 036415, 036418,
036419, 036421
Duthion, Jacques 013718
Dutilleux, Léon 003615
Dutoit, Albert 041760
Dutra, James 054229
Dutton, Sandra 046517
Duval Reese, Becky 048508
Duval, Dennis 047762
Duvall, John S. 048718
Duvernois, Bruno 013227
Duvivier, Christophe . 014959, 014960
Duvoisin, Madeleine 042171
Duvoisin, Olivier 041767
Duvosquel, Jean-Marie 004132
Duwel, Emely 053701
Duyckaerts, E. 003831
Duysters, K. 032027
Dvir, Orah 025137
Dvořáková, Dr. Hana 009468
Dwyer, Christine M. 046352
Dybbro, Inge 010157
Dybing, Leif 033575, 033576
Dyck, Angela 007372
Dyckmans, Heinrich 020627

Dye, Dr. Joseph M. 052565
Dye, Norman 050750
Dyer, David 050964
Dyer, Deborah 046783
Dyer, John T. 054169
Dyer, Michael P. 053285
Dyer, Mike 001422
Dykhuis, Peter 005944
Dymarski, Lech 035139
Dyment, Dave 007189
Dynak, Sharon 047792
Džanaeva, Alla Akimovna 038022
Dzeko, Lebiba 004283
Dzeranov, Timur Efimovič 038026
Dzhalilov, I. 042621
Dziechciaruk-Maj, Bogna 034875
Dzięciołowski, Edwin 034754
Dziedzic, Bogusław 034630
Dziembowski, Bettina von 020700
Dziewior, Yilmaz 019002
Dziurzyński, Tadeusz . 034871, 034872
Dzjadik, Jevgenija Mychajlivna . 043253
Džuma, Galina Ivanivna 043235
Eade, Coila 049346
Eadie, Bridget 045015
Eager, Barry W. 046944
Eakin, Eric 046836
Eames, Fred 001663
Earenfight, Dr. Phillip J. 047428
Earl, L.M. 001244
Earle, Edward 051406
Earle, Susan 050187
Early, J. W. 033142
Earnest, E. Peter 054126
Earnest, Ola May 052203
Earnst, Arlene E. 048346
Eastgate, S. 044729
Eastleigh, Lord 031716
Easton-Moore, Barbara 048040
Easton, Lois 049936
Eastwood, Stuart 043872
Eaton, Alex 046018
Eaton, Kenneth M. 049523
Eaton, Mary Candace 052021
Eaton, Nellie 006608
Eatts, M. 001189
Eayrs, Walter F. 047245
Ebbinghaus, R. 017155
Ēbedžans, Svetlana Georgievna . 037568
Eberhardt, Dr. Hans-Joachim . . 020576
Eberhardt, Herman 049649
Eberhardt, Ursula 016928
Eberhart, Rolf 020255
Eberle, Josef 030304
Eberle, Lidia 035329
Eberle, M. 049416
Eberle, Richard H. 051357
Eberle, Urs 041456, 041457
Eberli, Ulrich 042251
Eberold, Dr. Bernd 042202
Ebersberger, Eva 003091
Ebert, Glenn 007384
Eberwein, Dr. Roland 002192
Ėbinger, Margarita Astemirovna . 036965
Ebner, Dr. Lois 002306
Ebner, Mark 054234
Ebner, Silvia 002308
Ebneter, Curdin 041985
Ebrahimi, Nader 024777
Ebster, Dr. Silke 001766
Ebtehaj, V. 024743
Eccles, Tony 044312
Echeverría Llanos, Clemente . . . 008783
Ecker, Dr. Berthold 003037
Ecker, Reingard 002920
Eckerbom, Jonas 040739
Eckerson, Rosemary 052424
Eckert, Bob 050622
Eckert, Horst 016651
Eckert, Jochen 018866
Eckert, Prof.Dr. Rainer 020024
Eckhardt, Prof. Dr. Andreas . . . 017483
Eckhardt, Hans J. 041956
Eckhardt, Mária 023507

Eckhardt, Susan 049859
Eckhel, Nerina 009028
Eckhoff, Audun 033523, 033524, 033525
Eckhout, Debbra 046926
Eckloff, Dr. Wolfram 020161
Eckmann, Sabine 052809
Eckner, Harald 019032
Eckstrom, Carolyn 048579
Edblom, Mats 041011
Eddie, I. 033272
Édeleva, Inessa Aleksandrovna . 037101
Edelmann, Dr. Martina 022306
Eden, Dave 051002
Edenfield, W. Vernon 048954
Eder, Dr. Johanna 022056
Ederer, Walter 020779
Edgar, Angela 044244
Edgar jr., Arthur C. 050881
Edge, Yvonne 045215
Edgert, Elizabeth 051160
Edgren, Bengt 040677
Edhofer, Elisabeth M. 003053
Edidin, Stephen R. 051372
Edina, Mèri 023832
Edinger, Dorothy 033367
Edinger, Reuben 048543
Edison, Carol 052911
Edison, Robert D. 053822
Edland, Svein 050390
Edler, Dr. Doris 018629
Edler, Hans 002700
Edlinger, Matth. 002446
Edlund, D. 043958
Edmonds, Pam 007207
Edmonds, Tom 048824
Edmondson, J. 044869
Edmonson, Dr. Jim 047813
Edmunds, Allan L. 052035
Edo, Austin 034191
Edson Way, Dr. J. 053108
Edson, Gary 050517
Eduard, Ebner 001832
Edus, M.J. 032884
Edvardsen, Erik Henning 033836
Edwards, David 053651
Edwards, Denise 043936
Edwards, Geoffrey 001093
Edwards jr., Harford 048866
Edwards, Jeff 047936
Edwards, John 043350
Edwards, Karen 006525
Edwards, N. 043398
Edwards, Nancy E. 048904
Edwards, Nat 043397
Edwards, Pamela 051284
Edwards, Paula 049448
Edwards, Rita 049013
Edwards, Susan 051181
Edwards, William 044980
Eecher, Danilo 027584
Eekhof, H.R. 032945
Eekhout, L.L.M. 032564
Effenberger, Prof. Dr. Arne 017089
Efimova, Alla 046936
Efimovskij, A.F. 037162
Efrati, Y. 025224
Efsic, Dianne 047940
Eftekhari, M. 024774
Egan, Caroline 043705
Egan, Dominic 024829
Egawa, Tokusuke 029186
Egbert jr, Francis 049815
Egelseder, Berthold 002384
Eger, Peter 021300
Egermann, Gerhard . . . 002524, 002525
Egg, Prof. Dr. Markus 020233
Egge, Dr. Reimer 022247
Eggen, S. 003274
Egger, Dr. Franz . . . 041267, 041268
Egger, Norbert 002190
Eggers, Bernardo 007527
Eggers, Dr. José Carlos 004841
Egging, Dr. Björn 021248
Eggl, Christine 002764

Egglezos, Evangeline T. 048764
Eggli, Dr. Urs 042234
Egholm, Aage 010140
Egilsson, Kristjan 024119
Egipciaco, Wanda 051393
Egli, Markus 041730
Egli, Monika 042223
Egloffstein, Dr. Albrecht Graf von und
zu 021073
Egorova, Ljudmila Semenovna . 038007
Egoryčev, Viktor Vasiljevič 037236
Egounlety, Micheline 004184
Egret, Daniel 014624
Eguía, Dr. José 040425, 040429
Ehiro, Prof. Masayuki 029363
Ehis, Martin 043615
Ehling, Dr. Kay 020577
Ehlitzer, Bernd 018722
Ehls, Heiner 019292
Ehm, Rainer 021318
Ehrenberger, Mag. Anton 001926
Ehrenfellner, Dr. Karl 002637
Ehrenheim, Jacob von 040603
Ehrenhuber, Franz 001747
Ehrentraud, Adolf 002016
Ehrenwerth, Dr. Manfred 018905
Ehres, Maria 041138
Ehrhart, Bernhard 019937
Ehrmann-Schindlbeck,
Anna-Maria 019445
Ehry, Carl A. 048536
Eibl, Simon 019934
Eichelberger, Allison 050124
Eichenauer, Dr. Jürgen 020971
Eichhorn jun., Georg 002930
Eichhorn, Herbert 021386, 021387
Eichler, Dr. Anja 022610, 022612,
022613, 022618
Eichler, Heidrun 020289
Eichmann, Erland 042143
Eick, Dr. Simone 017605
Eichhorst, Dr. William S. 049889
Eickmann, Margaret 052303
Eickmann, Theodore 048437
Eidam, Hardy 018243, 018247
Eidemüller, Gisela 017390
Eidlin, Nathan 025185
Eidsaune, Thor Helge 033772
Eifel-Porter, Rhoda 051428
Eifrig, Jennifer S. 054356
Eigler, Gerd 019079
Eigner, Walter 001768
Eikelmann, Dr. Renate 019373, 019586,
019587, 020509, 020921, 020923,
020924
Eiken, Dr. Douglas 050054
Eikenoord, R.G. 054593
Eikermann-Moseberg M.A., Silke 017649
Eiland, William U. 046593
Eilat, Galit 025064
Eilertsen, Kate 053014
Eilertson, Orie 052348
Eillott, John 001517
Eilschou Holm, Niels 010193
Eimbeck 031837
Eimer, Prof. Gerhard 019360
Eimert, Dr. Dorothea . 018021, 018022
Einarsdóttir, Björg 024120
Einberger, E. 020527
Einreinhofer, Nancy 054250
Eipeldauer, Heike 041247
Eipper, Paul-Bernhard 041247
Eiras, Prof. Jorge 035939
Eirola, Martti 031849
Eiselt, Gerhard 020176
Eisen, M.A. Markus 020505
Eisenburger, Dr. C.M.S. 032062
Eisenhofer, Dr. Stefan 020579
Eisenmann, Mark 051255
Eisenreich, Jessica 053052
Eisenstein, Daniela 021690
Eisenstein, Dr. Daniela F. 018548
Eiskonen, Satu 011090
Eisl, Sigmund 002803

Eisler, Dr. János 023553
Eisley, Patricia 054034
Eissenhauer, Dr. Michael 016903, 017241,
017713, 019526, 019527, 019530,
019532, 019533, 019538, 019540,
019542, 019543, 019545, 019546,
019548, 019549, 019551, 019553
Eitam, David 025052
Eiynck, Dr. A. 020090
Eizenberg, Arie 025027
Ek-Nilsson, Katarina 041050
Ekberg, Anna 040888
Ekblom, Bengt 041116
Ekeberg, Jonas 033672
Ekelhart-Reinwetter, Dr. Christine 002931
Ekinci, H. Ali 042855
Eklund, Linda 040888
Eklund, Dr. Sigrid 041028
Ekpo, Violeta I. 033447
Ekström Söderlund, Birgitta 010801
Ekström, Börje 040804
Ekström, Kjell 011143, 011146
Eksturm, Nancy 052106
El-Amrani, Dr. Ahmed 031800
El-Aziz, Fatma Abd 010549
El-Hajj, Amal Abu 025074
El-Hajraoui, Mohamed
Abdelwahed 031788
El-Husseini, Dr. Mehrez 010521
El-Kasri, Houceine 031790
El-Moussli, Majed 042262
El-Nur, Prof. Osama A. 040493, 040496
El-Ouafi, Prof. Boubker 031797
El-Tawab, Garethe 045937
El-Younsi, Dr. Amin 031799
Eland, Steve 001395
Elas-Necel, Karol 034685
Elayyan, Khaled 034156
Elba La Gioiosa, Rosa 000379
Elbert, Dirk 021893, 021895
Elbracht-Iglhaut , M.A. Gisela . . 021898
Eldamaty, Dr. Mamdouh 010536
Elder, Bob 052376
Elder, E.J. 007416
Elder, Sarah M. 052788
Elderfield, John 051427
Eldredge, Bruce B. . . . 050304, 053442
Eldridge, Ruth 043949
Eleki, Dr. Krisztina 050412
Elen, A. 032826
Elerd, Udo 020991
Eletson, Helen 045136
Elgen, Fran 046982
Eliades, Chryso G. 009432
Elias Ortiz, Alejandrina 031509
Elias, Prof. Giacomo 028168
Elias, Margery M. 049798
Elias, Ramon Jan 049798
Eliasen, Kirstin 010242
Eliason, Sara 041176
Eliassen, Sven G. 033927
Eliëns, Prof. Dr. T.M. 032171
Eliot, D. 045952
Elk, Sara Jane 052042
Elkin, G.U. 037249
Elkington, Marian 043414
Elkington, Peter 043414
Elkins, Scott 051491
Elkjær, Trine 010006
Elkmann, Alain 028132
Elkovitch, Pat 046651
Elleberg Petersen, Mads 010004
Elleboudt, André 003554
Ellebrecht, Pat 049992
Ellen, Marc ter 032997
Ellenbroek, Dr. F.J.M. 032928
Ellender, Dan 050071
Ellensint Kremer, M. 022286
Ellerman, Deborah L. 046549
Ellermeier, Dr. Friedrich 019071
Ellingsen, Harry 033548, 033549
Ellingson, Eugene 052663
Ellington, Howard W. 054395
Elliott, Dana R. 048714

Elliott, J.H. 046610
Elliott, Keith 045241
Elliott, Lucy 048378
Elliott, Mary Gene 046610
Elliott, Patrick 044273
Elliott, Rachael 001130
Elliott, Richard 043775
Elliott, Simon 000937
Ellis-Schön, Jane 045727
Ellis, Anita 047739
Ellis, Caroline 046003
Ellis, Evelyn M. 050314
Ellis, Gillian 043626
Ellis, J.C. 050165
Ellis, Jane 044968
Ellis, Joe 047456
Ellis, Lori 005382
Ellis, Pam 047471
Ellis, Robert E. 049858
Ellis, Ruth 001458
Ellis, Vern 000791
Ellison, Dr. Curtis W. 051870
Elliston Weiner, Sarah 051426
Ellsworth, Frank L. 051411
Ellsworth, Lynn 051093
Ellwanger, Natalie 001794
Ellwanger, Ricard 054358
Ellwood, Tony 001279, 001503
Elmén Berg, Anna 040915
Elmer, Hans R. 041957
Elmes, Robert 047258
Elmore, Fletcher L. 050501
Elmore, George 050149
Elmqvist Söderlund, Inga 041029
Elms, Anthony 047661
Elorduy, Julieta Gil 030802
Elorza Guinea, Juan Carlos 039169
Eloy, Paul 003832
Elsberg, Carl J. 047037
Elsbury, Kathy 047330
Elsdon, Judith 045982
Elsen-Schwedler, Dr. Beate 019818,
021756
Elsen, Dr. Thomas . . 016654, 016665,
016665
Elsener, Walter G. 042155
Elser, Oliver 018411
Elsig, Patrick 041905
Elšin, Sergej Veniamovič 037132
Elsner, Dr. Tobias von 020207
Elsposch, Friedrich 018056
Elston, Tim 051976
Elsworth, Jo 043755
Elustondo, André 015377
Eluyemi, Dr. Omotoso . 033410, 033413
Elven, Dr. Reidar 033847
Elvers-Švamberk, Dr. Kathrin . . 021560
Elvira Barba, Miguel Angel 039629
Ely, Bruno 011560, 011565
Elyasiv, Isaac 025181
Elzinga, Dr. C. 032570
Emack Cambra, Jane 053688
Emanuel D., Hough 050397
Emanuel, Ludovit 009834
Emanuel, Nina 019968
Emanuelli, Dr. Giuseppe 027299
Ember, Dr. Ildikó 023553
Emberling, Geoff 047686
Emenegger, Ashley 054307
Emerick, Emily W. 054389
Emerson, Francis . . . 054561, 054562
Emerson, Julie 053248
Emerson, Lucius J. 024985
Emerson, Myra 050113
Emerson, Pam 053784
Emerson, Philip G. 054413
Emerson, Robert L. 054597
Emery, Dominique 013055
Emig, Ulli 017823
Emigholz, Dr. Björn 022318
Emiliani, Prof. Andrea 025574
Eminger, Dr. Jürgen 022175
Emmanuel, Victoria 043623
Emmendörffer, Dr. Christoph . . . 016662

Emmert, Dr. Claudia 018265
Emonson, Janet 007009
Emont Scott, Deborah 049888
Emptoz, Dr. F. 013937
Emsden, Katharine 050050
Ena, Jan 000898
Enache, Ştefan 036418
Encinas, Isabel 039647
Enciso Núñez, José Antonio ... 030893
Enckell Julliard, Julie 042108
End, Reinhard 018636
Ende, Arlyn 053276
Endenburg, Constanze 032820
Ender, Benjamin 041720
Enders, Don 052767
Enders, Donald L. 052915
Enders, John 050758
Endersbe, Susan U. 053210
Endert, D. van 016570
Endicott, John 043911
Endo, Koichi 029441
Endraß, Max 016599
Endslow, Ellen 054293
Endthoff, Gertrude 053411
Endzweig, Pamela 048622
Ene, Ioana 036645
Enei, Flavio 027822
Enei, Vicente 000489
Enevoldsen, V. 010199
Engberg, Siri 050948
Engdahl, Sarah 047787, 047788
Engel, Beate 041327
Engel, Edmée 030594
Engel, Erik 041101
Engel, Michael 050186
Engel, Nicolas 014689
Engel, Robert 047223
Engel, Silke 020842
Engel, Volker 017486
Engel, Dr. Walter 018045
Engelbosch, Christa 004028
Engelbrecht, Dr. C.M. 038655
Engelbrecht, Dean Mark. 046441
Engelhardt, Jeannine 022258
Engelhardt, Thomas 018264
Engell, Oluf 010305
Engelmann, Barbara 022511
Engelmann, Mag. Josef 001729
Engels, D. 009397
Engels, Kim 022631
Engels, Maria 016439
Engelsing, Dr. Tobias 019763
Engelskirchen, Lutz 018917
Engelsman, Dr. S.E. 032585
Engen, Jorunn 033580
Engen, Luc 003720, 033693
Engeseter, Aage 033694
Engfors, Christina 040990
Engh-Grady, Linda 049864
Enghoff, Dr. Henrik 010201
Engi, Dr. Andrea 042088
England, Ann 046620
England, Peter 047648
Engle, Duane E. 052485
Englebright, Steven E. 053549
Englert, Emil 022521
Englick, Gerd 016865
English, Catherine 006786
English, Charles L. 054470
English, Donise 052336
English, Louella F. 054470
English, Marie 051861
Engström, Urban 040836
Enguita, Nuria 039057
Enkhtsetseg, Dashdavaa 031733
Enne, Eva 002387
Enneberg, Lee 053967
Ennen, Dr. J. 022067
Enomoto, Toru 029426
Enriqueta del Castillo y Mejía,
  María 031225
Enriquez, Felonilla D. 034470
Enseki, Carol 047251
Ensign, Garald 049605

Ensslen, D. 005766
Entin, Daniel 051458
Entremont, Paul d' 007381
Entrudo, Ana Flores 035971
Entwistle, R. 044646
Enzler, Jerome A. ... 048375, 048376
Epp, Bear 005733
Eppel, Dr. Peter 003095
Eppensteiner, Heinz 002311
Eppich, Linda 052385
Eppinger, Virginia 047210
Epple, Sabine 019992
Eppler-Jehle, Ursula 016494
Epprecht, Dr. Katharina 042221
Epps, Renee 051418
Erasmus, Prof. Theuns 038846
Erašova, Galina 037571
Erb, Christian 018422
Erb, Dr. Martin 041719
Erbelding, Susanne 019505
Erbentraut, Dr. Regina 018886
Erbguth, Dr. Horst 021313, 021318
Ercolani, Fausto 028287
Erdem, Emin 042996
Erdenebat, P. 031737
Erdmann, Walter Johannes ... 018772
Erdős, Iudita 036682
Eremina, Natalija Aleksandrovna 037608
Ereoli, Mario 026910
Erf, Lisa K. 047658
Erffa, Dr. Axel von 019524
Erftemeijer, A.F.W. 032392
Ergin, Muhlis 042932
Ergino, Nathalie 016289
Erhart, Heinrich 002725
Erhart, Ingeborg 002138, 002140
Erhatic Širnik, Romana 038562
Erias Martinez, Alfredo 039134
Ericani, Dr. Giuliana ... 025470, 025473
Erice Lacabe, Romana .. 040447, 040450
Erich, Frasl 002586
Erichsen, Dr. Johannes 019826, 019831,
  019832
Erichson, Ulf 021408
Ericini, Carlo 025605
Erickson, Bruce R. 052852
Erickson, Florence 006272
Erickson, Marsha 049903
Erickson, Ruth 052004
Erickson, Willis D. 046725
Eriksson, Håkan 041118
Eriksson, Karl 040559
Eriksson, Maud 040795
Eriksson, Nadja 040732
Eriksson, Susan C. 047028
Eringen, Peri 052456
Eritsyan, Elena Sergeevna 037396
Erk, Werner 018709
Erkelens, Dr. A.M.L.E. 032013
Erkens, Michet 003704
Erkinger, Helmut 002537
Erkinova, Rimma Michajlovna .. 037020
Erlandsen, Roger 033996
Erler-Striebel, Petra 019508
Erlindo, Vittorio 026845
Erlinger, Fritz 002646
Ermacora, Dr. Beate 002123
Ermeling, Andreas 020607
Ermert, Axel 017178
Ermolaeva, Ljudmila Nikolaevna 037229
Ernst-Adolf, Kloth 019184
Ernst, Dr. Alexander B. 022817
Ernst, Alfred 020472
Ernst, Dr. Antje 021912
Ernst, Gene 053490
Ernst, Jürgen 020007
Ernst, Mathias 042034
Ernst, Dr. Rainer 018365
Ernstell, Eva-Sofi 040991
Erochin, Viktor Ivanovič 037985
Erol, Dr. Erdogan 043001, 043003,
  043004, 043007, 043008, 043009
Erramah, Mourad 042749
Errotabéréa, Claude 012147

Erskine, David 043456
Eršova, Valentina Petrovna ... 037666
Ertel, Dr. Christine 002375
Ērtjukova, Nadežda Semenovna 037053
Ertler, Karl 002682
Erven, Wilma 048235
Erwin, Sarah 053848
Eržen, Miro 038578
Escalas, Román 039081
Escat, Monique 015414, 015848
Escauriza, Edgardo 000644
Eschapasse, Anne 014750
Eschbach, Barbara 053702
Esche, Charles 032293, 040832
Eschebach, Dr. Erika .. 017552, 017566
Eschebach, Dr. Insa 018540
Eschenburg, Dr. Barbara 020580
Eschenburg, Wieland 021176
Eschment, Renate 018261
Escobar, Gloria 051776
Escobar, Ticio 034193
Escobar, Washington 046296
Escoffier, André 015711
Escolà, Dr. Olegrer 039077
Escriche Jaime, Carmen 040170, 040226
Escrivà, J. Ramón 040305
Escudero, Juan Miguel 034590
Esera, T. 038094
Esguerra, Liwliwa 034443
Eshenroder, Glenda 049564
Eshleman, Janet 051053
Eshun, Elton 045005
Eskelinen, Liisa 011112
Eskildsen, Prof. Ute 018306
Eskritt, Jodi Ann 007233
Esman, J. 032929
Espejo-Saavedra Romero, Luis . 040100
Espeland, Else 033827
Espinagosa Marsà, Jaume ... 040210
Espinola, Vera 052861
Espinós, Adela 040308
Espinosa Jiménez, Eleazar ... 009306
Espinosa Rodriguez, Antonio ... 030754,
  030767
Espinosa Ruiz, Antonio 040408
Espinosa Yglesias, Angeles ... 031375
Espinosa, Deborah 052408
Espinosa, Gina Agnes 034480
Espinoza Arrubarrena, Luis ... 031184
Espinoza Mella, Adelaida 008611
Espinoza Pérez, Egdar 033390
Espinoza, Carlos A. 007541
Espinoza, Felipe 031173
Espinoza, Dr. Isabel Flores ... 034348
Espírito Santo Bustorff Silva, Dr. Maria
  João 035797
Espíritu Gaston, Lynell 034596
Espointour, Elizabeth 046578
Esposito, Prof. Vincenzo 027001
Eßbach, Elke 019649, 022921
Esseiva, Renato 042018
Essen, Dr. Manfred von 020824
Esser, Dr. Hans-Joachim 020513
Esser, Dr. Werner 020882, 022057
Essl, Prof. Karlheinz 002207
Estall i Poles, Vicent . 039836, 039838
Esteban Darder, Vicente 040271
Estep, Connie 052528
Esterhuizen, Valerie 038863
Estes, Judith .. 049619, 049620, 049622,
  049623
Esteve, Y. 013262
Estham, Inger 041113
Estrada Sánchez, Ramiro 030831
Estrada, Cesar P. 034454
Estrada, Gerard Martí 039211
Eszter, Hajdu 023487
Etcheverry, Charles 003269
Eteläinen, Suvi-Mari 011138
Etgar, Raphie 029452
Ethelberg, Per 010092
Etter, Dr. Walter 041282
Etter, Willi 041445
Eubanks, Charles 051456

Eubanks, Ray 053431
Eugster, Katharina 042087
Eugster, Willi 041935
Euler-Schmidt, Dr. Michael ... 019693
Euler, Dr. Andrea 002330
Eure, Linda Jordan 048471
Eutsler, Therese 050313
Evangelinou, Rallou 023193
Evanich, Joan 054480
Evans-Cracknell, Victoria 004192
Evans, Bob 006452
Evans, Catherine 047936
Evans, Cathy 047552
Evans, Ceryl .. 044518, 044519, 044739
Evans, David . 045548, 045549, 045550,
  045559
Evans, Ed 005444
Evans, Eddie L. 046867
Evans, Edward 045450, 050713
Evans, Elaine A. 050011
Evans, G.H. 043877
Evans, Graham 044774
Evans, Greg 007331
Evans, Helen H. 048932
Evans, Jaime 051587
Evans, Jonathan 045095
Evans, Leroy 049647
Evans, Linda 053605
Evans, Lynn 050556
Evans, Margaret 049331
Evans, Mark 044800
Evans, Megan E. 049320
Evans, Mervyn 000655
Evans, Michael 046825
Evans, N. 033297
Evans, Neil 045168
Evans, Paul 044784, 045088
Evans, Steven 046841, 051378
Evdokimov, Vladimir
  Aleksandrovič 037241
Evdokimov, Vladimir Nikolaevič . 037002,
  037003
Eveleigh, David J. 043746, 043747
Eveleigh, Valerie 045288
Evelyn, Douglas E. 051448
Evensen, Anne Bjørg 033746
Everard, R.H.A.J. 032397
Everbrand, Lars-Åke 040554
Everett, Julienne 005408
Everett, Martin 043714
Everly, Nancy 048061
Evers, Cécile 003492
Evers, Richard 049461
Evers, Dr. Ulrika 021093
Eversberg, Dr. Gerd 019370
Eversole, Diane 050116
Evgenidou, Despina 023023
Evgeniev, Todor 004996
Evstigneeva, Irina V. .. 037819, 037820
Evtimov, Dejan 005213
Evtukova, Olga Aleksandrovna .. 037527
Ewald, Jill 051632
Ewbank, Nick 044355
Ewell, Matson 048680
Ewers, William 047688
Ewert, George H. 050974, 050979
Ewertowski, Zbigniew 035042
Ewigleben, Prof. Dr. Cornelia . 020127,
  022047
Ewigleben, Dr. Cornelia 020128
Ewin, Dr. Tim 043747
Ewing, Phyllis 049364
Ewing, Steve 051099
Ewing, William A. 041659
Ewington, Julie 001503
Ex, K.M.T. Sjarel 032826
Exley, Stuart 052574
Exnarová, Alena 009528
Eyb, Alexander Freiherr von ... 022722
Eyb, Silvia Freifrau von 022722
Eychner, George 049702
Eyland, Cliff 007444
Eyler, Carolyn 049164
Eynatten, K.C. 049594

Eyün Aksu, S. 043016
Ezquerra, Beatriz 040226
Ezsöl, Adolf 002744
Faass, Dr. Martin 017214
Fabbri, Fabio 027253
Fabbri, Dr. Franco 026322, 026323,
026324, 026325, 026327, 026330
Fabbri, Jennifer S. 038807
Fabeck, Sylvie 003907
Fabela, Pamela K. 053775
Fabényi, Dr. Júlia 023528
Faber, Alain 030594
Faber, Dr. Elfriede-Maria 002965
Faber, Lucien 003712
Faber, Dr. Marion 020877
Faber, Dr. Michael 020323
Faber, Dr. Monika 002931
Faber, O. 032052
Faber, Ole 010020, 010021, 010022,
010236, 010266, 010267, 010350,
010351, 010352
Faber, P. 031950
Faberman, Hilarie 053488
Fabian, Dr. Claudia 020507
Fabiani, Jérôme 012886
Fabiani, Rossella 028211
Fabiankowitsch, Gabriele 003033
Fabich, Uwe 019935
Fabing, Suzannah 051626
Fabiny, Dr. Tibor 023476
Fabisiak, Wojciech 035206
Fabra Carreras, Carlos 040384
Fabre, Alexia 016312
Fabre, Eric . . . 013158, 014708, 014709
Fabre, Jean-Louis 014968
Fabre, Regis 012836
Fabrega i Gallaguet, Marta 039760
Fabregues, M. de 014087
Fabrezi, Dr. Marissa 000556
Fabricius, Karen M. . . . 010000, 010001,
010002, 010322
Fabris, Dr. Corrado 025394
Fabritius, Mag. Art. Elisabeth . . 010305
Fabritius, Dr. Ruth 021395
Fabrizi, Elisabetta 044917
Fabrizio, Fernando di 027246
Fabyan, Dr. E. Joseph 053992
Facchinetti, Claudio 025761
Facchini, Prof. Fiorenzo 025552
Faccinto, Victor . . . 054487, 054496
Facciolongo, Sabino 025741
Fachrudin, H. 024582
Facklam, Gudrun 017709
Fadani, Dr. Andrea 022264
Fadden, John 051771
Fadden, Karen 048383
Fadden, Lee R. 048383
Fadeev, Igor 037250
Fae Kamm, Mary 048118
Fährig, Ute 018953
Faelmar-Abad, Mary 034456
Faenzi, Giovanna 026259
Färber, Steffen 019062
Færøy, Frode 033848
Faesler Bremer, Dr. Cristina . . . 031192
Fässler, Klaus 041226
Fagbohunmi, Johnson 033424
Fage, André 014653
Fagerli, Karl A. 033807
Faherty, William B. 049423
Fahey, Genevieve 001512
Fahey, John M. 054110
Fahey, Peggy 005987
Fahl-Dreger, Axel 022305
Fahl, Dr. Andreas 019053
Fahmy, Prof. Dr. Mamdouh Amin 010585
Fahrenbruch, Melanie 048018
Fahrer, Uwe 017572
Fahy, Everett 051424
Faietti, Dr. Marzia 025533, 026252
Faill, Carol E. 050126
Failla, Dr. Donatella 026400
Faille d'Huysse, B. Baron della . 004166
Faine, Susan 001470

Fair Sleeper, Sara 046374
Fair, Donald 046332
Fair, Eileen 053077
Fairbank, Lynn 048210
Fairbanks, Theresa 051266
Fairclough, Oliver 043863
Fairfax, Anne 051423
Fairfax, James A. 054169
Fairfield, Douglas 046358
Fairfull, Thomas M. 048802
Fairman, Elisabeth 051266
Faison, Elizabeth 053082
Faißner, Waltraud 002323
Faivre, Bernard 013438
Fajzullina, Galija Šauketovna . . 029797
Falabrino, Ugo 028214
Falaschi, Prof. Pierluigi 025711, 025717
Falcão, Prof. José Antonio 035514
Falco, João 004393
Falco, Joseph 016222
Falcon, Orlando Arturo 000238
Falconbridge, Prof. Brian 045089
Falconbridge, Liz 044704
Falgayrettes, Christiane 014694
Falk, Dr. Birgitta 018292
Falk, Dr. Brigitta 018312
Falke, Ulla 013335
Falkenberg, Edward 007165
Falkenberg, Dr. Regine 017299
Falkenberg, Dr. Uwe 017986
Falkenstien-Doyle, Cheri 053125
Falkinger, Johann 002552
Falkner, Dr. Gerd 021143
Fallet, Estelle 041532
Falletti, Dr. Franca 026258
Fallon, Steve 054022
Falls, John 048103
Falomir, Miguel 039676
Faltaits, Manos 023269
Faltings, Dr. Dina 019143
Falvey, Tom 047918
Falvy, Dr. Zoltán 023571
Falz, A. 021802
Falz, Andreas . . . 019145, 020245
Falz, Bernd 019211
Famà, Dr. Maria Luisa 028165
Famuyiwa, M.B. 033423
Fan, Shenglin 008427
Fan, Wanglin 008499
Fan, Xiyi 008017
Fan, Yousheng 008470
Fan, Zhiqian 007967
Fanciotto, Cristiana 027288
Fandrey, Dr. Carla . . . 019627, 022465
Fanelli, Doris 052044
Fang, Guangyao 008419
Fang, Guoquan 007948
Fang, Madeleine 046941
Fang, Yingsan 008142
Fang, Zaifu 008412
Fangarezzi, Dr. Riccardo 027049
Fanger, Dr. Henrik 021797
Fankhauser, Christina 041771
Fanni, Dr. Ignazio 028390
Fanning, Dale A. 052401
Fanony Fulgence 030652
Fansa, Prof. Dr. Mamoun 020988
Fanti, Mario 025559
Fanucci, Dennis 049978
Fanzun, Jon 042047
Farah, Ahmed 038642
Farahmanesh, Nazi 024710
Farar, Bill 048079
Farasin, Kurt 002724
Farbaky, Dr. Péter 023497
Farhad, Dr. Massumeh 054116
Faria, Dr. João Carlos Lázaro . . 035479
Faria, Sofia 035893
Farid Bin Abdul Jalal, Enoik
Ahmad 030721
Fariña Busto, Francisco 039857, 040000
Farina, Dr. Ermes 028326
Farina, Prof. Tina 019678
Farinelli, Dr. Leonardo 027212

Faringdon, Lord 044330
Faris, Douglas D. 053289
Faris, M.P. 024861
Farkas, Jesse 048603
Farley-Harger, Sara 048929
Farley, Annie 053919
Farley, G. 049416
Farley, Henry G. 053783
Farley, Dr. Linda E. 048254
Farmer, Dennis P. 047622
Farmer, Dorothy 005717
Farnden, Bob 001193
Farnell, C. 044527
Farnell, Cynthia 054033
Farnham, Priscilla 048684
Farnós y Bel, Alex 038971
Farooqi, Dr. Anish 024228
Farr, Amanda 045378
Farr, Fonda 051690
Farrajota Ataide Garcia, Dr. Maria
Madalena 035781
Farrand, Dr. William R. 046491
Farrar, H.D. 051721
Farrar, Robert A. 049474
Farrell, Anne 050040
Farrell, Kathleen 052619
Farrell, Laurie 050381
Farrell, Lesley 045833
Farrell, Thomas L. 052345
Farrell, Walter 052697
Farrenkopf, Dr. Michael 017433
Farrington, Charles W. 051603
Farrington, Dr. Gregory C. 052995
Farrow, Victoria 045045
Farver, Bonnie 047054
Farver, Jane 047364
Farwell-Gavin, Robin 053116
Farynyk, Diane 051390
Fasching, Josef 001819
Fasciati, Luciano 041487
Fasel, André 041501
Fash, Barbara 023399
Fash, Dr. William L. 047370
Fasold, Dr. Peter 018405
Fasold, Dr. Regina 019171
Fassati, Tomáš 009443
Fassbinder, Dr. S. 018808
Fasse, Dr. Norbert 017514
Fasser, Ernst 041930
Fast, Dr. Kirsten . . . 018319, 018321
Fastlicht, Dr. Samuel 031202
Fatah, Sayed Abdul 000004
Fateh, Suzanne 054084
Fatehi, F. 024700
Fatimah, T. 024539
Fatka, Jaroslav 009780, 009786, 009787,
009792, 009845
Faton, Louis 014631
Faubert, Claude 006576
Fauchille, Bernard 014235
Fauerso, Joey 052933
Faulds, W. Rod 047070
Faulkner, D. Lindsay 033316
Faulkner, John 047781
Faulkner, Peter 001576
Faull, Caroll 049550
Faull, Dr. Margaret 046031
Faure-Comte, Karen 016220
Faure, Philippe 013686
Fauria i Roma, Dr. Carme 039095
Faust, Alexis 053827
Faust, David T. 053821
Faust, Pierre-Paul 015174
Faust, Dr. Sabine 022202
Fautin, Daphne 050186
Fauver, Lynn N. 051095
Fava, Lourdes 031689
Favarato, Dr. Paolo 026603
Favaro, Gian Pietro 027429
Favata, Alexa 053666
Favell, Georg 049591
Favre, Lee M. 050356
Fawaz, Salma 030276
Fawdry, Eddy 045077

Fawzi, Ibrahim 025101
Faxon, Susan 046479
Fay, David 048694
Fay, Martin 049398
Fayard, Armand 013186
Faydali, Erol 043036
Fayeq, Abdul Zahir . . . 000002, 000003
Fayet, Dr. Roger 041950
Fazekas, Patricia L. 051246
Fazia, Gloria 026308
Fazzi, Kim 052573
Fazzone, Mary Ann 051253
Fearing, Chris 048529
Fearn, Peter 001212
Feather, Philip H. 050208
Fechter, E. 018927, 018928
Fechter, Dr. Sabine 018368
Fedele, Prof. F.G. 027003
Federico, Jean 046388
Federle Orr, Dr. Lynn . . 053009, 053033
Fedič, Vasil 038356, 038357
Fediw, Mary 052073
Fedorenko, Tamara Grigorjevna . 037778
Fëdorova, Ljubov Nikolaevna . . 037586
Fëdorova, Ljudmila Ivanovna . . 037093
Fëdorova, Natalja Viktorovna . . 037038
Fedorowicz, Władysław 034918
Fedyna, Holger 020662
Fedynsky, Andrew 047827
Feeley, Betsey 051323
Feely, Dr. Martin 024926
Feeney, Lawrence 047929
Feeney, Warren 033169
Feest, Prof. Dr. Christian 003046
Feffee, Claudette 052333
Fehér, Béla 023786
Fehér, Dr. György 023516, 023767
Fehle, Dr. Isabella 020229
Fehlemann, Dr. Sabine 022823
Fehr, Horst 020104
Fehrmann, Silvia 017159
Feichtlbauer, Brunhilde 001712
Feickert, Rudolf 022478
Feigel, Martine 014375
Feil, Fr. Carl 053764
Feinberg, Norman M. 047253
Fejös, Dr. Zoltán 023530
Fekeza, Lidija 004283
Felbauer, Josef 002276
Felbauer, Maria 002276
Felbermayer, Dietrich 002540
Feldberg, Dr. Michael 051355
Feldes, Angelika 021923
Feldkamp, Dr. Jörg 017740
Feldkamp, Ursula 017606
Feldman Bennet, Anita 045303
Feldman, Arthur M. 053851
Feldman, Kaywin 050783
Feldman, M.M 025082
Feldmane, Milda 030166
Feldmann, Dorothee 021170
Feldmer, Peter 019090
Felgenhauer-Schmiedt, Prof. Dr.
Sabine 002823
Felice, Silvio 030738
Felisati, Vittorio 026209
Fellay, André 041398
Felley, Jean-Paul 041512
Fellmeth, Prof. Dr. Ulrich 022052
Fellner, Fritz 001906
Fellner, Marianne 008709
Fellner, Reed 050945
Fellows, Leslie 053100
Felsher, Lynn 051433
Felshin, Nina 050863
Felsinger, Benedikt 001931
Feltz, Georges 030591
Felz, Joseph 048993
Fenech, Michael 030756
Feng, Baoxi 007861
Feng, Feng 008026
Feng, Shaozhu 008209
Feng, Wenxue 008021
Feng, Yonghao 008533

Feng, Zhaoxun 007750
Fenger, Lars 010146
Fennema, A.T. 032149
Fenner, Gloria J. 053841
Fenoglio, Norma 000514
Fenoglio, Norma C. 000515
Fenton, Dr. Frank M. 054421
Fenton, John 000922
Fenton, Nicole 053497
Fényi, Tibor 023550
Fenz, Dr. Werner 001978
Ferarresi, Roberto 026236
Ferasin, Luigi 013788
Ferchland, Andrea 017086
Ferdi, Sabah 000081
Ferebee, Peggy V. 049256
Ferez Kuri, Maria del Carmen . . 031253
Fergola, Dr. Lorenzo 028148
Fergurson, Boyd 049778
Ferguson, Francesca 041287
Ferguson, Ian 033339
Ferguson, Mary-Catherine 048605
Ferguson, Moira 044701
Ferguson, Randy 052364
Ferguson, Ray 052650
Ferguson, Robert S. 044412
Fergusson, Ian L.C. 045092
Ferigo, Dr. Giorgio 028106
Ferleger Brades, Susan 044994
Fernan, Marissa N. 034465
Fernandes Gomes, João José . . 035495
Fernandes, Isabel Maria 035730
Fernandes, João 035940
Fernández Bal, María Luisa 040000
Fernández-baldur, Fidel Torcida . 040048
Fernández Castañón Carrasco, José
  Antonio 039862
Fernández Cid, Miguel 040118
Fernández de Calderón, Cándida 031226
Fernández de Ulibarri, Rocío . . . 008854
Fernández García, Juan Alfonso 039437
Fernández Gómez, Fernando . . . 040168
Fernández Lóp, Olga 040347
Fernández Martinez, Dr. Miguel
  Ángel 040248
Fernandez Segorra, Natalia 040132
Fernández Ugalde, Antonio 039358
Fernández Vega, Pedro Ángel . . 040114
Fernández Villaverde, Duque
  Alvaro 040064
Fernández, Dr. Antonio 007517
Fernandez, Joanne 051484
Fernández, Jorge H. . . 039360, 039361
Fernandez, Jorge Osvaldo 000212
Fernández, Kathleen M. 047078
Fernandez, Luisito F. 034576
Fernández, Maria Angélica 000517
Fernandez, Nohema 049720
Fernández, Paloma Acuña 040244
Fernier, Jean-Jaques 014591
Fernoile, Kate 047252
Ferolla Frizzi, Dr. Vanda 025799
Ferran Sans, Marc 039998
Ferran, Domènec 040219, 040223
Ferrand, Jean 014256
Ferrandi, Dr. Giuseppe 028176
Ferrara, Giovanni 027020
Ferraresi, Roberto 026235
Ferrari, Alessio 026947
Ferrari, Amedeo 026099
Ferrari, Ciro 025856, 026703
Ferrari, Davide 026408
Ferrari, Michele 026408
Ferraris, Francesco 028336
Ferraris, Pietro 025261
Ferrazzini, Petra 041393
Ferré, Dr. Rosario 036141
Ferreira Brandão, Dr. Carlos
  Roberto 004890
Ferreira da Costa, Paulo 035808
Ferreira da Cunha, Sergio
  Henrique 004627
Ferreira, Claudia Marcia 004726
Ferreira, Ivette Senise 004869

Ferreira, João Carlos A. 004732
Ferreira, José 000114
Ferrell, John 047708
Ferren, Dick 054061
Ferrence, Cheryl 047146
Ferrer Ciari, Agustín 038968
Ferrer Dalgà, María Rosa 039449
Ferrer i Godoy, Joan 040089
Ferrer-Joly, Fabien 011781
Ferrer Soria, José Luis 039426
Ferrer jr., Dominador 034545
Ferrero, Marcelo 000152
Ferreyra, Prof. Carlos Alfredo . . 000404
Ferrin, Russ 050387
Ferrini, Giovanni 042009
Ferris, Alison 047285
Ferriso, Brian 053850
Ferro, Dr. Inês 036044
Ferryman, Charles 050788
Fersen, Vera von 010848
Fertig-Möller, Heidelore 022578
Fesik, Kateryna Borysivna 043283
Fest, Dr. Bärbel 017270
Festa, Mario 025230
Festanti, Dr. Maurizio 027502
Festini, Prof. Zvonko 009029
Fett, T. 033531
Fetter, Rosemary 046743
Fetzer, Fanni 041637
Fewkes, Nancy 047652
Fewster, Dr. Kevin . . . 001552, 001600
Fey, Cass 053824
Feyler, Gabrielle 015726, 015727,
                                   015728
Feys, Nicole 003266
Fiallo Hernández, Eduardo 010388
Fiaschetti, Maria Egizia 027624
Fibicher, Bernard 041656
Fibicher, Dr. Bernhard 041317
Fibiger, Susanne 010122
Fichner-Rathus, Prof. Dr. Lois . . 053795
Fichtenbaum, Judy 047602
Fick, Dr. Astrid 019366, 019367
Ficker, Peter 017786
Fickert, Hildegard 017689
Fideucci, Elaine 049397
Fiechter, Hermann 042074
Fiedler Streng, Evelyn 053263
Fiedler, Dr. Elisabeth 002856
Fiedler, Marek 035158
Fiedler, Thomas 018739
Fiedler, Uwe 017741
Fiege, Kirsten 019818, 021756
Fiehler, Tim 047079
Field, Hannah 043448
Field, Monica . 005666, 005667, 007255
Field, Peter 045389, 045391
Field, Dr. Richard 049675
Field, William 053117
Fielden, Dr. Kate 043831
Fielding, Andrew 045393
Fielding, James 007467
Fields, Edward D. 047075
Fields, Hoyt 053064
Fields, Laura 046841
Fields, Vilma S. 047590
Fierro Gossman, Julieta N. 031271
Fierz, Gaby 041276
Fietsam, Robert 046880
Fiette, Alexandre 041530
Fietz, Dr. Rudolf 020986
Fifield, Donald 053733
Fifiiţă, Mariana 036464, 036479
Figari, G.B. Roberto 025720
Figdor, Anna Maria 002584
Figiel, Flavia 019505
Figueiredo, Alexandrine 012691
Figueras i Casademont, Lluís . . 039049
Figueres, Roger 014088
Figueroa, Elisa 007555
Figueroa, Paul 053101
Figuerola i Pujol, Prof. Iris 039080
Figuti, Midory Kimura 004862
Fike, Nancy J. 053885

Filchner, Gerhard 020919
Filieri, Dr. Maria Teresa 026642, 026643
Filimonos, Dr. Melina 023247
Filina, Larissa 037322
Filip, Aurel 036781
Filipćić-Maligec, Vlatka 008915
Filipetti, Dr. Giuseppe 026572
Filipová, Maria 009765
Filipović, Marica 004283
Filipow, Dr. Krzystof 034644
Filipowicz, Zygmunt 035233
Filippi, Dr. Giorgio 054682
Filippo, Giuseppe 025414
Filippova, Éleonora Anatoljevna . 037013
Filipunas, Winifred 050100
Fillman, Joseph B. 046409
Filtvedt, Per Kr. 033884
Fimpeler-Philippen, Annette . . . 018043
Finaldi, Dr. Gabriele 039676
Finamore, Dr. Daniel 052880
Finamore, Prof. Nino 027862
Finat, Joelle 013667
Finch-Crisp, Sarah 045938
Finch-Morales, Nikii 053056
Finch, Jon 044833, 044834
Finch, Michelle 047222
Finch, Rick 053911
Finch, Sarah 043828
Finckh, Dr. Gerhard 022825
Finding, Gerhard 002186
Findley, Alvin 007052
Findley, Bill Brown 052450
Fingernagel, Dr. Andreas 003053
Fings, Dr. Karola 019707
Fink, Courtney 053027
Fink, David G. 046330
Fink, Roland 003037
Fink, Susanne 001794
Fink, Dr. Verena 019011
Fink, Prof. William L. 046493
Finke, Karl 018200
Finkele, Diana 020444
Finkelstein, Jacqueline A. 004753
Finkelstein, Lori 051430
Finks, Harvey 048343
Finkster, Paul 049431
Finlay, James 046225
Finlayson, Prof. J.C. 022941
Finley, Janet 049149
Finley, Valerie 006231
Finnegan, Paul M. 051880
Finneiser, Klaus 017065
Finnemore, Donnice 054221
Finnemore, Harry 054221
Finnerty, Sinéad 054307
Finney, Libby 044801
Finnigan, Penny 001355
Finnigan, Stephen 049284
Finotti, Franco 027683
Finweg, Sabine 022886
Finžgar, Maja 038549
Finzi, Dr. Ranny 025189
Finzi, Dr. Sonia Guetta 028303
Fionik, Doroteusz 034648
Fiorentino, Alessandro 027993
Fiorenza, Joseph 049031
Fiori, Dennis 046731, 047120
Fiori, Dennis A. 046764
Fiori, Dr. Pietro 026819
Fiorillo, Anthony 048116
Fiorio, Dr. Maria Teresa 026785, 026787,
                                   026789, 026802
Fiouri, E. 009414
Firl, Wolfgang 022571
Firmenich, Andrea 016786
Firmin, Sandra 047310
Firnges, Jörg 019266
Firnhaber, Waltraud 020753
Firstenberg, Jean 050406
Firszt, Stanisław 034795
Firth, Neil 045914
Fischbacher, Dr. Marianne 041603
Fischer-Elfert, Prof. Dr. Hans-W. 019972

Fischer Jonge, Ingrid . 010183, 010186,
                                   010262
Fischer, Andrea 021599
Fischer, Angelika 018933
Fischer, Barbara 006339
Fischer, Dr. Bernhard 022494
Fischer, Christian 010300
Fischer, D. 019516
Fischer, Dr. Eckhard 022741
Fischer, Felice 052081
Fischer, Dr. Fritz 016677, 021525,
                                   022047
Fischer, Germann 019400
Fischer, Gottlieb 020301
Fischer, Hannelore 019690
Fischer, Hans 019400
Fischer, Dr. Hartwig 018306
Fischer, Helmut 017745, 017746
Fischer, Dr. h.c. Hermann von . . 041609
Fischer, Ilse 018993
Fischer, Karl-Günther 017551
Fischer, Klaus 017857
Fischer, Magdalena 003033
Fischer, Manfred W.K. 002462
Fischer, Dr. Manuela 017112
Fischer, Mark 049423
Fischer, Prof. Dr. Martin S. 019454
Fischer, Martina 020349
Fischer, Peter 002669, 041708
Fischer, Reinhold 020886
Fischer, Ron 047542
Fischer, Steffen 018721
Fischer, Susanne 018381
Fischer, Tatjana 017733
Fischer, Thomas 017029
Fischer, Uwe 018131
Fischer, Prof. Dr. Volker 018435
Fischer, Dr. Werner 020388
Fischetti, Amy 048751
Fischhaber, Martin 020404
Fischlin, Jean-Mario 041242
Fischman, Lisa 053839
Fisecker, August 002099
Fish, Paul 053823
Fish, Suzanne 053823
Fishback, Pat 052560
Fisher Sterling, Dr. Susan 054149
Fisher, Alice 048045
Fisher, Claire 045976
Fisher, Clive 045531
Fisher, D.L. 033236
Fisher, Debra 047219
Fisher, Debra S. 050885
Fisher, Elaine 045812
Fisher, F.S. 047537
Fisher, Frederick J. 054119
Fisher, Jay 046733
Fisher, John R. 053694
Fisher, Kathleen 051043
Fisher, Kathryn 006993
Fisher, Kay 054450
Fisher, Kevin M. 050272
Fisher, Lawrence A. 046785
Fisher, Lilian 053980
Fisher, M. 048421
Fisher, Pat 044278
Fishman, Bernard P. 046410
Fishmann, Bernard B. 052395
Fiske, David A. 053516
Fiskesjö, Dr. Magnus 041030
Fissiaux, Jean-Claude 004062
Fisson, Bruno 012975
Fitch, Arwen 045696, 045700
Fitch, Blake 054464
Fitch, James A. 049083
Fitts, Catherine 054163
Fitz, Prof. Erwin 001795
Fitz, Péter 023497
Fitzgerald, Annamary 051755
Fitzgerald, Dr. Edward 052437
Fitzgerald, Elizabeth 052808
Fitzgerald, Emma 032469
Fitzgerald, Erika 053456
Fitzgerald, Jane 050907

Fitzgerald, Lawrence 044409
Fitzgerald, Mary Ann 050595
Fitzgerald, Maryann 053159
Fitzgerald, Paul 051601
Fitzgerald, Shannon 052796
Fitzhugh, Robin 048676
Fitzpatrick, Gareth 044684
Fitzpatrick, M. 054847
Fitzpatrick, Mike 024956
Fitzpatrick, Pat 007486
Fitzpatrick, Robert 047676
Fitzpatrick, Sheila E. 054052
Fitzpatrick, Tom 025099
Fitzsimmons, Rita 049863
Fitzsimons, Dr. J. Michael 046819
Fiumi, Prof. Enrico 028426
Fivez, Dr. Liesbeth 003311
Fjellhøi 033483
Fjermedal, Torkell 034075
Flacke, Prof. Dr. Monika 017299
Flacke, Robert 050078
Flaherty, Mary A. 046473
Flaksman, Sandy 050492
Flanagan, Anne 001535
Flanagan, Brendan 044744
Flanagan, I.D. 044197
Flanagan, Michael 054378
Flanagan, Shelia M. 050979
Flanety, Lynda 050030
Flanigan, Brian 052124
Flannery, Kent V. 046491
Flannery, Kerry 049741
Flannery, Prof. Tim 000771
Flask, David 050577
Flasza, Jan 034656
Flaud, Jackie 016066
Flavin, Samantha 044777
Flechas Corredor, Jorge Arturo . 008819
Fleck, Robert 014411
Fleck, Dr. Robert 017494, 018979
Fleckenstein, Jutta 020534
Fleet, Prof. A. 045061
Fleet, Judy 049062
Flegel-Wantier, Klaus 021826
Flegel, Andreas 018127
Fleischanderl, Rudolf 001870
Fleischer, Dr. Martina 002995
Fleischhauer, Dr. Carsten 021668
Fleischman, Stephen 050598
Fleischmann-Heck, Dr. Isa ... 019784
Fleissner, Alfred 002364
Fleming-Pawelek, Brenda 046873
Fleming, Dr. David 044869
Fleming, Elvis 052691
Fleming, Geoffrey 053423
Fleming, Jeff 048282
Fleming, Leslie 051678
Fleming, Marty 054179
Fleming, Sam 043477
Flensburg, Birgitta 040652
Flensburg, Brigitta 040879
Flentje, Rachel Daly 047052
Flesch, Edward 049560
Flesche, Susanne 019562
Flesher, Karen 053909
Fletcher, Ann 033209
Fletcher, Carrol 050715
Fletcher, Charles 051140
Fletcher, Corinne 054452
Fletling, Walter 018466
Flett, Dr. Alistair M. 043363
Flett, Harry 044514
Fleurke, Else 033113
Fleury, Nathalie 041437
Fleury, Dr. Robert 011708
Flick, André 014544
Fliegel, Stephen N. 047808
Flieger, Erwin 020243
Flinsch, Stefanie 002637
Flinspach, Joan L. 048896
Flint, Jody 047533
Flintoff, Paul 043641, 043642
Flodin, Göran 040989
Flögel-Kostić, Lela 038184

Flögel, Dr. Evelyn 022395
Flößer, Dr. Reinhard 016751
Flon, André 015403
Flon, Dominique 014403, 014407
Flood, Ellen 050809
Flood, Leon 051311
Flood, Richard 050948
Flood, Rick 054575
Flood, Robert 050809
Flood, Timothy 047694
Floquet, Prof. Jean 016117
Florance, Robyn 001375
Florea, Ileana 036716
Florentino, M. Raquel 035768
Florenzano, Clorinda 026977
Florer, Michael R. 049089
Flores, Frank 048609
Flores, Juan C. 054764
Florescu, Aurelia 036415
Flórez de Grajales, Aura 008809
Flórez, Helmer de Jesús 008801
Florian, Carmen 036311
Floridi, Dr. Massimiliano 026424
Florineth, Rico 041971
Floris, Prof. Giovanni 025681
Floris, Lene 010132
Florizoone, Jan 003921
Flory, Maurice 013906
Floryan, Dr. Margrethe 010198
Flotow, Julia von 006526
Flour, Joke 004099
Flourentzos, Dr. Pavlos 009414
Flowers, Betty Sue 046696
Flowers, Randy 048005
Floyd, David 046821
Floyd, Roger 053260
Fluck, Dr. Andreas 020715
Flucker, Turry 049757
Flude, Kevin 045068
Flückiger, Dr. Peter F. 041824
Flügge, Matthias 017168
Fluke, Pam 045788
Fluke, Phil 045788
Flukinger, Roy 046692
Fluksi, Josip 009025
Flumm, S. 021123
Flunger, Mag. (FH) Julia 002953
Fluri, Annette 041289
Flury-Steiner, Barbara 041336
Flury, Dorothea 041289
Flutsch, Laurent 041663
Flynn Johnson, Robert 053009, 053033
Flynt, Suzanne 048219
Foa, Piers 000841
Foa, Robyn 000841
Foanaota, Lawrence 038641
Fobbs, Archie 054146
Focht, John 046843
Focke, Sue 052853
Fodor, Mihály 023574
Foe, Peter 053666
Foelix, Dr. Rainer F. 041189
Foelsche, Gudrun .. 002402, 002405
Førde, Anne-Marie 034088
Först, Dr. Elke 018995
Foerster, Dr. Cornelia 017376
Förster, Monique 018245
Förster, Roswitha 021272
Förster, Steffen 020349
Föttinger, Franz 002779
Fogarasi, Zsuzsa 023723
Fogarty, Lori 053170
Fogle, Douglas 050948
Fogliazza, Virginio 026617
Foister, Dr. Susan 045057
Fojtík, Miroslav 009741, 009952
Folch, María Jesús 040305
Folch, Stella 039059
Foldøy, Oddveig 033972
Foley, Cindy M. 052280
Foley, J.A. 033321
Foley, Jim 047621
Foley, Suzanne 047585
Folga-Januszewska, Dorota ... 035352

Folger, Willibald 001804
Folgić-Korjak, Angelina 038168
Folie, Dr. Sabine 002996
Folk, Luther 047177
Folkins, Alice 005713
Follbert, Léon 011742
Folsach, Dr. Kjeld von 010173
Foltin, Viktorija Tyborivna 043312
Folwell Krueger, Hilary 048891
Fonclare, Guillaume de 014835
Fong, Lawrence 048618
Fonio, Prof. Ezio 028129
Fonsmark, Anne-Birgitte 010029
Font, Walter 048892
Font Sentias, Josep 039475
Fontaine, Axel 012544
Fontaine, Christian 015979
Fontaine, M. 013753
Fontaine, Patrick 014897
Fontaine, Susanne ... 017284, 017286
Fontana, Agostino 028278
Fontana, Armon 041487
Fontana, Prof. Dr. Eszter 019993
Fontana, Lilia 050833
Fontanarossa, Dr. Raffaella .. 025952
Fontaneda, Cristina 038972
Fontenioue, Henri du 013015
Fontenote-Jamerson, Belinda . 050454
Fontenoy, Paul 046847
Foote-Smith, Christy 050286
Foote, W.T. 001504
Fopp, Dr. Michael A. 045087
Foppa, Christian 041414
Forand, Edward J. 054305
Forand, Peggy 053417
Foray-Carlier, Anne 014733
Foray, J.-M. 012044
Foray, Jean-Michel 014487, 016104
Forber, Oskar 016827
Forbes, Duncan 044274
Forbes, Ian 046071
Forbes, Jamie 007231
Forbes, Dr. John M. 046723
Forcada Serrano, Miguel 039973
Ford, Chris 048214
Ford, David 005896
Ford, Peggy A. 049222
Ford, Richard I. 046491
Ford, Dr. S.D. 052150
Ford, Sandra 048755
Ford jr., Thomas 034478
Fordham, Gerald J. 052876
Fordham, Michael 044498
Fordyce, Prof. Dr. R. Ewan ... 033191
Forellad, Dolors 040027
Foreman, Martin 044715
Forero, Dr. Enrique 008605
Foresman, Dr. Kerry 050964
Forest, Barbara 012294
Forest, Dominique 014733
Forest, Gilles 013255
Forest, Marie-Cécile 014762
Forgan, Dr. Sophie 046114
Forgang, David M. 054596
Forgeng, Dr. Jeffrey 054544
Forhaug, Helén 040780
Forkl, Dr. Hermann 022048
Forman, Weldon 047971
Formanek, Verena 042217
Formanová, Markéta 009761
Formery, Valerie 003765
Formicone, Luigi 027053
Fornals Villalonga, Francisco ... 039244
Fornara, Livio 041520, 041526
Fornasari, Giuseppe 026967
Forner Caballero, Ester 040420
Fornés i Garcia, Prof. Josep ... 039095
Forney, Jack D. 048263
Forni, Prof. Gaetano 027835
Forno, Dr. Carla 025405
Forrer, Dr. M. 032585
Forrest, Richard 006233
Forrester, Gillian 051266
Forrester, Hugh 043562

Forrester, Jim 045206
Forsberg, Anders 033972
Forsberg, Diane 049386
Forsberg, Helena 041106
Forsberg, Virpi 010795
Forschler, Dr. Anne 047005
Forssberg, Anna Maria 040991
Forssblad, Marianne 053241
Forster, Helen 005354
Forster, Judith 042147
Forster, Michael 016552
Forsythe, Dee 050184
Fort, Thomas A. 048478
Forte, Richard V. 046990
Fortenberry, Kay 051197
Forti, Fabio 025598
Forti, Dr. Micol 054682
Fortier, Jana 050710
Fortin, James 006201
Fortin, Verna 050578
Fortino, Sally 016808
Fortson, Judith 048408
Fortun, Pascal 004060
Fortuna, Mary 052609
Fortunat Černilogar, Damjana .. 038622
Fortune, C.F. .. 038762, 038765, 038766,
                              038767
Fortune, Collin 038872
Foschi, Dr. Pier Luigi .. 027521, 027523
Foss, Birgit 033888
Foss, Marybeth 053660
Fossati, Alessandro 041696
Fosse, Martine 012294
Fossey, John M. 006391
Fossheim, Eivind 034074
Fossum, Finn 033521
Fostekew, Ken 046186
Foster, Alasdair 001390
Foster, Dr. Allan 007487
Foster, Clarence 005600
Foster, Cynthia 052754
Foster, Di 001245
Foster, Jeanne K. 051859
Foster, Kathleen 052081
Foster, Keith 050813
Foster, Kenneth 053032
Foster, Laura 051719
Foster, Marusia 007471
Foster, Preston 047971
Foster, Robert 051153
Foster, Stephen 045826
Foster, Velma 005473
Fotchi Chian, Mansoor 024738
Fotoghi, M. 024728
Foucart, Bruno 012133
Foucher, Jean-Pascal 015749
Foulds, Kris 005301
Foulon, Françoise 003710
Foulon, Pierre-Jean 003890
Fountouli, Mary 023023
Fouray, Monique 015173
Fourie, Dr. H. 038860
Fouriscot, Mick 013693
Fournet, Claude 014475
Fournier, Johanne 006274
Fournier, Dr. M. 032581
Fournier, Monique 013939
Fournier, Vincent 012500
Fourny-Dargere, Sophie 013679, 016278
Foust, Dorothy 050579
Fowkes, Dudley 045618
Fowler Shearer, Christine ... 050737
Fowler, Martin 043899
Fowler, Sophie 044285
Fowler, Stephen 000923
Fowler, Victoria 054340
Fox, Andrew 001106, 043423
Fox, Catherine 043965
Fox, Charles 053383
Fox, Christopher D. 053743
Fox, Emily 053466
Fox, Eugene 049272
Fox, Dr. Georgia 047712
Fox, Georgia 049746

Fox, Howard	050444	Franklin, Joyce	046466
Fox, Jill	045850	Franklin, Karen S.	047231
Fox, Julie	045993	Franklin, Kelly	050720
Fox, Lisa	047322	Franko, Blanche	007063
Fox, Maureen	033227	Franková, Božena	009882, 009920
Fox, Patric	047886	Franković, Dr. Bernard	008888
Fox, Peter	046022	Franks, Carolyn J.	046526
Fox, Rodney	001106	Franks, Clara	050714
Fox, Dr. Ronald	050824	Franks, Curtis J.	047541
Fox, Susan	043525	Franks, R.J.	006435
Foyer, Marle	013621	Fransen, Linda	054468
Fraaye, Dr. R.H.B.	032096	Fransen, Martin	006787
Fraboni, Anne	005631	Fransson, Elisabeth	040793
Fraccaro, Prof. Marco	027236	Frantz, Lewis	049991
Fradregada, Bruno	028168	Frantzen-Heger, Gaby	030606
Frady, Jackie L.	052508	Frantzen, Ole J.	010199
Fraga Vila, Dr. Mª Isabel	040125	Franulic, Markita	009048
Fraipont, J. de	003897	Franz, Dr. Corinna	016791
Fraisset, M.M.	011558	Franz, Dr. Erich	020601
Frakes, Rita	050749	Franz, Prof. Dr. G.	022204
Frame, Ron	047514	Franz, Dr. Manfred	018076
Framke, Dr. Gisela	017955, 017962	Franzel, Frank E.	050746
Framnes, Svein	033768	Franzén, Ann	041025
Franc, Karl	002820	Franzen, Dr. Brigitte	016440, 020601
França, Prof. Alexandre	004944	Franzen, Darlyne S.	054532
Francell, Larry	046421	Franzén, Eva-Lotta	040721
Francés, Fernando	039697	Franzen, Helga	019203
Francès, Henry	012619	Franzen, Merry	049117
Frances, Mary	048546	Franzke, Dr. Jürgen	020846
Franceschi, Branko	008977	Franzolin Trevisan, Amélia	004904
Franceschini, Luca	026726	Frapporti, Giovanni	025075
Franceschini, Maria Anna	026263	Fraser, Leslee	052933
Francesconi, Gino	051462	Fraser, Dr. Nicholas C.	050727
Francia, Ennio	054693	Fraser, Nonie	005591, 005780, 007097
Franciolli, Marco	041695	Fratelli, Dr. Maria	026780, 026818
Franciosi, Dr. Ario	025930	Fraueneder, Dr. Hildegard	002621
Francis, Dr. Carl	047367	Fraux, Sébastien	012234
Francis, Dermot	044093	Frawley, Sylvia	024898
Francis, Elaine	043804, 044304	Frayling, Prof. Christopher	045091
Francis, James H.	048628	Frazer, John	005406
Francis, Jesse	047621	Frazer, Kay J.	050018
Francis, Kathy	047117	Frazier, Ronald F.	048209
Francis, Dr. R.	044360	Freaney, Linda	054530
Francis, R.W.	046080	Frechen, Anne	020793
Francis, Rachelle	048423	Frechette, Myles R.R.	051329
Francis, Ross	006580	Frechou, M.	012064
Francis, Tim	053479	Freddi, Dr. Loris	025886
Francis, Wayne	050704	Frederici, Joan	047193
Francisco, Lisa	050772	Frederick, Frances	052155
Francisco, Paul S.	004175	Frederick, Karen	053189
Franck, Dr. Frederick	054075	Frederick, Margaretta S.	054433
Franck, Marketta	011400	Fredericks, Marie	051428
Franclieu, Comte de	015329	Fredj, Jacques	014676
Franco, Barbara	054118	Freece, David W.	049916
Franco, Leoncio Verdera	039323	Freed, Douglass	053258
Franco, Dr. Maria Nobre	036046	Freed, Peter	048708
Francois, Gary	050086	Freed, R. Mitchell	048576
François, Lionel	012921	Freed, Rita E.	047123
Franconville	016158	Freedman, Richard	038674
Francou, Dr. Carlo	025831	Freedman, Robert A.	047795
Francová, Zuzana	038310	Freegman, David	045480
Frâncu, Simona	041536	Freeland, Clara	047377
Frandsen, Lene B.	010021, 010022,	Freeland, George W.	050067
	010236, 010266, 010267, 010350,	Freeman, A.	045628
	010351, 010352	Freeman, B.J.	046594
Frangos, Ioannis	023319	Freeman, Craig C.	050186
Frank, Claude	012629	Freeman, D. Carl	048314
Frank, Dr. Günter	017621	Freeman, Dottie	054340
Frank, Heinrich	004647	Freeman, Dr. George	050081
Frank, Peggy	051161	Freeman, Lenette	051115
Frank, Richard J.	047066	Freeman, Marina	050376
Frank, Richard W.	048831	Freeman, Michael	043372
Frank, Stuart M.	053285	Freeman, Rusty	051174
Frank, Thomas	019405	Freemantle, Andrew	045528
Frank, Thomas H.	019821	Freemantle, Barbara	038759
Frank, Mag. Veronika	002014	Freer, Elene J.	005452, 005453
Franke, Bernd	018860	Freeze, Nancy	050792
Franke, Dr. Ute	017242	Frehner, Dr. Matthias	041318, 041319
Frankel Nathanson, Marjorie	047841	Frei, Prof. Dr. Alfred Georg	022784
Frankenberger, Anette	020539	Frei, Thomas	041666, 041975
Frankfurt, Judith	047253	Freiberg, Dr. A.	025214
Frankignoul, Daniel J.	003455	Freidel, Karl	018573
Franklin, Ailesia	048397	Freimark, Dan	046945
Franklin, Dr. David	006594	Freises, M.	015787

Freitag-Mair, Claudia	021737, 021738,	Friese, Dr. Elisabeth	019582, 019583
	021739, 021740	Friesen, Steve	049144
Freitag, Dr. Anne	041655	Friesenbichler, Gerhard	002573
Freitag, Matthias	021324	Friesenecker, Josef	002685
Freitag, Winfried	018082	Friesinger, Prof. Dr. Herwig	002827
Freitas, Dr. Henrique de	004474	Frieß, Jörg	017299
Freixinet, Mercè	039575	Friis-Hansen, Dana	046685
Freland, Colin	033151	Friis, Birte	010125, 010126
Frémaux, Thierry	013953	Frinchillucci, Dr. Gianluca	026199,
Frembgen, Dr. Jürgen W.	020579		026201
Fremd, T.	049969	Frinsko, Linda M.	049163
French, Ann	045219	Frisch Lehrer, Susan	053533
French, B.G.	046306	Frisch, Patricia L.	053375
French, Brenda	044074	Fritsch, Martin	017187
French, David	044949	Fritts, Teresa	048392
French, Mary	047901	Fritz, Beatrice	019139
Freng, Prof. Martin J. Earwicker	045103	Fritz, Dr. Ingomar	001984
Frenkley, Helen	025151	Fritz, Dr. Norbert	019139
Frennesson Westblad, Karin	040954	Fritz, Apl. Prof. Dr. U.	018008
Frennet, Raymond	003273	Fritzell, Charles	054093
Frenssen, Dr. B.	018808	Fritzsche, Elmar	017611
Frenz, Achim	018428	Frizzi, Dr. Vanda	025798
Frenz, Bonnie	047789	Frlan, Damodar	009028, 009028
Frenz, Dr. Hans G.	020233	Froberville, de	012664
Frerichs, Dr. Klaus	017709	Froehlich, Conrad G.	047524, 047525,
Freschi Conti, Dr. Graziella	025604		047526, 047527
Frese, Dr. Annette	019137	Fröhlich, Günter	003095
Fresk, Klas	040970	Frohmann, Dr. Sieglinde	002595
Fresneau, Estelle	013589, 013590	Frohriep, Monika	021584
Freudenberg, Dr. Mechtild	021663	Froitzheim, Dr. Eva-Marina	017452,
Freud, Franzika	002907		017455
Freund, Leslie	046941	Frolcova, Dr. Milada	009931
Freunschlag, Jörg	002188	Frolec, Dr. Ivo	009931
Frew, Dr. David	048600	Frolova, Aleksandra Anatoljevna	036969
Frey, Barbara	047948	Frolova, T.V.	038070
Frey, Dietrich	022557	From-Brown, Kristy	049632
Frey, Dr. Eberhard	019518	Frommhage-Davaz,	
Frey, Dr. Paul	003026	Gudrun-Sophie	022623
Frey, Pierre	041645	Fromont, Chantal	011939
Frey, Thomas	019011	Fromont, Dr. Jane	001403
Freyberg, Wolfgang	018154	Froning, Prof. Dr. H.	020266
Freyburger, J.-Claude	014410	Frontalini, Costantino	025987
Freytag, H.	017888	Fronty, Colette	045703
Friary, Donald R.	048218	Frorud, Harald	033946, 033947
Frías Acosta, Alvaro Francisco	008830	Frost-Pennington, Peter	045581
Frias, Jennifer	052597	Frost, Cheryl A.	049657
Friberg, Gösta	040591	Frost, Nada	053793
Friborg, Flemming	010188	Frost, Natalie	045349
Frick, William R.	050917	Frost, Sarah	017559
Fricke, Evelyn	019001	Frotscher, Matthias	019926
Fricke, Inge	019876, 019880	Frottier, Dr. Elisabeth	003074
Fricke, Dr. Ronald	022056	Froud, Alan	000937
Frickman, Linda	048798	Frumușelu, Teodor	036232
Fridfinnsson, Jóhann	024118	Fry, Judith	001299
Frieb, M.T.	016949	Fry, Stephen	005964
Friebel, V.	019936	Fryberger, Betsy	053488
Fried, S.	041000	Frỳda, Dr. František	009760, 009763
Friedberg, Erin H.	049006	Frye, J. Paul	052142
Friedberg, Rachel	049300	Frye, Todd	052402
Friedel, Prof. Dr. Helmut	020580	Fryer, John	043715
Friedhofen, Barbara	017025	Fryer, Jonathan	045768
Friedl, Gerlinde	016777	Fryer, Molly	048673
Friedli, Michael	041218	Fu, Songliang	008395
Friedli, Susanne	041318	Fu, Suanbao	008198
Friedman, Dr. Alan J.	048767	Fu, Dr. Tse Tak	007882
Friedman, Ann	051756	Fu, Yinan	007914
Friedman, Maxine	053498, 053504	Fu, Zhongxian	008131
Friedman, Robert	047899, 047900	Fuchs, Anni	002983
Friedrich-Sander, Harald	019290	Fuchs, E.	034963
Friedrich, Hanns	016799	Fuchs, Dr. Friedrich	021314, 021317
Friedrich, Heinz-Jürgen	022864	Fuchs, Friedrich von	020086
Friedrich, Holger	021882	Fuchs, Hofrat DI Georg	002513
Friedrich, Julia	019705	Fuchs, Dr. Gerald	002252
Friedrich, Dr. Sven	016988, 016991,	Fuchs, Dr. Ingrid	002997
	017001	Fuchs, Matthias	002699
Friedrich, Thomas	017734	Fuchs, Max	041458
Friel, Dr. I.	043940, 043943	Fuchs, Monique	015884
Frielinghaus, Volker	017439	Fuchs, Dr. Rainer	003050
Friemerding, Wolfgang	017810	Fuchs, Ronald	002419
Friend, Amy	051642	Fuchs, Tanja	002097
Friend, Glenda	054083	Fuchs, Thomas	041593
Frierson, Amy H.	047594	Fuchssteiner, Ingrid	002093
Fries, Dee	046917	Füköh, Dr. Levente	023653
		Füldner, Dr. Kai	019544

Fülep, Dr. Katalin 023539
Füllner, Dr. Karin 018030
Fülöp, Dr. Gyula 023825, 023932,
  023933, 023934, 023935, 023938,
  023939, 023941, 023942, 023990
Fünfschilling, Sylvia 041243
Fuente Andrés, Félix de la 040374
Fuentes, Isabel 038923
Fürészné-Molnár, Anikó 024003
Fürle, Gisela 022317
Fürst, Sabine 019517
Füvessy, Anikó 024011
Fugami, Tracey 051379
Fugaro, Anthony F. 053959
Fugate, Charlotte 051783
Fugate, Dody 053114
Fugate, Jane 053928
Fugazza, Prof. Stefano 027294
Fugazzola Delpino, Maria
  Antonietta 027634
Fuge, Rex 000961
Fugger-Babenhausen, Markus
  Graf 016685
Fuglie, Gordon L. 050441
Fuglsang, Harald 010074
Fuhr, Dr. Michael 003029
Fuhrman, Loretta 052226
Fuhry, Eva 019608
Fujii, Kazuo 028631
Fujii, Miyako 028924
Fujii, Yoshitugu 028924
Fujii, Zenzaburou 028924
Fujisaki, Aya 028652
Fujita, Chikako 029241
Fujita, Hirohiko 029169
Fujita, Yukio 028541
Fukagawa, Masafumi 028808
Fukuda, Hiroko 028652
Fukuda, Mamoru 028725
Fukuhara, Yoshiharu 029616
Fukuji, Eiji 028616
Fulconis, Marie-Madeleine . . . 015306,
  015307, 015308
Fulfer, Glenn 051107
Fulga, Ligia . 036272, 036669, 036670
Fulghum, Joyce 048962
Fuller, Benjamin A. G. 053227
Fuller, C.D. 032316
Fuller, Candy 048822
Fuller, Elaine L. 048086
Fuller, Elizabeth E. 052091
Fuller, Janae 050257
Fullerton, Jane 006893
Fulop, Lajos 036425
Fulton, Dr. R. 049514
Fulton, William 045922
Fumero, Patricia 033401
Funck, Eberhard 017560
Funck, Ludwig 017624
Funk, Dr. Frank E. 054449
Funk, Michael 021107
Funk, Tamara 050931
Funk, Uli 017123
Funke, Gerhard 019564
Funsho Adedayo, Oluremi 033441
Furber, Rich 052407
Furey, Linda 051637
Furger, Dr. Alex 041243
Furgol, Dr. Edward 054154
Furiesi, Dr. Alessandro 028427
Furingsten, Agne 040758
Furlong, James 051417
Furlow, Elizabeth 053240
Furman, Robert 053470
Furness, Jane 044273
Furnival, Gloria 051606
Furr, Kathlyn 053299
Furrer, Gerti 001828
Furrer, Dr. H. 042227
Furrer, Heinz 041803
Furry, Nancy 053632
Fursdon, E.D. 043821
Furshong, Peg 048698
Fursova, Irina Alekseevna 037585

Furtner, Alois 016557, 016559
Furtwengler, U. 016923
Fuscarino, Elise 053386
Fusi, Pietro 025853
Fuso, Dr. Silvio 026985, 028316
Fussey, Dr. George D. 044306
Fustig, Manuela 018685
Futivić, Vlasta 009057
Futoransky, Donna J. 048688
Futter, Christina 049888
Futter, Ellen V. 051327
Fux, Franz 001935
Fyfe, R. 033168
Fylypčuk, Dr. Olga Ivanivna . . . 043126
Ga, Dr. Paolo 027412
Gaaff, G. 032145
Gaál, Dr. Attila 023946, 023950
Gaálné, Dr. Merva Mária 023645
Gaand, Jacques 003844
Gaasch, Cynnie 048966
Gabarra, Edward 051704
Gabarra, Josep Feliu 038990
Gabay, Oshra 025158
Gabbard, Dennis 048902
Gabbitas, Victoria 044073
Gaber, Vizeleutnant Josef 002800
Gabet, Francis 041662
Gabet, Oliver 014743
Gabici, Dr. Franco 027487
Gable, Edward 047036
Gabler, Dr. Josephine 021083
Gábor, Komáromi 024046
Gabra, Dr. Gwadat 010534
Gabriel, Carol A. 049328
Gabriel, Prof. Reinhold 002206
Gabriel, Siegfried 018454
Gabrielson, Bengt 040733
Gabrielsson, Carl-Eric 040988
Gabršek Prosenc, Meta 038573, 038574
Gabyševa, Asja Lvovna 037057
Gachet, Louis-Jean 012421
Gachet, Louis Jean 012418
Gačková, Dr. Lýdia 038400
Gadd, Richard W. 051013
Gaddis, Eugene R. 049390
Gaddy, Kenneth 053860
Gadebusch, Raffael D. 017239
Gądor, Józef 035156
Gäbler, Kristin 018012
Gaede, Carla 020520
Gaede, Carola . 020510, 020532, 020533,
  020547
Gädeke, Dr. Thomas 021664
Gähler, Walter 041966
Gaensheimer, Dr. Susanne 018438,
  020580
Gärtner, Monika 002133
Gäßler, Prof. Dr. Ewald 020985, 020991
Gaete Harzenetter, Rodrigo . . . 039517
Gafert, Alfred 019194
Gaffino Rossi, Rodolfo 028122
Gaffney, T.J. 052251
Gagarina, Dr. Elena Jurjevna . . 037233,
  037234, 037235, 037248, 037287,
  037486, 037490, 037504
Gage, Mary Margaret 046459
Gage, Mike 049225
Gagea, Pavelina 036663
Gagge, Carl-Magnus 041140
Gagnaison, Jean-Claude 015729
Gagneur, Didier 011704
Gagneux, Dominique 014697
Gagneux, Yves 014669
Gagnon, Kathryn 005729
Gagnon, Louis 006035
Gagro, Ivan 009025
Gaiber, Maxine 054434
Gaida, Dr. Maria 017112
Gaier, Dr. Claude 003801
Gaigalat, Michael 020898
Gaillard, Dr. Karin 032575
Gain, Prof. D.B. 038727
Gaines, Robert 048567
Gair, Michael 005379

Gais Muller, Deborah del 052387
Gaisbauer, Ernst 001734
Gaisberger, Karl 001734
Gaissert Jackson, Mary 046383
Gaitán de Santamaria,
  Magdalena 008672
Gaither, Barry 047125
Gajárszky, Vera 023710
Gajda, Odette 013076
Gajdukova, Nadežda Michajlovna 037105
Gajewski, Dr. Pawel 026742
Gajski, Zora 009036
Galan, Brenda de 003649
Galanova, Raisa Aleksandrovna . 037950
Galanti Papola, Mary 052064
Galassi, Peter 051427
Galassi, Susan Grace 051390
Galasso, Prof. Elio 025486
Galasso, Giampiero 025299
Galasso, Dr. Giovanna 025481
Galasso, Prof. Maria 000361
Galati, Dr. Roberto 025497
Galbert, Antoine de 014675
Galbraith, Marie 054199
Galčenko, Sergij A. 043222
Gale, Marilyn 054334
Galea, Victoria 030770
Galeazzi, Dr. Claudio 027405
Gałecki, Zdzisław 035345
Galendo, Dr. Pedro G. 034548
Gales, Melinda 052557
Gałęza, Jarosław 035078
Galicer, Jorgelina 000446
Galicki, Piotr 034635
Galiègue, Josette 011956
Gálig, Dr. Zoltán 023985
Galili, Avner 025149
Galimberti, Dr. Paolo 026829
Galindo Rosales, Hortensia . . . 031162
Galindo Steffens, Margarita . . . 008578
Galindo, Luis 054726
Galizia, Anna Lisa 041302
Galkina, Inna Konstantinovna . . 036883
Galkina, Tatjana Ivanovna 037675
Gall, Bert 047677
Gall, Dr. Wolfgang M. 020976
Gallagher, Charles R. 051235
Gallagher, Marlys M. 046922
Gallagher, Marsha V. 051765
Gallagher, Mary 024927
Gallagher, Sheila 024927
Gallant, Aprile 051626
Gallant, Cécile 006334
Gallant, Greg 005598
Gallardo, Luis 031264
Gallasch, Harold E. 001131
Gallatin, Morgan Dean 054069
Gallavam, Teresa 047609
Galleano, Dr. Franco 026373
Gallego Londoño, María Elena . . 008774
Gallegos Franco, Francisco . . . 031540,
  031541
Gallemí, Dr. Jaume 039077
Galler, Peter 041881
Gallert, Andrea 022783
Galli, Dr. Cinzia 026101
Galli, Fiorenzo 026820
Galli, Mario 041731, 041732
Gallico, Prof. Claudio 026700
Gallie, Jim 043535
Gallifant, David W. 046092
Galliffet, P. de 013741
Galliker-Tönz, Robert 041333
Gallimore, Janet 054230
Galliot, Gérard 012004
Gallivan, Dr. P.A. 001485
Galljamova, Dolaris G. . 036996, 036998
Gallo, Prof. Pietro 027604
Gallois, J.L. 011622
Galloway, Dr. Ian 001501
Galloway, Mary Lou 054382
Galluzzi, Prof. Paolo 026264
Gallwitz, Prof. Klaus 021357
Galočkina, Svetlana Jurjevna . . 038018

Galonska, Ann 053554
Galopim de Carvalho, Prof. A.M. 035809
Galscott, Joseph 001111
Galtier, Jacques 014499
Galusha, Emily 050946
Galvani, Bill 049961
Galvin, John E. 050583
Galvin, Maureen 047297
Galvis, Hermana Martha Lucía . 008718
Galway, Mary 051917
Galyckyj, Mikola Oleksandrovyč . 043179
Gamache 012647
Gamazo Barrveco, Mercedes . . 039653
Gambetta Uría, Fredy 034404
Gambin, Kenneth 030765
Gambiroza, Roland 009052
Gamboa, Lyn 034596
Gamboa, Marilyn B. 034428
Gameiro, Dr. Ana Cristina 035593
Gamet, Prosper 016062
Gammersvik, Ågot . . . 033608, 033742,
  034059
Gamolin, Vladimir Danilovič . . . 037616
Gamperer, Franz 002372
Gamsjäger, Erna 001715
Gamsjäger, Rudolf 002052
Gamsjäger, Siegfried 002469
Gamwell, Lynn 046999
Gamzatova, Patimat Saidovna . 037208,
  037209
Gan, Jianbo 008546
Gancarski, Jan 034922
Gandee, Cynthia 053668
Gandhi, K.M. 024472
Gandhi, M. 024401, 024517
Gandino, Guido 026397
Gandioso, Carmen Roman 004926
Gandon, Gérard 011593
Gandy, Doris 049395
Gane, Constance 046948
Ganebnaja, Nina Evgenjevna . . 036977
Gangi, Sabine 016187
Gangkofner, M. 022915
Gangloff, Roland 048662
Gangopadhyay, Paula 049195
Gangwere, Stanley K. 048314
Ganiev, A. 054646
Ganijon, Alijonov 054634
Ganino, Luigi 025340
Ganje, Melissa 048561
Gannett, Michael R. 047997
Gannon, Thomas 048385
Ganong, Dr. Overton G. 052746
Ganozova, Radoslava 005183
Ganpat, Ganesh 046435
Gansohr-Meinel, Dr. Heidi 017499
Gansz, Kevin 053357
Gant, Bob 047628
Gantenberg, Anette 020898
Gantner, Bernard 013385
Gantt, Judy M. 046622
Gantt, Nancy M. 053267
Ganz, James A. 054419
Ganzelewski, Dr. Michael 017433
Ganzer, Dr. Gilberto . . . 027410, 027411
Ganzkow, Martin 019318
Gao, Prof. Chong-Wen 007589
Gao, Hua 007933
Gao, Jun 008254
Gao, Liren 007928
Gao, Maojie 008504
Gao, Qiang 008358
Gao, Xia 008402
Gao, Xianzhong 007711
Gao, Zenglin 008329
Gao, Zongyu 007991
Gappmayer, Helga 002377
Garanne Modise, Rogoff 004300
Garaudy, Roger 039310
Garavaglia, Gianluigi 026113
Garavaglia, Giuseppe 026778
Garay, Nora Cecilia 008572
Gârbacea, Adriana Rodica 036577
Garbarini Islas, Dr. Guillermo . . 000274

Garbe, Dr. Detlef 019003
Garbes, Pam 033233
Garbrecht, Dr. Jörg 017265
Garbutt, Kirstie 043602
Gárces Julve, Vicente 039326
Garces, Donna M.P. 034492
Garcia-Anton, Katya 041517
García Arévalo, Manuel Antonio . 010403
García Chacón, Daniela 004208
García de Del Valle, Prof. Delia . 054765
García de la Rosa, Ignacio 039542
Garcia Forner, Anna 039175
García-Frias Checa, Carmen . . . 040351
García Fuertes, Pilar 040390
García Limón, Fernando 031394
García López, Dr. J.R. 039453
Garcia-Lorca de los Ríos, Laura . 039463
García Martínez, Carlos 040130
García Menárguez, Antonio 039480
García Moreno, Beatriz 008595
García Mota, Francisco 039700
García Muriano, Felipe O. 039126
Garcia-Nakata, Lorraine 053011
García Quiros, Paz 039438
García Rozas, Rosario 040440
García Ruíz, María Esther 039428
García Serrano, Rafael 040229, 040235,
. . . . 040237, 040238
Garcia Serrano, Rafael 040199
García Soto, Rubén 034382
García-Tuñon, Guillermo 039440
García Vanegas, Rubén Darío . . 008722
García Velasco, Francisco 039382
Garcia, Anna 047013
Garcia, Damian 052412
Garcia, Denis 016122
Garcia, Elizabeth F. 050570
García, Dr. Eulàlia 039077
García, Joaquín Emilio 008782
García, José 050836
García, Julian 039180
García, Lola 040308
Garcia, Magdalena A. 051766
Garcia, Miki 053090
Garcia, Noah 052938
Garcia, Rene R. 034605
García, Susana 040447
Garcia, Susana 040450
Garcías y Truyols, Joan 039000
Gard, Jean-Noël 012210, 014488,
. . . . . . 014775, 014970, 015120, 015123,
. . . . 015982
Gardent, Bruno 013429
Gardi Dupré, Maria Grazia 026227
Gardi, Dr. Bernhard 041276
Gardiner, Mary N. 046570
Gardner Broske, Janet 051494, 051495
Gardner Gates, Mimi . . 053248, 053249
Gardner, David 001411
Gardner, John 049756
Gardner jr., William F. 050581
Garduño Pulido, Blanca 031170, 031223
Garduño Ruíz, Dr. Everardo . . . 031141
Garey, Rose 053631
Garfield, Leonard 053240
Gari Lacruz, Dr. Angel 038883
Garibaldi, Dr. Patrizia . 026397, 026413
Gariboldi, Leopoldo 026972
Garifdžanova, Roza G. 036960
Garimorth, Julia 014697
Garland, Jeff 049281
Garland, Joan 050601
Garland, Kathleen 049888
Garland, Steve 043668
Garlinski, M. 041533
Garlits, Donald G. 051696
Garncarzová, Hana 009487
Garner, I.I.D. 044764
Garner, Jennifer 048262
Garner, Les 051103
Garner, Sheila 001597
Garnham, Peter 007035
Garniche, Marie José 013387
Garnier 015401

Garnier, J.F. 016275
Garnier, Laurent 012571
Garnier, Maurice 014684
Garnier, Nicole 012445
Garnier, Renée 013065
Garnier, Sylvie 012762
Garniewicz, Rex 049688
Garrels, Gary 051427
Garret, Patty 047349
Garrett, Anne 007057
Garrett, Jim 050535
Garrett, Jural J. 050462
Garrett, Rose M. 046950
Garrido, Emilio M. 034533
Garriga, Francesc 040085
Garrigan, Curt 051180
Garrigue, Philippe 015647
Garrison, Elizabeth 052302
Garrity, Lois 007145
Garrott, Martha 052571
Garrut, Dr. J.M. 039054
Garside, Sioux 001554, 001556
Garside, Vivienne 038711
Garstenauer, Karl 002583
Gartner, Joachim L. 003019
Garton, Diane 006618
Garua, Marina Prieto 038914
Garuti, Alfonso 025779
Garweg, Udo 022825
Garwood, Susan 048700
Gary, Grace 054440
Garza Camino, Dr. Mercedes de
  la 031239
Garza, Michael A. 048839
Gasc, Michel 014176
Gasco, Yves 011933
Gascon, France 006051
Gash, Lyn 044648
Gąsior, Stanisław 035101
Gąska, Wiesław 035186
Gaskell, Ivan 047362
Gaspar, Burghard 001844
Gaspar, Filorena 035474
Gaspari, Prof. Gianmarco 026817
Gaspari, Kristin H. 054316
Gasparini, Gauchesco Luis 000565
Gašparović, Miroslav . 009047, 009054
Gasperment, Dr. Philippe 015617
Gass, Monika 019280
Gassen, Dr. Richard W. 020140
Gasser, Dr. Christoph 025975
Gasser, Ferdinand 025976
Gasser, Fredi 042074
Gasser, Dr. Martin 042149
Gasset y Salafranca, Josep M. . 039576
Gaßner, Prof. Dr. Hubertus 018991
Gassner, Therese 003087
Gast, Dr. Otto 020853
Gastal, Ney 004644
Gastaut, Amélie 014713
Gaston, Elizabeth 050921
Gaston, François-Régis 015988
Gatchell, Melanie 049202
Gateau, Fabienne 013787
Gatellier, Michel 013018
Gately, Barbara 047886
Gatenbröcker, Dr. Silke 017558
Gates, Charles 046674
Gates, Daphnea 033230
Gates, David 043839
Gates, Jay 054157
Gates, Jim 047975
Gates, Sue 046299
Gates, William 047941
Gatewood, Bill 050337
Gattemair-Farthofer, Ingrid 018494
Gatti, Dr. Sandra 027192
Gau, Sönke 042240
Gaubert, Prof. Jean 031723
Gaubert, Philippe 013466
Gaudet-Blavignac, Richard 041422
Gaudiau, Edmond 012884
Gaudichon, Bruno 015162
Gaudio, Jennifer 050538

Gaudlitz, Winfried 021658
Gaudnek, Prof. Dr. Walter 016561
Gaudzinski-Windheuser, Dr.
  Sabine 020233
Gaudzinski-Windheuser, Prof. Dr.
  Sabine 020776
Gauffin, Sten 040902
Gaugele, Roland 018731
Gauhar, Hasan 034148
Gaul, Roland 030574
Gauld, Gail 000901
Gauldine, K.C. 053232
Gaulke, Dr. Karsten 019527
Gault, Sam 043881
Gaulupeau, Yves 015172
Gaumé, Jean-Charles 014354
Gausling, Helga 016474
Gaustad, Randi 033840
Gauthier, Catherine 015732
Gauthier, Jacques A. 051265
Gauthier, Virgile 004048, 004049
Gautier, Jean 004033
Gautschi, Karl 041781
Gautschin, Hanspeter 041388
Gava, Hsueh 042477
Gavagan, Mary 044754
Gavazzoci, Dr. Maria Laura
  Tomea 027063
Gavín Arnal, Alfredo 039546
Gavin, Glenda 053749
Gavin, John 047311
Gaviria, Alba 008644
Gavish, Galia 025084
Gavran, Dr. Ignacije 004291
Gavranic, Cynthia 042217
Gavril, Victoria 036301
Gavrilaki, Eirini 023038
Gavrilova, Marija Jurjevna 037555
Gavrilović, Miladin 038251
Gawlick, Henry 018921
Gay, Dr. Hebe Dina 000326
Gay, Millicent 052653
Gay, Pierre-Raymond 012614
Gayler, Billie Sue 053846
Gayot, Lise 014731
Gayraud, Gilbert 015001
Gazengel, Alcide 014171
Gazlay, Cheryl 051640
Gazzaway, Ofiera 054538
Ge, Daigui 008477
Gea Calatayud, Manuel de 040013
Gealt, Adelheid M. 047047
Gear, Dr. Gilian 044913
Gearinger, Robert E. 048947
Geary, Adam 044697
Gebert, Walter 022253
Gebhardt, Hans 016987
Gebhardt, Matt 052583, 052584
Gebhardt, Dr. Ute 019518
Gebühr, Dr. Michael 021663
Gečeva, Magdalena 005225
Geddes, Jim 033210
Geddes, Victoria J. 044627
Geden, Dennis 006509
Gedeon, Luc H. 053969
Gedgaudienė, Eleonora 030557
Gedvilas, Algirdas 030504
Gee, Lisa 046197
Geens, Catherine 039622
Geeraerts, Liene 004012
Geerdinck, J.J.M. 032598
Geeroms, Luc 003257
Geert, Olieu 003864
Geerts, Dr. T. 032310
Geeslin, Phyllis 049697
Gefahun, Dr. Abebe 010726
Geffers, Dr. Andrea 020101
Geft, Liebe 050461
Gehör, Dr. Seppo 011219
Gehri, Emil 041568, 041570
Gehrig, Anette 041325
Gehrig, Bruno 041344
Gehrmann, Rudi 017429
Gehrt, Gertrude 046418

Geibig, Dr. Alfred 017752
Geidner, Oskar 022735
Geier, Bob 051715
Geier, Katrin 018988
Geiger 020917
Geiger, Gisela 021100
Geiger, H. 019800, 019801, 019802
Geiger, Malies 019434
Geiger, Maurice 012184
Geiger, Uwe 018079
Geiges, Michael 042212
Geil, Bernd 019849
Geiselman, Anette 048795
Geisenhofer, Paul 021476
Geiser, Anne 041661
Geisler, David A. 051121
Geisler, John 051121
Geisler, Dr. Jonathan 053507
Geißel, Helmut 022670
Geisser, Dr. Hannes 041492
Geisser, Hans 041227
Geissl, Gerhard 003105
Geißler, G. 021296
Geissler, Josef 002455
Geißler, Ramona 021429
Geist, Dr. Joseph E. 048713
Geitner, Mary Elizabeth 049458
Gelao, Clara 025463
Gelburd, Dr. Gail 050211
Geldenhuys, M. 038660
Geldin, Sherri 047945
Gelfi, Yves 014149
Gelin, Pär 040529
Gélinas, Cécile 007025
Gelis, Joseph 003557
Gelius, Dr. William 010198
Gelny, Alexandra 003033
Gelo, Josip 004266
Gelsthorpe, D. 045210
Gelszinnis, Hans Joachim 021456
Gelzer, Bev 051571
Gembroll, Robert 049753
Gemmell, Michael 007077
Gemmill, Moira 045123
Gemmingen, Freiherr von 019093
Gencer, Necdet 042998
Gencsi, Zoltán 023697
Gendre, Cathérine 016182
Gendron, Christian . . . 014506, 014508,
. . . . 014509
Generalli, Genevieve 047832
Genev, Hristo 005055
Genger, Angela 018040
Gengler, Guy 030593
Gennari, Massimo 025909
Geno, Jaiy 054244
Genorio, Rado 038565
Genoud, Jean-Claude 041660
Gens, María L. Alonso 039791
Gentele, Glen P. 052807
Gentet, Alfred 041910
Gentil, M. Bruno 014668
Gentile, Dr. Gerardo 027273
Gentile, Dr. Elio 028292
Gentili, Lamberto 028004, 028008
Gentilini, Dr. Anna Rosa 026174
Gentis, Thierry 047217
Gentle, Bill . . . 047629, 048345, 048789
Gentry, Nola 046827
Gentzel, Heiko 018132
Genys, Dr. Jonas 030392
Geoffroy, Pierre 015952, 015953
Geominy, Dr. Wilfred 017477
Geoppa, Judith 048150
Georg, Rolf 021908
George, Dave 052934
George, Dr. Donny 024804
George, Elizabeth 050488, 050497
George, Hardy 051734
George, Kerrianne 001557
George, M.R. 001215
George, Prof. Ron 046030
George, Sarah B. 052919
George, Tamsen E. 047132

George, Terri 048389
Georgel, Chantal 014743
Georgescu, Valeriu 036558
Georgeson, Melinda 053537
Georget, Luc 014053
Georghiades, A. 009414
Georghiou, G. 009414
Georgieva, Elene 005014
Georgieva, Tanička 005154
Georgsson, Agust 024112
Geppert, K. 021521, 021522
Geppert, Dr. Stefan 021905
Geraci, Ester 026048
Gérard, Peter 022519
Gerasimčuk, Olga 037801
Gerasimidi, Elena Ivanovna .. 036878
Gerbauckas, Maryanne 054315
Gerber, Barbara 041325
Gerber, Jan 048049
Gerber, Olive 053298
Gerber, Dr. Peter R. 042241
Gerber, Theodore 047104
Gerbino, Allen J. 050369
Gerchow, Dr. Jan 018428
Gercke, Irmgard 016440
Gerdes, Edward 048625
Gerdes, Peggy 053250
Gerdur-Gunnardóttir, Gudny .. 024092
Gereau, Rebecca 049122
Gergel, Belinda 047911
Gergely, Dr. Katalin 023484
Gerhards, Dr. Eva 018475
Gerhardt, Barbara 041364
Gerhardt, Dr. Gunther 022412
Gerhardy, Wilhelm 018074
Gerharz, Michael 021277
Gerichhausen, Heinz 022452
Gericke, Dr. Hans-Joachim ... 022146
Gerini, Fiorenzo 025269
Gerke, Lucien 004136
Germain, Christiane 015784
Germain, P. 012832
Germann-Bauer, Dr. Peter 021322,
                    021324, 021329
Germann, Edwin 041579
Germano, Anna 028286
Germanova, Galina 005191
Germiny, Comte Gabriel de .. 014955
Gerner, Dr. Cornelia 017165
Gerrard, Olga 050456
Gerrietts, Frank 046854
Gerritsen, J. C. 032068
Gershon, Stacey 051358
Gershuny, Justin 050428
Gerstein, Dr. Alexandra 044951
Gerstheimer, Christian 048508
Gertjejansen, Doyle 051290
Gertsch, Fritz 041664
Gertsch, Liliane 041913
Gervais, Alex 005313
Gervais, Louise 048096
Gesemann, Dr. Björn 020233
Gęsicki, Waldemar 034661
Gesing, Dr. Martin 017010
Gesquière, Dominique . 013661, 014120
Gess, Dr. F. W. 038723
Gessener, Andrew 051374
Gesser, Susanna 018428, 018431
Gessie, Dwight 007142
Gessl, Paul 002233
Gestin, Jean-Pierre 015561
Gestring, Tryggve 010771
Gestring, Ron 052490
Gestsson, Magnús 024071
Getchell, Katherine 047123
Gether, Christian 010151
Getman, Z.D. 037789
Getson, Ralph 006228
Gette, Timothy J. 050727
Gettinger, Maria 002231
Gettler, Peter 020983
Getzner, Manfred A. 001882
Geurts, Joe 005871, 006582
Geuze, M.A. 032871

Gevas, Sophia 048667
Gex, Catherine 042104
Geyer, Prof. Dr. A. 019443
Geyer, Prof. Dr. Angelika 019456
Geyer, Jeannette 038733
Geyer, Nancy 047138
Geyer, Dr. Roderich 002848
Gfeller, Irma E. 050312
Gfrereis, Dr. Heike ... 020260, 020262
Ghaeni, Farzaneh 024743
Ghambaryan, Armida 000691
Ghandgar, J. 024739
Gharrity, Mary Lou 050891
Ghedini, Prof. E. Francesca ... 027152
Gherardi, Dr. Pierluigi 027317
Ghesquières, L.-E. 006648
Ghez, Prof. Claude 041541
Ghez, Susanne 047694
Ghiara, Prof. Maria Rosaria ... 027022
Ghidini, Elena 027068, 027606
Ghini, Dr. Giuseppina 027027
Ghini, John 050612
Ghiorsi-Hart, Carol 053369
Ghiotto, Dr. Roberto 026926
Ghirlanda, Silvia 041390
Ghiron, Yves 013983
Ghislain, Bourdon 004064
Ghiță, Ioana 036340
Ghoniem, Atef 010543
Ghorbani, M. 024791
Ghosh, Indrani 024366
Giachino, Dr. Roberto Sandri .. 028136
Giacometti, Emmanuelle 041648
Giacometti, Florence 013418
Giacomini, Dr. Giovanni 025835
Giacomini, Dr. Pietro 025983
Giaimo, Catherine L. 052065
Giaj, Ezio 027336
Giamarchi, Paulette 013488
Giambuti, Lia 028392
Giampaolo, Francesca 026604
Gianadda, Léonard ... 041723, 041726,
                           041727
Gianaroli, Onofrio 025570
Gianelli, Ida 027537
Gianfranco, Armando 025616
Gianfreda, Dr. Sandra 042152
Gianmoena, Willi 021008
Giannetti, Prof. Antonio 027528, 027529,
                           027533
Giannetti, Elio 027114
Gianni, Dr. Ginsberti 025552
Giannoulia, Eirini 023096
Gianora, Giulietta 041978
Giaquinto, Michael 048243
Giarmoleo, Giuseppe 027631
Gibb Roff, Samantha 048380
Gibbard, Dr. Russell 033269
Gibbon, Imogen 044274
Gibbon, Wilda V. 007074
Gibbons, Cindy 006755
Gibbons, Michael L. 046730
Gibbs, Ivor 045641
Gibbs, Margaret 048527
Gibbs, Marie 024984
Gibbs, Maurice E. 051158
Gibbs, Roger 050055
Gibernau, Dr. Jesús Gil 040248
Gibert-Balboni, Françoise ... 016056
Gibson Garvey, Susan 005948
Gibson, Charlotte 053809
Gibson, Hugh 044086
Gibson, Keith E. ... 050270, 051283
Gibson, Ken 050602
Gibson, Paula 043349
Gibson, Ralph 054397
Gicà, Secție-Bâeștean 036681
Gichere, Hjuguna 029864
Giebelhaus, L. 007310
Gieben, Servus 027621
Gier, Dr. Helmut 016648, 016668
Gier, Dr. Klaus 018976
Giere, Prof. em. Dr. Wolfgang .. 018416
Gierisch, Nita 048031

Gierkinger, Gerda 001688
Gieschen, Margarete 018720
Giese, Dr. Dale F. 053339
Giesen, Alison 050148
Giesenschlag, Bill 047348
Giesert, Ted 049977
Giesler, Dr. Jochen 017499
Giesler, Dr. Ulrike 019711
Giffault, Michèle 011780
Giffeler, Elke 017276
Gigandet, Eric 041865
Giglierano, Geoff 054200
Giglio, Alessia 041360, 041698
Gigou, Laure 013756, 014859
Giguet, Aurore 050172
Gil de los Reyes, Soledad 039232
Gil de Prado, Eva 039548
Gil Imaz, Maria Cristina 040452
Gil Rodríggez, Juan Juis 039185
Gil, Dr. Euzebiusz 034929
Gilardi Polar, Hernando 008589
Gilberger, Ruth 018064
Gilbert, André 006747
Gilbert, Barbara 050468
Gilbert, Courtney 049947
Gilbert, David 045639
Gilbert, Estelle 046078
Gilbert, Frank 054130
Gilbert, Jennifer 050511
Gilbert, Sylvie 006376
Gilbertson, Elsa 046319
Gilbertson, Laurann 048208
Gilch, Eva 017692
Gilchrist, Rod 052996
Giles, Mel 006377
Giles, Peter 053045
Giles, Sue 043747
Gilfillan, Richard M. 054270
Gilg, Karen 046666
Gilgen, Ruth 041318
Gili, Antonio 041395, 041700
Gill, Angela 043769
Gill, B. 033142
Gill, Cecil 050844
Gill, J. D. 043497
Gill, Peter 001332
Gill, Sandra 006836
Gill, Stephen J. 030288
Gill, Vince 051175
Gill, Wendy 044391
Gille, Martha 031852
Gillerist, Christopher H. 053959
Gilles, Dr. Karl-Josef 022202
Gillespie, Gordon 052372
Gillespie, Margaret 047746
Gillespie, Maria 043674
Gillespie, Marilyn 050166
Gillespie, Rosemarie G. 046934
Gillespie, Sarah 024852
Gillespie, Shane 024929
Gillett, Marnie 053021
Gillette, Dr. David D. 048740
Gilliam-Beckett, Elizabeth 049906,
                           050274
Gilliam, Jan 054408
Gilliam, Kevin 053839
Gilliland, Michel 041835
Gillieaux, L. 003472
Gillies, A. 046134
Gillis, Jean 053073
Gillis, Nancy S. 046773
Gillis, Stuart 044483
Gillman, Derek 052078
Gillon, Patrick 016344
Gilman, Robert G. 049434
Gilman, Sondra 051491
Gilmore-House, Gloria 050399
Gilmore, Jerry 047140
Gilmour, Rosemary ... 043490, 043943
Gilpin, Melinda 050693
Giltaij, J. 032826
Gimborn, C.H. von 018188
Giménez Carrazana, Dr. Manuel 004242,
                           004244

Gimenez, Carmen 051470
Gimeno Salueña, Joaquín ... 039414
Ginalski, Jerzy 035191
Ginette, Eric 047340
Ginger, Greg 046887
Gingerich, Jim 053586
Ginlund, Hans 010762
Ginter, Donald W. 051605
Ginton, Ellen 025212
Gioacchini, Delfo 027110
Giobbi, Maria Paola 025389
Giombini, Dr. L. 026004
Gionio, Maria Teresa 026828
Giordano, Prof. Christian 041498
Giordano, Claudio 025759
Giorgetti, Pia 041696
Giorgi, Dr. Angelo 025623
Giorgi, Benedetta 041695
Giorgi, Dr. Marisa 027627
Giorolano, Fabris 028189
Giovanazzi, Carol 051132
Giovannoni, Günther 041394
Giovetti, Dr. Paola 025538
Giovine, Dr. Francesco Paolo .. 028284
Gipson Adams, Shelley 052355
Giradeau, Ted 005532
Girard, Alain . 011871, 011872, 014944,
                    014945, 016268
Girard, Hélène 005363
Girardin, Daniel 041659
Girardon, Eric 012776
Giraud, Milou 014825
Girgenti, Cecilia 000144
Gironde, Herberto 000294
Girouard, Terry 047891
Giroud, Magali 015663
Girousse, Laurent 011954
Giroux, Patrice 006134
Girsberger, Jürg 041455
Girst, Roswitha 018427
Girvan-Brown, Chris 001083
Girveau, Bruno 014642
Gisbert Santonja, Josep Antoni . 039345
Gisler-Jauch, Dr. Rolf 041210
Gisler, H. 041996
Gittelman, Steve 047499
Giubilei, Dr. Maria Flora 026389, 026419,
                           026429
Giuliano, Prof. Carlo 028144
Giuliano, Dr. Laura 027627
Giuntini, Trey 050786
Giurescu, Dinu C. 036320, 036496
Giuricco, Judith 051474
Giusti, Dr. Annamaria 026278
Giusti, Carla 027018
Giusti, Dr. Giovanna 026256
Givnish, Gerry 052077
Gizatulin, Marat R. 037635
Gjaurov, Toško 054364
Gjedsted, Kirsten 010214, 010215,
                           010216
Gjerdi, Trond 033829
Gjipali, Ilir 000031
Gjmacan 040217
Gjurinovic Canevaro, Pedro ... 034329
Gladstone, Prof. M. 024199
Gläßel, Adolf 022401
Glaister, Dr. Olivier 041655
Glaizot, Dr. Olivier 044667
Glanzer, Harvey W. 052529
Glanzer, Michelle 047244
Glaser-Schnabel, Dr. Silvia ... 020853
Glaser, Prof. Dr. Franz . 001937, 002192,
                           002289
Glasgow, Istiharoh 050408
Glaskova, Vera Evgenjevna ... 038029
Glass, Brent D. 048611, 053753, 054144
Glass, Catherine 050130
Glass, Christian 022258
Glass, Frank 006827
Glass, Jennifer 048891
Glass, Patrice 047591
Glassel, Jim 048772
Glasser, Dale 051487

Glasser, Susan 052550, 052831
Glasser, W. 001571
Glassman, Elizabeth .. 013115, 047702
Glassman, Susan 052102
Glassmoyer, Jeniffer 050388
Glassner, Prof. Dr. Gottfried 002381
Glasson, Michael 046042
Glasswell, Samantha 045605
Glaubinger, Jane 047808
Glauser, Daniel 041914
Glauser, Julien 041797
Glavočić, Daina 008977
Glaz, Kazimir 007216
Glaze, George 047030
Gleason, Mat 050410
Gleaton, Sonja 051210
Gleeson, Elizabeth 001025
Gleeson, Val 001619
Gleffe, Waltraud 016586
Glehr, Dr. Reinhold 002058
Gleirscher, Dr. Paul 002192
Gleissner, Stephen 054394
Gleixner, Johann 022360
Gleizer, Raquel 004907
Glenholme, Ian 044786
Glenn, Rosalyn 053691
Glenn, Sherrain 050099
Glevenko, Elena Jurjevna 037755
Glick, Dr. Shmuel 025104
Glidden, Joy 047256
Gligor, Ierusalima 036789
Glimme, Dr. Hans Peter 018179
Gliozheni, Lorenc 000021
Glistak, John 001452
Gliszczyńska, Elżbieta 034626
Glock, Walter 021425
Glocker, Dr. Jürgen ... 017505, 018788,
                                021589
Glöckner, Prof. Harry 002963
Gloor, Dr. Lukas 042233
Gloria, Dr. Heidi K. 034476
Gloria, Rogelia 034506
Glößner, F.X. 020723
Glósz, Dr. József 023950
Glot, Claudine 012651
Glotzhober, Robert 047941
Gloudemans, M. 032709
Glover, Jack 053609
Glover, Mike 043808
Glover, Sarah Elizabeth 005403
Glover, Susan 047106
Glover, Thomas 049325
Gloza-Rausch, Florian 016877
Gluchova, Tatjana Vladimirovna . 037049
Gluck, Daniel 051440
Glüber, Dr. Wolfgang 017822
Glückert, Ewald 019932
Glueckert, Stephen 050960
Glušacov, Viktor Ivanovyč 043275
Glušenkova, Ljudmila Nikolaevna 036918
Gluskova, Marina 037133
Glyda, Krzysztof 035133
Glynne-Percy, Colin 045635
Gmelch, Christa 021385
Gnecco, Prof. María Julia 000588
Gnidovec, Dale 047942
Gnilšak, Ida 038595
Go, Sugil 029907
Gobbi, Elmar 026761, 026763
Gobel, Mary 049856
Goberna, María Victoria 040305
Goble, Carole 047329
Goble, Rebecca 048204
Gochenour, Anne 050334
Gockel-Böhner, Christoph 021064
Gockerell, Dr. Nina 020509
Gockov, Ilija 004989
Gócsáné-Móró, Csilla 023992
Gocz, Teodor 035294
Godart, Freddy 003527
Godbey, Martin 053150
Godbout, Dominique 006139
Goddard, Stephen 050187
Godefroy, Hubert 015424

Godelet, Andre 003721
Godenschweger, Ulrich 018229
Godfrey, Honor 045137
Godfrey, Judith 046307
Godfrey, Laurie 046452
Godgenger, Jean-Marc 012970
Godoy Muñoz, Maria Antonieta . 023362
Godoy, Arnaldo Augusto 004348
Godoy, Eliana 051335
Godoy, José 041530
Godwin, Carmen 048724
Godwin, Mary 045542
Godwin, Sharon 007146
Goebel, Susanne ... 016492, 016493,
                016494, 016496, 016497, 016498
Göbel, Prof. Dr. Dr. Ulf 017273
Göçmen, Ali Bülent 043086
Goedbloed, J. 031885
Goedicke, Dr. Christian 017272
Goehring, Alicia L. 050046
Gökten, Serdar 042818
Göldner, Andrea 022531
Göller, Luitgar 016944
Göpfrich, Jutta 020970
Görcke, Bärbel 020752
Gördüren, Petra 019606
Goerg, Ute 019376
Görgner, Ernst 017866
Görlich, Manfred 003097
Görner, Dr. Irina 019553
Görner, Klaus 018438
Görner, Dr. Rolf 022672
Görner, Veit 019054
Goertzen, Peter 005912
Goes, Drs. André W.A. van der . 017989
Goes, B.R. 031979
Goes, Dr. Georg 016969
Goess-Enzenberg, Ulrich Graf . 002147
Goessen, J. 032631
Gößwald, Udo 017251
Göthberg, Elisabeth 040844
Goethert, Dr. Klaus-Peter 022201
Götte, Dr. Gisela 017597
Göttner, Dr. Jens 018962
Götz, Angela 020873
Götz, Matthias 041278
Götz, Dr. Norbert 020551
Götz, Reinhard 002831
Götz, Rudolf 042118
Goetz, Scott 053413
Götz, Dr. Ulrike 018501
Götze, Barbara 017320
Götze, Dr. Bettina 021287
Goetze, Lisel 048613
Götze, Robby Joachim 018711
Goetzman, Richard 049729
Götzmann, Dr. Jutta 021198
Goff, Nick 046101
Goffe, Gwendolyn H. 049588
Goffinet, Bernard 003844
Goffman Cutler, Judy 051536
Goger, Gottfried 001912
Goguen, Georges 006347
Goheen, Marion 052527
Gohlke, Gerrit 021179
Goj , Margarete 021326
Golay, Laurent 041660
Golay, Muriel 041982
Golbin, Pamela 014705
Gold-Levi, Vicki 046642
Gold, Dr. Helmut 018436
Gold, Jack Anthony 052392
Gold, Ruth C. 053381
Goldbacher, Alexander 002942
Goldberg, Beth 053061
Goldberg, Beverly 053896
Goldberg, Elizabeth 053273
Goldberg, Marsea 054309
Goldberg, Paul 053502
Goldberg, Shalom 047069
Goldblum, Bonnie 051296
Golden, John 054447
Golden, Rorey 051353
Golden, Thelma 051476

Goldenkova, Vera Anatoljevna .. 037839
Goldenvejzer, Elena Ivanovna .. 037410
Golder, Sheri 047993
Goldfarb, Hilliard 006391
Goldfinger, I. 025041
Goldfinger, Yaffa 025212
Goldfuhs, Ingeborg 001841
Golding, Norman 033289
Goldman, Steven 051808
Goldnagl, Friedrich 001818
Goldner, George R. 051424
Goldowsky, Alexander 054543
Goldschmidt, Lawrence P. 054152
Goldsmith, Jeremy 045860
Goldsmith, John 044616
Goldsmith, Robert B. 051390
Goldstein, Ann 031986
Goldstein, Gabriel M. 051493
Goldstein, Jon 053544
Goldstein, Lisa 048579
Goldstein, Sid 052818
Goldwasser, Edwin 054145
Goleas, Janet 048438
Goley, Mary Anne 054111
Golf, Anamari 049793, 049794
Golias, Allison 046857
Golikova, Ljudmyla Vasylivna .. 043294
Golinelli, Jean 016148
Golinski, Dr. Hans Günter 019096,
                019097, 019100, 019101, 019103
Golisek, Lisa 049848
Goll, Dr. Jürg 041399
Goll, Dr. Roland 001660
Gollmann, Sandra 003036
Golloway, Zoe C. 052429
Golofast, Larisa 043293
Gologranc Zakonjšek, Bronislava 038501
Golonka, Jan 034708
Golovčenko, Ljudmila 043293
Golshan, J. 024761
Goltz, R. von der 017211
Gołubiew, Zofia 034897
Gomá Lanzón, Javier 039623
Gombert, Florence 013807
Gombotz, Christian 001685
Gomel, Luc 014503
Gomendio Kindelán, Montserrat . 039674
Gomes de Pinho, António 035940
Gomes Leite, Gisélia Antonia ... 004799,
                                004800
Gomes, José Antonio Alves 004552
Gómez-Alba, Dr. Julio 039077
Gómez Arnanz, Miguel 039027
Gómez Gómez, Alicia 039633
Gómez Hildawa, Isidro 034562, 034565
Gómez Macías, Juan Antonio ... 040073
Gómez, Arley 008707
Gomez, Carla 053106
Gómez, Carlo Humberto 008609
Gómez, José Luis 000357
Gómez, Nadine 012773
Gómez, Roque Manuel 000465, 000558,
                                000559
Gomez, Victor 048502
Gomig, Dr. Leo 001822
Gomis, Joan Carles 039881
Gomozkova, Marina 037288
Gompertz, P. 045857
Gomringer, Prof. Eugen 021335
Gomringer, Dr. Nortrud 021335
Gonçalves, Cristina Martins 035557
Goncalves, Fabian 054090
Gonçalves, Maria Manuel 035893
Gonçalves, Prof. Dr. Maria Teresa
        Serra Silva 035621
Gonçalves, Vereador Renato ... 035869
Gončar, Petro Ivanovyč 043206
Gončarenko, Larisa Fedorivna .. 043163
Gončarova, Elena Alekseevna .. 037037
Goncerut Estèbe, Véronique ... 041513
Gong, Dazhong 008058
Gong, Ping 008297
Gong, Stephen 046933
Gong, Xiaoan 008410

Gong, Yunxue 008282
Gonos, Arthur 048883
Gonseth, Marc-Olivier 041797
Gonzaga Bastidas Ortíz, Rodrigo 008680
Gonzales Baldoui, Mariano 040432
Gonzales-Belleza, Erlinda 034493
Gonzales, Ben 034559
Gonzales, Michael E. 051726
González Alonso, Alberto 039653
González Amezúa, Mercedes 039650
González Antón, Dr. Rafael 040098
González Aranda, Lucilia 008638
Gonzalez Azcoaga, Miguel
        Fernando 000348
González Cevallos, María A. ... 000179
Gonzalez-Cid, Malena 052937
González de Aspuru Hidalgo,
        Sara 040427
González de Durana, Javier ... 040422
González de León, Antonio 031169
Gonzalez de Santiago, Ignacio
        María 039170
González Entidad, María
        Inocencia 008780
González González, Ignacio ... 031564
González Gonzalez, Mildgros ... 039721
González Gozalo, Dr. Elvira ... 039380
González-Hontoria y Allendesalazar,
        Guadalupe 039637
González Lorente, Margarita ... 009137
González Padrón, Antonio María 040216
González Parra, Samuel 008575
González Penagos, Claudio 008564
Gonzalez Pons, Esteban 039506
González Rodríguez, Rosalía ... 039528
González Romero, Santiago 040066
González-Sancho, Eva 012780
Gonzalez, Anita 052847
Gonzalez, Arturo Gonzalez 031425
González, Aurelio 039551
Gonzalez, Christine 013029
González, Doña Ligia 033398
González, Gilbert 048976
González, Jaime 008803
González, Jorge 008582
Gonzalez, Dr. Juan G. 036135
González, Julieta 054724
Gonzalez, Manuel 051358
González, Milagros 054724
Gonzalez, Sylvie 015285
Gonzalez, Victor 036169
Gonzalo Andrade, Luis 008814
Good, Les 043747
Good, Liz 051303
Good, Patricia 050280
Goodale, Thomas 048640
Goodband, Lara 046230
Goodchild, Anne 045772
Goode, Dr. David 007087
Goodell, Greg 049090
Goodger, Margaret 043810
Goodin, Kitty 049847
Gooding, Brian 044843
Goodlett, Chris 050492
Goodman, A. 005456
Goodman, Barbara 049769
Goodman, Carl 046581
Goodman, Gwen 052075
Goodman, Norman 046128
Goodman, Susan 051412
Goodman, Wallace 053737
Goodnight, Gwenyth 051850
Goodrich, Dr. James W. 047903
Goodrich, Judith 050793
Goodridge, Janet 044580
Goodridge, Wendy 045933
Goodsell, Leo J. 046947
Goodwin, John 048584, 048586
Goodwin, Marsi 052005
Goodwin, Murray 033207
Goodwin, Sue 053374
Goodwinl, Patricia 051144
Goody, Dick 052607
Goodyear jr., Frank H. 052118

Goos, D.	017647	
Goossens, Dr. E.J.H.P.	031953	
Goossens, Gerda	004151	
Gor-Arie, D.	025168	
Gorbenko, Anatolij Aleksandrovič	036879	
Gorbey, Ken	033360	
Gorbunova, Éleonora Grigorjevna	037040	
Gorczyca, Krzyśtof	035022	
Gorczynski, Bohdan	047691	
Gordenker, Emilie	044255	
Gordon-Smith, Peter	044108	
Gordon, Alec	044145	
Gordon, Alexander	043487	
Gordon, B.	038690	
Gordon, David	046718, 050916	
Gordon, Dr. Dillian	045057	
Gordon, Don	007003	
Gordon, Donnarae	052515	
Gordon, Iain	043979	
Gordon, Lindsay	043360	
Gordon, Nick	044800	
Gordon, Sharon	047904	
Gordon, Stanley	054045	
Gordon, Sue	043930	
Gorduza, Victor	036220	
Gordy Edwards, Esther	048312	
Gore, Michele	050089	
Goren, Dr. M.	025214	
Goren, Nili	025212	
Gores, Mary	049762	
Goreta, Aleksandra	008925	
Goretti, Ndayarinze M.	005280	
Gorewitz, Shalom	050605	
Gorham, Laura T.	004191	
Gorham, Linda	046724	
Gorham, Marilyn R.	050294	
Gori, Beatrice	025887, 025888	
Gorini Esmeraldo, Eugênia	004879	
Gorini, Prof. Giovanni	027146	
Gorjão, Sérgio	035875, 035876, 035877	
Gorman, Jill	046890	
Gorman, Dr. Martyn	043351	
Gormley, John	053819	
Gormley, Michelle	047134	
Gormley, Nina Z.	053425	
Gormsen, Gudrun	010310	
Gorny, Dagmar	021064	
Goroch, Tansija Vladimirovna	037955	
Gorochov, Vladimir Aleksandrovič	037179	
Gorodnij, Aleksandr Ivanovič	038030	
Gorove, Margaret	051863	
Gorr, Claudia	020880	
Gorroño de Crespo, Maria del Carmen	000406	
Górski, Mirosław	034841	
Gorski, Stefan	035033	
Gorsline, Marie	012258	
Gosden, Dr. Chris	045459	
Gosell, Julie	049661	
Goshen, R.	025134	
Goshow, Nancy	051429	
Gosling, Susan	045763	
Goss, Fred	049251	
Gosse, Rita	053280	
Gosselin, Claude	006366	
Gossy, Reinhard	002676	
Gostick, Misha	007259	
Gostisha, Don	046417	
Gostomski, Michelle L.	047339	
Gostwick, Martin	044045	
Goswami, Raj Ratna	024440	
Goszczyńska, Grażyna	034726	
Gotgelf, Lev K.	036846	
Gothe, Nina	019518	
Goto, Hidehiko	029680	
Goto, Ikuo	028687	
Gottfried, Claudia	021290	
Gottfried, Heinz	002415	
Gottfried, Dr. Michael D.	048441	
Gottlieb, Jaromir	009602	
Gottlieb, Paul	051350	
Gottskálksdóttir, Júliana	024098	
Goudry, Gail	005455	
Goudy, David	051654	

Goueiva Santos, Dr. Antonio Alberto	035946	
Gough, David	048308	
Gough, Joan	000927	
Gouinguenet, Hélène	011703	
Goujat, Bérangère	011706	
Goula, Gemma	039095	
Goulaine, Marquis de	013236	
Goulandris, Niki Angelos	023128	
Gould, Alac	054593	
Gould, Alec	049792	
Gould, Claudia	052058	
Gould, Gail	007184	
Gould, Jonathan H.	054273	
Gould, Marilyn	054458	
Gould, Richard	052517	
Gould, William	049145	
Goulden, M.R.R.	044550	
Goumand, Bruno	014088	
Gounari, Dr. Emmanuela	023295	
Gourdin, Michel	003390	
Gouriotis, Lena	023150	
Gourlin, Thierry	011578	
Gout, Jacques	012503	
Goutayer, Virginie	012608	
Goutorche, Christian	012743	
Gouvea, Alexandre Patrício	035960	
Gouveia, Tiago Patricio	035597	
Gouy-Gilbert, Cécile	013544	
Govan, Michael	046841, 051378	
Gove, Domingos Zefanias	031810	
Govindaraj, C.	024441	
Gowing, Laura	048651	
Goycea, Ana Maria	000415	
Gozak, Roman	043238	
Gozani, Tal	050468	
Gozdowski, Krzysztof	035235	
Gozzi, Dr. Fausto	025895, 025897	
Graaf, F.M. de	033058	
Graas, Lucette	003322	
Grabar, Olga	043293	
Grąbczewski, Wiktoryn	035358	
Grabenstatter	015037	
Graber, Cleon	048968	
Graber, Marcel	041346	
Grabert, Walter	018881	
Grabham, Chris	045168	
Grabher, Gerhard	001794	
Grabka, Dr. Marion	017829	
Grabner, FOL Adolf	002014	
Grabner, Dr. Sabine	002953	
Grabovec'kyj, Volodymyr Vasil'ovyč	043158	
Grabow, Astrid	017875	
Grabowska-Chałka, Janina	035262	
Grabowska, Genowefa	035180	
Grabowski, Dr. Jörn	017320	
Graça, Filipe	036019	
Grachos, Louis	047298	
Gracyalny, Devan	054286	
Gradin, Eva	040804	
Gradova, Dina Aleksandrovna	037727	
Gradstein, Prof. Felix	033847	
Gradwohl, Hans	001762	
Graells, Eulàlia	039054	
Grämiger, Hans	041920	
Graen, Dr. Dennis	019456	
Graen, Dr. Monika	021058, 021064	
Graepler, Dr. Daniel	018757	
Gräser, Dr.	022558	
Grässer, Irina	020237	
Graesslé, Isabelle	041536	
Grässlin, Thomas	021594	
Graeve-Ingelmann, Dr. Inka	020508	
Graf, Dr. Angela	019011	
Graf, Bernhard	041882	
Graf, Prof. Dr. Bernhard	017178	
Graf, Brigitte	017305	
Graf, Diane	054577	
Graf, Felix	042214	
Graf, Dr. Rudolf	041208	
Graf, Stefan	001794	
Grafe, Steve	051732	
Grafenauer, Michael	001735	

Graff, Allison	048015	
Graff, Nicholas	052750	
Graff, Terry	006977	
Graffagnino, J. Kevin	051036	
Graffam, Olive	054101	
Grafton, Duke of	045968	
Graham LeBlanc	005556	
Graham Wade, Karen	047753	
Graham, Lady	045621	
Graham, Angela	045882	
Graham, Barry	044963	
Graham, Bruce	052844	
Graham, Douglass J.M.	053152	
Graham, Fiona	044801	
Graham, J.	044284	
Graham, Sir James	045621	
Graham, John	001465	
Graham, John R.	050569	
Graham, Kenneth	024976	
Graham, Michael S.	051257	
Graham, Shelby	053100	
Graham, Tom	047539	
Grahan, Dr. John	047302	
Grahn, Göran	041044	
Grahn, Martin	044019	
Gram, Dr. Magdalena	041028	
Gramatikov, Georgi	005004	
Gramatikov, Jančo	005133	
Grambye, Lars	040828	
Gramer, Kevin	048251, 048265	
Gramm, Jean-Jacques	041885	
Grammel, Sören	001973	
Grammenos, Dr. D.	023286	
Grammer, Mag. Manfred	002606	
Gramolina, N.N.	037902	
Grams, Greg	054009	
Grančarov, Michail	005118, 005120, 005121	
Grančarova, Kamelia	004990	
Grancea, Maria	036210	
Grand, Stanley I.	047418	
Grande, Prof.Dr. Nuno	036014	
Grander, Johann	002149	
Grandidier, Daniel	015292	
Grandjean, Gilles	013692, 014643	
Grandmont, Jean-François	006144	
Grando, Constance	046721	
Grandrud, Dr. Reba	048343	
Grandy, Patrick	047173	
Grandy, Peter	042000	
Grange, Bill	044082	
Grangé, Claire	011579	
Grange, Sylvie	012375, 012377	
Granger, Brenda	048481	
Granier, Jean-Marie	014771	
Granitto, Andrew	054566	
Granitz, Gerhard	002409	
Granlund, Mark E.	052305	
Grannan, Jill	048300	
Grannis, Sue Ellen	050745	
Granomont, Josée	007247	
Granqvist, Eirik	011273, 015181	
Grant Morrisey, Marena	051807	
Grant, A.E.	043422	
Grant, Anne	001430	
Grant, Bronwyn	033285	
Grant, Carolyn	047431	
Grant, Christin	052623	
Grant, Elinah	004298	
Grant, Frank F.	006195	
Grant, Gavin	045676	
Grant, Gillian	033133	
Grant, Hugh	048266	
Grant, J.	043473	
Grant, Jaqueline	047999	
Grant, Laurence	005932, 043453	
Grant, Rachel	044427	
Grant, Ronald	044943	
Grantčarov, Michail	005037	
Granville, Lee Z.	053364	
Grapin, Claude	011599	
Gras, Mag. Claudia	002981	
Grasberger-Lorenz, H.	022440	
Graskoski, Debra	051647	

Grasmann, Lambert	022343	
Grasmug, Dr. Rudolf	001878	
Grass, Yasmin	000766	
Grasse, Carola	018183	
Grasse, Marie-Christine	013156, 013161, 013162	
Grassi, Alfredo	000264	
Grassi, Dr. Renzo	027955	
Grassmayr, Johannes	002124	
Grasso, Giacomo	026411	
Graswinkel, J.	032156	
Gratacós i Teixidor, Georgina	039049	
Grathwohl-Scheffel, Christiane	018483	
Gratl, Dr. Franz	002141	
Grattan, Patricia	006897	
Gratte, Stan	001614	
Gratz, Dr. Reinhard	002619	
Gratzer, Heinz	003046	
Grau Lobo, Luís A.	039557, 039558	
Grau, Salvador	040396	
Graudina, Anna	052652	
Grauer, Michael R.	047414	
Graumlich, Anne	053957	
Grauwiller-Straub, Kathrin	041674	
Gravatt, Kristin	046985	
Grave, Marion	046019	
Gravel, Mabel	005384	
Graves-Brown, Carolyn	045933	
Graves, Tolley	050239	
Gray, A.	045210	
Gray, Alan	033287	
Gray, Dr. Anna	000937	
Gray, Bonnie	006256	
Gray, Brian	050456	
Gray, Campbell B.	052405	
Gray, Cherry	043374	
Gray, Gabrielle	051852	
Gray, Ian	005528	
Gray, Jo	044318, 045737, 046039	
Gray, John	033336	
Gray, John L.	050455	
Gray, K.	046153	
Gray, Les	001192	
Gray, Mary H.	047943	
Gray, Muriel	001192	
Gray, Peter	001481, 045569	
Gray, Richard	043999	
Gray, Sam	051748	
Gray, Sara	052511	
Gray, Sarah	043456	
Gray, Dr. Sharon R.	053478	
Gray, Susan Inskeep	048665	
Gray, Susie	045079	
Gray, Xanthi	052309	
Gray, Y.	044679	
Grayburn, Patricia	044489	
Grayson, Neil	051371	
Graze, Sue	046682	
Grąziewicz-Chludzińska, Barbara	035386	
Graziola, Prof. Claudio	027305	
Grazzini, Patricia J.	050944	
Grdanjski, Dipl.-Ing. Mirjana	020505	
Greaves, Debbie	045961	
Greaves, Sarah	043846	
Grébert, Eliane	015606	
Grečenko, Ljudmila Pavlovna	037170	
Grech, Frances	030763	
Greco-Kaufmann, Prof. Heidy	041324	
Greeeson, Laura	047589	
Greeff, I.B.	038675	
Green, Adrian	045727	
Green, Alison	043806	
Green jr., Byron E.	050977	
Green, Cheryl	051445	
Green, Chris	043575, 043576, 045664, 045668, 045672	
Green, Dr. D.	045210	
Green, Erica	000761	
Green, James	045565	
Green, Jeremy	001403	
Green, John F.	053432	
Green, Prof. Jonathan	052591	
Green, Louise	053306	
Green, Mike	004180	

Green, Nancy 049739
Green, Pat 048345
Green, R.M. 001068
Green, Richard 046230
Green, Tina 047337
Green, Dr. William 046905
Green, William D. 046833
Green, William Davis 052712
Green, Dr. William E. 047414
Greenan, Althea 045140
Greenbaum, Elizabeth R. .. 047052
Greenberg, Judy A. 054128
Greenberg, Suzy 050947
Greene, A. Wilson 052013
Greene, Gerard 044625
Greene, Marilyn 044992
Greene, Dr. Patrick 000947
Greene, Virginia 052100
Greener, Peter 043449
Greenfield, Mark 050446
Greenhill, John 045956
Greenhow, Dr. Desna 043784
Greenlaw, William 005958
Greenlee, Hal 048021
Greenlee, Nina 050692
Greenlee, Sandy 047925
Greeno, Judy 007355
Greenway, Lord 045356
Greenwood, Gail .. 000765, 000836
Greenwood, Mike 045776
Greenwood, Tom 049949
Greer, Alberta 052616
Greer, Judy 052466
Greer, Kim E. 052900
Greer, Rina 007220
Greer, William 007187
Greffrath, Dr. Bettina .. 022739, 022742,
                            022746
Grefstad, Ola 034038, 034038
Gregersen, Bo 010298
Gregersen, R. 018980
Gregersen, Thomas 048232
Gregg, David 047217
Gregg, David W. 054351
Gregg, Rosalie 048206
Gregl, Dr. Zoran 009025
Gregoire, Jean 011699
Gregoire, Noël 003778
Gregoor, Yves 031989
Gregori i Berenguer, Joan .. 040328
Gregors, Tommy 046339
Gregory, Gail 007204
Gregory, Joseph .. 046939, 052040
Gregory, Linda 036172
Gregory, Susan 046938
Gregory, Tabitha 053925
Gregson, Brian 001127
Greguski, Eva 053548
Greif, B. 017774
Greif, Steve 051593
Greig, John 044052
Greimel, Susanne .. 002630, 002631
Grein, Gerd J. 021055
Greinwald, Dr. Elisabeth .. 001741
Greisenegger, Dr. Vana ... 003059
Grelier, Joseph 015408
Grella, Consalvo 025423
Grella, Prof. P. 027157
Grellier, Gérard 013633
Gremillet, E. 015903
Grémont, Pascale 015546
Grenda, Edward 006100
Grendi Ilharreborde, Paola .. 007530
Grenon, Prof. Ernesto A. .. 000633
Grenter, Steve 046206
Grenville, John H. 006096
Grépier, Gabriel 012756
Greppi, Graciela B. 000541
Grepstad, Ottar 033675
Greschny, Nicolaï 014035
Greshner, Barb 006668
Greve, Jürgen 019029
Grever, T.F. 031964

Grevisse, Edgard 003917
Grew, Dr. Priscilla C. ... 050302
Grewenig, Dr. Meinrad Maria .. 022349
Grexa, Dr. Susanne 017202
Grezet, Cédric 041243
Grezinger, August 020079
Gribble, Michael 033247
Gribbon, Deborah 050436
Grice, Jean 043704
Gricenko, Vladimir Petrovič .. 037960
Griebel, Dr. Rolf 020507
Griebler, Reinhold 002590
Grieder, Susanne 041788
Grierson, Dr. Manfred 022740
Griepink, Dr. B. 032755
Grießmair, Dr. Hans .. 025650, 025689
Griff, Hanna 051381
Griffeuille, Marie F. 015353
Griffeuille, Marie-Françoise .. 015351
Griffith Gregory, Diane .. 049579
Griffin, Dave 048257
Griffin, David J. 024888
Griffin, Elizabeth P. 049589
Griffin, Ian 045212
Griffin, James 047435
Griffin, Jennifer 048349
Griffin, Jonathan 044326
Griffin, Kjerstin 033972
Griffin, O.T. 044831
Griffioen, B. K. P. 032158
Griffis, Larry L. 050316
Griffith, Amanda 046548
Griffith, Ken 046715
Griffith, Lee Ellen 048967
Griffiths, A. 044923
Griffiths, Bill 045274
Griffiths, Caitlin 046141
Griffiths, Coralie 001205
Griffiths, D. 005657
Griffiths, D.F. 044882
Griffiths, David 009518
Griffiths, Dr. G. 043749
Griffiths, Jayne 043719
Griffiths, Jean 043997
Griffiths, Mary 045219
Griffiths, Pauline 045313
Griffiths, Peter 045395
Griffiths, R. John 045563
Griffo, Dr. Pietro 026843
Griggs, David W. 047438
Griggs, K. 001263
Grigo, Dr. M. 019685
Grigore, Nina 036276
Grigorian, Anelka 000733
Grigorian, Prof. Dr. Gurgen G. .. 037495
Grigorjeva, Tamara Andriïvna .. 043183
Grigoryan, Avedik 000752
Grigsby, Roger 047993
Gril du Guern, Cajetan ... 002605
Grimaldi, Dr. David 051327
Grimaldi, Floriano .. 026626, 026627
Grimaldi, Gloria 023403
Grimaldi, Louis A. 048451
Grimaldi, Stéphane 012275
Grimaud, Patricia 014484
Grimes, John 052880
Grimes, Richard 001326
Grimm, Albert 042122
Grimm, Dr. Alfred 020578
Grimm, Prof. Dr. Günter .. 022201
Grimm, Jakob 041819
Grimm, Tess C. 048645
Grimmer, Dietgard 002623
Grimod, Dr. Ivana 027702
Grimshaw, Sir Nicholas ... 045086
Grimsrud, Halvard 033894
Grinage, Jeanine 046344
Grinbaum, Blanche 012551
Grincevičiené, Zita 030369
Grinell, Sheila 052114
Grinenko, Lucy 043293
Griněv, Nikolaj N. 038016
Grinnell, Debbie 051161
Grinnell, Nancy 051538

Grinsted, Stephen 045102
Gripari 023161
Griset, Suzanne 053823
Grišina, Ekaterina Vasiljevna .. 037672
Griss, Victor 000917
Grissom, Don 050551
Griswold, Sarah 054519
Gritt, Stephen 006594
Grivegnee, M.R. 004009
Grjaznova, Ann 037675, 037788
Grob, Dr. B.W.J. 032581
Grobler, I. 038830
Groce, Prof. Aldo A. 000619
Groce, Dr. W. Todd 053176
Groden, Anke 021777
Grodskova, Tamara 037841
Grodwohl, Marc 016070
Grodzicki, Prof. Andrzej .. 035421
Gröger, Walter 002255
Groen, Ir.B. 032013
Groen, M.F. 032808
Groenendijk, Drs. G.H. ... 032157
Groeneveld, A. 032830
Groeneveld, A.C. 032859
Groenewald, A. 038660
Groenewald, M. 038773
Groeneweg, H. 032570, 032577
Groenewold, Rod 054005
Groenewoud, G.J.M.M. 032187
Grøngaard Jeppesen, Torben .. 010011,
            010253, 010254, 010257, 010258,
                                 010259
Grönholm, Kirsti 010832
Grønn Iversen, Trine 033813
Grønnegaard Jeppesen, Torben . 010260
Grønvold, Ulf 033825
Grøtan, Geir A. 033796
Grötsch, Dr. Kurt 040177
Groff, John M. 052104
Grohnert, René 018291
Grohs, Kurt 018075
Groiß, Dr. Albert 001702
Groleau, Michel 006874
Gromer, Kim 053707
Gromova, Ljudmila Ivanovna .. 037756
Grondman, Dr. A.C.M. 032216
Grondman, Drs. A.C.M. 032449, 033126
Gronert, Dr. Stefan 017495
Groninga, Janet 046305
Gronne, Claus 010188
Gronychová, Marie 009898
Groom, Simon 044865
Groos, Dr. Ulrike 018036
Grooss, Dr. K.S. 032581
Groot, E. de 032826
Groot, J.P. de 032386
Groot, Rolf de 020910
Groot, Dr. Susan 032575
Grootveld-Parrée, C. 032800
Grootveld, Cor 031990
Gros, C. 041533
Gros, Dr. Patrick 002625
Grosboll, Sue 047478
Groschner, Dr. Gabriele .. 002636
Grosdidier de Matons, Benoît .. 013134
Grosfeld, Dr. J.F. 032101
Grosfeld, J.F. 032100
Groß, Helmut 018218
Gross, Jennifer 051267
Groß, Klaus M. 021431
Gross, Martial 041728, 042070
Gross, Miriam 051356
Gross, Dr. Raphael ... 018430, 018442
Gross, Renee 051356
Groß, Simone 022386
Groß, Thea 020326
Grossbard, Elayne 046936
Grosshans, Dr. Rainald ... 017147
Großjohann, Heidrun 017380
Großkinsky, Dr. Manfred .. 018441
Grossman, Cissy 051367
Grossman, David J. 047877
Grossman, Grace Cohen 050468

Großmann, Prof. Dr. G. Ulrich .. 020853,
                           020854, 020875
Grossmann, Robert 015863
Grossmann, Wilhelm 019669
Grosspietsch, J. 022087
Grote, Dr. Dietmar 021017
Grote, Dr. Hans-Henning .. 022733
Grote, Dr. Udo . 020594, 020595, 022839
Grotkamp-Schepers, Dr. Barbara . 021897
Grotz, Helmut 022325
Grouwet, Carl 018046
Grove, David 052806
Grove, Jeffrey 047808
Grover, Ruth 047593
Grover, Ryan 048348
Groves, Bonnie Sue 048339
Grower, Steve 000929
Grozin, Vladimir 037289
Gruat, Philippe . 013534, 014313, 015793
Grubb, Tom 048720
Grube, Prof. Dr. NiKolai. . 017478
Grube, Susanne 017493
Grube, Walter 020620
Gruben, Dr.des Eva-Maria . 018047
Gruber, F. 002253
Gruber, Franz 002013, 002017
Gruber, Dr. Gabriele 017822
Gruber, Herbert 001789
Gruber, Dr. Herbert 018067
Gruber, Petra 021085
Gruber, Rosina 002012
Gruber, Sonja 002945
Gruber, Wolfgang 001801
Grubessi, Prof. Dr. Odino . 027612
Grubler, Mitchell 048766
Grubola, James 050498
Grubola, Prof. James 050485
Grüger, Dr. Andreas 022010
Grüner, Almut 044787
Grünewald, Dr. Mathilde .. 022764
Grünwald, Karin ... 020207, 020213
Grünwald, Manfred 002542
Grünwald, Prof. Michael .. 001915
Grünzweil, K. 045391
Grütter, Daniel 041950
Grütters, Prof. Monika ... 017223
Gruetzmacher, Mark 053671
Gruetzner, Sara Jane 005520
Gruffydd, Alun 043542, 044885
Gruhl, Ulrike 021183, 021200
Gruhs, Ernie 006015
Grujić, Dragana 038187
Grunberg, Evelina 004660
Grund, M. Maria 017139
Grund, Dr. Rainer 017998
Grundler, Raymond 013556
Grundmann, Dr. Kornelia .. 020275
Grunenberg, Christoph 044865
Grunz, Karin 017126
Grunzweil, Karen 045168
Grusman, V.M. 037816
Gruzdeva, Tatjana Aleksandrovna . 037904
Gruzska, Larysa Petrivna . 043329
Grygiel, Prof. Ryszard ... 034965
Grygorenko, Iryna Vasylivna . 043134
Gryka, Jan 034990
Grynszpan, Anna 034790
Grynsztejn, Madeleine 053025
Grys, Dr. Iwona 035361
Gryseels, Guido 004057
Gryspeert, Hans 003654
Grzegorczyk, Dr. Detlef .. 020602
Grzesiak, Marion 049819
Grześkowiak, Andrzej 034772
Grzymkowski, Andrzej 035018
Gschwantler, Dr. Kurt . 002982, 003026
Gschwendtner, Stefan 002330
Gstrein, Josef 002114
Gu, Feng 008468
Gu, Jianxiang 008112
Gu, Wenxia 008294
Guadarrama Guevara, Jorge . 031187
Gual Dolmau, Oriol 039065
Gualdoni, Flaminio 028276

Guardia, Gabriel de la 034176
Guardiola, Juan 039073
Guarin, Claudia 003036
Guarino, Dr. Ariberto 027636
Guarnieri, M. Chiara 026170
Guati Rojo, Prof. Alfredo 031245
Guay, Serge 006806
Guayasamín, Pablo 010486
Gubb, A.A. 038710
Gubbels, J.H.M. 032543
Gubčevskaja, Ljudmila
 Aleksandrovna 037889
Gube, Gundula 019468
Gubel, Eric 003492
Gubin, V.V. 037283
Gubler, Bernhard 041838
Gubser, Paul 042121
Gudelis, V. 030550
Guderzo, Dr. Mario 027429
Gudina, Jurij Volodymyrovyč ... 043248
Gudmundsdottir, Anna Lísa 024092
Gudnadóttir, Kristin 024097
Gudnason, Thorarinn 024079
Gudonytė, Alma 030533
Guedj, Gabrielle 013841
Gügel, Dominik 041916
Güiraldes, Consuelo 000473
Gül, Sebahat 042963
Guelbert de Rosenthal, Eva 000470
Güler, M. Güven 042971
Güllüce, Hamza 042903
Güllüoğlu, F. Nevin 043052
Günaydın, Celal 042876
Gündogdu, Veysel 042871
Gündüz, Yaşar 043017
Guenette, Irene 007089
Guennec, Aude Le 012570
Günter, Werner 020940
Güntert, Prof. Dr. Marcel 041321
Günther, Albert J. 018122
Guenther, Bruce 052302
Günther, Clas 040998
Günther, Michael 019761
Günther, Rolf 018502
Günther, Dr. Sigrid 001855
Guenther, Todd 050128
Günther, Wolfgang 021899
Günzler, Lilo 018426
Guerbabi, Ali 000080
Guerber, Steve 047076
Guerci, Prof. Antonio 026409
Guerdy Lissade, Joseph 023395
Guerin, Charles A. 053839
Guerin, Kerry 001385
Guermont, Mauricette 015548
Guerra Pizòn, Olga 010453
Guerra, Jairo 008823
Guerra, Yolanda 008823
Guerrero Cestero, Felipe 039455
Guerrero, Doris 008652
Guerrero, Javier 052979
Guerreschi, Prof. Antonio 027912
Guerri, Dr. Roberto ... 026808, 026811,
                               026813
Guersen, Pascale 014383
Gürtler, Dr. Eleonore 002141
Gürtler, Dr. W. 001861
Güse, Dr. Ernst-Gerhard 016972, 017942,
 018852, 019389, 021020, 022489,
 022491, 022495, 022496, 022497,
 022500, 022502, 022504, 022506,
 022507, 022508, 022509, 022510,
                               022512
Guest, Fiona 044737
Guest, Raechel 054199
Guevara Hernández, Jorge 031572
Guex, Stéphanie 041686
Güzel, Selahattin 042901
Guezi, Dr. Angela 028074
Guffee, Eddie 052190
Guffee, Patti 052190
Gugelberger, Rachel 051489
Guggenberger, Hannes 002194
Guggenberger, Hans 002229

Gugier, Pierre 012009
Guheen, Elizabeth 047793, 051096
Guhl, Klaus 041443
Guibal, Jean 011695
Guibert, Etienne 011890
Guichard, Christiane 015457
Guichard, Marie-Claire 013296
Guichard, Vincent 014240
Guichaumon, Jean-Louis 014007
Guichon, Françoise 011567
Guidon, Suzanne 012679
Guidotti, Prof. Francesco 026760
Guidotti, Nives 025663
Guidry, Keith 051788
Guiducci, Dr. Anna Maria 027961
Guiffre, Florence 048773
Guigo, Elsa 014697
Guigon, Emmanuel 015874
Guihaumé, Antoine 015460
Guiheneuf, Bernrd 015431
Guilardian, D. 003467
Guilbaud, Laurence 016154
Guilbert, Manon 005871
Guilbot 013324
Guilbride, Kurt 006150
Guild, John 053504
Guild, John W. 053498
Guilhot, Jean-Olivier 012898
Guillard, Alain 013842
Guillard, Maire-Joe 013378
Guillaume 004140
Guillaut, Laurent 012288
Guillemeteaud, François 012108
Guillemin, Daniel 011725, 013215
Guillemin, Jean-Claude 011993
Guilleminot, Jacques 012446
Guillemot, Hélène 013606
Guillén Jiménez, Marilú 031628
Guillen, Dr. Carlos B. 010598
Guillerme, Lucienne 014367
Guillermo, Joseph 034424
Guilliem Arroyo, Salvador 031206
Guilliod, Raymond 023354
Guillot 013677
Guimarães, Prof. Dr. Edi Mendes 004360
Guinand, Jean 041585
Guinid, Rosario 034490
Guinness, Alison 048608
Guinness, Robert 024990
Guiotto, Prof. Dr. Mario 028022
guiraud, Louis 016226
Guisolan, Dr. M. 042035
Gulbina, Anna Alekseevna 038041
Gulbrandson, Alvhild 033745
Guldemond, J. 032826
Guljaev, Vladimir Andreevič ... 037506
Gullan, Penny J. 048163
Gullickson, Nancy 048378
Gullickson, Rachel 053305
Gullsvåg, Per Morten 034077
Gulyásné Gömöri, Anikó 023482
Gulyes, Andreas 003056
Gumă, Nicoleta 036355, 036356
Gumnit, Daniel 050939
Gumpert, Lynn 051399
Gunarsa, Dr. Nyoman 024605
Gundacker, Reinhard 003145
Gundel, Dr. Marc 019157, 019160,
         019162, 019163, 019164
Gunderson, Catherine F. 006979
Gunderson, Dan 048183
Gundry, Beverly F. 049343
Gunér, Tullan . 040738, 040739, 040742
Gunjić, Ranko 008909
Gunn, Lorna 006070, 006071
Gunnarsson, Ann-Marie 040620
Gunnarsson, Annika 041021
Gunnarsson, Torsten 041024
Gunning, Judy 001503
Gunnlaugsdóttir, Elín 024065
Gunther, Uwe 038862
Gunz, Karin 017122
Guo, Erfu 007938
Guo, Kang 008311

Guo, Qinghua 008082
Guo, Xiankun 007667
Guo, Xingjian 007705
Guo, Yu'an 008354
Guo, Zhenghong 008289
Guohua, Gu 008115
Gupta Wiggers, Namita 052296
Gupta, Chitta Ranjandas 024180
Gupta, Preeti 048767
Gupta, V.N. 024344
Gupta, Dr. Vibha 024520
Guralnik, Nehama 025212
Guratzsch, Prof. Dr. Herwig ... 017662,
          018827, 021664, 021667
Gurba, Norma 050110, 050111
Gurevič, Dr. E. 037448
Gurkina, Tatjana Aleksandrovna 037218
Gurley, Diana 005377
Gurney-Pringle, Chantal 000802
Gurney, George 054160
Gurney, Helen 044579
Gurney, Marilyn 005954
Gursky, André 018958
Gusarov, Ass. Prof. Vladimir .. 033847
Guse, Ann 050748
Gusenleitner, Fritz 002316
Gusev, Vladimir 037743, 037744
Gushulak, Ruth 005809
Gusjeva, Svitlana Ivanivna 043262
Gust, Donald F. 046886
Gustafson, Donna 047841
Gustafson, Kelli 053365
Gustafsson Ekberg, Kent 040541
Gustafsson, Malin 040652
Gustafsson, Pierre 041143
Gustavson, Carrie 047013
Gustavson, Todd 052617
Gustin, Marcelo 000521
Gut, Andreas 018158
Gut, Heinrich 042134
Gut, Ursula 041960
Gut, Werner 042068
Gutbrod, Dr. Helga 020678
Gutek, Dr. František 038290
Guth-Dreyfus, Dr. Katia 041269
Guthardt, Dr. Wolfgang 022745
Guthier, Mark 050602
Guthrie, Alyce N. 049087
Guthrie, Giles 045193
Guthrie, Jeanne 047349
Guțică-Florescu, Laurențiu ... 036227,
   036369, 036616, 036720, 036721
Gutiérrez Alba, D. Julio 040178
Gutiérrez Campos, Yolanda
 Milagros 009390
Gutiérrez Carcia, Inés 009208
Gutiérrez Coto, Fernando 008849
Gutiérrez García, María Angeles 039795
Gutiérrez Salamanca, Carlos
 Julio 008727
Gutierrez-Solana, Carlos 047264
Gutiérrez Velásquez, Amado .. 008641
Gutiérrez, Ana 039676
Gutiérrez, Guillermo Giraldo .. 008712
Gutiérrez, José A. 000206
Gutierrez, Mario 052729
Gutiérrez, Raquel 040305
Gutiérrez, Reynaldo 004222
Gutker de Geus, Sven E. 032745
Gutknecht, Eugene 047024
Gutman, Kevel Jane 048363
Gutmann, Gabriela 002946
Guttormsen, Sissel 034036
Gutzmann, Dr. Ulrike 022740
Guy, Dr. A.J. 045056
Guy, Frances 043945
Guy, Grant 007433
Guy, Troger 014141
Guyader, Raymond 015012
Guye, Francis 041911
Guyonneau, Olivier 013441
Guyot, Philippe 015012
Guzman, Antonio 016095
Guzmán, Antonio 014401

Guzmán, Manuel Gil 031185
Guzman, Rene de 053032
Guzzi, B. Anthony 049453
Guzzo, Enrico Maria 028346
Guzzo, Prof. Pietro Giovanni .. 025826,
          026157, 027393, 028148
Gvišiani, G.M. 016405
Gwan, Pak Yong 029881
Gwara, Czesław 035208
Gweki Booi, Vuyani 038644
Gwiazda, Bronislaw 035104
Gwilim, Ian 043346
Gwyn, Brenda 050696
Gye, Peggy 045231
Gyergyádesz, László 023722
Gygax, Christian 041238
Gygax, Raphael 042211
Gyger, Patrick 042172
Gyger, Robert 042098
Gyllenberg Pernevi, Bodil 040660
Gyllenhaal, C. Edward 047291
Gyllkrans, Monika 041078
Gyöngyössy, Márton 023517
Györe, Zsófia 023405
Györgyné, Dr. Varga László .. 023593
Gyorody, Ninette 006087
Gysin, Jacques 041785, 041787
Gyuchané Veres, Krisztina ... 023442
Gyula, Dr. Ernyey 023514
Gyulay, Endre 023923
Ha, Hong-Chien 042474
Ha, Paul 052796
Haack, Jan 019006
Haag, Angela 053768
Haag, Olinde S. 051313
Haag, Dr. Sabine 003026
Haag, Tilla 019818
Haak, Dr. Bill 049131
Haakanson, Sven 050016
Haake, Susan M. 053452
Haakestad, Jorunn 033538
Haaland, Anders 033537
Haamer, Kaia 010682
Haan, David de 044649
Haan, Maartje de 032186
Haan, Drs. Peter de 032957
Haapala, Anu 011514
Haar, Dr. Frauke von der 017585
Haar, Dr. Frauke von der 017594
Haas, Gordon 048662
Haas, Prof. Dr. Helmuth 002977
Haas, James R. 048215, 048216,
                               048217
Haas, Josef 009541
Haas, Margarete 001991, 001992
Haas, Dr. Mechthild 017822
Haas, Dr. Richard 017112
Haas, Ruth 054445
Haas, Sabine 019873
Haas, Dr. U. 019765
Haase, Dr. Evelin 017566
Haase, Werner 022799
Haavaldsen, Per 033972
Habartova, Dr. Romana 009931
Habditch, Max 045174
Habegger, Dr. Ueli 041712
Habenicht, Wilfried 016711
Haber, Michael 003036
Haberhauer, Günther ... 016907, 016908,
                               016910
Haberman, Patty 050811, 053706
Habermann, Dr. Sylvia 016989
Habermann, Yvonne 021552
Habersatter, Dr. Thomas 002636
Haberstroh, Sarah ... 050911, 050925
Habes, Scott 047869
Habeshaw, Julia 045489
Habichler 002781
Habrich, Prof. Dr. Dr. Christa . 019408
Habsburg, Francesca von 003091
Hacala, Mark T. 054171
Hachet, Michel 015975
Hachmer, Hendrik 032972
Hack, Dr. C. 032488

Hack, Janice 050077
Hack, Shery N. 047405
Hackenschmidt, Sebastian ... 003033
Hacker, Heinrich 018368
Hackett, Brian 048172
Hackett, I. O. M. 045421
Hackford, Clive 044095
Hackl, Prof. Dr. A. 002909
Hackler, Cornelia 021788
Hackman, John 011440
Hacksley, M.M. 038725
Hacksley, Malcolm 038694
Hackstein, Hermann 018912
Haddad, Regina Clélia 004942
Hadders, Rieks 031939
Haddington, Earl of 044448
Haddock, Brent 052358
Hadjicosti, Dr. M. 009414
Hadjisavvas, Dr. Sophocles ... 009407,
          009417, 009424, 009426
Hadjusarnas, S. 009406
Hadland, Brian 006720
Hadland, Tony 046050
Hadley, Diana 053823
Hadley, V.J. 044022
Hadsbjerg, Lene 010281
Haduch, Henryka 034623
Häberli, Hansjörg 041220
Haeck, Jules 003599
Häder, Dr. Ulf 017674
Haedo, Dr. José Antonio 000598
Häffner, Martin 022210
Häffliger, Toni 041703
Haegen, Erik van der 004148
Häggman, Gunvor 011207
Häggström, Ralf 010800
Häggström, Stig 010962
Haenel, Johanna 001748
Haenel, Dr. Matthias 020507
Hänggi, Dr. Ambros 041282
Hänggi, Dr. Christoph 041976
Hänggi, René 041370
Hänisch, Sabine 020469
Hänsel, Dr. Alix 017245
Hänssler, Erica 042235
Häring, Dr. Friedhelm . 018689, 018691,
                      018694
Häring, Simona 022731
Haese, Josette d' 012321
Haese, Jürgen 018713
Haesler, Andreas 022898
Häuser, Dr. Christoph 022056
Häusle, Magdalena 001794
Häusler, Anita 041508
Häußler, Theodor 016686
Hafen, M.A. Thomas 018901
Hafenecker, Ingo 018077
Hafer, Mary K. 053978
Haffenden, Peter 001255
Haffner, Dr. Marianne 042225
Haffner, Zsoltné 023666
Hafner, Dr. Mark S. 046819
Hafok, Barbara 003089
Hagan, Annemarie ... 006338, 006340
Hagan, Marylee 053718, 053721
Hagberg, Magnus ... 040533, 041004,
                      041015
Hagdorn, Dr. Hans 019401
Hagedorn Olsen, Claus 010142
Hagedorn-Saupe, Prof. Monika . 017178
Hagel, Frank von 017178
Hagemeier, Anna Mae 053519
Hagen, Dr. Harold H. 053219
Hagen, John 049314
Hagen, Rien 031939
Hagendorfer, Dr. Johann 002092
Hagenmaier, Peter 019254
Hager, H. 038732
Hager, Helmuth 021269
Hager, Joyce 051116
Hagerup, Asbjørn 033797
Haggart, M.W. 006280
Haggenmiller, Sharon 049439
Haggerty, John 045001

Haglund, Elisabet 040573
Hagmark-Cooper, Dr. Hanna ... 011144
Hagnau, Dr. Carola 019706
Hagström, Lars 040784
Hague, Alan 044351
Hahling, Albert 041202
Hahn Pedersen, Morten 010045
Hahn, A.K. 017941
Hahn, Dr. Andreas 017479
Hahn, Cheryl H. 054565
Hahn, Prof. Dr. Dietbert 022800
Hahn, Donna 052218
Hahn, Dr. Sylvia 018500
Hahner, Cathi 047084
Hahnl, Prof. Dr. Adolf 002373
Haider, Adolf 003147, 003148
Haider, Prof. Gertrude 002837
Haider, Johann 003122
Haigh, Brian 043530
Haight, Pete 054479
Haiko, Pertti 011103
Haila, Leena 011033
Haimerl, Elfriede 017338
Hain, Dr. Beatrix 003089
Hainard, Dr. Jacques 041533
Haine, Heiko 018146
Haine, Prof. Dr. Malou 003474
Haine, Marie-Louise 003492
Haines, Alan P. 047171
Haines, Jeniffer 046653
Hainfellner, Gerhard 002818
Hair, Denny G. 049582
Hair, Julia 049370
Haja, Simion 036820, 036821
Hajdamach, Charles 044587
Hajdamach, Charles R. 044147
Hajduk, Stjepan 009015
Hájek, A. 009827
Hájíček, Dr. Jaroslav 038339
Hakamata, Akihiro 028600
Hakansson, Roger 040663
Hakasalo, Kalevi 011167, 011168
Hake, Helga 021129
Hakes, Jay E. 046628
Hakins, Judy 046461
Hakobyan, Lilit 000749
Hakoda, Chiho 029020
Halaas, Kristin 033788
Halamish, Uzi 025138
Halberstadt, Benjamin 017974
Halbmayerl, Prof. Dr. Ernst ... 020277
Halbreich, Kathy 050948
Haldemann, Dr. Anita 041273
Haldemann, Dr. Matthias ... 042250
Haldimann, Marc-André ... 041530
Haldner, Bruno 041278
Haldo, Soeren 033752
Haldy, Lanny 046432
Hale, Donald G. 052117
Hale, James K. 044559
Hale, Pat 048392
Halén, Widar 033840
Halevy, Maya 025071
Haley, Helen 006371
Haley, Melissa 051485
Halfpenny, Geoff 054436
Hall Miller, Thelma 052563
Hall-Patton, Mark P. 049441
Hall, Alma 054511
Hall, Anne M. 053031
Hall, Brian 043795
Hall, Charles W. 049595
Hall, Eli 034035
Hall, Fay 001634
Hall, Gail 050258
Hall, Gary 044017
Hall, Herbert 052320
Hall, Jenny 043531
Hall, Jim 048833
Hall, Dr. Joe B. 050320
Hall, Mark 045486, 050731
Hall, Marlene 047378
Hall, Melissa 045334
Hall, Michael D. 050829

Hall, Michael S. 049560
Hall, Nancy 054465
Hall, Peter J. 038751
Hall, Philip 045468
Hall, Robert 044613
Hall, Robin 045237
Halla, Dr. Jaana 010819
Hallam, Doug 007002
Hallan, Knut Christian ... 033849
Hallas-Murula, Karin 010667
Hallatt, Kim 052618
Hallberg, Sara 051691
Halldorsson, Gudni ... 024082, 024084
Hallé, Antoinette 015795
Halle, Dr. Uta 017585
Haller, Prof. Dr. Heinrich . 042177
Haller, Dr. Kerstin 017602
Haller, P. 016484
Hallett, Vic 045971
Hallgrimsdóttir, Margrét ... 024107
Hallin, Jens 010138
Halling, Finn 033960
Hallman, Charles V. 053899
Hallman, Christina 041026
Hallmark, Dr. Donald P. .. 053450
Hallock, Robert 048047
Hallonová, K. 038486
Hallová, Dr. Markéta . 009650, 009714
Halls, E.J. 043569
Halma, Sidney 051555
Halmágyi, Dr. Pál 023768, 023769
Halmová, Dr. Mária ... 038389, 038394
Halmyradow, Ahmed 043092
Halonen, Maija-Liisa 011388
Halota, Marianna 034904
Halouzková, Marcela 009965
Halsey Gardiner, Margaret .. 051423
Halstead, Graham 043620
Halter, Marc 013276
Halusa, Gerhard 003056
Halvorsen, Prof. Rune ... 033847
Ham, Mrs. Robert 048601
Hama, Atsushi 028516
Hamada, Naotsugu 029361
Hamaguchi, Dan 053005
Hamanishi, Koji 029622
Hamann, Dr. Klaus 019044
Hamano, Tokuei 029370
Hamberger, M.A. Heike ... 019660
Hambolu, Dr. M.O. 033437
Hamby, Elaine 054427
Hamed, Dr. Dawi Musa ... 040498
Hamed, Mohammed 040494
Hamel, Bettina 041774
Hamel, W.A. 032202
Hamelin, Joël 014888
Hamelmann, M.A. Cordula . 017907
Hamer, Diana 000534
Hamer, Fritz 047918
Hamil, Lisa 049306
Hamilton-Smith, Katherine .. 054230
Hamilton, Duke of 044493
Hamilton, Barbara 045486
Hamilton, Catherine 044955
Hamilton, Dan 048117
Hamilton, David 047547
Hamilton, Elaine 033302
Hamilton, James 043613, 043630
Hamilton, Dr. James ... 043619
Hamilton, John 000902
Hamilton, Kathleen 052979
Hamilton, Kelly 050601
Hamilton, Laura 044395
Hamilton, Mary 047422
Hamilton, Mary Ellin ... 046789
Hamilton, Mort 046525
Hamilton, Olivia 045526
Hamilton, R.V. 001215
Hamilton, Roy 050422
Hamilton, W. 033212
Hammell, Peter 053157
Hammer, Anders 040826
Hammer, Bettina 017829
Hammer, Dr. Karen-Elisabeth .. 022771

Hammer, Dr. Karen Elisabeth .. 021027,
                021028, 022769
Hammer, Victoria 002125
Hammer, William R. 052630
Hammerl, M. A. Tobias 016457
Hammerle-McGuire, Peggy ... 051447
Hammermann, Dr. Gabriele ... 017797
Hammerschmidt, Elmar 019426
Hammerschmied, Gerhard ... 002846
Hammersla, Keith E. 050725
Hammerstrom, Kirsten 052395
Hammond, Barbara M. ... 047740
Hammond, Eleanor A. 047149
Hammond, Gary R. 046721
Hammond, Jennifer . 044591, 049454
Hammond, Karen L. 049012
Hammond, Dr. Michael ... 051895
Hammond, Niall 043635
Hammond, Stacie 051875
Hammond, Theresa N. ... 049252
Hammonds, Kit 045108
Hamoniaux, Jacky 012794
Hamooya, C. 054840
Hampe, Henrike 022258
Hampshire, Camilla 044312
Hampson, Louise 046231
Hampton-Porter, Joan ... 053467
Hampton, Jan 033176
Hampton, Joel L. 048023
Hampton, John 046183
Hamraeva, G. 054656
Hamran, Ulf 033500, 033502
Hamre, Harald 033985
Hamrin, Örjan 040620
Hamshere, John 045769, 046573
Hamzabegović, Amra ... 004283
Hamzaev, A. 054632
Han Keuk 029912
Han, Guangyun 008045
Han, Jinke 007773
Han, Liji 008360
Han, Yong 007600, 007646
Han, Zhenfei 007788
Han, Zongxiang 007567
Hanau, E. 038852
Hancock, Robert 052557
Hand Evans, Dana ... 052566
Handel, Elke 003074
Handler, Jeremy 047826
Handley, Jerry 047926
Handley, Neil 044924
Handley, Pamela 006508
Handlos, Helga 020084
Hando, Justin 042627
Handy, Ilke 021678
Hanebutt-Benz, Dr. Eva-Maria . 020226
Hanejko, Eugeniusz ... 035274
Hanemann, Dr. Regina . 016948, 016955,
                      016959
Hanemza, Jarett 052710
Hanes, Alice C. 052323, 052326
Hanfstengel, E. 018123
Hangl 022438
Hanhardt, John 051470
Hanhart, Karin 007075
Hanisak, Dr. Dennis ... 048856
Hanisch, Heinz 016440
Hankel, Lillian 000967
Hanken, James 047369
Hankewych, Jaroslaw J. .. 047704
Hankins, Evelyn 047328
Hankins, Roger ... 048161, 054545
Hanley, Joanne 048706
Hanley, Robin ... 044705, 044707
Hanley, Theresa ... 051781
Hanna, Annette Adrian .. 048278
Hanna, Dan 048568
Hanna, Emily 047005
Hanna, George 046828
Hanna, John 054592
Hanna, Kathleen ... 053014
Hanna, Martha ... 006577, 006594
Hanna, Michael ... 049985
Hanna, Richard 051451

Hannah, Dr. Michael 033342
Hannan, Sarah 047862
Hannay, Alf 043423
Hannay, Rosemary 045470
Hannecart, Jean 012582
Hanner, Frank 048780
Hannesen, Dr. Hans Gerhard . . . 017066
Hannibal, Dr. Joseph 047809
Hannick, P. . . . 003321, 003322, 003323
Hanniffy, Constance 024823
Hanniffy, Kieran 024823
Hannula, Betsy 048738
Hannus, L. Adrien 053354
Hanor, Stephanie 050040
Hanot, Nicole 003697
Hanow, Orazmyrat 043091
Hanrahan, Kim 051534
Hanreich, Georg 002530, 002532
Hanreich, Liselotte 002530
Hansch, Dr. Wolfgang . 019162, 019164
Hansel, Anna-Elisabeth 018841
Hanselman, David S. 048809
Hansen, Arlette 048188
Hansen, Arne 005881
Hansen, Barbara 052360
Hansen, David K. 053947
Hansen, Don 052004
Hansen, Dr. Dorothee 017592
Hansen, E.P.H. 004028
Hansen, Emma 047855
Hansen, Eva Maria 052108
Hansen, Frank 005422
Hansen, Dr. Harriet M. 010232
Hansen, Jesper Vang 010194
Hansen, Jörg 018131
Hansen, Klaus Georg 023340
Hansen, Larry 049140
Hansen, Laura R. 048439
Hansen, Maiken 010280, 010345
Hansen, Mogens 010010, 010086,
                    010087, 010087, 010363
Hansen, Moya 048255
Hansen, Palle Birk 010233
Hansen, Prof. Pearl 054249
Hansen, Shannon 052245
Hansen, Sidsel 033921, 033922
Hansen, Dr. Steven D. 053349
Hansen, Thor Gunnar 033765
Hansen, Prof. Dr. Dr. Volkmar . . 018029
Hansen, Wayne 047628
Hansford, Michele 047452
Hansing, Anke 020985
Hansley, Kevin 051389
Hanslok, Andreas 017917
Hanson, Brent 052770
Hanson, Christopher 049945
Hanson, Collen W. 053273
Hanson, Dee 046648
Hanson, Georgia 046580
Hanson, Gloria 048490
Hanson, Dr. John 049505
Hanson, Luther D. 048834
Hanson, Stephen 047551
Hanssen-Bauer, Françoise 033845
Hanssen, L.H. 032830
Hanssens, Marjolaine 004007
Hansson, Lars 041149
Hanta, Dr. Rajoharison 030650
Hantschmann, Dr. Katharina . . . 020509,
                                        020921
Hanus-Möller, Christl 021633
Hanzel-Wachter, Dr. Lieselotte . . 002724
Hao, Liangzhen 007835
Hao, Liguo 007951
Hapkemeyer, Andreas 025580
Happe, Michael 019302
Haque, Dr. Enamul 003235
Hara, Fumihiko 029129
Hara, Toshio 029371, 029512
Harada, Heisaku 029020
Harada, Minoru 028642
Harbige, Sarah 043454
Harbin, Melanie M. 049137
Harck, Lorraine A. 050057

Hard, Michael W. 048368
Hardacre, John 046157
Hardardottir, Asa S. 024069
Hardberger, Linda 052940
Harden, Bob 053759
Harden, Mike 046561
Harden, R. 043521
Harden, Stan R. 049475
Harder, H. 032532
Harder, Katie 005707
Harder, Dr. Matthias . . 017169, 018719
Harder, Dr. Ruth E. 041956
Hardesty, Sarah 053826
Hardiati, Dr. Endang 024581
Hardie, Andrea 044542
Hardiman, Thomás 024925
Hardiman, Tom 052314
Hardin, Jennifer 052863
Harding-Davis, Elise 005332
Harding, Keith 045392
Harding, Maureen 054291
Harding, Shelley 006671
Hardingham, Dennis 044188
Hardjonagoro, K.R.T. 024660
Hardman, Genee 052690
Hardt, Nis 017814
Hardung, Klaus 021850
Hardwick, Tammy 005664
Hardy-Marais, Laurence 014083
Hardy, Alwyn 046053
Hardy, Anne 046179
Hardy, Channing C. 048526
Hardy, Dr. Florent 046816
Hare, Sandra J.. 049941
Harent, Sophie 014405
Harger, Sara 048931
Hargittai, Emil 023495
Harinck, Prof. Dr. G. 031944
Harinen, Dot 006693
Harju, Jouni 011042
Harker, Robert 053297
Harkins, Michael J. 046797
Harkiw, Michael 044058
Harleman, Kathleen 046887
Harlen, Willie 052878
Harling, Nick 043641, 043642
Harlow, Ray 045736
Harltow, Harvey 007058
Harman, Fred C. 051889
Harman, Malcolm 043604
Harman, Robert L.B. 048692
Harmon, Darla 049036
Harmon, Horace E. 050262
Harmon, Dr. J. Scott 046503
Harmon, Mark S. 048361
Harms, Annette 022821
Harms, Dr. Eduard 019518
Harmsworth, Lord 044960
Harn, Alan D. 050235
Harnesk, Helena 041118
Harni, Heli 011453
Harnly, Marc 050436
Harnoncourt, Anne d' . 052081, 052090
Haro, Jambo 029869
Harold, Steve 050644
Harper Lee, Laura 047157
Harper, Avis 046802
Harper, Mary Kay 052653
Harper, Robert W. 052743, 052745
Harpley, Melissa 001401
Harr, Jemima 048627
Harrand, Peggy 052004
Harre Hérin, A. 003866
Harreither, Dr. Reinhardt 001870
Harreld Love, Josephine 048307
Harrell, Deloris 050658
Harrer, Anton 002382
Harrer, Nicole 003056
Harrewijn, H. 032347
Harriett, Rebecca 049358
Harrington, Bill 001300
Harrington, Caren Anne 046874
Harrington, Dennis 052589
Harrington, Gary 049719

Harrington, Katie 049428
Harrington, Sian 044659
Harris-Fernandez, Al 053348
Harris, Adam Duncan 049764
Harris, Amy S. 046488
Harris, Dr. C. 045459
Harris, Carter 047904
Harris, Dee 052898
Harris, Donna 052455
Harris, Dr. E.C. 004193
Harris, Ed 046876
Harris, Florence E. 053490
Harris, Gene E. 047506
Harris, George 006717
Harris, Ivan 007167
Harris, James T. 046786
Harris, Jane 051330
Harris, Dr. Jennifer 045219
Harris, Joe 049474
Harris, Julie 051885
Harris, Kara 052740
Harris, Kaye 001132
Harris, Kevin 052259
Harris, L. 045210
Harris, Mac R. 050155
Harris, Miriam B. 048794
Harris, Ottawa 048250
Harris, René 048509
Harris, Rhian 044972
Harris, Richard 043947
Harris, Sandra 046941
Harris, Scott H. 051283
Harris, Tina B. 049796
Harris, Tom 048919
Harrison, Camilla 046226
Harrison, Christina 045302
Harrison, Don 051831
Harrison, Elizabeth 051112
Harrison, Gordon 046414
Harrison, Harry 052024
Harrison, Dr. J. 038860
Harrison, Jefferson 051573
Harrison, John 045995
Harrison, Keith 044803
Harrison, Loretta 051528
Harrison, Lynette 001565
Harrison, Marie 033320
Harrison, Michael 043843
Harrison, Monica 050601
Harrison, Myra 047246, 047247
Harrison, Stephen . . . 025006, 025007,
        043472, 043889, 043891, 043892,
        043893, 044769, 045471, 045472,
                            045534, 045535
Harrison, Susan . . . 044597, 045617
Harriss, Joan 007115
Hârşan, Aurel 036462
Harsch, René 041236
Hart Hennessey, Maureen 053537
Hart, Brooks 006512
Hart, John 046344
Hart, Karen 049569
Hart, Katherine 049353
Hart, Macy B. 049754
Hart, Nancy 048922
Hart, Peter R. 044195
Hart, Robert L. 048620
Hart, Susan E. 054444
Hart, William 052793
Harte, Jeremy 044309
Harte, Miriam 043538
Hartelman, Bart 032556
Hartemink, A.E. 033020
Hartfield, Libby 049752
Harth, Arno 022349
Harthorn, Sandy 047073
Hartke, Becki 052810
Hartl, Christa 021084
Hartl, Georg L. 017341
Hartl, Dr. Nina 017098
Hartleb, Silvia Waltraud 003139
Hartleib, Dr. Rudolf 018439
Hartley, C. 043842
Hartley, David B. 052129

Hartley, Keith 044273
Hartley, Kristin 052733
Hartman, Bruce 051846
Hartman, C. 038778
Hartman, Diane 053839
Hartman, Janice 046425
Hartmann, Andy 041936
Hartmann, Bonnie 053584
Hartmann, Christine 020871
Hartmann, Emil 041953
Hartmann, Felix 041936
Hartmann, Joseph 018360
Hartmann, K. 021472
Hartmann, Dr. Ralf F. 017135
Hartnigk-Kümmel, Dr. Christa . . 019507
Harto, Edward S. 050900
Hartog, Dr. Arie 017587
Hartshorn, Gary 052293
Hartshorn, Willis 051406
Hartung, Lee 049126
Hartweger, Hans 003056
Hartwell, Carroll T. 050944
Hartwell, Jane 045039
Hartwich, Bernd 021106
Hartwick, Sue 048779
Hartwig, Laurie 052140
Hartwig, Paul B. 036154
Hartz, Jill 047585
Hartz, Dr. Sönke 021663
Hartz, Susanne 010146
Hartzell, Matthew D. 048241
Hartzold, Susan 047044
Harun, Rahimie 030683
Harvey, A.M. 038788
Harvey, Liz 050376
Harvey, Dr. Mark 001403
Harvey, Matt 045712
Harvey, Paul 047971
Harvey, Phyllis 051435
Harville, Vickie 050526
Harwell, Marcelle 051446
Harwood, Barry 047253
Haryu, Tsutomu 028915
Haschke, Heiner 021830
Haşdeu, N. Titus 036271
Hasegawa, Koichi 028764
Hasegawa, Mariko 029513
Hasegawa, Saburo 029358
Hasegawa, Sakae 029571
Hasegawa, Tei 029305
Hasegawa, Tokushichi 028790
Hasert, Eva-Maria 022585
Hashagen, Joanna 043495
Hashimoto, Akinobu 028469
Hashimoto, Kentarou 029569
Hashimoto, Masao 028915
Hashimoto, Ryutaro 029604
Hashimov, Bahodir Djuraevič . . 054631
Hashitomi, Hiroki 028911
Hashtroodi, A. 024739
Haskell, Barbara 051491
Haskell, Dr. Eric T. 047755
Haskell, Heather 053460, 053461
Håskoll Haugen, Liv . . 033704, 033706,
                                        033708
Haslauer, Dr. Wilfried 002626
Hashlehner, Hubert 002505
Hasler, Margrit 041554
Hasler, Norbert W. 030303, 030308
Hasprunar, Jill 048935
Hasque, Jürgen de 020847
Haß, M.A. Heike 019132
Hassan, Syed Jamal . 024219, 024223
Hassbecker, Egon 019141
Hassdenteufel, Dr. Erwin 020225
Hasse, Dr. Claus-Peter 020207
Hasselbalch, Kurt 047368
Hassinen, Esa 011073
Hassler, Donna 053813
Hassler, Hermann 030309
Hassmann, Dr. Elisabeth 003094
Hassol, Milton D. 051451
Hasson, Rachel 025089
Hassrick, Peter H. 048258

Hastaba, Dr. Ellen 002141
Hastedt, Catherine A. 047873
Hastenteufel, Michel 013576
Hastings, Lord 045751
Hastings, Bill 050340
Hastreiter-Lamb, Lisa 053839
Hasvold, Carol 048208
Hata, Nobuyuki 028629
Hatchell, Steven J. ... 047888, 047890
Hatcher, Alexandra 006842
Hatcher, Alison 047043
Hatcher, Jennifer 047374
Hatcher, Stephen 044039
Hatfield, P. 046164
Hathaway, Merle 001154
Hatibu, Hadji 042637
Hatlen, Jean 005771
Hatoum, Hassan 042264
Hatschek, Christoph 003004
Hattam, Edward 001219
Hattingh, J. 038664
Hatton Davidson, J. 045379
Hatzitaki-Kapsomenou, Prof. Ch. 023290
Hau, Hal 049989
Hauck, Richard 051718
Hauck, Dr. Wolfgang 017652
Haudiquet, Annette 013646
Haudot, Charles 013450, 013451
Hauer, Armin 018400
Hauer, Karl 002966
Hauer, Ulrich 018943
Hauerberg, Louiswe 010146
Hauerová, Dr. Stanislava 009923
Hauff, Rolf Bernhard 019327
Haufschild, Kurt 021831
Haug, Prof. Dr. Andreas 018262
Hauge, Tore .. 033776, 033778, 033779,
                         033915, 033958
Haugen, Charleen 052678
Hauger, Therese 033593
Haugg, Hans 017708
Hauglid, Anders Ole .. 033608, 033742
Hauk, Roman 035072
Haumesser, Giselher 019494
Haunerland, Prof. Dr. W. 020542
Haupenthal, Dr. Uwe .. 019369, 020788
Haupt, Eva 019561
Haupt, Dr. Herbert 003026
Hauptmann, Prof. Dr. Andreas .. 017433
Hauregard, Léon J. 003272
Haurie, Michel 014593
Haus, Anny-Claire 015862
Hausammann, Hans Peter 042027
Hauschild, Dr. Jan-Christoph .. 018030
Hauschke, Dr. Norbert 018960
Hausding, Michaela 017988
Hauser-Schäublin, Prof. Dr.
  Brigitta 018747
Hauser, Claire 042038
Hauser, Emil 041601
Hauser, Erin 054025
Hauser, Hansgeorg 019785
Hauser, Larry 049444
Hauser, Monika 021478
Hauser, Peter 042203
Hauser, Robert 051225
Hauser, Rudolf 041803
Hauser, Urban 041363
Hauser, Dr. Zoltán 023615
Hausmann, Doris 051933
Hauss, M. 013309
Haussonville, Comte d' 041426
Haustein-Bartsch, Dr. Eva 021306
Hauswald, Dr. Kenneth 046562
Hautekeete, Stefaan 003459
Havard, Carine 003938
Havas, Sandra H. 051713
Havasi, Bálint 023737
Havassy, Péter 023672
Havassy, Dr. Péter ... 023670, 023671,
  023675, 023676, 023677, 023678,
                         023680, 023681
Havel, Joseph 049588

Havelková, Miloslava .. 009796, 009824
Havelund, Jørgen 010279
Havens, Patricia 053345
Havens, Ron 049244
Haveri-Nieminen, Marja-Liisa .. 011349
Haverkamp, Frode 033845
Havlíček, Jaroslav 009884
Havlíčková, Dr. Miloslava 009692
Havlik, Betty 049481
Havlík, Dr. Vlastimil 009502
Havron, Rina 025025
Haw, Jane 051943
Hawadi, T. 054849
Hawelka, Dr. Walter 018655
Hawell, Dr. Stevan 053229
Hawes, Irene 053887
Hawk, Alan 054146
Hawke-Smith, Cameron 043838
Hawke, Jenny 038712, 038713
Hawkey, Diane 053701
Hawkins, Andrea 045219
Hawkins, Lynda 048041
Hawkins, Marilyn 050548
Hawkins, Renee 050392
Hawkins, Dr. Ruth 049839
Hawkins, Dr. Seth C. 052842
Hawksworth, Susan 052898
Hawley, Anne 047117
Hawley, Greg 049893
Hawley, Liza 052043
Hawley, Pam 005817
Haworth, Brenda 049467
Hawvermale, Hettie G. 046943
Haxhiu, Prof. Dr. Idriz 000033
Haxton, Jason 049994
Hay, Irma 044911
Hay, Malcolm 045070
Hay, Mary 044656
Hay, Patricia 046847
Hay, Susan 052387
Haya, Dan 053673
Hayashi, Hirotumi 029539
Hayashi, Kiichiro 028708
Hayashi, Masayoshi 029740
Hayashi, Misa 029507
Hayashi, Toshihiko 029639
Hayashibara, Ken 029212
Haycock, Dr. Lorna 044096
Hayel, Janice 046057
Hayes, C. Daniel 048655
Hayes, David 043880
Hayes, Emily 048633
Hayes, Fiona 044411
Hayes, Neila C. 053333
Hayes, Randall 051761
Hayes, Sally 043583
Hayes, Virginia B. 053490
Hayhow, Simon 044349
Haynes, Bruce 005985
Haynes, Jackie 046482
Haynes, John H. 045846
Haynes, Peter 000935, 001572
Haynes, Tom 044314
Haynes, Valerie 052234
Haynie, Ron 054176
Hayoz, Roger 041768
Hays, Bonnie J. ... 051898, 051900,
                                 051901
Hays, George W.S. 052886
Haysom, David 045931
Hayter, Stephen 005463
Hayward, Joanna 045689
Hayward, M.R. 044646
Hayward, Ron 033138
Hayward, Susan P. 045916
Haywood, Dr. C. 024909
Hayworth, Ross 043362
Hazelrigg, Scott 046565
Hažír, Dušan 038431
He, Fanf 007596
He, Hongyuan 007956
He, Kezhou 008087
He, Nonglin 007730
He, Yinzhong 008291

He, Yu 007742
He, Zhongyong 008204
Headley, Dr. David 052249
Heady, Karen 051832
Healey, Mary 053465
Heaman, Ken .005968, 005971, 005978
Hearn, Brian 051734
Hearn, Redell 050411
Hearn, Robert 049078
Heath, Betty 053518
Heath, Hether 007062
Heath, Sophie 043977
Heathcote, Richard ... 001059, 001513
Heather, Robert 001237
Heaton, Alice 046788
Heaton, Colin 045235
Heaume, R.L. 044358
Heaver, Stephen G. 050528
Heazlewood, Brenton 001636
Hebblethwaite, Christine 044097
Hebden, M.J. 044473
Hebestreit, Prof. Dr.-Ing. Andreas 019978
Hebestreit, Dr. Franz 020572
Hebler, Ingo 017644
Hebron, Victoria 046159
Hecher, Markus 022184
Hecht, Dr. Winfried 021529
Hećimović, Branko 009044
Heck, M.A. Brigitte 017631
Heck, Brigitte 019505
Heckenbenner, Dominique 015695
Heckenberg, Dr. Norman 001476
Heckert, Dr. Uwe 022376
Heckes, Jürgen 017433
Heckl, Prof. Dr. Wolfgang M. .. 017487,
                  020516, 020517, 020919
Heckscher, Morrison H. 051424
Hectors, Jan 003583, 003584
Heddon, Gordon 044890
Heddy, Suzanne 048179
Hedegaard, Esben ... 010332, 010334,
                         010336, 010337
Hedel-Samson, Brigitte 012044
Hedelin, Jean-Louis 011790
Heder, Terence 050806
Hedgecock, Deborah 044927
Hedger, Michael 000930
Hedin, Marika 041027
Hedin, Svante 040920
Hedinger, Prof. Dr. Bärbel 018994,
                                 018997
Hedlund, John 047441
Hedrick, Nancy 050237
Heege, Dr. Elke 018130
Heeley, H.F. 045331
Heer, Ed de 031960
Heerebout, G.R. 032538
Heerebout, Dr. G.R. 032355
Heerink, B. 032388
Heermann, Dr. Ingrid 022048
Heese, Dr. Thorsten 021015
Hefermehl, Norbert 017371
Hefferhan, Tom 001293
Heffern, Marion 006965
Heffley, Scott 049888
Heffner, Jinger 050409
Heffner, Sarah 049359
Heflin, Kirk 048849
Heflin, Tori 052979
Hefner, L. 020905
Hefti, Dr. Ernst 042191
Hegan, M. 043479
Hegel, Andrea von 017299
Hegeman, Carol A. 049089
Heger, Gabriele 002783
Heger, Wolfgang 018969
Heggde, Dr. D. Veerendra 024245
Heggen, Bert 032964
Hegidio, Margaret 053984
Hegnar von Ubisch, Frederikke . 033677,
                                 033678
Hegseth Garberg, Ann Siri ... 034038
Heguerte de Soria, María del
  Carmen 046266

Hegyi, Lorand 015303
Hehr, Dr. John 048717
Heidbrink, Dr. Ingo 017606
Heide, Dr. Birgit 020229
Heide, Diana 009028
Heidecke, Dr. D. 018971
Heideman, William N. 050808
Heidemann, Dr. Stefan 019453
Heidenreich, Barbara 034694
Heidrich, Bill 049511
Heidrich, Prof. Dr. Hermann .. 020451
Heij, Dr. J.J. 032035
Heijmans, Dr. H.G. 032155, 032164
Heil, Gerd 021490
Heilbrun, Françoise 014743
Heilbrunn, Margaret 051455
Heiligenstein, Louis 015743
Heiliger, Don 049972
Heiligmann, Dr. Jörg 019757
Heilman, Anna Bea 050550
Heilman, Patrick 046841
Heilmann, Dr. Angela 020083
Heiman, Mary-Ellen 005470
Heimann-Jelinek, Felicitas 003011
Heimann, Dr. Curry 040958
Heimann, Sybille 022419, 022420
Heimerdinger, Luc 041530
Heimühle, Bernd 016813
Hein, Carolyn 006272
Hein, Ivar-Kristjan 010668
Hein, Jørgen 010165
Hein, Lisbet 010131
Hein, Louise 049059
Heinämies, Ms. Kati 010829
Heinänen, Seija 010938
Heine, Sabine 017504
Heinecke, Friedhelm 021580
Heinemann, Dr. Christiane 020211
Heinemann, Ingo F. 019352
Heinemann, Klaus-Maria 017872
Heinerud, Jans 041106
Heinicke, Gabriele 016536
Heinien, Jouko 011080
Heinimaa, Mia 011398
Heinisch, Prof. Dr. H. . 018959, 018960
Heinisch, Dr. Severin 002233
Heinlein, Margitta 022057
Heinlein, Roland 016740
Heinlein, Dr. Stefan 021560
Heinmiller, Lee D. 049314
Heinonen, Jouko 011071, 011072,
        011073, 011074, 011075, 011077,
                                 011079
Heinrich, Dr. Chirstoph 018991
Heinrich, Christoph 048258
Heinrich, Paul 030583
Heins, Dr. 033171
Heintzelman, Scott A. 053604
Heintzman, John 051986
Heinz, Dr. Marianne 019545
Heinz, Stefan . 017181, 020152, 020153
Heinz, Ulrich von 017174
Heinzelmann, Dr. Markus 020058
Heinzelmann, Prof. Michael ... 041309
Heinzelmann, P. 019154
Heinzl, Richard 003085
Heischmann, Dr. Günter 020581
Heise, Bernd 017994
Heise, Dr. Karin 020374
Heiser, Jerry 046847
Heising, Cornelia ... 018032, 018033
Heisner, Ron 052122
Heiss, Alanna 050382
Heiss, Herbert 041471
Heitele, Thomas 019810
Heitlinger, Maria Theresia 019818,
                                 021756
Heitmann, Dr. Claudia 019664
Heitz, Henri 015727
Heizmann, Dr. Elmar 022056
Hejdová, Dagmar 009543
Heks, Silke 020673
Helavuori, Hanna-Leena 010873
Helbig, Gabriele 018335

Helbig, Heike 021543
Helbling Gerster, Regine 042025, 042026
Held, Peter 053700
Heldt, Helma 021404
Helfenstein, Marie-Therese ... 041979
Helfrich, Kurt 053094
Helg, Lukas 041460
Helgesson, Michael 040739
Helke, Dr. Gabriele 003026
Hellandsjö, Karin 033658
Helleland, Allis 010195
Hellemans, Marijke 003295
Hellenkemper, Prof. Dr. Hansgerd 019711
Heller-Karneth, Dr. Eva 016568
Heller, Barbara 048301
Heller, Dr. Erna 051437
Heller, Laura 052420
Heller, Margot 045108
Hellerstein, Bruce S. 047495
Hellfaier, Detlev 017875
Hellmann, Anouk 041764
Hellmann, Birgitt 019459
Hellmund, Dr. M. 018959
Hellrung, Reinhard ... 018628, 018629
Hellstern, Dr. John R. 046377
Helm, Charles 047945
Helm, Cherie 053607
Helm, Erika 003133
Helm, Lorna 047764
Helm, Dr. Reinhardt 022485
Helmberger, Dr. Werner 016624, 016625,
                        016626, 016954
Helme, Mehis 010633
Helmecke, Gisela 017242
Helmerding, Willy 016839
Helmerson, Klas ... 041039, 041055
Helmig, Loeda 050366
Helminen, Hilkka 010908
Helminen, Liisa 010841
Helminen, Ritva 011367
Helms, Cheryl V. 051540
Helms, Hans 020467
Helms, Dr. Leo 009398
Helmus, L.M. 032944
Helstrom, Linnea 051315
Helton, Debra M. 053325
Helzer, Richard 047167
Héman, S. 031986
Hemdal, Jay F. 053759
Hemdorff, Olle 033972
Hemm, Marsha 052373
Hemm, Michael 001789
Hemmendorff, Ove 040902
Hemmer, J. 032099
Hemmerle, Andreas 020511
Hemmerlein, Tamara 048054
Hemmes, H.H. 032945
Hemmi-Ragaz, Christian .. 042069
Hemsley, Jacqueline 000893
Hemsworth, Sandra 048973
Henar, A.F. 032622
Henchoz, Jean-Frédéric .. 041404
Hendel, Dr. Margarete ... 020694
Hendeles, Ydessa 007226
Henderby, D. 044154
Henderson Fahnestock, Andrea . 051443
Henderson, Cherel 050010
Henderson, Diana 043438
Henderson, E. Ray 047751
Henderson, Graeme 001086
Henderson, Jai 050411
Henderson, Justin 053920
Henderson, Kevin 033278
Henderson, Margaret ... 001224
Henderson, Marie 006783
Henderson, Robbin 046932
Henderson, Stan 007033
Hendler, Karla 017179
Hendren, Eileen 048162
Hendricks, Albert J. .. 053780
Hendricks, Dr. Alfred . 020602
Hendricksh, M. 038785
Hendrickson, Dean A. . 046701
Hendricx, Hilde 004025

Hendrikman, Dr. Lars 032621
Hendrikx, R. 033119
Hendrix, Lee 050436
Hendrix, Mittie B. 052868
Hendry, R. 033333
Hendryx, Michael 054605
Heneberry, June L. 051308
Hengelhaupt, Regine 020068
Hengstenberg, Thomas 021839
Hengster, Sabine 019039, 019041
Hengstman, G. 032204
Henke, James 047822
Henke, Jerry D. 048913
Henkel, Jens 021519
Henkel, Katharina 018176
Henkel, Dr. Matthias 020877
Henkens-Lewis, Helen 043915
Henmi, Yoshiharu 029180
Henn, Catherine E. C. 047127
Henn, Char 052492
Hennaut, Eric 003458
Hennebert, Diane 003431, 003764
Hennedy, Jane M. 053688
Henneman, Inge 003287
Hennessey, Dr. William 051573, 051578,
                        051583, 053993
Hennessy, Thomas F. 053722
Hennig, Alexandra 003091
Hennig, Lutz 020478
Hennig, Dr. Michael 020601
Hennig, Ulrich 021361
Henning, Darrell D. 048208
Henning, Dirk 021553
Hennze, Dr. Joachim ... 019163
Henrich, Sarah 051129
Henricksen, Paul 050711
Henriksen, Kenneth Roy .. 010107
Henrikson, Steve 049848
Henrikssen, Tracey 007146
Henrion, Pierre 003810
Henriot, Elisabeth ... 012683
Henriques, Paulo 035810
Henry-Barnaudière, Elise 016360
Henry, Barbara 051472, 051681
Henry, Cassandra L. ... 050518
Henry, Chris 044371
Henry, James Pepper .. 046470
Henry, John B. 048748
Henry, Karen A. 005502
Henry, Patricia 047715
Henry, Patrick 044862
Henry, Dr. Sarah ... 051443
Henrykowska, Anna .. 035157
Henrysson, Harald ... 040581
Hens, Pierre 003570
Henschel, Sylvia ... 022132
Hensel, C. 018151
Hensel, Diethardt .. 020751
Henseler, Petra ... 020724
Henseler, Rainer .. 021548
Henshall, Barbara E. .. 047526
Henshall, W.A. 046227
Henshaw Jones, Susan .. 051443
Hensinger, Irene D. ... 052739
Hensley, Dr. Dannie ... 036135
Henson, Leonora 047308
Henson, Trevis 046748
Henstra, Klaas R. ... 032127
Hentschel, Dr. Martin .. 019786
Hentz, Dr. Károly ... 023647
Henwood, Peter C. ... 033231
Heon, Laura 051590
Hepburn, Lorna ... 044422
Hepp, Dr. Frieder .. 019137, 019147
Heralus, Dr. Stefanie .. 017381
Héran, Emmanuelle ... 014743
Heraud, François ... 015250
Herb, Reinhard 022490
Herbaugh, Karen ... 050508
Herber, Bob 053894
Herberg, Mayde ... 053088
Herbert, Jacky ... 015762
Herbert, Lynn M. ... 049574
Herbert, Sharon C. .. 046490

Herbst, Andreas 017144
Herbst, Detlev 022290
Herbst, Dr. Helmut 022375
Herbst, John 049683
Herbuveaux, Arlette 015918
Herce, Ana I. 040226
Herda, Isabel 018483
Herdits, H. 001861
Heredia Cortés, Gabriel .. 031676
Herger-Kaufmann, Josef 042014
Herholdt, Nancy 047564
Herje, Torunn 033901, 033902
Herkenhoff, Paulo 004755
Herkt, Silvia 003074
Herlihy, Berry H. 050428
Hermalyn, Dr. Gary D. . 047228, 047234
Herman, Judy 049406
Herman, Monica 048908
Herman, Richard 048856
Hermand, Michel 014574
Hermand, Nathalie 003777
Hermann-Szabó, Sara ... 010387
Hermann, Adolf 022191
Hermann, Dr. Claudia .. 041715
Hermann, Daniel 044273
Hermann, Reyh 021651
Hermann, Dr. Siegfried W. .. 002520
Hermansdorfer, Mariusz 035426, 035430
Hermansson, Jan 041099
Hermellin, G. 015765
Hernández Carralón, Gema 039645
Hernández Carrión, Emiliano .. 039536
Hernandez-Cruz, Luis ... 052577
Hernández Gutiérrez, Jorge .. 008639
Hernández Hervás, Emilia .. 040035
Hernández Nieves, Román .. 039036
Hernàndez Pérez, Mauro .. 038947
Hernandez, Annie 015605
Hernández, Carlos Arturo .. 008593
Hernández, Jo Farb 053040
Hernandez, John 050195
Hernández, Rodolfo ... 031257
Hernandez, Sofia 051329
Herold, G. 020874
Herold, Dr. Inge 020249
Herold, Joyce L. 048261
Herold, Prof. Roland .. 002957
Herold, Prof. Roland-Peter .. 002163
Herr, Daniela 017137
Herrán, Alfonso Carcero .. 008698
Herranen, Merja 011275
Herrbach-Schmidt, Dr. Brigitte . 019505
Herren, Hans 041352
Herren, Walter 041590
Herrera Acosta, Elvira .. 031258
Herrera de La Muela, Ignacio .. 040019
Herrera Salazar, Germán .. 008679
Herrero, Prof. Eutimio .. 039484
Herreshoff, Halsey C. ... 047218
Herrick O'Hara, Virginia .. 047506
Herrick, Phyllis M. 049716
Herrin, Kari 053180
Herring, Madeline ... 046805
Herring, Sarah 045057
Herrity, Emma 044225
Herrmann Fiore, Dr. Kristina . 027622
Herrmann, Dr. Elisabeth ... 003026
Herrmann, Gerd-Ulrich .. 021818
Herrmann, Josef 020115
Herrmann, Jürgen ... 022405
Herrmann, Dr. Klaus .. 022027
Herrmann, Thomas ... 021363
Herrmann, Uwe 019497
Herron, Christopher .. 048266
Herron, James 051930
Herron, Mary 049652
Herselman, D.E. 038837
Hersey, Irwin 051356
Hershfield, Shauna .. 051016
Herskowitz, Sylvia A. . 051493
Herst, Dr. Katarina .. 019505
Herta, Ana 036195
Hertel, Friderike ... 018965
Hertel, Kathleen .. 046470

Hertford, Marquess of 043379
Hertig, HP. 041812
Hertiš, Iris 038501
Hertling, Nele 017098
Hertling, Nele 017066
Hertner, Max 041961
Hertwig, Dr. Stefan 041321
Hertz, Betti-Sue 052978
Hertzlieb, Gregg 053938
Herz, Carrie 047048
Herz, Silke 022019
Herzer, Wolfgang 022460
Herzmanatus, Klaus .. 018627
Herzog, M.A. Claudia .. 021337
Herzog, M. A. Claudia .. 021336
Herzog, Dr. Georges .. 041609
Herzog, Hannes 041856
Herzog, Dr. Hans-Michael . 004681,
                        042192, 042193
Herzog, John E. 051434
Herzog, Melinda 050619, 054020
Herzog, Monika 018540
Herzog, Rainer 018091
Herzogenrath, Prof. Dr. Wulf . 017592
Hesche, Eva-Maria 002097
Hesketh-Campbell, Lorraine . 045254
Heslewood, Emma 045565
Hesling, Mary 044521
Hess, Carsten 010153
Hess, Dr. Daniel 020853
Hess, Roman 042222
Hessaby, Mohandes Iraj ... 024759
Hesse-Mohr, Dr. Petra .. 020255
Hesse, Gisela 019432
Hesse, Dr. Sabine .. 020127, 022047
Hessel, Rita 042119
Hesselbom, Ted 040660
Hessellund, Birgit .. 009999
Heston, Dana 050057, 050308
Heth, Charlotte 051448
Hetterich, Jochen .. 022045
Hettinga, Mary Jane .. 054237
Heubaum, Dr. Regine .. 020828
Heuberger, Vera ... 041813
Heuck-Neelsen, Dietrich .. 021605
Heuckeroth, Erwin .. 020340
Heuer, Bernd 017815
Heuer, Christoph .. 021826
Heufelder, Jochen .. 019681
Heukemes, Dr. Berndmark .. 019845
Heunemann 021954
Heuschen, Imelola .. 003839
Heusden, G. van ... 032404
Heuser, Prof. Dr. August 018414, 020080
Heuser, Douglas ... 050630
Heuser, Franz-Josef .. 017997
Heuser, Frederick J. .. 052086
Heuser, Dr. Mechthild . 041323
Heußner, Karen ... 022306
Heußner, Christiane .. 017377
Heuwinkel, Dr. Hubert .. 018047
Heuwinkel, Martin .. 017776
Hevers, Dr. J. 017565
Hevia, Agustin ... 039863
Hew Wing, Micheal .. 051438
Hewel, Christine ... 018435
Hewelt, James 045544
Hewitt-Dower, Bill .. 000980
Hewitt-Soar, Hugh David .. 045436
Hewitt, Karen 047519
Hewitt, Marion ... 045214
Hewitt, Susan 001609
Hewitt, Timothy M. .. 051710
Hewson, Alan 043370
Hewson, Karen E. .. 006534
Hewson, Rose 033267
Hewson, Shirley O. . 001656
Heyburn, Tim 044646
Heyd, Paul 020952, 020955
Heyden, Dr. Thomas .. 020873
Heyen, Lora 046715
Heyes, John 044249
Heyl, Cornelius A. von .. 022765
Heyl, Dr. Matthias 018540

Heym, Dr. Sabine 020565, 020566,
  020571
Heymans, Dr. Hubert 003834
Heymer, Kay 018046
Heyneman, C. 003419
Heynen, Arnold 042136
Heynen, Dr. Julian 018033
Heyning, Dr. John 050462
Heystek, A.C.W. 032484
Heyward, Carroll 047907
Heywood, Dawn 045814
Hibbits, Paul E. 054060
Hibel, Edna 049852
Hibler, Dolores 002861
Hickey, Janet 050645
Hickey, Leo J. 051265
Hickey, Michael 005534
Hicklin, Karen 047715
Hickman, Carole 046939
Hickman, Warren 054065
Hickmann, Manfred 021140
Hickock, Paul 049565
Hicks-Connors, Robin 053577
Hicks, Christine 014987
Hicks, Hugh Francis 046763
Hicks, Louis 046382
Hicks, Oberstabsfeldwebel Paul . 018341
Hidasi, Prof. José 004473
Hidber, Alfred 041248
Hideo, Kurita 028488
Hidook, Tatjana 043154
Hiebel, Kaye 050705
Hiebler, Prof. Dr. Herbert ... 002880
Hieblinger, Faith 006123
Hiekisch-Picard, Sepp . 017436, 019103
Hiesinger, Kathryn B. 052081
Hiesmayr, Herbert 002701
Higashi, Shunro 029653
Higginbotham, Evangeline J. 052726
Higgins, James 053745
Higgins, Kate 047287
Higgins, Sarah 046832
Higgs, Jane 044228
High, Dennis 047434
High, Laura 046528
High, Steven S. 052510
Hightower, Anne Marie 050061
Hightower, John B. 051546
Higman, David 044875
Hignette, Michel 014777
Higuchi, Dr. Takayasu 028980
Higuchi, Yoshikazu 028737
Hiis Søftestad, Hanne 033492
Hijiwara, Takshi 029230
Hilaire, Michel 014292, 014298
Hilarides, John 047574
Hilberg, Thomas D. 050763
Hilberg, Dr. Volker 021663
Hilbun, Dorinda 046818
Hildebradtová, Barbora ... 009875
Hildebran, Julia 047849
Hildebrand, Dr. Josephine ... 017205
Hildebrand, Walter 001923
Hildebrandt, Alexandra ... 017221
Hildebrandt, Dr. Frank ... 019011
Hildebrandt, Sandra ... 051516
Hildenbrand, Manfred ... 019086
Hiles, Wayne 054228
Hilger, H. 022067
Hilgers, T. 032659
Hilkenbach, Rainer ... 019599
Hilkhuijsen, J. 032157
Hill Carter, Charles ... 047536
Hill-Festa, Lisa 053241
Hill Perrell, Franklin ... 052685
Hill, Allan 045123
Hill, Andrew 051265
Hill, Andrew F. 043488
Hill, Becky 048976
Hill, Brian O. 053801
Hill, Carla 046806
Hill, Carolyn 051734
Hill, Charles 006594
Hill, Christine C. 049377

Hill, Claire 045864
Hill, Heidi L. 047802
Hill, Dr. John 054471
Hill, John 051737
Hill, Joyce 048646
Hill, June 044499
Hill, Karen 048341
Hill, Lindsay 006678
Hill, N.B. 045199
Hill, Peter 044611
Hill, Ray 050347
Hill, Robert D. 049371
Hill, Robin 044690
Hill, Sergej Ivanovič .. 036843
Hill, Shannen 048275
Hill, Steve 046559
Hillard, Elbert R. ... 049749
Hillebrand, Dr. Melanie .. 038807
Hiller von Gaertringen, Dr. Julia 017875
Hiller von Gaertringen, Dr. Rudolf 019971,
  019977
Hiller, Fritz 021855
Hiller, Dr. Hubertus .. 021716
Hiller, N. 033168
Hiller, Robert J. ... 047882
Hillermann, Barbara . 003090
Hillesheim, Dr. Jürgen 016648
Hillhouse, Susan ... 053097
Hilliard, Elbert R. .. 049750
Hillier, Kenneth ... 043440
Hillis, Florence E. .. 050007
Hillsman, Matthew .. 052269
Hillstrom, Richard L. . 050942
Hilpert Stuber, Susanne ... 041754
Hilpert, Dr. Wilhelm .. 020507
Hiltermann, Elisabeth 016845, 017041
Hilti, Joseph 041556
Hiltula, Leena 011494
Himanen, Jaakko ... 011406
Himber, René 012367
Himmelbauer, Robert . 002085
Himmelsbach, Sabine . 020984
Himmelstrand, Erik .. 041160
Hincapié Santa Maria, Jaime ... 008746
Hincapié, Alberto ... 008690
Hinchliffe, Robin ... 053059
Hind, Charles 054084
Hindelang, Prof. Eduard . 019897
Hindman, James ... 050406
Hinds, Roger 046957
Hine, Charles Henri .. 052865
Hine, Marjorie 007252
Hine, N.A. 043931
Hinerman, Lenore .. 049103
Hines, Christine ... 007013
Hines, Tommy 053418
Hingst, Eckhard ... 019279
Hinkley, Marilyn J. . 054570
Hinman, Alison ... 046477
Hinman, Marjory .. 049367
Hinnant, Vandorn .. 049255
Hinrichsen, Dr. Torkild . 019017
Hinrichsen, Prof. Dr. Torkild . 018976
Hinsch, Werner ... 019930
Hinsching, Susanne . 020827
Hinson, Tom E. .. 047808
Hinton, Dr. Brian .. 044372
Hinton, William ... 050483
Hinz, Dr. Hans-Martin . 017299
Hinz, Sally 001294
Hiott, William D. .. 047798, 047799
Hippe, Margarethe . 002600
Hipps, Will 046759
Hipwell, John ... 007344
Hír, Dr. János ... 023834
Hirabayashi, Kunio . 029223
Hirai, Takuya ... 028998
Hirakawa, Minami . 029325
Hirako, Akihiko .. 028572
Hiramatsu, Hideki . 028520
Hirano, Irene ... 050437
Hirano, Itaru ... 029305
Hirata, Shozo ... 029457
Hirata, Yutaka .. 029725

Hirayama, Miyako 029305
Hirci, Andrej 038552
Hird, Maureen 045160
Hirner, Dr. René 019152
Hiroshi, Doi 028643
Hiroshi, Sugaya 028719
Hiroya, Yukiatu 029276
Hirsch, Hans Joachim . 020250
Hirsch, Helen 042055
Hirsch, Ilana 052428
Hirsch, Jan 053322
Hirsch, John 046694
Hirsch, Marjorie .. 053945
Hirsch, Dr. Vanessa . 018976
Hirschbiegel, Dieter . 019783
Hirschel, Anthony G. . 047697
Hirschler, Christa .. 021914
Hirschlieb, Brandy . 052608
Hirst, M. 038770
Hirst, Marlys ... 046829
Hirth, Dr. Antoine . 012033
Hirvonen, Anna-Riikka . 010996, 010997
Hisa, Asuka 053135
Hiscock, Donna .. 006896
Hiscock, E. 043640
Hiscock, Rosella . 006996
Hiscox, Margaret . 001421
Hiss, Elfi 041773
Hitchmough, Dr. Wendy . 044346
Hite, Andy 052146
Hite, Jessie Otto . 046693
Hitov, Peicho .. 005167
Hittaller, Hans Martin . 001775
Hitz-Walser, Cordula . 041616
Hitzing, Jörg .. 021377
Hiukka, Toini .. 011213
Hjartarson, Sigu'236ur . 024083
Hjermind, Jesper . 010361
Hjorth, Carl-Gustav . 041070
Hladky, Michael .. 051425
Hladky, Sylvia .. 020517
Hlaváčková, Dr. Miroslava . 009865
Hlôška, L. 038486
Hlusko, Leslea .. 046939
Hnateyko, Olha .. 051486
Ho Ho, Ying ... 038268
Ho, Keimei ... 028908
Ho, Lynn 046613
Hoag, Everett H. . 051226
Hoagland, Glenn D. . 049048
Hoard, Constance G. . 047283
Hoare, Adrian D. . 046210
Hoare, Nell ... 046155
Hobaugh, J.H. .. 053166
Hobbs, M.D. ... 044300
Hobbs, Patricia A. . 050266
Hobein, Beate .. 018914
Hoberg, Dr. Annegret . 020580
Hobin, Ellen .. 005970
Hobolth, Nina . 009999
Hobson, Marlena . 053511
Hoch, Manfred .. 021636
Hochberg, Dr. F.G. . 053093
Hochberger, Ernst . 019509
Hochdörfer, Achim . 003050
Hoche, Frank .. 019115, 021023
Hochleitner, Dr. Martin . 002323, 002328
Hochreiter, Otto .. 001972, 001986
Hochscheid, Christian . 018047
Hochstrasser, Markus . 042008
Hochuli-Gysel, Dr. Anne . 041245
Hochuli, Dr. P. .. 042227
Hockett, Esther . 048342
Hockman, Winnie . 006809
Hodel, Daniel E. . 041701
Hodel, R. 032570
Hodge, Anne .. 024900
Hodge, Dr. G. Jeannette . 051287
Hodge, Katherine . 049497
Hodges, Greg A. . 052558
Hodges, Rhoda . 048499
Hodgins-Wollan, Sonya . 006043
Hodgson, Beverly . 049321
Hodgson, Clarence . 005810

Hodgson, Kenneth O. 051567
Hodgson, Polly 044289
Hodgson, Tom 043988, 043990
Hodkinson, Wendy 045210
Hodslavská, Lenka 009638, 009639
Hodson, Janice 047365
Höbler, Dr. Hans-Joachim ... 020001
Höck, Anton 002141
Höck, Dr. Christoph .. 019503
Høeg, Asger 010100
Höfchen, Dr. Heinz .. 019480
Hoefer, Denise ... 053877
Höfer, Dr. Hubert .. 019518
Höfert, Dr. Dorothee . 020249
Höffer, Dr. Volker .. 021488
Höfler, Werner Oskar . 019177
Höfling, Eckart Hermann ... 004720
Hoefnagels, J. 003397
Hoegen, Friedrich . 020369
Høgestøl, Mari .. 033972
Högl, Dr. Günther . 017959
Höglander, Agneta . 040888
Höglmeier, Barbara . 020692
Höhl, Dr. Claudia . 019247
Höhne, Annett .. 018818
Høibo, Roy ... 033654
Hôibo, Roy ... 033918
Høie, Bjørn K. .. 033792
Højer Petersen, Anne . 010229
Højgaard, Johnny . 010061
Hoeksema, B.W. .. 032582
Hoekstra-van Vuurde, C.J. . 032177
Hoekstra, Drs. Ridsert . 032795
Hoel, Arild ... 033819
Hoel, Randall . 053286
Höller, Anette . 041287
Höller, Silvia . 002135
Hoelscher, Dr. Monika . 016519
Hölscher, Dr. Petra . 020519
Höltl, Georg . 016771, 021081, 022161
Hoeltzel, Susan . 047232
Hölz, Marjatta . 022233
Hölzel, Josef .. 001763
Hoelzeman, Buddy . 051059
Hölzer, Dr. Adam . 019518
Hölzinger, Nora . 017829
Hölzl, Dr. Christian . 003026
Hölzl, Martin . 017144
Hölzl, Dr. Regina . 003026
Höneisen, Markus .. 041950, 041956
Hönggi, Ambras . 041805
Hoenig, Thomas M. . 049887, 049890
Hönigmann-Tempelmayr, Mag.
  Alexandra .. 002947, 002949, 002950,
   003003, 003010, 003081, 003082
Hönigmann, Andrea . 003095
Höper, Dr. Corinna . 022057
Höpfner, Martin 016616, 016621, 016627,
   016628
Höppl, Birgit .. 020299
Hoerig, Dr. Karl . 048778
Hörker, Alois . 002101
Hörmann, Barbara . 019229
Hörmann, Fritz . 002925
Hörner García, Gabriel . 031410
Hoerschelmann, Dr. Antonia . 002931
Hösch, Jürgen . 018439
Höser, Dr. Norbert . 016535
Hoesgen, Jennifer . 006020
Hoesli, Hans Rudolf . 041362
Host-Madsen, Lene . 010178
Høvsgaard, Thomas . 010327
Høyersten, Erlend G. . 033717
Höylä, Risto .. 011006
Hofbauer, Erwin . 001897
Hofbauer, John D. . 046972
Hofer, Elke ... 002062
Hofer, Ernst .. 041952
Hofer, Hans .. 041940
Hofer, Manfred . 002262
Hofer, Rudolf . 002262
Hofer, Verena . 041692
Hoff Jørgensen, Jan . 033585
Hoff, Dr. Marlise . 017199

Index of Persons

Hoff, Prof. Dr. Ralf von den
– Horwath, Alexander

Hoff, Prof. Dr. Ralf von den 018476
Hoff, Svein Olav 033741
Hoffberger, Rebecca 046727
Hoffer, Andreas 002207
Hofferer, Robert 002349
Hoffert, Tonia 053744
Hoffman, Angela 052461
Hoffman, Dan 006462
Hoffman, Jill 054013
Hoffman, Dr. Kerstin 018059
Hoffman, Marilyn 053290
Hoffman, Michael E. 051350
Hoffman, Dr. Richard S. 050727
Hoffman, Steeven A. 053358
Hoffman, Thomas 053078
Hoffmann, Almut 017245
Hoffmann, Dr. Andreas 018977
Hoffmann, Anja 017958, 019098
Hoffmann, Audrey 001627
Hoffmann, Christa 022782
Hoffmann, Christian 011450
Hoffmann, Dr. H. 017915
Hoffmann, Helmut W. 020760
Hoffmann, Joel M. 050840
Hoffmann, John 047680
Hoffmann, Paul 002120
Hoffmann, Dr. Per 017606
Hoffmann, Dr. Silke 018029
Hoffmann, Susanne 021077
Hoffmann, Tobias 019412
Hoffmann, Walter 021471
Hoffmann, Will 030601
Hoffpauer, Diane 048078
Hofgartner, Mag. Heimo 001968
Hofmacher, Anna 002548
Hofman, Andreas 018619
Hofman, Jack 050185
Hofman, O. 025168
Hofmann 021173
Hofmann-Kastner, Dr. Iris 022902
Hofmann, Dr. Alexander 017239
Hofmann, Dr. B. 041321
Hofmann, Beate Alice . 019039, 019040
Hofmann, Bernard 047163
Hofmann, C. 031858
Hofmann, Christa 003053
Hofmann, Gertrud 019563
Hofmann, Helmut 002361
Hofmann, Irene 051544
Hofmann, Jürgen 019385
Hofmann, Lois 048150
Hofmann, Martina 020480
Hofmann, M.A. Rainer 021208
Hofmann, Ramona 022901
Hofmann, Susanne 041987
Hofmeister, Walter 019307
Hofschen, Dr. Heinz-Gerd 017585
Hofstad, Turid 034082
Hofstetter-Sprunger, Heidi ... 041625
Hofstetter, Adolf 021085
Hofstetter, Willy 042175
Hofwolt, Gerald 049548
Hogan, Jacqueline 046999
Hogben, Barbara 001103
Hogben, Ian 001103
Hoge, Robert 051328
Hogen, Tim 047144
Hogenhaug, Gunnar 033997
Hogg, C.G.O. 043577
Hogue, Kate 049989
Hoguet, Anne 014721
Hohbaum, R. 018653
Hoheisel-Huxmann, Reinhard . 017606
Hoheisel, Tim 054217
Hohenberg, Anita Fürstin von . 001722
Hohenlohe-Bartenstein, Ferdinand
Fürst zu 021741
Hohenlohe-Langenburg, Fürst zu 019904,
019908
Hohenlohe-Oehringen, Katharina
Fürstin zu 020704
Hohenlohe, Prinz Constantin von 021652
Hohensee, Falko 017938
Hohl, Erhard 001775

Hohl, J. 022517
Hohmann, H.-J. 019207
Hohmann, Markus 020111
Hohrath, Daniel 017299
Hohri, Sasha 051342
Hohut, Khrystyna 005763
Hoisington, Rena 049390
Hokanson, Randee 051060
Hol Haugen, Bjørn Sverre 033582
Holahan, Mary F. 054433
Holahan, Paula M. 050600
Holben Ellis, Margaret 051428
Holbrook, Christopher 047974
Holbrook, Maureen 001292
Holcomb, Grant 052618
Holcomb, Julie 048010
Holcomb, Michael 048624
Holcombe, David C. 047810
Holcombe, Holly 047825
Holdcraft, Rose 047370
Holden, Charles 044471
Holden, Florence 047960
Holden, Rebecca 053271
Holden, Sally 043853
Holden, Steve 049020
Holden, Thomas 048384
Holdenbusch, Heinrich 021538
Holderbaum, Linda 046824
Holding, John 007314
Holdren, Greg 046968
Holdstock, Barry C. 044607
Holdsworth, Jane A. 051543
Hole, Frank 051265
Holečková, Dana 009544
Holeman, Claude A. 048167
Holen, Dr. Stephen 048261
Holeva, Daniel 050878
Holian, Sarah 049191
Holland, J. 001543
Holland, Kim 051631
Holland, M. 038737
Holland, Neil 043373
Holland, Rudolf 022903
Hollander, Barbara J. 046426
Hollander, Stacy . 051325, 051383
Hollander, Volker 017844
Hollein, Max . 018446, 018448, 018449
Hollender, G. 042236
Holler-Schuster, Mag. Günther . 001985
Holler, Prof. Dr. Wolfgang 017992
Hollerbach, Dr. R. 019685
Hollerová, Petra 009444
Holleschorsky, Catherine und
Thomas 020300
Holliday, Anne 051881
Holliday, Dr. John 043536
Holliger, C. 041369
Hollinetz, Alfred 002876
Hollingworth, Dr. Brad 052980
Hollingworth, John 044928
Hollingworth, Mar 050411
Hollister, Robert 050456
Hollitzer, Tobias 019990
Holló, Dr. Szilvia Andrea 023997, 023998,
023999
Hollomon, James W. 049341
Hollosy, Gyuri 050803
Holloway, Donald 049190
Holloway, James 044274
Hollows, Victoria 044398
Hollstein, Dr. Wilhelm ... 017998
Holm Sørensen, Bodil 010160
Holm, Axel 051570
Hólm, Geir 024074
Holm, Torben 010330
Holman, Graham 007378
Holman, Kurt 051997
Holman, Robin 050231
Holman, William M. 048341
Holmberg, Lori 050970
Holmes Helgren, Heidi 047402
Holmes, Carolyn C. 047580
Holmes, Dugald 001349

Holmes, Meryle 053036
Holmes, Saira 043620
Holmes, Selean 047656
Holmes, Willard 049390
Holmesland, Liv 033952
Holmgrain, Ardith 051323
Holmgren, Bengt 040912
Holmlund, Bertil 011207
Holmsted Olesen, Chritian ... 010172
Holmstrand-Krueger, Lena 040957
Holo, Dr. Selma 050421
Holoch, Jürgen 041065
Holodňáková, Dr. Radmila 009970
Holowacz, Eric V. 046849
Holozet, Andrée 016376
Holpainen, Anja 010983
Holsbeke, Mireille 003284
Holst, Dr. Daniela 020776
Holst, Gerd 022781
Holst, Lars 010109
Holstein, Lars 041106
Holsten, Christina 040681
Holsten, Dr. Siegmar ... 019517
Holt, Daniel D. 046303
Holt, Frank 051805, 051809
Holt, Loucinda 047825
Holt, Sharon 051472
Holtappels, Heiner ... 031971
Holtby, J.Mick 044463
Holte, Bettye 047778
Holten, Arnfinn J. 033510
Holten, Johan 019134
Holten, Lars 010217
Holthaus, Dr. Katrin 021107
Holthof, Marc 003286
Holthuis, Dr. Gabriele . 021748, 021749,
021750
Holtman, Marge 053400
Holton, Linda 049159
Holtry, Cara 047427
Holtz, Jerzy 041040
Holtz, Wolfgang ... 017298
Holtzapple, John C. 047919
Holverson, John H. 053253
Holvēt, Brigitte 013397
Holz, Francisc 036525
Holz, Uwe 017405, 022729
Holzach, Cornelie 021121
Holzer-Weber, Ruth 042245
Holzer, Dr. Elmar 022481
Holzhey, Magdalena 018036
Holzinger, Hans 020568
Holzleitner, Mag. Laurin ... 001870
Holzman, Rae 051680
Holzner, Dr. Anna ... 002050, 002051
Homa-Rożek, Ewa 034709
Homann, Dr. Klaus 022278
Homberger, Lorenz ... 042221
Homburg, Cornelia 052818
Homen, Prof. Zoran 008941
Homer, Russell 051899
Homering, Liselotte ... 020255
Hommel, Michel 013853
Homolova, Dr. Alexandra 038328
Homs-Chabbert, Arlette 014186
Honadel, Elroy 051664
Honaker, Corney A. 053692
Honcoopova, Dr. Helena ... 009790
Honda, Masamitsu 028770
Honda, Shizuo 029023
Honerød, Anne Grete 033908
Hong, Trinh Van 054803
Honkanen, Pekka 010867
Honour, Peggy 045861
Hoo, Gloria Chun ... 047391, 047392
Hoober, David 052115
Hood, Betty 047443, 054195
Hood, Michael 049676
Hood, Sadie Ann 054449
Hoogenboom, W. 032060
Hoogendonk, Mabel 032393
Hook, I.D. 043921
Hook, Lu 049083
Hooker, Saralinda 047397

Hooks, Rosie Lee 050472
Hooks, Terri 047968
Hoon, Wilhelm 016716
Hooper, Dr. John 001501
Hoopes, John 050185
Hoops, Erik 017606
Hoorn Bakker, A. 032836
Hoover, Bessie 052487
Hoover, Rhonda J. 054068
Hooymayers, Dr. J.M.M. ... 032581
Hoozee, Dr. Robert 003636
Hopfinger, Rebecca 047784
Hopfner, Dr. Rudolf 003026
Hopfner, HR Dr. Rudolf 003072
Hopgood, Dr. James F. 049470
Hopkins Coughlin, Joan 054269
Hopkins, Dave 006837
Hopkins, S.K. 045732
Hopkins, Terri M. ... 050728
Hopkinson, Susan 045747, 045748
Hopmann, Gerd 018197
Hopp, Dr. Andrea 021710
Hopp, Doris 018421
Hoppe, Barbara G. 050021
Hoppe, Michael 017965
Hoppe, Dr. Werner ... 016920
Hoppe, Werner 041276
Hoptman, Laura 052157
Horacio Aranguren, Juan José . 000581
Horat, H. 041702
Horat, Dr. Heinz ... 041707
Hordyński, Piotr 034907
Horeftosz, Konstantin 023731
Horgan, Judy 052172
Horgen, Randi 033677
Horký, Milan 009501
Horman, Trevor ... 000774
Horn, Damian 045708
Horn, Elizabeth 052369
Horn, Elizabeth G. 006722
Horn, Gabriele ... 017210
Horn, Dr. Karsten 018235
Horn, Ken 050433
Horn, Murray 052179
Horn, Dr. Reinhard ... 020507
Horn, Robert 047682
Horn, Tom 021363
Hornbostel, Prof. Dr. Wilhelm .. 019011
Hornby, D. Mary ... 043969
Horndasch, Mickey ... 046542
Horne, Daphne 001129
Horne, Gene 033367
Horne, Graham 001087
Horne, Jerry G. 052135
Horner, Gerald ... 048266
Horner, John R. ... 047166
Hornig, Dieter ... 022777
Hornik, Dr. Heidi ... 054017
Hornstrup, Nils ... 010100
Hornung, Herb 052955
Hornung, Prof. Dr. Maria .. 003058
Horoiu, Otilia ... 036211
Horpeniak, Dr. Vladimír 009625, 009899
Horray, Henrietta ... 049165
Horrigan, William 047945
Horsfield, Louise 043985
Horsley, Juliet 045920, 045926
Horsnæs , Helle W. ... 010179
Horst, Randy ... 048330
Horstik, Deborah 050140
Horsting, Archana 046937
Horstrup, Ellen ... 021931
Horton, Constance ... 049929
Horton, Lon 047492
Horton, Melissa 051136
Hortz, Laura ... 053357
Horvath, Barbara ... 003091
Horváth, Dr. Béla ... 023633
Horváth, Eugen ... 002337
Horváth, György ... 023518
Horváth, Dr. István ... 023627, 023628
Horváth, Dr. László ... 023804
Horváth, Ursula ... 002420
Horwath, Alexander ... 003054

Horwood, Michelle 033335
Horz, Ulrich Chr. 021362
Horžen, Vesna 038600
Hosack, George 054404
Hosaka, Kenjiro 029608
Hosar, Kåre 033742
Hoschek, Mag. Eleonore 002667
Hose, Henry 053796
Hoskin, Rebecca 000861
Hoskins, Rebecca 051062
Hoskins, Susan 052145
Hoskins, William J. 053357
Hoskinson, Susan 052194
Hoskisson, Don 050993
Hošková, Jitka 009773, 009905, 009922
Hosley, William 049383
Hosoe, Eikoh 029440
Hosokawa, Morisada 029502
Hosokawa, Toshiro 029545
Hossain, Shamsul 003212
Hossaka, Luiz 004879
Hoštálek, Petr 009505
Hostetter, David 050062
Hostyn, Norbert 003927, 003933, 003934
Hotchner, Holly 051324, 051436
Hoti, Prof. Afrim 000013
Hotz, Benedict 041710
Hotz, Claus-Dieter 018797
Hotz, Dr. Gerhard 041282
Hou, Fuxin 007682
Hou, Jiming 008411
Hou, Liancheng 007976
Hou, Wenyu 007659
Hou, Yancheng 008353
Hou, Yumei 008024
Houby-Nielsen, Dr. Sanne ... 041019
Houdamer, Chloé 012160
Hough, Douglas R. 051908
Hough, Katherine 051896
Hough, Melissa E. 052036
Houghtling, Kim 007110
Houghton-Bowen, Linda Kay ... 050087
Houghton, J. 051137
Hougue, Veronique de la 014733
Houkjaer, Ulla 010172
Houlden, Penny 006907
Houle, Joy 046724
Houlihan, E.T. 050246
Houliston, Laura 045014
Hoult, R. 033175
House, Emma 043495
House, George M. 046334
House, Kirk 049333
House, Patricia 048995
Houseago, Avril 045402
Housego, Bill 001027
Housen, Jean 003810
Houssinot, Jean-Claude 015228
Houston, David 047800
Houston, Joe 047038
Houston, John 052576
Houston, L.S. 050827
Houston, Dr. Margaret 050183
Houyez, Marcel 003806
Houzet, Bernard 012731
Hovagimian, Sarkis Vahan 003039
Hovell, Bob 044163
Hoveng, Dot 046544
Hovens, P. 032585
Hovers, Dr. Erella 025091
Hovi, Jouko 010757
Howard, Betty L. 054332
Howard, Donna J. 050098
Howard, Helen 045057
Howard, J. 038645
Howard, James 051127
Howard, Jan 052387
Howard, Jane 044999
Howard, Jean 045157
Howard, Jerome 006549
Howard, Jerry 050813
Howard, Julia 043915
Howard, Louise 051161
Howard, Nora 046708

Howard, Paul 046234
Howard, Stephen R. 047624
Howarth, Crispin 000937
Howarth, Frank 001537
Howdle, Joanne 044634
Howe, Angela 045673
Howe, Eileen 052471
Howe, Henry D. 048819
Howe, Katherine S. 049588
Howe, Kathleen ... 046136, 047758
Howe, Philip 033312
Howell, Alan C. 045712
Howell, Beverly 050150
Howell, Busser 053814
Howell, Helena 052351
Howell, Karen 045068
Howell, Kathy 054004
Howell, Melva 052514
Howells, David 045906
Howes, Stephen 048092
Howland, Robert 053402
Howoldt, Dr. Jenns E. 018991
Hoyd, Ed 047999
Hoyer, Dr. Eva Maria 019992
Hoyer, Klaus 016617
Hoyer, Lindy 051768
Hoyos Galarza, Melvin 010453
Hoyos, Cristina 040177
Hoyos, Hans 002609
Hoyt, Christine 047730
Hoyt, Marilyn 048767
Hoyum, Wanda 046911
Hozumi, Toshiaki 029334
Hrabánková, Dr. Světlana 009655,
                              009657
Hrebíčková, Marta 038432
Hrisenková, Nataša 038399
Hristov, Dr. Martin 005191
Hristova-Radoeva, Vessela .. 005203
Hristova, Mariana 005193
Hroch, David 009879
Hron, Vincent 047055
Hrtánková, Valéria 038328
Hrushetska, Maryna 010545
Hruška, Václav ... 009640, 009856
Hsiao, Tsung-huang 042541
Hsieh, Chia-Liang 042407
Hsieh, Kun-Shan 030302
Hsu, Claire 007870
Hu, Daxin 008448
Hu, Jiqiao 007566
Hu, Qingyun 008287
Hu, Shiliang 008550
Hu, Xiaojiang 008007
Hu, Yaoqing 007842
Hu, Yuanhong 008446
Hu, Zhisheng 007584
Huan, André 015087
Huang, Chongyue 008266
Huang, Chun-Min 042307
Huang, Dong 007863
Huang, Fengquan 007794
Huang, Gaoqian 007613
Huang, Huaixin 008219
Huang, Jianbang 007735
Huang, Juming 008102
Huang, Kuang-Nan ... 042539
Huang, Liang 007972
Huang, Ning 008242
Huang, Peiqi 007822
Huang, Qiang 008090
Huang, Qishan 008121
Huang, Tsai-Lang ... 042551
Huang, Weixiong ... 007701
Huang, Xizheng ... 007673
Huang, Yaohua ... 008177
Huang, Yongbang ... 007580
Huang, Yunzhong ... 008123
Huart, Micheline 007005
Huart, Nicole d' 003476
Hubach, Lynn 046680
Hubbard Mecham, Denny ... 053225
Hubbard, Dr. Charles M. ... 049382
Hubbard, Dr. Jane 046170

Hubbard, Jeremy 033155
Hubbs, Heather 047701
Hubbs, Jenice 005326
Huber-Bitzer, Anne 022057
Huber, Arthur 041365
Huber, Dr. Bernhard 017504
Huber, Christoph Anselm 003446
Huber, Franz 002866
Huber, Markus 041949
Huber, Mary 047477
Huber, Dr. Waldemar 001928
Huber, Walter 041464
Huber, MMag. Wolfgang 002209
Huberman, Anthony 014794
Hubert-Jeannin 012417
Hubert, François 003416, 015063
Hubinger, Horst 012022
Huck, Thomas 018774, 018776
Hudák, Julianna 023656
Huddleston, David 007015, 007016
Hudeczek, Dr. E. 002751
Hudgins, Don 053231
Hudson, Briony 045053
Hudson, Dr. Julian D. 047779
Hudson, Leslie 046339
Hudson, Louise 052462
Hudson, Mark S. 048945
Hudson, Dr. Neville 033148
Hue, Michel .. 012655, 012850, 013721,
          013833, 014178, 014180, 016094
Hueber, Régis 012628
Hübl, Prof. Dr. Richard 002848
Hübner, Heinz 001682
Hübner, Dr. John 001990
Hübner, Dr. Ute 018582
Hübsch, Norbert 016982
Hückel, Angela 020551
Hückstädt, Jutta 021214
Hügel, Dr. Peter 036204
Hüls, Johannes 017428
Hülsewig-Johnen, Dr. Jutta .. 017377
Hülsmann, Gaby 021396
Huemer, Dr. Peter 002132, 002141
Hümme, Dr. Julia 018329
Hünninghaus-Böckmann, Ute .. 002789
Hürkey, Dr. Edgar J. 018395
Huerta, Benito 046533
Huerta, Enrike 039714
Hürzeler, Adelheid 042016
Hüsam, Gernot 017760
Huescar, Duque de 039625
Hüsers-Doehmen, Dr. Katharina 018802
Hüsken, Wim ..003858, 003859, 003860
Hüßner, Reinhard 019418
Huet, Dr. A. 013937
Huete, Alberto 034187
Hueter, Diana 048125
Hueter, Ken 046368
Hütt, Dr. Michael ... 022334, 022335
Hüttel, Dr. Richard 020003
Hütter, Prof. Dr. Hans Walter .. 017490
Hüttermann, Prof. Dr. Armin .. 020265
Hüttner, Rudolf 002878
Huff, Richard E. 049724
Huffaker, Shirley 050812
Huffman, Joan 049380
Huffman, John 053558
Huffman, Lorilee C. 047425
Huffman, Vincent 006281
Hufnagl, Prof. Dr. Florian .. 020519
Hufnagle, Sue 047501
Hufschmid, Thomas ... 041243
Hufschmidt, Dr. Anke ... 018917
Hufschmied, Richard ... 003004
Hug, Guy 013780
Hug, Hannes 041262
Hug, Hans 041618
Hug, Shamsul 003234
Hug, Dr. Thomas 041507, 041871
Hugel, André .013350, 015103, 015105
Hugentobler, Hanspeter .. 041607
Huggert, Anders 041106
Hughbanks, Bette 052524
Hughes, Dr. David 054390

Hughes, Debra 054437, 054438
Hughes, Gareth 045659, 045802
Hughes, Gary 006893
Hughes, Helga 044443
Hughes, James T. .. 054209, 054211
Hughes, John 044886
Hughes, K.C. 046163
Hughes, Marianne 051251
Hughes, Mark 045371
Hughes, Michael 044719
Hughes, P. 001150
Hughes, Richard Ll. ... 044878
Hughes, Sheila 043455
Hughes, Susan ... 007213, 043933
Hughes, Thomas A. ... 051513
Hugo 038671
Hugomiot, Jean-Yves ... 012022
Hugonich, J. Yves ... 015634, 015635,
                    015636, 015637
Huguet, Jean-Pierre .. 041793
Huhtamäki, Tuula 011524
Huhtamäki, Ulla 010862
Hui, Hwang Sun 029878
Huijs Janssen, Dr. Paul ... 032472
Huijts, Stijn 032879
Huizi, María Elena 054724
Huizing, Pat 051087
Hujala, Susanna 011453
Hull, Prof. Andrea 001510
Hull, Dale 052020
Hull, Howard 044001
Hull, Karen 045590, 045592
Hull, Priscilla 043783
Hulló, István 038250
Hulse, E. 043709
Hulse, Sir Edward 043708
Hulse, Robert 044929
Hulser, Kathleen 051455
Hulshof, G. 032034
Hulston, Nancy 049873
Hultsch, Ute 018862
Humbel, Frédéric 030601
Humbert, Jean-Baptiste .. 034166
Humbert, Vincent 013522
Hume, J. Wilbert 005567
Hume, Ted 005873
Humer, Franz .001736, 002502, 002503
Humes, Alex B. 052053
Hummel, Alfred 020315
Hummer, Hermann ... 003057
Humphrey, L. 043837
Humphrey, M. 001543
Humphrey, Sylvia ... 045347
Humphreys, Mary ... 043715
Humphreys, Richard .. 045115
Humphries, Alan ... 044787
Humphries, Martin .. 044943
Humpl, Christine ... 002207
Hundertmark, Cordula ... 018540
Hundt, W.-D. Graf von ... 016652
Hunger, Mattli 041903
Hunt Chamberlain, Carola ... 053179
Hunt Miller, Lee ... 053009
Hunt, Alice 052500
Hunt, C. 045836
Hunt, Clare 045832
Hunt, Dr. D.A. 044821
Hunt, David 051791
Hunt, Dr. Donald W. ... 046601
Hunt, Eugene A. 049908
Hunt, Harrison 052686
Hunt, Jacqueline ... 045454
Hunt, Michael 048452
Hunt, Perdita 043998
Hunt, Phillip 047105
Hunt, Rhonda ... 001332
Hunt, Susan 050797
Hunt, Tony 033206
Hunt, Wilma ... 048023
Hunter Carsch, C. Morag ... 045375
Hunter Larsen, Jessica ... 049219
Hunter-Stiebel, Penelope .. 052302
Hunter, Charles 054614

Hunter, Elizabeth 005326, 006646, 043430
Hunter, Elizabeth Ives 048243
Hunter, Fredericka 050671
Hunter, Jamie 006304
Hunter, Jim 043957
Hunter, Judith 046167
Hunter, Ken 045851
Hunter, Lance 051851
Hunter, Michael 044201
Hunter, Paul 051382
Hunter, Richard 051791
Hunter, Terry 049402
Hunter, Thomas 053637
Hunter, Valerie 044255
Huntoon, Katherine 051582
Huntzinger, John 050890
Hunziker, Armando T. 000320
Hunziker, John 052611
Huo, Qiming 008030
Huo, Prof. Wei 007697
Hupe, Dr. Joachim 022202
Hupponen, Elina 011325
Huret, J. 015932
Hurezan Pascu, Georghe 036206, 036535, 036536, 036568, 036716, 036833
Hurkmans, Riek 032339
Hurler, R. 016484
Hurley, Arthur L. 047565
Hurry, S.P. 044101
Hursey, Dick 050276
Hurst, Henry 043845
Hurst, Rachel 044202
Hurtado Alfaro, Miguel Ángel .. 039911
Hurtado Londoño, Gloria 008602
Hurum, Prof. Jørn H. 033847
Hurvitz, Uri 025216
Hušáková, Ivana 009882
Husband, John 033227
Husband, Michael B. 049612
Husband, Timothy 051364
Huseboe, Arthur R. 053351
Husedžinović, Meliha 004282
Huskisson, Christine 049079
Husmann, Karin C.E. 032701
Husmeier-Schirlitz , Dr. Uta ... 020743
Huss, John 050624
Hussain, Syed Mohammad 024720
Hussein, Dató Ibrahim 030722
Hussein, Elena 036829
Husseini, Frederic 030268
Husseini, Frédéric 030275
Hussenot, Eric 012211
Hussey-Arntson, Kathy 054430
Hussey, Noelene 033235
Husslein-Arco, Dr. Agnes .. 002951, 002952, 002953, 002954
Husted, Charlene 049059
Husty, Mag. Peter 002637
Huszár, Zoltán .. 023836, 023837, 023838, 023841, 023842, 023846, 023847, 023849, 023851, 023852, 023857, 023858, 023859, 023860
Hutchcrot, Sally 050002
Hutchings, Lamar 050216
Hutchins Bobyrk, Dawn 053347
Hutchins, Patricia 047462
Hutchinson, J. Dwight 050865
Hutchinson, John 024873
Hutchinson, Susan 001499
Huterberger, Renate 002429
Huth, Nancy 051113
Huthwohl, Joël 014623
Hutin, Albert 047530
Hutt, Marilyn 053519
Hutt, Richard B. 001559
Hutt, Sarah 047131
Hutter, Dr. Ernestine .. 002637, 002641
Hutterer, Dr. Karl 053093
Hutterer, Dr. Rainer 017504
Huttlinger, John B. 050086
Hutto, Dr. Richard 050964
Hutton, Susan 053340
Huttunen, Marianna 011054

Hutz, Dr. Ferdinand 001800
Huuskola, Arto 011392
Huuskonen, Kari 010916
Huvenne, Patrick 003703
Huvenne, Dr. Paul 003289
Huwer, Elisabeth 019132
Huwyler, Dr. Edwin 041364
Hux, P. Meinrado 000432
Huxley, Geralyn H. 052153
Huybrechts, R. 004067
Huyge, Dirk 003492
Huygens, Ch.E.C. 032830
Huynh, Michel 014780
Huysza, Hans 003116
Huzel, Karl 019963
Hwan Kim, Myung 029900
Hwang, Pae Gang 029988
Hyams, Prof. Dr. Helge-Ulrike .. 020272
Hýbl, Dr. František 009850
Hyde, Hilary 045380
Hyde, J.R. 050787
Hyde, Lex 024939
Hyer, Warren W. 054553
Hygate, Nadine 044033
Hyland, Alice 048414
Hyland, Douglas 051240
Hylmarsová, Eva 009688
Hyman, Dr. Paul 045168
Hyman, Ryan 051064
Hynes, Patricia 051324
Hynes, William Q. 053315
Hyross, Dr. Peter 038298, 038302, 038303, 038304, 038308, 038309, 038310, 038313, 038314, 038324, 038330
Hysell, Barbara 046974
Hyšková, Veřa 009569
Hyson, Dr. John M. 046743
Hyss, Dr. Lothar 020611
Hytha, Prof. Dr. Robert K. 002472
Hyvönen, Marja-Liisa 010937
Hyytiäinen, Maija .. 011368, 011369
Iaconis, Héctor José 000115
Iacopi, Dr. Irene 027554, 027638
Iancu, Emilia . 036556, 036627, 036719
Iancu, Pavel 036490
Iannelli, Dr. Maria Teresa 028361
Iannucci, Heather 054578
Ibanez, Jean-Louis 012357
Ibarra, Jorge A. 023369
Ibbertson, Prof. H.K. 033145
Ibbotson, Michael P.S. 033234
Ibing-Harth, Krista 017003
Ibolya, Dr. Szathmári 023593
Ibragimov, Token Smailovič ... 029811
Ibragimov, Ulykbek Šarachynovič 029811
Ibragimova, Gulnar 029825
Ibrahim 029863
Ibrahim, Abbas 030733
Ibrahim, Dr. Asma 034123
Ibsen, Tim 021663
Icart, Françoise 011775
Ice, Dr. Joyce 053115, 053116
Ichijyo, Akiko 029608
Ichim, Dorinel 036214
Ickowicz, Pierre 012767
Idänheimo, Raija 011138
Ide, Evan 047248
Ide, Richard 021720
Idemitsu, Shosuke 029517
Idleman, Paul 047517
Ido, Makoto 029384
Idris, Hassan Hussein 040497
Idshinnorov, S. 031738
Idzikowski, Waldemar 035115
Iemura, Katsuyuki 029680
Iercoşan, Neţa 036357
Ieronimidis, M. 023325
Ieronymidou, M. 009414
Ievleva, Nina Vasiljevna 037706
Ifkovits, Kurt 003059
Ifland, Dorothee 017087
Igel, Dr. Walter 018475
Igitian, Henry 000739

Iglesias Grande, Manuel 039262
Igna, Ion 036388
Ignatev, Viktor Jakovlevič 037155
Ignatjeva, Tatjana Ivanovna 037360
Igrakova, Natalja Ivanovna 037657
Ihrenberger, Elisabeth . 002630, 002631
Iida, Kishiro 028688
Ikeda, Hosai 028640
Ikeda, Sakae 029465
Ikefuji, Kiyoshi 029578
Ikejima, Kazuo 029685
Ikibaeva, Roza Eržanovna 029833
Ikitian, H. 000725
Iles, Chissie 051491
Ilg, Werner 041623
Ilg, Dr. Wolfgang 001881
Ilić, Dragica 038168
Ilich-Klančnik, Breda 038549
Iliescu, Prof. Constantin 036690
Iliev, Christo 005001
Iliev, Petăr 005101
Ilieva, Marija 005009
Ilisch, Dr. Peter 020601
Iljasova, Ryšžan Iljasovna 029791
Iljin, Vadim Jurjevič 036922
Iljina, Ljudmila Ivanovna 036870
Ilkjær, Marianne 010124
Ilkosz, Jerzy 035419
Ille-Kopp, Regina 017385
Illich, Dr. Inge 002625
Illner, Dr. Eberhard 022820
Illuzzi, Dr. Michael C. 053215
Ilming, Thomas 003004
Ilse-Neuman, Ursula 051324
Ilshöfer, B. 022298
Imai, Yoko 029610
Imanse, G. 031986
Imbeault, Nisk 006348
Imber, Sofia 054722
Imbernor, Laurence 015067
Imbert, Christiane 012341, 012342, 012344
Imboden, Dr. Monika 042202
Imhof, Dirk 003295
Imhoff, Bert C. 006944
Imiela, Edward 035218
Immel, Dr. Isabell 021897
Immel, Dr. Wolfgang 017841
Immenkamp, Dr. Andreas ... 017958, 019846
Immerzeel, A.A.G. 032500
Immisch, T.O. 018969
Immler, Verena 020534
Immonen, Kari 011450
Immonen, Olli 011091, 011092
Imponente, Anna 026562
Impraim, Mary 022925
Imura, Yoshihiro 028932
Imychelova, E.C. 037990
Inaba, Hisao 029183
Inaba, Yoshinori 029305
Inai, Keijirô 029516
Inauen, Roland 041224
Inazuka, Hiroko 029639
Inbar, Janet 025212
Inbar, Yehudit 025080, 025116
Incardona, Dr. Angela 025256
İnce, R. Sefa 043064
Inch, John 033183
Inckmann, Lisa 020600
Indahl, Trond M. 033538
Indāns, Juris 030193
Indeck, Dr. Jeff 047414
Indra, Dr. W.I. Pandji 024571
Ineba, Pilar 040308
Infante, Victor Manuel 000323
Ingber, Abie 047745
Ingber, Barbara 051338
Ingebretsen, Gry 033681
Ingebrigtsen, Dr. Kristian 033856
Ingelheim, gen. Echter von Mespelbrunn, Graf zu ... 020384
Inghels, Cyriel 003303
Ingilby, Sir Thomas C.W. 045619

Ingle, R. 043912
Inglis, Elspeth 049860
Inglis, Marion L. 007254
Inglis, Robin 006518, 006520
Inglis, Stephen 005871
Ingmarsson, Niklas 040809
Ingo, Emi 010778
Ingold, Ed 048733
Ingraham, Joannie 054339
Ingram, Cathy 054115
Ingram, Daniel 049832, 049833
Ingram, Doris 047618
Ingram, J. 043510
Ingram, Jack 033305
Ingram, Jack E. 048843
Ingvordsen, Jens 010006
Iniesta González, Montserrat ... 040385
Innes, J.F. 006563
Innes, Kathryn V. 052498
Innes, Moira 044782
Innocenti, Pierluigi 025327
Ino, Koushin 029309
Inojatov, K.Ch. 054654
Inostroza Saavedra, Jorge E. .. 007554
Inoue, Kouki 028548
Inoue, Tadashi 028674
Insam, Peter 021321
Inselmann, Andrea 049739
Inserra, Margaret 053714
Insley, Debra 049277
Intas, Vaclovas 030424
Internoscia, David 050684
Intihar Ferjan, Jana 038549
Inui, Kazuo 028705
Inversini, Eva 041565
Io, Lau Si 008075
Iolen, Cadi 044873
Ion, Prof. Dr. I. 036512
Ion, Toma 036807
Ionescu, Ruxandra ... 036343, 036625
Ionesov, Anatoly 054645
Ioniţă, Ionel . 036284, 036285, 036292, 036293, 036294, 036299, 036301, 036303, 036306
Iotta, Ivana 026100
Iovănuţ, Camelia ... 036775, 036776, 036777
Iovinelli, Maria Teresa 027654
Iprajian, Avedis 025103
Ipsen, Dr. Anne 016877
Irazú, José María 000568
Ireland, Lynne 050549
Ireland, Shasta 052239
Ireland, Terese 048316
Irén, Kállai 023446
Iriarte, Prof. Elba Esther 000566
Iribarren Avilés, Rodrigo 007558
Irie, Yoshiro 029609
Irie, Yoshiyuki 028753
Irinina, Elena . 004977, 004978, 004979
Irish Jr., Mahlon 049522
Irle, Lisa 054071
Irlich, Franz 021808
Irrgang, Judith 016926
Irša, Dr. Rudolf 038386
Irvin, Mary A. 046968
Irvin, Stan 046690
Irving-Dusting, L. 001440
Irving, Dorien 043578
Irving, Gillian 044019
Irving, James 045100
Irving, Mary L. 053709
Irwin Meyer, Elaine 051589
Irwin, Charles W. 047086, 051712
Irwin, Ruth E. 050131
Irwinnson, Charles W. 047088
Isaac, Gwyneira 053701
Isaac, Preston 043953
Isakova, Ksenija Viktorivna ... 043151
Isaksen, Knud Fink 010093
Isawumi, Prof. Dr. M.A. 033426
Isbrandt, Lester 051273
Isbruch, Ursula 021477
Isdale, John 033310

Iseki, Masaaki	029617	Ivory, Ken	033331	Jacquat, Marcel S.	041408	James, Louise	054539
Isensee, Dr. Sabine	020991	Ivory, Margaret	033268	Jacqué, B.	015108	James, Marion E.	048402
Iser, Wunibald	020639	Iwaki, Ken-ichi	028959	Jacquemin, Hervé	015300	James, Rodney	001313
Isernhagen, Karin	042223	Iwakiri, Masanori	029044	Jacquemyn, Erik	003861	James, Sigrid	005811
Isert, Ingo Rüdiger	022037	Iwamoto, Mitsuo	028693	Jacques, Claude	013052	Jamieson, Louise	001628
Işfânoni, Doina	036310	Iwasaki, Yohachiro	028667	Jacques, R.A.	015890	Jamieson, Malcom E.	047534
Ishag, Ali	034147	Iwata, Tadao	028690	Jacquesson, Alain	041514, 041537	Jamieson, P.M.	006424
Ishibashi, Kentaro	028652	Izaqirre, Manu	038898	Jacquet, Alain	016016	Jaminon, Martine	003797
Ishiga, Harue	028639	Izdebska, Barbara	035366, 035368	jacquinot, Marc	015036	Jamison, Wayne	048242
Ishihama, Beniko	029252	Izquerdo Pascual, José María	039641	Jacquot, Dominique	015882	Jamkhedkar, Dr. A.P.	024465
Ishii, Hisataka	028737, 029600	Izsatovska, Dora	004984	Jadid-al-Eslam, Rasoul	024742	Jammot, Dr. Dominique	014587
Ishii, I.	028585	Izui, Hidekazu	029305	Jadoda, Rachel	050447	Jampen, Ulrich	042169
Ishikawa, Chiyo	053248	Jaacks, Prof. Dr. Gisela	019009	Jae, Jun Young	050459	Jamro, Ron D.	051162
Ishikawa, Koichi	028879	Jaakkola, Markku	010797	Jaeger, Anders	010023	Jamščikova, Natalja Borisovna	038020
Ishikawa, M.	028589	Jaatinen, Aija	011054	Jäger, Dr. Hagen	018133	Jan, Darja	038501
Ishikawa, Tetsuko	028652	Jaatinen, Carina	010771	Jäger, Hans	002477	Janaloo H., Hough	050397
Ishimaru, Shoun	028624	Jaatinen, Toimi	011393, 011394, 011394, 011407	Jäger, Dr. Joachim	017154	Janas, Stanisław	035095
Ishizaki, Katsumoto	029653		Jäger, Dr. M.	017970	Jančár, Dr. Ivan	038306
Ishizuka, Kaname	028596	Jaber Al-Jaber, Ibrahim	036181	Jaeger, Steven D.	050924	Jančeková, Viera	038474, 038475
Ishtiaq Khan, Muhammad	034121	Jablonská, Dr. Beáta	038328	Jäger, Veronika	019292	Jancsy, Peter	038284
Isifidis, Kaplanis	023124	Jabłońska, Teresa	034686, 035445, 035446, 035449, 035451	Jaegere, Isabelle De	003756	Janda, Don	050647
Iske, Gary	046884		Jaeggi, Dr. Annemarie	017077	Janda, Martin	002743
Isken, Suzanne	050456	Jablovska, Gaida	030215	Jaehner, Inge	021013, 021015	Jandl, Ferdinand	001719
Iskender, Prof. Kemal	042953	Jacanović, Dragan	038226	Jämbäck, Juha	011275	Jandorová, Monika	009975
Iskrzycka, Milena	021289	Jachson, Haidee	045418	Jäppinen, Jere	010826	Jané, Joseph	038982
Islamova, Tatjana Michajlovna	037946	Jacklitch, Prof. Paul	046929	Järliden, Barbro	040744	Janecek, James	052378
Isler, Andreas	042241	Jackowska, Anna	035282	Järvinen, Marja-Liisa	011333	Janes, Todd	005755
Isler, Ernst	021426	Jackson-Forsberg, Eric	051560	Järvinen, Vesa	010804	Janev, Georgi	005209
Isler, Jacques	042167	Jackson, Angela	049765	Jaffe, David	045057	Jangoux, Prof. M.	003493
Ismai, Reza	024782	Jackson, Barry	043898	Jaffe, Linda	046776	Janics, Dr. Hunert	002019
Isman, Eileen	007484	Jackson, Bee	005873, 007000	Jaffe, Steven	051455	Janik, Michael	002192
Isogai, Hideo	028558	Jackson, Bev	046353	Jagenteufel, Hermann	003138	Janisch, Mag. P. August	002585
Ison, Judy C.	052413	Jackson, Bryan	005820	Jager, Chuck	046924	Janjić, Gordan	038210
Ison, Susan P.	050506	Jackson, Henry	044336	Jager, Drs. G.W.M.	032864	Janke, Volker	021793
Isoz, Nicolas	041200, 041201	Jackson, Jerry	052658	Jaggi, Fred	041734	Janker, B.	020723
Israel, Sue	050687	Jackson, Kenneth T.	051455	Jaggi, Martin	041508	Janki, Ruth	042123
Israfilovglu, I.R.	003174	Jackson, Margaret	047986	Jagodin, Karen	010667	Jankóné Pajor, Ildikó	023555
Isshi, Kaihei	028664	Jackson, Paul	038089	Jahan-Giry, Dr. Ali	024715	Jankov, Dr. Peter	022265
Istas, Dr. Yvonne	022000	Jackson, Sailor	046816	Jahn, Thomas	019559	Jankova Kazakova, Rumjana	005010
Istha, Marieke	031971	Jackson, Steven B.	047166	Jahn, Dr. Wolfgang	018179	Jankova, Julia	005038
István, Dr. Berta	023502	Jackson, Toby	045116	Jahnigen, Rick	047740	Jankovskyj, Anatolij Andrijovyč	043133
Itai, Josef	025130	Jackson, Tony	046160	Jahnke-Lowis, Ilse	020670	Jankowski, Andrzej	034834
Itakura, Fumiaki	029609	Jacob-Friesen , Dr. Holger	019517	Jahnke, Dr. Elke	021931	Janků, Miloslav	009522
Iten, André	041521	Jacob, Dr. Christina	019157, 019164	Jahns, Achim	018771	Jankulovski, Plamen	005116
Iten, Brigitte	041714, 041980	Jacob, Eric	013280	Jaimes Camargo, Jairo Antonio	008742	Janneau, O.	012259
Ito, Masakazu	029211	Jacob, François	041525	Jain, Dr. Jyotindra	024222	Jannek, Norbert	019475, 019655
Ito, Tomoyuki	029226	Jacob, Dr. João Manuel Neto	035578	Jaindl, Franz	002860	Jannet, Monique	014485
Itoh, Hiroji	029170	Jacob, Milner	033313	Jajčević, Zdenko	009039	Jannke, William	054219
Itter, Wolfgang	019734	Jacob, Pascal	003330	Jakeuchi, Yoichi	028857	Janović, Ognjana	008956
Ittmann, John	052081	Jacob, Walter	052151	Jakeway, Robert	052640	Janovy, Karen	050300
Iuga, Georgeta Maria	036219	Jacobacci, Emilia	027624	Jaki, Dr. Barbara	038552	Janovyč, Oleksandr Vasyljovyč	043281
Iurman, Emilio	027214	Jacobi, Dr. Fritz	017262	Jakibčuk, Mykola Mychajlovyč	043156	Janscó, Istvan	004852
Ivančević, Nataša	008977	Jacobs, Carin	046936	Jakimińska, Grazyna	034994	Janse, K.P.J.	031978
Ivanenko, Pjotr Ivanovič	043177	Jacobs, Charles	050742	Jakober, Yannick	038932	Jansen, F.W.	032495
Ivannikova, Ljudmila	037367	Jacobs, Edwin	032737	Jakobi, Paul	020415	Jansen, G.	032826
Ivanov, Aleksandr Aleksandrovič	037215	Jacobs, F.J.	038664	Jakobs, Dr. Ingrid	020382	Jansen, Dr. Gregor	019522
Ivanov, Ivan	005141, 005148	Jacobs, Flora Gill	054174	Jakobsen, Björn M.	040701	Jansen jr, John D.	047906
Ivanov, Stefan	005075	Jacobs, John	049229	Jakobsen, Egil Olav	033594	Jansen, Kristian	033604
Ivanov, Vladimir Alexandrovič	037690	Jacobs, Joseph	051499	Jakobsen, S.E.	010116	Jansen, Leo	031996
Ivanova, Elena	047038	Jacobs, Karen	053080	Jakobsons, Imants	030094	Jansén, Maija	011438
Ivanova, Galina Olegovna	037127	Jacobs, Lisa D.	050139	Jakomin, Dušan	028208	Jansen, Siri	033526, 033534
Ivanova, Katja	005195	Jacobs, Marty	051443	Jakoubĕová, Dr. Vladimíra	009925	Janser, Andres	042217
Ivanova, Maja I.	037691	Jacobs, Dr. Stephanie	019981	Jakovleva, G.A.	037432	Janssen-Reinen, M.	032658
Ivanova, Natalja Vasiljevna	036937	Jacobs, Ted Seth	013746	Jakovleva, Marina Dmitrievna	037818	Janßen, B.	016743
Ivanova, Silvia	046999	Jacobs, Warwick	044452	Jakovleva, Nadežda Petrovna	038008	Janssen, Dr. Elske	003281
Ivanova, Tatjana Nikolaevna	038078	Jacobsen, Anne Marie	010092	Jakovljević, Goran	008884	Janssen, Dr. J.A.M.M.	032894
Ivanovskaja, Dr. Irina Nikolaevna	037328	Jacobsen, Bent Sonne	010234	Jakubowski, Dr. Krzysztof	035375	Janssen, Laura	054328
Ivanuš, Rhea	009036	Jacobsen, Gaute	033742	Jakunskienė, Audronė	030312	Janssen, Peter G.	019812
Ivask, Krista	010668	Jacobsen, Harald	033972	Jakuševa, Valentina Maksimovna	038077	Janssens, Dominique-Charles	011815
Ivaškina, Ljudmila	037995	Jacobsen, Robert	050944	Jalal, Masoody	024793	Janssens, Jan	003909
Ivčević, Sanja	008993	Jacobsen, Signe	010354	Jalandoni, Carlos L.	034480	Janssens, Sandra	003411
Ivelic, Milan	007546	Jacobson, Harold	049919	Jalava, Mari	010952, 011464, 011465, 011466	Jansson, Björn	040978
Ivens, R.	038765	Jacobson, L.	038765		Jansson, Tarja	011101
Iversen, Mette	010361	Jacobson, LouAnn	048337	Jalilian, S.J.	024764	Jantsch, Karl	002031
Iverson, Dr. John B.	052534	Jacobson, Sally	048589	Jalon, Rita	003304	Janus, Jerzy	034691
Iverson, Stephen C.	050375	Jacobson, Thomas N.	051992	Jalšić Ernečić, Draženka	008931	Janzen, Thomas	019786
Ives, Alicia	046376	Jacobson, Thora E.	052095	Jambon, Marcel	013361, 013362	Janzweert, Jan	018926
Ives, Dr. William	048591	Jacobsson, Ingela	041087	James-Sarazin, Ariane	014722	Jaouen, Jean-Pierre	011997
Iveson, Jon	044130	Jacobsson, Yvonne	041008	James Jr., Alton	051445	Japardi, Tilly	050039
Ivić, Prof. Vinko	009043	Jacoby, Christina	030305	James, Beth	048497	Jappy, Lorna	044557
Ivica, Križ	038585	Jacoupy, Charles	015812	James, Betty	005793	Jara, Dr. Holguer	010471
Ivičić, Ratomir	008887	Jacqmard, Jules	052169	James, Debbie	043518	Jarábik, Gabriella	038318
Ivinskis, Antanas	030500	Jacqmard, Philippe	003767	James, Frank	045037	Jaramillo Agudelo, Darío	008604
Ivkanec, Ivanka	009028	Jacquat, Jeannine	041844	James, Jon G.	054510	Jaramillo de Carrión, María	

Mercedes 010485
Jaramillo Salazar, Juan Andrés . 008764,
008768
Jarchow, Jens 022430
Jareon, Cécile 015295
Jarh, Dr. Orest 038494, 038562
Jaritz, Josef 002703
Jarmoljuk, Ol'ga Mikolaivna . . . 043239
Jarocka-Bieniek, Jolanta 035245
Jaros, John R. 046671
Jaros, Urszula Maria 035093
Jarosch, Jessica 001742
Jaroševskaja, Valentina
Michajlovna 037165
Jaroszewicz, Mieczysław 035203
Jarova, Galina Ivanivna 043203
Jarquín Pacheco, Ana María . . 031123
Jarratt, Ardell 052362
Jarratt, Dervla 045599
Jarrige, Jean-François . 014660, 014776
Jarron, Matthew 044168
Jarvie, Robert 045486
Jarvis, Anthea 045208
Jarvis, Antony 044830
Jarzembowski, Prof. Dr. E. 045193
Jaschke, Mag. Beatrice 002209
Jašić, Ivana 008900
Jasim, Dr. Sabah 043340
Jasiuk, Jerzy 035195, 035364
Jasiūnas, Dr. Henrikas 030526
Jaskanis, Paweł Olaf 035354
Jaskowiak, Jennifer 050421
Jaskulski, Janusz 035128
Jastram-Porsche, Sabine 016480
Jastrebova, Raisa 037573
Jastrzębski, Jerzy 034982
Jatti Bredekamp, Prof. Henry C. 038672,
038673, 038679
Jaukkuri, Maaretta 010836
Javed, Raja Muhammad 034138
Jávor, Dr. Anna 023518
Jawlensky-Bianconi, Angelica . . 041682
Jaworowicz, Hanna 034798
Jaworowska-Mazur, Ewa 035034
Jay, Hilary 052041
Jaysane, A. Patricia 050190
Jean, Louis 043384
Jeanbrau, Héléne 013742
Jeandron, Angela 049960
Jeanes, K.I. 038719
Jeanmart, Jacques 003900
Jeannot, Bernard . . . 014717, 014795
Jeanselme, Brigitte 016021
Jeansson Kock, Brgitta 041040
Jebreil Zadeh, A. 024727
Jebsen, Mary 053113
Jecha, James 050027
Jecht, Heidrun 019505
Jeck, Siegrun 019350
Jeckel, J. 032945
Jedda McNab, Barbara 052815
Jędrzejewski, Antoni 035257
Jędrzejowski, Grzegorz 034820
Jefferies, Dr. Chris 048391
Jefferies, Susan 007176
Jeffers, Christine 048094
Jeffery, Gordon 007253
Jeffries, Bill 005506, 007293
Jeffries, Bob 045117
Jeffries, Donna 000880
Jegorov, Boris Michajlovič 036859
Jehin, Paul 004041
Jehl, Dr. Rainer 019422
Jeitler, Johannes 002887
Jekov, Vasko 005192
Jelavić, Duška 009028
Jeletich, Dr. F. 032294
Jelfs, Ann 033293
Jeličić, Janko 009036
Jelínek, Petr 009510
Jelinek, Dr. Zdeněk 009779
Jelinková, Jaroslava . . 009980, 009981
Jeliseeva, Tatjana 043227
Jelvakova, Irena A. 037324

Jemison, Gerald Peter 053979
Jenal, Christian 041919
Jenal, Ludwig 041919
Jenied, Carol 053102
Jenisch, R. Dietlinde 018730
Jenkins, Adrian 043495
Jenkins, Carol 052008
Jenkins, Dr. Deborah 045012
Jenkins, Donald 052302
Jenkins, Frances Susong 049080
Jenkins, Sabra 053360
Jenkins, Susan 044904
Jenkner, Ingrid 005957
Jenko, Helena 038554
Jenko, Lilija Dmytrivna 043155
Jenko, Mojca 038552
Jenks, Tracey 046511
Jenne, Hans-Jörg 018184
Jenneman, Eugene A. 053791
Jenni, Rudi 000966
Jennings, Anne B. 049436
Jennings, Bill 049436
Jennings, David 043667
Jennings, Lucille 051689
Jensch, Dr. Rainer 022422
Jensen, B. 033303
Jensen, Ellen Elisabeth 010348
Jensen, Erik 010191
Jensen, Herbert 019378
Jensen, James 049527, 049528
Jensen, Jennifer R. 051977
Jensen, Johan M. 010178
Jensen, Kathleen A. 053473
Jensen, Ken 050707
Jensen, Leslie D. 054322
Jensen, M. Wayne 053746
Jensen, Mona 010005
Jensen, Niels Erik 010209
Jensen, Ole G. 023341
Jensen, Otto 010347
Jensen, Pam 048221
Jensen, Sterling 006463
Jensen, Tom 033828
Jensen, Vivi 010206
Jensen, William 051564
Jensner, Magnus 040690
Jenssen, Ragnar 033559
Jentjens, Kathrin 019692
Jenvold, Birgit 010206
Jenzen, Dr. Igor A. 018000
Jepbarov, Rejep 043096
Jeppesen, Hans 010107
Jeppsson, Kristina 040578
Jepson, David 045423
Jerg, Alfred 019939
Jerin, Robert M. 048453
Jerković, Dragan 038185
Jerlich, Walter 001738
Jerome, Jeff 046744
Jerris, Rand 048694
Jerrome, Peter 045498
Jervois, W. 038724
Jésman, Prof. Czesław 034973
Jesretić, Miroslav 038245
Jessat, Mike 016535
Jessewitsch, Dr. Rolf 021898
Jessop, L. 045347
Jessup, Wesley 051938
Jeszenszky, Dr. Sándor 023511
Jett, Cheryl 047878
Jette, Carol 046460
Jetzer, Gianni 041922, 051477
Jeufroy, Pierre 013101
Jevdokimova, Jemilija
Mykolaivna 043161
Jevsejeva, Olga Jevgenivna . . . 043136
Jevtović, Mila 038168
Jewell, Anne 050496
Jewell, Brian 044520
Jewell, Robert 051131
Jezierski, Richard 014566
Jeziorowski, Tadeusz 035140

Jezler, Peter 041315, 041475
Jezmann, Meike 022131, 022133
Ježowitzová, Jiřina 009591
Jha, Dr. Uma Raman 024519
Ji, Gon Gil 029942
Ji, Liancheng 007581
Ji, Qiang 007619
Ji, Zhongliang 007944
Jian, Li 048170
Jiang, Chuanhao 007786
Jiang, Guangyou 007816
Jiang, Hua 008494
Jiang, Huimin 008046
Jiang, Shidong 007744
Jiang, Tingyu 008122
Jiang, Wei 007685
Jiang, Weiming 008378
Jiang, Xukui 008084
Jiang, Yaonan 007724
Jiang, Yibing 008558
Jiang, Yuanjin 008040
Jiang, Zhili 007955
Jianu, Magda 036509
Jiao, Desen 007960
Jibah, Haji Matassim Bin Haji . . 004960,
004962, 004964
Jicărean, Monica 036535
Jie, Youjun 008064
Jilhjalmsson, Larus 024105
Jiliá, Pablo 038957
Jillings, Sarah 045011
Jillis, Nellie 038733
Jiménez Cadena, Gustavo 008749
Jiménez Chávez, Alma Rosa . . 031266
Jiménez Espejo, Francisco J. . . 039783
Jiménez Losantos, Encarna . . . 040305
Jiménez Martínez, Dr. Rafael . . 039502
Jiménez, Angel 040086
Jimeñez, Mar Villafranca 039459,
039468
Jimenez, Mercedes 031285
Jiménez, Pedro López 039317
Jimeno Santoyo, Myriam 008629
Jimison, Tom 051124
Jimmie, Charles 049314
Jin, Baidong 008366
Jin, Dechang 008344
Jin, Hongde 008350
Jin, Jon Moon 029876
Jin, Yongfeng 008001
Jin, Yongtian 007579
Jing Weng, Liao 007651
Jing, Guo 007992
Jing, Zhanyuan 008148
Jirmusová, Hana 009509
Joans, Dr. Barbara 051678
Jobke, Ryszard 035026
Jobst-Rieder, Marianne 003053
Jobst, Franz 017137
Joch, Dr. Peter 017826
Jochem, Marlene 019482
Jochims, Jennifer 053552
Jochims, Richard . . . 018605, 019175
Jochum, Franz 002459
Jockel, Nils 019011
Jodłowski, Prof. Antoni 035391
Joeckel, Reecanne 046532
Joenniemi, Sirpa . . . 011408, 011409
Jönsson, Bertil 040611
Jönsson, Tomas 040744, 040876
Jørgensen, Anja 010209
Jørgensen, Erik 010092
Jørgensen, Mogens 010188
Jöris, Dr. Olaf 020776
Joffre, Roland 013691
Joger, Prof. Dr. Ulrich 017565
Johannes, Janet 049413
Johannessen, Arild 033602
Johannessen, Kåre 010319
Johannessen, Preben L. 033656
Johannesson, Birgitte 010176
Johannsen, Frerk 019911
Johannsen, Ilona 022686
Johannsen, Karin 018065

Johannsen, Ruby 049641
Johannson, Sigfus 006227
Johansen, Annette 010307
Johansen, Jens 010138
Johansen, Michael 010123
Johansen, Rikke 010361
Johanson, Bruce 051784
Johansson, Börje 040510
Johansson, Conny 040809
Johansson, Eric J. 047233
Johansson, Hjördis 041066
Johansson, Jan 040796
Johansson, Kennet . . . 040824, 040833
Johansson, Ketty . . . 010240, 010241
Johansson, Kjell 040758, 041017
Johansson, Lars G. 040781
Johansson, Marie . . . 041080, 041098
Johansson, Sonny 041089
Johansson, Sven-Eve 040792
Johansson, Tina 041066
Johansson, Ulf 011208
Johansson, Ulrica 041106
John-Willeke, Bettina . 017820, 017828
John, Holverson 053414
John, J. 009751
Johncock, M. 001179
Johnes, John V. 046609
Johns, Daniel 049616
Johns, Kevin L. 050419
Johns, Sharleen 048392
Johns, Valerie 049181
Johnsen, Aksel 033999
Johnsen, Berit Åse 033693
Johnsen, Ron 052469
Johnson Bowles, Kay 048711
Johnson, Alfred E. 050185
Johnson, Amy Jo 052223
Johnson, Andrew 044232
Johnson, Anne 054089
Johnson, Arden 049405
Johnson, Bert 048450
Johnson, Brooks 051573
Johnson, Byron 051127
Johnson, Byron A. 054021
Johnson, C. Clayton 052319
Johnson, Carl 006807
Johnson, Christina 050419
Johnson, Christine 033270
Johnson, Craig 052835, 052841
Johnson, D. Thomas 046826
Johnson, Dale 043484
Johnson, David 052163
Johnson, David A. 053863
Johnson, Diane 048091
Johnson, Douglas 047043
Johnson, Drew 051681
Johnson, Edward 053505
Johnson, Dr. Eileen 050517
Johnson, Frank 038757
Johnson, Gerardo 023400
Johnson, Gina 053530
Johnson, Grace 052979
Johnson, Harold 053185
Johnson, Dr. Heide-Marie 050813
Johnson, James 047610
Johnson, Jean 047574
Johnson, Jeanne 053076
Johnson, Jennifer L. 047935
Johnson, Jeremy 000817, 000819
Johnson, Jerome Martin 050064
Johnson, Joan 046004
Johnson, Dr. John R. 053093
Johnson, Judi 047318
Johnson, Judy 047315, 047317
Johnson, Karen 053138
Johnson, Kathleen E. 053685
Johnson, Kaytie 049235, 049236,
049237
Johnson, Kenneth R. 049022
Johnson, Dr. Kirk R. 048261
Johnson, Laurel 000816
Johnson, LaVern M. 050539
Johnson, Libba 047930
Johnson, Lisa 048703

Johnson, Lydia 047473
Johnson, Mark M. 051025
Johnson, Marlo L. 049907
Johnson, Mary A. 052914
Johnson, Maurianna 050195
Johnson, Melinda 051136
Johnson, Mirma A. 046337
Johnson, Nancy T. 050999
Johnson, Nichola 045406
Johnson, Patricia 001133, 052731
Johnson, Pearl 049445
Johnson, Penny 044981
Johnson, Polly 049997
Johnson, Rebecca A. 048715
Johnson, Dr. Richard E. 052414
Johnson, Robert S. 048842
Johnson, Robin 044019
Johnson, Dr. Ron 007364
Johnson, Sandra L. 051982
Johnson, Sharon 048104
Johnson, Sheila A. 007491
Johnson, Steven 048208
Johnson, Stuart 051126
Johnson, Suni 046756
Johnson, Terry 050821
Johnson, Thomas B. 054584
Johnson, Tom 054582
Johnson, Willard 051731
Johnson, William C. 049062
Johnston, Curtis 050906
Johnston, Eileen 054122
Johnston, Phillip M. 053092
Johnston, Renée 005979
Johnston, Sally 046769
Johnston, Sona 046733
Johnston, Stephen 045455
Johnston, Virginia 048327
Join-Diéterle, Catherine 014759
Joiner, Janet 053984, 053985
Joiner, S. 005732
Jojo, Eva 050953
Jokio, Mikko 010754, 010755
Jokulo Cooper, Jimmie 046751
Jolliff, Dawna 049440
Jolly, Dr. Anna 041883
Jolly, Dianne 001298
Jolly, Linda Lee 053815
Jolly, Renee 053116
Joly, M. 015127
Jomppanen, Tarmo 010904
Jonaitis, Aldona 048662
Jonaitis, J. 030477
Jones Cox, Beverly 054150
Jones Harper, Suzanne 050571
Jones, Alan 045827
Jones, Amanda C. 050668
Jones, Amy 043846
Jones, Andy B. 054383
Jones, Ann 044244
Jones, Anne 044335
Jones, Barbara L. 049259
Jones, Betty Lou 050805
Jones, C.R. 047972, 047973
Jones, Dr. Cheri A. 048261
Jones, Clive 043944
Jones, Cynthia L. 049315
Jones, D. 043826
Jones, D. A. 046211
Jones, D.A. 043471
Jones, David 044646, 047635
Jones, Diana 001403
Jones, Donna P. 053376
Jones, Dr. Douglas S. 049001
Jones, Duffie 049825
Jones, E. Michael 004197
Jones, E.B. 033325
Jones, Edward 048992
Jones, Eric 043956
Jones, Erin 049267
Jones, Glenn 049213
Jones, Harvey 051681
Jones, Hugh R. 050529
Jones, Huw 044019
Jones, Jeanne 045150

Jones, Jeff 051853
Jones, Jill 053238
Jones, Joanna 045838
Jones, Josephine 044187
Jones, Joy 006235
Jones, Julie 051424
Jones, Kathy 005547, 048466
Jones, Keith 001032, 044442
Jones, Kevin 001416, 046143
Jones, Larry A. 050480
Jones, Lester 043944
Jones, Lial A. 052725
Jones, Linda 047504
Jones, Lisa 000889
Jones, Lowell N. 047722
Jones, Marilyn 049749, 052341, 053614
Jones, Mark 045123
Jones, Nia 045283
Jones, P. 044198
Jones, R. 006422
Jones, Rhian 046362
Jones, Richard 043866
Jones, Robert 000825
Jones, Dr. Robert 049752
Jones, Sharain 007138
Jones, Sian 045830
Jones, Stephen 046557, 052227
Jones, Sue 050037
Jones, Suzi 046470
Jones, Thomas 050878
Jones, Prof. Tom 043617
Jones, Tom 044332
Jones, Trevor 044132, 049229
Jones, W.J. 033322
Jonet, Sylviane 013095
Jong-Wook, Prof. Bae 029898
Jong, A. de 032957
Jong, Andy de 005737
Jong, C.C. de 032510
Jong, Dr. H. de 032002
Jong, Dr. M.A. de 032195
Jongh, Marja de 032781
Jonghe, Mieke de 003532
Jongs, Carel 050819
Jongstra, Jaap 032575
Jonin, Marie-Thérèse 012510
Jonker, Drs. A. 032209
Jonker, Drs. M. 031952
Jonsson, Prof. Goran 040953
Jonsson, Ingrid 040973
Jónsson, Petur 024070
Jonušienė, Eleonora 030430
Jonušienė, Vitalija Teresė 030431
Joo-Sik, Kim 029962
Joohee, Kim 030008
Joos, Cristian 042181
Joos, Erwin 003285
Joos, Marleen 003265
Jooss, Dr. Birgit 020853
Joppien, Dr. Rüdiger 019011
Joralmon, Mary 052989
Joraslafsky, Helen 000790
Jordaens, Lode 003583, 003584
Jordan Wass, Catherine H. . . . 051573
Jordan, Brenda 050028, 050030, 050031
Jordán, Dr. František 009963
Jordan, Jai 053682
Jordan, Prof. Dr. Lothar 018399
Jordan, Mary Alice . . . 053066, 053067
Jordan, Peter 042008
Jordan, Robert 005678, 018814
Jordan, Thomas E. 049096
Jordanov, Ivan 005134
Jordanović, Branislava 038178
Jordon, G. 000920
Jordon, J. Greg 048579
Jorgensen, Roy 048736
Jorikson, Håkan 040667
Joris, Philippe 003793
Joris, Yvònne G.J.M. 032476
Jorquera Caro, Sonia Alexandra 031338
Jorrand, C. 013569
Jorrín García, J.A. 039808
Josanu, Vitalie 036769

Joselu, Sergio 025596
Joseph-Debaque, Sandrine . . . 012558
Joseph, Carolyn 049408
Joseph, Jean-Louis 011699
Joseph, Lorraine 047974
Joseph, Sarah 049193
Josephson, Eva 040887
Josephson, Magnus 040888
Joshi, Prof. B.D. 024289
Joshi, Mallickarjun 024515
Joshi, Vishwanath N. 024427
Josić, Zoran 038124
Joslin, Stewart 053762
Jossen, Dr. Erwin 041997
Josset, Dr. Patrice 014755
Jossifova, Maria 005191
Jost-Prześlakowska, Maria 035264
Jost, Gerda 018623
Jost, Loren 052599
Jost, Stephan 051679
Jósvai, Dr. Katalin Dankó 023871
Joubert, Alain . 013432, 013433, 014068,
. . . . 014526
Jouen, Jean-Pierre 013392
Joüon des Longrais, F. 015062
Jouffroy 011983
Joulin, Guy 014310
Jourdan, Aleth 011594, 011595
Jourdan, Jean-Pierre 051346
Jourdan, Patrick . . . 014326, 014327
Journiac, Virginie . . . 012282, 012285
Jous, Léon 003568
Jouve, Gerard 012501
Jouzeau, Marie-Hélène 014419
Jovanov, Dr. Jasna 038219
Jovanović, Igor 038187
Jovel Castro, Walter E. 008862
Jovelić, Marijana 038185
Jover Maestre, Francisco Javier 039942
Jović, Vukašin 038187
Joyce, Dan 049927
Joye, Jean 003452
Joyeux, Y. 003391
Joyner, Elizabeth . . . 053973, 053976
Joyner, Roger D. 049256
Ju, Calvin 042288
Ju, Jin-Soon 029943
Ju, Wancang 008192
Juárez Nieto, M. Magdalena
Sofía 031393
Juaye, Comte Pierre du Manoir
de 012429
Juchli, Cornelia M. 042013
Juchniewich, Daniel 050648
Juck, Marcel 038320
Jucuti, Ioan 036338
Judd, Michael 046316
Judge, Joseph M. 051575
Judson, E. 038853
Judson, Etta 038820
Judson, Jack 052941
Judson, William D. 052157
Juelich, Clarence 054367
Jülich, Dr. Theo 017822
Jüllig, Carola 017299
Jürries, Wolfgang 019823
Jufa, Marija 037651
Juffinger, Dr. Roswitha 002636
Jugie, Sophie 012787
Juhász, Gyula 023780
Juhl, Kirsten 033972
Jul, Ole 010343
Jules Kouloufoua, Jean Gilbert . 008844
Julia y Macias, Manuel 039761
Julia, I. 014782
Julien-Laferriére, Alain 016014
Julin, Richard 041016
Julio, Roberto 025734
Juliusburger, Thomas 047203
Julkunen, Pirjo 011160
Jullian, Nathalie 014637
Jumalon, Julian N. 034439
Jumalon, Prof. Julian N. 034462
Jumarä, Dan 036506

Jumelet, J. 032113
Jun, Yong-Bok 029054
Juncosa, Enrique 024891
Juneau, Madeleine 006407
Junek, David . 009770, 009771, 009772
Jung, H. 032705
Jung, Ilse 003026
Jung, Ludwig 021423
Jung, M. 032436
Jung, Dr. Michael 027627
Jung, Seung Sun 029943
Jung, Dr. Werner 019707
Jungblut, Maria-Paule 030592
Junge-Stevnsborg, Kirse 040690
Junge, Brigitte 016681
Junge, Cornelia 019977
Junge, Joachim 022901
Junge, Dr. Michael 019314
Junge, Dr. Peter 017112
Junginger, Ernst . . . 019899, 019901
Junglas, Uta 022511
Jungmann, Judith 003033
Junhem, Inga 041053
Junier, Caroline 041796
Junior, Peter 019214
Junker, Patricia 048898
Junno, Johanna . . . 011225, 011515,
. . . . 011516, 011517
Junod, Barbara 042217, 042219
Juntikka, Ilse 011222
Juola, Emmy 011287
Juranek, Dr. Christian 022585
Jurca, Emanuela 036193
Jurčenko, Oleg Grygorovyč . . . 043182
Jurdana, Ela 009036
Jurdyga, Judy 006978
Jurecka, Małgorzata 035166
Jurevičius, Juozapas 030378
Jurgemeyer, Marne 048848
Jurić, Radomir 008953, 009020
Jurić, Spomenka 008991
Jurica, Varina 008915
Jurkäne, Anna 030232
Jurkiewicz, Jan Gustaw 035442
Jurriaans-Helle, Dr. G. 031923
Jussila, Neil 046986
Just, Arthur 003056
Just, Helmut 002959
Just, Rüdiger 018970
Justice, Ed jr. 048371
Justice, Gloria 052251
Justus, Dirk 018983
Justus, Mary Beth 046592
Jusupova, Dr. Majram
Ažibekovna 030074
Juton, Alain 014177
Jutras, Michel 007242
Juul Holm, Michael 010146
Juva, Antti 011257, 011258
Juvigny, Sophie de 015271
Juvonen-Eskola, Annamari 010771
Juzwenko, Dr. Adolf 035435
Kaak, Dr. Joachim 020508
Kaarto, Sari 011305
Kaasen, Svanhild Alver 033614
Kaba, Sory 023379
Kabić, Dušan 038187
Kábrtova, Jiřina 009743
Kabrud, Jack A. 050937
Kačalova, Alisa D. 036912
Kačírek, Ľuboš 038313
Kacso, Carol 036217
Kaczarová, Iveta 038385
Kaczmarczyk, Witold 035226
Kaczmarek, Drs. J. 032660
Kadake, Henrich 049855
Kader, Dr. Ingeborg 020554
Kadlec, Dr. František . 009783, 009799,
. . . . . 009800, 009803, 009807, 009830,
. . . . 009840
Kadowaki, Akira 029631
Kadurina, Elena 005149
Kadyrov, A.S. 042616
Käär, Pekka 011447

Käch, Robert 041867
Käck, Fredrik 040686
Kähler, Kai 017608
Kaelcke, Wolfgang . . . 021074, 021075
Kaelin, Léon 041595
Kämpf, Michael 019838
Kämpfner, Gabriele 017520
Kaenel, Gilbert 041653
Känsälä, Risto 010978
Käppeli, Bruno 041743
Kärgling, Dr. Karlheinz 020207, 020210
Kärki, Sirpa 040873
Kärnefelt, Prof. Ingvar 040806
Käs, Rudolf 020859
Kaeser, Dr. Bert 020575
Kaeser, Prof. Dr. Marc-Antoine . 041586
Kaeser, Olivier 041512
Kaesler, Roger L. 050186
Käss, Rudolf 020878
Kästner, Ursula 017072
Kästner, Dr. Volker 017072
Käyhkö, Unto 010985
Kafetsi, Anna 023020
Kagioka, Masanori 029214
Kahan, Mitchell 046329
Kahl, Elizabeth 051840
Kahle, Patricia 051275
Kahle, Shawn M. 048315
Kahlen, Prof. Wolf 017327
Kahler, Caroline 050318
Kahn, David M. 049384
Kahn, Leo 053122
Kahn, Tammie 049573
Kaijt, W.P.A.M. 032000
Kailashnathan 024123
Kaim, M.-L. 020527
Kaindl, Heimo 001969
Kaiqu, Liu 007608
Kaire, Inese 030176
Kaiser-Reka, Berol 017541
Kaiser, F.W. 032171
Kaiser, Friederike 020502
Kaiser, H. 018846
Kaiser, Dr. Hermann 017747
Kaiser, Michael 020851
Kaiser, Miriam 031256
Kaiser, Peter 041822
Kaiser, Philipp 041273
Kaiser, Dr. Stephan 021289
Kaißer, Paul 022366
Kaizuka, Tsuyoshi 029490
Kaján, Imre 023629
Kajanus, Pirkko 010892
Kajikawa, Yoshitomo 028934
Kakaras, Gunaras 030400
Kakko, Leena 011398
Kakourou-Chroni, Dr. Georgia . . 023272
Kakuda, Arata 028652
Kakusei, Miki 028795
Kal, W.H. 031950
Kalaber, Ladislau 036651
Kalaitzis, Myron 023243
Kalajainen, Jason 049507
Kalantzes, Isabelle 052918
Kalašnik, Tatjana Nikolajevna . . 037046
Kalašnikov, Nikolaj Ivanovič 037892
Kalčev, Kamen 005032
Kalčev, Petar 005211
Kalchthaler, Peter 018478
Kaldewei, Prof. Dr. Gerhard . . . 017851
Kaldor, Cynthia 050747
Kaldy, Robert 003045
Kalela-Brundin, Maarit 040816
Kalenberg, Ägel 046272
Kalenberg, Angel 046273
Kalenitchenko, Liliane 015189
Kalicki, Jerzy 034853
Kaligin, Pavel N. 037834
Kalina-Vidić, Ivana 038168
Kalinová, Dr. Alena 009468
Kalinsky, Nicola 044274
Kaljanac, Muhiba 004275
Kaljnyj, Dmytro Pavlovyč 043287
Kalkhoff, Gabriele 020989

Kalkmann, Hans-Werner 016858
Kallas, Nick 053884
Kállay, György 023575, 023734
Kallberg, Ulla 011435
Kallicas, Antonia 009399, 009400
Kalligas, P. 022975
Kallio, Kalle 011412
Kallio, Maija-Riitta 011075
Kalmár, Ágnes 023733
Kalmikova, Elena 037425
Kalmite, Leide Alida 050851
Kalmring, Sven 021663
Kalnins, Indulis 052734
Kalogeros-Chattan, Charlie 050403
Kalokerinos, Alexis 023096
Kaloutsakis, Agisilaos 023096
Kalpakova, Akgul 043100
Kalpickaja, Elena Jakovlevna . . . 037745
Kalšan, Prof. Vladimir 008891
Kalt, Daniel 003091
Kalt, Ewald 042040
Kaltenberger, Johann 002678
Kaltenbrunner, Regina 002638
Kaltofen, Dr. Andrea 020368
Kaltsas, Dr. Nikolaos 022997
Kaltved, David 048615
Kalus, Dr. Jaromír 009734, 009736,
                                         009738
Kamal, Hassan 042259
Kamansky, David 051936
Kamarun, Prof. Ham Rabeah . . . 030728
Kamber, Ljubomir 038204
Kamber, Pia 041268
Kamber, Robert 041642
Kamber, Urs 041196
Kamburg, Petra 019102
Kamel, Linda A. 054124
Kameneva, Marina Rimovna . . . 038039
Kamer, S.J. 032668
Kamerling, Leonard 048662
Kames, Holger 019983
Kamii, Monsho 029675
Kamijo, Kazuo 029390
Kamino, Yoshiharu 028867
Kaminski, Jane 017080
Kamiya, Yukio 028610
Kamm, David 048207
Kamm, James J. 050777
Kammann, Rolf 021229
Kammel, Dr. Frank Matthias 020853
Kammerhoff, Alexander 020567
Kamp 021542
Kamp, Dr. Diane D. 053689
Kamp, Michael 020089
Kampe, Allen 051315
Kampe, Katrine 009997
Kamper, Gerard de 038845
Kamprath, Audrey 052201
Kamras, Barbro 040579
Kamras, Borge 040579
Kamsky, Virginia A. 051362
Kamuhangire, Dr. E 043108
Kamyšova, Antonina Abramovna 036926
Kan, G. 033065
Kan, Marina 030210
Kana Anderson, Gail 051586
Kanabe, Albert 005779
Kanai, Takao 028734
Kanapkiené, Vida 030397
Kanaseki, Hiroshi 029189
Kanavrov, Darin 004975
Kandeler-Fritsch, Martina 003033
Kandl, Veronica 049195
Kane, Allen 054151
Kane, Debra 033307
Kane, Erin 053212
Kane, Gordon 045255
Kane, Katherine 049385
Kane, Linda 005672
Kane, Mary Ann 048014
Kane, Patricia 051267
Kaneichi, Ninomiya 028682
Kaneko, Kenji 029610
Kaneko, Yuriko 029305

Kanetaka, Kaoru 029730
Kanetuki, Takeshi 029005
Kanev, Ivajlo 005191
Kang, Prof. Joeng-Won 030040
Kang, Utchang 029943
Kangas, Faye 049604
Kania-Schütz, Monika 018866
Kania, Rolf 020898
Kankaala, Katri 011166
Kann, Lisa von 052778
Kann, Yanhua 007851
Kannan, Dr. R. 024197
Kannu, T. 038271
Kano, Kenji 028841
Kano, Takumi 029625
Kant Monteith, Joanita 050971
Kanter, Dr. Karin 020207
Kanter, Laurence B. 051424
Kantvilas, G. 001150
Kao, Deborah M. 047362
Kao, Li-Chen 042268
Kao, Shu-Kwei 042496
Kapches, Dr. Mima 007200
Kapfer, Otto 002456
Kapfer, Rudolf 002852
Kapinan, Catherine 054100
Kapinčeva, Dr. Iskra 005242
Kaplan, Dave 050331
Kaplan, Ilee 050376
Kaplan, Ludĕk 009566
Kaplan, Dr. Susan A. 047286
Kaplan, Wendy 050444
Kapler, Joseph 050601
Kappelmark, Ragnhild 041071
Kapronczay, Dr. Károly 023457, 023551
Kaptain, H.J. 010035
Kapteijn, Christine 044334
Kaptein, Helen 001532
Kapusta, Jan 009710
Kapustina, Nadežda Ivanovna . 043146,
                                  043148, 043149, 043150
Karačevcev, Vladimir F. 037015
Karadi, Gabriella 007187
Karadontis, Fleurette 022960
Karajotov, Dr. Ivan 004994
Karakas, Steve 052287
Karamaliki, Nota 023038
Karamanou, Dr. Aglaia 023055
Karamitrou-Mentessidi, Dr.
Georgia 022952
Karamut, Dr. İsmail . . . 042930, 042935,
                                               042941
Karamzin, Alex 053015
Karanitsch, Peter 002402
Karantzali, Efthymia 023147
Kararadeva, Svetozara 005168
Karas, Lourdes 052171
Karasawa, Masahiro 029610
Karaseva, Alevtina Pavlovna . . . 036872
Karaska, Dušan 038346
Karastamati, Dr. Helen 022995
Karbacher, Ursula 041927
Karbaum, Roswitha 020793
Kardinal, Royce 054399
Kardon, Nina 048888
Kardong, Don 053440
Kardosh, Judy 007288
Karegar, M. 024768
Karel, Johannes 003037
Karetsou, Al 023038
Karey, Carol 054191
Kargar 024763, 024776
Kargere, Lucretia 051364
Karhoff, Brent 048476
Karhunen, Lars 010817
Karibo, Lou 048928, 048932
Karinen, Helge 040983
Karingi, Lina 029867
Karis, Snježana 038587
Karis, Sirje 010701
Kärklina, Rūta 030187
Karlinger, Friedrich 002858
Karlsdottir, Anna Stella 033912, 033913
Karlsson, Anders 041106

Karlsson, Christer 040852
Karlsson, Eva-Lena 040841
Karlsson, Lars Olov 040514
Karlsson, Stig 040551
Karlstedt, Harry 011207
Karman Ohlberger, Christina . . . 040884
Karmazin, N.I. . 037200, 037202, 037204
Karnath, Sonja 021920
Karnauchova, Galina Georgievna 036873
Karner, Dr. Regina 003095
Karnes, Andrea 048906
Karneth, Dr. Rainer 016568
Karojeva, Ljudmyla Rostislavivna 043321
Károly, Simon 023490
Karon, Stephen A. 053635
Karosas, Algis 030537
Karosas, Gintaras 030512
Karov, Zdravko 005129
Karp, M. 051319, 051349
Karpf, Kurt 002610
Karpińska, Zofia 035280
Karpova, M.I. 037853
Karppinen, Eira 011031
Karr, Ed F. 051954
Karsch, Sylvia 020191
Karshens, Xander 032393
Karskens, Xander 032392
Karsmizki, Ken 048130
Karstadt, Bruce N. 050932
Karsten, Joan 052579
Karstens, E. 038822
Kartio, Kai 005129
Karttunen, Matti 010991, 011159,
                           011161, 011162, 011163
Karu, Mihkel 010667
Karwatzki, Dr. Barbara 021384
Karwowski, Lech . . . 035247, 035251,
                                  035252, 035253
Kasal, Václav 009561
Kasbi, Yasmine 003587
Kaschkat, Dr. Hannes 020920
Kashima, Teishin 029746
Kashiwabara, Shinji 028723
Kasilag, Emma Y. 034576
Kaširin, Oleg Semënovič 037959, 037965,
                                               037969
Kasparavičius, Stasys 030488
Kasperovič, K.K. 043113
Kasperska, Helena 034924
Kaspruk, Stepan Dmytrovyč . . . 043159
Kass, Amalie 047120
Kass, Emily S. 053672
Kassal-Mikula, Dr. Renate 003095
Kaštelionovaité, Zofija 030530
Kastler, Daniel 011882
Kastler, Dr. Raimund 002637
Kastriot, Dervishi 000010, 000011
Kasumov, Nazym Kasumovič . . 036967
Kasym, Sofija Vasiljevna 037762
Kaszak, Carole 045642
Kasziołka, Stanisław 035464
Kaszubkiewicz, Andrzej 034938
Katagishi, Shoji 029639
Kátai, Dr. László 023753
Katakami, Syusaku 029301
Katalin, Dr. Keserü 023474
Katanekwa, K.V. 034836
Katanekwa, N.M. . . . 054837, 054838
Katayama, Takashi 029435
Katerinkino, Lidija P. 037197
Katevenidou, Tassoula 023038
Kathane, M. Y. 024419
Katili, Dr. Ekki Husein 024565
Katins, Anne-Marie 018046
Kato, Bruce 049848, 053361
Kato, Jatsuo 029436
Kato, Sadao 029034, 029547
Kató, Zoltán 036237
Kato, Yuji 028808
Katona, Dr. András . . 023502, 023526
Katona, Dr. Stephen 046784
Katona, Dr. Tamás 023510
Katsalaki, Georgia 023096
Katsiff, Bruce 048366

Katsimpalis, Tom 050506
Katsura, Nobuo 029337
Katz, Alain . . . 012568, 012569, 012570
Katz, Paul 046436
Katzenbach, Dr. Annette 018179
Katzensteiner, Peter 002929
Katzinger, Dr. Willibald . . 002325, 002326
Kauder-Steiniger, Dr. Rita 020608
Kauder, Hans-Peter 002208
Kauders, Audrey S. 049910
Kauer, Dr. Andrea 041417
Kauffmann, André 011700, 011701
Kauffmann, Bernd 020705
Kaufhold, Dr. Tobias 020493
Kaufman, Donald G. 051867
Kaufmann, Bert Antonius 016926
Kaufmann, Burkhard 017403
Kaufmann, Carol 038687
Kaufmann, Christoph 020021
Kaufmann, Don 047961
Kaufmann, Hans 041802
Kaufmann, Dr. José Luis 000408
Kaufmann, Dipl.-Ing. Mechthild . . 020505
Kaufmann, Patrick 041553
Kaufmann, Dr. Sylke 019488
Kaukiainen, Hannu 010798
Kaukiainen, Heli 011034, 011035,
011037
Kaukoreit, Doz. Dr. Volker 003053
Kaulbach, PD Dr. Hans-Martin . . 022057
Kaumanns, Kurt 017023
Kaupang, Brit 033840
Kaupinis, Juozas . . . 030421, 030422
Kaus, Dr. K. 002754
Kauschat, Gerda 018205
Kauth, K.O. 049641
Kava, Ritva 010802
Kavanagh, Martha 005549
Kavasch, E. Barrie 054193
Kawabe, Masami 029188
Kawaguchi, Kei 029110
Kawaguchi, Kimio 029536
Kawaguchi, Masako 029536
Kawai, Masaki 029646
Kawai, Suyako 028937
Kawakami, Kunihiko 028792
Kawamithu, Sigeo 028718
Kawanabe, Hiroya 028912
Kawanabe, Kusumi 029694
Kawano, Katsuhiko 029002
Kawasaki, Matsuju 028658
Kawasaki, Shoichiro 029496
Kawashima, Akira 029721
Kawashima, Etsuko 029254
Kawiorski, Wacław . . 034925, 035047,
035050, 035052, 035108
Kay, Hannah 043639
Kay, Jeremy 053130
Kay, Joann 051453
Kaya, Settar 042984
Kayaselcuk, Mary 051547
Kaye, Carole 050413
Kaye, Jennifer 007210
Kayem, Abdul 000009
Kayombo, Dr. Norbert Andrew . . 042630
Kaysel, Dr. Roger 041256
Kayser, Christine 013908
Kayser, David 052881
Kayser, Gisela 017128
Kayser, Horst 020757
Kayumov, Bekmurad Djumayević . 043088
Kazbekov, F.P. 038023
Kazlienė, Viktorija 030423
Kaźmierczyk, Adam 034777
Kazus, Igor A. 037300
Kazutoshi, Kato 028524
Kazychanova, Marisa Grigorjevna 038047
Ke, Fengmei 008165
Ke, Shengwu 008460
Ke, Shilun 008094
Keall, Dr. Edward 007200
Kean, John 053889
Kean, Roger 052361
Kearney, Jarod 052588

Kearney, John 047653
Kearney, Lynn 047653
Keasal, Debra L. 045268
Keasler, Suzy 050672
Keates, Eve 005314
Keating, Neal B. 049599
Keaveney, Raymond 024900
Keck, Gabriele 041315
Keck, John 049290, 053556
Keckeisen, Robert J. 053773
Keddie, Robert 005798
Kędzierski, Sławomir 035369
Kędziora, Prof. Dr. Andrzej 035137
Keeble, Phyllis 044014
Keech, Brenda 049839
Keefer, Dr. Erwin 022047
Keegan, Daniel T. 053043
Keehner, Jill 053773
Keel, Othmar 041496
Keen, Lewis 048889
Keena, Katherine 053177
Keene Fields, Catherine 050324
Keene Muhlert, Jan 053899
Keene, John 038758
Keene, Rochelle 038740
Keery, John 045912
Keesecker, Martin 050725
Keeve, Valencia 050127
Kefer, Kons.OStR. Mag. Wilfrid . . 001743
Kehrer, Helma 016565
Kehrli, lic. phil. Manuel 041609
Keil, Franka 019773
Keil, Katherine 049804
Keil, Toralf 016536
Keim, Dr. Helmut 016573
Keinänen, Timo 010866
Keinan, Ofra 025011
Keinath, Rudi 016497
Keiper, Elke 022398, 022399
Keiper, Dr. Joe 047809
Keisuke, Iwamoto 028494
Keitel, Anja 017207
Keitemeier, Bernd 016433
Keith, Barbara W. 048830
Keith, Dorothy 050615
Keith, Jack 046643
Keith, Larry 045057
Keith, Uwe 022359
Kejariwal, Dr. O. P. 024235
Kejzar, Josef 002703
Kekäle, Leena 011303
Kekeke, Daniel O. 033421
Kekic, Dr. M. 001545
Kéler, Yves 006514
Kell, Dr. Klaus 019335
Kellams, Marty 048522
Kellar, Jane 049984
Kellein, Dr. Thomas 017377
Kellein, Madonna M. 050094
Keller, Dana 050823
Keller, Esther 041268
Keller, Jaroslav 009512
Keller, Linn 049360, 049361
Keller, Dr. Peter 002619
Keller, Pierre 041649
Keller, Sylvia M. 052158
Keller, Tim 053520
Kellerer, Anna 002029
Kellerhoff, Reinhard 022575
Kellett, John 046732
Kelley, Joyce J. 052477
Kelley, Robert P. 048139
Kelley, Samantha 047005
Kelley, Timothy 051860
Kellgren, Ola 041106
Kelling, Paisley 051963
Kellner, Bernd 018186
Kellner, Heike 002423
Kellokumpu, Mariina 011090
Kelly Elder, Catherine 049514
Kelly, Chris 047306
Kelly, Dr. Christopher 043844
Kelly, D. 005694
Kelly, Eamonn 024897

Kelly, Elaine 046997
Kelly, Gemey 006840
Kelly, Gillian 044177
Kelly, Jim 053236
Kelly, Karen 051378
Kelly, Linda 007014
Kelly, Michelle 043413
Kelly, Nancy 051769
Kelly, Patricia 052864
Kelly, R.J. 043387
Kelly, Samantha 051025
Kelly, Sheila 006782
Kelly, Vonda 047053
Kelly, William 046344
Kelm, Dr. Bonnie G. 053094
Kelm, Dr. Rüdiger 016489
Kelnberger, Raimund 021454
Kelsey Foley, Kathy 054236
Kelsey, Brian 049098, 053580
Kelso, Karen 051471
Kelso, Kirsty 045453
Kelso, Marcie 047575
Keltti, Pia 010760
Keltz, Ursula 020580
Kemdal, Anne Louise 041051
Kemkes, Martin 019449
Kemkes, Dr. Martin . . . 020917, 021525
Kemm, B. 041334
Kemmerer, Allison 046479
Kemp, Alan 045644
Kemp, Blanch 050552
Kemp, Dawn 044277
Kemp, Jane 048207
Kemp, Jo 045413
Kemp, N.R. 001150
Kemp, Richard 045266
Kemp, Dr. T.S. 045457
Kemp, W. Bradley M. 050186
Kemper Fields, Laura 049879
Kemper, Heinrich 018059
Kemper, Juli 049804
Kemper, Julie Maio 052799
Kemperdick, Dr. Stephan 041273
Kempf, Jürgen 011498
Kempf, Klaus 020507
Kempf, Peter 021861
Kempkes, Susen 018032
Kempler, Cheryl A. 054095
Kempling, Steven 005718
Kenagy, Dr. James 053229
Kendall, Andrew 046575
Kendall, Andy 048093
Kendall, Clifford P. 048710
Kendall, LeeAnn 049747
Kendall, Shirley J. 047199, 053654,
053655
Kendell, Shirley J. 053816
Kendrigan, Ernie 001204
Kenen, Susan 051242
Kenichi, Yuyama 029150
Kenichiro, Makino 029108
Kenig, Piotr 034652
Kennedy Harty, Leanne 053553
Kennedy, Arlene 006214
Kennedy, Caroline 052241
Kennedy, Cheryl 050604
Kennedy, Clay W. 047282
Kennedy, Dr. Craig 044245
Kennedy, Elaine 044153
Kennedy, Elizabeth 047700
Kennedy, Dr. George L. 052963
Kennedy, Greg 054292
Kennedy, Ian 049888
Kennedy, Mary 046545
Kennedy, Nancy 050018
Kennedy, R.H. 007197
Kennedy, Dr. Robert 051155
Kennedy, S.B. 043564
Kennedy, Sally E. 052526
Kennedy, Prof. W.J. 045457
Kennedy, William 048168
Kenney, Kimberley 047413
Kenney, Stephen 047110
Kennis 032063

Kenny, Jessica 043993
Kenny, Kevin 048773
Kenny, Michael 024898
Kensler, Terry 036151
Kent, David 045728
Kent, Jerry 052808
Kent, Joan 052266
Kent, Liz 049657
Kent, Rachel 001544
Kenworthy, M. 043768
Kenwright, Carole 046097
Kenwright, Peter 045298
Kenyon, Colleen 054529
Kenyon, Sir George 044049
Kenyon, Kathleen 054529
Kenyon, Sarah 045661
Kenyon, Therese 001249
Keo, Pich 005286
Keogh, Anthony J. 038807
Keough, Beverly 049921
Keough, Donald R. 051326
Keough, Tracey 052388
Kepford, Kris S. 052039
Kephart, Betsy 051918
Keppert, Jozef 038350
Kerbourc'h, Emmanuel 011615
Kerefov, Betal Muradinovič 037536
Kerekesh, Michael 052325
Keren, S. 025060
Kerényi, Péter 023572
Kerimova, L.L. 054628
Kerkhoff, Dr. Jacqueline A.M. . . 032439
Kerkhoven, Dr. Ronald 032181
Kerkmeijer, W. 032384
Kermode, Deborah 043620
Kern Paster, Gail 054112
Kern, Emil S. 041600
Kern, James E. 053931
Kern, Robert 016847
Kern, Steven 052978
Kern, Ursula 018428
Kernan, Thomas J. 050332
Kerndler, Werner 001734
Kerner, Johann 019902
Kernhan, John 006481
Kerr-Wilson, Ian 005972
Kerr, Andrea J. 045466
Kerr, Fred 001044
Kerr, John 044223
Kerr, Paul K. 046904
Kerr, Peter 001378
Kerschbaum, Rupert 003095
Kerschbaumer, Erich 002554
Kersing-Blok, Drs. I. 032153
Kerssenbrock-Krosigk, Dedo von 047994
Kersten, Hendrik 022349
Kersting, Dr. Hannelore 020439
Kert, Christian 011559
Kervern, Jeannine 012041
Kervran, Sophie 015451
Kervyn, Benoît 003424
Kes, Vicki 050545
Kesaeva, E.A. 038027
Kessel, Jane 044327
Kesselman, Steve 053163
Kessler, Adolf 041810
Kessler, Agnes 041596
Kessler, Earl 024238
Kessler, M. A. Horst 016658
Kessler, Dr. Michael 041283
Kessler, Monika 017827
Kessler, Nava 025112
Kessler, Dr. Ulrich 017089
Kessler, Werner 041580
Kestel, Michael 021991
Kesteman, Prof. Emile 003453
Kestenbaum, Stuart J. 048213
Kester, Marilyn 047372
Ketelsen, Cheryl 047765
Ketelsen, Dr. Thomas 017992
Ketner, Joseph D. 054051
Ketteler, Georg 017430
Kettmann, Dr. Otto . . 019807, 019808
Kettula, Suvi 010771

Ketzer, Dr. Hans-Jürgen 017521
Keulen, W.J. 031918
Keunecke, Dr. Hans-Otto 018260
Keurs, P. ter 032585
Keuter, J.H.G. 032047
Keutgens, Eric 003358
Keweloh, Hans-Walter 017606
Key, Eric 054388
Keyaerts, Geneniève 015471
Keyaerts, Magali 015471
Keyes, George S. 048301
Keyse, Val 001425
Keyser, Ignace de 003474
Keyser, Ralph 047466
Keyworth, Gretchen 047220
Keyworth, John 044909
Kezich, Giovanni 027783
Khabibullaev, Nozim 054670
Khaelss-Khaelssberg, Alexander 002944
Khalid, Syed Ali 030688
Khan Jatoi, Saeed 034140
Khan, Arif 053108
Khan, Aurangzeb 034145
Khan, Gulzar Muhammad 034153
Khan, Muhammad Afzal 034131
Khan, Nazir Ahmad 034151
Khanhoushi, Mustapha 042748
Kharissov, Ildar 017132
Khasanov, S.R. 054650
Khasawneh, Alia 029780
Khatib Eghdami, Roshanak ... 024756
Khazam, Sandra 001470
Khevenhüller-Metsch, Maximilian 002275
Khewhok, Sanit 049532
Khom, Helmut 001909
Khotko, P.S. 003248
Khreis, Dr. Khalid 029773
Khulani-Mkhize, Leo 038736
Khuldashev, Tulkin .. 054623, 054625,
                054626, 054627
Kibritova, Vanja 005106
Kicherer, Michael 018583
Kida, Junichiro 029716
Kida, Takuya 029610
Kidd, Alan 052973
Kidd, Jane 045466
Kiebert, Susan 053069
Kieblinger, Hans 002129
Kiedel, Klaus-Peter 017606
Kiefer, Theresia 020140
Kieffer-Olsen, Jakob .. 010275, 010277
Kiehl, David 051491
Kiel, Dennis 047739
Kielak, Bernard 035084
Kielholz, Walter B. 042207
Kielland, Morten 034083
Kielmayer, Oliver 042151
Kielwein, Dr. Matthias 017980
Kiely, Ann 046085
Kiely, David 046085
Kiemele, E. 018273
Kiendl, Anthony 005381
Kiener-Hofer, Trudi 042059
Kienholz, Kathleen 051323
Kieninger, Ernst 002940
Kienitz, Gerlinde 020773
Kientz, J.J. 012243
Kienzle, Karen 053096
Kiereś, Małgorzata 035394
Kiermeier-Debre, Prof. Dr.
   Joseph 020362
Kiers, J. 031962
Kiest, Rhonda 051645
Kietlińska-Michalik, Barbara .. 034886
Kieweg 002781
Kift, Dr. Dagmar 017442, 017958
Kiger, Robert W. 052165
Kihiyo, Jackson M. 042631
Kiiski-Finel, Päivi 011453
Kijima, Prof. Shunsuke 029445
Kikić, Prof. Vesna 008955
Kilaghbian, Elia 028306
Kilani, Huda 029771, 029772
Kilby, Virginia 049829

Kildee, Magee 047355
Kilger, Dr. Gerhard 017954
Kilham, D.D. 049366
Kilian, Dr. H. 022067
Kilian, Dr. Ruth 020217
Kiliti, Zsuzsanna 023748
Killguss, Hans-Peter 019707
Killias, Augustin 041633
Killoran, Elizabeth 051955
Kilner, Josefin 040660
Kilroy, Frank 045181
Kilsdonk, Betty 048613
Kim, Chi-Eun 030039
Kim, Christine Y. 051476
Kim, Hongnam 029995
Kim, Jin-Hyuck 029915
Kim, Jong Chull 029913
Kim, Judy 047253
Kim, Jung Soo 029981
Kim, Dr. Ke Chung 053897
Kim, Kumja Paik 052992
Kim, Sun-Jung 029944, 029983
Kim, Sung-Rae 029967
Kim, Sunggu 029963
Kim, Wook-Han 029944
Kim, Yong-Ha 029943
Kim, Yongsan 030055
Kim, Yun Soon 030061
Kimball, Cathy 053042
Kimball, Cindy 051803
Kimball, Kathryn A. 051189
Kimfoko-Madoungou, Joseph .. 008846
Kimmel, Dominik 020233
Kimsey, Dr. Lynn S. 048163
Kimura, Hidenori 029475
Kin, Wook-Han 029983
Kincaid, Jean E. 005364
Kinderen, Ronny de 003867
Kindgren, Hans 041098
Kindlen, Helen 036531
Kindler, Dr. Gabriele 019505
Kindrachuk, William 007350
Kindred, Amy 049686, 054617
Kindschy, Errol 054323, 054324
Kindt, Chantal 003655
Kindt, Diane 050982
King, Alan 051653
King, Andy 043747, 043748
King, Beverly 050808
King, Brad 044997
King, Carolyn 024839, 047633
King, Charles 047387
King, Douglas R. 052821
King, Dr. Duane H. 050469
King, Edward .043686, 044674, 044676
King, Elspeth 045874
King, Evelyn 001354
King, G.J. 001215
King, George G. 053107
King, Graham 044011
King, Harry 050014
King, Hugh R. 048433
King, John 033196
King, Jonathan 044923
King, Karma S. 050014
King, Lyndel 050936
King, Margaret H. 044359
King, Margery 052153
King, Martha 048877
King, Michael D. 044137
King, Muriel 044392
King, Tamara 050959
King, William C. 050795
Kingan, Beryl A. 046201
Kingdon, Charles 006326
Kingery, James 049120, 051503
Kingsbury, Bernard 033182
Kingsford Hale, James 044560
Kingsley, April 048440
Kingston, Jean T. 054453
Kinii, Shoko 029639
Kinnear, J. 045485
Kinno, Seiichi 029052
Kinoshita, Kyuichiro 029511

Kinoshita, Masafumi 029011
Kinsell, Tracey 050791
Kinsey, Danna 047893
Kinsman, Jane 000937
Kinsolving, Lucie 051444
Kinson, Alan 000809
Kintisch, Christine 002946
Kintzel, Jürgen 021838
Kinyon, Catherine 053750
Kinzelbach, Prof. Dr. R. 021496
Kinzinger, Nancy 053590
Kipiani, M.A. 016403
Kipkorir, Daniel 029857
Kipnis, Jeffrey 047945
Kippes, Dr. Wolfgang . 003012, 003079
Kipping, Jens 016535
Kiprop, Nancy 029858
Kiràly, Elvira 023529
Kirby, C. 044506
Kirby, Jo 045057
Kirby, John 005918
Kirby, Kris 051594
Kirby, Michael 054416
Kircher, Gunter 019629
Kirchhauser, Hannes 019518
Kirchmayer, Christian 002236
Kirchmeier Andersen, J.C. 010199
Kirchner, Dr. Ferdinand 002634
Kirchner, Gertrud 002093
Kiričenko, E.I. 023737
Kiričenko, Galina Vladimirovna . 037163
Kirigin, Branko 008993
Kirilova, Lena 005016, 005017
Kirk, Natalie 054058
Kirk, Rachel 044477
Kirk, Ruth 047920
Kirk, Wendy 044976
Kirkbride, Cinoy 005786
Kirkby, Patrick 046152
Kirkeby Andersen, Dorte 010301
Kirkham, Thomas G. 043561
Kirklin, Deborah 053143
Kirkman, John 001064
Kirkpatrick, Andrea 006893
Kirkpatrick, David 044050
Kirkpatrick, Eleanor 051730
Kirkpatrick, Dr. Gail B. 020592
Kirkpatrick, John 051730
Kirkpatrick, May 043477
Kirkpatrick, Roger 054191
Kirkpatrick, Stephanie 045614
Kirkwood, Michael 048512
Kirmeß, Max 020922, 020925
Kirnbauer, Dr. Martin .. 041268, 041281
Kirr, Katharina 019763
Kirrane, Siobhan 045168
Kirsch, Eva 052957
Kirsch, Dr. Jeffrey W. 052969
Kirsch, Dr. John A.W. 050600
Kirsch, Dr. Steven 050124
Kirschbaum, Barbara 019707
Kirschberg-Grossman, Reva G. . 051367
Kirschstein, Jörg 021002
Kirstein, Michaela 020068
Kirsten, Alex F. 006305
Kisapáti, István 023451
Kisëv, Ērdan 005161
Kishioka, Yukio 028482
Kisjova, Vera 005128
Kisler, Mary 033141
Kisler, Shuli 025212
Kisling, Ing. Andreas 002802
Kisman, A.K. 032233, 032235
Kiss, Andrei 036784
Kiss, K. 044954
Kissack, Terence 053001
Kisser, Gerhard 001932
Kissi, Flavio 027122
Kissinger, Scot 006516
Kissov, Kostadin 005124
Kistner, Hans-Jürgen 019487

Kitagawa, Kazuo 028974
Kitagawa, Seika 029182
Kitamura, Hitomi 029610
Kitamura, Sadao 028911
Kitamura, Yoshiro 029251
Kitamura, Yutaka 029622
Kitauo, Tsuguto 029072
Kitazawa, Toshio 029416
Kitchen, Dr. M. 044186
Kitchin, Cameron 053996
Kitchin, Simon 043940, 043943
Kitching, Matthew 038746
Kite Powell, Rodney 053671
Kittas, Prof. Dr.. Chr. 023085
Kittel, Gerti 002958
Kittelmann, Udo 017068, 017230,
                       017259
Kittleson, Rosalie 046725
Kittner, Ed 049560
Kitto, Suzanne 044784
Kittredge, Cindy 049218
Kiturka, Jury 003244
Kityo, Robert Martin 043104
Kitz, Hilary 053848
Kitzler, Gottfried 002452
Kitzlinger, Dr. Christine 019011
Kitzmann, Andrew 053632
Kiuchi, Kazuhiro 028486
Kivimäe, Juta 010668
Kivimäki, Kati 011304, 011307
Kiyoaki, Morita 029036
Kiyohiko, Sakurai 029530
Kiziltan, Zeynep 042941
Kjeldbæk, Esben 010174
Kjeldgaard, Inge Merete 010043
Kjeldsberg, K. 033670
Kjeldsberg, Peter Andreas ... 034036
Kjeldsen, Kjeld 010146
Kjellberg, Anne 033840
Kjems, Folke 010135, 010137
Kjerström Sjölin, Eva 040809
Kjupčupov, Georgi 005169
Kładna, Dr. Aleksandra 035250
Klages, Dr. Claudia 017499
Klameth, Wolfgang 018420
Klang Eriksson, Eva 040809
Klanicová, Dr. Evženie . 009456, 009457
Klapczyński, Stanisław 034683
Klar, Alexander 019840
Klar, Dr. Alexander 018916
Klar, Annika .. 040738, 040739, 040742
Klar, Herma 020636
Klarić, Josip 009049
Klassnik, Robin 045034
Klasson, Gunnar 041098
Klatt, David ..006420, 006506, 006968,
             006986, 007500, 007501
Klatt, Dr. Ingaburgh 020158
Klattenberg, Hinrich 020707
Klaus, Dietmar 016535
Klaus, Franz Josef 002311
Klaus, Judy 033248
Klaus, Werner 041598
Klauser, K.D. 003998
Kleckerová, Marcela 009927
Klee, Sonja 019818, 021756
Kleeblatt, Norman 051412
Kleefisch, Stephen J. . 052195, 052196
Kleeman, Drs. E. 031911
Kleemann, Christoph 021488
Kleemann, Kurt 021359
Kleespies, Gavin 051100
Klei, Mary 050207
Kleiger, Christian 051681
Kleijn, G. de 031916
Klein-Langner, Wolfgang 022802
Klein, A. 032563
Klein, Adaire 050461
Klein, Dr. Angela 017553
Klein, Anton 003032
Klein, Beatrixe 022631
Klein, Carolyn 048553
Klein, Christophe 014204
Klein, Dr. Emanuel 051170

Klein, Mag. phil. Eveline 003108
Klein, Holger A. 047808
Klein, Janice B. 048639, 048642
Klein, Jean-Claude 015248
Klein, Jean-Raymond 011636
Klein, Matthias 016610
Klein, Dr. Michael 020229
Klein, Peter 053603
Klein, Thomas 017908, 017909
Klein, Valérie 014185
Kleinbongart, M.A. Sigrid 018044
Kleindorfer-Marx, Dr. Bärbel ... 022389
Kleine, Dr. Joachim 022871
Kleiner, Gerald 051003
Kleiner, Michael 022312
Kleinert, Dr. Annette 019372
Kleinhans, M. .038868, 038869, 038870
Kleinová, Olga 038424
Kleinschmidt-Altpeter, Dr. Irene . 017495
Kleinschmidt, Torsten 019443
Kleinschrot, Dr. Dorothée 022793
Kleinsteuber, Wolfgang Friedrich
R. 017275
Kleipaß, Dr. Herbert 018191
Kleisner, Ted J. 054375
Kleist, Monika von 016730
Klejman, N.I. 037397
Klemm, Dr. Christian 042207
Klemm, Federico 000160
Klemm, Hélène 042039
Klemm, Randy 049226
Klemm, Susanne 018357
Klencke, Lippold von 018194
Klenner, Ingrid 022043
Klenow, Rosemary E. 048448
Klepač, Koraljka 008979
Klerk, Prof. Dr. L.F.W. de 032470
Klesman, Kees 031954
Klettner, Emmi 001960
Kley, Dr. Stefan 020868
Kliege, Dr. Melitta 020873
Klim, T.B. 037283
Klimaszewski, Cathy 049739
Klimek, Larry 050965
Klimenova, Tamara
Aleksandrovna 036953
Klimesch, Dr. Gertraud 001856
Klimm, Erich 021211
Klimov, Vladislav Petrovič 036938
Klimova, Anna Danilovna 036966,
038079, 038080
Klimova, Elena Michajlovna 037931
Klimpel, Dr. Paul 017240
Kline, Dr. John 048292
Kline, Katy 047285
Kline, Margaret 048943
Kline, Phil 047576, 047577
Kling, Burkhard 021974, 021976
Kling, Dr. Gudrun 041320
Klingelhöfer, Claudia 021848
Klinger, Dr. H.C. 038686
Klinghardt, G. 038686
Klinghardt, Gerald 038685
Klingman, Catherine 048300
Klingsöhr-Leroy, Cathrin 019669
Klink, Alfonso Caballero 039288
Klink, Peter 052139
Klink, R. 022213
Klinka, Isabelle 014586, 014588
Klinkhardt, Bernd 022734
Klinkow, Michaela 020986
Kljajić, Damijan 038629
Kljueva, Valentina Ivanovna .. 036975
Klobe, Tom 049546
Klobučníková, Magdaléna 038416
Kloch, Ove 010359
Klochko, Deborah 052968
Kloiber, Augustin 002660
Kloosterboer, A.E. 032787
Klootwijk, L.P. 032790
Klooz, Donan 050978
Kloppenborg, David N. 048334
Klos, Günther 019325
Klose, Dr. Dietrich 020577

Klose, Simon 000847
Klosek, Melesa 047933
Klostermeier, David 052755
Klotzer, Dr. Georg 026762
Kloub, Ola 029785
Klucas, Kevin 048138
Kluchko, Halya 007223
Klučina, Petr 009805
Klüver, M.A. Dennis 021701
Klüver, Heinrich 020072
Klug, Dr. C. 042227
Kluge, Dr. Arnd 019287
Kluge, Prof. Arnold G. 046493
Kluge, Prof. Dr. Bernd 017088
Kluge, Joachim 022430
Kluge, Martin 041265
Kluge, W. 020292
Kluger, Reinhard 001805
Klukina, Anna I. 037250
Klumpp, Sibylle 014381
Klupp, Krzysztof 035138
Kluska, Adam 035227
Kluss, Dr. Maciej 035153
Kluver, William C. 053593
Kluza, Radosław 035007
Klydie, Thomas 052555
Klykova, Galina Ivanova .. 037611
Kmieck, Anne T. 047810
Kmínek, Petr 009445, 009978
Knackert, Bruce J. 050922
Knape, Gunilla 040649
Knapen, I.A.M. 032764
Knapp, Clark 048604
Knapp, D.R. 044887
Knapp, Marylin 046470
Knapp, S. 044887
Knapp, Trevor 046046
Knappertsbusch, Dr. Michael ... 041282
Knauer, Wilfried 022731
Knauss, Carol 046486
Knauss, Dr. Florian 020575
Knauss, Dipl. Geogr. Jürgen .. 017783
Knebelsberger, Caroline 046893
Knecht, Dr. Richard 053882
Kneubühl, Dr. Urs 041325
Kneuss, Martine 042010
Knezevich, Alan 054142
Knibiely, Philippe 015426, 015427
Knick, Dr. Stanley 051969
Kniefeld, Lothar 022129
Knieriem, Dr. Michael 022819
Knierriem, Peter 020025, 021438
Kniesz, Jürgen 022430
Knigge, Prof. Dr. Volkhard .. 022492
Knight, Brian L. 050632
Knight, Cynthia 054518
Knight, David J. 049471
Knight, Gregory 047649
Knight, Irene 006537
Knight, Jeremy 044606
Knight, Jim 047918
Knight, Katie 049432
Knight, Michael 052992
Knight, Robert 046989
Knight, V.M. 044988
Knights, Mary 000770
Knill, Phyllis 048554
Knipe, Peter 051912
Knipp, Otto 019363
Knipprath, Ute 017499
Kniprath Smith, JaNeen 047853
Knittel, Pierre 016351
Knittel, W. 020711
Knížák, Dr. Milan 009823
Knížák, Milan 009844
Knoch, Francis 030584
Knoche, Norma 048340
Knocke, Erich 018303
Knode, Marilu 050915, 053212
Knöchlein, Dr. Bernhard 021865
Knödel, Dr. Susanne 019012
Knödler-Kagoshima, Brigitte .. 019503
Knöpfel, Sylvia 018596
Knofler, Dr. Monika 003027

Knogler, Helmut 002564
Knol, Egge 032378
Knol, M. 032944
Knoll, Andrew H. 047358
Knoll, Annegret 019769
Knolle, Dr. P. 032310
Knop Kallenberger, Christine .. 053850
Knop, Matthias 018027
Knop, Monika 021000
Knoppers, K. 032098
Knorowski, Mariusz 035080
Knorre, Dr. Alexander von ... 019217
Knott de Arechaga, Deborah ... 052541
Knowles, A.M. 043935
Knowles, Elizabeth 045481
Knowles, Elizabeth P. ... 050479
Knowles, Dr. P. 024843
Knowlson, Paul 043701
Knox, J. 043563
Knox, Judy 046775, 047178
Knox, Keith 006774
Knox, Lu 051624
Knox, R. 044923
Knox, Tim 045107
Knudsen, Ann Vibeke . 010090, 010285,
010286
Knudsen, Henning 010168
Knudsen, Herluf 010268
Knudsen, Karsten 010312
Knudsen, Marie Hedegaard ... 043445
Knudsen, Svend Aage 010028
Knüppel, Stefan 018345
Knuth, Dr. Michael ... 017089, 017149
Knutsen, Jim 053435
Knutson-Hedetun, Anna ... 040962
Knysh, Leann 005308
Ko, David C. 042580
Ko, Seong-Jun 029169
Kob, Karin 041243
Kobayashi, Akira 029399
Kobayashi, Fumio 029623
Kobayashi, Kazuhiko ... 028579
Kobayashi, Kiroku 029330
Kobayashi, Koichi 029472
Kobayashi, Saburo 029549
Kobayashi, Shigeko ... 029583
Kobayashi, Tadashi ... 028508
Kobayashi, Dr. Tatsuo ... 029081
Kobayashi, Toshinobu ... 028781
Kobayashi, Yasuyuki ... 029169
Kobayashi, Yotaro ... 029567
Kobe, Michael 018962
Koblmiller, Gertraud ... 002062
Kobosil, Karel 009462
Kobusiewicz, Zofia ... 035016
Koç, Sema 042820
Koca, Osman 043026
Koççoban, Cahit ... 042975
Kocebenkova, Vera Ivanovna .. 037081
Kočeřženko, Jevgenija Ivanivna . 043304
Kočevar, Sanda ... 008925
Koch, Dr. Alexander ... 021931
Koch, Claudia 002995
Koch, Cynthia M. 049649
Koch, Eva 017433
Koch, Fred 050525
Koch, Friederike ... 019489
Koch, Dr. Lars-Christian ... 017112
Koch, MA Matthias ... 018047
Koch, Dr. Michael ... 020509
Koch, Sven 018422
Koch, Werner ... 017448, 019061
Koch, Wilhelm ... 020372
Koch, Will 053098
Kochančuk, Vitalij Mykolajovyč .. 043174
Kochanoeicz, Roman ... 035073
Kochanova, Ljudmila Petrovna .. 038009
Kochendörfer, Reinhard ... 019938
Kocher, O.Univ.Prof. Dr. DDr. h.c.
Gernot 001975
Koches, Susan ... 046755
Kočík, Zdeněk . 009511, 009623, 009820,
009821

Kock, G.M. de 032923, 032925
Kock, Prof. Hans 019603
Kock, R.D. de 038729
Kocsis, Dr. Gyula 023580
Koczur, Anton 002001
Koda, Harold 051424
Kodani, Susan 049536
Köb, Edelbert 003050
Koebbel, Wolfgang 017566
Köberl, Prof. Mag. Rudolf 002507
Köberl, Sieglinde 001734
Koechlin, Bernard 041535
Koefoed, Lizette 010212
Koegel, Alice 022057
Köger, Dr. M.A. Annette ... 019962
Kögl, Reinhard 021688
Kögler, Johannes 018517
Köhler, Birte 019011
Köhler, Dr. Georg 002912
Köhler, Dr. Hubert 018917
Köhler, Dr. Jörn 017822
Köhler, Dr. Juliana 022664
Köhler, Neeltje 032392
Köhler, Otto 019304
Köhler, R. 019065
Köhler, Thomas 022429
Köhlmeier, Ernst 002055
Köhn, Regine 017853
Köhne, Dr. Eckart 022202
Koehnline, Lyn 047528
Kölbl-Ebert, Dr. Martina ... 018120
Kölbl, S. 017416
Köll, Helmut 002707
Köllhofer, Dr. Thomas 020249
Koelliger, Marcus 041446
Kölmel, Dr. Reinhard 016941
Könenkamp, Dr. Wolf . 020351, 020352
König, Cornelia 022165, 022166, 022167,
022168, 022169
König, G. 017401
König, Dr. Gabriele 018563
König, H. 016884
König, Hans 042141
König, Heinz 019182
König, Helmut 019292
König, Josef 002967
König, Prof. Kasper 019705
König, Norbert 017168
König, Reginald 022045
Koenig, Robert J. 053712
König, Prof. Dr. Viola .. 017112, 017186
Koenigsmarková, Dr. Helena ... 009843,
009926
Köntös, László 023830
Köpke, Ingeborg 018529
Köpke, Prof. Dr. Wulf 019012
Köppen, Thomas 022625
Köppl, Günther 001909
Koerber, Susannah 052601
Koerhuis, E. 033114
Környei, Dr. Attila ... 023802, 023890,
023903
Koesoemowarni, Endang 024638
Köster, Dr. Beate 020593
Köster, Hein 017277
Koevend, Mary Helen de 051201
Koffi, Bilé Yapo 008875
Kofler, Claudia 003033
Kofoeb, Jakob 010123
Kofuku, Akira 029536
Koga, Kim 050458
Kogan, Lee 051325, 051383
Kogler, Leopold 002908
Kogure, Susumu 028676
Koh, Dr. Eugen 001397
Kohan, Christopher 046431
Kohl, Barbara 021515
Kohlberg, Ilona 018507
Kohler, Ruth 053296
Kohlschmidt, Ulrich 020708
Kohlstruck, Gerd 021459
Kohn Loncarica, Prof. Dr. Alfredo
G. 000194, 000247
Kohnhorst, Bridgette 051192

Index of Persons

Kohnke, Dr. Hans-Georg
– Kowalski, The Very Reverend James A.

Kohnke, Dr. Hans-Georg 017546, 017547
Kohút, Dr. Teodor 038358
Koike, Tomio 029129
Koivisto, Mikko 011014
Koja, Dr. Stephan 002953
Kojaev, M. 029818
Koji, Yoneda 029096
Kojić, Liljana 038194
Kok, Harry de 004100, 004101
Kokki, Kari-Paavo 010809, 010810
Kokkotidis, Dr. Klaus Georg .. 022047
Kokošínek, Nataša 038520
Kokoska, Monika 018285
Kokott, Dr. Jeanette 019012
Kokroko, Grace Juliet 022930
Kokuryo, Masao 029622
Kolarić, Dr. Juraj 009027
Kolb, Jennifer 050601
Kolb, Nancy D. 052085
Kolb, Rainer 016617
Kolb, Sarah 002945
Kolbach, B. 032446
Kolbasina, Larisa Aleksandrovna 037577
Kolbe, Joachim 019893
Kold, Anders 010146
Kolde, H. 019476
Kolenberg, Hendrik 001535
Koles, L.A. 001398
Kolesnikova, Ljudmila Arsenjevna 037071
Kolesnikova, Ljudmila Petrovna . 037284
Kolesov, S.N. 037906
Koletsky, Sue 047826
Koletsky, Susan 046839
Kolev, Cvetan 005227, 005230
Koli, Pamela 018446
Koll, Kersti 010663, 010668
Kolland, Dr. Dorothea 017200
Kollar, Mary 050314
Kollár, Dr. Štefan 038280
Kollath, Jeffrey 050603
Kollbaum-Weber, Jutta 022836
Kollbaum, Marc 052802, 052805
Koller-Jones, Kimberly B. 051253
Koller, Karl 002974, 003101
Koller, Margaret H. 048228
Kollintzas Drikakis, Elaine 047663
Kollmann-Caillet, Virginie 011722
Kollmann, Dr. Karl ... 018278, 020043
Kolm, Herbert 002517
Kolmos, Dr. Alfred S. 054379
Kolner, Brian A. 047176
Kolodzei, Natalia 051911
Kolodziejak, Adeline 006291
Kołodziejski, Dr. Adam 035238
Kolomceva, Raisa Michajlovna . 037362
Kolonas, L. ... 022950, 022966, 023215
Kolonia, Rozina 023059
Kolonoas, Lazaros 023168
Koloskova, T.G. 037420
Kolosov, Viktor 036962
Kolosova, Ekaterina Matveevna . 037779
Kolsteren, Steven 032378
Kolt, Ingrid 007102
Kolt, Sheryl 007469
Kolter, Horst 019095
Koltz, Jean-Luc 030593
Koltzer, Ulrich 021176
Koludrović, Ivana 009016
Kolveši, Željka 009043
Kolyesnik, Ljudmila Michajlovna . 037026
Komanecky, Michael .. 048170, 052120
Komarek, Beverly 049215
Komarov, Ilja Iljič 037350
Komatu, Nario 029319
Kombu, Anatolij Stepanovič 037191,
                                037192
Komenan, Robert Yao 008877
Komick, J. 007310
Komissarov, Eha 010668
Komnenović, Ilija 038215
Komor, Valerie 051455
Komzak, Wolfgang 002467
Komzyuk, Vera 043185
Konan, Louise Constante

N'Guesssan 008879
Končakovs'kyj, Volodymyr
  Petrovyč 043201
Koncevaja, Tatjana Leonidovna . 037875,
                                037877
Kondakova, Nina Viktorovna ... 037962
Kondo, Tetsuo 028662
Kondratenko, Tatjana Michailovna 037158
Kondziela, Mariusz 035215
Konečná, Katarína 038413
Konecny, Dr. Andreas 001736
Konevskaja, Tatjana Ivanovna .. 037696
Konieczny, Marian 035065
Konieczny, Władysław 035439
Koning, Elizabeth 053608
Konishi, Masafumi 029145
Konishi, Tamao 029169
Konkle, Scott 005335
Konnikov, Boris Aleksandrovič .. 037595
Konno, Hirofumi 028631
Kono, Osamu 028890
Kononov, Alexander 038774
Kononova, Natalia 038774
Konopka, Lawrence 048808
Konova, Ljubava 005191
Konst, Thomas 047440
Konstam, Nigel 025802
Konstantinos, Manioudakis 023041
Konstantinus, Laharakis 023306
Konstantios, Dr. Dimitris 022981
Konukiewitz, Insa 017629
Konuma, Hirokazu 028873
Konvalinka, Danilo 054507
Konz, Klaus 020045
Kooiman, R. 032966
Kooiskra, L. 032361
Kooistra-Manning, Kevin 046988
Koolkin, Kenneth 025027
Koontz, Darlene 048214
Koop, Kathy 051318
Koopmans-Witt, C.E. 032073
Koopmans, J.J.G. 032828
Koopmans, Y. 032027
Koos, Greg 047044
Kopálová, Dr. Danka 038419
Kopbosinova, R.T. 029801
Kopco, Mary A. 048188, 048189
Kopczak, Dr. Chuck 050412
Kopec, Diane 046781
Kopecky, Carle J. 053198
Kopeinig, Josef 002806
Koperkiewicz, Adam . 034730, 034743
Koperski, Kate 051560
Kopf, Sibylle 017005
Kopf, Vicki C. 054495
Kopil, Olga Andriïvna 043327
Kopis, Mildred 052003
Kopówka, Edward 034852
Kopp, Klaus 022630
Kopp, Dr. Marianne 016832
Kopp, Siegfried 002576
Kopp, Werner 017271
Koppel de Ramírez, Elsa 008627
Koppel, Greta 010668
Koppel, John 010014
Koppel, Merike 010668, 010682
Koppelman, Dorothy 051480
Koppenhöfer, Dr. Johanna ... 018525
Koppensteiner, Dr. Erhard 002637
Kopplin, Bärbel 020531
Kopplin, Dr. Monika 020605
Koppy, Ann 051593
Kopydłowski, Janusz 034676
Koralova, Ilina 019986
Korb, Uwe 020210
Korčagin, E.N. 037726
Korčevnikova, I.L. 037429
Kordina, Julia 002234
Kordiš, Ivan 038527
Kordos, Dr. László 023536
Kordowska, Barbara 035405
Koref, Barbara 016698
Koren, Elisabeth 033507
Koren, Elisabeth S. 033852

Koreny, János 023558
Korhonen, Paavo 011396
Korhoven, Teppo 010830
Korina, Olga Aleksandrovna ... 037325
Korir, Daniel 029859
Korkmasova, Maimusat
  Arslan-Alievna 037211
Kormano, Riitta 011453
Korn, Brigitte 016454, 016455
Korn, Prof. Dr. Lorenz 016952
Korn, Dr. Werner 017753
Kornacki, Krzysztof 034804
Kornatek, Anna 035306
Kornblum, Aaron T. 046936
Kornejeva, Galina Viktorivna .. 043245
Kornetchuk, Elena 053279
Kornhofer, Elisabeth 002884
Korobar, Klime 030628
Kőrös, Dr. Endre 023684
Kőrösi, Krisztina 023807
Koroly, József 023691
Korotaeva, Ljudmila Dmitrievna . 038058,
                                 038059
Korotkij, Aleksandr Nikolaevič .. 037205
Korovina, Vera Vasiljevna 037662
Korppi, Svea 011269
Korpusik-Jelonkowa, Zofia 035465
Korres, Manolis 022991
Koršikov, Anatolij Seměnovič .. 037612
Korsmeier-Humbert, Helga ... 020599
Kort, Betty 052491
Kort, Drs. Toon 031908
Korten, Noel 050446
Korteweg, Drs. A. J. 032176
Kortländer, Prof. Dr. Bernd ... 018030
Kortlander, Christopher 049057
Kortschak, Dr. Werner 002928
Kortunova, M. 037771
Korz, Johannes-Rudi 018422
Korzeniewska, Iwona 034883
Kos, Dr. Peter 038553
Kos, Stanisław 035188
Kos, Dr. Wolfgang ... 003006, 003060,
   003061, 003066, 003092, 003095
Košacký, Petr 009600
Kosaka, Hirokazu 050424
Kosanji, Kozo 029368
Kosasih, Dikdik 024501
Koschatko, Alois 002875
Koschnik, Dr. Leonore 017299
Kościńska, Halina 033664
Kościołko, Eugeniusz 034993
Kosica, Marilyn K. 047172
Kosicka, Anna . 035277, 035283, 035285,
                    035287, 035288
Kosinski, Dr. Dorothy 048115
Kosinski, Joan 047691
Koski, Ann 050601
Koski, Elizabeth 046576
Koski, Janne 011306
Koskijoki, Maria 010774
Koskimies-Envall, Marianne ... 011471
Koskinen, Pekka 011487
Kośko, Maria 035125
Kosler, Barbara 020535
Kosmadaki, Polina 022960
Kosok, Prof. Dr. Lisa 019006
Kosonen, Elina 011284
Kosonen, Juhani 011120
Kosova, Parim 038228
Kossatz, Dr. Tilman 022792
Kossmann, M.A. Joachim 018998
Kosť, Ljubov Vasylivna 043237
Kosta, Bernard 049387
Kostadinova, Snežana 005146
Kostanian, Rachel 030519
Kostede, Ulrich 019486
Kostelny, Elizabeth ... 052553, 053368,
              053618, 053995, 053998

Koster, Richard de 054321
Kostet, Juhani . 011226, 011433, 011437,
                  011447, 011448, 011449
Kostina, N.V. 037910
Kostková, Magda 038458
Kostołowski, Andrzej 035460
Kostov, Georgi 005252
Kostova, Penka 004973
Kostrowicki, Andrzej 035111
Kostyál, Dr. László 024052
Kostynowicz, Roman 035824
Kosunen, Mirja 011202, 011348
Kosvincev, Boris Jurjevič 036901
Kosvinceva, Tatjana Dmitrievna . 036900
Koszkul, Jan 035222
Kosznik, Danuta 034827
Koszutski, Oskar 035109
Kotai-Ewers, Trisha 001533
Kotasiak, Wojciech 035086
Koteł, Ebenezer 054438
Kothe, O. 019959
Kothe, Robert J. 049097
Kotilainen, Eija-Maija 010839
Kotilainen, Kari 011331
Kotilainen, Simo 010942
Kotitonenko, Nadija Oleksiïvna . 043303
Kotivuori, Hannu 011321
Kotkin, Lauren 046316
Kotljakova, N.V. 037220
Kotoň, Friedrike 021786
Kotooka, Mitsushige 028888
Kotouč, Friederike 021781
Kotrošan, Dražen 004283
Kotrouzinis, Dr. Chrysanthi ... 020746
Kotsch, Natalie 049630
Kotschwar, Sandra 002323
Kotsiubynsky, Igor 043120
Kottcamp, Edward 048655
Kottenbrink, Magdalene 021264
Kotterer, Michael 021326
Kottmann, Ingeborg 022341
Kotula, Krystyna 035403
Kotzan, Peter 020067
Kotzur, Dr. H.-J. 020224
Kouakou, Thierry Konan 008876
Kouba, Jess 052504
Kouchi, Osami 029640
Koukal, Petr 009735
Koukal, Dr. Vitezlaw 009871
Koumantakis, Prof. I. 023326
Kourinou, Eleni 022997
Kourouma, Moussa 023382
Koutoudis, Irina 001794
Koutsomallis, Kyriakos 022960
Kouvaras, Fotini 023276
Kouvoutsakis, Panagiotis 023129
Kouzin, E. 037089
Kouzmin, Alexander 037737
Kovačev, Dimitar 004971
Kovačević, Dr. Bojan3a 038159
Kovačić, Goranka 008915, 008939
Kovačić, Marcelo 008979
Kovačić, Vladimir 008968
Kovács, Dr. Anna ... 023587, 023869,
                                 023870
Kovács, Irén 023544
Kovács, Józsefné 023576
Kovács, Dr. Tibor ... 023519, 023630
Kovácsné Bircher, Dr. Erzsébet . 023893
Kovácsné Ládi, Katalin 023781
Kovacsovics, Dr. Wilfried 002618, 002637
Kovalenko, Donna 053234
Kovalskij, Sergej Viktorovič ... 037718
Kovanen, Marja-Leena 011046
Kovarik, Jiri 009626
Kovrakis, Stavros 023201
Kovtun, Vladimir Vasiljevič ... 038085
Kowalczyk, Longin 035005
Kowalczyk, Wiktor 034825
Kowald, Dr. Margaret 001501
Kowalewska, Dr. Agnieszka ... 035402
Kowalski, Beth 050601
Kowalski, The Very Reverend James
  A. 051352

Kowalski, Michael 019408
Kowalski, Zygfryd 035020
Kowarik, Prof. Wilfried 022381
Kowark, Dr. Hannsjörg 022067
Kowerko, Oleh 047703
Kowski, Robert 049254
Kox, Jo 030585
Koyama, Atsushi 029004
Koyama, Miturn 028538
Kôyama, Noboru 029210
Koyama, Sherry 053008
Kožachmet, Chasen 029808
Kozáková, K. 038486
Kozen, Hiroshi 028958
Kozerski, Paweł 034662
Kozij, Galyna Ivanivna 043279
Kožiškova, Lenka 009696
Kozlova, T.D. 037730
Kozlova, Vera Danilovna 036914
Kozlovskaja, Nadežda Borisovna 036919
Kozlowski, Kazimiera 047404
Kozubek, Gwalbert 034713
Kožucharov, Dimitar 005046
Kozyra, Peter H. 005405
Kozyreff, Chantal 003492
Kraber, Rosa 053587
Kraciak, Aniela 035152
Kräftner, Dr. Johann 003031
Krämer, M.A. Anja 022066
Krämer, Dr. Annette 022048
Krämer, Dirk 018063
Krämer, Frank 022349
Krämer, Gerd 021295
Kraemer, Jörg 017512
Krafczyk, Alois 019087
Krafl, Martin 017308
Kraft, Gro C. 024104
Kraft, Joe C. 051217
Kraft, Karl Ludwig 016798
Kraft, Dr. Perdita von 017766
Kragelund, Patrick 010170
Kragh, Mogens 010349
Kraglund, Ivar 033848
Kragness, Kurt K. 046865
Krahn, Susan 005874
Krahn, Dr. Volker 017089
Krajasisch, Dr. P. 002276
Krajewski, Sara 050598
Krajneva, Anna Anatoljevna 037224
Krajniak, Paul J. 050912
Krakauer, Dr. Thomas H. 048412
Kral, Dr. Martina 041713
Kralevska, Julijana 030630
Kraljevič, Marija 008933
Kraljni, Ariana 009032
Kramatschek, Christopher 020506
Kramer, Prof. Cheryl 049738
Kramer, Dr. Dieter . . 002509, 018434
Kramer, Dr. E. 032570
Kramer, Finn Erik 010120
Kramer, Hermann 002844
Kramer, Lovetta 050373
Kramer, M. 021120
Kramer, Dr. Mario 018438
Kramer, Trudy C. 053420
Kramme, Dr. Mike 051094
Kramp, Dr. Mario 019664
Kranjc, Igor 038549
Kranjc, Katja 038549
Kranl, Christine 001769
Kranner, Romana 002068
Kranz, Betsy 049678
Kranz, Prof. Dr. P. 018258
Kranz, Tomasz 035003
Kranzelbinder, Norbert 001833
Krapež, Mateja 038552
Krapf, Eva 002948
Krapf, Dr. Michael 002953
Krasilnikov, Grigorij Vladimirovič . 036839
Kraske, Bernd M. 021349, 021350
Kraskin, Sandra 051467
Krasnokutskaja, Lidiya Ivanovna . 037659
Krasovskaja, E.S. 037255
Krasovskaja, Nonna 043293

Krasowska, Dorothy 051374
Krasteva, Dimitijka 005191
Krästeva, Natasha 005051
Krátká, Drahomíra 009878
Kratochvil, Jiří 009733
Kratochvilová, Alena 009859
Kratochvilová, Irena 009567
Kratochwil, Franz 002967
Kraus, E. 026895
Kraus, Gerhard 016505
Kraus, Jeannie W. 046847
Kraus, Dr. Jens 021208
Kraus, Dr. Jürgen 019407
Kraus, Karola 016928
Kraus, Dr. Michael 002723
Kraus, Peter 022269
Kraus, Dr. Stefan 019694
Krause-Willenberg, Horst 018902
Krause, Ines 020132
Krause, Joanne 049186
Krause, Dr. Jürgen 020601
Krause, Manfred 021903
Krause, Dr. Markus 018210
Krause, Melissa 051233
Krause, Dr. Siegfried 020853
Krause, Walt 052523
Krauß M.A., Marianne 018261
Krauß, M. 019467
Krauss, Dr. Rolf 017245
Krautgartner, Toni 002535
Krautschneider, Astrid Britt 001273
Krautwurst, Dr. Miriam . 018235, 018236
Kravcov, Igor Nikolaevič 037898
Kravcova, Ljudmila Petrovna . . . 037124
Kraxner, Ing. Hubert 002703
Krazmien, Mindy 049783
Kreamer, Todd A. 050979
Krebernik, Prof. Dr. Manfred . . . 019448
Krech, Shepard 047217
Kreem, Tiina-Mall 010668
Kregar, Tone 038501
Krehe, Lynette 047708
Kreibich, Matthias 018109
Kreijn, Drs. L. 032739
Kreil, Lisa 002944
Kreim, Dr. Isabella 019410
Kreindl, Richard 002273
Kreiser-Francis, Deborah 047953
Kreitmeier, Sabine 022791
Krejči, Irena 009942
Krejci, Jim 054585
Kremer, Dr. Roberta 007301
Krempel, Leih 020529
Krempel, Prof. Dr. Ulrich 019066
Kremsreiter, Hans 021116
Kren, Thomas 050436
Krenkel, Ljudmila Ėrnestovna . . . 037415
Krenn, Mag. Katharina 002841
Krens, Thomas 051400, 051470
Krenzlin, Kathleen 017137
Kreplin, Charles 054263
Kreps, Dr. Christina 048169
Kress, Brady 047376, 048169
Kreß, Hannelore 018853
Kress, Dr. Theo 032007
Krestin, Steffen 017768, 017769
Kret, Robert A. 047595, 051868
Kreth, PhD Rasmus 010141
Krethlow-Benziger, Dr. Donata . . 041704
Kretschmer, Karl 020473
Kretser, David de 045842
Kretzmer-Diepold, Monika 022244
Kretzschmar, Bettina . . 022828, 022829
Kretzschmar, Ulrike 017299
Kreuger, S.C.P. 032661
Kreul, Dr. Andreas 017592
Kreulen, Ivast 032892
Kreunen, Jean 032463
Kreutzer, Teresa 049402
Kreuzberger, Robert 053022
Kreuzer, Dr. Barbara K. 053407
Kreuzer, Hermine 003028
Kreuzer, Dr. Stefanie 020058
Krichman, Michael 052964

Krick, Robert K. 048952
Kriebner, Prof. Dr. 021256
Krieg, M. 011900
Krieger, Edino 004715
Krieger, Stefan 017499
Kriegs, Dr. Jan Ole 020602
Kriegseisen, Dr. Jacek 034731
Kriegsman, L. 032582, 032582
Kriek, Johan 038670
Kriemler, Andrea 041276
Krier, Jean 030593
Krier, Patricia 048622
Krijnen, Marloes 031940
Kriller, Dr. Beatrix 003026
Krings, Dr. Michael 020561
Krippner, Regina 021691
Krischel, Dr. Roland 019715
Krischer, René 018845
Krischka, Karl 002285
Krishnappa, Dr. M.V. 024150
Krishtalka, Dr. Leonard 050186
Kristament, Werner 002375
Kristan, Dr. Markus 002931
Kristensen, Hans H. 010323
Kristensen, Tenna 010092
Kristfinnsson, Örlygur 024115
Kristiansen, Peter 010165
Kristic, Dr. Danilo 023962
Kristinsdottir, Gudrun 024064
Kristjansdottir, Gudbjorg 024087
Krištof, Andreas 003033
Kristoffersen, Åge 033607
Kristy, Ben 048897
Kritzas, Charalambos 022985
Kritzman, George 050469
Krivoshejev, Vladimir 038254
Krivova, N.N. 037561
Krizanac, Milica 038168
Krizner, L.J. 051455
Krjakunova, Nadežda
PProkofjevna 036980
Krjukov, Anatolij Charlampievič . . 037016
Kroča, Květoslav 009619, 009620,
009988
Kroczak, Czesław 034677
Krody, Sumru 054164
Kröger, Dr. Michael 019203
Kroeger, Ted 047092
Kröhnert, Gesine 021793
Krönke, Rudolf 019737
Krönsjö, Marianne 040923
Kroeper, Dr. Karla 017065
Krösche, Klaus 018047
Kroeze, Hans 032007
Krogedal, David 048085
Krogh, Leila 010071
Krogseng, Gari 053167
Krogstad, Bjørn R. 033771
Krohm-Steinberg, Sabine 018047
Krohn-David, Robin 046992
Król, Adam 035094
Krolevec', Sergyj Pavlovyč 043216
Kroll, Stefanie 002944
Kromann, Erik 010230
Kronberger, Dr. Michaela 002933,
003015, 003052, 003070, 003071,
003095
Kronbichler, Dr. Johann 002693
Kronenfeld, Daniel 051393
Kronenwett, M.A. Heike 016929
Kronig, Karl 041320
Kronkright, Dale 053116
Kronquist, Bo 011200
Kroodsma, Donald E. 046452
Kroon, Rita 010668
Kropf, Dr. Christian 041321
Kropff, Heike 018306
Kropivka, Georgij Andrijovyč . . . 043323
Kroqh, Finn E. 033980
Krotky, Gene 050321
Krouthen, Mats 034036
Kroy, Eadwierd 051767
Krpan, Sonja 009025
Krstić, Nina 038194

Kruckÿ, Marian 009933
Kruczyński, Dr. Andrzej 035363
Kruegel, Rosalie 053445
Krüger, Frank 019179
Krüger, Dr. Gundolf 018747
Krüger, Hans 021535
Krueger, Hilary 054192
Krueger, Dr. Ingeborg 017499
Krüger, Prof. Dr. Kersten 021489
Krüger, Dr. M. 038860
Krüger, Martin 020168
Krueger, Marylin 051941
Krueger, Phyllis 054036
Krueger, Thomas 018541
Krüger, U. 020129
Krüger, Ulrich 019627
Krueger, Ute 020943
Krüper, Jochen 018313
Krug, Prof. Dr. Dr. Klaus 020373
Krug, Martina 019048
Krug, Wolfgang 002697
Kruger, Gert 031859
Kruitwagen, W.H. 033099
Kruk, Karol 035407
Krukowski, Wojciech 035311
Krul, James 050348
Krull, Regine 018212
Krummenacker, Carolyne 015676
Krumova, Božura 005185
Krumpöck, Dr. Ilse 003004
Krupa, Alain-Gérard 003804
Krupa, Dr. Vladimír 038422
Krupp, M.A. Dr. Ingrid 022513
Kruppa, Gail 053781
Kruschen, Franziska 007482
Kruse, Anette 010290
Kruse, Hans 010162
Kruse, Prof. Dr. Joseph A. 018030
Kruseman, Pauline W. . 031924, 031967
Krusenstiern, Konstantin von . . . 047182
Kruska, Peter 019604, 019610
Kruszynski, Dr. Anette 018032
Kruter, Stefan 018952
Krutina, Dr. Ivan 009453
Krutisch, Dr. Petra 020853
Krutova, Marina Anatoljevna . . . 036847
Kruus, Ülle 010663, 010668
Kruytz, Rick 048977
Kružel, Krzysztof 034865
Krylov, V.M. 037809, 037824
Krylova, Raisa Nikolaevna 037888
Krystof, Doris 018033
Krzemiński, Jerzy 035189
Krzemiński, Prof. Wiesław 034900
Krzesinska, Caroline 045774
Krzeszowiak, Aleksander 035228
Krzmarzick, Pam 051317
Krznarić, Stjepan 008882
Krzykowska, Zofia 035271
Kseniak, Dr. Mieczysław 035001
Kuang, Maosheng 008186
Kuasnicka, Mellanee 052491
Kubáček, Jiří 038316
Kubach-Reutter, Dr. Ursula 020852
Kubadinow, Irina 003026
Kubassek, Dr. János 023626
Kube, Dr. Alfred 017607, 017609
Kube, Jan K. 016914, 022074
Kubiak, Anna 034632
Kubíčková, Marcela 009584
Kubizek, Mathilde 001747
Kublanovskaja, Natalija
Alekseevna 037570
Kubli, Dr. Renate 019931
Kubo, Hiroji 029173
Kuboschek, Andreas 018770
Kuboshima, Seiichiro 029669
Kubota, Gaylord 052422
Kubota, Dr. Tadashi 029406
Kubrak, Zygmunt 034985
Kučerová, Marie 009532
Kucharski, Karen 048565
Kuchenbecker, Antje 016766
Kuchorenko, Konstantin

Gavrilovič 037086
Kuchyňka, Zdeněk 009627
Kučinskij, Stanislav A. 037740
Kuckertz, Josefine 017065
Kučuk, Julija Valerjevna 037203
Kuczek, Irena 035025
Kuczyńska, Urszula 035044
Kudielka, Prof. Dr. Robert .. 017066
Kudo, Masahide 028644
Kudrjavcev, Aleksandr Abakarovič 037901
Kudrjavceva, Olga 037728
Kübler, Christof 042230
Kübler, Dr. Gabriele 021388
Kübler, Sabine 016830
Küchler, Max 041496
Kuecker, Susan 047483
Kuefstein, Elisabeth Gräfin .. 002601
Kügler, Dr. Martin 018737
Kühebacher, Dr. Egon 027733
Kühhas, Margareta 002884
Kuehhas, Thomas A. 051875
Kuehl, Tracey K. 046965
Kühle, Dr. Sandra 020879
Kühling, Karin 020021
Kühn, Christine ... 017202, 017241
Kühnel, Dr. Anita 017202
Kuehnert, Leonard A. 046422
Kuelpman, Darrell 050568
Kümin, Dr. Beatrice 042241
Kümmel, Dr. Birgit ... 016701, 016702,
                          016703, 016704
Künnapuu, Kalver 010708
Künnecke, Otto 017443
Künnemann, Thorsten-D. 042162
Kueny, Michel 015213
Kuenzel, Kurt 049052
Künzig, Anne 022057
Künzl, Uta 016536
Künzle, Fredy 041670
Künzler, Walter 042006
Kürten, Hans Peter 021358
Küster, Babette 019992
Küster, Prof. Dr. Bernd 020987
Küster, Dr. Ulf 041879
Küttner, Sybille 018995
Kufel, Tadeusz 035071
Kufeld, Klaus 020133
Kugel, Natalija Arkadjevna ... 037629
Kugina, Marija Borisovna 036888
Kugler, Andrea 020816
Kugler, Hans 018114
Kugler, Dr. Lieselotte 017243
Kugler, Dr. Walter 041447
Kuhfeld, Ellen 050934
Kuhl, Hartmut 018908
Kuhl, Dr. Uta 021664
Kuhler, Astrid 021663
Kuhlmann-Hodick, Dr. Petra . 017992
Kuhn, Prof. Dr. Axel 020044
Kuhn, Dieter 042167
Kuhn, Hans Jörg 041380
Kuhn, Marko 020021
Kuhn, Peter 042223
Kuhn, Reinhilde 019503
Kuhn, Richard 041569
Kuhn, Thomas G. 047721
Kuhn, Werner 020663
Kuhnt-Saptodewo, Dr. Jani ... 003046
Kuijers, Arie 038657
Kuiper, A. 032509
Kuiper, J. 032166
Kuipers, J. H. 032362
Kuipers, T. 032946
Kuitunen, Antero 010923
Kujukina, Tatjana Savvateevna . 037974
Kujundžić, Zilka 004283
Kukk, Inge 010705
Kukla, Gabriele 017134
Kukla, Dr. Jon 047271
Kuklik, Mirosław 035154
Kuklová, Zdenka 009986
Kukovalska, Nelya M. 043224
Kukuruzova, Nina V. 037747

Kula, Krysztof 034813
Kulakov, Viktor Evgenevič 036956
Kuldna, Kersti 010668
Kulęgowski, Mieczyslaw 035102
Kulemeyer, Jorge Alberto 000609
Kulesa, Chester 046572, 048470, 053218
Kulešov, Viktor Ivanovič 003254
Kulichikchin, Vladimir 043147
Kulik, Gary 054502
Kulik, Zbigniew 034806
Kulinčenko, Galina
   Aleksandrovna 037206
Kulinič, V'jačeslav Jakovyč ... 043194
Kuljachtina, Ljubov Fëdorovna .. 037012
Kulkarni, B.S. 024246
Kulke, Willi 019846
Kullenberg, Ulrika 040925
Kulling, Catherine 041660
Kullnigg, Johann 002879
Kulmala, Matti 011459
Kulnyté, Birutė 030528
Kulsarina, Alma 030075
Kulygina, Svetlana Sergeevna .. 037592
Kumada, Tsukasa 029267
Kumar, Jitendra 024387
Kumar, Niraj 024232
Kumar, Samarendra 024373
Kumar, Dr. Sarvesh 024213
Kumasaka, Yoshihiro 028831
Kumazawa, Hiroshi 029509
Kume, Atsuko 029541
Kumela, Marjut 010812
Kumirai, Albert 054844, 054847
Kumler, Deloris M. 047780
Kumler, Phyllis 046891
Kummer, Edelbert 041878
Kummer, Prof. Dr. Stefan 022792,
                                022801
Kunac, Prof. Ana 008946
Kunaev, Diar Askarovič 029798
Kunar, Dr. Arun 024212
Kunar, Cynthia A. 051238
Kundert, Andreas 041440
Kunert, Mechthild 021013
Kung-Haga, Vivian 050777
Kung Haga, Vivian 052968
Kuniholm, Thor H. 031794
Kunst, Beate 017080
Kunštek, Vesna 008938, 008940
Kunst, Edward 047971, 051845
Kunz, Hartmut 002593
Kunz, John 054216
Kunz, Jürgen 020507
Kunz, Martin 041235
Kunz, Mary Ellyn 051071
Kunz, Richard 041276
Kunz, Veronika 002593
Kunze, Prof. Dr. Max 021995
Kunze, Dr. Ronald 018962
Kunze, Prof. Dr. Walter 002413
Kunzelmann, Dr. Steffen 019316
Kunzemann, Dr. Thomas 021219
Kuononoka, Américo António .. 000112
Kuosmanen, Hannes 010893
Kupec, Dr. Tomas 009826
Kupiec, Janina 035393
Kuprijanov, Aleksandr Alekseevič 037426
Kuragina, Ljudmila Anatoljevna 037136
Kuraishi, Dr. Mahmud 003217
Kuraku, Yoshiyuki 029274
Kurant, Dr. Bogdan 035344
Kuraya, Mika 029608
Kurban, Michael J. ... 047674, 050169
Kurbel, Michaela 017822
Kurčakova, Olga Mykolaïvna .. 043282
Kurdiovsky, Horst W. 001903
Kurella, Dr. Doris 022048
Kuri, Prof. Vilson 000032
Kuribara, Masami 029307
Kuring, Alfons 021069
Kuring, Gertrud 021069
Kuroda, Hideo 029446
Kuroda, Raiji 028546
Kuronen, Darcy 047123

Kurosawa, Hiroshi 029549
Kuroyanagi, Tetsuko 029494
Kurpik, Maria 035355
Kurre 019236
Kurt, Halim 042814
Kurtz Schaefer, Guillermo S. ... 039034
Kurtz, Jeffrey J. 051133
Kurtz, Dr. John K. 047295
Kurućev, Dragana 038257
Kurz, A. 021531
Kurz, Birgit 022045
Kurz, Elmar 002368
Kurz, Friedhelm 020482
Kurz, Harry 016885
Kurz, Helma 018359
Kurz, Jürgen 018216
Kurz, Manfred 022647
Kurz, Toni 002104
Kurze, Gisela 020022
Kurzel-Runtscheiner, Dr. Monica 003026,
                                003094
Kurzenberger, Jim 053385, 053387
Kushwaha, M.R.S. 024384
Kusnecova, Anna Grigorevna .. 036840
Kussé, Kristel 003571
Kusserow, Karl 048050
Kussi, Kurt 022856
Kussmann-Hochhalter, Andreas . 020907,
                                020908
Kustelny, Elizabeth 053617
Kuster, Emma Maria 016772
Kusuma, Agung 024672
Kusumajaya, Dr. Made 024559
Kutcher, Ellen M. 054489
Kutluay, Sevgi 042963
Kutsioğlu, F. Suzan 042841
Kutter, Fritz 001741
Kutuev, Anas Abdrahmanovič . 036899
Kutz, Artur 016951
Kuuioka, Yasuo 029632
Kuusk, Sven 001185
Kuuskemaa, Jüri 010668
Kuutsa, Tommi 010963
Kuvšinskaja, Ljudmila
   Anatoljevna 037072
Kuyama, Masaru 028901
Kuzeev, Rail Gumerovič 037981
Kuzina, Marina Nikolaevna ... 037968
Kužma, Pavel 038338
Kužma, Teresa 034830
Kuzmanić Pierotić, Ivo 007514
Kuzmenkova, Natalija Vasylivna . 043328
Kuzmić, Zdenko 009043
Kuźmicki, Dr. Miecysław 034969
Kuzmin, Aleksej 037823
Kuzmin, Anatolij Ivanovyč ... 043117
Kuzmin, Kirill Stanislavovič .. 037751
Kuzmits, Dr. Wolfgang . 001859, 001860
Kuznecov, Nikolaj 037573
Kuznecova, Ljudmila
   Valentinovna 037716
Kuznecova, Natalija 037766
Kvalø , Frode 033852
Kvaran, Gunnar B. 033826
Kvaran, Dr. Olafur 024099, 024108
Kvisle, Colleen 007414
Kvist, Kari 010795
Kwack, Dong Suk 029936
Kwadwo, Osei 022935
Kwakernaak, A.D. 032787
Kwami, Dr. Ataa 022934
Kwang-ho, Dr. Hahn 030004
Kwashigroh, David 050059
Kwaśniewicz, Dr. Włodzimierz . 034944
Kwasnik, Elizabeth I. .. 043461, 043462
Kwiadas-Wierzbicka, Ludwika . 034805
Kwiatkowski, Prof. Marek ... 035346
Kwiatkowski, Phillip C. 050138
Kwiatkowski, Zygmunt 034766
Kwiecinski, William 048502
Kwok, Dr. Kenson 038266
Kwok, Kian Chow 038274
Kyas, Dr. Vojtěch 009480
Kyek, Mag. Martin 002625

Kyle, Peter 045105
Kyler, Duane 047762
Kyllönen, Eeva 010771
Kyø Hermansen, Cathrine ... 010062
Kyparissi-Apostolika, Dr.
   Aikaterini 022974, 023147
Kyriazi, Nelli 023004
Kyriazopoulos, Prof. Basil ... 023181
Kyriazopoulou, Katerina 023038
Kyroudi, Prof. Dr. A. 023085
Kytr, Hobe 049659
La Bianca, Domenico 026152
La Fayette, Aubert 016321
La Gamba, Prof. Filippo 025248
La Malfa, Francesca 026749
La Morgia, Dr. Jasmine 026640
La Porte, John Thomas 050114
La Regina, Prof. Adriano 027635
La Roche, A.J. 033273
La Rue, Eva 053768
La Tour, Bruno de 011604
La Vove, Eleanor 053131
Laabs, Dr. Annegret 020208
Laajo, Eero 010899
Laakso, Kathy 053611
Laaksonen, Tapio 011051, 011052
Laas, Anne 010667
Labaig Gallardo, Esther 039851
Labaníč, Eugen 038366, 038368
Labarre, Angélique de 012360
Labaste, Jacqueline 014622
Labat, Béatrice 016212
Labaugh, Sally 050497
Labaune, Patrick 016091
L'Abbate, Vito 026061
Labbé 013588
Labbe, Armand 053086
Labbe, Bruno 015237
Labbe, Daniel 016021
Laber, Philip 050730
LaBier, Brenda 053996
Labonne, Michel 011943
Laborderie, Gérard 015902
LaBorwit, Melanie 050175
LaBounty, Robert 053611
Labourdette, Anne 016177
Labovvić Marinković, Liljana .. 004255
Labrador, Ana Maria Theresa P. 034583
Labrails, Marie-Dominique ... 015999
Labrosse, Gilles 012776
Labuda, Dr. Jozef 038283, 038284,
             038285, 038286, 038287, 038288,
                                038289
Labuschagne, L. 038786
Lacabanne, Marie 005651
Lacabe, Jean-Marc 015984
Lacasse, Johanne 007020
Lacaussade, James de 011915
Lacchi, G. 026483
Lâce, Mara 030222
Lace, Simon 045193, 045195
Lacena, Daniel L. 034472
Lacerna, Daniel L. 034485
Lacey, Eileen A. 046940
Lacey, Michael 001052
Lacey, Susan J. 046799
Lachatenere, Diana 051466
Lacher, Gerhard 003080
Lacher, Dr. Reimar 018938
Lachish, Z. 025196
Lachmann, Dr. Stefan 022225
Lachmuth, Uwe 018096
Łaciuk, Adam 034660
Lackner, Dr. Helmut 003089
LaComb, Gerald 051656
Lacovara, Dr. Peter 046626
Lâcrâmioara, Marin 036215
Lacy, Kyleen 049701
Lacy, Tandy 053909
Laczko, Valerie 048681
Ladd, Audrey 051517
Ladenburger, Dr. Michael ... 017483
Ladier, E. 014225

Ladleif, Dr. Christiane . 016616, 016619, 016627, 016628
Ladleif, Dr- Christiane 016621
Ladlow, Nick . . 044791, 044792, 044794, 044801
Ladouceur, H. 006883
LaDouceur, Philip Alan 054614
Laenen, Marc 004026
Länsivaara, Ilkka 010751
Läufer, Egbert 019013
Laevskaja, Galina Leonidovna . 037583
Lafargue, Guy 012107
Lafer, Celso 004856
Laffon, Gèraud de 013121
Laffon, Juliette 014687, 014697, 014697
Lafon, Alain 013874
Lafont-Couturier, François 012112
Lafontant Vallotton, Chantal . . 041795, 041796
Laforet, Andrea 005871
Laforge, Rosanne 005702
LaFountain, W. Weston 048403
Lafragette, Florence 015596
LaFrance, Liselle 046341
Lagabrielle, Sophie 014780
Lagercrantz, Jacob 040922
Lagercrantz, Kristian 040998
Lagerkrans, Jacob . . . 040593, 040594, 041152
Lagerkvist, Cajsa 040663
Lagerwaard, Drs. Cornelieke . . 021613
Lagés, Hélène 016274
Lagier, Valérie . 016306, 016308, 016309, 016310
Łagiewski, Dr. Maciej 035422
Lagiglia, Dr. Humberto A. 000606
Lagler, Dr. Annette 016440
Lagnel, Cosette 041661
Lagnel, Valdy 041761
Lagoria, Georgiana . . . 049527, 049528
Lagos, Marta E. 048293
Lagos, Patricio Acevedo 007526
Lagoutte, Claudine 011855
LaGrave, Michelle 050622
Lagree, Dr. Kevin 049698
Laguiche, Comte R. de 011727
Lagumdžija, Ermana 004283
Lagvik, Elisabeth . . . 040607, 040608, 040610
Lahcéne, Zghidi Mohamed 000061
Laher, Herman 001703
Lahikainen, Dean 052880
Lahn, Tore 033585
Lahs-Gonzales, Olivia 052823
Lahtinen, Maija 011356
Lai, Charles 051438
Laï, Christine 014702
Lai, Guoyun 008043
Lai, Hsiang-Ling 042531
Lai, Piero 026313
Lai, Selina 017118
Lai, Weng-Hsian 042432
Lai, Yingxin 007575
Laidlaw, William K. 047941
Laing, D.W. 043817
Laing, Henry 004196
Laing, Lloyd 045420
Lainhoff, Thomas A. 050736
Lair, George 053863
Laird, Ruth 045955
Lais, Bruno 021697
Laitinen, Suvi 010816
Laitman, Nanette 051324
Lajevs'kij, Sergij Lazarevič 043122
Lajos, Asperjan 023705
Lajos, Dargay 023705
Lajoux, Jacques 013108
Lake, Paul 045879
Lake, Simon 044800
Lake, Susan 054120
Laken, Lorraine 051617
Lakhani, Prof. S.R. 001143
Lakin, Ian 043357
Lakin, W.G. 045474

Lakson, Stephen 047144
Lakstigala, Aivars 030235
Lal Gaur, Banwari 024221
Lalande, Bernard 015001
Lalaounis, Ioanna 022992
Lalev, Ivan . . 005043, 005085, 005086
Lalevic, Olga 004283
Lallemant, Marie-Françoise 012363
Lally, Dr. Janice 000804
Lally, Michael H. 050513
Lalošević, Aleksander 031761
Laluc, Christine 014707
Laluc 014567
Lam, Andrew 007883, 007884
Lam, Anita 050606
Lam, Joseph 046494
Lam, Prof. Peter Y.K. 007869
Lam, Tiffany 007875
Lama, Luís 034308
Lamaestre, Laurence de 013102
Lamb, Donna 046652
Lamb, Dorothy 044871
Lamb, Nancy 053206
Lamb, Rebecca B. 046557
Lamb, Susan 045696, 045700
Lamb, Vaughn 053055
Lambacher, Lothar 017205
Lambaša, Gojko 008988
Lambel, T. de 012993
Lambers, Dr. P. 032957
Lambert, Al 050627
Lambert, Anne 050593
Lambert, Dan 046722
Lambert, Françoise 042107
Lambert, Dr. Frederick A. 054368
Lambert, Jean Marie 016128
Lambert, John 044184
Lambert, Kirby 049433
Lambert, Marianne 046638
Lambert, Marie-Jeanne 012433, 013866
Lambert, Dr. Miles 045202
Lambert, Phyllis 006361
Lambert, Ruth 045189
Lambert, Thierry 003840, 003841
Lambert, Véronique 003757
Lambertson, John Mark 049670
Lambertye, Jean de 012662
Lambeth, Linda 005640
Lambić, Sunčica 038261
Lambilly, Comtesse Louis de . . 015627
Lambon, Corinne 015665
Lambraki-Plaka, Prof. Dr. Marina 022986
Lambrecht, Miriam 003492
Lambrigger, Josef 041778
Lambropoulos, Iraklis 023230
Lamesch, Chantal 012595
Laminou, Souley 033409
Lammassaari, Taina 010790
Lammertse, F. 032826
Lamonaca, Marianne 050849
Lamontagne, Nicole 005867
Lamort, Jean-Pierre 003652
Lamouret, Prof. Alfredo Daniel . 000157
Lampard, D. 044646
Lampe, Dr. Karl-Heinz 017504
Lampl, Johann 002880
Lampl, Dr. Sixtus 022300
Lamprecht, Anett 019992
Lamprecht, M.S. 038809, 038810
Lamprecht, Wolfgang 002944
Lamprechter, Cornelia . 002234, 002711
Lamschus, Dr. Christian 020178
Lamy, Frank 016312
Lamy, Laurence 016060
Lan, Kaigeng 008159
Lan, Nguyên Khả 054812
Lan, Zhongzhao 008377
Lancaster, Bill 053308
Lancelot, Antoine 015218
Lancmanis, Imants . 030132, 030169
Landauer, Susan 053043
Lande Moe, Tore 033994
Landen, A. Wayne 006604

Landert, Markus 042124, 042125
Landes, Carolyn 050814
Landis, Prof. Dr. Anna Maria . . . 026500
Landis, Mary Ann 046437
Landman, Dr. Neil H. 051327
Landmann, Jörg 016644
Landmann, Jürgen 020706
Landois, Marie-Ange 012068
Landová, Blanka 038406
Landress, Helen T. 046805
Landrum-Bittle, Jenita 046600
Landry, Edmond 005909
Landry, Frédéric 005993, 006028
Landry, Johanne 006374
Landry, Mario 005630
Landsbergen, Michael 001600
Landsmann, Hannah 003049
Lane, Dr. Arne H. 007336
Lane, H.E. 043847
Lane, Joshua W. 048218
Lane, Margaret Mary 049626
Lane, Martin 006589
Lane, Richard B. 052761
Lane, Steve 044019
Lanfermann, Dr. des. Petra 017386
Lang Pierce, Beverly 054120
Lang Walker, Barbra 005513
Lang, Alexandra 020880
Lang, Anton 002960
Lang, Brian J. 052169
Lang, Elisabeth 015887
Lang, Ewald 018857
Lang, Frank 022047, 022385
Lang, Gabriele 021136
Lang, Irma 038240
Lang, Merike 010674
Lang, Dr. Odo 041459
Lang, Paul 041530
Lang, Sabine 020590
Lang, Sherryl 053177
Lang, Sherryl P. 052430
Lang, Sonia 041894
Lang, Tom 053085
Lang, Ulrich 018438
Langauer, Dipl.-Päd. Fritz 002889
Langbein, Wolfram 021529
Langberg, Dr. Harald 010200
Langberg, Signe 010123
Lange Malinverni, Jonas R. 004525
Lange, Dr. Albert de 020966
Lange, Allynne 049983
Lange, Amanda E. 048218
Lange, Andrzej 035199
Lange, Carmen 022709
Lange, Dr. Chrisatine 020541
Lange, Christian 020366
Lange, Cortina 041001
Lange, Dietmar 022435
Lange, Jane 053061
Lange, Dr. Justus 017566
Lange, Karl Wilhelm 018535
Lange, Martina 017582
Lange, Monika 020255
Lange, Nicolas 003346
Lange, Rich 053823
Langelaan, M.M.C. . . 032524, 032525
Langelier, Jacques 015720
Langen-Crasemann, Sabine . . . 020746
Langenthal, Antje 016877
Langer, Agnes von 023904
Langer, Axel 042221
Langer, Brian 001560
Langer, Dr. Brigitte . . 017687, 019882, 019886, 020690
Langer, Dr. Heiderose . 021527, 022309
Langer, Margit 018901
Langford, Jan 049109
Langhammer, Peter 002225
Langhammerová, Dr. Jiřina 009828
Langhof, Jörgen 040624
Langhoff, Helmut 022591
Langhoff, Thomas 017066
Langhorne, Richard 045889
Langkafel, Sonja 019204

Langland, Leo 050813
Langley-Ware, Lyntia 047967
Langley, Lloyd 043836
Langlois, Daniel 006930
Langman, David 033353
Langner, Kerstin 019996
Langos, Barbara 016526
Langrell, Opal 007488
Langsbury, Hoani 033197
Langton jr., Sara 053039
Långvik-Huomo, Marianne 010771
Lanigan Wood, Helen 044295
Lankford, Éamon 024987
Lanko, Valeri 048051
Lannerbro Norell, Maria 040620
Lanning, Bruce 053215
Lannon, John 047102
Lannoo, Luc 003492
Lannoye, Henri 003381
Lanphier, Barbara 053884
Lanser, Dr. Klaus Peter 020602
Lansky, Aaron 046453
Lanszweert, Willem 003923
Lantos, Titus 002252, 002509
Lantz, Jerry A. 049025
Lantz, Lynn 048222
Lantz, Marcel 015425
Lanyon, Scott M. 050940
Lanz, Dr. Hanspeter 042246
Lanz, Vreni 041735
Lanzarini, Viviana . . . 025548, 025553, 025566, 025571
Lanzinger, Dr. Michele . 026876, 028177
Lapa, Dr. Pedro 035789
Lapalus, Marie 013958, 013959
Lape, Dr. Peter 053229
Lapénienė, Vitalija 030515
Lapetra, Damaso 038974
Lapeyre, Jean-François 015999
Lapham, Jennifer 047045
Lapi Ballerini, Dr. Isabella 026293
Łapiak, Dorota 034699, 034700
Lapious, Sydney 051355
Laplana, Josep de C. 039769
Lappalainen, Ari 011316
LaPrelle, Robert 048109
Laqua, Prof. Dr. Wolfgang 018692
Laqueste, Marc 012740
Laquindanum, Felicidad 034571
Lara, Maria 053918
Lara, Rita 048185
Laranjeira de Schroeder, Prof. Viviana Nuñes 000385
Larcher, Pedro Bruno 004522
Larco de Alvarez Calderon, Isabel 034323
Larey, Frank 044139
Large, Carol E. 051752
Large, Christine 043658
Large, Joan T. 052070
Large, Stephen 001418
Larguem, Salima 000075
Lark, Thomas C. 046358
Larken Royds, Meriel Victoria . 034392
Larkin, Annie 047013
Larkin, Graham 006594
Larkins, Catherine 054245
Larl, Rupert 002122
Larocque, Peter 006893
LaRose, Helen 006168
LaRowe, Annette 049260
Larraburu, Margo 000665
Larrañaga, Juan Ramón 039551
Larrañaga, Maria Isabel de 000185
Larry, Billings 047290
Lars, Roger 013546
Larsch, Georg 018367
Larsen, Carsten U. . . 010167, 010187
Larsen, Karen 054344
Larsen, Lynne 050336
Larson, Dr. Andre P. 053960
Larson, Barbara 054061
Larson, Bradley 051829

Larson, Eric 050213, 053522
Larson, John 049557
Larson, Dr. Judy L. 052601
Larson, Julie 052888
Larson, Larry 051021
Larson, Michael E. 047522
Larson, Nathan 050338
Larson, Neal L. 049474
Larson, Sara 053653
Larson, Sidney 047903
Larson, Silvia 007393
Larsson Modin, Anita 041075
Larsson, Åke 040535
Larsson, Göran 040830
Larsson, Hans-Ove 041001
Larsson, Hugo 040888
Larsson, Lars-Erik 040993
Larsson, Thorbjörn 040997
Larsson, Ulf 041027
LaRue, Dr. Héléna 045449
LaRue, Dr. Héléne 045459
Lárusdóttir, Unnur 024092
Lary, Christian de 014092
Lasansky, Jeannette 050227
Lasansky, Leonardo 052840
Lasfargues, Jacques .. 013947, 015554
Lash, Michael 047683
Lasher, Leigh 052659
Lasheras Corruchaga, José
  Antonio 040140
Lasić, Dr. Stanko 008912
Lasić, Višnja 008925
Lasius, Dr. Angelika 020345
Laska, Dr. Vera 054350
Laskey, Jane 045233
Laskowski, Rainer ... 019625, 019626
Lasmane, Daina 030230
Lasowa, Teresa 035385
Lassen, Thomas W. .. 010227, 010228
Lassere, Odile 014454
Lasson, Dr. E. 020968
Lassotta, Dr. Arnold .. 017431, 017958
Lastic, Henri de 015223
Laštovková, Dr. Věra 009666
Lastra Belgrano, Prof. M. Cristina 000264
Lastra, Cristina 000195
Lastu, Didier 016024
Laszlo, Želimir 009048
Latady, Bill 047145
Latanza, Dr. Antonio 027628
Latchford, Becky 043920
Latchford, Norma 001176
Lategan, L. 038660
Latham, Charles 053733
Latham, Stanley A. 054015
Latimer, Dr. Bruce 047809
Latondress, Paul 005633
Latour, Patrick 014622
Latourelle, Philippe 015544
Latsay, Brenda 006310
Latschar, Dr. John A. .. 049089, 049090
Lattanzi, Dr. Elena .. 026108, 026109,
                        027497, 028361
Lattanzi, Gregory 053799
Lattanzi, Vito 027634
Lattey, Gordon 046347
Latuconsina, Dr. Achmad .. 024590
Latus, Urs 020877
Latvala, Anja 011327
Lau, Nina 021663
Lauand, Débora 004879
Laub, Dr. Peter 002637
Laub, Dr. Richard S. 047302
Laubach, Simone 019438
Laubard, Charlotte 012108
Laube, Kurt 020580
Laube, Robert 019098
Laubie, M. 016127
Laubier, Lucien 014630
Lauble, Andreas 021965
Laudat, František 009829
Lauderdale, David 049767
Laue, Christoph 019204

Lauenborg, Michael 010042
Lauer, Dr. Bernhard 019528
Lauer, Keith G. 046607
Lauermann, Dr. Ernst 001729
Lauermann, Frank 050685
Laughlin, Patricia J. 053517
Laughlin, Phyllis 048496
Laughton, Rodney 053188
Laugier, Pierre-Henri 011681
Laukkanen, Rauni ... 011481, 011482
Laurence, Sue 044688
Laurent, Dominique 005868
Laurent, Elizabeth M. 052096
Laurenza, Joseph 049168
Laurette, Sandra 046851
Lauridsen, Henning Ringgaard .. 010358
Lauridsen, Inger ... 010129, 010342
Laurière, Claudine ... 014347, 014350
Laurinavičius, Vidmantas 030523
Lauring, Kåre 010107
Lauritzen, Ruth 049231
Lauro, Dr. Maria Giuseppina .. 027595
Laus, Harry 004454
Lausson, Mirja 041056
Lautanen-Raleigh, Marcia ... 048562
Lauter, Dr. Marlene 022796
Lautz, Thomas 019684
Lauwers, J. 003305
Laux, Regula 041643
Lauxerois, Roger 016203, 016204,
                              016206
Lauzière, Yves 007024
Lavalle Argudín, Mario 031232
Lavanchy, Jean-Claude 041320
Lavanchy, John 048343
Lavandera López, José 039903
Lavandeyra-Schöffer, Eléonore de 014618
Lavarnway, Carrie 047244
Laveck, Jocelyn 007409
LaVelle, Nancy 048619
Lavelle, Phyllis 047174
Lavender, Prof. Tony 045831
LaVere, Jeanette 051871
LaVerne, Thomas 053332
Laverty, Bruce 052030
Lavery, Peter 000890
Lavery, Sally Ann 046444
Lavie, Pierre 013784
Lavigue, Emma 014707
Lavin, Lucienne 054193
Lavino, Jane 049764
Laviolette, Suzanna ... 052827, 052828
Lavis, Tom 005873
Lavoie, Suzanne 006884
Lavoy, Carl L. 005605
Lavoyer, Martin 012240
Lavrenjuk, Benedykt Antonovyč . 043309
Lawitschka, Valérie 022216
Lawler, Jo 044449
Lawless, Ann 054472
Lawlor, Seamus 024833
Lawrance, Sarah 044008
Lawrason, Helena 005675
Lawrence, Brad 006768
Lawrence, D.R. 001438
Lawrence, John ... 050034, 050563
Lawrence, John H. 051294
Lawrence, Kirk 006130
Lawrence, Sarah 049282
Lawrence, Wade 047546
Lawrimore, Katrina P. 049082
Lawson, H. William 054598
Lawson, Jane 045123
Lawson, Julie 044274
Lawson, Karol 050533
Lawson, Pam 001159
Lawson, Scott J. 052428
Lawton, Gregory 049642
Lawton, Peter 045560
Lawton, Rebecca 048898
Lawton, Sheryl 049107
Lawton, Simon R. ... 045963, 045964
Lay, Chan Kim 030703

Layher, Hermann 021879, 021935
Layne, Janice 049028
Layne, Valmont Edward ... 038676
Layton, Dwight 046976
Lazareva, Lilija Georgievna ... 036954
Lázaro López, Agustín . 039168, 039171
Lazárová, Eva 038294
Lazarus, Fred 046759
Lazer, Laurie 053010
Lažetić, Predrag 038185
Lazier, Isabelle 013183
Lázničková, Dr. Ilona 009479
Lazov, Gavrail 005191
Lăzureanu, Emil 036240
Lazzarini, Dr. Antonio 027356
Le Bail, Jean-Marc 014892
Le Balc'h, Jean 012209
Le Blanc, Lisette 033047
Le Blanc, Tony 043383
Le Boulanger, Yann 015593
Le Brun, Foni 030593
Le Cieux, L. 012935
Le Cieux, Laurence 036184
Le Dinh, Diana 041660
Le Foll, Typhaine ... 014322, 014323
Le Gall, Régine 014413
Le Gendre, Bernard 013591
Le Grelle, Daniël Graaf 003309
Le Harivel, Adrian 024900
Le Leyzour, Philippe .. 011854, 016017
Le Loup, Willy 003422, 003425
Le Maguet, Jean Paul 015066
Le Maître, Claude 012134
Le Nouëne, Patrick ... 011653, 011654,
                011655, 011658, 011659
Le Pelley, Paul 043894
Le Restif, Claire 013304
Le Rouzic-Giovanneli, Gisèle .. 013847
Le Saux, Marie-Françoise ... 016120,
                               016121
Le Stum, Philippe 015021
Le Tissier, Paul 045712
Le Van Ho, Mireille 014622
Le Vay, Jonathan 045284
Le, Ma Luong 054790
Le, Trung 054811
Lea, Sandra 044913
Lea, Vanessa 045454
Leach, Mark 047576, 047577
Leach, Patricia ... 046895, 049453
Leach, Sally 046692
Leader, Prof. Mary 024908
Leafdale, Norris 049368
League, Alice 048325
Leah, Mervyn 044807
Leahu, Victor 036505
Leahy, K.A. 045749
Leahy, Mary 051037
Leal Spengler, Eusebio 009175
Leal, José H. 053084
Leamon, David L. 053769
Lear, Julie 051132
Lear, Paul 051838
Leard Coolidge, Lindsay ... 047359
Leatherman, Marvin 054238
Leavesley, Dr. James 001527
Leavey, Jane 046639
Leavitt, Tracey 006314
Lebailly, Florence 003606
Lebas, Frédéric 019011
Lebeau, Monique 014385
Lebedeva, Alla Markovna ... 037554
Lebegern, Robert 022164
LeBihan, Olivier 012115
LeBlanc, Denise 046316
LeBlanc, Guy R. 053588
Leblanc, Jacqueline 015547
Leblanc, Pascal ... 016054, 016055
LeBlanc, Suzanne 050168
Lebouteiller, Claire 016323
LeBouthillier, Louis 005557
Lebrasseur, Marie-Josée .. 006616
Lebrero Stals, José 040165
Lebrun, Jacques 016174

Lebrun, Jean-Marc 003576
Lebsack, Ron 006756
Lecasseur, Marie 015671
Lečev, Dečko . 005088, 005123, 005212,
            005214, 005215, 005216, 005218
Lechner, Georg 018845
Lechner, Dr. Gregor Martin 001915
Lechner, Martin 002825
Lechowski, Andrzej 034641
Lechuga, Dr. Ruth 031171
Leclair, Alain 013540
Leclair, Tom 006561
Leclef, Louis 003298
LeClaire Taylor, Gay 050901
Leclerc, Claudette 007453
Leclerc, Marie-Andrée 006353
Leclerc, Patrick 012191
Leclerc, Paul 051456
Leclerc, Paul-André 006137
Leclercq, Jean-Paul 014705
Leclert, Hélène 015143
LeCompte, Sarah 046716
Leconte, Philippe 003820
L'Ecotais, Emmanuelle de .. 014697
Lecouturier, Yves 012276
Lecuillier, Michel 011918
Lecuyer, Françoise 014514
Ledbetter, William C. 051125
Leder, Alan 048637
Leder, Carolyn R. 045069
Lederer, Carrie 054043
Lederer, Irene 019233
Lederer, Nan 047144
Ledermair, Martin 002742
Ledet, Claudette 006733
Ledford, Thomas G. 050532
Ledoux, François 006252
Leduc, Dawn 005322
Lee Reid, Katharine 047808
Lee Song, C.J. 032203
Lee, Bok Hyung 029966
Lee, Chris 045466
Lee, Christopher 047063
Lee, Colin 044823
Lee, Cynthia 051438
Lee, Dae-Sup 029943
Lee, David L. 052918
Lee, Eric M. 051586
Lee, Hee-Jae 030044
Lee, Jackie 054072
Lee, Jane 051159
Lee, Jean 051259
Lee, Jim 048969
Lee, Prof. Jong-Sang 030040
Lee, Karen 006500
Lee, Linda 044823
Lee, Michel 043522
Lee, Molly 048662
Lee, Ra Hee Hong 030062
Lee, Robert 051342
Lee, Seung-Koo 029921
Lee, Steven 046738, 046749
Lee, William D. 047244
Lee, Wuh-Kuen 042477
Leebrick, Gilbert 049264
Leech, James R. 048357
Leech, William J. 047064
Leeder, P.G. 045966
Leedy, Harriet 049060
Leedy, Sherry 049885
Leen, Frederik 003460
Leenes, Tony 032594
Leerintveld, A. 033004
Lees, Diane 045120
Lees, Traci 051738
Leeth, Sara 053772
Leeuw, G. de 032259
Leeuw, Prof. Drs. R. de ... 031982
Lefarth, Ute 022744
Lefebvre-Carter, Michelle .. 047763
Lefebvre, Eric 014689
Lefebvre, Pascal 003796
Leferman, Mariana 036414
Lefever, Joel 049508

LeFevre Stratton, Leslie 049047, 051308
Lefevre, Gérard 014305
Lefevre, John 047892
Lefèvre, Prof. Patrick 003491
Leff, Cathy 050849
Leff, Deborah 047118
Lefftz, M. 003831
Lefki, Michaelidou 009419, 009420
Lefour, Pierre 015967
Lefrançois, Michèle ... 012136, 012137
Lefrançois, Thierry ... 013474, 013476,
                                013478
Lefrancq, Janette 003492
Legac, Mirjana 008979
Legarda, Eduardo 034508
Legarda, Rosalita R. 034508
Legendre, Françoise .. 012317, 015171
Legêne, Dr. S. 031950
Legg, Dr. G. 043734
Legget, J.A. 044797
Legleiter, Kim 051022
Legout, Gérard 011747
Łęgowska-Radosz, Dr. Anna ... 035207,
                          035333, 035379
Legrand-Rossi, Sylvie 014785
Legrand, Sybille 003390
Legrand, Véronique 015537
Legro, Patrice 054132
Legros, Charles 004110
Lehberger, Prof. Dr. Reiner .. 018992
Lehman, Arnold L. 047253
Lehmann, Gérard 013777
Lehmann, Dr. Hans-Ulrich ... 017992
Lehmann, Hilmar 022812
Lehmann, Kai 021678
Lehmann, Dr. Karin 019208
Lehmann, Klaus 019267
Lehmann, Markus 041801
Lehmann, Randy 049067
Lehmann, Dr. Stefan 018968
Lehmann, Dr. Ulrike 020140
Lehmann, Ursula 022577
Lehnemann, Dr. Wingolf ... 020185
Lehner, Adam 051379
Lehner, Astrid 002995
Lehnert, Marie Josée 030598
Lehnherr, Dr. Yvonne 041497
Lehodey, Colette 015058
Lehr, William 049373
Lehtinen, Dr. Martti 010819
Lei, Anjun 007562
Lei, Weixin 008171
Lei, Xianzhong 008475
Leiber, Christian ... 018869, 018870
Leibnitz, Dr. Thomas 003053
Leich, Jeffrey R. 048920
Leicht, Walter 021479
Leichty, Erle 052100
Leidl, Mag. Bettina 003025
Leidner, Udo 020763
Leidy, Susan 050626
Leighton, John 032186, 032196, 044255,
                               044273
Leijen, A.J.M. 032065
Leijerzapf, Dr. I.Th. 032583
Leijten, C.J.M. 032322
Leilund, Helle 010166
Leim, Nico 022152
Leimu, Tuula 011489
Lein, Sharon 048239
Leiner, Dr. Srećko 009037
Leinfelder, Prof. Dr. Reinhold .. 017244
Leinicke, Kris G. 052631
Leinonen, Paula 011393
Leinweber, Dr. Ulf 019548
Leinz, Dr. Gottlieb 018064
Leiseth, Sylvia 054223
Leismann, Burkhard 016476
Leisner, Silke 019388
Leistenschneider, Dr. Eva ... 022268
Leister, Claudia F. 048350
Leister, Michael D. 048357
Leistner, Dr. Gerhard 021326

Leite, Ana Cristina ... 035760, 035768,
                               035817
Leite, Prof. Michael 047510
Leitgeb, Dr. Franz 001878
Leitgeb, Otto 002200
Leitinger, Prof. Josef .. 002853, 002854
Leitmeyer, Wolfgang 021931
Leitnaker, Connie 050116
Leitner-Böchzelt, Susanne ... 002291,
                               002293
Leitner, Andreas 002843
Leitner, Cynthia 048587
Leitner, Dr. Friedrich W. 002192
Leitner, Mag Irene 001695
Leitner, Norbert 002709
Leitz, Prof. Dr. Christian 022212
LeJambre, L.A. 047096
Lejeune, Louis 003321
Lejnieks, Dr. Janis 030190
Lekavičienė, Angelė 030374
Lekens, Luc 003541
Lekholm, Kerstin 040660
Lelaurain 013639
Leloup, Michel 003439, 012949
Lemaire, Jean-Michel 012663
Lemaire, Michèle 012163
Lemaire, Philippe 015792
Lemal Mengeot, Chantal 003513
Lemańska, Stefania 034748
LeMarie, M.J. 043404
Lemay, Brian 047128
Lembke, Dr. Katja ... 019250, 019251
Lemcke, J. Peter 017729
Lemelle, François 014621
Lemercier, Christian 012126
Lemeunier, A. 003722
Lemeunier, Albert 003793
Lemežienė, Laima 030448
LeMieux, Linda D. 047891
Leming, C.F. 045832
Lemire, Marie 006398
Lemky, Barbara 005669
Lemle, Robert 049045
Lemmens, Claude 003432
Lemmon, Dr. Alfred 051294
LeMoine, Dr. Geneviève 047286
Lemoine, Serge 014743
Lemonedes, Heather 047808
Lemons, C.R. 048822
Lemos, Nikolas S. 023204
Lemothe, Peter Thomas 051264
Lempart, Desirée 018729
Lenaers, J.H. 032275
Lenaghan, Patrick 051404
Lenarduzzi, Nelson ... 000393, 000394
Lenart-Juszczewska, Krystyna .. 035077
Lenarz, Michael 018430
Lencova, Valentina Bronislavovna 037258
Lencquesaing, May-Eliane de .. 014811
Lenglet, Marie-Cornélie 013803
Lenglin, Catherine 013154
Lengyel-Kiss, Katalin 023534
Lenhart, Ralf 002224
Lenherr, Niklaus 041466
Lenk, Elfriede 002526
Lenkiewicz, Jolanta ... 034856, 034895
Lennard, John Barret 001001
Lennard, Mark 001097
Lennig, Dr. Petra 017080
Lennoy, Kim 049143
Lensen, Jean-Pierre .. 004117, 004118
Lenski, Jean-Pierre 015746
Lenssen, Dr. Jürgen .. 022356, 022788
Lent, Eric 047389
Lenten, Joke 031965
Lentino, Miguel 054711
Lentz, Ingo 018038
Lentz, Lamar 052693
Lentz, Thomas W. 047362
Lenz-Weber, Dr. Diana 019029
Lenz, Berta 018706
Lenz, M.A. Gerhard 017513
Lenz, Helmut 018418

Lenz, Iris 022042, 022043
Lenz, Mary Jane 051448
Lenz, Prof. Dr. Norbert 019518
Lenz, Vera 018751
Lenze, Susanne 017292
León Rivera, Jorge de 031163
León Rivera, Melba Rocío 008619
León, Imelda de 030893
Leon, Lamgen 051438
Leonard, Amy 046316
Leonard, Ben 052855, 052856
Leonard, Claude 014275
Leonard, Douglas 005382
Léonard, Georges 003603
Leonard, Glen M. 052915
Leonard, Jessie 050909
Leonard, Mark 050436
Leonard, Pascal 003271
Leonard, Robert 050616
Leonardo-Finger, Jascin 051156
Leoncini, Dr. Luca 026410
Leone, Carol 054058
Leonenko, Gajane ... 030332, 030544
Leonhardt, Günter 019837
Leonhardt, M.A. Monika 042239
Leonhardt, Rainer 020423
Leonhardt, Dr. Ute 019458
Leonori, Dr. Maria Chiara 026197,
                               026202
Leontiev, Andrej 037693
Leontjeva, Sofija Jakovlevna .. 037588
Leopardi, Anna 027488
Léotoing, Comte de 016007
Lepădatu, Ioana 036621
Lepage, Jean . 014428, 014430, 014431
Lepard, David 049072
Lepdor, Catherine 041656
Lepe, Marcella 031407
Lepičić, Janko 038192
Lépine, André 003509
Lépine, Brigitte 011800
Lépine, Luc A. 006847
Leplat, R. 003376
Lepley, John G. 048781, 048782
Leppala, Vesa 010805, 011344
Lepper, Beate 022488
Lepper, Dr. Katharina B. 018064
Leppiniemi-Järvinen, Anita ... 010969
Leproust, Evelyne 015023
Lepší, Miroslav 009617
Leray, Victor 013905
Lerch, Michaela 014683
Lerebours, Lionel 023393
Leresche, Anne 014870
Léri, Jean-Marc 014645, 014688
Leritt, Sarah 044792
Lernaut, Françoise 011632
Lernbaß, H. 019822
Lerner, Peter 047235
Lernout, Françoise 011633
Lerolle, Aggy 014749
Leron-Lesur, Pierre 015538
Lerouge-Benard, Jean-Pierre .. 013160,
                               014481
Leroux, Doug 006626
Leroux, Jean-François 016230
Lerov, Trifon 005040
LeRoy, Angelina 047454
Leroy, Gilbert 013862
Leroy, Jacques 011816
LeRoy, Louis 054611
Leroy, Vincent 001952
Lerrick, Dr. Yacob 024606
Lervy, Annick 014173
Lesaffre, Jean-Marie 013808
Lesák, Dr. Barbara 003059
Leschhorn, Prof. Dr. Wolfgang . 017558
Lescombes, Germain 012289
Lescroat-Cazenave, Elizabeth .. 014571
Lesdorf, Svetlana Pavlivna ... 043295
Leser, Rupert 016897
Lesher, Ronald E. 052834
Leshikar, Howard B. 053708

Lesko, Diane B. 053179, 053182
Leskova, Marina 030630
Leskovar, Jutta 002330
Leskovar, Peter 038614
Leskovec, Ivana 038510, 038516
Leslie, James 043481
Leslie, Mike 051732
Leśniewska, Józefa Ewa 034936
Lesouef, Jean Yves 012204
Lesser, Ann 054067
Lesseur, Catherine 011655
Leßmann, Dr. Sabina 017495
Lester, Alan K. 047119
Lester, Ellen 050017
Letenayova, Dr. Zdenka 038323
Letendre, Julie 048210
Letertre, J.B. 015898
Leth, Gwen I. 052528
Lethen, Paulien 047260
Letourneur, Céline 012947
Letschert, Edmund 022479
Lettany, Lieve 003858
Letts, A. 038825
Letz, Andrea 019655
Leu-Barthel, Angelika 018979
Leuba, Marion 011949, 011950, 011951
Leumann, Alexander 041490
Leung, Corinne 049608
Leuschner, Dr. Jörg 021575
Leutenegger-Kröni, Tanja 041564
Leutenegger, Dr. Marco 042002
Leuzinger, Dr. Urs 041491
Lev-Ari, Y. 025125
Levá, Kimmo 010955
Levallet, Didier 014232
Levander, Karin 041138
LeVangie, Patrick 048938
Levčenko, Vitalij Vasiljevič ... 037187
Levee, Catherine 049831
Levens, Dwight R. 048296
Levental, Olivia 017822
Leverton, Raymond 049729
Lévesque, André 006952
Levey, Georges 015727
Levi, A. 025128
Levièvre, Francine 006413
Levin, Annika 014663
Levin, Dalia 025062
Levin, Mai 010668
Levin, Nikolaj Iljič 037886
Levin, Pamela 053213
Levine, Beverly 048888
Levine, Janet 051474
Levine, Kenneth 051468
Levine, Dr. Louis 051439
Levine, Phyllis 051406
Levinho, José Carlos 004736
Levitt, S. 044795, 044797
Levkin, Jaroslav Borisovič ... 037656
Levy, David C. 054098, 054099
Levy, Harry A. 050831
Levy, Sue M. 052036
Lew, Dr. Irena 035059
Lewallen, Constance 046933
Lewandowski, Leszek 034822
Lewandowski, Marian 035122
Lewandowski, Ryszard 035037
Lewenhaupt, Dr. Inga 040869
Lewicki, Ireneusz 035010
Lewin, S.J. 043907
Lewis, Amy E. 048126
Lewis, Bethan 043822
Lewis, Carolyn Chinn 050187
Lewis, Frank C. 046514
Lewis, Greg 053903
Lewis, Karen 046853
Lewis, Larry 054529
Lewis, Lesley 007193
Lewis, Louise 051638
Lewis, Maisie 000835
Lewis, Mario 034176
Lewis, Merri 046366
Lewis, Oli 050163
Lewis, Sally-Anne 001589

Lewis, Stephanie 053446
Lewis, Vel A. 042733
Lewis, William R. 052109
Lewitzki, Wolfgang 019674
Lewry, Joan 046177
Lewyllie, Ingrid 003646
Ley, Dr. Astrid 021000
Leyge, François 014170
Leyoudec, Maud 015745
Leypold, Denis 015883
Leys, K. 003433
Leys, Kathleen 003447
Leysen, Veerle 003826
Leyten, C.J.M. 032939
Leyva, Maria 054090
Lezarrdiere, A. de 013384
Lhagvasüren, I. 031736
L'Hostis, Denez 012822
Li, Bingzhong 008488
Li, Chaozhen 007710
Li, Chia-Wei 042476
Li, Chun 008158
Li, Chunfang 008224
Li, Chunrong 008182
Li, Denggao 008143
Li, Deyang 008506
Li, Fuxin 008479
Li, Gengshan 008293
Li, Guicheng 008214
Li, Hancai 007915
Li, He 007918
Li, Hongri 007901
Li, Jiahua 007713
Li, Jianeng 008363
Li, Jianxin 008188
Li, Jichun 008142
Li, Jinghan 008325
Li, Jinglin 008552
Li, Jubao 007777
Li, Junping 008556
Li, Juzheng 008507
Li, Kexiang 008509
Li, Kuanhong 008276
Li, Kunsheng 007992
Li, Lifu 007666
Li, Linna 007811
Li, Peiyuan 008210
Li, Qifu 008216
Li, Qiliang 007564
Li, Qingrong 007577
Li, Rongji 007953
Li, Shangshu 007691
Li, Shecan 008435
Li, Shunjian 007860
Li, Shutao 007702
Li, Tai 008544
Li, Te'er 007736
Li, Tianbao 008415
Li, Weidong 008439
Li, Weigui 008518
Li, Weiguo 008536
Li, Wensheng 007945
Li, Xi 008012
Li, Xia 008144
Li, Xianheng 007803
Li, Xiliang 007977
Li, Xinling 007924
Li, Xiongkun 008373
Li, Xuewen 008409
Li, Xumei 008544
Li, Yan 008126, 008127
Li, Yangke 008549
Li, Yanming 008319
Li, Yecheng 007569
Li, Yinde 007817
Li, Yinsheng 008155
Li, Youcheng 008447
Li, Yousheng 008489
Li, Yuanqing 008447
Li, Zegao 008543
Li, Zhongshan 008369
Li, Zuyao 008053
Liabeuf, Brigitte 015934
Liagouras, Angelos 023168

Liang, Binggui 007973
Liang, Hengtang 008364
Liang, Huimin 007999
Liang, Jimin 007806
Liang, Juxiu 008408
Liang, Nongxi 008076
Liang, Sichong 008532
Liang, Tingbao 008103
Liang, Xiaomei 008157
Liang, Yaoji 007737
Liang, Zhaobi 007927
Liang, Zuozhen 008193
Liaño, Marta 054724
Liao, Jinxiong 008277
Liao, Kui-Yin 042502
Liapis, Kostas 022947
Liavas, L. 023001
Libbus Boon, Edna 005722
Liberatore, Laurie 048565
Libert, Laura L. 052065
Liby, Håkan 041110, 041118, 041118, 041123
Lică, Paul 036352, 036353
Lichnerova, Rut 038285
Licht, Ruth 041232
Lichte, Dr. Claudia 022791
Lichtenberg, James J. 052480
Lichter, Dr. Clemens 019505
Lichtin, Christoph 041708
Lichtlé, Francis 013350
Lichty, Robert C. 047410
Lickleder, Dr. Hermann 019575
Lickteig, Dorothy L. 049053
Licky, Maurice 007506
Liczkov, Sergej Nikolajevič 036840
Liddell, Dave 054320
Liddle, Peter 044113
Liddle, Dr. Peter 044604
Lideń, Magnus 041114
Lidequist, Dr. Christina 040965
Lidman, Roger 052123
Lidner, Andreas 019983
Lidtke, Thomas D. 054289
Līduma, Ilze 030165
Lieb, Hartmut 018907
Liebelt, Glenn T. 046375
Liebenberg, O.J. 038660
Lieber, Vincent 041807
Lieberman, Christina 053840
Lieberman, Richard K. 050380
Lieberman, William S. 051424
Liebgott, Dorion 007161
Liebgott, Niels-Knud 010165
Liebhart, Prof. Dr. Wilhelm 016562, 016563
Liebich, Haimo 020536
Liebich, Ursula 042184
Liebling, Carolina 041658
Liebscher, Sabine 020207
Liechtenstein, Friedrich 002598
Liechti, Werner G. 042019
Liechty, Ronald 052208
Liedekerke Beaufort, Comte Hadelin de 003508
Lieder, Cecilia 046912
Liedtke, Peter 018309
Liekonen, Marita 010988
Lien, Mary 052221
Lienau, Jennifer 053103
Lienerth, J. 018320
Lienhard, Kurt 041290
Lienhard, Ulrich 042185
Liesbrock, Dr. Heinz 017523
Liesch, F. 020405
Liesens, Liliane 003458
Lieser, Wolf 017106
Lieth, Elke-Luise von der 017195, 017317
Lif, Malka 025190
Lifjeld, Prof. Jan T. 033847
Lifshitz, S. 025143
Light, Brad 054476
Light, Brian 045679

Light, Dr. Helen 001470
Light, Natalie de 020843
Lightbourn, Brian 004189
Lightfoot, Ricky L. 048011
Lighthall, Penny L. 007252
Ligne, Michel Prince de 003356
Lignori, Prof. Laura 028121
Ligon, Doris H. 047896
Ligon, Dr. J. David 046365
Ligúde, Ginsa 030134
Liikkanen, Hilkka 011324
Liken, Paula 050813
Liknes, Fern 005797
Lilienthal, Dr. Georg 018906
Lillehammer, Grete 033972
Lilley, Clare 046086
Lillington, Sophie 045081
Lilly, Bruce 052847
Lilly, Stephanie 047564
Lilly, Tina 050582
Lillystone, Jane 043925
Lim-Yuson, Dr. Nina 034538
Lim, Hyo-Taek 029896
Lim, Sarah 050801
Lim, Yujin 050440
Lima Greene, Alison de 049588
Lima, Dr. Francisco 035530
Limat, Marc 041678
Limbacher, Dr. Gabor 023921
Limbree, Christian 003972
Limburg-Stirum, Comtesse F. de 015954
Limme, Luc 003492
Limón Delgado, Antonio 040172
Limpo Pérez, Luis Alfonso 039827
Lin, Chien-Ling 042505
Lin, Chunxin 007678
Lin, Huanzhen 008091
Lin, Jiafu 008119
Lin, Lee Chor 038272
Lin, Maoyuan 007952
Lin, Shenggui 008140
Lin, Weiqi 008125
Lin, Wen-Ren 042528
Lin, Xiaoming 008239
Lin, Xing'er 007926
Lin, Zhaoren 007923
Lin, Zhifang 008375
Lin, Zuliang 008165
Linaeque, Dr. P.M. 031846
Liñán, Prof. Eladio 040455
Linciano, Dr. Antonio 026362
Lincoln, Lee 048226
Lincoln, Louise 047654
Lind-Sinanian, Gary 054213
Lind-Sinanian, Susan 054213
Lind, Axel 010304
Lind, Dr. Christoph 020255
Lind, Ed 052405
Lind, Evgenij Alekseevič 037763
Lindahl, Per 040694, 040696
Lindauer, H. 033300
Lindauist, Christopher 054336
Lindberg, Annika 041106
Lindberg, David R. 046939
Lindberg, Janelle 051761
Lindberg, Kathy M. 048723
Lindblom, Michelle 047019
Lindbom, tina 040801
Lindeman, Jean 054352
Lindemann, Prof. Dr. Bernd Wolfgang 017147
Lindemann, Marilyn 048634
Linden, Grace 053349
Linden, Jaap ter 032814
Lindén, Lena M. 040711
Linden, Peppy G. 047586
Lindenhofer, Harald 003056
Linder, Karin 040674
Linder, Maria 040744
Linder, Marja-Liisa 011409
Linderer, Steve 050149
Linderman, Mike 048643
Lindert, Dietmar 002837
Lindh, Anders 040875

Lindh, Marianne 041157
Lindhartseni, Ole 034070
Lindholm, Dan 010765
Lindhorst, André 021016
Lindley, Ralph 045622, 045623
Lindmayr, Burgi 002626
Lindmier, Tom 050146
Lindner, Gerd 016765
Lindner, Dr. Helmut 018013
Lindner, Mathias 017738
Lindner, Dr. Robert 002625
Lindner, Walter 018632
Lindquist, Beth 048035
Lindquist, Everett 053952
Lindquist, Sven-Olof 041174
Lindqvist, Catharina 041027
Lindqvist, Per-Inge 040740
Lindqvist, Svante 041027
Lindsay Baker, Dr. T. 049479
Lindsay, Frances 001279
Lindsay, Judy 044116
Lindsay, Dr. Laura F. 046817
Lindsay, Mike 046587
Lindsay, Sally 053212
Lindsberger, Josef F. 002005
Lindseth, Knuth 033763
Lindsey, David 044660
Lindsey, Jack 052067, 052081
Lindsey, Jane 049849
Lindsey, Dr. Rob 050059
Lindström, Anneli 010872
Lindström, Inger 040789
Lindström, Leif 040890
Lindstrom, Janet 051248
Lindvall, Karin 040888
Lineros, Ricardo 039233
Lines, Sue 044079
Ling, Bo 008112
Ling, Rencai 007836
Lingier, Rafael 003542
Lingstadt, Kirsty 044509
Linhares, Philip 051681
Linhart, Dr. Eva 018435
Lininger, Kris 053907
Link, Wayne 005377
Linke, Irmgard 002515
Linkov, Krasimir 005127
Linkowski, Zdzislaw 034771
Linley, Jim 044636
Linn, Joe Dale 052363
Linn, Dr. M. 051588
Linna, Markku 011421
Linnehan, Sharon 048326
Linnel, Eric 047057
Linnen, Robert 007363
Linnertz, Rudolf 017550
Linničenko, Antonina 043226
Linnie, Martyn J. 024915
Linnikova, Nina Jakovlevna 037796
Lino, Elizalde 040464
Linou, Marie-Josée 015096, 015097
Lins, P. Andrew 052081
Linschinger, Josef 001954
Linsky, Carol 052646
Linsmann, Dr. Maria 022208
Linsmayer, Thomas 020562
Linthorne, Marguerite 005437
Linton, Dr. M. 001080
Linton, Meg 050409
Lintott, B. 033172
Linville, Tony 053532
Liot, David 015052
Liotta, Jorge 000603
Lipchinsky, Zelek L. 046927
Lipfert, Nathan 046809
Lipka, Dieter 016782
Lipkens, Myriam 003682
Lipo, Frank 051667
Lipońska-Sajdak, Jadwiga 034816
Lipovanski, Georgi 005201
Lippens, Kathleen 003473
Lippert, M.A. Anja 016616, 016621, 016627, 016628
Lippi, Angelo 026638

Lippi, Dr. Emilio 028183, 028184
Lippiett, John 045555
Lippincott, Louise 052157
Lippman, Judith 046761
Lippmann, Renate 017759
Lippold, Björn 040956
Lipps, Jere 046939
Lips-Liechti, Ursula 041721
Lipschutz, Jeff 051826
Lipscombe, Bruce 054016
Liptuga, Tatjana Ivanivna 043276
Lira, Dr. Raúl 000326
Lisai, Virginia 054344
Lisberg Jensen, Ole 010190
Liščáková, Daniela 009716
Lisenko, Mychajlo Lukyč 043284, 043285
Lisicyn, Viktor Borisovič 038055
Lisk-Hurst, Corrie 049250
Lisk, Susan J. 049308
Liška, Peteras 038465
L'Isle, Viscount de 045479
Lisnjak, Katerina Jevgenivna ... 043180
Lisowski, Silvia 022580
Liss, David 007191
Lissague, Bernard 012186
Lissens, Paul 003462
List, Manfred 001974
Listen, Simon 044199
Lister, Alison 045007
Listoe, Philip 006976
Litak, Anna 034902
Lithén, Hans-Olof 011470
Litochleb, Dr. Jiří 009837
Litteral, Kate 050813
Little, Ann 050198
Little, Faye 043925
Little, Stephen 049534
Little, Susan 048724
Little, Tricia 049457
Litvak, Silvia 000342
Litwin, Dr. Jerzy .. 034742, 034737,
      034742, 034753, 034778, 035273
Litwin, Józefa 035448
Litz, Dr. Christine 019705
Litzenburg, Thomas V. 050267
Liu, Bailing 008059
Liu, Chuan 008010
Liu, Dejun 008520
Liu, Fo'an 007943
Liu, Guang Rui 007706
Liu, Guang Tang 007718
Liu, Guangyu 008011
Liu, Guangzhong 008016
Liu, Guojin 008190
Liu, Haichao 007780
Liu, Hong 008422
Liu, Hongbo 007862
Liu, Hongshi 008008
Liu, Hui 008303
Liu, Ji 008189
Liu, Jianli 008247
Liu, Jianzhong 008524
Liu, Jionghai 007850
Liu, Liben 008195
Liu, Lingyun 007759
Liu, Ning 008271
Liu, Qi'an 008480
Liu, Qingyan 007795
Liu, Shun 007573
Liu, Suiqun 007970
Liu, Wenquan 008243
Liu, Wie 007657, 008542
Liu, Wujun 008555
Liu, Yan 008145
Liu, Yi 007963
Liu, Ying 007571
Liu, Yirong 008167
Liu, Yiwen 007959
Liu, Yuchuan 007707
Liu, Yulin 008153
Liu, Zhiquan 007776
Liu, Zhiyi 007704
Liu, Zhongxin 008146
Liu, Zubei 007966

Lium, Randi N. 034039
Lively, Carter C. 046501
Liverani, Dr. Paolo 054682, 054683,
      054686, 054689
Livernette 013766
Livezey, Dr. Bradley C. 052158
Livingston, Glenda A. 054024
Livingston, Jan 052536
Livingston, Dr. Valerie 053265
Livingstone, Judy A. .. 005696, 006383
Livland, Haakon 033707, 034018
Livneh, M. 025125
Livsey, Karen E. 049786
Lixfeld, Gisela 021733
Lixl, Dr. Christine 003089
Lizarazo, Plata 008576
Lizon, Sylvian 014951
Lizzadro, John S. 048563
Ljiljak, Aleksandar 004278
Ljøkjell, J. 033691
Ljubaščenko, Anna Ivanovna .. 037928
Ljubimov, Vladimir Alekseevič .. 036860
Ljubović, Prof. Blaženka ... 008984
Ljungberg, Louise 040938
Ljunggren, Lieutenant colonel
      Ingmar 041185
Ljungström, Lars 041040
Llabrés Martorell, Pere-Joan .. 039670
Llabrés, Pere-Joan 039894
Llagostera Martínez, Dr. Agustín 007534
Llanas, Joan 040396
Llanes-Myers, Susan 049578
Llanes, Manuel 040231
Llewellyn, Mike 044545
Llobet, Guy 012619
Llongarriu, Marissa 039831
Llorente, Tina 049588
Lloveras, Fernando 036137
Lloyd Harvey, Peter 045520
Lloyd, Alan 001535
Lloyd, Deanna S. 050663
Lloyd, Kristin B. 046394
Lloyd, Stephen 044274
Lluch, Montserrat 040364
Lmas, Jean de 012901
Lobanova, Galina Pavlovna ... 036978,
      036986
Lobaõ, Dr. Paulo 035530
Lobas, Aleksandr 029856
Lobb, Muriel 050309
Lobb, Sylvia 049786
Lobjakas, Kai 010671
Lobkareva, Antonina Vasilevna . 037992
Lobner, Walter 002787
Lobo, Dr. Wibke 017112
Lobos Lápera, Victor 007520
Loburev, Vladimir Vasiljevič .. 037366
Lochman, Dr. Tomas .. 041261, 041291
Lochmann, Reinhart 016466
Lock-Weir, Tracey 000762
Lock, Victoria A. 050046
Locke, Dr. H. 006467
Locke, Michelle 048002
Lockett, Daniel 045165
Lockwood, D.C. 033318
Lockwood, Dave 048358
Lockwood, David 044153, 044155,
      044156, 044157
Lockwood, Dr. Yvonne R. .. 048441
Locmelis, Prof.em. Dr. Jazeps . 030188
Loderer, Klaus J. 016697
Lodewijk, W. 032410
Lodge, Penny 052172
Lodowski, Dr. Jerzy 035417
Löb, Heidrun 042223
Löbel, Andrea 003023
Loeben, Dr. Christian E. ... 019060
Löber, Eva 020196
Löbert, Dr. Horst W. 022071
Löbke, Dr. Matthia 019161
Löckemann, Karsten 020567
Löcken, Monika 016638
Löffler, Dr. Erzsébet . 023614, 023616
Löffler, Dr. Michael 022910

Loeffler, Robert J. 051046
Löffler, Ruth 018057
Löfgren Ek, Anita 040789
Löfgren, Pia 040598
Löfström, Lena. A 041063
Löher, Jochen 017962, 017966
Löhr, Dr. Alfred 017585
Löhr, H. 018968
Löhr, Dr. Hartwig 022202
Loehr, Scott W. 046657
Løken, Trond 033972
Löliger-Henggeler, Elisabeth .. 041849
Lønborg Andersen, Vivi 010166
Lönnebo, Lovisa 041021
Lönnström, Yvonne 011479
Loera Chávez y Peniche, Dr.
      Margarita 031244
Lösch, Bernhard 027733
Lösche, Wolfgang 020526
Lötsch, Prof. Dr. Bernhard .. 003051
Lövgren, Amanda 040966
Loew, Benedikt 021565
Löwenkopf, A.Z. 025031
Loewy, Dr. Hanno 002097
Lofgren, Donald 047759
Loft, Elaine P. 049556
Lofthus, Brian 049177
Loftin, Jack 046523
Loftin, Jim 048999
Loftis, Lynnea 049747
Loftus, John 021799
Loftus, Sydney L. 051772
Logan, Brad 050185
Logan, Burt 047137
Logan, Joyce 006670
Logan, Judy 049265
Loge, Øystein 033572
Logger, R. 031981
Logghe, Marc .004077, 004078, 004079
Logunov, Dr. D. 045210
Lohberg, Dr. Gabriele ... 022199
Lohiniva, Leena 011324
Lohman, Prof. Jack 045041
Lohman, Jack 045048
Lohmann, J.H.A. 032427
Lohmann, Uwe 022514
Lohner, Beate 018866
Lohrenz, Mary 049750
Loichemol, Hervé 012968
Loichinger, Hannes ... 020179
Loidl, Josef 001752
Loikkanen, Kirsi 011206
Loisy, Antoine de 011709
Loizou Hadjigavriel, Loukia .. 009418
Lojdová, Ivona 009896, 009903
Lokensgard, Dr. Lynne ... 046855
Lokin, Dr. D.H.A.C. ... 032160, 032161,
      032162
Lokki, Juhani 010845
Lomaheftewa-Slock, Tatiana . 053110
Lomakov, Dmitrij Aleksandrovič . 037318
Loman, D.E. 032377
Lomatter, Alice 041901
Lomawaima, Hartman H. 053823
Lomax, James 044785
Lomax, Jenni 044932
Lombardi, Gloria R. ... 051549
Lombardi, Michele 013800
Lombardo, Candice ... 053515
Lombardo, Dr. Giovanna . 027627
Lombardo, Marta Alicia ... 000377
Lombardo, Romualda ... 034173
Lombino, Mary-Kay 050376
Łonak, Wojciech 035063
London, Elizabeth 051360
Londoño, Jaime Manzur .. 008608
Londos, Eva 040719
Londt, Dr. Jason 038801
Lonergan, Gillian 045627
Long, Charles 052942
Long, Greg 054542
Long, J. Grahame ... 047543
Long, Jena 053703
Long, John D. 052895

Long, Linda 047921
Long, Philip 044273
Long, Phillip C. 047749
Long, Steve 051418
Long, Xibin 008275
Longaven, Dorris 048691
Longbourne, Dr. D. ... 045248
Longcore, Robert R. .. 051554
Longenecker, Martha W. 048607, 052966
Longenecker, Vickie .. 048835
Longevialle, Ghislain de . 013122
Longeway, Tricia 006032
Longhauser, Elsa 053135
Longhi, Alberto 027692
Longhi, Janice 043762
Longhi, L.A. 043759
Longhi, Rolf 042249
Longley, Dione 050864
Longuein, Paul 015583
Longwell, Alicia 053420
Lonnberg, Thomas R. .. 048644
Loodts, Isabelle 004005
Look, Johannes 017011
Loomeijer, F.R. 032809
Looper, Gail S. 053370
Loos, Margit 018131
Loose, Annegret ... 018938
Loose, Helmut 016699
Loosli, Jean-Pierre .. 041755
Lootens, Michel 003654
Lopata, Robert 046961
Łopatkiewicz, Zdzisław . 034792
Lopatkin, Michail Vasiljevič . 037882
Lopes, Dr. Joaquim José de
      Castro 035826
Lopes, Manuel Teixeira . 036101
Lopes, Manuel José Ferreira .. 035965
López Álvarez, Joaquín ... 039439
López de Arriba, Mercedes . 039647
López Fernández, Francisco
      Javier 039292
López Franco, Amparo .. 008705
López Hervás, Maria Victoria ... 039653
López Pimentel, María del
      Carmen 009138
López Portillo Romano, Carmen 031203
Lopez Rodriguez, Jose Ramón . 040169
López-Vidriero, Luisa ... 039689
López Wehrli, Silvia 040421
López, Al Ignatius G. .. 034579
López, Ana María 040240
Lopez, Donald 054138
López, Edgar 008661
López, Javier Fernandez . 039438
López, Jorge 031238
Lopez, Julie Anne ... 053106
López, Magdalena ... 039647
López, María Jose Muñoz . 039307
López, Mónica Alejandra . 000146
López, Sebastián ... 004681
Lopiene, Regina 030531
Loppa, Linda 003293
Lopreato, Paola 026011
Loquenz, Norbert ... 002588
Loránd, Klára 023721
Lorbeer, Marie 017217
Lorblanchet, Hélène . 014288
Lorck, Bjorn 031855
Lord, Anne 001592
Lord, Gary T. 051635
Lord, Victoria 049945
Loren, Dr. Diane ... 047370
Lorenčak, Marja ... 038552
Lorenz, Dr. Angelika . 020601
Lorenz, Barbara ... 001770
Lorenz, Dr. Harald . 018763
Lorenz, Jens D. ... 017311
Lorenz, Margaret .. 044709
Lorenz, Dr. Ulrike . 018649, 018650,
      020249
Lorenzen, Dr. Heidrun . 021493
Lorenzi, Franco ... 026790
Lorenzi, Georges .. 015817
Lorenzo Rosales, Elaine ... 009241

Lorenzo Villamor, Felícitas 039434
Lorenzoni, Dr. Mario 027957
Loret, John 050635
Lorian, F. 016660
Lorimer, Wayne 033213
Loring Brace, C. 046491
Lorquin, Olivier 014770
Lorys, Jan 047691
Łos, Ewa 034996
Los, Dr. W. 032002
Losada, Santiago 039647
Losavio, Sam 046814
Losch, Michael 054341
LoSchiavo, Joseph A. 052753
Loser, Alois 041751
Loshi, Ygal 025166
Losonczy, Attila 023493
Lotfullin, Galim F. 037587
Lotorev, N.D. 037445
Lott-Büttiker, Dr. Elisabeth ... 041614
Lottermann, W. 016743
Lotze, Manfred 018610
Lotze, Ulrich 017045
Lou, Xia 008313
Louati, Ali 042782
Loud, Patricia C. 048904
Louédec, Patricia 013547
Lõugas, Anne 010668
Loughridge, Sally 054506
Loughton, Gregory 007199
Louis, Dr. Eleonora ... 002630, 002631
Louis, Hubert 012144, 012145
Louma, Mark 051118
Lounibous, Leroy 053137
Lounsbury, Kathryn 046304
Lourdes Barreto, Maria de 004449
Louring, Aage 010081
Loustau, Alain 013772
Louvencourt, Amaury de 012384
Louvi, Dr. Aspasia 023061, 023270,
                          023274, 023296
Louw, Prof. D.J. 038835
Louw, J.P. 038794
Louw, P. 038772
Louwagie-Neyens, Linda 006937
Louwman, E.V.N. 032774
Lovász, Stephanie 041276
Lovdzieva, Marija 005181
Love, Jackie 051598
Love, Mindi 048473
Lovell, Albert 033244
Lovell, Carol 050278, 053570
Lovell, Charles M. 053676
Lovell jr., Russell A. 053076
Lovén, Lucie 033866
Lovenz, Marianne 050164
Lovero, Elizabeth 053090
Lovett, Margaret 050278
Lovick, Emily 048872
Loving, Charles R. 051657
Lovis, Dr. William A. 048441
Low, Dr. James 006109
Low, William 050232
Lowagie, Luc 003591
Lowe, Carole 052230
Lowe, Dr. C. 045072
Lowe, Janet 005628
Lowe jr., John 048370
Lowe, Joseph 047939
Lowe, Susan 030769
Lowe, Vivian J. 050281
Lowell, Molly W. 048367
Lowenfeld, David 047819
Lowkunsombat, Chavalit 042694
Lowly, Tim 047643
Lowrey, Annie 054393
Lowrey, Carol 051445
Lowrey, Timothy K. 046365
Lowry, Dale 005562
Lowry, Glenn D. 051427
Loxa, Juan de 039418
Loyau, Anne 011931, 011932
Loynes, G. 046184
Loyrette, Henri 014749

Lozada Arias, José 008566
Lozado de Solla, Noemi 000120
Lozano Peña, Carlos 008827
Lozano, Jacob 050323
Lozano, Prof. Luis-Martín 031180
Lozuk, Josip 008992
Lsayeva, Olga Nikolaevna 043173
Lu, Benshan 007913
Lu, Cangxin 008019
Lu, Chenghan 008162
Lu, Chimu 008466
Lu, Dongzhi 007622
Lu, Wensheng 007959
Lu, Xiaodang 008359
Lu, Xincai 008484
Lu, Yinghong 007626
Lubac, Marcel 013057
Lubbe, T.A. ... 038662, 038666, 038798,
             038831, 038877, 038879
Lubic, James E. 049375
Lubimova, L.M. 037297, 037298
Lubina, Michail 035178
Lubitz, Dr. Telse 019127
Lubowsky Talbott, Susan 048282
Luca, Claudia 036802
Lucas, Annemarie 020596
Lucas, Cynthia L. 049240, 049241
Lucas, Daniel .011897, 012347, 012348,
                          014152
Lucas, Enriqueta T. 034608
Lucas, Judith S. 047747
Lucas, Martyn 044804
Lucas, Mike 043539
Lucas, Palma B. 052080
Lucas, Ritchie 047984
Lucchese, Dr. Angiolo del ... 028332
Lucchesi Ragni, Dr. Elena 025629,
                  025637, 025638
Lucchini, Mario 041545
Luce, Debbie 048001
Luce, Donald T. 050940
Lucewicz, Cher 051727
Lucey, Denis 024941
Luchetti, Dr. Glauco 026233
Luchter-Krupińska, Lidia 034712
Luciani, Luciano 027646
Lucio, Cilia 019268
Lucke, Dr. Arne 019268
Luckeneder, Walter 002715
Luckey, George 044513
Luckhardt, Prof. Dr. Jochen ... 017553
Luckhardt, Prof. Dr. Jochen ... 017558
Luckhardt, Dr. Ulrich 018991
Luckhurst, Ailene 047770
Luckow, Dr. Dirk 019606
Łuczkowski, Jan 035074
Luderowski, Barbara 052168
Ludescher, Inge 018294, 018297
Ludewig, Dr. Thomas 020743
Ludington, Link 050585
Ludvigsen, Lucia 023343
Ludvigsen, Peter 010166
Ludwig-Weber, Birgit 020842
Ludwig, Andreas 018143
Ludwig, David 053407
Ludwig, Dr. Jochen 018483
Ludwig, Jürgen 017903
Ludwig, Dr. Renate 019137
Ludwig, Rudolf 002174
Ludwig, Walter 051504
Ludyga, Johann 016859
Lübbers-Ukena, A. 017873
Luebbers, Dr. Leslie 050772
Lübcke, Peter 018603
Lübeck, Elisabet 040735
Lübeck, Karla-Kristine 021793
Lübke, Steffen 020324
Lüchinger, Paul 042086
Lüdi, Heidi 041679
Lüdtke, A. 018481
Lüdtke, Prof. Dr. Hartwig 020251
Lüdtke, Dr. Hartwig 020254
Lueg-Werner, Dr. Gabriele ... 019700
Lührs, Günter 018503

Lüken, Dr. Sven 017299
Lüll, M. 022067
Lüpkes, Dr. Vera 020031
Lüps, Dr. Peter 042089
Lüscher, Hans 042244
Lüscher, Regula 041927
Luesma, Teresa 039498
Lütgens, Dr. Annelie 022744
Lüth, Dr. Friedrich 021618, 021996
Lueth, Virgil W. 053375
Lüthi, Werner 041384, 041387
Lüthy, Jürg 041827
Lütkemeyer, Ursula 021396
Lütken, Ilona 022053, 022057
Lüttichau, Dr. Mario-Andreas von 018306
Luft Dávalos, Rafaela 031365
Lugger, Leopold 002357
Luginbill, Troy 050536
Luginbühl-Wirz, Regula 041315
Lugo, Flavia Marichal 036169
Lugosi, Dr. József 023720
Ługowski, Ryszard 035322
Lugtenburg, Anneke 038848
Luh, Willi 017665
Luhta, Caroline 047817
Luijten, G. 031983
Luisi-Potts, Billy 053270
Luisi, Vincent 048396
Lukas, Clemens 016984
Lukasch, Dr. Bernd 016587
Lukasiewicz, Nancy 046595
Łukaszewski, Andrzej 035202
Lukaszuk-Ritter, Joanna 002932
Lukatis, Dr. Christiane 019533
Luke, Gregorio 050372
Luke, Suzanne 007366
Lukeš, Dr. Michal 009825
Lukesch, I. 003005
Lukić, Ivan 038189
Lukičeva, Alla Stepanovna 037294
Lukina, Inna Nikolaevna 037670
Lukowczyk, Evelyn 018969
Lukšić, Hilar 008981
Luksic, Tugomir 009045
Lulav, S. 025031
Lum, Shawn 053919
Luma, Ramon E.S. 034578
Lumb, Dorothy 048007
Lumley, SimonWhite 044911
Lummel, Dr. Peter ... 017116, 017300
Luna, Juan J. 039676
Lunardi, Prof. Ernesto 026396
Lund, Arnstein 034038
Lund, Elaine H. 050708
Lund, Mikael 010155
Lund, Tora Jervell 033585
Lundberg, Åsa 041106
Lundberg, Carl-Anders 011041
Lundbergh, Hans 041128
Lundblad, Lena 040663
Lunde, Oeivind 034032
Lundgreen, James 050651
Lundgren Nielsen, Hanne 010112
Lundgren, Britta 041106
Lundin, Kennet 040653
Lundkvist, Anders 040622, 040623
Lundrigan, Dr. Barbara 048441
Lundström, Sven 040682, 040683
Lundtofte, Anne Sofie 010044
Lungarotti, Dr. Maria Grazia ... 028107
Lunger-Valentini, Eva 002261
Lunger, Brian 006040
Lunghi, Enrico 030585
Lungu, Alexandru ... 036698, 036702,
                  036703, 036704, 036706
Lungu, Cornel 036702
Lungu, Hariton 036812
Lunn, Will 051777
Lunning, Elizabeth 049585
Lunsford, P. Celina 018417
Lunsingh Scheurleer, Dr. R.A. .. 031923
Lunt, Sara ..044476, 044476, 044827,
                          046208
Luo, Anhe 008041

Luo, Biaoyuan 007818
Luo, Jianzhi 007725
Luo, Tingru 007751
Luo, Xizhang 007771
Luo, Xuewang 008000
Luo, Xuzhong 008278
Luo, Yisheng 008525
Luo, Zhangui 007630
Luojus, Susanna 011450
Luong, Tran 054807
Luoto, Jukka 011090, 011093
Lupal, Kimberly A. 055501
Luperini, Prof. Ilario 027348
Lupiy, Dr. Tetiana 043252
Luppino, Tony 005750
Luptáková, Eva 038345, 038407
Lupton, Ellen 051368
Lupton, Jo 000898
Lupton, John M. 054354
Lupu, Sabina 036756
Lupu, Valentin 036216
Lurie, Doron J. 025212
Lusa Monforte, Guillermo 039068
Lusardy, Martine 014662
Lusby, D.A. 033298
Luse, Vel 049939
Lushaba, Yvonne K. 038739
Lushear, Jen 051303
Lussi, Kurt 041480, 041707
Lussky, Richard 048180
Luštinec, Jan 009606
Luterbach-Renggli, Hansjörg ... 041963
Luth, Jan 050148
Luther, Charles 047989
Luti, Claudio 027071
Lutnes, Kjell 033550
Lutnesky, Dr. Marvin 052272
Luttropp, Bernhard 040663
Lutyens-Humfrey, A. 043918
Lutz, Dr. Albert 042221
Lutz, Gottlob 041584
Lutz, Marian 049588
Lutz, Peter 022645
Lutz, Tina 053853
Lutz, Dr. Volker 002784
Lutz, William 048891
Lux, D. Ruth 051824
Lux, Simonetta 027624
Lux, Ulla 018037
Luxford, Bob 033274
Luyken, Dr. Gunda 019606
Luz Paula Marques, Maria da .. 035932
Luz Ramírez Estrada, Maria de la 009165
Luz, David W. 051979
Luz, Henrique 004879
Luzin, Sergej Stepanovič 037139
Luzina, Lubomir 038283
Lyall, Clare 046000
Lyall, Dr. D.C. 022067
Lydakis, Prof. Dr. Stelios 023016
Lydamore, Chris 044513
Lydon, Andrea 024900
Lydon, Kate 052176
Lyke, Linda 050473
Lyke, Charles T. 049056
Lyle, Janice 051896
Lyles, Pelham 054481
Lynam, Seamus 024898
Lynas, Donna 043679
Lynch, Cindy 047927
Lynch, Geraldine 024964
Lynch, Harry P. 046331
Lynch, Kevin 044827
Lynch, Sheila 054381
Lynch, Thomas F. 047289
Lynch, Thomas J. 047164
Lyndaker, Vernon 048069
Lynesmith, P. 045287
Lynn, Bob 033137
Lynn, Devell 048827
Lynn, Judy G. 048038
Lynn, Kitty 048488
Lynn, Michael Anne 050269
Lynn, Ralph 006304

Lynx, David	054566	McCallum, Cheryl	049573
Lyon, David Peter	001381	MacCallum, Malcom	045567
Lyon, Katharine	050333	McCallum, Sue	052007
Lyon, Richard	049127	McCalmot, Faye	053335
Lyons, Don	048359	McCammon, Alice	050478
Lyons, Jana	054162	McCamphill, Debbie	045382
Lyons, Joan	048359	McCandless, Barbara	048898
Lyons, Dr. John D.	050600	McCann, Mac	047344
Lyons, Marilyn	051702	McCarroll, Stacey	047108
Lyons, Michael	046125	McCarter, John W.	047657
Lyons, Michele	046956	McCarthy, Brian	006332
Lyons, Rex	048923	McCarthy, Christine	052400
Lyra, Fernando Soares	004667	McCarthy, Darlene	049142
Lysdahl, Per	010123	McCarthy, Frances	047558
Lysenko, Nikolaj Stepanovič	029836	McCarthy, Indi	049720
Lysnjak, Kateryna Jevgenivna	043184	McCarthy, J.	001653
Lysogórski, Krzysztof	034755	McCarthy, John J.	053116
Lyssaris, Agelos	023266	McCarthy, Judith	024952
Lytjen, Lokey	049762	McCarthy, K.	033168
Lytle Lipton, Susan	051412	McCarthy, Michael J.	046033
Lytton Cobbold, Martha	044740	McCarthy, Dr. Stephen	052839
Lytwyn, Stephen	007480	McCarthy, Wayne	054052
Lyu, Ma-ri	029972	McCartney, Dr. Nancy G.	048717
M(/a)cLain, Brenda	046803	McCarty, Cara	052818
Ma, Fazhu	007616	McCarty, Gayle	052478
Ma, Hong	007989	McCarty, Kimberly	051071
Ma, Honglin	008459	McCarty, Laura	048918
Ma, Jiayu	007696	McCarty, R. Paul	048808
Ma, Kuijun	007865	McCaskey, Jane	051034
Ma, Liangyun	007699	McCauley, John F.	047099
Ma, Long	008335	McCay, Bill	049800
Ma, Peilin	008442	Macchi, M.	015824
Ma, Wei-du	007620	Macciotta, Prof. Giampaolo	025679,
Ma, Weidou	007621		025682
Ma, Weiyi	007849	McCleave, John	006790
Ma, Wenwei	007741	McClelland, Jesse	047672
Ma, Zhengyuan	007946	McClenahan, Jane	051724
Maanit, G.	015825	McClenny, Bart	046428
Maarek, Lucien	015825	McCloskey, Anne M.	050350
Maas, Dr. A.J.P.	032581	McClung-Oakes, Katherine	043806
Maas, Klaus	018063	McClung, Elizabeth	050868
Maas, Wilhelm	020717	McClure, Donna	051030
Maass, R. Andrew	052860	McClure, Kent	051030
Maassen, Dr. Walter	016444	McClure, Nancy	053309
Maat, Erik	032348	McClure, Rod	052308
Maaz, Dr. Bernhard	017068	McCluskey, Holly H.	054368
Mabee, Kim	053196	McClusky, Pamela	053248
Mabungule Chissano, Alberto	031811	McCollough, Pam	051009
Mac Phee, S.D.	005787	McCollum, Jay	047881
MacAdam, Barbara J.	049353	McComb, Joseph	051058
McAfee, Michael J.	054322	MacConnel, Cal	048777
McAffee, Dionne	007229	McConnell, Bridget	044405
McAinsh, Deirdre S.	046005, 046006	McConnell, Glenn	045317
Macalister, Helen	052629	MacConnell, Raymond	046930
McAlister, Judith	047149, 048693	McConnell, Ruth	006229
McAllister, Lowell	051719	McCooey, Patrick	005900
Macalpine Foshay, Susan	005956	McCord-Rogers, Suzette	049468
McAlpine, Dr. Donald	006893	McCord, Dr. Robert	050813
McAnally, Fiona	050015	McCormack, Anne	044721
McAneney, Helen	045285	McCormack, Peter	044732
McArdle, Jean	043616	McCormick, Donald	051456
McArthur, Meher	051936	McCormick, Edgar L.	049931
McArthur, Randy	050816	McCormick, Todd	046878
MCarty, Kenneth W.	052065	McCorquodale, Ross	001518
Macat, Andreas	020491	McCosh, John	033204
MacAulay, Donald	005369	McCourt, Frank A.	051834
MacAulay, Margot	005369	McCourt, Steve	000832
McAuley, Joyce	046714	McCowan, Kurt	052677
McAuley, Kathleen A.	047228, 047234	McCown, Donna	052205
McAuliffe, Dr. Chris	001275	McCoy, Bill	054260
McBay, Arthur	045505	McCoy, James P.	052026
McBean, Jude	001118	McCoy, Jeff	052707, 052709
MacBeath, Doug	044254	McCoy, Kathy	051004
McBratney-Stapleton, Deborah	046474	McCoy, Nancy	050395
MacBride, Frank	000885	McCracken, Katherine	005933
McBride, Heather	046309	McCreary, Kathy	049518
McBride, Joe	046464	McCree, Timothy	046351
McBride, Michael A.	049291	McCue, Debbie	046974
McBride, Peggy	052438	McCue, Donald	052495
McCabe, Jennifer	053017	McCue, Kaz	053720
McCafferty, Estelle	053473	McCuistion, Mary Ann	053964
McCall, Janet L.	052176	McCullick, Nancy	050743
McCall, Louise	043955	MacCulloch, Alexandra	043456
McCalla, Mary Bea	049027	McCulloch, Bruce	033262
McCulloch, Judith	049133	McGhie, H.	045210
McCullough, Catherine	043431	McGill, Forrest	052992
McCully, Kathleen J.	047371	McGill, Shirley	048352
McCurtis, Barbara S.	052962	McGillivray, Allan	007263
McCutcheon, Caroline	046166	McGillivray, Don	006456
McDaniel, George W.	047546	McGillivray, Dr. W. Bruce	045760
McDaniel, Robert	048399	McGinley, Dr. Jack	024890
McDaniels, Garvin	051152	McGinnis, Howard	048048
McDaniels, Warren E.	051301	McGinnis, Melissa S.	053642
McDermott, Eugene	048115	McGinnity, Betsy L.	054214
McDianmid, Gordon	005459	McGirr, Margo	053513
McDonald Crowley, Amanda	051384	McGivern, Andrew J.	054236
McDonald Pang, Toni	049721	McGlumphy, D.	053849
MacDonald, Andrew	044036	McGlynn, Robert	051456
McDonald, Bonnie	046505	McGovern, Robert M.	047464
MacDonald, Catherine	044775	McGovern, Thomas	050446
MacDonald, Colin	045518, 046035	McGowan, Jeanie L.	047426
MacDonald, Colinn	043886	McGowan, Rebecca	007207
MacDonald, Craig	007019	McGrady, Dr. Patrick J.	053899
McDonald, Donald M.	001315	McGrath, Deborah	001225
McDonald, Doris	006570	McGrath, Joe	001266
McDonald, Douglass W.	047741	McGrath, Steph	054363
MacDonald, Elizabeth H.	052084	McGreevy, Candace	053573, 053574,
McDonald, Garry	005716		053576, 054447
MacDonald, Dr. George	053229	Macgregor, Elizabeth Ann	001544
MacDonald, Gerald J.	051404	MacGregor, Neil	044923
Macdonald, I.H.	045866	McGrew, Rebecca	047758
McDonald, J.P.	050523	McGuffey, Patrick	049035
MacDonald, Jessie	006932	McGuire, Donna	006543
MacDonald, Judith A.	052210	McGuire, Janice	048914
MacDonald, Kate	045946	McGuire, Dr. Jim A.	046819
McDonald, Dr. L.	001456	McGuire, Jimmy A.	046940
McDonald, Marty	047319	McGuire, Raymond J.	051476
MacDonald, Mary V.	048192, 048193	McGuire, Ross	052722
McDonald, Michael	048370	MacGunnigle, Bruce C.	048427, 048429
MacDonald, Ned	006037	Mach, Jiří	009537
McDonald, Paul	046653	Machacek, R.J.	048111
MacDonald, Peter	044735	Machan, Dallas	044727
McDonald, Ruth	043724	Macheret, Elisabeth	041421
MacDonald, S.	045073	Machhammer, Helmut	001853
MacDonald, Stanley	006551	Machinek, Robert	021852
McDonald, Sussanne	005814	Machneva, Margarita A.	037129
McDonnell, Dr. Patricia	053649	Machnicki, Monika	022263
McDonough, Stephanie	046310	Machon, Abdurremane	031802
McDonough, Stephen	046444	McHugh, Eileen	046651
MacDougal, Bruce	052932, 052947,	McHugh, Joseph	007300
	052956	McHughs, David	048916
McDougall, Jamie	005443	McChuny, V.B.	038719, 038760
McDougall, Kim	005871	Macías Ramos, María Teresa	040358
McDowell, Anna	045765	Maciejewski, Andrzej	035340
McDowell, Ina	053725	Maciejunes, Nannette V.	047936
MacDowell, Dr. Marsha	048441	MacIlreith, Rob	005785
McDowell, Susanne	017724	McInerney, Gay	046831
McDowell, Suzanne	048087	McInerney, Michaela	048921
McEachern, Mark	053781	MacInnis, Karen	050667
Macedo, Maria de Fátima	035952	McIntosh, Jean	006671
Macedonio, Prof. Giovanni	026156	McIntosh, Mildred	050744
Macek, Stanislav	009547	McIntosh, Muriel	044428
McElhone, William K.	047012	McIntyre, Art	006816
McElroy, L.J.	051166	McIntyre, Barbara	006924
McEneaney, Eamonn	024999	McIntyre, Chris	044535, 045667
Maceroni, Prof. Giovanni	027517	McIntyre, Stella	043656
McEwen, Gloria	049731	Mačionis, Z.	030552
McEwin, Florence	052634	MacIsaac, Kim	051957
McFadden, David	051324, 051436	McIver, J.H.	047183
McFadden, Lila	005639	Mack, Angela	047548
McFadden, P.	038695	Mack, Gary	048123
McFalls, W.J.	005982	Mack, Jeanne P.	049112
McFarland, Seth	047735	Mack, Vanessa	001543
MacFarlane, Aynsley	005367	McKague, Margot	006066
MacFarlane, Nathalie	007045	McKain, Martha	050108
Macfarlane, Sheena	045435	McKale, William	048865
McFeely, Marie	024900	McKamey, Sheldon	047166
McFerrin, Jean	052925	MacKay, Allan	006125
McGahen, DeAnn	046403	Mackay, Andrew	045810
McGarrie, Mary	044283	McKay, B.	044954
McGarry, John H.	051135, 051136	McKay, Barrie	049669
McGarry, Susan	051836	McKay, Dr. David	050908
McGary, Nancy	051412	Mackay, Jim	052666
McGaugh, Houston	054190	Mackay, John L.	047571, 047572
McGavin, Dr. G.	045457	McKay, Kelly	005851
McGee, Prof. Mike	048994	MacKay, Kevin	005592
McGee, Patrick	024882	McKay, Maria	053304
McGehee, Thomas C.	053728	McKay, Mary Alice	051451

McKay, Patrick J.	052610	McLellan, Richard A.	047507
MacKay, Robert B.	047867	Maclennan, F.	044745
MacKay, Sarah	043357	Macleod, Chris	044692
Mackay, Thomas	045978	MacLeod, Cynthia	052559
Mackay, W.A.F.	006776	McLeod, Cynthia	052555
McKean, Virginia	050973	McLeod, Deborah	050567
Mackechnie, John	044399	Macleod, Dr. Ian	001403
McKee, Barry	005977	MacLeod, Karen	050213
McKee, Dorothy	050565	MacLeod, Kathleen	005553
McKee, Francis	044393	MacLeod, N.	045061
McKee, Kevin	046591	MacLeod, Patricia	052883
McKee, Marshall	044599	McLeod, Roddy	044408
MacKeil, David	006650	McLeod, Susan M.	048463
MacKellar, Dr. John	001308	McLerran, Jennifer	046599
McKellar, Stacey	005469	MacLoy, David	000865
McKenna, George L.	049888	McMahan, Jackie	053131
McKenna, James	051745	McMahon, Cynthia	047693
McKenna, Sharon L.	051957	McMahon, James	049454
McKenrick, Fremont	048466	McMahon, James E.	049094
Mackensen, Dr. Götz	017601	MacMahon, Rose	044891
McKenzie Smith, Dr. Ian	044268	McMahon, Stephen	053535, 053536
McKenzie, Cameron	044144	McMains, John	053055
McKenzie, Colin	044346	McManus Zurko, Kathleen	054540
McKenzie, Molly	047342	McManus, Greg W.	033296
Mackenzie, Philippa	044502	McManus, Marcia	048818
MacKenzie, Stuart C.	047725	McMasters, Thomas O.	048868
McKenzie, Wendy	006058	McMeekin, G.	046147
McKeown, Jack J.	049139	McMeekin, Jim	048026
McKernan, Robert L.	047724, 052457,	McMeekin, Rebeeca B.	052464
	052497, 054607, 054608	McMenamin, Brendan	045143
Mackey, Brian	044844	McMichael, Sindy	048236
Mackey, David M.	049553	MacMillan, Arthur	044757
Mackey, Marjorie R.	046587	McMillan, Edward	047839
Mackie, Louise W.	047808	McMillan, Erin	048318
McKie, S. Zurie	049781	McMillan, Patsy H.	049999
Mackie, Thomas	048023	McMillan, Dr. R. Bruce	053451
McKillop, Beth	045123	McMullen, Darlene	006550
McKillop, Joan	045437	McMurtrie, I.	044251
McKinlay, Dawn	007340	McNabb, Debra	007073
McKinley, Nancy R.	050877	McNabb, Liz	043671
McKinney-Tandy, Ruth	048595	McNair, Cynthia	050853
McKinney, Evans	049483	McNally, Ginger	054597
McKinney, Henry C.	054171	McNamara, C. J.	048750
McKinney, Jack N.	054014	McNamara, Carole	046492
McKinney, Janice	054392	McNamara, Mary E.	051484
McKinney, Rick	048716	McNaught, Lewis	045031
MacKinnon, Ann. M	044658	McNealy, Terry	051273
McKinnon, Karl L.	054456	McNeely, Don	053487
McKitterick, Michael	010421, 053262	MacNeil, Malcolm	044654
Mackow, Leigh	050122	McNeill, Anne E.	043695
Maćkowiak, Dr. Janryk	034846	McNeill, C.	001598
Mackowiak, Reinhild	017388, 020174	McNelly, Kath	000975
Maclagan, Dr. Helen	046060, 046063	McNelly, Dennis	052094
McLain, Charles	049962	McNett, Sarah	049589
McLain, Guy	053459	McNichol, Susan	006632, 007049
McLallen, Helen M.	049970	McNichols, Patricia	051673
McLaren, Hazel	046792	McNicol, Jean	007380
McLaren, M.	001210	McNicol, Morag	045268
McLauchlan, Mary Nell	047293	McNulty, E.C.	048630
McLaughlin, David	050940	McNulty, F.M.	048930
McLaughlin, Heather Jo	049456	McNulty, J.L.	048180
McLaughlin, Penny	049828	Macon, Robert	051981
McLaughlin, Robert	053475	McOuat, Gordon	006319
McLaughlin, Stacey B.	052679	McOwan, Daniel	001135
McLaughlin, William F.	048311	MacPhee, Dana	044842
McLaughlir, Dr. John	005838	MacPhee, Dr. Ross D.E.	051327
McLaurin, Genette	051466	McPherson, Beth Ann	051931
McLea	052910	McPherson, Dan	048099
McLean Ward, Dr. Barbara	052313	MacPherson, Kathy	050409
McLean, Claire	054415	McPherson, Loretta	001067
McLean, Dr. Janice V.	046420	McPherson, Tommy	051931
MacLean, John	047784	McQuaid, Matilda	051368
McLean, Kathleen	006523	Macquart-Moulin, X.	014467
McLean, Libby	044175	Macquart, Bruno	014773
McLean, Linda E.	049602	McQueen, Jack	053052, 053054
Maclean, Marge	005924	McQuillis Iscrupe, Shirley	050277
McLean, Rita	043615	McRae, Jennifer	049251
MacLean, Roderick	005945	McReynolds, Anne	043559
MacLean, Rusty	045649	McShane, Michael	044054
McLean, Steve	045347	McVay, Darice	053942
MacLeish, Bruce	051541	McVey, Dorothy	053693
McLellan, Darlene	048482	McWayne, Barry	048662
MacLellan, Heather	006656	McWilliams, Mary	047362
MacLellan, Iain	050625	Macyszyn, Dr. Jacek	035371
Mczakez, Franz	002958	Magidson, Phyllis	051443
Maczulis, Norbert	034809	Magill, Tina M.	052240
Madanian, M.	024725	Magilton, Jack	006105
Madawick, Paula	048238	Magini, Gianluca	026964
Madden, Jennifer	053073	Magiru, Maria	036396
Madden, Pamela	047424	Maglica, Nadja	009028
Maddox, Iva Jean	048857	Magnan, Oscar	049820
Maddox, Margaret	050009	Magnetti, Daniela	028142
Madec	014901	Magnifico, Marco	028279
Madeline, Laurence	014743	Magnin	011866
Mader, Dr. Gerald	002770	Magnin-Gonze, J.	041657
Maderek, Dorota	035176	Magnin, Bénédicte	011867
Madert, Johannes	020381	Magnini, Prof. Marsilio	026130
Madesta, Dr. Andrea	002196, 021326	Magnussen Svare, M.	010185
Madin, John	044312	Magnusson Staaf, Björn	040809
Madisia, Joeseph	031862	Mago, Norma A.	034617
Madjarov, Mitko	005005	Magolfi, Dr. Miria	026314
Madkour, Christopher	050633	Magomedova, Seir Seitovna	037933
Madonia, Ann C.	054414	Magrane, Nigle	046222
Madonsela, Dudu	038876	Magras, M.	023350
Madrid Rojas, Juan José	008779	Magrath, Philip	044328
Madrid, Candy L.	052659	Magré, Piet	032521
Madsen, Inge	010091	Magrini, Boris	041749
Madsen, Jan Skamby	010127	Magro, Antonio	025355
Madsen, John	053823	Magro, Pasquale	025402
Madsen, Lennart S.	010092	Magsaysay, Dr. Eulogio Amang	034530
Madsen, Orla	010092	Maguindanao, Macmod	034550
Madsen, Per Kristian	010355	Maguire, Brian	044112
Madshus, Eva	033825	Maguire, Cathy	053157
Madurini, Dr. Giulio	026795	Maguire, Dr. Eunice	046728
Madžarac, Nada	038549	Maguire, Nancy	047380
Maeda, Kenichiro	029148	Magun', Marija Karpivna	043242
Maeda, Koichiro	028651	Magurany, J.D.	052802, 052805
Maeda, Shoso	028570	Mahamadou, Kelessé	033408
Maeder, Andreas	041803	Mahé, Lucien	013203
Maeder, Edward F.	048218	Mahecha Useche, Ernesto	008693,
Maeght, Adrien	014348		008694
Mähl, Andrea	022729	Mahéo, Noël	011633
Mäkelä, Asko	010868	Maher, Kathleen	047200
Mäkelä, Sirkku	011136, 011137	Maher, Kristie	052130
Mäkeläinen, Reijo	011158	Maher, Tina	043393
Mäki, Riitta	011089	Mahida, Bhamini A.	024487
Mäkinen, Antti	010948	Mahilom, Mila M.	034469
Mäkinen, Marketta	010935	Mahler, Oskar	018423
Mael, Stephen	044808	Mahlknecht, Dr. Eduard	002187
Mämmetnurow, Öwezmuhammet	043089	Mahlman, Jennifer	051446
Mänd, Anu	010682	Mahlstedt, Thomas	054032
Maeng, Injae	029964	Mahmoud, Assad	042260
Männchen, Prof. Dr. Julia	018807	Mahmudov, N.	054644
Mäntele, Dr. Martin	022268	Mahmutović, Džafer	004258
Mänttäri, Roy	010866	Mahon, Rose	045607
Maergerman, Roger	004092	Mahran, Younes	010540
Maerten, Michel	015237	Mahy, Ivan	003788
Maes de Wit, Yvan	003862	Mai Hung, Prof. Dr. Pham	054799
Maes, Jean-Michel	033386	Mai, Do Thi Tuyet	054807
Maes, Leopold	004016	Mai, Hai	008218
Maesen, Mathieu	003963	Maia e Castro, Catarina	035952
Maessen, R.	032922	Maicki, Richard J.	047806
Maestro Guerrero, Dr. Fernando	039212	Maidt, Petra	022791
Mäuser, Dr. Matthias	016953	Maier-Schöler, Barbara	022373
Maeyama, Mikio	029637	Maier, Daniela	018324, 018325
Maeyama, Yuji	029305	Maier, Erhard	002679
Maffia, Prof. Cesare	026930	Maier, Jakob	016574
Maffioli, Dr. Monica	026294	Maier, Marco	042030
Mafre, Christian	011586	Maier, Stefan	022014
Magacz, Joanna	017987	Maier, Thomas	025098
Magajne, Miljojka	038516	Maierhofer, Fritz	002216
Magalhães, Carlos	004905	Maikweki, James	029868
Magalhães, Dr. Ivone	035660	Mailer, Allison	007277
Magaña Cota, Gloria E.	031024	Mailhot, Philippe R.	006857
Magdich, Bob	053759	Maillard, Anne	014107
Magee, Eileen M.	052030	Maillé, Jacqueline	015283
Magee, Eugene	026225	Maillot, Suzanne	036182
Magee, Henry J.	052048	Mailloux, Eleanor F.	049435
Magendans, J.R.	032584	Main, Sally	051304
Mager, Gerald Paul	021526, 021530	Minds, Paul E.	044566
Mager, Dr. Karl	002117	Mainella, Fran P.	046509
Maggi, Ersilio	027462	Mainieri Piedade, Leticia	004713
Maggi, Gianfranco	025265	Mainland, Katrina	043633
Maggi, Roberto	025953	Mlntosh, George	052620
Maggini, Roberto	042102	Mainzer, Prof. Dr. Udo	020458
Maggioli, Andrea	025558	Mair, Walter	001948
Magglo-Serra, Dr. Rosanna	006975	Maire, Marcel	014573
Maggs, Richard	045576	Mairesse, François	003890

Mairinger, Jutta 002927
Mairot, Philippe 015651
Maisak, Dr. Petra 018421
Maisano, Jack 051362
Maisch, Reiner 020844
Maisch, Ursula 020844
Maischberger, Dr. Martin 017072
Maisel, Jeff 016996
Maistrello, Roberto 026684
Maitilasso, Mario 028218
Maitland, Richard 053621
Maitre-Allain, Olivier 014515
Maiwald, Dr. Christine 019011
Maiwald, Gabriele 020291
Maiwald, Simone 021525
Maixner, Dr. Birgit 021663
Majača, Antonia 009030
Majahad, Carol 051591
Majboroda, Ljudmyla Oleksíïvna . . 043123
Majburova, Irina Borisovna 036988
Majcen, Jože 038537
Majewska, Grażyna 035079
Majka, Mary 006811
Major, Jürgen 022007
Major, Richard 052342
Majorov, Aleksandr Gennadjevič . . 037022
Majstorovic, Prof. Božo 008995
Majstorović, Ljuba 038185
Majstorski, Georgi 005248
Majtanov, Bachytžan
  Kauaskanovič 029798
Majumdar, Prof. Swapan 024467
Makabe, Tadahiko 028903
Makarewicz, Anna 035296
Makariou, Sophie 014749
Makarov, Nikolaj Michajlovič . . . 036925
Makarova, N.B. 037283
Makeeva, I.F. 038001
Maketz-Anderle, Sabine 002244
Makholm, Kristin 050943
Maki, Linda 047776
Makin, Jean 053700
Makkonen, Esko 011319
Makoev, Feliks H. 037858
Makoni, P. 054847
Makovičková, Marie 009707
Makowiecki, Wojciech 035118
Maksić, Nemanja 038245
Maksić, Slobodan 038245
Maksimova, L.D. 037689
Makwana, R.K. 024507
Malahoff, Dr. Alex 033224
Malak, Gregory 047763
Malakar, Diana 049066
Malao, N. 038853
Malaspina, Bartolomeo 028133
Malatare, Marilyn 053775
Malatesta, Giuseppe 025385
Malausssena, Paul-Louis 014478
Malavssi, Paola 019705
Malceva, Ljudmila Dmitrievna . . 037694
Malcolm, John 034095
Malcolm, Megan 033296
Malcolm, N. 033260
Malden, John 043579
Maldent, Dominique 012130
Maldon Fitch, M. Derek 045196
Maldonado, Eduard M. 047651
Maldonado, Laural R. 052664
Maleev, Dr. Michail 005196
Malek, Vincent 002795
Malenfant, Robert 006805
Malenke, Patti 048022
Malet Ybern, Rosa María 039058
Malgeri, Dina G. 050610
Malgin, Andrej . . . 043299, 043300
Malgorzewicz, Anna . . 001021, 001274
Malherbe, C. 038860
Malicky, Michael 002316
Maliet, Vincent 012013
Malik, Sajid Anwar 034134
Malikova, D. 038486
Malinar, Branka 048453
Malinauskas, V. 030334

Malinová, Miluše 009754
Malinverni, Pierluigi 026652
Mališeva, Valentina Pavlovna . . . 037148
Maliszewski, Juliusz Michał 019988
Maljković, Rada 038139
Malkočev, Angel . . . 004981, 004982
Małkowski, Dr. Krzysztof 035331
Mallalieu, William P. 003237
Mallamo, J. Lance 047499
Mallard, Michael 049760
Malley, Diane 044141
Malliet, Muriel 048677
Mallin, Risa 052112
Mallinson, Justin 043802
Mallo Area, Alberto 039728
Malloy, Dean 007172
Malmberg, Brita 041168
Malme, Arnfinn 033733
Malme, Heikki 010813
Malmlöf, Maria 040729
Malmquist, Hilmar 024088
Malnati, Dr. Luigi 026721
Malnica, Andrej 038584
Malone, Gail 006912
Malone, Mary 024830
Malone, Peter 047250
Malonek, Maryla 020647
Malonly, Rachelle 053870
Malov, Nikolaj Michajlovič 037848
Malsch, Dr. Friedemann 030305
Maltais, Marie E. 005849
Maltby, Janet . 007113, 007114, 007115,
  007116
Maltezou, Prof. Chryssa 028309
Malura, Andrew 022276
Malushi, Petrit 000016
Małuszyńska, Krystyna 034707
Maluțan, Cornelia 036434
Malý, Dr. František 009480
Malý, Karel 009604, 009904
Malý, Reinhold 002063
Malýševa, Tatjana Vasiljevna . . . 037128
Mamaev, Aleksandr A. 036874
Mamalakis, Kostis 023096
Mamaserikova, Šyryn Bukeevna . . 029827
Mambetova, D. 030080
Mambrini, Prof. Caterina 026363
Mamer, Helen 007394
Mammadov, Emin 003175
Mamuro, Yoshitake 029614
Man, Grigore 036219, 036221
Man, Xueyuan 007561
Manabe, Masahiko 028996
Manakidou, Dr. Eleni 023295
Manakou, Paraskevi 023293
MNaught, William 051800
Mancaş, Ioan . 036431, 036439, 036726,
  036819
Manceau, Sophie 013676
Manchak, Barry 005765
Mancinelli, Dr. Fabrizio 054678
Mancini, Prof. Francesca 028095
Manconi, Dr. Dorica 027258
Manconi, Francesca 027417
Mancusi-Ungaro, Carol 051491
Mancuso, Francesco 027178
Mandanas Esquito, Joven 034534
Mandeville, Jeffrey 050261
Mandeville, Joyce 051035
Mandic, Dr. Davor 008974
Mandis, Alexander . . . 022965, 022971
Mândran, Ion 036679
Mândru, Doina 036572
Manera, Brigida 014838
Manes, John 048849
Maness, David C. 051549
Manfield, Nigel 045573
Manfredini, Prof. Alessandra . . . 027601
Manfrina, Mario 041604
Mangado, Denis 012971
Mangan, Dianne 001618
Mangan, Richard 045032
Mangani, Elisabetta 027634
Mangels, W. Gilbert 052231

Mangia, Dr. Paola 027622
Mangiagalli, B. 038844, 038880
Mangiagalli, Tizzie 038844
Mangion, Eric 014041, 014493
Mangnus, G. 032690
Mango, PD Dr. E. 042188
Mangold, Dr. Meret 041310
Mangucci, Celso 035674
Manguso, John M. 048867
Manhart, Patty 048974
Manhès, Bruno 013510
Mania, Dr. Thomas 018832
Manibusan, Rose 023359
Manicastri, Dr. Claudio 027582
Maniez, Erica S. 049735
Maniguet, Thierry 014707
Maninger, Peter 016382
Manini, Paula . 053804, 053805, 053807
Maniurka, Dr. Piotr 035076
Manka, Riitta 011230
Manke, Elke 021057
Mankin Kornhauser, Elizabeth . . 049390
Mańkowska, Ewa 034923
Mankus, Edward 047640
Mann, David 054353
Mann, Esther 045026
Mann, George E. 051785
Mann, Harald J. 021265
Mann, Dr. James 050164
Mann, Jim 049352
Mann, Judith 052818
Mann, LaDonna 054243
Mann, Dr. Stephan 018721
Mann, Theresa 053431
Mann, Vivian 051412
Mannberg Wretin, Karin 041118
Manne, Robert 053204
Manneby, Dr. Hans 041098
Mannerheim, Lowissa 040635
Manni, Dr. Riccardo 027613
Manning, Anne 046693
Manning, Dr. P. 045210
Mannini, Maria Pia . . 027447, 027452
Mannion, Dr. Susan 024822
Manns, John 001657
Mano, Livio 026116
Manojlović, Miroljub 038226
Manoli, Britt 013457
Manring, Yvonne 054335
Mansell, Jefferson 046850
Manser, Jurg 041702
Mansfield, Dr. Bernard M. 049026
Mansfield, Lisa 051372
Mansfield, Nicholas 045216
Manske, Beate 017604
Manske, Prof. Dr. Hans-Joachim . . 017539
Manson, John 043397
Manson, Terry 033327
Manson, Wayne 049998
Mansour Naguib, Sylvia 046628
Mansperger, Linda W. 051833
Manssen, W.J. 032049
Mantel, Ken 045055
Mantovani, Prof. Giuseppe 028153
Manuel, R.S. 006628
Manuelas, Paul 050934
Manuhutu, W. 032948
Manus, Dieter 019404
Manyon, Ellen 054365
Manzanares, Rick 048810
Manzanilla, Dr. Linda 031213
Manzato, Prof. Eugenio 028186
Manzi, Prof. Giorgio 027606
Manzini Gazzani, Giorgio Mario . 008755
Mao, Liqiang 037937
Mao, Peiqi 007632
Mao, Yanwen 008018
Mao, Yuehong 008432
Mapfarakora, Jacques 005281
Maphet, L. Ernestine 049299
Maple, Dr. William . . 051155, 051156
Maples, Philip G. 052614
Maquet, Nadine 003804
Maquiso, Dr. Teresita R. 034554

Mara, Kanfela 023386
Mara, Kevin 048137
Mara, Maria 013132
Maračić, Antun 008913
Marafioti, Dr. Sergio . . 027196, 027197,
  027198
Marais, Caroline 045080
Marais, Caroline de 048444
Maráky, Dr. Peter 038329
Marandel, J. Patrice 050444
Marandu, Gallus 042628
Marangon, Maria 023245
Maraniello, Prof. Gianfranco . . . 025542
Marapana, P.B. 040486, 040487
Maraszek, Dr. Regine 018965
Maraval, P. 011963
Maravelaki, Noni 023038
Marazzi, Laura 028277
Marbach, Wilhelm 017744
Marble, Lori L. 051215
Marc, Denis 016082
Marc, Dorel 036793
Marcaccini, Ignazio 026167
Marcantel, Gregory 049812
Marcato, Prof. Paolo Stefano . . . 027136
Marcel, Louis 013241, 013570
Marcet, Roger 039099
March and Kinrara, Earl of 043942
Marchado, Philippe 013520
Marchand, Charlotte 016194
Marchand, Lionel 041912
Marchand, Scott 053834
Marchant Smith, Dick 046139
Marchant, Dr. R. 041654
Marchante, Carlos M. 009184
Marchbank, Frances A. 049958
Marchbank, James 007094
Marchenko, Leonid V. 043293
Marchesi, M. 032826
Marchesi, Dr. Marinella 025538
Marchesseau, Daniel 014716
Marcheville, de 013581
Marchini, Gian Paolo 028352
Marchione, Kenneth 053484
Marchois, Bernard 014756
Marciniak, Leszek 035438
Marció, Pietro 026964
Marco, Sixio 039366
Marcon, Gaby 014260
Marcos V., Emilio 039862
Marcos, Dr. Dieter 019664
Marcos, Ludovic 014704
Marcos, Mariano 015914
Marcoux, Charles 041840
Marcovici, Vladimir 036780
Marcu, Aurora 036471
Marcu, Maria 036354
Marcum-Estes, Leah 051670
Marcum, Donald 048429
Marcus, Evelyn 049353
Marcus, Jed 051468
Marcus, Joyce 046491
Marcus, Mette 010146
Marcus, Susan 047698
Maréchal, Dominique 003460
Marek, Danuta 034767
Marek, Stanisław 035054
Marenco, Jean-Jacques 015161
Marengo, Simeone 028146
Mareos Ruiz, P. 039020
Marepo Eoe, Soroi 034189
Mares, Dr. Michael A. 051588
Marešić, Jagoda 008993
Maret, J.F. 032640
Maretzky, Dieter 019707
Marfà i Riera, C. . . . 039734, 039736
Marghieri, Roland 014490
Margita, Laura 006981
Margkudilaga, Sufwandi 024591
Margles, Judy 052299
Margos, Rina 003848
Margrey, Peter 047422
Margulies, Stephen N. 047585
Mariani, Prof. Franco 026166

Mariani, Giuseppe Antonio 025970
Máriássy, Eva-Maria von 018811
Mariátegui Ezeta, José Carlos .. 034304
Marić, Dušan 031746
Marichal, Rémy 007235
Maričić, Dušanka 038185
Maricich, Andrea 049458
Maricki-Ostojić, Vesna 038245
Maries, John 044020
Marijan, Jassmina 038552
Marín Torres, Dr. Maria Teresa . 039805
Marín Vázquez, Gilberto 031161
Marin, Prof. Dr. Emilio . 008949, 008993
Marin, Jean-Yves 012277
Marin, Jeanne 015191
Marin, Lăcrămioara 036326
Marinac, Bogdana 038587
Marinazzo, Dr. Angela 025640
Marincig, Harald 003100
Marincola, Michele 051364
Marinelli, Dr. Giampiero Gabrio . 026386
Marínez García, Luisa 038926
Maringer, Christine 003095
Marini, Gabriele 026560
Marini, Dr. Giorgio 028350
Marini, Mariagrazia 027886
Marini, Marinella 027886
Marini, Dr. Paola .. 028348, 028350
Marinis, Dr. Giuliano de 025319
Marino, Christy 046860
Marino, Mari 004882
Marino, Ruta 052753
Marinoiu, Vasile 036329, 036492,
      036756, 036757, 036758, 036759,
            036826, 036827
Marinov, Evstati 005261
Marinska, Dr. Rouja 005184
Mariotti, Prof. G. 027610
Marits, Ülle 010617
Mark R., Thompson 053415
Mark, Guido 002873, 002898
Mark, John 048530
Markel, Stephen A. 050444
Markelova, Nina Konstantinovna 037529
Marker, Peggy 050953
Markert, Leo 018865
Markey, Teresa 048495
Markham, Philip 033352
Markiewicz, Katarzyna 035149
Markina, Natalja Petrovna 037929
Markishtum, Merlee 053246
Markko, Helvi 010907
Markland, Eleanor 045168
Markley, Penny 049319
Marko, Dr. Eva 001979
Markoe, Glenn 047739
Markoulaki, Stavroula 023038
Markovčič, Albin 038592
Marković, Mirjana 038239
Marković, Prof. Slavica 009040
Marković, Srdjan 038208
Markowitz, John 051984
Markowitz, Zev 047254
Marks, Dede 046232
Marks, Edward 053524
Marks, Jacqueline 022607
Marks, John 047663
Marks, John C. 049068
Marks, Katherine Duer 050319
Marks, Michael 022607
Marks, Michael J. 049485
Marks, William 048824
Marley, Rita 028439
Marlow, Martha 048377
Marlowe, Dean 053411
Marloyannis, Georgi 023221
Marmolejo, Martha Lucía 008676
Marnetté-Kühl, Dr. Beatrice 017566
Maroević, Igor 009016, 009026
Marøy Hansen, Arild 033528
Marović, Metodija 009025
Marquardt, Claudia 020873
Marquardt, Jean F. 053679
Marquardt, Dr. Markus 016616

Marquardt, Regine 021790
Marquart, Dr. Markus 016627
Marquart, Dr. Markus . 016621, 016628
Marques, António Fernando ... 035676
Marques, Luiz 004879
Marquet, Helmut 003270
Marquet, Jean-Claude . 013635, 014072
Marquet, Sybil 015299
Marquete, Ronaldo 004702
Márquez Martínez, Teresa 031250
Márquez Rojas, Margoth 008710
Marquis, Jean M. 041391
Marr, Louise 045745
Marr, Matthias 019597
Marre-Muls, Anne 003507
Marrin, Jim 053936
Marriott, James 043514
Marriott, Wayne P. 033253
Marsac, Michel 015350
Marsan, Geneviève 013903
Marsch, Peter 033362
Marschall, Jolanta 035171
Marsden-Atlass, Lynn 052078
Marsden, Dale 050553
Marsden, Melinda 049311
Marsden, Peter 044534
Marsden, Susan 006718, 006719
Marseille, Jeremias 025218
Marsh-Matthews, Dr. F. 051588
Marsh, Joanna 049390
Marsh, John S. 052002
Marsh, Natalie 047934
Marsh, Tracy 005637
Marsh, Warwick 000786
Marshall, Alan 013936
Marshall, Ann E. 052118
Marshall, Betty 047614
Marshall, Cari 050371
Marshall, Dr. Cora 051239
Marshall, Danny 052775
Marshall, Francis 045565
Marshall, Jerry 051141
Marshall, Karen 050025
Marshall, Dr. Larry 050813
Marshall, Robert A. 049304
Marshall, Ron 046434
Marshall, Steven 045175
Marshburn, Jeff 048000
Mársico de Larribité, Viviana ... 000137
Marson, Luigi 028414
Marsoun-Kaindl, Sandra 002047
Marston, Elizabeth 048521
Marsure-Rosso, Danila 025468
Marszalec, Roman 034917
Marszałek, Zygmunt 035334
Marszewski, Tadeusz 035173
Martel, Philippe 012424
Marteljanova, Elisaveta
      Vladimirovna 037621
Martens, Caspar 032378
Martens, H.E.D. 032027
Martens, Jean 003373
Martens, Melissa 046753
Martens, Ralph 041508
Martens, Warburg 016481
Martensen, Robert 046956
Martensen, Robert L. 049873
Marth, Dr. Regine 017558
Marthalie, Furber 046400
Martí Aixalà, Joseph 040204
Martí Armengol, Teresa 039058
Martí Bonet, José María 039092
Martí i Llambrich, Miquel 039871
Martí, Maria 000105
Martig, Prof. Dr. Peter 041609
Martín Flores, Alfonso 039653
Martín Merás, Luisa 039677
Martín Pascual , Crescencio ... 039741
Martin-Vignes, Nicole 011562
Martin, Albert 006877
Martin, Alvia D. 050739, 051746
Martin, Amanda 045705
Martín, Andrés Peláez 038953
Martin, Andy 001639

Martin, Anna 051439
Martin, Anthea 038696
Martin, Antoinette 049868
Martín, Aurora 039444, 040284
Martin, Carol 007162
Martin, Danny 045327
Martin, Darryl 044264
Martin, Dirk 020140
Martin, Erin L. 051254
Martin, Ètienne 015881
Martin, Eva 020873
Martin, Fabian 036779
Martin, François 012478, 012479,
      013439, 015210, 016130
Martin, Frank 051795
Martin, Glynn B. 050448
Martin, Jacques 005769
Martin, James E. 052472
Martin, Jean-Claude 011592
Martin, Jenni 053037
Martin, Jennifer 051065
Martin, Jerry 054520
Martin, JoAnne 005346
Martin, Dr. Joanne M. 046748
Martin, Juliet 052118
Martin, Larry 050186
Martin, Loretta 053806
Martin, Dr. Lutz 017315
Martin, M. 044798
Martin, Macarena Alés 039567
Martin, Marilyn 038687
Martin, Mary 048582
Martin, Maurice 013282
Martin, Michael 051772
Martin, Dr. Michael 019874
Martin, Otto-Walter 017354
Martin, Pascal 014317
Martin, Paul 048793
Martin, Dr. Paula 043900
Martin, Dr. Peter-Hugo 019505
Martin, Pierre de 015861
Martin, Rebecca J. .. 051196, 053773
Martin, Richard 048237
Martin, Rick 048582
Martin, Robert 045739
Martin, Roger 005426
Martin, Ros 033364
Martin, Sandra 045208
Martin, Dr. Shéonagh 044242
Martin, Stéphane 014751
Martin, Dr. Sylvia 019786
Martin, Terry 048997
Martin, Theo 041495
Martin, William 054350
Martin, William J. 052561
Martin, Wilson 052920
Martina, Dr. Alessandra 026458
Martindale, Bruce 007132
Martindale, Wendy 006759
Martinek, Jaroslav ... 009598, 009705,
      009919
Martinelli, Massimo 026535
Martinelli, Dr. Roberta . 027421, 027422
Martinet, Jean-Louis 013523
Martínez Alvárez, Dr. José María 039749
Martínez Arvizu, Jesus 031335
Martínez Cañabate, Juan Carlos 039373
Martínez Carril, Manuel 046252
Martínez Cortés, Ariel 008787
Martínez de Aguilar, Dr. Ana ... 039669
Martínez Fuentes, Dr. Antonio
      J.F. 009149
Martínez Mestre, Miguel 031224
Martínez Rodríguez, Andrés ... 039595
Martínez Sarria, Victoria 008767
Martínez Vargas, Enrique 031123
Martínez, Dr. Armando Luis ... 031207
Martínez, Carlos 048978
Martínez, Connie 053037
Martínez, Daniel 049547
Martínez, Gonzalo 008686
Martínez, Jean-Luc 014749
Martínez, José Luis 031217
Martínez, José Roberto 036156

Martínez, Margarita 054711
Martínez, Pita 048269
Martínez, Ramiro 031260
Martinez, Ray 046312
Martínez, Ricardo Ariel 008761
Martini, Prof. Dr. Erlend 021171
Martiniani-Reber, Marielle 041530
Martiniz-Lemke, Tania 050442
Martínková, Jana 009704
Martino, Giuseppe 026136
Martino, Mike 052656
Martino, Dr. Ronald L. 049625
Martinović, Jovan 031757
Martins Carneiro, José Manuel . 036043
Martins Catharino, Henriqueta .. 004803
Martins e Silva, José 035784
Martins Fernandes, Francisco R. 035728
Martins, Josephine 049910
Martins, Michael 048687
Martinson, Gordon 047186
Martinson, Helle-Reet 010673
Martis, Ela-Heigi 010703
Martis, Val 047640
Martonová, T. 009563
Martorell, Jaume 039879
Marty-Francart, Marie-Jo 015347
Marty, Hanspeter 042207
Marty, John F. 051256
Marty, Theodor 041818
Martykánová, Marie 009931
Martynjuk, Oleg Ivanovič 029837
Martynova, Nina Petrovna 037131
Martynova, Raisa Fёdorovna .. 037048
Maruschke, Thomas 017539
Marusic, Irena 038562
Maruszeczka, Dave 006010
Maruyama, Mariko 028562
Maruyama, Masatake 029605
Maruyama, Tamiko 029639
Marvin, Anne 053292
Marwell, Dr. David G. 051439
Marwick, Fiona C. 044366
Marx, Dr. Erich 002618, 002637, 002639,
      002641
Marx, Günther 022349
Marx, Prof. Dr. Harald 017984
Marx, Dr. Petra 020601
Marx, Prof. R. 038772
Mary Bess K., Paluzzi 046402
Marzahn, Dr. Joachim 017315
Marzatico, Dr. Franco . 025502, 028014,
      028169
Marzio, Frances 049588
Marzio, Peter C. 049572, 049588
Marzolf, John G. 046669
Más Vilaplana, Antonio 039570
Mas, Monique 014342
Masa, Karin 041969
Masaaki, Ueda 028945
Masafumi, Shimooka 028865
Masahide, Maekawa 029326
Masako, Sakashita 028880
Masanek, Wolfgang 021262
Masanori, Komiya 028811
Masaraci, Yasemin 042957
Masato, Ueno 028992
Maštátová, Ivana 009448
Masau Dan, Dr. Maria 028198
Masayuki, Ogata 028528
Mascarenhas, Dr. Aubrey A. .. 024404
Mascherbauer, Michael 001717
Mascia, Sara 053687
Masé, Guido 041678, 041678
Maselli Scotti, Dr. Franca 025345,
      025346, 028190
Masemore, Dr. Ira 050351
Masen, Larz 050943
Mashhadi, Syed Gulzar 034132
Masing, Maia 010682
Masino, H. 022382
Masjoan, Prof. Estela 000273
Maske, Andrew 052880
Maslić, Fatima 004285
Maslov, V.I. 037482

Mason-Chaney, Lisa	048551	Matheny, Paul	047918	Matthews, Marsha	052298	Maxwell, Robert N.	049654
Mason, B.	001143	Mathes, Wayne	051675	Matthews, Nancy	054135	Maxwell, Robyn	000937
Mason, Bonnie N.	051868	Matheson, Susan	051267	Matthews, Ronald	049986	Maxwell, Dr. Thomas	051516
Mason, Carl	046510	Matheson, Trueman	006217	Matthiesen, Kjell Marius	033742	May Dobbins, Melissa	048678
Mason, Darielle	052081	Mathews, George K.	051549	Matthiessen, Joy A.	048290	May, Denis	001573
Mason, Dorothy	048689	Mathews, Greg	051269	Matthieu, Bernard	015709	May, Dr. Elisabeth	018919
Mason, Erik	050390	Mathews, Nancy	049927	Matti, Arnold	042253	May, Guy	030603
Mason, G.	045347	Mathews, Susan	044291	Matticks, Rebecca	048498	May, Jane	044800, 044801
Mason, Gina	052780	Mathewson, Jon	050857	Mattila, Karen	006434	May, Jörg	017561
Mason, Julia	043717	Mathez, Dr. Edmond A.	051327	Mattioni, Fernando	027329	May, John M.	047887
Mason, Dr. Kevin	043661	Mathias, Gérard	012961	Mattl-Wurm, Dr. Sylvia	003095, 003096	May, Sally	001403
Mason, Laurie	006418	Mathiassen, Tove E.	010006	Mattle, Anton	001922	May, Thérèse	030603
Mason, Michael E.	048488	Mathie, William	048477	Mattlin, Håkan	010766	Mayberry, Matthew	047884
Mason, Ngahirika	033141	Mathieu, Caroline	014743	Matton, A.	003676, 003677	Mayeda, Cynthia	047253
Mason, R.P.	044267	Mathieu, Frederic	014046	Matton, Ann	003674	Mayer, August	002726
Mason, Richard	046093	Mathieu, Dr. Kai R.	018622, 021835	Mattox, Boyd	053694	Mayer, Dr. Gregory C.	050600
Mason, Wally	046779, 051817	Mathis, Bernhard	041465	Mattsson, Christina	041028	Mayer, Günter	002917
Mason, William J.	049305	Mathis, Brandi	047139	Matukova, Marija Bajramovna	037073	Mayer, Hedwig	001829
Masoodi, Omara Khan	000005	Mathis, Walti	041707	Matuschka, Dagmar Gräfin von	021622	Mayer, Herbert	002986, 003053
Masriera, Dr. Alícia	039077	Mathison, Judy K.	054003	Matuska, Mike von	041006	Mayer, Jannie	014745
Masriera, Alicia	039078	Mathusa, Parker D.	053266	Matuss, Dr. Márta	023610	Mayer, M.	020361
Massada, Satê	029779	Matijaca, Ivo	008936	Matuszczyk, Andrzej	034908	Mayer, Marc	006385
Massagra, Dr. Maria Grazia	027592	Matijašič, Prof. Sonja	008966	Matuszewski, Nina	019707	Mayer, Olafa	020313
Massari, Dr. Stefania	027630	Matijević, Jožef	038585	Matveeva, Tatjana Sergeevna	036863	Mayer, Ronald	053402
Massaro, Marilyn	052379, 052388	Matile, Michael	042196	Matvenov, Evgenij Rostitovič	037006	Mayer, Walter	047300
Massart, Claire	003492	Matilla, José Manuel	039676	Matyniak, Margaret	054074	Mayerhofer, M.A. Susanne	017709
Massart, Roger	003367	Matilsky, Barbara	047528	Matz, Darrell	053650	Mayes, Catherine	048414
Massas, Natacha	003492	Matino, Dr. Lucia	026825	Matzke, Klaus	021657	Mayes, James	052649
Massé, Eric	015186	Matiskainen, Dr. Heikki	011313, 011315	Matzke, Dr. Michael	041268	Mayfield, Signe	051903
Massenbach, Elisabeth von	020550	Matits, Ferenc	023422	Matzon, Ea	010325	Mayhew, Kim	051509
Massey, Kristine	053831	Matland, Susan	033740	Maubano, M.	013643	Mayland-Mason, Paul	044796
Massey, Lewis	046633	Mato, Nancy	051893	Mauch, Michael	041270	Mayle, A.H.	044974
Massey, Rachel	045743	Matos Moctezuma, María		Maucq, R.	003567	Maynard, Helen B.	049081
Massey, Timothy	047219	Fernanda	030999	Maucuer, Michel	014689	Maynard, Marie-Noëlle	012331
Massicotte, Jean	006249	Matos Reis, Dr. António	036076	Maudet, Serge	030302	Maynard, Rex	046907
Massie Lane, Rebecca	053625	Matoušková, Martina	009849	Mauer, Dr. Alexander	025727	Mayo, Dr. Margaret Ellen	052565
Massier, John	047305	Matrica, Vasilica	036197	Mauerhan, Joëlle	012005	Mayo, Martha	050510
Massimbe, Julieta	031808	Matschiner, Dieter	002702	Mauersberger, Dr. Klaus	018608	Mayo, Marti	049574
Massó Carballido, Jaume	039997	Matskási, Dr. István	023522	Mauersich, Angelika	002813, 002814	Mayor, N.	012158
Masso, Kaisa	010667	Matson, Jaff	049552	Maugain, Gregory	014340	Mayorga Sanchón, D. Juan	040239
Massobrio, Dr. Giulio	025285	Matson, Dr. Timothy	047809	Maugham, Joan	000872	Mayou, Roger	041535
Masson Rusinowski, Fabienne	003698	Matsubayashi, Hosai	029674	Maul, Dr. Bärbel	021540	Mayr-Oehring, Dr. Erika	002636
Måsson, Anette	010281	Matsuda, Hiroshi	028652	Mauldin, Dr. Barbara	053115	Mayr, Jürgen	018891
Masson, Colette	006147	Matsuda, Motoi	029217	Mauldin, Barbara	053116	Mayr, Karin	018672
Massotti, Jeffery R.	052391	Matsuda, Yoshiho	028573	Mauler, Norbert	001993	Mayr, Manfred	017889
Massy, Jean-Baptiste	042113	Matsumoto, Takeshi	029070	Maulsby, Richard	046541	Mayr, Michael	020877
Mastemaker, Lorie	049336	Matsumoto, Tohru	029608	Maumont, A.	015999	Mayr, Peter	002925
Master-Karnik, Paul	050285	Matsuo, Nobuhiro	029255	Maung, U Min Khin	031820	Mayr, Thomas	021877
Masterson, Dorothy	048120	Matsuoka, Masao	029100	Mauoloux, Herman	004047	Mayr, Walter	002563
Masterson, Judy	048649	Matsuoka, Mieko	029544	Maurer, Eleonore	002083	Mayrhofer, Martin	002752
Mastorgio, Carlo	025379	Matsura, Akira	028634	Maurer, Prof. Dr. Friedemann	019111	Mayrhofer, Meinrad	002531
Mastorito, Susana	000362	Matsushita, Kensuke	028464	Maurer, Gudrun	039676	Mayrhofer, Monika E.	003135
Mastrogiuseppe, Joy	052415	Matsuura, Kunio	029399	Maurer, Lena	002953	Mayrhuber, Petra & Gerti	002892
Mastrovic, Mikica	009031	Matsuzaki, Yukiko	029509	Maurer, Dr. Philipp	003016	Mayrock, Roger	019585
Masuda, Hideaki	028617	Matsuzawa, Shigefumi	029718	Maurer, Dr. Rudolf	001773	Mays, N.	000807
Masuda, Rei	029608	Matsuzono, Makio	029410	Maurer, Sherry C.	052627	Maysonnave, J.	016362
Masuda, Yoko S.	029165	Matt, Edvin	050194	Maurer, Simon	042200	Maytk, Paul	054258
Mašulovic, Branimir	038242	Matt, Dr. Gerald	003024, 003025	Mauri, Dr. Gemma	026316	Maywald, Dr. Claus	020226
Masuo, Tano	028856	Mattausch-Schirmbeck, Dr.		Maurice-Chabard, Brigitte	011810	Maza, Roberto Ramos	031627
Masuyer, François	012748	Roswitha	016788	Maurin, Ange	015007	Mazeika, Dr. Rasa	006342
Mata González, Juan C. de la	039689	Matteotti, Mario	027535	Maurin, Bernard	015675	Mazepov, Jurij Ivanovyč	043292
Mata, Dr. Aida Maria Reis da	035857	Matter, K.	041812	Maurin, Ray	049719	Maziarka, Włodzimierz	034784
Mata, Rene Luis S.	034545	Matter, Kuno	041958	Maurirere, Naomi	033208	Mazibuko, Elizabeth	038712, 038713
Matajis, Andrijana	038168	Matter, Victor	041669	Maurizio, Dr. Remo	042020, 042021	Mazitova, Svetlana Vladimirovna	037157
Matakieva, Prof. Dr. Teofana	005191	Mattes, Glenda	048352	Mauro, Dr. Marta	027166	Mazlish, Anne	053388
Matamoros, Joséphine	012386, 012621	Matteson, Judith M.	048817	Mauron, Christophe	041382	Mazlumoğlu, Hasan	042894
Matarazzo, Eva	000264	Mattet, Laurence	041529	Mautner-Markhof, Dr. Marietta	002931	Maznev, Kostadin	005060
Matasci, Mario	042049	Matteucci, Prof. Ruggero	027613	Mauzi Martí, Alfred	039431	Mazonowicz, Douglas	051394
Matat, Leonid Nykonorovyč	043192	Matthäi, Bernhard	019694	Mavric, Kaye	000984	Mazor, Robin	049899
Mataušić, Nataša	009036	Matthaes, Gottfried	026801	Mavrodiev, Zdravko	005202	Mazourek, Paul	001640
Mateescu, Ilie	036724	Matthes, Lilian	010092	Mavrotas, Takis	023084	Mazow, Dr. Leo	053899
Mátéffy, Dr. Balázs	023467	Matthes, Olaf	019021	Mavuso, Makana R.	040507	Mazur, Leonid Viktorovič	038073
Matei, Rodica	036312	Matthew, A.T.	038714	Maw, Dr. Martin	045458	Mazur, Tracey	053095
Matejková, Adriana	038288	Matthew, Malcom	052950	Mawdsley, Simon	033572	Mazza, Germana	025789
Matenga, Edward	054856	Matthewman, Su	046226	Mawe, Theodore	034190	Mazza, Thomas C.	051365
Mater, B.	032035	Matthews Hoffmann, Catherine	050259	Mawleke, M.	038760	Mazzanti, Laura	048776
Maternati-Baldouy, Danielle	014060	Matthews, Angus	007034	Maxfield, Dorothy	054454	Mazzari, Roberto	025525
Materón Salcedo, Edgar	008654	Matthews, Dr. B.	051588	Maxfield, Mike	045211	Mazzei, Dr. Lapo	026276
Maters, Drs. Petra	032972	Matthews, Dr. B.E.	044240	Maxham, R.S.	048866	Mazzei, Marina	026693
Matešic, Suzanne	021663	Matthews, Bert	051177	Maxheimer, Greta	048221	Mazzeo, Dr. Donatella	027627
Mateu y Miró, Josep	039246	Matthews, Christy	048295	Maxholmes, Thomas	001677	Mazzerioli, Paolo	027260
Mateva, Borjana	005039	Matthews, Claire	051145	Maxim, Dr. Elena	036296	Mazzini, Dr. Laura	026507
Matharan, Paul	012112	Matthews, Julia	007200	Maxwell Stuart, Catherine	044633	Mazziotti Gillan, Maria	051943
Matheis, Adam	020221	Matthews, Keith	046220	Maxwell, Jennifer	045910	Mazzocchi, G.	027691

Mazzolti, Prof. Mario 027483
Mazzoni de Bravo, Helga 000286
Mazzoni, Anne-Hélène 015145
Mazzotta, Gabriele 026792
Mazzotti, Dr. Stefano 026207
Mazzotti, Tullio 025276
Mbe, John C. Jacques 045764
Mbodji, Abdoulaye 038122
Mbokashanga, Nyimi Kuete . . . 008842
McCusker, Carol 052968
McDonnell, Dr. Patricia 054384
McGarry, James R. 049959
McIntosh, Jane 048749
McIntyre, Dr. Nancy 050517
McLean, Al 045786
Meabon, Linda 047173
Meachum, Gil 052277
Mead, Tina 054175
Meade, C. Wade 052713
Meade, Diane 048423
Meade, John 050199
Meador-Woodruff, Robin 046490
Meadors, William 050364
Meadows, Larry G. 047781
Meagher, George E. 047540
Meagher, Kathy 005846
Meagher, Tom 006223
Mealing, David 033226, 033354
Mealings, Bob 044453
Mears, Peter 046692
Mearwah, Reena 024282
Mecham, Denny 046863
Mechan, Dallas 043777, 043796, 044728
Mechell, Barbara 054018
Mechler, Sabine 019144
Mechler, Dr. Wolf Dieter 019053
Mecklin, Kai 011421
Mecredy, Stephen D. 006099
Medaković, Anica 038257
Medard-Blondel, Anne 014062
Medeiros, Francisco 035744
Medek, Prof. Robert 002973
Meder, Katrin 017559
Mederski, Mark 052127
Medhi, Prof. Birinchi K. 024270
Medhurst, Alex 045786
Medica, Dr. Massimo 025541
Medici, Felix Nazar 000134
Medicus, Mag. Christine 002625
Medina i Montes, Susana 040224
Medina, Dennis H.J. 046303
Medley, Mark 051175
Medlock, Leon 052236
Medlycott, Sir Mervyn 045731
Medri, Prof. Gianluca 026177
Medri, Dr. Litta 026262, 027381
Medvedev, Nikolaj Jurjevič . . . 037718
Medvedow, Jill 047116
Medzihradská, Dr. Elena 038382
Mee, Dr. Jane 045556
Meehan, Brian 006209, 006215
Meehan, Joseph M. 048739
Meek, Kenneth 046801, 046801
Meek, Noel 033203
Meeke, Dr. Noel 044577
Meeker, Amanda 052722
Meekers, Christa 003684
Meena, K.L. 024328
Mees, Dr. Allard 020231
Meester de Heyndonck, B. de . . 003332
Meeusen, M. 004075
Meeuwsen, Jeffrey 049194
Meewis, Menno 003291
Meglio, Luciano di 026516
Mehaska, Dr. Walter 049372
Mehler, Sarah 042240
Mehmann, Moritz 041190
Mehra, Sandeep 024233
Mehren Gorev, Leslie 052837
Mehring, Albert G. 049372
Mehring, Karen 048770
Mehta, Jawahar G. 024140
Mei, Runxin 007683
Meiborg, E. 032652

Meidell, Bjarne A. 033536
Meidenbauer, Roy 051228
Meidinger, Ronald J. 046574
Meier, Andreas 041837
Meier, Anton C. 042170
Meier, Bernhard 041367
Meier, Ernest 041672
Meier, Heinz 041463
Meier, Iris 041288
Meier, Jakob 042132
Meier, Jürgen 021295
Meier, Michelle 051489
Meier, Rupert 041295
Meier, Stefan 041265
Meier, Wolfgang 018380
Meighörner , Dr. Wolfgang 002141
Meijer, M. 032027
Meiler, Olaf 018791
Meillat, Eric 011608
Meillat, Joël 014261
Meillon, Anne 013196
Meilutiené, Gražina 030457
Meinderts, Drs. A.J.M. 032172, 032184
Meindre, Colette 013878, 015193
Meinelschmidt, C. 019295
Meiners, Dr. Jochen 017719
Meiners, Prof. Dr. Uwe 017747
Meinig, U. 016674, 016675
Meinrad-Meier, Dr. Anton 041566
Meir, Cecilia 025202
Meireles, Alexa Bastos Gambogi 004390
Meiring, Barbara 048863
Meiring, Jacob 041419
Meirose, Jan-Dirk 018094
Meïrovics, Prof. Imauts 030196
Meisch, Jim 030594
Meisser, Fritz 041718
Meisser, Dr. N. 041654
Meissimilly-Rosin, Martine 016188
Meissner, Dipl.-Ing. Irene 020505
Meissner, Roman 035120
Meißner, Uwe 018970
Meissonnier, J. 012870
Meister, Arthur 041375
Meister, Mark 048168, 048175
Meister, Nicolette 046905
Meister, Pamela 047570
Meisterman, Ingrid 040559
Meiszies, Dr. Winrich 018048
Meixner, Brigitte 021247
Mejdahl, Ulla 010044
Mejer, Robert Lee 052431
Mejía Maya, Rafael 008822
Mejia, Javier 008786, 008793
Mekas, Tauras 030364
Méla, Prof. Dr. Charles 041421
Melamed, Nuril 025203
Meland, Marte 033598
Melandri, Leonardo 025500
Melandri, Lisa 053135
Melcher, Horst 020458
Melcher, Dr. Ralph 021560
Melchert, Dr. Monika 017069
Melchionne, Kevin 052098
Melchior, Dr. Christoph 022817
Meldgaard, Bjarne 010113
Meleško, Dr. V.P. 037792
Melgarejo Cabrera, Elizabet . . . 009325
Meli, Pietro 005594
Meline, Anne 052763
Melisi, Prof. Francesco 026997
Melissa, Talachy 053119
Melke, Gunta 030114
Melki, Dr. Loutfalla 030269
Melkonyan, Hasmik 000724
Mella, Joseph S. 051191
Mellander, Barbo 040515
Mellander, Barbro 040760
Mellby, Julie 051244
Mellbye, Maren-Otte 034025
Mellemsether, Hanna 034038
Meller, Dr. Harald 018965
Mellinger, A. Clair 049377
Mello, Luiz Carlos 004727

Mellon, Lori A. 048274
Mellor, Jackie 045876
Mellott, Tina 047213
Mellow, Ralph 005639
Melly, Anne-Lyse 042113
Melly, R.G. 044979
Melman, Herman 054011
Melničuk, Andrej Fëdorovič . . . 037642
Melnikova, Iraida Fëdorovna . . 029843
Melnikova, Natalija Ivanovna . . 037032
Melnikova, Tamara Michajlovna . 037193
Mel'nyk, Anatolij Ivanovič 043215
Melo Miranda, Selma 004782
Melo, Jorge Orlando . . 008588, 008590
Melo, Dr. Manuela 035935
Melody, Paul 033239
Melomina, Olga Alekseevna . . . 036875
Melone, Dr. Francesco 027070
Meloni, Alberto C. 052788, 054193
Melrose, A. 001150
Melton, Babs B. 053565, 053566,
. . . . 053567, 053568
Melton, R. Neal 053275
Meltzer, J.L. 038688
Meltzer, Lalou 038685
Meltzer, Ellen 018810
Meltzer, Gustav 002514
Melzer, Dr. Miloš 009703, 009935,
. . . . 009968
Melzer, Dr. Walter 021891
Memoire, Natalie 012120
Mena Segarra, Prof. Enrique . . 046265
Mena, Manuela 039676
Menaker Rothschild, Deborah . . 054420
Menalo, Romana 008899
Ménard, Jean 015452
Menard, Jean-Daniel . 015961, 015962
Menardi, Dr. Herlinde 002142
Menasse-Wiesbauer, Dr.
 Elisabeth 003102
Menato, Prof. Giuliano 028251
Menconeri, Kate 054529
Mencucci, Angelo 027918
Mendel, Shlomoh 025178
Mendelssohn, Dr. Gabriele . . . 019403
Mendelssohn, Prof. H. 025214
Mendenhall, Winnifred 047853
Mendes Bürgi, Dr. Bernhard . . 041273,
. . . . 041277
Mendes Gonçalves, Maria Arlete 004731
Mendes Marques, Dr. Maria da
 Concerção 035733
Mendes, Albano 023388
Méndez Lecanda, Dr. Elia
 Josefina 031261
Méndez López, Nohra Isabel . . 008671
Mendez, Ivan 036169
Méndez, Dr. Marion 018826
Mendizabal, Tomás 034174
Mendoza Ramírez, Patricia . . . 031172
Menefee, Antonio 047111
Menegazzi, Dr. Alessandra . . . 027152
Meneguzzo, Giobatta 026683
Meneses, Adriana 054735
Menezes Gomes, Geraldo de . . 004887
Meng, Fancai 007986
Meng, Jianyao 001150
Meng, Qingjin 007716
Meng, Ung Vai 008068
Mengden, Dr. Lida von 020140
Menge, Heide 017601
Mengede, Dr. Bernd . . 018294, 018297
Mengele, Mark 053597
Mengis, Tom 050543
Menis, Gian Carlo 028228
Menk, Jeanette 021097
Menke, Dr. Annette 022357
Menke, Hubertus 019471
Menke, Wilhelm 019073
Menna, Edith 052914
Menne, Gilbert 003613
Mennel, Suzanne 041315
Mennicke-Schwarz, Christiane . . 017991
Mennicken, Ralph 003959

Mennucci, Dr. Antonello 027751, 027755
Mens, Drs. I.J.A. 033098
Mentel, Anette 021242
Menthoonex, Jean 041430
Mentzelopoulos, Constantine . . 023301
Menu, Daniel 016257
Menu, Louis 003868
Menz-Bächle, Anke 022057
Menz, Dr. Cäsar 041513, 041516,
 041520, 041526, 041528, 041530,
. . . . 041532, 041538
Menz, Christopher 000762
Menz, Martin 021826
Menzel, Joachim 021942
Menzel, Steffen 020285
Menzi, Renate 042217, 042219
Menzie, Mary 052184
Menzies, Donald 054517
Menzies, G.M. 043364
Menzies, Jackie 001535
Menzies, Sheila A. 049425
Merabet, Née 000044
Meral, A. Seda 042815
Merc, Viktor Karlovič 029835
Mercadal i Fernández, Oriol . . . 039987
Mercado Lopez, Eugenio 031311
Mercado, Stephanie 050410
Mercandino Jucker, Gabriella . . 006848
Mercede, Nevin 054576
Mercer, Anna 045706
Mercer, Bill 052302
Mercer, Régis 016294
Mercer, Valerie 048301
Mercier, Poppi 047067
Mercuri, Joan B. 047659, 051665
Mercuri, Mariano 027291
Merdjanov, Plamen 005076
Merdžanov, Plamen 005089
Meredith, Jill 046451, 046451
Meredith, Mary Ellen 053652
Merhib, David 047243
Merhiou, Olivier de 014128
Mérilleau 011886
Merino Acosta, Dr. Luz 009201
Mériot, Christian 012737
Mériot, Christian 012117
Merk, Jan 020499
Merk, Zsuzsa 023418
Merkel, Dr. Michael 018995
Merker, Matthias 022505
Merkert, Prof. Jörn 017085
Merland, Bénédicte 003795
Merlin-Anglade, Véronique 014829
Merlin, Georges 011853
Mermet, Christian 014367
Merode-Westerloo, Princess de . 015335
Meronen, Mikko 011435
Merrell Domingo, Merle 034496
Merrell, Ed 048525
Merriam, Robert W. 048428
Merrill-Mirsky, Dr. Carol 050429
Merrill, Kimberly 046334
Merrill, Michael 053341
Merriman, Larry 049393
Merriman, Dr. Nicholas 045210
Merrin, Keith 044662
Merrit, Carrie 049129
Merritt, Jody A. 046641
Merritt, Roxanne M. 048788
Mersseman, Denise 004108
Merten, Jürgen 022202
Mertens, Beate 017086
Mertens, Carlos Frederico 000445
Mertens, Dr. Eberhard 019075
Mertens, Dr. Rainer 020846
Mertens, Dr. Sabine 020229
Mertens, Dr. Veronika 016491
Mertes, Lorie 052052
Mertin, Dr. Michael 019258
Merveldt, Catherine Gräfin von . . 017948
Merwe, Somine van der 038771
Mery, Estelle 016334
Merz, Werner 041817
Merzenich, Hildegard 020521

Merzweiler, Heinz H. 041293
Mesa, José de 004209, 004211, 004215, 004217
Mesa, Laura 050985
Mesbahi, Sh. 024752
Meščerina, Nadežda Alekseevna 037615
Meschede, Dr. Friedrich ... 017098
Meschede, Prof. Dr. Martin ... 018806
Mesejo Maestre, Eduardo 009140
Mešerjakov, Vladimir 037769
Meshke, R. 005661
Meskin, Maria 005115
Meslin, Chantal 013822
Mesnil du Buisson, Laurent du .. 016233
Mesquida García, Dr. Mercedes . 039919
Mesquita, Ivo 004881
Messel, Nils 033845
Messenger, Jane 000762
Messenger, Janet M. 051916
Messenger, Margaret 005555
Messer, Jennifer 052632
Messer, Michael 016667
Messer, Renate 020935
Messer, Richard 050463
Messerli, Jakob 041320
Messerli, N. 041677
Messerli, Ursula 041967
Messiaen, Marcel 003267, 003736
Messika, Natalie 025179
Messmer, Dorothee 042125
Messner, Kathrin 003048
Messner, Reinhold ... 025816, 025980
Meštan, Prof. Pavol 038325
Mészáros, Dr. Ferenc 023522
Mészáros, Júlia 023659, 023663, 023667, 023668
Metallinou, Garyfalia 023051
Metcalf, Charles D. 054557
Metcalf, Craig 006557
Metcalf, Elizabeth 051379
Metcalf, Mike 053336
Metcalfe, Edward C. 050702
Metcalfe, John 044122
Metcalfe, Robin 005963
Meteku Zekeke, Endalew 010728
Metherall, John 045172
Methuen-Campbell, James ... 044012
Methven, John 001483
Metropolis, William C. 047367
Mets, Esta 010711
Mettala, Teri 051723
Metthews, Ray 050238
Mettjes, Dr. Gerd 021945, 021946, 021947, 021949
Metz, Donna 052781
Metz, Katharina ... 017732, 017733
Metz, Lance 048457
Metz, Thomas 019662
Metz, Wolfgang 020019
Metzger, Amy 049312
Metzger, Claudine 041318
Metzger, Frederick W. 007272
Metzler, Jeannot 030593
Metzler, Dr. Sally 047671
Metzner, Paula 049860
Meulemeester, Jean Luc 003942
Meulenbeld, B.C. 031950
Meurant, Anne 003474
Meuter, Günter 020748
Meûter, Ingrid de 003492
Mewborn, Frankie 054057
Mewes, Claus 019000
Mewton-Hynds, Sally 046039
Meyer 021226, 021230, 021232
Meyer Bühler, Monica 041338
Meyer-Büser, Dr. Susanne .. 019066
Meyer-Friese, Dr. Boye 018976
Meyer-Stoll, Christiane 030305
Meyer Utne, Jeannette 033741
Meyer zu Bargholz, Wolf ... 041715
Meyer zu Schwabedissen, Doris 020653
Meyer, Alain 014333
Meÿer, Bob 031949
Meyer, Charles G. 046718

Meyer, Dr. Christian .. 002404, 002937
Meyer, Dr. Christian A. 041282
Meyer, Ellen L. 046605
Meyer, Fifi 038689
Meyer, Franz 016861
Meyer, G. 021767
Meyer, Hans Rudolf 041741
Meyer, Helen 050198
Meyer, Hélène 041530
Meyer, Hubert 015756
Meyer, Irin von 019011
Meyer, J.B. 032428
Meyer, Jackie-Ruth 011581
Meyer, Jochanan 025023
Meyer, Karl 020667
Meyer, Katharina 014386
Meyer, Kimberli 050451, 054308
Meyer, Kurt 021227
Meyer, Lioba 020991
Meyer, M. 032021
Meyer, Marc 030594
Meyer, Marie-Emmanuelle ... 016166
Meyer, Maurice 011636
Meyer, Niels 010006
Meyer, Rachel 051904
Meyer, Dr. Reiner 021887
Meyer, Robert 051438
Meyer, Simone 022468
Meyer, Stefan 021437
Meyer, Steve 049598
Meyer, Thomas 042190
Meyer, Werner 018729
Meyerdierks, Judith 047459
Meyere, J.A.L de 032953
Meyers, Dr. Albert. 017478
Meyers, Amy 051266, 053053
Meyerson, Barbara 050810
Meyerson, Randy 053759
Meyrick, Robert 043373
Meysner 022103
Mezan Barrera, Jesús Manuel .. 031145
Mezger, Prof Klaus 020603
Mezhlumyan, G.B. 000730
Mezil, Eric 011832
Mežinski, Jelena 038139
Mezvinski-Smithgall, Elsa .. 054157
Mezzadri, Prof. Maria Grazia .. 027216, 027217, 027218
Mhire, Herman 050063
Mhute, Josiah Rungano 054854
Miano, Ederick C. 034461
Miao, Duo'an 008194
Miaoulis, Ioannis 047124
Michael A., Flannery 047002
Michael, Dean 001201
Michael, Dr. Eckhard 020182
Michael, Dr. Elisabeth 020507
Michael, Helene 042181
Michael, Janet 043829
Michael, William H. 049660
Michaelidon, Maria 023248
Michaelis, Dr. Rainer 017147
Michaels, Bonni-Dara 051493
Michaelsen, Miah 046647, 051636, 053857
Michailov, Christo 004985
Michailov, Igor 037774
Michajlov, A.B. 037956
Michajlov, Delčoe 030634
Michajlova, Ljudmila Vasiljevna . 037777
Michajlova, Zdravka ... 005082, 005217
Michajlovna, Jelena 037288
Michalaki-Kollia, Maria ... 022973
Michalevskij, Aleksandsandr Vlaimirovič 029829
Michalka, Matthias 003050
Michard, Dr. S.G. 014659
Michaud, Florine 041987
Michaud, François 014697
Michaud, Jean-Marc .. 013630, 015125
Michaud, Jocelyne 006879
Michaud, Raynald 006405
Michavila, Joaquín 040404
Michel, Dr. Ch. 003790

Michel, D. 014297
Michel, Ernest W. 051382
Michel, Gundega 030198
Michel, Helmut 022456
Michel, James 053312
Michel, Jean-François . 011568, 012517, 012518, 012519
Michel, Kaspar 041968
Michel, Samuel 041567
Michel, Prof. Dr. Thomas .. 022048
Michelangelo, Dr. 026056
Michelet, Annick 015002
Michelina, Prof. Dr. 025424
Michellod, Jean-Pierre 042111
Michelmann, Nina 019039
Michelon, Olivier 015118
Michelotti, Dr. Francesca .. 038096, 038099, 038103, 038106
Michels, Dr. Anette 022215
Michels, Dr. Hubertus 017877
Michels, Dr. Norbert 017862
Michélsen, Pia 040813
Michie, Thomas 052387
Michieli, Catalina Teresa ... 000118
Michoutouchkine, N. 054675
Mickenberg, David 054266
Mickle-Smaczny, Lynne 048290
Micklich, Judy 052896
Micklich, Dr. Norbert 017822
Micko, Ingeborg 002914
Miclescu, Constanţa 036815
Micol, P.M. 013003
Mičurová, M. 038486
Micynska, Tamara Petrivna . 043143
Middag, B.M. 032562
Middelbasch, Max 032772
Middlemost, Thomas A. 001610
Middleton, J. 001150
Middleton, J.W. 043852
Middleton, S.I. 043852
Miedema, B. 032077
Mieles, André 018412
Mielke, Dr. Heinz-Peter ... 018802
Mielsch, Prof. Dr. Harald . 017477
Mierkiewicz, Edward 035359
Mierzwiński, Mariusz 035006
Miesbeck, Peter 021474
Miesenberger, Martin 002444
Miessgang, Dr. Thomas 003024, 003025
Mieß, Peter 002191
Mieth, Dr. Matias 019450
Mieth, Thomas 017513
Miethe, Susann 016481
Mietke, Dr. Gabriele 017089
Miettinen, Sinikka 011500
Miggitsch, Jörg 002399
Miglbauer, Dr. Renate . 002913, 002914, 002918, 002919
Miglinas, Skirmante 047670
Migliori, Gabriele 026947
Mignot, D. 031986
Migritone Gondwe, Michael . 030656
Miguet, Danièle 012457, 012458
Mihai, Raul 036266
Mihail, Benoît 003500
Mihăilă, Sanda 036431
Mihailović, Milica 038145
Mihalić, Jana 008925
Mihelić, Sanjin 009025
Mihevc, Bibijana 038565
Miho, Terao 028750
Mihovilić, Kristina 008971
Mihut, Elena 036323
Mijić, Suzanalav 038187
Mijnlieff, E.M. 032365, 032366
Mikaelian, Kajik 000700
Mikaelsen, Laila 023338
Mikelson, Mary 007060
Mikes, Johannes H. 001931
Miki, Akiko 014794
Mikkalo, June 047962
Mikkelsen, Dr. Egil . 033831, 033835, 033844, 033869, 033872
Mikkelsen, Falk 010080

Mikkola Huk, Eira 011122
Mikkola, Ari 011172
Mikołajska, Ewa 035230
Mikulaštík, Dr. Tomáš 009962
Mikus, Joan 048829
Mikuž, Marjeta 038550
Milani, Arch 051634
Milani, Dr. Domenico 027219
Milano, Barbara R. 053155
Milano, Dean 048564
Milano, Prof. Ernesto 026850
Milbank, Bill 033334
Milburn, Alan 044380
Milchram, Gerhard 003049
Milde, Brigitta 017733
Mildner, Dr. Paul 002192
Miles, Blanche 052915
Miles, Carolyn P. 052790
Miles, Carolyn W. 046323
Miles, Christine M. 046340
Miles, David A. 054532
Miles, Ed 046995
Miles, Ellen 054150
Miletić, Jelena 038187
Mileusnić, Slobodan 038171
Mileusnić, Snježana Radovanlija 009048
Milev, Svetoslav 005170
Milhau, Muriel 015347
Milhazes, Dr. Maria Cláudia C. . 035547
Miliadzidou-Ioannou, Dr. E. .. 023289
Milian, Mihai 036365, 036797
Milivojević, Radmila 038201
Mill, Dennis J. 051797
Millan, William I. 043529
Millar, C. 044164
Millar, Jim 006693
Millar, Joyce 006665
Millard, Ellen 007127
Millard, John 044869
Millard, Rhoda 050700
Millberg, Per-Olof ... 040718, 040719
Mille, Dr. Diana Dimodica . 048670
Millen, A.J. 005353
Millenbruch, Gary L. 046408
Millender, Lynn R. 046458
Miller-Betts, Christine ... 046661
Miller, Alan 000895
Miller, Dr. Albrecht . 018155, 020862
Miller, Alison 043501, 050189
Miller, Andree . 046410, 047464, 048492
Miller, Angus 044464
Miller, Ann 050739
Miller, Archie W. 006473
Miller, Bob 052593
Miller, C.J. 049296
Miller, Carl R. 050030
Miller, Carloyn 051896
Miller, Charles 051818, 051820
Miller, Dana 051491
Miller, David 053160
Miller, David M. 050802
Miller, Denise 047652
Miller, Douglas 051071
Miller, E.M.J. 043504, 045179
Miller, Erika 049127
Miller, Prof. George 049418
Miller, George 052200
Miller, Gerald J. 053047
Miller, Gordon 047678
Miller, Harriet 051508
Miller, Heidi 046384
Miller, Howard L. 049388
Miller, John 045958
Miller, John F. 049283
Miller, Dr. Kathy Ann 053644
Miller, Kevin N. 048181
Miller, Kristin 046393
Miller, Lenore D. 054104
Miller, Leueen 049247
Miller, Linda L. 048537
Miller, Lisa 046408
Miller, Lois 049587
Miller, Margaret A. 053666
Miller, Marguerite 046405

Miller, Dr. Markus 018117
Miller, Mary 006971, 050493
Miller, Michael . . . 052289, 053419
Miller, Mike 046684
Miller, Nancy Lee 047141
Miller, Pamela 052357
Miller, Paul F. 051539
Miller, R. 045574
Miller, Randy 006893
Miller, Dr. Rick 046335
Miller, Rosemary 001148
Miller, Sammy 045325
Miller, Sandy 053562
Miller, Sarah 001368
Miller, Stephen 050079
Miller, Steve 051065
Miller, Steven 045651
Miller, Terry J. 049018
Miller, Tom 005878
Miller, Tracy 046930
Miller, Walter 053636
Miller, Warren 052352
Millereau, Marie-Pierre 012091
Millet, Florence 013582
Millet, Teresa 040305
Milli, Konrad 019466
Milligan, Dr. Frank D. 051157
Milligan, Robert B. 051530
Millin, Laura J. 050960
Milller, Marcus 006598
Millman, David 050170
Millroth, Thomas 041184
Mills, Brendon 050085
Mills, Dan 050224
Mills, Dinah 053982
Mills, Ellsworth 049472
Mills, John 044280
Mills, John K. 052371
Mills, Josephine 006188
Mills, R.G. 046059
Mills, Susan 043396
Milne, Jane 044671
Milne, Julie 045349
Milne, Lee 001134
Milne, V. 000775
Milner, Dr. M.J. 045672
Milner, Renate 053926
Milnor, Nancy 049034
Milojković, Jelica 038225
Milonjić, Miroslav 008983
Milosavljević, Branka 038185
Milošević-'320erek, Svjetlana . . 008901
Milošević, Prof. Ante 009000
Milosevich, Joe B. 049836
Milotd, Jocelyne 006740
Miloux, Yannick 013814
Milovanović, Dušan 038168
Milović, Olivera 038226
Milstreu, Dr. Gerhard 041074
Milton-Tee, Ann 033208
Milton, Lester R. 046791
Milton, Peter 045410
Milyov, Simeon 005191
Min, Choi Hyong 029874
Min, Susette 051379
Mina, Dr. Gianna A. 041680
Minar, Rena 053661
Minard, Barbara 049659
Minard jr., Richard A. 047958
Minarini, Dr. Laura 025538
Minassian, Levon 024699
Minchinton, J. 045391
Mindell, Dr. David 052995
Mindell, Prof. David P. . . 046493, 046493
Minder, Nicole 041848, 042108
Mindhoff, Otto 021803
Mine, William K. 043488
Mineau, Marc 007056
Mineeva, Valentina Alekseevna . 036932
Minelli, Donato 025369
Miner, Curt 049372
Minerva, Michael 051439
Minetti, Dr. Alessandra 027863
Minighetti, Clément 030588

Minish, Marveen 050794
Minke, Prof. Dr. Alfred 003590
Minkenberg, Dr. Georg 016435
Minkkinen, Aimo 011398
Minmeister, J.C. 012823
Minney, Michael J 050120
Minns, Dr. Jonathan E. 044609
Mino, Dr. Yutaka 029266
Minoja, Dr. Marco 025393
Minse, C. 032401
Minshew, Cheryl 048494
Minta, Laura 043806
Minto, Don 049791
Minto, Heather 049791
Minton-Elmore, Kara 051622
Minton, Dwayne 023359
Mintz, Ann 054293
Mintz, Deborah Martin 048719
Minx, Jim 051735
Mion-Jones, Esta 045828
Mir-khalaf, A. 024700
Mir Sancho, Joaquín 039731
Mir, Dr. Ronen 046676
Mira Bonomi, Angelo Vittorio . . 026366,
. . . . 027937
Mira, Vegar 008883
Mirabdolbaghi, Zia 016150
Mirabella Roberti, Prof. Mario . 028280
Miranda, Gaudenzio 027087
Mirelez, Cathy 049100
Mires, Cherry 046717
Mirmoghtadaie, S. 024701
Mirnik, Dr. Ivan 009025
Miró i Cuberes, Enric 039095
Mirolo, Diana 041236
Miron, Dr. Andrei 021558
Miron, Dr. O. 025134
Mironneau, Paul 014810
Mironowicz-Panek, Maria 035031
Miroslavljević, Dragisa 038252
Mirror, Louise 051455
Mirzejewska, Maria 035326
Mirzoyan, Pharaon 000743
Misago, Dr. Kanimba 038086
Misailova, Valentina Ivanovna . . 037722
Miscault, René de 013571
Mischke, Petra 019167
Mischkewitz, Franz 020683
Mišečkina, Marija V. 037023
Mishima, Tadao 029196
Mishima, Yashuhisa 029230
Mishima, Yoshinori 029230
Mishne, Merle H. 050764
Mishra, H. P 024173
Mišina, Ljudmila Aleksandrovna . 037684
Misjuga, Bogdan Vasil'ovič . . . 043255
Miskovic, Svetlana 007425
Miskowski, Anthony 053799
Mislej, Irene 038492
Misner, Mary 005537
Misočnik, Svetlana Michajlovna . 037880
Misof, Dr. Bernhard 017504
Miss, Dr. Stig 010198
Mistrot, Vincent 012112
Mitanov, Petăr 005204
Mitar, Mladen 008943
Mitchell, Archie 053887
Mitchell, Charles 051978
Mitchell, D.I. 045836
Mitchell, Freda 047187
Mitchell, Harold 000948
Mitchell, James 047564
Mitchell, Jena 052679
Mitchell, Joan 001048
Mitchell, Joanne 046395
Mitchell, John 050217
Mitchell, Leah 049789, 051501
Mitchell, Marjorie 052596
Mitchell, Paul 005659
Mitchell, Dr. Peter 045459
Mitchell, Renee 049974, 049975
Mitchell, Scott 051697
Mitchell, Starr 050335
Mitchell, Sue 001564

Mitchinson, David 045303
Mitev, P. 005150
Miteva, Kremena 005019
Mitkevič, Maja 036861, 036862
Mitkovs'ka, Tetjana Sergiïvna . 043264
Mitrev, Dimităr 005107, 005109, 005164
Mitsides, Dr. Andreas 009411
Mitten, David G. 047362
Mittendorfer, Gabriele 002683
Mittendorfer, Katja 002631
Mitterhöfer, Dr. Jakob 002400
Mitterwenger, Dr. Christine . . . 002494,
. . . . 002495, 002496, 002497
Mittmann, Dr. Hans-Walter . . . 019518
Mittwoch, Wolfgang 018355
Mitu, Aneta 036591
Mitzel, Horst 022553
Mitzel, June 005796
Mitzscherlich, Dr. Birgit 016976
Miwa, Kenjin 029608
Miyagawa, Ellen 051902
Miyaji, Akira 029404
Miyake, Akiko 028826
Miyamoto, Akio 028739
Miyamoto, Makiko 028652
Miyaoka, Kiyomi 029702
Miyatani, Masayuki 029431
Miyazaki, Takashi 029149
Miyazawa, Akimichi 029122
Miyazima, Hisao 029245
Mizerová, Hana 000777
Mizerski, Dr. Włodzimierz 035335
Mizina, Ljudmila V. 037911
Mizinga, Flexon Moono 054839, 054842
Mizukami, Iwao 028502
Mizukami, Takeshi 029481
Mizuno, Keizaburo 029080
Mizuno, Takashi 029305
Mizuta, Dr. Akira 029146
Mizutani, Takeshi 029608
Mizzan, Dr. Luca 028312
Mjärdner, Stellan 040658
Mjasičševa, Olena Georgïïvna . 043311
Mjøs, Lise 033871
Mlecz, Richard S. 052526
Mlecz, Stella C. 052526
Mleczko, Radosław 035360
Mlynarčík, Dr. Jozef 038391
Mlynarski, Ryszard 053181
Mnasaganyan, Susanna 000703
Mndaweni, Aurelian . 038712, 038713
Mo, Charles 047576
Mo, Guofu 007681
Mo, Huamin 007984
Moacir Maia, Prof. Pedro 004795
Moberg, Jack 050681
Mobley, Ree 053115
Moc, Joan S. 050749
Mochi Onori, Dr. Lorenza 028238
Mochola, Tatiana 034797
Mochoruk, Mike 006497
Mochova, Jelena Fillipovna . . . 036840
Mock, Janelle 048267
Mock, Rayonna 046418
Modegi, Masaaki 029598
Modersohn, Antje 021046
Modest, Wayne 028443
Modi, Amrutbhai 024124
Modonesi, Dr. Denise 028351
Modotti, Luigi 026012
Modrzejewska, Dr. Irena 034631
Moe, Barbara 049961
Moe, Helene 033808, 033809
Möbius, Ingrid 020468
Moed, G.J. 032689
Moeferdt, Dr. Horst 017911
Möhlenkamp, Dr. Klaus 020440
Møhlmann- de Grÿs, Laura . . . 032016
Moehlmann, Nathan 051555
Møhlmann, Rob 032016
Möhlmeyer, Erich 021025
Möhring, Barbara 019046
Möhring, Dr. Helmut 021510
Moehring, Markus 020098

Möhrle, Reinhard 041345
Möller-Arnsberg, Birgit 020540
Møller-Hansen, Keld 010364
Møller Jensen, Johnny 010006
Møller Pedersen, Kurt 010009
Möller, Axel 017149
Møller, Dan E. 010361
Möller, Hilke Gesine 020173
Möller, Dr. Kathrin 022694
Möller, Kristina 040969
Møller, Lars Kærulf 010089
Moeller, Prof.Dr. M. 017095
Möller, Marie von . . . 033523, 033524,
. . . . 033525
Möller, Peter 041098
Møller, Stig B. 009998
Møller, Thorkild 010056
Möllmann, Dr.Dr. Elke 017963
Moen, Eivind 034021
Mönig, Dr. Roland 019644
Mönkkönen, Mikko 011218
Moens, Dominique . . . 003928, 004023,
. . . . 004157
Moens, Karel 003316
Möntl, Karin 003059
Mörgeli, Prof. Dr. Christoph . . . 042209
Mörgeli, Markus 041770
Mörike, Dr. Doris 022056
Moering, Dr. Renate 018421
Moers, Rainer 018952
Mörter, Hermann 042210
Mörtinger, Franz 001787
Mörzer Bruyns, W.F.J. 031972
Moes, K. 032006
Moesgaard, Jens Christian . . . 010179
Moeske, Ulrich 017964
Moesser, Suzanne 050742
Mössinger, Ingrid 017733
Moffat, Jill 048980
Moffat, John 043412, 044546
Mofford, Juliet 046480
Moga, Jack 049226
Mogensen, Ove 010343
Moget, Layla 015808
Mohamadi, Naghme 024731
Mohammadrezaie, P. 024709
Mohan, S. 040504
Mohl, Dr. Max 016961
Mohr, Jeanette 022824
Mohr, Julie 045218
Mohr, Margit 020850
Mohr, Dr. Peter 045218
Mohr, Vibeke 033945
Mohrenberger, Ingeborg 001974
Mohring, Marilyn 052650
Moigne, Roland de 012168
Moilanen, Arja 011406
Moinet, Eric 014403, 014407
Moinet, Josette 012112
Moise, Leonida 036397
Moisen Lechuga, Patricia 031236
Moises, Loly Keith 034491
Moisio, Roberto 028139
Moissaides, Christina 048432
Mojica, Brenda 034488
Mojseeva, T.M. 037791
Mojzer, Miklós 023553
Mojzes, Ildikó 023545
Mojžíšová, Dr. Olga . . 009587, 009814
Mokarram Hossain, Mahmud . . 003228
Mokre, Jan 002999, 003053
Mol 004084
Mol, I. 032145
Mol, W. 032103
Moland, Arnfinn 033848
Molas, Gustau 039095
Moldenhauer, Susan 050145
Moldon, John 006283
Molen, Dr. J.R. ter 032013
Molenaar, Corine 032857
Molesworth-Saint Aubyn, Lady
Iona 043666
Molesworth, Kris 050024
Molina Cabal, Alvaro José 008675

Molina Pico, Horacio 000646
Molina, Prof. María Fernanda .. 000530
Molinari, Danielle 014672, 045713
Molinari, Giovanni 026438
Molinaro, Ettore 025619
Molinaro, Marcene 047498
Molist, Miguel 039071
Mollano, Prof. Olga 000486
Mollay, Catherine 006265
Moller-Racke, Heidi 020228
Molles, Dr. Manuel C. 046365
Mollet, Fran 053799
Molifulleda i Borrell, Joaquin .. 038988
Mollo, Boris 044230
Molloy, Dr. Peter M. 052117
Molloy, Tristan 044671
Molnár-Sipos, Judit 023517
Molnar, August J. 051246
Molnar, Cindy Lou 054150
Molnár, Dr. Erzsébet .. 023744, 023745
Molnár, Ildikó 023434, 023694
Molnar, Michael 044973
Molnár, Sándor 024028
Molnar, Veronika 021067
Molnos, Alison 044103
Mols, P. 032979
Molsbergen, Bert 032181
Molteni, A. 013747
Molteni, Dr. Gionfranco 025658
Moltesen, Mette 010188
Moltke-Hansen, David 052055
Molvær, Edvard 033486
Molzberger, Ansgar 019674
Momar Toboso, Manuel 040025
Momas, Jacques 012195
Mombrun, C. 023345
Momčilov, Dimčo 005050
Momin, Shamim 051491
Momoito Astoria, Iratxe 039433
Momonkulov, Džumaaly
  Momonkulovič 030072
Monachon, Danuta 014650
Monaco, Dr. Giuseppe 027577
Monaghan, A. 045722
Monaghan, Bryan 005753
Monaghan, Katherine 053833
Monaghan, Kathleen 048979
Monaghan, Nigel 024899
Monahan, Gloria 006240
Monbron, Comte de 013435
Monch, Manfred 021538
Moncrief, Dr. Nancy D. 050727
Moncur, Shannon 005435
Mond, Ulrike 017489
Mondini, Massimo 025620
Mondragón, Lourdes 031152
Mondragón, Martín Antonio ... 031532
Monds, Kathy 052242
Monedoñedo Espinoza, Juan ... 034365
Monego-Venzin, Barbara 041973
Monery, Jean-Paul 015577
Mones, Prof. Alvaro 046274
Money, Sarah 045656
Monfort, Catherine 003961
Mongan, Barbara 047955
Monge, Janet 052100
Monge, Maria de Jesus 036108, 036109,
                              036110
Mongellaz, Jacqueline 015713
Mongon, Alain 011740
Moniak, Eugeniusz 035470
Monigan, Michael F. 054169
Monk, Suny 053624
Monkhouse, Christopher 050944
Monmousseau, Patrice 015712
Monnani, Walter 001741
Monnet, C. 013805
Monney, Paul 041906
Monnier, Georges 041982
Monnier, Jean-Laurent 014824
Monod, Frédéric 014192
Monok, István 023539
Monot, Michel 016296
Monrad, Kasper 010195

Monroe, Dan L. 052880
Monroe, Dr. Sam 052245
Monschaw, Elline von 024900
Monsef, Gity 044967
Montagne, Dr. Cor 032181
Montagu of Beaulieu, Lord ... 043764
Montaldo, Dr. Anna Maria 025672,
                              025676
Montalembert , Roxane de ... 014731
Montaner, Jaume 039082
Montañés García, Carmen ... 040445
Montanti, Deborah 049840
Montaut 015456
Montebello, Philippe de 051424
Monteil, Esclarmonde 012212
Montel, Philippe 015488
Montenegro Carneiro, Dr. Maria
  Margarida V. 035841
Montérémal, R. 013701
Montero Agüerz, Ildefonso ... 039309
Montero de Povarché, Elena ... 000277
Montero, Alberto E. 007512
Montero, Prof. Paula 004876
Montesano, Frank 048006
Montesdeoca García-Sáenz, Antonio
  Daniel 039905
Montet-White, Anta 050185
Montezoma, Anna de 005688
Montfort, Helen G. 049302
Montgomery, A. 044635
Montgomery, Alexander 007176
Montgomery, Brian 046048
Montgomery, Cathie 045897
Montgomery, Dr. John 052269
Montgomery, Küllike 041109
Montgomery, N. 038660
Montgomery, S.E. 046061
Montgomery, Dr. Sally 043565
Montgomery, Prof. W.I. 043558
Monticelli, Dr. Paolo 027061
Montijo, Mark 052975
Montileaux, Paulette 052473
Montin, Kim 011443
Montjalin, Olaf de 015724
Montmany, Marta 039083
Montoya Soto, Sonia 008665
Monzó, Josep Vicent 040305
Moodie, Kathleen 053101
Moodie, Sandra 028447
Moody, Dr. Barry 005340
Moody, Eric N. 052509
Moody, Tammy 048334
Moogan, Kerri 044393
Mooij, Dr. Charles de 032472
Moombolah-Goagoses, Esther U. 031860,
                    031863, 031864
Moon, Diane 045030
Moons, L.J.M. 032440
Moor, Rebecca 054056
Moore Booker, Margaret 051154
Moore Ede, Minna 045057
Moore Stucker, Kathryn 052138
Moore, Anne 045297
Moore, Barbara 051857
Moore, Bennett 049337
Moore, Bill 005871, 044385
Moore, Carleton B. 053705
Moore jr., Dan 046603
Moore, Dan 046603
Moore, David 043712, 046847
Moore, Dona 006074
Moore, Dorothy 049552
Moore, Duane 053843
Moore, Edra 050109
Moore, Elke aus dem 022043
Moore, Ellen K. 052692
Moore, Fran 005691
Moore, Gary F. 053848
Moore, Geoff 001041
Moore, George 048831
Moore, Harold D. 053359
Moore, J. Kenneth 051424
Moore, J. Roderick 048727
Moore, James C. 046358

Moore, Jennifer 049249
Moore, John 047568
Moore, Juanita 049875
Moore, Kevin 045567
Moore, Linda 051142
Moore, Michael R. 052540
Moore, Richard 001213
Moore, Richard H. 049110
Moore, Rick 046474
Moore, Rod 043983
Moore, Rowan 044905
Moore, S. 005740
Moore, Susan J. 049559
Moore, Thomas 047483
Moore, Prof. Thomas E. ... 046493
Moore, Dr. Viola 051910
Moore, Wayne 047081
Moore, William S. 048314
Moorefield, Amy G. 052543
Moores-Franklin, Loísa 052191
Moos, Norbert 019680
Moosbauer, Christine 016772
Moppett, Ronald 005523
Moqimi-zadeh, Hamid 024707
Mor, Judith 025222
Mora García, Carlos 031654
Mora, Alejandra 031333
Mora, Dr. Rafael A. 000203
Morabito, Dr. Leo 026402
Moraes Branco, Pércio de 004646
Moraes Sophia, Eunice 004879
Moragas, Cora Romy 007524
Moraitou-Apostolopoulou, Dr.
  Maria 023327
Morales-Coll, Eduardo 036150
Morales Gómez, Antonio ... 031320
Morales Gómez, David 031321
Morales Marenco, Eliázar .. 033379
Morales Pérez, Eduardo ... 031624
Morales Ramón, José María ... 039722
Morales, Dr. Michael 048578
Moran-Zerda, Leticia .. 034543, 034552
Moran, Maryann 053214
Moran, Paula G. 046332
Moran, Vivienne 001588
Moraña, Prof. Oscar 054717
Morand, J. 005735
Morand, Marie Claude 041992, 041993,
                    041994, 041995
Morandini, Dr. Carlo 028229
Morando, Paul 047259
Morante López, Dr. Rubén
  Bernardo 031667
Morash, Terrence 047130
Morath-Vogel, Dr. Wolfram .. 018235,
              018236, 018237
Moravčík, J. 038486
Moravec, Tomáš 009742
Morawietz, Doris 019583
Morawińska, Agnieszka 035381
Mórawski, Karol 035372
Morawski, Mike 049952
Morbach, Philippe 014769
Morbis, Daniela 025469
Morcom, G. 038709
Morcos Gayyed, Mervat ... 010570
Mordal, Christen A. 033586
Mordi, Pietro 026542
More-Molyneux, Michael ... 044490
Moreau, Michel 013253
Morehouse, Dorothy V. 050311
Moreira, Ana Arbués 036039
Moreira, João Arbués 036039
Morel-Deledalle, Myriame ... 014056
Morel, Michel 006862
Morel, Pascal 015291
Morel, Sylvie 005871
Morelli, Gianni 027485
Morello, Dr. Giovanni .. 054691, 054692
Morelock, Dr. Jerry 048998
Morena López, José Antonio ... 039040
Moreño de Medellín, Patricia .. 031639
Moreno Garrido, Maria Jesús .. 039305
Moreno Pérez, D. Jorge ... 039891

Moreno Ulloa, Graciela 008853
Moreno, Anne Marie 034598
Moreno, Barry 051474
Moreno, Enrique 039647
Moreno, Fred 051320
Moreno, María Cristina 008706
Moreno, Mercedes 040043
Moreno, William 053011
Moret, J.-L. 041657
Moret, Jean-Charles 041401
Moretti, Dr. A. Maria 027633
Moretti, Dr. Giovanni 026168
Moretti, Dr. Mario .. 025914, 028407
Moretti, Prof. Mario .. 027798, 027800
Morgan-Welch, Beverly 047122
Morgan, B.V. 006474
Morgan, Chris . 043819, 045328, 048811
Morgan, Dahlia 050826
Morgan, Dave 050979
Morgan, David 044100
Morgan, Dina 050428
Morgan, Dr. Gary 001084
Morgan, Dr. J.D. 052230
Morgan, Jeremy 006778
Morgan, K. 033141
Morgan, Martin 046521
Morgan, Dr. Michele 047370
Morgan, Peter J. 045804
Morgan, Peter M. 051178
Morgan, Ron 044016
Morgany, Jacquie 006519
Morgenegg, Hans 041436
Morgenroth, Heiner 021409
Morgenstern, Undine 016535
Morgner, Bernd 019246
Morhun, Sue 005821
Mori, Ichiro 029653
Mori, Kei 029362
Mori, Masatoshi 028666
Moriarty, M.J. 043939
Morier-Genoud, Gabriel .. 041403
Morigi Govi, Dr. Cristiana .. 025538
Morillo-Cajiao, Pablo 008754
Morillo León, Anabel 040166
Morimoto, Prof. Mitsuo ... 029032
Morimura, Hideyuki 029732
Morin, Robert G. 051134
Morin, Yvan 013685
Morinelli, Dom Leone 025881
Morishigue, Akifumi 029235
Morisod, Dominique 041885
Morita, Hajime 029307
Morita, Takashi 029095
Moritsch, Dr. Otmar 003089
Moritz, Craig 046940
Moritz, Franz 019725
Moritz, Heidi 018089
Moritz, Dr. Marina 018246
Moritz, Per 040804
Moriyama, Mayumi 029564
Moriyama, Masanori 029610
Mork, Eileen 049405
Morl, Dr. Claudia 026314
Morley, Stephen H. 047291
Morlok, Dr. Wolfgang 021727
Morman, C. 032115
Morna, Dr. Teresa 035785
Morneburg, Otto 016980
Moro, Alberto E. 000402
Moro, Luigi 025240
Moroder, Robert 027111
Morofushi, Masami 028809
Moroyama, Masanori 029610
Moroz, Vira Dmytrivna .. 043331
Morozov, Ivan Ivanovič .. 037171
Morozov, Vladimir Vasilevič .. 037491
Morozova, Ludmila Ivanovna .. 037658
Morozova, T.V. 037090
Morral, Eulàlia 040220
Morrall, Dr. Andrew 051344
Morran, Gerald 005916
Morré, Jörg 017231
Morretta, Rosemary 052570
Morrill, Hope 054268

Morrin, Peter	050502	
Morris Hambourg, Maria	051424	
Morris, C.I.	044433	
Morris, Charles	047564	
Morris, Christopher	044432	
Morris, Clive	043867	
Morris, David	045219, 052521	
Morris, Henry	044898	
Morris, Jack	006067	
Morris, Jeffory	052399	
Morris, Joan	044762	
Morris, Johnathan	001058	
Morris, Lynda	045403	
Morris, Nell	050075	
Morris, Pamela	052302	
Morris, Patricia A.	052871	
Morris, Robbie	046543	
Morris, Valerie	045991	
Morrison, Alan	045228	
Morrison, Andrew	043574, 045948, 046117	
Morrison, Charlotte K.	048236	
Morrison, Cindi	050122	
Morrison, D.	001496	
Morrison, David A.	005871	
Morrison, Garth	044492	
Morrison, James F.	049831	
Morrison, John	052229	
Morrison, Kim	049490	
Morrison, Lynn	045661	
Morrison, Lynne	050117	
Morrison, Sandra	051922	
Morrissey, Dr. Kristine	048441	
Morrissey, Tom	050307	
Morrow, David	046501	
Morrow, Frances	046378	
Morsch, Prof. Dr. Günter	021000, 021003	
Morsch, Nathalie	005431	
Morsch, Sigrid	020934	
Morscheck, Charles	052069	
Morscher, Dr. Lukas	002130, 002137	
Morse, Carla	005974	
Morse, F.E.	047734	
Morse, Paula	049608	
Morseth, Kari	033188	
Morsiani, Paola	049574	
Mort, Mark	050186	
Moś, Ignacy	035129	
Mosbrugger, Prof. Dr. Volker	018447	
Moscatelli, Daniel	000521	
Moschig, Günther	002182	
Mosco, Dr. Marilena	026270, 026275	
Mosconi, Prof. Andrea	026098, 026103	
Moscoso Möller, M.A. Fernando	023372	
Mosdal, Brita	010084	
Moseley, Bruce M.	051773	
Mosena, David R.	047681	
Mosenfelder, Mrs. Robert	048602	
Mosenstine, Ken	047388	
Moser-Gautrand, Claire	012235	
Moser, Audrey	048370	
Moser, Dr. Bernd	001984	
Moser, Hans-Rudolf	041781	
Moser, Dr. Juliane	018331	
Moser, Kenneth	047253	
Moses Nesmith, Cynthia	047914	
Moses, Stephen A.	051292, 051293	
Moses, Vincent	052596	
Moshenson, Edna	025212	
Mosica, Phillippe	015496	
Mosier, Milton	053575	

Moskovaja, Ljudmila Ivanovna	037613, 037614	
Moslavac, Slavica	008943	
Mosler, Dr. Bettina	019693	
Mosneagu, Marian	036401	
Mosquera, Lic. Maria	000264	
Moss, Karen	051544	
Moss, Kenneth	051429	
Moss, Michael E.	054322	
Moss, Rick	051674	
Moss, Dr. Roger W.	052030	
Mossman, Kate	005851	
Mostegl, Sabine	002945	
Moster-Hoos, Dr. Jutta	020985	
Mostert, Antoinette	031840	
Mothe, Florence	014974, 014975	
Mothe, Hubert	013594	
Motlotle, Ntikwe Pene	004298	
Motornaja, Larisa Ivanovna	037548	
Motoyama, Yoshiao	029529	
Motre, Doina	036332	
Mott, Diane	053009, 053033	
Mott, Joyce	050229	
Mottart, Eugène	003938	
Motte, Vic	004056	
Mottram, Graham	046218	
Mottrecht, Uta	022828	
Motts, Warren E.	049288	
Motyka, Jerzy	034697	
Motyka, Maria	034921	
Mouche, Alexandre	041542	
Moudry, Mary Lou	047175	
Mouely-Koumba, Marie-Agnès	016382	
Mougel, René	013353	
Moughdad, S.	042255	
Mougin, Denis	013641	
Mould, Laurel	005997	
Moulias, Christos	023217	
Moulin, Hélène	016093	
Moulin, Pierre	011796	
Moult, Lynne	044810	
Mounier, Bernard	015900	
Mountfort, Dick	046984	
Moura Delphim, Angelo Alberto de	004533	
Moura, Fernando	035779	
Mourão, R.	004600	
Mourlet, Gérard	015624	
Mousley, Patricia K.	049562	
Moussa Nene, Djamil	007509	
Mousseigne, Alain	015987	
Moussellard, Renaud Pascal	012340	
Mousset, Jean-Luc	030593	
Moutashar, Michèle	011731	
Moutsopoulos, Thanassis	023030	
Mouw, Christine	054291	
Movillon, Mario M.	034498	
Movsisyan, Vergine	000734	
Mowat, Dr. P.	045438	
Mowll Mathews, Nancy	054420	
Mowrey, Debbie	046553	
Mowry, Robert	047362	
Moxeme, Oscar O.	052282	
Moxley, Richard W.	052541	
Moy, Henry	049652	
Moye, L.M.	050527	
Moye, Matthew M.	050527	
Moyes, Nick	044082	
Moyle, N.	045512	
Moyle, Nicola	045511, 046215	
Moynes, Debbie	005327	
Moyroud, Raymond	012200	
Mpouna Coline, Samuel	023385	
Mrabet, Ridha	042765	
Mrachacz, Maria	019712	
Mravlincić, Ivan	009006	
Mraz, Dr. Gottfried	003001	
Mroczkowski, Dennis P.	048847	
Mroski, Anne	017118	
Mrozinski, Mary Anne	049782	
Mrugalska-Banaszak, Magdalena	035102	
Mrva, Dr. Marián	038486	
Mrva, Dr. Marian	038341	
Mrvová, J.	038486	

Msemwa, Dr. Paul J.	042630	
Msimanga, A.	054847	
Mtshiza, Zola	038687	
Mu-Liepmann, Véronique	016380	
Muceus, Cheryl	046313	
Muceus, Cheryl L.	046311	
Muchametčina, Lejla Munirovna	037153	
Muchanov, Gennadij Stepanovič	037119	
Muchitsch, Hr. Dr. Wolfgang	001981	
Mucnjak, Prof. Alois	002964	
Mudd, Douglas	047889	
Mudd, Margaret	050961, 050962	
Mudd, Michael	049629	
Muddiman, Susi	001331, 001611	
Mühlbacher, Anton	001961	
Mühlbacher, Peter	021331	
Mühlbauer, Heinrich	019953	
Mühlberger, Dr. Kurt	002936	
Mühldorfer-Vogt, Ch.	021088	
Mühlegger-Henhapel, Dr. Christiane	003059	
Mühlemann, Lorenz	042065	
Mühlemann, Yves	041417	
Mühlenbrock, Dr. Josef	019219	
Mühlhäußer, Werner	018898, 018899	
Mühlhausen, Priv.-Doz. Dr. Walter	019142	
Muehlig, Linda	051626	
Mühne, Dr. Christian	021139	
Mührenberg, Dr. Anke	019515	
Mülhaupt, Freya	017085	
Müllejans-Dickmann, Dr. Rita	018605, 019175	
Müller-Bahlke, Dr. Thomas	018954	
Müller-Gaebele, Erich H.	018527	
Müller-Gerbl, Prof. Dr. med. Magdalena	041260	
Müller-Späth, Esther	019292	
Müller-Tamm, Dr. Pia	018032, 019517	
Müller-Westermann, Dr. Iris	041021	
Müller, Alois	041582	
Müller, Andreas	018492	
Müller, Dr. Anette	022911	
Müller, Arndt	016598	
Müller, Astrid B.	017178	
Müller, Berit	022773	
Müller, Brigitte	041746	
Müller, Bruno	001831	
Müller, Dr. Christian	041273, 041274	
Müller, Dr. Claudius	020579	
Müller, Conrad	018533	
Müller, Erwin	020401	
Müller, Eugen	002312	
Müller, Prof. Dr. Felix	041315	
Müller, Mag. Florian Martin	002118	
Müller, Gabriele	017601	
Müller, Gerda	018896	
Müller, Giesela	018462	
Müller, Dr. Gino	041657	
Mueller, Dr. Gordon H.	051299	
Müller, H.	017515	
Müller, Hans	002926	
Müller, Hans-Jörg	022734	
Müller, Dr. Hans-Peter	019974	
Müller, Heike	020993	
Müller, Heinz-Willi	016707	
Müller, Hermann	019827, 019829, 019830, 041349	
Müller, Prof. Dr. Irmgard	017438	
Mueller, Jane	047296	
Mueller, Jo	049842	
Mueller, John	050544	
Mueller, Judith	051401	
Müller, Dr. Jürg P.	041413	
Müller, Dr. Jürgen	019707	
Müller, M. A. Julia	022051	
Müller, K.	041812	
Müller, Karsten	018985	
Müller, Klaus	017533	
Müller, Dr. Klaus-Peter	020986	
Müller, Dr. Maria	018032	
Müller, Prof. Dr. Markus	020598	
Mueller, Prof. Marlene	054249	

Müller, Dr. Martin	018646, 018651	
Müller , Dr. Martin	022837	
Müller, Matthias	019299	
Müller, Dr. Norbert	002585	
Müller, Dr. Peter	023540	
Müller, Peter	022091, 022093	
Müller, Roland	041467, 041730	
Müller, Rudolf	042133	
Müller, Sabine	018464, 019061	
Mueller, Shirley	054590	
Müller, Dr. Siegfried	017433, 020987	
Müller, Stefan	020470, 020474	
Müller, M.A. Stefanie	020249	
Müller, Prof. Telmo Lauro	004841	
Müller, Dr. Thomas	019407	
Müller, Thomas	020014, 042186	
Müller, M. A. Thomas T.	020483, 020484, 020485, 020486, 020488	
Müller, Dr. Volker	022330	
Müller, Walter	016462	
Müller, Werner	018032	
Müller, Prof. Dr. Werner	021489	
Müller, Wolfgang	017949	
Müller, Zsuzsanna	040998	
Müllner, Dr. Jenő	023485	
Muench Navarro, Pablo	031621, 031624, 031628	
Münch-Severin, Kornelia	017119	
Münch, Dipl.Lehrerin Annegret	016588	
Münch, Eugen	022229	
Münch, Dr. Roger	022363	
Muenkel, Harold	050553	
Münzner, Harald	019485	
Muertegui, Charles	034457	
Müsch, Dr. Imgard	022047	
Muffet, A.	045163	
Mufti, Uxi	034118	
Mugavero, C.J.	046407	
Muggeridge, Rosalind	051430	
Muheeb Zada, Mohammad Yahia	000005	
Muhl, Dr. Arnold	018965	
Muhlbauer, Mic	052274	
Muir, Eleanor	045872	
Muir, J. Duncan	047737	
Muirhead, Frank H.	044047	
Muirhead, George	045787	
Mujanović, Prof. Fehim	004289	
Mujta, Dr. Józef Stanisław	034842	
Mukai, Tetsuya	029455	
Mukai, Tomoko	029509	
Mukanova, Najlja Arystanovna	029799	
Mukasa, Josephine	024149	
Mukaševa, Šajzada Tlegenovna	029791	
Mukela, Manyando	054835	
Mukulu	031848	
Mulcahy, Michael	024947	
Mulder, Drs. D.	032503	
Mulder, Dr. E.W.A.	032213	
Mulder, Jenny	032147	
Mulholland, Hugh	043560	
Mulinas, Cristina	040305	
Mulk, Dr. Inga-Maria	040723	
Mulkey, Terry	051077, 051078	
Mull, Astrid	015102	
Mullally, Matt	052700	
Mullen, Beverly J.	047720	
Mullen, Karen	052514	
Mullen, Kathleen	052266	
Muller, Donald	047211	
Muller, Jean-Paul	014378	
Muller, Joe	049566	
Muller, John	048833	
Muller, Judith	051395	
Mullers Bast, G.J.	032409	
Mulligan, Dermot	024840	
Mullin, Amanda	052041	
Mullin, D.	043570	
Mullin, Diane	050936	
Mullin, Marsha A.	049453	
Mullins, James	024935	
Mullins, Lashe	052538	
Mullins, Margaret	024923	
Mullins, Rose	043672	
Mullins, Sam	045029	

Mullis, Connie 051029
Mulqueen, Grace 024950
Mulvaney, Rebecca 046300
Mulvany, Dr. John 047652
Mulvay, Roger 033250
Mummery, Brian 044630
Mun-lee, Lin 042540
Mun, Kyong Hyun 029914
Munakata, Pariji 028760
Munakata, Shinsaku 028830
Munalov, Temir 029789
Munch, Jytte 010338
Munch, Paul Bernard 013275
Munck, Jacqueline 014697
Munday, Graham K. 043703
Munday, Val 044293
Munder, Heike 042211
Mundorff, Angelika 018543
Mundus, Doris 020021
Mundwiller, Mollye C. 049452
Mundy, James 052338
Munk Pedersen, Hans . 010067, 010068,
                                    010070
Munk, Jan 009913
Munk, Dr. Jan 009908
Munksgaard, Jan Henrik 033718
Munktell, Ing-Marie 041115
Munns, Judith 053363
Muno Jordan, Iris 048481
Muno, Ed 051732
Muñoz-Alonso, Laura 017098
Muñoz, Adriana 040663
Muñoz, Concha Carmona 039726
Munro, Dr. David M. 044725
Munro, Gale 054170
Munro, J.A. 043842
Munro, Lewis 044612
Munroe, Alexandra 051411
Munshi, Prof. Supriya 024158
Munsick, Lee R. 051906
Munson, Larry 053301
Munson, Robin 049033
Munstermann, Leonard 051265
Munt, Martin 045734
Muntean, Ioan 036696
Muntean, Vasile 036542
Munteanu, Simona . . 036380, 036386
Munteanu, Vasile 036513
Munthe, Adam 045109
Muntildeoz Rivera, Lorena 008620
Munyer Branyan, Cheryl 054183
Munyikwa, D. 054847
Munzenrieder, Claire 053116
Muolo, Vicenzo 026890
Mur, Pilar 039149
Mura-Smith, Collin 051489
Mura Sommella, Dr. Anna 027553,
                    027573, 027637, 027658
Murai, Yasuhiko 028961
Murakami, Isamu 028652
Muramatsu, Masafumi 029559
Murariu, Dr. Dumitru 036315
Murase, Masao 028543, 029590
Murat de Chasseloup-Laubat,
   Prince 014016
Murat, Francine 023391
Murata, Tetsuro 028991
Muratore, P. Umberto 028024
Murauchi, Michimasa 028581
Murch, Cliff 006158
Murchie, John 006841
Murchison, Morag 045687
Murdin, Alexander 043684
Murdoch, John 053053
Murdock, Gordon R. 050940
Murdock, Helen 001620
Murenina, Galina Platonovna . . 037845
Murgueítio de Montoya, Lucy . . 008660
Murillo Toro, Edificio 008633
Murken, Prof. Dr. Jan 021049
Murko, Matthias 020871
Murlasits, Mag. Elke 001968
Murphey, F. Warren 049808
Murphy, Alistair 044046

Murphy, Anne 052261, 052819
Murphy, Arlene 051050
Murphy, Beth 050934
Murphy, D. 044267
Murphy, Emily 052881
Murphy, Dr. Jack A. 048261
Murphy, Joe 024958
Murphy, John 044118
Murphy, Kenny 046745
Murphy, Patricia 051692
Murphy, Patrick 046909, 047956
Murphy, Patrick T. 024867
Murphy, Paula 046427
Murphy, Randy 054601
Murphy, Rosalinde 024898
Murphy, Russell 001529
Murphy, Shawna 047714
Murphy, Sue 046692
Murphy, Tony 044431
Murphy, Tricha J. 024860
Murphy, Valary 043548, 045022
Murra-Regner, Georg 017945
Murray, Alisdair 046108
Murray, Ann H. 051642
Murray, Graeme 043353
Murray, Janice 046226
Murray, Jeffrey 049234
Murray, Johanne 005344
Murray, Karin G. 053927
Murray, Kay 024912
Murray, Linda 049319
Murray, M. 005955
Murray, Mandy 043355
Murray, Mary 051013, 053916
Murray, Dr. Peter 000789
Murray, Peter 024858, 046086
Murray, Roddy 045888
Musa, George 005491
Musa, Peter 005294
Musameh, Abdulrahman S. . . . 003210
Musango, Simon Peter 043108
Muscă, Luciana 036751
Muscă, Mihai 036218
Muschwitz, Tanja 017059, 022331
Muscolino, Celly 027486
Mušeţeanu, Dr. Crişan 036314
Musgrave, Lee 053629
Musil-Jantjes, Angela 033840
Mušírovski, Volodymyr
   Makarovyč 043288
Muškalov, Sergej Michajlovič . . 037182
Musketa, Dr. Konstanze 018967
Mussari, Antonio 027012
Musser, Bernice 048234
Musser, Hugo 000316
Musset, Daniele 013987
Mustaine, Beverly 050091
Musto, David F. 051265
Musty, Tony . . 044069, 044331, 044495,
                    045464, 045545
Muswasi, P. 040507
Musy-Ramseyer, Sylviane 041407
Muszała-Ciałowicz, Joanna . . . 035163
Muth, Dr. Wolfgang 020156
Muxeneder, Therese 002937
Muya Wa Bitanko, Dr. Kamuanga 008840
Muzii, Dr. Rossana 027015
Mužík, Jiří 009594
Mveng, Engelbert 005297
Mvinjelwa, A. 038858
Mya, U San 031828
Mychajlunow, Lorraine 005745
Mycock, Chris 045264
Myers Jr., Barry W. 052745
Myers, Dr. Arnold 044264
Myers, Arnold 044270
Myers, Bob 046369
Myers, Dr. Charles W. 051327
Myers, Jane 048898
Myers, Kenneth 054116
Myers, Mark 044522
Myers, Maryann 048421
Myers, Mickey 053560

Myers, Pam 050105, 050106
Myers, Pamela L. 046550
Myers, Prof. Philip 046493
Myette, Ellen M. 054158
Myhrvold, Sissel 033840
Myllyharju, Taina . . . 011217, 011409,
                                    011410
Myllylä, Pälvi 011480
Myllymaa, Marju 010768
Mylona, Alex 023005
Mylona, Zoe A. 023321
Myltygaševa, Ljudmila Petrovna 036838
Mylyadi, Dr. Wulan Rujiati 024564
Myon, Kim Ik 029873
Mytton-Famié, Elisabeth 012829
Myzgina, Valentina Vasylivna . . 043132
Myzina, Larisa Ivanovna 037125
N., N. 011138, 017601, 018647, 019700,
      020021, 020577, 020853, 020898,
      039708, 051434, 051434, 053962
N.N. . 019693, 020273, 047253, 047253,
                                    047253
Naab, Michael 049945, 049946
Naar, Harry I. 050192
Nabačikov, Dr. V.A. 037379
Nabaura, Trudy 025076
Nabot, Anna 025397
Nachama, Prof. Dr. Andreas . . 017307
Nachtgäller, Roland . . 019203, 020833
Nacimento Brito, M.F. do 004721
Nacke, Josef 018688
Nacking, Åsa 040811
Nad-Jerković, Erika 008891
Nadaillac, Jean de 014978
Nadan, José 013631
Nadel, D. 025058
Nader, Georges S. 023394
Nadkarni, Dnyaneshwar 024405
Nadler Thomas, Genny 046820
Nadler, Dr. Rajaa 018271
Nadyrova, Galja O. 029793
Nadzeika, Bonnie-Lynn 051061
Naef, Weston 050436
Naegel, Marie-Luise 041928
Nägele, Dr. R. 022067
Näse, Li 011004
Näslund, Erik 040996
Naess, Jenny-Rita 033972
Naeßl-Doms, Vinzenz 020327
Näther, M.A. Ulrike 017631
Nätscher, Dr. Wolfgang 021077
Nafziger, Helen 046528
Nagai, Akio 028652
Nagai, Taizan 028591
Nagai, Yoshitaka 028627
Nagasaki, Motohiro 029211
Nagashima, Kosuke 028748
Nagel, Prof. Günter 020544
Nager, Hans-Werner 041221
Naghawy, Aida 029770
Nagiełło, Zbigniew 029459
Nagl, Michaela 002938, 003037
Nagorskij, Nikolaj Viktorovič . . 037746,
                                    037799
Nagueira, Dr. Lenita W. Mendes 004383
Nagy-Molnár, Dr. Miklós 023716
Nagy, Dr. András 023459
Nagy, Dr. Aniko S. 023515
Nagy, Dr. Árpád 023553
Nagy, Endre 023577
Nagy, Dr. István 023553
Nagy, László 023815
Nagy, Michael 046632
Nagy, Rebecca 049004
Nagy, Veronika 023688, 023692
Nagy, Dr. Zoltán 023752
Nagyné-Martyin, Emilia 023673
Naheb, Abél 011632
Nahon, Pierre 016151
Nahser, Dr. Siegmar 017112
Nairne, Andrew 045453
Nairne, Sandy 045060
Naish, Carolyn 045018
Naito, Makio 029418

Najafi, M. 024737
Naji Sari, Ahmed 054831
Nakabayashi, Kazuo 029608
Nakabe, Yoshitaka 029146
Nakabo, Tetsuji 028953
Nakagawa, Hajime 029369
Nakagawa, Misao 029639
Nakahara, Yusuke 028843
Nakajima, Prof. Jimi Naoki . . . 004940
Nakamura, Hiroshi 029311
Nakamura, Hitoshi . . . 028988, 029246
Nakamura, Junsuke 029151
Nakamura, Kenjiro 029593
Nakamura, Makoto 029305
Nakamura, Minoru 029565
Nakamura, Nobuo 028826
Nakamura, Reiko 029608
Nakamura, Saihachi 028505
Nakamura, Satoru 028507
Nakano, Masaki 028557
Nakano, Seijiro 029474
Nakano, Yuzo 046937
Nakashima, Kunio 028475
Nakashima, Michiko 029121
Nakazawa, Eigorou 028842
Nakrem, Prof. Hans Arne 033847
Naksuetrong, Somporn 042707
Namiki, Masashi 029129
Nan, Baosheng 008180
Nance, Krista 051191
Nance, Paul J. 050367
Nanfeldt, Philippa 040652
Nanni, Dr. Romano 028398
Nanni, Romano 028400
Nannup, Brett 000999
Nanu, Dan Basarab 036472
Nanzer, Angel Omar 000590
Naoki, Togawa 029511
Napier, Ellen B. 052047
Napier, John H. 051024
Náplavová, Dr. Marie 009931
Nápoli, Prof. Dr. José 000216
Napombejra, Sandy 053836
Naqshbandi, Ghulam 000001
Naqvi, Prof. Asif A. 024133
Naramore, Bruce E. 047802
Narbonne, Peter J. 052311
Narbutaite, Gražina 030468
Nardelli, Lucia 025585
Nardi, Christopher J. 048686
Nardin, A. 014719
Narinskaja, Natalija Efimovna . . 037702
Narkiewicz-Laine, Christian K. . 053192
Narr, Claudia 021678
Narro, Alicia 034229
Narzt, Karl 002684
Nasalski, Zygmunt 034997
Nasarov, N. 042624
Nasby, Judith M. 005934
Nascimento, Leo 004499
Nasi, Stefano 026928
Naski-Multanen, Marke 011458
Naski-Multanen, Marki 011460
Nason Davis, Robert 052643
Nason, Dr. James 053297
Naspleda i Arxer, Joan 039451
Nasset, Yvette 048696
Nassi, Talma 025039
Nasu, Takayoshi 029264
Nasybullin, Émir Salimovič . . . 037116
Natali, Dr. Antonio 026256
Nater, Sandra 042031
Nathan-Dorn, Anja 019692
Nathan, Gail 047225
Nathan, Jacqueline S 047161
Nathans, David 047843
Nathanson, Morris 052380
Nathiri, N. Y. 048461
Natier, Michel 013909
Natter, Dr. Tobias G. 001794
Naturel, Michèle 012505
Natuschke, Peter 016767
Natzer, Dr. E.-M. 020527
Naudascher, Josef 020216

Nauderer, Ursula Katharina 017795
Naukkarinen, Tiina 010840
Nault, Gilles 006948
Naumann-Steckner, Dr.
  Friederike 019711
Naumann, Gerd 021150
Naumann, Heike 017162
Naumann, Johannes 022152
Naumann, Peter 020843
Naumann, Rolf 019461
Naumann, M.A. Thomas 022407
Naumer, Sabine 019542
Naumova, Nadija Grygorivna .. 043202
Naupp, Thomas 002874
Nause, Jean Pierre 012301
Nauta, Dr. J.F 032557
Naval Mas, Antonio 039501
Navarro García, Ramón 039655
Navarro Guitart, Jesús 039581
Navarro Linares, Antonio 039460
Navarro, Anna Dolores 038962
Navarro, Daisy A. 034536
Navarro, Edward 050466
Navarro, Elías Terés 039021
Navarro, Dr. Fausta ... 026272, 026274
Navarro, Dr. Fausto 026260
Navarro, Mariana Varela 008600
Navas Gutiérrez, Mariano 008772
Navascués Benlloch, Pilar 039162
Navascués Benlloch, Pilar de .. 039630
Navascués, Pedro 039671
Nave, Dr. Francine de 003295
Naveau, Jacques 013336
Naveaux, Ralph 051000
Navia, Olga Lucía 008649
Nawafleh, Sami 029783
Nawka, Tomasz 016978
Nawrocki, Czesław 034619
Nawrot, Anna 034990
Nawroth, Dr. Manfred 017245
Naylor, Dr. Bruce G. 005719
Naylor, Eleanor 053523
Nazarov, Roman Grygorovyč .. 043144
Nazarov, Temirchan 029852
Naziebło, Dr. Jerzy 035472
Nazionale, Nina 051455
Ndacyayisenga, Albert 038088
Ndege, Conchita F. 049255
Neacşu, Epraxia 036234
Neagu, Gheorghe 036413
Neagu, Marian 036341
Neal, George 049379
Neal, Dr. John R. 047716
Neale, Susan 006641
Nealis, Sharon 048089
Néar, Marie-José 016033
Neault, Caroline 047928
Neave, R. 044164
Nebel, Dr. Martin 022056
Nebes, Prof. Dr. Norbert 019453
Nebl, Rudolf 002720
Nebuloni, Dr. Gabriela 026589
Nečajeva, Praskovija Mychajlivna 043131
Nečasová, Eva 009449
Necker, Hans 016809
Necker, Louis 015947
Nedelčeva, Cvetanka 005205
Nédélec-Bellevenue, Diane 014142
Nedeljković, Dragana 038245
Nedelkovska-Dimitrovska, Liljana 030630
Nédellec, François 014229
Nedeva, Galja 005268
Nedoma, Petr 009795
Neeb, Vanessa 053759
Needham, Dr. Claude 051976
Needham, Melanie 007405
Neel, David 049211
Neely, Alastair 006211
Neergaard, Mogens de 010204
Nees Holcomb, Sara 050242
Neff, Emily 049588
Neff, Jean W. 046455
Neff, Jerry W. 050327
Nefzi, Tarak 042736

Negrisin, Dr. Barbara 026983
Negro Cobo, Marta 039169
Negro, Dr. Angela 027566
Neich, R. 033142
Neidert, David 046475
Neidhart, Herbert 002519
Neidigh, Ronald 048484
Neighbors, Dale 051123
Neighbours, Dr. K.F. 051519
Neil, Erik 051304
Neil, Ph.D. Erik H. 049621
Neill, Teda W. 048801
Neilsen, Henry L. 054076
Neinhuis, Prof. Dr. Ch. 021138
NeJame, Art 048233
Nejdl, Josef 009540
Neka, Suteja 024561
Nekrasov, Aleksandr Michailovič 037677
Nekrasov, Sergej Michailovič .. 037826
Nel, M. 038660
Nelgares Raye, José 039522
Nelhams, Mike 045997
Nelis, Laurence 003613
Nelišerová, Dr. Eva 038327
Nelje, Kerstin 040804
Nelley, Betty 033244
Nellis, Mike 046683
Nelly, Catherin 014934
Nelsen, Rick 053695
Nelson-Balcer, Mona 054428
Nelson, Amanda 049715
Nelson, Bob 052633
Nelson, Bryan 005925
Nelson, Cindy J. 049926
Nelson, Cortland 047145
Nelson, Daniel A. 050143
Nelson, Darrel 048188
Nelson, Diane Bassette 049703
Nelson, Doug 048486
Nelson, Ellie 045447
Nelson, Eric 054601
Nelson, Esther 047972, 047973
Nelson, Geoff 046316
Nelson, Gwen 048334
Nelson, Harold B. 050371
Nelson, Henry 045318
Nelson, James R. 048008
Nelson, John 044678, 049310
Nelson, Julie D. 052433
Nelson, Dr. Larry 051996
Nelson, Lee Ann 049244
Nelson, Lori 006078
Nelson, Mike 053136
Nelson, Mindy Honey 051506
Nelson, Natalie 048162
Nelson, Philip 044678
Nelson, Ray 048071
Nelson, Rebecca 053656
Nelson, Robert 050419
Nelson, Sarah E. 046410
Nelson, Stephen R. 053058
Nelson, Susan 050204
Nelson, Taylor 047556
Nelsson, Sigvard 040850
Němeček, Jaroslav 009535
Nemes, András 023896
Nemes, András 023888, 023892, 023894,
         023895, 023897, 023901
Nemeth Ehrlich, Dorica 009025
Nemeth, Alice 007426
Nemeth, Elaine 007144
Németh, Dr. Péter 023812
Nemetz, Norbert 022757
Nenciu, Paul 036242
Nenno, Dr. Rosita 020970
Nenonen, Yrjo 011445
Nenov, Dr. Nikolai 005152, 005153,
                     005155
Nenquin, Prof. Dr. J. 045900
Nentwig, Dr. Franziska 017116, 017151,
        017218, 017236, 017248, 017249,
        017252, 017253, 017254, 017260,
                     017293, 017302
Nenz, Cornelia 021967

Nenzioni, Gabriele 027764
Nepi, Dr. Chiara 026280
Nepočatych, Viktor Alekseevič . 036958
Nepomuceno, Susan 046864
Nepputh, Fritz 018129
Néray, Katalin 023508
Nerdinger, Prof. Dr. Winfried .. 016646,
                     020505
Neretniece, Anastasija 030151
Nerheim, Gunnar 033854
Neri, Joanne 046206
Nero, Dr. Donnie 054059
Nesbitt, Dr. Elizabeth 053229
Nesbitt, Melissa 053724
Neshushtan, M. 025122
Nesin, Jeffrey D. 050784
Nesje Bjørlo, Aslaug 033919
Nesme, H. 013501
Nesmith, Michael 053111
Nespoli, T. 032002
Ness, Dr. Arlin E. 046356
Ness, Gary C. 050678, 050682
Ness, Dr. Gary C. 051310
Nessell Colglazier, Gail 050628
Nesselrath, Dr. Arnold . 054679, 054682,
                     054695
Nessi, Daniela 041748
Nesta, Prof. Paolo 025973
Nestler-Zapp, Dr. Angela 016805, 016806
Nestorova, Ani 005002
Neszmelyi, Adrienne 048366
Neto, Luis 036028
Netta, Irene 020580
Nettekoven, Katharina 018032
Netter, Esther 050475
Netterlund, Karin 040975
Netuschil, C.K. 017825
Netzer, Dr. Nancy 047625
Netzer, Dr. Susanne 017205
Neu, Noreen 006766
Neubauer, Alexandra 018429
Neubert, André 019311
Neubert, Christhard-Georg 017280
Neubert, George W. .. 052949, 052949
Neubert, Hermann 020408
Neubert, Dr. Michaela 022801
Neuburger, Dr. Susanne 003050
Neuendorf, Petra 020778
Neuer, D. 019135
Neufeld, Alice 007395
Neufeld, Carol E. 053643
Neufeld, Helmut 005362
Neugebauer-Maresch, Dr. J.-W. 002835
Neugebauer, Rud. 016594
Neugent, Susan E. .. 046615, 046616
Neuhaus, A.F. 032552
Neuhaus, Daniel 041423, 041424
Neuhaus, Hanspeter 019583
Neuhoff, Stephan 019687
Neuhuber, Prof. Dr. W.L. 018257
Neuland-Kitzerow, Dr. Dagmar . 017237
Neulinger, Adolf 001904
Neulinger, P. Maximilian 002259
Neuman, David 046862
Neuman, Thom 046862
Neumann-Eisele, Petra 019573
Neumann, B. 045808
Neumann, Christoph 020461
Neumann, Dr. Dieter 002862
Neumann, Herbert 003059
Neumann, Timothy C. 048219
Neumann, Dr. U. 022222
Neumann, Ursula 051436
Neumayer, Dr. Heino 017245
Neumayer, Sonja 002793
Neumayer, W. 001709
Neuner, Ernst 003140
Neuner, Meinhard 002131, 002141
Neuner, Mag. Wolfgang 002132
Neuner, Wolfgang 002141
Neunzert, Hartfrid 019877, 019879
Neuwirth, Karin 043059
Neve, André 003519
Nevermann, Ulrike 019061

Neves, Julio 004879
Neves, Maria da Soledade S. .. 035728
Neveux, Yannick 014316
Neville Blakey, W.H. 043730
Nevin, Michael E. 054002
Nevraumont, Marcel 003648
Nevskaja, Galina N. 036912
Nevskij, Aleksej 037288
Newbery, Chris 045557
Newbery, David 045554
Newbery, Maria 045753
Newbury, Barbara 043678, 043799
Newcomb, Nancy 051455
Newhard-Farrar, Nancy 047833
Newhouse, Larry 054602
Newkirk, Leslie 053449
Newkirk, Sally 051219
Newland, Judy 046906
Newman, Bruce 048941
Newman, David 048309
Newman, Ed 005977
Newman, Lilly 052537
Newman, Megan M. 050269
Newman, R. D. 051319, 051349
Newnham, Kate 043747
Newsome, Mary 048397
Newsome, Steven C. 054087
Newton Gibson, Yasabel 051757
Newton, Bette 048105, 049609
Newton, Christina 052544
Newton, Gael 000937
Newton, Paula H. 048542
Nex, Jenny 045047
Neyer, Dr. Hans Joachim 019069
Neylon, Nicholas 051947
Neyses-Eiden, Mechthild 022202
Neysters, Dr. Silvia 018046
Nezu, Kohichi 029561
Nga, Dang Viet 054783
N'Gele, Edward O. 030289
Nghien, Nguyen Thi 054800
Ngoubeli, Joseph 008844
Ngozi, Niobe 049573
Nguyen, Phi 051405
Nguyên, T.P. 041511
Nhlane, Dr. Med 030654, 030610
Ni, Ming 008112
Nias, Peter 043697
Nica, Cristina Mihaela 036752
Nica, Viorica 036389
Nicandri, David L. 053651
Nicasius, Sandu 042636
Nichitoiu, Tinca 036603
Nicholas, Darcy 033286
Nicholas, Jeremy 046040
Nicholis, Alexandra 050737
Nicholls, David 046470
Nichols, Charlotte 049221, 053412
Nichols, Elaine 047918
Nichols, John 053136
Nichols, Lawrence W. 053760
Nichols, Madeleine 051456
Nichols, Marian 045168
Nichols, Sarah 052157
Nichols, William O. 053976
Nicholson, Claudia J. 051615
Nicholson, Joseph 052103
Nicholson, Marilyn L. 054007
Nicholson, Robert 024894, 024914
Nicht, Dr. Christoph 016654
Nichterlein, Peter 022581
Nick, Lloyd 046629
Nickel, Douglas 053824
Nickel, Manfred 021463
Nickell, Jeff 046719
Nickens, Harry 052604
Nickerson, Cindy 048030
Nickl, Dr. B.E. 017321
Nicklaus, Jörg 016589
Nickles, Karen 048480
Nicod, Annelise 041530
Nicodemus, Dr. Victoria 031844
Nicola, A. 032992
Nicolae, Nina 036624

Nicolai, Dr. Bernd 018939
Nicolai, David 051682
Nicolaisen, C. 010318
Nicolaisen, Dr. Jan 020003
Nicolas, Alain 014052
Nicolas, Brigitte 014969
Nicolas, Jean 011674
Nicolas, Jean-Marc ... 041752, 041756
Nicolau, Antoni 039090
Nicolau, Irina 036320
Nicolau, Sanda Gabriela 036808
Nicolaÿ, Louis-Jean de 013652
Nicolaysen, Katja 033684
Nicolescu, Costion 036320
Nicoletta, Catherine 012631
Nicolini, Michèle 016147
Nicolini, Renato 027656
Nicoll, Jessica 052283
Nicosia, Angelo 025348
Nicosia, Prof. Dr. Francesco ... 027869
Nicu, Mircea 036773
Nie, Fuhe 008251
Niederalt, Michael 021775
Niederer, Robert 041592
Niederhöfer, Hans-Jörg 022056
Niedermaier, Robert 021165
Niedzielenko, Andrzej . 034940, 034941
Niehaus, Dr. Andrea 017487
Niehoff, Dr. Franz ... 019883, 019884
Niehoff, Matthias 017314
Nield, Michael 045342
Nield, Thenetta 053356
Niell-LLabrés, Francisca 039882
Nielo Alcaide, D. Victor 039650
Nielsen, Anna Kathrine 010034
Nielsen, Anne-Marie 010188
Nielsen, Christin 040773
Nielsen, David R. 046392
Nielsen, Debbie 006852
Nielsen, Erik 010331
Nielsen, Ervin 010256
Nielsen, Hans 010303
Nielsen, Jan 051711
Nielsen, Johannes 010282
Nielsen, Jytte 010148, 010340
Nielsen, Keld 010019
Nielsen, Kim 054154
Nielsen, Kori L. 048034
Nielsen, Niels Jul 010178
Nielsen, Ole 010092
Nielsen, Poul Bryde 010320
Nielsen, Teresa 010353
Nielsen, Wendy 006799
Niemann, Dr. Carsten 019067
Niemann, Willi 020604
Niemeier, Dr. Jörg-Peter 021086
Niemeyer, Kirk 051728
Niemi, Anneli 011451
Niemi, Marja 011305
Nier, Eberhard 016942
Nieraad, Karla 022267
Nierengarten-Smith, Dr. Beej ... 046359
Niermann, Katrin 017524
Niesel, Bernd 018565
Niesner, Dr. Tomáš ... 009487, 009652
Nieswandt, Dr. Helge 020591
Nieszczerzewska, Elżbieta 035060
Nieuwdorp, Hans 030294
Nieuwstraten, Frits 032391
Nievera, Dr. Eduardo 034542
Nievergelt, Frank 042139
Nieves, Marysol 051329
Nieweg, J. 032507
Niewegłowski, Stanisław 035217
Niewerth, Herbert 022414
Niezabitowski, Michał 034890
Nigbur, MR Werner 017641
Niggemann, Ralf 022618
Niggemann, Dr. Stefan 019426
Niggli, Stefan 041571
Nigratschka, Kurt 019359
Nigro, Dr. Lorenzo 054682
Niinistö-Alanne, Arja 011245
Niizeki, Kimiko 029509

Nijazbekova, Sajrapan
  Jakubbekovna 030077
Nijkamp, J.H. 033010
Nijsten, J. 032299
Nikbakhsh, M. 024727
Nikiforov, Valerij Ivanovič 037484
Nikiforov, Viktor S. 043277
Nikita, Eleni S. 009422
Nikitović, Lidija 038188
Niklasch, Wilfried 017834
Nikočević, Lidija 008965
Nikolaev, V.A. 037719
Nikolaeva, Irina Alekseevna ... 037439
Nikolaeva, Lidija Dmitrievna ... 037541
Nikolaidis, J. 031769
Nikolajenko, Galina 043293
Nikolaou, Yorka 023023
Nikolasch, Dr. Franz 002389
Nikoleišvili, M.V. 016391
Nikolić, Dejan 038168
Nikolić, Dragan 038185
Nikolić, Dušica 038187
Nikolić, Jadranka 008902
Nikolić, Tatjana 038143
Nikolov, Prof. Dr. Vasil 005172
Nikolova, Asja 005044
Nikolova, Prof. Dr. Diana 005191
Nikolova, Erika 005110
Nikolova, Petja 005191
Nikoltsios, Vasilios 023300
Nikonov, Aleksandr K. 037244
Nill, Annegreth T. 047936
Nilsen, Rolf Erik 033624
Nilsen, Tore L. 033528
Nilson, Göran 040653
Nilsson, Bo 041014
Nilsson, Christer 033586
Nilsson, Christin 040739
Nilsson, Håkan ... 041181, 041183
Nilsson, Harald ... 041121, 041121
Nilsson, Isabella 040778
Nilsson, Karin ... 040739, 040742
Nilsson, Marianne 041062
Nilsson, Rune 040699
Nilsson, Torsten 040875
Nimac, Dr. Dragan 009018
Nimmich, Dr. Gunter 017133
Nimri, Mohamed 042785
Nimz, Tim 050340
Ning, Yinchang 008527
Ning, Yongqing 008086
Nini, Dr. Roberto ... 027024, 027025,
                              027026
Niniou-Kindeli, Vanna 023038
Nirša, Alla Mychajlivna 043269
Nisbet, A. 046214
Nisbet, Peter 047362
Nishi, Dr. Genjirou 029400
Nishi, Ken-ichiro 028549
Nishi, Yukitaka 028915
Nishida, Aquio 029025
Nishida, Dr. Hiroko 029561
Nishidera, Masaya 029427
Nishikawa, Kyotaro 029715
Nishimura, Atsushi 029703
Nishitakaesuji, Nobuyoshi 028522
Nishitani, Noboru 029572
Nishizawa, Jun-ichi 029359
Nishizawa, Keiichi 029209
Niskz, Siiri 011309
Nisman, Dr. V. 032002
Nissen, Carl 020537
Nissen, Dan 010171
Nissen, M.A.J.M. 032291
Nit, A. 032419
Nițǎ, Gabriela 036754
Nitsch, Hubert 002318
Nitsch, Dr. Ulla M. 017598
Nitta, Hideo 029021
Nitter, Marianne 033972
Nittmann, Karl 001874
Nittve, Lars 041021
Nitzsche, Dr. Gunther 022122
Nitzschke, Katrin 017976

Niu, Lisuo 008212
Niven, Catharine 044641
Nivière, Marie-Dominique 011539
Nix-Hauck, Nicole 020731
Nix, Dr. J.C. 032825
Nix, M. 044522
Nixon, Chris 048354
Nixon, Isabelle 006161
Nixon, James D. 048354
Nixon, Prof. Jim 001142
Nixon, Nigel 043377
Nixon, W.A. 001200
Nizamiev, Muzip Tazievič 037108
Nizegorodova, Elena
  Vladimirovna 036896
Nko'o, Hulio 005298
Noack, W. 017888
Noack, Wita 017225
Noah, Randy 053208
Nobis, Dr. Norbert 019066
Noble, Jo 047352
Noble, Phil 048789
Nóbrega, Enrique 054724
Noda Gómez, Onaysi 009146
Noda, Renate 003046
Nodges, David 044591
Nodine, Jane 053433
Noe, Adrianne 054146
Noe, Per ... 010246, 010247, 010248,
                              010265
Noeding, Tom 053675
Nöhammer, Josef 001928
Nöhmayer, Andreas 002299
Noel, Albert 003524
Noël, Étienne 015061
Noël, Jacques 014414
Nørager Pedersen, Erik 010005
Nørbach, Lars Chr. 010239
Nørbach, Lars Christian 009998
Nørballe, Inge 010065
Nörenberg, Bernd 016504
Nöth, Dr. Stephan 016956
Noetinger, Robert-Mare 014841
Noever, Prof. Peter 002998
Noever, Peter 003033
Noguchi, Mitsunari 028582
Noguchi, Reiichi 029509
Nogueira da Costa, Maria Inés . 031807
Nogueira, Abrelino 033144
Nogueira, Sandy 033144
Nohsoh, Toshio 028556
Noirot, André 012148
Noko, Oarabilke J. 004297
Nolan Jones, Nancy 047805
Nolan, Edward V. 053651
Nolan, John 049267
Nolan, Julia 044010
Nolan, Patrick B. 049637
Nolan, Peg 050670
Noland, Dr. Martin R. 052795
Nold, Carl ... 047115, 048652, 049132,
        051276, 052310, 052315, 053168,
                      053398, 053399
Noli, Gino 031841
Noll, Linda M. 049411
Noll, Roger 046862
Nollar, Philip 053919
Nolle, Karl 017995
Nollée, Armand 003264
Nollert, Dr. Angelika 020873
Nolley, Lance 050336
Nolte-Fischer, Dr. Hans-Georg .. 017829
Nolte, Joseph 047491
Nombluez, Pieter 003690
Nome, Alexander 046412
Nommensen, Dr. Sven 017558
Nomura, Harukata 028973
Nonesteid, Mark 048424
Nong, Wentian 047969
Nonnemacher, Eugen 017457, 017458
Noon, Patrick 050944
Noon, Rosemary 054348
Nooter Roberts, Mary 050422
Noppe, Catherine 003890

Norberg, Deborah 053043
Norby, Lory 050592
Nordal, Bera 040949
Nordberg, Henri 011430
Nordberg, Olle 040805
Nordbrock, Carol 046673
Norden, Linda 047362
Nordgren, Sune 033843, 033845, 033846
Nordheim, Lutz von 022861
Nordiguian, Levon 030266
Nordin, Maxine 006003
Nordine, Dr. Malki 000071
Nordlinder, Gunnar 041032
Nordlund, Ivar 011361
Nordmo, Per E. 033636
Nordmo, Sverre 033622
Nordon, Deborah 051433
Nordqvist, Jørgen 010024
Nordstrand, Polly 048258
Nordström, Lena 040900
Nordstrom, Alison 052617
Nordstrom, Ralf 011469
Nordtorp-Madson, Shelly ... 052853
Norell, Dr. Mark A. 051327
Norgaard Larsen, Peter 010195
Norheim, Kari 033630
Norheim, Svein 033630
Noriega, Natividad 034586
Noring, Ann-Sofi 041021
Norkute, J. 030346
Norland, Lissie 033618
Norley, Mark 049864
Norlund, Yvonne 040977
Norman, Dr. David B. 043848
Norman, Heather L. 049021
Norman, Jan 049286
Norman, Margaret Jane 054358
Normann, Mark 045448
Normann, Øystein 033640
Normark, Ben 046452
Norovtseren, Oyuntegsh ... 031744
Norrington, S. 001543
Norris, B. 033168
Norris, Baden N. 033173
Norris, Johnny 047177
Norris, Dr. Patrick 049860
Norseng, Per 033712
Norseng, Per G. 033852
Northam, J. 006753
Northcote, P.S. 033319
Northcutl, Ginia 050391
Northrop Adamo, Amy 051389
Northwood, John 001296
Norting, Jack 049201
Nortmann, Dr. Hans 022202
Norton Moffatt, Laurie 033537
Norton, Elaine 047382
Norton, Fran 053595
Norton, Frances 045127
Norton, George 054351
Norton, I.S. 001215
Norton, Jessica A. 046455
Norton, Kelly 049114
Norton, Lyle 050110
Norton, Maria 051132
Norton, Mary Lynn 052625
Norton jr., Richard W. 053323
Norton, Scott L. 048837
Norton, Susan E.S. 054110
Norvaišas, Ignas 030506
Norvell, Edward 052905
Norwood, Nancy 052618
Nory, Jean-Marc 015203
Noryńska, Barbara 035200
Nose, Dr. Martin 020561
Nosek, Elizabeth 050100
Nosič, Dr. Stipe 008907
Noskova, Alla Anatoljevna ... 037131
Nostrala, Justin 049698
Nota, Dr. Maria Teresa 027578
Notari, Bernard 013013
Notet, Jean-Claude 014241
Nothnagl, Hannes 002422
Nothnagl, Reinhold 002767

Notin, Véronique 013816, 013821
Notley, Ben 045506
Nottage, James H. 049681
Notteghem, Patrice 013624
Notter, Eveline 041519
Nougarède, Martine 041504
Nougué, Cendrine 014251
Nouvel, Odile 014733
Nouyrit, H. 012265
Novak, Diane 052501
Novak, Gabriela 002446
Novák, Dr. László 023806
Novák, Dr. Pavel 009610
Novak, Stanislav 008903
Nováková, Dr. Lenka 009468
Novakovska, Nada 030622
Novellis, Mark de 046011
Novershtern, Prof. Avraam ... 025211
Novgorodova, Prof. M.I. 037343
Novick, Leslie 048506
Novikov, Timur 037798
Novo, Marla 053101
Novoselova, Rimma
  Vjačeslavovna 037639
Novotná, Dr. Mária 038379
Novotná, Mariana 038374
Novrie, Robert 049468
Nowack, Meg 054178
Nowacki, Janusz 035115
Nowacki, Marek 035239
Nowak, Cornelia 018235
Nowak, Holger 019449, 019451, 019459
Nowak, Joanna 035172, 035236
Nowak, Romuald 035434
Nowakyniak, Dr. Edmund 035075
Nowobilski, Dr. Józef Andrej ... 034878
Nowotny, Erich 002247
Noyes, Julia 050299
Nozaka, Motoyoshi 028614
Nozet, Hervé 016068
Ntisana, N. 038792
Ntsema Khitsane, Julia 030287
Nu, Li 008112
Nuber, Dr. W. 020619
Nuculov, Georgi 005138
Nudelman, Robert 050432
Nüesch, Ernst 041258
Nüesch, Rosmarie 042051
Nützel, Dr. Alexander 020561
Nugent, M. 005833
Nuki, Tatuto 028756
Nummelin, Esko 011265
Nummelin, Rauni 011390
Numonova, Gulchra Tojitinovna 042618
Nunes de Vasconcelos junior,
  Walter 004745
Nunes de Oliveira, Adilson ... 004446
Nunes, Dave 052248
Nunes, Dr. Graça Soares 036090
Nunes, Isabella 004681
Núñez Centella, Ramón 039314, 039316
Núñez, César Marku 040385
Nunge 012488
Nunley, John 052818
Nunn, Chris 045927
Nunn, Dr. Tey Marianna 053115, 053116
Nunziati, Anna 025775
Nupen, Nikki 031850
Nurabullaeva, Svetlana 054635
Nurdan Drignath 017451
Nurfeisova, Nurshamal
  Bekvossunovna 029801
Nurger, Heli 010627
Nuri, Kedir 010730
Nurmi-Nielsen, Anna 011305
Nurmi, Virpi 011305
Nursan, Alimbaj 029796
Nurtaev, Kaken Bejsenbekovič . 029849
Nurullayev, Mansur 054622
Nusbaum, Pat 048575
Nussbaum, Felix 021013
Nussbaum, Patricia .. 041769, 041823
Nussbaum, Prof. R.A. 046493
Nussbaumer, Christiana 002822

Nußbaummüller, Winfried 001793
Nutkins, Geoff 045789
Nutt, Ruth J. 053248
Nuttall, R. J. 038661
Nuttall, Rick 038663
Nutting, Ryan 052720
Nutzi, Hans 041706
Nuyttens, M. 003423
Nwaneri, Justin C. 033436
Nyakas, Dr. Miklós 023682
Nyamabondo, Constantinus M. . 042625
Nyamenya, Peter 029861
Nyblod, Ruth Ann 046541
Nybom, Jeny 011450
Nyce, Doug 054429
Nye, Steve 045628
Nye, Tom 047227
Nyegaard, Georg 023337
Nyein, U Kyaw 031819, 031823
Nyerges, Alexander Lee 048170
Nyffenegger, Katharina 041318
Nygaard, Jon Erik 033811
Nygaard, Knut M. 033596
Nygårds, Eva 040652
Nygren, Kris 005830
Nykänen, Veijo 011278
Nykolyshyn, Z. 007447
Nylén, Lars 041129
Nyman, Ola 041013
Nyman, Sven-Eric 041104
Nys, Wim 003319
Nystad, Marianne 034015
Nyström, Hazze 010165
Nysveen, Aagot 049476
Nyunt, Kyaw 031835
O'Connor, David 000762
O'Barr, James B. 048059
O'Connor, Michael J. 048585
O'Neill, Heidi 051379
O'Reilly, Margaret 053799
Oaklander, Christine L. 046408
Oakley, Lisa 050010
Oakley, Steve 047031
Oaks, Martha 049136
Oalka-Konieczko, Małgorzata . 034903
Oandasan, Crescencion 034486
Oanh, Phan 054808
Oates, Sarah 046580
Oathout, Kathleen C. 052861
Oatley, Val 045993
Oatman, Michelle 051236
Obal, Franc 038581
O'Bannion, Tami 050874
O'Baoill, Eilis 045381
Obedat, Nassim 029758
Oberchristl, Monika 002323
Oberer, Brigitte 002419
Oberfell, J. 022067
Oberfrank, Wilhelm 002347
Oberg, Paul Chancy 033479
Oberhänsli, Silvia 041475
Oberhänsli, Dr. This 041715
Oberhaus, Betty 048064
Oberhausen, B. 022067
Oberholenzer, Gunther 002207
Oberlin Steel, Virginia 047380
Obermark-Stiller, Dr. Folke ... 022824
Obermeier, Christoph 020732
Obermeyer Simmons, Helen .. 048738
Obernosterer, Beatrix . 002185, 002198
Obhof, Dr. Ute 019503
Oblin, Annie 013807
Obojes-Rubatscher, Johanna ... 002464
O'Bomsawin, Nicole 006539
Obradović, Slobodan 031747
Obrascova, Maja Aleksandrovna 037708
Obrecht, Elsa 042006
O'Brian, John 048792
O'Brien, Barry 041486
O'Brien, David 024969
O'Brien, Eileen 050628
O'Brien, Freda 033368
O'Brien, James 054338
O'Brien, Kevin 050060

O'Brien, Maureen 052387
O'Brien, Michael J. 047901
O'Brien, Patrick J. 053955
O'Brien, Ruairí 017975, 017979
Obrist, Hans-Ulrich 014697
O'Bryen, Antoinette 024862
Ocaña i Goma, M. Teresa 039102
Ocaña, Maria Teresa 039103
O'Carroll, Helen 024994
Ocello, Claudia 051498
Ochi, Catherine Megumi 029500
Ochi, Yujiro 028843
Ochoa, Cristina 050467
Ochonko, Nikolaj Anatoljevič . 037899
Ochsenmeier, Peter 018566
Ochsner, Henrik 010005
Očirova, Tsiren-Handa V. 037988
O'Connell, Daniel M. 052181
O'Connell, Michael 024851
O'Connell, Shirley 052106
O'Connor, Prof. Barry M. 046493
O'Connor, Emma 044819
O'Connor, Karen 048779
O'Connor, M. 033282
O'Connor, Thomas P. 024876
O'Connor, Tony 046045
O'Conor Nash, Marguerite ... 024846
Odavić, Djordjo 004286
Odawara, Kaname 028914
Odden, Tor 033895
Odding, A. 032568
Ode, Robert 012351
Odegaard, Nancy 053823
Odehnal, Petr 009939
O'Dell, Kathryn 054285
Odendaal, E.A. 038838
Odenhoven, Dr. S.L. 032914
Oder, Karla 038598, 038599
Odermatt, Eduard 041195
Odess, Daniel 048662
Odin, Samy 014711
Odoj, Romuald 035224
O'Donnell, Carolyn 053103
O'Donnell, Janice 052391
O'Donnell, John 048216
O'Donnell, Molly 049805
O'Donnell, P. 045438
O'Donnell, Robert M. 050649
Odrowąż-Pieniążek, Janusz ... 035347
Odum, R. Andrew 053759
Oechslin, Dr. Ludwig 041409
Oegema, D. 038856
Öhlinger, Walter 003095
Öhlknecht, Rupert 002756
Öhman, Nina 041052
Oehms, Günter 020670
Øien, Elisabeth Adeleide 033589
Öjemyr, Hans 041047
Økland, Olaug N. 033575
Ökten, Haydar 043031
Oellers, Dr. Adam C. .. 016442, 016445
Oelschlägel, Dr. Petra 017057
Oelschlaeger-Garvey, Barbara .. 050604
Oerichbauer, Edgar S. 049704
Oerlemans, Magi 003989
Örsi, Dr. Julianna 024020
Ørskou, Gitte 010005
Oertel, Michael 020002
Oertlin, Jean-Paul 014549
Oesch, Joseph 041496
Oeschger, Dr. Bernhard ... 019505
Østargård, Niels Jøan 010316
Östberg, Maths 041169
Östberg, Wilhelm 040998
Oester, Hanspeter 041194
Østerås, Bodil 033988
Österholm-Granqvist, Sonja .. 011470
Österlund, Petter 041064
Oesterreicher, Claudia ... 003095
Østervig, Torsten 010221
Oesterwind, Dr. Bernd C. .. 020321
Oetterli, Christoph 042128

Oettingen-Wallerstein, Fürst zu . 017509,
    019070, 022410
Oettinger, Dr. Marion 052949
Øvstebø, Solveig 033522
Oexle, Dr. Judith 017993
Özbey, Şahin 042895
Özcan, Hilmi 042993
Özçelik, Serpil 042938
Özdemir, Abdulhalik 042845
Özerdem, Hürrem 042807
Özeren, Öcal 043028
Özgen, Emine 042848
Özgerey, Mehmet 042958
Özkütük, Zafer 042854
Özsağır, Ahmet 042976
Öztürk, Erdal 042965
Offers, Donald 041243
Offringa, D.C. 038856
Offroy, Martine 015985
Offutt, Cathy 049671
O'Flaherty, Rosemary ... 005732
O'Floinn, Raghnall 024898
O'Foighil, Prof. Diarmaid ... 046493
Ogata, Hisanobu 029660
Ogawa, Toshiro 029484
Ogden, Dale 049688
Oggenfuss, Daniel 020580
Ogilvie, Peter 045722
Ogle, Brian 043756
Ogle, Nigel 033220
Ogle, Susan 054432
Ogliari, Prof. Francesco ... 027475
Ogny, Jean d' 013380
O'Gorman, Dr. Jodie A. ... 048441
O'Gorman, Tim 048834
O'Grady, Brian 051018
Ogris, Mag. Kerstin ... 002421
Ogrodniczak, Tomasz ... 035124
O'Grody, Jeannine 047005
O'Halloran, Jim 033308
O'Hanlon, Dr. Michael ... 045459
O'Hara, Bruce 053782
O'Hara, David 007180
O'Hara, Michael 046052
Ohashi, Koji 028480
Ohashi, Yasushi 028896
Ohe, Dr. Werner von der . 017723
O'Hear, John 050956
O'Hern, John D. 048565
Ohgane, Takuya 029277
Ohl, Thomas 022591
Ohlmann, Dan 013948
Ohlsén, Marie 040789
Ohlsen, Dr. Nils 018176
Ohlsson, Birger 040563
Ohm, Dr. Mathias 022047
Ohmann, Agathe 010365
Ohmura, Etumasa ... 029055
Ohngemach, Dr. Ludwig . 018106
Ohno, Akira 029717
Ohno, Masaji 029155
Ohrt, Karsten 010261
Ohrt, Nils 010235
Ohta, Yoshinori 028571
Ohtsubo, Kenji 029639
Oilliamson, Comtesse d' ... 013009
Oinas, Tapani 011322
Ojalvo Prieto, Roberto .. 008725
Ojeda Sato, Alonso ... 008729
Ojha, M.B. 024282
Okabe, Nobuyuki ... 029696
Okada, Akiko 028677
Okada, Hidenori ... 029609
Okada, Hirashi 029125
Okada, Itsuki 028592
Okajima, Hisashi ... 029609
Okániková, M. 038486
Okano, Kiheita ... 029401
Okano, Shun-ichiro . 029519
Okawa, Eiji 028618
Okaya, Michiko ... 048456
Okazaki, Kyoichi .. 029696
Okazaki, Dr. Masaki . 029300
Okazawa, Kunro ... 029191

O'Keefe, Barbara K. 047987
O'Keefe, John F. 052019
O'Keefe, Raymond 051418
O'Keeffe, Tony 044020
O'Kelly, Barry 024868
Oki, Shinichi 029218
Okimoto, Hiroshi 028649
Okkenhaug, Eli 033523, 033524, 033525
Okonogi, Miyoko 028818
Okoshi, Hisako 029305
Okubo, Takishi 028823
Okubo, Tokiko 029706
Okumura, Yoshitsugu 029049
Okuoka, Shigeo 029335
Okura, Haruhiko 028927
Olachea, Elena 031155
Olaczola, Juanjo 039032
Olalla Mercade, Leandro 039706
Olano Trujillo, Manuel José ... 008766
Ólason, Vésteinn 024113
Olausson, Magnus ... 040841, 041024
Olbrantz, John 052889
Ołdakowski, Jan 035357
Olde Dubbelink, R.A. 032712
Olden, Sam 054574
Oldenburg, Rie 023333
Oldham, Terry L. 052783
Oldknow, Tina 047994
O'Leary, Daniel E. 052283
Olejniczak, Józef 035099
Oleščina'ka, Alla Pavlivna ... 043115
Olfra, Michelle 050995
Olijnyk, Michael 052168
Oliva, Francisco Miguel González de
la 039458
Olivarelli, Sebastiano .. 027196, 027197,
027198
Olivares, Isabel 034211
Olivarez, Fátima 004222
Olivas, Art 053116
Olivas, Arthur 053118
Olive Garrard, Jacqueline 000652
Olivé Serret, Prof. Enric 040207
Oliveira Caetano, Dr. Joaquim .. 035678
Oliveira Mesquita Spränger, Paulo
Roberto 004756
Oliveira Rodrigues, Adélia E. de 004339
Oliveira, Jorge de 035849
Oliveira, Dr. José Carlos 035554
Oliveira, Natércia 035893
Oliveira, Nicolas de 045046
Oliveira, Pedro 035893
Oliver, Dr. Graham 043863
Oliver, Katherine C. 050574
Oliver, Marvin 053229
Oliver, Mary E. 054523
Oliver, Miguel 039880
Oliver, Robyn 001161
Oliver, Virginia 048720
Oliveri del Castillo, Patrizia ... 028404
Oliveri, Meg 049044
Olivero García, Lucía 023371
Olivié, Jean-Luc 014733
Olivier, André 012432
Olivier, Louis 015409
Olivo, Sandra 050858
Ollenburger, Jeff 049645
Ollikainen, Anja 011499
Olmedo Patiño, Dolores 031220
Olmedo, Freddy 010444
Olmer, Dr. Beate 022273
Olmo Fernández, Jesús del ... 039173
Olmo Ilarta, Francisco Javier del 040116
Olmstead, Dr. Richard 053229
Olmsted, Betty 051932
Olofsson, Jan 040590
O'Looghaire, Dr. Niamh 007225
Olschowski, Petra von 022046
Olsen, Dr. Arnold 049433
Olsen, Betsy 048759
Olsen, Ed 049744
Olsen, Johanne 023339
Olsen, John 033738
Olsen, Phyllis J. 049153

Olsen, Sandra H. 047309, 047310
Olsen, Tammery 048168
Olsnes, Aanund 033659
Olson Peters, Ruth 050150
Olson, Alice 047469
Olson, David A. 051553
Olson, Dennis 053828
Olson, Don 053640
Olson, Dona A. 052185, 052716
Olson, Dorothy 046619
Olson, Janis 046895
Olson, Joan 053332
Olson, Kathleen 050264
Olson, Kathryne 051848
Olson, Kathy 051853
Olson, Kristina ... 051056, 051057
Olson, Link 048662
Olson, Marilyn 052004
Olson, Nicky L. 048106
Olson, Oliver 053490
Olson, Roberta 053164
Olson, Ruth Ann 050184
Olson, Suzanne 052278
Olson, Thomas H. 052716
Olson, William 006659
Olsson, Daniels Sven 040621
Olsson, Murray 000760
Olsson, Susanne 040744
Olsson, Tommy 040571
Olszak, Eliane 013059
Olszanowski, Zdzisław 035431
Oltean, Dr. Vasile 036274
Olthoff, Gabriele 022576
Oltmanns, Kay 047142
Oltmans, Kay 047141
Olucha Montíns, F. 039249
O'Malley, Dennis M. 052382
O'Malley, Jeanette 051939
O'Malley, Tom 054546
Oman, Dr. Hiltrud 002656
Oman, Richard G. 052767
Omar, Hadmar 042643
Omar, Hamad H. ... 042629, 042640,
042641, 042642
Omar, Jana 009885
Omar, R.H. ... 038701, 038703, 038704,
038705
O'Mara, John F. 006900
O'Meara, John 046577
O'Meara, Nora 024973
Omedes, Dr. Anna 039077
Omeñaca, Jesus 039909
Omer, Gil 025067
Omer, Prof. Mordechai 025212
Ometto, Giulio 028127
Omi, Hideo 029169
Omieczyński, Witold 034942
Ommen, Dr. Eilert ... 020807, 020808
Omtvedt, Petter M. 033852
Omurbekova, Ainagul 030073
Onaga, Stephen 052422
Ondevilla, Alex 034437
Ondish, Andrea 053895
Ondrejčeková, Zora 038328
Ondrušeková, Dr. Anna 038428
Onea, Nicolae 036265
O'Neill, Anthony 044529
O'Neill, Colleen 005647
O'Neill, John 051404
O'Neill, John J. 046736
O'Neill, Mark 044404
O'Neill, Peter 001654
O'Neill, Vicki 001383
Ono, Masaharu 029281
Onodera, Hiroshi 029199
Onodera, Reiko 029509
Onorevoli, Dr. Giuseppe 028080
Onuki, Dr. Yoshi 034397
Onuma, Sunao 029491
Onyejekwe, Dr. Umebe N. 033446
Onyejiako, Joy 044928
Oost, T. ... 004020, 004021, 004022
Oost, Tony 004019
Oostdijk, Dr. Alexandra E. 033005

Oosterbaan, A. 032145
Oosterhof, H. 032929
Oostra, Dr. R.J. 031966
Ootukasa, Mitukuni 029230
Ooyama, Masaaki 029028
Op de Beeck, Bart 003435
Op de Beeck, Jan 003854
Op de Beeck, Jozef 003855
Op de Beek, R. 032112
Opaleye, A.O. 033427
Opaleye, Yemi 033427
Opdenberg, Laurence 003442
Operta, Mevlida 004283
Ophir, Daniel 025192
Opinel, Annick 015399
Opolã, Valentina 030177
Opp, Edmund 048629
Oppenhimer, Ann 052549
Opperman, G. 038865
Oppizzi, Rosita V. 042104
Oprea, Emanuela 036438
Opstad, Jan-L. 034034
Oquendo, Dante 034589
Oquiche, Antonio 034374
Orač-Stipperger, Dr. Roswitha .. 001989
Oram, Dr. Richard W. 046692
Oratowska, Barbara 034998
Oražem, Vito 018310
Orbons, Dr. A. 032495
Orchard, Dr. Karin 019066
Ordine, Mauriziano 028028
Ordoñez Delgado, Napoleón .. 008613
Ordoš, Dr. Ján 038430
Ore, Phyllis 006817
Oredsson, Prof. Lars-Göran .. 040810
Orefici, Giuseppe 034376
Oreglia, Luciano 025485
Orehovec, Martina 038562
O'Reilly-Lawrence, Priscilla ... 051294
Oremus, Frederick L. 051213
Oren, Dan 025022
Orfali, Dalila 000046
Orgeix, Charles-Louis d' 013879
Orgill, Peter 043803
Orgonikidze, I. 016401
Oriol, Calvo 038991
O'Riordan, E. 024962
O'Riordan, Ian 044233
Orkin, Frederic I. 049472
Orlandi, Giuseppe 028418
Orlando González, Jorge 008591
Orlando, Fran 051556
Orlić, Alfons 009024
Orloski, Julie 053215
Orlov, Evgenij Michajlovič ... 037797
Orlov, Viktor 036898
Orłowski, Jarosław 034787
Orlson Agbo, Raymond 022927
Ormenyi, Steven 050428
Ormerod, Robin 006120
Ormhaug, Knut 033523, 033524, 033525
Ormond, Dr. Rupert 045269
Ormonde, Dr. Helena 035530
Ornsby, Philip 044531
Oroianu, Silvia 036765
Oropeza Segura, Manuel 031412
Oroveanu, Mihai 036313
Orozco Móron, Edna Maria ... 031247
Orozco, Prof. Margarita V. de .. 030995
Orozco, Sylvia 046697
Orphal, Christina 020151
Orr-Cahall, Christina 054317
Orr, Brenda P. 006350
Orr, Charlene 050817
Orr, Dr. Patrick J. 024926
Orrick, Mary Norton 051966
Orro, Leena 010929
Orsi, Dr. Oriana 026507
Orsini, Ph. 015978
Orski, Marek 035212
Orsler, Christine 045622
Ortal, Dr. R. 025081
Ortega DeSantis, Diana 053118
Ortega Rivera, Enric 040396

Ortega, Enric 040396
Ortega, Luis Angel 040073
Ortiz Crespo, Alfonso 010478
Ortiz de Zevallos, Felipe 034371
Ortíz Lan, José E. 030854
Ortíz Muñoz, Idania 008682
Ortiz Volio, Melania 008863
Ortiz, Deborah V. 052729
Ortiz, Gustavo A. 008597
Ortiz, Rosa 039088
Ortmann, Matthias 041393
Ortmeier, Dr. Martin .. 020316, 020320
Ortner, Josef 003048
Ortner, M. Christian 003004
Orton, Anna 046082
Orton, Donald 053477
Oruch, Sam 051735
Orzsi, Zoltán 023435
Osajima, Rachel 053014
Osawa, Tomomichi 029111
Osborne, Bernard 054470
Osborne, Duncan 043394
Osborne, Lois I. 047829
Osborne, Lori 048638
Osborne, Peter 052253
Osborne, Richard 048984
Osborne, S. 044863
Osborne, Victoria 046230
Oscarsson, Ulla 040902
Oselies, David 005309
Oset, Hans Petter 033671
O'Shaughnessy, Margaret 024924,
051389
O'Shea, Deborah H. 052674
O'Shea, John 046491
O'Shea, Patrick B. 024949
O'Shea, William A. 006223
Oshel, Laurie 049042
Oshima, Seiji 029585
Osimitz, Stefanie 041399, 041780
Oslen, John C. 005744
Oslund, Christopher 005940
Osman, Prof. Ali Mohamed .. 040491
Osmann, Gudrun 018134
Osmond, Lynn ... 047645, 047646
Ospina Cubillos, Carlos Enrique . 008628
Ospina de Muñoz, Esperanza .. 008741
Ospina Navia, Francisco 008790
Ospino de la Rosa, José 008677
Osses, Dietmar 017442
Ostapova, Svitlana Ivanivna .. 043271
Ostašenko, E.J. 037504
Ostaszewska, Stanisława 035258
Osten, Dr. Till 022056
Ostendorf, Eleanor 046440
Ostendorf, Dr. Thomas .. 022120, 022121
Ostenrieder, Dr. Petra 020967
Osterhus, Cynthia B. 052904
Ostermiller, Jerry 046583
Osterrith, Martine 003914
Osticresi, Dr. Patrizio 026277
Ostlund, Ralph 051315
Ostojić, Milan 038257
Ostrenko, W. 053669
Ostritz, Dr. S. 021457, 022501
Ostrosky, Amelia 054318
Ostrouchova, Marina Georgievna . 037194
Ostrovskaja, Raissa Iljinična .. 037256
Ostrow, Mindy 051841
Ostrow, Victor 053211
Ostrowski, Prof. Jan 034913
Ostrum, Meg 050859
Ostwald, Thomas 017555
O'Suilleabhain, Sean 024824
O'Sullivan, Dr. A.M. 025002
O'Sullivan, Keith M.C. 043855
Osyra, Tadeusz 034854
Ota, Diane 047106
Ôta, Motoji 029574
Otáhal, Ivo 009631
Otani, Kazuhiko 029560
Otani, Shogo 029608
Otavová, Marie 009969
Otero, Álvaro 008662

Otomo, Tsutomu 029106
O'Toole, Joseph 050541
O'Toole, Judith H. 049259
Otsuka, Masahito 029158
Otsuka, Ronald 048258
Otsuka, Yoshio 029657
Ott, Hanspeter 041899
Ott, Irene 018468
Ott, John H. 050256
Ott, Lili R. 051747
Ott, Tanja 021176
Ott, Wolfgang 022547
Otte, Prof. 003800
Otte, Stacey A. 046706
Otte, Wulf 017553
Ottenbacher, V. 017361, 017363
Otterspeer, Prof. Dr. W. 032578
Ottillinger, Dr. Eva 003008
Ottinger, Bénédicte . . . 015767, 015768,
                                015769
Ottini, Alicia 000571
Ottmann, Helmut 020847
Ottmann, Rosemarie 020878
Ottnad, Clemens 021382
Otto-Hörbrand, Martin 022048
Otto-Mller, Jackie 051804
Otto, Henry S. 051280
Otto, Kristin 022851, 022852
Otto, Martha 047941
Otto, Dr. Sigrid 017086
Otto, Veronika 021584
Otto, Werner Th. 021441
Ottomeyer, Prof. Dr. Hans 017299
Otton, Dr. William G. 048002
Ottosson, Jon Gunnar 024103
Ottou Onana, Irène . . . 005292, 005295
Ottowitz, Arthur 001786
Ottrubay, Dr. Stefan 001895
Otttaviani, Jean-Claude 011591
Otzen, Dr. Peter 018717
Ou, Prof. Shan-Hui 042484
Ouarzazi, Abdul Wahed 039190
Oudet, Benoît 014470
Ouellet, Nicole 006728, 006745
Ouellet, Raynald 006356
Ourecky, Irma 054401
Ourmieres, Jean-Guy 015575
Outerbridge, Amanda 004199
Outridge Field, Stephanie 001077
Ovalle Neira, Pedro Ubert 008784
Ovaskainen, Heikki 011486
Ovečkinaova, Olga 037770
Overdick, Thomas 018371
Overdiep, Dr. G. 032227
Overgaard, Ibsen 010360
Overland, Christian 048195
Overlie, Jill 050390
Overmyer, Lorraine 048398
Overstreet, Laura 046395
Overton, Jonathan 043527
Oviedo de Coria, Beatriz 000585
Ovsjankina, Ljudmila
  Vladimirovna 037947
Owczarski, Marian 051796
Oweis, Iman 029782
Owen, Ann 044889
Owen, Brian 046013
Owen, Gwilym 044889
Owen, Paula 052943, 052952
Owen, Thomas S. 053784
Owens-Celli, Morgyn 050368
Owens, David 045681
Owens, Heather 051102
Owens, Marshall 049547
Owens, Sue Ellen 048199
Owens, Suzanne 051729
Owers, Lorna 044171
Owingayak, Winnie 005374
Owyoung, Steven D. 052818
Oxley, Nicola 045046
Oya, Mina 029536
Oygur, Dr. Sule 051499

Oyuela, Raul M. 047985, 050834
Ozaki, Masaaki 029608, 029609, 029610
Ozanam, M. Yves 014744
Ozay, Alev 042963
Ožegova, Ljudmila Ivanovnă 037138
Ozola, Agrita 030242
Ozola, Arija 030125
Ozolina, Vita 030177
Ozolins, Diana 049742
Paap, Hans-Uwe 019917
Paarlberg, Bill 053409
Paarlberg, S. 032246
Paas, Dr. Sigrun 018092, 020229
Paaverud, Merl 047018, 050765, 054424
Paaverud, Merlan E. . . 047020, 047022,
                                051967
Pabst, Fieke 031996
Pabst, Stewart J. 047968
Paccoud, Stéphane 011531
Pace, Dr. Lorenzo 053905
Pacere, Tifinga Frédéric 005275
Pačev, Nikolaj 005079
Pache, Pierre 042173
Pacheco, Anelise 004718
Pacheco, Gustavo 004664
Pacholik, Kazmimierz 035389
Pachomova, Naděžda Petrovna . . 036987
Pachter, Marc 054150
Pacini, Marina 050775, 050783
Pacjukov, Vladimir V. 037207
Packard, Andrea 053623
Packard, Elinor 049148
Packard, Robert W. 051881
Packard, Vance 047971
Packauskas, R. 049416
Packer, Dr. Claude 028445
Packer, Nancy 052041
Packheiser, Michael . . 020433, 020434
Paczkowski, Dr. Jörg 022588
Paczos, Andrzej 034796
Padalino, Nick 031936
Padalkina, Olga Viktorovna 036882
Paddock, Eric 048255, 048258
Paddock, Dr. J.M 043962
Paderni, Loretta 027634
Padfield, Joe 045057
Padgett, Michael 052368, 052585
Padgett, Sharon 048080
Padhy, Dr. K.P. 024337, 024370, 024451
Padian, Kevin 046939
Padovan, Dr. Stefanie 025972
Padovani, Dr. Serena . 026260, 026274
Paduraru, Nicolae 036320
Paemelaere, Jean-Pierre 003948
Pärdi, Heiki 010700
Paesano, Dr. Ciro 027002
Paetzold, Angelika 021029
Pagac, Gerald J. 050968
Pagano, Dr. Denise Maria 027013
Pagano, Ezio 025433
Pagano, Sergio 054677
Page Harrington, A. 054159
Page, Eric 049874
Page, John 043526
Page, Jutta Annette 053760
Page, Loa 050283
Pagé, Suzanne 014697
Pagella, Dr. Enrica 028118
Pagès, Solange 013673
Paget-Woods, David 045511
Pagliardi, M. Nicoletta 027635
Pagnotta, Ph. 012409, 012411
Pagot, Serge 014427
Pagsinohin, May M. 034552
Pagulayan, Teresita L. 034532
Pahl, Dr. Henning 021280
Pahl, James R. 051705
Pahlke, Dr. Rosemarie E. 017961
Pahlow, Martina 022178
Paiardini, Lino 027340
Paidasch, Christoph 003026
Paige Linn, Nancy 052363
Paillard, Ph. 013941

Pailler, Henri 012915, 014238
Paillet, Antoine 015857
Paillet, Patrick 015444
Paillot, Noël 015160
Pain, Theresia 018866
Paine, Fiona 044282
Paine, Tony 033167
Paine, Wesley M. 051183
Païni, Dominique 015495
Paino, Prof. Giuseppe 027002
Painter, Bill 049858
Païpetú, Andreas 023142
Pais de Brito, Joaquim 035808
Pais Pinto, Joaquim 000106
Pais, Teresa 035702
Paisie, Cinar 036761
Paiva Cavalanti, Lauro Augusto
  de 004765
Paiva, Ricardo Luís 035544
Paizis-Paradelis, Constantin 023224
Pajović, Maja 038199
Pajskrová, Dr. Jana 009558
Pakesch, Peter 001980
Pakkala, Taru 010750
Pakoma, Katariina 010931
Pakuningrat, Maulana 024554
Pakzad, Dr. Amir 019064
Pal, Raj 046073
Pal, S.N. 024359
Pala, Valter 027749, 027752
Palacios i Manuel, Josep . . . 039200,
                                039202
Palacpac, Remedios 034502
Palacz, Tomasz 035080
Palágyi, Dr. Sylvia 024040
Palas, Jan 009788
Palavonov, D. 042614
Palella, Dr. Graciela Elizabeth . . . 000337
Paletta, Céline 011994
Pálffy, András 003083
Palgi, Adva 025147
Palisse, A. 015958
Páll, Dr. István 023905
Pall, Martina 001974
Palladino-Craig, Allys 053657
Pallan, Norbert 002541
Pallanca, Dr. Claude 031722
Pallavicini, Maria Camilla . . . 027559,
                                027562
Pallestrang, Kathrin 003057
Pallmert, Sigrid 042228
Pallottino, Paola 026211
Pallu, Jocelyn 022349
Pálmason, Ólafur 024102
Palmberg Eriksson, Stina 040888
Palmeiri, John 047218
Palmenaer, Els de 003284
Palmer, Lord 044180
Palmer, Cheryl 047576
Palmer, Lindsay 052124
Palmer, Mike 043456
Palmer, Oonagh . . . 044329, 044451
Palmer, Dr. Oonagh 044537
Palmer, Peregrine 046161
Palmer, Ros 045988
Palmer, Sharon S. 049970
Palmer, Stephen 044323
Palmer, Sue 046120
Palmgren, Christine 040663
Palmgren, Kaj 010767
Palmgren, Ola 040739, 040742
Palmiero, Franesco 026145
Palmquist, Janice 052890
Palmquist, Dr. Lena 041028
Palmquist, Linda 047961
Palmy-Brandstätter, Lotte 042137
Palo-oja, Ritva 011404
Paloheimo, George B. 053106
Palokangas, Markku . 010847, 010863,
                                010864
Palombo, Aldo 051621
Palombo, Ionah 023083
Palomero, Santiago 040240
Palomo, Antonio 023355

Palop, José Alcántara 038997
Palou Sampol, Joana M. 039892
Palov, Dr. József 023917
Palsa, Raija . . 011117, 011118, 011119
Pálsson, Lýður 024075, 024076
Paltrinieri, Roberto 025529
Palzerová, Hana 009553, 009554
Pamart, Jean-Luc 014528
Pambrun, Jean-Marc 016380
Pamperin, David L. 047059
Pamuk, Ömer 043051
Pan, Giuseppe 027105
Pan, Junxiang 008233
Pan, Yue 007784
Pan, Zhaolin 007998
Pan, Zheng 008236
Pan, Zhenzhou 007636
Panabière, Louis 031154
Panafieu, Bruno de 015390
Panagiotopoulos, Prof. Dr.
  Diamantis 019130
Panamaroff, Emilia 006978
Pañares, Tonette 034464
Panaskin, Vladimir Vasiljevič 036921
Panawatta, Senarath 040485
Panayotova, Dr. S. 043842
Panazza, Dr. Gaetano 026620
Pančeva, Petrana 004972
Pančíková, Eva 009450
Pančuk, Fedir Vasil'ovyč 043318
Panczenko, Dr. Russell 050593
Panda, Prof. C. 024369
Pandey, A.K. 024421
Pandey, Dr. S.N. 024177, 024244
Pandilovski, Melentie 000764
Pandolfi, Sylvia 031263
Pandžić, Ankica 009036
Pane, Josh 052729
Panea, Traian 036803, 036804
Panek, Kornelia 018210
Paner, Henryk 034728
Panese, Prof. Dr. Francesco 041658
Pangburn, D.A. 052752
Pange, Roland de 014610
Panhorst, Michael W. 053533
Panić, Djordje 038244
Paniccia, Antonio 028341
Panizza, Marco 028047
Panjuškin, A.D. 037281
Pankok, Eva 019361
Pannabecker, Rachel K. 051609
Panneton, Jean 007249
Pannewitz, Otto 021871
Panni, Frédéric 013216
Pansi, Dr. Heinz 002076
Pant, Kurt 003076
Pant, Dr. Madhu 024227
Panteghini, Ivo 025634
Pantellini, Giovanni 041366
Panter, Dr. Armin 021751
Panter, John 052974
Pantermalis, Prof. Demetrios 023299
Pantoja, Dianne 047384
Pantoja, Hernán Gil 008720
Panţuru, Prof. N. Ioan 036309
Panvini, Dr. Rosalba 025701
Pany, Franz Norbert 020681
Panzanaro, Mauro 026561
Panzig, Dr. Christel 020197
Paoletta, Dr. Donald 050289
Paoli, Arturo 025707
Paoli, Feliciano 028234
Paolillo, Dr. Antonio 026106
Paolillo, Giuseppe 025465
Paolozzi Strozzi, Dr. Beatrice . . . 026295
Papadakis, N. 022949
Papadakis, Dr. Nikos 023265
Papademetriou, Dr. Eleni 009412
Papadopoulos, A. 009414
Papadopoulos, Prof. N. 023007
Papadopoulou-Pangou, K. 023131
Papadopoulou, Eleni 023038
Papageorgiou, Ioannis 023154
Papagno, Dr. Francesco Saverio . . 027262

Papait, Jean	012451	Parker, Tim	006479, 006486
Papakonstantinou, Maria-Photini	023147	Parker, Dr. W.A.	045145
Papaldo, Dr. Serenita	027555	Parkes, Susan	024906
Papangeli, Kalliopi	023069	Parkhurst, Jo	048753
Papantoniou, Ioanna	023194	Parkin, Peg	000846
Paparella Treccia, Raffaelle	027277	Parkinson, C.D.	043388
Paparone, Guiseppe	028049	Parkinson, Glenn	048920
Papazian, Aline	051943	Parkinson, Gordon	045977
Papazoglou, Lena	022997	Parkinson, Melissa	052568
Papco, Ján	038296	Parks McClain, Patti	050057, 050308
Papendieck, Dr. Christine	022416	Parks, Dennis	053239
Papendorf, Dorothee C.	021718	Parks, Gary W.	050226
Papet, Edouard	014743	Parks, Heather	051490
Papetti, Prof. Stefano	025390, 027842	Parks, Patty	047596
Papike, Dr. James J.	046362	Parks, Robert E.	051428
Papillard, Marie Laure	016131	Parks, Ron	048036
Papin-Drastik, Y.	012707	Parks, Shirley	054462
Papin, Gregory J.	047356	Parlic-Barisic, Slobodanka	030630
Papone, Elisabetta	026404	Parma, Raphaël	014882
Papounaud, Benoît-Henry	015321	Parma, Dr. Rita	027555
Papp, Dr. Claudia	021567	Parmal, Pamela A.	047123
Pappi, Evangelia	023192	Parmalee, Dr. Paul	050011
Papst, Helga	002156	Parman, Susan	048996
Papuc, Gheorghe	036524, 036790	Parmentier, J.-P.	013804
Papuš, Ivan Agafonovyč	043260	Parmentier, Stéphane	003602
Paquelot, Françoise	015418	Parnell, Dr. Geoffrey	045088
Paquette, Adrian	051953	Parnes, Stuart	048608
Paradis, Michèle	006495, 007250	Parola, Publio	000513
Paradisopoulos, Sofronis	043272	Parpagiola, Dr. Giuseppe	027077
Paramo Sureda, Ernesto	039471	Parr, Jesse	050744
Páramo, Amanda Lucía	008567	Parr, Melissa	051758
Paramonov, Vladimir		Parra Casamun, Gerardo	031234
Aleksandrovič	037967	Parra Santos, Alirio	008669
Paraschos, Georgios	023228	Parra, Raúl	031262
Paraskevas-Gizis, Evaggelos	023302,	Parreiras Horta, Maria de	
	023304, 023305	Lourdes	004628
Parberg, Dee	052367	Parreño Rivera, Prof. Manuel	030302
Parchman, Bryn	046766	Parrick, Allen	000898
Pardilla, Helena	008812	Parrilla de Alvarez, Estela	030968
Pardini, Prof. Edoardo	026279	Parrinello, Prof. Nicolo	027185
Pardo, Alfredo	008594	Parris, Benjamin J.	049046
Pardo, Isabel	039647	Parrish, M.	044673
Pardo, Rengifo	008586	Parrish, Ron	048342
Pardue, Diana	052118	Parrott, Robert W.	050729
Pardue, Diana R.	051474	Parrott, Shirley	052695
Pare, Glenn S.	053073	Parrow, Enid	053967
Paredis, Albert	003399	Parry, Wendy	045648
Parent, Béatrice	014697	Parsch, Marianne	020567
Parent, Geraldine	006627	Parsekiau, Robert	054213
Parent, Dr. Thomas	017958, 017967	Paršina, Lidija Semënovna	036910
Paret, John J.	046669	Paršina, Ljudmila Vladimirovna	038015
Pargner, Dr. Birgit	020518	Parsons, Brian	000994
Parham, Herbert S.	049350	Parsons, Ernie	006224
Pariente, Myriam	013820	Parsons, Jeffrey R.	046491
Parikka, Anne	010787	Parsons, K.	001541
Parillo, James D.	053159	Parsons, Louise	051725
París Pérez, Dr. Joaquín	039118	Parsons, Ralph	045854
Parisch, Manfred	002528	Parsy, Paul-Hervé	014554
Parise, Dr. Roberta	027146	Partanen, Päivi M.	011090, 011093,
Parisella, Prof. Antonio	027647		011095
Pariset, Jean-Daniel	014745	Partin, Marc	054558
Parish, David W.	049065	Parton, Neville	001149
Parisi, Chiara	011945	Partridge, Victoria	043551
Park Curry, Dr. David	052565	Partyka, Józef	035061
Park Leggett, Barbara	048422	Parusel, Janina	035455
Park, Janette A.	044734	Parušev, Todor	005244
Park, Mee-Jung	030052	Pârvan, Lucia	036253, 036254, 036255,
Park, You Chul	029903		036256, 036257, 036442, 036443,
Parke, Julie	050494		036444, 036538, 036567, 036686,
Parker, Angela	044910		036728
Parker, David	048805	Parviainen, Helena	011071
Parker, Donna	047157	Pârvulescu, Dorina	036783
Parker, Elizabeth	052794	Paša, Natko	009028
Parker, Eris	033164	Pascal, Huston	052450
Parker, Harry S.	053009, 053033	Pascall, Chris	046039
Parker, Herschell	051176	Paschen, Dr. Joachim	018987
Parker, John	000943	Paschke, Ilona	020167
Parker, Margaret	052232	Pascual, Jordi	033407
Parker, Niles	051157	Pascual, José	039506
Parker, P.	045161	Pascucci, Marisa	051025
Parker, Ray	046885	Pâsculescu, Gică	036602
Parker, Robert	049055	Pasetti, Elena	027513
Parker, Sandi	046866	Pasiciel, Stanisław	034764
		Paskova, Spaska	005034

Pasquali, Prof. Flavia	027945	Paul, Dr. April	051356
Pasquet	013374	Paul, C.D.	045416
Pasquier, Thierry du	014714	Paul, Eli	049886
Pasquine, Ruth	052137	Paul, Frédéric	012039
Pasquini-Barisi, Isabella	028093	Paul, Gary A.	049003
Pass, Robert	046648	Paul, Gayle	052322
Passaglia, Elio	026859	Paul, Renee	033296
Passarin, Dr. Mauro	028368	Paul, Ryan	047474
Passe, Eugene	052487	Paul, Shimrath	033194
Passi, Leena	010752	Paul, Yvonne	045718
Paßlick, Tonio	022468, 022471	Păuleanu, Dr. Doina	036395, 036399,
Passoja, Merja	011018		036561, 036791
Pasté de Rochefort, Aymeri	012653	Pauletić, Mirjana	008890
Pastellas, Dominiue	014765	Pauley, Ed	051928
Paster, Carol	052236	Pauley, Edward E.	048699
Pasternak, Natalija Anisimovna	037634	Paulicka-Nowak, Dr. Lucja	034847
Pasti, Sara	051309	Paulo, Dália	035687
Pastor Vázquez, Prof. J.F.	040335	Paulocik, Linda	007399
Pastor, Paloma	040060	Paulová, Dr. Eva	009781
Pastore, Heather	048561	Paulsen, John Q.	048699
Pastore, Prof. Michele	028056	Paulsen, Nils	033598
Pastorek, Ivan	038353	Paulson, Christopher	052441
Pastori, Jean-Pierre	041753	Paulson, Dr. Dennis R.	053644
Pastoriza, Juan A.	036132	Paulson, Warren	005357
Pasture, B.	003882	Paulsrud, Geir	033591
Paszylka-Glaza, Małgorzata	034740	Paulter, Charles D.	050329
Patà, Franco	042009	Pauly, Dorothy	047512
Pataki, Dezsö	023606	Pausch, Gottfried	001838
Patané, Dr. Rosario	025898	Pausch, Hans-Jürgen	017576
Patay, Dr. Franz	003020	Pause, Dr. Carl	020743
Patchen, Dr. Jeffrey	049679	Paust, Dr. Bettina	017011
Pátek, Jiří	009853	Pautet, C.	014097
Patek, Sheila	046939	Pautrizel, Françoise	012028
Paten, Richard	045490	Pavaloi, Dr. Margareta	019149
Patenaude, Val	006263	Pavelčík, Dr. Jiří	009931
Pater, Dr. Józef	035418	Pavelcikova, Monika	038340, 038456,
Pater, Philippe	003828		038457
Paterson, Anne	045286	Pavelec, Karen Marie	048621
Paterson, John	031845	Pavelka, Dr. Karel	009940, 009962
Pathak, Dr. Bindeshwar	024238	Pavia, Prof. Giulio	028128
Pathak, G.M.	024507	Pavic, Stjepan	004291
Patin, Sylvie	014743	Pavić, V.	008963
Patnode, J. Scott	053441	Pavić, Vladimira	009048
Paton, W.J.	033179, 033180	Pavičić, Snježana	009036
Patorska, Joanna	035009	Pavlát, Dr. Leo	009847
Patrick, Bill	001367	Pavletić, Mira	008887
Patrick, Brian	033194	Pavlides, Paul	012420
Patrick, J.A.	043624	Pavliňák, Dr. Petr	009744
Patrick, Pat	053803	Pavljuk, Vasil Mykolajovyč	043129
Patrick, Ruth	052023	Pavlov, Aleksandr Anatoljevič	036915
Patrick, Stephen E.	047151, 047152,	Pavlov, Slavi	005080
	047153, 047154, 047155	Pavlova, Irina Vasiljevna	037998
Patry, J. Michel	050233	Pavlu, Viktor	003018
Patry, Sylvie	014743	Pavlunina, Ljudmila Fëdorovna	038072
Patterschal, Donald	047602	Pavúková, Olga	038401
Patterson-O'Regan, Prof. J.J.	043603	Pavuljak, Jaroslav Ivanovyč	043307
Patterson jr., Aubrey B.	053852	Pavuza, Dr. Rudolf	003014
Patterson, David	045258	Pawlowska, Krystyna Iwona	035400
Patterson, Ken	005433	Pawłowski, Dr. Antoni J.	034932
Patterson, Lillian	046382	Pawłowski, Sebastian	009809
Patterson, Dr. Mark A.	047312	Paxton, Janet	044459
Patterson, Richard	053800	Pay, Dr. Peter	003089
Patterson, Robert B.	052862	Payet, Hugues	036183
Patterson, V.A.	049748	Payling, Catherine	027570
Patterson, Vivian	054420	Payne, Prof. Anthony	044407
Patteson, Rita	054012	Payne, Elizabeth	005658
Patti, Lynn	054209	Payne, James	054038
Pattie, M.	033294	Payne, Mary	050207
Pattinier, Philippe	011954	Payne, R.G.	045836
Pattinson, G.	045735	Payne, Prof. Robert B.	046493
Pattison, Alison	044439	Payne, Susan	045486
Patton, Becky	052122	Payrich, Wolfgang	002081
Patton, James L.	046940	Payton, Charles	045578
Patton, LuAnn W.	046380	Payton, Sidney	048271
Patton, Robert	053814	Paz Morales Osorio, Maria de la	031183
Patton, S.M.	046554, 046555	Paz, Dr. Boaz	017272
Patwa, Elizabeth	012002	Paz, Diego Alonso	008763
Patzak, Dr. B.	003064	Paz, Olivier	014145
Patzer, George	005982	Paz, Valeria	004222
Patzig, Eberhard	019992	Pazmandy Horvath, Margaret	052255
Paugh, James J.	054549	Pazoutova, Katerina	009609
Pauka, T.R.	032012	Pčelinceva, Nelli Sergeevna	037150
Paukert, Karel	047808	Peabody Turnbaugh, Sarah	051956
Paukšte, Regina	030096	Peabody, Ann F.	049928

Peach Lynch, Ann 052674
Peacock, Pat 053586
Peacop, James 045301
Peak, Marianne 052435
Peake, Jane 054356
Pearce, Andrew 045626
Pearce, Barry 001535
Pearce, John N. 048956
Pearce, Pelham G. 047503
Pearce, Robert J. 006212
Pearsall, Barbara 051606
Pearson, A. Faye 005981
Pearson, Dave 046583
Pearson, Deidre 051132
Pearson, Fiona 044273
Pearson, Katherine 045921
Pearson, Marion 053858
Pearson, Rudy 006701
Pearsons, Don 007434
Peashock, Kristine 053825
Peaucelle, Denis 011667
Peavy, John M. 049770
Pečara, Dr. Zdravka 038155
Peces Rata, Felipe Gil . 040182, 040183
Pech, Edwin 034808
Pech, Dr. Jürgen 017643
Pecháček, Ivo 009646
Pechacek, Wilson 051686
Pechal, Dorothy 053708
Pecharromán y Morales, Ricardo 039917
Pechová, Dr. Jarmila 009468
Pecht, Ulrike 019137
Peci, Innocenzo 027532
Pečiukonienė, Mariona 030330
Peck Bracy, Tralice 050989
Peck, James F. 053850
Peck, Katie 053515
Peck, Scott 048111
Peckham, Cynthia A. 052647
Pecoraro, Joseph P. 047634
Peddell, Lesley 001578
Pedder, John 045177
Peden, Harold R. 053902
Pedersen, Birgit 010005
Pedersen, Jørgen 010006
Pedersen, K.S. 010199
Pedersen, Kenno 010048, 010108,
010341
Pedersen, Knud 010177
Pedersen, Lisbeth 010156
Pedersen, Lykke 010208
Pedersen, Morten 009998
Pedersen, T. 010199
Pederson, Curt 050932
Pederson, Jennifer 047718
Pedevillano, Diane 052340
Pedlar, Arthur 054285
Pedneault, Richard 007338
Pedretti, Cornelia 041932
Pedrini, Dr. Claudia . . . 026501, 026502,
026505, 026506, 026507, 026508
Pedrocco, Dr. Filippo . 028294, 028303
Pedrosa, Adriano 052964
Peduzzi, Francis 012292
Peebles, Cheryl 007374
Peek, Andre 053324
Peel, Prof. John S. 041111, 041112,
041124
Peele, Anna M. 048885
Peeperkorn-van Donselaar, Dr.
L.A. 032799
Peer, Dr. Peter 001985
Peers-Gloyer, Axel 021210
Peers, Elida 007054
Peers, Dr. Laura 045459
Peet, Catherine 043571
Peeters Saenen, E. 003549
Peeters, Herman 003545
Peeters, L. 003318
Peeters, Ludo 003547
Peffer, Susanne 017210
Peggie, David 045057
Peić Caldarović, Dubravka 009036
Peiffer, Jacques 013256

Peik, Karl-Heinz 019006
Peikštenis, Eugenius 030516
Pein, Max Gerrit von 022031
Peine, Arnold 016950
Peiry, Lucienne 041646
Peisch, Dr. Sándor 017131
Peitler, Karl 001982
Pejčinova, Ekaterina 004980
Pejković, Božidar 008927
Peković, Mirko 038185
Peláez Arbeláez, Leopoldo 008626
Peláez Tremols, Dr. Lucía 039436,
039441
Pélatan, Martial 011547
Pełczyk, Antoni 034939
Pelegrini, Dr. Béatrice 041534
Péligry, Christian 014622
Pelikánová, Jaroslava 009721
Pélissié, Gérald 012537
Pelkey, Ann 052186
Pella, Nicholas W. 051759
Pellard, Sandra-Diana 012490
Pellegrinetti, Jerri 007154
Pellegrino, Fabrizio 026145
Pelleman, Lode 003744
Pellen, Robin 044056
Pellengahr, Dr. Astrid 019565
Pellew, Robin 044620
Pellmann, Dr. Hans 020211
Pelon-Ribeiro, Laurence 012735
Pelong, Dany 012030
Pelosio, Prof. Giuseppe 027220
Pelser, R. 038873
Pelsers, Dr. L. 032310
Peltier, Cynthia 050224
Pelz, H. 017634
Pemberton, E. 001150
Peña Castillo, Agustín 031139
Peña Javier, Estrellita 034582
Peña Reigosa, Maria Cristina . 009127
Peña, Pedro R. Elizabeth 047302
Peñalver Gómez, Henriqueta . . 054765
Penard, Michel 016299
Penati, Fabio 026976
Pence, Debra 049410
Pence, Debra L. 047895
Pence, Noel 049606
Penčev, Prof. Dr. Vladimir 005191
Penčeva, Radka 005254
Penda, Octavian Ion 036307
Pender, Branko 008993
Pendleton, Chris 048850
Penent, Jean 015996, 015997
Peng, Fang 008559
Peng, Guirong 008101
Peng, Qi 007787
Peng, Shifan 008095
Peng, Yunqiu 008156
Pengzeng, Yuguan 007815
Pénicaud, Pierre 012599
Penick, George 049368
Penick, Steven 052763
Penin, David R. 053665
Pénisson, Elisabeth 014830
Penka, Bradley 050026
Penketh, Sandra 045536
Penkina, Elena Ivanovna 037082
Penkov, Petio . 005249, 005250, 005251,
005258
Penn, Joyce 047995
Penn, Michael 045846
Penna, Liisa 046585
Penna, Dr. Vasso 022982
Pennant 004084
Penndorf, Jutta 016534
Penner, E. Paul 007361
Penner, Jim 007072
Penner, Liisa 046582, 046586
Penney, David 048301
Penney, S.F.J. 006453
Penning, Dr. Mark 038706
Pennington, Claudia L. 049951, 049955,
049956
Pennino, Gaetano 027170

Penno-Burmeister, Karin 019842
Pennock, David 049653
Penny, Nicholas 045057
Pensabene, Dr. Giovanni 026640
Pensom, Carole 006104
Pentangelo, John 046771
Pentrella, Dr. Ruggero 027632
Penvose, Jane 053766
Penz, Marta 028252
Penzin, Viktor 037433
Pèon, Marie-Thérèse 013956
Peoples, Jo 000769
Peora, Prof. Oscar Carlos 000262
Pepall, Rosalind 006391
Pepić, Eva 030300
Pepich, Bruce W. 052440
Pepin, Didier 013512
Pepion, Loretta 047276
Pepper, Myles 044348
Pepper, Victoria 045717
Peppler, Tina 005492
Perac, Jelena 038168
Perales Piqueres, Rosa 039183
Peralta Juárez, Juan 038905
Peralta Vidal, Gabriel 007528
Peralta, Juan 034308
Peralta, Prof. Tole 007521
Perchinske, Marlene 047902
Percival, Arthur 044338
Percival, Sandra 053017
Perco, Daniela 025935
Percy, Ann B. 052081
Percy, David O. 047560
Perego, Graziella 026464
Pereira Cabral, Augusto J. 031805
Pereira de Araujo, J. Hermes . . 004750
Pereira Herrera, David M. 004205
Pereira Neto, Prof. João 035798
Pereira, Maria Isabel de Sousa . 035538
Pereira, Prof. Sonia Gomes . . . 004743
Pereireira Reis, Brasiliano 004497
Perelló, Antonia M. 039073
Perepelkina, Aleksandra
Fëdorovna 037881
Peretti, Pierre de 014290, 014296
Peretto, Prof. Raffaele 027688
Pereyra Salva de Impini, Prof.
Nydia 000620
Pérez Avilés, José Javier 040293,
040296
Pérez Basurto, Alejandro 031188
Pérez Camps, Josep 039715
Perez Castro, Pedro 040039
Pérez Espinosa, Juan Fidel . . . 031381
Perez Gollan, Dr. José Antonio . 000230
Pérez Gónzález, Melba 009319
Pérez Gutiérrez, Arturo 031208
Pérez Maldonado, Raúl 023365
Pérez Mejía, Dr. Angela 008587
Pérez Mesuro, Dolores 040256
Pérez Moreno, Rafael 039628
Pérez Navarro, Amalia 039653
Pérez Ornes, Nora 010385
Pérez Outeiriño, Bieito 040122
Pérez Quintana, José Andrés . . 009145,
009176
Pérez Recio, Manuel 039556
Pérez Rodríguez, Francisco
Javier 039873
Pérez Ruán, Liliana 008824
Pérez Villarreal, Pedro 008798
Pérez, Fernando Alberto 008760
Pérez, Juan Fernando 010493
Pérez, Júlia 039095
Perez, Patricia 034459
Pérez, Rosa María 023375
Perg, Franz 001801
Pergament, Allen 053381
Pergola, Prof. Philippe 028331
Pería, Alejandra 031216
Périer-d'Ieteren, C. 003441
Perillat, Shirley 005723
Perin, Giovanni 027936
Périn, Dr. Patrick 015342

Périnet, Francine 006535, 006536
Perini, Mino 027508
Perisho, Sally L. 048252
Perišin, Stane 008893
Perissinotto, Antonio 027305
Perkins, Bruce C. 054502
Perkins, Deborah 053268
Perkins, Diane 045919
Perkins, Karen 045168
Perkins, Martin 048417
Perkins, Stephen 049227
Perkko, Mariliina 010771, 010773
Perl, Alexander 017658
Perlee, A. 031728, 031735, 031742
Perlein, Gilbert 014479
Perline, Carol 053079
Perlis, Natasha 052998
Perlman, Patsy 046741
Perlov, Dr. Diane 050412
Pernas Ojeda, Maria Emma . . . 031182
Pernicka, Prof. Dr. Ernst 020255
Pernot, M.M. 053802
Pero, Linda 053537
Péron, Christian 014286
Peron, Claude 014879
Perot, Jacques 012071, 012645
Perra, Dr. Carla 025769
Perram, Richard 000828
Perraudin, Marguerite 041690
Perrault, Lise 007267
Perreault, Michel 006038
Perrefort, Dr. Maria 019029
Perren, Stephan 042176
Perrenoud, Raymond 041420
Perret-Sgualdo, Janine 041794
Perret, Gilles 041796
Perret, Jean-Claude 013755
Perret, Pascale 041658
Perrier, Dr. Danièle 016759
Perrigo, Thomas E. 001676
Perrine-Dubost 012391
Perron, Jean L. 006914
Perron, Michelle M. 048294
Perrot, Annick 014786
Perrot, C. 014356
Perrot, Mollie 051842
Perrott, Trenton 005914
Perroud, Paul 013396
Perry, Andy 050487
Perry, Dr. Barbara 047576
Perry, Bob 046774
Perry, Carole 051661
Perry, Ed 053968
Perry, Gregory P. 051244
Perry, James 053925
Perry, John 054271
Perry, Karen E. 051576
Perry, Michael L. 049182
Perry, Nancy . 049876, 050567, 052322
Perry, Peter 000951
Perry, Rianne 047601
Perry, Sheila 050735
Persan, Marquis de 013125
Perse, Marcell 019472, 019473
Persengieva, Marina 005098
Peršin, Vladimir 008925
Persson, Ann-Charlotte 040875
Persson, Bertil 040803
Persson, Curt 040804
Persson, Dr. Per-Edvin 011491
Persson, Sven 040659
Persson, Thomas 040739
Pertin, B. 024316
Pertola, Esko 011014
Péru, Dr. Laurent 014408
Peruga, Iris 054724
Pesarini, Dr. Fausto 026207
Pescaru-Rusu, Adriana . 036423, 036437,
036474
Peschel-Wacha, Claudia 003057
Peschel, Tina 017237
Peschler, Malgorzata 042222
Pescia, Sergio 042017
Pešek, Dr. Ladislav 009668

Pesenecker, Marita 018819
Pesi, Dr. Rossana 026640
Pešková, K. 038486
Peso, Jean-Michel del 015430
Pesolillo, Carlo 041739
Pesqueira de Esesarte, Prof.
  Alicia 031332
Pesquer, Dr. Ismael Angulo 038899
Pesqueur, Jean-Yves . . 012297, 012298
Pessa, Joanna 051429
Pessa, Loredana 026419
Pessemier 's Gravendries, Paul
  de 003769
Pessey-Lux, Aude 011590
Pessoa, Miguel 035639
Pessy, Emanuelle 015740
Pestalozzi, Dr. Martin 041191
Pestel, Michael 052161
Pestiaux, Pierre 003843
Petchey, Tim 046145
Peter, Carolyn 050426
Peter, Christopher 038826, 038827
Peter, Elmar 002171
Peter, Dr. Frank-Manuel 019675
Peter, Jack 052752
Péter, Dr. László 023831
Peter, Dr. Markus 041243
Peter, Dr. Michael 041883
Petercsák, Dr. Tivadar . 023613, 023620,
                         023621, 023622
Petermann, Jörg 018141
Petermann, Ralf 016640
Peterposten, Carlo 041204
Peters Bowron, Edgar 049588
Peters-Schildgen, Dr. Susanne . . 021289
Peters, Dr. Dirk J. 017606
Peters, Georg 019051
Peters, Dr. Gustav 017504
Peters, Dr. Hermann 018169
Peters, Irmtrude 018047
Peters, Karen 044949
Peters, M. 032246
Peters, Michael 022698
Peters, Ted H. 052191
Peters, Dr. Ursula 020853
Peters, W. 005518
Peterseim, Dr. Siegfried 020602
Petersen, Christopher 045992
Petersen, Flemming 010019
Petersen, Jim 046945
Petersen, John 052984
Petersen, Michele 006824
Petersens, Magnus Af 041021
Peterson-Moussaid, Kathleen . . . 046698
Peterson, Alan 053803
Peterson, Allan 051985
Peterson, Allan E. 052988
Peterson, Brent 053529
Peterson, Brian 048366
Peterson, Dr. Cheryl D. 049801
Peterson, Dennis 053439
Peterson, Glenn 051433
Peterson, James L. 052852
Peterson, John 052921
Peterson, Karin 049930
Peterson, Leslie 049473
Peterson, Mark F. 054482, 054483
Peterson, Merlin 049130
Peterson, Mildred 046903
Peterson, Pam 050665, 050667
Peterson, Paula J. 051604
Peterson, Roger 053072
Peterson, Thomas 047036
Peterson, Tom 052767, 053820
Peterson, Townsend 050186
Peterson, William 051146
Petey-Hache, Philippe 015949
Petit, Jean-Paul 012073
Petit, Nathalie 003916
Petit, René 012372
Petit, Prof. Th. 015879
Petke, Debra 049386
Petkov, Dimitär 005186

Petkov, Pavel 005045, 005046
Petkov, Vladimir 005157
Petout, Philippe 015438, 015439, 015440
Petr-Marčec, Prof. Smiljana 008891
Petranov, Borislav 003002
Petre, John L. 033283
Petrelli, Dr. Flavia 027014
Petrenko, Aleksej Stanislavovič . 037851
Petrenko, Vladimír Grigorjevič . . 037214
Petri, Gerda 010165
Petri, Susanne 020003
Petrica, Vasile 036654
Petridis, Constantine 047808
Petrie, Brian 045515
Petrie, Hugh 044941
Petrins, Aivars 030206
Petrinski, Ivan 005191
Petritsch, Dr. Ernst 002987
Petritsch, Mag. Tanja 002637
Petrone, P. Paolo 027003
Petroskey, Dale A. 047975
Petrov, Angel 004988
Petrov, Evgenij 043114
Petrov, Najden 005159
Petrova, Lidija 005237
Petrova, Nadežda 037060
Petrova, Velička 005053
Petrović-Raić, Marijana 038168
Petrović, Dejan 038188
Petrović, Dragutin 038257
Petrović, Marijana 038168
Petrović, Vladimír 038354
Petrovnina, Galina Michajlovna . . 037636
Petrovski, Zoran 030630
Petrovszky, Dr. Richard 021931
Petruccelli, Lic. M. Rosa 000264
Petrucci, Stefano 027283
Petruchno, Alla Sergeevna 036844
Petrus, Jerzy T. 034913
Petrus, Magdolna 024050
Petrusevičienė, Vilmantė 030313
Petrussen, Thomas 023331
Petryta, Bliderişanu 036662
Petry, Michael 045046
Petryshyn, M. 006023
Petschar, Dr. Hans 003053
Petschek-Sommer, Birgitta 017839,
                               017840
Petsini, Penelope 023030
Pett, Joachim 016633
Pettengill, Shirley 051923
Pettersson, Jan 040726
Pettersson, Kerstin 041058
Pettersson, Lennart 040652
Pettibone, Jody 054088
Pettibone, John W. 049134
Pettit, Marilyn 047252
Pettit, Steve 053677
Petty-Johnson, Gayle 054588
Petty, Dr. Edgar 051739
Petty, Ian 046174
Petty, Laura 053874
Petuchov, Jurij Michajlovič 037303
Pétursdóttir, Brynhildur 024066
Pétursson, Björn 024077
Pétursson, Halldór 024065
Petz, Reinhold 001693
Petzäll, Leif 041101
Petzal, Monica 044971
Petzold, A. 017919
Petzold, Lorraine 006197
Peuckert, Valérie 003323, 003324
Peukert, Jörg 018508
Peuzin, Jean 015033
Pevereit, S. 045391
Pevernagie 003499
Pewitt, Dr. J. Dudley 047008
Peyer, Ricard de 045186
Peyer, Richard de 045185
Peyer, Samuel 042184
Peylhard, Anne-Marie 011834
Peyre, P. 014122
Peyton-Jones, Julia 045104
Pezarro, D. 031905

Pezzino, Prof. Antonio 025867
Pfaff, Bettina 022418
Pfaffe, Andreas 021619
Pfaffenberger, Dr. Gary 052272
Pfaffenhuemer, Dr. Johannes . . . 002385
Pfanner, Dario 026644
Pfanner, Ute 001794
Pfanz, Donald C. 048952
Pfarr, Christa 003014
Pfau, F.D. 021671
Pfau, Gisela 004540
Pfeifer, Anton 001842
Pfeifer, Sandra 003091
Pfeiffenberger, Deborah 051241
Pfeiffer, Brian 047132
Pfeiffer, Günter 004002
Pfeiffer, Dr. Ingrid 018446
Pfeiffer, Rainer 021824
Pfeifhofer, Elisabeth 001945
Pfeifhofer, Helmut 001945
Pfeil, Elke 017069, 017092
Pfennings, Edgar 007550
Pferdehirt, Dr. Barbara 020233
Pfister, Emil 041418
Pfister, Sara 041776
Pflanz, Paul H. 053637
Pfleger, Prof. Dr. Susanne 022747
Pflüger, Helmut 022682
Pflug, Dr. Hermann 019130
Pflumm, Manfred 022338
Pfnorr, Reinhard 020781
Pfnuer, Linda 017037
Pfosser, Dr. Martin 002316
Pfotenhauer, Louise 049229
Pfretzschner, Prof. Dr.
  Hans-Ulrich 022217
Pfrunder, Dr. Peter 042149
Phagan, Patria 052338
Phares, Jane 050879
Phayana, Leah 038652
Phelan, Wynne 049588
Phéline, Jean-Michel 015666
Phelps, Kathleen D. 054711
Phelps, Martin C. 050761
Pheney, Kathleen 046812
Pheysey, Dawn 052405
Phil, D. 024369
Philbin, Ann 050427
Philbrick, Harry 052569
Philbrick, Nathaniel 051154
Philbrook, Wyman 053987
Philipon, Peter 007093
Philipp, Werner 017151
Philippart, Jean-Paul . . 003793, 003799,
                              003809
Philippe, Pierre 041437
Philippon, Annie 015131
Philippot 014227
Philipson, Jean 001336
Philipson, Peter 045030
Phillip, D.M. 033181
Phillips Olmedo, Carlos 031221, 031228
Phillips, Amanda 044779
Phillips, Christopher 051406
Phillips, Colin 001506
Phillips, Dale 053988
Phillips, Elizabeth 003021
Phillips, G.R. 038780
Phillips, George 006714
Phillips, J. Michael 043545
Phillips, James 047418
Phillips, John 043660
Phillips, Joseph R. 046667
Phillips, Dr. Kenneth E. 050412
Phillips, Lisa 051452
Phillips, Prof. Ruth B. 007290
Phillips, Dr. Sandra S. 053025
Phillips, Steven 048937
Phillips, Tim 048815
Phillipson, Anthony 042952
Phillipson, Prof. David W. 043849
Phillilps, Jane 045935
Phillp, John 043669
Philo, P. 045594

Philo, Phil 045259
Philotheou, G. 009414
Philp, B.J. 044134
Philp, Brian 045441
Philpot, Chris 043600
Philpott, Colin 043696
Philpott, Pat 006569
Philpott, Vandean 053019
Philson, William C. 050802
Phinney Kelley, Sarah 050225
Phinney, Chad 053707
Phinney, Gail 052461
Phipps, Bbetty K. 049251
Phipps, Rebecca J. 050189
Phleps, Dr. Roland 018488
Phuong, Pham Xuân 054820
Phuong, Tran Thi Thuy 054810
Piacentini, Dr. Paola 027627
Pianca, Dr. Vittorino 028413
Pianea, Dr. Elena 027716
Piasecka, Grażyna 035292
Piastrella, Dr. Carlo 026096
Piatek, Bruce J. 051894
Piątek, Jacek 035141
Piatt, Margaret 054312
Piatti, Verena 002342
Piaty, Karl 002891
Picard, Dr. Bertold 018225
Picard, Dennis 054327
Picard, Michel 014877
Picard, Thomas 051463
Piccand, Marcel 015725
Piccardi, Dr. Marco 026279
Picci, Prof. Dr. Piero 025532
Picciano, Eric 047721
Piccioli, Prof. Rossana 026540
Piccirilli, Guido 028035
Piccirillo, M. 025094
Piccoli, Prof. Adalberto 025886
Picelj, Zdenko 038585
Picha, Krzysztof Vincenc 034711
Picher, Claude 005587
Pichler, Dr. Isfried H. 001692
Pichler, Jutta M. 002233
Pichler, Lucia 001427
Pichorner, Dr. Franz 003026
Pickard, Kelli L. 048905
Pickering, Mona 050661
Pickford, Stephanie 044960
Pickin, John 045898, 045899
Pickles, Owen 033329
Pickup, A. 044368
Picollet, Auguste 011579
Picon, Carlos 051424
Picot 011952
Picozza, Prof. Paolo 027558
Picpican, Isikias T. 034433
Pidault, Monique 014865
Pido, Asuncion Q. 034520
Pidpala, Nila Andriïvna 043195
Piek, W. 033061
Pieke, Dr. Gabriele 017476
Pieke, Hannelore 016657
Pienimaa, Vilho 010914
Pieplow, Jane 048690
Pierce, Dr. Alexandria 005975
Pierce, Dr. Charles E. 051428
Pierce, Donald L. 052885
Pierce, Dr. Donna 053106
Pierce, Donna 048258
Pierce, Sally 047102
Piergollini, Alberto 027082
Pierides, Demetrios Z. . 009401, 009409
Pierlot, Claire 016039
Pierottavio, Daniele 027038
Pierrangeli, Prof. Carlo 027564
Pierré, Fernand 013507
Pierret, Philippe 003487
Pierrot-Bults, Dr. A.C. 032002
Piers, John C. 052298
Piersantini, Dr. Michele 028242
Pieschacón González, Gabriel . . 008816
Pieschel, Janet 005517
Pieters, Dr. Georges 003632

Pieterson, A. 032465
Pietilä-Juntura, Katriina 011428
Pietri de Caldera, Alicia 054731
Pietropaolo, Dr. Lisa 026645
Pietrzak, Stanisław 035289
Pietrzak, Ted 047303
Pietrzyk, Prof. Zdisław 034907
Pietsch, Dr. Martina 018737
Pietsch, Dr. Ted 053229
Pietsch, Dr. Ulrich 018004
Pietscher, Charles 049938
Pietschmann, Wolfgang 019320
Piette, Jacques 014516
Piezzi, Aron 041396
Pigeat, Jean-Paul 012532
Pigott, Tayloe 049763
Pihema, Joseph 033208
Pihlak, Jaak 010718
Pihlgren, Kjell 040673
Piirainen, Sari 010771
Piiroinen, Päivikki 011040
Pijbes, W. 032807
Pijeaux, Dr. Lawrence J. 047004
Pijl, B.H. 032560
Pijl, J. 031917
Pika-Biolzi, Dr. M. 042195
Pikal, Dr. Jaroslav 009608
Pike, Alan 038650
Pike, J.P. 006112
Pike, Julia 005686
Pike, Stephen J. 050785
Pikett, Vivian 048423
Pilar Muñoz Jiménez, María del 008614
Pilar, Jacquelin 048978
Pilati von Thassul-Filo della Torre,
    Francesca Gräfin 002056, 002597
Pilato, Dominique 015614
Pilchowski, Władysław 035013
Pilcz, Prof. Dr. Karlheinz 002402
Pileckas, Atanas 030417
Pili, Leleala E. 000084
Pilides, Dr. D. 009414
Pilioko, A. 054675
Pilkilton, Sue 053864
Pilkinton, Hunter M. 054240
Pillaert, E. Elizabeth 050600
Piller, Ferdinand 041326
Piller, Dr. Gudrun 041268
Piller, Josienne N. 052178
Pillichshammer, Johann 002439
Pillsbury, Dr. Edmund P. 048119
Pilotti, Bianca Maria .. 026016, 026078
Pils, Walter 002333
Pilven, Denis 012167
Pilz, Anette 022755
Pilz, Winfried 016441
Pimentel, Cristina 035942
Pina, Nora 051132
Pinate, Marc 053000
Pincemin, Cecilia 006092
Pindinelli, Prof. Elio 026368
Pineau, Jacky 013758
Piñeiro, Alberto Gabriel 000236
Pinelli, Prof. Antonio 027352
Pinette, Matthieu 015760
Pingel, Carol 049246
Pingeot, Anne 014743
Pingree, Shelda 048987
Pinheiro-Fankhauser, Marie-Anne 042046
Pinheiro, Maria Joao 035779
Pini, Lucia 026798
Pinilla, José 008689
Pinner, Keith 045670
Pinnix, Cleve 054376, 054477
Pino Morán, Mónica 007560
Pinon, Mary 049127
Pinsker, Dr. Bernhard 017822
Pinsky, Dave 053142
Pinson, Dr. Patricia 051702
Pintarić, Adam 009006
Pintarić, Snježana 009046
Pintarić, Tanja 009016
Pintea, June 048598
Pintea, Rodica 036476

Pinto de Matos, Maria Antonia . 035753
Pinto Simon, Elena 051344
Pinto, Dr. Bianca Alessandra ... 027568
Pinto, Dr. Ettore 025868
Pinto, Matt 051468
Pinzl, Christoph 022756
Pinzón, Carlos 008828
Pinzón, Rito 008805
Pionk, Richard 051463
Piot, Marc-Etienne 041453
Piotrovski, Michail 037734
Piotrowski, Jerzy 034654
Pipek, K. 009552
Piper, Catharina 041136
Piper, William 052799
Pipino, Dr. Giuseppe 027455
Pipkin, Roy 053600
Piquera, F. 013654
Piquereddu, Dr. Paolo . 027075, 027076
Piquer López, Blanca 039650
Pirenne, R. 004095
Pires Martins, Prof. Calebe 004957
Piriyakul, Yukol 042677
Pirker-Aurenhammer, Dr.
    Veronika 002953
Pirker, Christina 019434
Pirker, Rudolf 002864
Pirkmajer, Darja 038502
Pirkonen, Pekka 011062
Pirlot, Anne-Marie 003458
Pirnat, Miha 038552
Pirnie, M. 045509
Pirotte, Daniel 003964
Pisa, L.J. 032907
Pisa, Regina M. 053384
Pisani, Camille 003494
Pisani, Étienne du 038721
Pisano, Dr. Jane G. 050462
Pisarev, S.V. 037109
Pişiren, Tuncay 043030
Piskal, Irina Leonidovna 037894
Piskur, Bojana 038549
Piso, Prof. Dr. Ioan 036383
Pistono Grand, Mario 036420
Pištora, Jiří 009503
Pistorius, Dr. Agnes 003059
Pitcher, Stephen 054543
Piterskaja, Larisa Karlovna 037731
Piteša, Ante 008993
Pithie, Helen 000783
Pitkänen, Erkki 010798
Pitman, Bonni 048115
Pitoň, Helena 035445
Pitot 004083
Pitre, Marc 006146
Pitrey, Thierry 015655
Pitruk, Dmitrij L. 038038
Pitscheider, Erich 027972
Pitsios, Prof. Dr. Th. K. 023085
Pitt Jones, Priscilla 048351
Pitt, James 023344
Pittner, K. Heinz 021950
Pittock, Tina 043385
Pitts, Priscilla 033189, 033195
Pitts, Terence 047485
Pittz, Franeis 047882
Pitxot, Antoni 039401
Pitzen, Jutta 022331
Pitzen, Marianne 017489
Pitzer, Robert W. 054355
Piuma, Dr. Fabio 028331
Pixley, Patricia 051763
Pixner, Albin 027768
Pizitz, Jackie 051516
Pizzigoni, Prof. Gianni 028334
Pizzinini, Dr. Meinrad . 002131, 002141
Pizziolo, Marina 026815
Pizzo, Tony 053183
Pizzollo, Sissy 052068
Pjankova, Svetlana Anatoljevna . 036939
Pła, Josefina 034196
Plaat, Jaanus 010691
Plada, Mabel 046242
Plaen, Guy de 004051

Plaismone, Ehza Cilio 026382
Plan, Dr. Lukas 003014
Planas Marcé, Silvia 039448
Planas, Marta 000104
Planas, Ramón 039086
Planchard, Ch. 015212
Planchon, Jacques 012766
Planck, Prof. Dr. Dieter 019757
Planisek, Dr. Sandra 050558
Planjer, W.E.J.M. 031909
Plank, Benedikt 002671
Plánka, Dr. Ivan 009984
Plankensteiner, Dr. Barbara 003046
Plantalamor, Luis 039694
Plante, Stella 006219
Plantin, Colonel Kjell 041185
Planzer-Steiner, Rita Lisa 041969
Plasser, Dr. Gerhard 002637
Plaßmann, Dr. Otmar 021682
Plaßmeyer, Dr. Peter 017996
Plata Vega, Edith 008818
Platen, Dr. Michael ... 018953, 019447,
                        019457
Platero, Ramon 039863
Plath, Dr. Carina 020610
Platonova, T.V. 037007
Platow, Raphaela 054051
Platt jr., Geoffrey 052556
Platt, Ron 047005
Platt, Ronald 049257
Platteter, Janet 050056
Plattner, Fritz 002777
Platz, Daniela 022773
Platzgummer, Hanno 001828
Plavan, Tamara 043293
Plavec-Butković, Lidija 008891
Plazotta, Carol 045057
Pledienė, Daina 030409, 030410
Plenković, Ivo 008886
Plenter, Dr. H.A. ... 032993, 032994,
                      032995
Pleská, Leona 009877
Plessis, H. du 038856
Plessis, M. du 038819
Pletnjov, Valentin 005247
Plitek, Dr. Karl-Heinz 022809
Ploch, Dr. Gregor 021289
Plößl, Dr. Elisabeth ... 018680, 020693
Ploetz, Franziska 019656
Plomp, Dr. M. 032399
Płonka, Krzysztof 035130
Plotkowiak, Wiesław 034785
Plotoagă, Gabriela ... 036393, 036400
Ploug, Mariann 010021, 010022, 010236,
        010266, 010267, 010350, 010351,
                                  010352
Plouin, Suzanne 012034
Ploum, Y. 031907
Plourde, Nelie 046704
Plowright, Georgina . 043954, 044008,
                                  044608
Pluciński, Dr. Józef 035243
Pluhařová, Jana 009953
Plukavec, Lina 009049
Plume, Thierry 012900
Plummer, Dr. Ellen 050334
Plummer, Marguerite R. 053321
Plummer, Mary Ann 048935
Plummer, Sue J. 046665
Plummer, Val 046179
Plump, Darlene 052004
Plunkett, Stephanie 053537
Pluntke, Margitta 016535
Plutat-Zeiner 018980
Pluymaekers, Anne 003848
Pochman, Kathryn K. 051155
Pochwała, Stanisław 034892
Pocius, D. 001373
Počivalova, Elena Ivanovna 037045
Počivavšek, Prof. Marija 038501
Pocsainé Eperjesi, Eszter 023874
Podany, Jerry 050436
Podedworny, Carol 007367
Podestá, Horacio 000275

Podlesnik, Mateja 038547
Podmaniczky, Christine B. 047506
Podnar, Franc 038609
Podos, Lisa Beth 051344
Pögelt, Erika 018953
Poel, Dr. Peter te 032621
Pölcher, Bertold 018139
Pöll, Ing. Helmut 001922
Pöllmann, Rotraud 021420
Poelman, Shannon 007409
Poelmans, Marc 004116
Poenaru, Constantin 036722
Pöppelmann, Dr. Heike 020207
Pöpping, Johannes 020171
Pörnbacher, Prof. Dr. Hans .. 021985
Pörnbacher, Dr. Karl 019557
Pöschel, Helmut 022784
Pöss, Ondrej 038317
Poester, Teresa 004653
Pöstinger, Eberhard 002755
Poette, Simonne 016135
Pöttler, E. 002804
Pöttler, Prof. DDr. V.H. 002804
Pötzsch, Herbert 020263
Poggi, Patrizia 027485
Poggi, Dr. Roberto 026398
Poggiani-Keller, Raffaella 025753
Pogliani, Dr. Marisa ... 028246, 028247
Pogonec', Olena Dmytrivna 043207
Pogorzelska, Renata 034702
Pogt, Dr. Herbert 022825
Pohjakallio, Lauri 010781
Pohl, Burkhart 053731
Pohl, G. 017565
Pohl, Gerhard 020361
Pohl, Dr. Klaus-Dieter 017822
Pohl, Dr. Norman 018471
Pohlman, Lynette ... 046439, 046440
Pohowsky, Carolyn 051019
Pointal, Jean-Louis 012082
Poirel, Evelyne 015165
Poirier, Daphne 006989
Poirier, Diane 005610
Poissant, Margaret 049571
Poisson, Philippe 013702
Poitevin, B. 013855
Poivet, Clémence ... 016010, 016011
Pokorná, Blanka 009658
Pokorná, Libuše 009709
Pokorny-Nagel, Kathrin 003033
Pokorný, Jaroslav 009941
Pokorný, Marek 009473, 009474, 009475
Pokorný, Zdeněk 009556
Pokorski, Janice 050512
Poláčková, Dr. Dagmar 038328
Polacsek, John 048300
Polacsek, John F. 048303
Polak, Dr. J.J.H. 032981
Polak, Dr. Jerzy 034653
Polakovic, Dr. Vincent 038300
Polanco de Laverde, Cecilia ... 008696
Polanía Ramírez, Tiberio 008664
Polanich, Judith K. ... 047199, 053654,
                        053655, 053816
Polanská, Zuzana 009778
Polapink, Tom 052522
Polat, Cemal 042891
Poldermans, Dr. M. 032390
Poldervaart, Dr. A.A. 032991
Pole, Merle 001388
Polenov, Juri A. 036994
Poleske, Lee E. 053277
Poletti, Catherine 012427
Polfer, Michel 030593
Poli, Cristina Tessari 025474
Poliakoff, Miranda 045044
Polich, Debra 046487
Polichetti, Maria Luisa 027917
Polichetti, Dr. Massimilano A. .. 027627
Polikarpova, Marina Alekseevna 037075
Polioudovarda, Anna M. 023034
Poliščuk, Natalija 043273
Poliščuk, Svetlana Alekseevna . 036905
Politopoulos, Catherine 050477

Polivka, Marcus 049526
Poljakova, Valentina Alekseevna 038051
Poljanceva, I.B. 037370
Poll, Lothar C. 017207
Poll, Walter 020684
Pollák, Dr. Róbert 038370
Polland, Annie 051381
Pollard, Ian 045781
Pollard, S. 033168
Pollard, Shari E. 050177
Pollard, Ted 054254
Pollei, Dane F. 050770
Polley, Amy E. 046534
Polli, Kadi 010668, 010676
Polli, Dr. Vittorio 028433
Pollini, Lucia 041696
Pollock, Mallory 047768
Polnick, Bill 006990
Polo, Laura W. 050751
Polonovski, Max 014739
Polott, E. Leslie 049603
Poloznev, Dmitrij F. 037095
Polozova, Natalja Ivanovna . . . 038045
Pols, Dr. Ivor 038803
Polverari, Dr. Michele 025323
Polvi, Arvo 010892
Polyarnaya, Janna 037733
Polychronopoulos, Mari Elleneca 023200
Polymerou-Kamilakis, Dr. Aik. . . 022990
Polzine, Roxann L. 054552
Pomar, Maria Teresa 030893
Pomarede, Vincent 014749
Pomeroy Draper, Stacy 053813
Pomeroy, Dan E. 051187
Pomeroy, Marilyn 054528
Pommeranz, Dr. Johannes 020853
Pommies, Marie-Anne 012111
Pompa, Vince 046735
Pomponi, Luigi 027999
Ponce de León, Carolina 053000
Poncho, Roland A. 050346
Ponciş, Gheorghe 036608
Pond, Jennifer 052620
Ponder, Gary 052224
Pongracz, Dr. Patricia 051437
Ponomareva, Galina B. 037409
Ponomareva, Tatjana 037795
Pons-Pons, Fermí 040029
Pons, Marc 040396
Ponsonnard, Christian 013660
Ponsoy, Teresita 034614
Pont, Jeanne 041530
Ponte, Dr. António 036084
Ponte, Emily 049580
Ponte, Dr. Susanne de 020518
Ponton d'Amecourt, Jean de . . . 013344
Pontual, Sylvia 004666
Pontvianne, Jean-Paul 013520
Poodt, J.G.M. 032476
Poolaw, Linda 046463
Poole, Helen 044497
Poole, Helen E. 045790
Poole, J.E. 043842
Poole, Malcolm 045197
Pooran Yosef, Simin 024756
Poore, Henry 053509
Poorter, Nora de 003313
Poot, J.A. 031986
Pop, Valer 036762, 036763, 036765
Pop, Dr. Valer 036764
Popa, Gabriela 036713
Popa, Nicolae 036766
Popa, Onusia 036823
Pope, Clare 046173
Pope, Linda 053099, 054234
Popejoy, Tammy L. 050647
Popelier, Betty 006644
Popelka, Dr. Pavel 009932
Popescu, Dr. A. 043842
Popescu, Camelia 036728
Popescu, Doina 036813
Popescu, Ilie 036552
Popescu, Ioana 036320
Popescu, Viorel 036785

Popivoda, Mileva 038204
Poplack, Robert 046899
Popoiu, Dr. Paulina 036310
Popolla, Don Gianluca 028040
Popov, Aleksandr Nikolaevič . . . 038037
Popov, Gennadij Viktorovič . . . 037239
Popova, L.B. 037609
Popova, Nina Ivanovna 037736
Popovčak, Borivoj 009050
Popović, Bojana 038168
Popovic, Miroslav 030630
Popović, Pavle 038245
Popovic, Tina 009036
Popovici, Elena 036387
Popovici, Georgeta . . 036211, 036212,
                         036214, 036638
Poppa, Dr. Rudolf 017900
Poppi, Dr. Cesare 027434
Popski, Dragoljub 038180
Porada, Jerzy 035043
Porenta, Tita 038627
Porezag, Karsten 021908
Porignaux, Jonathan 003685
Porinchak, Mary Ann 054066
Pornin, Claudie 016054
Poroszlai, Dr. Ildikó 023920
Porr, Dr. Martin 021243, 021245, 021249
Porra-Kuteni, Valérie 011967
Porras, César Augusto 008732
Porrelli, D.M. 045339
Porsay, François 012108
Porsmose, Erland 010159
Porstmann, Dr. Gisbert 018011, 018012
Port, Jane L. 052211
Portal, Jane 047123
Portanje, B.K. 032117
Portatius, Botho von 018482
Porte, Patrick 014781
Portell, Rosa 053484
Porteous, Ken 007464
Porter, Ann 049174
Porter, Barry 053744
Porter, Frances J. 050705
Porter, Herman 066682
Porter, John R. 006742
Porter, Dr. Joseph 052451
Porter, Julian 043586
Porter, Marty 049872
Porter, Paul 052888
Porter, Paul R. 052793
Porter, Phil 050556, 050559
Porter, Phillip 005704
Porter, Richard T. 051780
Porter, Russ 001576
Porter, Suzanne 033258, 048409
Portet, Angel 039957
Portiabohn, Ivece 051460
Portigliatti Barbos, Prof. Mario . 028125
Portillo, Monica 030588
Porto, Prof. Dr. Nuno 035620
Porttier, W. 003300
Portugal-Humeau, Marie-Joëlle . 013190
Portús, Javier 039676
Portz, Willi 021609
Porzenheim, Maria 017883
Posada, Teresa 039676
Posani, Dr. Miguel 054726
Posch, Eva 018461
Poser, Adriana 051340
Poser, Steffen 020021, 020023
Pošiené, Irena 030559
Posinjak, Jurij Michajlovič 037955
Pospišilova, Dr. Alena 009641
Poßegger, Dr. Iris 017441
Possehl, Gregory L. 052100
Posselt, Dr. Martin 019559
Possémé, Evelyne 014733
Possenti, Dr. Elisa 027078
Post, Dr. K. 033052
Post, Melissa G. 047577
Post, Dr. William 047543
Postek, Roman 034950
Postel, Michel 012027

Postel, Mitchell P. . . . 051883, 052499
Postier, Wilfried 017201
Postles, Suella 045410
Poston, Jonathan 047551
Poston, Roger D. 047914
Potapour, Roald Leonidovič . . . 037832
Poteau, Gérard 013035
Poteete, Timothy 050195
Potet, Jean-Pierre 016287
Potgieter, M.J. 038867, 038867
Pothan, Scott 033365
Potocka, Maria Anna 035036
Potorczyn, Pawel 045076
Potrak, Rainer 020281
Potsch 001780
Pott, Karin 017157
Pott, Dr. Ute 018938
Potter, Byrn 050469
Potter, Edmund D. 053514
Potter, John 046014, 051033
Potter, Ted 052543
Pottier, Marie Hélène . . 012979, 012980
Potts, Albert 052530
Potts, Prof. Daniel T. 001548
Potts, Jennifer 047176
Potts, Timothy 048904
Potucek, Delvin 054063
Potužáková, Dr. Jana 009762
Poueymirou, Frank 037432
Pouget, Daniel 012977, 015404
Poulain, Ivan 013198
Poulet, Anne L. 051390
Poulis, Helen 006093
Poullain, Christine 013181
Poulliot, Elizabeth A. 051575
Poulos, Ann 053707
Poulos, Paul 053879
Poulsen, Hanne 010107
Poulter, Gill 044167, 044170
Pounis Guntavid, Joseph 030675
Pourtalès, Laure de 011609
Pourtalet, P. 005960
Pourtaud, François 013347
Povilionienė, Vita 030444
Poviliúnas, Virgilijus 030491
Povšič, Sabina 038549
Powar, N.V. 024469
Powders, Dr. Vernon 046430
Powell Harold, Fran 053177
Powell, Bobbie 052595
Powell, Derek 046075
Powell, Earl A. 054141
Powell, JoAnne 046847
Powell, Maurice F. 019539
Powell, Michael 045041
Powell, Robert 044724, 045376
Powell, Shana 049011
Powell, Shona 045411
Powell, William V. 050004
Power, Cathy 044940, 046008
Power, Christopher 001351
Power, Deirdre 024845
Power, Dennis 051681, 051684
Power, Nicola 045722
Powers, Carolyn T. 048693
Powers, Catherine 046564
Powers, Claire 006510
Powers, Dale 052742
Powers, James L. 049633
Powers, Leland 049415
Powers, Linda 050044
Powers, Richard 050097
Poyner, Noelle 001197
Pozilijonov, Avazbek 054627
Pozsonyi, József 023431
Pozzi, Dr. Raffaella 026183
Pozzuoli, Prof. María Cristina . . 000412
Prackwieser, Walter 026223
Prada Pérez, Isaac 008684
Prader, Angela 041435
Prader, Jürg 041435
Prado, César O. 034272
Prätzel, Maritta 021429
Prag, Dr. A.J.N.W. 045210

Prahase, Mircea 036245, 036407,
         036539, 036580, 036657, 036677
Prakapcov, Vladimir I. 003253
Prall, Steve 046565
Pralong-Salamin, Rachel 041984
Pralormo, Marcella 028145
Prammer, Dr. Johannes 022014
Prang, Werner 020467
Pranza, Denise M. . . . 052531, 052735
Prasch, Ing. Baldur 002703
Prasch, Dr. Hartmut 002766
Prasolova, Elena Ivanovna 037787
Prasser, Gabriele 020872
Praßl, Johann 001878, 001879
Praszelik-Kocjan, Grażyna 035064
Pratesi, Dr. Giovanni 026286
Prati, Dr. Luciana 026322, 026323,
         026324, 026325, 026327, 026330
Pratt Brown, Alice 051491
Pratt, John 045388
Prause, W. 017805
Prayitno, Dr. R. Joko 024613
Praz, Jean-Claude 041994
Praženilová, Dr. Viera 038429
Préaux, Alain 015168
Prebushewsky Danyliuk, Janet . 006984
Preciado, Kimberly 052974
Preece, Geoff 044109
Preininger, Kristina 038552
Preising, Dr. Dagmar . 016434, 016445
Preiss, Prof. Heinz 002241
Preiss, Slawomir 035219
Preiße, Erika 019718
Preissler, Leonilda Maria 004490
Premack, Ellen 048267
Premerl, Nada 009043
Premoli Silva, Adolfo 027304
Premoli Silva, Dr. Isabella 026812
Premović-Aleksić, Dragica 038213
Prendergast, Mary B. 024844
Přenosilová, Dr. Vera 009797
Prentiss, Margo 052200
Presgrave, A.D. 001109
Presl, Igor 038587
Presnell-Weidner, Denise 052217
Pressl, Hannes 001734
Pressley, Sheila 053033
Presson, Virginia 054248
Prest, Véronique 014697
Prestaus, J.M. 011743
Preston, Bill 049219
Preston, Bryan 047489
Preston, Carol 053018
Preston, Dr. Charles R. 047855
Preston, P. Anthony 033170
Preston, Dr. P.M. 044240
Preston, Tantum 047005
Preszler, Robert E. 052471
Pretola, John P. 053464
Pretorius, A.F. 038769
Pretorius, Mimi 038656
Pretsch, Dr. Peter 019519
Pretty, Carol Anne 005536
Pretzel, Elke 020682
Pretzl, Manuel 020515
Preuß, Hartmut 018145
Preuß, Helmut 019757
Preuß, M.A. Patricia 022083
Prévost-Bault, Marie-Pascale . . 014835
Prévost-Bouré, Pascal 014494
Prevosti, Marta 039972
Prevratil, Joseph F. 050373
Prew, R. 043908
Prewer, Steven 043506
Prewett, Vickie 047732
Preyer, Josef 002527
Pría, Melba 031242
Přibáňová, Dr. Svatava 009480
Pribbernow, Marcus 020211
Přibil, Vladimír 009746
Přibyla, Radomír 009578
Price Shanis, Carole 052079
Price, Alan 045435

Price, Anita 005681, 005682
Price, John 050631
Price, Kari 047048
Price, Margaret 052333
Price, Dr. Marla J. 048906
Price, Marti 047642
Price, Paula T. 052579
Price, Renée 051450
Price, Sally Irwin 046834
Price, Terri 050979
Price, Thomas E. 047919
Prickett, Marcus 031838
Prickett, N.J. 033142
Priddle, Th. 046236
Priddy, Dr. Bennie 022362
Prie, Jean-Yves 014890
Priego Fernández del Campo,
   Carmen 039621, 039668
Priem, Dr. R. 032687
Prière, Marc 011850
Prieto, Gabriel 008631
Prieur, Abel 016291
Prieur, P. 041358, 041561
Prikolota, Olga Vasylivna 043232
Přikrylová, Ludmila 009707
Priljeva, Srdjan 038260
Prillinger, Judith 002416
Primozic, Melanija . . . 038625, 038627
Prince, David 053634, 053638
Prince, J.R. 045198
Prince, Nigel 043620
Prince, Shannon 006511
Prince, Todd 047474
Pringle, Roger . 045901, 045902, 045903,
            045904, 045905, 045907
Prinz-Büchl, Mag. Margarethe . . 001907
Prinz, Gerhard 022047
Prinz, Dr. Ina 017480
Prinz, Peter 047009
Prinz, Dr. Ursula 017085
Prinzky, Manfred 022043
Prinzler, Hans Helmut 017066
Prioli, Dr. Ausilio 025752
Prior, Annette 040660
Prior, Claudine 005497
Prior, Terrence M. 051839
Pripačkin, I.A. 037186
Prister, Boris 009036
Prister, Lada 009036
Pritchard, Frances 045219
Pritchard, Philip 007182
Pritchard, Susan 045311
Pritchett, Leslie 052994
Pritikin, Renny 048164
Prittwitz und Gaffron, Dr. Hans-Hoyer
   von 017499
Pritzker, Elisa 049469
Privette, Carol B. 054048
Privette, David 050321
Prix, Jörg 001889
Prjamikova, Vera Pavlovna 037156
Probst, Heinz 020661
Probst, Peter 041999
Probst, Dr. Rudolf 041323
Probst, Dr. Volker 018883
Probst, Volker 018884
Procacci, Ugo 026273
Procelli, Dr. Enrico 027474
Prochaska, James J. 047504
Prochazka, Denise 048496
Prochina, Irina Evgenjevna 037444
Procopé Ilmonen, Susanne 011142
Procopiou, E. 009414
Proctor, Jacob 046492
Proctor, Rebecca 043838, 043842
Proctor, Vicki 048517
Próder, István 024033
Prod'Hom, Chantal 041652
Proebst, Walter 016470
Prøhl, Henning 041074
Prøsch Danielsen, Lisbeth 033972
Prösch, Dr. Christoph 022842

Prokisch, Dr. Bernhard   001870, 002328,
            002330, 002333
Prokop, J. 009726
Prol, Elbertus 052575
Proll, Jochen 016966
Pronti, Dr. Stefano 027296
Prophet, R. 046192
Proschwitz, Ted von 040653
Proskynitopoulou, Rosa 022997
Proß, Christian 020685
Prosser, Jo 045123
Prost, Darin 052434
Prota, Prof. Romolo 005574
Protektore, Evilena 030210
Prothero-Brooks, C.C. 007028
Proto Pisani, Dr. Rosanna
   Caterina . . . 026246, 026248, 026282,
                        026301
Protz, Edward L. 049037
Proud, Hayden 038687
Proud, Hayden Russell 038681
Proust, Georges 014703
Prouty, Elisha 049170
Provencio, Bobbie 050154
Provick, A.M. 005790
Providi, Dr. Ioanna 023018
Provo, Dan 051737
Provorov, Viktor Petrovič 037364
Prpa Stojanac, Ika 008993
Prudhomme, James 051198
Prudhomme, Mary Louise 046816
Prudhomme, Suzanne M. 050998
Pruett, Herbert E. 054427
Prugh, Burton 047316
Pruitt, John E. 053443
Prum, Richard O. 050186
Prunet, Pierre 015386
Prusinovsky, Rupert 021047
Pruska, Mariola 034935
Pryor, Eric 053599
Prys-Jones, Dr. R. 045998
Przeczewski, P. Marek 026345
Przekazinski, Andrzej 035327
Przigoda, Dr. Stefan 017433
Przybył, Grazyna 034757
Przybylek, Leslie 048874
Przybylek, Stephanie 053197
Przybylska, Grażyna 035179
Przybysz, Jane 053044
Przychowski, Alexandra von . . . 042221
Przygoda, Udo 018419
Przypkowski, Piotr Maciej 034793
Psaropoulou, Betty 022982
Psenner, Dr. Barbara 002122
Psota, Dr. Thomas 041315
Pstrag, Renata 012119
Pucciarelli, Bepi 026457
Puccini, Simonetta 028151
Puchner, Manfred 002327
Puckett, Jakobus 019375
Puckett, James 016683
Puckett, Lori 053792
Puckett, Robert A. 054396
Puckett, Tom 048572
Puckitt, Sarah 053038
Puddy, Audrey 043715
Pudelska, Aleksandra 035122
Puech, Laurent 013715
Puentes de Valderrama, Ana
   Rosa 008796
Pürro, Gertrud 041372
Puerta, Prof. J. 039612
Puertas Tricas, Rafael . 039702, 039703
Puerto, Dr. F. Javier 039648
Püütz, Ursula 021060
Püüa, Endel . . 010628, 010630, 010631
Puga, Rosa D. 007537
Pugačevskaja, Alla
   Aleksandrovna 037456
Puget, Catherine 014931
Puggaard, Ole 010144
Pugi, Guido 027448
Puharich, Nick 033184

Puhle, Prof. Dr. Matthias 020207, 020210
Puhlmann, Dr. Helga 020509
Puig, Felipe 046264
Pujol, Albert 039097
Pujol, Jochen 000094
Puklowski, Tracy 033240
Pule, D.E. 004296
Pule, Tickey 004295
Pulinka, Steven M. 051708
Pulle, Thomas 002698
Pullen, Bernice 051815
Pullen, Reg 046775
Pullen, Sue 044626
Pullin, Robert 044229
Pulnar, Janusz 035164
Pultnones, Lucio 034470
Pultz, John 050187
Pulvenis de Séligny,
   Marie-Thérèse 014486
Pumbaya, Ismael 034550
Pumphrey, Jim 047437
Puncher, Sebastian 044079
Pungerčar, Ivana 038585
Puntman, Paul 033056
Pupăză, Grigore 036207
Pupo de Mogallón, Yolanda . . . 008657
Purba, Suruhen 024615
Purcell, Beth 047881
Purcell, John K. 047436
Purdy, Dorothea 047330
Purdy, John 048822
Purdy, Margaret 049676
Pure Via, Luis 040386
Purin, Bernhard 020534
Purini, Sergio 003492
Purmal, Kathie 049996
Purpus, Dr. Elke . . . 019696, 019710
Pursley, David W. 052769
Purves, W.M. 044252
Puscher 019753
Puschnig, Dr. André 041282
Pushman, Tracie 000999
Pusineri Scala, Prof. Carlos
   Alberto 034199
Pusineri, Prof. Adelina 034201
Pust, Dr. H.-C. 022067
Pusterla-Cambin, Patrizia 041689
Putin, Olga 036780
Putnam, Christy 051491
Putnam, Michael 046786
Putnam, Polly 044785
Putney, Carolyn 053760
Puts, Claude 003772
Putsch, Dr. Jochem 021903
Puttock, Dr. Sonia 001475
Putyńska, Elżbieta 034663
Putz, Karl 022082
Putze, Alfred 019387
Putzert, MA Nadja 018047
Puzanova, N.A. 029801
Puzik, Olga Vitalijovna 043189
Puzo, Joseph 048012
Puzrina, Oksana Jurjevna 037231
Py, Cyille 015179
Py, Jacques 015904
Py, Jean-Samuel 041914
Pyatt, John 033221
Pycke, J. 004088
Pyhrr, Stuart 051424
Pyka, Henryk 034815
Pykäläinen, Mari 011424
Pylkkänen, Heikki 010993
Pym-Hember, Peter 006573
Pyrrou-Drivalou, Ourania 022984
Pysher, Ray 050755
Pytlik, Wincenty 035096
Qaavigaq, David 023336
Qёndro, Gёzim 000030
Qi, Junjie 008355
Qian, Xiaoping 008301
Qiao, Delong 007829
Qiao, Yuguang 007868
Qiao, Zheng'an 008032
Qin, Xinghan 007612

Qiu, Aidong 007896
Qiu, Damin 007828
Qiu, Fang 007595
Qiu, Jinkun 008444
Qiu, Yongsheng 008456
Qu, Liping 008490
Quaas, Achim 019006
Quackenbush, Bill 007375
Quackenbush, Laura J. 050217
Quackenbush, Molly 047308
Quackenbush, W.G. 005383
Quaghebeur, Marc 003430
Quail, Sarah 045551
Quairiaux, Yves 003890
Qualizza, Dr. Marino 028227
Qualmann, Kenneth 046158
Quam, Jeffery 050718
Quan, Benning 008161
Quan, Zhongde 008006
Quann, Byron 049374
Quarcooporne, Nii 051499
Quarles Van Ufford, Dr. C.C.G. . 032960
Quartero, Dr. Luigi 028383
Quarterone, Dr. Anna 025869
Quatember, Dr. Wolfgang 001836
Quattrocchi, A. 006632
Quattrociocchi, Maria Pia 027285
Quayle, Frederick M. 053616
Quéau, Jean-François 012299
Queckenstedt, Dr. Hermann . . . 021011
Quellenberg, Björn 042207
Queree, Jennifer 033168
Querejazu Calvo, Dr. Jorge . . . 004243
Quermann, Carolin 018012
Quero Castro, Salvador 039653
Quetard, Patrice 011808
Quevedo, Emilio 008603
Quevedo, Jacinto 039904
Quick, Betsy D. 050422
Quien, Prof. Guido 008996
Quijano, Marcela 022260
Quijano, Ritchie 034519
Quilitzsch, Uwe 017868
Quinata, Tom 023358
Quinlan, Karen 000849
Quinlan, M. 043567
Quinn, Alexandra 053144
Quinn, Betty 024866
Quinn, Thomas 024934
Quinn, Vivienne 043816
Quinquis, Pierre 016223
Quinsonas, M. de 013705
Quint, Eduard 032516
Quintana Reyes, Remedios de . 040054
Quintanilla, Faustino 046838
Quintanilla, Mimi 052954
Quintanilla, Sonya 052978
Quintas de Sousa, Dr. Cipriano . 000109
Quintelier, Johan 003649
Quintens, Patricia 003429
Quintens, Roger 003785
Quintero Durán, Miguel 008738
Quirce, Sylvie 013286
Quirke, Stephen 045073
Quist, Bob M. 048531
Quist, Kris N. 051017
Quoc Binh, Prof. Dr. Truong . . 054798
Quoc Quan, Phạm 054797
Raab, DI Walter 002703
Raabe, Dr. Christiane 020510
Raabe, Dr. Christiane . 020510, 020520,
            020532, 020547
Raabe, Dr. Eva Ch. 018434
Raahauge, Kristine 023332
Raassen Kruimel, E.J.C. 032562
Raastad, Ass. Prof. Jan Emil . . 033847
Raat, Drs. R. 032957
Raath, A. 038847
Rabal Merola, Victoria 039225
Rabano, Isabel 039664
Rabben, Anne Lise 033845
Rabbidge, Ted 033369
Rabeder, Dr. G. 001734
Rabeling, C.M. 032231

Raber, Christie	047162	Rahlff, Reiner	018342, 018343	Rampazzi, Filippo	041696	Rasson, Camille	003329
Rabie, L.N.	038715	Rahman Chuchu, Abdul	004969	Rampl, Dr. Franziska	001772	Rassweiler, Janet	051498
Rabin Siegal, Robert L.	036175	Rahman-Steinert, Uta	017239	Ramsauer, Dr. Gabriele	002628, 002629	Rašticová, Dr. Blanka	009931
Rabino Massa, Prof. Emma	028126	Rahmani, Dr. A.R.	024407	Ramsay, Gael	000815	Rastorguyeva, Olga	036850
Rabinovitch, Victor	005871	Rahmatulaev, I.	054639	Ramsay, Peter	001319	Rataiczyk, Matthias	018964
Rabinowitz, Rivka	025176	Rahoult, Lena	040990	Ramsay, Susan	007076	Ratajczak, Rafał	035139
Rabner Cymerman, Mauricio	031002	Raič, Kristijan Dragan	008906	Ramsden, Alan R.	006459	Ratcliffe, Shirley	043987
Rabner, Mario	031255	Raiger, Johann	020511	Ramsden, William	053573, 053576	Rath, Andreas	003098
Rabner, Mauricio	031015, 031270	Raihle, Jan	040620	Ramseier, Dr. Markus	041673	Rath, H.	021653
Rabold, Dr. Joachim Martin	017002	Raimond, Jean-Bernard	011971	Ramsey, Ron	001355	Rath, I.	021653
Raboud-Schüle, Isabelle	041382	Raimondi, Dr. Ludovico	026449	Ramsey, Steve	051596	Rath, L.	019295
Raby, Dr. Julian	054091, 054116	Rainbird, Sean	022057	Ramseyer, Alfred	041834	Rathbone, Eliza	054157
Race, Anne Katrin	016774	Rainbow, Prof. P.	045061	Ramseyer, Dr. Denis	041586	Rathburn, Robert R.	047007
Rachbauer-Lehenauer, Mag. Karin	002620	Rainer, Mag. Gerhard Florian	003046	Ramskjaer, Liv	033854	Rathgeber, Paul	017714, 017716
Rachewiltz, Dr. Siegfried de	028089, 028090	Raineri, Maguerite	050032	Ramskov, Conni	010206	Rathke, Dr. Christian	020970
Rachimova, Akajša Rachimovna	029787	Raines, James L.	052915	Ramsour, Bruce	046679	Rathore, Dr. Madhulika	024168
Rachinger, Dr. Johanna	003053	Raines, K.	006264	Ramus, Lourdes I.	036163	Ratier, Hugo	000480
Rachman Zeth, H.R.A.	024627	Rainger, Don	045209	Ramušćak, Prof. Ljubica	008891	Ratner, Phillip	025223
Rachmankulov, Ilgiz Šamiljevič	037170	Rainsbury, Anne	043927	Ramuschi, Maria Cristina	028166	Ratnikova, Marina Seměnovna	038074
Racine, Bruno	014627, 014773	Rais, Dagmar	006438	Ramut, Eliza	034622	Rattemeyer, Dr. Volker	022641
Racine, Ellen	051628	Raisor, Jerry	052537	Rana, A.K.	024214	Rattenbury, Richard	051732
Racine, Yolande	005615	Raiss, Gerhard	018274	Ranade, Dr. Hari Govind	024450	Ratti, Dr. Enrico	028312
Rack, Theo	018666	Rait, Eleanor	049437	Ranaldi, Prof. Franco	027430	Ratti, Dr. Marzia	026538
Rackow, Marcia	051480	Raith, Robert	017930	Rand, Michael	054224	Ratti, Pietro	026535
Racky, Kurt	021007	Raithel, Andreas	020630	Rand, Patrick	047848	Rattigan, John	024919
Raczka, Laurel	052077	Rajaram, K.	024438	Rand, Richard	054419	Ratzenböck, Hans-Jörg	002317
Raczka, Robert	050752	Rajčeva, Vasilka	005145	Randall, M.	038861	Ratzky, Dr. Rita	023492, 023543
Rada, Judith K.	053754	Rajchevska, Tsvetana	005191	Randall, Patricia J.	053517	Rau, Daniel	041619
Radam, Brian	045840	Rajić, Delfina	038188	Randall, R. Dennis	049601	Rau, Gudrun	019212
Radawetz, Natalia	007466	Rajkinski, Dr. Ivan	005264, 005265	Randall, Ross	046387, 046398	Rau, Lorraine	048728
Radčenko, Olga Ivanovna	037714	Rajkowic-Kamara, Dr. Julia	018434	Randazzo, Giuseppe	027186	Rau, Dr. Patrick	019757
Raddatz, Birgit	017730	Rajkumar, Kimcha	005919	Randić Barlek, Mirjana	009028	Rauch, Elvira	022550
Radday, Helmut	017744	Rak, Natalija Egorovna	037539	Randow, Renate von	022626	Rauch, L.	041971
Radecki, Martin J.	049692	Rakonjac, Srdjan	038168	Raney, Floyd	050341	Rauch, Margot	002136
Radel, Chuck	052432	Rakotoarisoa, Jean-Aimé	030644	Rangaraju, Dr. N.S.	024417	Rauch, Peter	041972
Rademacher, Henning	019019	Rakowitz, Dietlinde	001731	Rangström, Lena	041015	Rauch, Werner	014592
Radford, Ron	000937	Raku XV, Kichizaemon	028977	Raninen-Siiskonen, Tarja	010917, 010918	Rauchbauer, Judith von	016571
Radić, Danka	009009	Ral'čenko, Iryna Grygorivna	043119			Rauchenwald, Alexandra	001736
Radić, Mladen	008962	Ralčev, Prof. Georgi	005199	Ranjbaran, Sa'eid	024746	Raud, Anu	010613
Radicke, Rüdiger	021702	Ralli, Efterpi	023023	Rank-Beauchamp, Elizabeth	053286	Raudsepa, Ingrida	030177
Radigois, Bertrand	013809, 013810	Ralls, Nick	043610	Rank, Karin	018470	Raudsepp, Renita	010668
Radigonda, Ron	051733	Ralph, Honeycutt	047350	Rankadi, Daniel	038668	Rauer, Helmut	018198
Radin, Saul	047238	Ralston, B.	033142	Rankin, Gary	043884	Rauf, Barb	048063
Radins, A.	030207	Ramachandran, V.S.	024372	Rankin, John L.	005980	Rauhut, Dr. Oliver	020561
Radivojec, Milenko	004270	Ramaekers, M.T.	004042, 004043	Rankin, Marty	046919	Raulff, Prof. Dr. Ulrich	020260, 020262
Radivojević, Milan	038186	Ramalho, Prof. Dr. Miguel Magalhães	035801	Rankin jr, P.R.	048550	Raunio, Erkki	010758
Radley, Seán	024966	Raman, Anna	042937	Rankin, Ray	044945	Raue, Kåre William	033597
Radman-Livaja, Ivan	009025	Rambelli, Faustolo	026714	Rankin, Shan	048478	Raus, Edmund	050620
Radojičić, Milica	004256	Rambourg, Bernard	015047	Ransom, Karen	051680	Rausch, Amalia	003142
Radovanović, Dragan	031764	Rambow, Charles	048844	Ransome Wallis, Rosemary	044979	Rausch, Felicitas	042193
Radović, Anđelija	038185	Ramesh, J.	024125	Ranšová, Eva	009725	Rausch, Guy	030590
Radović, Roman Bogdanovyč	043249	Ramesmayer, Catrin	002630	Rantanen, Reijo	011512	Rausch, Jürgen	021094
Radrizzani, Dominique	042108	Ramet, Guy	004127	Ranweiler, Margaret	052611	Rauschenberg, Chris	052288
Radschinski, Margitta	022888	Ramet, Robert	003856	Rao, D.V.	024456	Rauscher, August	002535
Radtke, Edward	034947	Ramey, Marilyn	048002	Rapetti, Rodolphe	014784	Rauscher, Hannelore	022529
Radulescu, Speranta	036320	Ramharter, Dr. Johannes	002848	Rapisarda, Alfredo Pablo	000673	Rauschmann, PD Dr. Michael	018413
Radulović, Ksenija	038167	Ramhorn, Amber	049758	Rapkin, Grace	051412	Rausis, Jean-Pierre	041905
Radut, Ion	036416	Ramirez Casali, María Eliana	007547	Rapmund, J.	032826	Raussmüller-Sauer, Christel	041947
Radwan, Krzysztof	034896	Ramírez de Lara, Ligia	008668	Raposo, Luís	035806	Raussmüller, Urs	041947
Radycki, Dr. Diane	046964	Ramírez Hernández, Jose Luis	031529	Rappe-Weber, Dr. Susanne	022710	Rautjoki, Ville-Matti	011075
Radzak, Lee	053871	Ramírez, Angel Antonio	001882	Raptou, E.	009414	Rauzi, Prof. Estebar	000483
Radzina, K.	030221	Ramirez, Jan	051455	Raptou, Dr. Eustathios	009433	Ravanelli Guidotti, Dr. Carmen	026175
Raemdonck, André	003481	Ramirez, Joe	047456	Rardin Johnson, Sara	052319	Ravaonatoandro, Aldine	030646
Raetzer, Maren	019038	Ramírez, José Juan	040212	Ras-Dürschner, Dr. Kerstin	017607	Ravasi, Prof. Gianfranco	026827
Rafanelli, Dr. Simona	028356	Ramírez, Marcela E.	031168	Rasajski, Javor	038257	Ravasio, Dr. Tiziana	025477
Rafey, Souhad	051323	Ramírez, Mari Carmen	049588	Rasario, Dr. Giovanna	026285	Rave, Dr. August B.	022057
Raffaelli, Prof. Mauro	026280	Ramírez, Mari	054283	Rasbury, Patricia	051187	Raved, Noga	025026
Raffel, Suhanya	001503	Ramírez, Suny	031005	Rasch, Elisabeth	041106	Raveendran, K.	024500
Rafferty, Pauline	007327, 007335	Rammer, Gottfried	002219	Rasch, Simone	021211	Raveh, Kurt	025160
Rafuse, Linda	006203, 006204	Ramón Vilarasau, Sandra	054717	Rasch, Tone	033854	Ravel, Christian	013998
Ragan, Charlotte	047407	Ramon, René	044149	Rasche, Dr. Adelheid	017202	Ravelo Virgen, Margarita	031628
Ragbir, Lise	051044	Ramond, Serge	016175	Rash, Roswitha M.	052015, 052018	Ravenal, John	052565
Ragionieri, Dr. Pina	026243	Ramond, Sylvie	013940	Rasjid, A.	024610	Raverty, Norm	001049
Ragni, Dr. Elena	025629	Ramoneda, Josep	039056	Rask, Sven	010082	Ravid, Zami	025155
Ragon, Jacques	015419	Ramoneda, Prof. Luis Alberto	000386	Rasmussen, Alan H.	010154	Raviv, Efrahim	025173
Ragsdale, Tammie	050562	Ramos Fernández, Dr. Rafael	039365	Rasmussen, Else	010179	Ravnik Toman, Barbara	038496, 038497, 038531, 038532, 038533, 038621
Raguindin, Ethelinda A.	034430	Ramos Horta, Dr. José	010408	Rasmussen, Frank Allan	010182	
Rahaman, F.	024482	Ramos Pérez, Froilán	031123	Rasmussen, Keith	050033	Ravoire, Olivier	014583
Rahe, Diane	049517	Ramos Ruíz, Martha	008730	Rasmussen, Dr. Pamela C.	048441	Ravonatoandro, Aldine	030645
Rahe, Gebhard	021588	Ramos, Dr. Manuel Francisco Castelo	036031	Rasmussen, Robert	051981	Raw, Jo	043443
Rahe, Dr. Thomas	020105	Ramos, Margarida	035815	Rasmussen, William	052564	Rawdon, David D.	038782
Rahim, Abdul Gaum Abdul	000006	Ramos, Salome D.	034604	Rasmussen, Allan	041182	Rawe, Dr. Kai	020492
				Raspail, Thierry	013934	Rawle, Caroline	046084
				Rasseaux, Anne-Françoise	003890	Rawles III, S. Waite	052557

Rawlinson, Philippa 045496
Rawluk-Raus, Bożena 035170
Rawson, Geoff 001574
Rawson, Helen 045678
Rawstern, Sherri 046299
Rawsthorn, Alice 044958
Ray, Dulce Ivette 048905
Ray, Dr. Ella Maria 048261
Ray, Jeffrey 052031
Ray, Justice A.N. 024360
Ray, Randy W. 049583
Ray, Romita 046593
Ray, Sandra 046568
Rayburn, Ella S. 053219
Raymond, André 003960
Raymond, Dr. C. 048597
Raymond, Nancy 007311
Raymond, Richard 054593
Raymond, Yves 006865
Raymond, Yvette 006062
Rayne, Angela 048648
Rayworth, Shelagh F. 007067
Razafindratsima, Clarisse . . . 030649
Razzaq Moaz, Dr. Abdal 042258
Read, Albert J. 051778
Read, Bob 050176
Read, Judith 000996
Read, Liz 024898
Read, Peter 033195
Read, Robert F. 048098
Read, Rolly 024898
Real, Nelson A. 000372
Reale, Giuseppe 026988
Reale, Dr. Isabella 028226
Reale, Sara W. 049786
Reap, Mary 053217
Reason, Robert 000762
Reaves, Dr. Stacy 053065
Rebbin, Karin 019658
Rebehn, Lars 018000
Reber, Dr. Alfred 041692
Reber, Eugene 047855
Reber, Paul 054492
Reber, Paul C. 054491
Rebhan, Dr. Helga 020507
Rebiere, Georges 011982
Rebman, Dr. Jon 052980
Rebmann, Margarete 016459
Réboli, Eliana Moro 044441
Rebolledo Godoy, Dr. Alvaro . . 018986
Rębosz, Ireneusz 035058
Rebuschat, Karlheinz 018626
Recalt, Gustavo 000289
Recanati, Harry 007549, 039723, 046290
Recanatini, Cesare 025321
Reccia, Ludovic 003890
Reček, Antonín 009776, 009848
Rečev, Vasil 005166
Rechberger, Erwin 002061
Rechlik, Dr. Karel 009463
Rechter, Jerril 001074
Recio Crespo, Miguel Angel . . 040064,
                                  040250
Reck, Pat 046361
Reckert, Dr. Annett 018729
Reckitt, Helena 046606
Recklinghausen, Daniel von . . 022212
Recouvrot, M. 015543
Rector, P. 039029
Recupero, Jacopo 025333
Reczek, Krzysztof 034855
Reddeker, Lioba 002945
Reddell, James R. 046701
Redding Jr., Dr. Forest 054059
Redding, Mary Anne 050156
Reden, Mauritz von 021087
Redenbach, Marj 006290
Redlberger, Melanie 018919
Redlich, Daniela 018429
Redlin, Jane 017237
Redman, Guy 038699
Redmann, Barbara 042120
Redmer, Axel 017394
Redmond, Dr. Brian 047809

Redol, Pedro 035633
Redpath, Donna 005829
Redvale, Jolene 052497
Reece Hardy, Saralyn 052897
Reed, C. 045836
Reed, Dorothy 050760
Reed, Gerard 000759
Reed, Gretchen 049465
Reed, Ian 044288
Reed, Karla 051641
Reed, Leon B. 048032
Reed, Mark 052293
Reed, Mary 051075, 051203
Reed, Michael 049574
Reed, Philip 044942, 045003
Reed, Scott 052849
Reed, Ursula 052063
Reedie, K.G.H. 043857, 043858
Reedie, Kenneth 046127
Reekie, Clara 051139
Reel, David M. 054322
Reel, Paula 046477
Reenberg, Holger 010114
Rees, Diane 049410
Rees, Elfan ap 046100
Rees, James C. 051104
Rees, Jim 046226
Rees, Monica 043584
Reese, Dr. Beate . . . 020496, 022796
Reese, David L. 051398
Reese, Gary 053650
Reese, Julie 051971
Reeskamp, E. 032944
Reeter, Rose 048970
Reeve, Christopher 043787
Reeve, Tori 045130
Reeve, William F. 006685
Reeves, Don 051732
Reeves, Kate A. 051130
Reeves, Lee 046462
Reeves, Dr. Nicholas 044307
Refice, Dr. Paola 025360
Refsum, Siri 033871
Regan, Brian 049689, 051428
Regan, Mary 049042
Regêncio Macedo, Carlos
  António 035630
Regenvanu, Ralph 054676
Reggiori, Albino 026578
Regibus, Raymond de 041945
Regier, Charles 051609
Reginster, Françoise 014468
Regli, Kurt 041581
Regnauer, Dr. A. 021812
Regnier, Gérard 014787
Regoli, Dr. Edina 027675
Regull, Enrique 040386
Reh, Tazalika te 018063
Řeháková, Dr. Hana . . 009748, 009749
Rehberg, Dr. Rolf 021224
Rehberg, Vivian 014697
Rehberger, Prof. Dr. Dr. Karl . . 002652
Rehde, Anders 010053, 010055
Reher, Charles A. 050144
Reher, Dr. Uwe 019551
Rehfeld, Max 017952
Rehler, Martha 051234
Rehm, Dr. Heinrich 021004
Rehor, Ophelia 016977
Rei, M. 038800
Reich, Dr. Annette 019480
Reich, Christopher J. 048155
Reich, Dindy 048479
Reich, Dr. Ines 021000
Reich, Jacqueline 015584
Reich, Dr. Mike 018748
Reichard, Tim 053759
Reichardt, Marina 022511
Reichel, Antje 019115
Reichel, Maik 020187, 020188
Reichel, Reinhard 018707
Reichelt, Norman 021534
Reichelt, Petra 021806
Reichen, Quirinus 041315

Reichenbach, Dr. María Cecilia
  von 000411
Reichenbach, Svenja Gräfin von 017102
Reichert, Klara 020311
Reichl, Leo 002161
Reichling, Dr. Norbert 017947
Reichmann, Dr. Christoph 019787
Reichmann, Hans-Peter 018412
Reichmann, Kurt 021007
Reickart, Harry 047791
Reid-Bavis, Glenda 005658
Reid-Cunningham, James 047102
Reid Hardman, Mimi 050090
Reid-Wallace, Dr. Carolyn 051179
Reid Williamson, J. . . . 047376, 049695
Reid, Alan 044062
Reid, David 044152
Reid, Dennis 007157
Reid, Henry 044128
Reid, Jackie 054329
Reid, Jamie 048379
Reid, Dr. John S. 043352
Reid, Kim 006641
Reid, Marie-Claude 006528
Reid, Megan 054610
Reid, Richard 005905
Reid, Scott 045250, 045251
Reid, Prof. Seona 044400
Reid, Vivien 044152
Reidel, Dr. Hermann . . 021314, 021317
Reidy, Susan 053904
Reifenscheid-Ronnisch, Dr. Beate 019663
Reijmers, M. 031950
Reijnders, Dr. W. 032067
Reilly, Gerry 054369
Reilly, John W. 048417
Reilly, Margaret T. 044403
Reilly, Robin 005358
Reilly, Valerie A. 045466
Reimann, Urs 041277
Reime, Albrecht 019049
Reimer, Andreas 020608
Reimer, Caroline 019012
Reimer, Wayne 006661
Reimets, Ülle 010672
Reimold, Gesine 018771
Reimonenq, L. 023353
Rein, Dr. Anette 018434
Rein, Elfriede 016822
Reinbacher, Adolf 001908
Reinbold, Friederike 017039
Reinbold, Dr. Michael 020987
Reinckens, Sharon 054087
Reinders, Drs. Carin E.M. 032009
Reinecke, Klaus 019492
Reinert, Frank 030593
Reinert, Hans-Walter 020899
Reingold, Daniel 047231
Reinhard-Felice, Dr. Mariantonia 042033,
                                  042160
Reinhard, Adelheid 021424
Reinhard, Erika 019766
Reinhard, Rolf 019766
Reinhardt, Beatrix K. 052894
Reinhardt, Dr. Brigitte . 022260, 022268
Reinhardt, David 049807
Reinhardt, Dr. Georg 020058
Reinhardt, Hannelore 022246
Reinhardt, Tanja 038700
Reinhart, Susi 041482
Reinhold, Mag. Bernadette . . . 003074
Reinhold, Katrin 017501
Reinhold, Renat 016536
Reinholdt, Bruce 054084
Reinholdt, Helle 010308
Reinig, Margot 018999
Reinikka, Satu 011453
Reininghaus, Ruth 051463
Reinisch, Christine 002654
Reinisch, Dr. Tobias 003062
Reinke, Doris M. 048538
Reinsch, Prof. Dr. Dietmar . . . 017562
Reinthaler, August 002217
Reis, Armindo 035484

Reis, Heitor 004794
Reis, Jennifer 051051
Reis, Dr. José 035635
Reis, Paulo 035893
Reisacher, Anton 041364
Reisetbauer, Hans 002651
Reisinger, Anton 002411
Reisinger, Iva Karolina 051370
Reisinger, Dr. Nikolaus 002421
Reisner, Peter 021466
Reiss, Johannes 001862
Reisse, Hans Peter 019535
Reißer, Gudrun 017933, 017935
Reißer, M.A. Gudrun . 017931, 017932,
                                  017934, 017937
Reiter, Alois 002705
Reiter, Carmen 001788
Reiter, Dr. Fabian 017065
Reiter, Günther 002279
Reiter, Heribert 021112
Reitere, Elina 030217
Reiterer, Bruno 002637
Reithinger, Gilles 014639
Reitner, Prof. Dr. J. 018748
Reitz, Thomas A. 006122
Reitzammer, Margit 021245
Reitzler, Anne 048416
Reiyu, Mori 028531
Rejn, J. Surya Narayana 024269
Rejune Adams, Gloria 048231
Rekawek, Theresa 047774
Reker, Les 052740
Rekker, L. 032976
Rell, Zofia 035112
Rem, Dr. P.H. 032013
Remaud 013763
Remen, Geir 033769
Remig, Nicola 019739
Remington, Charlotte 050366
Rémon, Régine 003791
Remsen, Dr. J.V. 046819
Remy 004087
Ren, Shude 007799
Ren, Xiangxian 007772
Renaux, Jean-Pierre 015860
Renda, Dr. Gerhard 017376
Render, Lorne E. 050640
Render, Shirley 007475
Renders, Bram 003319
Rendić-Miočević, Prof. Ante . . 009025
Renfrew, Katherine A. 051574
Renfro, Francis 048347
Renfro, Gladys 049607
Renftle, Dr. Barbara 017357
Renger, F. 044473
Rengnath, Rudi 017891
Renie, Pierre-Lin 012112
Renken, Dr. Lisa 050362
Renn, Wendelin 022340
Renner, Prof. Dr. Susanne S. . . 020513
Renner, Zsuzsanna . . . 023488, 023516
Rennie, Ian 044155
Rennie, J. 044933
Rennie, Jennifer A. 043376
Rennie, Keely 048035
Rennie, Mairi C. 054012
Rennox, Munuma 031842
Renold, Karin 042213
Renouf, Nicholas 051268
Renouf, Phil 001436
Renslow, Chuck 047669
Renson, Bernoît 004056
Rentfrow, Donna 053628
Renting, Dr. A.D. 032013
Renton, Alan 045483
Renton, Helen 045408
Renton, John 045396
Rentsch, Ursula 041469
Rentzhog, Sten 040902, 040903
Renvall, Dr. Pertti 011053
Renz, Dorothea 017977
Renz, Joachim 020672
Renz, Ruth 053165
Renzl, Ludwig 002501

Renzo, Luigi 027679
Repede, Titus 036837
Repetti, Gregory 049835
Repetto, Arnaldo 005526
Rephann, Richard 051268
Reppe, C. 021236
Requejo, Angel 040406
Rérat-Oeuvray, Géraldine . . . 041843
Reres, Sara 053220
Reris, L.S.D. 040480
Resina, Laura 039708
Resl, Jaroslav 009723
Resnick, Alyssa 049116
Rességuier, Bernadette de . . . 014365
Ressiter, Jeremy 046069
Ressmeyer, Richard 047564
Restall, Ralph 054711
Restle, Prof. Dr. Conny 017258
Resurreccion jr., Teodoro R. . . 034500
Retallick, Leslie 045989
Réthly, Ákos 023554
Retière, Ch. 012796
Retière, Marie-Noëlle 014537
Retka, Sharon 051083
Retno, Dr. Sulistianingsih S. . . 024575
Rettenbeck, Georg 017902
Rettenmund, Jürg 041602
Rettig, Wolfram 019007
Retzlaff, Detlef 020683
Reumer, Prof. Dr. J.W.F. 032815
Reuß, Carsten 020419
Reuße, Dr. Felix 018478
Reustle, M. Charles 048006
Reustle, Dr. S. 022683
Reuter, Dr. Astrid 019517
Reuter, Cecelia 048821
Reuter, Laurel J. 049177
Reuter, Lisa 021289
Reuter, Pat 048517
Reuter, Sally 052268
Reuter, Simone 016881
Reuther, Dr. Manfred 020715
Reutimann, Daniel 042084
Revelard, Michel 003366
Révész, József 023762
Révész, Dr. T. Mihály 023644
Rew, Christine 043348, 043361
Rex Svensson, Karin . . 041130, 041131,
041132
Rey-Delqué, Monique 015994
Rey-Vodoz, Véronique 041808
Rey, Jean-Pierre 013875
Rey, Kurt 042083
Reyer, Manfred 017004
Reyes Flórez, Graciela 008663
Reyes Mota, Dr. Alicia 031166
Reyes, Manuel 023363
Reyes, Patricia 051428
Reymann, Patrick H. 047828
Reynaert, Bieke 004145
Reynies, Béatrice de 014761
Reynolds Brown, C. 054446
Reynolds, Catherine 052180
Reynolds, Craig 049624
Reynolds, Earl 051094
Reynolds, Edmond 052771
Reynolds, Edwin 050894
Reynolds, Elsie 006801
Reynolds, Fiona 044142
Reynolds, Frank 007014
Reynolds, James T. 048196
Reynolds, Janelle 006943
Reynolds, Jock 051267
Reynolds, Kimberly 047106
Reynolds, Nancy 007208
Reynolds, R.G. 043878
Reynolds, Stanley G. 007392
Reynolds, Susan 001665
Reynolds, Dr. Terry 050157
Reynolds, Valrae 051499
Reys, M. 032639
Rezai, M. 024791
Rezeanu, Prof. Paul 036417
Rezler, Aleš 009656

Rezník, Branislav 038281
Rheault, Martine 007177
Rhees, David J. 050934
Rhein, Dr. Stefan 020193, 020194,
020198, 020199
Rhi, Chong-Sun 030060
Rhi, Prof. Ju-Hyung 030040
Rhie, Dr. Jong-chul 030019
Rhoades, John S. 050401
Rhoback, Kristine 046455
Rhode, Michael 054146
Rhodes, Dr. Michael 045989
Rhodes, Ulysses S. 047273
Rhodes, William 051329
Rhomberg, Anja 001794
Rhys-Michael, Cellan 043744
Riaño Bernal, Wolfram Armando 008625
Riaño Lozano, Fernando 040421
Ribas San Emeterio, Neus 038987
Ribbert, Dr. Margret . . 041267, 041268
Ribé Monge, Genis 040030
Ribeiro de Faria, Manuel José
Marques 035804
Ribeiro de Magalhães, Thales . . 004729
Ribeiro Martins, António 035728
Ribeiro, Agostinho 035747
Ribeiro, Francisca 036028
Ribemont, Francis 015067
Riben, Hans 041023
Ribera, Agustí 039841
Ribero, Michel 016102
Ribes, Sonia 036185
Ribeton, Olivier 011924
Ribey, Barbara 007061
Riboreau, Brigitte 012169
Ricafrente, Bernardino 034455
Ricart, Dolors 039058
Riccardi, Dr. Ada 027696
Ricci, Massimo 027810
Ricci, Pat 051288
Ricci, Dr. Stefania 026296
Ricciardi, Prof. Mario 028135
Riccòmini, Prof. Eugenio 025539, 025541,
025569
Rice Irwin, John 051589
Rice, Dr. Albert R. 047756
Rice, Annette 048579
Rice, Calvin 047375
Rice, Danielle 054433
Rice, David 044432
Rice, Donald 046970
Rice, Jack 052582
Rice, Larry 053776
Rice, Nancy N. 052811
Rice, Sean 051265
Rich, Andrea L. 050444
Rich, Beverly 053343, 053344
Rich, Hans 018489
Rich, Margaret 000815
Rich, Merle 046547
Rich, Nancy 051626
Rich, Tamara S. 051145
Richard, Christopher 051681
Richard, Jack 048101
Richard, M.L. 006559
Richard, Michel 013924, 014652
Richards, Beth 052045
Richards, Chuck 052300
Richards, Dolores 047077
Richards, Jane 046722
Richards, Janet 046490
Richards, L. Jane 050959
Richards, Nancy 049548
Richards, Nancy E. 051231
Richards, Pat 053162
Richards, Paul 045778
Richards, Ron 049688
Richards, W.R. 043590
Richardson Banks, Melissa . . . 050465
Richardson, Barbara 047729
Richardson, Catherine 044646
Richardson, Don H. 053521
Richardson, E. 044646
Richardson, Edgar 053203

Richardson, Gary 044143
Richardson, I. 001143
Richardson, Dr. James B. 052158
Richardson, Jeanette 033328
Richardson, Katherine 053601
Richardson, Larry D. 046835
Richardson, Michael 048873
Richardson, Paul 053203
Richardson, Roy 033211
Richardson, Sarah 045349
Richardson, Steve 053925
Richardson, Thom 044784
Richardson, Tim 005847
Richardson, Vivian 054070
Richart, Christian 046553
Richartz, Christoffer 017086
Richaud, Maurice 016254
Riche-Descamps, Jeannine . . . 003961
Richelson, Paul W. 050978
Richerson, James J. 051987
Richey, Susan 047920
Richi, Manuela 025933
Richmond-Rex, Phyllis E. 033448
Richmond, Alan 046452
Richmond, Lauren 051968
Richmond, Robert 044166
Richner, Trudi 042080
Richon, Marco 041339
Richoux, Sylvie 014049
Richter, Bettina 042217, 042218
Richter, Bob 006175
Richter, Donald 048144
Richter, Dr. Erhard 018812
Richter, Gert 018967
Richter, Gunnar 018902
Richter, Helmut 022537, 022538, 022539
Richter, Jenett 007452
Richter, Jörg 018936
Richter, Paul 001846
Richter, Dr. Susanne 020004
Richter, Susie 050047
Richter, Dr. Thomas . 016616, 016621,
016622, 016627, 016628
Richter, Ursula 021481, 021482
Richthofen, Dr. Jasper von . . . 018735,
018736
Richtsfeld, Dr. Bruno 020579
Richwagen, Suzette 047288
Rick, Jane 045110
Rickards, Dr. R.B. 043848
Rickenbach, Bertha von 041393
Rickenbach, Judith 042221
Ricker, Maria 007490
Ricketts, Paul W. 054024
Ríćko, Oka 008916
Rico Perrat, Gabriela 030858
Riddell, R.M. 045487
Ridder, M. A. Thomas 017947
Riddington, Kate 043933
Riddle, Janet 033367
Ridet, Christophe 013373
Ridgway, Ginger 051895
Ridler, Gerda 022386
Ridley, David J. 043963
Ridley, Jackie D. 044115, 044117
Ridley, Jacqeline 044120
Ridley, Dr. Michael 044121
Ridpath, Marjorie 047372
Rieche, Christiane 044620
Rieck, Silvana 016744, 016745
Ried, Martha 019804
Riedel, Dr. Alexander 019518
Riedel, Andrea 020959
Riedel, Angela 024898
Riedel, Christiane 019522
Riedel, David 016928
Riedel, Erik 018430
Riedel, Frank 022832
Riedel, Gerd-Rainer 018249
Riedel, Uwe 020439
Riedel, Walter G. 051791, 051792
Rieder, Alfred 042099
Riedesel, Rikarde 016726

Riedmiller, Thomas . . . 018552, 018554
Rief, Julia 016440
Rieger-Jähner, Prof. Dr. Brigitte . 018400,
018401
Rieger, Sigrun 020509
Riegler, Frank 002990
Riegler, Ingrid 003108
Riego González, Angel 039217
Riehl, Ken 005496
Riek, Peter 018222, 018223
Riek, Rolf 019862
Rieken, Duane 049714
Riel, M. van 015229
Riemann, A. 017941
Riemersma, Tiny 032206
Riepe, Ulrike 019077
Riepl, Heiner 021763
Riepniece, Ineta 030123
Riepula, Anne 011047
Ries, Christian 030594
Riese, Dr. Ute 018690
Riesen, Dr. Jeorge M. 041429
Riesgo, Juan M. 039634
Riestra de Cuevas, Bertha . . . 031235
Riether, Dr. Achim 020576
Riethmüller, Dr. Marianne 018561
Riethmuller, Bruce 001562
Rietmann, Urs 041316
Rietvelt, Carol 049756
Rietzke, I. 017033
Rife, Sandra B. 054417
Riffel, Carola 018171
Riffle, Tim 051502, 051505
Rifkin, Edwin 054088
Riganti, Riccardo 028181
Rigau, Adlin Rios 036158
Rigby, S. 043641, 043642
Rigdon, Lois 047316
Rigdon, Marge 048392
Rigeade, Pascal 011965
Rigert, Markus 042150
Rigg, Frank 047118
Riggall, R.A. 001461
Riggins-Ezzell, Lois S. 051187
Riggio, Leonard 051378
Riggs, Carol 050524
Riggs, Gerry 047885
Riggs, Mary Fran 048583
Riggs, Richard 050230
Riggs, Timothy A. 047528
Righetti, Olimpia L. 000630
Righolt, Niels 040690
Rigoli, Adriano 028248
Rigoli, Roxanne 001494
Rigon, Lidia 027686
Řihová, Dr. Ivana 009892
Riibak, Renate 010678
Riker, Janet 047265
Rikkoert, A. 032866
Rikkonen, Mikko 011483
Riley, Angelika 019011
Riley, Grace 045106
Riley, Ian 044857
Riley jr., Joseph P. 047545
Riley, Karen 049700
Riley, Norman 046654
Riley, Pat 048907
Riley, Peter 045723, 045724
Riley, Terence 051427
Rill, Martin 022258
Rimer, Greame 044784
Rimmö, Christina 041138
Rimoldi, Luciano 027171
Rimon, Ofra 025057
Rinaldi, Paolo 028079
Rinaldini, Dr. Francesco 025879
Rinberger, Sabine 020582
Rinck, Aksel 020815
Rindfleisch, Jan 048095
Rindlisbacher, Rolf 041773
Rindom Madsen, Finn 010236
Rindsfüsser, Annemarie 016825
Rinetti, Dr. Luigi 028357
Ringborg, Eva 040789

Ringgaard Lauridsen, Henning . . 010361
Ringler, Sarah 054287
Ringold, Judith 052078
Ringquist, Alden 048413
Rink, Christine 019985
Rinker, Dr. Dagmar 022260
Rintala, Liisa 011395
Rintoul, Dr. Gordon . . 044253, 044256,
044265, 045322, 045385, 045770,
045772, 045774
Rinuy, Anne 041530
Rionda, Patricia Elena 000637
Rionnet, Florence 012793
Rionnet, Dr. Florence 012792
Riopelle, Christopher 045057
Riordan, Mary Ann 049965
Riordon, Bernard 005837
Ríos de Molina, Martha 031194
Ríos Montero, Sylvia 007540
Ríos Reyes, Dr. Carlos Alberto . . 008640
Rios-Samaniego, Milita 052726
Riottot El-Habib, Béatrice 014797
Rioux, Alfred 006172
Ripatti, Marja-Liisa 010792
Ripley, John W. 054133
Ripoll Isern, Luis Díaz 040152
Ripoll, Rafael Romá 038930
Ripperová, Dr. Marie 009946
Rippstein, Laurence 042108
Rippy, Francine 053126
Rippy, Linda 052208
Risani Doran, Dr. Carola 026065
Risch, Laurie 048042
Riseman, Henry 047249
Rish, Sarah 050279
Rishel, Joseph 052059
Rishel, Joseph J. 052081, 052090
Risk, Shannon M. 047865
Risser, Mary 047209
Risskov Sørensen, Kurt 010158
Rister, Faye 006009
Risthein, Helena 010668
Ritch, Irene 006936
Ritchie, Christina 007282
Ritchie, Timothy 047006
Ritschard, Claude-Janine 041530
Ritschard, V. 041812
Ritson, Liz 045348
Rittenour, Maralyn 048430, 048431,
048434, 048435
Ritter-Lutz, Dr. Susanne 041784
Ritter, Beate 017733
Ritter, Martin 020931
Ritter, Nicole 017642
Ritter, Ruedi 041389
Ritter, Shelley 047771
Rittmann, Josef 002602
Rittwage, Roy 005479
Ritzmann, Dietrich-Wilhelm 021692
Ritzmann, Hans 042142
Riva, Davide 025744
Riva, Maurizio 025744
Rivard, Céline 006832
Rivas Octavio, José A. 039658
Rivas, Francisco 039900
Rivas, Helen 052279
Rivatton, Bernard 015305
Rivé, Isabelle 013931
Rivera, Dianna 047184
Rivera, George 053097
Rivera, Gigi 034591
Rivera, Mario 049652
Rivero Borrell, Héctor 031227
Rivero, Dr. María Eloisa 046288, 046289
Rivers, Gabrielle 045564
Rivière, Danielle 013952
Rivière, Dr. Dominique 014863, 016167
Rivoira, Prof. Giuseppe 027866
Rivollet, Karin 041391
Rivolta, Barbara 053765
Rix, Peggy 047480
Rixhon, Maryse 003596
Rizzo, Antonio A. 000518
Rjabcev, Igor Mykolajovyč 043181

Rjabov, Gennadij Evgenjevič . . . 037917
Rjazanov, Aleksandr Michajlovič 036907
Rjazanova, Lilija Aleksandrovna . 036974
Rjuken, Kjetil 033586
Roach, Ed 006178
Roaldset, Prof. Dr. Elen 033847
Robb-Dietzen, Jenny 052996
Robb, Doug 045404
Robb, Lisa 051963
Robbins, Anna 000840
Robbins, Anne 045057
Robbins, Carol 048115
Robbins, Carolyn 053212
Robbins, Daniel 045019
Robbins, Neil 044593
Robe, Sofia 036642
Robel, Sigrid 017767
Robelen, William G. 048230
Roberson, Jeanette W. 050081
Roberson, W. Allen 047915
Robert, Henri 012714
Roberti, Dou Giuseppe 027833
Roberts-Douglass, Kathy 052649
Roberts, Becky 049232
Roberts, Bette 050741
Roberts, Brady 052120
Roberts, C. J. 052239
Roberts, Chad 053416
Roberts, Coles 051597
Roberts, Dr. Dafydd 044873
Roberts, David 045776
Roberts, Day 005689
Roberts, Esther 043490
Roberts, Gary 047617
Roberts, Greg 050499
Roberts, Hugh 045096
Roberts, James D. 053463
Roberts, Jan 047727
Roberts, Jane V. 046320
Roberts, Jill 051112
Roberts, John 043781
Roberts, Martin 044019
Roberts, Peg 046923
Roberts, Perri L. 047986
Roberts, Sir Samuel 043986
Roberts, Shona 033159
Roberts, Teresa 048012
Róbertsdóttir, Gerdur 024092
Robertson Hughes, Neville 001101
Robertson, A. 044164
Robertson, Beatrice L. 051980
Robertson, Beverly 050787
Robertson, Bruce 050444
Robertson, Byron 049071
Robertson, Dongwol Kim 048003
Robertson, Dr. H.G. 038686
Robertson, I.G. 045732
Robertson, Lynn 047912
Robertson, Pamela B. 044401
Robertson, Pat 049633
Robertson, Preston 006439
Robertson, Rosalyn 052637
Robertson, Susan 054050
Robidoux, R.R. 052099
Robins, R. . . . 000776, 000777, 000778
Robinson, Alistair 045923
Robinson, Beulah 047710
Robinson, Bonnell 047100
Robinson, Bruce 001121
Robinson, Christine 053530
Robinson, Cliff 000926
Robinson, Cynthia 050256
Robinson, D.N. 045157
Robinson, Dan 043933
Robinson, David 053958
Robinson, Deborah 046041
Robinson, Don 046468
Robinson, Duncan 043842
Robinson, Franklin W. 049739
Robinson, Dr. George W. 049568
Robinson, Graham 033301
Robinson, Herb 006048
Robinson, John A. 000845
Robinson, Dr. Joyce 053899

Robinson, Julie 000762
Robinson, Katharine S. 047551, 047559
Robinson, Louise 048910
Robinson, Malcolm J. 053819
Robinson, Michael 005521
Robinson, Michelle 048154
Robinson, Dr. Paul 044096
Robinson, Phaedra 048299
Robinson, Ruth 054426
Robinson, Sue Ann 050371
Robinson, Susan 050336
Robinson, Tara 048301
Robinson, Thomas A. 048519
Robinson, Travis 050243
Robinson, Wendy 045282
Robinson, William H. 047808
Robinson, William W. 047362
Robles, Constantino 039561
Robles, José Angel 031456
Robles, Sylvia de 034176
Roblin, Laurent 012658
Robson, Diana 000867
Robson, Judy 046576
Robson, Keith 046002
Robson, Pamela 044023
Robson, Ruth 043495
Roby, Teresa 043426
Roca, Teresa 039585
Rocchi, Dr. Silvano 027784
Roch-Stübler, Dr. Heidi 018767
Roch, Dr. Heidi 018765
Roch, Neal 046370
Rocha Moreno, Leonor 008632
Rocha Virmond, Eduardo da 004433
Rocha, Dr. Olívio 035530
Rochard, Dr. Patricia 019402
Roche, Fr.S. 024494
Roche, James 047406
Rochefort, Hervé de 012715
Rochette, Christelle 013467
Rochette, Didier 014483
Rochford, John 051173
Rochlitz, Heike 018876
Rochmad, Edi 024524
Rock, James T. 054604
Rockefeller, David 051329
Rockenbach, Susanne 019537
Rockhill, King 051074
Rockhold, Jean 052691
Rockstein, Margitta 016733
Rockweiler, Susanne 020255
Roda, Dr. Burkard von 041264, 041267,
041268, 041275, 041281
Roda, Dr. Hortensia von 041950
Rodák, Dr. Jozef 038461
Rodarie, M. 013930
Rodd, H.C. 046142
Roddenberry, Heather 051983
Roddick, J.H. 007339
Roddie, Ken 053869
Roddis, William H. 050501
Rodekamp, Dr. Volker . 019973, 020000,
020010, 020015, 020019, 020020,
020021, 020023
Rodepeter, Dr. Ralf 020511
Roder, Dr. Hartmut 017601
Roderick, Myron 053532
Rodeschini Galati, Dr. Maria
Cristina 025492
Rodger, David 045947
Rodger, Robin 045486
Rodgers, Forrest B. 046914
Rodgers, James 050047
Rodgers, Jeff 047416
Rodgers, Kenneth G. 048411
Rodgers, Maureen 005460
Rodgers, Michael W. 048812
Rodgers, Michele 047447
Rodgers, Rachel A. 047420
Rodgers, Susan 001353
Rodgers, Tommie 050180
Rodgers, Z. 024913

Rodica, Radu 036412, 036427
Rodiek, Dr. Thorsten . . 020159, 020160,
020162
Rodimceva, I.A. 037248
Rodin, Matis 010667
Rodina, Lucia-Letiția 036194
Rodionov, V.A. 037277
Rodman, Vickie 052921
Rodney, Clayton 023389
Rodolfi, Rosanna 026319
Rodon, Francesc 040399
Rodrigues Meyer, Fernando 004638
Rodrigues, Dr. Dalila 035807
Rodrigues, Henrique A.N. 035517
Rodrigues, Dr. Paulo 004724
Rodríguez Angarita, María Esther 008584
Rodríguez-Boette, Taryn 052748
Rodríguez Culebras, Dr. Ramón . 040150
Rodríguez de Montes, María
Luisa 008565
Rodríguez del Camino, Marco . . 054724
Rodríguez Delucchi, Jorge 046268
Rodríguez García, Serafín 040067
Rodríguez Martín, Aurora 039616
Rodríguez Molina, Cap. D.
Antonio 039746
Rodríguez Ruza, Concepción . . . 039333
Rodriguez, Carlos Armando 008645
Rodriguez, Carmen 047984
Rodríguez, Dionisio 033389
Rodriguez, Geno 051322
Rodríguez, José Alfonso 008650
Rodríguez, Dr. Marco Antonio . . 010481,
010484
Rodriguez, Roberto M. 047566
Rodríguez, Rosa 040308
Rodriguez, Teliza V. 050292
Rodriguez, Teresa 053905
Rodríguez, Tomás 054724
Rodríguez Culebras, Ramón . . . 040404
Rodríquez, Jennie 053012
Roduit, Abt Joseph 041909
Rodway, Em 047926
Roe, Diane 044078
Röber, Dr. Ralph 019757
Röck, Erwin 002546
Roeckl-Niesen, Dr. Sabine 020519
Rödel, Klaus 010069
Roeder, Corinna 020986
Röder, Heinz 020835
Roeder, Paul 048062
Röder, Dr. Sabine 019786
Rödger, Ralf 016764
Röding, Lutz 017419
Roegholt, Alice 031961
Roegiers, Prof. Dr. Jan 003784
Röhl, Johannes 016727
Roehmer, Dr. Marion 020819
Röhrig, Prof. Dr. Dr. Floridus . . . 002209
Röhring, Micha 019292
Roelink, Géke 031939
Roeloffs, Abigail 047068
Roels, André 004121
Rømer Sandberg, Ole 033630
Römer, Prof. Dr. Cornelia 003063,
003053
Römer, Kurt 001753
Römer, Neria 041696
Römhild, Michael 019244
Rönn-Kollmann, Claudia 018488
Röper, Dr. Martin 021910
Rörig, Irene 020077
Røsand, Toril 033720, 033721
Rösch, Dr. Libuse 019760
Rösch, Mathias 020876
Roesink, Macha 031901
Röske, Dr. Thomas 019144
Roesler jr., Henry E. 051005
Rösner, Dr. Corinna 020519
Rössl, Dr. Joachim 002907
Rössler-Köhler, Prof.Dr. Ursula . . 017476
Rößler, Dr. Hans 020698
Rössler, Heiner 020353
Rössler, Prof. Leopold 003000

Rössler, Dr. R.	017736	Rolley, Jean-Pierre	011596
Rössler, Ralph	041191	Rollig, Stella	002324
Rößner, Dr. Alf	022499, 022511	Rollin, Thierry	011975
Rößner, Dr. Gertrud	020561	Rollings, James R.	054319
Roest-den Hartogh, G.C.	032481	Rollins, Avon William	050004
Retting, Dorrit	010225	Rollins, Fred H.	054215
Röver-Kann, Dr. Anne	017592	Rollins, Mike	049814
Roffia, Dr. Elisabetta	026133, 027970	Rollko, Daniel	016755
Rogalski, Dr. Andrzej	041699	Rolo Fajardo jr., Gaudioso	034592
Roganti, Gabriella	026848	Roloff, Prof. Dr. A.	022144
Rogantini, Franco	041694	Romagnoli, Martha Beatriz	000677
Rogdeberg, Guttorm A.	033793	Romagnoli, Prof. Myriam Ruth	000526
Rogé, Monique	016034	Romagnolo, Dr. Antonio	027689
Rogers, Alexander K.	052568	Romagnosi, Angelo	028159
Rogers, Brett	045074	Román Berrelleza, Juan Alberto	031219
Rogers, Dianne	054512	Roman de Zurek, Teresita	008656
Rogers, Frank	051946	Román, Enrique	000342
Rogers, Gerry	046537	Romanelli, Prof. Giandomenico	025659,
Rogers, Ginny	050010		026985, 028293, 028294, 028295,
Rogers, James	048527		028303, 028310, 028316, 028320,
Rogers, Malcolm	047123		028328
Rogers, Peggy	048529	Romanenko, Maja Petrovna	037534
Rogers, Rachael	043366	Romanenko, Vera Vladimirovna	036887
Rogers, Richard	051323	Romaniuc, Elvira	036487
Rogers, Robert	048558	Romano, Simona	027071
Rogers jr, Roy Dusty	047181	Romano, Socorro	039356
Rogers, Ted	053437	Romanova, Raisa Michajlovna	036927
Rogers, Thomas	054526	Romanowska, Marta	034901
Rogerson, J.-C.	044133	Romard, Judith	006223
Rogge, Marc	004103, 004168	Romatet, J.C.	016026
Rogger, Prof. Iginio	028174	Rombauer, Tamás	023411, 023412
Roggero, Prof. Mario Federico	025902	Romdhane, Khaled Ben	042734
Roggero, Sue	044488	Romeo, Paolo	026776
Roggmann, Bettina	016473	Romero Ardila, Dídimo Ernesto	008703
Roggow, Barbara	022752, 022753	Romero, Dr. Edgardo J.	000168
Rogić, Milena	008985	Romero, Mark	052273
Rogina, Bojana	038549	Romero, Virginia de	034178
Roglán-Kelly, Dr. Mark	048119	Romero, Yolanda	039461
Rogne, Odd	034003	Roming, Lorenz	021730
Rogozińska, Ewa	034628	Romińska, Elina	035057
Roháček, Dr. Jindřich	009737	Romito, Dr. Matilde	027041, 027164,
Rohan, Antoinette Duchesse de	013327		027708, 027709, 028379
Rohde-Enslin, Dr. Stefan	017178	Rommé, Dr. Barbara	020608
Rohde-Hehr, Patricia	019011	Rommel, Dr. Gabriele	022624
Rohdenburg, Dr. Ernest A.	047588	Rommel, Dr. Ludwig	019238
Rohlfs, Stefan	018254	Rommers, Sabine	003872
Rohmer, Hans Jürgen	018801	Romney, George G.	049667
Rohner, John R.	047144	Romo R., Hermana Aura Elisa	008751
Rohner, Rudolf	041588, 041589	Romo, Tere	053011
Rohovit, Dr. Ron	050412	Romportlová, Simona	009480
Rohr, Dr. Gerlinde	020019, 020021	Romsics, Imre	023706
Rohrbach, John	048898	Romualdi, Dr. Antonella	026256
Rohrer, Denis	042104	Rona, Zeynep	042909
Rohrer, Susan	051758	Ronaghan, Allen	006046
Rohwer, Dr. Sievert	053229	Ronald, William	001392
Roiberts, David Alan	048641	Ronchi, Dr. Benedetto	028163
Roidl, Iris	018866	Roncière, Florence de la	013976
Roig Toqués, Francisco	040389	Rončkevičová, Antónia	038450
Roige, Annabella	052097	Róncoli, Mónica	000116
Roiß, Dr. Hubert	003118, 003119,	Rondeau, James	047636
	003122, 003123	Ronfort, J.N.	014631
Roitman, Dr. Adolfo D.	025106	Rong, Enqi	007754
Rojanova, Galina Iljinična	037091	Ronig, Prof. Dr. Dr. Franz	022197
Rojas de Moreno, María Eugenia	008815	Ronin, Christophe	014749
Rojas Iragorri, Ximena	008607	Ronning, Peggy	050697
Rojas Toro, Elizabeth	004244	Ronowicz, Małgorzata	034983
Rojas y Cruzat, Oscar M. de	009095	Roode, I. de	031986
Rojo Dominguez, Ana Cristina	039183	Roodenburg, Teun	033078
Rola, Laura	035893	Roodt, Frans	038806
Rola, Stefan	035081	Roohfar, Z.	024768
Roland, Dr. Berthold	016720	Rooizpekar, Mandana	041987
Roland, Debbie U.	052833	Rooney, Dr. Brendan	024900
Roland, Greg	049243	Rooney, Jan	051322
Roland, Heather	047930	Rooney, Mary	033222
Roles, John	043365, 044779, 044781,	Rooney, Steve	051406
	044783	Roos, Pieter N.	051541
Roll, Dr. Carmen	018500	Roosa, Jan	053971
Roll, Kempton	046553	Roose, B.	003423
Rolla, Maureen	052157	Roosen, Jean-Paul	012254
Rolland, C.	012368	Roosevelt, Christopher	050519
Rolland, Raymond	014872	Roosing, Jan Jaap	022873, 022874
Rollenitz, Maria	002179	Root, Deane L.	052160
Roller, Scott	050601	Root, Margaret	046490
Roller, Dr. Stefan	018449	Roots, Gerrard	044941

Roper, Donna K.	051973	Rosjat, Dr. Erika	016979
Roper, Mark	043906	Roškiewicz, Wiesław	035237
Ropers, Emmanuel	013304	Rosljakov, Sergij Mykolajovič	043265
Ropp, Ann	049824, 049827	Rosoff, Nancy	047253
Ropp, Harry	048370	Rosoff, Susan	051807
Roppelt, Dr. Tanja	017706	Ross, Alan	052149
Ropponen, Liisa	010771	Ross, Andrea	044273
Roque Gameiro, Maria Alzira	035854	Ross, Dr. Cathy	045048
Roquebert, Anne	014743	Ross, David A.	048784
Roques, Thierry	006247	Ross, Donald A.	001561
Rorem, Ned	051323	Ross, Dugald Alexander	045849
Rorschach, Dr. Kimerly	048410	Ross, Ian D.	005499
Rosa, Jocelyn P. de la	034452	Ross, Jerry	048623
Rosa, Joseph	052157, 053025	Ross, Mari B.	046735
Rosa, Mercedes	004788	Ross, Marian	053777
Rosales Huatuco, Dr. Odón	034280	Ross, Pam	051143
Rosales Vergas, Manuel	031200	Ross, Phill	049496
Rosales, Felipe	034435	Ross, Richard	051926
Rosanbo, Richard de	013567	Ross, Sandria B.	046499
Rosandich, Dr. Thomas P.	048149	Ross, Stewart	000995
Rosas, Moisés	031148	Ross, Wendy	049901
Rosas, Dr. Silvia	034357	Rossellini, Laurence	014061
Roschanzamir, Dr. Béatrice	017489	Rosser, Merlin	007096
Rose, Ben	054201	Rosset-Magnat, Chantal	015138
Rose, Coral	049316	Rossetti de Scander, Dr. Antonio	028207
Rose, Kate	046200	Roßfeldt, Heike	017137
Rose, Mary	050954	Rossi Brunori, Ignazio	026941
Rose, Randall A.	053692	Rossi, Cristina	028021
Rose, Sally	044225, 044226	Rossi, Filli	026013
Rose, Susan	006647	Rossi, Dr. Francesco	025498
Rose, Wilhelm	017019	Rossi, Guido	026397
Rosefelds, Andrew	001150	Rossi, John S.	048710
Roselló, Núria	040396	Rossi, Rochelle	047396
Rosemann, Günther	020102	Rossiter, Shannon	049976
Rosemann, Thomas	042207	Rossiter, Sybil	047453
Rosen Poley, Rita	048540	Rossner-Dietz, Petra	022437
Rosen, Ferial Horesh	025172	Rossolimo, Dr. Olga L.	037525
Rosen, Rhoda	047696, 047698	Rost, Sven	018856
Rosenbaum, Joan	051412	Rostami	024769
Rosenbeck, Dr. Hans	017036	Rostek, Charlotte	044555
Rosenberg, Dr. H.I.	005535	Rostek, L.	001587
Rosenberg, John	052356	Rostholm, Hans	010115
Rosenberg, Dr. Stefan	020853	Rostworowska, Dr. Magdalena	035420
Rosenberger, Pat	047775	Rosu, Georgeta	036320
Rosenblattl, Christina	001819	Roșu, Steliana	036391
Rosenbleeth, Herb	054145	Roszko, Grzegorz	034765
Rosenblum Martin, Amy	047224	Rota, Anna Francesca	025958
Rosenblum, Robert	051470	Rota, Dr. Laura	025746
Rosenbluth, Betsy	047326	Rotaru, Costel și Elena	036743
Rosenbruch-Hammer, Angie	052768	Rotaru, Georgeta Monica	036547
Rosendahl, Paul	010046	Rotenberg, Nir	025038
Rosenfeld, Daniel	054220	Roth-Oberth, Dr. Erna	018354
Rosenfield Lafo, Rachel	050285	Roth, Dr. Anja-Maria	019137
Rosengart, Dr. h.c. Angela	041713	Roth, Dan	046598
Rosengren, Anna	040661	Roth, Eric	051308
Rosengren, Hans	040730	Roth, Flemming	010083
Rosengren, Jim	051693	Roth, Gary G.	052449
Rosengren, Kerstin	040642	Roth, Isabelle	041987
Rosenkranc, Milan	009534	Roth, John	047249
Rosenkrantz, Baron Erik Christian	010139	Roth, Josef	002865
Rosenkranz, Roger	048202	Roth, Jules	014447
Rosenmeier, Edith Marie	010328	Roth, Linda H.	049390
Rosenquist, Perry E.	048072	Roth, Marc-André	041945
Rosenstock, F.	016447	Roth, Prof. Dr. Martin	018010
Rosensweig, Larry	048232	Roth, Max	041192
Rosenthal, Beryl	047366	Roth, Dr. Michael	017209
Rosenthal, Deborah	053584	Roth, Neal	052241
Rosenthal, Dina R.	053503	Roth, Dr. Peter	038427
Rosenthal, Ellen	048735	Roth, Ronald C.	052484
Rosenthal, Turpin	017527	Roth, Dr. Thomas	019707
Rosenzveig, Charles	048709	Roth, Dr. Ursula	018030
Rosenzweig, Dr. Rainer	020880	Rothe-Wörner, Heidrun	018656
Rosero Solarte, Luciano	008750	Rothenberg, Peter	050590
Rosewitz, Mindy	048846	Rothenberger, Manfred	020861
Rosher, Fiona	044541	Rotheneder, Martin	002381
Rosiak, Andrzej	035388, 035469	Rother, Dietmar	020376
Rosich i Salvó, Mireia	040392	Rother, Dr. Rainer	017240
Rosiek, Barbara	035473	Rothermel, Barbara	050530
Rosing, Emil	023334	Rothfuss, Joan	050948
Rosing, M.	010196	Rothfuss, Dr. Otto	016459
Rosini, Prof. Dr. Corrado	025359	Rothmund, Cornelia	001794
Rosinski, Dr. Rosa	017373	Rothove, Billi R.S.	049061
Rosioara, Iustin	036685	Roths, Jaylene	051085, 051086, 053388
Rosito, Massimiliano G.	026265	Rothschild, Eric de	014676

Rothschild, Jan 051491
Rothschild, Baroness Philippine
de 014814
Rotkale, Rita 030179
Rotovnik, Mitja 038542
Rotsart, John A. 052976
Rott, Dr. Herbert W. 020508
Rottermund, Prof. Andrzej ... 035382
Rottmann, Carmen 021492
Rottmann, Dr. Joseph 016793
Rottmann, Lee 052816
Rouchaud, Alain 015054
Roughley, Dr. R.E. 007448
Rouijel, Souâd 031780
Rouijel, Souad 031793
Rouleau, Reynald 005618
Roulston, Peggy 050613
Roumell, Lisa 051452
Rountree, Robin K. 053589
Rouquereol, N. 011799
Rouquet, Chantal 016047, 016052,
                      016053, 016054
Rourke, Ralph 047230
Roush, Margaret 049615
Rousseau-Arlardi, Daniele 014729
Roussel, Dominique .. 015826, 016282
Roussel, Louis 013978, 015790
Rousset-Rouard, Yves 014123
Roussia, Pierre 013386
Roussier, François 016320
Routhier, Nathalie 005445
Routledge, M.A. 045926
Roux, Bertrand 012638, 015551
Roux, Jean-Pierre ... 012943, 013159
Roux, Nathalie 012596
Rouyer, Morgan 016194
Rouyer, William 016194
Rovati, Clementina 027234
Rowan, Edward 050627
Rowe-Shields, Michele 053153
Rowe, Alan L. 033264
Rowe, Ann P. 054164
Rowe, Barbara 054445
Rowe, C. 038805
Rowe, Donald 051753
Rowe, Matthew 044220
Rowe, Victoria 050357
Rowinski, Dr. Stanislaw 019011
Rowland Jones, Mark 043602
Rowland, David B. 049811
Rowley, Jennifer 043763
Rowley, Roger H.D. 052416
Rowold, Kathleen L. 047046
Rowsell, Wayde 006918
Rowsom, Christopher 046771
Rox-Schulz, Heinz 021554
Roxas Lim, Prof. Aurora 034577
Roxburghe, Duke of 044672
Roy, Dr. Archana 024356
Roy, Dr. Ashok 045057
Roy, Carolyn 052316
Roy, David T. 043460
Roy, Jan-Bernard 014442
Roy, Jean-Marc 005614
Roy, Kenneth J. 051197
Roy, Nathalie 015165
Roy, Noëlle 014790
Royce, Ralph 049036
Royer, Claude 013410
Royez, Robert 003954
Royle, Ted 043917
Royo, Dr. Ana Carmen 038889
Rozanov, Aleksej Jurjevič ... 037489
Rozenberg, Dr. Silvia 025100
Rozendaal, Prof. A. 038834
Rozenfeld, Boris Matveevič .. 037142
Rozewicz, Beverly L. 049434
Rozłowski, Tadeusz ... 034824, 034826
Roznoy, Cynthia 053485
Rózsa, Zoltán 023819
Rozsošyns'ka, Nina Leonidivna . 043193
Rozzo, Dr. Ugo 028158
Rúa, Rosa Edith 008770
Ruan, Yingqi 008531

Ruas dos Santos, Dr. Margarida 035765
Rub, Jürg 041312
Rub, Markus 041316
Rub, Timothy 047739
Ruban, Nikolaj Ivanovič 036949
Ruben Cuñeo, Dr. Nestor 000651
Rubenstein, Earl 050407
Rubidge, Prof. Bruce 038744
Rubin, Jane 051412
Rubinstein, Adam 052460
Rubinstein, Ernest 051484
Rubinstein, Helena 051491
Rubio Pediata, Luis 040465
Rubio Sánchez, Pedro 030957
Rubio, J. David 008691
Rubnikowicz, Marek 035286
Rucellai, Oliva 027939
Ruch, Pam 052366
Ruchhöft, Bernd 021146
Rucka, Barbara 035023
Ruckel, Jay G. 051396
Rucker, Denise 049766
Rucker, Henrike 022543
Rucker, Ken 053342
Ruckstuhl, Dieter 041510
Ruczynski, F.C.K. 032102
Rudaz, Patrick 041402
Rudd, Karen 045725
Rudd, Robert 048457
Rudd, Shirley J. 051965
Rudder, Jennifer 007084
Ruddle, John 024837
Rudi, Dr. Thomas 019992
Rudigier, Dr. Andreas . 001929, 002731
Rudin, Vladimir Vasiljevič 037884
Rudinger, Torsten 021185
Rudisill, Dr. Richard .. 053116, 053118
Rudkin, David J. 043941
Rudman, Karen 031851
Rudnickaite, Eugenija 030517
Rudnik, Galina Anatolevna ... 037860
Rudolp, Dr. Andrea 021101
Rudolph, Beth 046373
Rudolph, Jeffrey N. 050412
Rudolph, Johanna 021252
Rudolph, Ursula 021247
Rudzińska, Wanda M. 035316
Rübesam, Manfred 020012
Rückel, Robert 017100
Rücker, Dr. Thomas 002625
Rückert, Dr. Birgit 021572
Rückert, Hans 016467
Rueda, Horacio 010494
Rüdel, Dr. Holger 021665, 021666
Rüdiger, Bernd J. 018732
Ruedin, Dr. Pascal 041995
Ruedrich, Mark 048785
Rüger, Axel 031996
Ruegg, Dr. François 041498
Rühl, Karin 018459
Rühlig, Cornelia 020442, 020443
Rühmer, Erich 019309
Rührl, Helmut 017575
Ruell, David 046568
Rümbeli, Monica 041506
Rümmelein, Andreas 020512
Rünger, Dr. Gabriele 018328
Ruesch, Carol 054616
Ruess, Andreas 016862, 016863
Rueß, Dr. Karl-Heinz . 018727, 018728,
                      018733, 018734
Ruet, Karen 005844
Rüther, Dr. Martin 019707
Rütsche, Dr. Claudia 042204
Ruette, Monique de 003492
Rütti, Dr. Beat 041243
Rüttimann, Franziska 042213
Ruf, Beatrix 042206
Ruf, Dietmar 020509
Ruf, Peter 020134, 020139
Ruf, Susanne 022057
Rufe, Laurie J. 053837
Ruff, Eric J. 007495
Ruff, Joshua 053548

Ruff, T.J. 046115
Ruffo, Joseph M. 050290
Rugama, Miriam 033374
Ruggeri, Prof. Alessandro 025547
Ruggiero, Dr. Laurence J. 054497
Rugoff, Ralph 053007
Rugunanau, S. 038731
Rugus, Dot 052556
Ruhe, P. Wilhelm 016713, 016714
Ruhe, Rikke 010209
Ruhl, Erwin 019200
Ruhl, Hans-Eberhard 017831
Ruhl, Horst 022461
Ruhlig-Lühnen, Dagmar 019393
Ruhmann, Dr. Christiane 021060
Ruhmer, H.-H. 018823
Ruhrberg, Dr. Bettina 019445
Ruhstaller, Tod 053171, 053541
Ruiter, Dr. Marjan 032281
Ruivo, José 035639
Ruiz de Azua Murua, Asun ... 040436
Ruiz de Fischler, Dr. Carmen T. .. 036160
Ruiz-Martinez, José D. 023328
Ruiz Sastre, Elena 039362
Ruiz-Torres, Maricarmen 051781
Ruíz, Dimas Fernández-Galiano .. 039476
Ruiz, Jauier Luna 031438
Ruíz, Leticia 039676
Rukavina, Vjeročka 009043
Rukmini Devi, Srimatha 024498
Rules, John 044785
Ruley, Dennis 052679
Rulíšek, Dr. Hynek 009560
Rullmann, Anne 022418
Rum, Piera 027478
Rumbaugh, Catherine 052776
Rumbaugh, Darlene 052108
Rumi, Satoh 029511
Rumiche, Prof. Antonio 034383
Rumjanceva, Valentina Pavlovna . 037664
Rummel, Dr. Michael 016664
Rump, Elisabeth 047505
Rump, Elizabeth 052375
Rumrill, Alan F. 049911
Rumsby, J.H. 044614
Rumscheid, Prof. Dr. Frank ... 019601
Runde, Dr. Sabine 018435
Rundhovde, Eric 033529
Rundle, Stella 004294
Rundqvist, Dmitry V. 037285
Runey, Dennis I. 054514
Runia, E. 032180
Runte, Markus 021064
Runton, Bourke A. 053829
Ruohonen, Risto 010880
Rupert Enriquez, Carola 046719
Rupert, Dr. Manfred 002182
Rupp, Hansruedi 041983
Rupp, Jane E. 047717, 050694
Ruppeiner, Paul 041213
Rupprecht, Dr. Klaus 016956
Rupprechter, Hanns-Michael .. 022063
Ruprechtsberger, Prof. Dr. Erwin
M. 002326
Rus, Valer 036270
Ruseckaitė, Aldona 030368
Rusev, Rusi 004976, 005175
Rush, Don 048865
Rush, Dr. Michael 050093
Rushton, Ray 045028
Rusibaev, D.S. 054665
Rusina, Prof. Ivan 038328
Rusinová, Dr. Zora 038328
Ruske, Wolfgang 016615
Ruskule, Skaidrite 030191
Rusnell, Wesley A. 052692
Russe, Christel 018648
Russel-Smith, Dr. Lilla 017239
Russel, Jack 000850
Russell Howe, Dr. Stanley ... 046955
Russell, Barry 005395
Russell, Brenda 005808
Russell, Prof. Bruce 044722
Russell, Douglas 048017

Russell, Fiona 046447
Russell, J. Fiona 051627
Russell, John I. 047239
Russell, Laurence J. 050886
Russell, Marilyn M. 052157
Russell, Pamela 048737
Russell, Paul 047710
Russell, Willie 053237
Russi, Graciela 000367
Russmann, Edna 047253
Russo, Prof. Antonio . 026851, 026860
Russo, Pam 054527
Rust, Wilfried W. H. 048836
Rusterholz, Sabine 041549, 042005
Rustia, Augusto P. 034527
Rustichelli, Arlette 041240
Rustige, Rona 005414
Rusu, Olga 036508
Ruszczyc, Ferdynand B. 035352
Ruszczyc, Grazyna 035321
Ruszel, Dr. Krzysztof 035183
Ruszkowska-Macur, Małgorzata 034741
Ruta Serafini, Dr. Angela 026163
Rutberg, Carl 053495
Rutherford Conrad, Marth ... 005449
Rutherford Conrad, Martha ... 006118
Rutherford, John 001288
Rutherford, Dr. Richard 045451
Rutishauser, Werner 041950
Rutkauskas, Dr. Petras 030547
Rutkowski, Czesław 035196
Rutledge, Bethany 053076
Rutsch, Franz 021110
Rutschmann, Hans 042127
Rutte, Helgard 021678
Rutten, B. 032476
Rutter, A. 044528
Rutter, Michael 006068
Rutter, Mike 043934
Ruttinger, Jacquelyn 049857
Ruttner, Nancy 052166
Ruuska, Aulis 011463
Rux, Sandra L. 049293
Ruy Van Beest Holle, C.J. du .. 032832
Ruys, Christoph 003287
Ruyssinck, Micheline 003492
Ružič, Ivan 009043
Ruzo, Victor 041757
Ryan, Amy 047106
Ryan, Dan 053133
Ryan, Daniel J. 053834
Ryan, Debora 053633
Ryan, Dr. E. 048805
Ryan, Graham 001372
Ryan, Jack Dingo 051182
Ryan, Lynne 046869
Ryan, Meryl 000867
Ryan, Dr. Michael 047809
Ryan, Michael 024871
Ryan, Paul 053511
Ryan, Robert 047093
Ryan, Thomas R. 050121
Ryba, Kurt 027202
Rybakov, Ivan Alekseevič 036840
Rybakova, Tamara Viktorovna . 037340
Rybka, Walter P. 048599
Rybnikova, Larisa 037304, 037309
Rychener, Dr. Jürg 041243
Rychlik, Dr. Otmar 002458
Rychliková, Marie 009929
Rychtarik, John 047243
Ryckman, Heather 006527
Rycquart, Walter 003403, 003405,
              003406, 003407, 003408, 003409,
              003410, 003411, 003412, 003413,
              003414, 003422, 003425
Rydel, Maria 034911
Ryder-Davis, P. 046182
Rydergård, Evy 040888
Rydzewski, Dr. Jacek . 034876, 034877
Rykels, Sam 051298
Ryland, James 051000
Rylands, Philip 028322
Ryman, Susan 000867

Rynd, Chase 054139
Rynierse, H. 032301
Ryoichi, Minegishi 028992
Ryol, Thae Pyong 029879
Rypar, Astrid 003037
Rypka, Tom 046681
Ryšánková, Dr. Milada 009924
Ryščanova, Urkija
  Muchamedjarovna 029823
Ryser, Madeleine 041640
Rysin, Oleksandr Mykolajovyč . 043270
Ryska, Norbert 021061
Rystedt, Prof. Eva 040812
Ryusawa, Aya 029129
Ryzdynski, Mark 049440
Ryžov, Stanislav 043293
Ryžova, Tamara Aleksandrovna . 037546,
                            037554
Rzepecka, Mirosława 035087
Rzepiela, Urzula 035142
Rzeszotarski, Jan 035198
Sá Marques, Ana Rita 035808
Saabye, Marianne 010176
Saadé, William 011671, 011672
Saage, Wallace 049577
Saager-Bourret, Stephanie .. 052195,
                            052196
Saal, Dr. Wolfgang 020229
Saalmann, Ute 018061
Saar, A. 010664
Saaret, Tarmo 010668, 010682
Saarimaa, Hanna-Maija 011173
Saarinen, Jarmo 011435
Saarist, Tiiu 010649
Saarnio, Robert 046746, 046750
Saathoff, Joachim 021089
Saatsi, Niko 011197
Saavedra, Vicente Peña 040133
Šabalin, Rostislav Viktorovič . 037638
Sabaté i Piqué, Gemma 040366
Sabaté, Antonio 040386
Saber, Mohammad Davood
  Mohammad 000007
Sabin, Carleen 050869
Sabin, Owen 050869
Sabina, Choclán 039566
Sabloff, Dr. Jeremy 052100
Sabogal, Bernardo 008678
Sabolic, Prof. Dubravka 009017
Sabran-Ponteves, E. Comte de . 014993
Sabroe, Charlotte 010326
Šačalova, Natalja Vladimirovna . 037455
Sacarés Mas, Antoni 040191
Sacco, Louise R. 048211
Saccocci, Prof. Andrea 027146
Sach, Martin 045025
Šachalova, Natalja Vladimirovna 037288
Sachanbiński, Prof. Michał ... 035425
Sachs, Angeli 042217
Sachs, Charles L. 047264
Sachs, Sid 052092
Sachse, Jutta 020033
Sachse, Mike 022544
Sack, Dr. Birgit 017983
Sack, Georg 022632
Sackett, Margot Magee 006850
Sackstetter, Susanna 017772
Sada, Ricardo J. 039526
Sadion, Martine 012885
Sadjukov, Nikolaj I. 036929
Sadler, Dr. Donna 048198
Sadler, Patsy 051964
Sádlo, Dr. Václav 009711
Sadok, Nedjah 000062
Sadowski, Władysław 034881
Sadowsky, Dr. Thorsten 016509
Al-Saeed, Susan 034097
Saeed, Tahir 034126
Saegrov, Magnhild 033598
Säkkinen, Senja 011231
Sæland, Frode 033706
Sälejoki-Hiekkanen, Leena .. 010875
Sänger, Dr. Reinhard 019505
Sæther, Per 033486

Sæthre, Ellen 033982
Sættem, Nils 010152
Saeverud, Dag H. 033951
Safarnia, P. 024716
Saffiotti Dale, Maria 050593
Safford, Pete 048186
Safin, Françoise 003802
Safina, Zulfira Midechatovna .. 037531
Safoulis, Maxine 046979
Šafr, Pavel 009506
Safronova, V.V. 037601
Safsafi, Nourddine 031793
Saft, Marcia 051412
Saganic, Livio 050588
Sagar, Scott 054417
Sagdalen, Betty Jo 049963
Sage, J. 044164
Sager, Judy 048899
Saghaar-Bessiére, Francoise . 015486
Saghi, Mohamed 031792
Sági, Dr. Zoltán 023548
Sagin, Mike 044562
Sagmeister, Dr. Rudolf 001793
Sagov, Magomet Zakrievič ... 037542
Sagrera, Enric 039450
Sagristà, Ferran 040396
Sahakyan, Kamo 000718
Saheki, Ryoken 028679
Sahlstrand, James 048548
Said, Sanib 030701
Saies-Jones, Frank 005530
Sailard, Olivier 014705
Sailer, Andreas 002706
Sain, Annette J. 052224
Saines, Chris 033141
Sainio, Tuulu 011433
Sainsbury, Derek 044306
Saint-Affrique, Olga de 013483
Saint-Bris, François 011620
Saint-Clair, Anne-Laure 012247
Saint Hippolyte, Mary 053373
Saint-Pierre, Adrienne 048669
Saint-Pierre, François 013720
Saint-Pierre, Marie Odil de .. 012307
Saint-Pol, Pierre P. 012834
Saint Romain, P.W. 052189
Sainte-Fare-Garnot, Nicolas . 014765
Sáinz González, Elena 031165
Saisí, Enrique 000481
Saito, Satoshi 028802
Saja, Dr. David 047809
Sajenko, Dr. Milan 038447
Saji, Nobutada 029595
Sakaeva, Guzel Rinatovna .. 037113
Sakai, Makoto 029218
Sakai, Mark 006795
Sakai, Nobuo 029012
Sakai, Tadayasu 028615
Sakai, Tetsuo 028555
Sakai, Yasuo 029334
Sakamoto, Bishop 029444
Sakari, Maija 010836
Sakazume, Hideichi 029586
Sakli, Dr. Mourad 042772
Sakmirda, Marta 021698
Šakova, Marjana Kušbievna . 037535
Sala, Susan M. 034464
Salabert, Patricia 017601
Salachova, P.Ch. 037997
Säläjean, Ioan 036328
Salajić, Prof. Silvija 009017
Salameh, Khader 025085
Salamon, Sabina 008944
Salapi, Constantin 036527
Salardon, Catherine 011639
Salas Burguera, Magdalena . 039713
Salas Vázquez, Eduardo ... 039653
Salas, Jaime de 040275
Salata, Jim 006556
Salatino, Kevin 050444
Salaverria, Ana 039350
Salazar Soto, José Aníbal .. 008569,
                            008570
Salazar, Alvaro 006055

Salazar, Ernesto 010490
Salazar, Lucila A. 034561
Salcedo Zambrano, Alonso ... 008752
Salcher, Hubert 003114
Salcher, Johann 002357
Saldo, Josip Ante 008990
Šale, Majda 008881
Salé, Marie-Pierre 014743
Saleh, Ali Ibrahim 010561
Salehi, M. 024778
Salerno, Osvaldo 034193
Sales Marques, José Luís de . 008066
Salesch, Dr. Martin 022648
Saletti, Prof. Cesare 006302
Salgo, Nicolos 052265
Salib, Dr. Mahar 010534
Saliklis, Ruta 046408
Salin, Anne-Maj 011468, 011472
Salisbury, Joyce 051095
Salisbury, Ruth 045720
Salje, Prof. Dr. Beate 017315
Salkin, Asen 005054
Salkoranta, Marja 011449
Sall, Joan C. 052082
Salleh, Dr. Badriyah .. 030708, 030709,
          030710, 030711, 030712, 030713,
          030714, 030715, 030716, 030717
Sallen, Sheila 048841
Salmen, Dr. Brigitte 020625
Salmen, Nora 047922
Salmi, Lyle 048201
Salmon, Béatrice 014705, 014713,
                    014733, 014785
Salmon, Irit 025110
Salmon, Patricia 053505
Salmona, Riccardo 051383
Salmons-Perez, Carl 053163
Šalna, Rimantas 030551
Salo, Prof. J. 011436
Salome, Elizabeth 049674
Salomé, Laurent 015164, 015166,
                            015169
Salomeia, Paul 036501
Salomón Zugbi, Alex Faruk .. 034370
Salomon, Joel 052641
Salomone, Mario R. 000158
Salomon, Jouni 011254
Saloniemi, Marjo-Riita 011394
Salsi, Dr. Claudio .. 026781, 026783,
                            026788
Saltamanikas, P. 023210
Salter, Ann C. 052624
Salter, Vanessa 044720
Saltiel, Solita 023291
Salton, Jenny 044614
Saltonstall, Patrick 050016
Saltveit, Ass. Prof. Svein. J. . 033847
Salu, Luc 003287
Salus, Ernst R. 025124
Saluz, Prof. Eduard 018571
Salva, Gary 053330
Salvador Zazurca, Alejandro .. 040442,
           040444, 040456, 040457
Salvador, Fred 005588
Salvador, Josep 040305
Salvador, Mari Lyn 046364
Salvador, Marilyn 052979
Salvatierra, Julio 040404
Salvator, Markus 001749
Salvatore, Ferraro 025242
Salvatori, Prof. Franco 027594
Salvesen, Britt 053824
Salvesen, Ole Bjørn 033887
Salveson, Doug 048732
Salvig, Jacob 010335
Salvioni, Mario Casanova ... 041299
Salzgeber, Christoph 041863
Salzillo, Bill 047843
Salzman, Alexej 009701
Salzmann, Prof. Dr. Dieter .. 020591
Samaniego, Lydia 034564, 034566
Samarati, Dr. Luigi 026614
Samardžić, Draženko 008928
Samardžić, Nadežda 038185

Samejima, Yasutoyo 029190
Samek, Tomasz 020608
Samet, Jennifer 051354
Samford, Patricia 046810
Sami-Azar, Dr. A.R. 024760
Samietz, Rainer 018775
Saminger, Herbert 002905
Sammallahti, Leena 011268
Sammons, Mark J. 052317
Samochvalova, Ljudmila
  Nikolaevna 037130
Samodurov, Jurij Vadimovič .. 037377
Samon, Josep 040396
Samoškina, Lilija Vjačeslavovnas 037567
Samoy, Antenor R. 034429
Sampaio, Heli de Almeida ... 004802
Sampaio, Madalena 036030
Sampaolo, Prof. Valeria 026992
Samper, Cristian 054147
Sampil, Mamadou 023381
Sampson, Bill 050706
Sampson, Michael 048565
Sams, Maria 001751
Samson, Dr. Bill 044165
Samson, Fr. Antonio S. 034449
Samson, James 033326
Samson, Rémy 012512
Samsonova, Élvira Aleksandrovna 037185
Samuels, Clifford 052825
Samuels, Philip 052825
Samuelson, Laura 051571
San Agustin, Leonora P. ... 034431
San Augustin-Filaretos, Pilar ... 012112
San Diego, Rodolfo 034557
San Quirce, Sylvie 013285
San Román, Ángel García ... 040274
Sana, Costantino 025296
Sanborn, Andrea 005312
Sanborn, Colin J. 047761
Sanborn, Michael 054431
Sánchez Aguilera, Marco 007519
Sánchez de Velandia, Myriam . 008561
Sánchez del Barrio, Antonio .. 039741
Sánchez Fuertes, Caayetano .. 038986
Sanchez García, Javier 031297
Sánchez Hernández, Américo . 031237
Sánchez Mastranzo, Nazario A. . 031487
Sánchez Murguía, Socorro ... 030893
Sánchez Peña, José Miguel .. 039190
Sánchez Pérez, Luz María ... 031208
Sánchez Ramírez, Angeles ... 040461
Sánchez Recio, Miguel Angel .. 039626,
                            039627
Sánchez Ruiz, José Fernando . 038920
Sánchez Sancho, Dr. Luis María 040446
Sánchez Trujillano, Teresa ... 039594
Sánchez-Villagra, Prof. Dr. M. . 042227
Sanchez, Artero 004884
Sanchez, Basilio 004883
Sanchez, Dr. Bettina 003089
Sanchez, Marc 014794
Sanchez, Maria José 040110
Sánchez, Osvaldo 031178
Sanchez, Roger 052474
Sanchezová, Hanaěk 009507
Sanchis, Frank Emile 051431
Sancho, Emanuel A. 036009
Sanclemente Girón, Américo
  Alfonso 008687, 008688
Sand, Christiane 013388
Sand, Gabriele 019066
Sand, Viki 052031
Sandachi, Paul 036288
Sandahl, Jette 040663
Sandberg, Fredrik 040620
Sandberg, Jane 010101, 010207
Sandbichler, Veronika 002136
Sandell, Joanna 041092
Sandell, Margaret 046047
Sandell, Stephen 050938
Sanden, Jarle 033781, 033783
Sander, Dr. Antje 019463, 019464
Sander, Dr. Birgit 018441
Sander, Dr. Dietulf 020003

Sander, E. 018664
Sanders, Dr. Albert E. 047543
Sanders, Boyd 053873
Sanders, Bronwen 050884
Sanders, Debra 046549
Sanders, Emma 041522
Sanders, G. 032013
Sanders, James A. 049664
Sanders, Jennifer 001600
Sanders, Mildred 054226
Sanderson, Cecil 048789
Sanderson, Dr. Dewey D. 049625
Sanderson, Natalie 053094
Sanderson, R. 007028
Sandford, G. 045688
Sandin, Per 041015
Sandland, Bronwen 000755
Sandler, Ilan 005947
Sandlofer, Michael 047236
Sandmann, Oskar 019226
Sandmeyer, Marilyn 051075
Sandmo, Sigurd S. 033535
Sando, Rainer 017558
Šandová, Dr. Miroslava 009536, 009863
Sandoval, Josefina de 054707
Sandoval, Raquel 008577
Sandoval, Sonia 010487
Sandra, Galley 048559
Sandrini, Dr. Francesca 027207
Sandrock, Dr. Oliver 017822
Sandström, Paul 011207
Sandstrom, Ake 040705
Sandvoß, Hans-Rainer 017144
Saner, Gerhard 042037
Sanfelice di Bagnoli, Fabio 027563
Sanford, Dr. Beverly S. 054494
Sanford, Lamonica J. 050896
Sanftner, Phyllis 051067
Sang, Guozhen 007572
Sângeorzen, Monica 036746
Sangl, Dr. Sigrid 020509
Sangmanee, Kitti Cha 014753
Sanguinetti, María Isabel 000505
Sangvichien, Dr. Sanjai 042654
Sangvichien, Prof. Sood M.D. . . 042679
Sangwill, Richard 040606
Sani, Michele 026966
Sankowski, Janusz 035343
Sanna, Maria Antonia 028066
Sannibale, Dr. Maurizio 054682, 054685
Sano, Dr. Emily J. 052992
Sans, Bartolomé 038985
Sansano Costa, Lina 040104
Sansbury, Joan 054121
Sansom, Andrew 052354
Sansoy, Jaqueline 014627
Sant, Victoria P. 054141
Santa Ana Alvárez Ossorio, Florencio
de 039684
Santa Elena, Barrio 008697
Santaca, Dr. Susanna 010827
Santamaria, Jose 054371
Santamaria, Prof. Ulderico 054682
Santamarina, Guillermo 031158
Santana Quintana, Luis Carlos . 031160
Santander, Diego 010469
Santandreu, Edmundo 000223
Santaniello, Giovanni 027044
Santay, B. Mundade 024249
Sante Serpa, Prof. Balbín 034284
Santelli, Bob 053233
Santer, Rolf 021300
Santesso, Flavia 000607
Santiago, Isabel 035779
Santiago, Mark 046334
Santiago, Rubio 040295
Santinacci-Boitelle, Françoise . . 013492
Santini, Roberto Mario 004821
Santiot, Jacques 014423
Santis, Jorge 048827
Santomauro, Angelica M. 049317
Santoni, Mirta Elsa 000554
Santonja, Josep A. Gisbert 039346
Santoro, Geraldine 051474

Santoro, Dr. Paola 026672
Santos Correia Victor, Ana Paula
dos 000113
Santos Ramírez, Marco Antonio 031675
Santos Sátz Gomes, Ilmo 039331
Santos Simões, Joaquim A. . . . 035728
Santos, Ana Paula 035952
Santos, Prof. Eliane Aparecida . 004941
Santos, Maria José Ferreira dos 035893
Santos, Maria Soares dos 015854,
015855
Santos, Osmar 046291, 046292
Santos, Paula Mesquita 035952
Santos, Sara 035893
Santos, Turibio 004763
Santrot, Jacques 014418
Santzevláriz, David 039831
Santzevláriz, David 039830
Sanz Gamo, Rubí 038904
Sanz Sánchez, Erundino 039048
Sanz, Eduardo 040404
Sanz, Javier 039654
Sanz, Joaquim 039718
Sanzi, Dr. Maria Rita 025969
Šapošnikova, Ljudmila Vasiljevna 037342
Sapouna-Sakellaraki, Dr. Efi . . . 023158
Šapovalov, Georgij Ivanovyč . . . 043325
Sarabhai, Gira 024122
Sarajlić, Čazim 004288
Saran, P.K. 024280
Šarapa, Olga Mychajlivna 043118
Sarashiya, Kazuko 029442
Sarauw, Torben 009998
Saraver, Ida 006018
Saravia Garrido, Pablo 039786
Sarkisjan, David A. 037300
Saray, Michael 044596
Sarda, Marie-Anne 012153
Sardella, Filomena 027008
Sardo, Delfin 035782
Sarelius, Seppo 011435
Sargent, William 052880
Sargsyan, Zaven 000745
Sargus, David 046435
Sarholz, Dr. Hans-Jürgen 016760
Sari, Dr. Alireza 024787
Sarie, Pierre 013893
Sariego, Manuel Lucío Julio . . . 008733
Sarioglu, M. Adnan 042846
Sariola, Helmiriitta 010813
Sarkisjan, David A. 037300
Sarkissian, Karen 045101
Sarl, P.M. 045716
Sarl, Peter M. . 045710, 045712, 045714,
045715
Sarner Williams, Amy 052038
Sarosácz, György 023795
Saroyan, Ione 051455
Sarpong, Kwame 022929
Sarrail, Cécile 012337
Sarret, Françoise 013824
Sarsfield, Stanley 044063
Sarteanesi, Prof. Nemo 026000, 026001
Sartes, Minna . . . 011432, 011434
Sarti, Fabiola 027935
Sarto, Pietro 041759
Sartori, Dean 048266
Sartori, Dr. Michel 041655
Sartorio Pisani, Dr. Giuseppina . 027593
Sartorio, Matteo 026824
Sartorius, K. 017463
Sartorius, Tara 051025
Saruchanian, Saro 000746
Sary, Monique 014152
Saryan, Rouzan 000727
Saržickiené, Biruté 030532
Sas, Adrian 051334
Sasagawa, Takami 028997
Sasaki, Hideya 029051
Sasaki, Masamine . . 029533, 029534
Sasaki, Yoneyuki 028930
Sasayama, Kazutoshi 028850
Şaşmaz, Abdülkadir 042794
Sasse, Julie 053837, 053840
Sassen, J.H. 031986

Sasso, Rosalie 048370
Sassorossi, Larry 052467
Sastri, C.L.N. 024267
Sastry, Dr. V.V. Krishna 024295, 024457
Sasymanov, Y.A. 030082
Sather, Dean 051047
Sather, Stanley 007354
Šatkauskas, Gediminas 030489
Sato, Keichi 028892
Sato, Mitsuo 028778
Sato, Naoki 029536
Sato, Takeshi 028483
Sato, Teiichi 029607
Sato, Toyozo 029129
Sato, Yoshio 028909
Sato, Yumika 029334
Satoko, Maemura 028750
Satsuma, Masato 029509
Sattelmair, Hartwig 018232
Satterfield, Barbara 047966
Satterfield, Foad 053063
Satterthwaitt, Dr. Leann 001474
Sattlecker, Dr. Franz . . 003012, 003079
Satut, Jorge 046266
Satvé Oliván, Enrique 040032
Satyan, M.S. 024266
Satzer, Wolfgang 002255
Saubel, Katherine 046780
Sauciuc, Benedict 036817
Sauder David, D. 046522
Sauer, Dr. Albrecht 017606
Sauer, Daniel 004697
Sauer, Marc 011537
Sauer, Dr. Marina 016491
Sauerborn, Ulrich 016449, 016452
Sauermann, Bernhard 019828
Sauers, Dr. Richard A. 053612
Sauerwein, Herbert 002284
Sauge, Brigitte 033825
Saul, Dainiel 006059
Saul, Lydia 044714
Saulieu, Pascal de 012745
Saulnier, Lydwine 015771
Saulsbery, Jeff 047042
Saumade, Jeanine 013278
Saunders, Andrew 044031
Saunders, Angela 006228
Saunders, Dr. David 044923
Saunders, Dr. Gail 003201
Saunders, Jean 043981
Saunders, Kristin 045994
Saunders, Richard 050722
Saunders, Richard H. 050858
Saunders, Stuart 000857
Saunders, Dr. Timothy 020821
Saunier, Bernard 016231
Saurage, Hank 046820
Sautai, Daniel 012257
Sauter, Pierre 041200
Sauter, Wolfgang 019565
Sautov, Ivan Petrovič . 037671, 037673
Sautter, Dr. Heribert 018349
Sauvebeuf, B. de 012856
Savage, Harland 054459
Savage, Pamela 054459
Savage, Paula 033344
Savage, Ronald 046375
Savard, Denise 007429
Savary, Lee 046372
Savčenko, Andrej Grigorjevič . . 036936
Savčenko, Nikolaj Jakovlevič . . 037932
Savel, Irena 038582
Savelja, Oleg 043293
Savellon, Romula 034460
Savelsberg, Dr. Wolfgang 017868
Savery, Suzanne T. . . 052010, 052011,
052012, 052015, 052017, 052018
Savi, Dr. Dario 028252
Savickaja, Ljudmila
Aleksandrovna 037184
Savidge, Amanda 046193
Savidis, Prof. G. 023181
Savill, Rosalind J. 045125
Saville, Ansley 052670

Saville, Reg 044759
Savini Nicci, Marco 025855
Savini, Thomas M. 051357
Savinič, Irina Borisovna 037830
Savko, Iraida Lvovna 037961
Savolainen, Kari 011072
Savouet, Silvia 039108
Savu, Elisabeta 036622
Sawa, Ryuken 028921
Sawada, Keizo 029169
Sawadogo, Alimata 005279
Sawauchi, Takashi 029550
Sawchuk, Barry 007342
Sawchuk, Sherry 051097
Sawchyn, Linda 006074
Sawdey, Michael P. 046675
Sawelo, Victor 007459
Sawicka-Oleksy, Daniela 034695
Sawkins, Annemarie 050923
Sawyer, Carol 052565
Sawyer, Doug 052113
Sawyer, Ed 052737
Sawyer, Kenneth C. 001006
Sawyer, Sean 047269
Sawyer, Tom 048704
Saxinger, Karl 022396
Sayer, Christopher 045133
Sayers, Andrew 000939
Sayler, John G. 050188
Sayles, Mary 048485
Sayner, Mary 053165
Sayunain, Selamat 030690
Saywell, Edward 047123
Sazdova, Katerina 030643
Sazonov, Valerij Petrovič 037630, 037632
Sazonova, T.V. 037225
Scafetta, Stefano 054160
Scaggs, Tom 054188
Scaglia, Orlando 000447
Scalamandre Bitter, Marc J. . . . 050383
Scalamandre Bitter, Robert F. . . 050383
Scalera Liaci, Prof. Lidia 025457
Scalia, Dr. Fiorenza 026261
Scalini, Mario 025907
Scalora, Salvatore 053555
Scandola, Dr. Pietro 041340
Scanlon, Maureen 050200
Scannapieco, Carlos A. 000433, 000434,
000435
Scapellati, Claire 051812
Scapigliati, Juan 000591
Scărlătescu, Fl. 036429
Scarpetti, Ivo 026729
Scarratt, David 044248
Ščeboleva, Galina Fedorovna . . 037254
Scelsa, Dr. Joseph V. 051410
Ščerbakova, I.T. 037099
Ščerban, Aleksandr Nikolaevič . 036903
Schaaf, Ania 003989
Schaaff, Dr. Holger 020233
Schaal, Gerald 047056
Schaar, Dana 047455
Schaar, Elke 020021
Schaatsenberg, W.J.M. 032254
Schaber, Manfred 016496
Schachl-Raber, MMag. Dr. Ursula 002640
Schachtner, Dr. Sabine 017052
Schachtsiek, Richard 050282
Schack von Wittenau, Dr. Clementine
Gräfin 017752, 021451
Schade-Didschies, Peter 021407
Schächter, Willi 019106
Schädle, Bernarda 016915
Schädler, Ulrich 042064
Schäfer, Dr. Alfred 019711
Schäfer, Andreas 042006
Schäfer, Bernd 018777
Schaefer, Dr. Claudia 018052
Schaefer, D.E. 038764
Schäfer, Denis 020534
Schäfer, Dr. Dorit 019517
Schaefer, Elizabeth Meg 047905
Schäfer, M.A. Gerd 019425

Schäfer, Gerd 019427, 019429
Schäfer, Günter 002749
Schäfer, Prof. Dr. H.-P. 018634
Schäfer, Dr. Hans-Michael 018978
Schäfer, Dr. Hans-Thomas 022057
Schaefer, Dr. Hartmut 020507
Schaefer, Johann 047714
Schäfer, Julia 019986
Schäfer, Klaus 020489
Schäfer, Dr. Klaus 016580
Schäfer, Leane 018629
Schaefer, Patricia M. 051278
Schaefer, Rainer 019523
Schaefer, Scott 050436
Schaefer, Dr. Stacy 047712
Schäfer, Dr. Wolfgang 022291
Schäffel, Brigitte 002759
Schäffel, Karl 002759
Schaeffer, Alvina 046885
Schaeffer, Astrida 048403
Schäffer, Gisela 016434
Schäffer, Prof. Dr. Dr. Johann .. 019068
Schael, Alfredo 054734
Schaeps, J. 032583
Schär, Peter 003142
Schärer, Dr. Martin R. 042104
Schaetzel, Désiré 012086
Schafer, Barbara L. ... 046960, 046962
Schafer, Carl 046408
Schafer, Colin 044698
Schaffer-Hartmann, Richard .. 019039,
                              019040, 019042
Schaffer, Dale 048100
Schaffer, Dr. Herbert 002478
Schaffer, Dr. Nikolaus 002637
Schaffers Bodenhausen, Dr. K. . 032174
Schaffgotsch, Marie-Elise Gräfin 002453
Schaffner, Ingrid 052058
Schafhausen, Nicolaus 032831
Schafroth, Colleen 049151
Schagen, C. 032498
Schaible, Larry 053747
Schaible, Silvia 020490
Schaiek, Zvi 025219
Schakel, F. 031895
Schalhorn, Dr. Andreas 017209
Schalke, Prof. Dr. Thomas 017080
Schall Agnew, Ellen 050533
Schall, Jan 049888
Schallenkamp, Kay 048580
Schaller, Anne K. 051200
Schaller, Hydee 046500
Schallert, Dr. Elmar 001788
Schalles, Dr. Hans-Joachim ... 022838
Schallmayer, Prof. Dr. Egon ... 016789
Schaltenbrand, Therese 041678
Schaluschke, Anja 019542
Schamberger, Rolf 018556
Schamfuß, Axel 021459
Schampers, K.J.J. 032392, 032393
Schanche, Kjersti 034069
Schantl, Alexandra 002697
Schantz, Michael W. 052103
Schanz, K.H. 020886
Schaper, Henning 022744
Schapp, Rebecca M. 053096
Schardin, Anne 001331
Scharfenberg, Bernd 018523
Scharff, Prof. Wolfram 021347
Scharloo, Dr. M. 032399
Scharmann, Detlev 022611
Scharmann, Rudolf 017288
Scharmann, Rudolf G. .. 017079, 017222,
                         017264, 017283
Scharnowski, Joachim 017167
Scharnweber, Karl 017920
Schaschl-Cooper, Sabine 041786
Schaschl, Sabine 041493
Schatorje, Dr. Jos 032982
Schatteman, Alice 051008
Schatten, Gereon 016906
Schatz, Duane 048523
Schatz, Günther 002395
Schaub, M.A. Cajetan 022080

Schaub, Didier 005290
Schaub, James 053832
Schauber, Dr. Gregor 002582
Schauer, Karl 002365
Schaufelberger-Breguet, Simone 041925
Schaukal, Dr. Barbara 001968
Schaumann, Prof. Walther 002221,
                              002222
Schaumburg-Lippe, Wolfgang Prinz
zu 002315
Schaunaman, Lora 046299
Schaus, Dick 049858
Schawaller, Dr. W. 022056
Schawe, Dr. Martin ... 016669, 016958,
                       020508
Scheau, Prof. Dr. Mihail 036760
Schebesta, Dr. Kurt 002943
Scheck, Klaus Peter 021768
Scheckle, Marian 031841
Schedin, Pernilla 040744
Schedler, Christian 020409, 020410,
                   020411, 020412, 020413, 020414
Scheef, Vera 017876
Scheele-Schmidt, Gisela 031843
Scheele, Edward 046304
Scheele, Dr. F. 018179
Scheelen, J. 003687
Scheen, Rolf 033833
Schefcik, Jerry A. 050159, 050161
Scheffler, Ina 017694
Scheffler, Jürgen 020029, 020030
Scheflen, Mark 051414
Schefzik, Dr. Michael 018965
Schegg, Jakob 041329
Scheibe, Joachim 022075
Scheiblechner, Margit 041243
Scheiblechner, Mag. Petra 001968
Scheicher, Dr. Harald 001786
Scheicher, Harald 002870
Scheid, Dr. Eva 019293
Scheidt, Diana 049356
Scheifele Giger, Carmen 041573
Scheiffele, Hans 019900
Scheifinger, Otto 041786
Schein, Valentin 038591
Scheinost, Vojtěch 009831
Schelde-Jensen, Bodil 010064
Schelfhaut, Andrée 003668
Schellack, Dr. Fritz 021868
Schelle, Elise 032961
Schels, Bertin 019574
Schelshorn, E. 022656
Schelter, Dr. Alfred ... 020145, 020360
Schenck, Jean-Luc 015236
Schenck, Marvin 053878
Schenini, Elio 041695
Schenk Freiherr von Stauffenberg,
Franz 019910
Schenk-Weininger, Dr. Isabell .. 017386
Schenk, L.J.B. 031950
Schenk, Matthias 022642
Schenk, Ueli 041320
Schenkenberger, Katja 020586
Schenker, Prof. Christoph 042208
Schepers, Heiner 020091
Schepers, Dr. Wolfgang 019060
Scheppink, Baukje 033063
Scher, Anne 051412
Scherbakov, Dr. Lev A. 037275
Scherbel, Dr. G. 020864
Scherer, Dr. Bernd M. 017159
Scherer, Ernst 041964
Scherer, H. 041964
Scherer, Jim 052219
Scherer, Dr. Kristine .. 019137, 019147
Scherer, Kurt 041250
Scherer, Manfred 041964
Scherer, Mark A. 049663
Scherer, Rodney 001316
Scherf, Angeline 014697
Scherf, Avner 025120
Schermuck-Ziesché, Margit 017862
Scherner, Dr. Antje .. 016903, 019546
Scherney, Dr. Ingrid 001926

Scherp, Dr. Astrid 020509
Scherrieble, Dr. Joachim 020281
Scherzer, Beatrice 019355
Schettino, Edvige 027006
Scheu, David R. 054442
Scheuerbrandt, Dr. J. 021024
Scheuerecker, Franz X. 021082
Scheuermann, Gernot 021163
Scheuers, Vera 020670
Scheuren, Elmar 019741
Scheurer, Bruno 041441
Scheurer, Heidi 041314
Schiborra, Ute 018938
Schick, Dr. Karin 041434
Schick, Martin 016689
Schick, Dr. Ulrike 021043
Schicklgruber, Dr. Christian ... 003046
Schidlowsky, V. 012410
Schieber, Dr. Michael 020814
Schieck, Dr. Martin 018401
Schiek, Friedolf 022043
Schiemann, Larry 046569
Schiepers, Sue 004097
Schierle, Dr. Sonja 022048
Schiers, Jörge 016850
Schierscher, Renè 030305
Schierz, Prof.Dr. Kai Uwe 018244
Schießl, Werner 018098
Schiestl, Dr. Michael 002151
Schietere de Lophem, A. de .. 003421
Schietinger, James 048201
Schietz, Ottar 033962
Schiff, Michael 048114
Schiffer, Tim 053957
Schiffino, Prospero 025721
Schiffmann, Dr. René 041490
Schiffner, Wolfgang 022734
Schilbach, Michael 020960
Schild, Hans 042041
Schild, Margit 018027
Schilder, A.M.Th. 031995
Schildkrout, Dr. Enid 051327
Schill, Jörg 021815
Schillemans, R. 031962
Schiller, Joyce K. 054433
Schilli, K. 018635
Schilling, Wolfgang 021833
Schiltz, Kelly 049225
Schimek, Dr. Michael . 016879, 018725
Schimmel, Dan 052046
Schimmel, Paul 050456
Schimmelpfennig, Michael 022238
Schimpf-Reinhardt, Dr. Hans ... 016937,
                          016938, 016939
Schimpf, Dr. Simone 022045
Schindel, Morton 054347
Schindlbeck, Dr. Ingrid 017112
Schindlbeck, Dr. Markus 017112
Schindler-Kaudelka, Dr. Elen ... 002508
Schindler, Annette 041284
Schindler, Claudia 017784
Schindler, Harald 022919
Schindler, Dr. Margot 003057
Schink, Hendrik 021401
Schinkel, Dr. Eckhard . 017958, 022414
Schinnerl, Heimo 002193
Schinnerling, Mechthild 022869
Schiøtz, Ottar 033961
Schipper, P. 033086
Schipper, Drs. Peter W. 032141, 032923,
                              032925
Schirmböck, Thomas . 020255, 020256
Schisler jr, Lee C. 047214
Schitinger, Jim 048203
Schiwek, Thomas 022764
Schizas, Dr. Nikolaos 036135
Schladebach, Tilman 017015
Schläfli, Bruno 041820
Schlaepfer, Ralph 041782
Schlag, Dr. Gerald .. 002287, 002465,
                         002754, 002776
Schlagkamp, Dieter 021843
Schlass, Gerhard 003047
Schlatter, Alexander 041279

Schlatter, Dr. Rudolf 020011
Schlatter, Stephan 041951
Schlautmann, Rainer 018309
Schlechter, Gunther 017791
Schlegel, Adolf 022462
Schlegel, Amy Ingrid 050754
Schleger, Debra 001113
Schleher, Linda 047786
Schleich, Jean 003490
Schleider, Holger 017590
Schleif, Tom 049926
Schlemmer, C. Raman 027086
Schlenk, Christoph 002134
Schleper, Dr. Thomas 020898
Schlereth, Bernhard 019637
Schleyerbach-Breu, Christiane . 020407
Schlicher, Fred J. 050753
Schlicht, Esther 018446
Schlicht, Dr. Udo 017376
Schlichting, Art 052257
Schlichting, Heike 016600, 016602
Schlicting, Eunice 048155
Schlieben, Katharina 042240
Schliebs, Des 001235
Schliefsteiner, Herbert 002432
Schliekau, Jürgen 016732
Schliemann, Ulrich 017613
Schlieper, Hans 021911
Schlimmgen-Ehmke, Katharina . 017877
Schlinkert, Dr. Dirk 022740
Schlinkmann, Jean 053194
Schlitter, Balz 041597
Schlombs, Dr. Adele 019702
Schloßbauer, Kurt 021279
Schlosser, Charles ... 013727, 013728
Schlosser, Christa 020632
Schlosser, Christoph 041488
Schlosser, Prof. Katesa 019609
Schloßer, Rosemarie 017834
Schloter, John E. 052863
Schluchter, Dr. André 041483
Schlüter, Dr. Andreas 022056
Schlüter, Dr. des. Arnulf 020578
Schlüter, D. 032213
Schlüter, Günter 019941
Schlüter, Dr. J. 019004
Schlüter, Maik 017563
Schlueter, Sue 050908
Schlumpf, Fernand 042088
Schlup, Daniel 041715
Schlup, Michel 041792
Schmälzle, Reinhard 021809
Schmager, Martin 022542, 022544
Schmahl, Prof. Dr. Wolfgang W. 020557
Schmalbrock, R.F. 019452
Schmalfuß, Dr. Helmut 022056
Schmalwaßer, Karl-Friedrich ... 017127
Schmalz, Lydia H. 051296
Schmandt, Dr. Matthias 017389
Schmauder, Dr. Michael 017499
Schmeh, Thomas J. 050003
Schmeja, Dr. Lia 001774
Schmelz, Prof. Dr. Bernd 019012
Schmelzer, Christian 022784
Schmetz, Marcel 004061
Schmid-Beurmann, Dr. Peter ... 020597,
                              020603
Schmid-Birri, Ingrid 041472
Schmid-Nafz, Gerlinde 020328
Schmid, A. 016870
Schmid, Albert 041876
Schmid, Dr. Anna 041276
Schmid, Debora 041243
Schmid, Elisabeth 042226
Schmid, Dr. Elmar D. 020546
Schmid, G. 021504
Schmid, Gerhard 042057
Schmid, Hanspeter 041927
Schmid, Jakob 042099
Schmid, John 041635
Schmid, Dr. Manfred .. 022036, 022058,
                              022059
Schmid, Dr. Margit 001824
Schmid, Dr. Peter 042216

Schmid, Reinhard 022323
Schmid, St. 014293
Schmid, Dr. Ulla K. 021722
Schmid, Ute 051437
Schmidl, Dr. Hermann 001870
Schmidl, Karin 017086
Schmidl, Matthias 002070
Schmidt-Bergmann, Prof. Dr.
  Hansgeorg 019513
Schmidt-Czaia, Dr. Bettina 019688
Schmidt-Esters, Gudrun 018465
Schmidt-Glintzer, Prof. Dr. Helwig 022732
Schmidt-Herwig, Dr. Angelika .. 018405
Schmidt-Lawrenz, Dr. Stefan .. 019123
Schmidt-Möbus, Dr. Friederike . 017789
Schmidt-Rutsch, Dr. Olaf 017958,
                              022698
Schmidt-Surdez, Maryse 041792
Schmidt, Aiko 018179
Schmidt, Andreas 022875
Schmidt, B. 022289
Schmidt, C. J. 049416
Schmidt, Dagmar 018962
Schmidt, Dwight 049141
Schmidt, Erwin 019847
Schmidt, Dr. Eva 021853
Schmidt, Franz Xaver 022056
Schmidt, Franziska 019058
Schmidt, Georg 018703, 018704
Schmidt, Dr. Hans-Günter 022785
Schmidt, Dr. Hans-Werner 020003
Schmidt, Harro 019056
Schmidt, Johannes 018012
Schmidt, Karl 017763
Schmidt, Kirsten 010138
Schmidt, Klaus 022806
Schmidt, Prof. Dr. Klaus 018266
Schmidt, Lawrence R. 051706
Schmidt, Dr. Mária 023559
Schmidt, Martin 018952
Schmidt, Muriel H. 048631
Schmidt, Peter 042074
Schmidt, Rolf 017329, 017330
Schmidt, Roman 020123
Schmidt, Dr. Rüdiger 019503
Schmidt, Dr. Sabine Maria 018306
Schmidt, Sebastian 019275
Schmidt, Teresa 053939
Schmidt, Dr. Theun-M. 017089
Schmidt, Thomas 017898, 017899
Schmidt, Volker 017589
Schmidt, Willi 021673
Schmidts, Dr. Thomas 020231
Schmied, Dr. Hartmut 021487
Schmiegel, Karol A. 048348
Schmieglitz-Otten, Juliane 017725
Schmigalle, Dr. Günther 019503
Schmith, Andreas 016631
Schmitt, Erika 022349
Schmitt, Horst 022198
Schmitt, Prof. Dr. Michael 017504
Schmitt, Robert 015758
Schmitt, Dr. Ulrich 018752
Schmitz, Dr. Bettina 019250
Schmitz, Dipl. phil. Brigitte 017259
Schmitz, Brigitte 017068
Schmitz, Dr. Britta 017154
Schmitz, Engelbert 021904
Schmitz, Hans-Hermann 017046
Schmitz, Dr. Jörg 020493
Schmitz, Margret 021904
Schmitz, Dr. Michael 020232
Schmitz, N. 021496
Schmitz, Silvia 019061
Schmitz, Uwe 017868
Schmocker, Susan 001535
Schmöhl, Thomas 017022
Schmoldt, Robin 050602
Schmolka, Gerhard 003095
Schmoller, Mag. Andreas 001836
Schmollgruber, Friedrich 002782
Schmook, Dr. R. 016768
Schmook, Dr. Reinhard 016769
Schmuck, Carolin 020812

Schmuhl, Boje 017409, 020048
Schmuttermeier, Elisabeth 003033
Schmutz, Christine 041432
Schmutz, Daniel 041315
Schmutz, Daniel 041315
Schmutz, Dr. Hans Konrad 042165
Schmutz, Prof. Dr. Hans-Konrad 042159
Schnabel, Helga 020648
Schnakenburg, Dr. Sabine 018979
Schneble, Dr. H. 019568
Schneider-Kempf, Barbara 017294
Schneider, Dr. Alfred 016576
Schneider, Prof. Dr. Angela .. 017154,
                              017262
Schneider, Anja 019818
Schneider, Dr. Carola 018299
Schneider, Claire 047298
Schneider, Dr. Cornelia 020226
Schneider, Dr. Ditmar 020206
Schneider, Don 051857
Schneider, Eckhard 001793
Schneider, Elke 019010
Schneider, Dr. Erich . 021781, 021782,
                      021784, 021786
Schneider, Ester 020393
Schneider, Dr. G.C. 031861
Schneider, H. 019521
Schneider, Hansjörg 019757
Schneider, Hélène 014928
Schneider, Dr. K. 018971
Schneider, Karl-Heinz 021508
Schneider, Dr. Katja 018969
Schneider, Kerry 052163
Schneider, Prof. Dr. Klaus 019709
Schneider, Klaus 018078
Schneider, Malou 015872
Schneider, Minna 051430
Schneider, Nadine 041187
Schneider, Paul 050417
Schneider, Philippe 015572
Schneider, Reinhild 021915
Schneider, Reisa Smiley 007296
Schneider, René 012255
Schneider, Dr. Thomas 021012
Schneider, Prof. Dr. Ulrich .. 016446,
                              018435
Schneider, Dr. Ulrich 021664
Schneider, Prof. Dr. Ulrich
  Johannes 019980
Schneider, Virgil 046519
Schneider, Dr. Wolfgang 017822, 018703
Schneidler, Dr. Herbert 021330
Schneidmiller, Kenneth A. 049256
Schnell, Christian 012051, 012052
Schnell, Dr. G. 051588
Schnell, Gertrude H. 048942
Schnellhammer-Ocist, Norbert .. 002775
Schnepp, Patrick 013482
Schneyer, Rosemary 053538
Schnieder, Andreas 016474
Schnitzer, Dr. Claudia 017992
Schnitzler, Bernadette 015873
Schnurbein, Dr. Rüdiger Freiherr
  von 017542
Schnurr, Marie 048423
Schober, Otto 001951
Schoberleitner, Martina 002412
Schoberwalter, Anton . 002203, 002204
Schobinger, Dr. Juan 000452
Schoch, Dr. R. 022056
Schöb, Kathrin 041872
Schöbel, Dr. G. 022256
Schoeller, Danielle 016266
Schöllmann, Dr. Lothar 020602
Schoemaker-Weeszenberg, M.P. 032160
Schoen, Kenneth 054025
Schön, Matthias D. .. 016709, 018920
Schönau, Dieter 020375
Schönberger, Dr. Angela 017205
Schöne, Dr. Anja .. 022120, 022121
Schöne, Jasmin 018721
Schoene, Lynn 021736
Schöne, Dr. Sally 018031, 018042
Schoeneck, Edith 020913

Schönenberg, Walter 019031
Schoenenberger, Nicola 041696
Schönewald, Dr. Beatrix 019406, 019413,
                               019414
Schönfelder, Ramona 018084
Schönfellner, Dr. Franz 002235, 002239
Schönhagen, Dr. Benigna 016671
Schönig, Dagmar 041826
Schoenke, Marilyn 049541
Schönsweiter, Prof. Karlheinz .. 002655
Schönweger, Astrid 026764
Schøtt, Vibehe 010059
Schofield, Arnold 048870
Schofield, Dorothy 053075
Schokkenbroek, J. 031972
Scholbier, Noël 004032
Scholl, Prof. Dr. Andreas 017072
Scholl, Prof. Dr. Lars U. 017606
Schollenberger, Werner 020886
Scholler, Dr. Markus 019518
Schollian, Gerd 019124
Schollmeier, Dr. Axel 020608
Schollmeyer, Ullrich 016977
Scholtemeijer, G. 038856
Scholten, Lisa 047243
Scholten, Uta 018038
Scholz-Bauer, U. 021416
Scholz-Strasser, Inge 003084
Scholz, Angelika 021204
Scholz, Dr. Antony 001987
Scholz, C. 022067
Scholz, Dr. Dieter 017230, 017262
Scholz, Dr. Heike 022047
Scholz, Lothar 017473
Scholz, Margitta 017545
Scholze, Frank 017086
Schomers, Dr. Florian 002139
Schommers, Dr. Annette 020509, 022031
Schonewille, Dr. P. 032035
Schonlau, Ree 051760
Schooler, Jim 054609
Schoon, P.J. 032246, 032251
Schoonhoven Okken, W. 032138
Schoonover, Larry 053442
Schoonover, Mary Ellen 046670
Schopen, Klaus H. 019674
Schopp, N. 000812, 000815
Schopper, Dr. Franz 017540
Schopping, H. 032944
Schorta, Dr. Regula 041883
Schoske, Dr. Sylvia .. 020578, 021813
Schott-Smith, Christine 023356
Schotten, Christoph 022310
Schoutens, P. 031962
Schovel, Erin 048342
Schrade, Dorothea 022272
Schrade, Ewald 018104
Schrader, Kathleen 052565
Schrader, Wolfgang 021150
Schräer, Dr. Heinz 018290
Schramek, Maria-Antonia 002488
Schramer-Lorang, Fernande .. 030582
Schramm, Dr. Ágnes 038328
Schramm, Jeanne 054313
Schramm, Robert W. 054313
Schrank, Gustav 019273
Schratter, Rudolf 002106, 002107,
                      002108, 002109
Schraudner, Anita 020871
Schrecengost, Tammy 046980
Schreiber, Daniel J. 022220
Schreier, Dr. Christoph 017495
Schreilechner, Waltraud 002125
Schrein, Catherine 046804
Schreiner, Manuela 019757
Schreiter, Bernd 016591
Schrem, Wilfried 020680
Schretter, Reinhard 002871
Schrettl, Reinhold 002865
Schreuders-Derks, Annie 032263
Schrick, Annette 017052
Schriver, Stephanie 053102
Schroder, Dr. Anne 048410
Schroeck, Petra 017094

Schrøder Vesterkjær, Siri 033990
Schroeder, Allen 047459, 050931
Schroeder, Asher 050662
Schröder, Christoph .. 016530, 016532
Schroeder, Chuck 051732
Schroeder, Darlene 049141
Schroeder, Dorothy 047636
Schröder, Frank 017510
Schröder, Dr. Heike ... 016893, 020127,
                               022047
Schröder, Dr. Klaus Albrecht .. 002931
Schrøder, Lise Lotte 010121
Schroeder, Patricia 048729
Schröder, Ralf-Peter 018090
Schroeder, Stephan 039676
Schröder, Till 019006
Schröder, Dr. U. 018808
Schröter, Jürgen 022157
Schröter, Wolfgang 021833
Schroll, Ludwig 018202
Schroth, Dr. Adella 052497
Schroth, Dr. Sarah W. 048410
Schroth, Thomas 051949
Schrub, Antoine 014862
Schube, Inka 019066
Schubert, Dr. Alexander 020255
Schubert, Armin 017544
Schubert, Beate 021007
Schubert, Jason 047735
Schubert, Dr. Kristina 017080
Schubert, Werner 002960
Schubert, Wolfgang 019999
Schubiger, Letizia 041678
Schubiger, Pia 041969
Schuch, Eny 004653
Schuch, Herbert 019499
Schuchardt, Günter 018137
Schuchmann, Dr. Karl-Ludwig .. 017504
Schuelke, Chris 048723
Schülke, Ilsabe 018034
Schüllenbach, Konrad 031866
Schülly, Dr. Maria 018478
Schünemann, Dr. Gustav 016967
Schünemann, Karl 016502
Schünemann, Ulla 021805
Schüpbach, Beat 041318
Schüpbach, Ulrich 041343
Schürch, Theo 042071
Schürer 018112
Schürer, Dr. Ralf 020853
Schuerholz, Peter 053681
Schürmann, Dr. Brigitte 016668
Schürmann, Dr. Kay U. 020274
Schütt-Hohenstein, Dr. A. 022067
Schuette, Bobbie 049224
Schuette, Gerd 018302
Schütte, Dr. Margret 021664
Schütte, Rainer 032039
Schütz, Dr. Christian 022346
Schütz, Ilse 038885
Schütz, Jan 009523
Schütz, Jean-Luc 003793
Schütz, Dr. Karl 003026, 003026
Schütz, Volker 017397
Schütz, Wolfgang 022474
Schuh, Walter 002448
Schuhwerk, Dr. Franz 020513
Schuldt, Doren 020992
Schulenburg, Dr. Rosa von der . 017066
Schulenburg, Dr. Stephan von
  der 018435
Schulenburg, Susanne 019505
Schuler, Ewald 041197
Schuler, Romana 003029
Schuler, Shirley 052193
Schulkin, Jerrold L. 050275
Schuller, Kurt 001878
Schuller, Tom 049948
Schullerus, M. 016541
Schulson, Henry H. 047592
Schulte-Hobein, Dr. Jürgen .. 016608
Schulte-Kemper, Hubert 020305
Schulte-Wülwer, Prof. Dr. Ulrich 018372

Schulte, Dr. Birgit –
Sekerák, Jiří
Index of Persons

Schulte, Dr. Birgit 018915, 018919
Schulte, Elisabeth 016659
Schulte, Matthew 051595
Schulte, Wolfgang 003693
Schulten, Ton 032729
Schulters, Johannes 016946
Schultes, Alfred 002145
Schultes, Dr. Lothar .. 002323, 002330
Schultes, Peter 016562, 016563
Schultka, Prof. Dr. R. 018951
Schultz, Brian 049397
Schultz, Dr. Franz Joachim 016994
Schultz, Prof. Dr. Hartwig 018421
Schultz, Horst 016550
Schultz, J. Bernard 051057
Schultz, Jeffrey 049234
Schultz, Peter 041021
Schultze, Dr. Joachim 021663
Schulz-Hoffmann, Prof. Dr. Carla 020508
Schulz-Weddigen, Dr. Ingo 019758
Schulz, Andreas W. 020537
Schulz, Dr. Ch. 041881
Schulz, Emily L. 054161
Schulz, Hagen 016977
Schulz, Dr. Isabell 019066
Schulz, Lotte 034205
Schulz, Marion 021344
Schulz, Max 050421
Schulz, Ralph J. 051177
Schulz, Rebekah 001160
Schulz, Volker 022595
Schulz, Dr. Walter 018175
Schulz, Warren 047016
Schulze Altcappenberg, Prof. Dr. Hein-Th. 017209
Schulze, Armin 018941
Schulze, Dietmar 020015
Schulze, Falk 019751
Schulze, Prof. Dr. Sabine 019011
Schulze, T. 016673
Schumacher, Prof. Dr. Peter ... 041239
Schumacher, Dr. Renate 017497
Schumacher, Dr. Ulrich 018916
Schumacker, Mare 004104
Schumann, Kerstin 017535
Schumann, Maurice 013294
Schumann, Peter 049138
Schumann, Robert 001885
Schumann, Romain 030586
Schumard, Ann 054150
Schummel, Helle 010329
Schuppli, Madeleine .. 041186, 042060
Schur, Chantel 006858
Schure, Edeltraut 022716
Schurig, Michael 018412
Schurig, Dr. Roland 016450
Schurkamp, Trish 047085
Schuster-Cordone, Dr. Caroline . 041500
Schuster, Anna 001886
Schuster, Dr. Erika 001946
Schuster, Erwin 002784
Schuster, Eva 022631
Schuster, Prof. Dr. Gerhard 022503
Schuster, Kenneth L. 046976
Schuster, Marina 018046
Schuster, Prof. Dr. Peter-Klaus . 017154, 017262
Schuster, Robin 050550
Schuster, Vicki 047242
Schutte, Maria Christina 000121
Schutz, Carine 014861
Schutz, David 051037
Schutz, Jean-Luc 003799, 003809
Schutzbier, Heribert 002354
Schwab-Dorfman, Debbie 051412
Schwab, Bettina 019011
Schwab, K. 016927
Schwab, Sara 043367
Schwab, Sigrid 017830
Schwab, Yvan 041759
Schwabach, Dr. Thomas 022524, 022525, 022526, 022527
Schwabe, Uwe 019976
Schwager, Michael 052666
Schwaiger, Dr. Axel 018641
Schwaiger, Wilhelm 001930
Schwalenberg, Gregory 046730
Schwaller, M. 014511
Schwalm, Dr. Hans-Jürgen 021308, 021309
Schwalm, Helmut 021761
Schwan, Sandy L. 052739
Schwank, Prof. Dr. Benedikt ... 017350
Schwanzar, Dr. Christine 002330
Schwappach, Lisa 053041
Schwark, Dr. Thomas 019053
Schwarm, Larry W. 048580
Schwarts, Eleanore 053607
Schwartz, Constance 052685
Schwartz, David 046581
Schwartz, Deborah 047252
Schwartz, Elizabeth 049872
Schwartz, Gary 053742
Schwartz, Janet 007186
Schwartz, Jay 053799
Schwartz, Judith 007179, 007183
Schwartz, Michael 047811
Schwartzbaum, Paul 051470
Schwarz 021595
Schwarz, Bärbel 041619
Schwarz, Chris 034873
Schwarz, Dr. Dieter 042152
Schwarz, Ernst 002096
Schwarz, Dr. Godehard 002958
Schwarz, Heike 018093
Schwarz, Dr. Helmut 020877
Schwarz, Dr. Isabelle 019066
Schwarz, Prof. Dr. Karl 002987
Schwarz, Klaus 019455
Schwarz, Prof. Dr. Michael 017563
Schwarz, Peter 017392
Schwarz, Dr. Wolfgang 018965
Schwarzbauer, Dr. Franz 021302
Schwarzjirg, Dr. Helmuth 002267, 002269
Schweer, Dr. Helmut 022349
Schwegler Peyer, Barbara 041206
Schweiger, Andrea 041289
Schweigert, Dr. Günter 022056
Schweiggl, Wolfgang 026079
Schwein, Florence 048504
Schweizer-Makowska, Bòzena . 034718
Schweizer, Nicole 041656
Schweizer, Dr. Paul D. 053916
Schweizer, Dr. Rolf 020626
Schwelle, Dr. Franz 002392
Schwender, Judy 051884
Schwendner, Franz ... 016616, 016621, 016627, 016628
Schwenk, Dr. Bernhart 020508
Schwer, Dr. 022813
Schwering, Dr. Burkhard 019598
Schwerkolt, Charles L. 000806
Schwertner, Dr. Johann 002192, 002358
Schwiderowski, Dr. Peter 018287
Schwind, Stefan 017990
Schwind, Wilmont M. 052285
Schwinden, Lothar 022202
Schwinn, Wolfgang 021341
Schwintek, Monika 021679
Schwinzer, Dr. Ellen 019029
Schwitter, Josef 041548
Schwoeffermann, Catherine ... 046998
Schwolger, David 050135
Schymalla, Joachim 018336
Sci, LaVerne 048174
Sciallano, Martine 013302
Scichilone, Dr. Giovanni 025967
Scirè Nepi, Dr. Giovanna 028299
Scivier, J. 045553
Scoates, Christopher 050376
Scoccimarro, Fabio 028201
Scocos, John 050603
Scofield, P. 033168
Sconci, Maria Selene 027629
Scopelites, Celeste 007091
Scott-Childress, Katie 049233
Scott-Garrett, Pauline . 043736, 044610
Scott Garrity, Noreen 047380
Scott-Garrlett, Pauline 043740
Scott, Adrienne 047712
Scott, Amy 050455
Scott, Andrew 046226
Scott, Angela 053089
Scott, B. 001523
Scott, Barbara 051521
Scott, Carol E. 051934
Scott, Donna A. 051448
Scott, Elva 048418
Scott, Dr. Gerry 052949
Scott, Jane 001638
Scott, John 050263
Scott, Julie 053041
Scott, Kitty 006594
Scott, Marilyn 046176
Scott, Michael 045932, 050020
Scott, Preston 005431
Scott, Rodney 044607
Scott, Ron 007138
Scott, Rose 049205
Scott, Terri 048499
Scott, Timothy 052128
Scott, Vane S. 051522
Scott, Vanya 054391
Scott, Wendy 044065
Scott, William W. 053690
Scottez-de Wambrechies, Annie 013807
Scotti, Dr. Roland 041223
Scotti, Dr. Roland 041225
Scotton, Dr. Flavia 028293
Scoville, Pat 051165
Scrase, D.E. 043842
Scribner, John C.L. 046702
Scriven, Brian 047461
Scrobotă, Paul 036196
Scruggs, Bill 052276
Scudder, G.G.E. 007297
Scudero, Domenico 027624
Scudieri, Dr. Magnolia 026285
Ščukina, Irina Anatolivna 043209
Sculley, John 045722
Scully, Cammie V. 054208
Scully, Robert J. 054405
Seabold, Thomas 049937
Seage Person, Robin 049274
Seager, Pamela 050374
Seald-Chudzinski, Romi 052333
Seale, Sir David 003241
Searcy, Karen B. 046452
Searcy, Manera S. 053267
Searl, Majorie 052618
Searle, Ross 000891
Šeba, Vasyl Stanislavovyč 043313
Sebastián, Dr. Amparo 039673
Sebastian, Padmini 001276
Sebayang, Nas 024614
Šebek, Dr. F. 009750
Šebesta, Dr. Pavel 009515
Sebire, Heather 045712
Secher Jensen, Dr. Thomas 010008
Seck, Dr. Amadou Abdoulaye .. 038119
Seco, Fernando 039647
Sedano, Pilar 039676
Sedberry, Rosemary 046813
Seddon, Jane 043940, 043943
Seddon, Joan 006177
Sedikova, Larisa 043293
Sedilek, F. 038486
Sedinko, Svetlana Alekseevna .. 037572
Sedioli, Giovanni 025543
Sedláček, Dr. Richard 009945
Sedláček, Zbyněk 009643
Sedlazek, Dr. Dietmar 020467
Sedler, Dr. Irmgard ... 019771, 019772
Sedman, Ken 045260
Sedoi, Gabriela 005187
Sedova, Galina Michajlovna 037788
Sedova, Irina V. 037300
Seear, Lynne 001503
Seebeck, Eibe 017968
Seeber, Dr. Ekkehard 022301
Seeberg, Peter 002031
Seele, Ralf-Michael 020343
Seeley, Daniel 050557
Seelig, Thomas 042148
Seeliger, Dr. Matthias 019328
Seemann, Anja 020245
Seemann, Esther 021371
Seemann, Hellmut Th. 022498
Seemann, R. 022429
Seewald, Zahava 003487
Seewaldt, Dr. Peter 022202
Şefănescu, Radu 036267
Šefčík, Dr. Erich 009649
Sefcik, James F. 051298
Seffinga, M.J. 032889
Segal, Dror 025164
Segal, Merav 025180
Segalstad, Prof. Tom V. 033847
Segarra, Guido Barletta 036149
Segarra, Ninfa 051454
Segda, Jerzy 034914
Seger, Joe 050956
Segers, Prof. Dr. Dann 003633
Segessenmann, Vreni . 041829, 041830
Segger, Martin 007330
Seghatoleslami, Parvine Sadre . 024757
Seghers, George D. 046391
Segieth, Dr. Clelia 017338
Segl, Maria-Luise 018339
Seglie, Prof. Dario 027332, 027334, 027339
Segni Pulvirenti, Francesca 025683
Segni, Holly 052497
Segovia Barrientos, Francisco .. 039100
Segovia, Emilia de 039565
Séguin, Louise 006917
Sèguin, Mélanie 006868
Segura Martí, J.M. 038928
Segura, Angela 040372
Segura, Avelino 048116
Seibert, Dr. Elke 022469
Seibert, Georg 022179
Seibert, Peter S. 050118
Seidel, Elizabeth 048031
Seidel, W.A. 019843
Seidenberg, Ulrich 020399
Seider, Diana L. 050344
Seidl, Alfred 002039
Seidl, Andreas 002049
Seidl, Manfred 001857
Seidl, R. Bryce 053245
Seidl, Reinhard 021322
Seif el Din, Dr. Mervat 010515
Seifermann, Ellen 020865
Seifert, Dr. Andreas 020341
Seifert, M.A. Holger 021289
Seifert, Jörg 016590
Seifert, Sabine 018254
Seiffert, Claudia 019505
Seiffert, Prof. E. 045457
Seifriedsberger, Anna 002536
Seifriedsberger, Georg 002536
Seiji, Kamei 029365
Seilacher, Adolf 051265
Seiler, Oskar 041351
Seiler, R. 041613
Seillier, Bruno 013971
Seim, M.A. Andreas 017631
Sein, Joni L. 048483
Seip, Debra A. 053575
Seip, Ivar 033666
Seisbøll, Lise 010231
Seissl, Maria 003093
Seitter, John R. 047379
Seitz-Weinzierl, Beate 022644
Seitz, Erich 022441
Seitz, Gisela 017098
Seitz, Oliver 016647
Seitz, Walter 019332
Sejček, Zdeněk 009864
Sejdner Knudsen, Jesper 010012
Šejleva, Gergana 005033
Sekera, Dr. Jan 009782
Sekerák, Jiří 009471

Sekete, A. 038664
Sekgwele, David 038740
Sela-Zur, Bina 025152
Selbach, Josee 032378
Selbe, Nance 052705
Selbekk, Ass. Prof. Rune 033847
Selbert, Susanne 019291
Selborne, Dr. Ioanna 044951
Seleli, Prof. Y. 038697
Self, Dana 049884, 050012
Selfridge, Anna B. 050279
Selheim, Dr. Claudia 020853
Selig, Gary D. 005475, 005476
Seligman, Norman 001017
Seligman, Rachel 053195
Seligman, Thomas K. 053488
Seligmann, Susan M. 054105
Seligmann, Tine 010288
Selin, Laura 011265
Selirand, Urmas 010622
Sell, Manfred 022704
Sella, Lodovico 025509
Sellers, Mary B. 048287
Sellers, P.E. 031914
Selles, Didier 014749
Sellés, José Pascual 038930
Selley, Andrea 044124
Sellick, Justine 050723
Sellick, Roger 043803
Selliere-Charon 003347
Sellinger, Dr. Günter 002796
Sellink, Manfred 003403, 003405,
003406, 003407, 003408, 003409,
003410, 003411, 003412, 003413,
003414, 003422, 003425
Sello, Thomas 018991
Selmer Olsen, Knud 033873
Selmer, Jørgen 010178
Selmić, Leposava 038214
Selph, Judy 048247
Selsing, Lotte 033972
Selvorious, Clement 026387
Selway, Robert 053087
Selwyn, Sue 044322
Selz, Nan 050336
Selzer, Erika 001734
Selzer, Helmut 003095
Semakova, Tatjana Gennadjevna 037132
Semenova, Galina 037436
Semënova, Tatjana Andreevna . 037042
Semff, Dr. Michael 020576
Semočkin, Aleksandr
Aleksandrovič 037697
Semowich, Charles 046343
Semple, Eve 033449
Sen Gupta, Pradip Kumar 003221
Senadheera, S. 040482
Senanster, Hvon de 014900
Senchylo, Viktor Feodosiyvych . . 043330
Sencova, Ljubov Vladimirovna . . 036991
Sendra, Gemma 039073
Senften, Penny 050639
Seng, Dr. Joachim 018421
Sengele, J.-M. 014373
Sengewald, Angela 022254
Senior, Caroline 024997
Seniuk, Jake 052244
Sen'ko, Rostislav Oleksandrovyč 043211
Senn, Carol J. 048146
Senn, Dr. Matthias 041509
Sennequier, Geneviève 015165
Sens, Dr. Angelie 031977
Sensen, Stephan 016524, 016525
Seo, Chiaki 029441
Seoighe, Mainchín 024836
Seokgi, Kim 029920
Sepioł, Afred 034788
Seppänen, Anne 040620
Sequeira, Dora 008857, 008864
Sequeira, Dora María 008858
Serafin, Mary Beth 050683
Serafini, Flavio 026511
Serafini, Dr. Isabella 027589
Serafino, Gladys 005532

Serafińska-Domańska, Maria . . 035056
Serageldin, Dr. Ismail 010523
Seraja, Marina Leonidovna 038084
Serani Elliott, Daniela 007529
Šeravić, Ivana 009012
Şerban, Violeta 036611
Şerbănescu, Dr. Done 036592
Şerbănescu, Done 036553
Šerbec, Jože 038526
Šerbelj, Dr. Ferdinand 038552
Serboli, Laura 025710
Serdar, Elizabeta 009038
Serdarević, Mevlida . . 004276, 004277
Serena, Dominique 011732
Serexhe, Dr. Bernhard 019522
Serge, Pascal 016049
Sergeenko, Valentin Ivanovič . . 037649
Sergeeva, Elena Nikolaevna . . . 038002
Sergeeva, I.N. 037000
Sergent Nemeth, Dana 046581
Seright, Susan 045634
Séris, Anne-Jean 013477
Serkownek, Edith 050798
Sermeno, Rhodora J. 034584
Serna Muñoz, Martha Judith . . . 008674
Serota, Sir Nicholas . . 045115, 045116
Serpe, Louis 015690, 015691
Serra de Manresa, Valentí 039094
Serra Ferreira, Dr. Ana Margarida 035086
Serra, Florenci 040396
Serra, Sophie 013557
Serradj-Remili, Nedjma 000042
Serré-Van Wijk, Brenda 031954
Sersha, Tom 048647
Serson, Christine 003390
Sertznig, R. 030580
Sérullaz, Arlette 014783
Servais Lebrun, Françoise 003869
Servan, René 015083
Service, Dr. Alexandra 053729
Service, Pam 048626
Servitje de Lerdo de Tejada,
Marinela 031215
Serwer, Jacquelyn D. 054099
Sesić, Maria 038165
Šestakova, Tamara Nikolaevna . 037926
Sestan, Ivan 009028
Šestopalova, Galina Pavlovna . . 029834
Sethuraman, K. 024231
Setinský, Jiří 009488
Setouchi, Betsuji 028952
Šetrakova, Svetlana N. 037348
Sett, Dr. Saradesh K. 024292
Setter, Michele 054044
Settle Cooney, Mary 053865
Setzer, Heinz 016930
Seuffert, Dr. Ottmar 017936
Seugwoo, Nam 030008
Sevastjanova, Lidija Ivanovna . . 037096
Ševčenko, Ljudmyla Mychajlivna 043297
Ševčenko, Ol'ga Ivanivna 043196
Sevda, Dadashewa 003179
Ševeček, Dr. Ludvík 009983
Sevellec, Jacques 012205
Sever, Nancy 000756, 000757
Severance, Christopher C. 005597
Severens, Martha R. 049268
Ševereva, Tamara Anatoljevna . . 037569
Severino, Tenente Gerardo 027646
Severstad, Jamie H. 051208
Sévery, Sébastien 015059, 015060
Sevi, Uriel J. 000249
Sevier, James 048229
Sevigney, Nichola 054077
Sevilla, Guillermo A. 000282
Sevilla, Miguel J. 039638
Sevinç, Nurten 042866
Ševkomud, Igor Dmitrievič 036950
Ševyrin, Sergej Andreevič 037643
Seward Hannam, Kim 006126
Sewell, Dennita 052120
Sewell, Jack 001002
Sewell, Jane S. 054343
Sexton, Nick 044034

Seybold, Russell G. 051600
Seybold, Silke 017601
Seyfried, Dr. Friederike 019972
Seymour, C. 001468
Seyrl, Harald . . 002713, 002714, 003099
Sezen, Nuri 042834
Sezgin, Mustafa 043074
Seznec, A. 015020
Sforzi, Dr. Andrea 026475
Sfrappini, Dr. Alessandra 026659,
026661, 026663
Sgritta, Richard A. 051944
Sgromo, V. Joseph 052721
Sgubin, Dr. Raffaella . . 026453, 026454,
026455
Sgubini Moretti, Dr. Anna Maria 025739
Shaath, Shawqi M. 042254
Shacham, Tzvi 025186
Shackelford, George T.M. 047123
Shacter, Joseph 047690
Shady Solís, Dr. Ruth 034328
Shafer, Geri 048374
Shaffer Nahmias, Dr. Susan . . . 050813
Shaffer, Leslie 046740
Shaffer, Marcy 046345
Shaffer, Randy 046508
Shaffer, Vivienne 053448
Shafts, Karen 047106
Shagonaby, Joyce 046454
Shah, Hiralal P. 024140, 024141
Shahi, M. 024694
Shahzad, Tasneem 034146
Shaiboub, Dr. Abdullah 030291, 030292,
030294, 030295, 030296, 030298,
030299
Shaiman, Jason 047912
Shaker, H. 024690
Shakespeare, Valerie 051321
Shalev, Avner 025080, 025116
Shalgosky, Sarah 044021
Shamis, Bob 051443
Shanahan, Carl 049064
Shanahan, J.P. 001079
Shane, Dr. Orrin 052852
Shank, Nicholas B. 050941
Shankle, Kent 054208
Shanley, Ellen 051433
Shanley, Grace 051643
Shanmugam, Dr. P. 024194
Shannon, George W. 049401
Shannon, Dr. Helen 053799
Shannon, Steve 044188
Shao, Guotian 007576
Shao, Jiuhua 008517
Shao, Qinglong 007867
Shao, Zitong 008215
Shapiro, Anna 047107
Shapiro, Avra 050461
Shapiro, Denise 050160
Shapiro, Lynda 047978
Shapiro, Michael 046623
Shapiro, Dr. Michael E. 046618
Shapiro, Selma 051669
Sharan, Michael 052049
Sharer, Robert 052100
Sharma, Dr. O. P. 024154
Sharma, R.D. 024131
Sharon, Douglas 046941
Sharp, Corwin 054533
Sharp, James 050481
Sharp, Lewis I. 048258
Sharp, Lindsay 046207
Sharp, Prof. M. 005757
Sharp, Megan 046679
Sharp, R.W.D. 046020
Sharp, Randall 048544
Sharps, David 047268
Sharrock, D.J. 044679
Sharygin, G.V. 037494
Shastri, N.H. 024512
Shastri, Nandan 024511
Shatanawi, M. 031950
Shaudys, Vincent 053397
Shaughnessy, Beth 006115

Shaul, Nissim 025169
Shave, Sue 043901
Shaw, Alvin 047338
Shaw, Dr. Anita 041385
Shaw, D. 001453
Shaw, Dan 050893
Shaw, Deirdre 054300
Shaw, Elizabeth . . . 043670, 046085
Shaw, Harry 044772
Shaw, J. 043668
Shaw, Karen 048438
Shaw, Lana 006572
Shaw, Lytle 051379
Shaw, R. Douglas 054325
Shaw, R.W. 044846
Shaw, Tad 048650
Shawhan, Jeff 050799
Shawn Barber, Barbara 052022
Shawn, Josephine 052022
Shawver, David 053645
Shea, Carol 048133
Shea, John C. 049915
Shea, Josephine 049283
Sheads, Scott 046747
Sheaffer, Ed 053316
Shean-Hammond, Edie 047742
Shearer, Linda 047742
Shearer, Thelma 001057
Shedd, Linda 046842
Sheedy, Gilles 005404
Sheehan, Liz 050232
Sheehan, Ron 049822
Sheesley, Tim 051775
Sheets, Arian 053960
Sheets, H.L. 050703
Sheets, Jeff 046302
Sheets, John W. 054070
Sheffield, Sir Reginald 045928
Shefsiek, Kenneth . . 049068, 051308
Sheku, Nail 000035
Sheley, David 049404
Shelhouse, Marvin 050889
Shelkrot, Elliot L. 052087
Shell, Lynn 052277
Shelley, Elise 049926
Shelley, Roger 044087
Shelly, Diane 047168
Shelly, Roger 044082
Shelswell White, E.R.G. 024832
Shelton, George 046644
Shelton, L.T. 049805
Shen, Bohong 008341
Shen, Changming 008092
Shen, Guoying 008463
Shen, Hsue-Man 053249
Shen, Jiangbo 008362
Shen, Junshan 008052
Shen, Dr. Lindsay 052839
Shen, Lumin 008304
Shendar, Yehudit 025116
Sheng, Yi Jia 008106
Sheng, Zhenggang 007840
Shengold, Ann 051006
Shepard, Betty J. 050434
Shepard, Charles A. 048894
Shepard, Joel 053032
Shepard, Rebecca 048978
Shepherd, Dr. Brian 001402
Shepherd, Rowena 044075
Shepley, Emma 045051
Shepp, James G. 050607
Sheppard, Audrey 054159
Sheppard, Beverly K. 053579
Sheppard, Jude 047725
Sheppy, Richard 043699
Sherer, Aaron 051830
Sheridan, Joan 054052
Sherin, Pamela V. 052365
Sherman, Alan 025163
Sherman, Cathy R. 047990
Sherman, Mimi 051423
Sherman, Sean 045065
Shermann, Robert L. 047216
Sherrill, Sarah B. 051344

Sherring, Brenda 007499
Sherwood, Frederick 048437
Sherwood, Penelope 053982
Sherwood, Susan I. 046998
Shestack, Alan 054141
Shetler, John 053789
Shetty, Prof. N.R. 024226
Shexnayder, Charlotte 048388
Shi, Jiliang 007864
Shi, Jinhai 007798
Shi, Jungui 008345
Shi, Lei 007903
Shi, Meilong 008497
Shi, Mingrong 007931
Shi, Qinyang 008336
Shi, Tingyuan 008047
Shi, Tuankuo 008413
Shibuya, Taku 028818
Shickles, Tim 048749
Shieh, Suewhei 053786
Shields, Ann 053898
Shields, R. Michael 007295
Shields, Scott 052725
Shiels, Margaret M. . . 048369, 054504
Shiffman, Carol 052258
Shiftins, Liz 045596
Shiga, Taro 029129
Shigaki, Betty 052612
Shih, Chia-Chun 048904
Shikada, Norimitsu 029195
Shikata, Miho 029674
Shilliday, Charles 007442
Shilo-Cohen, Nurit 025102
Shimada, Kazutaka 029549
Shimada, Kunio 029405
Shimada, Dr. Norio 028869
Shimasaki, Susumu 028772
Shimazu, Misato 029509
Shimizu, Minoru 029399
Shimo, Elko 029653
Shimoda, Norio 029490
Shimp, Dr. Robert E. 047855
Shin-tsu Tai, Susan 053092
Shin, Chul-Won 029915
Shin, Prof. Kwang Cho 029959
Shin, Tak Keun 029248
Shin, Young Koog 030042, 030058
Shinkle, Bradford 050945
Shinn, Deborah 051368
Shino, Masahiro 028858
Shinoda, Aishin 028812
Shiozawa, Yukio 029514
Shipley, Jim 044548
Shipman, Douglas R. 054357
Shipp, Kim 006210
Shippee Lambert, Lisa 047414
Shirahara, Yukiko 053248
Shirahira, Yukiko 053249
Shiraishi, Masami 029653
Shirbacheh, Tahereh 024747
Shireman, Joyce A. 051202
Shirley, Donna 053247
Shirley, Pippa 046027
Shirley, Rosie 045828
Shives, Jenny 050507
Shivvers, Mel 048284
Shlomo, Eli Ben 025156
Shlomo, Eliyahu-Ben 025037
Sho, Masaki 029248
Shoaff, Jeanne 048799
Shock, Helen 049617
Shockley, Richard 006185
Shockro, Megan 053086
Shoemaker, Innis H. 052081
Shoemaker, John 050074
Shoemaker, Rene D. 046596
Shokoofi, M.R. 024786
Sholes, John 046913
Shôno, Prof. Susumu 029422
Shook, Ray 048173
Shore, Virginia 054089
Shoreman, Mark 007357
Shorey, Prof. C.D. 001541
Short, John D. 049696

Short, Juliet 044291
Short, William 053057
Shoshani, Jesheskel 048314
Shotelier, Jean-François 014412
Showalter, Natasha 046548
Shozo, Tamura 028750
Shrago, Esther 048159
Shrago, Mindy 048159
Shrake, Peter 046787
Shrigley, Ruth 045204, 045208
Shriver, James M. 054343
Shriver, Karen K. 049554
Shriver, Thomas L. 046506
Shrum, Kenneth 052680
Shu, Zhimei 008367
Shubert, Gabrielle 047264
Shubinski, Raymond 052556
Shugart, Dr. Gary 053644
Shulman, Steven E. 054086
Shuman, Greg 052702
Shuman, Lynne 050679
Shunsuke, Ogata 029428
Shupeck, Fern 050910
Shury, Jane 005400
Shust, Maria 051486
Shuttleworth, John 049618
Shydlovskyy, Ihor 043258
Si, Luoping 008389
Siachoono, Stanford Mudenda . . 054843
Siampanopoulos, Konstantinos . . 023144
Siano, Mary Ann 047232
Sibanda, D. 054851
Sibbalds, Mary H. 052647
Siber, Hans Jakob 041193
Sibeth, Achim 018434
Sibille, Marie-Hélène 011729
Sibiril, Marceline 014115
Sibley, S.F. 046163
Sibley, Shelia M. 051553
Sibson, Tracy 033307
Sibutha, Addelis 054846
Sicard, D. 015473
Sichelstiel, Ludwig 020852
Sicilia, Fiorella 027509
Sidaway, Joan 044301
Sidebottom, D.J. 044458
Sidén, Karin 041024
Siders, Phyllis 048308
Sidibé, Samuel 030734
Sidiropoulos, Kleanthis 023096
Sidorov, Konstantin Nikolaevič . . 036935
Sidorova, Elena Iljinična 037052
Siebenberg, Theo 025107
Siebeneicker, Dr. Arnulf 019251
Siebenmorgen, Prof. Dr. Harald . 017631,
    017715, 019505, 019512, 019514,
                                       021964
Sieber, Christian 041864
Sieber, Ellen 047050
Sieberer, Dr. Wido 002182
Siebert, Andreas 016812
Siebert, Dr. Anne Viola 019060
Siebert, Tim 053525
Sieblist, Kerstin 020021
Siebol, Michael 054566
Siebrand, Michel 004139
Siebrecht, Silke 019653
Sieburgh, J.W. 031982
Siedell, Daniel 050300
Siefert, Chris 052172
Siefert, Dr. Helge 020508, 021048
Siegal, Matthew 047123
Siegel, Bernard 053314
Siegel, Bob 050800
Siegel, Eric 048767
Siegel, Gail 051076
Siegel, Gerlinde 018861
Siegel, Nancy 049613
Siegenthaler, Gustavo Bernardo . 000625
Siegfried, Clifford A. 046344
Siegman, Joe 025163
Siegrist, Urs 042185
Sielaff, M.A. Ingo 017513
Siembor, Natalie 051839

Siemen, Palle 010044
Siemen, Wilhelm 016613, 019301,
                                       021833
Siemers, Dr. Sven-Hinrich 019369,
                                       022734
Siena, Pier Luigi 025580
Siener, William 047300
Sienkiewicz, Joanna 035169
Siepmann, Werner 021731
Sierpowski, Prof. Dr. Stanisław . 034848
Sierra Bello, Prof. Luz 031179
Sierra de la Calle, Blas 040346
Sierra Jones, Alvaro 008721
Sietz, Phillip 052039
Sievernich, Gereon 017220
Sievers-Flägel, Dr. Gudrun 020842
Sievert, Annette 018979
Siewert, Simone 018941
Sifkovits, Dr. Adam . . . 001750, 001751
Šiftar, Davor 009052
Sigal, Bronia 031231
Sigal, Laurence 014695
Sigaud, Anne 013910
Sigerist, Rudolf 022421
Sigmon, Nancy 050089
Sigmond, Dr. J.P. 031982
Signorello, Fortunato Orazio . . . 025870
Signorini, Prof. Rodolfo 026701
Sigo, Charles 053614
Sigurdardottir, Aldis 024116
Sigurdardóttir, Elin S. 024068
Sigurdardóttir, Maria Karen 024101
Sigurdardóttir, Sigridur 024117
Sigurdarson, Erlingur 024067
Sigurdson, Annchristine 040751
Sigurdsson, Helgi 024092
Sigurjonsdottir, Iris Olöf 024072
Sigurjónsson Jonsdottir, Sigrun
Asta 024091
Sigwing, Marty 054385
Sihm, Marcel Agang-Ang 034441
Sijarić, Mirsad 004283
Šikić, Vojko 008905
Sikorska, Janina 034782
Sikorski, Marek 034674
Sikorskyj, Mychajlo Ivanovyč . . . 043280
Šikšniené, Roza 030478
Sila, Dr. Roland 002141
Silber, Evelyn 044402
Silber, H. 022218
Silek, Suzanne 048989
Sill, Robert 053451
Silla, Dr. Chiara 026251
Silla, Dr. Chiaretta . . . 026250, 026268,
                         026284, 026289, 026294
Sillanpää, Pirkko 010771
Silliman, Thomas 051020
Sillner, Helmut 001764
Šilnikova, Inna Evgenjevna 037976
Šilov, Lev Alekseevič 037633
Šilov, Viktor Valentinovič 037859
Šilova, Tatjana Ēduardovna 036849
Silva de Oliveira, António Manuel 035920
Silva Montauntildea, Margarita . 008799
Silva Silva, Párroco Isaac 008728
Silva, Adelino Baptista e 035895
Silva, António Marques da 035893
Silva, Cyro Advincula da 004717
Silva, Dan 049128
Silva, Eduardo Pereira da 035676
Silva, Dr. Heliodoro 035530
Silva, Henrique 036094
Silva, Isabel 035574
Silva, Ivan Mariano 004574
Silva, John 034453
Silva, Dr. Leonardo D. 004668
Silva, Dr. Maria Rosa . 035506, 035507
Silva, Dr. P.H.D.H. de . 040484, 040488
Silva, Pilar 039676
Silva, Urbain Karim Elisio da . . . 004186
Silveira Godinho, Isabel 035816
Silveira, Prof. Maria Auxiliadora . 044589
Silverio, Prof. Franco di 027310

Silverio, Dr. Loca di 027310
Silverman, David 052100
Silverman, Lanny 047649
Silvester, Suzanne 051731
Silvestre, Ines K. 034566
Silze, Peter 018983
Sim, Angela 038273
Sim, Lee 030016
Simak, Ellen 047595
Simanjuntak, Pareli 024526
Simard, Claire 006736
Simard, Guylaine 006131
Simard, Julie 006946
Simavi, Çiğdem 042937
Simcock, Dr. J.P. 033145
Simerly, Enixx 050732
Simić, Prof. Jasna 038245
Simina, Nicolae-Marcel 036689
Simion, Luca 036701
Simister, Kirsten 044712
Simkins, Barbara 006154
Simková, Dr. Anežka 009891
Simma, Barbara 003091
Simmendinger, Jürgen 017701
Simmonds, Paul 045702
Simmons, Arlene 050210
Simmons, Ian 045400
Simmons, Jamie 053724
Simmons, Joyce 048722
Simmons, Julie M. 007098
Simmons, Mark 045486
Simmons, Ruth 052386
Simmons, Susan 048438
Simms, Kate 044445
Simms, M.H. 046144
Simon-Hamel, Jeanne-Yvonne . . 015242
Simon, Dr. Barbara 018229
Simon, C.M. 052139
Simon, Christiane 004040
Simon, Gerlinde 016661
Simon, Hal 048121
Simon, Katalin 023469
Simon, Dr. László . . . 023952, 023953,
        023954, 023955, 023956, 023957,
                         023959, 023961, 023963
Simon, Martina 020825
Simon, Nancy 054448
Simon, Patrick 031717
Simon, Robert 048285
Simon, Rolf 016737
Simon, Ronald 051441
Simon, Prof. Dr. Stefan 017272
Simon, Susanne 018075
Simonato Rabim, Dr. Chiaza . . . 028307
Simončič, Alenka 038552
Simončič, Vlastja 038513
Simoneau, Christiane 007244
Simonelli, Lynn 048168
Simonett, Dr. Jürg 041417
Simonetti, Dr. Farida 026394
Simoni, Kate 009025
Simonnet, Philippe 015050
Simons, Andrew 050940
Simons, Ulrich 021030
Simonson, George 050236
Simpel, Francis de 004132
Simpson Niven, Shelly 005546
Simpson, Ann 044273
Simpson, Colin 043608
Simpson, Helen J. 044196
Simpson, J.L. 045942
Simpson, John W. 006220
Simpson, Ken 043463
Simpson, M.A. 007384
Simpson, Paul 045655
Simpson, Shannon 054242
Simpson, Tracy 046205
Simpson, Wendell 053854
Simpson, William G. 045207
Sims-Burch, Wylene 047897
Sims, Judith 046685, 046686
Sims, Patterson 051007
Sims, Richard 052352
Simson, Susanna von 019203

Simui, Muyoyeta 054841
Simundson, Peter 007196
Sin, Jang Jong 029877
Sinaga, T. Moesa 024634
Sinanovitch, Laura 041286
Sinchak, Cindy 052779
Sinčić, Branko 009058
Sinclair, Dr. Bradley 017504
Sinclair, F. 044164
Sinclair, H.M. 045926
Sinclair, Mark E. 049457
Sindermann, Sabine 021499
Sinding Steinsvik, Tone 033493
Singer, Debra 051491
Singer, Ilana 025059
Singer, June 046298
Singer, Karl 020034
Singer, Dr. Marijane 051789
Singer, Robert 050444
Singer, Ruth 041950
Singh, Dr. Bhanu Pratap 024342
Singh, Himmat 024319
Singh, Prof. I.B. 024381
Singh, S. 024181
Singh, S. Amolak 024142
Singh, Vidya Nand 024191
Singhvi, Dr. L.M. 024226
Singleton, G. David 053850
Singleton, Robert W. 053074
Singley, Eric 050418
Sinick, Chris 052946
Sinisalo, Henna 010829
Šinjaev, Nikolaj Ulanovič 037004
Sinn, Margaret 054080
Sinn, Prof. Dr. Ulrich 022792
Sinnig-Haas, Christiane 012486
Sinnott, George 006653
Sinopoli, Carla 046491
Sinopoli, P.A. 000546
Sinsheimer, Karen 053092
Sintes, Claude 011730
Sinyard, Elizabeth 006476
Siokalo, Zoriana 048366
Šípková, Želmíra 038489
Sipkovsky, Frank 050396
Siqueira Celidonio, Helio
    Fernando 004362
Sirak, James 051091
Sirech, Alexander 009180
Siri, Marta 027221
Širin, Jurij 037573
Širkanova, Maja Konstantinovna . 037596
Širmat, Mirna 009048
Širová, Marie 009976
Sirugue, Christophe 014589
Sisdal, Odelbjørn 034001
Sisi, Dr. Carlo 026255, 026292
Šišigin, Egor Spiridonovič 037055
Šiškin, Anatolij Aleksandrovič . . . 037738
Šiškova, Živka 005012
Sisley, Alan 001386
Sitar, Mag. Gerfried 002690
Sitarski, Steve 052044
Sitch, B. 045210
Sithole, R. 054847
Sitjar, Corazon 034511
Sitkowska, Maryla 035325
Sitt, Dr. Martina 018991
Sitwell, Sir Reresby 045604
Siudziński, Kazimierz 034749
Siváček, Dr. Vladimír 038402
Šivačev, Stefan 005130, 005131
Sivan, Pnina 025045
Sivcova, Natalja Viktorovna 037540
Sivec, Nina 009052
Sivertson, Al 005307
Sivkov, Dr. Janaki 005246
Sivkova, Svetlana G. 037085
Sivonen, Esa 011098
Sivrikaya, İsmet 043044
Sivula, Leena 011186
Sivusaari, Tiina 011176
Sivyer, M.J. 001241
Siwek, Alberyk 035297

Siwek, Jan 034620
Siximons, Les 001088
Sizemore, Bernice 046872
Sjöberg-Pietarinen, Solveig 011439
Sjöblom, Seth 041156
Sjöö, Robert 040638
Sjöskog, Per 011116
Sjöström, Anne 011178
Sjöström, Mats 011449
Sjösvärd, Lars 040637
Skaarup, J.C. 010287
Skaarup, Jørgen 010296
Skala, William 050201
Skamby Madsen, Jan 010128
Skanis, M. 030103
Skar, Anne Kari 033972
Škarić, Mila 009025
Skarpia-Heupel, Prof. Dr.
    Xanthippi 023292
Skarsbø, Einar 033876
Skarstein, Sigrid 034062
Skarstøl, Gunnar 033573
Skartsounis Schwab, Diane 041410
Skatov, Nikolaj Nikolaevič 037752
Skavhaug, Kjersti 033668
Skeide, Cecilie 033741
Skelly, David 051265
Skenbäck, Urban 041028
Skerswetat, F. 018268
Skibbe, Uwe 017645
Škiljan, Maja 009036
Skinner, J.F. 045836
Skinner, John 045834, 045835
Skip, R. 048256
Skippings, Lin 044934
Skirtz, Stefan A. 047736
Skiver, Sharon 053732
Skjoldal, Bjørg 034014
Sklar, E. 005880
Sklenářová, Vilma 009947
Skobeleva, T.P. 037853
Škoberne, Želimir 009043
Skog, Fredik 041027
Skoghäll, Lars 040989
Skoglund, Eric Y.W. 040712
Skoglund, Ivar 033520
Skoglund, Urban 040955
Skonecki, Leonard 048912
Skopec, Dr. Manfred 003034
Skorikova, Svetlana Nikolaevna . . 036952
Skorobogatova, Nina Vasiljevna . . 037274
Skoruls'ka, Oksana Mykytivna . . 043210
Skory, Gary F. 050872
Skotheim, Hege 033701
Skott, Inga 011467
Skougaard, Mette 010119
Skov Kristensen, Henrik 010271
Skov, Torben 010134, 010136
Skovgaard, Margaret A. 048804
Skozen, Connie 051120
Skriebeleit, Jörg 018376
Škrinjarić, Srećko 009025
Skrobov, Sergej 036992
Skudal, Tina M. 033556
Skuggen, Guri 033871
Skuja, Andrejs 030204
Škulj, Božica 009051
Skura, Christopher 051399
Škurko, Dr. A.I. 037286, 037497
Skutecki, Andreas 022546
Škvařilová, Božena 009798
Skvarla, Diane K. 054172
Škvarnová, M. 038486
Skye, Stephen. 048082
Slabá, Jaroslava 009588
Slabakova, Ganka . . . 005007, 005008
Slade, Norman 050186
Slader, John 048790
Sladky, Vaclav 003068
Slamenová, Dr. Zuzanah 038321
Slaney, Deb 046358
Slanina, Súsanne 021953
Slapinski, Tim 047343
Slastunin, A.A. 037120

Slattery, Fiona 045486
Slavik, Jan 040663
Slavin, Ruth 046492
Sławski, Marek 035174
Slear, Gary 050227
Śledź, Edward 034725
Slehofer, Vera 041261
Slej, Dr. Karen 041019
Slettjord, Lars 033801
Sleurs, Jean-Pierre 003670
Slichter, Jennifer 050506
Sliggers, B.C. 032399
Slim, Soumaya 031259
Slimon, Gary 006583
Slinger, Frederico 046284
Śliwiński, Eugeniusz 034849
Sloan, Alan P. 053147
Sloan, Catherine 045158
Sloan, Mark 047549
Sloan, Nancy 050158, 050171
Sloan, Sarah 030931
Sloar, Steve 052973
Slobodzian, John 005685
Slocum, Kay 047439
Sloof Vermeij, P. 032849
Sloss, Tencha 047280
Slotta, Prof. Dr. Rainer 017433
Slottved, Lis 009997
Slover, Louise 049443
Slovin, Rochelle 046581
Slowe, V.A.J. 044002
Složenikina, Valentina Ivanovna . . 037663
Sluis, R. ter 032385
Sluiter, Drs. G. 031973
Slyngborg, Mette 010044
Slythe, Margaret 045443
Smadar, Harpazi 025213
Smailes, Helen 044255
Smal, M. 038859
Small, Colleen 053974
Small, Jim 049239
Small, Lisa 051372
Small, Nicole G. 048116
Smallman, Anne 047870
Smart, Diana 046200
Smart, Dragana 043968
Šmatov, Victor 003252
Smedt, Raphaël de . . 003435, 003482
Smeets, H. 032948
Smerdel, Inja 038558
Smerling, Walter 018060
Smidt-Jensen, Jörgen 010273
Smidt, D.A.M. 032585
Smidt, Erzsébet 023984
Smietana, Evelyn 050147
Smiley, Kristine 050822
Smiley, Mary-Ellen 046966
Smilova, Dimitrina 004999
Smirnov, Valentin Vladimirovič . . 037424
Smirnova, N.V. 036964
Smirnova, Natalja Petrovna 037856
Smit, K. 032273
Smit, Kim 024900
Smit, Sandra 038651
Smit, Toril 033871
Smith-Christmas, Kenneth L. . . . 052425
Smith de Tarnowsky, Andrea . . . 051172
Smith-Ferri, Sherrie 053878
Smith-Speelman, Dr. Khristina . . 047838
Smith Turner, Brier 050778
Smith Walley, Robyn 000999
Smith Woods, Karen 046798
Smith, Adam 045749
Smith, Adam. 051827
Smith, Alan 001047
Smith, Albert 048455
Smith, Allan 033141
Smith, Allison A. 050600
Smith, Amanda 053717
Smith, Dr. Amy C. 045593
Smith, Ann . . 043964, 045781, 050151,
                                054199
Smith, Antony 046089
Smith, Art 006784

Smith, Arthur 044427
Smith, Dr. B.J. 001215
Smith, Bea 048948
Smith, Bertyne 046796
Smith, Betsy 048367
Smith, Beverly 046471
Smith, Bill 046587
Smith, Bonnie H. 049715
Smith, Bradley 053564
Smith, Brian 050195
Smith, Brooke 046938
Smith, Bruce 047924
Smith, Calvin B. 054020
Smith, Carol 048417
Smith, Chris 033199
Smith, Christopher 047620
Smith, Chuck 052162
Smith, Cindy 047520
Smith, Clive 033232
Smith, D. Ryan 046861
Smith, Daniel J. 054288
Smith, Darryl 053010
Smith, David 006251, 050621
Smith, David K. 050623
Smith, Deborah A. 047287
Smith, Diane 053389
Smith, Dr. Donald 044246
Smith, Donna 005788
Smith, Donna M. 005392
Smith, Dottie 052493
Smith, Douglas G. 046452
Smith, Duane 049611
Smith, Dwight 048305
Smith, Earl 054246
Smith, Elizabeth . . . 047676, 048659,
                                053436
Smith, Elizabeth A. 051042
Smith, Francis 055432
Smith, Fred T. 049932
Smith, Frederick G. 047600
Smith, Dr. G.T. 046953
Smith, Garrett 053156
Smith, Gary 049286
Smith, Gary N. 048121
Smith, Gary T. 052899
Smith, George 053361
Smith, Graham 043386
Smith, Dr. Gregory 048581
Smith, Dr. H. Duane 052404
Smith, Heather 006421
Smith, Helen 050897
Smith, Ian 049244
Smith, J. 033168, 046009
Smith, J.B. 051948
Smith, Dr. James N.M. 007283
Smith, Jan 050648
Smith, Janet 006196
Smith, Janice 005967
Smith, Jaqueline 044086
Smith, Jay 049646
Smith, Jean 006455
Smith, Jerry 054173
Smith, Jill 051222
Smith, Joanne 043975
Smith, Joel 052338
Smith, John 046065
Smith, Julie 043485
Smith, Kelly 053814
Smith, Ken 050306
Smith, Kent 047665
Smith, Kevin P. 047217
Smith, Kym 000851
Smith, Lisa Deanne 007229
Smith, M. 006926
Smith, Prof.Dr. M.P. 043621
Smith, Mariann 005323
Smith, Marilyn 006187, 006613
Smith, Mark 044969, 045276
Smith, Mary 006603, 006605
Smith, Maureen M. 052367
Smith, Maurice 006111
Smith, Melvin R. 051736
Smith, Michael 051116
Smith, Michael J. 050508

Smith, Michael O. 048288
Smith, Michele L. 053944
Smith, Michelle 001445
Smith, Dr. Moya 001403
Smith, Nancy 024991
Smith, Nelson 048306
Smith, Norman 050903
Smith, Owen 046669
Smith, Pamela 053118
Smith, Pat A. 047845
Smith, Patricia F. . . . 050279, 054513
Smith, Patsie 050634
Smith, Pauline M. 046607
Smith, Peggy 050115
Smith, Peter W. 048567
Smith, R. 005522
Smith, R.E.G. . 006708, 006709, 006712,
006713
Smith, Dr. R.M.H. 038686
Smith, Raechell 049882
Smith, Randi 053434
Smith, Rebecca A. 050830
Smith, Robert 044205, 045909
Smith, Robert J. 050185
Smith, Rose J. 051736
Smith, Sam 048055
Smith, Samantha 045721
Smith, Scott 053178
Smith, Shana 052248
Smith, Sharon 046017
Smith, Shirley 048988
Smith, Stephanie 047697
Smith, Steve 052516, 053403
Smith, Sue 001457
Smith, Susan 053307
Smith, Terri 051174
Smith, Thomas 051623, 052182
Smith, Tim 049609
Smith, Tobi 053132
Smith, Valerie 017159
Smith, Verna 051658
Smith, Vivian DuRant 051501
Smith, W.D. 043879
Smith, Warwick 033261
Smith, Wayne C. 050107
Smitherman, Joy 053221
Smithline, Howard 025145
Smits, Eduard 004155
Smitten, Al 005922
Smogór, Anna 034629
Smolders, F.J. 032721
Smolders, R. 032214
Smole Gavrilovic, Margita 004283
Smolek, Michael A. 052789
Smolík, Luboš 009629
Smoljaninova, Iryna Mykolaïvna 043199
Smolkin, Robert 046971
Smorenburg, D. 032966
Smorodinsky, Marina 050418
Smothers, Ann E. 049709
Smrcka, Mariann 050756
Smrekar, Dr. Andrej 054613
Smucker, Linda 054613
Smulian, Rob 046606
Smurthwaite, D. 045056
Smutná, Silva 009943
Smylski, Brian 006483
Smyth, Anthony 006575
Smyth, Deirdre 044639
Smyth, R. 006216
Smyth, Victor 051466
Snape, Dr. Steven R. 044853
Snars, Frank 001252
Sneddon, Russell C. 048491
Šneiter, G.G. 037261
Snel, C. 038794
Snell, Dr. Gesa 019027
Snell, Dr. Howard L. 046365
Snell, Prof. Ted 000852
Snellen, Wayne 051415
Snelling, Graham 045360
Snichelotto, Paolo 027805
Snider, Katherine 051418
Snider, Phyllis 046535

Śniegocki, Michał 035374
Śniokienė, Nijolė 030455
Snoddy, Stephen 046041
Snoddy, Suzie 053592
Snodgrass, Dianna H. 049304
Snørteland, Dr. Målfrid 033794
Snørteland, Målfrid 033795
Snørtzland, Dr. Målfrid 033799
Snoeys, Johan 003308
Snotun, Trygve 033598
Snow, David H. 053118
Snowden, Frances 043687
Snowden, Gilda 048302
Snowden, Karen 045744, 045746
Snydal, Dave 005466
Snyder, Amy 051081
Snyder, Barbara C. 050755
Snyder, Craig 046390
Snyder, Hills 052945
Snyder, Howard 005564
Snyder, James 025087, 052468
Snyder, James S. 025100
Snyder, Jill 047818
Snyder, Jim 054596
Snyder, Marion 054234
Snyder, Rob 052416
Snyder, Scott 052638
So, Nguyen 054805
Soans, Howard 010668
Soares de Souza, Prof. Jonas . 004503
Soares, Elisa 035952
Sobčenko, Viktor Kindratovič . . 043116
Sobčuk, Galina 043319
Sobczyk, Pawel 035401
Sobezyk, Stéphanie 012838
Sobieraj, Edward 035103
Sobieszek, Robert 050444
Sobiesznak-Marcimak, Margaret 035350
Sobita Devi, Dr. K. 024310
Soboleva, L.L. 037154
Sobotka, Joseph 050181
Sobrepeña, Enrique A. 034603
Sobrequés i Callió, Jaume 039089
Sobreques, Dr. Jaume 040117
Socarrás, Carlos 008785
Soccoro R. Samson, Marlene . . 034468
Socorro Mallorca, Maria 034481
Sodagar, L. 024781
Soden, Jack 050779
Soden, Joanna 044269
Sodermanns, Christina 016440
Søborg, Hans Christian 033489
Söderlind, Lars 040899
Söderlind, Prof. Solfrid 041024
Söderlund, Göran 041033
Soeharto, Ki 024535
Søholt, Petter . 033770, 034038, 034038
Soejanto, B.U.W. 024535
Soekiman, Dr. Djoko 024683
Sölder, Wolfgang 002141
Söll-Tauchert, Dr. Sabine 041267,
041268
Söllner, Jürgen 021744
Sömer, Hermann 016746
Søndergaard, Ph. d. Morten
Karnøe 010306
Søndergaard, Per 010270
Søndergaard, Sidsel 010188
Sönmez, Dr. Necmi 018305
Sönnichsen, Hans F. 020787
Söptei, István 023876
Sørensen, Anne Birgitte 010092
Sørensen, Anny 010058
Sørensen, Einar 033572
Sørensen, Jens 010264
Sørensen, Jens Erik 010005
Sørensen, Marianne 010354
Sørensen, Ove 010252
Sørensen, Steinar 033585
Sørheim, Helge 033972
Sörries, Christine 017388, 020174
Sörries, Prof. Dr. Reiner 019541
Soesilo Hadi, Dr. Hidayat 024674
Soetjipto, Dr. 024655

Softić, Dr. Aiša 004283
Softic, Ajsa 004283
Soganci, Dr. Marion 020637
Soggin, Gian Piero 027096
Sogoba, Djaié 030736
Sohal, Neena 045822
Sohinki, Arnie 052651
Sohl, Katrin 020021
Sohlern, Gilbert von 018743
Soileau, Melba 054182
Soja, Henryk 034840
Sojka, Nancy 048301
Sojková, Dr. Jana 009958
Šoka, Milan 038282
Sokač-Štimac, Prof. Dubravka . . 008969
Soklevski, Valentin 030610
Soklove, Deborah 054100
Sokól-Augustyńska, Maria 035161
Sokol Jr., William 047233
Sokol, Vladimir 008986
Sokolov, Dr. Aleksej 037721
Sokolov, Dr. Nikolaj N. 037260
Sokołowska, Renata 035260
Sokolowski, Thomas 052153
Sol i Bachs, Josep M. 039098
Solà Turell, Ramón 039952
Solana, Guillermo 039686
Solander, Terry J. 049054
Solano Quirós, Mayela 008848
Solańska-Alexandroff, Krystyna . 034903
Solberg, Per 033530
Šolc, Dr. Josef 009694
Soldaini, Julio E. 000179
Soldatova, Nina Ivanovna 037855
Soldatova, Tamara Mikolaïvna . 043200
Soldini, Dr. Simone 041740
Sole, Allen 031847
Sole, Laura M. 044662
Solé, Maties 039773, 039774
Solender, Katherine 051691
Soler García, José María 040416
Soler, Notary Enrique Oscar . . 000266
Soles, Mary Ellen 052450
Solheim Pedersen, Einar 033972
Solias i Aris, Josep 039495
Šolić, Dr. Marija Edita 008947
Soliman, R. 028233
Solimano, Sandra 026399
Solimon, Ron 046361
Solis, Claudia 049571
Solis, Isabel 007559
Solms, Fürst zu 017549
Šolochov, Aleksandr Michajlovič 038019
Solomatenko, Marg 006705
Solomko, Juri 037335
Solomon Kiefer, Carol 046451
Solomon, Dara 046933
Solomon, Elizabeth 052104
Solomon, Valerica 036360
Solórzano Pérez, Diana 031007
Solovev, Aleksandr Borisovič . . 037739
Solovjeva, Ljubov Pavlovna . . . 037044
Soloyeni, Joanna 023188
Soltek, Dr. Stefan 020972
Solymosi, Rosalia 023462
Soma, Akio 029333
Sombathyova, Magdalena 038286
Some, Porosper 005273
Somers, David 005457
Somers, Dominique 003287
Somers, Stephen G. 047998
Somerville, Ann 033296
Somma, Dr. Thomas P. . 048957, 048959
Sommani, Dr. Donatella 028154
Sommer Ribeiro, José 035757
Sommer, Dr. Achim 017643
Sommer, Gertrud 016992
Sommer, Jeffrey 052739
Sommer, Lawrence J. 050296
Sommer, Mogens 010047
Sommer, Dr. Susanne 018057
Sommerfeld, Renate 021699
Sommerfeldt Gikas, Carol 046814
Sommerhuber, Christian 002786

Sommers, Joyce A. 049690
Sommers, Larry 050301
Somoza, Maria E. 036159
Somsuda, Khun 042649
Sonderegger, Cristina . 041360, 041698
Sondergaard, Marten 010288
Song, Aizhen 008031
Song, Bing 007831
Song, Dehui 007578
Song, Jianping 008400
Song, Jiashan 007721
Song, Prof. Ki-Ho 030040
Song, Misook 030050
Sonke, Denton 047014
Sonnberger, Gerwald 051450
Sonnen, David G. 054515
Sonnenberger, Dr. Franz 020844, 020858,
020859, 020871, 020878, 020879
Sonnenburg, Hubert von 051424
Sonnenschein, Isabell 021663
Sonnleitner, Dr. E.T. 051828
Sonnleitner, Mag. Elena 003053
Sonntag, Dr. Dina 022047
Sonntag, Gabrielle 013922
Sonntag, Johannes 018178
Sonntag, Dr. Johannes-Hendrik . 018675,
018676, 018677, 018678
Sonntag, Marilyn 051478
Sonoyama, Hiroshi 029521
Soós, Nándor 023410
Soosloff, Philip 052639
Sopóliga, Dr. Miróslav 038462
Sorber, Frieda 003293
Soregaroli, Dr. Art 005483
Soren, Joshua 053223
Soriano, Alain 011643
Soriano, Dr. Corazon 034553
Soriano, Dr. Dolors 039095
Sorin, Gretchen 048446
Soroka, Halyna 043220
Sorokin, Igor Vladimirovič 037842
Sorokina, N.I. 037374
Sorsa, Jaakko 011455
Sosić, Sarival 038547
Sošin, Aleksandr Pavlovič 037069
Šošmin, Aleksandr Aleksandrovič 037814
Sosnovs'ka, Tatjana Viktorivna . 043197
Sosnovskaja, Marina Jurjevna . . 036981
Sotheran, Cheryll 033360
Soto Bastida, Nicolás 031211
Soto, Jules M.R. 004492
Soto, Susana 039354
Součková, Dr. Jana 009822
Souichi, Takahata 029273
Souindoula, Simão 000111
Souka Séko, Cisse Nèe 004185
Soule, Robert M. 047598
Souplet, Michel 013539
Sourakli, Judy 053235
Sousa, Maria José Rego 035603
Sousa, Suzanne 050555
Souter, Teri 006022
South, Allison 052916
South, Will 049257
Southard, Edna Carter 051868
Southerland, Cindy 049073
Southern, John 044105
Southern, Kathy Dwyer 054140
Southwell, Carey 048259
Southworth, Edmund . 043793, 044749,
044750, 044751, 044752, 044753,
044755, 044869, 045566
Soutif, Prof. Daniel 027445
Sovaudjieva, Natalia 005020
Sovluk, Viktor Ivanovič 037108
Sowada, Dr. Karin 001548
Sowder, Marvin 048132
Sowers, Russ 049544
Sowinski, Larry 051263
Soyhan, Cihat 042963
Soylu, Nilüfer 043005
Sozio, Janet L. 051089
Space, Parker 053620
Špaček, Jaroslav 009497

Spackman, A.J. 045033
Spadavecchia, Dr. Fiorella 028304
Spadea, Dr. Roberto .. 026108, 026109
Späth, Annette 016734, 016735, 016736
Spagnolo, Francesco 046936
Spahn, Merrill 050119
Spahr, Elisabeth 041486
Spahr, P. Andrew 050626
Spalding, Jeffrey 005946
Spalek, Dr. Ladislav .. 038387, 038454
Spang-Babou, François 015198
Spang, Veronika 022057
Spangenberg, Kristin 047739
Spangenberg, L. 038772
Spanke, Dr. Daniel 022045
Spanos Nordan, Antoinette ... 047010
Spânu, Mihail 036738
Spargo, J. 044561
Sparks, Brian 050345
Sparks, Jim 005743
Sparks, Kevin S. 054451
Sparks, Laurel 054267
Sparre, Comte de 016001
Sparrow, John 045493
Sparrow, Judy 045493
Spas'ka, Larisa Ivanivna 043247
Spatafora, Dr. Francesca 027814, 028073
Spathari, Elizabeth .. 022957, 022964,
                                 023099
Spaulding, Daniel 046504
Spear, June 005816
Spear, Robert N. 049624
Spearman, Grant 046467
Spears, Carolyn 051149
Spears, Dolores 049812
Spears, Kimberly 046476
Spears, Susan 050716
Specht Corsi, Martha 028015
Speciale, Giovanni 025703
Speciale, Michael P. 054474
Speckels, Gabriele 021056
Specter, David 051451
Spector, Nancy 051470
Spedding, Janaki 044682
Spedding, Michael 033208
Speed, Bonnie 046626
Speicher, Merry 052675
Speierl, Dr. Charles 047842
Spelak, Lucyna 035040
Speler, Dr. Ralf-Torsten 018963
Speliers, Thierry 004131
Spellazza, William Salvador ... 048643
Speller, Margaret 044135
Spence, Cathryn 043517
Spence, David 045041
Spence, Millie 006675
Spencer-Davies, Jenni 045934
Spencer Forsythe, Laurel 051281
Spencer-Longhurst, Dr. P. 043612
Spencer, Elizabeth .. 044082, 044084
Spencer, Georgia 049752
Spencer, Jeff 051976
Spencer, Jim 047781
Spencer, Leonard H. 047340
Spencer, Ted 047975
Spencer, Wendy .. 052826, 053659
Spendel, Dr. Cölestin 002689
Spender, Michael 044209
Sperath, Albert 051864
Sperber, Dr. G. 005749
Sperber, Dr. Lothar 021931
Sperebet, Yilmaz 033474
Sperling, Jörg 017765
Speth, Cornelia 020081
Speth, John D. 046491
Spevec, Branko 009014
Spezia, Michael J. 049460
Spicer, Joaneath 046772
Spicer, Margaret E. 049353
Spichtig, Klara 041939
Špička, Dr. Dušan 009654, 009907
Spiegel, Dr. Beate 018680
Spiegel, Dr. Beate 020693
Spiegel, Elisabeth 019711

Spiegel, Marty 053244
Spiegelberg, Dr. U. 019265
Spiegelman, Deborah 050837
Spielmann, Andreas 041969
Spies, Dr. Bernhard 017494
Spies, Dr. Britta 020744
Spies, Dr. Heike 018029
Spiessens, Frans 003382
Spigo, Dr. Umberto 026600
Spilka, Abby R. 051439
Spilker, Rolf 021018
Spillemaecker, Chantal 013399
Spiller, Christine 017748
Spiller, Harriet 047299
Spiller, Mortimer 047299
Spillman, Jane S. 047994
Spillmann, Dr. Andreas 041509, 042230
Spindle, Carol 052331
Spindler, Gabriele 002323
Spindler, Jean-Charles 012087
Spinelli-Flesch, Marie 012001
Spinner, Helmut 019109
Spinola, Dr. Giandomenico ... 054682,
                                 054690
Spiridon, Maica 036201
Spiridonov, Jurij Ivanovič 037178
Spiro, Stephen B. 051657
Spitler, Bing G. 053287
Spitzbart, Ingrid 001953
Spivak, Lisa 048273
Spivey, Towana 048871
Splaine, Shirley 046567
Šplíchalová, Vladimíra 009964
Spliethoff, Dr. M. 032013
Splitt, Jerzy Aleksander 034801
Splitter, Dr. Rüdiger 019526
Splonskowski, Joann 048318
Spock, Dan 052846
Spohn, Clarence 048590
Spooner, Peter 048387
Spoorman, R.L.F. 031950
Sporer-Heis, Dr. Claudia 002131, 002141
Sporleder, Janet 046518
Sporn, Dr. Katja 018476
Sportiello, Anne 014059
Spottorno Díaz-Caro, Rafael .. 039691
Spotts, George 049230
Spowart, Cliff 001435
Spoz, Andrzej 035307
Sprague, John 052590
Sprain, D. 054339
Sprankling, Miriam 051516
Sprecher, Dr. Eva 041282
Sprecher, Dr. Dr. Thomas 042236
Spree, Tommy 017071
Spreer, Elmar 018281
Sprem Lovrić, Branka 008942
Spriggs, Dr. Lynne 049219
Spring, Marika 045057
Springer, Bill 047198
Springer, Ernst 002095
Springer, Jonathan 047813
Springer, Kathleen 052497
Springer, Dr. Tobias 020853
Springhorn, Prof. Dr. Rainer .. 017876
Sprink, Claus-Dieter 017163
Sproat, Dr. John G. .. 047910, 047913
Sproat, Michael 049637
Sprouls, Ellen 050136
Spruill Redford, Dorothy 048065
Spruill, Sarah C. 047603
Spruit, R.J. 032510
Sprung, Y. 025031
Spüasić-Djurić, Dragana 038226
Spur, Birgitta 024100
Spurgeon, Lucy 054541
Spurney, Dr. Robert 053284
Spychała, Jerzy 034966
Squier, P.H.M. 043946
Squiers, Carol 051406
Squire, S. 001428
Squires, Ann 049813
Sretenovic, Vesela 052381
Srimal, Sudip 024364

Srinivasan, Shanta 024237
Srisuchat, Amara 042721
Sroka, Steven D. 053963
Sroka, Teresa 034773
St. Onge, Dave 006097
Staarman, A. 032157
Staat, D. 032157
Staber, Deborah 049486
Stabrovskij, Aleksandr S. 037641
Stachel, G. 019905
Stachel, M. 019905
Stack, Joan 047902
Stack, Lotus 050944
Stack, R. 001543
Stackpole, Matthew 048472
Stackpole, Renny A. 053734
Stacy, Kaye 001498
Stade, Anette 041454
Stadelmann, Kurt 041320
Stadler, Mag. Eva Maria 002953
Stadler, Hilar 041629
Stadler, Dr. Peter 002402
Stadler, Wilhelm 002614
Stadlin, Bernhard 041742
Stadler, Helmut P. 003042
Stadtman, Kenneth L. 052403
Staebell, Sandra 047157
Staebler, Claire 014794
Staeck, Prof. Klaus 017066
Städtler, Leopold 001969
Stäheli, Charles E. 041606
Stähli, Alois 041929
Stählin, Eduard 041898
Stämpfli, Markus 041353
Stäuble, Dr. Nicole 042104
Staffen, Dr. Zdeněk 009519
Stafford, Michael D. 050918
Stage, Dr. Elisabeth K. 046938
Stager, Prof. Lawrence 047363
Stagg, Barbara 052701
Stagleder, Linda 048701
Stagno, Dr. Laura 026424
Stahel, Urs 042148
Stahl Gretsch, Laurence-Isaline 041534
Stahl, E. 021752
Stahl, Judith R. 046823
Stahl, Kathleen P. 047386
Stahl, Konrad 020465
Stahl, Marguerite 013791
Stahl, Patricia 018428, 018445
Ståhlberg, Nina 011326
Stahlecker, Robert C.G.A. ... 032188
Stahn, Frank 018295
Stahnke, Peter 018047
Staikova, Evdokia 005191
Stainback, Charles 053161
Staker, Mark L. 052915
Staker, Ron 048682
Stakulienė, Silvija 030553
Stalcup, Wanda 051128
Stalder, Dieter 041675
Staley, Dr. Thomas F. 046692
Stalf, Oscar 022406
Stallings, Diane G. 049792
Stallwitz, Kurt 048389
Stalskaja, Lidija Mirzejusofovna . 036869
Stalvey, Dorrance 050444
Stalvey, Rip 053667
Stamarski, Sheri 005782
Stamas, Christine 048298
Stambach, Dr. Rudi 041892
Stambrau, Dr. Doris 018434
Štambuk, Slobodan 008918
Stamey, Emily 054384
Stamhuis, I.M. 032936
Stamm, Jörg 042053
Stamm, Prof. Dr. Rainer 017593
Stamm, Wolfgang 021820
Stammers, Michael 044859
Stammers, Mike 044861
Stan, Dr. Ana-Maria 036384
Stancati, Antonella 027622
Stanciu, Ştefan 036408, 036473
Stancu, Dr. Radu 036619

Stancu, Radu .036347, 036532, 036618,
                                 036727
Standen, Roger 044500
Standenat, Peter 002401
Stander, Tom 049328
Standley, Janet 046171
Standring, Timothy 048258
Stanek, Barbara 035066
Stanek, Dr. Christian 003043
Stanevičienė, M. 030485
Stângă, Ion ... 036451, 036452, 036453,
                                 036718
Stange, Lorraine 052872
Stangl, Denise 046961
Stangl, Reinhard 020124
Stangler, Prof. Dr. Ferdinand .. 002402
Staniczek, Dr. Arnold 022056
Stanislaus, Richard 053218
Stanislav, Dr. Ján 038279
Stanislavska, Ljudmyla Pavlivna 043166,
                                 043167
Štankina, Irina Valeríïvna 043121
Stankov, Vesna 038257
Stanley, B. 054106
Stanley, Diane 048574
Stanley, Eric 053144
Stanley, Michael 043620, 045273
Stanley, Tom 052626
Stansberry, Lucy 048528
Stanton, Carole 047082
Stanton, Dr. John E. 000999
Stanton, Meridith 054125
Stapel, Drs. H.K. 032209
Stapf, Arnulf 020809
Stappaerts, Greet 003291
Starcky, Emmanuel ... 012646, 012789
Starham, J.T. 045236
Stark, Dr. Barbara 019764
Stark, Bonnie 049356
Stark, Douglas 051533
Stark, Dr. Frank 047716
Stark, Julie 054606
Stark, Linda 006982
Stark, Ole 022447
Stark, Peter 051707
Stark, Steve 054298
Stark, William 049416
Starke, Jörg 017303
Starker, Deborah 050590
Starkey, Marsha B. 049365
Starlin, Mildred 046328
Starling, Kate 045048
Starling, Peter 045307
Starodubcev, Gennadij Jurjevič . 037188
Starr Hart, Leslie 049829
Starr, Lori 050468
Stasch, Dr. Gregor ... 018560, 018564
Stasinopoulou, Elisavet 022997
Stasiuk, Savelia 007299
Staski, Edward 050157
Stastna, Dr. Eva 038328
Statham, Cheryl 053575
Staton, James D. 053594
Stattler, Rick 052395
Staub, Bettina 042045
Staub, Edith 041246
Štauber, Dr. Bedřich .. 009683, 009684
Stauber, Richard 041770
Staubli, Gallus 041320
Staubli, Thomas 041496
Stauder, Heiner 020082
Staudinger, Katrin 001974
Staudinger, Wilhelm 018409
Staudt, Benno 018498
Stauffacher, Ueli 041632
Stauffer, Nancy 052703
Stauß, Christian 020507
Stave, Pari 051343
Stavitsky, Gail 051007
Stavropoulos, Savvas 023199
Stavrou, Dr. Patroclos 023097
Stawarczyk, Prof. Tadeusz ... 035428
Stawarz, Dr. Andrzej 035353
Stayton, Kevin 047253

Stažić, Andjelko	009041	Steiner, Dr. Erich	002697, 002697	Stephenson, Sally	052731	Stiassny, Dr. Melanie L.J.	051327

Stažić, Andjelko 009041
Steacy, Archie M. 007276
Stearns, Emily 054282
Steatham, Helen 046178
Stebbins, Theodore E. 047362
Steblecki, Edith 047129
Stecher, Anton 002115
Stecher, Dr. Hans Peter 002320
Steck, Dr. Volker 019506
Stecker, Heidi 019986
Stedje, Ass. Prof. Brita 033847
Steehler, Dr. Kirk W. 048596
Steel, David H. 052450
Steel, Gregory 046495
Steel, Martin 051408
Steele, Brian 046322
Steele, Curtis 049838
Steele, Jeffrey 049450
Steele, Joe 006499
Steele, Julia 045613
Steele, Steven 048273
Steele, Dr. Valerie 051433
Steeley, Audrey 043554
Steen Jacobsen, Jansten 010250
Steen, Johan 032326
Steen, Dr. Jürgen 018428
Steenbeek, Dr. T.G.D. 033006
Steenbruggen, Han 032378
Steenhaut, Dr. Woutez 003627
Steer, Christine 017164
Ştefan-Nicolin, Mihail 036304
Ştefančíková, Dr. Alica 009682
Stefanelli, Claudio 027282, 027284
Stefanescu, Dan 036320
Ştefănescu, Radu 036273
Ştefănescu, Radu 036268
Stefani, Claude 015124
Stefani, L. 023311
Stefanidou-Tiveriou, Prof. Dr. Th. 023295
Stefanko, V.M. 043188
Stefano, Prof. 025424
Stefanov, Georgi 005026
Stefanova, Morena 005191
Stefánsdóttir, Ragnheidur 024062
Stefansson, Stig 040797
Steffan, Ferdinand 022439
Steffen, Carlos 004641
Steffen, Geneviève 048421
Steffen, Paul 041958
Steffian, Amy 050016
Ştefură, Elena Mihaela 036442
Stegall, Ron 051816
Stege, Frank 022847
Stegeman, E. 032098
Stegemann, Marco 016535
Steger, Adrian 042146
Stegmann, Dr. Markus 041950
Stegner, Dr. Cornelia 017751
Steguweit, Dr. Leif 018266
Steguweit, Dr. Wolfgang 017088
Stehlin, Vera 041279
Steidl, Erwin 002183
Steidl, Dr. Thomas 002372
Steiger, Ivan 009816, 020574
Steigerwald, Mark 054224
Steigerwald, Steve 049801
Steigman, Kenneth 050564
Stein, Arno 041385
Stein, Longest F. 017129
Stein, Sebastian 020545
Stein, Simcha 025152
Stein, Susan R. 047583
Stein, Dr. Wolfgang 020579
Steinacher, Bettina 018309
Steinau, Marlies 021774
Steinberg, Dr. David 047275
Steinbrenner, Theophil 021913
Steinbring, Wally 005384
Steinegger, Franz 041715
Steiner, Barbara 019986
Steiner, Charles K. 054394
Steiner, Chris 046358
Steiner, Dietmar 002935

Steiner, Dr. Erich 002697, 002697
Steiner, Franz 051685
Steiner, Dr. Juri 041328
Steiner, Ralph 045663
Steiner, Riccardo 041789
Steinhart, PD Dr. Matthias 020575
Steinhauer, Dr. George 022942, 023220
Steinhaus, Federico 026765
Steinhauser, H. J 021814
Steinhauser, Norbert 018527
Steinhof, Dr. Monika 017601
Steininger, Prof. Erich 002696
Steininger, Florian 002694
Steinkeller, Prof. Piotr 047363
Steinklauber, Dr. Ulla . 001965, 002882
Steinkrüger, Uwe 020898
Steinle, Dr. Christa 001985
Steinle, E. 021757
Steinle, Karl-Heinz 017290
Steinle, Dr. Robert Fin 017338
Steinmann, Dr. Axel 003046
Steinmann, Dr. Marc 019694
Steinmetz, Robert 012014
Steinmetz, Dr. Walter 002585
Steinmetz, Dr. Wiebke 017852
Steinmetz, Wolf-Dieter 017553
Steinmetz, Wolf Dieter 022730
Steinwendtner, Hans .. 002033, 002034
Stejskal, Václav 009930
Stejskalová, Eliška 009747
Štekar-Vidic, Verena .. 038493, 038536,
               038595, 038596, 038597
Steldinger, M. 017155
Stella, Dr. Clara 025638
Steller, Tom 051681
Stellmacher, Bernhard 021858
Stellwes, Carla 052931
Stelzhamer, Ernst 002921
Štem, Galina Michajlovna 036913
Stemmer, Dr. Klaus 017064
Stemmler, Dr. Sabine 017159
Stemmrich, Dr. Daniel 020898
Stenberg, Gunilla 040709
Stender, D. 018327
Steneberg, Suzanne 041106
Steneker, J. 032086
Stengel, Karin 019531
Stengel, Tom 050972
Stengert, Monika 054522
Steninger, Kim 048541
Stenkamp, Dr. Hermann Josef . 017431
Stenkula, Anna 041027
Stenlake, Frances 044053
Stenman, Paul 011017, 011020
Stenning, Howard 043411
Stensholt, Morten Chr. 034004
Stenström, Anders 040963
Stenstrop, Georg 010075
Stenz, Michael 053379
Stepančič, Lilijana 038546
Stepanova, Anna Alekseevna . 037935
Stepanova, Ljudmila Vasiljevna . 036955
Stepanova, Maria 037011
Stepens, Nicolas 045304
Stephan-Maaser, Dr. Reinhild . 022273
Stephan, Eric 019459
Stephan, Erik 019450
Stephan, H. 021256
Stephan, Peter 022321
Stephan, Ralph 021876
Stephen Weppner, Mary Dessoie 046983
Stephen, Charla 047068
Stephens Kruize, Pricilla G. 050827
Stephens, Dawn 047244
Stephens, Gordon 000962
Stephens, Dr. Howard 044102
Stephens, John R. 049169
Stephens, R. 045152
Stephens, Retha 049924
Stephens, Ruby A. 052874
Stephens, Spencer 053326
Stephenson, Holly 049857
Stephenson, Marcus 046482
Stephenson, Roy 045048

Stephenson, Sally 052731
Stephenson, Tim 044035
Stepic, Barbara 047270
Šter, Janez 038627
Stergar, Branka 008964
Šterk, Slavko 009043
Sterly, Marita 019186, 019188
Stermer, Jenifer 050245
Stermitz, Martin 002192
Stern, Barry 047275
Stern, Gail F. 052367
Stern, Günter 003069
Stern, Hans 004746
Stern, Jean 049722
Stern, V. 032457
Sternath-Schuppanz, Dr.
Marie-Luise 002931
Sternback, Mike 048033
Sternecker, Dieter 016740
Sternschulte, Agnes 017877
Sternweiler, Dr. A. 017290
Sterrett, Jill 053025
Stertz, Dr. Stephen 047234
Stets, Mary Anne 051146
Stetson, Daniel E. 050097
Stetzenbach, Dr. Klaus 050172
Steuer, Harald 022147
Steuer, Matthias 018162
Steuer, Yvette 016748
Steurbaut, W.W.G. 032011
Stevan Bathoorn, H. 033030
Steven, Jackie 048160
Stevens Hardeman, Dr. H.J.M. . 031999
Stevens, Andrew 050593
Stevens, B. 045698
Stevens, Dr. Crosby 044108
Stevens, F. Scott 054584
Stevens, H. 032157
Stevens, Jane 047665
Stevens, Jenny 043416, 043417, 044356
Stevens, Lawrence 045517
Stevens, N. Lee 049372
Stevens, Norman S. 049869
Stevens, Scott 054579, 054580, 054581,
               054582, 054583
Stevenson, Candace 006437
Stevenson, Colin 033315
Stevenson, Esma 033315
Stevenson, Fenners W. 047087
Stevenson, James 007316
Stevenson, Moira 045208
Stevenson, Sara 044274
Stevenson, Victor 007270
Ševko, Anton 038410
Steward, James 046492
Steward, Maureen 005941
Steward, Travis 051455
Stewardson, Joanne 044787
Stewart-Abernathy, Judith C. .. 052711
Stewart-Leach, Sheila 050572
Stewart-Young, J. 044164
Stewart, Ann 024900
Stewart, Barbara 044650
Stewart, Beth A. 052215
Stewart, Brian 044235
Stewart, Dan 050597
Stewart, David 052928
Stewart, Doug 006592
Stewart, Fiona 000882
Stewart, Dr. Ian 043354
Stewart, J. Jason 046348
Stewart, Lorelei 047661
Stewart, Pamela 053549
Stewart, Richard 046883
Stewart, Rick 048898
Stewart, Sarah 045974
Stewart, Steven 047776
Stewart, W. Lyn 007386
Stewart, Wilma 006242
Steyer, Dr. Bernd 022308
Steymans, Hans Ulrich 041496
Steyskal, Josef 002966
Stgiverson, Greg 046502
Sthanapati, Dr. Jayanta 024357

Stiassny, Dr. Melanie L.J. 051327
Stiawa, Wolfgang 002904
Štíbr, Dr. Jan 009671, 009673
Štíbrová, Eva 009672
Stichel, Dr. Rudolf 017829
Stidolph, Diana 033242
Stieb, Ernst 006489
Stief, Gerhard 017905
Stiefelmayer, Ulrike 018412
Stiefmiller, Helen ... 049298, 049299
Stiegel, Dr. Achim 017205
Stiegemann, Prof. Dr. Christoph 021060
Stiegler, Hans 002703
Stieglitz, Leo von 022047, 022385
Stiehler, Ursula 016788
Stiepani, Ute 017144
Stieper, Jürgen 017680
Stier, David 053464
Stieve, Mary Farrell 046787
Stiewe, Dr. Heinrich 017877
Stiffler, David L. 046315
Stifter-Faust, Marianne 020858
Stijns, H.J.M.R. 032767
Štika, Dr. Jaroslav 009871
Stiles, Victoria 053250
Stilje, Emma Kristina 041180
Still, Todd 048394
Stille Neto, Bernardo 004733
Stiller, Dr. Bernd 022110
Stilling, Niels Peter 010138
Stillström, Gunilla 041066
Stilo, Cathy 007175
Stimmel, Amy 052532
Stine, Jack 052000
Stipling, Trevor 043665
Stipperger, Prof. Walter 002064
Stippich, Wolfram 019592
Stirnratt, Betsy 047049
Stitely, David 050759
Stiwich, Undine 017812
Stochr, Bernard 012630
Stock, S. 001305
Stock, Wolfram 019446
Stockar, Rudolf 041696
Stockdale, David 046029
Stockdale, Glynn 044741
Stockebrand, Marianne 050671
Stocker, Gerfried 002314
Stockhammer, Prof. Dr. Helmut . 002195
Stockhaus, Heike 022447
Stockhausen, Dr. Tilmann von .. 018478
Stockhoff, Val 053890
Stockman, Edgard 003375
Stockman, Dr. René 003631
Stockmann, Robert 019371
Stoddard, Leah 047579
Stoddard, Sheena 043747
Stoddart, Simeon 047855
Stöcker, Jens 019482
Stöckler, Elisabeth 002090
Stöckli, Edi 041282
Stöcklin, Konstantin 041479
Stöckmann, Birgit 017154
Stöhr, Michael 017856
Stöhr, Dr. Oliver 002625
Stöllner, Dr. Thomas 017433
Stölting, Dr. Siegfried 017606
Stölzel, Eckhard 022920
Stoerring, Hanne 010195
Stößl, Dr. Marianne ... 020923, 020924
Stöttinger, Christoph 002259
Støvern, Håvard 033882
Stövesandt, Birka 022700
Stoff, Joshua 049043
Stofferan, Jan 053329
Stoga, Dr. Andrzej 035324
Stoica, Maria 036262, 036263
Stojanov, Georgi 005090
Stojanov, Stojan 005016, 005017,
                        005020
Stojanović, Nada 038127
Stojič, Jelena 038152
Stojkovih, Dragana 038152
Stojneva, Nadežda 005013

Stokes Sims, Lowery 051476
Stokes, David 005832
Stokes, Mike 045798
Stokesbury, Bria 006083
Stolberg-Wernigerode, Alexander Graf
   zu 021006
Stolk, Ton 032989
Stoll, Alexander 017738
Stolle, Dr. Walter 017822
Stoller, Blanche 051031
Stollfuß, Mandy 022882
Stollman, S. 033168
Stolte, Roland 017269
Stolwerk, Rick 033324
Stolwijk, Chris 031996
Stolyarov, Gwen 051399
Stolzenburg, Dr. Andreas 018991
Stone, Donna K. 049555
Stone, Gaynell 047947
Stone, Karen 046571
Stone, Karen E. 053377
Stone, Linda 046801
Stone, Margaret 043732
Stone, Mike 043950
Stone, Nathan 052443
Stone, Pam 049943
Stone, Paul 006534
Stone, Robert E. 051616
Stone, Thelma 048903
Stone, Walker S. 048405
Stone, Walter 052393
Stoner, Sheila 053586
Stoops, Susan 054548
Stooss, Toni 002630
Stooss, Dr. Toni 002631
Stoppioni, Maria Luisa 025880
Stopps, N. 043741
Storch, Dr. Ursula 003095
Storch, Prof. Dr. Dr. Volker ... 019150
Storck, Dr. Ekkehard 022116
Storer, Rick 047669
Storer, Prof. Robert W. 046493
Storer, Russell 001544
Stork, Dr. Simone 018072
Storm, Alex 006224
Stormark, Hege 033977
Storto, Fred 017895
Storvik, Jehans 034063
Storz, Dr. Dieter 019407
Stoscheck, Dr. Jeannette 020003
Stošić, Jovica 038194
Stoter, Dr. M. 032570
Stothoff, Richard H. 048746
Stottrop, Ulrike 018304, 018311
Stoughton, Kathleen 050042
Stoullig, Claire 012003, 014405
Stoumann, Ingrid 010044
Stovall, Linda Joy 054360
Stover, James 046461
Stovold, Fran 044218
Stow, Dr. Stephen 051668
Stoyanov, Stoyan 005191
Stoyas, Yannis 023023
Stoye, M.A. Martina 017239
Stoye, Wilfried 022909
Strachova, Alevtina Gennadjevna 037619,
                              037620
Strackerjan, Wilfried 019353
Straczek, Dr. Tomasz 035365
Stradeski, Leigh-Anne 044501
Stradiotti, Dr. Renata . 025629, 025631,
                     025637, 025638
Straehly, Christiane 011607
Strässler, Walter 042100
Strandberg Olofsson, Prof.
   Margareta 040648
Strander, Esa 011305
Strandgaard, Kirsten 023329
Strang, Alice 044273
Strang, T.H. 054306
Strang, Tim 053328
Strange, Shirley 047834
Stránsky, Jiří 009527
Straßburger, Jens 021195

Strassen, Hermann zur 019805
Straßer, Dr. Josef 020519
Straßer, Wilfried 020694
Strathy, Mark 051239
Stratis, Katie 051475
Stratmann, Dr. Wilhelm 020440
Stratton, Ceilia M. 048813
Stratton, David C. 052886
Stratton, Frances R. 049730
Straukamp, Werner 020832
Strauss, Bryson 050455
Strauss, Dorothea 042197
Strauss, Dr. Frank B. 053206
Strauss, Dr. Karl W. 017743
Strauss, Linda 050455
Strauß, Michael 017670, 017671, 017672
Strayer, Jenny 052173
Strazar, Dr. Marie D. 049484
Streatfield, George C. 044436
Street, Dr. Martin 020776
Streetman, John W. 048644
Streets, Kim 045778
Strege, Gayle 047938
Strehl, Dieter 003086
Strehlen, Martina 018287
Streicher, Dr. Gebhard 016556
Streidt, Gert 017770
Streiff, Elsie 041858
Streiff, Dr. Hans Jakob 041439
Streihammer, Dr. Rudolf 003141
Streisselberger, Josef 002436
Streitberger, Christine 002345
Streitt, Ute 002330
Strekalova, Valentina Michajlovna 037915
Strelnitski, Vladimir 051156
Stremel, Ted 005778
Strémoouhoff, Catherine 015917
Strempfl, Dr. Heimo 002197
Stremsterfer, Joanne 052819
Strenge, Bernhard von 020716
Stretton, Jenny 038698
Strick, Jeremy 050456
Strickland, Alice 049152
Strickland, D.H. 047190
Strickland, Dan 007412
Strickland, Leigh V. 048975
Strickland jr., William 052167
Strickler, Susan E. 050626
Strieder, Dr. Barbara 017011
Striefler, Christian ... 019751, 020468
Striehl, Dr. Georg 019942
Strigl, Alfred 002721
Strinati, Dr. Claudio 027629
Strine, Linda T. 052483
Strittmatter, Gabi 041644
Stritzinger, Wolfgang 003089
Stritzler-Levine, Nina 051344
Strižnikova, Svetlana S. 037292
Strnadová, Z. 009815
Strobbe, Johan 003976
Strobel, Alexandra 041707
Strobel, Barbara 017058
Strobel, Margaret 047667
Strobel, Sigruth 021032
Strobel, Sonja 020877
Strobl, Dr. Andreas 020576
Strobl, M.A. Hilde 020505
Strode, Molly 049803
Strode, Wendell 047158
Stroebel, G. 038660
Ströbele, Dr. Werner .. 021381, 021383
Ströle-Jegge, Dr. Ingeborg ... 021284
Strömbäck, Stig 040727
Strömberg, Ulla 010197
Strömfors, Göran 011141
Strömsdörfer, Thomas 021773
Strohfeldt, Guido 018545
Strohhammer, Andreas 002326
Strohl, Katrin 017100
Strohler, Peter 041521
Strohmaier, Franz 002341
Strom, George 048573
Strom, Peter 048601, 048602
Stromberg, Ann 040653

Strombo, Cathryn J. 053610
Strommer, Gabriele 003037
Stromskaja, Tatjana
   Aleksandrovna 029846
Strong, Charles 046899
Strong, Dan 049279
Strong, Michael 000921
Stroobants, Dr. A. ... 003537, 003538,
                     003539, 003540
Stroobants, Bart ... 003858, 003859,
                              003860
Stroppel, Alice 053256
Strote, Joel 050167
Stroud, Michael 030755
Strsembski, Stephan .. 017481, 017485
Struchtrup, Mechthild 020608
Struck, Juliusz 034789
Struhs, David B. 047481
Štrumej, Lara 038549
Strumia, Prof. Franco 025687
Strutwolf, Prof. Dr. Holger 020593
Stryker Lewis, Evelyn 051216
Stryker, Richard R. 048004
Stuart Elliott, David 029556
Stuart, Ann 048246
Stuart, Jacqueline 005361
Stuart, Jan 054116
Stubbe Østergaard, Jan 010188
Stubbe, Els 031989
Stubbendieck, James 050291
Stubbs, Ruth 006751
Štubova, Elena Valentinovna .. 036985
Stuchlík, Ladislav 009660, 009661
Stucke, Dr. Frank 020281
Stucker, Noah Lon 052138
Stucker, Richard 051654
Stuckert, Carolin M. 047904
Stuckey, Robert 053942
Stucky, Anton 041938
Studencka, Jolanta 034837
Studencki, Zbigniew 035213
Studenroth, Zachary 051644
Studenroth, Zachary N. 052738
Studer, Dr. Daniel 041921
Studer, Winfried 020696
Studley, David 052922
Stueber, Nancy 052299
Stuempfle, Dr. Steve 050830
Stüning, Dr. Dieter 017504
Stütz, Michael 019722, 019723
Stüwe, Hartmut 019572
Stuhec, Mojca 038549
Stuhlman, Theresa 052089
Stulić, Anselmo 008975
Stull, Ellen D. 047415
Stults, Jeannine 051552
Stumpe, L. 044869
Stumpf, Jean 012957
Stuppner, Dr. Alois 003009
Šturc, Libor 009842
Sturdy, Tom C. 006272
Sturgell, Ron 049214
Sturgeon, Ava 005907
Sturgeon, James 047061
Sturgeon, Keni 050994
Sturgis, Alexander 043519
Sturgis jr, Ewell G. 047917
Sturgis, Randy 053272
Sturm, Martin 002329
Sturm, Ulli 002191
Sturtz, Rodney 053210
Stussi, Edmond 016334
Stutzer, Dr. Beat ... 041412, 041932
Stutzinger, Dr. Dagmar 018405
Stuvøy, Ingunn 033942
Štvánová, Jana 009715
Štýbrová, Miroslava 009985
Stylianou, Petros 009413
Štyrkalo, Jaroslav Jevgenovyč . 043157
Styron, Thomas W. 049268
Styskal, Brenda 049203
Su, Guijie 007825
Su, Nan 007664
Su, Runshu 008255

Su, Ziming 007920
Suanda, Wayan 024641
Suárez Zuloaga, María Rosa .. 040467
Suárez, José 008775
Suárez, Margarita 009170
Subler, Craig A. 049897
Šubová, Dana 038383
Suchan, Laura 006565
Suchanov, Michail Konstantinovič 037499
Suchinina, Svetlana Jurjevna . 037838,
                              037840
Suchocki, Wojciech 035132
Suchy, Ladislao 026917
Suchy, Theodore D. 050322
Šučkaja, Galina Konstantinovna . 037497
Sudarmadji, Dr. 024589
Suddaby, R.W.A. 045003
Sudduth, William M. 046617
Sudwidja, Ketut 024648
Südekum, Dr. Karl 022785
Süel, Mustafa 042811
Sünderhauf, Dr. Esther 017299
Sünkel, Werner 021459
Sünwoldt, Sabine 021745
Sürth, Astrid 019707
Süsserova, Jarmila 009937
Sütterlin, Hans 041243
Sützl, Werner 001681
Sueur, Catherine 014749
Sueur, Catherine le 038682
Suezawa, Shoichi 029622
Suffolk, Randall 049123
Sugawara, Eikai 029184
Sugden, K. 045210
Suggitt, Mark . 043690, 043691, 043692
Suggs, Patricia 046846
Sugihara, Dr. Shigeo 029548
Sugimoto, Lisa Ann 049895
Sugino, Hideki 029639
Sugonjako, Mykola Mykolajovič . 043268
Sugrue, Christine 051840
Suhr, Dr. Norbert 020229
Suhrbier, Dr. Birgit 018434
Suhre, Terry 052800
Sui, Dr. Claude 020255
Suk, Joo-sun 030046
Sukpramool, Patcharin 042690
Sukumaran, Dr. C. 024176
Sulaiman, Dr. Nasruddin 024527
Sulayman, Dr. Tawfiq 042265
Šulce, Baiba 030102
Suldina, Elena Nikolaevna .. 037993
Şulea, Ioan 036763
Sulebust, Jarle 049467
Šuleř, Dr. Petr 009464, 009476
Suliga, Dr. Jan Witold 035378
Sullivan Sturgeon, Catherine .. 047711
Sullivan, B. 001215
Sullivan, Carole 046537
Sullivan, Christine M. 050904
Sullivan, Cyndi 052633
Sullivan, Irene 047236
Sullivan, Joanne 049407
Sullivan, John 050344
Sullivan, Kathryn D. 047937
Sullivan, Larry 051783
Sullivan, Dr. Lynne 050011
Sullivan, Martin E. 052832
Sullivan, Mary Louise 052026
Sullivan, Michael 050975
Sullivan, Dr. Robert M. 049372
Sully, Georges 013561
Šulskytė, Aldona 030407
Sultan, Samir M. 010529
Sultan, Terrie 049596, 054008
Sum, Dr. Chow Weng 030664
Šumacher, Larisa Nikiforovna . 029840
Sumalvico, Ulrike 022562
Sumawilaga, H. Djamhir 024651
Sumberg, Dr. Barbara 053115
Sumberg, Jack 047277
Sumers, Anne R. 051913
Sumi, Naoko 028873
Sumikawa, Kiichi 029002

Summerfield, Jan 043860, 045980	Suvitie, Osmo 011285	Sy, Joaquin 034522
Summers, Blake 053020	Suvorova, Tatjana 037084	Syahali, Teuku Sahir 024592
Summers, Carl 050349	Suwanasai, Avudh 042723	Šychlinskaja, Sabina 003155
Summers, Jacquie 043729	Suwentra, I. Ketut 024557	Sydow, Barbara 018832
Summers, John 007187, 047784	Suzuki, Katsuo 029608	Sykes, Diana A. 045674
Summers, Marcia J. .. 051151, 053111	Suzuki, Keizo 028765	Sykes, Russell 049510
Summers, Reece 050291	Suzuki, Koji 028496	Syljuk, Anatolij Michajlovyč .. 043228
Summerwill, Kristin 047988	Suzuki, Makoto 029324	Sylvester, Morten 010346
Šumnaja, Dr. T.G. 037282	Suzuki, Norio 029543	Symes, Dr. Robert F. 045801
Sumner, Kent 048020	Suzuki, T. 029755	Symmes, Edwin C. 046630
Sumoza, Rosa María 008666	Suzuki, Takatoshi 029618	Symmes, Marilyn 051368
Sumrall, Robert F. 046503	Suzuki, Yusai 029224	Symons, Allan 005692
Šumskaja, Ljubov Vladimirovna . 037228	Svachová, Jarmila 009632	Symons, Prof. T.H.B. 006640
Sun, Changqing 007847	Svae, Karin 040888	Synder, Judith 047379
Sun, Chengfu 008207	Svanascini, Prof. Osvaldo 000258	Syndram, Prof. Dr. Dirk 017985, 018005
Sun, Dr. Dennis 007885	Svedberg, Berit 041048	Synofzik, Dr. Thomas 022911
Sun, Desheng 007565	Svedfelt, Bjorn 040964	Syre, Dr. Cornelia 020508
Sun, Huijun 008044	Svedlow, Andrew Jay 050629	Syré, Dr. Ludger 019503
Sun, Jixin 007689	Švedov, Valerij Gennadevič ... 037717	Syrjänen, Anneli 011223
Sun, Ming 007668	Švedova, Inga 043293	Syrjänen, Inkeri 011415
Sun, Mingxiao 007668	Svens, John-Erik 011519	Šyškina, Raïsa Petrivna 043162
Sun, Minxin 008217	Svenshon, Dr. Helga 017829	Syson, Luke 045057
Sun, Qingfei 008495	Svensson, Else Marie 040545	Systad, Gunnhild 033605
Sun, Yigang 007739	Sverrir Árnason, Gísli 024081	Sytek, Wioletta 035119
Sun, Yinglong 007747	Svetina, Ivo 038559	Šytikov, Viktor Mychajlovyč ... 043152
Sun, Yun 008128	Sviblova, Olga 037347	Syverson, Duane 048593
Sun, Zhongguo 007782	Svirčević, Rosa 038222	Syvertsen, Per Ole 033772
Sunday Oyegbile, Olayinka ... 033411	Sviridova, Tetjana Mykolaïvna .. 043243	Szabo, Dr. Agneza 009043
Sundblad, Stefan 040968, 040971	Svitkova, Ani 005176	Szabo, Andras 036545, 036565
Sundborg, Peter 041066	Swaan, Trudy 006716	Szabó, Dr. Botond Gáborjáni ... 023597
Sundelin, Åsa 041027	Swadley, Ben H. 053208	Szabo, Ernst 002857
Sundell, Eva 040771	Swagemakers, Walter 031939	Szabó, Imre 023406
Sundell, Karin 041066	Swager, Karen 047912	Szabó, Dr. J. József 023965
Sundher, Chanchil 050431	Swain, Adrian 051052	Szabó, Lajos 023560, 023561
Sundin, Sven-Olov 040907	Swain, David 053953	Szabó, Lilla 023567
Sunding, Prof. Per 033847	Swain, Kristin 047995	Szabóné, Dr. Juhász Katalin ... 023455
Sundström, Susanne 041106	Swallow, Dr. Deborah 044996	Szabóné, Lázár Ibolya 023739
Suné, Prof. J.M. 039080	Swalve, Prof. Dr. 018966	Szadziewski, Prof. Dr. Ryszard . 034746
Suominen, Tapio 011409	Swan, Dr. Daniel 050773	Szakacs, Dennis 051544
Supamena, M.S. 010407	Swan, Daniel C. 053848	Szakmary, Joanna 051167
Superior, P. 038781	Swanagan, Don Jo 048050	Szála, Dr. Erzsébet 023900
Suppes, Paul D. 048235	Swanee, Lois 049512	Szameit, Prof. Dr. Erik 001926, 002827
Supriyatno, Y. 024677	Swanepoel, S. 038766	Szaniawska, Wanda 035330
Supry, Fred 049658	Swank, Scott T. 047405	Szaraniec, Dr. Lech 034819
Surace, Dr. Angela 025849	Swann, Dan 051822	Szathmáry, László 023684
Surberg-Röhr, Silke 020368	Swann, Paula 005524	Szatmári, Dr. Imre 023444, 023447
Šurdić, Borislav 038164	Swanson, Alan 054552	Szczebak, Dr. Władysław ... 035266
Sureau, Jean-Louis ... 012563, 015789	Swanson, Ann 052675	Szczepański, Marek 035244
Sureda Vallespir, Catalina 039150	Swanson, Brook 054233	Szczesiak, Rainer 020683
Surmann, Dr. Ulrike 019694	Swanson, Erik 048066	Szczypiorski, Krzysztof 035015
Surrett, Janet 053281	Swanson, Robert 005494, 050744	Székely, Dr. Ferenc 023618
Surroca i Sens, Joan 039445	Swanson, Tom 048261	Szekely, Linda 053537
Surup, Axel ... 022711, 022712, 022714	Swanson, Dr. Vern G. 053478	Szekeres, Viv 000767
Suryana, Otong 024532	Swantek, John E. 054222	Szente, Béla 023443
Surzyńska-Błaszak, A. 035129	Swartout, Dennis W. 049179	Szentkuti, Károly 023799
Susanna, Dr. Alfonso 039104	Swartz, Leslie 047103	Szentpéteri, Dr. Erzsebet 023537
Suščenko, Evgenija Michajlovna 037576	Swartzlander, Martha 052676	Szép, Dr. Zoltán 023470
Sušić, Željka 009052	Swatosch, Michael 003045	Szerdahelyi, Dr. Károly 023489
Susoj, Elena Grigorjevna 037083	Swatschina, Wilhelm 002095	Szevera, Dr. Walter 003089
Sussman, Elisabeth 051491	Sweasy, Lisa 048845	Szewczuk, Mieczysław 035165
Šuštar, Branko 038560	Sweeney Price, Mary Sue 051499	Szewczyk, Kazimierz 035303
Šutalo, Kata 009035	Sweeney, Alan 053214, 054206	Szewczyk, Monika 034636
Suter, Daniel 041243, 041715	Sweeney, Colin 046103	Szinai, Stephan 003047
Suter, Dawn 001423	Sweeney, Dennis F. 051390	Szinvavölgyi, Oszkár 023787
Suter, Dick 000871	Sweeney, John 047248	Szirányi, János 023461
Suter, Hans and Marlis 041811	Sweeney, Karen 052518	Szkop, Henryk 034720
Suter, Josef Ignaz 041331	Sweets, Henry 049350	Szmolicza, Józef 023698
Suter, Mary 048717	Sweha, Dr. Fawzi 010547	Szocs, Francisc 036658
Suter, Paul 041484	Sweitzer, Dr. Rick 049178	Szöke, Dr. Mátyás 024047
Sutherland, Cara A. 052502	Świda, Piotr 035348	Szőnyi, Dr. Eszter 023669
Sutherland, Gordon 006206	Swiderski, Christine 007192	Szpakiewicz, Alina 034729
Sutherland, Maire-Claire 003988	Swift, M. 044646	Szpakowska, Ewa 035155
Sutherland, Maureen 006178	Swift, R.G. 046010	Szram, Dr. Antoni 034975
Sutherland, Rene 001037	Swigart, Arthur L. Stewart 053935	Szücs-Irén, B. 023438
Sutherland, Dr. Stuart 007287	Swigart, Stacey A. 053934	Szücs, Dr. Judit 023590
Suthers, Terence 044511	Swim, Ann 047403	Szura, Andrzej 034835
Sutoyo, R. 024656	Swingle, James 052684	Szuromi, Béla 023868
Sutschek, Felix 017507	Swingle, Mark 054000	Szygenda, Lynn 044209
Sutter, Christoph 041888	Swinkels, Dr. L.J.F. 032687	Szymańska, Dobrochna 035305
Sutter, M. A. Julia Johanna 021160	Śwircz, Kazimierz 034920	Szymańska, Ewa 034724
Sutton-Vane, Angela 044310	Świst, Lucyna 034838	Szymczyk, Adam 041272
Sutton, Cynthia 049866	Swoboda-Körner , Christine ... 019292	Szymon Gwerchnik, Robert ... 027492
Sutton, Darren 000837	Swody, Russell 050123	Szymusiak, Dominique 013617
Sutton, Peter C. 049270	Swords, Carol 045075	Szyndzielorz, Albert 034760
		Taaffe, Renee 050960
		Taarup, Frank 010192
		Taawo, Kjell 040888
		Tabacu, Jenica 036342
		Tabah, Béatrice 016054
		Tabaka, M. 006480
		Tabakoff, Sheil 051451
		Tabakov, Michail Ivanovič 038048
		Tabbernor, Christine 007455
		Tabbert, Mark 050256
		Tabeaud, Marcel 012940
		Tabel-Gerster, Margit 019011
		Taber, David 050552
		Tabkoff, Sheila K. 051649
		Taboada Vázquez, Rafael 039320
		Tabor, Andy 051132
		Taborga Rivas, Matilde 004251
		Taburet-Delahaye, Elisabeth .. 014780
		Tac An, Dr. Nguyen 054816
		Tackett, Natalie 049802
		Tadahisa, Sakai 029659
		Tadiello, Dr. Roberto 028315
		Taeckens, Michel 013808
		Taegert, Prof. Dr. Werner 016957
		Täube, Dr. Dagmar 019706
		Täubrich, Hans-Christian 020847
		Tafilica, Zamir 000028
		Taft Andrews, Nathalie 050500
		Taft Terry, David 046767
		Taga, Yoshiya 028791
		Tagaev, Muchtar 030081
		Tagage, L.M. 032629
		Taggart, Travis 049416
		Taghiyeva, Roya 003167
		Taghon, David 048577
		Tagiri, Takashi 029523
		Tagliasacchi, Dr. Paolo 026069
		Taguchi, Yasuo 028721
		Taha, Dr. Haji Adi bin Haji ... 030719
		Taher, Salah E. 010550
		Taimagambetov, Prof.Dr. Zhaken 029800
		Tainturier, Cécile 014635
		Tajedin-Noureldaim, Yousif .. 010602
		Tajima, Susumu 029684
		Takács, Dr. Imre 023553
		Takács, Mária 023881
		Takahara, Mayako 028515
		Takahashi, Akiya 029536
		Takahashi, Prof. Susumu 029619
		Takakusa, Shigeru 028515
		Takanashi, Mitsumasa 029536
		Takano Yoshizawa, Hirohisa .. 028515
		Takano, Masahiro 028515
		Takase, Tamotsu 029641
		Takashi, Tuda 028880
		Takashina, Shuji 028899, 028905
		Takata, Yoshikazu 028824
		Takeda, Sharon 050444
		Takehana, Rintaro 028470
		Takemoto, Hiroshi 029744
		Takemoto, Mikio 029620
		Takenaka, Yasuhiko 029692
		Takeuchi, Jun-ichi 029509
		Takeuchi, Makoto 028872, 029501
		Takeuchi, Tadashi 029124
		Takita, Minoru 029375
		Taksas, Prof. Čuner 037764
		Taktak, Fehmi 042917
		Tal, Felix 031998
		Tal, Ilan 025121
		Tal, Dr. Nissim 025042, 025051, 025055
		Talaga, Sally 049525
		Talalay, Lauren 046490
		Talamo-Kemiläinen, Maarit .. 011449
		Talamo, Emilia 027581
		Talamo, Prof. Giuseppe 027580
		Tális, Dr. László 023977
		Talazov, Veselin 005077
		Talbot-Stanaway, Susan 051137
		Talbot, Dennis 045213
		Talbott, Tim 051743
		Talerman, Margaretha 052027
		Tales, Sofia 027153
		Tall, J.J. 044453

Talla, Primitiva C. 034609
Talland, Valentine 047117
Talley, Dan R. 050023
Talsky de Ronchi, Prof. Alicia .. 000617
Talvio, Tuukka 010855
Talvitie, Tarja 011054
Tamari, Dr. Dov 025194
Tămaş, Mihaela 036246
Tamassia, Dr. Marilena 026253
Tamba, Dumitru Gheorghe 036834,
036835
Tambini, Michael 043683
Tamburini, Pietro 025577
Tamindarova, Miljauša Amirovna 037104
Tamir, Peleg 025200
Tamis, Catherine M. 052686
Tamla, Toomas 010666
Tammen, Mike 050591
Tammiksaar, Erki 010694
Tamminen, Jukka A. 011121
Tamminen, Marketta .. 011271, 011272,
011275, 011276
Tamminen, Osmo 010756
Tamplin, Illi-Maria 006634
Tamura, Haruhiko 029437
Tanabaeva, Sara Eskairovna .. 029831
Tanabe, Emi 029196
Tanaka, Akira 029461
Tanaka, Dr. Hideo 029726
Tanaka, Masayuki 029536
Tanaka, Morio 029622
Tanaka, Takeyoshi 028479
Tanaka, Toshiaki 029507
Tanaka, Toshiya 028587
Tanaka, Yoshiaki 029653
Tănăsescu, Gabriela 036629
Tanata, Charles 049788
Tanatarova, Sara Šaripbaevna .. 029815
Tanchyk, Alla 017182
Tandefelt, Petra 054556
Tandy, Virginia 045204, 045208, 045220
Taneva, Valentina 005158
Tang, Hanzhang 007939
Tang, Shulan 007954
Tang, Weizhe 008202
Tangco, Manny 034509
Tangen, Eigil 033883
Tangen, Kirsten 033560
Tani, Gohei 028827
Taniguchi, Dennis 053006
Taniguchi, Harumichi 029424
Taniichi, Takashi 029216
Tanino, Dr. Akira 029711
Tanko, Majda 038585
Tankoano, Ounteni Ernest 005273
Tannehill, Suzette 046818
Tannen, Jason 047713
Tannenbaum, Barbara 046329
Tannenbaum, Judith 052387
Tanner, Julia 044530
Tanner, Matthew 043754
Tanner, Paul 042196
Tannert, Christoph 017198
Tanninen-Mattila, Maija 010813, 010827
Tannous, John 052353
Tannous, Youssef 030285
Tantaquidgeon, Gladys 053883
Țânțăreanu, Ecaterina 036200
Tantoco, José P. 034513
Tao, Hsin 042607
Tao, Tao 008318
Tapias Moreno, Rito Antonio ... 008817
Tapié, Alain 012279, 013807
Tàpies, Miquel 039057
Tapken, Dr. Kai Uwe 021286
Tapliss, Sue 043511, 044221
Tappe-Pollmann, Dr. Imke 017876
Tappe, Dr. Heinrich .. 017953, 017962
Tapsell, P. 033142
Tapsirul Dahum, Langca 034484
Taragin, davira S. 052440
Taran, Galyna Mychajlivna 043169
Taran, Jurij Jurijovyč 043170
Tarasek, Władysław 034714

Tarasenko, Tatjana Ivanovna ... 038005
Tarasov, Aleksandr Vladimirovič . 037749
Tarasova, Nadežda Petrovna ... 036983
Tarawneh, Khaled 029759
Tarbell, Mike 049599
Tarcalova, Dr. Ludmila 009931
Tardiem, Robert 013928
Targan, Alfons 017617
Taricco, Dr. Bruno 025943
Tarmann, Dr. Gerhard . 002132, 002141
Tarnawesky, Prof. M. 007474
Tarnovskaja, Tatjana Nikolaevna . 037579
Tarnow Ingvardson , Gitte 010179
Tarnowski, A. 044085
Taron, Dr. Douglas 047690
Tarr, Blair D. 053773
Tarrant, Josh 000979
Tarrats Bou, Francesc . 040208, 040209
Tarrús Galter, Josep 039048
Tarshish, Noa 025054
Tartaglio, Michael 047204
Tartuferi, Dr. Angelo 026256
Tarus, Käpy 011347
Taruvinska, Pascall 054850
Tasanen, Terni 011426
Taschenmacher, Peter 016794
Tashi, Khenpo Phuntsok 004200
Tashmuchamedova, Ainur 054655
Tasick, Julian 050653
Tasker, Diana 048544
Tašlaj, Ljudmyla Borysivna 043263
Tašlialan, Dr. Mehmet . 042973, 043083
Tasma, G.L. 032330
Tassell, C.B. 001215
Tassi, Prof. Franco ... 025519, 026020,
028397
Tassianari, Duilio 025394
Tassignon, Jean-Marie 003506
Tassini, Francesco 025998
Tatangelo, Mireille 016156
Tataru, Marius J. 018893
Tate, Barbara L. 051420
Tate, Belinda 054488
Tatewaki, M. 028458
Tatlican, Zerefşan 043061
Tátrai, Dr. Vilmos 023553
Tatsch, Madeleine 017761
Tatter, Paul 046360
Tatton Sykes, Baronet 044140
Taub, Liba 034851
Taube, Dr. Angelika 019736
Taubhorn, Ingo 018979
Tauerová, Dr. Jarmila 009813
Taufi 031857
Taulbee, Ann 051869
Taulé i Delor, Mariàngela 039093
Taupe, Christina 039508
Taupe, Georg & Ingrid 002863
Taut, Dr. Konrad 019990, 020202
Tautorat, Ina 017186
Tavares Garcia, Sonia Maria ... 004620
Tavassoli 024785
Tavčar, Dr. Lidija 038552
Taveneaux, Benoît 015479
Tavernier, Bertrand 013953
Tavernier, Marc 015141
Tavlakis, Ioannis 022945
Tawalba, Dia 029780
Tawfiq Gharib, Essam El-Din .. 010546
Tayler, Nicola 044377
Taylor Bennett, Cheryl 053393
Taylor Dixon, Tricia 048907
Taylor, Allison 044019
Taylor, Anne 043711, 054526
Taylor, Barbara 043809
Taylor, Barbara E. 049466
Taylor, Catherine A. 052724
Taylor, Darrell 047476
Taylor, David G. 006966
Taylor, Deborah 049635
Taylor, Donald B. 048915
Taylor, Edith 050186
Taylor, Eric B. 007298
Taylor, Gloria 006324

Taylor, Hollis 049110
Taylor, Howard 052923, 052927
Taylor, J. Lawrence 049759
Taylor, J. William 047603
Taylor, James C. 049617
Taylor, Jeanette 005540
Taylor, Jeff 048749
Taylor, Jennifer 044156
Taylor, Jill E. 046016
Taylor, Joseph A. 049425
Taylor, June 048518
Taylor, Karen 047451
Taylor, Kathryn T. 054331
Taylor, Kathy 045066
Taylor, Lillian 049334
Taylor, Linda 046462
Taylor, Lloyd 044668
Taylor, M.A. 045486
Taylor, Martin 045436
Taylor, Mary Rose 046624
Taylor, Michael 045485
Taylor, Mike 033282
Taylor, Richard 045398, 046765
Taylor, Robert L. 051844
Taylor, Robertine 051456
Taylor, Ron 007153
Taylor, Ruth 051532, 051534
Taylor, S.K. 001260
Taylor, Saddler 047912
Taylor, Sue Ellen 047864
Taylor, Susan H. 051272
Taylor, Susan M. 052368
Taylor, Thomas C. 053571
Taylor, Tom 050186
Tazlari, Marianne 022587
Tazzoli, Vittorio 027232, 027233
Tckcli, Gonul ....°... 042963
Teach, Joce 048423
Teague, John 001139
Teahan, John 049390
Teatero, Barbara E. 005501
Tebani, Shahzor Ali 034114
Tebbe, Dr. Karin 019137
Tebbutt, Michael L. 045240
Tecce, Dr. Angela 026987, 026999
Tederick, Lydia 054177
Tedeschi, Dr. Carlo .. 028042, 028044
Tedesco, Joseph 030759
Tedesco, Paul H. 048351, 048352
Tedesko, Frank W. 048757
Tedford, Catherine L. 047408
Teeling, Eugenia 011523
Teerimäki, Matti 010750
Teeson, Douglas H. 051146
Teeter, Wendy 050422
Teets, James W. 053715
Teets, Robert G. 053715
Teets, Ruth E. 053715
Teeuwisse, Dr. J.B.J. 032182
Tegawa, Tetsuro 029268
Tegelberg, Jaana 010829
Tegin, Taylan 042833
Tegou, Eva 023038
Țeicu, Dumitru 036574, 036600, 036601
Teinilä, Soile 011523
Teinonen, Markku 010887
Teisen, Michael 044715
Teitelbaum, Matthew 007157
Teixeira, Joaquina B. 052393
Teixeira Jiménez, Natalia 048715
Tejada, Susan 050425
Tejada, Susana 047298
Tekampe, Dr. Ludger 021931
Tekatlian, Dominique 012256
Teleanu, Eugenia 036331
Telek, Ulrike 016977
Teles Dos Santos, Prof. Dr.
Jocelio 004787
Telford, D.B. 033161
Telford, Helen 033254
Telford, Pamela 046121
Tellenbach, Dr. Michael 020255
Telleria de Gallardo, Prof. María
E. 000395

Tellería Greiger, Dr. José Luís .. 004226
Tellgren, Anna 041021
Tellier, Arnaud 041540
Tellier, Dr. Cassandra 047944
Tellini, Dr. Alberto 028275
Tello Figueroa, Jimmer Jesús .. 008636
Telsemeyer, Ingrid .. 017958, 022698
Tělupilová, Marie 009669
Temkin, Ann 052081
Temmer, James D. .. 050911, 050925
Temming, Dr. S.N. 032760
Tempal, Denison 049950
Tempel, Benno 031996
Tempelaars, P.A. 032322
Temperli, R. 042238
Tempesti, Giorgio 025427
Tempestini, Luciano .. 027363, 027365
Temple, Ann 024865
Temple, Patty 045333
Temple, Steve 048999
Templeman, Jack 007477
Templer, Peggy 050793
Templin, M. A. Brigitte 020165
Tempone, Mario C. 046269
Tena, Asunción 040308
Tenabe, Gabriel S. 046752
Tenbergen, Dr. Bernd 020602
Tenenbaum, Ann 051378
Teneul, Thierry 014004
Teneva, Dr. Nadežda 005190
Teng, Mingrui 008078
Tengblad, Stefan 033366
Tengku, Abdul Hamid 030695
Tennant, Robert 043609
Tennent, Elaine 049545
Tennent, Mike 052869
Tennent, William 052259
Tenorio Vera, Ricardo 039466
Tentoni, Dr. Damiana 027373
Teodoru, Violeta Veturia 036502
Teofilov, Teofil 005256
Tēplyj, Nikolaj Ivanovič 037732
Tera, Mitsuhiko 029652
Teran, Reba Jo 048886
Terasima, Yoko 029536
Terblanche, E. 038660
Tercan, Şenel 042904
Terčon, Nadja 038587
Terechov, Anatolij Vladimirovič . 037648
Terechov, Vladimir Michajlovič . 037545
Terentev, I.S. 037945
Terentjev, Vladyslav Sergijovyč . 043306
Terežčuk, Volodymyr 043310
Tergun, Dr. Alimuddin 024576
Terhune, Joyce 052490
Terlouw, Dr. Willem 032575
Terlutter, Dr. Heiner 020602
Terman Frederiksen, Finn 010274
Termure, Taisia 036566
Ternova, Natalja 037446
Teropšič, Tomaž 038499
Terpening, Virginia L. 049687
Terra, Prof. Pedro 004559
Terrail, Claude 014798
Terrasse, Jean-Marc 014749
Terreehorst, P.J. 032944
Terrel, Emmanuelle 014484
Terrell, Walter D. 053190
Terrier, Claude-Cathérine ... 012346
Terrier, France 042173
Terry, Blanche S. 053975
Terry, Cathy 045408
Terry, Joe 051026
Terry, Ray 052677
Tersewe, Sylvie 023348
Teruzzi, Dr. Giorgio 025501
Tervahauta-Kauppala, Liisa .. 010976
Terziev, Penjo 005056
Terzo, Dr. Fatima 028366
Tesch, Ilse 021979
Tesch, Sten 040940, 040941
Teschner, Adelheid 020479
Tesfatsion, Costentinos 010724
Teshigahara, Akane 029591

Teske, Christe 017569
Teske, Robert T. 050917, 054239
Tessenow, Inge 018883, 018884
Tesseraux, Hartmut 019272
Tessman, Norm 052352
Testoni, Marcus 000952
Testoni, Theresa 000952
Tétaz, Marie Noëlle 012593
Tête, Michel 012769
Tetlow, David 045203, 045610
Tetterode Ravestein, Elke 032315
Tetzlaff, Reinhild 018014
Teuber, Dr. Dirk 016928
Teubler, M. 020527
Teufel, Tina 002630, 002631
Teunissen, P. 032826
Tew, Julie 044383
Texel, Mark 051062
Teysen, Kenneth 050561
Thacker, Murray 033263
Thacker, Shirley 047719
þah̃yn, Seracettin 042963
Thakeray, Dr. Francis 038860
Thakkar, V.M. 024507
Thaler, Dr. Herfried 002326
Thaler, Janice M. 049027
Thaler, Klaudia 001828
Thalheimer, Patricia 048180
Thaller, Toni 042179
Thalmann, Jacqueline 045451
Thalmann, Rolf 041289
Thalmann, Simone 041270
Thames, Todd 052721
Thangwa, Josephine 029866
Tharp, Dr. Brent W. 053507
Thaw, Michael 043365
Thaxton Soule, Debora 052632
Thaxton, Kate 045405
Théberge, Pierre 006594
Theile, Michael 019752
Theiler, Theo 041444
Thein, Andreas 018027
Thein, U Soe 031818
Theis, Heinz-Joachim 017189
Theis, Wolfgang 017290
Theisen, Lee Scott 053197
Theiss, Nancy 050035
Theissen, A. 017295
Theißen, Dr. Peter 017907
Thelen, Klaus 021288
Thelle, Mikkel 010208
Theml, Dipl. KuWi Christine 019457
Thenoux, Jean-Marc 014881
Theodorescu, Roxana . 036291, 036297,
       036298, 036300, 036312, 036324
Theoret, Yves 006594
Thériault, Michèle 006375
Thériault, Serge 006594
Theriot, Edward C. 046701
Therre, Bernd 022349
Therre, Marie-Line 016199
Therrien, Richard 005770
Thestrup Truelsen, Susanne 010051
Thestrup, Poul 010255
Theune-Großkopf, Dr. Barbara . . 019757
Theunis, Karin 003492
Thévenin, M. Claude 015268
Thevenon, Luc F. . . . 014478, 014492
Thewes, Guy 030592
Theys, Guido 003754
Thi Nu, Ha 054817
Thi, Nguyen Toan 054804
Thiagarajan, Dr. Deborah 024412
Thibault, Barbara 047114
Thibault, Pascale 015003
Thibodeau, Tony 053114
Thiébaut, Philippe 014743
Thiel, Dr. Ulrich 018472, 018474
Thiele, Dr. Klaus 018798
Thielen, Daniel P. 047445
Thiels-Mertens, Lydia 003548
Thieme, Helga 018883, 018884
Thieme, Thomas 041154
Thieme, Dr. Uwe-Ekkehart 018773

Thiemler, Dagmar 021903
Thier, Dr. Bernd 020608
Thier, Wolfgang 002086
Thierbach, Melanie 016650
Thierolf, Dr. Corinna . . 016670, 020508
Thierry, Olivier 012350
Thiery, Robert 014279
Thiesen, Barbara A. 051610
Thiesen, Bill 050649
Thiesen, John D. 051610
Thieven, Mike 053203
Thijsse, L. 032163
Thill, Edmond 030593
Thingvold, Terje 033781
Thoben, Peter 032288
Thobois, Elisabeth 002931
Thøgersen, Mogens 010123
Thømt, Torill 033592, 033803
Thöny, Christof 001808
Thörn, Eva 041076
Thörnberg, Magnus 040794
Thörne, Michael 040979
Thörner, Ilka 017535
Thogersen, T. 033610
Thole, Dr. Bernward 020270
Thomanek, Kurt 001737
Thomann, Marcel 014024
Thomas-Ziegler, Sabine 020323
Thomas, Angela 043668
Thomas, Anne 006594
Thomas, B. 038792
Thomas, Barry 001350
Thomas, Becci 046413
Thomas, Dr. Beth 045686
Thomas, Brad 048158
Thomas, Celestine 028443
Thomas, Charles 047037
Thomas, Charles C. 049943
Thomas, David H. 048419
Thomas, David T. 053544
Thomas, Dean V. 044000
Thomas, Don 054501
Thomas, Eleanor 045170
Thomas, Dr. Floyd R. 054402
Thomas, Gillian 050838
Thomas, Harold Pratt 047559
Thomas, Heidi 054037
Thomas, Joe 047767
Thomas, Kathryn 052806
Thomas, Larry 053030
Thomas, Linda 047293
Thomas, Linda L. 046446
Thomas, M. Simon 032826
Thomas, Maureen . . . 052891, 052892
Thomas, Nancy K. 050444
Thomas, Nicholas 045853
Thomas, Raymond 011754
Thomas, Prof. Dr. Renate 019711
Thomas, Robert G. 045169
Thomas, Rosalyn 045941
Thomas, Staf 004069
Thomas, Steve 052591
Thomas, Susan 049219
Thomas, Thelma 046490
Thomas, Tim 051011
Thomas, Valerie 014404
Thomas, Valérie 014409
Thomas, William T. 047879
Thomas, Y. 013390
Thomason, Barbara 052496
Thomasson, J.R. 049416
Thome, Teresa L. 049192
Thomenius, Kristina 011055
Thomes, Russell 053423
Thomina-Berrada, Alice 014743
Thommen, Lynn 051412
Thommesen, Nils H. 033499
Thompson-Klein, Mark 050395
Thompson, Ann 048360
Thompson, Anna 041027
Thompson, Anna M. 052782

Thompson, Barb 005328
Thompson, Barbara . . . 049353, 053488
Thompson, Becke 048212
Thompson, Bernard J. 007498
Thompson, Bernell 053224
Thompson, Brian 045862
Thompson, Dale 049447
Thompson, David 052636
Thompson, Deborah . . 041119, 053169
Thompson, Dixie 047513
Thompson, Franklin 050247
Thompson, Gwen 050248
Thompson, Ida 049586
Thompson, Isobel 045941
Thompson, J. 047422
Thompson, James 049057
Thompson, Jennifer 043978
Thompson, Jo 054439
Thompson, Joe H. 053184
Thompson, John 047335, 053169
Thompson, Joseph 051590
Thompson, Kelly 050019
Thompson, Livia 049727
Thompson, Marc 048507
Thompson, Margaret 044570
Thompson, Mervyn 044102
Thompson, Dr. P. 044586
Thompson, Paul 033351, 044019
Thompson, Pauline 007303
Thompson, Peter 006717, 045973
Thompson, Ronda 054277
Thompson, Roy 006419
Thompson, Scott 048372
Thompson, Steve 045749
Thompson, Susan 051268
Thompson, Trish 048181
Thompson, Wilma 048089
Thomsen, Douglas 048425
Thomsen, Sancie 048425
Thomson, Beth 045579
Thomson, Bruce 005855
Thomson, Ceri 043649
Thomson, David 044150, 044394,
                                           044652
Thomson, Frances 001590
Thomson, Frank E. 046550
Thomson, Dr. Philip 001505
Thonet, Anke-Maria 018388
Thongthew, Suvaporn 042715
Thorarinsson, Einar 024089
Thorburn, Ken 043766
Thorkildsen, Åsmund . 033571, 033572
Thorlacius, Ethel 054063, 054064
Thorláksson, Eiríkur 024093
Thorláksson, Eiríkur . . 024095, 024096
Thormann, Dr. Olaf 019992
Thorn, Brett 043456
Thorndahl, Jytte 010019
Thorne, Lynne 004193
Thornton, Hetty 045929
Thornton, Pam 051865, 051866
Thoroddsen, Asdís 024090
Thorp, John P. 046978
Thorpe, Christine 043888
Thorpe, J. Derle 050273
Thorpe, Linda 044003
Thorpe, Timothy 044705
Thorsell, William 007200
Thorsen, Sue 053362
Thorsteinsson, Arne 010731
Thoulouze, Daniel 014734
Thouret, Nicole 011767, 011768
Thowsen, Atle 033528
Thrailkill, Helen 053492
Thrane, Helle 010123
Thresher, Leon 049357
Throckmorton, C. 043378
Throl, Dr. Hans-Joachim 022743
Thron, Rudolf 002602
Thümmler, Dr. Sabine . 017713, 019530,
       019538, 019540, 019546, 019549
Thünenkötter, Sabine 018832
Thüring, Michéle 041320
Thüring, Silvan 042006

Thuesen, Søren T. 023335
Thum, Agnes 053207
Thum, Marvin 053207
Thuma, Kathryn 051379
Thuring, Basil 041282
Thurmann, Dr. Peter 019606
Thurmer, Robert 047811
Thurn, Ulrike 020511
Thurner, Ludwig 002428
Thurnm, Maggie 045965
Thúroczy, Dr. Csaba 023983
Thurow, Chuck 047664
Thuy, X. de 014303
Thwaites, Dr. Peter 043834
Thygesen, Anker 010092
Thykier, Claus 010339
Thyrring, Ulla . . 010111, 010115, 010117
Thys, Joseph 004147
Tiacci, Giovanni Battista 027259
Tian, Likun 007687
Tian, Mingxin 007740
Tian, Tao 007917
Tian, Xiao 007979
Tian, Xiuping 008424
Tibbs Thurlow jr., E. 054109
Tiberi, Prof. Romolo 027118
Tiberio, Dr. Artioli 027327
Tichenor, Daryl L. 054001
Tichonov, Dr. Igor L. 037780
Tichonov, Vladimir Viktorovič . . . 037024
Tiderman, Conny 040634
Tidy, Josh 044815
Tiedemann, Klaus 021293
Tiedemann, Dr. Nicole 018976
Tiefenbach, Dr. J. 001861
Tiefenbach, Oskar 001878
Tielman, Prof. R.A.P. 032986
Tiendrebeogo, Prosper . 005272, 005277
Tierney, Janet 044446, 045808
Tierney, Lennox 052966
Tiesing, F. 032698
Tietjen, Hinrich 020071
Tietken, Antje 020985
Tietmeyer, Dr. Elisabeth 017237
Tietz, Annette 017136
Tietz, Jürgen 021893, 021895
Tietz, R.M. 038710
Tietzel, Prof. Dr. Brigitte 019784
Tiffany, Harold 049422
Tiffany, Sharon 053527
Tifft, Douglas C. 048566
Tigert, Jennifer 005583
Tiitsar, Ketli 010671
Tijdink, W. 032081
Til, Barbara 018046
Tilarinos, George 052014
Tilbrook, Val 000965
Tiley, Sian L. 038818
Till, Christopher 038678, 038742
Till, Dr. Wolfgang 020551
Tillequin, Prof. François 014727
Tiller, Helmut 003038
Tilley, Nicola 045699
Tilley, Peter 044900
Tilliet, Marie-Françoise 003890
Tillmann, Dr. Doris 019605
Tillmann, Ulrich 019705
Tillmann, Walter 020670
Tillmanns, Eric 003580
Tilmant, France 003845
Tímár, Gabriella 023471
Timbrook, Mark 050952
Timdal, Ass. Prof. Einar 033847
Timlia, Robert 047794
Timlin, Aaron 048298
Timm, Kristal 048758
Timm, Robert 050186
Timmelmayer, Andreas 003056
Timmerman, B.W. 032950
Timmermann, Dale 053949
Timmermann, Dietrich 010184
Timmermans, L. 032620
Timmins, John 033195
Timmons, P.J. 046058

Timms, Peter 048737
Timofeeva, Nina Vasiljevna 037891
Timotijevic, Milos 038188
Timpano, Anne 047743
Tinčev, Valentin 005028
Tindale, Bruce 001310
Tindall, Anne 044694
Tindel, Raymond D. 047686
Tindell, Anne 043409
Ting, Dr. Joseph 007881
Ting, Dr. Joseph S. P. 007880
Ting, Joseph S.P. 007874
Tinguely, Rofer 013762
Tinguy de la Giroulière, Dominique
de 013554
Tinius, Joanna 053172
Tinjum, Larry 052344
Tinker, Dr. Keith 003208
Tinney, Donal 024989
Tinney, Harle H. 051531, 051542
Tinte, Francisco 000649
Tintori, Dr. Andrea 026513
Tió-Fernandez, Dr. Teresa ... 036119,
036120, 036131, 036157, 036158,
036162, 036170
Tip, J.C. 032731
Tippitt, Dr. V. Ann 049058
Tira, Masa Morioka 049542
Tirado, Vincent 050841
Tirelli, Dr. Margherita 027463
Tirkalova, Nina Oleksandrivna . 043261
Tischer, Achim 017591
Tisdale, Brenda 048514
Tisdale, Shelby J. 053677
Tisken, Jörg 022209
Tisserand, Gérard 012597
Tissier de Mallerais, Martine .. 014001
Tissier, Gève 012634
Tissot, Susan 053946
Titley, Dr. Spencer R. 053838
Titova, Larisa Michajlovna 037146
Titova, Valentina Vasiljevna ... 037605
Tittoni Monti, Dr. Elisa . 007465, 027626
Tittoni, Dr. Maria Elisa . 027560, 027614,
027615
Titz, Susanne 020439
Tivadar, Petercsák 023617
Tixier, J.M. 014929
Tiziani, Andrea 042157
Tjan, R. 032116
Tjan, Ranti 032365, 032366
Tjandra, Gunawan 024547
Tjomlid, Steinar 034016
Tjossem, Susan 053920
Tjupina, Lidija Afanasjevna ... 036906
Tjutčeva, Valentina Sergeevna . 037735
Tkačenko, Katerina Grygorivna . 043298
Tkačeva, Irina Vladimirovna ... 037843
Tkach, Steven 054184
Tkačuk, Jaroslava Juriïvna 043187
Tkešelašvili, A.V. 016426
T'Kint de Roodenbeke, Juan
Graaf 003535
Tleuberdina, Piruza Ablaevna .. 029809
Toase, Charles 045138
Tobe, Carol B. 050486
Tobe, Eiko 028521
Tober, Barbara 051324
Tobias, Lara 052250
Tobin, Shelley 043758
Tobler, Beatrice 041320
Tobler, Mathilde 042248
Tobón Arbeláez, Elvira 008778
Tocev, Petär 005031
Tochigi, Akira 029609
Toczewski, Dr. Andrzej 035463
Tod, Chris 045867
Toda, Masayasu 029230
Todd de Iramain, Margarita Tula 000600
Todde, Nicola 025769
Todiraş, Silviu 036252
Todorov, Todor 005147
Todorova-Stojanova, Kresa ... 005112
Todorova, Ani 005006

Todorova, Pepa 005048
Todts, Herwig 003289
Toedtemeier, Terry 052302
Toejner, Poul Erik 010146
Tölle, Ch. 021392
Toelle, Wolfgang 018799
Tønneberg, Ingrid 033970
Toepfer, Lutz 016801
Törmälä, Satu 010771
Törnblom, Görel 040738
Törnig-Struck, Jutta 020363
Török, Tibor 023740
Toerper, Delaine A. 054586
Toftegaard, Kirsten 010172
Toftgaard Poulsen, Søren 010115
Togay, János Can 017097
Tognali, Dino Marino 028401
Togneri, Carol 051935
Tohjoh, Kohkoh 028913
Tohno, Haruyuki 028492
Toit, Elizabeth du 038735
Toivanen, Pekka 010910
Toivanen, Yrjö 011493
Toka, Arnita 030199
Tokarczyk, William 049070
Tokarski, Longin 034678
Tokiwa, Yutaka 028652
Tokombaeva, Tamara Alievna . 030075
Tokuda, Toshiaki 028774
Tokugawa, Narimasa 029038
Tokugawa, Yoshitaka 029129
Toledo, Eva 034510
Tolentino, Onofre R. 034568
Tolhurst, Robert 043760
Tollefson, Roy 006433
Tollet, André 012436
Tolmačeva, Valentina Michajlovna 037069,
037070
Tolman, Keith 048882
Tolmoff, Mati 010692
Tolnick, Judith E. 049988
Tolonen, Tatiana 011453
Tolson, Peter 053759
Tolstoj, Vladimir I. 037068
Tolu, Giuseppe 025484
Toluse, Joe 047076
Toma, Shiichi 029131
Tomamichel, Leonhard 041354
Tomar, Y.P.S. 024445
Tomaria Cortes, Manuel 040325
Tomaschett, Dr. Pius 042072
Tomašík, Gennadij Aleksandrovič 036902
Tomasina, Marie-Cécile 011814
Tomasović, Nives 008917, 008919
Tomassetti, Larissa 001943
Tomassini, Gabriella 028078
Tómasson, Dr. Þórður 024085
Tomaszewski, Robert 035028
Tomczak, Bogdan 035436
Tomea Gavazzoli, Dr. Maria
Laura 025645, 027058, 027059,
027062
Tomey, J.K. 043972
Tomić, Vladeta 038226
Tomiyasu, Reiko 028515
Tomkins, John 052197
Tomkins, Ken 044573
Tomlinson, Janis A. 051495
Tomlinson, Dr. Janis A. 054094
Tomlinson, Lori 052417
Tomlison, Lyn 051305
Tomljenović, Zvjezdana 009025
Tommassen, W. 032652
Tomor, Michael 049834
Tomor, Dr. Michael 050398
Tomov, Lazar 005191
Tomova, Krasimira 005243
Tompkins, Jack D. 052495
Tomplak, Darinka 038624
Toms, Donald D. 050199
Toner, Dominga N. 046952
Toner, Eden R. 052498
Tong, Guorui 008529
Tong, Zerong 008290

Tonini, Dr. Camillo 028303
Toniolo, Elia 028344
Tonkin, Steven 000924
Tonković, Snježana 008921
Tonneau-Ryckelynck, Dominique 013169
Tonon, Dr. Marco 025633
Tony, Mollica 051561
Tooby, Michael 043863
Toogood, Peter 000870
Toohey, Jeanette 049768
Tooker, David P. 053615
Toom, Maire 010668
Toomet, Tiia 010702
Toome, Keith 005514
Toperzer, Tom R. 046433
Toplovich, Ann 051186
Topoleanu, Florian 036799
Topoleanu, Florin 036365, 036455,
036796, 036797, 036798, 036800
Topolinsky, Verna 005337
Toraason, Mary C. 053913
Torataro, Yoneyama 029580
Torchia, Richard 049124
Torchia, Robert W. 050777
Torchio, Maurizio 028110
Torfs, Freya 000802
Torgersen, Ulf E. 033800
Torjusen, Bente 050205
Torki, Nasser 024696
Torm, Helge 010210
Tormo, Fernando 038906
Torno, Janet E. 047011
Torp, Lisbet 010185
Torra, Eduardo 040441
Torra, Emilio 026527
Torralba, Casiana C. 034447
Torrance, Joshua C. 048560
Torre Bonachera, José Luis la . 040282
Torre González, Norberto de la . 031462
Torre, Antonio 027643
Torre, Prof. Danilo 026281
Torre, Fuensanta García de la . 039306
Torre, Graciela de la 031241
Torre, Dr. Paola 027627
Torremocha Silva, Antonio 038942
Torrence, Gaylord 049888
Torrent í Bagudà, Enric 040266
Torrents, Carme 039402
Torrenzieri, Claudio 028080
Torres Correa de Migliorini,
Beatriz 000601
Torres Gonzàlez, Begoña 039682
Torres Valderrama, Hernando .. 008622
Torres, Elizabeth 004223
Torres, Prof. Emmanuel 034578
Torres, Francesc X. 039363
Torres, Jesse 054612
Torres, John . 051607, 053114, 053116
Torres, Julio 039647
Torres, Pedro Pablo Alberto 008757
Torsteinsrud 042904
Tortell, Victor E. 030748
Tortolero, Carlos 047673
Torunsky, Dr. Vera 017499
Tosatto, Guy 013181
Tošev, Miroslav 005099
Toškova, Svetla 005174
Tosi, Dr. Maria Teresa 027366
Tosini Pizzetti, Simona 026686
Tosswill, Richard 005421
Tostevin, Patrick 045608
Toteva, Teodora 004991
Totevski, Totjo 005532
Tóth, Hajnalka 024015
Tóth, János 023701, 024053
Toti, Dr. Enrico 027951
Toti, Dr. Odoardo 025295
Tottis, Jim 048301
Touche, Hugues de la . 014130, 014131,
014132, 014133, 014134
Touhey, Paula 049927
Toulgoat, Morgane 012682
Toulmond, Dr. André 015155
Toumit, François 012280

Touret, Lydie 014728
Tourneux, Alain 012456
Tourta, A. 023294
Tourtillotte, William 053394
Tous, Rafael 039067
Toussaint, Christophe 012806
Toussaint, Jacques 003898
Toussaint, Yolaine 006358
Touw, Susanne 020567
Towe, Andrew C. 048215, 048217
Towell, Debbie 007321
Towell, Gerry 045310
Towl, Betsy 047530
Town, Geri 054573
Townley, Kim 033345
Townsend, Clifford 048474
Townsend, Mary Ann 047533
Townsend, Richard P. 046800
Townsend, Simon 043484
Townsend, Wayne 006826
Townsend, Wilma T. 047398
Toyama, Kazuo 028871
Toyama, Tohru 029550
Trabitsch, Dr. Thomas 003059
Trabold, Dr. Rudolf 017299
Trace, Howard C. 049682
Tracey, D. 043572
Trachsler, Ernst 042118
Trachtenberg, Daniel S. 047866
Tracy, Conrad 047880
Tracy, Duane 051168
Tracy, Susan 007162
Träger, Hans-Heinrich 020501
Träris, Susanne 020347
Tragent, Henry 050861
Trahan, Eric 047395
Traikasem, Saengchan . 042692, 042696
Trainor, Eugene N. 053202
Trakospoulou, Helen 023130
Trâmbaciu, Şefan 036346
Trâmbaciu, Stefan 036345
Tramer, Nancy 048336
Tramonte, Pietro 027461
Tramp, Lana 053519
Tramposch, Dr. Bill 033360
Tranié, Pierre-Tarcisius 016044
Transier, Dr. Werner 021931
Trapnell, Dr. Simon H. 044435
Trapness, Thor 033676
Trapp, Graf Johannes 027971
Trappeniers, Dr. Maureen ... 032472
Trask, David 054499
Trask, Richard B. 048141
Trauffer, Walter 041364
Traugott, Joseph ... 053113, 053116
Traunecker, Claude 015865
Trautmann, Charles H. 049741
Trauzettel, Ludwig 017868
Travella, Eugenio A. 000545
Travers, John 045268
Travis Boley, F. 052787
Travis, Betsy L. 052332
Travis, P. 033323
Trazegnies, Marquis de 003523
Trebbi, Marco 033533
Trebilcock, Evelyn 049602
Trebilcock, Samantha 043747
Treble, B. 005668
Tréboit, Marcel 014085
Trecht, Daniela 041374
Trefnỳ, Martin 009866
Trefor Jones, Iwan 044042
Trefzger, U. 018481
Tregaskis, Kate 044276
Tréguier, Frank 012772
Trekell, Kirk 046429
Tremain, Robert 005899
Trembath, Lynne 044907
Tremblay, Diane 006833
Tremblay, Ermenegildo 027413
Tremblay, M. Joland 006389
Treml, Robert 022401
Tremp, Prof. Dr. Ernst 041924
Tremper Hanover, Lisa 047875

Trentin-Meyer, Maike	016822	Trowill, John	050219	Tucker, Anne W.	049588	Turner, Peter	045580
Trepesch, Dr. Christof	016649, 016654,	Trowles, Peter	044400	Tucker, Brian	051937	Turner, Sam	048131
	016658, 016663	Trox, Dr. Eckhard	020170	Tucker, Gary	047113	Turner, Sheridan	049127
Tresch, Karl	041974	Troy, Gloria	051431	Tucker, Mark S.	052081	Turner, Terry	044883
Treskow, Dr. Sieglinde von	017543	Trubicyna, Tatjana	037557	Tucker, Michael S.	052723, 052732	Turner, Dr. Thomas F.	046365
Trešlová, Romana	009622	Truc, Sylvie	013178	Tucker, Mike	052727	Turner, William H.	051007
Tresner, Cliff	050997	Truchina, Galina Vasiljevna	038062	Tucker, Norman P.	049980	Turnham, Stephanie	046908
Tresseras, Miquel	040375	Trucksis, Jane	054554	Tucker, Prof. Priscilla K.	046493	Turnquist, Jan	047954
Tretheway, Linda	048459	Trudel, Marianne	005334	Tucker, Renee	048902	Turovski, Jevgenij	043293
Trethewey, Graham	001400	Trudgeon, Roger	000817	Tucker, Sara	051378	Turpin, Susan	052660, 053429, 054524
Tretjakov, Nicolaj S.	037622	True, Marion	050436	Tucker, Sue	048733	Turton, Alan	043508
Tretjakova, Vera Dmitrievna	036971	Trueb, Linda	050186	Tucker, Susan A.	051207, 051028	Turunen, Eija	011293
Treude, Dr. Elke	017876	Trücher, Karl	001816	Tuckett, Noreen	000877	Turza, Witold	034874
Treuherz, Julian	044864, 044867	Trueman, Debbie	006445, 006450	Tucoo-Chala, Jean	014364	Tuschell, Peter	003075
Trevisani, Dr. Enrico	026207	Trümper, H.-J.	017274	Tudeer, E.	011375	Tut, Dr. Ye	031834
Trevisani, Dr. Filippo	026702, 026862	Trümpler, Dr. Charlotte	018288	Tudor, Kim	045852	Tutorov, Jasmina	038261
Trevitt, Peter	043869	Trümpler, Rico	041257	Tudose, Mihail	036223	Tuttle, Barbara	048815
Trevor, Tom	043743	Trümpler, Dr. Stefan	041889	Tuegel, Michele	052859	Tuttle, Lyle	053028
Trezzi, Daniela	028120	Trufanov, E.I.	038075	Tuftrow, Kate	053026	Tutty, Lauren	006202
Triandaphyllos, Diamandis	023135	Trufini, Antonio	026916	Tugarina, Polina Innokentjevna	037031	Tuyttens, Deborah	007382
Triantafyllidis, Jutta	018261	Trulock, Sue	052136	Ţuglui, Traian	036494, 036495, 036550	Tuzar, Dr. Johannes M.	001843
Triantis, Alice	023168	Trulock, Trent	047409	Tuhumwire, Joshua	043102	Tvedten, Lenny	048677
Triapitsyn, Dr. Serguei V.	052592	Truissen, Torbjorn	034028	Tulaeva, Marina Aleksandrovna	037964	Tvers'ka, Ljudmila Volodymyrivna	043190
Tribouillois, Brigitte	015529	Truman, Nell	046858	Tulekova, Margarita Vasiljevna	036916	Twalba, Dia'a	029786
Trice, Robbert	051976	Trumble, Angus	051266	Tuleškov, Vencislav	005081	Tween, Trevor	043815, 045168
Trickett, Jessica D.	054598	Trumble, Edward P.	047143	Tulin, Vasil Ivanovyč	043175	Twellman, Ron	051827
Trickett, Peter	043716	Trumble, Thomas E.	047143	Tulku, Jorma	011288	Tweneboa-Kodua, Maxwell
Tridente, Tommaso	026872	Trummer, Manfred	003033	Tulku, Doboom	024239	Ohene	022933
Triebel, Dr. Dagmar	020513	Trumpie, Dr. Ank	032575	Tull, Earl B.	053222	Twiehaus, Christiane	018548, 021690
Triem, Katleen Heike	049093	Trumpler, Julie	051958	Tulloch, Stuart	043644	Twist, William	006155
Trier, Dr. Marcus	019711	Trumps, Patricia	047945	Tumanova, Anna Andreevna	037836	Twitchell, Linda	006317
Triesch, Dr. Carl	019012	Trumpy, Sigrid	046503	Tumasyan, Tatyana	043135	Tworek, Dr. Elisabeth	020548
Triffaux, J.M.	003321, 003322	Trupp, Adam	053770	Tumidei, Stefano	025539	Twyford, Mark	048443, 051098
Trifiletti, Dee	050396	Truque, Angeles Penas	039321	Tungarova, Saran-Térèl		Txabarri, Myriam Gonzalez de	040459
Trifonova, Mariana	005267	Trusch, Dr. Robert	019518	Sultimovna	037989	Tyack, Gerry V.	045295
Trifut, Viorica	036209	Truškin, Michail Danilovič	037176	Tunheim, Arne	033619	Tyagi, Prof. K.N.	024397
Trimbacher, Hans-Peter	002821	Truss, Bernard	045980	Tunis, Dr. Angelika	017112	Tyamzarne, G.W.T.	054858
Trimmer, Jason	052753	Trzeciak, Andrzej	035223	Tunnell, Bill	050981	Tybring, Tina	040862
Trimpe, Pam	049712	Tsafrir, Prof. Yoram	025069	Tunnicliff, Iris	046112	Tych, Felix	035376
Tringali, Mario	028265	Tsai, Eugenie	047253	Tunsch, Dr. Thomas	017242	Tychenko, Volodymyr
Trinidad, Miguel Angel	031301	Tsaneva, Svetla	005191	Tuomi, Marja-Liisa	010844	Oleksandrovyč	043208
Tripanier, Romain	006856	Tsang, Gerard C.C.	007873	Tuomi, Timo	010866	Tychobayeva, Halina	043251
Tripathi, Dr. S.K.M.	024377	Tsang, Kathleen	007403	Tuomi, Tuulia	010777, 010788, 010794	Tydesley, Carole	046135
Trippi, Peter	051372	Tsaoussis, Vassilis	023258	Tuominen, Laura	010771	Tykkyläinen, Mla	011450
Trivelli, Marifrances	053060	Tsatsos, Irene	050443	Tuomola, Satu-Miia	011305	Tyler, Dr. Alan	045442
Trividic, Claude	014933	Tschabold, M.	041812	Tupan, Dr. H.R.	032035	Tyler, Fielding L.	053999
Trnek, Dr. Helmut	003026	Tschachotin, Peter	022057	Tupper, Christine	006318	Tyler, Frances P.B.	047535
Trnek, Dr. Renate	002995	Tschamber, Théo	014563	Tupper, Elton	051920	Tyler, Grant	007446, 007457
Troché, Bruno	013389	Tscherter, Erwin	016108	Tupper, J.	045572	Tyler, Gregory	054471
Tröster, Dr. G.	018759	Tschirner, Manfred	017262	Tupper, Jon	005593	Tyler, Ian	044680
Trofimova, Nadežda Nikolaevna	037862	Tschoeke, Dr. Jutta	020844, 020855	Tura, Jordi	038983	Tyler, Jean	049168
Trogemann, Hans-Dieter	017499	Tscholl, Erich u. Elisabeth	041780	Turán, Róbert B.	023524	Tyler, Karen	050191
Troiani, Dr. Stefano	027874, 027875,	Tschopp, Walter	041796	Turbanisch, Gérard	014361	Tyler, Kay Montgomery	047535
	027876, 027878	Tschorsnig, Dr. Hans-Peter	022056	Turcanu, Dr. Senica	036511	Tyler, Kim	000826
Trojak, Therese	048729	Tschudi, Gabriele	019144	Turci, Dr. Mario	027839	Tyllack, Thomas	017075
Troll, Tim	048328	Tschudi, H. U.	041550	Turdo, Mark A.	051205	Tylleskär, Annika	040737
Trollerud, Elin	033987	Tse Bartholomew, Terese	052992	Turek, Stanisław	034684	Tyre, William	054232
Troman, Louise	046137	Tselekas, Panayotis	023023	Turekanov, E.T.	038085	Tyrefelt, Ronny	040628
Tromantine, Margaret	046811	Tselos, George	051474	Turekian, Karl K.	051265	Tysdal, Bobbie Jo	051520
Tromme, François	004119	Tsenor, Carey	005964	Turgeon, Christine	006739	Tyson, Rose	052979
Trommer, Markus	020559	Tsereteli, Zurab	037351	Turicyna, T.K.	037999	Tzekova, Katia	005191
Tromp, C.P.	032842	Tsipopoulou, Dr. M.	022949	Turja, Teppo	011413	Tziafalias, Athanasios	023074, 023149
Tromp, Dr. G.H.M.	032775	Tsokas, Andreas	023077, 023079	Turkington, M.	007013	Tzikalov, Borislav	002141
Tron, Prof. Federico	026944	Tsomondo, Tarisai	054848	Turkov, Shura	025024	Úbeda, Andrés	039676
Tronchetti, Dr. Carlo	025673	Tsoukalas, P.	023297	Turkowski, Guntram	021668	Ubierna, César	039420
Tronem, Donatella	026831	Tsoutas, Nicholas	001661	Turley, Chuck	052399	Ubl, Prof. Dr. Hansjörg	002895
Troniou, Marie-France	013965	Tsuchiya, Shuzo	029277	Turnbow, Chris	053114	Uboh, Chioba Francisca	033442
Tronrud, Thorold	007148	Tsuji, Shigebumi	029188	Turnbull, A.	000973	Uburğe, Baiba	030177
Trop, Sandra	053633	Tsujii, Dr. Tadashi	028463	Turner, A.K.	033228	Uccello, Vincenza	054304
Tropeano, Placido	026770	Tsujimoto, Isamu	028474	Turner, Annette	001228	Uchida, Akihiko	029546
Trosper, Robert	046516	Tsujimoto, Karen	051681	Turner, Barbara L.	054521	Uchida, Hiroyasu	029723
Trost, Maxine	047110	Tsujimura, Tetsuo	029608, 029609,	Turner, Brenda	051971	Uda, Makoto	028647
Trost, Dr. V.	022067		029610	Turner, David	048618	Udø, Ovin G.	033714
Trotnow, Dr. Helmut	017067	Tsunoyama, Sakae	029314	Turner, E.	001150	Udovičić, Ivana	004282
Trott, Eileen	005464	Tsurumi, Kaori	029608	Turner, Dr. Elaine	020776	Uduwara, J.S.A.	040471
Trott, Martin	017940	Tsutsumi, Seiji	028786	Turner, Elizabeth Hutton	054157	Udvary, Ildikó	023481, 023547
Trotter, Mark A.	048674	Tsuzuki, Chieko	029608	Turner, Grady T.	051440	Üçbaylar, Enis	043054
Trotter, Ruth	045349	Tu, Zhigang	008385	Turner, J. Rigbie	051428	Ueda, Ichiro	029328
Trotti-Bertoni, Prof. Anna	026177	Tubaja, Roman	035284	Turner, Kathy	049705	Ueda, Koji	019689
Troubat, Catherine	012990	Tucci, Franco	026469	Turner, Dr. Kenneth R.	049160	Ueda, Shinya	029463
Trouplin, Vladimir	014726	Tuček, Helga	020122	Turner, Laura	044712	Üiepmeyer, Prof. Dr. L.	018572
Trout, Amy L.	051264	Tuch, Gabriele	011264	Turner, Linda	053726	Ueki, Hiroshi	028595
Trout, William	050790	Tuchel, Dr. Johannes	017144, 017146	Turner, Nathan	049298, 049299	Ueland, Hanne Beate	033826
Trouw, Bernward	019249	Tucher, Bernhard Freiherr von	019484	Turner, Neil	045416		

Uelsberg, Dr. Gabriele . 017499, 020495
Ueltschi, Hans-Peter 042028
Ueno, Katami 029596
Ueno, Kazuhiko 028591
Ueno, Takaaki 029699
Ueno, Tetsukazu 028491
Ürgová, Dana 008813
Uerscheln, Gabriele . . 018047, 018047
Uetz, Kitty 047744
Uffmann, Rüdiger 017376, 017380
Ufimcev, Valeri K. 036993
Ugarte-Peña, Alfredo 007544
Ugat, Pedro R. 034427
Ugawa, Masahiro 029317
Ugo, Prof. Gianni 000438
Ugren, Dragomir 038217
Uhen, Mark 047040
Uhlemann, Dr. Silvia 017829
Uhlenbrauck, Craig 046511
Uhlig, Rainer 017844
Uhlmann, Klaus 017121
Uhlmann, Reiner Jens . 021175, 021194,
                                     021197
Uhrig, Dr. Sandra 020625
Ujházi, György 023578
Ujiie, Seiichiro 029615
Új, Dr. Írisz 023531
Ujváry, Dr. Péter 023553
Ukdah, Radi 042261
Ukena, H. 017873
Ul-Haq, Saleem 034131
Ulak, Dr. James 054116
Ulbert-Wild, Corinna 017692
Ulbrich, Marorie 046357
Ulbrich, Thomas 030004
Ulbricht, Dr. Ingrid 021663
Ulčar, Mirko 038495
Ulenberg, Dr. S.A. 032002
Uliančić-Vekić, Prof. Dr. Elena . . 008968
Ulieru, Nina 036617
Ulin, Dr. Jonna 040849
Ulinskyte, Vilija 030465
Ulitin, Valerij G. 037825
Ulkenbajuly, Nariman 029813
Ullmann, Silke 020280
Ulloa Rojos, Sergio 007551
Ulloa, Gina 031294
Ullock, Jack 006331
Ullrich, Prof. Dr. Ferdinand . . . 021306,
                                     021309
Ullrich, H. R. 020998
Ullrich, Karin 018580
Ullrich, Tyrena 021547
Ulm zu Erbach, Franz Freiherr
  von 018228
Ulm, Dagmar 002330
Ulmann, Dr. Arnulf von 020853
Ulmann, Jana 041678
Ulmis, Janis 030157
Ulrich-Hicks, Barbara 049646
Ulrich, Jean 015013
Ulrich, Matthias 018446
Ulrichs, Dr. 021413
Ulrichs, Dr. Friederike 020556
Ulrichs, Dr. Ulrike 019277
Ultan, Prof. Lloyd 047234
Umberger, Eugene 049229
Umberger, Leslie 053296
Umemoto, Michio 028657
Umezu, Gen 029305
Uminowicz, Glenn 048455
Umney, Nick 045123
Umorbekov, Baytursun Esjanovič 029801
Unbehaun, Dr. Lutz . . 021532, 021533,
                                     021771
Underwood, David 049806
Underwood, Katie 050844
Undesser, Gottfried 003013
Ungar, Ann 006362
Unge-Sörling, Suzanne 041019
Unger, Dr. Achim 017272
Unger, Gerhard 020214
Unger, P. Karl 002681
Ungerfeld, M. 025191

Unitt, Phil 052980
Uno, Tomohisa 029579
Unrau, Peter 006430
Unruh, Dr. Frank 022202
Unsain, José María 039356
Untch, Katharine 052565
Unterberger, Dr. Gerald 001686
Unterdörfer, Dr. Michaela 041591
Unternährer, Nathalie 042024
Unterreiner, Katrin 003012
Unterschultz, Judy 007079
Untersteiner, Eva 003144
Unverfehrt, Dr. Gerd 018750
Unzueta Reyes, Prof. Ana
  Geraldina 031175
Uoti, Maunu 011238
Uotila, Dr. Pertti 010834
Updegraff, Kurt 047331
Upeniece, Daiga 030177
Uphoff, Johanna 019434
Upp, Janeanne A. 053649
Upshaw, Lisa 052898
Upton, Graham 044215
Upton, Jan 044215
Upward, Geoff 052608
Uracz, Andrzej 035062
Urakova, Ljudmila Evgenjevna . . 037047
Uranić, Igor 009025
Urban, Dr. Andreas 019053
Urban, Mag. Bettina 002954
Urban, Elke 020016
Urban, Erin 053500
Urban, Heinz 020335
Urban, Robin 046987
Urban, Wolfgang 021520
Urbaneck, Rotraud 022582
Urbanek, Erich 001962
Urbanek, Dr. Wilhelm 000926
Urbanelli, Lora 052387
Urbani, Vittorio 028319
Urbanova, V. 038486
Urbański, Andrzej 035453
Urbanski, Prof. Dr. Krysztof . . . 034833
Urbas, Herbert 001888
Urbascheck, Dr. Stephan 020567
Urey-Weeks, Burdie 030290
Urgesi, Prof. Domenico 026772
Uribe Ramón, Jorge 008808
Uribe, Gloria Zea de . . 008598, 008599
Urio, Leonarda 041391
Urioste Avana, Jaime 004244
Urmas, Olli 011030
Urquhart, Margaret 045943
Urquina Llanos, Henry 008681
Urquiola, Beatriz S.A. 034531
Urrea Fernández, Jesús 040345
Urrea, Ana Isabel 008826
Urrutibehety, Clément 015490
Urs, Ilarion 036674
Ursachi, Vasile 036661
Ursch, Jacqueline 012771
Urschel, Joe 046540
Ursel, Robert d' 003569
Ursić, Desanka 007531
Ursina, Marija Ivanovna 037373
Ursoi, Ana 036258
Ursu, Mihai 031711
Urteaga, PhD Mertxe 039515
Urumov, Vanko 005236
Urumova, Rumjana 004997
Urushadze, Prof. Levan 016416
Ušakov, Prof. Sergej
  Aleksandrovič 037480
Ušakova, Tamara Ivanovna 037164
Usanov, Evgenij Nikolaevič 036845
Usaquén Ramos, Martha Lucía . . 008573
Uschalt, Susanne 020880
Uschner, Ralf 016817
Usik, Boris Grigorjevič 038049
Usler, Dr. Alexander 018686
Ussing Olsen, Peter 010224
Ustinov, Vladimir Jurjevič 038069
Ustvedt, Østein 033864
Ustvedt, Øyтеin 033845

Usundag, Nell 000999
Utech, Dr. Frederick H. 052158
Utermöhlen, Bernd 017709
Uteulina, Ajsha 029790
Uthardt, Lorenz 011140
Uthemann, Ernest W. 021561
Utili, Dr. Mariella 027004
Utriainen, Onerva 011401
Utry, Dr. Attila 023788
Utter, Dr. Brad L. 054202
Utz, Karen 047007
Uvarova, Elena Anatolevna 037974
Uzunov, Nikolai 005029
Vaccaro, Dr. Luis Fernandez . . . 000199
Vaccaro, Prof. Mario 025699
Vache, Bernd 018869, 018870
Vačiullin, Gali Fajzrachmanovič . 037983
Väisänen, Raija 010934
Väkeväinen, Lea 010874
Välicki, Antti 010919
Väljas, Mait 010667
Vaelske, Urd 002637
Vaerten, Frank 003353
Värtinen, Lea . 010753, 010759, 010895,
        010896, 011064, 011131, 011196,
                                     011252, 011462
Vaessen, J.A.M.F 032028
Väterlein, Dr. Christian 016677
Vagriené, Birutė 030546
Vahey, John 047104
Vahl, Heidemarie 018030
Vahling, Ray 053723
Vai, Prof. G.B. 025565
Vaičeskauskienė, Aldona 030365
Vaida, Evghenia 036632
Vaida, Lucian 036579
Vaillancourt, Sandra 053104
Vaillant, L. le 033142
Vaimbois, Laurent 014406
Vainorius, Julius 030475
Vaireaux, Bernard 015126
Vairo, Carlos Pedro 000661
Vaitkute, Karile 047640
Vajnerman, Viktor Solomonovič . 037594
Vajšnorajte, I.A. 037971
Valadés Sierra, Juan M. 039183
Valaison, Marie-Claude 014842
Valat, Georges 012669
Välčev, Aleksandǎr 005197
Valcke, Ludo 003351
Valda Martínez, Edgar Armando . 004234
Valdeny da Rocha, Manoel 004917
Valderas, Elena Ruíz 039237
Valderrama de Díaz, Maritza . . . 008734
Valdés Figueroa, Eugenio 004681
Valdés Gonzalez Salas, Ana Luisa 031147
Valdes, Nieves 034440
Valdez, Ana Luisa 031190
Valdez, E.A. 050053
Valdez, Juan Carlos 031351
Vale, Jessica 043721
Vale, Prof. Dr. Maria da Graça Pratas
  do 035625, 035635
Valencia de Redondo, María
  Eugenia 008762
Valencia Ospina, José Aníbal . . . 008562,
                                     008563
Valencia, Alvaro Pío 008769
Valencia, Maria L. 036119
Valencia, Dr. Niels 034337
Valente, Maurizio 041392
Valentien, Prof. Donata 017066
Valentin, J.C.P. 032321
Valentin, Jocelyne 015853
Valentin, Laetitia 011639
Valentin, Noel 051432
Valentine, Eunice E. 054044
Valentine, James 046939
Valentine, Kathryn 045268
Valentine, P. 038667
Valentini, Paolo 028102
Valentová, Dr. Jarmila . 009633, 009634,
                          009635, 009642
Valenzuela Ríos, José Ignacio . . 039175

Valerio, Michele 025460
Valette, Jean 013063
Valeza, Luisa S. 034450
Vališková, Vladimíra 009959
Valk, Marika . . 010663, 010668, 010676,
                          010678, 010680, 010682
Vall Petitpierre, Carmina 039075
Valla, T. 015963
Vallan, Dr. Denis 041710
Vallarino, Adriana L. de 034176
Vallat, Pierre 012196
Valle Pérez, Dr. José Carlos . . . 039960
Valle, Carlos 039639
Valle, Dr. Marco 025494
Vallee, Dianne 006004
Vallee, Dianne 039988
Vallejo Laso, Manuel 039988
Vallejo Triano, Antonio 039303
Vallely, Dana S. 051597
Vallery, Tullio 028325
Valles-Bled, M. 013851
Valles, Dave Anthony 034469
Vallette, Philippe 012141
Valley, Derek R. 051758
Vallianos, Dr. Christopher 023317
Valliant, John R. 052834
Vallo, Brian D. 052412
Vallvieille, Michel de 015620
Valor, E. 000642
Valorinta, Riitta 011402
Valowitz, Rhoda 053742
Valter, Dr. Claudia 021783
Valtere, Māra 030118
Valuska, Dr. David 050022
Valverde Canoil, Mercedes 039308
Valz, Jean-Paul 011566
Vametova, Farida H. 036999
Van Adrichem, J. 031986
Van Allen, Victor 005627
Van Alphen, Jan 003284
Van Amerongen, R.A.M. 032155
Van Arsdel, Sara 051806
Van Beek, Dr. R. 031923
Van Beekhofff, D. 031906
Van Belle, Eddy 003661
Van Belle, Jean-Louis 003394
Van Berge-Gerbaud, M. 014635
Van Berge Henegouwen, Dr. Arno 032181
Van Boghout, Erik 003686
Van Boheemen, P. . . . 032217, 032218
Van Boorne, Geert 003642
Van Bossche, André 003669
Van Boven, Dr. M.W. 032191
Van Boxtel, E. 032826
Van Brakel, J.H. 031950
Van Brakel, J.W. 033019
Van Broekhoven, L. 032585
Van Caloen, Veronique 003829
Van Campenhout, Nico 003898
Van Cauteren, J. 033031, 033032
Van Cleempoel, Dr. Roland 003296
Van Cleve, David 047098
Van Daele, Micke 003924
Van Daele, Dr. Rik 003683
Van Damme 003752
Van Damme, I. 003463
Van de Beuque, Guy 004705
Van de Guchte, Dr. Maarten . . . 049768
Van de Maele, Maurice 007530
Van de Perre, M. 003256
Van de Poel, T.L. 032238
Van de Schoor, Dr. F.C.M. 032687
Van de Steeg, Dan 049191
Van de Velde, Hildegard 003312
Van de Voorde, Veronique 003892
Van de Wiele, Johan 003629
Van Delen, Wilfried 038685
Van Delft, Dr. D. 032581
Van den Abeele, Jean-Pol 004035
Van den Akker, Ad 049509
Van den Akker, M. 032414
Van den Berg, S. 032769
Van den Bogaart-Vugts, Dr. H. . . 032709
Van den Bosch, Paula 032621
Van den Branden, Walter 004124

Van den Brink, Dr. Peter 016445
Van den Broek, J.P.G. 032145
Van den Broeke, J. 032977
Van den Bussche, Willy 003737, 003931
Van den Driessche, B. 003831
Van den Eijnde, Jerden 033014
Van den Elzen, Dr. Renate 017504
Van den Heede, Joszef 004017
Van den Hole, Edmond 003650, 003651
Van den Hout, Drs. A.H.P.J. ... 032949
Van den Linden, Louis 003659
Van den Meiracker, C. 032830
Van den Steen de Jehay, Comte
Guy 003740
Van den Veen, J. 032704
Van der Auwera, M. 003688
Van der Beele, P. 032512
Van der Burgh, Dr. J. 032954
Van der Coelen, P. 032826
Van der Donckt, Marie-Claire . . 003940
Van der Goot, Welmoed 032565
Van der Ham, Pieter 032967
Van der Heever, Chast 038821
Van Der Heijden, Merijn 049209
Van der Holst, A.P.J. 032860
Van der Horst, L. 031997
Van der Kemp, Gerald 013114
Van der Laan, Anke 032392
Van der Laar, P. 032804
Van der Lelie, H. 032580
Van der Linden, Judy 048360
Van der Lugt, Dr. P. 032435
van der Meiden, Kees 032382
Van der Meijden, Dr. Hellebora . 041261
Van der Merwe, Jake 005383
Van der Merwe, T. 038691
Van der Merwe, Thijs 038685
Van der Meulen, G. 032926
Van der Meulen, G.P. 032319
Van der Mieden, O.J.E. 003007
Van der Ploeg, J. 033106
Van der Ploeg, P. 032180
Van der Spek, Dr. T.M. 032581
Van der Stoep, F. 032807
Van der Veer, Hugh G. 053571
Van der Wall, Dr. Frauke 022791
Van der Werdt, E. 032944
Van der Westhuizen, V. 038810
Van der Zalm, Leo 032352
Van der Zijpp, Sue-An 032378
Van der Zwaag, M. 031953
Van Deun, Charles 004015
Van Deuren, Guy 003320
Van Deurs, Lone 010030
Van Deventer, Anriët 038806
Van Dijck, Leen 003279
Van Dijk, E.A. 033128
Van Dijk, J. 031950
Van Dongen, A.G.A. 032826
Van Dongen, P.L.F. 032585
Van Dooren, Rita 003294
Van Doorne, Geert 003639
Van Doorne, Veerle 003536
Van Dorst, Drs. M.H.J. 032984
Van Droogenbroeck, M. 003435
Van Duuren, D.A.P. 031950
Van Dyck, Ann 003287
Van Dyk, Pieter 001415
Van Dyk, Stephen H. 051368
Van Dyke, Dore 046645
Van Dyke, Jonathon 049373
Van Egmond, C. 032458
Van Endert, Dr. Dorothea 020413
Van Faasen, Drs. S.A.J. 032176
Van Frank, M. 049478
Van Geluwe, Johan 004130
Van Gennip, J. 032398
Van Genuchten, S. 033021
Van Gerwen, Dr. Ch. 032969
Van Gijseghem, A.M. 003637
Van Goethem, Wendy 003292
Van Gompel, F. 003873
Van Grasdorff, Maria 004014
Van Grevenstein, Alexander 032621

Van Groeningen, Drs. J.J.W.P. . . 032271
Van Grootheest, Joke 032181
Van Haaften, Joel 050360, 050361
van Haaren, M.A. Christiane . . . 018189
Van Haaren, Joke 032392
Van Haaren, Drs. S.E. 032949
Van Haarlem, Dr. W.M. 031923
Van Haecke, Bert 003533
Van Hammeren, W. 032308
Van Harinxma, Barones K.L. . . . 032976
Van Harten-Boers, H. 032884
Van Hattem, M. 032157
Van Heerden, H. 038772
Van Heerden, Martie 031859
van Helmond, Mary 043854
Van Hengstum, R. 032582
Van Heugten, Sjraar 031996
Van Hoof, Werner 003317
Van Hoofstat, Jan 004001
Van Hoonacker, Albin 003779
Van Hoonacker, Marie-Henriette . 003779
Van Hoorn, Dr. M.A.M. 032399
Van Horn, Walter 046470
Van Hout, I.C. 031950
Van Houten, P. 032619
Van Huis, Edwin 032486
Van Huy, Prof. Dr. Hguyen 054789
Van Impe, Dr. D. 003301
Van Kampen, Dr. Iefke Johanna 026332
Van Kerkhoven, Geertrin 003941
Van Kerkhoven, Laurence 003411
Van Klinken, M.F. 032017
Van Koetsveld, Rik 031996
Van Kooten, Drs. C. 032739
Van Krimpen, W. 032171
Van Krimpen, Wim 032169
Van Krugten, Dr. D. 019436
Van Kruistum, G. 032540
Van Laar, Barb 053355
Van Laarhoven, Dr. Jan 032442, 032472
Van Laer, Marcel 003598
Van Landschoot, G. 003259
Van Leeuwen, P.C. 032049
Van Leeuwen, P.P.H. 033077
Van Lerberghe, Anja 003492
Van Leverink, J. 032642
Van Liebergen, Dr. L.C.B.M. . . . 032933
Van Lier, F.J. 032124
Van Lieshout, Jan 032053
Van Lingen, C. 032807
Van Loo, Rene 003688
Van Loon, V. 032001
Van Looy, Walter 003967
Van Maarseveen, Dr. M.P. 032035
Van Marion, A.J.H. 032952
Van Meerbeeck, Patrick 003761
Van Meerten, M.M. 033102, 033104
Van Melsen, Frances 038707
Van Mensch, Dr. 032485
Van Mol, Bruno 003881
Van Mourik, Justine 001318
Van Mullekom, Elisabeth 001155
Van Ness, Sally D. 048921
Van Nieuwenhuijzen, M. 031986
Van Nieuwenhyse, Ann 003635
Van Noort, Dr. S. 038686
Van Nüss, Berndt 022665
Van Oijen, M.J.P. 032582
Van Oord, Aaike 032181
Van Oordt, H.W. 031907
Van Opstal, Piet 003918
Van Peteghem, Dr. Sylvia 003643
Van Pottelberghe, Michel 004132
Van Praët, Michel 014659
Van Putten, J.H.J. 040506
Van Raalte, Christopher 053003
Van Raalte, Ronald C. 052681
Van Raay, Stefan 043945
Van Raemdonck, Mieke 046342
Van Ravensteyn, Fr. 003610
Van Regteren, R. 032692
Van Remoortel, P. 003391
Van Rensburg, F.S. 006629
Van Rhyn, Jaqueline 052088

Van Rijn, Gert-Jan 032181
Van Rijnberk, Jan Willem 032941
Van Rijnswall, Dr. M. 032619
Van Rijswick, Dr. A. 032913
Van Roosmalen, Arno 032200
Van Rosmeilen 031853
Van Roy, J.P. 003451
Van Royen, Dr. P.L. 033118
Van Schaik, Prof. C. 042216
Van Schalkwyk, Frieda 038854
Van Schalkwyk, Thea 038854
Van Schie, A. 032945
Van Seagraues 049776
Van Shields 050547
Van Shields, W. 052625
Van Sinderen, Wim 032170
Van Soest, Dr. R.W.M. 032002
Van Sprang, Sabine 003459
Van Straaten, Dr. E.J. 032739
Van Stuivenberg, F. 032309
Van Suchtelen, A. 032180
Van 't Klooster, Dr. Aguita 032575
Van 't Veen, Dr. Cees 032575
Van Tichelen, Fernand 003310
Van Tiggelen, R. 003450
Van Tilburg, A. 032970
Van Tilburg, B. 032754
Van Tilburg, C. 032434
Van Tol, J. 032582
Van Tooren, M.M. 032159
Van Tuyl, Prof. Gijs 031986
Van Tuyll, Carel 014749
Van Twist, Kees 032378
Van Valkenhoef, Drs. L. 032739
Van Veen, J.W. 032383, 032184
Van Vilsteren, Dr. V.T. 032035
Van Vleet, Sue 052749
Van Vlijmen, Dr. P.M.L. 032952
Van Vollenharen, A.C. 038828
Van Volsem, Cindy 006671
Van Voorhis, Joan 046840
Van Voorst Van Beest, R. 032399
Van Waterschoot, J. 031958
Van Wesemael, Willy 004058
Van Westerop, W.C.J. 032860
Van Wetten, Katja 022057
Van Wichelen, Peter 003360
Van Wijk, M. 032926
Van Zante, Gary 047368
Van Zijl, J.J.R. 038658
Van Zyl, S. 038808
Van Zÿverden, J. 032808
Váña, R. 009452
Vanackere, Luc 003608
Vanadzigs, A. 030111
Vanagiene, Birute 030435
Vanausdall, John 049681
Vanautgaerden, Alexandre 003433,
                              003447
Vance, Alex D. 051208
Vanco, John L. 048596
Vančugov, V.P. 043278
Vandamme, Heidi 031996
Vandaule, Marianne 003506
Vande Wetering, Joe 047354
Vandecan, Suzanne 003904
Vandekerchove, Veronique 003787
Vandelannoote, Jean-André 015366
Vanden Berghe, E. 003418
Vanden Bout, Veronica 053071
Vanden Broeck, Bavo 003559
Vandenberghe, Stéphane 003411
Vandenbulcke, Anne .. 003436, 003448
Vandenven, Marc 003313
Vandepitte, Dr. Francisca 003457,
                              003459
Vandeput, Ann 003679, 003683
Vanderbeck, LeeAnne 052459
Vanderhenst, Clara 003297
Vandermaesen, M. 003423
Vanderpijpen, Willy 003435
Vanderschooten, Alain 015630
Vandersteen, Alain 003645
Vanderstraeten, Carine 004073

Vandervellen, Pascale . 003474, 003492
Vanderway, Richard 046895
Vanderwerff, Laverne 046544
Vandever, William 048882
Vandevivere, I. 003831
Vándor, Dr. László 024051, 024052
Vandulek, Márta 023500
Vandura, Djuro 009050
VanDyke, Trudi 046397
Vanegas López, Walberto 008695
Vanerck, Patricia 003515
Vanhaelen, M. 003475
Vanhalewijn, Rudy 003750
Vanhemelryck, Kris 003354
Vanheusden, Margot 003306
Vanja, Prof. Dr. Christina 018930
Vanja, Prof. Dr. Konrad 017237
Vanlatum, Anne 015698
Vanloocke, Danny 003922
Vannini, Prof. Marco 026287
Vannoni, Filippo 026271
Vanns, Michael 043773, 044661
Vannson, Françoise 011613
Vannurden, James 050297
Váňová, Kristina 009889
Váňová, Dr. Libuše 009442
Vanreuerm-Rossetto, Roxane . . . 012846
Van't Veen, Dr. Cees 032574
Vantrimpont, Jean-Denis 003770
Vaoteté, Joseph 016374, 016379
Vaoteté, Joseph 016381
Varadinis, Mirjam 042207
Varaksina, Tamara Ivanovna . . 036884
Varaldo, Prof. Carlo .. 025267, 025268,
                     025270, 025592, 025593
Varani, Alberto 025413
Vardanjan, Irina Valerjevna 037600
Vardy, Anton 001479
Vare, Talis 010608
Varela, Katia 009200
Varela, Rudi 000130
Varenius, Linnéa 040179
Varga, Dr. Kálmán 023535
Varga, Tamás 023889
Varga, Vincent J. 006129
Vargas Cuervo, Roberto 008735
Vargas de Martínez, Evelia 008739
Vargas Guadarrama, Dr. Luis
Alberto 031265
Vargas, Heriberto 008592
Vargas, María Cristina 008592
Vargas, Mercedes L. 034569
Vargha, Mihály 036692, 036694
Vargoz, Sylvine 014710
Vargyas, Herbert 002611
Varnam, Emma 043446
Vârnav, Prof. Remus . 036581, 036582
Varno, Edward 047398
Varrak, Maruta 010687
Varsami, George 036402
Varsányi, Dr. András 020551
Vartiainen, Ulla 011164
Váry, Dr. László 023563
Vasconcelos, Maria João 035929,
                              035932
Vaseva Dimeska, Viktorija 030635
Vasher-Dean, April 051260
Vasić, Dr. Voislav 038179
Vasicek, Werner 001843
Vasile, Ioan Romulus 036498
Vasile, Prof. Radu 036376
Vasileva, Dineta 005226
Vasileva, Ljudmila 037446
Vasileva, Rumjana 005188
Vasilevič, Georgij N. 037681
Vasiliauskas, Ernestas 030340
Vasiliu, Lucian 036491, 036503, 036504,
             036505, 036506, 036507, 036508,
             036509, 036514, 036570
Vasiliu, Magdalena 036212
Vasilui, Lucian 036515, 036518, 036520,
                              036569
Vaslin, Pierre 015587
Vassallo de Cettour, Cristina . . . 000315

Vassallo, Dr. Adriana 027611
Vasseur, Bernard 015227
Vasseur, Dominique 053471
Vasseur, Marie-Madeleine 013549
Vassiliadis, Dimitrios 023136
Vasström, Annette 010208
Vasudeva Bhatta, K. 024152
Vašut, Karel 009944
Vates-Roth, Brigitte 020852
Vatksy, Sharon 048769
Vattulainen, Päivi 011094, 011496
Vatutin, Konstantin Ivanovič 037582
Vaucher, Laurence 041765
Vaughan, Ellen 047581
Vaughan, Dr. Gerard 001279
Vaughan, James 054153
Vaughan, Robert 046673
Vaughan, Roger 043464, 044019
Vaughn, Ann L. 051080
Vaughn, Greg 048111
Vauth, Henry 018032
Vautrin, Germain 012221, 012222
Vavilova, Anna Pavlovna 037660
Vavilova, N.I. 037655
Vavilova, Natalja Veniaminovna . 037413
Vavra, Jan 054383
Vázquez Agraz, Joaquín 030893
Vázquez Otero, Milagros 017207
Vázquez Ramírez, Nelson 030831
Vazquez Zuleta, Sixto 000387
Vázquez, Lucía 039708
Vazzana, Dr. Angelo 027498
Vdovin, Dr. Gennadij Viktorovič . 037352
Vecarić, Nenad 008893
Vecchio, Chris 052076
Vecchio, Marji 047637
Večerková, Dr. Eva 009468
Vedaa, Lisa 051045
Vedernikova, Galina I. 037387
Vedjaeva, Natalja Anatoljevna .. 037628
Védrine, Laurent 015322
Védrine, Mireille 012419
Vedsted, Jakob 010037
Veen, Dr. Cees van 't 032570
Veening, Hans 051639
Veenvliet, J. 032213
Veerabhadra Rao, K. 024268
Veerman, Hein 033001
Vees, L. 019121
Vega, Carmen L. 036164
Vegelin Van Claerbergen, Dr.
Ernst 044951
Vegesack, Alexander von 022472
Vegesack, Baron Henrik von ... 041035
Vehma, M. 021001
Vehniä, Harri 011057, 011058
Vehns, Johannes 019034
Veisz, Ottó 023770
Veit, Johann 001691
Veit, Walter 020396
Veit, Prof. Dr. Willibald 017239
Veitch Alexander, Mary 050815
Veith, Karl-Heinrich 022349
Vejle, Jan 010031, 010032, 010033
Vekemans, B. 032097
Vela de la Torre, Antonio 040097
Velado Craña, Bernardo 039009
Velani, Dr. Livia 025353
Velaochaga Rey, Irene 034332
Velasquez Alquizaleth, Julio
Cesar 004221
Velasquez, Dr. Geraldine 050101
Vélasquez, Dr. M. Manuel 035530
Velázquez Martínez del Campo,
Roxana 031251
Velázquez, Agustín 039749
Velázquez, Teresa 040347
Velder, Linda 051523
Veldhuÿzen, H. A. 032369
Velers, J.B.A. 032714
Veleva, Snejana 005260
Vélez, Jesús Hernán 008708
Velez, Pedro 047701
Vélez, Pilar 039096

Velfl, Dr. Josef 009852
Velghe, Brita 003449, 003460
Velghe, Elviera 003411
Velhagen, Dr. Rudolf 041255
Velička, Gintautas 030336
Velikanova, Olga Alekseevna ... 037794
Velikova, Elena 005151
Velikovskaja, Galina Viktorovna . 037418
Velilla Moreno, Pilar 008714
Vella, Antonino 046034
Vella, John J. 030741
Vellacott, Robert B. 001026
Vellekoop, Marije 031996
Veller, Prof. M. 038747
Velluto, Giovanna 028217
Velnic, Melchisedec 036637
Veloccia Rinaldi, Dr. Maria Luisa 026333,
028091, 028092
Veloso, Abraã9 036080
Veltheim, M. von 019185
Veltzke, Dr. Veit 020419, 022591
Venable, Charles L. 047808
Venborg Pedersen, Dr. Mikkel .. 010209
Venderby, Lone 010123
Vendeuvre, Elyane de 016153
Vendeuvre, Elyane de 016152
Venizelos, Eleftherios 023017
Venkataraman, R. 024390
Venkatesh Jois, Dr. K.G. 024343
Venner, Peta 045462
Ventisette, Renzo 026568
Vento, Minna 010771
Ventosa, Silvia 039106
Ventsch, Warren 053933
Venturo Rubino, Donata 025298
Vera, Gina C. de 034438
Vera, Teresa N. de 034499
Verastegui, Maura 039708
Verbanck Piérard, Annie 003890
Verbeeck, Luc 003379
Verbrugge, Vincent 004146
Verbrugghen, Johanna M. 051924
Vercauteren, Drs. R.A.M. 032983
Verckens, Jozef 003657
Verdaasdonk, Dr. Kees 033116
Verdaasdonk, P. 032179
Verderame, Dr. Lori 046411
Verdi, Prof. Richard 043612
Verdú Alonso, María Rosa 039371
Verdugo Santos, Javier 040141
Veremeeva, Irina Lukivna 043214
Veres, Dr. László 023783
Vereščagina, Tatjana Fëdorovna 038003
Verey, A.P. 046165
Vergara Rodríguez, Ancízar ... 008568
Vergara Torres, David 007548
Vergara, Alejandro 039676
Verge, Laurie 047837
Verger, Sylvestre 014750
Vergho, Dr. Helge ... 017020, 019615
Vergne, Philippe 050948
Vergnetta, Angela 027874
Vergun, Neonila Rodionivna ... 043254
Verhaagh, Dr. Manfred 019518
Verhaard, A.J.G. 032715
Verheggen, Mariëtte 032225
Verheyen, Ivo 003873
Verheyen, Roland 019643
Verhoeff, R.N.S. 032402
Verhoeven, Carolina 032014
Verhoeven, P.J.C. 032026
Verhofstadt, Dr. Karl 003691
Verhulst, Sabine 004046
Verkest, Idesbalo 003260
Verlinden, Frédérique 013078
Verlingue, Bernard Jules 015019
Vermeulen, Drs. C.M.A. 031923
Vermeulen, Richard 003337
Vermeulen, S. 038660
Vermeulen, Drs. W. 032739
Vermillion, Linda Marie 054556
Verna, Gaëtane 006181
Verner Hamilton, Elizabeth ... 047547
Verneuil, Yves 016050

Verney, Antoine 011919
Vernon, C.J. 038709
Vernon, G.N. 038709, 038710
Vernou, Christian 012782
Verone, Amy 051877
Verrando, Joyce 049399
Verrill, John H. 051770
Verroen, Dr. Th.L.J. 032240
Verschelden, G. 004055
Verschoor, J.C. 031920
Verschoor, M. Lynn 047243
Verschuur, P. 032277
Versluis, P. 032027
Verspeazt, V. 003435
Verster, J.L. 032797
Vertenten, A. 003572
Verthamon, Henri de 011638
Verwey, Drs. A.D.M. 031959
Verwiebe, Dr. Birgit 017068
Verževikina, Marina 037925
Verzosa, Fe Angela 034529
Verzuh, Valerie 053114
Veskovic, Esad 004283
Vesselinova, Temenuschka ... 025260
Vest, Ladonna 047532
Vest, Windsor 053371
Vestergaard, Ole Schou 010087
Vestergaard, Rasmus 010309
Vestermanis, Margers 030210
Vetter, J.W.M. 032291
Vetter, Manfred 020761
Vetter, Maureen 050298
Vetter, Remo 042050
Veyradier, Henri 013678
Vezzosi, Alessandro 028399
Via, Marie 052618
Vial, Marie-Paule 014040, 014053
Vialla, Andre 013326
Vialla, Gérard 015015
Vialle, Christian 016272
Viana, Dr. Maria Teresa 035952
Viane, Françoise 003349
Vianin, Georges et Claire 042183
Viar, Javier 039139
Vibe, Ellen S. de 033828
Vice, Mary L. 052867
Vicini, Dr. Donata 027231, 027235,
027237
Vicini, Maria Lucrezia 027569
Vicino, Giuseppe 026237
Vick, Daniel 049652
Vickers, Irene 044852
Vicq Carvalho, Beatriz de ... 004747
Victor, S. 038770
Victor, Stephanie 038771
Vida, Štefan 038417
Vidal de Alba, Beatriz 031246
Vidal Huguet, Carme 039584
Vidal Inglès, Anton 039394
Vidal, Juan Teodoro 039647
Vidal, Nélida 031177
Vidal, Pierre 014625
Vidal, Silvia M. 054760
Vidali, Dr. Gloria 028313
Vidarte, Juan Ignacio 039144
Vidović, Prof. Josip 008891
Vidulich, Tullio 028175
Viehmeier, Klaus 022652
Viehöfer, Dr. Erich 020131
Vienet, José 036117
Vienet, Bernard 013174
Viennct, Christine 012021
Vienonen, Inga 011208
Viera, Ricardo 046961
Vierck, Sigrid 022413
Vièrille, Dominique 014154
Vierk, Viola 019020
Vierlein, Ralf 020511
Vierny, Dina 014770
Vierock, Ronald 017238
Vietti, Paola 027515
Vieste, Arwed 022888
Vieux, Jacky 013119
Viéville, Dominique 014682

Viflanueva, Rudy 034458
Vigarani, Guido 026861
Vigié-Chevalier, Béatrice 014056
Vigilante, Amy 049007
Vigna Taglianti, Prof. Augusto . 027617
Vigorelli, Prof. Giancarlo 026817
Vigreux, Jean 015244
Viguier, Florence 014226
Vihovde, Anne Brit 033530
Viinikainen, Kari 011109
Vikan, Dr. Gary 046772
Vikki, Lily 033871
Vikström, Lena 040848
Viktorova, Jelena 030177
Vikulova, V.P. 037257
Vila Álvarez, Nuria 039438
Vilà Noguera, Francesc 039717
Vilà Planas, Xavier 040396
Vila, Dominique 012508, 012509
Vila, Emma R. 000135
Vilać, Pavica 008904
Vilademunt, Josep 040396
Vilain, André 003781
Vilaplana, Susana 040308
Vilas, Felipe Arias 039254
Vílchez Vílchez, Carlos 039465
Viler, Darko 038516
Viletová, Dagmar .. 009972, 009973
Vilhu, Raimo 010786
Viljoen, H. 038849
Viljoen, L. 038600
Vilks, Dr. Ilgonis 030181, 030182
Vilkuna, Janne 010936, 010937
Villa-Lacson, Angela de 034504, 034505
Villa Santa, Alberto 027808
Villa Santa, Emanuele F. 027808
Villa Torá, Geronimo 039282
Villa, Azucena . 034593, 034594, 034606
Villadier, Francis 014153
Villagran Valdespino, Manuel
Angel 031354
Villain, Agnio 015526
Villalon, Michel 012680
Villamayor Coto, José 031209
Villamizar Duarte, Eduardo Adolfo 008744
Villamizar Lamus, Eduardo ... 008745
Villanueva, Gustavo M. 033380
Villarroya i Font, Joan 039038
Ville, Roy V. de 046379
Villela, Milu 004881
Villeneove, Daniel 006191
Villeon, H. de la 013752
Villiers, Jaline de 038795
Villiger, Dr. Verena 041500
Vilman, Vladimir 038562
Vilsker, Bella 047231
Vilsmeier, Cäcilie 020774
Vilus, Ljiljana 009028
Vilutienė, Aldona 030418
Vimpari, Maritta 011171
Vimpère, Jean 013377
Vincent, Ariel 015607
Vincent, Danielle ... 012852
Vincent, François ... 006221
Vincent, Glenda 053596
Vincent, Hélène ... 013181
Vincent, Karie C. .. 051224
Vincent, Patricia ... 051871
Vincentelli, Moira 043371, 043373
Vincenti, Dr. Patrick 049410
Vincenzi, Marco 026149
Vinduška, Jan 009767
Viner, David 043960
Vingtain, Dominique .. 011838
Vinje, Tor 033789
Vinke, Albert 021393
Vinogradov, Z. 025031
Vinogradova, Vira Illivna ... 043212
Vinsen, Bob 033255
Vinyet Estebanell, Josep .. 039589
Vinzens, Brigitte ... 042163
Vio, Ettore 028311
Virasoro, Carlos Alberto .. 000621
Virgili, Bianca ... 025767

Virgin, Louise	054548	Vogel, Wilhelm Dieter	018450
Virieu, A. de	013002	Vogelaar, Dr. C.J.J.	032586
Virole, Agnès	015340	Vogelbacher, Martin	022160
Virtanen, Risto	011220	Vogeler, Dr. Hildegard	020159, 020163
Vis-Best, A.	032654	Vogels, J.J.	032365, 032366
Visbach, Alida	006721	Vogelsang, Harald	017434
Vischer, Dr. Theodora	041775	Vogelsang, W.J.	032585
Visentin, Prof. Dr. Mario	026039	Vogelstein, John L.	051412
Visini, Patrick	016075	Vogl, Elisabeth	022084
Viskari, Leena	011113	Vogler-Zimmerli, Brigitta	041318
Višnevskaja, Julija	037472	Vogler, Franz	002958
Vispi, Maria	026493	Vogt, Barbara	030306
Visschedijk, Dr. R.A.M.	032818	Vogt, Dr. Christine	020896
Visser, Dr. Anna Maria	026203	Vogt, Daniel	041285
Visser, H.	031986	Vogt, George L.	054437
Visser, Johann	020822	Vogt, Gisela	018131
Visser, Dr. S. H.	032044	Vogt, Dr. Renate	017503
Visser, Susan R.	053394	Vogt, Sharon L.	051048
Viswanadham, M.	024480	Vogt, Wolfgang	020412
Vít-Trčka, Jan	009862	Vogüé, de	012639
Vitáček, Zdeněk	009499, 009538	Vogüé, Comte B. de	014556
Vitagliano, Carol Edith	000240	Vogüé, Patrice de	013972
Vital, Christophe	015852	Voicu, Mariana	036623
Vital, M.	013016, 014216	Voigt Andersen, Mogens	023330
Vitali, Christoph	041879	Voigt, Dr. Kirsten	019517
Vitali, Dr. Samuel	041318	Voigt, Sigrun	019305
Vitali, Tiziana	026966	Voigt, Stefan	017981
Vitek, Dr. Gerhard	003036	Voigt, Dr. Wolfgang	018411
Viti, Paolo	028321	Voiron, Henri	011577
Vitie, Heikki	010851	Voisin-Thiberge, Claire	014421
Vitković Zikić, Milena	038168	Voisin, Bernard	012546
Vitmayer, Janet	045000	Voithofer, Waltraud	001791
Vitt, Dr. L.J.	051588	Vojtenko, Vladimir Georgievič	037564
Vitte, Bruno	014937	Vokotopoulou, Dr. Julia	023062
Vittereli, Margo	049540	Volker, Joye	000937
Vittorelli, Dr. Pietro	025807	Volkert, James	051448
Vitulskis, Evaldas	030350, 030356, 030357	Volkmann, Jürgen	018834
Vivanco Vega, Manuel	008624	Volkmann, Swantje	022258
Viver, F. Xavier	040253	Volkov, Sergej Pavlovič	037559
Vivies, Bertrand de	013072, 013073, 013074	Volkova, Valentina Borisovna	037094
Vivirito, Jackie	047471	Vollbrecht, Dr. Jürgen	016977
Vizi, Ondrej	031767	Vollert, Adalbert	018425
Vizzini, Salvatore	025704	Vollgraaff, Helene	038685
Vjalikov, P.F.	037238	Vollmer, David	053719
Vjalych, Ljudmila Makarovna	037189	Vollmer, Jennifer	048459
Vješnic, Borko	008993	Vollmer, Stephen	053837
Vlack, Donald J.	051456	Vollnhals, Rudolf	022907
Vlad, Florin	036548, 036591	Volmar, Michael	049398
Vladimirescu Băcanu, Angela	036826	Voločkova, Olga Kuzminična	037669
Vladimirova, Galina A.	037806	Volosenkova, Nadežda	
Vladu, Ioana	036714	Evstafievna	037868, 037871, 037872
Vlaemynck, Joris	003551	Volosova, Galina Pavlovna	037828
Vlasáková, Beata	038460	Vološyna, Ljubov Vasylivna	043236
Vlasic, Valentina	019644	Volovenskaja, Ljudmila	
Vlasova, Tetjana Mykolaïvna	043233	Vladimirovna	037041
Vlcek, Dr. Tomas	009785	Volpe, Jorge	000163
Vliegenthart, Dr. A.W.	032011	Volskij, Svjatoslav O.	037578
Vlieghe, Prof. Dr. Hans	003282	Volstorf, Bryan	053741
Vlk, Antonin	009621	Volz, Günther	016721
Vlnas, Dr. Vit	009801, 009802	Volz, Robert L.	054418
Vlnas, Dr. Vit	009841	Volz, Ulrich	031919
Vlok, Rochelle	038875	Vomm, Dr. Wolfgang	017051, 017057
Vögele, Dr. Christoph	042005	Vomsattel, Julian	042112
Voegele, Günter	021189, 021203	von Kerssenbrock-Krosigk, Dr.	
Voeks, Gretchen L.	053841	Dedo	018046
Völcker-Janssen, Dr. Wilhelm	019767	von Pape, Dr. Cora	016928
Voélin-Dubey, Monique	041807	Vondras, Barbara	050258
Voelkel, David B.	009543	Vonhof, M.J.	032739
Völker, Angela	033033	Vonier, Mag. (FH) Johannes	001827
Völker, Heike	017086	Vonk-Hennig, Daniel	031888
Völki, Rosemarie	021313	Vonscheidt, Dr. Wolfgang	017660
Voelkle, William M.	051428	Vonville, Auguste	015428
Vörös, Dr. Gabriella	023926, 023928	Voogt, Leo	032185
Vörös, Dr. Gabriella	023931	Vookles, Laura L.	054577
Voets, Michael	018037	Voordendaf, A.G.	032909
Vötter-Dankl, Susanna	002438	Voorheis, Lynn	052596
Vogeding, Dr. Ralf	022090	von Voorst tot Voorst	032494
Vogel, Gail	050771	Vorachek, Pamela	052887
Vogel, Prof. Klaus	017978	Vorano, Tullio	008945
Vogel, Dr. Michael	017040	Vorel, Otakar	009607
Vogel, Dr. Sabine	020726	Vorländer, Dr. Hermann	020697
Vogel, Tanja	020255	Vormaier, Andreas	003089
		Vorobjova, Solomija	
		Oleksandrivna	043250

Voropanov, Vladimir Valentinovič	038057	Wachter, Dr. Markus	002502, 002503
Vorres, George	023206	Wachtmeister, Ingegerd	040888
Vorres, Ian	023206	Wackerfuß, Winfried	018366
Vorsteher, Dr. Dieter	017299	Wacs, Darin	051904
Vort-Ronald, Michael P.	001172	Wada, Kunibo	029434
Vorwig, Dr. Carsten	020323	Wada, Norio	029020
Vos, H.N.	038838	Wadas, John W.	052965, 052974, 052981, 052986
Vos, K.	032585	Wade Thomas, Randall	048972
Vos, Margaret	052760	Wade, Blake	049427, 049524
Vosa, Dace	030185	Wade, Don	053606
Voss, Dr. Björn	020602	Wade, Hilary	043874, 043875
Voss, Kurt	049039	Wadensten, Hillevi	041106
Voß, Dr. Rolf	020683	Wadsworth, David H.	047859, 047860
Voss, Willem J.	048902	Wadum, Jorgen	010195
Vosseler, Peter	022078	Wäckerlin, Martin	041799
Vossenkuhl, Katharina	020567	Wägele, Prof. Dr. Wolfgang	017504
Vostrikov, Vladimir Ivanovič	037686	Wäldchen, Günter	021907
Votaw, John F.	054364	Wälder, Manfred	020064
Votroubková, Dr. Iva	009515	Wærdahl, Tove	033680
Vouk, Dr. Morana	009052	Wäre, Dr. Ritva	010865
Voute, Dr. A.M.	032957	Waern, Karl Erik	040520
Vovk Čepič, Taja	038548	Wäspe, Roland	041923
Voyadzis, G.	023016	Wagemakers, C.	031937
Voyame, Christian	041320	Wagemakers, T.	032929
Voyles, Robert S.	048806	Wageman, Patty	032378
Voza, Dr. Giuseppe	005434, 027968	Wagener, Danièle	030592, 030596, 030597
Voznickyj, Boris Grygorovyč	043240	Wagesreiter, Viktoria	003095
Voznjuk, Dr. Volodymyr		Waggoner, Lynda	050895
Oksentijovyč	043127	Wagini, Dr. Susanne	020576
Vrabec, Vesna	009043	Wagner-Autesserre, Pierre	013579
Vračević, Marica	038194	Wagner-Gnan, Angela	020883
Vraesos, Alexis	011942	Wagner, Connie	052706
Vrat, Prof. Prem	024462	Wagner, Dieter	018020
Vratislavskỳ, Ivo	009895	Wagner, Ethel Mae	049801
Vrba, Elisabeth	051265	Wagner, Gerhard A.	017548
Vrbanus, Milan	008951	Wagner, Günther	021137
Vrčic, Vjeko	008922	Wagner, Prof. Herbert	001905
Vrebos, M.	003436, 003448	Wagner, Hermann	001884
Vrellis, Paul P.	023093	Wagner, Hilke	017559
Vribe, Francesc	039077	Wagner, Dr. Jens-Christian	020828
Vriend, Anita	031996	Wagner, Kay	052961
Vries, D.P. de	032576	Wagner, Lynn	049919
Vries, J. de	032236	Wagner, Prof. Marek	034953
Vries, Janekke de	017588	Wagner, Nora	020587
Vries, Dr. M.E.A. de	031894	Wagner, Patrice	046932
Vries, P.A. de	032004	Wagner, Renate	019270
Vrieze, John	031919	Wagner, René	021257
Vroede, Erik de	003705	Wagner, Rita	019693
Vrousos, Alexis	013985, 015368	Wagner, Ruth	053298
Vrtiska, Floyd	053639	Wagner, Sandra	017513
Vrugteman, Gerda	032378	Wagner, Dr. Siegfried	020646, 020647
Vuadens, G.	042097	Wagner, Thomas K.	048464
Vuille, Roger	041904	Wagner, Tracy	052121
Vuilloud, R.	041750	Wagner, Dr. Velten	018211
Vujčić, Davorin	008927	Wagner, Vivian	047750
Vujičić, Jasminka	009028	Wagrodzki, Jerzy	035373
Vujnović, Andrej	038144	Wagstaff, Sheena	045115, 045116
Vuk, Vili	038572	Wagstaffe, Stephen	044299
Vukadin, Hela	009043	Wahdan, Tracey	045737
Vukotić, Biljana	038168	Wahl-Pletscher, Emil	041371
Vuković, Prof. Radovan	009053	Wahl, Hans-Rudolf	042178
Vukovich, Franz	002367	Wahl, Johann	002895
Vulf, Galina Gerasimovna	037173, 037174, 037175	Wahl, Sonja	047785
Vuori, Elina	011090, 011095	Wahler, Marc-Olivier	014794
Vuorinen, Pirjo	010932, 010937	Wahlers, Jack	046945
Vusoniwailala, Kate	010748, 033216	Wåhlin, Sidsel	010123
Vutela, Dr. Anneli	010819	Wahlöö, Claes	040809
Vuyk, Dr. K.	031992	Wahlquist, Håkan	040998
Vydra, Dr. František	009525	Wahlroos, Tuija	010772
Vydrina, Ljudmila Sergeevna	037783	Waiboer, Adriaan	024900
Vykoukal, Dr. Jiří	009516	Wain, Lisa	001005
Vykydal, Vlastimil	009482	Waintraub, Aron	025033
Vyntu-Terycanu, Dr. Olena		Wainwright, Paige	047007
Dmytrivna	043130	Waisbrot, Ann	050719
Vyskocil, Hubert	003075	Waite, Dawn D.	047468
Vysockij, Nikita Vladimirovič	037317	Waites, Robin	047911
Waage, Einy	033810	Wajima, Kunio	028706
Waal, M. de	038660	Wakae, Haruko	029739
Wachna, Pamela	007188	Wakamatsu, Motoi	029639
Wachs, Diane C.	050243	Wake, William	044183
Wachsmund, Ulrike	018199	Wakeling, Melissa	005402
Wachten, Dr. Johannes	018430	Wakili, Ousmanou-Mouhamadou	007507
Wachter, Fritz	041870		

Wakita, Osamu 029262
Wakitani, Masami 029454
Wakiyama, Yoshitaka 028593
Walaszczyk, Anna 035461
Walaszczyk, Antoni 034783
Walburg, Dr. Reinhold 018422
Walch, Christof 002348
Walch, Heribert 001865
Walch, Timoth 054291
Wald, Albrecht 018369
Wald, Jane 046448
Waldbott-Bassenheim, Baronin
   Marietheres 002045
Walde, Prof. Dr. Elisabeth 001822,
   002278
Walden, Barbara B. 049995
Walden, Ian N. 044146
Walder, Emma 046200
Walder, Hans Kaspar 042135
Waldherr, Johann 002499
Waldherr, Kaspar 020037
Waldkirch, Bernhard von 042207
Waldman, Carolyn M. 052065
Waldmann-Glaser, Dr. Erich . . . 002805
Waldmann, Burta 042062
Waldmeier, Kurt 041448
Waldner, Daniel 041643
Waldren, Dr. William 039342
Waldron, Richard 052027
Walek, Karl 002043
Walentowicz, T. 043918
Walgamott, Libby 052498
Walgenbach, H. 032804
Walgrave, René 003785
Waligurski, Carolyn M. 049638
Waligurski, Stephen S. 049638
Walinga, Ingeborg 032376
Walk, Stephen G.W. 051540
Walkensteiner, Wolfgang 002191
Walker-Coultice, Lauren 007365
Walker Mecky, Debra 049271
Walker, Alexandra 045565
Walker, Anita 048288
Walker, Barry 049588
Walker, Carol 045383
Walker, Catherine 050951
Walker, Celia 051174
Walker, Christopher 043788
Walker, Daniel 051424
Walker, David A. 044425
Walker, Dean 052081
Walker, Diane 045319
Walker, Edwin G. 048200
Walker, Eleanor 006545
Walker, Gavin 045646
Walker, Gerald 052701
Walker, Gwendolyn 050173
Walker, Helen 045075
Walker, John 046224
Walker, June 049977
Walker, Kate 001603
Walker, Lulen 054117
Walker, Neil 045411
Walker, Nicola 045219
Walker, Pam 054509
Walker, Peter 043649
Walker, Rebecca 051909
Walker, Rob 005378
Walker, Sam 044918
Walker, Sonia 046714
Walker, Stephen 045811
Walker, Stephen D. 054358
Walker, Steward 047106
Walker, Tim 033223
Walker, Tom 053275
Walker, William 050624
Wall, Dora 006244
Wall, Gordon 048465
Wall, Ian 046074
Wall, Kathleen 005932, 005933
Wall, Dr. Michael 052980
Wallace, A. 007396
Wallace, Amanda 045208

Wallace, Anaret 038659
Wallace, Brian 051309, 052062
Wallace, George 051637
Wallace, Hilt 007359
Wallace, I. 044869
Wallace, J. Don 050876
Wallace, James A. 050788
Wallace, Lynda 033132
Wallace, Meghan 046456
Wallace, Mike 050927, 051994
Wallace, Dr. Patrick F. . 024845, 024897,
   024898, 024899
Wallace, Teresa 047952
Wallace, William D. . . 054547, 054549
Wallach, Nancy 049902
Wallbrecht, Dr. Andreas 017628
Wallcraft, Wilma K. 006138
Waller, Bret 049680
Waller, Josip 008951
Waller, Richard 052551
Waller, Susan R. 052904
Wallers-Bulot, Danielle 012584
Wallet, Marie 012315, 012316
Walley, G.P. 045416
Walling, Mary 051412
Wallis, Andrew 044987
Wallis, Brian 051406
Wallis, Jean 049478
Wallis, Jonathan 044082
Wallis, Simon 044939
Wallner Bärbulescu, Luminiţa . . 036541
Wallner-Bärbulescu, Luminiţa . . 036543
Wallner, Dr. Thomas 042004
Wallon, Jean-Marie 003844
Walloschek, Arnold 021607
Walls, Nancy 049852
Walmsley, Frances 006444
Walrath, Gary 052142
Walseman, Laurie 048069
Walser, Stefan 042087
Walseth, J. 010199
Walsh-Piper, Kathleen 050252
Walsh, Alice 052384
Walsh, Brian 024917
Walsh, Debbie 024854
Walsh, John 001528
Walsh, Larry 024957
Walsh, Mary 051073
Walsh, Nigel 044779
Walsh, Noa Martin 024849
Walt, Barbara 041942
Waltenspül, Walter 041213
Walter-Brown, Nancy F. 050950
Walter, A. 018142
Walter, Alfred 017910
Walter, Bakk. phil. Barbara 002637
Walter, Dr. D. 022501
Walter, Ernst-Ulrich 018723
Walter, Friederike C. 019522
Walter, H. 021957
Walter, Dr. Karin 022660
Walter, Katharina 002141
Walter, Michael 021879
Walter, Patricia 046881
Walter, Philippe 014315
Walter, Susanne 020759
Walter, Prof. Dr. Ulrich 018354
Waltereit, Manfred 018173
Waltermire, Joan 051654
Walters, Andra 054618
Walters, Cheryl 053736
Walters, Cliff 043656
Walters, Janet 007221
Walters, John A. 053512
Walters, Margaret 054403
Walters, Nerina 038775
Walterskirchen, Prof. Dr. Gerhard 002626
Walther, Alois 017998
Walther, Andreas 017735
Walther, Gerd 018549
Walther, Hubert 016354
Walther, Jim 046366
Walton, Ann 048883
Walton, John 001050

Walton, Julia 043604
Walton, Karin . 043747, 043750, 043753
Walton, Les 045570
Waltoś, Prof. Dr. Stanisław 034905
Waltz, James 047420
Walujo, Dr. Ekoparoto 024545
Walz, Dr. Alfred 017558
Walz, Carol 051090
Walz, Glenn 007068
Walz, Jack 051090
Walzer, Herwig 002418
Wamelink, Garrit 047814
Wamers, Prof. Dr. Egon 018405
Wamser, G. 022108
Wamser, Prof. Dr. Ludwig 018874,
   020504
Wamulungwe, Patrick Lisina . . . 054837
Wan Ahmad, Wan Ali Bin 030674
Wan, Jamaluddin 030688
Wan, Norhiyati Ibrahim 030689
Wan, Robert 016377
Wan, Yusoff 030688
Wanachiwanawin, Prof. Darawan 042673
Wanamaker, Mark 050432
Wand-Seyer, Dr. Gabriele 019217,
   019220
Wander, Evelyne 015276
Wandl, Monika 002958
Wandrei, Ellen A. 046872
Wandschneider, Dr. Andrea . . . 021058,
   021064, 021067, 021068
Wang, Bingsen 007935
Wang, Bingwan 008351
Wang, Bohua 008181
Wang, Chang'an 007855
Wang, Chen 007726
Wang, Chongshun 007897
Wang, Conghui 007895
Wang, Cunxiao 007975
Wang, Desong 008361
Wang, Dezhi 008002
Wang, Fengqing 008307
Wang, Fengxian 007911
Wang, Guojie 008438
Wang, Guoquan 008340
Wang, Hebing 008425
Wang, Huaqing 008187
Wang, Jiaheng 008135
Wang, Jianhua 007932, 007993
Wang, Jinbao 008232
Wang, Jinhou 008258
Wang, Jinxiang 008284
Wang, Juguo 008541
Wang, Kuiyi 007663
Wang, Kwei-Ian 042566
Wang, Lanzhou 008163, 008170
Wang, Lianlan 008306
Wang, Prof. Lianmao 008197
Wang, Libin 007908, 008169
Wang, Linchang 008033
Wang, Luheng 008269
Wang, Pengfei 007727
Wang, Qi 007846
Wang, Qihui 007648
Wang, Qingshi 007728
Wang, Qinling 007671
Wang, Ragnhild 033866
Wang, Runyuan 008302
Wang, Shaoming 007894
Wang, Shenzhang 007768
Wang, Shijie 008226
Wang, Shouyin 008133
Wang, Shujin 008417
Wang, Taiming 008515
Wang, Wei-Hsien 042285
Wang, Weiyin 008316
Wang, Wenjing 008526
Wang, Wenwu 008474
Wang, Xi Rong 008230
Wang, Xiangyu 008320
Wang, Xianyou 007930
Wang, Xiaoping 007857
Wang, Xinchang 008132
Wang, Xiping 008472

Wang, Xiu 008060
Wang, Yanliang 008322
Wang, Yanqing 008516
Wang, Yonggen 007919
Wang, Yufu 008015
Wang, Yuxian 008454
Wang, Yuzhen 008414
Wang, Zheng 008051
Wang, Zhenghua 008279
Wang, Zhijie 008433
Wang, Zhiping 007598
Wang, Zhiqiong 008478
Wang, Zhiwen 007832
Wang, Zongtao 008009
Wånge, Cecilia 040691
Waning, Alexander 032593
Wankel, James 005775
Wanner, Prof. Dr. E. 019821
Wanner, Monika 041606
Wanriek, Benedicta 021070
Wappenschmidt, Dr. Kathrin . . . 021467
Wappis, Erich 002192
Wappler, Dr. Friederike 017437, 017441
Warburton, Stuart 046044
Warchałowski, Walerian 034984
Ward Grubb, Barbara 052032
Ward Povse, Sandra 053217
Ward Shannon, Dr. George 046847
Ward, Alberta 046634
Ward, Barry J. 050048
Ward, Barry K. 048390
Ward, Danette 051585
Ward, Esmé 045219
Ward, James 052117
Ward, Kathy 046826
Ward, Ken 046072
Ward, Kimberley D. 048390
Ward, Dr. Lauck W. 050727
Ward, Mark 046727
Ward, Mike 043391
Ward, Nancy 049834
Ward, Prof. Paul 042243
Ward, Phil 048163
Ward, Ralph E. 049381
Ward, Robert 050997
Ward, S. 043547
Warda, Rebecca M. 047616
Warde, John St. Andrew 046099
Warden, Canon 043873
Warden, Donald 053334
Warden, P. 042626
Wardius, Janine 033523, 033524,
   033525
Wardlaw, Alivia 049588
Wardle, Michelle 049931
Wardropper, Ian 051424
Ware, Alan 049940
Ware, Brechelle 049425
Ware, Donna M. 048475
Ware, Dr. John A. 048368
Ware, Katherine 052081
Ware, Mark 053383
Ware, Robert 046363
Warfel, Stephen G. 049372
Warger, Julia M. 053551
Wargo, Richard 047080
Warhurst, Margaret 044869
Warigai, Tsutomu 028658
Waring, H.B.H. 045194
Warinsie Kankpeyeng, Benjamin 032629
Warkiewicz Laine, Christian K. . 047647
Warlitner, Crystal C. 051548
Warming, Dagmar 010220
Warne, Pam 038687
Warnecke, Jan-Christian 022047
Warner, Henry 044497
Warner, Jan 050330
Warner, Joanne 051368
Warner, Judith 053199
Warner, Malcolm 048904
Warner, R. 043564
Warnes, Dr. Carl E. 051114
Warnes, Robert 044478
Warnier, Valérie 003521

Warnke, Dr. Ursula 017606
Warp, Harold G. 050928
Warren, David B. 049588
Warren, Eleanor 047769
Warren, Elizabeth 051383
Warren, Emma 043929
Warren, George W. 050268
Warren, Harry 054377
Warren, Jack D. 054161
Warren, Sally 006641
Warsdale, Godfrey 045261
Warstler, Pasqua 048603
Warten, Roland 020871
Wartenberg, Dr. Ute 051328
Warth, Siegfried 020463
Warth, Werner 042140
Warther, David R. 048355
Wartke, Dr. Ralf-Bernhard 017315
Wartna, Fred 032829
Warton, Sue 045796
Warwick Thompson, Paul 051368
Warzecha, Newton M. 049156
Waschka, Adelheid 016883
Wasescha, Elisabeth 041943
Washbrook, Harry 005741
Washington Goyaud, Rolando . . 000391
Washington jr, W. Curtis 047561
Waskowic, Lorraine 006702
Wass, Glenys 045489
Wassermann, Fred 051412
Wassing, A. 032830
Wassmann, Harry 046915
Wassmer, Edouard 041503
Wastian, Mag. Josef 002703
Wastiau, B. 041533
Wat, Kathryn 046329
Watanabe, Hisashi 029077
Watanabe, Junko 028647
Watanabe, Kiriko 029639
Watanabe, Kousho 029320
Watanabe, Osamu 029466
Watanabe, Seiichiro 028471
Watanabe, Shinsuke 029536
Watanabe, Taeko 029030
Watanabe, Takao 028976
Watanabe, Yasuhiro 029510
Watari, Shizuko 029621
Watelet, Pierre 014417
Waterlow Oam, Nick 001391
Waterman, Amy E. 051381
Waters Yarsinske, Amy 051581
Waters, Dr. D. 045457
Waters, Davis 048404
Waters, Deborah D. 051443
Waters, Fergus 045374
Waters, Ruth D. 047773
Watkin, Jeffrey 044773
Watkins, Jonathan 043620
Watkins, Kathleen 045697
Watkins, Rod 050914
Watkins, Sarah 047137
Watne, Eva 034072
Watson-Armstrong, Francis 043482
Watson, Andy 047845
Watson, Bob 044447
Watson, Claudia 048172
Watson, Daryl 049015
Watson, Doreen 043553
Watson, Edwin W. 048953
Watson, Jamie 046203
Watson, John 001220, 001531
Watson, Julian 044986
Watson, Julie 045349
Watson, Kristy 047071
Watson, Mark 044060
Watson, Matt 050758
Watson, Maureen 046129
Watson, Roger 044746
Watson, Ronald 048909
Watson, Sara C. 049273
Watson, Scott 007289
Watson, Sheena 045579
Watson, Dr. Sheila 044474
Watson, Dr. Wendy 053406

Watson, Wendy M. 053405
Watson, William S. 044275
Watt, David 044237
Watt, James C.Y. 051424
Watt, Keith 044378
Watt, Marie 052297
Watt, Robert 005329
Watt, T. 044179, 044811, 044812
Wattenberg García, Eloisa 040340
Watters, Frank X. 052170
Watters, Jane 001549
Watterson, Norma 048460
Watteyne, Damien 003994, 003996
Watton, Graham C. 033271
Watton, Janet 052464
Watts McKinney, Sylvia 052079
Watts, Kelly 049516
Watts, L. 001434
Watts, Michael 047538
Watts, Pete 054573
Watts, Peter . . 001056, 001374, 001459,
                001463, 001538, 001539, 001546,
                001551, 001607, 001613
Watts, Steve 044816
Waufle, Alan D. 052959
Waugh, Carmen 007542
Wautier, Jacques 003733
Wawrzyk, Henryk . . . 035045, 035046
Wawrzyniak, Andrzej 035328
Way-Jones, M.F. 038724
Way, Catherine A. 049784
Way, Paul 044224
Wayne, Cynthia 046726
Wayne, Ph.D. Kenneth 049621
Wayne, Kenneth 047298
Wayne, Nan 049563
Wcześny, Marek 035048, 035049
Weadick, Nathalie 024938
Weaner, Arthur 049088
Wear, Lisa 052904
Weatherby, Ken 006330
Weathers, Allen 050804
Weaver, Henry 006013
Weaver, Janet 049966
Weaver, Leslie J. 052116
Weaver, Nigel 045660
Weaver, Pat 052870
Weaver, R. 044646
Weaver, Susan 049442
Weaver, Suzanne 048115
Weaver, Vey O. 046304
Webb, Belinda 001555
Webb, Betsy 049521
Webb, Eddie 052329
Webb, Heather 006635
Webb, Judy 052654
Webb, Keith 046056
Webb, Lisa 044096
Webb, Vaughan 048727
Webber, Barry 051572
Webber, Chris 045148
Webber, Eddie 000869
Webber, Maurie 045433
Webel, Sophie 014651, 014827
Weber-Brosamer, Dr. Bernhard . 018055
Weber-Lehmann, Dr. Cornelia . . 017437
Weber Soros, Dr. Susan 051344
Weber-Stöber, Dr. Christianne . . 019035
Weber, Brigitta 022892
Weber, C. Sylvia 019818, 021756
Weber, C.S. 019816
Weber, Dr. Erich 022227, 042001
Weber, Francis J. 050954
Weber, Dr. Gerhard 019588
Weber, Prof. Dr. Gregor J.M. . . 019532
Weber, Dr. Hans 041219
Weber, Hans Ruedi 042207
Weber, Heidi 042198
Weber, Heiko 020427
Weber, Jean 030608
Weber, Jens 019015
Weber, Johanne 022073
Weber, John 053025
Weber, K.R. 053128

Weber, Karl . . . 016790, 017035, 022480
Weber, Dr. Klaus 017077
Weber, Dr. Marion 008723
Weber, Markus 022731
Weber, Michael 031855
Weber, Otto 020886
Weber, Dr. Peter 036562
Weber, Robin 006719
Weber, Dr. Stefan 017242
Weber, Stephan W. 001692
Weber, Timothy 053370
Weber, Ute 020947
Weber, Drs. W.I.M. 032999
Weber, Dr. W.I.M. 032788
Weber, Prof. Dr. Winfried 022196
Weber, Yvette 014000
Weberg, William 052372
Webersberger, Josef 002372
Webley, Dr. L. 038723
Webley, Dr. W. 038724
Webster, Barbara 047333
Webster, Harvey 047809
Webster, L. 044923
Webster, R.M.O. 046154
Webster, Robert 053759
Webster, Roger J. 043662
Webster, Susan 049251
Webster, Prof. William 001540
Wechner, Wilma 002141
Wechs, Thomas 016785
Weckbacher, Vernon 050544
Weckström, Gunnar 011490
Weddegierde, Per 033486
Weddle, Sam 050394
Wedlake, Roger 046068
Wedler, Jacque 046848
Weeda, K.A. 032305
Weedmark, Lillian 007046
Weekley, Carolyn 054408
Weekley, Carolyn J. . . 054409, 054410
Weekly, Nancy 047303
Weeks, Jason 047360
Weems, Marcia 053863
Weenink, G.J. 032599
Weerdt, G.A. de 032918
Weese, Josh De 049430
Wefers, Dr. Sabine 019460
Wegener-Welte, Kathrin 016712
Wegge, Aslak 034080
Wegmann, Dr. Peter . . 042154, 042158
Wegner, Gene 034451
Wegner, Hartmut 017408
Wegner, Manfred 020551
Wehdorn, Armine 002994
Wehgartner, Dr. Irma 022792
Wehling, Arno 004737
Wehmeyer, Egon 016710
Wehmeyer, Pete 046691
Wehrberger, Gerhard 002377
Wehrberger, Kurt 022268
Wehrle, Karl 021342
Wehrli, Martin 041359
Wei, Biao 007661
Wei, Haibo 007656
Wei, Hsing-Ru 042484
Wei, Kangquan 007988
Wei, Mengling 007833
Wei, Zhongying 008022
Weibel, Prof. Peter 019522
Weibel, Dr. h.c. Peter 001985
Weibel, Univ. Prof. Dr. h.c. Peter 009819
Weibel, Dr. Urs 041950
Weibezahn, Dr. Ingrid 017584
Weibull, Ph.D. Dr. Nina 041049
Weichselbaum, Dr. Hans 002624
Weichselbaum, Karl 002463
Weick, Wolfgang E. . . 017955, 017962
Weidemann, Dr. Friedegund . . . 017154
Weidemüller, Karola 018235
Weidenschläger, Christine 017205
Weidinger, Mag. Dr. Johannes T. 001953
Weidisch, Peter 016797
Weidl, Beverly 049553
Weidlich, Ariane 018866

Weidman, Mary C. 051651
Weidner, Dr. Thomas 020551
Weidner, Timothy 049122
Weidner, William 052217
Weiffenbach, Jeannie 052692
Weigel, Jay 051289
Weigel, Dr. Viola 022661
Weight, A. 045003
Weigl, Stephan 002316
Weigle, Petra 020866
Weigt, Karin 017867
Weil-Seigeot, Claude 012583
Weil, Benjamin E. 053025
Weil, Gilbert 012181
Weil, Dr. Mark S. 052809
Weiland, Linde 018562
Weileder, Rainer 020812
Weilepp, Bruce 053397
Weiler-Streichsbier, Dr. Doris . . 020987
Weiler, Gerhard 022307, 022789, 022798
Weimert, Dr. Helmut 019155
Weinberg, Adam D. 051491
Weinberg, H. Barbara 051424
Weinelt, Hannes 001999
Weinelt, Wendeline 018555
Weinen, Willy 022361
Weinert, Renate 016766
Weingarten, Ulfried 021798
Weingartner, Heinrich 003044
Weinhäusel, Mag. Bernhard . . . 001858
Weinhart, Dr. Martina 018446
Weinheber-Janota, Brigitte 002178
Weinig, Prof. Dr. 022028
Weinke, Jane 054236
Weinkein, John 048248
Weinland, Dr. Martina 017248
Weinmayr, M.A. Stefanje 019885
Weinshrott, William J. 048772
Weinstein, Michael 051381
Weintraub, Aviva 051412
Weinzapfel, Connie 051259
Weipert, Eve 049466
Weir, Kathryn 001503
Weir, Margaret 045268
Weir, Margaret S. 053581
Weise, Roland 019658
Weise, Sigrid 017086
Weisenberg, Dr. Angelika 017068
Weiserová, Dr. Jana . . 009581, 009812
Weiskircher, Arrand 015845
Weisman, Billie Milam 050423
Weisman, Bruce 049964
Weiss, Al 050070
Weiß, Bernd 021113
Weiß, Berthold 001919
Weiss, Brigitte 017728
Weiss, Christian 019617
Weiß, E. 021661
Weiß, Ferdinand 001875
Weiß, Frank 021150
Weiss, Dr. Gabriele 003046
Weiss, Ján 038333
Weiss, Lois 053190
Weiss, Martin 048767
Weiss, Prof. Dr. Rainer-Maria . . 018995
Weiss, Sabine 022448
Weiss, Dr. Thomas . . . 017868, 017869,
                017870, 020999, 022717, 022718
Weiß, Dr. Ulrike 019053
Weiss, Wolfgang 002955
Weissend, Patrick 046807
Weissengruber, Dr. Thekla 002330
Weisser, Dr. Bernhard 017088
Weit, Janina 017499
Weitschat, Dr. W. 018989
Weitzendorfer, Dr. Alfred 002290
Weium, Frode 033854
Welch, Cheryl 053970
Welch, Richard 048705
Weldon, Eleanor 006349
Welfele, Odile 014779
Welin, Kjell-Arne 040692
Wellens, Luc 003319

Weller, Bernhard 016904, 016905
Weller, Dennis P. 052450
Weller, Susan J. 050940
Wellicome, Lavinia 046175
Wellington, Amy 049747
Wellington, Judith L. 047562
Wellman, Lesley 049353
Wellman, Dr. Marc 017148
Wellmann, Nóra 023509
Wells-Cole, Anthony 044785
Wells Fitzgerald, Barbra 049619, 049620,
    049622, 049623
Wells Zaia, Kay 051652
Wells, Carol A. 050414
Wells, Edna 048291
Wells, Gerald 045124
Wells, Keith 052416
Wells, Kenneth D. 047163
Wells, Licia A. 053016
Wells, Nicole 051455
Wells, Penny 005902, 005903
Wells, Stephen D. 048836
Wells, Trevor C. 030302
Wellspeak, Charles 048133
Wellspeake, Raymond 054326
Wellwood, Bridget 033280
Welsh, Caroline 047059
Welsh, Dr. Peter 049115
Welsh, Tracy 044472
Welter, Barbara 041252
Welter, Matt 049229
Welti, Georges 041481
Weltmaier, Marion 020518
Welty, Claude 014377
Welu, James A. 054548
Welzig, Ingrid 020837
Wemhoff, Prof. Dr. Matthias .. 017245,
    021065
Wemmlinger, Raymond 051402
Wen, Yanyuan 007766
Wendel, Birgit 019505
Wendel, Candice 049701
Wendel, Michael G. 047016
Wendel, Siegfried 021536, 021539
Wendl, Dr. Tobias 016990
Wendling, Britt 040905
Wendling, Pia 013223, 013224
Wendling, Urs 041679
Wendorf, Richard 047102
Wendowski-Schünemann,
    Andreas 017791
Wendt, M.A. Liane 019144
Wenger, Rhona 005928
Wenham, Julia 046141
Weniger, Prof. Dr. Gerd-C. .. 020395
Weniger, Dr. Matthias 020509
Weninger, Gehard 002499
Wenn, Diane 007373
Wenn, Norbert 017648
Wentworth, Peggy 046778
Wenzel, Dr. Astrid 019505
Wenzel, Cornelia 017450
Wenzel, Gundula 022534
Wenzel, Kai 018736
Weon-Kettenhofen, Hae-Yon .. 017504
Wepfer, Dr. Hans-Ulrich ... 041627
Weppelmann, Dr. Stefan 017147
Werd, Drs. Guido de .. 019644, 019645
Were-Were, Liking 008870
Werf, Hans van der 015888
Werkmäster, Daniel .. 040881, 040882
Werkmäster, Lisskulla 040620
Werkner, Prof. Dr. Patrick ... 003074
Werlich, Dr. Uta 022048
Wern, Christine 018430
Werne, Gordon 052959
Werneburg, Dr. Ralf 021670
Werner, Dr. Anne-Marie 021560
Werner, Bettina 051464
Werner, Constanze 020888
Werner, Dietrich 052683
Werner, Dr. Edwin 018967
Werner, Hans-Hubert 020376
Werner, Harald 002762

Werner, Heinz 019231
Werner, Jane 052172
Werner, Manuela 022418
Werner, Petra 019301
Werner, Dr. Sigrid 016963
Werner, Dr. Winfried .. 020527, 020561
Wernevik, Lena 040888
Werneyer, Rita 022744
Wernz-Kaiser, Heike 016834
Werring, Ole A. 033832
Wert, Nancy L. 050898
Werth, Linda 054219
Wertkin, Gerard C. .. 051325, 051383
Wéry, Benoît 004112
Wery, Bernoît 003678
Weschenfelder, Dr. Klaus 017752,
    021451
Wescott, Dusty 052454
Weski, Thomas 020529
Wesling, Ferdinand 021338
Wesney, Russell 033265
Wessel, Dr. Alexandra 041276
Wessels, E.M. 038664
Wessels, Reinhard 019074
West, Charlotte 010263
West, Chris 043748
West, Gerald C. 047723
West, J. Martin 050277
West, Janis 050615
West, Dr. Patricia 049971
West, Penelope A. 050277
West, Richard 052719
West, W. Richard .. 051448, 054148
West, Wanda 048745
Westad, Lynn 048329
Westbrook, Nicholas 053743
Westbrook, T.G. 044148, 046193
Westby, Carole 046877
Wester la Torre, Carlos 034301
Westerlund, Jens 010321
Westerlund, Kasper 011435
Westermann-Angerhausen, Prof. Dr.
    Hiltrud 019706
Westers, Drs. Abraham 032673
Westfall, Deborah 047033
Westfehling, Dr. Uwe 019715
Westheider, Dr. Ortrud 018977
Westheider, Dr. Rolf 018889
Westholm, Gun 041174
Westin Berg, Elisabeth 041015
Westland, Alan 007051
Westling, Stefan 040761
Westlund, Frances 006287
Westmacott, Jean .. 049008, 049010
Westman, Anne 040547
Westmoreland, Debra 051061
Westmorland, Lamont 050420
Westoby, Carolyn 005336
Weston-Lewis, Aidan 044255
Weston, Catherine 043456
Weston, Elizabeth 045221
Weston, Gill 045243
Westphal, Florian 021663
Westphal, Jean 013449
Westphal, Dr. Martin 021373
Westphal, Ursula 017861
Westra, J.F. 031926
Westrop, Dr. S. 051588
Westwood, Ros. 043820
Wetenhall, John 053149
Wetherell, Virginia 048725
Wetherington, Mark V. 050490
Wetmore, Shirley 053838
Wettengl, Dr. Kurt 017961
Wetter, Ellen de 006699
Wetter, Dr. Evelin 041883
Wetterau, Glenadore 048859
Wetton, Allan 000800
Wettre, Håkan 040652
Wetzel, Brigitte 018827
Wetzel, Dietrich 021149
Wetzel, James K. 048234
Wetzel, Roland 041280
Wetzel, Walter 041377

Weuts, Sylvain 003782
Wex, Dr. Norbert 021893, 021895
Wexell, Dale 049561
Wey-Eiholzer, Josef 041875
Weydemann, Ingrid 002443
Weyerhaeuser, Charles A. .. 048414
Weyers, E. 038810
Weygandt, Virginia 053470
Weyher Dawson, V. 048721
Weyns, Marie 003812
Weyns, Sara 003291
Weytjens, Laurent 003371
Whalen, Connie 052125
Whaley, Martha 049826
Whallon, Robert 046491
Whang, Prof. Won-Koo 030054
Wharton, David 054197
Whatmoor, D.J. 045447
Wheaton, Marilyn L. 053895
Wheaton, R. 001446
Wheeler, A. 045061
Wheeler, Geoffrey 050797
Wheeler, Karla 046905
Wheeler, Ken 051885
Wheeler, Lawrence J. 052450
Wheeler, Mary 049896
Wheeler, Matt 045394
Wheeler, Robert K.L. 050886
Wheeler, Verna Anne 048073
Wheeley, T.R. 044617
Wheldon, G.H. 044670
Whidden, John 007483
Whissell, Raymond 006845
Whitaker, Monique 038824
White, Adam 044785
White, Babs 048150
White, Barbara 053322
White, Barry 024873
White, Betsy K. 046309
White, Blair H. 049823
White, C.S. 045558
White, Carol 046858
White, Christopher 046013
White, David .. 007239, 045118, 050313
White, Derrick 046536
White, Donald 052100
White, Donald F. 053917
White, Donny 006284
White, Elisa 051322
White, Gail W. 053761
White, Sir George 044947
White, James J. 052165
White, Joe L. 049967
White, Julia 049534, 054462
White, Karen 054531
White, Kathleen 052538
White, Kathryn 043530
White, Kristy 047992
White, Marcelle 048132
White, Mark A. 053531
White, Mary 045798
White, Matthew 051602
White, Maylon 054000
White, Matthew 046859
White, Pam 046099
White, Patsy 051878
White, Richard S. 053831
White, Sally 044817
White, Susanna 047843
White, Timothy J. 047629
White, Warren O. 048493
Whitehead, F. 043823
Whitehead, James T. 053282
Whitehead, Jane 046228
Whitehead, Leah 046925
Whitehouse, Adrienne 044553
Whitehouse, C.M. 043629
Whitehouse, David B. 047994
Whitelaw, R. Lynn 053683
Whiteley, Jim 050195
Whiter, Robert 001051
Whitey, Matthew 044779
Whitfield, Carol 006223
Whiting, Dr. Cliff 033360

Whiting, Elizabeth K. 048440
Whitlock, Pamela 001102
Whitlow, Paula 006542
Whitman-Grice, Lisa 052530
Whitmore, Damien 045123
Whitmore, Dennis 046756
Whitney, Lisa 050427
Whitney, Richard 048322
Whitt, Teresa C. 053401
Whittaker, Dr. J. 043897
Whittall, Ann 044880
Whittington, Michael 047576
Whittle, Stephen 045435
Whitworth, Kent 048926
Whitworth, Peggy 047484
Wholey, B.J. 043739
Whorlow, David 043814
Whyel, Rosalie 046890
Whyte, Alan 007360
Wiant, Dr. Michael D. 050235
Wiblé, François 041727
Wible, Dr. John R. 052158
Wichert, Dr. Susanne 017494
Wichrowski, Zbigniew . 034915, 034916
Wick Reaves, Wendy 054150
Wick, Susan B. 053880
Wickenden, N.P. 043918, 043921
Wickers, Tine 010195
Wickham, Ann 044486
Wickham, W.C. 049354
Wicklander, Dan 054081
Wicklund, Doug 048666
Wickman, Viktor 030771
Wickramasinghe, Dr. Nanda .. 040475,
    040479
Wicks, Dr. Fred 007200
Widdowson, Prof. J.D.A. ... 045777
Widemann, Norbert 018475
Widén, Britt-Marie 040652
Widenheim, Cecilia 041021
Widjeskog, Susanna 011275
Widmann, Dr. Marion 017499
Widmer-Heuberger, Kurt .. 041619
Widmer, Walter 042078
Widrick, Melissa 054215
Widulin, Navena 017080
Wie, Guangzhen 008183
Wie, Na 008061
Wie, Xia 008464
Wiebe, Charles M. 053279
Wiebel, Dr. Christiane 017752
Wiech, Stefanie 020157
Wiechert, Ralph D. 052824
Wiechmann, Dr. A.H.C.A. .. 032581
Wieczorek, Prof. Dr. Alfried .. 020255
Wieczorek, Jan 035136
Wied, Fürst zu 021551
Wied, M. Prinz zu 021551
Wiedenhofer, Dr. Josef 002370
Wieg, Cornelia 018969
Wiegand, Otto 020358
Wiegand, Tanya 005471
Wiegand, Winfried 020341, 020342,
    020344
Wiegartz, Dr. Veronika 017587
Wiegmann, Dr. Karlheinz 022221, 022223
Wiehager, Dr. Renate 017099
Wiehart, Michael 003047
Wielek, Jan 034948
Wiell, Stine 010092
Wiemann, Dr. Elsbeth 022057
Wiemer, Regine 020095
Wienand, Paul 022477
Wiench, Barbara 018046
Wiene, Inger 010178
Wienecke, Joachim 019197
Wieninger, Johannes 003033
Wiens, Dr. John J. 052158
Wier, Nan 051788
Wierbicki, Wilma 052264
Wiermann, Dr. Roland 017333
Wiermann, Susanne 016741
Wierschowski, Dr. Myriam .. 020092
Wiersema, J. 032793

Wierzbicka, Alina 035021
Wierzbowska, Ilona 034670
Wiese, Dr. André 041261
Wiese, Joyce 053756
Wiese, Kerstin 019979
Wiese, Larry 050814
Wiese, Prof. Dr. Rolf .. 020310, 020449,
    020450, 021473, 022564
Wiese, Sharon A. 046544
Wiese, Dr. W. 021282
Wiesel, Yael 025018
Wieseman, Betsy 045057, 047739
Wiesen, Barbara 049108
Wiesen, Richard A. 047300
Wiesenhofer, Franz .. 002547, 002549
Wiesenmüller, H. 022067
Wieser, Dr. Christian 002192
Wieser, Gertrud 003095
Wiesner, Cary 047630
Wiesner, Tomáš 009936
Wiethege, Dr. Katrin 001793
Wiethoff, Charlotte 032392
Wiethorn, Hans 021936
Wieting, Margaret 049710
Wiewelhove, Dr. Hildegard 017378
Wiget, Dr. Josef 041968, 041970
Wiggins, Denise 050601
Wight, Anthony 047660
Wigmore, John R. 007320
Wihebrink, Peggy 050358
Wiig, Prof. Øystein 033847
Wijers, Gaby 031971
Wijesooriya, Bandusena 040470
Wijn, Drs. J.J.A. 032881
Wikström-Haugen, Inger 040657
Wikström, Anne-Maj 011418
Wilcox, Chris 054448
Wilcox, Dr. David R. 048740
Wilcox, John 047971
Wilcox, Mike 048483
Wilcox, Richard 050699
Wilcox, Scott 051266
Wilcox, Susan 048739
Wilcox, Thomas R. 046809
Wilcox, Tim 045208
Wilczak, Susan 052781
Wild, Dr. Beate 017237
Wild, Jane 006641
Wild, Maxine F. 001060
Wild, Teresa 045999
Wildberger, Anne-Käthi 041261
Wilde, Cathleen 052231
Wilde, Kevan A. 033330
Wilde, Dr. Manfred ... 017846, 017847
Wildeman, G.J.D. 031972
Wilder, Jacalin W. 047240, 047241
Wilder, Oliver L. 052645
Wilder, Sofie 003306
Wildgans, Ilse 002398
Wildgust, Deborah 045521
Wildgust, Patrick 044027
Wildman, Janet 044454
Wildman, S.G. 044756
Wildung, Prof. Dr. Dietrich ... 017065
Wilenius-Rantala, Carita 010817
Wiley, E.O. 050186
Wiley, Florence 048339
Wiley, John 006316, 046190
Wilezek, Alexander 002855
Wilfong, Terry 046490
Wilhelm, Ada 047922
Wilhelm, Elliot 048301
Wilhelm, Franz 018107
Wilhelm, Martin 001901
Wilhelm, Robert 049414
Wilhelmer, Erika 016911
Wilhelmer, Hans-Sieghart 001716
Wilhelmson, Pia 011127
Wilhelmy, Dr. W. 020224
Wilhite, Janet 048110
Wilk, Christopher 045123
Wilk, Zbigniew 034744
Wilke, Erich 017611, 022668
Wilkens, Dr. Barbara 026640

Wilker, Klaus-J. 017916
Wilkie, Jo-Anne 051082
Wilkins, Prof. David G. 052178
Wilkins, Margaret 044344
Wilkinson, Dr. D.J. 044903
Wilkinson, MaryAnn 048301
Wilkinson, Patty 047060
Wilks, Harry T. 049329
Willaert, Saskia 003492
Willander, Dr. Alfred 002604
Willard-Wright, Rowena 044129, 044131
Willard, David 049061
Willburger, Dr. Nina 022047
Willeboordse, A. 032888
Willeboordse, A.C.J. 031885
Willeboordse, Arco 032870
Willems, Alfons 004149
Willems, Carmen 004074
Willems, Jean 003853
Willems, Roger 003340
Willemsen, M. 032157
Willer, Dr. Susanne 017499
Willer, Thérèse 015886
Willert, Dr. Barbara 022386
Willesme, Jean-Pierre 014628
Willi, Dawn 051135
Willi, Jürg 041877
Williams-Davies, John 045686
Williams, Alice 049082
Williams, Barb 052498
Williams, Bart 049634
Williams, Bill 049330
Williams, Clay 049756
Williams, Dr. D. 044923
Williams, Debby 053294
Williams, Don ... 048142, 051393
Williams, Edna 050643
Williams, Edward 046729
Williams, Eric K. 051566
Williams, Fred 047979
Williams, Gareth H. 043824
Williams, Dr. George 054330
Williams, Gloria 051935
Williams, Dr. Gloria 052839
Williams, Hilary 044104
Williams, James 049389
Williams, Janice W. 053602
Williams, Jay ... 046663, 047912
Williams, Jon 054438
Williams, Judy 033355
Williams, Karen 033304
Williams, Kathy 048873
Williams, Kay P. 051231
Williams, Dr. Keith 054281
Williams, Kevin 050940
Williams, Laura 044990
Williams, Linda 045654
Williams, Lisa 050541
Williams, Dr. Lorraine 053799
Williams, Lyle 052940
Williams, Mary-Louise 001016
Williams, Matt 043515
Williams, Matthew 043864
Williams, Michael ... 045410, 045419
Williams, Nadezhda 051389
Williams, Pamela 047154
Williams, Paul 052917
Williams, Peter J. 001642
Williams, Richard 044563
Williams, Robert J. 053283
Williams, Roger 052379
Williams, Sheldon 046875
Williams, Stacy 052913
Williams, Stephen R. 052251
Williams, Teri . 046604, 046620, 046631
Williams, Thomas 050664
Williams, Victoria ... 044532, 044533
Williamson, Eleanor 053480
Williamson, J. Reid 049685
Williamson, Jane 048726
Williamson, Norma 048373
Williamson, Paul 045123
Williamson, Roberta 048743
Willis, Carol 051468

Willis, Kelly B. 050240
Willis, Leanne 000917, 001495
Willis, Michael 052238
Willis, Pamela 045050
Willis, Patricia 044902
Willis, William 052350
Willits, Mary Lou 049208
Willmann, Prof. Dr. R. 018759
Willmitch, Thomas R. 047060
Willmont, Joanne 050234
Willms, Dr. Christoph 018405
Willms, Karl 017612
Willomitzer, Gerry 007408
Willoughby, Alan 051044
Willoughby, Brenda 048145
Willoughby, Laura 052011
Willour, Clint 049032
Willow, Hai Chang 051362
Willows, John 046110
Wills, Douglas 005914
Willscheid, Bernd 020777
Willson, Michael 052182
Wilmes, Ulrich 020529
Wilmes, Dr. Ulrich 019705
Wilmink, Drs. G.J.G. 032137
Wilmot, Jonel 050143
Wilmot, Patricia J. 049394
Wilmot, Robert 044245
Wilms, Kurt 016722
Wilschewski, Dr. Frank 020769
Wilsdon, D.E. 001511
Wilsey, Bart 048705
Wilson Davine, Blossom 052374
Wilson Jenkins, Kay 049279
Wilson-McKee, Marie 047632
Wilson, A. 045343
Wilson, Dr. Alan 043447
Wilson, Amy H. 048566
Wilson, Audrey 006235
Wilson, B. 045915
Wilson, Beth 047338
Wilson, Bob 047338
Wilson, Carolyn 054499
Wilson, Carrie 051480
Wilson, Chris . 024951, 043347, 044874,
    054455
Wilson, Christine 044055, 044086
Wilson, Courtney B. 046729
Wilson, Dale 048498
Wilson, Dan 049207
Wilson, David 048088
Wilson, Diane 046882
Wilson, Dorothy 043622
Wilson, Eric 045776
Wilson, Floyd S. 047569
Wilson, Gillian 050436
Wilson, Gordon 052744
Wilson, Hugh Allen 047080
Wilson, I.A. 043489
Wilson, Inge 006012
Wilson, J. Keith 050444
Wilson, Jane 051854
Wilson, Jerry 048166
Wilson, Jim 049339
Wilson, John 043564
Wilson, Jonathan 045511
Wilson, Kathy 006892
Wilson, Kay 049280
Wilson, Kevin 001570
Wilson, Lamar 053302
Wilson, Lesley 044906
Wilson, Marc F. .. 049880, 049888
Wilson, Mark 032378
Wilson, Martha 047257
Wilson, Nancy 050523
Wilson, Peter 045115, 045116
Wilson, Pierre 006394
Wilson, Robert 047311
Wilson, Robert C. 053556
Wilson, Robin 045497
Wilson, Dr. Rodney 033142
Wilson, Roger N. 051552
Wilson, Roy 045507
Wilson, Scott 048211

Wilson, Sophia 043925
Wilson, Steven 049382
Wilson, Dr. Thomas 050813
Wilson, Thomas H. 046906
Wilson, Tom 054494
Wilson, Viv 045341
Wilson, Wayne 006070, 006075
Wilson, William 045178
Wilsony, Thomas A. 053116
Wiltfang, Helga 020818
Wiltzius, Ruth 052004
Wiluś, Bogusława 035092
Wilusz, Mike 053543
Wilz, Mick 047274
Wimberger, Dr. Peter 053644
Wimmer, Franz 002514
Wimsett, Brian 043772
Win, U Kyaw 031834
Winans, Cathie 053284
Winberg, Hélène 041107
Winbladh, Marie-Louise 041019
Winch, Barbara 053408
Winch, Dinah 045435
Winchell, Janey 052880
Winchester, Dr. Judi A. 047855
Wincke, Claus-Jürgen 016679
Wind, D.A. 032856
Winderickx-Joos, Myriam ... 003263
Winders, Dr. Bruce 052928
Windham, Michael 052919
Windholz, Sascha 002836
Winding, Dr. Norbert 002625
Windle, Ann S. 050586
Windsor, Deirdre 050508
Windsor, Noel 000854
Wine, Dr. Humphrey 045057
Winegrad, Dr. Dilys V. 052029
Wines, Claudia 048541
Wines, Richard 052588
Winfield, Rosemarie 050129
Winford, Laura 050073
Wingate, Lloyd R. 038770
Wingfield, Carolyn 045661
Winglhofer, Rudolf 001926
Wingo, Calvin 048616
Winkelmayer, Sylvia 002945
Winkelmeyer, Iris 020580
Winker, Kevin 048662
Winkler-Hermaden, Burkhardt .. 002157
Winkler, Anke 018711
Winkler, Axel 022045
Winkler, Bernhard 022408
Winkler, Dr. Ferenc ... 023426, 023711,
    023713, 023886
Winkler, Dr. G. 001861
Winkler, Dr. Gerhard 001860
Winkler, Gerhard 001740
Winkler, Howard 046530
Winkler, Karin 050878
Winkler, Paul 049575
Winkler, Dr. Raffael 041282
Winkler, Steffen 018711
Winkler, Susanne 003095
Winkler, Ursula 019584
Winkworth, David R. 043984
Winn, Allan 046165
Winnekes, Dr. Katharina 019694
Winner, Scott 047058
Winnicki, Tomasz 035295
Winning, Fiona 001069
Winslow, Bruce B. 050871
Winsryg, Marian 053134
Wint, Dennis M. 052050
Winteler-Isoz, Florence 041199
Winteler, Florence 041201
Winter-Berger, Robert 051402
Winter, Dr. Amy 048765
Winter, Dr. B. 005505
Winter, C. 033190
Winter, Cordula 016535
Winter, Emil 019241
Winter, Gerhard 019106
Winter, Hilda 045156
Winter, Joan G. 000907

Winter, Judith 045261
Winter, Małgorzata 034669
Winter, Petra 017320
Winterhalter, Greg 046921
Winterich, Douglas E. 047322
Winters, Laurie 050916
Wintgens, Dr. D. 032586
Wintz, Klaudius 002240
Winzeler, Marius 018735, 022888
Winzingerode, Margit Freifrau
von 021209
Wiotte-Franz, Dr. Claudia 021564
Wipaki, Gail 000977
Wippermann, Hartmut 020354
Wippert, Bernd 019795
Wirenfeldt Asmussen, Marianne 010297
Wirnsberger, Mag. Karlheinz ... 002773,
                                002774
Wirrell, Wynne 001024
Wirrmann, Fanny 011752
Wirsperger, Dr. Peter 002377
Wirtanen, Lyle 048027
Wirth, Dr. Klaus 020255
Wirth, Paul 041350
Wirth, Peter 053121
Wirthgen, G. 019975
Wirtler, Dr. Ulrike 019664
Wirtz Eybl, Dr. Irmgard 041323
Wirtz, Gudrun 020507
Wirtz, Wes 007348
Wirxel, Julia 016928
Wirz, Arno 016960
Wisbey, Peter A. 046653
Wise, Carol 049862
Wise, Philip 043988
Wise, Dr. Stephen R. 051929
Wise, Tess 050606
Wiséhn, Ian 041010
Wiseman, Christopher . 049841, 049843
Wishart, E. 001150
Wismar, Franz Werner von ... 021364
Wismeijer, P.J. 032247
Wismer, Beat 018046
Wisowaty, Elaine 050343
Wißkirchen, Prof. Dr. Hans 020155
Wissler-Thomas, Carrie 049369
Wissler, Richard 047935
Wistar, Caroline 052060
Wistoft, Birgitte 010191
Wisuri, Marlene 047847
Wit, B.H. de 032082
Wit, Prof. Dr. G.W. de 032817
Witcherly, Zak 054370
Witcox, David E. 051569
Witenstein, Michael 050370
Witham, Jim 048610
Withrow, Sharon 046557
Withycombe, C. 043842
Witkofsky, Adrien 049269
Witmer, Linda F. 047427
Witschey, Walter R.T. . 052560, 052562
Witschi, Heinz 041609
Witschi, Dr. Peter 041593
Witt-Dörring, Dr. Christian ... 002998
Witt, David 053676
Wittbold, Dr. Annette 018883
Wittboldt, Dr. Annette 018884
Witte de Haelen, Didier de ... 003663
Witte, Amy 047289
Witte, David 053352
Witte, Hubert De 003409
Witte, Dr. Reinhard 016585
Wittekindt, Heiner 018387
Wittenbauer, Christine 020063
Witter, Thomas 019654
Wittig, Peer 016587
Wittigo Keller, Dr. 002955
Wittington, Dr. Stephen 054490
Wittkopp, Gregory M. 047038
Wittman, Jörg 022889
Wittmann, Dr. Helmut 002625
Wittmann, Dr. Reinhard G. .. 020543
Wittmer, Heike 021134
Wittmer, Marcilene 047986

Wittmer, Stephan 041709
Wittwer, Felix 041803
Witzke, Gitta 019218
Witzmann, Nora 003057
Wlattnig, Robert 002192, 002870
Włodarczyk, Tadeusz 035429
Woehlk Hansen, MA Lola 010141
Wöhrer, Hannes 001739
Wöhrer, Josef 002367
Woelfl-Graff, Elisabeth 003095
Woelk, Dr. Moritz 018009
Wörlen, Hanns Egon 021083
Wörtz, Walter . 019577, 020665, 021462
Wörz, Dr. Arno 022056
Wöß, Gerhard 002851
Woestenburg, K. 032765
Wohlgemuth, Caroline 016457
Wohlgemuth, Werner 018499
Wojciechowski, Jerzy 034828
Wojciechowski, Zbigniew 034747
Wójcik, Adam 035265
Wójcik, Andrej J. 035441
Wojewoda, Maciej 035209
Wojtowicz, Wit Karol 034934
Wojtusiak, Prof. Janusz 034906
Wojtysiak, Władysława 034988
Wolanin, Dr. Barbara A. 054166
Wolde, H.B. ten 032529
Wolenitz, Dorit 025174
Wolf, Alexandra 002837
Wolf, Dr. Barbara 016646
Wolf, Dr. Beate 022057
Wolf, Claus 042094
Wolf, Connie 052998
Wolf, Dieter 017707
Wolf, Dr. Eberhard 021426
Wolf, H.-J. 022008
Wolf, Dr. Helmut ... 016641, 019815
Wolf, Martha Leigh 052366
Wolf, Maurice de 003936
Wolf, Reinmar 002402
Wolf, René 031939
Wolf, Dr. Rüdiger 002723
Wolf, Sara 047576
Wolf, Sarah 052636
Wolf, Sylvia 051491
Wolf, Vicki 052982
Wolf, Werner 001983
Wolfe, Adam 001098
Wolfe, Douglas 050813
Wolfe, Jeraldine J. 052489
Wolfe, L. 006471
Wolfe, Sara 051350
Wolfe, Sarah 054060
Wolfe, Susan 049121
Wolfensberger, Dr. Rolf 041320
Wolfer, Johannes 001819
Wolfer, Dr. Martin 001819
Wolff, Prof. Heinz 043797
Wolff, Thomas 018814
Wolff, Dr. Tobias 017602
Wolff, W. 019951
Wolford, Craig 047782
Wolfová, Dr. Eva 009832
Wolfs, R. 032826
Wolfs, Rein 019536
Wolfschlag, Robert 050584
Wolfson, R.D. 032650
Wolins, Inez S. 049216
Woliung, Dorothy 046891
Wolkow, Gary 048187
Woll, Gunda . 022232, 022235, 022236
Wollan, Alexa 006043
Wolle, Dr. Stefan 017100
Wollert, Christian 017076
Wolleswinkel, E.J. 032779
Wollkopf, Peter 019763
Wollmann, Dr. Klaus 021066
Wollner, Dr. Bernd 020398
Wołodko, Wiesław 034733
Wolsdorff, Dr. Christian ... 017077
Wolsk, Nancy 050249
Wolstenholme, L. 045210
Wolston, Bill 053416

Wolter, Rüdiger 017586
Woltereck, Jane D. 046762
Wolters, R. 032110
Wolters, Prof. Dr. Reinhard ... 022224
Wolterstorff, Robert 052285
Wolthoorn, Dr. R. 032586
Wolts, Silke 018309
Wolynec, Dr. Renata 054203
Wolz, Robert J. 049954
Wolzogen, Wolf von 018428
Wong-Uparaj, Acharn Damrong . 042652
Wong, Danny 007888
Wong, Michele 051399
Wonisch-Langenfelder, Dr.
Renate 002637, 002639
Wooby, William 054136
Wood, Brian 005468
Wood, Bryen 043814
Wood, Carolyn 047528
Wood, Charles A. 052425
Wood, David 047951
Wood, Dr. Donald A. 047005
Wood, Eileen 045603
Wood, Elaine 047573
Wood, James N. 047636
Wood, Jeff 050874
Wood, Jeremy 045046
Wood, Katherine 043989
Wood, Kaye 000862
Wood, Lydia 047895
Wood, Mara-Helen 045354
Wood, Margaret 052004
Wood, Margaret K. 049165
Wood, Nancy 001426
Wood, P. 043905
Wood, Ronald M. 050521
Wood, Sue 043443
Wood, Tim 001173
Wood, Virginia 053971
Wood, Willis 054259
Wood, Wilma 006186
Woodcock, Vivi 052020
Woodings, David J. ... 033229
Woodrow, James ... 043544
Woods, Christine .. 045219
Woods, Claude N. .. 049426
Woods, Don 052141
Woods, James C. .. 053867
Woods, Jean 049312
Woods, John S. .. 051577
Woods, Margaret M. . 047166
Woods, Mark H. .. 050860
Woods, Wally ... 054354
Woodside, April .. 053357
Woodson, Celia .. 048310
Woodson, Yoko .. 052992
Woodward, Alan . 033347
Woodward, Geoff . 046038
Woodward, John . 043685
Woodward, Mac . 049637
Woodward, Peter . 044116
Woodward, Richard . 052565
Woody Cywink, Linda . 053110
Woof, Dr. Robert 044465
Woog, Dr. Friederike . 022056
Woolcott, Isabelle S. . 048025
Woolf, Derrick 045321
Woolley, Brandon .. 047207
Woolley, F. 045193
Woolliscroft, Pam .. 045885
Woolmore, J.D. ... 043948
Woolrich, Gill 045778
Woolsey, Sam 048129
Worcester, Mike .. 047863
Worden, Gretchen . 052071
Worden, John 052594
Workman, Robert . 048898
Workmann, Tom .. 049861
Worm, Dr. Heinz-Lothar . 020085
Worrall, Adam ... 000937
Worsech, Kathrin . 016535
Worschech, Thomas . 018412
Worsdale, Godfrey . 045260
Worsley, Lucy ... 044210

Worswick, Clark 052880
Worth, Claudia R. 054361
Worth, Cyril 049510
Worth, Tim 007439
Worthen jr., William B. 050335
Worthington, Gary 052198
Worthington, Robert 049287
Wortmann, Susanne 022791
Worton, Graham 044147
Worton, James 052816
Worts, Martin 033642
Woster, Don 049567
Wotovic, Audrey 012966
Wouters, Lisbeth ... 003278
Woyda, Christina .. 035143
Woyke, Karin 018027
Woytowicz, Barbara . 051406
Woźny, Dr. Alexander .. 034769
Wpadhyays, Dr. Chandrakant B. 024128
Wragg, Gary 043509
Wray, JaLayne 054391
Wrazidlo, Dr. Isolde .. 017383
Wrębiak, Celestyn ... 034634
Wrenn, Dr. W.J. ... 049178
Wrey, Dr. James D. . 051976
Wright-Parsons, Ann . 048223
Wright, Amina 043519
Wright, Anthony ... 033168
Wright, Dr. Barnaby . 044951
Wright, Belinda ... 000913
Wright, Carolyn .. 048128
Wright, David M. . 054514
Wright, George .. 044353
Wright, Geri ... 052118
Wright, Henry T. . 046491
Wright, John .. 005580, 049476
Wright, Lesley .. 049279
Wright, Lynne . 005301
Wright, Margot . 043359
Wright, Michael . 049756
Wright, Raymond . 054251
Wright, Rita ... 033136
Wright, Dr. Robin K. . 053229
Wright, Dr. Sheila .. 045416
Wright, Sherri ... 051122
Wright, Simon ... 000888
Wright, Twyla ... 046808
Wrigley, John ... 000928
Wrobel, Horst ... 018696
Wroczyńska, Ewa Maria . 035293
Wroz, Winfried .. 019566
Wu, Cheng-Shang . 042534
Wu, Ching-Feng . 042596
Wu, Chu Pang .. 008070
Wu, Douming .. 008519
Wu, Gangzhong . 007792
Wu, Gaobin .. 008491
Wu, Hongzhou . 008231
Wu, Hung Yuen . 007733
Wu, Jin .. 008429
Wu, Jingchao .. 007722
Wu, Kai .. 008423
Wu, Ke Shun .. 008268
Wu, Li-Chu .. 042490
Wu, Monical .. 042431
Wu, Qingying . 008191
Wu, Qisheng .. 008149
Wu, Rongguang . 007757
Wu, Shengtian . 007837
Wu, Shiliang . 007899
Wu, Tianming . 007907
Wu, Wenxin . 007995
Wu, Xiaosong . 007905
Wu, Xuebin .. 008164
Wu, Xuhui .. 007922
Wu, Yulu .. 008440
Wu, Zhengguang . 007823
Wudel, Tom .. 050969
Wüest, Charles . 041628
Wünsche, Prof. Dr. Raimund . 020575
Würsch, Paul .. 041962
Würzelberger, Jörg . 002845
Wüst, Léonard .. 042043
Wüthrich, Dr. Lucas 041866

Wulf, Heino	016877	Xu, Xinwan	008437
Wulf, Karin	017474, 017475	Xu, Xueshu	007695
Wulf, Rüdiger	017962	Xu, Xuyao	008038
Wulfsohn, Iris R.	000381	Xu, Yiqun	008458
Wulkan, Reba	051493	Xu, Yiyi	008112
Wullen, Dr. Moritz	017241	Xu, Yuli	008443
Wullen, Dr. Moritz	017202	Xu, Zhile	007934
Wullus, Linda	003492	Xue, Deng	007694
Wunderer, Dr. Hansjörg	021328	Xue, Dongxing	008338
Wunderlich, Dr.		Xue, Fusheng	008005
Christian-Heinrich	018965	Xylander , M. A. Marlen von	020824
Wunderlin, Dominik	041276	Xylander, Prof. Dr. Willi	018738
Wunsch, G.	018378	Y, Dr. Rainer	020128, 022047
Wurda, Andreas	020200	Yaacobi, Effi	025163
Wurfel, Max	001407, 001408, 001409	Yabacıoğlu, Ömer	042872
Wurmfeld, Sanford	051405	Yabe, Yoshiaki	028885
Wurtele, Gail	051207	Yaeger, Dan	054049
Wurzer, Günther	002192	Yager, Bob	050646
Wurzinger, Erika	002031	Yaghobi	024784
Wussow, Indra	021278	Yagid, Meira	025212
Wuthenau von Pietsch, Carlota		Yajima, Loretta	049530
von	031513	Yakar, Saffet	042915
Wyatt, Frederica	049846	Yalakki, C.S.	024490, 024491
Wyatt, Steve	050758	Yaldız, Fethiye	042911
Wyckoff, Dr. D.	051588	Yaldiz, Mehmet	043066
Wyczółkowski, Mariusz	034829	Yamada, Rikuzou	029295
Wydler, Henry	041715	Yamada, Shimako	029307
Wye, Deborah	051427	Yamada, Yasuhisa	029698
Wykersloath, Baron E. de	003395	Yamada, Yasushi	029693
Wyld, Lesley	045234	Yamaga, Motoko	028632
Wyld, Martin	045057	Yamagishi, Youichi	029229
Wylie, Charles	048115	Yamaguchi, Kichiko	028492
Wyman, James	046599	Yamaguchi, Seiichi	029120
Wymbs, Norm	048333	Yamakawa, Kazutoshi	029366
Wÿnandts, Dr. H.	032571	Yamamoto, Akira	028472
Wyniger, Dr. Denise	041710	Yamamoto, Atsuo	029069
Wynn, Rachel	052267	Yamamoto, Chikyo	028889
Wynne, M.V	043952	Yamamoto, Kazumaro	029043
Wyrsch, Dr. Jürg	042116	Yamamoto, Dr. Masao	029074
Wyse Jackson, Dr. Patrick N.	024883	Yamamoto, Minoru	029638
Wyss, M.	041812	Yamamoto, Shuji	029692
Wyss, Matthias von	042066	Yamamoto, Yasukazu	029129
Xamena, Miguel A.	000285	Yamamoto, Yoshiko	052990
Xanke, Uwe	017897	Yamanaka, Osamu	028841
Xavier, Andrew	044937	Yamazaki, Taeko	029624
Xerri, Alda	030768	Yan, Hongbin	007585
Xi, Jiping	008228	Yan, Lijin	008177
Xi, Mude	008339	Yan, Yiyuan	008221
Xia, Meizhen	008467	Yanagi, Munemichi	029411
Xia, Suixiang	008310	Yanagihara, Masaki	029639
Xia, Xiangjun	008503	Yanamoto, Akira	029485
Xia, Zhaoli	007983	Yanari, Sachi	048894
Xian, Shihai	007586	Yancey, Shirley S.	054241
Xiang, Qukui	008352	Yande, R.	003321, 003322
Xiao, Changjin	007928	Yandle, Ashley	047560
Xiao, Gaoshe	008476	Yanega, Dr. Doug A.	052592
Xiao, Kanghua	008368	Yáñez Anllo, Lucila	039602
Xiao, Qi	008048	Yang, Daoyi	007790
Xiao, Runjun	008545	Yang, Huifu	007949
Xie, Daohua	007942	Yang, Jianmin	008381
Xie, Daoxin	007714	Yang, Jinghua	008257
Xie, Kaihua	008184	Yang, Jingnan	008299
Xie, Qihuang	007611	Yang, Jun	008227
Xie, Shanguang	007965	Yang, Mingzhu	008511
Xie, Suilian	008539	Yang, Shi-Rong	042484
Xie, Ying-Cong	042495	Yang, Xiaoneng	049888, 053488
Xie, Yuan'an	007898	Yang, Xin	008534
Xie, Zhijie	008482	Yang, Yanqiang	007859
Xiejun, Chen	008234	Yang, Yingji	008035
Xin, Fa	007974	Yang, Yongfang	008450
Xin, Xue Jian	047249	Yang, Yuanlong	007985
Xinmiao, Zheng	007639	Yang, Zishan	008547
Xiong, Bufa	008418	Yanik, Kadir	043067
Xiong, Guoyao	008332	Yaniv, Shosh	025111
Xiong, Jianping	007981	Yank, Paul	047492
Xiong, Keda	007978	Yankov, Dr. Angel	025126
Xiong, Qiu Shi	008253	Yano Bretón, Dr. Sabino	031588
Xu, Dali	007582	Yao, Jingzhou	008160
Xu, Fan	007760	Yap, Priscilla M.	034573
Xu, Huping	008112	Yapelli, Tina	052985
Xu, Qiaohua	008513	Yarlow, Loretta	046454, 047266, 051461
Xu, Rongmin	008342	Yarmosh, Linda	047094
Xu, Sanjian	008023	Yarsinske, Amy Waters	051580
Xu, Xiaozhong	007902	Yaşar, Hasan	043006
Yashimoto, Mami	029307	Yoshikawa, Miho	029129
Yasin, Dr. Walid	043332	Yoshinaka, Yasumaro	028653
Yassin, Robert A.	052461	Yoshioka, Akemi	029561
Yasuda, Haruki	029561	Yoshioka, Yoji	028497
Yasui, Teruo	028900	Yoshitaka, Abe	028720
Yasumura, Toshinbu	029518	Yoshiyuki, Murakawa	029707
Yasuo, Hasimoto	029215	Yotsutsuji, Hideki	029129
Yates, Jane M.	047544	You, Mee Chung	030024
Yates, R.	038686	Youck, Ingrid	007083
Yates, Royden	038685	Young Lee, Nan	029945
Yates, Sam	050006	Young-Sanchez, Margaret	048258
Yates, Dr. Steve	053116	Young-Vignola, Eilee	052049
Yates, Dr. Steven A.	053113	Young, Anna	046979
Yates, Dr. Terry L.	046365	Young, Bill	051106
Yatim, Prof. Dr. Othman	030692	Young, Bobbie	048009
Yatras-Dehove, Elena	022978	Young, Brian	050334
Yawazaki, Takashi	029076	Young, C.	044164
Yaworski, Eileen	006978	Young, Dan	005951
Yazbeck, Janine	030278	Young, David C.	046650
Yde Iversen, Gerrud	010223	Young, Dede	052418
Ydstie, Ingebørg	033842	Young, Ernest	049905
Ye, Dingxiang	008288	Young, Gary	006973
Ye, Faling	008428	Young, Holly	052123
Ye, Jianhong	008286	Young, Jocelyn	046470
Ye, Qinglin	007570	Young, John	045142
Ye, Rongguang	008063	Young, John J.	007147
Ye, Xiangrong	008448	Young, Joyce	051764
Yeager, Shelley	051611	Young, Judi	007165
Yeager, William	047042	Young, Karen S.	053680
Yeatts, David L.	054532	Young, Kyle	051175
Yébenes, M. Sierra	039310	Young, Laurie	005831
Yee de Cury, Ana	010394	Young, Dr. Marcia	047042
Yee, Lydia	047224	Young, Marvin R.	053321
Yeffeth, Laura	051489	Young, Mary Jane	046916
Yelavich, Susan	051368	Young, Sir Nicholas	044926
Yeler, Nihat	042809	Young, Patience	053488
Yellow Horn, Jo-Ann	005486	Young, Richard	053410
Yelton, Chris	053004	Young, Richard L.	047455
Yener, Emin	042839	Young, Sherilyn B.	047959
Yerdon, Lawrence	052316	Young, Steve	048669
Yerdon, Lawrence J.	052183	Youngken, M. Joan	051534
Yerhot, Judith	047052	Youngs, Christopher	052469
Yerkovich, Dr. Sally	051498	Younker, Dr. Randall W.	046948
Yernaux, Danielle	003738	Younkin, Elva	052568
Yeroulanos, Marinos	023002	Yount, Frank	050271
Yersin, Jean-Marc	042109	Youssoufian, Jerry	006492
Yesalis, Dr. Charles	047064	Yperman, Jan	003865
Yetkin, Mehmet	042793	Ysel, H.J.D.	038814
Yeudall, Robert	005752	Ysenburg und Bündingen, Casimir
Yeung, C.T.	007893	Erbprinz zu	021468
Yi, Jiasheng	008117	Yu, Eric Howyi	042654
Yi, Shu	007637	Yu, Jianling	008208
Yi, Xuelin	008298	Yu, Jianyong	007844
Yi, Zhu	007615	Yu, Jun	008380, 008431
Yıldız, Adem	042880	Yu, Lanshuan	008274
Yıldız, Hacı Ahmet	042792	Yu, Louis	007885
Yılmaz, Ayla	042943	Yu, Shenghua	008514
Yin, Bilan	008037	Yu, Sun	007627
Yin, Hexian	008538	Yu, Teng Kuei	042543
Yin, Liming	008025	Yu, Weimin	007731
Yin, Qingsong	007994	Yu, Zemin	007779
Yin, Xueqiang	008485	Yuan, Daojun	008540
Yip, Chee-Kuen	007877	Yuan, Hua	007738
Ylioja, Merja	011331	Yuan, Zhongyi	008029
Ylvisåker, Anne Britt	033538	Yüce, Ahmet	042808
Yocarini, T.	032189	Yüksek, Mustafa	042987
Yocham, Doris R.	053145	Yuergniaux, Danièle	014979
Yochim, Jordan	050186	Yüki, Ryōgo	029085
Yoder, Harold E.	052482	Yukio, Miyashita	028880
Yoel, Ben-Yosef	025127	Yuko, Isazawa	029511
Yokomizo, Hiroko	029509	Yun, Jang Sub	030003
Yokoyama, Saki	029536	Yunus bin Zakaria, Dr. Mohd	030724
Yoneya, Tetsuo	029338	Yusupov, Muhammad	054626
Yoneyama, Torataro	029581	Yut, Joyce	047413
Yoon, Prof. Se-Young	030018	Yutaka, Minami	028880
York, Beverly	054423	Yzereef, Dr. Gerard	031910, 031912
York, Brian	046565	Zaadnoordijk, Hans	032590
Yosef, A.	025204	Zabarovskij, Vladimir Ivanovič	037242
Yoshida, Hidekazu	029037	Zabehlicky-Scheffenegger, Dr.
Yoshida, Kazuo	029331	Susanne	002508
Yoshida, Shoji	029633	Zabel, Anne	017762
Yoshihara, Lisa	049533	Zabel, Barbara	034843, 034844
Yoshihisa, Suto	028489	Zabel, Jürgen	022918
Yoshikata, Ishihara	029478	Zabērs, Miervaldis	030163

Zabka, Reinhard 019836
Zabrodin, A.V. 037332
Zabusch, Prof. Stephanie 002962
Zaccaguini, Dr. Marisa 027566
Zaccome, Dr. Gian Maria 028121
Zacepina, Olga Vasiljevna 037849
Zacharias, Elke 021574
Zachariasse, Dr. F. 033045
Zachariou, E. 009414
Zacharkiw, Irene 054062
Zacharuk, Dr. Richard 018429
Zache, Dirk 017958, 017967
Zachman, Jon 049251
Zachman, Patty 052663
Zachorowska, Maria 034884
Zachos, Dr. K. 023089
Zachou, Eleni 022974, 023147
Zack, Wendy 005636
Zackel, Mag. Christine 003046
Zadgorska, Valentina 005191
Zadka, Ruth 025068
Zadoropžna, Liudmila 043141
Zadorožna, Nadija Fedorivna . . 043290
Zäch, Benedikt 042153
Zaffarano, Antonio 028373
Zaffino, Antonio 027925
Zaforgia, Elena 026668
Zafortesza Calvet, José 039164
Zafran, Eric 049390
Zagala, Maria 000762
Žagar, Zora 038587
Zaglmann, Klaus 020583
Zagnoli, Roberto 054682, 054687
Zagrina, Natalja Aleksandrovna . 037688
Žaguń, Wanda 035150
Zahaniciuc, Marcel 036348
Žahirović, Prof. Šaban 004269
Zahn-Biemüller, Dr. Eva 022791
Zahn, Donald 047752
Zahrádka, Jiří 009480
Zahradník, Dr. Zdeněk . 009577, 009961
Zaicev, Nikolai Alekseevič 029828
Zając, Zdzisława 035447
Zajączkowski, Wiesław 034721
Zajceva, Natalija Leonidovna . . 037134
Zajic, Václav 009628
Zajtsev, Oleg 043315
Zakatova, Anastasija Vasiljevna . 036920
Zakian, Michael 050611
Zakin, Helen 051841
Zakrevskaja, G.P. 037725
Zakrzewski, R.J. 049416
Zakythinos, Prof. D. 023181
Zalba, Pat E. 047512
Zaldivar, Brenda Projas 031517
Zaleski, Pierre 014678
Zaliene, Olga 030391
Zalpienė, Danutė 030382
Zalto, Kurt 002427
Zamarchi Grassi, Dr. P. 025358
Zambrano de Gentile, Flor 054718
Zamora Bonet, Imma 000093
Zamora Canellada, Alonso 040157,
                                  040160
Zamudio Díaz, Hermano Luis
  Alberto 008716
Zanche, Prof. Vittorio de 027150
Zander-Seidel, Dr. Jutta 020853
Zander, Charlotte 017460
Zander, Dr. Gisela 020209
Zandler, Richard 050178
Zandstra, Anne 032181
Zanella Manara, Dr. Emma 026365
Zanella, Prof. S. 026364
Zanelli, Dr. Gianluca 026394
Zanetta, Michèle 041259
Zang, Gongmei 008445
Zangenfeind, Hermann 002649
Zankl, Mag. Eva 002889
Zankman, Patricia 049185
Zannas, D.K. 023300
Zanni, Dr. Annalisa 026821
Zannieri, Nina 047129
Zanola, Oscar Pablo 000660

Zanoli, Emilio 026370
Zansitis, Peter Paul 047670
Zant, Herbert 002265
Zantfrisco, Liana 003693
Žantudieva, Elizaveta
  Magamedovna 037532
Zapałowa, Jadwiga 035035
Zapałowa, Kazimiera 034832
Zapata, Marian 023396
Zapolis, Frank 047640
Zaporožskaja, Valentina
  Aleksandrovna 036951
Zappa, Giancarlo 041431
Zaracki, Kathleen 051008
Zaragüeta, Vicente 039355
Zarb, Michael 007334
Zaremba, Ewa 035147
Zaremba, Tadeusz 035105
Zarev, Dr. Kosio 005057, 005058,
                                  005061
Zarimi, Dominique 015731
Zarina, Elena Gennadjevna 037652
Zarina, Ronald 051270
Žarmagambetov, Nurlan
  Kanikeevič 029802
Zarp, Frank 021664
Zarri, Dr. Enrico 026640
Zarrilli, Dr. Carla 027956
Zarzecka, Natalia 034860
Zarzuelo Villaverde, Manuel . . . 039916
Zarzycka, Joanna 035308
Zasur, Dahas 004711
Zatloukal, Prof. Dr. Pavel 009731
Zatulovska, Olena Petrivna 043125
Zau, Daw Nu Mya 031826
Zausch, Bärbel 018969
Zavada, Jeanne L. 051953
Zavala, Ignacio M. 039030
Zavattaro, Dr. M. 026279
Zavjalova, Irina Vasiljevna 037107
Zawacki, Andrew 050335
Zawiła, Gabriela 034803
Zawisza, Norbert 034952
Zawol, Zofia 035301
Zawrel, Dr. Peter 002978
Zayani, Abdelhak 031784
Zazo, Jennifer 051314
Zbarnea, Tudor 031710
Żbikowska, Janina 035144
Zbikowski, Dr. Dörte 019606
Zborowski, Janusz . 035011, 035012
Zdesenko, Vjačeslav Iosifovič . 037019
Zdravkovski, Dr. Dragiša 030629
Zea, Philip 049791, 051515
Zec, Prof. Dr. Peter 018310
Zecca, Walter 028230
Zech, Karen 048054
Zecha, Horst 021874
Zechlin, René 019057
Zeder, O. 014298
Zee, Dr. Pauline van der 003623
Zegadłowicz-Rudel, Atessa 035299
Zegher, Catherine de 051379
Zegna Rata, Olivier 011611
Zeh, Julia 016440
Zeh, Thomas 002066
Zehavi, Zvi 025143
Zehethofer, Gerhard 001887
Zehnder, Andreas 041559
Zehnder, Angela 038687
Zehnder, Walter 042096
Zeidler, Ingrid 021258
Zeiger, Antje 022708
Zeigerer, Wolfgang . 019604, 019610
Zeik, Travis 054368
Zeiner, Ulrich 018955
Zeipel, Agneta von 041098
Zeisler, Stefan 003026
Zeitler, Dr. Kurt 020576
Zeitlin, Marilyn 053700
Zeitlin, Dr. Richard 050603
Zeitschel, Charlotte 019989
Zekaria, Ahmed 010722
Zekorn, Beate 018452

Zeković, Marina 038185
Zeldenrust, Irene 032181
Zelenović, Radoslav 038162
Želev, Ivan 005097
Železnik, Adela 038549
Zelfel, Prof. Dr. Hans Peter 001858
Zelfel, Dr. Hans Peter 002402
Zelie, M. 038866
Zella, Robbin 047202
Zeller, Anselm 002874
Zeller, Peter van 045583
Zeller, Regine 018029
Zeller, Dr. Ursula 017177, 018528
Zellner, Donna 051619
Zemanová, Jana 009881
Zemanová, Zita 009917
Zembala, Dr. Dennis 048300
Zembik, Andrzej 034706
Zemer, Avshalom 025055
Zemovičs, Elmārs 030211
Zemter, Dr. Wolfgang 022697
Zeńczak, Anna 034862
Zender, Rick 047552
Zenegas-Millan, Dorenda 051960
Zeng, Dehai 008206
Zenks, Ralph 022388
Zentgraf, Henrike 020847
Zentz d'Alnois, Philippe 012174
Zepa, Līvija 030160
Zepa, Līvija 030159
Zepeczaner, Eugen 036590
Zeppenfeld, Dr. Burkhard . 020897,
                                  020898
Zerah, Dov 014706
Zerdina, V. 030200
Zernig, Kurt 001984
Zerr, Peter 020871
Zerrudo, Eric 034563
Žertogova, Izabella Jakovlevna . 036946
Zeschick, Dr. Johannes 021464
Zeskinskaja, Nelly 043274
Zessnik, Monika 017086
Zettel, Ernest 001958
Zetterström, Margareta 041009
Zettl, Charleen 047188
Zettler, Richard 052100
Zeuli, Prof. Errico 027608
Zeuli, Maria Francesca 027624
Zevallos Quinones, Dr. Jorge . . . 034410
Zevi, Dr. Anna Gallina . 027126, 027127,
                                  027128
Ževrnja, Prof. Nediljko 009001
Zgaga, Višnja 009048
Zhai, Dang 008441
Zhai, Tunjian 007909
Zhan, Xiangsheng 008388
Zhang, Benren 008453
Zhang, Cheng Zhi 008237
Zhang, Chengzhong 008496
Zhang, Dengxian 008141
Zhang, Dianren 008317
Zhang, Guangshan 007962
Zhang, Guiyun 007775
Zhang, Guowen 008334
Zhang, Huaide 008508
Zhang, Jia 007941
Zhang, Jiaji 007809
Zhang, Jian 008333
Zhang, Jianting 007762
Zhang, Jianwen 007987
Zhang, Jidong 007672
Zhang, Jinchao 008168
Zhang, Jing 008088
Zhang, Junshao 007793
Zhang, Linxiang 008172
Zhang, Liushun 008203
Zhang, Lizhi 008401
Zhang, Lizhu 008273
Zhang, Min 008112
Zhang, Ming 008098
Zhang, Mingjun 007755
Zhang, Muga 007834
Zhang, Ning 007587, 007591
Zhang, Peilian 008510

Zhang, Qiguo 008471
Zhang, Qiqing 007921
Zhang, San 008500
Zhang, Shuliang 007743
Zhang, Songrong 007700
Zhang, Wanxing 008461
Zhang, Wei 007666
Zhang, Wen Long 008343
Zhang, Wenbao 008370
Zhang, Wenju 008152
Zhang, Wenjun 008537
Zhang, Wu'ai 008249
Zhang, Xi'ai 008154
Zhang, Xiaojun 008130
Zhang, Xinli 007982
Zhang, Xuqiu 007968
Zhang, Yingchao 008455
Zhang, Yingjun 008365
Zhang, Yongxi 008436
Zhang, Yongzhong 007848
Zhang, Yuandong 007834
Zhang, Yushan 008134
Zhang, Zeguo 008337
Zhang, Zhaosheng 008049
Zhang, Zhigang 007703
Zhang, Zhipan 008036
Zhao, Changshu 007658
Zhao, Fusheng 007650
Zhao, Genxi 008426
Zhao, Guoxiang 008392
Zhao, Mingyong 008434
Zhao, Shuzhong 008083
Zhao, Sichao 008088
Zhao, Wannian 008056
Zhao, Yuan 008349
Zhao, Yuquan 008458
Zhao, Ziming 007693
Zhen, Qingfeng 008452
Zheng, Bingyan 008530
Zheng, Binhai 008280
Zheng, Chongjin 008365
Zheng, Junsheng 007858
Zheng, Mingjin 008308
Zheng, Peisheng 007783
Zheng, Siguang 008096
Zhi, Yunting 008259
Zhong, Chuangjian 007745
Zhong, Chuanya 007769
Zhong, Fayong 008557
Zhong, Fengcai 008100
Zhong, Guangcai 008013
Zhong, Mingqi 008013
Zhong, Yuqing 008521
Zhou, Chengshan 007971
Zhou, Chuike 008211
Zhou, Chunru 008528
Zhou, Dehua 008416
Zhou, Delin 008128
Zhou, Diguang 008080
Zhou, Diren 007734
Zhou, Enwei 007789
Zhou, Hao 007900
Zhou, Hengbin 008088
Zhou, Huajun 008079
Zhou, Jiaju 008062
Zhou, Jiakang 007758
Zhou, Jichang 007912
Zhou, Jingyi 008420
Zhou, Jinkang 007764
Zhou, Jinkui 008151
Zhou, Jisheng 007688
Zhou, Junzhi 008356
Zhou, Rongben 007827
Zhou, Ronglin 007964
Zhou, Shengrui 008089
Zhou, Shiwu 007708
Zhou, Tianlin 007947
Zhou, Tianyou 008399
Zhou, Weilin 008398
Zhou, Xinhua 008379, 008486
Zhou, Zhao 007692
Zhou, Zhongqing 008465
Zhu, Jiping 008235
Zhu, Mingye 007568

Zhu, Xianjin 008245
Zhuang, Zhongneng 008387
Zhuo, Daocheng 008407
Ziąbka, Leszek 034802
Ziauddin, Muhammad 003219
Zibelo, Fisseha 010727
Zibler, Barbara 017166
Zich, Dr. Bernd 018965
Zidić, Igor 009042
Zidlick, Pauline 046795
Zieglar, Ken 052107
Ziéglé, Anne 012112
Ziegler, Dr. Reinhard 022056
Ziegler, Dr. Susanne 017112
Ziegler, Thomas 022065
Zieglgänsberger, Dr. Roman . . 021326
Ziehe, Dr. Irene 017237
Zielinski, Melissa L. 046520
Zielonka, Mariusz 035204
Zieman, Judge Fred 050088
Ziemann-Heitkämper, Sonja . . 019204
Zierden, Martha 047543
Zierke, Franciszek 035456
Ziesche, Ferdinand 018975
Zieselman, Ellen 053113
Ziesemer, Günter 018359
Ziessow, Dr. Karl-Heinz 017747
Zięzio, Ryszard 034836
Ziffer, Dr. Irit 025193
Zifferero, Dr. Andrea 028105
Žiganova, Nina IljiniČna 037114
Zigmunde, Alida 030220
Zijlmans, Dr. J. 032586
Zijp, R.P. 032305
Zika, Paul 001147
Zilianti, Dr. Romualdo 025560
Žiljaeva, Natalja Evenjevna . . . 029826
Žiljanina, Tatjana Vladimirovna . 037077
Zillenbiller, Prof. Dr. Erwin 022319,
                                022320
Zillner, Gerd 002945
Zimina, I.D. 037973
Zimmer, Georg 020055
Zimmer, Jim L. 050353
Zimmer, Patricia 005365
Zimmerer, Kathy 047442
Zimmerman, Anne 047482
Zimmerman, Bob 048344
Zimmerman, Cindy 048344
Zimmerman, Daniel W. 048803
Zimmerman, Diane 052655
Zimmerman, Ryan 047909
Zimmerman, Sara C. 051511
Zimmerman, Vickie 048392
Zimmermann, Andreas 016926
Zimmermann, Friedrich 001869

Zimmermann, Harvey 054078
Zimmermann, Prof. Dr. Konrad . 021486
Zimmermann, Paul 019297
Zimmermann, Petra 019522
Zimmermann, Prof. Udo 017066
Zimmern, Emily F. 047574
Zimmern, Friederike 017822
Zimprich, Shirley 048575
Zindler, Debra 050796
Zingarelli, Thomas V. 048670
Zingg, Ernst 041825
Zink, Bob 050940
Zinke, Dr. Detlef 018478
Zinnkann, Dr. Heidrun 018435
Zint, Günter 019018
Zintz, Daniel 015807
Ziolkowski, Anne 048057
Ziolkowski, Jan 054108
Zipfel, Dr. Bernhard 038744
Zipursky, Gerry 007296
Žirgulis, Rimantas 030381
Zirnitis, P. 030214
Žitko, Duška 038587
Zitko, Salvator 038529
Živankić, Radovan 038261
Živković, Vesna 038134
Zizman, Ron 006859
Zjablicki, Aleksej Ivanovič 029845
Zjeżdżałka, Ireneusz 035113
Zjumčenko, Ljudmila Petrovna . 037876
Zlatar Stambuck, José 034319
Zlatar, Milena 038612
Zlobina, Nina Nikolaevna 037135
Žmuda, Anna 034987
Žmuda, Maurycy 034931
Znamenov, V.V. 037812
Žnidarič, Dr. Marjan 038571
Znojek, Jerzy 035100
Zoe, Vivian F. 051650, 054303
Zöller, Achim 021836, 021837
Zoer, J.A. 032245
Zoete, Johan de 032395
Žogina, Viktorija Stepanovna . . 037768
Zoli, Renato Giancarlo 028080
Zoller, Tamara 046677
Zolli, Barbara T. 052207
Zollinger Dettwiler, lic. phil.
   Yvonne 041673
Zolotareva, Tatjana Nikolaevna . 037661
Zolotina, I.A. 037753
Zolotinkina, Dr. Larissa Igorevna 037758
Zolotova, Anastasija 037353
Zoltán, Kató 036695
Zombori, Dr. István 023924
Zombork, Márta 024025

Zomborka Szűcsné, Márta 024027
Zommer, Raya 025036
Zona, Louis A. 054599
Zong, Changhai 008136
Zonneveld-Kouters, M.C. 032751
Zopf, August . . 001910, 002411, 002414
Zorić, Ivanka 038168
Zorn, Dr. Bettina 003046
Zorn, Dr. Olivia 017065
Zorzi, Dr. Marino 028324
Zorzoliu, Traian 036446
Zou, Lianhe 008330
Zou, Zhifang 007791
Zoubek, Ernst 002849
Zoubek, Thomas 053483
Zrebiec, Alice 048258
Zrelov, Vladimir Andreevič 037705
Zselyi, István 023425
Zsidi, Dr. Paula 023456, 023480
Zsigmond, Attila 023463
Zsutty, Andrea 002944
Zsutty, Dr. Gerhard 003101
Zsuzsa, Dr. Fodor 024040
Zubajdullaeva, S. 054643
Zuber, Michaela 018385
Zuberer, Peter 022467
Zubiaurre, Pablo 000131
Zubov, Aleksandr Nikolaevič . . . 037991
Zubrij, Elena Stanislavovna . . . 037025
Zucal, Wendy R. 048244
Zuccali, Jean-Pierre 013865
Zucchi, Cristina 026834
Zuckerman Jacobson, Heidi . . . 046579
Zuckerman, B. Michael 047419
Zuckriegl, Dr. Margit . . 002630, 002631
Zünd, Marcel 042031
Zünd, Peter 041815
Zürcher, Barbara 041209
Züst, Ernst 042168
Zueva, Ljubov Vasiljevna 037921
Zufferey, Anne-D. . . . 041917, 041986
Žugan, Vitalij Ivanovič 037934
Zugaza , Miguel 039676
Zugazagoitia, Julian 051432
Zuidervaart, Dr. Huib J. 032727
Zuiderwish, Dr. A.C. 032002
Žuk, Oksana Jaroslavivna 043234
Żuk, Stanisław 034657
Zukauskiené, Genovaite 030474
Žukauskiené, Genovaite 030463, 030470
Žukauskiené, Rita 030425
Zukiwski, Wilson 007008
Zukulis, Ivars 030216
Zuleski, Jim 049747
Zuleta, Pilar 008612
Zumaeta Zúñiga, Hector 007552

Žumalieva, Tamara
   Kalmurzaevna 029788
Zumbini, Vincenzo 026087
Zumpe, Ulrike . 021180, 021181, 021199,
                                021202
Zumwalt, May Louise 051884
Zuna-Kratky, Dr. Gabriele 003089
Zunk, Bodo 021351
Žunusov, Remistokol Tuskenovič 029855
Zuo, Zijun 008220
Zupan, Michelle 049148
Zuparko, Robert 046934
Žuravlev, M.N. 003243
Žuravleva, Tatjana Jurjevna . . . 037537
Zurawski, Jarosław 035462
Zurbrugg-Tonezzer, Lucia 041399
Zurbuchen, Max 041667
Zurcher, Sarah 041499
Zurfluh, DDr. Anselm 041851
Zurfluh, Hubert 012514
Zuris, Donald P. 048004
Zurowski, Dr. Stanis'Elaw 035346
Zuskinová, Dr. Iveta 038380
Zutautas, Sofia 047649
Žutikov, Aleksandr Ivanovič . . . 037603
Zuver, W. Marc 054113
Zvelebil, Miloš 009869
Zviguilsky, Dr. Alexandre 012131
Zvirin, Prof. Yoram 025056
Zwaenepoel, Dr. J. 003703
Zwahlen, Ulrich 041622
Zwaka, Petra 017184
Zwanger, Meryl 051379
Zwart-Van der Gaag, A. 032522
Zwart, Joel 049188, 049189
Zweifel, Gwen 005852
Zwez, Annelise 042073
Zwicky, P. 041335
Zwiers, S. 033129
Zwierzyna, John 049372
Zwink, Dr. E. 022067
Zwolak, Adam 035498
Zybok, Prof. Dr. Oliver 021363
Zydlik, Stanisław 035304
Zykaner, Michaela 003034
Zykov, Andrej Viktorovič 037123
Zyman, Daniela 003091

Index of Institutions and Companies
Register der Institutionen und Firmen

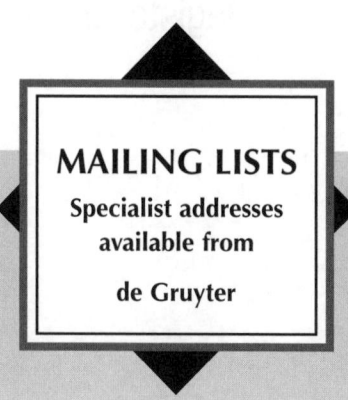
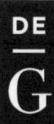

008 Antiquités, Paris 070853
02 Gallery, Akureyri 109215
1 2 3. . . Art Now, Bruxelles . . 100541
1 Artshop, Frechen 075836
1 Over 1a Gallery, Denver . . . 120677
I. Türkiye Büyük Millet Meclisi Binası,
Ankara Kurtuluş Savaşı Müzesi,
Ankara 042811
1. Deutsches Bockbier-Museum,
Taucha 022109
1. Deutsches Bratwurstmuseum,
Wachsenburggemeinde 022359
100 Titres, Bruxelles 100542
100% Natural, Cancún 132357
1000 et 1 Collections, Montluçon 070264
1000 Objekte, Zürich 087925
De 1001 Dingenhal, Alkmaar . . 082614
100th Bomb Group Memorial Museum,
Diss 044101
The 100th Meridian Museum,
Cozad 048048
103ans Antik och Kuriosa,
Sävedalen 086919
103rd Ohio Volunteer Infantry Civil
War Museum, Sheffield Lake . 053298
104 Antiques & Collectables,
Regina 064544
106 South Division Gallery, Calvin
College, Grand Rapids 049188
107 Workshop Gallery, Shaw . . 119075
1078 Gallery, Chico 047708
108 Fine Art, Harrogate 117912
10th Avenue Antiques, Calgary . 064244
111 Place Gift Antiques, Tulsa . 097704
11th Avenue Grafic Visuals,
Regina 101525
12 Sedie, Torino 081639
12 Vancouver Service Battalion
Museum, Richmond 006791
121 Photography Gallery, San
Antonio 124001
1220 Productions, Los Angeles . 121420
123 Watts Gallery, New York . . . 122260
1301 PE, Los Angeles 121421
13th Street Gallery, Omaha 123300
14–1 Galerie, Stuttgart 108684
14 Sculptors Gallery, New York . 122261
1405 Art Gallery, New York 122262
145 Antiques, New York 095184
14th Street Painters, New York . 122263,
122264
150 ans de Couture, Aix-les-
Bains 011566
15th Field Artillery Regiment Museum,
Vancouver 007270
15th Street Gallery, Salt Lake
City 123941
16th Street Gallery, Salt Lake
City 123942
17 Reasons, San Francisco 124199
17 Wing Heritage Collection,
Winnipeg 007432
1708 Gallery, Richmond 123715
1719 Galerie, Alkmaar 112099
1796 Bantry French Armada Exhibition
Centre, Bantry 024831
The 1811 Old Lincoln County Jail
and 1839 Jailer's House Museum,
Wiscasset 054504
1822-Forum Frankfurter Sparkasse,
Frankfurt am Main 018403
1859 Jail-Marshal's Home and
Museum, Independence 049662
1874 House Antiques, Portland . 096347
1890 House-Museum and Center for
Arts, Cortland 048012
18m Galerie für Zahlenwerte,
Berlin 106140
19 Vek, Sankt-Peterburg 085227
19. Sajandi Tartu Linnakodaniku
Muuseum, Tartu 010688
1900–1970, Bruxelles 063294
1908 Cash Register, Tampa . . . 097532
1910 Boomtown, Western

Development Museum,
Saskatoon 006968
1914 Plant City High School
Community Exhibition, Plant
City 052192
1918 Artspace, Shanghai 102167
198 Gallery, London 044897
1997 Avrupa'da Yılın Müzesi, Anadolu
Medeniyetleri Müzesi, Atpazarı 042836
19th Century Shop, Baltimore . . 144979
19th Century Willowbrook Village,
Newfield 051524
1blick – Minigalerie im Vorhaus,
Hallein 100031
2–28 Memorial Museum, Taipei . 042495
2 Effe, Trento 111024
2 Erre, Modena 131555
2 Ravens, Kirkby Stephen 144507
II. Türkiye Büyük Millet Meclisi Binası,
Cumhuriyet Müzesi, Ankara . . 042812
2+3D.Grafika plus produkt,
Kraków 138949
20 North Gallery, Toledo 124680
20/20, Dublin 138779
20/21 British Art Fair, London . . 098347
20/21 International Art Fair,
London 098348
2000 & Novecento, Reggio
Emilia 110705
2000 Art Galleries, Sydney 099758
2000 Cadres & Idées, Paris . . . 128885
2000 Ciutat Vella, Barcelona . . 133090
2023 Antiques, Louisville 094512
2040 Gallery, Austin 119634
20first Century Art, Godalming . 117865
20th Century Antiques,
Vancouver 064786
20th Century Antiques,
Wellington 084150
20th Century Consortium, Kansas
City 127273
20th Century Gallery, Baltimore . 092283
20th Street Art Gallery,
Sacramento 123752
21 Grand Gallery, Oakland 123239
21 Notre Dame O, Montréal . . . 064356
21-seiki Gendai Bijutsukan,
Kanazawa 028769
21_21 Design Sight, Tokyo 029486
217 Art Contemporain, La Croix-de-
Rozon 116234
21st Ctenuary Gallery, Bungay . 111443
21st Century Antics, Salisbury . 091218
21st Century Bob, San Diego . . 097047
221 Antiques, Forest 127225
222 Westlake Antique Mall,
Seattle 097385
23rd Street Antique Mall, Oklahoma
City 095922
24 Carat, Damascus 088025
24 Hour Museum, The National Virtual
Museum, Brighton 139118
24HR Art, Northern Territory Centre for
Contemporary Art, Parap 001395
26 Martyrs Museum, Nagasaki . 029085
26th Field Regiment RCA/XII Manitoba
Dragoon Museum, Brandon . . 005458
274 Gallery, Northbridge, Western
Australia 099482
275 Antiques, London 089876
Le 29, L'Isle-sur-la-Sorgue 069548
29 Palms, Fresno 093488
291 Antiques, London 089877
2AH, Grenoble 068707
2bis Editions, Paris 136846
2X de Afstand, Zaltbommel 112719
3 – Fil, Bari 079756
3 Arts Studio, Honolulu 093519
3 Day Gallery, Dallas . 120531, 120532
3 Falk Gallery, Tucson 124704
3 Monkeys, Portland 096348
3 Ten Haustudio, Atlanta 119529
3 Vue Gallery, Chicago 120019
303 Gallery, New York 122265

313 Antiques, London 089878
33rd Street Galleria, New York . 122266,
122267
34., Beograd 085326
364 Hayes Street Gallery, San
Francisco 124200
3A Antiques, Norwich 090864
3AR, Versailles 129337
3Arts, Three Art Clubs of Chicago,
Chicago 120020
3d Gallery, Bristol 117434
3F, Versailles 074037
Le 3L, Toulon (Var) 129282
3M Collection of Contemporary
Western and Americana Art, Long
Lake 050386
4 Art, Chicago 120021
4 Cats Gallery, Abbotsford 098606
4-H Schoolhouse Museum,
Clarion 047765
4 Ladies Art Gallery, Chicago . . 120022
4 Star Gallery, Indianapolis 121160
40 George Street Contemporary Art,
Mount Eden 084002
41 Artecontemporanea, Torino . 110946
A la 42ème Ligne, Paris 141014
43rd Street Gallery, Richmond . 123716
448th Bomb Group Memorial Museum,
Seething 045752
4599 Antiques, Miami 094673
45th Infantry Division Museum,
Oklahoma City 051726
4731 Gallery, Detroit 120796
48th Highlands Museum, Toronto 007155
The 4th Floor Gallery, Surry Hills 099741
4th Infantry Division Museum, Fort
Hood 048813
5 + 5, New York 122268
5 Art, Tampa 124622
5 Clock Gallery, Lausanne 116407
Le 5 Elément, Paris 141015
5 Elemente Museum, Waidhofen an
der Ybbs 002889
5 Green Boxes, Denver 120678
5-Star Collectibles, Honolulu . . 093520
50 Penn Place Art Gallery, Oklahoma
City 123270
500 X Gallery, Dallas 120533
50er Jahre Erlebniswelt,
Zusmarshausen 022903
50er-Jahre-Museum, Büdingen . 017663
511 Gallery, Lake Placid 121327
511 Gallery, New York 122269
515, Phoenix 123475
51st Avenue Art, Calgary 101210
539 Gallery, New Orleans 122110
54 the Gallery, London 118141
56 Gallery, Milano 110228
57th Street Antique Mall,
Sacramento 096589
5BE Gallery, New York 122270
5th and J Antiques, San Diego . 097048
5th Avenue Books, San Diego . . 145514
6 B Antik og Brocante,
København 065819
600 Jahre Kunst – Studiensammlung
der Universität Leipzig, Kustodie der
Universität Leipzig, Leipzig . . . 019971
626 Art Gallery, Los Angeles . . . 121422
63 Gallery, Seoul 111609
64 Steps, Toronto 101555
66 Cuadros, Valencia 115472
7 Galerie, Athinai 109014
El 7 Peus, Barcelona 085713
7. Antykwariat, Warszawa 143511
779, Paris 104441
799 Art Gallery, New York 122271
7th Avenue Antique Mall, Saint
Paul 096798
80 Washington Square East Gallery,
New York University, New York 051319
82 Republic, Hong Kong 102045
824 Gallery, Jacksonville 121228
825 Art, Denver 120679

825 Gallery, San José 102476
82nd Airborne Division War Memorial
Museum, Fort Bragg 048787
82nd Street Beanies,
Indianapolis 093844
84th Street Antiques Corporation, New
York 095185
871 Fine Arts, San Francisco . 124201,
145533
901 Gallery, Fort Worth 120848
918 Antique Gallery, Los Angeles 094236
921 Earthquake Museum, National
Museum of Natural Science,
Kengkou 042375
99' Art Center, Taipei 116900
A & A, Osnabrück 077956
A & A, Praha 065455
A & A Antique Restoration, Long
Beach 135690
A & A Antiques, Milwaukee . . . 094823,
127323
A & A Antiques, Pittsburgh 096248
A & A Art, Denver 120680
A & A Art and Frame,
Washington 124843
A & A Upholstery Company, San
Antonio 136244
A & A Upholstry, New Orleans . . 135853
A & B Antique Furniture,
Albuquerque 092018
A & B Antique Furniture, San
Antonio 136250
A & B Antiques, Hampton 061587
A & B World Coins and Antiques,
Chicago 092643
A & C, Zeuthen 079141
A & C Black, London 138180
A & C Black Publishers, London 138181
A & C Gallery, London 118142
A & C Upholstery, San Francisco 136302
A & D Antiquity, Birmingham . . . 092379
A & D Gallery, London 118143
A & E Armitage Antique Gallery,
Chicago 092644, 135397
A & E Downtown Antiques, San
Antonio 096932
A & F Galleries, Tweed Heads
South 099844
A & H Antiques, Milwaukee 094684
A & J Book World, Speedway . . 145610
A & J Glas-Porslinslagning,
Stockholm 133563
A & M Antique, Albuquerque . . . 092019
A & N Art Gallery, New York . . . 122272
A & P, Palermo 081013
A & S Art, New York 122273
A & T Kunsthandel, Plaidt 078053
A-1 Antiques and Auction House,
Houston 093580
A-1 Furniture, San Jose 136353
A-1 Furniture Refinishing, Los
Angeles 135697, 135698
A-1 Jewelry and Coin, Chicago . 092645
A 1 Upholstery, San Antonio . . . 136251
A-3, Moskva 114189
A A Antiques, Christchurch 083846
A à Z Antiquités, Lannion 069181
A Ages Ago, Denver 093236
A Althée, Marseille 140938
A-Antikvariaatti, Pori 140741
A-Antique, Long Beach 135691
A Antiques, New Orleans 095043
A Art, Grimmen 076095
A Arte Studio Invernizzi, Milano . 110229
A-Auction by Jer, Reading 127416
A Badacsonyi Állami Gazdaság Üzemi
Gyűjteménye, Badacsony 023413
A Barn Full of Brass Beds,
Conisholme 088968
A Better Antique Buyer,
Baltimore 092284
A Big Difference, Eagle Heights . 098992
A Bit of the Past, Austin 092200

A Book A Brac Shop, Miami
Beach 145259
A Book Hunter, Saskatoon 140385
A Booksearch, Nelson 143332
A Cappella Book, Atlanta 144959
A Child's Heart, Pittsburgh . . 096249
A City Jinshangjing History Museum,
Acheng 007561
A Classical Record, New York . . 122274
A Coin Gallery, Jacksonville . . 093943
A Collector's Bookshop, Saint
Louis 145495
A Contrario, Limoges 104022
A Country Place Antiques, Saint
Louis 096658
A Creative Eye, Goulburn 099093
A Damn Fine Artist, Austin . . 119635
A Different Angle, Oakland . . 123240
A Doll Affaire, Sacramento . . . 096590
A Dong Gallery, Ha Noi 125051, 125052
A l'Espace des Arts, Lyon . . 104058
A Field of Dreams, Oklahoma
City 095923
A Gallery, Göteborg 115623
A Gallery, London 118144
A Gallery, Salt Lake City 123943
A Gallery for Fine Photography, New
Orleans 122111
A Gift for All Seasons, Honolulu 120917
A Gift of Art, Tampa 124623
A Glas of Beauty, Amsterdam . 082672
A Glory to Behold, Tucson . . . 097621
A Good Buy, Los Angeles . . . 094237
A Healy Furniture Restorers,
Dublin 130844
A Joy Forever Antiques, Brighton,
Victoria 061197
A la Belle Epoque, Caen 067553
A la Belle Epoque, Giromagny . . 068667
A la Belle Epoque, Provins . . 072315
A la Belle Epoque, Sallanches . 073383,
129249
A la Belle Epoque,
Sarreguemines 073417
A la Belle Epoque, Vevey . . . 087873
A la Bonne Brocante, Paris . . . 070854
A la Bourse d'Or, Paris 070854
A la Brocante de l'Eglise, Marcq-en-
Ostrevent 069868
A la Brocante du Musoir, Le
Tréport 069366
A La Capucine, Québec 064517
A la Carte Antik, Stockholm . . 086955
A la Cour de France, Montreuil-sur-
Mer 070328
A la Galerie Placier, Paris . . . 104442
A la Grange, Sains-en-Amienois 072658
A la Grange du Léman, Morges 087683
A la Maison, Sammeron 073392
A la Manière, Rouen 105465
A la Mûrisserie, Le Mans . . . 069299
A la Porte de Sologne, Argent-sur-
Sauldre 066708
A la Porte Notre Dame, Cambrai 067613
A la Poussière d'Or, Marin . . . 069876
A la P'tite Vitrine, Lausanne . . 087558
A la Recherche du Temps Perdu,
Cabourg 067544
A la Recherche du Temps Perdu, Le
Puy-en-Velay 069341
A la Reine Margot, Paris 070856
A la Vieille Cité, Paris 070857
A la Vieille France, Metz 070082, 128796
A la Vieille Russie, New York . . 095186
A l'Age de Bronze, Murat
(Cantal) 070399
A Lamp & Fixture Shoppe,
Chicago 135398
A l'Ancien Temps, Ajaccio . . 066490,
066491
A l'Arbre de Jessé, Sens 073495
A l'Art Antique, Marly-le-Roi . . 069878
A l'Art Notre, Périgueux 105333
A l'Enseigne du Commissaire Maigret,

Liège 140239
A l'Espace des Arts, Saint-Rémy-de-
Provence 105606
A l'Espace des Particuliers,
Clamart 067998
A l'Estampille, Metz 070083
A Likely Story, Alexandria . . . 144955
A l'Imagerie, Paris . . 070858, 104443
A Literary Saloon, Seattle . . . 145574
A Little Bit of Kiwi, Roxburgh . . 113002
A Logger's Life Provincial Museum,
Grand Falls-Windsor 005901
A Lost Era, Chicago 092646
A Marie Stuart, Paris 070859
A Muse Gallery, Columbus . . 120468
The A Museum of Art, Haruhiko
Tanaka Bikutsukan, Takane . . 029438
A New Doll in Town, Miami . . 094674
A Niche in Time, Chicago . . . 092647
A Nitch in Tyme, Norfolk 095839
A-One Gallery, New York . . . 122275
A Piece of History, Launceston . 061723
A Place in Time, Jacksonville . 093944
A Pleasant Cove, Jacksonville . 093945
A Plus Auctions, Delavan . . . 127200
A Plus Furniture, Houston . . . 093581
A Quaint Little Shoppe, Atlanta . 092096
A Refinishing Touch, Jacksonville 135641
A Renaissance Book Shop,
Milwaukee 145263
A Repeat Performance, New York 095187
A Restoration Centre, Dún
Laoghaire 079601, 130864
A Room With a Vieux, Boston . . 092439
A Room With A View, Kortrijk . . 083714
A Rug Importer, Tucson 097622
A Ruth Creation, Cincinnati . . 120311
A Saint-Benoît-des-Prés, Paris . 141016
A Second Look Books, Kitchener 140353
A Secret Closet, Chicago 092648
A Select Few, New York 095188, 135868
A Shoal of Mackerel, Glasgow . 138166
A Simpler Time, Minneapolis . . 121960
A Soldier's Life, Newcastle-upon-
Tyne 045343
A Tegnap Autói, Tatabánya . . 024001
A Touch of Asia, San Francisco . 097153
A Touch of France, Boston . . . 119825
A Touch of Mystery, Norfolk . . 095840
A Touch of Nostalgia, Edmonton 064281
A Touch of the Past, Maryborough,
Queensland 061796
A Unique Antique, Chicago . . . 092649
A Unique Blend, Salt Lake City 096869
A Vintage Boutique, Saint Louis 096659
A Votre Servies, 's-
Hertogenbosch 083380
A-Z, Nürnberg 077808
A! Diseño, Tlalpan 138900
A&A Galeria Sztuki Dawnej,
Poznań 084635, 113715
A&B, Roma 081188
A&E Graphics Complex, Houston 120981
A&I Stripping & Refinishing,
Albuquerque 135275
A&M Galerie, Binz, Ostseebad . 106502
A. Giordano Gallery, Dowling College,
Dept of Visual Art, Oakdale . . 051672
A. Kutateladze State Art Academy,
Tbilisi 055431
A. Paul Weber-Museum,
Ratzeburg 021292
A. Schiiwesch Musée Rural,
Landmuseum A Schiewesch,
Binsfeld 030565
A. Shervashidze Art School,
Tbilisi 055432
A. Sohos Museum, Tinos 023302
A.A. Antik, Göteborg 086689
A.A. Gallerie, Greve 065711
A.A. Gallery, Ho Chi Minh City . 125108
A.A. Ringsted Antik, Ringsted . 066073
A.A. Gallery, Bruxelles 063295
A.A.T.A. Journal, Alexandria . . 139233

A.B. Antiquités, Bruxelles 063296
A.B. Works, Pittsburgh 096250, 136126
A.B.A. Bookhouse, Saint Kilda,
Victoria 139903
A.B.C. Antiquités, Marseille . . 069890
A.B.C.D, Châteauneuf 067886
A.B.C.D., Louviers 069649
A.B.D.S., Anglet 066619
A.B.D.S., Pau 072030
A.B.L, Chassy 067866
A.B.L.E. Tyson China Repairs, South
Yarra 127677
A.B.O, Plouénan 072198
A.C., Roma 081189
A.C. Gilbert's Discovery Village,
Salem 052887
A.C.O. – Art de Chine & d'Orient,
Paris 070860
A.C.R., Paris 128886
A.D. Buck Museum of Natural History
and Science, Nothern Oklahoma
College, Tonkawa 053767
A.D.O.M., Paris 070861
A.D.S., Léguevin 069382
A.E. Seaman Mineral Museum,
Michigan Technological University,
Houghton 049568
A.F. Antic, Constanţa 143623
A.F. Antiquités, Bruxelles 063297
A.G. Ar., Roma 081190
A.H. Tammsaare Muuseum,
Vetepere 010715
A.J. Antiquités, Paris 070862
A.J.'s Antiques, El Paso 093382
A.J.'s Auction and Antiques, Oklahoma
City 095924, 127376
A.J.'s Posters and Prints,
Chicago 120023
A.J.'s Toys, El Paso 093383
A.K.A. Gallery, Saskatoon . . . 006969
A.K.R., Quirnbach 078113
A.L. Fetterman Educational Museum,
Fort Valley 048883
A.L.F.A, Paris 104444
A.M. de Jonghuis, Nieuw
Vossemeer 032670
A.M.J.P. Antiquités, Montlouis-sur-
Loire 070263
A.M.Y. Japan, Yokohama . . . 082222
A.N.D., Paris 070863
A.N.Z. Gallery Fine Art,
Brunswick 098812
A.P Art, Lille 103997
A.R. Bowman Memorial Museum,
Prineville 052372
A.R. Mitchell Memorial Museum of
Western Art, Trinidad 053803
A.R. Ullmann Ltd., London . . . 089879,
134576
A.R.C. Gallery, Chicago 047633
A.R.C.O., Abano Terme 109733
A.R.CA.M.S., Autun 128454
A.R.T Mimo, Marseille 128739
A.S. Antique Galleries,
Manchester 090687, 134829
A.S. Starožitnosti, Pardubice . . 065441
A.S. Starožitnosti, Praha 065456
A.S.I.E., Marseille 069891
A.Shiwi A.Wan Museum, Zuni . . 054621
A.V. Recycle Center, Tokyo . . 082129
A.V.M., Paris 104445
A.W. Campbell House Museum,
Alvinston 005325
A.W.A. Electronic-Communication
Museum, Bloomfield 047036
a+mbookstore, Milano 142706
A1 Antik Snurrepiben, Nyborg . . 066008
A1 Antiques, Baltimore 092285
A1 Furniture Restoration,
Wellington 132755
A1 Jukebox and Nostalgia,
Baltimore 092286
A1A Antiques, Ulverston 091688, 135169
AA Antik-Schatzkammer,

München 077442
Aaaz Scris, Liège 063736
AAB Antiquités, Bruxelles 063298
Aabenhus Århus Antikvariat,
Århus 140548
Aabenraa Museum, Sønderjyllands
Søfartsmuseum, Aabenraa . . 009995
AABP, Bruxelles 063299
Aachen, A., Miami 094675
Aachener Kunst- und
Antiquitätenhändlerverband 1976,
Aachen 058702
Aachener Kunstblätter, Köln . . 138666
Aachener Printenmuseum,
Aachen 016431
AAF Museum, Washington . . . 054085
Aagaard & Co, Robert,
Knaresborough 089721
Aakar Art Gallery, Mumbai . . . 109411
Aal-Jays, Las Vegas 094105
Aalbers & Israëls, Enschede . . 083245
Aalborg Historiske Museum,
Aalborg 009998
Aalborggaarden, Kragerø 084264
Aalderink, Amsterdam 082673
Aalders, Evelyn, La Garde-Freinet 103890
Aalesunds Kunstforening i Perken
Kulturhus, Ålesund 059652
Aalholm Automobil Museum,
Nysted 010251
Aaltense Oorlogs en Verzetscollectie
1940–1945, Aalten 031882
AAM, Tübingen 078719
Aamiri, Berlin 074737
Aan de Oude Delft, Delft 083025
Aanblick Galerie, Groningen . . 112486
Aankoop Alle Antiek, Antwerpen 063180
Aankoop en Verkoop Antiek,
Antwerpen 063181
De Aanschouw, Rotterdam . . 112627,
112628
AAntikvariat, Glostrup 140576
AAOARTS, Paris 070864
Aaor, Denver 093237
Aarde, Almere 082639
Aardewerk, A., Den Haag . . . 083044
Aardewerk & Zoon, Betty, Den
Haag 083045
Aardewerkvermuseum Petrus Pegout,
Noordbroek 032695
Aardvark, Fyshwick 061492
Aardvark, Milwaukee 145264
Aardvark, Moorak 061882
Aardvark, San Francisco 145534
Aardvarks, Christchurch 083847, 083848
Aareena, Castle Hill 061294
Aarehüsi, Thun 116668
Aargauer Kunsthaus, Aarau . . 041186
Aargus, Frankfurt am Main . . 075767
Aarhus Kunsthandel, Århus . . 065593
Aaron, Chicago 120024
Aaron Bros., Fort Worth 120850
Aaron, Fort Worth . . . 120849, 120851
Aaron Bros., Houston . . 120982, 120983
Aaron Gallery, London 089880
Aaron, New Orleans 095044
Aaron, Santiago de Chile 064940
Aaron, Tunbridge Wells 091644, 144891
Aaron, A., Singapore 114469
Aaron, Anton, New York 095189, 135869
Aaron, Didier, London . 089881, 118145
Aaron, Didier, New York 095190
Aaron, Didier, Paris . . 070865, 070866,
104446
Aaron, Didier, Washington . . . 124844
Aaron, John, Denver 120681
Aaron's Art, Indianapolis 121161
Aarons Galerie, Art & Living,
Hamburg 076178, 107216
Aaronson House, Zichron Yaakov 025224
Aarquebus Antiques, Saint Leonards-
on-Sea 091195
Aars, Buenos Aires 139541
A'Arte, Saint-Ouen 072998

Aarwun, Ngunnawal 099455
Aas, Hannelore & Norbert,
Bayreuth 074689
Aasiaat Katersugaasiviat Museum,
Aasiaat Museum, Aasiaat 023328
Aasland Museum Taxidermy,
Cranbrook 005662
Aatma, Helsinki 140688
Aauld Antiques, Tauranga 084097
Aavikute Majamuuseum, Saaremaa
Muuseum, Kuressaare 010630
Aavishkar, Kolkata 109392
AB Antik, Praha 065457, 065458
AB Antiquités, Toulon (Var) 073696
AB Antiquités, Versailles 074038
AB Bogmarked, København 140588,
140589
AB Books, Croydon 139707
AB Furniture, Caulfield South .. 061305
ABA Centenary Antiquarian Book Fair,
London 098349
ABA Chelsea Book Fair, London 098350
ABA Militaria, Saint-Fargeau-
Ponthierry 072780
Abaad Antigüedades Boedo, Buenos
Aires 060844
Ababord Antiquités, Marseille .. 069892
Abac – Librairie Gilles Wolles,
Bruxelles 140163
Abaca, Bordeaux 067298
Abaca, L'Aigle 069143
Abaca, Libourne 069457
Abaca, Marseille 069893
Abaca Conservation-Restauration,
Paris 128887
Abacadabra, Nice 070543
Abacadabra, Nîmes 070682
Abacanthe, Pornic 072278
Abacar Antics, Valencia 086494
Abaco, Valencia 086495
Abaco Borgobello, Julio M., Florida
Oeste 060944
Abacule, Aix-en-Provence 066449
Abacus, Lowick 118671
Abacus, Roma 131790, 131791
Abacus Antiques, Louisville 094513
Abacus Art, Ashwood 127561
Abacus, Michel, Montréal 064357
Abacusgomberi, Milano 080377, 142707
Abad Duato, Antonio, Zaragoza . 133485
Abada, Felix, New York 122276
Abadie, Alain, Arles 066751
Abadir – Accademia di Belle Arti e di
Restauro, Palermo 055941
Abadir-Accademia di Belle Arti e
Restauro, Sant'Agata li Battiati 055977
Abafe, Saint-Thibéry 073312
Abakus Galeria 3, Wrocław 084759,
113915
Abal Tiempo's, Buenos Aires ... 060845,
127505
Abalache, Oklahoma City 095925,
145391
Abandoned Planet Bookstore, San
Francisco 145535
Abanico del Arte, Bogotá 102267
Abanicos Publicitarios Neder,
Barcelona 085714
Abante, Portland 123585
Abante, Stockholm 115824
Abantos Restauracion, Madrid .. 133284
Abaris Books, Norwalk 138435
Abascal, Ana & Patricia Medina,
Sevilla 086433
Abashiri Kyodo Bijutsukan,
Abashiri 028457
Abashiri Kyodo Hakubutsukan,
Abashiri 028458
Abashiri Prison Museum,
Abashiri 028459
Abásolo, Bilbao 133179
Abaúj Antikvárium, Budapest ... 079229,
126147
Abaziou, Hélène, Marseille 069894

Abbacus, Den Haag 083046
Abbadie, Marie-Claude, Bordeaux 067299
Abbakuse, Malmö 086842
Al-Abbar, Al-Kamara 117171
Al-Abbar, Dubai 117173, 117174
Abbas Antiques, Sherborne 091301
Abbas, Amer, Wien 100224
Abbasi Mosa, Torino 132082
The Abbasid Palace Museum,
Baghdad 024799
A Abbatia, Lisboa 084811, 143551
Abbaye Antiquité, Romainville .. 072518,
141316
Abbaye Antiquités, Montivilliers . 070258
Abbaye de Beaulieu en Rouergue,
Centre d'Art Contemporain, Centre
des Monuments Nationaux,
Ginals 013110
Abbaye de Bonnefont, Proupiary 015001
Abbaye de Cluny, Musée d'Art et
d'Archeologie, Cluny 012608
Abbaye de Flaran, Valence-sur-
Baïse 016094
Abbaye de Saint-Amand de Coly,
Saint-Amand-de-Coly 015209
Abbaye de Vauclair, Bouconville-
Vauclair 012129
Abbaye des Moines, Bruxelles .. 063300,
063301
Abbaye du Val Richer, Saint-Ouen-le-
Pin 015487
Abbaye Royale du Moncel,
Pontpoint 014961
Abbaye Saint-André, Villeneuve-lès-
Avignon 016267
Abbe Museum at Downtown, Bar
Harbor 046781
Abbe Museum at Sieur de Monts
Spring, Bar Harbor 046782
Abbé, Christophe, Seynod 073518
Abbeville, Paris 136847
Abbeville Press, New York 138370
Abbey and Stones Museum,
Margam 045224
Abbey Antiquarium,
Knaresborough 144509
Abbey Antiques, Cahir 079466
Abbey Antiques, Hemel
Hempstead 089494, 134451
Abbey Antiques, Norwood 062009
Abbey Antiques, Ramsey,
Cambridgeshire 091067
Abbey Antiques, Romford 134978
Abbey Antiques, Stirling 091428, 135083
Abbey Antiques, Winnipeg 064908
Abbey Antiques & Art, Auckland 083746
Abbey Artifacts, Duns 117700
Abbey Bookshop, Paris 141017
Abbey Clock Clinic, Austin 135318
Abbey Fine Arts, Ballina 130825
Abbey Galleries, Bath . 088375, 134071
Abbey Green, Utrecht 083632
Abbey House, Derby . 089052, 134321
Abbey House Museum, Leeds
Museums and Galleries, Leeds 044777
Abbey Museum of Art and
Archaeology, Caboolture 000921
Abbey Owney Upholstery,
Limerick 130879
Abbey Pumping Station Museum,
Leicester 044791
Abbey, Ron, Bowral 139652
Abbeydale Industrial Hamlet,
Sheffield 045769
Abbeygate Gallery and Antiques,
Grimsby 089390
Abbeys Auctions, Box Hill 125181
Abbona, Alessandro, Genova ... 080230
Abbot Hall Art Gallery, Kendal .. 044674
Abbot's, Kansas City 135673
Abbotsford House, Arbroath 043425
Abbotsford House, Melrose 045244
Abbott, Woodbridge 127098

Abbott & Holder, London 089882, 118146
Abbott, A., Ipswich 089643
Abbott, Dick, Burwell 117490
Abbott, Nicholas, Hartley Wintney 089448
Abboud & Haidar, Berlin 074738
Abby Aldrich Rockefeller Folk Art
Museum, Williamsburg 054408
ABC, Amsterdam 143008
ABC Antik, København 065820
ABC-Antiquariat, Hamburg 141853
ABC Antiquariat, Zürich 144155
ABC Antique, Bonn 075059
ABC Antiques, Knokke-Heist ... 063693
ABC Antiques, Books and Collectables,
Port Macquarie 062094, 139876
ABC Architectuurcentrum Haarlem,
Haarlem 032389
ABC Book Store, Toronto 140389
ABC Books and Posters, Denver 120682,
145121
ABC Carpet, New York 095191
ABC Chalonnaises, Chalon-sur-
Saône 067769
ABC Coin Company, Cincinnati . 092821
ABC Coin Company, Indianapolis 093845
ABC Coinarama, San Francisco . 097154
ABC Collectibles, Tucson 097623
ABC Collections, Tonnay-
Charente 073687
ABC de la Carte Postale Ancienne,
Lyon 069677
ABC Depôt, Tinqueux 073683
ABC Dépôt Vente, La Valette-du-
Var 069111
ABC Galerie, Bruxelles 100543
ABC Galleriet, Bergen 113123
ABC Gallery, Poznań . 113716, 113717
ABC-Kirja, Turku 140748
ABC Libri, Torino 142845
Abc Pascal, Clamart 067999
ABC-Solna Antikvariat, Solna ... 143922
Abcdaire, Angers 066594
ABCollectables, Lobethal 061744
Abd El-Fatah El-Sayed Soliman,
Alexandria 066215
Abdeen Palace Museum, Cairo . 010528
Abderhalden, Emil, Forch 138044
Abdis, Maulette 069998
Abdis, Sorel-Moussel 073539
Abdou, Casablanca ... 082537, 132371
Abdül Antik, İstanbul 088096
Al-Abdul Wahab, Riyadh 085311
Abe, Calgary 064245
Abé, A., Aalst 063170
Abecassis, Maxime, Mitry-Mory 070134
Abecassis, Paul, Saint-Ouen .. 072999
Abeceda Antiquariat, München . 142137
l'Abécédaire, Nice 140992
Abegg-Stiftung, Riggisberg 041883
Abeille, Argenteuil 066712
Abeille Antiquité, Saint-Ouen .. 073000
Abeille Débarras, Auzits 066858
Abel, Los Angeles 135699
Abel, Oslo 084289
Abel, Saint-Ouen 073001
Abel & Armand, Bruxelles 063302
Abel Tasman Kabinet, Lutjegast 032610
Abel, A., Paris 104447
Abelard, Toronto 140390
Abelein-Windsheimer & Windsheimer,
Blaufelden 075016
Abell, A.N., Los Angeles 127291
Abella, Antibes-Juan-les-Pins .. 066661
Abella, Barcelona 085715
Abella, Monaco 082495
Abels, Enschede 143138
Abels Kunsthandel, Oslo 084290
Abelsson, Mikael, Lidköping ... 086808
Abencerage, Paris 070867, 141018
Abend, Denver 120683
Abendstern, Wien 100225
Abeni Guerra, Brescia 131014
Abenteuermuseum Saarbrücken –
Sammlung Heinz Rox-Schulz,

Saarbrücken 021554
Aberbach, New York 122277
Aberbach, Susan, New York 122278
Abercauph, Carole, Philadelphia 136072
Aberconwy House, Conwy 044003
Aberdeen Art Gallery, Aberdeen . 043348
Aberdeen Arts Centre, Aberdeen 043349
Aberdeen Maritime Museum,
Aberdeen 043350
Aberdeen University Natural History
Museum, Aberdeen 043351
Aberdeen University Natural
Philosophy Museum, Aberdeen 043352
Aberdeen University Pathology and
Forensic Medicine Collection,
Aberdeen 043353
Aberdeenshire Farming Museum,
Mintlaw 045279
Aberdona Gallery, Alloa 043395
Aberfeldy Gallery, Aberfeldy ... 117214
Åberg, Kenneth, Göteborg 115624
Åberg, Lennart, Gamleby 086684
Abergavenny Art Gallery, Toodyay 099829
Abergavenny Museum,
Abergavenny 043366
Åbergs Museum, Bålsta 040556
Aberlé, Bruxelles 125304
Abernethy Museum, Abernethy . 043368
Abernethy Nature Heritage Museum,
Abernethy 005303
Aberseer Lipphaus, Strobl 002803
Abertillery Museum, Abertillery . 043369
Aberystwyth Arts Centre,
Aberystwyth 043370
Abe's Antique Attic, Indianapolis 093846
Abe's Antiques, Atlanta 092097
Abe's Antiques, New York 095192
Abe's House Clearance, Dublin . 079518
ABGG's Gallery, Makati 113377
Abgineh va Sofalineh, Tehrän .. 024743
Abguss Sammlung antiker Plastik,
Berlin 017064
Abhai Smarak Panchal Sangrahalaya,
Bareilly 024154
Abhey Arts, Bromley 117462
Abid, Ezzedine, Saint-Nazaire (Loire-
Atlantique) 072987
Abigael's Antiques, Annerley .. 061000
Abiko-shiritsu Kinchou Hakubutsukan,
Abiko 028461
Abim-Pljus, Sankt-Peterburg ... 085228
Abingdon Museum, Abingdon .. 043374
Abinger Bazaar, Abinger Hammer 088232
Abington, Finningham . 089278, 144426
Abington, Philadelphia 096069
Abington Art Center, Jenkintown 049810
Abington Park Museum,
Northampton 045389
Abis, Bonaria Giuseppa, Cagliari 079970
Abitarte, München 077443
Abitat, Ascona 116032
Abitzsch, R., Merseburg 077349
Abiuso, Vladimiro, Roma 131792
Abkar Museum, Sa'd-Abad Museum,
Tehrän 024124
Able, Houston 093582
Able, Saint Paul 096799
Able Art, Rotorua 112997
Ableart, Houston 120984
ABM, Saint-Ouen 141335
ABM-Utvikling, Oslo 059653
ABN Numismatique, Colmar ... 068080
ABN Numismatique, Saint-Louis 072897
Abner, Tampa 097533
Aboa Vetus – Arkeologis-historiallinen
Museo, Turku 011432
Abode, Providence 123692
Abon, Louis, Dijon 068310
Abony Lajos Falumúzeum, Abony 023405
abooks.de Detlef Thursch,
Düsseldorf 141712
Aboriginacorp, Milano 110230
Aboriginal and Pacific Art,
Ivanhoe 099156

Aboriginal and Pacific Art,
 Sydney 099759
Aboriginal and Tribal Art Centre,
 Sydney 099760
Aboriginal Art, Echuca 061435
Aboriginal Art and Culture Centre,
 Alice Springs 098626
Aboriginal Art Centres, Paddington,
 New South Wales 099508
Aboriginal Art Centres, Sydney . 099761
Aboriginal Art Galleries of Australia,
 Melbourne 099295
Aboriginal Art Gallery Concept Store,
 Sydney 062274, 099762
Aboriginal Art Museum, Utrecht . 032941
Aboriginal Art Online, West End . 062420
Aboriginal Art Prints, Darlinghurst 098941
Aboriginal Artists of Australia,
 Newcastle 061955
Aboriginal Arts, Vancouver 101730
Aboriginal Arts Adelaide,
 Adelaide 060964
Aboriginal Desert Art Gallery, Alice
 Springs 098627
Aboriginal Dreamings Gallery
 Ginninderra, Nicholls 099456
Aboriginal Fine Arts Gallery,
 Darwin 098947
Aboriginal Gallery of Dreamings,
 Melbourne 061811, 099296
Aboriginalartworks.com, Perth . . 062073
Aboriginart, Toronto 101556
Abos, Nérac 070501
Abott's, Turramurra 062353
Abou & Mireille Benibre, Thérèse,
 Toulouse 073731
Abou Dabou, Berlin 141465
Abou Jurab, Isa, Damascus 088026
Aboucaya, Georges, Paris 070868
Aboucaya, Martine, Paris 104448
Aboudaram, Christian, La
 Roquebrussanne 069097
About Antiques, Westbourne Park 062423
About Studio Gallery, Bangkok . 116944
About Time for Clocks,
 Greenwich 061558
Abovecloud, Los Angeles 094238,
 145208
Abovo Antik, Gnesta 086687
ABP, Vincennes 129357
ABPi, Lausanne 116408
Abracadabra, Bordeaux 067300
Abracadabra Bookshop, Rockfort 145480
Abracadabroc, Vernon 074034
Abraham Gallery, Southsea 119127
Abraham Lincoln Birthplace,
 Hodgenville 049498
Abraham Lincoln Book Shop,
 Chicago 145029
Abraham Lincoln Museum, Lincoln
 Memorial University, Harrogate 049382
Abraham Ojanperän Museo,
 Liminka 011117
Abraham, Eleanor, New York . . . 095193
Abraham & Co., Jackie, Miami . 094676
Abraham, Harry N., New York . . . 138371
Abraham, Joel, New York 135870
Abraham, Mike, Lichfield 144537
Abrahams, Hamburg 107217
Abraham's Mosterdmakerij,
 Eenrum 032273
Abrahams, Christine, Richmond,
 Victoria 099629
Abral Artproduct, Bucureşti 132990
Abrams, Harry N., New York . . . 138371
Al-Abrar, Sharjah 088193
Abrash, Dubai 117175
Abraxas Studio, San Diego 136282
Abreu, Belo Horizonte 063875
Abreu, Sacramento 123753
Abreu, Francisco F. Plácido, Porto 114071
Abreu, Iara, Belo Horizonte 100873
Abri du Cap-Blanc, Marquay . . . 014209
Abri du Passé, Castres 067720, 140817
Abri du Passé, Marseille 069895

Abri Préhistorique de Laugerie Basse,
 Gisement Préhistorique, Les Eyzies-
 de-Tayac-Sireuil 013747
Abrial, Saint-Tropez 105619
Abrocadabra-Passé par Ici, Neuilly-sur-
 Seine 070512
Abrocantique Antiquités Brocantes,
 Chalon-sur-Saône 067770
Abs, Hans-Joachim, Essen 106904
ABSA Group Museum,
 Johannesburg 038739
Absalom's Gallery, Broken Hill . . 098792
Absenta, León 115105
Abside Restauraciones, Madrid . 133285
Absidial, Plaudren 105355
Absinthe, Angers 066595
Absinthe Antiquité, Angers 066596
Absolu, Notre-Dame-des-Landes 070751
Absolute Art, Nottingham 118836
Absolute Art Company, Mermaid
 Beach 099331
Absolute Art Gallery, Knokke-
 Heist 100719
Absolute Art Gallery, New
 Orleans 122112
Absolutely, Toronto 064574
Absolutely North, Toronto 064575
Abstein, Atlanta 119530
Abstract Books, Indianapolis . . . 145176
Abstrait, Toulouse 105720
Abstrakt Creations, Chicago . . . 120025
Absynthe Antiquité, Angers 066597
ABTart, Stuttgart 108685
Abtei Rommersdorf, Neuwied . . . 020775
Abteilung für Asiatische und
 Islamische Kunstgeschichte, Institut
 für Orient- und Asienwissenschaften,
 Bonn 055489
Abteilung für Christliche Archäologie
 und Byzantinische Kunstgeschichte,
 Institut für Archäologische
 Wissenschaften (IAW), Freiburg im
 Breisgau 055543
Abteilung für Kunstgeschichte
 Ostasiens, Kunsthistorisches Institut
 Universität Zürich, Zürich 056441
Abteilung Handschriften
 und Alte Kunstdrucke der
 Universitätsbibliothek,
 Universitätsbibliothek Würzburg,
 Würzburg 022785
Abteilung Handwerk und dörfliches
 Leben des Heimatmuseums Neu-Ulm,
 Neu-Ulm 020676
Abu Ahmed, Dubai 088166
Abu Ahmed, Ibrahim, Manama . 063130
Abu Ali Ibn Sina Yodgorlik Muzeyi,
 Afchona 054622
Abu Dhabi Antiquarian Book Fair, Abu
 Dhabi 098343
Abu Al-Mudhyaf, Ruwi 084406
Abu Qais, Madinat Al-Sultan
 Qaboos 113297
El Abuelito Arturo, Bogotá 065167
Abuelo Antigüedades, Valparaíso 064991
Abuk Staraja Kniga, Sankt-
 Peterburg 143658
Aburaya, Sendai 082109
ABV Antik-Servis, Velikij
 Novgorod 085279
Abyss Marine, Liège 063737
Abysse, Morges 116516
Abysses, Donville-les-Bains 068383
AC – Sperka, Bopfingen 137119
AC Vent, La Roche-sur-Yon 069069
ACA Galleries, New York 122279
Acacia Antiques, Berwick 061151
Acacia Out Of Print, Adelaide . . 139609
Acacia Tree, Singapore 085353
Academia – Antiquariat an
 der Universität, Freiburg im
 Breisgau 141795
Academia Anticuario Euroartes,

Bogotá 065168, 128257
Academia Antiquariate an
 der Universität, Freiburg im
 Breisgau 141796
Academia Brasileira de Arte, São
 Paulo 004846, 055099
Academia Brasileira de Letras, Rio de
 Janeiro 004678
Academia Chilena de Bellas Artes,
 Santiago de Chile 055183
Academia de Arte George Enescu,
 Iaşi 056263
Academia de Arte Vizuale Ioan
 Andreescu, Cluj-Napoca 056261
Academia de Artes Plásticas San
 Alejandro, Marianao 055287
Academia de Artes Visuales,
 México 056142
Academia de Bellas Artes Martín Soria,
 Vitacura 055186
Academia de Dibujo y Pintura Mirart
 Mirasierras, Madrid 056367
Academia de Letras e Artes do
 Planalto, Luziânia 055083
Academia de San Carlos, Universidad
 Nacional Autónoma de México,
 México 056143
Academia do Saber, Rio de
 Janeiro 140287
Academia Gallery of Heffers Virgin
 Academy, Jerewan 098593
Academia Music, Tokyo 142918
Academia Nacional de Bellas Artes,
 Buenos Aires 054862
Academia Nacional de Bellas Artes
 Hernando Siles, La Paz 055061
Academia Nacional de Bellas Artes
 San Alejandro, La Habana 055285
Academia Naçional de Bellas Artes
 Santiago del Estero, Santiago del
 Estero 054877
Academia Provincial de Bellas Artes,
 Cádiz 056360
Academic File International News and
 Photo Syndication, London . . . 139119
Academician Norayr Sisakyan House-
 Museum, Ashtarak 000690
Academie Beeldende Kunsten,
 Oudenaarde 055045
Academie Beeldende Kunsten
 Maastricht, Hogeschool Zuyd,
 Maastricht 056178
Academie Brasschaat,
 Brasschaat 055000
Académie d'Auguste, Le Puy-en-
 Velay 103964
Academie des Arts de Woluwe-Saint-
 Pierre, Bruxelles . . . 055002, 136626
Academie des Arts et Métiers
 Constantin Meunier, Bruxelles . 055003
Académie des Beaux-Arts,
 Kinshasa 055272
Académie des Beaux-Arts,
 Verviers 055055
Académie des Beaux-Arts Alphonse
 Darville, Charleroi 055015
Académie des Beaux-Arts de la Ville
 de Tournai – École Supiérieure des
 Arts de Type, Tournai 055053
Academie des Beaux-Arts de Namur,
 Namur 055043
Académie des Beaux-Arts de Saint-
 Josse-ten-Noode, Bruxelles . . . 055004
Académie des Beaux-Arts et Arts
 Décoratifs, Binche 054999
Académie des Beaux-Arts Saint-Gilles,
 Bruxelles 055005
Académie des Beaux-Arts, Institut de
 France, Paris 055392
Académie des Inscriptions et des
 Belles-Lettres, Paris 055393
Académie des Sciences, Arts et Belles-
 Lettres de Dijon, Dijon 055367, 058585
Académie des Sciences, Belles Lettres

et Arts, Lyon 055377
Académie en Vaktekenschool,
 Temse 055051
Académie Internationale de la
 Céramique, Genève 059917
Académie Internationale des Arts et
 Collections, Paris 058586
Académie Libanaise des Beaux-Arts,
 Université de Balamand, Beirut 056113
Académie Libanaise des Beaux-Arts,
 Université de Balamand, Deir El-
 Balamand 056117
Académie R. Defossez, Spa 055050
Académie Royale des Beaux-Arts,
 Liège 055035
Académie Royale des Beaux-Arts de
 Bruxelles, École Supérieure des Arts,
 Bruxelles 055006
Académie Royale des Sciences, des
 Lettres et des Beaux-Arts de
 Belgique, Bruxelles 055007
Academie Salesienne Conservatoire
 d'Art et d'Histoire, Annecy . . . 011670
Academie van Bouwkunst,
 Groningen 056174
Academie van Bouwkunst Amsterdam,
 Amsterdamse Hogeschool voor de
 Kunsten, Amsterdam 056166
Academie voor Beeldende Kunsten
 Mol, Mol 055041
Academie voor Beeldende Vorming,
 Amsterdamse Hogeschool voor de
 Kunsten, Amsterdam 056167
Academie voor Schone Kunsten,
 Brugge 055001
Academie voor Schone Kunsten,
 Turnhout 055054
Academisch Historisch Museum van
 de Universiteit Leiden, Leiden . 032578
Academy Art Museum, Easton . 048454
Academy Billiard Company, West
 Byfleet 091784
Academy Fine Books, Las Vegas 145203
Academy Fireplaces, Dublin . . . 079519
Academy Galleries, Wellington . . 033340
The Academy Gallery, Beverly
 Hills 046971
Academy Gallery of Art, New
 York 051320
Academy Gallery, New Orleans
 Academy of Fine Arts, New
 Orleans 051285
Academy Hall Museum of the Rocky
 Hill Historical Society, Rocky
 Hill 052656
Academy of Art University, San
 Francisco 057852
Academy of Arts of the Ukraine,
 Kyïv 056490
Academy of Arts of Uzbekistan,
 Toshkent 058146
Academy of Fine Arts, Kolkata . 055809,
 109393
Academy of Fine Arts, Tirupati . 055839
Academy of Fine Arts and Literature,
 Delhi 109299
Academy of Fine Arts Museum,
 Kolkata 024352
Academy of Natural Sciences
 of Philadelphia Museum,
 Philadelphia 052023
Academy of Spherical Arts,
 Toronto 007156
Academy of Wall Artistry,
 Alexandria 056640, 119525
Academy Picture Restorations,
 Sydney 127683
Academy Upholstery, Hull 134484
Acadia Art & Rare Books,
 Toronto 140391
Acadia University Art Gallery,
 Wolfville 007482
L'Acajou, Beirut 082269
Acajou, Liège 128071

Acajou Bambou, Paris 070869
Acala Gallery, Bangkok 116945
Acanthe, Ambert 066547
Acanthe, Auch 066796
Acanthe, La Varenne-Saint-
Hilaire 069116
Acanthe, Le Vésinet 069373
Acanthe, Nice 070544
Acanthe, Saint-Raphaël 073285
Acanthe, Vannes (Morbihan) . . 073964
Acanthe, Versailles 074039
Acanthe Antiquités, Grenoble . . 068708
Acanthus, Cincinnati 092822
Acanthus, Nottingham . 090888, 134897
Acanthus, Twello 132644
Acanthus, Utrecht 143239
Acanthus & Reed, New York . . . 122280
Acanthus Ancient Art, New York 095194
Acanthus Leaf Antiques, Houston 093583
Acanthus Press, New York 138372
Acanto, Almeria 114835
Acanto, Donostia-San Sebastián 133239
Acanto, Roma 081191
Acantus Antik, Liseleje 065990
A'capella Galleri, Köpingsvik . . . 115712
Acapulco Arte Galería, Bogotá . 102268
ACC Galerie Weimar, Weimar . . 108860
Accademia Albertina di Belle Arti,
Torino 055981
Accademia Angelica Costantiniana di
Lettere Arti e Scienze, Roma . 055958
Accademia Artistica Internazionale,
Napoli 055933
Accademia Carrara di Belle Arti,
Bergamo 055874
Accademia dei Concordi, Rovigo 055975
Accademia dei Virtuosi, Roma . . 142795
Accademia del Libro Usato San Gallo,
Firenze 142661
Accademia di Architettura,
Università della Svizzera Italiana,
Mendrisio 056437
Accademia di Belle Arti, Bologna 055876
Accademia di Belle Arti, Carrara 055880
Accademia di Belle Arti,
Catanzaro 055885
Accademia di Belle Arti,
Frosinone 055906
Accademia di Belle Arti Aldo Galli,
Como 055886
Accademia di Belle Arti Cassino,
Cassino 055881
Accademia di Belle Arti Cuneo,
Cuneo 055887
Accademia di Belle Arti dell'Aquila,
L'Aquila 055912
Accademia di Belle Arti di Bari,
Bari 055873
Accademia di Belle Arti di Brera,
Milano 055918
Accademia di Belle Arti di Catania,
Catania 055883
Accademia di Belle Arti di Firenze,
Firenze 055891
Accademia di Belle Arti di Foggia,
Foggia 055903
Accademia di Belle Arti di Lecce,
Lecce 055913
Accademia di Belle Arti di Livorno,
Livorno 055916
Accademia di Belle Arti di Macerata,
Macerata 055917
Accademia di Belle Arti di Napoli,
Napoli 055934
Accademia di Belle Arti di Palermo,
Palermo 055942
Accademia di Belle Arti di Ravenna,
Ravenna 055954
Accademia di Belle Arti di Reggio
Calabria, Reggio Calabria . . . 055955
Accademia di Belle Arti di Roma,
Roma 055959
Accademia di Belle Arti di San Remo,
Istituto I. Duncan, Pavia 055947

Accademia di Belle Arti di Sassari,
Sassari 055978
Accademia di Belle Arti di Urbino,
Urbino 055984
Accademia di Belle Arti di Venezia,
Venezia 055988
Accademia di Belle Arti Fidia di
Stefanaconi, Stefanaconi 055980
Accademia di Belle Arti G.B. Cignaroli,
Verona 055995
Accademia di Belle Arti Lorenzo da
Viterbo – ABAV, Viterbo 055997
Accademia di Belle Arti Michelangelo
di Agrigento, Agrigento 055869
Accademia di Belle Arti Nike di
Catania, Catania 055884
Accademia di Belle Arti Picasso di
Palermo, Palermo 055943
Accademia di Belle Arti Pietro
Vannucci, Perugia 055949
Accademia di Belle Arti Santa Giulia di
Brescia, Brescia 055877
Accademia Giacinto Gigante,
Napoli 110515
Accademia Internazionale d'Arte
Moderna, Roma 055960
Accademia Internazionale d'Arti
Marvelli, Forlì 055904
Accademia Italiana di Restauro,
Milano 055919
Accademia Ligustica di Belle Arti di
Genova, Genova 055908
Accademia Nazionale dei Lincei,
Roma 055961
Accademia Nazionale di Belle Arti,
Parma 055945
Accademia Petrarca di Lettere, Arte e
Scienze, Arezzo 055870
Accademia Tedesca Roma Villa
Massimo, Roma 055962
Accart, Paris 104449
Accent Antiques, Austin 092201
Accent Arts, Minneapolis 121961
Accent Asia, Chicago 092650
Accent Fine Art, Phoenix 123476
Accent Fine Gifts, Phoenix 123477
Accent Gallery, Wembley 099873
Accent Office Art, Dallas 120534
Accent on, Portland 096349
Accent on Antiques, Pittsburgh . 096251
Accent on Books, Wellington . . . 143358
Accent on Design, Philadelphia . 096070
Accent Your Home, Jacksonville 093946
Accents and Beyond Consignment,
Minneapolis 094889
Accents Antiques, Miami 094677
Accents Toniques, Marseille . . . 140939
Accès Bijoux, Montréal 064358
Access Art, Auckland 112743
Access Art, West Perth 127697
Access Artisans, New York 122281
Access to Arts, Dublin 109569
The Accokeek Foundation,
Accokeek 046314
Accord Galéria, Budapest 109124
Accordeon en Harmonica Museum,
Malden 032633
Accou, Pascal, Seichamps 073477
Accra, Ille-sur-Têt 068865
l'Accroche-Coeur-Ducloy, Mons . 063737
L'Accrosonge, Paris 104450
Accton Arts Foundation, Hsinchu 042328
Accumulations, Birmingham . . . 092380
Accurate Antiques, Columbus . . 092983
Accurate Trading, London 089883
Ace Art, Winnipeg 007433, 101863
Ace Coin Shop, Salt Lake City . . 096870
Ace Furniture Refinishers, San
Antonio 136252
Ace Gallery, Los Angeles 121423
Ace Gallery, New York 122282
Ace International, Boronia 061175
Ace of Arts, Letterkenny 109663
Ace of Clubs House, Texarkana . 053724

Ace Traders, Palmerston North . 084053
Ace Upholstery, Albuquerque . . . 135276
Acero, Giuseppe, Roma 131793
Acervo Antiguidade, Belo
Horizonte 128097
Acervo Artístico Cultural dos Palácios
do Governo, São Paulo 004847
Acervo Didático de Invertebrados I
e II, Instituto de Biociências, São
Paulo 004848
Aces High Aviation Gallery,
Wendover 119320
Acetres, Sevilla 086434
Acevedo, Alicante 085690
Acevedo, William, Cali . 065238, 128282
ACF, Montpellier 070286
ACF les Sculptures, Yverdon-les-
Bains 116729
Achallé, Marc, Fontenay-le-
Comte 068572
Achard, Dominique, Bligny-sur-
Ouche 103400
Acharya Jogesh Chandra Purakirti
Bhavan, Bangiya Sahitya Parisat
Bishnupur Branch, Bishnupur . 024180
Achat Antiquité Brocante, Athis-
Mons 066767
Achat Antiquités, Nice 070545
Achdijan, Berdj, Paris 070870
Achenbach + Costa, Düsseldorf 137153
Achenseer Museumswelt, Maurach am
Achensee 002374
Achète Tout, Bruxelles 063303
Achète Tout Antiques, Bruxelles 063304
Achille, Nice 070546
Achilli & Rossi, Milano 131362
Achkal, Jounieh 082319
Achnamian, Pierre, Le Rouret . . 069351
Acht P!, Bonn 106522
Achtelik, Peter-Michael,
Gelsenkirchen 075966
Achter de Zuilen, Overveen 032761
Achterhoeks Museum 1940–1945,
Hengelo, Gelderland 032463
Achterkamp, H., Stanggaß 142373
Achternkamp, Petra, Darmstadt . 129667
Achternkamp, Stefan, Obernzenn 130481
Achtertalbahn, Ottenhöfen im
Schwarzwald 021037
Achtiani, Mehdi, Tours 073832
Acker, Rolf, Freiburg im Breisgau 075845,
075846
Ackerbürgermuseum, Gartz . . . 018594
Ackerbürgermuseum, Reichenbach,
Oberlausitz 021345
Ackerbürgermuseum Haus Leck,
Grebenstein 018799
Ackerl, Walter, Graz 127741
Ackerman, Waitara 062399
Ackermann Kunstverlag,
München 137384
Ackermann, Friedrich Adolf,
Ellenberg 137175
Ackermann, Wilhelm, Wuppertal 079099
Ackersgott, A., Duisburg 075546
Ackland Art Museum, University of
North Carolina at Chapel Hill, Chapel
Hill 047528
Acland Coal Mine Museum,
Acland 000754
Acme, Boston 135358
ACME – Accademia di Belle Arti
Europea dei Media, Milano . . . 055920
ACME – Accademia di Belle Arti
Europea dei Media, Novara . . . 055939
ACME Art Gallery, Columbus . . . 047933
ACME Fine Art, Boston 119826
ACME Gallery, Los Angeles 121424
Acme Gallery, New Orleans 122113
ACME Studios, London 060021
Aconerre Arte Conservazione Restauri,
Milano 131363
Acorn, Armadale, Victoria 061007
Acorn, Sheffield 091278

Acorn Antiques, Deganwy 089049
Acorn Antiques, Dulverton 089111
Acorn Antiques, London 089884
Acorn Antiques, Worthing 091952
Acorn Books, San Francisco . . . 145536
Acorn Bookshop, Columbus . . . 145093
Acorn Furniture, Atlanta 135296
Acorn Galleries, Billingshurst . . . 117364
Acorn Hall House Museum,
Morristown 051061
Acorn Lane Antiques, Seattle . . 097386
Acosta & Jiménez, Santiago de
Chile 064941
L'Acquario, Forlì 110078
L'Acquario, Roma 110720
Acquario Associazione Culturale,
Roma 059389
Acquavella Galleries Inc., New
York 122283
Acquire, Arnhem 112325
Acquisitions, Atlanta 092098
Acquisitions, Charlotte 092571
Acquisitions, London 089885
Acquisitions, Los Angeles 094239
Acquisitions, Prahran 062105
ACR Antiquités, Rouen 072554
ACR Atelier de Conservation et de
Restauration, Dannegger, Köninger &
Stebler, Fribourg 133740
ACR Edition Internationale,
Courbevoie 136810
ACR Gallery, Yoxford 119414
Acri Ciro, Napoli 080830
Acropolis, Nice 070547
Acrostiche, Saint-Étienne 072744
ACS Editions, Milano 137666
Act Conservation Services,
Baltimore 135335
Acta Historiae Artium/Academiae
Scientiarium Hungaricae,
Budapest 138769
Acta International, Roma 110721
Actana, Paris 070871
Act'Art, Bouglon 128512
Acte 2 Galerie, Paris 104451
Actée, Charenton-le-Pont 103572
Actes Sud, Arles 136798
Action, Indianapolis 135624
Action Auction Bendigo, Bendigo 125179
Action Auktioner i Östjämtland,
Hammarstrand 126595
Action Images, Lincolnwood . . . 127284
Action Sports Cards, Cleveland . 145080
Active Imagination, Albuquerque . 092020
Activité Bois, Marseille 128740
Acton Scott Historic Working Farm,
Acton Scott 043377
Actor, Tokyo 082130
Actua, L'Isle-sur-la-Sorgue 069549
Actual Art Foundation, New York 051321
Actuel Art, Paris 104452
Actuell & Antiques, Flensburg . . 075738
Acuarela, Badajoz 114460
Açude Museum, Museu Castro Maya,
Rio de Janeiro 004679
Acuna & Hansen, Los Angeles . 121425
Acuña, José Antonio, Bogotá . . . 102269
Acworth, Semaphore 062200
Acxelle Antique, Hyères 068848
AD Antiques, Johannesburg . . . 085643
AD Antiques, Stone 091455
AD Antiquités, Labastide-
d'Armagnac 069123
Ad-Art Decor, İstanbul 139110
ad artem, Antiquariat Galerie
Kunstbuchhandlung Verlag,
München . . . 108041, 137385, 142138
Ad-Dukkan, Sharjah 144210
AD Gallery, Athinai 109015
Ad Infinitum Galleries, San Juan 113622
Ad Invest, Napoli 110516
Ad Libitum Antikvitetshandel,
Hellerup 065728
Ad Plus, Praha 140499

Ad Sumos Antiguidades, Rio de
Janeiro 064004
Adacarte, Modena 110490
Adachi Bijutsukan, Yasugi 029706
Adachi Institute of Woodcut Prints,
Tokyo 111395
ADAH Museum, Alabama
Department of Archives and History,
Montgomery 051023
Adair County Historical Society
Museum, Kirksville 049992
Adalbert-Stifter-Haus,
Oberösterreichisches
Literaturmuseum und Adalbert-Stifter-
Gedenkraum, Linz 002313
Adam, Bath 117296
Adam, London 118147
Adam, Luxembourg 082357
Adam, Melbourne 061812, 099297
Adam & Dave's Group, Buffalo . 092504
Adam & Eve, London 144549
Adam & Eve Antiques, Dallas . . 093058
Adam & Ziege, Kleinmachnow . 107565
Adam Art Gallery, Victoria University of
Wellington, Wellington 033341
Adam Auctions, Burnham-on-Sea 126840
Adam East Museum Art Center, Moses
Lake 051077
Adam Gallery, Penarth 118885
Ádám Jenő Emlékház,
Szigetszentmiklós 023970
Adam-Ries-Museum, Adam-Ries-Bund
e.V, Annaberg-Buchholz . . . 016588
Adam Thoroughgood House, Virginia
Beach 053993
Adam, Alice, Chicago 120026
Adam, Costas, Athinai 137571
Adam, David, Lindfield 089836
Adam, Henry, Tarbes 125819
Adam, Jacques, Vannes
(Morbihan) 105816
Adam, James, Dublin 126167
Adam, Philippe, Pompièrre 072231
Ádama Alkšņa Memoriālais Muzejs,
Rūjiena 030226
Adamar, Miami 121719
Adamec, Eugen, Rožňava 085586
Adameck, Ted, Honolulu 093521
Adami & Pagano, Milano 131364
Adamiec & Schmidt, Hannover . 076357,
130002
Adamo, Las Vegas 121328
Adamo Petrausko Muziejus, Kupiškio
Etnografijos Muziejus, Uoginiai 030498
Adam's, Brescia 079907
Adams, Chester 088854, 134250
Adams, Milwaukee 094825
Adams, München . . . 077444, 130320
Adams, Richmond 096519
Adams Antiques, Nantwich 090793
Adams Avenue Bookstore, San
Diego 145515
Adams Blackrock, Blackrock . . . 126166
Adams County Historical Museum,
Decatur 048204
Adams County Historical Society
Museum, Gettysburg 049088
Adams County Historical Society
Museum, Lind 050312
Adams County Museum, Brighton 047210
Adams Davidson, Washington . . 097799,
124846
Adams Museum, Deadwood . . . 048188
Adams Museum and Historic Adams
House, Deadwood 048189
Adams Old Stone Grist Mill, Bellows
Falls 046896
Adams Room Antiques, Dorking 089076
Adams State College Luther Bean
Museum, Alamosa 046336
Adams, Ansel, Fresno 120890
Adams, B.J., Washington 124845
Adams, George, New York 122284
Adams, Jacqueline, Atlanta 092099

Adams, John, London 118148
Adams, John, New York 122285
Adams, Mark, London 118149
Adams, Mary, Birmingham 092381
Adams, Maurice, Miami 121720
Adams, Norman, London 089886
Adams, Pamela, Birchgrove . . . 061155
Adams, Robert Henry, Chicago . 120027
Adams, Rodney, Pwllheli 091061
Adams, Tony, Ipswich 089644
Adams, V., Frome 144432
Adams, Yvonne, Stow-on-the-
Wold 091459
Adamski, Stephan, Aachen 105927
Adamson Armoury, Skipton 091346
Adamson-Duvannes, Los Angeles 121426
Adamson-Eric Museum, Eesti
Kunstimuuseum, Tallinn 010663
Adamson, David, Washington . . 124847,
138478
Adamson, Nigel, Harrogate 089421,
134434
Adamstuen, Oslo 143392
Adamstuen Antikvariat,
Trondheim 143426
Adan E. Treganza Anthropology
Museum, San Francisco State
University, San Francisco 052990
Adana Arkeoloji Müzesi, Adana
Müzeleri, Adana 042788
Adana Atatürk Müzesi, Adana Müzesi,
Adana 042789
Adana Devlet Güzel Sanatlar Galerisi,
Adana 042790
Adana Etnografya Müzesi, Adana
Müzeleri, Adana 042791
Adaoust, Sylvie, Sanary-sur-Mer 105647
Adare Heritage Centre, Adare . . 024816
Ada's Place, Millthorpe 099344
Aday, Pazo de, Alicante 085691
Adda, Robert, Saint-Ouen 073002
Addams, Amanda, Bulleen 125183
Addenda Art Expressions, Lyon . 069678
Addenda Espace Image, Lyon . . 069679
Addenda Galerie d'Art, Lyon . . . 104059
Addi, Las Vegas 121329
Addington – Ryde House of Heritage,
Ryde 001466
Addison, Oakland 095879
Addison & Assoc., San Francisco 124202
Addison, Toronto 064576
Addison & Ripley, Washington . . 124848
Addison Fine Art, London 089887,
134577
Addison Gallery of American Art,
Phillips Academy, Andover . . . 046479
Addison Ross, London 118150
Addison, Richard, Wien 127898
Addisow, Ronald, Krefeld 076948
ADEC, London 089888
Adega, Houston 093584
Adel, Tomas, Berlin 106141
Adelaide Antiquarian Books,
Adelaide 139610
Adelaide Artistic Glass, Glenelg . 099081
Adelaide Arts Society, North
Adelaide 058182, 099465
Adelaide Booksellers, Adelaide . 139611
Adelaide Central Gallery,
Norwood 001372
Adelaide Central School of Art,
Norwood 054915, 099486
Adelaide Centre for the ARTS,
Adelaide 054890
Adelaide City Council Civic Collection,
Adelaide 000759
Adelaide Exchange, Adelaide . . 060965
Adelaide Gaol, Thebarton 001573
Adelaide House Museum, Alice
Springs 000786
Adelaide Hunter Hoodless Homestead,
Saint George 006870
Adelaide Junction Antiques,
Wellington 084151

Adélaïde le Grenier, Pontlevoy . . 072271
Adelaide Lithuanian Museum,
Norwood 001373
Adelaide Masonic Centre Museum,
Adelaide 000760
Adelaide Old Books, Kent Town . 139791
Adelaide River Railway Heritage
Precinct & Motorcycle Haven,
Adelaide River 000774
Adelaide's Antiques and Gallery,
Wentworth Falls 062419
Adelantado, Luis, Valencia 115473
Adelca, Pointe-à-Pitre 137579
Adele's Gallery of Fine Arts,
Canterbury 098859
Adelés ir Pauliaus Galaunių namai,
Nacionalinis M.K. Čiurlionio Dailės
Muziejus, Kaunas 030347
Adelhausermuseum, Natur- und
Völkerkunde, Städtische Museen
Freiburg, Freiburg im Breisgau 018475
Adelin, Antwerpen 063182
Adell McMillan Gallery, University of
Oregon, Eugene 048617
Adelman, J., San Antonio 096933,
124002
Adelmann von Adelmannsfelden, Jörg
Graf, Abtsgmünd 125846
Adelphi, Milano 137667
Adelphi, Redlynch Salisbury . . . 118962
Adelsö Konservatorsateljé, Adelsö 133518
Adelson, New York 122286
Adelson, Paul, Dallas 120535
Aden Society of Arts, Aden 060828
Aden, Jacqueline, Zürich 087926
Adena State Memorial, Home of
Thomas Worthington, Chillicothe 047717
Ader, Françoise, Arcachon 066700
Ader, Rémi, Paris 125671
Aderton, Stockholm 143925
Adès, Ginette, Saint-Ouen 073003
Adet, Jean-Louis, Longnes 069615
Adhoc, Vigo 115531
Adi-Art, Łódź 113656, 137897
ADI Art Gallery, Kathmandu 112060
Adi Art Gallery, Łódź 113657
Adia-M, Sankt-Peterburg 137928
Adiex, Paris 070872
Adin, Antwerpen 063183
Adirondack History Center,
Elizabethtown 048527
Adirondack Museum, Blue Mountain
Lake 047059
Adiyaman Müzesi, Adiyaman . . . 042793
Adjutant General's Corps Museum,
Winchester 046147
ADK – Actuele Kunst,
Amsterdam 112133
Adkins, Houston 093585
Adkins, Gary, Minneapolis 094890
Adkins, Keith, Launceston 139799
Adler, Gstaad 116364
Adler, Paris 104453
Adler & Co., San Francisco 124203
Adler, Wien 062881
Adler & Conkright, New York . . . 122287
Adler Kit, Los Angeles 094240
Adler Museum of Medicine,
Johannesburg 038740
Adler, Arno, Lübeck 142084
Adler, B., New York 095195
Adler, Carl, Dresden 141690
Adler, Julian, Cape Town 085629
Adler, Kurt S., Tampa 097534
Adler, Ulrike, Frankfurt am Main 106966
Adlerskij Kraevedčeskij Muzej – Muzej
Istorii Adlerskogo Rajona, Soči 037875
ADM Art Gallery, Philadelphia . . 123346
Admella, Mount Gambier 139841
Admiral Digby Museum, Digby . 005708
Admiral Nimitz National Museum of
the Pacific War, Fredericksburg 048949
Admiral Upholstery, Tulsa 136425
Admiral Vernon Antiques Market,

London 089889
Admiralty Antiques, Crows Nest, New
South Wales 061374
Admiralty House Antiques,
Seattle 097387
Admiralty House Museum and
Archives, Mount Pearl 006436
Admirer, Sofia 101138
ADN, Luxembourg 082358
Ado, Gdańsk 113453
Adobe, San Francisco 145537
Adolf-Bender-Zentrum, Sankt
Wendel 021609
Adolf Dietrich-Haus, Berlingen . . 041308
Adolf, Stephanie, Wiesbaden . . . 130754
Ádolfa Alunāna Memoriālais Muzejs,
Jelgava 030129
Adolfs, M., Münster 077651
Adolfsson, Segeltorp 115807
Adonis Antiquités, Saint-Parres-aux-
Tertres 073239
Adopt-A-Chair, Ulverstone 062371
Ador, Toulouse 129292
Adornments, Paddington,
Queensland 062047
Adornments Contemporain,
Albuquerque 135277
Adornos Antiguos, Buenos Aires 060846
Adornos Arañas Antiguas, Buenos
Aires 060847
Adracare, Busselton 139673
Adrénaline Antiquités, Marseille . 069896
Adri-Art-16, Barcelona 114844
Adriaan, Den Haag 083047
Adriaan's Antiques, New Orleans 095045
Adrian, David, Knokke-Heist . . . 100720
Adriana, San Vito Al Tagliamento 110902
Adrianno's Fine Art Gallery,
Forrestfield 099046
Adrian's Ashgrove Antiques,
Ashgrove 061065
Adriatica Editrice Divisione Arte,
Bari 109767, 137618, 142620
Adrien Numismatic, Dijon 068311
Adro de Santos, Lisboa 084812
ADS Anticuarios, Buenos Aires . 060848
ADS Antiques, Miami 094678
Adscriptio Iberica Librariae
Antiquariorum, Palma de
Mallorca 059825
Adsera Carvajal, Maria Teresa,
Barcelona 085716
Adsmore Museum, Princeton . . 052362
Advanced Coatings, San Diego . 136283
Advanced Graphics London,
London 118151
Advanced Refinishing, Nashville 135842
Advanced Wood Finishing, San
Diego 136284
Advances in Art, Kansas City . . 121276
Advertising and Graphic Design
Department, Southwestern
Community College, Sylva . . . 057937
Advertising Art Department,
Chattanooga State Technical
Community College,
Chattanooga 056862
Advocate Art, London 118152
Ady Endre Emlékmúzeum, Petőfi
Irodalmi Múzeum, Budapest . . 023453
Adyar, Sydney 139921
Adyar Art Club, Chennai 059354, 109272
Adygejskij Respublikanskij
Kraevedčeskij Muzej, Majkop . 037216
Adze Gallery, York 119403
Äänekosken Kaupunginmuseo,
Äänekoski 010749
Æbelholt Klostermuseum,
Hillerød 010118
Aebischer, Elisabeth,
Schwarzenburg 116635
Aebischer, Erwin, Basel 087212
Aebischer, Erwin, Frenkendorf . . 087430,
133739

AeDis Kieferle Reiner Schmid GbR,
Hochdorf 130050
Die Ägayrischen Gewölbe,
Viechtach 022323
Aegean Maritime Museum,
Mykonos 023179
Aêgerter, Bernard, Marseille 069897
Aegerter, Bernard, Nice 070548
Aegerter, Claude, Chêne-Bourg . 087364
Ægidius, Randers 102955
Ægidius, Anne Grethe & Ken,
Randers 066051
Aegis, Ulm 142440
Æglageret Udstillingslokaler,
Holbæk 102845
Ägyptische Sammlung der
Universität / Museum Schloß
Hohentübingen, Tübingen 022212
Ägyptischer Kulturladen, Berlin . 106142
Ägyptisches Museum der Universität
Bonn, Bonn 017476
Ägyptisches Museum der Universität
Leipzig – Interim, Leipzig 019972
Ägyptisches Museum und
Papyrussammlung, Staatliche
Museen zu Berlin – Stiftung
Preußischer Kulturbesitz, Berlin 017065
Ähtärin Kotiseutumuseo, Ähtäri . 010750
AEKO Veilingen, Hasselt 125339
Aelbrecht Galerie, Rotterdam . . 112629
Äldre Ting, Aakersberga 086624
Ängelholms Antikvariat,
Ängelholm 143819
Aenigma, Amsterdam 143009
L'Æntre, Neuchâtel 116525
Aequitas, Lyon 069680
AER Verlag, Bolzano 137626
Aera Memorial Museum, San
Pablo 034590
Aerial House Antiques, Long
Melford 090615
Aero, Helsinki 066293
Aero Nautical Museum-Armadillo,
Mona Vale 001303
Aero Space Museum of Calgary,
Calgary 005514
Ærø Museum, Ærøskøbing 010000
Ærøskøbing Antik, Ærøskøbing . 102770
Aeronautics Museum, Nanjing . . 008104
Aeronauticum – Deutsches Luftschiff-
und Marinefliegermuseum Nordholz,
Nordholz 020830
Aeroplastics Contemporary,
Bruxelles 100544
Aeroseum, Göteborg 040647
Aérospatiale, Toulouse 015983
Aerts, S., Bruxelles 063305
AES Gallery, Chicago 120028
Aeschbach, Max, Derendingen . 087395
Aesthetic Eye, Chicago 120029
AF Galerie, Karlsruhe . 107507, 125996
AFAS Journal, Automotive Fine Arts
Society, Brooksville 139234
AFC Edition, Wiesbaden 108897, 137552,
142483
Afdeling Nuttige en Fraaie
Handwerken, Openluchtmuseum Het
Hoogeland, Uithuizermeeden . . 032937
Affaire d'Antique, Eindhoven . . 083205
Affaire d'Eau, Amsterdam 082674
Affairs Art Gallery, Mumbai . . . 109412
Affairs of the Art, Toronto 101557
Affaitati, Antonio, Napoli 080831
Affaitati, Giuseppe, Napoli 080832
Affari d'Epoca, Firenze 080073
Affenberger, Graz 062574
L'Affiche, Milano 110231, 110232,
137668
L'Affiche, Nantes 104327
Affiche, Palermo 110585
Affiche Museum Hoorn, Hoorn . 032505
Affiches Anciennes, Paris 070873
The Affinity Art Gallery, Midsomer
Norton 118744

Affinity's Antiques, Regina 064545
Affirmation Arts, New York 122288
Afford-a-Bulls, Bulls 083843
Affordable Aboriginal Arts,
Indooroopilly 061648
Affordable Antiques, Hale 089407,
134431
Affordable Art, Dallas 120536
Affordable Art, London 118153
Affordable Art, Oklahoma City . 123271
Affordable Art, Virginia Beach . 124819
Affordable Art Fair Bristol, Bristol 098351
Affordable Art Fair London,
London 098352
Affordable Art Fair New York City, New
York 098410
Affordable Art Gallery, Chicago . 120030,
120031
Affordable Black Art, Jacksonville 121229
Affordable Books and Collectables,
Phoenix 096177, 145423
Affordable Posh, Mittagong . . . 061865
Affresco di Piero della Francesca,
Monterchi 026953
Affrica-African Art, Washington . 124849
Afghanisches Museum, Hamburg 018974
Afif Abad Military Museum,
Shiraz 024734
Afinsa, Madrid 126525
Afkari, Richard, New York 095196
Afnan, Cairo . 103011, 103012, 103013,
103014
Aforisma, Milano 142708
Africa, Cairo 066224
Africa Calls, Dublin 109570
Africa Cultural Gallery, Nairobi . 111557
Africa Dreaming, Gawler Belt . . 061505
Africa Exclusive, Delft 083026
Africa Just, Armadale, Victoria . 098652
Africa Museum, Seoul 029982
Africa Visions Two Thousand One, Las
Vegas 121330
Africa Within, Cincinnati 120312
African Accents, San Diego . . . 124085
African Accents Hawaii, Honolulu 093522
African American Art, Charlotte . 119972
African American Art Gallery, New
York 122289
African American Historical Museum,
Cedar Rapids 047483
African American Movie Posters, Los
Angeles 094241
African American Museum,
Cleveland 047805
African American Museum,
Dallas 048108
African-American Museum at Oakland,
Oakland 051674
The African American Museum in
Philadelphia, Philadelphia . . . 052024
African American Museums
Association, Chicago 060282
African and Tribial Art Markus Gunti,
Berlin 106143
African Art, Düren 075443
African Art, Saint-Denis 114115
African Art and Designs, Miami . 121721
African Art Center, Houston . . . 120985
African Art Centre, Durban 038696
African Art Collection, San Diego Mesa
College, San Diego 052960
African Art Gallery, Tampa 124624
African Art Museum of Maryland,
Columbia 047896
African Art Museum of the S.M.A.
Fathers, Tenafly 053712
African Art Room, Houston 120986
African Arts, Los Angeles 139235
African Arts, Tucson 124705
African Arts Cultural Center, New
York 122290
African Attractions, Virginia
Beach 124820
African Connection, Saint Ives,

Cornwall 119016
African Cultural Art Forum,
Philadelphia 052025
African Gallery, Bramsche 075114
African Gems, Denver 120684
African Hands, Washington 124850
African Hemingway Gallery, New
York 122291
African Heritage, Amsterdam . . 082675
African Heritage, Nairobi 111558
African Heritage Gallery, Dallas . 120537
African Imports, Dallas 120538
African Imports, Fort Worth . . . 120852
African Legacy Gallery, Kirup . . 099200
African Muse Gallery, Paris 104454
African Mystical Arts, Balwyn . . 098701
African Mystical Arts, Forest Hill 099044
African Odyssey, San Antonio . . 124003
African Palace, Indianapolis . . . 121162
African Safari Imports, Chicago . 120032
African Tribal Art, Haarlem 112507
African Village, Detroit 120797
Africana, Zürich 116740
Africana Museum, Mariannhill . 038781
Africana Museum, Monrovia . . . 030289
AfricanColours, Press Centre,
Nairobi 137799
Afric'Art, Château-Thierry 067876
Africart Motherland, New York . 122292
Afrika Centrum, Cadier en Keer . 032130
Afrika Galeri, Berlin 106144
Afrika Museum, Berg en Dal . . . 032062
Afrika-Museum, Bergen, Kreis
Celle 017045
Afrika-Museum, Zug 042247
Afrika Múzeum, Balatonederics . 023423
Afrikaanse Taalmuseum, Paarl . 038794
Afrikahaus Museum und Nold-Namibia-
Bibliothek, Sebnitz 021807
Afrikamuseum, Gallneukirchen . 001920
Afrikana Gallery, South
Melbourne 099699
Afrique Arts Premiers, Nice . . . 104356
Afrique Congo, Liège 063738
Afrique Congo Sculptures, Liège 063739
Afro-American Cultural Center,
Charlotte 047568
Afro-American Historical Society
Museum, Jersey City 049816
Afrocrafts, Manama 100409
Afrodite Antik, Stockholm 086956
Afrosiyob – Samarqand Thariki Muzeyi,
Samarqand 054643
After Noah, London . . . 089890, 089891,
134578
After the Paint, Saint Louis 096660
After-Words, Chicago 145030
Afterall, London 139120
Afterall, Valencia 139236
Afterimage, Rochester 139237
Afterimage Gallery, Dallas 120539
Aftermodern, San Francisco . . . 124204
Afton, Totland Bay 119246
Afval-Museum, Zwolle 033124
Afyon Devlet Güzel Sanatlar Galerisi,
Afyon 042794
Afyon Etnografya Müzesi, Afyon . 042795
Afyon Müzesi, Bolvadin Müzesi, Zafer
Müzesi ve Dumlupınar Müzesi,
Afyon 042796
Aga Khan Museum of Islamic Arts,
Marawi 034550
Agabio, Cagliari 079971
Agale Antiques, London 089892
Agaméde, Paris 104455
Agamennone, Savona 110915
Agape, Albuquerque 092021
Agape, Cleveland 120393
Agapi, Bordeaux 103410
Agar, Rosie, Saintfield 091212
Agardh & Tornvall, Stockholm . . 115825
AgArt, Warszawa 137903
Agartha, Paris 104456

Agartha, Rio de Janeiro 064005, 140288
Agassiz-Harrison Historical Society
Museum, Agassiz 005305
Agatha Antikvariat, Täby 143977
Agathe, Saint-Ouen 073004
Agathi, Athinai 109016
Agathon Oudheden, Antwerpen . 063184
Agatunet, Nå 033793
Agave, Valencia 133444
Agder Naturmuseum,
Kristiansand 033714
L'Age d'Or, Charleroi 140214
L'Age d'Or, Parma 081086
Age of Elegance, Whitchurch,
Shropshire 091829
Age Of Grace Antiques,
Christchurch 083849
Age of Jazz, Whitby 091823
Age of Steam Railroad Museum,
Dallas 048109
Agecroft Hall, Richmond 052541
The Aged Art Jia Association,
Guangzhou 058486
The Aged Art Society, Anshan . 101927
The Aged Art Society, Fuxin . . . 102021
Aged on High, Mount Waverley . 061925
Ageing Fancies, Seattle 097388
Ageless Treasures, Saskatoon . . 064559
Agem Produtora Cultural, São
Paulo 101079
Agen Philatélie, Agen 066421
Agence Alsacienne de l'Or et des Méraux
précieux, Strasbourg . 073561, 125818
Agence Alsacienne Or et des Méraux
Précieux, Nancy 070409
Agence Art Process, Paris 104457
Agence C.B.S., Boulogne-
Billancourt 103446
Agência Arte, Porto Alegre 100963
Agencia Cultural Oboe Financeira,
Fortaleza 100932
Agência de Leilões Xira, Lisboa . 126437
Agensteinhaus und Simmentaler
Hausweg, Museum der Alten
Landschaft Niedersimmental,
Erlenbach im Simmental 041470
Agentur Wort und Kunst, Berlin . 106145
Ageo Bustillo, Milagrosa, Sanlúcar de
Barrameda 133421
Ages Ago, Vancouver 064787
Agesilas, Claude, Fontaine-
Lavaganne 068560
Agharkar Museum, Botany Department
Museum, Allahabad 024134
Agharta, Żurawski & Walas,
Warszawa 113794
Aghetta, Giovanni, Torino 132083
Agia Triada Folklore Museum, Agia
Triada 022944
Agiart, Sarajevo 063866
Aginskij Okružnoj Kraevedčeskij Muzej
im. G. Cybikova, Aginskoe . . . 036841
Agios Andreas Convent, Kammeni
Chora 023109
Agios Nikolaos Archaeological
Museum, Agios Nikolaos 022949
Agios Stephanos Basilica,
Kastoria 023119
Agis-Verlag, Baden-Baden 137039
A Glaia, Porto 085005
AGM, Zagreb 136752
Agneray, Josette, Paris 128888
Agnes Etherington Art Centre,
Kingston 006095
Agnes Gallery, Birmingham . . . 047001
Agnes Jamieson Gallery, Minden 006321
Agnes-Miegel-Haus, Bad
Nenndorf 016832
Agnes Photography and Book Arts,
Birmingham 119783, 144986
Agnès, M., Rueil-Malmaison . . 072636
Agnew's, London 089893, 118154
Agniel, Aigues-Vives 066438
Àgora, Barcelona 133091

Agora, Honfleur 068825
Agora, Lyon 069681, 140877
Agora, New Orleans 095046
Agora, New York 122293
Agorá, Prato 081165, 131769
Agora, Wien 062882
Agora 3 Art Gallery, Sitges 115460
Agora 3 Art Gallery Show Room,
  Barcelona 114845
Agora Açıkhava Müzesi, İzmir . . 042970
Agora Antiguedades, Salamanca 086405
Agora Antiqua, Schwäbisch Hall 078408
Agora Gallery, Chelsea 120018
Agora Restauraciones de Arte,
  Madrid 133286
Agora Verlag, Berlin 137046
Agoraï Mozzanico Leckie, Manuela,
  Carouge 087345, 133711
Agostinelli, Berardo, Roma 110722,
  126220
Agostinelli, Roland, Marseille . . . 128741
Agostini, Arredo, Roma 081192, 126221
Agostini, Giuseppe, Roma 081193
Agostini, Nino, Roma 131794
Agostino, New York . . . 095197, 095198
Agostinucci, Andrée, Nice 128844
Agouridis, Nik., Thessaloniki . . . 079203
Agouridou, Chrisoula,
  Thessaloniki 079204
Agra-Art, Warszawa 126417
Agra, María Teresa & Vicente,
  Valladolid 086540, 143810
Agrargeschichtliche Sammlung,
  Stadtmuseum Wels-Burg, Wels 002913
Agrarhistorisches Aktionsmuseum,
  Trendelburg 022186
Agrarhistorisches Museum,
  Domsühl 017928
Agrarhistorisches Museum,
  Schlepzig 021662
Agrarisch en Wagenmuseum, De
  Waal 032150
Agrarisch Museum Westerhem,
  Middenbeemster 032653
Agrarmuseum Wandlitz, Wandlitz 022416
Agrármüszaki Emlékek Gyüjteménye,
  Zirc 024059
Agre Møllens Antik, Rønne 066081
Agre, Michel, Toulouse 129293
Agri Musée-Musée de l'Agriculture du
  Temps Passé, Rochehaut 003973
Agricultural Association Museum of
  Jhonghe, Jhonghe 042342
Agricultural Exhibition Hall, National
  Taiwan University, Taipei 042496
Agricultural Heritage Museum,
  Boerne 047071
The Agricultural Memories Museum,
  Penn Yan 051977
Agricultural Museum, Faisalabad 034110
Agricultural Museum, Jasin 030666
Agricultural Museum, Yangon . . 031829
Agricultural Museum and Cotton
  Museum, Cairo 010529
Agricultural Museum at Stone Mills,
  La Fargeville 050032
Agricultural Museum Brook,
  Brook 043772
Agricultural Museum of Jeollanam-do,
  Samho 029979
Agricultural Museum of New
  Brunswick, Sussex 007104
L'Agriculture au Fil du Temps,
  Maniquerville 013988
L'Agriculture Bressane Musée,
  Ecomusée de la Bresse
  Bourguignonne, Saint-Germain-du-
  Bois 015339
Agriculture Museum, Inakadate . 024305
Agropolis Museum, Montpellier . 014286
Agrotechnorama, Agrartechnische
  Entwicklungsschau,
  Ettenhausen 041478
Agrupació de Aquarel.listes de

Catalunya, Barcelona 059826
Agua Caliente Cultural Museum, Palm
  Springs 051895
Aguado, Logroño 115120
Agüera-Madrid, Juan, Bourges . 067421
Aguéli, Stockholm 115826
Aguélimuseet, Sala 040931
Aguer, Hélène, Saint-Jean-de-
  Luz 072840
Aguiar, Henri de, Beaune 067009
Aguilar, Granada 133263
Aguilar, Leo, Houston 093586
Aguilera, Madrid 133287
Aguilera Puntas, Córdoba 085939
Aguilera y Restauracion, Madrid 133288
Aguinaleo Narvaez, Fernando,
  Málaga 086284
Agung Rai Museum of Art, Ubud 024668
Agurcho Iruretagoiena, Pamplona 086376
Aguttes, Claude, Neuilly-sur-
  Seine 070513, 125660
Aguttes, Claude, Paris 070874
AH Antique Mall, Las Vegas . . . 094106
Ah Sam, San Francisco 136303
Ah-Tah-Thi-Ki Museum, Big
  Cypress 046975
Ah, Theo von, Sachseln 087767
AHA Art History Abroad, Bury Saint
  Edmunds 060022
Ahlat Müzesi, Ahlat 042798
Ahlberg, Lund 086827
Ahlbergshallen, Östersund 115795
Kurt Ahlbrecht & Dr. Gerhard Röper,
  Hamburg 107218
Ahler Krâm – Volkskundliche
  Sammlung, Partenstein 021077
Ahlers, Göttingen 107144
Ahlgrim, Harald, Gaienhofen . . . 075937
Ahlström, B., Stockholm 086957, 126666
Ahmadi, Manama 100410
Ahmed Pervez Art Gallery,
  Karachi 034120
Ahnert, Knut, Berlin 141466
Ahnlund, Mariann, Umeå 115960
Aholan Talomuseo, Keitele 010981
Ahorn Antik, Århus 065594
Ahrens, Varel 078778
Ahrens & Hamacher, Düsseldorf 141713
Ahrer, Weyer 062880
Ahsan Manzil Museum, Bangladesh
  National Museum, Dhaka 003215
Åhus Museum, Åhus 040515
Åhuset, Køge 102917
Ai, Nara 111334
Ai Baffi, Oberwil-Lieli 116551
Ai due Santi, Padova 142771
Aiassa, Daniela, Torino 132084
Aibido, Tokyo 082131
AIC News, Washington 139238
Aichberger, Helga, Neunkirchen . 062747
Aichberger, Peter, Großrust . . . 127761
Aichelbaum, Léon, Paris 141019
Aichele, Frieder, Stuttgart 078589
Aichelmann & Hawelka,
  Oberursel 077884
Aichholzer, Manfred, Pörtschach am
  Wörther See 100139
Aichi-ken Bijutsukan, Nagoya . . 029108
Aichi-ken Toji Siryokan, Seto . . 029365
Aichi-kenritsu Geijutsu Ongaku
  Daigaku, Nagakute 056019
Aichinger, Bernhard & Comp.,
  Wien 140055
Aichmaier, Johann, Salzburg . . . 062783
Aicon Gallery, London 118155
Aicon Gallery, New York 122294
Aidan Gallery, Moskva 114190
Aidan Phelan, Nadia, Kilkenny . . 079642
L'Aidjolate, Saint-Ouen 073005
Aiello, Denver 120685
Aiello, Roberto, Roma 131795
Aifacs Gallery, Delhi 109300
Aigaia School of Art and Design,
  Lefkosia 055291

Aigantighe Art Museum, Timaru . 033311
Aigenbauer-Schindler, Maria,
  Krems 062685
L'Aigle Impériale, Marseille 069898
L'Aiglon, Polliat 072227
Aiglonnes du Moulin de Mérouvel,
  Alençon 066519
Aigner, Christoph Wilhelm,
  Salzburg 136547
Aigner, Johannes,
  Deutschlandsberg 062542
Aiguier, Jean-Louis, Le Pradet . . 069337
Aiguière de Cuivre, Montlhéry . . 070261
Aikawa Folk Museum and Exhibition
  Hall of Folk Crafts, Sado 029290
Aikawa Kokokan, Isesaki 028699
Aiken Antiques and Auction,
  Aiken 092017, 127114
Aiken County Historical Museum,
  Aiken 046323
Aiken Thoroughbred Racing Hall of
  Fame and Museum, Aiken 046324
Aileen's Antiques, Vancouver . . . 064788
Ailene's Antiques and Collectibles,
  Albuquerque 092022
Ailes d'Anges, Genech 068647
Les Ailes du Temps, Béziers . . . 067163
D'Ailleurs, Beirut 082270
Ailsa Gallery, Twickenham 091672
AIM Enterprises, Cleveland 120394
Al-Ain Gallery, Cairo 103015
Ainažu Jürskolas Memoriālais Muzejs,
  Ainaži 030095
Aineen Taidemuseo, Tornio 011428
Aineh va Roshananal Museum,
  Yazd 024792
Ainger, E.J., Richmond, Victoria . 125229
Ainhoa Ziganda Apaolaza, Donostia-
  San Sebastián 133240
Ainola, Järvenpää 010908
Ainscough Contemporary Art,
  London 118156
Ainsley House, Campbell 047391
Ainsley, Sandra, Toronto 101558
Ainslie, Perth 090975
Ainslie, Alan, Tshwane 114818
Aintzinatekarien Azoka, Bilbao . . 098293
Ainu Museum, Hokkaido Daigaku
  Shokubutsuen Hakubutsukan,
  Sapporo 029328
Air Classic Museum of Aviation, Sugar
  Grove 053591
Air de Paris, Paris 104458
L'Air du Temps, Brie-Comte-
  Robert 067491
Air du Temps, Gironville-sur-
  Essonne 068669
Air du Temps, La Ferté-Alais . . . 069025
Air du Temps, Nantes 070439
Air du Temps, Schoelcher 082457
L'Air du Temps, Septfonds 073505
Air du Temps, Varces-Allieres-et-
  Risset 073977
Air Force Armament Museum, Eglin
  Air Force Base 048491
Air Force Flight Test Center Museum,
  Edwards 048486
Air Force Heritage Museum and Air
  Park, Winnipeg 007434
Air Force Museum, Christchurch . 033150
Air Force Museum, Delhi 024216
Air Force Museum, Ha Noi 054787
Air Force Museum, Tehrän 024745
Air Galleries, New York 122295
Air Gallery, London 118157
Air Mobility Command Museum, Dover,
  Air Force Base 048357
Air of Art Museum, Bologna . . . 025529
Air Power Park and Museum,
  Hampton 049339
Air Victory Museum, Lumberton 050525
Air Victory Museum, Medford . . 050755
Airault, Jean-Marc, Brigueuil . . . 067499
Airborne and Special Operations

Museum, Fayetteville 048718
Airborne Forces Museum,
  Aldershot 043385
Airborne Museum Hartenstein,
  Oosterbeek 032721
Airbrush Action, Allenwood 139239
Airbrush Art School, Werribee . . 099876
Aird, John B., Toronto 101559
Aireloom, Honolulu 135576
Airmen Memorial Museum,
  Suitland 053594
Airoldi, Isabella, Breganzona . . . 133703
L'Airone, Messina 110217
Airport Gallery 1 & 2, Frankfurt am
  Main 018404
Airport Museum, Cairo 010530
Airpower Museum, Ottumwa . . . 051844
Airvault, Philippe, Châtelaillon-
  Plage 067901
Aisling Antiques, Tralee 079706
Aisling Gallery,
  Newtownmountkennedy 109680
Aisne Collections, Bucy-le-Long 067531
Aisthesis, Jérôme Cordie, Paris . 128889
Aithousa Skoufa, Athinai 109017
Aitihasic Puratatva Sangrahalaya,
  Delhi 024217
Aitkin County Museum, Aitkin . . 046326
Aitonevan Turvemuseo, Kihniö . . 010999
Aître Saint-Saturnin, Blois 012076
Aittouarès, Paris 104459
Aiwen e Rifat Museum and Art Gallery,
  Karachi 034121
AIX, New York 095199
Aix Antiquités, Saint-Cannat . . . 072702
Aix-en-Provence Antiquités, Aix-en-
  Provence 066450
Aixia, Las Rozas de Madrid . . . 133412
AixLibris, Aachen 141393
Aizenman, Buenos Aires 139542
Aizkraukles Vēstures un Mākslas
  Muzejs, Aizkraukle 030096
Aizpuru, Juana de, Madrid 115129
Aizpuru, Juana de, Sevilla 115442
AJ Net, Montigny-les-Cormeilles 070253
Ajak-Nostalgie, Halberstadt 076157
Ajanpatina, Helsinki 066294
Ajassa, Torino 081640, 110942
Ajasse, Lyon 140878
Ajax Museum, Amsterdam 031922
AJEX Military Museum, London . 044898
Ajia-no Mingeikan, Noda 029193
Ajisai, Kobe 111278
AJM, New York 122296
AJNA, Amiens 066560
Ájtte – Svenskt Fjäll- och
  Samemuseum, Jokkmokk 040723
Ajucla, Salamanca 115400
AK Akt e.V., Berlin 058703
AK Artist Run Centre, Saskatoon 101538
AK-Express, Essen 138667
Aka-Renga Kokusaikan,
  Yokohama 029710
Aka Salon Antykow i Galeria,
  Kraków 084546, 113523
Akadamija za Umetnosti 'Braco Karic',
  Beograd 056320
Akademia e Arteve e Tiranes,
  Tiranë 054859
Akademia Sztuk Pięknych,
  Gdańsk 056237
Akademia Sztuk Pięknych,
  Poznań 056237
Akademia Sztuk Pięknych,
  Warszawa 056242
Akademia Sztuk Pięknych,
  Wrocław 056245
Akademia Sztuk Pięknych im. Jana
  Matejki, Kraków 056232
Akademia Sztuk Pięknych im.
  Władysława Strzemińskiego,
  Łódź 056235
Akadémia Umení v Banskej Bystrici,
  Banská Bystrica 056332

Akademiai Kiadó, Budapest 137583
Akademie der Bildenden Künste,
München 055644
Akademie der Bildenden Künste in
Nürnberg, Nürnberg 055668
Akademie der bildenden Künste Wien,
Wien 054964
Akademie der Künste, Berlin ... 017066,
058704
Akademie Franz-Hitze-Haus,
Münster 055662
Akademie für Bildende Kunst
Vulkaneifel, Galerie Norbert Blaeser
GmbH & Co. KG, Steffeln 055711
Akademie für Eurythmische Kunst
Baselland, Aesch 056405
Akademie Mode und Design (AMD),
Hamburg 055563
Akademie Schloss Solitude,
Stuttgart 055712
Akademie Verlag, Berlin 137047
Akademie voor Beeldende Kunsten en
Vormgeving, Akademie Minerva,
Groningen 056175
Akademie voor Kunst en Vormgeving,
's-Hertogenbosch 056176
Akademie Výtvarných Umění,
Praha 055298
Akademija, Vilnius 111794
Akademija Likovnih Umetnosti
Sarajevo, Sarajevo 055064
Akademija Likovnih Umjetnosti,
Sveučilište u Zagrebu, Zagreb . 055281
Akademija za Primenjene Umetnosti,
Beograd 056321
Akademische Druck- und
Verlagsanstalt, Graz 136505
Akademisches Kunstmuseum,
Antikensammlung der Universität,
Bonn 017477
Akademisches Münzkabinett, Institut
für Altertumswissenschaften der
Friedrich-Schiller-Universität Jena,
Jena 019443
Akademisk Antikvariat, Odense . 140631
Akademkniga, Moskva 143628, 143629,
143630,   143631
Akademkniga Magazin N1, Sankt-
Peterburg 143659
Akademkniga Magazin N3, Sankt-
Peterburg 143660
Akademkniga Magazin N4, Sankt-
Peterburg 143661
Akademkniga Magazin N5, Sankt-
Peterburg 143662
Akademski Slikar, Koper 114683
Akadimia Athinon, Athinai 055749
Akadomari-mura Folk Museum,
Sado 029291
Akai, Nanmeido, Nara 082067
Akainyah, Chicago 120033
Akalo, Bordeaux 067301
Akama-jingu Homotsukan,
Shimonoseki 029382
Akamon Bijutsukan, Sendai 029356
Akanat Art Gallery, Jerewan ... 098594
Åkande, Holstebro 065779
Akane, Nagoya 111310
Akant, Praha 065459, 102644
Akant, Szczecin 143504
Akant, Wrocław 084760, 132911
Akanthos, Antwerpen 063185
Akanthus, Erftstadt 075630
akanthus Restaurierungen, Berlin 129462
Akantus Antikt, Visby 087161
Akar Prakar, Jaipur, Rajasthan . 109378
Akariya, Tokyo 082132
Åkarps Antikvariat, Lund 143885
Akasaka Green Gallery, Tokyo .. 111396
Akateeminen Kirjakauppa,
Helsinki 140689
AKB – Angewandte Kunst Bremen,
Scholen 058705
Akbank Bahariye Sanat Galerisi,

İstanbul 116992
Akbank Bebek Sanat Galerisi,
İstanbul 116993
Akbank Beylerbeyi Sanat Galerisi,
İstanbul 116994
Akbarzadeh-Glatthard, A.,
Oberdiessbach 116543
Akbarzadeh-Glatthard, Antoinette,
Bern 116116
Akcent, Kwidzyń 113653
Akcent, Warszawa 113795
Akcija, Moskva 085154, 085155, 126471,
143632
Akd-Forum, Basel 087213
Akemi, Singapore 085354
Åkers Hembygdsmuseum,
Strängnäs 041057
Akers, Timothy, Beckenham ... 134086
Akershus Fylkesmuseum,
Strømmen 033996
Akesu Zone Museum, Akesu .. 007562
AKF Trading, New York 095200
AKI-Akademie voor Beeldende Kunst
en Vormgeving, Enschede 056173
Aki Gallery, Taipei 116901
Akiba, Miami 094679
Akiba Anti-ku Okimono Bijutsukan,
Nasu 029160
Akimbek, Azat, Almaty 082236, 111548
Akimoto, Nagoya 082055
Akin, Boston 092440
Akin, A.R., Birmingham 092382
Akinci, Amsterdam 112134
Akira Ikeda Gallery, Yokosuka . 111546
Akira Ikeda Gallery/ Berlin, Berlin 116146
Akita Daigaku Kogyo Hakubutsukan,
Akita 028468
Akita-kenritsu Bijutsukan, Akita . 028469
Akita-kenritsu Hakubutsukan,
Akita 028470
Akita-kenritsu Kindai Bijutsukan,
Yokote 029741
Akita Senshu Bijutsukan, Kenzo Okada
Kinenkan, Akita 028471
Akiyama, Tokyo 111397
Akiyoshi-dai Kagaku Hakubutsukan,
Shuuhou 029408
Akizuki Hakubutsukan, Amagi .. 028473
Akko Gallery, Bangkok 116946
Akko Municipal Museum, Akko . 025013
Akkolades, Den Haag 083048
Aklundh, Magnus, Malmö 115739
Akmal Ikramov nomidagi Tarikhi
Kültüri va San''atshunoslik
Özbekiston Khalqlari Davlat Müzeyi,
Samarqand 054644
Akomena, San Zaccaria 132078
Akon, Warszawa 084686
Akpol Museum, Semarang ... 024643
Akránum, Budapest 079230
Akriti, Delhi 109301
Akron Art Museum, Akron ... 046329
Akropolis, Brescia 079908
Akros Gallery, Bilbao 114977
AKS Holographie-Galerie, Essen 106905
Aksanat Resim Galerisi, İstanbul 116995
Aksaray Müzesi ve Bağli Örenyerleri,
Aksaray 042799
Akşehir Müzesi, Akşehir 042800
Aksel Bakuntz House-Museum,
Goris 000700
Aksent 167, Leuven 100774
Akshaya Kumar Maitreya Museum,
Siliguri 024482
Aksum Archaeology Museum,
Aksum 010727
Akta, Gabrovo 101125
Akta Lakota Museum,
Chamberlain 047513
Aktaion, Zaandam 143267
Die Aktgalerie, Berlin 106147
Aktion Museum M, Mistelbach an der
Zaya 002392
Aktionsgalerie Bern, Wohlen

(Bern) 116726
Aktives Museum Spiegelgasse für
Deutsch-Jüdische Geschichte in
Wiesbaden, Wiesbaden 022629
Aktives Museum Südwestfalen,
Siegen 021851
Aktivmuseum Ländliches Brauchtum,
Breitungen an der Werra 017581
AKUS Gallery, Eastern Connecticut
State University, Willimantic .. 054422
Akvamarín, Znojmo 102753
Akvarel', Kyïv 117115
Akvarius, Chimki 114177
Akvarius, Moskva 114191
Akwesasne Museum,
Hogansburg 049499
Akzente, Bremen 141641
Akzente, Mettmann 077364
Akzente, Nettetal 077704
Al-Ain Museum, Al-Ain 043332
AL Antik, Næstved 066002
Ål Auksjon, Ål 126335
Al Bassir, Hiba, Damascus ... 134007
Ål Bygdemuseum, Ål 033482
Al Desvan Antigüedades, Lima . 084424
Al Gallery, Podgorica 112025
Al-Harithy, İzmir 117100
Al Ibriz Orcadre, Rabat 082595, 112046
Al-Itekal, Chicago 145031
Al-Karak Museum for Archaeology and
Folklore, Al-Karak 029759
Al Kassaba Museum Thafer al Masri
Foundation, Nablus 034167
Al-Khor Museum, Al-Khor 036176
Al Mada'in Museum, Baghdad .. 024800
Al-Mahatah – Sharjah Aviation
Museum, Sharjah 043338
Al-Mansora Museum Dar-ibn-Luqman,
Daqahlia 010566
Al Manyal Palace and Nilometer,
Cairo 010531
Al Mawsil Museum, Nineveh ... 024813
Al-Mukalla Museum, Al-Mukalla 054826
Al-Muntaza Palace Museum,
Alexandria 010512
Al Portico, Verona ... 081886, 132264
Al Pozzo, Venezia 081816
Al-Rawi, Idris, Paris 104460
Al Shaheed Monument and Museum,
Baghdad 024801
Al Sulaimaniya Museum,
Sulaimaniya 024815
Al Tezzon, Camposampiero ... 109930
Al-Toyan Tower Trading, Ruwi .. 084407
Al-Wakra Museum, Al-Wakra .. 036177
Al Zubara Fort, Doha 036178
Al.ba.tro's, Toulouse 073732
Ala Moana, Honolulu 093523
Ala, Salvatore, Milano 110233
Alabama Agricultural Museum,
Dothan 048341
Alabama Art Agency, Klagenfurt 100068
Alabama Arts Magazine,
Montgomery 139240
Alabama Auction Room,
Birmingham 127142
Alabama-Coushatta Indian Museum,
Livingston 050346
Alabama Furniture, Houston ... 093587
Alabama Mining Museum, Dora 048339
Alabama Museum of Natural
History, University of Alabama,
Tuscaloosa 053856
Alabama Museum of the Health
Science, Birmingham 047002
Alabama Museums Association,
Tuscaloosa 060283
Alabama Music Hall of Fame,
Tuscumbia 053863
Alabama Pastel Society,
Birmingham 060284
Alabama River Museum, Franklin 048934
Alabama State Council on the Art,
Montgomery 060285

Alabarces Jimenez, Jose, Palma de
Mallorca 086334
Alabaster Bookshop, New York . 145318
Alabastri, Anne, Paris 128890
Alacahöyük Müzesi, Alaca 042804
La Alacena, Málaga 086285
Aladdin Antiques, Montmorency 061873,
127642
Aladdin Antiques, Wellington ... 084152
Aladdin Book Shoppe, Oklahoma
City 145392
Aladdin Finishers, Portland 136145
Aladdin Gallery, Houston 093588
Aladdin's Antique Alley, Saint
Paul 096800
Aladdin's Cave, Freshwater 089307,
144431
Aladdin's Cave, Leeds 089766
Aladdins Cave, Nenagh 079682
Aladdin's Cave, Swansea 091526
Aladdins Cave Antiques, Berry . 061143
Aladin, Dijon 068312
Aladin, Grenoble 068709
Aladin, Guérande 068758
Aladin, Lille 069471
Aladin, Versailles 138625
Aladin Antik, Praha 065460
Aladin Antik, Zürich .. 087927, 133973
Aladine, Paris 070875
Aladjem, Sabitaï, Aubervilliers .. 103283
Alaerts, J.P., Bruxelles . 063306, 100545
Alagarda Irriberri, Maria,
Barcelona 085717
Alaimo, Providence 123693
Alaimo, Michele, Milano 080378
Alain Encadreur, La Rochelle ... 103899
Alain, Bernard, Fontenay-sous-
Bois 068576, 140842
Alain, Levent, Athies-sous-Laon 066764
Alain, Michel, Bruxelles 063307
Alal, Paris 128891
El-Alam El-Arabi, Cairo 066225
Alamance Battleground,
Burlington 047325
Alamanda, Saint-François 109111
Alamannenmuseum, Weingarten, Kreis
Ravensburg 022514
Alamannenmuseum Ellwangen,
Ellwangen 018158
Alamo, Montpellier 070287
The Alamo, San Antonio 052928
Alamo Antique Mall, San Antonio 096934
Alamo City Galleries, San Antonio 124004,
124005,   124006,   124007
Alamo Military Collectables, Saint
Louis 096661
Alamo-Mo Book Shop, San Diego 145516
Alamo Township Museum – John
E. Gray Memorial Museum,
Kalamazoo 049856
Alampo, Orazio, Catania 080018, 131078
The Alan Cristea Gallery, London 118158
Alan Rutherford's Studio Gallery,
Toronto 101502
Alan Wofsy Fine Arts, San
Francisco 124205, 145538
Alan, Adrian, London 089894
Alana, Seattle 097389
Ålands Jakt och Fiskemuseum,
Eckerö 010764
Ålands Konstmuseum,
Mariehamn 011142
Ålands Museum, Mariehamn .. 011143
Ålands Sjöfartsmuseum,
Mariehamn 011144
Alaniesse, Jacques, Lyons-la-
Forêt 069786
Alankrita Antiques, Bangalore . 079341
Alankritha Arts Gallery,
Hyderabad 109350
Alan's Antique, Woking 135250
Alan's Antiques, Leatherhead ... 089758,
134529
Alan's Antiques, Omaha 096020

Alan's Collectables, Gosford ... 061544
Alanteaks, Jacksonville 093947
Alantic, Lyon 069682
Alanya Müzeleri ve Örenyerleri,
Arkeoloji Müzesi, Kızılkule Etnografya
Müzesi ve Atatürk Evi Müzesi,
Alanya 042805
Alapítvány Érc és Ásványbányászati
Múzeum, Rudabánya 023868
Alarce, Madrid 133289
Alarcon, Pedro Lopez, Madrid .. 086089
Alaska Aviation Museum,
Anchorage 046468
Alaska Centennial Center for the
Arts, Fairbanks Arts Association,
Fairbanks 048657
Alaska Indian Arts, Haines ... 049314
Alaska Museum of Natural History,
Anchorage 046469
Alaska Northwest Books,
Portland 138441
Alaska On Madison Gallery, New
York 122297
Alaska Shop of Chicago Gallery,
Chicago 120034
Alaska State Council on the Arts
Communique, Anchorage 139241
Alaska State Museum, Juneau . 049848
Alaska Watercolor Society,
Anchorage 060286
Alaskaland Pioneer Air Museum,
Fairbanks 048658
Alastar Crawford, New York ... 095201
Alatornion Pitäjämuseo, Tornio . 011429
Alaux, Jean, Paris 070876
Alauzet, Yvon, Millau .. 070119, 140957
Alazarine, Pierre, Gordes 068682
Alazraki, Daphne, New York 095202
ALB Antiquités, Paris 070877
Alba, Ferrara 110017, 137641
Alba, Madrid 086090
Alba Art Gallery, San José 102477
Alba Burguillos, Antonio, Bilbao . 133180
Alba Cabrera, Valencia 115474
Alba Editorial, Barcelona 137949
Alba Longa, Madrid 086091
Alba, Iulia, Palma de Mallorca . 086335
Albadia Museum, Arab Bedouin
Heritage Center, Osafia 025167
Alban, Cristin, Noce 070722
Albanese, Giovanni, Roma 131796
Albano, Jean, Chicago 120035
Albany Antiques, Hindhead 089532
Albany Antiques, Malvern,
Victoria 061776
Albany Civil Rights Movement
Museum, Albany 046337
Albany Gallery, Cardiff 117521
Albany Gallery, London 089895, 118159
Albany Hill, Albany ... 083744, 132679
Albany Historical Society,
Sabetha 052718
Albany Institute of History and Art,
Albany 046340
Albany Museum, Grahamstown . 083723
Albany Regional Museum, Albany 046349
Albany Residency Museum,
Albany 000775
Albarium, Zaragoza ... 133486, 133487
Albarran, Margarita & Rodríguez
Armijo, Sevilla 115443
Albatros, Milano 110234
Albatros, Moskva ... 114192, 114193
Albatros de Pitcairn, Etables-sur-
Mer 068477
Albatros Flugmuseum, Stuttgart . 022025
Albatrossmuseet, Romakloster . 040920
Albayalde, Donostia-San
Sebastián 133241
Albemarle, Scottsville 127452
Albemarle Gallery, London 118160
Alber, Herbert, Bregenz 099965
Alber, Martin, Haidershofen 062623

Albergo, Chicago 120036
Albergo-Museo Atelier sul Mare,
Castel di Tusa 025813
Albergo, Lorenzo, Palermo 131680
Alberni Valley Museum, Port
Alberni 006671
L'Albero del Ricchio, Milano ... 142709
Albers, Memphis 121679
Albers, Manning, San Francisco 124206
Albert, Atlanta 092100
Albert & Partner, Palma de
Mallorca 086336
Albert & Herbert's Antik og
Bruktbutikk, Torp 084377
Albert and Tove Boyajian Exhibition
Hall, Jerewan 098595
Albert County Museum, Hopewell
Cape 006014
Albert Cuijp Antiquiteiten,
Amsterdam 082676
Albert Edelfeltin Ateljeemuseo,
Porvoo 011270
Albert Engström-Museerna,
Grisslehamn 040673
Albert Kersten GeoCentre, Broken
Hill 000892
Albert-König-Museum, Unterlüß . 022278
Albert Lane Antiques, Berry ... 067784
Albert Park Gallery, Albert Park . 098621
Albert-Sammt-Zeppelin-Museum,
Niederstetten 020798
Albert-Schweitzer-Gedenk- und
Begegnungsstätte, Weimar ... 022488
Albert Schweitzer Haus, Königsfeld im
Schwarzwald 019728
Albert, Daniel, Sedan 073466
Albert, Iris, Landau in der Pfalz . 107784
Albert, Susan Carol, São Paulo . 064137
Alberta Art Foundation Collection,
Edmonton 005744
Alberta Association of Registered
Nurses Museum, Edmonton . 005745
Alberta Aviation Museum,
Edmonton 005746
Alberta Beach and District Museum,
Archives Society, Alberta Beach 005308
Alberta Central Railway Museum,
Wetaskiwin 007389
Alberta College of Art & Design,
Calgary 055119
Alberta Forestry Service Museum,
Hinton 006008
Alberta Hospital Museum,
Ponoka 006667
Alberta Museums Association,
Edmonton 058414
Alberta Print Makers Society,
Calgary 058415, 101211
Alberta Railway Museum,
Edmonton 005747
Alberta Society of Artists,
Calgary 058416
Alberta Sports Hall of Fame and
Museum, Red Deer 006756
Albertella, Genova 131285
Albertfalvai Helytörténeti Gyűjtemény
és Iskolamúzeum, Budapest . 023454
Alberth, Marina, Bad
Mergentheim 074549
Alberti, Bilbao 133181
Alberti, Nice 070549
Alberti Gallery, London 118161
Alberti, Bianca Maria, Milano . 131365
Albertiina, Helsinki 066295
Albertina, Wien 002931
Albertis, Hanau 107334, 125983, 130001,
141905
Albertland and Districts Museum,
Wellsford 033362
Alberto Galerija, Vilnius 111795
Alberto Jose, Salamanca 086406,
133415
Alberto Levi Gallery, Milano ... 080379
Alberton, Auckland 033139

Alberton Museum, Alberton 005310
Alberts, Almelo 082624
Alberts, Jacksonville .. 093948, 135642
Albert's Antiques Restorations and
Refinishing, San Mateo 136362
Albert's Bookshop, Guyra 139758
Alberts-Langdon, Boston 119827
Alberts-Van Smirren, Amsterdam 082677
Albigeois Antiquités, Albi 066503
The Albin O. Kuhn Gallery, University
of Maryland-Baltimore County,
Baltimore 046726
Albino Books, Birmingham 144249
Albino Tableaux, Paris 104461
Albino, William, New York 095203
Albins Antik, Linköping 086812
Albinson-Kurz, Eutin 075721
Albion, London 118162
Albion Academy Historical Museum,
Edgerton 048474
Albion Antique Auction Centre,
Albion 125175
Albion Books, Vancouver 140421
Albion Clocks, London 134579
Albion Gallery, Saint David's .. 119012
Albisetti, Alain, Cuzac 068262
Albo Antiques, Hamilton 064312
Albonetti, Domestici Mara, Roma 110723
Albrecht, Dortmund 075381
Albrecht-Daniel-Thaer-Gedenkstätte-
Landesausstellung, Reichenow-
Möglin 021346
Albrecht-Dürer-Gesellschaft, Kunstverein
Nürnberg, Nürnberg . 020843, 058706
Albrecht-Dürer-Haus, Museen der
Stadt Nürnberg, Nürnberg 020844
Albrecht-Kemper Museum of Art, Saint
Joseph 052783
Albrecht, A., Leipzig 077052
Albrecht, Andreas, Zinnowitz,
Ostseebad 079144
Albrecht, C.P., Nürnberg 077809
Albrecht, Donata, Burgdorf 129646
Albrecht, E., Augsburg 074449
Albrecht, Frank, Schriesheim .. 137500,
142350
Albrecht, Harro, Büsum 075220
Albrecht, Hubert, Sargans 144132
Albrecht, Mario, Aue .. 074446, 074447
Albrecht, Martina, Isernhagen .. 107472
Albrechtsburg Meißen, Meißen . 020345
Albrici & Paderni, Firenze 080074
Albrici, Antonio, Brescia 079909
Albright-Knox, Buffalo 119946
Albright-Knox Art Gallery, Buffalo 047298
Albrink, Torben, Birkerød 128378
Albrow & Sons, Norwich 090865
Albry, Berrima 061139
Albu Mõisahoone, Albu 010603
Album, Paris 141020
Album-Het Nieuwe Reclamemuseum,
Bruxelles 003427
Albums Books, New York 135871,
145319
Albuquerque & Sousa, Lisboa .. 084813
Albuquerque Art Business Association,
Albuquerque 060287
Albuquerque Coins, Albuquerque 092023
Albuquerque Furniture,
Albuquerque 135278
The Albuquerque Museum of Art and
History, Albuquerque 046358
Albuquerque, Celma, Belo
Horizonte 100874
Albuquerque, Michael & Jane de,
Wallingford 091710, 135181
Alburda, Lyon 140879
Albury Antiques, Albury 060981
Albury Regional Art Gallery,
Albury 000782
Albury Regional Museum, Albury 000783
Alby, Cagliari 131058
Alcade, Lyon 140880
Alcal, Houston 135582

Alcalá, Madrid 126526
Alcan Museum and Archives,
Montréal 006357
Alcarria, Sevilla 133428
Alcaston Gallery, Fitzroy 099019
Alcázar de Segovia, Segovia ... 040152
Alchemy: The Group for Collections
Management in Yorkshire and
Humberside, Halifax 060023
Alcheringa, Bodegraven 143093
Alcheringa ou Le Temps du Rêve,
Bordeaux 103411
Alchimia, Roma 081194, 131797
Alchimistenmuseum, Kirchberg am
Wagram 002170
Alcina, La Rochelle 103900
Alcobas Fernandez, Juan Miguel,
Cartagena 115005
Alcoforado, Washington 097800
Alcolado, Marcos Diez, Madrid . 115130
Alcolea, Madrid 086092, 115131
Alcolea, Fernando, Barcelona .. 085718,
114846
Alcolea, Jorge, Madrid 115132
Alcon, Houston 135583
Alcot, Los Angeles 121427
Alcuin, Scottsdale 145573
Alcuvilla, Jérôme, Rivieres 072481
La Aldaba, A Coruña 085946
El Aldabon, Bogotá 065169
Aldag, Eva, Buxtehude 125912
Aldborough Roman Site,
Aldborough 043380
Aldea del Tiempo, Bucaramanga 065233
Aldea Gallery, Buenos Aires ... 098545
Aldebert, Alain, Neuville-de-
Poitou 070525
Aldebert, Comdor, Castres 067721
Aldeburgh Museum, Aldeburgh . 043381
Aldeia Museu de João Franco,
Mafra 035839
Alden, Denver 093238, 135523
Alden B. Dow Creativity Center,
Northwood University, Midland 057441
Alden B. Dow Museum of Science and
Art, Midland Center for the Arts,
Midland 050870
Alden Hamilton, Richmond 096520
Alden Historical Society Museum,
Alden 046375
Alden House, Duxbury 048413
Alden, Michael, New Orleans .. 122114
Aldenhoven, Hans-Heiner, Kerken 141986
Alder, Ha Noi 125053
Alder Cocinas, Alicante 133079
Aldermaston Antiques,
Aldermaston 088243
Alderney Railway – Braye Road
Station, Alderney 043383
Alderney Society Museum,
Alderney 043384
Aldersey Hall, Chester 088855
Aldershot Military Museum and
Rushmoor Local History Gallery,
Aldershot 043386
Alderson, C.J.R., Tetbury 091562
Aldessan Gallery, Glasgow 117828
Aldfaers Erf Route, Museumdorp
Allingawier, Allingawier 031895
Aldheidskeamer Uldrik Bottema,
Aldeboarn 031889
Aldidea, León 086059, 086060
Aldine, Los Angeles 145209
Aldonga Auctions, Wodonga ... 125246
Aldous-Cook, David, Sutton ... 138286
Aldredge House, Dallas 145106
Aldrich Museum of Contemporary Art,
Ridgefield 052569
Aldridge, Bath 126810
Aldridge, F.W., Takeley 135117
Aldus, Brașov 143609
Aldus, Frederiksberg 140566
Ale-Kirja, Oulu 140738
Alecrim 50, Lisboa 113978

Alefeld, A.W., Bonn 075060
Alegria, Mandaluyong 113391
Alegria, San Antonio . . 096935, 136253
Alegria, Luis, Porto . . 085006, 114072
Aleksandra Čaka Memoriālais
Dzīvoklis, Rīga 030175
Aleksandrovskaja Sloboda –
Gosudarstvennyj Istoriko-
architekturnyj i Chudožestvennyj
Muzej-zapovednik, Aleksandrov 036844
Aleksandrovskij Chudožestvennyj
Muzej, Aleksandrov 036845
Aleksandrovskij Dvorec –
Gosudarstvennyj Muzej Carskoje Selo,
Puškin 037671
Aleksandrovskij Literaturno-
chudožestvennyj Muzej
Mariny i Anastasii Cvetaevych,
Aleksandrov 036846
Aleksandrovskij Narodnyj Muzej M.E.
Pjatnickogo, Aleksandrovka . . . 036847
Aleksiev, Rajko, Sofia 101139
Aleksis Kiven Syntymäkoti,
Palojoki 011230
Ålen Bygdetun, Ålen 033484
Alencastro, Porto Alegre 100964
Aleph, Bruxelles 100546
Aleph Books, Utrecht 143240
Aleppo National Museum, Aleppo 042254
Alert Bay Museum, Alert Bay . . 005311
Aléry, Annecy 066633
Ales, Saint Paul 136210
Alésina, Eric, Clairvaux-les-Lacs 067997
Alésina, Eric, Saint-Étienne . . . 072745,
                                     129223
Alessandri, Roma 131798
Alessandri, Roberto, Roma . . . 110724
Alessandrino, Carmelo, Milano . 131366
Alessi, Warszawa 084687
Alessia Librairie, Paris 141021
Ålesunds Museum, Ålesund . . . 033485
Alex Brown Cycle History Museum,
Thornhill, Dumfriesshire . . . 045975
Alex Gallery, Washington 124851
Alex Youck School Museum,
Regina 006764
Alex, Hans–Jürgen, Lüneburg . . 077205
Alex, Susanne, Geseke 075990
Alexander, Alassio 109740
Alexander, Haarlem 132513
Alexander, Henfield 089502
Alexander, Houston 093589
Alexander, New York . . 095204, 095205
Alexander, San Francisco 097155
Alexander, Sofia 101140
Alexander & Bonin, New York . . 122299
Alexander & David Fax, Kristen,
Canberra 139684
Alexander & Thurlow, London . . 089897,
                                     134580
Alexander & Victor, New Orleans 122115
Alexander and Baldwin Sugar
Museum, Puunene 052422
Alexander Antiques, Portrush . . 091042
Alexander Antonow Verlag, Dr.
Alexander Antonow, Frankfurt am
Main 137193
Alexander Brest Museum,
Jacksonville 049767
Alexander Fleming Laboratory
Museum, London 044899
Alexander Gallery, Bristol 088642,
                                     117435
Alexander Graham Bell, Baddeck 005367
Alexander Keiller Museum,
Avebury 043452
Alexander-Mack-Museum, Bad
Berleburg 016725
Alexander McKay Geological Museum,
Wellington 033342
Alexander–Morgan, Dee, Epping 117759
Alexander Ochs Galleries Berlin /
Beijing, Berlin 106148
Alexander Ramsey House, Saint

Paul 052835
Alexander Spendiaryan House
Museum, Yerevan 000720
Alexander Verlag Berlin, Berlin . 137048
Alexander, Julian, London 089896
Alexander, Keith, Vancouver . . 101731
Alexander, Marion, New York . . 122298
Alexander's, Oklahoma City . . . 095926
Alexander's, San Francisco . . . 097156
Alexanders, Titchfield . 091602, 135137
Alexanders Afsyrede Møbler,
Frederiksberg 065655
Alexander's Antiques, Richmond 096521,
                                     127418
Alexander's Antiques, Saint Louis 096662
Alexander's Galerie, Telgte . . . 108761
Alexander's Gallery, Saint Louis . 123793
Alexander's Sculptural Service, New
York 122300
Alexandra Antiquité, Groslay . . 068752
Alexandra Galleries, Beckenham 117322
Alexandra Museum, Alexandra . 033133
Alexandra Timber Tramway and
Museum, Alexandra 000784
Alexandre, Fort-de-France 142974
Alexandre, Namur 063805
Alexandre, New York 122301
Alexandre, Toronto 140392
Alexandre d'Or, Paris 070878
Alexandre Hogue Gallery, University of
Tulsa, School of Art, Tulsa . . 053847
Alexandre, Charles, Montréal . . 101353
Alexandre, Jean-Christophe,
Bavilliers 066968
Alexandria Antiques, Brighton . . 088611
Alexandria Archaeology Museum,
Alexandria 046381
Alexandria Black History Resource
Center, Alexandria 046382
Alexandria Livros, Lisboa 143552
Alexandria Museum of Art,
Alexandria 046378
Alexandria National Museum,
Alexandria 010513
Alexandria Press, London 138182
Alexandris, Konst., Patras 079198
Alexandroff, Arnaud, Bénodet . . 067090
Alexandros Papanastassiou Museum,
Levidion 023153
Alexandrovics, Ivanhoe 099157
Alexa's Treasures, Ballarat . . . 061080
Alexej von Jawlensky-Archiv,
Locarno 041682
Alexia Gallery, Washington . . . 124852
Alexia M Galerie, Lille 103998
Alex's Art Galerie, Klosters . . . 116390
Alex's Now & Then Collectibles, New
York 095206
Aleyne, Noëlle, Paris 104462
Alfa, Brno 140459
Alfa – Galerie Dila, Praha 102645
Alfa 2, Gdynia 084516
Alfa Antikvariat, Stockholm . . . 143926
Alfa Antikvariat Väst, Stockholm 143927
Alfa-Art, Moskva 085156, 126472
Alfa Art, Sankt-Peterburg 126476
Alfa Böcker, Malmö 143892
Alfa Haganum, Den Haag 143106
Alfa-Omega, Karlovy Vary 102603
Alfa Restauri, Roma 131799
Alfa Studio Editoriale, Bologna . 137622
Alfabeta, Stockholm 138010
Alfano, Domenico, Bologna . . . 130928
Alfano, Simone, Palermo 131681
Alfarrábios, Lisboa 143553
Alfarrabista Candelabro, Porto . 143599
Alfart, Napoli 131595
Alfazema, Porto Alegre 063953
Alfea Rare Books, Milano 142710
Alfia, Cluj-Napoca 143619
Alfican, Bruxelles 100547
Alfie's Antique Market, London . 089898
Alföldfásítási Gyüjtemény, Bugac 023574
Alföldi Galéria,

Hódmezővásárhely 109195
Alföldi Galéria, Tornyai János Múzeum,
Hódmezővásárhely 023688
The Alfoldy Gallery, Erikson . . . 005789
Alfombras Ghali, Alicante 133080
Alfombras Miguel, Zaragoza . . 133488
Alfombras y Kilims, Madrid . . . 086093,
                                     133290
Alfons Graber-Museum, Steinach 002777
Alfonsin, Marinete Rejane Zanetti,
Porto Alegre 063954
Alford Heritage Centre, Alford,
Aberdeenshire 043390
Alford House-Anderson Fine Arts
Center, Anderson 046474
Alford Manor House Museum, Alford,
Lincolnshire 043392
Alford Studio and Gallery,
Jacksonville 121230
Alfort, J.M., Stockholm 086958
Alfred Dunhill Museum, London 044900
Alfred East Art Gallery, Kettering 044683
Alfred Ost Museum, Zwijndrecht 004170
Alfred-Vogel-Museum, Teufen . 042050
Alfred Wegener Museum,
Rheinsberg 021400
Alfredo Jonušo Žemaitiška-japoniška
Sodyba-muziejus, Telšiai . . . 030487
Alfreds Antik, Sandnes 084350
Alfred's Antiques, Burpengary . . 061236
Alfreton Antiques Centre, Alfreton 088247
Alfriston Antiques, Alfriston . . . 088249
Alf's Antiques, Toronto 064577
Ålgården Konstnärernas Galleri,
Borås 115584
Alger, Minneapolis 094891
Alger County Heritage Center,
Munising 051117
Alghalaf, Saleh, Manama 100411
Algonkin, Braunschweig 141626
Algonquin College Museum,
Nepean 006460
Algonquin Culture and Heritage Centre,
Golden Lake 005891
Algonquin Logging Museum,
Whitney 007412
L'Alguer, La Bisbal d'Empordà . 085898,
                                     133207
Alhambra, Maracaibo 097971
Alhambra Antigüedades, Buenos
Aires 060849
Alhamra Art Gallery, Lahore . . 113308
Alhelm, Hameln 076344
Alhena, Fontainebleau 068564
Ali Baba, Le Havre 069271
A Aliança Livros, Lisboa 143554
Alianza, Boston 119828
Alianza Artistica, Barranquilla . . 102261
Alianza Editorial, Madrid 137980
Alianza Gallery, Louisville 121628
Alias, Paris . . 070879, 136848, 141022
Alibek, Hamburg 076179
Alibert, Daniel, Albi 128432
Alibert, Lisette, Paris 104463
Alibrandi, Firenze 110025
Alice Austen House Museum, Staten
Island 053495
Alice C. Sabatini Gallery, Topeka and
Shawnee County Public Library,
Topeka 053769
Alice in The Wonderland, Venezia 111077
Alice Springs RSL War Museum, Alice
Springs 000787
Alice T. Miner Colonial Collection,
Chazy 047600
Alice's, London 089899
Alice's Bookshop, Beechworth . 139638
Alice's Bookshop, Carlton North 139689
Alice's Old Furniture, Vancouver 064789
Aliceville Museum, Aliceville . . 046402
Alicke, Paul J. W., Frankfurt am
Main 141767
Alien Libros, Buenos Aires . . . 139543

Alienor, Albi 066504
Aliénor 2000, Duras . . 068418, 103709
Alig, K., Schwenda 108594
Alikartano, Numminen 011196
Alimentarium, Musée de
l'Alimentation – Une Fondation
Nestlé, Vevey 042104
Fratelli Alinari, Firenze 110026, 137643
Aline, Fryde, Saint-Ouen 073006
Alinea, Knokke-Heist 063694
Aline's Antiques, Louisville . . . 094514
Alingsås Antik, Alingsås 086626
Alingsås Konsthall, Konsthall och
Konstpedagogisk verksamhet,
Alingsås 040517, 056395
Alingsås Museum, Alingsås . . . 040518
Aliprandi, Gianantonio, Milano . 080380
Alis, Bucureşti 126469
Alisan Fine Arts, Hong Kong . . 102046
Alisea Art & Object Design,
Bologna 109810
Alisgårdenin Kotiseutumuseo,
Lapinkylä 011088
Alisher Navoii nomidagi Davlat
Adabiyot Muzeyi, Toshkent . . 054650
Alison, Chennai 079367, 079368
Alison Jacques, London 118163
Alisson RP 42, Saint-Étienne . . 072746
Alivertti, René, Vésenaz 087870
Aliwal North Museum and Church
Plein Museum, Aliwal North . . 038645
Alix Wagon Wheelmuseum, Alix 005314
Alix, Jean-Bernard, Paris 128892
Aliya, Atlanta 119531
Alizadeh, Saboor Ghadir,
Speltenbach 127876
Alizarin, San Francisco 124207
Alizarine, Albi 058587, 103213
Alizarine, Auvers-sur-Oise 103294
Alizés, Ronnet 072533
Aljira Center for Contemporary Art,
Newark 051496
Alkader, Antwerpen . . . 100479, 100480,
                                     127991
Alkader, Mechelen 100785
Alkame, Milano 080381
Alkent Actuel Art, İstanbul 116996
Alkgård, Peter, Gnesta 086688
Alkgårds Auktionskammare,
Södertälje 126665
Alkmaar Galeries voor Hedendaagse
Kunst, Alkmaar 112100
Alkonost, Sankt-Peterburg . . . 143663
Alkoven, Basel 087214
All Aboard, Wallington 062404
All About Antiques, Oklahoma
City 095927
All About Antiques, Wagga
Wagga 062395
All About Art, Albuquerque . . . 119452
All About Coins, Salt Lake City . 096871
All About Pictures, Cincinnati . . 120313
All American Vogue, Denver . . . 093327
All-Antik, Stockholm 086959
All Antique Art Restoration & Gilding,
Karel Kellner, Coorparoo . . . 127589
All Antique Glass, Norfolk 095841
All Art & Antique Restorer,
Kardinya 127621
All Art 1900, Bruxelles 063308
All Art Gallery, Houston 120987
all arts : alles Kunst, Stoltebüll . 108676
All Arts Bookshop, Woollahra . . 139964
All Booked Up, Jacksonville . . . 145181
All Booked Up, Louisville 145240
All Books Used & Rare, Houston 145158
All-Brite, Philadelphia 136073
All Goods Bought and Sold,
Belgooley 079456
All Hallows Undercroft Museum,
London 044901
All in Luck, Baltimore 092287
All India Annual Art Exhibition,
Mumbai 098173

All-India Fine Arts and Crafts Society,
Delhi 059355
All Most Antiques, Toronto 064578
All My Relations, Indianapolis . . 093847
All Our Yesterday, Sandwich . . . 091246
All Our Yesterdays, Glasgow . . . 089327
All Star Collectibles, Cleveland . 092910
All the Best Antiques, Buffalo . . 092505
All the Right Angles, Seattle . . 124477
All The Way Militaria, Rotterdam 083548
All Things Considered, Columbus 092984
All Things Old and Interesting,
Yackandandah 062493
All Uppköpet, Stockholm 086960
Alla Corte del Re, Cagliari 079972
Alla Corte Interiors, Padova . . . 080956
Alla Prima Arts, Portland 123586
Alla Prima Studios, Dallas 120540
Alla Stazione, Genova 110084
Alla Tiders Moderna och Antika,
Stockholm 086961
Alladin Antik, Roskilde 066083
Allahabad Museum, Allahabad . 024135
Allain, Françoise, Plurien 072211
Allam, Newbridge-on-Wye 090818,
134873
Allan Macpherson House,
Napanee 006453
Allan Shivers Museum, Woodville 054537
Allan, N.S., New York 095207
All'Angolo, Torino 081641
All'Angolo dell Antichità, Torino . 132085
Allanix, Les Acacias 116010
Allan's Antiques and Reproductions,
Cheadle Hulme 088825
Allard, Saint Ignatius 127425
Allard Pierson Museum Amsterdam,
Archeologisch Museum der
Universiteit van Amsterdam,
Amsterdam 031923
Allard, Isabelle, Annecy 128445
Allard, Serge, Montbenoit 070214
Allardi, Jean-Philippe, Paris . . . 125672
Allary, Irène, Pradelles 072302
Allbooks, Leidschendam 143189
Allcroft, Michael, Disley 089059
Allcroft, Michael, Hayfield 089480
Alldecor, Kirchberg (Sankt Gallen) 116387
Alle Antiche Armi, Milano 080382
Allecto, Tallinn 140676
Allée du Roy, Willgottheim 074298
Allegan County Historical and Old
Jail Museum, Old Jail Museum,
Allegan 046405
Allegany County Historical Museum,
Cumberland 048089
Allegany County Museum,
Belmont 046900
Alleghany Highlands Arts and Crafts
Center, Clifton Forge 047833
Allegheny Portage Railroad,
Gallitzin 049029
Allegoria, Aspelt 111815
Allegreto, Lisboa 113979
Allegretti Arte, Mantova 110199
Allegri, Claude, Flendruz 087426
Allegri, Claude, Gstaad 087517
Allegrini, Wiesbaden 078968
Allegro, Barcelona 133092
Allegro – Galerie Dila, Praha . . 102646
Allemand, Philippe, Issoire 128627
Allemandi & Beltrametti, Brescia 079910
Allemandi, Umberto, Torino . . . 137744
Allemann, Jean-Pierre, Valdoie . 129326
Allemann, Pierre, Evette-Salbert 128598
Allen, Pittsburgh 096252
Allen County-Fort Wayne Historical
Society Museum, Fort Wayne . 048892
Allen County Historical Society,
Iola 049706
Allen County Museum, Lima . . . 050279
Allen Gallery, Alton 043406
Allen Memorial Art Museum, Oberlin
College, Oberlin 051691

Allen Priebe Gallery, University of
Wisconsin Oshkosh, Oshkosh . 051826
Allen R. Hite Art Institute, University of
Louisville, Louisville 050485
Allen Sapp Gallery, North
Battleford 006503
Allen, Anthony, New Mills 134870
Allen, J., San Francisco 097157
Allen, John, Norwich 118824
Allen, M., Four Oaks . . 089300, 134394
Allen, Martin, Prahran 062106
Allen, Michael, East Brisbane . . 061428
Allen, Robert, San Francisco . . 124208
Allen, Tricia, Nungurner 099495
Allena, Costanzo Giovenale,
Torino 132086
Allende & Cia., Roberto, Santiago de
Chile 064942
Alleno, Janine, Amboise 066553
Allen's, Branksome 088573
Allen's Antiques, North Shore . . 084027
Allenspach, Kurt, Azmoos 087206
Allentown Art Festival, Buffalo . . 119947
Allentown Art Museum,
Allentown 046408
Allentuck, Gary, Washington . . 124853
Aller Art-Kunst von Frauen, Bad
Oldesloe 074568
Allerhand, Ansbach 105994
Allerhande, Hippolytushoef . . . 032489
Allerton, Mornington 099375
Allerton Books, Leeds 144518
Allerts, København 065821
Alles Altes, Neuenbürg 130445
Alles Antik, Bremen 075143
Alles-Kitsch Curiosa, Den Haag . 083049
Alletiders Brukt og Antik,
Lillehammer 084274
Alletru, Frédéric, Noyant 070757
Alley Cat, Saint Louis 096663
Allgäu-Museum, Kempten 019584
Allgäuer Auktionshaus, Kempten 125998
Allgäuer Burgenmuseum,
Kempten 019585
Allgäuer, Robert, Vaduz 142959
Allgas-Volkmann, Winden am
See 063118
Allhallows Museum of Lace and
Antiquities, Honiton 044600
L'Alliance, Toulon (Var) 141355
Alliance and District Museum Society,
Alliance 005315
Alliance Biblique des Antilles
Françoises, Baie Mahault 142547
Alliance Européenne des Experts, La
Guéroulde 058588
Alliance Francaise, San José . . . 102478
Alliance Française Galerie, Ha
Noi 125054
Alliance of Artists' Communities,
Providence 060288
Allianz deutscher Designer e.V.,
Braunschweig 058707
Allie Griffin Art Gallery, Weyburn 007394
Allied Air Force Museum,
Allentown 046409
Allied Artists of America, New
York 060289
Allied Arts Association, Richland 060290
Allied Arts Center and Gallery,
Richland 052527
Allied Arts Council of Lethbridge,
Bowman Arts Center,
Lethbridge 058417
Allied Arts of Seattle, Seattle . . 060291
Allier Antiquités, Vichy 074102
Les Alliés, Paris 104464, 104465
Alligri, Dominique, Bischheim . . 067229
Alliierten-Museum Berlin, Berlin 017067
Alling, München 077445
Alling Coverlet Museum, Palmyra 051898
All'Insegna del Sole, Genova . . 080231
Alliot Morillon, Brigitte, Lyon . . 104060
Allison, Miami 094680

Allison-Antrim Museum,
Greencastle 049238
Allison, Marie, Nashville 094961
Allisonville Road Antique Mall,
Indianapolis 093848
Allix, Brix 067515
Allmänna Galleriet 925,
Stockholm 115827
Allmendinger, Rolf, Vaihingen an der
Enz 130718
Allmer, Helmut, Graz 127742
Allmogegården Gulläkra,
Staffanstorp 086952
Allo, Robert, Menton 070057
Allobroc, Savigny-sur-Orge . . . 073452
Allodi, Elena, Brescia 131015
Allongue, Michel, Fayence 068516
Allongue, Michel, Tourrettes . . . 073829
Allora Historical Museum, Allora 000793
Allora Treasures, Allora 060994
Alloy, Los Angeles 135700
Alpine, Hythe 134495
Allpress Antiques, Malvern,
Victoria 061777, 127631
Allschwiler Kunst-Zentrum,
Allschwil 041207
Allschwiler Kunstverein, Allschwil 059918
Allsop, Duncan M., Warwick . . . 144907
Allsop, Jonathan, Mouilleron-en-
Pareds 070365
Alltime Antiques, Bairnsdale . . . 061074
Alluard, Régis, Andéchy 128441
Alluchon, Bernard, Paris 070880
Allure, Brugge 063258
Allure Antiques, Perth 062074
Alluto, Annalisa, Torino 132087
Allwardt, R., Amberg 074402
Allwood Antiques, Nowra North . 062019
Allwood Furniture Repair, Calgary 128143
Allwood House, Hurstbridge . . . 001159
Ally, Saint Paul 145500
ALM Déco, Le Pré-Saint-Gervais 128668
Alma, Montpellier 104270
Alma, Praha 065461
Alma, Saint-Jean-de-Luz 105531
Alma Antique Accent, Chicago . 092651
Alma Firehouse and Mining Museum,
Alma 046417
El Almacen, Fuengirola 085997
Almacén a la Bina, Bogotá 065170
Almacén Anticuario Antigüedades
Bazaar, Bogotá 065171
Almacén Bazar la Subasta,
Bogotá 065172
El Almacén del Arte, Lima 113340
Almacén Galeria de Arte, Rio de
Janeiro 101005
Almacén Jalan Dahlan, Bogotá . 065173
Almacén Jalan Dhalan,
Cartagena 065257
Almacenes el Hogar, Santa Cruz de
Tenerife 115417
Almaden, San Jose 136354
Alman Arkeoloji Enstitüsü,
İstanbul 056477
Almanar, Casablanca 112030
Almandine, Paris 070881
Almar, Conilhac-Corbières 068138
Almar, C. & A., Bucureşti 085075
Almarza, Santiago de Chile . . . 064943,
128244
Almaz-Lombard, Balakovo 085144
Almaznyj Fond, Gosudarstvennyj
muzej-zapovednik Moskovskij Kreml,
Moskva 037233
Almazovo, Moskva 114194
Almecap-Arte-Creacion, Bogotá . 102270
Almeda Antique Mall, Houston . 093590,
093591
Almediheem, Khalid, Riyadh . . . 085312
Almehagen, Anita, Firenze 110027
Almeida Franco, Lisboa 084815
Almeida Fuentes, Alejandra,
Zaragoza 133489

Almeida, Arlette de, Ville-sur-
Jarnioux 074135
Almeida, Monica Filgueiras de, São
Paulo 101080
Almeida, Vitória Silva, Lisboa . . 084814
Almelek, İstanbul 116997
Almenningen Galleri, Bergen . . 113124
Alméras, Jacky, Anduze 066591
Almici, Bologna 130929
Almira Restauradores, Madrid . . 133291
Almirah Antiques, Bangalore . . . 079342
Almoina, Palma de Mallorca . . . 086337
Almond Valley Heritage Centre,
Livingston 044870
Almoneda, Madrid . . . 086094, 133292
Almoneda, Valencia 086496
Almoneda Abando, Bilbao 085862
Almoneda Altamira, Vigo 086552, 126561
Almoneda Angelo, Málaga 086286
Almoneda Arienza, Madrid 086095
Almoneda Bastero, Madrid 086096,
126527, 143747
Almoneda la Fuentecilla, Madrid 086097
Almoneda Trastienda, Pontevedra 086398
Almoneda Verona, Madrid 086098
Almonedas Muñecas, Madrid . 086099
Almonry Heritage Centre,
Evesham 044308
Almorial, Paris 104466
Almost All New Zealand Books,
Oamaru 143339
Almost Heaven, Tucson 097624
Almqvist Antik och Auktionshandel,
Burträsk 086655, 126572
Almshouses Arcade, Chichester 088870
Almuehuset, Esbjerg 128380
Almuen Antik, Aalborg 065578
Almusbah, Shuaa, Jeddah 085296
Almviks Tegelbruksmuseum,
Västervik 041142
Alnot, Colette, Saint-Ouen 073007
Alnwick Antiques, Alexandra . . . 060989
Aloha, Brugheas 067520
Aloha Antiques, Honolulu 093524
Aloha Images, Honolulu 120918
Aloha Wood Art, Honolulu 120919
Along the Way, Baltimore 092288
Alonso Castrillo, Isabel & Jose Ramon,
Madrid 086100
Alonso Novoa, Málaga 115295
Alonso, Jean-Claude, Aillant-sur-
Tholon 066440
Alonso, Jose, Miami . . 121722, 121723
Alonzo's, Salt Lake City 136229
Alopaşalı, Kütahya 088152
Alosery, Almere 112113
Alostum, Mathias, Aalst 063171
't Aloude, Utrecht 083633
Alouette, San Diego 097049
Alp' Antiquités, Moutiers-
Tarentaise 070380
ALP Galleri Peter Bergmann,
Stockholm 115828
ALP Galleries, New York 122302
Alp-Museum Fasons, Seewis
Dorf 041977
Alp'Antic, Goncelin 068675
Alpen-Adria-Galerie im Stadthaus,
Klagenfurt 002185
Alpenländische Galerie, Zweigmuseum
des Bayerischen Nationalmuseums,
Kempten 019586
Alpenrail Swiss Model Village and
Railway, Claremont, Tasmania . 000966
Alpenrose, Moskva 114195
Alper, Maurice, Terrasson-
Lavilledieu 073635
Alper, Sherri, Nashville 094962
Alpha Antiques, Baltimore 092289
Alpha Antiquites, Saint-Ouen . . 073008
Alpha Arts Enchères, Lyon 125619
Alpha Collection, Paris 070882
Alpha Fine Art, Cape Town 114756
Alpha Gallery, Boston 119829

Alpha House Gallery, Sherborne 119090
Alpha Konsthandel, Halmstad . . 086729
Alpha Museum, Bangalore 024148
Alpha Numis, Lagor 069141
Alpha-Presse, Sulzbach 137531
Alpha Space, Sapporo 111373
Alphacentauri, Parma 110625
Alphadelta Gallery, Athinai 109018
Alpha's Dolls, Dallas 093059
Alphaville, Rio de Janeiro 101006,
                                        125359
Alphecca, Paris 104467
Alpin- und Heimatmuseum, Hohe
  Wand-Stollhof 002094
alpin+art+gallery – Schatzhaus der
  Natur, Ebene Reichenau 001833
Alpinarium Galtür, Galtür 001922
Alpine Antiques, Astley Bridge . . 088316
Alpine Art, Salt Lake City 123944
Alpine Coin Company, Portland . 096350
Alpine Curiosa, Stilfs 028018
Alpine Design, New York 095208
Alpines Ballonsport-Museum,
  Mürren 041779
Alpines Museum, Zermatt 042175
Alpines Museum des Deutschen
  Alpenvereins, München 020502
Alpineum, Hinterstoder 002084
Alpineum, 3D-Alpenpanorama/Diorama,
  Luzern 041701
Alpinmuseum Dachstein, Ramsau am
  Dachstein 002566
Alpinmuseum, Zweigmuseum des
  Bayerischen Nationalmuseums,
  Kempten 019587
Alpmuseum Riederalp, Riederalp 041878
Alp'Ours, Annecy 066634
Alps, Atlanta 119532
Alpsennereimuseum, Hittisau . . 002088
Alquimia, Puntarenas 102474
Alresford Gallery, Alresford 117229
Alro, Fieberbrunn 127732
Alro's Antiques, Woody Point . . 062455
Al's Collectables and Antiques,
  Denver 093240
Al's Fine Art, Nashville 122060
Al's Gallery, Nashville 122061
Als Hjemstavnsmuseum,
  Hadsund 010095
Alsace, Strasbourg 073562
Alsacienne-de-l'Or et des Métaux
  Précieux, Metz 070084
Alsaticarta, Mulhouse . 070382, 140972
Alsatroc, Strasbourg 073563
Alšova Jihočeská Galerie, Hluboká nad
  Vltavou 009560
Alsters Herrgård, Karlstad 040743
Alstertal-Museum, Hamburg . . . 018975
Alstin's Antique Gallery, Hamilton,
  Victoria 061582, 125197, 127609
Alston Antiques, Alston 088263
Alt, Hamilton 128172
Alt & Antik, Buxtehude 075237
Alt-Arnstorf-Haus, Arnstorf 016611
Alt-Freden-Sammlung, Freden . . 018466
Alt-Ginnheim-Antik, Frankfurt am
  Main 075768
Alt-Gut, Bad Nauheim 141430
ALT Production, Paris . 104468, 136849
Alt-Refrath, Bergisch Gladbach . . 106134
Alt-Rothenburger Handwerkerhaus,
  Rothenburg ob der Tauber . . 021505
Alt-Segeberger Bürgerhaus, Bad
  Segeberg 016876
Alt-Stade im Baumhaus, Stade . . 021944
Alt Wien Kunst, Wien . 062883, 125268
Alt, Konrad, Schillingsfürst 078364
Alt, Manfred, Saarbrücken 078313
L'Alta Epoca, Brescia 079911
Alta Mira, Eivissa 115057
Alta Museum, Alta 033489
Alta, R., Tilburg 083611
Altadena Antiques, Birmingham . 092383
Altai, Milano 110235

Altair, Palma de Mallorca 115333
Altajskij Gosudarstvennyj Institut
  Iskusstv i Kultury, Barnaul 056264
Altajskij Gosudarstvennyj
  Kraevedčeskij Muzej, Barnaul . . 036882
Altajskij Respublikanskij Kraevedčeskij
  Muzej im. A.V. Anochina, Gorno-
  Altajsk 037020
Altamira, Caracas 125038
Altamira, Warszawa 084688
Altamiradeco, Bournemouth 088530
Altamont Press, Asheville 138299
Altamura, Altamura 138785
ALTANA Kulturstiftung, Bad Homburg
  v.d.Höhe 016786
Alte Apotheke, Stadtgeschichtliches
  Museum Ottweiler, Ottweiler . . 021050
Alte Arche, Furtwangen im
  Schwarzwald 018570
Alte Börse, Stadtgeschichtliches
  Museum Leipzig, Leipzig 019973
Alte Bücher-Stube, Steyr 140042
Das Alte Bureau, Berlin 074739
Alte Cuesterey, Kultur-Historischer
  Verein Borbeck e.V, Essen 018286
Alte Dorfmühle, Monzelfeld 020459
Alte Dorfschmiede,
  Waldbreitbach 022380
Alte Feste, National Museum of
  Namibia, Windhoek 031860
Alte Feuerwache Loschwitz e.V., Kunst-
  und Kulturverein, Dresden . . 058708,
                                        106714
Alte Galerie am Landesmuseum
  Joanneum, Graz 001964
Alte Galerie Enge, Zürich 087928
Alte Huf- und Wagenschmiede,
  Dietmanns 001817
Alte Kunst, Wien 062884
Alte Kunst & Militaria, Wien . . . 062885,
                                        062886
Alte Lederfabrik Güttgemanns, Halle /
  Westfalen 018948
Alte Marktschmiede, Lasberg . . 002271
Alte Mühle Donsbrüggen, Kleve . 019641
Alte Mühle Schrattenberg,
  Schrattenberg 002728
Alte Nationalgalerie, Staatliche
  Museen zu Berlin – Stiftung
  Preußischer Kulturbesitz, Berlin 017068
Alte Oberamtei, Gerabronn 018653
Alte Pinakothek, Bayerische
  Staatsgemäldesammlungen,
  München 020503
Alte Saline Bad Reichenhall, Bad
  Reichenhall 016845
Alte Schmiede, Aurich 074476
Alte Schmiede, Bedesbach 017012
Alte Schmiede, Bennhausen . . . 017032
Alte Schmiede, Mühlbeck 142133
Alte Schmiede, Schwaikheim . . 021758
Alte Schmiede – Kunstverein Wien,
  Wien 002932
Alte Schule mit Clara-Zetkin-
  Gedächtnisstätte, Königshain-
  Wiederau 019733
Alte Spedition, Gladbeck 107133
Alte Synagoge, Essen 018287
Alte Synagoge, Hechingen 019121
Alte Tradition, Hameln 076345
Alte Wache, Dresden 075406
Alte Wache, Groitzsch 018828
Alte Wassermühle, Splietsdorf . . 021938
Alte Welt Museum, Nußbach . . . 020884
Alte Zeiten, Neustadt in Holstein 142228
Das Alte Zollhaus Heimatmuseum,
  Hitzacker 019267
Altea, Madrid 115133
Altea Gallery, London 144550
Altenburg, Berlin 125866
Altendorf, Irmeli & Wolfgang,
  Freudenstadt 107069, 137212
Alter, New York 095209
Alter Dinge Hof, Wels 062873

Alter Ego, Lagrasse 103913
Alter Ego, Parma 081087
Alter native, Diesse 116247
Alter, Anneliese, Hamburg 076180
Alter, Hélène, Malakoff 069819
Alterio, Pasquale, Roma 131800
Altermann, New York . 122303, 127340
Altermann, Santa Fe . 124465, 127449
Alterna Kontor für antike Öfen,
  Berlin 074740, 129463
L'Alternance, Arras 103274
Alternariato, Torino 081642
Alternativa, San José 102479
L'Alternative, Paris 070883
Alternative Antiques,
  Warrnambool 062411
Alternative Ink, San Antonio . . . 124008
Alternative Museum, New York . . 051322
Alternatives, Roma 110725
Alternatives, Tucson 124706
Altero, Nicole, Paris 070884
Alterosa Arte Comtemporanea e
  Antiga, São Paulo 101081
Alterthümer, München 077446
Alterthümer, Wien 062887
Altertumsammlung der
  Benediktinerabtei, Engelberg . . 041465
Altes Amtshaus, Karl-Pollender-
  Stadtmuseum, Werne 022578
Altes Bücherkabinett, Linz 140008
Altes Bürgerhaus, Stadtmuseum
  Waidhofen, Waidhofen an der
  Thaya 002885
Altes en Vogue, Basel 133630
Altes Hafenamt, Ständige Ausstellung
  für Hafen und Schiffahrt,
  Dortmund 017950
Altes Rathaus, Lüneburg 020176
Altes Rathaus – Potsdam Forum,
  Potsdam 021174
Altes Rathaus Wilster, Wilster . . 022673
Altes Schloss, Abteilung
  Gemäldegalerie und Kunsthandwerk,
  Oberhessisches Museum und
  Gail'sche Sammlungen, Gießen 018689
Altes und Neues Schloss Eremitage,
  Bayerische Verwaltung der
  staatlichen Schlösser, Gärten und
  Seen, Bayreuth 016981
Altesse, Nendeln 111777
Altfield, Hong Kong 065006
Altfriesisches Haus, Sylt 022091
Althorp Museum, Althorp 043405
Altia Oyj, Rajamäen Tehdasmuseo,
  Rajamäki 011297
Altiero, Thierry, Le Thor 069355
El Altillo, Zaragoza 086586
Altius, Warszawa 084689, 113796,
                                        126418
Altkirch Antiquités, Mulhouse . . 070383
Altkunst Josefstadt, Wien 062888
Altländer, Jork 107484
Altlandtechnik-Museum, Klein
  Upahl 019639
Altmärkische Antiquariats &
  Kunsthandlung, Gardelegen . . 075944,
                                        141821
Altmärkisches Museum, Stendal 021993
Altman, Robert, New York 095210
Altmann, Aspremont 066763
Altmann, Gerhard, Tulln 100206
Altmeyer, R., Paderborn 078002
Altmühltaler Mühlenmuseum, Dietfurt
  an der Altmühl 017891
Altnöder, Ferdinand, Salzburg . . 100153
Alto Garda, Arco 109754
Altomani & Sons, Milano 080383
Altomani & Sons, Pesaro 081147
Alton Museum of History and Art,
  Alton 046423
Altona Antikviteter, Malmö 086843
Altonaer Antiquariat, Hamburg . . 141854
Altonaer Antiquitätenhof,

  Hamburg 076181, 129944
Altonaer Museum für Kunst und
  Kulturgeschichte, Stiftung Historische
  Museen Hamburg, Hamburg . . 018976
Altonaer Silberwerkstatt,
  Hamburg 076182, 129945
Altos de Chavon Escuela de Diseño,
  La Romana 055315
L'Altra Antichità, Lugano 087602
Alträucher, Köln 076843
Altrama, Madrid 133293
Altri Tempi, Taranto 081629
Aitschäfl, Claus M.A.,
  Pfarrkirchen 142301
Altstadt Antiquariat, Erfurt 141742
Altstadt Antiquariat, Freiburg im
  Breisgau 141797
Altstadt Antiquariat, Rapperswil . 144121
Altstadt-Antiquariat, Zürich 144156
Altstadt-Galerie, Mainz 107944
Altstadt Galerie, Remscheid . . . 108475
Altstadt Galerie Hall, Hall in Tirol 100028
Altstadtgalerie, Regensburg 108455
Altstadtgalerie, Rheda-
  Wiedenbrück 108487
Altstadthaus, Bad Bentheim . . . 141423
Altstadtrathaus, Städtisches Museum
  Braunschweig, Braunschweig . 017552
Altube Barandiaran, Fernando,
  Donostia-San Sebastián 115043
Altun Ha Archaeological Site, Altun
  Ha 004171
Altunizade Kültür Merkezi,
  İstanbul 116998
Alturath, Manama 063131
Altus, Berlin 074741, 125867
Altus, San Isidro 060955
Altxerri, Donostia-San Sebastián 115044
Alūksnes Novadpētniecības un
  Mākslas Muzejs, Alūksne 030098
Alumni Museum, Catonsville . . . 047467
Alumny Creek Museum, Grafton 011117
Alun, Pierre d', Bruxelles 136627
Aluna Gallery, Washington 124854
Alupka Deržavnyj Palac ta Park
  Zapovidnyk, Alupka 043113
Alutiiq Museum and Archaeological
  Repository, Kodiak 050016
Alva DeMars Megan Chapel Art
  Center, Saint Anselm College,
  Manchester 050625
Alvar Aalto Museo, Jyväskylä . . 010931
Alvarenga, Eliana, Rio de Janeiro 064006
Alvarez, New York 135872
Alvarez, Orléans 070804
Alvarez, Porto 114073
Álvarez García, Félix, León 086061
Alvarez Rodrigo, Maria Pilar,
  Zaragoza 133490
Álvarez Rojo, Berango 133178
Alvarez Soleil, Levroux 069449
Alvarez White, María Cecilia,
  Bogotá 102271, 128258
Alvárez, Ernesto, Buenos Aires . . 060850
Alvarez, John Jairo, Medellín . . . 102421
Alvárez, Julietta, Medellín 102422
Alvarez, Robert, Bayonville-sur-
  Mad 103346
Alvaro Galeria, Sevilla 115444
Alvdal Museum, Alvdal 033490
Alvear, Washington 124855
Alvear Antiguedades, Buenos
  Aires 060851
Alvear Art Gallery, Buenos Aires 060852
Alvear, Helga de, Madrid 115134
Alvensleben, München 108042
Karen Alvermann & Heiko Geier,
  Bensheim 074719
Alves & Marques, Lisboa 084816
Alves & Rodrigues, Coimbra . . . 084786
Alves-Albich, Helenice, Friedberg,
  Hessen 107070
Alves Beato, Manuel, Lisboa . . . 084817
Alves, Carlos, Miami 121724

Alves, Leonardo Donizete, São
Paulo 064138
Alves, Maria Fernanda, Coimbra 143548,
143549
Alvida-Antiken, Halmstad 086730
Alwane, Casablanca 112031
Alwane, Sarba Kaslik 082328
Always Buying Antiques,
Jacksonville 093949
Always Collectible, Austin 092202,
144975
Alwin, Goult 068686
Alwun, Phoenix 123478
Alyea's Jewellers, Ottawa 064490
Alyed, Marrakech 082587
Alys Antiques 1906, Waikato . . 084120
Alyskewycz, Jorge, Paris 104469
Alytaus Kraštotyros Muziejus,
Alytus 030312
Alyth Museum, Perth Museum and Art
Gallery, Alyth 043410
Alzgalerie, Truchtlaching 108783
Alzon, Yves, Hirsingue 068821
Am Dragos, Madrid 115135
AM Galeria de Artes, Belo
Horizonte 100875
AM-Galerie, Bad Schwartau . . . 106091
Am Quadros, São Paulo 064139
AM Trading, Aldgate 060987
AM Visual Arts Management, Elizabeth
Bay 061439
AM.PM Gallery, Hong Kong . . . 102047
AMA, Rödermark 078257
Amacla, Toulouse 105721
Amadei, Maria Luigia, Bologna . 079796
Amadeus, Wien 062889
Amadeus Antikues, Bucureşti . . 085076
Amadeus Antiques, Tunbridge
Wells 091645
Amadeus Antiquités, Paris 070885
Amadeus Maroc, Casablanca . . 082538
Amador County Museum,
Jackson 049746
Amador de los Rios, Madrid . . . 115136
Amager Antikstue, København . . 065822
Amagerbro Antik, København . . 065823,
140590
Amagermuseet, Dragør 010031
L'Amalgame, Alès 066526
Amalienborg, Danske Kongers
Kronologiske Samling,
København 010165
Åmåls Hembygdsmuseum, Åmål 040520
Åmåls Järnvägsmuseum, Åmål . 040521
Åmåls Konsthall, Åmål 040522
Amalthea Signum Verlag, Wien . 136571
Amana, Aix-en-Provence 103188
Amanda's Collectibles, Houston . 093592
Amanda's of Mogo, Mogo 099358
Amanda's Secondhand Hypermarket,
Newport, Shropshire 090837
Amandi, Ars, Berlin 141467
L'Amandier Artémis, Auxerre . . 066841
Amandine, Bouguenais 067383
Amandine – d'Ostensjhe
leslandvisjherie, Oostende . . . 003924
Amann, Berlin 074742
Amann, Sandra, New York 135873
Amano, Kazuo, Osaka 111345
Amapola Gallery, Albuquerque . . 119453
Amar Nadège, Lioux 069540
Amaral, Adriano, Porto 132963
Amarant Konst och Antikviteter,
Tystberga 087108
Amarante, Grenoble 068710
Amarante, La Rochelle 069078
Amarante, Marcq-en-Barœul . . 069859,
128738
Amarante, Montrouge 070333
Amarante & Zebrano, Paris 128893
Amararte, Lugo 086082, 115123
Amara, Paul, Paris 104470
Amarilli, Torino 081643
Amarillo Art Association, Amarillo 060292

Amarillo Museum of Art, Amarillo 046433
Amarna, Maastricht 112566
Amaryllis, Bruxelles 100548
Amasra Müzesi ve Tarihçesi,
Amasra 042806
Amaste Comunicacion, Bilbao . . 114978
Amasya Devlet Güzel Sanatlar Galerisi,
Amasya 042807
Amasya Müzesi, Kral Mezarlan,
Hazeranlar Konaği ve Alpaslan
Müzesi, Amasya 042808
L'Amateur, Beirut 111770
L'Amateur, Buenos Aires 139544
L'Amateur, Lyon 140881
L'Amateur d'Art, Beirut 111771
Amateurs d'Art, Perros-Guirec . . 072099
Amathole Museum, King William's
Town 038770
Amato, Nice 070550
Amato, Thomas M., New York . . 135874
Amatto, Hasparren 103808
Amaury Goyet Galerie, Aix-en-
Provence 066451
aMAZE Cultural Lab, Milano . . . 059390
Amazing Adventures Collectibles, San
Francisco 097158
Amazing Art and Prints, Chicago 092652
Amazing Grates, London 089900, 134581
Amazonite, Casablanca 082539
't Ambacht, Enschede 132501
Ambalama, Amstelveen 082661
Ambalangoda Mask Museum,
Ambalangoda 040470
Ambassador Antiques, Albion . . 060977
Amber, Nijmegen 083506
Amber Antiques, Auckland 083747
Amber Antiques, Southampton . 091369,
135063
Amber Arts, Edinburgh 117713
Amber, B., Seattle 097390
Amberes, Antwerpen 125292
Amberg, Hardy, Schleusingen . . 078370,
130624
Ambergris Museum and Cultural
Centre, Ambergis Caye 004172
Amberley Working Museum,
Amberley 043411
Ambiance Antiques, San
Francisco 097159
Ambiance Antiquités, Montréal . 064359,
128179
Ambiance Art and Antiques, Los
Angeles 094242, 135701
Ambiance Deco, La Rochelle . . . 069079
Ambiance et Discernement Antiquités,
Montréal 064360
Ambience, Kolkata 109394
Ambience Antiques, New Orleans 095047
Ambient Art, Amsterdam 112135
Ambiente, Bad Arolsen 074489
Ambiente, Cuxhaven 075283
Ambiente, Heringsdorf, Seebad . 107420
Ambiente 51, Heidelberg 076431
Ambiente Antik, Aschaffenburg . 074428
Ambiente Antik, Bad Münstereifel 074553
Ambiente de Antaño, Medellín . . 065261
Ambiente Kunst, Bayreuth 074690
Ambienti, Casablanca 082540
Ambijenta, Beograd . . 085327, 114393
Ambika, Paris 141023
Ambit Find, New Orleans 095048
ÀMBIT Galeria d'Art, Barcelona . 114847
l'Amboine, Bruxelles 063309
Ambra, Milano 131367
Ambra Restauri, Napoli 131596
Ambre, Besançon 067117
Ambre, Cenon 067744
Ambre, Compiègne 068112
Ambre Antiquités, Casablanca . . 082541
Ambre Congo, Bruxelles 063310
Ambre de Chaldée, Strasbourg . 073564
Ambridge Ferazzutti, Toronto . . 064579
Ambrogi, Roma 131801
Ambrosanio, Nello, Roma 131802

Ambrose, Cleveland 092911
Ambrose & Assoc., Columbus . . 092985
Ambrosia, Los Angeles 145210
Ambrosia Books, Balmain 139634
L'Ambrosiana, Milano 110236
Ambrosini & Marinari, Livorno . . 080313
Ambrossino, Miami 121725
Ambrozik, Henryk, Warszawa . . 084690
Ambühl, Hans-Rudolf, Thalwil . . 144142
Ambühl, Marlies, Luzern 116465, 133828
Ambulante Galerie, Passau 108360
Ame Couleur, Fleurier 116264
Ame du Bois, Béziers 067164
Amedeo Sacchetto, Brigitta,
Zürich 087929
Amelang, J., Berlin 129464
Amelia Earhart Birthplace Museum,
Atchison 046588
Amelia Island Museum of History,
Fernandina Beach 048724
Amelia's, Ashburton 083745
Amelia's Antique, Wagga Wagga 062396
Amelia's Retro-Vogue and Relics,
Austin 092203
Ameliasburgh Historical Museum,
Ameliasburgh 005326
Amélie Antiquités, Paris 070886
Ameline, Paule, Paris 128894
Amell, Verner, London . 089901, 118164
Åmell, Verner, Stockholm 086962,
115829
Ameln, Eilika, Koblenz 076830
Amend & Assoc., E.M., Dallas . 120541
Amendolara, Robert, Providence 096496
América Antigüedades, Buenos
Aires 060853, 127506
América Móvel, Lisboa 084818
America Oh Yes, Washington . . 124856
America West Primitive and Modern
Art Gallery, Tucson 124707
Americabilia, Las Vegas 094107
American Abstract Artists, New
York 060293
American Academy in Rome,
Roma 055963
American Academy of Appraisers of
Art, Antiques and Rug, Upper Saint
Clair 127483
American Academy of Art,
Cambridge 056821
American Academy of Arts and
Letters, New York 057538
American Academy of Arts and Letters
Art Museum, New York 051323
American Advertising Museum,
Portland 052287
American Airlines C.R. Smith Museum,
Fort Worth 048897
American Airpower Heritage Museum,
Midland 050874
American Antiquarian Society,
Worcester 060294, 060295
American Antique Mall, Tucson . 097625
American Antique Restoration &
Reconstruction, Cumberland . . 135488
American Antiques, Cleveland . . 092912
American Antiques Auction, New
Brunswick 127332
American Art, Baltimore 119715
American Art, Miami 138357
American Art Journal, Smithsonian
American Art Museum,
Washington 139242
American Art Resources, Houston 120988
American Art Review, Kansas
City 139243
American Art Therapy Association,
Alexandria 060296
American Artisan, Nashville 122062
American Artist, New York 139244
American Artists Professional League,
New York 060297
American Arts Alliance,
Washington 060298

American Arts Antiques Appraisals and
Associates, Chicago . 092653, 092654
American Association for Museum
Volunteers, Denver 060299
American Association for State and
Local History, Nashville 060300
American Association of Museums,
Washington 060301
American Baptist – Samuel Colgate
Historical Library, Rochester . . 052613
American Bicycle and Cycling
Museum, Santa Fe 053104
American Bronzes, Las Vegas . . 121331
American Catholic Historical Society
Museum, Philadelphia 052026
American Center, Sofia 101141
American Ceramic Society,
Westerville 060302
American Classical II, Nashville . 094963
American Classical Music Hall of
Fame and Museum, Cincinnati 047736
American Classics, Minneapolis . 094892
American Clock and Watch Museum,
Bristol 047211
American Coin and Currency,
Omaha 096021
American Coin and Jewelry, Oklahoma
City 095928
American Coin Collectibles,
Columbus 092986
American Coin Company,
Cleveland 092913
American Coin Exchange, Fort
Worth 093422
American Coin Exchange, Miami 094681
American Coins, Portland 096351
American Collectibles, Las Vegas 094108
American Collectibles, Norfolk . . 095842
American Color Print Society,
Princeton 060303
American Contemporary Art Gallery,
München 108043
American Council for the Arts, New
York 060304
American Craft, New York 139245
American Craft Museum, New
York 051324
American Craftsmen, Portland . . 136146
American Estates, Milwaukee . . 094826
American Estates and Antiques,
Portland 096352
American-European Fine Art, New
York 122304
American Federal Rare Coins,
Phoenix 096178
American Federation of Arts, New
York 060305
American Film Institute, Los
Angeles 050406
American Fine Art and Frame,
Dallas 120542
American Fine Art Editions,
Scottsdale 124476, 139246
American Fine Art Show,
Philadelphia 098411
American Folk Art Gallery, Zürich 087930
American Folk Art Museum, New
York 051325
American Furniture Stripper,
Melbourne 135772
American Gallery, Sylvania 053630
American Garage, Los Angeles . . 094243
American Helicopter Museum &
Education Center, West Chester 054292
American Heritage "Big Red" Fire
Museum, Louisville 050505
The American Historical Foundation
Museum, Richmond 052542
American Historical Print
Collectors Society. Newsletter,
Farmingdale 139247
American Illustration, New York . 139248
American Illustrators Gallery, New
York 122305

American Independence Museum,
Exeter 048651
American Indian Art Magazine,
Scottsdale 139249
American Indian Contemporary Arts,
San Francisco 052991
American Institute for Conservation
of Historic and Artistic Works,
Washington 060306
American Institute for Conservation of
Historic and Artistic Works. Journal,
Washington 139250
American Institute of Architects,
Washington 060307
American Institute of Graphic Arts,
New York 060308
American Interiors, Norfolk .. 136029
American Irish Historical Society
Museum, New York 051326
American Jazz Museum, Kansas
City 049675
American Jewelers and Coin Company,
Chicago 092655
American Journal of Archaeology,
Boston 139251
American Labor Museum,
Haledon 049317
American Legacy Gallery, Kansas
City 121277
American Legation Museum,
Tanger 031794
American Maple Museum,
Croghan 048069
American Memorabilia, Las
Vegas 094109
American Merchant Marine Museum,
Kings Point 049978
American Military Museum,
Charleston 047540
American Motorcycle Museum,
Raalte 032772
American Motorcycle Museum,
Unterweitersdorf 002859
American Museum in Britain,
Bath 043512
American Museum of Asmat Art, Saint
Paul 052836
American Museum of Cartoon Art,
Santa Monica 053130
American Museum of Fire Fighting,
Hudson 049601
The American Museum of Fly Fishing,
Manchester 050631
American Museum of Magic,
Marshall 050708
American Museum of Natural History,
New York 051327
American Museum of Nursing,
Tempe 053698
American Museum of Radio and
Eletricity, Bellingham ... 046892
American Museum of Science and
Energy, Oak Ridge 051668
American Museum of Straw Art, Long
Beach 050368
American Museum of the Miniature
Arts, Dallas 048110
American Museum of the Moving
Image, Astoria 046581
American Numismatic Association,
Colorado Springs 060309
American Numismatic Society, New
York 060310
American Numismatic Society
Museum, New York 051328
American Photography, New York 139252
American Plating, Phoenix 136110
American Police Center and Museum,
Chicago 047634
American Precious Metals, San
Jose 097353
American Precision Museum,
Windsor 054472
American Primitive, Columbus .. 092987

American Primitive Gallery, New
York 095211, 135875
American Printing House for the Blind,
Callahan Museum, Louisville .. 050486
American Quarter Horse Heritage
Center and Museum, Amarillo . 046434
American Rare Coin and Collectibles,
Minneapolis 094893
American Rare Coin and Jewelry,
Kansas City 094055
American Rare Coins, Charlotte . 092572
American Red Cross Museum,
Washington 054086
American Refinishing,
Jacksonville 135643
American Renaissance for the 21st
Century, New York 060311
The American Revolution Center,
Wayne 054253
American Royal Museum, Kansas
City 049876
American Saddlebred Horse Museum,
Mexico 050822
American Saddlebred Museum,
Lexington 050239
American School of Classical Studies
at Athens, Athinai 055750
American School of Classical Studies at
Athens, Princeton .. 057721, 138445
American Showcase, New York . 138373
American Society of Appraisers,
Houston 060312
American Society of Appraisers,
Kansas City 060313
American Society of Appraisers,
Portland 060314
American Society of Portrait Artists –
ASOPA, Montgomery 060315
American Sport Art Museum,
Daphne 048149
American Studio Plus,
Washington 124857
American Swedish Historical Museum,
Philadelphia 052027
American Swedish Institute,
Minneapolis 050932
American Table Design, Tampa . 097535
American Textile History Museum,
Lowell 050508
American Trading Company,
Cincinnati 092823
American Truck Historical Society,
Kansas City 049877
American Visionary Art Museum,
Baltimore 046727
American Visions, Saint Louis .. 123794
American Watchmakers-Clockmakers
Institute, Harrison 049375
American Watercolor Society, New
York 060316
American Watercolor Society
Newsletter, New York 139253
American West Heritage Center,
Wellsville 054277
American Work Horse Museum,
Lexington 050263
Americana Antiques, Miami ... 094682
Americana House, Chicago 145032
Americana Manse, Whitney-Halsey
Home, Belmont 046901
Americana Museum, El Paso ... 048502
Americana Museum, Terra Alta . 053715
Americans for the Arts,
Washington 139254
America's Black Holocaust Museum,
Milwaukee 050909
America's Ice Cream and Dairy
Museum, Medina 050762
America's National Music Museum,
Vermillion 053960
Americas Society Art Gallery, New
York 051329
America's Stonehenge, North
Salem 051616

Amerigo Tot Múzeum, Janus
Pannonius Múzeum, Pécs 023836
Amerika Museum Nederland,
Cuijk 032138
Amerind Foundation, Dragoon .. 048368
Ameringer & Yohe, New York .. 122306
Amerongs Historisch Museum/
Tabaksmuseum, Amerongen . 031905
Amersham Auction Rooms,
Amersham 126798
Amersham Museum, Amersham . 043415
Ames, Jacksonville 121231
The Ames Gallery, San Francisco 124209
Amethystwelt Maissau, Maissau . 002350
Ameublement Comme Neuf Antiquités,
Montréal 064361
Ameye, Croix 103647
Ameziane Museum, Nador ... 031786
Amfiaraion Archaeological Museum,
Kalamos 023103
Amfora, Constanţa 114165
Amfora, Oostende 063811
Amgad, Elnily, Fgura 082413
Amgueddfa Cymru – National Museum
of Wales, Cardiff 043863
Amgueddfa Pontypool Museums,
Pontypool 045521
Amherst College Museum of Natural
History, Amherst 046446
Amherst County Museum,
Amherst 046456
Amherst History Museum,
Amherst 046447
Amherst Museum, Amherst 046455
l'Ami de Musée, Paris 138626
L'Ami des Arts, Paris 104471
L'Ami des Lettres, Bordeaux .. 103412
Ami Voyage, Avignon 140770
Amiano, Juan, Pamplona 115383
Amiaux, Christiane, Sainte-
Ruffine 073365
Amicale des Brocanteurs du Poitou
Charentes, Niort 058589
Amici Antiques, Nashville ... 094964
Amici della Bonda, Mezzana
Mortigliengo 059391
Amici, Francesco, Venezia ... 111078
Amicis, Alberto de, Rio de
Janeiro 064007
Amico dell'Arte Cristiana, Milano 138786
Amico, Francesco Enrico Giuseppe,
Meda 131351
Amics de les Arts i Joventuts Musicals,
Terrassa 059827, 115466
Amics de los Museos de Cataluña,
Barcelona 059828
Amiel, Karen, New York 122307
Amiens & Asta-Richard, Lyon .. 128695
Amieva Mexico, Madrid 086101, 143748
Amighini, Enzo, Verona 081887
Amigos de los Museos, Madrid . 059829
Amigos del Arte, Asunción ... 059684
Amigos del Arte, Santiago de
Chile 058482
Los Amigos del Libro,
Cochabamba 136664
Amilien Attic, San Diego 097050
Aminta, Skopje 082396
Amir, San José 102480
Amiras, Daniele, Milano 080384
Amire and Co, Ménerbes 070048
Amiri, A. & A., Berlin 074743
Amirouche, Réda, El Mouradi . 060838
Amirouche, Réda, Staoueli ... 060842
Amiry, Los Angeles 121428
Amis des Arts, Aix-en-Provence . 103189
Amis des Arts, Neuchâtel 116526
Amis du Château de Pau. Bulletin,
Pau 138627
Amis du Vieux Marche le Brol,
Bruxelles 063311
Amis Galerie d'Art Bagatelle, Aix-les-
Bains 103206
Amistad, San Antonio 124009

Amities, Minneapolis 121962
Amity and Woodbridge Historical
Society Museum, Woodbridge . 054517
Amity Gallery, Albany 098618
Amity Studio, New York 122308
Amivi, Castello de la Plana 115011
AMK Kunsthandel, Grafing bei
München 076075
AMM Newsletter, Winnipeg 138567
Amma, Antwerpen 100481
Ammann, Pöcking 108391
Ammann & Co., Zürich 138102, 144157
Ammann, M., Locarno 116444
Ammann, Thomas, Zürich 116741
Ammendolea, Luigi Fortunato,
Messina 080358
Ammering, Ernst, Ried im
Innkreis 100148
Ammirata, Alessandro, Palermo . 110586
Ammitzbøll, Anne, København .. 065824
Ammoniet, Hasselt 100711
Amodio, Antonio, Verona 132265
Amon Carter Museum, Fort
Worth 048898
Amor, Albert, London 089902
Amorati, Mauro, Bologna 079797
Amorim, Pedro, São Paulo 064140
Amorosi, Laura, Firenze 131105
Amorotti, Bernard, Nice 070551, 128845
Amory Regional Museum, Amory 046458
Amos Andersonin Taidemuseo,
Helsinki 010811
Ámos-Anna Gyűjtemény, Ferenczy
Múzeum, Szentendre 023952
Amos Eno Gallery, New York .. 051330
Amos Herr House, Landisville .. 050130
Amos Seaman School Museum,
Minudie 006329
Amos, Diana & Eric, Saint
George's 100836
Åmot Antik og Auksjon, Åmot .. 084209,
                            126336
L'Amour de l'Art, Paris 104472
Amourette, Daniel, Rouen 105466
Amourette, Gérard, Ivry-la-
Bataille 068887
Amouyal, Zouzou, Lyon 069683
Amparo, Madrid 133294
Amparo Barbero Ferrandiz,
Alicante 114827
Amparo Cabrera, Guadalajara . 111938
Amparo Caro Gancedo, Maria,
Madrid 133295
Ampersand, San Francisco 124210
Amphora Galleries, Balfron .. 088342
Ampir, Beograd 085328
Ampir, Sankt-Peterburg 133018
Ampir limited, Moskva 085157
Ampthill Antiques, Ampthill .. 088273
Ampthill Emporium, Ampthill .. 088274
Amrein, Werner J., Luzern ... 087628
Amreli Archaeological and Art
Museum, Amreli 024140
Amrit Art Gallery, Kanpur ... 109391
Amro, Abdou-Ibrahim, Paris .. 070887
Amrumer Museum und Windmühle,
Nebel 020649
Amsler & Ruthardt, Berlin ... 074744
AMSTE Arte Contemporanea,
Lissone 110171
Amstein, Philippe, Basel 087215, 133631
Amstel Antiques en Curiosa,
Amsterdam 082678
Amsteltuin, Amstelveen 112129
Amsterdam Antiquarian Book, Map
and Print Fair, Amsterdam ... 098252
Amsterdam Antiques, Denver .. 093241
Amsterdam Art Gallery,
Amsterdam 082679
Amsterdam Art Gallery, New York 122310
Amsterdam Avenue Antiques,
Atlanta 092101
Amsterdam University Press,
Amsterdam 137824

Amsterdam, Whitney, New York   122309
Het Amsterdams Automuseum,
   Zwanenburg 033121
Amsterdams Beeldhouwers Kollektief,
   Amsterdam 059590, 112136
Amsterdams Historisch Museum,
   Amsterdam 031924
Amsterdams Openbaarvervoer
   Museum, Amsterdam 031925
Amsterdamsch Venduhuis,
   Amsterdam 126270
Amstutz, Edi, Wil 133963
AMT, Rouen 072555
Amtshofmuseum, Gernsbach . . . 018659
Amtsrichterhaus, Stiftung
   Herzogtum Lauenburg, Mölln,
   Schwarzenbek 021772
Amtsturm-Museum, Wendländischer
   Altertumsverein, Lüchow 020168
Amüseum – das Saarburger Museum
   am Wasserfall, Saarburg 021562
L'Amuleto, Perugia 142784
Amuletum, Perugia 142785
Amundsen, Inderøy 126354
Amuri Historical Museum, Waihu 033319
Amurin Työläismuseokortteli,
   Tampereen Museot, Tampere . 011393
Amurskij Gorodskoj Kraevedčeskij
   Muzej, Amursk 036849
Amurskij Oblastnoj Kraevedčeskij
   Muzej im. G.S. Novikova-Daurskogo,
   Blagoveščensk 036902
L'Amusette, Musée Interactif pour
   Enfants, Mesvin 003869
AMVC-Letterenhuis, Antwerpen . 003279
Amyot, Paris 104473
Amy's Antiques, Portland 096353
Amy's Attic, Oklahoma City 095929
Amz Artz, Okato 112977
An Antique Affaire, Los Angeles 094244
An Antique Store, Chicago 092656
An Dun Transport and Heritage
   Museum, Ballinahown 024823
An Iodnlann, Baugh 043536
An Lanntair Art Centre,
   Stornoway 045888
AN Magazine, Newcastle-upon-
   Tyne 139121
An Siopa, Waterford 079718
An, Thanh, Ha Noi 125055
Ana Chiclana, Madrid 086102
Ana Ros, Barcelona 085719
ANA World's Fair of Money,
   Baltimore 098412
Anabas-Verlag, Frankfurt am
   Main 137194
Anabel's Galerie, Barbizon 103326
Anacleto, Joaquim, Lisboa 084893
Anacortes Museum, Anacortes . 046461
Anacostia Museum, Washington 054087
Anadarko Philomathic Museum,
   Anadarko 046462
Anadolu Medeniyetleri Müzesi,
   Ankara 042813
Anadoluhisan Müzesi, İstanbul Hisarlar
   Müzeleri, İstanbul 042919
The Anaesthesia Heritage Centre, The
   Charles King Collection, London 044902
Anaesthetic Museum, London . . 044903
Anaf, Saint-Priest (Rhône) 125807
Anaf, Jean-Claude, Lyon 125620
Anagama, Saint-Germain-en-
   Laye 105518
Anagramme, Montpellier 070288
Anaheim Museum, Anaheim . . . 046466
Anahy Alta Joyeria, Cancún . . . 082458
Anakita Oriente, Burgos 085907
Analecta, Versailles 141379
Analepse, Paris 070888
Anales de Historia del Arte,
   Madrid 139024
Analogon, Hannover 141906
Analogue, Nantes 104328
Anamnese, Jarnac 068893

Anamnesis, Balaclava 061076
Anamorfose, Kortrijk 100770
Anamorfosis, Barcelona 085720
Anamur Kültür Merkezi, Anamur 042809
Anamur Müzesi ve Örenyerleri,
   Anamur 042810
Ananda, Baltimore 119716
Ananda, Paris 070889
Ananda Niketan Kirtishala,
   Bagnan 024145
Anandhaa Nilayam, Chennai . . . 142588
Ananse, Houston 120989, 120990
Anapskij Archeologičeskij Muzej-
   zapovednik, Anapa 036851
Anares, Bremen 141642
Anars, Ljubljana 114691
Ana's Attic, Omaha 096022
ANA's National Money Show,
   Phoenix 098413
Anasazi, Phoenix 145424
Anasazi, Saint-Tropez 105620
Anasazi Heritage Center, Dolores 048337
Anasazi State Park Museum,
   Boulder 047145
Anastacia's Antiques,
   Philadelphia 096071
Anastasis, Gent 063623
Anatoliaus Tyčinos Paveikslų
   Ekspozicija, Šiaulių rajono literatūros
   muziejus, Gruzdžiai 030338
Anatoli's Restoration, New York . 135876
Anatomical Museum, Chulalongkorn
   University, Bangkok 042646
Anatomisch Museum, Groningen 032375
Anatomisch Museum, Utrecht . . 032942
Anatomisch Museum Nijmegen,
   Nijmegen 032684
Anatomisch Museum van de
   Rijksuniversiteit Leiden, Leiden 032579
Anatomische Sammlung, Friedrich-
   Alexander-Universität Erlangen-
   Nürnberg, Erlangen 018257
Anatomische Sammlungen des Institut
   für Anatomie und Zellbiologie, Martin-
   Luther-Universität Halle-Wittenberg,
   Halle, Saale 018951
Anatomisches Museum,
   Greifswald 018805
Anatomisches Museum Basel,
   Basel 041260
Anatomisches Museum, Institut
   für Anatomie der Medizinischen
   Universität Innsbruck, Innsbruck 002117
Anatomy Museum, Aberdeen . . 043354
Anatomy Museum, Chennai . . . 024193
Anatomy Museum, Medical College,
   Mumbai 024400
Anatomy Museum, University of
   Queensland, Saint Lucia 001473
AnBau 35 – Projektraum und Edition,
   Bonn 106523, 137113
Anbriosa, Antwerpen 063186
Ancelo, Helsinki 128407
Ancestor House Antiques, Toledo 097596
Ancestors Alive, Philadelphia . . 145404
Ancestral, Lisboa 132928
Ancestrale, Saint-Lô 072895
Ancestry, Mumbai 079399
Ancestry Antiques, Rozelle 062172
Anchor, London 089903
Anchor Art Gallery, Saint-
   Barthélemy 109109
Anchor Fine Arts, Christchurch . 112824
Anchor Graphics, Chicago 047635
Anchor House Antiques, Honolulu 093525
Anchorage Gallery, Port Chalmers 112993
Anchorage Museum at Rasmuson
   Center, Anchorage 046470
Ancien Couvent, Treigny 016038
Ancien Couvent de Vagnas,
   Vagnas 073928
Ancien Palais de Bruxelles, Site
   Archéologique et Musée du du
   Coudenberg, Bruxelles 003428

Ancien Palais de Justice – Centre
   d'Art et d'Histoire, Kamouraska 006061
Ancienne École de Médecine Navale,
   Rochefort (Charente-Maritime) . 015120
Ancienne Mairie-Musée Annexe,
   Lillebonne 013811
Ancienne-Mans, Le Mans 069300
Ancienne Salle de Ventes, Le
   Neubourg 069319
Anciennement Vôtre, Gondreville 068679
Anciennement Vôtre, Nancy . . . 070410
Ancient Agora Museum, Athinai 022975
Ancient and Honorable Artillery
   Company Museum, Boston . . . 047099
Ancient and Modern, Blackburn,
   Lancashire 088493, 134117
Ancient and Oriental, Grimston . 089392
Ancient Antique Furniture,
   Singapore 085355
Ancient Antiques, Indianapolis . . 093849
Ancient Architecture Museum,
   Beijing 007587
Ancient Art and Coins, London . 089904
Ancient Art of the New World, New
   York 095212
Ancient Artefacts and Egyptian
   Antiquities, Saint Kilda, Victoria 062179
Ancient Arts, Long Beach 094187
Ancient Asia Gallery, Tulsa 097705
Ancient Basics, Los Angeles . . . 094245
Ancient City, Samut Prakan 042715
Ancient Coins Museum, Beijing . 007588
Ancient Craft, Singapore 085356
Ancient Earth Indigenous Art Gallery,
   Cairns 061245
Ancient Gallery, Ha Noi 097979
Ancient Gallery, Ho Chi Minh City 097982
Ancient Greek and Roman Coins,
   Portland 096354
Ancient Grounds, Seattle 097391
Ancient High House, Stafford . . 045850
Ancient House Museum, Thetford 045967
Ancient Mariner Art Effects,
   Singapore 085357
Ancient Monuments Society,
   London 060024
Ancient of Days, Oklahoma City 095930
Ancient Order of Foresters,
   Southampton 045823
Ancient Pearl, Las Vegas 094111
Ancient Spanish Monastery of Saint
   Bernard de Clairvaux Cloisters,
   Miami 050824
Ancient Traditions Gallery,
   Albuquerque 092024
Ancient Treasure, Sacramento . 096591
Ancient Valley, Singapore 085358
Ancora Ancora, San Jose 097354
Ancre Aldine, Lyon 140882
Ancre de Miséricorde, Saint-Malo (Ille-
   et-Vilaine) 072911
Anczok, Maximilian, Stuttgart . . 078590
Anczykowski, C. & P., Münster . 108211
And Antiques, Calgary 064246
AND Gallery, Saint Paul 123888
AND Gallery, Trnava 114669
And So To Bed, London 089905, 134582
Andancas, Rio de Janeiro 101007
Andanti, Schwäbisch Gmünd . . 142352
Andelos, Amsterdam 132389
AndenHånden Køge Antikvariat,
   Køge 140619
Andenken, Denver 120686
Andera, Maastricht 083453
Anderberg, Christin, Lund 133546
Die Andere Seite, Nürnberg . . . 142237
Andere Welten, Hamburg 141855
Anders Galerie, Düsseldorf 106759,
   137154
Anders Svors Museet, Hornindal 033669
Anders, Bengt, Landskrona 115717
Andersch, Nürnberg 077810
Andersen, Aage, København . . . 065825
Andersen, Anders, Ribe 066059

Andersen, Annamarie M., Zürich 087931,
   116742
Andersen, Bjarne, Aalborg 065579
Andersen, Hanne, Helsingør . . . 140580
Andersen, Jan, Vedbæk 066184
Andersen, Mikael, Berlin 106149
Andersen, Mikael, København . . 102859
Andersen, Per, København 065826
Andersen, Siv, Sandnes 113253
Anderson, Burford 117474
Anderson, Cremorne 139705
Anderson, Markranstädt 077299
Anderson, Oakland 095880
Anderson, Saint Paul . 096801, 136211
Anderson, San Francisco 097160
Anderson & Garland, Westerhope 127084
Anderson & Paul, Dunedin 083918,
   132715
Anderson Alistair, Glasgow 117829
Anderson County Arts Center,
   Anderson 046476
Anderson County Arts Council,
   Anderson 060317
Anderson County Historical Museum,
   Garnett 049053
Anderson County Museum,
   Anderson 046477
Anderson Farm Museum, Madoc 006239
Anderson Farm Museum, Greater
   Sudbury Heritage Museum,
   Lively 006201
Anderson Gallery, Richmond . . . 052543
Anderson Hill, Kingston-upon-
   Thames 118051
Anderson Museum of Contemporary
   Art, Roswell 052689
Anderson O'Day, London 118165
Anderson O'Brien, Omaha 123301
Anderson Park Art Gallery,
   Invercargill 033227, 112921
Anderson Slater, Settle 091263, 135015
Anderson, A., Minneapolis 094894
Anderson, Alistair, Glasgow . . . 134403
Anderson, Evelyn, Nashville . . . 094965
Anderson & Son, F.E., Welshpool 091776
Anderson, H. & A., Stanwell Park 062241
Anderson, Jan, New York 122311
Anderson, Judith, Auckland . . . 112744
Anderson, Vivien, Caulfield North 098878
Anderson/Abruzzo Albuquerque
   International Balloon Museum,
   Albuquerque 046359
Andersons, A., Clayfield 061329
Andersonville Prison,
   Andersonville 046478
Anderssen, Trondheim . 084383, 126394
Andersson, Kalmar 086765
Andersson, Karlskrona 143875
Andersson, Börje G., Stockholm 115830
Andersson, F., Vevey 087874
Andersson, Gunnar, Lund 143886
Andersson, Herman, Tollarp . . . 133590
Andersson, Stefan, Umeå 115961
Andersson, Thomas, Sala 143912
Andersson, Thomas, Uppsala . . 143987
Anderssonhall, London 118166
Anderwereld, Groningen 112487
Andhra Academy of Arts,
   Vijayawada 055846
Andhra Pradesh State Museum,
   Hyderabad 024295
Andhra Sahitya Parishat Government
   Museum, Kakinada 024335
Andia Blanco, Javier, Murcia . . . 086309
Andiamo, Milwaukee 121889
Andijon Viloyat Adabiyot va San'at
   Muzeyi, Andijon 054623
Andijon Viloyat Ülkashunoslik Muzeyi,
   Andijon 054624
Andina, Miami 121726
Andino, Tshwane 085669
Andino, J.P. de, Washington . . . 124858
Andipa Gallery, London 089906, 118167,
   134583

Andi's Gallery, Jakarta 109505
Andjić, Beograd 114394
Andler & Co., A.R., Saint Paul . . 123889
Ando, Sendai 082110
Andó, Fabio, Roma 081195
Andoe, Tulsa 124788
Andøymuseet, Risøyhamn 033892
Andong Folk Museum, Andong . 029884
Andor Semsey Múzeum,
   Balmazujváros 023431
Andorra Romànica Centre
   d'Interpretació, La Massana . . 000095
Andover Gallery, New York 122312
Andover Historical Society Museum,
   Andover 046480
Andover Museum, Andover 043416
Andra, Antwerpen 063187
Andrade, Gecy C., Belo Horizonte 063876
Andrade, José, Lisboa 084820
Andrä, Ernest Leopold, Krems . . 062686
Andraeas, Renate, Hamburg . . . 107219
Andrašiovská Obrazáreň, Banícke
   Múzeum Rožňava, Krásnohorské
   Podhradie 038372
Andraz Antique, Bucureşti 085077
André, Paris 128895
André, Roma 110726
André Antiquités, Belfort-du-
   Quercy 067071
Andre Emile, Vancouver 101732
Andre Garitte Foundation,
   Berchem 003357
André Kertész Fotómúzeum,
   Szigetbecse 023969
André Schlechtriem Contemporary,
   New York 122313
André, Anthony, Saint-Pierre-des-
   Corps 073256
André, Jacques, Etampes 068478
André, Michel, Coutevroult 068211
Andrea, Dallas 135489
Andrea Arte Contemporanea,
   Vicenza 111201
Andrea Robbi Museum, Andrea Robbi-
   Stiftung, Sils Maria 041988
Andréani, Jean-Claude, L'Isle-sur-la-
   Sorgue 069550
Andreani, Luciano, Bern 087270
Andrea's Antiques, San Francisco 097161,
                                     136304
Andrea's Antiques, Tauranga . . . 084098
Andreas Grimm München,
   München 108044
Andreas, S.C., Timişoara 085127
Andreasbutiken, Falköping 115608
Andreasen.Art, Højbjerg 102842
Andreasson, Järpås 143865
Andréemuseet, Stiftelsen
   Grännamuseerna, Gränna 040667
Andreescu-Bernhardt, Despina,
   Saarbrücken 078314
Andreevskij spusk 3, Kyïv 117116
Andréewitch, Stephan, Wien . . . 062890
Andréhn-Schiptjenko, Stockholm 115831
Andreini, Milano 080385
Andreja Pumpura Muzejs,
   Lielvārde 030151
Andreja Upīša Memoriālmāja,
   Skrīveri 030234
Andreja Upītša Memoriālais Muzejs,
   Rīga 030176
Andreotti, Raffaele, Bologna . . . 130930
Andre's Art Gallery, San Diego . 124086
Andres Gonzalez, Jose Angel,
   Zaragoza 115548
Andres Särevi Kortermuuseum,
   Eesti Teatri- ja Muusikamuuseum,
   Tallinn 010664
Andres, Marie-Christine, Mutzig . 070404
Andresen, Berlin 141468
Andresen, Elke, Geesthacht . . . 075956
Andrew, Zurrieq 082453, 132347
Andrew and District Local History
   Museum, Andrew 005337

Andrew Carnegie Birthplace Museum,
   Dunfermline 044171
Andrew County Museum,
   Savannah 053185
Andrew J. Blackbird Museum, Harbor
   Springs 049355
Andrew Johnson National Historic Site,
   Greeneville 049239
Andrew Logan Museum of Sculpture,
   Welshpool 046081
Andrew Ross Museum, Kangaroo
   Ground 001178
Andrew & Assoc., Brian, Chicago 092657
Andrew, Hilditch & Son,
   Sandbach 127032
Andrew, William, Norwich 138436
Andrews & Andrews, New York . 095213
Andrew's Collectibles, Buffalo . . 092506
Andrew's Pueblo Pottery and Art
   Gallery, Albuquerque 119454
Andrew's Refinishing, Dallas . . . 135490
Andrews, L., München 077447
Andrews, Marilyn, Nelson 112954
Andrews, Meg, London 089907
Andrews, Michael, Milford,
   Surrey 090759
Andrews, Stephen, Saint Peter
   Port 091203
Andrien Antiquités, Dammpart . . 068268
Andriesse, Paul, Amsterdam . . . 112137
Andrieu-Bourelly, Edmonde,
   Fumel 068618
Andrighetti Glasswork, Vancouver 101733
Andrin, Joël, Antibes-Juan-les-
   Pins 066662
Andriola, Thom, Houston 120991
Andriot, Bernard, München 077448
Andriotto, Fabio, Roma 131803
Andrischewski, Heinz Joachim,
   Düsseldorf 075450
Andrle, V., Praha 065462
Andrlová, Marie, Praha 065463
Andrøymuseet, Polar- og
   Fiskerimuseet, Andenes 033497
Android's Dungeon, Oklahoma
   City 095931
Andromeda Books, Athinai 142535
Androna degli Orti, Trieste 081781
Androscoggin Historical Society
   Museum, Auburn 046650
Androt, Pierre, Beaulieu-sur-Mer 066997
Androutsopoulou, Evangelia,
   Chalkida 109060
Androx, Vigo 115532
Andruchowicz, Dietlinde,
   Bodenheim 106516
Andy Warhol Art Authentication Board,
   Andy Warhol Foundation, New
   York 060318
The Andy Warhol Museum,
   Pittsburgh 052153
Andy's Antique Restoration,
   Killala 130876
Andy's Antiquitäten, Karlsruhe . . 076681
Andy's Corner, Norwood 062010
Andy's Curio Shoppe, Gisborne . 083939
Andzelm Galeria, Lublin 113684
Ané, Patrick, Foix . . . 103746, 128600
Anekojikan – Kyoto Mangekyo
   Hakabutsukan, Kyoto 028918
Aneks, Kraków 113524
Anelim, Lisboa 084821
Anelli, Enrico, Genova 131286
Anet Antiquités, Anet 066593
Anew Fine Consignment,
   Columbus 092988
Anfield Museum, Liverpool 044849
Anfiteatro, Milano 079849
Angarskij Gorodskoj Vystavočnyj Zal,
   Angarsk 036852
Angarskij Muzej Časov, Angarsk 036853
Angarskij Muzej Mineralov,
   Angarsk 036854
Angaston Abbey Trading,

Angaston 060998
Änge Zoologiska Museum, Änge 040529
Angel Añor, Tarragona 115462
Angel Antiques, Petworth 090986
Angel Antiques, Redditch 091080
Angel Antiquités, Marseille 069899
Angel Arcade, London 089908
Angel Art Sanctuary, Rathdrum . 109685
The Angel Bookshop, North
   Walsham 144746
L'Angel d'Or, Palma de Mallorca 086338
Angel Gento, Valencia 086497
Angel Mounds, Evansville 048643
Angel Orensanz Foundation, New
   York 060319, 122314, 138374
Angel Row Gallery, Nottingham . 045409
Angel Town, Albuquerque 092025
Angel, David, Miami 094683
Angela & J.'s Gallery, New York . 095214
Angela & J's Gallery, New York . 095215
Angela Reitz Art Forum, Galerie 45,
   Riehen 116585
Angele, Montendre 070234
Angelfall Studios, Baltimore . . . 119717
Angelfish, Denver 120687
Angélic'Art, Genève 087442
Angelika Kauffmann Museum –
   Gemeindemuseum, Schwarzenberg
   in Vorarlberg 002737
Angelini, Alessandra, Roma 081196
Angelique's Cottage, El Paso . . 093384
Angeliter Antik, Quern 078111
Angell Gallery, Toronto 101561
Angell Street Curiosities,
   Providence 096497
Angelo d'Oro, Bologna 130931
Angelo D'Oro, Parma 081088
Angelo Restauracion, México . . 132362
Angelos and Angeliki Giallina Gallery,
   Corfu 023050
Angelou, Michel, Saint-Étienne . 072747
Angelou, Michel, Saint-Genest-
   Malifaux 072792
Angel's Antiques, Toronto 064580
Angels Attic Museum, Santa
   Monica 053131
Angel's Auctions, Atkinson 127124
Angels Divine, Louisville 094515
Angels Eagles Dolls, Sacramento 096592
Angel's on the Walk, Mount
   Tamborine 061921
L'Angelus, Paris 070890
Angenard, Pierre, Beaucé 066992
Angerer, Vomperbach 100215
Angermuseum, Erfurt 018235
Angermuseum- Margaretha-Reichardt-
   Haus, Erfurt 018236
Angers Hof, Wernigerode 108880,
                                     142477
Angevaren – Antiquairs, Den
   Haag 083050
Angie, Singapore 114470
Angiletta Antichità, Ponte Taro . 081163
Angius, Pietro, Venezia 132225
The Angkor Conservation, Siem
   Reap 005288
Angkorwat Art of Cambodia, Long
   Beach 094188
Angle, Saint-Paul-Trois-Châteaux 105602
Angle, Denver 120688
Angles of Art, Theale 119225
Anglian Art, King's Lynn 118044
Anglican Cathedral Museum, Saint
   John's 006896
Anglicana Aste di F. Giorardini,
   Roma 081197, 126222
Anglo American Antiques,
   Albuquerque 092026, 135279
Anglo Antiques, Invercargill . . . 083962
Angolo Al Sole, Massagno 087663
L'Angolo dé Buttieri, Bologna . . 079798
L'Angolo del Collezionista, Milano 080386
L'Angolo del Collezionista, Reggio
   Calabria 081183

L'Angolo del Collezionista, Roma 142796
L'Angolo del Lume, Roma 081198
L'Angolo del Passato, Venezia . . 081817
Angono National Museum,
   Angono 034424
Angost, Paris 070891
Angourie Road Antiques, Yamba 062497
Angren Viloyat Ülkashunoslik Muzeyi,
   Angren 054628
Angsana, Bangkok 088071
Angstrom, Dallas 120543
Anguera Maura, Salvador, Palma de
   Mallorca 086339
Angulo, Oviedo 115323
Angus & Co, Toronto 064581
Angus Antique Furniture, Richmond,
   Victoria 062144, 125230
Angus Antiques, Dundee 089114
Angus Fine Art, Kirriemuir 118060
Angus Folk Museum, Glamis . . . 044389
Anguz, Saint-Aygulf 072684
Angyalföldi Helytörténeti Gyűjtemény,
   Budapest 023455
Anhaltische Gemäldegalerie, Schloß
   Georgium, Dessau 017862
Anhaltischer Kunstverein Dessau e.V.,
   Dessau 058709
Anhava, Helsinki 103130
Anhui Agriculture Museum, Hefei 007851
Anhui Fine Arts Publishing House,
   Hefei 136716
Anhui Museum, Hefei 007852
Anhui Sheng Bo Wu Guan, Hefei 007853
Ani C, Musée de la Faïence,
   Longfosse 013858
Anica Art, Leura 099228
Anicennes Mines de Grandfontaine,
   Grandfontaine 013149
Anich-Hueber-Museum,
   Oberperfuss 002464
Anicroche, Eze 103726
Aniela, Kangaroo Valley 099172
Aniha, Delhi 109302
Anila Arts and Crafts, Chennai . 079369
Anima, Bergamo 138787
Animal and Fish Museum, Evros 023073
Animal Arts, Canterbury 117516
Animal Ecology and Dinosaur Museum,
   Taipei 042497
Animal Museum, Amman 029761
Animalaine, Bastogne 003341
Animas Museum, Durango 048399
Animation Art, Amsterdam 112138
Animation Art Gallery, Honolulu . 120920
Animation Art Gallery, London . . 118168
Animation Connection, Toronto . 101562
Animation USA, Seattle 124478
Animazing Gallery, New York . . . 122315
Aninat, Isabel, Santiago de Chile 101900
Anita Déco, Marseille 069900
Anita Gallery, San Antonio 034588
Anita Shapolsky Gallery, New
   York 122316
Anita's Corner Mall, Oklahoma
   City 095932
Antkabir-Atatürk ve Kurtuluş Müzesi,
   Ankara 042814
Anitlar Müzeler, Ankara 060005
Anixis, Baden 116046
Anjalan Kartanomuseo,
   Anjalankoski 010753
Anjali, Sangeeta, Hyderabad . . . 109351
Anjely, Austin 092204
Anji Museum, Anji 007563
Anjou Antiquités, Avrillé 066894
Anjou Antiquités, Craon 068218
Anjou, Bernard d', Rouen 125792
Ankang Zone Museum, Ankang . 007564
Ankara Devlet Güzel Sanatlar Galerisi,
   Ankara 042815
Ankara Sanat Fuarı, Ankara . . . 098339
Ankara Sanat ve El Sanatları Fuarı,
   Ankara 098340
Den Anker, Brugge 125301

Het Anker, Dordrecht 112431
de Ankerburg Wagner, Wacław,
  Kraków 132830
Anker's House Museum, Chester-le-
  Street 043934
Ankersen, Værløse 066171
Ankh Antiquarian Books,
  Oakleigh 139862
Ankkapurhan Teollisuusmuseo,
  Anjalankoski 010754
Anli Art and Culture Museum,
  Shengang 042448
Anlu Museum, Anlu 007565
Anmoder, Madrid 086103
Ann, Nagoya 082056
Ann & Ed's Antiques, Oklahoma
  City 095933
Ann & Neil Antiques, Phoenix . 096179,
                          127396
Ann Arbor Art Center, Ann Arbor 046485
Ann Arbor Hands-on Museum, Ann
  Arbor 046486
Ann Bryant Art Gallery, East
  London 038708
Ann-Morris Antiques, New York . 095216
Ann-Teeks, Toronto 064582
Ann-Tiques, Townsville 062337
Ann-Tiques and More,
  Jacksonville 093950
Anna, Den Haag 083051
Anna Amalia, Berlin 074745
Anna Antiek, Leeuwarden 083412
Anna Antikvitás, Budapest 079231,
                          142551
Anna Art Studio and Gallery, Bellevue
  Hill 098729, 127563
Anna Bemis Palmer Museum,
  York 054585
Anna Eaton Stout Memorial Art Gallery,
  Owensboro 051851
Anna Fresko, İstanbul 088097
Anna Leonowens Gallery, Halifax 005944
Anna Miller Museum, Newcastle 051520
Anna-Seghers-Gedenkstätte, Akademie
  der Künste, Berlin 017069
Anna, Ollia & Maude, Omaha .. 096023
Annabelle, Seattle 097392
Annabelle Antiques, Los Angeles 094246
Annabelle's Antiques, Subiaco .. 062253
Annabelle's Attic, Cincinnati .. 092824
Annabellinski, Orange 062029
Annabell's Attic, Tucson 097626
Annaberger Kunst- und Kulturverein e.V.,
  Annaberg-Buchholz .. 058710, 137029
Annadale Upholstery, Tralee .. 079707
Annaka Fine arts and Antiques,
  Annaka 081958
Annalan Kotiseutumuseo,
  Pyhäjoki 011289
Annales d'Ethiopie, Addis Ababa 138619
Annali dell'Istituto Italiano di
  Numismatica, Palazzo Barberini,
  Roma 138788
Annalina, Briançon 103466
Annals Antique, Singapore ... 085359
Annam, Bangkok 088072
Annam Héritage, Paris 127396
Annamoe Antiques, Glandalough 079629
Annan Gallery, Glasgow 117830
Annandale Galleries, Annandale 098649
Annandale, National Historic Site,
  Tillsonburg 007150
Annapolis Valley Macdonald Museum,
  Middleton 006303
Annas Brigaderes Memoriālais Muzejs,
  Tērvete 030237
Anna's Old Wares, Morwell .. 061892
Anne, Paris 070893
Anne & Gordon Samstag Museum of
  Art, Adelaide 000761
Anne & Pénélopé, Paris 070895
Anne & Thierry Antiquités, Lyon 069684
Anne Antiquités, La Baule-
  Escoublac 068944

Anne Antiquités, Le Mans 069301
Anne Frank Center USA, New
  York 051331
Anne Frank Huis, Amsterdam . 031926
Anne Frank Zentrum, Berlin ... 017070
Anne Grimdalens Minne, Dalen i
  Telemark 033567
Anne Hathaway's Cottage,
  Shakespeare Birthplace Trust,
  Stratford-upon-Avon 045901
Anne Hvides Gård, Svendborg .. 010332
Anne Julie, Læsø 102924
Anne-Maj's Antik och Kuriosa,
  Fritsla 086679
Anne of Cleves House Museum,
  Lewes 044816
Anne of Green Gables Museum,
  Kensington 006080
The Anne Spencer Memorial
  Foundation, Lynchburg 050529
Anne, Jean-François, Paris 070894
Anneberg-Samlingerne, Nykøbing
  Sjælland 010249
Année Galerie, Haarlem 112508
Années 1900, Bétaille 067147
Les Années de l'Or, Nîmes 070683
Années Folles, Villenave-d'Ornon 074157
Annelore Westerheyde, Antiquariat &
  Kunst, Oldenswort . 077947, 142272
Annemarie Verna Galerie, Zürich 116743
Anne's Antiques, Cape Town .. 085630
Anne's Attic, Tampa 097536
Anne's Books, Annerley 139626
Annett Antique, Vanløse 066178
Annette Antiques & Collectables,
  Lincoln 089827, 134558
Annette's Antiques, San Jose .. 097355
Annette's of Rue Royal, New
  Orleans 095049
Annetts, Horsham 118002
The Annex, New York . 122317, 122318
Annex, Portland 096355
Annex 14, Bern 116117
Annex Books, Toronto 140393
The Annex Galleries, Santa Rosa 124474
Annex Gallery, San Diego 124087
Annia, Salamanca 115401
Annick Vidal, Versailles 074040
Annick's Antiques, Sorell 062216
Annie E. Woodman Institute,
  Dover 048353
Annie Herold et Philippe, Sens . 073496
Annie Riggs Memorial Museum, Fort
  Stockton 048877
Annie's, Cleveland 092914
Annie's Affordable Art, Seattle . 124479
Annie's Antiques, Cooden 088973
Annie's Antiques, Farnham ... 089258
Annie's Antiques and Collectibles,
  Jacksonville 093951
Annie's Attic, Louisville 094516, 145241
Annie's Book Company, Calgary 140332
Annie's Book Stop, Portland .. 145441
Annie's Gallery, Norbreck 118810
Annie's Old Things, Camp Hill . 061260
Annie's Petroleum, Columbus . 092989
Annie's Sales, Milwaukee 094827
Annikas Antikt och Kuriosa, Eslöv 086669
Annima, Helsinki 066296
Annina Nosei Gallery, New York 122319
Anniston Museum of Natural History,
  Anniston 046504
Annmary Brown Memorial,
  Providence 052377
Anno 1666, Zwolle 083729
Anno 1800 Antikhandel,
  Västervik 087144
Anno 1850, Groningen 083274
Anno 85, Breda 082970
Anno Aktuell, Rheda-
  Wiedenbrück 078218
Anno Antik, Bonn 075061
Anno Antik, Otterdingen 077913
Anno-Antikes Aller Art, Berlin . 074746

Anno Antikvitás, Sopron 079322
Anno Antiquitas, Budakeszi ... 079228
Anno Bijutsukan, Tsuwano 029662
Anno Dazumal, Celle 075245
Anno Dazumal, Hattstedt 076419
Anno Dazumal, Wolfenbüttel .. 079050
Anno Domini, Meißen 077331
Anno-Domini, Osnabrück 077957
Anno-Domini, Wissen 079029
Anno Domini Antiques, London . 089909
Anno Tubak, Luxembourg 082359
L'Annonciade, Musée de Saint-Tropez,
  Saint-Tropez 015577
Ann's Antiques, Houston 093593
Ann's Antiques, Stoke-on-Trent . 091451
Ann's Art, Groningen 112488
Ann's Caning Service, Baltimore 135336
Annterior Antiques, Plymouth . 091020
Anntiques, Dallas 093060
Annual Antique Ethnographic Art Show,
  Santa Fe 098414
Annual August Antiques Show,
  Nantucket 098415
Annual Invitational Antique Indian Art
  Show, Santa Fe 098416
Annual Spring Show, San
  Francisco 098417
Anny's Antik, Hjørring 065761
Ano Atelier, Lübeck 107909
Ano Moni Panagias Xenias,
  Almyros 022954
Anoka County Museum, Anoka . 046505
Anola and District Museum,
  Anola 005343
Anomaly, Portland 096356
Anonimo e d'Autore, Roma 081199
Años Luz Antiguedades,
  Concepción 064936
Anotarte, Santiago de
  Compostela 115437
Another Age Antiques, Austin .. 092205
Another Antique Shop, Louisville 094517
Another Chance, Philadelphia .. 096072
Another Man's Treasure, El Paso 093385
Another Period in Time,
  Baltimore 092290
Another Time, San Francisco .. 097162
Another Time, Seattle 097393
Another Time Antiques,
  Christchurch 083850
Another Time Antiques, Detroit . 093357
Anouk Miga, Maastricht 083454
Anova Books, London 138183
Anqing Museum, Anqing 007566
Anqiu Museum, Anqiu 007567
Anquez, Etaples 068480
Anrep, Sabina, Milano 080387
An's Rommelzoo, Haarlem ... 083313
Ansamblul Muzeal şi Monumental Dr.
  Vasile Lucaciu, Şişeşti 036717
L'Anse aux Meadows National Historic
  Park, Saint-Lunaire-Griquet .. 006922
Anse-Blanchette, Gaspé 005861
Anselme, Philippe, Saint-Bonnet-du-
  Gard 072692
Anselmi, Guido Alessandro,
  Firenze 131106
Anselmo Matučio Memorialinis
  Muziejus, Alytus 030313
Ansermet, Pierre-Jean, Fribourg 133741
Anshan Museum, Anshan 007568
Anshun Zone Museum, Anshun . 007569
Ansley, Ruby, Birmingham ... 092384
Anson, Tampa 124625
Anson-Cartwright, Hugh, Toronto 140394
Anson County Historical Society
  Museum, Wadesboro 054024
Ansorena, Madrid .. 086104, 115137,
            115138, 126528, 126529
Anstead, Mary, San Diego ... 127439
Anstead, Victoria, New York .. 122320
Anstensen, Kim, Göteborg ... 115625
Anstey, Harrogate 117913
Ant Hill Gallery, Broken Hill .. 098793

Ant Hill Gallery, Wynnum 099925
Antal & Társa, Budapest 079232
Antalya Devlet Güzel Sanatlar Galerisi,
  Antalya 042834
Antalya Müzesi, Antalya 042835
Antalya Sanat ve El Sanatları Fuarı,
  Antalya 098341
Antami Libros, Buenos Aires .. 139545
Antan, Luxembourg 082360
Antaño, Buenos Aires 139546
Antaño, Estepona 085991
Antaño Antigüedades, Buenos
  Aires 060854
Antano Baranausko ir Antano
  Vienuolio-Žukausko Memorialinis
  Muziejus, Anykščiai 030315
Antano ir Jono Jušku Etninēs
  Kultūros Muziejus, Vilkija 030508
Antano Jonyno Memorialinis Muziejus,
  Alytus 030314
Antano Mončio Namai-Muziejus,
  Palanga 030436
Antano Žmuidzinavičiaus Memorialinis
  Muziejus, Nacionalinis M.K. Čiurlionio
  Dailēs Muziejus, Kaunas ... 030348
Antar, Latifeh, Zghorta 082330
Anteak Interiör, Stockholm ... 086963
Antéart, Dijon 068313
Antebellum, Blois 067248
Antebellum Antiques, New
  Orleans 095050
Antelope County Historical Museum,
  Neligh 051211
Antelope Valley Indian Museum,
  Lancaster 050109
Anten-Gräfsnäs Järnväg,
  Alingsås 040519
L'Antenato, Milano 080388
Antenne de l'Imprimerie de la Presse,
  Louhans 013883
Anterieur, Leiden 083423
Antes, Den Haag 083052
Anthea, Quimper 072339
Anthea's, Mitchelton 061864
Anthea's Antiques, Auckland .. 083748
Antheaume, Christiane, Rouen . 072556
Anthéaume, Fabrice,
  Foussemagne 068603
Anthemion, Bruxelles 063312
Anthemion, Cartmel 088804
Anthemion Gallery, Toronto ... 064583
Anthéos, Bordeaux 067302
Anthèse, Arcueil 136797
Anthía, Charleville-Mézières .. 067842
Anthill, Richmond 096522
Anthinéa Brocante, Saint-Quentin
  (Aisne) 073273
Anthology Film Archives, New
  York 051332
Anthony, Dublin 079520
Anthony Antigüedades,
  Fuengirola 085998
Anthony Henday Museum,
  Delburne 005695
Anthony, Ivan, Auckland 083749
Anthony, John, Bletchingley .. 088506
Anthony, John, Kitchener 064334
Anthony's, Minneapolis 135823
Anthony's Antiques and Fine Arts, Salt
  Lake City 096872
Anthropological Museum, Xiamen 080393
Anthropological Museum, Department
  of Anthropology, Guwahati University,
  Guwahati 024270
Anthropological Museum, Manipur
  University, Canchipur 024187
The Anthropological National Folklore
  Museum, Amman 029762
Anthropologische Sammlung der
  Universität Göttingen, Göttingen 018746
Anthropology at Creswick,
  Creswick 061372
Anthropology Museum, DeKalb . 048223
Anthropology Museum, Delhi ... 024218

Register der Institutionen und Firmen

Anthropology Museum, Johannesburg – Antiguedades Juan Antonio, Madrid

Anthropology Museum, Johannesburg 038741
Anthropology Museum, Ranchi . 024459
Anthropology Museum, Winnipeg 007435
Anthropology Museum, Southwest State University, Marshall 050710
Anthropology Museum, Sri Jai Narain Degree College, Lucknow 024374
Anthropology Museum, University of Queensland, Saint Lucia 001474
Anthropos, Fribourg 139086
Anti-Aircraft Museum, Ha Noi . . 054788
Anti-Keer, Hengelo, Overijssel . . 143170
Anti-Kriegs-Museum, Berlin 017071
Anti Reflets, Nantes . . 104329, 128824
Anti-Saloon League Museum, Westerville 054333
Antiaquariat FBV, Köln 142003
Antic, Arras 066736
Antic – Starožitnosti, Ústí nad Labem 065563
Antic . Arte, Roma 081200
Antic 190, Barcelona 085721
Antic 1900, Villeurbanne 074183
Antic 2000, Roquebrune-Cap-Martin 072537
Antic 80, Abbeville 066408
Antic Art, Toulon (Var) 073697
Antic Art Connexion, Bezons . . . 067183
Antic Art et Déco, Lege-Cap-Ferret 069379
Antic Arts and Crafts, Atlanta . . 092102
Antic Atelier Cresta, Brienz (Graubünden) 087331
Antic Atelier Cresta, Surava . . . 087837
Antic Bisbal, La Bisbal d'Empordá 085899, 085900
Antic Boutique, Luzern 087629
Antic Brocante, Barcelona 126502
Antic Center Informatique, Nice . 070552
Antic Comimpex, Bucureşti 085078
Antic Daro, La Bisbal d'Empordá 085901, 133208
Antic Deco, Grenoble 068711
Antic Déco International, Loupian 069645
Antic Design, Saint-Étienne 072748
Antic Dolls, Nice 070553
Antic Dolls-Toys, Lyon . 069685, 128696
Antic Ex Libris, Bucureşti 143610
Antic Julie & Renato, Toulon (Var) 073698
Antic-Kaufhaus Falkenried, Hamburg 076183, 129946
Antic-Land, Saint-Étienne 072749
Antic-Line, Perpignan 072069
Antic Passion, Belmont-sur-Lausanne 087268
Antic Renova Art, Savognin 087800, 133917
Antic Restauro, Barcelona 133093
Antic Saint-Maclou, Rouen 072557
The Antic Shop, Berlin 074747, 129465
Antic Shop, Neuilly-sur-Seine . . 070514
Antic Sud Loire, Saint-Melaine-sur-Aubance 072975
Antic-Tac, Paris 070896, 128896
Antica, Angers 066598
Antica, Braunschweig 075117
Antica, Gex 068657
Antica, Marines 069877
Antica, Olomouc 065428
Antica, Paris 070897
Antica, Tarbes 073623
Antica, Vitry-aux-Loges 074257
Antica Antiques, Claremont 061320
Antica GS, Sarreguemines 073418
Antica Libreria Cappelli, Bologna 142624
Antica Libreria Cappelli, San Gregorio di Catania 142843
ANTICA Namur, Namur 098009
Antica Opera Romana, Bari 109768
Antica Persia, Napoli 080833
Anticaja, Roma 081201, 131804
Anticaja & Petrella, Roma 081202, 131805
Anticamente, Roma 081203
Antic'Ange, Saint-Cassien 072706
Anticariat, Craiova 143624
Anticariat Curtea Veche, Bucureşti 143611
Anticariat Galeria Halelor, Bucureşti 114122, 143612
Anticariat Nr. 2, Bucureşti 143613
Anticariat Nr. 4, Bucureşti 143614
Anticariat Nr. 9, Bucureşti 143615
Antic'Art, Ajaccio 066492
Antic'art, Bouere 067374
Antic'Arte, Genova 110085
Antic'Arts, L'Absie 069130
Antic'Arts, Parthenay 072025
Antic'Brocante, Bayonne 066971
Antic'Diffusion, Clermont-Ferrand 068009
Antiche Belle Cose, Bologna . . . 079799
Antichi Maestri Pittori, Torino . . 110948
Antichi Ricordi, Milano 131368
Antichitá, Firenze 080075
Antichita, Glebe 061523
Antichità, Lignano Sabbiadoro . . 110169
Antichità, Padova 080957, 131651
Antichità, Parma 081089, 131715
Antichitá Al Collegio, Bologna . . 079800
Antichità Al Pozzo, Asolo 079752
Antichità al Teatro di Duse, Bologna 079801
Antichità Alla Porta, Bologna . . . 079802
Antichità alle Grazie, Milano . . . 080389
Antichità Ardeatina, Roma 131806
Antichità As, Messina 080359
Antichità Barberia, Bologna 079803
Antichità Belsito, Roma 081204
Antichità Biedermeier, Milano . . 080390
Antichità Brunello, Treviso 081779, 081780
Antichità Carlo III, Napoli 080834
Antichità Casolini, Milano 131369
Antichità Chiapuzzi, Alassio 079738
Antichità dal Luzzo, Bologna . . . 079804
Antichità Dante, Palermo 081014
Antichità De Giosa, Milano 080391
Antichità Dei Bardi, Firenze 080076
Antichitá del Corso, Bologna . . . 079805
Antichità della Moscova, Milano 080392
Antichità di Nobili Alessio, Canonica Lambro 080008
Antichità due Torri, Bologna . . . 079806
Antichità Due Torri, Verona 081888
Antichità e Arte, Milano 080393
Antichità e Restauro, Firenze . . . 131107
Antichità Estense, Modena 080770
Antichità Federico II, Bari 079757
Antichità G. N., Milano 080394
Antichita Gattamelata, Padova . . 080958
Antichità Gatti, Crema 080059
Antichità Giardini, Modena 131556
Antichità Giulia, Roma 081205
Antichità Gotha, Milano 080395
Antichità Grand Tour, Roma 081206
Antichità Il Cherubino, Cagliari . 131059
Antichitá Il Leone, Bologna 079807
Antichità Isabella, Milano 080396
Antichita M.M., Napoli 080835
Antichità Mona Lisa, Ascona . . . 087196
Antichità Nirone 5678, Milano . . 080397
Antichità Nova, Milano 080398
Antichità Piantarose, Perugia . . . 081137
Antichità Piselli Balzano, Firenze 080077
Antichità Porta Venezia, Milano . 080399
Antichità Restauro, Roma 131807
Antichità Rialto, Bologna 079808
Antichità Rinascimento, Roma . . 081207
Antichità Ripetta, Roma 081208
Antichità Roberto Cocozza, Roma 081209
Antichità San Basso, Venezia . . 081818
Antichità San Federico, Torino . . 081644
Antichità San Salvatore, Bologna 079809
Antichità San Samuele, Venezia 081819
Antichitá San Vitale, Bologna . . 079810
Antichitá Santa Giulia di Borelli, Brescia 079912
Antichità Santoro, Bologna 079811
Antichità Sforza, Milano 080400
Antichità Sodani, Roma 081210
Antichità Sturni, Roma 081211
Antichità Tanca, Eredi Tanca, Roma 081212
Antichità Unicorno, Cortina d'Ampezzo 109979
Antichitá Vecci, Firenze 080078
Antichitá Volta, Milano 080401, 142711
Antichitati Lu & Ad, Cluj-Napoca 085112
Antic'in, Le Grand-Saconnex . . . 087514
Antic'Inn, Beauvoisin 067058
Antická Gerulata v Rusovciach, Mestské Múzeum v Bratislave, Bratislava 038298
Anticmag, Clermont (Oise) 068008
Antico e Antico, Roma 081213
Antico Frantoio, Massa Marittima 026731
Antico Mania, Modena 131557
Antico Mondo, Köln 076844
Antico per Antico, Milano 080402
Antico Restauro, Roma 131808
L'Antico Viaggio Intorno Ad Un Oggetto, Roma 081214
Antico, Marlene, Paddington, New South Wales 099509
Antic'Oro, Brescia 079913
Antics, Paris 070898
Antics, Saint-Martin-d'Heres . . . 072946
Antics Lleida, Lleida 086068
Anticthermal, Nancy 125651
Anticua, Barranquilla 102262
Anticuaria, Oviedo 086326
Anticuaria de la Casa Verde, Guatemala 079220
Anticuaria del Retiro, Buenos Aires 139547
Anticuaria en Muebles, Medellín 065262
El Anticuario, Medellín 065263
Anticuario, Montevideo 091987
El Anticuario, Las Palmas de Gran Canaria 086369
El Anticuario, Vigo 086553
Anticuario Cooperativa, Buenos Aires 060855
Anticuario Cotanda, Madrid 086105
Anticuario El Arcón, Cartagena . 065258
Anticuario el Fin del Afán, Medellín 065264
Anticuario El Museo de Santa Severa, Bogotá 065174
El Anticuario Francés, Cali 065239
Anticuario Frances, Cochabamba 063860, 128095
Anticuario H & S, Bogotá 065175
Anticuario La Candelaria, Bogotá 065176
Anticuario Las Palmas, Medellín 065265
Anticuario Miriñaque, Buenos Aires 060856
Anticuario Pallarols, Buenos Aires 060857, 127507
Anticuario San Antonio, Cartagena 065259
Anticuarios 2000, Buenos Aires 060858
Anticuarios Agrupacion Profesional de Anticuarios de las Rea, Barcelona 059830
Anticuarios Arte Decorativo, Medellín 065266
Anticuaris, Asunción 084417
Anticums, Les Anglès 069397
Anticus, Russell 084072
Anticus, Warszawa 084691
Anticuus, Roma 081215, 081216
Antiek, Gouda 083262
Antiek & Curiosa, Amsterdam . . 082680
Antiek & Design Centre, Eindhoven 083206
Antiek Brocante, Antwerpen . . . 063188
Antiek Brocante Royal, Brugge . 063259
Antiek Care, Boxtel 132451
Antiek Care, Bussum 132462
Antiek de Weihoek, Roosendaal 083539, 132616
De Antiek-Express, Zuidwolde . . 143271
Antiek Franky, Hasselt 063678
Antiek Garden en Pine, Mechelen 063788
Antiek Oud Aelst, Aalst 063172, 127988
Antiek Paris Provence, Brugge . 063260
Antiek Roosendaal, Roosendaal . 083540
Antiek Show Case, Antwerpen . 063189
Antiek Stockhouse, Gent 063624
Antiekbeurs A'dam 700, Amsterdam 082681
Antiekcenter, Kortrijk 063715
De Antieke Schuit, Dordrecht . . 083175
De Antiekhoek, Almelo 082625
De Antiekhoek, Leiden 083424
't Antiekhuis, Breda 082971
Antiekimport Oss, Oss 083521
Antiekmarkt Leiden, Leiden 083425
Antietam National Battlefield, Sharpsburg 053288
Antifaschistische Mahn- und Gedenkstätte, Lieberose 020067
L'Antiga, Denia 085961, 133237
Antiga Libros, Palma de Mallorca 086340, 143789
Antigo, A Coruña 085947, 133231
Antigo Lavradio, Rio de Janeiro . 064008
Antigona, Ciutadella de Menorca 085936
Antigonish Heritage Museum, Antigonish 005345
Antigua, Cheptainville 067958
Antigua Antichità Restauro, Verona 081889, 132266
Antigua Cárcel de Jaruco, Jaruco 009227
Antigua Casa de los Marqueses de Campo Florido, San Antonio de los Baños 009311
Antigua Casa Histórica, Buenos Aires 060859
Antigua Clinica de Muñecas, Buenos Aires 127508
Antigua de Mexico, Tucson 097627
Antigua Librería, Buenos Aires . 139548
Antigua Upholstery, San Diego . 136285
Antigualla, Santa Cruz de Tenerife 086418
Antiguedades, Valencia 086498
Antiguedades Abraham, Palma de Mallorca 086341
Antiguedades Alfonso XIII, Madrid 115139, 115140
Antiguedades Antaño, Jerez de la Frontera 086054
Antiguedades Antiqua, Donostia-San Sebastián 085965
Antiguedades Atenas, Granada . 086033
Antiguedades Atrio, Valencia . . . 086499
Antiguedades Barros, Valladolid . 086541
Antiguedades Capileida, Santa Cruz de Tenerife 086419, 133423
Antiguedades Carlota, Barcelona 085722
Antiguedades Castellarnau, Tarragona 086481
Antiguedades Consuelo Roman, Madrid 086106
Antiguedades Dato 12, Zaragoza 086587
Antiguedades del Sureste, Cancún 082459
Antiguedades Don Bosco, Santiago de Chile 064944
Antiguedades el Hallago, Monterrey 082485
Antiguedades el Sol, Guadalajara 082461
Antiguedades Emilio, Madrid . . . 086107
Antiguedades Esperanza, Oviedo 086327
Antiguedades Jónico, Santiago de Chile 064945
Antiguedades Jorge, Punta del Este 092015
Antiguedades Juan Antonio, Madrid 086108

Antiguedades la Flora, Burgos .. 085908
Antiguedades la Peregrina, Las
Palmas de Gran Canaria ... 086370
Antiguedades la Salamandra,
Monterrey 082486
Antiguedades Linares, Toledo .. 086490
Antiguedades Luna, Sevilla 086435
Antiguedades Macor, Madrid 086109
Antiguedades Maite, Santa Cruz de
Tenerife 086420
Antiguedades Martin, Madrid .. 086110
Antiguedades Miguel Angel,
Madrid 086111
Antiguedades Montelongo, Sevilla 086436
Antiguedades Montenegro, Vigo 086554
Antiguedades Palacios, Madrid . 086112
Antiguedades Paris, Palma de
Mallorca 086342
Antiguedades Perojo, Las Palmas de
Gran Canaria 086371
Antiguedades Querol, Valencia . 086500
Antiguedades Rey Don Pedro,
Burgos 085909
Antiguedades Romero, Madrid . 086113
Antiguedades Santa Teresa,
Sevilla 086437
Antiguedades Santiago Marti,
Barcelona 085723
Antiguedades Segundo, Sevilla . 086438
Antiguedades Siloe, Valencia ... 086501
Antiguedades Talavera, Valencia 086502
Antiguedades Tandem, Barcelona 085724
Antiguedades Union, Guadalajara 082462
Antiguedades Viera, Las Palmas de
Gran Canaria 086372
Antigüedades, Bogotá 065177
Antigüedades Abraxas, Buenos
Aires 060860
Antigüedades Alabastro, Cali .. 065240
Antigüedades Alfonso XII, Jaén . 086052
Antigüedades Almoneda, Lima . 084425
Antigüedades Antiques, Bogotá . 065178
Antigüedades Arredondo, Santiago de
Chile 064946
Antigüedades Atelier Inroma,
Lima 084426
Antigüedades Barón, Madrid ... 086114,
143749
Antigüedades Bohemia, Buenos
Aires 060861
Antigüedades Britania, Las Palmas de
Gran Canaria 086373
Antigüedades C C, Santiago de
Chile 064947
Antigüedades Cambalache,
Badajoz 085708
Antigüedades Capdepont, Buenos
Aires 060862
Antigüedades Carlos III,
Pamplona 086377
Antigüedades Challer Palace,
Barranquilla 065166
Antigüedades Cuca Romero,
Madrid 086115
Antigüedades de la Fuente,
Bilbao 085863, 133182
Antigüedades de los Abuelos,
Medellín 065267
Antigüedades El Arcón, Lima .. 084427
Antigüedades El Coleccionista, Viña
del Mar 064994
Antigüedades el Museo, Sevilla . 086439
Antigüedades el Patio, Sevilla .. 086440
Antigüedades El Rastro, Lima .. 084428
Antigüedades El Rastro de San Telmo,
Buenos Aires 060863
Antigüedades El Rodeo, Buenos
Aires 060864
Antigüedades Fernando, Buenos
Aires 060865
Antigüedades Guemes, Buenos
Aires 060866
Antigüedades Juan, Buenos Aires 060867

Antigüedades Karlos, Madrid ... 086116,
143750
Antigüedades La Casa Azul, Lima 084429
Antigüedades la Caverna, Murcia 086310,
143787
Antigüedades La Dehesa, Santiago de
Chile 064948
Antigüedades La Glorieta, Lima . 084430
Antigüedades La Mandinga, Buenos
Aires 060868
Antigüedades la Ocasión,
Barcelona 085725
Antigüedades la Portada de Prado
Centro, Medellín 065268
Antigüedades La Rueda, Buenos
Aires 060869, 127509
Antigüedades la Telaraña,
Medellín 065269, 065270
Antigüedades la Verja, Oviedo .. 086328
Antigüedades La Yapa, Buenos
Aires 060870
Antigüedades Laura, Medellín .. 065271
Antigüedades Los Reyes, Cali .. 065241
Antigüedades Macor, Madrid ... 086117
Antigüedades Maria Teresa,
Medellín 065272
Antigüedades Matusalén, Castello de
la Plana 085929
Antigüedades Miguel Angel, Lima 084431
Antigüedades Milagros, Lima .. 084432
Antigüedades Nuria, Murcia ... 086311
Antigüedades Ondarbide, Donostia-San
Sebastián 085966
Antigüedades Pantin, Buenos
Aires 060871
Antigüedades Porto Bello, Buenos
Aires 060872
Antigüedades Rodas, Vigo 086555
Antigüedades San Expedito,
Medellín 065273, 128297
Antigüedades San Lorenzo,
Medellín 065274
Antigüedades San Telmo Chico,
Buenos Aires 060873
Antigüedades Santa Clara,
Cáceres 085918
Antigüedades Santander, Bogotá 065179
Antigüedades Seleccion, Buenos
Aires 060874, 127510
Antigüedades Siglo Pasado, Lima 084433
Antigüedades Siglo XVII, Lima .. 084434
Antigüedades Sonia, Cali 065242
Antigüedades Tadeo, Lima 084435
Antigüedades Tres Morillas, Jaén 086053
Antigüedades Trueque, Pamplona 086379
Antigüedades Valladolid 2000,
Valladolid 086542
Antigüedades Venus, Valencia .. 097977
Antigüedades Vernissage, Lima .. 084436
Antigüedades y Arte San Antonio,
Bogotá 065180
Antigüedades y Artesanias La 5a,
Cali 065243
Antigüedades y Avaluos, Medellín 065275
Antigüedades y Decoración 1900,
Concepción 064937
Antigüedades y Decoración San Telmo,
Bogotá 065181
Antigüedades y Objetos de Arte,
Arequipa 084419, 084420
Antigüedades y Segundas,
Medellín 065276
Antigüedades Z.S., Pamplona .. 086379
Antiguidades Doll's, Lisboa 084822
Antiguidades e Velharias Fim do
Século, Funchal 084802
Antiguidades Escritorio de Artes, Rio
de Janeiro 064009
Antiguidades Estima, São Paulo 064141
Antiguidades Francisco, São
Paulo 064142
Antiguidades Galera São Pedro,
Funchal 084803
Antiguidades la Gemme, Rio de

Janeiro 064010
Antiguidades Laura, São Paulo . 064143
Antiguidades Mayflower, Coimbra 084787
Antiguidades Minha Avo Tinha, São
Paulo 064144
Antiguidades Mobel, Funchal .. 084804
Antiguidades Opera, São Paulo . 064145
Antiguidades Ouro Velho, Belo
Horizonte 063877
Antiguidades Sampaio Rocha, São
Paulo 064146
Antiguidades São Sebastião, Rio de
Janeiro 064011
Antiguidades Serralves, Porto .. 085007,
132964
Antiguitats Capri, Tarragona ... 086482
Antiguo, Alicante 085692, 085693
Antiguo Ayuntamiento y Cárcel de
Bejucal, Bejucal 009077
Antiik, Tallinn 066279, 066280
Antiik & Käsitöö, Tallinn 066281
Antiikki & Nykyaika, Helsinki ... 066297
Antiikki Erik, Helsinki 066298
Antiikki ja Entisöinti, Jokela .. 066376
Antiikki Jokela, Pori 066392
Antiikki PP, Turku 066400
Antiikki Trianon, Helsinki 066299
Antiikkihalli, Helsinki 066300
Antiikkilinja, Helsinki 066301
Antik, Aalborg 065580
Antik, Beuzeville 067155
Antik, Castres 067722
Antik, České Budějovice 065366
Antik, Esbjerg 065636
Antik, Holice v Čechách 065388
Antik, Horsens 065789
Antik, İstanbul . 126789, 138131, 144207
Antik, La Garenne-Colombes .. 069047
Antik, Milano 080403
Antik, Moskva 085158
Antik, Náchod 065424
Antik, New York 095217
Antik, Odense 066026
Antik, Pardubice 065442
Antik, Pécs 142580
Antik, Plzeň 065449
Antik, Praha 065464
Antik, Přerov 065546
Antik, Samara 085223
Antik, Sohland an der Spree .. 078492
Antik, Stadthagen 078531
Antik, Znojmo 065569
Antik – Den Gamla Skole,
Svendborg 066155
Antik – Gammel, Viborg 066203
Antik- u. Sammlerstube, Dresden 075407
Antik- und Kunstkabinett, Eva-Maria
Schnez, Bonn 075062, 106524
Antik- und Trödelstube, Feldberg 075730
Antik & Antikvariat MH, Ústí nad
Labem 065564, 140542
Antik & Art, Karlsruhe 076682
Antik & Auksjon, Sandnes 084351,
126385
Antik & Chic, Halstenbek 076175
Antik & Dekor, İstanbul 139111
Antik & Design, Konstanz 076934
Antik & Design, Mainz . 077231, 130264
Antik & Kunst, Wien 062891
Antik & Mehr, Potsdam 078068
Antik & Militaria, Stockholm .. 086964
Antik & Modern, Budapest 079233
Antik & Schmuck, Flensburg ... 075739
Antik & Trends, Luzern 087630
Antik & Trends, Wangen im
Allgäu 078861
Antik 141, Frederiksberg 065656
Antik 161, Berlin 074749
Antik 18, København 065827
Antik 19, Frederiksberg 065657
Antik 1900, Berlin 074750
Antik 2000, Lahr 076994
Antik 2000, Sankt Leon-Rot .. 078351
Antik 22, Karlsruhe 076683

Antik 25, Bremen 075144
Antik 6200, Haderslev 065716
Antik 77, København 065828
Antik A+J, Mariánské Lázně ... 140484
Antik Acente, Köln 076845
Antik Äpplet, Norrköping 086879
Antik am Bahnhof, Berlin 074751
Antik am Deutschhof, Heilbronn 076467
Antik am Marstall, Delingsdorf . 075329
Antik am Meer, Bremen 075145
Antik am Zoo, Wuppertal 079100
Antik Ambiente, Winnenden .. 079021
Antik and More, Immenhausen . 076602
Antik-Antikvariat, Praha 065465, 140500
Antik Art, Moskva 085159
Antik Art Centrum, Trnava 085595
Antik-Art Deco, Bremen 075146
Antik Art Lüneburg, Lüneburg .. 077206
Antik-Atelier, Balje 074656
Antik Atrium, Göttingen 076039, 141833
Antik Baden, Karlsruhe 076684
Antik Bansin, Bansin, Seebad .. 074678
Antik Basar, Nittel 077790
Antik-Basar, Waghäusel 078823
Antik-Bauernmöbel, Nürnberg .. 077811
Antik Bazaar, Toronto 064584
Antik Bazar, Košice 085563
Antik Bazar, Leissigen 087589
Antik Bazar, Nancy 070411
Antik Bazar, Svitavy 065558
Antik Birkely, Gislinge 065702
Antik Blå, Odense 066027
Antik Blå och Kuriosa, Umeå ... 087110
Antik Böngészde, Budapest ... 142552
Antik-Bolstrún, Reykjavík 079331
Antik Brocante, Odense 066028
Antik Bútor – Régiség, Pécs ... 079316
Antik-Center, Karstädt 076709
Antik Center, Köln 076846
Antik-Center-Wiesenburg,
Wildenfels 079009
Antik-Centr, Sankt-Peterburg ... 085229
Antik Centrum, Bratislava 085539
Antik Centrum, Dvůr Králové nad
Labem 065384
Antik Cleversulzbach, Neuenstadt am
Kocher 077724
Antik D+D, Znojmo 065570
Antik Daniel, Fürth 075924
Antik-Deele, Fröndenberg 075918
Antik Depó, Enese 079307
Antik-Depot, Wien 062892
Antik-Diele, Heede 076425
Antik e Nuovo, Horw 087529
Antik-Eck, Hannover 076358
Antik-Eck, Lemgo 077073
Antik-Eggli, Luzern 087631
Antik Elvira, København 065829
Antik-Eppi, Jena 076641
Antik Fascination, Timişoara .. 085128
Antik-Forum, Zürich 087932
Antik France, Marseille 069901
Antik Galéria, Zalaegerszeg ... 079330
Antik-Galerie, Broderstorf 075197
Antik-Galerie, Potsdam 078069
Antik-Galerie, Vaduz . 082337, 111779
Antik Galerie Braunschweig,
Braunschweig 075118
Antik Galerie im Lichtental, Wien 062893
Antik Gallerí 95, Reykjavík ... 079332
Antik Galleriet, Roskilde 066084
Antik-Gården, Hörby 086747
Antik Glas Kunst, Hamburg ... 076184
Antik-Häusla, Cadolzburg 075240
Antik-Hall, Maguenau 068774
Antik Halle, Kassel 076711
Antik Håndsnedkeri,
Frederiksberg 128381
Antik-Harry, Uddevalla 087109
Antik-Haus, Butzbach 075236
Antik-Haus Eisenach, Eisenach . 075590
Antik Haus Sankt Leon-Rot, Sankt
Leon-Rot 078352
Antik HBS, Halberstadt 076158

Antik Heiligenstedten,
Heiligenstedten 076479
Antik Hillerød, Hillerød 098062
Antik Hjørnet, Nykøbing Falster . 066012
Antik Hof, Delmenhorst 075331
Antik Hof, Ganderkesee 075939, 129885
Antik-Hof, Oberteuringen 077881
Antik-Hof, Schwedt / Oder . . . 078420
Antik-Hof, Wokuhl-Dabelow 079049
Antik i Bandhagen, Bandhagen . 086637
Antik im Hinterhof, Waiblingen . 078825
Antik im Hof, Osnabrück 077958
Antik Im Rudert, Marburg 077283
Antik im Zwinger, Sankt Wendel 078354
Antik JN, Hellerup 065729
Antik Kælderen, Tønder 066166
Antik Kate, Frederiksberg 065658
Antik-Klasik, Praha 065466
Antik Konservatorsateljé,
Göteborg 133527
Antik-Kontor, Lüneburg 077207
Antik-Kunst, Innsbruck 062642
Antik Kuriosa, Reykjavík 079333
Antik Laden, Bad Hofgastein . . . 062524
Antik Lageret, København 065830
Antik-Lagret, Mölndal 086869
Antik Landemærket, København 065831
Antik Lar, Moskva 085160
Antik Loppen, Kolding 065973
Antik Luma, Praha 065467
Antik-Magazinet, Tranås 087099
Antik-Markt, Helgoland 076486
Antik Mignon, Karlovy Vary 065402
Antik Modärn, Sundsvall 087088
Antik-Möbel-Restaurierung, Bad
Homburg v.d.Höhe 129410
Antik-Möbelhaus, Kapellen-
Drusweiler 076677
Antik Møgeltønder, Tønder 066167
Antik Mühely – Antikvitás, Pécs 079317,
130817
Antik Nordjylland, Løkken 065991
Antik Nordwest, Frankfurt am
Main 106967
Antik Nostalgie, Lingen 077127
Antik Nr 9, Rostock 078280
Antik och Dekoration,
Sundbyberg 087086
Antik och Kuriosa, Helsingborg . 086736
Antik och Kuriosa, Skellefteå . . . 086932
Antik och Kuriosa Hörnan, Skara 086928
Antik och Kuriosa i Silvergruvan,
Sala 086921, 126657
Antik och Kuriosa Utsigten,
Karlskrona 086773
Antik Örestad, Malmö 086844
Antik Öst, Jönköping 086759
Antik of the Future, Stockholm . 086965
Antik og Design, København . . . 065832
Antik og Gammelt, Stavanger . . 084362
Antik og Genbrug, Frederiksberg 065659
Antik og Genbrug, Silkeborg . . . 066106
Antik og Hobby, Holstebro 065780
Antik og Kunst, København 065833
Antik og Kunst, Viborg 066204
Antik og Kunst i Svaneke,
Svaneke 066149
Antik og Lamper, København . . . 065834
Antik og Livsstil, Kastrup 065809
Antik og Lopper, København . . . 065835
Antik og Ting, Århus 065595
Antik Oriental, Tampa 097537
Antik Oskar, Helsinki 066302
Antik-Palette, Freising 075897
Antik Pjot, København 065836
Antik-Régiségbolt, Pécs 079318
Antik Restaurator., Budapest . . . 130810
Antik Rørbæk, Frederikssund . . . 065842
Antik-Rosenheim, Rosenheim . . 078272
Antik Russo, Nürnberg 077812, 130462
Antik-Sammelsurium, Limburg an der
Lahn 077109
Antik Sanat Galerisi, İstanbul . . 088098
Antik Sankt Rupert, Salzburg . . 062784

Antik Scheune, Mülheim an der
Ruhr 077431
Antik Scheune, Nettetal 077705
Antik-Schorsch – G. Joas,
Laufen 077032
Antik Schuppen, Bad Segeberg . 074596
Antik Seefeldmühle, Adnet 062513
Antik Shoppen, København 065837
Antik Stadel, Wörth, Kreis Erding 079047
Antik-Stadl, Bad Leonfelden . . . 062528
Antik-Stadl, Donzdorf 075374
Antik-Stadl, Krumpendorf 062696
Antik Starožitnosti, Mladá
Boleslav 065422
Antik-Store, Alteglofsheim 074390
Antik-Stube, Hinzert-Pölert 076539
Antik-Stube, Hirzenhain 076542
Antik-Stübchen, Erfurt . 075633, 075634,
129799
Antik-Stübchen, Friedrichstadt . 075916
Antik Style, Hannover 076359
Antik Style, Mainz 077232
Antik Svitidla, Praha 065468
Antik-Tage Singen, Singen 098117
Antik Textil och Fynd, Ludvika . 086824
Antik Torningmølle, Kolding . . . 065974
Antik Treff, Rheda-Wiedenbrück 078219
Antik Troja, Mersch 082379
Antik Truhe, Isernhagen 076622
Antik Ulla, Gilleleje 065691
Antik und Art, Hannover 076360
Antik und Kunst, Cottbus 075271
Antik & Kunst, Sindelfingen . . . 098118
Antik und Kunst Am Wilden Mann,
Dresden 075408
Antik und Kunsthandel,
Tangermünde 078666, 142408
Antik Vertiko, Walsrode 078855
Antik-Welt, Hartmannsdorf 076408
Antik West, Göteborg 086690
Antik West, London 089910
Antik Winkeltje, Aurich 074477
Antik Wohndiele, Fassberg 075728
Antik-Zentrum, Göppingen 076028
Antik Žiro, Trjavna 064242, 140328
Antik zum scharfen Eck, Uttwil . 087866,
133948
Antik, Jürgen Findeisen, Berlin . 074748
Antika, Graz 062575
Antika, Seattle 097394
Antika 1, Lyon 069686
Antika Kranj, Kranj 085601
Antika Lyon Terreaux, Lyon . . . 140883
Antika Tizian, Ljubljana 085602
Antika Vedspisar och Kaminer,
Sigtuna 086924
Antikælderen, Viborg 066205
Antikaffären Blå Huset, Kalix . . 086763
Antikaffären Rothboden,
Åtvidaberg 086634
Antikaffären Spegeln, Örebro . 086895
Antikarat, Luxembourg 082361
Antikarnas, İstanbul 088099
Antik'Art et Déco, Sète 073513
Antikas, Kaslik 082325, 132317
Antikavriata Veikals, Rīga 142953
Antikayes, Le Touquet-Paris-
Plage 069358
Antikblick, Henstedt-Ulzburg . . 076494
Antikboden, Nynäshamn 086893
Antikboden, Oslo 084291, 084292
Antikboden, Vordingborg 066212, 066213
Antikboden i Västertorp,
Hägersten 086726
Antikboden Trollstskåpet, Lund . 086828
Antikbúðin, Reykjavík 079334
Antikbyrån i Skara, Skara 086929
Antikcentret, Odder 066021
Antike-Fundgrube, Gronau,
Westfalen 076097
Antike Kunst, Basel 139087
Antike Puppen, Nürnberg 077813
Antike Welt, Mainz 138668
Antikea, Törökbálint 079328

Antiken- und Abgußsammlung
des Archäologischen Seminars,
Marburg 020266
Antiken-Kabinett, Frankfurt am
Main 075769
Antikenmuseum Basel und Sammlung
Ludwig, Basel 041261
Antikenmuseum der Universität
Leipzig, Leipzig 019974
Antikenmuseum und Abgusssammlung
des Instituts für Klassische
Archäologie Universität Heidelberg,
Heidelberg 019130
Antikensammlung, Bern 041309
Antikensammlung der Friedrich-
Alexander-Universität Erlangen-
Nürnberg, Erlangen 018258
Antikensammlung, Kunsthalle zu Kiel,
Kiel 019601
Antikensammlung, Museumslandschaft
Hessen Kassel, Kassel 019526
Antikensammlung, Staatliche Museen
zu Berlin – Stiftung Preußischer
Kulturbesitz, Berlin 017072
Antikes, Freiburg im Breisgau . 075847
Antikes am Deich, Hamburg . . . 076185
Antikes an der Burg, Unna 078761
Antikes Dütt und Datt, Dörverden 075368
Antikes im Glashaus, Graz 062576
Antikes im Torhaus, Altenkrempe 074394
Antikes in Nymphenburg,
München 077449
Antikes Kabinett, Graz 062577
Antikes vom Land, Graz 062578
Antikes Wohnen, Klagenfurt . . . 062667
Antikes Wohnmuseum, Weinheim 022519
Antikfönstret Mariestad,
Mariestad 086866
Antikgården, Lindesberg 086810
Antikgården, Mariefred 086862
Antikgården i Salomonstorp,
Köpingsvik 086784
Antikhallarna, Göteborg 086691
Antikhallen, København 065838
Antikhallen Nygammalt, Ystad . . 087167
Antikhandel, Bautzen 074688
Antikhandel Bonn, Königswinter 076927
Antikhandel Cottbus, Cottbus . . 075272
Antikhandeln Tennsoldaten,
Vadstena 087130
Antikhandl, Náchod 065425
Antikhandlarn, Västervik 087145
Antikhandlung Gohlis, Leipzig . . 077053
Antikhane, İstanbul 088100
Antikhaus, Hilden 076526
Antikhaus, Velbert 078784
Antikhaus Blomberg, Blomberg, Kreis
Wittmund 075025
Antikhjørnet, Århus 065596
Antikhjørnet, København 065839
Antikhjørnet Bagsværd, Gentofte 065688
Antikhof, Lingen 077128
Antikhof Kloschwitz, Wettin . . . 078956
Antikhof Kropp, Kropp 076975
Antikhof Matthias Figl,
Michelhausen 062735
Antikhof Mellendorf, Wedemark . 078878
Antikhof Sachsenhausen,
Waldeck 078833
Antikhus, København 065840
Antikhuset, Eslöv 086670
Antikhuset, Gilleleje 065692
Antikhuset, Motala 086873
Antikhuset, Nykvarn 086892
Antikhuset Barnap, Jönköping . . 086760
Antikhuset i Olofström, Olofström 086911
Antikhúsgögn, Reykjavík 079335
Antikhúsið, Reykjavík 079336
Antiki Aro, Sofia 064227, 140319
Antikia i Växjö, Växjö 087148
Antikimport, Montréal 064362
Antikis Interiør, Trondheim 084384
Antikk Møbelrestaurering,
Gjettum 132775

Antikk og Brukt, Fauske 084233
Antikk og Brukt, Mosjøen 084280
Antikk Restaurering, Trøgstad . . 084378,
132798,   132799
Antikk salg, Molde 084278
Antikkeller, Saarbrücken 078315
Antikkhuset, Levanger 084273
Antikkhuset, Mysen 084283
Antikkhuset Hamar, Furnes 126348
Antikkloftet, Kleppe 084260
Antikkstua, Harstad 084244
Antikladen, Horsens 065790
Antikladen, Klagenfurt . 062668, 062669
Antikladen am Wasserturm,
Oldenburg 077924
Antiklädchen, Rauschenberg . . . 078145
Das Antiklager, Warmsen 078870
Antiklagret, Örebro 086896
Antiklyckan, Markaryd 086867
Antikmagasinet-Lampverkstad,
Stockholm 086966
Antikmarknaden i Gustavsberg,
Gustavsberg 086725
Antikmarkt am Alex, Berlin 074752
Antikmarkt Basel, Basel 087216, 133632
Antikmarkt Opfertshofen,
Opfertshofen 087725
Antikmöbelhof Lohmen, Lohmen,
Sachsen 077149
Antikmunir, Reykjavík . 079337, 126159
Antikmuseet, Göteborg 040648
Antikname, İstanbul 088101
Antiknet, Fana 084232
Antikören Mynthandel, Göteborg 086692
Antikoteca, Montevideo 127108
Antikpalast, Egnach 087403
Antikregisághaus Sievershausen,
Lehrte 019955
Antikrundan Classic Antik,
Göteborg 086693
Antiks, Sankt-Peterburg 085230
Antik's, Thomery 073662
Antiks Design, Lisboa 084823
Antiksalan, Reykjavík 079338
Antiksawa, Szczecin 084670
Antikscheuer, Heidelberg 076432
Antikscheune Elmschenhagen,
Kiel 076770
Antikschopf, Rheinmünster 078230
Antiksnedkeriet, Haderslev 128388
Antikspeicher Plau, Plau am See 078054
Antikstället, Slöinge 086940
Die Antikstube, Wiefelstede . . . 078963
Antikstudio im Dritten, Wien . . . 062894
Antikstübli, Sankt Gallen 087773
Antikt, Surry Hills 062262
Antikt Kalleguta, Köpingsvik . . . 086785
Antikt och Antikvariskt,
Vimmerby 087159, 144004
Antikt och Design, Båstad 086639
Antikt och Diverse, Kristinehamn 086791
Antikt och Inredningar, Avesta . . 086636
Antikt och Möbler, Rättvik 086916
Antikt och Nytt i Norrtälje,
Norrtälje 086886
Antikt Tryckt och Präglat,
Karlshamn 086772
Antikt, Gammalt och Nytt,
Stockholm 086967
Antikt, Konsthantverk och Kuriosa,
Göteborg 086694
Antikum, Mainz 107945
Antikurmakaren, Malmö 086845
Antikus, Bydgoszcz 084493
Antikva, Frýdek-Místek 065385
Antikva – Starožitnosti,
Bratislava 085540
Antikva-Art, Brno 065341
Antikva Mikos, Třebíč 065561
Antikva Nova Kodl, Praha 065469,
125417,   140501
Antikva Nova Praga, Praha 125418
Antikva Radio Praha, Praha . . . 065470
Antikva René, Praha 065471

Antikvaar, Tallinn 066282, 140677
Antikvaarinen Kirjakauppa,
Helsinki 140690
Antikværkstedet, Hillerød 065755,
128389
Antikvar, Irkutsk 085146
Antikvár, Miskolc 142578
Antikvar, Nižnij Novgorod 085219,
143656
Antikvar, Sankt-Peterburg 085231
Antikvar-Metropol, Moskva 085161
Antikvar na Mjasnickoj, Moskva   085162
Antikvaria Bökkeri, Helsinki . . . 140691
Antikvaria Poopa Joe, Espoo . . 140684
Antikvariaat, Tallinn 140678
Antikvariaatit, Helsinki . 140692, 140693
Antikvariaatti Anno Domini,
Vantaa 140756
Antikvariaatti Brigitta, Helsinki . 140694
Antikvariaatti Grafika, Tuusula . . 140753
Antikvariaatti Lumikannas,
Helsinki 140695
Antikvariaatti Syvä uni, Helsinki 066303,
140696
Antikvariaatti Tyrwää, Vammala . 140755
Antikvariaatti Vistan Divari,
Paimo 140740
Antikvariat & Co, Halmstad 143857
Antikvariat, Kaliningrad 143627
Antikvariat, Kroměříž 140478
Antikvariat, Murska Sobota 085622,
143699
Antikvariát, Olomouc . . 140488, 140489
Antikvariát, Praha 140502, 140503,
140504,
140505
Antikvariát, Prešov 143692
Antikvariat, Rostov-na-Donu . . . 085222,
143657
Antikvariat, Samara 085224
Antikvariat, Sankt-Peterburg . . . 085232
Antikvariát, Sofia 140320
Antikvariát, Třebíč 140540
Antikvariát, Zlín 140544
Antikvariat – Starožitnosti,
Domažlice 065381, 140470
Antikvariat & Galleri Marsjö,
Borgholm 115590, 143823
Antikvariat 2B, København 140591
Antikvariát Akropolis, Uddevalla . 143982
Antikvariát Antiqua, Stockholm . 143928
Antikvariát Arcanum, Třebíč . . . 140541
Antikvariát Argo, Täby 143978
Antikvariat Asken, Mariestad . . . 143903
Antikvariat Åsnan, Uppsala 143988
Antikvariat Athenaeum,
Stockholm 143929
Antikvariat August, Stockholm . . 143930
Antikvariat Banvaktstugan,
Mellösa 143904
Antikvariat Bělehradská, Praha . 140506
Antikvariat Bibliomanen, Uppsala 143989
Antikvariat Bode Carlshamn,
Karlshamn 143874
Antikvariat Böcker och Kuriosa,
Lund 143887
Antikvariat Boken, Stockholm . . 143931
Antikvariat Bokhyllan, Falun . . . 143834
Antikvariat Bokstödet, Enskede . 143829
Antikvariat Classica, Växjö 144001
Antikvariat Classica, Vetlanda . . 144003
Antikvariát Dali, Bratislava 143686
Antikvariat Drotten, Visby 144005
Antikvariat Era, Stockholm 143932
Antikvariat Erato, Stockholm . . . 143933
Antikvariat Felix Jenewein, Kutná
Hora 140479
Antikvariát Globus, Liberec 140480
Antikvariát Goggles, Örebro . . . 143909
Antikvariat Hafnia, København . . 140592
Antikvariat Helenes Bokhylla,
Kristianstad 143876
Antikvariat Hum, Göteborg 143839
Antikvariat Islevbro, Rødovre . . . 140647
Antikvariat Källaren, Eslöv 143830

Antikvariat Kant, Praha 140507
Antikvariat Kungsgatan,
Halmstad 143858
Antikvariát-Kuriozity, Praha 140508
Antikvariat Läslustan, Lund 143888
Antikvariat Lectura, Jönköping . . 143867
Antikvariat Mullvaden, Stockholm 143934
Antikvariat Můstek, Praha 140509
Antikvariat na Bronnoj, Moskva . 085163
Antikvariat na Manežnoj, Moskva 085164
Antikvariát Nazca, Žilina 143694
Antikvariat Nordbok, Skellefteå . 143919
Antikvariat Nordkalotten, Boden . 143821
Antikvariat Oasen, Hvidovre . . . 140587
Antikvariat Oceanen, Stockholm 143935
Antikvariat Østerbro, København 140593
Antikvariat Pan, Göteborg 143840
Antikvariat Paradiso, Vanløse . . 140669
Antikvariát Pařížská, Praha 140510
Antikvariát Pegasus, Bratislava . 143687
Antikvariat Perssonligt, Kalix . . . 143870
Antikvariát Pod Svícnem, Frýdek-
Místek 140471
Antikvariát Ptolomaeus, Praha . . 140511
Antikvariat Röde Orm, Göteborg 143841
Antikvariát Runa, Jablonec nad
Nisou 140475
Antikvariát Š, Chrudim 140468
Antikvariat Sør, Kristiansand . . . 143388
Antikvariát Steiner, Bratislava . . 143688
Antikvariat Strängnäs, Strängnäs 143972
Antikvariat Strix, Stockholm . . . 143936
Antikvariát Tälje, Stockholm . . . 143937
Antikvariát v Suterénu, Ostrava . 140492
Antikvariat Valentinská, Praha . . 140512
Antikvariat Viking, Luleå 143884
Antikvariát Votobia, Olomouc . . 140490
Antikvariát Votobia, Přerov 140536
Antikvariat X, Vänersborg 143998
Antikvariát za Věží, České
Budějovice 140463
Antikvariatet, Herlev 140581
Antikvariatet, Herning 140582
Antikvariatet, Kalmar 143871
Antikvariatet, København 140594, 140595
Antikvariatet Akasia, Århus 140549
Antikvariatet Bokmalen, Göteborg 143842
Antikvariatet i Norrköping,
Norrköping 143907
Antikvariatet i Odense, Odense . 140632
Antikvariatet i Skärhamn,
Skärhamn 143917
Antikvariatet i Skövde, Skövde . 143920
Antikvariatet i Växjö, Växjö 144002
Antikvarijat Danica, Beograd . . . 085329
Antikvarijat Jesenski i Turk,
Zagreb 140454, 140455, 140456
Antikvarijat Tin Ujević, Zagreb . . 140457
Antikvarisk Restaurering, Stange 132796
Antikvariska Bokhandeln,
Helsingborg 143859
Antikvariska Bokhandeln,
Linköping 143879
Den Antikvariske Samling, Ribe 010275
Antikvárium, Makó 142576
Antikvárium, Szeged 142581
Antikvárium – EX–Libris,
Budapest 142553
Antikvárium – Font, Budapest . . 142554
Antikvarnaja Galereja, Moskva . . 085165
Antikvarnaja Galereja, Sankt-
Peterburg 085233
Antikvarnaja Galereja Archipova, Sankt-
Peterburg 085234
Antikvarnaja lavka, Moskva 085166
Antikvarnaja Mebel, Moskva . . . 085167
Antikvarnica Galerija Mirjana,
Beograd 085330, 114395
Antikvarnica i Galerija, Sombor . 085352,
114463
Antikvarnica Konzul, Beograd . . 085331
Antikvarno-Bukinističeskij Salon, Sankt-
Peterburg 085235, 143664
Antikvarno-Numizmatičeskij Magazin,

Sankt-Peterburg 085236
Antikvarno-restavracionnaja
Masterskaja, Jaroslavl 085147
Antikvarnyj Aukcion, Sankt-
Peterburg 126477
Antikvarnyj Centr, Sankt-
Peterburg 085237
Antikvarnyj Juvelirnyj Magazin,
Moskva 085168
Antikvarnyj Magazin, Moskva . . 085169
Antikvilla Ludwigshöhe,
Rückersdorf 078305
Antikvita, Nymburk 065427
Antikvita-Starožitnosti, Praha . . 065472
Antikvitás, Budapest . . 079234, 079235
Antikvitás Régiségbolt, Miskolc . 079313
Antikvitete MK, Maribor 085614
Antikviteten, Sandefjord 084349
Antikviteten, Skodsborg 066127
Antikviteter Fjordgården,
Haderslev 065717
Antikviteter i Forum og Internationale
Møntbørs, Frederiksberg 098063
Antikviteter og Kuriosa,
København 065841
Antikviteter på Børsen,
København 098064
Antikvity, České Budějovice . . . 065367
Antikvity Art Aukce, Brno 065342,
125413
Antikwest, Helsinki 066304
Antillaise, Fort-de-France 142975
Antillaise, Le Lamentin 142980
Antillaise, Rivière-Salée 142984
l'Antillaise Art Gallery, Austin . . 119636
Antilles Médailles, Chevilly-Larue 067968
Antilope, Casablanca 082542
L'Antilope, Lyon 140061
Antim, Bruxelles 063313
Antimo Bertolino, Lugano 087603,
133818
Antin, Brenda, Los Angeles 094247
Antin, Jean-Pierre, Lavérune . . . 069210
Antiorario a Ritroso nel Tempo,
Modena 080771
Antipodean Gallery, Nedlands . . 139849
Antipodean Map and Print Gallery,
Subiaco 139916
Antiq' Art, Saint-Raphaël 073286
Antiq Déco, Quiberon 072334
Antiq Photo Gallery, Toulouse . . 073733
Antiq'Ardenne, Libramont 098010
Antiqart, Košice 085564, 143691
Antiq'Arts, Courthiezy 068200
Antiqe & Art Restoration by Wiebold,
Terrace Park 136402
Antiqe och Qriosa, Stockholm . . 086968
Antiqládeli, Frauenfeld 087429
Antiqua, Amsterdam 143010
Antiqua, Athinai 079158
Antiqua, Emmerich am Rhein . . 075611
Antiqua, Firenze 080079
Antiqua, Göteborg 086695
Antiqua, Hradec Králové 065389
Antiqua, Jihlava 065397
Antiqua, Qormi 082424, 132339
Antiqua, Santiago de Chile 064949
Antiqua, Verona 081890, 081891
Antiqua, Warszawa 143512
Antiqua – Starožitnik, Jihlava . . 065398
Antiqua Coppedè, Rimini 081186, 131789
Antiqua Domus, Palermo 081015
Antiqua Galleria d'Arte, Trento . . 111025
Antiqua Novis, Funchal 084805
Antiqua Res, Roma 081217
Antiqua Russos, Setúbal 085061
L'Antiquaille, Lausanne 087559
L'Antiquaire, Albi 066505
L'Antiquaire, Benfeld 067085
L'Antiquaire, Fribourg 087431
L'Antiquaire, Montréal 064363
Antiquaire, Provins 072316
L'Antiquaire, Sainte-Clotilde . . . 085070
Antiquaire, San Francisco 097163

L'Antiquaire & the Connoisseur, New
York 095218
Antiquaire de Quinsac, Quinsac . 072356
Antiquaire des Gâtines Rouges, La
Chaussée-d'Ivry 068989
L'Antiquaire du Fourneau,
Balbigny 066924
L'Antiquaire du Tilleul, Fribourg . 087432
Antiquaire du Village, Saint-Leu-la-
Forêt 072893
Antiquaire d'Yerres, Yerres 074305
l'Antiquaire Galerie, Lefkosia . . . 065320
L'Antiquaire Joyal, Montréal . . . 064364
Antiquaire Suisse, Bruxelles . . . 063314
Les Antiquaires, Milano 080404
Antiquaires de la Geôle,
Versailles 074041
Antiquaires de la Pauline, La
Garde 069042
Antiquaires de Lignane,
Puyricard 072330
Antiquaires de Sant-Vincens,
Perpignan 072070
Les Antiquaires de Stockel,
Bruxelles 063315
Antiquaires Drouot, Paris 070899
Les Antiquaires du Château,
Plougoumelen 072201
Antiquaires du Rhin, Strasbourg 073565
Antiqualha, Porto 085008, 132965
Antiqualhas Brasileiras Livraria, Rio de
Janeiro 140289
Antiquália, Lisboa 084824
Antiqualia Salinha de Arte, São
Paulo 064147
AntiquaNova, Wiesbaden 142484
Antiquaren Boekenbeurs Utrecht,
Utrecht 098253
Antiquária, Lisboa 126438
Antiquária, Sliema . . . 082432, 142973
Antiquaria, Tours 073833
Antiquaria Bok- & Bildantikvariat,
Göteborg 143843
Antiquaria di La Rosa & C.,
Catania 080019
Antiquaria-English Antiques,
Tours 073834
Antiquaria Padova, Padova 098180
Antiquaria Sant'Angelo, Roma . . 081218,
142797
Antiquaria Veronese di Piazza S.
Anastasia, Verona 081892
Antiquariaat De Bilt, De Bilt . . . 143104
Antiquariaat De Koppelpoort,
Amersfoort 143005
Antiquariaat De Oude Wereld,
Antwerpen 140140
Antiquariaat Friesland,
Leeuwarden 143179
Antiquariaat Sint Joris Maritiem,
Amstelveen . . 082662, 132386, 143007
Antiquariaat Smedeman, Nieuw
Vennep 143205
Antiquarian, Plattsburgh 139255
Antiquarian, San Diego 097051
Antiquarian & Trading, Kingston 081953,
142892
Antiquarian and Fine Books,
Belfast 144236
Antiquarian and Landmarks Society
Museum, Hartford 049383
Antiquarian and Pre-Owned Books,
Miami 145256
Antiquarian Antiques, London . . 064341
Antiquarian Book Fair, Malvern,
Victoria 097990
Antiquarian Book Mart, Alamo
Heights 144950
Antiquarian Bookcrafts, Dublin . 130845,
142598
Antiquarian Booksellers, Glebe . 139749
Antiquarian Booksellers' Association,
London 060025
Antiquarian Booksellers Association of

Register der Institutionen und Firmen

Antiquarian Booksellers Association of Canada, Victoria – Antiqu'Art-Caritas, Genève

America, New York 060320
Antiquarian Booksellers Association of Canada, Victoria 058418
Antiquarian Booksellers Association of Japan, Tokyo 059488
Antiquarian Booksellers Association of Korea, Seoul 059526
Antiquarian Horological Society, Ticehurst 060026
Antiquarian Horologists, New York 095219, 135877
Antiquarian Maps and Prints, Potts Point 139879
Antiquarian of Dallas, Dallas . . . 145107
Antiquarian Print Gallery, Unley . 127693, 139944
Antiquarian Restorers, New York 135878
Antiquarian's Delight, Philadelphia 096073, 145405
Antiquariat, Bowral 139653
Antiquariat "Unter der Muren", Velbert 142448
Antiquariat 44, Berlin . 141469, 141470
Antiquariat Aix-La-Chapelle, Aachen 141394
Antiquariat am Bachhaus, Eisenach 141736
Antiquariat am Bäckerbrunnen, Wiesbaden 142485
Antiquariat Am Ballplatz, Mainz . 142100
Antiquariat am Bayerischen Platz, Berlin 141471
Antiquariat am Dom, Trier 142417
Antiquariat am Domshof, Bremen 141643
Antiquariat am Fischtor, Mainz . 142101
Antiquariat am Gendarmenmarkt, Berlin 141472
Antiquariat am Hermannplatz, Berlin 141473
Antiquariat am Kräherwald, Kirchheim unter Teck 141998
Antiquariat am Marktplatz, Solothurn 144136
Antiquariat am Moritzberg, Hildesheim 141944
Antiquariat am Prater, Ihlow . . . 141956
Antiquariat Am Prater, Woltersdorf 142505
Antiquariat am Reileck, Halle, Saale 141849
Antiquariat am Roßacker, Rosenheim 142335
Antiquariat am Soonwald, Sponheim 142371
Antiquariat am Stadtbach, Memmingen 142124
Antiquariat am Vareler Hafen, Varel 142446
Antiquariat am Westwall, Krefeld 142035
Antiquariat an der Donau, Neuburg an der Donau 142223
Antiquariat an der Herderkirche, Weimar 142469
Antiquariat an der Moritzburg, Halle, Saale 141850
Antiquariat an der Nikolaikirche, Leipzig 142053
Antiquariat an der Stiftskirche, Bad Waldsee 141438
Antiquariat an der Universität, Freiburg im Breisgau 141798
Antiquariat Athenaeum, Stockholm 143938
Antiquariat Atlas, Hamburg 141856
Antiquariat Aurora, Hannover . . . 141907
Antiquariat Beim Steinernen Kreuz, Bremen 141644
Antiquariat Bibermühle, Ramsen 144120
Antiquariat Blechtrommel, Jena . 141962
Antiquariat Buchfänger am Pferdemarkt, Oldenburg 142264
Antiquariat Buchhaltestellle, Berlin 141474
Antiquariat Buchseite, Wien . . . 140056
Antiquariat Buchstabei, Oldenburg 142265
Antiquariat Carpe Diem, Koblenz 142002
Antiquariat Claraplatz, Basel . . . 144013
Antiquariat Curiosum, Benz, Kreis Nordwestmecklenburg 141463
Antiquariat der Literatur, Zürich . 144158
Antiquariat Fasan, Wien 140057
Antiquariat Goethe & Co., Heidelberg 141923
Antiquariat Grimbart, Burgdorf . . 141666
Antiquariat Hagenbrücke, Braunschweig 141627
Antiquariat Heureka, Bad Staffelstein 141436
Antiquariat Hierana, Erfurt 141743
Antiquariat im Baldreit, Baden-Baden 141440
Antiquariat im Hopfengarten, Braunschweig 141628
Antiquariat im Hufelandhaus, Berlin 141475
Antiquariat im Lenninger Tal, Lenningen 142071
Antiquariat im Marienpalais, Neustrelitz 142229
Antiquariat im Schnoor, Bremen 141645
Antiquariat Im Seefeld, Zürich . . 144159
Antiquariat im Willy-Brandt-Haus, Berlin 141476
Antiquariat in der Goltzstraße, Berlin 141477
Antiquariat Karla, Lüneburg 142092
Antiquariat Les-Art, Gerlingen . . 141827
Antiquariat Lesbar, Graz 139984
Antiquariat Libretto, Soest 142367
Antiquariat Löcker, Wien 136572, 140058
Antiquariat Magister Tinius, Falkensee 141761
Antiquariat Maxvorstadt I, München 142139
Antiquariat Mephisto im Bücherzentrum, Unna 142444
Antiquariat Mercurius, Köln 142004
Antiquariat Messidor, Bamberg . 141448
Antiquariat Montfort, Feldkirch . 062558, 139980
Antiquariat Numero 45, München 142140
Antiquariat Oktav, Zürich 144160
Antiquariat-Papierier, Sint Jansteen 143229
Antiquariat-Puls, Fredersdorf . . . 141794
Antiquariat Querido, Düsseldorf . 141714
Antiquariat Rabenschwarz, Braunschweig 141629
Antiquariat Rosenstrasse, Braunschweig 141630
Antiquariat Roter Stern, Marburg 142111
Antiquariat Stargarder Land, Burg Stargard 141665
Antiquariat Suleika, Weimar . . . 142470
Antiquariat Tilmann Riemenschneider, Osterode am Harz 142284
Antiquariat Tresor am Römer, Frankfurt am Main 075770, 141768
Antiquariat und Buchhandel, Halle, Saale 141851
Antiquariat und Galerie im Rathausdurchgang, Winterthur 116720, 144152
Antiquariat zum Rathaus, Bern . 144044
Antiquariat Zwiebelfisch, Weimar 142471
Antiquariato, Milano 138789
Antiquariato, Parma 081090
Antiquariato, Pisa 081151
Antiquariato 1800 1900 e dintorni, Venezia 081820
Antiquariato 800, Roma 081219
Antiquariato Chiale, Racconigi . . 081180
Antiquariato Europeo, Roma . . . 081220
Antiquariato Florida, Napoli 080836, 142756
Antiquariato Indie, Firenze 080080
Antiquariato Librario Bado e Mart, Padova 080959, 142772
Antiquariato Librario Bisello, Milano 080405, 142712
Antiquariato Navale, Milano 080406
Antiquariato Petrucci, Bologna . . 079812
Antiquariats, Rīga 142954
Antiquariats-Messe Zürich, Zürich 098325
Antiquariatsgesellschaft, Berlin . 141478
Antiquaries Journal, London . . . 139122
Antiquarimercanti, Chiesuol del Fosso 098181
Antiquario, San Francisco 097164
Antiquario, Wellington . 084153, 132756
Antiquario da Atlantica, Rio de Janeiro 064012
Antiquário do Chiado, Lisboa . . . 143555
Antiquario DPL Numismatica, Rio de Janeiro 064013
Antiquario Epocas, São Paulo . . 064148
L'Antiquario Gemmologo, Padova 080960
Antiquario Genesi, Rio de Janeiro 064014
Antiquário HR, Fortaleza 063940
Antiquario Retalhos do Tempo, Rio de Janeiro 064015
Antiquário Sé, Faro 084798
Antiquario Soho, Rio de Janeiro 064016
Antiquarios, San Francisco 097165
Antiquaris Barcelona, Barcelona 098294
Antiquarische Fundgrube bei der Volksoper, Wien 140059
Antiquarische Gesellschaft in Zürich, Zürich 059919
Antiquarium, Agropoli 025255
Antiquarium, Ariano Irpino 025366
Antiquarium, Avella 025422
Antiquarium, Borgia 025597
Antiquarium, Brescello 025625
Antiquarium, Buccino 025652
Antiquarium, Caldarola 025690
Antiquarium, Cesenatico 025932
Antiquarium, Cimitile 025985
Das Antiquarium, Erlangen 141744
Antiquarium, Filadelfia 026232
Antiquarium, Golasecca 026451
Antiquarium, Houston 145159
Antiquarium, Loreto Aprutino . . . 026628
Antiquarium, Lugnano in Teverina 026649
Antiquarium, Manduria 026690
Antiquarium, Massarosa 026741
Antiquarium, Mergozzo 026771
Antiquarium, Minturno 026843
Antiquarium, Monasterace 026881
Antiquarium, New York 095220
Antiquarium, Nonantola 027048
Antiquarium, Numana 027072
Antiquarium, Palazzolo Acreide . 027169
Antiquarium, Palmi 027195
Antiquarium, Partinico 027222
Antiquarium, Patti 027225
Antiquarium, Porto Torres 027417
Antiquarium, Prato 027442
Antiquarium, San Marzano sul Sarno 027779
Antiquarium, Santa Flavia 027814
Antiquarium, Serravalle Scrivia . 027931
Antiquarium, Tindari 028086
Antiquarium, Trieste 028190
Antiquarium – Villa Maritima, Minori 026842
Antiquarium Cantianense, San Canzian d'Isonzo 027734
Antiquarium Civico, Bagnolo San Vito 025441
Antiquarium Civico, Vazzano . . . 028285
Antiquarium Civico, Vico del Gargano 028373
Antiquarium Comunale, Colleferro 026035
Antiquarium Comunale, Contessa Entellina 026059
Antiquarium Comunale, Milena . 026837
Antiquarium Comunale, Monte Romano 026912
Antiquarium Comunale, Roma . . 027553
Antiquarium Comunale, Sezze . . 027945
Antiquarium Comunale, Sutri . . . 028042
Antiquarium Comunale, Tiriolo . . 028088
Antiquarium Comunale N. Pansoni, Cossignano 026091
Antiquarium del Castello Eurialo, Siracusa 027966
Antiquarium del Parco della Forza, Ispica 026523
Antiquarium del Seminario Vescovile, Nola 027044
Antiquarium del Serapeo, Villa Adriana, Tivoli 028091
Antiquarium del Teatro Greco-Romano, Taormina 028054
Antiquarium del Varignano, Portovenere 027426
Antiquarium delle Grotte di Catullo, Sirmione 027970
Antiquarium di Agrigento Paleocristiana Casa Pace, Valle dei Templi, Agrigento 025250
Antiquarium di Canne, Barletta . 025464
Antiquarium di Himera, Termini Imerese 028073
Antiquarium di Megara Hyblaea, Augusta 025419
Antiquarium di Monte Cronio, Sciacca 027908
Antiquarium di Nervia, Ventimiglia 028330
Antiquarium di Poggio Civitate, Murlo 026986
Antiquarium di Sant'Appiano, Barberino Val d'Elsa 025447
Antiquarium di Tesis, Vivaro . . . 028416
Antiquarium di Torre Nao, Crotone 026108
Antiquarium di Villa Romana, Patti 027226
Antiquarium e Archeologica Museo, Cuglieri 026114
Antiquarium e Mosaico Romano, Bevagna 025504
Antiquarium e Zona Archeologica, Lugagnano Val d'Arda 026648
Antiquarium Forense, Roma . . . 027554
Antiquarium Hungaricum, Budapest 126148, 142555
Antiquarium Ipogeo dei Volumni, Perugia 027254
Antiquarium Jetino, San Cipirello 027738
Antiquarium Lucio Salvio Quintiano, Ossuccio 027125
Antiquarium Schönwalde, Wandlitz 078860, 142462
Antiquarium Sebatum, Sankt Lorenzen 027807
Antiquarium Sestinale, Sestino . 027935
Antiquarium Tellinum, Teglio . . . 028065
Antiquarium Torre Cimalonga, Scalea 027895
Antiquarium, Anfiteatro Romano, Santa Maria Capua Vetere 027818
Antiquarium, Chiostro di San Francesco, Fondi 026317
Antiquarium, Museo Archeologico, Sant'Antioco 027838
Antiquarium, Uomo e Ambiente nel Territorio Vesuviano, Boscoreale 025612
Antiquarium, Villa Rufolo, Ravello 027480
Antiquarius, Kraków 137892
Antiquarius, London . . 089911, 144551
Antiquarius, North Shore 143334
Antiquarius, Roma . . . 081221, 142798
Antiquarius, San Francisco 097166
Antiquarius Antiques Show, Greenwich 098418
Antiquarius Center, Los Angeles 094248
Antiquarius of Ampthill, Ampthill 088275, 134030
Antiqu'art, Berck-sur-Mer 067095, 103363
Antiqu'Art-Caritas, Genève 087443

Antiquary, Nelson 084013
Antiquary, Tulsa 097706
L'Antiquasi, Torino 081645, 132088
Antiquasia, Arts d'extrême orient,
Paris 070900
Antiquated, Petworth 090987
Antiquax, Madrid 086118
Antiquax-Victoria, Auckland 132680
Antique, Levoča 085574
Antique, Lisboa 084825
Antique, Nová Baňa 088581
Antique, Plzeň 065450
Antique, Spišská Nová Ves 085591
Antique & Ambiente, Bielefeld .. 074971
Antique & Ancient Coins, Beijing 064997
Antique & Art, Warszawa 084692
Antique & Contemporary Leasing,
Washington 097801
Antique & Decorative Warehouse,
Cookham 088974, 134296
Antique & Modern Jewellery Xchange,
Auckland 083750
Antique & Unique Treasures,
Indianapolis 093850
Antique 2000, Los Angeles 094249
Antique 2000, Paris 070901
Antique 21, Reims 072375
Antique 89, Brønshøj 065617
Antique Accents, Salt Lake City 096873
Antique Accents, Tulsa 097707
Antique Acquisitions, New
Orleans 095051
Antique Addict, Oklahoma City . 095934
Antique Adoption, Long Beach . 094189
Antique Affaire, Kings Park 061692
Antique Affaire, Wanganui 084142
Antique Age, Los Angeles 135702
Antique Alcove, Denver 093242
Antique Alley, Auckland 083751
Antique Alley, Honolulu 093526
Antique Alley, Phoenix 096180
Antique Alley, Portland 096357
Antique Alley, Seattle 097395
Antique and Art Around Florida,
Keystone Heights 139256
Antique and Art Exchange, San
Francisco 097167
Antique and Collectable Restoration,
Omaha 136056
Antique and Collectible Dolls of
Seattle, Seattle 097396
Antique and Collectibles Marketplace,
Sacramento 096593
Antique and Collector's Books,
Southport 139911
Antique and Collector's Gallery, Las
Vegas 094111
Antique and Concept Gallery,
Miami 094684
Antique and Country Pine, Crewe 089002
Antique and Design, Canterbury 088770,
134203
Antique and Doll Affaire Shoppe,
Sacramento 096594
Antique and Modern Caners,
Morden 134857
Antique and Resale Shoppe,
Chicago 092658
Antique Angle, Dallas 093061
Antique Antiques, Singapore ... 085360
Antique Appraisers Association of
America, Garden Grove 060321
Antique Arcade, Cleveland 092915
Antique Architectural Decor, Saint
Louis 096664
Antique Arms and Armour,
Nowergup 062018
L'Antique-Art, Montréal 064365
Antique-Art, Saint-Jean-de-Luz . 072841
Antique Art, Salon-de-Provence . 073387
Antique Art and Arms,
Minneapolis 094895
Antique Art Gallery, Oklahoma
City 095935

Antique Asylum, Toronto 064585
Antique Auction Hall,
Constableville 127187
Antique Auto and Race Car Museum,
Bedford 046867
Antique Auvergne, Ambert 066548
Antique Bad Company, Emsworth 089205
Antique Barn, Toledo 097597
The Antique Barometer Company,
London 089912, 134584
Antique Bed Shop, Halstead ... 089416
Antique Bed Specialists, Tyabb . 062358
Antique Beds, Hove .. 089594, 134477
Antique Bishamon, Hiroshima .. 081973
Antique Boat Museum, Clayton . 047784
Antique Bookshop, Buenos Aires 139549
Antique Bookshop, London 144552
Antique Bottle and Glass Collector,
Boyertown 139257
Antique Boutique, Edmonton ... 064282
Antique Boutique, New York ... 095221
Antique Boutique, Richmond ... 096523
L'Antique Bric a Brac, Prato .. 131770
Antique Broadway Broker,
Denver 093243
Antique Buildings, Dunsfold ... 089120
Antique Buyers of Melbourne,
Yarraville 062505
Antique Buying, Columbus 092990
Antique Cache, New York 095222
Antique Caméras, Paris 070902
Antique Castings, San Diego .. 097052
Antique Cellar, Regina 064546
Antique Center, Néa Kifissia .. 079197
Antique Center, San Antonio .. 096936,
136254
Antique Center Antiquarios, Rio de
Janeiro 140290
Antique Center at Federal Hill,
Baltimore 092291
Antique Center-Milwaukee,
Milwaukee 094828
Antique Center of Texas, Houston 093594
Antique Center of Tucson,
Tucson 097628
Antique Center on Broadway,
Denver 093244
Antique Center-Walker'S Point,
Milwaukee 094829
Antique Centre, Cleobury
Mortimer 088928
The Antique Centre,
Kidderminster 089678, 134510
Antique Centre, Nenagh 079683
Antique Centre, Oakland 095881
Antique Centre, Oklahoma City . 095936,
095937
Antique Centre, Preston 091053
Antique Centre Gloucester,
Gloucester 089338
Antique Centre of Devonport, North
Shore 084028
Antique Centre of Hawthorn,
Hawthorn 061598
Antique Centre of Stonnington,
Armadale, Victoria 061008
The Antique Centre on the Quay,
Exeter 089228, 144417
Antique Chairs and Museum,
Launceston 089749
Antique Chapelle, La Chapelle-
d'Armentières 068971
Antique Church Furnishings, Walton-
on-Thames 091720
Antique Circus, Baltimore 092293
The Antique Circus, Baltimore .. 092292
Antique-City, Atlanta 092103
Antique-City, Knokke-Heist 063695
Antique City, Rozelle 062173
The Antique Clique, Kilkenny .. 079643
Antique Clock Gallery, Long
Beach 094190, 135692
Antique Clock Repair, Dallas ... 135491
Antique Clock Repair and Antiques,

Bowral 061179, 127567
Antique Clock Restoration,
Columbus 135473
Antique Clock Shop, Saint Paul . 096802,
136212
Antique Clocks, Harston 089446
Antique Clocks – Pendulastic, Saint-
Blaise 087768, 133897
Antique Clocks by Mike, Oklahoma
City 095938
Antique Coins, Saint Paul 096803
Antique Collecting, Woodbridge . 139123
The Antique Collector, London .. 064342
Antique Collector's Club,
Easthampton 138332
Antique Collectors' Club,
Woodbridge 138294, 144932
Antique Company, Houston 093595
The Antique Company,
Sacramento 096595, 096596
Antique Connection, San Antonio 096937
Antique Connections, Fort Worth 093423
Antique Conservation, New York 135879
Antique Consignment, Los
Angeles 094250
Antique Cooperative,
Albuquerque 092027
Antique Cooperative, Oklahoma
City 095939
Antique Cooperative and Auction
House, Malden 127302
Antique Corner, Bristol 088643, 134157
Antique Corner, Fort Worth 093424
Antique Corner, Narrandera 061945,
125218
Antique Cottage, Albuquerque .. 092028
Antique Cottage, San Antonio .. 096938
Antique Cottage, San Diego ... 097053
Antique Cottage at Estes Park,
Omaha 096024
Antique Country Furniture and
Interiors, Sandgate 091232
Antique de la Radio, Madrid ... 086119
Antique Dealer and Collectors' Guide,
London 139124
Antique Dealers, Shepparton ... 062206
Antique Dealers' Association of Poland,
Kraków 059691
Antique Dealers Association of San
Anselmo, San Anselmo 060322
Antique Dealers Fair, Dublin .. 098174
Antique Decor, Armadale, Victoria 061009
Antique Decoration, Cluj-Napoca 085113,
133004
Antique Decoration Art Noveau,
Braşov 085071
Antique Department Store, Glebe 061524
Antique Depot, Richmond 096524
Antique Design, Philadelphia ... 096074
Antique Design Center, Norfolk . 095843
Antique Detours, Sacramento .. 096597
The Antique Dining Room, Totnes 091619,
135142
Antique Doctor, Tucson 097629, 136414
Antique Doll Restorers, Auckland 132681
Antique Dolls and Bears, Torbay 084114
Antique East, Singapore 085361
Antique Effects, Ballarat 061081
Antique Elements, Leichhardt . 061731
Antique Emporium, Albuquerque 092029
Antique Emporium, Dallas 093062
Antique Emporium, Healesville . 061611
Antique Emporium, Nashville ... 094966
Antique en Stock, Rouen 072558
Antique English Windsor Chairs,
Chipping Norton 088886
Antique Engravings,
Camperdown 061265
Antique Estate Jewelry Gallery, New
Orleans 095052
Antique Europa, Houston 093596
Antique Exchange, Baltimore .. 092294
Antique Exchange, Denver 093245,
127202

Antique Exchange, Winnipeg ... 064909
Antique Eyes, Stockbridge 091430
Antique Firearms,
Dommershausen 075371
Antique Fireplace Centre,
Hartlepool 089447
Antique Fireplace Restoration,
Dublin 079521, 130846
Antique Fireplaces, Liverpool .. 089853
Antique Fireplaces, Manchester . 090688
Antique Forum, Newcastle-under-
Lyme 090822
Antique Furniture, Baltimore ... 092295
Antique Furniture, Brno 065343
Antique Furniture, Sherwood .. 127674
Antique Furniture, Toledo 136403
Antique Furniture Services,
Miami 094685, 135787
Antique Furniture Warehouse,
Houston 093597, 135584
Antique Furniture Warehouse,
Stockport 091439
Antique Galerie, Düsseldorf 075451
Antique Gallery, Atlanta 092104
Antique Gallery, Cleveland 092916
Antique Gallery, Los Angeles .. 094251
Antique Gallery, Memphis 094610
Antique Gallery, Nairobi 082242
Antique Gallery, New York 095223
Antique Gallery, Philadelphia .. 096075
Antique Gallery, Phoenix 096181
The Antique Gallery, Sutton ... 119184
Antique Garage, Yokohama ... 082223
Antique Garden, Saint Paul ... 096804
Antique Gas and Steam Engine
Museum, Vista 054005
Antique Gatherings, Phoenix ... 096182
Antique General Store, Narrabeen 061943
Antique Guild, Denver 093246
Antique Guild, Los Angeles ... 094252,
135703
Antique Guild Japan, Tokyo 059489
Antique Guild United Kingdom,
Burnley 060027
Antique Hardware Store, Oklahoma
City 095940
Antique Heaven, Miami 094686
Antique Helper, Indianapolis ... 127258
Antique Home Improvement, Los
Angeles 094253
Antique House, Cairo 066226
Antique House, Göteborg 086696
Antique House, Honolulu 093527
Antique House, Jacksonville ... 093952,
135644
Antique House, Oklahoma City . 095941
Antique House, Waikato 084121
Antique Imports, Hutt City 083958
Antique Inspirations, Houston .. 093598
Antique Interiors, Charlotte 092573
Antique Interiors, Twickenham . 091673
Antique Iron Market, Nashville .. 094967
Antique Jewelry, Saint Paul 096805
Antique John's Auction, Clarinda 127180
Antique Junction, Columbus ... 092991
Antique Junque Shoppe, Walcha 062401
Antique Kaprova, Praha 065473
Antique Kingdom, Charlotte ... 092574
The Antique Lamp Shop, London 089913
Antique Lamp Shop, Maylands,
Western Australia 061806
Antique Legacy, Sacramento ... 096598
Antique Light, Abbotsford 060963
Antique Lighthouse, Philadelphia 096076,
136074
Antique Lighting, Ottawa 064491
Antique Lighting Company,
Seattle 097397
Antique Liquidators, Seattle ... 097398
The Antique Loft, Clarecastle .. 079481
Antique Mall, Edmonton 064283
The Antique Mall, Las Vegas ... 094112
Antique Mall, Regina 064547
Antique Mall, San Diego 097054

The Antique Mall, Tucson 097630
Antique Mall of Austin, Austin .. 092206
Antique Mall of Creve Coeur, Saint
Louis 096665
Antique Mall of Green Hills,
Nashville 094968
Antique Mall of Irvington,
Indianapolis 093851
Antique Mall of Midtown,
Memphis 094611
Antique Mall of Palma Ceia,
Tampa 097538
Antique Mall of West Seattle,
Seattle 097399
Antique Mall on Clark, Chicago . 092659
Antique Mall on Montgomery Street,
Fort Worth 093425
Antique Mall Village, Houston .. 093599
The Antique Man, Baltimore .. 092296
Antique Man, Buffalo 092507
Antique Manor, La Richardais .. 069062
Antique Manor, Trémereuc 073879
Antique Map & Book Shop,
Dorchester 144370
Antique Map & Print Gallery,
Hallow 144450
Antique Maps, Belmont 139639
Antique Maps and Fine Prints,
Dallas 145108, 145109
Antique Maps Room, Toronto .. 064586,
140395
Antique Margurite, Ringe 066063
Antique Market, Adelaide 060966
Antique Market, Denver 093247
Antique Market, Hay-on-Wye .. 089478
Antique Market, Phoenix 096183
Antique Market, Sacramento .. 096599
Antique Market, Vancouver ... 064790,
064791
Antique Market of Warrnambool,
Warrnambool 062412
Antique Marketplace, Austin ... 092207
Antique Mart, Saint Paul 096806
Antique Master, Nairobi 082243
Antique Mattress, Houston ... 093600
Antique Max, Praha 065474
Antique Memories, San Jose .. 097356
Antique Menagerie, Saint Louis . 096666
Antique Merchants Mall,
Nashville 094969
Antique Mini-Mall, Tucson ... 097631
Antique Moderne, København .. 065842
Antique Odyssey, Seattle 097400
Antique Oise, Saint-Samson-la-
Poterie 073298
Antique Oriental Rug Buyers,
Miami 094687
Antique Oriental Rugs, Pittsburgh 096253,
136127
Antique Oriental Rugs Center, Las
Vegas 094113
Antique Outdoor Market, Long
Beach 094191
Antique Outpost, Phoenix 096184
Antique Panache, Houston 093601
Antique Park, Bordeaux 067303
Antique Parlor, San Diego 097055
Antique Passion, Aulnay-sous-
Bois 066812
Antique Pavilion, Houston 093602
Antique Pendulum, Washington . 136451
Antique Pine & Country Furniture Shop,
Driffield 089105, 134342
Antique Place That, Houston ... 093603
Antique Plus, Las Vegas 094114
Antique Plus, Los Angeles 094254
Antique Pohořelec, Praha 065475
Antique Presidio, Tucson 097632
Antique Print, Main Beach ... 061759
Antique Print Gallery,
Christchurch 083851, 143300
Antique Print Room, Sydney ... 062275,
139922
Antique Print Shop, East

Grinstead ... 089132, 134346, 144378
Antique Print Shop, Redbourn .. 091076
Antique Prints, Dublin 142599
Antique Prints, Hawthorn 061599
Antique Properties, Nairobi ... 082244
Antique Radio and Audio Repair, San
Diego 136286
Antique Radio Store, San Diego 097056
Antique Refinishers, San Diego . 136287
Antique Refinishing, Virginia
Beach 136441
Antique Repair, Cincinnati 135431
Antique Repair Madeira Junction,
Cincinnati 135432
Antique Resources, Chicago ... 092660
Antique Resources, Tulsa 097708
Antique Restoration, El Paso .. 135548
Antique Restoration 45 Rose, San
Francisco 097168, 136305
Antique Restoration Atelier,
Washington 136452
The Antique Restoration Centre,
Cheltenham 134231
Antique Restoration Service,
Nashville 135843
Antique Restoration Services, Box
Hill 127568
Antique Restoration Services, East
Norriton 135546
Antique Restoration Studio,
Denver 135524
Antique Restoration Studio,
Stafford 135069
Antique Restoration Tralee, Tralee 079708,
130893
Antique Restorations,
Christchurch 083852, 132701
Antique Restorations, Hawkes
Bay 132721
Antique Restorations, Salt Lake
City 136230
Antique Restorations, Tulsa ... 136426
Antique Restorations of Unley,
Unley 127694
Antique Restoring, Nairobi .. 082245,
132314
Antique Riviera, Helsinki 066305
Antique Rooms, Maldon 090678
Antique Row, Dallas 093063
Antique Row on South Broadway,
Denver 093248
Antique Rugs as Art, New York . 095224
Antique Salad, Vancouver 064792
Antique Salvage, Dublin 079522
Antique Sampler Shoppes, Kansas
City 094056
Antique Sampler Shoppes, Las
Vegas 094115
Antique Services of Texas, San
Antonio 096939
Antique Shop, Blaenau Ffestiniog 088501
Antique Shop, Bruton 088691
Antique Shop, Dunshaughlin .. 079613
Antique Shop, Godalming 089339
Antique Shop, Hamilton, New South
Wales 061576
The Antique Shop, Kirkcudbright 089717
Antique Shop, Kjustendil 064222
The Antique Shop, Langholm . 089745
Antique Shop, Lille 069472
The Antique Shop, Newtonmore 089042
Antique Shop, Omaha 096025
The Antique Shop, Saint-Nom-la-
Breteche 072993
Antique Shop, Sutton Bridge .. 091515
Antique Shop Saborna, Plovdiv . 064226
Antique Shoppe, Ottawa 064492
Antique Shoppe, Salt Lake City . 096874
Antique Shoppe, Tauranga ... 084099
Antique Show, Dallas 093065
Antique Showcase, Philadelphia 096077
The Antique Silver Company,
Balaclava 061077, 127562
Antique Specialty Mall,

Albuquerque 092030
Antique Station, Tulsa 097709
Antique Stove Heaven, Los
Angeles 094255, 135704
Antique Styles, Loganholme ... 061747
The Antique Superstore,
Hamilton 064313
Antique Swan, Austin 092208
Antique Textiles & Lighting, Bath 088376
Antique Thrift Shop, San Diego . 097057
Antique Touch, Seattle 097401
Antique Toy and Board Collector, Salt
Lake City 096875
Antique Toy World, Chicago ... 139258
Antique Toys & Dolls, Milwaukee 094830,
135810
Antique Tractor Museum and Frontier
Village, Maple Creek 006257
Antique Trade Center, Dallas ... 093065
Antique Trader, Denver 093249
Antique Trader, Iola 139259
Antique Traders, San Francisco . 097169
Antique Trading, Dublin 079523
Antique Trading Company,
Wellington 084154
The Antique Travelshop, Bussum 083014
Antique Treasure Traders,
Toronto 064587
Antique Treasurers, Phoenix .. 096185
Antique Tribal Art Dealers Association,
Albuquerque 060323
Antique und Kunstmesse
Herrenhausen, Hannover 098119
Antique Valdemars Slot-
Møbelrestaurering, Svendborg . 066156,
128402
Antique Village, Manchester ... 090689,
134830
Antique Village, Toronto 064588
Antique Warehaus, Houston ... 093604
Antique Warehouse, Baltimore . 092297
Antique Warehouse, Claremont . 061321
Antique Warehouse, Fort Worth . 093426
Antique Warehouse, London .. 089914
Antique Warehouse, Memphis .. 094612
The Antique Warehouse,
Vancouver 064793
Antique Warehouse, Winnipeg .. 064910
Antique Warehouse, Worcester . 091939
Antique Watch and Clock Repair,
Minneapolis 135824
Antique Way, Los Angeles 094256,
135705
Antique Week, Knightstown ... 139260
Antique Wholesalers, Miami ... 094688
Antique Wholesalers, Salt Lake
City 096876
Antique Wooden Horse,
Jacksonville 093953, 135645
Antique Workshop, Cincinnati . 135433
Antique Workshop, Detroit 135538
Antique World, Calgary 064247
Antique World, Surrey . 064572, 128205
Antique World, Winnipeg 064911
Antique y Geneve, Santiago de
Chile 064950
Antique Yamamoto Shoten, Tokyo 082133
Antique Zoo, Denver 093250
Antiquedirect, København 065843
Antiquehaus Marienthal, Kersten,
Hamminkeln 076351
Antiqueland USA, Austin 092209
Antiqueland USA, Dallas 093066
Antiqueland USA, Fort Worth .. 093427
Antiqueland USA, San Antonio . 096940
Antiqueline, New York 095225
Antiqueria, New York 095226
Antiquerie Médoc-Antiquités-de-
Lescapon, Queyrac 072333
Antiquers, Toronto 064589
Antiques, Bucureşti 085079
Les Antiques, Dallas 093067
Antiques, Fochabers 089287
Antiques, Kingston 081954

Antiques, Marazion 090700
Antiques, Menston 090748
Antiques, Minsk 063165, 063166,
063167
Antiques, Saint-Florent-sur-Cher 072785
Antiques, Szczecin 084671
Antiques, Toronto 064590
Antiques, Vancouver 064794
Antiques – Antigüedades, La Paz 063861
Antiques & Art, Frankfurt am
Main 075771
Antiques & Art Around Florida,
Keystone Heights 139261
Antiques & Arts, Toronto 064591, 101563
Antiques & Co, Nantes 070440, 128825
Antiques & Collectables,
Auckland 083752
Antiques & Collecting Magazine,
Chicago 139262
Antiques & Curios Centre, Cork . 079491
Antiques & Curiosities, Auckland 083753
Antiques & Design, Minden 077372
Antiques & Dreams, Birmingham 092385
Antiques & Eccentricities,
Columbus 092992
Antiques & Elegant Junk,
Wanganui 084143
Antiques & Fine Art at Park Avenue
Armory, New York 098419
Antiques & Fine Furniture,
Sevenoaks 091267, 135016
Antiques & Funk, Winnipeg 064912
Antiques & Gallery, Praha 065476
Antiques & Gardens, Birmingham 092386
Antiques & Interiors, Genova ... 080232
Antiques & Interiors, Tralee 079709
Antiques & Knick Knacks, Nairobi 082246
Antiques & Militaria, Toronto ... 064592
Antiques & Stuff, Houston 093605
Antiques @ The George, Burford 088701,
144288
Antiques @ Tirau, Waikato ... 084122
Antiques 1071, Napier 084003
Antiques 4 USA, San Diego ... 097058
Antiques AAA, Indianapolis 093852
Antiques across the World,
Nottingham 090889
Antiques Ad Hoc, Ashburton ... 088305
Antiques Adventure, Coventry ... 088983
Antiques Allentown, Buffalo ... 092508
Antiques Amended, Memphis . 094613,
135773
Antiques and Angels, North
Shore 084029
Antiques and Art, Hobart 061625
Antiques and Art, Miami 094689, 135788
Antiques and Art, Seattle 097402
Antiques and Art in Queensland, Bondi
Junction 138490
Antiques and Art in Victoria, Bondi
Junction 138491
Antiques and Arts, Dubai ... 088167,
117176
Antiques and Auction News, Mount
Joy 139263
Antiques and Beyond, Atlanta .. 092106
Antiques and Bricabrac, Miami . 094690
Antiques and Bygones,
Chichester 088871
Antiques and Bygones, Durban . 088561
Antiques and Bygones, Rochdale 091124
Antiques and Classical Interiors Fair,
Naxxar 098243
Antiques and Collectables,
Coolangatta 061360
Antiques and Collectables,
Geelong 061506
Antiques and Collectables,
Honolulu 093258
Antiques and Collectables, North
Fremantle 061983
Antiques and Collectables,
Scunthorpe 091254, 135013
Antiques and Collectables Centre,

Ballarat 061082
Antiques and Collectables Fair,
London 098354
Antiques and Collectors Centre,
Diss 089061
Antiques and Country Living,
Frome 089313
Antiques and Country Pine,
Crewkerne 089003
Antiques and Crafts Centre,
Chorley 088901
Antiques and Curios,
Peterborough 090981, 134935
Antiques and Curios, Worcester 091940,
135257
Antiques and Decorative, Tampa 097539
Antiques and Decorative Art,
Tampa 097540
Antiques and Everything,
Triabunna 062346
Antiques and Fine Art, Omaha . 096026
Antiques and Fine Art, Penzance 090962,
134930
Antiques and Fine Art,
Watertown 139264
Antiques and Fine Art Fair,
Dublin 098175
Antiques and Furnishings,
Bournemouth 088531
Antiques and Gifts, Aberdour . . . 088227,
134012
Antiques and Gifts, Downham
Market 089103
Antiques and Interiors, Nashville 094970
Antiques and Interiors, Norwich 090866
Antiques and Interiors,
Philadelphia 096078
Antiques and Interiors on Selwyn,
Charlotte 092575
Antiques and Meubles, North
Fremantle 061984
Antiques and More, Indianapolis 093853
Antiques and More, Las Vegas . 094116
Antiques and More, Long Beach 094192
Antiques and More, Miami 094691
Antiques and More, Tulsa 097710
Antiques and Oddities, Portland . 096358
Antiques and Pearl, Austin 092210
Antiques and Piano Restoration,
London 134585
Antiques and Restoration by Irina, San
Francisco 097170, 136306
Antiques and Stuff, Denver 093251
Antiques and Stuff, San Diego . 097059
Antiques and Stuff by Ruth, San
Diego 097060
Antiques and Things, New
Orleans 095053
Antiques and Treats,
Albuquerque 092031
Antiques Anonymous,
Washington 097802
Antiques Antiques, San Francisco 097171
Antiques Anywhere, Baltimore . . 092298
Antiques Appraisers and Auctioneers,
Middletown 127317
Antiques Art, Rotterdam 083549
Antiques Arts & Crafts, Rathcoole 079692
Antiques As Well, Heidelberg . . 061614
Antiques at 115 Broadway, San
Antonio 096941
Antiques at 80 Charles, Boston . 092441
Antiques at 99 Charles, Boston . 092442
Antiques at Budleigh House, East
Budleigh 089131
Antiques at Dunlavy, Houston . . 093606
Antiques at Forge Cottage,
Gargrave 089319
Antiques at Glen Harrow,
Belgrave 061120
Antiques at Heritage, Woodstock 091926
Antiques at Hyde Park, London . 064343
Antiques at India Point,
Providence 096498

Antiques at Pike Place, Seattle . 097403
Antiques at Redbank, Redbank . 062140
Antiques at Rummel Creek,
Houston 093607
Antiques at Saint Lawrence Market,
Toronto 064593
Antiques at Spruce Meadows,
Calgary 098030
Antiques at the Beach, San
Diego 097061
Antiques at the Market, Las
Vegas 094117
Antiques at Weldwood, London . 064344
Antiques Auction House,
Cleveland 127184
Antiques Avignon, Hackney . . . 061571
Antiques Avignon, Moret-sur-
Loing 070344
Antiques Barometers, Ramsey,
Cambridgeshire . . . 091068, 134958
Antiques Books, Woolloongabba 139966
Antiques Brass Lantern,
Hiroshima 081974
Antiques by Billy, Toronto 064594
Antiques by George, Tucson . . . 136415
Antiques by Lillian, Miami 094692
Antiques by Loraine, Milwaukee 094831
Antiques by Ralph, Miami 094693
Antiques by Rebecca, Miami . . . 094694
Antiques by Rudy, Virginia Beach 097770
Antiques by Sandy, Miami 094695
Antiques by the Wishing Corner,
Portland 096359
Antiques by Three Lions,
Jacksonville 093954
Antiques Bye Wells, Toronto . . . 064595
Antiques Canada, Toronto 098031
L'Antiques Center, Cherbourg . . 067960
The Antiques Centre, Trowbridge 091639
Antiques Centre at Olney, Olney 090915
Antiques Centre of Wellington,
Wellington 084155
Antiques Centre York, York 091970
Antiques Collections, New York . 095227
Antiques Colony, San Jose . . . 097357
Antiques Consortium,
Albuquerque 092032
The Antiques Corner, Lower
Dingle 079666
Antiques Corner, New York . . . 095228
The Antiques Council, Warren . . 060324
Antiques de France, Montréal . . 064366
Antiques de Provence, New
Orleans 095054
Antiques Decor and More, Saint
Paul 096807
Antiques Down Under, Coffs
Harbour 061349
Antique's du Monde, Los Angeles 094257
Antiques Etc, Auckland 083754
Antiques Etc., Atlanta 092107
Antiques Etc., Boston 092443
Antique's Etc., Denver 093252
Antiques Etc., Sacramento 096600
Antiques Etc., Stretton 091498
Antiques Etcetera, El Paso 093386
Antiques Fair, Naxxar 098244
Antiquès Faubourg Saint-Gilles,
Liège 063740, 128072
Antiques Fleamarket, San Diego 097062
Antiques for All, Miami 094696
Antiques for Entertaining, New
York 095229
Antiques for Everyone Spring Fair,
Birmingham 098355
Antiques for Everyone Summer Fair,
Birmingham 098356
Antiques for Everyone Winter Fair,
Birmingham 098357
Antiques Forever, Phoenix 096186,
136111
Antiques Hall, Fyshwick 061493
Antique's House, Thessaloniki . . 079205

Antiques Houston, Houston 093608,
145160
Antiques Idées, Chelles 067954
Antiques Imp-Exp, Midwoud . . . 132587
Antiques Import Export,
Stockport 091440, 135087
Antiques in Ashbourne,
Ashbourne 088295
Antiques in Indy, Indianapolis . . 093854
Antiques in Sassafras, Sassafras 062192
Antiques in the Bank, Cleveland 092917
Antiques in the Rough, San
Francisco 097172
Antiques in Time, Cleveland . . . 092918,
135455
Antiques in Wales, Kidwelly . . . 089680
Antiques International, Houston . 093609
Antiques International, Miami . 094697
Antiques International, West
Hollywood 097870
Antiques Limited, Chicago 092661
Antiques Magazine, Birmingham 139125
Antiques Magazine, New Orleans 095055
Antiques Marie-Claire, Naarden . 083505
Antiques Market, Knokke-Heist . 063696
Antiques Memories, Qormi 082425
Antiques Minnesota Midway, Saint
Paul 096808
Antiques 'n Stuff, Oklahoma City 095942
Antiques 'n Stuff, Tampa 097541
Antiques Natalie, Lima 084437
The Antiques Network, Cincinnati 092825
Antiques Notre Dame, Montréal 064367
Antiques of Brentwood, Nashville 094971
Antiques of Cool Springs,
Nashville 094972
Antiques of Denver, Denver . . . 093253
Antiques Of Epsom, Auckland . 083755
Antiques of Georgetown,
Washington 097803
Antiques of Penrith, Penrith . . . 090955
Antiques of Shadyside,
Pittsburgh 096255
Antiques of Sherborne,
Sherborne 091302, 135032
Antiques of Southside,
Birmingham 092387
Antiques of the Orient, Singapore 085362,
085363, 143684
Antiques of Tomorrow, Tshwane 085670
Antiques of Western NY, Buffalo 092509,
135369
Antiques of Woodstock,
Woodstock 091927, 135256
Antiques on Broadway,
Vancouver 064795
Antiques on Ellsworth, Pittsburgh 096256
Antiques on High, Armadale,
Victoria 061010
Antiques on High, Oxford 090929,
134915
Antiques on Hildebrand, San
Antonio 096942
Antiques on Lloyd, Moe 061869
Antiques on Locke, Hamilton . . 064314
Antiques on Macquarie, Hobart . 061626
Antiques On Manukau, Auckland 083756
Antiques on Memorial, Houston . 093610
Antiques on Mount Pleasant,
Toronto 064596
Antiques on Nineteenth, Houston 093611
Antiques on Queen, Toronto . . . 064597
Antiques on Railway, Boonah . . 061173
Antiques on the River, Detroit . . 093358
Antiques on the Square,
Indianapolis 093855
Antiques on View, Bendigo . . . 061127
Antiques-Point Chev Plaza,
Auckland 083757
Antiques Productions, Houston . 093612
Antiques Provencal, New York . 095230,
135880
Antiques Puces Libres, Montréal 064368,
128180

Antiques Restoration, New York . 135881
Antiques Rich Man Poor Man,
Montréal 064369
Antiques Riverwalk, Minneapolis 094896
Antiques Services, Bruxelles . . . 063316
Antiques@Chicago, Chicago
Minneapolis 098420
Antiques Starozitnosti, Bratislava 085541
Antiques Super Mall, Houston . . 093613
Antiques to Go, Winnipeg 064913
Antiques Tour, Auckland 083758
Antiques Trade Gazette, London 138184,
139126
Antiques Unique, Dallas 093068
Antiques Warehouse, Bath 088377
The Antiques Warehouse, Bristol 088644,
134158
Antiques Warehouse, Buxton . . . 088747,
134195
The Antiques Warehouse,
Farnham 089259, 134384
Antiques Warehouse, Glasgow . 089328
Antiques Warehouse, Marlesford 090727
Antiques Warehouse, San
Antonio 096943
Antiques Warehouse, Uxbridge . 091700
Antiques Warehouse, Whangarei 084202,
132768
Antiques White Bear, Saint Paul 096809
Antiques with Grace, New York . 095231
Antiques Within, Leek . 089774, 134538
Antiques Within, Memphis 094614
Antiques Workshop, San Antonio 136255
Antiques World, Brugge 063261
Antiques, Art and Investment Fair,
Seinäjoki 098067
Antiques, Books, Ephemera and
Collectibles, Atlanta . 092105, 144960
Antiques, Fine Art & Decorative
Furnishings Fair, Harrogate . . . 098353
Antiques, Tiques & Toys,
Pittsburgh 096254
Antiques@Chicago, Chicago . . . 092662
Antiques'Atelier, Camblain-l'Abbé 067611
Antiquest-Verity, Toronto 064598
Antiquetrade, Roma 081222
AntiqueWeek, Knightstown 139265
AntiqueWest, Knightstown 139266
Antiquitäten "Im alten Zollhaus",
Weida 078885
Antiquitäten 1900, Viktring 062868
Antiquitäten 96, Berlin 074753
Antiquitäten am Alten Hof,
München 077450, 130321
Antiquitäten am Belvedere,
Weimar 078894
Antiquitäten am Berliner Platz,
Kronberg im Taunus 076971
Antiquitäten am Ellinger Tor,
Weißenburg in Bayern 078915
Antiquitäten am Franziskanerplatz,
Würzburg 079073, 142509
Antiquitäten am Hafenplatz, Bad
Karlshafen 074534
Antiquitäten am Klosterplatz,
Solothurn 087819
Antiquitäten am Kosttor,
München 077451
Antiquitäten am Markt, Olpe . . . 077948
Antiquitäten am Rathaus,
Würzburg 079074, 126143
Antiquitäten am Schloß, Kiel . . 076771,
125999
Antiquitäten am Schloß, Weimar 078895
Antiquitäten am Schloß Moyland,
Bedburg-Hau 074707
Antiquitäten am Schloß, Inh. Ingo Henjes,
Torgau 078686, 142414
Antiquitäten am Schlossberg,
Pegnitz 078029
Antiquitäten am Schloßpark, Bad
Homburg v.d.Höhe 074518
Antiquitäten am Schwabentor, Freiburg
im Breisgau 075848

Antiquitäten am Schwanenwall,
Dortmund 075382
Antiquitäten am See, Rottach-
Egern 078293
Antiquitäten am Thielenplatz,
Hannover 076361
Antiquitäten am Zielemp, Olten . 087724
Antiquitäten am Zielemp, Rickenbach
(Solothurn) 133892
Antiquitäten an der Heiliggeistkirche,
Landshut 077011
Antiquitäten Diele, Wiefelstede . 078964
Antiquitäten-Galerie, Abtwil . . 087180,
133614
Antiquitäten-Galerie, Erfurt 075635
Antiquitäten Hofgut, Mannheim . 077250
Antiquitäten im alten Zollhaus,
Wallerfangen 078852
Antiquitäten im Bachgau,
Großostheim 076110
Antiquitäten im Beethovenhaus, Baden
bei Wien 062530
Antiquitäten im Bennohaus,
Meißen 077332
Antiquitäten im Hagenstal,
Hagenbuch 087521, 133767
Antiquitäten im Hinterhaus,
Lörrach 077141
Antiquitäten im Hof, Krefeld . . 076949
Antiquitäten im Holländerviertel,
Potsdam 078070
Antiquitäten im Kolerschloß, Lauf an
der Pegnitz 077029
Antiquitäten im Wolfgang-Borchert-
Haus, Hamburg 076186
Antiquitäten in Altdorf, Altdorf bei
Nürnberg 074388
Antiquitäten in Herrenhausen,
Hannover 076362
Antiquitäten Kaufhaus, Leipzig . 077054,
077055
Antiquitäten Miklautz, Bad
Kleinkirchheim 062527, 127712
Antiquitäten-Münzen,
Quedlinburg 078106
Antiquitäten Pavillon, Andrea,
Stuttgart 078591
Antiquitäten Pöll, Kufstein 062700
Antiquitäten Spatenmühle,
Wöllstein 079043
Antiquitäten und Kunsttage, Bad
Säckingen 098122
Antiquitäten und Kunsttage,
Regensburg 098120
Antiquitäten und Kunsttage,
Wasserburg 098121
Antiquitäten vom Schneeberg,
Wattendorf 078877
Antiquitäten Vorstadt,
Schwarzenberg 078417
Antiquitäten zum Bühlhof, Iffwil . 087534
Antiquitäten zum Heiligkreuz, Sankt
Gallen 087774
Antiquitäten zum Rosshof, Basel 087217
Antiquitäten zum Schlössli,
Zweisimmen 088024
Antiquitäten zur Zinglermühle,
Viktring 062869
Antiquitäten.Design.Raum,
Stuttgart 098123
Antiquitätengalerie, Potsdam . . . 078071
Antiquitätengalerie Oben, Halle,
Saale 076161
Antiquitätenhalle Schüttorf,
Schüttorf 078401
Antiquitätenhandel, Nußdorf am
Attersee 062751
Antiquitätenhaus, Unterpleichfeld 078765
Antiquitätenwinkel, Hall in Tirol . 062626
Antiquitätenzentrum, München . 077452,
130322
Antiquitás, Budapest . . 079236, 079237
Antiquitás, Györ 079308
Antiquitat, Balatonfüred 079226

Antiquité, Cahors 067591
Antiquite, Dallas 093069
Antiquité, Québec 064518
Antiquite 56, Saint-Ouen 073009
Antiquité AJR, Saint-Maur-des-
Fossés 072960
Antiquité Art Ancien, Pérouges . 072065
Antiquité Art Ancien, Vaujours . 073986
Antiquité Art Contemporain, Nice 070554
Antiquité au Chiabout, Luppy . 069667
Antiquité au Temps Jadis,
Carcassonne 067665
Antiquité Au Temps Qui Passe,
Paris 070903
Antiquité Bonheur D'Autrefois,
Montréal 064370
Antiquité Brancion, Paris 070904
Antiquité Brocante, Coux 068214
Antiquité Brocante A, Reims . . 072376
Antiquité Brocante du Lac, L'Isle-
Jourdain 069545
Antiquité Brocante du Laonnois, Mons-
en-Laonnois 070171
Antiquité Brocante Guibert,
Narbonne 070487
L'Antiquité Classique, Louvain-la-
Neuve 138541
L'Antiquité Curiosité, Montréal . . 064371
Antiquité de Geneston, Pont-Saint-
Martin 072262
Antiquité de la Chapelle,
Plouigneau 072207
Antiquité de la Couronne,
Québec 064519
Antiquité de la Place, Codognan 068056
Antiquité de la Prefecture,
Orléans 070805
Antiquité de la Reine, Bourg-la-
Reine 067414
Antiquité de l'Abbaye, Hambye . 068783
Antiquité de l'Almanarre, Hyères 068849
Antiquité de l'Eglise, Sainte-
Geneviève 073351
Antiquité Décor, Dieppe 068295
Antiquité Décoration, Montville . 070341
Antiquité des Marais, Guérande 068759
Antiquité des Templiers, La Baule-
Escoublac 068945, 125599
Antiquité du Chalet, Caluire-et-
Cuire 067606
Antiquité du Château,
Rambouillet 072362
Antiquité du Hanau,
Philippsbourg 072150
Antiquité du Lac, Saulxures-sur-
Moselotte 073431
Antiquité du Musée, Troyes . . . 073896
Antiquité du Rest, Pleuven 072187
Antiquité du Village Fleuri,
Rémérangles 072410
Antiquité Ehrmann, Metz 070085
Antiquité Fontaine aux Bretons,
Pornic 072279
Antiquité Franco Anglaise, Paris 070905
Antiquité Franco Anglaise, Villiers-le-
Bel 074235
Antiquité la Barbotine, Alençon . 066520
Antiquité La Méridienne, Le
Pellerin 069324
Antiquité la Vigie, Quiberon . . . 072335
Antiquité le Grenier de Saint-Lié,
Charleville-Mézières . 067843, 067844
Antiquité les Deux Chênes,
Coutiches 068212
Antiquité Marie Rue Subes, Dax 068277
Antiquité Saint-Maur, Martel . . 069979
Antiquité Saint-Prix, Saint-Prix . 073270
Antiquité TDC, Castres 067723
Antiquitas, Brockenhurst 088681
Antiques, Aubigny (Falaise) . . . 066789
Antiqués, Bougival 067375
Antiques, Kraków 084547
Antiques 165 vh, Paris 070906
Antiqués 1900–1950, Cannes . 067626

Antiquités 2000, Chalon-sur-
Saône 067771
Antiquités 2000, Juvisy-sur-Orge 068930
Antiquités 29, Brest 067461
Antiquités 34, Nancy 070412
Antiquités 44, Vertou 074074
Antiquités 56, Marolles-en-Brie . 069888
Antiquités 80, Saint-Ouen 073010
Antiquités 87, Limoges 069513
Antiquités 88, Paris 070907
Antiquités 89, Essey-lès-Nancy . 068474
Antiquités 95, Deuil-la-Barre . . . 068290
Antiquités A la Clef d'Or, Genève 133753
Antiquités A&G, Montréal 064372
Antiquités Ada, Montréal 064373
Antiquités Allais, Honfleur 068826
Antiquités and Co, Nantes 128826
Antiquités Antonio, Boisseuilh . 067276,
128475
La Antiquités Art, Strasbourg . . 073566
Antiquités Art et Grenier, Nevers 070530
Antiquités Arts Artisanats, Lons-
Montmorot 069628
Antiquités Arts et Antiques,
Beaulieu 066995
Antiquités au Cadran Solaire,
Mathieu 069994
Antiquités au Palais, Rennes (Ille-et-
Vilaine) 072417
Antiquités au Palais du Grand Yaka,
Bouillac 067384
Antiquités au Saint-Hubert,
Rieulay 072462
Antiquités au Sourire du Passé,
Marboué 069848, 128737
Antiquités Auberchicourt,
Auberchicourt 066783
Antiquités aux Tonneaux, Méounes-lès-
Montrieux 070065
Antiquités B, Biscarrosse 067228
Antiquités B.G., Rennes (Ille-et-
Vilaine) 072418
Antiquités Bagatelle, Thonon-les-
Bains 073664
Antiquités Blancoises, Le Blanc . 069218
Antiquités Broc Troc, Narbonne . 070488
Antiquités Brocante & Lustincone,
Aléria 066525
Antiquités Brocante au Fond
du Chemin, Notre-Dame-de-
Sanilhac 070750
Antiquités Brocante Barbéris,
Nice 070555
Antiquités-Brocante des Vertus, Saint-
Aubin-sur-Scie 072680
Antiquités Brocante du Murier,
Angoulême 066625
Antiquités Brocante Riff,
Rohrwiller 129209
Antiquités Brocantes, Valbonne . 073932
Antiquités Brocantes du Faubourg,
Troyes 073897
Antiquités Brocantes le Phare,
Sète 073514
Antiquités Brocant'Stock,
Bergerac 067097
Antiquités Broquantes, Puttelange-aux-
Lacs 072327
Antiquités Caroline, Montréal . . 064374
Antiquités Claude, Puteaux . . . 072326
Antiquités Colombette, Toulouse 073734
Antiquités Contemporain, Binic . 067224
Antiquités Croix Rousse, Lyon . 069687
Antiquités d'Alet, Alet-les-Bains 066533
Antiquités d'Assise, Québec . . . 064520
Antiquités de Bailly, Bailly 066919
Antiquités de Beure, Beure 067152
Antiquités de France,
Hommarting 068824
Antiques de Grand-Papa,
Vetheuil 074079
Antiquités de la Baie, Perros-
Guirec 072100
Antiquités de la Bonne Etoile,

Confolens 068134
Antiquités de la Carderie, Saint-Martin-
Labouval 072953
Antiquités de la Cathédrale,
Auxerre 066842
Antiquités de la Cave au Grenier,
Moings 070137
Antiquités de la Choisille, Saint-Cyr-
sur-Loire 072726
Antiquités de la Cité, Troyes . . . 073898
Antiquités de la Dronne, Aubeterre-sur-
Dronne 066786
Antiquités de la Garde, La Garde 069043
Antiquités de la Gare,
Beaurepaire 067052
Antiquités de la Landriais, Dinard 068353
Antiquités de la Madeleine, Verneuil-
sur-Avre 074027
Antiquités de la Maison Carrée,
Nîmes 070684
Antiquités de la Place, Épinal . . 068452
Antiquités de la Poste, Cavaillon 067734
Antiquités de la Rabine, Landaul 069162
Antiquités de la Renaissance,
Hem 068805
Antiquités de la Sucrerie,
Chavenay 067947
Antiquités de l'Ancienne Poste, Sainte-
Geneviève-des-Bois 073353
Antiquités de l'Angoumois, Gond-
Pontouvre 068676
Antiquités-de-l'Autie, Saint-Ouen 073011
Antiquités de l'Eglise, Etréchy . 068486
Antiquités de l'Estel, Avignon . 066870
Antiquités de l'Ile Saint-Louis,
Paris 070908
Antiquités de l'Ill, Strasbourg . . 073567
Antiquités de l'Océane, Carnac . 067685
Antiquités de Locquemeau, Tredrez-
Lacquemeau 073873
Antiquités de Mauperthuis,
Mauperthuis 070006
Antiquités de Meaux, Meaux . . 070026
Antiquités de Printemps, Troyes 098069
Antiquités de Saint-Michel en Grève,
Saint-Michel-en-Grève 072980
Antiquités de Sainte-Agathe,
Longuyon 069622
Antiquités de Thueyts, Thueyts . 073677
Antiquités de Vaux, Cesson-
Sévigné 067754
Antiquités de Verneuil, Verneuil-sur-
Seine 074033
Antiquités de Volubilis, Moulay-Idriss
Zerhoun 031785
Antiquités Déco, Binic 067225
Antiquités Décor et Confort de France,
Montréal 064375
Antiquités Décoration, Sausset-les-
Pins 073440
Antiquités des Bateliers, Les Rosiers-
sur-Loire 069422
Antiquités des Bateliers, Saumur 073432
Antiquités des Bords du Cher, Saint-
Aignan (Loir-et-Cher) 072660
Antiquités des Carmes, Avignon 066871
Antiquités des Cordeliers, Dinan 068347
Antiquités des Corps Saints,
Avignon 066872
Antiquités des Emmurées, Rouen 072559
Antiquités des Escoulettes,
Rocbaron 072492
Antiquités des Halles, Airvault . 066448
Antiquités des Mercoeurs, Murat
(Cantal) 070400
Antiquités des Mesnuls, Les
Mesnuls 069416
Antiquités des Meulières, La Ferté-
sous-Jouarre 069032
Antiquités des Plantades, La
Garde 069044
Antiquités des Provinces de France,
Pézenas 072126
Antiquités des Quatre Fontaines,

Narbonne 070489
Antiquités des Rochettes, Nantes 070441
Antiquités des Souvenets, Le
Lude 069298
Antiquités des Ternes, Paris ... 070909
Antiquités du 12, Paris 070910
Antiquités du Barry, Vaujours .. 073987
Antiquités du Bourg, Martigny .. 087659
Antiquités du Canal, Troyes ... 073899
Antiquités du Château, Aubenas 066777
Les Antiquités du Château, Cagnes-
sur-Mer 067587
Antiquités du Château, Le Château-
d'Oleron 069247
Antiquités du Cloître, Jonzac .. 068909
Antiquités du Grand Bras, Saint-
Georges-sur-Loire 072802
Antiquités du Lavoir, Labosse .. 069129
Antiquités du Manoir de Bellanton,
Thouare-sur-Loire 073673
Antiquités du Marais, Challans . 067765
Antiquités du Marché, Levallois-
Perret 069443
Antiquités du Mas Chatelan,
Mallemort 069829
Antiquités du Matelot, Québec . 064521
Antiquités du Meneyer, Bénodet 067091
Antiquités du Moulin, Clamart .. 068000
Antiquités du Moulin, Saint-
Gelais 072791, 129230
Antiquités du Palais, Albi 066506
Antiquités du Palais, Niort 070709
Antiquités du Parc de l'Europe, Saint-
Étienne 072750
Antiquités du Parc de l'Europe,
Villeurbanne 074184
Antiquités du Pont, Bruxelles .. 063317
Antiquités du Pont, Monnières .. 070166
Antiquités du Pont des Arches,
Liège 063741
Antiquités du Pont Neuf, Paris . 070911
Antiquités du Port, Antibes-Juan-les-
Pins 066663
Antiquités du Poulbenn, Crach . 068217
Antiquités du Roy, Pacé 070833
Antiquités du Théatre, Noyon .. 070760
Antiquités du Théatre, Valence
(Drôme) 073937
Antiquités du Val de Loir, Bazouges-
sur-le-Loir 066990
Antiquités du Vieux Bourg,
Villersexel 074181
Antiquités du XXe Siècle, Genève 087444
Antiquités en Top, Montréal ... 064376
Antiquités et Brocante Glainoise, La
Chapelle-Glain 068975
Antiquités et Curiosités, Monaco 082496
Antiquités et Puces, Annecy ... 066635
Antiquités Félibien, Nantes 070442
Antiquités Ferme Sainte-Marie,
Doncourt-lès-Conflans 068382
Antiquités Finkwiller, Strasbourg 073568
Antiquités Galerie d'Art,
Guerlesquin 068764
Antiquités Le Galion, Paris 070912
Antiquités Guy, Montréal 064377, 101354
Antiquités H. Füllemann,
Lausanne 087560
Antiquités Hors Château, Liège . 063742
Antiquités Jasper, Aubenas ... 066778
Antiquités Jean Boyvin, Dole ... 068365
Antiquités Jean-Claude, Montréal 064378
Antiquités Jean Macé, Nantes .. 070443
Antiquités JPS, Montréal 064379
Antiquités la Bigarderie, Blaye . 067241
Antiquités la Chine du Port,
Capbreton 067658
Antiquités la Comtoise,
Montauroux 070205
Antiquités la Comtoise Fleurie,
Narbonne 070490
Antiquités la Farandole, Allauch 066536
Antiquités la Grange, Châteauneuf-Val-
Saint-Donat 067897

Antiquités la Grange, Chinon ... 067976
Antiquités la Grange, Gourdon .. 068688
Antiquités la Grange, La Vallée-au-
Blé 069115
Antiquités La Maladière, Bourg-de-
Péage 067409
Antiquités La Musardière,
Saintes 073372
Antiquités la Pinsonnière,
Chaville 067948
Antiquités la Vieuserie, Le
Quesnel 069348
Antiquités Le Cerisiers, Quillan . 072337
Antiquités le Damiette, Rouen .. 072560
Antiques le Grenier Olarguais,
Olargues 070776
Antiquités le Guillou, Brest 067462
Antiquités le Louis XV,
Magagnosc 069799
Antiquités le Marchand d'Oublis,
Bondues 067282
Antiquités le Padoue, Molesmes 070147
Antiquités Le Plessis, Saint-
Urbain 073323
Antiquités le Presbytère, Limoges 069514
Antiquités le Presbytère, Rilhac-
Lastours 072466
Antiquités le Prieuré, Varades .. 073976
Antiquités le Roi Soleil, Sanary-sur-
Mer 073396
Antiquités le Vertugadin, Villefranche-
sur-Mer 074148
Antiquités le Village,
Wambrechies 074287
Antiques le Voltaire, Narbonne . 070491
Antiquités l'Ecritoire, Nantes ... 070444
Antiquités les 3 Fontaines, Lancon-
Provence 069159
Antiquités les Cinq, Thonon-les-
Bains 073665
Antiquités les Puces Angevines,
Angers 066599
Antiquités les Remparts, Vannes
(Morbihan) 073965
Antiquités les Vieilleries, Vannes
(Morbihan) 073966
Antiquités Longchamp, Paris ... 070913
Antiquités Lormaye, Lormaye .. 069641
Antiquités Lyonnaises, Vaulx-en-
Velin 073988
Antiquités Magasin, Montréal .. 064380
Antiquités Marilyn, Lyon 069688
Antiquités Martine, Pons 072233
Antiquités Milleroses, Pully 087742
Antiquités Milord, Montréal 064381
Antiquités Montcalm, Paris 070914
Antiquités Montréal, Montréal .. 064382
Antiquités Montréal Universels,
Montréal 064383
Antiquités Nazareth, Toulouse .. 073735
Antiquités Parcheminerie, Nevers 070531
Antiquités Passion, Cognac 068057
Antiquités Passion, Javrezac ... 068901
Antiquités Plestin, Plestin-les-
Grèves 072181
Antiquités pour la Table,
Montréal 064384
Antiquités Privilège, Québec ... 064522
Antiquités Renaissance, Montréal 064385
Antiquités Rhône Alpes,
Villeurbanne 074185
Antiquités Rond Point Elysées 7,
Paris 070915
Antiquités Saint-Antoine,
Lucciana 069654
Antiquités Saint-Christophe, Le
Merlerault 069315
Antiquités Saint-Eloi, Clermont-
Ferrand 068010
Antiquités Saint-Honoré, Arnay-le-
Duc 066728
Antiquités Saint-Louis, Versailles 074042
Antiquités Saint-Maclou, Pontoise 072272
Antiquités Saint-Nicolas,

Lunéville 069666
Antiquités Saint-Spire, Corbeil-
Essonnes 068142
Antiquités Sainte-Anne, Paris .. 070916
Antiquités Salon de Thé,
Bordeaux 067304
Antiquités Scientifiques, Genève 087445
Antiquités Seconde Chance,
Montréal 064386
Antiquités Ségal, Basel 087218, 126705
Antiquités sous Lu Barri, Saint-Paul-
de-Vence 073241
Antiquités Transformation Ebénisterie,
Baudreville 066966
Antiquites Trésor d'Ivoire,
Montréal 064387
Antiquités Vence, Vence 074001
Antiquités Viaud, Château-Arnoux-
Saint-Auban 067867
Antiquités Voltaire, Nantes 070445
Antiquities, Auckland 083759
Antiquities, Wolverhampton ... 091902
Antiquities Dealers' Association,
London 060028
Antiquities International, Las
Vegas 094118
Antiquities Kerikeri, Kerikeri ... 083974
Antiquities Museum of Tel Aviv-Yafo,
Eretz Israel Museum, Tel Aviv . 025186
Antiquities Museum, Bibliotheca
Alexandrina, Alexandria 010514
Antiquities Museum, University of
Queensland, Saint Lucia 001475
Antiquities of Nevada, Las Vegas 094119
The Antiquities Shoppe, London . 064345
Antiquitum, Québec 064523
Antiquity, Cambridge 139127
Antiquity Furniture, Indianapolis 135625
Antiquity Hands of the Hills,
Singapore 085364
Antiquity Shop, Nairobi 082247
Antiquity Workshop, Austin 135319
Antiquorum, Roma 081223
Antiquorum Auctioneers, Hong
Kong 125402
Antiquorum L. A., Beverly Hills . 127137
Antiquorum Milan, Milano 126201
Antiquorum Moscow, Moskva .. 126473
Antiquorum Munich, München ... 126040
Antiquorum Paris, Paris 125673
Antiquorum SA, Genève 126721
Antiquorum Shanghai, Shanghai 125407
Antiquorum Tokyo, Tokyo 126244
Antiquorum USA, New York 127341
Antiqués du XXe siècle,
Lausanne 087561
Antiquus, Christchurch 083853
Antiquus, London 089915
Antiquus, Olsztyn 084632, 143492
Antiquus, Vauhallan 073985
Antiquus, Venezia 081821, 081822,
081823
Anti's, Milano 080407
Antis Jonáš, Chrast u Chrudimě 065378
Antix Brocante, Clermont-Ferrand 068011
Antix e Artes, Rio de Janeiro .. 064017
Antix-Stadl, Dornbirn 062545
Antler River Historical Society
Museum, Melita 006288
Antoine, Claude, Liège 100778
Antoine, Michel, Bruxelles 063318
Antoinette, San Diego 097063
Antoinette Westphal College of Media
Arts & Design, Drexel University,
Philadelphia 057664
Antoinette's, Salt Lake City ... 096877
Antologia di Belle Arti, Torino .. 138790
Antom, Stockinish 135086
Anton, Paris 070917
Anton Bruckner-Museum,
Oberösterreichische Landesmuseen,
Ansfelden 001717
Anton-Buchner-Heimatmuseum,
Betzenstein 017348

Anton Hanak-Museum,
Langenzersdorf 002267
Anton Hiersemann KG, Stuttgart 137514
Anton Pieck Museum, Hattem .. 032419
Anton van Wouw House, University of
Pretoria, Brooklyn 038668
Anton Wildgans-Haus, Mödling . 002398
Anton, Henri, Montreuil-sur-Mer 125649
Antona Austriņa Memoriālais Muzejs,
Vecpiebalga 030249
Antonacci, Francesca, Roma .. 081224
Antonacci, Paolo, Roma 081225
Antonelli Institute of Art and
Photography, Cincinnati 056895
Antonelli, Felice, Genova 080233
Antonelli & C., Franco, Torino .. 132089
Antonelli, Massimo, Roma ... 081226
Helga Antoni & Karl-Heinz Hüter,
Zellertal 079136
Antoni, Philippe, Mézy-sur-Seine 070118
Antoni, Pietro, Bologna 130932
Antonia's Antiques, Washington . 097804
Antonietti, François, Nantes ... 125654
Antonija, Rīga 111752
Antonijev Rov – Rudnik Živega Srebra,
Idrija 038515
Antonini, Antoinette, Monaco ... 082497
Antonini, Maria Maddalena,
Milano 080408
Antônio Antiguidades, São Paulo 064149
Antonio Rinaldini & Figlie, Napoli 110517
Antonio's Antique, San Francisco 097173
Antonio's Antiques, San
Francisco 097174, 136307
Antonopouleion Archaeological
Museum, Pylos 023237
Antonopoulou, Ang., Athinai ... 109019
Antonovich, François, Paris 070918
Antons & Co, Kim, København . 065844
Antonucci, Cosima, Milano 080409
Antony & Woodgate, Brighton,
Victoria 061198
Antony, Daniel, Fleurance 068544
Antony, Henrietta, Montréal ... 064388
Antony's, Oberstaufen . 077873, 126088
Antsla Gümnaasiumi Muuseum,
Antsla 010604
Antunes, Lisboa 143556
Antunes, Angela, Belo Horizonte 063878
Antunes, Manuel, Lisboa 084826
Antwerp Antiques Markt,
Antwerpen 063190
Anty-Raj, Warszawa 084693
Antyk, Gdańsk 084499, 132813, 143446
Antyk, Toruń 084680
Antyk & Art, Kraków 084548
Antyk Antykwariat, Szczecin .. 084672,
143505
Antyk Bielski Salon Sztuki, Bielsko-
Biała 084489
Antyk Bukowski, Szczecin 084673
Antyk Lombard, Kraków 084549
Antyk Salon Sztuki Dawnej,
Lublin 084622
Antyki, Białystok 084486
Antyki, Dąbrowa Górnicza 084498
Antyki, Gdańsk 084500
Antyki, Katowice 084529
Antyki, Kęty 084540, 132826
Antyki, Kobyłka 132827
Antyki, Kraków 084550, 084551, 132831
Antyki, Lublin 084623, 084624
Antyki, Sopot 084661
Antyki, Warszawa 084694, 084695
Antyki – Styl, Katowice 084530
Antyki Ars Silesiae, Wrocław ... 084761,
143537
Antyki Biżuteria Salon Sztuki,
Kraków 084552
Antyki i Dzieła Sztuki Grażyna Niezgoda,
Przemyśl ... 084658, 113755, 143501
Antyki Meda, Bielsko-Biała 084490
Antyki-Militaria, Kraków 084553
Antyki przy Farnej, Bydgoszcz .. 084494

Antyki Salon Sztuki Dawnej,
Lublin 084625
Antykwariat, Bielsko-Biała 143429
Antykwariat, Bydgoszcz 143434
Antykwariat, Jelenia Góra 084526,
143457
Antykwariat, Kalisz 084527
Antykwariat, Kołobrzeg 143463
Antykwariat, Kraków 143464
Antykwariat, Łódź . . . 084607, 143486
Antykwariat, Lublin 143490
Antykwariat, Warszawa 084696, 143513
Antykwariat, Wrocław . 084762, 143538
Antykwariat AB – Galeria Hetmanska,
Kraków 084554
Antykwariat Argos, Dąbrowa
Górnicza 143445
Antykwariat Atest, Kraków 084555,
143465
Antykwariat Barok, Poznań 084636,
132857
Antykwariat Beskidzki, Bielsko-
Biała 143430
Antykwariat Bibelocik, Kraków . 084556,
084557
Antykwariat Czas, Szczecin . . . 084674,
143506
Antykwariat DAES, Wrocław . . . 084763
Antykwariat Fraszka,
Częstochowa 143440
Antykwariat Gdański Salon Sztuki
Dawnej, Gdańsk . . . 084501, 143447
Antykwariat Górski Filar, Kielce . 143461
Antykwariat Julia, Bydgoszcz . . 143435
Antykwariat Kama, Bielsko-Biała 143431
Antykwariat Kolekcjoner, Toruń . 084681,
143509
Antykwariat Księgarski, Kraków 143466
Antykwariat Księgarski Oficyna,
Katowice 143458
Antykwariat Lamus, Częstochowa 143444
Antykwariat Meblowy, Sopot . . . 084662
Antykwariat Meda, Bielsko-Biała 143432
Antykwariat Naukowy, Łódź . . . 143487
Antykwariat Naukowy, Wrocław . 143539
Antykwariat Naukowy im. A.
Krawczyńskiego, Gdańsk 143448
Antykwariat Naukowy im. J.K.
Żupańskiego, Wrocław 143540
Antykwariat Naukowy im.
Żupańskiego, Poznań 143493
Antykwariat Naukowy na Długim
Pobrzeżu, Gdańsk 143449
Antykwariat Naukowy Wojtowicz,
Kraków 143467
Antykwariat pod Szermierzem,
Wrocław 143541
Antykwariat Pomorski, Szczecin 143507
Antykwariat Przy Ratuszu,
Poznań 143494
Antykwariat Rara Avis, Kraków . 143468
Antykwariat Retro, Bydgoszcz . 143436
Antykwariat Skrypt, Kraków . . . 143469
Antykwariat Staromiejski,
Warszawa 143514
Antykwariat Starożytności, Kraków 084558,
143470
Antykwariat Szkolny, Warszawa 143515
Antykwariat Tezaurus, Katowice 143459
Antykwariat Verbum, Kraków . . . 143471
Antykwariat Wolumen,
Częstochowa 143442
Antykwariat Wroc#lawski,
Wrocław 084764
Antykwariat Współczesny,
Częstochowa 143443
Antykwarnia Księgarnia Niezależna,
Częstochowa 143444
Antykwaryczny, Częstochowa . . 084497
ANU Antiques and More,
Albuquerque 092033
ANU School of Art Gallery, Acton 000755
ANU School of Art, Australian National
University, Canberra 054900

Anubis Ancient Art, Rotterdam . 083550
Anubis Antikvariat, Bergen 143380
l'Anverre-Atelier, Antwerpen . . . 100482
Anvil Antiques, Leek 089775
Anvill Antiques, Hartley Wintney 089449,
134440
Das Anwesen and Pharmacy Museum,
New Braunfels 051232
Anxi Museum, Anxi, Fujian . . . 007570
Anxi Museum, Anxi, Gansu 007571
Any Amount of Books, London . 144553
Anyang Glaswegianjian Art Museum,
Anyang 007572
Anyang Museum, Anyang 007573
Anyang Yinxu Bowu Yuan,
Anyang 007574
Anything Goes, London 089916
Anything Goes, Tucson 097633
Anything Goes On Bay, Toronto . 064599
Anyuan Museum, Anyuan 097575
Anyž, Komárov 065408
ANZ Banking Museum,
Melbourne 001264
Anza-Borrego Desert Park Museum,
Borrego Springs 047098
Anzac Cottage, Mount Hawthorn 001319
ANZAC Memorial, Sydney 001534
Anzali Military Museum, Bandar-
Anzali 024687
Anzenaumühle, Bad Goisern . . . 001742
Anzidei, Roberto, Roma 131809
Anzilotti & Martelli, Roma 131811
Anzilotti, David & Stefano, Roma 081227,
131810
AOC, Lyon 069689
Aohan Qi Museum, Aohan Qi . . 007576
Aoiya, Kyoto 082004
Aoki, Fukuoka 081963
Aoki, Yoshiichiro & Yoko,
Fujinomiya 111245
Aomori Chuo Gakuin Daigaku,
Tokyo 056036
Aomori-kenritsu Kyodokan,
Aomori 028475
Aoraki Polytechnic Art and Design
Department, Timaru 056197
Aouizerats, Alain, Job 068902
Aoyama Yanagawa, Tokyo 082134
AP-Art, Knokke-Heist 100721
AP4-Art, Lugano 116452
Apa, Nagoya 111311
Apa Tani, Paris 070919, 128897
APAA Newsletter, Summerville . 139267
Apache Coin, Minneapolis 094897
Aparatus Som e Imagem, Rio de
Janeiro 101008
Apart, London 118169
Aparté, Pézenas 141304
Apartheid Museum,
Johannesburg 038742
The Apartment Gallery, Athinai . 105020
Apax Libri, Firenze 142662
APD Singapore, Singapore 137935
Apeiron, Palermo 110587, 131682
Apellaniz, Vitoria-Gasteiz 086568, 115540
Apellaniz Saez de Ibarra, Juan Antonio,
Vitoria-Gasteiz 086569
Apelle, Paris 128898
Apelstein, Katherine, Saint-Ouen 073012
Aperture, Denville 139268
Aperture Foundation, New York . 138375
Apestaart, de, Enschede 083246
Apesteguy, Anne, Deauville 103657
APEX Gallery, Washington 124859
Apex Kunstverein Pro Art e.V.,
Göttingen 058711, 107145
Apex Museum, Atlanta 046603
apexart, New York 051333
Apexim '93, Bucureşti 114123
Aphaia Verlag, Berlin . 137049, 141479
Aphold, Roland, Allschwil 116015
Aphold, Roland, Weil am Rhein . 100854
Aphrodisias Müzesi, Aydın 042837
Apiary, Denver 093254

Apice Art, São Paulo 064150
Apis, Pézenas 072127
Apitz, Joachim, Babenhausen, Kreis
Unterallgäu 074483
Apitz, Manfred, Hamburg 076187
Apocalypse Boutique, Denver . . 120689
Apogee, Seattle 097404
Apolinario Mabini Birthplace Museum,
Tanauan 034606
Apolinario Mabini Shrine and Museum,
Manila 034520
Apollo, Bucureşti 114124
Apollo, Helsinki 125475
Apollo, Indianapolis 121163
Apollo Antik och Konsthandel,
Stockholm 086969
Apollo Antique Galleries,
Westerham 091796
Apollo Antiques, Warwick 091740
Apollo Antiques, Würzburg 079075
Apollo Apparao Galleries,
Mumbai 109413
Apollo Art Books, Den Haag . . . 143107
Apollo Gallery, Dublin . 109571, 109572
Apollo Gallery, Taipei 116902
Apollo Raamatuaja, Tallinn 140679
Apollo, The International Magazine for
Collectors, London . . 138185, 139128
Apollon Galeria, Bielsko-Biała . . 113428
Apollon Galerie, Frei-
Laubersheim 107050
Apolloni, W., Roma 081228
Apollo's Art, Doetinchem 083167
Apolonia, Denver 093255
APOM Informações, Actas dos
cólóquios, Lisboa 138997
Apostles Island National Lakeshore
Maritime Museum, Bayfield . . 046837
Apostolos, Lemesos 102544
l'Apostrophe, Montréal 064389
The Apothecary, Chicago 120037
Apotheken-Museum, Bad
Münstereifel 016826
Apotheken-Museum, Schiltach . 021653
Apothekenmuseum, Hofgeismar 019290
Apothicairerie, Pont-de-Veyle . . 014935
Apothicairerie de l'Hôpital, Moutiers-
Saint-Jean 014368
Apothicairerie de l'Hôtel Dieu-le-
Comte, Troyes 016047
Apothicairerie, Centre Hospitalier de
Gonesse, Gonesse 013127
Appalachian Center for Crafts,
Smithville 053370
Appalachian Spring, Washington 124860
Appaloosa Horse Club of Canada
Museum, Claresholm 005622
The Appaloosa Museum and Heritage
Center, Moscow 051074
Appap, Paris 070920
Apparences, Villeneuve-lès-
Avignon 105871
L'Appart Art Gallery, Les Sables-
d'Olonne 103986
Appay, Gairoard & Besch,
Cannes 125544
Appel, Frankfurt am Main 106968
Appel d'Air, L'Isle-sur-la-Sorgue 069551
Appel, Dr. Sabine, Berlin 141480
Appel, Eric, New York 095232
Appel, Hélène, Paris 104474
Appel, Vera-M. & Anton, Weßling 142482
Appelbaum & Himmelstein, New
York 135882
Appelfeld, New York 145320
Appenzeller Brauchtumsmuseum,
Urnäsch 042087
Appenzeller Schaukäserei, Stein 042030
Appenzeller Volkskunde-Museum,
Stein 042031
Appia Hérbert, Grenoble 103782
Appiani Arte 32, Milano 110238
Apple Barrel, Honolulu 120921
Apple Barrel Antique Mall, Oklahoma

City 095943
Apple Blossom Antique Mall,
Oklahoma City 095944
Apple Card Studios, Kingston-upon-
Hull 138173
Apple Gallery, Godalming 117866
Apple Market Stalls, London . . . 089917
Apple Orchard Antique Mall, Oklahoma
City 095945
Apple Press, London 138186
Apple Tree Gallery, Saundersfoot 119063
The Apple Trees Museum,
Burlington 047315
Apple, Annie, Tampa 097542
The Appledore Gallery, Appledore,
Devon 117242
AppleGum Antiques, Bega 061119
Apples & Pears Gallery, Tarleton 119206
Appleton Art Center, Appleton . . 046511
Appleton Galleries, Vancouver . . 101734,
125388
Appleton Museum of Art, Ocala . 051693
Appletree Press, Belfast 138138
Applewood, Charlotte 119973
Applewood – Shaver Homestead,
Etobicoke 005794
Applewood Gallery, Swords 109697
Applewood Studio Gallery, Bega 098724
Appleyard, Roger, Rotherham . . 091143
Appliance Plus, Tucson 097634
Applicat-Prazan, Paris 104475
Applied Art College, Rīga 056108
Applied Arts Magazine, North
York 138568
Applin, Jess, Cambridge 088758
Appolo Subastas, Bilbao 126517
Appomattox Court House,
Appomattox 046515
Appraisers Association of America,
New York 060325
Appraising Association,
Springfield 060326
L'Approdo, Sestri Levante 110929
Apricot Gallery, Ha Noi 125056
Apricot Gallery, Ho Chi Minh City 125109
Apriķu Novada Muzejs, Apriķi . . 030100
Aprilborn, Saint Paul 096810
April's Gallery, New York 122321
Apriori Antiques, San Francisco . 097175
Apropos Antiques, Richmond . . . 096525
Apropos Galleries, Saint Paul . . 123890
Apsara, Long Beach 121392
Apsara, Singapore 085365
Apsheron Art Gallery, Bakı 003149
Apskaitos Muziejus, Vilnius 030509
Apsley, Den Haag 083053
Apsley, Miami 094698
Apsley House, London 044904
APT Galerie d'Art, Houston 120992
Apta, Paris 070921
Apteekkimuseo, Kouvola 011034
Apteekkimuseo ja Qwenselin Talo,
Turun Maakuntamuseo, Turku . 011433
Apter-Fredericks, London 089918
Aptos Cruz, Stirling 099720
Aptucxet Trading Post Museum,
Bourne 047149
AQ-Verlag, Saarbrücken 137494
Adnan Al-Aqel & Khaled Al-Mallah,
Damascus 088027
Aqua, Koshigaya 082003
Aqua 20th Century Modern,
Austin 092211
Aqua Art Miami, Miami Beach . . 098421
Aqua Blues, Le Lavandou 103953
Aqua-Blues, Les Sables-d'Olonne 069425
Aqua-Design, Aichach 141401
Aquadrat Galerie, Kaiserslautern 107491
Aquafortis-ADC, Firenze 080081, 142663
Aqualeon Parc de la Natura,
Albinyana 038911
L'Aquamanile, Lyon 069690
Aquamarin Verlag, Grafing bei
München 137234

Aquarell Verlag, Lauda-Königshofen ... 137336
Aquarella, Les Trois-Ilets ... 111929
Aquarelle, Kortrijk ... 063716
Aquarelle Gallery, London ... 118170
Aquarelles de Loire, Angers ... 103234
Aquaria Vattenmuseum, Stockholm ... 040989
The Aquarium, London ... 118171
Aquarium-Musée de la Mer, Laboratoire Maritime de Dinard, Dinard ... 012796
Aquarium-Musée-Institut Paul Ricard, Fondation Océanographique, Six-Fours-les-Plages ... 015819
Aquarium-Muséum, Université de Liège, Liège ... 003790
Aquarius, Dubai ... 117177
Aquarius Antiques, Leeds ... 089767
Aquarius Rare Coins, Miami ... 094699
Aquarius Wassermuseum, Mülheim an der Ruhr ... 020491
Aquatic Hall of Fame and Museum of Canada, Winnipeg ... 007436
Aquatics Museum, Institute of Oceanography and Fisheries, Suez ... 010585
Aquazoo-Löbbecke Museum, Düsseldorf ... 018024
Aquila Books, Calgary ... 140333
Aquilanti, Buenos Aires ... 139550
Aquiles Adler, Richmond ... 123717
Aquincumi Múzeum, Budapesti Történeti Múzeum, Budapest ... 023456
AR-Group, Tbilisi ... 058698
Ar Stal, Brest ... 067463
Ar Stal, Concarneau ... 068117
Ar Stal, Pontivy ... 072270
Ar Stal, Quimper ... 072340
Ara, Lisboa ... 113980
ARA, Rekken ... 132612
Ara Sargsyan and Hakob Kojoyan House-Museum, Yerevan ... 000721
Arab International Services, Cairo 066227
Araba Fenice, Cagliari ... 079973
Arabesque, Beirut ... 082271
Arabesque, Port Aven ... 105392
Arabesque Galerie, Cairo ... 103016
Al-Arabesque Gallery, Manama ... 063132, 100412
Arabesques, Castres ... 067724
Arabia Museum-Gallery, Helsinki 010812
Arabian Trade House, Riyadh ... 085313
Arad Museum, Arad ... 025017
Arader, W. Graham, Houston ... 093614, 145161
Arader, W. Graham, New York ... 095233, 095234, 145321
Arader, W. Graham, Philadelphia 096079, 145406
Arader, W. Graham, San Francisco ... 097176, 145539
Aragão, Antónia, Lisboa ... 084827
Araghi, Genova ... 110086
Aragon, Krupina ... 085572
Aragon, Zvolen ... 085599
Aragon, Zvolenská Slatina ... 085600
Aragon Antiqués, Paris ... 070922
Aragon Museum, Ilupeju ... 033428
Aragón, López de, Madrid ... 086120
Aragon, Modeste, La Coucourde 069007
Aragon, Modeste, Savasse ... 073446
Arai Kinen Bijutsukan, Iwanai ... 028724
Arai-shiritsu Hakubutsukan, Arai 028477
Al-Araimi Gallery, Al-Qurum ... 113295
Āraišu Muzejparks, Ezerpils, Drabeši ... 030116
Araki Shuseikan Museum, Nagoya ... 029109
Aral, A Coruña ... 133232
Araldi, Albin, Chaumont (Haute-Marne) ... 067931
Araluen Centre, Alice Springs ... 098628
Aram, Den Haag ... 083054

Aram Khachaturyan Museum, Yerevan ... 000722
ARAM Kunstmagazin, Wien ... 138501
Arama, Maurice, Paris ... 070923
Arame Art Gallery, Jerewan ... 098596
Arami Galería, Bucaramanga ... 102375, 128281
Aran, Clonmel ... 079485
Arana, Getxo ... 086004
Aranaz Darras, Donostia-San Sebastián ... 085967
Arango, Guillermo, Medellín ... 065277
Arango, Maria, Las Vegas ... 121332
Aranha, Silvia de Souza, São Paulo ... 064151
Arany Egyszarvú Patikamúzeum, Kőszeg ... 023761
Arany János Irodalmi Emlékház, Geszt ... 023643
Arany János Múzeum, Nagykőrös 023806
Arany Sas Patika – Könyvtár és Levéltár, Semmelweis Orvostörténeti Múzeum, Budapest ... 023457
Arar, Florian, Bordeaux ... 067305
Ararat Regional Art Gallery, Ararat ... 000799
Ararat Trading Post, Ararat ... 061003
Arario Beijing, Beijing ... 101929
Arario Gallery, Cheonan ... 111578
Arario New York, New York ... 122322
Arario Seoul, Seoul ... 111610
Araripe, Nilda, Rio de Janeiro ... 101009
Aras, Ravensburg ... 108439, 137478
Aras, Patrick, Lamorlaye ... 069151
Arash, New York ... 122323
Arasi, Chennai ... 109273
Arati, Kathmandu ... 112061
Aratoi Wairarapa Museum of Art and History, Masterton ... 033240
Arauco, Nürnberg ... 108290
Araujo, Evly da Rosa, Porto Alegre ... 063955
Arauna, Oliva, Madrid ... 115141
Aravete Külamuuseum, Aravete . 010605
Arax Orientmattor, Stockholm .. 086970
Araya Vera, Sergio, Santiago de Chile ... 128245
Arbæjarsafn, Reykjavík Museum, Reykjavík ... 024092
Arbalet, Minsk ... 063168
Arbatskaja Iverija, Moskva ... 114196
Arbatskaja Nachodka, Moskva ... 085170, 143633
Arbaud, Aix-en-Provence ... 066452
Arbeia Roman Fort and Museum, South Shields ... 045819
Arbeidsgenot, Oudtshoorn ... 038791
Arbeiterfotografie Galerie, Köln . 107584
Arbeitsgemeinschaft "Historische Nahverkehrsmittel Leipzig" e.V., Leipzig ... 019975
Arbeitsgemeinschaft bildender Künstler am Mittelrhein e.V., Koblenz . 058712
Arbeitsgemeinschaft der Kunstbibliotheken, Berlin ... 058713
Arbeitsgemeinschaft der Museen in Bayern, Freising ... 058714
Arbeitsgemeinschaft des Kunsthandwerks Nordrhein-Westfalen, Recklinghausen ... 058715
Arbeitsgemeinschaft Deutscher Kunstvereine, Berlin ... 058716
Arbeitsgemeinschaft Diessener Kunst, Dießen am Ammersee ... 058717
Arbeitsgemeinschaft Friedhof und Denkmal, Stiftung Zentralinstitut und Museum für Sepulkralkultur, Kassel ... 058718
Arbeitsgemeinschaft kirchlicher Museen und Schatzkammern, Würzburg ... 058719
Arbeitsgemeinschaft Mittelfränkischer Münzensammler e.V., Dinkelsbühl ... 058720

Arbeitsgemeinschaft Schweizer Keramiker, Association Céramique Suisse, Sursee ... 059920
Arbeitsgemeinschaft Selbständige Industrie-Designer e.V., Beendorf ... 058721
Arbeitsgruppe für Theoretische und Angewandte Museologie – exponat, Wien ... 136573
Arbeitskreis 68, Künstlergemeinschaft Wasserburg am Inn, Wasserburg am Inn ... 058722
Arbeitskreis Bild- und Kunstbuchverlage im Börsenverein des Deutschen Buchhandels e.V., Frankfurt am Main ... 058723
Arbeitskreis Deutscher Kunsthandelsverbände, Frankfurt am Main ... 058724
Arbeitskreis Moderne Graphik, Frankenthal, Pfalz ... 058725
Arbeitskreis selbständiger Kulturinstitute, Bonn ... 058726
Arbejdermuseet, København ... 010166
Arbeo, Barcelona ... 085726
The Arbery Centre, Wantage ... 091724
Arbets Museum, Norrköping ... 040875
Arbil Museum, Arbil ... 024797
Arbin, Lars, Stockholm ... 086971
Arbiter, Wallasey ... 091708
Arboga Bryggeriumseum, Arboga 040532
Arboga Museum, Arboga ... 040533
D'Arbois, Lasne ... 128068
El Arbol de Dios, San Salvador . 103096
Árbols Skolmuseum, Åmål ... 040523
Arbor Antiques, Jacksonville ... 093955, 135646
Arbor Lodge, Nebraska City ... 051206
Arboré, Jean-François, Paley ... 070845
Arbouse Bastide, Patrice, Avallon ... 066860
Arbousse-Bastide, Patrice, Thory 073672
Arbras Gallery, London ... 118172
L'Arbre à Collections, Bazoche-Gouet ... 066988
L'Arbre à Pain, Grenoble ... 068712
L'Arbre aux Papiers, Le Mans .. 136825
Arbre Bleu, Gillonnay ... 068665
Arbre de Jessé, La Carité-sur-Loire ... 140861
Arbroath Art Gallery, Arbroath . 043426
Arbroath Museum, Arbroath ... 043427
Arbuthnot Museum, Peterhead .. 045492
Arc, Edinburgh ... 117714
Arc Abatement, Austin ... 092212
Arc Antique, Nantes ... 128827
Arc Communicatio, Subotica ... 114464
The Arc Contemporary Crafts, Chester ... 117569
Arc de Triomphe, Saintes ... 105638
Arc en Ciel, Dinard ... 103687
L'Arc en Seine, Paris ... 070924
Arc Linea, Milano ... 080410
Arc One Gallery, Melbourne ... 099298
Arc Prints, Saint Louis ... 123795
ARC Yinnar, Yinnar ... 099945
L'Arca, Brescia ... 079914
L'Arca, Genova ... 080234
L'Arca, Modena ... 131558
Arca de L'Àvia, Barcelona ... 085727
Arca de L'Àvia, Mahón ... 086283
Arcache, Beirut ... 082272
Arcache, Lucien, Paris ... 070925
Arcad, Annecy ... 103242
Arcad'art, Grenoble ... 068713
Arcadas São José, Lisboa ... 084828
Arcade, Apeldoorn ... 082917
Arcade, Linz ... 100097
Arcade 127, Paris ... 070926
Arcade Art, New Orleans ... 122116
Arcade Books, Belfast ... 144237
Arcade Books, Toronto ... 139933
Arcade Chausse-Coqs, Genève . 116275
Arcade Colette, Paris ... 104476

Arcade Historical Museum, Arcade ... 046517
Arcades, Rennes (Ille-et-Vilaine) 072419
L'Arcadia, Brescia ... 079915
Arcadia, Madrid ... 115142
Arcadia, Milano ... 080411
Arcadia, Montréal ... 064390
L'Arcadia, Napoli ... 080837
Arcadia, New York ... 122324
Arcadia, Oswestry ... 090922
Arcadia, Phoenix ... 096187
Arcadia Antichità, Bologna ... 079813
Arcadia Antiques, Athlone ... 079438
Arcadia Antiques, Bristol ... 088645
Arcadia Antiques, Tetbury ... 091563
Arcadia Antiques, Toronto ... 064600
Arcadia Books, Ballarat ... 139630
Arcadia Books, Mylor ... 139848
Arcadia Booksellers, Oxford ... 144764
Arcadia Galerie, Lyon ... 069691
Arcadia Historical Museum, Arcadia ... 046519
Arcadia Publishing, Mount Pleasant ... 138366
Arcadia University Art Gallery, Glenside ... 049124
Arcadian Antiques, Auckland ... 083760
Arcadian Books and Art Prints, New Orleans ... 145300
Arcadian Clock &CO, New York . 135883
Arcadie, Jouars-Pontchartrain .. 068914
Arcadium, Baden-Baden ... 074629
ARCAM Architectuurcentrum Amsterdam, Amsterdam ... 031927
Arcana, Colmar ... 068081
Arcana, Roquefort-les-Pins ... 072539
Arcana Artis, Warszawa ... 143516
Arcane, Toulon (Var) ... 105708
Arcane, Versailles ... 074043
Arcano Antigüedades, Buenos Aires ... 060875
L'Arcantica, Torino ... 081646
Arcas, Provins ... 105401
Arcaz, Madrid ... 133296
ARCE, Madrid ... 137981
Arce Subastas, Barcelona ... 126503
Arceneaux, Birmingham ... 119784
Arch Angel Antiques, Hong Kong 065007
Arch Angel Art, Hong Kong ... 102048
Arch Galleri, Kalmar ... 115697
Archaeological and Byzantine Museum Florina, Florina ... 023076
Archaeological and Historical Museum, Ramat Hashofet ... 025178
Archaeological and Land Study Collection, Alumot ... 025016
Archaeological Collection, Baltimore ... 046728
Archaeological Collection, Farsala 023074
Archaeological Collection, Filiatra 023075
Archaeological Collection, Kozani 023143
Archaeological Collection, Loutra Aidipsou ... 023158
Archaeological Collection, Molyvos ... 023175
Archaeological Collection, Symi . 023277
Archaeological Collection, Yehi'am ... 025220
Archaeological Collection of Agios Kirikos, Agios Kirikos ... 022948
Archaeological Collection of Amorgos, Amorgos ... 022956
Archaeological Collection of Pythagorion, Pythagorio ... 023241
Archaeological Collection of Youth Village Nitzanim, Doar-Na Evtach ... 025033
Archaeological Collection Panagia Kastrou, Platanos ... 023227
Archaeological Museum, Amman 029763
Archaeological Museum, Andros 022958
Archaeological Museum, Anuradhapura ... 040471
Archaeological Museum,

Argostolion 022966
Archaeological Museum, Arhaia
Olympia 022969
Archaeological Museum, Atalanti 022974
Archaeological Museum, Bijapur 024178
Archaeological Museum,
Bodhghaya 024181
Archaeological Museum,
Chalkeio 023036
Archaeological Museum,
Chalkida 023037
Archaeological Museum,
Colombo 040473
Archaeological Museum, Corfu . 023051
Archaeological Museum,
Dedigama 040481
Archaeological Museum, Delhi . 024219
Archaeological Museum, Delos . 023058
Archaeological Museum,
Gorakhpur 024262
Archaeological Museum, Guntur 024267
                                     024268
Archaeological Museum, Gwalior 024280
Archaeological Museum,
Halebidu 024284
Archaeological Museum, Hampi 024285
Archaeological Museum, Hardwar 024287
Archaeological Museum, Iráklion 023095
Archaeological Museum, Jaunpur 024323
Archaeological Museum,
Kamalapur 024337
Archaeological Museum, Karachi 034122
Archaeological Museum, Kavala 023123
Archaeological Museum,
Khajuraho 024344
Archaeological Museum, Kibbutz Ein
Dor 025126
Archaeological Museum, Kibbutz
Gat 025130
Archaeological Museum, Kilkis . 023130
Archaeological Museum, Konarak 024370
Archaeological Museum, Lamia . 023147
Archaeological Museum, Lárissa 023149
Archaeological Museum,
Lemesos 009423
Archaeological Museum,
Mahastangarh 003229
Archaeological Museum,
Mainamati 003230
Archaeological Museum, Merbok 030718
Archaeological Museum,
Mykonos 023180
Archaeological Museum,
Myohaung 031822
Archaeological Museum, Myrina 023182
Archaeological Museum, Nalanda 024421
Archaeological Museum, Néa
Anchialos 023198
Archaeological Museum,
Neishaboor 024726
Archaeological Museum, Pagan . 031823
Archaeological Museum,
Paharpur 003231
Archaeological Museum, Paros . 023212
Archaeological Museum, Pune . 024442
Archaeological Museum,
Rethymnon 023242
Archaeological Museum, Rhodos 023247
Archaeological Museum, Sanchi 024463
Archaeological Museum, Sarnath 024468
Archaeological Museum, Spanish
Town 028455
Archaeological Museum, Swat . 034152
Archaeological Museum, Taxila . 034153
Archaeological Museum, Tbilisi . 016395
Archaeological Museum, Trichur 024499
Archaeological Museum, Udaipur 024502
Archaeological Museum, Vaisali 024510
Archaeological Museum, Varanasi 024513
Archaeological Museum, Vravrón 023318
Archaeological Museum and Portrait
Gallery, Goa Velha 024261
Archaeological Museum Blitar,
Blitar 024544

Archaeological Museum Eressos,
Eressos 023070
Archaeological Museum Harappa,
Harappa 034111
Archaeological Museum Kediri,
Kediri 024601
Archaeological Museum Konarak,
Puri 024451
Archaeological Museum Moen-Jo-Daro,
Moen-Jo-Daro 034140
Archaeological Museum of Agrinion,
Agrinion 022950
Archaeological Museum of Aiani,
Aiani 022952
Archaeological Museum of Aigina,
Aegina 022942
Archaeological Museum of Ancient
Corinth, Ancient Corinth 022957
Archaeological Museum of Argos,
Argos 022965
Archaeological Museum of Astros
Kynourias, Astros 022972
Archaeological Museum of Astypalaia,
Astypalaia 022973
Archaeological Museum of Baluchistan
Quetta, Quetta 034147
Archaeological Museum of Chios,
Chios 023044
Archaeological Museum of Chora,
Messenia 023168
Archaeological Museum of Corinth,
Kórinthias 023139
Archaeological Museum of Delphi,
Delphi 023059
Archaeological Museum of Elefsina,
Elefssina 023069
Archaeological Museum of Epidaurus,
Asklepieion Lygourio 022971
Archaeological Museum of Eretria,
Eretria 023071
Archaeological Museum of Galaxidi,
Galaxidi 023081
Archaeological Museum of Gytheion,
Gythio 023088
Archaeological Museum of Ioannina,
Ioannina 023089
Archaeological Museum of Isthmia,
Isthmia 023099
Archaeological Museum of Kampos
Ikaria, Kampos 023110
Archaeological Museum of Karystos,
Giokaleion Cultural Foundation,
Karystos 023116
Archaeological Museum of Kea,
Kea 023126
The Archaeological Museum of Kfar
Saba, Kfar Saba 025119
Archaeological Museum of Kimolos,
Kimolos 023132
Archaeological Museum of Komotini,
Komotini 023135
Archaeological Museum of Kos,
Kos 023140
Archaeological Museum of Megara,
Megara 023165
Archaeological Museum of Mytilene,
Mytilini 023186
Archaeological Museum of Nafplion,
Nafplion 023192
Archaeological Museum of Naxos,
Naxos 023196
Archaeological Museum of Nemea,
Archaia Nemea 022964
Archaeological Museum of Nicopolis,
Preveza 023235
Archaeological Museum of Patiala,
Patiala 024430
Archaeological Museum of Patras,
Patras 023215
Archaeological Museum of Pella,
Pella 023218
Archaeological Museum of Piraeus,
Piraeus 023220
Archaeological Museum of Plaka,

Milos 023174
Archaeological Museum of Polygyros,
Polygyros 023229
Archaeological Museum of Samos,
Samos 023252
Archaeological Museum of Samothraki,
Paleopoli 023207
Archaeological Museum of Sifnos,
Kastro 023121
Archaeological Museum of Siteia,
Siteia 023265
Archaeological Museum of Spárti,
Spárti 023271
Archaeological Museum of Stavros,
Stavros Ithakis 023276
Archaeological Museum of Syros,
Ermoupolis 023072
Archaeological Museum of Tegea,
Museum of the Godness Athena,
Tegea 023279
Archaeological Museum of Thassos,
Thassos 023280
Archaeological Museum of Thebes,
Thebes 023282
Archaeological Museum of Thera,
Thera 023283
Archaeological Museum of Thermos,
Thermo 023285
Archaeological Museum of Tinos,
Tinos 023303
Archaeological Museum of Veroia,
Veria 023311
Archaeological Museum Peshawar,
Peshawar 034141
Archaeological Museum Poros,
Poros 023233
Archaeological Museum Red Fort
Delhi, Delhi 024220
Archaeological Museum Saidu Sharif,
Saidu Sharif 034151
Archaeological Museum Serres,
Serres 023256
Archaeological Museum Taj Mahal,
Agra 024121
Archaeological Museum Umarkot,
Umarkot 034154
Archaeological Museums of Chania
and Rethymnon, Chania 023038
Archaeological Preserve Nicopolis ad
Istrum, Veliko Tărnovo 005249
Archaeological Receipts Fund,
Athinai 059324
Archaeological Sit of Vitsa, Vitsa 023313
Archaeological Site, Paramythia 023211
Archaeological Site and Museum,
Sicyon 023261
Archaeological Site Avedat,
Avedat 025021
Archaeological Site Museum,
Alampur 024132
Archaeological Site Nimrud (Calah),
Nimrud 024812
Archaeological Site of Edessa,
Edessa 023066
Archaeological Site of Geraki,
Geraki 023083
Archaeological Site of Lindos,
Lindos 023155
Archaeological site of Megalopolis,
Megalopolis 023164
Archaeological Site of Perachora,
Perachora 023219
Archaeological Site Walraversijde
1465, Het Domein Raversijde,
Oostende 003925
Archaeological Survey of Alberta,
Edmonton 005748
Archäologie am Landesmuseum
Joanneum, Graz 001965
Archäologie-Museum Neuburg,
Archäologische Staatssammlung
München, Neuburg an der
Donau 020688
Archäologie-Museum, Städtische

Museen Heilbronn, Heilbronn . 019157
Archäologiemuseum im Fränkischen
Freilandmuseum, Archäologische
Staatssammlung München, Bad
Windsheim 016912
Archäologisch-Ethnographische Lehr-
und Studiensammlung des Instituts
für Altamerikanistik und Ethnologie
der Universität Bonn, Bonn . . . 017478
Archäologisch-Ökologisches Museum,
Albersdorf 016489
Archäologische Ausgrabung Borg,
Perl 021102
Archäologische Ausstellung in der
Schule, Nassenfels 020639
Archäologische Gedenkstätte,
Klosterneuburg 002206
Archäologische Sammlung
der Universität, Freiburg im
Breisgau 018476
Archäologische Sammlung der
Universität Fachbereich III, Trier 022195
Archäologische Sammlung der
Universität Graz, Graz 001966
Archäologische Sammlung der
Universität Rostock, Institut
für Altertumswissenschaften,
Rostock 021486
Archäologische Sammlung der
Universität Zürich, Zürich 042188
Archäologische Sammlung des
Historischen Museums, Luzern 041702
Archäologische Sammlung Keltenhaus,
Historisches Rathaus, Taufkirchen,
Kreis München 022111
Archäologische Sammlung mit
Heimatvertriebenenstube und Kruk-
Sammlung, Königsbrunn 019724
Archäologische Sammlung,
Ruhrlandmuseum, Essen 018288
Archäologische Sammlung,
Stadtmuseum Wels-Minoriten,
Wels 002914
Archäologische Staatssammlung
München, Museum für Vor- und
Frühgeschichte, München 020504
Archaeologische Verzameling der
Rijksuniversiteit, Utrecht 032943
Archäologischer Park Cambodunum –
APC, Kempten 019588
Archäologischer Park Carnuntum,
Petronell 002502
Archäologischer Park Carnuntum,
Amt der Niederösterreichischen
Landesregierung, Bad Deutsch-
Altenburg 001736
Archäologischer Park Magdalensberg,
Landesmuseum Kärnten,
Pischeldorf 002508
Archäologischer Park mit Museums-
Turm, Kellmünz an der Iller . . . 019577
Archäologischer Park Ostkastell,
Welzheim 022556
Archäologisches Freilichtmuseum
Oerlinghausen, Oerlinghausen . 020963
Archäologisches Grabungsfeld
Michaelerplatz, Wien Museum,
Wien 002933
Archäologisches Landesmuseum
Baden-Württemberg, Konstanz 019757
Archäologisches Landesmuseum
Brandenburg, Brandenburg an der
Havel 017540
Archäologisches Landesmuseum
Mecklenburg-Vorpommern,
Schwerin 021789
Archäologisches Landesmuseum,
Stiftung Schleswig-Holsteinische
Landesmuseen Schloß Gottorf,
Schleswig 021663
Archäologisches Museum,
Burgheim 017693
Archäologisches Museum,
Donauwörth 017931

Archäologisches Museum,
Rimpar 021433
Archäologisches Museum Amberg,
Archäologische Staatssammlung
München, Amberg 016570
Archäologisches Museum
Colombischlössle, Freiburg im
Breisgau 018477
Archäologisches Museum der Stadt
Kelheim, Kelheim 019573
Archäologisches Museum der Stadt
Weißenhorn, Weißenhorn 022546
Archäologisches Museum der
Universität Münster, Münster .. 020591
Archäologisches Museum des
Historischen Vereins für Oberfranken,
Bayreuth 016982
Archäologisches Museum Essenbach,
Essenbach 018315
Archäologisches Museum Frankfurt,
Frankfurt am Main 018405
Archäologisches Museum Gablingen,
Gablingen 018575
Archäologisches Museum in der
Feldmühle, Rennertshofen 021374
Archäologisches Museum Innsbruck,
Sammlung von Abgüssen und
Originalen der Universität Innsbruck,
Innsbruck 002118
Archäologisches Museum Lavant,
Lavant 002278
Archäologisches Museum Neu-Ulm,
Archäologische Staatssammlung
München, Neu-Ulm 020677
Archäologisches Museum Renningen,
Renningen 021375
Archäologisches Museum und
Dokumentationszentrum Via Claudia,
Fließ 001893
Archäologisches Museum, Verein
zur Förderung des Archäologischen
Museums Oldendorf/Luhe,
Oldendorf 020995
Archäologisches Pilgermuseum
Hemmaberg-Juenna, Globasnitz 001937
Archäologisches Zentrum Hitzacker,
Hitzacker 019268
Archaeology Centre, Bagshot . 043466
Archaeology Museum, Chennai . 024194
Archaeology Museum, Shamir .. 025184
Archaeology Museum Gozo,
Victoria 030760
Archaeos, Fitzroy 061466
Archaïa, Lyon 069692
Archambault, Guy, Vitry-le-
François 125843
L'Archange, Paris 104477
Archangel Antiques, New York . 095235
Archangelos Mihail İkon Müzesi,
Girne 033456
Archangelskij Gosudarstvennyj Muzej
Zodčestva i Narodnogo Ikusstva –
Malye Karely, Archangelsk ... 036858
Archangelskij Literaturnyj Muzej,
Archangelsk 036859
Archangelskij Oblastnoj Kraevedčeskij
Muzej, Archangelsk 036860
Archangelskij Oblastnoj Muzej
Izobrazitelnych Iskusstv, GMO
Chudožestvennaja Kultura Russkogo
Severa, Archangelsk 036861
Archangelskij Sobor, Gosudarstvennyj
muzej-zapovednik Moskovskij Kreml,
Moskva 037234
Archdiocesan Museum of Manila,
Manila 034521
Archè, Catania 131079
Arché, Milano 080412
L'Arche, Paris 136850
Arche de Noé, Orléans 070806
Arche Nebra, Die Himmelsscheibe
erleben, Wangen 022418
Archè Restauri, Parma 131716
Archea, Amsterdam 082682

Archea, Genova 131287
Archelaus Smith Museum, Cape Sable
Island 005555
Archenhold Sternwarte, Berlin .. 017073
Archéo- et Géo-Espaces Moulin du
Broukay, Eben-Emael 003564
Archéo-Musée du Kronos, Espace
Patrimoine, Albens 011577
Archeo Plan, Delft 132467
Archéo-Site Gaulois de Saint-Julien,
Saint-Julien 015394
L'Archéodrome de Bourgogne,
Meursault 014156
Archeologia Classica, Roma .. 138791
Archeologica di Sovana, Sorano 027991
Archeologičeska Ekspozicija v Džamia
Achmed Bej, Istoričeski Muzej,
Kjustendil 005062
Archeologičeska Muzej, Silistra . 005160
Archeologičeski Institut s Muzej,
Bălgarska Akademija na Naukite,
Sofia 005172
Archeologičeski Muzej, Burgas . 004994
Archeologičeski Muzej, Plovdiv . 005124
Archeologičeski Muzej, Radnevo 005142
Archeologičeski Muzej, Septemvri 005158
Archeologičeski Muzej, Sozopol . 005206
Archeologičeski Muzej, Varna .. 005235
Archeologičeski Rezervat Starijat
Grad Šumen i Šumenska Krepost,
Šumen 005212
Archeologičeskij Muzej, Biškek . 030070
Archeologičeskij Muzej,
Čerepovec 036940
Archeologičeskij Muzej Kazanskogo
Universiteta, Kazanskij
gosudarstvennyj universitet,
Kazan 037097
Archeologické Muzeum, Břeclav 009456
Archeologické Múzeum,
Národné Múzeum, Bratislava . 038299
Archeologické Pracoviště, Slezské
zemské muzeum, Opava 009734
Archéologie Monnaies,
Montpellier 070289
Archéologie Monnaies Médailles,
Montpellier 070290
Archeologiemuseum, Stein 032905
Archeologisch Centrum C.W. Bruinvis,
Alkmaar 031890
Archeologisch en Paleontologisch
Museum Hertogsgemaal,
Gewande 032345
Archeologisch Museum,
Harelbeke 003674
Archeologisch Museum,
Mechelen 003853
Archeologisch Museum Abdij,
Affligem 003260
Archeologisch Museum Haarlem,
Haarlem 032390
Archeologisch Museum Van Bogaert-
Wauters, Hamme 003669
Archeologisch Museum van de
Universiteit, Gent 003619
Archeologisch Site Museum,
Werken 004145
Archeologisch Streekmuseum Het
Labyrint, Kemmel-Heuvelland . 003748
Archeološki Grad Stobi, Negotino 030619
Archeološki Muzej Cherakleja,
Bitola 030609
Archeonet, Bruxelles 063319
Archeoparc, Schnals 027907
Archéoscope, Le Mont-Saint-
Michel 013668
Archéoscope du Pays de Salm,
Vielsalm 004109
Archéoscope Godefroid de Bouillon,
Bouillon 003388
Archéosite de l'Hospice des Pèlerins
de Saint-Jacques-de-Compostelle,
Pons 014927
Archéosite du Château et Village

Médiéval, Montségur 014314
Archéosite Gallo-Romain des Flaviers,
Mouzon 014369
Archéosite Préhistorique de Cambous,
Saint-Martin-de-Londre 015450
Archer, Sorell 062217
Archer County Museum, Archer
City 046523
Archer Gallery, Clark College,
Vancouver 053945
Archer Street Antiques and Gallery,
Longford 061749
Archer, Barbara, Atlanta 119533
Archer, R.G., Lavenham 144516
Archer, R.G., Sudbury, Suffolk . 144869
Les Arches, Paris 104478
Arches Museum of Pioneer Life,
Winona 054482
Archetto, Salvatore, Napoli 080838
Archétypes, Hyères 103847
Archeus Fine Art, London 118173
Archevêché, Rouen 072561
Archi-Noire, Paris 070927
ArchiArt, Port Douglas 099577
Archibald Graham Mcilwaine
House, The Petersburg Museums,
Petersburg 052010
Archibald Historical Museum, La
Rivière 006138
Archibalds, Wellington 113054
Archibook Verlag, Berlin 137050
Archicendolo, Napoli 080839
Archie Bray Foundation, Helena . 049430
Archief en Museum van het Vlaams
Leven te Brussel, Bruxelles ... 003429
Archief en Museum Zusters van Liefde
van Jezus en Maria, Gent 003620
Archie's, Chicago 092663
L'Archimede, Roma 110727
Archimede Staffolini, Lefkosia . 102534
Archimedes, Leipzig 142054
Archipel, Paris 070928, 128989
Archipel Antiek, Den Haag ... 083055
Archipelago, Sheffield 119076
Archipelago Art, Griffith 099098
Archipelago Press, Singapore . 137936
Archipelagos, Skiathos 079202, 109085
Archipelagos Cultural Center,
Athinai 022976
Architech Gallery, Chicago 120038
Architect Alexander Tamanyan
Research Institute-Museum,
Yerevan 000723
Architects' Journal, London ... 139129
Architectura, Berlin 138669
Architectura et Natura,
Amsterdam 137825
Architectural Antiques,
Albuquerque 092034
Architectural Antiques, Atlanta . 092108,
135297
Architectural Antiques, Bedford . 088423
Architectural Antiques, Denver . 093256
Architectural Antiques, Dublin .. 079524
Architectural Antiques, Kendal . 089668
Architectural Antiques, London . 089919,
134586
Architectural Antiques, Miami .. 094700
Architectural Antiques,
Minneapolis 094898
Architectural Antiques, Oklahoma
City 095946
Architectural Antiques, Ottawa . 064493
Architectural Antiques, San
Antonio 096944
Architectural Antiques, Vancouver 064796
Architectural Antiques Exchange,
Philadelphia 096080
Architectural Art Glass Studio,
Cincinnati 135434
Architectural Artifacts, Chicago . 092664
Architectural Artifacts, New York 095236

Architectural Artifacts, Toledo .. 097598,
127476,   136404
Architectural Association, London 060029,
118174
Architectural Association Publications,
London 138187
Architectural Colonial Antiques of
Indianapolis, Indianapolis 093856
Architectural Decor, Austin 092213
Architectural Design, London ... 139130
Architectural Heritage,
Birmingham 092388
Architectural Heritage,
Taddington 091537
Architectural Imports, Edmonton 064284
Architectural Recovery, Portland 096360
Architectural Relics, Phoenix ... 096188
The Architectural Review, London 139131
Architectural Salvage, Denver .. 093257
Architectural Salvage, Kansas
City 094057
Architectural Salvage, Louisville 094518
Architectural Salvage of San Diego,
San Diego 097064
Architectural Warehouse,
Auckland 083761
Architectural Warehouse Antiques, San
Antonio 096945
Architecturals, Philadelphia 096081
Architecture and Art School, Architect
Union of Armenia, Jerewan ... 054885
Architecture and Design Scotland,
Edinburgh 060030
Architecture Foundation, London 044905
Architecture in Greece, Athinai . 138763
Architektonski Fakultet, Univerzitet Sv.
Kiril i Metodij, Skopje 056125
Architektur- und Kunstgeschichte
20./21. Jahrhundert, Institut für
Geschichte und Theorie der
Architektur, Hannover 055576
Architektur Galerie Berlin, Berlin 106150
Architektur im Ringturm, Wien . 002934
Architektur in Progress, Wien . 058228
Architektura i Biznes, Kraków . 138950
Architektura-Murator, Warszawa 138951
Architekturforum Aedes, Berlin . 106151
Architekturforum Aedesland,
Berlin 106152
Architekturforum Zürich, Zürich . 042189
Architekturgalerie e.V., München 108045
Architekturgalerie Luzern,
Ausstellungsraum Partikel,
Luzern 041703
Architekturmuseum der Technischen
Universität München, München 020505
Architekturmuseum Schwaben,
Augsburg 016646
Architekturnaja Galereja, Moskva 114197
Architekturno-chudožestvennyj
Ansambl byvšego Spaso-
Jakovlevskogo Monastyrja,
Gosudarstvennyj muzej-zapovednik
Rostovskij Kreml, Rostov 037692
Architekturno-ètnografičeskij Muzej
Angarskaja Derevnja, Bratskij
gorodskoj objedinennyj muzej istorii
osvoenija Angary, Bratsk 036913
Architekturno-ètnografičeskij Muzej
Derevjannogo Zodčestva Chochlovka,
Perm 037637
Architekturno-etnografski Muzej Etăr,
Gabrovo 005028
Architekturno-istoriči Muzej-rezervat,
Boženci 004992
Architekturno-muzeen Rezervat
Arbanassi, Architectural Museum
Preserve Arbanassi, Veliko
Tărnovo 005250
Architekturno-muzeen Rezervat
Carevec, Veliko Tărnovo 005251
Architektūros Muziejus, Vilnius . 030510
Architekturzentrum Wien, Wien . 002935
Architus Antiques, Kirkby

Lonsdale 089713
Archiv Bürgerbewegung Leipzig e.V.,
  Leipzig 019976
Archiv der deutschen
  Jugendbewegung,
  Witzenhausen 022710
Archiv der Universität Wien,
  Schausammlung, Wien 002936
Archiv für Philatelie der
  Museumsstiftung Post und
  Telekommunikation, Bonn ... 017479
Archiv und Kollektion Paul Kaiser-
  Reka, Brandenburg an der
  Havel 017541
Archiv und Museum des
  Heimatkreises Leitmeritz, Fulda 018555
Archiv Verlag, Wien 136574
Archiv Werner Kittel, Zentralarchiv
  der Staatlichen Museen zu Berlin,
  Berlin 017074
Archival Conservation Center,
  Cincinnati 135435
Archive, Caen 067554
Archive, Toronto 101564
Archive Books and Music,
  London 144554
Archives, Paris 141024
Archives Antiques, Belfast 088431
Archives de la Construction Moderne,
  Lausanne 041645
Archives et Musée de la Littérature,
  Bruxelles 003430
Archives Fine Books, Brisbane . 139659
Archives Museum – Temple Mickve
  Israel, Savannah 053173
Archives of American Art Journal,
  Washington 139269
Archives of American Art, Smithsonian
  Institution, Washington 054088
Archives of Asian Art, New York 139270
Archivio Segreto Vaticano, Città del
  Vaticano 054677
Archivio Storico Fiat, Torino 028110
Archivo Español de Arte, Madrid 139025
Archivo-Museo Don Alvaro de Bazán,
  Palacio del Marqués de Santa Cruz,
  Viso del Marqués 040421
Archivo-Museo San Juan de Dios-
  Casa de los Pisas, Granada . 039455
L'Archivolto, Genova 080235
Archway Antiques, Glaslough .. 079630
Archway Books, Falmouth 144419
Archway Gallery, Aberdeen ... 117204
Archway Gallery, Houston 120993
Arcila Castaño, Alvaro, Medellín 065278
Arcimboldo, Bologna 079814
Arckx, Argenteuil 066713
El Arco, Gijón 086015
L'Arco, Lignano Sabbiadoro .. 110170
ARCO, Oslo 084293
Arco, Praha 140513
Arco & Haenle, München 108046,
                           137386
Arco Antico, Parma 081091
Arco Arredamenti, Noceto 080950
L'Arco del Tempo, Milano 080413,
                           131370
Arco-Zinneberg, Maximilian Graf von
  und zu, München 077453
Arcola Museum, Arcola 005346
Arcos Verlag, Ergolding 137179
The Arctic Bear, Toronto 101565
Arctic Corsair, Kingston-upon-
  Hull 044710
Arctic Experience Gallery,
  Hamilton 101310
d'Arcy, Montréal 064391
Arcy, Philippe d', Lyon 069693
Ardalan, Behzad, Saint-Ouen ... 073013
Ardavin, Anthony, Atlanta 119534
Ardéco, Avignon 103298
Ardeco, Dolch 083027
Ardeco, Fremantle 061484
Ardel, Bangkok 116947

Ardell Burchard, Saint Louis ... 123796
Ardellewa, Giza 103093
Ardelt, Frank, Dessau 129678
Arden, Boston 119830
Arden, West Hollywood 097871
Arden, Woollahra 099904
Arden Gallery, Buntingford 117473
Arden Gallery, Henley-in-Arden . 089504,
                           117954
Arden, C., Hereford 144471
Ardengo, Roma 142799
Ardennen Poteau '44 Museum, Sankt
  Vith 003997
Ardeta, Paris 070929, 128900
Ardey, Münster 137431
Ardies, Jacques Paul Andre, São
  Paulo 101082
Ardingly International Antiques and
  Collectors Fair, Ardingly 098358
Arditti, J.L., Christchurch 088904,
                           134270
Ardoisalle, Alle-sur-Semois ... 003261
Ardor, Auckland 112745
Adress Farmyard Museum,
  Annaghmore 043418
Ardrossan Historical Museum,
  Ardrossan 000801
Ards Art Centre, Newtownards . 045381
Ardshiel Gallery, Whiting Bay . 119340
Ardustyl, Paris 070930, 128901
Ardy, Bernadino, Genova 142691
Area, Houston 093615
Area de Anitua, Madrid 086121, 133297
Area Restauro, Venezia 132226
Ared Art, Tiranë 098514
Årekonstnärerna Ateljé och Galerie,
  Åre 115576
Arellano, Bolivar, New York ... 122325
Aremark Historielag, Ostfold ... 033873
Arena, Gent 063625, 140219
Arena, Sofia 101142, 101143
Arena Artists Studio, Arena Art and
  Design Association, Liverpool . 060031
Arena International, Taunton ... 119209
Arenas, Antonio, Trondheim ... 132800
d'Arenberg, Bruxelles 063320
Arend Dieperink Museum,
  Potgietersrus 038814
Arendal Auksjonsforretning,
  Arendal 126337
Arendal Bymuseum, Arendal ... 033499
Arendall, Memphis 094615
Arends, IJmuiden 132548
Arends, Venlo 083673
Arends & Tammes, Amsterdam . 112139
Arendt, Gerhard, Bielefeld 074972
L'Arengario, Gussago 142698
Arenhövel, Dr. Willmuth, Berlin . 137051
Arenhövel, M., Warendorf 108847
Arens, Anne, Freiburg im
  Breisgau 075849
Arens, Luis Fernando, München 077454
Arens, Renate, Berlin 074754
Arenthon, Paris 104479, 141025
Arentshuis, Musea Brugge,
  Brugge 003403
Arenz, Jürgen & Petra, Bermel . 074955
Ares Sanat Evi, İstanbul 116999
Ares Vapen & Antikt, Göteborg . 086697
Aresco, Breda 132452
Areta, Varna 101183
Areté, Milano 110239
Aretè, Roma 131812
Arethusa, Roma 131813
Arev Art Gallery, Jerewan 098597
Arévalo Rojas, Alberto, Lima ... 113341
Arévalo, Maresa, México 111943
Arfa, Knokke-Heist 063697
Arfaoui, Habib, Paris 141026
ARG Antique, Dubai 088168
Argar, Almeria 114836
Argar, Barcelona 085728
Arguaisch Kantonales
  Weinbaummuseum, Tegerfelden . 042048

Arge, Griffith 061561
arge kunst, Bolzano . 059392, 109873
Argence-Lecaillé, Aurore,
  Carcassonne 067666
Argenor Numismatique, Paris .. 070931
Argenson, Paule, Saint-Didier-au-Mont-
  d'Or 072735
Argent, Bronze & Cuivre,
  Bordeaux 128479
Argenteria Gioielleria Antica,
  Firenze 080082, 142664
Argenterie Abadie, Nice 070556
Argenterie de Turenne, Paris ... 070932
Argentium, Broughton 117467
Argentum, Double Bay 061403
Argenville, Louis d', La Rochelle 103901
Argeu, Ananias, Belo Horizonte . 128098
Argiero, Maria, Firenze 131108
Argile Gallery, London 118175
Argiles, Paris 070933
Argo, Athinai 109021
Argo, Knokke-Heist 100722
Argo, Lefkosia 102535
Argo, Warszawa 084697, 113797
Argo Gold Mine, Mill and Museum,
  Idaho Springs 049654
Argonaut Book Shop, San
  Francisco 145540
L'Argonaute, Cité des Sciences et de
  l'Industrie, Paris 014616
Les Argonautes, Paris . 070934, 141027
Argor, Valenciennes 073941
The Argory, Dungannon 044174
Argos, İstanbul 117000
Argosy, New York 145322
Argosy, Ottawa 140372
Argoud, Jean-Claude, Paris 128902
Argus, Brugge 100530
Argus, Bruxelles 063321
Argus Livres Anciens, Québec .. 140381
ARGVS Vereniging van Kunstexperts,
  Association d'Experts d'Art,
  Bruxelles 058352
Argyle, Goulburn 139755
Argyll and Sutherland Highlanders
  Regimental Museum, Stirling . 045869
Arhaiologiko Mouseio Thessaloniki,
  Thessaloniki 023286
Arheološka Zbirka, Nin 008953
Arheološka Zbirka, Stari Grad .. 009003
Arheološka Zbirka Crne Gore,
  Podgorica 031763
Arheološka Zbirka i Lapidarij Dr. Grga
  Novak, Hvar 008917
Arheološka Zbirka Osor, Nerezine 008952
Arheološka Zbirka, Arheološki Muzej u
  Splitu, Metković Split 008949
Arheološki i Prirodoslovni Muzej,
  Nacionalni Park Brijuni, Brijuni 008887
Arheološki Muzej, Budva 031748
Arheološki Muzej Đerdap,
  Kladovo 038197
Arheološki Muzej Dubrovnik,
  Dubrovnik 008899
Arheološki Muzej Istre, Pula ... 008971
Arheološki Muzej u Splitu, Split . 008993
Arheološki Muzej Zadar, Zadar . 009012
Arheološki Muzej Zagreb, Zagreb 009025
Arheološki Muzej, Tolminski Muzej,
  Most na Soči 038579
ARHIS, Paris 136851
Arhitekturni Muzej Ljubljana,
  Ljubljana 038539
Arhitext Design, Bucureşti 139001
Århus Bymuseum, Århus 010004
Aria, Saint-Ouen 073014
Aria Antiques, San Francisco .. 097177
L'Aria del Tempo, Milano 080414
Ariadne Galleries, New York ... 095237,
                           122326
Arians, Horst, Remels 130579
Aribaud, C., Cornebarrieu 068160
Aribel, Montevideo 119416
Arical, Barcelona 114848

Aricchi, Akié, Paris ... 104480, 141028
Aricò, Venezia 111079
Aricò, Kraków 113525
Ariel, Paris 104481
Ariëns Kappers, E.H., Amsterdam 143011
Ariental, Gent 063626, 128046
Arienti, Franco, Milano 080415
Aries, Dubai 088169
Arieta, Valerie, London 089920
L'Ariete, Padova 110555
Arigoni, Catherine, Paris 070935
Arild's Antikk, Oslo 084294
Arimatsu Narumi Shibori Kaikan,
  Nagoya 029110
Arimon, Järvenpää 140724
Ariodante, New Orleans 122117
Arion Press, San Francisco ... 138452
Arioso, Simrishamn 115809
Arioso, Stockholm 115832
ARIS, Lund 139078
Arise-Gallery, Washington 097805
Arise-Window to the East,
  Washington 097806
Arist Kunst, Garmisch-
  Partenkirchen 075945
Arista, Cannes 103694
Ariste, Santiago de Chile 128246
Arister, New York 122327
Aristeuros, Barcelona 143709
Aristippos Antikvariat, Höör ... 143861
Aristocrat Gallery, San Diego .. 124282
Aristoloches, Grenoble . 068714, 103783
Aristos, Getxo 086005, 086006
Aristos Broc's, Vannes (Morbihan) 073967
Arita-machi Minzoku Hakubutsukan,
  Arita 028478
Arita Toji Bijutsukan, Arita 028479
Arithmeum, rechnen einst und heute,
  Bonn 017480
Aritza, Bilbao 114979
Ariz, Tucson 124708
Ariz, Coree, Tucson 097635
Arizona Antiques, Tucson 097636
Arizona Artist Guild, Phoenix .. 060327
Arizona Coins, Phoenix 096189
Arizona Commission on the Arts,
  Phoenix 060328
Arizona Doll and Toy Museum, Rosson
  House Museum, Phoenix 052111
Arizona Gallery, Tucson 124709
Arizona Historical Cache, Phoenix 096190
Arizona Historical Museum,
  Tucson 053820
Arizona Historical Society Museum at
  Papago Park, Tempe 053699
Arizona Historical Society Museum-
  Downtown, Tucson 053821
Arizona Historical Society Pioneer
  Museum, Flagstaff 048739
Arizona Jewish Historical Exhibitions,
  Phoenix 052112
Arizona Mining and Mineral Museum,
  Phoenix 052113
Arizona Museum For Youth, Museo de
  Arizona para Jóvenes, Mesa .. 050810
Arizona Pastel Artists Association,
  Phoenix 060329
Arizona Science Center, Phoenix 052114
Arizona-Sonora Desert Museum,
  Tucson 053822
Arizona State Capitol Museum,
  Phoenix 052115
Arizona State Museum, University of
  Arizona, Tucson 053823
Arizona State University Art Museum,
  Tempe 053700
Arizona Watercolor Association, Sun
  Sity West 060330
Arizona Women Hall of Fame,
  Phoenix 052116
Arizzi, Giovanni, Parma 081092
Ark Antiques, Bishop's Castle .. 088486,
                           134115

Ark Art, Almaty 111549
Ark Kyoto, Kyoto 082005
Ark-La-Tex Antique and Classic
  Vehicle Museum, Shreveport . 053318
Ark Restoration, New York 135884
Arka, Paris 070936
Arka, Vilnius 111796
Arka, Warszawa 084698, 113798
ARKA Kulturwerkstatt e.V., Essen 018289,
  058727, 106906, 137180
Arkada, Gdańsk 113454
Arkade, Zemun 114465
Arkadi Monastery Collection,
  Rethymnon 023243
Arkadia, Gdańsk 113455
Arkadia, Palermo 110588
Arkaeologisk Museum, Korsør . . 010210
Arkansas Air Museum,
  Fayetteville 048716
The Arkansas Arts Center, Little
  Rock 050334
Arkansas County Doctor Museum,
  Lioncoln 050320
Arkansas Historic Wine Museum,
  Paris 051914
Arkansas Museum of Natural
  Resources, Smackover 053367
Arkansas Museums Association, North
  Little Rock 060331
Arkansas National Guard Museum,
  North Little Rock 051608
Arkansas Post Museum, Gillett . 049096
Arkansas State University Art Gallery,
  Jonesboro 049838
Arkansas State University Museum,
  Jonesboro 049839
Arkaringa Gallery, Black Rock . 098749
Arke Galerie, Dortmund 106694
Arkè Galleria Antiquaria e Moderna,
  Milano 080416
Arkea, Nice 070557
Arkell & Monk, Dorking 089077
Arken Museum of Modern Art,
  Ishøj 010151
Arkeologisk Museum i Stavanger,
  Museum of Archaeology Stavanger,
  Stavanger 033972
Arkeoloji Müzesi, Konya 043000
Arkeoloji Müzesi, Kütahya 043010
Arkeoloji ve Etnografya Müzesi,
  Elazığ 042887
Arkeos Arte Precolombino,
  Bogotá 102272
Arkesteijn, P.J.M., Delft 083028
Arkitektur Förlag, Stockholm . . . 138011
Arkitekturmuseet, Stockholm . . . 040990
Arkitekturmuseet, Nasjonalmuseet
  for Kunst, Arkitektur og Design,
  Oslo 033825
Arklio Muziejus, Horse Museum,
  Anykščiai 030316
Arklow Maritime Museum,
  Arklow 024818
Arkoloji ve Etnografya Müzesi,
  Edirne 042884
Arkona Lion's Museum, Arkona . 005350
Arkos, Clermont-Ferrand 103599
Arktisk Museum Nanoq,
  Jakobstad 010909
Arktouros, Chard 128546
Arkway, Richard B., New York . . 145323
The Arkwright Society, Cromford 044048
Arkwright Traders, Dunedin . . . 083919
L'Arlecchino, L'Isle-sur-la-Sorgue 069552
Arlequin, Buenos Aires 139551
Arlequin, Madrid 086122
Arlequin, Paris 070937
Arlequín Antigüedades, Don
  Torcuato 060940, 127542
Arleston, Flinders 061476
Arlet, Arzignano 109755
Arlet, Asiago 109756
Arlinghaus, H., Boxberg 075112
Arlington, Columbus 092993

Arlington Arts Center, Arlington . 046537
Arlington Books, Ottawa 140373
Arlington Court, Arlington 043429
Arlington Heights Historical Museum,
  Arlington Heights 046542
Arlington Historical Museum,
  Arlington 046538
Arlington Historical Society Collection,
  Arlington 046530
Arlington House – The Robert E. Lee
  Memorial, Arlington 046539
Arlington Mill Museum, Bibury . 043589
Arlington Museum, Birmingham . 047003
Arlington Museum of Art,
  Arlington 046532
Arlis News-sheet, London 139132
Arlo, Warszawa 084699
Arlotta, G., Winterthur 087906
Arlt, Maria, Dresden . . 106715, 129704
Arlt, Peter, Jena 076642
Armada Gallery, Plymouth 118917
Armadale Antique Centre, Armadale,
  Victoria 061011
Armadale Community Museum,
  Armadale 043430
Armadale Watch Gallery, Armadale,
  Victoria 061012
Armadillo's Pillow, Chicago 145033
Armádní Muzeum Žižkov, Praha 009779
Ármaga, León 115106
Armagh County Museum,
  Armagh 043431
Armand, Nice 070558
Armand, Pittsburgh 123520
Armand Frappier Museum, Laval 006169
Armand, Philippe, Enghien-les-
  Bains 068442
Ármándia Katonai & Hadtörténeti
  Régiségek Boltja, Budapest . . 079238
Armando Museum, Amersfoort . 031907
Armando Vitrales, Bogotá 128259
Armandos, Sacramento 136182
Armando's Antiques, Miami 094701
Armanet, Andrée & Marc, Bourgoin-
  Jallieu 067433
Armangaud, Josy, Paris 070938
Armchair Books, Edinburgh 144383
Armchair Bookstore, Portland . . 145446
Armed Forces Museum, Taipei . . 042498
Armed Forces of the Philippines
  Museum, Quezon City 034576
Armed Forces Survey Department
  Museum, Bangkok 042647
Armeemuseum 'Friedrich der
  Große', Bayerische Verwaltung der
  staatlichen Schlösser, Gärten und
  Seen, Kulmbach 019826
Armeepferdemuseum, Scherz . . 041952
Armel, Pleudaniel 072183
Armémuseum, Stockholm 040991
Armen, Toronto 101566
Armengau, Patrick, Avignon 125504
Armenia Editions, La Destrousse 136823
Armenian Artistic Union, Cairo . 058564
Armenian Cathedral and Museum,
  Isfahan 024699
Armenian Center for Contemporary
  Experimental Art, New York . . 122328
Armenian Center for Contemporary
  Experimental Art ACCEA,
  Jerewan 058179, 098598
Armenian Museum, Jerusalem . 034163
Armenian Museum of America,
  Watertown 054213
Armenjon, Michel, Appenai-sous-
  Bellême 066687
Armentani, Gaetano, Taranto . . 132080
Armeria, Mondavio 026884
Armeria Reale, Torino 028111
Arms & Traditions, Strasbourg . 073569
Les Armes Anciennes, Nice . . . 070559
Armes du Chevalier, Chevry-
  Cossigny 067973
Les Armes du Chevalier, Paris . . 070939

Armi e Bagagli, Torino 081647
Armidale Folk Museum, Armidale 000803
Armit, Jenny, Los Angeles 121429
Armit, Jenny, West Hollywood . . 124984
Armitage Auctions, Launceston . 125206
Armitano Editores, Caracas . . . 138484
Armoire Amoureuse, Bollène . . . 067279
Armoire Charolaise, Baron 066954
Armoires and More, Dallas 093070
Armondi, Brescia 079916
Armoni, İstanbul 117001
Armonia Galería de Arte, Lima . 113342
Armonie, Palermo 081016
Armonie, Paris 104482
Armonie, Roma 081229
Armor, Kate, Garsington 134401
Armor, Kate, Kington Langley . . 134518
Armory Art Gallery, Blacksburg . 047026
Armory Center for the Arts,
  Pasadena 051933
The Armory Show, International Fair of
  New Art, New York 098422
Armour Antiques, Indianapolis . . 093857
Armour-Winston, London 089921,
  134587
The Armoury of Saint James' Military
  Antiquarians, London 089922
Arms and Armor, Minneapolis . . 094899
The Arms and Armour Society,
  London 060032
Arms and Armour Society Journal,
  London 139133
Arms Collecting, Alexandria Bay 139271
Arms Family Museum of Local History,
  Youngstown 054598
Armsby, Martin, Hamburg 129947
Armson, Michael, Wilstone 091852
Armstark, Schärding 062834
Armstark, Alfred, Tettenweis . . . 078676
Armstrong, Coventry 144355
Armstrong, Fort Worth 093428
Armstrong, Harrogate 089422
Armstrong, Las Vegas 094120
Armstrong, Sandringham 062188
Armstrong Browning Library,
  Waco 054012
Armstrong-Prior, Phoenix 123479
Armstrong-Spallumcheen Museum and
  Art Gallery, Armstrong 005351
Armstrong, Richard Reed,
  Chicago 120039
Armurerie Caderot, Berre-l'Etang 067114
Army Medical Corps Centre Museum,
  Lucknow 024375
Army Medical Services Museum,
  Mytchett 045307
Army Museum, Halifax 005945
Army Museum and Fort Copacabana,
  Rio de Janeiro 004680
Army Museum Bandiana,
  Bandiana 000822
Army Museum of Infantry Center,
  Phachuab Khiri Khan 042709
Army Museum of New South Wales,
  Paddington, New South Wales 001389
Army Museum of South Australia,
  Keswick 001185
Army Museum of Western Australia,
  Fremantle 001082
Army Physical Training Corps Museum,
  Aldershot 043387
Arn, Willy, Lyss 133850
Arnaboldi & Fratelli, Mario, Roma 081230
Arnak, Buffalo 119948
Arnau, Barcelona 114849
Arnau, Miami 121727
Arnaud-Vannier, Langres 125607
Arnaud, Florence, Paris 070940, 104483,
  141029
Arnaud, François, Boulogne-
  Billancourt 103447
Arnault, Jean Claude, Stockholm 115833
Arndean, London 118176
Arndt & Partner, Berlin 106153

Arndt & Partner, Zürich 116744
Arndt, D., Graal-Müritz,
  Seeheilbad 076066
Arndt, Doris, Biberach an der Riß 106479
Arndt, Doris, Witten 130773
Arndt, Dr., Unterleinleiter 078764
Arndt, Gebr., Elmshorn 106874
Arnés y Röpke, Madrid 115143
Arnheim, Brighton, Victoria 098775
Arnhems Oorlogsmuseum 1940–45,
  Arnhem 032020
von Arnim – Willisch, Susanne,
  München 108047, 130323
Arno, Paris 128903
Arno, Raquel, São Paulo 064152
Arnol Blackhouse Museum, Arnol 043436
Arnold, Christchurch 143301
Arnold, Rochy-Condé 072507
Arnold Schönberg Center, Wien . 002937
Arnold, Peter, Prahran 139882
Arnold, Phyllis, Donaghadee . . . 089066,
  134327, 144369
Arnold, Phyllis, Greyabbey 089387
Arnold, Roy, Needham Market . . 090803,
  144737
Arnold, Sean, London 089923
Arnold & Co., Susan, Tulsa 097711
Arnold, Thomas, Völklingen 078811
Arnold, Willi, Uznach . . 087867, 133949
Arnoldi, Bruxelles 100549, 140164
Arnoldi-Livie, Angelika & Bruce,
  München 077455, 108048
Arnoldsche Verlagsanstalt,
  Stuttgart 137515
Arnolfini, Bristol 043743
Arnon, Monique, San Francisco . 097178
Arnot Art Museum, Elmira 048565
Arnot, Herbert, New York 122329
Arnoud, Liliane, Bourmont 067435
Arnoux, Paris 104484
Arnoux, Jean-Paul, Dun-sur-
  Auron 068412
Arnprior and District Museum,
  Arnprior 005352
Arnulf-Rainer-Museum, Baden bei
  Wien 001769
Arnus, Evaristo, Barcelona 085729
Arohonka, Anssi, Helsinki 140697
Arohonka, Mika, Hämeenlinna . . 140686
Aroma d.d Gallery, Ha Noi 125057
Aromda, Paris 104485
Aron Museum, Teple Emanu-El-Beth
  Sholom, Westmount 007384
Arona Numismatica, Arona 079750
Aronowitsch, Stockholm 115834
Aronson, Amsterdam 082683
Aronson, Chatswood 061313
Aronson, Frances, Atlanta 119535
Aros Aarhus Kunstmuseum,
  Århus 010005
Arossita, Sofia 101144
Around the Clock, Brixham 088669,
  134166
Around the Clock, Chicago 135909
Around the Corner, Albuquerque 119455
Around the World Books, Benicia 144996
Arovell-Kulturzeitschrift, Gosau . 138502
Arovell-Verlag, Gosau 136504
Aroya, Zaragoza 115549
ARP, Wrocław 113916
Arp Museum Bahnhof Rolandseck,
  Remagen 021357
Arpad Johan, Vezelay 074083
Arpe, Cannes 103499
Arquebus Krigshistorisk Museum,
  Haugesund 033641
Arquebuse, Genève 116276
Arqueología, Cuauhtémoc 138901
Arqueología Mexicana, Miguel
  Hidalgo 138902
Arquine, Miguel Hidalgo 138903
Arquitectura Viva, Madrid 139026
Arquitectura y Crítica, Álvaro
  Obregón 138904

Arquivo Histórico e Museu de Canonas
    Dr. Sezefredo Azambuja Vieira,
    Canoas 004400
Arran Heritage Museum, Brodick 043765
The Arrangement, San Antonio . 096946
Arrangerie, Antwerpen 127992
Arrantzaleen Museoa, Museo del
    Pescador, Bermeo 039131
Arranz, New York 122330, 122331,
    122332
Arranz Gozalo, Loreto, Madrid . . 086123,
    133298
Arreda, Simona Marletta, Catania 080020
Arredamento Mimarlık Tasarim Kulturu
    Dergisi, İstanbul 139112
Arrendamiento de Objetos y Bienes,
    Palma de Mallorca . . 086343, 126555
Arrêt sur Image, Bordeaux 103413
Arretini, Giuliano, Firenze 131109
Arretini, Marco, Firenze 080083
Arribas Parra, Santiago,
    Valladolid 086543, 133475
Arrighi, Bastelicaccia . . 066957, 128461
Arrivabene, Gilberto, Milano . . . 080417
Arrizabalaga Garcia, Manuel,
    Córdoba 115025
Arrojo Alfredo, Monica, A Coruña 085948
Arrow Rock State Historic Site, Arrow
    Rock 046546
Arrowhead Museum, Berkshire County
    Historical Society, Pittsfield . . 052180
Arrowmont School of Arts and Crafts,
    Gatlinburg 057127
Arrowmont School of Arts and Crafts
    Collection, Gatlinburg 049061
Arroyave, Fernando, Medellín . . 128298
Arruda, Miguel, Lisboa 084829
Arruda, Virgínia, Brasília 063929
ARS, Bratislava 139020
Ars, Český Krumlov 102582
Ars, Skopje 111878
Ars, Verona 081893
Ars, Wrocław 132912
Ars (Iskusstvo Petrograda), Sankt-
    Peterburg 139007
Ars 2000 – Galerija Krvina, Gorenja
    Vas 114680, 137941
Ars Aevi, Sarajevo 004271
Ars Amandi, Berlin 141481
Ars Amandi, Bologna 079815
Ars Antiqua, Bad Abbach 074487
Ars Antiqua, Genève 087446
Ars Antiqua, Genk 063614
Ars Antiqua, Madrid 133299
Ars Antiqua, Milano 080418
Ars Antiqua, Saint-Ghislain 063825
Ars Antiqua Restaurí, Padova . . 131652
Ars Antiquités, Ars-sur-Formans 066748
Ars Artis, Feldkirchen in Kärnten 099988
Ars Artis, Oxford 144765
Ars Artis – Espai Gorina, Sant Cugat
    del Vallès 115406
Ars Boutique, Tallinn 103109
Ars Ceramica, Glen Head 139272
Ars Domus, Düsseldorf 075452
Ars Domus, Krefeld 076950
ars Edition, München 137387
Ars Edizioni Informatiche, Milano 137669
Ars Electronica Center, Linz . . . 002314
Ars Felix, Etzgen 059921, 138042
Ars-Forum, Jaroslavl 114181
Ars Islamica, Bruxelles 063322
Ars Italica, Milano 110240
Ars Labor, Lugano . . . 087604, 133819
Ars Labor, Roma 131814
Ars Libraria, Milano 131371
Ars Libri, Boston 145005
Ars Libri, Windeck 142499
Ars Ligni, Münchwald 130422
Ars Liturgica, Buch- und Kunstverlag
    Maria Laach, Maria Laach . . . 137376,
    142117
Ars Longa, Kraków . . . 132832, 132833
Ars Longa, Milanówek 113704

Ars Longa, Paris 104486
Ars Longa, Rīga 111753
Ars Millennia, Roma 081231
Ars Mundi, Timişoara 085129
ars mundi, Edition Max Büchner,
    Hannover 137264
Ars Nobilis, Berlin 098124
Ars Nova, Breda 112350
Ars Nova, Lörrach 107884
Ars Nova, Natters 062744, 100128
Ars Nova – Nykytaiteen Museo,
    Turku 011434
Ars Publicita – Galerija Gasspar,
    Piran 114739
ARS Recup', Nancy 070413
Ars Sacra, Madrid 139027
Ars Temporis, Reifnitz 100147
Ars Temporis, Reims 072377
Ars Toto, Bari 130903
Arsbel Tendo O-tomachikku Gakki
    Hakubutsukan, Tendo 029473
d'Arschot & Cie, Bruxelles 063323
Arsenal Americas, Pittsburgh . . 096257
Arsenal Gallery, New York 051334
Arsenal Museum, Arsenal Football
    Club, London 044906
Arsenale, Bertonico 109806
Arsenale Editrice, San Giovanni
    Lupatoto 137740
Arsenic and Old Lace, Salt Lake
    City 096878
Arsinopia, Saint-Ouen 073015
Arslan, Bilge, Berlin 129466
Arstudio, Faenza 110009
Arstudio, Portomaggiore 110685
Arstudio B, Knokke-Heist 100723
Årsunda Viking Museum,
    Årsunda 040538
Art, Burgas 101118
Art, Chrudim 102589
Art, Eivissa 115058
Art, Genova 080236
Art, Leeds 118085
Art, Ljubljana . . 085603, 085604, 114692,
    114693,   133063,   133064
Art, Maribor 085615, 114726
Art, Mińsk Mazowiecki 113705
Art, Silsden 119099
Art, Venezia 111080
Art, Warszawa 113799
Art & All That Jazz, Saskatoon . 101539
Art & Antik, Saarbrücken 078316
Art & Antique Baden-Baden, Baden-
    Baden 098125
Art & Antique Essen, Essen . . . 098126
Art & Antique Wiesbaden,
    Wiesbaden 098127
Art & Antiques, Malibu 139274
Art & Antiques, New York 139273
Art & Antiques and Bridge Miniatures,
    Eton 089213, 134371
Art & Antiques Fair 's-Hertogenbosch,
    Kunst en Antiekbeurs, 's-
    Hertogenbosch 098254
Art & Architecture Journal,
    London 139134
Art & Archival, Queanbeyan . . . 127668
Art & Australia, Paddington, New South
    Wales 136485, 138492
Art & Book, Hamburg 107220
Art & Book Edition, Hamburg . . 137238
Art & Business. Sztuka i Antyki,
    Warszawa 138952
Art & Co, Torino 081648
Art & Co, Warszawa 084700
Art & Company, Amsterdam . . . 082684
Art & Concept, Frankfurt am
    Main 106969
Art & Cultural Alliance,
    Johannesburg 114787
Art & Deco, Warszawa 113800
Art & Decor, Lübeck 077169
Art & Decoration, Breda 082972
Art & Decoration, Venlo 083674

Art & Design Gallery, Kaikoura . 112927
Art & Form, Bucureşti 114125
Art & Frame, Apeldoorn 098255
Art & Frame, Christchurch 112825
Art & Frame, Kongens Lyngby . . 065980
Art & Frame, Malvern, South
    Australia 099268
Art & Framing, Greenside 114785
Art & Graphic Magazine,
    Spardorf 138671
Art & Home, Holywood 117988
Art & Image, Amsterdam 082685
Art & Industry, Berlin 074755
Art & Interior, Praha 098058
Art & Investissement, Paris 104488
Art & More, Raunheim 108438
Art & Ors, Paris 070941, 128904
Art & Photo, Faaborg 102796
Art & Presse Diffusion, Les
    Brenets 139088, 144057
Art & Public, Genève 116277
Art & Repro, Westerkappeln . . . 078949
Art & Soul Gallery, Singapore . . 114471
Art & Style, Höxter 107436
Art & Style, Néris-les-Bains . . . 070505
Art & Tapis, Beirut 082273
Art & Understanding Magazine,
    Albany 139275
Art @ Large, New York 122333
art + form, Dresden . . 106716, 137147,
    141691
Art + Frame, Wien 062895
Art + Vision, Bern . . . 116118, 138028
Art + Wave, Hamburg 107221
Art 11, Paris 104489
Art-1900, Berlin 074756
Art-2 Gallery, Singapore 114472
Art 204, Paris 104490, 136852
ART 21, Alfortville 103225
Art 3, Paris 128905
Art 3, Valence (Drôme) 105798
Art 36, Sofia 101145
Art 4 Association, Caen 103480
Art 4 d, Bangkok 139109
Art 4 Less, Tampa 124626
Art 4 U, Banbury 117279
Art 4 Uz, York 119404
Art-4-you, Kramsach 062684
Art 45, Montréal 101355
Art 47 Diffusion, Nogentel 104410
Art 5, Szczecin 113771
Art 5 Galerie, Oberembrach . . . 116546
Art 6, Richmond 123718
Art 75, Paris 070942, 104491
Art 77, Auxerre 066843
Art 77, Paris 104492
Art 80, Houston 135585
Art 86 Galerie Rivaud, Poitiers . 105362
L'Art à la Page, Paris 104493
Art à l'Affiche, Pessac 105346
Art a Motive, Minneapolis 121963
Art A2Z, Woollahra 099905
Art Ability, Dallas 120544
Art Academy, Dresden 106717
Art Access, Columbus 120469
Art Action (Art Galleries & Dealers),
    Singapore 114473
Art Actualité, Megève 070030
Art Actuel, Paris 104494
Art Advice Corporate, New York 122334
Art Advisory, Chicago 120040
Art Advisory Group, New York . . 122335
Art Advisory Services, New York 122336
Art Affair, Cannes 103500
Art Affair, Las Vegas 121333
Art Affair, Torquay 119241
Art Affaire, Sorrento 099691
Art Affairs, Amsterdam 112140
Art Agencies International, Den
    Haag 112372
art agents gallery, Hamburg . . . 107222
Art Agentur Köln, Köln 107585
art aktuell edition, Köln 137300

Art Allée Antiquités, Vence 074002,
    129333
Art Alley, Birmingham 119785
Art Alley, Houston 120994
Art Alliance, New York 122337
Art Almanac, Sydney 099763
Art Almanac – The Essential Guide to
    Australian Galleries, Sydney . 138493
Art Ambiente, Wathlingen 078876
Art-Amour, Le Bouchage 103926
Art Amsterdam, Amsterdam . . . 098256
Art Ancien, Basel 087219
L'Art Ancien, Bordeaux 067306
L'Art Ancien, Orléans 104426
L'Art Ancien, Québec 101485
l'Art Ancien, Rabat 082596
Art ancien, Roma 081232
Art and Antique Bank, Bad
    Wurzach 074621
Art and Antique Bank, Bruxelles 063324
Art and Antique Bank,
    Neuenbürg 077720
Art and Antique Centre, Bangkok 088073,
    126784
Art and Antique Dealers League of
    America, New York 060332
Art and Antique Furniture, San
    Francisco 097179
Art and Antique Planet, Los
    Angeles 094258
Art and Antique Restorations,
    Oklahoma City 136047
Art and Antique World, Los
    Angeles 094259
Art and Antiques Bank, München 077456
Art and Antiques Dubai, Dubai . 098344
Art and Applied Design Department,
    Lane Community College,
    Eugene 057059
Art and Architecture Department,
    Daley College, Chicago 056870
Art and Architecture Department,
    Oakton Community College, Des
    Plaines 057008
Art and Art, Dallas 120545
Art and Art Education Department,
    Alabama A & M University,
    Normal 057585
Art and Art History Department,
    College of Saint Catherine, Saint
    Paul 057818
Art and Art History Department, Drury
    University, Springfield 057940
Art and art History Department,
    Florida International University,
    Miami 057430
Art and Art History Department,
    Gustavus Adolphus College, Saint
    Peter 057826
Art and Art History Department,
    Kalamazoo College, Kalamazoo 057273
Art and Art History Department,
    Loyola Marymount University, Los
    Angeles 057363
Art and Art History Department, New
    England College, Henniker . . . 057195
Art and Art History Department,
    Providence College, Providence 052378,
    057724
Art and Art History Department, Saint
    Mary's College of Maryland, Saint
    Mary's City 057817
Art and Art History Department,
    University of California Davis,
    Davis 056982
Art and Art History Department,
    University of Richmond,
    Richmond 057758
Art and Art History Department,
    Virginia Polytechnic Institute and
    State University, Blacksburg . . 056747
Art and Arts Editeur, Bordeaux . 136803,
    140786
Art and Charity, San Antonio . . 124010

Art and Clay Designs, Tucson .. 124710
Art and Creation, Jaipur,
  Rajasthan 109379
Art and Culture, Dubai 088170
Art and Culture Center of Hollywood,
  Hollywood 049513
Art and Design Department, Adrian
  College, Adrian 056626
Art and Design Department, Brenau
  University, Gainesville 057120
Art and Design Department, Cerritos
  College, Norwalk 057598
Art and Design Department, Columbia
  College Chicago, Chicago 056871
Art and Design Department, Lynn
  University, Boca Raton 056759
Art and Design Gallery, Hatfield .. 044535
Art and Design Programs, Maryville
  University of Saint Louis, Saint
  Louis 057808
Art and Frame, İstanbul 117002
Art and Frame Direct, Dallas ... 120546
Art and Frame Direct, Los
  Angeles 121430
Art and Frame Exhibition, Jeddah 085297
Art and Frame Expo, Dallas ... 120547
Art and Frame Gallery, Kansas
  City 121278
Art and Frame Gallery, Saint
  Louis 123797, 123798
Art and Frame Gallery @ Creve Coeur,
  Saint Louis 123799
Art and Frame Gallery @ Manchester,
  Saint Louis 123800
Art and Frame Workes,
  Cincinnati 120314
Art and Frame World, Fort Worth 120853
Art and Framing Express,
  Washington 124861
Art and Framing Gallery, El Paso 120822
Art and Framing Solutions, San
  Diego 124089
Art and Fun, Cleveland 120395
Art and Gold, Haifa 109712
Art and Humanities Department,
  Kennedy-King College, Chicago 056872
Art and Humanities Department,
  Malcolm X College, Chicago .. 056873
Art and Humanities Department, Olive-
  Harvey College, Chicago 056874
Art and Humanities Department,
  Truman College, Chicago 056875
Art and Humanities Department,
  Wright College, Chicago 056876
Art and Industrie, Armadale,
  Victoria 098653
Art and Industry Antiques,
  Auckland 083762
Art and Interior Design Program,
  Lasell College, Newton 057575
Art and Languages Department,
  College of the Redwoods,
  Eureka 057062
Art and Market, Chengdu 138593
Art and More, Columbus 120470
Art and Photography Department,
  Hartnell College, Salinas 057832
Art and Picture Works Gallery,
  Bowral 098766
Art and Soul, Saint Louis 123801
Art and Soul Furniture,
  Gerringong 061518
Art and Soul Gallery, Milwaukee 121890
Art and Technology Department,
  University of Maryland Eastern Shore,
  Princess Anne 057720
Art and the Artist, Fresno 120891
Art and Wine Birmingham,
  Birmingham 119786
Art Angels, Amsterdam 112141
Art Angle, Johannesburg 085644
Art Antic, Sankt Moritz 087792
Art Antico, Köln 076847
Art Antiguidades, Belo Horizonte 063879

Art-Antique, Groß Niendorf 076100
Art Antique, Hamburg 076188
Art Antique, Košice 085565
Art Antique, Lindlar 077122
Art-Antique-Zumkley, Rhede ... 078220,
                              130586
Art-Antiquite, Bruxelles 063325
Art Antiquités Artisanat, Jugon-lès-
  Lacs 068924
Art Antyk, Warszawa 132871
Art Appraisal Company, Tulsa .. 124789
Art Approach, Kangaroo Ground 061667
Art Aqua, Wien 100226
Art Architecture Antiquité,
  Cannes 103501
Art Area Studio, Tiranë 098515
Art Arena Gallery, Wollongong .. 099893
Art Ariane, Honfleur 103815
Art Around Gallery, Philadelphia 123347
Art Artisanat de Mongolie, Rennes (Ille-
  et-Vilaine) 105428
Art Asap, New York 122338
Art Asiatique, Marseille 069902
Art Aspects Gallery, Lismore ... 099233
Art Assets LLC, New York 122339
Art Associates, Hawthorn 099113
Art Association of Harrisburg,
  Harrisburg 049369
Art Association of Jacksonville,
  Jacksonville 060333
Art Association of Richmond,
  Richmond 060334
Art at Buck Fizz, Suffolk Park .. 099731
Art at Home, Washington 124862
Art at Linden Gate, Yarra Glen . 099936
Art at Work, Greenford 117884
Art Atelier Aquatinta, Lenzburg . 116434
Art Atelier Galleries, Hawthorn . 099114,
                              127611
Art Athina, Athinai 098171
Art Attack, Cleveland 120396
The Art Attack, Jacksonville .. 121232
Art Attack, Manawatu 112936
Art Attack, Paraparaumu 112985
Art Attacks Too, New Orleans .. 122118
Art Attract, O'Connor 099497
Art au Carré, Angers 103235
L'Art au Zénith, Nyons 104417
Art Avenue, Vallauris 105804
Art Avenue Magazine, Vancouver 138569
Art Avenues, New York 122340
Art Bank, London 101331
Art Bank, Philadelphia 123348
Art Bank, Rosebery 062167
Art Bar, Albuquerque 119456
Art Bar, Calalzo di Cadore 109927
Art Baraka, Pamplona 115384
Art Barn, Washington 124863
Art Base, Singapore 114474
Art Base Book Research,
  Kitzingen 141999
Art Basel Miami Beach, Miami
  Beach 098423
Art Beat Gallery, Louisville ... 121629
Art Beat International, Boston . 119831
Art Beatus, Hong Kong 102049, 102050
Art Beatus, Vancouver 101735
Art Best, Saint-Ouen 073016
Art Bodensee, Dornbirn 097998
The Art Book, Oxford 139135
Art Book Company, Taipei 138127
Art Book Survey, London 138188
Art Books, San Jose 138463
Art Books International, London 138189
Art Books of London, London . 144555
Art Boutique, Braintree 117414
Art-Box, Århus 065597
Art Broker Bigli, Milano 080419, 126202
Art Brokering, Los Angeles ... 121431
Art Brush, Saint George's 100837
ART Brussels, International
  Contemporary Art Fair,
  Bruxelles 098011
Art Brut Center Gugging,

Klosterneuburg 058229
Art Bubamara, Beograd 114396
The Art Bug, Norwich 118825
Art-Buks, Moskva 143634
Art Bulletin, New York 139276
Art Bureau & Werkstatt,
  Lamspringe 107783, 137334
Art Bureau NZ, Auckland 083763
Art Business News, Seven Hills . 139277
Art Business Today, London .. 138190,
                              139136
Art by Design, Houston 120995
Art by God, Dallas 120548
Art by God, Miami 121729
Art by Nature, Tulsa 097712
Art by Querch, Seiersberg 100200
Art by the Sea, Auckland 112746
Art by Will, Philadelphia 123349
Art Caballero, Barcelona 085730
Art Cadre, Paris 104495
Art Cadre Dauphine, Paris 104496
Art Calendar, Orlando 139278
Art Calendar International,
  Dronningmølle 065635
Art Capital Group, New York .. 122341
Art Care, Singapore 133034
Art Care and Conservation,
  Dunedin 112872, 132716
Art Care Professional, Ellwangen 075597
Art Carnival, Toronto 064601
Art CC, Saint-Ouen 105564
Art Cellar Exchange, San Diego . 124010
The Art Center, Jeddah 085298
The Art Center, Mount Clemens 051082
d'Art Center, Norfolk 123224
Art Center, Porto Alegre 063956
Art Center at Fuller Lodge, Los
  Alamos 050399
Art Center Berlin Friedrichstrasse,
  Berlin 017075
Art Center College of Design,
  Pasadena 057648
Art Center Gallery, Central Missouri
  State University, Warrensburg . 054069
Art Center In Hargate, Saint Paul's
  School, Concord 047957
Art Center Lavradio, Rio de
  Janeiro 064018
Art Center of Battle Creek, Battle
  Creek 046824
Art Center of Corpus Christi, Corpus
  Christi 048001
Art Center of South Florida, Miami
  Beach 050845
Art Center of Waco, Waco 054013
Art Center Sarasota, Sarasota . 053147
Art Center, National Yunlin University
  of Science and Technology,
  Douliou 042307
Art Center, Queens College,
  Flushing 048763
Art Center, Tainan National University
  of the Arts, Tainan 042483
Art Centr Puškinskaja 10, Sankt-
  Peterburg 037718
Art Centre, Barcelona 114850
Art Centre, Haifa 025042
Art Centre, Kathmandu 112062
Art Cetera Gallery of Memphis,
  Memphis 121680
Art Channel, Sendai 111380
L'Art Chez Soi, Baerenthal ... 066906
Art Chi, Düsseldorf 075453
Art Chicago, Chicago 098424
Art China, Shanghai 138594
Art Chronika, Moskva 139008
Art Circle, Amsterdam . 082686, 112142
Art-Cite, La Chaux-de-Fonds .. 116205
Art Clectic, Reims 105422
Art-Clinique, Bruxelles 128013
Art Clipper, Helsinki .. 066306, 103131
Art Co-Ordinates, Subiaco 062254
Art Collection, Rotterdam 083551
Art Collection, San Francisco .. 124211

Art Collection, Stirling 045870
Art Collection Canada, Toronto . 101567
Art Collection International,
  Amsterdam 082687
Art Collection Roméo,
  Luxembourg 111831
Art Collection, Australian National
  University, Acton 000756
Art Collections, Tulsa 124790
Art Collections Outlet, Miami ... 121730
Art Collective, Minneapolis ... 121964
Art Collective Gallery, San
  Francisco 124212
Art Collector, Pisa 081152
Art Collector, San Diego 124091
Art Collector, Taupo 084094
Art College, Khorram-Abaad University,
  Khorram-Abaad 055850
Art Cologne, Internationaler
  Kunstmarkt, Köln 098128
Art Comes to Life, Milwaukee .. 121891
Art Company, Cincinnati 120315
Art Company, Eindhoven 112452
Art Company, Tshwane 085671
Art Company, Tucson 124711
Art Company, Virginia Beach .. 124821
Art Company the Art Gallery,
  Edmonton 101271
Art Compass, Wellington 113055
Art Complex Museum, Duxbury . 048414
Art Concept, Amsterdam 082688
Art Concept, Enghien-les-Bains . 068443
Art Concept, Nantes . 070446, 104330
Art Concept, Paris 104497
Art Concorde, Paris ... 104498, 136853
Art Conexion, Sabadell 115395
The Art Connection Eton, Eton . 117763
Art Connoisseurs International, New
  York 122342
Art Conseil, Paris 104499
Art Conseil & Expertise, Paris .. 070943
Art Conservation, Küsnacht ... 133785
Art Conservation, Vlaardingen .. 132658
Art Conservation Associates, New
  York 135885
Art Conservation Framers,
  Melbourne 127638
Art Conservation Resource Center,
  Boston 135359
Art Conservation Service, Boston 135360
Art Consultant, Los Angeles .. 121432
Art Consulting, Praha 102647
Art Contact, London 118177
Art Contact, Marlborough 118721
Art Contacts, Maitland 114806
Art Copenhagen, Nordens Kunstmesse,
  Frederiksberg 098065
Art Corner, Roskilde 102963
Art Corporation, London 118178
Art Cottage, Louisville 121630
Art Criticism, Stony Brook 139279
Art Critics' Association, Moskva . 059768
Art Cru Muséum, Bordeaux ... 012107
Art Cuadros, Lima 084438
The Art Cube, Düsseldorf 075454
Art d' Ameublement, Bielefeld .. 106484
Art d'Aimer, Castillon-la-Bataille 067717
L'Art d'Aimer, Craponne (Rhône) 103641
Art Dais, Hyderabad 109352
Art Dallas, Dallas 120549
Art d'Antan, Grenoble 068715
Art Data, London 138191
Art de Asia, San Francisco ... 097180
Art de Basse-Normandie, Caen . 138628
Art de Chine, Nice 070560
Art de la Chine et du Japon,
  Paris 070944
Art de Vivre, Carcassonne 067667,
                              067668
L'Art de Vivre, Hautefort 103809
L'Art de Vivre, L'Isle-sur-la-Sorgue 069553
l'Art de Vivre, New York 095238
L'Art de Vivre, Vevey 087875
Art Dealers Association of America,

Register der Institutionen und Firmen

Art Dealers Association of California, Los Angeles
– Art Department, Fine Arts Division, Concordia College, Saint

New York 060335
Art Dealers Association of California,
Los Angeles 060336
Art Dealers Association of Canada,
Association des marchands d'art du
Canada, Toronto 058419
Art Dealers Association of Chicago,
Chicago 060337
Art Dealers Association of Greater
Washington, Washington 060338
Art Dec, Bourg-Saint-Bernard . . 067417,
128516
Art Deco, Berlin 074757
Art Deco, Bonn 075063
Art Deco, Bruxelles 063326
Art Dèco, Kraków 084559, 084560,
132834
Art Deco, München . . . 077457, 077458
Art Deco, Poznań 113718
Art Deco, Torino 132090
Art-Deco, Warszawa 132872
Art Deco, Wien 062896, 062897
Art-Deco, Zürich 087933
Art Deco – Galerie Mili Vávrové,
Praha 065477
Art Deco – Starožitnosti, Praha . 065478
Art Deco & Jugendstil, Wien . . . 062898
Art Déco 1930–1950, Nice 070561
Art Deco-Art Nouveau, Bruxelles 063327
Art Deco Collections, Oakland . . 095882
Art Deco en zo, Apeldoorn 082918
Art Deco en Zo, Hilversum 083399
Art Deco Français, San Francisco 097181
Art Deco Gallery, Roma 081233
Art-Deco-House, Breda 082973
Art Deco Metropolis, Stuttgart . . 078592
Art Deco Society of New York, New
York 060339
L'Art Décor, Mereglise 070068
Art d'elite, Ortenberg, Baden . . . 077953
d'Art Demain, Broken Hill 098794
Art Department, Kwa Dlangezwa 056348
Art Department, San Diego 124092
Art Department 1072, University of
Puget Sound, Tacoma 057972
Art Department Gallery, Community
College of Rhode Island,
Warwick 054077
Art Department, Abilene Christian
University, Abilene 056621
Art Department, Acadia University,
Wolfville 055177
Art Department, Agnes Scott College,
Decatur 056991
Art Department, Alabama Southern
Community College, Monroeville 057471
Art Department, Alabama State
University, Montgomery 057477
Art Department, Albertus Magnus
College, New Haven 057521
Art Department, Alice Lloyd College,
Pippa Passes 057682
Art Department, Allegany Community
College, Cumberland 056967
Art Department, Allen County
Community College, Iola 057239
Art Department, Alverno College,
Milwaukee 057448
Art Department, Alvin Community
College, Alvin 056648
Art Department, Amarillo College,
Amarillo 056649
Art Department, Ancilla College,
Plymouth 057697
Art Department, Anderson University,
Anderson 056657
Art Department, Angelo State
University, San Angelo 057837
Art Department, Anoka Ramsey
Community College, Coon
Rapids 056952
Art Department, Aquinas College, East
Grand Rapids 057029
Art Department, Art Division, Pacific

University in Oregon, Forest
Grove 057094
Art Department, Ashland University,
Ashland 056670
Art Department, Augsburg College,
Minneapolis 057455
Art Department, Augustana College,
Sioux Falls 057919
Art Department, Aurora University,
Aurora 056690
Art Department, Averett College,
Danville 056979
Art Department, Baker University,
Baldwin City 056698
Art Department, Baldwin-Wallace
College, Berea 056728
Art Department, Barton College,
Wilson 058118
Art Department, Bay de Noc
Community College, Escanaba 057058
Art Department, Belhaven College,
Jackson 057248
Art Department, Belleville Area
College, Belleville 056718
Art Department, Bellevue Community
College, Bellevue 056719
Art Department, Benedictine College,
Atchison 056673
Art Department, Berea College,
Berea 056727
Art Department, Berry College, Mount
Berry 057489
Art Department, Bethany Lutheran
College, Mankato 057403
Art Department, Birmingham-Southern
College, Birmingham 056743
Art Department, Black Hawk College,
Moline 057467
Art Department, Black Hills State
University, Spearfish 057931
Art Department, Bluefield State
College, Bluefield 056758
Art Department, Boise State University,
Boise 056760
Art Department, Brazosport College,
Lake Jackson 057311
Art Department, Briar Cliff University,
Sioux City 057917
Art Department, Bridgewater College,
Bridgewater 056786
Art Department, Bridgewater State
College, Bridgewater 056785
Art Department, Bronx Community
College, Bronx 056789
Art Department, Brookdale Community
College, Lincroft 057348
Art Department, Brooklyn College,
Brooklyn 056796
Art Department, Caldwell College,
Caldwell 056819
Art Department, California Baptist
College, San Bernardino 057846
Art Department, California Lutheran
University, Thousand Oaks . . . 057993
Art Department, California Polytechnic
State University Pomona,
Pomona 057701
Art Department, California State
University, Carson . . 056839, 056840
Art Department, California State
University Hayward, Hayward . 057192
Art Department, California State
University Long Beach, Long
Beach 057357
Art Department, California State
University Los Angeles, Los
Angeles 057364
Art Department, California State
University, Chico, Chico 056892
Art Department, Calvin College,
Kentwood 057285
Art Department, Cameron University,
Lawton 057332
Art Department, Cape Cod Community

College, West Barnstable 058082
Art Department, Capitol Community
Technical College, Hartford . . . 057185
Art Department, Cardinal Stritch
College, Milwaukee 057449
Art Department, Carlow College,
Pittsburgh 057684
Art Department, Carroll College,
Waukesha 058072
Art Department, Carson-Newman
College, Jefferson City 057263
Art Department, Carthage College,
Kenosha 057281
Art Department, Castleton State
College, Castleton 056842
Art Department, Catonsville
Community College, Catonsville 056843
Art Department, Cedar Crest College,
Allentown 056642
Art Department, Central Missouri State
University, Warrensburg 058059
Art Department, Central Piedmont
Community College, Charlotte . 056858
Art Department, Central Wyoming
College, Riverton 057768
Art Department, Chaffey Community
College, Rancho Cucamonga . . 057747
Art Department, Chapman University,
Orange 057623
Art Department, Citrus College,
Glendora 057136
Art Department, City College of New
York, New York 057539
Art Department, Clackamas
Community College, Oregon
City 057627
Art Department, Clarion University of
Pennsylvania, Clarion 056905
Art Department, Cleveland State
University, Cleveland 056912
Art Department, Clinton Community
College, Plattsburgh 057693
Art Department, Cochise College,
Douglas 057017
Art Department, Coffeyville Community
College, Coffeyville 056922
Art Department, Coker College,
Hartsville 057187
Art Department, Colby College,
Waterville 058071
Art Department, Colby-Sawyer College,
New London 057527
Art Department, College of Arts and
Sciences, Governors State University,
University Park 058027
Art Department, College of Lake
County, Gages Lake 057117
Art Department, College of Mount
Saint Joseph, Cincinnati 056896
Art Department, College of Notre
Dame of Maryland, Baltimore . 056700
Art Department, College of Saint
Benedict, Saint Joseph 057806
Art Department, College of Saint
Elizabeth, Morristown 057487
Art Department, College of Saint Mary,
Omaha 057618
Art Department, College of Southern
Idaho, Twin Falls 058018
Art Department, College of the
Canyons, Valencia 058035
Art Department, College of the Desert,
Palm Desert 057642
Art Department, College of the
Siskiyous, Weed 058079
Art Department, Colorado State
University, Fort Collins 057095
Art Department, Columbia Basin
College, Pasco 057650
Art Department, Columbia College,
Columbia 056932
Art Department, Community College of
Allegheny County, Monroeville . 057472
Art Department, Compton Community

College, Compton 056950
Art Department, Concordia College,
Bronxville 056793
Art Department, Concordia College,
Moorhead 057482
Art Department, Concordia College,
Seward 057903
Art Department, Connecticut College,
New London 057526
Art Department, Cornell College,
Mount Vernon 057494
Art Department, County College of
Morris, Randolph 057748
Art Department, Cuesta College, San
Luis Obispo 057861
Art Department, Culver-Stockton
College, Canton 056830
Art Department, Daemen College,
Amherst 056655
Art Department, Dana College,
Blair 056749
Art Department, Davidson College,
Davidson 056981
Art Department, Delta College,
University Center 058018
Art Department, DePauw University,
Greencastle 057154
Art Department, Des Moines Area
Community College, Boone . . . 056762
Art Department, Division of
Communications and Fine Arts,
Grossmont College, El Cajon . . 057041
Art Department, Division of Creative
Arts, Monterey 057474
Art Department, Division of Fine Arts,
Antelope Valley College, Quartz
Hill 057737
Art Department, Division of Fine Arts,
Howard College, Vealmoor . . . 058041
Art Department, Division of Fine Arts,
Limestone College, Gaffney . . . 057116
Art Department, Division of Humanities
and Fine Arts, College of Saint
Francis, Shorewood 057911
Art Department, Doane College,
Crete 056965
Art Department, Dominican College of
Blauvelt, Blauvelt 056750
Art Department, Dominican University
of California, San Rafael 057865
Art Department, Drake University, Des
Moines 057006
Art Department, Earlham College,
Richmond 057756
Art Department, East Central College,
Union 058022
Art Department, East Central
University, Ada 056625
Art Department, East Los Angeles
College, Monterey Park 057475
Art Department, Eastern Arizona
College, Thatcher 057991
Art Department, Eastern Kentucky
University, Richmond 057757
Art Department, Eastern Michigan
University, Ypsilanti 058144
Art Department, Eastern Wyoming
College, Torrington 057998
Art Department, Eastfield College,
Mesquite 057429, 121718
Art Department, Eckerd College, Saint
Petersburg 057827
Art Department, Edinboro University of
Pennsylvania, Edinboro 057035
Art Department, Elmhurst College,
Elmhurst 057049
Art Department, Everett Community
College, Everett 057067
Art Department, Feather River College,
Quincy 057738
Art Department, Fine Arts Division,
College of the Sequoias, Visalia 058049
Art Department, Fine Arts Division,
Concordia College, Saint Paul . 057819

Art Department, Fine Arts Division, Shasta College, Redding 057751
Art Department, Florida Community College at Jacksonville, Jacksonville 057254
Art Department, Fontbonne University, Saint Louis 057809
Art Department, Fort Lewis College, Durango 057024
Art Department, Franklin College, Franklin 057103
Art Department, Franklin Pierce College, Rindge 057761
Art Department, Fresno City College, Fresno 057110
Art Department, Friends University, Wichita 058102
Art Department, Garden City Community College, Garden City 057124
Art Department, Gaston College, Dallas 056972
Art Department, Gavilan College, Gilroy 057133
Art Department, Georgetown College, Georgetown 057130
Art Department, Georgia College, Milledgeville 057446
Art Department, Goshen College, Goshen 057143
Art Department, Goucher College, Towson 058000
Art Department, Graceland University, Lamoni 057318
Art Department, Grambling State University, Grambling 057144
Art Department, Grand Canyon University, Phoenix 057676
Art Department, Grand Rapids Community College, Grand Rapids 057148
Art Department, Grand View College, Des Moines 057007
Art Department, Grayson County College, Denison 056998
Art Department, Green River Community College, Auburn .. 056687
Art Department, Greenfield Community College, Greenfield 057156
Art Department, Guilford College, Greensboro 057157
Art Department, Hagerstown Junior College, Hagerstown 057171
Art Department, Hannibal La Grange College, Hannibal 057179
Art Department, Harold Washington College, Chicago 056877
Art Department, Hartwick College, Oneonta 057621
Art Department, Haskell Indian Nations University, Lawrence 057327
Art Department, Hastings College, Hastings 057188
Art Department, Henry Ford Community College, Dearborn . 056989
Art Department, Hesston College, Hesston 057197
Art Department, Hibbing Community College, Hibbing 057198
Art Department, Highland Community College, Freeport 057109
Art Department, Hiram College, Hiram 057203
Art Department, Holy Names College, Oakland 057602
Art Department, Houghton College, Houghton 057211
Art Department, Housatonic Community and Technical College, Bridgeport 056784
Art Department, Hunter College, New York 057540
Art Department, Huntington College, Huntington 057221

Art Department, Imperial Valley College, Imperial 057230
Art Department, Indiana Wesleyan University, Marion 057411
Art Department, Iowa Wesleyan College, Mount Pleasant 057491
Art Department, Iowa Western Community College, Council Bluffs 056962
Art Department, Jacksonville State University, Jacksonville ... 057253
Art Department, James H. Faulkner Community College, Fairhope . 057073
Art Department, Jamestown College, Jamestown 057261
Art Department, Jefferson Community College, Watertown 058070
Art Department, Jefferson Davis Community College, Brewton . 056783
Art Department, Jersey City State College, Jersey City 057264
Art Department, John Brown University, Siloam Springs 057914
Art Department, Jones County Junior College, Ellisville 057048
Art Department, Kansas State University, Manhattan 057400
Art Department, Kansas Wesleyan University, Salina 057831
Art Department, Kentucky Wesleyan College, Owensboro 057637
Art Department, Kenyon College, Gambier 057123
Art Department, Kingsborough Community College, City University NY, Brooklyn 056797
Art Department, Knox College, Galesburg 057122
Art Department, La Grange College, La Grange 057298
Art Department, La Sierra University, Riverside 057765
Art Department, Lackawanna College, Scranton 057889
Art Department, Lake City Community College, Lake City 057308
Art Department, Lake Erie College, Painesville 057641
Art Department, Lake Tahoe Community College, South Lake Tahoe 057926
Art Department, Lamar University, Beaumont 056716
Art Department, Laney College, Oakland 057603
Art Department, Le Moyne College, Syracuse 057967
Art Department, Lehman College, Bronx 056790
Art Department, Lenoir Community College, Kinston 057289
Art Department, Lewis and Clark College, Portland 057707
Art Department, Lewis and Clark Community College, Godfrey . 057139
Art Department, Lewis-Clark State College, Lewiston 057336
Art Department, Lincoln College, Lincoln 057345
Art Department, Lindenwood University, Saint Charles 057803
Art Department, Linfield College, McMinnville 057381
Art Department, Linn Benton Community College, Albany ... 056634
Art Department, Long Island University – Brooklyn Campus, Brooklyn 056798
Art Department, Lorain County Community College, Elyria ... 057052
Art Department, Loras College, Dubuque 057021
Art Department, Los Angeles City College, Los Angeles 057365

Art Department, Los Angeles Harbor College, Wilmington 058114
Art Department, Los Angeles Valley College, Valley Glen 058039
Art Department, Lourdes College, Sylvania 057966
Art Department, Lower Columbia College, Longview 057359
Art Department, Luther College, Decorah 056993
Art Department, Lutheran Brethren Schools, Fergus Falls 057084
Art Department, Lynchburg College, Lynchburg 057379
Art Department, Macomb Community College, Warren 058058
Art Department, Manchester College, North Manchester 057589
Art Department, Mansfield University, Mansfield 057405
Art Department, Marian College, Fond Du Lac 057092
Art Department, Marietta College, Marietta 057407
Art Department, Mars Hill College, Mars Hill 057413
Art Department, Marylhurst University, Marylhurst 057416
Art Department, Marywood University, Scranton 057890
Art Department, McMurry University, Abilene 056622
Art Department, McPherson College, McPherson 057385
Art Department, Mercer University, Macon 057383
Art Department, Meredith College, Raleigh 057743
Art Department, Merritt College, Oakland 057604
Art Department, Mesa State College, Grand Junction 057146
Art Department, Methodist College, Fayetteville 057082
Art Department, Metropolitan State College, Denver 057001
Art Department, Miami University, Oxford 057639
Art Department, Middle Georgia College, Cochran 056920
Art Department, Midland College, Midland 057443
Art Department, Midway College, Midway 057444
Art Department, Millersville University, Millersville 057447
Art Department, Millikin University, Decatur 056992
Art Department, Minnesota State University, Mankato, Mankato . 057404
Art Department, Minot State University, Minot 057460
Art Department, Miracosta College, Oceanside 057609
Art Department, Mississippi State University, Mississippi State .. 057462
Art Department, Missouri Western State University, Saint Joseph . 057807
Art Department, Mohawk Valley Community College, Utica ... 058031
Art Department, Monmouth College, Monmouth 057468
Art Department, Monroe Community College, Rochester 057774
Art Department, Montana State University Billings, Billings .. 056739
Art Department, Montgomery County Community College, Blue Bell . 056756
Art Department, Morehead State University, Morehead 057485
Art Department, Morningside College, Sioux City 057918
Art Department, Mount Marty College, Yankton 058138

Art Department, Mount Mary College, Milwaukee 057450
Art Department, Mount Mercy College, Cedar Rapids 056847
Art Department, Mount Saint Mary's College, Los Angeles 057366
Art Department, Mount San Antonio College, Walnut 058056
Art Department, Mount San Jacinto College, San Jacinto 057858
Art Department, Mount Vernon Nazarene College, Mount Vernon 057495
Art Department, Muskingum College, New Concord 057520
Art Department, Nassau Community College, Garden City ... 057125
Art Department, Navarro College, Corsicana 056958
Art Department, Nazareth College of Rochester, Rochester 057775
Art Department, Nebraska Wesleyan University, Lincoln 057346
Art Department, New Mexico State University, Las Cruces 057323
Art Department, Normandale Community College, Bloomington 056754
Art Department, North Arkansas College, Harrison 057183
Art Department, North Carolina Agricultural and Technical State University, Greensboro 057158
Art Department, North Central Michigan College, Bay View .. 056714
Art Department, North Hennepin Community College, Brooklyn Park 056803
Art Department, North Idaho College, Coeur d'Alene 056921
Art Department, North Park University, Chicago 056878
Art Department, North Seattle Community College, Seattle .. 057893
Art Department, Northampton County Community College, Bethlehem 056733
Art Department, Northeastern Illinois University, Chicago 056879
Art Department, Northeastern Oklahoma A & M College, Miami 057436
Art Department, Northeastern State University, Tahlequah 057975
Art Department, Northern Kentucky University, Highland Heights .. 057201
Art Department, Northern State University, Aberdeen 056620
Art Department, Northern Virginia Community College, Annandale 056661
Art Department, Northwest College, Powell 057717
Art Department, Northwest Nazarene College, Nampa 057503
Art Department, Northwestern College, Orange City 057624
Art Department, Northwestern Michigan College, Traverse City 058002
Art Department, Notre Dame de Namur University, Belmont ... 056721
Art Department, Oklahoma State University, Stillwater 057954
Art Department, Olivet College, Olivet 057617
Art Department, Olympic College, Bremerton 056781
Art Department, Orange County Community College, Scotchtown 057885
Art Department, Paris Junior College, Paris 057645
Art Department, Park University, Parkville 057647
Art Department, Peace College, Raleigh 057744

Art Department, Penn Valley Community College, Kansas City 057276
Art Department, Peru State Colllege, Peru 057661
Art Department, Piedmont College, Demorset 056997
Art Department, Pierce College, Woodland Hills 058131
Art Department, Pittsburg State University, Pittsburg 057683
Art Department, Plattsburgh State University College, Plattsburgh 057694
Art Department, Plymouth State University, Plymouth 057698
Art Department, Portland Community College, Portland 057708
Art Department, Prairie State College, Sauk village 057880
Art Department, Pratt Community College, Pratt 057718
Art Department, Queens College, Flushing 057091
Art Department, Radford University, Radford 057741
Art Department, Rhode Island College, Providence 057725
Art Department, Richard Bland College, Petersburg 057662
Art Department, Ripon College, Ripon 057762
Art Department, Rivier College, Nashua 057506
Art Department, Roanoke College, Salem 057830
Art Department, Roberts Wesleyan College, Rochester 057776
Art Department, Rochester Community College, Rochester 057772
Art Department, Rockland Community College, Suffern 057960
Art Department, Rocky Mountain College, Billings 056740
Art Department, Roger Williams College, Providence 057726
Art Department, Rogers State College, Claremore 056904
Art Department, Rowan University, Glassboro 057134
Art Department, Sacramento City College, Sacramento 057799
Art Department, Saint Ambrose University, Davenport 056980
Art Department, Saint John's University, Collegeville 056929
Art Department, Saint Louis Community College at Meramec, Saint Louis 057810
Art Department, Saint Mary-of-the-Woods College, Saint Mary-of-the-Woods 057816
Art Department, Saint Mary's School, Raleigh 057745
Art Department, Saint Mary's University of Minnesota, Rochester 057773
Art Department, Saint Olaf College, Northfield 057594
Art Department, Saint Thomas Aquinas College, Sparkill 057929
Art Department, Salem College, Winston-Salem 058124
Art Department, Salem State College, Salem 057828
Art Department, Salisbury State University, Salisbury 057833
Art Department, Salve Regina University, Middletown 057440
Art Department, Sam Houston State University, Huntsville 057225
Art Department, Samford University, Birmingham 056744
Art Department, San Bernardino Valley College, San Bernardino 057847

Art Department, San Diego Mesa College, San Diego 057849
Art Department, San Francisco State University, San Francisco 057853
Art Department, San Jacinto College North, Houston 057212
Art Department, San Juan College, Farmington 057078
Art Department, Santa Ana College, Santa Ana 057867
Art Department, Santa Clara University, Santa Clara 057870
Art Department, Santa Rosa Junior College, Santa Rosa 057875
Art Department, Seattle Pacific University, Seattle 057894
Art Department, Shepherd College, Shepherdstown 057908
Art Department, Sheridan College, Sheridan 057909
Art Department, Shippensburg University, Shippensburg 057910
Art Department, Shorter College, Rome 057794
Art Department, Siena Heights College, Adrian 056627
Art Department, Sierra College, Rocklin 057788
Art Department, Silver Lake College, Manitowoc 057402
Art Department, Simpson College, Indianola 057236
Art Department, Snow College, Ephraim 057055
Art Department, South Carolina State University, Orangeburg 057625
Art Department, South Dakota State University, Brookings 056795
Art Department, South Suburban College, South Holland 057925
Art Department, Southeastern Oklahoma State University, Durant 057025
Art Department, Southern Nazarene University, Bethany 056731
Art Department, Southern Oregon State College, Ashland 056671
Art Department, Southern University in New Orleans, New Orleans ... 057528
Art Department, Southern Utah State College, Cedar City 056845
Art Department, Southern Virginia College, Buena Vista 056813
Art Department, Southwest Baptist University, Bolivar 056761
Art Department, Southwestern College, Winfield 058119
Art Department, Southwestern Oklahoma State University, Weatherford 058077
Art Department, Southwestern University, Georgetown 057131
Art Department, Spoon River College, Canton 056829
Art Department, Spring Arbor College, Spring Arbor 057936
Art Department, State University of New York at Albany, Albany .. 056631
Art Department, State University of New York at Stony Brook, Stony Brook 057958
Art Department, State University of New York College at Cortland, Cortland 056959
Art Department, State University of New York College at Oswego, Oswego 057633
Art Department, Stephen F. Austin University, Nacogdoches ... 057502
Art Department, Sterling College, Sterling 057950
Art Department, Stetson University, De Land 056987
Art Department, Surry Community

College, Dobson 057016
Art Department, Tacoma Community College, Tacoma 057971
Art Department, Taft College, Taft 057974
Art Department, Tallahassee Community College, Tallahassee 057976
Art Department, Tarrant County Junior College, Hurst 057227
Art Department, Taylor University, Upland 058029
Art Department, Temple College, Temple 057986
Art Department, Texarkana College, Texarkana 057990
Art Department, Texas A & M University-Kingsville, Kingsville 057288
Art Department, Thomas More College, Crestview Hills 056964
Art Department, Tougaloo College, Tougaloo 057999
Art Department, Treasure Valley Community College, Nyssa ... 057600
Art Department, Tulsa Junior College, Tulsa 058012
Art Department, Umpqua Community College, Roseburg 057795
Art Department, Union College, Barbourville 056709
Art Department, University of Alabama, Tuscaloosa 058016
Art Department, University of Arkansas, Fayetteville 057081
Art Department, University of Arkansas at Pine Bluff, Pine Bluff 057680
Art Department, University of Central Arkansas, Conway 056951
Art Department, University of Central Florida, Orlando 057628
Art Department, University of Evansville, Evansville 057065
Art Department, University of Findlay, Findlay 057086
Art Department, University of Indianapolis, Indianapolis 057233
Art Department, University of Massachusetts Boston, Boston 056765
Art Department, University of Memphis, Memphis 057421
Art Department, University of Missouri, Columbia 056933
Art Department, University of Nevada Reno, Reno 057753
Art Department, University of North Carolina at Chapel Hill, Chapel Hill 056853
Art Department, University of North Dakota, Grand Forks 057145
Art Department, University of Redlands, Redlands 057752
Art Department, University of Saint Thomas, Houston 057213
Art Department, University of South Carolina at Aiken, Aiken 056628
Art Department, University of Southern Indiana, Evansville 057066
Art Department, University of Southern Maine, Gorham 057142
Art Department, University of Texas at Arlington, Arlington 056667
Art Department, University of Texas at Brownsville and Southmost College, Brownsville 056805
Art Department, University of Texas Pan American, Edinburg ... 057036
Art Department, University of the District of Columbia, Washington 058061
Art Department, University of Vermont, Burlington 056818
Art Department, University of West Georgia, Carrollton 056838
Art Department, University of Wisconsin-Green Bay, Green Bay 057153

Art Department, University of Wisconsin Parkside, Kenosha . 057282
Art Department, University of Wisconsin River Falls, River Falls 057763
Art Department, University of Wisconsin Whitewater, Whitewater 058099
Art Department, University of Wisconsin, Marinette, Marinette 057409
Art Department, Valencia Community College, Orlando 057629
Art Department, Valley City State College, Valley City 058038
Art Department, Vassar College, Poughkeepsie 057714
Art Department, Vermilion Community College, Ely 057051
Art Department, Victor Valley College, Victorville 058045
Art Department, Villa Maria College of Buffalo, Buffalo 056814
Art Department, Vincennes University Junior College, Vincennes ... 058048
Art Department, Virginia Wesleyan College, Norfolk 057583
Art Department, Viterbo College, La Crosse 057294
Art Department, Wabash College, Crawfordsville 056963
Art Department, Waldorf College, Forest City 057093
Art Department, Walla Walla Community College, Walla Walla 058054
Art Department, Washington and Jefferson College, Washington 058069
Art Department, Washington and Lee University, Lexington 057340
Art Department, Wayne State College, Wayne 058074
Art Department, Webster University, Saint Louis 057811
Art Department, Wellesley College, Wellesley 058080
Art Department, Wenatchee Valley College, Wenatchee 058081
Art Department, Wesleyan College, Macon 057384
Art Department, West Liberty State College, West Liberty 058087
Art Department, West Valley College, Saratoga 057878
Art Department, West Virginia State College, Institute 057237
Art Department, West Virginia University at Parkersburg, Parkersburg 057646
Art Department, Western Connecticut State University, Danbury 056977
Art Department, Western Illinois University, Macomb 057382
Art Department, Western Kentucky University, Bowling Green ... 056776
Art Department, Western Montana College, Dillon 057015
Art Department, Western Oklahoma State College, Altus 056647
Art Department, Western Washington University, Bellingham 056720
Art Department, Western Wyoming College, Rock Springs 057785
Art Department, Westfield State College, Westfield 058091
Art Department, Westminster College, New Wilmington 057537
Art Department, Wharton County Junior College, Wharton 058095
Art Department, Wheaton College, Norton 057597
Art Department, Whitman College, Walla Walla 058055
Art Department, Whitworth College, Spokane 057933

Art Department, Wilberforce University,
Wilberforce 058106
Art Department, William Jewell
College, Liberty 057342
Art Department, William Paterson
College, Wayne 058075
Art Department, William Penn
University, Oskaloosa 057632
Art Department, William Woods
University, Fulton 057115
Art Department, Williams College,
Williamstown 058112
Art Department, Wilmington College,
Wilmington 058116
Art Department, Winston-Salem State
University, Winston-Salem 058125
Art Department, Wittenberg University,
Springfield 057942
Art Department, Worcester State
College, Worcester 058134
Art Department, York College,
York 058140
Art Depot, London 118179
Art Depot, Minneapolis 121965
Art Dépôt, Paris 070945, 104500
Art Design, Paimpol 070840
Art Design ans Multimedia Program,
Lansing Community College,
Lansing 057321
Art Design Service, Salt Lake
City 123945
Art Designs, Detroit 120798
Art Development, Detroit 120799
Art Développement ADT, Genève 138045
Art d'Gott, La Paz 100842
Art Diagonal, Internationaler Club
bildender Künstler, Wien 058230
Art Dialogue Gallery, Buffalo . . 119949
Art didacta, Internationale
Sommerakademie für bildende Kunst,
Musik und Theater, Igls 054939
Art Dimensions, Saint Louis . . . 123802
Art Direct, Oakland 123241
Art Direct of Saint Paul, Saint
Paul 123891
Art Directors Club, New York . . . 060340
Art Directors Club für Deutschland e.V.,
Berlin 058728
Art Division, Department of
Humanities, Avila College, Kansas
City 057277
Art Domain Gallery, Leipzig 107807
Art Domaru, Plzeň 102640
Art d'Orient, La Courneuve 069014
Art du Nord, Groningen 112489
Art du Nord, Vence 105822
Art du Temps, Cassis . 067701, 067702
Art du Temps, Cleon-d'Andran . 103597
Art du Temps, Cogolin 068061
L'Art du Temps, La Rochelle . . . 069080,
103902
Art du Temps, Mouriès 070374
L'Art du Temps, Paris . 070946, 104501
L'Art du Temps, Toulon (Var) . . . 073699
Art È, Torino 132091
Art e Dossier, Firenze 138792
Art-e-Fax, Holt, Norfolk 117983
Art e Mobili, Winterthur 133965
Art-East, Kyïv 117117
Art Education, Reston 139280
Art Education College, Jimei University,
Xiamen 055250
Art Education Department, Coahoma
Community College, Clarksdale 056906
Art Education Program, State
University of New York at New Paltz,
New Paltz 057535
Art-Edycja, Bydgoszcz . 113438, 143437
The Art Effect, Otley 118858
Art-Effects, Schwerin 078427
Art Egypte Centre, Casablanca . 082543
Art Eko, Košice 085566
Art Elements Gallery, Milwaukee 121892
Art Emporium, Louisville 121631

Art Emporium, Vancouver 101736
L'Art en Direct, Boulogne-
Billancourt 103448
Art en Maguscule, Montréal . . . 101356
Art en Marge, Bruxelles 100550
L'Art en Mouvement, Paris 104502
Art en Noir, Wien 062899
Art en Nosïalgie, Groningen . . . 083275
Art en Stalles, Pouzac 014983
Art Enables, Washington 124864
Art Encounter, Dallas 120550
Art Encounter, Las Vegas 121334
Art Enterprise, Toronto 128206
Art Equity, Sydney 062276
Art Eritorija, Kaunas 111783
Art Espace Allain-Dehergne,
Nantes 104331
Art Espace Galerie, Blagnac . . . 103399
Art Estampe, Gentilly . 103768, 136818,
140844
Art et Affiche, Nice 140993
Art et Antiques, Loches-en-
Touraine 069605
Art et Antiquités, Schuttrange . . 082387
Art et Architecture, Paris 104503
Art et Art, Marsa 116976
Art et Bois, La Possession 132985
Art et Cadre, Versoix 116689
Art et Cetera, Straelen 108677
Art et Collections, Montélimar . . 070231
Art et Comédie, Paris 136854
Art et Conservation, Toulouse . . 129294
Art et Cristal, Saint-Ouen 073017
Art et Culture, Carpentras 103535
Art et Decoration, Firenze 131110
Art et Décoration, Fougères (Ille-et-
Vilaine) 103751
Art et Découvertes, Saint-Vaast-la-
Hougue 105631
Art et Design, Paris 070947
Art et Elégance 1870,
Luxembourg 111832
Art Et Facts, Edinburgh 117715
Art et Innovation, Vic-Fezensac . 074090
L'Art et la Manière, Lattes 103922
L'Art et la Manière, Reims 072378,
129182
Art et Lithographies, Pontarlier . 105389
Art et Lumière, Digne-les-Bains 103671
Art et Lumières, Vannes
(Morbihan) 105817
Art et Nature, Decazeville 103660
Art et Passion, Hautmont 068798
Art et Passion, Limoges 104023
Art et Passions, Mortagne-au-
Perche 104295, 128816
Art et Patrimoine, Bourbon-
l'Archambault 067407
Art et Patrimoine, Challuy 067768
Art et Poésie, Pagny-sur-Meuse 138629
Art et Réalisme, Le Castellet . . . 103930
Art et Société, Presses Universitaires
de Rennes, Rennes (Ille-et-
Vilaine) 137000
Art et Style, Nice 104357
Art et Tendances, Rouen 105467
Art et Thé, Yèvre-le-Châtel 105884
Art et Troc, Sarrebourg 073415
Art et Veline, Dinard 103688
Art et Zen, Le Lavandou 103954
Art-Etage, Biel 116153
Art Etc, Cape Town 114757
Art Etcetera, Paris 123587
Art Etcetera, Singapore 114475
Art-Etche, Saint-Laurent-de-la-
Barrière 072882, 105541
Art Evasion, Hoenheim 068823
Art-Event Antwerpen, Antwerpen 098012
Art Exchange, Columbus 120471
Art Exchange, London 101332
Art Exchange, San Francisco . . . 124213
Art Exhibition House, Ha Noi . . . 125058
Art Exhibition Room, Yilan 042587
Art Expertise, Paris 070948

Art-Expo, Hattingen 107382
Art Expo, Parow 114808
Art Expo, Viby 103007
Art Expo-Galerie – Deprez-Bellorget,
Paris 104504
Art Export, Pézenas 072128
Art Exposure Gallery, Glasgow . . 117831
Art Express, Miami 121731
Art Expressions, Windsor, Ontario 101849
Art Expressions Gallery, San
Diego 124093
Art Factors, Portland 123588
Art Factory No. 1, Amersfoort . . 082649
Art Factory Outlet, Las Vegas . . 121335
Art Fair Tokyo, Tokyo 098225
Art Fairs International, New York 139281
Art Find Association, New York . 060341
art Finder Galerie, Mathias Güntner,
Hamburg 107223
Art First, London 118180
Art Focus, Singapore 114476
The Art Folio, Highgate, Western
Australia 127616
Art Folio, Kuala Lumpur 111903
Art for All, Chichester 117575
Art for All, Little Hallingbury . . . 118123
Art for Art's Sake, Christchurch . 083854,
112826
Art For Arts Sake, Wellington . . 113056,
113057, 113058
Art for Change, New York 051335
Art for Colour, Byron Bay 098828
Art for Eyes, Glasgow 117832
Art for Offices, London 118181
Art for Sale!, Wien 100227
Art for the Soul, New Orleans . . 122119
Art for You, Minneapolis 121966
Art for Your Office, London 118182
Art Formel, Paris 070949
Art Forum, Antwerpen 100483
Art Forum, Bangkok 116948
Art Forum, México 111944
Art Forum, Milngavie 118745
Art Forum, Singapore 114477
Art Forum – Kunstverein, Graz . 058231
Art Forum Berlin, Die Internationale
Kunstmesse, Berlin 098129
Art Forum Oldenburg, Atelier Volker
Kuhnert, Oldenburg 108334
Art Forum Würth, Capena 025749
Art Foundry Gallery, Sacramento 123754
Art Frame, Jeddah 085299
Art from Art, Gent 100686
Art From Russia, Dallas 120551
Art From the Heart, Portland . . . 123589
Art Front Gallery, Tokyo 111398
The Art Fund, London 060033
Art Fundus Biedermeiermöbel,
München 077459
Art Fusion Gallery, Miami 121732
Art-FX, Cootamundra 098926
Art-Galereja Dača, Moskva 114198
Art-Galereja Dača, Žukovka . . . 114379
Art Galeri Ere, Tiranë 098516
Art Galeri Impression, Tiranë . . . 098517
Art Galeria, Częstochowa 113448
Art Galéria, Liptovský Mikuláš . . 114462
Art-Galerie, Frankfurt am Main . 106970
Die Art Galerie, Hildesheim 107430
Art Galerie, Künzelsau 107771
Art Galerie, La Baule-Escoublac 103860
Art Galerie, Ludwigsburg 107895
Art Galerie, Wolfenbüttel 108957
Art Galerie 7, Köln 107586
Art Galerie 94, Nijmegen 083507
Art-Galerie-Kunstmarkt für Siegen,
Siegen 108615
Art Galerie Richter, Berlin 106154
Art-Galerie-Rudenz, Meiringen . . 116500
Art Galerie Spektrum, Darmstadt 106651
Art-Galerie, Verein zur Förderung der
Kunst e.V., Michaelbeuern . . . 058232
Art Galerija, Čačak 114447
Art Galery, Medellín 102423

Art Galleries and Ceramics, New
York 122343
Art Galleries Association Singapore,
Singapore 059796
Art Galleries at Nanyang Academy of
Fine Arts, Singapore 114478
Art Galleries at UAH, Huntsville . 049632
Art Galleries of Ramapo College,
Mahwah 050605
Art Galleries, California
State University, Northridge,
Northridge 051638
Art Gallery, Amsterdam 112143
Art Gallery, Asunción 113327
Art Gallery, Atlanta 046604
Art Gallery, Avignon 103299
Art Gallery, Bangalore . 079343, 109253
Art Gallery, Edmonton 101272
Art Gallery, Ha Noi 125059
Art Gallery, Haifa 025043
Art Gallery, Ho Chi Minh City . . 125110
Art Gallery, Hyderabad 109353
Art Gallery, Innsbruck 062643
Art Gallery, İstanbul 117003
The Art Gallery, Kuala Lumpur . 030681
Art Gallery, Miami 121733
Art Gallery, Milano 110342
Art Gallery, Mostar 100852
Art Gallery, Portland 123590
Art Gallery, Ruislip 118991
Art Gallery, Stockholm 115835
Art Gallery, Tiranë 098518
Art Gallery, Wiesbaden 108998
Art Gallery 2, Trieste 111042
Art Gallery 3, Singapore 114479
Art Gallery 78, Gdynia 113493
Art Gallery and Framing, Tulsa . 124791
Art Gallery at the Westin Copley,
Boston 119832
Art Gallery Collections, Surfers
Paradise 099733
Art Gallery de la Rivoire, Bourgoin-
Jallieu 103455
Art Gallery Groeninge, Brugge . . 100531
Art Gallery Hillsborough,
Hillsborough 117969
Art Gallery Irena, Ljubljana 114694
Art Gallery Kendall, Miami 050825
Art Gallery Lai, Lenzerheide . . . 116435
Art Gallery Lichtstrahl, Widen . . 116716
Art Gallery Marist College,
Poughkeepsie 052336
Art Gallery Natascha, Knokke-
Heist 100724
Art Gallery New Zealand,
Wellington 033343
Art Gallery Newcastle Region,
Newcastle 099430
Art Gallery of Algoma, Sault Sainte
Marie 006987
Art Gallery of Bancroft, Bancroft 005377
Art Gallery of Calgary, Calgary . 005515
Art Gallery of Fells Point,
Baltimore 119718
Art Gallery of Greater Victoria,
Victoria 007318
Art Gallery of Hamilton, Hamilton 005966,
101311
Art Gallery of Hungkuang University,
Shalu 042443
Art Gallery of Mississauga,
Mississauga 006337
Art Gallery of New South Wales,
Sydney 001535
Art Gallery of Newfoundland and
Labrador, The Rooms, Saint
John's 006897
Art Gallery of North York, North
York 006522
Art Gallery of Northumberland,
Cobourg 005632
Art Gallery of Nova Scotia,
Halifax 005946
Art Gallery of Ontario, Toronto . 007157

Art Gallery of Peterborough,
Peterborough 006634
Art Gallery of Prince Albert, Prince
Albert 006707
Art Gallery of South Australia,
Adelaide 000762
Art Gallery of Southwestern Manitoba,
Brandon 005459
Art Gallery of Sudbury, Sudbury 007091
Art Gallery of Swift Current National
Exhibition Centre, Swift Current 007110
Art Gallery of the College of Fine
and Applied Arts, Sudan University
of Science and Technology,
Khartoum 040491
Art Gallery of the Faculty of Painting,
Sculpture and Graphic Art,
Bangkok 042648
Art Gallery of the Society
for Macedonian Studies,
Thessaloniki 023287
Art Gallery of the South Okanagan,
Penticton 006627
Art Gallery of Western Australia,
Perth 001401
Art Gallery of Windsor, Windsor,
Ontario 007424
Art Gallery of York University, North
York 006523
Art Gallery Sint-Raphaël, Hasselt 100712
Art Gallery Wholesales,
Indianapolis 121164
Art Gallery, Doto Daigaku,
Kitahiroshima 028821
Art Gallery, James Prendergast Library,
Jamestown 049784
Art Gallery, Kingsborough Community
College, CUNY, Brooklyn . 047250
The Art Gallery, NIE, Nanyang
Technological University,
Singapore 038264
Art Gallery, North Hennepin
Community College,
Minneapolis 050933
Art Gallery, State University of
New York College at Oneonta,
Oneonta 051775
The Art Gallery, University of Maryland,
College Park 047869
Art Gallery, University of Texas at San
Antonio, San Antonio ... 052929
The Art Gallery, White Image,
Hillsborough 044585
Art Gallia, Paris 070950
Art Garaža, Beograd 114397
Art Garden Kawasaki, Kawasaki 111275
Art Gate, Manama ... 063133, 100413
Art Gilding, Rosebery 127671
Art Globe Scandinavia, Bromma 115594
Art Gluchowe, Glauchau 018710
Art Gotham, SoHo 124620
The Art Group, Las Vegas 121336
The Art Group, New York 122344
Art Guild of Burlington,
Burlington 060342
Art Guild of Burlington Gallery,
Burlington 047316
The Art Gym, Marylhurst University,
Marylhurst 050728
Art Hall Miraku, Ikoma 028678
Art-Haus, Bad Bertrich 106043
Art Haven, Crosshaven 109559
Art-Helsinki, Helsinki 103132
Art Heritage, Delhi 109303
Art History, Oxford 139137
Art History and Archaeology
Department, University of Missouri,
Columbia 056934
Art History Department, Bowdoin
College, Brunswick 056807
Art History Department, Emory
University, Atlanta 056677
Art History Department, Sweet Briar
College, Sweet Briar 057964

Art History Program, University
of Massachusetts Amherst,
Amherst 056652
Art Holdings, Minneapolis 121967
Art-Home, Singapore 114480
Art Home Déco, Bordeaux 103414
Art Horizon, Chicago 120041
Art House, Camborne 117501
Art-House, Dornbirn 099975
Art House, Hasselt 100713
Art House, Houston 120996
The Art House, Middlesbrough . 118743
Art House Gallery, Paget 100834
Art House Gallery, Peterborough 118906
Art House Gallery, Rushcutters
Bay 099658
Art House Reproductions,
Newstead 099440
Art House Studio, Oklahoma City 123272
Art Image, Atlanta 119536
Art Image, Limoges 104024
Art Image Gallery, Vancouver . 101737
Art Images, Dallas 120552
Art Images Gallery, Norwood .. 099487
Art Impex, Kunsthandel Nora, Baden-
Baden 074630
Art Impressions, Columbus ... 120472
Art in America, New York 139282
Art in Architecture, Marina del
Rey 057408
art in asia, Seoul 138896
Art in Bark Display, Toowoomba 099833
Art in Bulgaria, Sofia 138564
Art in Embassies Program,
U.S. Department of State,
Washington 054089
Art in Fiber, Washington 124865
Art in Florence, Firenze 110028
Art in General, New York 060343
Art in Progress, Courmayeur .. 109993
Art in Progress, Genova 110087
Art in Your Home, Great Coxwell 117879
Art Inc, London 118183
Art Incorporated, San Antonio . 124011
Art Incorporated, Southport ... 119124
Art Index, Amsterdam 112144
Art Index, Milano 138793
Art Index, Sazuka, Mie 142913
Art Indus, Delhi 109304
Art Initiative, Atlanta 119537
Art Inspiration and Polcraft,
Chelmsford 117559
Art Institute of Atlanta, Atlanta . 056678
Art Institute of Boston at Lesley
University, Boston 056766
The Art Institute of Boston Main
Gallery, Boston 047100
The Art Institute of Chicago,
Chicago 047636
Art Institute of Chicago, Publications
Department, Chicago 138316
Art Institute of Colorado, Denver 057002
Art Institute of Dallas, Dallas .. 056973
Art Institute of Fort Lauderdale, Fort
Lauderdale 057096
Art Institute of Houston, Houston 057214
Art Institute of Philadelphia,
Philadelphia 057665
Art Institute of Pittsburgh,
Pittsburgh 052154
Art Institute of Seattle, Seattle . 057895
Art Institute of Southern California,
Laguna Beach 057305
Art Instruction Schools,
Minneapolis 057456
Art Interiors, Toronto 101568
Les Art International,
Johannesburg 114788
Art International Publishers, Paris 136855
Art Internationale, Québec 101486
ART-Internationale Messe für
zeitgenössische Kunst Innsbruck,
Innsbruck 097999
Art Ireland Exhibition, Dublin .. 098176

Art Ireland Summer Exhibition,
Dublin 098177
Art-is, Sofia 101146
Art Is Heaven, Nashville 122063
Art Is In, Miami 121734
The Art is In, Salt Lake City ... 123946
Art is OK Gallery, Albuquerque . 119457
Art Is Vital Gallery, Calgary ... 101212
ART-isotope, Galerie Schöber,
Dortmund 106695
Art Issues, West Hollywood ... 139283
Art Izzi, Rumelange 111874
Art Jalene Galerie, Nantes 104332
Art Jingle, Paris 104505
Art Journal, New York 139284
Art-Junives, Lviv 117153
art kaleidoscope, Kunstmagazin für
Frankfurt und Rhein-Main, Frankfurt
am Main 138672
Art-Kapella, Schkeuditz 108563
art Karlsruhe, Internationale Messe
für Klassische Moderne und
Gegenwartskunst, Rheinstetten 098130
Art Kas, Paris 104506
Art Kennedy, Aalborg 106697
Art Kids, Amsterdam . 059591, 112145
Art Kitchen, Amsterdam 112146
Art-Kollegija, Sankt-Peterburg . 114329
Art-Komitet, Moskva 085173
Art Konstant, Lisboa 113981
Art Konsult, Delhi 109305
Art Kontakt, Český Krumlov ... 128323
Art Lab, Makati 113378
Art Land Indooroopilly,
Indooroopilly 099151
Art League, Indianapolis 060344, 121165
Art League of Houston, Houston 049571,
060345
Art League of Manatee County,
Bradenton 047168, 060346
Art League of Marco Island, Marco
Island 060347
Art League of Rhode Island,
Newport 060348
Art Legacy, Odiham 118848
Art Lending of Ottawa, Ottawa . 101460
Art Libraries Journal, London . 139138
Art Libraries Society of North America,
Kanata 058420
The Art Libraries Society/ UK &
Ireland, ARLIS, London 060034
Art Library, Leeds 044778
Art Libre Edition, Wien 136575
Art Life, Clyst Saint George ... 117605
Art Life Design, Ottawa 101461
Art Limited, Houston 120997
Art-Line, Bredstedt 106571
Art Line, London 139139
Art-Line, Mannheim 077251
Art Line, Singapore 114481
Art Line Gallery, Riyadh 114387
Art Link, Oslo 084295
Art Lithographies, Paris 104507
Art Littoral Galerie, Dinard ... 103689
Art Location, Hildesheim 107431
Art Loft, Port 116571
Art Loft Studio, Birmingham .. 119787
Art Loss Register, Köln 137301
Art Loss Register, London 089924
Art Loss Register, New York .. 138376
Art Loss Register, Sankt-
Peterburg 137929
The Art Lounge, Birmingham .. 117367
Art Lounge, Coogee 098915
The Art Lounge, Siem Reap ... 101207
Art Lovers Gallery, New York .. 122345
Art Lovers League, Baltimore . 060349,
119719
Art-M, Kraków 113526
Art M, Trjavna 101181
The Art Magazine, Néa Psychico 138764
Art-Mah, Genève 116278
Art Makers Studio Gallery,
Heidelberg 099124

Art Man Gallery, Cairo 103017
Art Market, Milwaukee 121893
Art-Market, Piaseczno . 084633, 132855
Art Market, Tulsa 124792
Art Market, Warszawa . 084701, 132873
Art Mart, Omaha 123302
Art Mart, Tampa ... 124627, 124628
Art-Mart.co.uk, Chelmsford ... 117560
Art Masters, Auckland 132682
Art Masters, El Paso 120823
Art Masters, Oklahoma City .. 123273
Art Matters, Makati 113379
Art Media Resources, Chicago . 138317,
145034
L'Art Meme, Bruxelles 138542
Art Metropole, Toronto 007158
Art Mexicano, Columbus 120473
Art Miami, Contemporary and Modern
Art, Miami Beach 098425
Art Miya, Tokyo 111399
Art MLS, Miami 121735
Art Mob, Battery Point 098713
Art Mode Gallery, Calgary 101213
Art Mode Gallery, Ottawa 101462
Art Monaco, Monaco 126266
Art Montage i Stockholm,
Stockholm 115836
Art Montfalcon, Barcelona 114851
Art Monthly, London 139140
Art Monthly, Nanjing 138595
Art MoRo, Neustadt, Wied 108265
Art-Most, Moskva 114199
Art Motif, Delhi 109306
Art Motors, Milano 110242
Art Móveis, Belo Horizonte ... 063880,
128099
The Art Movement, London ... 118184
Art Multi, Paris 136856
Art Multiple Décoration, Meyzieu 070116
Art Mûr, Montréal 101357
Art Museum,
Thiruvananthapuram 024490
The Art Museum at Florida
International University, Miami 050826
Art Museum of Dazu Rock Carvings in
Chongqing, Chongqing 007705
Art Museum of Greater Lafayette,
Lafayette 050059
Art Museum of Hsinchu County,
Jhubei 042347
Art Museum of Los Gatos, Los
Gatos 050477
Art Museum of South Texas, South
Texas Institute for the Arts, Corpus
Christi 048002
Art Museum of Southeast Texas,
Beaumont 046851
Art Museum of the Americas,
Washington 054090
Art Museum of the Chinese University
of Hong Kong, Hong Kong ... 007869
Art Museum of the University of
Memphis, Memphis 050772
Art Museum of Western Virginia,
Roanoke 052601
Art Music Antic, Nice 070562
Art Musings, Mumbai 109414
Art Mystique, Tucson 124712
Art 'n Frame Warehouse, Dallas 120553
Art 'n' Such, Bundaberg 098819
Art-Navat, Almaty 111550
Art Nest, Dubai 088171
Art New England, Brighton ... 139285
Art New Media, Warszawa 113801
Art New Zealand, Auckland ... 138941
Art Newc Project, Kyïv 117118
The Art Newspaper, London ... 138192,
139141
The Art Newspaper, New York . 139286
Art Next – Saga, Vigo 115533
Art Nexus, Bogotá 138606
Art Nexus, Miami 139287
Art Nice Gallery, Hong Kong .. 102051
Art Nocturne Knokke, Internationale

Kunst- en Antiekbeurs, Knokke 098013
Art Noir, Philadelphia 123350
Art North Gallery, Gordon 099090
Art Not Terminal Gallery, Seattle 124480
Art Notes, Santiago de
   Compostela 139028
Art Nova, El Paso 120824
Art-Nova, Toruń 113786
Art-Nova 1, Katowice 113509
Art-Nova 2, Katowice 113510
Art Nova Galeria, Bielsko-Biała . 113429
Art Noveaux, Cairo 103018
Art Novum, Warszawa 113802
Art Now Gallery, Göteborg 115626
Art Now International,
   Johannesburg 114789
Art Nuvo, Buderim 098815
Art och Form, Stockholm 115837
Art of Animation, Adelaide 098608
Art of Deco Galerie, Alkmaar .. 082615
Art of Eastern Europe, Denver .. 120690
Art of Facts, Houston 120998
Art of Hands, Long Beach 121393
Art of History, Cincinnati 120316
Art of Life, Lindau 107872
The Art of Michael Lees,
   Ellingham 117748
Art of Money, Portland 096361
Art of Persepolis, New York 095239
Art of the 20th Century, New
   York 098426
Art of the Past, New York 095240
Art of the Times, West Palm
   Beach 139288
Art of the West, Minnetonka .. 139289
Art of Velvet Gallery De Stierenstal,
   Montfoort 032657
Art-Office, Mouvaux 104310
Art Office – Art Lab, Dortmund . 106696,
                               137143
Art Omi, Ghent 049093
Art Omi, New York 051336
Art On 5th, Austin 119637
Art on Aplin, Cairns 098834
Art on Argyle, Picton 099566
Art on Cairncross, Maleny 099260
Art on Canvas, Saint Louis 123803
Art On Line Sociéty, Tours 058590
Art on Paper, New York 139290
Art on Paper, Praha 140514
Art on the Avenue, Detroit 120800
Art on the Edge, Philadelphia .. 123351
Art on the Edge, West Hollywood 124985
Art on the Wall, Paisley 118880
Art on Wheels, Göteborg 115627
Art One, Toronto 101569
Art One Gallery, Santa Monica . 124468
Art One Studio, Hong Kong 102052
Art Options, San Francisco 124214
Art or Not, Bath 117297
Art Original, Bristol 117436
Art Osaka, Osaka 098226
Art Outlet, Minneapolis 121968
Art Pace, San Antonio 052930
Art Pack, Makati 113380
Art Palm Beach, West Palm
   Beach 098427
Art-Pap, Gdańsk 113456
Art Papers, Atlanta 139291
Art Partner Galerie, Champigny-sur-
   Marne 103567
Art-Passepartout, Berlin 074758
Art Passion, Deauville 103658
Art Passion, Montpelier 104269
Art Passion, Mulhouse 070384
The Art Place, Rotorua 112998
Art' Place – Galerie d'Expression,
   Lille 103999
Art Place Gallery, Dorrigo 098974
Art Placement, Saskatoon 101540
Art Placement International, New
   York 122346
Art Plakat, Braunschweig 106557
Art-Plakat, Oldenburg 108335

Art Plaza, Osaka 111346
Art Pluralis, Paris 104508, 128906
Art Podvalčik, Chabarovsk 036948
art Position, Wien 100228
Art Poster Gallery, Zürich 116745
l'Art pour Art, Tucson 124713
Art Praha, Praha 102648
Art Premier, Bruxelles 100551
Art Premium, Revista de Arte
   Contemporáneo, San Juan . 139000
Art Présent, Cannes 103502
Art Présent, Paris 104509
Art Preservation Services, New
   York 135886
Art Preserve, Saint Paul 123892
Art Preserve, Saskatoon 101541
Art-Press, Hajdúszoboszló 138770
Art Press, Paris 136857
Art Print Gallery, Madrid 115144
Art Print Japan, Tokyo 111400, 111401
Art Print Verlag, Saalbach 136546
Art Prints, Mayland 118729
Art Program, Maui Community College,
   Kahului 057272
Art Program, Pfeiffer University,
   Misenheimer 057461
Art Program, Seton Hill University,
   Greensburg 057161
Art Program, Trinity College,
   Washington 058062
Art Program, Tyler Junior College,
   Tyler 058020
Art Program, University of Saint Mary,
   Leavenworth 057333
Art Projects, London 118185
Art Projects International, New
   York 122347
Art Prom PS 88, Alkmaar 112101
Art Promotion Gallery, Bruxelles 100552
Art Proofing Gallery, Paris 136858
Art-Propozycja, Poznań 113719
Art Provenance, Chicago 120042
Art Public Contemporain, Paris . 104510
Art-Public.com, La Varenne-Saint-
   Hilaire 138630
Art Quarterly, London 139142
Art-Ray, München 130324
Art Reach, Jacksonville 121233
Art Reference Collection,
   Westport 139292
Art Reference Services Quarterly,
   Binghamton 139293
Art Reflex, Den Haag 083056
Art Rent, Tromsø 084380
Art Rental & Sales, Winnipeg .. 101864
Art Republic, Bangkok 116949
Art Resource Gallery, Columbus 120474
Art Resources, Tulsa 124793
Art Resources, West Hollywood . 097872
Art Resources Gallery,
   Minneapolis 121969
Art Resources Gallery, Saint Paul 123893
Art Resources International,
   Washington 060350
Art Restauration Technique,
   Marseille 128742
Art-Restauro, Valencia 133445
Art Restoration, Quakers Hill .. 127667
The Art Restoration Studios,
   Haslemere 134443
Art Restorations, Dallas 135492
Art Retreat, A Private Museum,
   Singapore 038265
Art Review, London 139143
Art Room, Lengnau (Aargau) .. 116432
The Art Room, Nottingham 118837
The Art Room, Taunton 119210
Art Room Gallery & Print Bin, Hyde
   Park 099150
The Art Room in Beatties,
   Wolverhampton 119379
Art Room No. 9, Eastbourne .. 117709
The Art Room, Castle Galleries,
   Solihull 119108

Art Route Studio, Camps Bay .. 085628
D'Art Royale, Québec 101487
L'Art Rue dans l'Art, Beauvais .. 103357
Art Saint Louis, Saint Louis 123804
Art Sales Index, Egham 138158
Art Salon, Bangsar Baru 111900
Art Salon, Chiba 111241
Art Salon, Kaliningrad 114182
Art Salon, Long Beach 121394
Art Salon, Tbilisi 105895
Art Salon v Moskovskom Muzeje
   Sovremennogo Iskusstva,
   Moskva 114200
Art Sans Frontieres, Carlton,
   Victoria 061285
Art Sans Frontieres, Kew 061682
Art Santa Fe, An International
   Contemporary Art Fair, Santa Fe 098428
Art Sation, Stone Mountain 053544
Art Sauf, Lyon 069694
Art Savvy International, Fort
   Worth 120854
Art Scene China, Shanghai 102168
Art Scene Sacramento,
   Sacramento 123755
Art Scene Warehouse, Shanghai 102169
Art Scenic, Cagliari 131060
Art-Scenic, Le Faou 069265
Art School, Atlanta 056679
Art School Maa, Helsinki 055340
Art School, Education Alliance, New
   York 057541
Art Search, London 118186
Art Seasons, Beijing 101930
Art Seasons, Jakarta 109506
Art Seasons, Singapore 114482
Art Seek, Appledore, Devon ... 117243
Art Seiller, Saint-Paul-de-Vence . 105573,
                               105574
Art Sentral Asia, Singapore 059797
Art Seoul, Seoul 111611
Art Service, Paris 104511
Art Services International,
   Houston 060351
Art-Servis, Moskva 114201
The Art Shed Gallery, Red Hill .. 099611
The Art Shop, Abergavenny 117217
Art Show, Atlanta 119538
Art Signum, Žilina 114671
Art Singel 100, Amsterdam 112147
Art Sintez, Kyїv 117119
Art Site of Chai-Yi Railway Warehouse,
   Chai-Yi 042275
Art-Sites Press, San Francisco . 138453
Art Society Gallery, Queanbeyan 099602
Art Society of India, Mumbai ... 059356
Art Solutions, Tulsa 124794
Art Source, Denver 120691
The Art Source, Los Angeles ... 121433
Art Source, Phoenix 123480
Art Source, San Francisco 124215
Art Source, Tulsa 124795
Art Source and Design, Charlotte 119974
Art Source and Design, Oklahoma
   City 123274
Art Sources, Miami 121736
Art Space, Milano 110243
Art Space 111, Fort Worth 120855
Art Space Gallery, Saint Ives,
   Cornwall 119017
Art Space Seoul, Seoul 111612
Art Spectrum New York, New
   York 122348
Art-Spektor, Sankt-Peterburg ... 085238
Art Spirit, New York 122349
Art Stall Gallery, Seattle 124481
Art Stock Books, Stuttgart 137516,
                               142381
Art Stop, Denver 120692
Art-Store, Arnhem 082931
Art Store, Dublin 109573
Art Store, Indianapolis 121166
Art Store Gallery, Pittsburgh ... 123521
Art Stuc, Barcelona 133094

Art Student Showcase, New York 122350
Art Students League News, New
   York 139294
The Art Studio, Beaumont 046852
Art Studio, Mostar 100853
The Art Studio, Newcastle,
   Staffordshire 118781
Art Studio, Podgorica 112026
Art Studio, Tauranga 113020
Art Studio, Vaduz 111780
Art Studio, Westport 109706
Art Studio 1, Deinste 106663
Art Studio 56, Winnipeg 101865
Art Studio Magazine, San José . 138608
Art Studio Muca, Tiranë 098519
Art Studio Yzeiri, Tiranë 098520
Art Studios International,
   Hawthorn 099115
Art-Styl, Warszawa 084702
Art Sud, Toulouse 105722
Art Sun, Ha Noi 125060
Art Symbol, Paris 104512
Art Tableaux, Collonges-au-Mont-
   d'Or 068079
Art Tageblatt, Esch-sur-Alzette . 111825
Art Taipei, Taipei 098338
Art Take Away, Amriswil 116021
Art Talks Gallery, Singapore ... 114483
Art Taller la Glicina, Barcelona . 133095
Art Tank, New York 122351
Art-Tech Exhibitions, Silicon Valley
   Institute of Art Technology, San
   Jose 053036
Art Tecnica, Valencia 133446
Art Temple, San Francisco 097182
Art-Tendence, Kehl 076739
Art Territory, West Hollywood .. 124986
Art-Thé, Antibes-Juan-les-Pins . 103253
Art Thema Gallery, Bruxelles ... 100553
Art Times, Saugerties 139295
Art to Order, Dublin 109574
Art To Treasure, Willingham ... 119352
Art Today, Delhi 109307
Art-Top, Montreux 116509
Art Total, Köln 076848
Art Touch Collection, Greenwich 120915
Art Trade Center, Bucureşti 114126
Art Trading, Zaandam 083694
Art Trak, Dallas 120554
Art Treasures, Chicago 120043
Art Treasures Gallery, Hong Kong 065008
Art Trend, Salzburg 100154
Art Trend Gallery, Singapore ... 085366
Art Tribal, Paris 138631
Art Tribal Himalaya, Nevers 070532
The Art Trust, Woollahra 062456
The Art Trust – Fine Art Rental, South
   Melbourne 099700
Art U, Osaka 111347
Art U, Tokyo 111402
Art und Bijou, Köln 107587, 137302
Art und Form, Bad Soden-
   Salmünster 074602
Art-Uni, Lučenec 085577
Art Unique, Lichfield 118110
Art Unlimited, Alkmaar 137820
Art Unlimited, Neustadt an der
   Weinstraße 108255
Art Unlimited, New Orleans 122120
Art Unlimited / Friends of Art,
   Amsterdam 112148
Art Unlimited / Twee van de Kaart,
   Amsterdam 112149
Art Unlimited / Van de Kaart,
   Amsterdam 112150
Art Up Déco, Lyon 104062
Art Up Déco, Paris 104513, 104514
art US, Los Angeles 139296
Art USA, Dallas 120555
Art USA Gallery, Austin 119638, 119639
Art Valeur, Helsinge 102825
Art Vance, Cuajimalpa 138905
Art Vegas, Las Vegas 121337
Art Vento, Berlin 129467

Art Ventures, Charlotte 119975
Art Verneuil, Paris 104515
Art Verre, Eindhoven 112453
Art-Via Investment, Den Haag .. 112373
Art-Vie, Paris 104516
Art Vietnam Gallery, Ha Noi 125061
Art Village Gallery, Memphis 121681
Art Vision Enterprise, Singapore 114484
Art Visions, Las Vegas 121338
Art Vitam, Miami 121737
Art Vue, New York 122352
Art Walk, Mumbai 109415
Art Walking Berlin, Christel Zelinski,
Berlin 106155
Art Warehouse, Miami 121738
Art Werks, Portland 123591
Art West Gallery, Community Art
Association, Jackson 049761
Art Window, Kristianstad 086788
Art Wise, New York 122353
Art Wood, Reims 072379, 129183
Art Work, Nindorf 077789
Art Work, Zürich 116746
Art Work Australia, Hawthorn
East 061607
Art Workers Guild, London 060035
Art Works, Amsterdam 112151
Art Works, Christchurch 112827
Art Works, Minneapolis 121970
Art Works, Newcastle-upon-Tyne 118783
Art Works, Vancouver 101738
Art Works Gallery, Norfolk 123225
Art Works Paris, Paris 070951
Art Workshop, Lemesos 102545
Art World, Chennai 109274
Art World, Christchurch 112828
Art World, Edmonton .. 101273, 101274
Art World, Osborne Park 099505
Art World, San Diego 124094
Art X Art, Milano 110244
Art-X Galéria, Budapest 109125
Art y Antique Antiguedades,
Montevideo 091988
Art Zone, Dunedin 083920
Art Zone, København 065845
Art Zone, New Orleans 122121
ART(212) Contemporary Art Fair, New
York 098429
Art, Antiques and World Imports,
Cincinnati 092826
Art, Das Kunstmagazin, Hamburg 138670
Art, Hamilton College, Clinton .. 056918
Art, Media, Theatre Department,
Vanier College, Saint Laurent . 055154
Art, Music & Theatre Department,
Armstrong Atlantic State University,
Savannah 057881
l'Art, l'Objet, le Goût, Paris 104487
Art.es, Madrid 139029
art.fair21, Messe für aktuelle Kunst,
Köln 098131
Art.M., Warszawa 113803
Art.Metz, Exposition d'Art
Contemporain, Metz 098070
Art.tm, Inverness 118020
Art+Auction Magazine, New York 139427
Art2rent, Innsbruck 100039
Art3, Punta del Este 119437
art4business, London 118187
art4business, Philadelphia 123352
art4public, Elsbethen 099986
Artä, Bucureşti 139002
Arta-Mangon, Bruxelles 063328
Artacasa, Amsterdam 112152
Artache, Milano 110245
Artacraft, Minneapolis 121971
Artact Art, Artarmon 061056
Artadentro, Faro 113968
ArtAe, Leipzig 107808
ArtAffair, Regensburg 108456
Artaffect, Groningen 083276
D'Artagnan, Buenos Aires 139552
d'Artagnan, Eindhoven 112454
Artagor, Caen 103481

Artagra, Grimaud 068746
Artakut, Beverungen .. 106478, 137107
Artal, Strasbourg 105677
Artal Borruel, Jose, Zaragoza .. 133491
Artalia, Palermo 110589
Artamation, Milwaukee 121894
Art'an, Aix-en-Provence 066453
Artandgallery, Milano 059393
Artane Restoration, Dublin 130847
Artani, Stuttgart 078593
Artanne, Sittard 112675
Artantic, Montpellier 070291
Artantique, Brasília 063930
Artareal, Timişoara 085130
Artaria & Co., Wien 062900
Artarmon Galleries, Artarmon .. 000805
Artas Folklore Museum,
Bethlehem 034157
Artasia, Milwaukee .. 094832, 121895
Artasia, San Francisco 097183
Artasia Gallery, Newtown, New South
Wales 099448
ArtAsiaPacific, New York 139298
Artatelier, Bucureşti 139003
Artaud, Jean-Pierre, Marseille .. 069903
Artaut, Myriam, Albi 066507
Ar'Tawoo, Guérande 103801
Artax Kunsthandel, Düsseldorf .. 106760
Artazart, Paris 104517
Artbank, Rosebery 099652
Artbeat, Paraparaumu 112986
Artbibliographies Modern, Oxford 138269
Artbox, Mattersburg 002370
Artbox Frankfurt, Frankfurt am
Main 106971
Artbox Studio, Adelaide 098609
Artbubble, Sant Cugat del Vallès 115407
Artbuero Berlin-Galerie, Berlin .. 106156
The Artbunker, Auckland 112747
Artcade Gallery, Bruxelles 063329
Artcadre, Saint-Macaire 105544
Artcentr+, Moskva 114202
Artchet, Paris 128907
Archivum, Budapest 023458
Artcite Hot, Windsor, Ontario .. 101850
Artclaire, Paris 136859
Artclass, Kraków 132835
Artclub, Lyon 069695
ArtClub Błędne-Koło, Kraków ... 113527
ArtClub Gallery, Paris 104518
Artclub2, Roma 081234
Artco, Leeds 118086
Artco, Lima 113343
Artcol, Montrouge 070334
Artcom Puces de Reims & Nancy,
Reims 072380
Artcompaz, Frederiksberg 102802
Artconservation, Firenze 131111
Artcor Barcelona, Barcelona ... 133096
Artcore, Toronto 101570
Artcothèque, Granville 103777
Artcour, Paris 070952
Artcube, Paris 070953
Artcurial, Briest – Poulain – F.Tajan,
Paris 125674, 136860
Artdanish.dk, Frederiksberg 102803
Artdata Verlag, Ascheberg 137032
ArtDesign Feldkirch, Feldkirch . 098000
Artdirekt Galerie, Bern 116119
Artdomain, Glenside 099087
Arte, Budapest 109126, 126149
L'Arte, Catania 109946
Arte, Coimbra 113961
Arte, Milano 138794
Arte, Montevideo 119417
De Arte, Nantes 104333
Arté, Roma 081235
Arte, San Marino 133026
Arte, Tokyo 082135
De Arte, Valladolid 115521
Art'e', Villanova di Castenaso .. 137765,
138795
Arte, Witten 079030
Arte & Antiguidades, Porto Alegre 063957

Arte & Arte, Palermo 110590
Arte & Cornice, Milano 138796
Arte & Molduras, Fortaleza 100933
Arte & Vicewersa, Trento 111026
Arte 10, Madrid 115145, 137982
Arte 43, Madrid 115146
Arte 63, Bogotá 102273
Arte '800, Roma 081236
Arte 9, Padova 110556
Arte 92, Milano 110246
Arte Afrocaribeña, Amsterdam .. 082689
Arte Al Borgo, Palermo 110591
Arte al Día Informa, Buenos
Aires 138486
Arte al Día Internacional, Miami 139299
Arte al Limite, Vitacura 138592
Arte Americas, Fresno 120892
Arte Andino, Bogotá . 102274, 102275
Arte Antica, Buenos Aires 060876,
127511
L'Arte Antica, Livorno . 080314, 131345
Arte, Antica, Messina 080360
Arte Antica, Napoli 080840
Arte Antica Rufini, Roma 081237
Arte Antiga, Belo Horizonte ... 063881
Arte Antiga, Lisboa 084830
Arte Antiquariato Andora, Andora 079743,
079744
Arte Aperta, Mendrisio 087670, 144111
Arte Argentino, Buenos Aires . 098546
Arte Armanti, Varese 111075
Arte Bello, Cali 102401
Arte Bersani, Finale Ligure 110021
Arte Blanco, San José 102481
Arte Cafe, Medellín 102424
Arte Canarias, Las Palmas de Gran
Canaria 115372, 115373
Arte Capital, Courmayeur 109994
Arte Casa, Genova 110088
Arte Centro Due, Milano 110247
Arte Club, Genova 080237
Arte Colonial, Arequipa 084421
Arte Colonial, Bogotá 065182
El Arte Colonial, Lima 084439
Arte Colosseo, Roma 110728
Arte Contemporanea e Dintorni,
Como 109968
Arte Contemporanea Italiana,
Milano 110248
Arte Contemporanea Maiocchi,
Milano 110249
Arte Contemporanea Monopoli,
Pavia 110637
Arte Contemporaneo,
Bucaramanga 102376
Arte Creativo, Bogotá . 102276, 102277
Arte Cristiana, Rivista Internazionale di
Storia dell'Arte e di Arti Liturgiche,
Milano 138797
Arte Cuadro, Néa Erythraia ... 109076
Arte da Vinci, Allershausen 074378
Arte de Colección, Bogotá 102278
Arte de Relojes, Madrid 086124
Arte de Vivir, Barcelona 139031
L'Arte del Restauro, Chiasso ... 087367,
133720
Arte del Restauro, Roma 081238, 131815
Arte del Siglo XX, Madrid 115147
Arte do Objeto, Brasília 128103
Arte Domus, Kortrijk 100771
Arte DOS Gráfico, Bogotá 136748
Arte Due, Milano 110250
Arte e Antichitá, Verona 081894
Arte e Antiguidade, Rio de
Janeiro 064019
Arte e Antiquariato, Bologna ... 079816
Arte e Antiquariato Barbara Viola,
Castelfranco Veneto 109937
Arte e Archeologia. Studi e documenti,
Firenze 138798
Arte e Arte, Bologna 109811
Arte e Estúdio, Lisboa . 113982, 113983
Arte e Naturalezza, Palermo ... 110592

Arte e Personae, Firenze 110029, 110030
Arte e Restauro, Padova 131653
Arte e Restauro, Palermo 131683
Arte e Restauro del Vetro, Roma 131816
Arte e Stile, Locarno 087596
Arte e Tradizione, Catania 080021,
131080
Arte e Vetro, Venezia 111081
Arte em Dobro, Rio de Janeiro . 101010
Arte En Tela, México 111945
Arte Esposito, Torino 110949
Arte et Opera, Ajaccio . 066493, 140760
Arte Extremeña, Badajoz 114841
Arte Fato Galeria, Porto Alegre . 100965
Arte Fino Barroco, Guadalajara . 111939
Arte Florentia, Palermo 110593
Arte Galeria, Fortaleza 100934
Arte Galeria, Santa Cruz de
Tenerife 115418
Arte Galicia, Revista de Información
de las Artes Plásticas Gallegas,
Ferrol 139032
Arte Gallery, Cleveland 120397
Arte Garduño, Sevilla 115445
Arte Giani, Frankfurt am Main . 106972
Arte Gráfico en Madrid, Madrid . 139033
Arte Heraldica, Rio de Janeiro .. 064020,
064021
El Arte Hispano, Ciudad de
Panamá 113312
Arte Ignorante, Palma de
Mallorca 115334
Arte Imagen, A Coruña 115028
Arte in, Venezia 138799
Arte in Cornice, Roma 131817
Arte in Movimento, Roma 110729
Arte Incontro in Libreria, Milano 138800
Arte Inversión, Guatemala 109115
Arte Italiano, Bucaramanga ... 102377
Arte La Quimera, Caracas 097943
Arte Lancia, León 115107, 115108
Arte Las Americas, Houston ... 120999
Arte Lombarda, Cesano Maderno 138801
Arte Madrid, Madrid 133300
Arte Maya Tz'utuhill, San
Francisco 060352, 124216
Arte Medievale, Cinisello
Balsamo 138802
Arte Milano, Milano 080420
Arte moderna e contemporanea in
Fiera, immagina, Reggio Emilia 098182
L'Arte nel Tempo, Bari 130904
Arte nel Tempo, Bologna 079817
Arte Núcleo, México 111946
Arte Padova, Padova 098183
Arte Paris, Santiago de Chile ... 101901
Arte Periférica, Lisboa 113984
Arte Persiana, Milano 080421
Arte Pictoris, Medellín 102425
Arte Piú, Catania 109947
Arte Piú, Ferrara 110018
Arte Plus, Frankenthal, Pfalz ... 075762
Arte Povera, Reggio Calabria ... 131785
Arte Primitivo, New York 095241
Arte Publico Press, Houston ... 138339
Arte Religioso Venecia, Bogotá . 065183
Arte Religioso y Comtemporaneo,
Medellín 102426
Arte Reyes, Granada 086034
Arte Sacra Antiguidades, Belo
Horizonte 063882
Arte Salutari, Lisboa 084831
Arte Santana, Las Palmas de Gran
Canaria 115374, 115375
Arte Santerasmo, Milano 110251
Arte Serrano 207, Madrid 115148
Arte Sol, Solothurn 116647
Arte Studio, Verona 111175
Arte Til, Fortaleza 100935
Arte Tiless, Zaragoza 115550
Arte Topaz, Barcelona 114852
Arte Unico, Los Angeles 094260
Arte Universale, São Paulo 064153
Arte Vária, Coimbra 113962

Arte Veintiuno, Madrid 115149
Arte Veneta, Venezia 111082
Arte Verso, Genova 110089
Arte Villa, Milano ... 080422, 131372
Arte Vinci, Bucaramanga 102378
Arte Vita Stil', Moskva 114203
Arte Viva, Levallois-Perret 103994
Arte Viva Kunst und Objekt,
  Dietramszell 075356
Arte-Wuerzburg, Würzburg ... 108973,
                             137560
Arte X Arte, Buenos Aires 098547
Arte Xerea, Valencia 115475
Arte y Antigüedades, Lima 084440
Arte y Artes, Coyoacán 138906
Arte y Color, Bogotá 102279
Arte y Diseño, Málaga 115296
Arte y Folklore, Guatemala 109116
Arte y Formas, Bucaramanga ... 102379
Arte y Marquetería la 35,
  Medellín 102427
Arte y Parte, Santander 139034
Arte y R, Málaga 115297
Arte y Reliquia, México 132363
Arte y Representaciones, Málaga 115298
Arte y Restauración, Barcelona . 133097
Arte y Restauracion de Antigüedades,
  Bogotá 065184, 128260
Arte y Ritual, Bogotá 065185
Arte y Ritual, Madrid 115150
Arte, Individuo y Sociedad,
  Madrid 139030
L'Arte, Luigi, Napoli 131597
Artea Centr, Varna 101184
Arteacontrario, Modena 110491
Arte'Alp, Leysin 116438
Artearredo, Roma 081239
Arteasel, Winnipeg 101866
Arteatelier, São Paulo 101083
Arteba, Zürich 116747, 144161
Artec, Denia 085962
Artec Consulting, Bondi Junction 127564
Artecarta, Torino .. 081649, 132092,
                             142846
Artecasa, Ciudad Real . 085935, 115022
ArteClasica, Buenos Aires 097986
Arteclio, Pamplona ... 086380, 086381
artecomm, Ebbs 099981
Artecomm, Ebbs 062550
Artecontexto, Madrid .. 139035, 139036
Artecuadro, Madrid 139037
Artecum, Dardilly 103654
Arteder, Irún 137979
Arteemu, Helsinki 103133
ArteF-Galerie für Kunstfotografie,
  Zürich 116748
ARtéFACT, Cannes 103503
Artefact, Den Haag 112374
Artefact, Oosterbeek 132603
Artefact Gallery, Saint Kilda,
  Victoria 099665
Artefacts, Bangalore 079344
Artefacts, Hobart 099130
Artefacts Conservation, Auckland 132683
Artefactum, Wiesbaden 108899
Artefakt, Gdańsk 113457
Artefakt, Karlsruhe 076685
Artefakt, Linz 062712
Artefakt, Oslo 084296
Artefakt Design, Lyon . 069696, 069697
Artefatto, Monza 110512
Artefatto, Palermo 131684
Artefigura, Milano 110252
Artefino, Cleveland 120398
Artefiz – Kunsthalle Zeitgenössischer
  KüntlerInnen, Zürich 116749
Arteforum, Bologna 109812
Artegma, Ciudad de Panamá ... 113313
Artegrafica, Helsinki ... 103134, 128408,
                             136790
Artegrafica, Verona 137760
Artegrafica Silva, Collecchio .. 137639
Artegrafica Sociale, Cittadella .. 137638

Artegres, Cali 065244
Arteguia, Madrid 139038
Arteide, Palermo ... 081017, 131685
Arteidea, Milano 080423
Arteincontro, Cesena 109959
Arteincornice, Torino 110950
Arteingenua, Brescia 059394
Arteinvest, Roma 110730
Artekon, Helsinki 128409
Artel Galéria, Budapest 109127
Artelab, Roma 131818
Arteleku, Donostia-San Sebastián 056363
Artelheiras, Lisboa 113985
Das Artelier, Bremen 106573
Artelier, Milano 131373
Atelier d'Artistes Crillon, Avignon 103800
Atelier Tatort Spittelberg, Wien . 100229
Artelis, Rīga 111754
Artelite, Milano 110253
Artem, Paris 141030
Artem, Quimper 105411
Artem Galleria, Palermo 110594
Artemar, Nice 070563
Artemarco, Viña del Mar 064995
Artemare, Roma 081240
Artemartell, Bogotá ... 065186, 128261
Artemercato, Palermo 110595
Artemesia, Aix-en-Provence ... 103190
Artemesia, Alresford 088256
Artemida, Moschheim 077421
Artemide, Verona 081895
Artemis, Atlanta 092109
Artemis, Donostia-San Sebastián 133242
Artemis, Hong Kong 065009
Artemis, İstanbul 117004
Artemis, Kraków 113528
Artemis, Liverpool 118125
Artemis, London 118188
Artemis, Miami 121739
Artemis, München 077460
Artemis, Strasbourg 073570
Artemis Archery Collection,
  Oldland 045436
Artemis Gallery, Richmond 123719
Artemis Gallery, Seattle 124482
Artemisa – Arte Antica, Madrid . 086125
Artémise, Dinard 103690
Artemise, Napoli 080841
Artemisia, Buenos Aires 098548
Artemisia, Falconara Marittima . 059395,
                             110013
Artemisia, Madrid 115151
Artemisia, Paris 104519
Artemisia Gallery, Chicago 047637
Artemision e.V., Görlitz 058729
Artemon, Zeuthen ... 079142, 137568
Artemundi, Miami 121740
Artemus, Portland 096362
Artemusa, Bogotá ... 102280, 102281
Arten, Paris 104520
Arten Galéria, Budapest 109128
ArteNa, Waltenschwil 116711
ARTénim, Grenoble 098071
Artenostrum, Dieulefit . 068301, 103667
Artenova, Cossato 109992
Artense, Amersfoort 082650
Artenvielfalt, München 137388
Artenz, Waikato 113034
Arteoggi, Livorno 110172
L'Arteologo – Piera di Bellia,
  Roma 081241
Arteology, Colne 117620
ARTeon, Poznań 138953
Artepe, İstanbul 117005
Arteranzoni, Milano 111374
Arterestauro, Firenze . 131112, 131113
Arteria, Groningen 112490
Artéria Galéria, Szentendre ... 109214
Arterial Gallery, Parkside 099552
Arteries, Glasgow 117833
artery berlin, Gerswalde 138673
Artery Melbourne, Fitzroy 099020
Artes, Annecy 066636
Artes, Knokke-Heist 100725

Les Artes, San Diego 124095
Artes Beatriz, Cali 128283
Artes Cova, Murcia 115310
Artes de México, México 137816,
                             142988
Artes-Galerie, Merlischachen .. 116502
Artes Goya, Santiago de Chile .. 064951
Artes Monet, Santiago de Chile . 128247
Artes Platicas Vital, Brasília ... 100910
Artes y Antigüedades, Bogotá . 128262
Artes, Kunsthandelsgesellschaft mbH,
  Hannover 107342, 137265
Artes, Revista Especializada en Arte
  Caribeño, Santo Domingo ... 138618
La Artesa, Valencia 133447
Artesan, Barcelona ... 085731, 133098
Artésania, La Chaux-de-Fonds . 087357,
                             133716
Artesania Puerto, Valencia 133448
Artesania y Antiguedades, Rubielos de
  Mora 133413
Artesanias Bueno, Zaragoza ... 133492
Artesano Lugo, Madrid 086126
Artesanos Ex Casa Maple, Buenos
  Aires 127512
Artesanterasmo Fidia, Milano .. 110254
Artesia, Catania 109948
Artesia Historical Museum and Art
  Center, Artesia 046547
Artespacio, Bogotá 102282
Artespazio Dieci, Bologna 109813
Artestile, Roma 081242
Artestudio, Milano 110255
Artetecnica Cornici, Bologna ... 109814
Artetica, Roma 059396, 110731
Arteum, Ljubljana 114695, 114696
Artevalori, San Marino 133027
Artevariarte, Firenze 110031
Artevinum, Hamburg 076189
ArteViva, Bairawies 074651
Artexpo, Bruxelles 138543
Artexpo, Cluj-Napoca 098278
Artexpo Atlanta, Atlanta 098430
Artexpo New York, New York .. 098431
Artexport, La Paz 100843
Artexte, Centre d'Information en Art
  Contemporain, Montréal ... 136676
ArtEZ Institute of the Arts,
  Arnhem 056170
Artfit, Phoenix 123481
Artfocus Gallery, Brookvale ... 098803
Artfolio, Singapore 114485
Artforms Gallery Manuyunk,
  Philadelphia 052028
Artforms, Artists Association of New
  Mexico, Las Cruces 060353
Artforum, Hannover 107343
Artforum, New York 139400
Artforum, Ostrava 140093
Artforum Newgate, Seoul 111613
ArtFrame Gallery, Tavistock ... 119214
Artframed, Ormskirk 118854
Artfrance, Paris 104521
Artful Arts, Harrogate 117914
Artful Deposit Galleries,
  Allentown 046407
Artful Dodger, Charlotte 119976
Artful Dodger, Los Angeles 094261
Artful Dog, Nashville 122064
Artful Hand Gallery, Boston ... 119833
Artfull Enterprises Mobile Gallery,
  Christchurch 117589
Artfully, Auckland 083764
Artfully Yours, Angaston 098646
Art'G, Béziers 103373, 103374
Artgalerie am Schloss, Bad
  Bergzabern 106040
Artgalerie Kunstkopie.de,
  Hamburg 076190
Artgarbs Gallery, Saint Paul ... 123894
Artgarden, Grassington 117875
Artgate, Oslo 084297
ARTgerecht, Eberbach 106850
Artginal, Paris 104522

Artgold, Montréal 101358
Artgrafic, Barcelona .. 114853, 114854
Artgument, Esure-sur-Indre 103719
Arth Art and Crafts, Ahmedabad 109246
Artha, Alkmaar 132381
Arthaud, Paris 136861
Arthaud, Damien, Lyon 069698
ArtHaus, San Francisco 124217
Arthe, Musée de la Veilleuse-Théière,
  Dinan 012791
Arthea Galerie, Wiesloch 108935
Ar'them Gallery, Paris 070954
Arthèmes, Cormeray 068157
Arthemis, Mulhouse 070385
Arthès, Bruxelles 125305
Arthis Fine Art, Taichung 116892
Artho, Edwin, Chez-le-Bart ... 133719
Arthobler, Porto 114074
The Arthonse, Christchurch ... 083855,
                             112829
Arthotel Billie Strauss, Kirchheim unter
  Teck 107559
Arthouse, Frederiksberg 065660
Arthouse, Navan 109679
Arthouse, San Francisco 124218
Arthouse at the Jones Center,
  Austin 046682
ArtHouse Gallery, Sittingbourne . 119100
Arthouse Oakland, Oakland ... 123242
Arthur A. Houghton Jr. Gallery and the
  Great Hall Gallery, New York . 051337
Arthur Antiquités, Le Vicel 069375
Arthur Boskamp-Stiftung,
  Hohenlockstedt 019306
Arthur Cottage, Ballymena 043477
The Arthur J. Moore Methodist
  Museum, Saint Simons Island . 052867
Arthur M. Sackler Gallery,
  Washington 054091
Arthur M. Sackler Museum,
  Cambridge 047357
Arthur M. Sackler Museum of Art
  and Archaeology, Beijing University,
  Beijing 007589
Arthur Rackham Society. Journal,
  Grosse Pointe 139301
Arthur Ross Gallery, University of
  Pennsylvania, Philadelphia ... 052029
Arthur, Elizabeth, Hamilton,
  Victoria 061583
Arthur, John, Dorking 117682
Arthurdale Heritage Museum,
  Arthurdale 046548
Arthus, Zell am Harmersbach .. 079134
Arthus-Bertrand, Paris . 070955, 070956
Arthus Galéria, Pécs 109204
Arti, Alkmaar 137821
Le Arti, Ravenna 110700
Arti, Roma 131819
Arti Capelli II, 's-Hertogenbosch 112523
Le Arti Decorative, Parma 081093
Arti e Restauro, Associazione Culturale
  di Volontariato, Bologna 130933
Arti et Amicitiae, Amsterdam .. 031928,
                             059592
Arti Figurative, Udine . 111057, 111058
Arti Grafiche, San Francisco .. 138454
Arti Grafiche Friulane.
  Tavagnacco 137742
Arti Studio, Milano 110256
Arti-unique, Gilfach Goch ... 117823
Artia, Knokke-Heist 100726
Artia Russian Art, Toronto ... 064602,
                             101571
Artiaco, Alfonso, Napoli 110518
Artiarticle, Mablethorpe 118688
Artibus, Milano 131375
Artibus Asiae, Zürich .. 138103, 139089
Artibus et Historiae, Kraków ... 138954
Artice, Koondoola 061700
Article, Charlotte 119977
ARTicle, Köln 107588
Articles Fine Arts Gallery, Stanwell
  Park 099717

Artico, Wallern an der Trattnach 100217
Articulate Gallery, WVSA Arts
Connection, Washington 054092
Articule, Montréal 101359
ArtiDoTuum Gallery, Kraków ... 113529
Artiek, Groningen 083277
Artifact Book & Art Conservation,
Kyneton 127626
Artifacts, Accrington 117220
Artifacts, Bruxelles 063330
Artifacts, Nashville 094973
Artifacts Gallery, Indianapolis . 121167
Artifakt Gallery, Deloraine 098962
Artifex, Brugge 100532
Artifex, Katowice 143460
Artifex, Sutton Coldfield 119185
Artifex, Warszawa 138955
Artifex Antik, Stockholm 086972
Artifexgallery, Wellington 113059
Artificia Consorzio, Roma 131820
Artificial, San Francisco 124219
Artificio, Firenze 137644
L'Artificio, Torino 081650, 110951,
                                132093
Artificium, Berlin 106157, 141482
Artificium, Witten 108950
Artiges & Fils, C., Bruxelles 100555,
                                128014
Artigiana Cornici, Roma 110732
Artigiani del Cotto Restauri,
Roma 131821
Artigiano, Benningen, Kreis
Unterallgäu 106128
L'Artigiano, Genova 131288
Artigiano Gallery, Auckland 112748
Artigo, Leigh-on-Sea 118098
Artik, København 065846
Artik Galerie, Wien 062901
Artika, Zürich 087934
Artikel Editionen, Berlin 137052
Artillery Company of Newport Military
Museum, Newport 051529
Artillery Museum of Finland,
Hämeenlinna 010786
artimage, Graz 136506
Artimanha, Lisboa 084832
Artimer, Turku 066401
Artimex, Bruxelles 100556
Artin, Varna 101185
Artin Evolution, Phoenix 123482
Artina, Zug 116876
Artinf, Buenos Aires 138487
Artinfo, Moskva 137924
ARTINFO, New York 139302
Artinform, Kyïv 117120
L'Artingegno, Firenze 131114
Artinoe, Huelva 133276
Artinque, Buderim 061226
Artinteriors, Watford 119314
Artinvest, Thalwil 087843
Artio, Athinai 109022
Artion Gallery, Hong Kong 102053
Artique, Guatemala 079221
Artique, New York 122354
Artique, Tetbury 091564
Artique Fine Art Gallery,
Cremorne 098930
Artique Galleria, Minneapolis 094900
Artique Galleries, Leicester ... 118090
Artique Gallery, San Antonio ... 124012
Artique Restorations, Houston .. 135586
Artiquity of Orient, Armadale,
Victoria 061013
Artirex, Bucureşti 132991
Artis, 's-Hertogenbosch 112524
Artis, Lyon 104063
Artis, Paris 104523
Artis, Pont-Aven 105368
Artis, Salamanca 115402
Artis, Venezia 111083
Artis, Warszawa 084703, 113804
Artis Amactor, Essen 075668
Artis Galerie, Büren an der Aare 116175
Artis Gallery, Auckland 112749

Artis Geologisch Museum,
Amsterdam 031929
ARTISA – Salon des Métiers d'Art,
Grenoble 098072
Artisaani, Helsinki 103135
Artisan, Berlin 106158
Artisan, Botany 127565
Artisan, Concepción .. 064938, 128243
Artisan, Den Haag 132469
Artisan, Epping 117760
Artisan, New York 138377
Artisan Antiques, Cleveland 092919
Artisan Antiques, Miami 094702
Artisan Books, Fitzroy 139730
L'Artisan Ebéniste, Prunay 129180
L'Artisan Encadreur, Paris 128908
Artisan Furniture Refinishing,
Vancouver 128222
Artisan Galleries, Woodford
Green 119385
Artisan Gallery, Robigana 099644
Artisan House, Wilton 125035
Artisan Net, San Antonio 124013
Artisan Picture Gallery, Subiaco . 099724
l'Artisan Pratique, Lausanne .. 133790
Artisan Sanat Galerisi, Istanbul . 117006
Artisanat, Perth 118902
Artisania, Sale 112054
Artisans, Chennai 109275
Artisans 21 Gallery, Chicago ... 120044
Artisan's Alley, London 101333
Artisans Antiques Emporium,
Jacksonville 093956
Artisans Art, Washington 124866
Artisans d'Art, Honfleur 103816
Artisans du Bois, San Diego ... 136288
Artisans Gallery, Akaroa 112739
Artisans Gallery, Cardiff 117522
Artisan's Gallery, San Antonio . 124014
Artisan's Studio Gallery, Dallas . 120556
Artisart, Maastricht 112567
Artiscope, Bruxelles 100557
Artisin, Enschede 083247
Artisjok, Tjerkwerd 083631
Artison, Budapest 079239
ARTisSpectrum, New York 139303
Artissum, Den Haag 112375
Artist, Amsterdam 082690
The Artist, Tenterden . 138287, 139144
Artist Alley Gallery, Las Vegas . 121339
Artist Atelier of Atlanta, Atlanta . 119539
Artist Cooperative, Detroit 120801
Artist Design Group, New
Orleans 060354, 122122
Artist Gallery, Killarney 109650
d'Artist Gallery, Singapore 114486
artist Kunstmagazin, Bremen .. 138674
Artist Magazine, Taipei 139104
The Artist Market, New Orleans 122123
Artist Paintbrush, Sacramento . 123756
Artist Publishing, Taipei 138128
Artist Run, Amsterdam 112153
Artist Showcase, Denver 120693
Artist Studio, Hyderabad 109354, 109355,
                                109356
Artist Studio Gallery, Delhi ... 109308
Artist Trade, Ljubljana 114697
artist window, Bremen 138675
Artista, Firenze 138803
Artistas visuales asociados de Madrid,
AVAM, Madrid 059831
Artistbutiken, Kalmar 115698
L'Artiste, Chamalières 128537
Artiste Animalier, Aigues-Mortes 103175
Artistea, Tshwane 085672
Artistenmuseum, Marburg 020267
Artistes & Artisan du Quartier, Saint-
Étienne 105509
Artistes de France – Comité European
des Beaux Arts, Saint-Étienne . 058591
Artistes-de-Notre Temps, Rauville-la-
Bigot 072368
Les Artistes et la Mer, Saint-Malo (Ille-
et-Vilaine) 105545

Artistes sans Frontières, Paris .. 104524
Les Artistes Témoins, Nancy ... 104316
Artistic, San Antonio 136256
Artistic Accents, Regina 101526
Artistic Antiques, Miami 094703
Artistic Collections, Seattle ... 124483
Artistic Endeavors, San Antonio . 124015
Artistic Frame Works, Silverton . 085665
Artistic Galleries, Cincinnati ... 120317
Artistic Ideas, Toronto 101572
Artistic Imports, Atlanta 092110
Artistic Impressions, Miami ... 121741
Artistic Licence, Carlingford .. 109541
Artistic License Gallery, North
Adelaide 099466
Artistic Metal, Cincinnati 135436
Artistic Services Unlimited,
Washington 124867
L'Artistica, Girona 115076
Artistica Emilio, Buenos Aires .. 098549
Artistica Marmi, Catania 131081
Artistically Bent, Columbus 120475
Artistically Speaking Gallery & School
of Fine Art, Edmonton 101275
Artistik, Indianapolis 121168
Artistix, London 101334
Artistixx, New Orleans 122124
Artistry Galleries, Hawthorn 099116
Artists Abode, Semaphore 099684
Artists Alliance, New York 060355
Artists Alliance of California,
Clovis 060356
Artists and Framers, Baltimore . 119720
Artist's and Graphic Designer's Market,
Cincinnati 139304
Artists and Illustrators, London . 139145
Artists and Illustrators Exhibition,
London 098359
Artists' and Photographers' Press Ltd.,
Surrey 138285
Artists Association of Nantucket,
Nantucket 060357
Artists Association of Nantucket
Gallery, Nantucket 051153
Artists Bookworks, Forest Row . 138163
Artists' Camp Antiques,
Cremorne 061371
Artists Center and Foundation,
Makati 034503
Artists' Collaborative Gallery,
Sacramento 123757
Artists Consultant, Philadelphia . 123353
Artists' Cooperative Gallery,
Omaha 051759
Artists Creations, Las Vegas ... 121340
Artists Eye, Dee Why 098961
Artists' Fellowship, New York .. 060358
Artists' Forum, San Diego 124096
Artists' Gallery, Austin 119640
Artists Gallery, Parañaque ... 113410
Artists Gallery, Virginia Beach .. 124822
Artist's Gallery of Seattle, Seattle 124484
Artists' Gallery TMS, Tbilisi 105896
Artists Garden, Fitzroy 099021
Artists' General Benevolent Institution,
London 060036
Artist's Handicrafts Association,
Chennai 059357, 109276, 137591
Artists House Tel Aviv, Israel Painters
and Sculptors Association, Tel
Aviv 025187
The Artist's Magazine, Cincinnati 139305
The Artists Museum, New York . 051538
Artists on Santa Fe, Denver ... 120694
Artists Own Gallery, Brisbane .. 098778
Artists Realm, Gisborne 099077
Artists See Boston, Boston ... 119834
Artist's Space, New York 051339
Artists' Space, Ayala Museum,
Makati 034504
The Artists Studio, Pwllheli 118951
Artists to Watch, Saint Paul 123895
Artists' Union of Armenia,
Jerewan 058180

Artists' Union of Azerbaijan, Bakı 058340
Artists' Union of Georgia, Tbilisi 058699
Artists' Union of Kazakhstan,
Almaty 059519
Artists' Union of Latvia, Rīga ... 059541
Artists' Union of the Moldova Republic,
Chişinău 059575
Artists' Union of Turkmenistan,
Aşgabat 060016
Artists Unlimited, Bielefeld 106485
Artists Unlimited, Tampa 124629
The Artists Village, Singapore .. 059798,
                                114487
Artists Village Gallery, Beijing .. 101931
Artists's Union of Tajikistan,
Dušanbe 059997
Artistuff, Swindon 119197
Artitalia, Cortina d'Ampezzo .. 109980
Artitalia, Milano 110257
Artitalia, Riccione 110713
Artitude, Paris 104525
Artitudes, Memphis 121682
Artitudine, Genova 110090
Artium – artcollections, Essen,
Oldenburg 138676
Artium – Centro Museo Vasco de Arte
Contemporáneo, Vitoria-Gasteiz 040422
Artium – LuxArt, Essen,
Oldenburg 138677
Artium Art Galleries, Luxembourg 111833
Artium Art Gallery, Luxembourg . 111834
Artium Kunstmagazine,
Luxembourg 137809
Artizen Gallery, Bendigo 098733
Artjärven Kotiseutumuseo,
Artjärvi 010757
ArtJaz Gallery, Philadelphia ... 123354
Artkids, Tshwane 085673, 133076
Artkollegija, Moskva 114204
Artkraft Diffusion, Paris 104526
Artkunst-Ungar, Karlsruhe ... 076686
Artlab Australia, Adelaide 127549
Artland, Aachen 137019
Artland, Indooroopilly 127618
Artland, Mumbai 109416
Artland Gallery, Hong Kong ... 065010
Artlands, Hillington 117968
Artlantis Kunst und Design,
Stuttgart 108686
Artlibris, Paris 141031
Artlieb.com, Kalmar 086766
Artline, Amsterdam 112154
Artline, Berlin 074759
Artline, Heilbronn 107404
artline. Kunstmagazin, Freiburg im
Breisgau 138678
ArtLink, Virginia Beach 124823
Artlounge-Projects, Calvia 115004
Artlovers, Artarmon 061057
Art'm, Bordeaux 103415
Artmagazine, Wien 138503
Artmakers, Stockholm 115838
Artmakers Trust, Hamilton 112903
Artmaori, Taranaki 113011, 113012
Artmark Galerie, Spital am Pyhrn 100201
Artmark Galerie Wien, Wien ... 100230
ArtMbassy, Berlin 106159
Artmedia Art Gallery, Beograd .. 038128
Artmedia, Arte en Mesoamerica,
Moravia 138609
Artmetro, Cleveland 120399
Artmill, Plymouth 118918
Artmix, Poznań 138956
Artmobile, Newtown 051556
Art'monie, Nice 070564
Artmosphere, Graz 062579
Artmosphère, Vence 105823
Artmosphere Galerien, Wien ... 100231
Artmosphere Gallery, Seattle .. 124485
Artmost, Moskva 114205
Artnau Espai d'Art, Girona 115077
Artnet, Istanbul 117007
Artnet Magazin, Berlin 138679
Artnet Magazine, New York ... 139306

ArtNetwork, Nevada City 060359
Artnews, New York 139307
Artnex, San Diego 124097
Artocratia, Zürich 116750
Artology, Chicago 120045
Artom, Domat, Ems 116250
Artonef, Aix-en-Provence 103191
Artonivo, Brugge 003404
Artonnet, München 108049
Artonomy, Cincinnati 120318
Artonomy, Truro 119256
Artop, Lille 104000
Artop-Atelier Restauration Tableaux et
  Objets Polychromes, Héry 128621
Artos, Sofia 101147
Artoteket, Oslo 113190
Artothek der Zentral- und
  Landesbibliothek Berlin, Berlin 017076
Artothek-Galerie, Wien 002938
Artothek im Bonner Kunstverein,
  Bonn 017481
Artothek Worpswede, Worpswede 022769
Artothek, Niederösterreichisches
  Landesmuseum, Krems 002230
artothek, Raum für junge Kunst,
  Köln 019672
Artothèque, Angers 011647
Artothèque Antonin Artaud,
  Marseille 014039
Artothèque de Montréal,
  Montréal 101360
Artothèque Les Art au Mur,
  Pessac 105347
L'Artothèque-Sud, Nîmes 104397
Artothèque-Sud, Nîmes 014496
Artourage, 's-Hertogenbosch . . . 083381
Artous, Paul & Fils, Montauban . 070187
Artower, London 118189
Artpa e.V., Leipzig 058730, 107809
artPAPIER, Katowice 138957
Artpilot, Kristiansand 113167
Artplace, Perth 099560
Artpleasure, Paris 104527, 136862
Artpoint 222, Wien 100232
Artpoint Gallery and Studios Society,
  Calgary 005516, 058421
Artport, Auckland 112750
Artpreciation, Hong Kong 102054
Artprice.com, Saint-Romain-au-Mont-
  d'Or 137006
Artprint, Kaiapoi 112926
Artprint on Canvas, Miami 121742
ArtProducenten Galerie, Stuttgart 108687
Artprofil – Magazin für Kunst,
  Mannheim 138680
Artquest, Saint Louis 123805
Artra, Genova 110091
Artra, Milano 110258, 110259
Artrad Art et Traditions, Maisons-
  Laffitte 104145
Artrade, Heerlen 112521
Artradex, L'Union 104054
Artrageous, San Diego 124098, 124099
Artrageous La Jolla, San Diego . 124100
Artrain USA, Ann Arbor 046487
Artreasury, New York 095242
Artrent, Frankfurt am Main 106973
Artrepco Galerie, Zürich 116751
Artrepreneur, New York 122355
Artrium Kunstgalerie,
  Hombrechtikon 116377
Artrock, San Francisco 124200
Artrom Network, Roma 059397
Artroom, Peter Wildbolz, Cudrefin 116236
Artropica Innisfail Studio Gallery,
  Innisfail 099152
Artropolis, Chicago 098432
Artrujillo, Minneapolis 121972
ARTS, La Mézière 128642
Arts, Saint-Cyr-en-Val 072724
Art's, Sumy 117162
Les Arts, Versailles 105842
Art's, Wellington 113060

Arts & Autographes, Paris 136863,
  141032
Arts & Business, London 060037
Arts & Concept, Singapore 114488
Arts & Livres, Luxembourg . . . 082362,
  142972
Arts & Media Access Centre, Cape
  Town 056339, 137945
Arts & Mirrors, Toronto 101573
Arts & Tradition, Beirut 082274
Arts Academy, University of Ballarat,
  Ballarat 054893
Arts Alive Gallery, Oamaru 112976
Arts Alliance Center at Clear Lake,
  Nassau Bay 051194
Arts Anciens, Grasse 068701
Arts Anciens, Montalchez 126747
Arts and Antique, Hong Kong . . 065011
Arts and Artisans, Chicago . . . 120046,
  120047, 120048
Arts and Associates Gallery,
  Pasay 113412
Arts and Carpets, Manama 063134
Arts and Communications Department,
  Christopher Newport College,
  Newport News 057574
Arts and Crafts Association of Meriden,
  Meriden 060360
The Arts and Crafts Furniture Company,
  London 089925, 134588
Arts and Crafts Museum,
  Bhavnagar 024163
Arts and Design, Bishop Grosseteste
  University College, Lincoln . . 056547
Arts and Hobby Centrum,
  Warszawa 084704
Arts and Humanities Department,
  Red Rocks Community College,
  Lakewood 057314
Arts and Humanities Division,
  Jamestown Community College,
  Jamestown 057260
Arts and Humanities Division, Leeward
  Community College, Pearl City 057652
Arts and Letters Club of Toronto,
  Toronto 058422
Arts and Papers, Stuttgart 142382
Arts and Science Center for Southeast
  Arkansas, Pine Bluff 052134
Arts and Science Council,
  Charlotte 060361
Arts and Sciences Cluster, Central
  Community College, Columbus
  Campus, Columbus 056942
Arts and Sciences Division, New
  Mexico Junior College, Hobbs . 057204
Arts and The Islamic World,
  London 139146
Arts Antiquités, Bordeaux 067307
Arts Antiquités, Castres 067725
Arts Asiatiques, Paris 138632
Arts Bibliographic, London 144556
Arts Bridge, Chicago 060362
Arts Broc, Talence 073613
The Arts Center, Saint Petersburg 052857
The Arts Center of the Capital Region,
  Troy 053811
Arts Center of the Ozarks-Gallery,
  Springdale 053446
The Arts Center, Chulalongkorn
  University, Bangkok 042649
Arts Center, Yuan Ze University,
  Jhongli 042343
Arts Centre, Singapore 085367
Arts Centre, Washington 046066
Arts Centre (Far East), Singapore 085368
Arts Centre of Christchurch,
  Christchurch 033167
Arts Club, Montréal 101361
The Arts Club of Chicago,
  Chicago 047638
Arts Club of Washington,
  Washington 054093, 060363
Arts Collector Gallery, Singapore 114489

Arts College, Colorado Mountain
  College, Aspen 056672
Arts Company Nashville,
  Nashville 122065
The Arts Connection Corporation, New
  York 060364
Arts Council England, London . . 060038
Arts Council England, East,
  Cambridge 060039
Arts Council England, East Midlands,
  Nottingham 060040
Arts Council England, North West,
  Manchester 060041
Arts Council England, Northeast,
  Newcastle-upon-Tyne 060042
Arts Council England, South East,
  Brighton 060043
Arts Council England, South West,
  Exeter 060044
Arts Council England, West Midlands,
  Birmingham 060045
Arts Council England, Yorkshire,
  Dewsbury 060046
Arts Council Gallery, Silver City . 053335
Arts Council Nelson, Nelson . . 059629
Arts Council of Fayetteville/
  Cumberland County, Fayetteville 048719
Arts Council of Mendocino County,
  Fort Bragg 060365
Arts Council of New Orleans, New
  Orleans 060366
Arts Council of Northern Ireland,
  Belfast 060047
Arts Council of Pakistan, Karachi 059679
Arts Council of Southwestern Indiana,
  Evansville 060367
Arts Council of the Blue Ridge,
  Roanoke 060368
The Arts Council of Wales, Cyngor
  Celfyddydau Cymru, Cardiff . . 060048
The Arts Council of Wales, Mid and
  West Wales Office, Carmarthen 060049
Arts Council of Wales, North Wales
  Office, Colwyn Bay 060050
Arts Council of Wayne County –
  Museum, Goldsboro 049152
Arts Counsel, New York 122356
Arts d'Afrique, København 065847
Arts d'Autrefois, Paris 070957
Art's Déco Gallery, La Seyne-sur-
  Mer 103906
Les Arts Décoratifs, Bruxelles . . 063331
Arts Décoratifs, Düsseldorf 075455
Arts Decoratifs, West Hollywood 097873
Arts Décoration Extrême-Orient,
  Valence (Drôme) 073938
Arts Delight, Augsburg 106023
Arts Department, Framingham State
  College, Framingham 057102
Arts Department, Kellogg Community
  College, Battle Creek 056713
Arts Department, Qinghai Normal
  University, Xining 055254
Arts Department, Saint Clair County
  Community College, Port Huron 057703
Arts Department, University of
  Science and Arts of Oklahoma,
  Chickasha 056891
Arts Design, Tampa 124630
Arts Desire, Auckland . 083765, 112751
Arts Division, North Wilkesboro . 057592
Arts Division, Emma Willard School,
  Troy 058006
Arts Division, Fordham University, New
  York 057542
Arts Division, Merced College,
  Merced 057427
Arts Division, Rosemont College, Bryn
  Mawr 056809
Arts du Marbre, Palaiseau 128882
Arts du Monde, New York 122357
Arts du XXe Siècle, La
  Conversion 087381
Arts du XXème Siécle, L'Isle-sur-la-

Sorgue 069554
Arts Education Policy Review,
  Washington 139308
Arts et Antiquités d'Alisson,
  Montmorency 070276
Arts et Antiquités Kuriosum,
  Luxembourg 082363
Arts et Bijoux, Melun 128794
Arts et Boiseries, Paris 070958
Arts et Cimaises, Cordes-sur-Ciel 068148,
  103629
Arts et Civilizations, Quimper . . 072341,
  105412
Arts et Collections, La Charité-sur-
  Loire 068983
Arts et Création, Limoges 069515,
  104025
Arts et Lettres, Vevey 116691
Arts et Passions, Les Forges . . . 069408
Arts et Passions-Espace Collections,
  Blois 067249
Arts et Techniques, Paris 128909
Arts et Vestiges, Bouloc 067389, 103445,
  128514
Arts etc., Bakı 003150
Art's Events, Torrecuso 111019
Arts Extended Gallery, Detroit . . 120802
Arts Extension Service, Hedley . 060369,
  138337
The Arts Factory, Las Vegas . . . 121341
Arts for All, Los Angeles County Arts
  Commission, Los Angeles 060370
Arts for Health, Manchester 060051
The Arts Gallery, Hong Kong . . . 102055,
  102056
The Arts Gallery, University of Arts
  London, London 044907
Arts Gambetta, Montigny-lès-
  Metz 104266
Arts Graphiques d'Aquitaine – AGA,
  Libourne 136828
Arts Hawaii, Honolulu 093529
Arts in Harmony, Indianapolis . . 121169
Arts in Progress, Roxbury 060371
Arts in the Academy, Washington 054094
Arts India, New York 122358
Arts Industria, New York 122359
Arts Infinite, Hong Kong 102057
Arts League of Michigan, Detroit 060372
Arts Limited, Richmond 123720
Arts Litho – Atelier Stéphane Guilbaud,
  Paris 104528
Arts Magna, Miami 121743
Aux Arts Majeurs, Paris 141033
Arts Massilia, Marseille 104157, 128743
Arts Midland Galleries, Midland . 050871
Arts Midwest, Minneapolis 060373
Arts NDG, Montréal 101362
Arts Nouveaux, Verona 081896
Arts of Africa, New Brunswick . . 139309
Arts of Asia, Armadale, Victoria . 061014,
  098654
Arts of Asia, Ho Chi Minh City . 125111
Arts of Asia, Hong Kong 138596
Arts of Asia Publications, Hong
  Kong 136717
Arts of Iran, San Francisco 124221
Arts of Paradise, Honolulu 120922
Arts on the Green, Omaha 123303
Arts Options, Mendlesham Green 118739
Arts Palace Fine Art Gallery,
  Cairo 103019
Arts Partnership of Greater
  Spartanburg, Spartanburg 060374
Arts Pluriels, Fort-de-France . . . 111927
Arts Primitifs, Arnay-le-Duc . . . 066729
Les Arts Production, Bruxelles . . 100558
Arts Project Australia, Northcote 001370
Arts Quarterly, New Orleans . . . 139310
Arts-Recherche Photographiques,
  Bruxelles 100559
Arts Tamborine, Mount
  Tamborine 099404
Arts University at Bournemouth,

Poole 056596
Arts World, New York 122360
Arts, Antiques & Auctions, Gent 138544
Arts, Humanities and Communications
  Division, Modesto Junior College,
  Modesto 057466
Artscademia, Lutry 116460
Artscan, Houston 121000
Artscape, New York 139311
Artscape Gallery, Palmerston
  North 112981
Artscape West International, San
  Francisco 124222
Artsconnexion, Genval 058353
Artsconnexion, Mons 100792
Artscope, New York 122361
ArtScope.net, Chicago 139312
Artseal Gallery, San Francisco . 124223
Artsearch Databank, Cincinnati . 120319
Artseeandbuy.com, Remagen .. 078193
Artset, Limoges 069516, 104026
ArtsFocus, Colorado Springs .. 139313
Artsgora, Barcelona 114855
Artshare, København 102860
Artshaus.Co.Uk, Caister-on-Sea . 117496
Artshed, Ware 119303
Artshop 24, Hamburg 107224
Artshouse Gallery, Northbridge,
  Western Australia 099483
Artshow, Milano 138804
Artside Gallery, Seoul 111614
ArtSingapore, The Contemporary Asian
  Art Fair, Singapore 098291
Artsiphartsi, Tampa 124631
Artsland, Birmingham 119788
Artslink, Hong Kong 138597
Artsmart, Christchurch 112830
Artsonia, Chicago 060375
Artsonje Center, Seoul 029983
Artsonje Museum, Gyeongju .. 029944
Artsoum, Nice 104358
Artsource, Denver 120695
Artsource, London 118190
ArtSource Quarterly, Nevada City 139314
ArtSouth Gallery, Gore
  Invercargill 112900
Artspace, Auckland 033140
Artspace, Ha Noi 125062
Artspace, Peterborough 006635
Artspace, Raleigh 052444
Artspace, Richmond 052544
Artspace, Woolloomooloo 001661
Artspace-Lima, Lima 050280
Artspace Mackay, Mackay ... 001237
Artspeak Gallery, Vancouver . 007271
Artspeak Studio Gallery,
  Warrandyte 099859
ArtsPost, Hamilton 112904
ArtSquare, London 118191
Artstation, Légny 103979
Artstil, Moskva 143635
Artstudio, Lille 104001
Artstudio, München 077461
Artsud, Conques 103628
Artsud, Villeneuve-d'Aveyron .. 105870
Artsud Beyrouth, Beirut 098238
Artsud Dubai, Dubai 098345
Artsud Moscou, Moskva 098280
Artsun, École-Valentin 103710
Artsville, Nashville 122066
Artswatch, Louisville .. 050487, 121632
Artswest, Seattle 124486
Artsyndication, Düsseldorf ... 075456
ArtTactic, London 139147
Arttalk, Beacon 139315
Arttech, West Thurrock 119326
Arttemple, Hyderabad 109357
Artterra, Praha 102649
Artterra Int, Groningen 112491
Arttig, Milwaukee 121896
Artur Media, Bucureşti 085080
Artur, Ramon, Barcelona 085732
Artura, Humlebæk 065801
Artus Enchères, Paris 125675

Artveler, Paris 104529
Artview, Calgary 101214
Art'Vigne, La Chaux-de-Fonds .. 116206
Artvillage, Bad Aussee 058233
Artvue, Atlanta 119540
Artware, San Diego 124101
Artware Consulting, Wien 100233
Artware Fine Art, London 089926
Artweek, Palo Alto 139316
Artwindow, Barcelona 114856
Artwork Mexican, Phoenix 123483
Artwork Publishing, San
  Francisco 124224, 138455
Artworks, Austin 119641
Artworks, Bradford 117411
Artworks, Calgary 101215
Artworks, Cincinnati 120320
Artworks, Paris 070959
Artworks, Trenton 053794
Artworks 21, Sacramento 123758
Artworks Cornwall, Saint Austell 119009
Artworks Gallery, Dubai 117118
Artworks Gallery, Dunedin ... 112873
Artworks Gallery, Southport .. 099711
Artworks of Kansas City, Kansas
  City 121279
Artworks on Q, Mermaid Waters 099332
ArtWorkz, Lowestoft 118669
ArtWorld Hotline, Nevada City . 139317
ArtXchange, Seattle 124487
Arty Art Gallery, Singapore ... 114490
Arty Bee's Books, Wellington . 143359
Artyard Sculpture Gallery, Denver 120696
Artyfacts, Brighton 139148
Artyfacts, Bromsgrove 117465
Artyfacts, Padstow 118870
Artyfacts, Walebridge 119292
Artyp, Bruxelles 100560
Artz, Petersfield 118909
ArtZ FartZ, Milwaukee . 121897, 121898
Artz Plus, Edmonton 101276
Artz, Martina, Wetzlar 078957
Artzco, Vancouver 101739
Artzone, Glasbury 117827, 138165
Artzonia, Vancouver 101740
Artzu Gallery, Manchester ... 118702
Artzygifts, Chicago 120049
Aru, Maria Pia, Cagliari 131061
Aruid-Panajotovi, Ruse 101132
Arun, Lalitpur 112089
Arundel, Los Angeles 145211
Arundel Bookshop, Arundel .. 144220
Arundel Museum and Heritage Centre,
  Arundel 043437
Arundel Toy Museum Dolls House,
  Arundel 043438
Arusha Declaration Museum,
  Arusha 042625
Aruther Ackermann Ltd., London 089927,
                          118192
Arvada Center for the Arts and
  Humanities, Arvada 046549
Arval, Cali 128284
Arv'Art, Chamonix-Mont-Blanc . 103560
Arvel, Dainville 068263
Arvidsjaurs Hembygdsmuseum,
  Arvidsjaur 040539
Arvika Fordonmuseum, Arvika . 040541
Arvika Konstförening, Arvika .. 059879
Arvika Konsthall, Arvika 040542
Arvim, Roma 081243
Arvis, Gabrielle, Paris 070960
Arviz, Gabrielle, Paris 070960
Arxiu del Monestir, Sant Joan de les
  Abadess 040089
Ary Jan, Mathias, Paris 070961
Ary, Patricia F., Fortaleza ... 100936
Aryan, Delhi 137592
Aryana, Lyon 069699
Aryaseni, Singapore 114491
Arzamasskij Gosudarstvennyj
  Literaturno-Memorialnyj Muzej A.P.
  Gajdara, Arzamas 036867
Arzamasskij Istoriko-chudožestvennyj

Muzej, Arzamas 036868
Arzat, Paris 070962, 128910
Arzemju Mákslas Muzejs, Rīga . 030177
Arzenal, Praha 102650
Arzenál Galéria, Budapest ... 109129
Arzile, Jean-Claude, Bordeaux .. 128480
Arzneipflanzenmuseum, Universität
  des Saarlandes, Saarbrücken . 021555
Arzney Küche, Bönnigheim ... 017459
AS 220 Studio, Providence ... 123694
Les As de la Renovation, Les
  Abymes 130808
As Time Goes By, Edmonton ... 064285
As Time Goes By, Hamilton, New
  South Wales 061577
As Time Goes By, Louisville .. 094519
AS Verlag, Zürich 138104
Asa 2000, Hermosillo 138908
Asa Packer Mansion, Jim Thorpe 049822
Asagaya Bijutsu Senmon Gakko,
  Tokyo 056037
Asagaya Some Kenkyujo, Tokyo 056038
Asago Art Village, Asago 028481
Asahi, Sendai 111381
Asahi, Tokyo 082136, 111403
Asahi-yaki Kiln Museum, Uji,
  Kyoto 029674
Asahikawa Bijutsukan,
  Asahikawa 028483
Asahikawa Hakabutsukan,
  Asahikawa 028484
Asahikawa Youth Science Museum,
  Asahikawa 028485
Asakichi, San Francisco 097184
Asakura Isokichi Bijutsukan,
  Terai 029476
Asakura Sculpture Gallery, Tokyo 029487
As'Art, Paris 104530
Asashi Seimei Gallery, Tokyo ... 111404
Ásavallen, Postmuseum, Skolmuseum
  och Bruksmuseum, Edsbro ... 040596
Asbæk, København 102861
Asbury, Peter, Birmingham ... 088472,
                          134106
ASC, Aachen 105928
ASC Deco, Dallas 093071
Les Ascenceurs du Canal Historique
  du Centre, Houdeng-Goegnies . 003715
Ascencio, Alain, Toulon (Var) ... 073700
Ascencio, Robert, Maureilhan . 070009
Ascenseur Funiculaire pour Bateaux
  de Strépy-Thieu, Thieu 004060
Aschanska Gården, Eksjö 040597
Asche, Johann, Berlin 141483
H. Aschehoug & Co., H., Oslo .. 137884
Aschenbach & Hofland,
  Amsterdam 112155
Aschenbrenner, Ernst,
  Hohenwarth 076561
Aschendorff Verlag, Münster ... 137432
Aschoff & Petri, Jena 141963
Ascot, Saint Marys 139905
Ascot Antik, Monchy-Saint-Eloi . 070154
Ascot Antique, Wellington ... 084156
Ascot Antiques, Blackpool ... 088497
Ascot Antiques, Invercargill ... 083963
Ascott House, Leighton Buzzard 044806
Åsele Antik och Auktion, Åsele . 086633,
                          126567
Asele Institute Nimo, Nimo ... 056204
Asenbaum, London 089928
Asenbaum, Renate, Wien 140060
Aseespän Paja, Ilomantsi ... 010897
Aşezământul de Artă şi Cultură
  Religioasă, Maia 036548
Asfhan, Sharjah 088194
Ali Asghar-Assadi & Elke Schenck,
  Itzehoe 107475
Ash, New York 122362
Ash, Westcliff-on-Sea 119328
Ash Lawn-Highland, College of William
  and Mary, Charlottesville ... 047580
Ash Rare Books, London 144557
Ash, Edward, Orpington 118855

Ash, Jim & Pat, Llandeilo 089862
Ashanti, Atlanta 119541
Ashanti, Jacksonville 093957
Ashbeys Galleries, Cape Town .. 126490
Ashbourne Antiques, Ashbourne 088296
Ashbrook, Ottawa 064494
Ashburton Art Gallery, Ashburton 112741
Ashburton Aviation Museum,
  Ashburton 033135
Ashburton Books, Ashburton .. 144222
Ashburton Marbles, Ashburton . 088306
Ashburton Museum, Ashburton . 033136,
                          043439
Ashburton Vintage Car Club Museum,
  Tinwald 033313
Ashby, Queenscliff 062130
Ashby, San Diego 124102
Ashby-de-la-Zouch Museum, Ashby-
  de-la-Zouch 043440
Ashby-Hodge Gallery of American Art,
  Fayette 048713
Ashby Moody, M., Tampa 097543,
                          145613
Ashby, Vanessa, Armadale,
  Victoria 061015
Ashcombe Coach House,
  Henfield 089503
Ashcroft Museum, Ashcroft ... 005354
Ashe Antiques Warehouse, Wickham
  Market 091840, 135226
Asher, Portland 096363
Asher & Co., A., IJmuiden ... 143176
Ashern Pioneer Museum, Ashern 005355
Asher's Bazar, Caracas 097944
Ashes & Snow, New York 122363
Asheville Art Museum, Asheville 046550
Ashford Gallery, Dublin 024867
Ashford Museum, Ashford 043441
Ashgate Publishing, Aldershot . 138136
Ashgrove Clocks, Brighton,
  Victoria 061199, 127569
Ashikaga-shiritsu Bijutsukan,
  Ashikaga 028487
Ashish Balram Nagpal, Mumbai 109417
Ashitaba, Sendai 082111
Ashiya-shiritsu Bijutsu oyo Rekishi
  Hakubutsukan, Ashiya 028490
Ashiyana, Jaipur, Rajasthan ... 079395
Ashkenazy, Los Angeles 121434
Ashkenazy Gallery, Los Angeles 121435
Ashland, Cincinnati 092827
Ashland – Henry Clay Estate,
  Lexington 050240
Ashland Historical Society Museum,
  Ashland 046564, 046573
Ashland Logging Museum,
  Ashland 046563
Ashland Railroad Station Museum,
  Ashland 046566
Ashley, Cleveland 092920
Ashley, Houston 093616, 093617,
                          093618
Ashley, Jacksonville .. 093958, 127264
Ashley, Louisville 121633
Ashley, Miami 121744
Ashley, Omaha 096027
Ashley, Portland 096364
Ashley Jackson, Holmfirth ... 117977
Ashley Park Antiques, Waitotara 084140
Ashley Pine, Saintfield . 091213, 134997
Ashley, Diane, Philadelphia ... 133355
Ashman, Jacksonville 093959
Ashmolean Museum of Art and
  Archaeology, Oxford University,
  Oxford 045448
Ashmolean Museum Publications,
  Oxford 138270
Ashok Art & Interior, Amsterdam 082691
Ashtabula Arts Center, Ashtabula 046576
Ashtabula County Historical Society
  Museum, Geneva-on-the-Lake 049072
Ashtabula Plantation, Pendleton 051972
Ashtead Park Lane Fine Arts,
  Ashtead 117259

Ashton, Chicago 120050
Ashton, Saint Paul 096811
Ashton & Gower, Stow-on-the-
Wold 091460
Ashton Court Visitor Centre,
Bristol 043744
Ashton Gallery, Winnipeg 101867
Ashton Villa, Galveston 049030
Ashton, K., Gants Hill . 089318, 134399
Ashton's Antiques, Brighton 088612
Ashtree, Fresno 093489
Ashtree Arts, Northallerton 118814
Ashvita Art Object, Chennai ... 109277
Ashwell Gallery, Ashwell 117261
Ashwell Village Museum, Ashwell 043449
Ashwood, Hong Kong 065012
Ashworths Treasures of the Earth,
Home Hill 001152
Asia, Dohren 075370
Asia 2000, Hong Kong 136718
Asia-Africa Museum, Genève .. 041511
Asia America, Portland 096365
Asia Art Archive, Hong Kong .. 007870,
136719
Asia Art Center, Beijing 101932
Asia Art Center, Taipei . 116903, 144202
Asia Art Gallery, Mandaluyong .. 113392
Asia Arts, Bangkok 088074
Asia Arts and Crafts Syndikat, Kuala
Lumpur 082399
Asia-Australia Arts Centre,
Sydney 099764
Asia Bookroom, Macquarie 139814
Asia Contemporary Art, London .. 118193
Asia Crate, Oklahoma City 095947
Asia Fine Art, Hong Kong 102058
Asia House, London 118194
Asia International Arts and Antiques
Fair, Hong Kong 098041
The Asia Pacific Contemporary Art Fair,
Shanghai 098042
Asia-Pacific Culture & Arts,
Singapore 114492
Asia Society, New York 060376
Asia Society Museum, New York 051340
Asia Tide, Armadale, Victoria ... 061016
Asiamöbel, Nittendorf 077791
Asian American Arts Centre, New
York 051341, 122364
Asian-American Arts Centre, New
York 051342
Asian and Primitive Art Gallery,
Paddington, New South Wales 099510
Asian Antique Art Gallery,
Ulaanbaatar 082535
Asian Antique Gallery, Denver .. 093258
Asian Antiques, Bangkok 088075
Asian Antiques and Furniture, Los
Angeles 094262
Asian Antiques Center, Houston . 093619
Asian Art, Kathmandu 138925
Asian Art, Santa Fe 139318
Asian Art Center, Washington .. 136453
Asian Art Coordinating Council,
Denver 060377
Asian Art Gallery, London 089929
Asian Art Gallery, New York ... 122365
Asian Art Gallery, Woollahra ... 099906
Asian Art in London, London ... 118195
Asian Art Museum of San Francisco,
San Francisco 052992
Asian Art News, Hong Kong ... 136720,
138598
The Asian Art Newspaper,
London 139149
Asian Art Options, Singapore ... 114493
Asian Arts, New York 095243
Asian Arts and Culture Center, Towson
University, Towson 053786
Asian Center Museum, Quezon
City 034577
Asian Civilisations Museum,
National Museums of Singapore,
Singapore 038266

Asian Connection Galleries, Noosa
Heads 061976
Asian Cultures Museum, Corpus
Christi 048003
Asian Design, New York 095244
Asian Feeling, Bruxelles 063332
Asian Galleries, Art Gallery of New
South Wales, Sydney 001536
Asian House of Chicago, Chicago 092665
Asian Resource Gallery, Oakland 123243
Asian Style, Bolzano 079892
Asian Style, Seattle 097405
Asianart, North Newton 090849
Asiart Corner, Hong Kong 065013,
140441
Asiatic Arte, Buenos Aires 098550
Asiatic Fine Arts, Singapore ... 085369
Asiatic Society, Kolkata 059358
Asiatic Society Museum, Kolkata 024353
Asiatica-Buchhandlung, Zürich . 144162
Asiatica Musée d'Art Oriental,
Biarritz 012027
Asiattitude, Senlis 073491
Asibelle, Couesmes 068176
Asie Antique, Paris 070963
L'Asie Exotique, New York 095245
Asien Antik, Berlin 074760
Asikkalan Kotiseutumuseo,
Asikkala 010758
Asil, Damascus 088028
Asil, Manama 063135, 100414
Asila, Cairo 103020
El Asilo del Libro, Valencia 143800
Asín Remírez de Esparza, F.,
Zaragoza 143815
L'Asino, Livorno 080315
Asioli Martini, Giovanni, Imola .. 080305
Aşiyan Müzesi, İstanbul 042920
Aska Antik, Sorunda 086949
Askanders, Indianapolis 135626
Askeladden Antikk, Stavanger .. 084363
Asker Museum, Hvalstad 033677
Askeri Müze ve Kültür Sitesi
Komtuanlığı, İstanbul 042921
Askersunds Auktionskammare,
Askersund 126568
Aski Galerie, Biarritz 103380
Askim Museum, Askim 033512
Askon Museo, Lahti 011070
Aslamazyan Sisters Museum,
Gyumri 000702
Aslı Tunca, İstanbul 088102
Asmat Art Gallery,
Mönchengladbach 108013
Åsmundarsafn, Listasafn Reykjavíkur,
Reykjavík 024093
Asmundarsalur, Reykjavík 109229
Aso-gun Kazan Hakubutsukan-Orugoru
Kyouwakoku, Aso 028493
Asociace Starožitníků, Praha ... 058536
Asociación Amigos del Museo
Nacional de Arte Decorativo, Buenos
Aires 058170
Asociación Amigos del Museo
Nacional de Bellas Artes, Buenos
Aires 058171
Asociación Colombiana de Museos,
ACOM, Bogotá 058511
Asociacion de Amigos de los Museos
Militares, Madrid 059832
Asociación de Anticuarios de Aragón,
Zaragoza 059833
Asociación de Artistas Visuales de
Euskal Herria, CreaZ, Donostia-San
Sebastián 059834
Asociación de Escritores y Artistas,
Madrid 059835
Asociación de Museos de Guatemala,
AMG, Guatemala 059335
Asociación del Amigos de Museo
Nacional de Arte Decorativo, Buenos
Aires 058172
Asociación Española de Archiveros,
Bibliotecarios, Museólogos y

Documentalistas, ANABAD,
Madrid 059836
Asociación Española de Fabricantes,
Comerciantes Mayoristas de Marcos,
Molduras y Productos afines,
Artecuadro, Madrid 059837
Asociación Española de Museólogos,
Madrid 059838
Asociación Española de Pintores y
Escultores, Madrid 059839
Asociación Española de Profesionales
del Diseño, Madrid 059840
Asociación Fotográfica Cordobesa,
Córdoba 059841
Asociación Galega de Artistas Visuais,
A Coruña 059842
Asociación Internacional Mujeres en
las Artes, Madrid 059843
Asociacion Madrileña de Galerias de
Arte, Madrid 059844
Asociacion Nacional Profesional de
Galerias de Arte, Madrid 059845
Asociación Numismática Española,
Barcelona 059846
Asociación Pintores y Escultores de
Chile, APECH, Santiago de Chile 058483
Asociación Venezolana Amigos del
Arte Colonial, Caracas 060822
Asociatia Naţională a Muzeelor in aer
Liber – România, Bucureşti ... 059764
Aspaas, Trondheim 113285
Aspazijas Māja, Jūrmalas Pilsētas
Muzejs, Jūrmala 030135
Aspect Design, Hobart 099131
d'Aspect Jacques, Marseille ... 140940
Aspect, The Chronicle of New Media
Art, Boston 139319
Aspects Gallery, Blairgowrie ... 098754
Aspects Gallery in Appleby,
Nelson 112955
Aspects of Art, Leyburn 118109
Aspekt Galerie, Brno 102561
Aspen, Denver 093259
Aspen Art Museum, Aspen ... 046579
Aspen Historical Society Museum,
Aspen 046580
Asperger, Berlin 106160
Aspidistra Antiques, Woking ... 091899
Aspinall, Walter, Sabden 091165
Aspingtons, Stockholm 143939
Aspire, Cleveland 127182
ASPN Galerie, Leipzig 107810
Aspö-Tosterö Hembygdsmuseum,
Strängnäs 041058
Asppan, Rivas-Vaciamadrid ... 138000
Asquier, Marie-Pierre, Veigné .. 129331
ASR, Hamburg 129948
Asri Environmental Education Center,
Bristol 047214
Assab One, Milano 059398
Assadi, D., Nyon 116538
Assai al-Hamra (Red Castle),
Tripoli 030293
Assam Forest Museum, Guwahati 024271
Assam State Museum, Guwahati 024272
Assam, Ingeborg, Villach 100208
Åssamuséet, ÅSSA-Industri- och
Bilmuseum, Åtvidaberg 040551
Assay, Caroline d', Paris 128911
Asseff, Henry, Cali 065245
Assembly Rooms Market,
Lancaster 089737
Assets Antiques, Buffalo 092510
Assiginack Historical Museum
and Norisle Heritage Park,
Manitowaning 006251
Assila, Giza 103094
Assinck, Ruud, Lochem 132569
Assiniboia and District Museum,
Assiniboia 005356
Assiniboia Gallery, Regina 101527
Assioma, Prato 110687
Assmann, Franz Josef, Graz ... 062580
Associação Brasileira de

Conservadores-Restauradores de
Bens Culturais, Rio de Janeiro 058398
Associação Brasileira de Livreiros
Antiquarios, Petrópolis 058399
Associação dos Arquitectos de Macau,
Macau 058487
Associação Portugesa de Museologia,
Lisboa 059753
Associação Portuguesa dos
Antiquários, Lisboa 059754
Associacija Muzeev Rossii, Tula 059769
Associacija Muzej, Varna 058403
Associacio Arts, Palma de
Mallorca 059847
Associacio d'Amics de les Antiguitats,
Barcelona 059848
Associacio d'Artistas Visuals de
Catalunya, Barcelona 059849
Associació d'Artistas Visuals de
les Illes Baleares, Palma de
Mallorca 059850
Associació de Galeries Art Barcelona,
Barcelona 059851
Associació del Museu de la
Ciència i la Tècnica i
d'Arqueologia Industrial de Catalunya,
Terrassa 059852
Associaciò Difusora d'Obra Gràfica
Internacional, Barcelona 059853,
137950
Associated American Jewish Museum,
Pittsburg 052151
Associated Antique Dealers of
America, Monticello 060378
Associated Artists of Pittsburgh,
Pittsburgh 060379
Associated Artists of Pittsburgh Gallery,
Pittsburgh 052155
Associated Artists of Winston-Salem,
Winston-Salem 054486, 060380
Association Alberto et Annette
Giacometti, Paris 058592
Association Art Promotion, Liège 058354
Association Artcanal, Biel 059922
Association Artistique Général Gérard,
Ligny 058355
Association Arts de Faire,
Fribourg 059923
Association des Amis de Cholaïc,
Ayer 059924
Association des Amis des Arts et de
la Culture, Monaco 059577
Association des Amis du Musée Asia-
Africa, AMA, Genève 059925
Association des Antiquaires
Brocanteurs et Negociants
en Oeuvres d'Art du Poitou,
Poitiers 058593
Association des Antiquaires et
Brocanteurs de la Charente,
Angoulême 058594
Association des Antiquaires et
Brocanteurs de la Place du Foirail,
Pau 058595
Association des Antiquaires et
Brocanteurs de la Vendée, Les
Sables-d'Olonne 058596
Association des Antiquaires et
Brocanteurs de l'Aude, Conilhac-
Corbières 058597
Association des Antiquaires et
Brocanteurs des Deux-Sèvres,
Chauray 058598
Association des Artilleurs de la
Garnison, Lévis 006189
Association des Brocanteurs et
Antiquaires de Loire-Atlantique,
Nantes 058599
Association des Conservateurs
d'Antiquités, Athinai 059325
Association des Conservateurs et du
Personnel Scientifique des Musées
de la Ville de Paris, Paris 058600
Association des Galeries d'Art

Contemporain, Montréal 058423
Association des Galeries d'Art
Parisiennes, Paris 058601
Association des Musées Automobiles
de France, Paris 058602
Association des Musées et Centres
pour le Développement de la Culture
Scientifique, Technique et Industrielle,
AMCSTI, Dijon 058603
Association des Musées Locaux
d'Alsace, Wingen-sur-Moder .. 016350
Association des Spécialistes en Art
Asiatique, Paris 058604
Association du Patrimoine Artistique,
Bruxelles 058356
Association Européenne des Musées
d'Histoire des Sciences Médicales
(AEMHSM), Paris 058605
Association for Cultural Advancement
through Visual Art, London 060052
Association for Cultural Enterprises,
Worcester 060053
Association for Living History, Farm
and Agricultural Museums, North
Bloomfield 060381
Association for Studies in the
Conservation of Historic Buildings,
London 060054
Association for Support of Women
Artists and Art Critics of Kyrgyzstan,
Biškek 059536
Association for Visual Arts, Cape
Town 059814, 114758
Association Française des
Amateurs d'Horlogerie Ancienne,
Besançon 058606
Association Française des Directeurs
de Centres d'Art, Pougues-les-
Eaux 058607
Association Française des Musées
d'Agriculture et du Patrimoine Rural,
AFMA, Paris 058608
L' Association Francophone des
Musées de Belgique, Hornu .. 058357
Association Générale des
Conservateurs des Collections
Publiques de France, Paris ... 058609
Association Genevoise des Galeries
d'Art Moderne, Genève 059926
Association Internationale de
Management des Arts et de la
Culture, Montréal 058424
Association Internationale des Amis de
Nicolas Schöffer, Paris 014617, 058610
Association Internationale des
Critiques d'Art (AICA), Paris ... 058611
Association Internationale des Musées
d'Armes et d'Histoire Militaire,
Liège 058358
Association Internationale des Musées
d'Histoire, Paris 058612
Association Internationale des
Numismates Professionnels,
Bruxelles 058359
Association Mathieu Vivian, Saint-Pryvé
Saint-Mesmin 015524, 058613
Association Muséographique Médicale,
Hautefort 013238, 058614
Association Museum and Industries,
Paris 058615
Association Nationale des Artistes
Professionnels des Arts Plastiques –
ANAPAP, Ouagadougou 058407
Association Nationale des Conseillers
aux Arts Plastiques, Orléans .. 058616
Association Nayart, Nay 104352
Association Numismatique de Monaco,
Monaco 059578
Association Numismatique du Centre,
Orléans 058617
Association of American Editorial
Cartoonists, Harrisburg 060382
The Association of Art and Antique
Dealers – LAPADA, London ... 060055

Association of Art and Antique
Dealers of Saint Peterburg, Sankt-
Peterburg 059770
Association of Art Historians,
London 060056
Association of Art Historians Bulletin,
London 139150
Association of Art Museum Directors,
New York 060383
Association of Artists in Ireland,
Dublin 059375
Association of British Transport and
Engineering Museums, Gaydon 060057
Association of Children's Museums,
Washington 060384
Association of College and
University Museums and Galleries,
Norman 060385
Association of Community Arts
Agencies of Kansas, Salina ... 060386
Association of Czech Moravian and
Silesian Museums, Praha 058537
Association of European Open Air
Museums, Arnhem 059593
Association of Greek Antiquarians and
Art Dealers, Athinai 059326
Association of Hawaii Artists,
Honolulu 060387
Association of Hispanic Arts, New
York 060388
Association of Icelandic Curators,
Reykjavík 059344
Association of Icelandic Visual Artists,
Reykjavík 059345
Association of Illustrators,
London 060058
Association of Independent Colleges of
Art & Design, San Francisco .. 060389
Association of Independent Museums,
Bristol 060059
Association of Independent Museums
Bulletin, Cocking 139151
Association of Indian Museums,
Delhi 059359
Association of Indiana Museums,
Indianapolis 060390
Association of International
Photography Art Dealers,
Washington 060391
Association of Manitoba Museums,
Winnipeg 058425
Association of Master Upholsterers
and Soft Furnishers, Cardiff .. 060060
Association of Midwest Museums,
Saint Louis 060392
Association of Museum Professionals
of the Republic of Macedonia,
Skopje 059557
Association of Museums and Galleries
of Czech Republic, Praha 058538
Association of Museums of the Indian
Ocean, Victoria 059794
Association of Northern, Kimberley
and Arnhem Aboriginal Artists,
Darwin 058183
Association of Painters and Sculptors,
Tiranë 058165
Association of Railway Museums,
Washington 060393
Association of Science-Museum
Directors, San Diego 060394
Association of Science-Technology
Centers, Washington 060395
Association of South Dakota Museums,
Sioux Falls 060396
Association of Systematics Collections,
Washington 060397
Association of the State Literary
Museums of the Republic of Belarus,
Minsk 058349
Association Perce-Oreilles Productions,
Fontenay-le-Comte 058618
Association pour la Formation et le
Dévelopement des Arts Plastiques,

Paris 058619
Association pour la Recherche
Scientifique sur les Arts Graphiques,
ARSAG, Paris 058620
Association pour les Musées de
Marseille, Marseille 058621
Association Prof. Marcel Hoc pour
l'Encouragement des Recherches
en Numismatique, Louvain-la-
Neuve 058360
Association Professionnelle
de Restaurateurs d'Œvres
d'Art, Beroepsvereniging voor
Restaurateurs van Kunstvoorwerpen,
Bruxelles 058361
Associazione Amici del Museo Bagatti
Valsecchi, Milano 059399
Associazione Amici di Brera e dei
Musei Milanesi, Milano 059400
Associazione Antiquari d'Italia,
Firenze 059401
Associazione Artisti Bresciani,
Brescia 059402
Associazione Artisti Martino Dolci,
Brescia 059403, 059404
Associazione Culturale Arte e Ricambi,
Verona 059405, 111176
Associazione Culturale Azibul,
Milano 059406, 110260
Associazione Culturale Franco Conti,
Alpignano 059407
Associazione Illustratori, Milano . 059408
Associazione Interazioni, Locarno 059927
Associazione Internazionale Mosaicisti
Contemporanei, Ravenna ... 059409
Associazione Librai Antiquari d'Italia,
Firenze 059410
Associazione Musei Ecclesiastici
Italiani, Orvieto 059411
Associazione Napoletana Antiquari,
Napoli 059412
Associazione Nazionale dei Musei dei
Enti Locali ed Istituzionali, ANMLI,
Brescia 059413
Associazione Nazionale dei Musei
Italiani, Roma 059414
Associazione Nazionale Musei
Scientifici, ANMS, Firenze ... 059415
Associazione Piemontese Antiquari,
Torino 059416
Associazione Restauratori d'Italia,
Roma 059417
Associazione Romana Gallerie d'Arte
Moderna, Roma 059418
Assouline, Paris 136864
Assour & Sumer, Bordeaux ... 067308
Den Ast, Buggenhout 003502
Astamangala, Amsterdam ... 082692
Astan-e-Qods-e Razavi Museums,
Mashad 024720
Astano, Nice 070565
Astarte, Lugano 087605, 126739
Astarte, Madrid 115152
Astarte Antichitá, Palermo ... 081018
Astelle, Lyon 128697
Astier, Françis-Sébasti, Saint-
Chaffrey 072711
Astley-Cheetham Art Gallery,
Stalybridge 045855
Astley Hall Museum and Art Gallery,
Chorley 043955
Astley House, Moreton-in-Marsh 090775,
118754,   134858
Astley Street Gallery, Wynnum . 099926
Aston, Charters Towers 061312
Aston Hall, Birmingham Museums,
Birmingham 043610
Aston Manor Transport Museum,
Birmingham 043611
Aston Pine Antiques, Faringdon . 089255
Astor, Memphis 094616
Astor, New York 095246
Astor House Hotel Museum,
Golden 049143

Astori, Simona, Milano 131376
Astorre, Bologna 109815
Astors' Beechwood-Victorian Living
History Museum, Newport 051530
Astra House Antiques Centre,
Hemswell Cliff 089498
Astrachanskaja Gosudarstvennaja
Kartinnaja Galereja im. B.M.
Kustodieva, Astrachan 036870
Astrachanskij Gosudarstvennyj
Obedinennyj Istoriko- architekturnyj
Muzej-zapovednik, Astrachan . 036871
Astrachanskij Kraevedčeskij Muzej,
Astrachanskij muzej-zapovednik,
Astrachan 036872
Astrachanskij Kreml, Astrachanskij
muzej-zapovednik, Astrachan . 036873
Astraea, Leuth 143191
Astrapas, Niš 114451
Astree, Rouen 072562
Astro Antique Auction Gallery,
Easley 127210
Astro-Zombies, Albuquerque ... 092035
Åström, Skellefteå 086933, 126661
Astrolabe, Le Touquet-Paris-
Plage 069359
Astrolabe, Peymeinade 072122
Astrolabe Booksellers, Hobart .. 139772
Astrolabe Gallery, Ottawa 064495,
101463
Astrolavos Artlife, Athinai 109023,
142536
Astrolavos Dexameni, Athinai .. 109024,
137572
Astrolavos Piraeus, Piraeus ... 109080
Astrologo Eugenio, Roma 081244
Astronautics Museum, Beijing .. 007590
Astronomisch Compensatieuurwerk
Kamiel Festraets, Sint-Truiden . 004025
Astronomisch-Physikalisches Kabinett,
Museumslandschaft Hessen Kassel,
Kassel 019527
Astruc-Plijnaar, J.M.Th.,
Rotterdam 083552
Astruc, Stéphane & François-Xavier,
Bordeaux 067309
Astrup & Mogensen, Ulfborg ... 066170
Astrup Fearnley Museet for Moderne
Kunst, Oslo 033826
Astrup, Ingeborg, Oslo 084298
Astruptunet, Skei i Jølster 033940
Asturiol Estany, Andres,
Barcelona 085733, 133099
ASU Museum of Anthropology,
Tempe 053701
Asuka, Tokyo 111405
Asuka Shiryokan, Asuka 028494
Åsum Antik, Odense 066029
Asutosh Museum of Indian Art,
Kolkata 024354
Asuurkeraamika, Tallinn 103110
Ásványkiállítás, Dr. Benko Géza
magángyujteménye, Kaposvár . 023708
Ásványmúzeum, Siófok 023882
Asyl der Kunst, Häuslingen 018910
Asylum House Antiques, Bradford-on-
Avon 088559
At-Bristol, Bristol 043745
At First Sight, Soisy-sous-
Montmorency 073531
At Home Antiques, Toronto 064603
At Home Antiques, Vancouver .. 064797
At Home Galerie, Salins-les-
Bains 105644
At Home Gallery, Limón 102471
ATA – Česky Antikvariat, Praha . 140515
ATA Arts Centre, Singapore ... 114494
Atala, Marseille 069904
Atala & Chactas, Valencia 086503
Atalante, Saint-Aygulf 072685
Ataloa Lodge Museum,
Muskogee 051138
Ataq Museum, Ataq 054827
Atascadero Historical Society Museum,

Atascadero 046587
Atassi, Damascus 116890
Atatürk Evi, Ankara 042816
Atatürk Evi Müzesi, Erzurum .. 042892
Atatürk Evi Müzesi, Mersin ... 043024
Atatürk Evi ve Müzesi, Sakarya
  Müzesi, Sakarya 043045
Atatürk Kitaplığı Sanat Galerisi,
  İstanbul 117008
Atatürk Köşkü, Diyarbakır 042878
Atatürk Kongre ve Etnografya Müzesi,
  Sivas 043063
Atatürk Kültür Evi, İstanbul ... 117009
Atatürk Kültür Merkezi Sanat Galerisi,
  İstanbul 117010
Atatürk Müzesi, Bursa 042858
Atatürk Müzesi, İzmir 042971
Atatürk Müzesi, Konya 043001
Atatürk Müzesi, Erciyes Üniversitesi,
  Kayseri 042990
Atatürk ve Etnografya Müzesi,
  Akşehir 042801
Atatürk ve Etnografya Müzesi,
  Denizli 042875
Atatürk ve Etnografya Müzesi,
  Uşak 043077
Atatürkün Mekanı Müze Köşk,
  Ankara 042817
Atayal Museum, Ren-Ai 042426
Ataz, Minneapolis 121973
ATC Ediciones, Madrid 137983
ATC Gallery, Ho Chi Minh City .. 125112
Atchinson, Topeka and Santa Fe
  Depot, Alden 046374
Atchison, Birmingham 119789
Atchison County Historical Society,
  Atchison 046589
L'Atelier, Angers 066600, 103236
Atelier, Barcelona 133100
Atelier, Brive-la-Gaillarde 103473
Atelier, Calvenzano 109928
L'Atelier, Châteauroux 103580
Het Atelier, Deventer 112426
Atelier, Épinal 128592
O Atelier, Faro 132926
L'Atelier, Firenze 131115
Atelier, Köln 138681
L'Atelier, La Clayette 068995
Atelier, Lannion 069182
Atelier, Nîmes 070685
L'Atelier, Pignans 105354
Atelier, Portland 123592
El Atelier, San José 102482
L'Atelier, Toronto 064604
Atelier, Vélizy-Villacoublay 073998
l'Atelier, Washington 097807
Atelier & Rahmen-Werkstatt, Rainer
  Eulenschmidt, Hamburg ... 107225,
                          129949
Atelier 13, Ellwangen 018159
Atelier 13b, Bochum 106509
Atelier 14, Bottrop 106551
Atelier 142, San Francisco 124225
Atelier 25, Chinon 103592
Atelier 31 2500 First Av, Seattle 124488
Atelier 340 Muzeum, Bruxelles . 100561,
                          136628
Atelier 35, Bucureşti 114127
Atelier 353, Kötschach-Mauthen 100078
Atelier 3a, Wien 100234, 136576
Atelier 4, Courpière 068193
Atelier 4, Honolulu 120923
Atelier 43, Sankt Veit an der
  Glan 100190, 140038
Atelier 5+1, Unterseen 116676
Atelier 52, Saint-Maur-des-
  Fossés 072961
Atelier 67, Nantes 104334
Atelier 68 – Kreativzentrum
  Floridsdorf, Wien 100235
Atelier 71, Fareham 117787
Atelier 82, Ortschwaben 133883
Atelier 91, Paris 128912
Atelier 98, Galerie auf der Uhlenhorst,

Hamburg 107226
Atelier A, Basel 133633
Atelier A-E, New York 122366
Atelier A2, Saint-Paul-de-Vence 073242
Atelier A3, Eferding 099983
Atelier Alexander, Winterthur ... 116721
Atelier am Chelebüel, Muri 116522
Atelier am Glanzhügel, Neumarkt am
  Wallersee 100129
Atelier Am Tönniesberg,
  Hannover 107344
L'Atelier Anne-Marie Frère,
  Contern 082349, 132327
Atelier Antigona, Buenos Aires . 127513
Atelier Antik, Prien am Chiemsee 078093
Atelier Aquitain de Restauration, Etude
  Patrimoine Archéologique, Villeneuve-
  sur-Lot 129350
Atelier Arts, Plzeň 128335
Atelier Arts, Regina 101528
Atelier Asja, Pressbaum 100143
Atelier Augenmerk, Chemnitz .. 106631
Atelier B, Aarau 116003
Atelier B, Knokke-Heist 100727
Atelier B, Vence 105824
Atelier BerGer, Delmenhorst .. 106666
Atelier Block 16, Hannover 107345
Atelier Boab, Saint-Malo (Ille-et-
  Vilaine) 105546
Atelier Bois et Décoration,
  Bischwiller 067231
Atelier Bonaparte, Nice 128846
Atelier Brera, Milano 131377
Atelier Cantoisel, Joigny (Yonne) 103858
Ateliér Carta, Oslo 113191
Atelier Cezanne, Marseille 104158
Atelier Conseil, Nice 128847
L'Atelier Conseils, Lille 069473
Atelier Coperta, Bruxelles 063333,
  063334, 063335, 128015, 128016,
                          128017
L'Atelier Création d'Art, Nice ... 104359
Atelier Creazione Delin, Ascona . 116033
Atelier Damrémont, Paris 128913
Atelier d'Art, Zürich ... 116752, 133974
Atelier d'Art Damrémont 25,
  Paris 070964
Atelier d'Art Point-de-Vue, Saint-Paul-
  de-Vence 105575
Atelier d'Artistes Réunis, Paris . 070965
Atelier de Colombiere, Berlin . 106161
Atelier de Conservation-Restauration,
  Nantes 128828
L'Atelier de Georges Flanet, Saint-
  Tropez 105621
Atelier de la Contre Courbet,
  Châteauneuf-de-Gadagne 103578
Atelier de la Courtine, Saint-Paul-de-
  Vence 105576
Atelier de la Loue, Montgesoye . 104265
L'Atelier de la Ramejane, Sorèze 129261
L'Atelier de la Reinière, Nice ... 070566
Atelier de la Renaissance, Reims 129184
Atelier de la Tour Saint-Nicolas,
  Dijon 068314
Atelier de l'Argenteur, Nantes . 128829
Atelier de l'Ébéniste, La Baule-
  Escoublac 128636
L'Atelier de Maître Poupée, Paris 128914
L'Atelier de Nicolas Schöffer á la Villa
  des Arts, Paris 014618
Atelier de Pan, Tournes 073818
Atelier de Photographie,
  Besançon 103365
Atelier de Restauration, Nantes . 104335,
                          128830
Atelier de Restauration de Tableaux,
  Gévezé 128610
Atelier de Restauration de Tableaux,
  Saint-Germain-en-Laye 129232
Atelier de Restauro, Belo
  Horizonte 128100
Atelier de Tristan, Paris 104531
Atelier d'Ecalgrain, Auderville .. 103285

Atelier Décor, København 065848
Atelier Dedieu, Villeurbanne .. 129352
Atelier degli Artisti, Brescia 109895
Atelier del Restauro, Roma 081245,
                          131822
Atelier des Artistes, Knokke-Heist 100728
Atelier des Arts et Techniques,
  Paris 128915
Atelier des Bastides, Monpazier 014200
Atelier des Clercs, Vienne 105860
Atelier des Cyprès, Hillion 103813
Atelier des Cyprès, Port Aven .. 105393
Atelier des Grames, Gigondas .. 136819
Atelier des Guillemites, Paris ... 070966
Atelier des Marches, Paris 128916
Atelier des Ombres, Lyon 104064
Atelier des Peintres du Marais,
  Paris 104532
Atelier des Petits Champs, Paris 128917
Atelier des Produits Résineux Vidal,
  Ecomusée de la Grande Lande
  Marquèze, Luxey 013925
L'Atelier des Tableaux Carina,
  Nice 104360, 128848
Atelier d'Estampe Imago,
  Moncton 006345
Atelier d'Estienne, Pont-Scorff .. 105388
L'Atelier d'Eugène, Ollainville ... 070779
Atelier di Restauro, Parma 131717
Atelier DL, L'Isle-sur-la-Sorgue . 069555
Atelier do Papel, Porto Alegre .. 063958
Atelier d'Or, Brighton, Victoria .. 127570
Atelier d'Ornano, Bordeaux 128481
Atelier du 18ème, Nevers 070533,
                          128843
Atelier du Bois, Lodève 069608
Atelier du Bois Doré, Paris 128918
L'Atelier du Bronze, Pantin 128884
Atelier du Caire, Cairo 103021
Atelier du Change, Tours 105760
Atelier du Châpitre, Roquemaure 072541
Atelier du Cuir, Tréguier 073876, 129322
L' Atelier du Décorateur,
  Marseille 128744
Atelier du Doreur, Rennes (Ille-et-
  Vilaine) 072420, 129200
Atelier du Meuble, Cogolin 068062,
                          128565
Atelier du Meuble, Le Chambon
  Feugerolles 128655
l'Atelier du Meuble, Nice 128849
Atelier du Pain de Sucre, Saulxures-
  lès-Nancy 105659
L'Atelier du Peintre, Bastia ... 103337
Atelier du Peintre, Gap 103758
l'Atelier du Peintre, Lyon 104065
Atelier du Porche, Saint-Tropez . 105622
L'Atelier du Temps Passé, Paris 128919
Atelier du Val de Grace, Paris .. 128920
Atelier-Editions Fanal, Basel ... 116054
Atelier eins, Nürnberg 108291
Atelier Engramme, Québec 101488
Atelier Fernando Moreira, Paris . 128921
Atelier Franz Knapp, Pöchlarn .. 002511
Atelier Freies Glas, Haibach ... 100027
Atelier für Architektur und Kunst,
  Kassel 107528
Atelier für Buchbinderei und
  Einrahmungen – ABE, Sankt
  Gallen 133900
Atelier für Kunst und Handwerk,
  München 108050
Atelier für Kunst und Rahmen,
  Stuttgart 130664
Atelier für Restaurierung,
  Regensburg 130560
Atelier Fuxx, Walchensee 108836
Atelier G, Amiens 128435
Atelier Galeria Staromiejskiego
  Centrum Kultury, Kraków 113530
Atelier-Galerie, Andelfingen ... 116022
Atelier Galerie Bleu, Lille 104002
Atelier Galerie Carrat, Saint-

Tropez 105623, 129247
Atelier-Galerie de la Haulte Voie,
  Grandcamp-Maisy 103775
Atelier-Galerie de la Valsainte,
  Vevey 116692
Atelier Galerie du Richard Rak,
  Coutures 012709
Atelier-Galerie im A-Dürer-Hof, Sofia
  Fränkl, Nürnberg 108292
Atelier-Galerie Pigoza, Wuppertal 108982
Atelier-Galerie, Wicki Regula, Lohn-
  Ammannsegg 116449
Atelier-Gallerlet, Nyköbing
  Sjælland 102939
Atelier Gallery, Vancouver 101741
Atelier GF, Toronto 101574
Atelier Guidoux, Nantes 070447
Atelier Herenplaats, Rotterdam . 112630
Atelier HL, Den Haag 132470
Atelier Hoguet, Musée de l'Eventail,
  Paris 014619
Atelier A. Hug & B. Bührer,
  Zürich 133975
Atelier im Bärenhof, Zürich ... 087935
Atelier im Hof, Berlin 106162
Atelier im Landhaus, Bramsche . 106555
Atelier im Musikverein, Wien ... 127899
Atelier Irmak, Bern ... 087271, 133667
Atelier Ismet Mujezinović, Galerija
  Portreta Tuzla, Univerzitet u Tuzli,
  Tuzla 004287
Atelier Isolde Zipperer, Darmstadt 129668
Atelier Ivan Meštrović, Ivan Meštrović
  Foundation, Zagreb 009026
Atelier Jens Rusch, Brunsbüttel . 106605
Atelier Kea Homan, Assen 032033
Atelier Kjarval, Frederiksværk .. 102812
Atelier KL, Bordeaux 103416
Atelier Klex, Rheine 108496
Atelier la Giradière, Charentay .. 128547
Atelier La Patine, Nouan-le-
  Fuzelier 070754
L'Atelier Legault, Pouance 014976
Atelier Limited, Münster 108212, 137433
Atelier Livre & Pinturas Especiais,
  Fortaleza 063941
Atelier Lucas, Aigues-Mortes ... 103176
Atelier M.B., Verona 132267
Atelier Manoli-Musée et Jardin de
  Sculpture, La Richardais ... 013457
Atelier Memorial Paul Erdös, Satu
  Mare 036682
L'Atelier Metal, Martigues 104207
Atelier Mind Art, Neupölla 100131
Atelier Mistral, Hyères 103848
Atelier Municipal de Passementerie,
  Assoiation Soierie Vivante, Lyon 013930
Atelier Musée de la Soie,
  Montboucher-sur-Jabron 014236
Atelier-Musée de la Soie,
  Taulignan 015919
Atelier-Musée des Tisserands et de la
  Charentaise, Varaignes 016125
Atelier-Musée du Bâton Basque,
  Larressore 013578
Atelier-Musée du Chapeau, Chazelles-
  sur-Lyon 012540
Atelier-Musée du Sabotier,
  Lautrec 013586
Atelier Musée du Verre, Ancienne
  Verrerie Parant, Trélon 016041
Atelier-Musée du Vitrail et des
  Métiers d'Art du Verre, Gréoux-les-
  Bains 013187
Atelier-Musée Livre et Typographie,
  Grignan 013195
Atelier-Museu António Duarte, Caldas
  da Raínha 035584
Atelier-Museu João Fragoso, Caldas
  da Raínha 035585
Atelier Na-iem, Gees 032335
Atelier Claude Nicolet, Paris ... 070967
Atelier Nordfelder Reihe 13,
  Hannover 107346

Atelier Ocre d'Art, Châteauroux . 103581
Atelier Opus, Saumur 129254
Atelier Oriental Rugs, Louisville . 094520
Atelier Otto Niemeyer-Holstein, Die
Neue Galerie, Koserow 019773
Atelier-Outlet Berlin, Berlin 106163
Atelier Palette, Cairo 103022
Atelier Paul Cézanne, Aix-en-
Provence 011558
Atelier Pierre, Vals-les-Bains . . . 105815
Atelier Produzioni d'Arte, Milano 110261
Atelier Régional de Restauration,
Bignan 128469
Atelier Renaissance, Nantes . . . 128831
Atelier Repinto, São Paulo 064154
L'Atelier Restauracion de Muebles,
Madrid 133301
Atelier Robert, Barcelona 133101
Atelier Römerberg, Wiesbaden . 108900
Atelier Saint Antoine, Reims . . 072381,
129185
Atelier Saint-Michel, Le Havre . . 069272,
103942,   128660
Atelier Saint-Romain, Rouen . . . 105468
Atelier Sala d'Art, Barcelona . . . 114857
Atelier Salem, Salem bei
Ratzeburg 078341
Atelier-Sankt-Georg, Meiningen . 127817
Atelier Segantini, Maloja 041719
Atelier Spyros Vassiliou, Athinai 022977
Atelier Steendam, Gent 063627
Atelier Stéphane B, La Grande-
Motte 103893, 128641
Atelier Stern, Konstanz 107710
Atelier sur Cour, Rueil-Malmaison 105488
Ateliér Triton, Praha . . 065479, 128336
Atelier Verlag, Köln 137303
Atelier-Wohnhaus Othmar Jaindl,
Villach 002860
Atelier XVIIIème, Montbrison . . . 070217
Atelier Yang, Sandnes 113254
L'Atelier Z, Paris 104533
Atelier Züri-Breizh, Villaz-Saint-
Pierre 116702
Atelier zum Eck,
Speicherschwendi 087823
Atelier zum Wöschhuus D.,
Stallikon 087830
l'Ateliere, Liège 063743
Ateliergalerie In den Gerbgruben,
Neusiedl am See 100132
Atelierhaus Aachen, Aachen . . . 016432
Atelierhaus der Akademie der
bildenden Künste Wien, Wien . 002939
Atelierhaus Filderstrasse 34 e.V.,
Stuttgart 108688
Atelierhaus Philipp Harth,
Bayrischzell 017003
Atelierhof-Galerie, Bremen 106574
Ateliermuseum und Skulpturengarten
Erwin Rehmann, Laufenburg . . 041643
Ateliers Brugier, Paris . 070968, 128922
Ateliers Couleurs Sud, Toulouse 073736
Ateliers d'Art, Paris 128923
Ateliers d'Artistes de la Ville de
Marseille, Marseille 136833
Les Ateliers de la Mabilais, Noyal-sur-
Vilaine 128876
Ateliers de l'Abeille, Chavignon . 012538
Les Ateliers de Vence, Vence . . 105825
Ateliers du Parc, Champigny-sur-
Marne 103568
Les Ateliers du Prisme, Paris . . 070969
Les Ateliers du Roy, Le Haillan . 128659
Ateliers Férignac, Hautefort . . . 128619
Ateliers Gai, Lille 069474
Ateliers im Alten Schlachthof e.V.,
Sigmaringen 108622
Ateliersammlung Herbert Hiesmayr,
Sankt Thomas am Blasenstein 002701
Atelierul de Restaurari, Bucureşti 132992
Ateljé Ann-Lill, Finspång 115616
Ateljé Bästekille, Kivik 115709
Ateljé Ebéniste, Stockholm . . . 133564

Atelje Firul, Negotin 114450
Atelje-galerija Rebeka, Ljubljana 114698
Ateljé Nygammalt, Stockholm . . 086973,
133565
Ateljé Tingshuset, Eskilstuna . . . 115601
Atelje za Umetnost in Obrt,
Metlika 085621, 114735, 133069
Atena, Paris 070970
Atena, Warszawa 084705, 113805,
126419
Atena Gallery, Ha Noi 097980
Atenea, Punta del Este 119438
Ateneo Art Gallery, Ateneo de Manila
University, Quezon City 034578
Ateneo de Madrid, Madrid 059854
Ateneo de Valencia, Valencia . . . 054777
Ateneo Puertorriqueño, San Juan 036150
Ateneumin Taidemuseo, Valtion
Taidemuseo, Helsinki 010813
Atezart, Avignon 103301
Atha Cliath Paint Stripping, Dolphins
Barn 130839
Athabasca, Edmonton 140342
Athanassakeion Archaeological
Museum, Volos 023314
Athanassiadou, Zina,
Thessaloniki 109086
Athar, Casablanca . . . 082544, 112032
Athelhampton House, Dorchester 044114
Athelstan Museum, Malmesbury 045198
Athena, Bari 130905
Athéna, Binic 103388
Athena, Livorno 110173
Athena, München . . . 077462, 126043
Athena, Palermo 081019, 081020
Athena, Southampton 119118
Athena Antiquités, Casablanca . 082545
Athena Art Gallery, Kortrijk 100772
The Athenaeum, Alexandria . . . 046383
Athenaeum, Pavia 138805
Athenaeum of Philadelphia,
Philadelphia 052030
Athenars, Roma 081246
Athena's Antiques, Portland . . . 096366
Athena's Upholstery, San Diego . 136289
Athens School of Fine Arts,
Athinai 055751
Athens University Museum,
Athinai 022978
Atherton Antiques, Atherton . . . 061069
Athesia, Bolzano 137627, 142649
Athey, G.M., Alnwick . 088252, 134020
Athlone, Calgary 128144
Athlone Antiques, Athlone 079439
Athlone Castle Museum, Athlone 024819
Athol Murray College of Notre Dame
Museum, Wilcox 007414
Atholl Antiques, Aberdeen 088218
Atholl Country Life Museum, Blair
Atholl 043651
The Atholl Gallery, Dunkeld 117696
Athus et l'Acier, Athus 003330
Atica, Buenos Aires 098551
El Atico de Los Angeles,
Maracaibo 097972
Atikokan Centennial Museum,
Atikokan 005357
Atkins, Garry, London 089930
Atkins, T. J., Tolland 091607
Atkinson, Street 119170
Atkinson Art Gallery, Southport . 045838
Atlanta Antique Gallery, Atlanta . 092111
Atlanta Antiques and Auction,
Atlanta 092112, 127125
Atlanta Art Gallery, Atlanta 119542
Atlanta Art Services, Atlanta . . . 119543
Atlanta Book Exchange, Atlanta . 144961
Atlanta College of Art, Atlanta . . 056680
The Atlanta College of Art,
Atlanta 046605
Atlanta Contemporary Art Center,

Atlanta 046606
The Atlanta Cyclorama, Atlanta . 046607
Atlanta Furniture, Atlanta 135298
Atlanta Furniture, Chamblee . . . 135385
Atlanta Galleries, Atlanta 119544
Atlanta History Museum, Atlanta 046608
Atlanta International Museum of Art
and Design, Atlanta 046609
Atlanta Museum, Atlanta 046610
Atlanta Panorama, Atlanta 119545
Atlantic 1 Museum, Swampscott 053621
Atlantic Art, Galway 109634
Atlantic Art Gallery, Cape Town . 114759
Atlantic Bay Gallery, London . . . 089931
Atlantic Canada Aviation Museum,
Enfield 005785
Atlantic City Historical Museum,
Atlantic City 046642
Atlantic County Historical Society
Museum, Somers Point 053381
Atlantic Gallery, Kinsale 109656
Atlantic Gallery, New York 122367
Atlantic Gallery of Georgetown,
Washington 124868
Atlantic Pacific Time, New York . 095247
Atlantic Provinces Art Gallery
Association, Wolfville 058426
Atlantic Salmon Museum on the
Miramichi River, Doaktown . . . 005710
Atlantic Wall Museum, Het Domein
Raversijde, Oostende 003926
Atlantic Wildfowl Heritage Museum,
Virginia Beach 053994
Atlántica Internacional, Las Palmas de
Gran Canaria 139039
Atlantik Verlags- und
Mediengesellschaft, Bremen . . 137124
Atlantiquasie, Le Bono 069224
Atl'Antique, Rochefort (Charente-
Maritime) 072494
Atlantique City Fall Festival, Atlantic
City 098433
Atlantique City Spring Festival, Atlantic
City 098434
Atlantique Restores and
Conservationists, Devonshire . 128093
Atlantiques, Long Beach 094193
Atlantis, Houston 135587
Atlantis, Stockholm 138012
Atlantis Antiques, Louisville . . . 094521,
135749
Atlantis Bookshop, London 144558
Atlantis Discovered, San Antonio 096947
Atlantis Kindermuseum, Duisburg 018050
Atlas, London 138193, 139152
Atlas Antiquariaat, Rotterdam . . 143216
Atlas Antiquités, Montréal 064392
Atlas Coal Mine, East Coulee . . 005736
Atlas Galleries, Chicago 120051, 120052,
120053
Atlas Gallery, London 118196
Atlas Sztuki, Łódź 113658
Atlatl- National Native Arts Network,
Phoenix 060398
L'Atelier Galerie Otto, Wien . . . 062902
Atlin Historical Museum, Atlin . . 005359
ATM Gallery, New York 122368
Atmanmöbler, Malmö 086846
Atmosphere, Bruxelles 063336
Atmosphère, Dijon 068315
Atmosphere Antique and Unique
Interior, Göteborg 086698
Atmosphère Antiquités, Créteil . 068420
Atmosphère d'Ailleurs, Lorient . . 069631
Atmosphère d'Antan, Ménilles . . 070053
Atnabrufossen Vannbruksmuseum,
Atna 033513
Atom Museum, Haigerloch 018927
Atomic Passion, New York 095248
Atomik Pop 2, Oklahoma City . . 095948
Atomium, Bruxelles 003431
Atout Coeur, Calais 103491
Atout Poils, Saint-Gervais-les-
Bains 105522

Atrandom, Gifu 111254
Atrax, Přerov 102730
Atre, Lipsheim 069541
Atrema, Orléans 070807
Atreo, Zaragoza 086588
Atribut, Moskva 085174
Atrio, Santiago de Chile 064952
Atrion, Thessaloniki 109087
Atrium, Plzeň 102641
Atrium, Waibstadt 078827
Atrium 1, İstanbul 088103
Atrium 2, İstanbul 088104
Atrium an der Schleuse,
Brunsbüttel 017647
Atrium du Casino, Monaco 111993
Atrium Galerija, Beograd 114398
Atrium Gallery, Bath 117298
Atrium Gallery, Clifton Beach . . . 098898
Atrium Gallery, Fort Worth 048899
Atrium Gallery, Saint Louis 052790
Atrium Gallery, Salt Lake City . . 052909
Atrium Gallery, University of Southern
Main, Lewiston 050231
Atrium na Žižkově, Praha 102651
Atrium Showcase Gallery, San
Antonio 124016
Atrozos, Madrid 133302
Atsukeshi-machi Mingeikan,
Atsukeshi 028498
Atsuta-Jingu Museum, Nagoya . 029111
Attal, Jeffrey C., Austin 092214, 144976
Attali, Ange, Le Perreux-sur-
Marne 069326
Attali, Jean-Hubert, Paris 104534
Attenson, Mitchell, Cleveland . . 092921,
145081
Attention to Detail, Columbus . . 120476
Attese Biennale Ceramica Arte
Contemporanea, Milano 059419
Atti dei Convegni di Studi Sulla Magna
Grecia, Roma 138806
Attias, Annemarie, Zürich 087936
Attias, Fortunato Jose, Rio de
Janeiro 064022
Attic, Baldock 088338
The Attic, Delhi 079382
The Attic, Newton Abbot 090838
The Attic, San Francisco 097185
Attic Antiques, Blackburn South 061159
Attic Antiques, Indianapolis . . . 093858,
135627
Attic Antiques, Saintfield 091214
Attic Books, Cheltenham 144321
Attic Finds, Tucson 097637
The Attic Gallery, Leven 118107
Attic Gallery, Portland 123593
Attic Gallery, Swansea 119191
Attic Treasures, Louisville 094522
Attic Treasures, New Orleans . . 095056
Attic Treasures, Norfolk 095844
Attic Treasures, Vancouver 064524
Attica, Edmonton 064286, 128161
Attica Furniture Restoration,
Edmonton 128162
El Attico, Guatemala . . 079222, 109117
Atticus, Warszawa 143517
Atticus Books, Grays 144442
Atticus Books, Toronto 140396
Atticus Bookshop, Lancaster . . . 144512
Attienaj, Toronto 064670
Åttingen Hantverk och
Lantbruksmuseum, Aneby 040526
Attingham Park, Shrewsbury . . . 045793
Attique, Tulsa 097713
Attitude Exact, Washington 124869
l'Attitude Gallery, Boston 119835
Attitudes, Espace d'Arts Contemporains,
Genève 041512, 059928
Attleboro Area Industrial Museum,
Attleboro 046644
Attleboro Museum-Cerner for the Arts,
Attleboro 046645
Attmore-Oliver House, New Bern 051229
Atualidades Panoramica e Galeria de

Arte, Rio de Janeiro . 064023, 101011
Åtvidabergs Bruks- och Facitmuseum,
    Åtvidaberg 040552
Atw Antiques, Hamilton 064315
Atwater Kent Museum – The
    History Museum of Philadelphia,
    Philadelphia 052031
Atwell House Gallery, Melville
    Community Arts Centre, Alfred
    Cove 000785
Atwell-Wilson Motor Museum,
    Calne 043830
Atwood House Museum,
    Chatham 047588
Atwood Memorial Center Art Collection,
    Saint Cloud State University, Saint
    Cloud 052760
Atypique, Brignais 067494
Atyrauskij Oblastnoj Istoriko-
    Kraevedčeskij Muzej, Atyrau . 029813
Atyrauskij Oblastnoj Muzej
    Dekorativno-prikladnogo Iskusstva,
    Atyrau 029814
Atzig, Peter, Pforzheim 130516
Au Bleu de Longwy, Herserange 068814
Au Bon Diable, Le Conquet . . . 069253
Au Bon Maitre d'Armes, Le
    Castellet 069235
Au Bon Usage, Paris 070971
Au Bon Vieux Temps, Belvès . . 067083
Au Bon Vieux Temps, Chalon-sur-
    Saône 067772
Au Bon Vieux Temps,
    Luxembourg 082364
Au Bon Vieux Temps, Royan . . 072624
Au Bonheur d'Antan, Duclair . . 068410
Au Bonheur d'Ombeline,
    Toulouse 073737
Au Bonheur du Jour, Genève . . 087447
Au Bonheur du Jour, Lausanne . 087562
Au Bonheur du Jour, Neuchâtel . 087699,
                                                133869
Au Bonheur du Jour, Paris 104535
Au Bonheur du Jour, Valence-
    d'Agen 073940
Au Bonheur du Meuble, Mées . . 070029,
                                                128793
Au Bonheur du Tour, Douai 068391
Au Bouffon du Roy, Les Rosiers-sur-
    Loire 069423
Au Bouquiniste, Le Havre 140865
Au Bout de la Cour, Maulette . . 069999
Au Bric à Brac, Metz 070086
Au Capharnaüm, Perpignan . . . 072071
Au Carré de Jouy, Provins 072317
Au Charme d'Antan, Cerizay . . 067752
Au Charme d'Antan, Saint-Omer 072994
Au Charme d'Autrefois,
    Luxembourg 082365
Au Charme d'Autrefois, Pont-de-
    Briques-Saint-Etienne 072245
Au Charme du Logis, Bruxelles . 063337
Au Charme du Passé, Saint-
    Cloud 072716
Au Charme du Passé, Tel Aviv . 079732
Au Charme du Temps, Pont-
    Audemer 072240
Au Chat Botté, Basel 087220
Au Cherche-Midi, Bruxelles . . . 063338
Au Chevalier Bayard, Grenoble . 068716
Au Chevalier Noir, Lausanne . . 087563
Au Chiabout, Metz . . 070087, 128797
Au Chien Vert, Pézenas 072129
Au Chineur, Gardonne 068635
Au Chin'Eure, Saint-André-de-
    l'Eure 072669
Au Clair de la Lune, Dozule . . . 068402
Au Clair Obscur, Dinan 068348
Au Coeur de l'Ardoise, Bertrix . 003363
Au Coeur du Temps, Sarzeau . . 073423
Au coeur qui voit, Yverdon-les-
    Bains 116730
Au Coin des Ombres, Pau 072031
Au Coin du Meuble, Québec . . . 064524

Au Coin Littèraire, Strasbourg . . 141346
Au Col Bleu, Brest 067464
Au Collectionneur, Paris 070972
Au Collectionneur, Tournai 063838
Au Comptoir des Hallettes,
    Montivilliers 070259
Au Coq d'Inde, Neuchâtel 087700
Au Debotté, Paris 070973
Au Décor d'Autrefois, Strasbourg 073571
Au Dénicheur, Lomme 069611
Au Dénicheur, Saint-Ouen-du-
    Breuil 073235
Au Duc de Chartres, Paris 070974
Au Facteur Cheval, Paris 070975
Au Faubourg, Reims . . 072382, 129186
Au Feuillet Jauni, Waterloo . . . 140263
Au Fil des Ans, Bois-Colombes . 067266
Au Fil des Arts, Montivilliers . . . 104267
Au Fil des Siècles, Fauverney . . 068509
Au Fil des Temps, Saint-Ouen . 073018
Au fil du temps, Beirut 082275
Au Fil du Temps, Bruxelles 063339
Au Fil du Temps, Bruz 067526
Au Fil du Temps, Caen 067555
Au Fil du Temps, Cortaillod . . . 087384
Au Fil du Temps, Douai 068392
Au Fil du Temps, La Coucourde 069008
Au Fil du Temps, Montauban . . 070188
Au Fil du Temps, Paris 070976
Au Fil du Temps, Pézenas 072130
Au Fil du Temps, Vigeois 074120
Au Foyer d'Autrefois, Lausanne . 087564
Au France, New York 135887
Au Furet, Bois-Guillaume 067269
Au Garage, La Tour-de-Trême . . 087853,
                                                133939
Au Glaize Village, Defiance . . . 048222
Au Goût du Jour, Ajaccio 066494,
                                                103209
Au Gré du Temps, Albi 066508
Au Grenier, Champenoux 067812
Au Grenier, Vouvry 087887
Au Grenier d'Ango, Dieppe . . . 068296
Au Grenier de Beauvallon, Sainte-
    Maxime 073356
Au Grenier de Lucie, Saint-Ouen 073019,
                                                073020
Au Grenier de Monette, Gannat . 068627
Au Grenier d'Elise, Bar-le-Duc . 068670
Au Grenier du Louvre, Maulette 070000
Au Grenier Normand, Le Tréport 069367
Au Jardin Antique, Casablanca . 082546
Au Jeton d'Argent, Rennes (Ille-et-
    Vilaine) 072421
Au Jeu de Paume, Montpellier . 072502
Au Jouet d'Autrefois, Lyon 069700
Au Lac Art Gallery, Ho Chi Minh
    City 125113
Au Lion d'Argent, Gisors 068670
Au Lion des Neiges, Woollahra . 062457
Au Louis d'Or, Caen 067556
Au Mee Co, Doetinchem 083168
Au Mobilier Vendéen, Les Essards
    (Vendée) 069407, 128673
Au Mobilier Vendéen, Saint-Jean-de-
    Monts 072849
Au Nom de la Rose, Aubagne . . 066771
Au Palais du Meuble, Changy . . 067826
Au Papier Monnaie, Nantes . . . 070448
Au Passage Gambetta, Biarritz . 067185
Au Passé Entreposé, Lyon 069701
Au Passé Présent, Maureillas-las-
    Illas 070011, 140950
Au Passé Simple, Aix-les-Bains . 066477
Au Passé Simple, Amiens 066561
Au Passé Simple, Ardenay-sur-
    Mérize 066706
Au Passé Simple, Nogent-le-roi 070725
Au Passé Simple, Paris 070977
Au Passé Simple, Poitiers 072218,
                                                129171
Au Pays d'Ardenne – Original
    Museum, Bastogne 003342
Au Petit Bonheur, Beaumont

(Vienne) 067002
Au Petit Chose, Genève 087448
Au Petit Grenier, Cluses 068051
Au Petit Grenier, Nice 070567
Au Petit Paris on the Corner, Los
    Angeles 094263
Au Petit Passage, Chêne-
    Bougeries 087363
Au petit Trésor, Basel 087221
Au Picpus, Genève 087449
Au Pin de Campagne, Liège . . . 063744
Au Point du Jour, Genève 144070
Au Point du Jour, Troyes 129324
Au Prince d'Argos, Limoges . . . 014027
Au Reflets d'Antan, Aubigny-au-
    Bac 066790
Au Regard des Anges, Forgès . . 068589
Au Réveil du Temps, Saint-
    Nabord 072986
Au Roy Soleil, Antibes-Juan-les-
    Pins 066664
Au Siècle Passé, Bois-Colombes 067267
Au Singe Violet Antiquités, Paris 070978
Au Sol d'Aix, Aix-en-Provence . 066454
Au Soli, Megève 070031
Au Sou d'Or, Béziers 067165
Au Sourire d'Ariane, Saint-Ouen 073021
Au Temps Conte, Cely-en-Bière . 067743
Au Temps Jadis, Buironfosse . . 067533
Au Temps Jadis, Challans 067766,
                                                128536
Au Temps Jadis, Den Haag . . . 083057
Au Temps Jadis, Fontenay-sous-
    Bois 068577
Au Temps Jadis, Huisseau-sur-
    Cosson 068847
Au Temps Jadis, Montagnac . . . 070178
Au Temps Jadis, Rumegies . . . 072645
Au Temps Jadis, Sailly-lez-
    Lannoy 072654
Au Temps Jadis, Saint-Quentin
    (Aisne) 073274
Au Temps Jadis, Vétraz-
    Monthoux 074081
Au Temps Passé, Amiens 066562
Au Temps Passé, Mamers 069830
Au Temps qui Passe, Courville-sur-
    Eure 068201
Au Temps Retrouvé, Carouge . . 087346
Au Trésor de Paris-ATP, Paris . . 070979
Au Troc Cevenol, Florac 068549
Au Troc Malouin, Saint-Malo (Ille-et-
    Vilaine) 072912
Au Vase de Delft, Paris 070980
Au Verre à Pied, Le Mans 069302
Au Verre Luisant, Voiron 074269
Au Vieil Artisan Bressan,
    Confrançon 068136
Au Vieil Horloger, Genève 087450
Au Vieu Parchet, Flumet 068552
Au Vieux Bibelot, Charleroi . . . 063591
Au Vieux Bougeoir, Marseille . . 069905
Au Vieux Canon, Vésenaz 087871
Au Vieux Chaudron, Lannebert . 069180
Au Vieux Chêne, Metz 070088
Au Vieux Chevreuse, Chevreuse 067970
Au Vieux Chinois Antiquités,
    Paris 070981
Au Vieux Colombier, Aubencheul-au-
    Bac 066782
Au Vieux Dijon, Dijon 068316
Au Vieux Document, Paris 141034
Au Vieux Fiacre, Mont-Saint-
    Martin 070176
Au Vieux Grenier, Bergerac . . . 067098
Au Vieux Grenier, Fléac 068539
Au Vieux Logis, Ploisy 072191
Au Vieux Magasin, Bruxelles . . 063340
Au Vieux Manoir, Bapaume . . . 066933
Au Vieux Marché, Chinon 067977
Au Vieux Minervois, Rieux-
    Minervois 072464
Au Vieux Paris, Paris 070982
Au Vieux Paris, Tournai 063839

Au Vieux Quartier, Namur 140255
Au Vieux Rodez, Rodez 072509
Au Vieux Saint-Georges,
    Pérouges 072066
Au Vieux Saint-Pierre, Calais . . 067601
Au Vieux Saussaies, Paris 070983
Au Vieux Souvenir, Versailles . . 074044
Au Vieux Tournay, Tournai 063840
Au Vieux Tours, Tours 073835
Au Violon d'Ingres, Casablanca . 082547
Au Violon d'Ingres, L'Isle 087537
Au Voltaire, Châtillon-d'Azergues 067918
Au XXème Siècle, Paris 070984
l'Aubaine, Liège 063745
L'Auberge du Hussard, Brantôme 067448
Auberge Ganne, Musée Departemental
    de l'École de Barbizon,
    Barbizon 011891
Auberlehaus, Trossinger
    Heimatmuseum, Trossingen . . 022209
Aubert, Fabrègues 068501
Aubert, Jacques, Gerbéviller . . 068653
Auböck, Carl, Wien 100236
Auböck, Max, Hallwang 127765
Aubonnière, Chaille-sous-les-
    Ormeaux 012401
Aubour, Michel, Limoges 069517
Aubri, Christophe, Einvaux . . . 068436
Aubri, Michèle, Xeuilley 074303
Aubry, Alain, Ménil 070051
Aubry, Geneviève, Rosières-aux-
    Salines 072545
Aubry, Jacques, Pont-l'Abbé . . 072252
Aubry, Jean, Paris 070985
Aubry, Roger, Morges . 087684, 116517
Auburn Antiques, Auckland . . . 083766
Auburn-Cord-Duesenberg Museum,
    Auburn 046648
Auburn Gallery, Auburn 098675
Auburn University Gallery, Auburn 046647
Auchère-Merle, Marie-Hélène,
    Brantôme 103458, 128520
Auchinachie & Son, G.L.,
    Woollahra 062458
Auchindrain Museum, Inveraray 044634
Auckland Antique Shop,
    Auckland 083767
Auckland Art Gallery, Auckland . 033141
Auckland Collectors Centre,
    Auckland 083768
Auckland Museum and Auckland War
    Memorial Museum, Auckland . 033142
Auckland Photo Restorers, South
    Auckland 132739
Auction Block Company,
    Sacramento 127423
Auction Broker, Oakland 127374
Auction Company of Southern Oregon,
    Coos Bay 127188
Auction Gallery at the Middle of Main
    Antique Emporium, Front Royal 127231
The Auction Guild, Fayette 139320
Auction House of West Salem, West
    Salem 127493
Auction Partners, Königswinter . 126016
Auction Place, Welshpool 125244
Auction Rooms, Falkirk 126895
Auction Trader, Footscray 125192
Auctioneers Association, New
    York 060399
Auctioneers Association of Canada,
    Edmonton 058427
Audap, Pierre Emmanuel, Paris . 125676
Auden, Bad Homburg v.d.Höhe . 106060
Audeoud, Georges, Genève . . . 087451
Audi Fondation, Saïda 030280
Audi Museum Mobile, Ingolstadt 019405
Audibert, Paray-Vieille-Poste . . 104440
Audibert, François, Toulon (Var) . 129283
Audibert, Jean-Marie, Nîmes . . 070686
Audie Murphy/American Cotton
    Museum, Greenville 049269
Audiello, Massimo, New York . . 122369
Audiger, Christian, Breteuil-sur-

Iton 067483
Audiovisuelles Zentrum Wien,
  Wien 002940
Audley End House, Saffron
  Walden 045659
Audorfer Museum im Burgtor,
  Oberaudorf 020889
Audouard Franck, Pordic 072277
Audouze, Jean-François, Pau .. 072032
Audrain Historical Museum, Graceland,
  Mexico 050823
Audrey Natural Body, New York . 122370
Audreys Antiques, Auckland ... 083769
Audrey's Place, Indianapolis ... 093859,
                         135628
Audubon Antiques, New Orleans 095057
Audubon Artists, New York 060400
Audubon House, Key West 049949
Audubon State Historic Site, Saint
  Francisville 052766
Audubon's Auctioneers of Fine Art and
  Antiques, Audubon 127129
Aue Art Galerie, Dortmund 106697
Auer, Saint Louis 096667
Auer, Walsrode 078856
Auer von Welsbach Museum,
  Althofen 001707
Auersperg, Andreas, Goldegg .. 127739
Aufbau Antiquariat, Zürich 144163
Aufbau-Verlag, Berlin 137053
Auffenbauer, Dieter, Perleberg .. 130509
Auffret, Catherine, L'Isle-sur-la-
  Sorgue 069556
Auffret, Francois, Olonzac 070781
Aufgemöbelt, Bielefeld 074973
Aufhauser, Wien 127900
Auge, Jens, Obernkirchen 130480
Augen, Portland 123594
Augenius, Aschaffenburg 125850
Augenstein, F., Sinsheim 078481
Auger, Françoise, Nançay 070406
Auger, Roger, Québec 140382
AuGIAS, Juziers 068933
Augier, Albert, Carpentras 067692
Auglaize County Historical Society
  Museum, Saint Marys 052829
Augsburger Puppentheatermuseum
  Die Kiste, Augsburg 016647
Augsburger Spielzeugkiste,
  Augsburg 074450
August Antiques, Houston 093620
August-Becker-Museum,
  Klingenmünster 019648
August Deusser-Museum, Bad
  Zurzach 041247
August-Gottschalk-Haus, Gedenkstätte
  und Ausstellung zur neueren
  Geschichte der ostfriesischen
  Juden – Jüdisches Museum,
  Esens 018279
August-Holländer-Museum,
  Emsdetten 018198
August Horch Museum Zwickau,
  Zwickau 022907
August House Studio, Chicago . 120054
August Kitzbergi Tubamuuseum,
  Karksi 010625
August Macke Haus, Bonn 017482
Augusta, Indianapolis 093860
Augusta Antiquariat, Augsburg . 141413
Augusta Historical Museum,
  Augusta 000809, 046664
Augusta Museum of History,
  Augusta 046657
Augusta Raurica, Augst 041243
Augustana College Art Gallery, Rock
  Island 052627
Augustana College Center for Western
  Studies, Sioux Falls 053351
Auguste, Alençon 066521
Auguste Gallery, Houston 121001
Auguste, François, Saint-Nicolas-lez-
  Arras 072992

Auguste, François, Saint-Ouen . 073022
Augustin, Lübeck 077170
Augustin, Claude, Les Chères .. 069405
Augustin, N., Middelburg 083491
Augustin, Peter, Innsbruck 062644,
                         100040
Augustin, Peter, Wien 100237
Augustiner Chorherrenstift Neustift,
  Varna 028282
Augustinerkloster und Kirche Hl.
  Augustinus, Wien 002941
Augustinermuseum, Rattenberg,
  Inn 002574
Augustinermuseum, Städtische
  Museen Freiburg, Freiburg im
  Breisgau 018478
Augustines de la Miséricorde de Jésus
  de Roberval, Roberval 006815
Aujourd'hui Jadis, Le Luc 069295
Aukčná Spoločnost SOGA,
  Bratislava 085542, 126489
Aukční Síň Nuselská, Praha ... 125419
Aukrug Antik, Aukrug 074473
Auksjonshuset Østlandet,
  Elverum 126345
Auktioner, Everöd 126577
Auktionerverket, Eskilstuna 126576
Auktions-Magasinet, Grums ... 126589
Auktions-Shop.ch, Kriens 126735
Auktionsbyrån, Trollhättan 126691
Auktionsgården, Mönsterås ... 126644
Auktionsgården, Sankt Olof ... 126660
Auktionsgodsaffären i Billinge,
  Billinge 126570
Auktionshallen, Gävle 126583
Auktionshallen, Järbo 126606
Auktionshallen, Karlskoga 126613
Auktionshaus am Dom, Köln .. 126003
Auktionshaus Arnold, Frankfurt am
  Main 125943
Auktionshaus Baden-Baden, Baden-
  Baden 125859
Auktionshaus Die Eiche, Lübeck 126030
Auktionshaus Fine Art, Leipzig . 126023
Auktionshaus für Historica, Spiesen-
  Elversberg 126119
Auktionshaus Hanisch, Baden-
  Baden 125860
Auktionshaus Hansa, Bremen .. 125907
Auktionshaus Hess GbR, Alexander Hess,
  Katja Ledderhos, Neuenstein,
  Hessen 077726, 126079
Auktionshaus Höchst, Frankfurt am
  Main 125944
Auktionshaus im Schloß Potzneusiedl,
  Potzneusiedl 125264
Auktionshaus Karrenbauer,
  Konstanz 126017
Auktionshaus Königsallee,
  Düsseldorf 125924
Auktionshaus Landshuter Rüstkammer
  oHG, Tiefenbach, Kreis
  Landshut 078678, 126128
Auktionshaus Nitschareuth, Jena 125995
Auktionshaus Nymphenburg,
  München 077463, 126042
Auktionshaus OWL, Bielefeld .. 125894
Auktionshaus Rheinsberg,
  Rheinsberg 126105
Auktionshaus Sankt-Peter-Ording,
  Sankt Peter-Ording 126112
Auktionshaus Satow, Satow .. 126113
Auktionshaus Sigalas,
  Hildrizhausen 125992
Auktionshaus Viktualienmarkt,
  München 126043
Auktionshaus Walldorf, Walldorf 126137
Auktionshaus Zofingen, Zofingen 126760
Auktionshaus, Dr. Elisabeth
  Buschmans, Berlin 125868
Auktionshuset, Hørsholm 125446
Auktionshuset Anders Antik,
  Landskrona 126620
Auktionshuset i Sundbyberg,

Sundbyberg 126678
Auktionshuset Kåremo, Kalmar . 126610
Auktionshuset Metropol,
  Stockholm 126667
Auktionskammare Västby,
  Helgum 126598
Auktionskammaren Hallsberg,
  Hallsberg 126592
Auktionskammaren i Åmål, Åmål 126565
Auktionskammaren i Varberg,
  Kungsbacka 126618
Auktionskammaren i Varberg,
  Varberg 126698
Auktionskompaniet Göteborg,
  Göteborg 126585
Auktionskompaniet Malmö,
  Malmö 126635
Auktionskompaniet Stockholm,
  Stockholm 126668
Auktionslokalet og Auktionskontoret,
  Odense 125461
Auktionspreise im Kunstpreis-
  Jahrbuch, München 138682
Auktionsregister Gemälde,
  Ascheberg 138683
Auktionsverket Eda Gamla Skola,
  Charlottenberg 126573
Auktionsverket Gefle, Gävle ... 126584
Auktionsverket i Härnösand,
  Härnösand 126590
Auktionsverket i Karlskrona,
  Karlskrona 126614
Auktionsverket i Uddevalla,
  Uddevalla 126570
Aula Didáctica de Prehistoria de Tito
  Bustillo, Ribadesella 040002
Aula Didáctica del Castro de Coaña,
  Coaña 039296
Aulagnon, Soillès Pont 129260
Aulanier, Marc, Langeac 069166, 128652
Aularge-Danigo, Quimper 141308
AULD Alliance Art Gallery,
  Nashville 122067
The Auld Kirk Gallery, Aberdeen 117205
Auld Kirk Museum, East
  Dunbartonshire Museums,
  Kirkintilloch 044732
Auld Sköll, Fair Isle 044319
Auldearn Antiques, Auldearn ... 088325
Aule Mettmanner, Mettmann ... 020394
Aulestad, Karoline og Bjørnstjerne
  Bjørnsons Hjem, Follebu ... 033608
Auli Mølle, Årnes 033503
Aulnette, Saint-Nazaire (Loire-
  Atlantique) 072988, 125804
Aulock, Andrea von, Münster . 077652,
                         130423
Aulos Musica, Genève . 116280, 144071
AUM Gallery, Denver 120697
Aumann, Thomas, Hildesheim . 076533
Aumont Bernard, Paris 070986
Aunan, Ingvild, Oslo 084299
Aunette, Saint-Nazaire (Loire-
  Atlantique) 072988, 125804
Aunt Betty's Antiques, Dallas .. 093072
Aunt Edna's Treasures, Moonee
  Ponds 061876
Aunt Hattie's Attic, Miami 094704
Aunt Margaret's Museum of Childhood,
  Dufresne 005725
Aunt Ruth's Antiques, Salt Lake
  City 096879
Aunt Shelly's Treasures, Fresno . 093490
Auntie Q's, Memphis .. 094617, 127314
Auntie's Affic, Whyalla 062426
Aunties Attic, Portland 096367
Auprete Delagenest, Guy, Nice . 070568
Aura, Bromma 086653
Aura Antiques, Masham 090733
L'Aura des Arts, Mallemort 104148
Aura Galerías, México 111947
Aura Gallery, Hong Kong 102059
Aura Gallery, Shanghai 102170
Aurantique, Nice 070000
Aurea Charta, Parma 131718
Aurel, Marc, Augsburg 074451

Aurélie, Paris 070987
Aurelio Sevilla Alvero Library and
  Museum, Pasay 034561
Aurelius Ink, Nashville 094974
Áureo, Barcelona 126504
Aureola, Bucureşti 085081
Aurévane, Aixe-sur-Vienne 066488,
                         128430
Auriel, Toulouse 105723
Auriga, Barcelona 137951
L'Auriga, Firenze 080084
Aurland Bygdetun og Lensmannsstova,
  Aurland 033514
Aurnhammer-Sammlung,
  Treuchtlingen 022188
Aurobindo Gallery, Delhi 109309
Aurobora, San Francisco 138456
Aurora, Vancouver 101742
Aurora Antiques, Columbus ... 092994
Aurora Art Museum, Taipei ... 042499
Aurora Galerie, Knokke-Heist .. 100729
Aurora Historical Museum,
  Aurora 046671
Aurora Historical Museum, Aurora
  Historical Society, East Aurora . 048420
Aurora Historical Society Museum,
  Aurora 046680
Aurora History Museum, Aurora 046670
Aurora Museum, Aurora 005361
Aurora Public Art Commission,
  Aurora 046672
Aurskog-Høland Bygdetun, Hemnes i
  Høland 033647
Aurum Press, London 138194
Aurygem, Madrid 133303
Aus dem Antiquariat, Zeitschrift für
  Antiquare und Büchersammler,
  Frankfurt am Main 138684
Aus Forschung und Kunst,
  Klagenfurt 138504
Aus-Tex Coins, Austin 092215
Ausada Duran, León 115109
Ausart International, Cheltenham, New
  South Wales 061316
Ausems, Hilversum 132540
Ausgrabungen Wüstung Hard,
  Thaya 002823
Ausgrabungsdokumentation
  6000 Jahre Wohnberg Oberleis,
  Ernstbrunn 001874
Ausgrabungsstätte Schanzberg,
  Thunau 002827
AuSi Gallery, Tucson 124714
Auskjonshuset, Oslo 126365
Ausmeyer & Gerling, Bremen .. 075147
Aušros Alėjos Rūmai, Šiaulių Aušros
  Muziejus, Šiauliai 030461
Aussandon, Yves, Aubagne 066772
Aussen, Rob, Amsterdam 082693
Aussie Art Store, Sydney 099765
Aussiedlermuseum im Schüttkasten,
  Allentsteig 001696
Ausstellung der Geigenbauschule
  Brienz, Brienz (Bern) 041362
Ausstellung der Kantonalen
  Schnitzerschule, Ausstellung der
  Schule für Holzbildhauerei, Brienz
  (Bern) 041363
Ausstellung Faszination und Gewalt,
  Nürnberg 020847
Ausstellung Fontane und Hankels
  Ablage, Zeuthen 022871
Ausstellung für Kraftwagen und
  Krafträder, Satteldorf 021620
Ausstellung Haus Wysburg und Ruine
  Wysburg, Remptendorf 021360
Ausstellung in der
  Universitätsbibliothek Basel,
  Basel 041262
Ausstellung Juden in Buttenhausen,
  Münsingen 020587
Ausstellung Kirche Zillis, Zillis .. 042179
Ausstellung Max-Kommerell,
  Münsingen 020588

Ausstellung zur Erd- und Vorgeschichte, Breitscheid, Lahn-Dill-Kreis ... 017580
Ausstellungräume Schloss Gloggnitz, Gloggnitz ... 001938
Ausstellungsforum FOE 156, München ... 020506
AusstellungsHalle, Frankfurt am Main ... 018406
Ausstellungshalle Neues Rathaus, Kunstverein Bayreuth, Bayreuth 016983
Ausstellungshalle Zeitgenössische Kunst Münster, Münster ... 020592
Ausstellungsraum 25, Zürich ... 116753
Ausstellungsraum Dr. Erdel Verlag, Regensburg ... 108457
Ausstellungsraum Horst Stauber Glas, u. Zentrum unabhängiger ausstellungen, Passau ... 108361
Ausstellungsraum Klingental, Basel ... 041263
Ausstellungszentrum für die Archäologie des Emslandes, Meppen ... 020368
Ausstellungszentrum Heft, Hüttenberg ... 002106
Ausstellungszentrum Heiligenkreuzer Hof, Universität für Angewandte Kunst Wien, Wien ... 002942
Ausstellungszentrum Kloster Cismar, Stiftung Schleswig-Holsteinische Landesmuseen Schloß Gottorf, Grömitz ... 018827
Ausstellungszentrum Kroch-Haus, Leipzig ... 019977
Ausstellungszentrum Lokschuppen, Rosenheim ... 021474
Aust & Köckritz, Ortrand ... 130496
Aust-Agder Museet, Arendal ... 033500
Austefjord Museum, Austefjorden 033515
Austin, Paris ... 070988
Austin & Co, San Jose ... 097358
Austin & Desmond, London ... 118197
Austin Antiques, Austin ... 092216
Austin Artworks, Austin ... 119642
Austin Auction Company, Austin 127130
Austin Children's Museum, Austin ... 046683
Austin E. Knowlton School of Architecture, Ohio State University, Columbus ... 056943
Austin Galleries, Austin 092217, 119643
Austin History Center, Austin ... 046684
Austin Museum of Art, Austin ... 046685
Austin Museum of Art Laguna Gloria, Austin ... 046686
Austin Nature and Science Center, Austin ... 046687
Austin Pastel Society, Austin ... 060401
Austin Presence, Austin ... 119644
Austin Rare Coins, Austin ... 092218
Austin Spirit Gallery, Austin ... 119645
Austin, John, Colchester ... 138151
Austis Prodejní, Znojmo ... 102754
Australasian Golf Museum, Bothwell ... 000870
Australia Council for the Arts, Surry Hills ... 058184
Australia Dream Space, Lutry .. 116461
Australian Aboriginal Dreamtime Gallery, Alice Springs ... 098629
Australian Abstracts, Coburg .. 098903
Australian and New Zealand Arms Company, Box Hill ... 061184
Australian and New Zealand Association of Antiquarian Booksellers, Pokolbin ... 058185
Australian and New Zealand Association of Antiquarian Booksellers, Surry Hills ... 059630
Australian Antarctic Division Display, Kingston ... 001194
The Australian Antique & Art Dealers Association, Malvern, Victoria . 058186

Australian Antiques, Fyshwick .. 061494
Australian Antiques and Fine Art Fair, Moore Park ... 097991
Australian Arms Auctions, Doncaster East ... 125188
Australian Art Auctions, Kareela 125202
Australian Art Connection, Sydney ... 099766
Australian Art Informs, Sydney .. 099767
Australian Art Link, Greenwich . 061559
Australian Art Resources, Woollahra ... 062459, 099907
Australian Art Sales, Kareela ... 136481
Australian Art Sales Digest, Armadale, Victoria ... 136477
Australian Book Auctions, Prahran ... 125225
Australian Branch of China Relics, Ferny Creek ... 127597
Australian Centre for Contemporary Art, Southbank ... 001508
Australian Centre for Photography, Paddington, New South Wales 001390
Australian Children's Folklore Collection, Museum Victoria, Melbourne ... 001265
Australian Commercial Galleries Association, Collingwood ... 058187
Australian Council of University Art and Design Schools, Perth ... 058188
Australian Electric Transport Museum, Saint Kilda, South Australia ... 001468
Australian Exhibition Center, Chicago ... 047639
Australian Federation of Friends of Museums, Sydney ... 058189
Australian Flying Museum, Wallon ... 001616
Australian Freethought Heritage Library, Balwyn ... 000821
Australian Galleries, Collingwood 098909, 098910
Australian Galleries, Paddington, New South Wales .099511, 099512, 136486
Australian Gallery of Sport and Olympic Museum, East Melbourne ... 001043
Australian Gold Exchange, Melbourne ... 061813
Australian Golf Museum, South Melbourne ... 001506
Australian Guild of Realist Artists, Camberwell ... 058190, 098849
Australian Heritage Print Company, Randwick ... 062137
Australian Institute of Genealogical Studies, Blackburn ... 000863
Australian Mallee Art, Cheltenham, Victoria ... 061317
Australian Museum, Sydney ... 001537
Australian Naive Art, Paddington, New South Wales ... 099513
Australian National Maritime Museum, Darling Harbour ... 001016
Australian Originals, Portland .. 123595
Australian Pearling Exhibition, Darwin ... 001019
Australian Print Workshop, Fitzroy ... 099022
Australian Prints & Paintings, Blackwoode ... 061167
Australian Prospectors and Miners Hall of Fame, Kalgoorlie ... 001176
Australian Racing Museum and Hall of Fame, Melbourne ... 001266
The Australian Shop, Gent ... 100687
Australian Stockman's Hall of Fame and Outback Heritage Centre, Longreach ... 001229
Australian Toy Museum, Collingwood ... 000985
Australian War Memorial, Campbell ... 000929
Australian Wildlife Art Gallery,

Munro ... 099415
Australiana-Africana, Caloundra . 061254
The Australiana Aspect, Warrandyte ... 062407
Australis Gallery, Clonakilty ... 109549
Australis Old Wares, Caringbah . 061282
Austrått Fort, Brekstad ... 033559
Austråttborgen, Trondheim ... 034031
Austruy, Jean-Claude, Niort ... 070710
Auswanderermuseum, Oberalben 020887
aut. architektur und tirol, Innsbruck ... 002119, 136517
Aut. Pressaut., Königsdorf ... 136529
Autant, en Emporte le Temps, Paris ... 070989
Aut'Antic, Chigny ... 087368
Autantiquité, Lille ... 069475
Autegarden-Rapin, Anna, Bruxelles ... 063341
Auterbe, Jean-Louis, Menton .. 070058
Auth, G., Duisburg ... 075547
Authentic Aboriginal, Gembrook 061515
Authentic Aboriginal Art Centre, Sydney ... 099768
Authentic Africa, New York ... 122371
Authentic Africa, Philadelphia .. 123356
Authentic African Arts, Houston . 121002
Authentic City, Singapore ... 085370
Authentic French Antiques, Lievin 069463
Authentic Lighting, Red Hill ... 062139
Authentiek, Gent ... 063628
L'Authentique, Lyon ... 069702
Authier, Thierry, Périgueux ... 072056
Author Bookstore, Calgary ... 140334
Autiero, Antonio, Napoli ... 080842
Autistic Specialists, Baltimore .. 119721
Auto- und Motorrad-Museum, Öhringen ... 020952
Auto- und Motorradmuseum, Erlebnispark Ziegenhagen, Witzenhausen ... 022711
Auto & Technik Museum Sinsheim e.V, Sinsheim ... 021879
Auto Museum, Escondido ... 048604
Auto Museum Dr. Carl Benz, Ladenburg ... 019843
Auto-Museum Fritz B. Busch, Wolfegg ... 022727
Auto Museum Moncopuli, Osorno ... 007527
Auto Pinto, Lisboa ... 084833
Autobook World, Sydney ... 139923
Autochrome, Lyon ... 069703
Les Autodidactes, Paris ... 141035
L'Autographe, Genève ... 144072
Les Autographes, Paris ... 141036
Automates Avenue, Falaise ... 012947
Automatik-Museum, Leipzig ... 019978
Automative Fine Arts Society, Lake Orion ... 060402
Autómatuszálem Gyüjtemény, Bordány ... 023451
Automne, Bruxelles ... 100562
Automobil- und Motorradmuseum Austria, Mitterndorf an der Fischa ... 002396
Automobil-Museum, Asendorf .. 016633
Automobil Museum, Dortmund .. 017951
Automobil-Museum Siegfried Marcus, Stockerau ... 002795
Automobil-Veteranen-Salon, Gundelfingen an der Donau .. 018891
Automobile Driving Museum, Los Angeles ... 050407
Automobilia Transport Museum, Hebden Bridge ... 044551
Automobilmuseum, Aspang ... 001726
Automobilmuseum, Stainz ... 002772
Automobilmuseum Stainz, Graz . 001967
Automoción Costa, Valencia ... 133449
Automotive Hall of Fame, Dearborn ... 048191
Automusa, Bergeijk ... 032063
Automusée du Lyonnais,

Seyssuel ... 015806
Automuseum Deventer, Deventer 032215
Automuseum Engstingen, Engstingen ... 018213
Automuseum Histo-Mobil, Giethoorn ... 032348
Automuseum Old Timer, Lo-Reninge ... 003822
Automuseum Oldie Monschau, Monschau ... 020454
Automuseum Praga, Zbuzany .. 009974
Automuseum Schagen, Schagen 032843
Automuseum Störy, Bockenem . 017443
AutoMuseum Volkswagen, Wolfsburg ... 022737
Autor & Argentus, Västra Frölunda ... 087147
Autoren Galerie 1, München ... 108051
Autorska Galeria, Gdańsk ... 113458
Autosammlung, Linz ... 002315
AutoStadt ZeitHaus, Wolfsburg . 022738
Autotron, Rosmalen ... 032798
Autour de Bois, Aléas, Le Robert 132354
Autour de l'Art, Antibes-Juan-les-Pins ... 103254
Autoworld, Bruxelles ... 003432
Autre Chose, Paris ... 070990
L'Autre Galerie, Bruxelles ... 063342, 100563
Autre Temps, Bordeaux ... 067310
Autrefois, Annecy ... 066637
Autrefois, Chartronges ... 067862
Autrefois, Fayence ... 068517
Autrefois, Jounieh ... 082320
Autrefois, Martigues ... 069985
Autrefois, Paris ... 141037
Autrefois, Poussan ... 072301
Autrefois, Saint-Étienne ... 141328
Autrefois, Tours ... 073836
Autrefois, Vendôme ... 074008
Autrefois Antiques, Boston ... 092444
Autrefois en Gascogne, Vic-Fezensac ... 074091
Autumn Gallery Jewellery, Brighton, Victoria ... 061200
Autumn Leaves Books, Chicago 145035
Autumn London Park Lane Arms Fair, London ... 098360
Auvergne Médailles, Vichy ... 074103
Auvermann, Detlef, Glashütten . 141850
Auversoise de Décoration, Auvers Saint-Georges ... 066839
Auwärter-Museum, Landau an der Isar ... 019867
Aux 3 Aches, Lyon ... 104066
Aux Affaires Anciennes, Massy . 069993
Aux Armes d'Antan, Paris ... 070991
Aux Armes de Lorraine, Lerrain . 069392
Aux Armes de Lorraine, Mirecourt ... 070126
Aux Arts et Caetera, Saint-Brieuc 072695, 105494
Aux Arts Etc, Clermont-Ferrand . 068012
Aux Arts Etc, Tours ... 073837
Aux Arts. . . Etc, Oingt ... 104423
Aux Beaux Arts, Mantes-La Jolie 069840
Aux Beaux Arts, Saint-Tropez .. 105624
Aux Bons Livres, Nice ... 140994
Aux Cases à Bulles, Paris ... 141038
Aux Charmes d'Antan, Mulhouse 070386
Aux Collectionneurs, Caen ... 140809
Aux Compagnons d'Hastings, Merville-Franceville-Plage ... 070074
Aux Couleurs de la Campagne, Saint-Jean-sur-Richelieu ... 006882
Aux Coups de Coeur, Nevers ... 070534
Aux Déménageurs Héraultais, Montpellier ... 070293
Aux Deux Anges, Labastide-d'Armagnac ... 069424
Aux deux Arts, Saulieu ... 073427
Aux Drapeaux Dimu, Saint-Étienne ... 072751
Aux Fils du Temps, Paris ... 070992

Aux Fruits des Arts, Meudon . . . 104234
Aux Merveilles d'Asie, Paris . . . 070993,
128924
Aux Merveilles de l'Asie, Paris . 070994,
128925
Aux Meubles Anciens, Besançon 067118
Aux Monnaies de la Bourse,
Paris 070995
Aux Mystères de l'Art,
Marrakech 082588
Aux Naufrages du Temps, Saint-Malo
(Ille-et-Vilaine) 105547
Aux Neiges d'Antan, Font-Romeu-
Odeillo-Via 068556
Aux Occasions, Dijon 068317
Aux Portes de la Champagne,
Bernex 041330
Aux Portes du Temps Passé, Bézu-
Saint-Eloi 067184
Aux Puces de Plainfaing,
Plainfaing 072169
Aux Puces Lavalloises, Laval
(Mayenne) 069200
Aux Quartiers de France, Saint-
Ouen 073023
Aux Reflets du Passé, Asnières-sur-
Seine 066758
Aux Reflets du Passé, Paris . . . 070996
Aux Reliquaires du XVIIIème,
Rolle 087750
Aux Remparts, Belfort 067066
Aux Rêves d'Antan, Le Mans . . 069303
Aux Siécles Passés, Bordeaux . . 067311
Aux Siécles Passés, Le Lardin-Saint-
Lazare 069291
Aux Sièges d'Autrefois, Clermont-
Ferrand 068013
Aux Soldats d'Antan, Paris 070997
Aux Souvenirs d'Autrefois, Porsel 087741
Aux Temps Anciens, Grenoble . . 068717
Aux Tortues, Paris 070998
Aux Tresors Perdus, Saint-Ouen 073024
Aux Trois Clés, Villeurbanne . . . 074186
Aux Trois Marchés, Auxerre . . . 066844,
066845, 128455
Aux Trois Siècles, Chamonix-Mont-
Blanc 067799, 140819
Aux Trouvailles de Pasteur,
Coudekerque-Branche 068174
Aux Trouvailles Galerie, L'Isle-sur-la-
Sorgue 069557
Aux Vestiges Habis, Jounieh . . . 082321
Aux Vieilles Portes, Cogolin 068063
Aux Vieiseries, Halloy 068780
AV Art Gallery, New York 122372
AV Gallery, Limerick 024953
Av Goldteam, Paris 070999
AV Monografías, Madrid 139040
AV Proyectos, Madrid 139041
Ava Gallery, Bangor 117285
Ava Gallery and Art Center,
Lebanon 050205
Avail Art, New York 122373
Avalon, Christies Beach 139697
Avalon, Houston 135588
Avalon, Las Vegas 121342
Avalon, Portland 096368
Avalon, Saint Paul 123896
Avalook Antiques, Berrigan 061138
Avalun, München 142141
Avampato Discovery Museum,
Charleston 047562
Avanersuup Katersugaasivia, Thule
Museum, Qaanaaq 023336
Avanidis, Filip A., Athinai 079159
Avant Art, Lüdinghausen 137362
Avant Garde, Padstow 118871
Avant Garden Bookshop,
Daylesford 139711
Avante Galleries, Nashville 122068
Avantgarde Gallery, London . . . 118198
Avantgarde Gallery, Möhnesee . 108010
Avantgart, Helmbach . . 107413, 137276
La Avanzada, Valencia 086504

Avdelningen för Konstvetenskap, Lunds
Universitet, Lund 056397
Ave-Art, Šumperk 102736
Ave Galeria, Kraków 113531
Aveline, Paris 071000, 104536
Avelino, Pedro, Rio de Janeiro . 101012
Avenida Art Gallery, Calgary . . . 101216
Avent Designs, London 118199
Aventinum, Praha 136761
Aventinus, Abensberg 125845
Aventinus-Museum, Abensberg . 016457
Aventura, Miami 094705
Aventure Leclerc, Saint-Augustin-de-
Desmaures 006855
Aventurine, Bruxelles 063343
Avenue, Oostende 063812
Avenue 30, Cairo 066228
Avenue Antiques, Boston 092445
Avenue Antiques, Scarborough . 062195
Avenue Antiques, Tauranga . . . 084100
The Avenue Art Gallery, Calgary 101217
Avenue Gallery, Austin 119646
Avenue Gallery, Baltimore 119722
Avenue Victor Hugo Bookshop,
Boston 145006
Averøy Skolemuseum, Averøy . . 033516
Averoff Gallery, Athinai 022979
Averoff Gallery, Metsovo 023171
Averty, Clair, Nantes 128832
Avery County Historical Museum,
Newland 051527
Avery Fine Art, Atlanta 119546
Avery Historical Museum,
Smethwick 045813
Avery Research Center for African
American History and Culture,
Charleston 047541
Avery, Ashley, Denver 093260
Avery, Ashley, Sacramento 096601
Avetik Issahakyan House-Museum,
Yerevan 000724
Aviamy, Dallas 135493
Aviation Art Gallery, Bankstown . 098706
Aviation Art Gallery, Orange . . . 139321
Aviation Art Studio, Lincoln 118115
Aviation Gallery, Sacramento . . 123592
Aviation Heritage Museum of Western
Australia, RAAF Association, Bull
Creek 000904
Aviation Museum of Kentucky,
Lexington 050241
AVICOM – International Committee
for the Audiovisual and Image and
Sound New Technologies, Paris 058622
AVID, Wellington 084157
Avida Dollars, Milano 110262
Avier, Philippe, Bar-sur-Seine . . 066942
Avignon House Antiques, New
Orleans 095058
Avijacijos Pradininko Aleksandro
Griškevičiaus Muziejus, Mažeikių
Muziejus, Viekšniai 030504
Avila Gómez & Cia., Ricardo,
Bogotá 065187
Aviso, Washington 139322
Avison, Wellington 084158
Avivson, London 138195
AVME, Los Angeles 121436
Avner, Saint-Ouen 073025
Avoca Beach Antiques, Avoca
Beach 061071
Avogadro, Enrico, Milano 131378
Avog's Crash Museum, Lievelde 032598
Avoid the Middle Man, San
Diego 097065
Avola, Ulyana, Verona 111177
Avolio, Luigi, Napoli 131598
Avon Antiques, Bradford-on-Avon 088560
Avon Books, Bristol 144273
Avon Historical Society Museum,
Avon 046708
Avon House Antiques,
Kingsbridge 089694
Avon River Heritage Society Museum,

Hants County 005983
Avon Valley Art Society, Northam 058191,
099480
The Avonbridge Antiques and
Collectors Market, Salisbury . . 091219
Avoncroft Museum of Historic
Buildings, Bromsgrove 043769
Avondale Discovery Farm,
Beverley 000855
Avondale Galleries, Oklahoma
City 123275
Avondale House, Rathdrum 024979
Avonds, L., Antwerpen 127993
Avonds, Ludo, Antwerpen 063191
Avonlea, Jacksonville 093960
Avonlea, Young 139969
Avonlea and District Museum,
Avonlea 005364
Avonturen, Den Haag 083058
Avos, Los Angeles 135706
Avshalom, Eitan, New York 122374
Avtorskaja Galereja, Moskva . . . 114206
Avusturya Kültür Ofisi, İstanbul . 117011
Awa Camara, Bruxelles 063344, 128018
Award Antik, Stockholm 086974
Awarji, Joseph, Damascus 088029
Awashi, Sato, Kamakura 111267
Aweart, Bournemouth 117401
The AXA Gallery, New York 051343
Axana, Saint-Ouen 073026
Axe Néo-7 Art Contemporain,
Gatineau 005864
Axe, Richard, Harrogate 144453
Axel, West Hollywood 097874
Axel Ebbes Konsthall, Trelleborg 041086
Axel Stenross Maritime Museum, Port
Lincoln 001422
Axelle, Boston 119836
Axelle, New Orleans 122125
Axelle, New York 122375
Axelle, San Francisco 124226
Axet Antik, Halmstad 086731
Axia Art Consultants, London . . 118200
Axia Modern Art, Armadale,
Victoria 098655
Axial, Lyon 104067
Axiom, Liverpool 139808
Axiom Centre for Arts,
Cheltenham 043922
AXIS, Tokyo . . 111406, 137774, 138883
Axis – Visual Arts Exchange and
Information Services, Leeds . . . 053902
Axis Fine Art, Brno 102562
Axis Gallery, New York 122376
Axminster Museum, Axminster . . 043455
Axpe & Saillard, Périgueux 072057
Axpe-Chaplain, Eric, Château-
l'Eveque 067874
Aya İrini Kilisesi Müzesi, Ayasofya
Müzesi, İstanbul 042922
Ayabe-shiritsu Hakubutsukan,
Ayabe 028499
Ayacs, Amsterdam 112156
Ayagallery, London 118201
Ayala Museum, Makati 034505
Ayala Museum, Vigan 034612
Ayala Rodriguez, G., Madrid . . . 086127,
143751
Ayasofya Müzesi, İstanbul 042923
Ayasofya Müzesi, Trabzon 043073
Aydın Devlet Güzel Sanatlar Galerisi,
Aydın 042838
Aydın Müzesi, Aydın 042839
Aydın Müzesine Bağlı Örenyerleri,
Aydın 042840
Aydiner, Z., Berlin 111484
Aydogmus & Aydogmus-Burkhardt,
Bad Friedrichshall 074512
Ayer & Co., Bristol 065188
Ayer & Hoy, Caracas 097945
Ayera, Zaragoza 086589
Ayerego, Toronto 140397
Ayesse, Odette, Rennes (Ille-et-
Vilaine) 072422

Ayios Dimitrios Crypt Museum,
Thessaloniki 023288
Aylmer and District Museum,
Aylmer 005365
Aylmer Military Collection,
Brading 043700
Aylward, Wooroloo 099923
Ayme, Bernard, Marseille 128745
Aynalıkavak Kasrı, İstanbul 042924
Aynsley & Martel, Louisville . . . 094523
Les Ayres, Bordeaux 140787
Ayres, Fort Worth 135556
Ayres-Steinmetz, Ann, Chicago . 120055
Ayrshire Yeomanry Museum, Ayr 043458
Ayscoughfee Hall Museum,
Spalding 045843
Ayşe ve Ercümend Kalmık
Vakfı-Müzesi, Gümüşsuyu 042909
Ayuka, Stevenage 091420
Ayuntamiento de Almeria,
Almeria 114837
Ayutthaya, Atlanta 119547
AZ Gallery, Bakı 100407
AZ-Kunstprojekten, Nieuw
Beijerland 112600
AZ Location, Aubière 103284
AZ Project, Sapporo 082098
Azabu Kasumicho Gallery, Tokyo 111407
Azadi Museum, Tehrän 024746
Azaïs, Jean, Mougins 070360
Azalea House Antiques, Norfolk . 095845
Azani, Chicago 120056
Azar, Carl A., Atlanta 135299
Azarbayezan Museum, Tabriz . . 024739
Azcarate, Juan F.F. de, Buenos
Aires 098552
Azemar, Christophe, Nice 070570
Azen, Louis, Damascus 088030
Azerbaijan State Historical and
Architectural Museum Ichery Sheher,
Bakı 003151
Azerbaijan State Museum of
Independence, Bakı 003152
Azerbaycan Tarixi Muzeyi, Bakı . 003153
Azeredo, José Mota, Porto 132966
The Azeri Miniature Art Center,
Bakı 003154
Azevedo, António Manuel Guimarães,
Lisboa 084834
Azimont, Jean-Pierre, Baillargues 066917
Azimut, Torino 059420
Azimuth, Rousson 072623
Ažio, Kyïv 117121
Aziz, Tehrän 079420
Aziz, Abdat, Amsterdam 082694
Aziz, Antoun, Cairo 066229
Azizian, Milano 080424
Azoulay, Daniel, Miami 121745
Azovskij Kraevedčeskij Muzej,
Azov 036879
Azpitarte, Florence, Paris 071001
Aztalan Museum, Jefferson 049801
Aztec Graphics, San Diego 124103
Aztec Museum and Pioneer Village,
Aztec 046709
Aztec Ruins, Aztec 046710
Azuchi-do, Tokyo 142919
Azul, Padova 080961
Azul Cobalto, São Paulo 064155
Azum, Sylvane, Bagnères-de-
Luchon 066909
Azuma, Seattle 124489
Azur Antiquès et Occasions,
Fréjus 068605
Azur Démolition, Cannes 067627
Azusa, Tokyo 111408
Azzams Antik, Kristianstad 086789
Azzola, Max, Toulouse 073738
Azzopardi, Marseille 069906
Azzopardi & Sons, Valletta 082438
Azzopardi, C., San Giljan 082427
B & A, Tauranga 124553
B & B, Balerna 087211, 144012
B & B, Den Haag 083059

B & B Antiques, Stickney 091423
B & B Arte, Mantova 080344
B & B International Gallery, New
York 095249
B & C Art, Omaha 123304
B & C Carpet Company, New
York 095250
B & D, Milano 110263
B & D Furniture Repair, Saint
Louis 136195
B & H Antique Auction, Danville 127196
B & H Antiques, Miami 094706
B & H Clock, San Jose 136355
B & J Booksellers, Coopers
Plains 139703
B & M Clock Repair, Houston .. 135589
B & M Gallery, Meito 111307
B & N Verlag und Edition, Bad Homburg
v.d.Höhe 106061, 137034
The B & O Railroad Museum,
Smithsonian Institution,
Baltimore 046729
B & O Railroad Station Museum,
Ellicott City 048551
B & O Starožitnosti, Praha 065480
B & Q Art Gallery, Long Beach . 121395
B & R Fittings, Jolimont 061658
B & W Antik, København 065849
B & W Antiques, Seattle 097406
B-4 It Was Cool, New York 095251
B 42 Galerie, Frankfurt am Main 106974
B B Antiques, Zabbar . 082450, 132345
B-G Bok och Tidskriftsantikvariat,
Malmö 143893
B-Gallery, Richmond 123721
B J Antiquités, Marseille 069907
B-Lumen, Schiltigheim 073460
B-semi Learning System of
Contemporary Art, Yokohama . 056065
B&D Art Gallery, Genève 116281
B. & B. Art, Venezia .. 111084, 111085
B. & T. Antiques, London 089932
B. Bissoondoyal Memorial Centre,
Tyack 030796
B. Langston's Antiques,
Jacksonville 093961, 127265
B.A.M., Toulouse 105724
B.a.S., Venezia 111086, 111087
B.B.M. Coins, Kidderminster ... 089679,
134511
B.C. Fine Art Gallery, Sanctuary
Cove 099673
B.C. Galleries, Armadale, Victoria 061017
B.C. Galleries, Woollahra 062460
B.C. Metalcrafts, London 089933,
134589
B.D.E – Bureau des Estimations,
Paris 125677
B.D.E. – Bureau des Estimations,
Paris 071002
B.D.V, Saint-Ouen 073027
B.F. Larsen Gallery, Provo 052402
B.H.A., Bibliography of the History of
Art, Vandœuvre-lès-Nancy 138633
B.J. Hales Museum of Natural History,
Brandon 005460
B.J.F., Paris 071003
B.J.'s Antiques, Fort Worth 093429
B.J.'s Antiques, Jacksonville ... 093962
B.J.'s Antiques, Virginia Beach . 097771
B.J.'s Auctions, Edwardstown .. 125191
B.K.I. Galerie, Darmstadt 106652
B.L.G. Antichità, Padova 080962
B.M. Birla Science and Technological
Museum, Hyderabad 024296
B.N.C., Paris 071004
B.O. Furniture Refinishing, El
Paso 135549
B.R.M. Coins, Knutsford 089729
B'S Ballpark Museum, Centennial 047495
B4A Fine Arts 237, New York .. 122377
Ba Bi, Song, Shanghai 102171
Baader, Joachim, München ... 108052
De Baaien Rok, Bussum 083015

Baal, Bielefeld 106486
Baarle's Museum, Baarle-Hertog 032043
Baas, Seattle 124490
Baba & Társa, Budapest 079240
Baba Antiques, Mumbai 079400
Baba Nyonya Heritage Museum,
Melaka 030703
Babamúzeum, Tihany 024005
Babault, Alain, Gondrecourt-le-
Château 068677
Babcock, New York 122378
Babe Didrikson Zaharias Museum,
Beaumont 046853
Babe Ruth Birthplace & Museum,
Baltimore 046730
Babel, Murcia 115311
Babel Galeria, Gdańsk 113459
Babendererde, Peter, Lübeck . 142085
Babendererde, Rosemarie,
Stockelsdorf 078564
Babilônia Antiguidades, Porto
Alegre 063959
Babinot, Francine, Aigues-Mortes 066436
Babits Mihály Emlékház,
Wosinsky Mór Megyei Múseum,
Szekszárd 023946
Bábkiállítás, Bosta 023452
Bablon, Béatrice, Paris 141039
Babls Uhrensammlung,
Waffenbrunn 022367
Babol Museum, Babol 024686
Babon, Frédéric, Pézenas 072131
Bab's Collectibles, Jacksonville . 093963
Babuino, Roma 126223
Babur Yodgorlik Muzeyi, Andijon
Viloyat Adabiyot va San'at Muzeyi,
Andijon 054625
Babushka Antiques, Auckland .. 083770
Babylon Museum, Babylon 024798
Babylon Oriental Rug, Los
Angeles 094264
Babylone, Nice 070571
Babylonian Jewry Museum, Or
Yehuda 025166
Bac Art Studio, Venezia 111088
Bac Saint-Germain-Editions Callithos,
Paris 104537, 136865
Bac, Georges, Paris 071005
Baca House, Trinidad History Museum,
Trinidad 053804
Bacarelli, Riccardo, Firenze ... 080085
Les Bacchanales, Neuilly-sur-
Seine 070515
Bacchi, Antonio, Roma 081247
Bacchiglione, Vicenza 111202
Bacchus Antik, Stockholm 086975
Bacci di Capaci, Giovanna,
Livorno 080316
Bacci, Gregorio, Venezia 111089
Baccus Marsh Antiques, Bacchus
Marsh 061073
Bach-Gedenkstätte im Schloß Köthen,
Köthen 019747
Bach Møller, Randers 102956
Bach-Museum, Bach-Archiv Leipzig,
Leipzig 019979
Bach-Stammhaus Wechmar,
Günthersleben-Wechmar 018879
Bach, Detlef, Winterbach, Kreis Bad
Kreuznach 130771
Bach, Hans Jürgen, Düsseldorf . 125925
Bach, Inge, Maribo 102929
Bach, M., Roetgen 078266
Bachčisarajs'kyj Palac Kryms'kych
Chaniv, Bachčisaraj 043114
La Bacheca, Cagliari 109920
La Bacheca, Catania 080022
Bachelier, Saint-Ouen 073028
Bachelier-Glotin, Gaëlle, Nantes 070449,
140980
Bachelier, Chantal, Nice 128850
Bachellier, Monique, Revigny .. 072447
Bachelor's Club, Tarbolton 045947
Bacher Tillmanns, Véronique,

Faoug 133736
Bacher, Adam, Portland 123596
Bachet, Vouvray-sur-Loir 074281
Bachet, Pierrette, Château-du-
Loir 067869
Bachgau-Museum, Großostheim 018857
Bachhaus Eisenach, Eisenach .. 018131
Bachmann, Luzern .. 087632, 133829
Bachmann & Berthold,
Crimmitschau 075280
Bachmann & Eckenstein, Basel . 087222,
116055
Bachmann-Museum Bremervörde,
Ehemalige Erzbischöfliche Kanzlei,
Bremervörde 017613
Bachmann, D., Zürich 087937
Bachmann, Erika, Puchheim ... 078097
Bachmann, Fred, Essen 129812
Bachmann, Irène, Luzern 116466
Bachmann, Nora, Marquartstein 077304
Bachmann, Petra, Bornheim, Rhein-
Sieg-Kreis 129618
Bachofer, C., Frickenhausen, Kreis
Esslingen 129865
Bachrach, Richmond 096526
Bachritterburg Kanzach, Kanzach 019496
Bachschmid, Stephan,
Kaufbeuren 130107
Bacht, Essen 137181
Bachus, Warszawa 084706
Bacin de Plata, La Paz 063862
Back Alley Antiques, San Antonio 096948
Back Copies, Melbourne 139824
Back in Time-Antiques, Chicago 092666
Back of Beyond, Gisborne 083940
Back Porch, San Diego 124104
Back Roads Art, Austin 119647
Back to Africa Imports, Brooklyn 119938
Back to Back Galleries, Cooks
Hill 098916
Back to the Picture, San
Francisco 124227
Back, Michael, Bad Windsheim . 129429
The Backdoors Warehouse,
Washington 097808
Backens Antik, Skara 086930
Backert, Hilmar, Andernach ... 105991
Backhausen, Bernd,
Mönchengladbach 077385
Backker, Luc de, Bruxelles 063345
Backlot, Los Angeles 121437
Backman, Umeå 087111
Back'N'Time Antiques, Waihi ... 084117
Backspace, Portland 123597
Backstreet Antiques, Calgary ... 064248
Backstreet Gallery, Dallas 120557
Backus Heritage Village, Simcoe 007041
Backwoods Tin, Milwaukee 094833
Baclayon Church Museum,
Baclayon 034427
Bacon & Cie, Saujon 073426
Bacon, Philip, Fortitude Valley .. 099048
Baconnais, Jean, La Baule-
Escoublac 103861
Bacon's Castle, Surry County .. 053618
Bacot, Jacques, Paris 071006
Bacot, Odette, Chalon-sur-Saône 067773
Bacqueville, Paris 071007, 071008
Bacstreet, Paris 071009
Baculo, Madrid 115153
Bad Boys Gallery, Baltimore ... 092299
Bad Gasteiner Museum, Bad
Gastein 001741
Bad Schussenrieder Kutschen
Sammlung, Bad Schussenried . 016869
Bad Schwalbacher Kur-Stadt-
Apotheken-Museum, Bad
Schwalbach 016873
Badaling Art Cooperative, Beijing 101933
Badarelli, Albertville 066501
Badawang, Cleveland 120400
Badcock's Gallery, Newlyn 118795
Baddack, J., Hildburghausen ... 076525
Baddow Antique Centre, Great

Baddow 089363, 134419
Badè Museum of Biblical Archaeology,
Pacific School of Religion,
Berkeley 046931
Badeeb & Sons, Mohammed Ahmed,
Sanaa 097985
Baden-Powell House, London .. 044908
Badener Puppen- und
Spielzeugmuseum, Baden bei
Wien 001770
Bader, Eilenburg 106869
Bader, Tübingen 142423
Bader, Ulm 142441
Bader-Koller, Edith E., Massagno 087664
Bader, A., Neu-Isenburg 077707
Bader, Alexander, Berlin 074761
Bader, Alfred, Milwaukee 121899
Bader jun., Ekkehard, Golling an der
Salzach 062573
Bader, Patick, Eloie 068441
Bader, Peter R., Luzern 087633, 116467
Badest, Milwaukee 121900
Badger, Worthing 144938
Badger Creek Museum,
Cartwright 005573
Badger Mine and Museum,
Shullsburg 053328
Badgers Books, Buderim 139668
Badgerys Attic, Lawson 061729
Badhaus Museum und Galerie,
Kulmbach 019827
Badiali, Maria Grazia Marisa,
Modena 080772, 110492
Badin, Paris 071010
Badin de Montjoye, Dominique, Sucy-
en-Brie 073601
Badin, Giraud, Paris 141040
Badische Antiquitäten und Kunsttage,
Donaueschingen 098132
Badische Gesellschaft für Münzkunde,
Ettlingen 058731
Badische Landesbibliothek,
Karlsruhe 019503
Badischer Kunstverein, Karlsruhe 019504,
058732
Badisches Bäckereimuseum
und Erstes Deutsches
Zuckerbäckermuseum, Kraichtal 019776
Badisches Landesmuseum Karlsruhe,
Karlsruhe 019505
Badong Museum, Badong 007577
Badorfer, Brühl 141659
Badorrek, Manfred, Bad
Kissingen 141427
Badum, Beate, Bamberg 074658
Bae, Andrew, Chicago 120057
Bächler, Hubert, Zürich 116754
Baechtold, Brigitte, Lausanne .. 133791
Baecker, Inge, Köln 107589
Bäckerei- und Dorfgeschichtliches
Museum, Bremervörde 017614
Bäckereimuseum, Rimpar 021434
Bäckereimuseum Poppinga's alte
Bäckerei, Krummhörn 019811
Bäckereimuseum Vörstetten,
Vörstetten 022350
Bäckermuseum, Wien 002943
Bäcklund, Thomas, Skellefteå .. 086934
Bäkmann, Hans-Joachim, Rinteln 078243
Bänder, Klaus, Düsseldorf 106761
Bär, Arnold, Altnau 087191
Baer, Barbara, Jacksonville ... 093964
Baer, Patrick-Daniel, Dresden .. 106718
Bärchen, Heidi, Stuttgart 078594
Baere, de, Beaufort 061111
Bären Galerie, Birr 116160
Bärenmaler, Leipzig 107811
Bärner Heimatstube, Langgöns . 019919
Bärnstensmuseet, Höllviken ... 040700
Bärtschi, Bern 116120
Bärtschi, Guy, Genève 116282
Bärtschi, Tobias, Luzern 133830
Baerveldt, Joure 126292
Bärwaldt, Maastricht 083455

Bärwinkel, Elfriede, Langenbach 142048
Bästkustens Auktionstjänst,
Svenshögen 126685
Baetica, Málaga 139042
Baets, I. de, Gent 100688
Bättig, Armin W., Luzern 116468
Baetzner, Herlit, Hamburg 076191
Baeuerle, Jochen, Stuttgart 142383
Bäuerliches Gerätemuseum
und Venezianer-Gatter,
Innervillgraten 002116
Bäuerliches Heimatmuseum der
Landtechnik, Geiselhöring ... 018607
Bäuerliches Museum im Achentaler
Heimathaus, Rohrdorf 021466
Bäumler, Peter, Regensburg ... 108458,
137479
Baexlbooks, Bonn 141611
Baez-Yankovic, Dinko V.,
Luxembourg 132330
Bafico Facal, Luis Alberto,
Montevideo 135272
Bafkham, Limburg an der Lahn 077110
Bagamoyo Historical Museum,
Bagamoyo 042628
Bagatela, Belo Horizonte 063883
Bagatelle Antiques, Barton,
Cambridgeshire 088372, 134069
Bagatelle Brocante, Bègles ... 067063
Bagatti, Arturo, Milano 080425, 142713
Bagdad Gallery, Portland 096369, 127409
Bagdaschwili, Remscheid 078198
Baggins, Rochester 144803
Baggins, B., Saint Louis 096668
Baggot Framing, Dublin 109575
Baggott Church Street, Stow-on-the-
Wold 091462
Baggott, Duncan J., Stow-on-the-
Wold 091461
Baghai Sain Shapoor, Torino ... 081651
Baghcheh-Jaq Palace, Urmieh . 024788
Baghdad Museum, Baghdad ... 024802
Baghery's, Genève 087452, 116283
Baghgheh Chogh Palace, Maku . 024717
Baghus, Carsten, Helsinge 065736
Bagiau, Asnières-sur-Vègre 066762
Baginski, M., Großefehn 076107
Bagira, Moskva 114207
Bagland, Mauricette, Conches-en-
Ouche 068121
Bagn Bygdesamling, Bagn 033518
Bagnai, Alessandro, Firenze ... 110032
Bagno di Nerone, Pisa 110661
Bagnolati, Giancarlo, Bologna ... 079818
Bagnoréa, Annecy 103243
Bagnoud & Roppa, Crans-
Montana 087386, 133726
Bagnoud & Roppa, Crans-sur-
Sierre 087388
Bagpipe Music Museum, Ellicott
City 048552
Bagshaw Museum, Batley 043530
Bagshaw, G., Siddington 091340, 135049
Bagshaws, Uttoxeter 127073
Bagué, M.A., Girona 115078
Bagué, Santiago, Barcelona .. 133102
Baguio-Mountain Provinces Museum,
Baguio 034431
Bahadourian, Avedis, Paris 071011
Bahamas Historical Society Museum,
Nassau 003201
Bahamas Museum, Nassau 003202
Bahawalpur Museum, Bhawalpur 034107
Bahay Laguna Museum, Liliw . 034494
Bahay Nakpil-Bautista, Manila .. 113402
Bahay Tsinoy Museum, Manila . 034522
Bahlburg, Hamburg 076192
Bahns, H., Hannover 141908
Bahnzeit, Pferdeeisenbahnmuseum der
Südstrecke, Wels 002915
Baho-Gallery, Bern 116121
Bahrain Art Center, Manama ... 100415
Bahrain Arts Society, Manama . 058343
Bahrain Arts Society Gallery,

Manama 003209
Bahrain Contemporary Art Association,
Manama 058344
Bahrain National Museum,
Manama 003210
Bahraini Moosa Al Demistani,
Manama 100416
Bai, Beijing 101934
Baia, Tbilisi 105897
Baibo Hua Yuan, Qingdao 102157
Baicao, Urumchi 140445
Baicheng Museum, Baicheng .. 007578
Baie Verte Peninsula Miner's Museum,
Baie Verte 005373
Baier, Fred, Tucson 124715
Baierweinmuseum, Bach an der
Donau 016686
Baihan Museum, Baihan 054828
Baihua Literature and Art Publishing
House, Tianjin 136742
Baihuazhou Literature and Art
Publishing House, Nanchang . 136730
Baik, Honolulu 093530
Baik & Song, Seoul 111615
Baïkal, Paris 071012
Baiken, Kyoto 111289
Le Bail Weissert, Paris 141041
The Baildon Furniture, Baildon . 088331
Bailen la Trastienda, Ourense . 086319
Bailer, Brigitte, Dortmund 106698
Bailey, Austin 092219, 135320
Bailey, Honolulu 093531
Bailey, Murwillumbah 139847
Bailey & Co., Oklahoma City ... 095949
Bailey, Thirlmere 099819
Bailey & Nelson, Seattle 124491
Bailey House Museum, Mauri
Historical Society, Wailuku ... 054027
Bailey-Matthews Shell Museum,
Sanibel 053084
Bailey, Andrew G., Haverfordwest 117931
Bailey, Beezy, Cape Town 114760
Bailey, Jeff, New York 122379
Bailey, W. H., New York 122380
Baileys, Ross-on-Wye 091138
Bailian Dongdongkong Science
Museum, Liuzhou 008040
Bailiffgate Antique Pine, Alnwick 088253
Baillet-Lefrère, René, Tourcoing 073814
Bailleul & Nentas, Bayeux 125505
Bailleul, Didier, Le Mans 069304
Bailleul, Patrick, Saint-Hilaire-du-
Harcouët 115529
Baillie, London 118202
Bailly, Charles & André, Paris . 104538
Bailly, Jacques, Paris . 104539, 104540
Bailly, Jean-Jacques & Francois,
Paris 128926
Bailly, Thierry, La Chapelle-
Montlinard 068978
Bailo, Buenos Aires 127514
Bailo, B., Wuppertal 079101
Bailo, Luciano, Firenze 131116
Baimasi Han Wei Gucheng Display
Center, Luoyang 008057
Baimbridge Antiques, Hamilton,
Victoria 061584
Bainbridge & Co, A., Ruislip ... 127024
Baines, Henry, Southborough . 091374
Baines, James, Malanda 099258
Bains, Christchurch 083856
Les Bains d'Aphrodite, L'Isle-sur-la-
Sorgue 069558
Baiocco, Fabiola, Roma 081248
Bair, Salt Lake City 123947
Baird Institute Museum,
Cumnock 044058
Baird, Robert M., Salt Lake City 136231
Bairin Youqi Museum, Balin
Youqi 007581
Bairin Zuoqi Museum, Bairin
Zuoqi 007579
Bairnsdale Antiques, Bairnsdale 061075
Bairro Alto, Lisboa 132929

Baisden, Tampa 124632
Baishing, Singapore . 085371, 114495
Bait Al Zubair, Muscat 034096
Bait Muzna Gallery, Muscat ... 034097
Baitson, Gilbert, Kingston-upon-
Hull 126932
Baituna al Talhami Museum,
Bethlehem 034158
Bajaj Art Gallery, Mumbai 109418
Bajardi, Marion, Graz 127743
Bajcsy-Zsilinszky Endre Emlékmúzeum,
Kővágóörs-Pálköve 023765
Bajer, André, Oslo 143393
Bajocchi, Pietro, Cairo 066230
Bajolet, Léon, Montigny-lès-Metz 070254
Bajor Gizi Emlékmúzeum,
Színháztörténeti Múzeum,
Budapest 023459
Bajt Dan, Charkiv 117105
Bajuwarenmuseum Waging am See,
Waging am See 022370
Bajza József Emlékszoba, Szűcsi 023988
BAK 25, Almelo 059594
Bak Negara Muzium Matawang, Kuala
Lumpur 030682
BAK, Basis voor Actuele Kunst,
Utrecht 112690
Bakara, Biarritz 067186
Bakarat, Kraków 113532
Bakehouse, Ohaupo 084044
Bakehouse Gallery, Patonga ... 099555
Bakelants, Ivo, Deurne 063611, 136651
Bakelite Museum, Williton 046138
Het Bakken, Almelo 112109
Baker, Denver 127203
Baker, Paddington, New South
Wales 099514
Baker-Cederberg Museum, Rochester
General Hospital, Rochester .. 052614
Baker Truesdale, Fortitude Valley 061482
Baker, A.P. & R., Wigtown 144924
Baker, Edith, Dallas 120558
Baker, Frederick, Chicago 120058
Baker, G., Horncastle . 089580, 134474
Baker, Gordon F, London 064346
Baker, Gregg, London 089934
Baker, J. & J., Lavenham 089751
Baker, Jeffrey, Coffs Harbour .. 098905
Baker, Keith, Woking 091900
Baker, Norma, Fort Worth 093430
Baker, R.J., Canterbury 088771, 134204
Baker, T., Langford,
Nottinghamshire 089743
Baker, T.R., Stockbridge 091431
Bakers Dozen Antiques,
Vancouver 064799
Bakersfield Museum of Art,
Bakersfield 046718
Bakhruz Kangarli Museum,
Naxçivan 003184
Bakhtari, Mehdi, Paris 071013
Bakhtiyar, Stockbridge 091432
Bakibey Konagi, Burdur 042853
Bakke, Dagfinn, Svolvær 113280
Bakkemuseet, Frederiksberg .. 010065
The Bakken, Library and Museum of
Electricity in Life, Minneapolis 050934
Bakker – Leeskamer van Maastricht,
Tjerk, Maastricht 143197
Bakker, Ad, Dordrecht 083176
Bakker, Bartje, Amsterdam ... 082695
Bakker, Wim, Rotterdam 083553
Bakkerij Museum, Oosterhout . 032724
Bakkerijmuseum, Geel 003609
Bakkerijmuseum, Groot-
Bijgaarden 003661
Bakkerijmuseum, Huizen 032518
Bakkerijmuseum De Grenswachter,
Luijksgestel 032608
Bakkerijmuseum de Meelzolder,
Almelo 031896
Bakkerijmuseum Walter Plaetinck,
Veurne 004108
Bakonybéli Erdok Háza, Balatoni

Nemzeti Park, Bakonybél 023420
Bakonyi Ház, Laczkó Dezső Múzeum,
Veszprém 024039
Bakonyi Természettudományi Múzeum,
Zirc 024060
Bakri Gazzaz, Hussein, Riyadh . 085314
Baks, E., Den Haag 112376
Baksa, Prof. Dr. Jean Pierre von,
Wien 100238
Baktay Kortárs Képzőművészeti
Múzeum, Baktay Ervín Gimnázium,
Dunaharaszti 023607
Baku, Fukuoka 111247
Baku Art Center, Baku 003155
Baku Art Gallery, Bucureşti ... 114128
Baku Museum of Education, Bakı 003156
Bal-Antiques, Wien 062903
Bal Gandharva Rang Mandir,
Pune 109491
Bal Sangrahalaya, Lucknow ... 024376
Bala Lake Railway, Bala 043470
Balaam, Barcelona ... 133103, 143710
Balachninskij Kraevedčeskij Muzej,
Balachna 036880
Baladna, Amman 111547
Balai Seni Lukis Negara, Kuala
Lumpur 030683
Balaklava Museum – Centenary Hall,
Balaklava 000812
Balaklava Museum – Urlwin Park,
Balaklava 000813
Balalaika Arts, New York 122381
Baland, Torsten, Berlin 141485
Balandier, Michel, Saint-Étienne 129224
Balanghai Archaeological Site
Museum, Butuan 034442
Balao-Balao Gallery, Angono ... 113372
Balás Jenő Bauxitbányászati Múzeum,
Gánt 023642
Balas, Bernhard, Wien 127901
Balassa Bálint Múzeum,
Esztergom 023627
Balassa Bálint Múzeum – Babits
Mihály Emlékház, Esztergom . 023628
Balassi, Kiadó, Budapest 137584
Balastre, Jacqueline, Varzy ... 073982
Balatoni Múzeum, Keszthely ... 023737
Balay Negrense Museum, Silay . 034596
Balaÿ, Charles, Lyon 069704
Balboa 13, Madrid 115154
Balboa Art Conservation Center, San
Diego 136290
Balbon, Denise, Saint-Ouen ... 073029
Balboni, Strasbourg 129263
Balcaen, Ets Claude, Paris 071014,
128927
The Balch Institute for Ethnic Studies,
Philadelphia 052032
Balchašskij Gorodskoj Istoriko-
Kraevedčeskij Muzej, Balchaš . 029815
Balcli's, Barcelona 085734, 126505
Balcombe Fine Art, Croydon ... 127591
Balconi Ricciardi, Angela, Milano 080426
Balcony House, Nassau 003203
Baldacchino, Los Angeles 094265
Baldaia, Laurindo Pinto, Porto . 132967,
132968
Baldaia, Laurindo Pinto, São Mamede
de Infesta 132982
Baldari, Antonio, Roma 131823
Baldem, İstanbul 117012
Baldfaced Stag, Ashurst, Kent . 088315
Baldi, Padova 080963
Baldi, Lella, Verona 081897
Baldinger, Artuur, Amsterdam . 112157
Baldo e Veronese, Verona 081899
Baldo, Antonio Juan, Alicante . 133081
Baldo, Veronese, Verona 081898
Baldo, Vittorio, Brescia 131016
Baldomero Aguinaldo House,
Kawit 034488
Baldoni, Nicole, Pignans 072162
Baldwin, Vineyard Haven 145629
Baldwin County Heritage Museum,

Elberta 048518
Baldwin Historical Society Museum,
   Baldwin 046721
Baldwin Photo Gallery, Middle
   Tennessee State University,
   Murfreesboro 051124
Baldwin-Reynolds House Museum,
   Meadville 050751
Baldwin & Sons, A.H., London . 089935,
                                     126953
Baldwin, M. & M., Cleobury
   Mortimer 144343
Baldwin, M. & M., Kidderminster 144503
Baldwin, Peter, Mullumbimby . . 139845
Balestra, Louis, Marseille 069908
Balfet, Christian, Le Pradet 069338
Balfet, Olivier, Mazamet 070014
Balfour, Toronto 140398
Balfour Museum of Hampshire Red
   Cross History, Winchester . . 046148
Bâlgarka, Burgas 101119
Bǎlgarski Chudožnik, Sofia . . . 136670
Balgoma, Madrid 086128
Bali-Africa Tribal Art, Bruxelles . 063346
Bali Bay Trading, Tampa 097544
Balian, Sylvie et Hervé, Paris . . 071015
Balias, D., Serans . . . 073507, 105671
Balıkesir Devlet Güzel Sanatlar
   Galerisi, Balıkesir 042841
Baliki, Saint-Paul 085069
Balio ir Vandos Sruogų Namai-
   Muziejus, Maironio Lietuvių
   Literatūros Muziejus, Kaunas . 030349
Balke, Evelyn, Berlin 129468
Balke, H., Alpirsbach 074381
Balkh Provincial Museum, Mazar-
   i-Sharif 000008
Balkh Religious Museum, Mazar-
   i-Sharif 000009
Balkhausen, Weilerswist 078891
Balkhausen, H., Köln 076849
Balkwill, Ray, Exmouth 117778
Ball & Claw Antiques, Tetbury . 091565
Ball State University Museum of Art,
   Muncie 051113
Ball, O., Hüffenhardt 076581
Balla-Wein, Portarlington 062101
Ballack, F.H.H., München 077464
Ballan & Pannan, Armadale,
   Victoria 098656
Ballan Shire Historical Museum,
   Ballan 000814
Ballance House, Glenavy 044426
Ballangen Bygdemuseet,
   Ballangen 033520
Ballantine Publishing, New York 138378
Ballarat Clayfire Gallery, Magpie 099252
Ballarat Fine Art Gallery, Ballarat 000815
Ballarat Tramway Museum,
   Ballarat 000816
Ballard, Seattle 136363
Ballard & Fetherston, Seattle . . 124492
Ballarin Marin, Enrique, Zaragoza 086590
Ballarin, Giancarlo, Venezia . . . 132227
Ballarin, Teresa, Venezia 132228
Ballario, Georges, Le Cannet . . 069233
La Balle de Plomb, Eguilles . . . 068433
Ballerup Egnsmuseum, Ballerup 010018
Ballet, Eddy, Thorey-Lyautey . . 129280
Ballinamore Local Museum,
   Ballinamore 024824
Balling, Jan, Århus 065598
Ballon- en Luchtvaartmuseum Zep/
   Allon, Lelystad 032589
Ballonmuseum Gersthofen,
   Gersthofen 018672
Ballot, Saint-Étienne 125800
Ballot, Michel & Dany, Romenay 072524
Ballpark Gallery, San Francisco . 124228
Ball's Falls Historical Park,
   Jordan 006053
Ballufin, Charles, Grenoble . . . 068718
Bally, Miami 135789
Bally Schuhmuseum,

Schönenwerd 041960
Bally, Maurice & Gisèla, Colmar 068082
Ballybunion Heritage Museum,
   Ballybunion 024825
Ballycastle Museum, Ballycastle 043474
Ballyheige Maritime Centre,
   Ballyheige 024828
Ballymena Museum, Ballymena . 024826
Ballymoney Museum, Ballymoney 043480
Balmain Framers and Gallery,
   Balmain 098697
Balmelli, Michele, Bellinzona . . 116112
Balmer, H.R., Zug 144200
Balmès Richelieu, Pierre, Paris . 071016
Balmoral, Toronto 064606
Balmoral Art Galleries, Hamlyn
   Heights 099107
Balmoral Grist Mill Museum, Balmoral
   Mills 005375
Balnarring Gallery, Balnarring . . 098700
Balneologické Múzeum, Piešťany 038422
Balocco Arte, Torino 110952
Balocco, Vassilli, Milano 080427
Balogh, Erika, Perchtoldsdorf . . 062762
Il Balôn, Torino 081652
Il Balon 2, Roma 081249
Al-Baloushi, Khalid Rasool Buksh
   Osman, Muttrah 084401
Balroop, Susanna, Wien 127902
Balsalobre Oliva, Fernando,
   Cartagena 085925
Balsano, Rosario, Torino 081653
Balser Morris, Shirley, Hampton 145155
Balser jun., Helmuth, Gernsbach 075984,
                                     129900
Balsiger-Sax, Walter, Bad Ragaz 087207
Balsiger, Otto, Delémont 116242
Baltazar, Ostrava 065435
Balteau, Alain, Liège 063746
Balter, Steven F., Seattle 136364
Balters, Regensburg 078162
Baltes, Bernd, Holzwickede . . . 137282
Baltic Art Centre, Visby 041173
BALTIC Centre for Contemporary Art,
   Gateshead 117817
Baltic Crossroads, Los Angeles . 121438
Baltic Imports, Minneapolis . . . 121974
Balticsaktionsraum, Stralsund . . 108678
Baltimore Book Company,
   Baltimore 127132, 144980
Baltimore Civil War Museum,
   Baltimore 046731
Baltimore Clayworks, Baltimore . 119723
Baltimore County Historical Museum,
   Cockeysville 047852
Baltimore Gallery, Baltimore . . . 119724
Baltimore Gold & Silver Center,
   Baltimore 092300
Baltimore Maritime Museum,
   Baltimore 046732
The Baltimore Museum of Art,
   Baltimore 046733
Baltimore Museum of Industry,
   Baltimore 046734
Baltimore Public Works Museum,
   Baltimore 046735
Baltimore Streetcar Museum,
   Baltimore 046736
Baltimore's Black American Museum,
   Baltimore 046737
Bałucki Ośrodek Kultury, Łódź . 034951
Balugani, Aurelio, Zocca
   Montombraro 081952
Balvu Novada Muzejs, Balvi . . . 030101
Balzac, Toronto 101575
Balzaretti, Daniela, Milano 080428
Balzeau, Liliane, Tours 073838
Balzekas Museum of Lithuanian
   Culture, Chicago 047640
Balzo, Roland del, Grenoble . . . 068719
BAM – Beaux Arts Mons, Mons 003876
Bama Ngappi Ngappi Aboriginal
   Corporation, Yarrabah 062503
Bambach, Klaus, Wiesbaden . . . 078969

Bamberger, Alan, San Francisco 145541
Bamberger, Chantal, Strasbourg 105678
Bamberger, Peter, Nürnberg . . . 126083
Bamble Museum, Langesund . . 033729
Bamboo, Jacksonville 121234
Bamboo, Kathmandu 112063
Bamboo Museum, Lugu 042383
Bamburgh Castle, Bamburgh . . 043482
Bame, Chris, Dallas 135494
Bamfield Community Museum,
   Bamfield Community School
   Association, Bamfield 005376
Bamfords, Derby 126874
Bamfords, Matlock 126987
Bammert, Urs, Luzern . 087634, 133831
Bammler, Thomas, Traunstein . . 078691
Bamps, Véronique, Bruxelles . . 063347
Bampton Gallery, Bampton,
   Devon 088346, 134054
Bamses Auktioner, Svedala . . . 126684
Bamyan, Paris 071017
Ban Kao Prehistoric Museum,
   Kanchanaburi 042691
Banaemah, Jeddah 085300
Bananas, Wien 062904
Bananenmuseum, Sierksdorf . . 021858
Banas, Alessandro Pietro, Milano 131379
Banbury Cross Antiques,
   Sacramento 096602
Banbury Cross Fine Arts and Antiques,
   Ascot 061060
Banbury Fayre, London 099936
Banbury Museum, Banbury . . . 043484
Banc Public, Sainte-Cecile-
   d'Andorge 105636
La Bancarella del Novecento, Monte
   San Pietro 110506
Banchory Museum, Banchory . . 043486
Banco 22, Firenze 080086
Banco-Münzcabinett, Münster . . 077653
Bancroft, Cleveland 120401
Bancroft Historical Museum,
   Bancroft 005378
Bancroft Mill Engine Trust,
   Barnoldswick 043497
Bancroft's Decorative and Fine Arts
   Society, Loughton 060062
Band Museum, Pine Bluff 052135
Bandamora, Katoomba 099174
Bandaranaike Museum, Colombo 040474
La Bande des Six Nez, Bruxelles 100564,
                                     140165
Bandeira, Rosário, Setúbal 114111
Bandelier National Monument, Los
   Alamos 050400
Bandello, Rocco, Genk 063615
Bandigan, Sydney 099769
Bandinelli, Benito, Firenze 131117
Bandirma Arkeoloji Müzesi,
   Bandirma 042843
Bandkowski, Andrzej, Wrocław . 113917
Bandol Antiquités, Bandol 066928,
                                     128459
Bandwirkermuseum Ronsdorf,
   Wuppertal 022814
BaneGården Kunst og Kultur,
   Aabenraa 102756
Banegårdens Antik, Assens . . . 065612
Banes, Richmond 136164
The Banff Centre, Banff 055116
Banff Museum, Banff 043488
Banff Park Museum, Banff 005379
Bang, Cape Town 114761
Bang, Mikael, København 065850
Bangabandhu Shek Mujibur Rahman
   Museum, Dhaka 003216
Bangalow Antiques, Bangalow . 061099
Bangalow House Antiques,
   Lismore 061740
Bangbu Museum, Bangbu 007582
Banghaus Com, Hamburg 125965
Bangiya Sahitya Parisad Museum,
   Kolkata 024355
Bangkok University Gallery,

Bangkok 042650
Bangla Academy Folklore Museum,
   Dhaka 003217
Bangladesh National Museum,
   Dhaka 003218
Bangladesh Shamorik Jadughar,
   Dhaka 003219
Bangladesh Shilpakala Academy,
   Dhaka 054990
Bangor Historical Society Museum,
   Bangor 046776
Bangs, Christopher, London . . . 089937
Bangsbo Museum, Frederikshavn 010067
Bangunan Alat Kebesaran Diraja,
   Bandar Seri Begawan 004960
Banícke Múzeum, Gelnica 038351
Banícke Múzeum, Rožňava . . . 038437
Banícky Dom, Vyšná Boca 038482
Bank Austria Kunstforum Wien,
   Wien 002944
Bank Gallery, Darlinghurst 098942
Bank Gallery, Los Angeles 121439
The Bank House Gallery, Norwich 090867,
                                     134889
Bank House Gallery, Norwich . . 118826
Bank Leumi Museum, Tel Aviv . 025188
Bank Museum, Meeteetse Museums,
   Meeteetse 050767
Bank of America Galleries, San
   Francisco 052993
Bank of England Museum,
   London 044909
Bank of Ireland Arts Centre,
   Dublin 024868
Bank of Mississippi Art Collection,
   Tupelo 053852
Bank of Montreal Museum,
   Montréal 006358
Bank of Valletta Museum, Sliema 030748
Bank of Victoria Museum, Royal
   Historical Society Victoria,
   Yackandandah 001665
Bank One Fort Worth Collection, Fort
   Worth 048900
Bank Street Antiques, Ottawa . . 064496
Bank Street Gallery, Newquay . . 118822
Banker's Furniture, Phoenix . . . 096191
Bankfield Museum, Halifax 044499
Banking and Currency Museum,
   Kadina 001172
Bankmuseet, Stavanger 033973
Banko, Walter M., Montréal . . . 064393
Banks Peninsula Antiques,
   Akaroa 083741
Banks, M.G., Christchurch 083857,
                                     112831
Banks, Robert H ., Dallas 120559
Banks, Simon, Finedon 089274
Banks, Yvonne, Seattle 124493
Banksia Grove Gallery, Lake
   Cathie 099214
Bankside Gallery, London 044910
Banner County Historical Museum,
   Harrisburg 049368
Bannier, Yves, Saint-Brandan . . 072694
Banning Residence Museum,
   Wilmington 054431
Banning, J., New York 122382
Bannister, David, Cheltenham . . 134232,
                                     144322
Bannister, Penny, Fremantle . . . 139740
Bannock County Historical Museum,
   Pocatello 052221
Bannockburn Heritage Centre,
   Stirling 045871
Bannow, Günther, Lübeck 077171
Banovsky, Chomutov 065377
Banovsky, Michel, Bagneux . . . 066911
Banq, Eric, Castres 067726
Banque du Livre, Saint-Étienne . 141329
Banquels, J., Courville-sur-Eure 068202
Banqueting House, Historic Royal
   Palaces, London 044911
Bansbach, B., Heidelberg 076433

Banské Múzeum v Prírode, Slovenské Banské Múzeum, Banská Štiavnica 038283
Bantam Dell Publishing, New York 138379
Banting House, London 006207
Banting Interpretation Centre, Musgrave Harbour 006440
Bantin, Cornelius, Reutlingen .. 078209
Bantock House, Wolverhampton 046178
Bantry House, Bantry 024832
Bányászati Kiállítás, Gyöngyös . 023651
Bányászati Kiállítóhelye, Nógrádi Történeti Múzeum, Salgótarján 023869
Banys Àrabs, Palma de Mallorca 039878
Banzhaf, Michael, Tübingen 142424
Banzi, Barbara, Parma 131719
Banzon Antiques, Manila 084477
Bào Đại Palace Museum, Đà Lat 054782
Bao Tàng Dân Tôc Hoc Viêt Nam, Ha Noi 054789
Bao Tàng Hai Phong, Hai Phong 054815
Bao Tàng Quan Doi, Ha Noi .. 054790
Bao Tàng Văn Hóa Các Dân Tôc Viêt Nam, Thai Nguyen 054817
Bao, Ly, Hoi An 125157
Baobab, Amsterdam 082696
Baoji Museum, Baoji 007584
Baoji Xiàn Museum, Baoji 007585
Baoji Zhouyuan Museum, Fufeng 007771
Baoshan Museum, Baoshan 007586
Baptista, A.M. Salgueiro, Lisboa 084835
Bar Anticuario, Montevideo 091989
Bar Bookstore, Scarborough .. 144826
Bar David Museum, Kibbutz Bar Am 025123
Bar Harbor Historical Society Museum, Bar Harbor 046783
Bara Art, Lalitpur 112090
Barabas, Ladislav, Bratislava .. 085543
Barabe, Nicolas, Rennes (Ille-et-Vilaine) 072423
Barabinskij Kraevedčeskij, Novosibirskij oblastnoj kraevedčeskij muzej, Barabinsk 036881
De Baracquen Museum voor Bourtanger Bodemvondsten, Bourtange 032090
Baragova Rojstna Hiša, Dobrnič 038506
Barahona, Víctor, Santiago de Chile 064953
Barailler, Jean-Pierre, Nice 070572
Baraka, San Diego 124105
Baraka Klub, Kraków 113533
Barakat, Beit Lahm 034410
Barakat Gallery, London 089938, 118203
Barake, Herbert, São Paulo 064156
Barale, Patrick, Ezanville 068500
Barale, Paula, Moulins (Allier) .. 070368
Baramasa, Manila 113403
Baran de Bordeaux, Auckland .. 083771
Barandela Blancoq, Jaime, Saint-Maurice 072967
Barangay Museum, Mandaluyong 034515
Baranov Museum, Kodiak 050017
Baranowsky, Roma 081250
Barasoain Church Museum, Malolos 034511
Barat, Jean-Luc, Le Dorat 069262
Baratta, Maria Celeste, Messina 080361
La Baratte, Megève 070032
Baratte, Frédérique, Chauny 067942
Baratti Sordelli, Brescia 131017
Baratti, Giorgio, Milano 080429
Baraz, İstanbul 117013
Barazer, Guillaume, Roscoff ... 105459
Barb, Tucson 124716
Barbacana, Montevideo 091990
Barbáchano & Beny, Cercedilla . 133228
Barbados Museum, Saint Michael 003240
Barbagallo, Sebastiano, London 089939
Barbakan, Warszawa 113806
Barbanel, Claude, Paris 071018

Barbanel, Claude, Saint-Ouen .. 073030
Barbara and Steven Grossman Gallery, School of the Museum of Fine Arts, Boston 047101
Barbara F., Bremen 075148
Barbara Fritchie House Museum, Frederick 048943
Barbara Galeria, Warszawa 113807
Barbara Hepworth Museum and Sculpture Garden, Saint Ives, Cornwall 045696
Barbara Mathes Gallery, New York 122383
Barbara's Books, South Auckland 143342
Barbara's Dolls, Fort Worth 093431
Barbara's kleine Welt, Schorndorf 078391
Barbareschi, Valentina, Torino .. 132094
Barbarian Book Shop, Wheaton . 145644
Barbarians, Saint Paul 123897
Barbarisi, Elisabetta, Roma 081251
Barbarlık Müzesi, Lefkoşa 033470
Barbarot, Claude, Jaulny 068897
Barbaroux, Suzanne, Marseille . 069909
Barbatelli, Nicola, Napoli 080843
Barbaud, Jean-Pierre, Soisy-sous-Montmorency 073532
Barbaud, Jean-Yves, Nantes ... 070450
Barbe-Lafarge, Francis, La Calmette 068961
Barbe, Christophe, Caen 067557
Barbe, Ludovic, Toulouse 073739
Barbe, Ludovic, Vicdessos 074101
Barbe, Patrice, Paris 071019
Barbed Wire Museum, La Crosse 050026
Barber Institute of Fine Arts, Birmingham 043612
Barber Kennedy, Mat, Chicago . 120059
Barber, Alison, West Haddon .. 091785, 135205
Barber, John M., Richmond 123722
Barber, Ron, Wellington 113061
Barbera, Evelyne, Saint-Ouen .. 073031
Barbera, Olivia, Saint-Ouen 073032
Barberini, Antonio, Bari 130906
Barberousse, Jeanlin, Antugnac 103262
Barberousse, Michel, Paris 071020, 141042
Barbers, Market Drayton 126985
Barbers, Woking 127095
The Barber's Clock, Worcester . 091941, 135258
Barbers, Ashley, Redfern 062141
Barberton Museum, Barberton . 038649
Barbéry, Paris . 071021, 104541, 141043
Barbery, Lia, Montevideo 091991
Barbesol, Christian, Ladon 069133
Barbetti, Roberto, Genova 110092
Barbian, Horst, Wadern 078819
Barbian, Martin, Saarbrücken .. 142340
Barbican, Manama 100417
Barbican Antiques Centre, Plymouth 091021
Barbican Art Gallery, London .. 044912
Barbican Bookshop, York 144940
Barbié-Nogaret, Elisenda, Barcelona 114858
Barbier, Antoine, Blumeray 067264
Barbier, Gérard, Dordives 068386
Barbier, Gérard, Sermaise 073509
Barbier, Hélène, Marseille 128746
Barbier, Martine, Metz 070089
Barbier, Michel, Le Bizot 069217
Barbière & Beuche, Mazan 070018
Barbieri & C., Carlo, Modena .. 131559
Barbieri, Fabio, Firenze 131118
Barbieri, Lorenzo, Firenze 131119
Barbizon, Arnhem 082932
Barbizon, Paris 104542
Barbj Lavorazione, Firenze 080087
Barboni, Magda, Milano 080430
Barbotin, Marcilly-sur-Vienne .. 069853
Barbotin, Frédéric, Le Pouliguen 069336
Barbotine, Mont-Louis 070175
Barbotine, Saint-Martin-de-Ré .. 072935

Barbour's General Store Museum, Saint John 006889
Barboza, Punta del Este 092016
Barby Antiques, Tucson 097638
Barcala Diaz, Juan Ramon, Jerez de la Frontera 086055
Barcaldine and District Historical Museum, Barcaldine 000823
Barcellino, Nino R., Pittsburgh .. 096258
Barcellos, Livia de Alencar Chaucs, Porto Alegre 100966
Barcelo, Ignacio, Córdoba 115026
Barcínio & Araújo, Porto 114076
Barcino Art, Barcelona 114859
Barclay, Headington 089486
Barclay Lennie, Glasgow 117834
Barclay, M.J. & D., Friockheim . 089312
El Barco, Madrid 133304
Barco, Puma 065481
Barco Museo Yaraví, Puno 034392
Barcsay Gyűjtemény, Ferenczy Múzeum, Szentendre 023953
The Bard Graduate Center for Studies in the Decorative Arts, Design, and Culture, New York .. 051344, 057543
Bardas, Galați 143625
Bardawil, Eddy, London 089940, 134590
Bardean, Albuquerque 119458
Bardelaere Museum, Lembeke .. 003780
Barden House Antiques, Tonbridge 091608
Bardet, Jacques, Plougoumelen 072202
Bardez, Jean-François, Paris ... 128928
Bardhaman Science Centre, Bardhaman 024153
Bardiani, Luigi, Parma 081094
Bardini, Genova 142692
Bardith, New York 095252
Bardo, Pamela Pierrepont, Chicago 120060
Bardón Mesa, Luis, Madrid 143752
Bardon, Andres, Buenos Aires .. 098553
Bardoscia, Fulvio, Lenzburg ... 087590
Bardoux, Patrick, Grimaud 068747
Bardu Bygdetun, Salangsdalen . 033915
Bårdu Dzimtas Muzejs Rumbiņi, Katvaru 030145
Barducci, Alberto, Bologna 130934
Bare Cove Fire Museum, Hingham 049488
Bare-Delhaise, Liège 063747
Bare Feet Gallery & Ikebana Supportal, Dublin 109576
Bare Hands Gallery, Birmingham 119790
Bare Wood, Birmingham 135350
Barebones Art Space, Bangalow 098703
Barefoot, Colombo 086610
Barefoot Gallery, Christchurch .. 112832
Barekat, Torino 132095
Baren Woodcut, Aloha 060403
Bareo, Singapore 114496
Barès, Paris 104543
Bares, Nello, Milano 080431
Bareyre, Martine & Gérard, Agen 066422, 128424
Barfield, Donna J., Jacksonville 093965, 145182
Barfino, Barcelona 085735
Barfoot, Craig, East Hagbourne . 089133, 134347
Barfüßerkirche, Angermuseum, Erfurt 018237
Barfüsserkirche, Historisches Museum Basel, Basel 041264
Barfuss, M., Rheinau 108492
Bargac, Philadelphia 096082
Bargain Box, Luton 090655
Bargain Upholstery, Saint Paul . 136213
Barganhata Comercio de Usados, Rio de Janeiro 064024
Bargaro Beach Gallery, Bargara 098710
Barge, Annie, Ambierle 066550
Bargera, Kenda, Köln 107590
Barghini, Isa, Milano 080432

Bargrave Collection, Canterbury Cathedral Archives and Library, Canterbury 043855
Barham, London 089941
Baria, Tristan, Sées 073469
Barilleau, Bruno, Le Chesnay .. 069249
Barillet, Marie José, Milano 080433
Barilli Antichità, Cavriago 080051
Barin Carpets Restoration, London 134591
Barioz, Christian, Lyon 069705
Baripada Museum, Baripada ... 024155
Barjavel, Françoise, Arry 066743
Barjol, Michel, Malaucene 104146
Bark House Gallery, Newquay .. 118803
Barker, Buffalo 092511, 135370
Barker-Cawood Art Gallery, Olivet College, Art & Communication Dept., Olivet 051753
Barker & Son, F.J., Bayview ... 061110
Barkerville Historic Town, Barkerville 005383
Barkham Antique Centre, Barkham 088350, 134059
Barkham, Julian, Wellington ... 132757
Barkost, Geilo 132774
Barkowski, Rolf, Recklinghausen 078153
Barkwell, Kansas City 094058
Barlanglakás Emlékkiállítás, Budapest 023460
Barley Twist, Charlotte 092576
Barley, Robert, London 089942
Barleylands Farm Museum, Billericay 043600
Barlier, Francis, Paris 104544
Barling Fine Porcelain, Wickham Bishops 091839
Barlow, New Orleans 122126
Barlow, Tulsa 136427
Barlow Collection of Chinese Ceramics, Bronzes and Jades, Brighton . 043733
Barlow, Isiaha, Wanganui 113049
Barlow, PM & MA, Malanda ... 061762
Barmorr, San Diego 097066
Barmouth Sailors' Institute, Ty Gwyn and Ty Crwn, Barmouth 043494
The Barn, Nairobi 082248
The Barn, Petersfield 090984
Barn Antiques, Barnstaple 088361
Barn Antiques, Nantwich 090794
Barn Antiques Centre, Long Marston 090613
The Barn at Bislington, Bisington 088492
Barn Court Antiques, Templeton 091556
Barn Galleries, Henley-on-Thames 117957
Barn Gallery, West Cork 109704
Barn Museum and Rail Depot Museum, Troutdale Historical Society, Troutdale 053808
Barnacle Historic State Park, Coconut Grove 047854
Barnadas, Jordi, Barcelona ... 114860
Barnard & Nedin, Lund 086829
Barnard's Mill Art Museum, Glen Rose 049110
Barnas Antikvariat, Oslo 143394
Barnboksantikvariat Bokmagasinet, Stockholm 143940
Barnegat Bay Decoy and Baymen's Museum, Tuckerton 053819
Barnes & Noble, Cleveland ... 145082
Barnes Candace, San Francisco 097186
Barnes County Historical Museum, Valley City 053933
Barnes Foundation, Merion 050807
Barnes, Daisy, Hutton-le-Hole .. 118006
Barnes, Jane, Honiton 089557
Barnes, R.A., Bicester 088459
Barnet Cattanach, Anne-Marie, Borehamwood 088516, 134124
Barnet Historical Society, Barnet 046792
Barnet Museum, London 044913
Barnet, Bertrand, Gournay-en-

Bray 068689
Barnett, Jacksonville 121235
Barnett Cane and Rush Seating,
    Hathern 134447
Barnett, David, Milwaukee 121901
Barnett, Roger, Eton 089214
Barney, Daniel, New York 095253
Barnicott, Linda, Pittsburgh ... 123522
Barnizados & Lacados, Hontoria 133273
Barnizados J. & M. Martínez,
    L'Hospitalet de Llobregat 133274
Barnizados Mayoral, Madrid ... 133305
Barnizados y Lacados, Segovia . 086430,
                            133427
Barnola, Antonio de, Barcelona . 114861
Barnoud, Dijon 103672
Barnsdale Gallery, Yoxford 119415
Barnsdall Art Center, Los Angeles 121440
Barnsdall Arts Center and Junior Arts
    Center, Los Angeles 050408
Barnt Green Antiques, Barnt
    Green 088367, 134066
Barnum House Museum, Grafton 005897
Barnum Museum, Bridgeport .. 047200
Barnwell County Museum,
    Barnwell 046795
Barnyard International, Atlanta . 092113
Barockberichte, Informationsblätter
    des Salzburger Barockmuseums,
    Salzburg 138505
Barockjagdschloß Eckartsau,
    Eckartsau 001838
Barockschloss Mannheim, Staatliche
    Schlösser und Gärten Baden-
    Württemberg, Mannheim 020245
Barockschloß Riegersburg,
    Riegersburg, Niederösterreich . 002597
Baroco, Neuilly-sur-Seine 070516
Baroda Museum and Picture Gallery,
    Vadodara 024507
Barois, Philippe, Aire-sur-la-Lys . 066443
Barok, Prigorje Brdovĕko 128313
Barok, Warszawa 138958
Barokh, R.M., West Hollywood . 097875
Baroko, Louny 128330
De Barometer, Rotterdam 083554
Barometer Shop, Leominster ... 089802,
                            134546
Barometer World, Merton 090752,
                            134850
Barometer World Museum,
    Merton 045252
Le Baron, Casablanca 082548
Baron, Chester 088856, 134251
Baron, Miami 094707
Baron & Boisanté, New York ... 095254,
                            138380
Baron & Ribeyre, Paris 125678
Baron Art, Holt, Norfolk 089545
Baron Hermanos, Barcelona .. 085736
Baron, Bertrand, Melle 070043
Baron, Chris, London . 089943, 134592
Baron, H., London 144559
Baron, Josiane, Paris 128929
Barone, Marseille 128747
Barone's Books, Hamilton 140346
Barong, Padies 070837
Baroni, Jean-François, Paris 104545
Baroni, Jean-Luc, London 089944,
                            118204
Baroni, Loris, Venezia 081824
Baroni, Marcello, Roma 131824
Baroni, Maurizio, Parma 081095
Baroni, Maurizio, San Pancrazio . 081616
Baroni, Sergio, Milano 080434
Baroni, Silvia, Bologna 130935
Baronian-Francey, Bruxelles .. 100565
Baroniet Rosendal, Rosendal .. 033908
La Baroquante, Saint-Maime ... 072907
Baroque, Copponex 068141
Baroque Silversmith, Chicago .. 135400
Baroque Silversmith, Skokie .. 136381
Baros Books, Trowbridge 144888
Barossa, Angaston 139623

Barossa Valley Historical Museum,
    Tanunda 001561
Baroti, Klaipėda 111787
Barovier, Marino, Venezia 111090
Barozzino, Giovanna, Lisboa ... 084836
Barquet, Ramis, Monterrey 111987
Barquet, Ramis, New York 122384,
                            122385
Barr, Houston 121003
Barr & Davis, Lancaster 127277
Barr & Ochsner, Zürich 087938, 116755
Barr Colony Heritage Cultural Centre,
    Lloydminster 006205
Barrachina-Ramoneda, Barcelona 133104
Barrack Cottage, Hobart 061627
Barral & Mondiola, New York . 122386
Barral, Angel, Buenos Aires ... 060877
Barran, Julian, Castle Cary ... 117541
Barranco, Madrid 086129
Barrand, Grace, Nutfield 118484
Barras, David, Bulle . 087339, 133709
Barrat, Vichy 074104
Barraud & Bonnet, Le Crès 069256
Barraud, Jean-Marie, Castelnau-le-
    Lez 067709
Barraud, Patrick, Saint-Prex ... 087770
Barrault, Alain, Puisserguier ... 072323
Barrault, Anne, Paris 104546
Barre Historical Museum, Barre 046796
Barre, Claire, Bordeaux 067312, 128482
Barre, Laurent-Claude, Dampierre-en-
    Yvelines 068266
Barré, Nicolas, Orléans 070808
Barreau, Rennes (Ille-et-Vilaine) 072424
Barreau, Thorigné-Fouillard ... 073670
Barreau Le Maistre Art Gallery, Jersey
    Museum, Saint Helier 045690
Barreau, Patrick, Tours 073839
Barrel, Genova 110093
Barrere, Charles, Saint-Ouen .. 073033
Barrère, Jacques, Paris 071022
Barrett House, Forest Hall, New
    Ipswich 051276
Barrett-Jackson, Scottsdale ... 127451
Barrett Street Antique and Auction
    Center, Virginia Beach 097772, 127486
Barrette, Kelly, Boston 119837
Barrhead Centennial Museum,
    Barrhead 005384
Barridoff, Portland 127408
Barrie H Coleman, Great
    Somerford 117883
Barrié, Patrick, Paris 071023
Barrientos, Guatemala 079223
Barrie's Birthplace & Camera Obscura,
    Kirriemuir 044736
Barrington, Christchurch 143302
Barrington Area Historical Museum,
    Barrington 046797
Barrington-Doulby, Edward, Barnard
    Castle 088355, 134061
Barrington Galley of London, New
    York 122387
Barrington Living History Farm,
    Washington 054188
Barrington Woolen Mill Museum,
    Barrington 005388
Barrington, David, Brasted ... 088576
Barrington, Michael, Dorchester 134328
Barrister, New Orleans 122127
Barrister, San Antonio 145510
Barro, Sacramento 096603
Barro, Maos de, Belo Horizonte . 063884
Barrocco, Montfort-l'Amaury .. 104263
Barrocos Galería, Bucaramanga 102380
Barrón, Santiago de Chile 064954
Barron Arts Center and Museum,
    Woodbridge 054518
Barron County Historical Society's
    Pioneer Village Museum,
    Cameron 047388
Barros, Antonio M. Ferreira,
    Lisboa 084837
Barros, Dario, São Paulo 101084

Barros, Maria Beatriz, Porto ... 085009
Barrow, Loughborough 118664
Barrows, Martinborough 112944
Barry & Rare, Minehead 144733
Barry, Kevin, Los Angeles 121441
Barrymore, Albuquerque 092036
Barry's Antiques, Berrima 061140
Barry's Antiques, Great Yarmouth 089381
Barry's Auction, Albuquerque .. 127115
Le Bars, Montauban 070189
Barsa Műtárgykaptár, Budapest . 109130
Barsacq Schaefli, Sylvie, Biarritz 067187
Barsanti, G. & Figli, Pisa 110662
Barsley, Belmont 061123
Bart, Betty & Jane, New York .. 095255
Barta, Manfred, Salzburg 062785
Bartel, Irmgard, Konstanze & Christian,
    Bonn 075064
Bartel, Klaus, Osnabrück 077959
Bartel, Wolfgang, Leipzig 077056
Bartels, Gudrun, Friedland, Kreis
    Göttingen 137216
Bartelt, Giedre, Berlin 106164
Barter, Alnwick 144218
Barter, Johannesburg 085645
Bartfield, J.N., New York 095256, 135888,
                            145324
Barth, Hofheim in Unterfranken . 130056
Barth, H.-D., Köln 076850
Barth, Kathrin, Frankfurt am
    Main 075772
Barth, Ute, Zürich ... 116756, 138105
Barth Erben, Walter, Zug 088019
Bartha, Miklos & Margareta von,
    Basel 116056
Bartha, Niklas von, London ... 118205
Barthel & Tetzner, Berlin 106165, 137054
Barthelmess & Wischnewski,
    Berlin 074762, 106166
Bartholomäus-Schmucker-
    Heimatmuseum, Ruhpolding . 021546
Bartholomew County Historical
    Museum, Columbus 047925
Bartn Kültür Evi, Bartın 042844
Bartke, Waltraud, Mühnesee ... 108011
Bartleby & Co., Bruxelles 136629
Bartleby, Washington 145630
Bartlesville Area History Museum,
    Bartlesville 046798
Bartlett Museum, Amesbury ... 046443
Bartlett Street Antiques Centre,
    Bath 088378
Bartlett, Laura, London 118206
Bartlett, Michael, Saint-Ouen .. 073034
Bartlett, Nigel A., London 089945
Bartlett, Philipp, Nantes 070451
Bartlett's Bazaar, Dublin 079525
Bartlow, Peter, Chicago 120061
Bartók 32 Galéria, Budapest ... 109131
Bartók Béla Emlékház, Budapest 023461
Bartoli, Cincinnati 092828
Bartoli, Herminé & Saliceti,
    Ajaccio 066495
Bartoli, Rodolfo, Firenze 080088
Bartolo, Roma 131825
Bartolomé, Jordi, Barcelona ... 085737
Bartolomei-Roussel, Yvon,
    Corbières 068144
Bartolomeo, Roberto, Milano ... 080435,
                            131380
Bartolomeu Dias Museum Complex,
    Mossel Bay 038786
Bartolotta, Giuseppe, Firenze .. 131120
Bartolozzi, Guido, Firenze 080089
Barton, Sacramento 123760
Barton, Saint Louis 123806
Barton County Historical Society
    Village and Museum, Great
    Bend 049215
Barton Faist, Chicago 120062
Barton Museum, Bhavnagar ... 024164
Barton Museum, Wilson 054455
Bartow History Center,

Cartersville 047447
Bartow-Pell Mansion Museum,
    Bronx 047223
Bartrick, Steven D., Gloucester . 144438
Bart's Coins, Chicago 092667
Barts Wonderwinkel, Gent 128047
Bartsch & Chariau, München .. 108053
Bartsch, Heine Robert,
    København 065851
Bartsch, Helge, Immenstadt im
    Allgäu 130073
Bartsch, Monika, München 130325
Bartscht, Katrin & Windfried,
    Berlin 106167
Bartz, Kristine, Blåvand 065615
Baruc, São Paulo 064157
Baruch, New York 122388
Baruch, Anne & Jacques,
    Chicago 120063
Barunčina Škola, Česká Skalice 009500
Barwinski, Lothar, Meschede .. 130300
Barwon, Queenscliff 139889
Barwon Booksellers, Geelong . 139747
Barzel, Berlin 074763
Basaglia, Gianpietro, Livorno .. 080317
Basak, İstanbul 117014
Basalt, Amsterdam 112158
Basalt Antikvitetshandel,
    København 065852
Basaltpark Bad Marienberg, Bad
    Marienberg 016821
Basantapur, Kathmandu 112064
Bascourt, M., Antwerpen 063192
Base Art Gallery, Cincinnati ... 120321
Base Ball Cards, Columbus ... 092995
Base Borden Military Museum,
    Borden 005441
Base Gallery, Tokyo 111409
Base Space, Brookvale 098804
Baseball Card Corner, Cincinnati 145064
Basecraft, Yorklea 099946
Basedau, Waltraud, Hamburg . 076193
Basel Ancient Art Fair, Basel .. 098326
Baseline, Providence 123695
Basement Art, Rødovre 102961
Basement Gallery, Enniskillen .. 117754
Bases, Saint Paul 096812
Bases Loaded, Buffalo 092512
Bashar, Manama 100418
Bashayer and Om El Saad, Cairo 066231
Bashert, Albuquerque 119459
Bashford, Auckland 083772
Bashford, Dobson, Christchurch 112833
Basho Memorial Museum, Ueno 029671
Basho Museum Koto, Tokyo ... 029488
Basic Antic, La Fouillouse 096037
Basic, Luka, Wien 062905
Basile, Melbourne 061814
Basilian Fathers Museum,
    Mundare 006438
Basilica Collegiata di Santa Maria
    Assunta, San Gimignano ... 027749
Basilica de Nuestra Señora del Pilar,
    Zaragoza 040441
Basilica del Santo Niño Museum,
    Cebu 034455
Basilica Museum, Saint John's . 006898
Basilica Museum, Schatkamer van
    de Onze-Lieve-Vrouwbasiliek,
    Tongeren 004073
Basilique Royale Saint-Denis, Saint-
    Denis (Seine-Saint-Denis) 015283
Basilisk, Basel 144014
Basilisk Bookshop, Fitzroy 139731
Basilisken-Presse, im Verlag Natur
    & Text in Brandenburg GmbH,
    Rangsdorf 137476
Basin Head Fisheries Museum, Prince
    Edward Island Museum and Heritage
    Foundation, Souris 007523
Basın Müzesi, İstanbul 042925
Basing House, Basingstoke 043508
Basingstoke Canal Exhibition,
    Mytchett 045308

Basis Antiquariat, München 142142
Basis Wien, Kunst, Information und
  Archiv, Wien 002945
Basix, Singapore 085372
Baske, Harry, Kappeln 076678
Baskerville Antiques, Petworth . 090988
Baskerville Bindings, Holt,
  Norfolk 144483
Baskerville Books, Wadhurst . . . 144904
Baskerville, Lee, Richmond . . . 123723
Baskett, Mary, Cincinnati 120322
Baskin Visual Arts, University of
  California Santa Cruz, Santa
  Cruz 057871
Baškirskij Respublikanskij
  Chudožestvennyj Muzej im. M.V.
  Nesterova, Ufa 037977
Basler Kunstverein, Basel 059929
Basler Papiermühle, Schweizerisches
  Museum für Papier, Schrift und
  Druck, Basel 041265
Basnet Art, Kathmandu 112065
Basnier, François, Caen 067558
Basquin, Christine, Sains-du-
  Nord 072656
Basrah Museum, Basrah 024806
Bass Manor Museum, Malta . . . 050614
Bass Museum of Art, Miami
  Beach 050846
Bassali, Paris 071024
Bassam, Arès 066707
Bassamat, Casablanca 112033
Bassanese, Trieste 111043
Bassani, Aldemaro, Ravenna . . 081182
Bassani, Rossella, Verona 132268
La Basse Fontaine, Vence 105826,
  141373
Bassereau, Nicole, Paris 104547, 136866
Bassery, François, Lille   069476, 069477,
  128675
Basset, Florence, Flassans-sur-
  Issole 103741
Basset, Jean-Claude, Auray . . . 066815,
  103290
Bassetlaw Museum and Percy Laws
  Memorial Gallery, Retford . . . 045605
Bassett Hall, Williamsburg 054409
Bassi, Barbara, Cremona 080060
Bassi, Giuliano, Occhiobello . . 110549
Bassi, N., Memphis 094618
Bassler, Freiburg im Breisgau . 075850,
  107051
Bassmann, G., Herford 076499
Bassoi, São Paulo 128122
Bassola, Ettore, Firenze 131121
Basta, Lausanne 116409, 144092
Bastejs, Rīga 111755
Bastiaans, Boxmeer 112349
Bastian, Jacques, Strasbourg . . 073572
Bastianacci, Vittorio, Firenze . . 131122
Bastide Gallery, Tournon-
  d'Agenais 073819
Bastide, Catherine, Bruxelles . . 100566
Bastide, Gil, Orléans . . 070809, 104427
La Bastidiole, Paris 141044
Bastien, J., Bruxelles 100567
Bastier, Patrick, Coligny 068074
Bastilippo, Sevilla 086441
Bastings & Van Tuyl, Oss 083524
Bastings, Charles, Oss 083522
Bastings, Frans & Geryon, Oss . 083523
Bastion, Nanaimo 006445
Al Bastione, Palermo 081021
Al Bastione, Trieste 111044
Il Bastione Antichità, Cagliari . . 079974
Bastogne Historical Center, American
  Memorial Hill of the Mardasson,
  Bastogne 003343
Baszta Pranieńska, Muzeum Ziemi
  Chełmińskiej – Dział Powstania
  Styczniowego, Chełmno 034680
Baszta Prochowa, Muzeum Ziemi
  Chełmińskiej – Dział Archeologiczny,
  Chełmno 034681

BAT Artventure Collectie,
  Amstelveen 031918
BAT Artventure Collectie,
  Zevenaar 033097
BAT CampusGalerie, Bayreuth . . 016984
Bata Shoe Museum, Toronto . . . 007159
Batagianni, Vaso, Athinai 109025
Batalla, Valentín, Cáceres 085919
Batán Museo, Val de San
  Lorenzo 040289
Batanero, Stephane, Marseille . . 128748
Batavia Art, Singapore 114497
Batavia Depot Museum, Batavia 046806
Batavia Furniture – Art Gallery,
  Fermoy 109633
Bataviawerf-Nationaal
  Scheepshistorisch Centrum,
  Lelystad 032590
Batchworth Lock Canal Centre,
  Rickmansworth 045615
Bate Collection of Musical Instruments,
  Oxford 045449
Le Bateau Lavoir, Grenoble . . . 103784
Le Bâteau Livre, Montpellier . . 140963
Batel, Bruno, Fanjeaux 068505
Bateman Historical Museum,
  Bateman 005394
Bateman, J. & R., Chalford 088820,
  134226
Bateman, June, New York 122389
Bateman's, Etchingham 044304
Bateman's Kipling's House,
  Burwash 043804
bateria – Zeitschrift für Kunst und
  Literatur, Fürth 138685
Bates College Museum of Art,
  Lewiston 050232
Bates College of Art and Visual
  Culture Department, Lewiston . 057337
Bates-Scofield Homestead,
  Darien 048150
Bath Abbey Heritage Vaults, Bath 043513
Bath and Bradford on Avon Antique
  Dealers Association, Batheaston 060063
Bath Galleries, Bath 088379
Bath Old Books, Bath 144226
Bath Postal Museum, Bath 043514
Bath Royal Literary and Scientific
  Institution, Bath 043515
Bath School of Art and Design, Bath
  Spa University, Bath 056508
Bath Stamp and Coin Shop, Bath 088380
Bath Street Gallery, Auckland . . 112752
Batheaston, Corsham 134297
Báthory Júlia Üveggyűjtemény,
  Dömsöd 023603
Bathurst Agricultural Museum,
  Bathurst 038650
Bathurst Books, Hobart 139773
Bathurst Museum, Bathurst . . . 000827
Bathurst Regional Art Gallery,
  Bathurst 000828
Bathurst Street Antiques Centre,
  Hobart 061628
Bathurst Teddy Bear Shop,
  Bathurst 061104
Bathurst, Georgina, Paddington, New
  South Wales 062038
Batik Gemi Müzesi, Girne 033457
Batillat, La Ferriere-aux-Etangs . 069023
Batiran, Kamil, Berlin 074764
Batisse, Michel, Colombes 068099
Batista, Antônio S., Fortaleza . . 100937
Batley Art Gallery, Batley 043531
Batlle, Barcelona 143711
Batman Kültür Merkezi, Batman 042845
Batoche, National Historic Site,
  Batoche 005398
Baton Rouge Gallery, Baton
  Rouge 046812
Batra Art Gallery, Mumbai 109419
Båtsamlingarna på Bassholmen,
  Uddevalla 041097
Båtsmanskasernen, Karlskrona . 040738

Batsto Village, Hammonton 049335
Batstone, Hilary, London 089946
Batta, Gio, Brescia 109896
Battaglia-Merkle, Maria, Ettlingen 075717
Battaglia, Antonio, Milano 110264
Battaglia, Cristina, Torino 142847
Battaglia, Francesco, Troissereux 073888
Battaglia, Gaetano,
  Castelmaggiore 080017
Battaglia, Riccardo, Roma 131826
Battais, Fougères (Ille-et-Vilaine) 068593
Battais, Francis, Fougères (Ille-et-
  Vilaine) 068594
Battais, Gilles, Dinan 068349
Batteau, Yves, Tourcoing 073815
Battei, Parma 137713, 137714
Batten, John, Hong Kong 102060
Battenberg-Verlag, in der H. Gietl
  Verlag & Publikationsservice GmbH,
  Regenstauf 137483
Batten's Jewellers, Bridport . . . 088606,
  134147
Batters Cage, Indianapolis 093861
Battifoglio, Giacomo, Monaco . . 111994
Battistella, Aldo, Luneray 069664
Battistoni, Giorgio, Verona 081900
Battle Abbey, Battle 043532
Battle and Farm Museum,
  Naseby 045314
Battle Museum, Kibbutz Gesher 025131
Battle Museum of Local History,
  Battle 043533
Battle of Lexington State Historic Site,
  Lexington 050257
Battle Shop, Oostende 063813
Battle, David, Umberleigh 135170
Battlefield House Museum, Stoney
  Creek 007076
Battlefield Military Antiques,
  Minneapolis 094901
Battles Museums of Rural Life,
  Girard 049102
Battlesbridge Antique Centre,
  Battlesbridge 088408
Battlesbridge Motorcycle Museum,
  Battlesbridge 043535
Battleship Cove – Maritime Heritage
  Museums, Fall River 048686
Battleship North Carolina,
  Wilmington 054442
Battleship South Dakota Memorial,
  Sioux Falls 053352
Battleship Texas, La Porte 050048
Battolo, Luigi, Genève 087453
Battou, Remiremont 072412
Baturins'kyj Deržavnyj Istoričnyj
  Zapovidnyk Get'mans'ka Stolicja,
  Baturin 043115
Batzer & Co., Roskilde 140650
Bau, København 065853
Bau-Bau, Amsterdam 112159
Bau de Antiguidades, Rio de
  Janeiro 064025
Bau-Xi Gallery, Toronto 101576
Bau-Xi Gallery, Vancouver 101743
Baublis, Warszawa 084707
Bauch, Hannover 141909
Bauché, Dominique, Le Pont-Chrétien-
  Chabenet 069333
Bauchet, Christian, Clermont-
  Ferrand 068014
Baudach, Guido W., Berlin 106168
Baudelaire, Antwerpen 100484
Bauder, M., Hüffenhardt 076582
Baudet, Eric, Le Havre 103943
Baudet, Laure, Toulouse 073740
Baudiere, Max, Combaillaux . . 068108
Baudin, Jean-Paul, Gimeux . . . 068666
Baudouin, René, Chauconin-
  Neufmontiers 067929
Baudoux-Boitel, Francis,
  Compiègne 068113
Baudouy, Toulouse 129295
Baudry, Villeurbanne 074187

Bauer, Kevelaer 076767
Bauer, Portland 123598
Bauer, Sedan 125814
Bauer & Abraham, Béziers . . . 125516
Bauer & Leible, Trödelmarkt,
  Renchen 078204
Bauer-Bornemann, Bamberg . . 129438
Bauer, Christian, Schruns 127870
Bauer, Christophe, Paris 128930
Bauer, Dominique, Charleville-
  Mézières 125550
Bauer, E., Kleinostheim 076818
Bauer, Gregor, Leipzig 130220
Bauer, Gretel, München 142143
Bauer, Harald & Marie-Luise,
  Beltheim 074715
Bauer, Harry, Hamburg 107227
Bauer, Inge & Aloys, Landkern . 107787
Bauer, Jens-H., Hannover 076363,
  107347, 137266, 141910
Bauer, Max, Sankt Gallen 116607
Bauer, Michael, Wien 140061
Bauer, Peter, Gmund am
  Tegernsee 076012
Bauer, R.M., Itzstedt 076636
Bauer, Robyn, Paddington,
  Queensland 099541
Bauer, Wolfgang, Bad Aussee . . 062521
Bauerfeind, Hartmund, Berg, Kreis
  Ahrweiler 074724
Bauerle, Jimmy, København . . . 065854
Bauermeister, Olivier, Neuchâtel 087701
Bauern- und Handwerker-Museum,
  Malgersdorf 020240
Bauern- und Heimatmuseum
  Haigermoos, Sankt Pantaleon . 002687
Bauern- und Waldmuseum,
  Haidmühle 018924
Bauern-Technik-Museum, Steyr-
  Gleink 002788
Bauerngerätemuseum des
  Stadtmuseums, Stadtmuseum
  Ingolstadt, Ingolstadt 019406
Bauernhaus Habach, Eppelborn . 018217
Bauernhaus-Museum, Bielefeld . 017373
Bauernhaus-Museum, Lindberg . 020800
Bauernhaus-Museum, Muttenz . 041785
Bauernhaus-Museum,
  Ruhmannsfelden 021545
Bauernhaus-Museum, Wolfegg . 022728
Bauernhaus-Museum Bortfeld,
  Braunschweigisches Landesmuseum,
  Wendeburg 022559
Bauernhausmuseum, Grafenau, Kreis
  Freyung-Grafenau 018785
Bauernhausmuseum, Steinen . . 021983
Bauernhausmuseum Amerang des
  Bezirks Oberbayern, Amerang . 016573
Bauernhausmuseum des Landkreises
  Erding, Erding 018232
Bauernhausmuseum Gebersheim,
  Leonberg 020042
Bauernhausmuseum Hattingen,
  Hattingen 019097
Bauernhausmuseum Hof Haina,
  Biebertal 017369
Bauernhausmuseum Hohenstein,
  Hohenstein 019310
Bauernhausmuseum im Bruckmann-
  Haus, Bad Füssing 016771
Bauernhausmuseum Knechtenhofen,
  Oberstaufen 020928
Bauernhausmuseum Schniderlihof,
  Oberried 020916
Bauernhofmuseum, Wagrain . . . 002883
Bauernhofmuseum Hof,
  Kirchanschöring 019618
Bauernhofmuseum Jexhof,
  Schöngeising 021709
Bauernhofmuseum Schreiner,
  Buchhofen 017654
Bauernhausmuseum, Weinstadt 022524
Bauernkriegsmuseum Kornmarktkirche,
  Mühlhäuser Museen, Mühlhausen/

Thüringen 020483
Bauernmöbelmuseum Hirschbach,
Hirschbach 002085
Bauernmöbelmuseum Propstkeusche,
Malta 002351
Bauernmuseum, Dahlen . . . 017802
Bauernmuseum, Kalladorf 002154
Bauernmuseum, Simmerath . . 021866
Bauernmuseum, Zahna 022841
Bauernmuseum Althuus, Gurbrü 041575,
041576
Bauernmuseum Blankensee,
Trebbin 022178
Bauernmuseum Bundwerkstadel,
Haigermoos 002041
Bauernmuseum Großklein,
Großklein 002008
Bauernmuseum Grottenhof, Kaindorf
an der Sulm 002153
Bauernmuseum im Tschötscherhof,
Castelrotto 025847
Bauernmuseum Inzigkofen,
Inzigkofen 019417
Bauernmuseum Lanzenkirchen,
Lanzenkirchen 002270
Bauernmuseum Liebenau,
Geising 018610
Bauernmuseum Mühlhausen, Villingen-
Schwenningen 022333
Bauernmuseum Ostdorf, Balingen 016936
Bauernmuseum Osternach, Ort im
Innkreis 002480
Bauernmuseum Pfarrscheuer,
Bösingen 017465
Bauernmuseum Rothenburg,
Rothenburg/Oberlausitz 021513
Bauernmuseum Schwarzbach,
Museum für Volksarchitektur und
bäuerliche Kultur, Schwarzbach 021770
Bauernmuseum Selfklant,
Selfkant 021834
Bauernmuseum Sollingerhof, Maria
Schmolln 002359
Bauernmuseum Voadohuibn, Valle di
Casies 028261
Bauernmuseum Wittstock,
Nordwestuckermark 020835
Bauernmuseum Zabeltitz,
Zabeltitz 022840
Bauernstube Pfahlheim,
Ellwangen 018160
Baugeschichte der ehemaligen
Fürstbischöflichen Residenz Schloß
Neuhaus, Ausstellung im Remter des
Schlosses, Paderborn 021058
Baugeschichtliches Archiv der Stadt
Zürich, Haus am Rech, Zürich . 042190
Baugh, J.Earl, Louisville 135750
Bauhaus-Archiv, Museum für
Gestaltung, Berlin 017077
Bauhaus Art Gallery, Brisbane . 098779
Bauhaus Books, Seattle 145575
Bauhaus Design, Oklahoma City 095950
Bauhaus-Museum, Klassik Stiftung
Weimar, Weimar 022489
Bauhausbauten Dessau, Meisterhäuser,
Dessau 017863
Bauhofer, Prien am Chiemsee . 078094
Bauhofer, Monika, Prien am
Chiemsee 108414
Baukunst Galerie, Köln 107591
El Baul, Logroño 086071
El Baúl de La Abuela, Maracaibo 097973
El Baúl de Las Pulgas, Bogotá . 065189
El Baúl de los Recuerdos, A
Coruña 143739
El Baul de los Recuerdos, Sevilla 086442
La Baule Auction, La Baule-
Escoublac 125600
Il Baule Blu, Venezia 081825
Baum Gallery of Fine Art, University of
Central Arkansas, Conway . 047966
Baum, Frank, Meiningen 077326
Baum, Jan, Los Angeles 121442

Baum, Jayne H., New York 122390
Baum, P., Kirchzarten 076813
Baum, Timothy, New York 122391
Bauman, New York 145325
Bauman, D., Philadelphia 145407
Baumann, København 065855
Baumann & Städeli, Zürich 087939
Baumann, Dr. Wolfgang,
Regensburg 078163
Baumann Nachf., H., Varel 078779
Baumann, Hansjörg, Baar 133627
Baumann, Helmut, Trendelburg . 078697
Baumann, Irmtrud, Monschau . . 108028
Baumberger-Sandstein-Museum,
Havixbeck 019116
Baumbusch, E., Detmold 075338
Baume, Poitiers 072219
Baume, Hervé, Avignon 066873
Baumeister, Nürnberg . 077814, 108293
Baumeister, Lothar, Abenberg . . 074349
Baumgärtel, Mike, Schönheide . 142349
Baumgaertner, François & Nicole,
Inneneim 068871
Baumgart, Torsten, Ratingen . . . 078133
Baumgartl, Andreas, München . . 108054
Baumgartner, Chicago 135401
Baumgartner & Grodecki, Wien . 127903
Baumgartner & Verdross, New
York 095257
Baumgartner, Herbert, Klagenfurt 062670
Baumgartner, Katharina, Salzburg 062786
Baumkotter, Nicholas, London . . 089947,
118207, 134593
Baur & Baur, Graz 062581
Baur, Erich, Nidau 087710
Baur, M. Johannes, München . . 130326
Baury, Raymond, Saint-Hilaire-Saint-
Florent 072829
Bausback, Franz, Mannheim . . . 077252
Bauscher, Traudl, Potsdam 108395
Bause, S., Usedom 078770
Bauskas Novadpētniecības un Mākslas
Muzejs, Bauska 030102
Bauskas Pils Muzejs, Bauska . . 030103
Bauspar-Museum im Georg-Kropp-
Haus, Wüstenrot 022807
Bautahaugen Samlinger, Hedalen 033646
Baute, Daniel, Jurancon 068926
Baute, Daniel, Pau 072033
Bautista, Alba, Bogotá 102283
Bavastro, Montevideo 127109
Baveux Le Pennec, Claire & Didier,
Perriers-sur-Andelle 072098
Baviera, Silvio R., Zürich 116757
BAWAG Foundation, Wien 002946
Bawden, Keith, Cheltenham . . . 134233
BAX Antykwariat, Lubin 084621
Baxeras, Magdalena, Barcelona . 114862
Baxter, Archerfield 061004
Baxter, Paris 104548
Baxter House Museum, Gorham . 049163
Baxter Springs Heritage Center and
Museum, Baxter Springs 046830
Baxter Warner, Philadelphia . . . 096083
Baxter, B., Indianapolis 127259
Baxter, Chang & Patri, San
Francisco 124229
Baxter & Sons, H.C., London . . . 089948
Bay Area Discovery Museum,
Sausalito 053170
Bay Area Museum, Seabrook . . 053221
Bay Auctions, Tauranga 126325, 126326
Bay Books, Bay Village 144992
Bay Bookshop, Colwyn Bay . . . 144350,
144351
Bay City Books, Los Angeles . . . 145212
Bay of Islands Gallery, Paihia . . 112980
Bay of Spirits Gallery, Toronto . . 101577
Bay Sala, Bilbao 114980
Bay Tree Antiques, Wilton,
Salisbury 091853
Bay Tree Gallery, Bradford-on-
Avon 117412
Bay Tree Gallery, Holywood . . . 117989

Bay View Historical Museum, Bay
View 046833
Bay, Jørgen, Odense . . 066030, 128400
Bayanihan Folk Arts Museum,
Manila 034523
Bayard Antique Village,
Jacksonville 093966
Bayarré, Fabrice, Paris 104549, 141045
Bayat, Paris 071025, 128931
Bayat Inuit Gallery, Winnipeg . . 101868
Baycrafters, Bay Village 046834
Baycrest Heritage Museum,
Baycrest Centre for Geriatric Care,
Toronto 007160
Bayer, Bietigheim-Bissingen . . . 106500
Bayer & Bayer, Praha 102652
Bayer, Gerhard, Baden-Baden . . 074631
Bayerische Akademie der Schönen
Künste, München 055645
Bayerische Numismatische
Gesellschaft, München 058733
Bayerische Staatsbibliothek,
München 020507
Bayerische
Staatsgemäldesammlungen,
Direktion, München 020508
Bayerisches Armeemuseum,
Ingolstadt 019407
Bayerisches Brauerei- und
Bäckereimuseum Kulmbach,
Kulmbach 019828
Bayerisches Eisenbahnmuseum,
Nördlingen 020813
Bayerisches Moor- und Torfmuseum,
Torfbahnhof Rottau, Grassau . 018797
Bayerisches Münzkontor,
Waldaschaff 078832
Bayerisches Nationalmuseum,
München 020509
Bayerisches Schulmuseum,
Bayerisches Nationalmuseum
München, Ichenhausen 019373
Bayerisches Strafvollzugsmuseum,
Kaisheim 019483
Bayerl, Christoph, Mariazell . . . 127816
Bayet Aisha Fahmi (Mogamaa el-
Fonoon), Cairo 010532
Bayet, Max, Villeurbanne 074188
Bayin Guoleng Zhou Museum,
Kuerle 007990
Bayle Museum, Bridlington 043724
Baylis, Chris, Woodstock 091928
Bayliss, D., A.B. & N., Ludlow . . 090643
Baynton-Williams, Arundel 144221
Bayntun, George, Bath 134072, 144227
Bayo, Vitoria-Gasteiz 086570
Bayou Bend Collection, Houston 049572
Bayram, Ali, Singen 078476
Bayreuther FeuerwehrMuseum,
Bayreuth 016985
Bayreuther Münzhandlung,
Bayreuth 074691
Bayreuther, Michael, Wörth . . . 079046
Bayser, Bruno de, Paris 104550
Baysgarth House Museum, Barton-
upon-Humber 043505
Bayside Antiques, Oakland 095883
Bayside Antiques Centre,
Hampton 061588
Bayside Books, Wynnum 139967
Bayside Galleries, Fukuoka 111248
Bayside Galleries, Yokohama . . 111532
Bayside Treasures, Toledo 124681
Bayview Galleries, Mona Vale . . 099362,
127641
Bayview Gallery, San Francisco . 097187
Baz, Antoine Rovi, Paris 071026
Bazaar, Cairo 066232
Bazaar, Prato 081166
Bazaar, Whitby 091824
Bazalar, Lima 084441, 132804
Bazan, Dordrecht 112432
Bazar – Starožitnosti, Pardubice 065443
Bazar Antik, Praha 065452

Bazar Bizar, Groningen 083278
Bazar de Antiguedades Cuauhtemoc,
Monterrey 082487
Bazar de Belen, Guadalajara . . . 082463
Bazar de las Oportunidades, Cali 065246
Il Bazar di Milú, Cagliari 079975
Bazar el Escarabajo, México . . . 082471
Bazar King David, Rio de Janeiro 064026
Bazar Kny, Praha 065483
Bazar na Dvoře, Brno 065344
Bazar Odessa, Monterrey 082488
Bazar San Javier, Monterrey . . . 082489
Bazarez, Hiroshima 111258
Bazarka Antik, Brno 065345
Bazille, Anne-Marie, Bagnoles-de-
l'Orne 066912
Bazin, Nîmes 070687
Bazin, Paris 071027
Bazin, Jean-Marie, Jouy (Eure-et-
Loir) 068921
Bazin, Pascal, Brécey 067454
Baznytinio Meno Skyrius, Kauno
Arkivyskupijos Muziejus,
Kaunas 030350
Bazza, Anthony, New York 135889
Bazzani, Verona 142886
Bazzanti & Marinelli, Firenze . . . 110033
Bazzicalupo, Claudia, Parma . . . 131720
Bazzini, Maurizio, Firenze 131123
Bazzolo, Stefania, Torino 081654
Bazzoni, Angelo, Aosta 079745
BB International Fine Arts, Pfäffikon
(Schwyz) . . . 116562, 138080, 144119
BBC Books, London 138196
BBC Homes and Antiques, Bristol 139153
BBD Galerie, Amersfoort 112119
BBK-Galerie, Oldenburg 108336
BBK Torhaus-Galerie,
Braunschweig 058734, 106558
BBK Ulm e.V. – Künstlerhaus,
Ulm 108799
BBR Auctions, Barnsley, South
Yorkshire 126809
BC Antiques, Singapore 085373
BC Kunstverlag GmbH, Wuppertal 137565
BCI Fine Art, Johannesburg 085646
BCN Art Directe, Barcelona 137952
Be-Bops, San Antonio 096949
Be Framed, Woollen Mills 109711
Be Precedento, Vilnius . 082341, 142963
Be.Ma., Milano 080436
Béacco, Toni, Nazelles-Négron . 128842
Beach Antique, Toronto 064607, 140399
Beach Antiques, Clevedon 088929
Beach Art, Thirroul 099820
Beach Books, Jacksonville 145183
Beach Gallery, Virginia Beach . . 124824
Beach Street Gallery, San
Francisco 124230
Beaches Antique Gallery,
Jacksonville 093967
Beaches Art and Frame Gallery,
Jacksonville 121236
Beaches Arts Center,
Jacksonville 121237
Beaches Heritage Centre,
Eastport 005739
Beachhead Gallery of Contemporary
Art, Warrnambool 099863
Beachville District Museum,
Beachville 005402
Beacon Hill Antiques, Frankfurt am
Main 075773
Beacon Historical Society Museum,
Beacon 046840
Beacon Street Gallery, Chicago . 120064
The Beacon, Whitehaven,
Whitehaven 046120
Beaconsfield Historic House, Prince
Edward Island Museum and Heritage
Foundation, Charlottetown . . . 005591
Bead Art, Amsterdam 112160
The Bead Goes On, Toronto . . . 064608
Bead Harbor, Boston 092446

The Bead Museum, Glendale .. 049114
Beagdruck, Verlag Die Region,
   Emmenbrücke 138040
Beagle Books, Sway 144876
Beajaim, Pontevedra .. 086399, 086400
Beaker, Tampa 124633
Beakys, Tuggeranong 139936
Bealey, Christchurch 112834
Beall, Miami 094708
Beall Center for Art and Technology,
   Claire Trevor School of the Arts,
   University of California, Irvine . 049720
Béalu & Fils, J.M., Paris 071028
Bealu, Jean-Jacques, Blanzay .. 067238
Beam, Nijmegen 112601
Beam Ends Galleria, Hartley
   Wintney 117921
Beaman, Atlanta 092114
Beaminster Museum, Beaminster 043537
Beamish, the North of England Open
   Air Museum, Beamish 043538
Bean Row Galleria, Hathersage . 117930
Bean, S.A., San Antonio 096950
Beaners, Houston 093621
Beanie Babies Collector's Shop, Las
   Vegas 094121
Beanie Planet, El Paso 093387
Bear Butte State Park Visitors Center,
   Sturgis 053585
Bear Creek Glass, Birmingham . 119791
Bear Island Bokks, Cardiff 144308
Bear Mountain Trailside Museums
   Wildlife Center, Bear Mountain 046843
Bear Museum, Petersfield 045493
Bear Paths, Cleveland 120402
The Bear Press, Bayreuth 137043
Bear Steps Antiques, Shrewsbury 091323
Bearclaw Gallery, Edmonton ... 101277
Beard Gallery, Norton 051642
Beard Outlet, Portland 123599
Beardall, Clive, Maldon 090679, 134826
Beardsmore, London 118208
Beare, John & Arthur, London .. 089949,
                          134594
Bearly Collectibles, Milwaukee . 094834
Bearly Read Books, Portland ... 145447
Bearly Trading of London,
   London 089950
Béarn, Pau 072034
Bearne, Exeter 126892
Bear's Collection, Indianapolis .. 093862
Bears, A., Yackandandah 062494
Bea's Silver Hummingbird,
   Albuquerque 119460
Beasley, P.T., Farningham 089265
Beast Creations, Flinders 099040
Beat, Sacramento 123761
Beate Uhse Erotik-Museum,
   Berlin 017078
Beatles Museum, Halle, Saale . 018952
Beatles-Museum Little Cavern,
   Biebelnheim 017366
Beatles Museum Remember,
   Gemmerich 018633
The Beatles Story, Liverpool ... 044850
Beatrice M. Haggerty Gallery,
   University of Dallas, Irving .. 049723
Beatričes Grincevičiūtės Memorialinis
   Muziejus, Beatričes Namai,
   Vilnius 030511
Beatrix Potter Gallery, The National
   Trust, Hawkshead 044546
Beatriz Antiguedades, Getxo .. 086007
Beatties Historic Photograph Museum,
   Dodges Ferry 001030
Beatty, Darlinghurst 098943
Beatty Museum, Newport Pagnell 045367
Beau Fort Plantation Home,
   Bermuda 046946
Beau-Gest, L'Hôtellerie 069456
Beau Matthews, Philadelphia .. 096084
Beau Nash Antiques, Tunbridge
   Wells 091646
Beau Village Museum, Saint

Victor 006940
Beau, Didier, Sceau-Saint-Angel 073456
Beauarts Studios, Chicago 120065
Beaubourg, Vence 105827
Beaubourg, Gana, Paris 104551
Beaubreuil, Guy, Feytiat 068529
Beauchamp & Beauchamp,
   Québec 101489
Beauchamp Newman Museum,
   Elizabeth 048524
Beaucourt, Eric, Maulette 070001
Beaucousin, Catherine, Louviers 069650
Beaudesert Museum, Beaudesert 000835
Beaudet, Royan 072625
Beaudoin, Philippe, Romorantin-
   Lanthenay 072527
Beaudran, Pierre, Mirande 125644
Beaufield Mews Antiques,
   Stillorgan 079702
Beaufils, Serge, Thorigné-
   Fouillard 073671
Beaufort County Arts Council,
   Washington 060404
Beaufort Historic Site, Beaufort . 046846
Beaufort Museum, Beaufort 046848
Beaufort West Museum, Beaufort
   West 038651
Beaufort, Robert, Chauray 067944,
                          128553
Beaufour, Pierre-Louis, Dieppe . 068297
Beauger, Jacques, Moulins
   (Allier) 070369
Beaugrand Kulturkonzepte,
   Bielefeld 106487
Beaujean, Jean-Pierre, Chaumont-en-
   Vexin 067938
Beaujeu, Anne de, Bruxelles ... 063348
Beaulé, Montréal 064394
Beaulieu Abbey Exhibition of Monastic
   Life, Brockenhurst 043764
Beaulieu Fine Arts, Brockenhurst 117460
Beaumarie, Jean, Ville-d'Avray . 074128
Beaumaris Castle, Beaumaris .. 043541
Beaumaris Gaol and Courthouse,
   Beaumaris 043542
Beaumel, Gérard, Cajarc 067600
Beaumont, Antwerpen 063193
Beaumont, Montréal 064395
Beaumont Antique Furniture, Hamilton,
   New South Wales 061578
Beaumont Art League, Beaumont 046854,
                          060405
Beaumont Hall, Saint Albans .. 119002
Beaumont, Jean, Die 068294
Beaumont, Jean, Saumur 073433
Beaumont, Odile, Mussey-sur-
   Marne 070403
Beaumont, Robin de, London .. 144560
Beaumontpublic & Königbloc,
   Luxembourg 111835
Beauport-Sleeper-McCann House,
   Gloucester 049132
Beaupuis, Hubert de, Saint-Victor-
   l'Abbaye 073327
Beauquis, Philippe, Annecy 066638
Beauregard-Keyes House, New
   Orleans 051286
Beaussant & Lefèvre, Paris 125679
Les Beautés, Québec 064525
Beautiful Memories, Ocho Rios . 081957,
                          111237
Beautiful Plains Museum,
   Neepawa 006456
Beautiful Tibet, New York 095258
Beauty and the Beach,
   Mornington 099376
Beauty Antique, Singapore ... 085374
Beauty Galleria, Los Angeles .. 121443
Beauvais Carpets, New York ... 095259
Beauvais Heritage Center, Clark 047770
Beauvais, Philippe de, Paris ... 071029
Beauvais, Séverine,
   Montmorency 070277
Beauvoir, Moustiers-Sainte-Marie 070376

Beauvoir, Nice 070573
Beauvoir, Jefferson Davis Home and
   Presidential Library, Biloxi 046990
Beaux Arts, Bath 117299
Beaux-Arts, Dallas 093073, 145110
Beaux-Arts, London 118209
Beaux Arts Gallery, Miami 121746
Beaux Arts Magazine, Paris 138634
Les Beaux Jours, Rouen 072563
Beaver, Deakin 098958
Beaver Bros., San Francisco ... 097188
Beaver Book, Atlanta 144962
Beaver Coin Room, London ... 089951
Beaver Corner, Gent 063629
Beaver Island Historical Museum, Old
   Mormon Print Shop Museum, Marine
   Museum and Protar Home, Beaver
   Island 046863
Beaver River Museum, Beaverton 005408
Beaver Valley Military Museum,
   Clarksburg 005624
Beaverbrook Art Gallery,
   Fredericton 005837
Beaverhead County Museum,
   Dillon 048329
Bebb, Ludlow 118672, 134820
Bébert, F. Tabatabai Asbaghi,
   Rotterdam 137869
Bebes & Jouets, Edinburgh 089159,
                          134357
Bebob, Amsterdam 112161
Bebop Antik, Göteborg 086699
Beca Fine Arts, Wellington 084159
Bécassine, Arnhem 082933
Beccaletto, Cristiano, Milano .. 131381
Beccaria & C., Carlotta, Milano . 131382
Beccia, Fernanda, Milano 080437
Becciolini, Paolo, Bagno a Ripoli 130902
Beccles and District Museum,
   Beccles 043544
Beccles Gallery, Beccles 117321, 134083
Beceren, Zürich 087940
Bech, København 065856, 102862
Bechman, København 102863
Bechtold, Jeroen, Amsterdam .. 112162
Bechtold, U., Pforzheim 142302
Béchu, Jean-Baptiste, Nantes .. 128833
Bechu, Johan, Blois 067250
Beciani, A., Charleroi 100677
C.H. Beck, München 137389
Beck, Shepparton 139908
Beck, Tampa 124634
Beck & Eggeling International Fine Art,
   Düsseldorf 106762
Beck & Eggeling New Quarters,
   Düsseldorf 106763
Beck & Nalley, Sacramento 096604,
                          145482
Beck Cultural Exchange Center,
   Knoxville 050004
Beck Isle Museum of Rural Life,
   Pickering 045502
Beck jr., Arthello, Dallas 120560
Beck, Astrid & Manfred,
   Überherrn 078745
Beck, G., Lörrach 077142
Beck, Heinz, Meilen .. 087667, 133856
Beck, James, Fakenham 126894
Beck, K., Sontheim an der Brenz 108645
Beck, Markus, Kirchheim unter
   Teck 076806
Beck, Monika, Homburg 107446, 137283
Beck, Roland M., Nürnberg 077815
Beck, Thomas L., Jacksonville . 135647
Beckel-Odille-Boïcos, Paris 104552
Beckendorf, Losheim am See .. 107894
Beckenham Antique Market,
   Beckenham 088420
Becker, Biebrich 106482
Becker, Cannes 067628
Becker, Lüchow 077195
Becker, Reinbek 078185
Becker, Vanves 073974
Gebr. Becker, Zeist 083706

Becker & Scheloske, Bückeburg 141661
Becker Anno 1902, Haarlem ... 143156
Becker County Historical Museum,
   Detroit Lakes 048318
Becker-Kater, Vechta 108812
Becker, Albert Z., Saint Louis .. 123807
Becker, Christiane, Krefeld 130191
Becker, Dirk, Hamburg 076194
Becker, Emanuelle, Luxembourg 082366
Becker, Eugène, Paris 071030
Becker, Fred, Luxembourg 111836
Becker, Hans Robert,
   Westergellersen 078945
Becker, Horst, Bremen 075149
Becker, Ingeborg, Hannover ... 141911
Becker, J.H., Zeist 083707
Becker, J.S., Amsterdam 082697
Becker, Jürgen, Hamburg 107228
Becker, Larry, Philadelphia 123357
Becker, Michael, München 130327
Becker, Rita & Dagmar, Münster 077654
Becker, Robert, Badem 074627
Becker, Susan, Walton-on-
   Thames 091721
Becker, Thomas, Siek 078468
Becker, Ulrich, Würzburg 142510
Becker, Wolfgang, Pritzwalk ... 130542
Beckers, Eindhoven 083207
Beckers Books, Houston 145162
Beckers, E. & L., Burscheid ... 075233
Beckers, Harald, Jüchen 130085
Beckert, Wolfgang, Aschheim .. 074445
Beckett, Toronto 101578
Beckford's Tower and Museum,
   Bath 043516
Beckham, New Orleans 145301
Beckholmens Förlag, Stockholm 115839,
                          138013
Beckie's Antiques, Dallas 093074
Beckmann, Edemissen 075565, 129771
Beckmann, Toledo 124682
Beckmann, Erich, Hannover ... 076364
Beckmann, Marlies, Rheine 108497
Beckoning Hills Museum,
   Boissevain 005433
Beckschäfer, H., Arnsberg 074423
Beckstones, Greystoke 117890
Beckwith & Son, Hertford 089517,
                          134459
Becky's Old Wares, Rivett 062153
Becq, Pierre, Metz 070090
Becq, Pierre, Senon 073494
Becquer-Rastrillo, Madrid 086130
Becuadro, Madrid 115155
Bed Bazaar, Framlingham 089301
Al-Bedaa, Dubai 088172
Bedale Museum, Bedale 043546
Bedale, David, Mobberley 090763
Le Bedane, Bruxelles 128019
Bede Art Gallery, Mount Marty College,
   Yankton 054568
Bedel, Pleurtuit 072184
Bedel de Buzareingues, Bertrand,
   Paris 128932
Bede's World, Jarrow 044662
Bedford, Bilbao 085864
Bedford Central Library Gallery,
   Bedford 043549
Bedford City/County Museum,
   Bedford 046872
Bedford Gallery, Johannesburg . 114790
Bedford Gallery, Dean Lesher
   Regional Center for the Arts, Walnut
   Creek 054043
Bedford Historical Society Museum,
   Bedford 046870
Bedford Museum, Bedford 043550
Bedford, P.E.L., Bridport 088607
Bedfordshire and Hertfordshire
   Regiment Association Museum
   Collection, Wardown Park Museum,
   Luton 045166
Bedi, Margaret, Billingham 088464
Bédier, Claude, Carqueiranne . 067697

Bedingfield Inn Museum,
  Lumpkin 050526
Bednarczyk, C., Wien 062906
Bedos, Raphaël, Paris 071031
Bedot, André, Tarascon (Bouches-du-
  Rhône) 073619
Bedouin Market, Cairo 103023
Bedrijfsmuseum NUON-ENW
  Amsterdam, Amsterdam ... 031930
Bedriye Kazancıgil, İstanbul 088105
Bedsteads, Bath 088381, 134073
Bedwell, Peter, Prahran 062107
Bee, New Orleans 122128
Bee & Bear, Nagoya 142904
Bee Family Museum, Yuanshan . 042604
Bee, Kay, Boston 092447
Beech House, Sheffield 091279
Beeches & Rale, Hawthorn 061600
Beechey, Neville, Colac 061355
Beechmountain Art Gallery,
  Beechmont 098718
Beechwold, Columbus 145094
Beechwood Gallery, Newport,
  Gwent 118798
Beechworth Antiques,
  Beechworth 061112
Beechworth Curiosity Shop,
  Beechworth 061113
Beechworth Gallery, Beechworth 098719
Beechworth Historic Court House,
  Beechworth 000837
Beechworth Past and Present,
  Beechworth 061114
Beecroft Antiques, Beecroft 061117
Beecroft Art Gallery, Southend-on-
  Sea 045832
Beecroft Treasure House,
  Beecroft 061118
Beedham Antiques, Hungerford . 089600
Beehive, Salt Lake City 096880
Beehive Art Space, Albury 098624
Beehive House, Salt Lake City . 052910
Beehive Old Wares, Maldon 061764
Beek, Johan, Barneveld 143089
Beekhuizen, Jan, Amsterdam .. 082698
Beekmans en Waning, Haarlem 083314,
  132514,   143157
Beeld en Aambeeld Galerie,
  Enschede 112475
Beeld in Beeld, Nijmegen 112602
Beeldentuin Belling Garde,
  Bellingwolde 032056
De Beeldenwinkel, Amsterdam . 112163
Beeldlijn, Amersfoort 112120
Beenleigh and District Museum,
  Beenleigh 000842
Beer Collectables, Beer 088428
De Beer Studios, London 134595
Beer, Franz X., Auerbach, Kreis
  Deggendorf 129391
Beer, Franz X., München 077465,
  130328
Beer, Hans-Peter, Québec 064526,
  128194
Beer, Mike, Düsseldorf 129733
Beer, Mike, Köln 076851, 130140
Beer & Co., Monika, Wien 062907
Beerents, Anouk, Amsterdam . 082699,
  132390
Beerewein Museum, Eulenbis .. 018326
Beerhorst, Ursula, Karlsruhe . 130094
Be'eri Archaeological Collection,
  Be'eri 025025
Beers, Sacramento 145483
Bees Antiques, Woodstock 091929
Beesley & Pace, Salt Lake City . 145454
Beesley's, D.T., Paignton 127007
Beeston Castle, Tarporley 045948
Beethoven Emlékmúzeum,
  MTA Zenetudományi Intézete,
  Martonvásár 023770
Beethoven Eroicahaus, Wien Museum,
  Wien 002947
Beethoven-Gedenkstätte Floridsdorf,

Wien 002948
Beethoven-Haus Bonn, Verein
  Beethoven-Haus, Bonn 017483
Beethoven Heiligenstädter Testament,
  Wien Museum, Wien 002949
Beethoven Pasqualatihaus, Wien
  Museum, Wien 002950
Beethovenhalle, Bonn 017484
Beethovenhaus, Krems 002231
Beethovenhaus "Haus der Neunten",
  Baden bei Wien 001771
Beetles, Chris, London 089952, 118210
The Beetroot Tree, Draycott ... 117686
Beetz, Kurt, Lübeck .. 077172, 130253
Beeville Art Museum, Beeville .. 046873
Bee'Z Arts, Lauzerte 103923
Before My Time, Toronto 064609
Befreiungshalle Kelheim, Bayerische
  Verwaltung der staatlichen Schlösser,
  Gärten und Seen, Kelheim ... 019574
Le Befroi et le Musée des Frères
  Caudron, Rue 015187
Bega Valley Regional Gallery,
  Bega 000844
Began, São Paulo 064158
Begas, G., Heerlen 112522
Begegnungsstätte Alte Synagoge,
  Wuppertal 022815
Begemann, Karen, Hamburg ... 129950
Begg, Jan, Stanmore 127679
Begijnhofkerk, Provinciaal Centrum
  voor Cultureel Erfgoed, Sint-
  Truiden 004026
Begijnhofmuseum, Dendermonde 003537
Begijnhofmuseum, Herentals .. 003694
Begijnhofmuseum, Kortrijk 003755
Bégin Edition, Lévis 101330
Begleiter, Baltimore 135337
Bégné, Jean-Luc, Bordeaux ... 067313
Begoña Alcorta Egaña, Donostia-San
  Sebastián 085968, 085969
Begsteiger, Günther, Linz 062713
Béguelin, Vérena, Saint-Imier .. 087769
Begüm, İstanbul 088106, 088107
Beguet, Yves, Saintes 125811
The Beguiling, Toronto 140400
Béguinage d'Anderlecht,
  Bruxelles 003433
Begunk, J., Stendal 078559
Běhal, Ilja, Olomouc 128331
Behar, Renée, São Paulo 064159
Behavka, Karl, Wien 062908
Behind the Boxes – Art Deco,
  London 089953
Behm & Brock, Toronto 064610
Behncke-Brahmer, D., Hamburg 107229
Behnhaus/ Drägerhaus, Die Lübecker
  Museen, Lübeck 020154
Behnke, Frank, Ahrensburg .. 141400
Het Behouden Blik,
  Uithuizermeeden 032938
Behr, Örebro 086897
Behr, A.A., Kaiserslautern 107492
Behr, H.W., Augsburg 129393
Behr, Winfried, Zellingen 079137
Behrendt & Markus, Berlin 129469
Behrendt, Thomas, Erfurt 075636
Behrens, Hannover 130003
Behrens, Leck 077038
Behrens, Sorø 066142
Behrens, Thomas, Leck 077039
Behrens, Thomas, Rastede 078128
Behringer-Crawford Museum,
  Covington 048042
Behringer, Werner, Fürth 125954
Behrmann, Kurt von, Phoenix .. 123484
Behrmann, Patrick, Bremen .. 106575
Behruz, Prahran 062108, 127661
Behzad Museum, Sa'd-Abad Cultural-
  Historic Complex, Tehrän .. 024747
Bei Tomi, Bologna 142625
Beiarn Bygdetun, Moldfjord 033785
Beier, Belgershain 074712

Beier, Hartmut, Leipzig 077057, 142055
Beier, Ulrich & Monika, Berlin .. 125869
Beier, Viola, Köln 130141
Beifang Literature and Art Press,
  Harbin 136715
Beighton, Paul, Rotherham 127023
Beihl, Werner, Bremen 129628
Beijers, Utrecht 126300, 143241
Beijing 798, Chen Ling Hui
  Contemporary Space, Beijing . 101935
Beijing Antique City, Los Angeles 094266
Beijing Antiques Shop, Hong
  Kong 065014
Beijing Art Gallery, Beijing ... 101936,
  101937
Beijing Art Museum, Beijing ... 007591
Beijing Arts and Crafts Publishing
  House, Beijing 136693
Beijing Cao Chang Di, Chen Ling Hui
  Contemporary Space, Beijing . 101938
Beijing Changcheng Hua Ren Huai,
  Beijing 007592
Beijing China Nationality Museum,
  Beijing 007593
Beijing Classic Art, Beijing 064998
Beijing Cloisonne, San Francisco 097189
Beijing Cultural Relic Protection
  Foundation, Beijing 007594
Beijing Dabao Taixi Western Han Tomb
  Museum, Beijing 007595
Beijing Drawing House, Beijing . 101939
Beijing Fine Arts and Photography
  Publishing House, Beijing 136694
Beijing Folklore Museum, Beijing 007596
Beijing Gu Guanxiang Station,
  Beijing 007597
Beijing Imperial Art Academy,
  Beijing 055188
Beijing Katong Art Museum,
  Beijing 007598
Beijing Man Zhoukoudian Museum,
  Beijing 007599
Beijing Museum, Beijing 007600
Beijing Museum of Natural History,
  Beijing 007601
Beijing Telephone History Museum,
  Beijing 007602
Beijing Theater Museum, Beijing 007603
Beijing Wen Shi Art Gallery,
  Beijing 101940
Beijing World Art Museum,
  Beijing 007604
Beik Thano, Yangon 082608
Beil, H., Weilheim Kreis Waldshut 078893
Beiliu Museum, Beiliu 007655
Beim Antiquaire, Machtum 082376
Beiner-Büth, Silke, Hamburg .. 129951
Beirut Antiques Fair, Beirut 098239
Beisac, Hermien de, Wiesbaden 078970
Beisac, R. de, Wiesbaden 078971
Beisler, Weilheim in Oberbayern 142468
Beisler, W., Aschaffenburg 074429
Beissel, R., Bad Honnef 074527
Beissert, Hans-Werner,
  Wildeshausen 108937
Beit Abraham Krinizi, Ramat Gan 025172
Beit al-Ajaib Museum with House of
  Wonders Museum, Zanzibar . 042640
Beit-Alfa Ancient Synagogue, Kibbutz
  Chefzi-bah 025124
Beit Hagdudim, Avihail 025022
Beit-Hagefen Art Gallery, Haifa . 025044
Beit Hameiri, Zefat 025222
Beit Hankin, Kfar Yehoshua ... 025121
Beit Ha'Omanim, Jerusalem ... 025068
Beit Hatefutsoth, Tel Aviv 025189
Beit Lohamei Haghetaot-Yad Layeled,
  Lohamei-Hagetaot 025152
Beit Miriam-Museum, Kibbutz
  Palmachim 025143
Beit-Shean Museum, Beit Shean 025027
Beit Shturman Museum, Kibbutz Ein
  Harod 025128
Al-Beit, Dunia, Riyadh 085315

Beiteddine Museum, Shouf 030282
Beitou Hot Spring Museum,
  Taipei 042500
Beiträge zur Kunst des Christlichen
  Ostens, Recklinghausen ... 138686
Bejannin, Saint-Ouen . 073035, 073036
Bejšovec, Jan, Praha 128337
Beker, Buenos Aires 139553
Bekking & Blitz, Amersfoort ... 137822
Bekko Crafts Museum, Nagasaki 029086
Bekman, Amsterdam 082700
De Bekoring, Gent 063630
Bel Age, Las Vegas 094122
Bel Air Gallery, Bournemouth . 117402
Bel Air Gallery, Sheffield 119077
Bel-Aire Coin, Saint Paul 096813
Bel Art, Novi Sad 114456
Bel Art, Québec 101490
Bel Art Gallery, Vancouver 101744,
  101745
Bel-Arte, Lengnau (Bern) 116433
Bel-Canto Antiek, Hilversum ... 083400
Le Bel Ecrin, Nice 070574
Bel Etage, Wien 062909
Bel Etage, Zürich 087941
Bel Fine Art, New York 095260
Bel Mondo, Karlsruhe 107508
Belair, Villeurbanne 074189
Belair Mansion, Bowie 047151
Belair Stable Museum, Bowie . 047152
Belaja Galereja, Ekaterinburg . 114179
Belaja Luna, Odesa 117156
Béland, D.M., Montréal 101363
Belanger, Michael, Los Angeles . 094267
Belarussian State Academy of Arts,
  Minsk 054993
Belarussian State University of Culture
  & Arts, Minsk 054994
Belasting en Douane Museum,
  Rotterdam 032799
Belaych, Christian, Paris 071032
Belaž, Sankt-Peterburg 114330
Belbezier, Lille 069478
Belcarra Eviction Cottage,
  Castlebar 024844
Belcher Studios, San Francisco . 124231
Belcher, Robert, Droitwich 089109
Belcourt Castle, Newport 051531
Belda, Praha 102653
Beldam Gallery, Brunel University,
  Uxbridge 046025
Belenky Bros., New York 122392
Beleva, Kristina, Sofia 128139
Belfast Book, Belfast 144238
Belfast Historical Society, Belfast 005411
Belfast Traders, Christchurch . 083858
Belforte, Livorno 137661, 142699
Belga Traders, Sint-Niklaas ... 063829
Belgazou, Rouen 072564
Belger Arts Center, University of
  Missouri-Kansas City, Kansas
  City 049878
Belger, Frank, Annaberg-
  Buchholz 074413
Belgorodskij Gosudarstvennyj
  Chudožestvennyj Muzej, Belgorod Art
  Museum, Belgorod 036886
Belgorodskij Gosudarstvennyj Institut
  Iskusstv i Kultury, Belgorod . 056266
Belgorodskij Gosudarstvennyj Istoriko-
  Kraevedčeskij Muzej, Belgorod 036887
Belgrave Antique Centre, Darwen 089031
Belgrave Carpet Gallery, London 089954
Belgrave Gallery, London 118211
Belgrave Gallery, Saint Ives,
  Cornwall 119018
Belgrave Hall and Gardens,
  Leicester 044792
Belgravia, San Giljan 126259
Belgravia Gallery, London 118212
Belhaven Memorial Museum,
  Belhaven 046876
Les Béliers, Cellule 067742
Believe It Or Not Curiosity Museum,

Register der Institutionen und Firmen

Believers, Phoenix – Benchmark Antiques, Bridport

Tamsui 042569
Believers, Phoenix 096192
Belija Kvadrat, Sofia 101148
Belin, D., Detroit 120803
Belin, Étienne, Périgueux 072058
Belinki & DuPrey, Portland 123600
Belinskij Rajonnyj Kraevedčeskij Muzej, Penzenskij gosudarstvennyj obedinennyj kraevedčeskij muzej, Belinskij 036889
Belisle, Maurice E., Denver 093261, 135525
Belize Audubon Society, Belize City 004173
Belize Maritime Terminal and Museum, Belize City 004174
Belkhandi Museum, Belkhandi . 024159
Belkhayat, Casablanca 082549
Belkin, Ronald A., Long Beach . 094194
The Belknap Mill Museum, Laconia 050055
Belk's Refinishing, Charlotte . . . 135388
Bell, Berrima 098739
Bell, Chicago 120066
Bell, Latrobe 061718
Bell, Nashville 122069
Bell, Winchester 091866, 135236
Bell, Woodbridge 144933
Bell & Wyatt, Invercargill 083964
Bell Antiques, Fresno 093491
Bell Antiques, Grimsby 089391
Bell Antiques, Romsey 091136
Bell Antiques, Twyford, Berkshire 091680
Bell County Museum, Belton . . . 046908
The Bell Gallery, Belfast 088432, 117332, 134090
Bell Gallery, Storrington 119154
Bell Homestead Museum Complex, Brantford 005468
Bell House Antiques, Cambridge, Gloucestershire 088767
Bell Island Community Museum, Bell Island 005412
Bell Passage Antiques, Wickwar 091842, 135227
Bell Pettigrew Museum, Saint Andrews 045672
Bell-Roberts Contemporary, Cape Town 114762, 137946
Bell Rock Mill Museum, Bethany 005421
Bell, Anne, Kingston Southeast . 061697
Bell, Darrell, Saskatoon 101542
Bell, J. & H., Castletown 088807
Bell, Mary, Chicago 120067
Bell, Mike, Chicago . . . 092668, 092669
Bell, Paul-J., Falkensee 129831
Bell, Peter, Edinburgh 144384
Bell & Co., Robert, Horncastle . . 126925
Bell, Zillah, Thirsk 119227
Bella, Virginia Beach 124825
Bella Antique, New York 095261
Bella Arte Internacional, Medellín 102428
Di Bella Concetta, Catania 109949
La Bella Epoca, Ciudad de Panamá 084411, 132803
Bella G, Los Angeles 094268
Bella Luna Gallery, Tucson 124717
Bella Vista Gallery, Chicago 120068
Bella Vista Historical Museum, Bella Vista 046877
Bella Vita, Oakland 095884
Bellachioma, Roma . . . 081252, 110733
Bellacorium, Houston 121004
Belladestino, Eskilstuna 115602
Bellaghy Bawn, Magherafelt . . . 045190
Bellagio Gallery of Fine Art, Las Vegas 121343
Bellamy, Dunedin 112874
Bellamy, Fremantle 099060
Bellamy, Wellington 143360
Bellamy Mansion Museum of History and Design Arts, Wilmington . . 054443
Bellange, Stockholm 115840
Bellanger, Nantes 140981
Bellanger, Patrice, Paris 071033, 104553
Bellapais Manastiri, Girne 033458
Bell'Art Gallery, Bellville 114754
Bell'Arte, Maastricht 112568
Bellarte Gallery, Seoul 111616
Bellas Artes, Gijón 115068
Bellas Gallery, Fortitude Valley . 099049
Bella's Market, Saint Louis 096669
Bellavista Gallery, South Auckland 113006
Bellcourt Books, Hamilton, Victoria 139763
Belle & Belle, Paris 104554
Belle Arte, Kraków 084561
Belle Arte, West Hollywood 124987
Belle Chambre, Wallerfangen . . 078853
Belle Cheminée, Dublin 079526
La Belle Chimaera, Rio de Janeiro 064027
Belle Chose, Calvi 067608
Belle Cose, Napoli 080844
Belle Epoque, Alès . . . 066527, 128433
Belle Epoque, Belo Horizonte . . 063885
A la Belle Epoque, Biarritz 067188
Belle Epoque, Bougival 067376
Belle Epoque, Bourges 067422
Belle Epoque, Brest 067465
La Belle Epoque, Châteaudun . . 067879
La Belle Epoque, Jarmenil 068891
A la Belle Epoque, Le Havre . . . 069273
La Belle Epoque, Mainz 077233, 130265
Belle Epoque, Morlaix 070349
La Belle Epoque, München 077466
Belle Epoque, Münster 077655
La Belle Epoque, Néris-les-Bains 070506
La Belle Epoque, Ronchin 072531
La Belle Epoque, Santiago de Chile 064955
Belle Epoque, Toronto 064611
Belle Epoque, Traunstein 078692
Belle Epoque, Vatan 073983
Belle Framing and Gallery, Woollahra 099908
Belle Grove Plantation, Middletown 050868
Belle Ile-en-Mer, Béziers 067166
La Belle Maison, New York 095262
Belle Maison, Porto Alegre 063960
Belle Meade Gallery, Nashville . 122070
Belle Meade Interiors, Nashville 094975
Belle Meade Plantation, Nashville 051173
La Belle Nouvelle Orleans, New Orleans 095059
Belle Starr Antiques, Saint Louis 127426
Belle, H.F., Dieren 132487
Belleau, Christopher R., Providence 123696
Bellecour, San Francisco 097190
Bellecour, De, Lyon 104068
Bellemare, Roger, Montréal 101364
Bellemin, Patrick, Chambéry . . . 067792
Bellenger, Gérard, Elbeuf-sur-Seine 068437
Beller, Dettingen unter Teck . . . 075344
Bellerio Antichità, Locarno 087597, 133811
Belles Années, Saint-Erme-Outre-et-Ramecourt 072743
Belles Choses, Strasbourg 073573
Belles et Bonnes, Trans-en-Provence 073870
Belletti, Piero & Chiara, Torino . 081655
Belleville Area Museum, Belleville 046882
Belleville Public Library and Art Gallery, Belleville 005413
Bellevue Antique Mall, Nashville 094976
Bellevue Art Museum, Bellevue . 046887
Bellevue Gallery, Edinburgh 117716
Bellevue House, Kingston 006096
Bellflower Genealogical and Historical Society Museum, Bellflower . . 046891
Belli, Armando, Bologna 142626
Belliard, F., Rouen 072565
Bellier, André, Nantes 128834
Bellier, Jean-Claude & Yann, Paris 104555
Bellier, Luc, Paris 104556, 104557
Bellin, A. & A., Ueken 087862
Bellinato, Padova 142773
Bellinge, Odense 066031
Bellinger, Rommerskirchen 130595
Bellinger, Carl, Barnet 088359
Bellinger, Katrin, München 077467
Bellingrath Home, Theodore . . . 053728
Bellini, Luigi, Firenze 110034
Bellini, Philip, New York 095263, 135890
Bellini, Tino, Milano 080438
Bellino, Rodolphe, Nice 128851
Bellinzona, Oreste, Milano 110265
Bellissimo, Knutsford 118062
Bellman, C.M., Stockholm 115841
Bellman, John, Billingshurst . . . 126819
Il Bello Ritrovato, Roma 081253, 081254
Bello, Claudio, Roma 131827
Bello, Stefano, Bologna 130936
Belloc, Preignac 072307
Belloc Lowndes, Chicago 120069
Bellomo Gallery, Siracusa 110933
Bellon, Orlando, Trieste 132216
Bellon, Renaud, Luzillé 069675
Bellos, Gelson Carlos, Porto Alegre 063961
Bellotti, Magda, Madrid 115156
Bellou, Guy, Paris 071034, 128933
Bellport-Brookhaven Historical Society Museum, Bellport 046898
Bell's, Christchurch . . . 126310, 126311
Bells Antiques, Christchurch . . . 083859
Belltable Arts Centre, Limerick . 109666
Bellu, Emmanuel, La Châtre . . . 068987
Bellucci, Paolo, Firenze 131124
Belluco, Maurizio, Padova 080964
Bellwether, New York 122393
Belly Art, Brighton, Victoria 127571
Belmarco, Asunción 113328
Belmer Antiquitätenhof, Belm . . 074714
Belmont, Faversham 044337
Belmont & Landon, Richmond . . 096527
Belmont and District Museum, Belmont 005417
Belmont Antiques, Nashville . . . 094977
Belmont Antiques, Norfolk 095846
Belmont Booksellers, Portland . . 145448
Belmont County Victorian Mansion, Barnesville 046791
Belmont, Gari Melchers Estate and Memorial Gallery, Fredericksburg 048951
Belmundo, Zug 116877
Belmundo Galerie, Zürich 116758
Belo Belo, Braga 113955
Beloit Historical Society Museum, Beloit 046904
Belokranjski Muzej, Metliški grad, Metlika 038576
Belorusskij Gosudarstvennyj Muzej Istorii Velikoj Otečestvennoj Vojny, Minsk 003245
Belos Tempos Moveis de Decoracoes, Rio de Janeiro 064028
Belotti, Francesco, Milano 080439
Beloveżskaja Pušča Muzej, Beloveżskaja Pušča 003243
Below Stairs, Hungerford 089601
Belozerskij Istoriko-chudožestvennyj Muzej, Belozersk 036891
Belozerskij Muzej Narodnogo Dekorativno-prikladnogo Iskusstva, Belozersk 036892
Belriguardo, Ferrara 137642
B'Eltaller, Getxo 086008
Beltane, Taden 073611
Belton House, Grantham 044461
Beltramelli, Angeline, Lyon 069706
Beltrami County Historical Center, Bemidji 046911
Belussi, Gianangelo, Milano . . . 080440
Beluxurious, West Hollywood . . . 124988
Belvárosi Galéria, Szekszárd . . . 109213
Belvárosi Régiségbolt, Budapest 079241
Belvèdere, 's-Hertogenbosch . . . 083382
Belvedere, München 126044
Belvedere – 20er Haus, Wien . . 002951
Belvedere – Augarten Contemporary, Wien 002952
Belvedere – Oberes Belvedere, Wien 002953
Belvedere – Unteres Belvedere, Wien 002954
Belvedere – Zeitschrift für bildende Kunst, Wien 138506
Belvedere Antiques, New York . . 095264
Belvedere auf dem Klausberg, Stiftung Preußische Schlösser und Gärten Berlin-Brandenburg, Potsdam . 021175
Belvedere im Schlossgarten Charlottenburg, Stiftung Preußische Schlösser und Gärten Berlin-Brandenburg, Berlin 017079
Belvedere Reproductions, Ipswich 134500
Belvedere und Pomonatempel auf dem Pfingstberg, Stiftung Preußische Schlösser und Gärten Berlin-Brandenburg, Potsdam 021176
Belvidere Gallery, Aberdeen . . . 117206
Belvoir Gallery, Grantham 117871
Belye Palaty, Moskva 114208
Bélyegmúzeum, Budapest 023462
Belz, Rainer, Kassel 076712
Belz, T., Brühl 075204
Bemark Galeria, Raszyn 113757
Bembridge Windmill Museum, Mottistone 045300
Bement-Billings Farmstead, Newark Valley 051508
Bemis Center for Contemporary Arts, Omaha 051760
Bémon, Gerald, Nice 070575
Ben & Johnny's Attic, Los Angeles 094269
Ben-Gurion House, Tel Aviv 025190
Ben Maltz Gallery and Bolsky Galleries, Otis College of Art and Design, Los Angeles 050409
Ben Shahn Galleries, William Paterson University, Wayne 054250
Ben Uri Gallery, London Jewish Museum of Art, London 044914
Ben, Thanh, Ho Chi Minh City . . 125114
Benaco, Bardolino 109764
Benadi, Axel, La Varenne-Saint-Hilaire 140863
Benador, Jacques, Genève 116284
Benakeion Archaeological Museum, Kalamata 023100
Benalcazar, R. de, Paris 071035
Bénalcazar, Robert de, Issy-les-Moulineaux 068886
Benalla Art Gallery, Benalla 000847
Benalla Costume and Pioneer Museum, Benalla 000848
Benalla Market, Benalla 061124
Benamou, Albert, Paris 104558
Benappi, Vittorio, Torino 081656
Benardout, Raymond, Los Angeles 094270
Benards Gallery, Llandudno 089864, 118136, 134571
Benares Historic House, Mississauga 006338
Benato, Flavia Maria, Verona . . 132269
Benavente, Madrid 086131
Benavides, Madrid 086132
Bencés Apátsági Múzeum, Tihany 024006
Benchaïb, Paris 104559
Benchamabophit National Museum, Bangkok 042651
Benchmark Antiques, Bridport . . 088608, 134148

Benchmark Books, Salt Lake City 145505
Bencich, Buenos Aires 139554
Bencik, Pforzheim 078040
Bencteux MM, Falaise 068503
Benczúr, Budapest 109132
Bend in the River, Gainsborough 089315, 117814
Bendel, Gregor, Bad Neuenahr-Ahrweiler 106082
Bendel, J., Edesheim 075568
Bendel, Judit, Szada 130819
Benden & Klimczak, Viersen .. 108819
Bender, El Paso 093388
Bender, A., Mannheim 142104
Bender, Manfred, Völklingen .. 142453
Bender, Renate, München 108055
Bender, Stefanie, Freigericht .. 075891
Bender, Udo, Bremen 075150
Bendig, Hanfried, Bremen 075151
Bendigo Antique Centre, Bendigo 061128
Bendigo Art Gallery, Bendigo ... 000849
Bendin, Dalldorf 075294
Bendixen, Bo, Århus 102771
Bendixen, Bo, Hjørring 102835
Bendixen, Bo, Knebel 102856
Bendixen, Bo, Skagen 066115
Bendrat, Julia, Jesteburg 130083
Bendschneider, K., Demmin .. 075332, 129675
Bendt, Marilen, Freigericht ... 075892
Béné, Paris 128934
Benedek Péter Emlékház, Uszód 024023
Bénedet, Alain, Saint-Cybardeaux 072721
Bénedet, Alain, Saint-Jean-d'Angély 072835
Benedetti-Fahrenkrug, Lüneburg 077208
Benedetti Gallery, New York .. 122394
Benedetto, Saint Petersburg .. 127433
Bénédétto, Alain, Grimaud 068748
Benedick, Alain, Postroff 072292
Benedicta Arts Center, College of Saint Benedict, Saint Joseph .. 052782
Benedictine Monastery – Church of the Multiplication of Loaves and Fish Tabgha, Tiberias 025218
Benedictis, Henri de, Montpellier 070294
Benedikt, Mödling 062738
Benedikt-Nimser-Haus, Wilhelmsdorf, Kreis Ravensburg 022655
Benediktinerstift und Stiftssammlungen, Altenburg . 001702
Benefit Street Antiques, Providence 096499
Beneforti, Simone, Firenze .. 131125
Beneke, Detmold 106672
Benelli & Lascialfari, Pisa .. 131761
Bénelli, Jacques, Paris 141046
Beneš, Václav, Plzeň 140498
Beneska Galerija, San Pietro al Natisone 110898
Benetti, Antonella, Roma 081255
Beneun, Neuss 108250
Benezet, Dublin 079527
Bénézit, Henri, Paris 104560
Bénézit, M., Paris 104561
Benga Oral Historic Centre, Dandenong 001013
Bengal Gallery of Fine Arts, Dhaka 100459
Bengal Natural History Museum, Darjeeling 024209
A Bengala, Porto 085010
Bengler, Michael, Dingolfing .. 129687
Bengoechea Carriedo, Elena, Valencia 115476
Bengough and District Museum, Bengough 005419
Bengtsson, Ove, Eskilstuna .. 086664
Benguet Museum, La Trinidad . 034491
Benhabbou, Mohamed, Casablanca 082550
Benham, Seattle 124494
Benham, Naomi, Camberwell . 098850
I Beni Culturali, Parma 138807

Beni Studio, Tiranë 098521
Benichou, Edmond, Marseille . 069910
Benicia Historical Museum, Benicia 046915
Beningbrough Hall, York 046219
Benini, Gabriele, Cervia 109958
Benitez Gonzalez, Elba, Madrid . 115157
Benito, Juan, Aigues-Mortes .. 103177
Benjaman's Art Gallery, Buffalo . 119950
Benjamin and Dr. Edgar R. Cofeld Judaic Museum of Temple Beth Zion, Buffalo 047299
The Benjamin Banneker Museum, Baltimore 046738
Benjamin Books, Ottawa 140374
Benjamin Franklin House, London 044915
Benjamin N. Cardozzo School, New York 057544
Benjamin Patterson Inn Museum Complex, Corning 047993
Benjamin, A.A., Houston 093622
Benjamin, Marcus, Columbus .. 120477
Benjamin's Books, Queanbeyan . 139888
Benjamins, John, Amsterdam . 137826, 143012
Benkert, Ralf, Garmisch-Partenkirchen 141822
Benkovic, Josip-Bepo, Herceg-Novi 112021
Benli Arts de l'Islam, Paris .. 071036
Bennati, Michel, Beynac-et-Cazenac 067160, 067161
Bennazar, Paul, Saint-Ouen .. 073037
Bennazar, Pierre, Boulogne-Billancourt 067390
Bennemann, Markus, Köln ... 130142
Benner, Gloria, Baden-Baden .. 129432
Bennett, Bedale .. 088422, 134087
Bennett, Louisville 135751
Bennett, Nashville 122071
Bennett, Vancouver 064800
Bennett Mitten, New York ... 095265, 135891
Bennett Place, Durham 048404
Bennett Street Gallery, Atlanta . 119548
Bennett, Alan, Truro 091640
Bennett, Paul, London 089955
Bennetter, Anna, Lidingö 133545
Bennie Museum, Bathgate ... 043529
Bennington Battle Monument, Old Bennington 051744
Bennington Battlefield State Historic Site, Walloomsac 054042
The Bennington Museum, Bennington 046920
Benny Chia, Hong Kong 102061
Benny's, Attard .. 082407, 132333
Benoit, Buenos Aires 060878
Benoît-Latour, Dominique, Nevers 070535
Benoit, Adolf, Bindernheim ... 067223
Benoît, Anne, Paris 071037
Benoit, Jacky, Fontvieille 068583
Benoit, Jean-Loup, Beaumont (Vienne) 067003
Benoît, Pierre, Paris 071038
Benoit, Raymond, Saint-Jean-du-Corail 072853
Benoldi, Luca, Parma 081096
Benot, Cádiz 115002
Benrido, Kyoto 137767
Benriya Yuko, Settsu 082123
Benrubi, Bonni, New York ... 122395
Ben's Rocking Horses, Sacramento 096605
Bensi Santini, Antonella, Milano 080441
Bensi, Silvia, Firenze 131126
Benso, Alessandro, Firenze .. 080090
Benson, Paris 071039
Benson, Singapore 114498
Benson Veteran Cycle Museum, Benson 043568
Bensusan Museum of Photography, Johannesburg 038743

Bensussan, Gérard, Montgiscard 070247
Bent Cover, Phoenix 145425
Bent Henriksen, Dorthe, Odder . 066022
Bent Tree Gallery, Dallas 120561
Bent, Michael J, Mulgrave ... 125216
Benteli Verlags AG, Sulgen ... 138092
Bentleg, South Grafton 099698
Bentler, Bernd, Bonn 129606
Bentley, Calgary ... 064249, 128145
Bentley, Tshwane 114819
Bentley & Skinner, London ... 089956
Bentley for Books, Huonville .. 139781
Bentley Projects, Phoenix ... 123485
Bentley, William, Hungerford .. 089602, 134487
Benton County Historical Museum, Philomath 052109
Benton County Historical Museum, Prosser 052376
Benton Fine Art, Moreton-in-Marsh 090776, 118755
The Benton Homestead, Tolland 053761
Bentonville Battleground, Four Oaks 048915
Bent's Old Fort, La Junta 050043
Bentz, Peter, Milwaukee 094835
Benucci, Ponte San Giovanni ... 137720
Benucci, Roma 081256, 110734
Benue Polytechnic Art and Design Department, Ugbokolo 056207
Benxi Museum, Benxi 007656
Benz & Maier, Freiburg im Breisgau 107052
Benz, Gertrud & Marcus, Meilen 144113
Benz, Gertrud & Marcus, Zürich 144164
Benz, N., Damp 075295
Benz, R., Alpirsbach 129378
Benzacar, Ruth, Buenos Aires . 098554
Benzakein, Cape Town 085631
Benzenberg, A., Seeshaupt ... 078442
Benzenberg, Annemarie, Tutzing 078735
Benzie Area Historical Museum, Benzonia 046926
Beobide Lantoki Museoa – Museo-Taller de Julio Beobide, Zumaia 040466
Beolibel, Beograd 085332
Beothuck Village, Grand Falls-Windsor 005902
Beothuk Interpretation Centre, Boyd's Cove 005451
Beowulf Galleries, Woollahra .. 062461, 099909
Beppu Daigaku Fuzoku Hakubutsukan, Beppu 028503
Bep's Antiques, New Orleans . 050060
Ber, Claude le, Camaret-sur-Mer 103494
Beranoaguirre, Rosa, Donostia-San Sebastián 085970
Berardo, Silvia, Roma 081257
Beraud, Marc, Marclopt 069858
Berauer, Strasbourg 073574
Berbé, M., Bruxelles .. 063349, 063350
Berbere, Los Angeles 094271
Berbey, Janine, Seurre 073516
Berceo Librería, Madrid 143753
La Berceuse, Fort-de-France .. 132349
Berche & Pagis, Paris 141047
Berche, Maguy, Capdenac-Gare 013526
Berchem, Winfried, Essen ... 129813
Berchi, J. Fernandez, Madrid . 086133
Berchtold, M., Allschwil 087188
Bercker, Lucien, Aix-en-Provence 066455
Berden, Roermond 083533
Berdery, Jean-Pierre, Aire-sur-l'Adour 066445
Berdini, Andrea, Bologna 079819
Berea College Burroughs Geological Museum, Berea 046927
Berea College Doris Ulmann Galleries, Berea 046928
Beredskapsmuseet, Viken 041166
Beregi Múzeum, Vásárosnamény 024036
Berenberg, Boston 119838
Berend Lehmann Museum,

Halberstadt 018935
Berendes, Lüder, München 077468, 130329
Berengo, Amsterdam 112164
Berengo Fine Arts, Murano ... 110514
Bérenguer-Topacio, Makati ... 113381
Bérenguer, Joseph, Auch 066797, 128453
Berenkamp Verlag, Innsbruck . 136518
Berensen, Montréal 101365
Berensmann, Barbara, Stuttgart 078595
Berenson, Ellen, New York ... 095266
Berès, Paris . 071040, 071041, 104562, 104563
Berès, Pierre, Paris . 136867, 141048
L'Béret Vert, Toulon (Var) 073701
Beretti, Giuseppe, Milano 131383
Beretti, Sandro, Milano 131384
Béreyziat, Pierre, Tournus 073822
Bereznikovskij Istoriko-kraevedčeskij Muzej, Berezniki 036896
Berg, Carlingford 098865
Berg Bros., Cincinnati 092829
Berg, Paris 136868
Berg- und Stadtmuseum Obernkirchen, Obernkirchen 020910
Berg-Klönhammer, Ottobrunn . 130500
Berg-Kragerø Museum, Kragerø 033712
Berg, A., Toronto 128207
Berg, Annette, Hannover 130004
Berg, B., Sonnewalde 078506
Berg, Carl, Los Angeles 121444
Berg, Gary, Seattle 124495
Berg, J., Homberg (Efze) ... 076568
Berg, Lene, København 065857
Berg, Martina, Barntrup 141454
Berg, Otto, København 065858
Berg, Reinhold, Regensburg . 078164, 142322
Berg, U. & H.-P., Stühlingen . 078584
Bergama Arkeoloji Müzesi, Bergama 042846
Bergamini, Andrea, Modena .. 131560
Bergauer, Peter, Thun 087847
Bergbahnmuseum, Kitzbühel .. 002181
Bergbau- und Gotikmuseum Leogang, Leogang 002295
Bergbau- und Greifenstein-Museum, Ehrenfriedersdorf 018108
Bergbau- und Heimatmuseum, Gloggnitz 001939
Bergbau- und Heimatmuseum, Goldkronach 018761
Bergbau- und Heimatmuseum, Jochberg 002148
Bergbau- und Heimatmuseum im Paulushof, Essen 018290
Bergbau- und Industriemuseum Ostbayern, Kümmersbruck ... 019815
Bergbau- und Mineralienmuseum, Oberwolfach 020940
Bergbau- und Stadtmuseum Weilburg, Weilburg 022477
Bergbauernmuseum, Alpbach . 001697
Bergbauernmuseum, Telfs ... 002815
Bergbauernmuseum Hausstein, Frankenfels 001899
Bergbauernmuseum z' Bach, Wildschönau 003113
Bergbaumuseum, Aldenhoven . 016499
Bergbaumuseum, Altenberg .. 016530
Bergbaumuseum, Deuben ... 017880
Bergbaumuseum, Grünbach am Schneeberg 002017
Bergbaumuseum, Hall in Tirol . 002046
Bergbaumuseum, Hausham .. 019113
Bergbaumuseum, Horgen ... 041598
Bergbaumuseum, Langelsheim . 019883
Bergbaumuseum, Mechernich . 020322
Bergbaumuseum, Mühlbach am Hochkönig 002415
Bergbaumuseum, Niedermoschel 020796
Bergbaumuseum, Oelsnitz ... 020959
Bergbaumuseum Achthal,

Register der Institutionen und Firmen
Bergbaumuseum BUV Kleinzeche, Dortmund – Bernardoff, Braidwood

Teisendorf 022117
Bergbaumuseum BUV Kleinzeche, Dortmund 017952
Bergbaumuseum Erzpoche, Hausach 019107
Bergbaumuseum Goberling, Stadtschlaining 002769
Bergbaumuseum Graubünden mit Schaubergwerk, Davos Platz . . 041433
Bergbaumuseum Grube Anna II, Alsdorf 016517
Bergbaumuseum Hückelhoven, Hückelhoven 019354
Bergbaumuseum im Besucherbergwerk Tiefer Stollen, Aalen 016447
Bergbaumuseum Klagenfurt, Klagenfurt 002186
Bergbaumuseum Knappenhaus, Altenmarkt bei Sankt Gallen . . 001704
Bergbaumuseum mit Mineralienschau und Schaubergwerk, Hüttenberg 002107
Bergbaumuseum Peißenberg, Peißenberg 021095
Bergbaumuseum Pöllau, Neumarkt in Steiermark 002445
Bergbaumuseum Schachtanlage Knesebeck, Bad Grund 016778
Bergbaumuseum und Schaubergwerk Röhrigschacht, Wettelrode 022603
Bergbaumuseum und Schaustollen, Fohnsdorf 001894
Bergbauschaustollen, Pölfing-Brunn 002520
Bergdon, Toronto 064612
Bergé, Pierre, Paris 125680
Bergel, Gerolf, Berlin 141486
Bergemann, Berlin . . . 074765, 129470
Bergemann, Margot, Düsseldorf 075457
Bergen & Putman, New Orleans 122129
Bergen County Historical Society Museum, River Edge 052583
Bergen Kunst- og Antikvitetsformidling, Minde 084276
Bergen Kunsthall, Bergen 033522
Bergen Kunstmuseum – Lysverket, Bergen 033523
Bergen Kunstmuseum – Rasmus Meyers Samlinger, Bergen . . . 033524
Bergen Kunstmuseum – Stenersens Samling, Bergen 033525
The Bergen Museum of Art and Science, Paramus 051912
Bergen Museum of Art and Science, Paramus 051911
Bergen Museum of Local History, Bergen 046930
Bergen Museum, Universitetet i Bergen, Bergen 033526
Bergen Skolemuseum, Bergen . 033527
Bergen, de, Eindhoven 083208
Bergens Kunstforening, Bergen . 059654
Bergens Sjøfartsmuseum, Bergen 033528
Bergens Tekniske Museum, Bergen 033529
Bergensantikvariatet, Bergen . . . 143381
Berger, Augsburg 074452
Berger, Mulhouse 070387
Berger, Nieder-Olm 108271
Berger, Schwerin 108595
Berger & Kowalski, Potsdam . . . 078072
Berger & Marletta, Roma 081258
Berger & Söhne GmbH, Ferdinand, Horn 136515
Berger, Alain, Beaune 067010
Berger, André, Saint-Denis-en-Bugey 072732
Berger, Christian & Marion, Gahlenz 075936
Berger, Ferdinand, Wien 062910
Berger, François, Paris 071042
Berger, Max, Paris 071043
Berger, Michael, Benshausen . . 074718
Berger, Michael, Pittsburgh 123523
Berger, Peter, Stuttgart 130665, 130666
Berger, Reinhold, Vellberg 130720
Berger, Rudolf, Bern . . 087272, 133668
Berger, Simon, Wabern 133956
Bergères et Marquises, Paris . . 071044
Bergerhaus, Gumpoldskirchen . . 002021
Bergerie, Persac 072109
Bergerie, Ury 073916
Bergeron, Christian, Québec . . . 101491
Bergeron, D.L., Windsor, Ontario 101851
Bergeron, J.F., New Orleans . . . 122130
Bergeron, Jean-Claude, Ottawa . 101464
Berges, Jean-Louis, Saint-Pe-d'Ardet 073251
Bergführer-Museum, Heimat-, Bergführer- und Mineralienmuseum, Sankt Niklaus 041933
Berggericht – Mineralogická Expozícia, Slovenské Banské Múzeum, Banská Štiavnica 038284
Berggren, Anders, Stockholm . . 115842
Berghoff, Tampa 124635
Berghuis, Jan, Voorschoten 132661
Bergische Bücherstube, Overath 142289
Bergische Museumsbahnen e.V, Wuppertal 022816
Bergisches Freilichtmuseum für Ökologie und bäuerlich-handwerkliche Kultur, Lindlar . 020089
Bergisches Museum für Bergbau, Handwerk und Gewerbe, Bergisch Gladbach 017051
Bergisel Museum, Tiroler Kaiserjäger-Museum, Innsbruck 002120
De Bergkerk, Deventer 032216
Berglöf, Stockholm 086976
Berglöf, Lennart, Malmö 086847
Bergmännisches Traditionskabinett, Besucherbergwerk "Sankt Christoph", Breitenbrunn 017576
Bergman, Las Vegas 094123
Bergman, Sonja, Växjö 115989
Bergmann, Hamburg 076195
Bergmann sen., Franz, Schwarzenbach an der Saale . 078416
Bergmann, S., Köln 130143
Bergmann, Thomas, Erlangen . . 125938
Bergmanns-Bauernmuseum, Breitenbach, Kreis Kusel 017574
Bergmans, Amsterdam 143013
Bergner, Falk, München 130330
Bergner+Job, Mainz 107946
Bergoglio – Libri d'Epoca, Torino 142848
Bergoglio, Danile, Torino 132096
Bergoglio, Marco, Torino 081657, 132097
Bergrath, J. & A., Lingen 077129
Bergringmuseum, Stadtmuseum Teterow, Teterow 022131
Bergsjö Konsthall, Bergsjö 040562
Bergsma, Wiego, Utrecht 132645
Bergstens Konstförlag, Tyresö . . 138019
Bergström, Jönköping 115694
Bergström, Sture, Eskilstuna . . . 086665
Bergstrom-Mahler Museum, Neenah 051208
Bergstrom, Kalle, Paris 104564
Bergsturz-Museum Goldau, Goldau 041553
Bergtheil Museum, Westville . . . 038878
Le Bergueleven, Saint-Martin-des-Prés 072945
Bergwerkmuseum Grube Glasebach, Straßberg 022013
Bergwerksmuseum, Penzberg . . 021099
Bergwerksmuseum Schacht Mehren, Käbschütztal 019478
Bergwinkelmuseum, Schlüchtern 021674
Berheim, Sandnes 084352
Berk, Harlan J., Chicago 092670, 145036
Berkei, Ilona, Zeuthen 130801
Berkeley, Saint Louis 096670
Berkeley Antiques, Tewkesbury . 091589, 135130
Berkeley Antiques Market, Berkeley 088442
Berkeley Art Center, Berkeley . . 046932
Berkeley Art Museum, University of California, Berkeley 046933
Berkeley Castle, Berkeley 043569
Berkeley Civic Arts Program, Berkeley 060406
The Berkeley Costume and Toy Museum, New Ross 024974
Berkeley Editions, Woollahra . . . 099910, 136492
Berkeley Plantation, Charles City 047534
Berkelouw, Berrima 139643
Berkelouw, Paddington, New South Wales 139863
Berkelouw, H., Los Angeles . . . 145213
Berkelouw, Leo, Leichhardt 139803
Berkes, Peter, Basel 133634
Berkhout, P.Y.M., Den Haag 132471
Berkman, Philadelphia 096085
Berko, Patrick, Bruxelles 100568, 136630
Berko, Patrick, Knokke-Heist . . . 100730
Berks Art Alliance, Reading 060407
Berkshire Antiques, Windsor . . . 091876, 135241
Berkshire Antiques Centre, Midgham 090756
The Berkshire Artisans, Lichtenstein Center for the Arts, Pittsfield . . 052181
Berkshire Metal Finishers, Sandhurst, Berkshire 091244, 135007
Berkshire Mill Museum, Moora . 001307
Berkshire Museum, Pittsfield . . . 052182
Berkshire Scenic Railway Museum, Lenox 050219
Berkswell Village Museum, Berkswell 043572
Berland, Edinburgh 089160
Berland, Didier, Paris 071045
Berleux, Gilles, Boutenac-Touvent 067438
Berlevåg Havnemuseum, Berlevåg 033540
Berlijn, Ger, Amsterdam 082701
Berlijn, Paul, Haarlem 083315
Berlin-Antiquariat, Berlin 141487
Berlin Art and Historical Collections, Berlin 046944
Berlin Art Projects, Berlin 106169
Berlin Art Scouts, Berlin 106170
Berlin Connection Association, Luzern 138069
Berlin Design, Achim 105955
Berlin Historical Society Museum of Local History, Berlin 046945
Berline, Grenoble 068720
Berliner Auktionshaus für Geschichte, Berlin 125870
Berliner Bilder Galerie, Berlin . . 106171, 129471
Berliner Bilderzauber, Berlin . . . 129472
Berliner Büchertisch, Berlin 141488
Berliner Handpresse, Berlin 137055
Berliner Kunstsalon, Berlin 098133
Berliner Liste, Berlin 098134
Berliner Medizinhistorisches Museum der Charité, Berlin 017080
Berliner Münzauktion, Berlin . . . 125871
Berliner Planetarium und Sternwarte Wilhelm Foerster, Berlin 017081
Berliner S-Bahn-Museum, Potsdam 021177
Berliner Technische Kunsthochschule, Hochschule für Gestaltung, Berlin 055458
Berliner U-Bahn-Museum, Berlin 017082
Berliner Unterwelten-Museum, Berlin 017083
Berliner Waldmuseum mit Waldschule Grunewald, Berlin 017084
Berlinghof, Heidelberg 076434
Berlinghof, Leimen 126022
Berlinische Galerie, Landesmuseum für Moderne Kunst, Fotografie und Architektur, Berlin 017085
Berlins Gem and Historical Museum, Kingaroy 001190
Berlioz, Annonay 066655
Berlioz, Guy, Maraussan 069846
Berman, Dover 127208
Berman, Torino 110953
Berman Hall, Jerusalem 025069
Berman, Bob, Philadelphia 096086
Berman, D., Austin 119648
Berman &Co., Ted, Sacramento 136183
Bermann, Gisèle, Bruxelles 063351
Bermanne, Marsillargues 069974
Bermondsey, Madrid 086134
Bermuda Arts Centre, Dockyard 100820
Bermuda Editions Contemporary Art, Saint George's 100838
Bermuda Historical Society Museum, Hamilton 004190
Bermuda Maritime Museum, Mangrove Bay 004193
Bermuda Memories, Saint George's 100839
Bermuda National Gallery, Hamilton 004191
Bermuda National Trust Museum, Saint George's 004195
Bermuda Natural History Museum, Flatts 004189
Bermuda Society of Arts, Hamilton 004192, 058394
Bermúdez Antiques Art's, Maracaibo 097974
Berna, Denise e Beppe, Bologna 079820
Berna, Patrick, Le Mas 069313
Berna, Patrick, Lorgues 069630
Bernabei, Firenze 131127
Bernack, Worpswede 108963
Bernadac, Saint-Paul-de-Vence . 105577
Bernadette's Galleries, Vancouver 101746
Bernadottebiblioteket, Stockholm 040992
Bernaerts, Antwerpen 125293
Bernal S., José, Santiago de Chile 064956
Bernal Troncoso, Pedro, Santiago de Chile 101902
Bernal, Marie, Strasbourg 129264
Bernam, Bogotá 102284
Bernandino Jalandoni Ancestral House, Silay 034597
Bernard, Chicago 092671
Bernard, Macclesfield 118689
Bernard, Montréal 101366
Bernard Historical Museum, Delton 048237
Bernard Judaica Museum, Congregation Emanuel of the City of New York, New York 051345
Bernard Price Institute Paleontology Museum, Johannesburg 038744
Bernard, Alain, Meyzieu 070117
Bernard, Antoinette, Paris 071046
Bernard, Christian, Paris 071047
Bernard, Christine, Cheminot . . 067955
Bernard, Claude, Paris 104565, 136869
Bernard, Henri, Bordeaux 067314
Bernard, Jean-Claude, Montélimar 070232
Bernard, Marion, Champlitte . . . 103569
Bernard, Michel, Bougnon 128513
Bernard, Michel, Lyon 069707
Bernard, Michel, Nolay 070741
Bernard, Theo, Erlenbach 075656
Bernardi, San Marino 114381
Bernardi, Toronto 064613, 101579
Bernardi, A. de, Aachen 105929, 137020
Bernardi & Sons, Bea Tullio, Tshwane 126497
Bernardi, Giulio, Trieste 081782
Bernardi, Lorenzo, Firenze 131128
Bernardi, Marino, San Marino . . 085282
Bernardino & Soares, Lisboa . . . 084838
Bernardis, Sergio, Torino 142849
Bernardoff, Braidwood 061188

Bernarducci & Meisel, New York 122396
Bernardus Verlag, Grevenbroich 137235
Bernareggi, Federico Ferdinando,
Milano 080442
Bernasconi, Roger, Marseille . . 069911,
128749
Bernaudat, Guy, Sainte-Savine . 073366
Bernays, Dallas 120562
Bernberg Fashion Museum, Museum
Africa, Johannesburg 038745
Berndt, Köln 107592
Berndt Museum of Anthropology,
University of Western Australia,
Crawley 000999
Berndt, Brigitta, Solothurn 133926
Berndt, Pierre, Carouge 087347, 133712
Berndt, Pierre, Genève 087454, 133754
Bernecker, J., Bötzingen 141610
Berner, Deland 127199
Berner Kunst- und Antiquitätenmesse,
Bern 098327
Berner Kunstmitteilungen, Bern . 139090
Bernera, Urras Eachdraibh Sgire
Bhearnaraidh, Bernera 043573
Berners, Richard, Köln 107593
Bernesga, León 115110
Bernet-Neumann, Ulrike,
Erlenbach 133732
Bernett, F.A., Boston 145007
Bernhagen, Bad Münstereifel . . 074554
Bernhammer, Horst & Frank,
Bischoffen 075004
Bernhard & Vögeli-Müller,
Winterthur 087907, 133966
Bernhard-Haus, Ohlsdorf 002479
Bernhard-Heiliger-Stiftung, Berlin 058735
Bernhard, René, Zürich 087942
Bernhardt Christensen, Niels,
Kastrup 065810
Bernhardt, Torsten, Hamburg . . 141857
Bernheim, Ciudad de Panamá . . 084412,
113314
Bernheim-Jeune & Cie, Paris . . 104566,
104567, 136870, 141049, 141050
Bernheimer, Fine Old Masters,
München 077469
Berni, Carla, Firenze 080091
Berni, Lazzaro, Parma 081097
Bernicat, G., Roanne 072483
Bernice Pauahi Bishop Museum,
Honolulu 049526
Bernier & Eliades, Athinai 109026,
137573
Bernigaud, Fabrice, Saint-Martin-du-
Mont 072948
Bernini, Andrea, Livorno 080318
Bernini, Dario, Parma 131721
Bernische Stiftung für angewandte
Kunst und Gestaltung, Bern . . 041310
Berns, Christiane & Günter,
Neuwied 077772
Berns, Perry, Dallas 120563
Bernstein Huus, Esens 018280
Bernstein Museum Bad Füssing, Bad
Füssing 016772
Bernstein & Co. Ltd., A. Nathan, New
York 122397
Bernstein, David, New York 122398
Bernstein, M., Los Angeles 145214
Bernstein & Co., S., San
Francisco 097191, 145542
Bernsteinmuseum, Sellin,
Ostseebad 021838
Berntsen, Kaare, Oslo . 084300, 113192
Berntsson, Vetlanda 115997
Beronnet, Marcel, Villeurbanne . 074190
Berquat, Pascal, Marseille 125634
Berr, Karl, Tirschenreuth 078683
Berra, Clementina, Milano 080443,
080444
Berra, Paola, Milano 080445
Berre, Alain le, Kerlaz 068939
Berretta Lemoine, Veronique, Toulon
(Var) 073702

Berrima District Art Society Gallery,
Bowral 000873
Berrima Galleries, Berrima 098740
Berrima Museum, Berrima 000853
Berrino, Alassio 109741
Berrino, Monaco 111995
Berríos Valenzuela, Orlando, Santiago
de Chile 064957
Berry, Saint-Maur 072958
Berry & Crowther, Huddersfield . 134480
Berry Antiques, Moreton-in-
Marsh 090777
Berry Art Gallery, Berry 098741
Berry-Hill, New York 122399
Berry Museum, Berry 000854
Berry Street Bookshop,
Wolverhampton 144929
Berry, J.J., Windsor, Ontario . . . 064901
Berry, Lesley, Flamborough 089281
Berry, Ray, Maylands, Western
Australia 061807
Berry, Sandra, New Orleans . . . 122131
Berryware, Tucson 124718
Berselli, Milano 131385
Bersham Heritage Centre and
Ironworks, Wrexham 046202
Bersia Antiquariato, Milano 080446
Berson, Toronto 064614
Bersotti, Giuseppe, Brescia 131018
Bert, Providence 123697
Bert, Dominique, Paris 104568
Das Berta-Hummel-Museum im
Hummelhaus, Massing 020315
Bertacchi, Orfeo, Modena 080773
Bertacchini, Gabriella, Modena . 131561
Bertana, Isabel S. de, Buenos
Aires 060879
Bertarini, Aldo, Modena 080774
Bertaux, Baltimore 119725
Berte Museum – Livet på Landet,
Slöinge 040963
La Bertesca, Genova 110094
Bertha V.B. Lederer Arts Gallery,
Geneseo 049064
Bertha's Kunsthandel, Buchholz in der
Nordheide 075208
Berthelin, Manuel, Cognac 068058
Berthelot, Paris 128935
Berthelot, Saint-Denis-de-l'Hôtel 072729
Berthelot-Vinchon, Françoise,
Paris 071048
Berthelot, Bertrand, Remouillé . . 072415
Berthelot, Jean-Louis, Bourg-
Beaudouin 067408, 128515
Berthet, André, Lyon 069708
Berthet, Anne-Sophie, Saint-Pourcain-
sur-Sioule 073268
Berthier, Renée, Limoges 069518
Berthiller, Claude, Villeurbanne . 129353
Berthold, Gelsenkirchen 107109
Berthold-Auerbach-Museum, Horb am
Neckar 019338
Bertholier, Joël, Le Pont-de-
Beauvoisin 069334, 128667
Berthommier & Dominique Langlet,
Geneviève, Rivarennes 072478
Berthoud, Le Vésinet 069374
Berti, Firenze 080092
Berti, Adriana, Firenze 080093
Berti, Luciano, Bern 087273
Bertilsson, Falkenberg 115605
Bertin Toublanc, Paris 071049
Bertocchi, Bologna 142627
Bertogalli, Parma 081098
Bertola Lauro Antichità, Torino . 081658
Bertola, Carlo, Torino 132098
Bertola, Lauro, Torino 132099
Bertold Gallery, London 118213
Bertolino, Antimo, Lugano 087606,
133820
Bertolino, Francesco, Palermo . . 131686
Bertolotti, Luciano, Brescia 131019
Bertoncelj, Danilo, Ljubljana . . . 085605,
133065

Bertoncelli, Verona 132270
Bertoncelli, Maurizio, Bologna . . 079821,
142628
Bertone, Flavio, Torino 132100
Bertoni, Alberto, Venezia 142881
Bertoni, Andrea, Bologna 130937
Bertoni, Félix, Magagnosc 069800
Bertouy, Didier, Lyon 069709
Bertram, Bremen 075152
Bertram & Williams,
Williamsburg 145645
Bertram House Museum, Iziko
Museums of Cape Town, Cape
Town 038672
Bertram, Sigynn, Lüneburg 107929
Bertran, Jean-Pierre, Villeneuve-sur-
Lot 074163
Bertran, Michel, Rouen 072566, 105469
Bertran, Michèle & Etienne,
Rouen 072567, 141318
Bertrand, Lisboa 137916
Bertrand, Ribaute-les-Tavernes . 072450
Bertrand, Tassin-la-Demi-Lune . 141354
Bertrand, Alex, Vendôme 074009
Bertrand, Arthus, Palaiseau . . . 070844
Bertrand, Dominique, Cambrai . 125543
Bertrand, Michel, Noron-l'Abbaye 070745
Bertrandt, Dorota, Berlin 106172
Bertrandy, Philippe, Le Puy-en-
Velay 140867
Bert's Antiques, San Diego 097067
Bertus, Rotterdam 083555
Berufsfachschule für Holzbildhauerei
und Schreinerei des Landkreises
Berchtesgadener Land,
Berchtesgaden 055457
Berufsgruppe des Salzburger
Kunst- und Antiquitätenhandels,
Salzburg 058234
Berufsverband Bildender Künstler
Berlins e.V., Berlin 058737
Berufsverband Bildender Künstler
Frankfurt/Main, Frankfurt am
Main 058738, 106975
Berufsverband bildender Künstler
Hamburgs e.V, Hamburg 058739
Berufsverband Bildender Künstler
Landesverband Bayern e.V.,
München 058740
Berufsverband Bildender Künstler
München und Oberbayern,
München 058741
Berufsverband Bildender Künstler
Niederbayern e.V., Fürstenzell . 058742
Berufsverband Bildender Künstler
Niederbayern/Oberpfalz e.V,
Regensburg 058743
Berufsverband Bildender Künstler
Oberbayern-Nord und Ingolstadt e.V.,
Ingolstadt 058744
Berufsverband Bildender Künstler
Schwaben-Nord und Augsburg e.V.,
Augsburg 058745
Berufsverband Bildender Künstler
Schwaben-Süd, Kempten 058746
Berufsverband Bildender Künstler
Unterfranken, Würzburg 058747
Berufsverband Bildender Künstler
Westfalen e.V., Dortmund 058748
Berufsverband Bildender Künstler,
Bezirksverband Bergisch-Land,
Wuppertal 058736
Berufsverband Bildender Künstlerinnen
und Künstler NRW, Bezirksverband
Ostwestfalen-Lippe, Bielefeld . 058750
Berufsverband Bildender Künstlerinnen
und Künstler Nürnberg,
Nürnberg 058751
Berufsverband Bildender Künstlerinnen
und Künstler Oberfranken e.V.,
Bamberg 058752
Berufsverband Bildender Künstlerinnen
und Künstler Südbaden e.V., Freiburg
im Breisgau 058753

Berufsverband Bildender Künstlerinnen
und Künstler, Regionalverband
Wiesbaden, Wiesbaden 058749
Berufsverband österreichischer Kunst-
und Werkerzieher, Wien 058235
Berufsverband österreichischer Kunst-
und Werkerzieher, Landesgruppe
Niederösterreich, Ternitz 058236
Berufsvereinigung Bildender Künstler
Oberösterreich, Linz . 058237, 100098
Berufsvereinigung bildender Künstler
Salzburg, Salzburg 058238
Berufsvereinigung Bildender
Künstlerinnen und Künstler Vorarlbergs,
Bregenz 058239, 136496
Berufsvereinigung der bildenden
Künstler Österreichs, Bundes- Zentral
und Landesverband für Wien,
Niederösterreich und Burgenland,
Wien 058240
Berufsvereinigung der bildenden
Künstler Österreichs, Landesverband
Kärnten, Klagenfurt 058241
Beruto, Carlo, Albisola Superiore 079741
Berwald, New York 095267
Berwick, Officer 062025
Berwick Barracks, Berwick-upon-
Tweed 043574
Berwick Borough Museum and Art
Gallery, Berwick-upon-Tweed . 043575
Berwick Pakenham Historical Museum,
Pakenham 001392
Beryl's Boutique Gallery,
Bundoora 098820
Beryl's Old Charm Antiques,
Chatswood 061314
Berzel, Speyer 078518, 078519
Berżera, Warszawa 113808
Bērzgales Pagasta Vēstures Muzejs,
Bērzgale 030106
Berzolari, Andrea, Milano 131386
Berzsenyi Dániel Emlékmúzeum,
Egyházashetye 023623
Berzsenyi Dániel Emlékmúzeum,
Nikla 023810
Besættelsessamlingen 1940–45,
Grindsted 010086
Besançon, Jacques & Dominique,
Bordeaux 067315, 140788
Besançon, Philiberte, Aix-en-
Provence 066456
Besch, Constant, Mâcon 069790
Besch, Gérard, Juziers 068934
Besenmuseum & Galerie,
Ehingen 018104
Besenval, Jean-Pierre, Paris . . . 128936
Besichtigungsstätte der Mumie
des Ritter Kahlbutz, Neustadt,
Dosse 020764
Beside the Wave, Falmouth 117782,
138161
Beşiktaş Kültür Merkezi, İstanbul 117015
Besim, Adil, Innsbruck 062645
Besim, Adil, Salzburg 062787
Besim, Adil, Wien 062911
Besley, Beccles 134084, 144235
Besnard, Christian, Nançay 070407,
128819
Besnard, Michel, Mortagne-au-
Perche 070354
Besnard, Mireille, Orléans 070810
Beso Ur och Antik, Göteborg . . . 086700
Bespoke Gallery, New York 122400
Besrest, Jean-Jacques, Paris . . 071050
Bessard, Ghislaine, Velines 073997
Besseiche, Daniel, Courchevel . . 103634
Besseiche, Daniel, Dinard 103691
Besseiche, Daniel, Paris 104569
Besseiche, Daniel, Pont-Aven . . 103690
Besselaar, Utrecht 112691, 112692
Bessels, Hendrik-W., Oslo 084301
The Bessemer Gallery, Sheffield 119078
Bessemer Hall of History,
Bessemer 046952

Register der Institutionen und Firmen

Bessenyei György Emlékszoba, Jósa András Múzeum, Tiszabercel – Bharat Itihas Samshodhak Mandal, Pune

Bessenyei György Emlékszoba, Jósa András Múzeum, Tiszabercel . . 024009
Besser, Erika, Bad Orb 074569
Besser, Manfred, Hamburg 107230
Bessie Surtees House, Newcastle-upon-Tyne 045344
Bessière, Georges, Ambérieu-en-Bugey 066545
Bessire, Jythe, Maisoncelle-Saint-Pierre 069812
Bessler, Wilhelm, Augsburg 074453
Besson, London 118214
Besson, Jean-Louis, Aurillac . . . 066819
Best, Bad Oeynhausen 106084
Best, Chaskovo 101122
Best Art Restoration by Lallier, Plano 136144
Best Finishing Company, Houston 135590
Best Friends Antiques, Tampa . . 097545
Best of Brass, Hereford 117964
Best of the West Antique Expo, Abbotsford 098032
Best, Ray, Newton Tony 090841
Bestattungsmuseum der Bestattung Wien, Wien 002955
Besteckfabrik Hesse, Schmallenberg 021680
Bester, Béatrice & Alain, Vincennes 074238
Bestman, Delft 083029
Besucher-Bergwerk Hüttenstollen, Orts- und Bergwerksmuseum Osterwald, Salzhemmendorf . . 021578
Besucher-Dienste, Staatliche Museen zu Berlin – Stiftung Preußischer Kulturbesitz, Berlin 017086
Besucherbergwerk, Nothweiler . 020838
Besucherbergwerk, Windeck . . . 022675
Besucherbergwerk "Drei Kronen und Ehrt", Elbingerode, Harz 018149
Besucherbergwerk "Mühlenberger Stollen", Bleialf 017421
Besucherbergwerk Bad Friedrichshall-Kochendorf, Heilbronn 019158
Besucherbergwerk Fell, Fell . . . 018348
Besucherbergwerk Finstergrund, Wieden 022621
Besucherbergwerk Finstertal, Schmalkalden, Kurort 021677
Besucherbergwerk Frischglück, Neuenbürg 020695
Besucherbergwerk Grube Anna-Elisabeth, Schriesheim 021735
Besucherbergwerk Grube Bindweide, Steinebach 021981
Besucherbergwerk Grube Fortuna, Solms 021907
Besucherbergwerk Grube Gustav, Meißner 020350
Besucherbergwerk Grube Wohlfahrt, Hellenthal 019180
Besucherbergwerk Rischbachstollen, Sankt Ingbert 021597
Besucherbergwerk Schiefergrube Christine, Willingen 022667
Besucherbergwerk und Ehemaliges KZ Kochendorf, Bad Friedrichshall 016770
Besucherbergwerk Vereinigt Zwitterfeld zu Zinnwald, Tiefer-Bünau-Stollen, Altenberg 016531
Bet Gordon, Deganya Aleph . . . 025031
Bet Hashomer Museum, Kibbutz Kfar Giladi 025136
Bet Pinhas Museum of Nature, Haifa 025045
Beta Railway Station Historic Museum, Alpha 000794
Bétabois, La Charité-sur-Loire . . 068984
Betbeder Espacio de Arte, Tandil 098592
Betbeder, Catherine, Anglet 066620
Betbeder, Catherine, Biarritz . . . 067189
Betelli & Terlenghi, Brescia 131020
Betesaraki, Donostia-San Sebastián 133243
Beth Ahabah Museum and Archives, Richmond 052545
Beth Tzedec Reuben and Helene Dennis Museum, Toronto 007161
Bethany, Bendigo 098734
Bethany College Art Department, Mingenback Art Center, Lindsborg 057349
Bethel Historical Society's Regional History Center, Bethel 046955
Bethenod, Christine, Paris 104570
Bethesdas, København 140596
Bethge, Wolfgang, Erfurt 106892
Bethlehem Historical Association Museum, Selkirk 053266
Bethlehem Museum, Bethlehem 038653
Bethlem Royal Hospital Archives and Museum, Beckenham 043545
Bethmann, Jean de, Bordeaux . 067316
Béthon, Patrick, Paris 071051
Beth's Antiques, Meeniyan 061810
Bethune Memorial House, National Historic Site of Canada, Gravenhurst 005921
Bethunes, Wellington 126332
Betis, Jean-François, Versailles . 141380
Betonzeitschiene, Plattenbaumicromuseum, Dresden 017975
Betri kaup, Reykjavík 079339
Betsey Williams Cottage, Providence 052379
Betsy Ross House, Philadelphia . 052033
Bett, North Hobart 099469
Bettcher, Miami 121747
Bettencourt Frères, Bagnolet . . . 128458
Better Buy Books, Nerang 139851
Better Coins, Houston 093623
Better Days, Menai Bridge 090745
Better Days Antiques, Mornington 061886
Better Homes, San Antonio 096951
Better Times Antiques, New York 095268
Betteto, Ampelio, Montegrotto Terme 110509
Betti, Roger, Toulouse 105725
Bettina und Achim von Arnim Museum, Künstlerhaus Schloss Wiepersdorf, Deutsche Stiftung Denkmalschutz, Niederer Fläming 020793
Bettinardi, Giorgio, Milano 131387
Bettinger, Marc T., Portland 096370
Bettini, Firenze 131129
Bettini, A. & F., Karlsruhe 076687
Bettles, Ringwood 118977
Bettoni Hourcade, Sandra, Montevideo 091992
Bettoni, Arturo, Brescia 079917
Betts House Research Center, Cincinnati 047737
Bettuzzi, Giorgio, Ponte Taro . . . 081164
Betty Brinn Children's Museum, Milwaukee 050910
Betty Foy Sanders Department of Art, Georgia Southern University, Statesboro 057946
Bettyann & Jimbo, Dallas 093075
Betty's Antiques, Long Beach . . 094195
Betuws Fruitteelt Museum, Erichem 032317
Between Bridges, London 118215
Between Cultures, Seattle 124496
Betws-y-coed Motor Museum, Betws-y-coed 043581
Betygalos Muziejus, Raseinių krašto istorijos muziejus, Betygala . . . 030321
Betz, Houston 121005
Betz, Christa, Cadolzburg 075241
Betz, Holger, Bretzfeld 075193
Betz, Holger, Obersulm 077880
Beugnot, Claire & Xavier, Versailles 129338
Beulecke, Rebecca, Eichstätt . . 075581
Beuneker, Leeuwarden 083413
Beuroner Kunstverlag, Beuron . . 137106
Beurret, Nicolas, Zürich 087943
Beurrier, Dominique, Déville-lès-Rouen 068293
Beurs van Berlage Museum, Amsterdam 031931
Beurskens, Roermond 083534
Beurville, Claudine, Méry-sur-Oise 070075
Beuse, C., Amersfoort 132384
Beutel, Elka, Potsdam 130528
Beuthner, Manfred, La Roche-Bernard 103898
Beutler, E., Bremen 141646
Beutler, Felix, Berlin 129473
Beutler, Gerhard, Calw 075243
Bev & Dale, Oklahoma City 095951
Bevan, Rupert, London 134596
Beveled Edge, Baltimore 119726
The Beveled Edge, Richmond . . 123724
Beveled Edge, Tampa 124636
Beverage Containers Museum, Millersville 050899
Beverley, London 089957
Beverley Art Gallery, Beverley . . 043583
Beverly Art Center, Chicago . . . 047641
Beverly Hills Antiquarian, Los Angeles 094272, 145215
Beverly Hills Gallery, Chicago . . 120070
Beverly Historical Museum, Beverly 046966
Bévilacqua, Raphaël, Burlats . . . 067536
Bevo, Saint Louis 096671
Bevrijdingsmuseum van Boom, Boom 003378
Bewdley Antiques, Bewdley . . . 088456, 134100
Bewdley Museum, Bewdley . . . 043584
Bewer, Petra, Stuttgart 142384
Bewley's Café Museum, Dublin . 024869
Bexfield, Daniel, London 089958, 134597
Bexhill, Saint Leonards-on-Sea . 091196
Bexhill Museum, Bexhill-on-Sea 043586
Bexhill Museum of Costume and Social History, Bexhill-on-Sea . 043587
Bexley, Columbus 092996
Bexley Historical Society Museum, Bexley 046974
Bexley Museum, London 044916
Bey, Patrick, Cavaillon 067735
Beyeler, Ernst, Basel . . 116057, 138022, 144015
Beyer, Claudia, Vellberg 130721
Beyer, Harry, München 077470
Beyer, Stefan, Helferskirchen . . 076485
Beyer, Thomas, Wernigerode . . . 078934
De Beyerd, Centrum voor Beeldende Kunst, Breda 032098
Beylerbeyi Sarayı, İstanbul 042926
Beyoğlu Belediyesi Sanat Galerisi, İstanbul 117016
Beyoglu Sanat Merkezi, İstanbul 042927
Beyond Futons, Charlotte 092577
Beyond Leidsche Rijn, Utrecht . . 059595
Beyond the Fridge Door Childrens Art, London 101335
Beyond The Mount Fine Art Conservation, Castlemaine . . . 127580
Beyond the Wall, Chicago 120071
Beyond the Wall, New York 122401, 122402
Beyond the Wall, Philadelphia . . 123358
Beyond the Wall, Saint Louis . . 123808
Beyond the Wall, Seattle 124497
Beypazarı Kültür ve Tarih Müzesi, Beypazarı 042847
Beyreuther, Jürgen, Neuengönna 077721
Beyul, New York 095269
Bezalel Academy of Arts and Design Jerusalem, Jerusalem 055866
Bézard, Xavier, Maintenon 069809
Bezerédj Kastély, Győr 023658
Bezert, Jean-Pierre, Cosne-d'Allier 068166
Bézert, Jean-Pierre, Marseille . . 128750
Ter Beziens, Kolhorn 032549
Béziers, Dominique, Le Havre . . 128661
Bežigrajska Galerija 2, Mestna Galerija Ljubljana, Ljubljana 038541
Bežigrajska Galerija, Mestna Galerija Ljubljana, Ljubljana 038540
Bezirksheimatmuseum, Völkermarkt 002870
Bezirksheimatmuseum mit Zdarsky-Skimuseum, Lilienfeld 002311
Bezirksmuseum Alsergrund, Wien 002956
Bezirksmuseum Brigittenau, Wien 002957
Bezirksmuseum Buchen, Buchen 017652
Bezirksmuseum Dachau, Zweckverband Dachauer Galerien und Museen, Dachau 017795
Bezirksmuseum Döbling, Wien . 002958
Bezirksmuseum Donaustadt, Wien 002959
Bezirksmuseum Favoriten, Wien 002960
Bezirksmuseum Floridsdorf, Wien 002961
Bezirksmuseum Hernals, Wien . 002962
Bezirksmuseum Hietzing, Wien . 002963
Bezirksmuseum Höfli, Bad Zurzach 041248
Bezirksmuseum Innere Stadt, Wien 002964
Bezirksmuseum Josefstadt, Wien 002965
Bezirksmuseum Landstraße, Wien 002966
Bezirksmuseum Leopoldstadt, Wien 002967
Bezirksmuseum Liesing, Wien . . 002968
Bezirksmuseum Margareten, Wien 002969
Bezirksmuseum Mariahilf, Wien 002970
Bezirksmuseum Marzahn-Hellersdorf, Berlin 017087
Bezirksmuseum Meidling, Wien . 002971
Bezirksmuseum Neubau, Wien . 002972
Bezirksmuseum Ottakring, Wien 002973
Bezirksmuseum Penzing, Wien . 002974
Bezirksmuseum Rudolfsheim-Fünfhaus, Wien 002975
Bezirksmuseum Simmering, Wien 002976
Bezirksmuseum Stockerau, Stockerau 002796
Bezirksmuseum Währing, Wien . 002977
Bezirksmuseum Wieden, Wien . 002978
Bezirksverband Bildender Künstlerinnen und Künstler Karlsruhe e.V., Karlsruhe 058755
Bezirksverband Bildender Künstlerinnen und Künstler, Regional Mannheim e.V., Mannheim 058754
Bezisten, Skopje 082397, 111879
Bezoekerscentrum Binnenhof, Den Haag 032167
Bezoekerscentrum de Meinweg, Herkenbosch 032466
Bezoekerscentrum Mijl Op Zeven, Ospel 032734
Bezoekerscentrum Oortjespad, Kamerik 032531
BFAS Blondeau Fine Art Services, Genève 116285
BFI Southbank Gallery, London . 044917
BGC, Bernard Grassin Champernaud, Paris 071052
Bhagalpur Museum, Bhagalpur . 024160
Bhagavan Mahavir Government Museum, Cuddapah 024206
Bhak, Young Duk, Seoul 111617
Bhakta Bahadur Sreshtha, Kathmandu 082612
Bharany's Curio House, Delhi . . 079383
Bharat Bhawan Museum, Bhopal 024167
Bharat Itihas Samshodhak Mandal, Pune 024443

Bharat Kala Bhavan, Varanasi . . 024514
Bharatiya Adim Jati Sevak Sangh
Museum, Delhi 024221
Bharatiya Kala Prasarini Sabha,
Pune 055829
Bhartiya Vidya Bhavan,
Hyderabad 109358
Bhavan's Art Gallery, Bangalore 109254
Bhavya Art Gallery, Bangalore . . 079345
BHC Depot, Raon-l'Etape 072366
Bhirasri Museum of Modern Art,
Bangkok 042652
BHL, Soustons 073555
BHNY Fine Art, New York 122403
Bhuri Singh Museum, Chamba . . 024190
Bi Gakko, Tokyo 056039
Bi-Hand, Bristol 117437
Bia, Marta, Verona 081901
Biaggi, Alexandre, Paris 071053
Biaggi, Anne-Marie, Menton . . . 125643
Biagiarelli, Roma 081259
Biagiotti & Sarti, Firenze 131130
Biagiotti, Caterina Diana, Firenze 080094
Biagiotti, Roberta, Firenze 080095
Bialik-Museum, Tel Aviv 025191
Białostocki Towarzystwo Przyjaciół
Pięknych, Białystok 059692
Białostockie Muzeum WSI,
Państwowe Muzeum w Białymstoku,
Wasilków 035384
Bianca, Buenos Aires 127515
Biancarelli, François, Paris 071054
Bianchera, Mario, Zürich 133976
Bianchetti, Carla & Mario,
Brescia 079918
Bianchi, Dante, Roma 131828
Bianchi, Donatella, Milano 131388
Bianchi, Leonardo, Firenze 131131
Bianchi, Lucia, Milano 131389
Bianchi, Riccardo, Firenze 131132
Bianchin, Luisa Maria Laura,
Milano 080447
Bianchini, Roma 081260
Bianchini, David, Verona 132271
Bianco, Kathy, New York 122404
Biapo Artes, São Paulo 064160
Biard, Gaël, Niort 125665
Biarom Design, Bucureşti 085082
Biartea, Asociación Artesana de
Bizkaia-Bizkaiko Artisau Elkartea,
Bilbao 059855
Biasio, Padova 080965, 080966
Biasoli, Monique, Bazas 103347
Biasutti & Biasutti, Torino 110955
Biasutti, Giampiero, Torino 110954
Biau, Jean-Francois, Montpellier 070295
Biavati, Emma, Bologna 130938
Biazzi, Tina Cecilia & Leonarda
Vercelli, Torino 081659
Bibak Museum, Benguet, Ifugao
Bontoc, Apayao and Kalinga Museum
of Culture, Baguio 034432
Biban, Daniele, Venezia 081826
Bibbo, Canio, Roma 081261
Bibby, Anna, Auckland 112753
Bibel und Orient Museum,
Fribourg 041496
Bibeles Muzejs, Alūksne 030099
Bibelgalerie Meersburg,
Meersburg 020326
Bibelhaus Erlebnismuseum, Frankfurt
am Main 018407
Bibelmuseum, Beuron 017350
Bibelmuseum, Münster 020593
Bibelmuseum Stuttgart, Stuttgart 022026
Bibelot, 's-Hertogenbosch 083383
Bibelot, Honolulu 120924
Bibelot, Warszawa 143518, 143519
Bibelot Snuisterij, Heerlen 083369
Bibelots and Books, Seattle . . . 145576
Biber, Jeffery L., Philadelphia . . 096087
Biberger, Peter, Hallwang 062628
Bibi's, Neulengbach 062746

Biblarte, Lisboa 084839, 143557
Bible Lands Museum Jerusalem,
Jerusalem 025070
Bible Museum, Trilofo 023307
Bible Museum, Dizengof House, Tel
Aviv 025192
Biblia Rara, Würzburg 142511
Biblical Arts Center, Dallas 048111
Biblical Heritage Collection,
London 144561
Biblical Museum of Canada,
Vancouver 007272
Biblical Resource Center & Museum,
Collierville 047876
Biblical World Gallery, Budapest 079242
Biblio-Globus Torgovyj Dom,
Moskva 143636
Biblioludothèque, Maison de la Nature,
Paris 014620
Bibliomania, Oakland 145385
Le Biblion, Toulouse 105726
Biblion Antiquariat, Zürich 144165
Biblionia Books, Sendai 142914
Bibliophile, Paddington, New South
Wales 139864
Bibliophile Bookshop, Norfolk . . 145380
Bibliophilen-Gesellschaft Köln,
Köln 058756
Les Bibliophiles, La Baule-
Escoublac 125601, 140860
The Bibliopole, Surry Hills 139918
Bibliosfera, Moskva 114209
Biblioteca Antònio Perdigão, Museu e
Arquivo da Cidade, Conselheiro
Lafaiete 004414
Biblioteca Culinaria, Lodi 142700
Biblioteca d'Arte, Milano 026778
Biblioteca Luis-Angel Arango,
Bogotá 008582
Biblioteca Medicea Laurenziana,
Firenze 026240
Biblioteca-Museo Alessandro Longo,
Amantea 025311
Biblioteka Antikvárium, Budapest 142556
Biblioteka Czartoryskich, Muzeum
Narodowe, Kraków 034856
Biblioteka-muzej A.P. Gajdara,
Kaniv 043172
Biblioteka Narodowa, Warszawa 035308
Biblioteka Sejmowa, Wydzial
Muzealiów, Warszawa 035309
Biblioteka, Muzeum i Archiwum
Warszawskiego Towarzystwa
Muzycznego im. Stanisława
Moniuszki, Warszawa 035307
Biblioteksmuseet, Borås 040571
Bibliotheca Albertina,
Universitätsbibliothek der Universität
Leipzig, Leipzig 019980
Bibliotheca Gastronomica, Zürich 144166
Bibliotheca Rara, Münster 137434
Bibliotheca Wittockiana-Musée de la
Reliure et du Hochet, Woluwe-Saint-
Pierre 004154
Biblithème, Reims 141310
Bibliothèque Centrale de Prêt du
Saguenay Lac-Saint-Jean, Alma 005317
Bibliothèque d'Art ENSAV de la
Cambre, Fondation Isabelle Errera de
l'Ecole nationale des Arts Visuels,
Bruxelles 003434
Bibliothèque d'Art et d'Archéologie
des Musées d'Art et d'Histoire de la
Ville de Genève, Genève 041513
Bibliothèque d'Art et d'Archéologie
Jacques Doucet, Paris 014621
Bibliothèque de Genève, Espace Ami
Lullin, Genève 041514
Bibliothèque de l'Agglomération, Saint-
Omer 015481
Bibliothèque des Arts, Lausanne 138060
Bibliothèque d'Histoire de l'Art,
Genève 041515
Bibliothèque Marmottan, Boulogne-

Billancourt 012133
Bibliothèque Mazarine, Paris . . . 014622
Bibliothèque Musée, Clermont
(Oise) 012592
Bibliothèque-Musée de la Comédie-
Française, Paris 014623
Bibliothèque-Musée de l'Observatoire,
Paris 014624
Bibliothèque-Musée de l'Opéra,
Bibliothèque National de France,
Paris 014625
Bibliothèque-Musée Forney, Paris 014626
Bibliothèque-Musée Valéry Larbaud,
Vichy 016198
Bibliothèque Publique et Universitaire,
Salle Rousseau, Neuchâtel 041792
Bibliothèque Royale de Belgique-
Koninklijke Bibliotheek van België,
Bruxelles 003435
Biblos, Trieste 132217
Bibracte Musée, Bibracte Centre
Archéologique Européen, Glux-en-
Glenne 013123
Bibro, Denise, New York 122405, 135892
Bical, Lydia, Neuilly-sur-Seine . . 070517
Bican, Istanbul 088108
Bicentennial Art Center and Museum,
Paris 051916
Bicentennial Historical Museum,
Cunnamulla 001008
Bich, Cau, Ho Chi Minh City . . . 125115
Bichl, Miriana, Neuwied 108267
Bichlmeier, Annemarie,
Holzkirchen 076566
Bichon, Vendargues 074006
Bichon, Claude, Pornic 105391
Bichsel, Dr. P., Zürich 144167
Bickel, Minnette D., Pittsburgh . . 123524
Bickleigh Castle, Bickleigh 043591
Bicocca Antic, La Bisbal
d'Empordà 085902
Bicycle Culture Center, Tokyo . . 029489
Bicycle Museum, Three Oaks . . 053741
Bicycle Museum of America, New
Bremen 051237
Bida Loca, Albuquerque 119461
Bidasoa, Hondarribia 086047
Biddington, New York . 122406, 127342
Biddle & Webb, Birmingham . . . 124525
Biddulph, Peter, London 089959, 134598
Bideford Pottery, Bideford 117358
Bider, Lawrence 127280
Bidges, Birmingham 092389
Bidjara C.D.E.P., Charleville 061311
Bidonlivre, Montréal 140355
Bidwell House Museum,
Monterey 051018
Bidwell Mansion, Chico 047709
Biebergrund-Museum,
Biebergemünd 017367
Bieberle, Andreas, Magdeburg . 126032
Biebl, Jürgen, Bremen 125908
Biedenharn Museum and Gardens,
Emy-Lou Biedenharn Foundation,
Monroe 050996
Bieder, Hans, Liestal 087593
Biederberg, Suzanne, Amsterdam 112165
Biedermann-Restaurierung,
Stolpen 130662
Biedermann, Jean-Paul, Colmar 130424
Biedermann, R., Billigheim 074998
Biedermeier, Nijmegen 083508
Biedermeier, Roosendaal 083541
Biegas, Detroit 120804
Bielak, Kraków 113534
Bielefelder Kunstverein e.V.,
Bielefeld 058757
Bielitz, Volker, Leipzig 130221
Bielski Antykwariat Naukowy, Bielsko-
Biała 143433
Biemann & Müller, Bad Bevensen 074493
Bien Phong Museum, Ha Noi . . . 125842
Bienaimé, Bruxelles 063352
Bienaimé, Eric, Hazebrouck . . . 068802

Bienaimé, Pierre, Commercy . . . 068111
Bienal de Arte de Pontevedra,
Pontevedra 098295
Bienal de Arte Panamá, Ciudad de
Panamá 098269
Bienal de Artes Plásticas da Marinha
Grande, Marinha Grande 098273
Bienal Internacional de Arte
Contemporáneo de Sevilla,
Sevilla 098296
Bienefeld, Martin, Lüdenscheid . 107922
Bienefeld, Mirabel, Kempen . . . 076746
Bienen- und Wagnereimuseum,
Bezirksheimathaus, Tollet 002828
Bienenkundemuseum, Felsberg . 018352
Bienenkundemuseum, Münstertal 020612
Bienenmuseum, Duisburg 018051
Bienenmuseum, Plau am See . . 021145
Bienenmuseum des VDRB,
Alberswil 041205
Bienenstein, Ingeborg, Wien . . . 062912
Bieniek, Campbell Town 098857
Bieniek, Wuppertal 130792
Biennale d'Art Contemporain,
Bourges 098074
Biennale d'Art Contemporain,
Lyon 098073
Biennale des Antiquaires de Monaco,
Monaco 098249
Biennale des Antiquaires Paris,
Paris 098075
Biennale des Arts Plastiques,
Besançon 098076
La Biennale di Venezia, Venezia 098184
Bienstock, Herman, New York . . 095270
Bienvenu, Christian, Beaumont-sur-
Sarthe 067007
Bienvenue, Arnhem 082934
Bienville Gallery, New Orleans . . 095061
Bienwaldmuseum, Kandel 019493
Bier, Elmira, Washington 124870
Bierarium, 1. Burgenländisches
Brauereisouvenier-Museum,
Oberwart 002471
Bierbaum, R., Vlotho 078807
Bierhinkel, H., Baden-Baden . . . 106105
Bierhinkel, H., Rheinau 078221
Bierhuizen, Tilburg 132640
Bierkrug-Museum, Kaiserslautern 019479
Biermaier, B. H., Minneapolis . . 145271
Biermann, Dirk Benedikt,
Dortmund 075383
Bierreclamemuseum, Breda 032099
Bies, J.H.P., Eindhoven 132499
Bies, Marie-José, Eindhoven . . . 083209
Biesbosch Museum, Werkendam 033036
Biesenbruch, Wolfgang, Haan . . 076141
Biesenbruch, Wolfgang, Wittmund-
Farlage 079040
Bietz-Hoth, Berlin 106173, 129474
Bietz, Egon & Irene, Nieder-Olm 108272
Biffarino, Brescia 079919
Big Apple Antiques, New York . . 095271
Big Apple Art Gallery, New York 122407
Big Apple Art Gallery & Framing Inc.,
New York 122408
Big Bear Trails Museum, Loon
Lake 006220
Big Beaver Museum and Nature
Centre, Big Beaver 005423
Big Ben, Buenos Aires 060880
Big Bone Lick State Park Museum,
Union 053886
The Big Book, Glen Innes 139754
Big Book Store, Detroit 145137
Big Canvas, Freshwater 117812
Big Cat Press, Chicago 120072, 138318
Big Dipper, Portland . . 096371, 136143
Big Don's Sport Card Shop,
Columbus 092997
Big Hole National Battlefield,
Wisdom 054510
Big Horn County Historical Museum,
Hardin 049356

Big Jar, Philadelphia 145408
Big Orbit Gallery, Buffalo 119951
Big People Toys, Seattle 097407
The Big Picture, London 118216
The Big Picture, New York 122409
Big Pit National Coal Museum of
Wales, Blaenavon 043649
Big Red Barn Gallery, Clearmont 047792
Big River Art Gallery, Dargaville 112870
Big River Memorial Museum, Big
River 005424
Big Shed Antiques, Shepparton . 062207
Big Shell Museum, Tewantin ... 001568
Big Shot, Oakland 123244
Big Sky in Mind, Quezon City .. 113416
Big Springs Museum, Caledonia 047349
Big Thunder Gold Mine, Keystone 049962
Big Timbers Museum, Lamar .. 050104
Big Well, Greensburg 049258
BigArt, Grantham 117872
Bigden, Edward, Debenham ... 089042
Bigelow House Museum,
Olympia 051755
Biger, Yves-Marie, Pont-l'Abbé . 072253,
129175
Biggar Gasworks Museum,
Biggar 043594
Biggar Museum and Gallery,
Biggar 005425
Biggenden Museum, Biggenden 000856
Biggio, Giuseppina, Torino 081660,
142850
Biggör, Paul, Neukirch,
Bodenseekreis 130448
Biggs, Cahirciveen 079469
Biggs, Richmond 096528
Biggs Museum of American Art,
Dover 048348
Biggs Powell, Memphis 094619
Bigler, Dr. Robert R., Rüschlikon 087762
Bigler, Marc, Bern 133669
Bignell, Graham, London 134599
Bigorneau Langoureux, La Flotte 069035
bigprofi GmbH, Frankfurt am
Main 075774, 106976, 125945
Biguet, Paule, Antibes-Juan-les-
Pins 066665
Bigwood, Brasted 088577
Bigwood Auctioneers, Stratford-upon-
Avon 127059
Bihan Antiquités, Saint-Lunaire . 072901
Bihar Tribal Welfare Research Institute
Museum, Ranchi 024460
Bihari Múzeum, Berettyóújfalu .. 023446
Bihler, Maria-Ana, Magstatt-le-
Bas 069806
Bihun, Mieczysław, Wrocław ... 084765
Bijan, Bolzano 079893
Bijbels Museum, Amsterdam .. 031932
Bijbels Openluchtmuseum, Heilig
Landstichting 032442
De Bijenkorf, Nijmegen 083509
Bijenteeltmuseum De Bankorf,
Amen 031904
Bijenteeltmuseum Kalmthout,
Kalmthout 003744
Bijkerk, Willem Fredrik,
Enschede 112476
Bijl, A., Dordrecht 083177
Le Bijou Art Glass, San Diego .. 124106
Bijoux-Bijar, Charleroi 063592
Bijskij Kraevedčeskij Muzej V.V. Bianki,
Bijsk 036898
Bijutsu Kenkyu, Tokyo 138884
Bijutsu Shi, Tokyo 138885
Bijutsu-shi Gakkai, Tokyo 059490
Bijutsu Shuppan-sha, Tokyo .. 137775,
138886
Bijutsu Techo, Tokyo 138887
Bijutsushi Kenkyushitsu, Tohoku
Daigaku Bungaku-bu, Sendai . 056033
Bijzonder Goed, Deventer 083154
Bil & Teknikhistoriska Samlingarna,
Köping 040750

BIL-DAX, Düsseldorf 129734
Bilall, Myri, Berlin 125872
Bilanzuoli, Lucio & Buccio, Pasqualino,
Modena 080775
Bilbao Antigue, Bilbao 085865
Bilbao XXI Subastas, Bilbao ... 126518
Bilboko Arte Eder Museoa / Museo de
Bellas Artes de Bilbao, Bilbao . 039139
Bilboko Itsasadarra Itsas Museoa,
Museo Marítimo Ría de Bilbao,
Bilbao 039140
Bilboko Zezen Museoa, Museo Taurino
de Bilbao, Bilbao 039141
Bilby, Port Douglas 099578
Bild- und Tonarchiv, Landesmuseum
Joanneum, Graz 001968
Bild und Forum, Mühlheim am
Main 108035
Bild und Rahmen, Norderstedt . 108282
Bildarchiv Foto Marburg,
Marburg 020268
Bildegalleriet, Oslo 113193
Bildende Kunst/Kunstpädagogik,
Hochschule Vechta, Vechta ... 055735
Bilder Eigl GmbH & Co. KG, Linz 062714,
100099, 127806
Bilder-Galerie in der Kulturkunststatt
Prora, Prora 108415
Bilder Galerie Minden, Minden . 108003
Bilder im Hof, Flensburg 075740
Bilder Schaab, Zwickau 109008
Bilder till salu, Stockholm 115843
bilderagentur KUNSTTICK.com,
Potsdam 108396
Bilderbogen-Galerie, Neuruppin . 020737
Bilderdijkmuseum, Vrije Universiteit,
Amsterdam 031933
Bildergalerie, Solingen 108633
Bildergalerie Kovan, Duisburg . 106844
Bildergalerie, Stiftung Preußische
Schlösser und Gärten Berlin-
Brandenburg, Potsdam 021178
Das Bilderhaus, Frankfurt am
Main 106977
Bilderhaus Krämerbrücke, Erfurt 106893
Bilderrahmen & Postershop,
Salzburg 127843
Bildhauergalerie für Kleinplastik,
Berlin 106174
Bildhauermuseum Prof.
Wandschneider, Plau am See . 021146
Bildkunst Akademie Hamburg,
Berufsfachschule für
Illustrationsdesign, Hamburg . 055564
Bildmuseet, Umeå 041103
Bildt, Klaus, Krefeld 067951
Bilecik Devlet Güzel Sanatlar Galerisi,
Bilecik 042848
Bilger Breustedt Haus, Taufkirchen an
der Pram 002810
Bilhenry, Milwaukee 121902
Biliárd Múzeum, Sopron 023887
Bilinelli, L., Bruxelles 100569
Biljarskij Gosudarstvennyj Istoriko-
Archeologičeskij i Prirodnyj Muzej-
Zapovednik, Biljarsk 036899
Bilkin Arte Garaikidea, Bilbao . 114981
Bilkova vila, Galerie Hlavního Města
Prahy, Praha 009780
Billabong, Coolangatta 139701
Billamoz, Emmanuel, Chambéry 067793
Billantique, Yverdon-les-Bains .. 087917
Billard, Alain, Roquefort-les-Pins 072540
Billcliffe, Roger, Glasgow 117835
Billebaude, Avallon 066861
Billebaude, Chalon-sur-Saône .. 067774
Billed-Galleriet, Aalborg 102759
Billedhuggerskolen Charlottenborg,
Kongelige Danske Kunstakademi,
København 055303
Billedhuggerskolen Frederiksholms
Kanal, Kongelige Danske
Kunstakademi, København ... 055304
Billedkunstnernes Forbund,

København 058544
Billedkunstskoler, Kongelige Danske
Kunstakademi, København ... 055305
Biller, Georg, Erlstätt 075658
Billes Antik och Kuriosa,
Eskilstuna 086666
Billesberger, Siegfried,
Moosinning 108032
Billett, Allan, Kanahooka 061665
Billett, Darren, Kanahooka 061666
Billiard Room Antiques,
Chilcompton 088879
Billich, Sydney 099770
Billich at the Rocks, Sydney ... 062277
Billing, Gaby & Gert, Baar 116042
Billingham Art Gallery, Billingham 043602
Billings Estate Museum, Ottawa 006571
Billings Farm and Museum,
Woodstock 054532
Billingsley House Museum, Upper
Marlboro 053903
Billis, George, Los Angeles 121445
Billis, George, New York 122410
Billmann, Bert, Aachen 074323
Billnäsin Kirvesmuseo, Billnäs .. 010761
Billon, Montmarault ... 070273, 128805
Bill's Antiques, Miami 094709
Bill's Beanie Babies, Omaha ... 096028
Bill's Sports Collectibles, Denver 093262
Bill's Tobacco Antiques,
Cincinnati 092830
Bill's Upholstery, Tampa 136384
Bill's Used Furniture, Pittsburgh 096259
Billum Antik, Billum 065613
Billy-Andrieu, Montpellier 125647
Billy Bishop Heritage Museum, Owen
Sound 006603
Billy Graham Center Museum,
Wheaton 054362
Billy The Kid Museum, Fort
Sumner 048879
Billy, Dominique, Les Ulmes ... 069436
Billy, Gérard, Soudat 073544
Billy, Jean-Charles, Le Touquet-Paris-
Plage 133967
Billycan Emporium, Coolgardie . 061363
Bilmuseum, Vatnsoyrar 010746
Bilo-Jäger, Brigitte, Bayreuth ... 074692
Bilø, Jens, Brønshøj 065618
Biloko, Bordeaux 103417
Biloxi Art Association, Gulfport . 060408
Bilsak, İstanbul 117017
Bilston Craft Gallery,
Wolverhampton 046179
Biltmore Estate, Asheville 046551
Bilton, Vancouver 101747
Bily Clock Museum and Antonin
Dvorak Exhibition, Spillville .. 053438
Bilyara Gallery, Bournda 098763
Bimba-The Art Hut, Chennai ... 079370,
109278
Bimbo-Martinez, Madrid 133306
Bimbo-Martinez, Sevilla 133429
Bimbo-Martinez, Zaragoza 133493
Bimier, Michel, Villemomble ... 074154
Bimmerle, Louisville 121634
Bín Bán Gallery, Ballyenaught . 109528
Binaghi, A.-P., Fiez 087422
Binalong Motor Museum,
Binalong 000857
Binard, Éric, Venteuil-en-Halatte 074026
Binc, Oklahoma City 145393
Binchester Roman Fort, Bishop
Auckland 043635
Binck, Moerdyk 132588
Binder, Andreas, München 108056
Binder, Andreas, Pregarten 062771
Bindereimuseum Hofbräu
Kaltenhausen, Hallein 002049
Binders, Charlotte 119978
Bindi, Newcastle 061956
Bindi, Alessandro, Firenze 110035
Bineth Gallery, Tel Aviv 109722
Binette-Schroeder-Kabinett,

München 020510
Bing, Seoul 111618
Bingara Museum, Bingara 000858
Binger, København 065859
Binger, Heike, Flensburg 129838
Bingham, Salt Lake City 123948
Bingham & Vance, Cleveland . 092922,
120403
Bingham Gallery, University of
Missouri, Columbia 047898
Bingham, Tony, London 089960
Bingley Antiques, Haworth 089477
Bingley, Robert, Wing 091880
Bingold, Hans, Nürnberg 108294
Bini, Singapore 085375
Bini, Angelo, Cagliari 131062
Bini, Dr. Leo, Roma 131829
Bink, Portland 123601
Binnenhuis 1900, Den Haag ... 083060
't Binnenhuys, Antwerpen 063194
Binnenschifffahrts-Museum Oderberg,
Oderberg 020947
Binnenskamers, Gent 063631
Binner, Barbara, Wuppertal ... 108983
Binnington, Peter, Wareham ... 135185
Binoche & De Maredsous,
Orléans 125668
Binoche, Jean-Claude, Paris ... 125681
Binor, Balakovo 085145
Binscarth and District Gordon Orr
Memorial Museum, Binscarth . 005427
Binsch, Harald, Reutlingen 078210
Binsfeld, Ursula, Ras-Al-Khaimah 134008
Binst, Jean, Pont-l'Abbé 072254
Binz & Andrey, Sankt Antoni .. 133899
Binz & Krämer, Köln 107594
Biodiversity Museum, Dehradun 024212
Biodôme de Montréal, Montréal 006359
Biografmuséet, Säter 040928
Biohistoricum, Museum und
Forschungsarchiv für die Geschichte
der Biologie, Neuburg an der
Donau 020689
Biola University Art Gallery, La
Mirada 050045
Biolchini, Stefano, Milano 080448
Biological Museum, Imphal 024306
Biological Sciences Collection,
Birmingham 043613
Biologičeskij Muzej, Al-Farabi
Atyndagy Kazak Ylttyk Universiteti,
Almaty 029793
Biologiezentrum der
Oberösterreichischen Landesmuseen,
Linz 002316
Biologiska Museet, Örebro 040890
Biologiska Museet, Södertälje .. 040968
Biologiska Museet, Stiftelsen Skansen,
Stockholm 040993
Biology Department Teaching Museum,
Muncie 051114
Biology Museum, Amman 029764
Biomedical Museum, Institute
of Medical Research, Kuala
Lumpur 030684
Bios Terra, Roma 059421
Biosca Galería, Madrid 115158
Bipasha Ghosh, London 118217
Bira Gallery, Balaclava 098689
Biraben, Aurèle, Bergerac 125511
Biraben, Aurèle, Bordeaux 125520
Biraben, Aurèle, Sarlat-la-Canéda 125812
Birbal Sahni Institute of Palaeobotany
Museum, Lucknow 024377
Birch, København 102864
Birch & Libralto, Toronto 101581
Birch, Kenneth, Roskilde 066085
Birch, Robert, Toronto 101580
Bircham Gallery, Holt, Norfolk . 117984
Birchip Local History Museum,
Birchip 000859
Birchmeyer & Son, Paul,
Syracuse 127469
Birchwood Art Gallery, Winnipeg 101869

Bird in Hand, Washington 124871, 145631
Bird, Joanna, London 118218
Bird, John, Petworth 090989
Birdie, Zuid-Beijerland 126302
Birds Birds Birds, Faversham .. 117791
Bird's Nest Miniatures, El Paso . 093389
Birds of Paradise Lampshades, Toronto 128208
Birds of Vermont Museum, Huntington 049624
Birdsey, Paget 100835
Birdsville Working Museum, Birdsville 000860
Birdtail Country Museum, Birtle . 005428
Birendra Museum, Kathmandu . 031871
Birger Sandzén Memorial Gallery, Lindsborg 050316
Birgersson, Halmstad 115672
Biri Auksjonsforretning Ans, Biristrand 126340
Birimbao, Sevilla 115446
Birjand Museum, Birjand 024689
Birka Vikingastaden, Birkamuseet, Adelsö 040509
Birkdale Antiques, Southport .. 091378, 135065
Birkehuset Antik, Nykøbing Sjælland 066018
Birkenes Bygdemuseum, Birkeland 033542
Birkenhagen, D., Quedlinburg . 078107
Birkenhead Priory and Saint Mary's Tower, Birkenhead 043605
Birkholz, Dirk, Chemnitz 129651
Birks, Edmonton 101278
Birks Museum, Decatur 048200
Birla Academy of Art and Culture, Kolkata 055810
Birla Academy of Art and Culture Museum, Kolkata 024356
Birla Century Art Gallery, Mumbai 109420
Birla Industrial and Technological Museum, Kolkata 024357
Birla Museum, Bhopal 024168
Birla Museum, Pilani 024436
The Birman Gallery, Haywards Heath 117941
Birmingham Antique Centre, Birmingham 088473
Birmingham Antique Mall, Birmingham 092390
Birmingham Art Association, Birmingham 060409, 119792
Birmingham Arts Journal, Birmingham 139323
Birmingham Bloomfield Art Center, Birmingham 047011
Birmingham Civil Rights Institute Museum, Birmingham 047004
Birmingham Coin and Jewelry, Birmingham 092391
Birmingham Historical Museum & Park, Birmingham 047012
Birmingham Institute of Art and Design, Birmingham 043614
Birmingham Institute of Art and Design, Birmingham City University, Birmingham 056510
Birmingham Museum and Art Gallery, Birmingham 043615
Birmingham Museum of Art, Birmingham 047005
Birmingham Museums and Art Gallery, Publications Department, Birmingham 138139
Birnam & Wood, New York 122411
Biro & Sons, San Francisco ... 136308
Birocco, Cesare, Torino 142851
Biron 75ter, Saint-Ouen 073038
Birsfelder Museum, Birsfelden . 041347
Birštono Muziejus, Birštonas .. 030323

Birštonas 030324
Birthplace of John Wayne, Winterset 054499
Birties of Worcester, Worcester . 119392
Birtwhistle, David, Worcester .. 119393
Birzgales Novada Muzejs Rūķi, Birzgale 030107
Biržų Krašto Muziejus Sėla, Biržai 030325
Bis, Paris 128937
Bis-Art, København 065860
Bisanzio, Roma 110735
Bisaro, Spilimbergo 110936
Bisarri, Helsinki 066307
Bisbee Mill Museum, Chesterfield 047619
Bisbee Mining and Historical Museum, Bisbee 047013
La Biscaglina, Milano 080449
Biscarini, Paolo, Perugia 131743
Bischöfliches Diözesanmuseum für christliche Kunst, Münster .. 020594
Bischöfliches Dom- und Diözesanmuseum, Trier 022196
Bischöfliches Dom- und Diözesanmuseum Mainz, Mainz 020224
Bischof, Helmut, Sydney 099771
Bischof, Petra, Oberhausen ... 077861
Bischofberger, Bruno, Zürich .. 116759, 138106
Bischoff & Partner, Bernhard, Bern 116122
Il Biscione, Milano 110266
Biscuit Tin Museum, Duns 044180
Bisel, Marcos y Molduras, Cartagena 115006
Bishnu Museum, Assam Regimental Centre, Shillong 024473
Bishop Bonner's Cottage Museum, Dereham 044088
Bishop Hill Colony, Bishop Hill . 047015
Bishop Hill Heritage Museum, Bishop Hill 047016
Bishop Martin Museum, Natchitoches 051197
Bishop of Winchester's Palace Site, Witney 046172
Bishop Philpotts Gallery, Truro .. 119257
Bishop's Falls Heritage Museum, Bishop's Falls 005429
Bishop's Gallery, Bishop's Castle 117384
Bishops' House Museum, Sheffield 045770
Bishop's Lodge Museum, Hay ... 001140
The Bishop's Palace, Galveston . 049031
Bishop's Stortford Museum, Bishop's Stortford 043638
Bishops Waltham Gallery, Bishop's Waltham 117385
Bishops Waltham Museum, Bishop's Waltham 043640
Bishopsgate Antiques, York 091971
Bishopsgate Gallery, London ... 118219
Bishopston Books, Bristol 144274
Bisia Temple Museum, Yilan ... 042588
Biskupski Muzej i Katedralna Zbirka, Hvar 008918
Biskupský Dvůr, Moravské zemské muzeum, Brno 009460
Bislet, Oslo 143395
Bisman, Jean-Jacques, Rouen . 125793
Bismarck Antichità, Bologna ... 079822
Bismarck-Museum, Bad Kissingen 016797
Bismarck-Museum, Fürstlich von Bismarck'sche Kulturstiftung, Friedrichsruh 018529
Bismarck-Museum, Otto-von-Bismarck-Stiftung, Schönhausen 021710
Bisogno, Denis, Sanary-sur-Mer 073397
Il Bisonte, Scuola Internazionale di Arte Grafica, Firenze 055892, 110036
Het Bisschopshof, Utrecht 143242
Bissell, Sara H., Charlotte 092578
Bissogianni, Diana, Athinai 109027

Bissonnet, André, Paris 071055
Bistarelli, Eric, Nice 128852
Bistrot, Washington 097809
Bistrots d'Autrefois, Paris 071056
Bitar, Mohammed & Ahmad, Damascus 088031
Bitforms, New York 122412
Bitouzet, Annie, Moulins (Allier) . 070370
Bits & Pieces, Kitchener 064335
Bits & Pieces, Ta' Xbiex 082436
Bits & Pieces, Tshwane 085674
Bits and Pieces, Amsterdam ... 082702
Bits and Pieces, Broken Hill ... 061217
Bits and Pieces, Egkomi 065316
Bitschnau, Walter, Kufstein 062701, 062702
Bitté Franck, Marckolsheim 069855
Bittel, Oliver, München 108057
Bitter, Alain, Marckolsheim 069856
Bitterlich, Christine, Frauenberg 107048
Bitterne Local History Centre, Southampton 045824
Bittersweet Antiques, Atlanta .. 092115
Bittker, D. & J., New York 122413
Bittners, Louisville ... 094524, 135752
Biuletyn Historii Sztuki, Warszawa 138959
Biuletyn Informacyjny Konserwatorów Dzieł Sztuki, Łódź 138960
Biunno, Joseph, New York 095272, 135893
Biv, Roy G., Columbus 120478
Biville, Laurent, Rouen 072568
Le Bivouac, Cannes 103504
Biwa-ko Bunkakan, Kusatsu ... 028912
Biwa-ko Sosui Kinenkan, Kyoto . 028919
Bix, Austin 135321
Bix, Kansas City 135674
Bix, Louisville 135753
Bix, Milwaukee 135811
Bix, Tucson 136416
Bix, Tulsa 097714, 136428
Bixen, Nykøbing Falster 066013
Bixio, Manuela, Sestri Levante . 110930
Biyun Xuan, Hangzhou 102036
Biz Art, Shanghai 102172
Bizancio, Bogotá 102285, 128263
Bizarre Antiques, Bristol 088646
Bizarre Bazaar, New York 095273
Bizarre Decorative Arts North West, Altrincham 088265, 134026
Biz'Art, Paris 071057
Bize, Hervé, Nancy 104317, 140976
Bizet, Raymond, Beaulieu-sur-Mer 103348
Bizeul, Michel, Taverny 073631
Bizjak, Drago, Bochum 075031
Bizományi Műtárgybolt, Szeged . 079323
Bizy Bee's Books, Wellington .. 143361
Bizzarri, Daniele, Modena 131562
Bizzarro, Catania 080023
Bizzy Lizzy, Bournemouth 144252
BJ Antiques, Chicago 092672
BJ Antiques, Toronto 064615
BJ Oldies, Houston 093624
Bjärnums Museum, Hembygdsmuseet, Bjärnum 040563
Bjarkøy Bygdemuseum, Bjarkøy 033523
Bjarne, Edmonton 140343
Bjarne Ness Gallery, Fort Ransom 048861
Bjerager Fotogalleri, Odder ... 102943
Bjerget, København 065861
Bjerggaard, Bo, København ... 102865
Bjerre, Henrik, København 128394
Bjørgan Prestegård, Kvikne ... 033725
Björkholmen, Roger, Stockholm . 115844
Björklidens Gårdsmuseum, Ljusterö 040794
Björklund, Lars, Halmstad 133536
Bjørn West-Museet, Matredal .. 033767
Bjørnegaard, Viborg 066206
Bjørnlund, Slagelse 140659
Bjørnlund, Sussie, København .. 102866

Bjørn's Antik, København 065862
Björns Handelshus, Örebro ... 086898
Björnsson, Lund 126629
Bjorling & Grant, Minneapolis .. 094902
BJ's Curiosity Shop, Toledo ... 097599
Bjugn Bygdatun, Bjugn 033547
BKW-Museum, Mühleberg 041770
De Blå Antikvariater, Brønshøj . 140561
Blå Galleriet, Göteborg 115628
Det Blå Huset, Tranby 113282
Det Blå Marked, Randers 066052
Blå Pakhus, København 065863
Blå Stolen, Fellingsbro 115615
Blaabjerg Egnsmuseum, Varde Museum, Nørre Nebel 010236
Blaase, Rieseby 078238
Blaavandshuk Egnsmuseum, Varde Museum, Oksbøl 010266
Blache, Pierre, Grenoble 125592
Black, Winnipeg 064914, 140429
Black & Kausel, Paris 104571
Black & Pretty, Baltimore 119727
Black & White Art, Halmstad ... 115673
Black & White Gallery, New York 122414
Black Ace Books, Los Angeles . 145216
Black American West Museum and Heritage Center, Denver 048250
Black Art by Erni, Houston 121006
Black Art Collection, New Orleans 122132
Black Art Plus, Columbus 120479
Black Art Production, Oakland .. 123245
Black Arts, Houston 093625
Black Arts National Diaspora, New Orleans 051287
Black Artwork, Baltimore 119728
The Black Cat Bookshop, Leicester 144522
Black Chat Bookshop, Wendouree 139958
Black Cherry Blues, Paris 141051
Black Country Living Museum, Dudley 044146
Black Creek Pioneer Village, North York 006524
Black Cultural Archives, London 044918
Black Cultural Centre for Nova Scotia, Dartmouth 005679
Black Diamond Antiques, Bulli . 061230
Black Dog Antiques, Bungay ... 088697
Black Dog Gallery, Virginia Beach 097773
Black Dot Gallery, Cremorne ... 098931
Black Dragon Society, Los Angeles 060410, 121446
Black Fashion Museum, New York 051346
Black Gull, London 144562
Black Hawk State Historic Site, Hauberg Indian Museum, Rock Island 052628
Black Heritage Gallery, Houston 121007
Black Heritage Museum, Miami . 050827
Black Hills Mining Museum, Lead 050199
Black Hills Museum of Natural History, Hill City 049474
Black Horse Antiques Centre, Norwich 090868, 134890
Black Images Book Bazaar, Dallas 145111
Black Ink, Stow-on-the-Wold .. 119157
Black Kettle, Mosgiel Dunedin . 083999
Black Kettle Museum, Oklahoma Historical Society, Cheyenne . 047627
Black Laboratory, Seattle 124498
Black Legends of Professional Basketball Museum, Detroit . 048292
Black Letter Books, Winnipeg .. 140430
Black Light Gallery, Berlin 106175
Black Masks, Athens 139324
Black Mountain Circle @ Court Cupboard Gallery, Abergavenny 117218
Black Mountain College Museum and Arts Center, Asheville 046552

Black Nugget Museum,
Ladysmith 006150
The Black Palette Gallery of Modern
Art, Mandaluyong 113393
Black Peppers Gallery, Brisbane 098780
Black Phoenix Galleries, Atlanta 119549
Black Point Editions, Chicago . . 138319
Black River Academy Museum,
Ludlow 050522
Black River Historical Society of Lorain
Museum, Lorain 050396
Black Rock Arts Foundation, San
Francisco 052994
The Black Sheep, Tetbury 091566
Black Sheep Books, Jacksonville 145184
Black Sheep Gallery, Hawarden 117933
Black Sun Books, New York . . . 145326
Black Swamp Creek Antiques,
Tiaro 062320
Black Swan, London 118220
Black Swan, Piedmont 145432
Black Swan Gallery, Congleton . 117622
Black Tulip Antiques, Denver . . 093263
Black Watch of Canada Regimental
Memorial Museum, Montréal . . 006360
Black Woods Museum, Houbi . . 042327
Blackamoor Antiques, New
Orleans 095062
Blackberry Farm-Pioneer Village,
Aurora 046673
Blackberry Patch, Columbus . . 092998
Blackbird, San Francisco 124232
Blackbird Antiques, Naracoorte . 061939
Blackbird Pottery, Dunedin 112875
Blackbrook Antiques Village,
Weeford 091767
Blackburn Museum and Art Gallery,
Blackburn, Lancashire 043641
Blackburn, Charles H., Neuilly-sur-
Seine 070518, 140990
Blackburn, Norman, Stamford . . 091403
Blacket Book's, Edinburgh 144385
Blackfish Gallery, Portland 123602
Blackflash, Saskatoon . 134681, 138570
Blackfriars Bookshop, Victor
Harbor 139947
Blackgang Sawmill and Saint
Catherine's Quay, Blackgang . 043643
Blackhawk Museum, Danville . . 048142
Blackheath, London 144563
Blackheath Antiques, Mount
Victoria 061922, 139843
Blackheath Gallery, London . . . 118221
Blacklock, Egham 144405
Blackman Museum, Snohomish . 053371
Blackmon, James, San Francisco 097192
Blackmore, Cassandria, Seattle . 124499
Blackpool Mill, Narberth 045312
Blackridge Community Museum,
Blackridge 043646
Blackrock Art Gallery, Blockrock 109533
Blackrock Upholstery, Blackrock 130829
Blacksburg Regional Art Association,
Blacksburg 060411
Blacksmith Shop Museum, Dover-
Foxcroft 048358
Blacksmiths Antiques,
Blacksmiths 061166
Blacksmith's Cottage, Bacchus
Marsh 000810
Blackstone, Honolulu 093532
Blackstone Valley Historical Society
Museum, Lincoln 050306
Blackstone, Linda, Pinner 118913
Blackthorn Gallery, Riverton . . . 099642
Blackwater Draw Museum, Eastern
New Mexico University, Portales 052269
Blackwater Pine Antiques, Truro 091641,
135152
Blackwater Valley Museum,
Benburb 043566
Blackwell, London 144564
Blackwell – The Arts and
Crafts House, Bowness-on-

Windermere 043686
Blackwell Publishing, Oxford . . . 138271
Blackwell Street Center for the Arts,
Denville 048278
Blackwell's Rare Books, Oxford . 144766
Blackwood, Pittsburgh 123525
Blackwood Gallery, University
of Toronto at Mississauga,
Mississauga 006339
Blackwood, Simon, Hawick 134449
Blad-bixen, Århus 140550
Blades, Katrina L., Birmingham . 092392
Blaeschke, Darmstadt 141679
Blaess, Thomas, Bern 116123
Blätter für Grafik und Literatur,
Trier 138687
De Blaeue Acolye, Vlissingen . . 143258
Blågårds Plads Antikvariat,
København 140597
Blagbrough, Saint Louis 123809
Blagg, Daniel, Fort Worth 120856
Blagoveščenskij Sobor,
Gosudarstvennyj muzej-zapovednik
Moskovskij Kreml, Moskva . . . 037235
Blain, Claude, Montréal 064396
Blaine County Historical Museum,
Hailey 049313
Blaine County Museum, Chinook 047725
Blaine House, Augusta 046665
Blaine Lake Museum, Blaine
Lake 005430
Blair Antiques, Blair Atholl 088502
Blair Castle, Blair Atholl 043652
Blair House Antiques,
Washington 097810
Blair House Museum, Kentville . . 006082
Blair Museum of Lithophanes,
Toledo 053757
Blair, Brad, Los Angeles 094273
Blair, Robert, Richmond 096529
Blairman & Sons, H., London . . 089961
The Blairs Museum, Aberdeen . 043355
Blais, Michel, Vancouver 101748, 101749
Blais, Simon, Montréal 101367
Blaise Castle House Museum,
Bristol 043746
Blake, Woodbridge 144934
Blake & Palmer, Chicago 120073
The Blake Gallery, Crewkerne . . 117645
The Blake Gallery, York 119405
Blake Museum, Bridgwater . . . 043721
Blakesley Hall, Birmingham . . . 043616
Blakey, Colin, Nelson . 090806, 134867
Blakey & Sons, J.H., Brierfield . 088609
Blakstad Sykehus Museum,
Asker 033510
Blana, Matthias, Pforzen 130515
Blanc-Anselme, J. P., Paris 071058
Blanc Bertrand, Marssac-sur-
Tarn 069976
Blanc et Bleu, Jouy-en-Josas . . 068923
Blanc, Alain, Castres 067727
Blanc, André, Toulouse 073741
Blanc, Annie, Lyon 104069
Blanc, Peter, München 137390
Blanc, Philippe, La Canourgue . 068962
Blanchaert, Milano 080450
Blanchard, Froxfield 089314
Blanchard, London . . . 089962, 134600
Blanchard & Co., New Orleans . 095063
Blanchard, Nîmes 070688
Blanchard, Tourtour 105767
Blanchard, Albert, Paris 141052
Blanchart & Cie., Gerard,
Bruxelles 136631
Blanche & Co., Los Angeles . . . 094274
Blanchet, Deni, Montréal 064397
Blanchet, Monique, Ribérac . . . 072452
Blanchetière, Yves, L'Aigle 125606
Blanchetti, Jean-François de,
Paris 071059
Blanchon, J.P., Lisboa 126439
Blanchy & Lacombe, Bordeaux . 125521
Blanckaert, Alain, Reims 072383

Blanco Family Museum, Angono 034425
Blanco Seco, Iván, Gijón 086016
Blanco y Negro, Valencia 086505,
133450
Blanco, Mercedes, Pamplona . . 086382
Blancon, Gilles, Hannover 076365
Blancpain, Genève 116286
Le Bland, West Wyalong 099883
Bland District Historical Museum,
West Wyalong 001635
Bland, Gerald, New York 095274
Blanden Art Museum, Fort Dodge 048804
Blandford Forum Museum, Blandford
Forum 043654
Blandorama, Karlstad 086777
Blank Art, Kyïv 117122
Blank Canvass Art Gallery,
Houston 121008
Blank, Herbert, Stuttgart 142385
Blank, Marcus, Växjö 087149
Blank, Peter, Wilburgstetten . . . 079007
Blank, Reinhard, Bad Vilbel 125858
Blankenburger, Ingrid, Berlin . . . 141489
Blankschmiede Neimke, Dassel . 017832
Blaquière, Geneviève, Sorèze . . 073540
Blås & Knåda, Stockholm 115845
Blas Bruni Celli, Caracas 145648
Blas Garcia, Jose Maria,
Barcelona 133105
Blas Vega, Jose, Madrid 086135
Blásastova, Gøtu Fornminnissavn,
Norðagøta 010738
Blasberg, Doris & Hartmut,
Basberg 074687
Blasco, Monterrey 111988
Blasco Querol, Maria Antonia,
Sabadell 115396
Blasco, Francesco, Milano 080451
Blaser & Herrmann, Bern 087274,
133670
Blaser, R., Lahnstein 076993
Blasichc, Lima 084442, 132805
Blasingame, Fort Worth 093432
Blasinstrumenten-Sammlung,
Zimmerwald 042182
Blåsippans Antikt och Gammalt,
Söderhamn 086942
Blasius, Umeå 115962
Blaskovich Múzeum, Pest Megyei
Múzeumok Igazgatóságá,
Tápiószele 023992
Blasmusik-Museum, Ratten 002573
Blason, A Coruña 085949
Blassiaux, Richard, Paris 128938
Blast From the Past, Kansas City 094495,
135675
Blast From the Past, Milwaukee 094836
Blasthaus, San Francisco 124233
Blaszczyk, Michael, Bad Homburg
v.d.Höhe 106062
Blatchfords, San Diego 136291
Blatter, Hans-D., Münster 142199
Blau & Scheibl, Wien 062913
Blau-gelbe Viertelsgalerie und Kleine
Schlossgalerie, Bad Fischau-
Brunn 001739
Blau-gelbe Viertelsgalerie Weistrach,
Weistrach 002908
Blau-gelbe Viertelsgalerie, Weinviertel,
Mistelbach an der Zaya 002391
Blau, Daniel, München 108058
Blau, Dorothy, Saint Louis 123810
Blau, Vojtech, New York 095275
Die Blaue Eule, Essen . 106907, 137182
Blaues Schloss mit
Museumspädagogischer Werkstatt
Rotes Schloss, Obernzenn . . . 020913
Het Blauwe Huys, Melick 132580
De Blauwe Olifant, Maastricht . . 083456
De Blauwe Vlinder, Haarlem . . . 112509
Blåvand Museum, Varde Museum,
Blåvand 010020
Blåvand Redningsbådsmuseum, Varde
Museum, Blåvand 010021

Blawnox Antiques, Pittsburgh . . 096260
Blaze, Bristol 117438
Błażej, Piotr, Poznań 132858
Blazing Sun, Cincinnati 120323
Blazquez Barbero, Carmelo,
Madrid 086136
Blázquez Barbero, Guillermo,
Madrid 143755
Blazquez, Miguel, Madrid 143754
Bleachers, Cincinnati 092831
Bleak House, Broadstairs 043759
Blech, Wolfgang, Heilbronn 076468,
130030
Der Blecheimer, Düsseldorf . . . 075458
Blecher, George, San Francisco . 124234
Blechschmidt, Gotthard, Frankfurt am
Main 075775
Le Bled, Sophie, Requeil 072445
Bledel, Marcos F., Buenos Aires 098555
Bleeker Street, Chicago 092673
Bleeker, Otfried, Sylt 112870
Blehle, Karl, Seligenstadt 108608
Blei, Karl, Reiskirchen . 078190, 130577
Bleibtreu-Antik, Berlin 074766
Bleibtreu Galerie, Berlin 106176
Bleibtreu-Tomic, Clara, La Garenne-
Colombes 103892
Bleich-Rossi, Wien 100239
Bleier, Milton A., New York 122415
Bleijenberg, Amsterdam 112166
Bleikeller, Sankt Petri Domgemeinde,
Bremen 017582
Bleisch, Adrian, Arbon 116028
Blekinge Museum, Karlskrona . . 040739
Bleloch Museum, Johannesburg 038746
Blem, København 065864
Blench & Son, John, Saint Helier 144815
Blend, Barry, Collioure 103614
Blender, Paddington, New South
Wales 099515
BlendeVier, Oldenburg 108337
Blenheim, Fairford 089243
Blenheim Palace, Woodstock . . 046187
Blenko Museum of Seattle,
Seattle 053228
Blens & Hartmayer, Ofterdingen 077914
Blenwell, San Francisco 136309
Blésoise, Blois 067251, 128473
Bletchley Park, Bletchley 043658
Bleu 214, Toulon (Var) 073703
Bleu de Chine, Fleurier 116265
Bleu de Paris, Paris 071060
Le Bleu du Ciel, Lyon 069710
Bleu Galerie, Carouge . 087348, 116187
Bleu Line, Lisboa 132930
Bleu Passé Antiquités, Paris . . . 071061
Bleu Passion, Fréjus 103754
Bleu Sonatine, Dinan . 068350, 103683
Bleu Vert, Fort-de-France 111928
Bleu-Vert Antiquités, Paris 071062
Bleue Comme Une Orange,
Ajaccio 066496, 103210
Bleue Forêt, Salers 105641
Blewbury Antiques, Blewbury . . 088509
Blewett, E.K., Austin 092220
Blewitt-Harrison-Lee Museum,
Columbus 047928
Bleymaier, E.A., Egling 075578, 106864,
129774
Blicheregnens Museum, Kjellerup 010162
Blichermuseet på Herningsholm,
Herning 010111
Blichfeldt, Randers 140642
Blickensdorff, Berlin 106177
Blickling Hall, Norwich 045395
Blickpunkt, Hamburg 107231
Blidah, Hellerup 102824
Bligh Museum of Pacific Exploration,
Bruny Island 000902
Blikmuseum De Blikvanger,
Renesse 032778
Bliktrommen Antik, Odense 066032
Blind Art, London 060064
Blind Tiger Gallery, Quezon City 113417

Blindaž gde podpisan akt o
 kapituljacii, Oblastnoj istoriko-
 chudožestvennyj Muzej,
 Kaliningrad 037079
Blindehistorisk Museum, Hellerup 010099
Blink Gallery, London 118222
Blinky, Denver 093264
Blissett, Gary, Long Preston 118658
Blists Hill Victorian Town, Telford 045954
Blitz, Amsterdam 082703
Blob Art, Livorno 110174
Bloc Acier, Rouen 072569
Bloch, Oberhausen 077862
Bloch, Francis, Delémont 144063
Bloch, René, Frick 087436
Blochererschule, Berufsfachschule
 für Kommunikationsdesign &
 Innenarchitektur, München . . . 055646
Blochwitz, J., Hamburg 076196
Block 43 Studio Gallery,
 Singapore 114499
Block, René, Berlin . . . 106178, 137056
Blocker, Peggy, San Antonio . . . 096952
Blockhaus de la Rivière-Lacolle, Saint-
 Paul-de-l'Île-aux-Noix 006929
Blockhaus d'Éperlecques,
 Éperlecques 012874
Blockheads and Granary Antiques,
 Flore 089285
Blockhouse Museum, Merrickville 006293
Blocquel-Chapart, Françoise, Pau 072035,
 105327
Blodveimuseet, Bygdetunet, Saltnes,
 Rognan, Saltdal, Rognan 033903
Blöchle, Peter, Basel 087223
Blödorn & Ewert, Krummwisch . 137332
Blohm, Reiner, Freiburg im
 Breisgau 075851
Blois Antiquités, La Chaussée-Saint-
 Victor 068990
Blokken, Rotterdam 143217
Blokkodden Villmarksmuseum,
 Engerdal 033588
De Blokschaaf, Bussum 083016
Blokwachterswoning Waterhuizen,
 Waterhuizen 033029
Blom, Dordrecht 112433
Blome, Lars, Hannover 107348
Blomqvist, Bergen 084214
Blomqvist, Minde 084277
Blomqvist, Oslo 084302, 113194, 126366
Blomqvist & Co, Stockholm . . . 126669
Blond Fine Art, Saint James . . . 062178
Blondeau & Breton, Paris 071063,
 104572
Blondeel, Bernard, Antwerpen . . 063195
Blondel, Alain, Paris 104573
Blondel, Jean-Michel,
 Talmontiers 073617
Blondie's Antiques, Louisville . . 094525
Blonka, Stanisław, Lublin 113685
Bloodhound-Lenkwaffenstellung,
 Menzingen 041742
Bloom, San Diego 097068
Bloom Mansion, Trinidad History
 Museum, Trinidad 053805
Bloom & Son, N., London 089963
Bloomfield, Bungendore 098822
Bloomfield, Pittsburgh 096261
Bloomfield Academy Museum,
 Bloomfield 047037
Bloomfield Science Museum
 Jerusalem, Jerusalem 025071
Blooming Art Gallery, Los
 Angeles 121447
Bloomington Art Center, Greenberg
 Gallery & Atrium Gallery,
 Bloomington 047052
Bloomington Historical Museum,
 Bloomington 047053
Bloomsbury & Co, Ottawa 064497
Bloomsbury Antiques, Jolimont . 061659
Bloomsbury Auctions, London . . 126954
Blore, Nottingham 144755

Blosfield, Tampa 136385
Bloss, Merzhausen 126037
Blossom Street Gallery, Houston 121009
Blou, Dorte & Jesper, Holte 065782
Blouet, Pascal, Mayenne 125641
Bloume, Christian, Sainte-Marie 132989
Blount-Bridgers House/ Edgecombe
 County Art Museum, Tarboro . . 053682
Blount Mansion, Knoxville 005005
Blow de la Barra, London 118223
Blowfly Art & Design, Brunswick 127572
Blowing Rock, Charlotte 092579
Bloxham, London 118224
Bloxham Village Museum,
 Bloxham 043660
Bloxham, John, London 089964, 134601
Blu Cina, Milano 080452
Blu Old Sheffield, Roma 081262
Bluart, Padova 110557
Blue & White, Jerusalem 109714
Blue Antiques, Dallas 093076
Blue Antiques, Toronto 064616
Blue Beyond, Balemartine 117269
Blue Butterfly, Nice 070576
Blue Dolphin Gallery, Mooloolaba 099368
Blue Dot Gallery, Bristol 117439
The Blue Dot Gallery, Toronto . . 101582
Blue Earth County Museums, Heritage
 Center Museum & R. D. Hubbard
 House, Mankato 050651
Blue Flag Gallery, London 118225
Blue Fonz, Antwerpen 063196
Blue Gallery, Kansas City 121280
Blue Gallery, Richmond, New South
 Wales 099623
Blue Gallery, Wolverhampton . . 119380
The Blue Heron, Vancouver 064801
Blue Hills, Grose Vale 061564
Blue Hills Trailside Museum,
 Milton 050903
Blue House Antique, West
 Hollywood 097876
Blue Lake Fine Arts Camp, Twin
 Lake 058019
Blue Leaf Gallery, Dublin 109577
Blue Lias, Lyme Regis 118675
Blue Licks Battlefield Museum, Mount
 Olivet 051092
Blue Moon Antiques, Taranaki . . 084089
Blue Moon Books, Waikato 143351
Blue Moon Contemporary Art,
 Haslemere 117922
Blue Moon Contemporary Art, Henley-
 on-Thames 117958
Blue Moon Trading, Tampa 097546
Blue Mountain Gallery, New York 122416
Blue Mountains Antiques,
 Woodford 062450
Blue Nile Papyrus, Cairo 103024
Blue Oyster Gallery, Dunedin . . 033187
Blue Pegasus, Albuquerque . . . 092037
Blue Penguin, Russell 084073
Blue Pomegranate Studios,
 Omaha 123305
Blue Pony Gallery, Charlotte . . . 119979
Blue Pump Antiques, Toronto . . 064617
Blue Ridge Institute and Museum,
 Ferrum 048727
Blue Room Gallery, San
 Francisco 124235
Blue Room Gallery & Gift Shop,
 Bethany Beach 119781
Blue Ruin, Pittsburgh 123526
Blue Shore Gallery, Taurikura . . 113028
Blue Sky, Oregon Center for the
 Photographic Arts, Portland . . 052288
Blue Space Gallery, Ho Chi Minh
 City 125116
Blue Star Contemporary Art Center,
 San Antonio 052931
Blue Steel Art, Bern 087275
Blue Stone Gallery, Kenmare . . 109646
Blue Sun Gallery, Ha Noi 125063
Blue Tulip Gallery, Windsor 119363

Blue Umbrella Gallery, Listowel . 109672
Blue Untitled, Chicago . 120074, 120075
Blue Words, Lyon 140884
Bluebell Gallery, Killarney 109651
Bluebell Railway, Uckfield 046014
Blueberry Hill in Old Town,
 Sacramento 096606
Bluebird Art Gallery, Louisville . . 121635
Bluebird Studio, Indianapolis . . 121170
Bluebottle Art Gallery 415 E Pine St,
 Seattle 124500
Blüchermuseum, Kaub 019556
Bluelight Pictures, Halmstad . . . 115674
Blüm, Bad Dürkheim 074504
Blümel, Ratingen 078134
Blümel, Unterguggenberger & Mandl,
 Wien 127904
Bluemingdeal, Indianapolis 093863
Blümmel, C. & N., Hasselberg . . 076412
Bluemoon Gallery, Tunbridge
 Wells 119268
Bluestone Art, Lincoln 118116
Bluestone Gallery, Devizes 117676
Bluestone Museum, Hinton 049492
Bluestone Stables Gallery,
 Axedale 098680
Bluette, Christian, Bordeaux . . . 067317
Bluewood Gallery, Redfern 099618
Bluff Hall Museum, Demopolis . . 048240
Bluff Maritime Museum, Bluff . . 033163
Bluff Park Antiques, Birmingham 092393
Bluff Park Rare Books, Long
 Beach 145206
Bluff Road Upholstery,
 Indianapolis 135629
Blum, New York 095276
Blum & Poe, Los Angeles 121448
Blum, Charly, Antibes-Juan-les-
 Pins 066666
Blum, Egon, Staad 087825
Blum, Karlheinz, Spechbach . . . 078516,
 130647
Blum, Norbert, Marambat 069841
Blum, Otto, Steinach, Kreis Straubing-
 Bogen 078547
Blum, Peter, New York 122417
Bluma, Jed, Baltimore 092301
Blumberger Mühle, NABU
 Erlebniszentrum, Angermünde . 016582
Blume, Frederiksberg 140567
Blume, Birgitt, Hamburg 076197
Blumer, Rolf-Dieter, Schwäbisch
 Gmünd 130631
Blumka, Anthony, New York . . . 095277,
 122418
Blunden, Vanessa, Wimborne . . 119355
Blunderbuss, London 089965
Bly, John, London 089966
Bly, John, Tring 091633
Blyburgate Antiques, Beccles . . 088417,
 134085
Blyth, Manchester 118703
BM Çağdaş Sanat Merkezi,
 İstanbul 117018
BM Traders, Singapore 114500
BMB Galerie, Amsterdam 112167
BML Fine Arts, New York 122419
BMV Books, Toronto . . 140401, 140402
BMW Museum, München 020511
B'nai B'rith Klutznick National Jewish
 Museum, Washington 054095
Bô, Montpellier 104271
Bo Fang Shopping Arcade,
 Shenzhen 102217
Bo-Jo's Antique Mall, Memphis . 094620
Bo-Kaap Museum, Iziko Museums of
 Cape Town, Cape Town 038673
BO!, 20th Century Design, Art and
 Ethnographica, Stockholm . . . 086977
Boa, Flavio, Padova . . 080967, 131654
Boa, Gianni, Padova 131655
BOA2000, Wil 087903, 144151
Boada Saquero, Agustin,
 Barcelona 085738

BOAFair asbl, Bruxelles 063353
Bo'ai Museum, Bo'ai 007658
Boal Mansion Museum,
 Boalsburg 047063
Boam, Clare, Horncastle 089581
Boandik Cultural Museum,
 Beachport 000832
Boann Art, Drogheda 109566
Boardwalkart, Merimbula 099327
Boarman Arts Center,
 Martinsburg 050724
Boase Antiques, Exmouth 089238
Boat'art, Saint Paul 123898
Boatmen's National Bank Art
 Collection, Saint Louis 052791
Boato, Giannino, Venezia 111091
Boats 'n Pieces, Paradise Point . 062060
Boatwright, Franklin 127227
Bob Burns Museum and River Valley
 Museum, Van Buren 053941
Bob Campbell Geology Museum,
 Clemson University, Clemson . 047797
Bob Jones University Museum &
 Gallery, Inc., Greenville 049267
Bob Marley Museum, Kingston . 028439
Bobbins, Whitby 091825, 135221
Bobbitt Visual Arts Center, Albion
 College Print Collection and Art
 Exhibitions, Albion 046355
Bober, Vitoria-Gasteiz 086571
Bobet, Rennes (Ille-et-Vilaine) . . 072425,
 072426
Bobik, Dr. W. & M., Graz 062582
Bobin, Matthieu, Vannes
 (Morbihan) 105818
Bobinen, Örebro 086899
Bobo, Linz 100100
Bobritz, Rechberghausen 078152
Bob's Attic Antiques, Saint Louis 096672
Bob's Coin Store, Milwaukee . . 094837
Bob's Coins, Milwaukee 094838
Bob's Comics, Birmingham 092394
Bob's Custom Woodworking,
 Columbus 135474
Bob's Used Furniture, Providence 096500
Bobsien, Winfried, Hamburg . . . 076198
Boca Raton Historical Society Museum,
 Boca Raton 047065
Boca Raton Museum of Art, Boca
 Raton 047066
Bocca, Milano 137670, 142714
Bocca Tigris, Stockholm 086978
Boccador, Jacqueline, Paris . . . 071064
Boccara, Paris 071065
Boccard, Olivier, Villemoyenne . 074156
Boccarossa, Padova 080968
Bocchia, Ettore, Parma 081099
Boch, Eva, Kürnach 076985
Bochmann, Klaus, Limburg an der
 Lahn 077111
Bocholter Brouwerij Museum,
 Bocholt 003373
Bochon, Pascal, Vincennes 074239
Bochumer Künstlerbund e.V.,
 Bochum 058758
Bock, Augsburg 074454
Bock & Herchen, Bad Honnef . . 137035
De Bock, François, Lille 128676
Bock, Friedel-Paul, Wien 127905
Bock, Günter, Saarbrücken 078317
Bock, H.-P., Nörten-Hardenberg . 077793
Bockrath, Cleveland 120404
Bockwindmühlenmuseum, Doberlug-
 Kirchhain 017916
Bocquet, Emmanuel, Frevent . . 068615
Bocquier, Gérard, Nantes 140982
Bocquillon, André, Honfleur . . . 068687
Bocskai István Emlékkiállítás, Bocskai
 Kastélymúzeum, Nagykereki . . 023805
Bodacious Books, Nashville . . . 145290
Bodanna, New York 122420
Bodart & Valter, Sipplingen 108631
Bodaskogs Affärs- och Postmuseum,
 Aneby 040527

Boddington, Pat, Mount Glorious 099398
Bode, Karlsruhe 107509
Bode-Museum – Münzkabinett,
Staatliche Museen zu Berlin –
Stiftung Preußischer Kulturbesitz,
Berlin 017088
Bode-Museum – Skulpturensammlung
und Museum für Byzantinische Kunst,
Staatliche Museen zu Berlin –
Stiftung Preußischer Kulturbesitz,
Berlin 017089
Bode, Katrin, Ratingen 130556
Bode, Klaus, Borgholzhausen . . 075099
Bode, Klaus D., Nürnberg 108295,
. . . . 137449
Bode, Ralf, Bonn 075065
Bode, Wolfgang, Meckenheim,
Rheinland 077315
Bodeco, Paris 128939
La Bodeguita del Medio Obrero,
Medellín 102429
Bodek, Hannover 107349
Bodelwyddan Castle Museum,
Bodelwyddan 043661
Bodenheimer Heimatmuseum,
Bodenheim 017446
Bodenheimer, Heinz, Frankfurt am
Main 075776
Bodensee-Naturmuseum,
Konstanz 019758
Bodensee-Obstmuseum,
Frickingen 018512
Bodes & Bode, Den Haag 083061
Bodhi, Bergen op Zoom 082958
Bodhi, Delhi 109310
Bodhi, Mumbai 109421
Bodhi, New York 122421
Bodhi, Portland 096372
Bodhi, Singapore 114501
Bodhi Tree Bookstore, West
Hollywood 145642
Bodhouse Antiques, Birkenhead 088470
Bodiam Castle, Robertsbridge . . 045624
Bodie Park Museum, Bridgeport 047199
Bodijn, Jan, Leiderdorp 132564
Bodils Antik, Maribo 065998
Bodin, Strasbourg 073575
Bodin-Lemaire, Brigitte, Olonne-sur-
Mer 070780
Bodin, François, Tours 073840
Bodin, Sophie, Soussans 073554
Bodin, Thierry, Paris 141053
Bodis, Frank, Wiesbaden 130755
Bodmin and Wenford Railway,
Bodmin 043662
Bodmin Town Museum, Bodmin 043663
Bodnar, Joseph, Somerset 127460
Bodø Kunstforening, Bodø 059655
Bodoque, J., Madrid 133307
Bodrogközi Kastélymúzeum,
Pácin 023824
Bodrum Sualtı Arkeolojisi Müzesi,
Bodrum 042849
Bodt, Chicago 120076
Body Focus, Providence 096501
Body, Martin, London 134602
Bodybuilder & Sportsman Gallery,
Chicago 120077
Bø Museum, Bø i Telemark 033551
Boe, Patrick, Marennes 069870
Böblinger Kunstverein, Kunstverein für
Stadt und Landkreis Böblingen e.V.,
Böblingen 058759
Böck, Anton, Wien . . 062914, 127906
Boeck, de, Antwerpen 136618
De Boeck, Paule, Gent 100689
Böck, Wilhelm & Sophie,
Langenbach 130211
Böckelmann, Uwe, Bretnig-
Hauswalde 129638
Böcking, Heinrich, Ralingen . . 130553
Böckli, August, Aathal-Seegräben 087179,
. . . . 133613
Böcksteiner Montanmuseum Hohe

Tauern, Böckstein 001789
Böd of Gremista Museum,
Lerwick 044811
Böddeker & Schlichting,
Paderborn 130503
Bödeker, Erich, Hannover 107350
Böer, Claudia, Hannover 107351
Bøgebjerg Antik, København . . . 065865
Bøger og Antik, Silkeborg 066107,
. . . . 140653
Bøger og Kuriosa, København . . 065866,
. . . . 140598
Bögge, Alfred, Hamm, Westfalen 076347
Böggemeyer, W., Hasbergen . . . 076410
Boegly, Serge, Rupt-sur-Moselle 072649
Böhlau Verlag, Köln 137304
Böhlau Verlag GmbH & Co. KG,
Wien 136577
Böhler, Julius, Starnberg 078536
Böhler, Lothar, Oppenweiler . . 077952
Boehm Gallery, San Marcos . . . 053051
Böhm, Andreas, Wien 062915
Böhm, Bernhard, Freilassing . . . 129864
Böhm, Harald, Berlin . . 074767, 141490
Böhm, J., Bielefeld 074974
Böhme, Husum 141955
Böhme, Sibylle, Berlin 141491
Böhme, Thomas W., Dresden . . 129705
Böhmer, Marita, Ostritz 142286
Böhmerwaldmuseum Wien und
Erzgebirger Heimatstube, Wien 002979
Böhmisch, Passau 078016
Böhmisch, Eva, Passau 078017
Böhmisch, Hans L., Passau . . . 078018
Böhmländer, Ch., Herrenberg . . 076508
Böhringer, Georg, Düsseldorf . . 075459,
. . . . 129735
Böhringer, Heinrich, Wunsiedel . 142519
Boeijink & Boekel, Haarlem . . . 132515
Boek & Glas, Amsterdam 143014
Het Boekdorado, Den Haag . . . 143108
Böke Museum, Leer 019949
Böke, Christiane, Neulingen,
Enzkreis 130449
Boekenantiquairaat de Lezenaar,
Hasselt 140229
De Boekenbeurs, Middelburg . . 143203
De Boekenboet, Alkmaar 143000
De Boekenboot, Hallum 143165
Boekendaal, Gouda 143146
De Boekenhof Emmen, Emmen . 143137
't Boekenschuurtje, Warffum . . . 143262
De Boekenvriend, Albergen . . . 142999
Böker, Bayreuth 074693
De Boekerij, Baarn 143088
Boekhout, Phoenix 096193
Boekie Woekie – Books by Artists,
Amsterdam 137827, 143015
Boekie Woekie, Books by Artists,
Amsterdam 112168
't Boektiekske, Holsbeek 140231
BÖKWE, Bildnerische Erziehung +
Technisches Werken + Textiles
Gestalten, Wien 138507
't Boekwerk Antiquariaat,
Nijmegen 143208
Boelens, Marian, Zwolle 083730
Bölicke, F., Schortens 078393
Böll, Klaus, Würzburg 079076
Böll, R., Flintbek 075752
Böllner, Herbert, Lauter 130218
Boema, Praha 102654
Boemboem, Utrecht 112693
Boer Tala Zhou Museum and Gallery,
Bole 007659
Boer, H. de, Groningen 083279
Boer, M.L. de, Amsterdam 082704
Boer, Theo de, Zandpol 143269
Börde-Museum Burg Ummendorf,
Ummendorf Bördekreis 022270
Das Bördemuseum, Lamstedt . . 019866
Bördenheimatmuseum,
Heeslingen 019125
Boerderij Klein Hulze, Almen . . 031865

Boerderij Museum Duurswold,
Slochteren 032883
Boerderijmuseum De Bovenstreek,
Oldebroek 032711
Boerderijmuseum De Lebbenbrugge,
Borculo 032080
Boerderijmuseum Zelhem,
Zelhem 033095
Boerenantiek, Eindhoven 083210
Boerenbondsmuseum, Gemert . 032339
Boerenkrijgmuseum, Overmere . 003946
Boerenwagenmuseum, Buren,
Gelderland 032122
Børgesen, Bent, København . . . 065867
Börjeson, Malmö 115740
Börjesson, Göteborg 115629
Børnenens Museum, Nationalmuseet,
København 010167
Boerner, C.G., Düsseldorf 075460,
. . . . 141715
Boerner, C.G., New York 095278
Börner, Claudia, Radeburg 078114
Börnert, Armin, Eichenau 129779
Boerth, Richard, Seattle 097408, 136365
Bø's Jakob Antikk, Voss 084396
Bösch, Walter, Bayreuth 141456
Boesen, Hans, Mindelheim 108002
Boesendorfer, New York 095279
Boesky, Marianne, New York . . . 122422
Boesmans, M., Köln 076852
Boesner Künstlerbedarf, Wien . . 140662
Bössmann, Osnabrück 142274
Bøtø Nor GI. Pumpestation,
Væggerløse 010347
Boettcher, Gerrit, Hannover . . . 076366
Böttcher, H., Heidelberg 076435
Boetto, Genova 080238, 126194
Bœuf, Frederic, Toulon (Var) . . 129284
Bœuf, Robert, Toulon (Var) 105709
Bofors Industrimuseum,
Karlskoga 040734
Bog-art, Holbæk 140585
Bog-Børsen, København 140599
Bogan, Spartanburg 127462
Bogan House Antiques, Totnes . 091620
Bogantikken, Roskilde 140651
Bogapott, Tallinn 103111
Bogart, Groningen 083280
Bogart, Bremmer & Bradley,
Seattle 097409, 136366
Bogat, Baltimore 092302, 135338
Boğazköy Müzesi, Boğazkale . . 042850
Bogcentralen, Viborg 140672
Bogcentrum, Esbjerg 140564
Bogd Khaan Winter Palace Museum,
Ulaanbaatar 031727
Bogdanoff, New Orleans 122133
Bogena, Saint-Paul 105572
Boggiali, Milano 080453
Boggiano, H. & H., Buenos Aires 098556
Boghytten, København 140600
Bogkælderen, Hillerød 140583
Boglio, Jean-Louis & Janet,
Currumbin 139709
Bogner, E., Landshut 077012
Bogner, Willi, München 130331
Bognor Regis Museum, Bognor
Regis 043667
Bogormen, Esbjerg 065637
Bogormen, København 140601
Bogorodickij Dvorec-muzej i Park,
Bogorodick 036905
Bogselskabet, København 140562
Bogshoppen, Århus 140551
Bogshoppen, Holstebro 140586
Bogstad Gård, Oslo 033827
Bogstadveien Antikk, Oslo 084303
Bogtrykmuseet, Esbjerg 010042
Bogwood Sculpture Artists,
Newtowncashel 024975
Bogyoke Aung San Museum,
Yangon 031830
Bohard, Marie, Auch 066798
Bohec, Alain, Nantes 070452

La Bohème, Gisborne 143317
Bohemia, Bromsgrove 117466
Bohemia Antiques, Vancouver . . 064802
Bohemia Art, Ústí nad Labem . . 102745
Bohemia Galleries, Beverley . . . 117350
Bohemia Paper, Praha . 102655, 128338
The Bohen Foundation, New York 051347
Bohlmann, Dr. Carolin, Berlin . . 129475
Bohm, Malmköping . . . 086839, 126633
Bohm, Skogstorp 086939, 126664
Bohm, Siegfried, Göttingen . . . 076040
Bohm, Stephanie, Göttingen . . . 076041
Bohman, Lars, Stockholm 115846
Bohn, Lörrach 107885
Bohn, Carl, København 065868
Bohn, Henry, Liverpool 144544
Bohnhoff & Pluntke, München . . 077471
Bohnsack, Hans-Jörg, Strausberg 078582,
. . . . 078583
Bohol Museum, Balay Hipusanan
sa Mgs Bilihong Butang sa Bohol,
Tagbilaran 034604
Bohr, Meldorf 077335
Bohr, Ingrid, Graz 062583
Bohun, Henley-on-Thames 117959
Bohusgalleriet, Uddevalla 115958
Bohusläns Antikvariat, Skärhamn 143918
Bohusläns Museum, Uddevalla . 041098
Bohuslav, Tomaškovič, Svätý Jur 085593
Boi, Tran, Hue 125159
Boidet, Philippe, Chambéry . . . 103854
Boige, Annie, Villennes-sur-Seine 105874
Boiler House Steam and Engine
Museum, Kurwongbah 001199
Boille, Béatrice, Grasse 068702
Boillon-Tajan, Marie-Thérèse,
Paris 104574
Boillot Maison, Rouen 072570
Boira & Barclay, Sant Cugat del
Vallès 133422
Bois d'Acanthe, Liège 063748
Le Bois Debout, Lyon . 069711, 104070,
. . . . 140885
Bois Doré, Paris 104575
Bois l'Epicier, Maulette 070002
Bois Rose, Toulouse 073742
Bois Sacré, Genève 116287
Bois Secula, Saint-Paul 132988
Bois, Peter C. du, New York . . . 122423
Bois, Robin des, Levallois-Perret 069444
Boisard, Pierre, Barfleur 066948, 066949
Bois'Arts, Villeurbanne 074191, 129354
Boisbaudry, Loïc du, Paris 071066,
. . . . 125682
Boise Art Museum, Boise 047073
Boisgirard & Ass., Paris 125683
Boisguérin, Annick, Guérande . . 068760
Boisnard, Jean Luc, Rouen 072571
Boisserée, Köln 107595, 142005
Boisseron Antiquités, Boisseron . 067273
Boissière, Nicole, Gazeran 068642
Boissieu, Frédéric de, Rouen . . . 072572
Boisson, Thierry, Sarcey 073409
Boissonnas SA, Pierre, Lucien &
Valentin Boissonnas, Zürich . . 133977
Boîte à Jouets, La Rochelle 069081
La Boite à Musique, Paris 071067
La Boite à Sons, Villeurbanne . . 105877
Boix, Barcelona 085739
Boj Restauri, Genova . 131289, 131290
Bojangles, Betty, Dublin 079528
Bojara & Kellinghaus, Osnabrück 142275
Bôk, Leeuwarden 143180
Bok, Dr. Vladimír, Praha 128339
Bókasafn Menntaskólans á Ísafirði,
Ísafirði 024086
Bokcheon Municipal Museum,
Busan 029891
Boke, Nagoya 082057
Bokens Stad, Mellösa 143905
Bokfloran, Stockholm 143941
Bokhoven, Josine, Amsterdam . . 112169
Bokhulen, Oslo 143396
Bokhuset, Göteborg 143844

Bókin.is, Reykjavík 142585
Bokkistan, Göteborg 143845
Boklådan, Eskilstuna 086667
Bokn Byggdemuseet, Bokn 033555
Bokslussen Antikvariat, Sorunda 143924
Bokvalvet, Kalmar 143872
La Bola, Barcelona 085740
Bolander, Lars, New York 095280
Bolaño, Lamas, Barcelona 085741,
085742, 126506
Bolaños, Madrid 086137
Bolbro Antikvar, Odense 140633
Bold Art Gallery, Galway 109635
Bold in Gold, Maleny 099261
Bold, B., Carlton, Victoria ... 061286
Bolders, J.G., Roosendaal 132617
Boldon Auction Galleries, East
Boldon 126884
Boldrini, São Paulo 064161
Bolduc House Museum, Sainte
Genevieve 052872
Bolduc, Louis, Québec 064527
Bolduc, Marcel, Québec 064528
Boldyreff, Natalie, Paris 104576, 136871
Bolens, R., Bruxelles 063354
Bolerium, San Francisco 145543
Boletin de la ANABAD, Madrid . 139043
Boletin del Seminario de Estudios
de Arte y Arqueologia, Universidad,
Valladolid 139044
Bolger, Pat, Gowran ... 079634, 130872
El Boliche del Restaurador, Buenos
Aires 060881
Bolin, Östersund 086909, 115796
Bolin Wildlife Exhibit and Antique
Collection, McKinney 050562
Bolin, W.A., Stockholm 126670
Bolinao National Museum,
Bolinao 034438
Bolinas Museum, Bolinas 047077
Bolivar Gallery, Kingston 111212, 142893
Bolivar Hall, San Antonio 052932
Bolkart, Josef, Augsburg 074455
Bollag Galleries, Zürich 116760, 126762
Bollag, Claudia, Unterstammheim 116679
Bolland & Marotz, Bremen 075153,
125909, 141647
Bolland, Drs. Ton, Amsterdam .. 143016
Bollati, Tommaso, Milano 131390
Bolle, Pierre-Alain, Collex 087377
Bollen, Genk 063616
Boller, Kassel 141975
Bollerhey, Karl, Kassel 076713, 141976
Bollerhey, Karl, Wabern 078816, 142457
Bollig, J., Schliengen 078372
Bolliger, Marlies, Lausanne ... 116410
Bolling Hall Museum, Bradford . 043690
Bollinger Mill, Burfordville 047313
Bollmohr, Nienstädt 077786
Bollnäs Antik, Bollnäs 086644
Bollnäs Antikvariat, Bollnäs ... 086645
Bollnäs Konsthallen, Bollnäs ... 040569
Bolognini, Françoise, Yutz 105885
Boloqui Larraya, Mirian,
Zaragoza 086591
Boloré, Jacques, Mougins 104296
Bolot, Saint-Jean-de-Losne 072839
Bolour, Y. & B., Los Angeles . 094275
Bolsa, Madrid 086138
Bolsa de Arte, Rio de Janeiro . 101013
Bolsa de Arte Porto Alegre, Porto
Alegre 100967
Bolšemuraškinskij Gosudarstvennyj
Istoriko-chudožestvennyj Muzej,
Bolšoe Muraškino 036908
Bolšoj Dvorec, Muzej-zapovednik
Oranienbaum, Lomonosov 037199
Bolson, Harry N., New York ... 095281
Bolsoni, Bern 087276, 144045
Bolt, Brian R., Portballintrae .. 091037
Bolt, Rainer, Berlin 074768
Bolte, Nienstädt 077787
Bolte, Stadthagen 078532
Bolte, H., Drensteinfurt 106713

Boltens, Berne 074960
Bolton Historical Museum, Bolton 047079
Bolton Library, Cashel 024843
Bolton Museum and Art Gallery,
Bolton 043668
Bolton Steam Museum, Bolton . 043669
Bolton, Peter, Carnforth 117536
Bolton, Richard, Mangerton ... 134834
Boltz, Waltraud, Bayreuth 125864
Bolu Devlet Güzel Sanatlar Galerisi,
Bolu 042851
Bolu Müzesi, Bolu 042852
Bolwerk, Bredevoort 143097
Bolz Center for Arts Administration,
University of Wisconsin-Madison,
Madison 057390
Bolzani Cornici e Stampe, Milano 080454,
110267
Bolzau, J., Sankt Augustin 078348
Bomaderry Antiques, Bomaderry 061169
Bomann-Museum Celle, Celle .. 017719
Bomarsundmuseet i Prästö Lotsstuga,
Sund 011377
Bomb, Brooklyn 139325
Bombay & Co, Los Angeles ... 135707
Bombay Art Society, Mumbai .. 059360
Bombe Chest Antiques,
Washington 097811
Bombini, Leonardo, Milano ... 131391
Bommels, Amsterdam 082705
De Bommelzolder, Zoeterwoude 033107
Bommen Elvemuseum, Vennesla 034079
Bommer, Rolf, Basel 133635
Bommer, Werner, Zürich 116761, 126763
Bompiani, Milano 131671
Bomsholmen Museum, Røed, Øyestad,
Arendal, Arendal 033501
Bomsucesso, São Paulo 064162, 128123
Bomuldsfabriken, Arendal 113119
Bon, Kyoto 082006
Bon Accord, Edinburgh 144386
Bon Gallery, Seoul 111619
Bon Gloor, Henriette, Zürich .. 133978
Le Bon Petit Diable, Paris 104577
Au Bon Vieux Temps, Agen ... 066423
Bon, Ellen, Tucson 097639
Bona, Cincinnati 135437
Bona Thompson Memorial Center,
Indianapolis 049677
Bona-Wight, Vancouver 064803
De Bona, Valentina, Trouville-sur-
Mer 105773
Bonacci, Tivoli 142844
Bonaccorso, Paolo, Palermo ... 081022
Bonacina, Alessia Erminia Ales,
Milano 080455
Bonack, Berlin 129476
Bonade, Alexandre, Dinard 103692
Bonafous-Murat, Arsène, Paris 104578,
141054
Bonakdar, Tanya, New York ... 122424
Bonali & Fasser, Brescia 131021
Bonamici, Marcello, Roma 131830
Bonamici, Maurizio, Roma 131831
Bonanni, Chantal, Gournay-sur-
Aronde 068690
Bonanni, Chantal, Saint-Ouen .. 073039
Bonanno, Gloria, Messina 131354
Bonanova Subastas, Barcelona . 126507
Bonaparte Galleria d'Arte, Milano 110268
Bonard, Pierre, La Sarraz 133916
Bonardi, Fabrizio, Brescia 079920
Bonassi Antichità,
Montechiarugolo 080829
Bonatelli, Milano 080456
Bonavera Parodi, Cristina,
Genova 131291
Bonavista Museum, Bonavista . 005437
Bonavista North Regional Museum,
Wesleyville 007379
Bonazza, Milano 080457
Bonazzoli, Giuseppe, Brescia .. 131022
Das Bonbon-Museum, Vaihingen an
der Enz 022294

Boncompagni, Nicla, Roma 081263
Boncopagni, Nicola, Roma 081264
Bond Latin Gallery, San
Francisco 124236
Bond Store Museum, Heritage
Gateway, Maryborough,
Queensland 001257
Bond Street Antique Centre,
London 089967
Bond Street Antiques, Cromer .. 089011,
134308
Bond van Handelaren in Oude Boeken,
Rotterdam 059596
Bond, A.P., Unley Park 062381
Bond, J.V., San Diego 097069
Bond, Kent, Tampa 124637
Bond & Son, S., Colchester ... 088948,
134287
Bondarenko, Karel, Sosnowiec . 113767
Bondu, Dominique, Paris 125684
Bondu, Jean-Philippe, Paris ... 125685
Bonduelle & Lancry, Corbeil-
Essonnes 125561
Bondzio, Ulrich, Berlin 129477
The Bone Dude's, Christchurch . 112835
Bonechi, Firenze 137645
Bonelli Arte Contemporanea,
Mantova 110200
Bonello, Pierre, Reims 072384
Bo'ness Heritage Area, Bo'ness 043673
Bonfanti, Giuliano, Milano 142715
Bonfils, Olivier de, Bergerac .. 067099
Bonfini, Gabriel, Paris 128940
Bonfoey, Cleveland 120405
Bonfreschi, Stefano, Modena . 131563
Bong Bong Books, Bowral 139654
Bong Sung, Daegu 111580
Bongartz & Partner, Aachen ... 107352
Bongartz, Georg-Paul, Aachen .. 074324,
125844
Bongiorno, Paolo, Modena 142748
Bonhage jr., Joseph B., New
Orleans 122134
Bonhagen, Anke, Oldenburg ... 108338
Bonham, Waikato 084123
Bonhams, Banbury 126806
Bonhams, Bath 126811
Bonhams, Bedale 126813
Bonhams, Cape Town 126491
Bonhams, Cardiff 126845
Bonhams, Carlisle 126846
Bonhams, Chester 126855
Bonhams, Edinburgh 126887
Bonhams, Glasgow 126900
Bonhams, Guildford 126912
Bonhams, Hale 126915
Bonhams, Henley-on-Thames . 126922
Bonhams, Honiton 126924
Bonhams, Hove 126926
Bonhams, Leeds 126938
Bonhams, London 126955, 126956
Bonhams, Norwich 126998
Bonhams, Oxford 127005
Bonhams, Par 127008
Bonhams, Paris 125686
Bonhams, Ringwood 127020
Bonhams, Saint Helier 127027
Bonhams, Sevenoaks 127035
Bonhams, Sherborne 127041
Bonhams, Solihull 127047
Bonhams, Southport 127051
Bonhams, Tetbury 127067
Bonhams, Toronto 101583, 125379
Bonhams, Whitstable 127087
Bonhams, Winchester 127089
Bonhams & Butterfields, Los
Angeles 127292
Bonhams & Butterfields, New
York 127343
Bonhams & Butterfields, San
Francisco 127442
Bonhams & Goodman, Sydney . 125236
Bonhanss, Riga 111756
Bonheur, Saint Louis 123811

Bonheur des Jours, Perros-
Guirec 072101
Le Bonheur du Jour, Amiens ... 066563,
128436
Bonheur du Jour, Les Sables-
d'Olonne 069426
Bonheur du Jour, Montauban .. 070190
Le Bonheur du Jour, Sainte-
Fortunade 073347
Bonhof, Apeldoorn 082919
Bonhomme-Raynaud, Agnès,
Cheveuges 128556
Boni, Ilaria, Firenze 080096
Bonifatius-Verlag, Paderborn .. 137467
Bonillo, Albora F., Valencia ... 086506,
133451
Bonillo, Anselme, Rochegude .. 072504
Bonin, Carola, Neumünster ... 077739
Boningtons', Loughton 126979
Bonino, Fernanda, New York ... 122425
Bonino, Patrizia, Genova 080239
Bonioni Arte, Reggio Emilia .. 110706
Bonn, Lutz, Neuwied 077773
Bonnadier, Régis, Limoges 069519
Bonnafoux, Dolorès, Poitiers .. 105363
Bonnal, Mende 070045, 070046
Bonnar, Joseph, Edinburgh ... 089161
Bonnard, Christian, Saint-Lubin-des-
Joncherets 072900
Bonnat, Marie, Chennevières-sur-
Marne 067957
Bonne Occase, Annemasse 066652
A la Bonne Occaz, Lausanne .. 087565
Bonnechere Museum, Eganville . 005773
Bonnefantenmuseum, Maastricht 032621
Bonnefoi, Henri, Paris 141055
Bonnefont, N., Villefranche-de-
Rouergue 074141
Bonnefous, Lyon 128698
Bonnefoy, Jean-Charles, La
Coucourde 069009
Bonnefoy, Pascale, Saint-Étienne 072752
Bonnefoy, Richard, Chécy 067950
Bonnemoy, Elisabeth, Thiers ... 073650
Bonnemoy, Jean-Luc, Escoutoux 068467
Bonner Antiquitäten Centrum,
Bonn 075066
Bonner County Historical Museum,
Sandpoint 053069
Bonner Kunstverein, Bonn 017485,
058760
Bonnes Choses Antiquarios, Rio de
Janeiro 064029
Bonnet House, Fort Lauderdale . 048824
Bonnet, Christian, Faverolles
(Cantal) 068513
Bonnet, Jean-Pierre, Nîmes ... 070689
Bonnet, Marie-Claude, Toulouse 073743
Bonnetain, Paulette, Azé 066898
Bonneton 5060 Eric, Villeurbanne 074192
Bonneton, Eric, Brignais 067495
Bonnier, Genève 116288
Bonnier, Peder, New York 122426
Bonnin Serra, Jose, Barcelona . 085743
Bonnin, Bernard, Béziers 067167,
067168
Bonnin, Christian, Béziers 067169
Bonno, Michel, Plougoumelen . 072203
Bonn'occase, Bonne 067289
Bonnot, Thierry, Nice 070577
Bons Tempos, Porto 085011
Bonsai Museum, Seeboden 002747
Bonsignori, Roma 137726
Bonson, S., Greenbank 099095
Bonte-Sioen, Brugge 125302
De Bonte Sonoy, Alkmaar 082616
Bontoc Museum, Bontoc, Ifugao 034440
Bontoc Museum, Bontoc, Mountain
Province 034441
Bonum, Kraków 084562
Bonum, Solna 115821
Bonvin, Crans-sur-Sierre 087389
Bonvin, René, Sion 087814

Bonvini, Bruno, Milano 080458, 110269
Bonzano, Paolo, Roma 110736
Bonzi, Maurice, La Clusaz 068997
Boobala, Norfolk 095847
Boodja Aboriginal Art and Craft,
Fremantle 061485
Boodle & Dunthorne, Sydney .. 062278
Boody, Saint Louis 123812
Boogaard, Amsterdam 082706
Book & Art Shop, Sheffield .. 119079,
144834
Book Affair, Carlton, Victoria ... 139690
Book and Comic Exchange,
London 144565
The Book Bag, Fitzroy 139732
Book Bank, Memphis 145250
Book Bank, Toowong 139930
Book Barn, Olympia Heights 145040
Book Barn, Surry Hills 139919
Book Bazaar, Ottawa 140375
Book Bazaar Ballarat, Ballarat .. 139631
Book Bin, Lincoln Park 145205
Book Broker, San Diego 145517
Book Browser, Houston 145163
The Book Buff, Broadmeadow .. 139664
The Book Bug, Hitchin 144477
Book Cellar, Temple 145619
Book City Collectables, Los
Angeles 145217
Book Collector, Houston 145164
Book Collector, Parramatta 139869
Book Collector, Sacramento 145484
Book Connection, Dubbo 139720
Book Connection, Invermay 139783
Book Cultural Museum, Qujiang 008201
Book Den, Cessnock 139695
Book Den, San Antonio 145511
The Book Depot, London 144566
Book Discoveries, Nashville ... 145291
Book Discoveries, Seattle 145577
Book Emporium, Scarborough .. 144827
The Book End, Hastings 139765
Book Ends, Winnipeg 140431
The Book Exchange, Amsterdam 143018
Book Exchange, Amsterdam ... 143017
Book Exchange, Bullhead City .. 145024
Book Exchange, Colchester 144346
Book Exchange, Lakewood 145199
Book Finding Services, Saint
Andrews 144813
Book Gallery, El Paso 145142
Book Gallery, Jerusalem 142613
Book Gallery, Moss Vale 139839
Book Gallery, Phoenix 145426
Book Garden, San Diego 145518
Book Horizons, Coral Gables .. 145103
Book House, Kirkby Stephen ... 144508
Book House, Minneapolis 145272
The Book House, Ravenstonedale 144797
Book Inn, Whangarei 143376
The Book Loft, Masterton 143329
Book Lore, Lyneham 139812
Book Mark, Dallas 145112
Book Mark, Philadelphia 145409
Book Market, Bruxelles 140166
Book Market, Ottawa 140376
Book Now, Bendigo 139641
Book Off, Kaida 142895
Book Off, Kanie 142896
The Book Palace, London 144567
Book Peddler, Tatum 145618
Book People, Richmond 145473
Book Place, Memphis 145251
Book Rack, Anderson 144956
Book Rack, Bartlesville 144991
Book Rack, Birmingham 144998
Book Rack, Blue Ash 145003
Book Rack, Cincinnati 145065
Book Rack, Columbus 145095
Book Rack, Denver 145122
Book Rack, Edmond 145141
Book Rack, Indianapolis 145177
Book Rack, Lakewood 145200

Book Rack, Middleburg Heights 145261
Book Rack, Midwest City 145262
Book Rack, North Olmsted 145384
Book Rack, Oklahoma City 145394
Book Rack, Phoenix 145427
Book Rack, South Hills 145608
Book Rack, Springdale 145611
Book Rack, Tulsa 145626
Book Rack West End, Regency .. 145472
Book Re-View, Tigard 145620
Book Realm, Enoggera 139726
Book Rendezvous Downtown,
Baltimore 144981
Book Restorer, Christchurch .. 132702
Book Returns, Auckland 143277
Book Scene, Houston 145165
Book Shelf, Oklahoma City 145395
The Book Shop, Castle
Donington 144320
Book Shop, Fort Worth 145148
Book Shop, Hay-on-Wye 144457
Book Shop at the Art Gallery of
Ontario, Toronto 140403
Book Station, Caringbah 139687
Book Store, Cincinnati 145066
Book Swap, Laughlin 145204
Book Swap of Carrollwood,
Carrollwood 145027
The Book-Swop, Edinburgh ... 144387
Book Trader, Minneapolis 145273
Book Trader, Oklahoma City ... 145396
Book Trader, Philadelphia 145410
Book Traders, Memphis 145252
Book Traffic, Amsterdam 143019
Book Turn, Sendai 142915
Book Warren, Brixham 144286
Book World, Benoni 143701
Book World, Indianapolis 145178
Book Worm of N O E, New
Orleans 145302
Book Wormz, South Auckland .. 143343
Book, Stamp and Picture Gallery,
Malvern East 139819
Bookarts Canberra, Griffith .. 127606
Bookbarn, Bristol 144275
Bookbeat Byron, Byron Bay ... 139674
Bookbox, Cheltenham 144323
Bookcase, Carlisle 144313
Bookcase, South Brisbane 139909
The Bookcase Limited, Huntsville 140352
Booked Out, Randwick 139892
Booked Up, Washington 145632
Bookends, Emsworth 144410
Bookends, Fowey 144429
Bookends, Louisville 145242
Booker Lowe, Houston 121010
Booker T. Washington National
Monument, Hardy 049358
Bookery, El Paso 145143
Bookery, Fyshwick 139742
Booketab, Ringwood 139896
Bookfinders, Belfast 144239
Bookfound, Edge Hill 139724
Bookhaven, Philadelphia 145411
Bookings Taken, Rozelle 139898
Bookish, Waikato 143352
Bookleaves, New York 145327
Booklegger's Used Books,
Chicago 145037
Booklovers, Bath 144228
Bookman, Nashville 145292
The Bookman, Norwich 144748
Bookman's Corner, Chicago ... 145038
Bookmans Halt, Saint Leonards-on-
Sea 144818
The Bookmark, Brighton 144263
Bookmark, Falmouth 144420
Bookmark, Oakland 145386
Bookmine, Sacramento 145485
Bookmonger, San Francisco ... 145545
The Bookologist, Nowra 139861
Bookplate, Hornsby 139778
Bookrack Iroquois, Louisville .. 145243
The Book's, Cheltenham 144324

Books, Cromer 144363
Books & Bygones, Weymouth .. 091812,
144916
Books & More, Wettingen 144148
Books & Things, London 144568
Books Afloat, Weymouth 091813, 144917
Books and Art on Beaumont, Hamilton,
New South Wales 139760
Books and Collectables,
Underwood 139943
Books Around, Geelong 139748
Books by George, Birmingham . 144999
Books et Cetera, North Perth .. 139860
Books Etc., Oklahoma City 145397
Books for Collectors, Toddington 144883
Books for Cooks, Fitzroy 139733
Books for Thought, Tampa 145614
Books 'n Bitz, Nerang 139852
Books 'n Boxes Plus, Atacocita . 144958
Books 'n Tees, Del City 145120
Books 'n Things, Albany 139620
Books of Fifth Avenue, Palm
Beach 139867
Books Of Oceania, Whangarei . 143377
Books of Wonder, New York ... 145328
Books on Bronte, Waverley 139957
Books on Dean, Albury 139621
Books on High, Columbus 145096
Books Rare and New Zealand, Cargills
Corner 143299
Books Unlimited, Edmonton ... 140344
Books Unlimited, Indianapolis . 145179
Books, Cases and Prints, Atlanta 144963
Books, Maps & Prints, Brecon . 144260
Books@Devonport Wharf, North
Shore 143335
Bookscene, Nottingham 144756
Booksearch, Chichester 144334
Booksearch, Launceston 144515
Booksearch, Leicester 144523
Booksearch, Saint Paul 145501
The Bookshelf, Cairns 139676
Bookshelf, Winnipeg 140432
The Bookshop, Bournemouth .. 144253
Bookshop, Cambridge 144294
Bookshop, Conwy 144354
Booksmart, Sweetwater 145612
The Booksmith, Clayfield 139699
Books'n'Print, Invercargill 143324
Bookstacks, Wellington 143362
Bookstop Gallery, Taranaki 113013,
143346
Bookstop Plus, Barlett 144990
Bookstore on West Twenty Fifth,
Cleveland 145083
The Bookstore, National Gallery of
Canada, Ottawa 140377
Booktalk, Richmond, Victoria .. 139894
Booktrader, København 140603
Bookwise, Wellington 143363
Bookworks, Chicago 145039
Bookworm, Dayton 145119
The Bookworm, Edinburgh 144388
Bookworm, Evesham 144415
Bookworm, Holland-on-Sea 144481
Bookworm, Piqua 145433
Bookworm Book Store, Kansas
City 145193
Booleroo Steam and Traction Museum,
Booleroo Centre 000865
Boom, Cleveland 092923, 120406
Boomalli Aboriginal Artists,
Leichhardt 061732
Boomerang, Kansas City 094060
Boomerang, San Diego 097070
Boomerang Art, Glenelg 099082
Boomkwekerijmuseum, Boskoop 032088
Boon Gallery, Bruxelles 100570
Boon Gallery, Knokke-Heist 100731
Boondooma Homestead –
Museum and Heritage Complex,
Mundubbera 001326
Boone County Historical Center,

Boone 047086
Boone County Historical Society
Museum, Belvidere 046909
Boone, Mary, New York 122427
Boone's Lick Site, Boonesboro . 047090
Boongar Arts & Crafts, Kuranda 061705
Boonsborough Museum of History,
Boonsboro 047091
Boonshoft Museum of Discovery,
Dayton 048168
The Boorman House, Mauston . 050743
Boos Antic Sud Est, L'Isle-sur-la-
Sorgue 069559
Boos Antic Sud-Ouest, Condom 068126
Boos, Frank H., Troy 127479
Boos, T. & G., Landau in der
Pfalz 077001
Booser, Luc de, Knokke-Heist .. 100732
Boot Hill Gallery, Beaumaris .. 098717
Boot Hill Museum, Dodge City . 048334
Booth 13 Vintage, Oklahoma City 095952
Booth-Clibborn Editions, London 138197
Booth Museum of Natural History,
Brighton 043734
Booth & Son, A.E., Hook 089578, 134472
Booth, David, Wells 135199
Booth, Joanna, London 089968
Booth, Richard, Hay-on-Wye .. 144458,
144459
Boothbay Railway Village,
Boothbay 047093
Boothbay Region Art Foundation,
Boothbay Harbor 047094
Boothe, Carol, Dún Laoghaire .. 079602,
130865
Boquet, Damien, Paris 104579
Bor – Art, Roma 110737
Boralevi, Alberto, Firenze 080097
Boralevi, Daniele, Firenze 080098
Boranup Galleries, Margaret
River 099280
Borås Idrottsmuseum, Borås .. 040572
Borås Konstmuseum, Borås ... 040573
Borås Museum, Borås 040574
Borås och Sjuhäradsbygdens
Konstförening, Borås . 059880, 115585
Borca, Simona, Torino 132101
Borch Jensen, Niels, København 102867
Borchardt, Peter, Hamburg ... 107232
Borcherdt, K., Braunschweig ... 106559
Borchert, Olaf, Willisau 087904
Borck, Andrea, Hamburg 129952
Bord, Jacques, Saint-Étienne .. 072753,
072754
Bord, M., London 089969
Borde, Thierry, Cosne-Cours-sur-
Loire 068164, 128572
Borden, Janet, New York 122428
Bordenave, Jean, Anglet 066621
Bordentown Historical Society
Museum, Bordentown 047096
Border Regiment and King's Own
Royal Border Regiment Museum,
Carlisle 043872
Borderland Antiques, El Paso .. 093390
Bordes, Alexis, Paris .. 071068, 128941
Bordes, Michel, Villeneuve-sur-
Lot 074164
Bordes, Patrick, Marnes-la-
Coquette 069884
Bordet-Secrettand, Elisabeth,
Vitteaux 074263
Bordier, Alain, Colmar 068084
Bordighera, Bordighera 109891
Bordihn, München 077472
Bordin, Padova 080969
Bordin, Giuseppe, Torino 132102
Bords d'Eaux, Mers-les-Bains .. 070073
Borduas, Montréal 101368, 136677
Boréal Express, Paris 071069
Borej Art, Sankt-Peterburg 114331
Borella, Anna Maria, Venezia .. 111092
Borella, Lars, København 102868
Borelli, Colombes 103621

Borelli-Edwards, Pittsburgh 123527
Bores, Emile, Montauban 070191
Borg, Oslo 113195
Borg, Tønder 066168
Borg Verhildersum, Leens 032565
Borg, Bernard, Bagneaux-sur-
Loing 066907
Borg, Christina, New York ... 095282
Borg, Michel, Villeveyrac 074233
Borg, Paul, B'Kara 082408, 111912
Borg, Richard ter, Groningen ... 083281
Borgarsyssel Museum, Sarpsborg 033927
Borgazdasági Múzeum, Tolcsva . 024016
Borgeaud, Claude, Basel 087224, 133636
Borgen, Valby 136783
Borges Reis & Almeida, Lisboa . 084840
Borgheresi, Leonardo, Firenze . 131133
Borghese, Rio de Janeiro 101014
Borghesi, Roberto, Trieste 081783
Il Borghetto, Milano 080459
Borghorst, Steinfurt 078549
Borgmeier, Hagen 129933
Borgmuseum, Spøttrup 010327
Le Borgne, Daniel, Villeneuve-le-
Guyard 074161
Borgobello, Parma 081100
La Borgognona, Roma 110738
Borgsjö Hembygdsgård,
Erikslund 040605
Borinquen Art, Tampa 124638
Boris Antiquaire, Béziers 067170
Boris-Becker-Zimmer, Leimen .. 019958
Boriscothi Antiques, Chiang Mai 088089
Borisoglebskij Kraevedčeskij Muzej,
Borisoglebsk 036909
Borja, R., Valencia 086507, 115477
Bork, København 065869
Bork, Dr. U., Freiburg im
Breisgau 075852
Bork, Moshe, El Paso 093391
Borkas, Lima 113344
Borkener Galerie, Borken,
Hessen 106545
Borkowska, Monika, Nové Zámky 133055
Borkowski, Hannover 107353
Borla, Silvano, San Vittore 087772,
133898
Borlach-Museum, Bad
Dürrenberg 016753
Borlänge Antik och Kuriosa,
Borlänge 086651
Bormann, Herbert, Rodenbach . 078249,
087250
Bórmida, Hugo, Punta del Este . 119439
Bormshuis, Antwerpen 003280
Born, Barcelona 085744, 126508
Born Again, Baltimore . 092303, 135339
Born in the USSR, Houston 093626
Born-Néo, Montréal 101369
Born, Frank, San Francisco ... 124237
Bornemann, Lutz, Dresden 141692
Borner Forst- und Jagdmuseum, Born
am Darß 017517
Borner, Andreas, Rickenbach
(Solothurn) 133893
Borner, Pia Anna, Luzern 116469
Bornholms Kunstmuseum,
Gudhjem 010089
Bornholms Museum, Hjorths Fabrik
und Bornholmsk Keramikmuseum,
Rønne 010285
Bornholt, Detlef, Kiel 076772
Bornmann, M., Kassel 076714
Borno Museum Society, University of
Maiduguri, Maiduguri 059648
Borno State Museum, Maiduguri 033434
Borobudur, Jakarta 126161
Borobudur, Singapore 126484
Boros, R., Rietheim-Weilheim .. 078240
Borosch, Dortmund 129695
Borough Galleries, Eaglehawk . 061424,
098993, 127594
Borough Museum and Art Gallery,
Berwick-upon-Tweed 043576

Borow, Hamburg 076199
Borras Mari, Fermin, Barcelona . 085745
Borrelli, Armando, Napoli 131599
Borri, Luciano, Roma 131832
Borring, Tove & Ole, Vejle 066189
Borroco N Roll, São Paulo 064163
Borromini Arte Contemporanea,
Ozzano Monferrato 110554
La Borsa del Fumetto, Milano .. 142716
Borsani, Rovereto 110877
Borsos Miklós Múzeum, Városi
Muvészeti Múzeum, Győr 023659
Borstei Museum, München 020512
Borstel, Mario von,
Emmenbrücke 087407, 133730
Bortolotti, Carla e Alberto, Milano 080460
Bortolotti, Pierre, Albi 066509
Bortolozzi, Isabella, Berlin 106179
Bortorello, São Paulo 064164
Bortz, Philip, Cincinnati 092832
Borup Antik, Ringsted 066074
Borups Antik, København 065870
Borycz, Anna, Wrocław 113918
Boryeong Coal Museum,
Boryeong 029890
Borzo, Amsterdam 112170
Bos-de Groote, C.J.M.S., Baarn . 132431
BoS Antik och Kuriosa, Göteborg 086701
Bos-Otten, Amsterdam 082707
Bos, Marten, Leeuwarden 132557
Bosák, Luboš, Česká Lípa 065365
Bosc, Alain, Cannes 067629
Bosch, Barcelona 143712
Bosch, M., Düsseldorf 075461
Bosch, Rudolf, Erlenbach 087412,
133733
Bosch, Rudolf & Ruth, Zug 088020
Boschbach, M., Kerpen 107546
Boscher & Oriot, Morlaix 125650
Boscher Belleissue, Jacqueline, Saint-
Brieuc 072696
Boscher, Samuel, Cherbourg-
Octeville 125557
Boscher, Studer & Fromentin,
Paris 125687
Boscher, Yvan-Guillaume, Paris . 071070
Boschert, Eberhard, Landau in der
Pfalz 077002, 130206
Boschert, Eberhard, Neustadt an der
Weinstraße 130456
Boschert, Hermann,
Kaiserslautern 107493
Boschert, Hermann, Neustadt an der
Weinstraße 108256
Boschetti, Enrica, Milano 131392
Boschi, Alessio, Bologna 109816
Boschi, Andrea, Parma 131722
Boschi, Rudolf, Innsbruck 062646
Boscia, Alberto, Palermo 131687
Boscia, S., Hausach 076421
El Bosco, Godoy Cruz . 060945, 127544
Bosco, Luciano, Roma 081265
Bosco, Paul J., New York 095283
Boscobel House, Brewood 043716
Boscobel Restoration, Garrison . 049056
Boscombe Militaria,
Bournemouth 088532
Boscombe Models & Collectors Shop,
Bournemouth 088533
Bosdure Marzocchi, Bastia 066958,
103338
Bose Pacia, New York 122429
Bose, Anne-Dorothee, Wittlich . 108953,
137556
Bosetti, Henri & Michèle, Rueil-
Malmaison 072637
Bosi Artes, Roma 081266
Bosica, Ernani, Torino 132103
Bosio, Yves, Menton 070059
Bosjökloster Slot & Trädgården,
Höör 040702
Boškov, P., Kjustendil 064223
Bosmuseum, Amsterdam 031934
Bosmuseum, Tessenderlo 004058

Bosmuseum Jan van Ruusbroec,
Hoeilaart 003703
Bosmuseum Pijnven, Hechtel .. 003686
Bosmuseum Wildert, Essen 003582
Bosnyák Helytörténeti és Néprajzi
Szoba, Áta 023410
Bosque de Ciência, Manaus ... 004550
Bosque Galería Artdicré, México 111948
Boss & Co, Urtenen-Schönbühl . 138094
Boss, Sławomir, Łódź 113659
Boss, Thomas G., Boston 119839,
145008
Bossaert, Brugge 063262
Bossanova, Seattle 124501
Bosse, Kitzbühel 100066
Bosse, P., Osnabrück 077960
Bosse, Sigrid, Cottbus . 075273, 129660
Bosser, Fiederic, Paris . 104580, 136872
Bostock, J., London 118226
Boston Antiquarian Book & Ephemera
Fair, Boston 098435
Boston Antique Cooperative,
Boston 092448
Boston Antique Textile & Vintage
Fashion Show, Boston 098436
Boston Antiques Show, Boston . 098437
Boston Art, Boston 119840, 119841,
119842
Boston Art Dealers, Boston ... 119843
Boston Artists Gallery, Tulsa .. 124796
Boston Athenæum, Fine Arts
Collection and Library, Boston . 047102
Boston Book Annex, Boston ... 145009
Boston Book Company, Jamaica
Plain 145192
Boston Center for the Arts,
Boston 119844
Boston Children's Museum, Museum
Wharf, Boston 047103
Boston Corporate Art Collection,
Boston 119845
Boston Directors' Lab, Boston . 119846
Boston Fire Museum, Boston .. 047104
Boston Gallery, Quezon City .. 113418
Boston Guildhall Museum,
Boston 043677
Boston House, Toronto 064618
Boston National Historical Park,
Boston 047105
Boston Printmakers, Emmanuel
College, Boston 060412
Boston Public Library, Boston .. 047106
Boston Rare Maps, Boston 145010
Boston Sculptors, Boston 060413
Boston Sculptors Gallery, Boston 047107
Boston Towne Coin Exchange,
Boston 092449
Boston University Art Gallery,
Boston 047108
Boston, Derek, Salisbury 091220
Boston, Julia, London 089970
Boswell, Ashgrove 139628
Boswell & Davis, London 134603
Boswell-Crowe, Los Angeles .. 121449
Boswell Museum of Music, East
Springfield 048446
Bosworth Antiques, Market
Bosworth 090703
Bosworth Battlefield Heritage
Centre and Country Park, Market
Bosworth 045228
Bősze, Budapest 109133
Bot, Christiane de, Antwerpen .. 063197
Bot, G.P.L., Vlaardingen 132659
Botanic Gardens Museum,
Southport 045839
Botanical Museum, Rawalpindi . 034149
Botanical Museum of Harvard
University, Cambridge 047358
Botanical Museum of the National
Gardens, Athinai 022980
Botanical Survey Museum of India,
Pune 024444
Botanical Survey of India,

Dehradun 024213
Botaničeskij Muzej, Sankt-
Peterburg 037719
Botanické Oddělení Národního Muzea,
Průhonice 009858
Botanické Oddělení, Moravské zemské
muzeum, Brno 009461
Botanische Staatssammlung München,
München 020513
Botanisches Museum Berlin-Dahlem,
Berlin 017090
Botanisk Museum, København . 010168
Botaniska Museet, Lund 040806
Botany Museum, Faizabad 024264
Botany Museum, Gorakhpur ... 024263
Botany Museum, Jaunpur 024324
Botany Museum, Kanpur 024338
Botany Museum, Lucknow 024378
Botelho, Antonio Pedro Martinho,
Lisboa 084841
Botelho, Francisco Silva, Lisboa 084842
Botelho, José Silva, Lisboa 084843
Boterf, La Chapelle-Basse-Mer . 068969
De Botermolen, Keerbergen ... 003747
Botero, Barranquilla 102263
Botesazan, New York 122430
Both, T., Bonn 075067
Bothnia, Vännäs 087136, 143999
Bothorel, Marie, Pleyber-Christ . 072189
Bothorel, Yves, Berrien 067115
Bothwell-Zone Oil Museum,
Bothwell 005442
Botkyrka Konsthall och Xet-Museet,
Tumba 041092
Botswana National Museum and Art
Gallery, Gaborone 004295
Bottarel & Foi, Brescia 079921
Bottari, Torino 110956
Botte Dorée, Vitré 074254
Bottega, Bologna 130939
La Bottega, Carouge 116188
La Bottega, Gempen 087440
Bottega Artigiana, Milano 131393
Bottega d'Altri Tempi, Prato .. 131771
Bottega d'Arte, Bari 130907
Bottega d'Arte, Casablanca 082551
Bottega d'Arte, Firenze 080099, 110037
Bottega d'Arte, Livorno 110175
Bottega d'Arte, Roma 081267
La Bottega d'Arte Antica, Napoli 131600
Bottega d'Arte di Anelli Antonietta,
San Benedetto del Tronto ... 081613
Bottega d'Arte Minerva, Torino . 132104
Bottega del Bastione, Andora . 109751
La Bottega del Borgo Nuovo,
Torino 081661
La Bottega del Legno Antichità e
Restauri, Torino 132105
Bottega del Restauro, Bari 130908
Bottega del Restauro, Brescia . 131023
Bottega del Restauro, Milano .. 131394
Bottega del Restauro, Modena . 080776,
131564
La Bottega del Restauro, Torino 132106
Bottega del Restauro, Torino .. 081662
Bottega del Restauro, Verona . 132272
Bottega del Restauro, Vezia .. 087884,
133952
La Bottega del Vecchio Ceppo,
Milano 131395
La Bottega dell'Antiquario,
Lugano 087607, 144099
La Bottega delle Arti, Bordighera 109892
Bottega delle Occasioni, Palermo 081023
Bottega delle Stampe, Firenze . 142665
Bottega delle Stampe, Milano . 080461,
131396
Bottega delle Vecchie Cose, Pisa 081153,
131762
Bottega di Restauro Bramante,
Milano 131397
La Bottega di San Luca, Torino . 081663
Bottega Farnese, Palermo 081024,
131688

La Bottega Margutta, Roma ... 081268
La Bottega Restauri e Mobili su
Misura, Modena 131565
Bottega San Carlo, Locarno 087598
Bottega San Felice, Firenze 080100
La Bottega Solferino, Milano 080462
Bottegantica, Bologna 079823
La Bottegantica, Cagliari 079976
El Botteghin, Milano 080463
Botteguccia d'Arte, Aosta 109752
Botteguccia dell'Arte, Bari 130909
Bottéro, Maryse & Gérard, Nice 070578
Botticelli, Lima 084443
Botticelli Via Maggio, Firenze .. 080101
Bottin, Peter, Vachendorf 078776
Bottland Gallery of Modern Art, Surry
Hills 099742
The Bottle Kiln, West Hallam
Ilkeston 119323
Bottom Drawer Antique Centre,
Moorooduc 061883, 139633
Bottom Feeders Studio Gallery, New
York 122431
Bottom of the Harbour Nautical
Antiques, Sydney 062279
Bottom of the Ninth, Cleveland . 092924
Botton, Isidoro, Milano 080464
Bottroper Auktions- und Handelshaus,
Bottrop 125903
Bottura, Milano 080465
Botwood Heritage Centre,
Botwood 005444
Botzenhardt, Walburga, Kempten 076748
Bou, Marie-Hélène, Decazeville . 068285
Boucard, Michelle, Paris 128942
Boucaud, Paris 071071
Boucaud, Charles et Philippe,
Maraussan 069847
Boucaud, Philippe, Béziers 067171
Bouchardon, Odile, Langeais ... 069167
Bouchart, Françoise, Granville .. 068696,
140845
Bouchart, Philippe, Aire-sur-la-
Lys 066444
Bouche, Bernard, Paris 104581
Bouché, Lucette, Villeréal 074174
Bouché, Olivier, Monpazier 070167
Bouchelier Antiquité Brocante 71,
Montceau-les-Mines 070223
Bouchene, Nasreddine,
Bouzaréah 060835
Boucher, Amiens 066564
Boucher, Gérard, Le Havre 069274
Boucher, Kevin, Mount Evelyn . 099396
La Boucherie, Orvilliers 070826
La Boucherie Antiquités,
Maulette 070003
Bouchetard, Catherine, Saint-
Ouen 073040
Bouchiba, Strasbourg 129265
Le Bouclart, Maussane-les-
Alpilles 070012
Boucle d'Or, Tours 073841
Bouclet, Christian, Bessé-sur-
Braye 067141
Boucq, Aubervilliers 066785
Boucq, Paris 071072
Bouda, Petr, Brno 140460
Boudewijn, Ton, Nijmegen 112603
Boudin, Arthur, Honfleur 103817
Boudin, Gérard, Rouen 105470
Boudinar, Karim Michel, Berlin . 074769
Il Boudoir della Nonna, Palermo 081025
Boudon, Gilles, Ferrières-Saint-
Mary 068527
Boudoux, Bernard, Nointel 070733
Boudreau, Montréal 064398
Boudreaux, Houston 135591
La Bouee de Noe, Ceton 067757
Bouet, Jean-Loïc, Vaubadon ... 073984
Boughton House, Kettering 044684
Bougival Antiquités, Bougival .. 067377
Bougo, Loyc, Saint-Congard ... 072718
Bouguen, Bernard, Avignon 140771

Bouhin, Jean-Michel, Montbard . 070210
Bouhours, Thierry, Château-du-
Loir 067870, 128549
Bouhours, William, Châteaudun . 067880
Bouillet, Beaune 067011
Bouillon, Dominique, Paris 128943
Bouillon, Michel, Autun 066832
Bouillotte, Sainte-Anne-d'Auray . 073340
Bouillotte, Sannois 073406
Bouin & Gardon, Paris 128944
Boulade, Isabelle, Villeurbanne . 074193
Boulakia, Fabien & Daniel, Paris 104582
Boulan, Vigneux-sur-Seine 105861,
137015
Boulanger & Fils, Paris 128945
Boulanger, Patrice, Paris 071073
Boularand, Jacques, Pézenas .. 072132
Boulard, Laurent, Ailly-le-Haut-
Clocher 066441
Boulard, Laurent, Saint-Riquier . 073295
Boulder History Museum,
Boulder 047138
Boulder Museum of Contemporary Art,
Boulder 047139
Boulder Pastel Painters Association,
Boulder 060414
Boul'Dingue Occasions, Lyon .. 140886,
140887
Boule, Monaco 082498
Boulenger, Bernard, Berck-sur-
Mer 067096
Boulenger, Daniel, Hardelot-Plage 068788
Boules, Betton 067151
Boulestin, Patrick, Saivres 073377
Boulet, Tulle 073912
Boulet, Jean-Claude, Sébécourt 073463
Boulet, Michelle, Paris 104583
Boulevard Antique, Leicester .. 089788
Boulevard Arts Center, Chicago . 047642
Boulevard Book Rack,
Muskeogee 145289
Boulevard d'Unica, Utrecht ... 112694
Boulevard Emporium, Dallas ... 093077
Bouley, Alain, Villeurbanne ... 074194
Bouley, Josiane, Villeurbanne .. 074195
Bouley, Roger, Villeurbanne ... 074196
Boulia Stone House Museum,
Boulia 000871
El Boulle, Milano 110270
Boulle, Saint-Grégoire 072818
Boullé, Claude, Paris 104584
Boullé, Yvon, Moutiers 070379
Boulogne, Pont-de-Briques-Saint-
Etienne 072246
Boulonnais, Anne, Vandières ... 073960
Boulouk, Georges, Paris 071074
Boulton & Cooper, Malton 126983
Boulton, John, Broseley 088688
Bouma, R.J., Amsterdam 082708
Bouman-Buwalda, Amsterdam . 082709
Bound for Pleasure, Sacramento 145486
Bound To Be Found, Indianapolis 093864
Bound Words, Hampton 139764
Boundary Gallery, London 118227
Boundary Museum, Grand Forks 005904
Bounica, Genève 087455
Bouniol, Denise, Issoudun 103855
Bountiful Books, San Diego ... 145519
Bountiful of Berry, Bourges 061145
Bounty Antiques, Topsham 091612
Bouquet-Jacaillot, Colette, Condé-sur-
Vesgre 068125
Bouquinerie, Valence (Drôme) .. 141372
Bouquinerie Comtoise, Besançon 067119,
140782
Bouquinerie de l'Institut, Paris .. 141585,
141056
Bouquinerie des Carmes, Liège . 140240
Bouquinerie des Quais, Rennes (Ille-et-
Vilaine) 141311
Bouquinerie du Palais, Ajaccio . 140761
Bouquinerie Le Meridien,
Bruxelles 140167
Bouquiniste, Bayonne 140777

Bouquiniste, La Rochelle 140862
Bouquiniste, Saint Andrews 144814
Bouquiniste, Saint-Étienne 141330
Bourassin-Rancon, Christian,
Airaines 066442
Bourbaki-Panorama Luzern,
Luzern 041704
Bourbon-Hanby, London 089971
Bourbon Street Gallery, New
Orleans 122135
Bourbonnais, Jean-Louis, Saint-Rémy-
lès-Chevreuse 073294
Bourcy & Paulusch, Wien 140063
Bourdareau, Marcel, Chasseneuil-sur-
Bonnieure 067864
Bourdariat, Albert, Paris 128946
Bourdely & Roig Diffusion,
Grenoble 068721
Bourdette-Gorzkowski, Danielle,
Honfleur 103818
Bourdon-Smith, J.H., London .. 089972,
134604
Bourdon, Daniel, Viriat 074250
Bourdon, Louis, Guingamp 068771
Bourdon, M., Mons 063792
Bourdonnais Antiquités, Paris .. 071075
Boureau, Eric, Coutances 125564
Bourel, Jean-Paul, Annonay ... 066656,
066657
Bouretz, Béthune 067149
Bourey Caldarera, Marie-Christine,
Avallon 066862
Bourgeneuf 1900, Saint-Julien-
Chapteuil 015395
Bourgeois, Besançon 067120
Bourgeois Boheme, Los Angeles 094276
Bourgeois, Daniel, Châtillon-sur-
Seine 067923
Bourgeois, Marie, Vaupillon ... 073990
Bourgeois, Pierre, Lyon 069712
Le Bourget, Montréal 101370
Bourget, Yves, Sainte-Cathérine-de-
Fierbois 105635
Bourgey, Sabine, Paris 071076
Bourgogne Antiquités, Dunkerque 068413
Bourgogne Antiquités, Mâcon .. 069791
Bourgues, Christian, Cavaillon .. 067736
Bourguet, Gérard, Québec 064529,
101492
Bourke, F., Casthel 079473
Bourkel, Christian, Les Baix-de-
Breteuil 069401
Bourlet, Catherine, Genève 116289,
133755
Bourlot, Torino 142852
Bourn, Hugh, Burnham-on-
Crouch 117484
Bourne, Edinburgh ... 089162, 117717,
134358
Bourne Antiques and Arts,
Bourne 088526
Bourne End Antiques Centre, Bourne
End 088527
Bourne End Auction Rooms, Bourne
End 088528, 126827
Bourne End Pictures Plus, Bourne
End 117400
Bourne Gallery, Reigate 118963, 134963
Bourne Hall Museum, Ewell ... 044309
Bourne Mill Antiques, Farnham . 089260
Bourne, Michael, Columbus ... 135475
Bournemouth Natural Science Society
Museum, Bournemouth 043680
Bournemouth Transport Museum,
Bournemouth 043681
Bournville Centre for Visual Arts,
Birmingham 043617
Bourquard-Furster, Pierre, Semur-en-
Auxois 073485
Bourquin, Offenbach an der
Queich 077904
Bourquin, René, Bessoncourt . 128468
Bourquin, Yveline, Dijon 068318
Bourrasset, Daniel, Le Muy 069318

Bourreau, Ravier, Noirmoutier-en-
l'Ile 104411
Bourret, Bernard, Avignon 066874
La Bourrine à Rosalie, Sallertaine 015656
Bourrut, Lucienne, Vic-en-Bigorre 074088
The Bourse Antiquarian Print Gallery,
Brighton, Victoria 061201
Bourse aux Affaires, Les Sables-
d'Olonne 069427
Bourse aux Armes Anciennes,
Rungis 098077
La Bourse aux Collections, Melun 070044
Bourse aux Livres, Tournai 140261
Bourse aux Timbres, Lyon 069713
Bourse Collectionneur, Paris ... 071077
Bourse du Mobilier, Roanne ... 072484
Bourse et Collections, Toulon
(Var) 073704
Boury, Jean-Yves, Janzé 068890
Bouscary, Monique, Dijon 068319
Bouscasse, Alain, Carpentras .. 067693
Bouscayrol, Biarritz 103381
Bouscayrol, Pau 105328
Bousquet, Richard, Arcachon .. 066701
Boussat, Philippe, Felletin 068521
Boustead-Bland, Chester 088857
Bout-Croizé, Sandrine, Lieuran lès
Béziers 128674
Bout d'Bois, La Baule-Escoublac 068946,
068947
Boutang, René, Collonges-la-
Rouge 103616
Boutang, René, Paris 104586
Bouteloup, Huguette, Saint-Ouen 073041
Boutemy, Paris 071078, 128947
Boutersky, Jacques, Valberg ... 105793
Boutet Brousmiche, Poitiers ... 072220
Boutet, Lionel, Lumio . 069658, 104053,
140876
Boutin, Josiane, Chamoux 067804
The Boutique, Hamilton 101312
Boutique, Opio 070787
Boutique Antiquités, Beaune ... 067012
Boutique d'Antan, Montcenis ... 070227
Boutique de Francine, L'Isle-sur-la-
Sorgue 069560
Boutique de l'Antiquaire,
Lourmarin 069647
Boutique de l'Histoire, Paris .. 141057
Boutique-de-l'Or, Rennes (Ille-et-
Vilaine) 072427
Boutique dell'Arredamento,
Modena 080777
Boutique du Collectionneur, Saint-
Louis 072898
Boutique du Fou, Loix-en-Ré .. 069610
Boutique Dunia, Montréal 101371
Boutique Fantasque, London ... 089973
Boutique Folies, Québec 064530
Boutique Paris-Musées, Paris . 141058
Boutique Rosa, Casablanca ... 082552
Boutique Virgilio, Ciudad de
Panamá 113315
't Boutje, Breda 082974
Boutle & King, London 144569
Bouton, Heidi, Bernburg 106474
Boutreux, Patrick, Chouzé-sur-
Loire 067989
Boutros, Zaki, Cairo 066233
Boutwell Draper, Redfern 099619
Bouvarel, Cécile, Saint-Sauveur-le-
Vicomte 105616
Bouveret, Luc, Paris 071079
Bouvet Ladubay, Saumur 105660
Bouvier, Jacques, Paris 071080
Bouvier, Jean-Louis, Saint-Quentin-
Fallavier 073280
Bouvier, Michel, Paris 141059
Bouwen, Rina, Madrid 115159
Bouwhuis, G., Diever 132508
Bouwman, Ivo, Den Haag 083062
Bov Museum, Padborg 010269
Bovani, Sabrina, Firenze 131134
Bovenschutte, Gelselaar 132505

Bovien, Ursula, Aurigeno 116038
Bovill, Miles, Malvern, Victoria . 061778
Bovo, Giorgio, Parma 131723
Bow House, Hungerford 089603
Bow-well Antiques, Edinburgh . . 089163
Bow Windows Book Shop,
Lewes 144528
Bowback Chair Company, Atlanta 092116
Bowden Pioneer Museum,
Bowden 005447
Bowden, Bryan, Harrogate 089423,
134435
Bowden, David, London 118228
Bowden, Jonathan, Riverside . . 099641
Bowdery, M.J., Hindhead 089533
Bowdoin College Museum of Art,
Brunswick 047285
Bowdoin College. Museum of Art.
Occasional Papers, Brunswick . 139326
Bowdon, Clitheroe 144344
Bowen, Wellington 113062
Bowen House / Roseland Cottage,
Woodstock 054527
Bowen Island Museum, Bowen
Island 005448
Bowen, Annabel, Baltinglass . . 130827
Bower, Mount Molly 061919
Bower, Peakhurst 062069
The Bower Gallery, San Antonio 052933
Bower House Gallery, Milton . . 099346
Bowerbank Mill, Deloraine 098963
Bowers & Merena, Irvine 127263
Bowers Mansion, Carson City . 047443
Bowers Mansion, Washoe Valley 054195
The Bowers Museum of Cultural Art,
Santa Ana 053086
Bowers Science Museum, State
University of New York College at
Cortland, Cortland 048013
Bowers, Francis, Peterborough . 090982
Bowery, Baltimore 092304
Bowery Gallery, New York 051348
The Bowes Museum, Barnard
Castle 043495
Bowes Railway Heritage Museum,
Gateshead 044380
Bowhaus, Los Angeles 121450
Bowhill Collection, Selkirk 045755
Bowie & Hulbert, Hay-on-Wye . . 117934
Bowie on Bowie Art, Fort Worth 120857
Bowie Railroad Station and Huntington
Museum, Bowie 047153
Bowie's Heritage and Children's
Museum, Bowie 047154
Bowinkel, Uberto, Napoli 080845
Bowkett, Knaresborough 089722, 134521
Bowlby, Nicholas, Poundgate . . 118946
Bowlby, Nicholas, Tunbridge
Wells 119269, 135155
Bowles, Franklin, New York ... 122432
Bowles, Franklin, San Francisco 124238
Bowls New Zealand Museum, New
Plymouth 033255
Bowman, Megahan and Penelec
Galleries, Allegheny College,
Meadville 050752
Bowman, Robert, London 089974,
118229, 134605
Bowman, Russell, Chicago 120078
Bowmanville Museum,
Bowmanville 005449
Bowmoore, Marlborough 090719
Bown, Matthew, London 118230
Bowne House, Flushing 048764
Bowood House and Gardens,
Calne 043831
Bowral Antiques, Bowral 061180
Bowraville Folk and Military Museum,
Bowraville 000874
La Box, Bourges 012160
The Box, Torino 110957
Box Heart Expressions,
Pittsburgh 123528
Box Kraftverk, Bjärtrå 040564

Box of Porcelain, Dorchester ... 089068
Boxart, Verona 111178
Boxenstop Auto- und
Spielzeugmuseum, Tübingen .. 022213
Boxer, San Isidro 060956
Boxer, Henry, Richmond, Surrey 118968
Boxes and Musical Instruments,
London 089975, 134606
Boxfield Gallery, Stevenage 045863
Boxleitner, Ludwig, Passau 078019
Boxster, Nashville 094978
Boxy, P. & L., Knokke-Heist ... 063698
Boya, Shenzhen 102218
Boyang Museum, Boyang 007660
Boyce, Victoria, Tucson 124719
Boycott, Bruxelles 100571
Boyd, Annie, Stanmore 062239
Boyd, John, Castle Cary 134222
Boyd, John & Maureen, Eliot .. 127214
Boyd, Roy, Chicago 120079
Boyd, Vance, Memphis 094621
The Boydell Galleries, Liverpool . 089854,
118126
Boyer, Argenton-sur-Creuse ... 066716
Boyer-Labarre, Frédéric, Genève 116290
Boyer Museum, Wildwood 054405
Boyer, Dorien, Saint-Hilaire-de-
Brethmas 072821
Boyer, Serge, Paris . . . 071081, 128948
Boyertown Museum of Historic
Vehicles, Boyertown 047163
Boyes, Judy, Ambleside 117236
The Boyle House, Hot Spring County
Museum, Malvern 050615
Boyle, Lena, London 118231
Boyles, Clonakilty 079484
Boyles-Snyder & Co, Los
Angeles 135708
Boyne Art Studio – Church Gallery,
Bective 109529
Boyne Valley Historical Cottage,
Ubobo 001599
Boys From the Bush Antiques,
Woodside 062453
Boys Town Hall of History & Father
Flanagan House, Boys Town . . 047164
Boz, Hay-on-Wye 144460
Bozar Shop, Bruxelles 140168
Bozart, Philadelphia 138438
Bozatlı, İstanbul 117019
Bozhou Museum, Bozhou 007661
Bozsik, Budapest 079243
Bozsó Gyűjtemény, Kecskemét . 023721
Bozzeto, Bilbao 114982
Bozzetto, Pedro, Porto Alegre . . 063962
Bozzi, Ivan, Bolzano 131006
Bozzi, Ugo, Roma 137727
Bozzini, Sonia, Milano 131398
BPA Collectables, Los Angeles . 094277
BPS 22 – Espace de Creation
Contemporaine, Charleroi ... 003510
BQ, Köln 107596
Braak, ter, Zwolle 112730
Braakhuis, P., Maastricht 083457
Braam, J.P.B., Bruxelles 128020
Braat & de Pater, Leiden 083426
Braatz, Rüdiger, Langwedel ... 142050
Brabant Antiquariaat, 's-
Hertogenbosch 143172
Brabante, Madrid 133308
Brabants Etsenkabinet, Geldrop . 143143
Brabants Museum Oud Oosterhout,
Oosterhout 032725
Brac de la Perrière, Lyon 128699
Brac, Jean, Mouleydier 070366
Bracaglia, Bartolomeo, Roma . 131833
Bracci, Angelino, Roma 081269, 081270
Bracci, Claudio, Roma 081271
Bracher, Kathrin, Bern 133671
Brachert, Dr. Eva, Mainz 130266
Brachert, Dr. Thomas, Nürnberg 130463
Brachot, Gabriel, Bruxelles ... 100572
Bracht, Christa, Kassel 076715

Bracias Peñafiel, Sergio,
Zaragoza 086592
Bracken Hall Countryside Centre,
Baildon 043467
Brackendale Art Gallery Theatre
Teahouse, Brackendale 005454
Bracket Clock, Córdoba 085940
Brački Muzej, Supetar 009005
Brackley Antique Cellar, Brackley 088552
Brackley Antiques, Brackley ... 088553,
134132
Bracko Brothers, Calgary 128146
Bradbury Science Museum, Los
Alamos 050401
Bradbury & Son, Edward,
Cheltenham 088829, 144325
Bradbury, Roger, Coltishall 088960
Bradbury, Stuart, Paulton 134925
Bradford, Cairns 061246
Bradford, Nashville 094979
Bradford Brinton Memorial, Big
Horn 046976
Bradford County Historical Society
Museum, Towanda 053783
Bradford Creek Antiques,
Bendigo 061129
Bradford Historical Society,
Bradford 047174
Bradford Industrial Museum and
Horses at work, Bradford ... 043691
The Bradford Museum of Collector's
Plates, Niles 051562
Bradford Ohio Railroad Museum,
Bradford 047172
Bradford-on-Avon Museum, Bradford-
on-Avon 043698
Brading Roman Villa, Brading . 043701
Bradley, Houston 135592
Bradley, Portland 096373
Bradley, Richmond 096530
Bradley Brent, Portland 123603
Bradley Gallery of Art, Lakeland
College, Plymouth 052217
Bradley Gallery of Art, Lakeland
College, Sheboygan 053295
Bradley Museum, Mississauga . 046842
Bradley, A. & M., Petworth 090990
Bradley, Robert, Salisbury 091221
Bradran, Manama 100419
Bradshaw, Birmingham 092395
Bradshaw, Portland 123604
Bradstreet's Books, Hawthorn . 139766
Bradwell Antiques Centre,
Bradwell 088565, 134137
Bradwin, Elizabeth, London ... 089976
Brady, Philadelphia 136075
Brady & Co., W.M., New York . 122433
Brae-Mar Antiques, Woonona . 062492
Brae, Clarence, Paterson 062068
Bräcke, Helsingborg 086737
Braem & Minetti, Toronto 064619,
101584
Braemar, Springwood 099714
Braemar Antiques, London 089977
Brändl, Werner, Würzburg 079077
Brändström & Stene, Stockholm 115847
Brænne, Victoria, Oslo 132783
Bräunlein, Dieter, Spalt 078511
Braga, Martel 069980
Braga, Piacenza 110646
Bragardo, Liège ... 063749, 128073
Bragegården, Vaasa 011467
Bragg-Mitchell Mansion, Mobile 050973
Bragg, Jean, New Orleans 095064
Bragg, John, Lostwithiel 090633
Bragg, Kimberly, San Francisco . 092617
Bragge & Sons, Rye . . 091159, 134984
Braggiotti, Amsterdam 112171
Braghuset Antik, Køge 065970
Brahma, Pierre, Paris 071082
Brahms & Netski, New York ... 095284
Brahmshaus, Baden-Baden ... 016920
Brahmshaus, Heide 019126
Brahuer Joyeros, Barcelona ... 085746

Braidwood Museum, Braidwood 000879
Brail-Millet, Alzonne 066543
Brailly, Amiens 066565
Brain Trust, Tokyo 111410
Brain-Watkins House, Tauranga . 033305
Brainard, C.R., Cleveland 120407
Braintree District Museum,
Braintree 043704
Braintree Historical Society Museum,
Braintree 047176
Braith-Mali-Museum, Museum
Biberach, Biberach an der Riß 017356
Braithwaite Fine Arts Gallery, Cedar
City 047473
The Braithwaite Gallery, York . 119406
Braitman, Charlotte 119980
Brajkovic, Berlin ... 074770, 129478
Brakebusch, Börries, Düsseldorf 129736
Brakebusch, Kristina, Düsseldorf 129737
Bralorne Pioneer Museum,
Bralorne 005455
Brāļu Amtmaņu Muzejs, Valle . . 030246
Brāļu Jurjānu Memoriālais Muzejs
Meņģeļi, Ērgļi 030121
Brāļu Kaudzīšu Memoriālais Muzejs,
Kaudzīši 030146
Braly, Jean, New Orleans 122136
Bram, Hobro 102840
Brama, Breda 112351
Brama, Warszawa 113809
Brama Człuchowska, Muzeum
Historyczno-Etnograficzne,
Chojnice 034687
Bramah Tea and Coffee Museum,
London 044919
Bramall Hall, Bramhall 043705
Brambilla, Marco, Milano 131399
Brambridge Antiques,
Brambridge 088566, 134138
Brambuk Living Cultural Centre, Halls
Gap 001132
Brame & Lorenceau, Paris ... 104587,
136873
Bramley Antiques, Eye 089240, 134379
Brammertz, H., Stolberg,
Rheinland 078565
Brampton Picture Gallery, Brampton,
Cambridgeshire 117415
Bramsnæs Museum og Arkiv, Kirke
Hyllinge 010161
Bramtidsmuseet, Borlänge 040580
Branca, Alessandra, Chicago .. 092674
Brancaccio, Simonetta, Roma . 081272
Branch Gallery, Durham 120821
Brancheverening Art & Frame,
Utrecht 059597
Branchville Railroad Shrine and
Museum, Branchville 047177
Branciforti, Nicolino, Bardolino . 109765
Branco, Catarina Faria, Lisboa . 084844
Branco, Maria Rosário, Porto .. 085012
Branco, Pedro Aguiar, Porto .. 085013
Brancolini Grimaldi, Firenze .. 110038
Brancolini Grimaldi, Roma 110739
Brand, Den Haag 083663
Brand, Roches-lès-Blamont ... 072526
Brand & Rauch, Bonn 075068
Brand Library and Art Galleries,
Glendale 049116
Brand X Antiques, San Francisco 097194
Brand, Claudia, Klosterneuburg . 062678
Brand, Joachim, Bern 087277
Brand, Marianne, Carouge 116189
Brand, W., Punthorst 132610
Branda, Sergio, Perugia 131744
Brandão, São Paulo 140300
Brande, Halstenbek 107215
Brande Antikvariat, København . 140604
Brande Museum, Brande 010025
Brandeis, Norland 145381
Brandel, Katrin, Berlin 141492
Brandenburg-Preußen Museum,
Stiftung Ehrhardt Bödecker,
Wustrau 022832

Brandenburg, Thomas, Alt
 Duvenstedt 074387
Brandenburgischer Kunstverein Potsdam
 e.V., Potsdam 021179, 058761
Brandenburgischer Verband Bildender
 Künstler e.V, Potsdam 058762
Brandenburgisches Apothekenmuseum,
 Cottbus 017764
Brandenburgisches Forstmuseum
 Fürstenberg/Havel, Fürstenberg,
 Havel 018538
Brandenburgisches Freilichtmuseum
 Altranft, Bad Freienwalde .. 016767
Brandenburgisches Museum für Klein-
 und Privatbahnen, Gramzow .. 018793
Brandenburgisches Textilmuseum
 Forst, Forst, Lausitz 018385
Brander Museum, Huntly 044619
Brander-Rodriguez, Luis, Nice .. 128853
Brandi, Maurizio, Napoli 080846
Brandicourt, Colette, Paris 071083
Brandis, Agnes von, Hamburg .. 129953
Brandkårsmuseet, Simonstorp .. 040943
Brandl, Eduard, Wien 062916
Brandler, Brentwood 134142
Brandler Galleries, Brentwood .. 117420
Brandon, Tulsa 127480
Brandon Heritage Museum,
 Brandon 043706
Brandon Mental Health Centre
 Historical Society Museum and
 Archives, Brandon 005461
Brandon, Elizabeth, Nashville .. 122072
Brandstätter, Baden-Baden 141441
Brandstätter, Feistritz an der
 Drau 127729
Brandstätter, Dr. Christian, Wien 136578
Brandstätter, Günther, Bergheim 127717
Brandstätter, Horst & Ulrike,
 Öhningen 142260
Brandstorps Auktionsfirma,
 Brandstorp 126571
Brandstrup, Oslo 113196
Brandstrup, Stavanger 113262
Brandt, London 089978
Brandt-Jensen, Peter, København 065871
Brandt, C., Itzstedt 076637
Brandt, Christophe, Neuchâtel .. 133870
Brandt, H., Siegburg 078459
Brandt, Tilman, Görlitz 129908
Brandt, Walfried, Høvikodden .. 132779
Brandtmayer, Horst, Seefeld in
 Tirol 062841
Brandval Museum, Brandval .. 033558
Brandweermuseum, Wouwse
 Plantage 033070
Brandweermuseum en
 Stormrampmuseum, Borculo .. 032081
Brandweermuseum Wassenaar,
 Wassenaar 033025
Brandy, Michel, Saint-Martin-de-
 Ré 072936
Brandywine Battlefield, Chadds
 Ford 047505
Brandywine River Museum, Chadds
 Ford 047506
Brandywine River Museum Antiques
 Show, Chadds Ford 098438
La Branère, Labenne 069128
Brangakmenių Muziejus, Kaunas 030351
Branger, François, Angers 125488
Branigan Cultural Center, Las
 Cruces 050152
Brank Sore Antigüedades, San
 Isidro 060957
Branko Deškovič Art Gallery, Bol 008885
Branksome Antiques, Branksome 088574
Brans, Clive & John, Mosman
 Park 061898
Brans, Sylvaine, Paris 128949
Branscheid, U., Lüdenscheid ... 077198
Brant, New York 138381
Brant Museum, Brantford 005469

Brant, Laird, Dallas 093078
Brantebjerg, Nykøbing Sjælland . 102940
Brantes, Flore de, Paris 071084
Brantwood, Coniston 044001
Brantwood, Warwick 091741
Brasher-Rucker, Memphis 121683
Brasil 40, Madrid 133309
Brasil Europa Moveis e Antiguidades,
 Rio de Janeiro 064030
Brasilianisches Kulturinstitut in
 Deutschland – Galerie, Berlin . 017091
Brasilienmuseum im Kloster Bardel,
 Bad Bentheim 016713
Brask-Nielsen, Gerda, Viborg .. 103004
Brask, Erling, Hørsholm 102844
Brass Armadillo, Omaha 096029
Brass Armadillo, Phoenix 096194
Brass Bed Factory, Philadelphia 136076
Brass Buff, Tulsa ... 097715, 136429
Brass Eagle, Portland 096374
Brass Eagle Antiques, Cincinnati 092833
Brass Key Antiques, Salt Lake
 City 096881
The Brass Knob, Washington ... 097812
Brass Light Gallery, Milwaukee . 094839,
                                135812
Brass Menagerie, New Orleans . 095065
Brass Monkey, New Orleans ... 095066
Brass Rubbing Centre, Edinburgh 044231
Brass Works, Chicago . 092675, 135402
Brass, Lawrence, Bath . 088382, 134074
Brass & Son, Lawrence, London 134607
Brassac, J.L., Villefranche-de-
 Rouergue 074142
Brassel, Rüdiger, Siegen 108616
Brassel, Verena, Altstätten 116019
Brasseur, Martine, Reims 072385,
                                105423
Brassworks Company, Baltimore 092305,
                                135340
Braswell Chase, New York 127344
Braswell, F.L., Chicago 127608
Bratskij Chudožestvennyj Vystavočnyj
 Zal, Bratskij gorodskoj objedinennyj
 muzej istorii osvoenija Angary,
 Bratsk 036914
Bratskij Gorodskoj Objedinennyj Muzej
 Istorii Osvoenija Angary, Bratsk 036915
Brattle, Boston 145011
Brattle Farm Museum,
 Staplehurst 045862
Brattleboro Museum and Art Center,
 Brattleboro 047182
Bratz, Jörg P., Berlin 074771
Brauchtums- und
 Trachtenpuppenmuseum der Banater
 Schwaben e.V., Würzburg ... 022786
Brauer & Brauer, Rio de Janeiro 101015
Brauer Museum of Art,
 Valparaiso 053938
Brauer, Jörg, Rudolstadt 078303
Brauer, Ulf-Roman, Erfurt 129800
Brauerei-Kontor, Kleines Museum
 der Privatbrauerei Moritz Fliege,
 Bochum 017432
Brauerei-Museum, Haßfurt 019091
Brauerei-Museum, Nesselwang . 020667
Brauerei-Museum Dortmund,
 Dortmund 017953
Brauereichäller Kulturforum,
 Laufen 041641
Brauereikulturmuseum Gut
 Riedelsbach, Neureichenau .. 020734
Brauereimuseum, Aldersbach .. 016500
Brauereimuseum, Alpirsbach ... 016514
Brauereimuseum, Altomünster .. 016560
Brauereimuseum Barre's Brauwelt,
 Lübbecke 020148
Brauereimuseum Bräu im Moos,
 Tüßling 022229
Brauereimuseum im historischen
 Kronen-Brauhaus zu Lüneburg,
 Ostpreussisches Landesmuseum,
 Lüneburg 020177

Brauereimuseum Schneider,
 Weißenburg in Bayern 022536
Brauereimuseum Schöneck,
 Schöneck / Vogtland 021704
Brauhaus-Galerie, Kunst- und
 Kulturvereinigung im Brauhaus
 Freistadt, Freistadt 001905
Brauhausmuseum,
 Stadtlauringen 021955
Brauhinia Shire Bicentennial Art
 Gallery, Springsure 001514
Braulio 2, Castello de la Plana . 115012
Braumuseum, Wieselburg an der
 Erlauf 003109
Braumuseum – Wolff Giersing's
 Brauerbe, Stadt- und
 Kulturgeschichtliches Museum
 Torgau, Torgau 022165
Braumuseum, Gösser Brauerei,
 Leoben 002290
Braun, Saint Louis 096673
Braun, Stockholm 086979
Braun, Bernd, Gengenbach 141826
Braun, Gabriele, Mühlacker 130316
Braun, Hanns-Gert, Wuppertal .. 079102,
                                142521
Braun, Hans-Gert, Wuppertal .. 079103,
            108984,   142522
Braun, Helmut, Rinnthal 078242
Braun, Horst, Spalt 078512
Braun, I., Detmold 141685
Braun, Klaus, Stuttgart 108689
Braun, Markus, Pirmasens 130521
Braun, Martin, Bad Zwischenahn 074622
Braun, Martin, Rastede 078129
Braun, R., Freiburg im Breisgau 075853
Braun, Wolfgang, Blieskastel ... 075022
Braunauer Heimatmuseum, Forchheim,
 Oberfranken 018380
Braune, Brad, San Antonio 124017
Braune, Irmgard, Edling 075573
Braunmiller, Wilfried, München . 130332
Brauns, Thomas & Curt, Frechen 141793
Braunschweigisches Landesmuseum,
 Abteilung Ur- und Frühgeschichte,
 Wolfenbüttel 022730
Braunschweigisches Landesmuseum,
 Niedersächsische Landesmuseen
 Braunschweig, Braunschweig . 017553
Braunshör, Peter, Wien 062917
Braunstein & Quay, San
 Francisco 124239
Braunstein, Alain, Nice 070579
Braunton and District Museum,
 Braunton 043707
Brautsch, H. & M., Greding 129917
Bravard, Frédéric, Monaco 082499
Braverman, Pittsburgh 096262
Bravi, Zola Predosa 111211
Bravin, Tamara, Torino 081664
Bravo, Quezon City 084484
Bravo Cárcamo, Mauricio, Santiago de
 Chile 064958
Bravo Carrión, Paulino, Madrid . 086139
Bravo Heredia, Francisco,
 Málaga 086287
Bravo of Old Town, San Diego . 124107
Bravo Panadero, Oscar, Madrid . 086140
Bray Antique Restoration Service,
 Bray 130832
Bray Heritage Centre and Museum,
 Bray 024835
Brayer, André, Damblain 068264
Brayer, Pascal, Nancy 070414
Brazen, Burnham Market 088720,
                                144289
Brazier, Richmond 123725
Brazil Galeria, Sydenham 099757
Braziller, George, New York ... 138382
Brazoria County Historical Museum,
 Angleton 046482
Brazos Valley Antiques, Fort
 Worth 093433
Brazos Valley Museum of Natural

History, Bryan 047289
Brazosport Museum of Natural
 Science, Brazosport 047183
Brazosport Museum of Natural
 Science, Clute 047850
Brazzini, Dario, Firenze 131135
Brdský Památník – Městské Muzeum,
 Rožmitál pod Třemšínem 009869
Bread & Puppet Museum, Glover 049138
Bread Street Gallery, Penzance . 118893
Breadalbane Folklore Centre,
 Killin 044694
Breakspeare, Tetbury 091567
Breamore Countryside Museum,
 Breamore 043708
Breamore House Museum,
 Breamore 043709
Breard, Hermann, Rouen 105471
Breault & Sons, Dallas 135495
Brebner Surgical Museum,
 Johannesburg 038747
Breche, Sevilla ... 086443, 086444
Brechin Museum, Brechin 043710
Brechó, Elio & Clair, São Paulo . 128124
Brecht-Weigel-Gedenkstätte, Akademie
 der Künste, Berlin 017092
Brechthaus, Augsburg 016648
Brecknock Museum and Art Gallery,
 Brecon 043712
Brecon Mountain Railway, Merthyr
 Tydfil 045249
Breda Aude, Chantal de, Lyon .. 069714
Breda's Begijnhof Museum,
 Breda 032100
Breda's Museum, Breda 032101
Bredase, Breda ... 082975, 132453
De Brede, Maasbree 032617
Brede Værk, Kongens Lyngby .. 010208
Bredgade Kunsthandel,
 København ... 065872, 102869
Bredovský Letohrádek na Lemberku,
 Jablonné v Podještědí 009589
Bredy Farm Old Farming Collection,
 Burton Bradstock 043802
Bree, Ilka, Bordeaux 103418
Breebaart, I.L., Amsterdam 132391
Breede, B. & J., Bremen 141648
Breede, Ulf, Berlin 074772
The Breeder, Athinai 109028
Breewood, Richmond, New South
 Wales 099624
Breffni Ireland, Carrigallen 130834
Bregenzer Kunstverein, Bregenz 058242
Brégeot, Eric, Aire-sur-l'Adour . 066446
Breginjski Muzej, Kobarid 038525
Bregt Zijlmans, Vught 132665
Bréhal'Antiques, Brehal 067457
Bréhélin, Jean, Rennes (Ille-et-
 Vilaine) 141312
Bréhéret, Paris 071085, 104588, 136874
Bréhéret, Michèle, Angers 066601
Brehm, Krefeld 107730
Brehm-Gedenkstätte,
 Renthendorf 021377
Brehm, Gerda, Krefeld 076952
Breidablikk, Stavanger 033974
Breidenbach, Claude, Fribourg . 133742
Breier & Breier, Dettingen unter
 Teck 075346
Breier, Marianne, Dettingen unter
 Teck 075345
Breier, V., San Francisco 124240
Breitbach, Norbert, Boppard ... 125901
Breitenbach, Christa,
 Aschaffenburg 074430
Breitenbach, Christa,
 Laudenbach 077028
Breitenecker, Leopold, Wien ... 127907
Breitengraser, Berlin 016180
Breiter, Langendorf 087554
Breitfeld, Annaberg-Buchholz .. 074414
Breitman, Hugo, Buenos Aires .. 060882
Breitmann, Francis, Paris 071086,
                                104589

Breitschuh, Klaus, Quedlinburg . 078108,
126100
Breiz, Carnac 067686
Bremecke, Ratingen 108434
Bremens, Lyon 125621
Bremens, Jean-Marc, Lyon ... 125622
Bremer, Berlin 106181
Bremer, Hamburg 076200
Bremer Institut für Architektur,
Kunst und Städtische Kultur,
Universität Bremen, Fachbereich 09
Kulturwissenschaften, Bremen 055496
Bremer Numismatische Gesellschaft,
Bremen 058763
Bremer-Röhner, Dr. Uta, Plön . . 108388
Bremer Rundfunkmuseum,
Bremen 017583
Bremer Verband Bildender
Künstlerinnen und Künstler,
Bremen 058764
Bremer, Beverly, Atlanta 092117
Bremer, R., Heidelberg 076436
Bremermann, New Orleans 095067
Bremerton Naval Museum,
Bremerton 047188
Bremilts, Jean-Louis, Lille 069479
Bremmer, Tilburg 112678
Bremmers, Maastricht . 132572, 143198
Bren Art, Toronto 101585
Brenau University Galleries,
Gainesville 049008
Brendlein, A., Heidelberg 141924
Brendow & Sohn, Joh., Moers . . 077405
Brennan and Geraghty's
Store Museum, Maryborough,
Queensland 001258
Brennan, Patricia, Roma 081273
Brennan, Robyn, Annerley 061001
Brennbergbányai Bányászati Emlékház,
Soproni Múzeum, Sopron ... 023888
Brennecke, Klaus-D., Berlin ... 106182
Brenner, Stein bei Nürnberg ... 078545
Brenner, Michel, Vesoul 074077
Brenner, Stefan, Stuttgart 078596
Brenner, Sylvia, Zwinge 079156
Brennereimuseum Haselünne,
Haselünne 019084
Brennig, Dr. Heribert, Bonn 141612
Brennpunkt, Schüpfen 116633
Brensbach Museum und Galerie im
alten Rathaus, Brensbach 017618
Brenske, Dr. Stefan, München ... 077473
Brenske, Dr. Stefan, Zürich 087944
Brent, Keighley 118032
Brent, Saint Louis 096674
Brentwood, Nashville 122073
Brentwood Antik Auctions,
Brentwood 126834
Brentwood Art Center, Los
Angeles 057367
The Brentwood Gallery,
Warrington 119309
Brentwood Museum, Brentwood 043715
Brenzett Aeronautical Museum Trust,
Romney Marsh 045632
Breon, Gagny 068621
Brepols Publishers, Turnhout ... 136663
Brereton, John, Dublin 079529
Brès, de, Amsterdam 112172
Brès, Martin, Marseille 069912, 104159,
140941
Bresaylor Heritage Museum,
Bresaylor 005473
Breslauer Sammlung Köln, Köln 019673
Bresler Eitel, Milwaukee 121903
Bressan, Renzo, Venezia 132229
Bresset & Fils, Gilles, Paris 071087,
071088
Bressingham Steam Museum,
Diss 044102
Bressoise Transactions, Le Crès 069257
Bressol, Barcelona 085747
Brest, Christophe, Épernay 068447
Bret, André, Faverges-de-la-Tour 068512

Bret, Michel, Lyon 128700
Bretagne Ancienne, Guérande .. 068761
Bretagne Antik, Brest 067466
Breteschè, Mamia, Paris 104590
Brethvad, København 102870
Bretin, Sophie, Lyon 128701
Bretón, Logroño 086072, 115121
Breton, Benoît, Bulgnéville 067534
Breton, Gérard, Saint-Léonard-de-
Noblat 072891
Breton, Sylvie, Dijon 068320
Bretschneider, Giorgio, Roma . . 142800
Brett Whiteley Studio Museum, Surry
Hills 001530
Brett & Co, A.J., London 134608
Brett, James, Norwich 090869
Bretts, Christchurch 143303
Bretz, Simone, München 130333
Breu, Bernadette, Portland 096375
Breu, Toni J., Bonstetten 116164
Breu, Zita, Wien 127908
Breuberg-Museum, Breuberg .. 017624
Breuer Moreno, Buenos Aires .. 060883
Breuer, Eva, Woollahra 099911
Breughel, Pieter, Amsterdam ... 082710
Breughel, Pieter, Venlo 083675
Breuil, Bellerive-sur-Allier 067078
Breul, Maria Christian von, Lyon 104071
Breunesse, Zeist 083708, 112723
Breunig, Eugen, Stuttgart 130667
Breuning, Ulrik, Borre 140560
Breuninger, Rolf, Ofterdingen .. 077915
Breuza & Larizza, Torino 081665
Brevard Museum, Cocoa 047853
Brevard Museum of Art and Science,
Melbourne 050770
Brevik Bymuseum, Brevik 033560
Brew & Sterling, Broadbeach .. 061214
Brewarrina Aboriginal Cultural
Museum, Brewarrina 000880
Brewer, Henry, West Bridgford . 119322
Brewery Art Association, Los
Angeles 050410, 060415
Brewery Chapel Museum,
Halstead 044507
The Brewhouse Gallery,
Grantham 117873
Brewhouse Yard Museum,
Nottingham 045410
Brewster & Brown, New York . . 095285
Brewster & Co., I., Philadelphia . 123359
Brexendorf, E., Worpswede 108964
Breyer, Laura, New York 095286
Brezina, Birmingham 135351
Brhel, Rudolf, Heidelberg 076437
Brian, Buffalo 119952
Brian & Anna's Collectibles,
Houston 093627
Brian Page Antiques, Brighton .. 088613
Brian, Robert, San Francisco ... 097195
Briance, Jean, Paris ... 104591, 141060
Briand, Daniel, Drémil-Lafage . 136814
Briand, Josiane, Notre-Dame-de-
Monts 070748
Brianes, Daniel, Gages 068620
Briani & Miotto, Cincinnati ... 120324
Brian's, Kaiapoi 083969
Brian's Books, Doubleview 139718
Brian's Books, Katoomba 139787
Brianza, Bruno, Kempten 076749
Briar Herb Factory Museum,
Clyde 033179
Briar Patch, Salt Lake City 096882
Briard, Alain, Marseille 128751
Briard, Claude, Marseille 128752
Briard, Thierry, Caen 067559
Briarpatch, Cincinnati 092834
Briarwood, San Jose 097359
Briata, Georges, Marseille 104160
Briatte, Williams, Jussy 068928
Briatte, Williams, Saint-Quentin
(Aisne) 073275
Briau, Jacques, Blois 067252
Bribie Island Community Arts Centre,

Bribie Island 000882
Bric à Brac, Orléans 070811
Bric-a-Brac, Padova 080970
Bric-a-Brac, Saint Ives, Cornwall 091191
Bric à Brac, Saint-Ouen 073042
Bric à Brac d'Armelle, Ris-
Orangis 072475
Bric à Broc, Cluny 068046
Bric Brac, Buenos Aires 060884
Bricabraque Mobrel, Lisboa ... 084845
Brice, Memphis 094622
Brice, Bruce, New Orleans 122137
Brice, Geneviève, Cagnes-sur-
Mer 103488
Brick House, Montgomery 051027
The Brick Row, Baltimore 092306,
135341
Brick Row Book Shop, San
Francisco 145546
Brick Store Museum, Kennebunk 049920
Brick Window Gallery, Naphill . 118773
Brickell Village Gallery, Miami .. 121748
Bricolage Móveis Antigos, Porto
Alegre 063963
Bricout, Patrice, Marcq-en-
Barœul 069860
Bricus, Beirut 082276
Bride, Patrick, La Cluse-et-
Mijoux 068999
Bridenne, Franck, Boulogne-sur-
Mer 067397
Bridewell Museum, Norwich ... 045396
Bridge & Art, Braunschweig ... 106560
Bridge Art Fair, Chicago, Chicago 098439
Bridge Art Fair, London, London 098361
Bridge Art Fair, Miami Beach, Miami
Beach 098440
The Bridge Art Gallery, Dublin . 109578
Bridge Books, Murray Bridge .. 139846
Bridge Center for Contemporary Art, El
Paso 048503
The Bridge Gallery, Northbridge, New
South Wales 099481
The Bridge Gallery, Southampton 119119
Bridge House Antiques,
Hungerford 089604, 134488
Bridge House Gallery, Saint
George's 100840
Bridge Magazine, Chicago 139327
Bridge Street Auction,
Plattsburgh 127406
Bridge, Christine, London 089979,
134609
Bridgehampton Historical Society
Museum, Bridgehampton ... 047196
Bridgeman, Taranaki 113014
Bridgeman Art Library, Berlin .. 137057
The Bridgeman Art Library,
London 138198
Bridgeman Art Library, New York 138383
Bridgeman Giraudon, Paris ... 136875
Bridgepoint Books, Mosman .. 139837
Bridgeport Antiques, Chicago .. 092676
BridgeSpotters Collective,
Pittsburgh 060416
BridgeSpotters Gallery, Pittsburgh 123529
Bridgestone Bijutsukan, Ishibashi
Zaidan, Tokyo 029490
Bridgetown Old Gaol Museum,
Bridgetown 000883
Bridgewater Books, Lewes 138175
Bridging Arts, London 118232
Bridgnorth Antiques Centre,
Bridgnorth 088597, 134144
Bridgnorth Northgate Museum,
Bridgnorth 043718
Bridgton Historical Museum,
Bridgton 047207
Bridgwater, David, Bath 088383
Bridlington Harbour Museum,
Bridlington 043725
Bridport Museum, Bridport ... 043727
Bridport Old Bookshop, Bridport 144262
Brié, Alain, Saint-Alban 072662

Briefmarkenmuseum im Kloster Bardel,
Bad Bentheim 016714
Briels, Breda 112352
Briercrest and District Museum,
Briercrest 005477
Brière, Michel, Sablé-sur-Sarthe 072651
Brieux, Alain, Paris ... 071089, 141061
Brigandi, Donald E., New York . 095287
Brigantini, C., Miami 121749
Briged Pasukan Polis Hutan,
Kuching 030700
Briggs-Robinson, New York ... 122434
Briggs, F.E.A., Brook 088687
Brigham City Museum-Gallery,
Brigham City 047208
Brigham Young University Earth
Science Museum, Provo ... 052403
Brigham Young's Winter Home, Saint
George 052767
Brighenti, Andrea, Modena ... 080778
Bright, Albuquerque 092038
Bright, Saint Kilda, Victoria ... 099666
Bright Antiques, Bright 061194
Bright Art Gallery, Bright 098774
Bright Henderson, J., San Diego 124108
Bright Metals of the Heights,
Houston 135593
Brighthouse, Wickham 119349
Brightlingsea Museum,
Brightlingsea 043732
Brighton, Dallas 093079
Brighton, Fort Worth 093434
Brighton, Houston 093628
Brighton, Kansas City 094061
Brighton, Murcia 086312
Brighton, Saint Louis 096675
Brighton, San Antonio 096953
Brighton, Tucson 097640
Brighton, Tulsa 097716
Brighton Antique Prints and Maps,
Brighton, Victoria 139656
Brighton Antiques, Brighton,
Victoria 061202
Brighton Architectural Salvage,
Brighton 088614
Brighton Books, Brighton 144264
Brighton Fishing Museum,
Brighton 043735
Brighton Galleries, Mortdale .. 099382
Brighton Lanes Antique Centre,
Brighton 088615
Brighton Museum and Art Gallery,
Brighton 043736
Brightwells, Leominster 126940
Brigl, G., Landsberg am Lech .. 077008,
130208
Brignell, Cambridge 134196
Brikettfabrik Hermannschacht Zeitz,
Zeitz 022850
Brilhante, Manuel, Porto 085014
Brill, Leiden 137864
Brill, Gary, Milton 061855
Brill, Normelio, Porto Alegre .. 063964
Brillantes Ancestral House,
Tayum 034608
Brilleman, Fabienne, Lyon 128702
Brillen-Museum, Homburg 019333
Brilliant Antiques, Saint Louis .. 096676
Brilliant Facets, Brisbane 098781
Il Brillocco, Roma 081274
Brim-Brim Gallery, Newtown,
Victoria 099454
Brimaud, Carole, Paris 104592
Brimo de Laroussilhe, Paris ... 071090
Brimstone Museum, Sulphur .. 053596
Brimstones, Lewes 144529
Brindle, T., Whalley 091818
Brindley Bank Pumping Station and
Museum, Rugeley 045650
Brings, Mülheim an der Ruhr .. 130319
Brini, Andrea, Bologna 079824
Brink & Van Dijk, Deventer ... 083155
Brink 7 Galerij, Yde 033071
Brinkens Museum, Malax 011140

Brinkhus, Heike, Bremen 141649
Brinkman Russell, Nelson 084014
Brinkman, A.P., Amsterdam ... 082711
Brinkman, H., Amsterdam 143020
Brinkmann & Verbeek, Köln ... 130145
Brinkmann, S., Köln 130144
Brinkmann, Volker, Mannheim .. 077253
Brinks & Partner Kunstberatung,
Köln 107597
Brinquin, André, Tours 073842
Brinsley Art Gallery, Dunedin .. 112876
Briolet-Lecaudey, Véronique, Saint-
Vigor-le-Grand 073328
Brion, Jean-Bernard, Rochefort-en-
Terre 072500
Brioni, Ernesto Antonio, Milano . 131400
Brioult, Lydie, Vernon 125832
Brisbane and Provincial Antique
Dealers Association, Tarragindi 058192
Brisbane Antique Market,
Clayfield 061330
Brisbane Auction Galleries, Bowen
Hills 125180
Brisbane City Gallery, Brisbane . 000885
Brisbane Galleries, Thornlands . 099823
Brisbane Tramway Museum, Ferny
Grove 001070
Brisbois, Marie-José, Paris 071091
Briscadieu, Alain, Bordeaux ... 125522
Brischen, Simone, Gleizé 068673
Brischigliaro & Figli, Giuseppe,
Padova 131656
Briscoe, Woolloongabba 062485
Au Brise Bise, Saint-Maur-des-
Fossés 129238
Brisigotti, London 089980
Brisólla, A., Schwetzingen 078434
Brissaud, Poitiers 141306
Brisset-François, Isabelle, Rennes (Ille-
et-Vilaine) 105429
Brissoni Marcellin, Audincourt .. 103288
Brissonneau, Paris 125688
Bristol Antique, Oslo 084304
The Bristol Arts Library, Bristol . 117440
Bristol Blue, Barcelona 133106
Bristol Books, Bristol 144276
Bristol Brocante, Bristol 088647
The Bristol Decorative and Fine Arts
Society, Compton Martin ... 060065
Bristol Guild of Applied Art,
Bristol 088648
Bristol Historical and Preservation
Society Museum, Bristol 047215
Bristol Trade Antiques, Bristol .. 088649
Bristol's City Museum and Art Gallery,
Bristol 043747
Bristol's Industrial Museum,
Bristol 043748
Bristow, Daniel, Tetbury 135127
Britain at War Experience,
London 044920
Britain's Heritage, Leicester ... 089789,
134542
Britannia, Petiville 072119
Britannia Antigüedades, Lima .. 084444
Britannia Antiques, Cagliari ... 079977
Britannia Antiques, Hahndorf .. 061572
Britannia Antiques, Houston ... 093629
Britannia Antiques, Stepney ... 062243
Britannia Antiques and Fine Art Gallery,
Nelson 084015, 112956
Britannia Art Publications,
London 138199
Britannia Gallery, Mellor Brook . 118734
Britannia Gallery, Ottawa 101465
Britannia Shipyard, Richmond .. 006792
British Academy Publications,
London 138200
British Antique Dealers' Association,
London 060066
British Antique Dealers' Association
Antiques and Fine Art Fair,
London 098362
British Antique Furniture Restorers'

Association, Dorchester 060067
British Antique Replicas, Burgess
Hill 088715
British Antiques, New Orleans .. 095068
British Arts Festivals Association,
London 060068
British Association of Friends of
Museums, Whitchurch 060069
British Association of Paintings
Conservator-Restorers, Blofield 060070
British Association of Paper Historians,
Longhoughton 060071
British Aviation Preservation Council,
Coventry 060072
British Balloon Museum,
Newbury 045340
British Brick Society, Stanmore . 060073
The British Button Society,
Tonbridge 060074
British Columbia Aviation Museum,
Sidney 007033
British Columbia Congenital Heart
Museum, Vancouver 007273
British Columbia Farm Machinery
and Agricultural Museum, Fort
Langley 005819
British Columbia Forest Discovery
Centre, Duncan 005728
British Columbia Forest Service
Museum, Victoria 007319
British Columbia Golf House Museum,
Vancouver 007274
British Columbia Museum of Medicine,
Vancouver 007275
British Columbia Museum of Mining,
Britannia Beach 005483
British Columbia Museums
Association, Victoria 058428
British Columbia Orchard Industry
Museum, Kelowna 006070
British Columbia Regiment Museum,
Vancouver 007276
British Columbia Sports Hall of Fame
and Museum, Vancouver 007277
British Columbia Sugar Museum,
Vancouver 007278
British Columbia Wine Museum,
Kelowna 006071
British Commercial Vehicle Museum,
Leyland 044820
British Council Arts Group,
London 060075, 138201
British Cycling Museum,
Camelford 043852
British Dental Association Museum,
London 044921
British Empire and Commonwealth
Museum, Bristol 043749
British Engineerium, Hove 044609
British Gallery Antiques, Neuilly-sur-
Seine 070519
British Gallery Antiques, Paris .. 071092
British Golf Museum, Saint
Andrews 045673
British Horological Institute, Newark-
on-Trent 060076
British Import Antiques, Paris .. 071093
British in India Museum, Nelson 045318
British Journal of Aesthetics,
Oxford 139154
British Lawnmower Museum,
Southport 045840
British Library, London 044922
British Library Bookshop, London 144570
The British Library Publishing,
London 138202
The British Museum, London ... 044923
British Museum Friends, London 060077
British Museum Magazine,
London 139155
British Museum Press, London . 138203
British Numismatic Trade Association,
Rye 060078
British Optical Association Museum,

College of Optometrists, London 044924
British Painting for Sale,
Sheffield 119080
British Photographic Museum,
Totnes 045992
The British Postal Museum & Archive,
London 044925
British Red Cross Museum and
Archives, London 044926
The British School at Rome, Accademia
Britannica, Roma ... 059422, 110740
British Schools Museums, Hitchin 044590
British Society of Aesthetics,
Eastbourne 060079
British Society of Master Glass
Painters, London 060080
British Society of Painters in Oil,
Pastels and Acrylics, Ilkley . 060081
British Trading Company, Dallas 093080
Brito, Antonio Carlos, São Paulo 064165
Brito, Jeannine de, Paris 071094
Britsch sen., Georg, Bad
Schussenried 074590
Britsch jun., Georg, Bad
Schussenried 074589
Britschgi, Melchior, Zürich 144168
Brittany Upholstering, Ottawa .. 128188
Britto, Miami 121750
Brittons, Clitheroe 088932
Brive Numismatique, Brive-la-
Gaillarde 067505
Briwax, Taranaki 132745
Brix Gallery, Makati 113382
Brixham Heritage Museum,
Brixham 043756
Brixia, Brescia 079922
Brjanskij Gosudarstvennyj Obedinennyj
Kraevedčeskij Muzej, Brjansk . 036917
Brjanskij Literaturnyj Muzej, Brjanskij
gosudarstvennyj obedinennyj
kraevedčeskij muzej, Brjansk . 036918
Brjanskij Oblastnoj Chudožestvennyj
Muzej, Brjansk 036919
Brněnská Antikva, Brno 065346
Broad Antiques International, New
York 095288
Broad Bay China, Dunedin 083921
Broad Ripple, Indianapolis 135630
Broad Ripple Vintage,
Indianapolis 093865
Broad Street Book Center,
Hereford 144472
Broadbent, London 118233
Broadfield House Glass Museum,
Kingswinford 044723
Broadfoot, Tony, Brighton 088616
Broadhurst & Co, C.K., Southport 144857
Broadstairs Antiques and Collectables,
Broadstairs 088671
Broadview Historical Museum,
Broadview 005485
Broadway 1602 Gallery, New
York 122435
Broadway and Ninth Antiques, San
Antonio 096954
Broadway Antique Market,
Chicago 092677
Broadway Antiques, Denver ... 093265
Broadway Ceramics, London ... 118234
Broadway Clock Shop, Seattle . 097410,
136367
Broadway Dolls and Bears,
Broadway 088672, 134167
Broadway Gallery, Houston ... 121011
Broadway Gallery, Tucson 124720
Broadway Gallery, Vancouver .. 101750
Broadway Magic Experience,
Broadway 043762
Broadway Modern, Broadway .. 117454
Broadway Windows, New York
University, New York 051349
Broadway, Andrew, Richmond,
Victoria 099630
Broan, Cynthia, New York 122436

Broberg, Lars, Stockholm 086980
Broc 3000, Elbeuf-sur-Seine ... 068438
Broc Amélie, Saint-Dizant-du-
Bois 072737
Broc-Antique, Paris 071095
Broc Antique, Thouars 073674
Broc Art, Basel 087225
Broc de Brenaz Sud, Sault-
Brenaz 073430
Broc en Stock, Lille 069480
Broc en Tout, Paris 071096
Broc et Puces, Le Havre 069275
Broc et Style, Beaune 067013
Broc et Troc, Arbus 066697
Broc Line Lainet, Sampigny ... 073395
Broc Rennes, Rennes (Ille-et-
Vilaine) 072428
Broc Shop, Bernay 067107
Broc Style, Claye Souilly 068006
Broc'a Boc', Voglans 074268
Broc'Adour, Biarritz 067190
La Brocalou, Les Alleuds 069393
Brocantaires, Megève 070033
Brocant'Art, Épône 068459
Brocante, Amsterdam 082712
La Brocante, Córdoba . 060936, 127541
Brocante, Heilbronn 076469
Brocante, Helsinki 066308
Brocante, Torino 081666
Brocante, Weybridge 091804
Brocante 51, Marseille . 069913, 128753
Brocante Abescat, Chaumont-sur-
Tharonne 067939
Brocante Aixoise, Aix-les-Bains . 066478
Brocante Almere, Almere 082640
El Brocante Andaluz, Málaga .. 086288
Brocante Antiek, Den Haag 083064
Brocante Antiques, Stavenhagen
Reuterstadt 078544
Brocante Antiques Centre,
Marlborough 090720, 134842
Brocante Antiquités, Neufchâteau 070509
Brocante Atmosphères, Paris .. 071097
Brocante au Bonheur du Jour,
Châteauneuf-du-Faou 067891
Brocante Aux 4 Sous, Meysse . 070115
La Brocante Baconnaise,
Baconnes 066902
Brocante d'Augan, Augan 066808
Brocante de Château, Château-
Salins 067875
Brocante de la Gare, Mouy 070381
Brocante de la Ronde, La Chaux-de-
Fonds 087358
Brocante de la Vallée de l'Arré,
Airion 066447
Brocante de l'Eglise, Arpajon .. 066732
Brocante de l'Epinay, Roncherolles-en-
Bray 072530
Brocante de Provence, Saint-Paulet-
de-Caisson 073248
Brocante des Chapelles, Archiac 066704
Brocante des Emmurées, Rouen 072573
Brocante des Fargues, Aurillac . 066823
Brocante des Isles, Pointe-à-Pitre 079218
Brocante des Quatre Moulins,
Cluny 068047
Brocante des Quatre Vents,
Bourges 067423
Brocante des Valois, Angoulême 066626
La Brocante du Batiment,
Azerailles 066899
Brocante du Bourget la Barbotine, Le
Bourget-du-Lac 069226
Brocante du Château, Apremont 066692
Brocante du Château, La
Clayette 068996
Brocante du Donjon, Domfront . 068372
Brocante du Lavoir, Saint-Pol-de-
Léon 073266
Brocante du Lude, Albi 066510
La Brocante du Mas, Pierrelatte 072158
Brocante du Plot Art Populaire
Savoyard, Groisy 068750

Brocante du Pont de Coise,
Coise 068073
Brocante du Pont de Paris, Laval
(Mayenne) 069201
La Brocante du Port, Antibes-Juan-les-
Pins 066667
Brocante du Rempart, Eguisheim 068435
Brocante du Rohu, Saint-Pierre-
Quiberon 073264, 129243
Brocante du Trieves, Mens 070056
Brocante du Valentin, Yverdon-les-
Bains 087918
Brocante la Petite Chine,
Meillonnas 070040
Brocante Merle, Pons 072234
Brocante Minaret, Rabat 082597
Brocante MV, Saint-Pierre-du-Val 073260
Brocante Normande, La Ferté-
Macé 069028
Brocante Réalmontaise,
Réalmont 072369
Brocante Saint-Eloi, Arbonne-la-
Forêt 066696
Brocante Saint-Nicolas, Moisdon-la-
Rivière 070140
Brocante Sainte-Cécile, Albi .. 066511
Brocante Service, Bellegarde-en-
Marche 067075
Brocante Vallée aux Loups, Le Plessis-
Robinson 069331
Brocante ZF, Alger 060833
De Brocanterie, Arnhem 082935
Brocanterie, Carcassonne 067669
Brocanterie, Honfleur 068828
La Brocanterie, Le Petit Quevilly 069328
Brocanterie, Palermo 131689
Brocanterie Annette, Haarlem . 083316
Brocanteur Koloniaal Teak, Venlo 083676
Brocanteurs, Amsterdam 082713
Brocanthèque, Chanceaux-sur-
Choisille 067822
Brocanthèque, Thionville 073654
Brocanthosaurus, Épinal 068453
Brocantic, Bouchain 067372
Brocantic, Le Havre 069276
Broc'antic, Lille 069481
Brocantic, Milano 110271
Brocantic, Nice 070580
Broc'Antic, Novalaise 070756
Brocantic, Oostende 063814
Brocantic, Saint-Quentin (Aisne) 073276
Broc'Antic'Collection, Courlaoux 068190
Broc'Antics, Etampes 068479
Brocantics, Torcy-le-Grand 073689
Brocantik, La Coucourde 069010
Brocantine, Corbigny 068145
Brocantine, Ecully 068428
Brocantine, Les Sables-d'Olonne 068429
Brocantine, Veneux-les-Sablons 074013
La Brocantine d'Autrecourt, Mesnières-
en-Bray 070078
Brocantique, Carcassonne 067670
Brocantique, Genève . 087456, 133756
Brocantique, Saint-Céré 072707
Brocantix, Sarrians 073419
Brocant'oc, Laloubière 069148
Brocaphil Keller, Colmar 068085
Brocapuce, Bern 087278
Broc'Armony, Langon 069170
Brocart, Girona 086025
Broc'Art, Lay 069211
Broc'Art des Remparts, Calvi . 067609,
                        140813
Brocastel, Vivier-Danger 074266
Broc'Avenue, Nice 070581
Brocchi, Roma 142801
Broccoletti, Remo, Roma 110741
Brochart, Jean-Paul & Brigitte,
Mazamet 070015
Brochu, Jean, Tour-en-Sologne 073813
Brock, Dülmen 075440
Brock & Art, Liebefeld 087591, 116440,
                        126738
Brock, J., København 065873

Brock, Martin, Dülmen 075441
Brock, Nick, Dallas 093081
Brock, P. de, Knokke-Heist .. 100733
Brockenstube, Brig 087332
Brockhaus Antiquarium,
Kornwestheim 142033
Brockmann, Daniela, Lugano . 133821
Brockmann, Georg, Mering ... 126036
Brockmann, H., Gescher 075988
Brockmanns, Dr. Karl-Josef,
Willich 108942
Brockshus, Bremen 129629
Brockville Museum, Brockville .. 005487
Broc's-en-Stock, Massillargues
Attuech 069992
Broc'Trente, Auxerre 068846
Brocza, Rudolf, Melk .. 062732, 062733,
                        127818
Brode, Dr. K., Bergisch Gladbach 074729
Brodergave, København 065874
Broderick, Portland 123605
Brodersen, Peter, København . 065875
Brodhag, Rosemarie, Berlin .. 106183
Brodick Castle, Brodick 043766
Brodie Castle, Forres 044361
Brodney, Boston 092450
Brodoux, Germaine, Agen ... 066424,
                        103172
Brodsworth Hall & Gardens,
Doncaster 044108
Brody, Dalton, Washington ... 097813
Broe-Andersen, Mogens,
Haderslev 125437
Brödlin, J., Effringen-Kirchen .. 075574
Brödlin, Johannes, Weil am
Rhein 078889
Bröhan-Museum, Landesmuseum
für Jugendstil, Art Deco und
Funktionalismus (1889–1939),
Berlin 017093
Broeke, Henk, Den Haag 083065
Bröker, Bremerhaven . 106594, 141658
Broekhoff, Nijmegen 083510
Broelmuseum, Stedelijke Musea
Kortrijk, Kortrijk 003756
Brøndsted, Jeanette, Kongens
Lyngby 065981
Brøndum Antikvariat, Skørping . 140658
Brønshøj Antikvariat, Brønshøj . 140562
Brønshøj Museum, Brønshøj .. 010027
Broes, K., Brugge 100533
Broess, 's-Hertogenbosch 083384
Brötje, G., Oldenburg 077925
Brötzmann & Linke, Uffenheim . 078751
Broeze, Edam 132494
Broggiato, Daniele, Padova ... 080971
Broggiato, Giancarlo,
Albignasego 079740
Broglin, Colmar 103617
Brohammer, Jürgen, Ettlingen . 129827
Broich, Christian, Iserlohn 141961
Brok, Barcelona 085748, 126509
Broken Art Restoration, Chicago 135403
Broken Art Restoration,
Momence 135840
Broken-Beau Pioneer Village Museum,
Beauséjour 005405
Broken Hill Regional Art Gallery,
Broken Hill 000893
Brokensha, Anne, Unley 062373
Brokidée, Marcq-en-Barœul .. 069861
Brol, Jacques, Bruxelles 063355
Brolga, Gundaroo 139757
Broman, Kalustamo, Helsinki .. 128410
Brome County Historical Society
Museum, Knowlton 006130
Bromer, Boston 145012
Bromfield, Boston 119847
Bromfield Art Gallery, Boston .. 047107
Bromfietsmuseum De Peel, Ospel 032735
Bromham Mill and Art Gallery,
Bromham 043768
Bromley Galleries, Bromley .. 117463
Hal Bromm – Art & Design, New

York 122437
Broms, Helsinki 103136
Bromsen, Maury A., Boston .. 145013
Bromsgrove Museum,
Bromsgrove 043770
Bromsten, Spånga 086950
Bromwell, Cincinnati 120325
Broncería Artistica, Buenos Aires 127516
Broncería Mateu, Las Heras .. 127545
Bronck Museum, Coxsackie .. 048047
Bronda, Helsinki 103137
Bronès Buivydaitès Memorialinis
Muziejus, Anykščiai 030317
Bronès ir Bronius Radeckų Keramikos
Muziejus, Kuršėnai 030403
Bronet, Bruxelles 063356
Bronidève, Viroflay 074252
Broniakowska, Ewa, Warszawa . 113810
Bronne, Henri, Monaco 111996
Bronner, L'Isle-sur-la-Sorgue .. 069561
Bronner, Astrid, Teningen ... 130701
Bronner, Christian, Coustellet . 068205
Brons, L., Amsterdam 082714
Bronsin, Bruxelles 063357
Bronte Gallery, Mapua 112942
Brontë Parsonage Museum,
Haworth 044547
Bronx Council on the Arts, Bronx 060417
Bronx Museum of the Arts,
Bronx 047224
Bronx River Art Center, Bronx . 047225
Bronz Stockholm, Stockholm .. 086981
Bronzart, Calgary 101218
Bronze and Brass Museum,
Bhaktapur 031867
Bronze et Lumière, Paris 128950
Bronzeplus, Birmingham 117368
Bronzes et Sculptures, Metz .. 070091
El Bronzino, Bogotá 065190
De Broodgalerie, Amsterdam .. 112173
Broodthaers, Marie-Puck,
Bruxelles 100573
Brook, Bad Homburg v.d.Höhe . 074519
The Brook Gallery, Budleigh
Salterton 117470
Brook Street Collectables,
Providence 096502
Brook, Michael, London 134610
Brookart Press, San Jose 138464
Brooke Gifford, Christchurch .. 083860
Brooke, Alexander, New York . 122438,
                        138384
Brookes-Smith, Forest Row ... 089294
Brookes, Diane, Tetney 119223
Brookes, Howard, Göteborg .. 115630
Brookfield, Healesville 061612
Brookfield Craft Center,
Brookfield 047239
Brookgreen Gardens, Murrells
Inlet 051131
Brookings Arts Council,
Brookings 047242
Brookings County Museum,
Volga 054008
Brookings & Co., J.J., San Jose 124445
Brookland, Washington 124872
Brooklands Museum, Weybridge 046105
Brookline Auction Gallery,
Brookline 127149
Brookline Historical Society Museum,
Brookline 047245
Brookline Village Antiques,
Boston 092451
Brooklyn Art Gallery, Cardiff . 117523
Brooklyn Arts Center,
Jacksonville 121238
Brooklyn Arts Council Gallery,
Bronx 047226
Brooklyn Arts Council News,
Brooklyn 139328
Brooklyn Children's Museum,
Brooklyn 047251
Brooklyn Historical Society Museum,

Brooklyn 047252, 047270
Brooklyn Museum, Brooklyn ... 047253
Brooklyn Sports Cards and Collectibles,
Cleveland 092925, 145084
Brooklyn Waterfront Artists,
Brooklyn 060418
Brooknead Modern, La Jolla .. 050039
Brooks, Den Haag 083066
Brooks, Dublin 079530
Brooks, Monaco 082500, 126267
Brooks, Sacramento 096607
Brooks, Washington 136454
Brooks Academy Museum,
Harwich 049400
Brooks and District Museum,
Brooks 005489
Brooksby, Angie Elizabeth,
Firenze 110039
Brookside Antiques, Kansas City 094062,
                        135676
Brookside Art Gallery, Kansas
City 121281
Brookside Museum, Ballston Spa 046724
Broome County Historical Society
Museum, Binghamton 046996
Broome Museum, Broome 000899
Broomehill Museum, Broomehill 000900
Broomer, A.R., New York 095289
Broqua, Christophe, Aire-sur-
l'Adour 128426
Bror Hjorths Hus, Uppsala ... 041109
Bros, Francoise, Bordeaux ... 128483
Brosa, Berthe, Tettnang 108765
Brose, Oliver, Hamburg 076201
Brose, Oliver, Kayhude 076737
Broßhauser Mühle, Solingen .. 078493
Brossy, Robert, Rolle 087751
Brot- und Mühlen-Lehrmuseum,
Dirnbachermühle, Gloggnitz .. 001940
Brot und Spiele Galerie, Berlin . 106184
Brotfabrik Galerie, Berlin 017094
Brother, Philadelphia 096088
Brotherton, Naomi, Dallas ... 120564
Brotlose Kunst,
Mönchengladbach 077386
Brotonne, Frédéric, Feron ... 068525
Brouard, Gilles, Gemages 068643
Brouard, Gilles, Saint-Germain-de-la-
Coudre 072806
Les Broucants, Verona 132273
Broughton, Berry 061146
Broughton Books, Edinburgh .. 144389
Broughton Gallery, Broughton . 117468
Broughton House and Garden,
Kirkcudbright 044729
Broughton House Gallery,
Cambridge 117503
Broughton, Bob, Taroona 139929
Broughty Castle Museum,
Dundee 044162
Broughty Ferry Art Society,
Dundee 060082
Brouillard, Patrick, Corcelles .. 133725
Brouns, Konrad, Maastricht ... 083458
Brous, Nancy, New York 095290
Brousse, Didier, Lournarin ... 104049
Broutta, Michèle, Paris 104593, 136876
Brouwershuis, Antwerpen 003281
Brouws Museum, Brouwershaven 032112
Brovot, Köln 076853
Broward County Historical Museum,
Fort Lauderdale 048825
Brower, David, London 089981
Brower, Robert, Cleveland ... 120408
Browgate Antiques, Baildon .. 088332
Brown & Co., Brigg 126838
Brown, Glasgow 089329
Brown, Kansas City 094063
Brown, Oklahoma City 095953
Brown, Seattle 097411
Brown, Tain 119203
Brown County Art Gallery and
Museum, Nashville 051170
Brown County Historical Society

Pioneer Museum, Nashville ... 051171
Brown County Museum, New
Ulm 051317
Brown House Museum,
Sandersville 053066
Brown Mansion, Coffeyville 047857
Brown Mountain Art and antique
Restoration, Dallas 135496
Brown, Alexander, Cincinnati ... 092835
Brown, B., Seattle 145578
Brown, Ben, London 118235
Brown, Cam, Eagle Farm 125190
Brown, David, Montréal 064399
Brown, David, Saint Andrews . 091177,
134990
Brown, Dina, Los Angeles 121451
Brown, Graeme, Toorak 062321, 127689
Brown, I. & J.L., Hereford 089514
Brown, I. & J.L., London 089982, 134611
Brown, James A., Charlotte 092580
Brown, Joseph, Prahran 099599
Brown, Juanita C., Oklahoma
City 095954
Brown, Karen, Darwin 098948
Brown, Linda L., San Antonio . 124018
Brown, Lou, Houston . 093630, 135594
Brown, Malcolm, Cleveland 120409
Brown, Martina, Berlin 074773
Brown, Murray, Pevensey 091012
Brown, P.J., Haslingden 089459
Brown, Paula, Toledo 124683
Brown, Reinhold, New York 122439
Brown, Robert, Washington . 097814,
124873
Brown, S.J., Jacksonville 135648
Brown, Sally, Stockport 119146
Brown, Sidney, San Francisco .. 124241
Browne & Co, Benny, Stoke-on-
Trent 119149
Browne & Assoc. Chuck, Tulsa . 097717
Browne, Gayle, Austin 092221
Browne, Martin, Potts Point ... 099595
Brownella Cottage, Galion 049025
Browning, G.R., Washington . 124874
Browning's of Wyoming,
Cincinnati 120326
Brownlee, Chicago 120081
Brownlow, Faldingworth 089249
Brownrigg, Petworth 090991
Browns, High Wycombe 089526, 134463
Brownsbank Cottage, Biggar 043595
Brownsboro Book Center, Saint
Matthews 145499
Brownstone, Chicago 092678
Brownstone, Houston 093631
Brownstone & Cie., Gilbert, Paris 104594
Brownsville Museum of Fine Art,
Brownsville 047280
Browntown Museum, Lake City . 050075
Brownvale North Peace Agricultural
Museum, Brownvale 005491
Brownville Historical Society Museum,
Brownville 047281
Browse & Darby, London 118236
Browse Antiques, Stafford . 091399,
135070
Browse Inn, Virginia 062392
Browser, Falmouth 144421
Browsers Bookshop, Epping 144411
Browsers Delight, Fremantle ... 061486
Browsers Nook, San Francisco . 097196
Browsers Secondhand Books,
Hamilton 143318
Browsers Secondhands Books,
Tauranga 143349
Browzers, Broken Hill 139665
Browzers, Manchester 144725
Broyles, Kansas City 094064
Brozzetti, Torino 081667, 132107
BRS Galerie, Ulrike Eisen,
Bickendorf 106480
Brubacher House Museum,
Waterloo 007361
Bruc Antic, Barcelona 133107

Bruce Castle Museum, London . 044927
Bruce County Museum,
Southampton 007061
Bruce Gallery of Art, Edinboro
University of Pennsylvania,
Edinboro 048477
Bruce Groner Museum,
Groundbirch 005931
Bruce Mines Museum, Bruce
Mines 005492
Bruce Museum of Arts and Science,
Greenwich 049270
Brucemore, Cedar Rapids 047484
Bruce's, El Paso 093392
Bruce's Antiques, San Jose 097360
Bruce's Auctions, Adelaide 125174
Bruchet-Méry, Jean-Luc, Rennes (Ille-
et-Vilaine) 072429
Bruchhausen, Heike, München . 108059
Brucke, Tokyo 111411
Brucknerzimmer, Kronstorf 002247
Brucoli, Mauro, Milano 080466
Brudavollen Bygdetun, Ørsta .. 033814
Bruder-Konrad-Museum, Altötting 016554
Die Brücke – Kärnten.Kunst.Kultur,
Klagenfurt 138508
Brücke Museum, Berlin 017095
Brückmann, Ursula, Berlin 106185
Brueckner Museum of Starr
Commonwealth, Albion 046356
Brückner-Zündorff, Jutta, Bad
Füssing 106056
Brückner, Hans, Schmelz 078377
Brückturm-Museum über der
'Steinernen Brücke', Arbeitskreis
Schiffahrtsmuseum Regensburg e.V,
Regensburg 021313
Brüder Grimm-Haus Steinau, Steinau
an der Straße 021974
Brüder Grimm-Museum Kassel,
Kassel 019528
Brüder-Grimm-Stube, Marburg . 020269
Brüggemann-Riemer, Simone,
Lübben 107905
Brüggemann, Ernst, Mömbris . 077383,
130308
Brüggen, Falk von der, Bremen . 129630
Brügger Antiquitäten, Brügge . 075203
Brügger, Christine, Bern 116124
Brühler Kunstverein e.V, Brühl . 058765
Brüll, Hans, Dorsten . 075380, 129694
Brüning, Lena, Berlin 106186
Brüning, Thomas, Ratingen 130557
Brünner Heimatmuseum, Schwäbisch
Gmünd 021747
Brünz & Rosowsky, Las Vegas . 121344
Bruet, Nicolas, Paris 071098
Brüxer und Komotauer Heimatstuben,
Palais Stutterheim, Erlangen . 018259
Bruford & Heming, S.J., London 089983,
134612
Bruford, W., Eastbourne 089143, 134351
Bruggemuseum – Archeologie, Musea
Brugge, Brugge 003405
Bruggemuseum – Belfort, Musea
Brugge, Brugge 003406
Bruggemuseum – Brugse Vrije, Musea
Brugge, Brugge 003407
Bruggemuseum – Gruuthuse, Musea
Brugge, Brugge 003408
Bruggemuseum – Onthaalkerk van
Onze-Lieve-Vrouw, Musea Brugge,
Brugge 003409
Bruggemuseum – Stadhuis, Musea
Brugge, Brugge 003410
Brugidou, Philippe, Paris 071099
Brugmann Antiquitès, Bruxelles . 063358
Brugs Antiekhuis, Brugge 063263
Brugwachterhuis, Dilsen-
Stokkem 003553
Bruhais, André, Le Loroux
Bottereau 128666
Bruhm, Serfling & Wollenschläger,
Jena 130082

Bruhns, Denver 093266, 127204
Bruijn & de Vaan, Nijmegen ... 143209
Bruijn, de, Maastricht 083459
Bruijn, Dick de, Amsterdam ... 112174
Bruijn, W.M. de, Groningen 083282
Bruijne, Ellen de, Amsterdam .. 112175
Bruijnzeels, Haarlem 083317
Bruini, Renzo e Stefano, Modena 080779
Bruk, Kevin, Miami 121751
Bruksmuseet Smedsgården,
Sandviken 040936
Bruksmuseum, Hudiksvall 040708
Brukt og Antikkhuset,
Kristiansand 084265
Bruktbu, Løvas, Stavanger ... 143421
Brullé, Pierre, Paris 104595
Brumado, Rio de Janeiro 064031
Brumaire, André, Pont-l'Évêque . 072258
Brument, Patrick, Rouen 072574
Brun Lea Antiques, Burnley ... 088728,
088729
Brun-Leglise, Dominique, Paris . 104596
Brun, Jean-Claude, Grenoble .. 068722
Brun, Jeanette G., Zürich 087945
Bruneau, Jean-Charles, Dinard . 068354
Bruneau, Philippe, Tours 073843
Bruneaux Graveurs, Caen 067560
Brunei Forestry Museum, Sungai
Liang 004969
Brunei Gallery, School of Oriental and
African Studies, University of London,
London 044928
Brunel, Cannes 067630
Brunel, Saint-Sever (Landes) ... 073307
Brunel Antiques and Collectors Fair,
Uxbridge 098363
The Brunel Museum, London .. 044929
Brunel, Didier, Digoin 068307
Brunelli, Fabio Marco, Bologna . 130940
Brunelli, Harold, Auxerre 066847
Brunello, Paolo, Bologna 130941
Brunello, Sergio, Venezia 111093
Brunet, Rio de Janeiro 064032
Brunet, Elisabeth, Rouen 141319
Brunet, Noemi, Gaillac 068622
Brunet, Patrick, Lyon . 069715, 128703
Brunetti, Silvia, Paris 128951
Brunia, Mascha, Breda 082976
Bruniková, Květa, Liberec 065412
Brunk & Sons, John, Phoenix .. 127397
Brunke, Berthold, Bochum 129598
Brunkeberg, Stockholm 133566
Brunn Museum, Nólsoy 010737
Brunnarius, Karl, Bad
Gandersheim 074514
Brunnen-Verlag, Basel 138023
Brunnenmuseum, Bad Vilbel ... 016895
Brunnenturm-Presse,
Regensburg 138084
Brunner, Höchst 139996
Brunner Heimathaus mit Rudolf-
Steiner-Gedenkstätte, Brunn am
Gebirge 001805
Brunner Mine Site, Dobson 033186
Brunner-Tafertshofer, Monika,
Wessobrunn 078943
Brunner-Wendel, Lieselotte,
Steißlingen 108671
Brunner, Andreas Christian,
Schwanberg 062838
Brunner, Beatrice, Bern 116125
Brunner, Elisabeth, Pfäffikon
(Zürich) 087735
Brunner, Fritz, Embrach 133729
Brunner, Katia, Honfleur 103819
Brunner, Klaus, Lörrach 077143
Brunner, Susi, Zürich . 116762, 116763
Brunnier Art Museum, Iowa State
University, Ames 046439
Bruno Weber Skulpturenpark
Spreitenbach, Dietikon 041442
Bruno & Fils, André, Paris 128952
Bruns, Thomas, Winsen, Luhe . 079023,
108945

Brunsch, Josephine, München . 077474
Brunskogs Hembygdsgård,
Brunskog 040585
Brunswick Antiques, Penrith ... 090956
Brunswick Railroad Museum,
Brunswick 047288
Brunswick Street Antiques,
Fitzroy 061467
Brunswick Town/Fort Anderson State
Historic Site, Winnabow 054478
Brunswick Valley Museum,
Mullumbimby 001325
Brunswiek, Braunschweig 075119
Bruntjen, Sven H.A., San
Francisco 124242
Brusasca, Torino 081668
Brusato, Tebaldo, Brescia 142651
Brusberg, Dieter, Berlin 106187
Bruschelli, Perugia 081138
Bruschi & Merry, Pittsburgh .. 123530
Bruschi, Paolo, Firenze 131136
Bruschweiler, F.G., Rayleigh .. 091073
Brusgaard Jensen, Erik, Hobro . 125444
Brush Art Gallery, Lowell 050509
Brush Country Museum, Cotulla 048031
Brush Gallery, Wingham 099891
Brushstrokes Art School, Dublin 055859
Brushwood Studio Gallery,
Sneem 109695
Brushworks, San Diego 124109
Brushworks, Tulsa 124797
Brushworks Gallery, Salt Lake
City 123949
Brusquet, Thierry, Sene 073490
Brussels Ancient Art Fair,
Bruxelles 098014
Brussels Antiques & Fine Arts Fair,
Bruxelles 098015
Brussels Non European Art Fair,
Bruxelles 098016
Brussels Oriental Art Fair,
Bruxelles 098017
Brusten, Annelie, Wuppertal .. 108985
Bruton, Holmfirth 117978
Bruton, New York 122440
Bruton Classic Furniture Company,
Bruton 134174
Brutschan, Katharina,
Schaffhausen 087801
Brutscher, H., Merseburg 142125
Bruun, Brørup 125432
Bruun, København 065876
Bruun Rasmussen, Århus 125430
Bruun Rasmussen, København . 065877,
125448, 125449
Bruun Rasmussen, Vejle 066190, 125471
Bruun, Lisse, København 102871
Bruunshåb Gamle Papfabrik,
Viborg 010357
Bruxela 1238, Musée de la Ville de
Bruxelles, Bruxelles 003436
Bruyere, Alain, Toulouse 073744
Bruyère, Catherine, Versailles . 105843
Bruyere, Jean, L'Etrat 103993
Bruyères, Serge, Marssac-sur-
Tarn 069977
Bruyn, Ch. de, Bruxelles 063359
Bruzaholms Bruksmuseum,
Bruzaholm 040586
Bry Gallery, Monroe 050997
Bry, François, Mortagne-au-
Perche 070355
Bryan, Douglas, Cranbrook ... 088993
Bryanston Graphics,
Johannesburg 114791
Bryant, Houston 121012
Bryant, Kansas City 135677
Bryant, New Orleans .. 122138, 122139
Bryant Cottage, Bement 046910
Bryant-Russ, Los Angeles 135709
Bryant, Jan, Gladesville 061520
Bryars, Tim, London 144571
Bryczkowska, Lena, Bydgoszcz . 113439
Brydone Jack Observatory Museum,

Fredericton 005838
Bryggens Museum, Bergen 033530
Bryggerhuset, Vollen 084394
Bryggeriets Antik, Nyköping . . . 086889,
133559
Brzezinski, Ewa, Rennes (Ille-et-
Vilaine) 129201
Brzostek, Phoenix 127398
Brzozowski, Kristof,
Mönchengladbach 077387
BS Antique Trading, Singapore . 085376
BSAT Gallery, British Sporting Art
Trust, Newmarket 045359
BSMD, Paris 104597
Bu Museum, Utne 034057
Buade, Québec 101493
Bubba's Baseball Bullpen,
Albuquerque 092039
Bubbly, Singapore 085377
Bubbly Hills Vintage, Long Beach 094196
Buby, Trenčín 085594
Bucaro, León 086062
Bucat, Pierre, Paris 128953
Buccaneer, 's-Hertogenbosch . . 083385
Buccaneer Heaven, Tampa 097547
Bucceri, Padova 080972
Bucchi, Federica, Roma 081275
Bucci, Enrico, Santiago de Chile 101903
Bucciarelli, Giovanni, Firenze . . 131137
Buccleuch Mansion, New
Brunswick 051242
Buccony, Buco, Bratislava 085544
Bucerius Kunst Forum, Hamburg 018977
Buceti, Carmelo, Roma 131834
Buceti, D & S, Roma 081276
Buceti, Domenico, Messina 080362
Buch- und Kunsthandlung der
Benediktinerabtei Sankt Michael,
Siegburg 142360
Buch & Boje, Haderslev 125438
Buch & Kunst, Braunschweig . . 106561,
141631
Buch & Schudrowitz, Berlin . . . 129479
Buch Druck Kunst e.V., Gesellschaft
zur Förderung zeitgenössischer
Buchkunst, Hamburg 058766
Buch + Kunst, Wesseling 142481
Buch-Kunstverlag Ettal, Ettal . . . 137186
Buch und Grafik Galerie der
Büchergilde Bonn, Bonn 106525
Buch, Claudia, Oberhausen 077863
Buch, Nils Ole Funder, Brøndby . 128573
Buchal, P., Groß-Umstadt 076104,
125961
Buchan, K.S., Leigh-on-Sea . . . 089795
Buchanan Center for the Arts,
Monmouth 050991
Buchanan, Duncan, Auckland . . 083773
Buchans Miners Museum,
Buchans 005493
Das Buchantiquariat, Ittigen . . . 144087
Buchantiquariat am Rhein, Basel 144016
Buchantiquariat Kloster Marienrode,
Hildesheim 141945
Buchart, Antibes-Juan-les-Pins . 066668
Buchdruckerei Museum und
Schauwerkstatt, Krakow am
See 019779
Buchdruckmuseum, Rohrbach in
Oberösterreich 002606
Buchdruckmuseum Druckwerk,
Dornbirn 001823
Bucheit, Rita, Chicago 092679
Bucheli, Rocio E., Bath 117300
Bucheon, San Francisco 124243
Bucher, Alois, Innsbruck 062647
Bucher, Herbert, Absam 062511
Bucher, Jeanne, Paris . 104598, 136877
Bucher, Stefan, Littau 133809
Buchfink, Brugg 144058
Buchfink, Lenzburg 144096
Der Buchfreund, Wien 140064
Buchgraitz, Gerda Louise & Frank,
Horsens 065791

Buchhandlung beim Lederhaus,
Ravensburg 142318
Buchhandlung Johannesstift,
Berlin 141493
Buchhandlung u. Galerie Böhler,
Bensheim . . 106129, 137044, 141461
Buchhandwerk Hans Peter Bühler,
Nürtingen 130473, 130474
Buchheim Museum, Sammlung
Buchheim – Museum der Phantasie,
Bernried 017338
Buchhold, Petra, Schrecksbach . 078398
Buchholz, Hannover 107354
Buchholz, Lisboa 143558
Buchholz & Pyroth, Hamburg . . 076202
Buchholz, Godula, Denklingen . . 106669
Buchholz, Karsten, Reinbek 142327
Buchholz, Susanne, Kassel 130103
s'Buchholze Hisli, Schuttertal . . 021742
Buchholzer Heimatmuseum, Buchholz
in der Nordheide 017655
Buchinger, Otto, Linz . . 062715, 100101
Der Buchladen, Saarbrücken . . . 142341
Buchloh, Helmut S. & Claudia,
Schnaitsee 078379
Buchmann, Agra 116013
Buchmann, Berlin 106188
Buchmann, Yvonne, Basel 133637
Buchmüller, J., Mülheim an der
Ruhr 077432
Buchmuseum, Solothurn 041999
Buchmuseum der Sächsischen
Landesbibliothek – Staats- und
Universitätsbibliothek Dresden,
Dresden 017976
Buchrestaurierung Leipzig,
Leipzig 130222
Buchs, Oskar, Gstaad 116365
Buchspecht, Saarbrücken 142342
Buchtip, Strasshof 140046
buchWeger, Frankfurt am Main . 141769
Buchwelten, Berlin 141494
Buchwerkstatt, Bremen 129631
Buck House, New York 095291
Buck House Antique Centre,
Beaconsfield 088411
Buck Mountain Treasures, Oklahoma
City 095955
Buck, Christopher, Sandgate . . . 091233,
135003
Buck, Daniel, Wickford 127495
Buck, J. de, Bruxelles 063360
Buck, Nicole, Strasbourg 105679
Buck, Siegfried & Hermine de,
Gent 100690, 136653, 140220
Buck, Sven, Neustadt-Glewe . . . 077766
Buckboard Antiques, Oklahoma
City 095956
Buckenham, Southwold 119129
Bucket Rider Gallery, Chicago . . 120082
Buckeye Furnace State Memorial,
Wellston 054274
Buckfast Abbey, Buckfastleigh . 043774
Buckham Gallery, Flint 048747
Buckhaven Museum, Fife
Council Libraries & Museums,
Buckhaven 043777
Buckhead, Atlanta 135300, 144964
Buckhorn Saloon & Museum, San
Antonio 052934
Buckhorn School of Fine Arts,
Buckhorn 055117
Buckie Drifter Fishing Heritage Centre,
Buckie 043778
Buckies, Cambridge . . 088759, 134197
Buckingham Antiques,
Buckingham 088694, 134175
Buckingham Movie Museum,
Buckingham 043780
Buckingham Palace, Royal Collections,
London 044930
Buckingham, C.W., Cadnam . . . 088752
Buckinghamshire County Museum,
Aylesbury 043456

Buckinghamshire Railway Centre,
Quainton 045574
Buckland Abbey, Yelverton . . . 046215
Buckland Gallery, Beechworth . . 098720
Buckland Southerst, Vancouver . 101751
Buckler's Hard Village Maritime
Museum, Beaulieu 043539
Buckley, Memphis 094623
Buckley, Penrith 062072
Buckley Center Gallery, University of
Portland, Portland 052289
Buckley Galleries, Sandycove . . 079696,
109688,   126178
Buckley-Kohler, Columbus 092999
Buckley, W., Congleton 088966
Buckleys Yesterday's World,
Battle 043534
Bucknall, Moya, Solihull 119109
Buck's Stove Palace, Portland . . 096376
Bucksport Historical Museum,
Bucksport 047294
Buckwitz, Sybille, Kronberg im
Taunus 107754
Bucx, Ben, 's-Hertogenbosch . . 132534
Bucyrus Historical Society Museum,
Bucyrus 047295
Bud & Bettys, Winnipeg 064915
Budai Antik, Budapest 079244
Budai Krónika, Budapest 142557
Budansky, A. Adriana, Buenos
Aires 127517
Budaörser Heimatmuseum,
Bretzfeld 017623
Budapest Art Fair, Budapest . . . 098172
Budapest Galéria Kiállítóháza,
Budapest 023463
Budapest Galéria Kiállítóterem,
Budapest 023464
Budapest Galéria Varga Imre Állandó
Kiállítóház, Budapest 023465
Budapesti Történeti Múzeum,
Vármúzeum, Budapest 023466
Budatétényi Galéria, Budapest . 109134
Budavári Mátyás Templom
Egyházművészeti Gyűjteménye,
Budapest 023467
Budawang Antiques, Milton . . . 061856
Budde, Hermann, Lemgo 077074
Budde, Klaus, Warendorf 078866
Buddelschiff-Museum,
Neuharlingersiel 020706
Buddelschiff-Museum,
Tangerhütte 022098
Buddelschiff-Museum, Wedel . . 022446
Buddenbrookhaus, Heinrich-und-
Thomas-Mann-Zentrum, Die
Lübecker Museen, Lübeck . . . 020155
Buddenbrooks, Boston 145014
Buddhawelt, Bremen 075154
The Buddhist Art Museum,
Yangon 031831
Buddle Arts Centre, Wallsend . . 046037
Budd's Beach Gallery, Surfers
Paradise 099734
Buddy Holly Center, Lubbock . . 050515
Buddy's Antique Auction, Arab . 127122
Bude-Stratton Museum, The Castle
Heritage Centre & Gallery, Bude 043782
Budenz-Ház – Ybl Gyűjtemény,
Szent István Király Múzeum,
Székesfehérvár 023932
Buderim Pioneer Cottage,
Buderim 000903
Buderim Rare Books, Buderim . 139669
Buderus Zentralheizungsmuseum,
Lollar 020111
Budget Antiques, Edmonton . . . 064287
BudiArt, Bussum 083017
Budil, S., Amsterdam 079580
Budinok-muzej A. Čechova, Sums'kyj
oblasnyj krajeznavčyj muzej,
Sumy 043303
Budinok-muzej Sevastopol's'ke
Komunistične Pidpilja 1942–1944,

Muzei geroïčnoï oborony
i vyzvolennja Sevastopolja,
Sevastopol 043290
Budrich & Schmidt, Friedeburg . 075909
Budrus, Renate, Stadtallendorf . 078529
Buechel, Louisville 094526
Buechel Memorial Lakota Museum,
Saint Francis 052765
Bücher-Brockenhaus, Bern 144046
Bücher-Brocky, Aarau 144007
Bücher-Brocky, Basel 144017
Bücher-Brocky, Luzern 144105
Bücher-Brocky, Zürich 144169
Bücher-Ernst, Wien 140065
Bücher-Fantasia, Zürich 144170
Bücher James, Bonn 141613
Bücher Oase, Wilhelmshaven . . 142498
Bücher Scheidegger, Affoltern am
Albis 144009
Bücherbogen am Savignyplatz,
Berlin 141495
Bücherbogen im Stülerbau,
Berlin 141496
Bücherbogen in der Knesebeckstraße,
Internationales Modernes Antiquariat,
Berlin 141497
Bücherbogen in der Nationalgalerie,
Berlin 141498
Büchergalerie Westend, München 142144
Büchergilde Art Club, Frankfurt am
Main 106978, 137195
Das Bücherhaus, Eldingen 141737
Bücherinsel, Leipzig 142056
Büchernarr, Wermelskirchen . . . 142476
Büchernische beim Neuen Dom,
Linz 140009
Büchers Best, Dresden 141693
Die Bücherstube, Bad
Kleinkirchheim 139974
Bücherstube am See, Konstanz . 142029
Der Bücherwurm, Dessau 141684
Bücherwurm, Kiel 141990
Bücherwurm, Schaffhausen . . . 144133
Büchi, Monaco 111997
Büchi, Max, Bern . . . 087279, 133672
Büchner, H., Sinsheim 078482
Büchner, Thomas, Berlin 129480
Büchnerhaus, Riedstadt 021420
Büchsenmacher- und Jagdmuseum,
Ferlach 001888
Bücken, Walter, Herzogenrath . . 107425,
130041
Bücker-Luftfahrt- und Europäisches
Eissegel-Museum, Rangsdorf . 021275
Bückmann, Monika, Krefeld . . . 076953
Büdenhölzer, Brigitte,
Emmendingen 075608
Büecheler-Vis, Kloten 041617
Bühler, Aschaffenburg 074431
Bühler, Irmtraud, Bremen 075155
Bühlmann, Astrid, Bern 133673
Bühlmayer, C., Wien . . 062918, 127909
Bühn, Josef, München 137391
Bühnemann, Brake 106554
Bühnemann, Michael, Berlin . . . 074774
Bühnemann, Rainer L., Hannover 076367
Bühning, Bohmte 075056
Bührer, Barbara, Neuhausen am
Rheinfall 133873
Buekenhout, Jean, Bruxelles . . . 063361
Büker, Eberhard, Hamburg 076203
Bükki Üveghuták Ipartörténeti
Múzeuma, Bükkszentkereszt . . 023573
Bülow, Holger, Berlin 129481
Bülow, Rose-Marie, København . 065878
Bültmann & Gerriets, Oldenburg 142266
El Buen Orden, Buenos Aires . . 060885
En Buen Uso, Logroño 086073
Buena Park Historical Society Museum,
Buena Park 047296
Buena Vista, Dordrecht 112434
Bündner Kunstmuseum, Chur . . 041412
Bündner Naturmuseum, Chur . . 041413
Büngener, H., Pegestorf 078028

Büren, Fritz von, Mettmenstetten 133860
Bürger- und Bauernmuseum,
Hilzingen 019257
Bürgerhausen, Reinhold, Aachen 074325
Bürgermeister-Müller-Museum,
Solnhofen 021910
Bürgermeister-Ringenhain-Haus, Stadt-
und Kulturgeschichtliches Museum
Torgau, Torgau 022166
Bürgerstiftung Kunst für Güglingen,
Güglingen 018877
Bürgi, Robert & Silvia, Lyss 087653
Bürki, Bernhard, Sigriswil 087813
Bürki, Bernhard, Tschingel ob
Gunten 087860
Bürkle, A., Freiburg im Breisgau 075854
Bürkle, Raimund, Luzern 087635
Bürkler, Urs Björn, Brugg 087334
Büro für Kunst, Dresden 106719
Büro für Kunst Michael G. Schüll,
Fürth 107083
Büro Otto Koch, Dessau 017864
Büromuseum, Mülheim an der
Ruhr 020492
Bürsten- und Heimatmuseum,
Schönheide 021711
Bürth, Karin, Greifswald 076084
Büsch, Großmaischeid 076109
Büsch, Karin, Berlin 106189
Bütow, Georg, Bad Oeynhausen 074564,
106085
Büttiker, Katharina, Zürich 087946
Büttner, Anke, Geisenhausen .. 129890
Büttner, W. & J., Hohenroth 076560
Büyük Saray Mozaik Müzesi,
İstanbul 042928
Buffa, Francesca, Palermo 081026
Buffalo and Erie County Historical
Society Museum, Buffalo 047300
Buffalo and Erie County Naval and
Military Park, Buffalo 047301
Buffalo Bill Cody Homestead,
Princeton 052361
Buffalo Bill Historical Center,
Cody 047855
Buffalo Bill Memorial Museum,
Golden 049144
Buffalo Bill Museum of Le Claire, Iowa,
Le Claire 050196
Buffalo Bill Ranch, North Platte . 051613
Buffalo Candle and Craft, Buffalo 092513
Buffalo Museum of Science,
Buffalo 047302
Buffalo Nations Luxton Museum,
Banff 005380
Buffalo Society of Artists,
Amherst 060419
Buffalo Stew Antiques, Tulsa .. 097718
Buffalo Trails Company, Fort
Worth 120858
Buffalo Trails Museum, Epping . 048593
Buffalo Wholesale Antiques,
Buffalo 092514
Buffet, Claude, Paris 141062
Buffs Museum, National Army
Museum, Canterbury 043856
Bugatti Trust, Cheltenham 043923
Bugdahn und Kaimer, Düsseldorf 106764
Bugden, Potts Point 139880
Bugeja, Victoria 082444
Bugera & Kmet, Edmonton 101279
Il Bugigattolo, Genova 080240
Bugmann, Sabine, Nunningen .. 133876
BuGok Railroad Exhibition Hall,
Railroad Museum, Uiwang 030058
Buguruslanskij Istoriko-kraevedčeskij
Muzej, Buguruslan 036923
Buhard, Jean-Yves, Saint-André-sur-
Vieux-Jonc 072673
Buhard, Jean-Yves, Villeurbanne 074197
La Buhardilla, Monterrey 082490
La Buhardilla, Pamplona 086383
Buhler, Elfriede, Milano 131401

El Búho, Cali 102402
Buhr, Carl, Hamburg 076204
Buhr, M. de, Dorum 075400
Buhr, Raimund, Wien 127910
Buia Gallery, New York 122441
Building Arts, Pittsburgh 096263
Building of Bath Museum, Bath . 043517
Buisine, Patrick, Lille 069482
Buisson, Claude, Seloncourt ... 073484
Buisson, Luc du, Cluny 068048
Buisson, Monique, Dijon 068321
Buisson, Roger, Paris 128954
Het Buitenhuis, Amstelveen ... 082663
Buitrago, Carlos Armando, Cali . 065247
Bujnakskij Istoriko-kraevedčeskij
Muzej, Dagestanskij gosudarstvennyj
objedinennyj istoriko-architekturnyj
muzej, Bujnaksk 036924
Bukara, Valencia 086508
Bukhara, Sharjah 088195
Bukinist, Moskva 143637
Der Bukinist, München 142145
Bukinist 7, Moskva 143638
Bukinist na Litejnom, Sankt-
Peterburg 143665
Bukowski, Göteborg 126586
Bukowski, Helsinki 125476
Bukowski, Malmö 126585
Bukowski, Oslo 084305, 126367
Bukowski, Stockholm 126671
Bukowski, JaceK, Szczecin 084675
Bukowski, Rafał, Kraków 132836
Bukyoung University Museum,
Busan 029892
Bul-bul Memorial Museum, Bakı 003157
Bul-bul Museum-Flat, Bakı 003158
Bulabula Arts, Winnellie 062440
Bulang & Zorn, Marburg 142112
BulArt, Varna 101186
Buldrini, Roberto, Roma 131835
El Bulevard dels Antiquaris,
Barcelona 085749
Bulgarelli, Fausto, Modena 080780
Bulgarelli, Renato, Torino 132108
Bulgarian Museum Chamber,
Sofia 058404
Bulger, Stephen, Toronto 101586
Bulimba Gallery, Bulimba 098816
Il Bulino, Milano 080467
Il Bulino, Roma 110742
Buljanskaja, Alla, Moskva 114210
Bulkley Valley Museum, Smithers 007046
Bull, Audrey, Carmarthen 088796
Bull, Audrey, Tenby 091557
Bull, John, London 089984
Bull & Son, Walter, Cirencester . 088913,
134271
Bullet, Saint Paul 096814
Bullet Space, New York 122442
Bulletin de l'Antiquaire et du
Brocanteur, Paris 138635
Bulletin de l'Institut Royal du
Patrimoine Artistique, Bruxelles 138545
Bulletin du Musée National de
Varsovie, Warszawa 138961
Bulletin of the Detroit Institute of Arts,
Detroit 139329
Bulletin van het Rijksmuseum,
Amsterdam 138926
Bullivant, Robert, Saint Louis .. 123813
Bulloch Hall, Roswell 052688
Bullock, Leslie, Haarlem 083318
Bullot, Patrick, New York 122443
Bullpen, Fresno 093492
Bullpen, Toledo 097600
Bullrich Gaona Wernicke, Buenos
Aires 125166
Bulls, Bulls 083844
Bulls in a China Chop, Burra ... 061237
Bullseye Connection Gallery,
Portland 123606
Bully's Fifties, Rotterdam 083556
Bullyworld Museum, Spraitbach 021939
Bulurru Australia, Marrickville .. 061795

Bulwagang Helena Benitez Gallery,
Manila 034524
Bulyanskaya, Alla, London 118237
Bulzoni, Roma 137728
Bum Steer, Fort Worth 093435
Bumb, H.-Jürgen, Contwig 075270
Bumble Hill Studio, Kulnura ... 099203
Bumbles, Ashtead ... 088313, 134042
Bumbongan Dua Belas, Bandar Seri
Begawan 004961
Bumper, Villenave-d'Ornon 074158
Bumpers Antique, Sandgate ... 062187
Bumpkin, Littlehampton 061743
Bumps, Los Angeles 094278
Bunbury Regional Art Galleries,
Bunbury 000906
Bunbury Watermill, Bunbury ... 043786
Bunch, Kongens Lyngby 065982, 102923
Bund Bildender Künstler Leipzig e.V.,
Leipzig 058767
Bund Bildender Künstler Lüneburg,
Strachau 058768
Bund Bildender Künstlerinnen und
Künstler, Braunschweig 058769
Bund Bildender Künstlerinnen und
Künstler für Niedersachsen e.V,
Hannover 058771
Bund Bildender Künstlerinnen und
Künstler für Niedersachsen, Gruppe
Hannover, Hannover 058770
Bund Bildender Künstlerinnen und
Künstler, Württemberg e. V., Stuttgart .. 058772
Bund der Kunsthandwerker Baden-
Württemberg e.V., Stuttgart .. 108690
Bund Deutscher Grafik-Designer e.V.,
Berlin 058773
Bund Deutscher Kunsterzieher e.V.,
Hannover 058774
Bund Freischaffender Foto-Designer
e.V., Stuttgart 058775
Bund mitteldeutscher Grafikdesigner
e.V., Leipzig 058776
Bunda, Veronica, Springwood .. 062237
Bunda, Veronica, Sydney 062280
Bundaberg Arts Centre,
Bundaberg 000907
Bundaberg Historical Museum,
Bundaberg 000908
Bundaberg Railway Museum,
Bundaberg 000909
Bundanon Art, Cambewarra ... 098853
Bundava, Sydney 062281
Bundelkhand Chhatrasal Museum,
Banda 024147
Bundesakademie für kulturelle Bildung
Wolfenbüttel, Wolfenbüttel ... 058777
Bundesbriefmuseum, Schwyz .. 041968
Bundesgremium des Juwelen-,
Uhren-, Kunst-, Antiquitäten und
Briefmarkenhandels, Wien 058243
Bundesverband Bild & Rahmen,
Rheinbach 058778
Bundesverband Bildender Künstler e.V.,
Landesverband NRW, Ortsverband
Bonn.Rhein-Sieg e.V., Bonn .. 058779
Bundesverband Bildender Künstler
Landesverband Hessen e.V.,
Darmstadt 058780
Bundesverband Bildender Künstler
Rheinland-Pfalz e.V., Mainz .. 058781
Bundesverband Bildender Künstler
Saarland e.V., Saarbrücken ... 058782
Bundesverband Bildender
Künstlerinnen und Künstler,
Bonn 058783
Bundesverband Bildender
Künstlerinnen und Künstler
Bezirksgruppe Stade-Cuxhaven,
Ottersberg 058797
Bundesverband Bildender
Künstlerinnen und Künstler
Bezirksverband Aachen/ Euregio,
Aachen 058798
Bundesverband Bildender

Künstlerinnen und Künstler
Bezirksverband Mittelhessen e.V,
Gießen 058799
Bundesverband Bildender
Künstlerinnen und Künstler für
Niedersachsen, Bezirksgruppe
Ostfriesland, Fehnhusen 058800
Bundesverband Bildender Künstlerinnen
und Künstler für Niedersachsen,
Landesgruppe Oldenburg,
Oldenburg 058801, 108339
Bundesverband Bildender
Künstlerinnen und Künstler
Landesverband NRW e.V., Köln 058802
Bundesverband Bildender
Künstlerinnen und Künstler
Niederbayern/Oberpfalz e.V.,
Regensburg 058803
Bundesverband Bildender
Künstlerinnen und Künstler Region
Bergischland, Wuppertal 058804
Bundesverband Bildender
Künstlerinnen und Künstler,
Bezirksgruppe Hildesheim,
Hildesheim 058789
Bundesverband Bildender
Künstlerinnen und Künstler,
Bezirksverband Düsseldorf e.V,
Düsseldorf 058790
Bundesverband Bildender
Künstlerinnen und Künstler,
Bezirksverband für Westfalen,
Dortmund 058791
Bundesverband Bildender
Künstlerinnen und Künstler,
Bezirksverband Heidelberg,
Heidelberg 058792
Bundesverband Bildender Künstlerinnen
und Künstler, Bezirksverband Köln e.V.,
Köln 058793, 107598
Bundesverband Bildender
Künstlerinnen und Künstler,
Bezirksverband Münsterland,
Münster 058784
Bundesverband Bildender
Künstlerinnen und Künstler,
Bezirksverband Unterfranken e.V,
Würzburg 058785
Bundesverband Bildender
Künstlerinnen und Künstler,
Landesgruppe Oldenburg,
Oldenburg 058786
Bundesverband Bildender
Künstlerinnen und Künstler,
Landesverband Schleswig-Holstein,
Kiel 058787
Bundesverband Bildender
Künstlerinnen und Künstler,
Regionalverband Celle, Celle .. 058794
Bundesverband Bildender
Künstlerinnen und Künstler,
Regionalverband Göttingen,
Göttingen 058795
Bundesverband Bildender
Künstlerinnen und Künstler,
Regionalverband Harz e.V.,
Goslar 058796
Bundesverband Bildender
Künstlerinnen und Künstler,
Regionalverband Kassel-Nordhessen,
Kassel 058788
Bundesverband der
Jugendkunstschulen und
Kulturpädagogischen Einrichtungen
e.V., Unna 058805
Bundesverband des Deutschen
Kunst- und Antiquitätenhandels e.V,
Drestedt 058806
Bundesverband Deutscher Galerien
e.V., Köln 058807
Bundesverband Deutscher
Kunstversteigerer e.V.,
Wiesbaden 058808
Bundesverband Kunsthandwerk,

Berufsverband Handwerk Kunst
Design e.V., Frankfurt am Main 058809
Bundesverband Museumspädagogik
e.V., Geschäftsstelle, München 058810
Bundis-Arndt, Maren, Hamburg . 129954
Bunëdi Farhangi Isfara, Isfara . . 059998
Bunëdi Farhangi Tojikiston,
Dušanbe 059999
Bunel, Patrick, L'Isle-sur-la-
Sorgue 069562
Bungalow Antiques, Oakland . . . 095885
Bungalow Gallery, Indianapolis . 121171
Bungalow Porslin, Stockholm . . 086982
Bungalow Upholstery, Auckland 132684
Bungay Museum, Bungay 043787
Bunglemarr Trading, Bundaberg 061232
Bunia, Sebastian-Jan, Krems . . 127794
H.K. Buniatyan Art Gallery, Gavar 000698
Buninyong Antiques, Buninyong 061235
Bunka Gakuen Fukushoku
Hakubutsukan, Tokyo 029491
Bunkamura Museum, Tokyo 029492
Bunker Hill Museum,
Charlestown 047565
Bunker Museet, Frederikshavn . 010068
Bunker Stalina, Samara 037704
Das Bunkermuseum, Emden . . . 018174
Bunkermuseum Oberhausen,
Oberhausen 020895
Bunkermuseum Wurzenpass,
Arnoldstein 001720
Bunkier Sztuki Wspólczesnej,
Kraków 113535
Bunnell Street Gallery, Homer . 049520
Bunratty Castle and Folkpark,
Bunratty 024837
Bunryu, Tokyo 142920
Buntin, Robyn, Honolulu 093533, 093534,
120925
Bunting, Peter, Bakewell 088333
Bunun Culture Exhibition Center,
Taoyuan 042573
Bunyevác Tájház, Türr István Múzeum,
Baja – Szentistván 023419
Buonasorte, Ugo, Milano 080468
Buono, Francesco, Roma 131836
Buonocore, Gennaro, Napoli . . . 131601
Buonocore, Mario, Napoli 080847,
131602
Buqaish, Sharjah 117195
Buque Museo Corbeta Uruguay,
Buenos Aires 000150
Buque Museo Ex BAP América,
Iquitos 034290
Buque-Museo Fragata Presidente
Sarmiento, Buenos Aires 000151
Bur, Marie-France, La Celle-Saint-
Cloud 068965
Buraczyński, Tomasz, Warszawa 132874
Burapha Museum of Art, Chon
Buri 042689
Burbage-Atter, Leeds 144519
Burbank, San Jose 097361
Burbiškio Dvaro Istorijos Ekspozicija,
Daugyvenés kultūros istorijos
muziejus-draustinis, Burbiškis . 030328
La Burbuja-Museo del Niño,
Hermosillo 031038
Burca, Eamon de, Blackrock . . 142595
Burcea, Ion, Bucureşti . 085083, 132993
Burchard, Saint Petersburg 127434
Burchard, Friedrich, Wuppertal . 142523
Burchard, Hans Martin, Hamburg 129955
Burchard, Harald,
Schwarzenberg 108591
Burchfield-Penney Art Center, Buffalo
State College, Buffalo 047303
Burckardt, Helga, Bielefeld 106488
Burden & Izett, New York 095292
Burden Gallery, Aperture Foundation,
New York 051350
Burden, Clive A., Rickmansworth 144800
Burdett, Melrose Park 061836
Burdett & Rawling, Hull 134485

Burdick, Washington 124875
Burdigal'Art, Bordeaux 103419
Burdur Devlet Güzel Sanatlar Galerisi,
Burdur 042854
Burdur Müzesi, Burdur 042855
Bureau County Historical Society
Museum, Princeton 052360
Bureau, Anne, Gaure 103764
Burebank Book's, Norwich 144749
Burel, Thierry, Paris 071100
Burel, Y., Luneray 069665
Buret, Roland, Paris 141063
Burford Antique Centre, Burford 088702,
134177
The Burford Gallery, Burford . . 117475
Burg- und Klosteranlage Oybin,
Oybin 021057
Burg- und Mühlenmuseum Pewsum,
Ostfriesisches Freilichtmuseum,
Krummhörn 019812
Burg- und Schloßmuseum Allstedt,
Allstedt 016513
Burg- und Schloßmuseum Jägersburg,
Homburg 019334
Burg- und Stadtmuseum Königstein,
Königstein im Taunus 019737
Burg-Antik, Nürnberg 077816
Burg Berwartstein, Erlenbach bei
Dahn 018269
Burg Eltz, Wierschem 022628
Burg Falkenberg, Falkenberg, Kreis
Tirschenreuth 018333
Burg-Galerie, Brüggen 106603
Burg Galerie, Fehmarn 106947
Burg Giebichenstein, Hochschule für
Kunst und Design Halle, Halle,
Saale 055551
Burg Gleichen, Schlossmuseum
Molsdorf, Wandersleben 022415
Burg Gnandstein, Staatsbetrieb
Staatliche Schlösser, Burgen und
Gärten Sachsen, Kohren-Sahlis 019751
Burg Gößweinstein, Gößweinstein 018743
Burg Greifenstein, Greifenstein . 001990
Burg Hagen, Hagen im
Bremischen 018920
Burg Herzberg, Breitenbach am
Herzberg 017573
Burg Hochosterwitz, Launsdorf . 002275
Burg Hohenzollern, Burg
Hohenzollern 017681
Burg Kreuzen, Bad Kreuzen . . . 001754
Burg Kreuzenstein, Korneuburg . 002224
Burg Kronberg, Kronberg im
Taunus 019804
Burg Lauenstein, Bayerische
Verwaltung der staatlichen Schlösser,
Gärten und Seen, Ludwigsstadt 020145
Burg Meersburg GmbH,
Meersburg 020327
Burg Mildenstein, Leisnig 020025
Burg-Museum, Parsberg 021076
Burg Neuhaus, Wolfsburg 022739
Burg Neustadt-Glewe, Neustadt-
Glewe 020767
Burg Pappenheim mit Naturmuseum
und Historischem Museum,
Pappenheim 021073
Burg Prunn, Bayerische Verwaltung
der staatlichen Schlösser, Gärten und
Seen, Riedenburg 021413
Burg Reichenstein,
Trechtingshausen 022182
Burg Sooneck, Niederheimbach 020795
Burg Stollberg, Stolberg,
Rheinland 022005
Burg Stolpen, Staatliche Schloesser,
Burgen und Gärten Sachsen,
Stolpen 022007
Burg Taufers, Campo Tures . . . 025727
Burg Thurant, Alken 016508
Burg Trausnitz, Bayerische Verwaltung
der staatlichen Schlösser, Gärten und
Seen, Landshut 019882

Burg Trifels, Annweiler am Trifels 016593
Burg Vischering, Münsterlandmuseum,
Lüdinghausen 020174
Burg zu Burghausen, Bayerische
Verwaltung der staatlichen Schlösser,
Gärten und Seen, Burghausen 017687
Burg Zug, Zug 042248
Burg Zwernitz, Bayerische Verwaltung
der staatlichen Schlösser, Gärten und
Seen, Wonsees 022758
Burgan, Claude, Paris 071101
burgart-presse, Rudolstadt 137493
Burgaski Muzej, Burgas 004995
Burgassi, Lucia, Firenze 110040
Burgate Antique Centre,
Canterbury 088772
Burgauer Galerie, Burgau 106617
Burgauer Tiermuseum, Burgau . 017684
Burge, Yale R. & Robert, New
York 095293
Burgen und Schlösser, Braubach 138688
Burgenkundliche Sammlung im
Burgenmuseum, Gossau (Sankt
Gallen) 041555
Burgenkundliches Museum Schloss
Alt-Kainach, Bärnbach 001774
Burgenländische Landesgalerie,
Eisenstadt 001856
Burgenländisches Brotmuseum, Bad
Tatzmannsdorf 001762
Burgenländisches Feuerwehrmuseum,
Eisenstadt 001857
Burgenmuseum, Dahn 017807
Burgenmuseum Eisenberg, Eisenberg,
Kreis Ostallgäu 018139
Burgeo Centennial Memorial Museum,
Burgeo 005495
Burger Weeshuis, Zierikzee . . . 033101
Burger, Anna-Maria, München . 108060
Burger, Christa, München 108061
Burger, Robert, Los Angeles . . . 135710
Burger, Veronika, Tübingen 108784
Burgerhuis, Stellenbosch 038833
Burgersdijk & Niermans, Leiden 143182
Burgersdorp Cultural Historical
Museum, Burgersdorp 038669
Burgert, Strasbourg . . 073576, 129266
Burgess Farm Antiques,
Morestead 090774
Burgess Farm Antiques,
Winchester 091867
Burgess, Derek J., Parkstone . . 090944,
134924
Burgess, J & J, Cambridge 144295
Burgevin, Denis, Gien 068660
Burggalerie, Magdeburg 107935
Burggalerie Laa, Laa an der
Thaya 002253
Burghead Museum, Burghead . . 043789
Burghley House, Stamford 045857
Burghofmuseum, Stadtgeschichtliches
Museum, Soest 021891
Burgi, Camille, Paris 071102
Burgin, Christine, New York . . . 122444
Burgio, Emanuele, Palermo . . . 081027
Burgio, Giuseppe, Palermo 081028
Burgmuseum, Güssing 002019
Burgmuseum, Seebenstein 002746
Burgmuseum Altnußberg,
Geiersthal 018604
Burgmuseum Archeo Norico,
Deutschlandsberg 001814
Burgmuseum Bad Bodenteich, Bad
Bodenteich 016737
Burgmuseum Burg Guttenberg,
Haßmersheim 019093
Burgmuseum Clam, Klam 002199
Burgmuseum der Wasserburg
Kapellendorf, Kapellendorf . . . 019497
Burgmuseum für Vor-
und Frühgeschichte,
Deutschlandsberg 001814
Burgmuseum Greifenstein,
Greifenstein 018803

Burgmuseum Grünwald,
Archäologische Staatssammlung
München, Grünwald 018874
Burgmuseum Kirschau, Kirschau 019634
Burgmuseum Krautheim,
Krautheim 019782
Burgmuseum Lenzen, Lenzen . . 020039
Burgmuseum Marksburg,
Deutsche Burgenvereinigung e.V,
Braubach 017548
Burgmuseum Nideggen,
Nideggen 020784
Burgmuseum Ortenberg, Ortenberg,
Hessen 021006
Burgmuseum Runneburg,
Weißensee 022548
Burgmuseum Schlossfreiheit,
Tangermünde 022099
Burgmuseum Schönfels,
Schönfels 021708
Burgmuseum Sulzberg, Sulzberg 022085
Burgmuseum Waldaist-Stöckl,
Tragwein 002830
Burgmuseum Waldeck, Waldeck 022383
Burgmuseum Wolfsegg, Wolfsegg 022748
Burgruine Montclair, Merzig . . . 020379
Burgruine Münzenberg,
Münzenberg 020615
Burgruine Teufelsberg, Überherrn 022241
Burgstaler, Markus, Templin . . . 078673
Burgturm Davert, Ascheberg . . . 016630
Burgues, Catherine, Gentilly . . . 128609
Burguet, Monique, Montrouge . . 070335
Burgverlag, Wien 140066
Burgwin-Wright Museum,
Wilmington 054444
Burica, New York 135894
Buried Treasures, Denver 093267
Buril, Madrid 086141
Buril, Sevilla 086445
Burin Heritage House, Burin . . . 005497
Burjatskij Respublikanskij
Chudožestvennyj Muzej im. C.S.
Sampilova, Ulan-Udé 037986
Burk, Jürgen J., Nürnberg 077817
Burkard, Gerhard & Viktoria,
Buttenwiesen 075235
Burkard, Peter, Luzern 116470, 126743
Burkart, Michael, Nürnberg 077818
Burke, Los Angeles 094279
Burke, Memphis 145253
Burke, New York 095294
Burke County Historical Powers Lake
Complex, Powers Lake 052344
Burke Museum of Natural History and
Culture, Seattle 053229
Burke News, Seattle 139330
Burke & Sons, John F., Cahir . . 079467
Burkert, George, San Francisco . 136310
Burkhardt, Vaihingen an der Enz 108807
Burkhardt-Spann, Gabriele,
München 130334
Burkhardt, Janet, Waramanga . . 139950
Burkhardt, Klaus-Joachim,
Stuttgart 078597
Burk's Falls and District Museum,
Burk's Falls 005498
Burlando, Mauro Silvio, Genova . 080241
Burle, Rio de Janeiro 101016
Burleson County Historical Museum,
Caldwell 047348
Burli, Pascal, Lucens 133816
Burlington, Denver 093268
Burlington, New York 095295
Burlington Art Centre, Burlington 005499
Burlington Bertie's, Auckland . . . 083774
Burlington County Historical Society
Museum, Burlington 047322
Burlington Historical Museum,
Burlington 047321
The Burlington Magazine,
London 139156
The Burlington Magazine Publications,
London 138204

Burlington Paintings, London . . . 089985,
118238, 134613
Burlington-Rock Island Railroad and
Historical Museum, Teague . . . 053693
Burlövs Gamla Prästgård, Arlöv . 040537
Burman, Stratford-upon-Avon . . 091491,
135100
Burman, C. & L., London 089986,
134614
Burmeister, Brigitte, Hasselberg 076413
Burmeister, Brigitte, Kiel 076773
Burmeister, J, Achim 105956
Burmese Fine Art, Singapore . . 085378
Burmester, James, Bristol 144277
Burnaby Art Gallery, Visual Arts
Burnaby, Burnaby 005502
Burnaby Village Museum,
Burnaby 005503
Burnay, Luís P., Lisboa 143559
Burne, W.G.T., London . 089987, 134615
Burnet, Amsterdam 143021
Burnet, Ross, Uralla 139945
Burnett, Richmond, Surrey . . . 118969
Burnett House and Myilly Point
Heritage Precinct, Darwin 001020
Burnham Deepdale Exhibitions,
Burnham Deepdale 117481
Burnham Model and Collectors Shop,
Burnham-on-Sea 088724
Burnham on Crouch and District
Museum, Burnham-on-Crouch 043790
Burnham Tavern, Hannah Weston
Chapter, National Society Daughters
of the American Revolution,
Machias 050554
Burnie Regional Art Gallery,
Burnie 000913
Burning Embers, Aberdeen 088219
Burning Issues Gallery,
Whangarei 113104
Burning Mountain Antiques,
Wingen 062439
The Burning Question Studio,
Vancouver 101752
Burns & Graham, Winchester . 091868
Burns Cottage Museum, Alloway 043397
Burns House Museum,
Mauchline 045238
Burns Visual Arts Society of Calgary,
Calgary 058429, 101219
Burns, Catherine, Oakland 123246
Burns, Greg, Oklahoma City . . . 123276
Burns, Stephanie, Deakin 098959
The Burnside, Brodick 117461
Burnside Plantation, Bethlehem . 046958
Burnt House Museum, Jerusalem 025072
Burntisland Edwardian Fair Museum,
Burntisland 043796
Buro Art, Edinburgh 117718
Burøsund Bygdemuseum,
Vannareid 034067
BuroK, Leeuwarden . . 112545, 137863
Burot, Y., Châtillon-sur-Seine . . 067924
Burpee Museum of Natural History,
Rockford 052635
Burr & Muir, Cape Town 085632
Burra Mine Open Air Museum,
Burra 000915
Burralga, Brisbane 098782
Burrangong Gallery, Young . . . 099947
Burrell Collection, Glasgow 044392
Burrell, Marceline, San Francisco 097197
Burren Centre, Kilfenora 024937
Burren College of Art,
Ballyvaughan 055858
Burri, Heinz & Therese,
Alchenflüh 087185
Burridge, John, Swanbourne . . 062273
Burritt on the Mountain, A Living
Museum, Huntsville 049633
Burrough of the Gleann Museum,
Glenboro 005882
Burrowes Mansion Museum,
Matawan 050738

Burrows, David E., Osgathorpe . 090921
Burrus, Claire, Paris 104599
Bursa Arkeoloji Müzesi, Bursa . . 042859
Bursa Devlet Güzel Sanatlar Galerisi,
Bursa 042860
Bursa Kent Müzesi, Bursa 042861
Bursa Türk-İslâm Eserleri Müzesi,
Bursa 042862
Bursledon Windmill, Bursledon . 043798
Burslem Antiques, Stoke-on-
Trent 091452
Burstert, Albrecht, Potsdam . . . 108397
Burston Strike School, Burston . 043800
Burstow & Hewett, Battle 126812
Bursztyn Galeria, Gdynia 113494
Burt County Museum, Tekamah 053695
Burté, Bernadette, Nancy 070415
Burtnieku Pagasta Briedes Krogs,
Burtnieki 030108
Burton, Invercargill 083965
Burton, Saint Paul 123899
Burton Agnes Hall, Burton Agnes 043801
Burton Antiques, Burton-upon-
Trent 088733
Burton Art Gallery and Museum,
Bideford 043592
Burton Ching, San Francisco . . . 097198
Burton Constable Hall, Kingston-upon-
Hull 044711
Burton Court, Eardisland 044196
Burton-Garbett, A., Morden . . . 144734
Burton Lysecki, Winnipeg 140433
Burton Marinkovich, Washington 124876
Burton, Ian, Auchterarder 088321
Burton, John, Ebrington 089155
Burton & Co., K.J., Balmain . . . 061096
Burtons' Colossal Collectables Fair,
Canterbury 097992
Burt's Antiques, Salt Lake City . 096883,
136233
Burwell-Morgan Mill, Millwood . 050902
Burwell Museum, Burwell 043805
Bury Art Gallery and Museum, Bury,
Lancashire 043806
Bury Saint Edmunds Art Gallery, Bury
Saint Edmunds 043809
Bury Transport Museum, Bury,
Lancashire 043807
Burzio, Vincenzo, Torino 132109
Burzym & Wolff, Kraków 113536
Busacchi, Daniela, Verona 132274
Busan Museum, Busan 029893
Busan University Museum, Busan 029894
Busan Woman's College Tea-ceremony
Museum, Busan 029895
Busch, Agnes, Krummennaab . . 076979
Busch, Bettina, Berlin 129482
Busch, Jürgen, München 108062,
130335
Busch, Klaus, Stakendorf 078535
Busche, Dr. Ernst A., Berlin . . . 106190
Buschlen-Mowatt, Vancouver . . 101753
Buschmann, Heinz & Peter,
Mönchengladbach 130309
Buschulte, Heinz, Oberlahr . . . 142254
Busck, Arnold, København 140605
Buscot House, Buscot 043813
Buscot Park House, Faringdon . . 044330
Buse, Peter-Jesko, Mainz 077234,
130267
Bush, New Orleans 095069
Bush and Beach Artwork, Barwon
Heads 098711
Bush Antiques, Saffron Walden . 091167
Bush House Museum and Bush Barn
Art Center, Salem 052888
Bush 'n Beach Gallery, Collaroy 098907
Bush 'n Beyond Studio Gallery, Broken
Hill 098795
Bush & Partners, A.E.,
Attleborough . . . 088319, 134043
Bush, Robert, London 089988
Bushehri, Manama 100420
Busher, Raad, Saarbrücken . . . 078318

Bushey Antique Centre, Bushey . 088744,
134193
Bushey Museum and Art Gallery,
Bushey 043814
Bushido Antiques, Honolulu . . . 093535
Bushuven, Düren 141709
Bushwhacker Museum, Nevada 051217
Bushwood, Redbourn 091077
Bushwood Books, Kew 138172
Bushy Park Ironworks, Dublin . . 130848
Bushy Run Battlefield Museum,
Jeannette 049795
Busi, Cristina, Chiavari 109963
Busiello, Pasquale, Napoli 131603
Busignani, Patricia, San Marino . 085283,
143679
Business and Creative Arts Division,
College of San Mateo, San
Mateo 057864
Business Committee for the Arts, Long
Island City 060420
Business Men's Art Institute,
Montrose 060421
Buskerud Auksjonsforretning,
Drammen 126343
Busnel, Jacques-Henry, Saint-Malo
(Ille-et-Vilaine) 072913
Busnel, Jean-François, Saint-Malo (Ille-
et-Vilaine) 072914
Buss, Jever 141966
Busse, Schleswig 108564
Busse, Holger, Hamburg 076205
Bussemakerhuis, Borne 032087
Bussien, Pierre, Martigny 087660
Bussière Jannel, Paris 104600
Bussolino, Roberto, Torino 132110
Busson, Bernard, Saint-Ouen . . 073043
Bussoni, Giuseppe, Milano 080469
Bussotti, München 130336
Al-Bustan, Manama 063136
Busters, Atlanta 092118, 135301
Busy Beaver Coins, Buffalo . . . 092515
Busy Bee Books, Mawson 139822
But Is It Art, London 101336
But is it Art, Reading 118958
Butchart & O'Hara, Long Beach 094197
Butchart International, Taipei . . 116904
Butchoff, Ian, London 089989
Bute Museum, Rothesay 045641
Buteau, Gérard, Clermont-
Ferrand 068015
Butera School of Art, Boston . . 056767
Butik 81, Odense 140634
Butik Antik, København 065879
Butik Antik, Kûblis 087547
Butler, Fort Worth 093436
The Butler & the Chef, San
Francisco 097199
Butler & Wilson, London 089990
Butler County Historical Museum,
Allison 046415
Butler County Museum, Hamilton 049328
Butler CountySociety Museum and
Kansas Oil Museum, El Dorado 048498
Butler Gallery, Kilkenny 024938
Butler Institute of American Art,
Youngstown 054599
Butler Street Gallery, Pittsburgh 096264
Butler, Brian D., Los Angeles . . 121452
Butler, Peter, Teralba 099815
Butler, Rita, Pangbourne 090943
Butler, Robin, Clare 088923
Butler, Roderick, Honiton 089558,
134469
Butler's Corner, Braintree 092307
Bútorbizományi, Budapest 079245
Butt, Lübeck 130254
Butt, Oldenburg 077926
Butt, Anthony, Baldock 088339
Butte County Pioneer Memorial
Museum, Oroville 051818
Butte Montmartre, Paris 104601
Butte Silver Bow Arts Chateau,
Butte 047334

Butter Pat Museum, Bigfork . . . 046983
Butterfass, Julie, Villegly 129346
Butterfield Trail Historical Museum,
Russell Springs 052710
Butterflies Gallery, Pokolbin . . . 099569
Butterfly Alphabet, Washington . 124877
Butternut and Blue, Baltimore . . 144982
Butters, Portland 123607
Butterworth, Sarah, Aboyne . . . 117219
Buttet, Pierre de, Poisy 129170
Buttin, Jean-Charles, Rumilly . . 072646
Buttin, Philippe, Annecy 066639
The Button Queen, London . . . 089991
Buttons Buttons Buttons, Sydney 062282
The Buttonwoods Museum,
Haverhill 049407
Butuan City Museum, Butuan . . 034443
Butuan Diocesan Museum,
Butuan 034444
Butuan National Museum,
Butuan 034445
Butzon & Bercker, Kevelaer . . . 107547,
137296
Buvat, Madeleine, Pionsat 072164
Bux, Gelsenkirchen 141825
Buxoro Qala Davlat Muzeyi
Qo'riqxonasi, Buxoro 054629
Buxtehude Museum für
Regionalgeschichte und Kunst,
Buxtehude 017709
Buxton Antiques Fair, Buxton . . 098364
Buxton Autumn Antiques and Fine Art
Fair, Buxton 098365
Buxton Museum, North Buxton . 006511
Buxton Museum and Art Gallery,
Buxton 043820
Buy and Sell Olde Wares,
Annandale 060999
Buy Design, Ancrum 088280
Buyenlarge, Philadelphia 126348
Buyeo National Museum, Buyeo 029902
Buyers, Toronto 064620
Buysscher, C. de, Ladeuze 125341
De Buytensael, Arnhem 112326
Buz White Screenprint, Portland 123608
Buzare, Hélène, Paris 071103
Buzeczki, M., Bruchsal 075200
Buzeczki, M., Ubstadt-Weiher . . 078741
Buzelim, Belo Horizonte 100876
Buzulukskij Gorodskoj Istoriko-
kraevedčeskij Muzej, Buzuluk . 036925
Buzzanca Stampe Antiche,
Padova 142774
Buzzi, Vittorio, Cagliari 131063
BV-Galerie, Klagenfurt 100069
BV Métaux Précieux, Paris 071104
By George, Parkhurst 085658
By George Antiques Centre, Saint
Albans 091172, 134987
By-Gone Days Antiques,
Charlotte 092581
By Many Hands, Minneapolis . . 121975
By the Book, Albury 139622
By the Book, Phoenix 145428
Bya Barros, São Paulo 064166
Byam Shaw School of Fine Art,
London 056551
Byard Art, Cambridge 117504
Byblos Art Gallery, Verona 111179
Byblos Bookshop, Mareeba . . . 139821
Bydee, Austin 119649
Bydgoski Antykwariat Naukowy,
Bydgoszcz 126399, 143438
Bydgoski Dom Aukcyjny Transfer,
Bydgoszcz 126400
Byegones, Zebbug 082451
Byens Antik, Århus 065599
Byers-Evans House Museum,
Denver 048251
Byethorpe, Barlow . . 088351, 134060
Byfield and District Museum,
Byfield 000919
Bygdarsavnið í Vestmanna,
Vestmanna 010747

Bygdetunet Jutulheimen, Vågå
  Historielag, Vågåmo 034063
Byggðasafn, Vestmannaeyjar .. 024118
Byggðasafn Akraness og Nærsveita,
  Akranes 024061
Byggðasafn Árnesinga,
  Eyrarbakki 024075
Byggðasafn Austur-Skaftafellssýslu,
  Höfn 024081
Byggðasafn Borgarfjarðar,
  Borgarnes 024069
Byggðasafn Dalamanna,
  Búdardalur 024071
Byggðasafn Dalvíkurbyggdar,
  Dalvík 024072
Byggðasafn Hafnarfjarðar,
  Hafnarfjörður 024077
Byggðasafn Húnvetninga og
  Strandarmanna, Brú 024070
Byggðasafn Reykjanesbaejar,
  Reykjanesbaer 024091
Byggðasafn Skagfirðinga,
  Varmahlid 024117
Byggðasafn Snæfellinga og
  Hnappdæla, Stykkishólmur .. 024116
Byggðasafnið Grenjadarstad,
  Húsavík 024082
Byggðasafnið í Skógum,
  Hvolsvöllur 024085
Byggförlaget, Stockholm 138014
Bygland Museum, Setesdalsmuseet,
  Bygland 033564
Bygone, Richmond 096531
Bygone Beautys, Leura 061735
Bygone Days, Naas 079679
Bygone Days, Ottawa 064498
Bygones, Angmering 088283
Bygones, Burford 088703, 134178
Bygones, Huntly 089621, 134493
Bygones, Limerick 079654
Bygones, Worcester 091942
Bygones & Antiques,
  Morningside 085652
Bygones by the Cathedral,
  Worcester 091943
Bygones Museum, Claydon 043965
Bygones Museum at Holkham, Wells-
  next-the-Sea 046079
Bygones of Ireland, Westport .. 079720
Byham, P.J. & J.F., Barooga ... 061102
Byland Abbey, Coxwold 044026
Byles, Patrick K., Kingston ... 111213
Byles, Robert, Bampton, Devon . 088347,
  134055
Bymodellen, Oslo 033828
Bymuseet, Skjern 010311
Bymuseet Møntergården, Odense 010253
Bymuseet, Oslo Museum, Oslo .. 033829
Byrd, Edward, Los Angeles ... 121453
Byrdie's, Jacksonville 093968
Byre Gallery, Stoneykirk 119153
Byrkley Books, Burton-upon-
  Trent 144290
Byrne, Bendigo 061130
Byrne, Christopher C., Dallas .. 120565
Byrnes, James B., Los Angeles . 121454
Byron, Taranto 081630
Byron Fine Art Gallery, Byron Bay 098829
Byron Historical Museum, Byron . 047338
Byron, Sandra, Woollahra 099912
By's Antigüedades, Caracas ... 097946
Byte Gallery, Northbridge, Western
  Australia 099484
Bytown Museum, Ottawa 006572
Bytte-Biksen, Struer 140662
Byttebørsen, Kolding 140620
Byttecentralen, Kolding 140621
Byttecentralen, Vejle 140670
Byzance Artes e Interiors, Rio de
  Janeiro 064033
Byzant, Prešov 085584
Byzantine and Christian Museum,
  Athinai 022981
Byzantine Museum, Mytilini ... 023187

Byzantine Museum, Paphos ... 009431
Byzantine Museum, Tinos 023304
Byzantine Museum and Art Galleries,
  Lefkosia 009411
Byzantine Museum of Antivouniotissa,
  Corfu 023052
Byzantine Museum of Ekatontapyliani,
  Paros 023213
Byzantine Museum of Naousa,
  Naousa 023195
Byzantium, Columbus 120480
C & C Arts Gallery, Singapore . 114502
C & C Coin, Phoenix 096195
C & D Sports, Columbus 093000
C & G Auctions, Rivett 125231
C & G Furniture, Houston 135595
C & R Gallery, Milwaukee 121904
C & T Antiques, Houston 093632
C 2, Montsoult 070338
C Fine Art, Wath-upon-Dearne . 119315
C-Galéria, Komárno 114661
C International Contemporary Art,
  Toronto 138571
C Konsthall, Upplands Väsby .. 041107
C Magazine, Toronto . 136682, 138572
C Photo Magazine, London 139157
C-Plus, Strasbourg 105680
C-Zone, Trieste 111045
C. Grier Beam Truck Museum,
  Cherryville 047611
C. Keller und Galerie Markt 21,
  Weimar 108861
C. Kermit Ewing Gallery of Art and
  Architecture, University of Tennessee,
  Knoxville 050006
C.A.D., Firenze 131138
The C.A.S.E Museum, Jersey City 049817
c.art, Dornbirn 099976, 136499
C.B. Restauración, Burgos ... 133209
C.B.O-Comptoir Belfortain de l'Or,
  Belfort 067067
C.C. Silver and Gold, Phoenix . 096196
c.d.n. Galeria, Warszawa 113811
C.E. Smith Museum of Anthropology,
  California State University, Hayward,
  Hayward 049418
C.F.C., Milano 080470
C.G.B., Paris 071105
C.G.B. Galerie, Honfleur 103820
C.G.F., Paris 071106
C.H. Moore Homestead, Clinton . 047834
C.H. Nash Museum-Chucalissa
  Archaeological Museum,
  Memphis 050773
C.J. Bucher Verlag, München .. 137392
C.J. Pictures, Hornchurch 117998
C.J.'s Upholstery, Albuquerque . 135280
C.L. Alexander Museum, Tumby
  Bay 001505
C.M. Bellman Museum,
  Stockholm 040994
C.M. Bookshop in Silverlake, Los
  Angeles 145218
C.M. Booth Collection of Historic
  Vehicles, Rolvenden 045631
C.M. Russell Museum, Great
  Falls 049216
C.M.B. Marseille Numismatique,
  Marseille 069914
C.M.C. – Comptoir Mondial de
  Création, Paris 104602
C.M.P, Paris 071107
C.O. Card Home, Cardston 005562
C.O. Müller-Galerie, Eichstätt .. 018118
C.P. Art Centre, Chennai 109279
C.P. Nel Museum, Oudtshoorn .. 038792
C.P.R. Antiques and Services,
  Barrhead 088369
C.R. Conservazione e Restauro,
  Milano 131402
C.R.A.C, Paris 104603
C.R.C. Restauri, Bologna 130942
C.R.O.A. Consultores, Madrid .. 133310
C.S.A.F., Paris 104604

C.S.A.O.-Galerie 3a, Paris 104605
C.T. Restauro, Bologna 130943
C.T.S., Roma 131837
C.T.S. France, Paris 128955
C.W. Niemeyer Buchverlage,
  Hameln 137260
C.Wichtendahl.Galerie, Berlin ... 106191
C/O Berlin, Kulturelles Forum für
  Fotografie, Berlin 017096
c/o Suti Galerie und Edition,
  Bern 116126, 138029
The C21 Shop, Mundesley 118765
C59 Productions, New York ... 122445
Ca' Bon Gallery, Motovun 008950
Cá d'Oro, Bologna 079825
Ca' d'Oro, Vicenza 111203
CA Galleri, Oslo 113197
Cá Venier dei Leoni, Venezia .. 111095
CAA Newsletter (New York), New
  York 139331
Caamano, Philippe, Saint-Martin 130809
Cabaia, Lisboa 084846
Cabalgata XVIII, Palma de
  Mallorca 086344
Caballero Vela Arturo, Acapulco 132355
Caballero, Inocencio Lopez,
  L'Hospitalet de Llobregat 086050
Caballin Muñiz, David Armando,
  Gijón 086017
Cabanel, Victoria, Paris 071108
Cabanis & Troschke, Esslingen . 129823
Cabanon de Le Corbusier, Roquebrune-
  Cap-Martin 015145
Cabat, Tucson 124721
Cabello, C.I.A., Sankt Margrethen 126751
Cabeza de Vaca, Mercedes,
  Madrid 086142
Cabib, P.F., Genova 080242
Cabinet, Berlin 129483
Cabinet d'Art Asiatique, Labarde 013506
Cabinet d'Art Contemporain,
  Bruxelles 100574
Cabinet de l'Amateur, Paris 071109
Cabinet des Curieux, Paris 071110
Cabinet des Estampes du Musée d'Art
  et d'Histoire, Genève 041516
Cabinet des Estampes et des Dessins,
  Liège 003791
Cabinet des Estampes et des Dessins,
  Strasbourg 015862
Cabinet des Médailles, Bruxelles 003437
Cabinet des Medailles et Antiques,
  Paris 014627
Cabinet des Monnais et Médailles
  du Revest-les-Eaux, Le Revest-les-
  Eaux 013695
Cabinet d'Expertise Tastet Souchon,
  Amiens 066566
Le Cabinet du Curiosité, Paris .. 071111
Cabinet Numismatique, Nantes . 070453
Cabinet Numismatique de Bretagne,
  Lorient 069632
Cabinet of Discoveries, Campbell 061261
Cabinet Quercy, Toulouse 073745
Cable Natural History Museum,
  Cable 047339
Cabo, Pascual de, Palma de
  Mallorca 115335
Cabom, Kalmar 086767
Cabon, Frédéric, Bordeaux 067318
Cabon, Frédéric, Libourne 069458
Caboodle, Edinburgh 089164
Caboolture Historical Village and
  Museum, Caboolture 000922
Cabot Historical Society, Cabot . 047340
Cabot's Pueblo Museum, Desert Hot
  Springs 048291
Cabotse, Michel, Paris 071112
Cabourdin, Annie, Lille 069483
Cabral, Rio de Janeiro 140291
Cabral Moncada, Lisboa 126440
Cabral, Alice, Rio de Janeiro ... 101017
Cabral, Gilberto E.M., Rio de
  Janeiro 101018

Cabral, Jorge, Montevideo 091993,
  144946
Cabrera Moran, M.I., Madrid ... 115160
Cabri and District Museum, Cabri 005510
Cabrol, Pierre, Toulouse 073746
Caburn Books, Lewes 144530
Cabuzel, Frédéric, Saint-Riquier 073296
Caccamo, Bertrand, Brignais ... 067496
Cacciabue-Drac, Delio, Château-
  Chinon 067868
Cacciatore, John, Albuquerque . 119462
Cacciola, J., New York 122446
Cache, Málaga 086289
Cache Antiques, West Hollywood 097877
Cachemire, Bessèges 067143
Cachet, Basel 087226, 133638
Los Cachorros, Buenos Aires ... 139555
Cachot, Den Haag 083067
Cachoux, Michel, Paris 104606
Cacialli, Giuseppe, Firenze ... 131139
Caco Velho, Porto 085015
Caco Velho, Rio de Janeiro ... 128110
Cad, Cincinnati 120327
Caddo-Pine Island Oil and Historical
  Society Museum, Oil City 051721
Caddoan Mounds, Cherokee
  County 047608
Cade, Denise, New York 122447
Cadeiral, Coimbra 084788
Cadena de Moteles, Madrid ... 115161
Cadenet, Martine, Perpignan ... 072072
Cades Cove Open-Air Museum,
  Townsend 053785
Cadet, Rousselle, Arles 066720
Cadet, Rousselle, L'Isle-sur-la-
  Sorgue 069563
Cadet, Rousselle, Saint-Germain-en-
  Laye 072807
Cadi, Barcelona 085750
Cadia Vintage, Boston 092452
Cadiac, Béziers 067172
Cadillac Museum, Cadillac 005512
Cadimage, Lyon 069716
Cadogan, London 118239
Cadot, Farideh, Paris 104607
Cadr'Art, Dijon 103673
Le Cadre, Cahors 103490
Cadre, Épinal 068454, 128593
Cadre à Part, Paris 128956
Cadre de Vie, Paris 104608
Au Cadre Doré, Genève 133757
Cadre Image, Saint-Cyprien ... 105499
Le Cadre Noir, Aulnay-sous-Bois 103289
Le Cadre Noir, Livry-Gargan ... 104042
Le Cadre Noir, Paris 104609
Le Cadre Noir, Saint-Denis ... 114116
Cadrerie, Le Puy-en-Velay 103965
Cadrerie, Marseille 104161, 128754
Cadres & Passions, Paris 103482
Cadres de Vie, Château Thierry . 103577
Cadres Décors, Aurillac 103292
Cadrissime, Troyes 105774
Cadrissimo, Clermont-Ferrand .. 103600
Cadrix, Montpellier 128807
Il Caduceo, Torino 081669
Caedmon House, Whitby 091826, 135222
Caelum Gallery, New York 122448
Caen Numisma, Caen 067561
Caernarfon Maritime Museum,
  Caernarfon 043823
Caerphilly Castle, Caerphilly ... 043827
Caesar, Benedict, Hamburg ... 107233
Caesarea Museum, Kibbutz Sdot
  Yam 025146
Café Aroma Photogalerie, Berlin 106192
Café Art Gallery, Hue 125160
Café Crème, Luxembourg 111837
Cafe de Matisse, San Jose ... 124446
Café des Arts, Ha Noi 125064
Le Café Français, Bruxelles ... 100575
Cafe Galeria, Brasília 100911
Cafe Gallery Projects, London .. 044931
Café Sting, Stavanger 113263
El Cafetal, Vitoria-Gasteiz 086572

Register der Institutionen und Firmen

Caffard, Guy, Colmar
– Cambridge and County Folk Museum, Cambridge

Caffard, Guy, Colmar 068086
Il Caffè, Milano 142717
Caffi, Franco, Brescia 079923
Cafiso, Milano 110272
Cafmeyer, Knokke-Heist 100734
Cagayan Museum, Tuguegarao . 034609
Çağdas Sanatlar Vakfi, Ankara . 060006
Cage, Robin, Richmond 123726
Cagliani, Luigi, Milano 080471, 131403
Cagnard, T, Charenton-le-Pont . 067837
Cagne, Edith, Lons-le-Saunier . . 069626
Cagny, Yves de, Paris 125689
Cahaba Cottage Antiques, Birmingham 092396
Cahaya, Singapore 085379
Les Cahiers de la Peinture, Paris 138636
Cahit Sıtkı Tarancı Müze-Evi, Diyarbakır 042879
Cahn, Jean-David, Basel 087227, 126706
Cahn, Jean-David, Sankt Moritz 087793, 126752
Cahokia Courthouse, Cahokia . . 047342
Cahokia Mounds, Collinsville . . . 047878
Cahoon Museum of American Art, Cotuit 048030
Cai Yuanpei Art College, Shaoxing University, Shaoxing 055237
Caiati Dipinti Antichi, Milano . . . 080472
Caiati, Lino, Milano 131404
Caidian District Museum, Wuhan 008369
Caifeng Culture Museum, Danei 042299
Caignard, Nantes 070454
Caignard, Erwan, Nantes 070455
Cailac, Paule, Paris 104610
Caillaud, Alain, Sainte-Reine-de-Bretagne 073364
Caillaud, Florence, Bologna 130944
Cailler, Nane, Pully 116573
Caillet, Gérard, Payerne 087731
Cailleux, Pierre, Paris 071113
Cailly, Geneviève, Hanches 068785
Cailun Ciwen Guan (Museum), Yangxian 008464
Cain, Corinne, Phoenix 123486
Caio, São Paulo 101085
Caird, Jane, Raglan 112995
Cairdean nan Taighean Tugha, Innerleithen 060083
Cairncross & Sons, Filey 089272
Cairns Museum, Cairns 000923
Cairns Regional Gallery, Cairns . 000924
Cairo Academy of Arts, Cairo . . 055328
Cairo Atelier, Cairo 103025
Cairo Berlin Art Gallery, Cairo . . 103026
Cairo Museum, Vicksburg 053973
Cais Gallery, Hong Kong 102062
Cais Gallery, Seoul 111620
Caistor Antiques, Caistor 088755
Caithness District Museum, Wick 046131
Caitlyn, Saint Louis 123814
Caixa Forum Madrid, Madrid . . . 039604
CaixaForum Barcelona, Barcelona 039053
CaixaForum Palma de Mallorca, Palma de Mallorca 039879
Caja de Castellon, Castello de la Plana 115013, 115014
Caja Madrid, Zaragoza 115551
La Caja Negra, Madrid 115162
Čajkovskaja Kartinnaja Galereja, Permskaja gosudarstvennaja chudožestvennaja galereja, Čajkovskij 036926
Čajkovskij Kraevedčeskij Muzej, Permskij oblastnoj kraevedčeskij muzej, Čajkovskij 036927
Cáki Pincesor – Szabadtéri Néprajzi Múzeum, Cák 023577
Çakirağa Konaği, Ödemiş 043037
Calabrò, Trevignano Romano . . . 142875
Calabró Antichitá 1968, Roma . . 081277
Calades, Avignon 103302
Calado, José Cândido, Lisboa . . 084847
Calala Cottage, Tamworth 001559
Calamassi, Simone, Firenze . . . 080102
Calame Lelievre, Michèle, Argilliers 066717
Calame, Marlyse, Vence 105828
Il Calamo, Torino 081670
Calamus, Mostar 100854
Calanca, Jean-Pierre, Paris 071114
Calandra Arte, Manarola 110197
Calandra, Sebastiano, Riomaggiore 110718
Calao, Vic-le-Comte 074093
Calard, François, Jouvencon . . . 068920
Calard, Jacques, Chénérailles . . 067956, 128555
Calart Actual, Genève 116291
Calaveras County Museum, San Andreas 052922
Calbourne Water Mill and Rural Museum, Newport, Isle of Wight 045363
Calçada da Estrela, Lisboa 084848
Calcografia Nacional, Real Academia de Bellas Artes de San Fernando, Madrid 039605
Calcografia, Istituto Nazionale per la Grafica, Roma 027555
Calcutta Painters, Kolkata 059361
Caldarella, Jack M., El Paso . . . 093393
Calderan, Torino 081671
Calderglen County Park, East Kilbride 044204
Caldicot Castle, Caldicot 043828
Caldon Gallery, Cheddleton 117558
Caldwell & Snyder, New York . . 122449
Caldwell & Snyder, San Francisco 124244
Caldwell, Ian, Walton-on-the-Hill 091722
Caldwell, J., Charlotte 092582
Caldwell, Tom, Belfast 117333
Caldwells, Elsternwick 061441
Caleb Pusey House, Upland . . . 053902
La Calebasse, Cabasse 067541, 067542
Caleca, Messina 080363
Calèche, Barneville-Carteret . . . 066953
Calèche, Le Luc 069296
Caledon Museum, Caledon 038670
Caledonia, Bilbao 085866, 114983
Caledonia Antiguedades, Madrid 086143
Caledonia Books, Glasgow 144433
Caledonian Antiques, Bristol . . . 088650, 134159
Caledonian Images, Banff 117281
Calegari, Dario, Milano 131405
Caleidos, Padova 110558
Caleidoscopio Arte, La Paz 100844
Calendas, Rio de Janeiro 064034
Calepinus, Paris 141064
La Calesa, Castello de la Plana . 085930
Caley, Auckland 112754
Calgary Coin & Antique Gallery, Calgary 064250
Calgary Contemporary Arts Society, Calgary 058430, 101220
Calgary Police Service Interpretive Centre, Calgary 005517
Calgary Science Centre, Calgary 005518
Calgary Summer Antique Show, Calgary 098033
Calgary Winter Antique Show, Calgary 098034
Calheiros, António M. Fernandes, Lisboa 084849
Calhoun County Museum, Port Lavaca 052254
Calhoun County Museum, Rockwell City 052654
Calhoun County Museum, Saint Matthews 052833
Caliban, Pittsburgh 145434
CàLibri, Landsberg am Lech . . . 142044
Calico, Essen 075669
Calico Cat, Chicago 092680
Calico Cat, Milwaukee 094840
The Calico Cat, Tampa 097548
Calico Museum of Textiles, Sarabhai Foundation, Ahmedabad 024122
Calico, X & F, Barcelona 085751
Califia, Birmingham 145000
California Academy of Sciences, San Francisco 052995
California African-American Museum, Los Angeles 050411
California Art Club Gallery, San Marino 053052
California Art Institute, Westlake Village 058092
California Art Investment, Kansas City 121282
California Art Investment, Pittsburg 136125
California Association of Museums, Santa Cruz 060422
California Center for the Arts, Escondido Museum, Escondido 048605
California Child, Los Angeles . . . 121455
California College of Arts and Crafts, Oakland 057605
California College of Arts and Crafts, San Francisco 057854
California Heritage Museum, Santa Monica 053132
California Indian Museum, Santa Rosa 053137
California Institute of the Arts, Valencia 058036
California Museum of Ancient Art, Beverly Hills 046972
California Museum of Photography, Riverside 052591
California Oil Museum, Santa Paula 053136
California Science Center, Los Angeles 050412
California State Capitol Museum, Sacramento 052721
California State History Museum, Sacramento 052722
California State Indian Museum, Sacramento 052723
California State Mining and Mineral Museum, Mariposa 050697
California State Railroad Museum, Sacramento 052724
California Upholstering Company, Sacramento 136184
California Watercolor Association, Walnut Creek 060423
Calik, Krystyna, Kraków 113537
Calil, São Paulo 140301
Calimala, Vessy 116690
La Calinière, Paris 071115
Calkins Field Museum, Iowa Falls 049714
Call of the Wild – Wildlife Museum, Gaylord 049062
Calla, Lily, Salt Lake City 123950
Callahan County Pioneer Museum, Baird 046714
Callan, New Orleans . . 095070, 122140, 138369
Callan, Thomas R., Ayr 126804
Callander Bay Heritage Museum, Callander 005536
Callant, Roland, Montay 070209
Callanwolde Fine Arts Center, Atlanta 046611
Callat & Cie., Paris 071116
Calle, Carmen de la, Madrid . . . 115163
Calle, Gerardo de la, Santiago de Chile 064959
Callegari, Dino, Padova 131657
Callegari, Roberto, Padova 080973
Calleja, Antwerpen 063198
Callejón Antiguo, Medellín 065279
Callejón Sanchez, José, Barcelona 133108
Callendar House, Falkirk 044322
Callewaert, Henri, Jaux 068899
Callies, Bettina, Regensburg . . . 108459
Callies, Gerhard, Dornhan 075377, 125920
Calligrammes, Ottawa . 101466, 128189
Callingham, Northchapel 090857
Callis, São Paulo 136665
Calloway, Susan, Washington . . 097815, 124878
Callsen, Gunter & Maximilian, Straßlach-Dingharting 078573
Calmels, Simone, Toulouse 073747
CalmelsCohen, Paris 125690
Calne Antiques, Calne 088757
Caloundra Fine Art Gallery, Caloundra 098847
Caloundra Regional Art Gallery, Caloundra 000925
Caltex Gallery, Biella 109808
Calton Gallery, Edinburgh 089165, 117719, 134359
Calumet County Historical Society Museum, New Holstein 051270
Calumet-Nuzinger, Ch., Heidelberg 107388
Calvari, Angelo, Roma 081278
Calvasina, Milano 110273
Calvente, Madrid 086144
Calverley Antiques, Tunbridge Wells 091647, 135156
Calvert, Nashville 094980
Calvert, Washington . . 097816, 124879
Calvert Marine Museum, Solomons 053377
Calvert, Karen, San Antonio . . . 124019
Calverton Folk Museum, Calverton 043832
Calvi e Volpi, Milano 080473
Calvin Coolidge Presidential Library and Museum, Northampton . . . 051624
Calvin Coolidge State Historic Site, Plymouth Notch 052220
Calvinia Regional Museum, Calvinia 038671
Calvo, Valencia 086509
Calvo Gonzalez, Pablo, Segovia . 086431
Calypso, Xaghra 082448
Calzada Salavedra, Arcadi, Barcelona 114863
Calzada, Marc, Barcelona 085752
Calzavara, Luca, Venezia 132230
Camafeo, Alicante 085694
Le Camaïeu, Tournai 063841
Cámara Santa de la Catedral de Oviedo, Oviedo 039860
Camara, Jose A., Madrid 086145
Camard, Jean-Pierre, Paris 071117
Camarev, Norton Summit 062008
Camargo, Ricardo Pagotto, São Paulo 101086
Camargue, Spit Junction 062234
Camaver Kunsthaus, Lecco 080309
Cambalache, Segovia 086432
Camberwell Antique Centre, Camberwell 061255
Camberwell Auctions, Hawthorn East 125199
Camberwell Books and Collectibles, Hawthorn East 061608, 139768
Camberwell College of Arts, University of the Arts London, London 056552, 056553
Cambi, Genova 080243
Cambiaso, Maria Barbara, Genova 131292
Cambieri, Marco, Milano 131406
Cambium, Madrid 133311
Cambon, Alain, Paris 141065
Camborne School of Mines Geological Museum and Art Gallery, Pool . 045525
Cambria Books, Blackheath . . . 139645
Cambria County Historical Society Museum, Ebensburg 048466
Cambria Historical Society Museum, New Providence 051312
Cambridge and County Folk Museum,

Cambridge 043838
Cambridge Antiques, Waikato .. 084124
Cambridge Art Association,
Cambridge 060424
Cambridge Auctioneering,
Waikato 126329
Cambridge Contemporary Art,
Cambridge 117505
Cambridge Cottage, Bolwarra .. 061168
Cambridge Fine Art, Cambridge 088760,
117506,   134198
Cambridge Fine Art, Little
Shelford 118124
Cambridge Glass Museum,
Cambridge 047373
Cambridge Historical Museum,
Cambridge 047359
Cambridge Museum, Cambridge 033164,
047372
Cambridge Museum of Technology,
Cambridge 043839
Cambridge Pine, Bottisham 088524
Cambridge University Collection of Air
Photographs, Cambridge ... 043840
Cambridge University Press,
Cambridge 138146
Cambridge University Press, New
York 138385
Camden, Pontevedra 086401
Camden Archives and Museum,
Camden 047383
Camden Arts Centre, London .. 044932
Camden Books, Bath 144229
Camden County Historical Society
Museum, Camden 047379
Camden Fine Art Gallery,
Camden 098854
Camden Haven Museum,
Laurieton 001216
Camden Museum, Camden 000928
Camden Passage Antique Market &
Pierrepont Arcade Antique Centre,
London 089992
Camden Society of Artists,
Camden 058193, 098855
Camden Valley Antiques,
Narellan 061942
Le Caméléon, Antibes-Juan-les-
Pins 103255
The Cameleon, Rotterdam 083557
The Camelford Gallery,
Camelford 117514
Camellia Grove Antiques, Eagle
Heights 061421
Camelot, Boston 119848
Camelot, Kraków 113538
Camelot Antiques, Toronto 064621
Camelot Castle Gallery, Tintagel 119236
Camelú, Roma 110743
Cameo, Columbus 120481
Cameo, Heilbronn 076470
Cameo, Mannheim 077254
Cameo, Tbilisi 105898
Cameo Antiques, Chester 088858
Cameo Framing and Gallery,
Vancouver 101754
Cameo Gallery, London 089993
Camera Austria, Graz .099997, 136507,
138509
Camera Magenta, Wien 058244
Camera Obscura, Hainichen .. 018932
Camera Obscura, Paris 104611
Camera Obscura and World of
Illusions, Edinburgh 044232
Camera Obscura Gallery, Denver 120698
Camera Obscura, Museum zur
Vorgeschichte des Films, Mülheim an
der Ruhr 020493
Camerawork, Berlin 106193
Camerawork, London 118240
Camerawork Books, San
Francisco 145547
Cameron, Portland 145449
Cameron & Hollis, Moffat 138266

Cameron Highlanders of Ottawa
Regimental Museum, Ottawa . 006573
Cameron, Donald, Kew 099186
Cameron, Patricia, Seattle 124502
Camet, Huguette, Saint-Ouen .. 073044
Camhi, Rebecca, Athinai 109029
Camiciotti, Ugo, Firenze 080103
Camilla's Bookshop, Eastbourne 144380
Camille, N., Bruxelles 063362
Camilleri, Paul, Balzan 132334
Camille's, Tucson 097641
Camin, Roberto, Bolzano 131007
Caminade, Corinne, Paris 104612
Caminade, Jean-Alain, Colmar . 068087,
140829
Caminha, Francisco B., Rio de
Janeiro 064035
Il Camino, Pordenone 110672
El Camino Gallery, Phoenix ... 123487
Càminul Artei, Bucureşti 114129
Çamlı Köşk, İstanbul 042929
Çamlıca Sanat Evi, İstanbul ... 117020
Cammarata, Sebastiano, Catania 131082
Cammarella, Loretta, Milano .. 110274
Cammies Country Living, Paringa 062061
CAMOC – International Committee
for the Collections and Activities of
Museums of Cities, Moskva .. 059771
Camões, Lisboa 143560
Camoin-Damachy, Paris 071118
Camouflage Brussels, Bruxelles 100576
Camp 6 Logging Museum,
Tacoma 053640
Camp Five Museum, Laona 050141
Camp Floyd, Fairfield 048674
Camp Gallery, Virginia Center for the
Creative Arts, Sweet Briar 053624
Campabadal y Torres, Barcelona 085753,
126510
La Campagne du Nord„
Rotterdam 112631
Campagne, Brigitte, Paris 071119
Campagne, Yves, Sainte-Cathérine-lès-
Arras 073341
Campagnola, Daniela, Verona . 081902,
132275
Campaign for Museums, London 060084
Campana, Guérin, Corte 068162
Campaña, Luis, Berlin 106194
Campanella, Michele, Genova .. 110095
Campbell, Baltimore 092308
Campbell, Cleveland 092926
Campbell, Dubai 117179
Campbell, Edmonton 101280
Campbell Carriage Factory Museum,
Middle Sackville 006302
Campbell Collections, Mashu Museum
of Ethnology and William Campbell
Furniture and Picture Collection,
Durban 038697
Campbell County Rockpile Museum,
Gillette 049097
Campbell Grant, Christchurch .. 083861
Campbell Hall Gallery, Western Oregon
University, Monmouth 050993
Campbell Historical Museum,
Campbell 047392
Campbell House Museum, Saint
Louis 052792
Campbell House Museum,
Toronto 007162
Campbell River and District Public Art
Gallery, Campbell River 005540
Campbell River Museum, Campbell
River 005541
Campbell River Optical Maritime
Museum, Campbell River 005542
Campbell Town Croft Antiques,
Campbell Town 061262
Campbell Whittlesey House, The
Landmark Society of Western New
York, Rochester 052615
Campbell Wilson, Brighton ... 088617
Campbell, Andrew, Jesmond ... 089657

Campbell, Duncan, London 118241
Campbell, Iain, Widnes 091844, 144922
Campbell & Assoc., J., Cincinnati 120328
Campbell, J.T., Dallas 093082
Campbell, John, Nashville 122074
Campbell, Lucy B., London 118242
Campbell, Mack, Toowoomba .. 139931
Campbell, Marcus, London ... 144572
Campbell, Meg, Southampton .. 091370
Campbell, Peter, Atworth 088320
Campbell, William, Fort Worth . 120859
Campbellford-Seymour Heritage
Centre, Campbellford 005544
Campbelltown Art Center,
Campbelltown 000930
Campbelltown Print Gallery,
Leumeah 099226
Campbelltown Museum,
Campbelltown 043854
Campden Gallery, Chipping
Campden 117583
Campe, Nürnberg ... 108296, 142238
Camperdown and District Museum,
Camperdown 000931
Camperdown Buggy Museum,
Camperdown 000932
Camperos, San Francisco 136311
Camphor-Hill, Baltimore 119729
Campi, Adriana, Milano 080474
Campidell, Walter, Feistritz an der
Drau 127730
Il Campiello, Milano 080475
Campillo Monforte, Carlos,
Barcelona 085754
Campion Gallery, Rye 118996
Campion, Peter, Cheltenham ... 134234
Campisi, Gabriele & Rodolfo,
Catania 080024
Campman, Zeist 083709
Campo, Antwerpen 125294
Campo & Campo, Berchem ... 100528,
125299
Campo Antolin, Fernando del,
Zaragoza 133494
Il Campo delle Fragole, Bologna 109817
Campo, Carine, Antwerpen 100485
Campo, Carmen del, Córdoba .. 115027
Campobasso, Roberto, Napoli .. 080848
Campobello Island Museum, Roosevelt
Campobello International Park,
Welshpool 007376
Campogiani, Claudia, Roma ... 080476
Campolmi, Alessandro, Firenze . 080104
Camponi, Enrico, Roma 081279
Campregher, J. & R., Steyr 062849
Campus Gallery, Georgian College
of Applied Arts and Technology,
Barrie 005385
Campus Martius Museum,
Marietta 050678
Camron-Stanford House, Oakland 051675
Camrose and District Centennial
Museum, Camrose 005546
Camstra, Eleonore, Lochem ... 132570
Camu, Th., Charleroi 063593
Camugli, Lyon 080479
Camus, Saint-Ouen ... 073045, 073046
Camus Ares, Ricardo, Santander 086421
Camus, Jean, Bordeaux 067319
Camuso, Armando, Prato 131772
Can Art Handworks, Detroit ... 135539
Canaan Historical Society, Falls
Village 048691
Canaan Historical Society Museum,
Canaan 047394
Canada 2000 Antiques,
Vancouver 064804
Canada Agriculture Museum,
Ottawa 006574
Canada Aviation Museum,
Ottawa 006575
Cañada Cordero, Antonio, Madrid 086146
Canada House Gallery, London . 044933
Canada Science and Technology

Museum, Ottawa 006576
Canada War Museum, Adegem . 003259
Canada West Antique Company,
Vancouver 064805
Canada's Aviation Hall of Fame,
Wetaskiwin 007390
Canada's Penitentiary Museum,
Kingston 006097
Canada's Sports Hall of Fame,
Toronto 007163
Canadeau, Québec 101494
Canadian Antique Dealers' Association,
Toronto 058431
Canadian Arctic Gallery, Basel . 116058
Canadian Art, Toronto 138573
Canadian Art Foundation, Toronto 058432
Canadian Art Museum Directors
Organization, Ottawa 058433
Canadian Artists' Representation,
CARFAC Ontario, Toronto 058434
Canadian Association of
Photographers & Illustrators In
Communications, Toronto ... 058435
Canadian Automotive Museum,
Oshawa 006563
Canadian Baseball Hall of Fame &
Museum, Saint Marys 006925
Canadian Broadcast Museum
Foundation, Toronto 007164
Canadian Bushplane Heritage Centre,
Sault Sainte Marie 006988
Canadian Canoe Museum,
Peterborough 006636
Canadian Centre for Architecture,
Montréal 006361
Canadian Clay and Glass Gallery,
Waterloo 007362
The Canadian Clock Museum, Deep
River 005692
Canadian County Historical Museum,
El Reno 048517
Canadian Embassy Art Gallery,
Washington 054096
Canadian Federation of Friends of
Museums, Toronto 058436
Canadian Fine Arts, Toronto ... 101587
Canadian FireFighter's Museum, Port
Hope 006689
Canadian Football Hall of Fame and
Museum, Hamilton 005967
Canadian Force Base Gagetown
Military Museum, Oromocto .. 006559
Canadian Golf Hall of Fame and
Museum, Oakville 006534
Canadian Guitar Players Association,
Lion's Head 006198
Canadian Lacrosse Hall of Fame, New
Westminster 006472
Canadian Medical Hall of Fame,
London 006208
Canadian Military Engineers Museum,
Oromocto 006560
Canadian Military Heritage Museum,
Branford 005467
Canadian Military Heritage Museum,
Saskatoon 006970
Canadian Military Studies Museum,
Limehouse 006195
Canadian Museum of Contemporary
Photography, National Gallery of
Canada, Ottawa 006577
Canadian Museum of Flight,
Langley 006160
Canadian Museum of Hindu
Civilization, Richmond Hills ... 006800
Canadian Museum of Nature,
Ottawa 006578
Canadian Museum of Rail Travel,
Cranbrook 005663
Canadian Museums Association,
Ottawa 058437
Canadian Olympic Hall of Fame,
Temple de la Renommée Olympique
du Canada, Ottawa 006579

Canadian Plains Gallery,
Winnipeg 101870
Canadian Scottish Regimental
Museum, Victoria 007320
Canadian Scouting Museum,
Ottawa 006580
Canadian Sculpture Centre,
Toronto 007165
Canadian Ski Museum, Musée
Canadien du Ski, Ottawa 006581
Canadian Society for Education
through Art, Société Canadienne
d'Éducation par l'Art, Thunder
Bay 058438
Canadian Society of Asian Arts,
Vancouver 007279
Canadian Society of Painters in Water
Colour, Société Canadienne de
Peintres en Aquarelle, Toronto . 058439
Canadian Transportation Museum and
Historic Village, Kingsville 006114
Canadian War Museum, Ottawa . 006582
Canadian Warplane Heritage Museum,
Mount Hope 006435
Canadian Wildlife and Wilderness Art
Museum, Ottawa 006583
Canajoharie Library and Art Gallery,
Canajoharie 047395
Çanakkale Anıtları ve Savaş Müzesi,
Çanakkale 042864
Çanakkale Devlet Güzel Sanatlar
Galerisi, Çanakkale 042865
Çanakkale Müzesi, Çanakkale . . 042866
Çanakkale Seramik Sanat Galerisi,
İstanbul 117021
Canal-de-Carillon, Saint-André-
d'Argenteuil 006846
Canal-de-Lachine – Lieu Historique
National du Canada, Lachine . 006145
Canal Fulton Heritage Society
Museum, Canal Fulton 047396
Canal Museum, Llangollen 044887
Canal Museum, Stoke Bruerne . 045880
Canale, Paris 071120
Canaletto, Warszawa 113812
Canalgrande 18, Modena 110493
Canals, Sant Cugat del Vallès . 115408
Canapé, Lisboa 084850
Canapé, Claque, Sindelfingen . . 078474
Can'Art, Toulouse 105727
Cañas, Barcelona 085755
Canasi, Ugo, Parma 081101
Canavèse, Catherine, Paris . . . 071121,
104613
Canavez Filho, Nuno, Porto . . . 143600
Canazza, Giuseppe, Padova . . . 080974
Canbaz, Faruk, Hannover 130005
Canberra Antiques Centre,
Fyshwick 061495
Canberra Bicycle Museum,
Dickson 001029
Canberra Contemporary Art Space,
Braddon 000878, 136479
Canberra Contemporary Art Space
Manuka, Manuka 001250
Canberra Museum and Gallery,
Canberra 000935
Canbulat Türbe ve Müzesi,
Gazimağusa 033451
Canby Depot Museum, Canby . . 047400
Cancelliere Candida, Messina . . 080364
Il Cancello, Firenze 080105
La Cancha de la Mamita, Santa
Cruz 063865
Canci, Lerici 110166
Cancino Chapinero, Bogotá . . . 065192
Cancino, María, Bogotá 065191
Candace, Chicago 092681
Candeias, Vítor, Lisboa 084851
De Candelaer, Rotterdam 083558
Candelaria Rug Galleries,
Albuquerque 092040
Candeloro, Carmela, Napoli . . . 131604

Candid Arts Trust, London 060085,
118243
Candido, Bilbao 085867
Cândido e Calado, José, Lisboa 084852
Candillier, André, Paris 104614
Candl, Virginia Beach 097774
Candle Lane Books, Shrewsbury 144845
Candlenut Gallery, Port Douglas 099579
Candlestick Books, Masterton . 143330
Candover Gallery, Alresford . . . 117230
Candyjacks Antique Store,
Louisville 094527
Cane & Able, Bury Saint
Edmunds 088739, 134190
Canelli, Milano 080477
Canepa, Sergio, Genova 080244
Caners & Upholders, Tarrant
Gunville 135118
Canesi, C., Milano 131407
Canessa Gallery, San Francisco 124245
Canesso, Maurizio, Paris 071122
Canestrelli, Claudia, Venezia . . 081827
Canevelli, Bernard, Genova 080245
Caney Valley Historical Society,
Caney 047401
Cangemi, Giuseppe, Roma 131838
Cango Caves Interpretive Centre,
Oudtshoorn 038793
Cangxian Fenghua Calligraphy and
Painting Gallery, Cangzhou . . 102008
Canicio, Heidelberg 141925
Canillo, Christian, Toulouse 073748,
129296
Canillo, Philippe, Toulouse 073749,
129297
Caninenberg, Horst, München . . 077475
Canivet, Jehan, Bruxelles 063363
Canivet, Stéphane, Le Mans . . . 069305
Cankarjev Dom, Ljubljana 038542
Cankarjeva Založba, Ljubljana . 137942,
143695
Çankırı Devlet Güzel Sanatlar Galerisi,
Çankırı 042868
Çankırı Müzesi, Çankırı 042869
Canmore Museum, Centennial
Museum Society of Canmore,
Canmore 005547
Cannage Antique, Peseux 087733
Cannaviello, Enzo, Milano 110275
Cannell, M.D., Raveningham . . 091072
Cannella, Catania 080025
Cannery Antiques, Nashville . . . 094981
Cannery Art Centre, Esperance . 099004
Canneva, Elio, Genova 131293
Canning A.R.T.S., Yarrawonga . 099042
Canning, Jo, Fairfield 099012
Cannington Centennial Museum,
Cannington 005548
Cannington Manor, Kenosee Lake 006079
Cannon Falls Area Historical Museum,
Cannon Falls 047402
Cannon Hall Museum, Barnsley, South
Yorkshire 043498
Cannon Village Visitor Center,
Kannapolis 049870
Cannon, Elizabeth, Colchester . . 088949
Cannonbury, Southwold 091389
Cannone, Domenico Antonio,
Bari 130910
Cano Granados, Manuel,
Granada 115083
Cano Puy, Miriam, Donostia-San
Sebastián 133244
Canolfan Y Plase, Bala 043471
Canolfan Y Plase, Y Bala 046211
Canon City Municipal Museum, Canon
City 047403
Canon España, Barcelona 114864
The Canon Gallery, Petworth . . 090992,
118910, 134938
La Canonica Antichità, Milano . . 080478
Canopia, Madrid 115164
Canova, Treviso 137754
Canova, Adriana, Martignacco . 110206

Canowindra Trading Post,
Canowindra 061276
Canso Museum, Canso 005549
Canta Attilio, Bari 079758
Cantaert, M., Brugge 063264
Cantalapiedra, Valencia 133452
Cantata Studio, Kuranda 099207
Cantelli, Prof. Patrizia, Bologna . 130945
Il Canterano di Romanazzi,
Torino 081672
Canterbury, Christchurch 112836
Canterbury, Los Angeles 121456
Canterbury Antiques, Canterbury 061277,
088773
Canterbury Antiques, Dallas . . . 093083
The Canterbury Auction Galleries,
Canterbury 126844
Canterbury Bookshop,
Canterbury 144304
Canterbury Fine Art Gallery,
Canterbury 098860
Canterbury Heritage Museum,
Canterbury 043857
Canterbury Museum,
Christchurch 033168
Canterbury Shaker Village,
Canterbury 047405
Canto, Umberto, Roma 081280
Canton Antiques, Baltimore . . . 092309
Canton Classic Car Museum,
Canton 047410
Canton Gallery, Baltimore 119730
Canton Historical Museum,
Canton 047406
The Canton Museum of Art,
Canton 047411
Canton, Agnes, Wettingen 087897
Cantor Arts Center Journal,
Stanford 139332
Cantor, Lawrence J., West
Hollywood 124989
Cantore Antichitá, Modena 080781
Cantore Galleria Antiquaria,
Modena 080782
Canty's Bookshop, Fyshwick . . . 139743
Canuit, Eindhoven 112455
Les Canularts, Lyon 144072
The Canvas Collection, Lincoln . 118117
Canvas Gallery, San Francisco . 124246
The Canvas Works, Dún
Laoghaire 109628
The Canvas Works, Wicklow . . . 109708
Canwood Museum, Canwood . . 005551
Canyan Antiques, New York . . . 095296
Canyon State Collectables,
Tucson 097642
Canzanella, Napoli 131605
Canzani Center Galleries, Columbus
College of Art and Design,
Columbus 047934
Cao Xue Qin Gallery, Liaoyang . 102145
Caoba, Almeria 085705
Caoba de Cuba, Valencia 133453
CAP Art Center and Don Sergio
Osmeña Memorabilia, Cebu . . 034456
Capa Esculturas, Bruxelles 100577
Capanema, José, Belo Horizonte 063886
Capanni, Mauro, Firenze 131140
Caparra Ruins Historical Museum,
Guaynabo 036131
Caparrini & Corzani, Firenze . . . 131141
CAPC Musée d'Art Contemporain,
Bordeaux 012108
Cape Ann Historical Museum,
Gloucester 049133
Cape Bonavista Lighthouse,
Bonavista 005438
Cape Breton Centre for Heritage and
Science, Sydney 007113
Cape Breton Miners' Museum, Glace
Bay 005878
Cape Clear Heritage Centre,
Skibbereen 024987
Cape Coast Castle Museum, Cape

Coast 022927
Cape Cod Clutter, San Diego . . . 145520
Cape Cod Museum of Natural History,
Brewster 047192
Cape Cod National Seashore Museum,
Wellfleet 054268
Cape Cod School of Art,
Provincetown 057730
Cape Fear Museum, Wilmington 054445
Cape Gallery, Byron Bay 098830
The Cape Gallery, Cape Town . . 114763
Cape Henry Lighthouse, Virginia
Beach 053995
Cape May County Historical Museum,
Cape May Court House 047420
Cape Museum of Fine Arts,
Dennis 048243
Cape of Good Hope Art Gallery,
Singapore 114503
Cape Ray Lightkeeper's House, Cape
Ray 005554
Cape Sable Historical Society Centre,
Barrington 005389
Cape Spear National Historic Site,
Saint John's 006899
Cape Town Holocaust Centre, Cape
Town 038674
Cape Vincent Historical Museum, Cape
Vincent 047422
Cape, Jonathan, London 138205
Čapek, Třebíč 102739
Čapek, Petr, Brno 128318
Y Capel, Llangollen 044888
Capel Mawr Collectors Centre,
Criccieth 144362
A Capela, São Paulo 064167
Capella Brancacci, Firenze 026241
Capelle, Nano, Cahors 067592, 067593
Capelletti, Mario, Meda 080350
Capellotti, Michele, Saluzzo . . . 081612
Ca'Pesaro Galleria Internazionale
d'Arte Moderna, Musei Civici
Veneziani, Venezia 028293
Capet, Patrick, Royan 072626
Capet, Paul, Le Gosier 130807
Capharnaum, Saint-Gilles-du-
Gard 072814
Capharnaum Ancient Synagogue, Kfar
Nahum 025118
Capia, Robert, Paris . . 071123, 128957
Capitain, Gisela, Köln 107599
Capital Antiques, San Antonio . . 096955
Capital Antiques, Washington . . 097817
Capital Art Corporation, Taipei . . 116905
Capital Bookshop, Cardiff 144309
Capital Children's Museum,
Washington 054140
Capital Corner, Bruxelles 100578
Capital Pawn & Gun, Regina . . . 064548
Capital Restoration, Saint Louis . 136196
La Capitale Galerie, Paris 104615
Capitani Art Gallery, Milano . . . 080479
Il Capitello, Torino 081673
Capitol Coin Company,
Washington 097818
Capitol Furnishings, New York . . 095297
Capitol Hill Art League,
Washington 060425
Capitol Hill Books, Denver 145123
Capitol Hill Books, Washington . 145633
Capitol Reef National Park Visitor
Center, Torrey 053780
Capitola Historical Museum,
Capitola 047423
Capitole Echange Estimation,
Toulouse 073750
Capitolium, Brescia 126187
Capitolium, Roma 081281, 081282
Capítulo Uno, Buenos Aires . . . 139556
Caplain-Matignon, Paris 104616
Caplan, Elliott City 127215
Caplow, Minneapolis 135825
Capo, Philadelphia 096089
Capo D'Arte, São Paulo 128125

Capodistria Museum, Koukouritsa,
Evropouli 023142
Capolavori, I., Seattle 124503
Capon, Patric, Bromley 088684, 134173
Capone, Pietro, Palermo 081029
Capozzi, Alberto, Genova 080246
Capozzi, Marco, Genova 080247
Cappelen, J.W., Oslo .. 137885, 143397
Cappelens, J.W., Oslo . 126368, 143398
Cappella degli Scrovegni, Padova 027142
Cappella Sistina, Sala e Gallerie
Affrescate, Musei Vaticani, Città del
Vaticano 054678
Cappellaro, Luca, Padova 131658
Cappelle Medicee, Firenze 026242
Cappelletti, Gabriele, Milano . . 110276
Cappelletti, Renzo, Perugia ... 131745
Cappelli, Bologna 142629
Cappellini Libreria, Firenze 142666
Cappellini, Leonardo, Firenze . . 131142
Cappon House Museum, Holland 049504
Capponi, Ettore & C., Brescia . 079924
Cappuzzo, Giuliano, Firenze 080106
Capra, Waldeck 108837
Caprara, Marisa, Bologna 130946
Capraro, Belisario, Milano 131408
Caprice, Bischwiller 067232
Caprice d'Antan, Saint-Ouen .. 073047
Caprices, Levallois-Perret 069445
Caprices, Paris 071124
Caprices du Temps, Angers 066602
Capricia, Le Blanc 069219
Capricious Magazine, Brooklyn . 139333
Capricorn, Smethwick 119107
Capricorn Antiques, Toronto 064622
Capricorn West, Kojonup 061699
Il Capricorno, Venezia 111096
Capricorno, Washington 124880
Capstick-Dale, Simon, New York 122450
Capsule Gallery, Cardiff 117524
CAPT Comics, Birmingham 092397
Captain Billys, Seevetal 078445
Captain Cook Birthplace Museum,
Middlesbrough 045259
Captain Cook Memorial Museum,
Whitby 046114
Captain Cook Schoolroom Museum,
Great Ayton 044469
Captain Forbes House Museum,
Milton 050904
Captain George Flavel House Museum,
Astoria 046582
Captain John Wilson House,
Cohasset 047859
Captain Nathaniel B. Palmer House,
Stonington 053546
Captain Phillips' Rangers Memorial,
Saxton 053186
Captain Salem Avery House Museum,
Shady Side 053281
Captain's Cottage Museum, Murray
Bridge 001327
Captain's Courtyard Antiques, New
Orleans 095071
Captain's House, Scarborough . . 119064
Captain's Saloon, München 077476
Captier-Barnes, Paris 071126
Captier, Bernard, Paris 071125
Captive Inspirations Photographic
Gallery, Brisbane 098783
Captured on Canvas Paintings of
Australia, Wembley Downs .. 099874
Captured Visions, Saint Paul ... 123900
Capuano, Emanuele, Napoli ... 080849
Capucine, Rochefort (Charente-
Maritime) 072495
Capus, Jean, Senningen 082388
Caputo, Bruno, Napoli 131606
Caputo, Lorenzo, Bari 079759
Caputo, Luciano, Padova 131659
Car and Carriage Museum,
Pittsburgh 052156
Car Life Museum, Charlottetown 005592
Car, Jean-Pierre, Gardanne 068634

Cara y Cruz, Madrid 086147
Carabaux, New Orleans 122141
Carabe, Luis, Sevilla 086446
Carabelli, Nino, Creazzo 109996
Carabinier, Montigny-lès-Metz . 070255
El Carabo, Valencia 143801
I Caracci Arte Antica, Casalecchio di
Reno 080012
Caracciolo, Italo, Genova 131294
Caracol Archaeological Site, Mayan
Ruins, Caracol 004176
Caractère, Paris 104617
Caractères, La Rochelle 069082
Caraffi, Maleny 099262
Carafizi, São Paulo 101087
Caral de Montety & Cornut, Aix-en-
Provence 066457
Caram, F., São Paulo 064168
Carambol, Basel 087228
El Carambolo, Sevilla 086447
Caramoor Center for Music and the
Arts, Katonah 049900
Carantec, Carantec 103528
Carápan, Monterrey 082491
Carasi, Massimo, Mantova 110201
Carat, Bordeaux 103420
Carat, Gdańsk 113460
Carati, Ambrogio, Milano 131409
Caravan, Tehrän 079421
Caravan Art, Toshkent 125037
Caravan Book Store, Los Angeles 145219
Caravanserai, Manama 100421
Caravanserai, München 077477
Caravanserail, Jounieh 082322
Caravansérail, Reims 072386
Caravati, Richmond 096532
Caravello, Beatrice, Palermo ... 110596
Caravello, Massimo & Giacomo,
Palermo 110597, 131690
Carayol, Bayonne 125506
Carayol, Biarritz 067191, 125517
Carbajosa, Valentin Fernandez,
León 086063
Carberry Plains Museum,
Carberry 005558
Carberry, Valerie, Chicago 120083
Carbethon Folk Museum and
Pioneer Village, Crows Nest,
Queensland 001004
Carbon Based Studios, Brisbane 098784
Carbon County Museum, Rawlins 052477
Carbonari, Giuseppe, Verona ... 081903
Carbonaro, Davide, Roma 081283
Carbone, Torino 081674, 132111
Carbone Rieger Fine Arts,
Toronto 101588
Carbone.to, Torino 110958
Carbonear Museum, Carbonear . 005560
Carboni, New York 122451
Carburantes Tarazona, Albacete 114825
The Carby Art Gallery, Aberdeen 117207
Carcoar Historic Village, Stoke
Stable Museum, Firearms
Technology Museum & Court House,
Carcoar 000945
Card Traders of Austin, Austin . . 144977
Cardamom Antiques, Buenos
Aires 060886
Cardana, Amersfoort 082651
Cardani, Daniel, Madrid 086148, 115165
Cardelli & Fontana, Sarzana ... 110910
Cardelli, Alessandro, Genova ... 131295
Cardellini, Roberto, Roma 131839,
131840
Cardellino, Emanuele, Torino . 081675,
137745
Cardenas Olaya, Enrique, Bogotá 065193
Cardi, Francis, L'Ile-Rousse 069469
Cardiff Antiques Centre, Cardiff . 088784
Cardiff Castle, Cardiff 043864
Cardiff Reclamation, Cardiff .. 088785
Cardiff School of Art and Design,
University of Wales Institute, Cardiff,
Cardiff 056517

Cardigan Heritage Centre,
Cardigan 043871
Cardigan Street Gallery, Carlton,
Victoria 098869
Cardilès, Jean, Onet-le-Château . 070785
Cardin, Pierre, Paris 104618
Cardinal Gallery, Annapolis 046498
Cardinal Ó Fiaich Heritage Centre,
Áras an Chairdiméil Ó Fiaich,
Cullyhanna 044054
Cardinal, Suzanne, Voiron 074270
Cardingmill Antiques, Church
Stretton 088909
Cardini, Andrea, Genova 080248
Cardini, Leon F., Los Angeles .. 135711
Cardona, Genova 080249
Cardona, Samwel, Zebbug 132346
Cardone, Roberta, Bari 079760
Cardoso, Amadeu, São Paulo .. 101088
Cardoso, Paulo Barreiros, Lisboa 132931
Cardwell Shire Museum, Tully .. 001594
Care of, Cusano Milanino 110000
Care of Collections Forum,
Grangemouth 060086
Carel, Miami 121752
Carel, Jean-Lou, Le Castellet .. 103931
Carello Frères, Nice 070582
Careof – Fabbrica del Vapore,
Milano 059423, 110277
Carette, Hervé, Linselles 069538
Caretto, Luigi, Torino 081676
Caretto, Patrizia, Torino 081677
Carew Manor and Dovecote,
Beddington 043548
Carey Museum, Serampore 024471
Ca'Rezzonico – Museo del Settecento
Veneziano, Musei Civici Veneziani,
Venezia 028294
Carfagna, Andrea, Roma 131841
Cargo, Hyères 068850
Cariano, Victoria 082445
Caribart, Amsterdam 112176
Caribbean Art Gallery, Montego
Bay 111229
Caribbean Creations, Chicago . . 120084
Caribbean Gallery, Tampa 124639
Caribbean Primate Research Center
Museum, San Juan 036151
Caricatura Galerie, Kassel 019529
Caricatura Museum Frankfurt,
Frankfurt am Main 018408
Caricyno – Gosudarstvennyj Istoriko-
architekturnyj, Chudožestvennyj i
Landšaftnyj Muzej-Zapovednik,
Moskva 037236
Carillon Antiques, Perth 062075
Carillon di Archita Calcante,
Taranto 081631
Carillon Historical Park, Dayton . 048169
Carina Prints, London 118244
Carina Straub, Bild + Rahmen, Galerie
im Künstlerhaus, Leonberg ... 107858
Carincotte, Delphine, Marseille . 104162
Carini, Pier Giuseppe, San Giovanni
Valdarno 110895
Carinthia Verlag, Klagenfurt ... 136525
Cariou, Bernard, Grenoble 140846
Cariou, Jean-Loup, Pont-Aven . 105370
Caris, Gordon, Alnwick 088254, 134021
Caris, Gordon, Hexham 089522, 134461
Carisbrooke Castle Museum, Newport,
Isle of Wight 045364
Carissa's Armoires and Antiques,
Nashville 094982
Caristi, Dominic F., Boston ... 135361
Caritas Galerie, Berlin 106195
Carivari, Leipzig 137338
Carl-Benz-Haus, Ladenburg ... 019844
Carl Bosch Museum, Heidelberg 019131
Carl Eldhs Ateljémuseum,
Stockholm 040995
Carl Freedman Gallery, London . 118245
Carl Goldmark-Gedenkhaus,
Deutschkreutz 001812

Carl-Henning Pedersen og Else Alfelts
Museum, Herning 010112
Carl-Hirnbein-Museum, Missen-
Wilhams 020421
Carl Jularbo Museum, Avesta .. 040554
Carl-Julius-Weber-Gedenkstube,
Langenburg 019903
Carl Larsson-Gården, Sundborn . 041062
Carl-Lohse-Galerie – Stadtmuseum
Bischofswerda, Bischofswerda 017399
Carl-Maria-von-Weber-Museum,
Stadtmuseum Dresden,
Dresden 017977
Carl Nielsen Museet, Odense .. 010254
Carl Nielsens Barndomshjem,
Árslev 010011
Carl Orff Museum, Dießen am
Ammersee 017888
Carl Sandburg Home, Flat Rock 048744
Carl Sandburg State Historic Site,
Galesburg 049020
Carl-Schweizer-Museum,
Murrhardt 020626
Carl Zeiss Ausstellungszentrum,
Optisches Museum Oberkochen,
Oberkochen 020901
Carl, Victor, New York 095298
Carlborg, Hans & Jennifer,
Woollahra 062462
Carle, Gerhard, Abstatt 129368
Carlén, Åmål 115575
Carlen, Ottawa 101467
Carlén, Stockholm 086983
Carlen House, Mobile 050974
Carleo, San Diego 136292
Carles, Bruno, Lunel 069660
Carleton College Art Gallery,
Northfield 051630
Carleton University Art Gallery,
Ottawa 006584
Carletti, Maurizio, Roma 131842
Carli, L'Isle-sur-la-Sorgue 069564
Carli-Livre-Brocante, Marseille . 069915,
140942
Carlier, Amiens 066567
Carlier, Saint-Germain-lès-Corbeil 105521
Carlier & Gebauer, Berlin 106196
Carlier & Imbert, Saint-Étienne . 125801
Carlier, Hélène, Condom 068127
Carlier, Jean-Edouard, Paris ... 071127,
141066
Carlimpex Reproduction, Paris . 104619,
136878
Carlin, Paris 104620
Carlina, Torino 110959
Carling Nichols, Washington ... 097819
Carlini, Gesuino, Cagliari 131064
Carlini, Marcel, Saint-Raphaël . 073287
Carlino, Las Vegas 094124
Carlisle & Finch, Cincinnati ... 092836,
135438
Carlisle Antique and Craft Centre,
Carlisle 088793
Carlisle Collection of Miniature Rooms,
Nunnington 045427
Carlo, Anita de, New York 095299
Carlos Gomes Museum, Centro
de Ciências, Letras and Artes,
Campinas 004383
Carlos P. Garcia Memorabilia,
Manila 034525
Carlos P. Romulo Collection,
Manila 034526
Carlos, Antonio, Fortaleza 100938
Carlotta, Roma 081284
Carlotte Weibulls Foklivscenter,
Åkarp 040516
Carlotti, Michel, Marseille 128755
Carlow County Museum, Carlow 024840
Carl's, Bruxelles 063364
Carl's Gallery, Killarney 109652
Carlsbad Caverns National Park,
Carlsbad 047432
Carlsbad Museum and Art Center,

Carlsbad 047433
Carlsberg Museum Valby, Valby 010349
Carlsen, Lymington . . . 090658, 118678
Carlsen, Bente, Svendborg . . . 066157
Carlskrona Antik och Konsthandel,
  Karlskrona 086774
Carlson Tower Gallery, Chicago . 047643
Carlson-Turner, Portland 145442
Carlsro Museum,
  Kristiinankaupunki 011040
Carlstads Konst och Antikhandel,
  Karlstad 086778
Carlton Antiques, Great Malvern 089370
Carlton Clocks, Amersham . . . 088270,
                                    134029
Carlton Clocks, Little Chalfont . . 089843
Carlton County Historical Museum and
  Heritage Center, Cloquet 047847
Carlton-Davidson, London . . . 089994,
                                    089995
Carlton Hobbs, London 089996
Carlton Hollis, Harrogate 089424
Carlton House Antiques,
  Wingham 091881
Carlton Rochell, New York 095300
Carlton-Smith, John, London . . 089997
Carlton Smith, M.D., London . . 118246
Carlucci, Roma 081285
Carlyle House, Alexandria 046384
Carlyle's House, London 044934
Carlyn, Dallas 120566
Carman House Museum, Iroquois 006042
Carmarthenshire County Museum,
  Carmarthen 043877
Carmel, Claremont 114779
Carmel, Constantia 114781
Carmel, Jerusalem 136185
Carmel Art Gallery, Ottawa . . . 101468
Carmelo, Santander 133424
Carmen Books, Granada 143745
Carmen Boreantiques, Vancouver 064806
Carmen Museo Max Moreau,
  Granada 039456
Carmicheal-Stewart House, New
  Glasgow 006466
Carmignani, Thierry, Marseille . . 069916
Carmine, Francis, Bern 087280
Carmo Ribeiro, Ana Maria,
  Lisboa 084853, 143561
Carmo, Marya, Porto Alegre . . . 100968
Carnaby Collection of Beetles and
  Butterflies, Boyup Brook 000877
Carnegie, New York 122452
Carnegie Art Center, Walla Walla 054037
Carnegie Art Center, Rotunda and
  Fireplace Galleries, Mankato . . 050652
Carnegie Art Museum, Oxnard . 051871
Carnegie Center for Art & History, New
  Albany 051219
Carnegie Hill Antiques, New York 095301
Carnegie Historical Museum,
  Fairfield 048671
Carnegie Magazine, Pittsburgh . 139334
Carnegie Museum, Inverurie . . 044643
Carnegie Museum of Art,
  Pittsburgh 052157
Carnegie Museum of Natural History,
  Pittsburgh 052158
Carnegie Science Center,
  Pittsburgh 052159
Carnegie, Chris, Yealmpton . . . 091968,
                                    135266
Carnegie, D., San Francisco . . 097200
El Carnero, Bogotá . . . 065194, 140446
Carnet, J.P.H., Villedieu-sur-Indre 074139
Carnevali, Carlo, Firenze 080107
Carney Gallery, Regis College Fine
  Arts Center, Weston 054348
Carney, K.M. & W.A., Eaglehawk 061425
Carnifex Ferry Battlefield State Park
  and Museum, Summersville . . 053597
The Carningli Centre, Newport,
  Dyfed 090832, 134880
Carniola Antiqua, Ljubljana . . . 085606

Carnot, Lorient 069633
Carnovale, Aurelio, Roma 081286
Carnton Plantation, Franklin . . 048939
Caro de Prielle, Bertrand,
  Bordeaux 128484
Caro, Yvan, Chézy-sur-Marne . . 067974
Carod, Santiago de Chile 101904
Caroje, Rennes (Ille-et-Vilaine) . 105430
The Carole and Barry Kaye Museum
  of Miniatures, Los Angeles . . 050413
Carolee, Saint Louis 096677
Caroli, Leonardo, Firenze 131143
Carolina Art Association,
  Charleston 060426
Carolina Augusta Art, São Paulo 064169
Carolina Böcker och Konst,
  Uppsala 087116, 143990
Carolinas Aviation Museum,
  Charlotte 047569
Caroline Wheels of Time Museum,
  Caroline 005570
Carolines Antikwaren, Nürnberg 077819
Carol's, Edmonton 064288
Carol's Antiques, Adare 079431
Carolus Rex Antikhandeln,
  Göteborg 086702
Caron, Vancouver 101755
Caron, Daniel, Montreuil-sous-
  Bois 070324
Caron, Guillaume, Lille 104003
Carosi, Aldo, Roma 131843
Carosio, Gianmatteo, Genova . . 131296
Carosso, New York 122453
Carowind Gallery, Portland . . . 123609
Carpanely, Badajoz 085709
Carpe Artem, Paris 104621
Carpe Diem, Léon 069388
Carpe Diem, Monaco 082501
Carpe Diem, Toulouse 105728
Carpe Diem Antiques,
  Philadelphia 096090
Carpe Diem Galleries, Bangkok . 116950
Carpe Diem Gallery, Wellington . 113063
Carpe Diem Konstklubb,
  Stockholm 115848
Carpenter Home Museum, Cle
  Elum 047787
Carpenter Museum of Antique
  Outboard Motors, Gilmanton . 049099
Carpenter, John, Llanelli 089868
Carpenter, Rosemary, Truro . . . 144889
Carpenter, Steve, Wisbech . . . 091884,
                                    135244
Carpentier, Paris 104622
Carpentier, Jean, Danze 068272
Carpentier, Jean, Paris 071128
Carpentier, Jean, Saint-Ouen . . 073048
Carpetbagger Aviation Museum,
  Harrington 044515
Carpi-Phil, Carpi 080011
Le Carpicorne, Saint-Paule-de-
  Vence 105603
Carpinteria Valley Museum of History,
  Carpinteria 047438
Carpinus, Bussum 143101
Carpio, Jacob, Escazú 102469
Carr, Adam, New Orleans 095072
Carr, Claudia, New York 122454
Carr, Harold J., Washington . . . 091755
Carr, John B., Saint Albans . . . 134988
Carr, Michael, Woollahra 062463
Carr, Ronald, Salisbury 091222
Carr, Virginia, Chicago 120085
Carra, Charlotte, Saint-Lattier . . 072879
Carrabba, Austin 135322
Carracci, I., Bologna 079826
Carracedo Carracedo, Benjamín,
  Barcelona 114865
Carraig, Blackrock 142596
Carranza, Madrid 086149
Carrara, Las Vegas 121345
Carrau, Gilbert, Cauneille 067733,
                                    128532
Carravieri, Giovanna, Milano . . 131410

Carré d'Ambre, Sarreguemines . 105656
Carré d'Art, Vinelz 116704
Carré d'Art, Musée d'Art
  Contemporain, Nîmes 014497
Carré d'Artistes, Aix-en-Provence 103192
Carré d'Artistes, Lyon 140073
Le Carre D'Or Antiguidades, Rio de
  Janeiro 064036
Carré Sainte-Anne, Montpellier . 104272
Le Carré-Scène Nationale, Château-
  Gontier 012480
Carré Verre, Obernai 104418
Carré, Louis, Paris 104623
Carrefour de la Création, Touvre 073869,
                                    105769
Carrefour du Meuble, Saint-Jean-
  d'Angély 072836
Carrefour du Vitrail, Lavaudieu . 013595
Carreiras, Carmen, Barcelona . . 085756
Carreño, Dunhe, Chicago 120086
Carrera, Bernard, Saint-Mariens 072928
Carrera, Federico, New York . . . 095302
Carrère, Xavier, Léon . 069389, 103980
Carrère, Xavier, Soustons 073556,
                                    105676
Carretero Garrido, Valencia . . . 133454
Carrì, Silvano, Firenze 131144
Carriage and Harness Museum of
  Murray Breweries Historic Museum,
  Beechworth 000838
Carriage Antiques, Jimboomba . 061657
Carriage House Antique,
  Jacksonville 093969
Carriage House Antiques,
  Houston 093633
Carriage House Antiques, Norfolk 095848
Carriage House Museum,
  Camden 047382
Carriage House Museum, Colorado
  Springs 047882
Carriage Museum of Bermuda, Saint
  George's 004196
Carriageway Gallery, New
  Orleans 122142
Carrick Hill Museum, Springfield 001513
Carrico, Memphis 135774
Carrie Chang Fine Arts Center,
  Tamkang University, Tamsui . . 042570
Carrie M. McLain Memorial Museum,
  Nome 051571
Carrie Weedon Natural Science
  Museum, Galesville 049023
Carrière-Lenglet, Philippe,
  Ganges 068626
Carrière Souterraine d'Aubigny,
  Taingy 015898
Carrigar, Houston 093634
Carrilhão, Belo Horizonte 063887
Carrillo Sanz, Julian, A Coruña . 115029
Carrington & Co., London 089998
Carrington-Gallagher, San
  Antonio 124020
Carrington House, Llanrwst . . . 089874
Carrion Bolivar, Carmen,
  Santander 115425
Carro, Gaetano, Modena 131566
Carroll, San Francisco 145548
Carroll County Farm Museum,
  Westminster 054340
Carroll Mansion Home – Leavenworth
  County Historical Society Museum,
  Leavenworth 050203
Carroll Reece Museum, East
  Tennessee State University, Johnson
  City 049823
Carroll, A., Saint Paul 136214
Carroll's Clocks, Saint Paul . . . 096815,
                                    136215
Carrollton Junque and Antiques, New
  Orleans 095073
Carromato de Max, Museo de
  Miniaturas, Mijas 039750
Le Carrousel, Tournai 063842
Carroza López, Jorge, Santiago de

Chile 101905
Carrozzo, Alfredo, Roma 131844
Carry A. Nation Home Memorial,
  Medicine Lodge 050760
Cars of the Stars Motor Museum,
  Keswick 044678
Carscadden's Museum, Plenty . . 006659
Carshalton Antique Galleries,
  Carshalton 088801
Carsley, Whetstone & Assoc,
  Montréal 064400
Carson, Dallas 120567
Carson, Vancouver 140422
Carson County Square House Museum,
  Panhandle 051910
Carson Pirie Scott, Chicago . . . 092682
Carson Valley Museum, Douglas
  County Historical Society,
  Gardnerville 049051
Carson, Sandy, Denver 120699
Carson, Thomas J., Denver . . . 120700
Carss Cottage Museum,
  Blakehurst 000864
Carswell, Rush & Berlin, New
  York 095303
Cart-Tanneur, Didier, Migé . . . 140956
La Carta Bianca, Genova 142693
Carte d'Arte, Bologna 138808
Cartel Fine Art, Granada 115084
Cartel Fine Arts, Málaga 115299
Carten & Assoc., Austin 119650
Carter, Atlanta 119550
Carter & Ormsby, Richmond,
  Victoria 062145, 139895
Carter Bowden Ford, Fort Worth . 093437
Carter-Coile Country Doctors Museum,
  Winterville 054503
Carter County Museum, Ekalaka 048493
The Carter House, Franklin . . . 048940
Carter, Katharine T., Kinderhook 121323
Carter, Mark, Fairford 089244
Carter, Stewart, Christchurch . . 112837
Carters, Hampton East 125198
Carter's Grove Visitor Center,
  Williamsburg 054410
Cartes Anciennes Ptolémée Plus,
  Montréal 140356
Cartesius, Trieste 111046
Cartexpo, Vulaines-sur-Seine . . 137018
Cartier-Brébeut, Lieu Historique
  National du Canada, Québec . . 006724
Cartier Co., Cairo 066234
Cartier, Paul, Zürich 116764
Cartiere Vannucci, Milano 110278
Il Cartiglio, Firenze 080108
Il Cartiglio, Torino 081678, 142853
Cartmill & Parish, Tauranga . . . 084101,
                                    132748
Carton, Karin, Versailles 074045
Carton, Michael, Dublin 130849
Carton, Taquin, Delroisse,
  Bruxelles 128021
Cartonumis, Orléans 070812
Cartoon Art Gallery, Houston . . 121014
Cartoon Art Museum, San
  Francisco 052996
Cartoon City, San Antonio . . . 096956
Cartoon Connection Studio,
  Brookvale 098805
Cartoon Factory, Salt Lake City . 123951
Cartoon Gallery, Sydney 099772, 099773
Cartoplus 21, Dijon 140831
Cartouche, Montrouge 070336
Cartref Taid, Trevelin 000654
Cartrefle Antiques, Mathry . . . 090736
Cartuja de la Asunción, Granada 039457
Cartwells of Tirau, Tirau 084113
Cartwheel, Mosta 082416
Cartwheel Antiques, Corrimal . . 061367
Cartwright Hall Art Gallery,
  Bradford 043692
Cartwright, Michael, Denmark . . 098965
Caruso, San Francisco 097201

Caruso, Antonio, Lausanne 087566, 133792
Caruso, Concetto, Catania 131083
Carvajal Cabello, Luis, Madrid .. 115166
Carvajal, Luis, Madrid 086150
Carvalho & Gil, Lisboa 084854
Carvalho, António Lima, Porto .. 085016
Carvalho, Antonio Tavares, Lisboa 143563
Carvalho, António Tavares, Lisboa 143562
Carvalho, José, Belo Horizonte . 063888
Carvalho, Julio C., Belo Horizonte 063889
Carvalho, Nilson Guimaraes, Rio de Janeiro 064037
Carver County Historical Society Museum, Waconia 054023
Carverne, Tampa 097549
Carvers & Gilders, London 134616
Carvings International, San Jose 124447
Carwick, Koroit 061702
Cary Cottage, Cincinnati 047738
Caryl House and Fisher Barn, Dover 048351
Carzaniga, Basel 116059
Casa, Bucureşti 085084
Casa Adobe de San Rafael, Glendale 049117
Casa Alegre de Sagrera, Museo de Terrassa, Terrassa 040218
Casa Amesti, Monterey 051009
Casa Anatta e Casa Selma, Percorso museale del Monte Verità, Ascona 041233
Casa Antano, Córdoba 060937
Casa Antica, Holte 065783
Casa Antica di Caruso, Catania .. 080026
Casa-Antik, Bad Honnef 074528
Casa-Antik, Kelkheim . 076742, 130109
Casa Arte Sacra, Braga 132922
Casa-Atelier Gabriel Popescu, Vulcana Pandele 036829
Casa-Atelier Gheorghe Petraşcu, Târgovişte 036746
Casa Aymat, Museu de Sant Cugat, Sant Cugat del Vallès 040082
Casa Belchior, Porto Alegre 063965
Casa Bella Collection, Valencia . 086510
Casa Berra, Buenos Aires 060887, 127518
Casa Blanca, Firenze 080109
Casa Bruzzone, Mar del Plata .. 000443
La Cas'à Bulles Martinique, Saint-Joseph 142985
Casa Buonarroti, Firenze 026243
Casa Cărţii, Iaşi 114170
Casa Ceramo, Belo Horizonte .. 063890
Casa Cercantes, Vélez Málaga . 040361
Casa Contadina, Forenza 026320
Casa Creativa, Moosburg an der Isar 077418
Casa Cuba, Miami 094710
Casa Culturale Ciseri, Ronco sopra Ascona 087758
Casa da Botica, Póvoa de Lanhoso 035963
Casa da Cerca, Centro de Arte Contemporânea, Almada 035503
Casa da Cultura, Machado 004548
Casa da Memória Arnaldo Estevão de Figueiredo, Campo Grande .. 004391
Casa Da Noal, Treviso 028183
Casa Daros, Daros-Latinamerica, Rio de Janeiro 004681
Casa d'Arte, Fortaleza 100939
La Casa d'Arte, Padova 110559, 126216
Casa d'Arte – Via dei Mercati, Vercelli 111174
Casa d'Arte Ascona, Ascona .. 116034
Casa das Artes, Brasilia 100912
Casa das Artes de Vigo, Vigo .. 040378
Casa das Artes-Galeria, São Paulo 101089

Casa d'Asta Della Rocca, Torino 081679, 126232
Casa d'Aste Bottega dell'Arte, Lugano 087608, 126740, 144100
Casa d'Aste Capitolium, Brescia 079925
Casa d'Aste Trionfante, Palermo 126217
Casa d'Aste Vincent, Napoli 126214
Casa de America, Madrid 115167
Casa de Andrés Eloy Blanco, Cumaná 054747
Casa de Antigüedades, Lima 084445
Casa de Antigüedades Los Bosquesillos, Caracas 097947
Casa de Antigüedades y Hierro Forjado, Medellín 065280
Casa de Cervantes, Esquivias .. 039385
Casa de Cora Coralina, Goiás .. 004476
Casa de Cultura Afrónio Peixoto, Lençóis 004535
Casa de Cultura João Ribeiro, Laranjeiras 004531
Casa de Deodoro, Rio de Janeiro 004682
Casa de Dulce María Loynaz, La Habana 009134
Casa de Garibaldi, Museo Histórico Nacional, Montevideo 046243
Casa de Giró, Museo Histórico Nacional, Montevideo 046244
Casa de Hidalgo, Dolores Hidalgo 030944
Casa de José Bonifácio, Rio de Janeiro 004683
Casa de José Jacinto Milanés, Matanzas 009250
Casa de la Constitución de 1814, Apatzingán 030827
Casa de la Cultura "Augusto Rivera Garces", Museo Simón Bolívar, Bolívar 008636
Casa de la Cultura Benjamín Carrión, Cuenca 010424
Casa de la Cultura, Museo Arqueológico, Museos Histórico y de la Ciudad, Curuzú Cuatiá 000351
Casa de la Panadería, Centro de Mesoneron Romanos, Madrid .. 039606
Casa de la Ribera, Peñafiel 039928
Casa de la Vall, Andorra la Vella 000085
Casa de las Américas, La Habana 058528
Casa de las Ciencias, A Coruña 039314
Casa de Lavalleja, Museo Histórico Nacional, Montevideo 046245
Casa de los Abuelos, Museo Folclórico, Sonsón 008800
Casa de los Arabes, La Habana 009135
Casa de los Cinteros, Villar del Arzobispo 040411
Casa de los Morlanes, Zaragoza 040442
Casa de los Tiros, Granada 039458
La Casa de los Tratados, Girón . 010441
Casa de Manuel Mujica Lainez, La Cumbre 000399
Casa de Marinos, Aranjuez 038979
Casa de Miguel Failde Pérez, Matanzas 009251
Casa de Oliveira Vianna, Niterói 004588
Casa de Osório, Rio de Janeiro . 004684
La Casa de Palermo, Buenos Aires 098557
La Casa de Pandora, Lima 084446
Casa de Pilatos, Sevilla 040164
Casa de Subastas de Madrid, Madrid 126530
Casa de Unidad, Detroit 048293
Casa de Usados, Medellín 065281
Casa de Vacas, Madrid 115168
Casa de Ximénez, Museo Histórico Nacional, Montevideo 046246
Casa del Águila y La Perla, Museo Regional, Santillana del Mar .. 040136
Casa del Arte de la Universidad de Concepción, Concepción 007517
La Casa del Balcón, San Isidro . 060958

Casa del Boccaccio, Certaldo .. 025909
La Casa del Coleccionista, Vitoria-Gasteiz 086573
La Casa del Collezionista, Roma 081287
Casa del Cordón, Vitoria-Gasteiz 040423
La Casa del Cuadro, Lima 084447, 084448, 132806, 132807
Casa del Escribano Real Don Juan de Vargas, Tunja 008811
Casa del Gatto, Brione 116167
Casa del Rey IV, Vitoria-Gasteiz . 086574
Casa della Nonna, Cairate 025684
Casa dell'Ariosto, Ferrara 026203
Casa Delmonte, Palma de Mallorca 086345
Casa Depa Antigüedades, Lima . 084449
Casa d'Esper, Miami 094711
Casa di Carlo Goldoni, Musei Civici Veneziani, Venezia 028295
Casa di Giorgione, Castelfranco Veneto 025824
Casa di Goethe, Roma 027556
Casa di Lazzaro Spallanzani, Scandiano 027897
Casa di Luigi Pirandello, Roma . 027557
Casa di Petrarca, Arquà Petrarca 025378
Casa di Pirandello, Agrigento .. 025251
Casa di Xilogravura, Campos de Jordão 004396
Casa do Brazão, Faro 084799
Casa do Corpo Santo, Museu do Barroco, Setúbal 036026
Casa do Escultor Ervin Curt Teichmann, Pomerode 004637
Casa do Livro Azul, São Paulo . 140302
Casa do Livro Usado, Fortaleza . 140278
Casa Domnească Brebu, Brebu . 036276
Casa Dorita Castel-Branco, Sintra 036035
Casa dos Livros Usados, Fortaleza 140279
Casa dos Patudos – Museu de Alpiarça, Alpiarça 035514
Casa dos Presentes, Lisboa ... 084855
Casa dos Sete Candeeiros, Salvador 004785
Casa Dr. José Celso Barbosa, Instituto de Cultura Puertorriqueña, Bayamón 036120
Casa e Chiesa, Prato 131773
La Casa Elizalde, Barcelona .. 114866
Casa España, Makati 113383
Casa Española, Buenos Aires .. 060888
Casa Galería Marcos J. Alegría, Dorado 036129
Casa Gorordo Museum, Cebu .. 034457
Casa Goyeneche, Colección del Banco de Crédito del Perú, Lima .. 034303
Casa Grande Ruins, Coolidge .. 047970
Casa Grande Valley History Museum, Casa Grande 047453
Casa International, Osaka 082080
Casa Jesús T. Piñero, Canóvanas 036126
Casa João Turin, Curitiba 004425
Casa Leopardi, Recanati 027488
Casa Lines, Buenos Aires 060889
Casa Loma, Toronto 007166
Casa Luis Barragán, México ... 031146
Casa Manana, Waikato 084125
Casa Manila Museum, Manila .. 034527
Casa Maria Esclasans, Barcelona 085757
Casa Martirilor de la Moisei, Moisei 036573
Casa Memória d' Alte, Alte 035515
Casa Memorială a Poetului Ion Şiugariu, Băiţa 036225
Casa Memorială Alexandru Sahia, Mănăstirea 036553
Casa Memorială Alexandru Vlahuţă, Agapia 036191
Casa Memorială Aloisie Tăutu, Valea Vinului 036810
Casa Memorială Anton Pann, Râmnicu Vâlcea 036643
Casa Memorială Benedek Elek, Bățanii

Mici 036237
Casa Memorială Cezar Petrescu, Buşteni 036331
Casa Memorială Constantin Joja, Bucureşti 036278
Casa Memorială Costache Negruzzi, Hermeziu 036491
Casa Memorială Crişan, Crişan . 036423
Casa Memorială D.P. Perpessicius, Brăila 036260
Casa Memorială Dinu Lipatti, Leordeni 036532
Casa Memorială Dr. Vasile Lucaciu, Apa 036202
Casa Memorială Ecaterina Teodoroiu, Târgu Jiu 036756
Casa Memorială Elena Cuza, Soleşti 036726
Casa Memorială Elena Farago, Craiova 036414
Casa Memorială Emil Racoviţă, Dăneşti 036431
Casa Memorială Eusebiu Camilar, Udeşti 036806
Casa Memorială Gala Galaction, Bucureşti 036279
Casa Memorială George & Agatha Bacovia, Bucureşti 036280
Casa Memorială George Bacovia, Bacău 036211
Casa Memorială George Enescu, Liveni 036538
Casa Memorială George Enescu, Sinaia 036711
Casa Memorială George Pop de Băseşti, Băseşti 036236
Casa Memorială George Topârceanu, Iaşi 036503
Casa Memorială Gheorghe Dima, Braşov 036267
Casa Memorială Gheorghe Ionescu-Şişeşti, Şişeşti 036718
Casa Memorială Gheorghe Rădulescu-Motru, Butoieşti 036333
Casa Memorială Gyarfas Jenö, Sfântu Gheorghe 036692
Casa Memorială I.C. Vissarion, Costeştii din Vale 036410
Casa Memorială I.G. Duca, Horezu 036494
Casa Memorială Ion Creangă, Târgu Neamţ 036766
Casa Memorială Ion Minulescu & Claudia Millian, Bucureşti 036281
Casa Memorială Ion Popescu, Voiteşti 036827
Casa Memorială Ionel Perlea, Ograda 036591
Casa Memorială Liviu & Fanny Rebreanu, Bucureşti 036282
Casa Memorială Liviu Rebreanu, Ştefăneşti 036727
Casa Memorială Lucian Blaga, Lancrăm 036527
Casa Memorială Mihai Codreanu, Iaşi 036504
Casa Memorială Mihail Sadoveanu, Fălticeni 036457
Casa Memorială Mihail Sadoveanu, Iaşi 036505
Casa Memorială Mihail Sadoveanu, Vânători-Neamţ 036816
Casa Memorială Moş Dumitru, Baldovineşti 036226
Casa Memorială Moş Ion Roată, Câmpuri 036350
Casa Memorială N.D. Cocea, Sighişoara 036709
Casa Memorială Nichita Stănescu, Ploieşti 036621
Casa Memorială Nicolae Gane, Muzeului Literaturii Române, Iaşi 036506
Casa Memorială Nicolae Iorga,

Muzeului Judeţean de Istorie
Botoşani, Botoşani 036253
Casa Memorială Nicolae Labiş,
Mălini 036551
Casa Memorială Nicu Enea,
Bacău 036212
Casa Memorială Nikolaus Lenau,
Lenauheim 036531
Casa Memorială Otilia Cazimir,
Iaşi 036507
Casa Memorială Panait Cerna,
Cerna 036365
Casa Memorială Panait Istrati,
Brăila 036261
Casa Memorială Simion Florea Marian,
Suceava 036733
Casa Memorială Tamasi Aron,
Lupeni 036545
Casa Memorială Theodor Bălăşel,
Ştefăneşti, Vâlcea 036729
Casa Memorială Traian Demetrescu,
Craiova 036415
Casa Memorială Tudor Arghezi,
Bucureşti 036283
Casa Memorială Tudor Vladimirescu,
Vladimir 036826
Casa Memorială Vasile Alecsandri,
Mireşti 036569
Casa Memorială Vasile Pogor,
Iaşi 036508
Casa Memorială Vasile Voiculescu,
Pârscov 036604
Casa Memorială Veronica Micle, Târgu
Neamţ 036767
Casa Mexicana, Dallas 120568
Casa Mia, Herrstein 076511
Casa Milchstrasse, Livorno ... 080319
Casa Mitropolitului Dosoftei, Muzeului
Literaturii Române, Iaşi ... 036509
Casa Munoa, Donostia-San
Sebastián 085971
Casa-Musei de Aquilino Ribeiro,
Peva 035913
Casa Museo 20 de Julio de 1810,
Bogotá 008583
Casa Museo A. Gramsci, Ghilarza 026435
Casa Museo A. Uccello, Palazzolo
Acreide 027170
Casa Museo Alberto Sols, Sax . 040149
Casa Museo Alfonso Ariza, La
Rambla 039992
Casa Museo Almirante Miguel Grau,
Piura 034383
Casa Museo Amado Nervo, Tepic 031549
Casa Museo Ángel Barrios,
Granada 039459
Casa-Museo Antonio Machado,
Segovia 040153
Casa Museo Antonio Nariñtildeo y
Álvarez, Villa de Leyva 008819
Casa Museo Antonio Padrón,
Gáldar 039420
Casa-Museo Azorín, Monóvar . 039770
Casa Museo Blasco Ibáñez,
Valencia 040299
Casa Museo Cadete Juan Escutia,
Tepic 031550
Casa Museo Carlos Pellicer Cámara,
Villahermosa 031656
Casa Museo Concha Piquer,
Valencia 040300
Casa Museo Coronel Gregorio Méndez
Magaña, Jalpa de Méndez .. 031078
Casa Museo Curros Enríquez,
Celanova 039261
Casa Museo da Fundação Camilo
Josè Cela, Padrón 039866
Casa Museo de Cervantes,
Valladolid 040332
Casa Museo de Colón, Valladolid 040333
Casa Museo de Goya, Museo del
Grabado, Fuendetodos 039414
Casa Museo de Juan Zorrilla de San
Martín, Museo Histórico Nacional,

Montevideo 046247
Casa-Museo de Lope de Vega,
Madrid 039607
Casa Museo de Rafael Alberti, El
Puerto de Santa María 039981
Casa Museo de Unamuno, Antigua
Casa Rectoral, Salamanca .. 040037
Casa Museo de Zorrilla,
Valladolid 040334
Casa-Museo de Zurbarán, Fuente de
Cantos 039416
Casa Museo dei Puccini,
Pescaglia 027271
Casa Museo del Banco Nacional,
Ciudad de Panamá 034172
Casa Museo del Cima,
Conegliano 026057
Casa Museo dell'Alta Valle del Cervo,
Rosazza 025933
Casa Museo Depero, Museo d'Arte
Moderna e Contemporanea di Trento
e Rovereto, Rovereto 027681
Casa Museo di A. Oriani, Casola
Valsenio 025801
Casa Museo di Monticello Amiata,
Cinigiano 025990
Casa Museo Emilia Pardo Bazán, Real
Academia Galega, A Coruña . 039315
Casa Museo F. Scaglione,
Sciacca 027909
Casa Museo Fantoni, Rovetta .. 027686
Casa Museo Federico García Lorca,
Fuente Vaqueros 039418
Casa Museo Fernando Fader,
Ischilín 000389
Casa Museo Francesco Baracca,
Lugo 026650
Casa Museo Francesco Torregrosa,
Muro de Alcoy 039811
Casa Museo G. de Chirico, Roma 027558
Casa Museo G. Mazzarino,
Pescina 027285
Casa Museo Gene Byron,
Guanajuato 031021
Casa Museo General San Martín,
Cervatos de la Cueza 039265
Casa Museo Giovanni Verga,
Catania 025866
Casa Museo Ignacio Zuloaga,
Zumaia 040467
Casa Museo Ivan Bruschi, Arezzo 025356
Casa Museo Joaquín Torrents Llado,
Palma de Mallorca 039880
Casa-Museo Jorge Eliécer Gaitán,
Bogotá 008584
Casa-Museo José Benlliure,
Valencia 040301
Casa Museo José Carlos Mariátegui,
Lima 034304
Casa Museo José Clemente Orozco,
Guadalajara 030995
Casa Museo José María de Cossío,
Casona de Tudanca, Tudanca . 040276
Casa Museo José Padilla, Madrid 039608
Casa Museo Julia Codesido,
Lima 034305
Casa Museo Julián Gayarre,
Roncal 040016
Casa Museo La Moreña, La
Barca 031095
Casa-Museo La Sebastiana, Fundación
Pablo Neruda, Valparaíso ... 007555
Casa Museo León y Castillo,
Telde 040216
Casa Museo Leonardo Valdez Esquer,
Etchojoa 030975
Casa Museo López Portillo,
Guadalajara 030996
Casa Museo M. Moretti,
Cesenatico 025933
Casa Museo Maestro Manuel
Altamirano, Tixtla de Guerrero . 031570
Casa-Museo Manuel de Falla,
Granada 039460

Casa Museo María Augusta Urrutia,
Quito 010469
Casa Museo Mario Urteaga,
Cajamarca 034227
Casa-Museo Menéndez Pelayo,
Santander 040112
Casa Museo Miguel Grau, Lima 034306
Casa Museo Miguel Hernández,
Orihuela 039843
Casa Museo Modernista, Novelda 039820
Casa Museo Nuñez del Prado, La
Paz 004208
Casa Museo Páez, Valencia 054778
Casa Museo Pavesiano, Santo Stefano
Belbo 027846
Casa-Museo Pedralba 2000,
Pedralba 039923
Casa Museo Pedro Muñez Seca, El
Puerto de Santa María 039982
Casa Museo Pérez Galdós, Las Palmas
de Gran Canaria 039899
Casa Museo Piccione d'Avola, Geraci
Siculo 026431
Casa Museo Pinazo, Godella ... 039452
Casa Museo Posada del Moro,
Torrecampo 040258
Casa Museo Quevedo Zornoza,
Museo Nacional de Colombia,
Zipaquirá 008829
Casa Museo Quinta de Bolívar,
Bogotá 008585
Casa Museo R. Bendandi,
Faenza 026169
Casa Museo R. Siviero, Firenze . 026244
Casa Museo Ricardo Gomez
Campuzano, Bogotá 008586
Casa Museo Ricardo Palma,
Lima 034307
Casa Museo Rosalía de Castro,
Padrón 039867
Casa Museo Sa Domu Antiga,
Santadi 027826
Casa Museo Signorini-Corsi,
L'Aquila 026559
Casa Museo Silvestre Rodríguez,
Nacozari de García 031318
Casa Museo Tiu Virgiliu, Suni . 028036
Casa-Museo Tomás Morales,
Moya 039791
Casa Museo V. Cermignani,
Giulianova 026447
Casa Museo Valle Inclán, La Pobra do
Caramiñal 039951
Casa Museo Vladimir Cora,
Acaponeta 030799
Casa Museo Zenobia y Juan Ramón
Jiménez, Moguer 039756
Casa Museu Abel Salazar da
Universidade do Porto, São Mamede
de Infesta 036014
Casa Museu Àngel Guimerà, El
Vendrell 040364
Casa Museu Atelier Henrique Medina,
Esposende 035658
Casa Museu Bissaya-Barreto,
Coimbra 035617
Casa-Museu Castell Gala Dalí, La
Pera 039938
Casa Museu Cristóvão Colombo, Porto
Santo 035962
Casa Museu da Freira, Penacova 035906
Casa Museu da Fundação Dionísio
Pinheiro e Alice Pinheiro,
Águeda 035476
Casa-Museu da Fundação Medeiros e
Almeida, Lisboa 035752
Casa Museu da Ordem Terceira de
São Francisco, Ovar 035889
Casa-Museu d'Alfàbia, Bunyola-
Mallorca 039164
Casa Museu de Aljustel, Fátima 035691
Casa-Museu de Almeida Moreira,
Viseu 036114
Casa-Museu de Camilo, Vila Nova de

Famalicão 036095
Casa-Museu de Francisco Ernesto
de Oliveira Martins, Angra do
Heroísmo 035528
Casa Museu de Monsenhor Ribeiro de
Almeida, Lamego 035746
Casa Museu de Penacova,
Penacova 035907
Casa-Museu de Salvador Dalí de
Cadaqués, Cadaqués 039186
Casa Museu de Sarah Beirão,
Tábua 036050
Casa-Museu del Poeta Verdaguer,
Folgueroles 039402
Casa Museu do Lorvão, Lorvão . 035827
Casa Museu Dom Hugo, Porto . 035927
Casa-Museu Dr. Anastácio Gonçalves,
Lisboa 035753
Casa-Museu Egas Moniz,
Estarreja 035661
Casa Museu Fernando de Castro,
Porto 035928
Casa Museu Fernando Namora,
Condeixa-a-Nova 035638
Casa-Museu Ferreira de Castro,
Ossela 035886
Casa Museu Frederico de Freitas,
Funchal 035700
Casa-Museu Gaudí, Barcelona . 039054
Casa Museu Guerra Junqueiro,
Porto 035929
Casa-Museu João Soares, Cortes 035643
Casa Museu José Furtado de
Mendonça, Santo Estevão ... 036007
Casa-Museu José Régio,
Portalegre 035923
Casa Museu José Régio, Vila do
Conde 036079
Casa Museu Manuel Nunes Corrêa,
Pedrógão Grande 035903
Casa-Museu Maria da Fontínha,
Gafanhão 035715
Casa Museu Marieta Solheiro
Madureira, Estarreja 035662
Casa Museu Marta Ortigão Sampaio,
Porto 035930
Casa Museu Mestre João da Silva,
Lisboa 035754
Casa Museu Pintor José Cercas,
Aljezur 035497
Casa-Museu Prat de la Riba,
Castellterçol 039252
Casa Museu Rafael Bordalo Pinheiro,
Life and work of R. Bordalo Pinheiro,
his ceramics, Caldas da Raínha 035586
Casa Museu Roque Gameiro,
Amadora 035520
Casa-Museu Sant Enric d'Ossó,
Jesús 039532
Casa Museu Segrelles, Albaida . 038906
Casa-Museu Teixeira Lopes, Vila Nova
de Gaia 036101
Casa-Museu Torres Amat, Sallent 040050
Casa Museu Vasco Lima Couto,
Constância 035640
Casa Museu Vieira Natividade,
Alcobaça 035482
Casa Museu Vitorino Ribeiro,
Porto 035931
Casa-Muzeu Constantin Brâncuşi,
Hobiţa 036062
Casa-Muzeu Cristian, Cristian . 036424
Casa-Muzeu Solca, Solca 036725
Casa Muzicii, Lugoj 036540
Casa Nacional de la Moneda,
Fundación Cultural del Banco Central
de Bolivia, Potosí 004234
Casa Nadiua de Pau Casals, El
Vendrell 040365
Casa Natal de Aquileo Parra,
Barichara 008574
Casa Natal de Jesús Montané
Oropesa, Nueva Gerona ... 009269
Casa Natal de José Antonio Ramos

Sucre, Cumaná 054748
Casa Natal de José María España, La
Guaira 054759
Casa Natal de Legazpi,
Zumárraga 040468
Casa Natal de Regino Botí,
Guantánamo 009129
Casa Natal de San Ignacio,
Azpeitia 039029
Casa Natal de Sant Antoni María
Claret, Sallent 040051
Casa Natal de Santa Teresa de Jesús,
Ávila 039019
Casa Natal de Santiago Ramón y
Cajal, Petilla de Aragón 039940
Casa Natal de Sarmiento, San
Juan 000585
Casa Natal del Hermano Francisco
Gárate, Azpeitia 039030
Casa Natal del Libertador Simón
Bolívar, Caracas 054707
Casa Natal del Pintor Wifredo Lam,
Santa Clara 009331
Casa Natale di A. Rosmini,
Rovereto 027682
Casa Natale di Giosuè Carducci,
Pietrasanta 027317
Casa Natale di Giuseppe Verdi,
Busseto 025662
Casa Natale di Leonardo da Vinci,
Vinci 028398
Casa Natale di Pascoli, San Mauro
Pascoli 027782
Casa Natale di Pio V, Bosco
Marengo 025610
Casa Natale di Raffaello, Urbino 028237
Casa Natale di Toscanini, Parma 027203
Casa Natale di V. Monti,
Alfonsine 025288
Casa Natale Gaetano Donizetti,
Bergamo 025491
Casa Navarro, San Antonio 052935
Casa-Oficina de António Carneiro,
Porto 035932
Casa Pedro Domingo Murillo, La
Paz 004209
Casa Personalităților, Lugoj 036541
Casa Piú, Caracas 097948
Casa Postal, Madrid 086151
Casa Quinta de José Batlle y
Ordóñez, Museo Histórico Nacional,
Montevideo 046248
Casa Quinta de Luis Alberto de
Herrera, Museo Histórico Nacional,
Montevideo 046249
Casa Ramón Power y Giralt, San
Juan 036152
Casa Real, São Paulo 064170
Casa Real Shrine, Malolos 034512
Casa Rocca Piccola, Valletta . . . 030750
Casa Roche Pombo, Morretes . . 004577
Casa Romei, Ferrara 026204
Casa Ros, Ourense 086320
Casa Salazar, San Antonio 124021
Casa Salsa, Cancún 132358
Casa San Isidro, Ciudad Bolívar 054743
Casa Santuario de San Pedro Claver,
Verdú 040370
Casa Schlatter, Firenze 080110
Casa Sheik, Porto Alegre 063966
Casa Siglo XIX, Fundación Sebastián,
Chihuahua 030874
Casa Tait, Porto 035933
Casa Taller de Arte, Medellín . . 102430
Casa Velha Antiguidades, Belo
Horizonte 063891
Casa Velha Antiguidades, Brasília 128104
La Casa Verde, Cancún 082460
Casa Vicente, Donostia-San
Sebastián 085972
Casa Victoria, Los Angeles 094280
Casa Wolf, Firenze 080111
Casa y Museo de El Greco,
Toledo 040231

Casablanca Antique Lamps, New
York 095304, 135895
Casablanca Bookshop, Kitchener 140354
Casadella, Ensie, Solliès-Pont . 073535
Casadio, Diletta, Roma 081288
Casado, José, Collioure 068076
Casagrande, Giampiero, Lugano 138065
Casagrande, Luciano, Roma . . . 081289
Casagrande, Mario,
Braunschweig 075120
Casagrande, Massimiliano, Roma 131845
Casagrande, Pino, Roma 081290
Casagwa National Museum,
Daraga 034474
Casaj, Oakland 095886
Casal Balaguer, Palma de
Mallorca 115336
Casal Solleric, Centre d'Exposicions i
Documentació de l'Art Contemporai,
Palma de Mallorca 039881
Il Casale dell'Antico, Roma 081291
Il Casale dell'Arte, Catania 080027,
131084
Casalini Libri, Fiesole 142660
La Casamaures, Saint-Martin-le-
Vinoux 015457
Casamonti, Roberto, Crans-
Montana 116231
Casanelly, Lima 084450
Casanova, Alcino F.M., Porto . . . 132969
Casanova, Alcino F.M., Vila Nova de
Gaia 132984
Casar Herrero, Santiago,
Santander 115426
Casarão Azul Antiguidades, Porto
Alegre 063967
Casares, Alberto, Buenos Aires . 139557
Casariego, Madrid 086152
Casartelli, Massimo, Modena . . . 080783
Casas Adobe's Collectibles,
Tucson 097643
Casas Estrade, Barcelona 114867
Casatelier Restauração e Arte, Rio de
Janeiro 128111
Casati, Alessandro, Milano 080480
Casati, Giorgio, Milano 080481
Casay, Anthony, San Diego 124110
Casbah, Amsterdam 112177
Cascabel, Oslo 132784
Cascade, Arnhem 082936
Cascade, Casablanca 082553
Cascade Blue Antiques,
Blackheath 061160
Cascapedia River Museum,
Cascapedia-Saint-Jules 005575
Cascar Art, Bremen 106576
Cascianelli, Roma 142802
Casciani, Firenze 131145
Casciello, Domenico, Paris 071129,
071130
Casciello, Domenico, Saint-Pryvé 073271,
129244
Cascina Museo della Civiltà Contadina,
Cremona 026097
Cascina San Vitale, Osmate . . . 110550
Cascione, Olimpia, Bologna 079827
Casconi, Mario, Roma 131846
Le Case d'Arte, Milano 110279, 110280
Casè, Fabio, Milano 131411
Casebow, Peter G., Worthing . . 135262
Casell, New Orleans 122143
Casella, Napoli 142757
Casella, Valencia 133455
Casella, West Hollywood 097878
Casella, Gianmaria, Brescia . . . 131024
Caselli, Elio, Napoli 080850
Caselli, Lucia, Firenze 080112
Casemate Museum, Fort Monroe 048847
Casements, Swinton 135115
Caseneuve, Christiane, Aurillac . 103293
Caseta de Balma, Castello de la
Plana 085931
Caseta de la Musica, Madrid . . . 086153
Casetti, Giuseppe, Roma 142803

Casetti, Hélène, Cannes 067631
Casey, Clonmel 130836
Casey, Jacksonville 093970
Casey, Las Vegas 094125
Casey & Assoc., San Francisco . 124247
Casey, Toledo 097601, 136405
Casey Farm, Saunderstown 053168
Casey Jones Home and Railroad
Museum, Jackson 049759
Casey Jones Museum, Vaughan 053954
Casey Kaplan, M., New York . . . 122455
Casey, William J., Warwick 091742
Cash America, Portland 096377
Cash Paid, El Paso 093394
Cashan, Sharjah 088196
Cashel Antiques, Cashel 079471
Cashi, Denver 093269
Cashmere Museum & Pioneer Village,
Cashmere 047454
Casho, Cairo 103027
Casillas 1 Art-Studio, El Paso . . 120825
Casimir, Tournai 063843
Casimir, Jack, London 089999
Casini, Philippe, Paris 104624
Casino, Saint-Céré 105497
Casino dell'Aurora Pallavicini,
Roma 027559
Casino Luxembourg – Forum d'Art
Contemporain, Luxembourg . . 030585
Casino Restorations, Saint Paul . 136216
Casita Antiques, New York 095305
Casita del Príncipe, Madrid 039609
Casnica, Cluj-Napoca 085114
Caso, København . . . 065880, 128395
Caso, Osaka 111348
Casón del Buen Retiro, Madrid . 039610
Caspar-David-Friedrich-Institut, Ernst-
Moritz-Arndt-Universität Greifswald,
Greifswald 055550
Casper, Borås 115586
Casperle, Bilbao 085868
Caspian Gallery, Paddington, New
South Wales 099516
Caspian Oriental Rugs, Chicago . 092683
Casque, Tokyo 111412
Casque and Gauntlet Militaria,
Farnham 089261, 134385
Cass County Historical Society
Museum, Logansport 050358
Cass County Historical Society
Museum, Plattsmouth 052200
Cass County History Museum at
Bonanzaville, West Fargo 054298
Cass County Museum, Walker . . 054036
Cass Freshford, Freshford 079621
Cass-Garr, New Orleans 095074
Cass Sculpture Foundation,
Goodwood 044445
Cass, Maria, Nayland . 090801, 134865
Cassagnaud, Joël, Auch 066799
Cassagne, Corinne, Romans-sur-
Isère 125791
Cassagne, Joël, Langon 069171
Cassagnes, Bernard, Brousses-et-
Villaret 067518
Cassandra Antiquités, Marseille . 069917
Cassandra's Art, Atlanta 119551
Cassandre, Saint-Raphaël 073288,
141340
Cassaro, Giovanni, Torino 132112
Cassetti, Dr. Giuliana, Pisa 131763
Il Cassetto, Taranto 081632
Il Cassetto della Nonna, Torino . 081680
Cassettone, Bologna 130947
Cassi, Marco, Firenze 080113
Cassia County Museum, Burley . 047314
Cassian de Vere Cole, London . . 090000
Cassidy, London 144573
Cassidy, Warminster . 091731, 135191
Cassidy, Barry, Sacramento 145487
Cassidy, P.J., Holbeach 144480
Cassimus, Angelo, New Orleans 095075
Cassina, Diego, Lugano 116453
Cassinis, Amedeo, Roma 131847

Cassiodor, Gauting 141823
Cassiopée, Paris 071131
Cassiopeia, Chicago 092684
Cassissium, Nuits-Saint-Georges 014536
Cassius Forum Bonn, Forum für religiöse
Kunst und Begegnung, Bonn . . 017486,
058811
Casson, Guillaume de, Paris . . . 104625
Castagner, Agathe, La Garenne-
Colombes 069048
Castaing, Frédéric, Paris 141067, 141068
Castaing, Hugues, La Rochelle . 069083
Castaing, Hugues, Le Bois-Plage-en-
Ré 069223
Castan, Echternach 111824
Castan, Perpignan 105335
Castan, Nellie, Melbourne 061815
Casteel-Linn House and Museum,
Princeton 052363
Castel Beseno, Besenello 025502
Castel Prösels, Fié 026223
Castel Sant'Elmo, Napoli 026987
Castel, Jeanne, Paris 104626
Castel'Art, Le Castellet 103932
Castelin & Lhote, Dunkerque . . . 068414
Castell Cartoixa Vallparadis, Museu de
Terrassa, Terrassa 040219
Castell y Waun, Chirk 043952
Castella-Ayral, Toulouse 073751
La Castellana, Palermo 081030
Castellana Serrano, Madrid 126531
Castellana Subasta, Madrid 126532
Castellana Subastas Barcelona,
Madrid 126533
Castellani Art Museum, Niagara
University, Niagara University . 051560
Castellani, Jean-Marie, Aix-en-
Provence 103193
Castellano Olmedo, Millan,
Santander 086422
Castellano, Claudio, Torino 081681
Castelli, Leo, New York 122456
Castellini, Brescia 079926
Il Castello, Adria 109737
Il Castello, Milano 110281
Castelló 120, Madrid 115169
Castello dei Burattini – Museo
Giordano Ferrari, Musei Civici di San
Paolo, Parma 027204
Castello dei Visconti di San Vito,
Somma Lombardo 027979
Castello del Buonconsiglio, Monumenti
e Collezioni Provinciali, Trento . 028169
Castello della Manta, Manta . . . 026695
Castello di Costigliole d'Asti – Museo
Nazionale del Territorio, Costigliole
d'Asti 026092
Castello di Donnafugata, Ragusa 027471
Castello di Giulio II, Ostia Antica 027126
Castello di Masnago, Musei Civici,
Varese 028276
Castello di Montebello, Museo
Archeologico e Civico,
Bellinzona 041297
Castello di Rivoli, Museo d'Arte
Contemporanea, Rivoli 027537
Castello di Sasso Corbaro,
Bellinzona 041298
Castello di Stenico, Stenico 028014
Castello d'Issogne, Issogne 026524
Castello Ducale, Agliè 025246
Castello e Museo Civico F. Baldinucci,
Lari 026526
Castello e Raccolta Archeologica,
Montechiarugolo 026928
Castello Fénis, Fénis 026194
Castello Malaspina, Bobbio 025524
Castello Malaspina, Massa 026725
Castello Pallotta, Caldarola 025691
Castello Principesco, Merano . . 026761
Castello Reale, Racconigi 027466
Castello Reale, Venaria 028291
Castello Reggia di Venaria Reale,
Torino 028112

Castello, G. & G. Cerato,
  Bevilacqua 109807
Castellon, Adora, Tresses 105771
Castells & Castells, Montevideo 127110
Castellum, Bovenden 075111
Castellvi, Maria Jose, Barcelona 114868
Castelnuovo, Tedesco Giovanni,
  Firenze 080114
Castelul Regal Săvârşin, Săvârşin 036685
Castera, Bordeaux 140789
Casteran, Eric, Toulouse 105729
Casterman, Bruxelles 136632
Castiaux, France, Mairieux 069811
Castiglione, Paris 071132
Castiglione, Saint-Paul-de-Vence 105578
Castiglione, Pasquale, Napoli . 131607
Castillo, Lima 113345
El Castillo, Santiago de Chile . 064960,
                                  128248
Castillo de Javier, Javier 039526
Castillo de Loarre, Loarre 039593
Castillo de los Tres Reyes del Morro,
  La Habana 009136
Castillo de San Felipe del Morro, San
  Juan 036153
Castillo de San Marcos Museum, Saint
  Augustine 052743
Castillo de Santa Catalina, Cádiz 039188
Castillo de Villasobroso,
  Mondariz 039763
Castillo, Aldo, Chicago 120087
Le Castillon, Toulon (Var) 129285
Castillos de Guayana la Vieja,
  Guayana 054756
Castine Scientific Society Museum,
  Castine 047462
The Castle, Marietta 050679
Castle Air Museum, Atwater ... 046646
Castle and Regimental Museum,
  Monmouth 045287
Castle Antiques, Burnham-on-
  Sea 088725
Castle Antiques, Deddington .. 089047
Castle Antiques, Deganwy 089050
Castle Antiques, Downham
  Market 089104
Castle Antiques, Edinburgh ... 089166
Castle Antiques, Gent 063632
Castle Antiques, Nottingham .. 090890
Castle Antiques, Orford 090918
Castle Antiques, Vancouver ... 064807
Castle Antiques, Waitara 062400
Castle Antiques, Warwick 091743
Castle Antiques Centre,
  Middleham 090754
Castle Antiques Centre,
  Westerham 091797, 135209
Castle Arte e Antiguidades, Rio de
  Janeiro 064038
Castle Arts, Canterbury 117517
Castle Ashby Gallery, Castle
  Ashby 088806, 117540, 134221
Castle Book's, Edinburgh 144390
Castle Bookshop, Colchester .. 144347
Castle Brand Visitor Centre,
  Nenagh 024972
Castle Cary and District Museum,
  Castle Cary 043882
Castle Clinton, New York 051351
Castle Close Antiques, Dornoch . 089098
Castle Combe Museum, Castle
  Combe 043883
Castle Coole, Enniskillen 044294
Castle Cornet Military and Maritime
  Museums, Saint Peter Port ... 045710
Castle Douglas Art Gallery, Castle
  Douglas 043885
Castle Fine Art, Birmingham ... 088474
Castle Frames, Caerphilly 117495
Castle Fraser, Sauchen 045740
Castle Galleries, Bath 117301
Castle Galleries, Birmingham .. 117369
Castle Galleries, Bluewater ... 117393
Castle Galleries, Bristol 117441

Castle Galleries, Cambridge ... 117507
Castle Galleries, Cardiff 117525
Castle Galleries, Chester 117570
Castle Galleries, Glasgow 117836
Castle Galleries, Leamington Spa 118081
Castle Galleries, London 118247
Castle Galleries, Manchester .. 118704
Castle Galleries, Meadowhall . 118730
Castle Galleries, Milton Keynes . 118746
Castle Galleries, Newcastle-upon-
  Tyne 118784
Castle Galleries, Norwich 118827
Castle Galleries, Nottingham .. 118838
Castle Galleries, Reading 118959
Castle Galleries, Salisbury 091223
Castle Galleries, Solihull 119110
Castle Galleries, Southampton . 119120
Castle Galleries, Stratford-upon-
  Avon 119167
Castle Galleries, Telford 119218
Castle Galleries, Windsor 119364
Castle Galleries, Wolverhampton 119381
Castle Galleries, York 119407
Castle Gallery, Inverness 118021
Castle Gallery, College of New
  Rochelle, New Rochelle 051314
Castle Gate Antiques Centre, Newark-
  on-Trent 090810
Castle Gate Gallery, Dunster ... 117701
Castle Hill Books, Kington 144506
Castle Hill Gallery, Reading ... 118960
Castle Hill National Historic Park,
  Placentia Bay 006656
Castle House Museum, Dunoon . 044178
Castle Keep Museum, Newcastle-upon-
  Tyne 045345
Castle Menzies, Aberfeldy 043364
Castle Military Museum, Cape
  Town 038675
Castle Museum, Clitheroe 043969
Castle Museum, Colchester ... 043988
Castle Museum, Gonder 010728
Castle Museum, York 046220
Castle Museum and Art Gallery,
  Norwich 045397
Castle Museum of Saginaw County
  History, Saginaw 052739
Castle of Antiquities, Hong Kong 065015
Castle of Saint John, Stranraer . 045898
Castle Park Arts Centre,
  Frodsham 044373
Castle Point Transport Museum,
  Canvey Island 043862
Castle Reclamation, Martock ... 090732,
                                  134846
Castle Restoration, Auchtertool . 134045
Castle Rushen, Manx National
  Heritage, Castletown 043889
Castle Story, Newark-on-Trent . 045329
Castle Tucker, Wiscasset 054505
Castle Ward, Downpatrick 044136
Castleford Museum Room,
  Castleford 043886
Castlegate House Gallery,
  Cockermouth 117607
Castlemaine Antiques,
  Castlemaine 061295
Castlemaine Art Gallery and Historical
  Museum, Castlemaine 000951
Castlemaine Collectors Centre,
  Castlemaine 061296
Castlemart, Bradford 061516
Castleton, Indianapolis 093866, 121172
Castleton Historical Society,
  Castleton 047463
Castleton Village Museum,
  Castleton 043888
Castletown Museum, Gerston .. 044178
Castletown, Teach Chaisleán an Bhaile,
  Celbridge 024847
Častnyj Muzej Figurskoj Kultury, Fond
  Azata Akimbeka, Almaty 029794
Častnyj Muzej Grammofonov i
  Fonografov, Sankt-Peterburg .. 037720

Casto, Gent 125333
Castor, Zug 088021
Castor and District Museum,
  Castor 005580
Castorano, Roma 131848
Castorrini, Renato, Firenze 131146
Castreau, Jean, Vernon 074035
Castrillo Libreria, Madrid 143756
Castro, Barcelona 085758
Castro e Rosa, Lizaura Marques de,
  Porto Alegre 128106
Castro e Silva, Pedro, Lisboa .. 143564
Castro, Amanda, Sacramento .. 096608
Castro, Antonio, Sevilla 143796
Castro, Juan Manuel de, Buenos
  Aires 139558
Castro, Meira, Rio de Janeiro .. 064039
Castro, Ricardo, Barcelona 085759,
                                  133109
Castropola, Pula 140452
Castrum, Eger 079303
Castrum Peregrini, Amsterdam . 137828,
                                  138927
Casttelar, Maria José, São Paulo 101090
Casubolo, Mariangela, Roma .. 110744
Casula Powerhouse Arts Centre,
  Casula 098877
Cat & Fiddle Gallery, Upper
  Coomera 099847
The Cat & the Fiddle, Regina .. 064549
The Cat & The Moon Gallery,
  Sligo 109692
Cat Alley, San Diego 124111
Cat, Marianne, Marseille 104163
Cat, Tuong, Ha Noi 125065
Catacombes, Paris 014628
Catala-Roca, Francesco,
  Barcelona 114869
Catala, Marie-Christine, Genève 087457
Catala, Pierre-Yves, Moulins
  (Allier) 070371
Catalane, Céret 103546, 128533
Cataldi, Giorgio, Marseille 069918
Catalina, Tucson 097644
Catalina Island Museum, Avalon 046706
Catalonia Grafic, Barcelona ... 114870
Catalonian International Independent
  Art Fair, Lleida 098297
Catalyst, Science Discovery Centre,
  Widnes 046133
Catanach, Armadale, Victoria .. 061018,
                                  127554
Catanach, Amanda, Broome ... 061220
Catarsis, Madrid 115170
Catawba County Museum of History,
  Newton 051555
Catawba Science Center, Hickory 049457
Catawissa Railroad Museum,
  Catawissa 047465
Catch-All, Salt Lake City 096884
Catch All Five, Los Angeles ... 094281
Cátedra Ediciones, Madrid 137984
Catedral Antiguidades, Rio de
  Janeiro 064040
Catelin, Gilles, Montluçon 070265
Catellani, Anna, Stockholm ... 133567
Catelli, Milano 110282
Cater Museum, Billericay 043601
Catera, Carlo, Roma 131849
Caters, Jacques-Emmanuel de,
  Bruxelles 063365, 063366
Cates & Co., Mary, Dallas 093084
Cathach, Dublin 142600
Catharina, Groningen 083283
Cathay, Paris 071133
Cathay, Singapore 085380
Cathcart, Valerie, San Francisco 124248
Cathedral Church of Saint John the
  Divine, New York 051352
Cathedral Gallery, Lichfield ... 118111
Cathedral Jewellers, Manchester 090690
Cathedral Museum, Mdina 030740
Cathedral Treasury, Chichester . 043939
Cathedral Treasury Museum,

Carlisle 043873
Cathédrale d'Images, Les Baux-de-
  Provence 013737
Cather's Furniture, Richmond .. 136165
Cathey, Nottingham 090891
Catholic Historical Society of the
  Roanoke Valley, Roanoke 052602
Cathrinesminde Teglværk, Museum
  Sønderjylland, Broager 010026
Cathy's Antiques, Dallas 093085
Catley, Brendon, Tauranga 084102,
                                  132749
The Catlin House Museum,
  Scranton 053214
Catlins Historical Museum,
  Owaka 033267
Cato, Lennox, Edenbridge 089156,
                                  134356
Caton, Hyères 068851
Caton, Peter, Brighton, Victoria . 098776
Catot, Olivier, Marseille 128756
Catrice, Wasquehal 074290
Catrina, Jeannette, Stäfa 116658
Cats Eye, Chicago 092685
Cats Paw, Cleveland 092927
Cats, Bernard, Knokke-Heist ... 100735
Catskill Fly Fishing Center and
  Museum, Livingston Manor ... 050348
Cattai, Michela, Corvara in Badia 080058
Cattai, Michela, Milano 080482
Cattaneo & C., Natalina, Genova 080250
Cattani, Gianfranco, Roma 131850
Cattaraugus Area Historical Center,
  Cattaraugus 047468
Cattedra di Archeologia e Storia
  dell'Arte Greca e Romana, Facoltà
  di Lettere e Filosofia, Libera
  Università Maria Santissima Assunta,
  Roma 055964
Cattelani, Modena 110494
Cattet, Alexandre, Paris 071134
Cattiaux, Chantal, Douchy-les-
  Mines 068395
Cattiaux, Didier, Josselin 068912
Cattle Raisers Museum, Fort
  Worth 048901
Cattle, Barbara, York 091972
Catto, Gillian, London 118248
Catucci, Enrico, Genova 131297
Catùs-Arte, Bologna 109818
Caudine, Laurent, Moncayolle-Larrory-
  Mendibi 128799
Caudron, Christian, Montreuil-sous-
  Bois 128814
Caudwell's Mill & Craft Centre,
  Matlock 045234
Cauet, Christian, Longueau 069620
Caula, Jacqueline, Pertuis 072113
Caulfield Gallery, Vancouver ... 101756
Caulfield Antique Warehouse, Caulfield
  South 061306, 127584
Caulfield Auction Centre,
  Caulfield 125186
La Causa, Milwaukee 121905
Causans, Etienne de, Paris 071135,
                                  104627
Causeway Collection, Tralee ... 079710
Causeway School Museum,
  Bushmills 043816
Causey Antique Shop, Gosforth, Tyne
  and Wear 089345
Causse, Georges, Saint-Sulpice-de-
  Favières 073310
Causse, Henri, Sainte-Enimie . 073346
Caussignac, Jean-Claude, Lavaur 069207
CAV Antiques, Providence 096503
Cavalcade of Costume, Blandford
  Forum 043655
Uchoa Cavalcanti, Rio de Janeiro 101019
Cavaleri, Carlo, Mendrisio 087671,
                                  133858
Cavalier, Chicago 092686
Cavalier, Geenwich 120913
Cavalier, Nantucket 122059

Cavalier Antiques, Glenelg 061530, 127604
Cavaliere, Giancarlo, Milano ... 131412
Cavalière, Henri, Toulon (Var) .. 073705
Cavalieri, Luciano, Roma 131851
Cavaliero, New York 122457
Il Cavalletto, Locarno 116445
Il Cavallino, Reggio Emilia 110707
Cavallino Edizioni, Venezia 111097, 137756
Cavallo, Carlo, Torino . 081682, 132113
Cavan County Museum, Ballyjamesduff 024829
Cavanacor Gallery, Lifford 109665
Cavani Antichità, Modena 080784
The Cave, Washington 124881
Cave au Grenier, Paris 071136
Cave au Grenier, Ségny 073473
La Cave aux Curiosités, Lyon .. 069717
La Cave aux Sculptures, Dénezé-sous-Doué 012759
La Cave D'Arts, Louviers 104052
Cave, F. & C.H., Northampton .. 090854
Cave & Sons, R.G., Ludlow 090644
Caveau, Bologna 079828
Caveau des Récollets, Château-du-Loir 067871, 103576
Caveau Muséalisé, Cave Coopérative, Cabrières 012266
Caveau-Musée Les Vignerons de Roueïre, Quarante 015014
Cavecanem, Sevilla 115447
Cavedagna, Stefano, Falerna ... 080068
Cavedagna, Stefano, Napoli ... 080851
Cavehill Gallery, Belfast 117334
Cavellini, Piero, Brescia 109897
Cavendish, Fremantle 061487
Cavendish Antiques, York 091973
Cavendish Fine Arts, Sonning-on-Thames 091362
Cavendish Laboratory, Cambridge 043841
Cavendish, Rupert, London 090001
La Caverne, Nangis 070435
Caverne, Tampa 097550
La Caverne d'Ali Baba, Cercles . 067748
La Caverne d'Ali Baba, Perpignan 072073
Caverne d'Antan, Hangest-en-Santerre 068787
La Caverne des Arts, Chantilly . 103571
La Caverne des Particuliers, Alfortville 066535
Caverne des Particuliers, Créteil 068239
Caverne des Particuliers Alma, Creil 068230
Cavernes de l'Abîme à Couvin, Couvin 003525
Les Caves des Minimes, Bruxelles 063367
Caves Muséalisées Byrrh, Thuir 015958
Les Caves Rupestres, Rochecorbon 015119
Cavestany, Madrid 086154
Cavicchi Vannoni, Prato 081167
Cavicchioli, Rossella, Modena .. 080785
Caviezel, Claudio, Bern 133674
Caviglia, Enrico, Maroggia 087658
Cavin & Morris, New York 122458
Cavite City Library and Museum, Cavite City 034452
Caviolo, Brescia 142652
Cawdor Castle, Cawdor 043896
Cawi, Elisabeth, Hamburg 076206
Cawood Antiques, Cawood 088809
Caxton, Dublin 079531
Cayfosa, Santa Perpetua de Mogoda 138002
Caylar, Guy, Bédarieux 067061
Caylus Anticuario, Madrid 086155
Cayman Arts, Miami 121753
Cayón, Madrid 126534
Cayon, Troyes 105775
Cayuga Museum, Auburn 046651

Cayuga-Owasco Lakes Historical Society Museum, Moravia 051050
Cazac, Abel, Perpignan 072074
Cazal, Patrick, Biarritz 067192
Cazals de Fabel, Paris 071137
Cazanove, Michelle, Le Gosier . 109108
Cazaux, Christian, Bordeaux .. 125523
Cazenave, Edwina, Biarritz 067193
Cazzulani, Massimiliano, Milano 131413
CBA Gallery, Tokyo 111413
CBC, Perugia 131746
CBC Museum, Toronto 007167
CBC Radio Museum, Saint John's 006900
CBK Amsterdam, Amsterdam .. 112178
CBO Antiquités, Roanne 072485
CC Arts Gallery, Vancouver ... 101757
CC Kunst, Stuttgart 108691
CCA-Centre for Contemporary Arts, Glasgow 044393, 138167
CCA Galleries Ltd., Beech Studio, Tilford 119234, 138288
CCAA Glasgalerie Köln, Köln ... 107600
CCC Weeks Gallery, Jamestown 049785
CCNOA, Bruxelles 003438
CDS Gallery, New York 122459
Čebelarski Muzej, Muzeji radovljiške občine, Radovljica 038595
Čeboksarskij Vystavočnyj Zal, Čeboksary 036928
Cebriá Saez, Vicente, Valencia . 133456
Cebrian, Miguel, Zaragoza ... 086593
Cebu Archdiodesan Museum, Cebu 034458
Cebu City Museum, Cebu ... 034459
Cebu City State College Museum, Cebu 034460
CECA International Committee for Education and Cultural Action of Museums, Apeldoorn 059598
Ceccarelli, Evelino, Perugia .. 131747
Ceccarelli, Glauco, Basel 133639
Ceccherelli, Renato, Genova .. 080251
Cecchin, Donatella, Roma 131852
Cecchini, Perugia 110639
Cecchini, Aldo, Roma 131853
Ceccotti, Carlo, Roma 131854
Cecelia Coker Bell Gallery, Coker College, Art Dept., Hartsville . 049393
Čech a Syn, Blansko 065340
Cech, Johannes, Wien 140067
Cechman, Franck, Versailles ... 074046
Cecil Higgins Art Gallery, Bedford 043551
Cecille R. Hunt Gallery, Webster University, Sankt Louis 053085
Cedar Antique Centre, Hartley Wintney 089450, 134441
Cedar Cliff Collectibles, Saint Paul 096816
Cedar Falls Historical Museum, Cedar Falls 047475
Cedar Grove Galleries, Morpeth . 061889
Cedar Grove Historical Society Museum, Cedar Grove 047479
Cedar Grove Mansion Inn, Vicksburg 053974
Cedar Grove Mansion, Philadelphia Museum of Art, Philadelphia .. 052034
Cedar House Antiques, Mullumbimby 061928
Cedar House Gallery, Ripley, Surrey 118980
Cedar Key Historical Society Museum, Cedar Key 047480
Cedar Key State Park Museum, Cedar Key 047481
Cedar Lodge Antiques, Newtown, Victoria 061967
Cedar Rapids Museum of Art, Cedar Rapids 047485
Cedar Ridge Creative Centre, Scaborough 006995
Cedar Root Gallery, Vancouver . 101758

Cedar Street Galleries, Honolulu 120926, 120927
Cedar Valley Memories, Osage . 051821
Cederwall, San Francisco 124249
CEDIAS Musée Social, Paris ... 014629
Cédille, Paris 114628
Los Cedros, Madrid .. 086156, 133312
Cee-Jay, Richmond ... 096533, 136166
Cefer Cabbarlının Ev-Muzeyi, Bakı 003159
Cefn Coed Colliery Museum, Neath 045315
Ceglédi Természetrajzi Vadászati Múzeum, Cegléd 023578
Cein, Julio & Fernando, Córdoba 060938
Ceiriog Memorial Institute, Glyn Ceiriog 044437
Cejudo Perez, Rafael, Córdoba . 085941
Celadon, Rio de Janeiro 064041
Celamuse – Porcelain Museum, Seoul 029984
Celander, Anna, Malmö 133555
Celant, Geneviève, Clermont-Ferrand 068016
Celantique, Paray-Vieille-Poste . 070850
Celbridge, Clane 130835
Celbridge Motor Museum, Celbridge 024848
Celebration Antiques, Philadelphia 096091
A Celebration of Immigration, London 044935
Celebrity Coin, Milwaukee 094841
Celebrity Sports Gallery, Houston 145166
Celest, Varna 101187
Celestat, Muzeum Historycznego m. Krakowa, Kraków 034857
Celeste & Co., Las Vegas 094126
Celestial Galleries, Fortitude Valley 099050
Celi, Maria Liliana, Taranto 081633
Celil Memmedquluzadenin Ev-Muzeyi, Bakı 003160
Celinda's Doll Shoppe, Minneapolis 135826
Céline Antiquaire, Montréal ... 064401
Čeljabinskaja Oblastnaja Kartinnaja Galereja, Čeljabinsk 036935
Čeljabinskij Oblastnoj Kraevedčeskij Muzej, Čeljabinsk 036936
Cellar Antiques, Hawes 089475
Cellar Antiques, Orroroo 062036
Cellar Coin, San Diego 097071
Cellar Gallery, Chippenham ... 117582
Cellar Stories, Providence 145466
Cellensia, Celle 141670
Celler Garnison-Museum, Celle . 017720
Celler Schützenmuseum, Celle . 017721
Cellier, Châtellerault 067903
Cellier, Jean-Pierre, Belfort ... 067068
Cellini, Bridgnorth 117422
Cellini, Leiden 112554
Cellini, Portland 123610
Cello Press, Witney 138293
Cellula Laboratorio, Brescia ... 131025
Celma, Firenze 110041
Celma Albuquerque Galeria de Arte, Belo Horizonte 100877
Celma Delgado, Maria Carmen, Barcelona 085760
Celotti, Vincent, Vevey 087876
Cels, Hamburg 107234
Celsius, Amsterdam 082716
Celsius Artes e Antiguidades, Brasília 063931
Celtic Antique, Firenze 080115
Cemal Reşit Rey Galerisi, İstanbul 117022
Cemara-6 Galeri, Jakarta 109507
Cementmuseet Slemmestad, Slemmestad 033952
Cenacchi, Maria, Bologna 079829
Il Cenacolo, Piacenza 110647
Il Cenacolo, Trento 111027

Cenacolo del Ghirlandaio, Firenze 026245
Cenacolo di Sant'Apollonia, Firenze 026246
Cenacolo Vinciano, Milano 026779
Cénatus, Pont-Pean 072259
Cencor, Lima 132808
Cendre, Jean-Luc, Prey 072310
Cennosti, Sofia 138565
Le Cent Garde, Nice 070583
Cent Noranta, Barcelona 085761
Cent8, Paris 104629
Centa-Damiano, Joseph, Nice .. 070584
Centar Savremene Umjetnosti Crne Gore, Podgorica 031764
Centar za Grafiku i Vizuelna Istraživanja, Beograd 114399
Centar za Sovremeni Umetnosti Sarajevo, Sarajevo 004272
Centar za Sovremeni Umetnosti Skopje, Skopje 030624
Centaur Boekhandel, Den Haag . 143109
Centaur Bookshop, Maylands, Western Australia 139823
Centaur Galleries, Las Vegas ... 121346
Centaurica, Amsterdam 112179
Il Centauro, Pisa 081154
Le Centenaire, Hydra 060839
Centenary Antique Centre, Newcastle 061957
Centenary College, Meadows Museum of Art, Shreveport 057912
Centennial Art Gallery, Calgary . 101221
The Centennial Museum at the University of Texas at El Paso, El Paso 048504
Centennial Museum of Sheguiandah, Sheguiandah 007014
Centennial Museum of the Nanaimo Regional General Hospital, Nanaimo 006446
Centennial Park 1910 Logging Museum, Thunder Bay 007142
Centennial Park Museum, Moose Factory 006419
Centennial Village Museum, Greeley 049220
Center Art Gallery, Calvin College, Grand Rapids 049189
Center Art Gallery, University of Georgia, Athens 046591
Center Art Studio, New York ... 135896
Center for Advanced Visual Studies, Massachusetts Institute of Technology, Cambridge 056822
Center for Agricultural Science and Heritage, Indianapolis 049678
Center for American Archeology, Kampsville 049866
Center for American History, Austin 046688
Center for American Music, Stephen Foster Memorial, Pittsburgh .. 052160
Center for Art and Design Studies, Cazenovia College, Cazenovia .. 056844
Center for Book Arts, New York 051353
Center for Contemporary Art, Kyïv 043191
Center for Contemporary Art, Sacramento 123762
Center for Contemporary Art Kitakyushu, Kitakyushu 028826
Center for Contemporary Arts of Santa Fe, Santa Fe 053105
Center for Contemporary Graphic Art and Tyler Graphics Archive Collection, Sukagawa 029412
Center for Creative Photography, University of Arizona, Tucson . 053824
Center for Curatorial Studies, Bard College, Annandale-on-Hudson 046497
Center for Emerging Art, Miami 121754
Center for Exploratory and Perceptual Art, Buffalo 047304
Center for Figurative Painting, New

York 051354
Center for Industrial Design, Induk
Institute of Technology, Seoul . 056093
Center for Intuitive and Outsider Art,
Chicago 047644
Center for Jewish Art, Hebrew University
of Jerusalem, Jerusalem 055867,
137612
Center for Jewish History, American
Jewish Historical Society, New
York 051355
Center for Maine Contemporary Art,
Rockport 052645
Center for Photography at Woodstock,
Woodstock 054529
Center for Puppetry Arts, Atlanta 046612
Center for the Arts, University of
Wisconsin La Crosse, La Crosse 057295
Center for the Arts, Wesleyan
University, Middletown 057438
Center for the Study of Political
Graphics, Los Angeles 050414
Center for the Visual Arts,
Denver 048252
Center for the Visual Arts, Brandywine
Workshop, Philadelphia 052035
Center for the Visual Arts,
Metropolitan State College of Denver,
Denver 048253
Center for Visual Communication,
Coral Gables 047981
Center for Wooden Boats, Seattle 053230
Center Galleries, College for Creative
Studies, Detroit 048294
Center Gallery, Old Church Cultural
Center School, Demarest 048238
Center Garo, Yokohama 111533
Center of Art, Design and Media,
Colchester Institute, Colchester 056522
Center of Arts, Cairo 103028
Center of Attention, Greensboro 139335
Center of Attention, Virginia
Beach 124826
Center of Nautical and Regional
Archaeology, Nahshonim 025160
Center of Photographic Art,
Carmel 047434
Center of the Earth Gallery,
Charlotte 119981
Center on Contemporary Art,
Seattle 053231
CenterCity Contemporary Arts,
Providence 123698
Centerville Historical Museum,
Centerville 047500
Cento Anni, Bruxelles . 063368, 063369
Cento Anni, Paris 071138
Centoira, Buenos Aires 098558
Centonze, Claudio, Firenze . . . 080116
Centorbi, Rossella, Catania . . . 080028
Centr Iskusstv 2000, Moskva . . 085175
Centr Istorii Aviacionnych Dvigatelej
im. N.D. Kuznecova, Samarskij
gosudarstvennyj aėrokosmičeskij
universitet, Samara 037705
Centr Sovremennogo Iskusstva,
Toshkent 054651
Centr Sovremennogo Iskusstva Sorosa,
Almaty 029795
Centraal Museum, Utrecht 032944
Central Academy of Arts and Design,
Beijing 055189
Central Antiques, Mount Lawley 061915
Central Antiques, Phoenix 096197
Central Archives for the History of the
Jewish People, Jerusalem 025073
Central Art Gallery, Ashton-under-
Lyne 043445
Central Australian Aviation Museum,
Alice Springs 000788
Central Avenue Strip Shop,
Toledo 136406
Central Book Exchange, Salt Lake
City 145506

Central Butte District Museum, Central
Butte 005585
Central Christian Supply, Phoenix 145429
Central City Coins and Antiquities,
Brisbane 061207
Central City Opera House Museum,
Central City 047503
Central Coin and Jewlery, Denver 093270
Central College Museum,
Bangalore 024149
Central Gallery, London 090002
Central Goldfields Art Gallery,
Maryborough, Victoria 099286
Central Hawke's Bay Settlers Museum,
Waipawa 033323
Central Institute of Fine Arts,
Beijing 055190
Central Iowa Art Association,
Marshalltown 060427
Central Iowa Art Museum,
Marshalltown 050717
Central Luzon State University
Museum, Muñoz 034554
Central Markets, Hutt City 126319
Central Mart, Colac 061356
Central Missouri State University
Archives and Museum,
Warrensburg 054070
Central Montana Historical Association,
Lewistown 050236
Central Museum, Bhopal 024169
Central Museum, Central 047502
Central Museum, Indore 024315
Central Museum, Nagpur 024419
Central Nation University Museum,
Beijing 007605
Central National Herbarium, Botanical
Survey of India, Haora 024286
Central Nevada Museum,
Tonopah 053768
Central New Brunswick Woodmen's
Museum, Boiestown 005432
Central Okanagan Museum,
Kelowna 006072
Central Park Gallery, Kansas City 121283
Central Park Gallery, Malvern
East 099266
Central Piedmont Community College
Art Gallery, Charlotte 119982
Central Saint Martin's College of Art
and Design, University of the Arts
London, London 056554
Central Sikh Museum, Amritsar . 024142
Central Street Antiques, Windsor 127497
Central Texas Area Museum,
Salado 052875
Central Texas Museum of Automotive
History, Rosanky 052677
Central Texas Pastel Society,
Temple 060428
Central Washington Agricultural
Museum, Union Gap 053894
Central Washington Gallery,
Ellensburg 048543
Centralantikvariatet, Stockholm . 143942
La Centrale dell'Arte, Roma . . . 110745
La Centrale Galerie Powerhouse,
Montréal 101372
Centrale Hydroélectrique du Val de
Poix, Saint-Hubert 083993
Centralivros, Lisboa 137917
Centralne Muzeum Jeńców Wojennych
w Łambinowicach -Opolu,
Opole 035075
Centralne Muzeum Morskie,
Gdańsk 034722
Centralne Muzeum Pożarnictwa,
Mysłowice 035027
Centralne Muzeum Włókiennictwa,
Łódź 034952
Centralnyj Dom Chudožnikov,
Moskva 037237
Centralnyj Geologorazvedočnyj Muzej
im. Akademika F.N. Černyševa,

Sankt-Peterburg 037721
Centralnyj Gosudarstvennyj Muzej
Kazachstana, Almaty 029796
Centralnyj Muzej Aviacii i
Kosmonavtiki, Moskva 037238
Centralnyj Muzej Drevnerusskoj
Kultury i Isskustva im. Andreja
Rubleva, Moskva 037239
Centralnyj Muzej Federalnoj
Pograničnoj Služby RF, Moskva 037240
Centralnyj Muzej MVD Rossii,
Moskva 037241
Centralnyj Muzej Oktjabrskoj Železnoj
Dorogi, Sankt-Peterburg 037722
Centralnyj Muzej Počvovedenija
im. V.V. Dokučaeva, Sankt-
Peterburg 037723
Centralnyj Muzej Svjazi im. A.S.
Popova, Sankt-Peterburg 037724
Centralnyj Muzej Velikoj Otečestvennoj
Vojny, Moskva 037242
Centralnyj Muzej Vnutrennych Vojsk
MVD Rossii, Moskva 037243
Centralnyj Muzej Voenno-Vozdušnych
Sil Rossijskoj Féderacii, Monino 037232
Centralnyj Muzej Vooružennych Sil,
Moskva 037244
Centralnyj Muzej Vooružennych Sil
Uzbekistana, Toshkent 054652
Centralnyj Muzej Vserossijskogo
Obščestva Slepych, Moskva . . 037245
Centralnyj Muzej Železnodorožnogo
Transporta, Sankt-Peterburg . . 037725
Centralnyj Voenno-morskoj Muzej,
Sankt-Peterburg 037726
Centralnyj Vystavočnyj Zal g. Perm,
Perm 037638
Centralnyj Vystavočnyj Zal Manež,
Moskva 037246
Centralnyj Vystavočnyj Zal Manež
g. Sankt-Peterburga, Sankt-
Peterburg 037727
Centre Arqueològic de Sagunt,
Sagunto 040034
Centre Artístic de Sant Lluc,
Barcelona 059856
Centre Artistique de la Vallée de la
Creuse, Fresselines 103756
Centre Artistique et Médiéval,
Richerenches 015087
Centre Artistique Jean Baffier,
Sancoins 015674
Centre Atlantique de la Photographie,
Brest 012202
Centre Belge de la Bande Dessinée,
Belgisch Centrum van het
Beeldverhaal, Bruxelles 003439
Centre Commémoratif de l'Holocauste
à Montréal, Montréal 006362
Centre County Historical Society
Museum, State College 053494
Centre County Library Historical
Museum, Bellefonte 046879
Centre Cultural de la Fundació La
Caixa, Vic 040372
Centre Culturel – Musée Montaut,
Saint-Sever (Landes) 015565
Centre Culturel Canadien, Paris . 104630
Centre Culturel Coréen, Paris . . 104631
Centre Culturel de la Memoire
Combattante, Musée des Costumes,
Carcassonne 012327
Centre Culturel de Lachute,
Lachute 006148
Centre Culturel de Verdun,
Verdun 007311
Centre Culturel des Roches,
Rochefort 003971
Centre Culturel et d'Interprétation
de Havre Saint-Pierre, Havre-Saint-
Pierre 005994
Centre Culturel et Patrimonial la
Poudrière de Windsor, Windsor,
Québec 007429

Centre Culturel Franco-Manitobain,
Winnipeg 007437
Centre Culturel Ipousteguy,
Doulcon 012827
Centre Culturel Marguerite Bourgeoys,
Troyes 016048
Centre Culturel Suédois, Paris . . 104632
Centre Culturel Suisse, Paris . . . 104633
Centre Culturel Ti Yarihuten,
Wendake 007377
Centre d'Animation Permanente du
Rail et de la Pierre, Jemelle . . 003741
Centre d'Archéologique et Historique,
Neuville-aux-Bois 014459
Centre d'Art, Baie-Saint-Paul . . 005370
Centre d'Art, Port-au-Prince . . . 023391
Centre d'Art Contemporain,
Alençon 011588
Centre d'Art Contemporain,
Bignan 012039
Centre d'Art Contemporain,
Bruxelles 003440
Centre d'Art Contemporain, Dijon 012777
Centre d'Art Contemporain, Istres 013301
Centre d'Art Contemporain, Lacoux-
Hauteville-Lompnes 013520
Centre d'Art Contemporain,
Meymac 014159
Centre d'Art Contemporain, Mont-de-
Marsan 014207
Centre d'Art Contemporain, Pougues-
les-Eaux 014979
Centre d'Art Contemporain, Revin 015083
Centre d'Art Contemporain,
Tanlay 015904
Centre d'Art Contemporain Bouvet-
Ladubay, Saumur 015712
Centre d'Art Contemporain de Basse-
Normandie, Hérouville-Saint-
Clair 013255
Centre d'Art Contemporain de Brétigny,
Brétigny-sur-Orge 012213
Centre d'Art Contemporain de Castres,
Castres 012363
Centre d'Art Contemporain du Creux
de l'Enfer, Thiers 105704
Centre d'Art Contemporain Genève,
Genève 041517
Centre d'Art Contemporain Georges
Pompidou, Cajarc 012290
Centre d'Art Contemporain la
Synagogue de Delme, Delme . 012751
Centre d'Art Contemporain le Quartier,
Quimper 015017
Centre d'Art Contemporain, École
d'Art du Rueil-Malmaison, Rueil-
Malmaison 015188
Centre d'Art Contemporain, Ferme du
Buisson, Marne-la-Vallée 014028
Centre d'Art de la Sarre, La
Sarre 006139
Centre d'Art de Saint-Georges, Saint-
Georges, Quebec 006872
Centre d'Art du Crestet, Crestet 012720
Centre d'Art en l'Ile, Genève . . 116292
Centre d'Art enfoc, Sant Cugat del
Vallès 115409
Centre d'Art et de Culture,
Soignies 004033
Centre d'Art et de Plaisanterie l'Allan,
Scène Nationale, Montbéliard . 014232
Centre d'Art et d'Exposition Jacques-
Henri Lartigue, L'Isle-Adam . . 013831
Centre d'Art et d'Histoire André
Auclair, Cruas 012727
Centre d'Art Graphique, Hameau Petit
Arran, Parly 014802
Centre d'Art le Quai de la Batterie,
Arras 103275
Centre d'Art Neuchâtel,
Neuchâtel 041793
Centre d'Art Nicolas de Staël, Braine-
l'Alleud 003391
Centre d'Art Passerelle, Brest . . 103460

Centre d'Art Présence Van Gogh,
Saint-Rémy-de-Provence 015544
Centre d'Art Sacré, Patrimoine
Oeuvres d'Art Roman et Baroque,
Ille-sur-Têt 013292
Centre d'Art Santa Mònica,
Barcelona 039055
Centre d'Art Sébastien, Saint-Cyr-sur-
Mer 015280
Centre d'Artistes Vaste et Vague,
Carleton 005565
Centre d'Arts Visuels de Saint-Hubert,
Saint-Hubert 006873
Centre de Création Contemporaine,
Tours 016014
Centre de Cultura Contemporània de
Barcelona, Barcelona 039056
Centre de Cultura Sa Nostra, Palma
de Mallorca 039882
Centre de Culture Catalane, C.D.A.C.C.,
Perpignan 014838
Centre de Culture Scientifique la
Rotonde Ecole des Mines, Saint-
Étienne 015300
Centre de Découvert du Maraîs Breton,
Le Barre-de-Monts 013603
Centre de Découverte de la Nature et
du Patrimoine Cynégètique, Sault
(Vaucluse) 015709
Centre de Découverte Martime,
Penmarch 014823
Centre de Design de l'UQAM,
Montréal 006363
Centre de Documentació i Museu
Têxtil, Terrassa 040220
Centre de la Gravure et de l'Image
Imprimée de la Communauté
Française de Belgique, La
Louvière 003763
Centre de la Marionnette de la
Communauté Française, Tournai 004080
Centre de la Mémoire d'Oradour,
Oradour-sur-Glane 014566
Centre de la Mer, Institut
Océanographique, Paris 014630
Centre de la Nature Montagnarde,
Sallanches 015652
Centre de la Photographie Genève,
Genève 041518
Centre de la Vieille Charité,
Marseille 014040
Centre de Natura de la Cortinada,
Ordino 000098
Centre de Paléontologie et Archéologie,
Valréas 016115
Centre de Patrimoine de Facture
Instrumentale, Le Mans 013653
Centre de Photographie de Lectoure,
Lectoure 013720
Centre de Recherches et d'Etudes
Technologiques des Arts Plastiques,
Bruxelles 003441
Centre de Recherches Historiques sur
les Maîtres Ebénistes, Paris .. 014631
Centre de Scriptophilie, Deinze . 140216
Centre de Secours Principal,
Tourcoing 016003
Centre de Vente entre Particuliers,
Angers 066603
Centre d'Edition Contemporaine,
Genève 041519
Centre Départemental d'Art Contemporain
Cimaise et Portique, Albi 011581,
136795
Centre des Antiquaires, Attin ... 066768
Centre des Antiquaires, Moirans 070138
Centre des Arts Contemporains du
Québec à Montréal, Montréal . 006364
Centre des Arts de Shawinigan,
Shawinigan 007011
Centre des Dossiers, Québec .. 006725
Centre d'Étude du Masque et
Mouvement, Walhain 004127
Centre d'Etudes Archéologiques,

Eyguières 012938
Centre d'Études Edmond Michelet,
Brive-la-Gaillarde 012234
Centre d'Études Historiques et
Archéologiques, Le Barroux ... 013604
Centre d'Exposition Archéologique de
l'Orangerie, Étiolles 012924
Centre d'Exposition Art-Image,
Gatineau 005865
Centre d'Exposition d'Amos,
Amos 005334
Centre d'Exposition de la Bibliothèque
Gabrielle Roy, Québec 006726
Centre d'Exposition de Rouyon-
Noranda, Rouyn-Noranda 006832
Centre d'Exposition de Saint-
Hyacinthe, Saint-Hyacinthe ... 006874
Centre d'Exposition de Val d'Or, Val-
d'Or 007265
Centre d'Exposition des Gouverneurs,
Sorel 007055
Centre d'Exposition du Costume,
Avallon 011828
Centre d'Exposition l'Imagier,
Gatineau 005866
Centre d'Exposition Mont-Laurier,
Mont-Laurier 006352
Centre d'Exposition Plein Sud,
Longueuil 006218
Centre d'Exposition Raymond-Lasnier,
Maison de la Culture de Trois-
Rivières, Trois-Rivières 007242
Centre d'Exposition sur l'Industrie des
Pâtes et Papiers, Trois-Rivières 007243
Centre d'Hébergement, d'Étude sur la
Nature et l'Environnement, Allouville-
Bellefosse 011603
Centre d'Histoire de la Résistance et
de la Déportation, Lyon 013931
Centre d'Histoire de Montréal,
Montréal 006365
Centre d'Histoire Locale,
Tourcoing 016004
Centre d'Histoire et d'Archéologie de
la Métabetchouane, Desbiens . 005702
Centre d'Iconographie Genevoise,
Genève 041520
Centre Didactique de l'Eau,
Seneffe 004006
Centre d'Interpretació del Castell de
Gilida, Gelida 039431
Centre d'Interpretació p.n.
d'Aigüestortes í Estany de Sant
Maurici, Espot 039384
Centre d'Interprétation Archéo Topo,
Grandes-Bergeronnes 005915
Centre d'Interprétation Archéologique
de la Gaspésie, La Martre 006135
Centre d'Interprétation de l'Agriculture
et de la Ruralité, Métabetchouan-Lac-
à-la-Croix 006296
Centre d'Interpretation de l'Ardoise,
Richmond 006793
Centre d'Interprétation de l'Histoire
de Sherbrooke, Société d'Histoire
de Sherbrooke, Sherbrooke,
Québec 007020
Centre d'Interprétation de l'Outil,
Braine-le-Château 003394
Centre d'Interprétation des Luys,
Brassempouy 012194
Centre d'Interprétation des
Mammifères Marins (CIMM),
Tadoussac 007121
Centre d'Interprétation des Mines
Minerais, Petit-Rocher 006643
Centre d'Interprétation du Bourg de
Pabos, Pabos Mills 006609
Centre d'Interprétation du Moulin
Bernier, Courcelles 005654
Centre d'Interpretation du Parc de l'Île-
Bonaventure-et-du Rocher-Percé,
Percé 006630
Centre d'Interprétation du Patrimoine

de Plaisance, Plaisance 006657
Centre d'Interprétation du Patrimoine
de Sorel, Sorel 007056
Centre d'interprétation du Phoque,
Grande-Entrée 005911
Centre d'Interprétation et de la
Recherche Philatélique du Canada,
Terrebonne 007133
Centre d'Interprétation Faunique du
Cap Tourmente, Saint-Joachim 006887
Centre d'Interprétration du
Site Archéologique Doulers
Tsiionhiakwahta, Saint-Anicet . 006853
Centre du Livre d'Artiste Contemporain,
Verderonne 016164, 141378
Centre Dürrenmatt, Neuchâtel .. 041794
Centre Européen d'Actions Artistiques
Contemporaines, Strasbourg .. 015863
Centre Européen de la Paix,
Souchez 015843
Centre Européen de l'Orgue,
Marmoutier 014024
Centre Européen des Métiers d'Art,
Givet 013118
Centre Européen du Résistant Déporté,
Site de l'ancien camp de Natzweiler,
Natzwiller 014433
Centre européen du résistant déporté,
Site de l'ancien camp de Natzweiler,
DIACVG Alsace, Natzwiller ... 014434
Centre Expo du Meuble,
Marseille 069919
Centre for African Studies Collection,
University of Cape Town,
Rondebosch 038824
Centre for Art Tapes, Halifax ... 005947
Centre for Black and African Art and
Civilization, Lagos 033433
Centre for Contemporary Photography,
Fitzroy 001071, 099023
Centre for Creative Communities,
London 060087
Centre for Creative Photography,
Stepney 099718
Centre for Exhibition and Art
Exchange, Ha Noi 125066
Centre for the Magic Arts,
London 044936
Centre for the Study of Greek
Traditional Pottery, Athinai 022982
Centre for Visual Art, University of
Kwazulu-Natal, Scottsville ... 056352
Centre France Estimation, Cognat-
Lyonne 068060
Centre Franco-Ontarien de Folklore,
Sudbury 007092
Centre Gallery, Miami 121755
Centre Gallery, Sunderland ... 119178
Centre Hill Museum, The Petersburg
Museums, Petersburg 052011
Centre Histoirque Médiéval,
Azincourt 011860
Centre Historique de la Résistance en
Drôme et de la Déportation, Romans-
sur-Isère 015138
Centre Historique Minier,
Lewarde 013785
Centre International d'Art
Contemporain – CIAC, Port
Aven 014963
Centre International d'Art
Contemporain de Montréal,
Montréal 006366
Centre International d'Art et du
Paysage, Beaumont-du-Lac .. 011945
Centre International d'Art-Raymond du
Puy, Le Poët-Laval 013685
Centre International des Artisans d'Art,
Bordeaux 128485
Centre International du Vitrail,
Chartres 012467
Centre Jean Giono, Manosque .. 013991
Centre Jean Rostand, Maison de la
Chasse, Roquefort 015148

Centre Jeanne d'Arc, Orléans .. 014581
Centre Juno Beach, Courseulles-sur-
Mer 012699
Centre Marguerite d'Youville,
Montréal 006367
Centre Minier de Faymoreau,
Faymoreau-les-Mines 012958
Centre Missionnaire Sainte-Therèse
Musée, Montréal 006368
Centre National d'Artisanat d'Art,
Ouagadougou 005277
Centre National de la Céramique d'Art
de Sidi Kacem Jelizi, Tunis .. 042780
Centre National des Arts Plastiques,
Paris 058623, 136879
Centre National d'Exposition,
Jonquière 006052
Centre National et Musée Jean Jaurès,
Castres 012364
Centre of Contemporary Art,
Christchurch 033169
Centre Philatélique de Polynésie
Française, Papeete 016376
Centre Photographique d'Ile-de-France,
Pontault-Combault 014951
Centre Pierre-Paul Riquet, Avignonet-
Lauragais 011844
Centre pour l'Image Contemporaine
Saint-Gervais, Genève 041521
Centre Provincial d'Hébergement Le
Caillou, Maison Emile Verhaeren,
Roisin 003979
Centre Régional d'Art Contemporain,
Fontenoy-en-Puisaye 013019
Centre Régional d'Art Contemporain,
Montbéliard 014233
Centre Régional d'Art Contemporain,
Sète 015785
Centre Régional de la Photographie,
Douchy-les-Mines 103706
Centre Régional de la Photographie Nord-
Pas-de-Calais, Douchy-les-
Mines 012824, 136812
Centre Régional de Restauration
et Conservation d'Œuvres d'Art,
Vesoul 129342
Centre Rhenan d'Art Contemporain,
Altkirch 011606
Centre Social i Cultural de la Fundació
La Caixa, Lleida 039575
Centre Social i Cultural de la Fundació
La Caixa, Tarragona 040203
Centre Thématique Fossilifère, Notre-
Dame-du-Nord 006531
Centre Touristique de la Laine & de la
Mode, Verviers 004105
Centricity, New York 095306
Centro Agost, Museo de Alfarería,
Agost 038885
Centro Ambientale Archeologico,
Archeologico Museo Civico,
Legnano 026588
Centro Andaluz de Flamenco, Jerez de
la Frontera 039527
Centro Andaluz de Arte
Contemporáneo, Sevilla 040165
Centro Andaluz de la Fotografía,
Almería 038957
Centro Andino de Tecnología
Tradicional y Cultura de la
Comunidades, Ollantaytambo . 034378
Centro Arte Contemporanea,
Napoli 110519
Centro Arte La Tesoriera, Torino 110960
Centro Arte Magenta, Pisa ... 110598
Centro Arte Moderna, Pisa 110663
Centro Atlántico de Arte Moderno, Las
Palmas de Gran Canaria 039900
Centro Camuno di Studi Preistorici,
Capo di Ponte 025751
Centro Ceremonial Indígena de
Caguana, Utuado 036172
Centro Comercio e Moveis Usados,
Rio de Janeiro 064042

Centro Cultural Banco do Brasil, Rio de Janeiro 004685
Centro Cultural Conde Duque, Madrid 039611
Centro Cultural Contemporaneo Pelaires, Palma de Mallorca .. 059857
Centro Cultural Corp Group, Caracas 054708
Centro Cultural Eduardo León Jimenes, Santiago de los Caballeros ... 010382
Centro Cultural España Córdoba, Córdoba 000317
Centro Cultural Isidro Fabela, México 031147
Centro Cultural Paraguayo-Americano, Asunción 034192
Centro Cultural Paraguayo-Argentino, Concepción 034206
Centro Cultural Polivalente, Bagnacavallo 025434
Centro d'Arte Contemporanea Ticino, Bellinzona 041299
Centro d'Arte La Maddalena, Genova 110096
Centro d'Arte Puccini, Firenze .. 110042
Centro d'Arte La Roggia, Conegliano 109974
Centro d'Arte S. Vidal, Venezia . 111098
Centro d'Arte Verrocchio, Casole d'Elsa 025802
Centro de Arqueologia Annette Laminge Emperaire, Lagoa Santa 004526
Centro de Arte Caja de Burgos, Burgos 039166
Centro de Arte Contemporáneo de Málaga, Málaga 039697
Centro de Arte Contemporáneo Fundación Chirivella Soriano, Valencia 040302
Centro de Arte Contemporáneo la Fábrica, Abarca de Campos .. 038882
Centro de Arte Interpretación de las Hoces del Río Duratón, Sepúlveda 040162
Centro de Arte Jaime Trigo, Pontevedra 086402
Centro de Arte José Guerrero de la Diputación de Granada, Granada 039461
Centro de Arte Joven, Madrid .. 115171
Centro de Arte La Cañuela, Caracas 054709
Centro de Arte La Estancia, Caracas 054710
Centro de Arte Moderna José de Azevedo Perdigão, Fundação Calouste Gulbenkian, Lisboa .. 035755
Centro de Arte Moderno y Contemporáneo Daniel Vázquez Díaz, Nerva 039818
Centro de Arte Visuales, Lima .. 034308
Centro de Arte y Antiguedades, Guadalajara 082464
Centro de Arte y Naturaleza. Fundación Balboa, Huesca 039498
Centro de Arte, Fundación Ortiz Gurdián, León 033382
Centro de Arte, Museo de Almería, Almería 038958
Centro de Artes e Comunicação, Universidade Federal de Pernambuco, Recife 055089
Centro de Artes Tradicionais, Antigo Museu do Artesanato Regional, Évora 035674
Centro de Artes UFF, Universidade Federal Fluminense, Rio de Janeiro 055090
Centro de Artes Visuais, Museo del Barro, Asunción 034193
Centro de Artes, Departamento de Artes Plásticas, Universidade do Estado de Santa Catarina,

Florianópolis 055074
Centro de Artistas Plásticos, Montevideo 046250
Centro de Bellas Artes, Maracaibo 054762
Centro de Biologia Marinha, São Sebastião 004911
Centro de Ciencia y Tecnología, Guadalajara 030997
Centro de Desarrollo de las Artes Visuales, La Habana 009137
Centro de Diseño Ambiental, La Habana 009138
Centro de Documentación Musical, Casa Museo Andrés Segovia, Linares 039565
Centro de Educação Ambiental Gralha Azul e Museu de História Natural, Cascavel 004405
Centro de Estudios Conservacionistas, Guatemala 023367
Centro de Estudios Folklóricos, Guatemala 023368
Centro de Estudos Murilo Mendes, Juiz de Fora 004518
Centro de Exhibiciones Marinas, Calzada de Amador 034170
Centro de Exposição Rural da Urzelina, Urzelina 036068
Centro de Exposiciónes de Geologia y Mineria, Buenos Aires 000152
Centro de Exposiciones La Casona de los Olivera, Buenos Aires 000153
Centro de Exposiciones-Ses Voltes Parc de la Mar, Palma de Mallorca 039883
Centro de Exposições, Arte e Multimédia, Sintra 036036
Centro de Fotografía, TEA, Santa Cruz de Tenerife 040097
Centro de Historia de Zaragoza, Zaragoza 040443
Centro de Información e Interpretación del Parque de Pagoeta, Aia 038897
Centro de Interpretação da Batalha de Aljubarrota, Porto de Mós 035960
Centro de Interpretación Baños de Montemayor, Baños de Montemayor 039045
Centro de Interpretación del Parque en San Martín, San Martín de Castañeda 040065
Centro de Investigaciones Culturales-Museo, Universidad Autónoma de Baja California, Mexicali 031141
Centro de Investigación de las Artes Escénicas, Miramar 055288
Centro de Museos de Buenos Aires, Buenos Aires 058173
Centro de Pesquisas Folclóricas, Rio de Janeiro 004686
Centro de Pesquisas Paleontológicas da Bacia do Araripe, Crato ... 004419
Centro de Restauración Artística, Santiago de Chile ... 064961, 128249
Centro del bel libro ascona, Ascona 133622
Centro del Diseño, Bilbao 039142
Centro del Restauro, Cagliari . 131065
Centro della Cornice, Milano 080483, 131414
Centro di Arte Contemporanea, Roma 110746
Centro di Conservazione e Restauro, Firenze 131147
Centro Di della Edifimi, Firenze . 137646
Centro di Documentazione delle Bonifiche Romane, Granze ... 026465
Centro di Documentazione Messapica, Oria 027098
Centro di Informazione e Documentazione Arti Visive, Archivio Biblioteca Arte Comtempoanea, Prato 027443

Centro di Porta Altinate, Padova 080975
Centro di Restauro, Firenze 131148
Centro di Restauro, Milano 131415
Centro di Restauro Artistico, Genova 131298
Centro di Scienze Naturali e Raccolta Natura ed Arte, Prato 027444
Centro di Scultura Contemporanea, Cagli 025669
Centro di Studi Salentini, Lecce 137660
Centro Escolar University Museum and Archives, Manila 034528
Centro Etnográfico Joaquín Díaz, Urueña 040287
Centro Ferroviário de Cultura, Volta Grande 004959
Centro Fotográfico Álvarez Bravo, Oaxaca 031329
Centro Galego de Arte Contemporáneo, Santiago de Compostela .. 040118
Centro Histórico-Cultural de Enfermagem Ibero- americana, Escola de Enfermagem / EE, São Paulo 004849
Centro Internazionale di Ceramica, Roma 055965
Centro Maggiori, Collezione di Disegni di Antichi Maestri, Monte San Giusto 026913
Centro Mundo Aborigen, Córdoba 000318
Centro Municipal de Exposiciones, Elche 039364
Centro Municipal de Exposiciones Subste, Montevideo 046251
Centro Museo Universitario di Storia Naturale e della Strumentazione Scientifica, Modena 026847
Centro Nacional de Artes, San Salvador 010592
Centro Nacional de Conservación, Restauración y Museología, La Habana 128316, 136759
Centro Nacional de las Artes, México 031148
Centro Numismatico, Ascona ... 087197
Centro Ordóñez Falcón de Fotografía, Donostia-San Sebastián 039349
Centro Palazzo Doria, Genova . 131299
Centro Parque Nacional Picos de Europa – Casa Dago, Cangas de Onis 039223
Centro per l'Arte Contemporanea Luigi Pecci, Prato 027445
Centro Pró-Memória do Club Athlético Paulistano, São Paulo 004850
Centro Pró-Memória Hans Nobiling do Esporte Clube Pinheiros, São Paulo 004851
Centro Regional de Arqueologia Ambiental, Piraju 004630
Centro Regional de Pesquisas Archeológicas Mário Neme, Piraju 004631
Centro Restauro Meranese, Merano 131352
Centro Restauro Mobili Antichi, Torino 081683, 132114
Centro Steccata, Parma 110626
Centro Studi Alfredo Muratori, Modena 137699
Centro Studi I. Silone, Pescina . 027286
Centro Studi Naturalistici del Pollino Il Nibbio, Morano Calabro 026974
Centro Universitário Belas Artes de São Paulo, São Paulo 055100
Centro Universitario Europeo per i Beni Culturali, Ravello 055953
Centro Visitantes Fabrica de Hielo, Sanlúcar de Barrameda 040078
Centrul de Cultură Palatele Brâncoveneşti, Mogoşoaia 036572
Centrum Arts and Creative Education, Port Townsend 052258
Centrum Beeldende Kunst,

Groningen 032376
Centrum Beeldende Kunst, Nijmegen 112604
Centrum Beeldende Kunst Dordrecht, Dordrecht 112435
Centrum Beeldende Kunst Emmen, Emmen 032301
Centrum Beeldende Kunst Utrecht, Amersfoort 031908
Centrum för Arbetarhistoria, Landskrona 040772
Centrum Fornikling, Oslo 084306
Centrum Hasičského Hnuti, Přibyslav 009853
Centrum Kultury Dwór Artusa, Toruń 035276
Centrum Kultury i Sztuki Afrykańskiej, Kraków 034858
Centrum Kultury Żydowskiej, Kraków 034859
Centrum Kunsthistorische Documentatie, Radboud Universiteit, Nijmegen 056180
Centrum Kunstlicht in de Kunst, Eindhoven 032281
Centrum Młodzieży, Kraków ... 113539
Centrum Pamięci Gen. Józefa Hallera, Władysławowo 035396
Centrum Rysunku i Grafiki im. Tadeuza Kulisiewicza, Muzeum Okręgowego Ziemi Kaliskiej w Kaliszu, Kalisz 034798
Centrum Rzeźby Polskiej, Orońsko 035080
Centrum Rzeźby Polskiej – Dział Wydawnictw, Orońsko 137899
Centrum Scenografii Polskiej – Galerija Anrzeja Majewskiego, Muzeum Śląskie, Katowice ... 034810
Centrum Sztuki Studio im S.I. Witkiewicza, Warszawa 035310
Centrum Sztuki Współczesnej #E Laźnia, Gdańsk 034723
Centrum Sztuki Współczesnej, Zamek Ujazdowski, Warszawa 035311
Centrum voor de Vlaamse Kunst van de 16de en de 17de Eeuw, Antwerpen 003282
Centrum voor Heiligenbeelden, Kranenburg 032556
Centrum voor Natuur en Landschap, Terschelling-West 032917
Centura House Historical Society, A.J. Snyder Estate, Rosendale 052683
Centuries, Birmingham 092398
Centuries, Minneapolis 094903
Centuries 46, Brno 065347
Centuries Old Maps and Prints, New Orleans 145303
Centurion Galleries, Chicago ... 120088
Century, Charlotte 092583
Century Antique Watch, Singapore 085381
Century Antiques, Cleveland ... 092928
Century Coins, Pittsburgh 096265
Century Fine Arts, Lymington .. 090659
Century Galleries, Henley-on-Thames 117960
Century Gallery, Slough 119104
Century Modern, Dallas 093666
Century of Stone Gallery, Louisville 121636
Century Shop, Louisville 094528
Century Village Museum, Burton 047331
Ceol – Irish Traditional Music Centre, Dublin 024870
CEP Sanat Galerisi, İstanbul .. 117023
CEPA Photographic Galleries, Buffalo 119953
CEPACI, Promotion de l'Artisanat de Côte d'Ivoire, Bordeaux 103421
Cepade, Paris 104634
Cepade Centro Paulista de Arte e Design, São Paulo 055101

Cepcolor, Mâcon 104141
Cepheus, Stockholm 115849
C'era Una Volta, Genova 080252
C'era Una Volta Il Mobile, Milano 131416
Ceralacca, Brescia . . . 079927, 142653
Ceram Antiek, Tilburg 083612
Cerami, Charleroi 063594
Ceramic Art Center, Mashiko . . 029000
Ceramic Art London, London . . 098366
Ceramic Collectors Society, North
Sydney 058194
Ceramic Restoration, New York . 135897
Ceramic Review, London 138206,
139158
Cerámica, Madrid 139045
Ceramica Arte 4, Cartagena . . . 128296
Ceramics Collector, Strasbourg . 073577
Ceramics Gallery, Aberystwyth . 043371
Ceramics Monthly Magazine,
Westerville 139336
Ceramics Museum, Tel Aviv . . . 025193
Céramiques du Château, Antibes-Juan-
les-Pins 103256
Cerclaje, Madrid 115172
Cercle des Beaux-Arts de Liège,
Liège 058362, 100779
Cercle des Beaux-Arts de Verviers,
Verviers 058363, 100819
Čerdynskij Kraevedčeskij Muzej A.S.
Puškina, Tcherdyn Regional Pushkin
Museum, Čerdyn 036939
Cereal Prairie Pioneer Museum,
Cereal 005586
Ceredigion Museum, Aberystwyth 043372
Cerenna, Mattydale 127311
Ceres, New York 122460
Ceres Antiques, Ceres 088810
Ceresa, Ferdinando, Maglio di
Colla 116489
Ceri, M., Firenze 080117, 080118
Cerino Canova, Laura, Torino . . 081684
Cerioli e Calvi, Milano 110283
Čerkas'ke Učylyšče Mystectv i Kul'tury
im. S. Gulaka-Artemovs'kogo,
Čerkasy 056483
Čerkas'kyj Chudožnij Muzej,
Čerkasy 043116
Čerkas'kyj Oblasnyj Krajeznavčyj
Muzej, Čerkasy 043117
Čerkechskij Muzej Jakutskaja
Političeskaja Ssylka, Jakutskij
gosudarstvennyj muzej istorii
i kultury narodov severa im.
Jaroslavskogo, Čerkech 036946
Cerkljanski Muzej, Mestni muzej Idrija,
Cerkno 038503
Cerkov Ionna Predteči, Jaroslavskij
gosudarstvennyj istoriko-
architekturnyj i chudožestvennyj
muzej-zapovednik, Jaroslavl . . 037058
Cerkov Pokrova v Filjach,
Moskva 037247
Cerkov Položenija riz Presvjatoj
Bogorodicy, Gosudarstvennyj muzej-
zapovednik Moskovskij Kreml,
Moskva 037248
Cerkov Roždestva Christova,
Jaroslavskij gosudarstvennyj istoriko-
architekturnyj muzej-zapovednik,
Jaroslavl 037059
Cerkov Troicy v Nikitnikach,
Gosudarstvennyj istoričeskij muzej,
Moskva 037249
CERN – Microcosm, Globe de
la Science et de l'Innovation,
Genève 041522
Cernan Earth and Space Center, River
Grove 052586
Cerne Antiques, Cerne Abbas . . 088812
Černigivs'kyj Oblasnyj Chudožnij
Muzej, Černigiv 043119
Černigivs'kyjy Literaturno-memorial'nyj
Muzej-zapovidnik Mychajlo
Kocjubynskogo, Černigiv 043120

Černivec'ke Učylyšče Mystectv
i Kul'tury im. S. Vorobkevyča,
Černivci 056484
Černivec'kyj Chudožnij Muzej,
Černivci 043124
Černivec'kyj Krajeznavčyj Muzej,
Černivci 043125
Černivec'kyj Literaturno-Memorial'nyj
Muzej im. Ju.A. Fedkoviča,
Černivec'kyj krajeznavčyj muzej,
Černivci 043126
Černivec'kyj Literaturno-Memorial'nyj
Muzej O. Kobyljans'koï, Černivec'kyj
krajeznavčyj muzej, Černivci . . 043127
Černivec'kyj Muzej Bukovins'koï
Diaspory, Černivci 043128
Černivec'kyj Muzej Narodnoï
Architektury ta Pobutu, Černivci 043129
Cernuschi, Ursula, Paris 104635
Cerny Inuit Collection, Bern . . . 041311
Cerny Inuit Gallery, Bern 116127
Černý Pavouk, Ostrava 140494
Cerny, P., Erlangen 075650
Ceros, Paris 141069
Cerruti, Massimo, Milano 131417
Cersósimo Picado, Emilia María, San
José 102483
Certainly Wood, Christchurch . . 132703
Cerutti & Miller, New York 122461
Cervantes Antiquités, Saint-Ouen 073049
Cervera, Jose, Valencia 086511
Cervera, Marta, Madrid 115173
CeSAC – Centro Sperimentale per le Arti
Contemporanee, Associazione Culturale
Marcovaldo, Caraglio 059424, 109932
El Cesar, Bogotá 102286
Cesaretti, Roma 142804
Cesaris, Luisa, Milano 131418
Cesaro, Giorgio, Padova 080976
Cesaroni, Brașov 085072
Cesati, Alessandro, Milano 080484
Il Cesello, Firenze 110043
Česká Koruna, Praha 065484
České Muzeum Hudby – Oddělení
Hudebních Nástrojů a Hudebně
Historické Oddělení, Národní
muzeum, Praha 009781
České Muzeum Hudby – Zámek
Hořovice, NM Praha – České
Muzeum Hudby, Hořovice 009571
České Muzeum Stříbra – Hrádek,
Kutná Hora 009655
České Muzeum Výtvarných Umění,
Praha 009782
Çeşme Müzesi, Çeşme 042870
Cessane, Bogotá 102287
C'est Chouette, L'Isle-sur-la-
Sorgue 069565
C'est la Vie Antiques, San Diego 097072
Cesteria Al Mare, Brescia 079928
Cēsu Izstāžu Nams, Cēsis 030110
Cēsu Vēstures un Mākslas Muzejs,
Cēsis 030111
Cetatea Brașov și Fortificațiile din Țara
Bârsei, Brașov 036284
Cetatea de Scaun a Sucevei,
Suceava 036734
Cetatea Devei, Deva 036436
Cetatea Neamțului, Târgu Neamț 036768
Cetatea Poienari, Căpățânenii
Pământeni 036351
Cetatea Sucidava, Corabia 036403
Cetatea Țărănească Râșnov,
Râșnov 036649
Cettina, Marino, Umag 102523
Cettour, Bernard, Trouville-sur-
Mer 073890
Ceuleers, Jan, Antwerpen 140141
Ceuleeners, Jan, Antwerpen . . . 140142
Cevenol, Saint-Martin-de-
Valgalgues 072942
Cévenols, Le Vigan (Gard) 069376
Ceviz, İstanbul 088109
Ceylon, Dallas 093087

Ceylon Society of Arts, Colombo 059876
CFB Esquimalt Naval Museum and
Military Museum, Victoria 007321
CFB Petawawa Military Museum,
Petawawa 006633
CFCC Webber Center, Ocala . . . 051694
CFM Gallery, New York 122462
CG Park, Saint-Paul-de-Vence . . 073243
CG's Curiosity Shop,
Cockermouth 088939
CH Art, Miami 121756
Ch. Frenkelio Rūmai, Šiaulių Aušros
Muziejus, Šiauliai 030462
Chaa Creek Natural History Centre,
San Ignacio 004180
Chabanian, Annecy 103244
Chabanian-le Baron, Annecy . . . 103245
Chabanian-le Baron, Honfleur . . 103821
Chabanian-le Baron, Saint-Paul-de-
Vence 105579, 105580
Chabanian-Perbet, Megève 104212
Chabanis, Roselyne, Bourg-Saint-
Andeol 067416
Chabanne, Michel, Pranles 072305
Chabannes, Cavaillon 067737
Chabarovskij Gosudarstvennyj Institut
Iskusstv i Kultury, Chabarovsk . 056268
Chabarovskij Gosudarstvennyj Muzej
Dalnego Vostoka im. N.I. Grodekova,
Chabarovsk 036949
Chabarovskij Muzej Archeologii im. A.P.
Okladnikova, Chabarovsk 036950
Chabloz, Vincent, Morges 133861
Chabolle, Jean-François, Paris . . 071139
Chabot Museum, Rotterdam . . . 032800
Chac & Mool, West Hollywood . 124990
Chácara do Céu Museum, Museu
Castro Maya, Rio de Janeiro . . 004687
Chacártegui Cirerol, Jorge, Palma de
Mallorca 115337
Chacewater Antiques,
Chacewater 088813
Chaco Culture National Historical Park,
Nageezi 051150
Chacón Tassara, Ana Catalina, San
José 102484
Chacon, Miguel, Vic-le-Comte . . 074094
Chacon, Rodriguez, Madrid 133313
Chacun, Christian, Paris 071140
Chadelaud, Michel-Guy, Paris . . 071141,
071142
Chadron State College Main Gallery,
Chadron 047508
Chadwell, Audrey & Jesse,
Buffalo 092516
Chadwick, Richmond 096534
Chadwick, Dolby, San Francisco 124250
Chadwick, Howard, Boston 092453
Chadz Antiques, Taranaki 084090
Chaffal, Roger, Morance 070342
Chaffee Art Center, Rutland . . . 052715
Chaffey, Pat, Tamworth 099803
Chagall-Galerie Wuyt,
Amsterdam 112180
Chagall House, Haifa 025046
Chagall Museum, Vitebsk 003255
Chagas, Funchal . . . 126435, 126436
Chagas, Onedina L., Belo
Horizonte 063892
Chagford Galleries, Chagford . . . 117548
Chagot, Francis, Tulle 073913
Chagrin Falls Historical Society
Museum, Chagrin Falls 047512
Chai Ma, Bangkok 088076
Chai-Yi 2–28 Memorial Park Museum,
Chai-Yi 042276
Chai-Yi Museum of Stone and Art,
Chai-Yi 042277
Chaieb, Daniel, Porto Alegre . . . 063968
Chailloux, Maria, Amsterdam . . 112181
The Chaim Gross Studio Museum,
New York 051356
The Chain, Amsterdam 082717
Chaineau, Jean-Marc, La Roche-

Clermault 069064
Chaini-zu Toji Bijutsukan, Usuki 029683
Chaintreuil, Georges, Ferney-
Voltaire 068522
Chair Doctor, Nashville 135844
Chair Re-Glue and Furniture Repair,
Saint Paul 136217
The Chair Set Antiques,
Woodstock 091930
La Chaise Antique, Stow-on-the-
Wold 091463, 135093
Chaise de Bois, Angers 066604
Chaix, Corine, Toulon (Var) 073706
Chaix, Corinne, Camps-la-Source 067625
Chaiya National Museum, Surat
Thani 042724
Chakasskij Respublikanskij
Kraevedčeskij Muzej, Abakan . 036838
Chakdara Museum, Chakdara . . 034108
Chak's, Hong Kong 065016
Chalabi Antik, İstanbul 088110
Chalais Brocante, Chalais 067764,
128535
Chalancon, Saint-Étienne 072755
Chalancon, Jacqueline, Le
Conquet 103939
Chalermnit, Bangkok 144203
Le Chalet, Allouis 066539
Chalet, Berlin 074775
Le Chalet, Villefranche-sur-Cher 074147
Châlet de l'Etambeau, Château-
d'Oex 041403
Chalet of the Golden Fleece, New
Glarus 051255
Chalifour, Gérald, Bessé-sur-
Braye 067142
Chalk House, Saint Ives,
Cornwall 119019
Chalkwell Auctions, Southend-on-
Sea 127050
Challiss, San Francisco 097202
Chalmeton, Chappes 067835
Chalmont, Christophe, Durtal . . 068421
Chalmont, Richard, Paris 071143
Chalon Collections, Chalon-sur-
Saône 067775
Chalon, Nick, Chard 088822
Chalrignac, Laurent, Paris 071144
Chalvet, Mado, Houston 121014
Chalvignac Isabelle, Versailles . . 074047
Chalvin, Yves, Lyon . . 069718, 104074
Chamarajendra Academy of Visual
Arts, Mysore 055824
Chamarande, Bruxelles 063370
Chamarilero, Madrid 086157
Chamberlain Furniture Restoration,
Denver 135526
Chamberlain Museum of Pathology,
Birmingham 043618
Chamberlayne School of Design,
Mount Ida College, Newton . . 057576
Chambers, London 118249
Chambers, Utrecht 083634
Chambers & Crosthwaite, Hamilton,
Queensland 061581
Chambers Fine Art, New York . . 122463
Chambers Gallery, Ipswich 118024
Chambers, Graham H., Lancaster 089738
Chambers, Ronald G., Petworth . 090993
Chambet, Jean-Paul, Saint-
Germain 072284
Chamblin Book Mine,
Jacksonville 145185
El Chambo, Vigo 086556
Chambon, C., Nogent-sur-Seine . 070730
Chambon, Claude, Favars 068510
Chambon, Claudie, Saint-Aubin . 072678
Chambon, Philippe, La Chapelle-la-
Reine 068872
Chambord, Virginia Beach 097775
Chamboux, Jean-François, Tours 073844
La Chambre, Rotterdam 083559
Chambre Belge des Experts en
Œuvres d'Art, Bruxelles 058364

La Chambre Blanche, Québec . . 101495
Chambre d'Amis, Amsterdam . . 082718
Chambre Européenne des Experts-
Conseils en Oeuvres d'Art,
Paris 058624
Chambre Nationale des Commissaires-
Priseurs Judiciaires, Paris 058625,
125691
Chambre Nationale des Experts
Spécialisés, Paris 058626
Chambre Professionnelle Belge de
la Librairie Ancienne et Moderne,
Bruxelles 058365
Chambre Royale des Antiquaires de
Belgique, Koninklijke Kamer van de
Antiquairs van Belgie, Bruxelles 058366
Chambre Syndicale de l'Estampe, du
Dessin et du Tableau, Paris . . 058627,
104636
Chambre, Claire, Annecy 103246
Chamberlin, Joëlle, Attin 066769
Chambrion, Martine, Lyon 069719
Chambure, Marie de, Paris 104637
Chameleon, Hong Kong 102063
Chameleon Antique Lighting, New
York 095307
Chameleon Gallery, Masterton . . 112946
Chaminade, Jacques, Lyon . . . 140889
Chamizal National Memorial, El
Paso 048505
Chammard, Monique de, Paris . 104638
Chamonal, Rodolphe, Paris 141070
Chamoroff, Bianca C., Beirut . . 082277
Chamouleau, Armand, Paris . . . 071145
Chamouleau, Jean-Claude,
Fouquebrune 068599
Champ de Mars, San Francisco 097203
Champ des Possibles, Noirmoutier-en-
l'Ile 104412
Champ Libre, Montréal 101373
Champagne dit Lambert, Ottawa 064499
Champaign County Historical Museum,
Champaign 047517
Champaign County Historical Museum,
Urbana 053910
Champaloux, Emile, Limoges . . 104028
Champavert, Toulouse 141360
Champeaux, Dominique, Thil . . 073653
Champetier, Michelle, Cannes . . 103505
Champignonnière-Musée du Saut aux
Loups, Montsoreau 014316
Champion & Kusel, Nîmes 125664
Champion Mill, Champion 047521
Champion, Alain, Nantes 070456
Champion, Alain, Nuille-le-Jalais 070764
Champion, Josy, La Sauvetat . . 069099
Champions Lighting, Houston . . 093635
Champion's Refinishing, Houston 135596
Champkins, Paul, London 090003,
134617
Champlain Trail Museum,
Pembroke 006623
Champs Chapel Museum, East
Hendred 044203
Champvallins, Jacqueline de,
Paris 104639, 136880
Chamsori Gramophone and Audio
Science Museum, Gangneung . 029925
Chan, San Francisco 097204
Chan Ngee, Singapore 085384
Chan, Kam Fuk, Hong Kong . . . 102064
Chan, P.F., Singapore 085382
Chan, Peter, Toronto 064623
Chan, Pui Kee, Singapore 085383,
133035
Chance Operations Gallery,
Denver 120701
Chancellor Robert R. Livingston
Masonic Library and Museum, New
York 051357
Chancenotte, Jacques, Dijon . . 128578
Chancery Lane Antiques,
Nashville 094983
Chandan Bulchand, Las Palmas de

Gran Canaria 086374
Chandelier, Maurice, Atlanta . . . 092119
Chandeliers, Clancy, Dublin . . . 079532
Chandler Fine Art, San Francisco 124251
Chandler Gallery, Randolph 052464
Chandler Museum, Chandler . . . 047522
Chandos Books, Bristol 144278
Chandradhari Museum,
Darbhanga 024208
Chang Da-Chien's Residence,
Taipei 042501
Chang Foundation Museum,
Taipei 042502
Chang Gung University Arts Center,
Gueishan 042319
Chang Wei Hwa, Taipei 088066
Chang, Bobby, Singapore 114504
Chang, Theo, Caracas 097949
Changcheng, Wuxi 102249
Changdao Museum, Changdao . 007667
Changdao Voyage Museum,
Changdao 007668
Change des Saints Pères, Paris . 071146
Change du Port, Marseille 069920
Change Plus, Biarritz 067194
Changement de Décor, Châlons-en-
Champagne 067785, 103557
Changement de Décor, Épernay 068448,
103716
Changi Village Antiques,
Singapore 085385
Changji Mulei County Museum,
Changji 007669
Changjizhou Museum, Changji . 007670
Changning District Revolutionary
Display Center, Shanghai 008222
Changping Yanjing Art Cooperative,
Beijing 101941
Changqing Museum, Changqing 007672
Changsha Museum, Changsha . 007673
Changshu Beike Museum,
Changshu 007676
Changshu Calligraphy and Painting
Gallery, Changshu 102010
Changshu Museum, Changshu . 007677
Changtai Museum, Changtai . . . 007678
Changting Museum, Changting . 007679
Changwu Museum, Changwu . . 007680
Changxing Museum, Changxing 007681
Changy, D. de, Bruxelles 063371
Changyangtu Sept Self-Rule Museum,
Yidu 008483
Changzhi Museum, Changzhi . . 007682
Changzhou Museum, Changzhou 007683
Chania Byzantine and Postbyzantine
Collection, Chania 023039
Chankit, Glenda, Toronto 064624
Le Channel – Galerie de l'Ancienne
Poste, Calais 012292
Channel Folk Museum, Lower
Snug 001234
Channel Islands Galleries, Saint Peter
Port 091204, 144819
Channel Islands Military Museum,
Saint Ouen 045708
Chanoit, Fréderic, Paris 071147, 104640
Chan's Ancient Jade and Art, Hong
Kong 065017
Chantal Muñoz, Rognac 129208
Chantal, Karli, Lausanne 133793
Chantala, Eric, Paris 071148
Chantebois, Chantilly . 067831, 128544
Chantek, Rumah, Singapore . . . 114505
Chanteloup, Marie de, Paris . . . 071149
Chantérac-Geiss, Alois,
Völklingen 130727
Chantharakasem National Museum,
Ayutthaya 042644
Chantier A.C. Davie, Lieu Listorique
national du Canada, Lévis 006190
Chantilly, Kelvin Grove 061675
Chantilly Museum, Geraardsbergse
Musea, Geraardsbergen 003646
Chantry Bookshop & Gallery,

Dartmouth 089030, 144365
Chantry House Gallery, Ripley, North
Yorkshire 118979
Chantry, Léon-Noël, Tours 073845
Chanu, Cedric, Narbonne 070492
Chanut, François, Paris 141071
Chao Hua Art Cooperative,
Beijing 101942
Chao Phraya Gallery, Washington 097820
Chaos Art Gallery, Budapest . . . 109135
Chaos Collectables, Christchurch 083862
Chaosamphraya National Museum,
Ayutthaya 042645
Chaotiangong Palace, Nanjing City
Museum, Nanjing 008105
Chaoyang Museum, Chaoyang . 007685
Chaoyang Museum, Chaoyang,
Liaoning 007687
Chaozhou Museum, Chaozhou . 007688
Chap Ku, Tong, Hong Kong . . . 065018
Chapeau, Antwerpen 100486
Chapeau Clacque, Berkel-
Enschot 143091
La Chapeauthèque, Anduze . . . 011644
Chapel, Pittsburgh 136128
Chapel Antiques, Barmouth . . . 088353
Chapel Antiques, Nantwich 090795,
134863
Chapel Antiques Centre,
Sheffield 091280
Chapel Art Gallery, Riseley 118981
Chapel Books, Chepstow 144328
Chapel Emporium Antique Centre,
Long Sutton, Lincolnshire 090628
Chapel Gallery, Bracebridge . . . 005452
Chapel Gallery, North Battleford 006504
Chapel Hill Museum, Chapel Hill 047529
Chapel Hill Museum, Shag
Harbour 007006
Chapel House Fireplaces,
Holmfirth 089542
The Chapel of Art, Criccieth . . . 117648
Chapel off Chapel, Prahran 099600
Chapel Place Antiques, Tunbridge
Wells 091648
Chapel Street Antiques Arcade,
Penzance 090963
Chapel Street Bazaar, Prahran . 062109
Chapelaine Antiquitités, La Chapelle-
sur-Erdre 068981
Chapelle de Bailly, Saint-Bris-le-
Vineux 015243
Chapelle de Capucins, Aigues-
Mortes 011545
Chapelle de Pénitents Blanc, Aigues-
Mortes 011546
Chapelle des Indiens, Tadoussac 007122
Chapelle des Visitandines, Fondation
François Brochet, Auxerre 011819
Chapelle Matisse, Vence 016149
Chapelle Saint-Calixte – Musée du
Château Comtal, Mons 003877
Chapelle Sainte-Radegonde, Musée du
Vieux Chinon, Chinon 012559
Chapellerie des Arcades, Brest . 067467
Chapellerie Houssin, Le Havre . 069277
Chaphariz, Brasilia 063932
Chapin Library of Rare Books,
Williams College, Williamstown 054418
Chapin, Bill, Toledo . . 097602, 136407
Chapitre, Paris 141072
Chapitre Un, Montréal 140357
Chapleau Centennial Museum,
Chapleau 005588
Chaplier, Ghislaine, Paris 071150
Chaplin-Archiv, Frankfurt am
Main 018409
Chapman, Milwaukee 112906
Chapman & Friedman, Louisville 121637
Chapman Art Center Gallery,
Cazenovia College, Cazenovia 047470
Chapman Gallery, Salford 045719
Chapman Historical Museum, Glens
Falls 049122

Chapman Museum, Brandon . . . 005462
Chapman, Chas, London 101337
Chapman, Lee, Keswick 118036
Chapman, Manuka, Griffith 099099
Chapman, Michael, Cley 117598
Chapo Formes Nouvelles, Gordes 103770
Chapotot, Claude & Sonia,
Nanterre 070436
Chapoulie, Agen 066425
Chappel Galleries, Chappel 117555
Chappell, New York 122464
Chappell & McCullar, San
Francisco 097205
Chappell Art Gallery, Chappell . . 047531
Chappell Hill Historical Society
Museum, Chappell Hill 047532
Chappell, Alice, Boston 119849
Chappell, Jordan, Houston 093636
Chappells Antique Centre,
Bakewell 088334
Chapron, Hubert, Saint-Marcel . 072927
Chaptard, Jeannine, Le Coteau . 066069
Chapter & Verse, Rye 144807
Chapter and Verse, Pune 142594
Chapter House Books,
Mornington 139835
Chapter House Books, Sherborne 135033,
144842
Chapter One, London . 090004, 144574
Chapters, Dublin 142601
Chapu, Marie-Françoise, Paris . 071151
Chapuis, Nice 140995
Chapuis, Frédéric, Saint-Ouen . 073050
Chapuis, Geneviève, Gacé 068619
Chapusot, Jean-Claude, Langres 069174
Chaput, Nicole, Tournus 073823
Characters Art Gallery, Bolton . . 117394
Charaire, Jean, Agen 066426
Charaire, Jean, Aubiac 066788
Charbit, Adolphe, Draguignan . . 125574
Charbo, Amsterdam 143022
Charbonneau, Boris, Cahors . . . 067594,
140811
Charbonneaux, Catherine, Paris . 125692
Charbonnel, Alain, Loches-en-
Touraine 069606
Charbonnet & Charbonnet, New
Orleans 095076
Charbonnier, Georges L., Genève 087458
Charbonnier, Michel, Marseille . 069921
Charbonnier, Noël, Paris 128958
Chard and District Museum,
Chard 043905
Chard Antique Centre, Chard . . 088823
Chard Coins, Blackpool 088498
Chardin, Tbilisi 105899
Chardon, Yves, Hauterives 068797
Charef, Guy, Douvres-la-
Delivrande 068401
Charèl, Amsterdam 082719
Charfa, Skopje 111880
Charil, Grégory, Nemours 070499
Charim Galerie, Wien 100241
Charim, Miryam, Wien 100240
Charing Cross Books, Annerley . 139627
Chariot, Earlwood 061427
Le Chariot d'Or, Lyon 069720
Charitonenko i Syn, Sumy 117163
Charivari, Knokke-Heist 100736
Charkivs'ka Deržavna Akademija
Kul'tury, Charkiv 056485
Charkivs'ke Chudožne Učylyšče,
Charkiv 056486
Charkivs'kyj Chudožnij Muzej,
Charkiv 043132
Charkivs'kyj Deržavnyj Institut Kul'tury
i Mystectv im. I. Kotljarevskogo,
Charkiv 056487
Charkivs'kyj Istoryčnyj Muzej,
Charkiv 043133
Charkivs'kyj Literaturnyj Muzej,
Charkiv 043134
Charlbury Museum, Charlbury . . 043908
Charles, Rouen 072575

Charles, Salt Lake City 136234
Charles, Whitwick 091838
Charles & Charles, Los Angeles 094282
Charles A. Grignon Mansion,
  Kaukauna 049904
Charles A. Lindbergh House, Little
  Falls 050329
Charles A. Weyerhaeuser Memorial
  Museum, Little Falls 050330
Charles A. Whitaker Auction Co,
  Philadelphia 127389
Charles A. Wustum Museum of
  Fine Arts, Racine Art Museum,
  Racine 052439
Charles Allis Art Museum,
  Milwaukee 050911
Charles B. Aycock Birthplace,
  Fremont 048975
Charles B. Goddard Center for Visual
  and Performing Arts, Ardmore . 046525
Charles Carroll House of Annapolis,
  Annapolis 046499
Charles Dickens Birthplace Museum,
  Portsmouth 045548
The Charles Dickens Museum,
  London 044937
Charles Gates Dawes House, Evanston
  Historical Society Museum,
  Evanston 048634
Charles H. MacNider Museum, Mason
  City 050735
Charles H. Scott Gallery,
  Vancouver 007280
Charles H. Taylor Arts Center,
  Hampton 049340
Charles H. Wright Museum of African
  American History, Detroit 048295
Charles-Harold Collection, Tampa 124640
The Charles Hosmer Morse Museum
  of American Art, Winter Park . 054497
Charles J. Beldon Museum of Western
  Photography, Meeteetse Museums,
  Meeteetse 050768
Charles M. Schulz Museum and
  Research Center, Santa Rosa . 053138
Charles R. Conner Museum,
  Pullman 052414
Charles Rennie Mackintosh Society,
  Glasgow 060088
Charles River Museum of Industry,
  Waltham 054049
Charles River Street Antiques,
  Boston 092454
Charles Sturt Museum, Grange . 001120
Charles Sturt University Art Collection,
  Wagga Wagga 001610
Charles Towne Landing 1670,
  Charleston 047542
Charles, Bernard, Pont-de-Briques-
  Saint-Etienne 072247
Charles, Daniel, Ashbourne .. 088297,
                            134039
Charles, Jean, Valenciennes ... 073942
Charles, Mary, Neuville-de-Poitou 070526
Charles, Oliver, Petworth 090994
Charles, Sylvain, Saint-Julien-les-
  Morey 072868
Charlesfort Antiques, Manakau . 083981
Charleston Heights Art Center Gallery,
  City of Las Vegas, Las Vegas . 050158
The Charleston Museum,
  Charleston 047543
Charleston Trust, Firle 044346
Charlesworth, Benny, Great Harwood,
  Lancashire 089367
Charleville Historic House Museum,
  Charlesville 000954
Charlie Finds Great Old Stuff,
  Seattle 097412
Charlier & Michiels, Bruxelles . 063372
Charlie's Art Studio and Gallery,
  Jacksonville 121239
Charlie's Arthurs Seat Auto Museum,
  Arthurs Seat 000806

Charlie's Restoration, Columbus 135476
Charlie's Used Furniture,
  Pittsburgh 096266
Charlot, Jack, Bessey-lès-Cîteaux 067144
Charlotte and Philip Hanes Art Gallery,
  Wake Forest University, Winston-
  Salem 054487
Charlotte Antiques and Collectibles
  Show, Charlotte 098441
Charlotte Berlins Museum, Ystad 041181
Charlotte County Historical Center,
  Charlotte Harbor 047578
Charlotte County Museum, Saint
  Stephen 006936
Charlotte Museum of History and
  Hezekiah Alexander Homesite,
  Charlotte 047570
Charlotte Nature Museum,
  Charlotte 047571
Charlotte's Web, Seattle 097413
Charlott's Antik, Stockholm ... 086984,
                            133568
Charlott's Antiques, San Antonio 096957
Charlton Hall Galleries, Columbia 127185
Charlton House Antiques, London 090005
Charm of Yesteryear, Houston . 093637
Charmasson, Danielle, Uzès ... 073921
Le Charme, Milano 080485
Charme Antique, Québec 064531
Charme d' Antan, Ribecourt-
  Dreslincourt 072451
Charme d'Antan, Beirut 082278
Charme d'Antan, Fresnes-sur-
  Escaut 068612
Charme du Passé, Le Havre ... 069278
Charmerie, Bourges 067424
La Charmerie, Saint-Denis 114117
Charming You, Jacksonville ... 093971
Charmoy, Xavier, Blois 140785
Charney Bassett Mill, Wantage . 046049
Charnley-Persky House Museum,
  Chicago 047645
Charnwood Antiques,
  Sawbridgeworth 091248, 135009
Charnwood Museum,
  Loughborough 045151
Charo y Pat, Ciutadella de
  Menorca 085937
Charpentier, Paris 104641
Charpentier, Denis, Metz 070092
Charpentier, Fabrice, Verdun
  (Meuse) 074018
Charpentier, Marc, La Rochelle . 069084
Charpentier, Marc, Metz 070093
Charpentier, Patrice, Evreux ... 068494
Charpille, Jean-Pierre, Saint-
  Étienne 072756
Charra Franck, Saint-Péray ... 073252
Charrette, Bernay 067108
Charrier-Villepreux, Brigitte, La
  Châtre 068988
Charrier, Fabrice, Bonneval ... 067291
CHArt Computers and the History of
  Art, School of History of Art, Film
  and Visual Media, London 060089
Chart House, Shenfield 091297
Charta, Bergamo 109791
Charta, Milano 137672
Chartaphilus, Milano 142718
Chartered Insurance Institute's
  Museum, London 044938
Chartered Society of Designers,
  London 060090
Charterhouse, Sherborne 127042
Charterhouse School Museum,
  Godalming 044438
Charton, Hubert, Salins-les-Bains 073381
Charton, Roland, Avignon ... 103303,
                            103304
Chartouni & Sehnaoui, Beirut .. 082279,
                            126255
Chartrain, Flers 068540
Chartres, Neuilly-sur-Seine ... 070520
Chartres Antiquités, Chartres .. 067858

Chartres, Stéphane, Ribérac ... 072453
Chartreuse International, Seattle 097414
Les Chartreux, Toulouse 141361
Chartwell, Westerham 046097
Charveriat, Jean, Lyon 104075
Charvet, Arnaud, Paris 104642
Charvet, Christian, Grenoble .. 103785
Charytonowicz, Zbigniew,
  Warszawa 084708
Chasamp, Cincinnati 120329
Chase and District Museum,
  Chase 005601
Chase Art Centre, Wadebridge . 119293
Chase Art Gallery, Wadebridge . 119294
Chase County Historical Society,
  Cottonwood Falls 048028
Chase Fine Art, Stafford 119132
Chase Gallery, Boston 119850
Chase Home Museum of Utah Folk Art,
  Salt Lake City 052911
The Chase Manhattan Bank Art
  Collections, New York 051358
Chase, Dolores, Salt Lake City . 123952
Chasen, Richmond 123727
Chass, Pittsburgh ... 096267, 136129
Chassagne, Annie & Jean-Pierre,
  Toulouse 073752
Chassaigne, Yvette de la, Pont-
  Aven 105371
Chassaing, Hervé, Toulouse ... 125823
Chassaniol-Bénard, Elisabeth,
  Vire 074247, 141391
Chassany, Georges, Le Broc ... 069230
Chassany, Raphaël, La Broc ... 069231
Chassay, Christiane, Montréal . 101374
La Chasse aux Trésors Antiques,
  Montréal 064402
Châsse Sainte-Bernadette et Musée,
  Nevers 014466
Chasset, Patrick, Rouen 072576
Chasseurs d'Horizons, Chamonix-Mont-
  Blanc 067800
The Chassidic Art Institute,
  Brooklyn 047254
Chassin, Joëlle, Saint-Ouen-
  d'Attez 073231
Chastagner, André, Lyon 069721
Chastagnier, Bernard, Bandol .. 066929
Chastenay, Florence de, Paris . 141073
Chatarras Luque, Alicante 085695
Château, Arlay 011727
Château, Assas 011760
Château, Duilhac-sous-
  Peyrepertuse 012840
Château, Le Bourg-Saint-Léonard 013610
Château, Prudhomat 015003
Château, Rousson 015180
Château, Saint-Ouen 073051
Château, La Sarraz 041941
Château, Vogue 016318
Château à Motte de la Haie Joulan,
  Saint-Sylvain-d'Anjou 015571
Château Beaufief, Mazeray ... 014100
Château Bisso, Voiron 074271
Chateau Chic, Phoenix 096198
Chateau d' Arts, Singapore ... 114506
Château d'Abbadie, French National
  Academy of Scienes, Hendaye 013248
Château d'Aigle, Aigle 041199
Château d'Aigremont, Seraing .. 004008
Château d'Angers, Angers 011648
Château d'Arcelot, Arceau 011709
Château d'Assier, Assier 011761
Château d'Aubenas, Aubenas . 011769
Château d'Auvers, Voyage au Temps
  des Impressionnistes, Auvers-sur-
  Oise 011814
Château d'Avignon, Saintes-Maries-de-
  la-Mer 015638
Château d'Azay-le-Ferron, Azay-le-
  Ferron 011854
Château de Beloeil, Beloeil .. 003356
Château de Belvoir, Belvoir .. 011983
Château de Biron, Monpazier . 014201

Château de Bonneval, Coussac-
  Bonneval 012706
Le Château de Bouchons, Saint-
  Loubès 105543
Château de Bouges, Bouges-le-
  Château 012130
Château de Bourdeilles,
  Bourdeilles 012150
Château de Boury-en-Vexin, Boury-en-
  Vexin 012174
Château de Bruniquel Semateg,
  Bruniquel 012246
Château de Busca-Naniban,
  Mansencôme 013994
Château de Buzay, La Jarne ... 013435
Château de Cadillac, Cadillac . 012270
Château de Castelnou, Castelnou 012362
Château de Cazeneuve, Préchac 014993
Château de Cénévières,
  Cénévières 012382
Château de Champchevrier, Cléré-les-
  Pins 012591
Château de Châteaudun,
  Châteaudun 012489
Château de Chevannes-Saint Racho,
  La Clayette 013393
Château de Chillon, Montreux .. 041753
Château de Clarbec, Paris 071152
Château de Coppet, Coppet ... 041426
Château de Courtanvaux, Bessé-sur-
  Braye 012011
Le Château de Craon, Craon .. 012713
Château de Crazannes,
  Crazannes 012715
Château de Deulin, Fronville .. 003607
Château de Fénelon, Sainte-
  Mondane 015628
Château de Fernelmont,
  Fernelmont 003595
Château de Fleurac, Nersac .. 014446
Château de Fleurigny, Thorigny-sur-
  Oreuse 015954
Château de Fléville, Fléville-devant-
  Nancy 012993
Château de Fougères-sur-Bièvre,
  Fougères-sur-Bièvre 013035
Château de Frolois, Frolois ... 013061
Château de Galard, Terraube .. 015925
Château de Gandels,
  Garrevaques 068637
Château de Germolles, Mellecey 014117
Château de Grandson, Grandson 041562
Château de Gratot, Exposition
  Permanente Historique Huit Siècles
  de Vie, Gratot 013165
Château de Gruyères, Gruyères . 041572
Château de Gy, Gy 013220
Château de Hautefort, Hautefort 013239
Château de Jannée, Pessoux .. 003949
Château de Kergrist, Ploubezre . 014900
Château de la Bastie d'Urfé, Saint-
  Étienne-le-Molard 015311
Château de la Baume, Prinsuéjols-
  Marvejols 014999
Chateau de la Compagne, Los
  Angeles 094283
Château de la Guerche, La
  Guerche 013430
Château de la Motte-Tilly, Centre des
  Monuments Nationaux, La Motte-
  Tilly 013444
Château de la Roche-Guyon, La
  Roche-Guyon 013463
Château de la Souche, Doyet .. 012833
Château de Lacypierre, Saint-Crépin-
  Carlucet 015273
Château de l'Alba-la-Romaine, Alba-la-
  Romaine 011574
Château de l'Angotière, Domjean 012805
Château de Larroque-Toirac, Larroque-
  Toirac 013579
Château de Lastours, Bazièe .. 066987
Château de Léhélec, Béganne . 011964
Chateau de l'Ermitage, Cercle

Historique et Archéologique de Wavre et de la Region, Wavre . 004139
Château de l'Hermitage, Condé-sur-l'Escaut 012654
Château de Lichtenberg, Lichtenberg 013792
Château de Losse, Thonac 015942
Château de Lourmarin, Lourmarin 013906
Château de Louvignies, Notre-Dame-de-Louvignies 003915
Château de Luynes, Luynes . . . 013926
Château de Maintenon, Maintenon 013973
Château de Maisons-Laffitte, Centre des Monuments Nationaux, Maisons-Laffitte 013976
Château de Meillant, Meillant . . 014113
Château de Messilhac, Raulhac 015039
Château de Meux, Jonzac 013324
Château de Mogère, Montpellier 014287
Château de Montaigut, Camares 012296
Château de Montaner, Montaner 014218
Château de Montcornet, Montcornet-en-Ardenne 014243
Château de Monte-Cristo, Le Port-Marly 013690
Château de Montignies-sur-Roc, Montignies-sur-Roc 003888
Château de Montméry et son Parc, Ambazac 011609
Château de Montmuran, Les Iffs 013752
Chateau de Mores State Historic Site, Medora 050765
Château de Moulineaux Robert-le-Diable, Moulineaux 014353
Château de Parentignat, Parentignat 014615
Château de Poudenas, Poudenas 014978
Château de Prades, Lafox 013524
Château de Puymartin, Marquay 014030
Château de Ratilly, Treigny 016039
Château de Reinhardstein, Robertville 003969
Château de Rixensart, Rixensart 003968
Château de Rochambeau, Thoré-la-Rochette 015950
Château de Rochefort-en-Terre, Rochefort-en-Terre 015125
Château de Roquedols, Meyrueis 014162
Château de Rosanbo, Lanvellec 013567
Château de Saint-Aubin-sur-Loire, Saint-Aubin-sur-Loire 015229
Château de Saint-Jean-de-Beauregard, Saint-Jean-de-Beauregard . . . 015372
Château de Sauvan, Mane 069836
Château de Solre-sur-Sambre, Solre-sur-Sambre 004036
Château de Sombreffe, Sombreffe 004037
Château de Sourdis, Gaujacq . . 013084
Château de Spontin, Yvoir 004159
Château de Sully-sur-Loire, Sully-sur-Loire 015892
Château de Talcy, Marchenoir . . 014001
Château de Tarascon, Musée du Château du Roy René, Tarascon (Bouches-du-Rhône) 015906
Château de Temniac, Sarlat-la-Canéda 015686
Château de Trazegnies, Musée Vie et Travail de nos Aïeux, Trazegnies 004089
Château de Trevarez, Saint-Goazec 015354
Château de Vascœuil, Centre d'Art Musée Michelet, Vascœuil 016131
Château de Vaurenard, Gleize . . 013122
Château de Véves, Celles 003508
Château de Villegongis, Villegongis 016251
Château de Villeneuve, Saint-Germain-Lembron 015346
Château de Villeneuve, Vence . . 016150
Château de Voltaire/Centre des

Monuments Nationaux, Ferney-Voltaire 012968
Le Château d'Eau, Toulouse . . . 015984
Le Château des Allymes, Ambérieu-en-Bugey 011611
Château des Pêcheurs, La Bussière 013370
Château des Princes de Chimay, Chimay 003520
Château des Rois Ducs, Sauveterre-la-Lémance 015722
Château des XIVe-XVIIe s, Époisses 012888
Château d'Estignols, Aurice 011798
Château d'Ételan, Saint-Maurice-d'Ételan 015462
Château d'Oiron, Oiron 014554
Château d'Oron, Oron-le-Châtel 041831
Château du Bousquet, Musée d'Art Ancien, Laguiole 013527
Château du Chassan, Faverolles (Cantal) 012955
Château du Clos de Vougeot, Vougeot 016326
Château du Clos-Lucé, Parc Léonard de Vinci, Amboise 011620
Château du Fosteau, Leers-et-Fosteau 003779
Château du Haras du Pin, Le Pin-au-Haras 013681
Château du Haut-Koenigsbourg, Orschwiller 014592
Château du Lude, Le Lude 013652
Château du Riau, Villeneuve-sur-Allier 016273
Château du Rivau, Lemere 013729
Château du Val-Saint-Lambert, Seraing 004009
Château du XVIe s, Ancy-le-Franc 011641
Château d'Ussé, Rigny-Ussé . . . 015092
Château et Musée de Filain-Fort, Filain 012981
Château et Musée de Valangin, Valangin 042092
Le Chateau Export, Tampa . . 097551, 097552
Château Féodal de Beynac, Beynac-et-Cazenac 012017
Château feodal de Corroy-le-Château, Monsieur le Marquis de Trazegnies, Corroy-le-Château 003523
Château Fort, Bouillon 003389
Château Franc-Waret, Franc-Waret 003605
Château Frontenac, Québec . . . 101496
Château Guillaume le Conquérant, Falaise 012948
Château Histoire de Tours, Tours 016015
Château l'Hospitalet, Narbonne . 014426
Château Logue, Maniwaki 006252
Château Malromé, Saint-André-du-Bois 015219
Château Médiéval de Tancarville, Tancarville 015903
Château Médiéval et Musée Delporte, Thy-le-Château 004066
Château Médiéval et Renaissance, Châteaubriant 012487
Château Morin Antiquités, Sablonnières 072652
Château-Musée, Aulan 011789
Château-Musée, Boën-sur-Lignon 012083
Château-Musée, Brissac-Quincé 012233
Château-Musée, Carrouges 012346
Château-Musée, Collonges-lès-Bévy 012625
Château-Musée, Cons-la-Grandville 012662
Château-Musée, Cormatin 012673
Château-Musée, Davayat 012745
Château-Musée, Dinan 012792
Château-Musée, Ferrières-en-

Brie 012972
Château-Musée, Haroué 013230
Château-Musée, Heudicourt 013262
Château-Musée, Lavardens 013594
Château-Musée, Lavoûte-sur-Loire 013601
Château-Musée, Le Plessis-Brion 013683
Château-Musée, Lenoncourt . . . 013730
Château-Musée, Maillebois 013969
Château-Musée, Menthon-Saint-Bernard 014128
Château-Musée, Meung-sur-Loire 014155
Château-Musée, Montmirail 014278
Château-Musée, Pange 014610
Château-Musée, Pirou 014876
Château-Musée, Roche-la-Molière 015115
Château-Musée, Saumur 015713
Château-Musée, Tourrette-Levens 016013
Château-Musée, Valprivas 016114
Château-Musée, Vendeuvre 016152
Château-Musée, Villandry 016217
Château-Musée, Virieu-sur-Bourbre 016305
Château-Musée Anjony, Tournemire 016007
Château-Musée d'Andurain, Mauléon-Licharre 014087
Château-Musée d'Anet, Anet . . . 011646
Château-Musée d'Art Populaire et Religieux, Clermont-en-Genevois 012593
Château-Musée d'Aulteribe, Sermentizon 015778
Château-Musée de Bois-Chevalier, Legé 013725
Château-Musée de Bonneville, Chamblac 012422
Château-Musée de Brie, Champagnac-la-Rivière 012429
Château-Musée de Bussy-Rabutin, Bussy-le-Grand 012257
Château-Musée de Cany, Cany-Barville 012318
Château-Musée de Château Dauphin, Pontgibaud 014955
Château-Musée de Chémery, Chémery 012544
Château-Musée de Chenonceau, Chenonceaux 012545
Château-Musée de Combourg, Combourg 012636
Château-Musée de Condé, Condé-en-Brie 012653
Château-Musée de Courson, Courson-Monteloup 012703
Château Musée de Cugney, Cugney 012731
Château-Musée de Culan, Culan 012733
Château-Musée de Dampierre-sur-Boutonne, Aulnay (Charente-Maritime) 011790
Château-Musée de Dieppe, Dieppe 012767
Château-Musée de Filières, Gommerville 013125
Château-Musée de Jussy, Jussy-Champagne 013344
Château-Musée de la Batisse, Chanonat 012444
Château-Musée de la Gataudière, Marennes 014016
Château-Musée de la Grange, Manom 013990
Château-Musée de la Nature, Bonnefamille 012099
Château-Musée de la Possonnière, Couture-sur-Loir 012708
Château-Musée de la Roche, Chaptuzat 012450
Château-Musée de la Roche-Racan, Saint-Paterne-Racan 015491

Château-Musée de la Rochelambert, Saint-Paulien 015502
Château-Musée de la Trémolière, Anglards-de-Salers 011660
Château-Musée de Lacassagne, Saint-Avit-Frandat 015230
Château-Musée de Lantheuil, Lantheuil 013566
Château-Musée de Maupas, Morogues 014335
Château-Musée de Miromesnil, Tourville-sur-Arques 016026
Château-Musée de Montaigu, Laneuveville-devant-Nancy . . . 013550
Château-Musée de Montal, Saint-Jean-Lespinasse 015390
Château-Musée de Montfleury, Avressieux 011850
Château-Musée de Montgeoffroy, Mazé 014099
ChÂteau Musée de Montmorin, Beillom 011966
Château-Musée de Panloy, Port-d'Envaux 014966
Château-Musée de Pont-Chevron, Ouzouer-sur-Trézée 014599
Château-Musée de Préhistoire, Belesta-de-la-Frontière 011967
Château-Musée de Puyguilhem, Villars (Dordogne) 016223
Château-Musée de Ravignan, Perquie 014845
Château-Musée de Ripaille, Thonon-les-Bains 015947
Château-Musée de Serrant, Saint-Georges-sur-Loire 015335
Château-Musée de Teillan, Aimargues 011553
Château-Musée de Terre-Neuve, Fontenay-le-Comte 013015
Château-Musée de Vaulaville, Tour-en-Bessin 016000
Château-Musée de Veauce, Ébreuil 012851
Château-Musée de Villeprévost, Tillay-le-Péneux 015965
Château-Musée de Villersexel, Villersexel 016287
Château-Musée de Villesavin, Tour-en-Sologne 016001
Château-Musée de Vollore, Vollore-Ville 016321
Château-Musée d'Effiat, Aigueperse 011543
Château-Musée des Ducs, Duras 012846
Château-Musée des Martinanches, Saint-Dier-d'Auvergne 015293
Château-Musée des XVIe-XVIIe s, Tanlay 015905
Château-Musée du Fort La Latte, Rennes (Ille-et-Vilaine) 015062
Château-Musée du Grand-Saussay, Ballancourt-sur-Essonne 011877
Château-Musée du Maréchal Lyautey, Thorey-Lyautey 015952
Château-Musée du Moulin, Lassay-sur-Croisne 013581
Château-Musée du Plessis-Bouré, Ecuillé 012856
Château-Musée du Plessis-lès-Tours, La Riche 013458
Château-Musée du Rhône, Tournon-sur-Rhône 016009
Château-Musée du Touvet, Le Touvet 013705
Château-Musée du Val, Lanobre 013561
Château-Musée du XVe et XVIIIe Siècle, Commarin 012639
Château-Musée du XVIIe Siècle, Fontaine-Française 013008
Château-Musée d'Urtubie, Saint-Jean-de-Luz 015376
Château-Musée et Ecuries de Chaumont-sur-Loire, Chaumont-sur-

Loire 012532
Château Musée Grimaldi, Musée de
l'Olivier, Musée d'Art Moderne et
Contemporain, Donation Suzy Solidor,
Cagnes-sur-Mer 012282
Château-Musée la Fayette, Chavaniac-
Lafayette 012536
Château-Musée le Denacre,
Wimille 016349
Château-Musée Listran, Jau-Dignac-et-
Loirac 013312
Château-Musée Mansart,
Sagonne 015198
Château-Musée Municipal,
Nemours 014441
Château-Musée Renaissance, Azay-le-
Rideau 011855
Château-Musée Turcan, Ansouis 011680
Château-Musée Verdus, Saint-Seurin-
de-Cadourne 015564
Château-Musée, Musée des Beaux-
Arts, Boulogne-sur-Mer 012138
Chateau Napoleon, Los Angeles 094284
Le Chateau of Antiques, Dallas . 093088
Château-Promenade des Parfums,
Chilleurs-aux-Bois 012558
Château Royal d'Amboise,
Amboise 011621
Château Royal, Musée de la Côte
Catalane, Collioure 012618
Château Saint-Hugues, Semur-en-
Brionnais 015761
Château, Bernard, Augnax 066809
Chateaubriand Antiques,
Burwash 088736
Chateauminois, Cyrille, Dardilly . 128576
Chateauminois, Marc, Toulon
(Var) 129286
Châtelain, Jeannette, Orbe 116556
Chatelet Chaume, Didier, Roanne 072486
Chatelherault, Ferniegair 044340
Chatelus, Alain, Lyon 128704
Chatham College Art Gallery,
Pittsburgh 052161
Chatham Islands Museum, Waitangi,
Chatham Islands 033329
Chatham-Kent Museum, Chatham,
Ontario 005604
Chatilez, Guy, Tourcoing 105756
Chatillon-DeMenil Mansion, Saint
Louis 052793
Châtre, Martine et Thibault de la,
Paris 104643
Chatsworth, Mamaroneck 127303
Chatsworth House, Bakewell . . . 043468
Chattahoochee Valley Art Museum, La
Grange 050033
Chattanooga African Museum, Bessie
Smith Hall, Chattanooga 047590
Chattanooga Regional History Museum,
Chattanooga 047591
Chattel, Dallas 135497
Chattel Changers, Milwaukee . . 094842
Chattels, Lija 082414
Chattels Antiques, Warrandyte . . 062408
Chattels Antiques, Wonga Park . 062448
Chattishgarh Mudra Sangrahalaya,
Katora Talab 024342
Chatto & Windus, London 105786
The Chatton Gallery, Chatton . . 117556
Chau, K.H., Hong Kong 065019
Chaubard, Dominique & Claude,
Fontanes 068569
Chaucer, London 090006
Chaucer Bookshop, Canterbury . 134205,
144305
Chaucer-Head Bookshop and Art
Gallery, Stratford-upon-Avon . . 144864
Chaudesaigues, Stéphane, Vaison-la-
Romaine 105786
Chaudoir, M.-C., Bruxelles 063373
Le Chaudron Magique, Brugnac . 012240
Chaufour, Chatellerault 067904

Chaume Gonzalez, Miguel Pablo,
Valencia 133457
Chaumeil Expo, Clermont-
Ferrand 103601
Chaumet, İstanbul 088111
La Chaumière, Paris 128959
Chaumière d'Antan, Mérignac
(Gironde) 070071
Chaumière Rustique, Saint-Jean-de-
Monts 072850
Chaussat, Dominique, Thairé . 073640,
129276
Chaussavoine-Delecroix, Paule,
Rennes (Ille-et-Vilaine) 072430
Chaussin, Jacques, Lyon 125623
Chaussinand, Yves, Villeurbanne 074198,
129355
Chauveau, Pierre, Hyères 068852
Chauvin Egaze Ghislaine, Longny-au-
Perche 069616
Chauvin, Didier, Chambéria 128539
Chauvin, J M, Rouen 072577
Chauzy, François, Tournus 073824
Chavanne, Henriette, Cormatin . 103631
Chavarría, Solano Heriberto,
Alajuela 102462
Chave, Alphonse, Vence 105829
Chaves Ferreira, Lisboa 137918
Chavey, Felix, New York 135898
Chavez, George, San Antonio . 136257
Chawkandi Art, Karachi 113302
Chawner, John, Birchington . . 088467,
134105
Chaya-machi Gallery, Osaka . . . 111349
Chaye, Honfleur 103822
Chayette & Cheval, Paris 125693
Chazanof, Ilene, New York 095308
Chazen Museum of Art, Madison 050593
Cheap Antiques, Kansas City . . 094065
Cheap Art, Athinai 109030
Chebellas, Trina, Sacramento . . 096609
Chebský Hrad, Krajské Muzeum Cheb,
Cheb 009514
Cheddar Man & the Cannibals,
Cheddar 043916
Cheddleton Flint Mill, Cheddleton 043917
Chedworth Roman Villa,
Yanworth 046212
Chedworth Roman Villa Museum,
Cheltenham 043924
Cheekwood Museum of Art,
Nashville 051174
Cheese Factory Studio Gallery,
Meadows 099292
Cheever, Alamo Heights 144951
Cheffertons, Antwerpen 063199
Cheffins, Cambridge 126843
Chégaray, Sylvie, Honfleur 068829
Chehel-Sotoon Museum, Qazvin 024727
Chehel-Sotoon Palace Museum,
Isfahan 024700
Chehel-Stoun Museum, Ghazvin 024691
Cheim & Read, New York 122465
Cheiro de Mofo, Belo Horizonte . 063893
Chelesa Art, Hong Kong 102065
Chelines, Hondarribia 086048
Chelini, A. & C., Firenze 080119
Chelini, Fabio, Livorno 080320
Chelley's Antiques, Dunedin . . . 083922
Chellini Silvano & C., Firenze . . 080120
Chelmer, Chelmsford 117561
Chelminski, London 090007
Chelmsford Historical Society Museum,
Chelmsford 047602
Chelmsford Museum, Chelmsford 043918
Chelsea, Laufen 116406
Chelsea, Philadelphia 136077
Chelsea Antichitá, Firenze 080121
Chelsea Antique Mirrors, London 090008,
134618
Chelsea Antiques, Los Angeles . 094285
Chelsea Antiques, Strandon
Taranaki 084086
Chelsea Antiques and Decorative Art

Centre, Albion 060978
Chelsea Antiques Fair, London . 098367
Chelsea Art Fair, London 098368
Chelsea Art Museum, Miotte
Foundation, New York 051359
Chelsea Book Shop, San
Francisco 145549
Chelsea Books, London 144575
Chelsea Ceramic Guild, New York 060429,
122466
Chelsea Clocks and Antiques,
London 090009
Chelsea College of Art and Design,
University of the Arts London,
London 056555
Chelsea Court House Museum,
Chelsea 000956
Chelsea Galleria, Miami 121757
Chelsea Galleries, Kingston 111214
Chelsea Gallery, London 144576
Chelsea Gallery, Tulsa 124798
Chelsea Gallery, Wynberg 114824
Chelsey, Metford 061845
Chelston Gallery, Torquay 119242
Cheltenham Antique Market,
Cheltenham 088830
Cheltenham Art Center,
Cheltenham 056865
Cheltenham Art Gallery and Museum,
Cheltenham 043925
Chelwood, Cheltenham 117562, 138150
Chemainus Valley Museum,
Chemainus 005609
Chemin Christian, Rivedoux-
Plage 072480
Chemin de Fer-Musée Blonay-Chamby,
Chaulin 041405
Chemin de la Soie, Molière-
Cavaillac 014193
Chemin des Arts, Paris 141074
Le Chemin du Fauvisme, Réseau
Culturel Terre Catalane,
Collioure 012619
Chemin, François, Paris 128960
La Cheminée Antiquitées, Paris . 071153
Chemini, Claudie, Darnieulles . . 068276
Chemins Privés, Hyères 068853
Chemnitzer Künstlerbund e.V.,
Chemnitz 058812
Chemnitzer Verlag, Chemnitz . . 137132
Chemould, Kolkata 109395
Chemould, Mumbai . . 109422, 109423
Chemstrip, Regina . . 064550, 128196
Chemung Valley History Museum,
Elmira 048566
Chen & Wiedemann, München . 077478
Chen Chung-Tsang Museum,
Dongshan 042304
Chen Feng-Hsien Miniature Art
Museum, Sindian 042457
Chen Jonghui Memorial Museum,
Yuanli 042599
Chen Peiqiang, Perugia 081139
Chen Sheung Ning, Hong Kong . 065020
Chen Xin Tang, Nanjing 102148
Chen, J.F., Los Angeles 094286
Chen, J.F., West Hollywood 097879
Chenal, Jacques, Grenoble 068723
Chenal, Jacques, Seyssinet-
Pariset 073519
Chenango County Historical Society
Museum, Norwich 051651
Le Chêne dans la Maison,
Tournai 063844
Le Chêne d'Or, Rouen 072578
Chenel, Alain, Paris 071154
Chenel, Alain-Pierre, Nice 070585
Chenevière, Antoine, London . . . 090010
Chenevière, Pierre, Antibes-Juan-les-
Pins 066669
Cheney, Sacramento 096610
Cheney Homestead, Manchester 050621
Cheney, Robert, Finedon 089275
Cheng Shuang Kuei, Hong Kong 065021

Cheng Ta Art Gallery, National Cheng
Kung University, Tainan 042484
Chengdu Museum, Chengdu . . . 007691
Chengdu Painting Gallery,
Chengdu 102012
Chenghai Museum, Chenghai . . 007701
Chengshiu University Art Center,
Niaosong 042410
Chenkier, Strasbourg . . 073578, 105681
Chenkier, M., Scharrachbergheim 073459
Chenoir, Lidie, Thouars 073675
Chenonceau, Washington 097821
Chenu, Jean, Lyon 125624
Chenu, Scrive & Bérard, Lyon . . 125625
Chenue à la Croix de Lorraine,
Clichy 068040
Cheo, Oakland 123247
Cheongdam Gallery, Seoul 111621
Cheongju Early Printing Museum,
Cheongju 029906
Cheongju National Museum,
Cheongju 029907
Cheongju University Museum,
Cheongju 029908
Chepstow Museum, Chepstow . . 043927
Cheraw Lyceum Museum,
Cheraw 047603
Chercheur d'Art, Metz 104221
Chéreau, Jacques, Sannerville . 105652,
129252
Cherif Fine Art, Sidi Bou Saïd . . 116982
Cheriff, Charles, New York 095309
Cherishables, Washington 097822
Cherner, Bruce, Boston 092455
Cherokee Coins and Jewelry,
Louisville 094529
Cherokee County Historical Museum,
Murphy 051128
Cherokee Courthouse Museum,
Gore 049162
Cherokee National Museum,
Tahlequah 053652
Cherokee Strip Land Rush Museum,
Arkansas City 046529
Cherokee Strip Museum, Alva . . 046429
Cherokee Strip Museum and Henry
S. Johnston Library, Oklahoma
Historical Society, Perry 051995
Chéron, Henry, Bernay 067109
Cheroutre, J., Oostende 128084
Cherrier, Francis, Agnac 066429
Cherrington, T.A., Teddington . . 144877
Cherruault, Isabelle, Verneuil-sur-
Avre 074028
Cherry, Washington 097823
Cherry Antiques, Hemel
Hempstead 089495
Cherry Art Studio, San Francisco 124252
Cherry County Historical Society
Museum, Valentine 053929
Cherry Holt Studio, Sutterton . . . 119183
Cherry Lane, Toronto 064625
Cherry Orchard Antiques,
Wellington 084160
Cherry Pelavin Fine Arts, New
York 138386
Cherry Ripe Antiques, Concord . 061358
Cherry Street Antiques,
Columbus 093001
Cherry Street Interiors, Kansas
City 094066
Cherry Tree Auctions, Gansevoort 127233
Cherry Valley Museum, Cherry
Valley 047610
Cherry, Peter, Hahndorf 061573
Cherrydelosreyes, Los Angeles . 121457
Cherrylane Fine Arts, Delgany . . 109563
Cherrys' Art Center, Columbus . 120482
Chertsey Antiques, Chertsey . . . 088849
Chertsey Museum, Chertsey . . . 043929
Cherub, Carshalton 088802
Cherub, Washington 097824
Cherwell Framing Gallery,
Kidlington 118041

Cheryl & Pete's, Yackandandah . 062495
Chesapeake and Ohio Canal Tavern
  Museum, Sharpsburg 053289
Chesapeake Bay Maritime Museum,
  Saint Michaels 052834
Chesapeake Beach Railway Museum,
  Chesapeake Beach 047612
Chesham, Singapore 114507
Chesham Graphics Gallery,
  Chesham 117566
Chesham Museum-The Stables,
  Chesham 043930
Cheshire Antiques and Fine Art Fair,
  Knutsford 098369
Cheshire Brick & Slate, Tarvin . 091541,
  135119
Cheshire County Museum, Keene 049911
Cheshire Military Museum,
  Chester 043931
Chesnais, Eric, Alençon 103220
Chesney, Atlanta 092120
Chesney, London . . . 090011, 090012
Chesney, New York 095310
Chess Antiques, Chesham 088851,
  134249
Chess Expo, 's-Hertogenbosch . 112525
Chessler, Baltimore 119731
Chest of Drawers, London 090013
The Chest of Drawers, Tetbury . 091568
Chester Antiques and Fine Art Show,
  Chester 098370
Chester Art Guild, Chester 047618
The Chester Beatty Library,
  Dublin 024871
Chester County Historical Society
  Museum, West Chester 054293
Chester County Museum,
  Chester 047617
Chester, Peter & Catherine, San
  Diego 097073, 136293
Chesterfield, Fresno 093493
Chesterfield Antiques,
  Birmingham 088475
Chesterfield Historical Society of
  Virginia, Chesterfield 047622
Chesterfield Museum and Art Gallery,
  Chesterfield 043935
Chester's Closet, Denver 120702
Chesterton, Margaret, Saint
  Austell 091180
Chesterwood Museum,
  Stockbridge 053533
Chesti, Valter, Roma 131855
Chestico Museum, Port Hood . . 006688
Chestnut Hill Antiques,
  Philadelphia 096092
Chestnut Hill Resale, Philadelphia 096093
Chet Krupski, Tampa 097553
Chethams's Hospital and Library,
  Manchester 045201
Chetopa Historical Museum,
  Chetopa 047626
Chetwynd Stapylton, Portland . . 123611
Cheung, Sacramento 096611
Cheung Kong School of Art and
  Design, Shantou University,
  Shantou 055236
Cheung, Shao Tieh, Milano 080486
Le Cheval Blanc, Ury 073917
Le Cheval de Sable, Paris 104644
Chevalerias, Stéphane, Dirac . . 068360
Chevalet d'Eon, Salon-de-
  Provence 105645
Chevalier-Bourdiol, Marie-Joëlle, Saint-
  Cannat 072704
Chevalier Conservation,
  Colombes 068100, 128567
Chevalier, Dominique, Paris . . . 071155
Chevalley, Annie, La Tour-de-
  Peilz 116671
Chevallier, Stéphane, Guillestre . 068770
Les Chevau-Légers, Lorient . . . 069634
Le Cheverny, Bruxelles 063374
Chevertons of Edenbridge,

Edenbridge 089157
Chevillon, Gilles, Corcelles-en-
  Beaujolais 068147
Cheviot, Cincinnati 092837
Cheviot, Los Angeles 135712
Cheviot Museum, Cheviot 033165
Chevrier Franck, Chaumont (Haute-
  Marne) 067932
Chevrier, Guy, Montréal 064403
Chevy Chase Antiques,
  Washington 097825
Chevy Chase Gallery, Washington 124882
Chew, Betek, Singapore 114508
Cheyenne Frontier Days Old West
  Museum, Cheyenne 047628
Cheyne House Antiques, Rye . . 091160
Cheyne, Patrick, Altrincham . . . 126797
Chez Annamarie, Praha 102656
Chez Antique, Sacramento 096612
Chez Arsène, Batz-sur-Mer . . . 066962
Chez Betty, Charleroi 063595
Chez Colombine, Auvers-sur-Oise 066840
Chez Daniel, Les Fourgs 069409
Chez Dany, Saint-Ouen 073052
Chez Desserey, Le Pailly 069321
Chez Grand'Mère, Cabestany . . 067543
Chez Jules de chez Smith,
  Gerpinnes 140228
Chez Maxine, Montréal 064404
Chez nos Anciens, Boulogne-sur-
  Mer 067398
Chez Pierrot, Notre-Dame-de-
  Riez 070749
Chez Swann, Antiquité – Décoration,
  Paris 071156
Chez Will, Paris 071157
Chez Willi, Biel 087314
Chez, Françoise, Annesse-et-
  Beaulieu 066653
Chez, Joanne, Piaseczno 113714
Chi City Museum, Chicheng . . . 007702
Chi Hong, Taipei 088067
Chi Mei Museum, Rende 042428
Chi-Wen Gallery, Taipei 116906
Chia Chi Lan Wine Museum,
  Yilan 042589
Chia Hsien Fossil Museum,
  Jiasian 042353
Chia Nan Museum of Culture and Art,
  Chia Nan University of Pharmacy and
  Science, Rende 042429
Il Chiacchierino, Bari 079761
Chiadò, Torino 132115
Chiais, Valentina, Venezia 081828
Chianelli, Rosario, Roma 131856
Chiang Kai-shek's Shihlin Residence,
  Taipei 042503
Chiang Mai Contemporary Art
  Museum, Chiang Mai 042685
Chiang Mai National Museum, Chiang
  Mai 042686
Chiang Saen National Museum,
  Chiang Rai 042688
Chiappa & Didonè, Brescia 131026
Chiappetta, Claude, Aix-en-
  Provence 066458
Chiara, Warszawa 113813
Chiara Antichità, Modena 080786
Chiari, Piero, Firenze . . 142667, 142668
Chiaroscuro, Chicago 120089
Chiaroscuro, Richmond 123728
Chiarugi, Simone, Firenze 131149
Chiavarini, Paola, Parma 081102
Chiave di Volta, Venezia 132231
Chiayi Distillery Museum,
  Minsiong 042400
Chiayi Natural History Museum,
  Singang 042459
Chiazzese, Pietro, Palermo 131691
Chiba-kenritsu Awa Hakubutsukan,
  Tateyama, Chiba 029470
Chiba-kenritsu Bijutsukan, Chiba 028506
Chiba-kenritsu Boso Fudoki-No-Oka,
  Sakae 029308

Chiba-kenritsu Boso-no-Mura
  Hakubutsukan, Sakae 029309
Chiba-kenritsu Chuo Hakubutsukan,
  Chiba 028507
Chiba-kenritsu Indasutori oyo
  Kagakukan, Ichikawa 028668
Chiba-kenritsu Kazusa Hakubutsukan,
  Kisarazu 028819
Chiba-kenritsu Sonan Hakubutsukan,
  Isumi 028704
Chiba-shi Bijutsukan, Chiba . . . 028508
Chicago Antique Centre, Chicago 092687
Chicago Antique Mall, Chicago . 092688
Chicago Antique Market, Chicago 092689
Chicago Architecture Foundation,
  Chicago 047646
Chicago Art Exchange, Chicago . 120090
Chicago Art Source, Chicago . . 120091
Chicago Artists' Coalition,
  Chicago 060430
Chicago Artists' News, Chicago . 139337
The Chicago Athenaeum – Europe,
  Néa Symirni 023202
The Chicago Athenaeum – Galena,
  Galena 049014
The Chicago Athenaeum, Museum
  of Architecture and Design,
  Chicago 047647
Chicago Center for the Print,
  Chicago 120092
Chicago Children's Museum,
  Chicago 047648
Chicago City Arts Gallery,
  Chicago 120093
Chicago Coin Company, Chicago 092690
Chicago Connection, San Antonio 096958
Chicago Conservation Center,
  Chicago 135404
Chicago Contemporary and Classic,
  Chicago 098442
Chicago Cultural Center, Chicago 047649
Chicago Fine Arts Exchange,
  Chicago 120094
Chicago Gallery News, Chicago . 139338
Chicago Gold Gallery, Chicago . 092691
Chicago Historical Society,
  Chicago 047650
Chicago International New Art Forms
  Exposition, Warren 139339
Chicago Printmakers Collaborative,
  Chicago 060431
Chicago Room Gallery Cafe,
  Chicago 120095
Chicago's Arts and Antique Center,
  Chicago 092692
Le Chicchere Cose Cosi,
  Manziana 080348
Chiché, Bertrand, Bouin 067385
Chichester Antiques Centre,
  Chichester . 088872, 134255, 144335
Chichester Bookshop, Chichester 144336
Chichester District Museum,
  Chichester 043940
Chichester Gallery, Chichester . 117576,
  144337
Chichi Antiquités, Montréal . . . 046405
Chichibunomiya Kinen Supohtsu
  Hakubutsukan, Tokyo 029493
Chickamauga-Chattanooga National
  Military Park, Fort Oglethorpe . 048854
Chickasaw Council House Museum,
  Tishomingo 053749
Chickasaw County Historical Society
  Museum, Bradford Village,
  Nashua 051168
Chicken Street, Leiden 083427
Chickies, Edmonton 064289
Chico Museum, Chico 047710
Chicochai, Vancouver 064808
Chicola, Gabriela, Buenos Aires . 127519
Chicory Gallery, Alkmaar 112102
Chiddingstone Castle,
  Chiddingstone 043949
Chide, François, Dinard 068355

Chido Hakubutsukan, Tsuruoka . 029659
Chidori, Seattle 097415
Chief John Ross House, Rossville 052687
Chief Oshkosh Native American Arts,
  Egg Harbor 048489
Chief Plenty Coups Museum,
  Pryor 052407
Chiefs & Spirits, Den Haag 112377
Chiefswood Museum, Ohsweken 006542
Chieftains Museum, Rome 052670
Chiellini, Livorno 110176
Chien Dai Sculpture Art Museum,
  Sanyi 042439
Le Chien d'Or, Québec 101497
Ch'ien Mu House, Taipei 042504
Chieng Huat Antique, Bangkok . 088077
Chiesa di Santa Maria Maddalena de
  Pazzi, Sala Capitolare Crocifissione
  del Perugino, Firenze 026247
Chiesa, Mario, Torino 132116
Chiesa, Valerio, Milano 080487
Chifeng Museum, Chifeng 007703
Chiffonnier, Jarmenil 068892
Chiffonnier d'Art, Beuzeville . . . 067156
Les Chiffonniers, Meschers-sur-
  Gironde 070076
Chifley Home, Bathurst 000829
Chigira Mingeikan, Katashina . . 028794
Chignoli, Patrizia, Milano 080488,
  110284
Chihaya-Akasaka Kyodo-shiryokan,
  Chihaya-Akasaka 028512
Chihiro Art Museum Azumino,
  Nagano 029070
Chihiro Art Museum Tokyo,
  Tokyo 029494
Chihuly, Seattle 124504
Chikatetsu Hakubutsukan, Tokyo 029495
Chikkyo Art Museum Kasaoka,
  Kasaoka 028791
Chikueido, Kyoto 082007
Chikuho-do, Yokohama 082224
Chikurin-ji Homotsukan, Kochi . 028855
Chikuryudo, Tokyo 082137
Chikusendo, Hiroshima 081975
The Child Museum, Cairo 010533
Childers Pharmaceutical Museum,
  Childers 000958
Childhood and Costume Museum,
  Bridgnorth 043719
Children Art Palace, Islamabad . 034115
Children's Museum of Southeastern
  Connecticut, Niantic 051561
Children's Science Center,
  Carbondale 047424
Children's Art Gallery, Children's
  Museum, Yerevan 000725
Children's Art Museum, San
  Angelo 052923
Children's Art Museum, Taipei . . 042505
Children's Discovery Museum of
  Central Illinois, Bloomington . . 047041
Children's Discovery Museum of North
  San Diego, Carlsbad 047430
Children's Discovery Museum of San
  Jose, San Jose 053037
Children's Discovery Museum of the
  Desert, Rancho Mirage 052459
The Children's Discovery Museum of
  Vermont, Burlington 047326
Children's Folklore and Folkgame
  Museum, Sikou 042453
Children's Gallery, Taichung . . . 042471
Children's Hands-On Museum,
  Tuscaloosa 053857
Children's Health Museum,
  Charlottesville 047581
Children's Heritage and Science
  Museum, Amman 029765
Children's Library Museum,
  Marondera 054855
The Children's Metamorphosis,
  Londonderry 050365
The Children's Metamorphosis,

Children's Museum, Derry 048279
Children's Museum, Bhavnagar . 024165
Children's Museum, Detroit 048296
Children's Museum, Holyoke .. 049518
Childrens Museum, Imphal 024307
Children's Museum, Quezon City 034579
Children's Museum, Qurm 034103
Children's Museum, Saint
  Petersburg 052858
The Children's Museum, Seattle . 053232
Children's Museum, Shanghai .. 008223
Children's Museum at La Habra, La
  Habra 050038
The Children's Museum at Saratoga,
  Saratoga Springs 053155
Children's Museum at Yunker Farm,
  Fargo 048696
Children's Museum Big Bang,
  Sakai 029310
Children's Museum in Dartmouth,
  South Dartmouth 053402
Children's Museum in Easton, North
  Easton 051604
Children's Museum of Art and
  Creativity, Kaohsiung 042356
Children's Museum of Boca Raton,
  Boca Raton 047067
The Children's Museum of Cleveland,
  Cleveland 047806
Children's Museum of Denver,
  Denver 048254
The Children's Museum of Green Bay,
  Green Bay 049224
Children's Museum of History, Natural
  History and Science at Utica, New
  York, Utica 053915
The Children's Museum of Houston,
  Houston 049573
The Children's Museum of
  Indianapolis, Indianapolis .. 049679
The Children's Museum of Kansas City,
  Kansas City 049872
Children's Museum of Lake Charles,
  Lake Charles 050071
Children's Museum of Los Angeles,
  Los Angeles 050415
Children's Museum of Maine,
  Portland 052278
Children's Museum of Manhattan, New
  York 051360
The Children's Museum of Memphis,
  Memphis 050774
Children's Museum of Oak Ridge, Oak
  Ridge 051669
Children's Museum of Portsmouth,
  Portsmouth 052309
Children's Museum of Richmond,
  Richmond 052546
The Children's Museum of Rose Hill
  Manor Park, Frederick ... 048944
Children's Museum of South Carolina,
  Myrtle Beach 051143
Children's Museum of Spokane,
  Spokane 053440
Children's Museum of Stockton,
  Stockton 053540
Children's Museum of Tampa – Kid
  City, Tampa 053665
Children's Museum of the Arts, New
  York 051361
Children's Museum of Transportation
  and Communication, Taipei . 042506
The Children's Museum of Utah, Salt
  Lake City 052912
Children's Museum of Virginia,
  Portsmouth Museums,
  Portsmouth 052321
Children's Museum, Museo de Los
  Niños, San Diego 052961
Children's Planet, Louisville 094530
The Children's Science Center, Cape
  Coral 047416
Childress County Heritage Museum,
  Childress 047714

Childs, Boston 119851
Childs, Graham, Newbury 134874
Childventure Museum, Fort
  Washington 048888
Chile & Tomate, Houston 121015
Chillagoe Historical Centre,
  Chillagoe 000960
Chillariz, Dnipropetrovs'k 117111
Chilliwack Museum, Chilliwack . 005616
Chiltern Athenaeum, Chiltern ... 000961
Chiltern Motor Museum, Chiltern 000962
Chiltern Open Air Museum, Chalfont
  Saint Giles 043901
Chilton, Miranda 061861
Chilton, Juliet, Shrewsbury ... 091324
Chimène, Meylan 070110
La Chimera, Roma 081292
Chimera Foundation, Mumbai . 109424
Chimera Oriental Art Books,
  Hurstpierpoint 144495
La Chimère, Cannes 067632
Chimes Textbook Exchange, New
  Orleans 145304
Chimmey Point, Addison 046319
Chimney Mill Galleries, Bury Saint
  Edmunds 088740
Chimneyville Crafts Gallery,
  Jackson 049748
Chin, Geraldton 061517
Chin Fung Tong Gallery, San
  Francisco 124253
China 2000, New York 122467
China Academy of Art, Hangzhou 055212
China Agricultural Museum,
  Beijing 007606
China Antique Furniture, New
  York 095311
China Antique Link, Shanghai .. 065128
China Antiques, Shanghai 065129
China Art, Aachen 074326
China Art Archives and Warehouse,
  Beijing 101943
China Art Exposition, Beijing .. 098043
China Art Gallery, Aschaffenburg 106005
China Art Gallery, Bruxelles ... 063375
China Art Objects, Los Angeles . 121458
China Arts, Kuala Lumpur 082400
China Arts, Paris 071158
China Arts Far East Trading,
  Beijing 064999
China Arts Hangzhou, Hangzhou 065005
China Arts Trade and Development,
  Beijing 065000
China Attic, Bruxelles 063376
China Avant-Garde, New York .. 122468
China Aviation Museum, Beijing 007607
China Bazaar, Honolulu 093536
China Book Centre, Auckland . 143278
China Caideng Museum, Zigong 008556
China Cat Antiques, Jacksonville 093972
China China, Toronto 064626
China Chine, Chatou 067927
China City, Al-Qurum 084398
China Clay Country Park Mining and
  Heritage Centre, Saint Austell . 045681
China Coal Museum, Taiyuan . 008311
China Collection, Luxembourg . 082367
China Cultural Relic Newspaper,
  Beijing 138599
China Curio Centre, Hong Kong . 065022
China Fine Art Institute,
  Hangzhou 102037
China Fine Arts Gallery, Beijing . 007608
China First History Archive Museum,
  Beijing 007609
China from China, Amstelveen . 082664
China Galerie, Hedingen 116374
China-Galerie, Lehrte 107802
China Gallery, New York 095312
China Great Wall Museum,
  Beijing 007610
China Guardian Auctions, Beijing 125396
China Guche Museum, Linzi ... 008034
China Handicraft Company,

Penang 082403
China Institute Gallery, New York 051362
China International Book Trading
  Corporation, Beijing 136695
China International Export Antique
  Furniture Fair, Shanghai 098044
China International Gallery Exposition,
  Beijing 098045
China Jia Long, Singapore ... 085386
China Jiawu Zhan Zhen Guan,
  Weihai 008355
The China Locker, Lamberhurst . 089736
China National Association of
  Literature and Arts, Taipei 059991
China Nationalities Photo-Arts
  Publishing House, Beijing .. 136696
China Nationality Museum,
  Beijing 007611
China Painter, Oklahoma City . 139340
China Penmanship Art Museum,
  Xian 008396
China People's Revolution Military
  Museum, Beijing 007612
China Photographic Publishing House,
  Beijing 136697
China Repair, Lincoln 134559
The China Repairers, Doughton . 134341
China Repairers, London 134619
China Revolution Museum,
  Beijing 007613
China Silk Museum, Hangzhou . 007836
China Society for the Study of Folk
  Literature and Art, Beijing 058488
China Sports Museum, Beijing . 007614
China Stamp Museum, Beijing . 007615
China Standard Cao Book Society,
  Nanjing 140444
China Suxiu Art Museum,
  Suzhou 008294
China Tea Museum, Hangzhou . 007837
China Tradition Medicine Museum,
  Chongqing 007706
China Works, Auckland 083775
China Xiamen Huaqiao Museum,
  Xiamen 008394
China Yin Xue Museum,
  Hangzhou 007838
Chinantiques, Basel 087229
Chinard, Blacourt 067235
Chinart Company, New York .. 095313
ChinaSquare, Beijing 101944
ChinaSquare, New York 122469
Chinati Foundation, Marfa 050671
Chinchilla and District Historical
  Society Museum, Chinchilla . 000964
Chinchilla, Guido, San José 102485
Chinchu Bijutsukan, Naoshima . 029142
Chinderah Bay Antiques,
  Chinderah 061318
La Chine, Amsterdam 082720
Chine Antique, Dijon 068322
Chine et Japon, Marseille 069922
Chine Gallery, Hong Kong ... 065023
La Chine Populaire, Uzès 073922
Chinerie, Riom 072469
Chines Antiques, Nantes 070457
Chinese-American Fine Art Society,
  Boston 060432
Chinese Ancient Architecture Museum,
  Yangmei 042586
Chinese Antique Repair and Appraisal,
  Oakland 136039
Chinese Antiques, Limoges ... 069520
Chinese Art Gallery, Dallas 120569
Chinese Art Gallery, New York .. 122470
Chinese Art Studio, Hong Kong . 102066
Chinese Artists' Association,
  Beijing 058489
Chinese Arts and Crafts, Hong
  Kong 065024, 065025, 065026
Chinese Arts Centre, Manchester 118705
Chinese Association of Museums,
  Taipei 059992
Chinese Association of Natural

Science Museums, Beijing ... 058490
Chinese Carpets, Sharjah 088197
Chinese Catholic Museum and
  University History Exhibition
  Gallery, Fu Jen Catholic University,
  Sinjhuang 042462
The Chinese Ceramic Society,
  Beijing 058491
Chinese Contemporary Art Gallery,
  London 118250
Chinese Contemporary Art Gallery,
  New York 122471
Chinese Cultural Centre of Vancouver,
  Vancouver 007281
Chinese Culture Center of San
  Francisco, San Francisco ... 052997
Chinese Decorative Art Studio,
  Singapore 085387
Chinese Dynasty Furnishings,
  Honolulu 093537
Chinese Fine Art, Manama ... 100422
Chinese Furniture Museum,
  Taoyuan 042574
Chinese Heritage, Singapore ... 085388
Chinese Historical Art Museum,
  Nagasaki 029087
Chinese House Museum, Echo . 048467
Chinese International Antique
  Decoration and Artworks Fair,
  Shanghai 098046
De Chinese Kunsthandel,
  Rotterdam 083560
Chinese Medicine Museum,
  Hangzhou 007839
Chinese Museum, Melbourne .. 001267
Chinese Museum of Military
  Weaponary, Tongzhou 008343
Chinese Opera Museum, National
  Taiwan Junior College of Performing
  Arts, Taipei 042507
Chinese Painting Academy,
  Beijing 055191
Chinese Painting and Calligraphy
  Institute, Wuhan 102247
Chinese Painting Gallery,
  Shanghai 102173
Chinese Painting Gallery, Wuxi . 102250
Chinese People's Political Confutative
  Conference Calligraphy and Painting
  Gallery, Ningbo 102154
Chinese Photographers Association,
  Beijing 058492
The Chinese Porcelain Company, New
  York 095314
Chinese Society of Museums,
  Beijing 058493
Chinesische Antiquitäten,
  Seeshaupt 078443
Chinesische Haus, Stiftung Preußische
  Schlösser und Gärten Berlin-
  Brandenburg, Potsdam 021180
Le Chineur du Marais, Paris ... 071159
Chineurs, Dieppe 068298
Ching Shiun, Taichung 126778
Ching Shiun, Taipei 126779
Ching Ting Chen, Hong Kong . 065027
Chingsi Culture Museum,
  Kaohsiung 042357
Chinikhaneh Museum, Sheikh Sati
  Mausoleum, Ardebil 024685
Chink in the Wall, Barton-upon-
  Humber 117293
Chino, Vancouver 101759
Chinoh, New York ... 122472, 122473
Chinqua-Penn Plantation, North
  Carolyna State University,
  Reidsville 052506
Chiocchini, Franca, Perugia 131748
Chionchio, Andrea, Milano ... 131419
Chios Museum of Byzantine and
  Postbyzantine Art, Chios 023045
Il Chiostro, Saronno 110909
Chiostro dello Scalzo, Firenze . 026248

Il Chiostro di Manfredi, Genova . 080253, 142694
Chiotakis, Ioannis, Athinai 142537
Chip n' Dale's Antique Mall, Memphis 094624
Chiparus, Parthenay 072026
Chippendale Eastlake, Houston . 093638
Chippendale Society, Leeds 060091
Chippenham Museum and Heritage Centre, Chippenham 043950
Chippewa County Museum, Montevideo 051021
Chippewa Valley Museum, Eau Claire 048463
Chipping Norton Antique Centre, Chipping Norton 088887
Chipping Norton Museum of Local History, Chipping Norton 043951
Chippokes Farm and Forestry Museum, Surry 053616
Chips, København 102672
Chirac, Chantal, Aubusson 066793
Chiran Museum, Chiran 028518
Chiran Tokko Heiwa Kaikan, Chiran 028519
Chirco, Toulon (Var) 129287
Chiron, Marie-Christine, Nantes . 070458
Chirone L'Arte nell' Antichità, Verona 081904
Chirurgiemuseum, Tuttlingen . . . 022230
Chisenhale Gallery, London 044939
Chishoken, Imai, Kyoto 082008
Chisholm, New York 095315
Chisholm & Larsson, New York . 095316
Chisholm Trail Heritage Center Museum, Duncan 048391
Chisholm Trail Historical Museum, Oklahoma Historical Society, Waurika 054235
Chisholm Trail Museum, Wellington 054270
Chisholm Trail Museum, Oklahoma Historical Society, Kingfisher . 049974
Chislehurst Antiques, Chislehurst 088896
Chista, New York 122474
Chistina's Cupboard, Toronto . . 064627
The Chiswick Fireplace Company, London 090014, 134620
Chiswick House, London 044940
Chita-shi Rekishi Minzoku Hakubutsukan, Chita 028520
Chitari Academy Of Fine Arts, Pune 109492
Chiti, Isabella, Prato 131774
Sri Chitra Art Gallery, Thiruvananthapuram 024491
Chitrakoot, Kolkata 109396
Chitram, Rekha, Kolkata 109397
Chitramandir School of Art, Belgaum 055781
Chittagong University Museum, Chittagong 003212
Chittenango Landing Canal Boat Museum, Chittenango 047729
Chiusa, Romano, Fidenza 080072
Chivers, Los Angeles 094287
Chiverton House Museum, Northampton 001367
Chivré, Claire de, Dieulefit 103668
Chiz, Pittsburgh 123531
Chizky, Jean, L'Isle-sur-la-Sorgue 069566
Chladecká, Zuzana, Žilina 085597
Chlebovec, Ondřej, Olomouc . . . 065429
Chloe Antiques, Worthing 091953
Chloupek, Zagreb 128314
Chmel'nyc'kyj Chudožnij Muzej, Chmel'nyc'kyj 043141
Chmel'nyc'kyj Oblasnyj Krajeznavčyj Muzej, Chmel'nyc'kyj 043142
Cho, Seoul 111622
Cho, Heidi, New York 122475
Chobot, Wien 100242
Chochon, Barré & Allardi, Paris . 125694
Chochou, Maria, Rhodos 109083

Chock, Déborah, Paris 104645
Choco-Musée Erico, Québec . . . 006727
Chocol Art Gallery, Ashburton . . 112742
Choctaw Museum, Hugo 049606
Chodorowski, Wiesław, Starogard Gdański 132869
Chodová, Alena, Příbram 065547
Chofu Museum, Shimonoseki . . 029383
Chohung Museum of Finance, Seoul 029985
Choice, Bergen op Zoom 082959
Choice, Delhi 109311
Choice Antiques, Warminster . . 091732
Choice Art Gallery, Dallas 120570
Choice Galeria de Arte, São Paulo 101091
Choice Gallery, Birmingham . . . 117370
Choisy Antiquités, Choisy-le-Roi 067982, 128557
Choisy, Christian, Saint-Paul-de-Vence 105581
Chojagahara Kokokan, Itoigawa . 028713
Chojin Lama Monastery Museum, Ulaanbaatar 031728
Chokaido Hakubutsukan, Yokkaichi 029709
Chokoku no Mori Bijutsukan, Hakone 028591
Cholet, Nicole, Busseol 103476
Chollet Vuaillat, Paris 071160
Chollet, Stéphane, Paris 128961
Cholmondeley, Peter, Stuart Park 062252
Choma Museum and Crafts Centré, Choma 054834
Chomarat, Gérard, Lyon 104076
Chomé, Bath 117302
Chomette, Michèle, Paris 104646
Chomsioglou, Maria, Thessaloniki 079206, 079207
Chonbuk University Museum, Jeonju 029959
Chongming Museum, Shanghai . 008224
Chongqing Chinese Painting Gallery, Chongqing 102017
Chongqing History Calligraphy and Painting Gallery, Chongqing . . 102018
Chongqing Museum, Chongqing 007707
Chongqing Natural History Museum, Chongqing 007708
Chongwen Zhai Art Cooperative, Jinan 102137
Chongyuan, Macau 125406
Chonigbaum, Antony 066683
Chonnam University Museum, Gwangju 029940
Choo Seng & Co., Yangon 082609
ChooChoo Train Depot, Houston 093639
Chopin, Bruxelles 063377
Choppin de Janvry et Asso., Paris 136881
Choquet, Caroline, Paris 071161
Choraku Homotsukan, Shimoda 029378
Chorfi, Casablanca 112035, 112624
Chorley-Burdett, Bournemouth . 088534
Chorus, Firenze 131150
Chorus, Roma 131857
Chorus, Torino 132117
Chorus-Verlag für Kunst und Wissenschaft, Mainz 137367
La Chose, Saint-Dizier 072738
Chosun, Pymble 062126
Chosun, Seoul 082256, 111623, 111624
Chosun-Ilbo, Seoul 111625
Chouakri, Mehdi, Berlin 106197
Chouchou, Magny-Lormes 069805
Choucroun, Richard, Houston . . 093640
Chouet-Darbois, Claude, Beaune 067014
La Chouette, Montolieu 140961
Chouinard, Bracebridge 101209
Chourlet, Vence 105830
Chourlet, Céline, Paris 104647
Chow, Carin, San Francisco . . . 097206
Chow, Raymond, Vancouver . . . 101760
Chowaiki & Mosionzhnik, New

York 122476
Choyce Gallery, Saint Albans . . 119003
Chr. Belser AG für Verlagsgeschäfte, Stuttgart 137517
Chrämai, Anduze 103233
Chrämai, Hamburg 107235
Chrétien, René, Paris 071162
Chrétienne, Le Lorrain 142981
Chrétienne La Foi, Fort-de-France 142976
Chris' Antiques, Healesville 061613
Chris Art & Antique, Bucureşti . 085085
Chris Newland Antiques, London 090015, 134621
Chris R, Paris 071163
Chris'Ary, Villeneuve-d'Ascq . . . 074159, 129349
Chrismas, Douglas, Los Angeles 121459
Chriss Antix, Roanne 072487
Chris's Crackers, Carhampton . . 088792
Chrissy, Auckland 083776
Christ, Bad Sooden-Allendorf . . 074603
Christ, Wülfrath 079071, 130782
Christ Boelens Jukeboxen Museum, Sint Oedenrode 032875
Christ Child Opportunity Shop, Washington 097826
Christ Church Cathedral Treasury, Oxford 045450
Christ Church Community Museum, Lakefield 006155
Christ Church Picture Gallery, Oxford 045451
Christ the King College Museum, Calbayog 034451
Christ, Elvira, Rohrbach, Pfalz . . 130594
Christ, Helmut, Raumbach 078143
Christ, Jürg, Erschwil 144065
Christ, Sabine, Kirchheim unter Teck 130126
Christa's Antiques, Chicago 092693
Christa's Fine Tribal & Modern Art Gallery, København 102873
Christchurch Art Gallery Te Puna o Waiwhetu, Christchurch 033170
Christchurch Auctions, Phillipstown 126323
Christchurch Carpets, Christchurch 088905
Christchurch Mansion, Ipswich . 044644
Christelijke Hogeschool voor de Kunsten Constantijn Huygens, Zwolle 056184
Christel's Antiek, Utrecht 083635
Christen-Fankhauser, Jetta, Reinach (Basel-Land) 087746
Christen, Mario, Luzern 133832
Christen, Richard K., Louisville . 094531
Christen, Rolf, Solothurn 116648
Christensen, Stavanger 132797
Christensen, Bent, Ahrensburg . 074360
Christensen, Erik, Nykøbing Sjælland 102941
Christensen, Hjartvar, Odense . . 140635
Christensen, Kenth, Varde 066181
Christensen, Lars, Gilleleje 065693
Christensen, Palle, Hellerup . . . 065730
Christensen, Reinhold, Frederiksberg 065661
Christensen, Rie & Jørgen, Gilleleje 065694
Christi, Neuss 077746
Christian Art Distributors, Vereeniging 137948
Christian Arts, London 060092
Christian C. Sanderson Museum, Chadds Ford 047507
Christian Daniel Rauch-Geburtshaus, Museum Bad Arolsen, Bad Arolsen 016701
Christian Daniel Rauch-Museum, Museum Bad Arolsen, Bad Arolsen 016702
Christian Gallery, Merano 110210

Christian Wagner Haus, Stadtmuseum Leonberg, Leonberg 020043
Christian-Wolff-Haus, Stadtmuseum Halle, Halle, Saale 018953
Christiania City Auktionsforretning, Oslo 126369
Christianity and the Arts, Chicago 139341
Christian's, Guatemala 109118
Christiansted National Historic Site, Saint Croix 052764
Christie, Cairo 066235
Christie, Midland 139829
Christie, Ottawa 064500
Christie Antique Show, Dundas . 098035
Christie Galerie, Mougins 104297
Christie, Robert, Carrickfergus . . 088800
Christie's, Amsterdam 126271
Christie's, Bangkok 126785
Christie's, Barcelona 126511
Christie's, Beijing 125397
Christie's, Beverly Hills 127138
Christie's, Bordeaux 125524
Christie's, Boston 127144
Christie's, Bruxelles 125306
Christie's, Buenos Aires 125167
Christie's, Chicago 127165
Christie's, Doublebay 125189
Christie's, Dubai 126792
Christie's, Dublin 126168
Christie's, Düsseldorf 125926
Christie's, Edinburgh 126888
Christie's, Firenze 126190
Christie's, Genève 126722
Christie's, Genova 126195
Christie's, Hamburg 125966
Christie's, Helsinki 125477
Christie's, Hong Kong 125403
Christie's, Jakarta 126162
Christie's, London 126957, 126958
Christie's, Madrid 126535
Christie's, Milano 126203
Christie's, Monaco 126268
Christie's, Moone 126177
Christie's, München 126045
Christie's, New York 127345
Christie's, Newport 127367
Christie's, Paris 125695
Christie's, Ramsey 126179
Christie's, Rio de Janeiro 125360
Christie's, San Francisco 127443
Christie's, Shanghai 125408
Christie's, South Yarra 125233
Christie's, Stuttgart 126120
Christie's, Taipei 126780
Christie's, Tel Aviv 126182
Christie's, Tokyo 126245
Christie's, Torino 126233
Christie's, Toronto 125380
Christie's, Venezia 126237
Christie's, Wien 125269
Christie's, Zürich 126764
Christie's Gallery, Dallas 120571
Christie's International Media Division, London 138208
Christie's Magazine, London . . . 135193
Christie's Publications, New York 138387
Christin Gallery, Toronto 101589
Christin, Jean, Genève 126723
Christinaz, Alain, Biel 087315
Christine Antiquité, Noisy-le-Grand 070739
Christine König Galerie, Wien . . 100243
Christinehof Slott, Brösarp 040583
Christinerose, New York 122477
Christine's Adapon Fine Arts, Makati 113384
Christine's Art Gallery, Baguio . 113375
Christine's Curiosity Shop, Tucson 097645
Christine's Time, Helsinki 066309
Christine's Vintage Specialty Shop, Cleveland 092929
Christmann, Wiesbaden 142486

Christmas Closet, Jacksonville . 093973
Christmas Morning, Dallas 093089
Christmas Museum Gallery, Horsley
Park 001155
Christmas Yorkshire Fine Art and
Antiques Fair, Harrogate 098371
Christodoulou, D.E., Amsterdam 143023
Christof, Franz, Schleusingen .. 078371
Christof, Franz, Weimar 078896
Christofski, Motueka 084000
Christoph & Co., Holger, Bonn .. 125898,
141614
Christophe, Clément, Embrun .. 103713
Christopher, Birmingham 092399, 135352
Christopher-Clark Fine Art, San
Francisco 124254
Christopher Henry Gallery, New
York 122478
Christopher, Alexander, Duncraig 127592
Christopher, Radko, Pittsburgh . 096268
Christopher's, Farnham 089262, 134386
Christopher's Collectibles, El
Paso 093395
Christopher's Discoveries, New
Orleans 095077
Christophs Friseur-Museum,
Leipheim 019969
Christos Capralos Museum,
Aegina 022943
Christos Capralos Museum,
Agrinion 022951
Christusbruderschaft Selbitz Buch-
und Kunstverlag, Selbitz 137506
Christy Bird, Dublin 079533
Christy's Antiques, Virginia Beach 097776
Chroma, Salt Lake City 123953
Chroma, San Francisco 124255
Chromatic Arts, Swindon 119198
Chromos, Belo Horizonte 100878,
100879
Chronicle Books, San Francisco 138457
Chronika, Böblingen 075052
Chronique d'Egypte, Bruxelles .. 138546
Chronis, Nikol., Iráklion 079195
Chronos, Brugge 063265
Chrostowski, Witold, Kraków ... 084563
Chrysalis, East Melbourne 098994,
136480
Chrysalis, Manchester 118706
The Chrysler Museum of Art,
Norfolk 051573
Chrysochoidis, Dimitrios, Athinai 079160
La Chrysolite, Auxerre 066848
Chrysolite, Paris 071164
Chryssikou, Angeliki G., Kifissia 109070
Chrystals Auctions, Ramsey 126180
Chrz, Jan, Žilna 085598
CHTJ Southard House Museum,
Richmond 052539
Chu Dong Tu Temple, Binh Minh 054781
Chu, Sammy, Los Angeles 121460
Chuan Cheng Art Center, Taipei . 042508
Chubu, Nagoya 111312
Chuck, San Francisco 097207
Chuck and Barb's Baseball Cards,
Columbus 093002
The Chuck Jones Studio Gallery, San
Diego 124112
Chudalla, Albert, München 130337
Chudarek, Rudolf, Wien 062919
Chudleigh Park of Sassafras,
Sassafras 099679
Chudožestvena Galerija, Balčik . 004973
Chudožestvena Galerija,
Blagoevgrad 004989
Chudožestvena Galerija, Burgas 004996
Chudožestvena Galerija,
Chaskovo 005002
Chudožestvena Galerija, Čirpan . 005007
Chudožestvena Galerija, Drjanovo 005021
Chudožestvena Galerija, Gabrovo 005029
Chudožestvena Galerija,
Gălăbovo 005033
Chudožestvena Galerija, Karlovo 005047

Chudožestvena Galerija, Kavarna 005052
Chudožestvena Galerija, Kazanlăk 005055
Chudožestvena Galerija, Levski . 005081
Chudožestvena Galerija, Loveč . 005084
Chudožestvena Galerija, Mezdra 005092
Chudožestvena Galerija, Pernik . 005110
Chudožestvena Galerija, Plovdiv 005125
Chudožestvena Galerija, Provadia 005138,
005139
Chudožestvena Galerija, Radnevo 005143
Chudožestvena Galerija, Silistra . 005161
Chudožestvena Galerija, Smoljan 101137
Chudožestvena Galerija, Sozopol 005207
Chudožestvena Galerija,
Velingrad 005259
Chudožestvena Galerija, Vidin .. 005261
Chudožestvena Galerija, Žeravna 005268
Chudožestvena Galerija "Petar
Persengiev", Novi Pazar 005098
Chudožestvena Galerija Boris Denev,
Veliko Tărnovo 005252
Chudožestvena Galerija Darenie
Kollekcia Svetlin Russev, Pleven 005115
Chudožestvena Galerija Dobrič,
Dobrič 005015
Chudožestvena Galerija Elena
Karamichajlova, Šumen 005213
Chudožestvena Galerija Georgi
Papazov, Jambol 005040
Chudožestvena Galerija Ilija Beškov,
Pleven 005116
Chudožestvena Galerija Kiril Petrov,
Montana 005094
Chudožestvena Galerija Nikola Marinov,
Targovište 005224
Chudožestvena Galerija Nikolaj
Pavlovič, Svištov 005219
Chudožestvena Galerija Prof. Ilija
Petrov, Razgrad 005147
Chudožestvena Galerija Ruse,
Ruse 005151
Chudožestvena Galerija Stanislav
Dospevski, Pazardžik 005105
Chudožestvena Galerija Varna,
Varna 005236
Chudožestvennaja Galereja –
Dom Kukly Tatjany Kalininoj,
Petrozavodsk 037650
Chudožestvenno-grafičeskij Fakultet,
Kubanskij Gosudarstvennyj
Universitet, Krasnodar 056279
Chudožestvenno-memorialnyj
Muzej K.S. Petrova-Vodkina,
Chvalynsk 036957
Chudožestvenno-pedagogičeskij Muzej
Igruški – Kulturno-delovoj Centr,
Sergiev Posad 037856
Chudožestvenno-vystavočnyj
Kompleks, Muzej Istorii i Kultury
Srednego Prikamja, Sarapul .. 037838
Chudožestvennyj Moskovskij
Gusudarstvennyj Akademičeskij
Institut im. V.I. Surikova,
Moskva 056283
Chudožestvennyj Muzej, Staryj
Oskol 037893
Chudožestvennyj Muzej, Valujki . 038009
Chudožestvennyj Muzej im. A.N.
Radiščeva, Saratov 037841
Chudožestvennyj Muzej im. M.S.
Tuganova, Vladikavkaz 038022
Chudožnij Muzej, Sevastopol .. 043291
Chudožnij Muzej, Žytomyr 043326
Chudožno-memorial'nyj Muzej Ivana
Truša, Nacionalnij Muzej u Lvovi,
Lviv 043234
Chudožno-memorial'nyj Muzej
Leopol'da Levyc'kogo, Nacionalnij
Muzej u Lvovi, Lviv 043235
Chudožno-memorial'nyj Muzej Oleksi
Novakivs'kogo, Nacionalnij Muzej u
Lvovi, Lviv 043236
Chudožno-memorial'nyj Muzej Oleni
Kul'čic'koï, Nacionalnij Muzej u Lvovi,

Lviv 043237
Chudožnyj Muzej im. O. Šovkunenka,
Cherson 043137
Chuen, Yue Pun Po, Hong Kong 065028
Chuey, Los Angeles 121461
Chugeisha, Takaoka 082125
Chugg, Keith, Swansea 091527
Chughtai Museum Trust, Lahore 034127
Chugoku, Kyoto 082009
Chugoku, Sapporo 111374
Chulpan Yodgorlik Muzeyi, Andijon
Viloyat Adabiyot va San'at Muzeyi,
Andijon 054626
Chumacero, Lourdes, México .. 111950
Chumbhot-Punthip Museum,
Bangkok 042653
Chun Chan, Beijing 101945
Chun & Co., Peggy, Honolulu .. 120928
Chunfeng Literature and Art Publishing
House, Shenyang 136738
Chung, Hong Kong 065029
Chung, Seoul 111626
Chung Art Center, Chung Ang
University, Seoul 029986
Chung-Cheng Art Gallery, Saint John's
University, Jamaica 049780
Chung Cheng Aviation Museum,
Taoyuan 042575
Chung Hua University Arts Center,
Hsinchu 042329
Chung Jark, Seoul 111627
Chung Li-Ho Memorial Museum,
Meinong 042395
Chung Yi Gallery, Danshuei ... 042301
Chung, Marguerite, Fontanil-
Cornillon 068570
Chungchern Art Museum
and Jhongjheng Art Center,
Pingtung 042414
Chungju Museum, Chungju 029909
Chungman Telecommunication
Museum, Daejeon 029916
Chungnam University Museum,
Daejeon 029917
Chuo Gallery, Nagoya 111313
Churburg, Sluderno 027971
Church Corner Antiques,
Christchurch 083863
Church Farm Museum, Skegness 045807
Church Farmhouse Museum,
London 044941
Church Gallery, Chagford 117549
Church Gallery, Claremont 098891
Church Hall Farm Antique,
Broxted 088690
Church Hill Antiques, Los
Angeles 094288
Church Hill Antiques Centre,
Lewes 089811
The Church Hill Gallery,
Knutsford 118063
Church House Antiques,
Weybridge 091805
Church House Design,
Congresbury 117624
Church Lane Antiques,
Claremont 061322
Church Lane Antiques,
Cranebrook 061370
Church Museum, Dimitsana ... 023060
Church of Scotland Committee
on Church Art and Architecture,
Edinburgh 060093
Church on the Lane Antiques,
Columbus 093003
Church Street Antiques,
Altrincham 088266, 134027
Church Street Antiques, London 090016,
134622
Church Street Antiques, Wells-next-
the-Sea 091774
Church Street Centre,
Woodbridge 091910
Church Street Gallery, Cromer . 117651

Church Street Surplus, New York 095317
Church-Waddel-Brumby House
Museum, Athens 046592
Churchgate Antiques,
Empingham 089204
Churchhill Clocks, Midhurst 090757,
134853
Churchill, North Fremantle 061985
Churchill & White, Boston 092456
Churchill County Museum and
Archives, Fallon 048690
Churchill House and Marine Memorial
Room Museum, Hantsport ... 005984
Churchill Island Agricultural Museum,
Newhaven 001357
Churchill Museum and Cabinet War
Rooms, London 044942
Churchilliana, Sacramento 145488
Churchills Gallery, Nazeing 118774
Churt Curiosity Shop, Churt ... 088911
Churzee Gallery, London 118251
Chusonji Sankozo – Fujiwara
Kinenkan, Hiraizumi 028635
Chuuko Tamba Yakimono
Hakubutsukan, Sasayama ... 029345
Chuxiongzhou Museum,
Chuxiong 007710
Chvalynskij Kraevedčeskij Muzej,
Chvalynsk 036958
Chybulski, Feldkirch 139981
Chybulski, Mauren 142957
Ci-Gong, Daegu 111581
CIA Museum, McLean 050566
Cia Rosa Art, São Paulo 064171
Ciäsa Granda, Stampa 042020
Cian Conservando el Arte,
Madrid 133314
Cianci, Tullio, Davos Platz 087390
Ciancimino, London 090017
Cianella, Pietro, Roma 081293
Ciani, Andrea, Genova 110097
Ciaralli, Mario, Roma 131858
Ciaravino, Carlo, Torino 132118
Ciarlo Brothers, San Francisco . 136312
Cicaloni, Massimo, Modena ... 080787
Ciccarelli, Lyon 069722
Ciccarelli, Nelly, Vaulx-en-Velin . 073989
Cicco, de, Lyon 069723
Cicconi, Alberto, Roma 131859
Cicerano, Alessandra, Napoli .. 142758
Cicero, Krummhörn 076980
Cicero-Galerie für politische Fotografie,
Berlin 106198
Cicero, Jan, Chicago 120096
Cicerone, Giuseppe, Roma 131860
Cicogna Casanova, Rolando,
Torino 081685
Cicogna, Michele, Venezia 132232
El Cid, Buenos Aires 060890
Le Cid Antiquités, Marseille ... 069923
Cider House Galleries,
Bletchingley 088507, 117390
Cider Museum and King Offa Distillery,
Hereford 044570
Cider Press Antiques Centre,
Taunton 091544, 135121
CIDOC International Committee
for Documentation of the ICOM,
Berlin 058813
Cielo, Portland 096378
Ciências Educação e Artes Luiz
de Queiráz, Escola Superior de
Agricultura Luiz de Queiroz / ESALQ,
Piracibaba 004629
Ciencias Naturales P. Ignacio Sala,
Valencia 040303
Cienfuego, Luxembourg 111838
Cieri, Gabriele, Roma 131861
Ciervo, Francesco, Napoli 080852
Cieslik, Marianne, Jülich 137289
Cietumu Vēstures Muzejs,
Jūrmala 030136
Ciferri, Armando, Brescia 079929
Cifola, Stefano, Roma 131862

Cifrapalota, Katona József Múzeum,
  Kecskemét 023722
Çifte Medrese Müzesi, Kayseri . 042991
Cigarpe, Cartagena ... 115007, 115008
Cigirac-Peter, Brigitte, Oberursel 108323
Cigliutti, Carole, Colombes 068101
Cigna Museum and Art Collection,
  Philadelphia 052036
Cigna, Antonino, Roma 081294
Il Cigno, Milano 080489
Ciguë, Saint-Étienne 141331
Cihan Antika, İstanbul 088112
Cilfynydd Art Society, Mountain
  Ash 060094
Cilicia Museum, Antélias 030262
Ciliento, Franco, Richwiller ... 072459
Cilindri, Alessio, Prato 081168
Cilla, Serge, Lyon 069724
CIMA Art Gallery, Kolkata 109398
La Cimaise, Besançon 103367
Cimaise, Besançon 103366
Cimal, Valencia 139046
Cimarron Valley Railroad Museum,
  Cushing 048098
CIMCIM – International Committee of
  Musical Instrument Museums and
  Collections, Comité International des
  Musées et Collections d'Instruments
  de Musique, Firenze 059425
Cimetière, Jean, Bourbon-Lancy 067406
Cimino, Luca, Genova 131300
CIMUSET – International Commitee of
  Museums of Science and Technology,
  Paris 058628
Cinà, Marcella, Palermo 131692
Cinabre, Paris 071165
Cincinnati Art Club, Cincinnati .. 060433
Cincinnati Art Galleries,
  Cincinnati 120330, 127171
Cincinnati Art Museum,
  Cincinnati 047739
Cincinnati Fire Museum,
  Cincinnati 047740
Cincinnati Gallery, Cincinnati .. 120331,
                           120332
Cincinnati Museum Center,
  Cincinnati 047741
Cincinnati Plating and Repair
  Company, Cincinnati 135439
Cindy's Dolls, Las Vegas 094127
Ciné-Images, Paris 104648
Cine-Qua-Non, Amsterdam ... 143024
Cinédoc, Paris 104649, 141075
Cinelli, Marzio, Firenze 080122
Cinema Books, Seattle 145579
Cinema Bookshop, London 144577
Cinema Collectors, Las Vegas .. 094128
Cinema Collectors, Los Angeles 094289,
                           145220
Cinema Graphics, Denver 120703
Cinema Museum, London 044943
La Cinémathèque de Toulouse,
  Toulouse 015985
La Cinémathèque Française,
  Paris 014632
Cinémathèque Québécoise – Musée
  du Cinéma, Montréal 006369
Ciney Antiquites, Ciney 098018
Cingyun Sword Craft Museum,
  Shalu 042444
Çini Müzesi, Kütahya 043011
Cini, Giovanni Battista, Genova . 080254
Çinili Köşk Müzesi, İstanbul 042930
Cinissimo, München 142146
Cinnamon Hill Antiques, Nashville 094984
Cinolter, Martin, Praha 065485
Cinquemani, Rosario, Palermo . 081031
Ciocca Arte Contemporanea,
  Milano 110285
Ciocșan, Octavian, Craiova ... 133005
La Ciotola, Verona 081905
Cipolat, Manduel 069835
Circa, Arles 066721
Circa, Fort Worth 093438

Circa, Indianapolis 093867
Circa 1900, Liverpool 089855
Circa 1904 Antiques, Portland . 096379
Circa 1910, Los Angeles 094290
Circa 610, New York 095318
Circa Ad, Portland 096380
Circa Antiques, Birmingham .. 092400
Circa Antiques, Jacksonville .. 093974
Circa Antiques, Los Angeles .. 094291
Circa Antiques, San Antonio .. 096959
Circa Antiques, Unley 062374
Circa Art Magazine, Dublin ... 137603,
                           138780
Circa Design Center, Dallas ... 093090
Circa Gallery, Minneapolis 121976
Circa Interiors and Antiques,
  Charlotte 092584
Circé, Strasbourg 073579
Circhirello, Carole, Prades-le-Lez 129179
Circle Art and Antiques, Miami . 094712
Circle Books, Bristol 144279
Circle Centre Gallery,
  Indianapolis 121173
Circle Craft, Vancouver 101761
Circle Press Publications, London 138209
Circolo Artistico di Arezzo,
  Arezzo 059426
Circolo degli Artisti, Torino 059427
Circuit des Enclus Paroissiaux en
  Miniatures, Commana 012637
Circuit Géologique des Boves,
  Arras 011743
Circuit, Association d'Art
  Contemporain, Lausanne ... 059930
Circulo Cinco, Belo Horizonte .. 063894
Círculo de Bellas Artes de Valencia,
  Valencia 059858
Círculo del Arte, Barcelona ... 114871
Circus City Festival Museum,
  Peru 052001
Circus Museum, Rapperswil .. 041858
Circus World Museum, Baraboo 046785
Circusmuseum "Bajasseum",
  Enkenbach-Alsenborn 018214
Cirencester Arcade, Cirencester 088914
Cirencester Art & Local Studies
  Collection, Bingham Library Trust Art
  Collection, Cirencester 043960
Cirencester Lock-Up, Cirencester 043961
Ciribino, Giuseppe, Roma 131863
Cirieco, Georges, Grenoble ... 068724
Cirilla, Kraków 143472
Cirillo, Giuseppe, Parma 131724
Ciris, Rīga 111757
Ciris, Tarrega 086486, 133443
Ciriza Vega, Jesus Carlos,
  Pamplona 115385
Cirkel Galerie, Almere 112114
Cirkevno-archeologické Múzeum,
  Nizná Mysľa 038412
Cirklaff, Dieter, Rodgau 108506
Cirkon, Sankt-Peterburg 085239
Cirone, Pierfrancesco, Roma ... 081295
Cirrus Gallery, Los Angeles ... 121462
Cirrus Gallery, Winnipeg 101871
Ciruíto Museale Urbano, Cascia . 025797
Ciry, Micheline, Paris 071166
Cisar-Holt, Tulsa 097719
Cisařová, Praha 128340
Cisařská Konírna, Prague Castle
  Gallery, Praha 009783
Čišinski-Gedenkstätte, Panschwitz-
  Kuckau 021069
Cislaghi Arredamenti,
  Abbiategrasso 079737
Cislaghi, Carlo, Milano 080490
Cissy, George, Memphis 121684
Cistercian Museum, Hailes 044495
The Citadel, Saint Helens 045687
The Citadel Archives and Museum,
  The Military College of South
  Carolina, Charleston 047544
Citadel Trading, Dubai 088173
Citadelles & Mazenod, Paris ... 136882

La Cité d'Art, Genève . 087459, 116293
Cité de la Mer Estran, Dieppe . 012768
Cité de l'Énergie, Shawinigan .. 007012
Cité Deco, Montréal 064406
La Cité des Antiquaires,
  Villeurbanne 074199
Cité des Métiers de Tradition, Saint-
  Laurent-de-la-Plaine 015408
Cité des Sciences et de l'Industrie,
  Paris 014633
Citiart Gallery, Santiago de Chile 101906
Citigold, Brisbane 061208
Čitinskij Oblastnoj Chudožestvennyj
  Muzej, Čita 036961
Čitinskij Oblastnoj Kraevedčeskij Muzej
  im A.K. Kuznecova, Čita 036962
Citiscapes, Seattle 124505
Citizens for the Arts in Pennsylvania,
  Harrisburg 060434
Citoyen de la Frontière de Markten,
  Bruxelles 100579
CITRIC, Brescia 109898
Città di Reggio, Reggio Emilia .. 110708
Cittadellarte – Fondazione Pistoletto,
  Biella 059428
City and County Museum,
  Lincoln 044829
City and Guilds of London Art School,
  London 056556
City Antik, København 065881
City Antik, Västerås 087138
City Antik, Valbo 087151
City-Antik, Wien 062920
City Antiques, Los Angeles ... 094292
City Antiques, Portland 096381
City Art, Birmingham 117371
City Art, Christchurch 112838
City Art, San Francisco 124256
City-Art, Tampere 066395
City Art Centre, Edinburgh ... 044233
City Art Gallery, Saint Paul ... 123901
City Art Gallery, Vancouver ... 101762
City Art Gallery, Leeds Museums and
  Galleries, Leeds 044779
City Art International, Piacenza . 110648
City Art Works, Atlanta 119552
City Arts, New York 051363
City Auction, Bern 126712
City Auction Rooms, Dunedin . 126314
City Auksjon, Oslo 126370
City Basement Books, Melbourne 139825
City Books, Pittsburgh 145435
City Clocks, London .. 090018, 134623
City County Pioneer Museum,
  Sweetwater 053627
City Galerie, Hamburg 076207
City-Galerie, Linz 100102
City Galerie, Regensburg 108460
City-Galerie, Wien .. 062921, 100244
The City Gallery, Leicester ... 044793
City Gallery, Nottingham 118839
City Gallery at Chastain, Atlanta 046613
City Gallery East, Atlanta 046614
City Gallery Wellington,
  Wellington 033344
City Hall Art Gallery, Ottawa ... 006585
City Hall Council Chamber Gallery,
  Charleston 047545
City Island Nautical Museum,
  Bronx 047227
City Kunst og Antik, Slagelse .. 066130
City Kunsthandel, København .. 065882
City-Mint, München 077479
City Museum, Ahmedabad 024123
City Museum, Makale 010729
City Museum, Saint Louis 052794
City Museum, Yirgalem 010730
City Museum and Art Gallery,
  Gloucester 044432
City Museum and Art Gallery,
  Plymouth 045511
City Museum and Records Office,
  Portsmouth 045549
City of Armadale History House

Museum, Armadale, Western
  Australia 000802
City of Belmont Museum,
  Belmont 000846
City of Brea Gallery, Brea 047184
City of Calgary-Civic Art Collection,
  Calgary 005519
City of Gilroy Museum, Gilroy .. 049100
City of Gosnells Museum,
  Gosnells 001115
City of Kingston Fire Department
  Museum, Kingston 006098
City of Las Vegas and Rough Riders
  Memorial Museum, Las Vegas 050175
City of Le Sueur Museum, Le
  Sueur 050198
City of London Phonograph
  & Grammophone Society,
  Edgmond 060095
City of London Phonograph &
  Grammophone Society – London
  Meetings, Croydon 060096
City of London Phonograph &
  Grammophone Society – Midlands
  Meetings, Wolverhampton ... 060097
City of London Phonograph &
  Grammophone Society – Northern
  Group, Barrow-in-Furness ... 060098
City of London Phonograph &
  Grammophone Society – Thames
  Valley Group, Charvil 060099
City of London Phonograph &
  Grammophone Society – West of
  England Meetings, Exeter ... 060100
City of London Police Museum,
  London 044944
City of Melbourne Collection,
  Melbourne 001268
City of Norwich Aviation Museum,
  Horsham Saint Faith 044607
City of Saint John Gallery, Saint
  John 006890
City of Unley Museum, Unley .. 001603
City of Wayne Historical Museum,
  Wayne 054248
City of York Museum, York 007498
City Park Radio Museum,
  Launceston 001211
City Pride, Canterbury 088774
City Salvage, Minneapolis 094904
City Streets Gallery, Toronto ... 101590
City Strippers Restoration,
  Wellington 132758
City Traders, Rockhampton ... 062160
City View Antique Mall, Dallas .. 093091
CityArt, Toronto 138574
Cityn kirja ja pap, Rovaniemi .. 140742
Ciudadela, Punta del Este 119440
Le Civette, Trieste 081784
Civic Bookshop, Newcastle ... 139854
Civica Galleria d'Arte Contemporanea,
  Lissone 026601
Civica Galleria d'Arte Contemporanea
  F. Scroppo, Torre Pellice ... 028153
Civica Galleria d'Arte G. Sciortino,
  Monreale 026893
Civica Galleria d'Arte Moderna,
  Gallarate 026364
Civica Galleria d'Arte Moderna,
  Genova 026389
Civica Galleria d'Arte Moderna,
  Milano 026780
Civica Galleria di Palazzo Rocca,
  Chiavari 025952
Civica Pinacoteca V. Crivelli,
  Sant'Elpidio a Mare 027842
Civica Raccolta d'Arte, Medole . 026754
Civica Raccolta d'Arte, Roseto degli
  Abruzzi 027673
Civica Raccolta d'Arte B. Biancolini,
  Potenza Picena 027431
Civica Raccolta delle Stampe Achille
  Bertarelli, Milano 026781
Civica Raccolta di Terraglia-Museo

Internazionale Design Cermico,
Laveno Mombello 026578
Civica Raccolta e Ceramiche
Rinascimentali, Camporgiano . 025737
Civiche Raccolte Archeologiche e
Numismatiche, Castello Sforzesco,
Milano 026782
Civiche Raccolte d'Arte, Busto
Arsizio 025665
Civiche Raccolte d'Arte Applicata,
Milano 026783
Civiche Raccolte d'Arte Applicata ed
Incisioni, Milano 026784
Civici Musei di Villa Paolina,
Pinacoteca L. Viani, Villa Paolina,
Viareggio 028358
Civici Musei e Gallerie di Storia ed
Arte, Udine 028225
Civico Gabinetto dei Disgeni, Civiche
Raccolte d'Arte, Milano 026785
Civico Museo Antropologico,
Balestrate 025444
Civico Museo Archeologico, Acqui
Terme 025240
Civico Museo Archeologico, Arsago
Seprio 025379
Civico Museo Archeologico,
Camogli 025720
Civico Museo Archeologico,
Milano 026786
Civico Museo Archeologico,
Ozieri 027135
Civico Museo Archeologico e di
Scienze Naturali Federico Eusebio,
Alba 025265
Civico Museo Archeologico L. Barni e
Pinacoteca C. Ottone, Vigevano 028380
Civico Museo Bibliografico Musicale,
Bologna 025530
Civico Museo Biblioteca dell'Attore del
Teatro, Genova 026390
Civico Museo d'Arte Contemporanea,
Milano 026787
Civico Museo d'Arte Moderne, Anticoli
Corrado 025333
Civico Museo d'Arte Orientale,
Trieste 028191
Civico Museo degli Strumenti Musicali,
Milano 026788
Civico Museo del Mare, Trieste . . 028192
Civico Museo del Merletto al Tombolo,
Villa Tigullio, Rapallo 027478
Civico Museo del Risorgimento e
Sacrario Oberdan, Trieste 028193
Civico Museo della Risiera di San
Sabba, Trieste 028194
Civico Museo di Guerra per la Pace
Diego de Henriquez, Trieste . . 028195
Civico Museo di Scienze Naturali,
Domodossola 026138
Civico Museo di Storia ed Arte ed Orto
Lapidario, Trieste 028196
Civico Museo di Storia Naturale della
Lombardia, Jerago con Orago . 026528
Civico Museo Dianese, Diano
Marina 026135
Civico Museo e Antiquarium
Comunale, Lanuvio 026556
Civico Museo Flaminio Massetano,
Massa Martana 026740
Civico Museo Frignanese, Pavullo nel
Frignano 027238
Civico Museo G. Galletti,
Domodossola 026139
Civico Museo G.V. Parisi-Valle,
Maccagno 026658
Civico Museo Ingauno, Palazzo
Vecchio del Comune, Albenga . 025267
Civico Museo Insubrico di Storia
Naturale, Induno Olona 026513
Civico Museo Morpurgo, Trieste . 028197
Civico Museo Naturalistico,
Ovada 027134
Civico Museo Naturalistico Ferruccio

Lombardi, Stradella 028023
Civico Museo Revoltella e Galleria
d'Arte Moderna, Trieste 028198
Civico Museo Sartorio, Trieste . . 028199
Civico Museo Setificio Monti, Abbadia
Lariana 025227
Civico Museo Storico, Palmanova 027194
Civico Museo Storico Archeologico,
Savona 027888
Civico Museo Teatrale di Fondazione
Carlo Schmidl, Trieste 028200
Civico Studio Museo Francesco
Messina, Trieste 026789
Civiero, Diano Castello 110002
Civil War Museum, Philadelphia . 052037
Civil War Museum of Lone Jack,
Jackson County, Lone Jack . . . 050366
Civil War Museum, McCook House,
Carrollton 047440
Civil War Store, New Orleans . . . 095078
Civile, Andrea, Roma 081296
Civilian Conservation Corps Museum,
Highlands Hammock State Park,
Sebring 053255
Civilization Art Gallery, Columbus 120483
Civitat, Zwolle 083731
Civitella Ranieri Center,
Umbertide 059429
Civran, Fiorenza, Venezia 132233
Cizeron, Marie-Claude, Le Crozet 069261
Cizeron, Marie-Noël, Romans-sur-
Isère 072521
Çizgi, İstanbul 117024
Čižková, Jitka, Česká Lípa 140462
CK Collection, Singapore 085389
Claassen-Schmal, Barbara,
Bremen 106577
Clac, Naucelles 104351
The Clachan Gallery, Glasgow . 117837
Clackmannshire Council Museum,
Alloa 043396
Claër, Eric, Brignais 067497
Claes, Sävedalen 086920
Claes, Jules, Genk 100682
Claeshallen, Malmö . . 086848, 126637
Clagahé, Lyon 140890
Clair Obscur, Montréal 064407
Claire Lane Antiques,
Jacksonville 093975
Claire Oliver Gallery, New York . 122479
Claire Trevor School of the Arts,
University of California Irvine,
Irvine 057243
Claire, Thomas le, Hamburg . . 076208,
141858
Clairefontaine, Luxembourg . . . 111839,
111840
Clairemont, San Diego 097074
Clairon-Labarthe, Paris 128962
Clampart, New York 122480
Clan Armstrong Museum,
Langholm 044758
Clan Cameron Museum, Spean
Bridge 045847
Clan Donnachaidh Museum,
Pitlochry 045505
Clan Grant Museum, Alness 043399
Clan Gunn Museum, Latheron . . 044763
Clan Macpherson House and Museum,
Newtonmore 045375
Clanart Gallery, Enniskillen 117755
Clancy, Bray 079463
Clandestine Immigration and Naval
Museum, Haifa 025047
Clapham Art Gallery, London . . 118252,
118253
Claphams, Lane Cove 061716
Claphams Clocks-The National Clock
Museum, Whangarei 033364
Clapson, Jenny, Kingscote 099196
Clara Barton Birthplace, North
Oxford 051611
Clara Barton National Historic Site,
Glen Echo 049106

Clara-Zetkin-Museum,
Birkenwerder 017395
Claradon, Saint Michael 063164
Claragh, Millstreet . . . 079675, 130881
Clara's Attic, Seattle 097416
Clare Ancient House Museum,
Clare 043963
Clare Antique Warehouse, Clare 088924,
134278
Clare Hall Company, Sudbury,
Suffolk 091502, 135103
Clare Heritage Centre, Corofin . 024862
Clare Museum, Ennis 024919
Clare National Trust Museum,
Clare 000965
Claremont Museum, Claremont . 047761
Claremont Museum, Claremont,
Western Australia 000967
Claremont, Antiques and Modern Art,
Sevenoaks 091268, 119069
Claren, Latrobe 061719
Clarenbridge Antiques,
Clarenbridge 079482
Clarence House Antiques,
Watchet 091757, 144909
Clarencetown and District Historical
Museum, Clarencetown 000969
Clarendon Books, Leicester . . 134543,
144524
Clarendon House, Evandale 001068
Claresholm Museum, Claresholm 005623
Clarfeld, Sabine, Seevetal 078446
Clarges, London 090019
Clarin Ancestral House, Loay . . 034497
Clarinbridge Antiques, Marton . . 083986
Claringbold, Berry 061147
Clark, Towcester 091625, 135144
Clark & Smith, Bordeaux 067320
Clark Art, Northwich 118821
Clark Atlanta University Art Galleries,
Atlanta 046615
Clark County Historical Museum,
Vancouver 053946
Clark County Museum,
Henderson 049440
Clark County Museum, Marshall 050707
Clark Galleries, Towcester 119251
Clark Hotel Museum, Van Horn . 053942
Clark House Museum Complex,
Wolfeboro 054512
Clark Humanities Museum, Scripps
College, Claremont 047755
Clark, Annaralla, Stow-on-the-
Wold 091464
Clark, Carson, Edinburgh 144391
Clark, Catharine, San Francisco 124257
Clark, Edward, Armadale,
Victoria 061019
Clark, Garth, New York 122481
Clark, Jonathan, London 118254
Clark, Peter, Birmingham 088476,
134107
Clark, Stephen L., Austin 119651
Clarke, Cape Town 143702
Clarke, Larchmont 127279
Clarke, Westbury 062424
Clarke County Historical Museum,
Berryville 046951
Clarke Gammon Wellers,
Guildford 126913
Clarke Hall, Wakefield 046029
Clarke Hall, J., Deal 144366
Clarke Historical Museum,
Eureka 048626
Clarke House Museum, Chicago 047651
Clarke Museum, Claringtons Museums,
Kirby 006118
Clarke, Christopher, Stow-on-the-
Wold 091465
Clarke, Graham, Maidstone . . 118696,
138262
Clarke, Janet, Bath 144230
Clarke, Janet, Freshford 144430
Clarke, Mary A., New York 122482

Clark's Antiques, Charlotte 092585
Clarkson Historical Museum,
Clarkson 047773
Clarksville, Austin . . . 119652, 119653
Clarksville-Montgomery County
Museum-Customs House Museum,
Clarksville 047776
Claro, Tucson 124722
Clary & Co., New York 095319
Clasic Stylle, Praha 065486
Clasic V & G, Bucureşti 114130
Clásica, Buenos Aires 139559
Clasicas Adriano, Sevilla 086448
Clasing, Münster 108213, 130424
Clasing & Langer, Osnabrück . . 077961
Classen Street Potters, Oklahoma
City 123277
Classensgade Antikvariat,
København 140606
Classic, Akaroa 083742
Classic, Cairo 066236
Classic, Taranaki 132746
Classic, Tbilisi 105900
Classic & Antique, Qormi 082426
Classic Age, West Hollywood . . 097880
Classic Antique, Freiburg im
Breisgau 075855
Classic Antique Wood, Charlotte 092586,
135389
Classic Antiques, Jacksonville . . 093976,
135649
Classic Antiques, Philadelphia . 096094,
136078
Classic Art, Albuquerque 092041
Classic Art, Attard 111911
Classic Art, Bromley 117464
Classic Art Gallery, Groningen . . 083284
Classic Artforms, West Hollywood 124991
Classic Attic, Charlotte 092587
Classic Bindings, London 134624,
144578
Classic Book Shop, Haymarket . 139769
Classic Cars International, Salt Lake
City 052913
Classic Century Square,
Albuquerque 092042
Classic Chapters, Broken Hill . . 139666
Classic Coins Hawaii, Honolulu . 093538
Classic Collectibles, Phoenix . . 096199
Classic Collection, Birmingham . 092401
Classic Collection, Zürich 087947
Classic Collections, San Antonio 096960
Classic Consignment Center,
Albuquerque 092043
Classic Custom Upholstery, San
Antonio 136258
Classic Desks, Wellington 084161,
132759
Classic Dreams 1960, Atlanta . . 092121
Classic Fine Arts, Mansfield . . . 118714
Classic Finish Art, San Jose . . . 136356
Classic Furniture, Milwaukee . . 094843
Classic Furniture Service, Salt Lake
City 136235
Classic Hall, Hellerup 065731
Classic Home, Budapest 079246
Classic Image, London 118255
Classic Jewellery, Christchurch . 083864
Classic Lighting Emporium,
Philadelphia 096095
Classic Picture Company, Dallas 120572
Classic Reproductions, Leicester 118091
Classic Reproductions, Pewsey . 134942
Classic Restoration Company,
Cleveland 135456
Classic Rocking Horses, Thirsk . 091596
The Classic Source, Mumbai . . . 109425
Classic Toy, Lyon 069725
Classic Traders, Auckland 083777
Classic Troc Cannes, Cannes . . 067633
Classic Upholstery, Philadelphia 136079
Classic Woods, Portland 136149
Classic World Imports, Dallas . . 093092
Clássica, Lisboa 084856

Classica, Sofia 101149
Classica Antikvariat, Oslo 143399
Classical Numismatic Group,
Lancaster 089739, 126936
Classical Numismatic Group,
London 090020, 126959
The Classical Quarterly,
Cambridge 139160
Classical Restorations and Design,
New York 135899
The Classical Review, Cambridge 139161
Classical Strings, Milwaukee . . . 094844
Classico, Binningen 087323
Classics Antiques, Miami 094713
Classics Kunstatelier, Brugge . . 063266
Classique, Lyon 140891
Classy, Houston 121016
Clastres, Jacques, Villeblevin . . . 129345
Clau, Rio de Janeiro 064043
Claude & Lima, Paris 071167
Claude Bernard Galerie,
Villedomer 074140, 105863
Claude Crayston Museum,
Glenora 005883
Claude, Daniel, Gampelen 087438
Claudel, Jean-Jacques,
Saussines 073442
Claudelands, Waikato 084126
Clausager Kunstvirke, Skjern . . . 066124
Clausen, København 102874
Clausen Memorial Museum,
Petersburg 052007
Clausen, Andreas, Sønderborg . 066136
Clausen & Sons, Arthur,
Edmonton 125365
Clausen, Grethe, København . . . 128396
Clauss, André, Strasbourg 129267
Claustro, Segovia 115439
Claustros del Pilar, Buenos Aires 000154
Clavel, Bonnaz & Lecomté,
Marseille 125635
Clavell & Morgades, Barcelona . 085762
Claveria, Emmanuel, Artigues-près-
Bordeaux 066750
Clavreuil, Bouchemaine 067373
Clavreuil, Jean, Paris 141076
Clawson Historical Museum,
Clawson 047780
Clay, Memphis 135775
Clay County Historical Museum,
Liberty 050272
Clay County Museum, Moorhead 051045
Clay Pan Gallery, Dubbo 098983
Clay Street Press, Cincinnati . . 120333,
138324
Clay Studio, Philadelphia 052038
Clay Tobacco Pipe Museum,
Broseley 043773
Clay, D.K., New Orleans 122144
Clay, John, London . . . 090021, 134625
Clayart Gallery, Manawatu . . . 112937
Claydon House, Middle Claydon 045257
Clayfield Antique Market,
Clayfield 061331
Claypool-Young Art Gallery, Morehead
State University, Morehead . . . 051051
Claypool, Esther, Seattle 124506
Clayton, Saint Louis 096678
Clayton, Tampa 124641
Clayton Brown, Matthew, New
Orleans 127334
Clayton Collection Museum,
Chollerford 043954
Clayton McLain Memorial Museum,
Cut Knife 005674
Clayton-Payne, Andrew, London 090022,
118256
Clayton, Tim, King's Lynn 089690,
134514
Clayways, Austin 119654
Cle Elum Telephone Museum, Cle
Elum 047788
Clear Creek Gallery, Tenterfield . 099814
Clear Lake Area Historical Museum,

Clear Lake 047790
Clear View Art Gallery, Hamilton
Parish 100833
Clearfield County Historical Society
Museum, Clearfield 047791
Clearing House, Charlotte 092588
Clearing House, Jacksonville . . . 093977
Clearwater County Museum,
Shevlin 053312
Clearwater Historical Museum,
Orofino 051815
Cleary, John, Houston 121017
Les Clédelles, Puy-l'Evêque . . . 105402
Clédic, Bar-sur-Aube 066941
Cleethorpes Discovery Centre,
Cleethorpes 043966
Cleeve Antiques, Bristol 088651
Cleeve Picture Framing, Bishop's
Cleeve 088488, 134116
La Clef, Uzès 073923
Clef du Passé, Reims 072387
The Cleft Art Gallery,
Donaghadee 117678
Clegg, Benedict, Tunbridge Wells 135157
Clegg, John, Ludlow 090645
Clemenceau, Pau 141301
Clemenceau Heritage Museum,
Cottonwood 048026
Clémençon Meillat, Martine,
Bosmoreau-les-Mines 067371
Clémençon Meillat, Martine,
Bourganeuf 067419
Clemens, Århus 140552
Clemens Antik, Lund 086830
Clemens-Galerien, Solingen . . . 108634
Clemens-Sels-Museum Neuss,
Neuss 020743
Clément, Culan 068257
Clément, Montluçon 070266
Clement, Catherine, Bonn 141615
Clement, Guy, Roubaix 072548
Clément, Louis, Cohade 068070
Clement, Piet, Amsterdam 112182,
137829
Clemente, Nice 070586
Clemente, Antonio, Torino 132119
Clemente, Luigi, Genova 110098
Clementine, New York 122483
Clements, Hixson 127244
Clendening History of Medicine Library
and Museum, University of Kansas
Medical Center, Kansas City . . 049873
Clent Books, Bewdley 144244
Cléo Antiquités, Saint-Dyé-sur-
Loire 072741
Cleo Redd Fisher Museum,
Loudonville 050481
Cléon, Michel, Rouen 072579
Cleopatra Bazar, Alexandria . . . 066216
Cleo's Black Cat, San Francisco 097208
Clepsidra, Bogotá 102288
Clepsidra, Lisboa 084857
Cléquin, Daniel, Villevallier . . . 074232
Cler Relojeros, Fernando, Madrid 086158
Clerc, Alain, Cour-Cheverny . . . 068183
Clerc, Catherine F., Lausanne . . 116411
Clerc, Pierre, Montpellier 140964
Clerck, F. de, Brugge 063267
Clergiot, Georges, Sainte-
Menehould 073359
Clergiot, Michel, Sainte-
Menehould 073360
Clergy House, Alfriston 043393
Clérigo Galeria, Lugo 115124
Clermont, Pézenas 072133
Clermont Collections, Clermont-
Ferrand 068017
Clermont State Historic Site,
Clermont 047802
Clerval, Xavier De, Colombes . . 068102
Clés de Saint-Pierre, Dax 068278
Clés de Saint-Pierre, Hossegor . 068837
La Clessidra, Messina 110218
La Clessidra, Trieste 081785

Cletienne, Olivia, Paris 128963
Cleve National Trust Museum,
Cleve 000973
Clevedon Court, Clevedon 043967
Clevedon Salerooms, Clevedon . 126863
Cleveland Art Association,
Cleveland 060435
Cleveland Artists Foundation at Beck
Center for the Arts, Lakewood 050102
Cleveland College of Art and Design,
University of Teeside, Isleworth 056543
Cleveland County Historical Museum,
Shelby 053302
Cleveland Institute of Art,
Cleveland 056913
Cleveland Institute of Art Gallery,
Cleveland 047807
Cleveland Ironstone Mining Museum,
Skinningrove 045809
Cleveland Living Arts Centre,
Dunedin 033188
The Cleveland Museum of Art,
Cleveland 047808
The Cleveland Museum of Natural
History, Cleveland 047809
Cleveland Police Museum,
Cleveland 047810
Cleveland Public Art, Cleveland . 120410
Cleveland State University Art Gallery,
Cleveland 047811
Cleveland Studies in the History of Art,
Cleveland 139342
Cleveland, Dwight, Chicago . . . 120097
Les Clévos, Paris 071168
Clewiston Museum, Clewiston . . 047830
Clichés du Passé, Marseille . . . 069924
Clifden Antiques, Clifden 079483
Clifden Art Gallery, Clifden . . . 109546
Cliff Street Collective, Fremantle 099061
Cliffe Antiques Centre, Lewes . . 089812
Cliffe Castle, Keighley 044667
Clifford, Dublin 079534
Clifford Cross Auctions, Wisbech 127092
Clifford Gallery, Olney 118851
Clifford-Smith, Boston 119852
Clifton, Bristol 088652
Clifton and District Historical Museum,
Clifton 000976
Clifton Art Gallery, Karachi 113303
Clifton Community Historical Society,
Clifton 047831
Clifton Gallery, Aberdeen 117208
The Clifton Gallery, Bristol 117442
Clifton Gallery, Clifton 117599
Clifton Park Museum, Rotherham 045638
Clifton, T. Joseph, Memphis . . . 121685
Climat, Les Sables-d'Olonne . . . 103987
Climax Community Museum,
Climax 005627
Clin d'Ailes, Musée de l'Aviation
militaire, Payerne 041834
Clin d'Oeil, Clermont-Ferrand . . 068018
Clin d'Oeil, Paris 071169
Clinco, Carlo, Torino . . 081686, 132120
Clínica de Arte, Cali . . 065248, 128285
Clínica de Muñecos Casas Reyes,
Bogotá 128264
Clínica de Porcelanas, Bogotá . 128265
Clinique de Poupées, Nice 128854
Clinique des Poupées, Lyon . . . 128705
Clinique des Poupées, Nancy . . 070416,
128820
Clinique du Meuble Ancien,
Abrest 066414, 128422
Clink Prison Museum, London . . 044945
Clinton, Jacksonville 127269
Clinton Academy Museum, East
Hampton Historical Society, East
Hampton 048430
Clinton Art Association, Clinton . 060436
Clinton County Historical Museum,
Plattsburgh 052197
Clinton County Historical Society
Museum, Wilmington 054450

Clinton County Museum, Clinton 047835
Clinton County Museum,
Frankfort 048922
Clinton Howell, New York 095320
Clinton Street Antique Centre,
Milwaukee 094845
Clinton Street Antiques, Portland 096382
Clintonville Antiques, Columbus . 093004
Clio, Paris 141077
Clio Antic, Bucureşti 085086
Clipper, Vannes (Morbihan) . . . 073968
Clipper Ship Cutty Sark, London 044946
Clisby, Bryan, Hartley Wintney . 089451,
134442
Clist, Gerry, London 134626
Clive House Museum,
Shrewsbury 045794
Cliveden House, Philadelphia . . . 052039
Cloarec, Didier, Pornichet 072280
Cloche d'Or, Foix 068553
Cloches de Corneville, Pont-
Audemer 014929
Clock 24, Blankenburg 075015
Clock and Watch Service Centre, Coffs
Harbour 061350
Clock Clinic, London . . 090023, 134627
Clock Doctor, Atlanta . 092122, 135302
Clock Gallery, London 134628
The Clock Gallery, North
Killingholme 134885
Clock-house, Aalst 063173
Clock House, Leavenheath 089759,
134530
Clock House Antiques, Keighley 089663
Clock Manor, Overland Park . . . 136070
Clock Shop, Hurstpierpoint . . . 089626,
134494
Clock Shop, Saint Albans 091173,
134989
The Clock Shop, Winchcombe . . 091861,
135235
Clock Shop of Richmond,
Richmond 096535, 136167
Clock Shop Weybridge,
Weybridge 091806, 135215
Clock Tower, Saint Albans 045662
Clock Tower Antiques,
Philadelphia 096096
Clock Tower Archive Centre,
Youghal 025005
Clock Wise Restoration, Henfield 134454
The Clock-Work-Shop,
Winchester 091869, 135237
Clock Workshop, Caversham . . . 088808,
134223
Clockmakers' Museum, London . 044947
Clockmaster, Saint Louis 136197
Clocks and Collectables, Eltham 061443
Clocks and Collectibles,
Jacksonville 093978, 135650
Clocks and More, Albuquerque . 092044,
135281
Clocks Unique Antique Mart,
Tampa 097554, 136386
Clocktower Museum, Elsham . . 044287
Clockwise, Emsworth . 089206, 134367
Clockwise, Launceston 061724
Cloisonne, Ando, Tokyo 082138
The Cloisters, North Sydney . . . 061999
Cloisters Antiques, Ely . 089199, 144406
The Cloisters, The Metropolitan
Museum of Art, New York 051364
Cloître de Cadouin, Le Buisson-de-
Cadouin 013615
Clola Antiques Centre, Clola-by-
Mintlaw 088935
Clonalis House, Castlerea 024846
Cloncurry and District Museum,
Cloncurry 000977
Clonfert Museum of Ecclesiastical Art,
Loughrea 024961
Clonmel Potters Gallery, Kingston 111215
Cloots, Christiane, Bruxelles . . . 100580
Clos Arqueologic Torre Llauder,

Mataró 039734
Clos Arsene Lupin, Maison Maurice
    Leblanc, Etretat 012928
Clos de la Fontaine, Nîmes 070690
Le Clos des Cimaises, Saint-Georges-
    du-Bois 072799
Le Clos Saint-Roch, Betoncourt-Saint-
    Pancras 067150
Close, David, Collyweston 117619
Close, Henry H., Abergavenny . . 088229
Closet, Sacramento . . . 096613, 145489
Closset, Jean-Jacques, Paris . . 128964
Closson, Cincinnati . . . 120334, 120335
Clot Bramsen et Georges, Paris 104650
Clothing and Textiles Hallway
    Museum, University of Manitoba,
    Winnipeg 007438
Clotworthy Arts Centre, Antrim . 043421
Cloud County Historical Society
    Museum, Concordia 047961
Clough-Hanson Gallery, Rhodes
    College, Memphis 050775
Clough, Sallie, Auckland 112755
Clovelly Collectables, Clovelly . . 061345
Cloverdale Antique Mall, Surrey 064573,
    125378
Clowes Fund Collection,
    Indianapolis 049680
Club degli Artisti, Foggia 059430
Il Club dell'Arte e della Cultura,
    Roma 110747
Clüver, Hanna, Loxstedt 077151
Clun Local History Museum, Clun 043972
Clune Galleries, Potts Point 099596
Clunes Museum, William Barkell
    Memorial Arts and Historic Centre,
    Clunes 000978
The Clunky Mantel, Edmonton . . 064290
Cluny, Edinburgh 144392
Cluny Auteur, Cluny 103612
Clusella Piñana, Carmen,
    Barcelona 114872
Cluster, Berlin 106199
Cluthé, Marie, Toronto 064628
Clutter House Antiques, Dallas . 093093
Clutterboxe Antiques, Cleveland 092930,
    135457
Cluttered Cupboard, Chicago . . 092694
Clutters, Toronto 064629
Cluva, Venezia 137757
Cluzel, Patrick, Cannes 128527
Clydach Antiques, Swansea . . . 091528
Clyde Antiques, Patrington 090948
Clyde Historical Museum, Clyde 033180
Clydebank Museum, Clydebank . 043973
Clydebuilt, Scottish Maritime Museum,
    Glasgow 044394
Clyde's Antiques, Oklahoma City 095957
Clymer Museum of Art,
    Ellensburg 048544
CM Antik, Uttenreuth 078773
CMA Gallery, Saint Louis 123815
cmagazine, Toronto 138575
CMC Art Museum, Taipei 042509
Cmuk, Wiesbaden 108901
CMZ-Verlag, Rheinbach 137486
CN Galeria, Bogotá 102290
CN Galeria, Bogotá 102289
CNAP, Lisboa 113986
CNCE, Aix-les-Bains 066479
Galerij Cnock, Knokke 125340
CNRS Editions, Paris 136883
Co-operative Storage Museum,
    Albany 000776
Co, Xhan, Ha Noi 125067
Co. Art, Genova 131301
CO.R.O.Art, Clermont-Ferrand . 128559
Coach House Antiques,
    Canterbury 088775
Coach House Antiques, Gomshall 089342,
    134410
Coach House Antiques, Hastings 089464,
    134446
Coach House Antiques, Moss

Vale 061901
Coach House Antiques, Stanley . 091413,
    135075
Coach House Antiques, Whitby . 091827
Coach House Books, Toronto . . . 136683
The Coach House Gallery,
    Eastwood 117711
Coach-House Gallery, Saint Peter
    Port 045711
Coachella Valley Museum, Indio 049699
Coachman, San Jose 097362
Coad, Jim, Albuquerque 092045
Coal Creek Heritage Village,
    Korumburra 001197
Coal Miners' Museum, Van Lear 053943
Coalfields Antiques, Greymouth . 083944
Coalfields Museum, Collie 000984
Coalition of Canadian Healthcare
    Museums and Archives, Toronto 058440
Coalmine Fotogalerie, Winterthur 042147
Coalport China Museum, Ironbridge
    Gorge Museums, Coalport . . . 043977
Coarte, A Coruña 115030
Coast, Fresno 093494
Coast & Country Browns Bay,
    Auckland 112756
Coast Gallery, Long Beach 121396
Coast Gallery, Mombasa 111556
Coast Salish Arts, Vancouver . . 101763
Coast to Coast, Tampa 124642
Coastal and Maritime Art,
    Narrabeen 099421
Coastal Center for the Arts, Saint
    Simons Island 052868
Coastal Curiosities, Opunake . . 084047
Coastal Discovery Museum, Hilton
    Head Island 049485
Coastal Heritage Museum, Crystal
    River 048080
Coastal Painting Gallery, Qingdao 102158
Coastal Peoples, Vancouver . . . 101764
Coatalem, Eric, Paris 104651
Coaticook Historical Society Museum,
    Coaticook 005629
Coats, A., London 090024
Cobalt Mining Museum, Cobalt . 005631
Cobb & Co., Box Hill 061185
Cobb & Co., Box Hill
    Toowoomba 001583
Cobb County Youth Museum,
    Marietta 050673
Cobb Institute of Archaeology
    Museum, Mississippi State . . . 050956
Cobb Lane Art Gallery,
    Birmingham 119793
Cobb Museum of Art, Marietta . 050674
Cobb, Clive & Lesley, Hove 134478
Cobb, Timothy, Milwaukee 121907
Cobbaton Combat Collection,
    Chittlehampton 043953
Cobbitty Arts, Crafts & Antiques,
    Cobbitty 061347
Cobble Hill, New York 095321
The Cobbled Yard, London 090025
Cobblers Garden Gallery,
    Millthorpe 099345
Cobblestone, Houston 093641
Cobblestone, Seattle 097417
Cobblestones, Portland 096383
Cobblestones Museum, Greytown 033215
The Cobb's Auctioneers,
    Peterborough 127388
Cobdogla Irrigation and Steam
    Museum, Cobdogla 000980
Cobh Museum, Cobh 024853
Cobham Bus Museum, The London
    Bus Preservation Trust, Cobham 043982
Cobisi, Giuseppe, Roma 131864
Cobland, Depden 089051, 134320
Cobo, Pepe, Madrid 115174
Cobra & Bellamy, London 090026
Cobra Museum voor Moderne Kunst,
    Amstelveen 031919
Coburg, Jacksonville 093979

Coburger Münzhandlung, Coburg 075262
Coburger Puppen-Museum,
    Coburg 017748
Coburn, Annandale 127551
The Coburn Gallery, Ashland
    College Arts & Humanities Gallery,
    Ashland 046569
Cobweb, Miami 135790
Cobweb Antiques, Cullompton . . 089022,
    134312
Cobweb Corners, Omaha 096030
Cobwebs, Galway 079624
Cobwebs, Salt Lake City 096885
Cobwebs, Southampton 091370, 144802
Cobwebs Collectables, Wanganui 084144
Coca-Cola Memorabilia Museum of
    Elizabethtown, Elizabethtown . 048526
The Coca-Cola Pavillion, Santa
    Cruz 034591
Coca-Nuts & Co., Oklahoma City 095958
Cocalico Valley Museum, Ephrata 048590
Cocardasse, Den Haag 083068
Cocchi, Nicola, Firenze 131151
Coccolo, Alessandro, Roma . . . 131865
Coccodrillo, Davos Platz 087391
Le Cocguen, Jean-Claude,
    Plouaret 072195
La Cochera de Pepe, Jerez de la
    Frontera 086056
Cochet, Dourdain 068399
Cochet, Paulette, Genève 087460
Cochin Ceramics Museum, Chai-
    Yi 042278
Cochise, Firenze 080123
Cochrane Railway and Pioneer
    Museum, Cochrane 005633
Cocinas Montañez, Málaga 086290
Cock-A-Too, Calgary 064251
Cockaigne, George, Lismore . . . 061741
Cockatoo Art Gallery, Cairns . . . 098835
Cockatoo Gallery, Broome 098807
Cockburn Museum, Edinburgh . 044234
Cockermouth Antiques,
    Cockermouth 088940
Cockermouth Antiques Market,
    Cockermouth 088941, 134284
Cockrell Antiques, Surbiton 091509
Coco-De-Mer, Singapore 085390
CoCo Kunst, Næstved 102932
Coco, Melissa, Catania 080029
Cocoa, Cheltenham . . . 088831, 134235
Coconut Grove Gallery, Miami . . 121758
Cocoon Art, Milano 080491
Coco's Galleria, Kansas City . . . 121284
Cocotte, Stuttgart 078598
Cocusse, Claudine, Reims 072388,
    129187,   129188
Coda Gallery, New York 122484
CODA Museum, Cultuur Onder Dak
    Apeldoorn, Apeldoorn 032009
Code Red Art Gallery, Claremont 098892
Codex, Stockholm 143943
Codinach, Arnaldo, Montevideo . 091994
Codington County Heritage Museum,
    Watertown 054217
Codman Estate – The Grange,
    Lincoln 050284
Côdô, Ha Noi 125068
Codosero, Luis, Madrid 086159, 115175
Cody Country Art League, Cody . 060437
Coe College, Cedar Rapids 056848
Coe College Collection of Art, Stewart
    Memorial Lbrary, Cedar Rapids 047486
Coe Hall, Oyster Bay 051874
Coed Hills Rural Artspace,
    Cowbridge 117628
Toca Coelhos, Lisboa 084858
Coen, Luciano, Roma . 081297, 137729
Coenen, Heide, Essen 075670
Coenobium, Asti 142619
Coens Galerie, Grevenbroich . . 107162
Coerten, Jacques, Le Paradou . 103960
Coeur d'Art, Brioude . . 067504, 103471
Coeur de Capitol Gallery,

Washington 124883
Coeur, Jacques, Bourges 067425
Coffee House Press, Minneapolis 138362
Coffin House, Newbury 051514
Coffin School Museum, Egan Institute
    of Maritime Studies, Nantucket 051154
Coffineau, Béatrice & Guy, La Roche-
    sur-Yon 069070
Coffman & Co., Boston 092457
Cofides, Saint-Ouen 073053
Cofion Books & Postcards, Tenby 144879
Cogges Manor Farm Museum,
    Witney 046173
Coggeshall Farm Museum,
    Bristol 047216
Coggeshall Heritage Centre and
    Museum, Coggeshall 043987
Coggins, Anthony, Longwell
    Green 134816
Coggins, Ray, Westbury 091792
Cogito, Setúbal 085062, 114112
Cognard, Christian, Pont
    Audemer 129174
Cognitive Arts, Boston 119853
Cohan & Leslie, New York 122486
Cohan, James, New York 122485,
    138388
Cohan, James, Shanghai 102174
Cohen, Saint Louis . 096679, 136198
Cohen & Cohen, Reigate 091088
Cohen Amador, New York 122489
Cohen, Byron C., Kansas City . . 121285
Cohen, Ethan, New York 122487
Cohen, James & Joanne, New
    York 095322
Cohen & Sons, James H. & Jerald L.,
    New Orleans 095079
Cohen, Michel, Paris 071170
Cohen, Philip, London 090027
Cohen, R., San Francisco 124258
Cohen, Rick, Ringwood 127670
Cohen, Ronny, New York 122488
Cohen, Valerie, Paddington, New
    South Wales 099517
Cohendy, Jean-Charles, Moissat 070146
Cohet & Feraud, Montauban . . . 125645
Cohn Edelstein, Rio de Janeiro . 101020
Cohn, E., Berlin 074776
Cohn, George & Peter, London . 090028,
    134629
Cohrs-Zirus, Ingrid, Worpswede 108965
Cohu, Stephen, Saint Ouen . . . 091202,
    134993
Cohuna Museum, Cohuna 000981
Coimbra, Rosa Martins, Lisboa . 132932
Coin, Pampigny 087728
Coin and Jewellery Shop,
    Accrington 088235
Coin and Stamp Gallery, Phoenix 096200
Coin and Stamp USA, Miami . . . 094714
Le Coin des Arts, Paris 071171
Coin du Miroir, Dijon 012778
Coin Exchange, Pittsburgh 096269
Coin Galleries of Seattle, Seattle 097418
Coin Invest Trust, Balzers 082331
Coin Mart, Kansas City 094067
Le Coin Perdu, Utrecht 083636
Coin Shop, Chicago 092695
Coin Shop, Cleveland 092931
Coin Shop, Houston 093642
Coin Shop, Indianapolis 093868
Coin Star, Chicago 092696
Coin Vault, New Orleans 095080
Coin, Gold and Silver Report, Las
    Vegas 094129
The Coinage Hall Antiques Centre,
    Truro 091642, 135153
Coinex II, Los Angeles 094293
Coins, Cincinnati 139343
Coins, Tel Aviv 079733
Coins and Curios Centre,
    Mumbai 079401
Coins and Currency, Miami 094715
Coins for Tomorrow, Dallas 093094

Coins Museum, Amman 029766
Coins Plus, Cincinnati . 092838, 092839
Coins Plus Gems, Milwaukee . . 094846
Coins, Michael, London 090029
Coisas & Coisas, Rio de Janeiro 064044
Coisas de Família, Lisboa 084859
Cojan, Anne-Marie, Vitry-sur-
Seine 074261
Cokato Museum and Akerlund
Photography Studio, Cokato . . 047863
Cokkie Snoei, Rotterdam 112632
Le Col Rouge, Aix-en-Provence . 066459
Cola Collection, Houston 093643
Colac Otway Shire Hammerton Bottle
Collection, Colac 000982
Colacicco, Giovanni, Roma . . 131866
Colacicco, Michele, Torino . . 132121
Colacino, Buenos Aires 127520
Colaert, Piete, Antwerpen 063200
Colangelo, Philippe, Saint-Ouen 073054
Luciano Colantonio, Brescia . . 109899
Colantonio, Luciano, Brescia . . 079930
Colanzi, Domenico, Milano . . 110286
Colas, Alain, Lyon 128706
Colas, Christine, Paris 104652
Colas, Marjorie, Brest 103461
Colasanti & Colasanti, Roma . . 081298
Colberg, Rønne 140649
Colbert, Paris 104653
Colboc, Laurent, Meylan 070111
Colborne Lodge, Toronto 007168
Colburn Gem and Mineral Museum,
Asheville 046553
Colby College Museum of Art,
Waterville 054220
Colby & Son, Ralph, Portland . 136149
Colchester Historical Museum,
Truro 007252
Colcţia Muzeală a Mănăstirii Lainici,
Bumbeşti-Jiu 036328
Cold Spring Harbor Whaling Museum,
Cold Spring Harbor 047866
Coldharbour Mill, Working Wool
Museum, Uffculme 046016
Coldstream Antiques, Coldstream 088954
Coldstream Gallery, Ulmarra . . . 099845
Coldstream Museum, Coldstream 043994
Coldwater Canadiana Heritage
Museum, Coldwater 005634
Cole & Contreras Books / Sylvan Cole
Gallery, Sitges 115461
Cole County Historical Museum,
Jefferson City 049802
Cole Harbour Heritage Farm Museum,
Cole Harbour 005635
Cole Land Transportation Museum,
Bangor 046777
Cole Museum of Zoology,
Reading 045588
Cole Pratt, New Orleans 122145
Cole, E.J., Houston 093644
Cole, John, Devonport 139715
Coleção Assis Pegado – Museu da
Moto, Fortaleza 004460
Coleção de Artes Visuais, São
Paulo 004852
Coleção do Palácio do Governo de
Campos do Jordão, Campos de
Jordão 004397
Coleção Entomológica de Referência,
Faculdade de Saúde Pública / FSP,
São Paulo 004853
Colecção Berardo, Lisboa 035756
Colecção Etnográfica da Câmara
Municipal de Sobral de Monte
Agraço, Sobral de Monte
Agraço 036047
Colecção Etnográfica da Direcção-
Geral Agrícola, Aljustrel 035501
Colecção Etnográfica da Paróquia de
São Mamede da Ventosa, Torres
Vedras 036062
Coleccion 21, Guatemala 079224
Colección Anatomía, Madrid . . . 039612

Colección Arqueológica, Génova 008685
Colección Arqueológica, Pijao . . 008760
Colección Benedito, Madrid . . . 039613
Colección Carlos Alberto Pusineri
Scala, Asunción 034194
Colección Confradía del Rollo,
Jumilla 039534
Colección de Anestesioloxía e
Reanimación, Santiago de
Compostela 040119
Colección de Arqueología y Etnografía
Americana, Madrid 039614
Colección de Arte y Museo Botero,
Banco de la República, Bogotá 008587
Colección de Escultura
Contemporánea, Madrid 039615
Colección de la Biblioteca Musical,
Madrid 039616
Colección de la Casa de Alba,
Epila 039377
Colección de la Real Academia de la
Historia, Madrid 039617
Colección de los Padres Escolapios,
Madrid 039618
Colección de Minerales y Arqueología,
Villanueva 008826
Colección de Mineralogía, Madrid 039619
Colección de Pintura de Olmeda
de las Fuentes, Olmeda de las
Fuentes 039829
Colección de Relojes Populares,
Salamanca 040038
Colección de Roberto Montenegro,
Instituto de las Artesanías
Jaliscienses, Guadalajara 030998
Colección del Banco de España,
Madrid 039620
Colección del Castillo de Mazanares El
Real, Manzanares el Real . . . 039720
Colección del Poeta Baudilio Montoya,
Clarcá 008665
Colección dos Gabinetes de Física,
Química e Historia Natural do Antigo,
Instituto de Santiago, Santiago de
Compostela 040120
Colección Enrico Poli, Lima . . . 034309
Colección Etnográfica, Colexió Público
Mosteiro de Caaveiro, Capela . 039224
Colección Maruja Dìaz, Villaviciosa de
Odón 040415
Colección Monográfica Ecomuseo de
Somiedo, Somiedo 040193
Colección Municipal, Granada . . 039462
Colección Municipal de Arte
Contemporáneo de Alcorcón,
Alcorcón 038925
Colección Museística de San Mateo,
Collecció Museogràfica Municipal de
Sant Mateu, Sant Mateu 040068
Colección Museo Parroquial del
Santuari de la Font de la Salut,
Traiguera 040272
Colección Museográfica Casa de la
Apicultura, Boal 039154
Colección Museográfica d'Cerámica de
Potríes, Potríes 039965
Colección Museográfica de la Torre
Musulmana, Silla 040184
Colección Museográfica d'Etnografia i
Costums de la Mancomunitat de
Municips de la Vall d'Albaida,
Benisoda 039128
Colección Museográfica Municipal de
Borriol, Borriol 039157
Colección Museográfica Municipal de
Castell de Castells, Esquina, Castell
de Castells 039245
Colección Museográfica Municipal de
Cirat, Cirat 039286
Colección Museográfica Museo de la
Baronía, Ribesalbes 040004
Colección Museográfica Permanente
Arte Bizantino, Monasterio Trinidad,
Tángel 040200

Colección Museográfica Permanente
de Biar, Museu Municipal de Biar,
Biar 039137
Colección Museográfica Permanente
de Buñol, Buñol 039163
Colección Museográfica Permanente
de Cinctorres, Cinctorres 039285
Colección Museográfica Permanente
de la Font de la Figuera, La Font de
la Figuera 039404
Colección Museográfica Permanente
de Municipal, Moncada 039762
Colección Museográfica Permanente
Luís García de Fuentes, Caudete de
las Fuentes 039257
Colección Museográfica Permanente
Parroquial de Jérica, Jérica . . 039530
Colección Museográfica Permenent
Municipal de Cocentania, Palau
Comptal, Cocentaina 039297
Colección Numismática, Banco de la
República, Bogotá 008588
Colección Ornitológica W.H. Phelps,
Caracas 054711
Colección Parroquial de Villahermosa
del Río, Villahermosa del Río . 040407
Colección Parroquial, Iglesia Santiago
Apóstol, Liétor 039564
Colección Taurina, Gerena 039432
Coleccionart, Santiago de Chile . 064962,
140439
El Coleccionista, Buenos Aires . 060891
El Coleccionista, Cuenca 085959
El Coleccionista, México 082472
Colecţia Arheologică, Zlatna . . . 036837
Colecţia Bibliotecii Seminarului
Teologic Ortodox, Caransebeş . 036354
Colecţia Boiţa, Tălmaciu 036744
Colecţia de Artă Plastică Al.
Bălintescu, Costeşti 036409
Colecţia de Artă Plastică Fr. Storck
şi Cecilia Cecilia Cuţescu Stock,
Bucureşti 036284
Colecţia de Artă Veche Bisericească,
Reşiţa 036654
Colecţia de Carte Veche şi Icoane
Ştefan Crişan, Lipova 036535
Colecţia de Ceramică Populară de pe
Valea Olteţului, Balş 036227
Colecţia de Etnografie şi Artă Populară,
Buzău 036335
Colecţia de Etnografie şi Artă Populară,
Chilia 036369
Colecţia de Istorie a Farmaciei Cluj-
Napoca, Cluj-Napoca 036378
Colecţia de Istorie şi Arheologie, Piatra
Olt 036616
Colecţia de Mineralogie Estetică a
Fierului Constantin Gruescu, Ocna de
Fier 036588
Colecţia de Obiecte Bisericeşti,
Agniţa 036193
Colecţia de Obiecte Bisericeşti,
Movile 036577
Colecţia Etnografică, Bărbăteşti . 036231
Colecţia Etnografică, Irimeşti . . 036523
Colecţia Etnografică, Poiana
Sibiului 036631
Colecţia Etnografică, Răşinari . . 036647
Colecţia Etnografică Zahacinshi,
Mihăileni 036567
Colecţia Eugenia şi Costache Burada,
Dăneşti 036432
Colecţia Mănăstirii Zamfira,
Lipăneşti 036534
Colecţia Maria şi Dr. George
Severeanu, Bucureşti 036285
Colecţia Memorială Maria Filotti,
Bucureşti 036286
Colecţia Memorială Octavian Goga,
Răşinari 036648
Colecţia Mitropolit Antonie Plămădeală,
Sâmbăta de Sus 036674
Colecţia Muzeală, Beliu 036240

Colecţia Muzeală, Colacu 036387
Colecţia Muzeală, Cotorca 036413
Colecţia Muzeală, Holod 036493
Colecţia Muzeală, Rimetea 036658
Colecţia Muzeală, Smeeni 036724
Colecţia Muzeală, Tărcaia 036745
Colecţia Muzeală a Arhiepiscopiei
Ortodoxe Cluj, Cluj-Napoca . . 036379
Colecţia Muzeală a Arhiepiscopiei
Ortodoxe Galaţi, Galaţi 036470
Colecţia Muzeală a Bisericii Ortodoxe
Sf. Nicolae, Bereşti 036241
Colecţia Muzeală a Episcopiei
Armeano-Gregoriene, Bucureşti 036287
Colecţia Muzeală a Episcopiei
Ortodoxe, Alba Iulia 036197
Colecţia Muzeală a Episcopiei
Ortodoxe, Buzău 036336
Colecţia Muzeală a Episcopiei
Ortodoxe, Oradea 036594
Colecţia Muzeală a Episcopiei
Ortodoxe Sârbe, Timişoara . . . 036780
Colecţia Muzeală a Mănăstirii
Adormirea Maicii Domnului, Bodrogu
Vechi 036249
Colecţia Muzeală a Mănăstirii Agapia,
Agapia 036192
Colecţia Muzeală a Mănăstirii Bistriţa,
Bistriţa, Vâlcea 036247
Colecţia Muzeală a Mănăstirii Cozia,
Căciulata 036339
Colecţia Muzeală a Mănăstirii Dintr-un
Lemn, Dezrobiţi 036438
Colecţia Muzeală a Mănăstirii
Dragomirna, Dragomirna 036449
Colecţia Muzeală a Mănăstirii Govora,
Mihăeşti 036566
Colecţia Muzeală a Mănăstirii Gura
Motrului, Butoieşti 036334
Colecţia Muzeală a Mănăstirii Hurezi,
Românii de Jos 036663
Colecţia Muzeală a Mănăstirii Jitianu,
Podari 036630
Colecţia Muzeală a Mănăstirii
Moldoviţa, Vatra Moldoviţei . . . 036822
Colecţia Muzeală a Mănăstirii Neamţ,
Vânători-Neamţ 036817
Colecţia Muzeală a Mănăstirii Nicula,
Fizeşu Gherlii 036463
Colecţia Muzeală a Mănăstirii
Ortodoxe Celic-Dere, Poşta . . 036633
Colecţia Muzeală a Mănăstirii
Ortodoxe Cernica, Cernica . . . 036368
Colecţia Muzeală a Mănăstirii
Ortodoxe Ciolanu, Tisău 036788
Colecţia Muzeală a Mănăstirii
Ortodoxe Cocoş, Niculiţel . . . 036586
Colecţia Muzeală a Mănăstirii
Ortodoxe Curtea de Argeş, Curtea de
Argeş 036428
Colecţia Muzeală a Mănăstirii
Ortodoxe Dealu, Aninoasa . . . 036201
Colecţia Muzeală a Mănăstirii
Ortodoxe Ghighiu, Bărcăneşti . 036234
Colecţia Muzeală a Mănăstirii
Ortodoxe Pasărea, Pasărea . . 036605
Colecţia Muzeală a Mănăstirii
Ortodoxe Sinaia, Sinaia 036712
Colecţia Muzeală a Mănăstirii
Ortodoxe Suzana, Măneciu . . 036555
Colecţia Muzeală a Mănăstirii
Ortodoxe Ţigăneşti, Ciolpani . 036372
Colecţia Muzeală a Mănăstirii
Ortodoxe Viforâta, Viforâta . . . 036823
Colecţia Muzeală a Mănăstirii
Polovragi, Polovragi 036632
Colecţia Muzeală a Mănăstirii Putna,
Putna 036637
Colecţia Muzeală a Mănăstirii Râmeţ,
Râmeţ 036641
Colecţia Muzeală a Mănăstirii Rohia,
Târgu Lăpuş 036761
Colecţia Muzeală a Mănăstirii Secu,
Mănăstirea Neamţ 036554

Colecţia Muzeală a Mănăstirii Sf. Ilie,
Topliţa 036792
Colecţia Muzeală a Mănăstirii Sf.
Simion-Stâlpnicul, Arad 036203
Colecţia Muzeală a Mănăstirii Strehaia,
Strehaia 036731
Colecţia Muzeală a Mănăstirii Suceviţa,
Suceviţa 036741
Colecţia Muzeală a Mănăstirii Tismana,
Tismana 036789
Colecţia Muzeală a Mănăstirii Văratec,
Văratec 036818
Colecţia Muzeală a Mitropoliei
Ortodoxe a Olteniei, Craiova . . 036416
Colecţia Muzeală a Protoieriei
Ortodoxe Lugoj, Lugoj 036542
Colecţia Muzeală Batthyaneum, Alba
Iulia 036198
Colecţia Muzeală Birchiş, Birchiş 036244
Colecţia Muzeală Colţi, Colţi . . . 036389
Colecţia Muzeală de Argintărie,
Braşov 036269
Colecţia Muzeală de Artă Bisericească
Veche a Mitropoliei Banatului,
Timişoara 036781
Colecţia Muzeală de Istorie,
Săcueni 036671
Colecţia Muzeală Geamia Medgidia,
Medgidia 036560
Colecţia Muzeală Sătească,
Cârligele 036358
Colecţia Muzeală Sătească,
Coteşti 036411
Colecţia Muzeală Sătească,
Gugeşti 036486
Colecţia Muzeală Sătească,
Paltin 036603
Colecţia Muzeală Sătească,
Vulcăneasa 036830
Colecţia Oologică Kalaber Ladislau,
Reghin 036651
Colecţia Sătească, Atid 036208
Colecţia Sătească, Breaza 036275
Colecţia Sătească, Cârţa 036359
Colecţia Sătească, Ciumani . . . 036377
Colecţia Sătească, Corund 036406
Colecţia Sătească, Dăneşti 036433
Colecţia Sătească, Gura Râului . 036488
Colecţia Sătească, Joseni 036526
Colecţia Sătească, Lăpuş 036528
Colecţia Sătească, Lăzarea 036529
Colecţia Sătească, Lunca de Sus 036544
Colecţia Sătească, Lupeni 036546
Colecţia Sătească, Mereşti 036564
Colecţia Sătească, Muşăteşti . . . 036578
Colecţia Sătească, Ocland 036587
Colecţia Sătească, Plăieşii de
Jos 036620
Colecţia Sătească, Praid 036634
Colecţia Sătească, Remetea . . . 036653
Colecţia Sătească, Sâncrăieni . . 036675
Colecţia Sătească, Sândominic . 036676
Colecţia Sătească, Sărmaş 036680
Colecţia Sătească, Subcetate . . 036732
Colecţia Sătească, Suseni 036742
Colecţia Sătească, Tulgheş 036801
Colecţia Sătească, Voşlăbeni . . . 036828
Colecţia Sătească, Zetea 036836
Colecţia Vasile Heisu, Răcăciuni 036638
Colecţias Muzeală a Mănăstirii
Ortodoxe Căldăruşani, Moara
Vlăsiei 036571
Colectivo Multimédia Perve,
Lisboa 113987
Colefax & Fowler, London 090030
Colegiata de Santa María la Mayor,
Toro 040255
Colegio de Arquitectura y Diseño de
Interiores, Universidad San Francisco
de Quito, Quito 055322
Colegio de Artes Liberales,
Especialización Historia del Arte,
Universidad San Francisco de Quito,
Quito 055323

Colegio de Santa Isabel Museum,
Naga 034556
Colegio del Arte Mayor de la Seda,
Valencia 040304
Colegio del Rey, Alcalá de
Henares 038912
Colegio Dominicano de Artes Plásticas,
Santo Domingo 055318
Coleham Pumping Station,
Shrewsbury 045795
Coleman House Museum,
Ainsworth 046325
Coleman, Garrick D., London . . 090031
Coleman, J., Cincinnati 145067
Coleman, Kathleen, Houston . . . 121018
Coleman, Lynda, Austin 119655
Coleman, Mary L., Pittsburgh . . 123532
Coleman, Simon, London 090032
Coleman, Spencer, Beverley . . . 117351
Coleman, Spencer, Lincoln 118118
Coleman, Spencer, Stamford . . . 119137
Coleman, Teresa, Hong Kong . . 065030
Colemore, Jolimont 061660
Colemore-Williams, Patricia,
Russell 084074
Coleraine Local History Museum,
Coleraine 000983
Coleridge Cottage, Nether
Stowey 045321
Coles, Ronald, Kenthurst 099183
Coles, Sadie, London 118257
Colesberg/Kemper Museum,
Colesberg 038690
Coleshill Antiques, Coleshill . . . 088959
Colesville and Windsor Museum
at Saint Luke's Church,
Harpursville 049367
Colette, Den Haag 143110
Coletti, Roma 142805
Coleurs d'Îles, Fort-de-France . . 137815
Coley Homestead and Barn Museum,
Weston 054345
Coliboeuf, Caroline, Pau 072036
Colicchio, Carlo, Napoli 080853
Colin, Cardiff 144310
A Colin Maillard, Paris . 071172, 104654
Colin, Jean-Pierre, Saint-Benigne 072688,
129217
Coliseum, Lugo 086083
Coliseum, Roma 142806
Coliseum Antiquariato, Belo
Horizonte 063895, 140266
Coliseum Antiques, Mayfield . . . 061802
Coll-Rotger, Paris 071173
Colla, Joseph, Montpellier 128808
Collado Aucejo, Carmen, Valencia 115478
Collage, Odense 102945
Collage 20th Century Classic,
Dallas 093095
Collage Gallery, San Francisco . 124259
Collalti, Filippo, Roma 131867
Collalti, Sergio, Roma 081299
Collard, Pontrieux 072275
Collard, Totnes 144885
Collay, Jean-Louis, Lyon 069726
Collecció la Porciuncula, L'Arenal 038985
Col·lecció Museogràfica d'Arqueologia
de Pedreguer, Pedreguer 039925
Col·lecció Museogràfica d'Arqueologia
Local, Riba-Roja de Turia 039999
Collecció Museogràfica Permanent de
l'Almassara, Muro de Alcoy . . 039812
Colección Carlos Ferreyros Díaz,
Bogotá 008589
Colleccíon de Instrumentos Musicales
"Monsenör José Ignacio Perdomo
Escobar", Bogotá 008590
Colección Museográfica Permanente,
Siete Aguas 040181
Colleció d'Autòbiis de Salvador Claret
Sargatal, Sils 040185
Colección de Antigüedades, San
Andrés de Sotavento 008783
Collecion du Louvre, Belo

Horizonte 063896
Colección Eugenio Fontaneda,
Ampudia 038972
Colleicons Municipals del Foment
Arqueològic Excursionista Sallentí,
Casal Municipal, Sallent 040052
Collect-Art, Amersfoort 112121
Collectabears, Portland 096384
Collectable Books, London 144579
Collectable Creations,
Toowoomba 062326
Collectables, Cowes 061369
Collectables, Honiton 089559
Collectables, Kingston-on-Spey . 089698
Collectables From Yesteryear, Greerton
Tauranga 083943
Collectables Trader, Bondi
Junction 138494
Collectamania, Waikato 084127
Collectania New York, New York 095323
Collectible Antiques, New
Orleans 095081
Collectible Art Gallery,
Philadelphia 123361
Collectible Caboose, Austin . . . 092222
Collectible Toys Unlimited,
Sacramento 096614
Collectible Trains and Toys,
Dallas 093096
Collectibles, Dubai 088174
Collectibles, El Paso . . 093396, 093397
Collectibles Etc., Oklahoma City 095959
Collectie A. Veltman, Bussum . . 032128
Collectie Anderson, Losser 032606
Collectif d'Expression Artistique-C.E.A,
Bayonne 103341
Collecting Doulton and Beswick
Collectors Club, Richmond . . . 139162
Collectio Navalis, Berlin 141499
The Collection, Columbus 093005
Collection, Denver 093271
The Collection, Estepona 085992
The Collection, Los Angeles . . . 094294
The Collection, Manama 063137
The Collection, Padstow 118872
The Collection, West Hollywood . 097881
Collection 2000, Saint-Étienne . 072757
Collection Archéologique,
Marmoutier 014025
Collection Archéologique, Villarzel-
Cabardes 016226
Collection Archéologique de la
Maison des Fouilles et Maison du
Memontois, Malain 013978
Collection Archéologique Nationale,
Tramont-Emy 016031
Colleción Banco de la República,
Bogotá 008591
Collection Capital, Toulouse . . . 073754
Collection Care, Dublin 130850
Collection Cérès Franco,
Lagrasse 103914
Collection Charles de l'Escalopier,
Amiens 011628
Collection d'Archéologie Biblique USH,
Strasbourg 015864
Collection d'Art, Pauillac 014811
Collection d'Art Céramique,
Longfosse 013859
Collection d'Art Religieux Ancien,
Prunay-en-Yvelines 015004
Collection d'Arts Décoratifs, Cazaux-
Saves 012378
Collection d'Arts Décoratifs,
Villevielle 016293
Collection d'Automates, Gilley . . 013106
Collection de la Guerre de 1870, Le
Bourget (Seine-Saint-Denis) . . 013611
Collection de l'Abbaye de l'Ile Chauvet,
Bois-de-Céné 012093
Collection de l'Art Brut,
Lausanne 041646
Collection de l'Art Brut Publications,
Lausanne 138061

Collection de l'Aventure Automobile à
Poissy, Carrieres-sous-Poissy . 012345
Collection de l'Institut d'Égyptologie,
Strasbourg 015865
Collection de Locomotives à Vapeur,
Obermodern 014541
Collection de l'Outillage de Tonnellerie,
Pauillac 014812
Collection de Matériel Ferroviaire,
Burnhaupt-le-Haut 012255
Collection de Minéraux de l'Université
Pierre et Marie Curie, Paris . . . 014634
Collection de Peintures, Cavaillon 012374
Collection de Peintures, Issac . . 013297
Collection de Tapisseries d'Aubusson
et des Flandres, eigene homepage,
Saint-Chamant 015250
Collection de Téléphones Anciens,
Saint-Jean-du-Gard 015387
Collection de Tire-Bouchons, Saint-
Christophe-des-Bardes 015257
Collection de Tonnellerie,
Boersch 012086
Collection de Vieux Pressoirs,
Paulx 014816
Collection de Voitures Anciennes
de S.A.S. Le Prince de Monaco,
Monaco 031715
Collection d'Etains, Sierre 041984
Collection Deyrolle, Paris 136884
Collection d'Orgues Mécaniques,
Palais des Orgues Mortier,
Herzeele 013257
Collection d'Outillage Ancien de
Tonnellerie, Cussac-Fort-Médoc 012735
Collection d'Outillage Viticole dans la
Cave, Saint-Estèphe 015299
Collection du Château d'Attre,
Attre 003332
Collection du Moulin du Cros,
Montagnac-sur-Lède 014214
Collection du Musée Archéologique,
Guise 013215
Collection Entomologique Ouellet-
Robert, Outremont 006602
Collection Fondation Zoubov,
Genève 041523
Collection Frits Lugt, Paris 014635
Collection Gallo-Romaine, La
Roquebrussanne 013488
Collection Historial Cognacq-Jay,
Paris 014636
Collection Historique de Fouets, Les
Micocouliers, Sorède 015835
Collection Historique de la Marqueterie
d'Art Spindler, Boersch 012087
Collection Historique Maritime,
Brest 012203
Collection Historique, Ossuaire Tour,
La Wavrille, Douaumont 012823
Collection Internationale d'Épinettes,
Dommartin-lès-Remiremont . . 012806
Collection J.M., Amiens 066568
Collection Jules Verne, Amiens . 011629
Collection Lambert en Avignon, Musée
d'Art, Avignon 011832
Collection Lapidaire, Gimont . . . 013107
Collection of Archaeology and Land
Study, Nahshonim 025161
Collection of Byzantine and
Postbyzantine Art, Kythira 023146
Collection of Icons and Church
Heirlooms at Pyrgos, Thera . . . 023284
Collection of Martinware Pottery,
Southall 045822
Collection of the Evangelismos tis
Theotokou, Skiathos 023266
Collection of the Mega Spileon
Monastery, Kalavryta 023104
Collection of the Saint John the Divine
Monastery, Patmos 023214
Collection Pierre Bouton, Berneuil 011995
Collection Préhistorique Pagès, Le
Moustier 013673

Collection Prevee, Miami 121759
Collection Rare de Tracteurs Viticoles,
Savigny-les-Beaune 015733
Collection Républicaine, Urimenil 016072
Collection Roux, Villefranche-sur-
Mer 016245
Collection Schlumpf, Musée National
de l'Automobile, Mulhouse ... 014374
The Collection Shop, Miami ... 094716
Collection Ski-Montagne, Maison
du Parc National des Ecrins,
Briançon 012214
Collection Vini-Viticole et de
Tonnellerie, Caveau J. Becker,
Riquewihr 015101
Collection Viti-Vinicole, Coopérative de
l'Étoile, Banyuls-sur-Mer 011887
Collection Viticole, Mardeuil ... 014014
Collectionantes, Nantes 070459
Collectionneur, Clermont-Ferrand 068019
Le Collectionneur, Middelburg .. 083492
Collectionneur, Nevers 070536
Collectionneur, Roanne 072488
Le Collectionneur Cevenol, Alès 066528
Collectionneur Impuni, Paris ... 141078
Le Collectionnisme, Wattignies . 074294
Collections, Christchurch 083865
Collection's, Saint-Ouen 073055
Collections 71, Autun 066833
Collections and Conservation Centre,
Toronto 007169
Collections Antiques, Indianapolis 093869
Collections Archéologiques,
Parmain 014803
Collections Artistiques de l'Université
de Liège, Liège 003792
Collections Baur-Musée des Arts
d'Extrême-Orient, Fondation Alfred et
Eugénie Baur-Duret, Genève .. 041524
Collections Brassicoles, Brasserie
Kronenbourg, Strasbourg 015866
Collections Claeyssens,
Wambrechies 016335
Collections Council of Australia,
Adelaide 058195
Collections d'Arts Populaires,
Eschau 012896
Collections de la Bibliothèque, La
Bernerie-en-Retz 013364
Collections de la Bibliothèque
Municipale, Auxerre 011820
Collections de la Bibliothèque Sainte-
Geneviève, Paris 014637
Collections de la Fondation de
Coubertin, Saint-Rémy-lès-
Chevreuse 015546
Collections de la Fondation Roger-van-
Rogger, Bandol 011882
Collections de la Maison de la R.A.T.P.,
Paris 014638
Collections de la Maison Henri IV,
Saint-Valery-en-Caux 015582
Collections de la Salle des Martyrs-
Chapelle-Crypte, Paris 014639
Collections de la Serrurerie Dantin,
Paris 014640
Collections de la Tour Eiffel,
Paris 014641
Collections de l'Abbaye Bénédictine
de Saint-Louis-du-Temple,
Vauhallan 016137
Collections de l'Auberge du Cygne,
Tôtes 015973
Collections de l'Ecole Nationale
Supérieure des Beaux-Arts,
Paris 014642
Collections de la Born, Saint-
Europe-de-Born 015312
Collections du Château de Breil de
Foin, Genneteil 013089
Collections du Château de Bridoré,
Loches-en-Touraine 013844
Collections du Château de Courances,
Courances 012695

Collections du Château de Gizeux,
Gizeux 013121
Collections du Château de la
Thibaudière, Juigné-Béné 013337
Collections du Château de Lantilly,
Cervon 012397
Collections du Château de Sury-le-
Comtal, Sury-le-Comtal 015897
Collections du Château des Rohan,
Pontivy 014957
Collections du Château du Bouchet,
Rosnay 015159
Collections du Château Médiéval des
Cars, Les Cars 013745
Collections du Donjon, Bazoges-en-
Pareds 011931
Collections du Haras, La Roche-sur-
Yon 013466
Collections du Restaurant Procope,
Paris 014643
Collections du Service Historique de
l'Armée de l'Air, SHAA, Armées 011734
Collections et Découvertes,
Montauban 070192
Collections II, New Orleans 095082
Collections in Situ de l'Unesco,
Paris 014644
Collections Internationales, Verdun
(Meuse) 074019
Collections Internationales de
Marionnettes, Charleville-
Mézières 012454
Collections Littéraires, Kolbsheim 013353
Collections Livres, Bruxelles ... 140169
Collections of Cypriot Painting,
Sculpture and Graphics,
Kolonaki 023134
Collectique, Fresno 093495
Collective Antiques, San
Francisco 097209
Collective Art Gallery,
Indianapolis 121174
Collectophile Librairie, Montréal 140358
The Collector, Barnard Castle .. 088356,
134062
The Collector, Cape Town 114764
The Collector, Clevedon 088930
The Collector, Haarlem 083319
Collector, La Verne 139344
The Collector, London 090033
The Collector, Murrumbeena .. 125217
The Collector Art Gallery,
Washington 124884
Collector Book and Print Gallery,
Cincinnati 120336, 145068
Collector Heaven, Atlanta 092123
Collector Items, Seattle 097419
Collectormaniac, Mount Gambier 061908
Collector's, Parma 081103
Collectors Art, San Diego 124113
Collectors Authority, San Antonio 096961
Collectors Books and Cards,
Calgary 140335
Collectors Cabinet, Aachen 141395
The Collectors Centre, Saint Peter
Port 091205, 144820
Collectors Centre – Antique City,
London 090034
Collector's Choice, Denver 093272
Collectors Choice, Modbury ... 090765
Collectors Choice, Omaha 096031
Collector's Choice, Saskatoon .. 101543
The Collectors Club, New York . 051365
Collectors Coin Exchange,
Phoenix 096201
Collectors Connection, Cincinnati 092840
Collector's Corner, Brüttisellen .. 087333
Collectors Corner, Burnie 139671
Collectors Corner, Calgary 064252
Collectors Corner, Carshalton .. 088803,
144317
Collectors Corner, Columbus ... 093006,
093007
Collectors Corner, Dún Laoghaire 079603

Collectors' Corner, Faversham .. 089267
Collector's Corner, Greenslopes . 061557
Collector's Corner, Grüningen .. 087516
Collectors Corner, Marieholm .. 086865
Collectors Corner, Napier 084004
Collectors Corner, Northallerton . 090853
The Collectors Corner, Oklahoma
City 095960
Collector's Corner, Omaha 096032
Collector's Corner, Toledo 097603
Collector's Corner, Truro 091643
Collector's Corner, Zürich 087948
Collector's Covey, Dallas 120573
Collectors Covey, Dallas 120574
The Collector's Den, Dublin ... 079535
Collectors Den, Indianapolis ... 093870
Collectors Den, Pittsburgh 096270
Collector's Den, Valletta 082439
Collector's Dream, Sacramento . 096615
Collectors Emporium, Buffalo .. 092517
Collector's Emporium, Los
Angeles 094295
Collectors Fine Art, Honolulu .. 120929
Collectors Firearms, Houston .. 093645
The Collectors Gallery, Calgary . 101222
Collector's Gallery, Kalamaria . 109066
Collectors Gallery, Kooyong ... 099201
Collectors' Gallery, Shrewsbury . 091325
Collector's Heaven, Dallas 093097
Collectors News, Des Moines .. 139345
Collectors News & the Antique Report,
Grundy Center 139346
Collectors Old Book Shop,
Richmond 145474
Collectors Old Toy Shop and Antiques,
Halifax 089412
Collectors Paradise, Albuquerque 092046
Collectors' Paradise, Leigh-on-
Sea 089796
Collector's Paradise Watches,
Chennai 079371
Collectors Path, Honolulu 093539
Collectors Pieces, Mount Eliza .. 061907
Collector's Place, Shrewsbury .. 091326
The Collectors Shop, Edinburgh 089167
Collector's Showcase,
Jacksonville 093980
Collectors Showcase, Winnipeg . 064916
Collector's Sports Books,
Earlwood 139721
The Collector's Store, Hamilton . 064316
Collectors Superstore, Virginia
Beach 097777
Collector's Toystop, Milwaukee . 094847
Collector's Warehouse, Cleveland 092932
Collectors World, Cromer 089012
Collectors World, Nottingham .. 090892
Collectors World, Pittsburgh ... 096271
Colleen's Art, New Orleans 122146
Colleen's Collectables, Columbus 093008
Colleen's Station Gallery, Miami 121760
College Art Association of America, New
York 060438, 139347
College Art Collections, London . 044948
College Art Gallery, New Paltz .. 051309
The College Art Gallery, College of
New Jersey, Ewing 048649
College Book Rack, Jacksonville 145186
College Book Rack – Vol 2,
Jacksonville 145187
College Bookstore, Houston ... 145167
Collège de France, Paris 055394
College Football Hall of Fame, South
Bend 053391
College Museum, Epsom 044300
College Museum, Maynooth ... 024965
College of Architecture and
Design, Daegu Haany University,
Gyeongsan 056089
College of Architecture and
Design, Kansas State University,
Manhattan 057401
College of Architecture and Design,
Lawrence Technological University,

Southfield 057928
College of Architecture and Fine
Arts, University of Santo Tomas,
Manila 056226
College of Architecture, Clemson
University, Clemson 056909
College of Architecture, Georgia
Institute of Technology, Atlanta 056681
College of Architecture, Illinois
Institute of Technology, Chicago 056880
College of Architecture, Texas A & M
University, College Station ... 056927
College of Architecture, University of
Illinois at Chicago, Chicago ... 056881
College of Art, Delhi 055789
College of Art and Architecture,
University of Idaho, Moscow .. 057488
College of Art and Communication,
Beijing Normal University,
Beijing 055192
College of Art and Design,
Philadelphia 057666
College of Art and Design, Center for
Creative Studies, Detroit 057009
College of Art and Design, Duksung
Women's University, Seoul ... 056094
College of Art and Design, Konkuk
University, Seoul 056095
College of Art and Humanities,
Humboldt State University,
Arcata 056665
College of Art and Science, Vanderbilt
University, Nashville 057507
College of Art Gallery, Kwame
Nkrumah University of Science and
Technology, Kumasi 022932
College of Art, Department of Art
Education and Design, Shenzhen
University, Shenzhen 055239
College of Art, Jilin University,
Changchun 055197
College of Art, Kwame Nkrumah
University of Science and Technology,
Kumasi 055748
College of Art, Yanbian University,
Yanji 055257
College of Art, Zhejiang University of
Technology, Hangzhou 055213
College of Arts and Architecture,
Pennsylvania State University,
University Park 058028
College of Arts and Crafts, Faculty
of Fine Arts, Lucknow University,
Lucknow 055816
College of Arts and Crafts, Patna
University, Patna 055827
College of Arts and Design, Ewha
Womens University, Seoul ... 056098
College of Arts and Design, School
of Fine Arts, Yeungnam University,
Gyeongsan 056090
College of Arts and Humanities,
Bangor University, Bangor 056507
College of Arts and Humanities,
University of Maine at Augusta,
Augusta 056689
College of Arts and Sciences
Gallery, University of Miami, Coral
Gables 047982
College of Arts and Sciences of
Beijing Union University, Beijing 055193
College of Arts and Sciences, Seton
Hall University, South Orange . 057927
College of Arts and Sciences,
University of Massachusetts Amherst,
Amherst 056653
College of Arts and Sciences,
University of Miami, Coral
Gables 056954
College of Arts and Sciences,
University of Tokyo, Tokyo 056041
College of Arts and Social Sciences,
Australian National University,
Canberra 054901

College of Arts Sichuan University,
Chengdu 055201
College of Arts, Chung Ang University,
Anseong 056079
College of Arts, Department of Fine
Art, Hebei University, Baoding . 055187
College of Arts, Department
of Fine Art, Jinan University,
Guangzhou 055208
College of Arts, Department of Fine
Arts, Chonnam National University,
Gwangju 056088
College of Arts, Department of Fine
Arts, Pusan National University,
Busan 056081
College of Arts, Dept. of Visual Arts,
Tamagawa Daigaku, Machida . 056017
College of Arts, Dong-A University,
Busan 056080
College of Arts, Dongduk Women's
University, Seoul 056096
College of Arts, Dongguk University,
Seoul 056097
College of Arts, Fine Art Department,
Changwon National University,
Changwon 056084
College of Arts, Fu-Jen Catholic
University, Hsinchuang 056452
College of Arts, Hue University,
Hue 058160
College of Arts, King Saud University,
Riyadh 056318
College of Arts, Nihon University,
Tokyo 056040
College of Arts, University of Baghdad,
Baghdad 055857
College of Arts, University of Nairobi,
Nairobi 056075
College of Creative Arts, Massey
University, Wellington 056200
College of Creative Arts, West Virginia
University, Morgantown 057486
College of Design, Arizona State
University, Tempe 057984
College of Design, North Carolina
State University, Raleigh 057746
College of Design, Shih Chien
University, Taipei 056458
College of Design, Shu-te University,
Yanchao 056461
College of Eastern Utah Prehistoric
Museum, Price 052357
College of Fine and Applied Arts,
Sudan University of Science and
Technology, Khartoum 056392
College of Fine Arts – School of Visual
Arts, Boston University, Boston 056768
College of Fine Arts & Music,
Chungnam National University,
Daejeon 056086
College of Fine Arts and Architecture,
Jawaharlal Nehru Technological
University, Hyderabad 055801
College of Fine Arts, Capital Normal
University, Beijing 055194
College of Fine Arts, Carnegie-Mellon
University, Pittsburgh 057685
College of Fine Arts, Chongqing
Normal University, Chongqing . 055203
College of Fine Arts, Daebul University,
Samho 056092
College of Fine Arts, Hongik University,
Seoul 056099
College of Fine Arts, Hunan Normal
University, Changsha 055198
College of Fine Arts, Marshall
University, Huntington 057222
College of Fine Arts, Seoul Women's
University, Seoul 056100
College of Fine Arts, Sookmyung
Women's University, Seoul ... 056101
College of Fine Arts, University of
Kerala, Triruvananthapuram .. 055842
College of Fine Arts, University of

Montevallo, Montevallo 057476
College of Fine Arts, University of New
Mexico, Albuquerque 056636
College of Fine Arts, University of New
South Wales, Paddington, New South
Wales 054916
College of Fine Arts, University of the
Philippines, Quezon City 056229
College of Fine Arts, Wichita State
University, Wichita 058103
College of Imaging Arts and Sciences,
Rochester Institute of Technology,
Rochester 057777
College of Industrial Art, Fuzhou
University, Fuzhou 055207
College of Liberal Arts and Education,
Tuskegee University, Tuskegee 058017
College of Liberal Arts and Sciences,
Iowa State University, Ames . 056651
College of Liberal Arts and Sciences,
University of Florida, Gainesville 057118
College of Liberal Arts, Azusa Pacific
University, Azusa 056696
College of Liberal Arts, Dakota State
University, Madison 057389
College of Marin Art Gallery,
Kentfield 049934
College of Music and Visual Arts,
Kyungpook National University,
Daegu 056085
College of Music, Visual Arts and
Theatre, James Cook University,
Townsville 054922
College of New Jersey Art Gallery,
Trenton 053795
College of New Rochelle, New
Rochelle 057536
College of Photographic Art and
Sciences, Rochester Institute of
Technology, Rochester 057778
College of the Arts, Ohio State
University, Columbus 056944
College of the Arts, University
of Southwestern Louisiana,
Lafayette 057304
College of Visual and Performing Arts,
Kutztown University, Kutztown . 057293
College of Visual and Performing Arts,
University of Mary Hardin-Baylor,
Belton 056722
College of Visual and Performing
Arts, University of Massachusetts
Dartmouth, North Dartmouth .. 057588
College of Visual Arts, Saint Paul 057820
College of Visual Arts Gallery, Saint
Paul 052837
The College of Wooster Art Museum,
Wooster 054540
College Park Aviation Museum,
College Park 047870
Collegi d'Arquitectes de Catalunya,
Barcelona 059859
Collegium Artisticum, Sarajevo . 100859
Collegium Hungaricum Berlin,
Berlin 017097
Collem Antiquités, Bruxelles ... 063378
Collet, Catherine, Saint-Mihiel .. 072984
Collet, Henri Bertrand, Saint-Denis-
d'Anjou 129220
Colleton Museum, Walterboro .. 054047
Colletti, Russell, Chicago 092697
Collezionando di Palazzo Giampiero,
Firenze 080124, 142669
Collezione A. Checchi, Palazzo Volta,
Fucecchio 026354
Collezione Archeologica e Foro
Romano, Assisi 025396
Collezione Borsari, Parma 027205
Collezione Calderara, Fondazinone
Calderara, Ameno 025315
Collezione Civica d'Arte, Pinerolo 027332
Collezione Contini-Bonacossi,
Firenze 026249
Collezione d'Arte della Banca Carige,

Genova 026391
Collezione d'Arte di Santa Maria di
Piazza, Ostra 027130
Collezione d'Arte Religiosa
Moderna, Musei Vaticani, Città del
Vaticano 054679
Collezione d'Arte Sacra,
Saludecio 027715
Collezione di Carrozze, Villa Barbaro,
Maser 026722
Collezione di Ceramiche Mostra
Permanente, Fiorano Modenese 026238
Collezione di Farfalle, Guardia
Sanframondi 026486
Collezione di Fisica e Medicina,
Pavia 027230
Collezione di Marionette Ferrari,
Istituto San Paolo, Parma ... 027206
Collezione di Minerali e Fossili,
Semione 041978
Collezione di Violini, Cremona . 026098
Collezione Ducati Fratelli Saltarelli,
Senigallia 027915
Collezione Etnografica, Forni
Avoltri 026334
Collezione G. Lorenzi, Milano .. 026790
Collezione Martini Carissimo,
Oria 027099
Collezione Memorie della Comunità,
Romano di Lombardia 027666
Collezione Minici Zotti-Un Museo di
Magiche Visioni, Palazzo Angeli,
Padova 027143
Collezione P. Mariani, Desio ... 026134
Collezione Permanente del Design
Italiano '45-'90, Milano 026791
Collezione Permanente di Valentino
Martinelli, Perugia 027255
Collezione Privata di Attrezzi ed
Utensili Contadini, Camerota .. 025718
Collezione Sarda L. Piloni,
Cagliari 025671
Collezione Scaglione, Locri 026611
Collezione Straka-Coppa, Spello 027998
Collezione Tessile della Tela Umbra,
Città di Castello 025999
Collezione Titta Ruffo, Pisa 027348
Collezione Wolfson, Genova ... 026392
Collezioni d'Armi e Ceramiche della
Rocca Sforzesca, Imola 026501
Collezioni d'Arte della Cassa di
Risparmio, Savona 027889
Collezioni d'Arte e di Storia della
Fondazione Cassa di Risparmio,
Bologna 025531
Collezioni del '900, Raccolta Alberto
della Ragione e Collezioni del
Novecento, Firenze 026250
Collezioni del Dipartimento di
Entomologia e Zoologia Agraria,
Università di Napoli, Portici ... 027413
Collezioni di Presepi del Convento
Muri-Gries, Bolzano 025579
Collezioni Diana, Roma 131868
Collezioni Egittologiche dell'Università
di Pisa, Università di Pisa, Pisa 027349
Il Collezionista Numismatica,
Navacchio 080949
Collier County Museum, Naples . 051162
Collier Memorial State Park and
Logging Museum, Chiloquin . 047722
Colliers, Newcastle-upon-Tyne . 118785
Colliette, Marie-Agnès, Semur-en-
Auxois 125815
Colligan, Dennis, Houston 093646
Collignon, Audun-le-Roman 066806
Collignon & Laurent, Laon 125608
Collignon & Laurent, Soissons . 125817
Collignon, Daniel, Mont-
Bonvillers 070172
Collignon, Daniel, Nancy 070417
Collignon, Gildas, Troyes 105776, 141371
Collignon, Sylvie, Paris 071174
Collin County Farm Museum,

McKinney 050563
Collin, Gillot, Paris 071175
Collin, Jean-François, Paris 071176
Collinge & Clark, London 144580
Collinge Antiques, Llandudno . 089865,
                                  134572
Collingwood, Bramley Gardens . 085627
Collingwood Gallery, Collingwood 098911
Collingwood Library and Museum on
Americanism, Alexandria 046385
Collingwood Museum,
Collingwood 005637
Collins, Houston 093647
Collins, München 130338
Collins, San Diego 097075
Collins & Chandler, Toronto ... 064630,
                                  101591
Collins & McCullough, Miami .. 094717
Collins & Paterson, Glasgow ... 126901
Collins Fine Art, New York 122490
Collins Gallery, Glasgow 044395
Collins Gallery, London 118258
Collins Street Gallery, Melbourne 099299
Collins, Bernadette, Minneapolis 121977
Collins, C.P., Leichhardt 139804
Collins, F.G. & C.,
Wheathampstead ... 091820, 135220
Collins, Frank, Mendlesham
Green 090747
Collins & Son, J., Bideford 088460,
                                  134103
Collins, Jim, Vancouver 101765
Collins, Louis, Seattle 145580
Collins, M.E., New York 095324
Collinsville Depot Museum,
Collinsville 047879
Colliton Antique Centre,
Dorchester 089069, 134329
Collot, Longué-Jumelles 069619
Collu, Giovanni, Cagliari 079978, 131066
Collyer-Bristow, London 118259
Collyer, Jean, Boughton 088525
Collyer, R., Birmingham 088477, 134108
Colman, Michael, Malvern,
Victoria 061779, 127632
Colmar, Marcel, La Garde 069045
Colmenares, Cartago 102466
Colnaghi & Co., P. & D., London 090035
Colnard, Marc, Pringy 072311
Colne Valley Museum, Golcar . 044440
Colobert & Letresor, Etampes . 125583
Cologne Fine Art & Antiques,
Köln 098135
Colomb, Alain, Chanac 067821
Colomb, Olivier, Saint-Estève .. 129222
Colombani, Dominique, Marseille 069925,
                                  128757
Colombano, Gianbeppe, Torino . 110961
La Colombe d'Or Art Gallery,
Houston 121019
Colombelle, Paris 071177
Colombi, Elda, Milano 080492
Colombi, Giuliana, Milano 131420
Colombine, Amsterdam 143025
Colombine, Montcaret 070221
Colombo National Museum,
Colombo 040475
Colombo, Alberto, Milano 131421
Colombo, Antonio, Milano 110287
Colombo, Maria Grazia, Bologna 079830
Colombo, Nicoletta, Milano 110288
Colomé, Raoul, Toulouse 129298
Colomes, Monique, Uzès 105783
Colomo, Lorenzo, Valladolid .. 115522
Colonel Ashley House, Ashley
Falls 046575
Colonel Davenport Historical
Foundation, Rock Island 052629
Colonel Stephens Railway Museum,
Kent and East Sussex Railway,
Tenterden 045958
Colonel William Jones House,
Gentryville 049075
Colonels' Choice, Milwaukee ... 094848

Coloneum, Regensburg 078165
La Colonia, Medellín 065282
Colonial Antiguidades, Rio de
  Janeiro 064045
Colonial Antiques, Tulsa 097720
Colonial Art Gallery, Oklahoma
  City 095961, 123278, 127377
Colonial Art Gallery, Russell ... 113003
Colonial Arts, San Francisco ... 097210
Colonial Attic Antiques,
  Spotswood 062236
Colonial Auctions, Philadelphia . 127390
Colonial Brass, Houston 135597
Colonial Burlington Foundation,
  Burlington 047323
Colonial Classics, Ascot Vale .. 127559
Colonial Concept, Saint-Ouen .. 073056
Colonial Cottage Antiques,
  Rockhampton 062161
Colonial Cottage Antiques, Virginia
  Beach 097778
Colonial Cottage Museum,
  Wellington 033345
Colonial Desks, Christchurch .. 083866
Colonial Discount Jewelers,
  Cleveland 092933
Colonial Estate Buyers,
  Minneapolis 094905
Colonial Fair Antiques,
  Dannevirke 083916
Colonial Frontiers, Tucson 097646
Colonial Industrial Quarter,
  Bethlehem 046959
Colonial Pemaquid Historical Site, New
  Harbor 051258
Colonial Pennsylvania Plantation,
  Media 050759
Colonial Restorations, Auckland 083778,
  132685
Colonial Times, Cross-in-Hand . 089015
Colonial Trading Company,
  Boston 092458
Colonial Williamsburg,
  Williamsburg 054411
Colonies Market, Oklahoma City 095962
Colonna, Bucureşti 114131
La Colonna, Lugano 087609
La Colonna, Roma 081300
Le Colonne, Firenze 080125
Le Colonne, Torino 142854
Colonnese, Napoli 142759
Colony Furniture Shops,
  Charlotte 092589
Colony House, Newport 051532
Colophonarte, Belluno 109788
Color Art, Saint Louis 123816
Color Association of the United States,
  New York 060439
Color Connection Gallery, Tulsa . 124799
Color Genesis, Atlanta 119553
Color Impressions, Charlotte .. 119983
Color y Forma, Medellín 102431
Colorado Gallery of the Arts, Araphoe
  Community College, Littleton . 050339
Colorado Historical Society Museum,
  Denver 048255
Colorado Photographic Arts Center,
  Denver 048256
Colorado Railroad Museum,
  Golden 049145
Colorado River Indian Tribes Museum,
  Parker 051927
Colorado School of Mines Geology
  Museum, Golden 049146
Colorado Ski & Snowboard Museum
  and Hall of Fame, Vail 053920
Colorado Springs Fine Arts Center,
  Taylor Museum, Colorado
  Springs 047883
Colorado Springs Museum, Colorado
  Springs 047884
Colorado University Art Galleries,
  Boulder 047140
Colorado University Heritage Center,

Boulder 047141
Colorado Watercolor Society,
  Denver 060440
Colorado-Wyoming Association of
  Museums, Sheridan 060441
Il Colore, Cagliari 109921
Colore Luna, Essen 106908
De Colores, Albuquerque 119463
Colorex, Norderstedt 077796
Colorido, Mumbai 109426
Coloris-Art, Villeneuve 116703
Colory, Berlin 106200
Colosseum Art Gallery, Haifa ... 025048
Colour Museum, Bradford 043693
Colourbox, Otley 118859
Colours, Cairo 103029
Colours, Edinburgh ... 117720, 134360
Colours, Masterton 112947
Colportages, Rouen 141320
Colter Bay Indian Arts Museum, Colter
  Bay Village 047893
Coltishall Antiques Centre,
  Coltishall 088961, 134294
Colton Area Museum, Colton ... 047894
Colton Hall Museum, Monterey . 051010
Colton, Gary, Kelvedon 089665, 134509
Coltsfoot, Leominster . 089803, 134547
Colucci, Fernando, Genova 110099
Coluccia, Lucrezia, Bari 079762
Coluccia, Piero, Bari ... 079763, 130911
Coluccini, Franco, Roma 131869
Columbia Antique Auction,
  Columbia 127186
Columbia City Gallery, Seattle .. 124507
Columbia Coins, Portland 096385
Columbia College Art Gallery,
  Chicago 047652
Columbia County Historical Museum,
  Lake City 050074
Columbia County Historical Society
  Museum, Kinderhook 049970
Columbia County Historical Society
  Museum, Saint Helens 052772
Columbia County Museum,
  Bloomsburg 047054
Columbia Fire Department Museum,
  Columbia 047906
Columbia George Interpretive Center,
  Stevenson 053527
Columbia Journal of Law & the Arts,
  New York 139348
Columbia Museum of Art,
  Columbia 047907
Columbia Rarities Group, El Paso 093398
Columbia River Exhibition of
  History, Science and Technology,
  Richland 052528
Columbia River Maritime Museum,
  Astoria 046583
The Columbian Museum and Art
  Center, Wamego 054054
Columbus, Groningen 083285
Columbus-Belmont Civil War Museum,
  Columbus 047927
Columbus College of Art and Design,
  Columbus 056945
Columbus Cultural Arts Center,
  Columbus 047935
The Columbus Museum,
  Columbus 047923
Columbus Museum of Art,
  Columbus 047936
Columbus Park Museum, Discovery
  Bay 028437
Columbus Sportscards,
  Columbus 093009
Column Works, New York 122491,
  122492
Columna 32, Zaragoza 086594
Las Columnas, San Salvador .. 103097
Colussi, Franco, Seven Hills .. 127673
Colville Lake Museum, Colville
  Lake 005638
Colville, Thomas, New York 122493

Colvin, Richard, Denver 135527
Colystock Antiques, Stockland . 091438
Colzium Museum, Kilsyth ... 044699
Com Arca, Bucureşti 085087
Com Art Graz, Verein zur Förderung
  von Computerkunst, Graz .. 058245
Com Arte Projetos Culturais, Belo
  Horizonte 100880
Comanche Crossing Museum,
  Strasburg 053562
Comarca, Cartagena 065260
Il Comaro, Iaşi 085122
Comatra, Colombes 068103
Comba, Marcus, Sankt Gallen . 133901
Combalbert, Françoise, Saint-Magne-
  de-Castillon 072906
Combastieix, Hervé & Robert, Le
  Grand-Lucé 069270
Combat Air Museum, Topeka .. 053770
Combe Martin Museum, Combe
  Martin 043997
Combe Mill Beam Engine and Working
  Museum, Long Hanborough .. 045145
Combe, Michel, Batz-sur-Mer .. 066963
Comber and District Historical Society
  Museum, Comber 005639
Comberton Antiques, Comberton 088965
Combes, Dominique, Toulouse . 105730
Combes, Michèle, Pouilly-sous-
  Charlieu 072298
Combi, Gian Luigi, Torino 081687
Combining Forces, London .. 090036,
  134630
Combret, Didier, Villefranche-de-
  Rouergue 074143
Combridge, Dublin 109579
Combs, Gary, New York 145329
Combs, Sharon, Tulsa 097721
Come in, Aalen 074347
Comeau, André, Montréal 064408
Comed, Milano 138809
Comenius-Antiquariat,
  Hilterfingen 144086
Comenius Museum, Naarden .. 032660
Comenos, George L., Boston .. 092459
Comer Museum, Morgantown .. 051055
Comércio de Arte Oriental Macau
  Antigo, Lisboa 084860
Comerford, Edward, Ballycallan . 079449
Comerlati, Didier, Paris 071178
Comes, Giovanna, Catania 131085
Comet Slices Gallery, Louisville . 121638
Cometa Arte Comercio, Belo
  Horizonte 063897
Comic Book World, Cincinnati .. 145069
Comic-Galerie Grober Unfug,
  Berlin 106201, 106202
Comic Warehouse, Albuquerque 144953
Comicladen-Kollektiv, Hamburg . 141859
Comicopia, Boston 092460
Comics Bazar, Roma .. 081301, 142807
Coming of Age Antiques,
  Withcott 062442
Comisión Nacional de Arqueología y
  Monumentos Historicos, Ciudad de
  Panamá 059682
Comisión Nacional de Museos y de
  Monumentos, Buenos Aires .. 058174
Comité Auguste Rodin, Paris ... 058629,
  136885
Comité Brasileiro do ICOM, São
  Paulo 058400
Comité Español del ICOM,
  Madrid 059860
Comité Nacional del Consejo
  Internacional de Museos, ICOM,
  Bogotá 058512
Comité National Français de l'ICOM,
  Paris 058630
Comité National Monégasque de
  l'Association Internationale des Arts
  Plastiques auprès de l'UNESCO,
  Monaco 059579
Comité Peruano del Consejo

Internacional de Museos, ICOM Perú,
  Lima 059687
Comité Professionnel des Galeries
  d'Art, Paris 058631
Comitée National de ICOM
  Centralafrique, Bangui 058480
Comitée National de ICOM de Algérie,
  Alger 058166
Comix Antikvariat, Oslo 143400
Comma 5, Milano 080493
Command Museum, San Diego . 052962
Commanda General Store Museum,
  Commanda 005641
Commandant's Quarters, Dearborn
  Historical Museum, Dearborn . 048192
La Commanderie d'Arville, Arville 011757
Commandery Museum,
  Worcester 046191
Comme Autrefois, Montauban .. 070193
Comme Avant, Ars-en-Ré 066744
Comme Ca NYC, New York 122494
Commedia, Münster 142200
Commedia Dell'Arte, Montpellier 104273
Commencement Art Gallery,
  Tacoma 053641
Commerce Bancshares Fine Art
  Collection, Kansas City 049879
Commerce Graphics, New York . 122495
Commercial and Industrial Museum,
  Kanpur 024339
Commercial and Technical Art
  Department, South Central College,
  North Mankato 057590
Commercial Art, Salt Lake City . 123954
Commercial Art Department, Honolulu
  Community College, Honolulu . 057207
Commercial Art Department, Luzerne
  County Community College,
  Nanticoke 057504
Commercial Arts Services, Dallas 120575
Commercial Museum, Gauhati
  University, Guwahati 024273
The Commercial Press, Hong
  Kong 136721
Commeter, Persiehl & Co.,
  Hamburg 107236
Commissaires-Priseurs de Chartrons,
  Bordeaux 125525
Commissaires Priseurs de Montpellier,
  Montpellier 125648
Commissariat House, Saint
  John's 006901
Commissariat Store Museum,
  Brisbane 000886
Commission a Craftsman,
  London 090037
Commission a Portrait, London . 118260
Common Ground Gallery, Windsor,
  Ontario 101852
Common Market, Memphis ... 094625
Common Space Center for Creativity,
  Toledo 124684
Common to Key Coin,
  Sacramento 096616
Commonwealth Air Training Plan
  Museum, Brandon 005463
Commonwealth Antiques, Boston 092461
Commonwealth Association of
  Museums, Calgary 058441
Commonwealth Books , Boston . 145015
Commonwealth Fine Art, Boston 119854
Commonwealth Institute, London 044949
Commonwealth Museum, Boston 047110
Commonwealth of the Northern
  Mariana Islands Museum of History
  and Culture, Saipan 033479
Commune du Pouancé, Pouancé 105399
Communication Arts, Menlo Park 138356,
  139320
Communication Arts Department,
  Madison Area Technical College,
  Madison 057391
Communications and History Museum
  of Sutton, Washington 054098

Communications Monument,
Seoul 029987
Communications Museum, Tbilisi 016396
Community Artists' Collective,
Houston 060442
Community Arts Centre,
Wanganui 033333
Community Arts Council of Richmond,
Richmond 058442
Community Arts Council of the Alberni
Valley, Port Alberni 006672
Community Arts Gallery, Wayne State
University, Dept. of Arts, Detroit 048297
Community Arts Project, Cape
Town 059815
Community College Art Gallery,
Baltimore 046739
Community Council for the Arts,
Kinston 060443
Community Fine Arts Center, Rock
Springs 052632
Community Gallery Spaces, CEMA Arts
Centre, Portland 001429
Community Historical Museum,
Maxwell 050744
Community Memorial Museum of
Sutter County, Yuba City 054606
Community Museums Association
of Prince Edward Island,
Charlottetown 058443
Community of Christ,
Independence 049663
Comocer, Castellón de la Plana . 133224
Comode, Simone, Autun 066834
Comoglio, Georges, Paris 071179
Comox Air Force Museum, Lazo 006175
Comox Archives & Museum,
Comox 005642
Comox Valley Centre for the Arts,
Courtenay 005655
Comox Valley Community Arts Council,
Courtenay 005656
Compagnia della Pietra, Verona 132276
Compagnia delle Indie, Firenze . 080126
Compagnie Avignonnaise de Change,
Avignon 066875
Compagnie de la Bourse, Paris . 071180
Compagnie de la Chine et des Indes,
Paris 071181
Compagnie de l'Or et des Métaux
Précieux, Thionville 073655
La Compagnie des Antiquaires,
Paris 058632
Compagnie des Arts, Strasbourg 073580,
141347
Compagnie des Commissaires
Priseurs de Paris, Paris 125696
Compagnie des Experts en
Ameublement, Objets d'Art et de
Collection près la Cour d'Appel et le
Tribunal de Grande Instance de Paris,
Paris 058633
Compagnie d'Expertise en Antiquités
et Objets d'Art, Paris 058634
Compagnie d'Experts Français en
Antiquités, Paris 058635
Compagnie Expertise en Antiquités et
Objets d'Art, Paris 058636
Compagnie Générale de l'Or,
Paris 071182
Compagnie Nationale des Experts,
Paris 058637
Compagnie XIX, Paris 071183
Compagnon Franck, Rouen . . . 072580
Compagnon, Emilie, Sains-du-
Nord 072657
Compagnons Ebénistes Associés,
Paris 128965
Compas, Les Mars 069414
Compas Antiquités, Les Mars . . 069415
Compass Gallery, Glasgow 117838
Compendium Gallery, Auckland . 112757
Complejo Arqueológico de Chavín,
Huari 034282

Complejo Histórico Chivilcoy,
Chivilcoy 000303
Complejo Museológico del Instituto
Félix Bernasconi, Buenos Aires 000155
Complejo Museológico La Estación,
Cruz Alta 000350
Complejo Museos Históricos Militares,
La Habana 009139
Complément d'Objet, Paris 071184
Complément d'Objets, Brantôme 067449
Complementy', Dallas 093098
Complesso Museale di Santa Chiara,
Napoli 026988
Complete Pictures, Houston . . . 121020
Complete Post, Los Angeles . . . 094296
Complete Traveller Antiquarian Bookstore,
New York 135900, 145330
Complex Muzeal de Ştiinţele Naturii,
Galaţi 036471
Complexe de Betange, Florange 012995
Complexul Memorial Nicolae Titulescu,
Nicolae Titulescu 036585
Complexul Monumental Curtea
Domnească, Târgovişte 036747
Complexul Muzeal Arad, Arad . . 036204
Complexul Muzeal Bucovina,
Suceava 036735
Complexul Muzeal de Stiinţe ale
Naturii Constanţa, Constanţa . . 036393
Complexul Muzeal de Stiintele Naturii
Ion Borcea, Bacău 036213
Complexul Muzeal Măldăreşti,
Măldăreşti 036550
Complexul Muzeal Naţional Goleşti,
Şefaneşti 036690
Complexul Muzeal Stelea,
Târgovişte 036748
Complexul Muzeistic Ciprian
Porumbescu, Ciprian
Porumbescu 036373
Complexul Muzeistic Gheorghe
Magheru, Râmnicu Vâlcea . . . 036644
Complexul Naţional Muzeal Astra,
Sibiu 036697
Complexul Naţional Muzeal Curtea
Domnească, Târgovişte 036749
Compliments, Buffalo 092518
Componere, Saint Louis 123817
Compostella, Mario, Napoli . . . 131608
Compra – Venta, Málaga 086291
Compra y Venta de Antigüedades
Belle Epoque, Bogotá 065195
Compra y Venta de Oro y Plata,
Madrid 086160
Compraventa Libros & Muebles,
Bogotá 065196, 140447
Compta Martinez, Antoni,
Barcelona 133110
Comptoir d'Achat Or et Argent,
Blois 067253
Comptoir d'Achat Or et Argent,
Tours 073846
Comptoir de Change, Toulon
(Var) 073707
Comptoir de la Chine, Mulhouse 070388
Comptoir de la Chine, Strasbourg 073581
Comptoir de la Main d'Or, Paris 071185
Le Comptoir de l'Aviation, Paris 071186
Comptoir de l'Occasion du Centre,
Aurillac 066821
Comptoir de Zélie, Chamalières 067790
Comptoir des Créateurs, Le
Pouliguen 103963
Comptoir des Isles, Fontevraud-
l'Abbaye 103749
Comptoir des Métaux Précieux,
Auxerre 066849
Comptoir des Monnaies Ancienne,
Lille 069484
Comptoir des Monnaies Anciennes,
Épernay 068449
Comptoir des Monnaies Anciennes,
Reims 072389
Comptoir des Quais, Reze 072449

Comptoir des Tuileries, Paris . . . 071187
Le Comptoir du Collectionneur,
Thionville 073656
Comptoir du Meuble Ancien,
Charleroi 063596
Comptoir du Troc, Saint-Melaine-sur-
Aubance 072976
Comptoir Français de l'Or, Paris 071188
Comptoir Marseillais de Bourse,
Marseille 069926
Comptoir Numismatique, Nevers 070537
Comptoir Numismatique, Orléans 070813
Comptoir Numismatique, Villeneuve-
sur-Lot 074165
Comptoir Numismatique du Roannais,
Roanne 072489
Comptoir Numismatique et
Philatélique, Bergerac 067100
Comptoir Numismatique la Semeuse,
Creutzwald 068241
Comptoir Numismatique Nantais,
Nantes 070460
Comptoir Numismatique Troyen,
Troyes 073900
Comptoir Parisien de l'Or, Le
Mans 069306
Comptoir Savoyard de Change,
Annecy 066640
Comptoirs du Chineur, Paris . . . 071189
Compton & Schuster, London . . 134631
Compton Cassey Gallery,
Withington 119371
Compton Castle, Paignton 045463
Compton County Historical Museum,
Eaton Corner 005740
Compton-Jones, Robert Victor,
London 090038
Compton Verney Collections, Compton
Verney 043999
Compton, Lucinda, Ripon 134971
Compulsive Collector, New York 145331
Computer Arts, Bath 139163
Computer Assisted Arts Association,
Zwijnaarde 058367
Computermuseum, Roeselare . . 003974
Computermuseum Aachen,
Aachen 016433
Computing Commons Gallery, Arizona
Sate University, Tempe 053702
Comstock House, Moorhead . . . 051046
Comte, Olivier, Meyrin 087676
Comte, Roland, Genève 116294
Comteroux, Perpignan 072075
Comtoise, Cannes 067634
Comtoise Uhren Museum,
Düsseldorf 018025
Con Arte, Ludwigsburg 077153
Con Edison Energy Museum, New
York 051366
Con Foster Museum, Traverse
City 053792
Cona, Nairobi 082249
Conan, Lyon 125626
Conart, Bucureşti 132994
Conart, Roma 131870
Conca d'Oro, Messina 110219
Concept Art, Granville 103778
Concept Art Gallery, Pittsburgh . 123533
Conception Bay Museum, Harbour
Grace 005987
Conceptions Framing Gallery Ltd,
Calgary 101223
Conceptions Southwest,
Albuquerque 139350
Concetta D-Gallery, Albuquerque 119464
Concha de Nazelle, Toulouse . . 105731
La Conchiglia, Sanremo 110903
Conchon, Ginette-Marie, Genève 087461
Concord Art Association, Concord 047950,
060444
Concord Heritage Museum,
Concord 000986
Concord Museum, Concord . . . 047951
Concordati, Enrica Maria Stella,

Lodi 080340
Concordia Historical Institute, Saint
Louis 052795
Concordia University-Wisconsin Art
Gallery, Mequon 050799
Concourse Aboriginal Gallery,
Winnipeg 101872
Condamin, Jacques, Nantes . . . 104336
Condamin, Jean-Paul, Saint-Laurent-
d'Agny 072881
Conde Gonzalez, Cesar, Madrid . 086161
Condé, Monica de, Vichy 141384
Conde, Susan, New York 122496
Condemine, Annie, Sens 073497
Condettan, Boulogne-sur-Mer . . 067399
Condettan, Le Touquet-Paris-
Plage 069360
Condor Ap Contract, Bucureşti . 085088
Condotti Fine Arts, Roma 081302,
081303
Condrau, Alexa, Disentis 116248
Conduit Gallery, Dallas 120576
Conejo Valley Art Museum, Thousand
Oaks 053739
Confederate Memorial Hall-Bleak
House, Knoxville 050007
Confederate Memorial State Historic
Site, Higginsville 049464
Confederate Museum,
Crawfordville 048056
Confederate Museum, New
Orleans 051288
Confederation Centre Art Gallery and
Museum, Confederation Centre of
the Arts, Charlottetown 005593
Confédération Française des Métiers
d'Art, Paris 058638
Confédération Internationale des
Négociants en Œuvres d'Art (CINOA),
Bruxelles 058368
ConFer Art Actuel, Nyon 116539
The Conference House, Staten
Island 053496
Confluences, Paris 058639
Le Confort des Etranges,
Toulouse 105732
Conforti, Firenze 142670
Confucianist Society of Japan,
Tokyo 059491
Congdom Anatomical Museum,
Bangkok 042654
Congdon-Clelford, Paul, Chandler's
Ford 134227
Congé, Michèle, Paris 128966
Congedo Editore, Galatina 137658
Congo Arts, Bruxelles 100581
La Congrega, Brescia 079931
Congregation Emanu-el Museum, New
York 051367
Congregation Shaar Hashomayim
Museum, Westmount 007385
Congressional Arts Caucus,
Washington 060445
Conifer Grove Antiques, South
Auckland 084077
Conil, Monique, Paris 071190
De Coninck-Minnaert, Kaprijke . 128065
Coningsby Gallery, London 118261
Coninx Museum, Zürich 042191
Conjunt Monumental de la Església
de Sant Pere, Museu de Terrassa,
Terrassa 040221
Conjunto Arqueológico de Baelo
Claudia, Tarifa 040202
Conjunto Arqueológico de Carmona,
Necrópolis y Anfiteatro Romano de
Carmona, Carmona 039232
Conjunto Arqueológico de Itálica,
Santiponce 040141
Conjunto Arqueológico de Madinat Al
Zahra, Córdoba 039303
Conjunto Cultural de Caixa, São
Paulo 004854
Conjunto de la Ferrería y Molinos de

Agorregi, Aia 038898
Conjunto Monumental Alcazaba de
  Almería, Almería 038959
Conklin Reed Organ and History
  Museum, Hanover 049352
Conley, Buffalo 092519
Conlibro Buch und Kunst
  Verlagsgesellschaft, Darmstadt 137134
Conlon, Dublin 079536
Conn, Jack M., Dallas 093099
Connaissance de l'Art, Lausanne 059931
Connaisseur, Kraków 084564
Connaisseur, Warszawa 084709, 084710
Connally, Oklahoma City 095963
Connaught Brown, London 090039,
  118262,   134632
Conneaut Railroad Museum,
  Conneaut 047963
Connecticut Audubon Birdcraft
  Museum, Fairfield 048668
Connecticut Avenue Collection,
  Washington 097827
Connecticut Fire Museum, East
  Windsor 048450
Connecticut Graphic Arts Center,
  Norwalk 051643
Connecticut Historical Society
  Museum, Hartford 049384
Connecticut Institute of Art,
  Pittsburgh 057686
Connecticut Museum Association,
  Roxbury 060446
Connecticut Pastel Society,
  Newtown 060447
Connecticut River Museum,
  Essex 048608
Connecticut State Museum of Natural
  History, Storrs 053553
Connecticut Trolley Museum, East
  Windsor 048451
Connecticut Valley Historical Museum,
  Springfield 053459
Connecticut Watercolor Society, West
  Simsbury 060448
Connell, Atlanta 119554
Connell, John, Wargrave 091728, 135189
Connell, Michael, Dublin 079537
Connelly, John, New York 122497
Connelly, Robert J. & Sally,
  Binghampton 127141
Connemara History and Heritage
  Centre, Clifden 024849
Conner, Washington 124885
Conner & Rosenkranz, New York 122498
Conner Prairie Living History Museum,
  Fishers 048735
Connex Caribe Mexico, Cancún . 132359
Conni Gordon Arts Foundation,
  Miami 121761, 138358
Connoiseur Antiques, Toronto . . 064631
The Connoisseur, Governor's
  Harbour 063128
Connoisseur, Toronto 064632
Connoisseur, Valletta 082440
Connoisseur Antiques, West
  Hollywood 097882
Connoisseur Art Gallery, Hong
  Kong 102067
Connoisseur Art Gallery,
  Singapore 114509
Connoisseur's Connection,
  Baltimore 092310
Connoisseurs Fine Antiques,
  Birmingham 092402
Connoisseur's Gallery, Paris . . . 104655
Connoisseurs' Store, Toorak . . . 062322
Connor, Barry, Moss Vale 061902
Conor Fennessy, San Francisco . 097211
La Conoscenza del Restauro,
  Verona 132277
Conquest House Antiques,
  Canterbury 088776
Conquest, Rosemary, London . . 090040
Conquet, Perpignan 072076

Conrad, Fresno 120893
Conrad, Merzig 077356
Conrad Mansion Museum,
  Kalispell 049863
Conrad Weiser Homestead,
  Womelsdorf 054515
Conrad, Arnold R., Tübingen . . . 078720
Conrad, Ch., Berlin 074777
Conrad, Hans, Alfter . . 105973, 137026
Conrad, René, Woippy 074301
Conrad, Ulrich, Göttingen 129910
Conrad, Wolfgang, Lutherstadt
  Eisleben 130259
Conrads, Düsseldorf . 106765, 137155
Conrads, Gerhard, Aachen 074327
Conrads, Mathias, Wuppertal . . 130793
Conrest2, München 130339
Conros, Sankt-Peterburg 126478
Conroy & Assoc., T.W.,
  Baldwinsville 127131
Cons Art, Craiova 133006
Consadori, Annamaria, Milano . . 080494
Consart, Helsinki 128411
Conseil Bruxellois de Musées,
  Bruxelles 058369
Conseil, Philippe, Antibes-Juan-les-
  Pins 125493
Consejo Nacional para la Cultura y las
  Artes, Cuauhtémoc 059571
Consejo Provincial de las Artes
  Plasticas, Santiago de Cuba . . 058529
Conservacion 3-a, Madrid 133315
Conservación y Restauración del
  Patrimonio Artístico, Valencia . 133459
Conservación y Restauración del
  Patrimonio Artístico, Valencia . 133458
Conservación y Restauración Siglo XXI,
  Madrid 133316
Conservation Center for Art and
  Historic Artifacts, Philadelphia . 136080
Conservation Centre Presentation,
  Liverpool 044851
Conservation Départementale Musées
  des Pays de l'Ain, Bourg-en-
  Bresse 012152
Conservation of Art, Hillsborough 135575
Conservation of Fine Art, Dublin 130851
Conservation Resource Centre,
  Glebe 001104
Conservation Studio, London . . . 134633
Conservation Studio, Thame . . . 135132
Conservation Studio for Art,
  Philadelphia 136081
Conservation, The Getty Conservation
  Institute Newsletter, Los
  Angeles 139351
Conservatoio, Milano . 080495, 110289,
                      131422
Conservatoire Botanique National de
  Brest, Brest 012204
Conservatoire de la Faune Sauvage de
  Sologne, Ménestreau-en-Villette 014125
Conservatoire de l'Air et de l'Espace
  d'Aquitaine, Mérignac (Gironde) 014139
Conservatoire de l'Estuaire de la
  Gironde, Blaye 012067
Conservatoire de l'Outil Maison de
  Pays des Crètes, Évigny 012933
Conservatoire des Machines de Guerre
  Médiévales, Tiffauges 015961
Conservatoire du Machinisme et de
  Pratiques Agricoles, Vouillé . . . 016327
Conservatoire du Patrimoine Religieux,
  Liessies 013793
Conservatoire Maritime Basque,
  Ciboure 012574
Conservatoire National des Arts et
  Métiers, Paris 055395
Conservatoire Vinicole, Plassac . 014883
Conservatoni du Musée,
  Gravelines 013168
Conservatorio di Santa Chiara, San
  Miniato 027784
Conservatory of Fine Arts, Ha

Noi 058153
Conservazione Beni Culturali,
  Roma 131871
Conservazione e Restauro,
  Brescia 131027
Consigli Arte, Parma 081105
Consigli, Igino, Parma 081104
Consiglieri, Sergio, Livorno 080321
Consignment 1st, Charlotte 092590
Consignment Cellar, Louisville . . 094532
Consignment Collection, Houston 093648
Consignment Collections, West
  Hollywood 097883
Consignment Company, Salt Lake
  City 096886
Consignment Connection,
  Richmond 096536
Consignment Cottage, Los
  Angeles 094297
Consignment Galleries, Dallas . . 093100
Consignment Galleries,
  Washington 097828
Consignment Gallery, Portland . . 096386
Consignment Interiors,
  Albuquerque 092047
Consignment Shop, Indianapolis 093871
Consignment Store, Dallas 093101
Consignment Treasures, Tulsa . . 097722
Consignments, Memphis 094626
Consignments 'n More, Houston 093649
Consignpointe, Memphis 094627
La Consolle, Palermo 081032
Consolo, Giuseppe, Genova . . . 131302
Consorti, Augusto, Roma 110748
Le Consortium, Dijon 012779
Consorzio Arkè, Roma 131872
Consorzio Capitolino, Roma . . . 131873
Consorzio L'Officina, Roma 131874
Constables Studio, Fowey 117807
Constant Reader Bookshop,
  Milwaukee 145265
Constantine & Mayer, Cheswick 127163
Constitution Convention Museum, Port
  Saint Joe 052256
Constitution Island – Warner House,
  West Point 054321
Constitution Museum, Tabriz . . . 024740
Constitution Square, Danville . . . 048145
Constitutional History Gallery, Bandar
  Seri Begawan 004962
Consuelo, Modena 080788
Consuelo Sierra Arranz, Madrid . 086162
Consulart Luca Lualdi, Milano . . 080496
Consulich, Chicago . . 092698, 135405
Contact Editions, Toronto 140404
Contact Gallery, Warszawa 113814
contact:c4, Forum internationaler
  Kunst aus Berlin, Berlin 058814
Contafili, Firenze 131152
Container Warehouse, Honolulu . 093540
Containers for Collectibles,
  Houston 093650
Contardo, A., Bruxelles 063379
Contavalli, Nino, Lima . 084451, 113346
Fratelli Conte, Bari 079764
Conte, Las Vegas 094130
Conte, Paris 071191
Conte, Gérard, Paris 071192
Conte, Guglielmo, Roma 131875
Conte, Silvana, Roma 131876
Contempary Temporary Art Space,
  Cardiff 117526
Contempo, Eindhoven 112456
Contempo, Rotterdam 112633
Contemporanea, Oberbillig 108314
Contemporanea, Pescara 110645
Contemporaneo, Venezia 111099
Contemporary, London 139164
Contemporary African Art, New
  York 122499
Contemporary Applied Arts,
  London 118263
Contemporary Art Australia, South

Yarra 058196, 099704
Contemporary Art Center of Virginia,
  Virginia Beach 053996
Contemporary Art Centre,
  Kingston 111216
Contemporary Art Centre of South
  Australia, Parkside 001396
The Contemporary Art Fair, New
  York 098443
Contemporary Art Gallery, Oklahoma
  City 123279
Contemporary Art Gallery,
  Vancouver 007282, 101766
Contemporary Art Gallery at Falkirk
  Cultural Center, San Rafael . . . 053061
Contemporary Art Institute – CAI
  School of Art, Sapporo 056030
Contemporary Art Museum,
  Raleigh 052445
Contemporary Art Museum of Saint
  Louis, Forum for Contemporary Art,
  Saint Louis 052796
Contemporary Art Museum, University
  of South Florida, Tampa 053666
Contemporary Art Services Tasmania,
  Hobart 099132
Contemporary Art Society,
  London 060101
The Contemporary Art Society for
  Wales, Cardiff 060102
Contemporary Art Workshop,
  Chicago 047653
Contemporary Arts and Crafts,
  Pune 109493
Contemporary Arts Center,
  Cincinnati 047742
Contemporary Arts Center, New
  Orleans 051289
Contemporary Arts Collective, Las
  Vegas 121347
Contemporary Arts Museum,
  Houston 049574
Contemporary Ceramics, London 118264
Contemporary Conservation, New
  York 135901
Contemporary Fine Art Gallery,
  Eton 117764
Contemporary Fine Arts, Berlin . 106203
Contemporary Fine Arts Center, Ha
  Noi 125069
Contemporary Gallery, Dallas . . 120577
Contemporary Graphic Art Dealer's
  Association of Japan, Tokyo . . . 059492
Contemporary History Museum,
  Tehrän 024748
Contemporary Impressions, Peachtree
  City 139352
The Contemporary Jewish Museum,
  San Francisco 052998
Contemporary Museum,
  Baltimore 046740
The Contemporary Museum,
  Honolulu 049527
The Contemporary Museum at First
  Hawaiian Center, Honolulu . . . 049528
Contemporary Studio Pottery,
  Chagford 117550
Contento, Michele, Bari 130912
Contes d'Orient, Grimaud 068749
Contes d'Orient, Hong Kong . . . 065031,
                       065032
Contessa, Arnhem 112327
Contessa Art Gallery, Austin . . . 119656
Contessa Gallery, Cleveland . . . 120411
Contessa, Michel, Miami 094718
Conti, Adriana & C., Catania . . . 109950
Conti, Alessandro, Firenze 131153
Conti, Giovanni, Firenze 080127
Conti, Nadia, Roma 131877
Conti, Pietro, Brescia 131028
Continental, Praha 102657
Continental Precious Metals, San
  Jose 097363
Continua, San Gimignano 110892

Contorno, Rio de Janeiro 101021
Contour Contemporary Art,
Newlyn 118796
Contrada Scura, Trieste 081786, 081787
A Contrario, Paris 141079
Contrast, Barcelona 114873
Contrast Gallery, Bruxelles ... 100582
Contrastarte, Lisboa 113988
Contraste e Antiguidades, Belo
Horizonte 063898
Contraste Rustico e Antigo, Belo
Horizonte 063899
Contrasten, Rotterdam 083561
Contrastes, Paris 071193
Contre Courant, Les Porrtes-en-
Ré 103985
Contreras, Tala, Mandaluyong .. 113394
Contretype, Bruxelles 100583
Contrisciani, Mara, Bari 079765
Controversial Bookstore, San
Diego 145521
Convent Gallery, Daylesford .. 098955
Convento de San Antonio el Real,
Segovia 040154
Convento de Santo Domingo el
Antiguo, Toledo 040232
Conversation Piece, Clayton
South 061342
Conway Fire Museum Aka Vintage Fire
Engines, New Albany 051220
Conway Scenic Railroad, North
Conway 051600
Conway, Susan, Washington .. 124886
Conwy Valley Railway Museum, Betws-
y-Coed 043582
Cony, Singapore 085391
Conzada, Milano 080497
Conzen-Sammlung, Düsseldorf . 018026
Conzen, F. G., Frankfurt am Main 106979
Conzen, F.G., Düsseldorf 106766
Coo-ee Aboriginal Art, Bondi
Beach 098757
Cooch & Son, Fred, Houston .. 135598
Coogan, Hobart 061629
Cooinda Gallery, Sawtell 099681
Cook, Boston 092462
Cook, New York 122500
Cook County Historical Museum,
Grand Marais 049185
Cook, David, Denver 120704
Cook, I.C., Exeter 117769
Cook, J., San Francisco 097212
Cook, L.J., Camberwell 061256, 127576
Cook, Martyn, Woollahra 062464
Cook, R.L., Sheringham 144843
Cook, Randal, Fort Worth 093439
Cook, Richard, Béziers 103375
Cook, Sheila, London 090041
Cook, Stephen, Broadway 088673,
                              134168
Cook, William, London 134634
Cook, William, Marlborough ... 090721,
                              134843
Cooke, New York 135902
Cooke, Mary, London . 090042, 134635
Cooke, Sandy, Long Melford ... 090616,
                              134808
Cooke, Sue, Wanganui 113050
Cookeville Depot Museum,
Cookeville 047969
Cookhouse Gallery, Ambleside . 117237
Cookhouse Gallery, Keswick ... 118037
Cook's Creek Heritage Museum,
Dugald 005726
Cooks Hill Galleries, Cooks Hill . 098917
Cookstown Antiques, Cookstown 088975
Cooktown School of Art Gallery,
Cooktown 000988
Cookworthy Museum of Rural Life in
South Devon, Kingsbridge .. 044709
Cool-Art Gallery, Coolum Beach 098920
Cool Saigon Company, Ho Chi Minh
City 125117
Cooljunk, Waimate 084139

Coolspring Power Museum,
Coolspring 047971
Coolstore Gallery, Mapua 112943
Coolyah, Shanghai 065130
Cooma Cottage, National Trust of
Australia (NSW), Yass 001672
Coomalie Culture Centre,
Batchelor 061103
Coombe Street Gallery, Lyme
Regis 118676
Coombes, J. & M., Dorking ... 089078
Cooner, Joel, Dallas 120578
Cooney Declan, Mullingar 130884
Cooper, Brasted 088578
Cooper, Cardiff 144311
Cooper, Phoenix 096202
Cooper & Tanner, Frome 126899
Cooper Gallery, Barnsley, South
Yorkshire 043499
The Cooper Gallery, Bideford .. 088461,
                              117359
Cooper Gallery, New York 122502
Cooper Hay Rare Books,
Glasgow 144434
Cooper-Hewitt National Design
Museum, Smithsonian Institution,
New York 051368
Cooper Hirst, Chelmsford 126851
Cooper Regional History Museum,
Upland 053900
The Cooper Union – School of Art,
New York 057545
Cooper Young Gallery, Memphis 121686
Cooper, Brad, Tampa 124643
Cooper, Bruno, London 090043
Cooper, Bruno, Norwich 090870
Cooper, Ian, Looe 118661
Cooper, John, Saint Helier 091181
Cooper, Melanee, Chicago 120098
Cooper, Paula, New York 122501
Cooper, Sonia, Aberfeldy 088228
Cooper, Suzanne, Kansas City . 094068
Cooper, Wendy, Chicago 120099
Cooperativa Archeologia, Firenze 131154
Cooperativa per il Restauro,
Milano 080498, 131423
Coopernook Antiques,
Coopernook 061364
Coopers, Epping 061449
Coopers of Ilkley, Ilkley 089634, 134497
Coopersmith, Ellen, Washington 097829
Cooperstown Art Association,
Cooperstown 060449
Coopersville Area Historical Society
Museum, Coopersville 047977
Coopmans, L., Venlo 143254
Coop'ren Duikhelm, Giethoorn . 032349
Cooran Art Gallery, Cooran ... 098923
Coorevits, A., Gent 063633
Coos Art Museum, Coos Bay .. 047978
Coos County Historical Society
Museum, North Bend 051593
Copacetic Comics, Pittsburgh . 123534
Cope Livros e Revistas, Brasília 140273
Copeland & de Soos, Woollahra 062465
Copeland Fine Art, Columbus . 120484
Copeland, Albert, Dallas 093102
Copelia Artes, São Paulo 101092
Copello, Alessandro, Milano ... 080499,
          080500,   131424,   131425
Copenhagen Art Gallery,
Charlottenlund 102784
Copercini & Giuseppin, Arquà
Petrarca 079751
Coperta, Bruxelles 063380
Copet, Christian, Bruxelles 128022
Copin, Michel, Paris 071194
Coplan-Hurowitz, Sharon, New
York 095325
Copley, Boston 119855
Copley Society of Art, Boston .. 060450
Coplu, Nevşehir 117102
Copnal, Crewe 144360
Coppelia Antiques, Plumley ... 091019

Coppens, Helene, Bruxelles 100584
Coppens, Josephine, Saint
Arnaud 099663
Copper Art Gallery, Woodburn . 099898
Copper Belt Museum, Whitehorse 007404
Copper Cliff Museum, Copper
Cliff 005645
Copper Country Antique Mall,
Tucson 097647
Copper Kettle Antiques,
Vancouver 064809
Copper King Mansion, Butte ... 047335
Copper Village Museum and Arts
Center, Anaconda 046460
Copperbelt Museum, Ndola ... 054843
Copperfields, Hawkes Bay 083950
Coppergate Gallery, York 119408
Coppertown U.S.A. Mining Museum,
Calumet 047353
Coppin, Jean-Patrick, Sougy-sur-
Loire 073548
Coppini Academy of Fine Arts, San
Antonio 057838
Coppini Academy of Fine Arts Gallery
and Museum, San Antonio ... 052936
Coppola, Alessandro, Roma ... 131878
Coppola, Mauro Roberto,
Montevideo 119418
Copshaholm House Museum and
Historic Oliver Gardens, South
Bend 053392
Coptic Museum, Cairo 010534
Copyright, Antwerpen 140143
Copyright, Paris 136886
Copyright Artbookshop, Gent .. 140221
Coquelet, Sophie, Bourges 067426
Coquery, Jean-Jacques, Paris . 128967
Coquet-Pourvis, Nyons 070766
Coquet, Nicole, Saint-Maurice-sur-
Eygues 072970
La Coquille, Haarlem 083320
Coquille River Museum, Brandon 047178
Coquille Valley Art Association,
Coquille 060451
Coquville River Museum, Bandon
Historical Society, Bandon 046775
Cor e Forma, Rio de Janeiro .. 101022
Cor Fontein, Zutphen 083716
Côr On Line, Lisboa 113989
Corak Rinesi, Karmen, Roma .. 131879
Coral Creation, the Cosmos of Buddha
Coral Museum, Su-Ao 042467
Coral Gables, Miami 094719
Coral Gables Gallery, Miami .. 121762
Coral Gables Merrick House, Coral
Gables 047983
Coral Garden, Pittsburgh 123535
Coral Springs Museum of Art, Coral
Springs 047987
Corau, Alain, Bordeaux 125526
Corazon, Houston 121021
Corbally Stourton, Mascot 061800
Corban Estate Arts Centre,
Auckland 033143
Corbelin, Sylvie, Saint-Ouen .. 073057
Le Corbellier, Paris 128968
Corberi Arte, Milano 110290
Corbet, Auguste, Monthuchon . 070249
Corbett, Cynthia, London 118265
Corbetta, Glória, Porto Alegre . 100969
Corbin, Tampa 097555
Corbin, Marc, Miami 094720
Corbridge Roman Site and Museum,
Corbridge 044008
Corcelle, Thierry, Paris 071195, 141080
Corcione, Rosaria, Napoli 110520
Corcoral, Guy, Pibrac 072153
Corcoran, Ashbourne 079436
Corcoran College of Art and Design,
Washington 058063
Corcoran Gallery of Art,
Washington 054099
Corcoran, James, Cleveland .. 092934,
                              120412

Corcoran, James, Los Angeles . 121463
Cordé, Amsterdam 082721
Cordeiros, Porto 114077
Cordelia Street Antique and Art Centre,
Brisbane 061209
Les Cordeliers, Montpellier ... 070296
Les Cordeliers, Penzenas 072054
Cordero, Alba 142616
Cordero Garcia, Monica, Madrid 133317
Cordero, Merce, New York 095326
Cordero, Michele, Giaveno 080303
Cordes, Kiel 141991
Cordier & Ekstrom, New York . 122503
Cordier-Dornat, Maurice,
Appoigny 066688
Cordier, Joël, Irancy 068874
Cordier, Patrice, Pontarlier ... 072267
Cordoba, Bucaramanga 102381
Cordoba, Francisco, Roma 110749
Cordouan, La Palmyre 103897
Cordova Historical Museum,
Cordova 047990
Cordy, John, Auckland 126303
Core, Denver 048257
Core New Art Space, Denver . 120705
Corenger, Den Haag 083069
Coresal, Madrid 133318
Corfe Castle Museum, Corfe
Castle 044009
Corfield Potashnick, London ... 090044
Corfu Byzantine Collection, Corfu 023053
Corfu Museum of Shells, Kastellanoi
Mesis 023117
Corge, Colette, Monflanquin ... 070163
The Corgi Heritage Centre,
Heywood 044581
Corgialenios Historical and Cultural
Museum, Argostolion 022967
Cori, Anna, Roma 131880
Corianne, Paris 104656
Corigliano, Giovanna, Milano .. 131426
Corimbo, Palermo 110599
Corin Maman Ashdod Museum,
Ashdod 025018
Corinium Museum, Cirencester . 043962
Corinto, Medellín 065283
Coriolan, Lyon 128707
Corita Art Center, Los Angeles . 050416
The Cork Art Fair, Cork 098178
Cork Art Society, Cork . 024855, 059376
Cork Brick Antiques, Bungay .. 088698
Cork Butter Museum, Cork ... 024856
Cork Public Museum, Cork 024857
Corkin, Jane, Toronto 101592
Corlett, Oliver, Los Angeles .. 094298
Corley, David, Dallas 093103
Corlouer, Gilbert, Paris 128969
Corman Arts, London 118266
Corn Maiden Gallery, Gordon
Park 099091
Corn Mill Antiques, Skipton ... 091347
Cornacchia, Madrid 133319
Cornall, Graham, Armadale,
Victoria 061020
La Corne d'Abondance, Fontaine-les-
Gres 068562
Corneli, Garmisch-Partenkirchen 107101
Cornelis, Rotterdam 083562
Corneliu, Florin, Iaşi 085123
Cornelius Low House/Middlesex
County Museum, Piscataway . 052147
Cornelius, L.A.U., Rotterdam .. 083563
Cornelius, Peter, Halle, Saale . 076162
Cornell Museum, Delray Beach . 048231
Cornell University Press, Ithaca . 138342
Cornell, Volker, Obertshausen . 108321
Corner Antiques, Las Vegas .. 094131
Corner Antiques, Windsor,
Ontario 064902
Corner Antiques, Worthing ... 091954
Corner Brook Museum & Archives,
Corner Brook 005646
Corner Collectables, Oamaru .. 084041
Corner Cupboard, Albuquerque . 092048

Corner Cupboard Curios,
Cirencester 088915
Corner Farm Antiques, Shifnal .. 091318
Corner of Arts, Singapore 114510
The Corner Portobello Antiques
Supermarket, London 090045, 134636
Corner Shop, Bradford 088556
Corner Shop, Oxford 090930
Corner Shop Antiques, Camelford 088768
Corner Store, Dallas 093104
Cornerhouse Publications,
Manchester 138263
Corners, Charlotte 119984
Cornerstone, Portland 096387
Cornerstone Antiques, Toronto .. 064633
Cornerstone Gallery, Saint Ives,
Cornwall 119020
Cornet à Dés, Quimper 072342
Cornette de Saint-Cyr, Pierre,
Paris 125697
La Cornice, Padova .. 080977, 080978
La Cornice, Roma 081304
Cornice Museum of Ornamental
Plasterwork, Peebles 045469
Cornici Antiche Enrico Ceci,
Formigine 080228
Cornier, Patrick, L'Isle-sur-la-
Sorgue 069567
Corning Museum of Glass,
Corning 047994
Cornish Antiques, Fort Worth ... 093440
The Cornish Collective, Penzance 118894
Cornish College of the Arts,
Seattle 057896
Cornish Mines, Engines and Cornwall
Industrial Discovery Centre,
Redruth 045597
Cornkammer, Kuddewörde 107767
Corno, Montréal 064409
Cornstalk, Glebe 139750
Cornstalk, Newtown, New South
Wales 139856
Cornu, Jean-Pierre, Balacet ... 066922
Cornucopia, London 090046
Cornucopia, Roma 081305
La Cornucopia, Torino 081688
Cornucopia Antiques, Hebden
Bridge 089489
Cornucopia Magazine, İstanbul . 138132,
139113
Cornucopia: Turkey for Connoisseurs,
Hawick 139165
Cornwall Galleries, Newquay ... 118804
The Cornwall Gallery, Kingston . 111217
Cornwall Geological Museum,
Penzance 045480
Cornwall Historical Museum,
Cornwall 047997
Cornwall Iron Furnace, Cornwall 047998
Cornwall Regional Art Gallery,
Cornwall 005648
Cornwall Traders, Auckland 083779
Cornwall, Heidi, Flensburg 075741
Cornwater Fine Art, Ravenshead 118957
Cornwell, Earl E., Indianapolis . 127260
Coroforte, Roberto, Roma 131881
Coroller & Gilbon, Paris 128970
Coromandel, Riec-sur-Belon ... 072460
Coromandel School of Mines Museum,
Coromandel 033181
La Corona, Ascona ... 087198, 133623
Corona, Spišská Nová Ves ... 143693
Corona Antiques, Montréal 064410
Corona Boned, Eivissa 085989
Coronach District Museum,
Coronach 005652
Coronado National Memorial Museum,
Hereford 049447
Coronado-Quivira Museum,
Lyons 050540
Coronado Studio, Austin 119657
Corpel, Jean-Guy, Nice 070587
Corporate Art, Atlanta 119555
Corporate Art, London 118267

Corporate Art, Norwood 062011
Corporate Art, West Burleigh ... 099878
Corporate Art Associates, New
York 122504
Corporate Art Directions, New
York 122505
Corporate Art Services, Kansas
City 121286
Corporate Art Source, Chicago . 120100
Corporate Art West, Phoenix ... 123488
Corporate Fine Art International,
London 118268
Corporate Image, Long Beach .. 121397
Corporate Images, Chicago 120101
Corporation Archiépiscopale C.R.
de Sherbrooke, Sherbrooke,
Québec 007021
Corpsmuseum van het Utrechtsch
Studenten Corps, Utrecht 032945
Corpus Christi Museum of Science
and History, Corpus Christi .. 048004
Corpus der Mittelalterlichen
Glasmalerei Österreichs, Wien . 138510
Corpus Vasorum Antiquorum. Italia,
Roma 138810
Corpus Vitrearum, Paris 138637
Corradi, Ettore, Polesine
Parmense 081162
Corradi, Giovanni, Parma 142780
Corradi, Paola, Genova 080255
Corradini, Enrico, Castelguelfo . 080016
Corrado, Mario, Roma 081306
Corraini, Maurizio, Mantova .. 110202,
137665
Corrales Miura, Angelina,
Cáceres 085920
Corran, Carmarthen 144315
Corrarello, Rosa, Roma 131882
Corre, François, Rennes (Ille-et-
Vilaine) 141313
Le Corre, Guy, Mougins 104298
Correa, Ruta, Freiburg im
Breisgau 107053
Correct Dental Arts, Tampa ... 124644
Corredor de Arte Antiguo, Bogotá 065199
Corredor, Gallery of the College
of Fine Arts, University of the
Philippines Diliman, Quezon
City 034580
Corredor, Luis Fernando, Bogotá 065197,
065198
Correia, Domingos da Silva,
Porto 132970
Correia, Gilbert, Hagen 107195
Correia, Maria C. Pereira, Braga 084777
Correia, Renato Silva, Lisboa .. 084861
Corrèze, Den Haag 083070
Il Corricolo, Napoli 080854
Corridor Gallery, Glenrothes ... 044430
Corridors of Colour, Bangalore . 109255
Corrie ten Boomhuis, Haarlem . 032391
Corrigan, Paisley 090942
Corrigan, Singleton 062214
Corrigin Pioneer Museum,
Corrigin 000994
Corrin, John, Douglas . 079727, 130898
Corris Railway Museum,
Machynlleth 045188
Corroyer-Signature, Lyon 104077,
128708
Corry Area Historical Society Museum,
Corry 048008
Corry's, Leicester .. 089790, 134544
Corsale, Maria Pia, Napoli 131609
Corsaren, København 140607
Corsarium, São Paulo 140303
Corsesca, Martínez 060948
Corsham Court, Corsham 044012
Corsi Arte Antiquaria, Milano .. 080501
Corsi Internazionali di Cultura
Sull'Arte Ravennate e Bizantina. Atti,
Ravenna 138811
Corsi, Clara, Roma 081307
Corsi, Mario, Altendorf 087189

Corsini, Piero, New York 122506
Corski, Bucureşti 132995
Corson, Yves, Plestin-les-Grèves 072182
Cortada, Xavier, Miami 121763
Cortade Art, Collioure 103615
Cortade, Jean Claude,
Montauban 070194
La Corte, Brescia 131029
La Corte Arte Contemporanea,
Firenze 110044
Cortemilia, Milano 131427
Cortes Casaña, Francisco Javier,
Castellón de la Plana 133225
Cortes, Armando Maria, Lisboa . 084862
Cortes, Carlos, San Antonio ... 124022
Cortés, Filip, Český Krumlov ... 065373
Cortese, Porto Alegre 063969
Cortese, Donato, Napoli 131610
Cortese, Raffaella, Milano 110291
Cortevecchia, Ferrara 110019
Cortex Athletico, Bordeaux ... 103422
Cortijo Bacardi, Málaga 039698
Le Cortil, Falaën 003592
Cortile Bisignano, Napoli 080855
Cortina Arte, Milano 110292
Cortland County Historical Society
Museum, Cortland 048014
Cortland-Leyten, Chicago 120102
Cortona, Enrico, Milano 080502
CortonAntiquaria, Cortona 098185
Çorum Devlet Güzel Sanatlar Galerisi,
Çorum 042872
Çorum Müzesi, Çorum 042873
Çorum Örenyerleri, Çorum 104274
Corum, Angle, Montpellier 104274
Coruzzi, Alessandra, Milano ... 131428
Corvaglia, Gianluigi, Torino ... 132122
Corvallis Art Center, Oregon State
University, Corvallis 048016
Corve Street Antiques, Ludlow . 090646
Corven Art, Howey 118003
Corvi, Alessandra, Roma 081308
Corvin János Múzeum, Gyula . 023670
Corvina Books, Budapest 137585
Corvinos, Pascal, Billom 067220
Corvol, Marie-France, Paris ... 104657
Cory, San Antonio 096962
Corzine & Co., Nashville 094985
Cosa, London 118269
Cosac & Naify, São Paulo 136666
Cosar HMT, Düsseldorf 106767
Cosas, Groningen 083286
Cosas de Entonces, Madrid ... 086163
Cose d'Altri Tempi, Sanremo .. 110904
Cose d'Altri Tempi, Venezia ... 132234
Le Cose del 900, Milano 080503,
080504
Cose del Passato, Firenze 080128
Cose del Passato, Lugano 087610,
133822
Cose del Passato, Torino 081689
Cose d'Epoca, Milano 080505
Cose d'Epoca, Roma 081309
Le Cose di Ieri, Perugia 081140, 131749
Cose di Vecchie Case, Padova . 080979
Cosediunaltromondo, Milano .. 131429
Cosentino, Los Angeles 094299
Cosgrove, Leila M ., Omaha .. 096033,
127378
COSI Columbus, Columbus ... 047937
Cosi Toledo, Toledo 053758
Coskun, London 118270
Cosma Elite, Médis 070028
Cosmic Galerie, Paris 104658
Cosmo, Tokyo 082139
CosmoCaixa Madrid, Museo de la
Ciencia, Alcobendas 038923
Cosmodôme, Laval 006170
Cosmopolitan Book Shop, Los
Angeles 145221
Cosmos Gallery, Ganghwa 111597
Cossa, Gabor, Cambridge 088761
Cossit House Museum, Cape Breton
Centre for Heritage and Science,

Sydney 007114
Cosson, Robert, Luxeuil-les-
Bains 069671
Costa, Nice 070588
Moreira da Costa, Porto 143601
Costa & Rodrigues, Sesimbra .. 085060
Costa Azul, Don Torcuato 060941
Costa Brocante, Riscle 072477
Costa de Diego, Oviedo 115324
Costa-Dupuy, Marie-Louise,
Toulouse 073755
Costa Guido, Torino 110962
Costa Martinez, Jorge, Barcelona 114874
Costa Rino, Valenza 111072
Costa, Alberto, Buenos Aires ... 139560
Costa, Alexander, Furth im Wald 107094
Costa, Andrea, Padova 131660
Costa, António, Lisboa 084863
Costa, Benito, Roma 081310
Costa & Cia, C.A., Brasília 063933
Costa, Elisabeth, Pontresina ... 116568
Costa, Eugenio, Genova 110100
Costa, Maria Pereira, Lisboa ... 084864
Costa, Pedro Felner, Lisboa 084865
Costa, Rodrigo Vieira, Porto ... 114078
Da Costa, Rosane Aguiar Fernandes,
Porto Alegre 100970
Costalunga Antique Restorations,
Malvern East 127630
Costantini, Giuseppe Maria,
Bologna 130948
Costantini, Paolo, Venezia 081829,
132235
Costantino, Giuseppe, Reggio
Calabria 131786
Costanzo, Alfio, Catania 080030, 131086
Costas, Milano 110293
Costela, Bucureşti 114132
Costelloe Kieran, Freshford 079622
Costentin, Michel, Brionne 067501
Coster, Laurens Jansz, Haarlem 143158
Costermans Antiques, Bruxelles 063381
Costes, Pierre & Sylvie, Plouha . 072205
Costey, Jean-Paul, Paris 071196
Costil, Jean-Jacques, Rouen ... 072581
Costta, Juçara, Belo Horizonte . 100881
Costume Museum of Canada,
Dugald 005727
Costume Museum of Shih Chien
University, Taipei 042510
Cosy Corner, Antwerpen 140144
Cosy Corner, Toulouse 073756
Cosy Home, Eguilles 068434
Cota, West Hollywood 097884
Cotabato National Museum,
Cotabato 034472
Cotanda Anticuario, Madrid ... 086164
Cote, Villeurbanne 074200
Côté Arts, Marseille 136834
Côte Basque, Bayonne 066972
Coté Cadre, Lure 104055
Côté Cave, Rueil-Malmaison .. 072638
Côté Cour, Chauny 067342
Côte d'Azur Antiques, Los
Angeles 094300
La Côte des Arts, Marseille ... 138638
Côté Galerie, Paris 071197
Côté Jardin Antiques,
Washington 097830
Côté Nord Antiek, Eindhoven .. 083211
Côte Ouest, Arcachon 066702
Côté Provence, Ascot Vale 061061
La Côte Rouge, Caen 103483
Côte Savoie, Val-d'Isère 105792
Côte Sud, Bruxelles 063382
Côté Versailles, Saint-Denis (Seine-
Saint-Denis) 072728
Coté, François, Montréal 140359
Cotehele House, Saint Dominick,
Cornwall 045683
Cotham Antiques, Bristol 088653
Cotham Galleries, Bristol 088654,
117443
Cotham Gallery 101, Kew 099187

Cothey Bottom Heritage Centre, Ryde,
Isle of Wight 045655
Cotinaud, Marie-Claire, Paris 071198
Cotiso, Bucureşti 085089
Coto, Rosy, Riom 072470
Cotswold, Cirencester 117591
Cotswold, New York 095327
Cotswold Antique Dealers Association,
Stow-on-the-Wold 060103
Cotswold Antiques, Winchcombe 091862
The Cotswold Auction Company,
Cheltenham 126853
The Cotswold Auction Company,
Cirencester 126859
The Cotswold Auction Company,
Gloucester 126904
Cotswold Galleries, Stow-on-the-
Wold 091466, 135094
Cotswold Motoring Museum and
Toy Collection, Bourton-on-the-
Water 043683
Cotswold Pine and Associates, Middle
Aston 090753, 134851
The Cottage, Melville 085651
Cottage, San Antonio 096963
Cottage, Tulsa 097723
Cottage & Castles, Enniskerry . . 079620
Cottage and Garden Antiques,
Portland 096388
Cottage Antiques, Houston 093651
Cottage Antiques, Los Angeles . 094301
Cottage Antiques, Margate 090701,
134836
Cottage Antiques, Nashville 094986
Cottage Antiques, Oklahoma City 095964
Cottage Antiques, Pontefract . . 091031,
134945
Cottage Antiques, Sanson 084075
Cottage Antiques, Stepney 062244
The Cottage Antiques, Tucson . . 097648
Cottage Antiques, Walsden 091718
Cottage Antiques, Zürich 087949
Cottage Blundell's, Canberra . . . 000936
Cottage Bookshop, Queenstown 139890
Cottage Collectables, Holt,
Norfolk 089546
Cottage Craft, Gifts & Fine Arts,
Calgary 064253, 101224
Cottage Curios, Allonby 088251
Cottage Door, Omaha 096034
Cottage Farm Antiques, Chipping
Campden 088882
Cottage Gallery, Huntingdon . . . 118005
Cottage Gallery, San Diego 124114
The Cottage Gallery, Skipton . . . 119101
Cottage Grove Museum, Cottage
Grove 048025
Cottage Hall, Wedemark 078879
Cottage Holdings, North
Fremantle 061986
Cottage House Mall, Memphis . 094628
Cottage Lawn, Madison County
Historical Society, Oneida 051772
Cottage Museum, Lancaster 044749
Cottage Rose, South Auckland . 084078,
132740
Cottage Style Antiques,
Rochester 091125
Cotte de Mailles, Avallon 066863
Cottees, Wareham 127076
Cotten, Martine, Pont-Aven 105372
Cottet, Jean-Claude, Lausanne . 087567,
133794
Cotthem, Bruxelles 100585
Cotthem Gallery, Barcelona 114875
Cottier, Sarah, Redfern 099620
Cottin, Frechen 075837, 125951
Cottin, Irene, Köln 076854, 126004
Cottingley Antiques, Bradford . . 088557,
134135
Cotton Country Antiques, Phoenix 096203
Cotton, Joan, West Bridgford . . 091780
Cotton, Lynda, Watchet 119313
Cotton, Mary, Houston 121022

Cotton, Nick, Watchet . 091758, 135193
Cotton, Yvonne, Vandoeuvres . . 133950
Cottone, Alessandra, Verona . . . 132278
Cottonlandia Museum,
Greenwood 049274
Cotton's Attic, Los Angeles 094302
Cottonwood County Historical Society,
Windom 054468
Cottonwood House, Quesnel . . . 006750
Cottreau, Thierry, Saint-Ouen . . 073058
Cottrell, Catherine, New Orleans 095083
Couch, Annette, Los Angeles . . 121464
Coudert, Michel, Meyrals 070114
Coudray-Haus, Bad Berka 016723
Coué, Jean-Yves, Nantes 070461
Coué, Philippe, Véretz 074023
Coueffé, Dominique, Lyon 140892
Couet, Martin de, Aachen 074328
Couffon de Trévois, Charles, Rennes
(Ille-et-Vilaine) 072431
Coughton Court, Alcester 043378
Coughton Galleries, Arthingworth 117250
Cougoule-Devergne, Sophie, Saint-
Ouen 073059
Coulborn & Sons, Thomas, Sutton
Coldfield 091518, 135108
Coulet & Faure, Paris 141082
Coulet, Laurent, Paris 141081
Couleur du Temps, Saint-Ouen . 073060
Couleur en Musique, Cherrueix . 013591
Couleur Konsthandel, Stockholm 115850
Couleur Sud, Le Thor 069356
Couleur Syracuse, Paris 071199
Les Couleurs d'Asie, Auray 066816
Les Couleurs de l'Artiste, Étoile-sur-
Rhône 103721
Couleurs du Siam, Marans 069842
Couleurs du Temps, Galerie d'art
contemporain, Grand Lancy . . 116358
Couleurs et Bois, Saint-Denis . . 085064
Coulibaly, Yousfouph, Montréal . 064411
Les Couloirs du Temps, Champigny-la-
Futelaye 067814
Les Couloirs du Temps, Pacy-sur-
Eure 070834
Coulon, Le Havre 069279
Coulter Robert, York . . 091974, 119409
Coultershaw Beam Pump,
Petworth 045497
Coumantaros Art Gallery of Sparta,
National Gallery, Spárti 023272
Council for the Prevention of Art Theft,
Warminster 060104
Council of American Jewish Museums,
New York 060452
Council of American Maritime
Museums, Bath 060453
Council of Australian Art Museum
Directors, Melbourne 058197
Council of Australian Museum
Directors, Haymarket 058198
Council, Michael, Columbus . . . 093010
Countermedia, Portland 145450
Counterpoint, El Paso 120826
Counterpoint Records and Books, Los
Angeles 145222
Countess von Buse, institute research
of conservation, Erfurt 129801
Country and Eastern, Norwich . . 090871
Country Antique Showroom, San
Francisco 097213
The Country Antiques, Antrim . . 088284,
134034
Country Antiques, Auckland . . . 083780
Country Antiques, Bagshot . . . 088330
Country Antiques, Kidwelly 089681
Country Antiques, Killearn 089685
Country Antiques, Long Melford . 090617
Country Antiques, Roscommon . 079693
Country Bear, Sacramento 096617
The Country Bedroom, Keswick . 089675
Country Charm, Masterton 083987
Country Charm Antiques, Tulsa . 097724
Country Chattels, North

Tamborine 062001
Country City, New York 095328
Country Clocks, Tring . 091634, 135149
Country Club Antiques, Denver . 093273
Country Clutter, Fresno 093496
Country Clutter, Phoenix 096204
Country Collector, Pickering . . . 091014
Country Cottage, Dallas 093105
Country Cottage Antiques,
Deniliquin 061397
Country Cottage Interiors,
Kingsley 089695
Country Cousins, San Diego . . . 097076
Country Cupboard Antiques,
Tanunda 062305
Country Depot, Fresno 093497
The Country Doctor Museum,
Bailey 046712
Country Eclectic Antiques,
Willoughby 062429
Country Emporium Antiques,
Tucson 097649
Country Fine Arts, Llanddeusant 118135
Country French Antiques, New
York 095329
Country Friends, Indianapolis . . 093497
The Country Gallery, Ballyclare . 117273
Country Gallery, Dereham 117675
Country Garden Antiques, Dallas 093106
Country Gentleman, Houston . . 093652
Country Heritage Park, Milton . . 006316
Country Home, New York 135903
Country Home, Salt Lake City . . 096887
Country House, Saint Mary 091201
Country House, Wien 062922
Country House and Antiques, São
Paulo 064172
Country House Antiques, Manuka 061790
Country House Antiques,
Seething 091261
Country in Time, San Antonio . . 096964
Country Life Antiques, Bushey . 088745
Country Life Antiques, Glen Iris . 061528,
127603
Country Life Antiques, Meadows 061809
Country Line Antiques, Denver . 093274
Country Living Antiques,
Callington 088756
Country Looks, Healesville 099120
Country Manor, Cincinnati 092841
Country Markets Antiques,
Chilton 088880, 134259, 144339
Country Music Hall of Fame and
Museum, Nashville 051175
Country Oak Antiques, Harrogate 089425
Country Peddlers, Miami 094721
Country Peddlers of America,
Jacksonville 093981
Country Pine Antiques, Market
Bosworth 090704
Country Pine Shop, West Haddon 091786
Country Roads, Ottawa 064501
Country Sailor, Torino 081690
Country Scene Art Gallery,
Molong 099361
The Country Seat, Henley on
Thames 089506
The Country Seat, Llandudno . . 089866,
134573
Country Seat Antiques, Newby
East 090821
Country Shoppe, Pittsburgh . . . 096272
Country Shops Antiques, Omaha 096035,
136057
Country Stuff Antiques, Bantry . 079453
Country Times, Rockhampton . . 062162
The Country Trader, Waterloo . . 062413
Country Treasures, Kerikeri 083975
Country Village, Palmerston
North 084054, 132736
The Countrymans Gallery,
Leicester 118092
Countryside Museum, East
Budleigh 044199

County Antiques, Ashford 088312
County Antiques, Ilminster 089636
County Antiques, Linlithgow . . . 089839
County Donegal Railway Heritage
Centre, Donegal 024865
County Hall Gallery, London . . . 044950
County Museum, Sligo 024989
County Museum Dundalk,
Dundalk 024917
County Museum Technical Centre,
Halton 044508
County of Los Angeles Century Gallery,
Sylmar 053629
County of Los Angeles Fire Museum,
Los Angeles 050417
Coup de Cour de Vesna, Sucy-en-
Brie 073602
Coup de Foudre, Vence 074003
Coupé, Michel, Cuvilly 068261
Coupes Languedoc Provence,
Avignon 066876
Coupure Galerie, Gent 100691
Couque, Janine, Saint-Ouen . . . 073061
Couque, Philippe, Paris 071200
Cour Beaudeval, Faverolles (Eure-et-
Loir) 068514
Cour Carrée, Paris . . . 071201, 104659
Cour d'Albane, Rouen 105472
La Cour d'Antan, Acquigny 066416,
128423
Cour de Rouen, Rouen 072582
A la Cour des Adieux,
Fontainebleau 068565
La Cour des Arts, Arles 103272
Cour des Arts, Belfort 103361
Cour des Miracles, Figeac 068530,
140837
La Cour des Miracles, Reignier . 072374
Cour des Miracles, Vannes
(Morbihan) 073969
Cour des Mirascles en Bourgogne,
Marcigny 069851
La Cour Renaissance, Antiquités,
Strasbourg 073582
Courant d'Art, Clermont-Ferrand 068020
Courant d'Art, Mulhouse 104311
Courant d'Art, Paris . . 104660, 141083
Courants d'Ere, Paris 071202
Courault, Noël, Gueugnon 068766
Courbis, Michel, Nice 070589
Courbois, Emile, Abbeville 066409
Courbois, R.M., Haarlem 083321, 083322
Courbon, Saint-Étienne 072758
Courbon, Armand, Saint-Étienne 072759
Courchet, Palloc & Fede, Nice . . 125662
Courchet, Thierry, Cagnes-sur-
Mer 125541
Courchinoux, Marie-Rose, Paris . 104661
Courcoux & Courcoux,
Stockbridge 119144
Courcy, Dodd Henry de, Killarney 079644
Court Gallery, Calne 117498
Court Gallery, Nether Stowey . . 090807,
134868
Court Hall Museum,
Sittingbourne 045803
Court House and Museum,
Pevensey 045500
Court House Antique Centre,
Sheffield 091281, 144835
Court House Arts Centre,
Busselton 098825
Court House Museum, Cardston 005563
Court House Museum, Stroud . . 001525
Court'Art, Paris 071203
Courtauld Institute Gallery, Courtauld
Institute of Art, London 044951
Courtauld Institute of Art, University of
London, London 056557
Courtenay and District Museum and
Paleontology Centre, Courtenay 005657
Courtenay's Fine Art,
Bournemouth 117403
Courthouse Gallery, Portsmouth

Museums, Portsmouth 052322
Courthouse Museum, Berrien
Springs 046947
Courthouse Museum, Exira 048654
Courthouse Museum,
Jacksonport 049765
Courthouse Square Museum,
Charlotte 047567
Courtiade, Bernard, Toulouse . 129299
Courtial, Maria & Jörg,
Darmstadt 106653
Courtier, Laurent, Pacy-sur-Eure 070835
Courtieux, Suresnes 105698
Courtin, Grégoire, Baugé 066967
Courtlands Restoration, Banstead 134057
Courtney and Lemmerman Galleries,
New Jersey City University, Jersey
City 049818
Courtois & Chauviré, Angers ... 125489
Courtois de Viçose, Germaine,
Toulouse 073757
Courtois, Georges, Massangis .. 069991
Courtoy, R., Bruxelles 063083
The Courtroom Gallery, Lampeter 118067
Courts Miscellany, Leominster .. 089804
Courtville Antiques, Dublin 079538
Courtyard, Brackley . 088554, 134133
Courtyard Antiques, Brasted .. 088579,
                               134140
Courtyard Antiques, Coxley ... 088991
Courtyard Antiques, Limerick .. 079655
Courtyard Antiques, Williton .. 091848
Courtyard Art Gallery, Austin .. 119658
The Courtyard Craft & Exhibition
Centre, Midleton 109675
Courtyard Gallery, Appleby-In-
Westmorland 117241
Courtyard Gallery, Beijing 101946
The Courtyard Gallery,
Monkstown 109677
The Courtyard Gallery, South
Crail 119115
The Courtyard Gallery,
Wokingham 119375
Courtyard Gallery, Washington Studio
School, Washington 054189
Courtyard Living Gallery, Portland 123612
Cousens, Peter, Newtown, New South
Wales 127652
Cousin, Jacqueline, Les Andelys 125613
Cousins & Son, E.W., Ixworth .. 089654,
                               134506
Cousland Smiddy, Cousland ... 044015
Coustal, Catherine, Villeseque-des-
Corbieres 105876
Coutant, Claude, Paris 071204
Coutau-Bégarie, Olivier, Paris .. 125698
Coutinho, Bussum 137850
Coutinho, Francisco Pereira,
Lisboa 113990
Coutinho, Salvador Ferreira,
Porto 085017
Coutinho, Sebastião L., Rio de
Janeiro 064046
Couton, Bertrand, Nantes 125655
Coutoula, Martine, Beynac-et-
Cazenac 067162
Coutoula, Martine, Sarlat-la-
Canéda 073410
Coutts Memorial Museum of Art, El
Dorado 048499
Couturier Gallery, Los Angeles . 121465
Couturier-Nicolay, Paris 125700
Couturier, Elizabeth, Bourgoin-
Jallieu 103456
Couturier, Eric, Paris 125699
Couven Museum, Aachen 116434
Couvent des Ursulines, Liège .. 128074
Couvrat Desvergnes, Paris 071205
Gebr. Couvreur, Breda 082978
Couvreur, Breda 082977
Couvreur, Oosterhout 132604
Couvreur, R., Oosterhout 132605
Cova, Cartagena 085926

Covatti, Assunta, Roma 081311
Cove Curios, Cove 088982
Cove Landing, New York 095330
Covent Garden Flea Market,
London 090047
Coventry Toy Museum, Coventry 044016
Coventry Transport Museum,
Coventry 044017
Coventry Watch Museum,
Coventry 044018
Covered Bridge Antique Mall,
Cincinnati 092842
Coves Quay Gallery, Salcombe . 119048
Covet Design Store & Gallery,
Kerikeri 112931
Covington, Tucson 124723
Covington Antique Market, Oklahoma
City 095965
Cow Hollow Woodworks, San
Francisco 136313
Cowall, Bernd, Den Haag 112378
Cowan Vertebrate Museum,
Vancouver 007283
Cowboy-Museum Fatsy, Linz .. 002317
Cowboys and Indian Antiques,
Albuquerque 092049
Cowbridge and District Museum,
Cowbridge 044023
Cowbridge Antique Centre,
Cowbridge 088985, 134300
Cowderoy, John, Eastbourne ... 089144,
                               134352
Cowdy Gallery, Newent 118793
Cowes Maritime Museum,
Cowes 044024
Cowfold Gallery, Cowfold 117634
Cowichan Bay Maritime Centre,
Cowichan Bay 005659
Cowichan Valley Museum,
Duncan 005729
Cowleigh Gallery, Great Malvern 117880
Cowles, Charles, New York 122507
The Cowley County Historical Museum,
Winfield 054476
Cowlitz County Historical Museum,
Kelso 049916
Coworth Gallery, Sunningdale .. 091508
Cowpens National Battlefield,
Chesnee 047613
Cowper and Newton Museum,
Olney 045437
Cowra Museums, Cowra 000997
Cowri, Seattle 097420
Cowtown Antique Mall, Fort
Worth 093441
Cowwarr Art Space, Cowwarr . 098928
Cox & Fils, Bruxelles 140170
Cox & Co., London 090048, 118271
Cox, Claude, Ipswich . 134501, 144500
Cox, Edd, Seattle 124508
Cox, Jon, Louisville 135754
Cox, Udo, Gerolzhofen 129901
Cox's Architectural Reclamation Yard,
Moreton-in-Marsh 090778
Coyaba Gardens and Awarak Museum,
Ocho Rios 028453
Coyle, Dublin 079539
Coyote Gallery, Butte-Glenn
Community College, Oroville . 051819
Coyote Point Museum for
Environmental Education, San
Mateo 053056
Cozze, Houston 093653
Cozzetto, Maxine, Portland 096389
Cozzolino, Floriana, Livorno ... 080322
CPH Restoration, Salt Lake City 136236
CPIT School of Art and Design,
Christchurch Polytechnic Institute of
Technology, Christchurch 056190
Crabbe, Peter Ian, Cambridge .. 088762
Crabble Corn Mill, Dover 044128
Crabnet, New Orleans 122147
CRAC – Centre de Recherche
et d'Action Culturelle, Valence

(Drôme) 105799
Cracker Country Museum, Tampa 053667
Cracker Factory Antique Center, San
Diego 097077
Cracker Trail Museum, Zolfo
Springs 054620
Crackerbox Palace, Cairns 098836
Crackerjack, Seattle 124509
Crackers, Portland 096390
Craddock, Kay, Melbourne ... 139826
Cradle of Aviation Museum, Garden
City 049043
Cradle of the Sun, San Francisco 124260
Cradlewell Antiques, Jesmond . 089658
Crafoord, Eslöv 143831
Crafoord, Lund 126630
Craft Alliance, Saint Louis 052797
The Craft Factor, Saskatoon ... 138576
Craft in the Bay, Cardiff 117527
Craft Potters Association, London 060105,
                               118272
Craftart & Gallery, Cape Town . 114765
Crafted Toys International, Salt Lake
City 096888
Crafthouse, Vancouver 101767
Craftiques, San Antonio 096965
Craftivist, New Orleans 122148
Crafts Council, London 139166
Crafts Council Gallery, London . 044952
Crafts Museum, Delhi 024222
Crafts Museum, Mequon 050800
Crafts Nouveau, London 090049, 134637
Craftselijke Zadelmakerij Museum,
Bellingwolde 032057
Craftsman Antiques, Cleveland . 135458
Craftsman Antiques, Oakland .. 095887
Craftsman Farms Foundation,
Parsippany 051931
Craftsman Furniture Service,
Sacramento 136186
Craftsman Gallery, Swansea .. 119192
The Craftsman Magazine,
Driffield 139167
Craftsmens Gallery, Woodstock . 119387
Craftspace Gallery, Winnipeg .. 101873
Crafty Olde Things, Rangiora .. 084068
Craggsdan, Altona 098645
Cragside House, Rothbury 045637
Craig, Tucson 124724
Craig Heritage Park Museum,
Parksville 006611
Craig, Edmund, Fort Worth 120860
Craigavon Museum, Craigavon . 044029
Craigdarroch Castle, Victoria ... 007322
Craighead-Green, Dallas 120579
Craiglea, Edinburgh ... 089168, 134361
Craig's, Buffalo 092520
Craik Community Archives & Oral
History Society, Craik 005660
Craik, Brian & Caroline, Bath .. 088384
Crail Museum, Crail 044030
Crailo State Historic Site,
Rensselaer 052515
Crake, Johannesburg 114792
Crals, Liège 100780
Cralyn, Manawatu 083982
Cramalot Inn Antiques, Stroud .. 062251
Cramer, Barbara, Bonn 106526
Cramer, Hans, Den Haag 083071
Cramer, Patrick, Genève 116295, 138047,
                               144073
Crames, W., Bergheim (Erft) ... 074728
Crampton Tower Museum,
Broadstairs 043760
Cranach-Haus, Cranach-Foundation,
Lutherstadt Wittenberg 020196
Cranage Lane, Phoenix 096205
Cranbrook Academy of Art, Bloomfield
Hills 056751
Cranbrook Antique Centre,
Cranbrook 088994
Cranbrook Art Museum, Bloomfield
Hills 047038

Cranbrook Gallery, Cranbrook .. 117639,
                               144359
Cranbrook House, Bloomfield
Hills 047039
Cranbrook Institute of Science –
Natural History Collections,
Bloomfield Hills 047040
Cranbrook Museum, Cranbrook . 044031
Cranbury Historical and Preservation
Society Museum, Cranbury ... 048050
Crane Collection, Magnolia ... 121678
Crane Gallery, Seattle . 097421, 124510
Crane Hill Publishers,
Birmingham 138306
Crane Kalman, London 118273
Crane Museum of Papermaking,
Dalton 048133
Crane Point Hammock Museum,
Marathon 050663
Crane, Cato, Liverpool 126950
Crane & Assoc., Stefan, Dallas . 120580
Cranford, Knutsford 089730
Cranglegate Antiques, Swaffham 091523
Crankos, Invercargill 083966
Cranmer, Dana, New York 135904
Crannog Gallery, Dingle 109564
Crantock Forge Gallery, Crantock 117640
Crantock Gallery, Crantock ... 117641
Crapie, Péronnas 072063
Crapper & Co., Thomas, Stratford-
upon-Avon 091492
Craque Allure, Haarlem 112510
Crary Art Gallery, Warren 054067
Craters of the Moon, Arco 046524
Crathes Castle, Banchory 043487
Craven, Freiburg im Breisgau .. 075856
Craven Gallery, London 090050
Craven Museum and Gallery,
Skipton 045810
Cravens House, Lookout
Mountain 050394
Cravinho, Vitorino P. Silva, Lisboa 084866
Craviolatti, Pier Carlo, Torino .. 132123
Crawford, Memphis 135776
Crawford Arts Centre, Saint
Andrews 045674
Crawford Auto-Aviation Museum,
Cleveland 047812
Crawford County Historical Museum,
Pittsburg 052129
Crawford Doyle, New York 145332
Crawford Municipal Art Gallery,
Cork 024858
Crawford W. Long Museum,
Jefferson 049796
Crawford, Matthew, London .. 134638
Crawford, Owen & Zachery,
Louisville 121639
Crawford, R.E., Richmond 096537
Crawfordjohn Heritage Venture,
Crawfordjohn 044032
Crawley & Asquith, London ... 090051,
                               144581
Crawley Museum Centre,
Crawley 044033
Le Crayon Gris, Sisteron 105675
Les Crayons Bleus, Marseille .. 104164
Crazy Collecting, Miami 094722
Crazy House Museum, Đà Lạt .. 054783
Crazy Mountain Museum, Big
Timber 046982
Crazymport, Milano 080506
Cre-Art-Ive Flair, Nelson 112957
Creaciones M.S., Bogotá 102291
Créahm, Région Wallonne, Liège 058370
Crealde School of Art, Winter
Park 058129
Crealine, Saint-Émilion 105505
Creamery Square Museum,
Tatamagouche 007126
Creange, Colette, Avignon 066877
Créartdis, Saint-Martin-Boulogne 072930
Crearte, Bogotá 102292
Creatabilitoys! – Museum of

Advertising Icons, Coral Gables 047984
Creation, Christchurch 112839
Création d'Ambiance, Gstaad . 087518
Creation Museum, Petersburg . . 052009
Creationists Konsortium, Denver 120706
Creations by Elaine, Indianapolis 121175
Créations RB, Marseille 069927, 104165
Créations Tesoro, Paris 071206
Creatistic Fine Art, Calgary ... 101225
Creativ Art, Lüneburg 077209
Creativa, San José 102486
Creative Art, London 101338
Creative Art, Philadelphia 123362
Creative Art Center, Van Nung
  University, Jhongli 042344
Creative Art Centre, Dubai 088175
Creative Art Concepts, Cincinnati 120337
Creative Art Conservation,
  Burwood 127574
Creative Art Galleries International,
  Miami 121764
Creative Art Gallery, Dallas 120581
Creative Art Gallery, Saint Louis 123818
Creative Arts, Cairo 103030
Creative Arts, Vancouver 101768
Creative Arts Center, Kansas City 049880
Creative Arts Center, Pontiac .. 052236
Creative Arts Department, West
  Virginia Institute of Technology,
  Montgomery 057480
Creative Arts Division, Western Oregon
  University, Monmouth 057469
Creative Arts Guild, Dalton .. 048131,
  060454
Creative Arts Maker, Los Angeles 135713
Creative Arts Program, Charles River
  School, Dover 057019
Creative Arts Studio, Louisville Visual
  Art Association, Louisville 050488
Creative Arts Unlimited,
  Durbanville 126493
Creative Burnings, Saint Louis . 096680
Creative Camera, London 139168
Creative Closets, Columbus . 135477
Creative Connections,
  Albuquerque 119465
Creative Discovery Museum,
  Chattanooga 047592
Creative Eye, Denver 120707
Creative Finishing Specialists,
  Vancouver 128223
Creative Furniture, Los Angeles . 094303
Creative Galleries, Los Angeles . 121466
Creative Galleries, Nashville .. 122075
Creative Growth Art Center,
  Oakland 051676
Creative Imagery, Columbus ... 120485
Creative Impressions Gallery,
  Sacramento 123763
Creative Insprations Gallery,
  Belmont 098731
Creative Mind, Germering ... 107117
Creative Napier, Hawkes Bay .. 112910
Creative Native, Fremantle .. 061488
Creative Native, Perth 062076
Creative Nature, San Diego ... 124115
Creative New Zealand, Arts Council of
  New Zealand Toi Aotearoa,
  Wellington 059631, 059632
Creative New Zealand, Toi Aotearoa
  Arts Council, Auckland 059633
Creative Palette, Columbus ... 120486
Creative Picture Framing, London 118274
Creative-shop, Ballwil 116052
Creative Solutions, Villa Park ... 058046
Creative Spirit Art Centre,
  Toronto 007170
Creative Spirit Gallery, Tucson . 124725
Creative Stüble, Bludenz 062539
Creativiteitscentrum Gouden Handen,
  's-Heerenberg 032427
Creativity Centre, Coorparoo . 098924
Creativity Explored, San
  Francisco 124261

Creativity Explored II, San
  Francisco 124262
Creazioni Fuoco, Roma 081312
Crédence, Issigeac 068880
La Crédence, Paris 071207
Credi Arte, Bucaramanga 102382
Crediorvi, Cartagena 102419
Crédit de l'Or, Lille 069485
Crédit de l'Or, Paris 071208
Crédit du Palet Médailles,
  Angoulême 066627
Creech, M. Ford, Memphis ... 094629
Creek Antiques, London 090052, 134639
Creek Council House Museum,
  Okmulgee 051741
Creek, Cherry, Oxted 118869
Creekhouse Gallery, Vancouver . 101769
Creekside Gallery, Hamilton ... 101313
Creetown Gem Rock Museum,
  Creetown 044035
Creetown Heritage Museum,
  Creetown 044036
Cregneash Folk Village,
  Cregneash 044037
Cregneash Village Folk Museum,
  Manx National Heritage, Port Saint
  Mary 045534
Creiche, Jean-Luc, Bazas 066985
Creighton-Davis, Mclean 094609, 121677,
  127301
Crema, Paola, Firenze 110045
Creman, Jurjen, Amsterdam ... 132392
Cremona Mdina Arts, Rabat ... 111918
CremonAntiquaria, Cremona .. 098186
Cremyll Antiques, Cremyll 089001,
  134304
Crescent Arts, Scarborough 045743
Crescent Bend/Armstrong-Lockett
  House and William P. Toms Memorial
  Gardens, Knoxville 050008
Crescent City Architecturals, New
  Orleans 095084
Crescent City Books, New
  Orleans 145305
Crescent Collection, Charlotte . 092591
Crescent Galleries, Bridgwater . 117428
Crescent Galleries, Taunton ... 119211
Crescent Gallery, New Orleans . 122149
Crescent Moon, Saint Paul ... 096817
Crescent Scents, Seattle 124511
Crescentia-Gedenkstätte,
  Kaufbeuren 019557
Crescentini, Remo, Roma 131883
Crescenzi, Antonio, Torino ... 132124
Creski Muzej, Cres 008895
Crespo, J., Madrid 143757
Cress Gallery of Art, Chattanooga 047593
Cressever, Ariane, Tarbes 073624
Cressida, Paris 071209
Cresswell, J D M, Turramurra . 062354
Cressy, Koroit 139798
Crest Collectables, Eastbourne . 089145
Crestar Bank Art Collection,
  Richmond 052547
Crested China Company, Driffield 089106
Crested Circle, Fetcham 139169
Cresten, Doherty, Dublin 130852
Crestet-Centre d'Art, Le Crestet 013623
Crestline Shunk Museum,
  Crestline 048061
Creston and District Museum,
  Creston 005664
Crestwood Antiques, Birmingham 092403
Crestwood Coins, Saint Louis . 096681
Creswell Crags Museum and
  Education Centre, Welbeck . 046074
Creswick Antiques, Malvern,
  Victoria 061780
Creswick Museum and Gold Battery,
  Creswick 001002
Creswick Station, Creswick ... 061373
Crété, Eric, Saint-Hélen 072820
Crétier, François, Vichy 074105
Le Creuset, Charleroi 063597

Creusy, J.P., Lyon 069727
Creusy, Joël, Lyon 069728
Creusy, Rémy, Saint-Étienne ... 072760
Creutzmann, Krostitz 076976
Le Creux de l'Enfer, Centre d'Art
  Contemporain, Thiers 015933
Creuze, Raymond, Paris 104662
Creuzevault, Colette, Paris 104663,
  141084
Crevoisier, San Antonio 096966
Crewe, Bob, Los Angeles 121467
Crewkerne and District Museum and
  Heritage Centre, Crewkerne .. 044041
Crewkerne Antique Centre,
  Crewkerne 089004, 134305
CRG Gallery, New York 122508
Cribier, Saint-Paul-de-Vence ... 105582
Criccieth Castle, Criccieth 044043
Criccieth Gallery, Criccieth 089007,
  134307
Crichton Museum, Dumfries ... 044151
Crick, M.E., London 090053
Cricket Book Collector,
  Blacktown 139647
Cricket Gallery, Atlanta 119556
Cricket Hill Gallery, New York .. 122509
Cricoteca-Galeria-Pracownia Tadeusza
  Kantora, Kraków 034860
Crid'Art, Amnéville-les-Thermes . 103230
Crimea, Madrid 086165
Criminology Museum, Abbasiya . 010508
Crimisos, Palermo 131693
Crimson, Bangalore 109256
Crinel, Yves, Paris 128971
Cringle, M. & A., Burnham
  Market 088721
Crippa Numismatica, Milano ... 080507
Cripple Creek District Museum,
  Cripple Creek 048066
Cripta e Museo di Sant'Anastasio,
  Asti 025404
Criptonit, Milano 080508
Crisce, Barcelona 085763
Crisp, J.W., London 134640
Crispi, Fabio, Napoli 131611
Crispin & Sons, J., London 134641
Crispin, Madeline, London 090054
Cristal, Jakarta 109508
Cristal d'Art, Aiguèze 103186
Cristalarte, Cádiz 115003
Cristall, Elissa, Vancouver 101770
Cristallamp, Milano 080509
Cristalleries Royale de Saint-Louis,
  Saint-Louis-les-Bitche 015429
Cristallo, Cortina d'Ampezzo ... 109981
Cristeph Gallery, Letterkenny . 109664
Cristiani & Foschini, Bologna ... 079831
Cristiani, Giancarlo, Torino ... 081691
Cristiano, Ferdinando, Quarto . 081179
Cristiano, Filippo, Ravenna ... 142790
Cristina's Bellas Artes, Ciudad de
  Panamá 084413
Cristo, Poissy 072216
Cristobal, Agustin, México 082473
Cristol, Jean-Marc, Labastide-Saint-
  Pierre 069127
Criswick, New York 122510
Criterion Auctioneers, London . 126960
Criterion Books, Vancouver ... 140423
Critica d'Arte, Firenze 138812
Critique d'Art, Châteaugiron ... 138639
Critiques Livres Distribution,
  Bagnolet 140776
Crivellaro, Lorenzo, Milano ... 131430
Croatian Heritage Museum,
  Eastlake 048453
Croce, Ugo, Torino 132125
Crocheted Afghans, Louisville . 094533
Crocker, Memphis 094630
Crocker Art Museum,
  Sacramento 052725
Crocker House, Mount Clemens 051083
Crockett County Museum, Ozona 051878
Crockett Tavern Museum,

Morristown 051069
Crockford, Sue, Auckland 112758
Crocodile, New York 095331
Crocq, Guy, Domagné 068369
Croes, Bernard, Paris 128972
Croës, Gisèle, Bruxelles 063384
The Croft, Calgary 101226
Croft, Portree 091041
Croft, Swindon 119199
Croft Castle, Leominster 044809
Croft Wingates, Market
  Harborough 118719
Crofton Roman Villa, Orpington . 045441
Crofts, Peter A., Wisbech 091885
Crohill Gallery, Bicheno 098746
La Croisée, Vezelay 105852
Croisée des Artistes, Maîche ... 104143
La Croisée des Arts, Albi 066512,
  103214
La Croisée des Temps, Carvin .. 067699
Croisette Antiques, Paris 071210
Croiseur Colbert, Bordeaux 012109
Croissant de Lune, Landerneau . 103068
Croissy, Bernard, Courbevoie ... 068185
Croizier, René, Craponne (Rhône) 068220
Croizit, Frédéric, Abzac 066415
Crolotte, Dominique, Nancy ... 128821
Crom Estate, Newtownbutler ... 045384
Croma Galleri, Stavanger 113264
Cromarty Antiques, Cromarty . 089010
Cromarty Courthouse Museum,
  Cromarty 044044
The Cromarty Gallery, Cromarty 117650
Crome Gallery & Frame Shop,
  Norwich 090872, 134891
Cromer Museum, Cromer 044046
Cromford Mill, Cromford 044049
Crompton, Paul H., London ... 138210
Cromwell Museum, Huntingdon . 044616
Crona, Dennis, Örebro 133560
Cronan, Sandra, London 090055
Crone, Berlin 106204
Cronin, Frank, Christchurch 083867
Cronomuseu de Queiriga,
  Queiriga 035969
Cronos, Milano 080510
Cronosaurio, Villa de Leyva ... 008820
Cronulla Art, Cronulla 098934
Crook County Museum and Art Gallery,
  Sundance 053605
Crooked House Gallery,
  Lavenham 118076
Crooked Tree Arts Center,
  Petoskey 052020
Crooked Tree Arts Council,
  Petoskey 060455
The Crooked Window, Dunster . 089121
Cropp, M., Kronberg im Taunus 130203
El Croquis, El Escorial . 137976, 139047
Cros, Anne, Pézenas 105348
Cros, J. Du, Petworth 090995
Crosby Art, Belfast 117335
Crosby County Pioneer Memorial
  Museum, Crosbyton 048073
Crosby Furniture, Cleveland ... 135459
Croset, Christian, Nogent-sur-
  Marne 104409
Cross Antiques, Watlington ... 091540
Cross Art and Books, Kings
  Cross 139796
Cross Art Projects, Kings Cross . 099184
Cross Gallery, Dublin 109580
Cross Gallery, Nagoya 111314
Cross Hayes, Chippenham 088881
Cross Keys Antiques, Baltimore . 092311
Cross Keys Jewellers, Devizes . 089056
Cross Plains Antiques Mall,
  Nashville 094987
Cross Street Gallery, London .. 118275
Cross, Edward, Weybridge 091807,
  119330, 135216
Cross, Michael, Dallas 120582
Crosseyed Moose Antique, Fort
  Worth 093442

Crossing Arte Club, Portogruaro 110683
Crosskeys Art Gallery, Perth ... 118903
Crossland & Hydra, Goomeri ... 061540
Crossland Furniture, Charlotte .. 092592, 135390
Crossley & Scott, Melbourne ... 099300
Crossman Art Gallery, University of Wisconsin Whitewater, Whitewater 054378
The Crossness Engines, London 044953
Crossover Gallery, Albuquerque . 119466
Crossroads Antiques, Prestwick 091058
Crossroads Art Center, Richmond 123729
Crossroads Bookshop and Gallery, Wellington 143364
Crossroads Gallery, Wellington . 113064
Crossroads Museum, Oyen 006608
Crossroads of America, Bethlehem 046957
Crosstown Arts, Boston 119856
Crouan, Marguerite, Montussan 070339
Crousel, Chantal, Paris 104664
Crouwel, Amsterdam 132393
Crouzat, Marie-Rose, Millau ... 070120
Crouzet, Jocelyne, Paris 071211
Crow & Co., Houston 093654
Crow Canyon Archaeological Center, Cortez 048011
Crow Wing County Historical Museum, Brainerd 047175
Crow, Dennis George, Hong Kong 065033
Crow, Dennis George, Los Angeles 094304
Crowder College-Longwell Museum and Camp Crowder Collection, Neosho 051215
Crowell, Providence 123699
Crowley Art Association, Crowley 060456
Crowley Art Association and Gallery, Crowley 048077
Crowley Museum, Sarasota 053148
Crowley, Yuill, Sydney 099774
Crown, New York 122511
Crown & Eagle, Columbus 093011
Crown Antique Furniture, Seoul . 082257
Crown Antiques, Pittsburgh 096273
Crown Arcade, London 090056
Crown City Hardware, Pasadena 136071
Crown Fine Arts, Ingleby Barwick 118017
Crown Gallery, Bruxelles 100586
Crown Gallery, Newcastle-upon-Tyne 118786
Crown Gardens Museum, Dalton 048132
Crown Hill, Seattle 097422
Crown Liquor Saloon, Belfast .. 043555
Crown Mercantile, Denver 093275
Crown Point Press, San Francisco 124263, 138458
Crown Point State Historic Site, Crown Point 048079
Crows Nest Books, Waikato ... 143353
Crowsnest Museum, Coleman .. 005636
Crowsnest Pass Ecomuseum, Bellevue 005415
Crowyn, P., Bruxelles 063385
Croxteth Hall, Liverpool 044852
Croydon Auction Rooms, Croydon 126872
Croydon Museum, Croydon 044051
Croydon Natural History and Scientific Society, Croydon 044052
Crozet, Pierre, Carmaux 067684
Crozier Antiques Restoration, Keady 130873
CRP, Brescia 079932
Le Cruchon, Haarlem 083323
Cruciani, Federica, Roma 131884
Crucible Gallery, Mandaluyong . 113395
Crucis, Gérard, Bruxelles 140171
Cruck House Antiques, Much Wenlock 090790
Crudelia?, Contemporary Art Magazine, Roma 138813
Crump, Lucien, Philadelphia .. 123363
Crusader Church, Abu Ghosh .. 025008

Cruttenden's, Burgess Hill 134183
Cruz del Sur, Buenos Aires 139561
La Cruz del Sur, Buenos Aires . 060892
Cruz Mateo, Julian, Madrid 133320
Cruz Reig Fito, Maria, Valencia . 086512
Cruz, Manuel Lopes, Lisboa ... 084867
Cruz, Mary, Bath 088385, 134075
Cry for the Moon, Guildford ... 089393, 134427
The Cry of the Gulls, Fowey ... 117808
The Crypt – Town Hall Plate Room, Oxford 045452
Crypte Archéologique du Parvis de Notre-Dame, Paris 014645
Crypte Carolingienne de l'Ancienne Abbaye, Flavigny-sur-Ozerain . 012990
Cryptoneum, Legenden-Museum, Rostock 021487
Crystal, Minneapolis 094906
Crystal Art Gallery, New York .. 122512
Crystal Arts Gallery, Kansas City 121287
The Crystal Ball, Annandale ... 127552
Crystal Brook Heritage Centre, Crystal Brook 001006
Crystal Caverns at Hupp's Hill, Strasburg 053565
The Crystal Ceiling, Haarlem ... 112511
Crystal City Community Museum, Crystal City 005668
Crystal Clinic, Phoenix 136112
Crystal D., Den Haag 112379
Crystal Dragon, Nashville 094988
Crystal Fine Art, Boston 119857
Crystal House, Manama 063138
Crystal Kingdom, Coonabarabran 000990
Crystal Lady, Omaha 096036
Crystal Lake Falls Historical Association, Barton 046802
Crystal Mountain – Crystal & Mineral Gallery & Museum, Auckland . 033144
Crystal Palace Antiques, London 090057
Crystal Palace Museum, London 044954
Crystal River State Archaeological Site, Crystal River 048081
Csaba Lengyel de Bagota, Budapest 137586
Csabai Tanya- és Gabonatörténeti Kiállítóhely, Munkácsy Mihály Múzeum, Békéscsaba 023441
Cselley Ház, Hansági Múzeum, Mosonmagyaróvár 023797
Csemadok, Dunajská Streda ... 143690
Csepel Galéria és Helytörténeti Gyűjtemény, Budapest 023468
Cserépfalvi Imre Emlékszoba, Cserépfalu 023583
Cserno, Amsterdam 082722
Csillár Bazár, Budapest 079247
Csiszar, Csaba, Hamburg 129956
Csohány Galerie, Pásztói Múzeum, Nógrádi Megyei Múzeum Szervezet, Pásztó 023833
Csók István Képtár, Szent István Király Múzeum, Székesfehérvár 023933
Csók, István, Budapest 079248, 109136
Csokay, Gols 099995
Csonth, Istvan, Bamberg 074659
Csontváry Múzeum, Janus Pannonius Múzeum, Pécs 023837
CSP Editions, Glasgow 117839, 138168
Csúcsi Fazekasház, Tornyai János Múzeum, Hódmezővásárhely .. 023689
CTS Turner Museum, Elkhart .. 048532
Čtvrtník, Dobroslav, Praha 128341
Cu Heritage Center, University of Colorado, Boulder 047142
Cu4rto Nivel Arte Contemporaneo, Bogotá 102293
Cuadrado, Teresa, Valladolid .. 115523
Cuadro, Cartagena 115009
Cuadro, Las Palmas de Gran Canaria 115376
Cuadros, Valencia 115479
Cuadros de Monterrey, Monterrey 111989

Cuadros Vivos de la Sabana, Corozal 008667
Cuare, México 138909
La Cuarenta, Madrid 086166
El Cuarto de San Alejo, Bucaramanga 065234
Cuartoscuro, Cuauhtémoc 138910
Cuatro Casas, Madrid . 086167, 115176
Cuba Aereo, Miami 094723
Cuban Art Space, New York ... 122513
Cuban Collection Fine Arts, Miami 121765
Cuban Masters Collection, Miami 121766
Cuban Museum of Arts and Culture, Miami 050828
Cubanostalgia, Miami 094724
Cubas Basterrechea, Carmen, Madrid 133321
Cubby Hole Antiques, Portland . 096391
Cube, Bowden 098764
Cube Gallery, Bristol 117444
Cubeta, C., Castlemaine 061297
Cubic Gallery Íteza, Kyoto 111290
Cubis Collection, North Sydney . 099475
Cubitt Artists, London 118276
Cubus Kunsthalle, Duisburg ... 018052
Cucchi, Paola, Roma 131885
Cuchifritos, New York 122514
The Cuckfield Gallery, Cuckfield 117658
Cuckfield Museum, Cuckfield .. 044053
Cuddle Pot Collectables, Cudal . 061382
Cudgewa Antique Shop, Cudgewa 061383
Cudrio Kystmuseum, Langesund 033730
Cudworth, Bath 117303
Cue Art Foundation, New York . 051369
Cuéllar, Arturo, Zürich 116765
Cuervo, Medellín 065284
Cueto Project, New York 122515
La Cueva de Caritas, Rosario . 060950
Cuivrerie de Cerdon, Cerdon .. 012383
Cukier, Raymond, Romainville . 072519
Cukormúzeum, Szerencs 023967
Culbertson Mansion, New Albany 051221
Culbertson Museum, Culbertson 048084
Culinair Historisch Kookmuseum de Vleer, Appelscha 032014
Culinair Museum, Amersfoort . 031909
Culinary Archives and Museum, Johnson and Wales University, Providence 052380
Cullity Gallery, Nedlands 099424
Cullman County Museum, Cullman 048086
Culloden House Antiques, Vancouver 064810
Culloden Visitor Centre, Inverness 044639
Cullompton Old Tannery Antiques, Cullompton 089023
Culo, Delphine, Lorient 069635
Culotta, Rosolino, Palermo ... 081033
Culross Palace, Culross 044055
Cult, Wien 100245
Cult Verein zur Förderung aktueller Kunst, Wien 058246
Cultura 2000 – Kunst im Kuhstall, Kirtorf 107564
Cultura Collection, Cleveland .. 120413
Cultural Center Hellenic Cosmos, Athinai 022983
The Cultural Center of Fine Arts, Parkersburg 051928
Cultural Center of the Philippines Museo, Pasay 034562
Cultural Centre de Dorval, Dorval 005714
Cultural Heritage Center Museum, Confederated Tribes and Bands of the Yakama Indian Nation, Toppenish 053775
Cultural Museum, Melaka 030704
Cultural Museum at Scarritt-Bennett Center's, Nashville 051176
Cultural Museum of Hemei,

Hemei 042323
Cultural Museum, State University Museum, Kabacan 034485
Cultural Research Institute, Kolkata 024358
Cultural Rights and Protection/Ute Indian Tribe, Fort Duchesne .. 048807
Culturati Art Gallery, San Diego . 124116
Culture & Economics, Minden .. 108004
Culture and Fine Arts Publishing House, Beijing 136698
Culture Cache Gallery, San Francisco 124264
Culture Exhibition Hall of Singang Fengtian Temple, Singang ... 042460
Cultured Pearl, Portland 096392
Cultureel Centrum van Gogh, Zundert 033116
Cultureel Maçonniek Centrum Prins Frederik, Den Haag 032168
Cultuur-Historisch Museum Sorgdrager, Hollum 032496
Cultuurhistorisch Museum Valkenswaard, Valkenswaard . 032968
Cultuurhistorisch Streek- en Handkarrenmuseum De Wemme, Zuidwolde, Drenthe 033113
Cultuurhistorisch Streekmuseum De Acht Zaligheden, Eersel 032275
Culzean Castle and Country Park, Maybole 045240
Cum Museum, Imphal 024308
Cum Tempore, Salzburg 062788
Cumalı Sanat Galerisi, İstanbul . 117025
Cumberland Art Conservation, Nashville 135845
Cumberland Art Society, Cookeville 060457
Cumberland County Historical Society Museum, Greenwich 049273
Cumberland County Museum, Amherst 005328
Cumberland Gallery, Chichester . 117577
Cumberland Gallery, Nashville .. 122076
Cumberland Gap National Historical Park, Middlesboro 050860
Cumberland Heritage Village Museum, Cumberland, Ontario 005670
Cumberland House, Regina 006765
Cumberland Museum, Cumberland, British Columbia 005669
Cumberland Pencil Museum, Keswick 044679
Cumberland Science Museum, Nashville 051177
Cumberland Theatre Arts Gallery, Cumberland 048090
Cumberland Township Historical Society, Ottawa 006586
Cumberland Toy and Model Museum, Cockermouth 043983
Cumbernauld Museum, Cumbernauld 044057
Cumbria Architectural Salvage, Raughton Head 091071, 134959
Cumbria Auction Rooms, Carlisle 126847
The Cumbrian Antiques Centre, Brampton, Cumbria 088569
Cumhuriyet Devri Müzesi, Ankara 042818
Cumhuriyet Egitim Müzesi, İstanbul 042931
Cumhuriyet Eğitim Müzesi, Karataş 042983
Cuming Museum, London 044955
Cummer Museum of Art, Jacksonville 049768
Cumming, A. & Y., Lewes 144531
Cumming's Corner, Kansas City . 094069
Cummings, J., Minneapolis 145274
Cummins, James, New York ... 145333
Cunderdin Municipal Museum, Cunderdin 001007
Cundiff, Bryan, Louisville 094304, 127295

Cuneo Museum, Vernon Hills .. 053966
Cunha, Rio de Janeiro 064047
Cunha, Carlos, Porto 085018
Cunha, Ruben, Lisboa 113991
Cunha, Vítor Manuel, Lisboa ... 143565
Cunha, Walter A., Rio de Janeiro 140292
Cuniberti, Beatrice, Firenze ... 131155
Cunin, Marie-Noëlle, Lacoste .. 069131
Cuningham, Betty, New York ... 122516
Cunnamore Galleries, Skibbereen 109690
Cunningham, New Orleans 122150
Cunningham, Portland 145443
Cunningham Dax Collection, Parkville,
  Victoria 001397
Cunnington, Lynette, Woollahra . 062466
Cuoccio, Antonio, Milano 080511
Cuoccio, Guiseppe, Milano 080512,
                              131431
Cuoghi, Paola, Modena 080789
Cuomo, Baltimore 092312
Cuore, Buono, Rio de Janeiro . 128112
Cupar and District Heritage Museum,
  Cupar 005671
Cupboard Antiques, Amersham . 088271
Cupboard Collectables,
  Minneapolis 094907
Cupertino Historical Museum,
  Cupertino 048094
Cupertino S, Lisboa 113992
Cupid World of Beauty, Ettalong
  Beach 061454
Cupid's Antiques, Oklahoma City 095966
Cupids Museum, Cupids 005672
Cupillard, Grenoble 103786
Cupola House, Egg Harbor 048490
Cuppacumbalong, Tharwa 099818
Cuppalot, Lyons 099241
Cupper, David, Norfolk 095849
Cupuri, Thannhausen 142411
Curá Antiques, London 090058
Curaçao Museum, Willemstad .. 009397
Curator, New York 139353
Curdnatta Art Gallery, Port
  Augusta 099575
Curfman Gallery and Duhesa Lounge,
  Fort Collins 048794
Le Curieux, Saint-Ouen 073062
Curiger, Carl-Rudolf, Uitikon
  Waldegg 087863, 126757
Curio Arts, Kathmandu 112066
Curio Books, Hahndorf 139759
Curio-City, Singapore 085392
Curio City, Southend-on-Sea .. 091376
Curio Corner, Tynemouth 091682
Curio Corner Books, Austin ... 144978
The Curio Shop, Dublin 079540
Curio Warehouse, Guildford ... 061565
Curios City, Vancouver 064811
The Curios House, Udaipur ... 079417
Curios Limited Antiques, Norfolk 095850
Curios of Chale, Chale, Isle of
  Wight 088817
Curiosa, Breda 082979
Curiosa de Markies Roosendaal,
  Roosendaal 083542
Curiosa, Meubelen & Hebbedingen,
  Den Haag 083072
Curiosamuseum Heraldiek en Folklore,
  Alveringem 003264
Curioser & Curioser, Bendigo .. 061131
Curiosidade's Guatemala, Oklahoma
  City 095967
Curiosité, Evian-les-Bains 068490
Curiosités, Besançon 067121
Curiosités, Sallanches 073384
Curiosités Marocaine, Rabat .. 082598
Curiosities, Highgate, Western
  Australia 061623
Curiosities Shop, Vancouver .. 064812
Curiosity Corner, Dallas 093107
Curiosity Corner, North Shore . 084030
Curiosity Shop, Hawkes Bay ... 083951
Curiosity Shop, Las Vegas 094132

Curiosity Shop, Nashville 094989
The Curiosity Shop, Saint
  Sampson 091209
The Curiosity Shop, South
  Shields 091367
The Curiosity Shop, Winnipeg .. 064917
Curiosity Shoppe, Jacksonville . 093982
Curiosum, Caslano 087353
Curious Art Co-Op, Chinderah .. 098885
Curious Arts, Singapore 114511
The Curious Kids' Museum, Saint
  Joseph 052780
Curious Sofa, Kansas City 094070
Currahee Military Museum,
  Virden 007344
Currency Museum of the Bank of
  Canada, Ottawa 006587
Currency Press, Redfern 136487
Currier Gallery of Art,
  Manchester 050626
Curry Historical Society Museum, Gold
  Beach 049142
Cursief, Brugge 125303, 140158
Curso Artes Visuais, Universidade
  do Extremo Sul Catarinense,
  Criciúma 055071
Curso de Artes Visuais, Universidade
  de Ijuí, Dept. DELAC, Ijuí 055078
Curso de Educação Artística,
  Universidade Estadual de Campinas,
  Campinas 055070
Curso de Educação Artística,
  Universidade Federal do Maranhão,
  São Luís 055098
Curso de Pintura, Fortaleza 100940
Curtessão, Porto Alegre 063970
Curti, Paolo, Milano 110294
Curtins, Sankt Moritz 116622
Curtis, Long Melford 090618
Curtis Museum, Alton 043407
Curtis, Edward S., Seattle 124512
Curtis, Lee J. & Eva, München . 142147
Curtiss, Louis F., San Diego ... 097078
Curtze, Dr. Heike, Wien 100246
Curulla, Marco, Cagnes-sur-Mer 067588
Curve, Sheffield 119081
Curwen, New Academy Gallery and
  Business Art Galleries, London 118277
Curwood Castle, Owosso 051856
Curzon, David, London 090059, 118278,
                             134642
Cusanus-Geburtshaus, Geburtshaus
  des Nikolaus von Kues, Bernkastel-
  Kues 017334
Cuscini, Guido, Bologna 079832
Cushing House Museum,
  Newburyport 051517
Cusimano, Giovanni, Brescia .. 079933
Cusset, Gérard, Clermont-
  Ferrand 103602, 128560
Cust Museum, Cust 033182
Custer Battlefield Museum,
  Garryowen 049057
Custer County 1881 Courthouse
  Museum, Custer 048099
Custer County Art Center, Miles
  City 050887
Custer County Historical Society
  Museum, Broken Bow 047221
Custer Museum, Clayton 047782
Custom Art Works, Houston .. 093655,
                             121023
Custom by Design, Surry Hills . 099743
Custom Collectables, Indianapolis 093873
Custom Crafted Upholstery, Saint
  Paul 136218
Custom Engraving, Baltimore .. 119732
Custom Frame Shop and Gallery,
  Philadelphia 123364
Custom House Maritime Museum,
  Newburyport 051518
Custom House Visitor Centre,
  Dublin 024872
Custom Made Furniture, Salt Lake

City 136237
Custom Oil Paintings Reproductions,
  Houston 121024
Custom Refinishing, Omaha .. 136058
Customer Claim Service, Phoenix 136113
Customs House Nautical Museum,
  Robe 001453
Customs Museum, Taipei 042511
Customs Wharf Gallery,
  Williamstown 099887
Cutelli, Roma 131886
Cutispoto, Domenico, Napoli .. 137702
Cutrini, Sabina, Roma 131887
Cutroneo, Giuseppe, Messina .. 080365
Cutts, Christopher, Toronto ... 101593
Čuvašskij Gosudarstvennyj
  Chudožestvennyj Muzej,
  Čeboksary 036929
Čuvašskij Gosudarstvennyj Institut
  Kultury i Iskusstv, Čeboksary . 056267
Čuvašskij Nacionalnyj Muzej,
  Čeboksary 036930
Cuwo 1801, Pamplona 086384
Cuxhavener Kunstverein,
  Cuxhaven 058815
Cuyahoga Valley Art Center, Cuyahoga
  Falls 056968, 120530
Cuyler Manor Museum,
  Uitenhage 038868
Cuyuna Range Museum, Crosby 048071
Cvetkov, Vladimir, Sofia 064228
CVW Piano Restoration,
  Jacksonville 135651
CW Exchange, Saint Paul 096818
CWM Galleries, Brisbane 098785,
                             098786
Cwmgwili Mill, Carmarthen ... 088797
CWU Anthropology Department
  Collection, Ellensburg 048545
Cy Twombly Gallery, The Menil
  Collection, Houston 049575
Cybèle, Paris .071212, 136887, 141085
Cybercoins, Pittsburgh 096274
Cybertery Memorials, Nashville . 122077
Cybertiques, Louisville 094535
Cyfarthfa Castle Museum and Art
  Gallery, Merthyr Tydfil 045250
Cygnaeuksen Galleria, Helsinki . 010814
Le Cygne Rosé, Paris 071213
Le Cygne Vert, Paris 071214
Cylburn Nature Museum,
  Baltimore 046741
La Cymaise, Paris 104665
Cymroza Art Gallery, Mumbai .. 119427
Cynon Valley Museum and Gallery,
  Aberdare 043347
Cynthia & Wim, Almere 082641
Cypher's & Kallander, Tacoma . 136383
Cypoly School of Art and Design,
  Aradippou 055289
Cypress College Fine Arts Gallery,
  Cypress 048102
Cyprus College of Art, Paphos . 055296
Cyprus College of Art, Larnaca,
  Larnaca 055290
Cyprus College of Art, Limassol,
  Lemesos 055292
Cyprus Federation of Associations of
  Friends of Museums, Lefkosia 058534
Cyprus Folk Art Museum,
  Lefkosia 009412
Cyprus Historical Museum,
  Lefkosia 009413
Cyprus Jewellers Museum, Laiki
  Geitonia 009405
Cyprus Medieval Museum,
  Lemesos 009424
Cyprus Museum, Lefkosia 009414
Cyprus National Struggle Museum,
  Lefkosia 009415
Cyprus Olympic Committee Museum,
  Lefkosia 009416
Cyr, Raymond Emile, Culver City 135487
Cyrette, Winnipeg 101874

Cyriel Verschaeve Museum,
  Alveringem 003265
Cyrlin, Philip, London 090060
Cyrny, H.-J., Moringen 077420
Cyrus Antiques, Petworth 090996
Cyrus H. McCormick Memorial
  Museum, Steeles Tavern .. 053516
Cytrangulo, Julio, Rio de Janeiro 064048
Cywaith Cymru. Artworks Wales,
  Cardiff 060106
Cyx Art, Paris 071215
Cyzer, E. & R., London 118279
Czaar Peterhuisje, Zaandam .. 033074
Czaja, Berlin 074778, 141500
Czajka, Nowy Dwór Mazowiecki 113706
Czajkowski & Son, Edmund, Woodhall
  Spa 135255
Czambor, Joachim, Erfurt 075637
Czarina, Monaco 082502
Czarny Butik, Warszawa 113815
Czarny, Veronika, München ... 077480
Czasostop Artystyczny, Gdynia . 113495
Czastka, H., Karlsruhe 076688
Czech Center New York, New
  York 051370
Czech Committee of ICOM, Brno 058539
Czech Heritage Museum and
  Genealogy Center, Temple .. 053708
Czell, Gertrud, Bamberg 074660
Czempik, Euskirchen 106940
Czernin, Albrecht, Wien 127911
Czerny, Sarzana 126231
Czerwonatis, Gerd, Hamburg .. 076209,
                             129957
Czibor Zoltán Emlékszoba, Klapka
  György Múzeum, Komárno .. 023754
Czisch, Georg, Aschaffenburg . 129387
Czóbel Múzeum, Ferenczy Múzeum,
  Szentendre 023954
Czonka, Piotr, Konstancin-
  Jeziorna 132829
Czotscher, Eduard & Susanna,
  Creglingen 106643
Czyganowski, Uwe, München .. 077481
Czymoch, Hans, Berlin 074779
D & B, Beograd 085333
D & B African Village Art,
  Houston 121025
D & B Antiques, Omaha 096037, 136059
D & D, Ljubljana 085607
D & D, Roma 081313
D & D Upholstery, San Francisco 136314
D & H Acrylic Designs,
  Indianapolis 093874
D & J Antik, Vanløse 066179
D & M Antiques, Portland 096393
D & S Antiques, Wien 125270
D & T Architectural Salvage,
  Birkenhead 088471
D-Art, Gdynia 113496
D-Art, Praha 102658
D-Day Museum and Overlord
  Embroidery, Portsmouth ... 045550
D-Day Omaha Musée, Vierville-sur-
  Mer 016208
D J B Woodworks, Austin 135323
D&D Silverware, Warszawa ... 084711,
                             113816
D. Art, Ostrava 102631
D. Baazov Georgian Jewish Historic
  and Ethnographic Museum,
  Tbilisi 016397
D. Ralph Hostetter Museum of Natural
  History, Harrisonburg 049377
D.A. Wurfel Grain Collection,
  Pinnaroo 001407
D.A.M., Corte 068163, 103632
D.C. Booth Historic National Fish
  Hatchery, Spearfish 053434
D.G. Asturias, Oviedo 086329
D.H. Lawrence Birthplace Museum,
  Eastwood 044225
D.M. Restorations, Weston-super-

Mare 091803, 135214
D.M.S, Saint-Martin-Boulogne .. 072931
d.o.c.s. Gallery, New Orleans .. 122151
D.P. Druck- und Publikations GmbH,
München 137393
D.P.M., Brescia 079934, 131030
D.R. Barker Historical Museum,
Fredonia 048964
D.R. Visual Art Center, Nakorn-
Chaisri 042705
D.S.L, Rennes (Ille-et-Vilaine) .. 072432
D.T.A. Galerie, Genk 100683
D137, Sankt-Peterburg 037728
D406, Modena 110495
Da Art Shop, Los Angeles 121468
Da Capo Music, Sydney 139924
Da Hunt, Las Vegas 094133
Da Huo, Shenzhen 102219
Da Jian Wen House, Shenzhen .. 102220
Da Lozzo, Sydney 099775
Da Luz, Rui, Lisboa 143566
Da Mosto, Venezia 081830
Da Silva, Villeneuve-d'Ascq ... 105869
Da Silva, Geneviève, Lazenay ... 069212
Da Silva, Isabel Lopes, Lisboa . 084868
Da Vera, Porto Alegre 100971
Da Vinci, Bucureşti 114133
Da Vinci Art Alliance,
Philadelphia 060458, 123365
Da Vinci Galería de Arte, Bogotá 102294
Da Vinci-Museum, Tongerlo 004075
Da Vinci Restauro, Madrid 133322
daadgalerie, Berlin 017098
Da'an Museum, Da'an 007711
DAAP Galleries, University of
Cincinnati, Cincinnati 047743
Daatselaar & Godhelp,
Amsterdam 082723
Daatselaar & Godhelp, Utrecht . 083637
Al-Dabagh, Jeddah 085301
Al-Dabbas, Mohammed Zuheir,
Damascus 088032
Al-Dabbas, Tawfiq, Damascus .. 088033
Dabbene Argenteria, Milano ... 080513
Dabbeni, Felice & Angela,
Lugano 116454, 138066
Dabdo, Croydon 117653
Dabek, Yolande, Le Mesnil-
Esnard 136826
Dablanc, Jack, Damville 068269
Dablemont, Daniel, Bruay-la-
Buissière 067519
Daborn, Patricia, Karrinyup ... 127623
Dabren, Lavington 099224
Dąbrowska, Ada, Wrocław 132913
Dacapo Arte Clasico, Alicante .. 085696
D'Acci, Antonio & Girgio, Torino 132126
Dach, Strasbourg 073583
Dacić, Ingrid, Tübingen 108785
Daco Verlag, Stuttgart 137518
D'Aco, Giuseppe, Palermo 131694
Dacobra, Tanilba Bay 139926
DaCoGos Leseratten,
Ganderkesee 141819
Dacorum Heritage, Berkhamsted 043571
Dacosta Noble, Nelly, Bordeaux . 067321
Dacotah Prairie Museum and Lamont
Art Gallery, Aberdeen 046299
Dacou Aboriginal Gallery,
Rosewater 099654
Dactyl Foundation, New York .. 051371
Dada, Beograd 085334
D'Adamo & Woltz, Seattle 124513
Dadaocheng Art Museum, Taipei 042512
Dade Battlefield, Bushnell 047333
Dade, Clifford & Roger, Thames
Ditton 091595
Dadian Gallery, Wesley Theological
Seminary Center for the Arts and
Religion, Washington 054100
Dado, Seoul 111628
Dad's Old Book Store, Nashville 145293
Daegu National Museum, Daegu 029911
Daegu National University of Education

Museum, Daegu 029912
Daegwallyeong Museum,
Gangneung 029926
Dähler, Niklaus, Sulgen 087835, 133936
Dähn, Dresden 075409
Daehn, H., Timmendorfer Strand 078680
Daehnhardt, Rainer, Lisboa 084869
Daejeon Museum of Art (DMA),
Daejeon 029918
Dändorf, Dierhagen, Ostseebad . 075352
Daensmuseum en Archief van de
Vlaamse Sociale Strijd, Aalst .. 003256
Daeppen, Guillaume, Basel 116060
Daewoo Art Gallery, Ha Noi 125070
DAF Automobiel Museum,
Eindhoven 032282
Daffodil, Skerries 109689
Dag en Zonne, Brugge 063268
Daga, Antonio, Cagliari 109922
Dagali Museum, Geilo 033616
Dagbladmuseum Abraham
Verhoevenhuis, Hoboken 003701
Dagda, Almere 112115
Dagestanskij Gosudarstvennyj
Objedinennyj Istoriko-architekturnyj
Muzej, Machačkala 037208
Dagestanskij Muzej Izobrazitelnych
Iskusstv, Machačkala 037209
Daggett, London 090061, 118280,
134643
Daglen, Toledo 136408
Dagmar, North Melbourne 099470
Dagommer, Chantal, Saint-Ouen 073063
Dagon Collection, Archaeological
Museum of Grain Handling in Israel,
Haifa 025049
Dagoneau, Gérard, Paris 128973
D'Agostino, Renato, Ospedaletti 080953
Dagot, Sylvie, Montluçon 125646
Dagu Paotai Yi Zhi Museum,
Tianjin 008321
Daguenet, Bernard, Niherne ... 070681
Daguenet, Berthe, Versailles ... 074048
Dagues-Fouché, Françoise,
Montpellier 104275
Dahe Cun Yi Zhi Guan,
Zhengzhou 008536
Dahesh Museum of Art, New
York 051372
Daheshist Publishing, New York 138389
Daheshite Museum, Beirut 030264
Dahiez, Dr. Anne, Zürich 116766
Dahl, Haugesund 113160
Dahl, København 065883
Dahl Arts Center, Rapid City ... 052469
Dahl Boisen, Søren, Odense ... 140636
Dahl, Volker, Ulmet 078759
Dahle, Halden 084243
Dahlem, Rita, Bispingen 106506
Dahlgård, Kristian, København . 065884
Dahling, Oscar, Croydon 089017
Dahlman, Mary, San Antonio .. 124023
Dahlonega Courthouse Gold Museum,
Dahlonega 048104
Dahm, Wolfgang, Mannheim ... 107963
Dahmann, Prof. Dr. Werner,
Berlin 137058
Dahmen, Xanten 079127
Dahms, Frank, Annaberg-
Buchholz 141410
Dai County Museum, Daixian .. 007712
Dai Ichi, New York 122517
Dai Nippon Printing, Tokyo 137776
Dai Token Ichi, Tokyo 098227
Daigaku, Osaka 111350
Daigaku, Tokyo 111414
Daigaku Bijutsukan, Kyoto-shiritsu
Bijutsu Daigaku, Kyoto 028920
Daigo Fukuryu Maru Exhibition Hall,
Tokyo 029496
Daigoji Reihokan, Kyoto 028921
Daihozoden Homotsukan, Ikoma 028923
Daiichi Bijutsukan, Nagoya 029112
Daikaku-ji Temple, Kyoto 028922

Daikan Yashiki oyo Setagaya Rekishi
Hakubutsukan, Tokyo 029497
Dailey, William, Los Angeles ... 138347,
145223
Dailininkės Veronikos Šleivytės
Paveikslų Galerija, Kupiškio
Etnografijos Muziejus,
Viktariskes 030507
Dailininko Prano Domšaičio Galerija ir
Kultūros Centras, Lietuvos Dailės
Muziejus, Klaipėda 030386
Daily-Bůl, La Louvière 136657
Daily Planet, Seattle 097423
Daimaru Museum Kobe, Kobe .. 028838
Daimaru Museum Kyoto, Kyoto . 028923
Daimaru Museum Sapporo,
Sapporo 029329
Daimaru Museum Shinsaibashi Osaka,
Osaka 029238
Daimaru Museum Tokyo, Tokyo . 029498
Daimaru Museum Umeda Osaka,
Osaka 029239
Daimarufujii, Sapporo 111375
Daimler Contemporary, Berlin .. 017099
Daimler Geburtshaus, Schorndorf 021725
Daimonion, Warszawa 143520
Daimyo Tokei Hakubutsukan,
Tokyo 029499
Dairy House Antiques, Semley . 091262
Dais, Charkiv 117106
Daishodo, Kyoto 082010
Daiso, Hiroshima 081976
Daisy Airgun Museum, Rogers . 052661
Daiter, Stephen, Chicago 120103
Dajmond Klassik, Sankt-
Peterburg 085240
Dajmont Art, Moskva 114211
Dakot, Oakland 123248
Dakota, Lausanne 087568
Dakota County Museum, South Saint
Paul 053416
Dakota Dinosaur Museum,
Dickinson 048325
Dakota Science Center, Grand
Forks 049174
Dakota Territorial Museum,
Yankton 054569
Dakotaland Museum, Huron ... 049641
Dakshina Chitra Museum,
Muttukadu 024412
Dal Bosco, Paolo, Rovereto ... 110878
Dal-Paso Museum, Lamesa 050107
Dala Antik, Krylbo 086794
Dala Antiktjänst, Stockholm ... 086985,
133569
Dalälvarnas Flottningsmuseum,
Sandviken 040937
Dalahästmuseum, Dala-Järna . 040588
Dalancon, Jean, Versailles 074049
Dalane Folkemuseum, Egersund 033575
Dalarnas Idrottmuseum, Falun .. 040619
Dalarnas Indianmuseum, Dala
Husby 040587
Dalarnas Museum, Falun 040620
Dalbeattie Museum, Dalbeattie . 044061
Dalbry Antiques, Canterbury ... 127578
Dalby Regional Gallery, Dalby .. 001011
Dalcici, Timişoara 085131
The Dale Gallery, Stanhope ... 119139
Dale House Antique, Moreton-in-
Marsh 090779
Dale, John, London 090072
Dalemain Historic House, Penrith 045477
Dalénium – Gustaf Daléns
Upptäckarcentrum, Stenstorp . 040987
Dalénmuseet, Stenstorp 040988
Dales Countryside Museum,
Hawes 044541
Daleside Antiques, Markington . 090718
Dalet, Lucienne, Soucirac 073542
Daleus, Miami 094725
Daley, Minneapolis 121978
Daley, Washington 124887

D'Alfama, Rosa, Lisboa 084870, 132933
Dalfling Gallery, Inverurie 118023
D'Alfonso, Maria Felicia, Perugia 081141
Dalgaard, Skjern 066125
Dalgaard, Jørgen L., København 065885
Dalgården Antikt och Gammalt,
Järfälla 086754
Dalgarven Mill, Museum of Ayrshire
Country Life and Costume,
Kilwinning 044701
Dalhousie Art Gallery, Halifax .. 005948
Dalhousie Station, Calgary 101227
Dali, Den Haag 083073
Dalí, Montevideo 091995
Dalí – Espace Montmartre, Paris 014646
Dali Museum, Dali 007713
Dalian Calligraphy and Painting Gallery,
Dalian 102019
Dalian Gallery, Dalian 007715
Dalian Natural History Museum,
Dalian 007716
Dalibor, Praha 065487
Dalin, Chicago 120104, 120105
Dalizhou Museum, Dali 007714
Dalkeith Arts Centre, Dalkeith .. 044062
Dalkeith Auctions, Bournemouth 126828
Dalkey Arts, Dalkey 109560
Dalla Venezia, Venezia 132236
Dall'Aglio, Mario, Bolzano 109874
Dallai, Genova 142695
Dallam-Hartley XIT Museum,
Dalhart 048106
Dallas Art and Design, Dallas .. 120583
Dallas Arts Studio, Dallas 120584
Dallas Auction Gallery, Dallas .. 127192
The Dallas Center for Contemporary
Art, Dallas 048112
Dallas Historical Society Museum,
Dallas 048113
Dallas Holocaust Museum, Dallas 048114
Dallas Museum of Art, Dallas .. 048115
Dallas Museum of Natural History,
Dallas 048116
Dallas Rare Coins, Dallas 093108
Dallendörfer, H. & M., Koppl ... 127790,
127791
Dalléra, Catherine, Banon 066932,
103324
The Dalles Art Association, The
Dalles 060459
The Dalles Art Association Gallery, The
Dalles 048128
Dallmann, Peter, Bad Feilnbach . 129408
Dall'Occa, Marco, Bologna 142630
Dall'Olio, Stefano, Sassuolo ... 081617
Dalmas, Richard, Nice 070590
Dalmau, Carles, Girona 137978
Dalmeny House, South
Queensferry 045816
Dalnavert Museum, Winnipeg .. 007439
Dalnevostočnyj Chudožestvennyj
Muzej, Chabarovsk 036951
Dalnevostočnyj Institut Iskusstva,
Vladivostok 056313
Dalpiaz, Renato, Roma 081314
Dalsgaard, Højslev .. 065771, 125445
Dalsland Antikvariat, Ed 143826
Dalslands konstmuseum,
Åsensbruk 040548
Dalta & Piccinini, Genova 131303
Dalton Centre Museum, Tignish
Cultural Centre Museum,
Tignish 007149
Dalton Defenders Museum,
Coffeyville 047858
Dalton Gallery, Decatur 048198
Dalton-Somaré, Milano 110295
Dalton, Midge, Palma de
Mallorca 095332
Dalva Bros., New York 095332
Dalvi's Art Institute, Kolhapur .. 055807
Daly House Museum, Brandon . 005464
Daly, Raymond, Geashill 079628, 130871
Dalype, Oslo 113198

Dam Mill Antiques, Codsall 088944
Dam Site Museum, Mangla 034139
Dam, Christian, København . . . 102875
Dam, Christian, Oslo 113199
Damais, Jean, Biarritz 067195
Damasco, Gino, Egnach 087404
Damascus Palace, Dubai 088176
Damask Asia, Bingley 117366
Damasquine, Bruxelles 100587
Damasse, Charles, Yerres 074306
Dambourges, Québec 064532
Dambra Laboratorio di Restauro,
  Firenze 131156
D'Ambra, Raffaele, Napoli 110521
Dambron, Arlette, Schiltigheim . . 073461
D'Ambrosio, Fabio, Roma 131888
Dambry, Antibes-Juan-les-Pins . 066670
La Dame de Malacca, Kuala
  Lumpur 082401
Dame du Castellet, Le Castellet 103933
Dame e Cavalieri, Milano 080514
D'Amelio & Terras, New York . . 122518
Damerla Rama Rao Art School,
  Rajahmundry 055830
Damerla Rama Rao Memorial Art
  Gallery, Rajahmundry 024456
Les Dames, Paris 071216
Damgaard-Lauritsen, Odense . . 066033
Damiano, Agostino, Torino 081692
Damiano, Lina, Cagnes-sur-Mer 067589
D'Amico Verena, Lamone-
  Cadempino 087552, 133789
Damidot, Patrick, Dijon 068323
Damiean, Pierre-Jean, Fabrègues 128599
Damien, B., Miami 121767
Damien, Jean-Philippe, Sancerre 073399
Damien, S., Mons 063793
Damioli, Sandro, Basel 087230
Damjanich János Múzeum,
  Szolnok 023977
Damm, Hamburg 076210
Damm, Oslo 143401
Damm, G., København 065886
Damm, Helmut, Gescher 075989
Damm, Henrik & Jeanette, Ringe 066064
Dammann, Detlef, Zwickau . . . 109009,
                                137570
Dammron, Claude, Genève 087462
D'Amodio, Antimo, Napoli 080856,
                                080857
Damokles Antiquariaat, Den
  Haag 143111
Damon, Poitiers 105364
Damon, Yvette, Saint-Martin-de-
  Ré 072937, 105556
Damongeot, Marie-Claude,
  Sampans 073394
D'Amore, Valeria e Andrea,
  Livorno 110177
Damoutte, Dominique, Vannes
  (Morbihan) 073970
Damoutte, Gérard, Vannes
  (Morbihan) 073971
Damp Frog Productions, New
  York 122519
Dampffährschiff "Stralsund" Wolgast,
  Wolgast 022751
Dampfkornbranntweinbrennerei-
  Museum, Wildeshausen 022652
Dampflok-Museum, Hermeskeil . 019211
Dampfmaschinenhaus-Moschee,
  Stiftung Preußische Schlösser
  und Gärten Berlin-Brandenburg,
  Potsdam 021181
Damß, E., Heidelberg 076438
Dan, Los Angeles 094305
Dan Andersson Museum,
  Ludvika 040800
Dan City Museum, Dancheng . . 007722
Dan-Cuthbert Sally, Rose Bay . . 099651
Dan Estuary Museum, Shiyan . . 008278
Dan Flavin Art Foundation Temporary
  Gallery, Dia Art Foundation, New
  York 051373

Dan Flavin Art Institute,
  Bridgehampton 047197
Dan Kook University Museum,
  Seoul 029988
Dan O'Laurie Canyon Country
  Museum, Moab 050972
Dan Quayle Center and Museum,
  Huntington 049616
Dan Winters House – Ancestral Home,
  Loughgall 045156
Dan-yr-Ogof Showcaves Museum,
  Abercrave 043346
Al-Dana, Manama 100423
Dana Kunsthus, Gudhjem 065713
The Dana-Thomas House,
  Springfield 053450
Danah, Manama 063139
De Danann Antique Centre,
  Dorchester 089070, 134330
Danant, Jean-Louis, Paris 071217
Danart, Bucureşti 085090, 132996
Dana's Art and Antiques, Braşov 085073
Danbury Antiques, Danbury . . . 089027,
                                134313
Danbury Museum, Danbury 048136
Dance Collection – Stiftung John
  Neumeier, Hamburg 018978
Dance Cottage Museum,
  Ravensthorpe 001444
Dancing Art, Nashville 122078
Dancing Geese Gallery, Tokyo . . 111415
Dancing Sheep, Cleveland 120414
Dandelion, Philadelphia 096097
Dando, Andrew, Bradford-on-
  Avon 088561
Dandois, Ariane, Paris . 071218, 071219
Dandois, Ariane, Saint-Ouen . . . 105565
Dandong Museum on the War to
  resist U.S. Aggression and Aid Korea,
  Dandong 007723
D'Andrea, Alberto, Roma 081315
D'Andrea, Antonietta, Borgo Val Di
  Taro 079906
Dane G. Hansen Memorial Museum,
  Logan 050356
Dane, Thomas, London 118281
Danechild, E.E., San Francisco . 124265
Danemberg, Arnaldo, Rio de
  Janeiro 064049
Danenberg, Valérie, Paris 071220
Danese, New York 122520
Danesi, Giacomo, Brescia 131031
Danet, Songeons 073538
Danevirke Museum, Sydslesvigsk
  Forening, Dannewerk 017814
Danford Collection of West African Art
  and Artefacts, Birmingham . . . 043619
Danforth Gallery, Portland 052279
Danforth Museum of Art,
  Framingham 048918
Dang Dynasty, New York 095333
Dang, Geomok, Daegu 082254
Dang, Samkook, Seoul 082258
Dang, Songa, Daegu 111582
Dangel, Erika, Murrhardt 108231
D'Angelo, Antonio, Roma 081316
D'Angelo, Gaetano, Napoli 131612
Danger, B., Vöhringen 078810
L'été Dangereux, Nîmes 141008
Dangol, Kathmandu 112067
Dangoor, Daniella, London 118282
Dangyang Museum, Dangyang . . 007724
Danhof, J., Brackenheim 075113
Danicourt, Benoît, Paris 071221
Danicourt, Georges, Toulouse . . 073758
Daniel, New York 095334
Daniel Boone Homestead,
  Birdsboro 047000
Daniel et Vincent, Paris 071223
Daniel W. Tantoco jr. Collection,
  Manila 034529
Daniel, Christiane, Paris 071222
Daniel, Christiane, Saint-Hilaire-du-

Harcouët 072825, 129233
Daniel, Michel, Saint-Nazaire (Loire-
  Atlantique) 072989
Danieli, Parma 081106
Danièlle, Zeist 083710
Daniels, Chicago 092699
Daniels County Museum and Pioneer
  Town, Scobey 053203
Daniel's Shop, Strovolos 065336
Danielsen, H., København 065887
Danielson-Kalmarin Huvila,
  Vääksy 011475
Danielsson, Anders, Ekshärad . 126574
Danihelka, Marita, Dorsten 106693
Daninos, Andrea, Firenze 080129
Daninos, Andrea, Milano 080515
Danish Canadian National Museum,
  Dickson 005705
Danish Country, Boston 092463
The Danish Cultural Institute,
  Edinburgh 044235
Danish Immigrant Museum, Elk
  Horn 048530
Danish Silver, København 065888
Dank Je de Koekoek, Utrecht . . 083638
Danker, H., Dublin 079541
Dankl, Ernst, Salzburg 127844
Danko, Dr. Waldemar, Lieboch . . 062710
Dankwardt, M., Leer 077041
Dánlann Yawl Art Gallery, Achill 109523
Danmar, Stockholm 133570
Danmarks Bogmuseum,
  København 010169
Danmarks Cykelmuseum,
  Ålestrup 010003
Danmarks Fotomuseum, Herning 010113
Danmarks Jernbanemuseum,
  Odense 010255
Danmarks Keramikmuseum,
  Middelfart 010231
Danmarks Kloster Museum, Ry . . 010298
Danmarks Kunstbibliotek,
  København 010170
Danmarks Mediemuseum,
  Odense 010256
Danmarks Tekniske Museum,
  Helsingør 010106
Dann Antiques, Melksham 090739
D'Anna Antiquariato, Roma 081317,
                                131889
Dannebrog, København 065889
Dannenberg, Berlin 125873
Dannenberg & Ryan, North
  Adelaide 061980
Danner, F., Appenweier 074422
Danowski, Marek, Zürich 087950,
                        138107,   144171
Dan's, Houston 135599
Dan's Antiques, Tucson 097650
Dan's Custom Woodworking,
  Houston 093656
Dans le Temps, Brugny-
  Vaudancourt 067521
Dans Mon Monde, Nancy 104318
Dan's Place, Albuquerque 092050
Dansk Billedhuggersamfund,
  Horsens 058545
Dansk Bogservice, København . 140608
Dansk Galerie, Paris 071224
Dansk ICOM, København 058546
Dansk Jagt- og Skovbrugsmuseum,
  Hørsholm 010130
Dansk Kunst og Antikvitetshandler
  Union, København 058547
Dansk Landbrugsmuseum,
  Auning 010016
Dansk Møbelkunst, København . 065890
Dansk-Retro, Nykøbing Mors . . 066015
Dansk Veteranflysamling, Skjern 010312
Den Danske
  Antikvarboghandlerforening,
  Frederiksberg 058548
Det Danske Filminstitut,
  København 010171

Det Danske Kunstindustrimuseeum,
  København 010172
Dansmuseet, Stockholm 040996
Danson Gallery, Orpington 118856
Dantan, Buenos Aires 060893
Dantas, Carlos, Noailles 070719
Dante, Cortina d'Ampezzo 080056
Dante, Padova 110560
Dante, Philadelphia 136082
Le Dantec, Bertrand, Paris 128974
Dantica, Amsterdam 112183
Dantikvariát, Praha 140516
Dantiques, Helsingør 065739
Danubiana, Meulensteen Art Museum,
  Bratislava 038300
Danusha Fine Arts, London 118283
Danvers Historical Society Exhibition,
  Danvers 048140
Danville Museum of Fine Arts and
  History, Danville 048147
Danville Science Center, Danville 048148
Dany Antic, Bellot 067080
Dany Art Déco, Saint-Ouen . . . 073064
Dany Keller Galerie, München . . 108063
Dany, Laure, Paris 104666
Daynsz, Magda, Paris 104667
Danzi, Angiolo & Cutri, Alessandro,
  Roma 081318
Danzinger, New York 122521
Daolio, Mario, Bologna 130949
Dapa International, Timişoara . . 085132
Dapa, Gean & Merry, Los
  Angeles 094306
Dapeng Gucheng Museum,
  Shenzhen 008261
Daphne du Maurier's Smugglers's at
  Jamaica Inn, Bolventor 043672
Daphne's Antiques, Penzance . . 090964
Dapple-Gray, Saint Louis 096682
Dappon, Cluj-Napoca 143620
Dapsens-Bauve, Françoise,
  Coulommiers 125563
Dapsens, Ludovic, Reims 125784
Dapu Museum, Dapu 007725
Daqian, Beijing 101947
Daqing Dumeng Museum,
  Daqing 007726
Dar Al-Bareh Gallery, Manama . 063140,
                        100424,   100425
Dar Al-Ghad Gallery, Manama . . 100426,
                                136617
Dar Al-Maaref, Cairo 136785
Dar-es-Salaam National Museum, Dar-
  es-Salaam 042630
Dar Galereja Naivnogo Iskusstva,
  Moskva 114212
Dar-Hervey Ely House, Rochester 052616
Dar Maâkal Az-Zaïm, Musée de
  l'Histoire du Mouvement National,
  Tunis 042781
Dar Museum First Ladies of Texas
  Historic Costumes Collection,
  Denton 048246
Al-Dara, Manama 100427
Daragnes, Pierre, Biarritz 067196
Daraphirom Royal Mansion
  Museum, Chulalongkorn University,
  Bangkok 042655
Daras & Gilbert, Düsseldorf . . . 141716
Darbarhall Museum, Junagadh . 024333
Darbre, Florence, Nyon 133877
Darby Houses, Rosehill and Dale
  House, Coalbrookdale 043974
Darchis, Fabian, Nancy 070418
Darcy's, Lucknow 061755
Dard, Jean, Orléans 125669
Dardamani, Anna, Thessaloniki . 079208
Dardengo, Andrea, Padova 131661
Dardenne, Bruxelles 063386, 063387
Dardo, Santiago de Compostela 139048
Dare-Dare, Centre de diffusion d'art
  multidisciplinaire de Montréal,
  Montréal 101375
Dare, I., Sydney 099776

Darga & Lansberg, Paris 104668
Dargate, Pittsburgh 127401
Dargate Auction Gallery,
  Pittsburgh 127402
Dargaville Museum, Dargaville . 033184
Daria, Hyderabad 109359
Daries, Valencia 133460
Darilo, Novo Mesto 085623
Darious, Atlanta 119557
Dárius, Budapest 079249
Darius, Galaţi 085121
Darius, New York 095335
Darjeeling, Manama 100428
Darkhorse Group Military Aviation
  Museum, Long Beach 050369
Darkoum, Lostorf 116450
Darl At Tifl Museum, Jerusalem 034164
Darlay, Olivier, La Frette-sur-
  Seine 069039
Darlie, Jette, København 065891
Darling, Stockholm 115851
Darling Down's Antique and Art
  Centre, Toowoomba 062327
Darling Street Books, Rozelle . 139899
Darling, Lindy, Yallah 099928
Darlingford School Heritage Museum,
  Darlingford 005678
Darlington Arts Centre and Myles
  Meehan Gallery, Darlington . 044065
Darlington Raceway Stock Car
  Museum, Darlington 048152
Darlington Railway Centre and
  Museum, Darlington 044066
Darl'ys Antiquités, Charleville-
  Mézières 067845
Darms, Gieri, Ilanz ... 087536, 133776
Darmstädter Münzhandlung,
  Darmstadt 075301
Darnall's Chance House Museum,
  Upper Marlboro 053904
Darnas, Daniel, Couzon-au-Mont-
  d'Or 068215
Darnault, Véronique, Paris 071225
Darni, Ulla, New York 122522
Darnley Fine Art, London 118284
Daros Exhibitions, Zürich 042192
Daros-Latinamerica AG, Zürich . 042193
D'Arpino, Rosa Bianca, Roma .. 081319
Darquenne, Jean, Krombach ... 142036
Darracq, Bray-sur-Seine 067452
Darre, Cosette, Villepinte 074173
Darreau, Norbert, Mâcon 140934
Darrell, Dean, Washington 097831
Darrow, George, New York 095336
Darru, Figtree 099017
D'Ars, Milano 059431, 138814
La Darsena, Modena 142749
Darshan, Ottawa 101469
Darß-Museum, Prerow,
  Ostseebad 021214
Darßer Arche, Wieck am Darß . 022620
Dart, Gdańsk 084502, 143450
Dart Antiques & Restoring,
  Gdańsk 132814
D'Art en Déco, Rodez 072510
The D'Art Gallery, Dartmouth .. 117663
D'Arte e Di Vetro, Bergamo ... 109792
D'Arte Mev, Gateshead 117818
Dartevelle, P., Bruxelles 063388
Dartford Borough Museum,
  Dartford 044068
Dartigoeyte, Richard, Castelnau-de-
  Médoc 067707
Dartington Crystal Glass Museum,
  Torrington 045990
Dartmoor Bookshop, Ashburton . 144223
Dartmoor District Coach House
  Museum, Dartmoor 001018
Dartmouth Castle, Dartmouth .. 044069
Dartmouth Fine Art, Totnes 119247
Dartmouth Heritage Museum,
  Dartmouth 005680
Dartmouth Museum, Dartmouth 044070

Dartmouth Street Gallery,
  Albuquerque 119467
Darvas József Irodalmi Emlékház és
  Kutatóépület, Szántó Kovács János
  Területi Múzeum, Orosháza ... 023817
Darvinovskij Muzej, Moskva ... 037250
Darwall, Hunnebostrand 133540
Darwin, Philadelphia 096098
Darwin Gallery, Darwin 088949
Darwisch, Rashad, Graz 062584
Dary's, Paris 071226
Darzalas, Varna 101188
Das Damianstor, Kunstverein Bruchsal
  e.V., Bruchsal 058816, 106600
DASA, Deutsche
  Arbeitsschutzausstellung der
  Bundesanstalt für Arbeitsschutz und
  Arbeitsmedizin, Dortmund 017954
Dasar, Forfar 117804
D'Ascanio, Anna, Roma 110750
D'Ascenzi, Gian Carlo, Roma ... 110751
Dåsettunet, Lyngdal 033759
Daspelgruber & Lötsch, Steyr-
  Gleink 062854
Dasse, Wilfried, Kiel 076774
Dassin, West Hollywood 077885
Daßler, Georg, Stuttgart 078599
Dassonvalle, Honfleur . 103823, 103824
Dastig, Marco, Dahme 075292
Dat Antiek, Den Haag 132472
Dat Lütt Museum, Vielank 022327
Datian Museum, Datian 007727
Datina, Prato 081169
Datong Museum, Datong 007728
Datrino, Marco, Torre Canavese 081775
D'Attili, Maria Novella, Roma ... 081320
Datu Fine Arts, Wishaw 119370
Datura, Chicago 120106
Daubet, Francis, Floirac 068548
Daubisse, Isabelle, Ribérac 072454
Daubitz & Son, Dallas 135498
Daubrée, Valérie, Paris 128975
Daudet, Alain, Toulouse 073759, 105733
Dauerausstellung im Stadtarchiv,
  Heilbronn 019159
Dauerausstellung Württembergische
  Schwarzwaldbahn, Althengstett 016546
Daugaard, Peter, Stege 140661
Daugava, Rīga 111758
Daugavas Muzejs, Salaspils ... 030230
Daugavpils Novadpētniecības un
  Mākslas Muzejs, Daugavpils . 030112
Daugavpils Universitātes Muzejs,
  Daugavpils 030113
Daugé, Claude, Richelieu 072457
Dauger, Yves, Paris 071227
Daughters of the American Revolution
  Museum, Washington 054101
Daughters of Utah Pioneers Pioneer
  Memorial Museum, Salt Lake
  City 052914
Daugyvenės Kultūros Istorijos
  Muziejus-draustinis, Radviliškis 030450
Daulaus, Nice 070591
Daum Museum of Contemporary Art,
  Sedalia 053258
Daumarie, Jacques, Amphion-les-
  Bains 103231
Daumas, Jean-Paul, Aigues-
  Mortes 066437
Daune, Alain, Paris 104669
Dauphin, Singapore 114512
Le Dauphin Bernard, Paris 071228
Dauphin-Descours, New York ... 095337
Dauphin Publishing, Oxford 138272
Dauphin, Cedric le, Bruxelles ... 063389
Dauphine Street Books, New
  Orleans 145306
Dauphinoise, Le Bouchage ... 103927
Daura Gallery, Lynchburg College,
  Lynchburg 050530
Dauriac, Jean-Louis, Dieupentale 068303
Dautrait, Daniel, Clermont-
  Ferrand 128561

Dauvillier, Middelburg 112591
Dauwe, Christel, Antwerpen ... 063201
Davaco, Doornspijk 137853
Davaki, Eleni, Iráklion 079196
Davao Museum, Davao 034476
Dave & Bill's Trading Post, Saint
  Louis 096683
Dave Antique, Paris 071229
Daveau, Jacqueline, Paris 071230
Davenham Antique Centre,
  Davenham 089035
Davenport House Museum,
  Savannah 053174
Davenport Museum of Art,
  Davenport 048154
Davenport, Karl & Claire, Enfield 130870
Daventry Museum, Daventry ... 044073
Daverio, Chrisgeni, Nice 128855
Daves, Tucson 136417
Dave's Antiques, Louisville 094536
Dave's World, Houston 093657
Davi, Timişoara 085133
David, Los Angeles 094307
David, Singapore 085393
David & Charles, Newton Abbot 138268
David & David, Philadelphia 123366
David & Laure Soustiel, Marie-
  Christine, Paris 071232
David Aaron Ancient Arts,
  London 090063, 134644
David Bradford House,
  Washington 054184
David Conklin Farmhouse, Huntington
  Historical Society, Huntington . 049619
David Crawford House,
  Newburgh 051512
The David Davis Mansion,
  Bloomington 047042
David de Sauzéa, Rambert, Nice 070592
David et ses Filles, Paris 071233
David L. Mason Children's Art
  Museum, Dunedin 048394
David Livingstone Centre,
  Blantyre 043657
David Palombo Museum,
  Jerusalem 025074
David S.r.l., Cortina d'Ampezzo . 080057
David S.r.l., Riccione 081185
David Stefánsson Memorial Museum,
  Akureyri 024062
David Strawn Art Gallery, Art
  Association of Jacksonville,
  Jacksonville 049778
David Winton Bell Gallery, Brown
  University, Providence 052381
David, Bernadette, Langoiran .. 069168
David, Charles, Providence 123700
David, Francine Marie,
  Beatenberg 116111
David, G., Cambridge 144296
David, Jean-Marc, La Trimouille 069108
David, Marie-Christine, Paris ... 071231
David, Philippe, Limey-
  Remenauville 069512
Davididi, Farmington 135554
Davidov, Lina, Paris 104670
Davidowski, F., Almelo 082626
David's, Brampton,
  Cambridgeshire 088568
David's, Miami 121768
David's, New Orleans 095085
David's Antiques, Chicago 092700
David's Art, Memphis 121687
Davids Book Exchange, Brighton 144265
David's Gallery, Pittsburgh 096275
Davids Samling, København ... 010173
David's Timeless Treasures,
  Toorak 062323
Davidsfonds, Leuven 136658
Davidson, Calgary 128147
Davidson, London 090064
Davidson, Newcastle-upon-Tyne 090824
Davidson, Saint-Ouen 073580

Davidson, Salt Lake City 096889, 136238
Davidson, Seattle 097424, 124514
Davidson County Historical Museum,
  Lexington 050259
Davidson County Museum of Art,
  Lexington 050260
Davidson, Greg, Seattle 097425
Davidson, Maxwell, New York .. 122523
Davie, Tickhill 119232
Davies & Tooth, London 118285
Davies Antiques, Ormond 062033
Davies, Daryl, Ringwood 118978, 138280
Davies & Son, Edmund, Whalley 091819,
                                135219
Davies, Hugh, London 090065
Davies, Jeffrey, San Francisco . 097214
Davies, John, Stow-on-the-Wold 091467,
                      119158,   135095
Davies, Phillip, Swansea 119193, 135110
Davila Camargo, Antonio Maria,
  Bogotá 065200, 128266, 140448
Davinci, Vancouver 064813
Davioud, André, Lille 128677
Davioud, Georges, Lille 069486
Davis, Austin 119659
Davis, Dallas 135499
Davis, Houston 093658
Davis, Long Beach 094198
Davis & Co., Los Angeles 094308
Davis, New Orleans 122152
Davis, Perry 127387
Davis, Worcester 138481
Davis & Davis, London 134645
Davis & Gardner, New York 095338
Davis & Langdale Company, New
  York 122524
Davis & Leron Galries, Brooklyn 119939
Davis Art Center, Bountiful 047147
Davis Art Center, Davis 048160
Davis Art Gallery, Columbia ... 047899
Davis Astrof Gallery, Montréal . 101376
Davis Dominguez, Tucson 124726
Davis Klemm Gallery, Frankfurt am
  Main 106980
Davis Mardine, Los Angeles ... 121469
Davis Museum, Wellesley College,
  Wellesley Hills 054266
Davis Place Antiques, Saint Louis 096684
Davis, A.B., London 090066
Davis, Barbara, Houston 121026
Davis, Caleb, Boston 135362
Davis, E., Nashville 122079
Davis, Gerald, Dublin 109581
Davis, Kenneth, London 090067
Davis, Leslie, Milwaukee 121908
Davis, Rachel, Cleveland 120415, 127183
Davis, Reginald, Oxford 090931, 134916
Davis, Roger A., Great Bookham 089364,
                                134420
Davis, Sharyl, San Antonio 096967
Davishire Interiors, Nashville ... 094990
Davison Art Center, Wesleyan
  University, Middletown 050862
Davison Gallery, Holywood 117990
Davoud-Salmani, Bruxelles 063390
Davoy, Coufouleux 068177
Davril, Bogotá 102295
Davut, Mizrahi, Wien 062923
Davy, Anthony Kent, Hong Kong 102068
Davy, Michel, Saumur 073434
Dawa, Kathmandu 112068
Dawes County Historical Society
  Museum, Chadron 047509
Dawes, Philip, Royston 091150, 134982
Dawes, Valentyne, Ludlow 090647,
                                134821
Dawlish Museum, Dawlish 044074
Dawne, Morgan, Vancouver ... 101771
Dawo, Udo, Saarbrücken 078319,
                                126109
Dawson, Jacksonville 121240
Dawson, Rotorua 112999
Dawson, Vancouver 128224
Dawson City Museum, Dawson

City 005686
Dawson County Historical Museum,
  Lexington 050258
Dawson Creek Art Gallery, Dawson
  Creek 005688
Dawson Creek Station Museum,
  Dawson Creek 005689
Dawson Folk Museum, Theodore 001575
Dawson of Stamford, Stamford . 091404
Dawson Springs Museum and Art
  Center, Dawson Springs 048167
Dawson, Douglas, Chicago . . . 120107
Dawson, Michael, Los Angeles . 145224
Dawson, P., Bentley . . 088440, 134093
Dawson, Phillip, Twickenham . . 144895
Dawu Museum, Dawu 007729
Daxer & Marschall, München . . 077482
Daxin Museum, Daxin 007730
Daxing Xing Du Art Cooperative,
  Beijing 101948
Day & Saber, London . 090068, 118286
Day Antiques, Tetbury 091569
Day, Alan, Edinburgh 089169
Day, Christopher, Paddington, New
  South Wales 099518
Day, Dean, Houston 121027
Day, Edward, Toronto 101594
Day, John, Eastbourne 089146, 134353
Dayan, Paris 071234
Dayan, Hubert, Saint-Jean-Cap-
  Ferrat 105530
Dayboro Shed, Samsonvale 062185
Daye Museum, Daye 007731
Dayi Liu's Manor-House Museum,
  Chengdu 007692
Daylesford and District Historical
  Museum, Daylesford 001022
Dayna's Antiques, Jacksonville . 093983
Days Gone By Antiques,
  Kitchener 064336
Days of Grace, Budleigh
  Salterton 088695
Days of Olde Antiques, Ballarat . 061083
Daystar Collectibles, Tucson . . . 097651
Dayton Art Institute, Dayton . . . 048170
Dayton Historical Depot Society
  Museum, Dayton 048178
Dayton Visual Arts Center,
  Dayton 048171
Daytona Gallery, Tauranga 113021
Daza, Francisco, Jerez de la
  Frontera 115102
Dazaifu Tenman-gu Hōmotsuden,
  Dazaifu 028522
Daze Gone By, Gympie 061568
Dazenire, Michel, Boulazac . . . 067387
Dazhong Siguzhong Museum,
  Beijing 077616
Dazibao, Centre de Photographies
  Actuelles, Montréal 006370
Dazy, Christian, Dijon 103674
Dazy, Christian, Megève 104213
Dazzle, Helmsley 117946
DB Museum, Deutsche Bahn,
  Nürnberg 020846
DBA California Route 66 Museum,
  Victorville 053983
DBL Diffusion, Nice 070593
dbv Deutscher Buchverlag GmbH,
  Oldenburg 137458
DC Antiques, Los Angeles 094309
DC Collection, Tucson 097652
DCI Gallery, Memphis 121688
DCKT Contemporary, New York . 122525
DD Starožitnosti, Praha 065488
DDR-Motorradmuseum, Borna . . 017518
DDR Museum, Berlin 017100
DDR-Museum im Filmpalast,
  Malchow 020237
DDR-Museum Radebeul,
  Radebeul 021256
DDS Antique Restoration, New
  York 135905
De, Beograd 085335

De Amicis Point, Milano 080516
De Angelis, Antonio, Napoli . . . 080858
De Angelis, Fabio, Roma 131890
De Antoni, Gaetano, Sarcedo . . 110908
De Antoni, Maria Luigia, Padova 080980
De Artis, Zug 116878
De Benedetti, Augusto, Roma . . 110752
De Bianchi, Luciano, Verona . . . 081906
De Bortoli, Alberto, Venezia . . . 132237
De Bretts, Moonee Ponds 061877
De Capola, Mario, Napoli 080859
De Cardenas, Monica, Milano . . 110296
De Carlo Studio, Milano 110298
De Carlo, Massimo, Milano 110297
De Carolis, Massimo, Modena . . 110496,
                                    131567
De Celles, Aliénor, Marseille . . . 104166
De Chene, Gene, Los Angeles . . 145225
De Chhitiz, Kathmandu 112069
De Choses en Choses, Reims . . 072390
De Cillia, Udine 111059
De Colellis, Urbano, Torino 132127
De Crescenzo, Roma 081321
De Crescenzo, Antonio, Roma . . 081322
De Domenico, Luigi, Messina . . 080366
De Donno, Rocco, Biel 087316, 133693
De Falco, Alberto, Napoli 080860
De Fazio, Indianapolis 121176
De Ferrari, Renata, Genova 080256,
                                    131304
De Filippis, Domenica, Messina . 080367
De Filippo, Gianluca, Napoli . . . 080861
De Giovanni Argenteria, Milano . 080517
De Girardi, Stefano, Roma 131891
De Goudron, Laura, Napoli 131613
De Lancastre, Sebastião, Lisboa 084871
De Land Museum of Art, De
  Land 048182
De Lazuli, Paris 128976
De Lee, Seoul 111629
De Locht – Streekmuseum, Nationaal
  Asperge- en Champignonmuseum,
  Horst 032511
De los Viejos Colonos, Bariloche 000143
De Luca, Adele, Roma 081323
De Luca, Francesco, Belluno . . 109789
De Lullo, Maurizio, Roma 110753
De Marchi, Giuliano, Venezia . . 132238
De Martini, Cagliari . . 079979, 142658
De Martini, G. & C., Cagliari . . 079980
De Matteis, Ciro, Napoli 131614
De Matteo, Rosella, Milano 080518
De Mayo, Cannes 067635
De Mayo, Paris 071235
De Mena Art Gallery, New York . 122526
De Mets, Bruxelles 128023
De Mon Temps, Chavanay 067945
De Mores State Historic Site,
  Bismarck 047018
De Morgan Centre, London . . . 044956
De Palma, Vincenzo, Roma 131892
De Palo, Michele, Rovigo 110885
De Pasquale, Germano, Bologna 130950
De Paz, Elisabetta, Livorno . . . 080323
De Pillis, Alberto, Roma 131893
De Pottenkarie, Dranouter 003558
De Renzo, Raffaele, Napoli 080862
De Ruvo, Franco, Milano 080519,
                                    131432
De Sanctis, Sergio, Roma 081324
De Santis, Fabrizio, Roma 131894
De Santis, Giuseppe, Torino . . . 132128
De School van Toen, Historisch Archief
  & Documentatiecentrum Stedelijk
  Onderwijs, Gent 003621
De Siena, Valter, Roma 131895
De Simone, Antonino, Roma . . . 131896
De Smet Depot Museum, De
  Smet 048187
De Stefani, Antonio, Genova . . . 080257
De tout un peu, Carouge 087349
De Valera Library and Museum,
  Ennis 024920
De Velodroom, Snellegem 004032

De Venere, Maria Gabriella,
  Catania 080031
De Vera, San Francisco 124266, 124267
De Vercelli, Danilo, Modena . . . 080790
De Vere, Woodhall Spa 119386
De Witt, Hobart 061630
De Witte Raaf/ The White Raven,
  Bruxelles 138547
De, Dis, Ex, London 139170
De, Rong, Beijing . . . 101949, 101950
Dea, Sofia 101150
Dea Galeri, Tiranë 098522
deA-Publishing Pool,
  Gumpoldskirchen 136513
Deaf Smith County Museum,
  Hereford 049448
Deák Ferenc Szülóháza, Söjtör . 023885
Deák Kúria, Kehidakustány 023736
Deakin, Montréal 064412
Deakin University Art Gallery, Deakin
  University, Burwood 000917
Deakin, Karen, Sydney 062283
Deal Castle, Deal 044075
Deal Maritime and Local History
  Museum, Deal 044076
Dealer's Choice, Saint Louis . . 096685
Dealer's Choice Antiques,
  Nashville 094991, 127329
Dealers Den, New York 095339
Dean, Minneapolis 121979
Dean Castle, Kilmarnock 044696
The Dean Clough Galleries,
  Halifax 044500
Dean Forest Railway Museum,
  Lydney 045172
Dean Gallery, Edinburgh 044236
Dean Heritage Museum, Soudley 045814
De'an Museum, De'an 007734
Dean Project, New York 122527
Dean Street Antiques, Albury . . 060982
Deann, Jeffrey, Philadelphia . . . 096099
Deann, Milwaukee 121909
Dean's Antiques, Spalding 091395
Deans, Thomas, Atlanta 119558
Dearborn Historical Museum,
  McFadden Ross House,
  Dearborn 048193
Dearing, Tyrone, Woollahra 062467
Deartre, Bologna 130951
Dease, Al, San Diego 097079
Death Valley National Park Museum,
  Death Valley 048196
Deaucourt, Pascal, Marseille . . 128758
Deauville Auction, Deauville . . . 125566
Debaigts, Jacques, Paris 104671
Debao Museum, Guansi 042318
Débard, André, Périgny 072055
Debard, Philippe, Mazet-Saint-
  Voy 070021
Débarras Bretons, Lorient 069636
Débarras d'Armor, Saint-Vran . . 073335
Débarras de l'Ouest, Lorient . . . 069637
Debat, Alain, Orthez 070824
El Debate, Buenos Aires 139562
Débatisse, Jean-Pierre, Riorges 072472
Debbas, Viviane, Beirut 082280
Debbia, Roma 081325
Debbrecht, B.,
  Georgsmarienhütte 075974
Dębczyński, Artur, Białystok . . . 084487
Debden Antiques, Debden 089041
Debelleyme, Paris 104672
Debély, Michel, Neuchâtel 087702
Debenham Antiques, Debenham 089043
Debeyne, A., Kortrijk 063717
Debladis, Paris 128977
DeBolt and District Pioneer Museum,
  DeBolt 005691
Debonnaire, Philippe, Bernay-
  Vilbert 067112
Debony, Singapore 114513
Debord, Claudette, Dignac 068305
Debord, Véronique, Tauriac . . . 129274
Debosu, Tokyo 142921

Deboulle, I., Bruxelles 063391
Debras-Bical, Bruxelles 100588
Debray, M., Coignières 068071
Debreceni Irodalmi Múzeum, Déri
  Múzeum, Debrecen 023591
Debruille, Luc, Paris 071236
Dębski, Tadeusz, Warszawa . . . 113817
Deburaux, Patrick, Paris 125701
Debusmann, Antje-Kathrin,
  Solingen 130645
Debutrie, Hugues de la, Marnay 069883
Decades Antique Shop, Strovolos 065337
Decades Antiques, Canning Vale 061273
Decades Antiques, Chicago . . . 092701
Decades Collectables, Hamilton . 064317
Décades et Lustres, Cruseilles . 068252
Décalage, Paris 071237
Decamp, François, Marseille . . . 069928
Décapage de la Maye, Crécy-en-
  Ponthieu 068225
Décapage Gil, Strasbourg 129268
Décap'Services, Mordelles 070343
Decárdenas, Gijón 086018
Decarlo, Houston 121028
Decart, Rivista di Arti Decorative,
  Firenze 138815
Decatur House Museum,
  Washington 054102
Decaux, Marc, Vandrimare 073963
Au Décervelé, Caluire-et-Cuire . 140812
Deceuninck, Nathalie, Saint-
  Omer 072995
Dechamby Horowicz, Caroline, Crans-
  sur-Sierre 116233
Decillo, New York 095340
Decina, Giovanni, Napoli 080863
Deck the Walls, Atlanta 119559
Deck the Walls, Austin 119660, 119661,
                                    119662
Deck the Walls, Birmingham . . . 119794
Deck the Walls, Cincinnati 120338
Deck the Walls, Fort Worth 120861
Deck the Walls, Houston 121029
Deck the Walls, Jacksonville . . . 121241,
                                    121242
Deck the Walls, Louisville 121640,
                                    121641
Deck the Walls, Minneapolis . . . 121980
Deck the Walls, Omaha 123306
Deck the Walls, Pittsburgh 123536
Deck the Walls, Richmond 123730
Deck the Walls, San Antonio . . 124024,
                    124025,   124026
Deck the Walls, Tampa 124645
Deck the Walls, Tucson 124727
Deck the Walls, Tulsa 124800
Decker-Heftler, Sylviane de, Paris 104673
Decker-Lehwald & Lehwald, Olpe 077949
Decker, Dr. Bernhard, Frankfurt am
  Main 075777
Decker, Meyerhoff and Pinkard
  Galleries, Maryland Institute, College
  of Art, Baltimore 046542
Decker, Mike, Luxembourg 082368,
                                    132331
Decker, Rolf, Schwarzheide . . . 130634
Decker, S., Achern 074351
Deckers, Gangelt 107097
Deckert, Inge & Bernd,
  Düsseldorf 075462
Deckmann, Wolfgang, Forchheim,
  Oberfranken 075755
Deckmann, Wolfgang, München 077483
Declare, Indianapolis 093875
Déclic, Luxembourg . . 111841, 111842
Deco, Knokke-Heist 063699
Decó, Verona 081907
Deco 2 Moderne, Saint Louis . . 096686
Déco 3003, Cogolin 068064
Deco Art, Dublin 079542
Deco Art 1900–1950, Salzburg . 062789
Deco Arte, Cali 102403
Déco Chatelet, Montluçon 070267

Deco Creations, Chicago 120108
Déco-d'Antan, Ars-en-Ré 066745
Deco Days, Vancouver 064814
Déco Deligths, Asnières-sur-
Seine 066759
Deco Deluxe, New York 095341
Deco Design, Stockholm 086986
Deco Etc., New York 095342
Déco-In, Cosne-Cours-sur-Loire 068165
Deco Jewels, New York 095343
Déco Saint-Georges, Paris 071238
Déco Silence, Quimper 072343
Decodence, Christchurch 083868
Decodence, San Francisco .. 097215
Decodream, Coulsdon 088980
Decollectables, Ranfurly 084067
Decolov, Amsterdam 082724
Decolux Comercio, Fortaleza ... 100941
Decomania, Milano 080520
Deconinck, Aalst 063174
Decor, Maryland Heights 139354
Décor, Miami 094726
Decor, Roma 081326
Le Décor Antic, Bussac 067537
Décor Antic, Perpignan 072077
Decor Antichità, Milano 080521
Decor Art, Dallas 120585
Decor Art Gallery, New York . 122528,
          122529,   122530
Decor Delights, Calgary 101228
Décor et Tradition, Cavaillon 067738
Decor Expo Atlanta, Atlanta 098444
Decor Expo New York, New York 098445
Décor Mobilier, Nice 128856
Decor Splendid, New Orleans .. 095086
Decorables and Antiques, Denver 093276
La Decoradora, Alicante 085697
Décorah, Ambilly 066551
Decorantigo, Porto 085019
Décorantique, Saint-Orens-de-
Gameville 072997
Décorantiques, Toulouse 073760
Decorart, Fortaleza 063942
Decorarte, Genova 131305
Decorarte, Rio de Janeiro 064050
Decoration Homewares,
Singapore 085394
Decorativa, Amsterdam 082725
Decorative and Fine Arts,
Oostende 063815
Decorative Antiques, Bishop's
Castle 088487
Decorative Antiques, London ... 090069,
                    134646
Decorative Antiques and Textiles Fair,
London 098372
The Decorative Arts Center of Ohio,
Lancaster 050114
Decorative Arts Collection, Museum of
Decorative Painting, Wichita .. 054383
Decorative Arts Museum of Rhodos,
Rhodos 023248
Decorative Arts Society,
Woodbridge 060107
Decorative Arts Society 1850 to the
Present. Journal, Woodbridge . 139171
Decorative Arts Society Newsletter,
Washington 139355
Decorative Expressions, Atlanta . 119560
Decorator Source, Tetbury 091570
DeCordova Museum and Sculpture
Park, Lincoln 050285
Decoremos Anticuario, Cali .. 065249
Decorglaz 2000, Madrid 115177
Decorum, Buffalo 092521
Décorum, Chevanceaux 067966
Decorum, San Francisco 097216
Decorum, Toronto 064634
Decostand, Mougins 104299
Decotiques Antique Refinishing, San
Francisco 136315
Decourcelles, Langres 069175
La Découverte, Nantes 104337, 128835
Découvertes Antiquités, Paray-le-

Monial 070849
Decruyenaere, L., Gent 063634
Dedalus, Madrid 143758
Deddington Antique Centre,
Deddington 089048
Dede, Roula, Thessaloniki 079209
Dedeystere, P., Sint-Niklaas 063830
Dedham Historical Museum,
Dedham 048209
Dedianous, Gérard de, Marseille 125636
Dedinje, Beograd 085336
Dediseño, Coyoacán 138911
Dedry, F., Liège 063750
Dee, Broken Hill 098796
Dee, Raceview 139891
Dee-Jay, Saint Louis 096687
Dee Why Books, Warriewood .. 139953
Dee, Atkinson & Harrison,
Driffield 126882
Dee, Elizabeth, New York 122531
Dee's Antique, Windsor 091877
Deele, Münster 077656
Deeler, Vancouver 064815
Deelers, Vancouver 064816
Deemer, B., Louisville 121642
Deep Art Gallery, Pune 109494
Deep Cove Heritage Society, North
Vancouver 006515
Deep Sea Adventure and Diving
Museum, Weymouth 046107
Deep South Pioneer Museum,
Ogema 006541
Deep Water Shark Gallery,
Tinonee 099827
Deepe, W., Beckeln 074702
Deepgroove-Antiquariat,
Freilassing 141814
Deer Isle-Stonington Historical Society
Museum, Deer Isle 048212
Deer Valley Rock Center,
Glendale 049115
Deere Museum, Moline 050987
Deerfield, Oakland 095888
Deerfield Beach Historical Society
Museum, Deerfield Beach .. 048220
Deerstalker Antiques, Whitchurch,
Buckinghamshire ... 091828, 135223
Deeside Books, Berkhamsted .. 144242
Defachelle, Michel, Les Attaques 069400
Defence Services Museum,
Adelaide 000763
Defence Services Museum,
Yangon 031832
Défense d'Afficher, Paris 104674
Defeo, L.B., Dordrecht 083178
Defet, Hansfried, Nürnberg 108297
Deffarges, Montvert 070340
Defiance Gallery, Newtown, New
South Wales 099449
Defina, Ennio, Venezia 132239
DeFinis, Kevin J., Philadelphia . 096100
Definition Press, New York 138390
Defrawi, Cairo 066237
Defrocourt, Cathy, Bresles 067459
Degan, Bruno, Venezia 111101
Degani, Giuliano, Verona 132279
Degas Pastel Society, New
Orleans 060460
Degenaar, Amsterdam 082726
Degener, Rolf M., Meerbusch .. 142120
Degenfeld, Regina von,
Waibstadt 078828
Degenhardt, Gertrude, Mainz .. 137368
Degenhart Paperweight and Glass
Museum, Cambridge 047374
Degenhart, J., Memmingen 077341
Degenkolbe, Leipzig 077058
Degeorge, Charles, Le Boulou . 128654
Degerfors Fotbollsmuseum,
Degerfors 040589
Degheldere, Willy, Brugge 140159
Degieux, Marc, Rots 105463
Degioanni, Léopold, Villeurbanne 074201
Degli Esposti, Paola Maria,

Bologna 130952
Deglise, Nicolas, Paris 128978
Degrange, Philippe, Charlieu ... 067850
DeGrazia Gallery in the Sun,
Tucson 053825
Degrémont, Madeleine, Clamecy 068005
Degrugillier, Hélène, Lagnes ... 069138
Degutsch, Ingrid, Stuttgart ... 142386
Degwitz, Uelzen 078749
Dehaan, Bankstown 098707
Dehennin, Guy, Cotignac 068168
Dehler, Remlingen 078196
Dehler, C., Raunheim 078144
Dehner jun., Georg, Bachhagel . 074485
Dehoogh, Philadelphia 096101
Dehua Ceramic Museum, Dehua 007735
Deià Museo Arqueológico, Deià . 039342
Deichmann, von, Schaan 132318
Deichtorhallen Hamburg, Haus der
Photographie – Aktuelle Kunst,
Hamburg 018979
Ulrich Deicke & Regina Himmel-Deicke,
Süderholz 078642, 130697
Deike, Horst, Konstanz 137325
Deile, Bodo, Dresden 075410
Deiller-Ducatel, Dijon 128579
Deillon, Robert, Villeneuve 087886
Deilmann, Dietrich & Maureen,
Herdecke 076497
Deinbacher, Andreas, Sankt
Pölten 140034
Deinert, U. & M., Herzebrock-
Clarholz 076516
Deines Cultural Center, Russell . 052705
Deinhard-Kellermuseum, Koblenz 019661
Deininger, Charlotte, Augsburg . 129394
Deir ez-Zor Museum, Deir ez-Zor 042260
Deitch, Jeffrey, New York 122552,
                    122533
Déja Vu, Gouda 083263
Déja Vu, Winnipeg 064918
Déjà Vu Antik, Uppsala 087117
Deja Vu Antiques, Leigh-on-Sea 089797
Deja Vu Antiques, Shrewsbury . 091327
Deja Vu Books, Hobart 139774
Dejans, P., Kortrijk 063718
Dejbjerg Jernalder, Skjern 010313
Dejima-machi Hakabutsukan,
Nagasaki 029088
Dějiny Královského Města Kolína,
Regionální Muzeum v Kolíně,
Kolín 009633
Deju Art, Carluke 117535
Dejwis Bis, Gdańsk 113461
DeKalb Historical Society Museum,
Decatur 048199
Dekanatsmuseum, Haus im
Ennstal 002064
Dekema State, Jelsum 032528
Deketelaere, Oostende 100799
Dekker, Amsterdam 082727
Dekker & Inkelaar, Amsterdam . 132394
Dekora, Baden 133628
Dekoratīvi Lietišķās Mākslas Muzejs,
Rīga 030178
Dekorativnoe Iskusstvo, Moskva 139009
Dekorator, Warszawa 084712
Del Bono, Domenico Savio,
Napoli 080864
Del Borgo, Giglio, Roma 081327
Del Cardal, Montevideo 119419
Del Cerro Gallery, San Diego .. 124117
Del Mestre, Umberto, Torino .. 132129
Del Monte, Renzo, Roma 131897
Del Norte County Historical Society
Museum, Crescent City 048058
Del Prado, Barcelona 102383
Del Recuerdo, Buenos Aires ... 139563
Delacour, Puymirol 072329
Delacour, Jean-Claude, La Palme 080420
Delacroix, Punta del Este 119441
Delacroix, Jean-Arnauld, Paris . 071239
Delacroix, Jean-Paul, Saint-Malo (Ille-
et-Vilaine) 072915

Delafontaine, Carouge 116190
Delafosse, Wormhout 074302
Delage, Philippe, Villeurbanne .. 074202
Delahaye, Didier, Pleurtuit ... 072185
Delahaye, Isabelle, Paris 071240
Delahaye, Olivier, Chaumont (Haute-
Marne) 067933
Delahenty, Olinda 099501
Delahoche, Gérard, Rémérangles 072411
Delaine, Rémi, Dole 068366
Delalande, Paris 071241
Delalande, Dominique & Eric,
Paris 071242
Delamain, Paris 141086
Delamarche, Thierry, Lecousse . 069378
Delamare, David, Portland 123613
Delan, Patrick, Pau 072037
Delaney, Los Angeles 094310
Delank, Claudia, Köln 107601
Delannoy, Xavier, La Garde-
Freinet 103891
Delano Heritage Park, Delano . 048229
Delany, Newport 079686
Delarce, Paris 071243
Delarue, Etretat 103722
Delarue-Turcat, Paris 128979
Delarue, Alain, Bordeaux 128486
Delarue, Dominique, Excideuil . 068498
Delaunay, Fabrice, Vitry-le-
François 074258
Delaune, Francis, Segré 073475
Delaune, Yves, Saumur 073435
Delaveau, Jean-Marc, Saint-
Mammès 072923
Delaware Agricultural Museum,
Dover 048349
Delaware and Hudson Canal Museum,
High Falls 049465
Delaware Art Museum,
Wilmington 054433
Delaware Center for the Contemporary
Arts, Wilmington 054434
Delaware Market, Tulsa 097725
Delaware Museum, Wilmington . 054435
Delaware Museum of Natural History,
Wilmington 054436
Delaware State Museums, Dover 048350
Delawood, Hunstanton 089617, 144492
Delbarre, Fabien, Lille 069487
Delbos, Frédéric, Chaumont (Haute-
Marne) 067934, 140826
Delbos, Jacques, Paris 071244
Delboy, Francoise, Vezin-le-
Coquet 074084
Delbridge Museum, Sioux Falls . 053353
Delcampo, Ricardo, Miami 094727
Delclaux, Myriam, Aurillac 066822
Déléage, Michel, Yssingeaux .. 074312
Deleau, Jacques, Carcassonne . 125546
Delecaut, M., Charleroi 063598
Delectable Collectibles,
Philadelphia 096102
Delectus Books, London 144582
Delehar, Peter, London 090070
DELEIKA, Dinkelsbühl 075359
DeLeon White, Toronto 101595
Delerive, Antoine, Lille 069488
Deletaille, Emile, Bruxelles 063392
Deleu, Vincent, Antwerpen 063202
Deleuil, Huguette, Marignane . 069874,
                    104156
Delfhavens, Rotterdam 112634
Delfi, Padova 080981
Delfina, London 044957
Delfino, Thierry, Marseille 128759
Delfitto, Alberto, Milano 131433
Delforge, M.-F., Bruxelles 063393
Delfour, Claude, Clermont-
Ferrand 068021
Delftenco, Bruxelles 063394
Delgatie Castle, Turriff 046004
Delhaye, Thierry, Craponne
(Rhône) 068221
Delhi Art Gallery, Delhi 109312

Delhi Ontario Tobacco Museum & Heritage Centre, Delhi 005696
Delic, Čačak 114448
Delicias Galeria, Madrid 115178
Delicias Restauración, Madrid . 133323
Delight Hamilton, Seattle 124515
Delignières, Jean-Marc, Sailly-sur-la-Lys 072655
Delikatessenhaus e.V., Leipzig . 058817, 107812
Delille, Francis, Paris 136888
Délire en Formation, Paris 104675
D'Eliso & Tomè, Trieste 132218
Delitz-Kater, Wilma, Hannover . 107355
Delius, Cornelia, Bielefeld 074975
Delizsánsz Kiállítóterem, Debrecen 023592
Dell, Baesweiler 074649
Dell, Buffalo 092522, 145020
Dell, Detroit 093359
Dell Mill, Isle-of-Lewis 044657
Dell, Klaus-M., Moers 077406
Della Corte, Vignola 111207
Della Porta, Roberto, Roma ... 131898
Della Toffola, Mauro, Venezia ... 081831
Della, Rondine, Ascona 144010
Dellacasa, Renata, Torino 081693
Dellachá Vaj, Maria Luisa, Torino 081694
Dellacherie, Alain, Bergerac ... 067101
Dellapiana Antichità, Alba 079739
Dell'Aquila, Alfredo, Torino 132130
Dell'Aria, Ignace, Marseille 069929, 104167
Delle Fave, Roberto, Roma 110754
Dellefant, Klaus, Puchheim 078098
Delley, François & Marina, Nyon 087713
Dellière, Christiane, Bellême ... 067076
Dello Ioio, Napoli 080865
Delloye Thoumyre, Anne-Marie, Châteauroux 128550
Dellus, Jacques, Noirmoutier-en-l'Ile 070735
Dellview, San Antonio 096968
Delmar, Hurst Green 089624
DelMar Gallery, Saskatoon 101544
Delmas, Bruguières 067522
Delmas, Solange, Paris 141087
Delmon, Manama 063141
Delmon, J F, Toulouse 073761
Delmore Gallery, Alice Springs . 136808
Delomosne & Son, North Wraxall 090852
Delon, Jean-Paul, Marigny-Saint-Marcel 140937
Delon, Michel, Malemort-sur-Corrèze 069824
Delor, Nîmes 070691, 070692
Delor, Léon, Nîmes 070693
Deloraine Museum, Deloraine ... 001023
DeLorenzo, New York . 095344, 095345
Delorme, Paris 104676
Delorme, Chantal, Milano 080522
Delorme, Daniele, Paris 104677
Delorme, Francis, Avèze (Sarthe) 066869
Delorme, Marc, Saint-Ouen 073066
Delorme, Michel, Eaubonne ... 068424
Delorme, Patricio, Santiago de Chile 064963
Delotte, Alain, Montrouge 070337
Delouche, Olivier, Saint-Malo (Ille-et-Vilaine) 072916
Delourmel, Nadine, Rennes (Ille-et-Vilaine) 105431
Delozier, D.L., Virginia Beach . 136442
Delpani, Luciano, Brescia 079935
Delperie, Adrien, Toulouse 073762
Delpero, Antonio, Genova 080258
Delphi Antiques, Dublin 079543
Delphica Restauri, Roma 131899
Delpierre, Philippe, Paris 071245
Delpire, Paris 136889
Delpla, Eric, Saint-Ouen 073067
Delpuech, Charles, Bonnieux ... 067294
Delray Beach Historical Society Antiques Show, Delray Beach . 098446

Delrey, Rio de Janeiro 064051
Delshan Art Gallery, Armadale, Victoria 098657
Delsing, Gerard, Maastricht 083460
Delta, Paris 104678
Delta, Rotterdam 112635
Delta – Arte, Donostia-San Sebastián 115045
Delta Antiques, İstanbul 088113
Delta Axis Contemporaray Arts Center, Memphis 050776
Delta Blues Museum, Clarksdale 047771
Delta Change, Antibes-Juan-les-Pins 066671
Delta County Historical Society Museum, Escanaba 048601
Delta County Museum, Delta ... 048234
Delta Cultural Center, Helena ... 049428
Delta Museum, Delta 005697
Delta-Phot, Limoges 104029
Delteil, Montréal 140360
Deltos, Firenze 131157
Deluermoz, Martine, Ussel 073919
Delusions of Grandeur, Ottawa . 101470
Deluxe Junk, Seattle 097426
Delvaille, Olivier, Paris 071246
Delvaulx, L., Bruxelles 128024
Delvaux, Jean-Marc, Paris 125702
Delvecchio, Erica, Aviano 130901
Delvigne, Hervé, Ham 068781
Delvoie, M., Maastricht 083461
Dely, Nice 070594
Delzongle, Jean-Paul, Drugeac . 068409
Demago, Maribor 085616
Demaison André, Restigné 015071
Demak, Sankt-Peterburg 085241
Deman, Nicolas, Paris 104679
Autographes Demarest, Paris .. 141088
Demary, Rheinbach 078222
Demazières, Sylviane, Courpière 068194
Dembinski & Bittner, Kassel 107529
Demenat, Schwabhausen 108584
Demenga, Hansruedi, Basel ... 116061, 133640
Demers, Columbus 135478
Demes, New York 145334
Demessieur, Thomas, Krefeld ... 126018
Demet, Berlin 141501
Demetriades Studio, Lemesos .. 102546
Demetriades, Maria, Athinai ... 109031
Demetrios, Bon Air 135357
Demetrios, Alphonso, New York 135906
Demetrius, Pittsburgh . 096276, 136130
Demetzy, London 144583
Demeulemester, Régis, Matignon 069995
La Demeure Antique, Colombe . 068098
Demeure des Prémontrés, Pagny-sur-Moselle 070838
La Demeure Médiévale, Beaune 067015
Demeyer, N., Gent 063635
Demharter, Walter, Krumbach .. 076978
DEMHIST-International Committee for Historic House Museums, Milano 059432
Demi-Lune, Bruxelles 063395
La Demi-Lune, Saint-Lunaire .. 072902
Demian, Antwerpen 140145
Demigod, Armadale, Victoria ... 061021
Demille, Melbourne 061816
Demilliere-Vergnais, Laurent, Lyon 128709
Deml, Brigitte & Heinrich, Aldersbach 074372
Demmer & Hohnig, Obertshausen 077883
Demmer, A., Karlsruhe 076689
Demmler, Schwerin 137504
Demner, M. & H., Wien 062924
Demolition Depot & Irreplaceable Artifacts, New York 095346
Démons & Merveilles, Paris 104680
Demontrond, Paul, Dannemarie . 103653
Demouron, Christine, Biarritz .. 067197
Demouron, Rémy, Beaune 067016

Demours Antiquités, Paris 071247
Demouzon, Michel, Barberaz ... 066943
Dempsey, Jacksonville 121243
Demtröder, Ulla, Dortmund 106699
Demun, Saint Louis 123819
Den Antiquaire, Tuntange 082394
Den of Antiquities, Yarra Glen .. 062500
Den of Antiquity, Calgary 064254, 128148
Den of Antiquity, Launceston ... 061725
Den of Antiquity, Los Angeles .. 094311
Den of Antiquity, Portland 096394
Denbigh, Vancouver 101772
Denbigh Castle, Denbigh 044080
Denbigh Museum and Gallery, Denbigh 044081
Denburn Antiques, Aberdeen ... 088220, 134009
Dene Museum, Fort Good Hope 005818
Denecke, S., Haibach 129935
Denenberg Fine Arts, Los Angeles 121470
Denesle, Christian, Rouen 125794
Deneuve, Marie-Odile, Ancretteville-sur-Mer 066588
Deng, Seattle 124516
Dengel-Galerie, Reutte 002593
Denger, E., Alsheim 074386
Dengis, Marc, Bruxelles 100589
Dengleant, Henry, Lunel-Viel ... 069662
Dengler, Erik, Esslingen 075708
Dengzhou Guchuan Museum, Penglai 008145
Denham, Warnham 127077
Dénichoir, Châteauneuf 067887
Denijs, P.A., Den Haag 083074
Denis Antiques, Montréal 064413
Denis, Francisco, Montigny-sur-Loing 070257
Denis, Jöel, Condé-sur-Noireau . 068124
Denis, Laurence, Caen 067562
Denise, Roma 081328
Denise Antiques, Cincinnati 092843
Denise, François-Henri, Saint-Jouin-de-Blavou 072859
Denise, Pierre, Yerres 074307
Deniset, René, Champagnole ... 128540
Denison, Toronto 101596
Denison University Art Gallery, Granville 049209
Deniston Cottage Antiques, Bungendore 061234
Deniz Müzesi, İstanbul 042932
Denizli Devlet Güzel Sanatlar Galerisi, Denizli 042876
Denki no Kagakukan, Nagoya .. 029113
Denkmal für die ermordeten Juden Europas, Berlin 017101
Denkmalhof Bechelsdorfer Schulzenhaus, Schönberg, Mecklenburg 021700
Denkmalhof Gererhof, Annaberg 001714
Denkmalhof Kösslerhäusl, Großarl 002002
Denkmalhof Maurerhaus, Zederhaus 003134
Denkmalhof Retschow, Retschow 021379
Die Denkmalpflege, München .. 138689
DenkStätte Weiße Rose, Weiße Rose Stiftung, Ludwig-Maximilians-Universität, München 020514
Denman Island Museum, Denman Island 005700
Denmark Historical Museum, Denmark 001024
Denneberg, Johann T., Hamburg 076211
Dennemans, Goch 076016
Denner, G., Maßbach 130292
Dennett, Clive, Norwich 090873
Denneweg 33 Antiek, Den Haag 083075
Denning, Guildford 089394
Denning & Son, Ron, Russell .. 132738

Dennis, Margaret, Beaudesert .. 098716
Dennis, Richard, Shepton Beauchamp 138283
Dennis, Ruud, Minneapolis 135827, 145275
Dennison Railroad Depot Museum, Dennison 048244
Dennler, Guy, Dulverton 089112
Dennos Museum Center of Northwestern Michigan College, Traverse City 053791
Denny Ship Model Experiment Tank, Scottish Maritime Museum, Dumbarton 044150
Denoel, Liège 063751
Denoël, Paris 136890
Denova-Tesar, Olga, Paris 128980
Denoyelle, Pascal, Le Cateau-Cambrésis 069242
Denoyer, Jean-Pierre, Sourzac . 073553
Densheway National Museum, Densheway Village 010567
Dental Museum, Edmonton 005749
Dentalhistorisches Museum, Zschadraß 022898
La Dentellière, Paris 071248
Denter, H., Bad Driburg 106052
Denter, M., Lünen 077213
Denter, Marlies, Lünen 077214
Denti, Mirella, Milano 080523
Denton, London 090071
Denton County Historical Museum, Denton 048247
Denton, Berenice, Nashville ... 094992
Denuo-A, Warszawa 113818
Denver Antique Shows and Sale, Denver 098447
Denver Art Dealers Association, Denver 060461
Denver Art Museum, Denver ... 048258
Denver Book Mall, Denver 145124
Denver Firefighters Museum, Denver 048259
Denver Museum of Miniatures, Dolls and Toys, Denver 048260
Denver Museum of Nature and Science, Denver 048261
Denver Woodwrights, Denver .. 135528
Deny, Bernard & Béatrice, Ervy-le-Chatel 068466
Denys, Philippe, Bruxelles 063396
Denzil Grant, Bury Saint Edmunds 088741, 134191
Denzinger, Herbert, Neustadt an der Weinstraße ... 077759, 126081, 130457
Denzler, W. & J., Weisslingen .. 087896
Deokpojin Educational Museum, Gimpo 029934
DeOro Galería, Bogotá 102296
Deosung Art Museum, National Museum of Contemporary Art, Seoul 029989
Dep Art, Milano 110299
Depaire, Alain, Sauveterre-de-Guyenne 073443
Depaoli, A., Höchst 062636
Departamento de Arquitectura y Diseño, Universidad de Los Lagos, Osorno 055180
Departamento de Arte Visual, Universidade Estadual de Londrina, Londrina 055082
Departamento de Artes e Comunicação, Universidade Federal de São Carlos, São Carlos ... 055097
Departamento de Artes e Design, Pontifícia Universidade Católica, Rio de Janeiro, Rio de Janeiro ... 055091
Departamento de Artes Plásticas, Facultad de Artes y Humanidades, Universidad de Caldas, Manizales 055267
Departamento de Artes Plásticas, Facultad de Artes, Universidad del

Cauca, Popayán 055268
Departamento de Artes Plásticas,
Facultad de Artes, Universidad
Nacional de Cuyo, Mendoza .. 054872
Departamento de Artes Plasticas,
Facultad de Humanidades y
Arte, Universidad de Concepción,
Concepción 055179
Departamento de Artes Visuales y
Estética, Universidad del Valle,
Facultad de Artes Integradas,
Cali 055265
Departamento de Artes Visuales,
Facultad de Artes, Universidad de
Nariño, San Juan de Pasto ... 055269
Departamento de Artes Visuales,
Instituto Universitario Nacional del
Arte, Buenos Aires 054863
Departamento de Artes Visuales,
Universidad de Chile, Facultad de
Arte, Santiago de Chile ... 055184
Departamento de Artes, Universidade
Federal de Rio Grande do Norte,
Natal 055085
Departamento de Bellas Artes,
Fundación Universitaria de Boyacá,
Tunja 055271
Departamento de Bellas Artes,
Universidad de Sonora,
Hermosillo 056140
Departamento de Escultura e Historia
del Arte, Universidad de Sevilla,
Facultad de bellas Artes, Sevilla 056382
Departamento de Historia del Arte
Diego Velázquez, Instituto de Historia
del CSIC, Madrid 056368
Departamento de Historia del
Arte, Universidad de Santiago
de Compostela, Santiago de
Compostela 056381
Departamento de Historia del
Arte, Universidad de Valencia,
Valencia 056385
Departamento de Humanidades y
Arte, Universidad de Los Lagos,
Osorno 055181
Departamento d'Historia de
l'Art, Universidad de Barcelona,
Barcelona 056355
Departement Archeologie,
Kunstwetenschap en Musicologie,
Katholieke Universiteit Leuven,
Leuven 055033
Departement Architectuur en
Beeldende Kunst, Provinciale
Hogeschool Limburg, Hasselt . 055025
Departement Architectuur,
Audiovisuele en Beeldende Kunsten
Academie, Hogeschool Gent,
Gent 055020
Département d'Archéologie et
d'Histoire de l'Art, Université
Catholique de Louvain, Louvain-la-
Neuve 055037
Departement des Arts Plastiques et
Sciences de l'Art, Université de
Provence, Aix-en-Provence ... 055347
Département des Arts Visuelles,
Université de Moncton,
Edmundston 055125
Département des Arts Visuels,
Université de Moncton,
Moncton 055138
Département d'Histoire de l'Art et de
Musicologie, Section de philosophie
et d'histoire, Genève 056425
Département d'Histoire de
l'Art, Université de Montréal,
Montréal 055139
Departement Kulturanalysen und
-Vermittlung, Zürcher Hochschule der
Künste, Zürich 056442
Departement Kunst & Medien, Zürcher
Hochschule der Künste, Zürich 056443

Department Architectuur Sint-Lucas,
Bruxelles 055008
Department Architectuur Sint-Lucas,
Gent 055021
Department Arts und Management,
Donau-Universität Krems,
Krems 054947
Department Design, Zürcher
Hochschule der Künste, Zürich 056444
Department für Kunstgeschichte und
Musikwissenschaft, Université du
Fribourg, Fribourg 056424
Department Gestaltung und mediale
Kommunikation, Fachhochschule
Vorarlberg GmbH, Dornbirn ... 054925
Department of Advertising Design &
Graphic Arts, The City University of
New York, New York City College of
Technology, Brooklyn 056799
Department of Architecture and
Landscape Architecture, North
Dakota State University, Fargo 057075
Department of Architecture, University
of Pennsylvania, Philadelphia . 057667
Department of Art, Eugene 057060
Department of Art, Raymond ... 057749
Department of Art & Art History,
Stanford University, Stanford .. 057943
Department of Art & Design, Harding
University, Searcy 057892
Department of Art and Archaeology,
Hood College, Frederick 057106
Department of Art and Archaeology,
Princeton University, Princeton 057722
Department of Art and Architecture,
Lehigh University, Bethlehem . 056735
Department of Art and Architecture,
Northeastern University, Boston 056769
Department of Art and Art Education,
Delaware State College, Dover 057018
Department of Art and Art Education,
Indiana University of Pennsylvania,
Indiana 057232
Department of Art and Art
History, Adelphi University, Garden City 057126
Department of Art and Art History,
Albion College, Albion 056635
Department of Art and Art History,
Bucknell University, Lewisburg 057335
Department of Art and Art History,
Carleton College, Northfield ... 057595
Department of Art and Art History,
Colgate University, Hamilton . 057173
Department of Art and Art History,
College of William and Mary,
Williamsburg 058110
Department of Art and Art History,
College of Wooster, Wooster . 058133
Department of Art and Art History,
DePaul University, Chicago ... 056882
Department of Art and Art History,
Dickinson College, Carlisle ... 056837
Department of Art and Art History,
Duke University, Durham 057028
Department of Art and Art History,
Florida Southern College,
Lakeland 057313
Department of Art and Art History,
Franklin and Marshall College,
Lancaster 057319
Department of Art and Art History,
George Mason University,
Fairfax 057070
Department of Art and Art History,
Lawrence University, Appleton . 056663
Department of Art and Art History,
Maple Woods Community College,
Kansas City 057278
Department of Art and Art History,
Mary Baldwin College, Staunton 057948
Department of Art and Art History,
Mary Washington College,
Fredericksburg 057107
Department of Art and Art History,

Mesa Community College,
Mesa 057428
Department of Art and art History,
Michigan State University, East
Lansing 057030
Department of Art and Art History,
Oakland University, Rochester . 057771
Department of Art and Art History,
Pomona College, Claremont .. 056900,
                              056901
Department of Art and Art History,
Scripps College, Claremont ... 056902
Department of Art and Art History,
Sonoma State University, Rohnert
Park 057793
Department of Art and Art History,
Texas Christian University, Fort
Worth 057101
Department of Art and Art History,
University of Alabama at Birmingham,
Birmingham 056745
Department of Art and Art History,
University of Alabama in Huntsville,
Huntsville 057224
Department of Art and Art History,
University of Connecticut, Storrs 057959
Department of Art and Art History,
University of Miami, Coral
Gables 056955
Department of Art and Art History,
University of Missouri Saint Louis,
Saint Louis 057813
Department of Art and Art History,
University of Nebraska, Lincoln 057347
Department of Art and Art History,
University of Nebraska at Omaha,
Omaha 057619
Department of Art and Art History,
University of New Mexico,
Albuquerque 056637
Department of Art and Art History,
University of Rhode Island,
Kingston 057287
Department of Art and Art
History, University of Rochester,
Rochester 057779
Department of Art and Art History,
University of Saskatchewan,
Saskatoon 055155
Department of Art and Art History,
University of South Alabama,
Mobile 057464
Department of Art and Art History,
University of Texas at Austin,
Austin 056692
Department of Art and Art History,
University of Texas at San Antonio,
San Antonio 057841
Department of Art and Art
History, University of the Pacific,
Stockton 057955
Department of Art and Art
History, University of the South,
Sewanee 057902
Department of Art and Art History,
University of Utah, Salt Lake
City 057835
Department of Art and Art History,
Vanderbilt University, Nashville 057509
Department of Art and Art History,
Wayne State University, Detroit 057010
Department of Art and Art History,
Western Maryland College,
Westminster 058093
Department of Art and Art History,
Wright State University, Dayton 056983
Department of Art and Art Professions,
New York University, New York 057547
Department of Art and Classics, Troy
State University, Troy 058005
Department of Art and Design
Arts, Barton Institute of TAFE,
Richmond 054918
Department of Art and Design,

California Polytechnic State University
at San Luis Obispo, San Luis
Obispo 057862
Department of Art and Design,
California State University Fresno,
College of Arts, Fresno 057111
Department of Art and Design,
California University of Pennsylvania,
California 056820
Department of Art and Design,
Charles Darwin University,
Darwin 054904
Department of Art and Design,
Chatham College, Pittsburgh .. 057687
Department of Art and Design,
Converse College, Spartanburg 057930
Department of Art and Design,
Crowder College, Neosho ... 057512
Department of Art and Design, East
Tennessee State University, Johnson
City 057267
Department of Art and Design, KvB
Institute of Technology, North
Sydney 054912
Department of Art and Design,
Missouri State Univeristy,
Springfield 057941
Department of Art and Design,
Montclair State University, Upper
Montclair 058030
Department of Art and Design,
Northern Michigan University,
Marquette 057412
Department of Art and Design,
Oklahoma Christian University,
Edmond 057038
Department of Art and Design, Pacific
Union College, Angwin 056658
Department of Art and Design, Red
Deer College, Red Deer 055151
Department of Art and Design,
Saginaw Valley State University,
Zilwaukee 058145
Department of Art and Design, Saint
Xavier University, Chicago 056883
Department of Art and Design,
Schoolcraft College, Livonia .. 057353
Department of Art and Design,
Southern Illinois University
Edwardsville, Edwardsville ... 057040
Department of Art and Design,
Southwest Texas State University,
San Marcos 057863
Department of Art and Design,
University of Alberta, Edmonton 055124
Department of Art and Design,
University of Charleston,
Charleston 056857
Department of Art and Design,
University of Minnesota Duluth,
Duluth 057023
Department of Art and Design,
University of Papua New Guinea,
Boroko 056220
Department of Art and Design,
University of Wisconsin Eau Claire,
Eau Claire 057034
Department of Art and Design,
University of Wisconsin Stevens
Point, Stevens Point 057951
Department of Art and Design,
University of Wisconsin Stout,
Menomonie 057425
Department of Art and Design,
Vancouver Community College,
Vancouver 055166
Department of Art and Design,
West Nottinghamshire College,
Mansfield 056582
Department of Art and Design,
Winthrop University, Rock Hill . 057783
Department of Art and Industrial
Technology, Western State College of
Colorado, Gunnison 057170

Department of Art and Music, Assumption College, Worcester 058135

Department of Art and Painting, SNDT Women's University, Faculty of Fine Arts, Mumbai 055818

Department of Art and Photography, Phoenix College, Phoenix 057677

Department of Art and Photography, Queensborough Community College, Bayside 056715

Department of Art as Applied to Medicine, Johns Hopkins University, Baltimore 056701

Department of Art History & Archaeology Visual Resources Collection, Columbia University, New York 051374

Department of Art History and Archaeology, New York 057549

Department of Art History and Art, Case Western Reserve University, Cleveland 056915

Department of Art History and Communication Studies, Faculty of Arts, McGill University, Montréal 055140

Department of Art History and Humanities, John Carroll University, University Heights 058026

Department of Art History and Theory, University of Otago, Dunedin . . 056193

Department of Art History and Theory, University of Sydney, Faculty of Arts, Camperdown 054899

Department of Art History and Visual Arts, Occidental College, Los Angeles 057369

Department of Art History, Columbia University, New York 057548

Department of Art History, Comenius University, Bratislava 056333

Department of Art History, Dartmouth College, Hanover 057181

Department of Art History, Northwestern University, Evanston 057063

Department of Art History, Rutgers State University of New Jersey, New Brunswick 057516

Department of Art History, State University of New York at Binghamton, Binghamton 056741

Department of Art History, University of Brighton, Brighton 056512

Department of Art History, University of Chicago, Chicago 056884

Department of Art History, University of Maryland, College Park 056926

Department of Art History, University of Minnesota, Minneapolis . . . 057457

Department of Art History, University of Nottingham, Nottingham . . . 056588

Department of Art History, University of Saint Thomas, Saint Paul . . 057823

Department of Art History, University of Wisconsin Madison, Madison 057393

Department of Art History, Villanova University, Villanova 058047

Department of Art History, Visual Art and Theory, University of British Columbia, Vancouver 055167

Department of Art History, Visual Arts and Musicology, University of South Africa, Tshwane 056353

Department of Art Practice, University of California Berkeley, Berkeley 056729

Department of Art Theory and Practice, Northwestern University Evanston, Evanston 057064

Department of Art, Albright College, Reading 057750

Department of Art, Alma College, Alma 056645

Department of Art, American River College, Sacramento 057800

Department of Art, American University, Washington 058064

Department of Art, Anna Maria College, Paxton 057651

Department of Art, Appalachian State University, Boone 056763

Department of Art, Arkansas State University, Jonesboro 057270

Department of Art, Arkansas Tech University, Russellville 057796

Department of Art, Art History and Design, Andrews University, Berrien Springs 056730

Department of Art, Art History and Design, University of Notre Dame, Notre Dame 057599

Department of Art, Art History and Film Studies Program, Mount Holyoke College, South Hadley 057924

Department of Art, Auburn University, Auburn 056686

Department of Art, Augustana College, Rock Island 057784

Department of Art, Austin Peay State University, Clarksville 056908

Department of Art, Ball State University, Muncie 057497

Department of Art, Barat College, Lake Forest 057309

Department of Art, Bay Path College, Longmeadow 057358

Department of Art, Baylor University, Waco 058051

Department of Art, Bethel College, North Newton 057591

Department of Art, Bethel University, Saint Paul 057821

Department of Art, Biola University, La Mirada 057301

Department of Art, Blackburn College, Carlinville 056836

Department of Art, Bloomsburg University, Bloomsburg 056755

Department of Art, Blue Mountain College, Blue Mountain 056757

Department of Art, Bowdoin College, Birch Island 056742

Department of Art, Bradley University, Peoria 057659

Department of Art, Brescia University, Owensboro 057638

Department of Art, Brigham Young University Idaho, Rexburg . . . 057754

Department of Art, California State University, Turlock 058015

Department of Art, California State University Bakersfield, 15 FA, Bakersfield 056697

Department of Art, California State University Northridge, Northridge 057596

Department of Art, California State University Sacramento, Sacramento 057801

Department of Art, California State University, Fullerton, Fullerton . 057113

Department of Art, Camden County College, Blackwood 056748

Department of Art, Catholic University, Washington 058065

Department of Art, Centenary College of Lousiana, Shreveport 057913

Department of Art, Central Connecticut State University, New Britain . 057515

Department of Art, Central Michigan University, Mount Pleasant . . 057492

Department of Art, Central Oregon Community College, Bend 056724

Department of Art, Central Washington University, Ellensburg 057047

Department of Art, Claflin College, Orangeburg 057626

Department of Art, Clarke College, Dubuque 057022

Department of Art, Cleveland State Community College, Cleveland 056916

Department of Art, College of Arts and Sciences, Gonzaga University, Spokane 057934

Department of Art, College of Liberal Arts, Drew University, Madison 057388

Department of Art, College of Marin, Kentfield 057284

Department of Art, College of New Jersey, Ewing 057068

Department of Art, College of Saint Rose, Albany 056632

Department of Art, College of Santa Fe, Santa Fe 057872

Department of Art, College of the Arts, University of Southern Mississippi, Hattiesburg 057189

Department of Art, College of the Ozarks, Point Lookout 057700

Department of Art, Colorado College, Colorado Springs 056930

Department of Art, Colorado State University – Pueblo, Pueblo . . 057732

Department of Art, Columbia College, Columbia 056936

Department of Art, Columbia State Community College, Columbia . 056939

Department of Art, Columbus State University, Columbus 056940

Department of Art, Community College of Philadelphia, Philadelphia . . 057668

Department of Art, Community College of Rhode Island, Warwick 058060

Department of Art, Contra Costa Community College, Martinez . 057415

Department of Art, Cornell University, Ithaca 057244

Department of Art, Cuyahoga Community College, Cleveland 056914

Department of Art, Dallas Baptist University, Dallas 056974

Department of Art, Davis and Elkins College, Elkins 057046

Department of Art, Del Mar College, Corpus Christi 056956

Department of Art, Delta State University, Cleveland 056911

Department of Art, Denison University, Granville 057150

Department of Art, Division of Fine Arts and Humanities, Dickinson College, Dickinson 057014

Department of Art, Division of Fine Arts, Malone College, Canton . 056832

Department of Art, Eastern Illinois University, Charleston 056854

Department of Art, Eastern New Mexico University, Portales . . 057704

Department of Art, Eastern Washington University, Cheney 056866

Department of Art, Elizabeth City State University, Elizabeth City 057044

Department of Art, Emporia State University, Emporia 057053

Department of Art, Fisk University, Nashville 057508

Department of Art, Florida State University, Tallahassee 057977

Department of Art, Folletts Lambuth College, Jackson 057251

Department of Art, Fort Hays State University, Hays 057191

Department of Art, Furman University, Greenville 057165

Department of Art, George Washington University, Washington 058066

Department of Art, Georgian Court College, Lakewood 057316

Department of Art, Goldsmith's College, London 056558

Department of Art, Grace College, Winona Lake 058123

Department of Art, Greensboro

College, Greensboro 057159

Department of Art, Greenville College, Greenville 057162

Department of Art, Grinnell College, Grinnell 057169

Department of Art, Hanover College, Hanover 057180

Department of Art, Hardin-Simmons University, Abilene 056623

Department of Art, Heidelberg College, Tiffin 057994

Department of Art, Hillsdale College, Hillsdale 057202

Department of Art, Hobart and William Smith Colleges, Geneva 057129

Department of Art, Hollins University, Roanoke 057769

Department of Art, Houston Baptist College, Houston 057215

Department of Art, Huntingdon College, Montgomery 057478

Department of Art, Idaho State University, Pocatello 057699

Department of Art, Indian Hills Community College, Centerville 056850

Department of Art, Indian Hills Community College, Ottumwa . 057635

Department of Art, Indiana State University, Terre Haute 057989

Department of Art, Iowa Central Community College, Sac Ciity . 057798

Department of Art, Iowa Lakes Community College, Spencer . 057932

Department of Art, Jackson State University, Jackson 057249

Department of Art, Jacksonville University, Jacksonville 057255

Department of Art, Juniata College, Huntingdon 057220

Department of Art, Kauai Community College, Lihue 057344

Department of Art, La Salle University, Philadelphia 057669

Department of Art, Lafayette College, Easton 057033

Department of Art, Lake Forest College, Lake Forest 057310

Department of Art, Lake Michigan College, Benton Harbor 056726

Department of Art, Lander University, Greenwood 057167

Department of Art, Lenoir Rhyne College, Hickory 057199

Department of Art, Letters and Social Sciences, Polk Community College, Winter Haven 058128

Department of Art, Longwood University, Farmville 057080

Department of Art, Louisiana College, Pineville 057681

Department of Art, Macalester College, Saint Paul 057822

Department of Art, MacMurray College, Jacksonville 057257

Department of Art, Manatee Community College, Bradenton 056780

Department of Art, Mercer County Community College, Trenton . 058003

Department of Art, Mercyhurst College, Erie 057057

Department of Art, Middle Tennessee State University, Murfreesboro 057499

Department of Art, Mills College, Oakland 057606

Department of Art, Millsaps College, Jackson 057250

Department of Art, Mississippi College, Clinton 056917

Department of Art, Missouri Southern State University, Joplin 057271

Department of Art, Monmouth University, West Long Branch . 058088

Department of Art, Montgomery College, Rockville 057791

Register der Institutionen und Firmen

**Department of Art, Moorhead State University, Moorhead
– Department of Display and Design, Langara College, Vancouver**

Department of Art, Moorhead State University, Moorhead 057483
Department of Art, Moravian College, Bethlehem 056734
Department of Art, Mount Olive College, Mount Olive 057490
Department of Art, Mount Union College, Alliance 056644
Department of Art, Muhlenberg College, Allentown 056643
Department of Art, Murray State University, Murray 057500
Department of Art, Music and Philosophy, John Jay College of Criminal Justice, New York ... 057546
Department of Art, Music and Theater, Georgetown University, Washington 058067
Department of Art, Newberry College, Newberry 057572
Department of Art, North Carolina Central University, Durham ... 057027
Department of Art, North Carolina Wesleyan University, Rocky Mount 057792
Department of Art, North Central College, Naperville 057505
Department of Art, North Dakota State College of Science, Wahpeton . 058053
Department of Art, North Florida Community College, Madison . 057387
Department of Art, North Iowa Area Community College, Mason City 057420
Department of Art, Northeast Mississippi Community College, Booneville 056764
Department of Art, Northeastern Junior College, Sterling 057949
Department of Art, Northwest Missouri State University, Maryville ... 057418
Department of Art, Oberlin College, Oberlin 057608
Department of Art, Ohio Dominican College, Columbus 056946
Department of Art, Ohio Northern University, Ada 056624
Department of Art, Ohio State University, Columbus 056947
Department of Art, Oklahoma Baptist University, Shawnee 057905
Department of Art, Oregon State University, Corvallis 056960
Department of Art, Ottawa University, Ottawa 057634
Department of Art, Ouachita Baptist University, Arkadelphia 056666
Department of Art, Our Lady of the Lake University, San Antonio .. 057839
Department of Art, Palm Beach Community College, Lake Worth 057312
Department of Art, Pitzer College, Claremont 056899
Department of Art, Portland State University, Portland 057709
Department of Art, Queen's University, Kingston 055134
Department of Art, Quincy University, Quincy 057739
Department of Art, Randolph-Macon Woman's College, Lynchburg . 057380
Department of Art, Reed College, Portland 057710
Department of Art, Regis College, Weston 058094
Department of Art, Rhodes College, Memphis 057422
Department of Art, Riverside Community College, Riverside . 057766
Department of Art, Rock Valley College, Rockford 057786
Department of Art, Rollins College, Winter Park 058130
Department of Art, Saint Cloud State University, Saint Cloud 057805

Department of Art, Saint Gregory's University, Shawnee 057906
Department of Art, Saint Louis Community College at Forest Park, Saint Louis 057812
Department of Art, Sarah Lawrence College, Bronxville 056794
Department of Art, School of Music and Fine Arts, Howard Payne University, Brownwood 056806
Department of Art, School of the Arts, Pacific Lutheran University, Tacoma 057973
Department of Art, Skagit Valley College, Mount Vernon 057496
Department of Art, Skidmore College, Saratoga Springs 057879
Department of Art, Smith College, Northampton 057593
Department of Art, Southeast Missouri State University, Cape Girardeau 056834
Department of Art, Southeastern Community College, Whiteville 058098
Department of Art, Southern Arkansas University, Magnolia 057394
Department of Art, Southern Connecticut State University, New Haven 057522
Department of Art, Springfield College in Illinois, Springfield 057937
Department of Art, State University of New York College at Brockport, Brockport 056788
Department of Art, State University of New York College at Geneseo, Geneseo 057128
Department of Art, State University of New York College at Oneonta, Oneonta 057622
Department of Art, Swarthmore College, Swarthmore 057963
Department of Art, Texas Tech University, Lubbock 057377
Department of Art, Thiel College, Greenville 057164
Department of Art, Towson State University, Towson 058001
Department of Art, Trinity University, San Antonio 057840
Department of Art, Union College, Schenectady 057884
Department of Art, Union University, College of Arts and Sciences, Jackson 057252
Department of Art, University at Buffalo, Buffalo 056815
Department of Art, University of Alaska Anchorage, Anchorage . 056656
Department of Art, University of Alaska Fairbanks, Fairbanks .. 057069
Department of Art, University of Arizona, Tucson 058011
Department of Art, University of Calgary, Calgary 055120
Department of Art, University of California Los Angeles, Los Angeles 057368
Department of Art, University of California Santa Barbara, Santa Barbara 057868
Department of Art, University of Delaware, Newark 057568
Department of Art, University of Hawaii at Manoa, Honolulu ... 057208
Department of Art, University of Houston, Houston 057216
Department of Art, University of Kansas, Lawrence 057328
Department of Art, University of Kentucky, Lexington 057338
Department of Art, University of La Verne, La Verne 057303
Department of Art, University of Maine,

Orono 057630
Department of Art, University of Maryland, College Park 056925
Department of Art, University of Massachusetts, Lowell 057376
Department of Art, University of Mississippi, University 058024
Department of Art, University of Montana, Missoula 057463
Department of Art, University of Nebraska Kearney, Kearney .. 057280
Department of Art, University of Nevada Las Vegas, Las Vegas . 057324
Department of Art, University of North Alabama, Florence 057089
Department of Art, University of North Carolina at Asheville, Asheville 056669
Department of Art, University of North Carolina at Greensboro, Greensboro 057160
Department of Art, University of North Carolina at Pembroke, Pembroke 057654
Department of Art, University of Northern Iowa, Cedar Falls ... 056846
Department of Art, University of San Diego, San Diego 057850
Department of Art, University of Sioux Falls, Sioux Falls 057920
Department of Art, University of South Carolina, Columbia 056937
Department of Art, University of South Dakota, Vermillion 058043
Department of Art, University of Tampa, Tampa 057979
Department of Art, University of Tennessee at Chattanooga, Chattanooga 056863
Department of Art, University of Texas of Permian Basin, Odessa 057610
Department of Art, University of the Cumberlands, Williamsburg ... 058109
Department of Art, University of the Ozarks, Clarksville 056907
Department of Art, University of Toledo, Toledo 057995
Department of Art, University of West Florida, Pensacola 057657
Department of Art, University of Wisconsin, Rice Lake 057755
Department of Art, University of Wisconsin Madison, Madison . 057392
Department of Art, University of Wisconsin Oshkosh, Oshkosh . 057631
Department of Art, University of Wyoming, Laramie 057322
Department of Art, Utah State University, Logan 057356
Department of Art, Valdosta State University, Valdosta 058034
Department of Art, Wake Forest University, Winston-Salem 058126
Department of Art, Wartburg College, Waverly 058073
Department of Art, Washburn University of Topeka, Topeka .. 057996
Department of Art, Waubonsee Community College, Sugar Grove 057961
Department of Art, Wayland Baptist University, Plainview 057691
Department of Art, Wayne Community College, Goldsboro 057140
Department of Art, Weatherford College, Weatherford 058078
Department of Art, Wells College, Aurora 056691
Department of Art, West Texas A & M University, Canyon 056833
Department of Art, Western Carolina University, Cullowhee 056966
Department of Art, Western Michigan University, Kalamazoo 057274
Department of Art, Westminster

College, Salt Lake City 057834
Department of Art, Wheaton College, Wheaton 058096
Department of Art, Whittier College, Whittier 058100
Department of Art, Wilkes University, Wilkes-Barre 058108
Department of Art, Williams Baptist College, Light 057343
Department of Art, Winona State University, Winona 058122
Department of Art, Wuhan University, Wuhan 055246
Department of Art, York College of Pennsylvania, York 058141
Department of Art, Young Harris College, Young Harris 058142
Department of Art, Youngstown State University, Youngstown 058143
Department of Arts & Sciences, Wilberfoce University, Wilberforce 058107
Department of Arts and Design, Guangzhou University, Guangzhou 055209
Department of Arts and Sciences, Fuji Joshi Daigaku, Sapporo 056031
Department of Arts, Chadron State College, Chadron 056851
Department of Arts, Fudan University, Shanghai 055229
Department of Arts, Goddard College, Plainsfield 057690
Department of Arts, Otero Junior College, La Junta 057300
Department of Arts, Rutgers University Newark, Newark 057569
Department of Arts, Seijo University, Faculty of Arts and Literature, Tokyo 056042
Department of Arts, Slippery Rock University of Pennsylvania, Slippery Rock 057921
Department of Arts, Southeast University, Nanjing 055224
Department of Arts, University of Central Oklahoma, Edmond ... 057039
Department of Arts, University of Minnesota, Minneapolis 057458
Department of Arts, University of New Hampshire, Durham 057026
Department of Byzantine Museums, Athinai 059327
Department of Communication and Fine Arts, Lubbock Christian College, Lubbock 057378
Department of Communication Design, Fine Art and Photography, Virginia Western Community College, Roanoke 057770
Department of Communications and Visual Arts, University of North Florida, Jacksonville 057256
Department of Creative Art, Purdue University, Hammond 057176
Department of Creative Arts, Muskegon Community College, Muskegon 057501
Department of Creative Arts, Siena College, Loudonville 057373
Department of Creative Studies, University of British Columbia Okanagan, Kelowna 055133
Department of Design and Technology, Loughborough University, Loughborough 056573
Department of Design, Housing and Apparel, University of Minnesota, Saint Paul 057824
Department of Design, Lowestoft College, Lowestoft 056574
Department of Design, University of Kansas, Lawrence 057329
Department of Display and Design,

Langara College, Vancouver .. 055168
Department of Expressive Arts,
Western New Mexico University,
Silver City 057915
Department of Fine and Applied
Art, Solano Community College,
Fairfield 057071
Department of Fine and Applied Arts,
University of Nigeria, Nsukka . 056205
Department of Fine and
Communication Art, Coppin State
College, Baltimore 056702
Department of Fine and Graphic Arts,
Northwestern State University of
Lousiana, Natchitoches 057511
Department of Fine and Performing
Arts, Baruch College, City University
of New York, New York 057550
Department of Fine and Performing
Arts, Johnson State College,
Johnson 057266
Department of Fine and Performing
Arts, Saint Joseph's University,
Philadelphia 057670
Department of Fine and Performing
Arts, Saint Louis University, Saint
Louis 057814
Department of Fine and Performing
Arts, Spring Hill College, Mobile 057465
Department of Fine Art, Chipola Junior
College, Marianna 057406
Department of Fine Art, Dongeui
University, College of Arts and
Physical Education, Busan ... 056082
Department of Fine Art, Sul Ross
State University, Alpine 056646
Department of Fine Art, University of
Chester, Chester 056521
Department of Fine Art, University of
Reading, Reading 056599
Department of Fine Art, University of
Toronto, Toronto 055158
Department of Fine Art, University of
Wisconson Platteville, Platteville 057692
Department of Fine Art, Washington
State University, Pullman 057734
Department of Fine Art, West Virginia
Wesleyan College, Buckhannon 056812
Department of Fine Arts,
Montgomery 057479
Department of Fine Arts,
Taichung 056457
Department of Fine Arts, Taipei . 056459
Department of Fine Arts and Design,
Louisiana State University at
Alexandria, Alexandria 056639
Department of Fine Arts Gallery, John
C. Calhoun State Community College,
Decatur 048197
Department of Fine Arts, Ahmadu
Bello University, Samaru 056206
Department of Fine Arts, Alcorn State
University, Lorman 057361
Department of Fine Arts, Allen R. Hite
Art Institute, University of Louisville,
Louisville 057374
Department of Fine Arts, Amherst
College, Amherst 056654
Department of Fine Arts, Andong
National University, Andong .. 056078
Department of Fine Arts, Anhui
Normal University, Wuhu 055249
Department of Fine Arts, Arcadia
University, Glenside 057137
Department of Fine Arts, Art History
and Humanities, Hofstra University,
Hempstead 057194
Department of Fine Arts, Augusta
State University, Augusta 056688
Department of Fine Arts, Baltimore
City Community College,
Baltimore 056703
Department of Fine Arts, Barry
University, Miami 057431

Department of Fine Arts, Berkshire
Community College, Pittsfield . 057689
Department of Fine Arts, Bethany
College, Bethany 056732
Department of Fine Arts, Bowie State
University, Bowie 056775
Department of Fine Arts, Brandeis
University, Waltham 058057
Department of Fine Arts, Butte
Community College, Pulga .. 057733
Department of Fine Arts, Cabrini
College, Radnor 057742
Department of Fine Arts, Calhoun
State Community College,
Decatur 056990
Department of Fine Arts, Cheyney
University of Pennsylvania,
Cheyney 056869
Department of Fine Arts, Claremont
Graduate School, Claremont . 056903
Department of Fine Arts, College of
our Lady of the Elms, Chicopee 056893
Department of Fine Arts, Concordia
University, Austin 056693
Department of Fine Arts, Daytona
Beach Community College, Daytona
Beach 056986
Department of Fine Arts, Delgado
College, New Orleans 057529
Department of Fine Arts, Ellsworth
Community College, Iowa Falls 057241
Department of Fine Arts, Francis
Marion University, Florence ... 057090
Department of Fine Arts, Georgia
Southwestern State University,
Americus 056650
Department of Fine Arts, Glenville
State College, Glenville 057138
Department of Fine Arts, Gordon
College, Barnesville 056710
Department of Fine Arts, Indiana-
Purdue University, Fort Wayne . 057099
Department of Fine Arts, Kangnung
National University, College of Arts,
Gangneung 056087
Department of Fine Arts, Kirkwood
Community College, Cedar
Rapids 056849
Department of Fine Arts, Kyoritsu
Joshi Daigaku, Tokyo 056043
Department of Fine Arts, Lincoln
University, Jefferson City 057262
Department of Fine Arts, Lock Haven
University, Lock Haven 057355
Department of Fine Arts, Maryville
College, Maryville 057419
Department of Fine Arts, Miami
University, Hamilton 057174
Department of Fine Arts, Miles
Community College, Miles City 057445
Department of Fine Arts, Mississippi
Delta Community College,
Moorhead 057484
Department of Fine Arts, Morgan State
University, Baltimore 056704
Department of Fine Arts, Mount
Allison University, Sackville ... 055153
Department of Fine Arts, Nagpur
University, Faculty of Arts,
Nagpur 055826
Department of Fine Arts, New Mexico
Highlands University, Las Vegas 057325
Department of Fine Arts, Panjab
University, Chandigarh 055785
Department of Fine Arts, Phillips
County Community College,
Helena 057193
Department of Fine Arts, Porterville
Community Youth College,
Porterville 057705
Department of Fine Arts, Rider
University, Lawrenceville 057331
Department of Fine Arts, Rockford
College, Rockford 057787

Department of Fine Arts, Rutgers
State University of New Jersey,
Camden 056826
Department of Fine Arts, Saint Anselm
College, Manchester 057398
Department of Fine Arts, Saint John's
University, Jamaica 057258
Department of Fine Arts, Saint Joseph
College, West Hartford 058083
Department of Fine Arts, Saint
Lawrence University, Canton .. 056831
Department of Fine Arts, Saint Mary's
University, San Antonio 057842
Department of Fine Arts, Saint Peter's
College, Jersey City 057265
Department of Fine Arts, Spokane
Falls Community College,
Spokane 057935
Department of Fine Arts, State
University of New York College at
Potsdam, Potsdam 057713
Department of Fine Arts, Syracuse
University, Syracuse 057968
Department of Fine Arts, Texas
Southern University, Houston .. 057217
Department of Fine Arts, Trinity
College, Hartford 057186
Department of Fine Arts, Troy
University Savannah Campus,
Savannah 057882
Department of Fine Arts, University of
Colorado Boulder, Boulder 056774
Department of Fine Arts, University of
Delhi, Faculty of Music & Fine Arts,
Delhi 055790
Department of Fine Arts, University of
Hong Kong, Hong Kong 055215
Department of Fine Arts, University of
New Haven, West Haven 058085
Department of Fine Arts, University
of New Orleans Lake Front, New
Orleans 057530
Department of Fine Arts, University
of North Carolina at Wilmington,
Wilmington 058115
Department of Fine Arts, University of
Pennsylvania, Philadelphia ... 057671
Department of Fine Arts, University
of Pittsburgh at Johnstown,
Johnstown 057268
Department of Fine Arts, University of
the Free State, Bloemfontein .. 056337
Department of Fine Arts, Waynesburg
College, Waynesburg 058076
Department of Fine Arts, Wiley
College, Marshall 057414
Department of Fine Arts, Xavier
University of Louisiana, New
Orleans 057531
Department of Graphic
Communcations, Pennsylvania
College of Technology,
Williamsport 058111
Department of Graphic Design, Saint
Lawrence College, Kingston .. 055135
Department of Graphic Design,
Technological Education Institute
Athens, Athinai 055752
Department of Graphic Design,
Woodbury University, Burbank . 056817
Department of History in Art,
University of Victoria, Victoria . 055171
Department of History of Art and
Architecture, Brown University,
Providence 057727
Department of History of Art and
Architecture, Harvard University,
Cambridge 056823
Department of History of Art and
Architecture, University of Pittsburgh,
Pittsburgh 057688
Department of History of Art, Bryn
Mawr College, Bryn Mawr ... 056810
Department of History of Art, North-

West University, Potchefstroom
Campus, Potchefstroom 056350
Department of History of Art, Ohio
State University, Columbus ... 056948
Department of History of Art,
University of Michigan, Ann
Arbor 056659
Department of Humanities-Art
Program, Bowling Green State
University, Huron 057226
Department of Humanities, Chemeketa
Community College, Salem ... 057829
Department of Language and Visual
Arts, Charleston 056855
Department of Liberal Arts and
Sciences, Nihon University,
Fujisawa 055998
Department of Literature, Language
and the Arts (Art History), Open
University of Israel, Ramat-Aviv 055868
Department of Modern Art and
Architecture, Beijing Dongfang
University, Beijing 055195
Department of Music Science and
Art, University of Macedonia,
Thessaloniki 055754
Department of Oriental Art and
Department of Art Design, Nankai
University, School of Humanities,
Tianjin 055244
Department of Performing and Visual
Arts, American University Cairo,
Cairo 055329
Department of Performing and Visual
Arts, Himachal Pradesh University,
Shimla 055834
Department of Studio Arts and Art
History, Hamline University, College
of Liberal Arts, Saint Paul 057825
Department of the Arts, Bucks County
Community College, Newton .. 057579
Department of the Arts, Rensselaer
Polytechnic Institute, Troy 058007
Department of the History of Art and
Architecture, Dublin 055860
Department of the History of Art,
Cornell University, Ithaca 057245
Department of the History of
Art, Johns Hopkins University,
Baltimore 056705
Department of the History of Art,
University of California Riverside,
Riverside 057767
Department of the History of Art, Yale
University, New Haven 057523
Department of the Treasury Museum,
Washington 054103
Department of Visual and Performing
Arts, GuanXi Normal University,
Guilin 055211
Department of Visual and Performing
Arts, Rutgers State University of New
Jersey, Newark 057570
Department of Visual and Performing
Arts, Springfield College,
Springfield 057939
Department of Visual and Performing
Arts, University of Colorado, Colorado
Springs 056931
Department of Visual Art, Frostburg
State University, Frostburg ... 057112
Department of Visual Art, Holyoke
Community College, Holyoke .. 057206
Department of Visual Art, Interlochen
Arts Academy, Interlochen 057238
Department of Visual Arts,
Worcester 058136
Department of Visual Arts and
Humanities, Florida A & M University,
Tallahassee 057978
Department of Visual Arts and New
Media, State University College at
Fredonia, Fredonia 057108
Department of Visual Arts and

Technology, San Antonio College, San Antonio 057843
Department of Visual Arts, Adams State College, Alamosa 056630
Department of Visual Arts, Asbury College, Wilmore 058117
Department of Visual Arts, Brigham Young University, Provo 057731
Department of Visual Arts, California State University San Bernardino, San Bernardino 057848
Department of Visual Arts, Casper College, Casper 056841
Department of Visual Arts, Chowan College, Murfreesboro 057498
Department of Visual Arts, Dowling College, Oakdale 057601
Department of Visual Arts, Dutchess Community College, Poughkeepsie 057715
Department of Visual Arts, Gettysburg College, Gettysburg 057132
Department of Visual Arts, Lakehead University, Thunder Bay 055157
Department of Visual Arts, Loyola University of New Orleans, New Orleans 057532
Department of Visual Arts, McNeese State University, Lake Charles . 057307
Department of Visual Arts, Otterbein College, Westerville 058090
Department of Visual Arts, Pensacola Junuior College, Pensacola .. 057658
Department of Visual Arts, School of Visual Arts, University of Northern Colorado, Greeley 057151
Department of Visual Arts, Southeastern Louisana University, Hammond 057177
Department of Visual Arts, Texas Lutheran University, Seguin .. 057901
Department of Visual Arts, Texas Woman's University, Denton .. 056999
Department of Visual Arts, Ulster County Community College, Stone Ridge 057957
Department of Visual Arts, University of Dayton, Dayton 056984
Department of Visual Arts, University of North Carolina at Charlotte, Charlotte 056859
Department of Visual Arts, University of Regina, Regina 055152
Department of Visual Arts, University of Victoria, Victoria 055172
Department of Visual Arts, University of Western Ontario, London .. 055137
Department of Visual Arts, Weber State College, Ogden 057611
Department of Visual Arts, Yakima Valley Community College, Yakima 058137
Department of Visual Arts, York University, North York 055147
Department of Zoology Museum, Kampala 043104
Departamento Arquitectura e Urbanismo, Universidade de Franca, Franca 055075
Departamento Artes Visuais, Universidade Federal de Paraiba, João Pessoa 055081
Departamento de Arquitectura y Diseño Gráfico, Universidad Cuauhtémoc, Aguascalientes 056136
Departamento de Artes, Universidad de las Américas Puebla, Puebla .. 056154
Departamento de Bellas Artes, Pontificia Universidad Católica de Puerto Rico, Ponce 056254
Departamento de Historia de Arte, Universidad de Murcia, Murcia 056377
Departamento de Historia del Arte, La Laguna 056364
Departamento Desenho Industrial,

Universidade de Franca, Franca 055076
Departure Lounge, Burleigh Heads 098824
DePaul University Art Gallery, Chicago 047654
DePauw University Anthropology Museum, Greencastle 049235
Depecker, Bernard, Hazebrouck 068803
Depejeteme, Mount Druitt 061906
Depen d'Ans, Den Haag 083076
Dépendance, Bruxelles 100590
Dephi Antiques, Montréal 064414
Depieds, Jean-Christophe, Paris 071249
D'Epoca, Bologna 079833
D'Epoca di Balmelli Renzo, Massagno 087665
Depoid, Concarneau 103622
Depósito del Automóvil, La Habana 009140
't Depôt, Maastricht 083462
Depot and Atelier, İstanbul 088114
Dépôt Archéologique, Saint-Thièbry 015573
Dépôt Archéologique du Fort Vauban, Alès 011592
Dépôt Art Contemporain, Sion .. 116640
The Depot Arts Space, Auckland 112759
Dépot Broc, Béziers 067173
Dépôt de Fouilles Gallo-Romaines et Préhistoriques de Montmaurin, Montmaurin 014274
Dépôt de Jadis, Montargis 070183
Dépôt des Artistes, Levallois-Perret 103995
Dépôt des Equipages, Le Conquet 069254
Dépot des Islettes, Les Islettes . 069412
Dépôt du Château, Vincennes .. 074240
Dépôt-Musée Diocésain, Metz .. 014150
Depot Museum and Fort Sedgwick Museum, Julesburg 049844
The Depot Museum Complex, Henderson 049442
Depot Museum Complex, Gilliam County Historical Society, Condon 047962
Depot Museum, Mammoth Spring State Park, Mammoth Spring . 050618
Depot on Chapel, Windsor, Victoria 062436
Depot Park Museum, Sonoma .. 053389
Dépôt Richelieu, Rueil-Malmaison 072639
Depot Square, Cincinnati 120339
Depot Street Antiques, Columbus 093012
Dépôt-Vente, Tarare 073618
Dépôt Vente Casino, Hautmont . 068799
Dépôt Vente de Coutances, Coutances 068207
Dépôt Vente de la Vallée, Chevreuse 067971
Dépôt Vente de Niort, Niort 070711
Dépôt Vente de Paris, Paris ... 071250
Dépot-Vente du 9ème, Paris ... 071251
Dépôt Vente du Particulier, Mondeville 070157
Dépôt Vente du Particulier, Paris 071252
Dépot-Vente la Fouine, Montargis 070184
Dépot Vente la Réserve, Nîmes . 070694
Dépôt Vente Particuliers, Valence (Drôme) 073939
Deppe, Mary, Hannover 076368
Depping, Gerit & Norbert, Münster 130425
Deppner, Market Drayton 090708
Depreciation Lands Museum, Allison Park 046416
DePree Art Center and Gallery, Hope College, Holland .. 049505, 057205
Depression Era, Saint Paul ... 096819
Depression Modern, New York .. 095347
Depretz, Bernard, Paris 104681, 125703, 128981
Dept of Applied Communication,

Stockholms Universitet, Stockholm 056398
Dept. Creative Industries and Technologies, York St. John University, York 056616
Depth of Field, Minneapolis 094908
Depuydt, Bruxelles 128025
Deqing Museum, Deqing, Guangdong 007737
Deqing Museum, Deqing, Zhejiang 007738
Der Ausguck, Stellwerk Gwt, Recklinghausen 108449
Der-Horng Art Gallery, Tainan .. 116897
Deracourt, Dominique, Bruxelles 063397
Deramo, Donald, Pittsburgh ... 096277
D'Eramo, Walter, Firenze 080130
D'Erasmo, Antonio, Bari 079766, 079767
Derby Arts, Derby 061399
Derby City Antique Mall, Louisville 094537
Derby Museum and Art Gallery, Derby 044082
Derby Park Traders Circle, Louisville 094538
Derbylius, Milano 110300, 137673, 142719
Derbyshire Clocks, Glossop ... 089337, 134407
D'Ercole, Renato, Trieste 132219
Dereks Militaria, Tønsberg 084375
Derenik Demirchyan House-Museum, Yerevan 000726
Derenzi, Castello de la Plana .. 115015
Deres, Uwe, Essen 075671
Derhaag, S.H., Maastricht 083463
Déri Múzeum, Debrecen 023593
DeRicci Gallery, Madison 050594
Derigny, Patrice, Saint-Nazaire (Loire-Atlantique) 125805
Dering-Schunk, Elisabeth, Aschaffenburg 106007
Dering, E., Aschaffenburg 106006
Derissi, Chiraz 109518
Derka, Igor, Bratislava 085545
Derks, Steven, Tucson 124728
Derksen, Arnhem ... 082937, 126276
Dernier Quartier Général de Napoléon, Genappe 003613
Dernuet, Testor & Codsi, Montréal 128181
Derom, Patrick, Bruxelles 100591
Derouard, Michel & Jacqueline, Antoigné 066682
Deroyan, Armand, Paris 071253, 071254, 128982, 128983
Derrey, Jérémie, Louviers 069651
Derribos Sales, Castello de la Plana 085932
Derrière la Dune, Le Touquet-Paris-Plage 133968, 103969
Derrière la porte, Toulon (Var) .. 073708
Derrière les Fagots, Clermont-Ferrand 068022
Derrive, Nicole, Hyères 068854
Derrymore House, Bessbrook .. 043580
Derrymane House, Caherdaniel . 024838
Dersigny, Patrice, Amiens 128437
Dersu, Buenos Aires 060894
Dert Restauratie, Hilversum ... 132541
Dervieux, Denis, Paris 071255
Dervieux, Ferdinand, Avignon .. 066878
Dervieux, Frédéric, Arles 066722
Dervish, Bellingen 098730
Dervish Pasha Mansion Ethnographical Museum, Lefkoşa 033471
Derwentside Antiques, Belper .. 088436
Deržavnyj Ajvazovskyj Kartynna Galereja, Feodosija 043154
Deržavnyj Istoriko-kul'turnyj Zapovidnyk, Kam'janec'-Podil's'kyj ... 043166
Deržavnyj Istoriko-kul'turnyj Zapovidnyk Bat'kivščyna Tarasa

Ševčenka, Ševčenkove 043297
Deržavnyj Istoryko-kul'turnyj Zapovidnyk Pole Poltavskoï Bytvy, Poltava 043281
Deržavnyj Muzej Teatral'noho, Muzyčnoho ta Kinomystectva Ukraïny, Kyïv 043192
Deržavnyj Muzej Ukraïns'koho Narodnoho Dekoratyvnoho Mystectva, Kyïv 043193
Derzon, David, Milwaukee 094849
Derzypolsky, Dijon 068324
Des 3 Vallées, Saint-Pierre-d'Albigny 073255
Des 4 Moulins, Vallet 073953
Des Arts et des Hommes, Ploërmen 105359
Des Hommes dans la Guerre de 1870, Ancienne Boulangerie de Citadelle, Bitche 012051
Des Journées Entières dans les Arbres, Gordes 103771
Des Lee Gallery, Washington University School of Art, Saint Louis 052798
Des Moines Art Center, Des Moines 048282
Des Plaines Historical Museum, Des Plaines 048290
Des Présents du Passé, Courtenay 068197
Desa, Białystok 084488
Desa, Bydgoszcz 084495
Desa, Kielce 113515, 126407
DESA-Dzieła Sztuki i Antyki, Gdynia 126403
DESA-Dzieła Sztuki i Antyki, Katowice 084531, 126404
DESA-Dzieła Sztuki i Antyki, Kraków 126408, 126409, 126410
DESA-Dzieła Sztuki i Antyki, Poznań 084637, 126415
Desa Unicum, Warszawa 084713
Desa Unicum Galeria Koszykowa, Warszawa 084714, 113819
Desa Unicum Galeria Staromiejska, Warszawa 084715, 143521
Desa Unicum Galerie Nowy Świat, Warszawa 084716
Desart, Alice Springs 060990
Desaute, Sylvie, Pont-à-Mousson 072237
Desbois, Christian, Paris 104682, 136891
Desbordes, Jean-Pierre, Huisseau-en-Beauce 068845
DesBrisay Museum and Exhibition Centre, Bridgewater 005475
Descamps, Paris 071256
Descamps, C., Mons 063794
Descamps, Claire, Toulon (Var) . 073709
Descant, Toronto 138577
Descarsin, Brienon-sur-Armancon 103469
Descateaux, Juany, La Rochelle 069085
Desch-Drexler, Pinkafeld 140021
Deschamps, Alain, Theminettes 073644
Deschamps, Jean-Pierre, Montreuil-Bellay 070323
Deschamps, Philippe, Paris ... 071257
Deschamps, Pierre, Gisors 068671
Deschamps, Sylvie, Harquency . 068794
Deschler, Marcus, Berlin 106205
Deschner, Chemnitz 141671
Deschutes County Historical Society Museum, Bend 046913
Descombes, Genève 144074
Descotis, Pierre-Marie, Vendôme 074010
Descours, Michel, Lyon 069729, 104078, 140893
Descours, Thierry, Nice 070595
Descubre Museo Interactivo de Ciencia y Tecnología, Aguascalientes 030807
Désert Aride, Saulieu 073428
Desert Art Distributors, Norwood 062012

Desert Art Shop, Kalgoorlie 099165
Desert Artisans Gallery, Tucson . 124729
Desert Caballeros Western Museum,
Wickenburg 054399
Desgans, Patrick, Villegouge .. 129347
Desha County Museum, Dumas 048388
Deshine Bijutsukan, Nagoya 029114
Design, Reichertshausen 138690
Design & Arts College of New Zealand,
Christchurch 056191
Design 18, New York 095348
Design 70, Nice 070596
Design Academy, Tbilisi 055433
Design Akademie Berlin, Hochschule
für Kommunikation und Design (FH),
Berlin 055459
Design and Art in Greece,
Athinai 138765
Design and Artists Copyright Society,
London 060108
Design and Arts, Isernhagen .. 076623,
125994
Design Arts Gallery, Drexel University,
College of Media Arts & Design,
Philadelphia 052040
Design Austria, Wien . 058247, 136579
Design Avenue, Almere 082642
The Design Center at Philadelphia
University, Philadelphia 052041
Design Centrum České Republiky,
Brno 009462
Design Circus Gallery,
Jacksonville 121244
Design Connections, Oakland .. 123249
Design Consortium, Cincinnati . 120340
Design Council, London 060109
Design de Museu, Rio de Janeiro 128113
Design Exchange, Toronto 007171
Design Factory, Oundle 118861
Design Gallery 1850–1950,
Westerham 091798, 135210
Design Gallery 52, Cannes ... 103506
The Design History Society,
Oxford 060110
Design Images, San Francisco . 124268
Design Issues, Pittsburgh 139356
Design Marlowe Galerie Ambiente,
Wien 062925
Design Modern, Los Angeles .. 094312
Design Museum, London 044958
Design Museum & Design Gallery,
International Design Center Nagoya,
Nagoya 029115
Design Museum Gent, Gent .. 003622
Design Network Switzerland,
Zürich 059932
Design Shop, Louisville 094539
Design Škola Zlín, Vysoká Škola
Uměleckoprůmyslová v Praze,
Zlín 055302
Design Utopia, Los Angeles .. 094313
Design Week Action Pack,
London 139172
Design Zentrum Bremen,
Bremen 106578
Designarts, Chicago 120109
Designated Hatter, Detroit ... 093360
Designer Antiques, Atlanta 092124
Designer Cargo Furniture, Los
Angeles 094314
Designer Renaissance, Nashville 094993
Designerinnen-Forum e.V.,
Hamburg 058818
Designers Craft, Tucson 097653
Designers Furniture and Finishing,
Pittsburgh 096278, 136131
Designer's Rugs, Los Angeles . 094315
Designing Block, Saint Louis .. 096688
Designmuseo, Helsinki 010815
Designs by Kathryn, Calgary ... 101229
Designs in Gold, Saint Louis .. 096689
Designs in Pine, Los Angeles .. 094316,
135714
Designs UK, Bristol 117445

Designscape Gallerie, Mumbai . 109428
Désilets, Pierre, Caen 067563
Desing Artifact Museum, Sinpu . 042463
Desiree, Inning am Ammersee . 107465
Desloubières, Jean-Marc, Paris . 071258,
128984
Desmettre, Nadir, Peronne ... 072064
Desmeules, Jean-François, Pully 087743
Desmond Castle, Kinsale 024946
Desmontils, Guy, Vitré 074255
Desmoulin at the Granary,
Newbury 118778
Desnoues, Bruno, Paris 128985
Desnoues, Jacques, Bagnolet .. 066915
Desormeaux, Fort-de-France ... 142977
Desoto National Memorial,
Bradenton 047169
Despalles, Mainz 137369
Despalles, Paris 136892
Despard, Hobart 099133
Desparrain, Christian, Clermont-
Ferrand 068023
Despesel, Port-à-Binson 072283
Despesel, Jean, Binson-et-
Orquigny 067226
Despića Kuća, Sarajevo 004273
Desprairies, Monique, Saint-Ouen 073068
Desroches, Bernard, Montréal .. 101377
Desroches, Brigitte, Montréal .. 101378
Desroches, Françoise, Paris ... 128986
Desrues, Gérard, Mont-Dol ... 070174,
128800
Dessain et Tolra, Paris 136893
Dessapt Editions, Viry-Chatillon . 105881,
137101
Dessarzin, Jean-Daniel, Colombier
(Neuchâtel) 087380
Dessaules, Robert, Paris 071259
Dessaut, Benoît, Chalon-sur-
Saone 125549
Dessein, Helsinki 066310
Dessemond, J.- P., Aubusson .. 066794,
128450
Dessers, Hasselt 100714
Dessinee Gallery, Toronto 101597
Destang, Moissac 070141
Destek Reasürans Sanat Galerisi,
İstanbul 117026
Desterne, Joëlle, Villeneuve-lès-
Avignon 105872
Destina, Eric, Philadelphia ... 136083
Destiny Studios, Pittsburgh ... 123537
Destito, Giovanni, Roma 131900, 131901
Destito, Saverio Umberto, Roma 131902
El Desvan, Barcelona 085764
El Desvan, Ourense 086321
El Desvan de Madrid, Madrid .. 086168
El Desvan de Olavide, Madrid .. 086169,
133324
El Desvan de Tendilla,
Guadalajara 086046
Det Hanseatiske Kvarter, Bergen 084215
Det Lille Galleri, Stavanger 084364
Details, Indianapolis 121177
Details Art, Tucson 124730
Detalhe, Lisboa 132934
Détections Loisir, Toulouse ... 073763
Detering, Houston 145168
Dethlefs, Hamburg 129958
Dętkoś, Aleksander, Bydgoszcz . 113440
Detlefsen-Museum, Glückstadt . 018718
Detling International Antiques and
Collectors Fair, Detling 098373
Au Détour d'une Promenade,
Bonnieux 067295
Détraz, Gérard, Basel 116062
Detre, Oakland 095889
Detroit Antiques Mall, Detroit .. 093361
Detroit Artists Market, Detroit .. 048298
Detroit Contemporary, Detroit . 048299
Detroit Focus, Royal Oak 060462
Detroit Focus Gallery, Detroit .. 120805
Detroit Historical Museum,
Detroit 048300

The Detroit Institute of Arts,
Detroit 048301
Detroit Repertory Theatre Gallery,
Detroit 048302
Detskaja Kartinnaja Galereja,
Samara 037706
Detskij Muzej, Čerepovec 036941
Detskij Muzej Buratino-Pinokkio,
Moskva 037251
Detskij Muzej Dom-Skazok Žili-byli,
Moskva 037252
Detskij Muzej Šantyrak, Almaty . 029797
Detten, Rupert von, Aufseß ... 129392
Dettinger Mayer, Lyon 104079
Dettori, Francesca, Genova ... 131306
Dettori, Renato, Milano 131434
Detursa, Madrid 115179
Deubner, Köln 137305
Deuchar & Son, A.S., Perth ... 090976
Deuerlein, Thomas, Nürnberg .. 142239
Deumens, Johan, Haarlem ... 083324,
137858, 143159
Deupoor, E., Bruxelles 140172
Deurbergue, Dominique,
Versailles 074050
Deuring, Gisela, Schnelldorf .. 078381,
078382
Deuschl, Horst, Furth bei
Göttweig 062564
Deuschl, Walter, Haag,
Niederösterreich 062622
Deusserhaus, Monheim am
Rhein 020453
Deutsch-Deutsches Museum
Mödlareuth, Museum zur Geschichte
der deutschen Teilung, Töpen . 022164
Deutsch-Französisches Puppen und
Spielzeugmuseum, Schweigen-
Rechtenbach 021780
Deutsch-Salsedo, Antoine, Paris 061242
Deutsch, André, London 138211
Deutsch, Berthold & Sigrid,
Jockgrim 076654
Deutsch, H. & W., London ... 090072
Deutsch, Paul Ernst, Wien ... 062926,
125271
Deutsch, R., Freiburg im
Breisgau 075857
Deutsch, Robert A., Saint Louis . 096690
Deutsche Barockgalerie,
Kunstsammlungen und Museen
Augsburg, Augsburg 016649
Deutsche Exlibris-Gesellschaft e.V.,
Berlin 058819
Deutsche Gesellschaft für Christliche
Kunst e.V., München 058820
Deutsche Gesellschaft für
Photographie e.V., Köln 058821
Deutsche Guggenheim, Berlin . . 017102
Deutsche Kinemathek – Museum
für Film und Fernsehen, Filmhaus,
Berlin 017103
Deutsche Numismatische
Gesellschaft – Verband der
Deutschen Münzvereine e.V.,
Speyer 058822
Deutsche Raumfahrtausstellung e.V.,
Morgenröthe-Rautenkranz .. 020465
Das Deutsche Stickmuster-Museum
Celle, Celle 017722
Deutsche Zeitungsmuseum, Stiftung
Saarländischer Kulturbesitz,
Wadgassen 022363
Deutscher Akademischer
Austauschdienst – Berliner
Künstlerprogramm, Berlin ... 058823
Deutscher Designertag e. V., Sektion
Design im Deutschen Kulturrat,
Hamburg 058824
Deutscher Künstlerbund e.V.,
Berlin 058825, 106206
Deutscher Kunsthandelsverband e.V.,
Berlin 058826
Deutscher Kunststudenten-Verband,

Köln 058827
Deutscher Kunstverlag GmbH, Berlin/
München, Berlin 137059
Deutscher-Menzies, Malvern,
Victoria 125209
Deutscher Museumsbund e.V. –
Bulletin, Berlin 138691
Deutscher Museumsbund e.V., Büro
Berlin, Berlin 058828
Deutscher Verein für Kunstwissenschaft
e.V., Berlin 058829, 138692
Deutscher Verlag für
Kunstwissenschaft, Berlin ... 137060
Deutscher Werkbund, Darmstadt 058830
Deutscher Werkbund Baden-
Würtenberg, Karlsruhe 058831
Deutscher Werkbund Hessen, Frankfurt
am Main 058832, 106981
Deutscher Werkbund Nordrhein-
Westfalen, Köln 058833
Deutscher Werkbund Rheinland-Pfalz
e.V., Mainz 058834
Deutscher Werkbund Saarland,
Saarbrücken 058835
Deutscher Werkbund Sachsen e.V.,
Dresden 058836
Deutsches Albert-Schweitzer-
Zentrum, Archiv, Museum und
Forschungsstätte, Frankfurt am
Main 018410
Deutsches Apotheken-Museum im
Heidelberger Schloß, Heidelberg 019132
Deutsches Architekturmuseum,
Frankfurt am Main 018411
Deutsches Auswandererhaus
Bremerhaven, Bremerhaven .. 017605
Deutsches Automuseum Schloss
Langenburg, Langenburg 019904
Deutsches Bauernkriegsmuseum,
Böblingen 017450
Deutsches Baumaschinen-
Modellmuseum, Weilburg ... 022478
Deutsches Bergbau-Museum,
Bochum 017433
Deutsches Bernsteinmuseum, Ribnitz-
Damgarten 021408
Deutsches Bienenmuseum,
Weimar 022490
Deutsches Blinden-Museum,
Berlin 017104
Deutsches Boxsport-Museum,
Sagard 021569
Deutsches Buch- und Schriftmuseum
der Deutschen Nationalbibliothek
Leipzig, Deutsche Nationalbibliothek,
Leipzig 019981
Deutsches Buchbindermuseum,
Mainz 020225
Deutsches Chemie Museum,
Merseburg 020373
Deutsches Damast- und
Frottiermuseum, Großschönau . 018862
Deutsches Dampflokomotiv-Museum,
Neuenmarkt 020701
Deutsches Drachenmuseum und
Stadtmuseum, Museum des
Heimatkreises Bischofteinitz/
Sudetenland, Furth im Wald . 018566
Deutsches Drahtmuseum, Altena 016524
Deutsches Edelsteinmuseum, Idar-
Oberstein 019376
Deutsches Eichendorff-Museum,
Wangen im Allgäu 022419
Deutsches Elfenbeinmuseum Erbach,
Erbach, Odenwaldkreis 018229
Deutsches Epilepsiemuseum Kork,
Kehl 019568
Deutsches Erdölmuseum Wietze,
Wietze 022648
Deutsches Fächer Museum, Barisch
Stiftung, Bielefeld 017374
Deutsches Fastnachtmuseum,
Kitzingen 019637
Deutsches Feld- und

Kleinbahnmuseum, Deinste ... 017844
Deutsches Feuerwehr-Museum,
Fulda 018556
Deutsches Film- und Fototechnik
Museum, Deidesheim 017841
Deutsches Filmmuseum, Deutsches
Filminstitut DIF e.V., Frankfurt am
Main 018412
Deutsches Fleischermuseum,
Böblingen 017451
Deutsches Freimaurermuseum,
Bayreuth 016986
Deutsches Gartenbaumuseum Erfurt,
Erfurt 018238
Deutsches Glasmalerei-Museum
Linnich, Linnich 020092
Deutsches Glockenmuseum,
Greifenstein 018804
Deutsches Goldmuseum,
Schalkau 021627
Deutsches Goldschmiedehaus,
Gesellschaft für Goldschmiedekunst,
Hanau 019035
Deutsches Harmonikamuseum,
Trossingen 022210
Deutsches Hirtenmuseum,
Hersbruck 019229
Deutsches Hopfenmuseum,
Wolnzach 022756
Deutsches Hugenotten-Museum, Bad
Karlshafen 016795
Deutsches Hygiene-Museum Dresden,
Dresden 017978
Deutsches Jagd- und
Fischereimuseum, München .. 020515
Deutsches Kaltwalzmuseum,
Hagen 018911
Deutsches Kartausen-Museum,
Buxheim 017708
Deutsches Kartoffelmuseum
Fußgönheim, Fußgönheim 018573
Deutsches Kinderwagenmuseum,
Zeitz 022851
Deutsches Kleingärtnermuseum,
Leipzig 019982
Deutsches Klingenmuseum Solingen,
Solingen 021897
Deutsches Knopfmuseum,
Bärnau 016933
Deutsches Kochbuchmuseum,
Dortmund 017955
Deutsches Korbmuseum, Michelau in
Oberfranken 020398
Deutsches Kunst- und Jagdkabinett,
München 077484
Deutsches Landwirtschaftsmuseum,
Stuttgart 022027
Deutsches Landwirtschaftsmuseum
Schloss Blankenhain,
Freilichtmuseum der ländlichen
Kultur, Technik und Arbeit
für Mitteldeutschland,
Crimmitschau 017783
Deutsches Märchen- und
Wesersagenmuseum, Bad
Oeynhausen 016837
Deutsches Maler- und Lackierer-
Museum, Hamburg 018980
Deutsches Marinemuseum,
Wilhelmshaven 022657
Deutsches Medizinhistorisches
Museum, Ingolstadt 019408
Deutsches Meeresmuseum, Museum
für Meereskunde und Fischerei –
Aquarium, Stralsund 022009
Deutsches Museum, München . 020516
Deutsches Museum Bonn, Bonn 017487
Deutsches Museum Flugwerft
Schleißheim, Oberschleißheim . 020919
Deutsches Museum für Schulkunst,
Hagen 018912
Deutsches Museum Verkehrszentrum,
München 020517
Deutsches Musikautomaten Museum,

Außenstelle des Badischen
Landesmuseums Karlsruhe,
Bruchsal 017631
Deutsches Nationalkomitee des
Internationalen Museumsrates ICOM,
Berlin 058837
Deutsches Ofenmuseum,
Burrweiler 017702
Deutsches Orthopädisches Geschichts-
und Forschungsmuseum, Frankfurt
am Main 018413
Deutsches Panzermuseum Munster,
Munster 020620
Deutsches Peitschenmuseum,
Burladingen-Killer 017701
Deutsches Pferdemuseum,
Hippologisches Institut, Verden 022317
Deutsches Phonomuseum und
Uhrensammlung, Sankt Georgen im
Schwarzwald 021591
Deutsches Pinsel- und
Bürstenmuseum, Bechhofen .. 017006
Deutsches Plakat Museum,
Essen 018291
Deutsches Polizeimuseum,
Salzkotten 021579
Deutsches Porzellanmuseum,
Porzellanikon in Hohenberg,
Hohenberg an der Eger ... 019301
Deutsches Röntgen-Museum,
Remscheid 021361
Deutsches Salzmuseum –
Industriedenkmal Saline Lüneburg,
Lüneburg 020178
Deutsches Schaustellermuseum,
Lambrecht 019864
Deutsches Schiefermuseum
Steinach/Thür., Steinach, Kreis
Sonneberg 021972
Deutsches Schifffahrtsmuseum,
Institut der Leibniz-Gemeinschaft,
Bremerhaven 017606
Deutsches Schilder- und Lichtreklame-
Museum, Lahr 019851
Deutsches Schloss-und
Beschlägemuseum, Velbert .. 022308
Deutsches Schreibmaschinenmuseum,
Bayreuth 016987
Deutsches Schuhmuseum Hauenstein,
Hauenstein 019106
Deutsches Schustermuseum,
Burgkunstadt 017695
Deutsches Schweinemuseum
Ruhlsdorf, Teltow 022122
Deutsches Segelflugmuseum mit
Modellflug, Gersfeld 018666
Deutsches Sielhafenmuseum in
Carolinensiel, Wittmund 022704
Deutsches Skimuseum, Planegg 021143
Deutsches Spiele-Archiv,
Marburg 020270
Deutsches SPIELEmuseum e.V,
Chemnitz 017729
Deutsches Spielkartenmuseum,
Landesmuseum Württemberg,
Leinfelden-Echterdingen 019962
Deutsches Spielzeugmuseum,
Sonneberg 021915
Deutsches Sport- und Olympia
Museum, Köln 019674
Deutsches Straßenmuseum,
Germersheim 018657
Deutsches Tabak- und
Zigarrenmuseum, Museum Bünde,
Bünde 017670
Deutsches Tabakpfeifenmuseum,
Oberelsbach 020892
Deutsches Tanzarchiv Köln, SK
Stiftung Kultur, Köln 019675
Deutsches Tapetenmuseum,
Museumslandschaft Hessen Kassel,
Kassel 019530
Deutsches Technikmuseum Berlin,
Berlin 017105

Deutsches Telefon-Museum,
Morbach 020463
Deutsches Textilmuseum Krefeld,
Krefeld 019784
Deutsches Theatermuseum,
München 020518
Deutsches Traktoren und Modellauto
Museum e.V., Paderborn .. 021059
Deutsches Uhrenmuseum, Furtwangen
im Schwarzwald 018571
Deutsches Verpackungs-Museum,
Heidelberg 019133
Deutsches Vulkanmuseum,
Mendig 020364
Deutsches Weinbaumuseum,
Oppenheim 020998
Deutsches Werkzeugmuseum,
Historisches Zentrum der Stadt
Remscheid, Remscheid ... 021362
Deutsches Zementmuseum,
Hemmoor 019190
Deutsches Zinnfigurenmuseum,
Kulmbach 019829
Deutsches Zollmuseum,
Hamburg 018981
Deutsches Zweirad- und NSU-
Museum, Neckarsulm 020656
Deutschheim State Historic Site,
Hermann 049451
Deutschmann, Michael, Kassel . 076716,
107530
Deutschordensmuseum Bad
Mergentheim, Bad Mergentheim 016822
Deux Anglès, Flers 103744
Les Deux Ebenistes, Marseille . 128760
Les Deux Illes, New York 135907
Les Deux-Mondes, Paris 141089
Les Deux Orphelines, Paris .. 071261
Deux Rue Antiquités, Caluire-et-
Cuire 067607
Deux Rue Antiquités,
Villeurbanne 074203
Deux, Henri, Delft 083030
Deuxièmement, Montréal 064415
Deva, Hartley Wintney 089452
Deva, Marseille 128761
Déva, New York 122534
DeValera Museum and Bruree
Heritage Centre, Bruree ... 024836
Devaney, Carol, Dublin 079544
Devar, Minneapolis 121981
Devauchelle, Alain, Paris 128987
Devaud, René, Sarras 073414
DeVaults, Tulsa 097726
Devaux, Alain, Camps-en-
Amienois 067624
Devaux, F.J., Bruxelles 128026
Devaux, Ghislaine, Lyon 069730
Devaux, J.F., Charleroi 063599
Devaux, Jean-Claude, Caen ... 067564
Devaux, Jean-Luc, Moulins
(Allier) 140970
Deveau, Robert, Toronto 101598, 125381
Deveaud, Laurent, Bois-le-Roi .. 067271
Deveaux, Los Angeles 094317
Deveko, Fourmies 068602
Devereaux, Emile, New Orleans 095087
Devereux Antiques, Linden Park 061738
Devic, Auch 066800
Devilé, Leiden 083428
Deville, Mâcon 069792
DeVille, New Orleans 145307
Deville, Saint-Étienne 072761
Deville, Bernard, Bernot-par-
Guise 067113
Devillé, J., Liège 063752
Devilleneuve, Philippe, Tonnerre 125821
Devillers, Bonnières 067293
Devillers, Liège 063753
Devillers, Charlie, Amiens ... 066569
Devil's Coulee Dinosaur Heritage
Museum, Royal Tyrrell Museum of
Paleontology, Warner 007355
Devils Tower Visitor Center, Devils

Tower 048319
Devin-Hürlimann, Barbara,
München 126046
Devisme, Bruno, Amiens 066570
Devisme, Bruno, Flesselles ... 068542
Devivi, Raymond, Saint-Paul-de-
Vence 105583
Devlay, Philippe, Cairo 066238
Devo, Brive-la-Gaillarde 067506
Devon and Cornwall Constabulary
Museum, Exeter 044310
Devon Gallery, Vancouver 101773
Devon Guild of Craftsmen, Bovey
Tracey Newton Abbot 060111
Devon Guild of Craftsmen Gallery,
Bovey Tracey 043684
Devon Pottery Collectors Group, Saint
Albans 060112
Devonia, Boston 092464
Devonport Antiques, North Shore 084031
Devonport Gallery and Arts Centre,
Devonport 001025
Devonport Maritime Museum,
Devonport 001026
Devons, Judith, Leicester 118093
DeVos Art Museum, Northern
Michigan University, Marquette 050704
Devotional Art, Los Angeles ... 094318
Devoto, Genova 110101
Devroe, J., Bruxelles 140173
Dew, Roderick, Eastbourne ... 144381
Dewa Roman Experience,
Chester 043932
Dewa San-Zan History Museum,
Tsuruoka 029660
Dewanoyuki Sake Museum,
Tsuruoka 029661
Dewards, Greenslopes 099097
Dewart, Bruxelles 100592
Dewasmes, Jean-Charles, Arras 066572
Dewazakura Bijutsukan, Tendo . 029474
Dewberry Valley Museum,
Dewberry 005704
Deweer Gallery, Otegem 100812
Dewenter, Runhild, Krefeld ... 130192
Dewey Hotel, Dewey 048320
Dewey Museum, Warminster .. 046053
Dewi Lewis, Stockport 138284
Dewindt, Bruxelles 063398
DeWitt County Historical Museum,
Cuero 048083
DeWitt Historical Society Museum,
Ithaca 049737
DeWitt Stetten Jr., Museum of Medical
Research, Bethesda 046956
DeWitt Wallace Decorative Arts
Museum, Williamsburg ... 054412
Dewsbury Museum, Dewsbury . 054401
Dewulf, Patrick, Froidestrees ... 068616
Dexing Museum, Dexing 007739
Dexter Area Museum, Dexter . 048323
Dexter Historical Society Museum,
Dexter 048322
Dextera, Bilbao 133163
Dey Mansion, Washington's
Headquarters Museum, Wayne 054251
Dey, J.C.A., Marnes-la-Coquette 069885
Deya, Buenos Aires 060895
Deyang Museum, Deyang 007740
Deygas, Bernard, Lyon 128710
Deyl, Wolfgang, Salzburg 062790
Dezign House, Jefferson 049798
Dezuttere, Brugge 063269
DF Gallery, Windsor, Ontario . 101853
D'Fab, Brunswick 061222
DG.Art, Paris 104683
DGEG Eisenbahnmuseum Würzburg,
Würzburg 022787
DGLM Gallery, Philadelphia .. 123367
Dhaka City Museum, Dhaka .. 003220
Dhaka College of Fine Arts,
Dhaka 054991
Dhakshini, Bangalore 079346
Dhamapala, Kathmandu 112070

Dharma, Cazadero 138315
Dhénin, Jean, Saint-Loup-de-
Naud 072899
Dhenkanal Science Centre,
Dhenkanal 024248
Dhennequin, Michèle, Paris ... 141090
D'Hier à Aujourd'hui, Génolhac . 068650
D'Hier à Aujourd'hui, Versailles . 074051
D'Hier et Aujourd'hui,
Fontainebleau 068566
D'Hier et d'Ailleurs, Annet-sur-
Marne 066654
D'Hier et d'Ailleurs, Lagny-sur-
Marne 069139
D'Hier et d'Ailleurs,
Wambrechies 074288
D'Hier et d'Aujourdhui, Sisteron . 073524
Dhikéos, Pierre, Lyon 104080
Dhoomimal Gallery, Delhi 109313
Dhouailly & Cie., Paris 071262, 141091
Dhyäna-Bernardin, Gap 103759
Di, Norfolk 095851
Di Antonietta, Roma 081329
Di Baio, Milano 137674
Di Bella, Pietro, Palermo 081034, 131695
Di Blanco, Avignon 103305
Di Blanco, Montélimar 104261
Di Blasi, Marcella, Altenahr 105976
Di Castro, Alberto, Roma 081330
Di Castro, Angelo, Roma 081331
Di Castro, Leone, Roma 081332
Di Castro, Richard, Roma 081333,
                          081334
Di Cave, Giuliana, Roma 142808
Di Clemente, Simone, Firenze .. 080131
Di Colman, Orange 099504
Di Consiglio, Gioia, Roma 081335
Di Domizio, Atlanta 092125
Di Elsden, Yandilla 099933
Di Gesaro, Angelo, Palermo 110600
Di Giacinto Romani, Tilde,
Bologna 130953
Di Giaimo, Saverio, Napoli 080866
Di Giugno, Catania 080032
Di Giugno, Angela, Catania 131087
Di Iacovo, Rosanna, Roma 081336
Di Ka, Changsha 102009
Di Ka, Liuyang 102147
Di Là Dal Fiume e Tra Gli Alberi,
Concesio 109973
Di Lorenzo, Vincenzo, Napoli ... 131615
Di Maccio, Laetitia, Sidi Bou
Saïd 116983
Di Maggio, Davide, Berlin 106207
Di Maio, Felice, Roma . 081337, 110755,
                          131903
Di Marco, Lidia, Roma 081338
Di Marco, Ugo, Roma 081339, 131904
Di Maria, Massimiliano, Palermo 081035,
                          131696
Di Matteo, Stefano, Roma 131905
Di Mauro, Emilio, Cava dei
Tirreni 137634
Di Meglio, Paloma, Milano 131435
Di Méo, Paris 104684
Di Napoli, Raffaele, Napoli 080867
Di Palma 1961, Napoli 080868, 131616
Di Paolo, Bologna 109819
Di Paolo, Domenico, Bologna .. 079834
Di Pasquale, Rocco, Bari 079768
Di Pasquale, Vincenzo, Bari ... 130913
Di Pietrantonio, Alfonso, Roma . 131906
Di Pietro, Giulia, Roma 131907
Di Pinto, Laura & Claudio, Trieste 081788,
                          142876
Di Pirro, Roma 131908
Di Pumpo, Matteo, Milano 131436
Di Rollo, Edinburgh 117721
Di Sabbato, Fausto, Roma 131909
Di Salvatore, Marseille 066930
Di Sciascio, Virginia, New York . 095349,
                          135908
Di Simone, Antonio, Palermo ... 131697
Di Stefano, Giuseppe, Palermo . 081036

Di Summa, Giovanni, Roma ... 081340,
                          110756
Di Teodoro, Dafne, Bari 079769
Di Tutto, Bogotá 065201
Di Vara, Isabella, Roma 081341
Di Veroli, Paris 104685
Dia Art Foundation, New York .. 051375
Dia Center for the Arts, New
York 051376, 051377
Dia Center for the Arts, Dia Art
Foundation, Quemado 052427
Dia:Beacon, Riggio Galleries, Dia Art
Foundation, Beacon 046841
Dia:Chelsea, Dia Art Foundation, New
York 051378
Diabira, Carole, Bernay-Saint-
Martin 067111
Le Diable Bouilli, Amiens 066571
Diablo Valley College Museum,
Pleasant Hill 052201
Diadema, Saratov 085277
Diagram, Nice 070597
Dial Gallery, Warkworth 119306
Dialog-Verlag, Reinbek 137485
Dialogues with the Art of Central
Eastern Europe, Trieste 138816
Diamante Pictures, New York .. 122535
Diamantmuseum, Antwerpen .. 003283
Diamantmuseum, Grobbendonk . 003659
Diamantschleifermuseum,
Brücken 017639
Diamond Coin, Indianapolis 093876
Diamond Gallery, Kobe 111279
Diamond Head Gallery, Honolulu 120930,
                          120931
Diamond Sport Cards, Austin .. 092223
Diamond & Son, Harry, Brighton 088618
Diamond, Mills & Co., Felixstowe 126896
Diamonds on Broadbeach,
Broadbeach 061215
Diana, Kroměříž 065409
Diana Antikvárium, Makó 142577
Diana Paul Galleries, Calgary ... 101230
Dianbai Museum, Dianbai 007741
Dianna's Vault, Saint Louis 096691
Diapason Gallery, Brooklyn 119940
Diard, Pierre-Olivier, Côte-Saint-
André 068167
Diarra, Françoise, Saint-Valery-en-
Caux 073324
Dias, Carlos, Lisboa 113993
Dias, Maria L. Gouveia, Lisboa . 084872
Dias, Paulo Cesar, Rio de Janeiro 064052
Diaspora Vibe Gallery, Miami .. 121769
Diaspro, Milano 080524
Diaz de Andres, Carlos, A
Coruña 126522
Diaz Perez, Baldomero, Palma de
Mallorca 115338
Díaz y Trancón, Cáceres 133218
Díaz, Gustavo, Cali 128286
Diaz, Salvador, Madrid 115180
Diba Antik, Jönköping 086761
Dibit, Norfolk 095852
Dibon, Jean-Luc, Amboise 128434
Dicembre, Carlo, Milano 131437
Dichter- und Stadtmuseum Liestal/
Herwegh-Archiv, Liestal 041673
Dichter bij Dordt, Dordrecht ... 143131
Dichtermuseum Joseph Maria Lutz,
Pfaffenhofen an der Ilm 021110
D'Ici et d'Ailleurs, Saint-Jean-de-
Luz 105532
Dick, Sylt 078656
Dick & Amby's Collectibles,
Jacksonville 093984
Dick-Art Design Galeria, Łódź .. 084608
Dick Institute Museum and Art Gallery,
Kilmarnock 044697
Dick, Hildegard, Recklinghausen . 078154
Dick, W., Lütisburg ... 116451, 138064
Dickel, Hanna, Wuppertal 078050
Dickel, Martin, Hamburg 076212, 129959
Dickens Curios, Frinton-on-Sea . 089309

Dickens House Museum Broadstairs,
Broadstairs 043761
Dickens Libros, Buenos Aires .. 139564
Dickens Museumtheater,
Bronkhorst 032111
Dickens, Alexandra, Saint Ives,
Cornwall 119021
Dickens, C., Atlanta 144965
Dickenson, Alastair, London ... 090073,
                          134647
Dickerson, Woollahra 099913
Dickhout, Alkmaar 082617
Dickins, H.C., Bloxham 088510, 117392
Dickins, Tom, Stretton Grandison 119172
Dickinson & Roundell, New York 095350,
                          122536
Dickinson County Heritage Center,
Abilene 046302
Dickinson, Bernard, Gargrave .. 089320,
                          134400
Dickinson, D. & B., Bath 088386
Dickinson, D.E., Mudgee 099457
Dickinson, Simon, Berlin 074780
Dickinson, Simon C., London .. 090074,
                     118287, 134648
Dickmann, Ingrid, Bad Nauheim 106079
Dicksmith, London 118288
Dickson & Rendall, London 090075
Dickson Mounds Museum,
Lewistown 050235
Dickson Store Museum, Dickson 005706
Dickson, José Manuel, Porto ... 085020
Dicon, Fuenlabrada 115063
Dicontal, Santiago de Chile 064964
Dicson, Alexandria 066217
Didaktische Ausstellung Urgeschichte,
Chur 041414
Didascalies, Aix-en-Provence .. 066460
Didcot Railway Centre, Didcot .. 044098
Diderichsen, Kastrup 065811
Didi Museum, Abuja 033412
Didier, Corbeil-Essonnes 068143
Didier, New Orleans 095088
Didier, Germain, Paris 104686
Didierjean, Stéphane, Lanvollon 069185
Didimatus, Bollène 067280
Dido Editions, Bristol 117446
Didrichsenin Taidemuseo,
Helsinki 010816
Didsbury and District Museum,
Didsbury 005707
Didžiulių Muziejus, Didziulis Family
Museum, Anykščiai 030318
Die Moebelwerkstatt,
Ebreichsdorf 127726
Die Neue Sammlung Design in
der Pinakothek der Moderne,
München 020519
Die Photographische Sammlung/SK
Stiftung Kultur, Köln 019676
Diebold, Juliette,
Niederhausbergen 070678
Diebold, Pascal, Semur-en-
Auxois 073486, 129256
Diecézní Muzeum, Brno 009463
Dieci Due, Milano 110301
Diederiks, Den Haag 126280
Dief- en Duifhuisje, Capelle aan den
IJssel 032133
Diefenbaker Canada Centre, University
of Saskatchewan, Saskatoon . 006971
Diefenbaker House Museum, Prince
Albert 006708
Diefenbunker-Canada's Cold War
Museum, Carp 005571
Diegoni, Christine, Paris 104687
Diego's Gallery, Toronto 101599
Diehl, Hanna, Bergisch Gladbach 106135
Diehl, Sibylle, Wettingen 116714, 133961
Diehl, Volker, Berlin 106208
Diela a Dielka Slovackych
Výtvarnákov, Banská Štiavnica 114647
Diele Galerii, Tallinn 103112
Dielo – The Art Center, Liptovský

Mikuláš 114664
Diem Gallery, Amsterdam 031935
Dienstbier sen., H. A., Lauf an der
Pegnitz 142051
Dienvidu Elektrisko Tilu Muzejs,
Jelgava 030130
Dieny, Jean-Pierre, Sauxemesnil 073444
Diepering Bijlsma, Bergen 082953
Dierich, H., Danstedt 075300
Dierick, J.-P., Bruxelles 063399
Dierken, Andrew, Los Angeles . 121471
Dierkes, Kevelaer 107548
Dierking, Köln 076855, 107602
Dierks, Leif, Risskov 066077
Dierre, Genova 080259
Dies & Das, Wien 062927
Diesel, Oakland 145387
Dieselmuseum, Stadtmuseum
Schramberg, Schramberg ... 021729
Diesenberg, W., Porta Westfalica 078065
Dietel, Miami 094728
Dietemann, Toronto 064635
Dieter Roth-Museum, Dieter Roth
Foundation, Hamburg 018982
Dieter, Norbert, Windesheim ... 079019
Dieterle, Dr. M., Schenkenzell . 078360
Dietgard Wosimsky, Online-Galerie für
Konkrete Kunst, Gießen 107126
Dietrich, Baden-Baden 129433
Dietrich, G., Billerbeck 074996, 125896
Dietrich, Horst, Berlin . 106209, 137061
Dietrichsteinský Palác, Moravské
zemské muzeum, Brno 009464
Dietro l'Angolo, Bologna 079835
Dietz, Heidelberg 076439
Dietz, Günter, Soyen 137509
Dietz, H ., Erlensee 075657
Dietzel, Berlin 074781, 125874
Dietzel, Wolfgang, Meiningen .. 142122
Dietzschold, Conny, Waterloo .. 099865
Dieutegard, Jean-Pierre, Paris .. 071263,
                          128988
Dieutegard, Marie-Noëlle, Paris . 071264,
                          128989
Dieutre, Christophe, Acquigny . 066417
Dieutre, Christophe, Le Mesnil-
Esnard 069316
Dieutre, Christophe, Rouen 072583
Diéz, Buenos Aires 098559
Diez Garcia, Maria Teresa,
Valencia 086513
Diez Monsalve, Eufemio, Madrid 086170
Diez Pérez, José Ramón, Burgos 085910,
                          133210
DiF Galerie-Artothèque,
Montbrison 104255
DIFC Gulf Art Fair, Jumeirah ... 098346
Diferença, Lisboa 113994
A Diferença, Lisboa 084873
Différences, Paris 104688
Different Strokes, Nashville ... 122080
DIG, York 046221
Dig Impex, Iași 143626
DiG Publishing, Warszawa 113919
Digard Maison des Ventes, Paris 125704
Digaro, Claude, Saint-Léonard-de-
Noblat 072892
Diggers Antiques, Winnipeg ... 064919
Diggins, Lauraine, Caulfield
North 061302
Diggs Gallery at Winston-Salem State
University, Winston-Salem ... 054488
Digha Science Centre & National
Science Camp, New Digha .. 024424
Digit Art, Santa Cruz de Tenerife 115419
Digital Art Lab, Holon 025064
Digital Art Museum, Berlin 017106
Digital Design Showcase,
Edinburgh 117722
Digterhuset, Farsø 110060
Diguet, Jean-Paul, Barfleur 066950
Dijana Centar za Preventivnu Zaštitu,
Beograd 133031
Dijeau, Olivier, Libourne 069459

Dijeau, Pierre, Les Billaux 069404
Dijeau, René, Rochefort (Charente-
  Maritime) 125789
Dijecezanski Muzej Zagrebačke
  Nadbiskupije, Zagreb 009027
De Dijk Te Kijk, Petten 032765
Dike, David, Dallas ... 093109, 120586,
                             127193
Dikemark Sykehus Museum,
  Asker 033511
Dikmayer, Joachim, Berlin ... 106210,
          106211,   137062
Dildarian, Louise C., New York . 095351
Dilemma, Leuven 100775
Dilemma Artworks, York 138296
Le Dilettante, Montolieu 140962
Le Dilettante, Paris ... 136894, 141092
Dilettante, Vex 116697
Diliberto, Palermo 081037
Dillard Mill State Historic Site,
  Davisville 048166
Dillaway-Thomas House, Roxbury
  Heritage State Park, Boston .. 047111
Dillée, Paris 071265
Dillée, Stéphane, Lisieux 069542
Diller, Johannes, München 077485
Dillingham & Co., New York ... 095352
Dillingham & Co., San Francisco 097217
Dillon, Dallas 093110
Dillon, New York 122537
Dillon, Pat, Toronto 064636
Dilmos, Milano 110302, 137675
Dilna Lochkov, Praha 065489
Dilo – Galerie, Brno 102563
Di'Lorenzo, London 118289
Dilworth, Charlotte 092593
Dima, Paris 104689
Dimadis, Stella, Collingwood .. 098912
Dimbola Lodge Museum, Freshwater
  Bay 044372
Dimboola Antiques Centre,
  Dimboola 061400
Dime a Dance, Milwaukee 094850
Dimension Painting Galeria, Delhi 109314
Dimension Plus, Montréal 101379
Dimensione Arte, Agnadello ... 109739
Dimensions, Taipei 116907
Dimensions Antiques, Houston . 093659
Dimensions Art Center, Taipei .. 059993,
                             116908
Dimitreion Oikima House, Museum of
  the Sculptor Nikolas, Agios Georgios
  Nilias 022947
Dimitris Pierides Museum of
  Contemporary Art, Glyfada ... 023084
Dimitrova, Z., Pleven 101129
Dimock Gallery, Washington ... 054104
Dimond & Son, Prospect 127664
Dimondstein, Joshua, Los
  Angeles 121472
Din Antik, Stockholm 086987
Dina, Los Angeles 121473
Dinaburg, Mary E., New York .. 122538
Dinajpur Museum, Dinajpur 003228
Dinallo, West Hollywood 097886
Din'art, Dinard 103693
Dinastia, Lisboa 084874, 126441,
                             138998
Dinbau, Flamatt 126720
Dinc, Abel, Sas van Gent 132627
Diner En Ville, Milano 080525
Diner, Geoffrey, Washington ... 097832
Dingles Steam Village – Fairground
  Museum, Lifton 044828
Dingley Dell Gallery, Ipswich .. 099154
Dingling Museum, Beijing 007617
Dinguidard, Alain, Tourdun 073817
Dingwall Museum, Dingwall ... 044099
Dingxi Zone Museum, Dingxi .. 007742
Dingzhou Museum, Dingzhou .. 007743
Dini, Daniela, Firenze . 080132, 131153
Dini, Stefano, Roma 081342
The Dining Room Shop, London 090076,
                             134649

Dink Antik, København 065892
Dinkelbergmuseum Minseln,
  Rheinfelden 021397
Dinkler, Marina, Berlin 106212
Dinky Toy, Enschede 083248
Dinkytown, Minneapolis 145276
Dinnebier, Gerald & Anja,
  Pleinfeld 078058
Dinnerware Contemporary Art Gallery,
  Tucson 053826
Dinolevi Antiques, Firenze 080133,
                             131159
Dinosaur Gardens Museum,
  Ossineke 051834
Dinosaur Isle, Sandown 045734
Dinosaur Museum, Dorchester . 044115
Dinosaur Museum, Malabon .. 034509
Dinosaur Museum, Mesalands
  Community College, Tucumcari 053842
Dinosaur Provincial Park, Royal Tyrrell
  Museum of Palaeontology Field
  Station, Patricia 006617
Dinosaur Valley State Park, Glen
  Rose 049111
Dinosauria – Musée des Dinosaures,
  Espéraza 012906
Dinosaurier-Freilichtmuseum
  Münchehagen, Rehburg-
  Loccum 021338
Dinosaurland, Lyme Regis 045173
Dinozzi, Daniela, Livorno 080324
Dinsmore Homestead, Burlington 047319
Dintenfass, Terry, New York ... 122539
Dinter, Jürgen, Köln 142006
Dinter, Jürgen, Schwerin 078428
Diodora, Milano 110303
Diözesanmuseum, Bamberg ... 016944
Diözesanmuseum, Klagenfurt .. 002187
Diözesanmuseum Eisenstadt,
  Eisenstadt 001858
Diözesanmuseum Freising,
  Freising 018500
Diözesanmuseum Graz, Das Museum
  der Steirischen Kirche, Graz .. 001969
Diözesanmuseum Linz, Linz ... 002318
Diözesanmuseum Rottenburg,
  Rottenburg am Neckar 021520
Diözesanmuseum Sankt Afra,
  Augsburg 016650
Diözesanmuseum Sankt Pölten, Sankt
  Pölten 002693
Diözesanmuseum Sankt Ulrich,
  Regensburg 021314
Diogenes, Nußdorf-Debant 062752
Diogenes Verlag, Zürich 138108
Diomedea, Palermo 110601
Dion Archaeological Museum,
  Dion 023062
Dion, Jean-Michel de, Bruxelles 063400
Dioni, Larnaka 102532
Dionie, Cholargos 109062
Dionis Bennassar, Madrid 115181
Dionisi Antichità, Susegana 081627
Dionizo Poškos Baubliai, Šiaulių
  Aušros Muziejus, Bijotai 030322
Dionne Quints Museum, North
  Bay 006507
Diop, Baba, Bordeaux 128487
Diorama "Bitva za Dnepr",
  Dnipropetrovskyj deržavnij
  istoričnij muzej im Javornic'kogo,
  Dnipropetrovs'k 043146
Diorama Arts, London 044959
Diorama Bethlehem, Einsiedeln . 041456
Diorama der Schlacht bei Rosbach,
  Reichardtswerben 021340
Diorama Kurskaja Bitva –
  Belgorodskoe Napravlenie,
  Belgorodskij Gosudarstvennyj
  Chudožestvennyj Muzej,
  Belgorod 036888
Diorama Saint-Bénilde, Saugues-en-
  Gévaudan 015702
Diorama Šturm Izmaïla, Izmaïl . 043160

Dioramenschau Altötting,
  Altötting 016555
Dios, Margarita de, Madrid ... 143759
Diósgyőri Papírgyár Zrt. Papíripari
  Múzeuma, Miskolc 023780
Diósgyőri Vármúzeum, Miskolc . 023781
Dip 'n' Strip, Calgary 128149
Dip-Strip, Dún Laoghaire 079604,
                             130866
Dipartimento Ambienta, Construzioni e
  Design (DACD, Scuola Universitaria
  Professionale della Svizzera Italiana
  (SUPSI), Canobbio 056422
Dipartimento di Discipline Artistiche,
  Musicali e dello Spettacolo,
  Università degli Studi di Torino,
  Torino 055982
Dipartimento di Scienze Archeologiche
  e Storico Artistiche, Università di
  Cagliari, Cagliari 055879
Dipartimento di Scienze dell' Antichità,
  Università di Padova, Padova . 055940
Dipartimento di Storia dell'Arte
  Medievale e Moderna, Università
  degli Studi di Roma, Roma ... 055967
Dipartimento di Storia dell'Arte,
  Facoltà di Lettere e Filosofia,
  Università degli Studi di Roma La
  Sapienza, Roma 055966
Dipartimento di Storia delle Arti e
  Concervazione dei Beni Artistici G.
  Mazzariol, Università Cà Foscari,
  Venezia 055989
Dipartimento di Storia delle Arti e
  dello Spettacolo, Università degli
  Studi di Firenze, Firenze 055893
Dipartimento di Storia delle Arti, delle
  Musica e dello Spettacolo, Università
  degli Studi di Milano, Milano .. 055921
Dipartimento di Storia delle Arti,
  Università di Pisa, Pisa 055950
Dipartimento di Storia e Storia
  dell'Arte, Università di Trieste,
  Trieste 055983
Dipinti Antichi, Bergamo 079791
Dipinti Antichi, Roma 081343
Dipinti, Enrico, Milano 110304
Diplomatic History Museum,
  Seoul 029990
Dipressa, Rosina, Torino 132131
Diquero, Yves, Malansac 069822
Diquéro, Yves, Rochefort-en-
  Terre 072501
Dirección General de Museos de la
  Nación, Asunción 059685
Direct Auction Galleries, Chicago 127166
De Directiekamer, Middelburg .. 112592
Direction de Musées de France,
  Paris 058640
Directiques, New York 095353
Direzione Generale per l'Architettura e
  l'Arte Contemporanee, Roma .. 059433
Diriwächter, Kurt, Luzern 087636
Dirkedaen, Wilnis 132669
Dirksen, Marion, Wiesbaden ... 078972
Dirkx, Mariska, Roermond 112619
Dirt Gallery, West Hollywood .. 124992
Dirven, Jan, Schoten . 063826, 100813
Disabled Person Art Cooperative,
  Wuxi 058494, 102251
Disagården, Upplandsmuseet,
  Uppsala 041110
Disaro, Roberto, Milano 142720
Dischinger Heimatmuseum,
  Dischingen 017911
Disciples of Christ Historical Society
  Museum, Nashville 051178
Discount Graphics, San
  Francisco 124269
Discover Houston County Visitors
  Center-Museum, Crockett ... 048068
Discover Nepal, Kathmandu ... 112071
Discoveries, Oklahoma City ... 095968
Discovery, Noosa Heads 061977

Discovery Antiques, Stirling 062246
Discovery Center, Fresno 048977
Discovery Center Museum,
  Rockford 052636
Discovery Center of Idaho, Boise 047074
Discovery Center of Springfield,
  Springfield 053466
Discovery Center of Taipei, Taipei 042513
Discovery Center of the Southern Tier,
  Binghamton 046997
Discovery Center Science Museum,
  Fort Collins 048795
Discovery Corner, Clayfield 061332
Discovery Creek Children's Museum of
  Washington DC, Washington .. 054105
Discovery Harbour,
  Penetanguishene 006625
The Discovery Museum,
  Bridgeport 047201
Discovery Museum, Newcastle-upon-
  Tyne 045346
Discovery Museum, Sacramento
  Museum of History, Science, Space
  and Technology, Sacramento .. 052726
The Discovery Museums, Acton . 046316
Discovery Park, Safford 052737
Discovery Place, Charlotte 047572
Discovery Science Center of Central
  Florida, Ocala 051695
Discovery Trail, Le Pin-au-Haras 013681
Discovery World – The James Lovell
  Museum of Science, Economics and
  Technology, Milwaukee 050912
Disella, Kolding 065975
Diseño Muebles Antiguos,
  Medellín 065285
Dishes From the Past, Fort Worth 093443
Dishman Art Gallery, Beaumont . 046855
Disjointed, Rotterdam 083564
Disparate, Bilbao 085869
Dispela, Los Angeles 094319
Display, Leipzig 107813
La Dispute, Paris 141540
Diss Antiques, Diss . 089062, 134324
Diss Museum, Diss 044103
Disselhoff, Los Angeles 094320
Dissmann, Jörg, Essen 106909
Distanza, Leeuwarden 112546
Distefani, Daniela, Bari 130914
Distill, Toronto 101600
Distilled Art, Bagnolet . 103322, 136801
Distillerie du Périgord, Sarlat-la-
  Canéda 015687
Distinctive African American Art, Salt
  Lake City 123955
Distinctive Antiques, Claremont . 061323
Distinctive Consignments,
  Richmond 096538
Distinctive Desks, Dunedin 083923
Distinctive Furnishings, New York 095354
Distribution Group, Bucureşti ... 114134
District Archaeological Museum,
  Dhar 024242
District Fine Arts, Washington .. 124888
District Gallery, Memphis 094631
District Museum, Barpeta 024157
District Museum, Darrang 024211
District Museum, Gulbarga 024266
District Museum, Guntur 024269
District Museum, Pillalamari ... 024437
District Museum, Shimoga 024478
District Museum, Shivpuri 024479
District of Columbia Arts Center,
  Washington 054106
District Science Centre,
  Dharampur 024243
District Science Centre, Purulia . 024452
District Science Centre,
  Tirunelveli 024495
District Six Museum, Cape Town 038676
Distrito Arte, A Coruña 115031
Distrito Cuatro, Donostia-San
  Sebastián 115046
Dit-Elle, Grignan 140853

Ditchburn & Boxshall, Armadale,
Victoria 061022
Ditchling Museum, Ditchling . . . 044104
Ditec, Lisboa 113995
Ditesheim, François, Neuchâtel . 116527
Dithmarscher Kunstauktionshaus,
Albersdorf 125849
Dithmarscher Landesmuseum,
Meldorf 020351
Ditjes en Datjes, Tilburg 083613
Dits, Hornu 138548
Ditt og Datt, Stavanger 084365
Ditterke, Gehrden bei Hannover 075959
Dittmann, A., Daun 075316
Dittmann, J.-U., Nordhorn 142235
Dittmar, Berlin 106213, 137063
Dittmar, Ratingen 078135
Dittmar Memorial Gallery, Nothwestern
University, Evanston 048635
Il Ditto Magico, Marseille 128762
Dittrich, Praha 128342
Dittrich, Klaus, Seligenstadt . . . 108609
Dittrick Museum of Medical History,
Case Western Reserve University,
Cleveland 047813
Dittus, Oberreichenbach 077872
Ditzoff, Bernard, Coppet 116225
Diva Designz, Oakland 123250
Diva Fine Art, Paris 071266
Diva Gallery, Bruxelles 100593
Divadelní Galerie, Uherské
Hradiště 102743
Divalco, Mechelen 128080
Divan Edebiyatı Müzesi, İstanbul 042933
Divari Kaleva, Helsinki 140698
Divary, Helsinki 140699
Diverse Works, Houston 049576
Diversified, Oklahoma City 136048
Diversified Collectibles,
Indianapolis 093877
The Diversion Gallery, Blenheim 083840
Diversity Books, Mordialloc . . . 139834
Diversity Gallery, Saint Louis . . 123820
Diverso, Amsterdam 082728
Diverssikauppa, Nummi 066390
Divet, Rennes (Ille-et-Vilaine) . 072433,
072434,   105432,   105433
Divide County Historical Society
Museum, Crosby 048072
La Divine Comédie, Aix-en-
Provence 103194
Divine Trash, Miami 094729
Divine Word University Museum,
Tacloban 034600
Divinely, Seattle 097427
Divisão de Museus, Patrimônio e
Arquivo Histórico, Taubaté . . . 004921
División Museo e Investigaciones
Históricas de la Policía Federal,
Buenos Aires 000156
Division of Art and Design, Purdue
University West Lafayette, West
Lafayette 058086
Division of Art, Design and
Architecture, Judson College,
Elgin 057042
Division of Art, Elmira College,
Elmira 057050
Division of Art, Florida College, Temple
Terrace 057987
Division of Art, Meadows School
of the Arts, Southern Methodist
University, Dallas 056975
Division of Art, School of Fine
Arts, Nicholls State University,
Thibodaux 057992
Division of Art, University of
Lethbridge, Lethbridge 055136
Division of Arts and Sciences,
Concordia University, Mequon . 057426
Division of Arts, Harrisburg Area
Community College, Harrisburg 057182
Division of Arts, Laramie County
Community College, Cheyenne 056868

Division of Arts, Russell Sage College,
Troy 058008
Division of Arts, Sage Junior College
of Albany, Albany 056633
Division of Arts, Western Nebraska
Community College, Scottsbluff 057886
Division of Communication and
Fine Arts, Unibetsedât Guahan,
Mangilao 055758
Division of Computer Graphics,
Graphic Design and Photography,
Sullivan County Community College,
Loch Sheldrake 057354
Division of Fine and Performing Arts,
Mississippi University for Women,
Columbus 056941
Division of Fine and Performing
Arts, Sinclair Community College,
Dayton 056985
Division of Fine Arts, Brevard College,
Brevard 056782
Division of Fine Arts, El Camino
College, Torrance 057997
Division of Fine Arts, Fullerton College,
Fullerton 057114
Division of Fine Arts, Gulf Coast
Community College, Panama
City 057643
Division of Fine Arts, Judson College,
Marion 057410
Division of Fine Arts, Midwestern
State University, Wichita Falls . 058105
Division of Fine Arts, North Central
Texas College, Gainesville . . . 057121
Division of Fine Arts, Oakland City
College, Oakland 057607
Division of Fine Arts, Orange Coast
College, Costa Mesa 056961
Division of Fine Arts, Truman State
University, Kirksville 057290
Division of Fine Arts, University of
West Alabama, Livingston 057352
Division of Fine Arts, Wingate
University, Wingate 058120
Division of Graphics, La Roche
College, Franklin Park 057105
Division of Humanities & Fine Arts,
Saint Norbert College, De Pere 056988
Division of Humanities and Fine Arts,
Illinois Valley Community College,
Oglesby 057612
Division of Humanities and Fine Arts,
West Shore Community College,
Scottville 057888
Division of Humanities, Fine and
Performing Arts, Okaloosa-Walton
College, Niceville 057581
Division of the Arts, Bard College,
Annandale-on-Hudson 056662
Divisional Kohima Museum, York 046222
Divison of Visual and Performing
Arts, Green Mountain College,
Poultney 057716
Divus, Praha 136762
Diwan Dar El Jeld, Tunis 088091
Diwania, Casablanca 112036
Dix, Neuffen 077728
Le Dix-Neuvième, Bruxelles . . . 063401
Dix-Sept, Framlingham 089302
Dixie Gun Works' Old Car Museum,
Union City 053893
Dixie's Doll House, Cincinnati . . 092844
Dixon, Midlothian 127318
Dixon & Dixon, Tell City 127473
Dixon & Harris, New Orleans . . 095089,
122153
The Dixon Gallery and Gardens,
Memphis 050777
Dixon, C.J. & A.J., Bridlington . . 088602
Dixon, John, Auckland 083781
Dixson, Denver 093277
Dixy, Antwerpen 125295
Diyarbakır Devlet Güzel Sanatlar

Galerisi, Diyarbakır 042880
Diyarbakır Müzeleri ve Tarihi Kültürel
Değerleri, Diyarbakır Arkeoloji
Müzesi ve Ziya Gökalp Müze-Evi,
Diyarbakır 042881
Diyarbakır Örenyerleri, Diyarbakır 042882
Dizajn Galerija, Beograd 114400
Djaburov, Bucureşti 085091
Djalil Mammadkulizade Home
Museum, Naxçivan 003185
Djalil Mammadkulizade Memorial
Museum, Nehram 003193
D'Jampel, Santiago de Chile . . . 064965
Djamshid Nakhchivansky Home
Museum, Naxçivan 003186
Djanman, Roosendaal 112624
Djanogly Art Gallery, University of
Nottingham, Nottingham 045411
DJT Fine Art Gallery, New York . 122540
Djuna, Denver 093278
Djursholm Möbelkonservator,
Djursholm 133520
Djursholms Antikvariat,
Djursholm 143825
Djurslands Museum – Dansk
Fiskerimuseum, Grenaa 010084
DK Project Art, Boston 119858
DKR China Repair Studio,
Auckland 132686
DLK Art, New Orleans 122154
DLM, Saint-Laurent-la-Conche . 072887
DLM – Deutsches Ledermuseum/
Schuhmuseum Offenbach,
Museum für Angewandte Kunst,
Ethnologisches Museum, Deutsches
Schuhmuseum und Kunstgalerie,
Offenbach am Main 020970
DLT, Philadelphia 136084
Dlum, Skopje 111881
Dmanisi Museum-Reserve of History
and Architecture, Dmanisi . . . 016388
DMC, Montereau-Fault-Yonne . . 070236
Dmitrijevič, Beograd 133032
D'n Diestkant, 's-Hertogenbosch 112526
DNA, Berlin 106214
DND, Rouen 105473
Dnipropetrovskyj Chudožnij Muzej,
Dnipropetrovs'k 043147
DNM Artgroup, Houston 121030
Do, Tokyo 111416
Do, Hyung The, Seoul 111630
Doa, Alain, Cannes 067636
Doak House Museum,
Greeneville 049240
Dobbs, John, Sacramento 136187
Dobeles Novadpētniecības Muzejs,
Dobele 030114
Dobergmuseum/Geologisches Museum
Ostwestfalen-Lippe, Museum Bünde,
Bünde 017671
Dobes, Heinrich, Petershagen . . 078037
Dobi, Marlies & Daniel, Bern . . 133675
Dobiaschofsky Auktionen, Bern . 126713
Dobkowitz, K., Neustrelitz 077769
Dobler, Renate, Essen 106910
Dobler, Uwe, Augsburg 074456
Dobloug, Oslo 113200
Dobó István Vármúzeum, Eger . 023613
Dobrá Kniha, Praha 140517
O Dobrão, Lisboa 084875
Dobrjanskij Istoriko-kraevedčeskij
Muzej, Permskij oblastnoj
kraevedčeskij muzej, Dobrjanka 036970
Dobry Styl, Wesoła 113913
Dobson Bashford, Christchurch . 112840
Dobson's, Abbots Langley 088216
Docal Gonzalez, Maria Estela,
Santander 115427
Dochartaigh, B., Dallas 120587
Docilia, Torino 132132
Dock- och Textilmuseum,
Katrineholm 040745
Dock Museum, Barrow-in-
Furness 043502

Dock och Leksaksmuseum,
Landsbro 040771
Dock One, Newtown, New South
Wales 061964
Dočkal, Dr., Hradec Králové . . . 140472
Docker Street Galleries, Wagga
Wagga 099855
Les Docks de la Rade, Brest . . 067468
Docks for Antiques, Antwerpen . 063203
Doc's Antique Auction,
Hammonton 127237
The Doctor Vigari Gallery,
Vancouver 101774
document Neupfarrplatz, Museen der
Stadt Regensburg, Regensburg 021315
document Schnupftabakfabrik,
Museen der Stadt Regensburg,
Regensburg 021316
documenta Archiv für die Kunst
des 20. und 21 Jahrhunderts,
Kassel 019531
Documentation Française, Lyon . 140894
Documentation Française, Paris . 136895
Dodd Cletus, P., Houston 121031
Dodd, Graeme, Lane Cove . . . 061717,
127627
Dodd, Michael L., Houston 135600
Doddington Hall, Lincoln 044830
Dodds, Mold 126989
Le Dodecaedre, Paris 141094
Dodengang en Oorlogsmuseum,
Diksmuide 003550
Dodge County Historical Museum,
Mantorville 050659
Dodge County Historical Society
Museum, Beaver Dam 046862
Dodge-Fjeld, New Orleans 095090
Dodier, Philippe, Le Val-Saint-
Pere 069370
Dodin, Roland, Erquy 068464
Dodin, Roland, Pléneuf-Val-André 072177
Dodington Antiques, Whitchurch,
Shropshire 091830
Dodis Shoppe, Vancouver 064817
Dodsland and District Museum,
Dodsland 005711
Dodson, Oklahoma City 123280
Dodsworth, N.J. & C.S.,
Spennithorne 091398
Döbele, Dettelbach . . . 106676, 125918
Döbele, Dresden 106720
Döbele-Brandensein, Hinterzarten 076537
Döderhultarmuseet, Oskarshamn 040908
Doehn, Dr. Uwe, Frankfurt am
Main 141770
Döllen, Bernhard von, Bad
Münstereifel 074555
Dölling & Galitz, München 137394
Dölling & Galitz Verlag, Hamburg 137239
Döllinger, Kurt, Markt Erlbach . . 077300
Dølmo Bygdetun, Tolga 034020
Dönmez, Galip Bülent, İstanbul . 088115
Dønna Bygdesamling, Dønna . . 033569
Döpkens, H.-J., Berlin 074782
Döppl, Robert, Oberlichtenau . . 130479
Dörfler, Taura 078669
Dörfler, F., Jena 076643
Dörflihaus-Museum, Spiringen . 042014
Doering, Abel, Berlin . . 141502, 141503
Döring, K., Barth 074683
Dørnsk, E., Sønderborg 066137
Dörpmuseum Münkeboe,
Südbrookmerland 022089
Dörr, Anna, München 077486
Doerr, Mary, Austin 119663
Dörrie-Priess Berlin, Berlin 106215
Dörrie-Priess Hamburg, Hamburg 107237,
137240
Dörschug, Harald, Augsburg . . . 074457
Doesburg, J., Utrecht 083639
Doesburgsch Mosterd- en
Azijnmuseum, Doesburg 032229
Dötlinger Antikhaus, Dötlingen . 075369
Dog & Pony Gallery, Birmingham 119795

Dog Collar Museum, Maidstone 045192
Dog Eared Books, San Francisco 145550
Dog Mushing Museum,
  Fairbanks 048659
Dogan, Schönecken 108572
Dogana d'Arte, Atripalda 025416
Il Doge, Bassano del Grappa . . . 109779
Il Doge, Marostica 110205
Dogenhaus Galerie, Leipzig . . . 107814
Doggerfisher, Edinburgh 117723
Doggett Street Studio, Newstead 099441
Doggett, Ronnie, Louisville 094540
The Doghouse, Walsall 091716
Dogny, Marc, Allaman 126704
Dogon, Roosendaal 112625
Dogra Art Museum, Jammu . . . 024321
Dogue Bleu, Chamonix-Mont-
  Blanc 067801
Doherty, Louis J., Sligo 079699
Dohm, Fritz & Silvia, Reutlingen 078211
Dohm, Michèle, Pont-à-Mousson 072238
Dohna, Graf zu, Barum 074684
Dohr, M., Lindlar 077123
Doi Orsi, Milano 080526, 080527
Doigts d'Or, Guebwiller 103800, 128616
Doison, Christine, Paris 128990
Doka, Tokyo 111417
Dokken, Hol i Hallingdal 033661
Al-Doks, Cairo 066239
Doku Sanat Galerisi, İstanbul . . 117027
Dokulil, Peter, Koblenz 076831
Dokumentačné Centrum Českej
  Kultúry Martin, Etnografické
  Múzeum – Slovenské Národné
  Múzeum, Martin 038388
Dokumentarna, Ljubljana 114699,
                                        133066
Dokumentation des ehemaligen
  Heilbades, Plastiken-Freilichtmuseum
  Prof. Knesl, Bad Pirawarth . . . 001758
Dokumentation Obersalzberg –
  Orts- und Zeitgeschichte,
  Berchtesgaden 017037
Dokumentations- und Gedenkstätte
  der BstU in der ehemaligen U-Haft
  der STASI, Rostock 021488
Dokumentations- und
  Informationszentrum,
  Stadtallendorf 021952
Dokumentations- und
  Informationszentrum Emslandlager
  (DIZ), Papenburg 021071
Dokumentationsarchiv des
  österreichischen Widerstandes,
  Wien 002980
Dokumentationsraum für staufische
  Geschichte, Göppingen 018727
Dokumentationsraum Georg Matthäus
  Vischer, Raabs an der Thaya . . 002554
Dokumentationsstätte im
  Mesnerhaus, Erzherzog
  Johann-Dokumentationsstätte,
  Scheiblingkirchen-Thernberg . . 002718
Dokumentationszentrum Alltagskultur
  der DDR, Eisenhüttenstadt . . . 018143
Dokumentationszentrum des Landes
  für die Opfer der Diktaturen in
  Deutschland, Schwerin 021790
Dokumentationszentrum des Landes
  für die Opfer deutscher Diktaturen,
  Rostock 021489
Dokumentationszentrum ehemaliges
  KZ Osthofen, Osthofen 021033
Dokumentationszentrum NS-
  Zwangsarbeit Berlin-Schöneweide,
  Berlin 017107
Dokumentationszentrum Oberer
  Kuhberg-Ulm KZ Gedenkstätte,
  Ulm 022257
Dokumentationszentrum Prof.
  Sepp Mayrhuber, Golling an der
  Erlauf 001961
Dokumentationszentrum
  Reichsparteitagsgelände, Museen der

Stadt Nürnberg, Nürnberg 020847
Dolan & Maxwell, Philadelphia . 123368
Dolder, Hans-Rudolf,
  Hombrechtikon 087527, 133770
Doldt, Stefan, Karlsruhe 076690
Dolenjski Muzej Novo Mesto, Novo
  Mesto 038585
Dolev, Pamplona 086385
Dolfen, G., Eschweiler 106902
Dolfin, Daniela, Venezia 081832
Dolfus, Paule, Villeurbanne . . . 074204
Dolge, B., Magdeburg 107936
D'Oliver, Palma de Mallorca . . . 086347
Doll, Louisville 138353
Doll, Seoul 111631
Doll and Bear Museum,
  Manunda 001251
The Doll and Toy Museum of NYC,
  Brooklyn 047255
Doll Gardner Art Gallery, Portland 052290
Doll Hospital New York, New
  York 095355, 135909
Dollar Museum, Dollar 044106
Dollar, Bob, Fort Worth 093444
Dolling, Michael, Glandford . . . 134402
Dollita, Quiberon 105404
Doll's & Bear's at Charlton Court, Long
  Beach 094199
Dolls 4 U, Louisville 094541
Dolls by Jerri, Charlotte 092594
Doll's House, Northleach 090859, 134887
Dolls, Bears & More, Fresno . . . 093498
Dolly Land, London . . . 090077, 134650
The Dolly Mixture, Langbank . . . 044757
Dolly's Antiques, Houston 093660
Dolma Thangka, Kathmandu . . . 112072
Dolmabahçe Sarayı Müzesi,
  İstanbul 042934
Dolnośląski Centrum Informacji
  Kulturalnej-Galeria, Wrocław . . 035413
Dolnośląski Inkubator Sztuki,
  Komorowice 059693
Dolnośląskie Centrum Fotografii
  Domek Romański-Galeria,
  Wrocław 035414
Dolnośląskie Stowarzyszenie
  Artystów Fotografików i Twórców
  Audiowizualnych, Wrocław . . . 059694
Dolo-Sallard, Annie, Alençon . . 066522
Dolomythos, San Candido 027732
Dolores Serier, Göteborg 143846
Dolors Junyent Galeria d'Art,
  Barcelona 114876
Dolphin Galleries, Honolulu . . . 120932
Dolphin House Gallery, Colyton . 117621
Dolphin Quay Antique Centre,
  Emsworth 089207, 134368
Dolphin Sailing Barge Museum,
  Sittingbourne 045804
Dolphin Square Antiques,
  Dorking 089079, 144372
Dolter, Raymond, Wangerland . . 130734
Dom- und Diözesanmuseum Wien,
  Wien 002981
Dom Artysta Plastyka, Warszawa 035312
Dom Aukcyjny Polswiss,
  Warszawa 126420
Dom Eskenów, Muzeum Okręgowe w
  Toruniu, Toruń 035277
Dom Fotografie v Poprade,
  Poprad 038426
Dom Galerie, Wiener Neustadt . 100400
Dom Gerharta Hauptmanna, Jelenia
  Góra 034794
Dom Hudby Mikuláša Schneidera-
  Trnavského, Západoslovenské
  Múzeum v Trnave, Trnava 038473
Dom I.S. Ostrouchova,
  Gosudarstvennyj Literaturnyj muzej,
  Moskva 037253
Dom Jana Matejki, Muzeum
  Narodowe w Krakowie, Kraków 034861
Dom Józefa Mehoffera, Muzeum
  Narodowe, Kraków 034862

Dom Knigi, Sankt-Peterburg . . . 143666
Dom Knigi v Sokolnikach,
  Moskva 143639
Dom Książki, Warszawa 143522
Dom Laraia, New York 095356
Dom L'udového Bývania,
  Matúškovo 038398
Dom L'udového Bývania, Šala . . 038442
Dom Marina Držića, Dubrovnik . 008900
Dom Miguel, Porto 085021
Dom Mikuláša Galandu, Turčianska
  Galéria Martin, Turčianske
  Teplice 038477
Dom-Museum, Fulda 018557
Dom-Museum Bremen, Bremen 017584
Dom-Museum Hildesheim,
  Hildesheim 019247
Dom-muzej A.P. Čechova, Jalta . 043164
Dom-muzej Akademika E.N.
  Pavlovskogo, Borisoglebskij
  kraevedčeskij muzej,
  Borisoglebsk 036910
Dom-muzej L.N. Andreeva,
  Gosudarstvennyj literaturnyj muzej
  im. I.S. Turgeneva, Orël 037597
Dom-muzej M. Auézova, Naučno-
  kulturnyj Centr, Almaty 029798
Dom-muzej S.M. Budënnogo,
  Proletarsk 037666
Dom-muzej A.P. Čechova,
  Gosudarstvennyj Literaturnyj muzej,
  Moskva 037254
Dom-muzej Velimira Chlebnikova,
  Astrachanskaja gosudarstvennaja
  kartinnaja galereja im. B.M.
  Kustodieva, Astrachan 036874
Dom-muzej N.N. Chochrjakova,
  Kirovskij oblastnoj chudožestvennyj
  muzej im. Viktora i Apollinarija
  Vasnecovych, Kirov 037128
Dom-muzej T.N. Chrenikova, Elec 037000
Dom-muzej G.V. Čičerina, Tambovskij
  oblastnoj kraevedčeskij muzej,
  Tambov 037925
Dom-muzej K.É. Ciolkovskogo,
  Kaluga 037087
Dom-muzej P.P. Čistjakova,
  Puškin 037672
Dom-muzej Čudomir, Kazanlǎk . 005056
Dom-muzej K.I. Čukovskogo,
  Peredelkino 037633
Dom-muzej M.I. Cvetaevoj,
  Moskva 037255
Dom-muzej Dekabristov, Kurganskij
  oblastnoj kraevedčeskij muzej,
  Kurgan 037183
Dom-muzej F.M. Dostoevskogo,
  Staraja Russa 037890
Dom-Muzej A.L. Durova, Voronežskij
  oblastnoj kraevedčeskij muzej,
  Voronež 038063
Dom-muzej A.A. Dydkina,
  Gosudarstvennyj muzej palechskogo
  iskusstva, Palech 037617
Dom-muzej Elin Pelin, Nacionalen
  Muzej na Bǎlgarskata Literatura,
  Bajlovo 004972
Dom-muzej M.N. Ermolovoj,
  Gosudarstvennyj centralnyj teatralnyj
  muzej im. Bachrušina, Moskva 037256
Dom-muzej M.V. Frunze, Samarskij
  oblastnoj istoriko-kraevedčeskij
  muzej im. P.V. Alabina, Samara 037707
Dom-muzej Gannibalov v Petrovskom,
  Muzej-zapovednik A.S. Puškina
  Michajlovskoe, Pskovskije Gory 037679
Dom-muzej Generala I.A. Plieva,
  Vladikavkaz 038023
Dom-muzej N.V. Gogolja, Gorodskaja
  biblioteka – kulturnyj centr N.V.
  Gogolja, Moskva 037257
Dom-muzej I.I. Golikova,
  Gosudarstvennyj muzej palechskogo
  iskusstva, Palech 037618

Dom-muzej M.B. Grekova,
  Muzej istorii Donskich kazakov,
  Novočerkassk 037571
Dom-muzej A.S. Grina, Kirov . . . 037129
Dom-muzej im. T.N. Granovskogo,
  Gosudarstvennyj literaturnyj muzej
  im. I.S. Turgeneva, Orël 037598
Dom-muzej Istorii Molodëžnogo
  Dviženija, Rjazan 037687
Dom-muzej Ivan Vazov, Sopot . . 005205
Dom-muzej E.M. Jaroslavskogo,
  Jakutskij gosudarstvennyj muzej
  istorii i kultury narodov severa im.
  Jaroslavskogo, Jakutsk 037052
Dom-Muzej P.D. Korina,
  Gosudarstvennyj muzej palechskogo
  iskusstva, Palech 037619
Dom-muzej V.G. Korolenko,
  Gelendžikskij istoriko-kraevedčeskij
  muzej, Gelendžik 037015
Dom-Muzej I.N. Kramskogo,
  Ostrogožskij istoriko-chudožestvennyj
  muzej im. Kramskogo,
  Ostrogožsk 037613
Dom-muzej P.A. Krasikova,
  Krasnojarskij kraevoj kraevedčeskij
  muzej, Krasnojarsk 037161
Dom-muzej Pavla Kuznecova,
  Chudožestvennyj muzej im. A.N.
  Radiščeva, Saratov 037842
Dom-muzej V.I. Lenina, Kazan . . 037098
Dom-muzej V.I. Lenina, Samarskij
  oblastnoj istoriko-kraevedčeskij
  muzej im. P.V. Alabina, Samara 037708
Dom-muzej M.Ju. Lermontova,
  Gosudarstvennyj Literaturnyj muzej,
  Moskva 037258
Dom-muzej N.S. Leskova,
  Gosudarstvennyj literaturnyj muzej
  im. I.S. Turgeneva, Orël 037599
Dom Muzej M. Saryana, Yerevan 000727
Dom-muzej M.K. Ammosova,
  Jakutskij gosudarstvennyj muzej
  istorii i kultury narodov severa im.
  Jaroslavskogo, Jakutsk 037053
Dom-muzej D.I. Mendeleeva,
  Boblovo 036904
Dom-muzej N.S. Muchina, Olikjal 037588
Dom-muzej N.S. Muchina, Nacionalnyj
  muzej respubliki Marij Ėl im. T.
  Evseeva, Joškar-Ola 037069
Dom-muzej Narodnogo Chudožnika
  V.A. Igoševa, Chanty-Mansijsk . 036952
Dom-muzej N.A. Nekrasova,
  Čudovo 036964
Dom-muzej I.S. Nikitina, Voronežskij
  oblastnoj literaturnyj muzej im I.S.
  Nikitina, Voronež 038064
Dom-muzej Osipovych-Vulf v
  Trigorskom, Muzej-zapovednik A.S.
  Puškina Michajlovskoe, Puškinskie
  Gory 037680
Dom-muzej A.N. Ostrovskogo,
  Gosudarstvennyj centralnyj teatralnyj
  muzej im. Bachrušina, Moskva 037259
Dom-muzej B.L. Pasternaka,
  Gosudarstvennyj literaturnyj muzej,
  Peredelkino 037634
Dom-muzej Penjo Penev,
  Dimitrovgrad 005012
Dom-muzej Petra Pervogo (Petrovskij
  Domik), Vologda 038055
Dom-muzej M.M. Prišvina,
  Gosudarstvennyj Literaturnyj muzej
  Moskva, Dunino 036974
Dom-muzej B.I. Prorokova, Ivanovskij
  oblastnoj chudožestvennyj muzej,
  Ivanovo 037037
Dom-muzej S.V. Rachmaninova,
  Ivanovka 037036
Dom-Muzej I.E. Repina, Samarskij
  oblastnoj chudožestvennyj muzej,
  Žigulevsk 038081
Dom-muzej V.A. Rusanova, Orlovskij

oblastnoj kraevedčeskij muzej, Orël 037600
Dom-muzej F.I. Šaljapina, Gosudarstvennyj centralnyj muzej muzykalnoj kultury im. M.I. Glinki, Moskva 037260
Dom-muzej M.E. Saltykova-Ščedrina, Kirov 037130
Dom-muzej Semji Cvetaevych, Novo-Talicy 037570
Dom-muzej Semji Lopatinych, Stavropol 037896
Dom-muzej Sestër Nevzorovych, Nižnij Novgorod 037544
Dom-muzej Skulptora A.S. Golubkinoj, Zarajskij Kreml, Zarajsk 038079
Dom-muzej K.S. Stanislavskogo, Muzej Moskovskogo Chudožestvennogo Akademičeskogo teatra, Moskva 037261
Dom-muzej Tukaevych, Nacionalnyj muzej Respubliki Tatarstan, Košlauč 037153
Dom-muzej G.I. Uspenskogo, Sjabrenicy 037861
Dom-muzej V.M. Vasnecova, Gosudarstvennaja Tretjakovskaja galereja, Moskva 037262
Dom-muzej Konstantin Veličkov, Pazardžik 005106
Dom-Muzej V.V. Veresaeva, Tula 037959
Dom-muzej N.M. Zinovjeva, Gosudarstvennyj muzej palechskogo iskusstva, Djagilevo 036968
Dom-Muzej N.N. Žukova, Gorodskoj kraevedčeskij muzej, Elec 037001
Dom na Humora i Satirata, Gabrovo 005030
Dom Naščokina, Moskva 037263
Dom Plastyka, Związek Polskich Artystów Plastyków, Toruń ... 035278
Dom Poëta A.N. Širjaevca – Muzej Krestjanskogo Byta, Žigulevsk . 038082
Dom Rodzinny Jana Kasprowicza, Inowrocław 034781
Dom Rodzinny Ojca Świętego Jana Pawła II, Wadowice 035298
Dom F.I. Šaljapina – Galereja, Moskva 114213
Dom Sztuki, Gdańsk 113462
Dom Sztuki, Wrocław 084766
Dom Tkacza, Muzeum w Bielsku-Białej, Bielsko-Biała 034649
Dom und Domschatz Halberstadt, Halberstadt 018936
Dom Uphagena, Muzeum Historyczne Miasta Gdańska, Gdańsk 034724
Dom Urodzenia Fryderyka Chopina, Sochaczew 035207
Doma Mungi Gallery, Chiltern . 098884
Domaine de l'Abbé Saunière, Rennes-le-Château 015068
Domaine MacKenzie King, Chelsea 005606
Domaine MB, Amsterdam 112184
Domaine National de Chambord, Chambord 012424
Domakowski, Lothar, Ostrach .. 077994
Domani, Christchurch 132704
Domański, W., Kraków 113540
Domarb & Döbele-Brandenstein, Hinterzarten 076538
Domati, Patrizio, Modena 131568
Domberger, Michael, Filderstadt 106951, 137190
Dombuchhandlung, Wiener Neustadt 140135
Dome Antiques, London 090078
Domein De Locht, Duffel 003561
Domein d'Oude Smelterij – Landbouwmuseum, Torhout . . 004076
Domek Bożeny Němcové, Červený Kostelec 009498
Domek Mauretański, Warszawa 113820

Domek Żeromskiego, Gdynia ... 034745
Doménat, Brive-la-Gaillarde ... 067507, 128523
Dômergue, Margaret, Clermont-Ferrand 068024
Domergue & Co., Robert, San Francisco 097218
Domgalerie, Merseburg 107996
Domgrabungsmuseum, Salzburg Museum, Salzburg 002618
Domicil, København 065893
Domicil, Mostar 100855
Domik Petra I, Sankt-Peterburg . 037729
Domingues da Silva, Marie-Claude, La Destrousse 069022
Dominguez Espinosa, Andres, Madrid 086171
Dominguez Guinaldo, Mario, Salamanca 115403
Dominguez Manteca, Felix Didimo, Getxo 115064
Domini Canes, Tallinn 103113
Dominici Art Edition, Leopoldsdorf bei Wien 136532
Dominicis, Paola, Milano 080528
Dominic's Art Gallery, Hong Kong 102069
Dominiiklaste Kloostri Muuseum, Tallinn 010665
Dominik, Königstein im Taunus . 130178
Dominikanerkloster Prenzlau, Kulturzentrum und Museum, Prenzlau 021212
Dominikanermuseum Rottweil, Zweigmuseum des Landesmuseum Württemberg Stuttgart und des Archäologischen Landesmuseums Konstanz, Rottweil 021525
Dominikonų Gatvė Antikvariatas, Vilnius 142964
Dominion Antiques, Toronto ... 064637
Dominion Books, Auckland ... 143279
Dominique, Bruxelles 063402
Dominique Antiquités, Lyon ... 069731
Dominique Antiquités, Saint-Jeoire-Prieure 072858
Dominique, C., Pont-Aven 105373
Domino Antik, Stockholm 086988
Domino Gallery, Liverpool 118127
Dominović, Zagreb ... 136753, 140458
Dominus, Bucureşti 114135
Dominus Flevit, Franciscan Custodia di Terra Sancta, Jerusalem ... 025075
Dominus Z Art, Bucureşti 114136
Domirai, Meersburg 107987
Domizil Antique, Esslingen ... 075709
Domkammer der Kathedralkirche Sankt Paulus zu Münster, Münster 020595
Dommanovich, Trudica, Großwarasdorf 062617
Dommuseum, Brandenburg an der Havel 017542
Dommuseum, Chur 041415
Dommuseum, Frankfurt am Main 018414
Dommuseum zu Salzburg, Salzburg 002619
Domnista Folklore and History Museum, Domnista 023063
Domo Art, Speyer 108650
Domo Verlag, Speyer 137510
Domobaal, London 118290
Domont, John, Indianapolis ... 121178
Domowniski muzej Dešno, Niedersorbisches Heimatmuseum Dissen, Dissen-Striesow 017913
Domschatz, Trier 022197
Domschatz, Würzburg 022788
Domschatz- und Diözesanmuseum, Passau 021080
Domschatz- und Diözesanmuseum Eichstätt, Eichstätt 018119
Domschatz der Stiftskirche St.Servatii, Quedlinburg 021242
Domschatz Sankt Ursen Kathedrale,

Solothurn 042000
Domschatz and Museum des Sankt Petri-Domes, Fritzlar 018533
Domschatzkammer, Aachen ... 016435
Domschatzkammer, Essen 018292
Domschatzkammer, Minden ... 020415
Domschatzkammer Sankt Petri, Bautzen 016976
Domschatzkammer und Diözesanmuseum, Osnabrück . 021011
Domschatzkammer, Metropolitan Kapitel der Hohen Domkirche, Köln 019677
Domschatzmuseum, Regensburg 021317
Domus, Napoli 080869
Domus, Rozzano 138817
Domus, Vancouver 101775
Domus – Casa del Hombre, A Coruña 039316
DoMuS – Museum und Galerie der Gemeinde Schaan, Schaan ... 030300
Domus Academy, Milano 055922, 059434
Domus Antiqua, Milano 080529
Domus Antiqua, Sassuolo 110913
Domus Antiqua-Città Dorica, Mostra Nazionale di Antiquariato, Ancona 098188
Domus Antiqua, Mostra Nazionale di Antiquariato, Civitanova Marche 098187
Domus Arte, Roma 110757
Domus Aurea, Catania 080033
Domus Galilaeana, Pisa 027350
Domus Mazziniana, Pisa 027351
Domus Nova, Parma 081107
Domus Romana, Rabat 030746
Domus Romana Archéosite, Aubechies 003333
Domus Starocie, Warszawa ... 084717
Domus Varius, Lisboa 113996
Domvorhalle, Goslar 018764
Domy Brocante, Erquinghem-le-Sec 068463
Don Bank Museum, North Sydney 001364
Don Books, Invercargill 143325
Don F. Pratt Memorial Museum, Fort Campbell 048792
Don Garlits Museum of Drag Racing, Ocala 051696
Don Harrington Discovery Center, Amarillo 046435
Don Murray's Museum of History, Neepawa 006457
Don Quijote Antiguedades, Montevideo 091996
Don River Railway Museum, Devonport 001027
Doña Antigualla, Madrid 086172
Dona Taraja, Lisboa 084876
Dona, Francesca, Los Angeles . 094321
Donabedian Koharig, Michel, Romans-sur-Isère 072522
Donacija Dr. Vladimir Malančec, Muzej Grada Koprivnice, Koprivnica . . 008929
Donado, Stefano, Milano 131438
Donald Agricultural Museum, Donald 001031
Donald G. Trayser Memorial Museum, Barnstable 046793
Donald, Julie, Coburg 098904
Donalda and District Museum, Donalda 005712
Donat-le-Drezen, Simone, Pont-l'Abbé 072255
Donat-Verlag, Bremen 137125
Donat, Hervé, Strasbourg 073584
Kurt Donath & Cornelia Szojda, Berlin 129484
Donath, Inge, Troisdorf 108781
Donáti, Budapest 079250
Donati, Pino & Class, Lugano . 087611
Donation Mario Prassinos, Saint-Rémy-de-Provence 105607

Donato, Colmar 068088
Donato, Niedermorschwihr ... 070679
Donatone, Lanfranco, Napoli .. 080870
Donatzsky, Torben, Gentofte ... 065689
Donau-Schiffahrts-Museum-Regensburg, Arbeitskreis Schiffahrtsmuseum Regensburg e.V, Regensburg 021318
Donauschwäbisches Zentralmuseum, Ulm 022258
Donay, London 090079
Donazione Putti Biblioteca, Raccolta dell'Istituto Ortopedico Rizzoli, Bologna 025532
Donazione Raffaele Vaccari, Brescello 025626
Donazione Sambo, Trieste 028201
Doncaster Museum and Art Gallery, Doncaster 044109
Doncaster Sales and Exchange, Doncaster 089067
Doncker, Thomas de, Chicago . 120110
Dondaine, Claude, Castres ... 067728
Donderbus, Rotterdam 083565
Dondo, Niš 114452
Dondtai Antique Market, Shanghai 065131
Done, Ken, Sydney ... 099777, 136491
Donec'ke Chudožne Učylyšče, Donec'k 056488
Donec'kij Oblasnyj Chudožnij Muzej, Donec'k 043152
Donec'kij Oblasnyj Krajeznavčyj Muzej, Donec'k 043153
Donegal Historical Society Museum, Rossnowlagh 024985
Donelly & Co., Alan, Perth 062077
Doneux & Soci, Torino 132133
Dong, Seoul 111632
Dong-Bang Gallery, Seoul 111634
Dong-eui University Museum, Busan 029896
Dong Fang, Shanghai 102175
Dong Nai College of Decorative Arts, Bien Hoa 058152
Dong Son, Ha Noi 125072
Dong Zuo Inkslab Art Museum, Ershuei 042309
Dong, Phong, Ha Noi 125071
Dong, Phuong, Ho Chi Minh City 125113
Dong, San Bang, Seoul 111633
Dong, Seo, Masan 111604
Donga University Museum, Busan 029897
Donggang Museum, Donggou . . 007744
Dongguan Museum, Dongguan . 007745
Dongguk University Museum, Seoul 029991
Dongha Museum, Dongen 032239
Dongho, Seoul 111635
Dongjin Irrigation Folk Museum, Gimje 029933
Dongshan County Artists Union, Tongling 058495
Dongshan Kite Museum, Dongshan 042305
Dongshan Museum, Dongshan . 007747
Dongtai Lu Antiques Market, Shanghai 065132
Dongtai Museum, Dongtai 007748
Dongtinghu Guan, Yuanjiang .. 008502
Dongwon Gallery, Daegu 111583
Dongying History Museum, Guangrao 007799
Dongyu Art Gallery, Shenyang . 102211
Dongzhang Calligraphy and Painting Gallery, Changshu 102011
Donhauser, Thomas, Sulzbach-Rosenberg 078649
Donhofer & Moser, Wien 140981
Donington Grand Prix Collection, Castle Donington 043884
Donjon, Ballon 011879
Donjon, Montréal 140361

Donjon-Château Carondolet,
Crupet 003526
Donjon Roman, Fondremand . . . 013000
Donker, Ad., Rotterdam 137870
Donkersloot, Peter, Amsterdam . 112185,
112186
Donkey Milk House, Key West . 049950
Donn, Henry, Whitefield 091833
Donn, Massimo, Torino 132134
Donna Beam Fine Art Gallery,
University of Nevada, Las Vegas, Las
Vegas 050159
Donna Gallery, Abu Dhabi 117165
Donnaint, Agincourt 128425
Donna's Not Just Antiques,
Indianapolis 093878
Donnelly, Mullingar 079678
Donnelly, Adrian, Shrewsbury . 091328,
135044
Donner, Helsinki 066311
Donner Memorial and Emigrant Trail
Museum, Truckee 053816
Donnini, Mauro, Firenze 131160
Donnybrook Art School, Dublin . 055861
Dono, Houston 093661
Donohoe, Martin & Michael,
Goresbridge 079631
Donohue & Bousquet, Ottawa . . 064502
Don's Antiques, Jacksonville . . . 093985
Don's Antiques, San Francisco . 097219
Don's Coin Center, Tulsa 097727
Don's Collectables, Seattle 097428
Don's Refinishing Service,
Omaha 136060
Don's Shop, Sylt 078657
Don's Village Coins, San
Francisco 097220
Donskoj Istoriko-kraevedčeskij Muzej,
Donskoe 036972
Donson, Theodore B., New York 122541
Donval, Yvesk, Port Aven 105394
Donvale Antique Clocks, Donvale 061402
Donzeau, Dax 125565
Donzella, New York 095357
Donzelli, Roma 137730
Donzelli, Ferdinando, Firenze . . 110046
Donzelli, Ferdinando, Milano . . . 110305
Doocollect, Madrid 126536
De Doofpot, Den Haag 083077
Doon Heritage Crossroads,
Kitchener 006122
Doon Valley Museum,
Dalmellington 044063
Dooneen, Crecora 079510
Door Art Gallery, Fresno 120894
Door County Maritime Museum, Gills
Rock 049098
Door County Maritime Museum,
Sturgeon Bay 053580
Door County Museum, Sturgeon
Bay 053581
Den Doorbraeck, Amsterdam . . 082729
Doorten, Henri, Meppel 132582
Doozoo Children's Museum, Grand
Junction 049181
Doppelkapelle Sankt Crucis,
Landsberg, Saalekreis 019880
Doppia V, Lugano . . 087612, 133823
Doppler-Brandis, Alexandra,
Salzburg 127845
Doppler, Denis, Rechesy 072370
Dor Antic, Nice 070598
Dor Art, Vancouver 101776
Doran, M.A., Tulsa 124801
Dorantiq di Patti Landini Simona,
Modena 080791
Dorarte Restauri, Perugia 131750
Dorate, Ali, Amsterdam 112187
Dorau, Gernot, Berlin . 074783, 125875
Dorchester Abbey Museum,
Dorchester-on-Thames 044122
Dorchester Antiques, Dorchester-on-
Thames 089074, 134333
The Dorchester Bookshop,

Dorchester 144371
Dorchester County Historical Society
Museum, Cambridge 047356
Dordrechts Museum, Dordrecht . 032246
Dordtse Antiekhandel, Dordrecht 083179
Dore, San Francisco 124270
Doré, Søndersø 102983, 136782
Le Doré Brocante, Paris 071267
Dorée, Jacques, Dole 128584
Doreth, Georg, Neustadt am
Kulm 130454
Dorety, John, Philadelphia 096103
Dorf- und Fahrradmuseum,
Pflückuff 021117
Dorf- und Heimatmuseum, Winterbach,
Rems-Murr-Kreis 022687
Dorf- und Kutschenmuseum,
Eisgarn 001864
Dorf- und Uhrenmuseum,
Gütenbach 018887
Dorfgemeinschaftshaus
Heimatmuseum Dörrenbach, Sankt
Wendel 021610
Dorfman, Cancún 111933
Dorfman Projects, New York . . 122542,
138391
Dorfmann, Patricia, Paris 104690
Dorfmuseum, Finkenstein 001890
Dorfmuseum, Günzach 018880
Dorfmuseum, Kritzendorf 002245
Dorfmuseum Ahnenhaus,
Pliezhausen 021158
Dorfmuseum Alter Forsthof,
Wetter 022605
Dorfmuseum Altkirchen,
Altkirchen 016549
Dorfmuseum Bad Alexandersbad, Bad
Alexandersbad 016700
Dorfmuseum Bellach, Bellach . 041296
Dorfmuseum Bennwil, Bennwil . 041305
Dorfmuseum Birmensdorf,
Birmensdorf 041345
Dorfmuseum Birr, Birr 041346
Dorfmuseum Bönigen, Bönigen . 041351
Dorfmuseum Bottmingen,
Bottmingen 041355
Dorfmuseum Buchenberg, Königsfeld
im Schwarzwald 019729
Dorfmuseum Daniel-Martin-Haus,
Rauschenberg 021298
Dorfmuseum Deckenpfronn,
Deckenpfronn 017836
Dorfmuseum Delligsen, Delligsen 017848
Dorfmuseum Dettingen und Illertal,
Dettingen an der Iller 017879
Dorfmuseum Dierscheid,
Dierscheid 017886
Dorfmuseum Dietersweiler,
Freudenstadt 018504
Dorfmuseum Dobin am See, Dobin am
See 017918
Dorfmuseum Düppenweiler,
Beckingen 017007
Dorfmuseum Eichstetten am
Kaiserstuhl, Eichstetten am
Kaiserstuhl 018124
Dorfmuseum Emmingen-Liptingen,
Emmingen-Liptingen 018195
Dorfmuseum Feldbrunnen,
Feldbrunnen 041482
Dorfmuseum Gahlenz, Gahlenz . 018580
Dorfmuseum Gebhardshain,
Gebhardshain 018598
Dorfmuseum Georgenthal,
Georgenthal 018642
Dorfmuseum Görzig, Görzig bei
Köthen 018741
Dorfmuseum Gontenschwil,
Gontenschwil 041554
Dorfmuseum Graberhaus,
Strengelbach 042036
Dorfmuseum Großgörschen,
Großgörschen 018849
Dorfmuseum Grube, Grube . . . 018868

Dorfmuseum Hanweiler, Städtisches
Museum, Winnenden 022681
Dorfmuseum Hausen im Wiesental,
Heimathaus des alemannischen
Dichters J. P. Hebel, Hausen im
Wiesental 019110
Dorfmuseum Haynrode,
Haynrode 019118
Dorfmuseum Herrnbaumgarten,
Herrnbaumgarten 002078
Dorfmuseum Hoffeld, Hoffeld . . 019289
Dorfmuseum Holzburg,
Schrecksbach 021734
Dorfmuseum Hüntwangen,
Hüntwangen 041601
Dorfmuseum im Greifenhaus, Hausen,
Oberfranken 019112
Dorfmuseum im Guggerhuus,
Ettingen 041479
Dorfmuseum im Kulturzentrum,
Fislisbach 041484
Dorfmuseum im Pulverturm,
Schwanden 041965
Dorfmuseum Jockenhof,
Simonswald 021869
Dorfmuseum Kelter, Marbach am
Neckar 020259
Dorfmuseum Kienberg, Kienberg 019613
Dorfmuseum Kirchbözberg,
Unterbözberg 042078
Dorfmuseum Konolfingen,
Konolfingen 041620
Dorfmuseum Langendorf,
Langendorf 041636
Dorfmuseum Langlingen,
Langlingen 019920
Dorfmuseum Leutesdorf,
Leutesdorf 020054
Dorfmuseum Linthal, Linthal . . 041681
Dorfmuseum Maschwanden,
Maschwanden 041729
Dorfmuseum Melihus, Möhlin . . 041746
Dorfmuseum mit Pfarrer-Lehmann
Stube im Bauernkrieghaus, Landau
in der Pfalz 019870
Dorfmuseum Mönchhof,
Mönchhof 002406
Dorfmuseum Niederbreitbach,
Niederbreitbach 020791
Dorfmuseum Niederlinxweiler, Sankt
Wendel 021611
Dorfmuseum Nikitsch, Sammlung
Anica, Nikitsch 002457
Dorfmuseum Nunningen,
Nunningen 041805
Dorfmuseum Ostheim, Nidderau 020782
Dorfmuseum Pfaffenweiler,
Pfaffenweiler 021114
Dorfmuseum Pöttelsdorf,
Pöttelsdorf 002521
Dorfmuseum Poppenweiler,
Ludwigsburg 020125
Dorfmuseum Ringewalde, Temmen-
Ringewalde 022124
Dorfmuseum Rohrbach, Rohrbach,
Pfalz 021465
Dorfmuseum Roiten,
Rappottenstein 002570
Dorfmuseum Rupperswil,
Rupperswil 041899
Dorfmuseum Schlossweid,
Ringgenberg 041884
Dorfmuseum Schönbach,
Großbothen 018840
Dorfmuseum Schönwalde, Schönwalde
am Bungsberg 021716
Dorfmuseum Seon, Seon 041983
Dorfmuseum Stove, Boiensdorf . 017468
Dorfmuseum Stricklerhuus,
Hombrechtikon 041597
Dorfmuseum Therwil, Therwil . . 042054
Dorfmuseum Tremmen, Tremmen 019595
Dorfmuseum Tündern, Hameln . 019026
Dorfmuseum Weckbach,

Weilbach 022476
Dorfmuseum Weinburg,
Weinburg 002903
Dorfmuseum Wiesen, Wiesen . . 042137
Dorfmuseum Zeihen, Zeihen . . . 042174
Dorfmuseum Zeissholz,
Bernsdorf 017342
Dorfmuseum Ziefen, Ziefen . . . 042178
Dorfmuseum Zierow, Zierow . . 022877
Dorfmuseum Zwingendorf,
Zwingendorf 003147
Dorfmuseum, Museumssölde,
Museumsstadel, Alte Schule,
Mertingen 020377
Dorfschmiede- und Heimatmuseum,
Empfingen 018196
Dorfschmiede Neckargröningen,
Remseck am Neckar 021365
Dorfschulmuseum, Ködnitz . . . 019670
Dorfstube Münchholzhausen,
Wetzlar 022606
Dorfstube Ötlingen, Weil am
Rhein 022467
Galerie Doria, Paris 104691
Doria, Pesaro 081148
Doria & Lino Santos, Franca,
Lisboa 084877
Dorian, Miami 121770
Doric Arts, King's Lynn 118045
Doric House, Flemington 048746
Dorigo, Richard, Zürich 087951, 133979
Doris' Place, Saint Louis 096692
Doriti, Maria, Lemesos 102547
Dorking and District Museum,
Dorking 044123
Dorking Desk Shop, Dorking . . 089080
Dorking House Antiques, Dorking 089081
Dorman Memorial Museum,
Middlesbrough 045260
Dormen, İstanbul 117028
Dorn, Wachtendonk 078818
Dorn, Gertrud, Stuttgart 108692
Dorn, Werner, Weisendorf 078913
Dornal, Golden Square 061536
Dornburg Galerie, Dornburg, Kreis
Limburg-Weilburg 106691
Dornburger Schlösser, Klassik Stiftung
Weimar, Dornburg, Saale 017942
Dornebusch, A., Düren 075444
Dorner, Wien 062928
Dorner, Anton, Darmstadt 141680
Dorney Court, Windsor 046161
Dornier, Vincent, Nancy 140977
Dorninger, Michael, Micheldorf . 127819
Dornoch Heritage Museum,
Dornoch 044126
Dorobanti, Bucureşti 114137
Dorobell, Castello de la Plana . . 115016
Doron, Esther, Jerusalem 079730
Dorothea B. Hoover Historical
Museum, Joplin 049840
Dorothea C. Witham Gallery,
Lewiston 050233
Dorotheum, Bruxelles 125307
Dorotheum, Dornbirn 125252
Dorotheum, Düsseldorf 125927
Dorotheum, Graz 125253, 125254
Dorotheum, Innsbruck 125257
Dorotheum, Klagenfurt 125259
Dorotheum, Linz 125260, 125261
Dorotheum, Milano 126204
Dorotheum, Praha . . . 065490, 125420
Dorotheum, Roma 126224
Dorotheum, Salzburg 125265
Dorotheum, Sankt Pölten 125266
Dorotheum, Verona 126241
Dorotheum, Wien 125272
Dorotheum, Wiener Neustadt . . 125289
Dorotheum, Zagreb 125415
Dorotheum GmbH & Co KG,
München 126047
Dorothy G. Page Museum,
Wasilla 054196
Dorothy's Antiques, Sheringham 091315

Dorothy's House Museum, Port
Hope 006690
Dorottya Galéria, Budapest 109137
Dorottya Galéria, Ernst Múzeum,
Budapest 023469
Dorozsmai Szélmalom, Szeged . 023922
Dorpsmuseum de Kleuskes,
Liempde 032597
Dorpsmuseum De Kluis, Eext .. 032276
Dorpsmuseum in de Drie Snoeken,
Oud-Gastel 032743
Dorr, Wiesbaden 078973
Dorridge Antiques, Dorridge ... 089100
Dorrigo Art of the Country,
Dorrigo 098975
Dorrigo Steam Railway and Museum,
Dorrigo 001032
Dorsaz, Christian, Vollèges ... 133955
Dorsch, Miami 121771
Dorsel, Friederike, Heimbach .. 107407
Dorset Coin Company, Parkstone 090945
Dorset County Museum,
Dorchester 044116
Dorset Rare Books, Shaftesbury 144831
Dorset Reclamation, Bere Regis 088441
Dorset Teddy Bear Museum,
Dorchester 044117
Dorsky, Carol, New York 122543
Dorsman, Titus, Amsterdam ... 143026
Dorta, Antonietta, Miami 121772
Dortmunder Kunstverein e.V,
Dortmund 058838
Dortmunder Museumsgesellschaft,
Dortmund 058839
Dorval, Régis, Lille 104004
Dory Shop Museum, Shelburne . 007015
Dos Coronas, Barcelona 085765
Dos Hermanos Viejo, Madrid ... 086173
Dosch, Peter P., Zürich 116767
Dose, Bernd, Lübeck 077173
Dose, Harry, Oldenburg 077927
Doskal, Montevideo 091997
Doß, Brigitte, Quickborn 142311
Dossat 2000, Madrid 137985
Dossev, Sofia 101151
Dossier de l'Art, Dijon 138640
Dossin Great Lakes Museum, Detroit
Historical Museum, Detroit ... 048303
Dostál, Praha 065491, 128343
Doster, K., Amstetten 105990
Doster, Thomas, Tübingen 078721,
                         142425
Dot, Genève 116296
Dot Concept, Singapore 114514
Dotart Galeria de Arte, Belo
Horizonte 100882
Dots to Nuts, Cleveland 120416
Dotti, Daniele, Roma 142809
Dotti, Paolo, Roma 131910
Dotzel & Sons, Michael J., New
York 135910
Dotzheimer Museum, Wiesbaden 022630
Dou, Gerard, Amsterdam 082730
Doual'art, Douala 005290
Double Bay Old Books, Double
Bay 139716
Double Dragon II, Honolulu ... 093541
Double Duck's, Eindhoven ... 083212
Double Vision Gallery, Los
Angeles 121474
Doublet, Avranches 103317
Doubletake Gallery, Minneapolis 121982
Doucet House, Farmers Bank of
Rustico, Rustico 006835
Douchet, François, Biéville-
Beuville 067218
Les Doucines, Lambres-les-
Douai 069150
Douet, Digosville 068309
Doug, Birmingham 092404
Doug Kerr Vintage Museum,
Oaklands 001378
Doughty, John, San Francisco . 097221
Douglas, Miami 094730

Douglas, South Deerfield 127461
Douglas & McIntyre, Vancouver . 136691
Douglas Art Association, Douglas 060463
Douglas-Baker, Minneapolis ... 121983
Douglas Brett – Modern Life Designs,
San Francisco 124271
Douglas County Historical Society
Museum, Superior 053611
Douglas County Museum Complex,
Armour 046544
Douglas County Museum of History
and Natural History, Roseburg . 052679
Douglas Design, Edmonton 101281
Douglas F. Cooley Memorial Art
Gallery, Reed College, Portland 052291
Douglas Hall, Brighton 088619
Douglas Heilman, Phillip, New
York 122544
Douglas Heritage Museum, Douglas,
Lanarkshire 044127
The Douglas Hyde Gallery, Dublin 024873
Douglas, Bobby, Comber 088964
Douglas, David, Indianapolis .. 121179
Douglas, Gavin, London 090080, 134651
Douglass, Dallas 120588
Douglass Historical Museum,
Douglass 048347
Douillard, Saint-Hilaire-de-Riez . 072823
Douillard, Frédéric, Saumur ... 073436
Doukhobor Village Museum,
Castlegar 005576
Doull, Winston, Haan 107192
Douma, Amsterdam .. 082731, 132395
Douma, A.R., Amsterdam 082732
Douma, M.H., Amsterdam 082733
Doumen County Fine Art Calligraphy
Society, Zhuhai 008551
Doumen Museum, Doumen 007750
Doumenc, Laurent, Saint-Medard-en-
Jalles 072974
Doussière, Marseille 128763
Doussot, Eze 103727
Doutrebente, Oliver, Paris 125705
Douwes, London 090081, 118291
Douwes, Sint Nicolaasga 132631
Douwes Fine Art, Amsterdam .. 082734,
           112188,   132396
Dove, San Diego 124118
Dove Cottage and the Wordsworth
Museum, Grasmere 044465
Dove House Gallery, Mornington 099377
Dover-Calais Antiques, Atlanta . 092126
Dover Castle, Dover 044129
Dover Museum, Dover 044130
Dover Publications, Mineola ... 138361
Dover Transport Museum,
Whitfield 046123
Doveridge House Antiques,
Wolverhampton .. 091903, 135252
Dovero, Gérard, Marseille 128764
Dovetail Antiques, Sheffield .. 091282,
                         135025
Dovetail Creations, Charlotte .. 135391
Dovetails Gallery, Ilfracombe .. 118010
Dovin Galéria, Budapest 109138
Dow, Fort Worth 120862
Dow Sheng Chinese Kuoshu Weapons
Museum, Tamsui 042571
Dowd Fine Arts Gallery, State
University of New York,
Cortland 048015
Dower House Antiques, Kendal . 089669
Dowerin District Museum,
Dowerin 001033
Dowers, Chester, Cincinnati ... 127172
Dowling, Tralee 079711
Dowling & Bray, Looe 090631
Dowling Street Gallery, Dunedin 083924,
                         112823
Dowling, Judith, Boston 092465, 119859
Down & Quilt Shop, New York . 095358
Down County Museum,
Downpatrick 044137
Down Home Antiques, Miami . 094731

Down Home Creations, Oklahoma
City 095969
Down Lane Hall Antiques, Tunbridge
Wells 091649, 135158
Down Memory Lane, Columbus 093013
Down Memory Lane, Fyshwick . 061496,
                         139744
Down the Lane, Hamilton 064318
Down Town, Paris 104692
The Downers Grove Park District
Museum, Downers Grove ... 048361
Downey Museum of Art, Downey 048362
Downey, Peter, Auckland 083782
Downeys Antique Shop,
Buncrana 079464
Downhill Castle, Castlerock ... 043887
Downie, Robert, Whitchurch,
Shropshire 144919
Downieville Museum, Downieville 048363
Downings, Ramsgate 118955
Downpatrick Railway Museum,
Downpatrick 044138
Downs Court Galleries,
Altrincham 117234
Downstairs Gallery, Vancouver . 101777
Downtown Antiques, West
Hollywood 097887
Downtown Antiques Mall,
Nashville 094994
Downtown Art Show Space, Los
Angeles 121475
Downtown Arts Projects, New
York 122545
Downtown Books, Milwaukee .. 145266
Downtown Gallery, Nanaimo ... 006447
Downtown Gallery, New Orleans 122155
Downtown Independent Gallery, Los
Angeles 121476
Downtown Model Trains,
Vancouver 064818
Dow's Books, Aberdeen 144211
The Dowse, Hutt City 033223
Dowse & Son, A.E., Sheffield .. 127040
DOX – Centrum současného umění,
Praha 009784
Doyen, Yannick, Vannes
(Morbihan) 105819
Doyle, Malvern East 061770
Doyle, New York 127346
Doyle, Rotorua 084069, 132737
Doyle & Doyle, New York 095359,
                         135911
Doyle, Stephen F., Providence . 096504
DPD Svoboda Tone Čufar,
Jesenice 114681
DPH Trading, Odense 066034
Dr. Arratta Memorial Museum,
Muttaburra 001334
Dr.-Arthur-Lindgens-Jagdmuseum,
Schönsee 021713
Dr. Batthyány-Strattmann László
Múzeum, Körmend 023752
Dr.-Bauer-Heimatmuseum, Bad
Rothenfelde 016849
Dr. Berger Heimatmuseum,
Forchtenberg 018382
Dr. Bhau Daji Lad Museum,
Mumbai 024401
Dr. Busso Peus Nachf., Frankfurt am
Main 075778, 125946
Dr.-Carl-Haeberlin-Friesen-Museum,
Wyk auf Föhr 022836
Dr. Daniel W. Kissam House,
Huntington Historical Society,
Huntington 049620
Dr. Do a Lot, Kumeu 132724
Dr. Eduard Benninger Heimathaus,
Bad Wimsbach-Neydharting . 001767
Dr. Eisenbarth- und Heimatmuseum,
Oberviechtach 020936
Dr. Engelbert Dollfuß-Museum,
Texing 002820
Dr. Fr.R. Kreutzwaldi
Memoriaalmuuseum, Võru . 010720

Dr. Francis Medical and Apothecary
Museum, Jacksonville 049766
Dr. Giri's Museum, Chennai 024195
Dr. Heins Classic Car Collection,
Christchurch 033171
Dr. Henry N. Payne Community
Museum, Cow Head 005658
Dr. I. Lehr Kunstauktionen GmbH,
Berlin 125876
Dr. Increase Mathews House,
Zanesville 054613
Dr. J.'S Baseball Cards,
Cleveland 092935, 145085
Dr. John Harris Dental Museum,
Bainbridge 046713
Dr. Johnson's House, London .. 044960
Dr. Jono Šliūpo Memorialinė Sodyba,
Lietuvos Nacionalinis Muziejus,
Palanga 030437
Dr. Juba Ferenc Magyar
Tengerészettörténeti Gyűjteménye,
Klapka György Múzeum,
Komárom 023755
Dr. Kui-Fang Lai Art Museum,
Singapore 038267
Dr. Lin Sian Tang Gallery, Ming Tai
High School, Wufeng 042582
Dr. Park Collection and Gallery,
Seoul 029992
Dr. Pepper Museum and Free
Enterprise Institute, Waco 054014
Dr. Raj Bali Pandey Puratatva
Sangrahalaya, Deoria 024240
Dr. Samuel D. Harris National Museum
of Dentistry, Baltimore 046743
Dr. Sun Yat Sen Memorial House,
Macau 008065
Dr. Vernon L. Watson Allied Arts
Centre, Dauphin 005683
Dr. Vinco Kudirkos Muziejus, Lietuvos
Nacionalinis Muziejus, Kudirkos
Naumiestis 030399
Dr. William Robinson Plantation,
Clark 047769
Dr. Woods House Museum,
Leduc 006180
Draaiorgelmuseum, Assen 032034
Draakoni, Tallinn 103114
Drabinsky, Toronto 101601
Drachenhöhle, Syrau 022094
Drachenloch Museum, Vättis .. 042091
Drachmanns Hus, Skagen 010303
Drackett, Cincinnati 092845
Draenert, Immenstaad am
Bodensee 107458
Dräxler, Dr. Bettina, Wien 127912
Dräxler, Thomas, Wien 127913
Dragesco-Cramoisan, Paris ... 071268,
                         104693
Dragets Galleri, Sollentuna 115820
Drago Artecontemporanea,
Bagheria 109763
Dragør Museum, Dragør 010032
Dragon, Ha Noi 125073
The Dragon, South Molton 091363
Dragon- og Frihedsmuseet,
Holstebro 010134
Dragon & Phoenix, Sunnybank . 062260
Dragon and Friends, Hong Kong 065034
The Dragon and the Phoenix, South
Molton 091364
Dragon Antique Furniture, Seoul 082259
Dragon Antiques, Harrogate ... 089426
Dragon Antiques, Kettering ... 089677
Dragon Antiques, New York ... 095360
Dragon Art, Portland .. 096395, 123614
Dragon Arts Gallery, Hong Kong 065035
Dragon Court Antiques,
Singapore 085395
Dragon Culture, Hong Kong .. 065036,
                         065037
Dragon Era, Shanghai 065133
Dragon Gate Antique, Hong Kong 065038
Dragon House, San Francisco .. 097222

Dragon House, Singapore 085396
Dragon Traders, Seattle 097429
La Dragona, Roma 081344
Dragonermuseum, Wels 002916
Dragonette, Los Angeles 094322
Dragonette, Patrick, West
  Hollywood 097888
Dragonflies & Daydreams,
  Pittsburgh 096279
Dragonfly Antiques, Louisville . 094542
Dragonfly Paintings, Amstelveen 112130
Dragonfly Paintings, Krommenie 112544
Draheim, Anna-Margarete,
  Rellingen 078192
Drahošová, Silvia, Rožňava ... 085587
Drai, Edmond, Le Pian-Medoc .. 069329
Drai, Guillaume, Bordeaux 067322
Drai, Simon, Bordeaux 067323
Draier Verlag, Friedberg, Hessen 137213
Draíocht, Dublin 109582
Draisienne, Saint-Étienne 072762
Drake House Museum, Plainfield 052188
Drake Well Museum, Titusville . 053753
Drake, Ashton, Wolfurt 063122
Drakensteijn, Amsterdam 082735
Drakesbrook, Waroona 062406
Drama Archaeological Museum,
  Drama 023064
Drammen Antikvariat, Drammen 143384
Drammens Museum for Kunst og
  Kulturhistorie, Fylkesmuseum og
  Galleri for Buskerud, Drammen 033572
Drammens Museum, Drammen
  Kunstforening, Drammen 033571
Drangedal Bygdetun, Drangedal 033573
Dranoff, New York 122546
Draper, Jim, Jacksonville 121245
Dráva Múzeum, Barcs 023435
Drawbridge, Saint Paul 096820
Drawe, G., Speyer 078520
Drawing Center, New York 051379
Drawing Gallery, Singapore ... 114515
The Drawing Room, Makati 113385
The Drawing Room, Pershore .. 090971
Drawing Society of Canada, Gabriola
  Island 058444
The Drawingroom, Chesham .. 117567
Drawn From Memory, London .. 118293
Dray Walk Gallery, London 118294
Drayton Hall, Charleston 047546
Drayton Valley Museum, Drayton
  Valley 005715
Drazen, Sydney, New York 095361,
                                127347
Dream a Theme, Detroit 120806
Dream Merchants, Denver 093279
Dream Toys, Bruxelles 063403
Dream, E., San Francisco 124272
Dreamcards, Las Vegas 094134
Dreamfactory, Degersheim 126718
Dreamhaven Books, Minneapolis 145277
Dreaming Art et Déco, Mantes-La
  Jolie 104152
The Dreaming Room, San
  Francisco 097223
Dreamland, Baltimore 092313
Dreamtime Cultural Centre,
  Rockhampton 001455
Drechsler- und Metalldrücker-Museum,
  Wendelstein 022560
Dreczko, M., Reutlingen 108479
Dred, Scott, Surry Hills 099744
Drees, Bruxelles 100594
Drees, Robert, Hannover 107356
Dreffler, H., Hishult 133537
Drefit, Kraków 132837
Dreher Sörmúzeum, Budapest .. 023470
Dreher, Anselm, Berlin 106216
Dreher, Konrad, Freiburg im
  Breisgau 141799
Drei-Brunnen-Verlag,
  Plüderhausen 137472
Drei-Königs-Antik, Reutlingen .. 098136

Drei-Seen-Galerie, Seeham 100199
Dreieich-Museum, Dreieich 017973
Dreier Verlag, Frankfurt am Main 137196
Dreier, Michaela, Burghausen .. 075226
Dreier, Wolfgang, Berlin 074784
Dreierlei, Wasserburg am Inn .. 078873
Dreimäderlhaus, Wien 062929
Dreimol, Uwe, Münster 077657
Dreipunkt Edition, München ... 137395
Dreiseitel, Helmut, Köln 107603
Dreispitz, Thunstetten 116670
Drenewydd Museum, Bute Town 043819
Drenth, Almelo 082627
Drents Museum, Assen 032035
De Drentse Glasbloazer, Dalen . 032142
Drescher, Olaf, Berlin 141504
Drescher, Stefan, Käbschütztal . 130086
Dresden Art Galleries, New York 122547
Dresdener-Antiquariat, Dresden . 141694
Dresdener Kunstauktionshaus,
  Dresden 125921
Dresdener Kunstblätter, Dresden 138693
Dresdner Münzhandlung,
  Dresden 075411
Dresdner Rahmenwerkstatt,
  Dresden 129706
Dresdner Sezession 89 e.V.,
  Dresden 058840
Dress and Textiles Specialists Group,
  Beamish 060113
Dreuille, Paris 071269, 128991
Drevet, Saint-Ouen 073069
Drew, Fort Worth 093445
Drew County Historical Museum,
  Monticello 051029
Drew, Simon, Dartmouth 117664
Drewdale, New York 145335
Dreweatt Neate, Bristol 126839
Dreweatt Neate, Eastbourne ... 126885
Dreweatt Neate, Godalming ... 126905
Dreweatt Neate, Tunbridge Wells 127070
Dreweatts 1759, Donnington .. 126878
Drews, Walter, Wien .. 062930, 140069
Drexler, Louisville 094543
Drey, Colmar 103618, 128566
Dreyer, Lillehammer 113174
Dreyer, H., Mönchengladbach .. 077388,
                               130310
Dreyer, Marc, Thann 073641
Dreyfus, Austin 092224, 135324
Dreyfus, Raymond, Paris 071270, 104694
Dreyfuss, Simone Michele, Bari . 079770,
                                079771
Dreyspring, Brigitte, Wiesbaden 130756
Dreysse, Roswitha, Quedlinburg 130545
Driant, Jany, Langres 069176
Dridan, McLaren Vale 099245
Drie Kronen, Alkmaar 082618
De Drie Uilen, Utrecht 083640
Driebe Freight Station,
  Stroudsburg 053573
Driels Oudheidkundig Museum,
  Kerkdriel 032542
Driemert, Hartmut, Damshagen . 075296
Driesen, Marianne, Taunusstein . 142410
Driessen & Schreurs, Genk 063617
Driessen, K., Selfkant 078451
Driessen, P.A.P., Maastricht 132573
Drieux, Pascale, Paris 071271
Drif, Wittenheim 074300
Driffold, Sutton Coldfield 091519, 119186
Driftwood, Cleveland 120417
Driftwood Gallery, Ilfracombe .. 118011
Drilandmuseum Gronau, Gronau,
  Westfalen 018831
Drill Hall Gallery, Australian National
  University, Acton 000757
Drill Hall Market, New Norfolk .. 061953
Driminagh Castle, Dublin 024874
Dring, C. & K.E., Lincoln 089828
Drinhausen, A., Nümbrecht 077807
Drishti, Nagpur 109487
Drisket, Anne-Marie, Liège 063754
Dritsoulas, Serafim, Neufahrn bei

Freising 130446
Drive-In Exchange, Baltimore .. 092314
Drjanovski Manastir Muzej,
  Drjanovo 005022
Droege, Rudolf, Düsseldorf 075463
Dröge, V., Bad Salzuflen 141434
Dröge, Volker, Bielefeld 141596
Dröll, Frank, Koblenz .. 076832, 130136
Dröscher, Elke, Hamburg 107238
Droin, Gérard, Auxonne 125503
Droit de Regard, Arcachon 103264
Droit de Regard, Sanguinet 073404,
                               105651
Droite, Den Haag 083078
Droitwich Spa Heritage Centre,
  Droitwich 044141
Dromana and District Museum,
  Dromana 001034
Dromana Antiques, Dromana .. 061407
Drome Magazine, Roma 138818
Dromos Restauri, Roma 131911
Dromson, Strasbourg . 073585, 105682
Dronfield Antiques, Sheffield .. 091283
Droogdok Jan Blanken,
  Hellevoetsluis 032456
De Droomfabriek, Amsterdam .. 082736
Drop Dial Antiques, Bolton 088512,
                               134121
Drori, Irene, Los Angeles 121477
Droschl, Max & Annette, Graz .. 099998,
                      136508,   139985
Drostdy Museum, Swellendam . 038844
Drostdy Museum, Uitenhage .. 038869
Droste-Museum, Münster 020596
Droste-Museum im Fürstenhäusle,
  Meersburg 020328
Droste-Museum, Burg Hülshoff,
  Havixbeck 019117
Droste Verlag, Düsseldorf 137156
Droste, Norbert, Hamm,
  Westfalen 076348
Droste, Rüdiger von, Andernach 074411
Drottningholms Slott,
  Drottningholm 040593
Drouart, Paris 104695
Drouet, Juliette, Paris 141095
Drouin, Audincourt 066805
Drouot de Richelieu, Paris 125706
Drouot Estimations, Paris 125707
Drouot Expertise, Paris 125708
Drouot Montaigne, Paris 125709
Drouot Nord, Paris 125710
Droysen-Keramikgalerie, Berlin . 106217
Droz, Genève 138048
Drtikola, Františka, Příbram 102731
Druckereimuseum, Freienbach . 041494
Druckereimuseum, Hatten 019095
Druckereimuseum, Kusel 058248
Druckereimuseum, Wildeshausen 022653
Druckereimuseum und Schaudepot
  der Museen der Stadt Erfurt im
  Benary-Speicher, Erfurt 018239
Druckhistorische Ausstellung,
  Kornwestheim 019770
Druckmuseum, Darmstadt 017818
Druckmuseum Graz, Graz 001970
Druckwerk, Dornbirn 058248
Druckwerkstatt im Bethanien,
  Berlin 106218
Drüner, Dr. Ulrich, Stuttgart ... 142387
Druivenmuseum, Overijse 003944
Drukkerijmuseum Meppel,
  Meppel 032646
Drum & Co., San Francisco ... 097224
Drum Barracks Civil War Museum,
  Wilmington 054432
Drum Castle, Drumoak 044145
Drum Museum Taikokan, Tokyo 029500
Drum, Denis, Malahide 126176
Drum, Paul, Killinick 079647
Drumheller and District Museum,
  Drumheller 005717
Drumlanrig Castle, Thornhill,
  Dumfriesshire 045976

Drumlanrig's Tower, Hawick ... 044542
Drumlin Farm, Lincoln 050286
Drummer, Matthias, Berlin 141505
Drummond, Hindhead 089534
Drummond, London 090082
Drummond Home, Oklahoma Historical
  Society, Hominy 049524
Drummond Island Historical Museum,
  Drummond Island 048370
Drummond, Nicholas, Odiham . 090909,
                               134904
Drummonds Antiques, Auckland 083783
Drumond, Fernanda & Gonzaga,
  Funchal 113973
Druon, Sylvie, Douchy-les-Mines 068396
Drusche, V., Bad Doberan 074503
Drusilla's Books, Baltimore 144983
Druskininkai Miesto Muziejus,
  Druskininkai 030331
Druskininkų Ž. Lipšico Žydų Muziejus,
  Valstybinis Vilniaus Gaono Žydų
  Muziejus, Druskininkai 030332
Društvo Istoricara Umetnosti Srbije,
  Beograd 059785
Društvo na likovnite umetnici za
  primeneta umetnost na Makedonija,
  Skopje 059558
Društvo Povjesnicara Umjetnosti
  Hrvatske, Zagreb 058520
Drutt, Helen, Philadelphia 123369
Druvienas Vecā Skola-Muzejs,
  Druviena 030117
Dry Falls Interpretive Center, Coulee
  City 048033
Dry Mounting, Brilon 075196
Dry, Dr. Graham, München 137396
Dryade, Paris 104696
Dryden, New York 122548
Dryden and District Museum,
  Dryden 005722
Dryden Galleries, Providence .. 123701
Dryphoto, Prato 110688
Drysdale, Bowen Hills 139651
Državni Muzej, Narodni Muzej Crne
  Gore, Cetinje 031750
Drzażdżyński, Marek, Warszawa 132875
D's Antiques, Kansas City 094071
DSA Bookart, Epping Green ... 138159,
                               144412
DsArt Galerie, Frankfurt am Main 106982
DSM, Saint-Maur-des-Fossés .. 072562
DSS Coin, Omaha 096038
DSV Kunstkontor, Deutscher
  Sparkasseenverlag, Stuttgart .. 108693
Du Page County Historical Museum,
  Wheaton 054363
Du Tam-Tam au Satellite, Musée
  Européen de la Communication,
  Angers 011649
Dual Galerija, Novi Sad 114457
Dualchas Heritage Service for Skye
  and Lochalsh, Portree 060114
Dualchas-Museum Bharraigh Agus
  Bhatarsaidh, Isle-of-Barra .. 044654
Dualibe, Fortaleza 100942
Duane, Anchorage 127120
Duane Reed, R., Saint Louis .. 123821
Duarte & Martins, Lisboa 126442
Dubach, Anna Louise,
  Murgenthal 087690
Dubai Museum, Dubai 043333
Dubard, Dijon 068325
Dubbel, Siegfried, Gevelsberg . 075991,
                               107121
De Dubbelde Palmboom, Historisch
  Museum Rotterdam, Rotterdam 032801
De Dubbele Deur Antiek,
  Rotterdam 083566, 132619
Dubbert, F., Nordhorn 077800
Dubbo Antiques, Dubbo 061410
Dubbo Military Museum, Dubbo 001036
Dubbo Regional Gallery, Dubbo . 001037
Dube, Saint-Ouen 073070
Dube-Heynig, Dr. Annemarie,

München 108064
Dubé, Laurier, Paris . . 104697, 136896
Dubernet de Garros, Agnès,
Cahors 067595
Dubey, Baltimore 092315
Dubinsky, Zürich 087952
Dubio, Den Haag 112380
Dubisar, Wilhelm, Sankt Michael im
Burgenland 127866
Dublin Art Metals, Dublin . . . 079545
Dublin Castle, Dublin 024875
Dublin Civic Museum, Dublin . 024876
Dublin-Kitzen, Miami 121773
Dublin-Laurens Museum, Dublin 048372
Dublin Print Museum, Dublin . . 024877
Dubly, Jacques, Le Castellet . . 069236
Duboc, Rodolphe, Andé 066589
Dubois Museum, Dubois 048373
Dubois Rémi, Colombes 068104
Dubois, Adrien, Pertuis 072114
Dubois, Colette, Paris 104698
Dubois, Dominique, Tilques . . . 073682
Dubois, Eric, Paris . . . 071272, 104699
Dubois, François, Saint-Cast-le-
Guildo 105496
Dubois, Jean-François, Paris . . 071273
Dubois, Jess E., Denver 120708
Dubois, Michel, Saint-Hilaire-du-
Harcouët 072826
Dubois, Nicole, Blas 067216
Dubois, Rémi, Suresnes 073606
Dubois, Rémi, Versailles 074052
Dubouilh, P., Langon 069172
Dubouilh, Patrick, La Réole . . . 069060
Dubouilh, Patrick, Verdelais . . . 074017
Dubourg, Jean-Pierre, Bordeaux 125527
Dubrovin Farm Museum, Yisod
Hama'ala 025221
Dubuque Museum of Art,
Dubuque 048374
Dubuy, Lucie, Paris 128992
Duc Minh Art Gallery, Ho Chi Minh
City 054802
Ducaat, Doetinchem 083169
El Ducado, Santander . 086423, 115428
Ducalerestauro, Venezia 081833, 081834,
132240,   132241
Ducasse, Patrick, Dax 103655
Ducastel, Avignon 103306
Ducate, John, Namur 063806
Ducci, Vittorio Veneto 111210
Ducellier, Hugues, Pithiviers-le-
Vieil 125779
Duch, Barcelona 085766
Duchang Museum, Douchang . . 007749
Duchange, Roger, Paris 071274
DuChâteau, Bernard, Saint-Ouen 073071
Duchâteau, Monique, Paris . . . 128993
Duchatel, Nicolas, Saxon-Sion . 073455
Duchayne, Nathalie, Saint-Tropez 105625
Duché d'Uzès, Uzès 016076
Duchemin, Guy, Rouen 072584, 072585
Duchêne, Toulouse 141362
Duchenet, Laurent, Caen 067565
Duchenne, Daniel, Compiègne . 128568
Ducher, François, Pouilly-sous-
Charlieu 072299
Ducher, Jean-Claude, Charlieu . 067851
Duchet, Jacques, Pervencheres 072115
Duchoze, Daniel, Rouen 105474
Duck Creek Antique Mall,
Cincinnati 092846
Duck Lake Regional Interpretive
Centre, Duck Lake 005723
Duckler, Alysia, Portland 123615
Ducky's Auctions, Hampstead . . 127238
Duclerc, André, Sennecey-le-
Grand 073492
Duclerc, Marie-Claude, Chalon-sur-
Saône 067776
Duclos la Chine, Clermont-
Ferrand 140827
Duco, Breda 112353
Ducos, Marc, Saint-Jean-Mirabel 072855

Ducoté, Frédéric-Charles, Lyon . 069732
Ducrocq, Emmanuel, Tournon-sur-
Rhône 073820
Ducroz, Michel, Chamonix-Mont-
Blanc 067802
Ducroz, Michel, Les Praz-de-
Chamonix 069420
Duda, Sonja, Steinheim am
Albuch 078553
Duda, W., Stegen 142376
Dudenhoeffer, Roger, Capestang 067662
Dudin, Skopje 111882
Duding, Daniel, Neirivue 087697
Dudit, Cholet 067983
Dudley and Mary Marks Lea Gallery,
Findlay 048732
Dudley Museum and Art Gallery,
Dudley 044147
Dudmaston House, Quatt 045575
Dudouit, François, Honfleur . . . 103825
Dudragne, Vincennes 141389
Le Due Emme, Bordighera 109893
Due Maestà, Scandiano 081620
Le Due Sicilie, Messina 080368
Le Due Spine, Rovereto 110879
Le Due Torri, Noceto 080951
Dührssen, A., Sprendlingen . . . 108657
Dülkener Narrenmuseum,
Viersen 022330
Duell, C., Boston 119860
Dülmer Kunstverein, Dülmen . . 058841
Dümmer-Museum, Lembruch . . 020028
Dueñas, Teodulo, Bogotá 065202,
128267
Dünnbier, Wolfgang,
Großschönau 129926
Dünne, R., Bad Salzuflen 074579
Dünnebacke, Gisela, Düsseldorf 106768
Duensing, Elfi, Aschaffenburg . . 141412
Dür, Rapperswil 116575
Dür, Wädenswil 116705
Dür, Isabella, Adliswil 116011
Dürbeck, Straubing 108681
Dürbeck, I., Deggendorf 106662
Dürer Terem, Corvin János Múzeum,
Gyula 023671
Düring, Leo, Bad Königshofen . . 074539
Dürmoser, P.J.H., Millstatt . . . 062736
Dürnberger, Franz, Ihringen . . . 076598
Dürnfelder, Silke, Hannover . . . 130006
Dürr, Worms 079064
Dürr, Bernd, München . 077487, 108065,
137397
Dürr, Christoph, München 108066
Dürr, Michel, Carouge 087350
Dürrbeck, Petra & Hubert, Höchstadt
an der Aisch 107435
Dürring, Kurt, Basel 116063
Dürrschmidt, M., Großbräschen . 129925
Dürsteler, Heinz, Zürich 087953
Dürstelerhaus, Ottikon 041833
Dürüst, A., Augsburg 074458
Düsseldorfer Auktionshaus,
Düsseldorf 125928
Düsseldorfer Museumsverein e.V.,
Düsseldorf 058842
Düster, J.B., Essen . . 075672, 129814
Düsterdieck, Jürgen, Vellmar . . 130722
Düsterwald, J., Hameln 107324
Düwal, Eckard, Berlin 141506
Dufey, Serge, Lausanne 133795
Duff Community Heritage Museum,
Duff 005724
Duff, George, Edinburgh 089170
Duffar, Nicole, Vichy 074106
Dufferin County Museum,
Rosemont 006826
Dufferin Historical Museum,
Carman 005568
Duffour, Pierre, Rochefort (Charente-
Maritime) 072496, 129207
Dufftown Whisky Museum,
Dufftown 044149

Duffy, Brian, London 134652
Duffy & Sons, Jimmy,
Philadelphia 123370
Duffy, John, Dunshaughlin 079614
Duffy, Michael, Dublin 079546
Dufillot, Georges, Bordeaux . . . 128488,
128489
Duflon & Racz, Bern 116128
Duflot, Jacques, Angers 066605
Dufor Quercy, Michel, Montauban 070195
Dufour, Amiens 103227
Dufour, Alain, Saint-Maur-des-
Fossés 072963
Dufour, Edmond, Chaumont-sur-
Tharonne 067940
Dufour, Eric, Trun 073911
Dufour, Roseline, Saint-Féliu-
d'Avall 072781
Dufreche, Gérard, Besançon . . . 125514
Dufresne Serre, Christine,
Genève 087463, 133758
Duftmuseum im Farina Haus,
Köln 019678
Dufu Caotang Museum, Chengdu 007693
Dufus, Nicolas, Piffonds 072161
Duggan-Cronin Gallery,
Kimberley 038762
Duggan, Ann, Adare 079432
Duggan, David, London 090083
Duggan, N.S., Baltimore 092316
Dugger Coal Museum, Dugger . 048377
Duggleby, David, Scarborough . 127033
Duggleby, David, Whitby 127085
Dugout, Indianapolis 093879
Dugovic, Fresno 093499
Dugrenot, Paris 071275
Dugue, François Xavier, Chatou . 067928
Duhamel, Guillaume, Paris 104700
Duhamel, Laurence, Montigny . . 070252
Duhme, Osnabrück 077962
Duhon-Benedetti, Sylvie, Toulon
(Var) 105710
Duinat, Isabella, Saint-Jean-de-Luz 072842
Duivenvoorden, P.H.A., Leiden . . 132558
Dujardin, Laurent, Marcq-en-
Barœul 069862
The Duka, Windsor, Queensland 062435
Dukan, Bangalore 079347
Dukat Galeria, Kraków 113541
Duke Homestead, Durham . . . 048405
Duke University Union Museum and
Brown Art Gallery, Durham . . 048406
Duke, Tom, Dallas 093111
Dukek, Ursula, Oberstaufen . . . 077874
Dukel, R.C.T., Leiden 083429
Dukelské Múzeum – Vojenské
Múzeum Svidník, Svidník . . . 038461
Duke's Antiques, Kitchener . . . 064337
Dukes Grove Auctions,
Dorchester 126879
Duke's of Dorchester, Dorchester 126880
Dukessa, Vitoria-Gasteiz 115541
Duksung Women's University Art
Museum, Seoul 029993
Dukue, New York 135912
Dukwon Museum of Art, Seoul . 029994
Dulce, El Paso 120827
Dulce & Filho, M.H., Porto Alegre 100972
Dulcey, Bucaramanga 102384
Dulitz, Joachim, Guben 107181
Dulkerian, Baltimore 092317
Dullaert, H.W.C., Amsterdam . . 082737
Dullak, Karin von, Freising 107067
Dulle, Martin, Alsdorf 074382
Dulle, Willi, Alsdorf 074383
Dullsville, New York 095362
Dulon, Bernard, Paris . 071276, 104701
Duluth Art Institute, Duluth . . . 048380
Duluth Children's Museum,
Duluth 048381
Dulwich Antiques, Dulwich 061413
Dulwich Decorative and Fine Arts
Society, London 060115
Dulwich Picture Gallery, London 044961

Dům Františka Bílka, Galerie Hlavního
Města Prahy, Chýnov 009531
Dům Gustava Mahlera, Jihlava . 009603
Dům Pánů z Kunštátu, Moravské
zemské muzeum, Brno 009465
Dům u Černé Matky Boží – Muzeum
kubismu, Národní Galerie v Praze,
Praha 009785
Dům u Jonáše, Východočeská Galerie,
Pardubice 009748
Dům u Kamenného Zvonu, Galerie
Hlavního Města Prahy, Praha . 009786
Dům u Rytířů, Litomyšl 102616
Dům u Zlatého Prstenu, Galerie
Hlavního Města Prahy, Praha . 009787
Dům Umění, České Budějovice . 009525
Dům Umění Města Brna, Brno . 009466
Dům Umění, Jihomoravské muzeum
ve Znojmě, Znojmo 009989
Dumac, London 064347
Dumain, Nadine, Puymoyen . . . 129181
Dumaine, Le Tronchet 069368
Dumaine-Lafare, Bandol 066930
Dumaine, Suzanne, Tulle 073914
Dumaresq, Rio de Janeiro 064053
Dumartin, Christian, Ligré 069465
Dumartin, Gilles, Chinon 067978
Dumas, Aix-les-Bains 066480
Dumas, La Monnerie-le-Montel . 128644
Dumas, François, Beaune 067017
Dumas, Gustave, Nice 128857
Dumas, Jean-Paul, Lyon 069733, 128711
Dumbarton House, Washington . 054107
Dumbarton Oaks Collections,
Trustees for Harvard University,
Washington 054108
Dumbo Arts Center, Brooklyn . . 047256
Dumesnil, Stéphane, Etretat . . 128597
Dumez Numismatique, Mulhouse 070389
Dumfries and Galloway Aviation
Museum, Dumfries 044152
Dumfries Museum & Camera Obscura,
Dumfries 044153
Dumins, Frank, Viersen 078798
Dumjahn, Horst-Werner, Mainz . 142102
Dummann, Rüdiger, Wien 062931
Dumont, Austin 119664
DuMont Buchverlag, Köln 137306
Dumont-Rivola, Caen 125540
Dumont, Bruno, Aouste 066686
Dumont, Eric, Troyes 105777
Dumont, F., Namur 063807
Dumont, Marcel, Aumale 066813
Dumont, Michel, Mainneville . . . 069808
Dumonteil, Andre, Montpellier . . 140965
Dumonteil, Pierre-Michel, Paris . 104702
DuMouchelle, Detroit . 093362, 127206
DuMouchelle, Toledo 124685
Dumoulin, Joël, Condrieu 068131
Dumoussaud, Paris 104703
Dumoussaud, Daniel, Avignon . . 066879
Dumousset, Patrick, Paris 125711
Dun Laoghaire Institute of Art, Design
and Technology, Dún Laoghaire 055864
Duna Galéria, Budapest 109139
Duna Múzeum – Magyar
Környezetvédelmi és Vízügyi
Múzeum, Esztergom 023629
Dunajský, Kežmarok 085561
Dunakapu Téri Régiségi Üzlet,
Győr 079309
Dunamelléki Református
Egyházkerület Ráday, Budapest 023471
Dunamelléki Református
Egyházkerület Ráday Múzeuma,
Kecskemét 023723
Dunántúli Képcsarnok, Pécs . . 079319,
109205
Dunántúli Református Egyházkerület
Tudományos Gyűjteményei Múzeum,
Pápa 023830
Dunaparti Aukciósház és Galéria,
Budapest 079251, 126150
Dunaskin Open Air Museum, Ayr 043459

Dunaújvárosi Papírgyár Üzemtörténeti
Gyűjteménye, Dunaújváros ... 023609
Dunbar Sloane, Auckland 083784,
126304
Dunbar Sloane, Wellington 084162,
126333
Dunbar Town House Museum,
Dunbar 044158
Dunbar, Marcus, South
Queensferry 119117
Dunblane Museum, Dunblane .. 044161
Duncan, Louisville 094544, 127296
Duncan & Reid, Edinburgh 089171
Duncan Baird Publishers, London 138212
Duncan Cottage Museum,
Metlakatla 050820
Duncan Gallery of Art, Stetson
University, De Land 048183
Duncan-McAshan Visual Arts Center,
Ingram 049702
Duncan of Jordanstone College of Art
and Design, Dundee 056526
Duncan, Ken, Matcham 099289
Duncan, Ken, Melbourne 099301
Duncan, Ken, Pokolbin 099570
Duncan, Ken, Sydney 099778
Dundalek, Peter-Moritz,
Göppingen 076029
Dundar Avenue Print Workshop,
Vancouver 101778
Dundas Historical Society Museum,
Dundas 005730
Dundas Valley School of Art,
Dundas 055123
Dundas Valley School of Art Gallery,
Dundas 005731
Dundee Art Galleries and Museum
Association, Dundee 060116
Dundee Art Society, Dundee ... 060117,
117692
Dundee Township Historical Society
Museum, Dundee 048393
Dundonian Studio and Gallery,
Dundee 098987
Dundurn Castle, Hamilton 005968
Dundy County Historical Society
Museum, Benkelman 046917
Dune Art, Latvia 103031
Dunedin Fine Arts Center,
Dunedin 048395
Dunedin Historical Society Museum,
Dunedin 048396
Dunedin Public Art Gallery,
Dunedin 033189
Dunedoo Country Antiques,
Dunedoo 061414
Duner, Göteborg 115631
Dunfermline Museum,
Dunfermline 044172
Dung Bui, Paris 104704
Dung, Peter & Petra, Köln 076856
Dungarvan Museum, Dungarvan 024918
Dungog Antiques, Dungog 061415
Dungog Historical Museum,
Dungog 001039
Dunguaire Castle, Kinvara 024948
Dunham Massey Hall, Altrincham 043408
Dunham Tavern Museum,
Cleveland 047814
Dunhill, Cleveland 092936, 135460
Dunhuang Glaswegiansu Guan,
Dunhuang 007753
Dunhuang Museum, Dunhuang . 007754
Dunilli Arts, Casuarina 061300
Dunkel, Helmut, Berlin 074785
Dunkeld Antiques, Dunkeld ... 089119
Dunkeld Cathedral Chapter House
Museum, Dunkeld 044176
Dunker & Nellissen, Greven ... 141842
Dunkers Kulturhus, Helsingborgs
Museer, Helsingborg 040690
Dunklin County Museum, Kennett 049925
Dunlop Art Gallery, Regina 006766

Dunluce Antiques, Bushmills ... 088746,
134194
Dunluce House Museum,
Kimberley 038763
Dunn & Brown, Dallas 120589
Dunn & Sonnier, New Orleans .. 095091
Dunn-Seiler Museum, Mississippi
State 050957
Dunn & Co, Capes, Manchester 126984
Dunn, Hamish, Wooler 091934, 144936
Dunn, John D. & Francis W., Malvern,
Victoria 061781
Dunn, Oliver, Washington 097833
Dunne, Mary, Portarlington 079689
Dunnes, San Jose 136357
Dunnet Pavilion, Castletown ... 043890
Dunolly Museum, Dunolly 001040
Dunree Military Museum,
Inishowen 024932
Dunrobin Castle Museum,
Golspie 044442
Duns Area Museum, Duns 044181
Dunsborough Gallery,
Dunsborough 098988
Dunsheath, Anah, Auckland ... 143280
Dunster Castle, Dunster 044183
Dunster, K.W., Staines 091401
Dunvegan Castle, Dunvegan ... 044184
Dunwell and Community Museum,
Weekes 007371
Dunzendorfer, Herwig,
Hellmonsödt 100033
Duo, Paris 071277
Duo Yun Xuan, Shanghai 102176
Duodecimo, Gent 128048
Il Duomo, New York 095363
Dupard, Yvonne, Pougues-les-
Eaux 072295
Dupasquier, Pierre-Alain, Bulle . 087340,
133710
Dupeux, L'Isle-d'Espagnac ... 140873
Dupeyron, Claude, Castelmoron-sur-
Lot 067706
Dupeyron, Dominique, Orléans . 070814
Dupin, Juliette, Paris 128994
Duplan, Frédéric, Lille 069489
Duplantier, Mireille, Hasparren . 068795
Duplessis, Jacques Vigoureux,
Vignory 074123, 141386
Dupont, Kortrijk 063719
Dupont, Washington 124889
DuPont-Antiquités, Saint-Ouen . 073072
Dupont Gallery, Lexington 050264
DuPont Historical Museum,
DuPont 048398
Dupont, Chantal, Troyes 073901
Dupont, Eric, Paris 104705
Dupont, Jean, Troyes 071782
Dupont, Marlène Edith, Montreuil-sous-
Bois 128815
Dupont, Roberte, Troyes 073903
Dupont, Sophie, Paris 071278
Dupont, Thérèse, Fribourg 133743
Dupoux & Co, Wien 127914
Duprat, Miguel, Saintes 073373
Dupraz, Jean-Baptiste, Villars-sur-
Glâne 133953
Dupré, Rob, Amsterdam 082738
Duprez, Monique, Paris 071279
Dupront, Salon-de-Provence ... 073388
Dupuis, Bernard, Nice 128858
Dupuis, Dominique, Saint-Quentin
(Aisne) 129245
Dupuis, Joël, Condette 103627
Dupuis, Joël, Hardelot-Plage ... 103806
Dupuis, Robert Alexander René,
Berlin 129485
Duputel, Paris 071280, 104706
Dupuy, Daniel, Pujols 072324
Dupuy, Eonnet, Paris 104707
Dupuy, Francis, Honfleur 125596
Dupuy, Michèle, Paris 071281
Dura-Strip of Santa Monica, Los
Angeles 135715

Dural Antiques, Dural 061419
Dural Galleries, Dural 098990
Durán, Madrid 086174, 115182, 126537
Duran, Exposiciones de Arte,
Madrid 115183
Durán, Fernando, Madrid 086175,
086176, 126538, 126539, 126540,
126541, 143760
Durán, Pedro, Madrid 126542
Durance, G., Nantes 140983
Durand, Calgary 101231
Durand-Réville, Françoise, Paris 071282
Durand-Ruel & Cie., Paris 104708
Durand, Brigitte, Saint-Maur-des-
Fossés 072964
Durand, Brigitte, Sens 073498
Durand, Gilles, Villeurbanne ... 074205
Durand, Guy, Montceau-les-
Mines 070224
Durand, Jean-Michel, Menil-la-
Tour 070052
Durandot, Denis, Dijon 128580
Durango, Carmen, Valladolid ... 115524
Durant & Kotek, Bruxelles 063404
Durant, Eliane, Saint-Laurent-de-la-
Salanque 072883
Durante, Giovanni, Roma 081345
Durations Gallery, Baltimore ... 119733
Durban Art, Westbridge 059816
Durban Art Gallery, Durban ... 038698
Durban House Heritage Centre,
Eastwood 044226
Durban Natural Science Museum,
Durban 038699
Durban Segnini, Caracas 125039
Durban Segnini, Miami 121774
Durbes Pils, Tukums 030238
Ďurčanský, Alexej, Trnava 133062
Durchblick-Galerie, Leipzig 107815
Durden, Memphis 121689
Dureau, Laurence, Bourges 067427
Durero, Gijón 115069
Duret Dujarric, Paris 104709
Durga's Art, Mumbai 109429
Durham Art Gallery, Durham ... 005733
Durham Art Guild, Durham 048407,
060464
Durham Cathedral Treasures of Saint
Cuthbert, Durham 044186
Durham Center Museum, East
Durham 048425
Durham Heritage Centre and Museum,
Durham 044187
Durham Historic Association Museum,
Durham 048402
Durham House Antiques Centre, Stow-
on-the-Wold 091468
Durham Light Infantry Museum and
Durham Art Gallery, Durham .. 044188
Durham Western Heritage Museum,
Omaha 051761
Durier, Limony 069536
Duriez, Alain, Grignan 068743
Durif, Rennes (Ille-et-Vilaine) ... 072435
Durjaven Kulturen Center Dvoreca,
Balčik 004974
Durkee, Michael, Wincanton ... 135233
Durlacher, Natasha, London ... 118295
Durot, Charles, Beaurepaire-en-
Bresse 067053
Durot, Gilles, Saint-Bris-le-Vineux 072700,
129218
Duroub, Cairo 103032
Durrell Museum, Durrell 005734
Durrenberger, Henri, Molieres .. 104245
Durst, H., Kirchheim unter Teck 076807
Durst, Jean, Paris 071283, 128995
Duru Artspace, Seoul 111636
Dury, Christian, Nice 104361
Dusable Museum of African-American
History, Chicago 047656
Dusan, Wollongong 062446
Dusan, Bartko, Halle, Saale ... 076163
Dusarduijn, Robert, Amsterdam . 082739

Dusart, Jean-Luc, Poitiers 072221
Dusch, Peter, Speyer 078521
Duse, Leonardo, Brescia 079936
Duseaux, Paris 071284
Duson, Los Angeles 094323
Duss, Martin, Niederrohrdorf ... 116536
Dussably, Marie-Thérèse,
Charolles 067857
Dussi, Paola, Milano 131439
Dussort, Maurice, Beaune 067018
Dussouchaud, Didier, Saint-
Junien 072871
Dust, Las Vegas 121348
Dust Jacket, Cincinnati 145070
Dutch Art, Rotterdam 112636
Dutch Dakota Association Exhibition,
Schiphol 032859
Dutch Iris, Baltimore 092318
Dutch Museum Amsterdam,
Amsterdam 112189, 137830
Dutch Period Museum, Colombo 040476
Dutch Touch Gallery, Philadelphia 123371
Dutch Wall Art, Hilversum 112536
Dutchess County Arts Council,
Poughkeepsie 060465
Dutchess County Historical Society,
Glebe House & Clinton House,
Poughkeepsie 052337
Dutel, Jean-Pierre, Paris 141096
Duteurtre, Philippe, Valognes .. 073954,
129328
Dutilleul, Albi 066513, 103215
Dutilleul, Noel & Giuseppe,
Firenze 080134
Dutko, Jean-Jacques, Paris 071285,
104710
Dutoit, Marcel, Unterentfelden .. 087864,
133945
Dutrieux, F., Bruxelles 063405
Dutta, Genève 087464, 116297
Duttenhofers Books, Cincinnati . 145071
Dutto, Riccardo, Torino 081695
Dutton, D., Los Angeles 145226
Duval, Jacksonville 093986
Duval-Wenta, Simone, Cabourg . 103477
Duval, Anne-Sophie, Paris 071286
Duval, Raymonde, Paris 104711
Duval, Robert & Myriam,
Équeurdreville-Hainneville ... 103717
Duvall, Atlanta 092127
Duvallier, L. & W., Dublin 079547
Duvauchelle, Frédéric, Bannalec 066931
Duvauchelle, Frédéric, Paris ... 071287
Duveau-Sage, Catherine & Philippe,
Paris 071288
Duvert, Martial, Crest 068235
Duvieu, Jacques, Villerville 074182
Duvillard, André, Beaune 067019
Duvillard, Guy, Beaune 067020
Duvillier, Arnaud, Chaumont (Haute-
Marne) 125556
Duvivier, Jacques & Anne, Châtillon
(Hauts-de-Seine) 067915
Duwisib Castle, Maltahohe 031844
Dux, Frank, Bath 088387
Duxbury Rural and Historical Society,
Duxbury 048415
Duyn, Joop, Kenmare 130874
Duys, Oss 083525
Dvärsätts Auktionsbyrå,
Östersund 126649
D'Vine House, Fredericksburg .. 127228
Dviračiu Muziejus, Šiauliu Aušros
Muziejus, Šiauliai 030463
Dvor Trakošcan, Trakošćan ... 009006
Dvorac u Čarevu, Čelarevo ... 038191
Dvořáček, Kolín 065405
Dvořák, České Budějovice 065368
Dvořák, Jan, Benešov 102559
Dvořáková, Jana, Náchod 065426
Dvořákovo Muzeum Pravěku,
Regionální Muzeum v Kolíně,
Kolín 009634
Dvorec-Muzej Betnava, Maribor 038570

Dvorec-Muzej Petra I, Sankt-
Peterburg 037730
Dvorec-muzej Petra III, Muzej-
zapovednik Oranienbaum,
Lomonosov 037200
Dvorec Petra I, Strelna 037903
Dvorec Štatenberg, Makole 038569
Dwight D. Eisenhower Library-
Museum, Abilene 046303
Dwight Frederick Boyden Gallery,
Saint Mary's College of Maryland,
Saint Mary's City 052831
Dwór Artusa, Muzeum Historyczne
Miasta Gdańska, Gdańsk 034725
Dwór w Dołędze, Muzeum
Okręgowego w Tarnowie,
Zaborów 035439
Dworek Jana Matejki,
Krzesławice 034927
Dworek Jana Matejki w Krzesławicach,
Kraków 034863
Dworek Wincentego Pola, Muzeum
Lubelskie, Lublin 034988
Dworski, Marijana, Hay-on-Wye . 144461
Dwyer, Paul, Melbourne 099302
Dwyer, Sammie, Austin 092225
Dyansen, Boston 119861
Dybeck, Stockholm 086989
Dycheling Antiques, Ditchling . 089063
Dyck International Art Collection,
Rosny-sous-Bois 105461
Dycke, Everhard F., München ... 142148
Dyckman Farmhouse Museum, New
York 051380
Dyehouse Gallery, Waterford ... 109702
Dyekjær & Kaae, Lemvig 065989
Dyer Memorial Library, Abington 046310
Dyer & Sons, W., Hayle 089482
Dyfed Antiques, Haverfordwest . 089472
Dyhrfjeld, Jørgen, Frederikssund 065684
Dyka, Corinne, Nice 070599
Dykällans Antik, Frändefors ... 086676
Dylan, Garches 068633
Dylan, Swansea 144873
Dylan Thomas Boathouse,
Laugharne 044765
Dylan's Bookstore, Swansea 144874
Dyma, Walter & Steffen, Rottweil 108532
Dyna, Dallas 135500
Dynamic Impressions, Calgary . 101232
Dynamic Traders, Milwaukee . 094851
Dynamo Arts, Vancouver 101779
Dynasties Antique and Art Gallery,
Singapore 085397, 114516
Dynasties Art, Montréal 064416
Dynasty Antiques Gallery, Hong
Kong 065039
Dynasty Art Gallery, Taipei . 116909,
138129
Dynasty Chinese Antiques,
Edmonton 064291
Dynasty Pacific, Göteborg 086703
Dynasty Pacific, Stockholm ... 086990
Dynasy, Warszawa 113821
Dynon, John, Silverton 099686
Dyrax, Frederikssund 065685
Dyrøy Bygdemuseum, Tennevoll 034011
Dyroff, Klaus-Peter,
Schmiedeberg 130627
Dyson & Son, Clare 126862
Dyson, D.W., Huddersfield 089597,
134461
Dyson, June, Calgary 064255
Dyte, C.W.E. & R.I., Highbridge 089529
Dyte, John & Colin, Burnham-on-
Sea 088726
Dyvelstens Flottningsmuseum,
Forshaga 040630
DZD Art, Roermond 112620
Dzeržinskij Gorodskoj Kraevedčeskij
Muzej, Dzeržinsk 036976
Dział Archeologii i Etnografii,
Muzeum Okręgowe im. Leona
Wyczółkowskiego, Bydgoszcz . 034665

Dział Zbiorów Specjalnych – Muzeum,
Biblioteka im. Stanisława Staszica w
Dzielnicy Bielany m.st. Warszawy,
Warszawa 035313
Dziatko, Maciej, Warszawa ... 113822
Dzieła i Interpretacje, Wrocław . 138962
Dziersk, Klaus, Söhlde 078486, 130644,
142366
Dzierzynski, Outreau 070830
Dzioba, Edmond, Genay 068644
Dziubas, Renate, Ludwigsburg . 077154
DZS, Ljubljana 137943, 143696
Dzubba, Stefan A., Hamburg ... 076213
E & C Sales, Tucson 097654
E & L Furniture, Las Vegas ... 135683
E & P Libros Antiguos, Madrid . 143761
E & S Art Gallery, Louisville ... 121643
E 31 Gallery, Athinai 109032
E Antik, Praha 065492
E-Gallery, London 118296
E-K:s Antik, Göteborg 086704
E-Konst, Falkenberg 115606
E-Miljö Antikt, Växjö 087150
E Street Gallery, Salt Lake City . 123956
E-Studio Galleria d'Arte, Milano . 110306
E-Werk Freiburg, Freiburg im
Breisgau 018479
E. & J.'s Antiques, Richmond . 096539
E. Akhvlediani Children's Picture
Gallery, Tbilisi 016398
E. Akhvlediani House Museum,
Tbilisi 016399
E. Bindstouw, Viborg 010358
E. Galéria, Kecskemét 109197
E. St. Julien Cox House, Saint
Peter 052855
E.C. Allison Research Center, San
Diego 052963
E.C.A. France, Paris . 104712, 136897
E.H. Starožitnosti, Praha 065493
E.K. Antiques, Finedon 089276, 134390
E.M. Violette Museum, Kirksville 049993
E.R. Burrowes School of Art,
Georgetown 055760
E.R.A., Roma 110758
E.T.A. Hoffmann-Haus, Bamberg 016945
E3 Gallery, New York 122549
EA Kanalbod, Trollhättan 087105
EAA AirVenture Museum,
Oshkosh 051827
Eacrett, Sean, Dublin 079548
Eagle Art, Saint Louis 123822
Eagle Bluff Lighthouse, Fish
Creek 048734
Eagle Consultants, Manama ... 100429
Eagle Gallery, Epsom 117761
Eagle Gallery, London 118297
Eagle Harbor Lighthouse and
Museums, Keweenaw County
Historical Society, Eagle Harbor 048419
Eagle Historical Society and Museums,
Eagle City 048418
Eagle Tavern Museum,
Watkinsville 054225
Eagle Valley Museum, Sicamous 007031
Eaglemont Antiques, Eaglemont 061426
Eagles Antiques, Dallas 093112
Eagle's Dream, Singapore 114517
Eagles Nest, El Paso 093399
The Eagle's Nest, Saint Louis .. 096693
Eagles Nest Fine Art Gallery, Aireys
Inlet 098617
Eagles Nest Odds and Ends, Eagle
Heights 061422
The Eakin Gallery, Belfast 117336
Ealing College Gallery, London . 044962
EAM-live-Museum Wasserkraftwerk
Lippoldsberg, Wahlsburg 022372
EAM-live-Museum Wasserkraftwerk
Merkenbach, Dillenburg 017896
EAM-live-Museum Wasserkraftwerk
Wülmersen, Trendelburg 022187
Eamonn's Bookshop, Sandycove 142612
Eandi, Torino 132135

Ear Falls District Museum, Ear
Falls 005735
Earist Museum, Manila 034530
Earl, New Orleans 122156
The Earle-Harrison House, Waco 054015
Earle Wightman Museum, Oyster Bay
Historical Society, Oyster Bay . 051875
Earle, John, Merewether 099326
Earles, Greg & Frances, Oklahoma
City 095970
Earls Barton Museum of Local Life,
Earls Barton 044197
Earls Court Gallery, Hamilton ... 101314
Earlsfield Bookshop, London ... 144584
Early American Museum,
Mahomet 050604
Early Auction, Milford 127320
Early Days Antiques, Paddington,
Queensland 062048
Early Halloween Vintage Clothing, New
York 095364
Early Oak, Knaresborough ... 089723
Early Works, Austin 119665
Earlystreet, Norman Park 001361
Earlywork, South Fremantle ... 099694
Earlyworks Museums, Huntsville 049634
It Earmhus en Oud Friese
Greidboerderij, Warten 033024
Earsham Hall Pine, Earsham ... 089125
Earth and Beyond, Salt Lake City 123957
Earth Art Studio and Gallery, Panton
Hill 099548
Earth Artspace, Auckland ... 112760
Earth Collectible Toy Mall,
Cincinnati 092847
Earth Link Art, Ballarat 098692
Earth Sciences Museum,
Waterloo 007363
Earth Spirit Gallery, Albuquerque 119468
Earth Water Sky, Portland ... 123616
Earthworks Gallery, Mount
Tamborine 099405
Easatis, Tehrän 079422
Easdale Island Folk Museum, Easdale
Island 044198
The Easel, Kingston 111218
The Easel Art Gallery, Chennai . 109280
Easiaart, Singapore 114518
East and Orient Company, Dallas 093113
East and Seaport Museum,
Greenport 049247
East and West, Roma 138819
East and West Art, Kew East . 061687,
099189
East Anglia Transport Museum,
Carlton Colville 043876
East Anglian Antique Dealers Fair,
Loddon 098374
East Anglian Film Archive, University
of East Anglia, Norwich 045398
East Anglian Railway Museum,
Chappel 043904
East Art Artworks, Guangzhou .. 102025
East Brunswick Museum, East
Brunswick 048424
East Campus Galleries, Valencia
Community College, Orlando .. 051804
East Carlton Steel Heritage Centre,
East Carlton 044200
East Coast Museum of Technology,
Gisborne 033207
East Coulee School Museum, East
Coulee 005737
East Ely Railroad Depot Museum,
Ely 048571
East End Arts & Humanities Council,
Riverhead 060466
East Essex Aviation Society and
Museum of the 40's, Saint
Osyth 045707
East Gallery, Taipei 116910
East Gippsland Museum,
Bairnsdale 000811
East Gippsland Star Auctions, Wy

Yung 125250
East Ham Nature Reserve,
London 044963
East Hampton Historical Society
Museum, East Hampton 048431
East Hants Historical Museum,
Maitland 006245
The East Hertfordshire Decorative and
Fine Arts Society, Knebworth . 060118
East Herts Council, Museum
Development, Hertford 060119
East Jersey Olde Towne,
Piscataway 052148
East Jordan Portside Art and Historical
Museum, East Jordan 048439
East Kentucky Center for Sciece,
Mathematics and Technology,
Prestonsburg 052356
East London Museum, East
London 038709
East Martello Museum, Key West 049951
East Meets West, Los Angeles . 094324
East Midlands Museums Service,
Colston Bassett 060120
East Point Antiques, Milwaukee 094852
East Poultney Museum, East
Poultney 048445
East Riddlesden Hall, Keighley . 044668
East Sacramento Gallery,
Sacramento 123764
East Side Antiques, Fort Worth . 093446
East Side Gallery, Milwaukee . 121910,
121911
East Side Upholstery, Buffalo . 135371
East Somerset Railway, Shepton
Mallet 045779
East Street Antiques, Rye 091161
The East Suffolk Decorative and Fine
Arts Society, Leiston 060121
East Surrey Museum, Caterham 043895
East Tennessee Discovery Center,
Knoxville 050009
East Tennessee Historical Society
Museum, Knoxville 050010
East Texas Antique Auction, Sulphur
Springs 127467
East Texas Oil Museum at Kilgore
College, Kilgore 049967
East Timorese Cultural Centre
Museum Collection, Dili 010408
East Troy Electric Railroad Museum,
East Troy 048449
East Village Books and Records, New
York 145336
East-West Center, Honolulu ... 049529
East West Designs, Denver ... 120709
East-West Gallery, Austin ... 119666
East West Gallery, London ... 118298
East Wing Art Gallery, Mount
Wachusett Community College,
Gardner 049049
Eastbourne Antiques Market,
Eastbourne 089147
Eastbourne Auction Rooms,
Eastbourne 126886
Eastbourne Engraving,
Eastbourne 134354
Eastbourne Heritage Centre,
Eastbourne 044214
Eastchester Historical Society Museum,
Bronxville 047238
Easterbrook, M.J., Saskatoon . 128200
Easterley Gallery, Waikato ... 113035
Eastern Arizona Museum, Pima . 052133
Eastern Art Company, Providence 123762
Eastern Art Gallery, Kranenburg 107729
Eastern Art Report, London ... 138213,
139173
Eastern Azerbaijan Province History
Museum, Tabriz 024741
Eastern Books, London 144585
Eastern California Museum,
Independence 049660
Eastern Curious Company,

Penang 082404
Eastern Discoveries Antiques,
  Singapore 085398, 133036
Eastern Edge Art Gallery, Saint
  John's 006902
Eastern Expression, Varsity
  Lakes 099851
Eastern Gallery, Chicago 092702
Eastern Living, Tucson 097655
The Eastern Museum, Derby . . 044083
Eastern Oregon Museum on the Old
  Oregon Trail, Haines 049316
Eastern Region Arts Centre,
  Midland 099337
Eastern Shore Art Association,
  Fairhope 060467
Eastern Shore Art Center,
  Fairhope 048676
Eastern Southland Gallery, Gore 033210
Eastern State Penitentiary Historic Site,
  Philadelphia 052042
Eastern Suburbs Auctions,
  Croydon 125187
Eastgate Antiques, Cowbridge . . 088986
Eastgate Gallery, Chichester . . . 117578
Eastgate Gallery, Cowbridge . . . 117629
Eastgate Gallery, Hawthorn 099117,
                                127612
Easthetics, Chicago 120111
Eastleigh Museum, Eastleigh . . . 044221
Eastney Industrial Museum,
  Portsmouth 045551
Easton Farm Park, Easton 044223
Easton, Andrea, West Hollywood 097889
Eastside Art Gallery, New York . . 122550
Eastside Gallery, Memphis 121690
Eastside Heritage Center,
  Bellevue 046888
Eastwood, Saint Paul 123902
Eastwood House, Glasgow 044396
Easy Art, London 118299
Easy Art, Mumbai 109430
Eaton, Austin 119667
Eaton-Buchan Gallery and Marvin
  Cone Gallery, Cedar Rapids . . 047487
Eaton Gallery, London 090084
Eatos Art, Wien 062932
L'Eau Forte, Bordeaux 103423
L'Eau Forte, Lyon 104081
Eavanston Ecology Center,
  Evanston 048636
Ebanistería Albaco, Madrid 133325
Ebanite, Albacete 085688
Ebbamâla Bruk, Kyrkhult 040768
Ebbas Hus, Malmö Museer,
  Malmö 040824
Ebeck, Aabenraa 102757
Ebeling, Bergisch Gladbach 074730
Ebeling, Köln 076857
Ebeltoft Antikvariat, Ebeltoft . . . 140563
Ebeltoft Kunstforening, Ebeltoft . 058549
Ebeltoft Museum, Ebeltoft 010037
Ebenbeck, Edmund O., München 130340
Ebene, Paris 071289
Ebène & Sycomore, Gex 068658
Ebène Original, Cannes 013507
Ebenezer Maxwell Mansion,
  Philadelphia 052043
Ebenhoch, Inge, Feldkirch 062559
Ebenhofer, Leopold, Perg 062765
L'ebéniste, Nice 128859
Ebeniste, D., New York 095365, 135913
Ebénisterie & Tradition, Ancenis 066587
Ebénistrie d'Art Campagna,
  Marseille 128765
Ebensperger, Patrick, Graz 099999
Éber Emlékház, Türr István Múzeum,
  Baja 023416
Eberhard, Solothurn 087820
Eberhard & Son, Harry A.,
  Philadelphia 096104, 136085
Eberhardt, Petra & Wolfgang,
  Münstermaifeld 108228
Eberhart, Karl, Ermatingen 087414

Eberhart, Robert,
  Schönholzerswilen 087807
Eberhart, Robert, Sulgen 087836
Eberl, H., Berlin 129486
Ebersberg, Erbach,
  Odenwaldkreis 075626
Ebersdorfer Schulmuseum,
  Chemnitz 017730
Ebert & Björk, Stockholm 115852
Ebert, Karl-Otto & Monika,
  Klingenberg am Main 076826
Eberwein, Freiburg im Breisgau . 075858
Eberwein, Roswitha, Göttingen . 076042
Ebes, Douwma, Melbourne . . . 061817,
                                099303
Ebeseder, Alois, Wien 140070
Ebetsu-shi Bijutsu Togeikan,
  Ebetsu 028525
Ebinger Heimatmuseum, Albstadt 016490
Ebinger, Jürgen, Bad Neuenahr-
  Ahrweiler 074559
Ebinger, Michael, Würzburg 079078
Ebisons & Harounian, New York . 095366
Ebisubashi, Osaka 111351
Ebiya, Tokyo 082140
Ebjay, Merbein 061842
Ebland, Verona 111180
Ebner von Eschenbach, Kristian,
  Berlin 074786
Ebner, Heinz Noé, Freiburg im
  Breisgau 075859, 125952
Ebony, Dallas 120590
Ebony Elevations, Indianapolis . . 121180
Ebony Framing Gallery, Limerick 109667
Ebran, Patrick, Le Havre 069280
Ebreuil, Royan 072627
Eburne, Paris 071290
ECA, Pézenas 072134
Ecce, Rotterdam 112637
Ecce Homo, Jerusalem 025076
Eccentricities, London 090085
Eccles Community Arts Center,
  Ogden 051713
Eccles Road Antiques, London . 090086
Ecclesiastical Museum, Komotini 023136
Ecclesiastical Museum, Mullingar 024970
Ecclesiastical Museum, Sifnos . . 023262
Ecclesiastical Museum
  of Alexandroupolis,
  Alexandroupolis 022953
Ecco Arte, Graz 062585
Eccola, Los Angeles 094325
Echancrure Galerie, Bruxelles . . 100595
Echandi, Enrique, San José 102487
Echange, Maurice, Brumath . . . 125539
Echarri, Pamplona 086386
Echauguette, Nevers 070538
Echauri, Fermin, Pamplona 086387,
                        115386,   115387
Echeberria Monteberria, J., Donostia-
  San Sebastián 115047
Echenard, Michel, Villars-sous-
  Mont 087885
Echeverri Valasquez, Eduardo,
  Medellín 102432
Echeverría, Zaragoza 086595
Echigo-Tsumari Art Triennial,
  Tokamachi 098228
Echizen Crab Museum, Echizen . 028526
Echizen Pottery Gallery, Echizen 028527
Echizen Take-ningyo-no-sato,
  Maruoka 028999
Echmiadzin Art Gallery,
  Echmiadzin 000692
Echmiadzin Cathedral Complex and
  Museum, Echmiadzin 000693
Echnaton Gallery, Cairo 103033
Echo, Columbus 093014
Echo, Toledo 097604
Echo Antiques, Reepham 091087,
                                134962
Echo Art Deco, Zürich 087954
Echo Chamber, Aptos 119527
Echo du Passé, Le Havre 140866

Echo Gallery, Chicago 120112
Echo Gallery, Washington 124890
Echo Historical Museum, Echo . 048468
Echo Home, Tulsa 097728
Echoes, Todmorden . . 091606, 135138
L'Echoppe du Cloître, Aigle . . . 087184
Echoppe Saint-Guénhaël, Vannes
  (Morbihan) 105820
Echos d'Art, Ollioules 104423
Echos From the Past, San
  Antonio 096969
Echuca Gem Club Collection,
  Echuca 001047
Echuca Museum, Echuca 001048
Eckardt, Dr. Hans-Georg, Berlin . 141507
Eckart, Hans, Altenmarkt im
  Pongau 127705
Eckart, Karl-Heinz, Ellwangen . . 075598
Ecker, Brigitte, Ingolstadt 076611
Ecker, Helmut, Spiegel 133929
Ecker, Michaela, Großkadolz . . . 062616
Eckert, Indianapolis 121181
Eckert & Nolde, Erfurt . 075638, 125937
Eckert, D., Bremen 141650
Eckert, Iris, Kampen 107501
Eckhardt, M., Aglasterhausen . . 074358
Ecklé, Bruno, Nogent-le-Rotrou . 125666
Eckley Miners' Village, Eckley . . 048470
Eclat d'Art, Colmar 103619
Eclectic, Abu Dhabi 088156
Eclectic Art Gallery, Pittsburgh . 123538
Eclectic Attic, Singapore 085399
Eclectica, London 090087
Eclectica, Saint Louis 096694
Eclectique, Exeter 089229
Eclectique Antiques, New
  Orleans 095092
Eclectiques, New York 095367
Eclectiques Antique Mall,
  Columbus 093015
Electors, Houston 093662
Eclips, Nijmegen 112605
Eclipse, Calgary 101233
Ecliptic Gallery, Armadale,
  Victoria 098658
Eco Bush Art Gallery, Surfers
  Paradise 099735
Eco d'Arte Moderna, Firenze . . . 059435,
                        137647,   138820
Éco-Musée de la Blanchisserie,
  Grezieu-la-Varenne 013193
Eco-Musée de la Noix, Castelnaud-la-
  Chapelle 012358
Eco-Musée des Etouvières,
  Thônes 015944
Éco Musée Vivant de Provence, La
  Gaude 013425
Ecodrome Zwolle, Zwolle 033125
Ecole Bohemiarte, Montréal . . . 055141
Ecole Camondo, Paris 055396
Ecole Cantonale d'Art de Lausanne –
  ECAL, Lausanne 056429
Ecole Cantonale d'Art du Valais,
  Sierre 056439
Ecole d'Architecture de Strasbourg,
  Strasbourg 055418
Ecole d'Art, La Chaux-de-Fonds 056423
École d'Art d'Avignon, Avignon . 055355
Ecole d'Art du Havre, Le Havre . 055373
École d'Art Gérard Jacot, Belfort 055356
Ecole d'Art Marseille-Luminy,
  Marseille 055379
Ecole d'Arts, Annecy 055352
École d'Arts de Rueil-Malmaison,
  Rueil-Malmaison 055415
École de l'Image d'Epinal, Epinal 055371
École de Recherche Graphique – ERG-
  ESA, Bruxelles 055009
École Dentellière, Bailleul 011874
Ecole des Arts, Bruxelles 055010
Ecole des Arts Appliqués, Haute
  Ecole Supérieure d'Arts Appliqués,
  Genève 056429
École des Arts et Céramique,

Tarbes 055421
Ecole des Arts Visuelles, Université
  Laval, Québec 055150
Ecole des Beaux-Arts, Wavre . . 055057
Ecole des Beaux-Arts de Bordeaux,
  Bordeaux 055358
École des Beaux-Arts de la Réunion,
  Le Port 056257
Ecole des Beaux-Arts, Galerie des
  Beaux-Arts, Cherbourg-Octeville 055365
Ecole des Buissonnets, Sierre . . 116637
Ecole du Louvre, Paris 055397
École du Rang II, Authier 005363
E.M.A.P., Nice 055388
École Municipal d'Art Plastique
Ecole Municipal d'Arts Plastiques,
  Saint-Jean-de-Védas 055417
Ecole Municipale d'Arts Plastiques,
  Menton 055381
Ecole Municipale d'Arts Plastiques,
  Riom 055413
Ecole Municipale des Beaux
  Arts, Collège Marcel Duchamp,
  Châteauroux 055364
Ecole-Musée de Champagny,
  Champagny 012434
Ecole Nationale d'Art Décoratif,
  Aubusson 055354
Ecole Nationale des Arts, Bangui 055178
Ecole Nationale des Arts d'Haiti, Port-
  au-Prince 055761
Ecole Nationale des Arts Douta Seck,
  Médina 056319
Ecole Nationale des Beaux-Arts de
  Lyon, Lyon 055378
Ecole Nationale des Chartes,
  Paris 055398
École Nationale Supérieure d'Art,
  Dijon 055368
Ecole Nationale Supérieure d'Art de
  Nancy, Nancy 055386
Ecole Nationale Supérieure d'Art
  Décoratif de Limoges-Aubusson,
  Limoges 055357
Ecole Nationale Supérieure d'Arts de
  Bourges, Bourges 055359
Ecole Nationale Supérieure d'Arts de
  Cergy-Pontoise, Cergy-Pontoise 055363
Ecole Nationale Supérieure d'Arts
  et Métiers, Université Française
  d'Egypte, El Shorouq 055336
École Nationale Supérieure de
  Création Industrielle ENSCI,
  Paris 055399
École Nationale Supérieure de la
  Photographie, Arles 055353
Ecole Nationale Supérieure des Arts
  Appliqués Duperré, Paris 055400
Ecole Nationale Supérieure des Arts
  Décoratifs (Ensad), Paris 055401
École Nationale Supérieure des Arts
  Visuels de la Cambre, Bruxelles 055011
Ecole Nationale Supérieure des Beaux-
  Arts, Paris 055402, 136496
École Régionale d'Artce,
  Dunkerque 055370
Ecole Régionale d'Arts Plastiques, Fort-
  de-France 056134
Ecole Régionale des Beaux-Arts,
  Nantes 055387
Ecole Régionale des Beaux-Arts,
  Rennes (Ille-et-Vilaine) 055412
Ecole Régionale des Beaux-Arts,
  Rouen 055414
Ecole Régionale des Beaux-Arts de
  Caen, Caen 055361
Ecole Régionale des Beaux-Arts de
  Valence, Valence 055428
Ecole Régionale des Beaux-Arts et
  Arts Appliqués, Besançon . . . 055357
Ecole Régionale Supérieure
  d'Expression Plastique,
  Tourcoing 055425
Ecole Saint-Luc, Ramegnies-Chin 055046

École Spéciale d'Architecture,
Paris 055403
Ecole Supérieur des Arts Saint-Luc
Liège, Liège 055036
Ecole Supérieure d-Art et de Design
de Saint-Étienne, Saint-Étienne 055416
Ecole Supérieure d'Art, Cambrai 055362
Ecole Supérieure d'Art, Metz . . 055382
École Supérieure d'Art d'Aix-en-
Provence, Aix-en-Provence . . . 055348
Ecole Supérieure d'Art de Clermont
Communauté, Clermont-Ferrand 055366
Ecole Supérieure d'Art de Grenoble,
Grenoble 055372
Ecole Supérieure d'Art de Toulon,
Toulon (Var) 055422
Ecole Supérieure d'Art et de Design
ESAD, Amiens 055349
Ecole Supérieure d'Art et Design,
Reims 055411
École Supérieure d'Arts de Brest,
Brest 055360
École Supérieure d'Arts de Lorient,
Lorient 055376
École Supérieure de l'Image,
Poitiers 055409
École Supérieure de l'Image ESI,
Angoulême 055351
Ecole Supérieure des Arts Décoratifs,
Strasbourg 055419
École Supérieure des Arts et de la
Communication de Pau, Pau . . 055407
École Supérieure des Arts Staint-Luc,
Bruxelles 055012
École Supérieure des Beaux-Arts,
Nîmes 055390
École Supérieure des Beaux-Arts,
Toulouse 055423
Ecole Supérieure des Beaux-Arts
Ahmed et Rabah-Salim Asselah
(L'ESBA), Alger 054860
École Supérieure des Beaux-Arts
d'Angers ESBA, Angers 055350
Ecole Supérieure des Beaux-Arts de
Casablanca, Casablanca 056162
École Supérieure des Beaux-Arts de
Cornouailles, Quimper 055410
Ecole Supérieure des Beaux-Arts
de Montpellier-Agglomération,
Montpellier 055383
École Supérieure des Beaux-Arts de
Tours, Tours 055427
Ecole Supérieure des Beaux-Arts de
Valenciennes, Valenciennes . . . 055429
École Supérieure des Beaux Arts du
Mans, Le Mans 055374
Ecoles d'Art Americaines de
Fontainebleau, Paris 055404
Ecolver, Barcelona 085767
Ecomare, Stichting Texels Museum,
De Koog 032145
Écomusée, Boigneville 012091
Écomusée, Saulge 015705
Ecomusée Agricole du Pays de
Fayence, Fayence 012957
Écomusée d'Alsace, Ungersheim 016070
Écomusée d'Alzen, Alzen 011608
Écomusée d'Anticosti, Port-
Menier 006692
Écomusée d'Arts et Traditions
Populaires, Ligny-le-Ribault . . . 013799
Écomusée de Fresnes, Fresnes (Val-
de-Marne) 013056
Écomusée de Hauteluce,
Hauteluce 013240
Écomusée de Hearst Ecomuseum,
Hearst 005999
Écomusée de Hull, Gatineau . . . 005867
Écomusée de la Basse-Seine, Notre-
Dame-de-Bliquetuit 014525
Écomusée de la Batellerie et de la
Voie d'Eau Le Thudo, Thuin . 004063
Écomusée de la Bourrine du Bois
Juquaud, Saint-Hilaire-de-Riez 015358

Écomusée de la Brenne, Le
Blanc 013606
Ecomusée de la Bresse
Bourguignonne, Pierre-de-
Bresse 014863
Écomusée de la Bresse
Bourguignonne, Le Vigneron et sa
Vigne, Cuiseaux 012732
Écomusée de la Cévenne, Saint-
Laurent-de-Trèves 015409
Écomusée de la Combe de Savoie,
Les Côteaux du Salin, Grésy-sur-
Isère 013191
Écomusée de la Communauté, Ciry-le-
Noble 012576
Écomusée de la Communauté Le
Creusot Montceau-les-Mines, Le
Creusot 013624
Écomusée de la Courneuve, La
Courneuve 013402
Écomusée de la Crau, Saint-Martin-de-
Crau 015449
Ecomusée de la Ferme de Kervazegan,
Pont-l'Abbé 014938
Écomusée de la Ferme de la Forêt,
Saint-Trivier-de-Courtes 015575
Écomusée de la Ferme du Chemin,
Parc Naturel Régional de Maine-
Normandie, Madré 013960
Écomusée de la Ferme du Moulin
Vanneau, Saints-en-Puisaye . . 015641
Écomusée de la Ferme et des Vieux
Métiers, Lizio 013842
Écomusée de la Forêt d'Orient,
Brienne-la-Vieille 012221
Ecomusée de la Forêt
Méditerranéenne, Gardanne . . 013079
Écomusée de la Forge, Auchy-la-
Montagne 011783
Écomusée de la Forge, Menton . 014129
Écomusée de la Grande Lande
Marquèze, Sabres 015195
Écomusée de la Haie, Châtillon-en-
Dunois 012515
Ecomusée de la Lomagne,
Miradoux 014178
Écomusée de la Margeride,
Loubaresse 013878
Écomusée de la Margeride, Ecole
de Clémence Fontille, Ruynes-en-
Margeride 015193
Écomusée de la Métairie,
Plouigneau 014911
Écomusée de la Montagne des
Reuns, La Maison du Bûcheron,
Germaine 013095
Écomusée de la Pêche, Thonon-les-
Bains 015948
Écomusée de la Pêche et de la Mer,
Capbreton 012320
Écomusée de la Pomme au Calvados,
Le Sap 013698
Écomusée de la Région de Fourmies
Trélon en Avesnois, Maison de la
Fagne, Trélon 016042
Écomusée de la Région du Viroin,
Treignes 004090
Écomusée de la Sainte-Baume, Plan-
d'Aups 014881
Écomusée de la Tradition Basque
Jean Vier, Saint-Jean-de-Luz . 015377
Ecomusée de la Truffe, Sorges . 015838
Écomusée de la Vallee d'Aspe,
Accous 011534
Écomusée de la Vallée de l'Aigre, La
Ferté-Villeneuil 013414
Écomusée de la Vallée du Galeizon,
Cendras 012381
Écomusée de la Vigne et du Vin,
Gradignan 013139
Écomusée de l'Abeille, Auriac-du-
Périgord 011797
Écomusée de l'Abeille, Grateloup 013164
Écomusée de l'Abeille, Le Faou 013629

Écomusée de l'Abeille et de la Flore
Locale, Faugères 012953
Écomusée de l'Agriculture, Aigues-
Vives 011550
Écomusée de l'Apiculture,
Montargis 014220
Écomusée de l'Armagnac, Labastide-
d'Armagnac 013508
Écomusée de l'Ile-de-Groix, Ile-de-
Groix 013285
Écomusée de l'Île d'Ouessant,
Maisons du Niou Huella, Ile-
d'Ouessant 013290
Écomusée de Marais Breton Vendéen,
La Barre-de-Monts 013359
Écomusée de Marie Galante, Musée
Schœlcher, Grand Bourg 023345
Écomusée de Montjean, Montjean-sur-
Loire 014268
Écomusée de Plein Air du Quercy,
Sauliac-sur-Célé 015706
Écomusée de Saint-Constant, Saint-
Constant 006863
Écomusée de Saint-Dégan,
Brec'h 012197
Écomusée de Saint-Joseph, Fargues-
sur-Ourbise 012952
Écomusée de Saint-Léger, Saint-Léger-
les-Mélèzes 015416
Écomusée de Saint-Nazaire, Saint-
Nazaire (Loire-Atlantique) . . . 015473
Écomusée de Salazie, Salazie . . 015644
Écomusée de Savigny-le-Temple,
Savigny-le-Temple 015732
Écomusée des Anciennes Carrières de
Pierre "Les Capucins", Paris . . 014647
Écomusée des Appeaux et de la
Faune, Saint-Didier 015289
Écomusée des Boutères-Pilat,
Riotord 015099
Écomusée des Bruneaux, Firminy 012984
Écomusée des Deux-Rives, Salaberry-
de-Valleyfield 006962
Écomusée des Goémoniers et de
l'Algue, Plouguerneau 014906
Écomusée des Monts d'Arrée,
Commana 012638
Écomusée des Monts d'Arrée, Saint-
Rivoal 015551
Écomusée des Pays de l'Oise,
Beauvais 011954
Écomusée des Pipes Ropp, Baume-les-
Dames 011911
Écomusée des Techniques, Maison
de l'Imprimerie & des Lettres de
Wallonie, Thuin 004004
Écomusée des Vallées du Paillon,
Contes 012663
Écomusée d'Hannonville, Hannonville-
sous-les-Côtes 013226
Écomusée d'Ichkeul, Bizerte . . . 042736
Écomusée du Ancien Moulinage,
Chirols 012564
Écomusée du Bocage, Sains-du-
Nord 015204
Écomusée du Bommelaers Wall,
Ghyvelde 013100
Écomusée du Carton, Mesnay . . 014146
Écomusée du Cheminot Veynois,
Veynes 016188
Écomusée du Cognac, Migron . . 014169
Écomusée du Commerce et de
l'Artisanat, Rochefort (Charente-
Maritime) 015121
Écomusée du Cuir de la Tannerie Nory,
Sainghin-en-Weppes 015203
Écomusée du Domaine de la Petite
Couère, Nyoiseau 014538
Écomusée du Fier Monde,
Montréal 006371
Écomusée du Haut Beaujolais,
Marnand 014027
Écomusée du Haut Pays de la Roya et
de la Bevera, Breil-sur-Roya . . 012199

Écomusée du Libournais,
Montagne 014215
Écomusée du Liège, Gonfaron . . 013128
Écomusée du Linge, Montcy-Notre-
Dame 014244
Ecomusée du Marais Salant, Loix-en-
Ré 013855
Écomusée du Mont-Lozère, Le Pont-
de-Montvert 013689
Écomusée du Montmorillon,
Montmorillon 014280
Écomusée du Moulin de Kerchuz,
Saint-Thurien 015574
Écomusée du Moulin de la Sée,
Maison de l'Eau et de la Rivière,
Brouains 012237
Écomusée du Moulin Saint-Martin,
Cany-Barville 012319
Écomusée du Pays de la Cerise de
Fougerolles, Fougerolles 013036
Écomusée du Pays de la Roudoule,
Puget-Rostang 015007
Écomusée du Pays de Montfort,
Montfort-sur-Meu 014260
Écomusée du Pays de Nied,
Gomelange 013124
Écomusée du Pays de Rennes, Ferme
de la Bintinais, Rennes (Ille-et-
Vilaine) 015063
Écomusée du Pays des Collines,
Lahamaide 003770
Écomusée du Paysan Gascon,
Toujouse 015974
Écomusée du Pèrche de Saint-Cyr-la-
Rosière, Bellême 011974
Écomusée du Phare de Trezien,
Plouarzel 014898
Écomusée du Port Ostréicole, Maison
du Marin et Traditions Populaires, La
Teste-de-Buch 013365
Écomusée du Roannais, Roanne 015109
Écomusée du Val de Saône,
Seurre 015790
Écomusée du Veron, Savigny-en-
Veron 015731
Écomusée du Verre de Biot, Biot 012042
Écomusée du Vignoble Drouais et des
Artisans, Dreux 012837
Écomusée du Village Breton de Poul-
Fetan, Quistinic 015026
Ecomusée Haldensleben-Hundisburg,
Haldensleben 018942
Écomusée Industriel de Lochrist-
Inzinzac, Musée des Forges, Lochrist-
Inzinzac 013847
Écomusée la Cité des Abeilles, Saint-
Faust 015315
Écomusée la Ferme du Hameau,
Bierre-les-Semur 012032
Écomusée La Vie d'Antan, Esparron-
de-Verdon 012905
Écomusée l'Abeille du Haut-Doubs,
Aubonne 011779
Écomusée Maison Michaud, Chapelle-
des-Bois 012448
Écomusée Picarvie, Saint-Valéry-sur-
Somme 015953
Ecomusée Rachana, Rachana . . 030279
Écomusée Régional du Centre,
Houdeng-Aimeries 003713
Écomusée Rural de Vigneux-La
Pauelais, Vigneux-de-Bretagne 016214
Ecomusée Sauvegarde du Vieil Auzon,
Auzon 011827
Ecomuseo, Tordehumos 040249
Ecomuseo – Molino de Zubieta,
Museo de Navarra, Zubieta . . 040464
Ecomuseo de Guinea, Frontera . 039413
Ecomuseo del Caserío Vasco,
Artea 039001
Ecomuseo del Habitat Subterráneo de
Rojales, Rojales 040012
Ecomuseo della Resistenza,
Coazze 026024

Ecomuseo della Val Nerina, Cerreto di
Spoleto 025906
Ecomuseo dell'Archeologia Industriale,
Schio 027906
Ecomuseo di Archeologia Industriale E.
Crumi, Villar Pellice 028395
Ecomuseo Los Torres, Lerdo .. 031110
Ecomuseo-Luogo della Memoria,
Lamezia Terme 026547
Ecomuseo Regional Maipú,
Maipú 000441
Ecomuseu da Fazenda Boa Vista,
Roseira 004780
Ecomuseo de Góis, Góis 035718
Ecomuseu de Itaipu, Foz do
Iguaçu 004465
Ecomuseu de les Valls d'Aneu, Esterri
d'Aneu 039391
Ecomuseu do Guadiana, Mértola 035850
Ecomuseu do Rabeirão do Ilha,
Florianópolis 004451
Ecomuseu, Centro Cultural de Santa
Cruz Dr. Antônio Nicolau Jorge, Rio
de Janeiro 004688
Ecomuséum, Sainte-Anne-de-
Bellvue 006947
Ecomuseum en Archief van de
Boomse Baksteen, Boom 003379
Ecomuseum Simplon, Simplon
Dorf 041990
Economuseum voor Papier, Karton en
Aandrijftechnieken, Alsemberg 003263
ECORE – Escola De Conservació i
Restauració d'Obres d'Art,
Barcelona 056356, 133111
Ecossaise, Saint-Ouen 073073
EcoTarium, Museum of Science and
Nature, Worcester 054543
Ecoteca-Museu João Lúcio,
Olhão 035880
Ecovillage Centre Africain Reconstitué,
Magny-Cours 013962
Écovillage de Lorraine, Ville-sur-
Yron 016232
Écovillage Gaulois, Pleumeur-
Bodou 014892
ECR Conservation and Restoration,
New York 135914
Ecra Restauraciones, Madrid .. 133326
L'Ecrin, Marseille 069931
Ecrin, Metz 070094
Ecrin de Clisson, Clisson 068043
Ecu de France, La Tremblade .. 069107
L'Ecuelle, Amboise 066554
Ecume des Jours, Bordeaux ... 067324
Ecume des Temps, Besançon .. 067122
Ecurie d'Antan, Orgeval 070798
Ecuries du Seigneur, Le Poët .. 069332
l'Ecuyer, Bruxelles 100596
Ed & Elaine's Custom Interiors,
Phoenix 096206
Ed Clark Museum of Missouri Geology,
Rolla 052667
Eda Skans Museum, Åmotfors . 040525
La Edad de la Madera y el Bronce,
Bogotá 065203
Edams Museum, Edam 032264
Edarcom Europa, Roma 110759
Eddie's Furniture, Mount Gambier 061909
Eddie's Furniture Refinishing,
Winnipeg 128237
Eddie's Shoppe, Indianapolis .. 093880
Eddy, Charenton-le-Pont 067838
Eddy's English Furniture, Brugge 063270
Ede, Charles, London 090088
Edel, Osaka 111352
Edelbauer, Adolf, Monheim am
Rhein 077412
Edelfelt-Vallgren Museo, Porvoon
Museo, Porvoo 011271
Edeling, København 065894, 102876
Edelman, Catherine, Chicago .. 120113
Edelman, H. Heather, New York 122551

Edelmann, Nürnberg .. 108298, 142240
Edelmann, N., Bensheim 074720
Edelstam, Stockholm 086991
Edelsteenslijperij de Steenarend,
Eindhoven 032283
Edelstein, Paul, Memphis 121691
Edeltrödel, Wettingen 087898
Edelweiss, Houston 093663
Edelweiss, Porto Alegre 100973
Eden, New York 122552
Eden, Olomouc 065430
Eden Antic, Cannes 067637
Eden Antiques, South Auckland . 084079,
                                  132741
Eden Camp, Modern History Theme
Museum, Malton 045199
Eden Court Art Gallery, Inverness 044640
Eden Killer Whale Museum, Eden 001051
Eden Valley Museum, Edenbridge 044228
Eden Wharf Antiques, Eden 061436
Eden, Matthew, Corsham 088979
Edenhurst, West Hollywood ... 124993
Edenside Gallery, Louisville ... 121644
Eder, Gregor, Wien ... 100247, 127915
Eder, Hans, Hermagor 139995
Eder, Herbert A., Wien 077488
Eder, Johann, Wiener Neustadt . 063113
Edera, Villeneuve-sur-Lot 074166,
                                  141388
Ederle, Isabelle, Grenoble ... 068725
Ederti, Bilbao 114984
Edet, Montreuil-sous-Bois 070325
Edet, Saint-Ouen 073074
Edgar Allan Poe Cottage, Bronx
County Historical Society, Bronx 047228
Edgar Allan Poe House,
Philadelphia 052044
Edgar Allan Poe House and Museum,
Baltimore 046744
Edgar Allan Poe Museum,
Richmond 052548
Edgar County Historical Museum,
Paris 051917
Edgar de Marchand d'Art, Paris 104713
Edgar Modern, Bath 117304
The Edge, London 139174
Edge City, Auckland 112761
Edge Gallery, Denver 120710
Edge of Glass Gallery, Seattle . 124517
Edge of the Cedars State Park
Museum, Blanding 047033
Edge, D., Memphis 121692
Edgerton and District Museum,
Edgerton 005743
Edgerton Explorit Center, Aurora 046678
Edgewater Antique Mall, Chicago 092703
Edgewood Coins, Jacksonville . 093987
Edgewood Station Antiques,
Pittsburgh 096280
Edgware Antiques, Edgware ... 089158
Ediarte, Napoli 110522
Edicarte, Lisboa 113997
Ediciones Encuentro, Madrid . 137986
Ediciones Omega, Barcelona .. 137953
Ediciones Polígrafa, Barcelona . 137954
Ediciones Serigráficas, Lima .. 137889
Ediciones Siruela, Madrid 137987
Ediciones Unión, La Habana .. 136760
Ediciones y Arte, Asunción ... 137888
Edicions de l'Eixample,
Barcelona 137955
Edies Bits & Pieces, Birmingham 092405
Edie's Antiques, Saint Louis .. 096695
Edieuropa Gallerie, Roma 110760
Edificiul Roman cu Mozaic, Muzeul
de Istorie Naţional şi Arheologie,
Constanţa 036394
Edilán – Ars Libris, San Fernando de
Henares 138001, 143793
Edinburgh Architectural Salvage Yards,
Edinburgh 089172
Edinburgh Coin Shop, Edinburgh 089173
Edinburgh College of Art,
Edinburgh 056527

The Edinburgh Gallery, Edinburgh 117724
Edinburgh House, Port Lincoln . 099587
Edinburgh Printmakers Workshop and
Gallery, Edinburgh 044237
Edinburgh Scout Museum,
Edinburgh 044238
Edinburgh Showroom, Edinburgh 117725
Edinburgh Square Heritage and
Cultural Centre, Caledonia .. 005513
Edinburgh University Anatomy
Museum, Edinburgh 044239
Edinburgh University Natural History
Collections, Edinburgh 044240
Edinburgh University Press,
Edinburgh 138154
Edirne Devlet Güzel Sanatlar Galerisi,
Edirne 042885
Edish, Houston 093664
Edish, Salt Lake City 096891
Edison and Ford Winter Estates, Fort
Myers 048850
Edison Community College Gallery of
Fine Art, Fort Myers 048851
Edison Museum, Beaumont ... 046856
Edison National Historic Site, West
Orange 054315
Edissi, Kitchener 101323, 125368
Edisto Island Historic Preservation
Society, Museum, Edisto Island 048480
Editart, D. Blanco, Chêne-
Bougeries 116210, 138035
Edith Elizabeth Antiques, Saint
Ives 062177
Edith-Ruß-Haus für Medienkunst,
Oldenburg 020984
Edithburgh Museum, Edithburgh 001053
Edith's Gallery, Hong Kong 102070
Editin H.-M. Heimath, Borken,
Westfalen 137120
Editio Maris, München 137398
Edition "Die Horen", Wirtschaftsverlag
NW GmbH, Bremerhaven ... 138694
Edition 5, Isa Teloeken,
Düsseldorf 137157
Edition A, Ried im Innkreis 136543
Edition Akanthus, Spröda 137511
Edition Alectri, Köln 137307
Edition Aragon, Moers 137383
Edition Artefides, Luzern 116471, 138070
Edition Augenweide, Bernburg . 137036
Edition ausart, Wien 136580
Edition Automobile,
Rangendingen 108428, 137475, 142315
Edition Axel Menges, Fellbach .. 137189
Edition Balance, Gotha 137229
Edition Braus, Im Wachter Verlag,
Heidelberg 107389, 137271
Edition Classic Art, Uster 138095
Edition Copenhagen, København 102877,
                                  136771
Edition Curt Visel, Inh. Jürgen
Schweitzer, Memmingen ... 137379
Edition des Archives d'Architecture
Moderne, Bruxelles 136633
Edition Dr. Heinrich Fuchs, Wien 136583
Edition du Ricochet, Nice 140996
Edition EIKON, Wien 136582
Edition Emma Kunz, Würenlos . 138100
Edition Grafic d'Or, Hindelbank . 138054
Edition Graphischer Zirkel,
Eichgraben 136502
Edition GS, Düsseldorf 137158
Edition Kilian, Großkarolinenfeld 107176,
                                  137237
Edition Kunst der Comics,
Sonneberg 137508
Edition Kunstpodium, München . 137399
Edition Lebensretter, Leipzig .. 137339
Edition Leipzig in der Seemann
Henschel GmbH, Leipzig 137340
Edition Libri Illustri, Raritäten früher
Buchkunst, Ludwigsburg ... 137355
Edition Lidiarte, Berlin 137064

Édition Lithelle, Ramatuelle 105421,
                                  136999
Edition Lito, Champigny-sur-
Marne 136807
Edition Maldoror, Berlin 137065
Edition Mull, Saarbrücken 137495
Edition Nautilus, Hamburg 137241
Edition Northwind, Sæby 136780
Edition Pittura & Graphica, Celle 137131
Edition Raetia, Bolzano 137628
Edition Reuss, Aschaffenburg . 137030
Edition Rovana, Deisenhofen .. 075324
Edition Schwarze Seite,
Frankenhardt 137192
Edition Spektrum, Regensberg . 138085
Edition Tertium, Stuttgart 137519
Edition Text + Kritik, München . 137400
Edition Tiessen, Neu-Isenburg . 137440
Edition Va Bene, Klosterneuburg 136528
Edition Waechterpappel,
Denkmalschmiede Höfgen,
Grimma 137236
Editions, Houston 121032
Editions, Liverpool 118128
Editions Alternatives, Paris .. 136899
Editions Art, Edmonton 101282
Editions Atlantica Séguier,
Biarritz 136802
Editions Atlantica Séguier, Paris 136900
Editions Atlas, Paris 136901
Editions Cahiers d'Art, Paris ... 136902
Editions Caraibes Marinique, Pointe-
à-Pitre 137580
Editions Cartes d'Art, Paris .. 104714,
                                  136903
Editions Cercle d'Art, Paris ... 136904
Editions de Grenelle, Paris ... 136905
Editions du Centre Pompidou,
Paris 136906
Editions du Chêne, Paris 136907
Editions du Félin, Paris . 136908, 141097
Editions du Griffon, Neuchâtel . 116528,
                                  138076
Les Editions du Latanier, Saint-
Barthélemy 137582
Editions du Marronnier, Montboucher-
sur-Jabron 104254
Editions du Regard, Paris 136909
Editions du Rocher, Monaco .. 142995
Editions du Seuil, Paris 136910
Editions Edisud, Aix-en-Provence 136794
Editions EFFA, Levallois-Perret . 136827
Editions F.B., Paris 136911
Editions Gadoury, Monaco 082503,
                                  137818
Editions Galilée, Paris 136912
Editions Gallery, Calgary 101234
Editions Gallery, Edmonton ... 101283
Editions Gallery, Thirroul 099821
Editions Gallery – Showroom,
Maastricht 112569
Editions Graphiques, London .. 090089,
                                  118300, 144816
Editions Heuwinkel, Carouge .. 138034
Editions Ides et Calendes,
Neuchâtel 138077
Editions Jasor, Pointe-à-Pitre . 137581
Editions La Part de l'Oeil,
Bruxelles 136634
Editions Limited, Indianapolis . 121182,
                                  121183
Editions Mango, Paris 136913
Editions Martinsart, Romorantin-
Lanthenay 137001
Editions Memoire Vivante, Paris 136914
Éditions Minkoff, Paris 136915
Éditions Minkoff – Reprint SA,
Genève 138049
Editions Privat, Toulouse 137010
Editions Revue Noire, Paris ... 136916
Editions Rivages, Fouesnant .. 136817
Editions Saint-Paul, Luxembourg 137810
Editions Scriptoria, Chasselay . 136808

Editions Stef, Pont-Aven 105374, 136997
Editions Textuel, Paris 136917
Editions Traversière, Paris 136918
Editions Zodiaque, Paris 136919
Editora da Universidade de São Paulo,
 São Paulo 136667
Editora Pensamento-Cultrix, São
 Paulo 136668
Editorial Casariego, Madrid 137988
Editorial Fuentemolinos, Madrid 137989
Editorial Gredos, Madrid 137990
Editorial Planeta, Barcelona 137956
Editorial Prensa Ibérica,
 Barcelona 137957
Editorial Teide, Barcelona 137958
Editoriale Jaca Books, Milano . . . 137676
Editoriale Umbra, Foligno 137655,
 142689
Editout, Paris 104715
Editrice Latina, Roma . 110761, 137731
Editrice Vaticana, Città del
 Vaticano 138483
Editti, Fabrizio, Milano 131440
Editura Meridiane, Bucureşti . . . 137922
Edizioni del Museo Luigi Pecci,
 Prato 137723
Edizioni dell'Acquario, Padova . . 110561
Edizioni dell'Arquata, Foligno . . . 137656
Edizioni Kappa, Roma 137732
Edizioni Mediterranee, Roma . . . 137733
Edizioni Poligrafiche Internazionali
 Cultura ed Arte, Palermo 137708
Edizioni Quattroventi, Urbino . . . 137755
Edizioni San Paolo, Cinisello
 Balsamo 137636
Edkins & Son, G.H., Bishop
 Auckland 126823
EDL & Assoc., Atlanta 119561
Edler, Edith, Wien 062933, 127916,
 127917
Edler, J., Porta Westfalica 078066
Edlin, Canberra 061270
Edlin, Andrew, New York 122553
Edlinger, Herta, Klagenfurt 100070
Edmond Antiguedades, Barcelona 085768
Edmond Historical Society Museum,
 Edmond 048481
Edmonds, Kent Town 139792
Edmonds Arts Festival Museum,
 Edmonds 048482
Edmonds South Snohomish County
 Historical Society Museum,
 Edmonds 048483
Edmonds, David, London 090090,
 134653
Edmonston House, Vails Gate . . 053921
Edmonton Antique Warehouse,
 Edmonton 064292
Edmonton Art Gallery, Edmonton 005750
Edmonton Furniture Upholsterers,
 Edmonton 128163
Edmonton Public Schools Museum,
 Edmonton 005751
Edmonton Telephone Historical Centre,
 Edmonton 005752
Edmund Adler-Galerie,
 Mannersdorf 002353
Edmunds, Andrew, London 090091,
 118301
Edmunds, Larry, Los Angeles . . 145227
Edna Historical Museum, Edna . 048484
Edo, Artarmon 061058
Edo Art, Limhamn 086809
Edo-Tokyo Museum, Tokyo 029501
Edo-Tokyo Tatemono-En, Koganei 028872
Edoardo Villa Museum, Tshwane 038845
Edoya, Tokyo 082141
Edrich, I.D., London 144587
Eds MC- och Motormuseum, Ed 040595
Edsberg, Oslo 084307
Edsel and Eleanor Ford House, Grosse
 Pointe 049283
Eduard Bargheer-Haus, Hamburg 018983
Eduarda Smilga Teātra Muzejs,

Rīga 030179
Eduarda Veidenbauma Memoriālais
 Muzejs, Liepa 030152
Eduardo Antik, København 065895
Educação Artística de Universidade
 Católica do Salvador, Salvador 055095
Educação Artística, Universidade de
 São Judas Tadeu, Mooca 055084
Educatieve Boerderij en Zijdemuseum
 de Schans, Oud-Gastel 032744
Educatión Artística, Miguel
 Hidalgo 138912
Education Museum, Hyderabad . 034112
Educational Science Museum,
 Kuwait 030067
Edurne, Madrid 115184
Eduskunnan Talvisotamuseo,
 Kauhajoki 010971
Edvard Grieg Museum, Paradis . 033875
Edventure, Columbia 047908
Edward and Helen Mardigian Museum,
 Jerusalem 025077
Edward and Marthann Samek
 Art Gallery, Bucknell University,
 Lewisburg 050224
Edward de Courcy Clarke Earth
 Science Museum, Crawley . . . 001000
Edward-Dean Museum, Cherry
 Valley 047609
Edward H. White II Memorial Museum,
 Brooks Air Force Base 047273
Edward Hopper House Art Center,
 Nyack 051661
Edward M. Bannister Gallery, Rhode
 Island College, Providence . . . 052382
Edward VIII Antiques, Tyabb . . . 062359
Edward Williams Gallery,
 Fairleigh Dickinson University,
 Hackensack 049300
Edwardian Shop, Ipswich 089645
Edwards, Cleveland 092937
Edward's, Johannesburg 114793
Edwards, Seattle 124518
Edwards County Historical Museum,
 Kinsley 049990
Edward's Jewellers, Liverpool . . 089856
Edwards Memorial Museum,
 Chesterfield 047620
Edwards, Charles, London 090092
Edwards, Christophe, London . . 090093
Edwards, Clint, Richmond 136168
Edwards, Francis, Hay-on-Wye . 144462
Edwards, Francis, London 144588
Edwards, Gary, Washington 124891
Edwards, Libby, Portsea 099594
Edwards, Libby, South Yarra . . . 099705
Edwards, Libby, Woollahra 099914
Edwards, Roy, Craigavon 117635
Edwards, Shaw, Hatherleigh . . . 089469
Edwin, Jakarta 109509
Edwin A. Ulrich Museum of Art,
 Wichita State University, Wichita 054384
Edwin Scharff Museum am
 Petrusplatz, Neu-Ulm 020678
Edwin Wolters Memorial Museum,
 Shiner 053314
Edwin Young Collection,
 Salisbury 045723
Edwin & Assoc., Clark,
 Rochedale 062156
Eek, Christian, Skien 126386
Eells-Stow House, Milford 050888
Eemu Tulik, Saaremaa Muuseum,
 Linnuse 010637
Een Twee Drie, Den Haag 109509
De Eenhoorn, Aalst . . . 063175, 127989
De Eenhoorn, Delft 083031
De Eenhoorn, Venlo 112714
Het Eerste Friese Schaatsmuseum,
 Hindeloopen 032487
Eerste Nederlandse Opel Automuseum,
 Tijnje 032926
Eesti Ajaloomuuseum, Tallinn . . 010666
Eesti Arhitektuurimuuseum,

Tallinn 010667
Eesti Kirjandusmuuseum, Tartu . 010689
Eesti Kunstimuuseum, Tallinn . . 010668
Eesti Loodusmuuseum, Tallinn . 010669
Eesti Meremuuseum, Tallinn . . . 010670
Eesti Muuseumiühing Tallinn,
 Tallinn 058568
Eesti Muuseumiühing Tartu, Tartu 058569
Eesti Muuseumraudtee,
 Lavassaare 010633
Eesti Piimandusmuuseum,
 Imavere 010618
Eesti Põllumajandusmuuseum,
 Tõrvandi 010708
Eesti Postimuuseum, Tartu 010690
Eesti Rahva Muuseum, Tartu . . . 010691
Eestimaa Vene Muuseum, Russkij
 Muzej Estonii, Tallinn 010675
Eesti Raudteemuuseum,
 Haapsalu 010607
Eesti Ringhäälingumuuseum, Türi 010709
Eesti Spordimuuseum, Tartu . . . 010692
Eesti Tarbekunsti- ja Disainimuuseum,
 Tallinn 010671
Eesti Teatri- ja Muusikamuuseum,
 Tallinn 010672
Eesti Tervishoiu Muuseum,
 Tallinn 010673
Eesti Vabaõhumuuseum, Tallinn 010674
Eewal Galerie, Leeuwarden 112547
EF Galéria & Régiségbolt,
 Budapest 079252, 109140
EFA Museum für Deutsche
 Automobilgeschichte, Amerang 016574
Efes Müzesi, Selçuk 043054
Effe-Erre, Napoli 110523
Effelsberg, K., Köln 076858
Effigy Mounds Museum, Harpers
 Ferry 049364
Effler, Elder, Cincinnati 092848
Efinger & Rippel, Berlin 074787
EFMA, Brighton 134149
Efron, Jean, Washington 124892
Efté, Paris 104716
EG Trading Post, Denver 093280
Egam, Madrid 115185
Egan, Sydney 099779
Egan, James, Bungaree 098821
Egawa Bijutsukan, Nishinomiya . 029186
Ege, Öpfingen 077899
Egea Sanchez, Miguel, Málaga . 086292
Egelkamp, T., Bornheim, Rhein-Sieg-
 Kreis 106550
Egelund, København 102878
Egelund, Kirsten & Jørgen, Holte 065784
Egense Antik, Otterup 066047
Egerländer-Elbogner Heimatstuben,
 Illertissen 019383
Egerland-Museum, Marktredwitz 020302
Egersund Fayancemuseum,
 Egersund 033576
Egeskov, København 065896
Egeskov Veteranmuseum,
 Kværndrup 010213
Egg Artistry Museum, Cyonglin . 042295
Egge Museum, Steinkjer 033988
Egge-Museum Altenbeken,
 Altenbeken 016529
Egged Museum of Passengers Traffic,
 Holon 025065
Egger, Innsbruck 100041
Egger, André, Fribourg 087433, 133744
Egger, Gerhard, Krumpendorf . . 062697
Eggers, M., Wedemark 078880
Eggers, Rüdiger, Lindewitt 107878
Eggers, Uwe, Bokel, Kreis Rendsburg-
 Eckernförde 075057
Eggert, Roswitha G., Hamburg . 076214
Eggimann, Heinz, Huttwil 087532
Eggler, Erika, Weingarten, Kreis
 Ravensburg 078906
Eggleston, Ben, Long Marton . . 090614
Egg's Farm and Museum,
 Ciaotou 042289

Egham Museum, Egham 044280
Egida Antiques, Roma 110762
l'Egide, Bruxelles 063406
Egidio, Ita, Innsbruck 127773
Egidis, Amsterdam 143027
Eğitim Özel Müzesi, Ankara . . . 042819
Eğitim ve Kültür Vakfı Özel Müzesi,
 Malkara 043019
d'Églantier, Amsterdam 082740
Eglantine, Beuzeville 067157
Eglise Saint-Rémy de Cuesmes,
 Cuesmes 003527
Egly, Franck, Lutzelhouse 128694
Egnsmuseet Jacob Michelsens Gård,
 Aabenraa 009996
EGO Publishers, Sankt-Peterburg 137930
Egon Schiele Art Centrum, Český
 Krumlov 009509
Egon Schiele-Museum, Tulln . . . 002844
Egri Galéria, Eger 109190
Egry József Emlékmúzeum,
 Badacsony 023414
Eguiguren, Jaime, Buenos Aires 060896
Egyetti, Taipei 116911
Egyházmegyei Múzeum és Kincstár,
 Szeged 023923
Egymalom, Debrecen 142575
Egypt Centre, Swansea 045933
The Egypt Exploration Society,
 London 060122
Egypt Papyrus Museum, Haram . 010576
Egyptian Antique Furniture and
 Decoration, Cairo 066240
Egyptian Archaeology, Bulletin of
 the Egypt Exploration Society,
 London 139175
Egyptian Art, Oostende 100800
Egyptian Art Gallery, Manama . . 100430
The Egyptian Auction Palace,
 Cairo 125474
Egyptian Center for International
 Cultural Cooperation Art (ECICC),
 Cairo 103034
The Egyptian Geological Museum,
 Cairo 010535
Egyptian Museum, Cairo 010536
Ehagaki Shiryokan, Kobe 028839
Ehbrecht, G. & T., Eicklingen . . 141735
Ehemalige Benediktinerabtei
 Seligenstadt, Seligenstadt . . . 021835
Ehemalige Synagoge, Freudental 018506
Ehemalige Synagoge, Odenbach 020946
Ehemalige Synagoge Ahrweiler, Bad
 Neuenahr-Ahrweiler 016833
Ehemalige Synagoge Erfelden,
 Riedstadt 021421
Ehemalige Synagoge mit Ausstellung
 Juden auf dem Lande, Beispiel
 Ichenhausen, Ichenhausen . . . 019374
Ehemalige Synagoge Sankt Pölten,
 Sankt Pölten 002694
Ehemalige Synagoge Sulzburg,
 Sulzburg 022086
Ehemaliges Jagdschloß Bebenhausen,
 Staatliche Schlösser und Gärten
 Baden-Württemberg, Tübingen 022214
Ehemaliges Postfuhramt, Berlin . 017108
Ehime Bunkakan, Imabari 028682
Ehime-ken Bijutsukan,
 Matsuyama 029020
Ehime-kenritsu Hakubutsukan,
 Matsuyama 029021
Ehime-kenritsu Kagakukan,
 Niihama 029179
Ehime Mingeikan, The Japan Folk
 Craft Association, Saijo 029300
Ehlen Marbe, Bert, Sittard 083605
Ehlenbeck, Dorothee, Düsseldorf 106769
Ehm, Lippstadt 077132
Ehm Welk- und Heimatmuseum,
 Angermünde 016583
Ehm Welk-Haus, Bad Doberan . 016744
Ehm, Manfred, Hamburg 141860
Ehmann, Rudolf, Nürnberg 130464

Ehmer, Erich, München 077489, 130341
Ehmke, Edgar, Werder 130747
Ehnbom, Leni, Lund 086831
Ehrat, Hanspeter, Baden 116047
Ehrbar, Othmar, Zürich 144172
Ehrensperger, Dr. Franz & Ingrid,
Zürich 116768
Ehrensvärd-Museo, Helsinki . . 010817
Ehrhardt, Gustave, Monthou-sur-
Bièvre 070248
Ehrhardt, H.-H. & B., Osterode am
Harz 077988
Ehrhardt, Hans, Künzelsau . . . 076984
Ehrhardt, Heinrich, Madrid . . . 115186
Ehrhardt, Konrad, München . . . 077490
Ehrhart Museum, Antwerp . . . 046508
Ehrhold, Loritta, Berlin 129487
Ehrl, Manfred, Greding 076083
Ehrler, Rosmarie, Zürich 087955
Ehrlich & Langwieser, München 126048
Ehrman Mansion, Tahoma 053654
Ehrmann, Fabien, Thionville . . . 073657
Ehrsam, J., Gershershausen . . . 075981
Ehrsam, Roland, Basel 133641
Ehrwalder Heimatmuseum,
Ehrwald 001846
Eibel, Münster 077658, 077659, 130426
Eibel, Burkhard, Münster 077660
Eiben, Hamburg 076215
Eicha-Museum, Archeologisch
Museum voor de Brabantse Kempen,
Bergeijk 032064
Eichberg, Helmut & Ursula,
Nickenich 077775
Eichelkraut, Potsdam . 078073, 126097
Eichelsdörfer, Jürgen & Christine,
Hallstadt 076174
Eichenberger, Dr. Walter, Beinwil am
See 144042
Eichendorff, Mannheim 077255, 130274
Eicher, Delhi 109315
Eicher, Fritz, Bad Dürkheim . . . 106054
Eichholz, Jürgen, Kassel 076717
Eichholzer, Werner, Sankt Gallen 087775,
133902
Eichinger, Georg, Berlin 106219
Eichinger, Max, Ansbach 074416
Eichler, Dietmar, Ludwigsburg . . 130247
Eichler, Helmut, Gössendorf . . . 127738
Eichler, Helmut,
Hausmannstätten 127766
Eichler, P., Berlin 141508
Eichsfelder Heimatmuseum,
Heiligenstadt, Heilbad 019170
Eickhoff, M. & D., Herford 107418
Eid & Sengle, Heidelberg 076440
Eid, Irmgard, Trostberg 078717
Eid, Jichi, Beirut 082281
Eid, Susanna-Isabel, München . 130342
Eidam, Chemnitz 141672
Eidam, Niederwiesa 077779
Eidam, Cornelia, Lichtenau . . . 142075
Eide-Dalrymple Gallery, Augustana
College, Sioux Falls 053354
Eidelstedter Heimatmuseum,
Hamburg 018984
Eiden, Werner, Hannover 076369
Eiderstedter Heimatmuseum, Sankt
Peter-Ording 021608
Eidos Immagini Contemporanee,
Asti 109761
Eidskog Museum, Eidskog 033580
Eidsvold Historical Society Museum,
Eidsvold 001054
Eidsvold Og Eidsvold, Enebakk . 126346
Eidsvoll 1814 – Rikspolitisk
Senter, Eidsvoll Bygningens
Nasjonalhistoriske Samlinger, Eidsvoll
Verk 033581
Eidsvoll Bygdemuseet, Eidsvoll
Verk 033582
Eifel Antik, Birreisborn 075002
Eifel-Heimatmuseum Bergweiler,
Bergweiler 017062

Eifel-Vulkanmuseum, Daun 017835
Eifeler Bauernhausmuseum,
Adenau 016468
Eifelmuseum Blankenheim,
Blankenheim 017410
Eifelmuseum mit Deutschem
Schiefergebergwerk, Mayen 020321
Eiffel Antique Centre, Singapore 085400
Eifler, Jörg, Potsdam 142306
Eigen Art, Dietzenbach 106679
Eigen-Art, Krefeld 107731
Eigenart-Verlag, Winterthur . . . 138099
The Eighteen Nineties Society,
London 060123
Eighth Avenue Antique Mall,
Nashville 094995
Eikaasgalleriet, Musea i Jølster, Skei i
Jølster 033941
Eikelmann, Burkhard H.,
Düsseldorf 106770
De Eikenhof, Meppen 132583
Eikens Antikvariat, Bergen 143382
Eikon, Villanova di Castenaso . . 138821
EIKON Gesellschaft der
Freunde der Ikonenkunst e.V.,
Recklinghausen 058843
Eikon Restauri, Roma 131912
EIKON, Internationale Zeitschrift für
Photographie und Medienkunst,
Wien 138511
Eikona, Venezia 132242
Eiktunet Kulturhistorisk Museum,
Gjøvik 033618
Eildon Hall Sibbald Memorial Museum,
Sutton, Ontario 007105
Eilean Donan Castle Museum,
Dornie 044125
Eileen of China, Seattle 097430
Eilers & Hondtong, Hengelo,
Overijssel 143171
Eilert, Evelyn, Ludwigsburg . . . 142081
Eilert, Gerhard, Dresden 141695
Eilert, Gerhard, Meißen 142123
Eilisha's Shoppe, New Farm . . . 061950
Eimer, Christopher, London . . . 090094
Einari Vuorelan Kirjailijakoti,
Keuruu 010995
Einaudi, Giulio, Torino 137746
Einbecker Schreibmaschinenmuseum,
Einbeck 018129
EinBlick-Dauerausstellung im
Missionswerk, Neuendettelsau 020697
Eindhovense Muntenhandel,
Eindhoven 083213
Eindiguer, Michel, Borg 087419
Einert, Steffen, Bleckede 075019
Einnolf, Friedeburg 075910
Einricher Heimatmuseum,
Katzenelnbogen 019555
Eins und . . . Zeitschrift für Kunst
und Gestaltung, Stuttgart . . . 138695
Einsiedel, Gräfin, Perchtoldsdorf 062763
Einsiedelei Erzherzog Maximilians des
Deutschmeisters, Innsbruck . . 002121
Einstein-Haus, Bern 041312
Einstein, G.W., New York 122554
Einzmann, Siegfried, Kassel . . . 076718
Eiraku, Kyoto 082011
Eisei Bunko Museum, Tokyo . . . 029502
Eisenacher Münzhandlung,
Eisenach 075591
Eisenbahn- und Heimatmuseum
Hochdahl, Erkrath 018256
Eisenbahn- und Technikmuseum
Rügen, Prora 021225
Eisenbahnmuseum Lokschuppen
Aumühle, Aumühle 016679
Eisenbahnmuseum, Blumberg . . 017425
Eisenbahnmuseum, Deutsch
Wagram 001809
Eisenbahnmuseum, Falkenberg,
Elster 018332
Eisenbahnmuseum, Gerstetten . 018669
Eisenbahnmuseum, Knittelfeld . . 002212

Eisenbahnmuseum, Losheim am
See 020120
Eisenbahnmuseum – Sammlung Frey,
Seifhennersdorf 021829
Eisenbahnmuseum Bayerischer
Bahnhof zu Leipzig e.V., Leipzig 019983
Eisenbahnmuseum Bochum-
Dahlhausen, Bochum 017434
Eisenbahnmuseum Darmstadt-
Kranichstein, Darmstadt 017819
Eisenbahnmuseum der Köln-Bonner-
Eisenbahn-Freunde, Wesseling 022596
Eisenbahnmuseum Historischer
Wasserturm, Bebra 017004
Eisenbahnmuseum Lehmann, Hofheim
in Unterfranken 019294
Eisenbahnmuseum Neustadt an
der Weinstraße, Neustadt an der
Weinstraße 020757
Eisenbahnmuseum Schwarzenberg,
Schwarzenberg 021773
Eisenbahnmuseum Strasshof – Das
Heizhaus, Strasshof 002802
Eisenbahnmuseum Vienenburg,
Vienenburg 022328
Eisenbahnverein "Hei Na Ganzlin",
Röbel/Müritz 021449
Eisenberg, Omaha 123307
Eisenerzer Krippenhaus, Eisenerz 001854
Eisenhammer Dorfchemnitz,
Dorfchemnitz 017939
Eisenhower Birthplace, Denison 048242
Eisenhower National Historic Site,
Gettysburg 049089
Eisenhüttenmuseum, Trippstadt . 022206
Eisenkunstgußmuseum, Erlebnispark
Ziegenhagen, Witzenhausen . . 022712
Eisenkunstgussmuseum, Stiftung
Schleswig-Holsteinische
Landesmuseen Schloß Gottorf,
Büdelsdorf 017662
Eisenlauer, Karl, Ichenhausen . . 076592
Eisenmuseum, Ainring 016487
Eisenmuseum, Murau 002423
Eisenmuseum Jünkerath,
Jünkerath 019474
Eisenuhren- und Antikuhrenmuseum,
Steyr 002782
Eisenverhüttungs-Schauraum,
Oberpullendorf 002465
Eisfabrik e.V., Hannover 058844, 107357
Eisho, Tokyo 082142
Eishockey-Museum, Hall of Fame
Deutschland e.V., Augsburg . . 016651
Eismann, Günter, Hamburg . . . 076216
Eisstockmuseum, Sankt Georgen im
Attergau 002657
Eiteljorg Museum of American Indians
and Western Art, Indianapolis . 049681
Eitzenberger, Hans, Hamburg . . 076217
Eizendo, Tokyo 111418
EJ Designs, Boston 119862
Eka Gallery, Valencia 115480
Ekain Arte Lanak, Donostia-San
Sebastián 115048
Ekaterina, Moskva 085176
Ekaterinburgskij Muzej Izobrazitelnych
Iskusstv, Ekaterinburg 036977
Ekaterininskij Dvorec –
Gosudarstvennyj Muzej Carskoje Selo,
Puškin 037673
Ekbatan, Firenze 131161
Eke, Sint-Niklaas 063831
Ekeby Qvarn Artspace, Uppsala . 115969,
133594
Ekegardh, P., Maastricht 112570
Ekehagens Forntidsby, Åsarp . . 040546
Ekenäs Museum, Ekenäs 010765
Ekenäs Slott, Linköping 040785
Ekenstam, Lena, Svedala 087093
Eketahuna and Districts Museum,
Eketahuna 033200
Eketorps Borg, Degerhamn . . . 040590
Ekhof-Theater, Stiftung Schloss

Friedenstein Gotha, Gotha 018774
Al-Ekhtiar, Tawfiq, Damascus . . 088034
Eklectic, Atlanta 119562
Eklektik, Montréal 064417
Eklektikos, Washington 124893
Eklund & Wallmark, Stockholm . 115853
Ekobio, Ekerö 115600
Ekomuseum Bergslagen,
Smedjebacken 040965
Ekomuseum Gränsland,
Strømstad 041060
Ekotechnické Museum, Stará
kanalisační čistírna, Praha . . . 009788
Ekron – The Philistine City and it's
Culture, Shiqmim 025185
Eksa, Sofia 128140
Eksi Şark Eserleri Müzesi,
İstanbul 042935
Eksjö Antikvariat, Eksjö 143827
Eksjö Auktionsverk, Eksjö 126575
Eksjö Museum, Eksjö 040598
Ekslibriso Muziejus, Šiaulių Apskrities
Povilo Višinskio Viešosios Bibliotekos,
Šiauliai 030464
Ekspo-88, Moskva 114214
Ėkspozicija "Čelovek. Duša.
Duchovnost.", Ivanovo 037038
Ėkspozicija "Ivanovskaja starina",
Ivanovo 037039
Ėkspozicija Kȇrpeeva Kašta,
Kotel 005074
Ekspozycija Rzemiosło w Białym
Spichrzu, Muzeum Okręgowe
im. Leona Wyczółkowskiego,
Bydgoszcz 034666
Ekström, Västerås 115980
Ekström, J. F., Uppsala 087118, 133595
Ektoft, Vittaryd 087166, 126703
EKZ-Museum Stromhaus Burenwisen,
Glattfelden 041550
El Arcón, Ciudad de Panamá . . . 084414
El-Art, Warszawa 113823
El-Barkai Museum, Khartoum . . 040492
El Baz, Patrick, Saint-Ouen . . . 073075
El Camino College Art Gallery,
Torrance 053779
El Campo Museum of Natural History,
El Campo 048496
El Castell-Ecomuseu Urbà, Rubí 040024
El Dorado County Historical Museum,
Placerville 052187
El Dorado Museum, Eldorado . . 001055
El Hamed, Samy Abdul, Cairo . . 128404
El-Hindi, Khaled, Ehrenberg . . . 129777
El-Ibied Regional Museum,
Khartoum 040493
El Monte Historical Society Museum,
El Monte 048500
El Morro National Monument,
Ramah 052456
El Naggar, Houston 093665
El Ouachni, Ali, Marvejols 069988
El Paso Arts, El Paso 120828
El Paso Coins and Collectible, El
Paso 093400
El Paso Holocaust Museum, El
Paso 048506
El Paso Import, Chicago 092704
El Paso Import, Dallas 093114
El Paso Museum of Archaeology at
Wilderness Park, El Paso 048507
El Paso Museum of Art, El Paso 048508
El Paso Museum of History, El
Paso 048509
El Presidio Gallery, Tucson 124731
El Pueblo History Museum,
Pueblo 052408
El Puente, Nordstemmen 137504
El Rancho de Las Golondrinas
Museum, Santa Fe 053006
El Shalzi, Walid Ramadan, Roma 081346
El-Souk El-Kabin, Alexandria . . . 066218
Elabidi, Nassereddine, Marrakech 082589
Elabužskij Gosudarstvennyj Istoriko-

architekturnyj i Chudožestvennnyj
Muzej-zapovednik, Elabuga ... 036996
Eläinmuseo, Luonnontieteellinen
Keskusmuseo, Helsinki 010818
Elaginoostrovskij Dvorec-Muzej
Russkogo Dekorativno-Prikladnogo
Iskusstva i Interjera XVIII–XX vv.,
Sankt-Peterburg 037731
Elahe Gallery, Tehrān 109519
Elaine Claire, Louisville 094545
Elaine L. Jacob Gallery, Wayne State
University, Dept. of Arts, Detroit 048304
Elaine Phillips Antiques,
Harrogate 089427
Elaine's Fine Art, New York 122555
Elaine's Jewel Box, Sacramento 096618
Elam School of Fine Arts,
The University of Auckland,
Auckland 056185
Elam, Carolyn, Nashville 094996
Elam & Assoc., Gary, Dallas ... 093115
Elan, Groningen 083287
Elan, New York 095368
Eland, Alkmaar 143001
Eland, Jonge, Amsterdam 132397
Elart, Praha 065494
Elas 95, Valladolid 086544
Elastic Galleri, Malmö 115741
Elazarova, Elena, Sofia 101152
Elaziğ Örenyerleri, Elaziğ 042888
Elba Arte y Antigüedades, Lima 084452
Elbaz, Frank, Paris 104717
Elbaz, Jacques, Paris 104718
Elbdörfer Galerie, Hamburg ... 107239
Elbé, Paris 141098
Elbers, Toronto 064638, 128209
Elbers, Ulrike, Kleve 076822
Elbert Hubbard-Roycroft Museum,
Aurora Historical Society, East
Aurora 048421
Elberton Granite Museum,
Elberton 048519
Elbogen, Malmö 143894
Elbow Museum, Elbow 005775
Elbschiffahrtsmuseum mit
stadtgeschichtlicher Sammlung,
Lauenburg 019930
Elbschloss Bleckede, Bleckede . 017419
Elbtal- Haus, Bleckede 017420
Elbtor Werben, Werben 022568
Elca London Gallery, Montréal .. 101380
Elcabas, Alain, Paris 104719
Elcho Island Art & Craft,
Galiwinku 061501
Elco, Bremen 075156
Elco, Wilhelmshaven 108939
Elden Antiques, Kirk Deighton .. 089712
Elder, Charlotte 119985
Elder, Saint Louis 096696
Elder Art Gallery, Rogers Center for
Fine Arts, Lincoln 050289
Elder Fine Art, North Adelaide . 061981,
125219
Elder, R., Nashville 145294
Eldh, Lysekil 133553
Eldnäsets Flottningsmuseum,
Ånge 040530
Eldon & Choukri, Casablanca . 082554,
132372
Eldon House, London 006209
Eldorado Antiquariat, Freiburg im
Breisgau 141800
Eldorado Antiques, San Antonio 096970
Eldre Norsk Kunst, Stavanger . 084366,
113265
Eldred, Robert C., East Dennis . 127211
Eldreds, Plymouth 127015
Eldridge Street Project, New York 051381
Eleanor Barbour Cook Museum of
Geology, Chadron State College,
Chadron 047510
Eleanor Roosevelt National Historic
Site, Hyde Park 049648
Eleb, Alain, Paris 071291

Eleckij Literaturno-memorialnyj Muzej
Pisatelja I.A. Bunina, Gorodskoj
kraevedčeskij muzej, Elec 037002
Electra Gallery, Waipukurau 113048
Electric Ladyland – The First Museum
of Fluorescent Art, Amsterdam 031936
Electric Mountain – Museum of North
Wales, Amgueddfa'r Gogledd,
Llanberis 044871
Electric Train Outlet, Saint Louis 096697
Electrical Engineering Museum,
Fredericton 005839
The Electrical Industry Museum,
Tehrān 024749
Electrium, Sainte-Julie 006955
Electro Antiques, New York 095369
Electro Rádio São João, Lisboa . 084878
Electrobroc, Broc-Fabrique 041368
Electroniquel, Gijón 133260
Electrum, Edmonton 101284
Electrum, London 118302
L'Elefante, Treviso 111036
El Elefante del Rastro, Madrid .. 086177,
143762
Elefantens Ateljé, Kumla 115714
Eleftherios K. Venizelos Museum,
Athinai 022984
Elégance, Mougins 104300
Elegance for Hire, Dallas 093116
Elegant Antiques, Paddington,
Queensland 062049
Elegant Art Painting Gallery, Hong
Kong 102071
Elegant Ceramic Mold, Cleveland 120418
Elegant Dollhouse, Sacramento . 096619
Elegant Extras, Cleveland 092938
Elegant Findings, Boston 092466
The Elegant Garage Sale, Toronto 064639
Elegant Junque Shop, Tucson .. 097656
Elegia, Strasbourg ... 073586, 141348
Elektriciteits- en Techniekmuseum,
Hoenderloo 032493
Elektrizitätsmuseum,
Münchenstein 041772
Elektrizitätszähler-Kabinett,
Braunschweig 017554
Elektro-Museum Rendsburg-EON
Hanse, Rendsburg 021371
Elektro Museum, Kraftwerk
Kappelerhof, Baden 041250
Elektromuseum Erfurt, Erfurt .. 018240
Elemente, Salt Lake City 096892
Elements, Palma de Mallorca .. 086348
Eléments, Paris 138641
Elements Glass, Portland 123617
Elements of Art, Columbus 120488
Eleouet, Cathérine, Grenoble ... 128614
Elephant & Castle, Den Haag .. 083080
Elephantine Island's Museum, Awan
Museum, Aswan 010524
Elephant's Toe, Richmond 096540,
096541
Elephantsgalore, Virginia Beach 097779
Élet az ártéren – kiállítás,
Szekszárd 023947
Elet és Irodalom, Journal Art
Alapítvány, Budapest 138771
Eleusis, Caen 067566
Elevation X, London 118303
Eley, Boston 126825
The Elf Shelf, Las Vegas 094135
Elfenbein-Museum, Michelstadt . 020399
Elfenbeinmuseum, Walldürn ... 022406
L'Elfo, Bologna 079836, 130954
Elford, Yelverton 119402
Elfred, Lee, San Diego 124119
Elfreth's Alley Museum,
Philadelphia 052045
The Elgar Birthplace Museum, Lower
Broadheath 045158
Elghanayan, Sina, New York ... 095370
Elgin Apple Museum, Grabouw . 038722
Elgin Area Museum, Elgin 048521
Elgin County Pioneer Museum, Saint

Thomas 006937
Elgin Museum, Elgin 044283
Elgin Public Museum, Elgin ... 048522
Elguabbas, Mohamed,
Casablanca 082555
Elham Valley Bookshop,
Canterbury 144306
Elham Valley Railway Museum,
Folkestone 044353
Eli Whitney Museum, Hamden . 049321
Elian, Lily, Beirut 082282
Elias, Miami 121775
Elias, New York 135915
Eliason, Carl Henrik, Tingstäde . 133588
Eliasu Dzimtas Muzejs Zīleni,
Platone 030170
eLightworks, Saint Paul 123903
Elijah Clark Memorial Museum,
Lincolnton 050310
Elimäen Kotiseutumuseo ja
Koulumuseo, Elimäki 010768
Elimäen Koulumuseo, Elimäki . 010769
Elingaard Museum, Fredrikstad
Museum, Gressvik 033623
Elins Eagles-Smith Gallery, 5th fl., San
Francisco 124273
Eliot, Toronto 140405
Eli's Gallery, Dallas 120591
Elisabet Ney Museum, Austin .. 046689
Elisabeta, Bucureşti 114138
Elisabeth & Sam, Bruxelles ... 063407
Elisabethen Antiquitäten, Basel . 087231
Elisabeth's, London 090095
Elisabeths Atelier, Trondheim .. 113286
Elisart, Châtenay-Malabry 103584
Élisej, Moskva 085177
Elisengalerie, Aachen 105930
Elitas, Pardubice 065444
Elite Antiques, San Diego 097080
Elite Autographs, Tucson 145622
Elite Gallery, Tbilisi 105901
Elite Upholstery, San Diego ... 136294
Elitzer, Ernst, Saarbrücken ... 108540
Eliza Cruce Hall Doll Museum,
Ardmore 046526
Eliza Gallery, Mornington 099378
Eliza Jane Antiques, Haberfield . 061570
Elizabeth & Son, Ulverston 091689
Elizabeth Anne's Antiques,
Gosford 061545
Elizabeth Antiques, Chatswood . 061315
Elizabeth Bay House, Elizabeth
Bay 001056
Elizabeth Castle, Saint Helier .. 045691
Elizabeth D. Walters Library,
Stroudsburg 053574
Elizabeth Farm, Rosehill 001459
Elizabeth Myers Mitchell Art Gallery,
Saint John's College, Annapolis 046500
Elizabeth O'Neill Verner Studio
Museum, Charleston 047547
Elizabeth P. Korn Gallery, Drew
University, Madison 050588
Elizabeth Perkins House, York .. 054579
Elizabeth R. Antiques, East
Molesey 089134, 134348
Elizabeth Rozier Gallery, Jefferson
City 049803
Elizabeth Sage Historic Costume
Collection, Bloomington 047046
Elizabeth Slocumb Galleries, Carrol
Reece Museum, Johnson City . 049824
Elizabeth Street Antiques and
Restoration, London 134654
Elizabethan House Museum, Great
Yarmouth 044474
Elizabethans, Fareham 089254
Elizabeth's Hallmark, New
Orleans 095093
Elizbarrutiko Arte Sakratuaren
Museoa/ Museo Diocesano de
Arte Sacro, Obispado de Vitoria,
Diputación Foral de Álava, Vitoria-
Gasteiz 040424

Elizium, Moskva 085178, 114215
Elk Grove Farmhouse Museum, Elk
Grove Village 048529
Elk River Area Arts Alliance, Elk
River 060468
Elka, Córdoba 139602
Elkhart County Historical Museum,
Bristol 047213
Elkjær Christensen, Bjarne,
Fredericia 065652
Elkoll, Set, Manama .. 063142, 100431
The Elkon Gallery, New York ... 122556
Ella Sharp Museum, Jackson .. 049747
Ellavale, Pymble 062127
Elleholm, Mörrum 086872, 126645
The Ellen Noel Art Museum of the
Permian Basin, Odessa 051709
Ellen Terry Memorial Museum,
Tenterden 045959
Ellenberg, Harrislee 076406
Ellenor Antiques, Otford 090923
Ellenroad Engine House, Milnrow 045271
Ellequadro Documenti, Genova . 110102
Eller, Derek, New York 122557
Eller, Peter, Albuquerque 119469
Ellermann, Heiko, Köln 130146
Ellermeiers Burgmannshof,
Baudenkmal und Geigenmuseum,
Hardegsen 019071
Ellgaard, Haderslev 065718
Elli Riehl Puppenwelt, Einöde . 001852
Ellinghaus, Affter 074374
Ellingson Car Museum, Rogers . 052663
Ellington White Project-Art Exhibitions,
Detroit 048305
Elliot Colliery Winding House, New
Tredegar 045328
Elliot Lake Nuclear and Mining
Museum, Elliot Lake 005779
Elliott, New Orleans 122157
Elliott, New York 095371
Elliott, Saint Paul 136219
Elliott, Singapore 085401
Elliott & Scholz, Eastbourne ... 089148
Elliott Avedon Museum and Archives
of Games, Waterloo 007364
Elliott Bay Antiques, Seattle .. 097431
Elliott Gallery, Braunton 117417
Elliott Museum, Stuart 053577
Elliott, Charlotte, Baltimore ... 092319
Elliott, Phyllis, New York 122558
Elliott, Samuel, Dublin 079549
Elliott's, Petworth 090997
Ellis County Historical Society
Museum, Hays 049413
Ellis County Museum,
Waxahachie 054242
Ellis, Anita, Avalon 098678
Ellis, J.R. & R.K., Norwich 144750
Ellis, Ruth, Bendigo 098735
Ellisland Farm, Auldgirth 043451
Ellison, Beaconsfield 088412
Ellmauthaler, Bischofshofen .. 062537
Ellsworth, Robert H., New York . 095372
Ellwood, Southampton 144853
Ellwood House Museum, DeKalb 048224
Elm Hill Contemporary Art,
Norwich 118828
Elm House Antiques, Merstham 090750,
134849
Elm Lodge, London 118304
Elm Museum, Yushu 008516
Elm Tree Antiques, Flaxton ... 089282
Elman W. Campbell Museum,
Newmarket 006476
Elmbridge Museum, Weybridge . 046106
Elmenwood, Potsdam 108398
Elmenzeh, Casablanca 082556
Elmhurst Art Museum, Elmhurst 048561
Elmhurst Historical Museum,
Elmhurst 048562
Elmina Castle Museum, Saint George's
Castle, Cape Coast 022928
Elmira, Saint Louis 096698

Elmira Railway Station – PEI Museum Branch, Prince Edward Island Museum and Heritage Foundation, Elmira 005780
Elmont Evergreen Firearms Museum, Belmont 005418
Elmore, Washington .. 097834, 136455
Elmore Old Wares, Elmore 061440
Elmore Progress Association Station Museum, Elmore 001057
Elmros Konst och Antikhandel, Umeå 087112
Elms Lesters, London . 090096, 118305
Elmuseet, Danmarks Museum for Elektricitetens Fysik, Teknologi og Kulturhistorie, Bjerringbro 010019
The Elmwood Gallery, Winnipeg 101875
Eloaitta-Kaavin Kotiseutumuseo, Kaavi 010945
Elobu, Apeldoorn 082920
Eloise Pickard Smith Gallery, University of California at Santa Cruz, Cowell College, Santa Cruz ... 053099
Eloquence, Philadelphia 096105
Elora Spring Antique Show, Elora 098036
Elovitz, Albert, Pittsburgh 096281
Eloy-Altmann, Jean-Sébastien, Nice 070600, 128860
Elrick-Manley, New York 122559
Elrose Heritage Museum, Elrose 005781
Els Encants, Palma de Mallorca 086349
Els Encants Terrenals, Sabadell . 086403
Els Llibres del Tirant, Barcelona 143713
Elsa Forde Galleries, Bismarck State College, Bismarck 047019
Elschker, K., Leverkusen 077092
Elsen & ses Fils, Jean, Bruxelles 063408, 125308
Elseneur, Paris 071292
Elsenhans, Veith, Freiburg im Breisgau 141801
Elsenzer Heimatstuben, Eppingen 018221
Elseroui, M., Graz 062586
Elsewhere Books, San Francisco 145551
Első Numizmatikai, Budapest .. 079253
Elspeth, Kangaroo Valley 061669
Elstir, Paris 104720
Elstow Moot Hall, Bedford 043552
Elt Custom Refinishing, Montgomery 135841
Elten & Elten, Zürich 116769
Eltgen, Dr. Ulrich, Königswinter . 130180
Eltham Gallery, Eltham 098997
Eltham Palace, London 044964
Elung Jensen, Søren, Skive 066118
ELV, Honfleur 103826
Elvarheim Museum, Åmli 033492
Elvarts, Ann, Chicago 092705
Elverket, Djursholm 115598
Elvick, San Francisco 136316
Elvina, Vence 105831
Elvira Antikt & Kuriosa, Halmstad 086732
Elvira Sanguino, Luís, Oropesa-del-Mar 086318
Elwert, H.-J., Baden-Baden 106106
Elwert, N.G., Marburg 142113
Elwood Haynes Museum, Kokomo 050018
ELX Koloniaal Meubel, Eindhoven 083214
Ely Books, Ely 144407
Ely Museum, Ely 044289
Elztalmuseum – Regionalgeschichte und Orgelbau, Waldkirch 022395
Em-Dash, Denver 120711
Emage, Los Angeles 121478
Email-Museum in der Internationalen Begegnungsstätte Alte Mühle, Großlittgen 018855
Emailmuseum Gertrude Stöhr, Vorchdorf 002876
Emal, San Diego 097081
Eman Fine Art, Hamilton 101315
Emani, München 077491
Emanouel Corporation, London . 090097

Emåns Ekomuseum, Bodafors .. 040566
Emanuel Vigeland Museum, Oslo 033830
Emar Sanat Galerisi, İstanbul .. 117029
Emaux d'Art de Longwy, Longwy 069623
EMB-Service, Adligenswil 138020
Emba Museum of Chinese Modern Art, Ashiya 028491
Embassy, Luzern 087637
Embden, Antony, Paris 071293
L'Embellie, Jouques 068919
Embellie, Orsay 070823
Embling, D.S., Alton 134024
Embom-Garpom Backstugusittarstuga, Liljendal 011114
Embregts, Breda 082980
Embroiderers Guild Museum, Mile End 001292
Embroiderers' Guild Museum Collection, East Molesey 044209
Embroiderers' Guild of America, Louisville 060469
Embry & Co., Portland 096396
Embryo, Leuven 100776
Embsay Bolton Abbey Steam Railway, Skipton 045811
Embu, São Paulo 064173
Emdes Hill, Malmö 086849
eMediaLoft, New York 122560
Emeishan Museum, Emei 007755
Emel, Bernard, Saint-Baudille-et-Pipet 072686
Emelies Antik och Kuriosa, Linghem 086811
The Emer Gallery, Belfast 117337
Emerald, Milwaukee 121912
Emerald Art, Magherafelt 118693
Emerald Art Center, Springfield . 053472
Emerald Art Gallery, Emerald .. 001060
Emerald Art Society, Emerald .. 058199, 099000
Emerald Bamboo, Seattle 124519
Emerald Bazaar, Emerald 061446
Emerald Empire Arts Association, Springfield 060470
Emerald Museum, Emerald ... 001061
l'Emeraude, Casablanca 082557
Emeraude, Le Touquet-Paris-Plage 103970
L'Emeraude, Paris 071294
Emeraude, Paris 071295, 071296
Emerenzia Antik, Sandviken .. 086923
Emerge Gallery, Wellington ... 113065
Emerging Artist Show, Scottsdale 098448
Emerson, Portland 145444
Emerson Gallery, McLean 050567
Emerson Gallery Berlin, Berlin .. 106220
Emerson Gallery, Hamilton College, Clinton 047843
Emery, Etoy 087420
Emery, Roger, München 077492
EMI Antik, Plzeň 065451
Emigrantmuseet, Kisa 040749
Emigre, Berkeley 139357
Emil A. Blackmore Museum of the American Legion, Indianapolis . 049682
Emil Cedercreutzin Museo, Harjavalta 010802
Emil Schumacher-Museum, Hagen 018913
Emil-von-Behring-Ausstellung, Marburg 020271
Emil Wikström Museo, Valkeakoski 011480
Emīla Dārziņa Muzejs, Jaunpiebalga 030125
Emīla Melngaiļa Memoriālais Muzejs, Vidriži 030255
Emile van Doren Museum, Stedelijk Museum, Genk 003616
Emile Verhaeren Museum, Sint-Amands 004012
Emiliani, Michèle, Dieulefit ... 103669
Emiliani, Michèle, La Bégude-de-Mazenc 103874

Emilie, Roskilde 066086
Emilijos Digrevičienės Kūrybos Muziejus, Tytuvénai 030492
Emilio Aguinaldo Shrine Museum, Kawit 034489
Emilio's Everthing Art, Miami .. 121776
Emily Carr House, Victoria 007323
Emily Carr Institute of Art and Design, Vancouver 055169
Emily Dickinson Museum, Amherst 046448
Emily et Nous, Brugge 063271
Emily'S Nook, Donostia-San Sebastián 085973
Emirates Fine Arts Society, Sharjah 060020
Emirates Very Special Arts, Sharjah 043339
Emiyu, Chiba 111242
Emma, Den Haag 083081
Emma Antiquités, Paris 071297
Emma Kunz-Museum, Würenlos 042170
Emmanuel Gallery, Auraria Campus, Denver 048262
Emmanuel Walderdorff Galerie, Köln 107604
Emmanuel, Camille-Bernard, Paris 141099
Emmanuelle, d', Liège 100781
Emmanuelli, Antoinette, L'Ile-Rousse 069470
Emmarp, Paris 071298
Emma's, West End 139959
Emmas Antikhandel, Oskarshamn 086914
Emma's Antiques, North Fremantle 061987
Emma's Arts Emporium, Pittsburgh 123539
Emmaville Mining Museum, Emmaville 001062
Emmebici – Metodologie di Indagine per i Beni Culturali, Roma ... 131913
Emmenegger, Franz, Luzern .. 133833
Emmenegger, Oskar, Zizers ... 133971
Emmett Kelly Historical Museum, Sedan 053260
Emmi, Firenze 137648
Emminghaus, Hans, Meerbusch 077317
Emócia, Žilina 114672
Emoldurarte, Porto Alegre 100974
Emonts-Holley, Gabriele, Ratingen 130558
Emperor Arts and Antiques, Manila 084478, 113404
Empire, Best 082967
Empire, Słupno 132868
Empire & Virgo, Toronto 101602
Empire Antikk, Porsgrunn 084343
Empire Antiques, New Orleans . 095094
Empire Area Heritage Group Museum, Empire 048577
Empire Art, Mainz 107947
Empire Auctions, Montréal ... 125369
Empire Auctions, Toronto 125382
The Empire Company, Alphington 060995
Empire Enterprises Antiques, San Diego 097082
Empire Exchange, Manchester . 090691
Empire Gallery, New York 122561
Empire State Aerosciences Museum, Glenville 049128
Empiric, Los Angeles 121479
Empirical Studies of the Arts, Amityville 139358
Empl, Alfons, Landshut 130209
Emporia State University Geology Museum, Emporia 048578
Emporio Brasileiro de Arte, São Paulo 064174
Emporium, Houston 093666
Emporium, Nashville 094997
The Emporium, Pontarddulais . 091030

Emporium Antique Centre, Lewes 089813, 144532
Emporium Antique Mall, Saskatoon 064560
Emporium Antique Shop, New York 095373
The Emporium Antiques, Welling 091768
Emporium at Warson Woods, Saint Louis 096699
Emporium Ex V, Nice 070601
Emporos, Greytown 083945
Empreinte, Lille 069490
Empresa de Conservacion y Restauracion, Elche 133258
Empress Scarlet, West Hollywood 097890
Emprin & Bersano, Torino 132136
Empty & Bur, Amsterdam 112190
Empty Walls, Dallas 120592
Empty Walls Dallas Galleria, Dallas 120593
EMR Imports, New York 095374
Emsbach, Jürgen, Koblenz 076833
Emser Bergbaumuseum, Bad Ems 016758
Emsland-Moormuseum, Geeste . 018600
Emslandmuseum Lingen, Lingen 020090
Emslandmuseum Schloss Clemenswerth, Sögel 021887
Emsworth Museum, Emsworth . 044292
Emu Park Historical Museum, Emu Park 001063
EN Japanese Arts, Los Angeles . 094326
En Passant, Amsterdam 082741
En Plein Air, Pinerolo 110659
En Suite, Apeldoorn 112321
ENAC Section d'Architecture, Lausanne 056430
Enamel Art Gallery, Tbilisi 105902
Enault, Bernard, Valenciennes . 125829
Enault, Michèle, Paris 128996
Enbaoui, Casablanca 082558
EnBW-Showroom Berlin, Berlin . 017109
Encadrements R.D, Reims ... 129189
Encadr'Heure, Bordeaux 103424
Encantada, San Francisco ... 124274
Los Encantes, Cádiz . 085922, 133221
Encantes, A Coruña 085950
Encanto da Freiria, Coimbra .. 084789
Los Encantos Granadinos, Granada 086035
Encapça, Girona 086026
Encarnacion Matran Bea, Cartagena 115010
Encart Chimy Gallery, Cairo ... 103035, 136786
Encaustique, Triel-sur-Seine .. 073884, 141370
Enchanted Aviary, Bury Saint Edmunds 088742
Enchanted Cottage, Cincinnati . 092849
Enchanted Light Galleries, Albuquerque 119470, 119471
Enchanted Mansion Doll Museum, Baton Rouge 046813
Enchanted Southwest, Albuquerque 119472
Enchanted What It's Shop, Jacksonville 093988
Enchanted World Doll Museum, Mitchell 050969
Enchantment Antiques, Kenthurst 061680
The Enchantry, Birkdale 061157
Enchères Champagne, Épernay . 125581
Encheres MSA – Martinot-Savignat-Antoine-Dumeynion, Pontoise . 125781
Encinas Ayuso, Alberto, Madrid . 133327
Encinas Merchan, Angel, Madrid 133328
Enck, Edwards, New York 122562
Encontro Marcado Locadora de Livros, Rio de Janeiro 140293
Encore, Wien 062934
Encre de Chine, Paris 104721
Encsy György Tokaj-Hegyalja Földtani Gyüjteménye, Tállya .. 023991

Encuadernacines Hernandez,
Zaragoza 133495
Encuentra, Jacques, Toulouse . . 129300
Encyclia, Paris 104722
Encyclopaedia Shop, Magill 139817
Encyclopedia World Book, Eaton 139722
End-O-Line Railroad Museum,
Currie 048096
End of History, New York 095375
End of Steel Heritage Museum, Hines
Creek 006007
End of the Row Antiques, Atlanta 092128,
135303
End, Karl-Heinz, Erwitte 075659
Endangered Books, Auckland . . 143281
Endangered Pieces, Portland . . . 096397
Endeavour Antiques, Dunedin . . 083925
Endeavour Museum, Wollombi . 001652
Endemann, Klaus, Kirchheim bei
München 130125
Enderby and District Museum,
Enderby 005784
Enders, G., Neuburg an der
Donau 077718
Enders, Jürgen, Fulda 129880
Endicott College Museum,
Beverly 046967
Endl, Ulrich, Bad Soden am
Taunus 074598
Endler, Dr. Franz, Kreuzwertheim 076967
Endlich, John, Haarlem 083325
Endo, Midori, Tokyo 082143
Endre Nemes Múzeum, Janus
Pannonius Múzeum, Pécs 023838
Endres, Detroit 135540
Endres, Helga & Dieter,
Uffenheim 108798
Endres, R., Bad Wildbad 074611
Endrődi Tájház, Gyomaendrőd . 023655
Endrös, Bernhard, Sparneck . . . 078515
Enebakk Auksjonsforretning,
Enebakk 126347
Enebakk Bygdetun, Trysil 034042
Enebakk Museet, Enebakk 033587
Enebybergs Antikbod, Eneryberg 086658
Enedina, Belo Horizonte 063900
Eneo, Maastricht 112571
EnergetidA, Museum voor
Energietechniek, Amsterdam . . 031937
Energie-Museum Berlin, Berlin . 017110
Energiemuseum,
Fürstenfeldbruck 018542
Energiemuseum, Rickenbach . . . 021411
Enes, J., Lisboa 126443
Enes, Z., Cruz Quebrada-Dafundo 126434
L'Enfance de l'Art, Lyon 104082
Enfield Historical Society Museum,
Enfield 048584
Enfield Shaker Museum, Enfield 048585
Enforcer Art Gallery, Philadelphia 123372
enformsak i Mälmö, Malmö . . . 115742
Eng, Ola, Uppsala 143991
Engadiner Museum, Sankt Moritz 041930
Engage, The National Association for
Gallery Education, London 060124
Engel, Jerusalem 109715
Engel, Leuven 128069
Engel & Co., Stuttgart 142388
Engel, Tel Aviv 109723
Engel & Hoff, Wien 062935
Engel Kunstauktionen, Koblenz . 126001
Engel, Fritz, Hamburg 141861
Engel, Peter, Wiesbaden 078974
Engel, Philippe, Courville-sur-
Eure 068203
Engelbertz, Gisela, Solingen . . . 108635
De Engelbewaarder, Utrecht . . . 083641
Engelbrecht, Alexander, München 077493,
130343
Engelen, 's-Hertogenbosch 083386
Engelen, Leuven 063730
Engelenburg House Art Collection,
Tshwane 038846
Engelhardt, Peter, Emersacker . . 129789

Engelhardt, U., Halle, Saale 076164
Engelhorn, Klaus, Wien 100248
Engeli, Dora, Biel 087317
Engelmann, A., Bellenberg 074713
Engelmann, W., Lahr 142041
Engeln Antik, Malmö 086850
Engelsmannen, Huskvarna 086751
O Engenho Antiguidades, Belo
Horizonte 100883
Engerbeau, Dominique, La Roche-sur-
Yon 069071
Engert, Eschweiler 106903
Engesser, Christabel, Thalwil . . . 116667
Engine House Project, Chelmsford
Museums, Chelmsford 043919
Engine House, Dunedin Gasworks
Museum Trust, Dunedin 033190
Engine 'n' Tender, London 090098
The Engine Shed, Paekakariki . . 033269
Eng'l, Kuchl 062698
England & Co., London 118306
England Antiques, Gräfelfing . . . 076067
England-Antiques, München 077494
Englands Gallery, Leek 089776, 134539
Engle, Indianapolis 093881
Englehart and Area Historical Museum,
Englehart 005786
Engler, Sevelen 087812
Englert, Seligenstadt 078452
Englert-Byrne, Amorbach 074409,
105988
Englert, O. jun., Walldürn 078851
Englich, Ulf, Salzburg 062791
Engling, Günter, Hamburg 076218
English Accent Antiques, Atlanta 092129
English and Continental Antiques,
Eye 089241
English Antiques, Antwerpen . . . 063204
English Antiques, El Paso 093401
English Antiques, Knokke-Heist . 063700
English Bridge Workshop,
Shrewsbury 119095
English-Center-Antiques, Zürich 087956
English Ceramic Circle,
Beckenham 060125
English Country Antiques,
Uffculme 091687
The English Craftsman,
Masterton 083988, 132729
English Heritage, Brighnorth . . . 088598
English Heritage, London 060126
English Heritage East Midlands Region,
Northampton 060127
English Heritage North West Region,
Manchester 060128
English Heritage South East Region,
Dover 044131
English Heritage South West Region,
Bristol 060129
English Heritage West Midlands
Region, Birmingham 060130
English Heritage Yorkshire Region,
York 060131
The English House, Marcus B. Pecher,
Aschau am Inn . . . 074439, 129389
English Interiors, Den Haag 083082
English Pine Company, Dallas . . 093117
The English Room, Charlotte . . . 092595
English Rose Antiques,
Coggeshall 088945, 134285
English Traditions, Cincinnati . . . 092850
English Workshop, Knokke-Heist 063701
English, Toby, Wallingford 144905
Englishman's Antiques, Charlotte 092596,
127158
Engman, Honolulu 120933
Engman, Kjell, Fågelmara 126579
Engping Museum, Enping 007757
Engrand, Chantal, Mouans-
Sartoux 070359
Engraving Art, Linxia 065125
Engraving Print Village Art Museum,
Sado 029292
Engs, Christianne, West

Hollywood 097891
Engsö Slott, Västerås 041136
Engström, Göran, Stockholm . . . 115854
Engströmsgården, Hult 040710
Enhjørningen, Oslo 084308
Enhörningen Antikvariat, Lidingö 143877
Enigma, Frome 117813
Enigma, Toronto 101603
Enigma Art Gallery, Delhi 109316
Enigma Art Studio, Denver 120712
Het Enigste Winkeltje, Eindhoven 083215
Enisejskij Kraevedčeskij Muzej,
Enisejsk 037008
Enjalbert, Ginette, Paris 071299
Enjoy, Roma 081347
Enkhuizer Almanak Museum,
Enkhuizen 032303
Enköpings Antikhandel, Enköping 086659
Enköpings Museum, Enköping . . 040602
Enlace, Miguel Hidalgo 138913
Les Enluminures, Paris 071300, 141100
Enmarcaciones Olivar, Tarragona 115463
Enmarcados Modernos, Ciudad de
Panamá 084415
Ennistuskoda Kanut, Tallinn . . . 128406
Ennsmuseum Weyer, Weyer . . . 002928
Ennsthaler, Steyr 140043
Eno, Amos, New York 122563
Enoagrarimuseum, Castelvetrano 025851
Enoch, New Orleans 122158
Enoch Turner Schoolhouse,
Toronto 007172
Enø Antik, Næstved 066003
Enoteca Regionale del Monferrato,
Vignale Monferrato 028383
Enoteca Regionale Permanente la
Serenissima, Gradisca d'Isonzo 026457
Enrico Gallerie d'Arte, Genova . . 110103
Enrico Gallerie d'Arte, Milano . . 080531,
110307
Enrico, Maria, Milano 080530
Enrile Corsini, C., Madrid 133329
Ensch, Werner, Münstermaifeld . 077681
Ensi Gallery, Český Krumlov . . . 102583
Ensival Historical Museum,
Ensival 003580
Enskede Antik, Enskede 086660
Ensoreanum, Oostende 058371
Ensorhuis, Oostende 003927
Enta, Warszawa 113824
Entente Cordiale, Warwick 091744
Enterieur Antikvitás, Budapest . . 079254
Enterprise Collectors Market,
Eastbourne 089149
Enterprises Antiques, Portland . . 096398,
096399
Entertainment Gallery, New
Orleans 122159
Enticott, B. J., Hampton 061589
Entlebucher Museum,
Schüpfheim 041964
Entler-Dvořáková, Praha 102659
Entomological Society Museum,
Cairo 010537
Entomologické Oddělení, Moravské
zemské muzeum, Brno 009467
Entomologické Oddělení, Národní
Muzeum, Praha 009789
Entomology Research Museum,
Riverside 052592
Entre-Deux, Bruxelles 063409
Entre Libros, Buenos Aires 139565,
139566
Entre Sable et Bruyère, Saint-Benoit-sur-
Loire 072690, 105493
Entre Temps, Bordeaux 067325
Entre-Temps, Bourges 067428
Entre Temps, Saint-Ouen 073076
Entre Vents et Marais, Batz-sur-
Mer 066964
Entrée des Artistes, Barbizon . . 103327
Entrée en Martière, Avignon . . . 103307
Entrée en Martière, Roussillon
(Vaucluse) 105481

L'Entrepôt, Carpentras 067694
L'Entrepôt des Chalets, Roncq . . 105458
L'Entrepôt du Cadre, Belbeuf . . 103360
Entrepôt du Troc, Mâcon 069793
Entropia, Wrocław 113919
Entropie, Paris 141101
Entschladen, Claudia, Bamberg . 129439
Entwistle, Lance, London 090099
Entwistle, Lance, Paris 071301
Entzmann & Sohn, Reinhold,
Wien 062936, 062937, 100249, 100250,
140071
Envers Chapin, Toronto 101604
L'Envers du Décor, Azay-le-
Rideau 066896
Enviedart, Paris 104723
Environment Canada's Biosphère, Île-
Saint-Hélène 046026
Environmental Education Center,
Basking Ridge 046804
Environmentally Friendly Sculpture and
Fine Art Show, Beijing 098047
Enwald-Museo, Tuomelan Talo,
Maatalousmuseo, Liperi 011120
Enzinger, Helmut, Salzburg 127846
Enzinger, Sebastian, Salzburg . . 127847
Eötvös József Emlékmúzeum,
Ercsi 023624
Eötvös Kúria, Beregi Múzeum,
Vásárosnamény 024037
Eötvös Loránd Emlékkiállítás,
Budapest 023472
Eötvös Lorand Tudomanyegyetem,
Régészettudomanyi Intézet,
Budapest 055764
Eolia, Paris 104724
Eon's Fashion Antique, Pittsburgh 096282
Eos Verlag, Sankt Ottilien 137497
EP Galerie, Düsseldorf 106771
Epcot, Lake Buena Vista 050070
Epelde, Bilbao 114985
Epema-State, IJsbrechtum 032523
Ephelides, Auxerre 066850
Ephemera – Dr. Hereth,
München 137401, 142149
Ephesos-Museum, Kunsthistorisches
Museum Wien, Wien 002982
Ephraim Foundation Museums,
Thomas Goodletson House, Pioneer
School House, Anderson Store
Museum, Anderson Barn Museum,
Ephraim 048589
Ephrata Cloister, Ephrata 048591
Epicene, Chicago 092706
Epicentro, Porto 114079
Epicureo, Cagliari 079981
Epifani, Maurizio, Milano 080532
Epifanio, Giuseppe, Genova 131307
l'Epigraphe, Lyon 140895
Epigraphical Museum, Athinai . . 022985
Epigraphy Museum, Department of
Antiquities, Tripoli 030294
Epipla, Milano 080533
Época, Badajoz 085710, 133087
Epoca, Paris 071302, 104725
Epoca Antiquario & Galeria,
Fortaleza 063943
Epoca Italiana, Milano 065286
Epoca Objetos de Arte, Rio de
Janeiro 064054
Epoca-Regalos, Madrid 086178
Epocas Antiques, Caracas 097950
Epoche, Parma 081108
Epoha, Beograd 085337
Epok, Stockholm 086992
Epok Antik, Umeå 087113
Epokernas Antik och Möbelrenovering,
Töreboda 087097, 133589
Epokhe, Breda 082981
Epokia, Täby 087095
Eponyme, Tourrettes-sur-Loup . 105759
Epopea, Paris 104726
Epoque, Château-Thierry 067877
Epoque, Kortrijk 063720

l'Epoque, New York 095376
Epoque, Palermo 081038
l'Epoque Dorée, Bruxelles 063410
Epoques Bleues, Genève 087465
Epoques et Styles, Survilliers .. 073610
Epp, Franz, Konstanz 076935
Eppe, Christian, Paris 141102
Eppel, Donaueschingen 106688
Epping Forest District Museum,
  Waltham Abbey 046045
Eppler, Karlsruhe 141969
Eps, H.J., Michelstadt . 077370, 130304
Epsilon Fine Art, London 118307
Epsom Art Gallery, Auckland ... 112762
Eptek Art & Culture Centre, Centre
  des Arts et de la Culture – Prince
  Edward Island Museum and Heritage
  Foundation, Summerside 097097
Epworth Old Rectory, Epworth .. 044301
EQ Konservatorsatelié – LAB,
  Uppsala 133596
Equator Gallery, Boston 119863
Equatorial America, New York .. 122564
Equilibrium, Folkestone 117803
Equilibro, Solothurn 144137
Equine Images, Fort Dodge 139359
Equine Museum, Gwacheon ... 029938
Equine Museum of Japan & JRA
  Racing Museum, Yokohama .. 029711
Equinocial Antiguidades, Rio de
  Janeiro 064055
Equinox Gallery, Vancouver ... 101780
Equinox Studio Gallery, Enmore . 099002
Equinoxe Groupe Artisanat
  Contemporain, Yvoire 105886
Equipement Culturel Les Pénitents,
  Gap 068629
Equipo Patina, Madrid 133330
Equity Art Brokers, New York . 122565,
                                122566
Equrna Gallery, Ljubljana 114700
Era, Toronto 064640
Erard, Patrick, Sauxillanges 129255
Erasmus, Amsterdam 143028
Erasmus Naklada, Zagreb 136754
Erasmushaus, Basel . 126707, 144018
Erata Edition und Galerie, Leipzig 107816,
                                137341
Erba, Bari 079772
Erbar, Kurt, Cochem 075265
Erben, Kurt, Wien 062938
Erben, Thomas, New York 122567
Erbetta, Paolo, Foggia 110077
Erbguth, Eckhardt, Dabel 106647
Erbland, Pierre, Paris 141103
Erbs-Mailleux, E., Angers 128442
Erbstück, Stade 087527
Erckens, Fritz, Aachen 074329
Ercoa, Bilbao 133184
Ercsi Helytörténeti Gyüjtemény,
  Ercsi 023625
Erdai Art Hall, Magong 042391
Erddig, Wrexham 046203
Erdefunksstelle Aflenz, Au bei
  Aflenz 001732
Erdész, Szentendre 079327
Erdészeti & Szabadtéri Erdészeti
  Múzeum, Szilvásvárad 023974
Erdészeti-, Faipari- és
  Földméréstörténeti Gyüjtemény,
  Sopron 023889
Erdgeschichtliches Museum, Lauffen
  am Neckar 019935
Erdmann, Erika, Konstanz 142030
Erdőhegyi, Péter, Budapest ... 109141
Erdős Kamill Cigány Múzeum,
  Pécs 023839
Erdos Renée-Ház – Kiállítóterem,
  Budapest 023473
Erdos, John, Singapore 085402
L'Ere du Temps, Genève 087466
Ere du Temps, Sainte-Savine .. 073367
Erebouni Museum of the
  Establishment of Yerevan,

Yerevan 000728
Eredi Antonio Esposito, Firenze . 080135
Eredi Giuseppe Tanca, Roma ... 081348
Ereğli Müzesi, Ereğli 042890
Erelöf, Stefan, Stockholm 086993,
                                133571
Eremita, Mario, Ponzano Veneto 110671
Eremiten-Presse, Düsseldorf .. 137159
Eremus Antichità, Langhirano .. 080307
Eren, İstanbul 117030
Erenköy İş Sanat Galerisi,
  İstanbul 117031
Erents, Almere 082643
Eretz-Israel Museum Tel Aviv, Tel
  Aviv 025194
Erewash Museum, Ilkeston 044627
't Erf Antiek, Rotterdam 083567, 132620
Erfahrungswelt Entfaltung der Sinne,
  Essen 018293
Erfatal-Museum Hardheim,
  Hardheim 019072
Erfurter Kunstverein e.V., Erfurt . 058845
Ergi, Düsseldorf 106772
Ergon Verlag, Würzburg 137561
Eric Carle Museum of Picture Book
  Art, Amherst 046449
Eric Thomas Galley Museum,
  Queenstown 001441
Erica Antique, Uppsala 087119
Erich-Bödeker-Gesellschaft für naive
  Kunst e.V., Hannover 058846
Erich Dohrendorf
  Restaurierungswerkstatt,
  Hamburg 129960
Erich Kästner Museum Dresden,
  Micromuseum, Dresden 017979
Erich-Kästner-Zimmer, München 020520
Erich Mäder-Glasmuseum,
  Grünenplan 018869
Erich Maria Remarque-
  Friedenszentrum, Osnabrück .. 021012
Erichsen, Jørgen, Helsingør ... 065740
Erichsens Gaard, Bornholms Museum,
  Rønne 010286
Erichson, Jörn, Bad Gandersheim 129409
Erickson, Arthur W., Portland ... 096400
Erickson, Dan, Saint Paul 096821
Erickson, Elaine, Milwaukee ... 121913
Ericson, David, Salt Lake City .. 123958
Erie Art Museum, Erie 048596
Erie Canal Museum, Syracuse .. 053632
Erie Canal Village, Rome 052672
Erie History Center, Erie 048597
Erikas Puppenstube, Bamberg .. 074661
Erikdals- och Söderboden Antik,
  Stockholm 086994
Eriksbergs Museum, Tranås ... 041084
Eriksborg Vagnmuseum, Ystad . 041182
Eriksdale Creamery Museum,
  Eriksdale 005788
Eriksen, Trondheim 132801
Eriksen, Tone, Oslo 132785
Eriksson, Anita, Falun 143835
Eriksson, Olle, Grums 115668
Erin Huse Prints, Burn 117480
Erindale's Antiques, Erindale .. 061451
Erinnerungsstätte an die
  Zwangsarbeit auf dem Gelände des
  Volkswagenwerk, Wolfsburg .. 022740
Erinnerungsstätte Baltringer Haufen –
  Bauernkrieg in Oberschwaben,
  Mietingen 020405
Erinnerungsstätte für die
  Freiheitsbewegungen in der
  deutschen Geschichte, Rastatt 021280
Erinnerungsstätte Luftbrücke
  Fliegerhorst Fassberg, Technische
  Schule der Luftwaffe 3,
  Fassberg 018341
Erinnerungsstätte Notaufnahmelager
  Berlin-Marienfelde, Flucht im geteilten
  Deutschland, Berlin 017111
Erinnerungsstätte Ständehaus,
  Karlsruhe 019506

Erith Museum, Erith 044302
Eritrean Archaeological Museum,
  Asmara 010602
Erkebispegården, Trondheim ... 034032
Erkel Ferenc Emlékház, Corvin János
  Múzeum, Gyula 023672
Erkel Ferenc Múzeum, Gyula ... 023673
Erkenbert-Museum, Frankenthal,
  Pfalz 018395
Erker-Galerie, Sankt Gallen ... 116608,
                        138086,   144126
Erland Lee Museum, Stoney
  Creek 007077
Erlauftaler Feuerwehrmuseum,
  Purgstall an der Erlauf 002547
Erlbacher, Erika, Wagrain 062870
Erlebniswelt "Mensch und
  Landschaft", Hermeskeil 019212
Erlebnismuseum Anzenaumühle, Bad
  Goisern 001743
Erlebniswelt Blockflöte, Fulda .. 018558
Erlebniswelt Grundwasser,
  Langenau 019898
Erlebniswelt Renaissance –
  Schloß Bevern, Bevern, Kreis
  Holzminden 017351
Erlebniswelt Sipplingen,
  Sipplingen 021885
Erlebniswelt U-Boot, Sassnitz . 021617
Erlemann, Hartmut, Eutin 141760
Erlen, Steffisburg 087832
Erlengut, Steffisburg 116660
Erler, Berlin 106221
Erler, Hans Jörg, Wien 062939
"erlesenes" – Antiquariat &
  Buchhandlung, Mag. Joy Antoni,
  Wien 127918, 140072
Erlianhaote Saurian Museum,
  Erlianhaote 007758
Erlinger, Günter, Nürnberg ... 077820
L'Erma, Roma 138822
Ermasch-Presse-Service,
  Hamburg 137242
Ermatinger Old Stone House, Clergue
  National Historic Site, Sault Sainte
  Marie 006989
L'Ermete, Roma 110763
Ermgassen, H., Hermannsburg . 076505
Ermini, Bastia 066959
Ermita de San Antonio de la Florida
  y Museo Panteón de Goya,
  Madrid 039621
Ermita de San Baudelio de Berlanga,
  Caltojar 039209
Ermita de Sant Feliu, Xàtiva ... 040431
Érmitaž Antikvarno-juvelirnyj Salon,
  Nižnij Novgorod 085220
Erna och Victor Hasselblads
  Fotografiska Centrum, Göteborg 040649
Ernes, E., Liège 063755, 128075
Ernest, Cluj-Napoca .. 085115, 143621
Ernest and Marion Davis Medical
  History Museum, Auckland .. 033145
Ernest Hemingway House Museum,
  Key West 049952
The Ernest W. Michel Historical
  Judaica Collection, New York . 051382
Ernesta Birznieka-Upīša Muzejs
  Bisnieki, Pastariņa Muzejs,
  Zentene 030259
Ernhardt, Holger, Stuttgart 078600
Erni, Muttenz 133867
Erni & Fornara-Erni, Simone,
  Luzern 116472
Ernie Pyle House, Dana 048135
Ernie's Antiques, Jacksonville .. 093989
Ernst, Konstanz 107711
Ernst, Paderborn 108350
Ernst Barlach Atelierhaus und
  Ausstellungsforum-Graphikkabinett,
  Ernst Barlach Stiftung Güstrow,
  Güstrow 018883
Ernst Barlach Gesellschaft e.V.,
  Wedel 058847

Ernst Barlach Haus, Stiftung Hermann
  F. Reemtsma, Hamburg 018985
Ernst Barlach Museum,
  Ratzeburg 021293
Ernst Barlach Museum Wedel,
  Wedel 022447
Ernst-Bloch-Zentrum, Ludwigshafen
  am Rhein 020133
Ernst Fuchs Museum, Wien ... 002983
Ernst-Haeckel-Haus, Jena 019444
Ernst-Moritz-Arndt-Haus,
  Stadtmuseum Bonn, Bonn ... 017488
Ernst-Moritz-Arndt-Museum,
  Garz 018596
Ernst Múzeum, Budapest 023474
Ernst, Albert J., Saarbrücken .. 078320
Ernst, Manfred, Ronnenberg .. 130596
Ernst, Rolf-Gerhard, München .. 130344
Ernsts Samlinger, Assens 010014
Erol Kerim Aksoy Vakfı, İstanbul 060007
Erotic Art Dresden, Dresden ... 106721
Erotic Art Museum Hamburg,
  Hamburg 018986
Erotic Museum, Los Angeles ... 050418
Erotics Gallery, New York 122568
Erotisch Museum, Amsterdam . 031938
Errance, Paris 136920
Errico, Umberto, Napoli 080871
Errol Station Railway Heritage Centre,
  Blairgowrie 043653
Ersego, Carla, Brescia 142654
Érseki Gyüjteményi Központ,
  Eger 023614
Ersta Diakonimuseum,
  Stockholm 040997
Erstes Allgäu-Schwäbisches
  Dorfschulmuseum, Erkheim . 018253
Erstes Bayerisches Schulmuseum,
  Sulzbach-Rosenberg 022082
Erstes Circusmuseum in Deutschland,
  Preetz 021210
Erstes Deutsches Fliesenmuseum
  Boizenburg e.V., Boizenburg ... 017473
Erstes Deutsches Historic-Actien-
  Museum, Kürnbach 019821
Erstes Deutsches Motorroller-Museum,
  Rüdesheim am Rhein 021535
Erstes Deutsches Museum für
  mechanische Musikinstrumente,
  Rüdesheim am Rhein 021536
Erstes Deutsches Strumpfmuseum,
  Gelenau 018619
Erstes Deutsches Türmermuseum,
  Vilseck 022344
Erstes Europäisches Hundemuseum,
  Österreichischer Kynologenverband,
  Klostermarienberg 002203
Erstes Imaginäres Museum,
  Sammlung Günter Dietz, Wasserburg
  am Inn 022437
Erstes Kärntner Handwerksmuseum,
  Baldramsdorf 001776
Erstes Niederbayerisches
  Automobil- und Motorrad-Museum,
  Adlkofen 016470
Erstes Niederrheinisches Karneval-
  Museum, Duisburg 018053
Erstes Oberösterreichisches
  Schnapsmuseum, Sankt Oswald bei
  Freistadt 002683
Erstes Österreichisches Biermuseum,
  Laa an der Thaya 002254
Erstes Österreichisches
  Bodenschätzungsmuseum,
  Hollabrunn 002100
Erstes Österreichisches
  Computermuseum, Bad
  Pirawarth 001759
Erstes Österreichisches
  Friedensmuseum und Heimatstube,
  Wolfsegg am Hausruck 003129
Erstes Österreichisches Funk- und
  Radiomuseum, Wien 002984
Erstes Österreichisches Museum für

Alltagsgeschichte, Neupölla . . . 002449
Erstes Österreichisches
  Rettungsmuseum, Hohenems . 002096
Erstes Österreichisches Tischler-
  Museum, Pöchlarn 002512
Erstes Österreichisches
  Tonbandmuseum, Wien 002985
Erstes Rundfunkmuseum Rheinland-
  Pfalz, Münchweiler an der
  Alsenz 020585
Erstes Südburgenländisches
  Schnapsbrennereimuseum,
  Kukmirn 002251
Erstes Tiroler Feuerwehrmuseum,
  Schwaz 002738
Erstes Tiroler Holzmuseum,
  Wildschönau 003114
Erstes Urgeschichtliches
  Freilichtmuseum der Steiermark,
  Kulm bei Weiz 002252
Erstes Waldviertler Webereimuseum,
  Waidhofen an der Thaya 002886
Erstes Württembergisches
  Polizeimuseum, Deizisau 017845
Ertl, Hallstadt 107214
Erud Billedkunstner, Birkerød . . 065614
Ervand Kochar Museum, Yerevan 000729
Ervasti, K., Helsinki 140700
Ervin, S.H., Millers Point 099341
Erwana, Strasbourg 105683
Erwin Museum, Painted Post . . . 051891
Erwin von Kreibig-Museum, Erwin von
  Kreibig Stiftung, München 020521
Eryldene Heritage House, Gordon 001111
Erzbergbahn- und Wintersportmuseum,
  Vordernberg 002879
Erzbergbaumuseum und
  Besucherbergwerk, Bestwig . . 017346
Erzbischöfliches Diözesanmuseum
  Köln, Köln 019694
Erzbischöfliches Diözesanmuseum mit
  Domschatzkammer, Paderborn 021060
Erzgebirge Toys Museum & Museum
  of Picture Books, Erzgebirgisches
  Spielzeugmuseum Seiffen,
  Karuizawa 028783
Erzgebirgische Volkskunst Ausstellung,
  Grünhainichen 018871
Erzgebirgischer Spielzeugwinkel,
  Eisenheim 018142
Erzgebirgisches Freilichtmuseum,
  Seiffen / Erzgebirge Kurort . . . 021827
Erzgebirgisches Heimatmuseum
  Hospital Sankt Johannis, Sayda 021623
Erzgebirgisches Spielzeugmuseum,
  Liestal 041674
Erzgebirgisches Spielzeugmuseum,
  Seiffen / Erzgebirge Kurort . . . 021828
Erzgebirgsmuseum mit
  Besucherbergwerk, Annaberg-
  Buchholz 016589
Erzherzog Franz Ferdinand Museum,
  Artstetten 001722
Erzherzog Johann-
  Dokumentationsstätte,
  Thernberg 002825
Erzincan Kültür Merkezi, Erzincan 042891
Erzstollen Silbergründle, Seebach 021809
Erzurum Arkeoloji Müzesi,
  Erzurum 042893
Erzurum Kültür Merkezi, Erzurum 042894
Erzurum Resim ve Heykel Müzesi,
  Erzurum 042895
Es Baluard – Museu d'Art Modern i
  Contemporani de Palma, Palma de
  Mallorca 039884
Esaias Antikk, Oslo 084309
Esbjerg Kunstmuseum, Esbjerg . 010043
Esbjerg Museum, Esbjerg 010044
L'Escabel, Zaragoza 133496
L'Escabelle, Dijon 068326
Escabelle, New York . . 095377, 135916
Escalante, Valparaíso 101925
La Escalera, Cuenca 115038

L'Escalier d'Argent, Paris 071303
L'Escalier du Commandant, Brest 067469
Escande, Jean-Luc, Auch 066801
L'Escapade, Périgueux 105334
Escape Trading, Surfers Paradise 099736
L'Escarpolette, Arzon . . 066754, 103279
Escatllar Soler, Maria, Girona . . 133262
Escazaux, C., Boissy-Saint-Léger 067278
Esch, Osnabrück 077963
Esch, Heinz Josef, Düsseldorf . . 075464
Eschen Clausen, Henning,
  Værløse 066172
Eschenau Summer Press & Eschenau
  Editions, Knetzgau 137298
Eschenburg, Harald, Kiel 141992
Escher in Het Paleis, Den Haag 032169
Eschmann, Karl-Hermann & Gudrun,
  Kelkheim 141982
Eschmann, Michael, Griesheim . 141843
Escobar, Medellín 102433
Escobar Estrada, Guillermo,
  Medellín 102434
Escobar, Adriana, Medellín 128299
Escola d'Art i Diseny de la Diputació,
  Tarragona 056384
Escola de Artes, Brasília 100913
Escola de Artes Visuais, Rio de
  Janeiro 055092
Escola de Belas Artes, Universidade
  Federal da Bahia, Salvador . . . 055096
Escola de Belas Artes, Universidade
  Federal de Minas Gerais, Belo
  Horizonte 055068
Escola de Belas Artes, Universidade
  Federal do Espírito Santo,
  Vitória 055108
Escola de Belas Artes, Universidade
  Federal do Rio de Janeiro, Rio de
  Janeiro 055093
Escola de Comunicações e Artes,
  Departamento Artes Plásticas, São
  Paulo 055102
Escola de Joalheria Bell' Arte, São
  Paulo 055103
Escola de Música e Belas Artes do
  Paraná, Curitiba 055072
Escola Massana, Centre d'Art i
  Disseny, Barcelona 056357
Escola Municipal de Belles Arts,
  Lleida 056366
Escola Profissional e Artística
  da Marinha Grande, Marinha
  Grande 056251
Escola Superior de Artes Decorativas,
  Lisboa 056247
Escola Superior de Bellas Artes de
  Lisboa, Lisboa 056248
Escola Superior de Disseny,
  Barcelona 056358
Escola Superior de Disseny de les
  Illes Balears, Palma de Mallorca 056378
Escoural, Lisboa 113998
Escrihuela Adelantado, Jose,
  Valencia 133461
Escritório de Arte Alto da Bronze,
  Porto Alegre 063971
Escudero, Madrid 086179, 126543
Escudor, Allez-et-Cazeneuve . . . 066537
Escuela de Arte Francisco Alcántara,
  Madrid 056369
Escuela de Arte y Antiguedades,
  Madrid 133331
Escuela de Arte y Antigüedades,
  Madrid 056370
Escuela de Arte y Comunicación
  Visual, Universidad Nacional,
  Heredia 055275
Escuela de Arte, Pontificia Universidad
  Católica de Chile, Facultad de Artes,
  Providencia 055182
Escuela de Artes Aplicadas,
  Camagüey 055284
Escuela de Artes Aplicadas,
  Madrid 056371

Escuela de Artes Plasticas de Puerto
  Rico, San Juan 056256
Escuela de Artes Plasticas y
  Arquitectura, Universidad Tecnologico
  de Santiago, Santiago de los
  Caballeros 055317
Escuela de Artes Plásticas y de Bellas
  Artes, Tuxtla Gutiérrez 056158
Escuela de Artes Plásticas,
  Universidad de Costa Rica, Facultad
  de Bellas Artes, San José 055276
Escuela de Artes Visuales, Universidad
  Autónoma de Nuevo León,
  Monterrey 056148
Escuela de Artes Visuales, Universidad
  de Cuenca, Facultad de Artes,
  Cuenca 055321
Escuela de Artes, Facultad de
  Humanidades y Educación,
  Universidad Central de Venezuela,
  Caracas 058150
Escuela de Artes, Universidad
  Nacional de Córdoba, Facultad
  de Filosofía y Humanidades,
  Córdoba 054867
Escuela de Bellas Artes,
  Asunción 056221
Escuela de Bellas Artes, San
  Francisco de Macorís 055316
Escuela de Bellas Artes de Azul,
  Azul 054861
Escuela de Bellas Artes General
  Urquiza, Buenos Aires 054864
Escuela de Bellas Artes y Artes
  Decorativas e Industriales,
  Tucumán 054878
Escuela de Bellas Artes, Facultad de
  Humanidades y Artes, Universidad
  Nacional de Rosario, Rosario . 054875
Escuela de Bellas Artes, Universidad
  Benito Juarez de Oaxaca,
  Oaxaca 056151
Escuela de Bellas Artes, Universidad
  de Querétaro, Querétaro 056155
Escuela de Bellas Artes, Universidad
  Juarez, Durango 056139
Escuela de Bellas Artes, Universidad
  Michoacana de San Nicolás de
  Hidalgo, Morelia 056149
Escuela de Dibujo y Pintura,
  Pamplona 056379
Escuela de Diseño y Artesanías,
  Morelia 056150
Escuela de Diseño y Escuela de
  Arquitectura, Universidad de la Salle
  Bajío, León 056141
Escuela de Flandes, Granada . . 086036,
                                    133264
Escuela de Fotografía Video e Imagen
  Digital, Centro Argentino Fotografico,
  Quilmes 054874
Escuela de Pintura y Artes
  Plásticas, Universidad del Atlántico,
  Barranquilla 055259
Escuela Mexicana de Arquitectura
  y Diseño, Universidad La Salle
  Pachuca, Pachuca 056153
Escuela Museo de Bellas Artes
  General Urquiza, Buenos Aires 000157
Escuela Nacional de Artes Plásticas
  Rafael Rodríguez Padilla,
  Guatemala 055759
Escuela Nacional de Artes Plásticas,
  Universidad Nacional Autónoma de
  México, México 056144
Escuela Nacional de Bellas Artes,
  Comayagüela 055762
Escuela Nacional de Bellas Artes,
  Santo Domingo 055319
Escuela Nacional de Bellas Artes,
  Universidad del República Uruguay,
  Montevideo 056618
Escuela Nacional de Conservación,
  Restauración y Museografía Manuel

del Castillo Negrete, México . . 056145
Escuela Nacional Superior Autonoma
  de Bellas Artes del Perú, Lima 056224
Escuela Provincial de Bellas Arte,
  Córdoba 054868
Escuela Provincial de Bellas Artes,
  Salta 054876
Escuela Provincial de Bellas Artes
  Fernando Fader, Córdoba 054869
Escuela Superior Autónoma de Bellas
  Artes Diego Quispe Tito, Cusco 056223
Escuela Superior de Arte Carlos Baca
  Flor, Arequipa 056222
Escuela Superior de Arte Neptali
  Rincón, Maracaibo 058151
Escuela Superior de Bellas Artes
  Ernesto de la Cárcova, Buenos
  Aires 054865
Escuela Taller de Artes Plásticas y
  Visuales, Maldonado 056617
Escuelas de Arquitectura, Bellas Artes
  y Diseño Publicitario, Universidad
  Autónoma de Centro América, San
  José 055277
Esculturas y Moldes, Medellín . . 128300
Esders, Viviane, Paris 104727
Eselstein, Zürich 116770
Esempi del Novecento, Milano . 080534
Eseral, Rio de Janeiro 101023
Esiew, Singapore 114519
Esinemuseo ja Koulumuseo,
  Polvijärvi 011261
Esk House Arts, Grosmont 117892
Eskær, Frederikshavn 065681
Eskape In Time, Esk 061453
Eskbank House, Lithgow 001225
Eskdale Antiques, Sleights 091352
Eskenazi, London 090100
Eskenazi, John, London 090101
Eski İstanbul, Ankara 088092
Eskilstuna Konstmuseum,
  Eskilstuna 040606
Eskilstuna Stadsmuseum, Faktoriet,
  Eskilstuna 040607
Eskimo Art Gallery, Toronto 101605
Eskimo Museum, Churchill 005618
Eskişehir Atatürk ve Kültür Müzesi,
  Eskişehir 042897
Eskişehir Devlet Güzel Sanatlar
  Galerisi, Eskişehir 042898
Eskişehir Müzeleri, Eskişehir . . . 042899
Eskolan Taide- ja Kehysliike,
  Lahti 066380
Eskoriatzako Museo Eskola,
  Eskoriatza 039381
Eslite GAllery, Taipei 116912
Eslövs Stadsmuseum, Eslöv . . . 040611
ESMA, El Madania 098539
Esmaili, Dallas 093118, 135501
Esman, Jacksonville 121246
Esman, Rosa, New York 122569
Esmonpe, Madrid 133332
Espace, Amsterdam 112191
Espace, Montréal 138578
Espace 161, Bruxelles 063411
Espace 36, Galerie d'Art
  Contemporain, Saint-Omer . . . 105563
Espace 4, Paris 071304
Espace Aïn, Kram 116973
Espace Antiquaires, Clermont-
  Ferrand 068025
Espace Antiquités, Le Puy-en-
  Velay 069342
Espace Apart, Luxembourg 111843
Espace Archéologique Forez-Jarez,
  Saint-Étienne 015301
Espace Architectural Normand,
  Crévecœur-en-Auge 012321
Espace Arlaud, Lausanne 041647
Espace Art Brenne, Concremiers 012652
Espace Arts Plastiques, Maison du
  Peuple, Vénissieux 016156
Espace Automobiles Matra,
  Romorantin-Lanthenay 015141

Espace Baroque Maurienne, Lanslebourg-Mont-Cenis 013565
Espace Baroque Tarentaise, Séez 015750
Espace BD3D, Bordeaux 103425, 140790
Espace Beaurepaire, Paris 104728
Espace Bétemps, Bons-en-Chablais 103409
Espace Blanc, Bourges 103452, 128517
Espace Brocs'Antiks, Roanne . 072490
Espace Castillon, Toulon (Var) .. 105711
Espace Châteaudun, Ivry-sur-Seine 103857
Espace Chevalier, Paris 071305, 128997
L'Espace Collections, Lyon 069734
Espace Collections, Quimper ... 072344
Espace Collections Evreux, Evreux 068495
Espace Commines, Paris 104729
Espace Contemporain, Beaune . 103350
Espace Courant d'Art, Chevenez 116211
Espace CREAS – C. Leckie, Carouge 116191
Espace Créativa, Montpellier ... 104276
Espace Cro-Magnon, Thonac .. 015943
Espace Crous, Rennes (Ille-et-Vilaine) 105434
Espace Culture Paul Doucet, Château-Thierry 012485
Espace Culturel Assens, Assens 041237
Espace Culturel La Douve, Langeais 013553
Espace Culturel Lucien Prigent, Landivisiau 013547
Espace d'Art Contemporain, Demigny 103664
Espace d'Art Contemporain, Marseille 014041
Espace d'Art Contemporain Camille-Lambert, Juvisy-sur-Orge . 013347
Espace d'Art Contemporain, André Malraux de Colmar, Colmar ... 012626
Espace d'Art Mille Feuilles, Marsa Plage 116979
Espace de Découvert du Cheval, Hennebont 013249
Espace de l'Art Concret, Mouans-Sartoux 014343
Espace-de-l'Or, Metz 070095
L'Espace-de-Nesle, Paris 104730
Espace Delille, Nice 104362
L'Espace des Arts, Furiani 013757
Espace des Arts, Le Boulou .. 013608
Espace des Arts Plastiques, Saint-Dié-des-Vosges 013854
Espace des Inventions, Lausanne 041648
Espace des Saveurs, Herve 089734
Espace des Sciences, Rennes (Ille-et-Vilaine) 015064
Espace des Vieux Metiers, Ayen 011852
Espace d'Exposition Minal, Héricourt 013252
Espace du Cheminot, Musée du Chemin de Fer, Nîmes 014498
Espace du Dedans, Lille 104005
L'Espace Enchère, Lyon 125627
Espace Enchères, Agen 125484
Espace et Traditions, Martigny . 087661
Espace Evasion, Carpentras .. 103536
Espace Exposition, Centre Culturel Jacques Brel, Thionville ... 015936
Espace Founoun, Alger 098535
Espace Gallo-Romain, Ath 003326
Espace Gare de l'Est, La Chaux-de-Fonds 116207
Espace Géologique "Les Frissons de la Terre", Château-Queyras ... 012483
Espace Giono, Lalley 013536
Espace Guisan, Pully 041853
Espace Historique de Sword Beach, Hermanville-sur-Mer 013254
Espace Horloger de la Vallée de Joux, Le Sentier 041982
Espace Jean de Joigny, Joigny (Yonne) 013318

Espace Jean Tinguely – Niki de Saint-Phalle, Fribourg 041497
Espace Joseph Besset, Vanosc . 016123
Espace-Jouets, Ecomusée des Bruneaux, Firminy 012985
Espace Journal, Gammarth 116970
Espace Karim Francis – Contemporary Art Gallery, Cairo 103036
Espace l'Aurore, Sorens 116655
Espace Lausannoise d'Art Contemporain, Lausanne 041649
Espace Marais d'Anchais, Maillezais 013970
Espace Maritime et Portuaire des Docks Vauban, Le Havre ... 013642
Espace MDC, La Baule-Escoublac 103862
Espace Meyer Zafra, Paris 104731
Espace Mica, Saint-Grégoire ... 105528
Espace Minéralogique Minéraux Couleur Nature, Eymoutiers ... 012940
Espace Mira Phalaina, Montreuil-sous-Bois 014304
Espace Montfort, Vaison-la-Romaine 105787
Espace Montoise, Le Mans 013955
Espace Multi Collections, Brehal 067458
Espace Muséal, Rosheim 015158
Espace Muséal de la Maison du Berger, Châtillon-sur-Saône . 012517
Espace Muséal de la Maison du Cordonnier, Châtillon-sur-Saône 012518
Espace Muséal de la Route Romane d'Alsace, Neuwiller-Saverne 014465
Espace Muséal Les Courtinals, Moureze 014361
Espace Musée Sainte-Madeleine, Besançon 011999
Espace Musée, Centre d'Information, Freissinières 013046
Espace Nael, Biarritz 013382
Espace Neige et Montagne, Bonneval-sur-Arc 012102
Espace Paragon, Luxembourg . 111844
Espace Paroles d'Objets Nègres, Musée Municipal, Blangy-sur-Bresle 012057
Espace Patrimoine le Cœur, Tignes-le-Lac 015963
Espace Paul Rebeyrolle, Eymoutiers 012941
Espace Pepin, Montréal 101381
Espace Pierre Bonnard, Le Cannet 013616
Espace Pierres Folles, Saint-Jean-des-Vignes 015386
Espace Point Barre, Lille 104006
Espace PR36, Neuchâtel 116529
Espace Riquet, Béziers 012020
Espace Saint-Eloi, Valbonne ... 073933
Espace Saint-François, Lausanne 116412
Espace Saint-Martin, Cognac .. 103613
Espace Sol, Seoul 111637
Espace Sophonisbe, Carthage Hannibal 116969
Espace Suisse, Strasbourg ... 105684
Espace Temps, Aix-en-Provence 103195
L'Espace Temps, Gex 068659
Espace Temps, Paris 128998
Espace Terre et Materiaux de la Faculté Polytechnique, Mons .. 003878
L'Espace Troc 57, Saint-Avold .. 072682
Espace Vauban, Besançon 012000
Espace Véga, Saint-Ouen 105566
Espace Verre, Vannes-le-Châtel 016122
Espace Virtuel, Chicoutimi 005614
Espace Wafabank, Casablanca . 112037
Espach, Greifensee 116361
Espacio 61, Asunción 113329
Espacio Amalfi, Punta del Este . 119442
Espacio Distrito Cu4tro, Madrid . 115187
Espacio Fortuny, Granada 115085
Espacio Lady Anne, Madrid 086180
Espacio Liquido, Gijón 115070

Espacio Marzana, Bilbao 114986
Espacio Minimo, Madrid 115188
Espacio y Restauro, Zaragoza . 133497
Espaço Cultural da Marinha, Rio de Janeiro 004689
Espaco Cultural e Galeria Retiro das Artes, Rio de Janeiro 101024
Espaço de São Bento, Lisboa .. 084879
Espaço dos Anjos, Leopoldina . 004536
Espaço Elinga, Luanda 098540
Espaço Lucio Costa, Brasilia ... 004356
Espaço Museu da Vida, Rio de Janeiro 004690
Espaço UFF de Fotografia, Universidade Federal Fluminense, Rio de Janeiro 004691
Espadim, Lisboa 084880
L'Espadon, Baudot-Visser, Paris . 071306
Espai Lluis Ribas, Sant Cugat del Vallès 115410
Espai Lucas, Valencia 115481
Espanner, Edmund A., Roskilde . 066087
La Española, Buenos Aires 060897
Españoleto, Madrid 133333
Espaxios – Arte Actual, Guatemala 109119
Espelho da Moda, Luanda 098541
Espelho Design Antiguidades, São Paulo 064175
Esperance Municipal Museum, Esperance 001065
Esperanto-Museum der Österreichischen Nationalbibliothek und Sammlung für Plansprachen, Wien 002986
Esperluette, Castres 103542
Espersit Ház, József Attila Múzeum, Makó 023768
Espinet, Carmen, Barcelona .. 114877
Espinosa, José Garcia, Rio de Janeiro 128114
Espinossa, Michel, La Couvertoirade 069016
Espinoza, Oakland 136040
El Espíritu del Mueble, Don Torcuato 060942, 127543
L'Espit, Paris 071307
Espoon Automuseo, Espoo ... 010770
Espoon Kaupunginmuseo, Espoo 010771
Esposito, Antonio, Napoli 080872
Esposito, Giuseppe, Napoli 080873, 080874, 080875, 080876
Esposito, Vincenzo, Napoli ... 131617
Esposizione Forte Airolo, Airolo . 041203
l'Esprit de l'Art, Antwerpen ... 063205
Esprit d'Artiste, Honfleur 103827
Esprit Decor Gallery, Phoenix .. 123489
l'Esprit des Lieux, Tournai 100818
Esprit d'Objet les Objets, Saint-Ouen 073077
Esprit d'Uzes, Uzès 073924
Esquerre, Raymond, Toulouse . 073764
La Esquina, Buenos Aires 060898, 127521
Esquina, Porto 143602
Esquire Appraisals, New York .. 095378
Esquisse et Lavis, Pontrieux .. 105390
Esra, Apeldoorn 082921
Esrum Antik, Græsted 065705
Essa, Hassan Ali, Muttrah 084402
Essaraya, Tunis 116988
Esseen & Co, Växjö 115990
Essem Art Gallery, Mumbai ... 109431
Essence of Time, Lichfield 089821
Essences de Siècles, Bruxelles . 063412
Essences Primitives, Langoiran . 013919
Essential Furniture, Los Angeles 094327
Éssentukskij Gorodskoj Kraevedčeskij Muzej im. V.P. Špakovskogo, Essentuki 037010
Esser & Esser, Köln 076860
Esser, Günther, Düsseldorf ... 106773, 129738
Esser, Marlene, Wachenheim an der

Weinstraße 108828
Esser, Norbert, Köln 076859
Essers, Neckargemünd 077695
Essex, Heathmont 139770
Essex Antiques, Winnipeg 064920
Essex Art Association, Essex ... 060471, 120846
Essex Fine Art, Ocean Grove ... 099496
Essex Historical Museum, Essex 048609
Essex Police Museum, Chelmsford 043920
Essex Regiment Museum, Chelmsford 043921
Essex Shipbuilding Museum, Essex 048610
Essex, Richard, Langford, Somerset 089744
Essie, London 090102, 134655
Essig Museum of Entomology, University of California, Berkeley, Berkeley 046934
Essl Museum – Kunst der Gegenwart, Klosterneuburg 002207
Essley-Noble Museum, Mercer County Historical Society, Aledo 046376
Esslinger Kunstverein, Esslingen 058848
Esslinger, W., Tuningen 078733
Esso, New York 122570
Est Europa Center, Bucureşti .. 085092
Est-Ouest Auctions, Tokyo 126246
El Establo, Gonzalez Catán 060946
Estação Ciência / EST CIENC, Reitoria da Universidade de São Paulo / RUSP, São Paulo 004855
Estação da Arte, Fortaleza 100943
Estación Enológica de Haro y Museo del Vino, Haro 039485
Estación Experimental de Zonas Áridas, Almería 038960
Estación Ferrocarril Bejucal, Bejucal 009078
Estades, Michel, Lyon . 069735, 104083
Estades, Michel, Toulon (Var) .. 105712
Estadieu, Portet-sur-Garonne .. 072287
Estadieu, Toulouse 073765
Estagnié, Solliès-Pont 073536
L'Estampe, Tournai 140262
Estampe Ancienne, Grenoble .. 068726
Estamperia Castells, Barcelona . 114878
Estampille, Les Sables-d'Olonne 069429
Estampille, Lille 104007
L'Estampille – L'Objet d'Art, Dijon 138642
Estampille d'Alpage, Briançon .. 067486
Estate 93, Caslano 116200
Estate Antiques, Las Vegas 094136
Estate Auction, Denver 093281, 127205
Estate Furnishings, Tucson ... 097657
Estate Furniture, Tulsa 097729
Estate Silver Company, New York 095379
Estates Galleries, New York ... 122571
Este, Singapore 085403
Esteban Pascual, Maria Mercedes, Bilbao 133185
Esteban, Jbanez, Barcelona ... 133112
Estela, Tarbes 105701
Estelle, Saint-Rémy-de-Provence 105608
Estelle's Antiques, Paddington, Queensland 062050
Estelle's Consignment, Nashville 094998
Esterhammer, Innsbruck 100042
Esterházy Community Museum, Esterhazy 005790
Esterházy-Ausstellung, Eisenstadt 001859
Esterházy Kastélymúzeum, Fertőd 023635
Esterházyo Palác, Slovenská Národná Galéria, Bratislava 038301
Ester's Gallery, Lockrose 099236
Estes Park Area Historical Museum, Estes Park 048613
Estes Plating, Atlanta . 092130, 135304
Estes-Simmons, Atlanta 135305
Estes-Winn Antique Automobile

Museum, Asheville 046554
Esteta, Porto 114080
Estetika/Aesthetics, Praha .. 138612
Estevan Art Gallery and Museum,
Estevan 005792
Estève, Brigitte, Genillé 128608
Esteve, C., Kintzheim 068940
Esteve, Denis, Revel 105444
Estevez, Austin 119668
Esther M. Klein Art Gallery,
Philadelphia 052046
Esther Prangley Rice Gallery,
Westminster 054341
The Esther Thomas Atkinson Museum,
Hampden-Sydney 049336
Esthete, Bruxelles 100597
Estiarte, Madrid 115189
Estonian Academy of Arts,
Tallinn 055339
Estonian Museums Association,
Tartu 058570
Estorick Collection of Modern Italian
Art, London 044965
Estrada Antiguedades, Barcelona 085769
Estrada Studios, Dallas 120594
Estrada, Félix, A Coruña 085951
Estrade Cédric, Saint-Robert ... 073297
Estrade, Michel, Bordeaux 103426
Estrany de la Mota, Barcelona . 114879
Estrella, Max, Madrid 115190
Estro, Padova 110562
Estuary Gallery, Leigh-on-Sea .. 118099
Estucolor, Barcelona 133113
Estudi B2, Barcelona 133114
L'Estudi i Onec, Barcelona 114880
Estudio Anticuario Hom, Madrid 086181,
133334
Estudio Arrieta, Bilbao 133186
Estudio la Casilla, Bilbao 133187
Estudio del Angel, México 132364
Estudio Métodos de la Restauración,
Moncada 133407, 137997
Estudio Museo Ramoneda,
Humahuaca 000386
Estudio San Angel, México 132365
Estudio Urgull, Donostia-San
Sebastián 115049
Estudio Urquijo, Bilbao 133188
Estudios de Restauracion Alcazaba,
Sevilla 133430
Esztergomi Muvészek Céhének
Galériája, Esztergom 109191
Eszterházy Károly Föiskola
Csillagászati Múzeuma, Eger .. 023615
Et Caetera, Paris 071308
Et Zetera, Neuhausen auf den
Fildern 077731
Etage 2, Montreal 108533
Il Etait Une Fois, Sainte-Savine . 073368
Etat d'Ame, Saint-Jean-de-Monts 105537
Etat d'Origine, Paris .. 071309, 071310
etc., Tokyo 138888
Etcetera, Kinsale 079649
Etcetera Aesthetic Additions,
Chennai 109281
Etcetera Antiques, Denver 093282,
135529
Etcetera Etc Antiques, Seaton . 091257
Etching, Pordenone 110673
Etchings Art Gallery, Balaclava . 061078,
098690
Ete-Bols Prylar, Visby 087162
Etelä-Karjalan museo,
Lappeenranta 011090
Etelä-Karjalan Taidemuseo,
Lappeenranta 011091
Etelä-Pohjanmaan Maakuntamuseo,
Seinäjoki 011349
Etele XI. Kerületi Helytörténeti Kör
Kiállítóhelye, Budapest 023475
Etellin, Jean-Claude, Clermont-
Ferrand 068026
Etemad, Brian, San Francisco . 097225
L'Etendard, Boulogne-sur-Mer .. 067400

Eterna, Varna 101189
Eternity Antiques, Tuggerah ... 062349
Eternity Museum, Wannian 008351
Ethel-20th Century Living,
Toronto 064641
Ethel, Anne, Hawera 083949
Ethel's, Toronto 064642
Etherton, Tucson 124732
Ethiopian American Jewish Art, Los
Angeles 060472
Ethnic Africa, Nairobi 111559
Ethnic Art, Bruxelles 063413
Ethnic Art Gallery, Kathmandu . 112073
Ethnic Arts, Calgary 101235
Ethnic Discovery, Singapore .. 085404
Ethnic Doll and Toy Museum,
Whitstable 046126
Ethnic Elegance, Jacksonville . 121247
Ethnic Extravaganza, Tampa .. 124646
Ethnic Homes, Singapore 085405
Ethniki Pinakothiki & Mouseio
Alexander Soytzoy, Athinai .. 022986
Ethnix, New York 122572
Ethno-Art Galerie, Basel 116064
Ethno Arts, Elwood 098999
Ethno-Textil-Galerie, Bremen ... 106579
Ethnoarte, Milano ... 080535, 110308
Ethnographic Museum, Addis
Ababa 010722
Ethnographic Museum, Avgorou 009401
Ethnographic Museum, Port
Harcourt 033444
Ethnographic Museum, Assam Institut
Guwahati, Guwahati 024274
Ethnographical Historical Museum of
Lárissa, Lárissa 023150
Ethnographical Museum, Aden .. 054822
Ethnographical Museum, Doha .. 036179
Ethnographical Museum, Paphos 009432
Ethnographical Museum, Sudan
National Museum, Khartoum .. 040494
Ethnographische Sammlung,
Fribourg 041498
Ethnographisches Museum Pöykkölä,
Rovaniemi 011320
Ethnographisches Museum Schloss
Kittsee, Kittsee 002180
Ethnography Museum, Department of
Antiquities, Tripoli 030295
Ethnological Museum, Cairo .. 010538
Ethnological Museum, Chittagong 003213
Ethnological Museum, Pune ... 024445
Ethnological Museum, Tehrän .. 024750
Ethnological Museum of Pyrsogianni
Stonemasons, Konitsa 023138
Ethnological Museum, Sa'd-Abad
Museum, Tehrän 024751
Ethnologische Sammlung der
Universität Göttingen, Institut für
Ethnologie, Göttingen 018747
Ethnologisches Museum, Staatliche
Museen zu Berlin – Stiftung
Preußischer Kulturbesitz, Berlin 017112
Ethos, Bologna 109820
Etibank Sanat Galerisi, İstanbul . 117032
Etienbled & Jean-Claude Cazenave,
Gérard, Paris 071311
Etienne & Van den Doel,
Oisterwijk 112615
Etienne, Danielle, Anières 133619
Etienne, Emmanuel, Riaz 133891
Etique, Den Haag 132473
Etkin, Argyll, London .. 090103, 144589
Etkin, Jay, Memphis 121693
Etna, Pittsburgh 096283
Etnarref Galerie, Saint-Amand-les-
Eaux 105492
Etnica Galleria, Milano 080536, 110309
Etnie, Verona 081908
Etnogalerija Skrina, Ljubljana .. 114701
Etnografi Museoa, Arcienaga ... 038984
Ètnografičeskij Muzej Kazanskogo
Universiteta, Kazanskij
gosudarstvennyj universitet,

Kazan 037099
Ètnografičeskij Muzej Narodov
Zabajkalja, Ulan-Udé 037987
Ètnografičeskij Muzej pod Otkrytym
Nebom Torum Maa, Chanty-
Mansijsk 036953
Etnografičeskij Muzej Tavrika,
Krymski Kraevedčeskij Muzej,
Simferopol' 043299
Etnografické Múzeum, Slovenské
Národné Múzeum, Martin ... 038389
Etnografický Ústav, Moravské zemské
muzeum, Brno 038520
Etnografiki Muzeý, Türkmenbaşi 043100
Etnografikoa Museoa Artziniega,
(Museo Etnográfico de Artziniega),
Artziniega 039005
Etnografinė Prano Dzūko Sodyba-
Muziejus, Lazdijų krašto muziejus,
Delnicai 030330
Etnografisch Museum,
Antwerpen 003284
Etnografische Collecties van de
Universiteit Gent, Gent 003623
Etnografisk Museum, Universitetets
Kulturhistoriske Museer, Oslo . 033831
Etnografiska Museet, Statens Museer
för Världskultur, Stockholm ... 040998
Etnografski Kompleks na Istoričeski
Muzej, Vraca 005264
Etnografska Ekspozicija,
Pazardžik 005107
Etnografska Ekspozicija, Svištov 005220
Etnografska Ekspozicija Stojova Kăšta,
Radomir 005144
Etnografska Zbírka Františkánský
Klášter, Zaostrog 009055
Etnografska Zbirka i Lapidarij
Franjevačke Gimnazije, Franciscan
Secondary School at Visoko,
Visoko 004291
Etnografski Kompleks Starijat Dobrič,
Dobrič 005016
Etnografski Muzej, Berkovica .. 004984
Etnografski Muzej, Dobrič 005017
Etnografski Muzej, Elchovo 005025
Etnografski Muzej, Plovdiv 005126
Etnografski Muzej, Silistra 005162
Etnografski Muzej, Varna 005237
Etnografski Muzej, Zagreb 009028
Etnografski Muzej Dunavski Ribolov i
Lodkostroenie, Tutrakan 005233
Etnografski Muzej Istre, Pazin .. 008965
Etnografski Muzej Rupe,
Dubrovnik 008901
Etnografski Muzej Split, Split .. 008994
Etnografski Muzej Srbije,
Beograd 038129
Etnografski muzej, Burgaski Muzej,
Burgas 004997
Etnografski-văzroždenski Kompleks,
Kavarna 005053
Etnografya Müzesi, Ankara 042820
Etnografya Müzesi, Konya 043002
Etnološki Muzej, Budva 031749
Etnološki Muzej, Ribnica na
Pohorju 038601
Etnološko društvo Srbije,
Beograd 059786
Etnomuseo dei Monti Lepini,
Roccagorga 027544
Etobicoke's Museum – Montgomery's
Inn, Etobicoke 005795
L'Etoile, Montceau-les-Mines .. 104258
Etoile des Arts, Moustiers-Sainte-
Marie 104308
L'Etoile d'Ishtar, Paris 071312
l'Etoile Royale, New York 095380
Eton Antique Bookshop, Eton .. 144414
Eton Antiques Partnership, Eton 089215
Eton Art Gallery, Eton 117765
Eton College Natural History Museum,
Eton 044306
L'Etourdi, Paris 141104

Etowah Indian Mounds,
Cartersville 047448
Etruria, Adria 109738
Etruria Industrial Museum, Stoke-on-
Trent 045881
Ets Chevalier, Paris 071313
Etta, Claremont 061324
Ettaswir, Tunis 116989
Ettinger, Eleanor, New York 122573
Ettlin, Urs, Sankt Moritz 087794, 133914
Etxea, Saint-Jean-de-Luz 072843
Etzel Museum, Tel Aviv 025195
Etzer, Peter A., Zell am See ... 100405
Etzikom Museum, Canadian
national Historic Windmill Centre,
Etzikom 005796
Etzold, S., Leonberg 077088
Eubie Blake National Jazz Museum
and Cultural Center, Baltimore 046745
Eugeen van Mieghem Museum,
Antwerpen 003285
Eugene and Marilyn Glick Indiana
History Center, Indianapolis .. 049683
Eugene Field House and Saint Louis
Toy Museum, Saint Louis 052799
Eugene Furniture, Cleveland .. 135461
Eugene Galleries, Boston 092467
Eugene O'Neill Home, Danville . 048143
Eugene V. Debs Home, Terre
Haute 053716
Eugène, Louis-Marie, Le François 132353
Eugenie-Victoria, Paris 071314
Eugenijaus ir Leonardo Matuzevičių
Memorialinis Muziejus, Pasvalio
krašto muziejus, Krinčinas ... 030398
Eugenio Collectta Carrozze Serreria,
Caronno Pertusella 025776
Eugster & Urs Brand, Lothar,
Appenzell 116023
Eugubium Studio Numismatico,
Gubbio 080304
Eulberg, Bernd, Dornburg, Kreis
Limburg-Weilburg 106692
Eule-Art, Davos Platz 116239
Eule-Art, Flims Waldhaus 116266
Eule-Art, Lenzerheide 116436
Eule-Art, Samedan 116605
Die Eulenburg, Historisches Museum
Rinteln, Rinteln 021437
Eulenspiegel Kunstbuchhandlung,
Göttingen 141834
Eulenspiegel Museum, Mölln .. 020433
Eulenthurm-Museum,
Weißenthurm 022549
Euler & Lösch, Saarbrücken ... 078321
Eulo Telegraph House Gallery,
Eulo 099007
Eumundi Museum, Eumundi ... 001066
Eunomia, Roma 110764
Euphrat Museum of Art, De Anza
College, Cupertino 048095
EURANTICA Brussels, Fine Art and
Antiques Fair, Bruxelles 098019
Eurantik, Woolloongabba 062486
Eurasia Antiques, Unley 062355
Eurasian Interiors, San Francisco 097226
Euregio Foto-College, Prager
Fotoschule Österreich, Neumarkt im
Mühlkreis 054958
Euregio-Kunstkreis e.V., Bocholt 058849
Eureka, Bruxelles 063414
Eureka, Cairo 066241
Eureka – Parque de la Ciencia,
Mendoza 000450
Eureka Antiques, Bowdon 088548
Eureka Antiques, London 090104
Eureka Antiques & Gallery,
Norwood 062013, 099488
Eureka Antwerp, Antwerpen ... 063206
Eureka Editions, Utrecht 143243
Eureka Fire Museum, Milltown . 050900
Eureka Pioneer Museum of McPherson
County, Eureka 048629
Eureka School House, Springfield 053474

Register der Institutionen und Firmen

Eureka! The Museum for Children, Halifax
– Excalibur, Los Angeles

Eureka! The Museum for Children,
Halifax 044501
Eurest, Barcelona 133115
Euricka-Antiques Trade Services,
Dublin 079550, 130853
Euro Antik, Haguenau 068775
Euro Antik, Västerås 087139
Euro Antiques, New York 095381,
135917
Euro Antiquités, Juziers 068935
Euro Art, Bushey 117494
Euro-Art-Club, Innsbruck 058249
Euro Arte, Vigo 086557
Euro Artes, Bogotá 065204, 128268
Euro Classics, Columbus 093016
Euro-Collectibles, Hilversum ... 112537
The Euro Collection, Saint Louis 096700
Euro Dépot-Vente Broc'Antique, Notre-
Dame-de-Bondeville 070747
Euro Exports, Boston 092468
Euro Galeria, Porto 114081
Euro-Galerie KIS, Riedt-Erlen ... 116584
EuRo-Kunst, Saarbrücken ... 108541
Euro Orient Arts, Saint-Ouen ... 073078
Euro-Pacific, Venezia 081835
Euro Pine, Houston 093667
Euro Tecnica Losheim, Hellenthal 019181
Euro UK Antiques, Wembley ... 062417
Euroa Fine Books, Euroa 139727
Euroantic, Molesmes 070148
Euroantik, Innsbruck 098001
Euroart, Iaşi 085124, 133009
Euroarte, Boston 119864
Eurobroc, Bayonvillers 066983
Eurocoupes, Illzach 068869
Eurocraft, Chicago 135406
Eurodecor Gallery, Hong Kong ... 102072
Eurogalerías de Arte Llamas,
Bilbao 114987, 133189
Eurogoldor, Bruxelles 063415
Eurographics, Barbing 137042
Euromint, Bochum 075032
Euronumix, Düsseldorf 075465
Europa Antiques, Birmingham ... 092406
Europa Books, London 144590
Európa Galéria – Lex antiquitás,
Budapest 079255
Europa Galerie, Hannover 107358
Europa House Antiques, London 090105
Europäische Glasmalerei- und
Krippenausstellung, Schramberg 021730
Europäische Kunstakademie,
Trier 055729
Europäische Messe für Denkmalpflege,
Restaurierung und Altbausanierung,
Leipzig 098137
Europäische Vereinigung bildender
Künstler aus Eifel und Ardennen,
Prüm 058850
Europäische Werkstatt für Kunst und
Kultur im Festspielhaus Hellerau,
Dresden 106722
Europäischer Skulpturenpark,
Willebadessen 022665
Europäisches Brotmuseum,
Ebergötzen 018074
Europäisches Frauenforum für Kunst
und Kultur, Leutkirch im Allgäu 107860
Europäisches Industriemuseum für
Porzellan und Europäisches Museum
für Technische Keramik, Rosenthal
Museum, Porzellanikon in Selb,
Selb 021833
Europäisches Klempner- und
Kupferschmiede-Museum,
Karlstadt 019523
Europäisches Museum für Frieden,
Stadtschlaining 002770
Europäisches Museum für Modernes
Glas, Kunstsammlungen der Veste
Coburg, Rödental 021451

Europäisches Spargelmuseum,
Schrobenhausen 021737
Europalia International Festival of the
Arts, Bruxelles 058372
Europ'Art, Aigues-Mortes 103178
Europ'Art, Dordrecht 112436
Europart, San Diego 124120
Europ'Art – Foire Internationale d'Art,
Genève 098328
Europe Art, Milano ... 080537, 126205
Europe Art Sélection, Suresnes . 105699
Europe in the Village, Houston . 093668
European Accent, Warminster .. 091733
European Antique Market,
Louisville 094546, 135755
European Antique Market-Wholesale,
Louisville 094547
European Antiques, Auckland .. 083785
European Antiques, Minneapolis 094909
European Antiques, Seattle 097432
European Antiques Restoration,
Chicago 135407
European Art Garage,
Jacksonville 121248
European Art Importers, Houston 121033
European Association of Open Air
Museums, Szentendre 059338
European Classic, Cincinnati ... 092851
European Collaborative for Science,
Industry and Technology Exhibitions,
Bruxelles 058373
European Confederation of
Conservator-Restorers Organisations
(ECCO), Bruxelles 058374
European Country Antieks, Saint
Louis 096701
European Decorative Arts, New
York 095382
European Express, Kansas City . 094072
European Federation of Associations
of Industrial and Technical Heritage,
E-FAITH, Terrassa 059861
European Federation of Museum and
Tourist Railways, Haaksbergen 059599
European Fine Arts, Preston ... 091054
European Leage of Institutes of the
Arts, Amsterdam 059600
European Makers Gallery,
Amsterdam 112192
European Museum Forum, Bristol 060132
European Photography, Berlin .. 137066,
138696
European Pine Imports, Phoenix 096207
European Restorations, New
Orleans 135854
European Textile Network,
Hannover 058851
European Treasures, Pittsburgh . 096284
Européenne de Ventes, Bourg-en-
Bresse 125534
Européenne des Arts et
Connaissances E.A.C, Mios ... 104238
Europese Bibliotheek,
Zaltbommel 137878
Europhil, Borgo San Dalmazzo . 079905
Europ'Occase, Saint-Pierre-des-
Corps 073257
Europos Parkas, Vilnius 030512
Eurotrocantic 87, Isle 068876
Eury, Nancy 070419
Eusebi, Romolo, Fano 080069
Eusebia's Antiques, Cleveland .. 092939
Euskal Abertzaletasunaren Museoa,
Museo del Nacionalismo Vasco,
Artea 039002
Euskal Arkeologia, Etnografia eta
Kondaira Museoa, Bilbao 039143
Euskal Gastronomi Museoa,
Llodio 039590
Euskal Poliziaren Museoa, Museo de
la Policía Vasca, Arkaute 038992
Eustatia, Utrecht 083642
Eustis Historical Museum, Eustis 048633
Euston Hall, Thetford 045968

Euzen, Nicole, Pont-l'Abbé 072256
Eva and Morris Feld Gallery, American
Folk Art Museum, New York .. 051383
Eva Brook Donly Museum,
Simcoe 007042
Eva Faschaunerin Heimatmuseum,
Gmünd, Kärnten 001942
Evan, Lugano 087613, 116455
Evandale Antiques, Evandale .. 061456
Evangélikus Országos Múzeum,
Budapest 023476
Evangelisches Bibelwerk im
Rheinland – Bibelzentrum,
Wuppertal 022817
Evangelisches Diözesanmuseum,
Fresach 001907
Evangelisches Diözesanmuseum,
Murau 002424
Evangelisches Kirchenmuseum
Pleidelsheim, Pleidelsheim ... 021151
Evangelisches Museum in Wien,
Wien 002987
Evangelisches Museum Oberösterreich,
Rutzenmoos 002614
Evangelista Costantino, Napoli . 080877
Evans, Miami 121777
Evans, New Orleans 122160
Evans, Oamaru 084042
Evans & Evans, Alresford 088257
Evans & Gerst, West Hollywood 097892
Evans Clarke National, Woodville
North 125248
Evans-Tibbs Collection,
Washington 054109
Evans, Ann, Valley 091701
Evans, Bob, Honiton 089560
Evans, D. D., Mönchengladbach 077389
Evans, Dennis, Saskatoon 101545
Evans, Geoff, Hay-on-Wye 117935
Evans, Larry, San Francisco ... 124275
Evans, Lionel, Eagle Heights ... 061423
Evans, Steven, Treorchy 091630
Evans, W., Bremen 075157
Evans, Wendy S., New York ... 122574
Evanston Art Center, Evanston . 048637
Evanstonia, Chicago .. 092707, 092708,
135408, 135409
Evansville Museum of Arts and
Science, Evansville 048644
Evarb, Houston 121034
Evelino Antiquario, Rio de
Janeiro 064056
Evellin Orfèvrerie, Rennes (Ille-et-
Vilaine) 129202
Evelyn Hone College of Applied Arts &
Commerce, Lusaka 058161
Evelyn Payne Hatcher Museum of
Anthropology, Saint Cloud ... 052761
Evelyn's Antique Chinese Furnitures, San
Francisco 097227, 136317
Evelyn's Antiques, Washington . 097835
Evénementielle Richelieu,
Montréal 101382
Evening Moon, Minneapolis ... 121984
Évenkijskij Okružnoj Kraevedčeskij
Muzej, Tura 037971
Eveno, Pascal, Yerres 074308
Evensen & Silverstolpe,
Stockholm 086995
Evensen, Klaus, Trondheim ... 084385,
126395
Evensen, Thor, Harstad 132778
Eventi e Interventi, Ravenna ... 138823
Ever Arts, San Francisco 097228
Ever English, Gerrards Cross .. 117822
Everard, Rosebank 114810
Everarts, Paris 104732
Everett Fine Art, West Buckland 091782,
135203
Everglades Fine Art Gallery, Port
Macquarie 099588
Everglades Gardens Gallery,
Leura 099229
Evergreen, Sint-Niklaas 100815

Evergreen 2 Antiques, New York 095383
Evergreen Antique, Budapest .. 079256
Evergreen Antiques, Newtown,
Victoria 061968
Evergreen Antiques, Redruth ... 091085
Evergreen Aviation Museum,
McMinnville 050568
Evergreen Books, Fort Worth ... 145149
Evergreen Books, North Shore . 143336
Evergreen Galleries, Evergreen State
College, Olympia 051756
Evergreen Gallery, Daventry ... 117669
Evergreen Gallery, Salt Lake City 123959
Evergreen House, Dartmouth ... 005681
Evergreen House, Johns Hopkins
University, Baltimore 046746
Evergreene, Genève 116298
Everhart Museum of Natural History,
Science and Art, Scranton ... 053215
Everitt & Rogers, Tiptree 135136
Everlasting Heritage, New York . 135918
Everley Antiques, Eversley 089224
Evers, Mark, Pittsburgh 096285
Evers, Olaf, Taunusstein 078668
Everson Museum of Art,
Syracuse 053633
Everson Museum of Art Bulletin,
Syracuse 139360
Everts, F., Baarn 132432
Every, Castel San Pietro 080015
Every Now & Then, Cincinnati .. 092852
Everybody's Book Shop, Hawkes
Bay 143320
Everything Goes, Phoenix 096208
Evetian & Beck-Arron, Doncaster 117680
Evje og Hornnes Museum, Evje . 033590
Evleen's Art Studio, Los Angeles 121480
Evnur Insaat, İstanbul 088116
Evo, Madrid 133335
Evocation, Saint-Paul-de-Vence . 105584
Evolog, Singapore 085406
Evolution – Prehistoric Art Gallery,
Singapore 085407
Evolution Art, Adelaide 098610
Evolution of Education Museum,
Prince Albert 006709
Evolutionsmuseet, Paleontologi,
Uppsala Universitet, Uppsala .. 041112
Evolutionsmuseet, Uppsala Universitet,
Uppsala 041111
Evolving Art, Chicago 120114
Evolving Art Gallery @INJA, San
Francisco 124276
Evraiko Mouseio tis Ellados,
Athinai 022987
Evro-Art, Moskva 114216
EW Gallery, San Diego 097083
Ewa Walinska Galerie, Arnhem . 112328
Ewald, Georg, Frankfurt am Main 141771
Ewald, Gustav, Würzburg 079079
Ewart Gallery AUC, Cairo 103037
Ewbank Clarke Gammon Wellers
Auctioneers, Woking 127096
Ewe, Kiel 141993
The Ewe Art Gallery & Sculpture
Garden, Glengafriff 109640
Ewe, Waltraud, Duisburg 106845
Ewelme Cottage, Auckland 033146
Ewha Yoja Taehakkyo Pakmulgwan,
Seoul 029995
Ewing, Kathleen, Washington .. 124894
Ex, Amsterdam 082742
Ex-Clou, Neuss 077747, 108251
Ex Convento de Santo Domingo de
Guzman, Oaxtepec 031338
Ex-Libris, Gdynia 143455
Ex-Libris, Oyonnax 104433
Ex Libris, Porto Alegre 142810
Ex Libris, Roma 142810
Ex Libris Books, Montréal 140362
Exbrayat Guadeloupe, Baie
Mahault 142548
Excalibur, Los Angeles 094328

Excalibur, Solcio di Lesa 081626, 110934
Excalibur Books and Comics,
Portland 145451
Excel Art, New York . . 122575, 122576
Excel Fine Art, New York 122577
Excel Fine Art Creative Custom
Framing, New York 122578
Excel Shop, Louisville 135756
Excell Blacksmith and Engineering
Workshop Museum, Tumby Bay 001596
Excelplus Antique, Singapore . . 085408
Excelsior Coin Gallery,
Sacramento 096620
Excelsior Gallery, Columbus . . . 120489
Excelsior House, Jefferson 049799
Excelsior-Lake Minnetonka Historical
Museum, Excelsior 048650
Excentury, Milano 080538
L'Exception, Poitiers 105365
Exchange, Fortaleza 100944
The Exchange, Totnes . 091621, 144886
Exchange Books, Liverpool 144545
Exchange Bookshop, Dalkey . . . 142597
Exchange Hotel Civil War Museum,
Gordonsville 049161
Exclusief Toegepaste Kunst,
Alkmaar 112103
Exclusiv Antik und Art,
Rorschach 087759
Exclusivas Europeas, Managua . 084207
Exclusive Fine Art, Burton-upon-
Trent 117489
Exclusive Fine Art, Houston 121035
Exclusive Lighting Gallery,
Milwaukee 094853, 135813
Exclusive Sanat Merkezi, İstanbul 117033
Exclusive Treasures, Las Vegas . 094137
Exe Gallery, Exeter 117770
Executive Book, Southampton . . 144854
The Executive Mansion, Frankfort 048923
Exedra Books, London 144591
Exeter Rare Books, Exeter 144418
Exhibit 231, Philadelphia 123373
Exhibit Museum of Natural History,
University of Michigan, Ann
Arbor 046488
The Exhibition Alliance, Hamilton 060473
Exhibition Center at the Artist Union,
Bakı 003161
Exhibition Center, National Taiwan
Craft Research Institute, Caotun 042274
Exhibition Gallery of the Islamic
Dáwah Centre, Bandar Seri
Begawan 004963
Exhibition Hall, Ha Noi 054792
Exhibition Hall of Agricultural
Implements, Neipu 042408
Exhibition Hall of Geology, Beijing
University, Beijing 007618
Exhibition of the Norwegian-American
Historical Association, Northfield 051631
Exhibition Studies Center,
Chicago 120115
Exhibitionist Gallery of Art, North
Perth 099474
Exhibitions @ Soap Factory,
Minneapolis 050935
Exhibitions International, Leuven 136659
Exhibits USA, Kansas City 060474
Exibart, Firenze 138824
Exile, Richmond 096542
Exit, Madrid 139049
Exit Art, New York . . . 060475, 120573
Exit Express, Madrid 139050
Exit Media, Madrid 137991
Exit To Art, Amstelveen 082665
Exit, Nowa Sztuka w Polsce,
Warszawa 138963
Exler & Co., Frankfurt am Main . 106983
Exlibris, Gdynia 143456
Exlibris, Praha 140518
Exlibris Polish Book Gallery,
Chicago 145040

voor Exlibris en andere Kleingrafiek,
Groenekan 059601
Exmoor Gallery, Dulverton 117689
Exmouth Gallery, Exmouth 117779
Exmouth Museum, Exmouth . . . 044314
Exner, R., Oberstdorf 077878
Exner, W., Wien 100251
Exodus, Naxxar 082421
Exota, Rotterdam 083568
Exotic Art, Itzstedt 076638
Exotic Art Gallery, Mumbai 109432
De Expeditie, Amsterdam 112193
Expeditionsmuseum Werner Freund,
Merzig 020380
Experegol, Oviedo 115325
Experience Children's Museum,
Erie 048598
Experience Music Project, Seattle 053233
Experimenta, Madrid 139051
Experimental Art Foundation,
Adelaide 000764, 139612
Experimentarium, Hellerup 010100
Expert Clock Repair, Houston . . 135601
Expert Restorations, Atlanta . . . 135306
Exploding Head Gallery,
Sacramento 123765
Explora Museum, Frankfurt am
Main 018415
Explora Science Center and Children's
Museum, Albuquerque 046360
Exploration Arts, Calgary 101236
Exploration House, Toronto 064643,
101606, 140406
Exploration Place, Wichita 054385
Exploration Station, A Children's
Museum, Bourbonnais 047148
Explorations, Tucson 097658
Exploratorio, Centro Cientifico
Tecnológico Interactivo, San
Isidro 000580
The Exploratorium, San Francisco 052999
Exploratorium-Kindermuseum Stuttgart
und Region, Stuttgart 022028
Explore & More . . . a Children's
Museum, East Aurora 048422
Explorers Hall, National Geographic
Society, Washington 054110
Explorion, Limburgs Science Centrum,
Heerlen 032433
Exploris, Raleigh 052448
Explorium of Lexington,
Lexington 050242
Explosion! The Museum of Naval
Firepower, Gosport 044449
Explosivos, Lugo 115125
Expo Ardos, Hoffelt 111830
Expo-Arte, Madrid 115191
Expo Haringvliet, Stellendam . . 032906
Expo-Musée Renault, Boulogne-
Billancourt 012134
Expoarte Uadi, Getxo 115065
Expodec, Madrid 115192
Expoint, Budapest 109142
Expol Arte Antiguidade, São
Paulo 064176
Expol-Sources Gallery, Taipei . . 116913
Exponatec Cologne, Internationale
Fachmesse für Museen,
Konservierung und Kulturerbe,
Köln 098138
Exponera Fotogalleri i Malmö,
Malmö 115743
Expopetro – Museo del Petróleo,
Lima 034310
Exporail, Musée Ferroviaire Canadien,
Saint-Constant 006864
Expose, Amsterdam 112194
Exposé, Orléans 136844, 138643
Exposición de Gregorio de Laferrere,
Gregorio de Laferrere 000380
Exposición Permanente Castro
Celtibéricon de el Ceremeño,
Herrería 039489
Exposición Permanente de Cerámica

Indígena, Manizales 008705
Exposición Roberto J. Mouras, Carlos
Casares 000293
Exposiciones y Salones Monograficos,
Madrid 115193
Expositie 40–45, Blitterswijk . . . 032072
Expositie Versteend Leven,
Drouwen 032256
Exposition Archéologique de la Maison
de la Cour, Nécropole Mérovingienne,
Givrand 013120
Exposition Archéologique Gallo-
Romaine, Mandeure 013986
Exposition Artistique et Viticole,
Aigues-Mortes 011547
Exposition au Château, Les
Ternes 013768
Exposition Bernardini, Obernai . 014542
Exposition Caillebotte, Yerres . . . 016359
Exposition d'Archéologie Médiévale,
Antignac 011690
Exposition d'Art et d'Histoire,
Pionsat 014874
Exposition de Coiffes et Broderies
Bretonnes, Guiclan 013211
Exposition de Hôtel de Ville,
Vevey 042105
Exposition de la Medievale, Evol 012934
Exposition de la Préhistoire, Mazières-
en-Mauges 014106
Exposition de la Tour de Castellane,
Musée de la Cave des Champagnes
de Castellane, Épernay 012875
Exposition de Patrimoine Meylanais,
Meylan 014158
Exposition de Tapisseries,
Felletin 012965
Exposition des Éditions Tarmeye,
Mazet-Saint-Voy 014105
Exposition d'Saint-Romain de la
Préhistoire, Saint-Romain . . . 015552
Exposition du Château, Cadillac-en-
Fronsadais 012272
Exposition du Château Fort du
Fleckenstein, Lembach 013727
Exposition du Cloître des Cordeliers,
Tarascon (Bouches-du-Rhône) . 015907
Exposition du Fort de l'Estissac, Port-
Cros 014964
Exposition du Fort Sainte-Agathe,
Porquerolles 014962
Exposition du Jouet en Bois, Arvieux-
en-Queyras 011754
Exposition du Mémorial Capétien,
Manoir de la Luzerne, Bernières-sur-
Mer 011996
Exposition du Moulin à Huile, La
Cadière-d'Azur 013371
Exposition du Petit Monde de Pagnol,
Musée Marcel Pagnol, Aubagne 011763
Exposition Historique, Cahors . 012286
Exposition Historique du Château, La
Roche-Maurice 013464
Exposition Historique du Monastère,
Sauxillanges 015725
Exposition Jadis, La Roche-
Derrien 013462
Exposition La Guerre de Cent Ans,
Saint-Mesmin 015465
Exposition Permanente,
Aigueperse 011544
Exposition Permanente, Thanvillé 015929
Exposition Permanente "Valence à
la Croisée des Chemirs", Valence
(Drôme) 016091
Exposition Permanente d'Archéologie,
Ainvelle 011555
Exposition Permanente d'Art,
Brasparts 012191
Exposition Permanente d'Art,
Éguilly 012859
Exposition Permanente d'Art
Contemporain, Bonnemazon . 012100
Exposition Permanente d'Arts

Plastiques, Brioude 012229
Exposition Permanente Pierre Miquel,
Maurs 014093
Exposition Permanente sur François
Pétrarque, Lombez 013856
Exposition Permanente sur la
Déportation, Charmes 012460
Exposition Permanente sur les
Églises Fortifiées de la Thiérache,
Vervins 016184
Exposition Préhistorique,
Choranche 012571
Exposition Ruchon Morin, Le
Merlerault 013662
Exposition sur la Parfumerie, Eze 012943
Exposition sur l'Art Cistercien, Les
Abbayes Cisterciennes en Limousin,
Bétête 012012
Exposition Universelle des Vins et
Spiritueux, Bandol 011883
Expositions de Chapelle Sainte-Thècle,
Seguret 015753
Expositions Pipe et Diamant, Saint-
Claude 015266
Expostion Grotte Chauvet Pont d'Arc,
Vallon-Pont-d'Arc 016107
Exposto, Maria Teresa, Lisboa . 084881
Exposure Gallery, Vancouver . . . 101781
Expozice Asijského Umění Národní
Galerie v Praze, Praha 009790
Expozice Čelákovického Košíkářství,
Čelákovice 009496
Expozice Dominikánský Klášter,
Jablonné v Podještědí 009590
Expozice Městského Vlastivědného
Mudruňkovo Muzeuma na Městském
úřadu, Říčany u Prahy 009861
Expozice Oddělení Stařších Českých
Dějin, NM – Historické muzeum,
Vrchotovy Janovice 009960
Expozice Řemesel, Moravské
Budějovice 009705
Expozícia Čičmany, Čičmany . . . 038341
Expozícia Historických Hodín Kaštiel'
v Budimíre, Slovenské Technické
Múzeum Košice, Budimír 038335
Expozícia Historických Hodín, Mestské
Múzeum v Bratislave, Bratislava 038302
Expozícia L'udovej Architektúry
a Bývania (Skanzen), Vihorlatské
Múzeum, Humenné 038356
Expozícia L'udovej Archtektúry a
Bývania, Martovce 038397
Expozícia Meštianskej Bytovej
Kultúry, Múzeum v Kežmarku,
Kežmarok 038360
Expozícia Zbraní a Stredovekého
Mestkého Opevnenia, Mestské
Múzeum v Bratislave, Bratislava 038303
Expoziţia Arheologica Tibiscum,
Caransebeş 036355
Expoziţia Curtea Domnească, Piatra-
Neamţ 036609
Expoziţia de Arheologie şi Istorie
Locală, Deta 036435
Expoziţia de Artă Feudală,
Domneşti 036441
Expoziţia de Artă Populară Iulia Folea
Troceanu, Buziaş 036338
Expoziţia de Etnografie şi Artă
Populară, Ciacova 036370
Expoziţia de Etnografie şi Artă
Populară, Pucioasa 036636
Expoziţia de Istoria Farmaciei,
Sibiu 036698
Expoziţia de Istorie a Culturii
Nămăeşti, Câmpulung 036344
Expoziţia de Istorie Locală,
Ciacova 036371
Expoziţia de Pictură Românească
Contemporană, Teremia Mare 036775
Expoziţia de Viticultură, Teremia
Mare 036776
Expoziţia Etnografică, Horezu . . 036495

Expoziţia Foto-documentară Ion
Ionescu de la Brad, Negri 036583
Expoziţia Foto-documentară Vasile
Pârvan, Huruieşti 036499
Expoziţia Hidrocentrală Porţile de Fier,
Drobeta-Turnu Severin 036451
Expoziţia Memorială Adam Müller-
Guttenbrunn, Zăbrani 036833
Expoziţia Memorială Béla Bartók,
Sânnicolau Mare 036679
Expoziţia Memorială Dimitrie Cantemir,
Dimitrie Cantemir 036439
Expoziţia Memorială Ion Borcea,
Racova 036639
Expoziţia Memorială Silviu Dragomir,
Gurasada 036489
Expoziţia Memorială Ştefan Jäger,
Jimbolia 036525
Expoziţia Memorială Traian Vuia,
Traian Vuia 036795
Expoziţia Muzeală, Turnu Roşu . 036805
Expoziţia Muzeală de Etnografie,
Cisnădie 036374
Expoziţia Muzeală I.L. Caragiale,
Ghirdoveni 036477
Expoziţia Muzeală Istoricul Industriei
Textile, Cisnădie 036375
Expoziţia Muzeală Vasile Dobrian,
Rod 036659
Expoziţia Nichita Stănescu, Teremia
Mare 036777
Expoziţia Permanentă de Artă
Plastică Contemporană, Ştefăneşti,
Botoşani 036728
Expoziţia Permanentă de Etnografie,
Pietroşiţa 036617
Expoziţia Permanentă de Istoria
Farmaciei Montanistice Bănăţene,
Oraviţa 036600
Expoziţia Teatrului, Colecţia Iosif
Sârbut, Arad 036205
Expoziţie Sătească, Stejaru 036730
Expoziţie Sătească, Lisa 036537
Expoziţie Sătească, Sericu 036551
Expresiva, San José 102488
Expression Gallery of Fine Art,
Chicago 120116
Expressions, Paris 141105
Expressions, Shrewsbury 091329
Expressions Art Gallery, Oakland 123251
Expressions Black Art Gallery,
Houston 121036
Expressions Gallery, Armadale,
Victoria 098659
Expressions in Art, Pittsburgh .. 123540
Expressions of Culture, Chicago 120117
Expressions of Soul, Bowral .. 061181
Expressive Therapies Program of
Lesley University, Graduate School of
Arts and Science, Cambridge . 056824
Exquisite Art, Tshwane 085675
Extence Antiques, Teignmouth .. 091551
Extence, Leigh C., Shaldon 091274
Exteri-Art, Hawkes Bay 112911
Extra Buch, Berlin 141509
Extrapool, Nijmegen 112606
Extrato Galeria de Arte, Porto
Alegre 100975
L'Extreme Orient, Hong Kong .. 065040
Extrêmes, Bruxelles 063416
Eyam Museum, Eyam 044316
Eybergen, Wassenaar 083692
Eye Blink Studios, Indianapolis . 121184
Eye Candy, Sheffield 119082
Eye Jammie, New York 122580
Eye Level Gallery, Halifax 005949
Eye Lounge, Phoenix 123490
Eye On the Square, New Orleans 122161
Eye Spy, Toronto 064644
Eye, Candy, Cleveland 120419
Eye, Candy, Denver 120713
Eye. The International Review of
Graphic Design, London 139176
Eyebeam, New York 051384

Eyedeal Imports, Minneapolis .. 121985
Eyeline, Kelvin Grove . 136482, 138495
Eyemouth Museum, Eyemouth . 044317
Eyeneek Youneek, Bobs Farm .. 098756
Eyes Gallery, Philadelphia 123374
Eyes of the Wild Art Gallery,
Minneapolis 121986
Eyes of the Wild Art Gallery, Saint
Paul 123904
Eyesong, Yokohama 111534
Eyestorm-Britart Gallery, Leeds . 118087
Eyestorm-Britart Gallery, London 118308
Eyestorm-Britart Gallery, Warwick 119311
EygenArt Verlag, Hofheim am
Taunus 137279
Eykelberg, Jan, Antwerpen 100487
Eylül Sanat Galerisi, İstanbul .. 117034
D'Eyncourt, Chertsey .. 088850, 134247
Eyraud, Emmanuel, Joigny
(Yonne) 068903
Eyre & Greig, London 118309
Eyrie Galleries, Mansfield 118715
Eyvind Johnson-Stugan, Boden . 040567
Ezair, New York 122581
Ezekiel Harris House, Augusta .. 046658
Ezhou Museum, Ezhou 007759
Ezra and Cecile Zilkha Gallery,
Middletown 050863
Ezra Meeker Mansion, Puyallup 052423
Ezry Gallery, Jerusalem 109716
Ёžvinskij Literaturnyj Muzej,
Syktyvkar 037907
F & C, Warszawa 113825
F & F Art, Jihlava 102598
F & W Publications, Cincinnati .. 138325
F 33, Praha 102660
F Rame's Gallery, Knokke-Heist 100737
F. & F., Perugia 081142
F.A. Antik, Michalovce 085579
F.A.N. Gallery, Philadelphia 123375
F.A.T.E. in Review, Charleston .. 139361
F.H. Restaurering, Aabenraa ... 128374
F.I.A., Dardilly 068274
F.L Arts, Villefranche-sur-Mer .. 105864
F.R. Szecesszió, Budapest 079257
F.R.I. Museums, Dehradun 024214
F.S. Antiques, Sheffield 091284, 135026
F.W.Raiffeisen Museum, Hamm,
Sieg 019028
F.Y.B. Arts, Long Beach 121398
F8 Fine Art Gallery, Austin 119669
FA Projects, London 118310
Fa. Stilke, Magdeburg 077218
Faaborg Arrest, Faaborg 010049
Faaborg Kulturhistoriske Museer,
Faaborg 010050
Faaborg Museum for Fynsk
Malerkunst, Faaborg 010051
Faba Film, Bucureşti 085093
Fabbrica Casa Museo G. Mazzotti
1903, Albissola Marina 025276
Fabbrica EOS, Milano 110310
Fabbrini, Carlo, Roma . 081349, 081350
Fabbrini, Marcello, Roma 081351
Faber & Faber, Leipzig 137342
Faber & Faber, London 138214
Faber Gotic, Barcelona 114881
Faber, Friedrich und Agi de,
Zülpich 079149, 130803
Faber, Johannes, Wien 100252
Faber, Thilo, Rudolstadt 078304
Fabert & Raoult, Meaux 070027, 128792
Fabian, Vancouver 128225
Fabian & Claude Walter Galerie,
Zürich 116771
Fabian & Samuels, Stäfa 087828,
                              133931
Fabian, F., Stuttgart 130668
Fabiani, Comacchio 109966
Fabien, Paris 071315
Fabien, Mathieu, Toulouse 105734
Fabing, Catherine, Toul 073695
Fabio & Max Antichità, Roma .. 131914
Fabio Paris Art Gallery, Brescia . 109900

Fabio's Upholstery, Los Angeles 135716
Fabius Frères, Paris .. 071316, 104733,
                        141106
Fabjbasaglia, Rimini 110715
Fables, South Fremantle 139910
Fables Antique, Auckland 083786,
                            132687
Fables Antiques, Los Angeles .. 094329
Fabre & Fils, Paris 071317
Fabre, Etienne, Peypin 129168
Fabre, Nicole, London 090106
Fabre, Régis, Rochemaure 072505
Fabri Antiquariat, Ulm 142442
The Fabric Workshop and Museum,
Philadelphia 052047
Fábrica, Asunción 113330
Fabrica, Brighton 043737
La Fabrica, Madrid 115194
Fabrica – Benetton's Communication
Research Center, Catena di
Villorba 059436
Fabricarestauri, Torino 132137
Fabricius-Ház, Soproni Múzeum,
Sopron 023890
Fabrik Fotoforum, Hamburg 058852,
                            107240
Fabrik.Galerie für Gegenwärtige Kunst,
Neustrelitz 108266
Fabrikmuseum der Leonischen
Industrie, Roth 021501
Fabrile, Armadale, Victoria ... 061023
Fabrile, Chicago 120118
Fabris, Giulia, Venezia 081836
Fabrizi, Lionel, Charleville-
Mézières 067846
Fabry, Paris 128999
Fabry, Forence, Talence 073614
Fabry, Olivier, Bruxelles 063417, 128027
Fabsits, Stinatz 062855
Fabula, Haren 141922
Fabulous Finds, Phoenix 096209
Fabulous Finds, Toronto 064645
Fabulous Frames and Art,
Cincinnati ... 120341, 120342, 120343
Fabyan Villa Museum and Dutch
Windmill, Geneva 049066
FAC, Kassel 076719
The Facade, London 090107
Daniela Facchinato Image Gallery,
Bologna 109821
Facchinetti, Brescia 079937
Facchini, Silvana, Miami 121778
Faccioli, Andrea, Verona 132280
Face aux Tilleuls, Gavray 068640
The Face Gallery, Sassari 110911
Face Painting Culture Museum,
Sanyi 042440
Facere, Seattle 097433
Facet Books, Bournemouth 144254
Fach, Sanem 087799
Fach, Joseph, Frankfurt am Main 075779,
                106984,   141772
Fachausstellung Arbeitsschutz/
Sicherheitstechnik, München . 020522
Fachbereich 01 Architektur, Hochschule
München, München 055647
Fachbereich 06: Architektur, Stadt-
und Landschaftsplanung, Universität
Kassel, Kassel 055606
Fachbereich 1: Architektur und
Innenarchitektur, Hochschule
Ostwestfalen-Lippe, Detmold .. 055510
Fachbereich 12 Design, Hochschule
München, München 055648
Fachbereich 15: Architektur,
Technische Universität Darmstadt,
Darmstadt 055505
Fachbereich 4: Fotografie, Industrial
Design, Kommunikationsdesign,
Folkwang Hochschule, Essen . 055531
Fachbereich Angewandte Kunst
Schneeberg, Westsächsische
Hochschule Zwickau (FH),

Schneeberg 055703
Fachbereich Architektur – Städtebau,
Universität Siegen, Siegen ... 055709
Fachbereich Architektur (AR), Georg-
Simon-Ohm-Fachhochschule
Nürnberg, Nürnberg 055669
Fachbereich Architektur
Hildesheim, Fachhochschule
Hildesheim/Holzminden/Göttingen,
Hildesheim 055586
Fachbereich Architektur Oldenburg,
Fachhochschule Oldenburg/
Ostfriesland/ Wilhelmshaven,
Oldenburg 055677
Fachbereich Architektur und
Bauingenieurwesen, Fachhochschule
Augsburg, Augsburg 055446
Fachbereich Architektur und
Bauingenieurwesen, Fachhochschule
Bielefeld, Minden 055642
Fachbereich Architektur und
Bauingenieurwesen, Fachhochschule
Wiesbaden, Wiesbaden 055739
Fachbereich Architektur und
Bauingenieurwesen,
Würzburg-Schweinfurt-Aschaffenburg,
Abteilung Würzburg, Würzburg 055742
Fachbereich Architektur und
Bauingenieurwesen, Fakultät Design,
Coburg 055501
Fachbereich Architektur und
Bauwesen (AB), Hochschule
Karlsruhe, Karlsruhe 055597
Fachbereich Architektur und
Gebäudetechnik, Beuth Hochschule
für Technik Berlin, Berlin 055460
Fachbereich Architektur und
Gestaltung, Fachhochschule
Konstanz, Konstanz 055620
Fachbereich Architektur und
Stadtplanung, Fachhochschule
Koblenz, Koblenz 055611
Fachbereich Architektur und
Städtebau, Fachgebiet Restaurierung,
Potsdam 055690
Fachbereich Architektur,
Bergische Universität Wuppertal,
Wuppertal 055745
Fachbereich Architektur,
Fachhochschule Aachen,
Aachen 055438
Fachbereich Architektur,
Fachhochschule Biberach, Biberach
an der Riß 055483
Fachbereich Architektur,
Fachhochschule Bochum,
Bochum 055486
Fachbereich Architektur,
Fachhochschule Dortmund,
Dortmund 055511
Fachbereich Architektur,
Fachhochschule Düsseldorf,
Düsseldorf 055520
Fachbereich Architektur,
Fachhochschule Frankfurt am Main,
Frankfurt am Main 055537
Fachbereich Architektur,
Fachhochschule Hannover,
Nienburg 055667
Fachbereich Architektur,
Fachhochschule Münster,
Münster 055663
Fachbereich Architektur,
Fachhochschule
Nordostniedersachsen,
Buxtehude 055500
Fachbereich Architektur,
Fachhochschule Regensburg,
Regensburg 055693
Fachbereich Architektur, Facility
Management und Geoinformation,
Fachhochschule Anhalt, Abteilung
Dessau, Dessau 055508
Fachbereich Architektur, Hochschule

Fachbereich Architektur, Hochschule für bildende Künste –
Faculty of Art and Design, Swansea Metropolitan University,

Index of Institutions and Companies

Darmstadt, Darmstadt 055506
Fachbereich Architektur, Hochschule
für bildende Künste Hamburg,
Hamburg 055565
Fachbereich Architektur, Hochschule
für Technik und Wirtschaft des
Saarlandes, Saarbrücken 055700
Fachbereich Architektur, SRH
Hochschule Heidelberg,
Heidelberg 055581
Fachbereich Architektur,
Westsächsische Hochschule Zwickau,
Bereich Reichenbach, Reichenbach
im Vogtland 055697
Fachbereich Architektur/Raum- und
Umweltplanung/Bauingenieurwesen,
Technische Universität Kaiserslautern,
Kaiserslautern 055595
Fachbereich Bauen und Gestalten,
Fachhochschule Kaiserslautern,
Kaiserslautern 055596
Fachbereich Bauingenieurwesen/
Architektur, Hochschule für
Technik und Wirtschaft Dresden,
Dresden 055514
Fachbereich Bildhauerei, Alanus
Hochschule, Alfter 055442
Fachbereich Design- und Kunst,
Bergische Universität Wuppertal,
Wuppertal 055746
Fachbereich Design (D), Georg-Simon-
Ohm-Fachhochschule Nürnberg,
Nürnberg 055670
Fachbereich Design und Medien,
Fachhochschule Hannover,
Hannover 055577
Fachbereich Design, Fachhochschule
Anhalt, Abteilung Dessau,
Dessau 055509
Fachbereich Design, Fachhochschule
Dortmund, Dortmund 055512
Fachbereich Design, Fachhochschule
Düsseldorf, Düsseldorf 055521
Fachbereich Design, Fachhochschule
Münster, Münster 055664
Fachbereich Design, Fachhochschule
Potsdam, Potsdam 055691
Fachbereich Design, Hochschule
für bildende Künste Hamburg,
Hamburg 055566
Fachbereich Design, Hochschule
Niederrhein, Krefeld 055621
Fachbereich Design, Informatik,
Medien, Fachhochschule Wiesbaden,
Wiesbaden 055740
Fachbereich Freie Kunst, Hochschule
für bildende Künste Hamburg,
Hamburg 055567
Fachbereich Gestaltung und Kunst,
Hochschule der Künste Bern,
Bern 056416
Fachbereich Gestaltung, Architektur,
Design und Innenarchitektur,
Wismar 055741
Fachbereich Gestaltung,
Fachhochschule Aachen,
Aachen 055439
Fachbereich Gestaltung,
Fachhochschule Bielefeld,
Bielefeld 055484
Fachbereich Gestaltung,
Fachhochschule für Technik und
Wirtschaft Berlin, Berlin 055461
Fachbereich Gestaltung,
Fachhochschule Mainz, Mainz . 055633
Fachbereich Gestaltung,
Fachhochschule Würzburg-
Schweinfurt, Abteilung Würzburg,
Würzburg 055743
Fachbereich Gestaltung, Hochschule
Darmstadt, Darmstadt 055507
Fachbereich Gestaltung, Hochschule
für Technik und Wirtschaft Dresden
(FH), Dresden 055515

Fachbereich Gestaltung, Studiengang
Edelstein- und Schmuckdesign,
Fachhochschule Trier, Idar-
Oberstein 055593
Fachbereich Innenarchitektur und
Integriertes Produktdesign, Fakultät
Design, Coburg 055502
Fachbereich Konservierung und
Restaurierung, Hochschule der
Künste Bern, Bern 056417
Fachbereich Künstlerische Therapie,
Alanus Hochschule, Alfter 055443
Fachbereich Kunst- und
Musikpädagogik, Universität Siegen,
Siegen 055710
Fachbereich Kunst und Design,
Hochschule für Künste Bremen,
Bremen 055497
Fachbereich Kunst und Design,
Universität Duisburg-Essen,
Essen 055532
Fachbereich Kunst, Kunstakademie
Düsseldorf, Düsseldorf 055522
Fachbereich Kunstbezogene
Wissenschaften, Kunstakademie
Düsseldorf, Düsseldorf 055523
Fachbereich Kunsterziehung,
Institut für Kunst und Musik,
Heidelberg 055582
Fachbereich Kunstgeschichte,
Universität Trier, Trier 055730
Fachbereich Kunstpädagogik
und Techniklehre, Hochschule
für bildende Künste Hamburg,
Hamburg 055568
Fachbereich Malerei, Alanus
Hochschule, Alfter 055444
Fachbereich Mode und
Textilwissenschaft, Institut für
Alltags- und Bewegungskultur,
Heidelberg 055583
Fachbereich Technik: Fachrichtung
Architektur, Fachhochschule Mainz,
Mainz 055634
Fachbereich Textil und Design,
Fachhochschule Reutlingen,
Reutlingen 055698
FachbereichTextil und Design,
Fachhochschule Hof,
Münchberg 055643
Fachgebiet Grafik und Malerei,
Philipps-Universität, Marburg . 055638
Fachgebiet Kunst, Universität Erfurt,
Erfurt 055527
Fachgruppe Allgemeine Künstlerische
Ausbildung und Werken, Staatliche
Akademie der Bildenden Künste,
Stuttgart 055713
Fachgruppe Architektur, Staatliche
Akademie der Bildenden Künste,
Stuttgart 055714
Fachgruppe Design, Staatliche
Akademie der Bildenden Künste,
Stuttgart 055715
Fachgruppe Kunst, Staatliche
Akademie der Bildenden Künste,
Stuttgart 055716
Fachgruppe Wissenschaft, Staatliche
Akademie der Bildenden Künste,
Stuttgart 055717
Fachhochschul-Studiengang
Medientechnik und -design,
Hagenberg 054938
Fachhochschul-Studiengang
MultiMediaArt (MMA), Puch/
Salzburg 054960
Fachhochschule für Gestaltung
Schwäbisch Gmünd, Schwäbisch
Gmünd 055704
Fachhochschule für Kunsttherapie
Nürtingen, Nürtingen 055672
Fachhochschule für Kunsttherapie und
Kunstpädagogik und Freie Bildende
Kunst, Ottersberg 055683

Fachhochschule Joanneum –
Ausstellungs- und Museumsdesign,
Graz 054927
Fachhochschule Joanneum –
Industrial Design, Graz 054928
Fachhochschule Joanneum –
Informationsdesign, Graz 054929
Fachhochschule Kunst Arnstadt,
Arnstadt 055445
Fachings- und Brauchtumsmuseum,
Knittelfeld 002213
Fachner, Sigrid, Lippstadt 142079
Fachrichtung Kunstgeschichte,
Universität des Saarlandes,
Saarbrücken 055701
Fachschule des Bezirks Oberbayern
für Schreiner und Holzbildhauer mit
Fachakademie für Holzgestaltung,
Garmisch-Partenkirchen 055546
Fachschule für Bildhauerei, HTL
Bau und Kunst – Innsbruck,
Innsbruck 054940
Fachschule für Malerei, HTL Bau und
Kunst – Innsbruck, Innsbruck . 054941
Das Fachwerk, Bad Salzuflen .. 016860
Fachwerkbaumuseum im Ständerbau,
Städtische Museen Quedlinburg,
Quedlinburg 021243
Fackler, Raimund, Traunstein .. 078693
Facoltà del Design, Politecnico di
Milano, Milano 055923
Facoltà di Design e Arti, Università
IUAV di Venezia, Venezia 055990
Facsimile Editions, London .. 144592
Facsimilia Art & Edition,
Darmstadt 106654, 137135
Factoría Del Barco, Sevilla .. 115448
Factory, Livorno 110178
Factory Fine Art, Modena 110497
Factory04, Padova 110563
Facts and Fiction, Rosebud 139897
Faculdade de Arquitectura, Artes e
Communicação, Universidade
Estadual Paulista Júlio de Mesquita
Filho, Bauru 055067
Faculdade de Artes do Paraná,
Curitiba 055073
Faculdade de Artes e Comunicação,
Universidade de Passo Fundo, Passo
Fundo 055086
Faculdade de Belas Artes da
Universidade do Porto, Porto .. 056252
Faculdade de Belas-Artes,
Universidade de Lisboa, Lisboa 056249
Faculdade de Comunicação e
Artes, Universidade Presbiteriana
Mackenzie, São Paulo 055104
Faculdade Santa Marcelina, Campus II,
Itaguera 055080
Facultad de Arquitectura y
Artes, Universidad Nacional
Pedro Henríquez Ureña, Santo
Domingo 055320
Facultad de Arquitectura, Universidad
de la República Uruguay,
Montevideo 056619
Facultad de Arquitectura, Urbanismo y
Diseño, Córdoba 054870
Facultad de Arte, Sección Artista
Visual, Universidad Mayor, Santiago
de Chile 055185
Facultad de Artes Plasticas,
Universidad Autónoma Tomás Frías,
Potosí 055063
Facultad de Artes Plásticas,
Universidad Veracruzana,
Xalapa 056159
Facultad de Artes y Diseño,
Mendoza 054873
Facultad de Artes, Area Artes
Plasticas, Universidad el Bosque,
Bogotá 055263
Facultad de Artes, Pontificia
Universidad Javeriana, Bogotá 055261

Facultad de Artes, Universidad
Autónoma del Estado de Morelos,
Cuernavaca 056138
Facultad de Artes, Universidad Central
del Ecquador, Quito 055324
Facultad de Artes, Universidad
Nacional de Colombia, Bogotá 055262
Facultad de Artes, Universidad
Nacional de Tucumán, Tucumán 054879
Facultad de Bellas Artes de Cuenca,
Universidad de Castilla, La Mancha,
Cuenca 056362
Facultad de Bellas Artes, Universidad
del Atlántico, Barranquilla 055260
Facultad de Bellas Artes, Universidad
Nacional de La Plata, La Plata . 054871
Facultade de Arquitectura, Universidad
Autónoma de Aguascalientes,
Aguascalientes 056137
Facultade de Artes Plásticas e Artes
Visuais, Universidade Federal de
Goiás Área de Lingüística, Letras e
Artes, Goiânia 055077
Facultade de Artes Plásticas,
Universidade da Região da
Campanha, Bagé 055066
Facultade de Artes y Humanidades,
Universidad del Noreste,
Tampico 056157
Facultade de Bellas Artes, Universidad
Complutense, Madrid 056372
Facultade de Bellas Artes, Universidad
de Panama, Ciudad de Panamá 056217
Facultade de Santa Marcelina,
Campus I, Perdizes 055088
Facultas Dombuchhandlung,
Melk 136536, 140017
Facultas-Dombuchhandlung,
Wien 140073
Facultat de Belles Arts, Universidad de
Barcelona, Barcelona 056359
Faculté de Beaux Art, Université de
Yaoundé I, Yaoundé 055114
Faculté des Arts, Université du
Québec à Montréal, Montréal . 055142
Faculté des Beaux-Arts et des Arts
Appliqués, Université Saint Esprit de
Kaslik, Juniyah 056119
Faculty Arts and Design, Thames
Valley University, Harrow 056538
Faculty Arts and Design, University of
East London, Croydon 056524
Faculty Arts and Design, University of
Greenwich, London 056559
Faculty Arts and Design, University of
Portsmourh, Portsmouth 056597
Faculty Arts, Design and Technology,
University of Derby, Derby 056525
Faculty of Applied and Creative Arts,
Universiti Malaysia Sarawak, Kota
Samarahan 056127
Faculty of Applied Arts and Sciences,
German University in Cairo, Al-
Khames 055325
Faculty of Applied Arts, 6th of October
University, Cairo 055330
Faculty of Architecture, Bangkok 056462
Faculty of Architecture and Arts,
University of Petra, Amman ... 056068
Faculty of Architecture and Planning,
Dalhousie University, Halifax .. 055130
Faculty of Architecture, Landscape
and Design, University of Toronto,
Toronto 055159
Faculty of Architecture, University of
Manitoba, Winnipeg 055175
Faculty of Art and Architecture, Shiraz
University, Shiraz 055851
Faculty of Art and Design,
Manchester Metropolitan University,
Manchester 056578
Faculty of Art and Design,
Swansea Metropolitan University,
Swansea 056611

Faculty of Art and Design, Universiti
Teknologi Mara, Shah Alam .. 056132
Faculty of Art, Alzahra University,
Tehrän 055853
Faculty of Art, Design and
Architecture, Technikon
Witwatersrand, Doornfontein .. 056343
Faculty of Arts & Education,
Benghazi 056121
Faculty of Arts and Architecture,
University of Brighton, Brighton 056513
Faculty of Arts and Communication,
Whitireia Community Polytechnic,
Porirua 056196
Faculty of Arts and Design, Dept. of
Visual Arts, University of Jordan,
Amman 056071
Faculty of Arts and Human Sciences,
Dept. of Graphic Design, Al-Ahliyya
Amman University, Amman ... 056072
Faculty of Arts and Humanities,
Jeddah 056317
Faculty of Arts and Letters, University
of Santo Tomas, Manila 056227
Faculty of Arts and Sciences,
American University of Beirut,
Beirut 056114
Faculty of Arts and Sciences, Edge Hill
University, Ormskirk 056591
Faculty of Arts Media and Social
Sciences, University of Salford,
Manchester 056579
Faculty of Arts, Ain Shams University,
Cairo 055331
Faculty of Arts, Alexandria University,
Alexandria 055326
Faculty of Arts, Annamalai University,
Annamalai Nagar 055777
Faculty of Arts, Assiut University,
Assiut 055327
Faculty of Arts, Australian National
University, Canberra 054902
Faculty of Arts, Cairo University,
Cairo 055332
Faculty of Arts, Department of Art
History, The University of Auckland,
Auckland 056186
Faculty of Arts, Dept. of Design,
Philadelphia University, Amman 056069
Faculty of Arts, Dept. of Fine Arts,
Applied Science Private University,
Amman 056070
Faculty of Arts, Design and
Media, University of Sunderland,
Sunderland 056610
Faculty of Arts, Dr. Ram Manohar
Lohia Avadh University,
Faizabad 055795
Faculty of Arts, Durban University of
Technology, Durban 056344
Faculty of Arts, Humanities and Social
Sciences, University of Central
Lancashire, Preston 056598
Faculty of Arts, Islamic University of
Omdurman, Omdurman 056394
Faculty of Arts, Makerere University,
Kampala 056481
Faculty of Arts, Media and Design,
Staffordshire University, Stoke on
Trent 056609
Faculty of Arts, Minufiya Univeristy,
Minufiya 055337
Faculty of Arts, Tshwane University of
Technology, Tshwane 056354
Faculty of Arts, University of Glasgow,
Glasgow 056535
Faculty of Arts, University of Khartoum,
Khartoum 056393
Faculty of Arts, University of Liverpool,
Liverpool 056549
Faculty of Arts, University of
Peradeniya, Peradeniya 056391
Faculty of Arts, University of Plymouth,
Plymouth 056595

Faculty of Arts, University of
Strathclyde, Glasgow 056536
Faculty of Arts, University of
Zimbabwe, Harare 058162
Faculty of Arty, Dept. of Creative
Art, University of Maiduguri,
Maduguri 056203
Faculty of Communication and Arts,
Ningbo University, Ningbo 055227
Faculty of Creative Arts, Technologies
and Science, University of
Bedfordshire, Luton 056575
Faculty of Decorative Arts, Silpakorn
University, Bangkok 056463
Faculty of Design and Visual Arts,
Cape Peninsula University of
Technology, Cape Town 056340
Faculty of Design, University of Essex,
Chelmsford 056519
Faculty of Fine and Applied
Arts, Chulalongkorn University,
Bangkok 056464
Faculty of Fine and Applied Arts,
University of Benin, Benin 056202
Faculty of Fine Art, Helwan University,
Cairo 055333
Faculty of Fine Art, Music and Home
Science, University of Calcutta,
Kolkata 055811
Faculty of Fine Arts, Chittagong 054989
Faculty of Fine Arts, Jaipur,
Rajasthan 055804
Faculty of Fine Arts and Architecture,
University of Art, Isfahan,
Isfahan 055849
Faculty of Fine Arts and Crafts, Bihar
University, Muzaffarpur 055823
Faculty of Fine Arts and Social
Sciences, Universiti Malaya, Kuala
Lumpur 056128
Faculty of Fine Arts, Aristoteleio
Panepistimio Thessalonikis,
Thessaloniki 055755
Faculty of Fine Arts, Chiang Mai
University, Chiang Mai 056469
Faculty of Fine Arts, Concordia
University, Montréal 055143
Faculty of Fine Arts, Damascus
University, Damascus 056450
Faculty of Fine Arts, Doto Daigaku,
Kitahiroshima 056007
Faculty of Fine Arts, Indira Kala Sangit
Vishwavidyalaya, Khairagarh .. 055806
Faculty of Fine Arts, Kyushu Sangyo
Daigaku, Fukuoka 055999
Faculty of Fine Arts, Maharaja
Sayaji Rao University of Baroda,
Vadodara 055844
Faculty of Fine Arts, Maharishi
Dayanand University, Rohtak .. 055832
Faculty of Fine Arts, Minia University,
El-Minia 055335
Faculty of Fine Arts, Momoyama
Gakuin Daigaku, Izumi, Osaka . 056003
Faculty of Fine Arts, Srinakharinwirot
University, Bangkok 056465
Faculty of Fine Arts, Universitas
Polonorum, München 055649
Faculty of Fine Arts, University of
Travoncore, Triruvananthapuram 055843
Faculty of Fine Arts, Yarmouk
University, Irbid 056073
Faculty of Media, Art and
Communications, University of
Gloucestershire, Cheltenham .. 056520
Faculty of Media, Arts and Society,
Southampton Solent University,
Southampton 056608
Faculty of Music and Art, Daugavpils
University, Daugavpils 056107
Faculty of Music and Fine Arts, Guru
Nanak Dev University College,
Amritsar 055775
Faculty of Music and Fine Arts,

University of Jammu, Jammu . 055805
Faculty of Painting, Sculpture and
Graphic Art, Bangkok 056466
Faculty of the Arts, Thames Valley
University, Reading 056600
Faculty of the Arts, University of
Cumbria, Carlisle 056518
Faculty of Visual Arts, Banaras Hindu
University, Varanasi 055845
Fad, Barcelona 059862
FAD Gallery, Melbourne 099304
Faded Fancy, Taranaki 084091
Faded Fines Antiques, Los
Angeles 094330
Il Fado, Bologna 130955, 130956
Fadó, Dublin 079551
Fächerstadtantiquariat, Karlsruhe 141970
Fädernegården i Rejpelt, Vörå . 011519
Fälschermuseum, Binz,
Ostseebad 017391
Fälschermuseum, Wien 002988
Fængselshistoriske Museum,
Horsens 010140
Faerber, Ingo, Ahrensburg 105965
Färbermuseum, Gutau 002027
Färgargården, Norrköpings
Stadsmuseum, Norrköping ... 040876
Färgfabriken, Stockholm 040999
Fässler & Söhne, Hermann,
Appenzell 087193, 133620
Fästningsmuseet Karlsborg,
Karlsborg 040732
Fäfängans Antik, Göteborg 086705
Fafouille & Grandchamps,
Charleroi 140215
Fagan, J & J, Dublin 130854
Fågelmuseet, Jönköping 040717
Fågelsjö Gammelgård Bortom Åa,
Los 040799
Faggionato Fine Art, London ... 118311
Faggruppe for Kunstvitenskap,
Tromsø 056214
Fagin, Crows Nest, New South
Wales 061375
Fagin n Books, Chicago 145041
Fagins Antiques, Exeter 089230
Fagioli, Donatella, Roma 131915
Fagus-Gropius-Ausstellung,
Alfeld 016502
FAH Galerie, Maastricht 112572
Faherty, Arty, Market Drayton .. 090709
Fahey & Klein, Los Angeles ... 121481
Fahl Kro, Hemmet 010110
Fahl, J. & Silvia, Freiburg im
Breisgau 141802
Fahlnaes, Göteborg 086706
Fahn, Jay, Chicago 120119
Fahnemann, Clemens, Berlin ... 106222
Fahr, Roland, Kranichfeld 130189
Fahrbach, Georg, Köln 076861
Fahrenheit, Denver 145125
Fahrenheit 451, Tilburg 143233
Fahrenson, Monika, München .. 077495
Fahrngruber Hammer mit
Köhlereimuseum, Ybbsitz 003132
Fahrni, Lünen 130257
Fahrradmuseum, Ybbs 003131
Fahrradmuseum der Fahrrad-
Veteranen-Freunde-Dresden 1990,
Dresden 017980
Fahrradmuseum Zumhaus
und Brauereimuseum,
Feuchtwangen 018356
Fahrzeugmuseum, Frankenberg,
Sachsen 018389
Fahrzeugmuseum Marxzell,
Marxzell 020311
Fahrzeugmuseum Suhl, Suhl ... 022075
Fahtz, H., Dessau 075335
Fahua Ta Tayuan Museum,
Shanghai 008225
Fahy, Gerard, Montenotte ... 130882
Fai Kee, Hong Kong 065041
Faïence d'Antan, Cloyes-sur-le-

Loir 068044
Faïencerie d'Art, Desvres 068289
Faïences et Emaux, Longwy .. 013863
Failer, Neuwied 108268
Fain, Eliane, Perpignan 072078
Fainting Couch, Los Angeles .. 094331
Fair for Conservation and Restauration
of Art and Architecture, Paris . 098078
Fair, Anthony, New York 095384
Fairbanks Arts Association,
Fairbanks 060476
Fairbanks Community Museum,
Fairbanks 048660
Fairbanks Gallery, Corvallis 048017
Fairbanks House, Dedham 048210
Fairbanks Museum and Planetarium,
Saint Johnsbury 052777
Fairbanks, H., Nashville 094999, 135846
Fairbridge Chapel Museum, Mill
Bay 006312
Fairburn Antique Mall, Atlanta .. 092131
Fairchild, Calistoga 127152
Fairfax Gallery, Burnham Market 117482
Fairfax Gallery, Jacksonville ... 121249
Fairfax Gallery, Tunbridge Wells . 119270
Fairfax House Museum, York ... 046223
Fairfax Museum and Visitor Center,
Fairfax 048665
Fairfax, Harriet, Langley Burrell . 089746
Fairfield Art Association, Fairfield 060477
Fairfield Art Museum, Fairfield . 048672
Fairfield Center for Contemporary Art,
Sturgeon Bay 053582
Fairfield City Museum and Gallery,
Smithfield 001499
Fairfield County Museum,
Winnsboro 054481
Fairfield Museum, Bothwell 005443
Fairfields Manor, Frampton
Cotterell 117811
Fairhurst Gallery, Norwich 090874,
134892
Fairlawn Mansion and Museum,
Superior 053612
Fairley, Malcolm, London 118312
Fairlie Antique Shop, Fairlie ... 089246
Fairlook, Seattle 097434, 145581
Fairlynch Museum, Budleigh
Salterton 043783
Fairmarket Antique Centre,
Lismore 061742
Fairmount Books, Markham ... 136675
Fairmount Park Art Association,
Philadelphia 060478
Fairport Historical Museum, Perinton
Historical Society, Fairport 048680
Fairport Marine Museum, Fairport
Harbor 048681
Fairview Museum of History and Art,
Fairview 048682
Fairview R.C.M.P. Centennial Museum,
Fairview 005799
Fairymead House Sugar Museum,
Bundaberg 000910
Fais Moi une Place dans ta Maison,
Biarritz 067198
Faisal, Al-Khobar 085292
Faisan, Strasbourg 105686
Faison, Kim, Richmond 096543
Faisons un Rêve, Bruxelles ... 063418
Le Fait de Savoir Cyber, Saint-
Martin 142550
Fait et Cause, Paris 104734
Faith Trumbull Chapter Museum,
Norwich 051647
Faithgrant, Victoria 064900
Faivre-Andrieux, M.-H., Paris .. 071318
Faivre-Reuille, Bruno, Paris 071319
Fakenham Antique Centre,
Fakenham 089247, 134381
Fakenham Museum of Gas and Local
History, Fakenham 044320
Fakkers, Marianne, Alkmaar ... 112104
Faksimile Verlag, Luzern 138071

Faktor-Verein zur Förderung künstlerischer Fotografie und –
Farmaceutická Expozícia, Mestské Múzeum v Bratislave,

Index of Institutions and Companies

Faktor-Verein zur Förderung künstlerischer Fotografie und artverwandter Künste, Graz ... 058250
Fakultät A: Architektur und Gestaltung, Hochschule für Technik, Stuttgart 055718
Fakultät Architektur, Bau und Umwelt, Hochschule Bremen, Bremen . 055498
Fakultät Architektur, Bauhaus-Universität Weimar, Weimar .. 055736
Fakultät Architektur, Bauingenieurwesen und Stadtplanung, Brandenburgische Technische Universität Cottbus, Cottbus 055503
Fakultät Architektur, Fachhochschule Erfurt, Erfurt 055528
Fakultät Bauingenieurwesen und Konservierung/Restaurierung, Fachhochschule Erfurt, Erfurt . 055529
Fakultät Bildende Kunst, Universität der Künste Berlin, Berlin 055462
Fakultät der Freien Künste, Freie Universität Teufen, Teufen 056440
Fakultät DMI, Studiendepartment Design, Hochschule für angewandte Wissenschaften, Hamburg 055569
Fakultät Erhaltung von Kulturgut, Fachhochschule Hildesheim/Holzminden/Göttingen, Hildesheim 055587
Fakultät für Architektur, Fachhochschule Köln, Köln .. 055613
Fakultät für Architektur, Universität Karlsruhe (TH), Karlsruhe ... 055598
Fakultät für Gestaltung, Hochschule Mannheim, Mannheim 055636
Fakultät für Gestaltung, Hochschule Pforzheim, Pforzheim 055688
Fakultät für Kulturwissenschaft-Institut Kunst, Musik, Gestaltung, Universität Paderborn, Paderborn 055685
Fakultät für Linguistik & Literaturwissenschaft – Abteilung Kunst & Musik, Universität Bielefeld, Bielefeld 055485
Fakultät Gestaltung, Bauhaus-Universität Weimar, Weimar .. 055737
Fakultät Gestaltung, Fachhochschule Augsburg, Augsburg 055447
Fakultät Gestaltung, Fachhochschule Hildesheim/Holzminden/Göttingen, Hildesheim 055588
Fakultät Gestaltung, Universität der Künste Berlin, Berlin 055463
Fakultas Seni Rupa dan Desain, Universitas Trisakti, Jakarta . 055847
Fakultet Architektury, Iskusstv i Stroitelstva, Novgorodskij Gosudarstvennyj Universitet im. Jaroslava Mudrogo, Velikij Novgorod 056311
Fakultet Chudožestvennogo Obrazovanija, Voronežskij Gosudarstvennyj Pedagogičeskij Universitet, Voronež 056316
Fakultet Iskusstv, Moskovskij Gosudarstvennyj Universitet im. M.V. Lomonosova, Moskva 056284
Fakultet Iskusstv, Omskij Gosudarstvennyj Pedagogičeskij Universitet, Omsk 056293
Fakultet Iskusstva i Dizajna, Severo-Ossetinskij gosudarstvennyj universitet im K.L. Chetagurova, Vladikavkaz 056312
Fakultet Iskusstva, Altajskij Gosudarstvennyj Universitet, Barnaul 056265
Fakultet Iskusstvovedenija i Kulturologii, Uralskij Gosudarstvennyj

Universitet, Ekaterinburg 056270
Fakultet Izobrazitelnogo Iskusstva, Rossijskij Gosudarstvennyj Pedagogičeskij Universitet im. A.I. Gercena, Sankt-Peterburg ... 056300
Fakultet Kultury i Iskusstv, Uljanovskij Gosudarstvennyj Universitet, Uljanovsk 056310
Fakultet Kultury i Iskusstva, Omskij Gosudarstvennyj Universitet im. F.M. Dostevskogo, Omsk 056294
Fakultet Likovnih Umjetnosti, Univerzitet Crne Gore, Cetinje . 056161
Fakultet Tehnologii i Dizajna, Vjatskij Gosudarstvennyj Gumanitarnyj Universitet, Kirov 056278
Fakultet za Likovni Umetnosti, Univerzitet Sv. Kiril i Metodij, Skopje 056126
Fakulti Rekabentuk dan Senibina, Universiti Putra Malaysia, Serdang 056131
Falabrègues, Pascal, Rodez 125790
Falanga, Roma 081352
Falanga Arte Antica, Napoli 080878
Falbygdens Antikvariat, Falköping 143833
Falbygdens Museum, Falköping . 040617
Falce, Philippe, Beaune 067021
Falch, Hermann, Sankt Gallen . 133903
Falchini, Simone, Firenze 080136
Falcieri, Armando, Milano 080539
Falcioni & Figlio, Moreno, Firenze 131162
Falck, Genève 116299
Falco, Breda 082982
Falcón, Buenos Aires 060899
The Falcon Gallery, Wortham .. 091951, 135261
Falcon, Mercedes, Jerez de la Frontera 086057, 115103
Falcone, San Jose 097364
Falconer Museum, Forres 044362
Falconi, Bellagio 109786
Falconnet, Pierre, Menton 070060, 140952
Faleide Skulemuseum, Stryn 033998
Falger-Museum, Elbigenalp 001865
Falia Art, Toronto 101607
Faligowska, Teresa, Wrocław ... 113920
Falinki, Alain, Orvault 128879
Falk & Falk, Zürich ... 087957, 126765, 144173
Falk Art Reference, New York .. 138392
Falk, Lars, Frederikshavn 102808
Falk, Ragnar, Lund 133547
Falke, Renate, Bleiburg 099962
Falkenberg, Petra, Düsseldorf .. 075466, 129739
Falkenbergs Antikvariat, Falkenberg 143832
Falkenbergs Museum, Falkenberg 040613
Falkenstein, Dr. Klaus, Dinslaken 141689
Falkland Palace and Garden, Falkland 044324
Falkner, Castlemaine 098875
Falknereimuseum Burg Hohenwerten, Werfen 002924
Falks Kuriosagård, Laholm 086797
Fall River County Historical Museum, Hot Springs 049565
Fall River Historical Society Museum, Fall River 048687
Falla Campos, Maria Eugenia, Madrid 086182
Fallani & Sons, Long Beach 135693
Fallani Best, Firenze 110047
Fallani, Alessandro, Firenze 142671
Fallani, Carlo Maria, Roma 081353
Falle, John, Saint Helier 091182
The Fallen Angels, Gent 063636, 140222
Faller, Dieter, Hannover 076370
Faller, Dieter, Riehen 133894
Falleur, Cambrai 103496
Fallingwater – Western Pennsylvania

Conservancy, Mill Run 050895
Fallourd, Claude, Saint-Vincent-Sterlanges 073333
Falls Gallery, Wentworth Falls .. 099875
Falls Road Antiques, Baltimore . 092320
Fallwell, Marshall, Nashville ... 145295
Falmouth Art Gallery, Falmouth . 044325
Falmouth Historical Museum, Falmouth 048693
Falon, Las Vegas 121349
Falret, Jérôme, Planioles 072175
Falsi d'Autore, Firenze 131163
Falstadminnet, The Falstad Memorial, Levanger 033739
Falstaff Antiques, Rolvenden ... 091132
Falsterbo Museum, Falsterbo .. 040618
Falteri, Fiorenza, Firenze 131164
Faltinsky, Rudolf, Braunschweig 075121
Falu Gruva Gruvmuseum, Falun . 040621
Falumúzeum, Hollókő 023694
Falumúzeum, Isaszeg 023698
Falumúzeum, Törökbálint 024014
Falvey, Boston 135363
Falzone, Angelo, Mannheim ... 107964
FAM Galerie, Nijmegen 112607
Fami, Hong Kong 102073
Famiglietti, Vittorio, Modena ... 131569
Familie Vrouwenhofje – Hofje van Aerden, Leerdam 032567
Familienmuseum, Schwerin ... 021791
Familienmuseum Bad Homburg, Bad Homburg v.d.Höhe 016787
Familienmuseum im Altbauernhaus, Göstling an der Ybbs 001958
Familienmuseum Schloss Braunfels, Braunfels 017549
Familistère de Guise, Guise 013216
Family Antiques, Manchester ... 090692
Family Art Gallery, Vancouver .. 101782
Family Broc, Caen 067567
Family Museum of Arts and Science, Bettendorf 046965
Family Ties, Portland 096401
Family Tree Art Gallery, Baltimore 119734
Family Values Antiques, Long Beach 094200
Famine Museum, Strokestown . 024992
Famous, Amsterdam 082743
Famous French Gallery, Tucson . 124733
Fan Collector, Clermont-Ferrand 068027
Fan Dali, Shanghai 102177
Fan Man, Dallas 135502
Fan Museum, London 044966
Fan Tower Museum, Fanshi ... 007762
Fancelli, Marco, Firenze 131165
Fancelli, Nello, Firenze 131166
Fanchang Museum, Fanchang .. 007760
Fancies, Toronto ... 064646, 064647
Fandos y Leonarte, Valencia ... 115492
Fandrich, Berlin 129488
Fandrich, Bernhard, Falkenhagen 129830
Fanelli, New York 095385, 135919
Fanelli, Piera, Roma 081354
Fanette, Paris 071320
Fang County Museum, Fangxian 007761
Fangmeyer-Ferrari, Eggenburg . 099984
Fangotto, Palermo 081039
Faninadahen Kosas Guahan, Guam Museum, Adelup 023355
Le Fanion, New York 095386
Fankhänel, Dietrich, Hildesheim 130044
Fanlac, Périgueux 136996
Fanny's Antiques, Reading 091074
Fanø Kunstmuseum, Fanø 010056
Fanø Museum, Fanø 010057
Fanø Skibsfarts- og Dragtsamling, Fanø 010058
Fans & Stoves of Five Points, Jacksonville 093990
Fanshawe Pioneer Village, London & Middlesex Heritage Museum, London 006210
La Fantasia, Roma ... 081355, 131916
Fantasias Artificiais, Braga 137915

Fantasmak, Paris 141107
Fantastic Art and Image, Hamburg 107241, 137243
Fantastic Fiction, Adelaide 139613
Fantastic Finds, San Antonio ... 096971
Fantastic French Furniture, San Diego 097084
Fantastic Modern Art Gallery, Beverly Hills 098745
Fantasy Crafts, Manama 063143
Il Fante di Quadri, Napoli 110524
Fantoni, Rosario 127547
Fanuele, Leonardo, Buenos Aires 098560
Fapa Ediciones, Barcelona 143714
Faqir Khana Museum, Lahore .. 034128
Faques, Brighton 088620
F'ar, Lausanne 041650
Far Corners, Los Angeles 094332
Far East Fine Arts, San Francisco 124277
Far East Gallery, Singapore ... 114520
Far Eastern Antiques, New York 095387, 135920
Far North Art, Alawa 060974
Far North Regional Museum, Kaitaia 033234
Fara, Maria Ernestina, Roma ... 110765
Clive Farahar & Sophie Dupré, Calne 144292
Farahbaksh, Houshang, Berlin .. 074788
Farahnick, Al, Bruxelles 063419, 063420
Farahnick, Philippe-John, Bruxelles 063421
Faraj, Ibrahim, Sharjah 088198
Farancz, Alan, New York 135921
La Farandole, Pont-Sainte-Maxence 072264
Faraoni, Luigi, Roma 131917
Farbene, Giorgio, Bolzano 079894
Farbene, S., Bolzano 079895
Farber, Indianapolis 121185
Farbror Fynd, Visby 087163
Farbrother, John, York 135269
Farce, Raleigh 139362
Fard, Reza, Milano 080540
Al-Fardan, Sharjah 117196
Al-Fardan, Ahmed Ibrahim, Manama 100432
Fardel, Le Touquet-Paris-Plage . 103971
Fardel, Demanie, Annecy 103247
Farella Franck, Marseille 069932
Farenc, Michel, Le Garric 069269, 128658
Fares, Saint Louis 136199
Farfouillette, Nice 070602
Farga Rossell Centre d'Interpretació del Ferro, La Massana 000096
Fargeaud, Jacques, Saint-Jean-de-Monts 072851
Fargel, Hameln 076346
Fargeton, Damien, Lyon 128712
Fargo Air Museum, Fargo 048697
Farg'ona Viloyat Ülkashunoslik Muzeyi, Farg'ona 054631
Farhi, Luc, Paris 071321, 125712
Farimagshuset, København 065897
Farina, Gilberto, Roma 131918
Farina, Prof. Tina, Köln 107605
Farinella, Comacchio 109967
Farinelli, San Francisco 097229
Faringdon, Arundel ... 088287, 134035
Fark, Friedhelm, Bissendorf 075007, 129555
Farkas, Växjö 115991
Farleigh House Antiques, Southwold 091390
Farleigh Hungerford Castle, Farleigh Hungerford 044331
Farlow, Peter, London 090108
The Farm, Sturgeon Bay 053583
Farm House Museum, Iowa State University, Ames 046440
Farm Shed Museum, Kadina .. 001173
Farmaceutická Expozícia, Mestské Múzeum v Bratislave, Bratislava 038304

Farmacevtsko-Medicinska Zbirka Bohuslava Lavičke, Ljubljana . 038543
Farmacia Museo Taquechel, La Habana 009141
Farmācijas Muzejs, Rīga 030180
Farmamerica, The Minnesota Agricultural Interpretive Center, Waseca 054079
Farmamuseo Sa Potecaria, Villacidro 028390
Farman, P., Paris 071322
Farmer Painting Gallery, Shanghai 102178
Farmer, Andrew, Armadale, Victoria 061024
Farmer's Arms Hotel Museum, Euroa 001067
Farmers' Bank of Rustico, Hunter River 006021
Farmers Bank, The Petersburg Museums, Petersburg 052012
Farmers Gallery, Leominster . . 089805, 118102, 144527
Farmer's House, Metulla 025154
The Farmers' Museum, Cooperstown 047972
Farmhouse Antiques, Bolton-by-Bowland 088514
Farmhouse Antiques, Chester . . 088859
Farmhouse Antiques, Great Houghton 089369
Farmhouse Antiques, Portland . . 096402
Farmhouse Antiques, Stoke Ferry 091450, 135092
Farmhouse Collections, Vancouver 064819
Farming and Fishing Museum of Sihu, Sihu 042452
Farming Through the Ages, Whyte Yarcowie 001641
Farming Village Museum, Cingshuei 042291
Farmington Historic Home, Louisville 050489
Farmington Museum, Farmington 048705
Farmland Museum and Denny Abbey, Waterbeach 046069
Farmleigh House, Dublin 024878
Farmlife Centre, Thornhill, Central 045974
Farnborough Gallery and Bookshop, Farnborough, Hampshire 144423
Farnese, Roma 081356, 081357
La Farnesiana Antichitá, Roma . 081358
Farnham Art Society, Farnham . 060133
Farnham Galleries, Simpson College, Indianola 049698
Farnhem Maltings Gallery, Farnham 044333
Farnier, Anne-Marie, Paris 104735
Farningham Pine, Farningham . . 089266
Lo Faro, Catania 131088
El Faro del Fin del Mundo, Buenos Aires 060900, 139567
Faro, Artur Lopo, Porto 085022
Faro, Maria José, Porto 114082
Farolfi, Francesco, Firenze 131167
Farolfi, Pierluigi, Firenze 131168
Farouq Corner Museum, Helwan 010578
Farr, Cincinnati 092853
Farradèche, Christian, Hambye . 068784
Farradèche, Janine, Chatonrupt-Sommermont 067926
Farragh House Art Gallery, Bunbrosna 109536
Farrando & Lemoine, Paris 125713
Farrar-Mansur House and Old Mill Museum, Weston 054352
Farrar, Straus & Giroux, New York 138393
Farré, Barcelona 143715
Farrel, Lisboa 113999
Farrelly, Tring 091635, 135150
Farrer, Singapore 085409

Farrington Gallery, Farrington Gurney 117790
Farrington, John, Dublin 079552
Farris, Diane, Vancouver 101783
Farron, Guy, Wien 062940
Farrs Gallery, Woking 119374
Farsettiarte, Cortina d'Ampezzo . 126188
Farsettiarte, Milano . . . 110311, 126206
Farsettiarte, Prato 110689
Farsh Museum, Tehrān 024752
Farshchian Museum, Sa'd-Abad Museum, Tehrān 024753
Farsi, Wien 062941
FART, Černigiv 117103
Farthings, Lynton 090668
Fartygsmagasinet i Stockholm AB, Stockholm . 086996, 126672, 143944
Faru, Saint Louis 096702, 123823
Farvan, Charlotte 092597
Farvergården, Ebeltoft 010038
Farvergården, Kerteminde 010158
Farzin, Dallas 093119
Fa's Antik, Skänninge 086925
Fasaani, Helsinki 066312
Fasaani, Muijala 066388
Fasan, Wien 140074
Fasanella, Margherita, Milano . . 080541
Fasano, Attilio, Napoli 080879
Fasciani, Leonardo, Giubiasco . . 133760
Fasciati, Luciano, Chur 087370, 116216, 138036
Fascination Street Gallery, Denver 093283
Fascino, Padova 131662
Fascino, Rotterdam 112638
Fascino Antico, Roma . 081359, 131919
Fase 7, Castello de la Plana . . . 115017
Fashion and Textile Museum, London 044967
Fashion Design Books, New York 145337
Fashion Institute of Technology-Fine Arts Faculty, State University of NY, New York 057551
Fashion Museum, Hasselt 003679
Fasnacht- und Heimatmuseum Noaflhaus, Telfs 002816
Fasnachtsmuseum, Imst 002112
Fasnachtsmuseum Schloß Langenstein, Orsingen-Nenzingen 021004
Fastnachtsmuseum, Bräunlingen 017531
Fastnachtsmuseum, Herbstein . 019200
Fastnachtsmuseum, Sankt Ingbert 021598
Fastnachtmuseum Fasenickl, Kipfenberg 019616
Faszination Zugspitze, Ehrwald . 001847
Faszinosum, Borna 017519
Fat Chance, Los Angeles 094333
Fat Elvis, Baltimore 092321
Fata Libelli, Paris 141108
Fata Morgana, Saint-Clément-la-Rivière 137002
Fatec, Iserlohn 076617
Fath, Gunter, Salies-de-Béarn . . 073379
Fath, Sebastian, Mannheim . . . 107965
Father Marquette National Memorial and Museum, Saint Ignace . . 052773
Father Pandosy Mission, Kelowna 006073
Father Time Antiques, Chicago . 092709, 135410
Father Weyland Gallery, Epworth 048594
Fathy Tantawi, Cairo 066243
Fathy, Ibrahim, Cairo 066242
Fatima College Fine Art Department,

Port-of-Spain 056470
Faton, Dijon 136811
Fattibene, Sabina, Roma 110766
Fatto, Amano, Garmisch-Partenkirchen 075946
Fattorini, Franco, Roma 131920
Fattorini, Massimo, Firenze 080137
Fattorini, P., Firenze 080138
Fattorini, Stefano, Firenze 131169
Fatucchi, Finisio, Milano 131441
Fatzer, B., Zug 144201
Faubourg, Amsterdam . 082744, 132398
Faubourg Diffusion, Paris 104736
Fauchart, Josette, Paris 071323
Fauchatre, Christophe, Campagne-les-Hesdin 067623
Faucher-Touvron, Brigitte & Raphaël, Sainte-Hermine 073354
Fauchet, Bruxelles 063422
Faucheux, Arnaud, Soulge-sur-Ouette 073550
Fauconberges, Beccles 088418
Faudou-Fayolle, Chapareillan . . 128545
Faudree, Charles, Tulsa 097730
Faulconer Gallery at Grinnell College, Grinnell 049279
Fauld Town and Country Furniture, Hereford 134457
Faulkner County Museum, Conway 047967
Faulkner House Books, New Orleans 145308
Faulkner, Ian, Abingdon 134015
Faulstich-Strümpfler, S., Bremen 075158
Fauroux, Philippe, Vic-Fezensac 074092, 129344
Faurschou, København 102879
Fauser, A., Dettingen unter Teck 075347
Fauser, Adolf, Dettingen unter Teck 075348
Fauske Museum, Fauske 033595
Faust-Museum, Knittlingen 019660
Faust, W., Freiburg, Elbe 075844
Faux-Mouvement, Metz 104222
Fauxtiques, Seattle 097435
Favand, Brigitte, Uzès 073925
Favardin, Patrick, Paris 071324, 104737
Favell Museum of Western Art and Indian Artifacts, Klamath Falls . 049999
Faverais, Josiane, Craon 068219
Favereaux, Lille 140869
Favero, Fiorenzo & Bellucco, Silvano, Padova 080982
Favorite Things, Denver 093284
Favre, Chaillé-les-Marais 067762
Favre Antiquité Brocante, Modane 070135
Favre-Boschung, Christine, Le Tholonet 128671
Favre, Christian, Royan 072628
Favre, Daniel, Lyon 069736
Favre, Frédéric, Marennes 069871
Favre, Monique, Marans 069843
Favreau, Lucie, Montréal 064418
Favrel, Didier, Vendôme 074011
Favrel, Jean-François, Plancoët . 072174
Fawbush, New York 138394
Fawick Art Gallery, Baldwin-Wallace College, Berea 046929
Fawkes, Keith, London 144593
Fay, Hans, Thannhausen 142412
Fay, Pammy, Indianapolis 093882
Fayard, Arthème, Paris 136921
Faye, Henriette, Carcassonne . . 067671
La Fayence d'Autrefois, Teting-sur-Nied 073639
Faye's Collectibles, Louisville . . 094548
Fayet, Reims 072391
Fayette Art Museum, Fayette . . 048712
Fayette County Historical Museum, West Union 054328
Fayette County Museum, Washington Court House 054191
Fayette Heritage Museum, La

Grange 050036
Fayette, Gilbert, Villeurbanne . . . 074206
Fayetteville Museum of Art, Fayetteville 048720
Fayla, Bruxelles 100598
Fayoum Art Center, Tunis, Fayoum . . 055338, 058565, 103095
Fazazz, Christchurch 083869
Fazekasház, Tihany 024007
Fazio, Maria Caterina, Roma . . . 081360
Fazzino, Hamburg 076219
FB, Auxerre 103295
FBW-Museum, Wetzikon 042132
FCC Phnom Penh Gallery, Phnom Penh 101203
Fearnley, Nambour 099418
Featherston Gallery, Masterton . 112948
Featherston Heritage Museum, Featherston 033201
Featherston Memorabilia Museum, Featherston 033202
Feats of Clay, Austin 119670
Feats of Clay, Columbus 120490
Feature Inc, New York 122582
Featured Wood Gallery, Australind 098677
Féau & Cie., Paris 071325
Febbraio, Franco, Napoli 080880
FeBland, Harriet, New York 122583
Febvre, Gilles, Lyon 104084
Fecarotta, Palermo 081040
Feddi, Abdellah, Casablanca . . . 082559
Fedeli, Andrea, Firenze 131170
Fedelich, Ghylaine, Riom 072471
Feder, Jörg, Germering 141828
Feder, Soraya, Paris 071326
Federação de Amigos de Museus do Brasil, São Paulo 058401
Federação de Amigos dos Museus de Portugal, Lisboa 059755
Federación Argentina de Amigos de Museos, Buenos Aires 058175
Federación Chilena de Amigos de los Museos, Santiago de Chile . . . 058484
Federación Española de Anticuarios, Madrid 059863
Federación Española de Asociaciones de Amigos de los Museos, Madrid 059864
Federación Mexicana de Asociaciones de Amigos de los Museos, México 059572
Federal Hall National Memorial, New York 051385
Federal Precious Metals, San Francisco 097230
Federal Reserve Bank of Boston Collection, Boston 047112
Federal Reserve Board Art Gallery, Washington 054111
Federatie Kunstuitleen (FKU), Den Haag 059602
Federation Antiques, Cincinnati . 092854
Fédération des Amis des Musées de Belgique, Bruxelles 058375
Fédération des Associations Européennes des Galeries d'Art, Paris 058641
Fédération des Ecomusées et des Musées de Société, Besançon 058642
Fédération des Maisons d'Ecrivain et des Patrimoines Littéraires, Bourges 058643
Fédération Européenne des Commissaires-Priseurs, Bruxelles 058376
Fédération Française des Conservateurs-Restaurateurs, Paris 058644
Fédération Française des Sociétés d'Amis de Musées, Paris 058645
Fédération Hellénique des Amis des Musées, Athinai 059328
Fédération Internationale des Musées

du Jouet et de l'Enfance, Moirans-
en-Montagne 058646
Federation Lighting and Antiques,
Canterbury 061278
Fédération Mondiale des Sociétés
d'Amis de Musées, Paris 058647
Fédération Nationale des Ateliers d'Art,
Guidel 058648
Fédération Nationale des Experts
Professionnels spéc. en Oeuvres
d'Art, Paris 058649
Federation of Art Societies,
Singapore . . . 038268, 059799, 137937
Federation of British Artists,
London 060134
Federation of Canadian Artists,
Vancouver 058445
Federation of Modern Painters and
Sculptors, New York 060479
Federation of Museums and Galleries
in Wales, Aberystwyth . . . 060135
Fédération Régionale des Amis des
Musées du Nord-Pas de Calais,
Dunkerque 058650
Federation Trading, Adelaide . . 060967
Federazione Italiana delle Associazioni
Amici dei Musei, Firenze 059437
Federazione Italiana Mercanti d'Arte,
Milano 059438
Federchen Verlag, Friedland . . 137215
Federseemuseum, Bad Buchau . 016741
Fedora Antiques, Winnipeg . . 064921
Fedrenes Minne, Hidrasund . . . 033652
Fedrizzi, Cles 109964
La Fée aux Miettes, Paris 071327
Feering and Kelvedon Local History
Museum, Kelevedon 044670
Feerique, Amsterdam 082745
Feet First: The Scholl Story, North
Chicago 051599
Feetham, Piers, London 118313
Fefeye, Liège 063756
Fegan, Dallas 135503
Feheley, Toronto 101608
Fehér Miklós Emlékmúzeum,
Győr 023660
Feher, Anne-Laure, Paris 129000
Fehérhajó Galéria, Budapest . . 109143
Fehérvári Babaház, Szent István Király
Múzeum, Székesfehérvár . . . 023934
Fehn- und Schiffahrtsmuseum,
Rhauderfehn 021389
Fehnmuseum Eiland, Großefehn 018844
Fehres, Arndt & Anja, Mülheim an der
Ruhr 077433
Fehringer, Gertrud, Bätterkinden 133629
Fehrle-Choms, C., Schwäbisch
Gmünd 108585
Feichter, F & CH, Bad Gastein . 139972
Feichtner, Lukas, Wien 100253
Feiden, Margo, New York 122584
Feierabend, W., Zürich 087958
Feiertag, Sieger 142361
Feige, Karin, Glückstadt 076011
Feigen, Richard L., New York . . 095388
Feil, Werner, Hallstadt 129941
Feilchenfeldt, Walter, Zürich . . 116772
Feilding & District Art Society,
Manawatu 059634, 112938
Feilding, Linnet, London 118314
Feiligrath Museum, Hotel Krone
Assmannshausen, Rüdesheim am
Rhein 021537
Feille, Clairac 103595
Feinberg, Alan, Washington . . . 124895
Feinberg, Robert, Milwaukee . . 121914
Feinig-Geisinger, Lucia, Altach . 099952
Feintes, Joël, Paris 071328
Feira de Arte Contemporânea,
Lisboa 098274
Feira de Artes Plásticas e
Antiguidades, Batalha 098275
Feiss, Muri 087692, 133866
Feitian, Anxi 064996

Fejkiel, Jan, Kraków 113542
Fekete-Ház, Móra Ferenc Museum
Szeged, Szeged 023924
Fekete István Emlékház, Gölle . 023648
Fekete Sas Patikamúzeum,
Szent István Király Múzeum,
Székesfehérvár 023935
Fekete Szerecseny Patikamúzeum,
Kőszeg 023762
Fel, Joseph, Aurillac 066823
Félagið Íslensk Grafík, Reykjavik 059346
Félags Íslenskra Safna og Safnmanna,
Hafnarfjörður 059347
Felberturmmuseum, Mittersill . . 002397
Felbrigg Hall, Norwich 045399
La Felce, Milano 110312
Feld- und Grubenbahnmuseum,
Solms 021908
Feld- und Industriebahnmuseum (FIM),
Freiland 001903
Feld, Lynn G., New York 095389
Feld, Stuart P., New York 122585
Feldbaumer, Norberg, Graz . . . 127744
Feldberg, Uwe, Mühlhausen/
Thüringen 077426
Feldès, Chantal, Ehnen 132328
Feldfoss & Stenstrøm,
Frederiksberg 065662
Ulrich Feldhaus & Martin Kaufmann,
Castrop-Rauxel 129649
Feldherrnhalle, Bayerische Verwaltung
der staatlichen Schlösser, Gärten und
Seen, München 020523
Feldman Gallery and Project Space,
Pacific Northwest College of Art,
Portland 052292
Feldman, Marc, Paris 071329
Feldman, Ronald, New York . . . 122586
Feldmann, Landshut 077013
Feldmann, Dietmar, Abensberg . 129367
Feledy-Gyűjtemény – Feledy Ház,
Miskolci Galéria, Miskolc . . . 023782
Felekidis, Konst., Kalamaria . . . 109067
Feletin, Thierry, Provins 125782
Felgitscher, Gottfried, Heiligenkreuz
am Waasen 062630
Felicity of Terrigal, Terrigal . . . 062314
Felisi, Sergio, Parma 081109
Felix, Sevilla 086449
Félix, Six-Fours-les-Plages 073525
Felix, Winterthur 133967
Félix-Arnaud Huet, Monique de,
Paris 071330
Felix-Müller-Museum im
Zehntspeicher, Neunkirchen am
Brand 020728
Felix-Nussbaum-Haus Osnabrück,
Osnabrück 021013
Felix Ravenna, Rivista di Antichita
Ravennate, Cristiane e Bizantine,
Ravenna 138825
Felix Valle State Historic Site, Sainte
Genevieve 052873
Félix, Patrick, Aubais 066776
Los Feliz Antiques, Los Angeles 094334
Fell Locomotive Museum,
Featherston 033203
Fell, Anthony, Holt, Norfolk . . . 089547
Fellay Urzua, María Cecilia, Santiago
de Chile 064966
Felle, Oberstaufen 077875
Felle, Dr. M., Kempten 107544
Fellenbergmühle Feinmechanisches
Museum, Merzig 020381
Fellenz, Saint Louis 096703
Fellini, Fresno 093500
Fellner von Feldegg, Marie-Luise,
Krefeld 107732
Fellner, Bernard, Morayfield . . . 127644
Fellner, Michael, Mattsee 062731
Fellows & Sons, Birmingham . . 126822
Fellows of Contemporary Art, Los
Angeles 060480

Fellows, Maurice, Birmingham . 088478,
134109
Fellowship of Makers and Researchers
of Historical Instruments,
London 060136
Fells Point Antique Mart,
Baltimore 092322
Felly, Haarlem 083326
Felman, Jacqueline, Paris 104738
Fels, W., Bocholt 075027
Felsberg-Museum, Lautertal . . . 019944
Felsenburg Neurathen, Lohmen,
Sachsen 020107
Felsendome Rabenstein,
Chemnitz 017731
Felsengarten Sanspareil mit
Morgenländischem Bau, Bayerische
Verwaltung der staatlichen Schlösser,
Gärten und Seen, Wonsees . . 022759
Felsenkeller Brauerei-Museum,
Monschau 020455
Felsenkeller-Labyrinth im Hirschberg –
Brauereimuseum, Beilngries . . 017019
Felsenmuseum, Bernstein 001780
Felsted Studio, Felsted 117795
Feltrinelli, Giangiacomo, Milano . 137677
Femme Rochas, Cairo 066244
Femörefortet, Oxelösund 040911
Fenaux, Brigitte, Lons-le-Saunier 125617
Fendalton Antiques, Christchurch 083870
Fendel, Vera, Gehrden bei
Hannover 129887
Fender Museum of Music and the
Arts, Corona 048000
Fene, Valenciennes 073943
Fenech, Adrian, San Giljan . . . 111919
Fenelon Falls Museum, Fenelon
Falls 005800
Fenes, Piraeus 109081
Fénestra, Eric, Condom 068128
Fenêtre-sur-Cour, Cour-Cheverny 068184
Feng Chia University Art Center,
Taichung 042472
Feng City Museum, Fengcheng . 007763
Fenghua Xikou Museum,
Fenghua 007764
Fengkai Museum, Fengkai 007765
Fenglin Shan Art, Guangzhou . . 102026
Fengshan Dr. Sun Yat-sen Memorial
Hall, Fengshan 042311
Fengshun Museum, Fengshun . . 007766
Fengxian Museum, Shanghai . . 008226
Fengxiang Museum, Fengxiang . 007767
Fengyuan Lacquer Art Museum,
Fengyuan 042312
La Fenice, Milano 080542
La Fenice, Roma 081361
La Fenice, Sassari 110912
La Fenice, Verona 132281
Fenice Arte, Castellanza 109939
La Fenice-Nottoli, Saint-Paul-de-
Vence 073244
Feniks, Ljubljana . . . 085608, 114702
Fenimore Art Museum, New York
State Historical Associaton,
Cooperstown 047973
Fenina, Hedi, Hammamet 116971
Fenix, Amsterdam 143029
Fenix, Göteborg 115632
Fenix, Miami 121779
Fenix, Stockholm 086997
Fenland and West Norfolk Aviation
Museum, West Walton 046093
Fenning, James, Dún Laoghaire 142606
Fenns Quay Antiques, Cork . . . 079492
Fenollar Olivert, José, Valencia . 086514,
133462
Fenomenalen, Visby 041174
Al-Fenoun Al-Gamila, Cairo . . . 066245
Fens, Amsterdam 082746
Fensch, Seremange-Erzange . . 073508
Fenske, Edmonton 125366
Fenske, Christine, Bernhardswald 129583
Fenton Gallery, Cork 109554

Fenton Glass Museum,
Williamstown 054421
Fenton History Center-Museum and
Library, Jamestown 049786
Fenton House, London 044968
Fenton Maclaren, Oakland 095890
Fenway, Boston . . . 119865, 135364
Fenway Gallery, Chicago 120120
Fenwick, Toledo . . . 124686, 124687
Fenwick & Fenwick, Broadway . 088674
Fenwick Gallery, Warkworth . . . 119307
Fenxi Museum, Linfen 008015
Fenyang Museum, Fenyang . . . 007768
Fenyi Museum, Fenyi 007769
Fenzi, Ermanno, Rovigo 110886
Fenzi, Gianpolo, Verona 132282
Feo Garcia, Maria Teresa,
Valencia 115483
Fer du Monde, Buenos Aires . . 098561
Feral, Jacqueline, Toulouse . . . 073766
Feraud, Patrick, Champtercier . 067820
Ferber, Daniel, Daun 129672
Ferber, Lothar, Wettenberg . . . 078953
Ferchaud, Monique, Bordeaux . 067326
Ferdinand Domela Nieuwenhuis
Museum, Heerenveen 032429
Ferdinando, Steven, Queen
Camel 144794
Ferdinandsen, Per, København . 065898
Fere, Gent 063637
Ferencvárosi Helytörténeti Gyűjtemény,
Budapest 023477
Ferenczy Múzeum, Szentendre . 023955
Ferens Art Gallery, Hull City Museum
& Art Gallery, Kingston-upon-
Hull 044712
Feres Archaeological Site,
Velestino 023310
Feret, Marc, Paris 129001
Ferfoje Eski, T.D., İstanbul 088117
Ferges, Düsseldorf 129740
Ferguson, Cincinnati 092855
The Ferguson Gallery, Ramelton 109683
Fergusson Gallery, Perth 045485
Feria de Anticuarios, Buenos
Aires 097987
Feria de Antigüedades, Valencia 098298
Feria de Antigüedades y Galerías de
Arte, Madrid 098299
Feria de Arte Contemporáneo, Buenos
Aires 097988
Feria de Arte Contemporáneo,
México 098245
Feria de Arte de Semana Santa, La
Cumbre 097989
Feria de Arte Moderno y
Contemporáneo de São Paulo, São
Paulo 098028
Feria de Arte y Antigüedades de
México, Mexico 098246
Feria de Ayer Antiguedades,
Montevideo 091998
Feria de las Artes Plásticas,
México 098247
Feria de Libros, Buenos Aires . . 139568
Feria de Videoarte en Barcelona,
Barcelona 098300
Feria del Arte Contrempráneo Sevilla,
Sevilla 098301
Feria Iberioamericana de Arte,
Caracas 098513
Feria Internacional de Arte
Contemporáneo, Madrid 098302
Feria Internacional de Arte
Contemporáneo de Marbella,
Marbella 098303
Feria Internacional de Arte y
Antigüedades, Madrid 098304
Feria Internacional de Fotografía
Contemporánea y Vídeo-Arte,
Donostia-San Sebastián 098305
Feria Internacional del Arte,
Valencia 098306
Feria Internacional del Diseño, Diseño,

Bogotá 098057
Feria Regional de Antigüedades,
  Albacete 098307
Ferihegy Repülögép Emlékpark,
  Budapest 023478
Feringa Sach – Ortsgeschichte
  und heimatkundliche Sammlung,
  Unterföhring 022277
Ferjan, Ljubljana 085609
Ferlelund Antiques, Gilleleje . . 065695
Ferlini, Bruno, Mons . . 063795, 140252
Ferlini, Renata, Bologna 130957
Ferma, Barcelona 085770
Fermanagh County Museum at
  Enniskillen Castle, Enniskillen . 044295
Ferme aux Abeilles, Saint-Brévin-les-
  Pins 015240
La Ferme Caussenarde d'Autrefois,
  Hures-la-Parade 013278
Ferme d'Antan, Plédéliac 014888
Ferme de Hot Springs, Mohon . 070136
Ferme de la Claison, Saint-Étienne-du-
  Bois 015309
Ferme de la Fontaine, Lausanne 087569,
  133796
Ferme de Paris, Paris 014648
Ferme des Bois Nois, Saint-Eloy La-
  Glaciere 015296
La Ferme des Étoiles, Mauroux . 014092
Ferme des Mangettes, Saint-Étienne-
  du-Bois 015310
Ferme d'Hier et d'Aujourd'hui,
  Aviré 011846
Ferme du Château, Beveuge . . 067158
Ferme du Château, Saint-
  Fargeau 015313
La Ferme Marine-Musée de l'Huître et
  du Coquillage, Cancale 012306
Ferme-Musée à Pans de Bois de
  Sougey, Montrevel-en-Bresse . 014307
Ferme-Musée de Bray, Sommery 015833
Ferme Musée de la Soyotte, Musée
  Arts et Traditions Populaires, Sainte-
  Marguerite 015616
Ferme-Musée de la Tonnelière,
  Courcerault 012697
Ferme-Musée des Frères Perrel,
  Moudeyres 014345
Ferme-Musée du Contentin, Sainte-
  Mère-Eglise 015625
Ferme-Musée du Léon,
  Tréflaouenan 016033
Ferme-Musée du Montagnon,
  Fournets-Luisans 013040
Fermes-Musée du Pays Horloger,
  Grand-Combe-Châteleu 013144
Fermette, Épreville 068460
Fermin Garcia, Felipe, Madrid . . 086183
Fermynwoods, Brigstock 117433
Fern, Natascha von, München . 108067
Fernandes & Costa, Carlos,
  Lisboa 084882
Fernandes, Paulo, Rio de Janeiro 101025
Fernandez, Zamora 086581
Fernández Alfaro, José Antonio,
  Alajuela 102463
Fernandez Antiguedades, Sevilla 086450
Fernandez Battagliese, Isaura,
  Vigo 086558
Fernández Blanco Libros Antiguos,
  Buenos Aires 139569
Fernandez de Bobadilla, Teresa,
  Madrid 133336
Fernandez-Escarzaga, Belen Roiz,
  Santander 086424
Fernández Gallart, Manuela,
  Valencia 086515
Fernandez Lopez, Maria Consuelo,
  Murcia 086313
Fernandez Muñiz, Maria Jesus,
  Gijón 115071
Fernández, J., Barcelona 085771
Fernández, Jesús Angel, León . . 086064
Fernández, José, Santiago de

Chile 064967
Fernandez, Miguel, Rosières . . . 072544
Fernandez, Salvador, Barcelona . 085772
Fernando, Luis, New York 135922
Fernando, Pedro, Belo Horizonte 100884
Fernbank Museum of Natural History,
  Atlanta 046616
Fernbank Science Center, Atlanta 046617
Fernbrook Gallery, Kurrajong
  Heights 099213
Ferndown Gallery, Ferndown . . . 117796
Ferner Galleries, Auckland 112763,
  112764
Ferner Gallery, Wellington 113066
Fernhill House, The People's Museum,
  Belfast 043556
Fernie and District Historical Society
  Museum, Fernie 005803
Fernlea Antiques, Manchester . 090693
Fernmeldemuseum Aachen,
  Aachen 016436
Fernwold, Pittsburgh 096286
Féron, Richard, Agon-Coutainville 066431
Ferra, Ray, Los Angeles 094335
Ferracini & Paglia, Bologna . . . 079837
Ferraglia, Parma 081110, 081111
Ferraguti, Parma 142781
Ferran Cano, Palma de Mallorca 115339
Ferrand, Michel, Dammarie . . . 086265
Ferrando Guerra, Santiago, Buenos
  Aires 060901
Ferrando, Andrea Armando,
  Genova 131308
Ferrando, Francesca, Genova . . 080260
Ferranti, Ugo, Roma 110767
Ferrara, Paris 071331
Ferrara, Jonathan, New Orleans 122162
Ferrara, Rosario, Catania 080034,
  131089
Ferrara, Stefano, Bologna 079838
Ferrari, Milano 131442
Ferrari, Saint Louis 096704
Ferrari, Treviglio 096703
Ferrari Antiquités, Martignargues 069984
Ferrari-Auer, Bolzano 142650
Ferrari Auer, Bolzano 137629
Ferrari Museum of Art, Gotemba 028573
Ferrari Rosetta de Antigüedades,
  Caracas 097951
Ferrari Sarzábal, Aldo,
  Montevideo 127111
Ferrari, Bernard, Nice 070603
Ferrari, Gian-Mario, Mendrisio . 087672
Ferrari, Giorgio, Roma . 081362, 131921
Ferrari, Neldo, Modena 131570
Ferrari, Olivier, Châlonvillars . . 067789
Ferrari, Pierluigi, Milano 131443
Ferraris, Paolo, Torino 132138
Ferraro, Roma 110768
Ferraro, Arturo, Napoli 131618
Ferraroni, Patrizia, Modena . . . 080792
Ferraton, Alain, Bruxelles 125309,
  140174
Ferrazza, Renato, Roma 081363
Ferrazzutti, Pam, Toronto 064648
Ferré, Jacqueline, Ingrandes-sur-
  Loire 068870
Ferré, Patrick, Saint-Mathurin-sur-
  Loire 072957
Ferreira, Bruxelles 063423
Ferreira, Oakland 136041
Ferreira & Manteigas, Lisboa . . 084885
Ferreira Vicente, José, Lisboa . . 143568
Ferreira Vicente, José, Paço de
  Arcos 143598
Ferreira, António Almeida, Braga 084778
Ferreira, Felismina, Porto 085023
Ferreira, Gilberto M. Teves, Ponta
  Delgada 132962
Ferreira, Helena Simões, Lisboa 084883
Ferreira, Henrique Meneses,
  Porto 085024
Ferreira, João, Cape Town 114766
Ferreira, Licínio S. Santos, Mafra 085002

Ferreira, Luísa & David, Porto . . 085025
Ferreira, Manuel, Porto 126458, 143603
Ferreira, Maria Lurdes, Lisboa . 084884
Ferreira, Maria Meira, Lisboa . . 114000
Ferreira, Matos, Lisboa 114001, 143567
Ferrer Carrer, Ramon, Barcelona 114882
Ferrer Marcs, Barcelona 114883
Ferrer Martí, Valencia 086516
Ferrería de Mirandaola, Lenbur Museo-
  Lurradea, Legazpia 039551
Ferrero, Nice 104363
Ferrero Zendrini, Carlo, Monaco 082504
Ferrero Zendrini, Carlo, Torino . 081696
Ferrers, Staunton Harold 119141
Ferret Bookshop, Wellington . . . 143365
Ferret, Georges, Villers-les-Guise 074179
Ferretti & Guerrini, Roma 081364
Ferretti, Gino, Maranello 080349
Ferretti, Sergio, Langhirano . . . 080308
Ferrey, Laurent, Pirey 072166
Ferri Associés, Paris 125714
Ferri, Daniele, Bologna 130958
Ferriani, Milano 131444
Ferrien-Marcognet, Eric, Paris . . 071332
Ferrier, Patrick, L'Isle-sur-la-
  Sorgue 069568
Ferriere de Lassus, André & Vonia,
  Chomerac 067988
Ferrigno, Gaetano, Napoli 131619
Ferris, Sacramento 096621
Ferris, Maurice, Oklahoma City . 095971,
  136049
Ferrito, Giancarlo, Palermo . . . 081041
Ferro, Antonino, Palermo 081042
Ferro, Nunzio, Messina 131355
Ferro, Paola, Milano 080543
Ferroni, Montréal 064419
Ferroni, Alice, Genova 131309
Ferrow, David, Great Yarmouth . 134425,
  144447
Ferrucci, Arnaldo, Roma 081365
Ferrum – Museum Kremayr,
  Ybbsitz 003133
Ferruzzi, Balbi Ferruccio, Venezia 081837
Ferry Antiques, Wargrave 091729,
  135190
Ferry Park Gallery, MacLean, New
  South Wales 099247
Ferry, Marcel, Rambervillers . . . 072361
Ferry, Nicole, Paris 104739
Ferrymead Heritage Park,
  Christchurch 033172
Fersen, Monaco 082505
Fersini, Andrea, Roma 131922
Fersini, Massimo, Roma 131923
Fersini, Rocco, Roma 131924
Ferstl, Robert, Karlsruhe 076691, 130095
Fert, Yvoire 105887
Fert-Rifaï, Nathalie, Nyons . . . 070767,
  141011
Fesel, Bernd, Sonsbeck 108644
Fesseler, K., Berlin 129489
Fester, Den Haag 112381
Festetics-Kastély, Dég 023599
Festetics-Kastély, Helikon
  Kastélymúzeum, Keszthely . . 023738
Festina Lente – italiAntiquariaat,
  Groningen 143148
Festival-Institute Museum, James Dick
  Foundation, Round Top 052693
Festival Internacional de Fotografía y
  Artes Visuales, Madrid 098308
Festung Kniepass, Unken 002853
Festung Königstein, Königstein . 019736
Festung Marienberg mit
  Fürstenbaumuseum, Bayerische
  Verwaltung der staatlichen Schlösser,
  Gärten und Seen, Würzburg . . 022789
Die Festung Rosenberg – Deutsches
  Festungsmuseum, Kronach . . . 019800
Festung Wilhelmstein im Steinhuder
  Meer, Wunstorf 022810
Festungs- und Heimatmuseum,

Kufstein 002250
Festungsanlage – Museum
  Senftenberg, Senftenberg 021841
Festungsmuseum Crestawald,
  Zillis 042180
Festungsmuseum Heldsberg, Sankt
  Margrethen 041929
Festungsmuseum Reuenthal, Fortress
  Museum, Reuenthal 041871
Festungsmuseum Salzburg,
  Salzburg 002620
Fethard Park and Folk Museum,
  Fethard 024923
Fethiye Cami Müzesi, Ayasofya
  Müzesi, İstanbul 042936
Fethiye Kültür Merkezi, Fethiye . 042901
Fethiye Müzesi, Fethiye 042902
Fetlar Interpretive Centre, Fetlar 044342
Fetoon, Sharjah 117197
Fetsund Lenser Fløtingsmuseum,
  Fetsund 033596
Fetterman, Peter, Santa Monica 124469
Fetting & Minx, Berlin 141510
Fety, F., Bruxelles 063424
Fety, Margaret, Bordeaux 140791
Fetzer, Sontheim an der Brenz . 108646
Fetzer, J., Ludwigsburg 142082
Fetzian & Kourd, Athinai 079161
Feucht, R., Allmendingen 141406
Feuchtmayer Museum, Salem,
  Baden 021570
Feuer, Zach, New York 122587
Feuerman, Carole A., New York . 122588
Feuerschiff, Wilhelmshaven . . . 022658
Feuerstätten-Ausstellung im Lausitzer
  Bergbaumuseum, Hoyerswerda 019348
Feuerstein, G. & E., Aachen . . . 074330
Feuerstein, Karl Heinz, Dornbirn 062546
Feuerstein, Thyl, Berlin 129490
Feuerstein, Wolfgang, Dornbirn . 127724
Feuerwehr- und Handwerkermuseum,
  Endingen 041464
Feuerwehr- und Technikmuseum,
  Städtisches Museum
  Eisenhüttenstadt 018144
Feuerwehr-Helm-Museum,
  Stuttgart 022029
Feuerwehr Museum, Lahr 019852
Feuerwehr-Museum Schloß
  Waldmannshofen, Creglingen . 017773
Feuerwehrhistorische Ausstellung,
  Dahlen 017803
Feuerwehrmuseum, Adligenswil 041195
Feuerwehrmuseum, Altenkirchen,
  Westerwald 016539
Feuerwehrmuseum, Asperg . . . 016635
Feuerwehrmuseum, Dobersberg 001818
Feuerwehrmuseum, Hermeskeil 019213
Feuerwehrmuseum, Kaufbeuren 019558
Feuerwehrmuseum, Kreuzlingen 041623
Feuerwehrmuseum, Lengenfeld . 020034
Feuerwehrmuseum, Niederwiesa 020804
Feuerwehrmuseum, Pegau . . . 021090
Feuerwehrmuseum,
  Perchtoldsdorf 002493
Feuerwehrmuseum, Rehau . . . 021334
Feuerwehrmuzeum, Roding . . . 021447
Feuerwehrmuseum,
  Sachsenheim 021566
Feuerwehrmuseum, Salem,
  Baden 021571
Feuerwehrmuseum, Sankt Leonhard
  am Forst 002657
Feuerwehrmuseum, Sayda . . . 021624
Feuerwehrmuseum, Spalt 021924
Feuerwehrmuseum,
  Stadtprozelten 021958
Feuerwehrmuseum, Stetten am kalten
  Markt 021999
Feuerwehrmuseum, Steyrermühl 002792
Feuerwehrmuseum, Türnitz . . . 002842
Feuerwehrmuseum, Tuttlingen . 022231
Feuerwehrmuseum, Wasserburg am
  Inn 022438

Feuerwehrmuseum Berlin, Berlin 017113
Feuerwehrmuseum des Landkreises
Harburg, Außenstelle des
Freilichtmuseums am Kiekeberg,
Marxen 020309
Feuerwehrmuseum Ellricher
Spritzenhaus, Ellrich 018156
Feuerwehrmuseum Gahlenz,
Gahlenz 018581
Feuerwehrmuseum Grethen,
Parthenstein 021078
Feuerwehrmuseum Hannover,
Hannover 019049
Feuerwehrmuseum Hordorf,
Cremlingen 017779
Feuerwehrmuseum Kalmbach,
Riedbach 021412
Feuerwehrmuseum Kradolf-
Schönenberg, Kradolf 041621
Feuerwehrmuseum Marxen,
Marxen 020310
Feuerwehrmuseum Musberg,
Leinfelden-Echterdingen 019963
Feuerwehrmuseum Nürnberg,
Nürnberg 020848
Feuerwehrmuseum Ravensburg,
Ravensburg 021300
Feuerwehrmuseum Salzbergen,
Salzbergen 021573
Feuerwehrmuseum Schleswig-Holstein,
Norderstedt 020823
Feuerwehrmuseum Schröttinghausen,
Preußisch Oldendorf 021219
Feuerwehrmuseum Winnenden,
Winnenden 022682
Feuerwehrmuseum Zeven, Zeven 022872
Feuille d'Acanthe, Jodoigne . . . 128063
Feuille d'Acanthe, Thionville . . . 073658,
129278
La Feuille d'Or, Allschwil 133616
Feuille d'Or et Mercure, Paris . . 129002
Feuilleraie, Vieux-Moulin 074119
Feuillets Libres, Toulouse 105735
Feurer, Dr. Reto, Obing 142257
Fevik Auksjonsforretning, Fevik . 084236
Février, André, Jouars-
Pontchartrain 068915
Février, Laurence, Marboue 069849
Few, Ted, London 090109
Fey-Weidner, Gudrun, Wiesbaden 130757
Fey, Dr. D., Eutin 075722
Fey, S., Karlsruhe 076692
Fey, Thomas, Seevetal 078447, 130636
Feycolor, Mäder 100115
Feyman, Wiesbaden . . . 108902, 108903
FFotogallery, Cardiff . . . 117528, 138148
FH Coin & Collectibles, New York 095390
Fi Donc, Alkmaar 112105
FIAF – Federazione Italiana
Associazioni Fotografiche,
Torino 059439
Fiakermuseum, Wien 002989
Fiala, Praha 065495, 128344
Fiale, Rosario, Milano 131445
Fiano, Fabrizio, Roma 081366
Fiartega, Vigo 115534
Fibber Mcgee, Seattle 097436
Fiberart Gallery, Saint Ives,
Cornwall 119022
Fiberarts Magazine, Loveland . . 139363
Fiberscene, San Francisco 124278
Fibre Arts Studio, Vancouver . . . 101784
La Fibule, Paris 071333
La Fibule, Rabat 082599
Ficara, Santo, Firenze 110049
Fich, Anne, København 102880
Fichera & Perkins, New York . . . 095391
Fichet, Jacques, Avèze (Puy-de-
Dôme) 066868
Fichtelgebirgsmuseum,
Wunsiedel 022809
Fichter, H.W., Frankfurt am Main 075780
Fick Fossil and History Museum,
Oakley 051689

Fick, Helene & Henrik,
Humlebæk 065802
Fickels, San Antonio 136259
Fidalgo et Fréres, Saint-Ouen . . 073079
Fiddes & Son, Cardiff 134212
Fiddys, Osterwald 077991
Fideicomisio de Conservación e
Historia de Vieques, Vieques . . 036174
Fidelo, Edinburgh 117726
Fides-Arte, Torino 110963
Fidesarte Italia, Pieve di Cadore 110657
Fidesarte Italia, Venezia 111102, 126238
Fidia, Lugano 138067
Fidia Arte Moderna, Roma 110769
FIDM Museum and Galleries,
Fashion Institute of Design and
Merchandising, Los Angeles . . 050419
Fidow-Fiddickow, A., Köln 076862
Fidus, Schwülper 142358
Fidus Antik, Hadsund 065722
Fiebach & Minninger, Köln 107606
Fiebig, S., Syke 078654
Fiechter, Münchenstein 133864
Fiedler, Wetzlar 078958
Fiedler Memorial Museum,
Seguin 053263
Fiedler, Karl, Waldstetten 130732
Fiedler, Ulrich, Köln 107607
Fiedler, Vladimir, Wien 062942
Field, San Francisco 145552
Field Museum of Natural History,
Chicago 047657
Field of Dreams, Dallas 093120, 145113
Field of Dreams, Houston 145169,
145170
Field of Dreams, Indianapolis . . 093883
Field of Dreams, Las Vegas 094138
Field of Dreams, Louisville 094549
Field of Dreams, Norfolk 095853
Field of Dreams, Saint Paul 096822
Field Staff Antiques, Rochester . 091126
Fielder, John, Denver 120714
Fielding L. Wright Art Center,
Cleveland 047804
Fielding Lecht, Austin 119671
Fieldings, Haslingden 089460
Fieldings Court Antiques, Kerry . 079641
Fields Galleries, Tulsa 124802
Fields of Valor, Vancouver 064820
Fields, Walter, New York 122589
Fieldstone, Dallas 120595
Fiera Internazionale d'Arte
Contemporanea, Bologna 098189
Fiera Internazionale d'Arte Moderna e
Contemporanea, Milano 098190
Fiera Piacentina di Numismatica,
Filatelia e Collezionismo,
Piacenza 098191
Fieros Fedros, N., Lefkosia 065321
Fiesta Galerie, Paris 071334
De Fietsenstalling, Den Haag . . . 112382
Fife Folk Museum, Ceres 043900
Fifh, Casablanca 082560
Fifteen Minutes, Minneapolis . . 094910
Fifteen Ten Antiques, Charlotte . 092598
Fifteenth Century Bookshop,
Lewes 144533
Fifth Avenue Antiques,
Birmingham 092407
Fifth Avenue Antiques,
Milwaukee 094854
Fifth Avenue Antiques, San Diego 097085
Fifth Avenue Galleries, Columbus 120491
Fifth Class, Atlanta 119563
Fifth Element Pottery, Portland . 123618
Fifth Maine Regiment Center, Peaks
Island 051957
Fifth Meeting House, Lancaster . 050112
Fifth Street Antiques, Kansas City 094073
Fifties-Sixties, Amsterdam 082747
Fifty Crows, San Francisco 124279
Fifty Five, Göttingen 076043
Fifty-One Antiques, Toronto 064649
Fifty One Fine Art Photography,

Antwerpen 100488
Fig Tree Gallery, Fresno 120895
Figari, Dr. Luisa, Cagliari 131067
Figgis, Moyard 142611
Fighter World, Williamtown 001644
Figl, Anton, Sankt Pölten 062828
Figl, Felice, Linz 100103
Figl, Matthias, Plankenberg 062767
Figueiredo, Carlos Alberto Veiga,
Lisboa 143569
Figueiredo, Carlos H.C., Rio de
Janeiro 064057
Figueiredo, Oswaldo, São Paulo 064177
Figueiredo, Rui, Lisboa 084886
Figuière, Geneviève, Lasalle . . . 069195
Figuline, Viroflay 074253
Figuline Antiquités, La Couarde-sur-
Mer 069006
Figur, Rainer, Ronnenberg 130597
Figurative Design Group, San
Francisco 097231
Figurencabinett Madame Lucrezia,
Rosegg 002608
Figures de Style, Paris 104740
Figurinen Antik och Kuriosa,
Vadstena 087131
Fiji Museum, Suva 010748
Le Fil d'Ariane, Antibes-Juan-les-
Pins 103257
Le Fil d'Ariane, Toulouse 141363
Fil De Fer Fransk Antik,
København 065899
Fil du Temps, Clermont-Ferrand 140828
Le Fil du Temps, Sanary-sur-Mer 073398,
129251
Fil-Money, Verbania 081885
Filambule, Lausanne 087570
Filaretos, Athinai 079162
Filatelia Numismatica Alex,
Mantova 080345
Filatelia Numismatica Sintoni,
Forlì 080226
Filatelica e Numismatica, Civitanova
Marche 098192
La Filature, Anduze 066592
Fileman, David R., Steyning . . . 091422,
135081
Filene, Washington 124896
Filentroc, Anceaumeville 066586
Filey Museum, Filey 044344
Filial Moskovskogo Gosudarstvennogo
Universiteta Kultury i Iskusstv,
Naberežnye Čelny 056288
Filial Moskovskogo Gosudarstvennogo
Universiteta Kultury i Iskusstv,
Norilsk 056291
Filidarte, Palermo 110602
Filigrana, Venezia 132243
Filipiak, Deborah, Buffalo 119954
Filipp & Rosbach, Leipzig 107817
Filippis, Konstantin, Athinai . . . 079163
Fillaire, Marie-Christine, Mantes-La
Jolie 125633
La Fille du Pirate, Paris 071335, 129003
Filleau, Joël, Belin-Béliet 067073
Filleau, Pascal, Bordeaux 103427
Filley, Christopher, Kansas City . 094074
Fillipi, B., Seattle 145582
Fillmore County Historical Center,
Fountain 048913
Fillmore Historical Museum,
Fillmore 048730
Filloux, Georges, Paris 141109
Film- und Fernsehmuseum Hamburg
e.V., Hamburg 018987
Film Art LA, Hollywood 120916
Film Forum – Film Archives,
Collingswood 047877
Film Line Gallery, New York 122590
Film Museum of Iran, Tehrän . . 024754
Filman Fine Art, Dorking 117683
Filmmuseet, Oslo 033832
Filmmuseet, Regionmuseet i Skåne,
Kristianstad 040758

Filmmuseum, Bendestorf 017023
Filmmuseum Foundation Nederlands,
Amsterdam 031939
Filmmuseum Landeshauptstadt
Düsseldorf, Düsseldorf 018027
Filmmuseum Potsdam, Potsdam 021182
Filmmuseum, Münchner Stadtmuseum,
München 020524
Filmmuseum, Museum van
Hedendaagse Kunst Antwerpen,
Antwerpen 003286
Filmworks USA, Atlanta 092132
Filnum, Parma 081112
Filoli Center, Woodside 054526
Fils, Françoise, Ecuras 068429
Filsinger, Claus, Neulußheim . . . 077736
The Filson Historical Society,
Louisville 050490
Filupeit, Harry, Frankfurt am
Main 106985
Fimmers-Van der Cruysse,
Brugge 063272
Fin-Art, Nova Gorica 114738
Fin de Siglo, Buenos Aires 139570
Fina, Gianfranco, Torino 081697
Finacchioli, Rita, Roma 131925
Final Chapter, Tuggeranong 139937
Final Lap, Las Vegas 094139
Final Touch, Indianapolis 121186
Final Touch-Up, Virginia Beach . 136443
Finale Art File, Makati 113386
Finanzgeschichtliche Sammlung der
Bundesfinanzakademie, Brühl . 017641
Finard, Alain, Paris 071336
Finarte, Milano 126207, 126208
Finarte, Roma 126225
Finarte, Venezia 126239
Finartis, Zug 116879
Finaz de Villaine, Manuela, Paris 071337
Finch, Sandgate . . . 091234, 135004
Finch Foundry Working Museum,
Okehampton 045432
Finch, Hector, London 090110
Finch, Simon, Holt, Norfolk 144484
Finch, Simon, London 144594
Finchcocks, Living Museum of Music,
Goudhurst 044456
Finches, Saint Louis 096705
Finchingfield Antiques Centre,
Finchingfield 089273
Finchley Fine Arts Galleries,
London 090111, 134656
Finck, Thomas, Rottenbach 078298
Finckenstein, Dresden 106723
Find, Los Angeles 094336
Findahl, K. & L., Norberg 086878
Finders Keepers, Long Beach . . 094201
Finders Keepers, Napier 084005
Finder's Keepers, Phoenix 096210
Finders Keepers, Pittsburgh . . . 096287
Finders Keepers, Toronto 064650
Finders Keepers Antiques,
Denver 093285
Finders Keepers Katikati, Katikati 083971
Findings, Philadelphia 096106
Findlater Museum, Dublin 024879
Findlay, Cynthia, Toronto 064651
Findlay jr., David, New York 122591
Findlay, David, New York 122592
Findley, Peter, New York 122593
Findley, Wally, New York 122594
Findorffhof, Grasberg 018796
Finds, Tucson 097659
Finds From the Far East, New
York 095392
Fine & Performing Arts, Oakland
Community College, Farmington 057077
Fine and Performing Arts Department,
Creighton University, Omaha . . 057620
Fine and Performing Arts Department,
Southwestern Michigan College,
Stevensville 057953
Fine and Performing Arts Department,
York College, City University of New

York, Jamaica 057259
Fine Antique Furniture, Los
Angeles 135717
Fine Antique Restoration, Denver 135530
Fine Antiques and More, Las
Vegas 094140
Fine Art, Granada 115086
Fine Art, Moskva 114217
Fine Art, Roermond 112621
Fine Art, Saint Louis 123824
Fine Art & Curios, North Shore . 084032
Fine Art & Frames, Lytham Saint
Annes 118686
Fine Art & Interiors, Norwich . . . 118829
Fine Art Acquisitions, New York . 122595
Fine Art and Antique, Colombo . 086611
Fine Art and Artists, Washington 124897
Fine Art and Framing,
Jacksonville 121250
Fine Art Associates, Honolulu . . 120934
Fine Art Association of Ho Chi Minh
City, Ho Chi Minh City 060824
Fine Art by Jane, Long Beach . . 121399
Fine Art Collectables, Lower
Portland 061753
Fine Art Collection & Gallery 1C03,
Winnipeg 007440
Fine Art College of Shanghai
University, Shanghai 055230
Fine Art Commissions, London . 118315
The Fine Art Company, Mumbai 109433
Fine Art Conservation Group, New
York 135923
Fine Art Conservation Services, New
York 135924
Fine Art Deichtorhallen, Hamburg 098139
Fine Art Department, Himachal
Academy of Arts, Culture &
Languages, Shimla 055835
Fine Art Department, Indian River
Community College, Fort Pierce 057098
Fine Art Department, Ithaca College,
Ithaca 057246
Fine Art Department, Joliet Junior
College, Joliet 057269
Fine Art Deptartment, Central
University of Technology,
Bloemfontein 056338
Fine Art di Zinutti, Udine 111060
Fine Art Distributors, Alphington 098643
Fine Art Division, Concord College,
Athens 056676
Fine Art Framing, Bridgetown . . 100460
Fine Art Gallery, Hong Kong . . . 102074
Fine Art Gallery, Metz 104223
Fine Art Gallery, Richmond,
Surrey 118970
The Fine Art Gallery, Rochester . 118984
Fine Art Gallery, Winnipeg 101876
Fine Art Gallery at Centro Cultural
Aztlan, San Antonio 052937
Fine Art Hawaii, Honolulu 120935
Fine Art Images, Richmond 123731
Fine Art Inn, Salt Lake City . . . 123960
Fine Art Limited, Chicago 120121
Fine Art Living, Chadstone 061309
Fine Art Paper Conservation,
Winnipeg 128238
Fine Art Photographics, Arbroath 117246
Fine Art Prints on Canvas, San
Diego 124121
Fine Art Program, Northwest Normal
University, Lanzhou 055222
Fine Art Rentals and Sales,
Sacramento 123766
Fine Art Research, Beijing 138600
The Fine Art Resource, Mumbai 109434
The Fine Art Society, London . . 090112,
. . . . 118316
Fine Art Studios, Eketahuna . . . 112897
Fine Art Trade Guild, London . . 060137
Fine Art University Exhibition Hall, Ha
Noi 054793
Fine Arts, Berlin 106223

Fine Arts, Oostende 100801
Fine Arts, Wien 062943
Fine Arts @ Cypress College,
Cypress 056970
Fine Arts and Communications
Division, Foothill College, Los
Altos 057362
Fine Arts and Humanities Department,
Fayetteville State University,
Fayetteville 057083
Fine Arts Appraisers, Chicago . . 120122
Fine Arts Associates, Los
Angeles 121482
Fine Arts Association, Willoughby 060481
Fine Arts by Todd, Atlanta 119564
Fine Arts Center for New River Valley,
Pulaski 052413
Fine Arts Center Galleries, Bowling
Green State University, Bowling
Green 047161
Fine Arts Center Gallery, Maryville
College, Maryville 050731
Fine Arts Center of Hot Springs, Hot
Springs 049563
Fine Arts Center of Kershaw County,
Camden 047384
Fine Arts Center, East Stroudsburg
University, East Stroudsburg . . 057032
Fine Arts Collection, Lucay-le-
Male 069653
Fine Arts Conservation & Restoration,
Wilmette 136473
Fine Arts Dealers Association,
Carmel 060482
Fine Arts Department Gallery, Xavier
University, Cincinnati 047744
Fine Arts Department, Academy of Art
College, San Francisco 057855
Fine Arts Department, Aichi Bunkyo
Daigaku, Komaki 056009
Fine Arts Department, Allan Hancock
College, Santa Maria 057874
Fine Arts Department, Benedict
College, Columbia 056938
Fine Arts Department, Benedictine
University, Lisle 057350
Fine Arts Department, Bismarck State
College, Bismarck 056746
Fine Arts Department, Blue Mountain
Community College, Pendleton 057655
Fine Arts Department, Boston College,
Newton 057577
Fine Arts Department, Burlington
County College, Pemberton . . 057653
Fine Arts Department, Capital
University, Columbus 056949
Fine Arts Department, City College of
San Francisco, San Francisco . 057856
Fine Arts Department, College of
Mount Saint Vincent, Bronx . . 056791
Fine Arts Department, Community
College of Allegheny County, West
Mifflin 058089
Fine Arts Department, Fairleigh
Dickinson University, Teaneck . 057983
Fine Arts Department, Gogebic
Community College, Ironwood . 057242
Fine Arts Department, Haverford
College, Haverford 057190
Fine Arts Department, High Point
College, High Point 057200
Fine Arts Department, Illinois Central
College, East Peoria 057031
Fine Arts Department, Indiana
University South Bend, South
Bend 057923
Fine Arts Department, Indiana
University Southeast, New
Albany 057514
Fine Arts Department, Jefferson
Community College, Louisville . 057375
Fine Arts Department, Kean University,
Union 058023
Fine Arts Department, Keystone

College, La Plume 057302
Fine Arts Department, Lakeland
Community College, Kirtland . 057291
Fine Arts Department, Loyola
University of Chicago, Chicago 056885
Fine Arts Department,
Manchester Community College,
Manchester 057397
Fine Arts Department, Marymount
Manhattan College, New York . 057552
Fine Arts Department, McLennan
Community College, Waco . . . 058052
Fine Arts Department, Mississippi
Valley State University, Itta
Bena 057247
Fine Arts Department, Niagara
University, Niagara Falls 057580
Fine Arts Department, North Georgia
College & State University,
Dahlonega 056971
Fine Arts Department, Northwestern
Connecticut Community College,
Pleasant Valley 057695
Fine Arts Department, Ohio Wesleyan
University, Delaware 056996
Fine Arts Department, Oral Roberts
University, Tulsa 058013
Fine Arts Department, Pace University,
Pleasantville 057696
Fine Arts Department, Pearl River
Community College, Poplarville 057702
Fine Arts Department, Pierce College
Fort Steilacoom, Lakewood . . . 057317
Fine Arts Department, Pikeville
College, Pikeville 057679
Fine Arts Department, Queens
University of Charlotte,
Charlotte 056860
Fine Arts Department, Quincy College,
Quincy 057740
Fine Arts Department, Regis University,
Denver 057003
Fine Arts Department, Saint Francis
University, Loretto 057360
Fine Arts Department, Saint Francis
Xavier University, Antigonish . . 055115
Fine Arts Department, Saint Michael's
College, Colchester 056924
Fine Arts Department, Santa Barbara
City College, Santa Barbara . . 057869
Fine Arts Department, Seattle
University, Seattle 057897
Fine Arts Department, South Plains
College, Levelland 057334
Fine Arts Department, State University
College at Buffalo, Buffalo . . . 056816
Fine Arts Department, Truett-
McConnell College, Cleveland . 056910
Fine Arts Department, Tusculum
College, Greeneville 057155
Fine Arts Department, University of
Arkansas at Little Rock, Little
Rock 057351
Fine Arts Department, University
of Arkansas at Monticello,
Monticello 057481
Fine Arts Department, University of
Waterloo, Waterloo 055173
Fine Arts Department, Victoria College,
Victoria 058044
Fine Arts Department, Virginia
Intermont College, Bristol 056787
Fine Arts Department, Virginia State
University, Petersburg 057663
Fine Arts Department, West Hills
Community College, Coalinga . 056919
Fine Arts Department, Xavier
University, Cincinnati 056897
Fine Arts Division, Brigham Young
University, Laie 057306
Fine Arts Division, Campbellsville
University, Campbellsville 056827
Fine Arts Division, Cottey College,
Nevada 057513

Fine Arts Division, Middlesex
Community College, Middletown 057439
Fine Arts Division, Niagara County
Community College, Sanborn . 057866
Fine Arts Division, Ventura College,
Ventura 058042
Fine Arts Division, Yuba College,
Marysville 057417
Fine Arts Exhibitions, Austin . . . 046690
Fine Arts Express, Dallas 120596
Fine Arts Fund, Cincinnati 120344
Fine Arts Gallery, Al-Khobar . . . 114385
Fine Arts Gallery, Breckenridge . 047185
Fine Arts Gallery, Columbus . . . 047929
Fine Arts Gallery, New Orleans . 051290
Fine Arts Gallery of New Orleans, New
Orleans 122163
Fine Arts Gallery, California State
University Los Angeles, Los
Angeles 050420
Fine Arts Gallery, Creighton University,
Omaha 051762
Fine Arts Gallery, University of Great
Falls, Great Falls 049217
Fine Arts Humanities Division, Ohio
University – Chillicothe Campus,
Chillicothe 056894
Fine Arts Imports, Columbus . . . 093017
Fine Arts Museum, Sa'd-Abad
Museum, Tehrän 024755
Fine Arts Press, Surrey Hills,
Victoria 136490
Fine Arts School of the Georgian
Architecture Institute, Tbilisi . . 055434
Fine Arts Section, University of Québec,
Trois-Rivières 055165, 101729
Fine Arts Services, Cincinnati . . 120345
Fine Arts Services, Los Angeles . 121483
The Fine Arts Studio, Hong Kong 102075
Fine Arts Warehouse, Houston . 121037
Fine Arts Work Center,
Provincetown 060483
Fine Arts, Columbia College,
Sonora 057922
Fine Books Classic, Croydon . . . 139708
Fine Books Oriental, London . . . 144595
Fine China Restoration, Malvern,
Victoria 127633
Fine European Arts Corporation,
Miami 121780
Fine Impressions, Seattle 124520
Fine Paintings, Bruxelles 100599
Fine Pine Antiques, Totnes 091622,
. . . . 135143
Fine Print Books, Newtown, New
South Wales 139857
Fine Refinishing, Albuquerque . . 135282
Fine Sporting Art, Lauder 118074
Fine Woodcraft Gallery,
Pemberton 099556
Fine Young Artists, Auckland . . . 083787
Fineart, Barsinghausen 074680
Finearts Contra, Berlin 106224, 106225
Finecraft Workshop, London . . . 090113,
. . . . 134657
Finedon Antiques, Finedon 089277
Fineline Gallery, Buffalo 092523
Finer Things, Nashville 122081
Finer Things of Life, Beechworth 061115
Finesilver Gallery, San Antonio . 124027
Finesse Fine Art, Dorchester . . . 089071
La Finestra, Amsterdam 082748
La Finestra Sull'Arte, Palermo . . 081043
Fingerhutmuseum, Creglingen . . 017774
Fingers, Auckland 112765
Finials, Washington 097836
Fini's Hoeve, Vriezenveen 033009
Finish Line, Las Vegas 135684
Finish Shop, Milwaukee 135814
Finished Look, Bartlett 135349
The Finishing Center, Miami . . . 135791
Finishing Touch, Atlanta 135307
The Finishing Touch, Hornchurch 117999

Finishing Touch, Milwaukee ... 094855, 135815
Finishing Touches, Derby 089053
Finishing Touches, Minneapolis . 094911, 135828
Finishing Traditions, Calgary .. 128150
De Finition Bois-Laque, Vimont . 128234
Finitude, Bordeaux 140792
Finizio, Rosario, Napoli 131620
Fink, André & Cecile, Woollahra 062468
Fink, Bernd, Heidelberg 130027
Fink, Josef, Lindenberg im
  Allgäu 137352
Fink, P., Altomünster 074400
Fink, Peter, Innsbruck 127774
Fink, Peter, München 077497
Fink, Ralf Andreas, Düsseldorf . 075467
Fink, Wilhelm, Paderborn 137468
Finlaggan Centre, Ballygrant ... 043476
Finlands Svenska Skolmuseum,
  Suomen Ruotsalainen Koulumuseo,
  Munsala 011169
Finlandssvenskt Konstcentrum, Pro
  Artibus Foundation, Ekenäs ... 010766
Finlay, K.W., Ciliau Aeron 088912
The Finley House, Wayne 054254
Finnantic, Helsinki 066313, 128412
Finnegan, Chicago 092710
Finnegan, Robin, Darlington ... 089028
Finnetunet, Grue Finnskog 033628
Finney County Kansas Historical
  Museum, Garden City 049042
Finney, Jason, Shelbyville 127458
Finney, Michael, London 090114, 144596
Finnkroken Bygdemuseum,
  Tromsdalen 034026
Finnskogsmuseet, Viksjöfors ... 034404
Fins de Siècles et Plus, Bruxelles 063425
Finsbay Gallery, Finsbay 117798
Finspångs Auktionstjänst,
  Finspång 126581
Finspångs Slott, Finspång 040626
Finstermacher, Bolzano 079896
Finta Múzeum, Túrkeve 024020
Fióka Art, Budapest 109144
Fiol, Xavier, Palma de Mallorca . 115340
Fiona & Co., Sacramento 096622
Fior di Loto, Cagliari 079982
Fior, Jake, London 144597
Fioravanti, Rachele, Genova .. 131310
Il Fiore, Bassano del Grappa .. 109780
Fiore, Angelo, Roma 131926
Fiore, Carmela, Torino 081698
Fiore, Maria Giuseppina, Roma . 081367
Fiore, Mario, Napoli 131621
Fiore, Mario, Roma 081368
Fiorentini, Enrico, Roma 081369
Fiorentino, Buenos Aires 139571
Fioretto, Firenze 080139
Il Fioretto, Padova 110564
Fiori, Willy, Ascona 087199
Fiori, Willy, Losone 133814
Fiorile Arte, Bologna 109822
Fiorillo, Marcos Eugenio, Bologna 130959, 130960
Fiorini, Firenze 080140
Fiorini – Rogai, Verona 081909
Fiorini, Susanna, San Lazzaro di
  Savena 081614
Fiorini, Umberto, Roma 131927
Il Fiorino, Pisa 081155
Fioroni, Giuseppina, Roma 110770
Fiorucci, Mario, Roma 131928
Fira d'Antiguitats de Mollerussa,
  Mollerussa 098310
Fira d'Antiguitats i Obres d'Art, Palma
  de Mallorca 098311
Fira d'Antiguitats, Coleccionisme i
  Obres d'Art, Mahón 098309
Fira d'Antiquaris i Brocanters,
  Girona 098312
Firat, Mehmet, Kassel 076720
Firbob & Peacock, Knutsford .. 118064
Fircks-Huth, Iris Freifrau von,

Fuchstal 107076
Firdusi, Chennai 079372
Fire Antiques, Miami 094732
Fire Brigades Museum, Reigate 045601
Fire Brigadesmen's Museum,
  Wellington 033346
Fire Fighters Museum of Winnipeg,
  Winnipeg 007441
Fire Fly Museum of Dongshi Forest
  Farm, Dongshi 042306
Fire Museum, Minsk 003246
Fire Museum of Maryland,
  Lutherville 050528
Fire Museum of Memphis,
  Memphis 050778
Fire Museum of Texas,
  Beaumont 046857
Fire Museum of York County,
  York 054586
Fire Police Museum, Sheffield .. 045771
Fire Safety Museum of Taipei Fire
  Department, Taipei 042514
Fire Services Museum of Victoria, East
  Melbourne 001044
De Fire Verdenshjørner, Vejle .. 066191
Fire Works @ Pioneer Square,
  Seattle 124521
Fire Works @ University Village,
  Seattle 124522
Fire Works @ Westlake Center,
  Seattle 124523
Fire-Works Gallery, Newstead . 099442
Firearms Technology Museum,
  Orange 001385
Fireborn, Pittsburgh 123541
Fireborne, Portland 096403
Firecracker, San Francisco 097232
Firefighters Museum of Nova Scotia,
  Yarmouth 007494
Firefly Antiques, Minneapolis .. 094912
Firehouse Antiques, Tucson ... 097660
Firehouse Art Center, Norman . 051585
Firehouse Art Gallery, Nassau
  Community College, Garden
  City 049044
Firehouse Gallery, Denver 120715
Firehouse Gallery, Fort Worth . 120863
Firehouse Gallery, Oak Bluffs . 051662
Firehouse Museum, Ellicott City 048553
Firehouse No. 1 Museum, Nevada
  City 051218
Firelands Association for the Visual
  Arts, Oberlin 060484
Firelands Historical Society Museum,
  Norwalk 051646
Firelight Museum, Leongatha .. 001219
Fireman's Hall Museum,
  Philadelphia 052048
Firenze Studio, Chicago 120123
The Fireplace, London . 090115, 134658
Firepower – Royal Artillery Museum,
  London 044969
Fireside Book Company,
  Columbus 145097
Firestone, Eric, Tucson 097661
Firewire Gallery, Totnes 119248
Firle Place, Firle 044347
Firmani, Giorgio, Roma 131929
Firmenmuseum Novatech,
  Reutlingen 021380
Firnkäs-Kuhnt, München 077498
First-Art, Amsterdam 082749
First Art Gallery, Lymington .. 118679
First Arts and Antiques, Chicago 092711
First at Bat, Buffalo 092524
First Austrian Scotch Whisky Museum,
  Steyregg 002790
First Cavalry Division Museum, Fort
  Hood 048814
First Class, Lohnsburg am
  Kobernaußerwald 127814
First Coast Trading Post,
  Jacksonville 093991
First Consolidated School in

Minnesota, Saum 053167
The First Division Museum at Cantigny,
  Wheaton 054364
First Draft, Surry Hills 099745
First Edition Collectibles, Houston 093669
First Floor Art Gallery, Jerewan . 098599
First Garden City Heritage Museum,
  Letchworth Garden City 044815
First Glas Galerie, München .. 108068
First Hussars Citizen Soldiers Museum,
  London 006211
First Light Gallery, Camborne . 117502
First Love, Singapore 114521
First Michigan Museum of Military
  History, Holland 049506
First Missouri State Capitol, Saint
  Charles 052755
First National Bank of Chicago Art
  Collection, Chicago 047658
First Nations Art, Albuquerque . 119473
First Oriental Rugs of Boston,
  Boston 092469
First Penny Farthing Museum,
  Knutsford 044741
First Raadsaal Museum,
  Bloemfontein 038655
First Street Antiques, Milwaukee 094856
First Street Gallery, Jacksonville 121251
First Street Gallery, New York .. 122596
First Territorial Capitol of Kansas, Fort
  Riley 048864
First-View, Stourton 119156
First White House of the Confederacy,
  Montgomery 051024
First Wing Historical Center
  (1WHC) The Golden Falcon,
  Beauvechain 003349
Firstsite, Colchester 060138
Firstsite at the Minories Art Gallery,
  Colchester 043989
Firt, Karel, Slany 065555
The Firth Gallery, Pickering ... 118911
Firth Tower Historical Reserve,
  Matamata 033245
Fiscal Prints, Ferny Creek 061463
Fisch, P., Regensburg 078166
Fischbach, New York 122597
Fischbacher, Potsdam 142307
Fischbacher, Anton, Schladming 062837
Fischel, Joan, New York 095393
Fischer, Berlin 106226, 137067
Fischer, Bogotá 065205
Fischer, Kaufbeuren 107539
Fischer, Worpswede 108966
Fischer- und Webermuseum
  Steinhude, Wunstorf 022811
Fischer & von Koeding-Fischer,
  Karlsruhe 076693
Fischer Fine Art, La Roquette-sur-
  Siagne 103905
Fischer-Menshausen, Angelika,
  Hamburg 129961
Fischer Trendy, Deventer 083156
Fischer-Zöller, Marlies,
  Düsseldorf 075468
Fischer, Albert, Kaufbeuren ... 076735
Fischer, Arno, Dießen am
  Ammersee 129683
Fischer, Bleker & Fischer, Freiburg im
  Breisgau 076693
Fischer, Carsten, Bonn 075069, 129607
Fischer, Christian, Berlin 129491
Fischer, Danielle, Saint-Ouen .. 073080
Fischer, Dora, Paris 071338
Fischer, Dr. Jürgen, Heilbronn . 076471, 125991
Fischer, E., Frickenhausen, Kreis
  Esslingen 075905
Fischer, E., Kirchenlamitz 076801
Fischer, E., Frank-Peter, Solingen . 078494
Fischer, Franz, Stuttgart 108694
Fischer, Fritz & Ruth,
  Gerhardshofen 075982
Fischer, Gabriele, Augsburg 106024

Fischer, Gebhard, Sankt Gallen . 133904
Fischer, Gerlinde, Hann. Münden 076355
Fischer, Gesine, Ludwigsburg . 077155
Fischer, Gottfried Michael, Wien 062944
Fischer, H.A., Bern 133676
Fischer, Heide, Oberstdorf ... 077879
Fischer, Hermann, Balingen 074653
Fischer, Jacques, Paris 071339, 104741
Fischer, Jeffrey W.T., Highgate, South
  Australia 061621
Fischer, Jochen, Bonn 075070
Fischer, Karin, Aachen 137021
Fischer, Karl, Wien 100254
Fischer, Klaus, Arnsberg 074424
Fischer, Klaus, Kirchhain 076802
Fischer, Konrad, Berlin 106227
Fischer, Konrad, Düsseldorf ... 075468
Fischer, Kuno, Luzern . 087638, 116473, 126744, 144106
Fischer, Martina, Schwäbisch
  Gmünd 130632
Fischer, Michael, Igling 137287
Fischer, Peter, Bratislava 085546
Fischer, Peter, München 077499
Fischer, R., Bad Buchau 074500
Fischer, R., Lienen 077102
Fischer, Ricarda, Berlin 141511
Fischer, Serge, Montpellier ... 070297
Fischer, Thomas, Würzburg ... 079080
Fischer, Walter, Aich . 062515, 127703
FischerBauKunst, Polling 021168
Fischerei-Museum, Zug 042249
Fischereimuseum,
  Wassertrüdingen 022442
Fischerplatz-Galerie, Ulm . 078753, 108800, 137539
Fischerstube, Mönkebude 020441
Fischetti, Carmine, Milano 080544
Fischinger Heimathaus mit
  Schimuseum, Fischen im Allgäu 018367
Fischinger, Angelika, Stuttgart . 078601
Fischinger, Markus, Ravensburg 108440
Fischinger, Peter, Stuttgart ... 078602
Fischl, Joseph, New York 122598
Fischlandhaus, Wustrow,
  Ostseebad 022833
Fischle, Erwin, Stuttgart 078603
Fischli, Markus, Appenzell 087194
Fish Art Center, Taipei 116514
Fish Museum Golden Angel,
  Damascus 042256
Fish Tales, Fish Creek 139729
Fishbourne Roman Palace and
  Museum, The Sussex Archaeological
  Society, Chichester 043941
Fisher, Christchurch . 112841, 132705
Fisher, Pittenweem 118915
Fisher, Rozelle 139900
Fisher & Sperr, London 134659, 144598
Fisher Gallery, University of Southern
  California, Los Angeles 050421
Fisher Grove Country School,
  Frankfort 048933
Fisher Museum of Forestry, Harvard
  University, Petersham 052019
Fisher, Barbara, Cammeray ... 139683
Fisheries Museum of the Atlantic,
  Lunenburg 006228
Fisherman's Cottage Museum,
  Tooradin 001582
Fishermen's Life Museum, Jeddore
  Oyster Ponds 006049
Fishermen's Museum, Hastings . 044531
Fishermen's Museum, Pemaquid
  Point 051966
Fishermen's Museum, Port-de-
  Grave 006682
Fishermen's Museum, Saint
  Vincent's 006942
Fishers, Auckland ... 083788, 112766, 132688
Fisher's Upholstery, Jacksonville 135652
Fisherton Mill, Salisbury 119053
Fishery Exhibition Hall, Lugang . 042380

Fishlake Antiques, Fishlake 089280
Fisk University Galleries, Aaron
Douglas Gallery and Carl Van
Vechten Gallery, Nashville 051179
Fiska- og Nåttúrngripasafn
Vestmannaeyja, Vestmannaeyjar 024119
Fiskars Museum, Fiskars 010778
Fiske Antiques, Armadale,
Victoria 061025
Fiskemuséet i Kaskö, Kaskinen . 010969
Fisken, Austin 092226
Fisker, Aksel, Randers 066053
Fiskeri- og Søfartsmuseet,
Esbjerg 010045
Fiskeriets Hus – Vestkyst Akvariet,
Fiskerimuseum, Hvide Sande . 010150
Fiskerimuseet på Hjertøya, Molde 033781
Fiskerimuseum i Måløy, Måløy . 033762
Fister, Paul, Recourt-le-Creux . 072371
Fitch-Febvrel Gallery, Croton-on-
Hudson 120527
Fitch, Michael, Sandgate 091235
Fitchburg Art Museum, Fitchburg 048737
Fitchburg Historical Society Museum,
Fitchburg 048738
Fitchett, Alan, Brighton 088621, 134150
Fitoussi, Lucien, La Chapelle-en-
Serval 068973
Fittkau, Stefan, Berlin 129492
Fitton Center for Creative Arts,
Hamilton 057175
Fitzgerald, Phoenix 096211
Fitzjohn, London 144599
Fitzkaplan, New York 135925
Fitzpatrick, Tony, Chicago 138320
Fitzroy, Prospect 062123
Fitzroy & Everest, Rotterdam .. 083569,
143218
Fitzroy Antiques, Ascot Vale .. 061062
Fitzroy Gallery, Fitzroy 099024
Fitzsimmons, Michael, Chicago . 092712
Fitzthum, Andreas, Griesheim .. 076092
The Fitzwilliam Museum,
Cambridge 043842
Fiumano, Farmilo, London 118317
Fiumicelli, Stefano, Bologna ... 079839
La Fiura, Gaetano, Palermo 081044
Ignez Fiuza, Fortaleza 100945
Fiúza, Ceres, Fortaleza 100946
Five Art & Images, Todmorden . 119239
Five Broads off Broadway, San
Antonio 096972
Five Civilized Tribes Museum,
Muskogee 051139
Five O Seven Antiques, Dallas .. 093121
Five Owls Bookshop, Richmond, North
Yorkshire 144798
Five Story Antique Market,
Shanghai 065134
Five Trees, Saint Paul 096823
Five Ways Galleries, Kalorama . 099171
Fizikos Muziejus, Vilniaus Universiteto,
Vilnius 030513
Fjällboden Antik, Järpen 086756
Fjärilshuset i Hagaparken,
Stockholm 041000
Fjeldgaard, Silkeborg . 066108, 102972
Fjeldgaards, Silkeborg 066109
Fjerritslev Bryggeri- og Egnsmuseum,
Fjerritslev 010063
Fjøsangersamlingene, Fjøsanger 033600
Fjøset Bondeantikk, Vollen 084395
Fjordane Forsvarsmuseum,
Nordfjordeid 033805
Fjordmuseet, Jyllinge 010154
Flaccovio, S.F., Palermo 137709, 142778
Flach, Henry, Auma 129399
Flach, Uwe, Gera 075976
Flach, Volker, München 130345
Flachard, Rémi, Paris 141110
Flachmann, Roland & Viera,
Dresden 129707
Flachsbrechhütte, Creglingen .. 017775
Flachsmann, Hildegard, Kronberg im

Taunus 076972, 107755
Flachsmuseum Wegberg-Beeck,
Wegberg 022452
Flacks, M.D., New York 095394
Flade, Jochen, Dresden 129708
Flagler Museum, Palm Beach . 051892
Flagstaff Antiques, Cowes 088989,
134301
Flagstaff Gallery, Auckland ... 112767
Flagstaff House Museum of Tea Ware,
Hong Kong 007871
Flagstones Pine, Melton
Mowbray 090743, 134848
Flahaut, Michel, Paris 129004
Flair, Oberstaufen 077876
La Flair, Ramstein-Miesenbach . 078123
Flais, Gérard, Le Touquet-Paris-
Plage 069361
Flajszer, Saint-Ouen 073081
Flakkebjerg Skolemuseum,
Slagelse 010318
Flamand, Bailleul 066918
Flamary, Jean-Jacques, Aurillac 066824
Flambards Experience, Helston . 044559
Flambards Victorian Village and
Gardens, Helston 044560
Flambro Head Heritage Museum,
Job's Cove 006050
Flame & Grate, Bristol 088655
Flame Lily Enterprises, Carindale 098864
Flame 'n' Grate, Barry 088371
Flame Trees Gallery, Nowa Nowa 099490
Flame!, The Gasworks Museum of
Ireland, Carrickfergus 043881
Flaménska Galleriet, Borås .. 115587
Flament, Paul-Michel, Brécey . 067455
Flamingo Gallery, Las Vegas .. 050160
Flamingo Street Antiques, Virginia
Beach 097780
Flammann, Winfried, Karlsruhe . 107510
Flammarion, Lyon 140896
Flammarion, Paris 136922, 141111
Flammarion Centre, Paris 141112
Flanagan, Seattle 097437
Flanagan Valley Campus Art Gallery,
Lincoln 050307
Flanagan, Christopher, Dublin . 079553,
130855
Flanagans of Mount Merrion, Mount
Merrion 079676, 130883
Flanders, Minneapolis 121987
Flanders, Yokohama 111535
Flanders Antiquity, Gent 063638
Flanders Center, Osaka 029240
Flanders Gallery, Knokke-Heist . 063702
Flandes Marcs, Barcelona 114884
Flandrau Science Center, Tucson 053827
Flandrin les Trois Grilles, Paris . 071340
Flaningan, J. Michael, Baltimore 092323
Flanquart, C., Agde 066419
Flao, Bernard, Les Ormes-sur-
Voulzie 069418
Flapper, Milwaukee 094857
Flash, Bordeaux 067327
Flash Art, Novi Sad 114458
Flash Art Gallery, Sydney 099780
Flash Art International, Milano .. 138826
Flash Art Italia, Milano 138827
Flash Color Imaging Center,
Richmond 123732
Flash Trash Antiques,
Camperdown 061266
Flashback, Memphis 094632
Flashback, Northcote 062003
Flashbacks, Las Vegas 094141
Flaske-Peters-Samling,
Ærøskøbing 010001
Flaster, Philippe, München 142150
The Flat, Milano 110313
Flat Cat Gallery, Lauder 118075
Flatanger Bygdemuseum,
Flatanger 033601
Flaten Art Museum, Saint Olaf College,
Northfield 051632

Flatfile Galleries, Chicago 120124
Flathead Indian Museum, Saint
Ignatius 052774
Flatland Gallery, Utrecht 112695, 137874
Flatowturm, Stiftung Preußische
Schlösser und Gärten Berlin-
Brandenburg, Potsdam 021183
Flatrock Museum, Flatrock 005806
Flaujac, New Orleans 095095
Flava, Bussum 112359
Flavoni, Roma 081370
Flax Art, Chicago 120125
Flaxbourne Settlers Museum,
Ward 033337
Flaxton Antique Gardens, Flaxton 089283
Flea Antique, Atlanta 092133
Flea Antique Depot, Atlanta ... 092134
The Fleamarket, London 090116, 134660
Fleck, Ernest, Barcelona 114885
Fleck, François, Marseille 125637
Fledderman, T.A., Cincinnati ... 092856
Fleet, Fort Worth 093447
Fleet Air Arm Museum – Concorde,
The National Museum of Science &
Industry, Yeovilton 046218
Fleet Air Arm Museum of New
Zealand, Auckland 033147
Fleet Galleries, Winnipeg 101877
Fleet Period Furniture, Tecoma . 062312
Fleetwood Museum, Fleetwood . 044349
Fleisher & Ollman, Philadelphia . 123376
Fleißig-Wächter, C., Krefeld ... 107733
Der Fleißige Biber, Griesheim .. 076093
Flekkefjord Museum, Flekkefjord 033602
Fleming Historical Museum,
Fleming 048745
Fleming, A., Portsmouth 091043, 134951
Fleming, Eris, Bakers Swamp . 098687
Flemming, L., Kenzingen 076756
Flensburger Schiffahrtsmuseum,
Flensburg 018371
Flesch, Wolfgang, Plettenberg . 078059
Flesner, Jan, Horsens 065792
Flessenscheepjesmuseum,
Enkhuizen 032304
Flesser, Helsingborg 115680
Fleßner, Rastede 078130
Fletcher, H.M., Much Hadham .. 144736
Fletcher, Leda, Carouge 116192
Fletcher, Leda, Shanghai 102179
Flett, Godalming 117867
La Fleur, Oakland 095891
Fleur de Lis Heritage Centre,
Faversham 044338
Fleuraux, Catherine, Peyrehorade 072125
Fleurieu Big Shed, Mount
Compass 099394
Le Fleuron, Ableiges 066413
Fleuron, Honfleur 103828
Fleurs du Mal, Nîmes 141009
Les Fleurs du Mal, Paris 141113
Fleurs du Sillon, Saint-Malo (Ille-et-
Vilaine) 072917
Fleury, Cahir 079468
Fleury, Dublin 079554
Fleury, Benoît, Dives-sur-Mer . 068361
Fleury, Michel, Thore-la-Rochette 073969
Flibusterie, Saint-Tropez 073313
Le Flibustier, Ars-en-Ré 066746
Flick, Stephan, Herxheim bei Landau
(Pfalz) 130040
Flicker, Portland 123619
Flieger Flab Museum, Dübendorf 041448
Fliegermuseum Altenrhein,
Altenrhein 041213
De Flierefluiter, Oss 083526
Flin Flon Museum, Flin Flon ... 005808
Flinders Fine Art Gallery, Flinders 099041
Flinders Gallery, Townsville ... 099837
Flinders Lane Gallery, Melbourne 099305
Flinders University Art Museum,
Bedford Park 000836
Flinders University City Gallery,
Adelaide 000765

Flinner, Craig, Baltimore 092324, 119735,
144984
Flint House, Scotia 053205
Flint Institute of Arts, Flint 048748
Flint Ridge State Memorial Museum,
Glenford 049120
Flint, Gregory, Auckland 112768
Flint, J.H., Amsterdam 143030
Flintham Museum, Flintham ... 044350
Flintlock Antiques, Stockport ... 091441
Flipje en Jam Museum Tiel, Tiel 032923
Flipo, Brigitte, Bondues 067283
Flipo, Emmanuel, Pézenas 105349
Flirey Art, Romans-sur-Isère ... 105456
Flirt, Lisboa 138999
Floating Gallery, Winnipeg 007442
Floating World Gallery, Chicago . 120126
Le Floch, Cyrillem, Saint-Martin-en-
Biere 072951
Flodin, Stockholm 143945
Flößer- und Heimatmuseum,
Wolfach 022724
Flößer- und Schiffermuseum, Kamp-
Bornhofen 019490
Flößerei- und Verkehrsmuseum,
Gengenbach 018635
Flößereimuseum, Lychen 020201
Flößermuseum Unterrodach,
Marktrodach 020303
Flohburg, Nordhausen 020826
Flok Arts, Haarlem 112512
Flomenhaft, Eleanor & Leonard, New
York 122599
Flood Museum, Johnstown 049832
Flooders Gallery, Ruthven 118992
Floors, Deventer 112427
Floors Castle, Kelso 044672
La Flor, Torino 081699
Flora Twort Gallery, Petersfield . 045494
Flora, Thomas, Innsbruck 100043,
136519
Floral Museum, Chiba 028509
Floravilla Art Gallery, Cow Bay . 098927
Flore, Elizabeth, New York 122600
Floréal, Evreux 068496, 140835
Florence Art Gallery, Dallas ... 120597
Florence Art Gallery, Firenze ... 110050
Florence Court House, Enniskillen 044296
Florence Griswold Museum, Old
Lyme 051749
Florence Hawley Ellis Museum of
Anthropology, Ruth Hall Museum of
Paleontology, Abiquiu 046311
Florence McLeod Hazard Museum,
Columbus 047930
Florence Mine Heritage Centre,
Egremont 044282
Florence Museum of Art, Science and
History, Florence 048759
Florence Nightingale Museum,
London 044970
Florence Price Pioneer Museum,
Florence 048757
Florence Ranch Homestead,
Mesquite 050817
Florence, Pascale, Le Port-Marly 069335
Florentin, Linselles ... 069539, 128661
Florentina Galleries, Glasgow .. 117840
Les Florentines, Apt 066693
Flores, Krefeld 076954
Flores, New York 095395
Flores & Iva, New York 095396
Florewood State Park,
Greenwood 049275
Florez Torrez, Guillermo, Ploudiry 072197
Floria Isensee GmbH, Oldenburg 137459
Florian, Houston 093670
Florian Sundheimer Kunsthandel
GmbH, München 077500
Florianerbahn, Sankt Florian .. 002648
Florida Adventure Museum, Punta
Gorda 052417
Florida Agricultural Museum, Palm
Coast 051894

Florida Art Museum Directors
Association, Vero Beach 060485
Florida Arts School, Tampa 057980
Florida Association of Museums,
Tallahassee 060486
Florida Coin and Jewelry, Tampa 097556
Florida Craftsmen Gallery, Saint
Petersburg 052859
Florida Fine Art Consignment Gallery,
Tampa 124647
Florida Frame and Art, Tampa .. 124648
Florida Gulf Coast Railroad Museum,
Parrish 051930
Florida History Museum, Jupiter 049851
Florida Holocaust Museum, Saint
Petersburg 052860
Florida International Museum, Saint
Petersburg 052861
The Florida Museum of Hispanic
and Latin American Art, Coral
Gables 047985
Florida Museum of Natural History,
Gainesville 049001
Florida Pastel Association,
Hollywood 060487
Florida Printmakers, Tarpon
Springs 060488
Florida State University Museum of
Fine Arts, Tallahassee 053657
Florida's Furniture Fixers,
Jacksonville 135653
Flories, K., Fort Worth 093448
Florijn, 's-Hertogenbosch 083387
Florilegium, Glebe 139751
Florin, Emmanuel, Le Taillan-
Médoc 128670
Florina Museum of Modern Art,
Florina 023077
Florissant Valley Historical Society,
Florissant 048760
FLORIX Multimedia, Firenze ... 137649
Florman, Yvonne, Stockholm .. 133572
Flormann, Waltraud, Mannheim . 077256
La Flosimed, Versailles 074053
Flossoms, Mosman 061893
Flotow, Ursula, Ottobrunn 142288
Flotsam Camera Collectables,
Bentley 061137
Flottmann-Hallen, Herne 019216
Floubuttek, Flou, Ettelbruck ... 082353
Flour Mill Gallery, Maryborough,
Victoria 099287
Flour Mill Museum, Sudbury ... 007093
Flow Contemporary Art Gallery,
London 118318
Flower House Antiques,
Tenterden 091559, 135124
Flowerdew Hundred Foundation,
Hopewell 049554
Flowers, London 118319
Flowers, Angela, London 118320
Flowers, Angela & Matthew, New
York 122601
George Floyd & Mark James,
London 090117
Floyd & Rita's, Toronto 064652
Floyd County Historical Museum,
Charles City 047533
Flucher, Tillmitsch 062858
Fluck, Eckhard, Berlin 141512
Flügel, M., Köln 107608
Flügen, Thomas, Lörzweiler ... 130245
Flühmann, Dr. Adrian, Zürich .. 144174
Flüthe, Reinhold, Telgte 078672
Flugausstellung Hermeskeil,
Hermeskeil 019214
Flughafen Modellschau, Hamburg 018988
Flugmuseum Aviaticum, Wiener
Neustadt 003104
Flugpionier-Gustav-Weißkopf-Museum,
Leutershausen-Wiedersbach .. 020051
Flumen Diez Martinez, Donostia-San
Sebastián 133245
Flur-Galerie im Kulturamt, Berlin 017114

Flurgalerie Eisenbart, Magdeburg 107937
Fluri-Bieri, Margrit, Zürich 087959
Flury & Co., Seattle . 097438, 124524,
145583
Flury, Matthias, Obereisenheim . 077859,
142251
Fluss & Charlesworth, London . 090118
Fluss, Niederösterreichische Initiative
für Foto- und Medienkunst,
Wolkersdorf im Weinviertel ... 058251
Fluweelengrot Valkenburg, Valkenburg,
Limburg 032961
Flux, Liège 100782
Fluxion, Omaha 123308
Fluxus-Freunde Wiesbaden e.V.,
Wiesbaden 058853
Fly by Nite Gallery, Chicago ... 092713
Flygande Huset, Sundsvall 115943
Flyhistorisk Museum, Sola 033965
Flying Arts, Brisbane 098787
Flying Colours Gallery, London . 118321
The Flying Cow Project,
Bruxelles 100600
Flying Cranes Antiques, New
York 095397
Flying Dutchman, Hondarribia .. 086049
Flying Fish Gallery, South
Melbourne 099701, 127676
Flying Goose, Beeston 117331
Flying Pig Precinct, Berry 098742
Flying Shuttles Studio,
Providence 123703
Flyingfrog, Jacksonville 093992
Flynderupgård Museet,
Espergærde 010048
Flynn, James, Warminster 119308
Flypast Museum of Australian Army
Flying, Oakey 001377
Flyvapenmuseum, Malmslätt ... 040834
Flyway Gallery, Albuquerque .. 119474
FMR, Paris 136923
FMR Art'e', Villanova di
Castenaso 137766
FMR Magazine, Villanova di
Castenaso 138828
FN Museet, Padborg 010270
Fo Guang Yuan Art Gallery and
Museum of Treasures, Dashu . 042302
Foam Fotografiemuseum Amsterdam,
Amsterdam 031940
Foam Lake Museum, Foam Lake 005809
Foata au Passe-Temps, Seine-
Port 073480
Focal Point Gallery, Southend-on-
Sea 045833
Focale-Photographie, Nyon ... 116540,
144117
Foch Antiquités, Le Creusot ... 069260
Foch, Desvigne, Lyon 140897
Fochabers Folk Museum,
Fochabers 044352
Fochem, Christian, Krefeld 107734
Focke-Museum, Bremer
Landesmuseum, Bremen 017585
Focks, Freren 075900
Foco Verde Noe Ciências Cultura e
Arte, Belo Horizonte 100885
Focus Galeria de Arte, Rio de
Janeiro 101026
Focus Gallery, Chennai 109282
Focus Gallery, Nottingham 118840
Focus Gallery, San Francisco . 124280
Focus on the Past, Bristol 088656
Fodera, New York 135926
Föglö Museet, Föglö 010779
Föegyházmegyei Könyvtár, Eger 023616
Foelkel, Th., Köln 076863
Föll, Münster 108214
Föll, Erhard, Spiegelberg 078525
Fölsing, Gröbenzell 107165
Förde Antik, Flensburg 075742, 129839
Förde Antiquariat, Flensburg ... 141764
Förde Galerie, Flensburg 106953, 129840
Fördergemeinschaft Kunst,

Karlsruhe 058854
Förderkreis Bildende Kunst e.V.,
Bernau bei Berlin 058855
Förderkreis der Städtischen Galerie
Karlsruhe, Karlsruhe 058856
Förderkreis Industrie- und
Technikgeschichte e.V., Frankfurt am
Main 018416
Förderkreis Schloss Plüschow e.V.,
Plüschow 058857
Förderkreis Stadtgalerie Saarbrücken,
Saarbrücken 058858
Förderkreis zeitgenössischer Kunst Kreis
Euskirchen e.V., Euskirchen .. 058859,
106941
Förderverein Aktuelle Kunst,
Münster 058860
Förderverein Keramik-Museum Berlin,
Berlin 058861
Förenade Antik- och Konsthandlare,
Linköping 059881
Förg, Roland, Singen 078477
Förgåtmigej i Åmål, Åmål 086628
Förnbacher, Kurt, Freiburg im
Breisgau 075861
Føroya Fornminnissavn, Hoyvík . 010731
Føroya Náttúrugripasavn,
Tórshavn 010742
Føroysk Myndlistafólk, Tórshavn 058572
Förr, Skövde 086936
Förster, Sabrina, Düsseldorf ... 075469
Foert, Frederik, Berlin 106228
Förtsch, Tobias & Karen, Kronach 076968
Föszékesegyházi Kincstár és Könyvtár,
Esztergom 023630
Foëx, Genève 116300
Fog, Dan, København 140609
Fogarty, Hojsgaard & Entwistle,
Auckland 112769
Fogarty, Terrence, Minneapolis . 121988
Fogelmarck, Stockholm 086998
Fogelsangh State, Veenklooster . 032976
Fogg, Sam, London ... 090119, 144600
Fogh Sørensen, Aksel,
Charlottenlund 065622
Fogh, Annemette, Græsted 065706
Fogle, Leigh, Jacksonville 121252
Fratelli Fogliato, Torino 110964
Fogola & Figli, Torino . 110965, 137747,
142855
Fogolino, Trento 111028
El Fogon de los Arrieros,
Resistencia 000521
La Foi, Fort-de-France 142978
Foire à la Brocante, Cogolin ... 068065
Foire d'Antiquités de Brocante,
Bussigny-près-Lausanne 087343
Foire d'Art Contemporain de
Strasbourg, Strasbourg 098079
Foire d'Art Moderne et Contemporain,
Paris 098080
Foire Internationale Brocante
Antiquités de Paris, Le Bourget 098081
Foire Internationale d'Art
Contemporain, Paris 098082
Foki's, Long Beach 135694
Folavril, Paris 071341
Foldager, Villy, Nykøbing Mors .. 066016
Foldalbruket, Kjøllefjord 033699
Foley, Columbus 120492
Foley, New York 122602
Foley House Antiques, Great
Malvern 089371
Foley l'Ecuyer, Montréal 064420
Foley, Kel, Miami 099333
Foley, Shannon C., New Orleans 122164
Folger Shakespeare Library,
Washington 054112
Folha D'Ouro, Lisboa 132935
La Folie, 's-Hertogenbosch 083388
La Folie des Arts, Montréal 101383

Folies Antiques, Torino 081700
Folini Artecontemporanea,
Chiasso 116213
Folio, Gronsveld 132511
Folio – The Graphic House,
Bruxelles ... 100601, 128028, 136635
Folio Verlag, Bolzano 137630
Folio Verlag, Wien 136583
Foliograph Gallery, Washington . 124898
Folk and Applied Arts Museum,
Tbilisi 016400
Folk and Country Antiques,
Ashbourne 088298
Folk Art and Photography Galleries,
High Museum of Art, Atlanta . 046618
Folk Art Finders, Essex 139364
Folk Art Institute, Lefkoşa 033472
Folk Art Messenger, Richmond . 139365
Folk Art Museum, Lemesos ... 009425
Folk Art Museum, Pyrgi 023238
Folk Art Museum, Yeroskipos .. 009434
Folk Art Society of America,
Richmond 052549, 060489
Folk Cultural Collection Museum,
Jhongpu 042346
Folk Custom Museum, Xiangfen 008408
Folk Experience Museum,
Dranouter 003559
Folk Heritage Museum, Thimphu 004201
Folk Museum, Anuradhapura .. 040472
Folk Museum, Câtel 043894
Folk Museum, Mbarara 043110
Folk Museum, Mevagissey 045255
Folk Museum, Millom 045267
Folk Museum, Soroti 043112
Folk Museum of Indian Immigration,
Moka 030788
Folkart Interiors, Vancouver ... 064821
Folkenborg Museum, Eidsberg og
Mysen Historielag, Mysen ... 033792
Folkes Antiques & Jewellers, Great
Yarmouth 089382
Folkestone Museum and Art Gallery,
Folkestone 044354
Folkets Hus, Hovmantorp 126604
Folklife Display, Lisnaskea 044845
Folklore and Ethnological Museum
of Macedonia and Thrace,
Thessaloniki 023289
Folklore Museum, Cairo 010539
Folklore Museum, Mysore 024415
Folklore Museum, Pogoniani ... 023228
Folklore Museum, Thessaloniki . 023290
Folklore Museum, Victoria 030761
Folklore Museum, Xanthi 023319
Folklore Museum and Municipal Art
Gallery, Karditsa 023111
Folklore Museum at Drama,
Drama 023065
Folklore Museum of Folegandros,
Folegandros 023080
Folklore Museum of Polygyros,
Polygyros 023230
Folklore Museum of the Culture Club
Florina, Florina 023078
Folkloremuseum Florishof,
Oostduinkerke 003921
Folkloristisch Museum, Swalmen 032913
Folkwang-Museumsverein e.V.,
Essen 058862
Folldal Bygdetun, Folldal 033606
Folldal Gruver, Folldal 033607
Follens, Hervé, Bapaume 066934
Follett House Museum, Sandusky 053070
Folliard, Tory, Milwaukee 121915
Follin, Italo, Venezia 111103
Follinge Antik och Kuriosa,
Skänninge 086621
Follo Museum, Drøbak 033574
Follot, Jeannine, Gevrey-
Chambertin 068655
Folly Antiques, Clitheroe 088655
Folly Antiques Centre, Petersfield 090985
Folsom History Museum, Folsom 048770

Folsom House Museum, Taylors
Falls 053690
Folsom Museum, Folsom 048771
Foltermuseum, Burghausen 017688
Foltermuseum, Museum für
mittelalterliche Rechtsgeschichte,
Wien 002990
Foluo Lun Sa, Shenzhen 102221
Fond Régional d'Art Contemporain,
Angoulême 011662
Fondaçao-Museu Chissano,
Matola 031811
Fondantico, Bologna 079840
Fondation André Hallet, Bruxelles 136636
Fondation Beyeler, Riehen 041879
Fondation Cartier pour l'Art
Contemporain, Paris 014649
Fondation Carzou, Manosque . . 013992
Fondation Charles de Gaulle,
Colombey-les-Deux-Eglises . . 012633
Fondation Cziffra, Senlis 015766
Fondation de Jau, Cases-de-
Pène 012349
Fondation de l'Automobile Marius-
Berliet, Lyon 013932
Fondation de l'Hermitage,
Lausanne 041651
Fondation Dosne-Thiers, Paris . . 014650
Fondation Dubuffet, Paris 014651
Fondation Dubuffet, Périgny-sur-
Yerres 014827
Fondation et Musée René Carcan,
Bruxelles 003442
Fondation Félicien Rops, Mettet 003870
Fondation Folon, La Hulpe 003762
Fondation Hergé ASBL, Bruxelles 136637
Fondation Herzog, Basel 041266
Fondation Jean Paul Blachère,
Abidjan 008868
Fondation Jean Paul Blachère,
Apt 011698
Fondation La Tène, Laténium, Parc et
Musée d'Archéologie de Neuchâtel,
Hauterive 041585
Fondation Le Corbusier, Paris . . 014652
Fondation L'Estrée, Espace Culturel,
Ropraz 041891
Fondation Louis Moret, Martigny 041722
Fondation Maeght, Saint-Paul-de-
Vence 015495
Fondation Marius Vazeilles,
Meymac 014160
Fondation Martin Bodmer,
Cologny 041421
Fondation Musée-Volti, Musée de la
Citadelle, Villefranche-sur-Mer 016246
Fondation Oscar et Janine Wiggli,
Muriaux 041783
Fondation Pierre Gianadda,
Martigny 041723
Fondation pour l'Architecture,
Bruxelles 003443
Fondation pour l'Art Contemporain
Claudine et Jean-Marc Salomon,
Alex 011597
Fondation Prouvost, Septentrion,
Marcq-en-Barœul 014010
Fondation Rainer Maria Rilke,
Sierre 041985
Fondation René Pous, Auterive . 011806
Fondation Saner, Stiftung für
Schweizer Kunst, Studen 042037
Fondation Van-Gogh Arles, Arles 011728
Fondation Vasarely, Aix-en-
Provence 011559
Fondazione Adriano Olivetti, Ivrea 059440
Fondazione Adriano Olivetti,
Roma 059441
Fondazione Antonio Mazzotta,
Milano 026792
Fondazione Antonio Ratti, Como 059442
Fondazione Arnaldo Pomodoro,
Milano 059443
Fondazione Bandera per l'Arte, Busto

Arsizio 025666
Fondazione Bevilacqua La Masa,
Venezia 059444
Fondazione Dazio Grande, Rodi-
Fiesso 041886
Fondazione delle Stelline, Milano 059445,
110314
Fondazione Europea del Disegno,
Meina 026755
Fondazione Galleria Gottardo,
Lugano 041694
Fondazione M. Minucci, Vittorio
Veneto 028412
Fondazione Magnani Rocca, Mamiano
di Traversetolo 026686
Fondazione March, Padova 059446
Fondazione Mazzucconi, Milano 059447
Fondazione Memmo, Roma 110771
Fondazione Merz, Torino 028113
Fondazione Michetti, Chieti 059448
Fondazione Musei Civici Veneziani.
Bollettino, Venezia 138829
Fondazione Museo di Fotografia
Contemporanea, Cinisello
Balsamo 059449
Fondazione Museo Glauco Lombardi,
Museo Lombardi – Napoleone e
Maria Luigia, Parma 027207
Fondazione Palazzo Albizzini
Collezione Burri, Città di
Castello 026000
Fondazione Palazzo Bricherasio,
Torino 059450
Fondazione Palazzo Coronini Cronberg,
Gorizia 026452
Fondazione per l'Arte Bartoli-Felter,
Cagliari 059451
Fondazione Prada, Milano 026793,
059452
Fondazione Rolf Gérard, Ascona 041234
Fondazione Romano nel Cenacolo di
Santo Spirito, Firenze 026251
Fondazione Sandretto Re Rebaudengo,
Torino 028114
Fondazione Spinola Banna per l'Arte,
Poirino 059453
Fondazione Studio Marangoni,
Firenze 059454
Fondazione T. Balestra, Longiano 026622
Fondazione Teseco per l'Arte,
Pisa 059455
Fondazione Trussardi, Milano . . 059456
Fonderie de Cloches, Robecourt 015111
Fondo del Sol, Washington 124899
Fondo Del Sol, Visual Art and Media
Center, Washington 054113
Fondo Internacional de Pintura,
Barcelona 114886
Fondon, Palma de Mallorca 115341
Fonds ancien de la Bibliothèque
cantonale jurassienne,
Porrentruy 041843
Fonds Ancien, Manuscrits du Mont
Saint-Michel, Avranches 011847
Fonds Régional d'Art Contemporain,
Saint-Paul 036187
Fonds Régionaux d'Art Contemporain
Bourgogne, Dijon 012780
Foney, E., Houston 121038
Fong, Los Angeles 094337
Fong & Yeung, Hong Kong 122076
Fong, D.P., San Jose 124448
Fong, Freddie, San Francisco . . 124281
Fong, Suzanna Mah, San
Francisco 124282
Al Fonoon, Kasr, Cairo 103038
Fons Blavus, Renningen 142329
Font Mascaro, Margarita,
Barcelona 133116
La Fontaine Antiquité, Paris 071342
La Fontaine Charlemagne, Paris 141114
La Fontaine d'Argent, Montluçon 070268
Fontaine des Arts, Herblay 103812
La Fontaine des Arts, Sanary-sur-

Mer 105648
La Fontaine du Temps, Fresnay-sur-
Sarthe 068609
Fontaine, Christophe, Colmar . . 068089
Fontaine, Danièle, Cherbourg-
Octeville 067961
Fontaine, Dominique, Caen 067568
Fontaliran, Jean-Marc,
Montignac 070251
La Fontana, Paris 104742
Fontana, Roma 131930, 131931
Fontana, Emi, Milano 110315
Fontana, Marco, Roma 081371
Fontana, Nazzareno, Roma 131932
Fontana, Nicola, Firenze 131171
Fontana, Nicola, Roma 131933
Fontanarosa, Riccardo, Cagliari . 131648
Fontane-Antiquariat, Berlin 141513
Fontane Marose 4, Genova 110104
La Fontanella, Carpi 109934
Fontanet, Hugues, Cherbourg-
Octeville 067962
Fontanez, Marc, Maisons-Laffitte 069814
Fontanini, Pietro, Roma 131934
Fontannaz, Serge, Lausanne . . . 133797
La Fonte, Muralto 087688
Fonte d'Abisso Arte, Milano . . . 110316
Fontenele, Fatima, Rio de
Janeiro 101027
Fontenelle, Marly-le-Roi 069879
Fonthill Museum of the Bucks County
Historical Society, Doylestown . 048364
Fontvieille Antiquités, Monaco . . 082506
Fontys Hogeschool voor de Kunsten,
Akademie voor Beeldende Vorming,
Tilburg 056183
Foo Yong Kong, Kuala Lumpur . 111904
Foochow, Hong Kong 065042
Foord-Brown, David, Cuckfield . 089020
Foot-Binding Culture Museum, Broad
River Hospital, Tucheng 042580
Foothills Art Center, Golden 049147
Footnote Books, San Diego 145522
Foot's Artworks, Hamilton Island 099106
Foot's Artworks, Main Beach . . . 099253
Footscray Community Arts Centre,
Footscray 001074
Foott, Andrew, Cheadle Hulme . 088826,
134228
For All Seasons, Columbus 093018
For Art Sake Galleries, Newcastle 099431
For Decorative Living, London . . 090120
For Love of the Earth, Esperance 099005
For Old Times Sake, Milwaukee 094858
For the Love of Dolls, Dallas . . . 093122
For the Record – City of London
Phonograph & Grammophone Society,
Edgmond 139177
For Your Walls, Louth 118667
Foragers, Christchurch 083871
Forain, Pierre & Suzanne, Annecy-le-
Vieux 066651
Foralosso, Catherine, Genève . . 087467
Foralosso, Jean-Louis, La
Murette 069055
Foralosso, Michel, Voiron 074272,
129359
Forante, Silvia, Verona 132283
Forbelsky, Johannes, Wien 127919
Forbes, Los Angeles 061479
Forbes Art Gallery, Forbes 099043
Forbes Collection, New York . . . 051386
Forbes Museum, Forbes 001076
Forbes Museum, Tanger 031795
Forbes, James, London 090121
Forbes, M.R., Sacramento 096623
The Forbidden City Museum,
Lujhou 042386
Forbidden Gallery, London 120598
Forbidden Planet, New York . . . 145338
Force, Debra, New York 122603
Ford County Historical Society
Museum, Paxton 051954
Ford End Watermill, Ivinghoe . . 044660

Ford Gallery, Eastern Michigan
University, Art Dept., Ypsilanti . 054602
Ford Green Hall, Stoke-on-Trent 045882
Ford Museum, Dumalag 034478
Ford, Rita, New York . . . 095398, 135927
Ford, Ruth, York 091975
Forde Abbey, Chard 043906
Fordholm, Flinders 061477
Ford's Theatre, Lincoln Museum,
Washington 054114
Fordsmith, Denver 093286
Fordyce Joiner's Visitor Centre,
Fordyce 044357
Forecast Public Artworks, Saint
Paul 138451
Foredown Tower, Portslade-by-
Sea 045547
Foreign Affairs, San Antonio . . . 124028,
124029
Foreman Art Gallery of Bishop's
University, Lennoxville 006181
Foreman Gallery, Hartwick College,
Oneonta 051776
Foreman, Skot, Atlanta 119565
Foreningen af Danske Kunstmuseer,
Nivå 058550
Foreningen af Danske
Museumsmaend, København . . 058551
Foreningen af Danske Naturhistoriske
Museer, Århus 058552
Foreningen Kunst på Arbejdspladsen,
København 058553
Foreningen til Norske
Fortidsminnesmerkers Bevaring,
Oslo 059656
Forensic Medicine Museum,
Bangkok 042656
Forero, Bogotá 102297
Fores Domus, Roma 081372
Foresight International, Omaha . 123309
Forest and Heritage Centre,
Geeveston 001096
Forest Art Gallery, Kuranda 099208
Forest Art Gallery, Stafford 119133
Forest Books, Chingford 144340
Forest Books, San Francisco . . . 145553
Forest Books, Uppingham 144901
Forest Capital State Museum,
Perry 051993
Forest City Gallery, London 101339
Forest Department Utilisation Division
Museum, Kampala 043105
Forest Discovery Center, Portland 052293
Forest Entomology Museum,
Bangkok 042657
Forest Gallery, Forestville 099045,
127599
Forest Gallery, Guildford 117894
Forest History Center, Grand
Rapids 049196
Forest-Lambton Museum, Forest 005810
Forest Lawn Museum, Glendale 049118
Forest Museum, Gyeonggi-do . . 029943
Forest Museum, Peshawar 034142
Forest Museum, Taungdwingyi . 031827
Forest Park Museum, Perry 051994
Forest, Laurent, Alès 066529
Forestburg and District Museum,
Forestburg 005811
Forestier, Antoine & Marie, Saint-
Martial-d'Albarède 072929
Forestier, Pierre, Pézenas 072135
Forestry Exhibition Hall,
Fengyuan 042313
Forestry Research Institute Malaysia
Museum, Kepong 030672
Forestwood Antique Mall, Dallas 093123
Forêt Verte, Paris 071343
Forêt, Patrick, Saint-Dizier 072739
Forever After Books, San
Francisco 145554
Forever Amber, Paddington,

Queensland 062051
Forever Finishing, Jacksonville . 135654
Forever Keepsakes, Saint Louis . 123825
Forever Plaza de Mexico,
  Reynosa 111990
La Forge, Grandchamp-des-
  Fontaines 068694, 128613
The Forge Antiques, Coleraine . . 088957
Forge Antiques, Lissarda 079664
Forge Antiques, Sandhurst, Kent 091245,
  135008
Forge Gallery, Robin Hoods Bay 118982
Forge-Menuiserie Cauchon, Rivière-
  Malbaie 006814
Forge Mill Needle Museum and
  Bordesley Abbey Visitor Centre,
  Redditch 045596
Forge-Musée d'Etueffont,
  Etueffont 012930
Forge Musée d'Ittre, Ittre 003731
Forge Museum, Much Hadham . 045302
Forge Studio Workshops,
  Manningtree 134835
The Forges and Manor of Ringwood,
  Ringwood 052575
Forges, Hubert des, New York . . 095399
Forget me not, Christchurch . . . 083872
Forget Me Not, San Jose 097365
Forget, Bernard, Poitiers 072222, 129172
Forgotten Shanghai, San
  Francisco 097233
Forgotten Times, Koorawatha . . 061701
Forichon, Gisèle, Bessines-sur-
  Gartempe 067146
Forlaget Cordelia, Skørping 136781
Form 1900, Mariefred 086863
Form Gallery, Albuquerque 092051
Form Gallery, Christchurch 112842
Form Gallery, Cork 109555
FORM London, Modern Contemporary
  Art + Design, London 098375
Form Vintage Modern, Oakland . 095892
Form, The Making of Design,
  Basel 139091
Forma, Moskva 114218
Forma Aktua, Groningen 112492
Forma-Centro Internazionale di
  Fotografia, Milano 026794
Formann, Wien 062945
Formanoir & Launoit, Bruxelles . 063426
Formargruppen, Malmö 115744
FormArt, Berlin 106229
Formas Galeria de Arte, Lima . . 113347
Formas Plásticas, Barcelona . . . 139052
Formas y Espacios,
  Bucaramanga 102385
Format, Kraków 113543
Format, Wrocław 138964
Format of Birmingham,
  Birmingham 088479
Formby, Jeffrey, Moreton-in-
  Marsh 090780, 144735
Forme d'Art, Las Vegas 121350
Forme Libre, Grenoble 068727
Formenti, Gianni, Padova 131663
Former Glory, Gosport . 089348, 134413
Former Glory Country Antiques,
  Masterton 083989
Former Governors' Mansion,
  Bismarck 047020
Former Hong Kong and Shanghai
  Bank Museum, Nagasaki 029089
Formes, Tokyo 111419, 111420
Formliner, Mannheim 077257
Fornaro, Rosalia, Bologna 079841
Fornasier, Alba, Wettolsheim . . 074295
Forncett Industrial Steam Museum,
  Forncett Saint Mary 044360
Forneck, Gerd Martin, Höhr-
  Grenzhausen 076544
Fornello, Nicola, Torino 110966
Fornès, Le Juch 069290
Forney Transportation Museum,
  Denver 048263

Forney, Stéphane, Cormondrèche 087383
Forni, Bologna 109823
Fornminnesgården, Eksjö 040599
Forno, Barry, Dunedin 126315
Fornoni, Eugenio, Verona 081910
Fornvännen, Journal of
  Swedish Antiquarian Research,
  Stockholm 139079
Foroni, Fausto, Milano 080545
Forré & Joisten, Engelskirchen . 075618
Forrest Scott Gallery, Millburn . . 121888
Forrester Gallery, Oamaru 033261
Forrest's Antiques, Fort Worth . . 093449
Forsand Bygdemuseum, Forsand 033609
Forschungsinstitut für Kunstgeschichte,
  Philipps-Universität, Marburg . 055639
Forssan Luonnonhistoriallinen Museo,
  Forssa 010780
Forst- und Köhlerhof, Rostock . . 021490
Forstbotanisches Museum, Technische
  Universität Dresden, Tharandt . 022144
Forster, Silvia, Friedenweiler . . 141816
Forstliche und Jagdkundliche
  Lehrschau Grillenburg, Tharandt 022146
Forstmuseum, Reichraming 002583
Forstmuseum Heringen, Heringen
  (Werra) 019206
Forstmuseum im Hochwildpark
  Karlsberg, Weikersheim 022464
Forsvarsmuseet, Oslo 033833
Forsvarsmuseet på Bornholm,
  Rønne 010287
Forsviks Bruk, Forsvik 040631
Forsyth, Dublin 079555
Forsythe, Frank & Dave,
  Cincinnati 127173
Fort Abercrombie Historic Site,
  Abercrombie 046298
Fort Amherst Heritage Museum,
  Chatham 043909
Fort Ancient Museum, Oregonia 051799
Fort Anne, Annapolis Royal 005339
Fort Atkinson State Historical Park,
  Fort Calhoun 048790
Fort Beaufort Museum, Fort
  Beaufort 038714
Fort Beauséjour, National Historic Site,
  Aulac 005360
Fort Bedford Museum, Bedford . 046871
Fort Belknap Museum,
  Newcastle 051519
Fort Bend Museum, Richmond . 052540
Fort Benton Museum of the Upper
  Missouri, Fort Benton 048781
Fort Bliss Museum, Fort Bliss . . 048783
Fort Boonesborough Museum,
  Richmond 052537
Fort Bowie, Bowie 047150
Fort Bridger State Museum, Fort
  Bridger 048789
Fort Brockhurst, Gosport 044450
Fort Buenaventura, Ogden 051714
Fort Buford, Williston 054424
Fort Calgary, Calgary 005520
Fort Carlton, Regina 006767
Fort Caroline Memorial Museum,
  Jacksonville 049769
Fort Carré, Antibes-Juan-les-Pins 011684
Fort Casey Interpretive Center,
  Coupeville 048037
Fort Caspar Museum, Casper . . 047455
Fort-Chambly, Chambly 005587
Fort Charles Maritime Museum,
  Kingston 028440
Fort Charlotte, Nassau 003204
Fort Chipewyan Bicentennial Museum,
  Fort Chipewyan 005814
Fort Christmas, Christmas 047732
Fort Clark Trading Post, Center . 047497
Fort Clatsop National Memorial,
  Astoria 046584
Fort Clinch, Fernandina Beach . 048725
Fort Collins Museum, Fort Collins 048796
Fort Columbia House Museum,

Chinook 047726
Fort Concho, San Angelo 052924
Fort Crawford Museum, Prairie du
  Chien 052345
Fort Croghan Museum, Burnet . 047329
Fort Crook Historical Museum, Fall
  River Mills 048689
Fort Dalles Museum, The Dalles 048129
Fort d'Artillerie A46, Champex . 041401
Fort d'Aubin-Neufchâteau et son
  Musée, Neufchâteau 003908
Fort Dauphin Museum, Dauphin 005684
Fort Davidson, Pilot Knob 052132
Fort Davis Museum, Fort Davis . 048800
Fort de Chartres Museum, Prairie du
  Rocher 052347
Fort de Seyne, Seyne-les-Alpes 015798
Fort de Vallorbe Pré-Giroud,
  Vallorbe 042095
Fort de Verny, Groupe Fortifié l'Aisne-
  Feste Wagner, Verny 016178
Fort Découvertes, Martigny . . . 041724
Fort Delaware, Narrowsburg . . . 051167
Fort Delaware Society Museum,
  Delaware City 048230
Fort Discovery, Augusta 046659
Fort Dix Museum, Fort Dix 048803
Fort Dobbs State Historic Site,
  Statesville 053508
Fort Dodge Historical Museum, Fort
  Dodge 048805
Fort Donelson National Battlefield
  Museum, Dover 048356
Fort Door, Edmonton 101285
Fort Dorsner, Giromagny 013111
Fort Douglas Military Museum, Fort
  Douglas 048806
Fort Edmonton Park, Edmonton . 005753
Fort Erie Historical Museum,
  Ridgeway 006802
Fort Erie Railroad Museum, Fort
  Erie 005815
Le Fort et son Musée, Flémalle . 003598
Fort Fetterman, Douglas 048345
Fort Fincastle, Nassau 003205
Fort Fisher, Kure Beach 050021
Fort Fleur d'Epée, Le Gosier . . . 023346
Fort Frances Museum, Fort
  Frances 005817
Fort Frederica, Saint Simons
  Island 052869
Fort Frederick, Big Pool 046977
Fort Garry Horse Regimental Museum,
  Winnipeg 007443
Fort George, Niagara-on-the-
  Lake 006488
Fort George and Buckingham House,
  Saint Paul 006927
Fort George G. Meade Museum, Fort
  Meade 048842
Fort Gibson Historic Site, Oklahoma
  Historical Society, Fort Gibson . 048811
Fort Glanville, Semaphore Park . 001490
Fort Grey and Shipwreck Museum,
  Saint Pierre du Bois 045716
Fort Hancock Museum, Sandy
  Hook 053078
Fort Harker Museum Complex,
  Kanopolis 049871
Fort Hartsuff, Burwell 047332
Fort Hays State Historic Site,
  Hays 049414
Fort Henrietta, Echo 048469
Fort Henry, Kingston 006099
Fort Hill – The John C. Calhoun House,
  Clemson University, Clemson . 047798
Fort Hill Museum, Hillsboro . . . 049477
Fort Huachuca Museum, Fort
  Huachuca 048815
Fort Hunter Mansion, Harrisburg 049370
Fort Ingall, Cabano 005509
Fort Inglish, Bonham 047081
Fort Jackson Museum, Fort
  Jackson 048817

Fort Jesus Museum, Mombasa . 029865
Fort Jones Museum, Fort Jones 048821
Fort Kearney Museum, Kearney 049907
Fort Kearny, Kearney 049908
Fort Kijkduin, Castricum 032135
Fort King George, Darien 048151
Fort Klock Historic Restoration, Saint
  Johnsville 052779
Fort Knox State Historic Site, Stockton
  Springs 053543
Fort La Reine Museum and Pioneer
  Village, Portage-la-Prairie 006698
Fort Langley, National Historic Site,
  Fort Langley 005820
Fort Laramie, Fort Laramie 048823
Fort Larned, Larned 050149
Fort Laurens State Memorial,
  Bolivar 047078
Fort Leaton, Presidio 052354
Fort Leboeuf, Waterford 054203
Fort l'Ecluse, Archéosite Militaire,
  Leaz 013718
Fort Lee Historic Park and Museum,
  Fort Lee 048833
Fort Lennox, Saint-Paul-de-l'Île-aux-
  Noix 006930
Fort Lewis Military Museum, Fort
  Lewis 048838
Fort Ligonier Museum, Ligonier . 050277
Fort Loudoun, Chambersburg . . 047515
Fort Lowell Furniture, Tucson . . 097662
Fort Lowell Museum, Tucson . . 053828
Fort McAllister, Richmond Hill . . 052567
Fort MacArthur Museum, San
  Pedro 053058
Fort McHenry, Baltimore 046747
Fort McKavett State Historic Park, Fort
  McKavett 048839
Fort Mackinac, Mackinac Island 050556
Fort Macon, Atlantic Beach . . . 046641
Fort Malden, Amherstburg 045329
Fort-Mardi Antiquités, Montévrain 070241
Fort Massac, Metropolis 050821
Fort Matanzas, Saint Augustine . 052744
Fort Meigs State Memorial,
  Perrysburg 051996
Fort Mifflin, Philadelphia 052049
Fort Montagu, Nassau 003206
Fort Morgan Museum, Fort
  Morgan 048848
Fort Morgan Museum, Gulf
  Shores 049294
Fort Morris, Midway 050881
Fort Mountain, Chattsworth . . . 047599
Fort Muséalisé d'Uxegney,
  Uxegney 016075
Fort N5 – Muzej Istorii Velikoj
  Otečestvennoj Vojny, Oblastnoj
  istoriko-chudožestvennyj Muzej,
  Kaliningrad 037080
Fort Namutoni, Tsumeb 031856
Fort Napoleon, Oostende 003928
Fort Nashborough, Nashville . . . 051180
Fort Necessity National Battlefield,
  Farmington 048706
Fort Nelson Heritage Museum, Fort
  Nelson 005826
Fort Nisqually Living History Museum,
  Tacoma 053642
Fort Normandeau, Red Deer . . . 006757
Fort-Numéro-Un de la Pointe-de-Lévy,
  Lévis 006191
Fort Okanogan, Pateros 051942
Fort Ontario, Oswego 051838
Fort Osage, Sibley 053330
Fort Ostell Museum, Ponoka . . 006668
Fort Peck Museum, Fort Peck . . 048855
Fort Pelley Livingston Museum,
  Pelly 006621
Fort Phil Kearny, Story 053556
Fort Pike, New Orleans 051291
Fort Pitt Museum, Pittsburgh . . 052162
Fort Plain Museum, Fort Plain . . 048859

Fort Point Museum, LaHave ... 006152
Fort Polk Military Museum, Fort
Polk 048860
Fort Pulaski, Savannah 053175
Fort Qu'appelle Museum, Fort
Qu'appelle 005827
Fort Queenscliff Museum,
Queenscliff 001438
Fort Raleigh, Manteo 050657
Fort Rammekens, Ritthem 032788
Fort Recovery Museum, Fort
Recovery 048863
Fort Richardson, Jacksboro 049745
Fort Robinson Museum,
Crawford 048052
Fort Rodd Hill, Victoria 007324
Fort Roosevelt Natural Science and
History Museum, Hanford 049349
Fort Sackville Foundation,
Bedford 005410
Fort Saint George Museum,
Chennai 024196
Fort Saint James, National Historic
Site, Fort Saint James 005828
Fort Saint Jean Baptiste,
Natchitoches 051198
Fort Saint John-North Peace Museum,
Fort Saint John 005829
Fort Saint Joseph, Richard's
Landing 006788
Fort Saint Joseph Museum, Niles 051563
Fort Sam Houston Museum, Fort Sam
Houston 048867
Fort San Pedro National Museum,
Cebu 034461
Fort Saskatchewan Museum, Fort
Saskatchewan 005830
Fort Savannah Museum,
Lewisburg 050228
Fort Scott National Historic Site, Fort
Scott 048869
Fort Scratchley, Newcastle 001352
Fort Selden State Monument, Radium
Springs 052443
Fort Selkirk, Yukon 007503
Fort Sidney Museum and Post
Commander's Home, Sidney .. 053332
Fort Sill Museum, Fort Sill 048871
Fort Siloso, Singapore 038269
Fort Simcoe, White Swan 054376
Fort Sisseton, Lake City 050076
The Fort Smith, Fort Smith 048742
Fort Smith Art Center, Fort Smith 048873
Fort Smith Museum of History, Fort
Smith 048874
Fort Snelling, Saint Paul 052838
Fort Spokane Visitor Center and
Museum, Davenport 048156
Fort Stanwix, Rome 052673
Fort Steele Heritage Town, Fort
Steele 005832
Fort Stewart Museum, Fort
Stewart 048876
Fort Sumner, Fort Sumner 048880
Fort Sumter National Monument,
Sullivan's Island 053595
Fort Tejon, Lebec 050209
Fort Témiscamingue, Lieu Historique
National, Ville-Marie 007341
Fort Ticonderoga, Ticonderoga .. 053743
Fort Totten State Historic Museum,
Pioneer Daughters Museum, Fort
Totten 048881
Fort Towson Military Park, Oklahoma
Historical Society, Fort Towson 048882
Fort Uncompahgre History Museum,
Delta 048235
Fort Union, Watrous 054227
Fort Union Trading Post, Williston 054425
Fort Vancouver, Vancouver 053947
Fort Vasquez Museum, Colorado
Historical Society Regional Museum,
Platteville 052194
Fort Verde, Camp Verde 047390

Fort Walla Walla Museum, Walla
Walla 054038
Fort Walsh, Maple Creek 006258
Fort Ward Museum and Historic Site,
Alexandria 046386
Fort Washington, Fort
Washington 048887
Fort Washington Museum, Fort
Washington 048889
Fort Washita, Oklahoma Historical
Society, Durant 048401
Fort Wayne Firefighters Museum, Fort
Wayne 048893
Fort Wayne Museum of Art, Fort
Wayne 048894
Fort Wellington, Prescott 006704
Fort Whoop-Up Centre,
Lethbridge 006185
Fort Widley, Portsmouth 045552
Fort Wilkins Historic Complex, Copper
Harbor 047980
Fort William Henry Museum, Lake
George 050078
Fort Winnebago Surgeons Quarters,
Portage 052267
Fort Worth Antiques, Fort Worth 093450
Fort Worth Coin Company, Fort
Worth 093451
Fort Worth Museum of Science and
History, Fort Worth 048902
Fort Worth Public Library Arts and
Humanities, Fort Worth 048903
Fort Yargo, Winder 054467
Fort York Armoury, Museum of
the Royal Regiment of Canada,
Toronto 007173
Fort Zachary Taylor, Key West .. 049953
Fortaleza de Maputo, Maputo .. 031803
Fortaleza de Nuestra Señora de los
Angeles de Jagua, Jagua 009225
Fortaleza de Sagres, Sagres ... 035986
Fortaleza San Cristóbal, San
Juan 036154
Forte Rinascimentale, San Leo . 027766
Fortebraccio, Firenze 142672
Fortelli, Alessandra, Pisa 131764
Fortenbacher, Klaus,
Crimmitschau 129661
Forteresse de Mornas, Mornas . 014334
Forteresse de Salses, Salses-le-
Château 015666
Forteresse Médiévale de Ducs,
Bourbon-l'Archambault 012146
Fortescue, Birdie, London 090122
Fortifications-de-Québec, Québec 006728
Fortín San Jerónimo del Boquerón,
San Juan 036155
Fortin, Catherine, Saint-Ouen .. 073082
Fortin, Evelyne, Paris 129005
Fortix Ltd, Sharjah 117198
Fortlands Antique, Charleville .. 079479
Fortner, Columbus 135479
Fortress of Louisbourg,
Louisbourg 006223
Forts Folle Avoine Historic Park,
Danbury 048138
Fortuijn, Amsterdam 082750
Fortuna, Merano 080351, 080352
Fortuna, Roma 142811
Fortuna Fine Arts, New York ... 095400
Fortunate Finds, South Auckland 084080
Fortune Antiques, New York ... 095401
Fortune Art Gallery, Pune 109495
Fortune Classics, Leeuwarden .. 083414
Fortune de Mer, Toulouse 073767
Fortune Green Books, London .. 144601
Fortune Line, New Orleans 095096,
.... 135855
Forty Fifty Sixty, New York 095402
Forty Hall Museum, Enfield 044293
Forty Nine, Manningtree 090698
Forty One Antiques, Heathfield . 089488
Forty River Street Antiques,
Boston 092470

Forum, 't Goy-Houten 143147
Forum, Toronto 064653
Forum – Chantal Grande,
Tarragona 115464
Forum Amalienpark, Berlin 106230
Forum Antik, København 065900
Forum Antiques, Cirencester ... 088916,
.... 134272
Forum Antiques, Row 079695
Forum Arte, Kötzting 107707
Forum Arte Contemporanea Merano,
Merano 110211
Forum Artis, Modena 138830
Forum Artis Museum, Montese . 026959
Forum Book Art, Hamburg 137244
Forum Cartes et Collections, Nieuil-
l'Espoir 141007
Forum d'Art Contemporain,
Sierre 116638
Forum de Artes Antiquario, Rio de
Janeiro 064058
Forum der Schweizer Geschichte,
Schwyz, Schweizerisches
Landesmuseum, Schwyz 041969
Forum des Antiquaires, Toulouse 073768
Forum des Arts, Aix-en-Provence 066461
Forum des Arts, Rouen 105475
Forum Flüh, Flüh 041489
Forum Fränkischer Hof, Bad
Rappenauer Museum, Bad
Rappenau 016844
Forum Frohner, Krems 002232
Forum für Angewandte Kunst
Nürnberg e.V, Nürnberg 020849
Forum für Fotografie, Köln 019680
Forum für Kunst, Heidelberg ... 107390
Forum für Medien und Gestaltung,
Bern 041316
Forum Gallery, Los Angeles 121484
Forum Gallery, New York 122604
Forum Hall, Handwerker- und
Heimatmuseum, Bad Hall 001747
Forum Internationale Photographie,
Mannheim 020246
Forum Konkrete Kunst Erfurt,
Erfurt 018241
Forum Kunst Rottweil e.V.,
Rottweil 058863, 108534
Forum Marinum, Turku 011435
Forum Museo, Omegna 027091
Forum Prudens, Praha 065496, 102661
Forum Schloßplatz, Aarau 041187
Forum Stadtscheune-Ortsmuseum und
Raum für Kunst, Mellingen ... 041738
Forum Vebikus, Schaffhausen .. 116629
Forum Würth, Arlesheim 041230
Forum+, Wrocław 113921
Forwarders' Museum, Prescott . 006705
Forwoods, Hay-on-Wye 117936
Forzani, Bruno, Milano 080546, 131446
Fos, Zürich 087960
Foscan, Hobart 099134
Foschini, Ivan, Genova 131311
Foshan Museum, Foshan 007770
Fosnes Bygdemuseum, Jøa 033688
Foss Street Galleries, Dartmouth 117665
Fossa Magna Museum, Itoigawa 028714
Fossa, Bruno, Capdenac-Gare .. 067660
Fossa, Enzo, Parma 131725
Fossalta, Bologna 109824
Fossard, Stéphane, Estry 068476
Fossati, Kilchberg (Basel) 116383
Fossati, Giuseppe, Firenze 131172
Fossati, Nerino, Brescia 079938
Fossati, Orazio, Firenze 131173
Fosse Gallery, Stow-on-the-Wold 091469,
.... 119159
Les Fosses d'Enfer, Maison des
Ressources Géologiques de
Normandie, Saint-Rémy-sur-
Orne 015547
Fossesholm Herregård,
Vestfossen 034084
Fossil Country Museum,

Kemmerer 049917
Fossil Decor, Cirencester 088917
Fossil Grove, Glasgow 044397
Fossil Museum, Fossil 048911
Fossil Station Museum, Russell . 052706
Fossilien- und Heimatmuseum,
Messel 020385
Fossilien- und Steindruck-Museum,
Gunzenhausen 018895
Fossilien-Museum, Fürth 018547
Fossilienmuseum im
Werkforum Rohrbach Zement,
Dotternhausen 017970
Fossilienschauraum, Ernstbrunn 001875
Foster & White, Seattle 124525, 124526
Foster and District Historical Society
Museum, Foster 001080
Foster Caddell's Art School,
Voluntown 058050
Foster Gallery, Eau Claire 048464
Foster-Gwin, San Francisco ... 097234
Foster Road Collectibles, Portland 096404
Foster, Kim, New York 122605
Foster, Marilyn, Jacksonville ... 093993
Foster, Matthew, London 090123
Foster, Michael, London 090124
Foster, Paul, London 144602
Foster, Peter, Sydney 062284
Foster, R.L., Denver 120716
Foster, Robert L. jr., Newcastle . 127365
Foster, Stephen, London 144603
Foster, W.A., London 144604
Fosterfields Living Historical Farm,
Morristown 051062
Foster's, Albuquerque 135283
Foster's, Rotherham 091144
Fostoria Area Historical Museum,
Fostoria 048912
Fota House, Carrigtwohill 024842
Fotart, Szczecin 113772
Fotevikens Museum, Höllviken . 040701
Foth, Freiburg im Breisgau 107054
Foth, Lothar, Osnabrück 077964, 130497
Fotheringham Gallery, Stirling .. 119143
Fotland Bygdemølle, Bryne 033561
Fotland Kraftverk, Bryne 033562
Foto-Galeria Związki Polskich Artystów
Fotografików, Kraków 113544
Foto-Grafix Books, San Francisco 145555
Foto Museet Sundvall – Bildens Hus,
Sundsvall 041064
Foto-Objektiv, Wien 138512
Foto-Production, Wien 136584
Fotoarchiv Sojuza Fotochudožnikov
Rossii, Moskva 037264
Fotofolio, New York 138395
Fotoforum, Luzern 116474
Fotoforum West, Innsbruck 002122
Fotogalereja im. Bratjev Ljumjer,
Moskva 037265
Fotogalina 1/3, Dom Muz – Miejska
Instytucja Kultury, Toruń 035279
Fotogalerie Friedrichshain, Berlin 017115
Fotogalerie in der Fabrik Heeder,
Krefeld 107735
Fotogalerie Lichtbild, Wiesbaden 108904
Fotogalerie No5, Leeuwarden .. 112548
Fotogalerie Wallraf, Köln 107609
Fotogalerie Weinberg, Kefermarkt 100064
Fotogalerie Wien, Verein zur Förderung
künstlerischer Fotografie, Wien 002991
Fotogalerie Wien, Verein zur Förderung
künstlerischer Fotografie und neuer
Medien, Wien 058252
Fotogalleriet, Oslo 113201
Fotografičeskij Muzej Dom Metenkova,
Ekaterinburg 036978
Fotograficzny Muzeum, Muzeumn
Regionalne, Mielec 035011
Fotografie Forum International,
Frankfurt am Main 018417
Fotografijos, Panevėžys 111790
Fotografijos Muziejus, Šiaulių Aušros
Muziejus, Šiauliai 030465

Fotoğrafın Tarihi, Osmaniye 043041
Fotografische Gesellschaft
Oberösterreich, Linz 058253
Fotohistorische Sammlung G. und W.
Pabst, Uhingen 022254
Fotohof, Salzburg 100155, 136548
Fotomuséet i Osby, Osby 040907
Fotomuseet Olympia, Falkenberg 040614
Fotomuseum, Motala 040856
Fotomuseum Den Haag, Den
Haag 032170
Fotomuseum Provincie Antwerpen,
Antwerpen 003287
Fotomuseum Winterthur,
Winterthur 042148
Fotomuzej Vlastja, Gorenja Vas . 038513
Fotoplastikon, Warszawa 035314
FotoSojuz – Galereja Sojuza
Fotochudožnikov Rossii, Moskva 114219
Fotostiftung Schweiz, Winterthur 042149
Fototapeta, Warszawa 138965
La Fototeca, Bologna 130961
Fototeca de Cuba, La Habana .. 009142
Fototeca Municipal de Reus, Institut
Municipal de Museus de Reus,
Reus 039996
Fototeka, Los Angeles 121485
Fotouhi, Renee, New Rochelle .. 122259
Fouad, Abdullah, Dammam 126482
Fouasse, Dominique, Toulon (Var) 129288
Foucault, Jeanne, Aubigny-sur-
Nère 066791
Foucher, Christian, Langoiran .. 069169
Foucher, Dominique, Le Pellerin 069325
Foucher, Jean-Claude, Bordeaux 067328,
140793
Foucher, Joël, Peujard 129167
Fouchet, Dijon 068327, 103675, 128581
Fougerolle, Christine, Paris ... 129006
Fougerouge, Essey 068473
Fougerouge, Christian, Dijon ... 068328
Fouilles Archéologiques, Jonzac 013325
Fouillis du Livre, Pessac 141303
Le Fouillis du Livre, Vendays
Montalivet 141376
Fouilloy, Monique, Clacy-et-
Thierret 067996
Le Fouineur, Bruxelles . 063427, 140175
Fouineur, Hossegor 068838
Fouineuse, Bayonne 066973
Found, Dallas 093124
Found Collections, Seattle 097439
Found It, Seattle 097440
Found Objects, Seattle 097441
Foundation for Hellenic Culture, New
York 051387
Foundation for Historic Christ Church,
Irvington 049728
Foundation for Women's Art,
London 044971
Founders' Hall, Canada's Birthplace
Pavilion, Charlottetown 005594
Founders Historic Park, Nelson . 033252
Founders Room, Museum of the
Pioneers, Raanana 025170
Foundling Museum, London ... 044972
Foundry Gallery, Washington ... 124900
The Foundry, Design Museum, Cape
Town 038677
Founès, Monique de, Lyon ... 128713
Fountain Antiques, Honiton 089561
Fountain Art Gallery, Memphis . 121694
Fountain Bookstore, Richmond . 145455
Fountain Gallery, East Molesey . 117708
Fountain Gallery, New York 051388
Fountain Gallery, Port Augusta . 099576
Fountain Press, Faringdon 138162
Fountain, Desmond, Hamilton . 100822
Fountainhead, Seattle 124527
Fouqué, Bernard, Signes 073521
Fouquer, Gérard, Lieurey 069462
Fouquet, Collégien 068075
Fouquet, Gino, Paris 071344
Fouquier, Olivier, Les Sièges .. 069432

Four Banal, Castillon-du-Gard .. 067715
Four Centuries, New Orleans ... 095097
Four Continents, Vancouver 064822
Four Corners Antiques, San
Francisco 097235
Four Corners Gallery, Kingston . 111219
Four Rivers Cultural Center,
Ontario 051783
Four Roses Kunst og Antik,
Charlottenlund 065623
Four Seasons, Indianapolis 093884
Four Seasons Gallery, Karama .. 117193
Four Seasons Gallery, Pune 109496
Four Sisters Gallery and the Robert
Lynch Collection, North Carolina
Wesleyan University, Rocky
Mount 052657
Four Solaire, Mont-Louis 014209
Four Star Galleries, Kansas City 121288
Four Star Galleries, Omaha ... 123310,
123311, 123312
Four Wheel Drive Foundation,
Clintonville 047846
Four Winds Gallery, Double Bay 098976
Four Winds Gallery, Pittsburgh . 096288,
096289, 123542, 123543
Four, Robert, Paris ... 071345, 129007
Fourbet, Martine, Saint-Jean-du-
Gard 072854
Fourcade, Jean-François, Paris . 141115
Fourcherot, Josselyn, Le Cellier 069244
Fourcherot, Josselyn, Vertou ... 074075
Fourchon, Serge, Luc-sur-Mer . 069652
Fourcroy, A., Bruxelles 063428
Fouré-Labrot & Bignon, Toulouse 125824
Fourment, Christian, Montblanc . 070216
Fourmond, Beaune 067022
Fournié Jean-Marc, Moissac ... 070142
Fournié-Olivié, Jean-Marc,
Toulouse 073769
Fournier, Lyon 140898
Fournier, Paris 071346
Fournier des Corats, Marie-Thérèse,
Villeneuve-sur-Lot 074167
Fournier-Guérin, Hélène, Paris .. 071347
Fournier, Françoise, Clermont-
Ferrand 068028
Fournier, Michel & Germaine, Pleine-
Fougères 072176
Fournier, Nicole, Dance 068271
Fournier, Yolande, Lambesc 069149
Fourquet, Patrick, Saint-Omer .. 125806
Fourquier, Stanislas, Paris 141116
Fours à Chaux, Antoing 003275
Fourteen, 's-Hertogenbosch 112527
Fourteen Sculptors, New York .. 122606
Fourtin, Patrick, Paris 071348
Foust, Charlotte 119986
Fouz Vega, Jesus Manuel,
Madrid 086184
Fővárosi Képtár, Kiscelli Múzeum,
Budapest 023479
Fovea, Bruxelles 063429
Fovea, Harrow 117920
Fowey River Gallery, Fowey ... 117809
Fowle, A. & J., London 090125
Fowle, Nicholas, Norwich 090875,
134893
Fowler, Lorne 139810
Fowler Museum of Cultural History,
University of California, Los Angeles,
Los Angeles 050422
Fowler, J.G., Auckland 083789, 132869
Fowler, Robin, Aylesby 088328, 134048
Fox, Brisbane 098788
Fox, Manuka 061791
Fox and Hen Antiques, Tampa .. 097557
Fox and Pheasant Antique Pine, White
Colne 091831, 135224
Fox Cottage Antiques, Stow-on-the-
Wold 091470
Fox Harbour Community Museum, Fox
Harbour 005834
Fox Island Historical Society Museum,

Fox Island 048916
Fox Lake Historical Museum, Fox
Lake 048917
Fox River Trolley Museum, South
Elgin 053404
Fox-Schröder, Ricarda, Mülheim an
der Ruhr 108038
Fox Talbot Museum, Lacock ... 044746
Fox, C., Queenscliff 062131
Fox, Jon, Moreton-in-Marsh ... 090781
Fox, Judy, London 090126
Fox, Leonard, New York 122607, 145339
Fox, Pearce, Philadelphia 096107
Foxfire Gallery, Albuquerque ... 119475
Foxford House Antiques, Mallow 079670
Foxgloves, Chepstow 088846
Foxhall Gallery, Washington ... 124901
Foxton Canal Museum, Foxton . 044367
Foxton Flax Stripper Museum,
Foxton 033205
Foxton Museum, Foxton 033206
Foxx Galerie, Zürich 116773
Foxx, Marc, Los Angeles 121486
Foxy, Denver 093287
Foxy Production, New York 122608
Foyer / James Hockey Gallery,
University College for the Creative
Arts, Farnham 044334
Foyer Rural, Musée du Bois et de
l'Artisanat, Jupilles 013343
Foyle, Londonderry ... 090610, 134806,
144716
Foynes Flying Boat Museum,
Foynes 024924
Fozzato, Lara, Verona 081911
Fra.Bo.No., Roma 081373
FRAC du Centre, Orléans 014582
FRAC Limousin, Fonds Régional d'Art
Contemporain, Limoges 013814
Fradécor, Gargenville . 068636, 128607
Frändeboden, Frändefors 086677
Fraenkel, San Francisco 124283
Fränkische Galerie, Zweigmuseum
des Bayerischen Nationalmuseums,
Kronach 019801
Fränkische Hopfenscheune mit
heimatkundlicher Sammlung,
Neunkirchen am Sand 020729
Fränkische Schweiz Museum,
Pottenstein 021208
Fränkisches Bauern- und
Handwerkermuseum,
Kirchenburgmuseum
Mönchsondheim, Iphofen ... 019418
Fränkisches Brauereimuseum,
Bamberg 016946
Fränkisches Freilandmuseum, Bad
Windsheim 016913
Fränkisches Freilandmuseum
Fladungen, Fladungen 018368
Fränkisches Museum,
Feuchtwangen 018357
Fränkisches Spielzeugmuseum,
Gößweinstein 018744
Fränkisches Spitalmuseum Aub,
Aub 016639
Fraeylemaborg, Slochteren 032884
Fragmenta Selecta, Amsterdam 143031
Fragments Editions, Paris 136924
Fragne, Paul, Angoulême 066628
Fragonard Anticuario, Valladolid 086545
Frahm, London 118322
Fraietta, Raffaele, Roma 131935
Fraise des Bois Antiquités, Plougastel-
Daoulas 072199
Fraisse-Jabot, Tours 125826
Fraisse, Pierre, Craponne-sur-
Arzon 068223
Fraisse, Pierre, Feurs 068528
Fraissinet, Pierre, Cahors 067596
Fraleigh, Toronto 064654
Framagraphic, Vancouver 101785
Framatic Art Gallery, Singapore . 114522
Framboyan, New Orleans 095098

Frame, Amsterdam 138928
Frame, Madrid 086185, 143763
Frame and Art, Dallas 120599
Frame and Art Center, Phoenix . 123491
Frame Art Gallery, Kingston 111220
Frame Art Magazine, Wien 138513
Frame Center and Gallery,
Milwaukee 121916
Frame Center Gallery, Kingston . 111221
Frame Concepts a Division of 3
Galleries, Toronto 128210
Frame Factory and Gallery,
Tampa 124649
Frame Gallery, Pittsburgh 052163
Frame Makers Gallery, Atlanta . 119566
Frame Merchant and Gallery,
Cleveland 120420
Frame 'n Art, Virginia Beach ... 124827
Frame of Mind, Edmonton 101286
Frame of Mind, Miami 121781
Frame of Mind, Winnipeg 101878
Frame Place and Gallery, El Paso 120829
Frame Place and Gallery, Tampa 124650
Frame Projects, Wien . 100255, 136585
Frame Shop and Art Gallery,
Jacksonville 121253
Frame Shop and Gallery, New
Orleans 122165
Framed Art Studios, Portland .. 123620
Framed Memories, Kitchener .. 101324
Framed to Perfection, Tucson .. 124734
Framein Place Gallery, Baltimore 119736
The Framery & Gallery, Waikato 113036
Framery Art Gallery, Sea Point . 114817
Frames and Fantasies,
Birmingham 119796
Frames by Rockwell, Ottawa ... 101471,
128190
Frames House, Tbilisi 105903
Frames USA and Art Gallery,
Miami 121782
Framework Gallery, Dún
Laoghaire 109629, 130867
Framework Gallery, London ... 118323
Frameworks, Killarney 079645
Frameworks Gallery, Charlotte . 119987
Frameworks Gallery, Hamilton . 100823
Framgi Dadabhoy Alpaiwalle
Museum, Parsi Panchayat Museum,
Mumbai 024402
Framing & Art Centre, Edmonton 101287
Framing & Art Centre, Hamilton 101316
Framing & Art Centre, Kitchener 101325
Framing & Art Centre, London . 101340
Framing & Art Centre, Vancouver 101786
Framing and Art, Charlotte ... 119988
Framing and Art Matters,
Randwick 099609
Framing Concepts and Gallery, El
Paso 120830
The Framing Gallery, Portlaoise . 109682
The Framing Machine & Gallery,
Swansea 119194
Framing Place and Gallery, Saint
Paul 123905, 123906
The Framing Workshop, Bath .. 117305,
134076
Framingham Historical Museum,
Framingham 048919
Frammuseet, Oslo 033834
Framtidsantiken Skrot & Korn,
Stockholm 086999
Fran Hill Gallery, Toronto 101609
Franc Segent, Marly-le-Roi ... 069880
Franc, Isabelle, Villeurbanne .. 074207
Franca, Rio de Janeiro 064059
Franča Trasuna Muzejs Kolnasäta,
Sakstagals 030228
Franca, Godofredo, Rio de
Janeiro 064060, 101028
Français, Jacky, Vittel 074265
La Franca, Tampa 097558
France Afrique, Bayonne 103342
France Création Style, Barbizon 066945

France Création Style, Les Baux-de-Provence ... 069402
France Expertises Enchères, Montastruc-la-Conseillère 104249
France Lejeune, Mechelen 100786
France Loisirs Antilles, Fort-de-France 142979
France Tradition, Los Angeles .. 094338
France, Sophie, São Paulo ... 064178
Francelet-Grosz, Jacqueline, Saint-Pierre-Quiberon 073265
Frances Burke Textile Resource Centre-Collections, RMIT University, School of Architecture and Design, Melbourne 001269
The Frances Lehman Loeb Art Center, Poughkeepsie 052338
Frances Willard House Museum, WCTU Museum, Evanston 048638
Frances, Lyliane, Chatelaillon-Plage 067902
Francés, Ricardo, Pamplona ... 086388
Francesc Mestre Art, Barcelona . 114887
Francesca's Collections, Houston 093671
Francesca's Italian Arts and Antiques, Sacramento 096624
Franceschi, Severino, Firenze .. 131174
Franceschini, Roberto, Brescia . 079939, 142655
Francesco della Gola, Scuol ... 087809
Francescoli, Rosemary, Tavannes 087841
Franchi, Angela, Bologna 079842
Franchi, Daniel, Paris 104743
Franchini, Firenze 080141, 142673
Franchini, Stefano, Bologna 130962
Franchino, Gherardo, Torino ... 132139
Franchitti, Natalina, Genova .. 131312
Franci, Mario, Roma 081374
Francia Antigüedades, Lima ... 084453
Francis Colburn Gallery, Burlington 047327
Francis Land House, Virginia Beach 053997
Francis Ledwidge Cottage and Museum, Slane 024988
Francis McCray Gallery, Western New Mexico University, Silver City . 053336
Francis, Annie & Martin, Jacksonville 121254
Francis, C., Providence 123704
Francis, Howard, Stockport ... 119147
Francis, Paolo, Croydon 117654
Francis, Peter, Carmarthen 126849
Francis, Peter, Saintfield 091215
Francisco-Deutch Fine Arts, New York 122609
Francisco Fort Museum, La Veta 050050
Francisco Romero, Julian, Valencia 133463
Franck & Jean-Pierre, Nice ... 128861
Franck Laigneau, Paris 071350
Franck, Georges, Paris 071349
Franck, Gérard & Lydia, Colmar 068090
Franck, Isabelle, Cahuzac-sur-Vere 067599
Franck, Patrick, Couches 068170
Die Franckeschen Stiftungen zu Halle, Halle, Saale 018954
Franco, Cannes 067638
Franco, El Paso 120831
Franco, Nice 070604
Franco Serrano, Maria del Rosario, Barcelona 114888
Franco, Cristian, Badajoz 114842
Francoarte, Medellín 102435
Francois & Catherine, Amsterdam 082751
François & Martine, Villers (Semeuse) 074175
François Antiquités, Sedan 073467
François-Miron, Paris 104744
Francois Pierre Antiquaire, Martel 069981
François, Denis, Paris 071351
François, Marie-Cecile, Bruxelles 063430

François, Roger, Gevigney-et-Mercey 068654
Francoise Cottage, Ulladulla ... 062367
Franconeri, Palermo 081045
Frandl, Johann, Gundertshausen 127762
Frandsen, Varde 102996
Franek, Wolfgang, Wien 140075
Frangos, Stratjs, Zürich 087961
Franic, M., Domažlice 065382
Franjevački Samostan Gorica, Livno 004266
Franjevački Samostan Jajce, Jajce 004263
Franjevački Samostan Kraljeva Sutjeska, Kraljeva Sutjeska .. 004264
Franjevački Samostan Krešovo, Krešovo 004265
Frank, Möglingen 077378
Frank & Dunya, Seattle 124528
Frank & Joe Gallery, New York . 122610
Frank Cameron Museum of Wynyard, Wynyard 007492
Frank H. McClung Museum, University of Tennessee, Knoxville ... 050011
Frank-Iniesta, Madrid 086186
Frank Lloyd Wright Home and Studio, Oak Park 051665
Frank Lloyd Wright Museum, Richland Center 052529
Frank Lloyd Wright's Pope-Leighey House, Alexandria 046387
Frank-Loebsches Haus, Landau in der Pfalz 019871
Frank Partridge V.C. Military Museum, Bowraville 000875
Frank Phillips Home, Oklahoma Historical Society, Bartlesville . 046799
Frank Pictures Gallery, Santa Monica 124470
Frank Slide Interpretive Centre, Crowsnest Pass 005666
Frank, Anna, Krün 076977
Frank, Hans-Peter, Elsbethen .. 127727
Frank, Jean, Mulhouse 070390
Frank, John, Phoenix 096212
Frank, Manfred, Mandern 077249
Frank, Michael, Waiblingen ... 078826
Frank, William, Dublin 109583
Frank, Zangl & Co., München .. 077501, 142151
Franke, München 077502
Franke, Stuttgart 108695, 142389
Franke & Seij, Rotterdam 143219
Franke, B., Großostheim 076111
Franke, Dr. J., Wien 062946
Franke, M., Berlin 074789
Franke, Monika, Münstertal ... 142214
Franke, R., Halle 076160
Frankel, E. & J., New York 095403
Franken, Dordrecht ... 083180, 083181
Franken & Driessen, Amsterdam 132399
Franken Galerie, Kronach 107749
Frankenberger Art Gallery, University of Charleston, Charleston 047563
Frankenburger Heimatstube, Frankenburg 001897
Frankenmuth Historical Museum, Frankenmuth 048921
Frankenstein, Peter, Stuttgart . 130669
Frankenwaldmuseum, Kronach . 019802
Frankfort Area Historical Museum, West Frankfort 054299
Frankfurter Äpfelwein-Museum, Frankfurt am Main 018418
Frankfurter Antiquariatsmesse, in der Frankfurter Buchmesse, Frankfurt am Main 098140
Frankfurter Feldbahnmuseum, Frankfurt am Main 018419
Frankfurter Künstlerclub e.V., Frankfurt am Main 058864
Frankfurter Kunstkabinett Hanna Bekker vom Rath, Frankfurt am Main 106986

Frankfurter Kunstverein e.V., Frankfurt (Oder) 058865
Frankfurter Kunstverein e.V., Frankfurt am Main 058866
Frankfurter Münzhandlung, Frankfurt am Main 075781
Frankfurter Numismatische Gesellschaft, Frankfurt am Main 058867
Frankfurter Sportmuseum, Frankfurt am Main 018420
Frankfurter Westend Galerie, Frankfurt am Main 106987
Frankie G. Weems Gallery & Rotunda Gallery, Meredith College, Raleigh 052447
Franklin 54 Gallery, New York .. 122611
Franklin Arts and Cultural Centre, Pukekohe 033290
Franklin College of Arts and Sciences, Department of Art, University of Georgia, Athens 056674
Franklin County Historical Museum, Pasco 051941
Franklin County Museum, Brookville 047274
Franklin County Old Jail Museum, Winchester 054465
Franklin D. Roosevelt Presidential Library-Museum, Hyde Park .. 049649
Franklin Furnace Archive, Brooklyn 047257
Franklin G. Burroughs-Simeon B. Chapin Art Museum, Myrtle Beach 051144
Franklin Historical Society Museum, Franklin 048936
Franklin House, Launceston ... 001212
The Franklin Institute, Philadelphia 052050
Franklin Mineral Museum, Franklin 048937
The Franklin Mint, Birmingham . 092408
Franklin Mint Museum, Franklin Center 048941
The Franklin Museum, Tucson .. 053829
Franklin Pierce Homestead, Hillsborough 049482
Franklin Street Gallery, Buffalo . 119955
Franklin Town Antiques, El Paso 093402
Franklin, David, Columbus 093019
Franklin, John, San Diego 097086
Franklin, Lynda, Hungerford ... 089605, 134489
Franklin, N. & I., London 090127
Frankly Speaking Antiques, Yackandandah 062496
Franks Antique Doll Museum, Marshall 050714
Frank's Auctions, Hilliard 127242
Frank's Coin Shop, Las Vegas .. 094142, 094143
Frank's Coins, Buffalo 092525
Frank's Upholstering Shop, Detroit 135541
Frank's Used Furniture & Antiques, Port Adelaide 062090
Frankston Auction Mart, Frankston 125194
Frankston Primary Old School Museum, Frankston 001081
Franky's Galerie, Langenfeld .. 107797
Fråno Antikbod, Helsingborg .. 086738
Fråno Antikbod, Kramfors 086787
Fran's Antiques, Virginia Beach . 097781
Fran's Fantasies Antiques, Norfolk 095854
Frans Hals Museum, Haarlem . 032392
Frans Walkate Archief/ SNS Historisch Archief, Kampen 032532
Franses, S., London ... 090128, 134661
Franses, Victor, London 090129, 134662
Fransız Kültür Merkezi, İstanbul 117035
Fransk Bokhandel l'Arc Librairie Française, Stockholm 143946

Franska Liljan, Västerås 087140
Fransson, Landvetter 086801
Fransu-tupa, Himanka 010884
Franta, G., Köln 076864
Frantz, Walther & Renate, München 077503
Franz Daurach-Sammlung, Hadres 002037
Franz Kafka Librería, Cuernavaca 142987
Franz Kafka museum, Praha .. 009791
Franz-Liszt-Museum der Stadt Bayreuth, Bayreuth 016988
Franz Liszt Zentrum, Burgenländische Kulturzentren, Raiding 002562
Franz Marc Museum, Kochel am See 019669
Franz Michael Felder-Stube, Schoppernau 002727
Franz Radziwill Haus, Varel 022301
Franz Schmidt Museum, Knappenhof, Perchtoldsdorf 002494
Franz-Stock-Museum, Franz-Stock-Komitee für Deutschland, Arnsberg 016606
Franz Traunfellner-Dokumentation, Pöggstall 002515
Franz Xaver Gruber-Gedächtnishaus, Ach 001681
Franz Xaver Gruber-Museum, Lamprechtshausen 002260
Franz, H., Cremlingen 075279
Franz, H., Stuttgart 142390
Franz, Volker, Michelau in Oberfranken 142126
Franzensburg, Laxenburg 002279
Franzin & C., Milano 080547
Franzini-Heinen, Arlette, Pont-Sainte-Maxence 072265
Franziskanermuseum, Villingen-Schwenningen 022334
Franzke, Wien 100256
Franztaler Heimatstube, Mondsee 002410
Frapin Beauge, Houston 121039
Frasassi-Le Grotte, Genga 026386
Frasca, Guglielmo, Roma 131936
Frasch Art Studio, Leipzig 107818
Frascione, Enrico, Firenze 080142
Frascione, Giulio, Firenze 080143
Fraser, Coldstream ... 088955, 134292
Fraser, Washington 124902
Fraser, Woodbridge 119384
Fraser-Fort George Regional Museum, Prince George 006714
Fraser Gifford, Tucson 136418
Fraser Lake Museum, Fraser Lake 005836
Fraser-Sinclair, South Godstone . 135060
Fraser, Daphne, Glasgow 134404
Fraserburgh 044370
Fraserorr, Robin, New Orleans .. 135856
Frasers, Inverness 126930
Frasnetti, Maurice, Cusset 068253
Frassine, Roberto, Brescia 131032
Fratelli, Casablanca 082561
Fratini, Marco, Roma 142812
Frato, Starnberg 078537
Frau Holle Expreß, Hessisch Lichtenau 019235
Frauchiger-Flückiger, Hans, Othmarsingen 087812
Frauen Kunst Wissenschaft, Marburg 138867
Frauen Museum, Bonn 017689
Frauen Museum Wiesbaden, Wiesbaden 022631
Frauenkultur, Leipzig 019984
Frauenmuseum, Hittisau 002089
Fraunces Tavern Museum, New York 051389
Fraux, Patrice, Lavergne 069209, 128653
Fray, Jean-Marc, Austin 092227
Frayling-Cork, Alistair J., Wallingford 135182

Fraysse, Paris –
French Country Imports, Memphis
Index of Institutions and Companies

Fraysse, Paris 125715
Frazer, Winnipeg 064922, 128239
Frazer's Museum, Beauval 005406
Freak-Ando, Bologna 079843
Frebel, Sylt 108745
Freckem, Elke, Bielefeld 074976
Fred, Leipzig 107819
Fred, London 118324
Fred, Philadelphia 123377
Fred Antiquités, Jandun 068889
Fred Dana Marsh Museum, Ormond Beach 051813
Fred Harman Art Museum, Pagosa Springs 051889
Fred J. Johnston House Museum, Friends of Historic Kingston, Kingston 049982
Fred Jones Jr. Museum of Art, University of Oklahoma, Norman 051586
Fred Light Museum, Battleford . 005399
Fred Wolf Jr. Gallery, Philadelphia 052051
Freddi, Fabio, Torino 142856
Freddy's Antiquitäten, Ubstadt-Weiher 078742
Frede, Den Haag 112383
Frede, Stefan, Münster 130427
Frederic Remington Art Museum, Ogdensburg 051719
Frederick, Milwaukee 121917
Frederick Antique, Fresno 135562
Frederick C. Robie House, Chicago 047659
Frederick Douglass, National Historic Site, Washington 054115
Frederick Gallery, Dublin 109584
Frederick Law Olmsted Collection, Brookline 047246
Frederick R. Weisman Art Foundation, Los Angeles 050423
Frederick R. Weisman Art Museum, University of Minnesota, Minneapolis 050936
Frederick R. Weisman Museum of Art, Malibu 050611
Fredericks & Freiser, Kristen, New York 122612
Fredericks & Son, C., London . . 090130, 134663
Fredericksburg and Spotsylvania National Military Park, Fredericksburg 048952
Fredericksburg Area Museum and Cultural Center, Fredericksburg 048953
Frédérik Antiquités, Arc-les-Gray 066698
Frederiksberg Antik, Frederiksberg 065663
Frederiksberg Antikvariat, Frederiksberg 140568
Frederiksberg Auktionshus, Rødovre 125464, 125465
Frederiksberg Kunsthandel, Frederiksberg 065664
Frederiksbog Antikvariat, Hillerød 140584
Frederikshavn Kunstmuseum- og Exlibrissamling, Frederikshavn 010069
Frederiksson, Kurt, Märsta 133554
Frederiksværk Bymuseum, Frederiksværk 010072
Fredhøj, Herning 065749
Fredman, Sydney 062285
Fredrick's Antiques and Art Gallery, Ettalong Beach 061455, 099006
Fredrik, Karl, Helsinki 066314
Fredriksdal Museer och Trädgårdar, Fredriksdal Museums and Gardens, Helsingborg 040691
Fredriksson, Ulla, Ludvika 126627
Fredrikstad Antikvariat, Fredrikstad 143385
Fredrikstad Museum, Fredrikstad 033610
Fred's, Baltimore 092325, 135342
Fred's Antique, Werne 078933
Fredtun Antikk, Hafrsfjord 084241
Fredy Brocante, Cabrieres 067548
Fredy's Mechanisches Musikmuseum, Lichtensteig 041670
Free Art School, Helsinki 055341
Free Atelier, Honfleur 103829
Free Painters and Sculptors, London 060139
Free Public Library Special Collections, New Bedford 051223
Free State Voortrekker Museum, Winburg 038879
Freeborn County Museum, Albert Lea 046353
Freedman Gallery, Albright College, Reading 052481
Freedman, Matthew, Hove 144490
Freeman, Birmingham 135353
Freeman, Hobart 099135
Freeman, Norwich 144751
Freeman, Roxwell 091149
Freeman & Lloyd, Sandgate . . . 091236
Freeman-Whitehurst, Phoenix . . 123492
Freeman, I. & J., Lefkosia 065322
Freeman, Peter C., New York . . 122613
Freeman, Tip, Miami . . 094733, 121783, 135792
Freeman, Vincent, London 090131
Freemans, Philadelphia 127391
Freemantle, Bernard, Romanel-sur-Lausanne 087757
Freeport Arts Center, Freeport . . 048969
Freeport Historical Society Museum, Freeport 048972
Freer, Chicago 135411
Freer Gallery of Art, Washington 054116
Freer Gallery of Art Occasional Papers, Washington 139366
Frees, Dirk, Marktschellenberg . 077302
Freess, Edmond, Roussillon (Vaucluse) 105482
Freestone County Historical Museum, Fairfield 048673
Fregatten Jylland, Ebeltoft 010039
Frégnac, Philippe, Paris 104745
Frégosi, Claire, Marseille 069933
Frei, Anton, Kißlegg 130130
Frei, Georges, Toffen 087851
Freiämter Strohmuseum, Wohlen 042167
Freibord, Wien 138514
Freiburger Antique & Art, Freiburg im Breisgau 098141
Freiburger Münzkabinett, Freiburg im Breisgau 075862
Freiburger Photohistorische Sammlung, Schallstadt 021630
Freie Akademie der Künste in Hamburg e.V., Hamburg 055570
Freie Akademie der Künste Rhein-Neckar e.V., Mannheim 055637
Freie Kunstakademie Basel, Basel 056406
Freie Kunstakademie Nürtingen, Nürtingen 055673
Freie Kunstakademie Rhein/Ruhr, Krefeld 055622
Freies Deutsches Hochstift/ Frankfurter Goethe-Museum mit Goethe-Haus, Frankfurt am Main 018421
Freight & Volume, New York . . . 122614, 139367
Freijo, Angustias & Manuel Gonzalez, Madrid 115195
Freiland-Grenzmuseum Sorge, Sorge 021922
Freilandmuseum Ammerländer Bauernhaus, Bad Zwischenahn 016917
Freilandmuseum Grassemann Naturpark-Infostelle, Warmensteinach 022432
Freilandmuseum Lehde, Lübbenau 020152
Freilicht- und Heimatmuseum Donaumoos, Karlshuld 019502
Freilichtanlage Gschlößl, Leithaprodersdorf 002287
Freilichtmuseum, Bad Tatzmannsdorf 001763
Freilichtmuseum, Diesdorf 017887
Freilichtmuseum, Herbertingen . 019195
Freilichtmuseum, Stade 021945
Freilichtmuseum, Wolfach 022725
Freilichtmuseum "Frelsdorfer Brink", Frelsdorf 018503
Freilichtmuseum "Mittelalterliche Bergstadt Bleiberg", Frankenberg, Sachsen 018390
Freilichtmuseum Alt Schwerin, Agrarhistorisches Museum, Alt Schwerin 016521
Freilichtmuseum am Kiekeberg, Rosengarten, Kreis Harburg . . 021473
Freilichtmuseum am Rätischen Limes, Rainau 021267
Freilichtmuseum Apriacher Stockmühlen, Heiligenblut 002070
Freilichtmuseum Beuren, Museum des Landkreises Esslingen für ländliche Kultur, Beuren 017349
Freilichtmuseum des Gebirgskrieges, Kötschach-Mauthen 002221
Freilichtmuseum Domäne Dahlem, Stiftung Stadtmuseum Berlin, Berlin 017116
Freilichtmuseum Ensemble Gerersdorf, Gerersdorf 001932
Freilichtmuseum Erdöl- und Erdgaslehrpfad, Prottes 002542
Freilichtmuseum Finsterau, Mauth 020320
Freilichtmuseum Fürstenhammer, Lasberg 002272
Freilichtmuseum Glentleiten, Großweil 018866
Freilichtmuseum Groß Raden, Landesamt für Kultur und Denkmalpflege-Archäologie und Denkmalpflege, Sternberg 021996
Freilichtmuseum Hackenschmiede, Bad Wimsbach-Neydharting . . 001768
Freilichtmuseum Handwerkerhaus Stegwagner, Windhaag bei Freistadt 003118
Freilichtmuseum Hayrl, Reichenthal 002581
Freilichtmuseum Hessenpark, Neu-Anspach 020672
Freilichtmuseum Kalte Kuchl, Rohr im Gebirge 002602
Freilichtmuseum Katzensteiner Mühle, Weyer 002929
Freilichtmuseum Klausenhof, Herrischried 019226
Freilichtmuseum Klockenhagen, Ribnitz-Damgarten 021409
Freilichtmuseum Ledermühle, Sankt Oswald bei Freistadt 002684
Freilichtmuseum Massing, Massing 020316
Freilichtmuseum Mondseer Rauchhaus, Mondsee 002411
Freilichtmuseum Neuhausen ob Eck, Neuhausen ob Eck 020711
Freilichtmuseum Oberlienz, Lienz 002306
Freilichtmuseum Ostenfelder Bauernhaus, Zweckverband Museumsverbund Nordfriesland, Husum 019366
Freilichtmuseum Pelmberg, Denkmalhof Mittermayer, Hellmonsödt 002074
Freilichtmuseum Petronell, Archäologischer Park Carnuntum, Petronell 002503
Freilichtmuseum Rätersiedlung Himmelreich, Museum Wattens, Volders 002873
Freilichtmuseum Rhöner Museumsdorf, Tann 022102
Freilichtmuseum Römerbad, Jagsthausen 019440
Freilichtmuseum Römersteinbruch, Sankt Margarethen 002676
Freilichtmuseum Römervilla in Brederis, Rankweil 002568
Freilichtmuseum Säge Buch, Buch 041375
Freilichtmuseum Scherzenmühle Weidenberg, Weidenberg 022461
Freilichtmuseum Schwebsingen, Bech-Kleinmacher 030562
Freilichtmuseum spätbronzezeitlicher Hügelgräber, Siegendorf 002754
Freilichtmuseum Stehrerhof – Dreschmaschinenmuseum, Neukirchen an der Vöckla 002439
Freilichtmuseum Sumerauerhof, Oberösterreichische Landesmuseen, Sankt Florian 002649
Freilichtmuseum Tiroler Bauernhöfe, Kramsach 002228
Freilichtmuseum Venetianersäge, Windhaag bei Freistadt 003119
Freilichtmuseum Vorau, Vorau . . 002875
Freimaurermuseum der Großen Landesloge der Freimaurer von Deutschland, Sankt Michaelisdonn 021605
Freipresse, Bludenz 136495
Freiräume, Bad Großpertholz . . . 136493
Freiraum, Heidelberg 076441
Freiraum – Verein für Kunst und Kultur e.V., Furth im Wald 058868
Freiria Aparicio, Maria Isabel, A Coruña 115032
Markus Freitag & Dotheeé Simmert, Kiel 130118
Freitag-Badenhausen, L., Reichshof 078184
Freitag-Lévy, Anne, Paris 071352
Freitag, Prof. Jörg, Potsdam . . . 130529
Freitag, Werner, Biebertal 129588
Freitag, Wolfgang, Berlin 074790
Freitagsgalerie, Solothurn 116649
Freitas, Bruno, Béziers 067174
Freitas, Francisco Gonçalves, Lisboa 143570
Freitas, Maria Carvalho, Lisboa . 114002
Freitas, Raphaël, Maureilhan . . . 070010
Freivogel-Sippel, Nicole, Pforzheim 130517
Frejamuseét, Kil 040747
Fréjaville, Jacques, Gigean 068662
Frelin, Saint-Paul-de-Fenouillet . 073240
Freller, Walter, Linz 062716
Fremantle Arts Centre, Fremantle 001083
Fremantle History Museum, Fremantle 001084
Fremantle Prison Precinct, Fremantle 001085
Frémeaux & Assoc, Vincennes . 074241
Fremont Antique Mall, Seattle . . 097442
Fremont County Pioneer Museum, Lander 050128
Frémont, Pascal, Le Havre 103944
Frémontier, Patrick, Paris 071353
French & Co., New York 095404
French Accents, Baltimore 092326
French Aisling, San José 102489
French Antique Corner, San Francisco 097236
French Antique Shop, New Orleans 095099
French Antiques, Menora 061838
French Antiques, Minneapolis . . 094913
French Art Colony, Gallipolis . . . 049027
French Art Network, New Orleans 122166
French Azilum, Towanda 053784
French Cable Station Museum, Orleans 051811
French Country Imports,

Memphis 094633
French Country Living, Mougins 070361
French Country Road, Cincinnati 092857
French Door Antiques, Norfolk . 095855
French Heritage Antiques,
Mentone 061839
French Heritage Antiques,
Ormond 062034
The French House, London 090132,
134665
The French House, York 091976
French House Decor, Abu Dhabi 088157
French Kings Antiques, West
Hollywood 097893
The French Legation Museum,
Austin 046691
French Provincial, Mogo 061871
French Provincial Upholstery, Los
Angeles 135718
French Quarter, San Jose 097366
French Style, Prahran 062110
French Trading, Queenstown . . 084063
French Twist, Saint Louis 096706
French, Lily, Saint Paul 096824
French, Peter, Hampton 061590
French, R. A., London 134664
French, Sherry, New York 122615
Frenchman Butte Museum, Frenchman
Butte 005852
Frenk, E., Braunschweig 075122
Frenoy, Guy, Flines-lez-Raches . 068547
Frenzel, Göppingen 076030
Frenzel, Heroldstatt 076507
Frenzel, Peter, München 142152
Fréret, Florent, Soudan 073543
Frers, Karl-Gerhard, Bremen . . 075159
Freschi, Renzo, Milano 080548
Fresen, Dr. Georg, München . . 108069
Fresh Art, Denver 120717
Fresh Art, New York . . 122616, 122617
Fresh Fruit Studio Gallery,
Portland 123621
Fresh Gallery, Dunedin 112878
Freshford House Museum,
Bloemfontein 038656
Fresnel, Sophie, Plumetot 072210
Fresno Art Museum, Fresno . . . 048978
Fresno Coin Gallery, Fresno . . . 093501
Fresno Flats Museum, Oakhurst 051673
Fresno Furniture, Fresno 135563
Fresno Metropolitan Museum,
Fresno 048979
Freter, Silke, Hagen 076147
Frettes, François de, Pomport . . 072232
Fretz, Alain, Fribourg 133745
Freud Museum, London 044973
Freudenberger, Herbert, München 108070
Freudenheim, Nina, Buffalo . . . 119956
Freudenthal, Peter, Hamburg . . 107242
Freudenthaler Sensenhammer,
Leverkusen 020056
Freund, Christian R., Mannheim 077258,
077259
Freund, Raija, Berlin 106231
Freunde aktueller Kunst Zwickau,
Reinsdorf bei Zwickau 058869
Freunde der Bildenden Kunst g.e.V,
Freiburg im Breisgau 058870
Freunde der Kunsthalle, Kunstverein in
Ostfriesland e.V, Emden 058872
Freunde der Pinakothek der Moderne
e.V., München 058873
Freunde der Schlumper e.V.,
Hamburg 058874
Freunde des Wallraf-Richartz-
Museums und des Museums Ludwig
e.V., Köln 058875
Freunde Mainfränkischer Kunst und
Geschichte e.V., Würzburg . . . 058876
Freundes- und Förderkreis der Burg
Giebichenstein Hochschule für Kunst
und Design e.V., Halle, Saale . 058877

Freundeskreis der Dresdner
Porzellansammlung im Zwinger e.V,
Dresden 058878
Freundeskreis des Stadtmuseums
Düsseldorf e.V., Düsseldorf . . 058879
Freundeskreis Grassi Museum für
Angewandte Kunst e.V., Leipzig 058880
Freundeskreis Kunstgewerbemuseum
Dresden e.V, Dresden 058881
Freundeskreis Max Klinger e.V.,
Leipzig 058882
Freundorfer, Sigrid, New York . . 122618
Freuville, J., Liège 063757
Frew, Robert, London . 090133, 144605
Frey, Wien 100257
Frey & Norris, San Francisco . . 124284
Frey, Anne-Sibylle, Basel 133642
Frey, Elke, Ehringshausen 106867
Frey, Helmut Leo, Montabaur . . 077416
Frey, Kayoko, Zürich 116774
Frey, M., Edenkoben 106861
Frey, M., Ulm 078754
Frey, Michel, Venelles 074012
Frey, Otto, Cuxhaven 075284
Frey, Simone & Günter,
Meiningen 077327
Freya, Melbourne 061818
Freya, Tacolneston . . 091536, 135116
Freyburger, Michel, Châtenois . . 067909
Freyer, Cornelius, Heilbronn . . 076472,
130031
Freyer, Ulli, Bern 133677
Freymond-Guth & Co., Jean-Claude,
Zürich 116775
Fri-Art, Centre d'Art Contemporain,
Fribourg 041499
Friargate Pine and Antiques Centre,
Derby 089054
Friary Art Gallery, Lichfield . . . 044822
Frias, Dallas 093125
Frias, Winnipeg 064923
The Frick Art Museum-Henry Clay
Frick Estate, Pittsburgh 052164
Frick Collection, New York 051390
Fricke, B., Hehlen 076428
Fricke, Marion & Roswitha,
Berlin 106232
Fricker, A., Meckenbeuren 077313
Fricker, Jürgen H., Dinkelsbühl . 075360,
106682
Frickers, Gordon, Plymouth . . . 118919
Fricktaler Museum, Rheinfelden 041872
Fričovo Muzeum, Lázně Bělohrad 009660
Frida & Fred – Grazer Kindermuseum,
Graz 001971
Friday Street Antique Centre, Henley-
on-Thames 089507
Frides Laméris, Amsterdam . . . 082752
Fridge Gallery, Wellington Institute of
Technology, Petone 033281
Fridrich, Gundelfingen, Breisgau 076133
Fridrich, Wiesau 130753
Fridrich János Fényírdája, Koszta
József Múzeum, Szentes 023964
Fridrich, Hans-Peter, Wiesau . . 078967
Fridriha Candera Kosmosa Muzejs,
Latvijas Universitātes, Rīga . . 030181
Fridriha Candera Memoriālā Māja,
Latvijas Universitātes, Rīga . . 030182
Friebe, Karin, Mannheim 107966
Friebes & Gruber, Graz 139987
Friebes, Wolfgang, Graz 139986
Friedelt, Roswitha, Düsseldorf . 129741
Friedemann, Gütersloh 107183
Friedenauer Antiquariat, Berlin . 141514
friedens räume, mehr als ein museum,
Lindau 020081
Friedenshistorisches Museum, Bad
Hindelang 016785
Friedensmuseum, Meeder 020324
Friedensmuseum, Remagen . . . 021358
Friedenszimmer im Schloß Altranstädt,
Großlehna 018854
Friederichsen &Co., Hamburg . 141862

Friedhuber, Eugen, Wels 140051
Friedl, Charlotte, Wildermieming 127984
Friedl, Oskar, Chicago 120127
Friedland & Rivault, Paris 104746
Friedlander, David R., Louisville . 094550,
145244
Friedler, N., Köln 076865
Friedli, Kurt, Netstal 087698
Friedman, Chicago 120128
Friedman & Vallois, New York . . 122621
Friedman, Anita, New York 122619
Friedman, Barry, New York 122620
Friedman, Phyllis, Montréal . . . 064421
Friedman, Stephen, London . . . 118325
Friedmann, Grit Karen, Leipzig . 130223
Friedmann, Nando-Martin, Wien 062947
Friedrich, Basel 116065
Friedrich-der-Große-Museum
im Museumshof Lerchennest,
Sinsheim 021880
Friedrich-Eckenfelder-Galerie,
Balingen 016937
Friedrich-Fröbel-Museum, Bad
Blankenburg 016733
Friedrich-Gerstäcker-Museum,
Braunschweig 017555
Friedrich-Ludwig-Jahn-Gedenkstätte,
Lanz 019921
Friedrich-Ludwig-Jahn-Museum,
Freyburg (Unstrut) 018507
Friedrich-Rückert-Gedächtnisstätte,
Coburg 017749
Friedrich-Wilhelm-Weber-Museum,
Bad Driburg 016746
Friedrich, Claus-Ulrich,
Niederhausen 142233
Friedrich, H., Freudenstadt . . . 075902
Friedrich, K., Freiburg im
Breisgau 075863
Friedrich, O. & U., Oldenburg . . 077928
Friedrich, R., Köln 076866
Friedrich, Regina, Wildenfels . . 079010
Friedrich, T., Mannheim 142105
Friedrichsgalerie, Kassel 076721
Friedrichson, Thomas, Berlin . . 129493
Friege, Heinz, Remscheid 108476
Friehs, Preding 062770
Friend Art, Tokyo 111421
Friend of Art, Amsterdam 112195,
137831
Friend or Faux, Bungay 088699, 134176
Friends Art, Barcelona 114889
Friends Art Gallery, Ha Noi . . . 125074
Friends in the Wild, Saint Louis 096707
Friends of Art, Kansas City . . . 121289
Friends of Bezalel Academy of Arts
and Designs, New York 060490
Friends of Firemen's Museum, New
Bern 051230
Friends of the Arts, Chicago . . . 120129
Friends of the Fan Museum,
London 060140
Friends of the Minneapolis Institute of
Arts, Minneapolis 060491
Friends of the Nukus Museum,
Nukus 060820
The Friends of the Royal Academy,
London 060141
The Friends of the V&A, London 060142
Friends' Preview, Glasgow 139178
Friends' Quarterly (Enfield),
Enfield 139368
Friendship Firehouse, Alexandria 046388
Friendship Hill, Point Marion . . 052225
Friendship Museum, Darhan . . 031726
Fries-Junak, Liesbeth, Bern . . . 087281
Fries Landbouw Museum,
Exmorra 032325
Fries Museum, Leeuwarden . . . 032570
Fries Natuurmuseum,
Leeuwarden 032571
Fries Scheepvaart Museum,
Sneek 032889
Fries, Ernst de, Hilden 076527

Fries, R., Kleinostheim 076819
Fries, Ulrich, Berlin 074791
Friesacher, Bolzano 109875
't Friesche Schathuis, Haarlem . 083327
Friese-Galerie, Wendländischer
Altertumsverein, Lüchow 020169
Friesen, Seattle 124529
Friesenecker, H., Lahstedt 076996
Friesenstube Honkenswarf,
Langeneß 019911
Friesia Cottage, Dunsborough . . 061418
Friesisches Heimatmuseum,
Freilichtmuseum, Niebüll 020786
Fries, Jean-Pierre, Joigny
(Yonne) . . . 068904, 068905, 128630,
128631
Frieß, Manfred, Augsburg 129395
Friessnig, Leo, Klagenfurt 062671
Frietkotmuseum, Antwerpen . . . 003288
Friez, Marie, Paris 071354
Frieze, London 139179
Frieze Art Fair, London 098376
Frihedsmuseet, Nationalmuseet,
København 010174
Frij, Ole, Dallas 120600
Frijns, Adriaan, Maastricht . . . 132574
Frilandsmuseet i Maribo, Maribo 010227
Frilandsmuseet, Nationalmuseet,
Kongens Lyngby 010209
Frilli, Firenze 110051
Friluftmuseet Hallandsgården,
Länsmuseet Halmstad,
Halmstad 040682
Friluftmuseet Skansen,
Stockholm 041001
Friluftsmuseet, Jönköping 040718
Friluftsmuseet Gammelgården,
Köping 040751
Friluftsmuseet i Apaladalen,
Värnamo 041134
Frings, Brunsbüttel 106600
Fripe c'est Chic, Fumay 068617
Frip'ou'net, Saint-Jean-de-Luz . 072844
Frippery, New Orleans 095100
Le Fripier, Napoli 080881
Frisch, Dieter, Hamburg 141863
Frisch, Gertrud, München 077504
Frisco Historical Society Museum,
Frisco 048985
Frisco Native American Museum and
Natural History Center, Frisco . 048986
Frisina, Francesco, Torino 132140
Frisso, Z., Brasília 100914
Frist Center for the Visual Arts,
Nashville 051181
Fritelli, Firenze 110052
Frith Street Gallery, London . . . 118326
Frith, Michael, Toowoomba . . . 062328
Fritsch, Georg, Wien . . 140076, 140077
Fritsch, Jean-Louis, Sand 073403
Fritschi, Rolf, Steinmaur 133934
Fritz, Mössingen 108024
Fritz & Fritz, Osnabrück 077965
Fritz-Best-Museum, Kronberg im
Taunus 019805
Fritz-Denneville, Hildegard,
München 108071
Fritz Prause's Erben GmbH u. Co KG,
Klagenfurt 062672
Fritz-Reuter-Literaturmuseum,
Stavenhagen Reuterstadt 021967
Fritz-Shaar, Pieterlen 116567
Fritz-von-Wille-Museum im Haus
Beda, Dr.-Hanns-Simon-Stiftung,
Bitburg 017402
Fritz-Winter-Haus, Museum und
Galerie für moderne Kunst,
Ahlen 016474
Fritz, G. & G., Regensburg 078167
Fritz, Maria Anna, Ilz 100037
Fritzøe Museum, Larvik 033732
Fritzsche, Ernst, Berlin 074792
Frobenius-Institut, Johann Wolfgang
Goethe-Universität, Frankfurt am

Main 055538
Frobisher Thresherman's Museum,
   Frobisher 005853
· Froden, Los Angeles 121487
Frode's Brukt og Antik, Oslo ... 084310
Frodsham & Co, Charles, London 090134
Froeba, Peter & Hedwig,
   Gunzenhausen 107188
Fröhlich, Linz 100104
Fröhlich, Hansueli, Sankt Gallen 087776
Fröhlich, Helmut, Wien 140078
Fröhlich, Patrik, Zürich 116776
Das fröhliche Wohnzimmer-Edition,
   Wien 136586
Frölich, Stuttgart 078604
Frölich & Kaufmann, Berlin 137068,
                               141515
Frölich, E.J. & M., Enschede ... 083249
Frölich, Hans-Peter, Stuttgart . 130670
Froelick Gallery, Portland 123622
Frösåkers Auktioner, Östhammar 126650
Froesch, Roger, Singrist 073523, 129259
Froese, Ralf, Wismar 130772
Frøslevlejrens Museum, Padborg 010271
Frösö Hembygdsmuseum, Frösön 040633
Froessel, Pascal, Strasbourg ... 105687
Frövifors Pappersbruksmuseum,
   Frövi 040634
Frøya Bygdemuseum, Sistranda 033936
The Frog & Mouse Picture,
   Leicester 118094
Frog Hollow Vermont State Craft
   Center, Middlebury 050856
Froger, Rochefort (Charente-
   Maritime) 072497
Frogmore House, Royal Collections,
   Windsor 046162
Frogtown Books, Toledo 145621
Frohberg, Ulrich, Flonheim ... 129843
From Around the World Vintage, New
   York 095405
From Holocaust to Revival Museum,
   Chof Ashkelon 025030
From Slaveship to Ownership,
   Denver 120718
From the Hip, Los Angeles 094339
From Times Past, Edmonton ... 064293
Fromagerie de Démonstration de
   Gruyères, La Maison du Gruyère,
   Pringy 041852
Fromaget, Régis, Paris 129008
Fromanger, Jean-Pierre & Veronique,
   Paris 071355, 125716
Fromantin, Dominique, Paris ... 071356
Frome Museum, Frome 044374
Froment, Michèle, Vallauris ... 105805
Fromm, Bruno, Parsberg 130505
Fromm, Dieter, Blieskastel ... 075023
Fromme, Suzanne H., Pörtschach am
   Wörther See 062768, 100140
Fronhouse Antiques, Barmouth . 088354
Front Gallery, Edmonton 101288
Front Gate, Indianapolis 093885
Front Porch Antiques, Atlanta ... 092135
Front Street Museum, Ogallala . 051711
The Front Window, San Jose . 097367
Frontera Audubon Society,
   Weslaco 054283
Fronti di Mare, Propriano 072313,
                               105400
Frontier Army Museum, Fort
   Leavenworth 048831
Frontier Culture Museum,
   Staunton 053510
Frontier Gallery, Denver 093288
Frontier Gallery, Fort Worth ... 120864
Frontier Gallery, Houston 121040
Frontier Gallery, Seattle 097443
Frontier Gateway Museum,
   Glendive 049119
Frontier Museum, Williston ... 054426
Frontier Times Museum, Bandera 046774
Frontier Village Museum,
   Jamestown 049788

Frontiers of Flight Museum,
   Dallas 048117
Frontline Books, Chicago 145042
Frontmuseet – Rintamamuseo,
   Karis 010962
Froschauergenossenschaft Züricher
   Antiquare, Zürich 059933
Froschmuseum, Münchenstein . 041773
Froseth, K.M., Minneapolis 094914
Frost, Monmouth 090768, 134855
Frost, Sydney 022286
Frost & Reed, London . 090135, 118327
Frost & Reed, New York 095406
The Frost Entomological Museum,
   University Park 053897
Frost. Procházková. GbR, Atelier für
   Konservierung, Restaurierung.
   Denkmalpflege, Frankfurt am
   Main 129846
Frosta Bygdemuseet, Frosta ... 033612
Froula, Barbara, Denver 120719
Fru Larsens Antikvariat, Hundvåg 143387
Fruch, Jean-Marie, Domremy-la-
   Pucelle 068381
Fruchtkasten, Museen der Stadt
   Tuttlingen, Tuttlingen 022232
Frübing, Ruth & Horst, Füssen . 075927
Früh-Jenner, Dr. Gustl, München 077505
Früh, Peter, Salem, Baden 130615
Frühauf, H., Flensburg 075743
Frühchristliche Kirchenanlage, Feistritz
   an der Drau 001877
Frühmittelaltermuseum Carantana,
   Rothenthurn 002610
Fruehsorge Contemporaray Drawings,
   Berlin 106234
Fruehsorge, Jan-Philipp, Berlin . 106233
Fruens Vilje, Humlebæk 065803
Fruitlands Museums, Harvard .. 049398
Fruitman, C., Bruxelles 063431
Fruitmarket Gallery, Edinburgh . 044241
Fruitmarket Gallery, Publishing
   Department, Edinburgh 138155
Fruitteeltmuseum, Kapelle 032538
Frumas, San Jose 097368
Frutabela, Lisboa 084887
Fry Art Gallery, Saffron Walden . 045660
Fry Model Railway, Demesne .. 024863
Fry Model Railway Museum,
   Malahide 024963
Fryč, Jaroslav, Liberec 140481
Frýda, Dan, Rybitví 065553
Fryde, Aline, Saint-Ouen 073083
Frydendal, Frederiksberg 065665
Frydman, Renée, Saint-Ouen .. 073084
Frye & Sohn, Münster 077661
Frye Art Museum, Seattle ... 053234
Frye, Ken, Washington 124903
The Fryeburg Fair Farm Museum,
   Fryeburg 048992
Fryer Memorial Museum,
   Munnsville 051119
Fryer, Fritz, Ross-on-Wye 091139,
                               134979
Frye's Measure Mill, Wilton ... 054459
Fryksdals-Auktioner, Sunne ... 126682
Fryoux, Elsie, New Orleans ... 095101,
                               135857
Fryś, Agnieszka, Kraków 084565
Frysk Skildershûs, Leeuwarden . 112549
Fryxell Geology Museum, Rock
   Island 052630
FS & AV Arte Contemporânea,
   Porto 114083
FTC Galerie 85, Chur 116217
Fu Hakubutsukan, Hijikawa ... 028623
Fu, Jian, Shenzhen 102222
Fu, Xia, Singapore 114523
Fúcares, Almagro 114834
Fúcares, Madrid 115196
Fuchs, Berlin 074793
Fuchs, Biesenthal 074995
Fuchs, Fulda . 107088, 129881, 141818
Fuchs, Oberhausen 108315

Fuchs & Reposo, Lugano 144101
Fuchs-Ochsner, Hanspeter & Rosmarie,
   Trachslau 087854, 133940
Fuchs, Alois, Mainburg 077227
Fuchs, Anton, Eisingen 129784
Fuchs, Günter, Bad Münder am
   Deister 141429
Fuchs, Jacques, Nice 128862
Fuchs, Maria, München 130346
Fuchs, Rolf, Aalen 129365
Fuchs, Wolfgang, Norderstedt .. 077797
Fuchsia, Arnhem 112329
Fuchsia Cottage Pottery,
   Cahirciveen 109537
Fuchsia Dentelle, Paris 071357
Fuchu-shi Kyodo-no-mori Ajisai
   Matsuri, Fuchu 028533
Fuchunjiang, Hangzhou 102038
Fuda, Suzhou 102238
Fudge, Charles, Crystal River . 127191
Fuego, Tilburg 083614
Führichův Dům, Městské Muzeum,
   Chrastava 009524
Füldner, Peter, Zemmer 079138
Füllhorn, Sommerhausen 108638
Fündling, Wilhelm, Stade 108659
Fünf-Giebel-Haus, Friedrichstadt 018530
Fuentes, San Antonio 136260
Fuentes Fortes, Vigo 086559
Fuentes-Rodriguez, Isidoro,
   Hamburg 076220
Fürdő Múzeum, Budapesti Történeti
   Múzeum, Aquincum Múzeum,
   Budapest 023480
Fürhacker, Robert, Graz 127745
Fürst Thurn und Taxis Zentralarchiv
   und Hofbibliothek, Universität
   Regensburg, Regensburg 021319
Fürstbischöfliches Museum,
   Meersburg 020331
Fürstenberger, Hans, Linz 140010
Fürstenberger Hof, Zell am
   Harmersbach 022853
Fürstengalerie, Fürstenwalde ... 107081
Fürstengruft mit russisch-orthodoxer
   Kirche, Klassik Stiftung Weimar,
   Weimar 022491
Fürstlich Esterházy'sche
   Sammlungen Burg Forchtenstein,
   Forchtenstein 001895
Fürstlich Fürstenbergische
   Sammlungen, Donaueschingen 017929
Fürstlich Hohenzollernsche
   Sammlungen, Sigmaringen .. 021861
Fürstlich Leiningensche
   Sammlungen – Heimatmuseum
   Amorbach, Amorbach 016577
Fürstlich Ysenburg- und
   Büdingensches Schloßmuseum,
   Büdingen 017664
Fürstliches Residenzschloß,
   Detmold 017874
Fürstliches Schloß Mausoleum,
   Bückeburg 017658
Fürstliches Schloß und Kreuzgang
   Sankt Emmeram, Fürst Thurn und
   Taxis Museen, Regensburg ... 021320
Fuerte de la Punta del Sauce, La
   Carlota 000398
Füssl & Jakob, München 108072
Fufeng Museum, Fufeng 007772
Fufeng, Zhuhai 065165
Fuhlrott-Museum, Wuppertal ... 027828
Fuhrer, Hack & Walther, Stuttgart 130671
Fuhrmann, Rostock 108517

Fuhrmann, J., Rostock 108518
Fuhrmann, Jürgen,
   Braunschweig 141632
Fuhrwerkswaage Kunstraum,
   Köln 019681
Fuiza, Belo Horizonte 063901
Fujairah Museum, Fujairah ... 043336
Fuji Art Craft, Kyoto 082012
Fuji Art Museum, Fujinomiya .. 028538
Fuji Fine Arts, Tokyo 082144
Fuji Photo Salon, Osaka 111353
Fuji-Torii, Tokyo 082145
Fuji Visitor Center, Kawaguchiko 028802
Fujian Fine Arts Publishing House,
   Fuzhou 136708
Fujian Museum, Fuzhou 007783
Fujii Bijutsu Mingeikan,
   Takayama 029451
Fujii Saisei-kai Yurinkan, Kyoto . 028924
Fujii, T., Honolulu 093542
Fujikawa Galleries, Osaka 111354
Fujikawa Galleries, Tokyo 111422
Fujikeido, Nara 142905
Fujimi-machi Rekishi Hakubutsukan,
   Fujimi 028536
Fujita Bijutsu, Fukui 028541
Fujita Bijutsukan, Osaka 029241
Fujita-Unkotei, Nara 082068
Fujiya, Tokyo 082069
Fujiyama Museum, Fuji-Yoshida 028535
Fujiyama Museum of Historical Data,
   Oonishi 029237
Fukahori Kaizukai Iseki Museum,
   Nagasaki 029090
Fukawa, Yokohama 082225
Fukuchiyama Shato-Mingeikan,
   Fukuchiyama 028539
Fukuchiyama-shi Tanba Seikatsui-kan,
   Fukuchiyama 028540
Fukuhara, Tokyo 111423
Fukui City Museum of Natural History,
   Fukui 028542
Fukui-ken Togeikan, Echizen ... 028528
Fukui-kenritsu Bijutsukan, Fukui 028543
Fukui-kenritsu Bunka Hakubutsukan,
   Fukui 028544
Fukui-kenritsu Kyoryu Hakubutsukan,
   Katsuyama 028796
Fukui-shiritsu Bijutsukan, Fukui 028545
Fukui, Asahido, Kyoto 082013
Fukuoka, Yamato-koriyama ... 082221
Fukuoka Ajian Bijutsukan, Fukuoka
   Asian Art Museum, Fukuoka ... 028546
Fukuoka-kenritsu Bijutsukan,
   Fukuoka 028547
Fukuoka-machi Shashinki
   Hakubutsukan, Fukuoka,
   Toyoma 028553
Fukuoka Nichi-do Gallery,
   Fukuoka 111249
Fukuoka-shi Bijutsukan, Fukuoka 028548
Fukuoka-shi Hakubutsukan,
   Fukuoka 028549
Fukushima Kaiyo Kagakukan,
   Iwaki 028720
Fukushima-ken Bunka Centre,
   Fukushima 028554
Fukushima-kenritsu Bijutsukan,
   Fukushima 028555
Fukushima-kenritsu Hakubutsukan,
   Aizuwakamatsu 028465
Fukuyama Auto and Clock Museum,
   Fukuyama 028556
Fukuyama Bijutsukan, Fukuyama 028557
Fukuyama Bungakukan, Fukuyama
   Gaijutsu Bunka Shinko Zaidan,
   Fukuyama 028558
Fukuyama Castle Museum,
   Fukuyama 028559
Fukuyama City Human Rights and
   Peace Museum, Fukuyama ... 028560
Fukuyama-renritsu Rekishi
   Hakubutsukan, Fukuyama ... 028561
Fukuzawa Ichiro Memorial Museum,

Nakatsu 029139
Fulcheri & Figli, Cuneo 080062, 131104
Fulcrand, Bessan 067139
Fulda Gallery, Salford . 119052, 134999
Fulda, Constance, Paris 104747
Fuldaer Münze, Ebersburg 075560
Fulham, London 090136
Fuling Museum, Fuling 007776
Full Circle, San Diego 097087
Full Circle Architectural Antiques,
Hawthorn 061601
Full Circle Art Consultancy,
London 118328
Fullam, Dublin 079556
Fuller Craft Museum, Brockton . 047220
Fuller, D'Arch & Smith, London . 144606
Fuller, Ernest, Denver 120720
Fuller, Jeffrey, Philadelphia ... 123378
Fullerton Museum Center,
Fullerton 048993
Fullyjie Zone Museum, Bijie ... 007657
Fulton, Fresno 093502
Fulton County Historical Museum,
Wauseon 054238
Fulton County Historical Society
Museum, Rochester 052606
Fulton County Museum,
Groversville 049289
Fulton Mansion, Rockport 052648
Fulton Street Art Gallery, New
York 122622
Fultz House Museum, Lower
Sackville 006226
Fulwood Gallery, Preston 118947
Fumagalli, Luigi, Honolulu 120936
Fumelli, Adele Maria, Roma 081375
Fumi, Okegawa 111343
Fumiki, San Francisco 097237
Fumusa, Vincenzo, Milano 131447
Funabashi Bijutsu Gakuin,
Funabashi 056001
Funaki, Melbourne 099306
Function + Art, Chicago 120130
Fund & Arte, Medellín 128301
Fundação Arpad Szenes e Vieira da
Silva, Lisboa 035757
Fundação Casa de Jorge Amado,
Salvador 004786
Fundação Clóuis Salgado – Palácio
das Artes, Belo Horizonte 004340
Fundação Cultural do Estado da Bahia,
Praça Thomé de Souza, Paláco Rio
Branco, Salvador 058402
Fundação Cultural Ema Gordon Klabin,
São Paulo 004856
Fundação Eça de Queiroz, Santa Cruz
do Douro 035992
Fundação Eva Klabin Rappaport, Rio
de Janeiro 004692
Fundação Hansen Bahia,
Cachoeira 004374
Fundação Maria Luisa e Oscar
Americano, São Paulo 004857
Fundação Memorial da América Latina,
São Paulo 004858
Fundação Museu do Zebu Edilson
Lamartine Mendes, Uberaba .. 004939
Fundaçao Oscar Niemeyer, Rio de
Janeiro 004693
Fundação Ricardo do Espírito Santo
Silva, Lisboa 035758
Fundacija Jože Ciuha, Bled ... 114675
Fundació Antoni Tàpies,
Barcelona 039057
Fundació Barceló, Palma de
Mallorca 039885
Fundació Casa Museu Llorenç
Villalonga, Binissalem 039150
Fundacio Espais, Girona 039442
Fundacio Fita, Girona 039443
Fundació Joan Miró, Barcelona . 039058
Fundació la Caixa, Revista el treua,
Barcelona 139053
Fundació Pilar i Joan Miró, Palma de

Mallorca 039886
Fundació Pública Institut d'Estudis
Ilerdencs, Fons Bibliogràfic i
Documental, Lleida 039576
Fundación Alfredo L. Palacios, Buenos
Aires 000158
Fundación Archivo Gráfico y Museo
Histórico de la Ciudad de San
Francisco y la Región, San
Francisco 000576
Fundación Banco Francés, Buenos
Aires 000159
Fundación Carlos de Amberes,
Madrid 039622
Fundación César Manrique,
Teguise 040212
Fundación Cultural Otero Pedrayo,
Amoeiro 038970
Fundación de Fomento a las Artes, La
Paz 004210
Fundación Federico Jorge Klemm,
Buenos Aires 000160
Fundación Folch, Barcelona 039059
Fundación Juan March, Madrid . 039623
Fundación Luis Muñoz Marín, San
Juan 036156
Fundación Luis Seoane, A
Coruña 039317
Fundación Museo de las Ferias,
Medina del Campo 039741
Fundación NMAC Montenmedio de
Arte Contemporáneo, Vejer de la
Frontera 040360
Fundación Pablo o Ruiz Picasso
Museo Casa Natal, Málaga 039699
Fundación Pi y Margall, Madrid . 059865,
115197
Fundacion Privada Espai Guinovart,
Agramunt 038886
Fundación Vela Zanetti, Casona de
Villapérez, León 039554
Fundación Yannick y Ben Jakober,
Finca Sa Bassa Blanca, Alcúdia
(Baleares) 038932
Fundacja Artgard, Warszawa ... 137905
Fundatia Galeria de Artă, Cluj-
Napoca 114158
Fundaziun Planta Samedan
Wohnmuseum, Samedan 041918
Funder, Dover 139719
Fundgrube, Glücksburg 076009
Fundgrube zur Laterne, Sursee . 087838
Fundus, Berlin 141516
Fundus, Sankt-Peterburg 114332
Fundus-Props for Yesterday, New
York 095407
Fundy Geological Museum,
Parrsboro 006612
Fune no Kagakukan, Tokyo 029503
Fung, Martin, Hong Kong 065043,
065044
Funk House, Las Vegas 094144
Funk, Robert, Saskatoon 064561, 128201
Funk, Werner, Grasbrunn 076079
Funkar Art Gallery, Karachi 113304
Funke, Peter, Sundern 078651
Funktionall Art, Minneapolis ... 121989
Funky Furnishings, Sacramento 096625
Funky Junktion, Whangamata .. 084201
Funnefoss Industriarbeidermuseum,
Årnes 033504
Funnix, Wittmund 079038
Funque's, Austin 092228
Funtiques, Seattle 097444
Fuoricentro, Roma 110772
Fur Museum, Fur 010075
Furas Levin, Sergio N.,
Montevideo 091999
Furdyna, Józef, Kraków 132838
Fureai Minatokan, Osaka 029242
Fureaikan – Kyoto Museum of
Traditional Crafts, Kyoto 028925
Fureso Museer, Farum 010062
Furiko, Fukuoka 081964

Furini Arte Contemporaena,
Arezzo 137616
Furini Arte Contemporaena,
Roma 110773
Furletti, Belo Horizonte 063902
The Furlong Gallery, University of
Wisconsin-Stout, Menomonie . 050797
Furlotti, Alessandra, Milano 080549,
131448, 142721
Furman jr., Donald, Buffalo ... 092526
Furn Davies, Holt, Wales 089553, 134468
Furnarture Gallery, Chicago 120131
Furnas-Gosper Historical Society and
Museum, Arapahoe 046516
Furneaux Museum, Flinders
Island 001073
Furness Abbey, Barrow-in-
Furness 043503
The Furnishing Touch, Tauranga 084103,
132750
Furnishings off Foster, Portland . 096405
Furniture and Refinishing,
Cincinnati 135440
Furniture and Upholstery,
Houston 135602
Furniture Artisans, Philadelphia . 136086
Furniture Artist Repair, Miami . 135793
Furniture by School Masters,
Milwaukee 135816
Furniture Cave, Aberystwyth ... 088231,
134014
The Furniture Cave, London ... 090137
Furniture Cave, Preston 127662
Furniture Conservatory,
Indianapolis 135631
Furniture Craft Company, Los
Angeles 135719
Furniture Craftsman, Cincinnati . 135441
Furniture Craftsman, San Antonio 136261
Furniture Craftsman, Tucson ... 136419
Furniture Depot, Denver 093289
The Furniture Doctor, Vestaburg 136439
Furniture Exchange, Atlanta ... 092136,
135308
Furniture Haven, Honolulu 093543
Furniture Heaven, Regina 064551
Furniture History Society, Haywards
Heath 060143
Furniture History Society, Surrey
Hills 058200
Furniture Manufacturing Museum in
Tainan, Rende 042430
Furniture Mart, Margate 090702, 134837
Furniture Medic, Cleveland 135462
Furniture Medic, Littleton 135689
Furniture Medic, Louisville 135757
Furniture Medic, Memphis 135777
Furniture Medic, Miami 135794
Furniture Medic, Phoenix 136114
Furniture Medic, Sacramento .. 136188
Furniture Medic, San Antonio .. 136261
Furniture Now and Then, Albany 060975
Furniture on Consignment,
Albuquerque 092052
Furniture Refinishing, Denver .. 135531
Furniture Renovations Company,
London 134666
Furniture Restoration, London .. 134667
Furniture Restoration Center, Sunland
Park 136382
Furniture Restorations, Houston 135603
Furniture Revisited, Brunswick . 061223
Furniture Revival, Hawthorn ... 061602
Furniture Service, Dunedin ... 132717
Furniture Shoppe, Norfolk 136030
Furniture Stripping, Kansas City 135678
Furniture Trading Company, Botley,
Hampshire 088523
Furniture Vault, London 090138
Furniture Wizards, Winnipeg .. 128240
Furniture Workshop, San
Francisco 136318
Furnitures Friend, Seattle 136368
Furonnet, Jean-Louis, Langon .. 069173

Furphy, John, Armadale, Victoria 061026
Furrer, Martin, Brig 133705
Fursdon House, Cadbury 043821
Furstenberg, Paris 104748
Furthmühle, Egenhofen 018097
Furthof, Heiligenberg 076477
Furtwengler, Uwe, Baden-Baden 074632
Furukawa Bijutsukan, Nagoya .. 029116
Fusa, Marta, Verona 081912
Fusay, Jean-Paul, Reims 129190
Fusban, Daniel, Utting am
Ammersee 078774
Fusco, Toni, Aarau .. 087171, 133607
Fuse Magazine, Toronto 138579
Fusebox, Washington 124904
Fushun Calligraphy and Painting
Gallery, Fushun 102020
Fushun Museum, Fushun 007777
Mario Fusi & Alberto Barigazzi,
Lugano 087614
Fusiliers London Volunteers' Museum,
London 044974
Fusiliers' Museum Lancashire, Bury,
Lancashire 043808
Fusiliers Museum of Northumberland,
Alnwick 043810
Fusion, Milwaukee 121918
Fusion, Portland 096406
Fusion Art Gallery, Torino 110967
Fusion Fine Arts, Mandurah ... 099272
Fusions Gallery, Fortitude Valley 001077
Fustat Rare Books, Cairo 140675
Fustier, Maxime, Paris 071358
Futselaar, Erna, Amsterdam ... 112196
Futter, Lothar, Sundern 078652
Futterknecht, Ralf, Stuttgart ... 130672
Futur Antérieur, Bruxelles 063432
Futur Anterieur, Le Touquet-Paris-
Plage 069362
Futur Antérieur, Paris 071359
Futur Antérieur, Saint-Ouen ... 073085
Futura, Paris 104749, 136925
Futurama, Los Angeles 094340
Futurart, Roma 081376, 131937
Futuras Antigüedades, Barcelona 085773
Future Antique Furniture,
Silverwater 062213
Future Antiques, Saint Louis .. 096708
Futures Antiques, Norfolk 095856
Futures Past, Dallas 131589
Futuro, Meisterschwanden 139092
Futuro – Associazione per l'Arte,
Roma 059457
Futurotroc, Chasseneuil-du-
Poitou 067863
Fuxin Calligraphy and Painting Gallery,
Fuxin 102022
Fuxin Chahai Ruins Museum,
Fuxin 007778
Fuxin Museum, Fuxin 007779
Fuyang Museum, Fuyang, Anhui 007780
Fuyang Museum, Fuyang,
Zhejiang 007781
Fuyou Lu Market, Shanghai ... 065135
Fuyu Museum, Fuyu 007782
Fuzhou Cangshan Museum,
Fuzhou 007784
Fuzhou City Hualin Si Museum,
Fuzhou 007785
Fuzhou East Art Cooperative,
Fuzhou 058496, 102023
Fuzhou Painting Gallery, Fuzhou 102024
Fuzoku Hakubutsukan, Kyoto .. 028926
FVSR Kunsthandel, München ... 077506
Fyansford Antiques, Fyansford . 061491
Fyfe, Toronto 064655, 128211
Fyndet Antik och Kuriosa,
Ängelholm 086625
Fyndgömmans antik, Lidingö .. 086806
Fyndhörnan, Luleå 086826
Fyndkällaren, Lund 086832
Fyns Antikvariat, Odense 140637
Fyns Grafiske Værksted, Odense 066035
Fyns Kunstmuseum, Odense ... 010257

Fynsantik, Ringe 066065
Den Fynske Landsby, Odense . . 010258
Fyresdal Bygdemuseet, Fyresdal 033613
Fyris Antik, Uppsala 087120
Fyshwick Antique Centre,
  Fyshwick 061497
Fyson, Jonathan, Burford 088704
Fyvie Castle, Fyvie 044375
G & G Furniture Restorations,
  Olney 134908
G & L Antiques, Wellington 084163
G & N Lebensart, Wien 100258
G & O Art, New York 122623
G & S, Zemun 114466
G + H Verlag Berlin, Berlin 137069
G 26, Bern 116129
G Fine Art, Washington 124905
G-Gallery, Lemesos 102548
G III Antiques, Brugge 063273
G-Loft, Bruxelles 100602
g-module, New York 122624
G-Module, Paris 104750
G&G Restaurateurs, 's-
  Hertogenbosch 132535
G&H Furniture Repairing,
  Cincinnati 135442
G. Leonidze State Literature Museum
  of Georgia, Tbilisi 016401
G. Tabidze House Museum,
  Tbilisi 016402
G.A.L.A, Gap 103760
G.A.M., Saint-Ouen 073086
G.B. Antiques, Lancaster 089740
G.C. Antik Design, København . 065901
G.G. Galerie, La Roche-sur-Yon 069072
G.G.Ice Impex, Bucureşti 114139
G.I. Katsigras Museum, Municipal
  Gallery of Larrisa, Lárissa . . . 023151
G.I.M., Torino 081701
G.K. Zhukov Museum,
  Ulaanbaatar 031729
G.N. Jha Kendriya Sanskrit
  Vidyapeetha, Allahabad 024136
G.P.F. Gallery, Newport, Gwent . 118799
G.R.C. Grupo Restauro Conservativo,
  Venezia 132244
G1 Galerie, Gütersloh 107184
G2K, Auckland 112770
G67 Antique, München 142153
GAA Museum, Dublin 024880
Gaál Imre Galéria, Pesterzsébeti
  Múzeum, Budapest 023481
De Gaarde, Utrecht 083643
Gaarde Schmidt, Horsens 102848
Gaasbeek El-Bar, Amsterdam . . 082753
Gaasch, Armand, Dudelange . . . 111820
Gaballa, Austin 092229
Gabara, Joan, Picton 084062
Gabarro Art, Sabadell 115397
Gabba Old Wares,
  Woolloongabba 062487
Il Gabbiano, La Spezia 110154
Il Gabbiano, Roma 110774
Gabel, Seattle 136369
Gaber, Heinrich, Trier 078700
Gabert, Pascal, Paris 104751
Gabillet, Richard, Tours 073847
Gabinet d'Assessors, Barberà del
  Vallés 126501
Gabinet de les Arts Gràfiques,
  Museu de les Arts Decoratives,
  Barcelona 039060
Gabinet Numismatyczny,
  Wrocław 084767
Gabinet Numizmaticzny, Gdańsk 084503
Gabinet Numizmatyczny Mennicy
  Państwowej, Warszawa 035315
Gabinet Pisarza Pawła Hulki-
  Lasowskiego, Muzeum Mazowsza
  Zachodniego, Żyrardów 035471
Gabinet Prof. Kazimierza
  Sosnowskiego, Kraków 034864
Gabinet Rycin Biblioteki
  Uniwersyteckiej w Warszawie,

Warszawa 035316
Gabinet Rycin Polskiej Akademii
  Umiejętności w Krakowie,
  Kraków 034865
Gabinete de Antigüedades,
  Madrid 039624
Gabinete de Arte Minas
  Contemporânea, Belo Horizonte 100886
Gabinete Numismático de Cataluña,
  Museu Nacional d'Art de Catalunya,
  Barcelona 039061
Gabinetto Comunale delle Stampe,
  Roma 027560
Gabinetto dei Disegni e delle Stampe,
  Pinacoteca Nazionale, Bologna 025533
Gabinetto Disegni e Stampe degli
  Uffizi, Firenze 026252
Gabinetto Disegni e Stampe,
  Università di Pisa, Pisa 027352
Gabinetto Fotografico, Firenze . 026253
Gabinetto Geologico e Botanico
  Piccinini, Pergola 027249
Gabinetto Nazionale delle Stampe,
  Roma 027561
Gables, Miami 094734
The Gables, Paarl 085654
Gablonzer Archiv Museum,
  Kaufbeuren 019559
Gablonzer Schmuckmuseum,
  Enns 001869
Gabone Art Gallery, Jerewan . . 098600
Gaboriaud, Marc, Saint-Ouen . . 073087
Gaboriaux, Frédéric, Brive-la-
  Gaillarde 067508
Gaboriaux, Frédéric, Chameyrat 067798
Gabory, de, Maule 069997
Gabriel, Footscray 099042
Gabriel, Christophe, Paris 104752
Gabriel, Erich, Wien 100259
Gabrielės Petkevičaitės-Bitės
  Literatūrinis Muziejus, Smilgiai 030482
Gabrielės Petkevičaitės Memorialinė
  Ekspozicija, Panevėžio kraštotyros
  muziejus, Panevėžys 030440
Gabrieli, Giovanni, Liebefeld . . 133807
Gabriëls, Den Haag 083083
Gabrielsen, E., Oslo 113202
Gabrijel, Sarajevo 100860
Gabriola Museum, Gabriola . . . 005854
Gabrius, Milano 110317
Gaby's Clocks and Things,
  Tenterden 091560, 135125
Gaby's Shoppe, Dallas 120601
La Gaceta, La Habana 138610
Gacheon Museum, Incheon . . . 029951
Gadabout, Toronto 064656
Gadaj, Barbara, Warszawa 132876
Gadda, Eugenio, Milano 131449
Gades Art, Roma 110775
Gadesmann, Dr. Detlev, Hannover 130007
GaDeWe – Galerie des Westens,
  Bremen 106580
Gadfly Gallery, Dalkeith 098939
Gadfly Online, Charlottesville . . 139369
Gadner & Reuven, Wien 062948
Gado, Kyoto 111291
Gadow, Hartmut, Itzehoe 076634
Gadow, Hartmut, Puls . 078104, 130544
Gadsby's Tavern Museum,
  Alexandria 046389
Gadsden Arts Center, Quincy . . 052429
Gadsden Museum, Mesilla 050815
Gadsden Museum of Fine Arts,
  Gadsden 048999
Gaedigk, Wolfgang, Rellingen . . 108471
Gaedtke, Hubert, Sonneberg . . 108643
Gähler-Maurer, Walter, Glarus . 144083
Gael Song, Seattle 097445
Gaelic College of Celtic Arts and
  Crafts, English Town 055126
Gällersta Forngård, Örebro . . . 040891
Gaemers, Voorburg 132660
Gäng, Alfred, Osnabrück 077966
Gärtner, Hilden 076528

Gärtner, Schlangenbad 078365
Gärtner- und Häckermuseum,
  Bamberg 016947
Gärtner, Erich, Garmisch-
  Partenkirchen 107102
Gärtner, Heinz, Hannover 141912
Gærup Skolemuseum og
  Brahetrolleborg Sogns
  Folkemindesamling, Faaborg . 010052
Gästrike Antikvariat, Sandviken . 143914
Gaetano, Antonella, Roma 081377
Gaetano, Bruno, Palermo 110603
Gätjens, Dieter, Hamburg 141864
Gäubodenmuseum, Straubing . . 022014
Gävle Konstcentrum, Gävle . . . 040636
Gaffé, Daniel, Dieppe 068299
Gaffelaer, Breda 112354
Gaffer, Hong Kong 102077
Gafor, Barcelona 133117
GAG Dépôt Vente, Poitiers 072223,
                                      141307
Gagarin, Antwerpen 138549
Gage County Historical Museum,
  Beatrice 046844
Gage, Deborah, London 090139
Gagé, Pierre, Cannes . 067639, 140814
Gage, Robert, Austin 092230
Gaggero & Corbo, Genova 131313
Gagliano, Salvatore, Firenze . . 131175
Gagliardi, London 118329
Gagliardi, San Gimignano 110893
Gagliardi, Taormina 110940
Gagliardi-Steinemann, Manuela,
  Ottoberg 116559
Gagliardi, Giovanni, Firenze . . . 131176
Gagnaire, Dallas 093127
Gagnaire, Saint-Étienne 129225
Gagnaire, Georges, Marseille . . 128766
Gagnaire, Leilani, Dallas 120602
Gagosian, London 118330, 118331
Gagosian, Los Angeles 121488
Gagosian, New York . . 122625, 122626
Gahlberg Gallery, College of DuPage,
  Glen Ellyn 049108
Gaia, Seoul 111638
GAIA Gallery of Art, Amsterdam 112197
Gaiarts, Marseille 104168
Gaia's Welt, Neumagen-Dhron . 108244
Gaidu un Skautu Muzejs, Ogre . 030166
Gaiffe, Daniel, Champagnole . . 067807
Gaignon, Claude, Saint-Ouen . . 073088
Gaillard, Agathe, Paris 104753
Gaillard, Christian, Genève . . . 087468
Gaillard, Christian, Paris 129009
Gaillard, Jean-Pierre, La Roche-sur-
  Yon 069073
Gaillard, Nathalie, Paris 104754
Gailtaler Heimatmuseum-Sammlung
  Georg Essl, Schloß Möderndorf,
  Hermagor 002076
Gaimusho, Gaiko Shiryokan,
  Tokyo 029504
Gaineswood, Demopolis 048241
Gainsborough, Oviedo . 086330, 115326
Gainsborough Galleries, Calgary 101237,
                                      101238
Gainsborough House Antiques,
  Tewkesbury 091590, 135131
Gainsborough Old Hall,
  Gainsborough 044376
Gainsborough's House, Sudbury,
  Suffolk 045919
Gainza, Miami 121784
Gairloch Heritage Museum,
  Gairloch 044377
Gaisbauer, Aschach an der
  Donau 127708
Gajac, Villeneuve-sur-Lot 105873
Gajah, Singapore 114524
Gajdarov, Sofia 140321
Gajendra, Kathmandu 112074
Gajewski, Dariusz, Warszawa . . 132875
Gajón Mozota, Antonio, Zaragoza 143816
Gajowy, Witold, Warszawa 132878

Gakken, Tokyo 137777
Gakki Hakubutsukan, Musashino
  Ongaku Daigaku, Tokyo 029505
Gakkigaku Shiryôkan, Tachikawa 029422
Gaku, New York 135928
Gakumon Kagaku Hakubutsukan,
  Tokyo-shi Kagagu Daigaku,
  Tokyo 029506
Gala, Paris 071360
Gala Galeria, Wrocław 113922
Gala Home Furnishings, San
  Diego 097088
Galán, José María, Zaragoza . . 133498
Galanos, Ioannis, Ilioupoli
  Athinon 109065
Galant, Catherine & Patrick, Vaison-la-
  Romaine 073930
Galantaris, Christian, Paris 141117
Galante Antiques, Tyabb 062360
Galante, Marie, Bénodet 067092
Galapagos Art Space, Brooklyn . 047258
Galaria Casa Postigliun, Disentis 116249
Galarie am Schloss, Koblenz . . 107569
Galarie Heimbürge, Arnstadt . . 106003
Gal'Art, Coimbra 113963
Galarthe, Saint-Emilion 105506
Galata Museo del Mare, Genova 026393
Galateau, Bernard, Paris 125717
Galathe, Nantes 104338
Galati Numismatica, Alessandria 079742
Galaverni, Reggio Emilia 110709
Galaxie des Arts, Barbizon . . . 103328
Galaxy, Dallas 093128
Galaxy Books, Dunedin 143314
Galbeada, A Coruña 085952
Galbraith, Austin 092231, 135325
La Galcante, Paris 141118
Galchon Mask Museum, Goseong 029937
Gale-Martin Fine Art, New York . 122627
Galea, C., Valletta 111923
Galeao Antiguidades, Rio de
  Janeiro 064061
Galearte, Bogotá 102298
Galearte, Sociedade de Leilões Lda,
  Lisboa 084888, 126444
Galen Lowe, Seattle 097446
Galena-Jo Daviess County Historical
  Museum, Galena 049015
Galena Mining and Historical Museum,
  Galena 049019
Galera Kombëtare e Arteve Figurative,
  Tiranë 000030
Galerei M, Vrchlabí 102750
Galereja 1000 Stupenej, Tjumen 114370
Galereja 102, Sankt-Peterburg . 114333
Galereja 2/2, Nižnij Novgorod . . 114323
Galereja 2000, Ekaterinburg . . 114180
Galereja 259, Moskva 114220
Galereja 36, Kyїv 117123
Galereja 8 Sov, Sankt-Peterburg 114334
Galereja A & T, Moskva 114221
Galereja A3, Moskva 114222
Galereja AJa, Kursk 114187
Galereja Ajmak, Ufa 114372
Galereja Akademii Chudožestv Rossii,
  Sankt-Peterburg 114335
Galereja Alternativna, Orenburg . 114328
Galereja Arbat, Moskva 114223
Galereja Asti, Moskva 114224
Galereja Astrum, Moskva 114225
Galereja Čertanovo, Moskva . . . 114226
Galereja Čingizchan, Ufa 114373
Galereja Elena, Moskva 114227
Galereja FF, Kursk 114188
Galereja Fonda Svobodnogo Russkogo
  Sovremennogo Iskusstva, Sankt-
  Peterburg 114336
Galereja Forma, Moskva 114228
Galereja Fortuna, Kyїv 117124
Galereja Forum, Sankt-Peterburg 037732
Galereja Fotoimage, Sankt-
  Peterburg 114337
Galereja Galina, Moskva 114229
Galereja Geoid, Moskva 114230

Galereja Gerceva, Moskva 037266
Galereja Getman, Lviv 117154
Galereja Goroskop, Moskva 114231
Galereja im. Bratjev Ljumjer,
  Moskva 114232
Galereja Irena, Kyïv 117125
Galereja Iskusstv Zuraba Cereteli,
  Rossijskaja Akademija Chudožestv,
  Moskva 037267
Galereja Jug, Krasnodar 114186
Galereja Kino, Moskva 114233
Galereja Kino-Foto-703, Sankt-
  Peterburg 114338
Galereja Kitaj-gorod, Moskva . . 114234
Galereja-klub v Saburovo,
  Moskva 037268
Galereja L, Moskva 114235
Galereja Le Vall, Novosibirsk . . 114324
Galereja Lužniki, Moskva 114236
Galereja M, Kyïv 117126
Galereja-masterskaja Chudožnika G.S.
  Rajševa, Chanty-Mansijsk . . 036954
Galereja Meždunarodnogo Universiteta,
  Moskva 037269
Galereja Milena, Kyïv 117127
Galereja Modern, Novosibirsk . . 114325
Galereja Morskaja, Odesa 117157
Galereja Moskovskogo Centra
  Iskusstva, Moskva 114237
Galereja Most, Odesa 117158
Galereja Mucha, Moskva 114238
Galereja na Bastionnoj, Pskov . . 037667
Galereja na Peschanoj, Moskva . . 114239
Galereja na Smolenskoj Naberežnoj,
  Moskva 085179, 114240
Galereja Navicula Artis, Sankt-
  Peterburg 114339
Galereja NU, Moskva 114241
Galereja Odoevskogo,
  Ekaterinburg 036979
Galereja O'K, Kyïv 117128
Galereja Olimp, Kyïv 117129
Galereja OR, Kyïv 117130
Galereja Pan-Ukrainskaja,
  Dnipropetrovs'k 117112
Galereja Panukrainskaja II,
  Dnipropetrovs'k 117113
Galereja Persona, Kyïv 117131
Galereja Pljus, Moskva 114242
Galereja Podsolnuch, Toljatti . . 114371
Galereja RA, Kyïv 117132
Galereja RusArta, Moskva 114243
Galereja Sacha-Art, Postojannoe
  Predstavitelstvo Respubliki Sacha pri
  Prezidente RF, Moskva 037270
Galereja-salon na Varvarke,
  Moskva 114244
Galereja Šatc'i, Moskva 114245
Galereja Selskaja Žizn, Sankt-
  Peterburg 114340
Galereja Sem Bruk, Kulturnyj Centr-
  muzej V.S. Vysockogo, Moskva 037271
Galereja Sem' Gvozdej, Moskva 114246
Galereja Sojuza Chudožnikov Rossii,
  Moskva 114247
Galereja Sojuza Tvorčestvo,
  Moskva 037272
Galereja Šon, Moskva 114248
Galereja Sovremennogo Iskusstva
  Arka, Vladivostok 038029
Galereja Sovremennogo Iskusstva
  Artëtaž, Vladivostok 038030
Galereja Sovremennoj Skulptury,
  Moskva 114249
Galereja Studija, Moskva 114250
Galereja Triptich, Kyïv 117133
Galereja UKV, Kyïv 117134
Galereja Ulica OGI, Moskva . . . 114251
Galereja Ural, Ufa 114374
Galereja Vizit, Moskva 114252
Galereja Vladimir, Charkiv 117107
Galereja Volodimir, Sevastopol . . 117161
Galereja Vrubel, Omsk 114326

Galereja XL, Moskva . . 114253, 114254
Galereja Zarubežnogo Iskusstva
  im. Professora M.F. Gabyševa,
  Jakutsk 037054
Galereja Žizn Stekla, Moskva . . 114255
Galereja Zoo, Moskva 114256
Galeri Alfa, İstanbul . . 088118, 144208
Galeri Artist, İstanbul 117036
Galeri Çerçeve, İstanbul 117037
Galeri Hanehøj, Hadsund 102822
Galeri Nev, İstanbul 117038
Galeri Oda, İstanbul 117039
Galeri Pago, İstanbul 117040
Galeri Suav ve Küsav, İstanbul . 042937
Galeri Ümit Çamaş, İstanbul . . . 117041
Galeri Valfisken, Simrishamn . . . 040944
Galeri Vinci, İstanbul 117042
La Galeria, A Coruña 085953
Galéria, Kecskemét 109198
Galeria, Krzeszowice 113652
La Galeria, Madrid 126544
Galeria, Miami 121785, 121786
Galeria, Nouméa 112737
Galéria, Poprad 114665
La Galeria, Stuttgart 078605
Galeria, Warszawa 113826
Galería 1–2–3, San Salvador . . . 103098
Galeria 10/10, México . 111951, 126262
Galeria 111, Lisboa 114003
Galeria 111, Porto 114084
Galeria 1199, São Paulo 064179
Galéria 13, Budapest 109145
Galeria 153, Porto 085026, 114085
Galeria 19, Sopot 113758
Galeria 2 Światy, Kraków 113545
Galeria 2000, Kraków 113546
Galeria 2000, Nürnberg 108299, 137450
Galeria 2000, Warszawa 113827
Galeria 24, Amberg 105984
Galeria 24 de Outubro, Porto
  Alegre 100976
Galeria 26, Wrocław 113923
Galeria 28, Kraków . . . 084566, 113547
Galeria 2piR, Wyższa Szkoła Nauk
  Humanistycznych i Dziennikarstwa,
  Poznań 035113
Galéria 3, Vaison-la-Romaine . . 105788
Galeria 3–2–0, Akademia Techniczno –
  Humanistyczna w Bielsku Białej,
  Bielsko-Biała 034650
Galeria 3 Punt, Barcelona 114890
Galeria 31, Łódź 113660
Galeria 32, Warszawa 113828
Galeria 34 Bis, Warszawa 113829
Galeria 5, Centrum Kultury i Sztuki,
  Kalisz 034799
Galeria 5, Górnośląskie Centrum
  Kultury, Katowice 034811
Galeria 65 A, Lisboa 114004
Galeria 85, Bydgoszcz 113441
Galeria 86, Łódź 113661
Galeria 88, Kraków 113548
Galeria 90, Łódź 113662
Galeria A – Der Kunstladen,
  Berlin 106235
Galeria A-5, Gdynia 113497
Galeria A. Cortina, Barcelona . . 085774,
                                  114891, 114892
Galería A. D. I., Montevideo . . . 119420
Galeria AB, Kraków 113549
Galeria Abakus – Artefe, Kraków 113550
Galería Acquavella, Caracas . . . 125040
Galería Actualidad, Bucaramanga 102386
Galeria AG, Kraków 113551
Galeria Akademicka, Łódź 113663
Galeria Akademii Sztuk Pięknych
  im. Władysława Strzemińskiego,
  Łódź 034953
Galeria Akcent, Kraków 113552
Galeria Akcji, Warszawa 113830
Galeria Alameda, Vigo 115535
Galeria Alfonso XIII, Madrid . . . 115198
Galeria Alianța Artelor, Cluj-
  Napoca 114159

Galeria Alicji, Gdańsk 113463
Galeria Alisina, Hamburg 107243
Galeria Alix, Bydgoszcz 113442
Galería Alzama, Madrid 115199
Galería Ambar, Bogotá 102299
Galería America, Caracas 125041
Galería America 92', Lima 113348
Galeria Amfilada, Pomorskie Centrum
  Edukacji, Szczecin 035246
Galeria Amsterdam, Lublin 113686
Galeria Aneri, Gdańsk 113464
Galeria Angel Romero, Madrid . . 115200
Galeria Angeles Penche, Madrid 115201
Galería Anmar, Madrid 113553
Galería Antika, Košice 085567
Galeria Antika Dorota, Bratislava 085547,
                                  085548
Galería Antíqua, Porto 085027
Galeria Antique, Kraków 084567
Galería Antiqvaria, Madrid 139054
Galería Antyczna, Nowy Targ . . 084631,
                                  113708
Galeria Antyk, Kraków 084568
Galeria Antykwaryczna Nadine,
  Warszawa 084718
Galería Apolo L, Punta del Este . 115042
Galería Aramayo, Montevideo . . 119421
Galeria Architektury SARP,
  Katowice 034812
Galería Architektury SARP,
  Kraków 034866
Galería Ardan, Bratislava 114648
Galeria Arka, Lublin 113687
Galería Arroyo, Buenos Aires . . 098562
Galería Ars Forum, Caracas . . . 125042
Galeria Ars Nova, Łódź 113664
Galería Arsena#I, Toruń 113787
Galeria Arsenal, Białystok 034636
Galeria Art-Deco, Barcelona . . . 085775
Galeria Art-Foto, Częstochowa . 034704
Galería Art Petritxol, Barcelona . 085776,
                                  114893
Galería Arte Actual Mexicano, Garza
  García 111937
Galería Arte Assinada, Lisboa . . 084889,
                                  114005
Galeria Arte d'Artes, Braga . . . 113956
Galeria Arte e Manifesto, Porto . 114086
Galeria Arte Nobre, Porto Alegre 100977
Galeria Arte Nuevo, Bogotá . . . 102300
Galeria Arte Originales, Medellín 102436
Galería Arte y Antigüedades,
  Maracaibo 097975
La Galeria Arte y Diseño,
  Granada 115087
Galeria Arteko, Donostia-San
  Sebastián 115050
Galeria Artela, Lisboa 114006
Galería Artemis, Bratislava 114649
Galeria Artemis, Kraków 113554
Galeria Artes Alba, Bucaramanga 102387
A Galeria Artes Antiquariato, São
  Paulo 064180
Galería Artesanal de Colombia,
  Bogotá 102301
Galería Artespacio, Santiago de
  Chile 101907
Galería Artfit, Lisboa 114007
Galería Artico, Medellín 102437
Galería Artieri, México 111952
Galeria Artificial, Madrid 115202
Galería Artwood, Bielsko-Biała . 113430
Galeria AT, Poznań 113720
Galeria Atlantica, A Coruña . . . 115033
Galeria Attyka, Kraków 084569
Galeria Autorska, Gdańsk 113465
Galeria Avima, Denia 115041
Galeria Azjatycka, Muzeum Azji i
  Pacyfiku, Warszawa 035317
Galeria Azul, Guadalajara 111940
Galeria B, Gdańsk 113466
Galeria Bałucka, Miejska Galeria
  Sztuki, Łódź 034954
Galeria Bar Santa Clara, Coimbra 113964

Galeria Barata, Lisboa 114008
Galeria Barcelona, Barcelona . . 114894
Galeria Barcelos, Vigo 115536
Galería BAT, Madrid 115203
Galeria BB, Kraków 113555
Galeria BB, Wrocław 113924
Galeria Beaskoa, Barcelona . . . 114895
Galería Belarca, Bogotá 102302
Galeria Belas Artes, Porto Alegre 100978
Galeria Belas Artes, Rio de
  Janeiro 101029
Galeria Benedito, Málaga 115300
Galería Benlliure, Valencia 086517,
                                  115484
Galeria Biała, Centrum Kultury w
  Lublinie, Lublin 034990
Galeria Białostockiego Towarzystwa
  Przyjaciół Pięknych, Białystok . 034637
Galeria Biblioteka, Patio Centrum
  Sztuki Wyższej Szkoły
  Humanistyczno-Ekonomicznej,
  Łódź 034955
Galeria Bielska BWA, Bielsko-
  Biała 113431
Galería Bienes Culturales, Victoria de
  las Tunas 009383
Galeria Bihodo, São Paulo 064181
Galeria Blue Moon, León 111942
Galeria Bogny Lewtak-Baczyńskiej,
  Kraków 113556
Galeria Bonino de Arte, Rio de
  Janeiro 101030
Galeria Brasil, Oxford 118863
Galeria Bukowski, Kraków 113557
Galeria BWA, Lublin 113688
Galeria BWA Awangarda,
  Wrocław 113925
Galeria BWA Design, Wrocław . . 113926
Galeria BWA Studio, Wrocław . . 113927
Galería Café Libro, Bogotá 102303
Galeria Camba, Palma de
  Mallorca 115342
Galeria Can Felipa, Barcelona . . 114896
Galería Canaria de Arte, Santa Cruz
  de Tenerife 115420
Galeria Canem, Castello de la
  Plana 115018
Galería Cano, Bogotá 102304
Galería Carlos MuRo, San Miguel de
  Allende 111991
Galería Carrión Vivar, Bogotá . . 102305
Galeria Casa Colon, Cancún . . . 111934
Galería Casa Cuadrada, Bogotá . 102306
Galería Casa Lamm, México . . . 111953
Galería Casa Negret, Bogotá . . . 102307
Galeria Casar, Alicante 114828
Galeria Catedral, Monterrey . . . 082492
Galería Cedritos, Bogotá 102308
Galeria Cenarius, Luanda 098542
Galeria Centralis, Budapest . . . 109146
Galeria Centre Civic Guinardo,
  Barcelona 114897
Galeria Centre Civic Pati Llimona,
  Barcelona 114898
Galería Centro, San José 102490
Galeria Centrum, Warszawa . . . 113831
Galeria Cervantes, Santander . . 115429
Galeria-Cervino, Augsburg 106025
Galería Charlie, Bogotá 102309
Galeria Charpa, Valencia 115485
Galeria Chodor, Wrocław 113928
Galeria Circulo Art, Palma de
  Mallorca 086350, 115343
Galería Clásica, Medellín 102438
Galería Clásico, Lima 084454
Galeria el Claustre, Girona 115079
Galeria Clave, Murcia 115312
Galería Clavia, Cartagena 102420
Galería Collage, Lima 115114
Galería Colón XVI, Bilbao 114988
Galería Comas, Barcelona 114899
Galería Contrast, Barcelona . . . 114900
Galería Cuatro Diecisiete, Madrid 115204
Galería Cultural de la Escuela Nacional

de Bellas Artes, Lima 034311
Galéria D, Trnava 114670
Galeria D.V., Donostia-San
  Sebastián 115051
Galeria da Arcada, Lisboa 084890
Galeria da Sé, Lisboa 084891
Galería Dacar, Medellín 102439
Galería d'Art Amat, Lleida 115115
Galería d'Art Cal Talaveró, Verdú 040371
Galeria d'Art Mar, Barcelona ... 039062
Galería d'Art Naif, Kraków 113558
Galería d'Art Terra Ferma, Lleida 039577
Galeria das Antas, Porto 085028
Galeria David, Oradea . 085125, 114172
Galeria de Artă, Braşov 114120
Galeria de Artă, Cluj-Napoca ... 114160
Galeria de Artă, Constanţa 114166
Galeria de Artă, Piteşti 036618
Galeria de Artă, Sibiu 114173
Galeria de Artă Ioan Sima, Muzeul
  Judeţean de Istorie şi Artă,
  Zalău 036834
Galeria de Artă Lascăr Viorel, Piatra-
  Neamţ 036610
La Galería de Arte, La Habana . 009143
Galería de Arte, Montevideo ... 119422,
                              119423
Galeria de Arte, Porto Alegre ... 100979
Galería de Arte 11–12, San José 102491
Galería de Arte 715, Lima 113349
Galeria de Arte Andre, São Paulo 064182
Galeria de Arte Augusto Augusta, São
  Paulo 101093
Galeria de Arte Brasileira, São
  Paulo 064183, 101094
Galeria de Arte CC22, Madrid .. 115205
Galería de Arte Colonial, San
  José 102492
Galería de Arte Crayons, San
  José 102493
Galería de Arte Dalí, Toledo ... 115468
Galería de Arte de la Escuela
  Armando Reverón, Barcelona . 054698
Galeria de Arte de la UCV,
  Caracas 054712
Galeria de Arte de Santi, Buenos
  Aires 098563
Galeria de Arte de São Bento,
  Lisboa 114009
Galería de Arte de Tabasco,
  Villahermosa 031657
Galeria de Arte del Aeropuerto,
  Instituto Nacional de Bellas Artes,
  México 031149
Galería de Arte Diners, Bogotá . 102310
Galería de Arte Dos V's, Universidad
  Femenina de'l Sagrado Corazon,
  Lima 034312
Galería de Arte El Callejón,
  Bogotá 102311
Galería de Arte el Taller, Medellín 102440
Galería de Arte Escuela Zeus,
  Medellín 102441
Galería de Arte Expo 80,
  Barranquilla 102264
Galeria de Arte Figuras, Cali ... 102404
Galeria de Arte Flagler, Miami .. 121787
Galeria de Arte Granada Capital,
  Granada 115088
Galeria de Arte Gravura, Porto
  Alegre 100980
Galeria de Arte Ipanema, Rio de
  Janeiro 101031
Galería de Arte Iztapalapa, UAM,
  México 031150
Galería de Arte Jaguar Despertado,
  Villahermosa 031658
Galería de Arte Klasse, Ciudad de
  Panamá 113316
Galería de Arte La Sala, Santiago de
  Chile 101908
Galería de Arte Leonardo da Vinci,
  Buenos Aires 098564
Galería de Arte Metal, Buenos

Aires 098565
Galería de Arte Mexicano,
  México 111954
Galería de Arte Moderna, Rio de
  Janeiro 101032
Galeria de Arte Moura Cosme,
  Lisboa 114010
Galería de Arte Nacional,
  Caracas 054713
Galería de Arte Nova, Málaga .. 115301
Galeria de Arte Porto Feio, Porto 114087
Galería de Arte Sadco,
  Bucaramanga 102388
Galeria de Arte T-Graba, Cali ... 102405
Galeria de Arte Tavira, Bilbao .. 085870,
                              114989
Galería de Arte Tetracktis,
  Bogotá 102312
Galería de Arte Travesia 4,
  Madrid 115206
Galería de Arte Trece, Santiago de
  Chile 101909
Galería de Arte UFF, Universidade
  Federal Fluminense, Rio de
  Janeiro 004694
Galeria de Arte UNICAMP,
  Campinas 004384
Galería de Arte UNIFE, Universidad
  Femenina de'l Sagrado Corazon,
  Lima 034313
Galería de Arte Universal, Santiago de
  Cuba 009342
Galería de Arte Valentina, Cali .. 102406
Galería de Arte Valerio Calles,
  Oruro 004228
Galeria de Arte Vertice,
  Guadalajara 111941
Galería de Arte Via Margutta,
  Aires 098566, 125168
Galería de Arte Via Margutta,
  Córdoba 098589, 125173
Galería de Arte Via Veneto,
  Lisboa 114011
Galeria de Arte Virreynal Cusco Art,
  Lima 113350
Galeria de Arte Ygrego, Lisboa . 114012
Galeria de Artes, Belo Horizonte 100887
Galeria de Artes Renoir, Fortaleza 100947
Galeria de Artes de São Paulo, São
  Paulo 101095
Galería de Artes Visuales, Universidad
  Ricardo Palma, Lima 034314
Galería de Catalanes Ilustres,
  Barcelona 039063
Galería de Cinemateca,
  Montevideo 046252
Galería de Escuela Nacional de Artes
  Plásticas, Universidad Nacional
  Autónoma de México, México . 031151
Galeria de Exposições Temporárias
  da CMMP, Macau 008066
Galería de Historia o Museo del
  Caracol, México 031152
Galería de la Asamblea Legislativa,
  Barcelona 054699
Galería de la Bahía, Montevideo 119424
Galería de la Ciudadela,
  Montevideo 119425
Galeria de la Escuela de Artes
  Plásticas, Universidad de Costa Rica,
  San José 008850
Galería de la Municipalidad de
  Miraflores, Lima 034315
Galería de la Oficina, Medellín .. 102442
Galería de la Paleta, Bogotá ... 102313
Galeria de la Raza, San
  Francisco 053000
Galería de Marcos, Guatemala . 109120
La Galería de San Isidro, Lima . 113351
Galeria Degas, Bratislava 085549
Galeria Degrau Arte, Porto 114088
Galería del Banco Continental,
  Lima 034316
Galería del Barrio, Cali 102407

Galeria del Bronce, Bogotá 102314
Galería del Castillo, San José .. 102494
Galeria del Centro, Boorowa ... 061174
Galería del Cerro, Santiago de
  Chile 101910
Galeria del Cisne, Madrid 115207
Galería del Instituto Cultural Peruano-
  Juan Pardo Heeren, Lima 034317
Galería del Instituto Cultural Peruano-
  Norteamericano, Lima 034318
Galeria del Medio Art, Miami .. 121788
Galeria del Museo Anzoátegui,
  Barcelona 054700
Galería del Pasador, Buenos
  Aires 098567
Galería del Paseo, Montevideo . 119426
Galeria del Sol, Miami 121789
Galeria del Sur, UAM Xochimilco,
  México 031153
Galeria Delgado, Madrid 115208
Galeria Der Reiter Kunstraum,
  Valencia 115486
Galería Di-Marco, Bogotá 102315
Galería Dick-Art, Łódź 084609
Galeria Diez, Madrid 115209
Galería DLA. ., Toruń 113788
Galería d'Museo, Caracas 125043
Galeria do Claustro, Lisboa 084892,
                              114013
Galeria Dom Aktora, Gdańsk ... 113467
Galeria Domek Ogrodnika, Patio
  Centrum Sztuki Wyższej Szkoły
  Humanistyczno-Ekonomicznej,
  Łódź 034956
Galeria Domplein, Utrecht 083644,
                              112696
Galeria Domu Kultury Włókniarzy,
  Bielsko-Biała 034651
Galeria Domus, Wrocław 113929
Galeria Dos Damas, San Diego . 124122
Galería Dr. Miguel Angel Gómez
  Ventura, Villahermosa 031659
Galería Duque Arango, Medellín 102443
Galería Eduarte, Medellín 102444
Galeria Ego, Poznań 113721
Galeria Ekslibrisu, Kraków 113559
Galeria el Cantil, Santander ... 115430
Galería el Cerrojo, Medellín ... 065287
Galeria el Circulo Azul, México . 111955
Galería El Fanal, Caracas 097952
Galería El Marco, Alcalá de
  Guadaíra 114826
Galería El Ojo Ajeno, Lima 034319
Galería El Socorro, Buenos Aires 098568
Galeria Empire, Oradea 085126
Galeria Engram, Górnośląskie Centrum
  Kultury, Katowice 034813
Galería Enlace Diseño, Bogotá . 102316
Galeria Enter, Poznań 113722
Galería Episcopal, Zacatecas .. 031686
Galeria Erotica-Museum für erotische
  Kunst, Köln 019682
Galeria ES Arte, Philadelphia .. 123379
Galería Espacio, San Salvador . 103099
Galería Espacio 36, Zamora ... 115547
Galería Espaço Arte, Brasília ... 100915
Galeria Estampa, Madrid 115210
Galeria Estandarte, Madrid 115211
Galería Estil, Valencia 115487
Galería Estilo, Bogotá 102317
Galería Estudio, Donostia-San
  Sebastián 115052
Galeria etniczna Ambo, Kraków . 113560
Galería Eude, Barcelona 114901
Galería Euroarte, Lisboa 114014
Galeria Evelyn Botella, Madrid . 115212
Galería Expreso del Arte, Bogotá 102318
Galería Extasis, Puntarenas ... 102475
Galeria Extravagance, Sosnowiec 113768
Galeria F5, Kraków 113561
Galeria Fabryka Kestenberga, Patio
  Centrum Sztukii Wyższej Szkoły
  Humanistyczno-Ekonomicznej,
  Łódź 034957

Galería Fábula, Bogotá 102319
Galería Félix, Caracas 125044
Galeria Ferran Cano, Barcelona . 114902
Galeria Ferso, Lisboa 114015
Galeria Filomena Soares, Lisboa 114016
Galeria Fizek, Poznań 113723
Galeria Foksal, Mazowieckie Centrum
  Kultury i Sztuki, Warszawa ... 035318
Galeria Fondo de Arte, Santa Cruz de
  Tenerife 115421
Galeria Forografii Prospero,
  Lublin 113689
Galería Forum, Lima 113353
Galeria Forum Fotografii, Łódź . 034958
Galeria Fotografii, Łódź 113665
Galeria Fotografii, Wrocław 113930
Galeria Fotografii B&B, Bielsko-
  Biała 113432
Galeria Fotografii Instytutu Sztuki
  Akademii Pedagogicznej,
  Kraków 034867
Galeria Fotografii Piwnica Wisłna 12,
  Kraków 113562
Galeria Fotografii Terminal,
  Kraków 113563
Galeria Freta 22, Warszawa ... 113832
Galeria FS, Poznań 113724
Galeria de la Fuente, Bilbao ... 114990
Galeria G de Artes, Rio de
  Janeiro 101033
Galeria Galeria Fotografii – BWA w
  Kielcach, Kielce 113516
Galería Galileu, Barcelona 085777,
                              143716
Galeria Galle, Szczecin 113773
Galería G'Arte, Braga 113957
Galería Gemis, Bogotá 102320
Galeria Gestual Artes, Porto
  Alegre 100981
Galeria Giart, Girona 115080
Galería Goya 2, Valencia 115488
Galeria Graça Brandão, Porto . 114089,
                              114090
Galeria Grafiki i Plakatu,
  Warszawa 113833
Galería Gran Avenida, Bogotá .. 102321
Galeria Greca, Barcelona 114903
Galería Grifus Negrus, Bogotá .. 102322
Galeria Grodzka, Lublin 113690
Galeria H2O, Barcelona 114904
Galeria Habitante, Ciudad de
  Panamá 113317
Galeria Hemar, Lublin 113691
Galeria Hemmerling, Gdańsk ... 113468
Galeria Henry, Lisboa 114017
Galeria Hermida, Bogotá 102323
Galeria Hetmańska, Kraków ... 084570,
                              113564
Galéria Horný Val, Žilina 114673
Galeria Humb-Humbi, Luanda . 098543
Galeria i Ośrodek Plastyczny
  Twórczości Dziecka, Toruń ... 035280
Galería IFAL, Instituto Francés de
  América Latina, México 031154
Galeria III, Porto 114091
Galeria Ikona Księży Młyn, Łódź 113666
Galeria Imperio, São Paulo ... 064184
Galeria Indigo, Lima 113354
Galeria Infantas, Madrid 115213
Galeria Instituto d'Estudis Fotografics
  da Catalunya, Barcelona 039064
Galería Italia, Alicante 114829
Galería Itinerante, Santiago de
  Chile 101911
Galería Itzcazú, San José 102495
Galeria J M, Málaga 115302
Galeria JAD, Kalisz 113506
Galéria Jána Koniarka v Trnave –
  Kopplov Kaštieľ, Trnava 038474
Galeria Janetas, Vedea, Lima .. 114018
Galeria Jatki, Małopolskie Biuro
  Wystaw Artystycznych, Nowy
  Targ 035053
Galeria Jaume III, Palma de

Mallorca 115344
Galeria Joani, Fortaleza 100948
Galeria Jorge Juan, Madrid 115214
Galeria Jorge Shirley, Lisboa ... 114019
Galería José María Velasco,
  Instituto Nacional de Bellas Artes,
  México 031155
Galéria Jozefa Kollára, Slovenské
  Banské Múzeum, Banská
  Štiavnica 038285
Galeria Kanon, Sosnowiec 113769
Galeria Kantorek, Bydgoszcz .. 113443
Galeria Kazimierz, Kraków 113565
Galería Kin, México 111956
Galeria Klub Kolekcionerów,
  Kraków 113566
Galeria Kolegium, Wrocław 113931
Galeria Kolumnada, Łódź 113667
Galéria KOMART, Bratislava ... 038305
Galeria Kombetare e Arteve,
  Tiranë 098523
Galeria Koneser, Lublin 084626, 113692
Galeria Kont, Lublin 113693
Galeria Kontrast, Kraków 084571,
                              113567
Galeria Kontur, Kraków 113568
Galeria Kontynenty, Warszawa .. 113834
Galeria Krakowskiej Fundacji Hamlet,
  Kraków 113569
Galeria Królewska, Lublin 084627,
                              113694
Galeria Krypta u Pijarów, Kraków 084572
Galeria Krytyków Pokaz, Mazowieckie
  Centrum Kultury i Sztuki,
  Warszawa 035319
Galeria la Aurora, Murcia 115313
Galería La Bagatela, Bogotá .. 102324
Galeria la Capella de l'Antic, Hospital
  de la Santa Creu i Sant Pau,
  Barcelona 039065
Galería la Cometa, Bogotá 102325
Galeria La Consolata, Medellín . 102445
Galería la Francia Arte, Medellín 102446
Galería La Quinta, Lima 113355
Galeria Labirynt 2 BWA, Lublin . 113695
Galería Las Pintadas, Estelí ... 033376
Galeria Latina, Montevideo ... 119427,
                              119428
Galería Latino Americana, La
  Habana 102529
Galería Lázaro, San José 102496
Galeria Leo, Lisboa 114020
La Galería l'Ermitage, Medellín . 102447
Galeria Lifeforart, Dąbrowa
  Górnicza 113452
Galeria Lipert, Kraków 113570, 113571
Galería Lonty-Petry, Częstochowa 113449
Galería Los Oficios, La Habana . 009144
Galéria Luca Marco, Prešov ... 114667
Galeria Lufcik, Warszawa 113835
Galeria Luhian 12, Bucureşti ... 114140
Galería Luluberlu, Limón 102472
Galeria M, Kraków 113572
Galeria M, Wrocław 113932
Galeria M Odwach, Wrocław ... 113933
Galéria M.A. Bazovského, Trenčín 038470
Galeria Maeght, Barcelona 114905
Galeria Margem, Faro 113969
Galeria Mariacka, Gdańsk 113470
Galeria Maes, Madrid 115215
Galeria Majolika, Gdańsk 113469
Galeria Malarstwa Akademii Sztuk
  Pięknych im. Jana Matejki,
  Kraków 034868
Galeria Mar Gaita, Palma de
  Mallorca 115345
Galéria Maráky, Bratislava 085550
Gáleria Marie Medvedckej, Oravská
  galéria, Tvrdošín 038478
Galeria Marii Ritter, Muzeum
  Okręgowe, Nowy Sącz 035048
Galeria Marlborough, Madrid ... 115216
Galeria Mat Mar T, Lublin 113696
Galería Mata, San José 102497

Galeria Matahari, Szczecin 113774
Galería Matiz, San José 102498
Galeria MBWA, Leszno 113655
Galería Medicci, Caracas 125045
Galeria Mediterrània, Palma de
  Mallorca 115346
Galeria Mesa Contemporary Arts,
  Mesa Arts Center, Mesa ... 050811
Galéria Mesta Bratislavy,
  Bratislava 038306
Galería Metropolitano, México .. 111957
Galeria Mexicana de Diseño,
  México 031156
Galeria Międzynarodowego Centrum
  Kultury, Kraków 034869
Galeria Miejska Arsenał, Poznań 137900
Galeria Miejska BWA, Bydgoszcz 113444
Galeria Miejsko, Gminny Ośrodek
  Kultury, Stary Sącz 035221
Galeria Millenium, Gdynia 113498
Galería Minimal, Porto 114092
Galería Minotauro, Caracas ... 125046
Galeria Miro, Bogotá 102326
Galería Miró, San José 102499
Galeria Mistral, Kraków 113573
Galería Mölln, Mölln 077379
Galeria Mokra, Poznań 084638, 113725
Galería Monaliza, San José ... 102500
Galeria Monet, Palma de
  Mallorca 115347, 115348
Galeria Monumental, Lisboa ... 114021
Galería Monumento Histórico Nacional
  a la Bandera, Rosario 000541
Galeria Multiarte, Fortaleza ... 100949
Galeria Mun, Madrid 115217
Galería Mundo, Bogotá 102327
Galería Mundo Arte, Medellín .. 102448
Galeria Municipal de Arte,
  Almada 035504
Galería Municipal de Arte Moderno,
  Barcelona 054701
Galeria Municipal Trem, Faro ... 035685
Galería Muntaner, Barcelona ... 114906
Galeria Muro, Valencia 115489
Galeria Muz, Dom Muz – Miejska
  Instytucja Kultury, Toruń ... 035281
Galeria na Freta, Warszawa ... 084719,
                              113836
Galeria Na Piętrze – BWA w Kielcach,
  Kielce 113517
Galeria na Piwnej, Warszawa .. 084720,
                              113837
Galeria na Strychu, Wrocław ... 113934
Galeria na Tarczyńskiej,
  Warszawa 113838
Galería Nacional de Arte,
  Tegucigalpa 023402
Galería Nacional de Arte
  Contemporáneo, San José ... 008851
Galería Nacional y Museo de los
  Niños, San José 008852
Galeria Nacional, Instituto de Cultura
  Puertorriqueña, San Juan .. 036157
Galeria Nadir, Valencia 115490
Galería Namu, San José 102501
Galeria Nave Diez, Valencia ... 115491
Galería Nazareth's, Porto 114093
Galeria Nova, Kraków 113574
Galeria Nowa, Łódź 034959
Galeria Nowa, Poznań 113726
Galeria Nowa Przestrzeń, Patio
  Centrum Sztuki Wyższej Szkoły
  Humanistyczno-Ekonomicznej,
  Łódź 034960
Galeria Nowohuckiego Centrum
  Kultury, Kraków 113575
Galeria Nr 1, Zbigniew,
  Warszawa 113839
Galeria Nuevoarte, Sevilla 115449
Galeria Nusantara, Muzeum Azji i
  Pacyfiku, Warszawa 035320
Galeria Oamenilor de Seamă,
  Fălticeni 036458
Galería Obsidiana, Lima 113356

Galeria Okendo, Vitoria-Gasteiz . 115542
Galeria Olimpus, Łódź 113668
Galeria Olympia, Kraków 113576
Galería OMR, México 111958
Galeria On, Poznań 113727
Galeria OPK Gaude Mater,
  Częstochowa 113450
Galeria Opus, Łódź 113669
Galería Orfila, Madrid 115218
Galería Oriente, Santiago de
  Cuba 009343
Galería Orizont, Bucureşti 114141
Galeria Osiedlowego Domu Kultury
  Pod Lipami, Poznań 035114
Galeria Osobliwości Este, Kraków 084573,
                              113577
Galeria Ostrołęka, Centrum Kultury,
  Ostrołęka 035082
Galeria PAC, Wien 100260
Galeria Palatina, Galerie + Edition,
  Heidelberg 107391, 137272
Galería Pandora, San Salvador . 103100
Galería Passe-Partout, Lisboa .. 114022,
                              114023
Galería Patio, Patio Centrum Sztuki
  Wyższej Szkoły Humanistyczno-
  Ekonomicznej, Łódź 034961
Galería Pazos, Bogotá 102328
Galería Pedro Cera, Lisboa ... 114024
Galeria Pegaz, Kraków 113578
Galeria Persja, Kraków 084574
Galéria Petra Michala Bohúňa,
  Liptovský Mikuláš 038381
Galería Picasso, Bogotá 102329
Galería Picasso, Medellín 102449
Galeria Piekary, Poznań 113728
Galeria Pies, Poznań 113729
Galeria Pilares, Cuenca 115039, 115040
Galeria Piotra Uznańskiego, Łódź 084610,
                              113670
Galería Piurarte, Piura 034384
Galeria Piwnice – BWA w Kielcach,
  Kielce 113519
Galería Pizarro, Valencia 115492, 115493
Galeria Plakatów, Warszawa ... 113840
Galeria Plakatu, Kraków 113579
Galeria Plakatu Astrid, Gdynia .. 113499
Galeria Plan B, Cluj-Napoca ... 114161
Galeria Plater, Warszawa 084721,
                              113841,   126421
Galería Plazuela del Artista,
  Medellín 102450
Galería Pluma, Bogotá 102330
Galería Pocitos, Montevideo ... 046253
Galeria Pod Arkadami, Białystok 113427
Galeria Pod Kopcem, Kraków .. 113580
Galeria pod Plafonem, Wrocław 113935
Galería Pod Podłogą, Lublin ... 113697
Galeria pod Rejentem, Kraków . 113581
Galeria Pod Sufitem, Katowice . 113511
Galeria Podlaska, Miejski Ośrodek
  Kultury, Biała Podlaska 034632
Galéria Poláková, Bratislava ... 114650
Galeria Polony, Poznań 113730
Galeria Polskiego Plakatu,
  Warszawa 113842
Galeria Ponto Artes, São Paulo . 064185
Galeria Ponto das Artes, Lisboa 114025
Galeria Por Amor à Arte, Porto . 114094
Galeria Potocka – Kolekcja Artystów,
  Kraków 113582
Galería Praxis, Lima 113357
Galería Praxis, México 111959
Galería Presença, Porto 114095
Galería Prestige, Caracas 097953
Galeria Pro Arte, Oświęcim 113713
Galería Pro-Arte, San José 102502
Galeria Prof. Sylvio Vasconcellos,
  Instituto Cultural Brasil-Estados
  Unidos, Belo Horizonte 004341
Galeria Profil, Poznań 113731
Galeria Profile, Gdynia 113500
Galeria Promocyjna, Warszawa . 113843
Galeria Przy Piastów, Szczecin . 113775

Galeria Przyssawka, Szczecin .. 113776
Galeria Puenta, Sopot 113759
Galeria Puerta de Alcala, Madrid 115219
Galeria Punto, Valencia 115494
Galería Punto Arte, Barcelona .. 114907
Galeria Quadrado Azul, Porto .. 114096
Galeria r, Poznań 113732
Galeria Radwan, Sopot 113760
Galería Rafael, Valladolid 115525
Galeria Ratton Cerâmicas, Lisboa 114026
Galería Rembrant, Cali 102408
Galeria Restelo, Lisboa 114027
Galeria Retxa, Ciutadella de
  Menorca 115023
Galería Reyes Católicos,
  Salamanca 115404
Galería Río de la Plata,
  Montevideo 119429
Galería Rodas, Bogotá 102331
Galeria Rogan, Barcelona 114908
Galería Rojo y Negro, Bogotá .. 102332
Galeria Romana, Bydgoszcz ... 113445
Galeria Rosengasse, Heide 107386
Galeria Rovira i Monti Se'Leccio,
  Sabadell 115398
Galeria Rynek, Miejski Osrodek Kultury,
  Olsztyn 035068
Galería Safia, Barcelona 114909
Galería Sam, Bucaramanga ... 102389
Galería San Antón, Pamplona .. 115388
Galería San Francisco, Caracas . 125047
Galería San Francisco, Lima ... 113358
Galeria San Ysidro, El Paso ... 093403
Galeria São Mamede, Lisboa .. 114028
Galería Sardon, León 115111
Galeria SARP, Wrocław 035415
Galeria Sen, Madrid 115220
Galeria Senatorska, Warszawa . 113844
Galería Sextante, Bogotá 102333
Galería Sibú, San José 102503
Galeria Siglo XXI, Lugo 115126, 115127
Galeria Signum, Poznań 113733
Galeria Sikoronja, Rosegg 100151
Galeria Sleńdzińskich, Białystok 034638
Galéria Slovenskej Sporitel'ne,
  Bratislava 038307
Galeria Słowiańska przy Parafii
  Prawosławnej, Kraków 034870
Galeria Sokoa, Madrid . 086187, 115221
Galeria SOSO – Arte Contemporânea,
  Luanda 098544
Galeria Stara, Łódź 034962
Galeria Starmach, Kraków 113583
Galeria Stary Browar, Poznań .. 113734
Galeria Stowarzyszenia Artystycznego
  Grupa Krakowska Krzysztofory,
  Kraków 113584
Galeria Stowarzyszenia Wspólnota
  Polska, Kraków 113585
Galeria Stowarzyszenie Moliere,
  Kraków 113586
Galería Studio, Buenos Aires ... 060902,
                              098545
Galeria Studio, Warszawa 113845
Galería Sur, Montevideo 119430
Galeria Surrealista, Barcelona .. 085778,
                              114910
Galeria Sztuki, Kraków 084575, 113587,
                              113588
Galeria Sztuki, Olsztyn 113709
Galeria Sztuki, Warszawa 113846
Galeria Sztuki Attis, Kraków ... 113589
Galeria Sztuki Dawna Synagoga,
  Muzeum Okręgowe, Nowy Sącz 035049
Galeria Sztuki Hand, Łódź 113671
Galeria Sztuki i Rzemiosła, Nowy
  Sącz 113707
Galeria Sztuki im. Włodzimierca i
  Jerzego Kulczyckich, Muzeum
  Tatrzańskie im. Tytusa
  Chałubińskiego w Zakopanem,
  Zakopane 035445
Galeria Sztuki Łowieckiej,
  Warszawa 084722

Galeria Sztuki Naiwnej & Ludowej,
Wrocław 113936
Galeria Sztuki Nowoczesnej PINXIT,
Toruń 113789
Galeria Sztuki Pasaż, Kraków . . 084576
Galeria Sztuki Polskiej XIX w. w
Sukiennicach, Muzeum Narodowe w
Krakowie, Kraków 034871
Galeria Sztuki Polskiej XIX wieku w
Sukiennicach, Muzeum Narodowego,
Kraków 034872
Galeria Sztuki Sakralnej,
Częstochowa 113451
Galeria Sztuki Secesja, Łódź . . 084611
Galeria Sztuki TAB, Warszawa . . 113847
Galeria Sztuki Wozownia, Toruń 035282
Galeria Sztuki Współczesnej,
Kołobrzeg 113521
Galeria Sztuki Współczesnej,
Przemyśl 113756
Galeria Sztuki Współczesnej BWA,
Kalisz 113507
Galeria Sztuki Współczesnej BWA,
Olsztyn 113710
Galeria Sztuki Współczesnej i
Pracownia Gobelinu, Poznań . . 113735
Galeria Sztuki Współczesnej Profil &
Galeria Fotografii pf, Centrum Kultury
Zamek, Poznań 035115
Galeria Sztuki Współczesnej
Winda, Kieleckie Centrum Kultury,
Kielce 034831
Galeria Sztuki Współczesnej,
Biuro Wystaw Artystycznych,
Katowice 034814
Galeria Sztuki Współczesnej, Galeria
Sztuki Współczesnej, Sopot . . 113761
Galeria Sztuki Współczesnej,
Muzeum Narodowe w Szczecinie,
Szczecin 035247
Galeria Sztuki Współczesnej,
Myślenicki Ośrodek Kultury,
Myślenice 035024
Galeria Sztuki Współczesnj,
Łomża 034981
Galeria Szyb Wilson, Katowice . . 113512
Galeria Szyperska, Związek Polskich
Artystów Plastyków, Poznań . . 113736
Galeria Talleres, Asunción 113331
Galeria Teatro Nacional Enrique
Echandi y Joaquín García Monge,
San José 008853
Galeria Tena'Art, Timişoara 114174
Galéria Teo, Banská Bystrica . . . 114646
Galeria Teraz, Szczecin 113777
Galería Terracota, Bogotá 102334
Galeria Terranova, Guadalajara . 082465
Galeria Test, Mazowieckie Centrum
Kultury i Sztuki, Warszawa . . . 035321
Galeria Testemunhos, Porto 114097
Galeria Thema, Valencia 115495
Galería Thema, Valencia 115496
Galeria Tkacka Na Jatkach,
Wrocław 113937
Galeria Tolmo, Toledo 115469
Galeria Torre de Anto,
Carapinheira 084785
Galeria Torre de Anto, Coimbra . 084790
Galeria Torre de Anto, Figueira da
Foz 084800, 084801
Galeria Towarzyszenia Historyków
Sztuki, Kraków 113590
Galeria Trama, Barcelona 114911
Galería Trapecio, Lima 113359
Galeria Trazo, México 111960
Galeria Trino Tortosa, Almería . . 114838
Galeria Ttopara, Hondarribia . . . 115097
Galeria Turleja, Kraków 113591
Galeria Tuset, Barcelona 114912
Galéria Typoars, Bratislava 114651
Galeria U-1, Warszawa 113848
Galeria u Teresy, Wrocław 113938
Galéria Umelcov Spiša, Spišská Nová
Ves 038453

Galéria Umenia, Nové Zámky . . 038416
Galeria Universitaria Artistos,
México 031157
Galeria Urbino, Madrid 086188
Galeria Valbom, Lisboa 114029
Galeria Valle Orti, Valencia 115497
Galeria van Gogh, Bilbao 114991
Galeria Veche, Cluj-Napoca 114162
Galeria Vena, Warszawa 113849
Galeria Venus, Bilbao 114992
Galeria Vértice, Cascais 113960
Galeria Vértice, Lisboa 114030
Galeria Viciana, Valencia 115498
Galeria Victoriei, Braşov 114121
Galeria Vigo, Vigo 115537
Galeria Virtual da Casa das Rosas,
São Paulo 004859
Galeria Visor, Valencia 115499
Galeria w Bramie, Bydgoszcz . 113446
Galeria w Forum, Kraków 113592
Galeria w Hallu, Centrum Kultury i
Sztuki, Kalisz 034800
Galeria w Pasażu, Wrocław . . . 113939
Galeria w Ratuszu, Bielsko-Biała 113433
Galeria w Willi Struwego,
Warszawa 084723
Galeria Weiss Nowina Konopka,
Kraków 113593
Galeria Wiktoria, Kościerzyna . . 113522
Galeria Willa, Miejska Galeria Sztuki,
Łódź 034963
Galeria Władysława Hasiora,
Muzeum Tatrzańskie im. Tytusa
Chałubińskiego w Zakopanem,
Zakopane 035446
Galeria Wnetrz Sanpol, Poznań . 113737
Galeria Wschodnia, Łódź 113672
Galeria Współczesnej Sztuki Sakralnej,
Kielce 113519
Galeria Wymiany, Łódź 113673
Galeria Wzgórze, Bielsko-Biała . 113434
Galeria Xeito, Madrid 115222
Galeria XXI, Mazowieckie Centrum
Kultury i Sztuki, Warszawa . . . 035322
Galería y Tienda de Regalos Indues
Ganges, Medellín 102451
Galéria Z, Bratislava 114652
Galeria z Rączką, Kraków 113594
Galería Zaguán, Valencia 115500
Galeria Żak, Gdańsk 113471
Galeria Zamek w Reszlu, Muzeum
Warmii i Mazur w Olsztynie,
Reszel 035171
Galeria Zamkowa, Bielsko-Biała 113435
Galeria Zapiecek, Warszawa . . . 113850
Galeria Zarząd Realności M. Bukowski,
Kraków 113595
Galeria Ziemi Pałuckiej Mariusza
Niewolskiego, Żnin 035467
Galeria Zona Corazon, Columbus 120493
Galeria ZPAP Sukiennice, Kraków 113596
Galeria Związku Polskich Artystów
Plastyków, Gdańsk 113472
Galeria Związku Polskich Artystów
Plastyków, Kielce 113520
Galeria Związku Polskich Artystów
Plastyków, Łódź 113674
Galeria Związku Polskich Artystów
Plastyków, Olsztyn 113711
Galeria Związku Polskich Artystów
Plastyków, Opole 113712
Galeria, Małopolskie Biuro Wystaw
Artystycznych, Nowy Sącz . . . 035047
The Galeria, Saint Mary's College,
Orchard Lake 051796
Galeria, Wojewódzki Ośrodek Kultury
w Lublinie, Lublin 034989
Galeriacuatro, Valencia 115501
Galerias Arther, Guadalajara . . . 082466
Galerías de Arte, San José 102504
Galerias de Arte Blubtes, Porto
Alegre 100982
Galerías del Oeste, San José . . 102505
Galerias Internacionales de Arte

Sebastian, Cancún 111935
Galerías La Granja, México 082474,
111961
Galerías Marquezrios, Caracas . 097954
Galerias Net, Amadora 132921
Galerías Pacífico de Centro Cultural
Borges, Buenos Aires 000161
Galerias Pina, Faro 113970
Galerias Turisticas, Córdoba . . 085942
Galerias Vandoma, Porto 085029
Galerias Yela, Lisboa 114031
Galericitra, Kuala Lumpur 111905
La Galerie, Ajaccio 103211
Die Galerie, Bünde 106613
La Galerie, Cork 109556
De Galerie, Den Haag 112384
Die Galerie, Detmold 106673
Die Galerie, Dötlingen 106686
Die Galerie, Frankfurt am Main . 106988,
137197
Die Galerie, Hannover 107359
Die Galerie, Hiddenhausen . . . 107428
Galerie, Hossegor . . . 103845, 103846
La Galerie, Les Trois-Ilets 069434
La Galerie, Limoges 104030
Die Galerie, Linz 127807
La Galerie, Luxembourg 111845
Die Galerie, Oldenburg 108340
Die Galerie, Paderborn 108351
La Galerie, Paris 104755
Die Galerie, Rheinbach 108493
La Galerie, Saint-Gilles-Croix-de-
Vie 105523
La Galerie, Seraincourt 073506
Galerie, Tourgéville 105757
La Galerie, Washington 124906
Die Galerie, Wilhelmshaven . . . 108940
Die Galerie, Worpswede 108967
Galerie – C, Benešov 102560
Galerie – Kunsthalle Hosp,
Nas25sereith 100127
Galerie "Bilder-Räume",
Kreuzlingen 116392
Galerie "Zu ebener Erde und im
ersten Stock", Korneuburg . . . 100080
Galerie & Edition Götz, Hohenfels 107441,
137280
Galerie & Edition Marlene Frei,
Zürich 116777, 138109
Galerie & Fotostudio, M. Fuhrmann,
Siegen 108617
Galerie & Kunsthandel Elisabeth den
Bieman de Haas, Amsterdam . 112198
Galerie 027, Sion 116641
Galerie 1, Basel 106066
Galerie 10, Wien 100261
Galerie 100, Berlin 017117
Galerie 100, Magdeburg 107938
Galerie 101, Ottawa 006588
Galerie 104, El Jadida 112042
Galerie-10er Haus, Gmunden . . 099991
Galerie 10er Haus, Gmunden . . 099992
Galerie 13, Esslingen 106933
Galerie 13, Freising 107068
Galerie 13, Neustadt bei Coburg 108264
Galerie 13, Sète 105672
Galerie 13 Rue Jacob, Paris . . . 071361
Galerie 130, Paris 071362
Galerie 14, Paris 104756
Galerie 14–16 Verneuil, Paris . . 104757
Galerie 15, Alès 103222
Galerie 1510, Saint-Jean Pied-de-
Port 105538
Galerie 16, Versailles 074054
Galerie 16, Wien 062949
Galerie 162, Essen 106911
Galerie 17, Salies-de-Béarn . . . 105642
Galerie 1900, Helsinki 066315
Galerie 1900–2000, Paris 104758
Galerie 1990, Eisenstadt 099985
Galerie 19ème, Limoges 069521
Galerie 1a, Husum 107453
Galerie 2 Fenster, Augsburg . . 106026
Galerie 2000, Bordeaux 067329

Galerie 2000, Montréal 101384, 101385
Galerie 201, Amsterdam 112199
Galerie 2016, Bruxelles 100603
Galerie 2016, Hauterive 116373
Galerie 21, Tours 105761
Galerie 21, Uzès 105784
Galerie 216, Bruxelles 063433
Galerie 22 Plus, Nancy 104319
Galerie 23, Lausanne 116413
Galerie 25, Siselen 116646
Galerie 251, Amsterdam 112200
Galerie 26, Paris 104759
Galerie 27, Paris 104760
Galerie 27, Viersen 078799
Galerie 29 Atalante, Evian-les-
Bains 103724
Galerie 2nd Floor, Schwabach . . 108582
Galerie 3, Athinai 109033
Galerie 3, Klagenfurt 100071
Galerie 3 – Vienna, Wien 100262
Galerie 3.2.1, Zug 116880
Galerie 31, Bastia 103339
Galerie 32, Luxembourg 082369
Galerie 32, Lyon 104085
Galerie 33, Aachen . . . 105931, 137022
Galerie 33, Essen 075673
Galerie 39, La Chaux-de-Fonds . 116208
Galerie 39 A, Limoges 104031
Galerie 3A, Kassel 107531
Galerie 3K, Rechberghausen . . 108447
Galerie 4, Cheb 102586
Galerie 40, Wiesbaden 108905
Galerie 41, Monaco . 082507, 111998
Galerie 419, Brisbane 098789
Galerie 422, Gmunden 099993
Galerie 43, Paris 104761
Galerie 44, Saint-Sulpice (Vaud) 116603
Galerie 48, Saarbrücken 108542
Galerie 49, Saumur 105661
Galerie 49 Ann'Cadres, Angers . 066606
La Galerie 5, Toulouse 105736
Galerie 5020, Salzburg 002621
Galerie 53, Paris 104762
Galerie 54, Paris 104763
Galerie 57, Biel 116154
Galerie 58, Magny-en-Vexin . . . 069802
Galerie 58, Montreux . 087680, 116510
Galerie 59, Wuppertal 108987
Galerie 5ünf Sinne, Halle, Saale . 107204
Galerie 6, Dijon 068329, 103676, 128582
Galerie 60, Feldkirch 099987
Galerie 64, Kriens 116395
Galerie 67, Bern 116130
Galerie 67, Québec 101498
Galerie 7, Riegersdorf 100150
Galerie 73, Bordeaux 067330
Galerie 78, Le Chesnay 103938
Galerie 7ème Avenue, Paris . . . 071363
Galerie 83, Zollikerberg 116736
Galerie '88, Hanau . . . 107335, 137263
Galerie 88, Kolkata 109399
Galerie 88, Lorsch 107892
Galerie 89, Aarwangen 116008
Galerie 9, Paris 071364, 104764
Galerie 92, Bruxelles 063434
Galerie 95, Luxembourg 111846
Galerie 96, Bourglinster 111819
Galerie A & A, Tokyo 111424
Galerie à la Grue, Gruyères . . . 116363
Galerie a Muzeum Litoměřické
Diecéze, Litoměřice 009671
Galerie A, Harry Ruhé,
Amsterdam 082754, 137832
Galerie A. & S., Husum 107454
Galerie A+0, Borken, Westfalen 106547
Galerie A16, Zürich 116778
Galerie AA, Clermont-Ferrand . 103603
Galerie AAA, Ostrava . . 065436, 102632,
125416
Galerie AB, Paris 071365, 104765,
141119
Galerie Abélard, Sens 105670
Galerie Academia, Salzburg . . . 100156
Galerie Accord, Marseille 104169

Galerie Action, Sierentz 105673
Galerie Adam, Paris 071366
Galerie Adina, Irene Stumpp,
Giebenbach 116347
Galerie Adlergasse, Kulturverein Riesa
Efau, Dresden 017981
Galerie Adolf Blaim, Messern . . 002383
Galerie Ägypten, Kufstein 100089
Galerie Afrique, Ramatuelle . . . 072360
Galerie Agnes Dutko, Atelier Bonaparte,
Paris 071367, 104766
Galerie Akzente, Minden 108005
Galerie Albane, Nantes 104339
Galerie Alberghina, Bad Homburg
v.d.Höhe 106063
Galerie Albert Baumgarten, Freiburg
im Breisgau 107055
Galerie Albert Prem1er, Bruxelles 100604
Galerie Albis, Ústí nad Labem . . 102746
Galerie Albrecht, München 108074
Galerie Albstadt, Städtische
Kunstsammlungen, Albstadt . . 016491
Galerie Alex Lachmann, Köln . . . 107610
Galerie Alif Ba, Casablanca 112038
Galerie-Alte-Kanzlei, Albstadt . . 074370
Galerie Alte Kunst, Baden-Baden 074633,
106107
Galerie Alte Meister, Köln 076867
Galerie Alte Mühle, Würenlos . . 116728
Galerie Alte Reichsvogtei, Städtische
Sammlungen Schweinfurt,
Schweinfurt 021781
Galerie Alte Schule, Bramstedt . 106556
Galerie Alte Schule Wittstedt,
Bramstedt 017537
Galerie Alte Technik, Berlin 074795
Galerie Alte und Moderne Kunst,
Koblenz 107570
Galerie Alte Wache, Cuxhaven . . 106644
Galerie Altes Kraftwerk, Borken,
Hessen 106546
Galerie Altstadt, Krems 062687, 100082
Galerie Altstadt, Speyer 108651
Galerie am Abend, München . . . 108075
Galerie am Alten Garten, Siegen 108618
Galerie am Alten Markt, Rostock 021491
Galerie am Alten Rathaus, Rupert
Kraus, Regensburg 108461
Galerie am Altpörtel, Speyer . . . 108652
Galerie am Bach, Neustadt an der
Weinstraße 108257
Galerie am Bach, Obererlinsbach 116547
Galerie am Berg, Oberwiesenthal,
Kurort 108326
Galerie am Berg,
Windischgarsten 100403
Galerie am Blauen Wunder,
Dresden 106724
Galerie am Brasselsberg, Kassel 107532
Galerie am Brunnen, Bergisch
Gladbach 106136
Galerie am Brunnental, Lauingen 107801
Galerie am Carlsplatz, Düsseldorf 075470
Galerie am Damm, Dresden . . . 106725
Galerie am Dillberg,
Marktheidenfeld 107981
Galerie am Dom, Aachen 105932
Galerie am Dom, Billerbeck . . . 074997
Galerie am Dom, Frankfurt am
Main 106989
Galerie am Dom, Wetzlar 022607
Galerie am Domhof, Zwickau . . 022908
Galerie am Domplatz, Halle,
Saale 018955
Galerie am Domplatz, Hamburg 076221
Galerie am Eichhof, Kiel 107552
Galerie Am Elisengarten, Aachen 105933
Galerie am Fluß, Pegestorf 108367
Galerie am Fürstenhof, Kempten 076750
Galerie am Gendarmenmarkt,
Berlin 106236
Galerie am Graben, Augsburg . . 016653
Galerie am Hafen, Glückstadt . . 107135
Galerie am Hafen, Ulsnis 108804

Galerie am Hauptplatz,
Fürstenfeldbruck 107078
Galerie am Haus der Kunst, Monika
Schmidt – Japankunst,
München . . . 077507, 108076, 142154
Galerie am Hechtplatz, Zürich . . 087962
Galerie am Hellweg, Werl 108873
Galerie am Hengstbrunnen, Bad
Tatzmannsdorf 099958
Galerie am Hengstbrunnen,
Sulzriegel 100204
Galerie am Holzmarkt,
Wolfenbüttel 079051
Galerie am Jakobsbrunnen,
Stuttgart 108696
Galerie am Johannisplatz, Jena 107478
Galerie am Judenberg, Augsburg 106027
Galerie am Kamp, Teterow . . . 022132
Galerie am Kietz, Schwedt /
Oder 108592
Galerie am Kirchplatz, Fiss 099989
Galerie am Kleinen Plan, Celle . 106628
Galerie am Kloster, Hausach . . . 076422
Galerie am Kloster, Lorsch 107893
Galerie am Kloster, München . . 108077
Galerie am Kreuzplatz,
Frauenfeld 116269
Galerie am Kröpke, Hannover . . 107360
Galerie am Kulm, Neustadt am
Kulm 108253
Galerie am Leewasser, Brunnen 116170
Galerie am Lindenplatz, Vaduz . 111781,
137803
Galerie am Lobkowitzplatz, Wien 100263
Galerie am Maintor,
Sommerhausen 108639
Galerie am Markt, Bad Düben . . 106053
Galerie am Markt, Bad Saulgau 016862
Galerie am Markt, Waldkirch . . . 108839
Galerie am Marktplatz, Büren an der
Aare 116176
Galerie am Michel, Hamburg . . . 107244
Galerie am Münster, Ulm 108801
Galerie am Museum, Flensburg . 106954
Galerie am Museum Glasgarten,
Frauenau 107047
Galerie am Neuen Palais,
Potsdam 108399
Galerie am Nikolaikirchhof,
Leipzig 107820
Galerie am Paradeplatz, Zürich . 116779
Galerie am Park, Kühlungsborn,
Ostseebad 107768
Galerie am Park, Sankt Gallen . 087777,
116609
Galerie am Park, Wien 100264
Galerie am Platz, Eglisau 116254
Galerie am Rathaus, Höxter . . . 104770
Galerie am Ratswall, Bitterfeld . 017404
Galerie am Rennweg, Wien 062950
Galerie am Rex, Wilnsdorf 108943
Galerie Am Rhein, Eglisau 087401
Galerie am Rotbrunnen, Schopp 108575
Galerie am Sachsenplatz, Leipzig 107821
Galerie am Salzgries, Wien 100265
Galerie am Schloss, Gottlieben . 116357
Galerie am Schloß, Heidelberg . 107392
Galerie am Schloß, Senftenberg 021842
Galerie am Schloßgarten,
Mötzingen 108027
Galerie am Schrotturm,
Schweinfurt 108593
Galerie am Spalengraben, Basel . 116067
Galerie am Springbrunnen,
Laßnitzhöhe 100091
Galerie am Stadthaus, Bonn . . . 106527
Galerie am Stein, Schärding . . . 100191
Galerie am Stern, Essen 106912
Galerie am Strichweg, Cuxhaven 106045
Galerie am Stubentor, Wien 100266,
136587
Galerie am Styg, Safnern 116600
Galerie am Tegernsee, Rottach-
Egern 108527

Galerie am Turm, Büren 106615
Galerie am Turm, Waiblingen . . 108832
Galerie am Undertor, Stein am
Rhein 087833
Galerie am Weinberg,
Mainstockheim 077230
Galerie am Weinkeller, Wien . . . 100267
Galerie am Ambiente, Flensburg . 106955
Galerie am Ambiente, Hamburg . 076222
Galerie am Ambiente, Kronach . . 107750,
107751
Galerie Amicorum, Paris 104767
Galerie Ammar Farhat, Sidi Bou
Saïd 116984
Galerie Amsterdam / Amsterdam
Outsider Art, Amsterdam 112201
Galerie Amyot, Paris 104768
Galerie an der Bleiche,
Ludwigslust 020141
Galerie an der Börse, Düsseldorf 106775
Galerie an der Brenz, Sontheim an der
Brenz 078507
Galerie an der Pinakothek der
Moderne, München 108078
Galerie an der Schirn, Frankfurt am
Main 141773
Galerie an der Sierichstraße,
Hamburg 076223
Galerie Anais, München 108079
Galerie Analix Forever, Genève . 116301
Galerie Anatome, Paris 104769
Galerie Anderswo, Berlin 106237
Galerie Ando, Tokyo 111425
Galerie Ange Antoine, Morlaix . . 104293
Galerie Anima, Waldshut-Tiengen 108840
Galerie Anita Beckers, Frankfurt am
Main 106990
Galerie Anke Zeisler, Berlin . . . 106238
Galerie Anna Paulownahuis,
Soest 032892
Galerie Anne de Villepoix, Paris . 104770
Galerie Anne Marie, København 102881
Galerie Annick Mancheron, Paris 071368
Galerie Anno 1589, Bensheim . . 106130
Galerie Anno 1900, Wien 062951
Galerie Ansichtssache, Lübeck . 107910
Galerie Antenna, Dakar 038113
Galerie Antic, Objat 070774
Galerie Antické Umění,
Hostinné 009574
Galerie Antilope, Düsseldorf . . . 106776
Galerie Antja Wachs, Berlin . . . 106239
Galerie Anton, Agen . . 066427, 103173
Galerie Antonigasse, Bremgarten
(Aargau) 116166
Galerie Antonína Chittussiho, Ronov
nad Doubravou 009864
Galerie Antonine Catzéflis, Paris 104771
Galerie Antonins, Nîmes 104398
Galerie Anzbach und Goldammer
Museum, Maria Anzbach 100116
Galerie AO, Emsdetten 106883
Galerie AP, Brno 102564
Galerie Apart, Bad Wimpfen . . . 106098
Galerie Apart, Pforzheim 108374
Galerie Aquarelle, Biel 116155
Galerie Aratora, Artern 106004
Galerie Arcade, Mödling 100120, 136537
Galerie Arcanes, Paris 071369
Galerie Arche, Hameln 107325
Galerie Architektur und Arbeit,
Gelsenkirchen 018624
Galerie Architektury Brno, Brno . 009469
Galerie Arcturus, Paris 104772
Galerie argus fotokunst, Berlin . 106237
Galerie Ariadne, Wien 100268
Galerie Ariane, Dijon 103677
Galerie Arnstein, Weismain . . . 108884
Galerie Ars, Brno 102565
Galerie ARS, Stuttgart 108697
Galerie Art, Biarritz 103383
Galerie Art, Dürnstein 099980
Galerie Art & Craft, Praha 102662

Galerie Art & Ride, Tholey 108768
Galerie Art 204, Brigitte Schmidt,
Düsseldorf 106777
Galerie Art Aurora, Zürich 116780
Galerie Art Céleste, Bordeaux . . 103428
Galerie Art Comparaison, Carnac 103533
Galerie Art Comparaison, La Baule-
Escoublac 103863
Galerie Art Comparaison, Paris . 104773
Galerie Art du Temps, Metz . . . 104224
Galerie Art et Culture, Montréal . 101386
Galerie Art et Patrimoine, Paris . 104774
Galerie Art G, Béziers 103376
Galerie Art-In, Meerane 107985
Galerie Art International, Nancy . 104320
Galerie Art Lympia, Nice 070605
Galerie art one, Zürich 042194
Galerie Art Stuttgart, Stuttgart . 108698
Galerie Art und Weise, Heide . . 107387
Galerie Art'7, Nice 104364
Galerie Artcadache, Vallorbe . . . 116685
Galerie Artès, Barbizon 103329
Galerie Arthéna, Eguisheim . . . 103712
Galerie Artificado, Hildesheim . . 107432
Galerie Artima, Paris 104774
Galerie Artina, Ebersdorf 099982
Galerie Artischocke, Bad
Schussenried 106090
Galerie Artisty, Hamburg 107245
Galerie Artitude, Montréal 101387
Galerie Artlantis, Bad Homburg
v.d.Höhe 106064
Galerie Artpark, Linz 100105
Galerie Arts Multiples, Metz . . . 104225
Galerie Asa, Praha 102663
Galerie Asakusa, Marseille 104170
Galerie Aspekte, Waibstadt . . . 078829
Galerie at Lincoln Center, New
York 051391
Galerie Ateliana 79, Zürich 116781
Galerie-Atelier III, Barmstedt . . 106121
Galerie Atelier Outlet, Berlin . . . 106241
Galerie Athanor, Marseille 104171
Galerie Athena, Bruxelles 100605,
125310
Galerie Atlantis, Cottbus 075274
Galerie Atrium ed Arte, Wien . . 100269
Galerie au Directoire, Paris 071370
Galerie Au Paon, Avenches 116040
Galerie auf dem Mühlenboden,
Teterow 108764
Galerie Auf dem Strich, Landsberg am
Lech 107788
Galerie auf der Spitze,
Biedenkopf 106483
Galerie auf der Treppe, Limburg an
der Lahn 107868
Galerie Aufsturz, Berlin 106242
Galerie Augenblicke, Berlin . . . 106243
Galerie Augenweide, Münster . . 108215
Galerie Aurea, Waiblingen 108833
Galerie Aurika, Frankfurt am
Main 075274
Galerie Aurore, Luxé 104057
Galerie Aurus, Paris 104776
Galerie Austria, Wien 100270
Galerie Avanti, Eze 103728
Galerie Azur, Saint-Paul-de-
Vence 105585
Galerie B, Port Aven 105395
Galerie B, Sinzheim 078485
Galerie B 15, München 108080
Galerie B-312, Montréal 101388
Galerie B des Frankfurter Kunstvereins,
Frankfurt (Oder) 018396
Galerie B. Vincent, Les Sables-
d'Olonne 103988
Galerie b2_, Leipzig 107822
Galerie B9, Neumarkt 107986
Galerie Bab Doukala, Marrakech 112045
Galerie Bab El Kébir, Rabat . . . 031581
Galerie Bab Rouah, Rabat 112047
Galerie Babylonia, Paris 071371
Galerie Baggio, Paris 071372

Galerie Bagnato, Konstanz 107712
Galerie Banpo, Paris 071373
Galerie Baptist Amandus, Bergen op
Zoom 112345
Galerie Bar zur Zinne,
Oberrieden 116548, 133880
Galerie Barbakane, Leipzig 107823
Galerie Barbara, Hradec Králové 102593
Galerie Barbara Weiss, Berlin .. 106244
Galerie Barbier, Eu 068487
Galerie Bassenge, Berlin 074796, 125877,
141517
Galerie Bauart, Dessau 106670
Galerie Bayard, Toulouse 105737
Galerie Beaubourg, Paris 104777
Galerie Beaubourg, Vence 016151
Galerie Beaumontoise, Beaumont-de-
Lomagne 067004
Galerie Beauséjour, Paris 071374
Galerie Beethovenstrasse,
Düsseldorf 106778
Galerie bei der Albertina, Wien . 062952,
100271
Galerie bei der Komödie,
Augsburg 106028
Galerie beim Rathaus, Wil 116717
Galerie beim Roten Turm,
Sommerhausen 108640
Galerie beim Stadthaus, Uster .. 116682
Galerie beim Steinertor, Krems . 100083
Galerie beim Uhrenmuseum,
Wien 062953
Galerie Bel-Etage, Amsterdam . 112202
Galerie Belphégor, Étaules 068482
Galerie Belvedere, Singapore .. 114525
Galerie Ben Kaufmann, Berlin .. 106245
Galerie Benedikta Rejta v Lounech,
Louny 009682
Galerie Bengelsträter, Düsseldorf 106779
Galerie Bengelsträter, Iserlohn . 107470
Galerie Benjamine, La Trinité-sur-
Mer 103907
Galerie Berinson, Berlin 106246
Galerie Berlin, Küttner Ebert GmbH,
Berlin 106247
Galerie Berlioz, Sausset-les-Pins 105662
Galerie Bernau des Förderkreises
Bildende Kunst e.V., Bernau bei
Berlin 017324
Galerie Bernd Klüser, München . 108081
Galerie Berthet-Aittouarès, Paris 104778
Galerie Bertram, Freiburg im
Breisgau 107056
Galerie Best, Bielefeld 106489
Galerie Best, Seoul 111639
Galerie Beyer, Galerie für
zeitgenössische Kunst, Dresden 106726
Galerie Bezirkshauptmannschaft Sankt
Pölten, Sankt Pölten 100188
Galerie Biebertal, Biebertal 017370
Galerie Biedermann, München .. 108082
Galerie Bild und Rahmen, Schwalbach
am Taunus 108588, 137501
Galerie Bilderwelt, Berlin 106248
Galerie Bildschön, Heidelberg .. 076442
Galerie Bildschön, Riesa 108501
Galerie bim Brunne, Dielsdorf .. 116245
Galerie Binnen, Amsterdam 082755
Galerie Birgit Ostermeier, Berlin 106249
Galerie Biró, Internationale
Schmuckkunst, München 108083
Galerie Bis Heute, Bern 116131
Galerie Bismarck, Bremen 106581
Galerie Blanche, Krefeld 107736
Galerie Blankenese, Hamburg .. 107246
Galerie Blaugelbe Zwettl, Zwettl 003143
Galerie Bleiche, Appenzell 116024
Galerie Bleue, Saint-Michel-en-
Grève 072981
Galerie Blickfang, Wuppertal ... 108988
Galerie Blickpunkte, München . 108084
Galerie Blondel, Ostrava 102633
Galerie Bodenseekreis,
Meersburg 020330

Galerie Bohéma, Olomouc 102622
Galerie Bohemiarte, Montréal .. 101389
Galerie Bohemica, Isernhagen . 107473,
137288
Galerie Bona Dea, Cherbourg-
Octeville 103588
Galerie Bordas, Venezia 111104
Galerie Bormes, Bormes-les-
Mimosas 103443
Galerie Bornéo, La Varenne-Saint-
Hilaire 069117
Galerie Borssenanger, Chemnitz 106632
Galerie Bovary, Musée d'Automates,
Ry 015194
Galerie Brandgässli, Luzern .. 116475
Galerie Bratři Špillarů, Domažlice 009539
Galerie Břetislava Tvarůžka,
Opava 102628
Galerie Briand, Royan 105485
Galerie Brigitte Schenk, Köln ... 107611
Galerie Brockstedt, Berlin 106250
Galerie Brockstedt, Hamburg ... 107247
Galerie Brötzinger Art e.V.,
Pforzheim 108375
Galerie Jean Brolly, Paris 104779
Galerie Bruder Baum, Linz 100106
Galerie Brûlée, Strasbourg 105688
Galerie Bruyer, Paris 104782
Galerie Buchwald, Übersee 108796
Galerie Bürgerhaus, Schwabach 108583
Galerie BurkARThof, Neukirch,
Egnach 116533
Galerie Buter, Paris ... 071375, 104781
Galerie BV, Amsterdam 112203
Galerie C, Amiens 066572, 103228
Galerie C, Piazzogna 116565
Galerie C, Sélestat 105666
Galerie C+C, Schwetzingen 108598
Galerie Cabotse, Paris 104782
Galerie Cadr'in, Nouméa 112738
Galerie Caesar, Olomouc 102623, 140491
Galerie Capazza, Musée Imaginaire du
Grand Meaulnes, Nançay ... 014398
Galerie Caprice Horn, Berlin ... 106251
Galerie Capriola, Großostheim .. 107177
Galerie Carinthia, Klagenfurt ... 100072
Galerie Cart, Besançon 103368, 140783
Galerie Casarte, Aschaffenburg . 106008
Galerie Castiglione, Paris 104783
Galerie Celsius, Gonderange ... 111828
Galerie České Plastiky, Praha .. 102664
Galerie Český Porcelán, Praha . 102665
Galerie Chaco, Hamburg 107248
Galerie Chagall, Brno 102566
Galerie Chapelle de la Salamandre,
Nîmes 104399
Galerie Chastel-Maréchal, Paris . 071376
Galerie Chez Anne, Bourg, Mosel 106616
Galerie Chez Marie-Rose, Kunst aus
Ha'iti, Oy-Mittelberg 108349
Galerie Chmelarová, Pardubice .. 065445
Galerie Choiseul, Paris 104784
Galerie Christa, Florstadt 106962
Galerie Christel, Limoges 069522,
104032
Galerie Christian Nagel, Köln ... 107612
Galerie Christine, Saint-Ouen .. 073089
Galerie Christoph Kühl, Hannover 107361
Galerie Cimaise, Genève 116302
Galerie Cimaise, Senlis 105668
Galerie Cimaise d'Or,
Luxembourg 111847
Galerie Circa, Rennes (Ille-et-
Vilaine) 072436
Galerie Clair-Obscur, Mornac-sur-
Seudre 104294
Galerie Claire, Aesch 116012
Galerie Clairefontaine,
Luxembourg 111848
Galerie Clairefontaine – Espace 2,
Luxembourg 111849
Galerie Clarence Gagnon,
Outremont 101484
Galerie Classic Art Deco 1930,

Düsseldorf 075472
Galerie Classico, Berlin 106252
Galerie Classicus, Bissen 111818
Galerie Claude, Köln . 076868, 130147
Galerie Claude Lafitte, Montréal 101390
Galerie Claude Samuel, Paris .. 104785
Galerie Claudia, Houston 121041
Galerie Claudiana, Innsbruck ... 100044
Galerie Claudie Alberio, Paris .. 071377
Galerie Claudius, Hamburg 107249
Galerie Claux, Nice 104786
Galerie Clément, Nice 104365
Galerie Clément, Fine Art Gallery,
Vevey 116693
Galerie Club der Begegnung, Verein
zur Förderung zeitgenössischer
Kunst, Linz 100107
Galerie Collage, Wangerooge,
Nordseebad 108845
Galerie Collections, Paris 071378,
104787,   129010
Galerie Colline, Edmundston .. 005769
Galerie Colombière, Fondettes .. 068555
Galerie Concorde, Mulhouse ... 104312
Galerie Condillac, Bordeaux 103429
Galerie Contact, Städtische Galerie,
Böblingen 017452
Galerie Conte, Perpignan 072079
Galerie Contemporaine, Paris .. 104788
Galerie Cora, Uzès 105785
Galerie Côté, Barbizon 103330
Galerie Coup d'Oeil, Saint-Cristol-lès-
Alès 072719, 129219
Galerie Coup d'Oeuil, Rimouski . 006804
Galerie Couturier, Paris 104789
Galerie CP, Wiesbaden 108906
Galerie Création Plus,
Luxembourg 111850
Galerie CROUS – Beaux-Arts,
Paris 104790
Galerie D, Brno 102567
Galerie D.H., Paris 071379
Galerie d'Art et d'Essai "Le Damier",
Mutuelleville 116981
Galerie D'Agata, Bruxelles 063435
Galerie Daguerre, Photo-Club du Val
de Bièvre, Paris 014653
Galerie d'Albertas, Gémenos ... 103765
Galerie-Damlos, Damlos 106650
Galerie Daniel Besseiche, Genève 116303
Galerie Daniel Buchholz OHG,
Köln 107613, 142007
Galerie d'Antan, Luxembourg .. 082370
Galerie d'Antrain, Rennes (Ille-et-
Vilaine) 105435
La Galerie d'Architecture, Paris . 104791
Galerie d'Arfi, Saint-Sulpice
(Vaud) 116604
Galerie d'Armorique, Quimper .. 105413
Galerie d'Art, Castelsarrasin ... 067714,
103541
Galerie d'Art, Nice 104366
Galerie d'Art, Paris 104792
Galerie d'Art, Saint-Gilles-Croix-de-
Vie 105524
Galerie d'Art & de Création B/C2,
Bettembourg 111816
Galerie d'Art Africain, Céret 067749
Galerie d'Art Ancien et Moderne,
Monaco 082508
Galerie d'Art and Craft, Pézenas 105350
Galerie d'Art Antinea, Marseille . 104172
Galerie d'Art Beauséjour, Bir Mourad
Rais 098538
Galerie d'Art Contemporain,
Chamalières 012414
Galerie d'Art Contemporain, La Seyne-
sur-Mer 013490
Galerie d'art contemporain,
Lausanne 116414
Galerie d'Art Contemporain,
Menton 014130
Galerie d'Art Contemporain, Rablay-
sur-Layon 015028

Galerie d'Art Contemporain Am Tunnel,
Luxembourg 030586
Galerie d'Art Contemporain du Cardo,
Reims 105424
Galerie d'Art Dar Hayet,
Hammamet 116972
Galerie d'Art d'Art, Saint-Martin-de-
Ré 105557
Galerie d'Art de Créteil, Créteil . 012721
Galerie d'Art de Matane, Matane 006273
Galerie d'Art d'Esch, Esch-sur-
Alzette 030581
Galerie d'Art du Centre Culturel,
Sherbrooke, Québec 007022
Galerie d'Art du Château,
Bettembourg 030564
Galerie d'Art du Château de Villeneuve,
Sablé-sur-Sarthe 105491
Galerie d'Art du Gray d'Albion,
Cannes 103508
Galerie d'Art du Parc, Trois-
Rivières 007244
Galerie d'Art du Prince Noir,
Dinard 103694
Galerie d'Art du Vieux Presbytère,
Sainte-Flavie 006950
Galerie d'Art du Wilson, La Ciotat 103882
Galerie d'Art Elysées, Paris 104793
Galerie d'Art Entrée des Artistes, Les
Sables-d'Olonne 103989
Galerie d'Art et d'Essais, Le
Portel 103962
Galerie d'Art Francais, New
Orleans 122167
Galerie d'Art International, Saint-
Julien-en-Genevois 105540
Galerie d'Art Les Amis de l'Art,
Monaco 111999
Galerie d'Art l'Union-Vie,
Drummondville 005720
Galerie d'Art Moderne, Passavant-la-
Rochère 014805
Galerie d'Art Municipale,
Oberkorn 030599
Galerie d'Art Municipale,
Rumelange 030600
Galerie d'Art Royale, Windsor,
Ontario 101854
Galerie d'Art Saint-Honoré, Paris 104794
Galerie d'Art Saint-Pierre,
Vézelay 105853
Galerie d'Art Stewart Hall, Pointe-
Claire 006665
Galerie d'Art Stewart Hall, Centre
Culturel de Pointe-Claire, Point-
Claire 006662
Galerie d'Art, Université de Moncton,
Moncton 006346
Galerie d'Artisanat d'Art Kalloné,
Crémieu 103643
Galerie d'Arts Contemporains,
Montréal 101391
Galerie d'Arts Decoratifs, Miami 094735
Galerie Das Da, Zug 116881
Galerie David, Bielefeld 106460
Galerie D'Avignon, Montréal ... 101392
La Galerie d'Alliance Française de
Manille, Makati 113387
Galerie de Beaune, Paris 071380,
129011
Galerie de Bel Air, Loudun 069642
Galerie de Bel Air, Verrue 074036
Galerie de Berne, Wabern 087888,
133957
Galerie de Bourglinster,
Bourglinster 030566
Galerie de Bretagne, Quimper . 105414
Galerie de Breyne-Lamy,
Bruxelles 063436
Galerie de Briord, Nantes 104340
Galerie de Buci, Paris 104795
Galerie de Cannes, Cannes ... 103509
Galerie de Chai du Terral, Saint-Jean-
de-Védas 015383

Galerie de Chartres, Chartres .. 125551
Galerie de Corbez, Saint-Germain-les-
Corbeil 072812
Galerie de Crécy, Crécy-la-
Chapelle 068227, 103642
Galerie de Crécy, Voulangis 074277
Galerie de Dampierre, Dampierre-en-
Yvelines 103652
Galerie De Di By, Paris 104796
Galerie de Florence, Paris 104797
Galerie de France, Paris 104798
Galerie de Grancy – Centre
International d'Art Contemporain,
Lausanne 116415
Galerie de Guise, Dijon 103678
Galerie de L'Isle, Montréal 101393
Galerie de Jonckheere, Paris ... 104799
Galerie de Joss Joss, Dijon ... 103679
Galerie de Kerbrest, Guidel 103804
Galerie de la Bièvre, Paris 104800
Galerie de la Bourdonnais, Paris 071381
Galerie de la Broye, Payerne ... 116560
Galerie de la Cathédrale, Aix-en-
Provence 066462
Galerie de la Cathédrale, Monaco 112000
Galerie de la Charité, Lyon ... 104086
Galerie de la Cité, Penne-
d'Agenais 105332
Galerie de la Colombe, Vallauris 105806
Galerie de la Corraterie, Genève 116304
Galerie de la Coste, Beaune 103351
Galerie de la Daurade, Toulouse 105738
Galerie de la Ferme de la Chapelle,
Grand-Lancy 116359
Galerie de la Fontaine, La Bresse 103876,
128638
Galerie de la Gare, Bayonne ... 103343
Galerie de la Gare, Molleges ... 070149
Galerie de la Gare, Nice 070606, 104367
Galerie de la Huchette, Paris ... 104801
Galerie de la Loge, Lyon 069737
Galerie de la Louve, Gland 087509,
116348
Galerie de la Madeleine, Béziers 067175,
103377
Galerie de la Maison, Champagne-
Vigny 103564
Galerie de la Marine, Cherbourg-
Octeville 103589
Galerie de la Marine, Nice 014474
Galerie de la Méditerranée, Saint-Paul-
de-Vence 105586
Galerie de la Mer, Paris 104802
Galerie de la Place, Lyon 104087
Galerie de la Place, Paris 104803
Galerie de la Place Ziem, Beaune 103352
Galerie de la Présidence, Paris . 104804
Galerie de la Rivière, Saint-
Savinien 105618
Galerie de la Rotonde, Paris ... 104805
Galerie de la Scala, Paris 104806
Galerie de la Schürra,
Pierrafortscha 116566
Galerie de la Sorbonne, Paris ... 104807
Galerie de la Tour, Bazens 011929
Galerie de la Vielle Fontaine,
Sion 087815, 116642
Galerie-de-l'Alpage I, Megève .. 104214
Galerie de l'Alpage II, Courchevel 103635
Galerie de l'Ancien Courrier,
Montpellier 104277
Galerie de l'Ancienne Douane,
Strasbourg 015867
Galerie de l'Ancienne Poste,
Toucy 073691, 105707
Galerie de Lange, Emmen 032302
La Galerie de Lans, Chambéry .. 103559
Galerie de Lara, Litomyšl 102617
Galerie de l'Arcade, Pézenas ... 072136
Galerie de l'Arche, Ecueille 068427
Galerie de l'Art Erotique,
Newstead 099443
Galerie de l'Aspic, Nîmes 104400
Galerie de l'Echarpe, Toulouse . 105739

Galerie de l'École d'Art, Marseille 014042
Galerie de l'École la Chaufferie, École
Superieure des arts décoratifs,
Strasbourg 015868
Galerie de l'École Nationale des
Beaux-Arts, Nancy 014401
Galerie de l'École Régionale des
Beaux-Arts, Nantes 014411
Galerie de l'Ecusson, Montpellier 104278
Galerie de l'Empreinte, Court .. 116230
Galerie de l'Encadrement,
Etampes 103720
Galerie de l'Ermitage, Le Touquet-
Paris-Plage 103972
Galerie de l'Estampe, Lyon 069738,
104088
Galerie de l'Estuaire, Honfleur .. 103830
Galerie de l'Europe, Paris 104808
Galerie de l'Evêché, Vence 105832
Galerie de l'Hibiscus, Pont-
l'Évêque 105386
Galerie de l'Hôtel de Saulx,
Beaune 103353
Galerie de l'Hôtel de Wicque,
Pézenas 105351
La Galerie de l'Impasse, Lyon .. 104089
Galerie de l'Iroise, Douarnenez . 103704
Galerie de l'Isle, Montpellier ... 070298
Galerie de l'Olympe, Lyon 104090
Galerie de l'Olympe, Perpignan . 072080,
105336
Galerie de l'Opéra, Vichy 105855
Galerie de l'Orne, Rémalard ... 072406,
105427
Galerie de l'Université, Lyon ... 104091
Galerie de l'UQAM, Montréal ... 006372
Galerie de Mai, Penang 111908
La Galerie de Marjac, Saint-Bauzille-
de-Putois 072687
Galerie de Médicis, Paris 104809
Galerie de Messine, Paris 104810
Galerie de Minéralogie, Paris .. 014654
Galerie de Nemours, Annecy ... 103248
Galerie de Nesle – Exposition,
Paris 014655
La Galerie de Noisy, Noisy-le-Sec 014522
Galerie de Normandie, Saint-Vaast-la-
Hougue 105632
Galerie de Paléobotanique, Paris 014656
Galerie de Pézenas, Pézenas ... 072137
Galerie de Pontaillac, Royan ... 105486
Galerie de Pouzauges,
Pouzauges 014984
Galerie de Prêt d'Œuvres d'Art,
Bruxelles 100606, 136638
Galerie de Prêt Ville d'Angers,
Angers 011650
Galerie de Rue, Rue 116596
Galerie De Ruimte, Eersel 112450
Galerie de SOUZY, Paris 071382, 104811
Galerie de Varenne-Gincourt,
Paris 104812, 136926
Galerie de Vigreyos, Noailhac .. 104408
Galerie de Villiers, Le Touquet-Paris-
Plage 103973
Galerie de Villiers, Paris 104813, 104814
Galerie de Vinci, Bruxelles 100607
Galerie De Vink, Utrecht 083645, 112697
Galerie de Winter, Paris 104815
La Galerie d'Egmont, Bruxelles . 063437
Galerie Deï Barri, Gassin 103762
Galerie Deichstrasse, Hamburg . 107250
Galerie Deichstrasse, Holm 107444
Galerie Déjà vu, Frankfurt am
Main 106991
Galerie Dekorat, Lübeck 107911
Galerie Dekorat, Neumünster .. 108246,
108247
Galerie del'Art, Dinan 103684
Galerie Delfi Form, Zwolle 112731
Galerie Delta 98, Den Haag ... 083084
Galerie Demay Debève, Le Touquet-
Paris-Plage 103974
Galerie d'Enfer, Bruxelles 100608

Galerie Depelmann,
Langenhagen 107798, 137335
Galerie der Berliner Graphikpresse,
Berlin 106253
Galerie der Bezirksbibliothek
Rheinhausen, Duisburg 018054
Galerie der Dinge, München ... 108085
Galerie der Fekete-Stiftung,
Südhessisches Handwerksmuseum,
Roßdorf 021481
Galerie der Heimat, Hamm, Sieg 107329
Galerie der Hochschule für Graphik
und Buchkunst Leipzig, Gallery of
the Academy of Visual Arts Leipzig,
Leipzig 019985
Galerie der Jenaoptik AG, Jena . 019445
Galerie der Kreissparkasse Esslingen-
Nürtingen, Esslingen 018316
Galerie der Künste, Berlin 017118
Galerie der Künstler, München . 108086
Galerie der Künstlergruppe Experiment
Stuttgart, Stuttgart 108699
Galerie der Kunststudenten, Wien 002992
Galerie der Kunstuniversität Linz,
Linz 002319
Galerie der Malschule See,
Lupburg 107933
Galerie der Mitte, Oberösterreichischer
Künstlerbund, Linz 002320
Galerie der Moderne, Mainz ... 107948,
137370
Galerie der Sinne, Erding 106886
Galerie Der Spiegel, Köln 107614,
137308
Galerie der Stadt, Bad Wimpfen 016906
Galerie der Stadt Backnang,
Backnang 016689
Galerie der Stadt Fellbach,
Fellbach 018349
Galerie der Stadt Plochingen,
Plochingen 021159
Galerie der Stadt Remscheid,
Remscheid 021363
Galerie der Stadt Salzburg im
Mirabellgarten, Salzburg 002622
Galerie der Stadt Sindelfingen,
Sindelfingen 021871
Galerie der Stadt Stuttgart,
Stuttgart 022045
Galerie der Stadt Traun, Traun . 002837
Galerie der Stadt Tuttlingen,
Tuttlingen 022233
Galerie der Stadt Vöcklabruck,
Vöcklabruck 002866
Galerie der Stadt Waiblingen
Kameralamt, Waiblingen 022375
Galerie der Stadt Wels, Wels ... 002917
Galerie der Stadt Wendlingen am
Neckar, Wendlingen am Neckar 022562
Galerie der Volksbank, Lindenberg im
Allgäu 107876
Galerie der Volksbank, Weinheim 108866
Galerie der Zeichner, München . 108087
Galerie der Zülow-Gruppe, Linz . 100108
Galerie Derby, Zürich . 087963, 116782,
144175
Galerie des 30, Mazamet 104209
Galerie des 4 Coins, Hauterives 103810
Galerie des Ambassadeurs,
Hyères 068855
Galerie des Antiquaires, Saint-
Ouen 073090
Galerie des Antiques, Saint-Rémy-de-
Provence 105609
Galerie des Arcades, Arcachon . 103265
Galerie des Arcades, Noyers-sur-
Serein 070759
Galerie des Arcs, Cortona 109988
Galerie des Artistes, Bergerac .. 067102
Galerie des Artistes l'Art dans la Rue,
Moulins (Allier) 104307
Galerie des Arts, Kaarst 105080
Galerie des Arts, Lailly-en-Val .. 103915
Galerie des Arts, New York 122628

Galerie des Arts, Nîmes 104401
Galerie des Arts, Trigance 105772
Galerie des Arts – Ouaiss Antiquités,
Paris 104816
Galerie des Augustins, Avignon . 103308,
140772
Galerie des Beaux Arts, Charleroi 063600
Galerie des Bildungshauses Sodalitas,
Tainach 002806
Galerie des Brotteaux, Lyon ... 104092
Galerie des Chartons, Bordeaux 067331
Galerie des Chartreuses, Gosnay 103773
Galerie des Dames, Musée de Cires,
Chenonceaux 012546
Galerie des Docks, Nice 104368
Galerie des Ducs, Nevers 104354
Galerie des Escaliers, Saint-
Émilion 105507
Galerie des Estampes, Lille 104008
Galerie des Estampes, Rouen ... 105163
Galerie des Expositions Pablo Picasso,
Denain 012753
Galerie des Franciscains, Saint-
Nazaire (Loire-Atlantique) ... 105561
Galerie des Glariers, Aigle 116014
Galerie des Grands-Augustins,
Paris 104817
Galerie des Iles d'Or, Hyères .. 068856,
128624
Galerie des Jenaer Kunstvereins im
Optischen Museum, Jena ... 019446
Galerie des Joueries, Saint-Germain-
en-Laye 072808
Galerie des Kultur- und
Festspielhauses Wittenberge,
Wittenberge 022699
Galerie des Kultur-Forum-Amthof,
Feldkirchen in Kärnten 001885
Galerie des Kunstvereins, Wetzlar 108891
Galerie des Kunstvereins Aalen e.V.,
Aalen 016448
Galerie des Kunstvereins Neustadt an
der Weinstraße, Neustadt an der
Weinstraße 020758
Galerie des Lampes, Paris 071383
Galerie des Landkreises Rügen,
Putbus 021237
Galerie des Laques, Paris 104818
La Galerie des Laques, Paris ... 071384
Galerie des Limonadiers, Vaison-la-
Romaine 105789
Galerie des Lumières, Saint-Paul-de-
Vence 073245
Galerie des Maîtres Contemporains,
Nice 104369
Galerie des Minimes, Bruxelles . 063438
Galerie des Modernes, Paris ... 104819
Galerie des Muses, Puily 116574
Galerie des Ocres, Roussillon
(Vaucluse) 105483
Galerie des Orfèvres-Carré Mevel,
Paris 104820
Galerie des Ormeaux, Challans . 103554
Galerie des Papes, Châteauneuf-du-
Pape 103579
Galerie des Peintres, Paris 104821
Galerie des Planches, Montreux 116511
Galerie des Polnischen Instituts
Düsseldorf, Düsseldorf 018028
Galerie des Ponchettes, Nice ... 014475
Galerie des Portes Chanac, Tulle 105781
Galerie des Portraitistes,
Toulouse 105740
Galerie des Quais, Plérin 072179
Galerie des Remparts, Bordeaux 103430
Galerie des Remparts, Hyères .. 103849,
128625
Galerie des Remparts, Le Mans 103956
Galerie des Remparts, Vaison-la-
Romaine 105790
Galerie des Sablons, Saint-Malo (Ille-
et-Vilaine) 105548
Galerie des Saints-Pères, Paris . 104822
Galerie des Talents, Sugiez 116664

Galerie des Tanneurs, Mulhouse 104313
Galerie des Tournelles, Paris 104823
Galerie des Trais A, Dieppe 103665
Galerie des Transports, Marseille 014043
Galerie des Trois Pignons, Dijon 068330
Galerie des Tropiques, Papeete . 105889
Galerie des Tuileries, Paris 104824
Galerie des Tuiliers, Lyon 104093
Galerie des Vélos Bleus, Saint-Malo
  (Ille-et-Vilaine) 105549
Galerie des Verbandes Bildender
  Künstler Thüringen e.V., Erfurt . 018242
Galerie des Vosges, Paris 104825
Galerie des Wantiers,
  Valenciennes 105803
Galerie Dethlefs, Ihr Bilderrahmer,
  Elmshorn 106875
Galerie d'Etraz, Lausanne 116416
Galerie Deux, Tokyo 111426
Galerie d'Exposition, Tours 105762
Galerie d'Exposition d'Art et d'Histoire,
  Villaries 016222
Galerie d'Exposition sur la Bourse,
  Paris 014657
Galerie Die schwarze Treppe,
  Haigerloch 107199
Galerie Die Treppe, Nürtingen . 108312
Galerie Dieter, Besigheim 074964
Galerie Dijk 1, Alkmaar 112106
Galerie Dina4 Projekte, München 108088
Galerie Djursholm, Djursholm .. 115599
Galerie Documents, Paris 104826
Galerie Dogon, Berlin 106254
Galerie Dolphin Court, Löwenberger
  Land 107887
Galerie Domi Nostrae, Lyon ... 104094
Galerie Domino, Paris 102666
Galerie d'Orangerie, Willemstad 102530
La Galerie Dorée, Lyon 104095
Galerie Dorée du Comte de Toulouse,
  Paris 014658
Galerie d'Orsay, Boston 119866
Galerie Down Town, Paris 071385,
  104827
Galerie Dr. Riedl, München 108089
Galerie Dream On, Paris 071386
Galerie Drei, Dresden 106727
Galerie Drei Eichen, Wieren .. 108896
Galerie Drei Könige, Zürich 087964
Galerie Dreyfus, Paris 071387
Galerie Driba, Marsa Plage ... 116980
Galerie Druck & Buch, Tübingen 108786,
  142426
Galerie Drucksache, Aichtal ... 105969
Galerie du BAC, Saint-Aubin-
  Sauges 116601
La Galerie du Bec, Le Bec-
  Hellouin 109215, 103925
Galerie du Bellay, Mont-Saint-
  Aignan 104248
Galerie du Berceau, Monaco .. 112001
Galerie du Bouchet, Châtelaine . 116204
Galerie du Canal, Limoges 104033
Galerie du Carlton, Cannes 103510
Galerie du Carré, Versailles ... 074055
Galerie du Carrois, Zürich 116783
Galerie du Cenacle, Develier .. 116244
Galerie du Centenaire, Carouge . 116193
Galerie du Centre, Paris 104828
Galerie du Château, Avenches . 116041
Galerie du Château, Biot 103390
Galerie du Château, Mougins . 104301
Galerie du Château, Vence 105833
Galerie du Château, Espace Graphique,
  Nice 014476, 104370
Galerie du Chatel, Senlis 105669
Galerie du Chêne, Lausanne .. 116417
Galerie du Cheval Blanc, Roanne 105450
Galerie du Chevalier et des
  Collectionneurs, Paris 104829
Galerie du Colombier, Corvol-
  d'Embernard 103633
Galerie du Conseil Général du Bas-
  Rhin, Strasbourg 015869

Galerie du Cotentin, Cherbourg-
  Octeville 103590
Galerie du Crescendo, Mougins . 104302
Galerie du Cygne, Ferney-Voltaire 068523,
  103736
Galerie du Cygne, Paris 104830
Galerie du Dauphin, Honfleur . 103831
Galerie du Dôme, Paris 104831
Galerie du Dourven, Tredrez-
  Locquemeau 105770
Galerie du Dragon, Mons 100793
Galerie du Duc Guillaume,
  Bayeux 103340
Galerie du Faucon, La Neuveville 116535
Galerie du Fleuve, Paris 104832
Galerie du Forum, Albi 103216
Galerie du Forum, Monaco 112002
Galerie du Hameau, Verbier .. 087868,
  116686
Galerie du Jas, Tourtour 105768
Galerie du Jorat, Mézières 116505
Galerie du Jour Agnès B, Paris . 104833
Galerie du Levant, Porto-Vecchio 105397
Galerie du Limas, Avignon ... 066880
Galerie du Lion d'Or, North Shore 112970
Galerie du Louvre, Montréal .. 064422,
  101394
Galerie du Luberon, Gordes .. 103772
Galerie du Manoir, Saint-Evarzec 105515
Galerie du Marché, Montreux . 116512
Galerie du Médoc, Bordeaux .. 103431
Galerie du Mercadial, Bonifacio . 103404
Galerie du Mercadial, Saint-Céré 072708,
  103405
Galerie du Monde, Hong Kong . 102078
Galerie du Multiple, Marseille . 069934
Galerie du Musée, Dijon 103680
Galerie du Musée, Les Baux-de-
  Provence 103983
Galerie du Musée, Paris 104834
Galerie du Musée Montebello,
  Trouville-sur-Mer 016045
Galerie du Palais des Papes,
  Avignon 103309
Galerie du Palais Gallien,
  Bordeaux 103432, 128490
Galerie du Panthéon, Paris ... 104835
Galerie du Paradis, Lisieux ... 104037
Galerie du Parc, Lyon 104096
Galerie du Parc, Vichy 105856
Galerie du Park Palace, Monaco 082509
Galerie du Passage, Moutier ... 116521
Galerie du Passage, Paris 071388
Galerie du Passé, Houdon 068840
Galerie du Pavillon, Bruxelles .. 063439
Galerie du Perron-Borel, Le Grand-
  Saconnex 116360
Galerie du Petit Champlain,
  Québec 101499
Galerie du Petit Michaux, Bar-le-
  Duc 103325
Galerie du Peyrou, Montpellier . 070299
Galerie du Pharos, Avignon 103310
Galerie du Pharos, Marseille ... 104173
Galerie du Pharos, Saint-Rémy-de-
  Provence 105610
Galerie du Phonographe, Dinan . 103685
Galerie du Phonographe, Dinard 103695
Galerie du Pistolet d'Or, Mons .. 063796
Galerie du Pont de Corot, Mantes-La
  Jolie 104153
Galerie du Port, Noirmoutier-en-
  l'Île 104413
Galerie du Port de Brigneau, Moëlan-
  sur-Mer 104243
Galerie du Pressoir, Ecublens .. 116253
Galerie du Prieuré, Carennac .. 103532
Galerie du Puy Joli, Brantôme .. 103459
Galerie du Quai, Strasbourg ... 073588,
  141349
Galerie du Rhône, Sion 116643, 126754
Galerie du Simplon, Vevey ... 087877,
  116694
Galerie du Soleil, Tampa 124651

Galerie du Soleil @ International Plaza,
  Tampa 124652
Galerie du Solstice,
  Treycovagnes 116673
Galerie du Sud, Kressbronn am
  Bodensee 107747
La Galerie du Sud, Marseille .. 104174
Galerie du Tableau, Marseille .. 104175
Galerie du Temple, Lourmarin .. 104050
Galerie du Théâtre, Orange ... 104424
Galerie du Triangle, Lyon 104097
Galerie du Verneur, Pont-Aven .. 105375
Galerie du Vert Galant, Paris .. 104836,
  104837
Galerie du Vexin, Vigny 105862
Galerie du Vieux Belfort, Belfort 103362
Galerie du Vieux Chapitre, Meaux 104210,
  140951
Galerie du Vieux Honfleur,
  Honfleur 103832
Galerie du Vieux Lyon, Lyon .. 104098
Galerie du Vieux Moulin, Clères 103598
Galerie du Vieux Pressoir,
  Onnens 116554
Galerie du Vieux Saint-Paul, Saint-
  Paul-de-Vence 105587
Galerie du Vitrail, Amiens 011630
Galerie du Voisinand, Lutry 116462
Galerie du XXe Siècle, Paris ... 104838
Galerie Due, Baden 087209
Galerie Duo, Bochum 106510
Galerie Dupont, Paris 071389
Galerie Dusseldorf, Mosman Park 099390
Galerie d'YS, Bruxelles 100609
Galerie E, Opfikon 116555
Galerie E-Damm 13, Hannover . 107362
Galerie Eclat, New Orleans ... 122168
Galerie Edelsteine, Bad
  Wörishofen 106101
Galerie Edition Copenhagen,
  København 136772
Galerie Edition Stalzer, Wien ... 100272,
  136588
Galerie Edmond Rostand,
  Marseille 069935
Galerie Eichenwand, Düsseldorf 106780
Galerie Eigen + Art, Berlin 106255
Galerie Eigen + Art, Leipzig ... 107824
Galerie EigenArt, Duisburg ... 107823
Galerie Eins!, Bad Münstereifel . 106076
Galerie Einzigartig, Altenmarkt im
  Pongau 099954
Galerie El Marsa, Marsa 116977
Galerie Elbchaussee, Hamburg . 107251
Galerie Elbhang, Dresden 106728
Galerie Elfenau, Biel 116156
Galerie Elle, Liberec 102613
Galerie Ellen Richard, Küsnacht . 116357
Galerie Emde, Grünberg 107178
Galerie Empreintes, Paris 104839
Galerie en Passant, Paris 106256
Galerie en Ré, Bois-Colombes .. 103403
Galerie Engelsaal,
  Tauberbischofsheim 108757
Galerie Entre Cadre, Montréal .. 101395
Galerie Ephémère, Colombier
  (Vaud) 116224
Galerie Epikur, Wuppertal ... 108989
Galerie Eric Devlin, Montréal .. 101396
Galerie Ernst Hilger, Wien 100273
Galerie eS, Praha 102667
Galerie Esca et Papiers Libres, Revue
  d'Art Contemporain, Milhaud . 104237
Galerie Eschberg, Saarbrücken . 108543
Galerie Espace, Thonon-les-
  Bains 105705
Galerie Espace de la Culture, El
  Jadida 112043
Galerie Espace, Morges 116518
Galerie Esprit Art, La Baule-
  Escoublac 103864
Galerie Eulenspiegel, Basel .. 087232,
  116068, 133643
Galerie Eurêka-Espace Montagne,

Chambéry 012417
Galerie Europeenne, Boston ... 119867
Galerie Europeenne de la Forêt et du
  Bois, Dompierre (Orne) 012807
Galerie Evelyne Canus, Basel . 116069
Galerie Evrard, Roissy-en-Brie .. 105455
Galerie Expression Libre – Les
  Chemins de l'Art, Paris 104840
Galerie Exquisit, Rechtmehring . 108448
Galerie F, Dresden 106729
Galerie F, Karditsa 109069
Galerie f 5,6, München 108090
Galerie f/8, Wels 100219
Galerie Faaturuma, Papeete .. 105890
Galerie Faider, Bruxelles 100610
Galerie Falkenbrunnen, Dresden 106730
Galerie Fallet, Genève 116305
Galerie Farb, Worb 116727
Galerie Feeling, Toulon (Var) .. 105713
Galerie Feille, Pézenas 105352
Galerie Felixe Jeneweina, Kutná
  Hora 009656
Galerie Fernand Léger – Le Crédac,
  Ivry-sur-Seine 013304
Galerie Fichtegasse, Wien 100274
Galerie Figure, Paris 104843
Galerie Finartis, Zug 116882
Galerie Fine Art, Eymet 103725
Galerie FL Arts, Villefranche-sur-
  Mer 105865
Galerie Flak, Paris ... 071390, 104842,
  136927, 141120
Galerie Flip-Flop, Hamburg ... 107252
Galerie Flo, Aigues-Mortes ... 103179
Galerie Flora, Paris 104843
Galerie Flore, Saint-Brieuc ... 105495
Galerie Florence-Basset, Flassans-sur-
  Issole 103742
Galerie Florian, Sceaux 073457
Galerie Florian Sundheimer,
  München 108091
Galerie Florian Trampler,
  München 108092
Galerie Fondation Vromans, IJlst 112543
Galerie Fonderie, Toulouse ... 105741
Galerie Fontaine, Paris 071391
Galerie Formart, Cuxhaven ... 106646
Galerie Forsblom, Helsinki ... 103138
Galerie Fortlaan 17, Tallieu & Tallieu,
  Gent 100692
Galerie Fortuna, Zürich 087965, 144176
Galerie Forum, Mainz 107949
Galerie Forum Lindenthal, Dr. Hans
  Thomas, Köln 107615
Galerie foto-forum, Bolzano ... 109876
Galerie Fototreppe 42, Hanau .. 107336
Galerie Framond, Paris 104844
Galerie Française, München .. 108093,
  137402
Galerie France des Arts, Lillers . 104021
Galerie Franck, Paris 104845
Galerie Frank, München 108094
Galerie Frank Schlag & Cie. GmbH,
  Essen 106913
Galerie Frankengasse, Zürich .. 116884
Galerie Fred, Lyon 069739
Galerie Freihausgasse, Galerie der
  Stadt Villach, Villach 002861
Galerie Freiraum, Lohfelden .. 107888
Galerie Fretags, Alzenau in
  Unterfranken 105982
Galerie Fuchstal, Fuchstal 107077
Galerie für angewandte Kunst,
  München 108095
Galerie für gegenständliche Kunst,
  Kirchheim unter Teck 107560
Galerie für Gegenwartskunst,
  Bonstetten 116165
Galerie für Landschaftskunst,
  Hamburg 107253
Galerie für Zeitgenössische Kunst
  Leipzig, Leipzig 019986
Galerie Fueterchrippe, Zuckenriet 116739
Galerie Fusion, Toulouse 105742

galerie futura, Berlin 106257
Galerie G, Heidelberg 107393
Galerie G, Judenburg 100062
Galerie G22, La Rochelle 103903
Galerie Gärtner, Berlin 106258
Galerie Gala, Montréal 101397
Galerie Gala Amalvy, Valence
(Drôme) 105800
Galerie Galax, Göteborg 115633
Galerie Galipette, Hamburg . . . 107254
Galerie Galya, Neuchâtel 116530
Galerie Gambra, Praha 102668
Galerie Gamma, Wien 062954
Galerie Ganesha, Mersch 082380
Galerie Georges-Goguen,
Moncton 006347
Galerie Gérard, Carcassonne . . . 067672
Galerie Gerbi, Grindelwald 116362
Galerie Gerersdorfer, Wien . . . 100275
Galerie Gerken, Berlin 106259
Galerie Gersag Emmen,
Emmenbrücke 041462
Galerie Gibel, Büron 116179
Galerie Gilgamesh, Paris 071392
Galerie Gillian Morris, Berlin . . . 106260
Galerie Giovanni, Paris 071393
Galerie Glacis, Graz 062587
Galerie Gladys Sani, Paris 071394
Galerie Glaswerk, Germersheim . 107118
Galerie Glockengasse, Köln . . . 019683
Galerie Gloggner Luzern, Luzern 087639,
116476, 126745
Galerie Gloria, Karlovy Vary 102604
Galerie Gluri-Suter-Huus,
Wettingen 042131
Galerie Gmünd, Gmünd, Kärnten 001943
Galerie Gmurzynska, Zürich . . . 116785
Galerie GNG, Paris 104846
Galerie GNG, Verneuil-sur-Avre . 105841
Galerie Godot, Praha 102669
Galerie Goethe 53, München . . . 020525
Galerie Göttlicher, Krems-Stein . 100086
Galerie Marie-Claude Goinard,
Paris 104847
Galerie Goldener Engl, Hall in
Tirol 100029
Galerie Goldturm, Sankt Gallen . 116610
Galerie Gong, Pardubice 102638
Galerie Gora, Montréal 101398
Galerie Gouverneur-Château
d'Homecourt, Homécourt . . . 103814,
136821
Galerie Graff, Montréal 101399
Galerie Grand-Rue, Genève . . 116306,
144075
Galerie Grand Siecle, Taipei . . . 116915
Galerie Grande Fontaine, Sion . 116644
Galerie Grand'Rue, Poitiers . . . 105366
Galerie Granero, Bern 116132
Galerie Grard, Fenin 116263
Galerie Grashey, Küsnacht 116398
Galerie Gražyna, Ústí nad Labem 102747
Galerie Grita Insam, Wien 100276
Galerie Groeneveld, Almelo . . . 112110
Galerie Gronert, München 108096
Galerie Grounded, Chemnitz . . . 106633
Galerie Grünstraße, Berlin 017119
Galerie Günther Frey, Radenthein 100145
Galerie Guillaume, Paris 104848
Galerie Guislain – États d'Art,
Paris 104849
Galerie Gunzenhausen, München 108097
Galerie Gunzenhausen, München 108098
Galerie Gut Gasteil, Prigglitz . . . 100144
Galerie H 17, Wien 100277
Galerie Hall du Mouggar, Alger . 098536
Galerie Hamelin, Honfleur 103833
Galerie Handwerk, München . . . 020526
Galerie Hans Steger, Zeulenroda-
Triebes 022869
Galerie Hans Strelow, Düsseldorf 106781
Galerie Hanssler, München . . . 108099
Galerie Haslinger, im Arik Brauer Haus,

Wien 100278
Galerie Haus 23, Cottbus 017765
Galerie Haus Hohenzollern,
Schlangenbad 078366
Galerie Heike Curtze, Berlin . . . 106261
Galerie Heimeshoff, Essen . . . 106914,
137183
Galerie Heinz Bossert, Köln . . . 107616
Galerie Heinz Teufel Multimedia,
Zingst, Ostseebad 022880
Galerie Heinz Teufel Schule des
Sehens, Zingst, Ostseebad . . . 022881
Galerie Heinz Teufel Villa Ruh, Zingst,
Ostseebad 022882
Galerie Heitz, Strasbourg 015870
Galerie Heja, Freiburg im
Breisgau 075864, 129859
Galerie Helbing, Hamm,
Westfalen 107330
Galerie Hell, München 108100
Galerie Henri Sie, Saint-Tropez . 105626
Galerie Henseler, München . . . 077508,
108101
Galerie Henze & Ketterer & Triebold,
Riehen 116586
Galerie Hilt, Basel 116070
Galerie Himmelreich, Magdeburg 107939
Galerie Himmelsweg, Tostedt . . 078689
Galerie Hinter dem Rathaus,
Gemeinschaft Wismarer Künstler und
Kunstfreunde e.V., Wismar . . 022692
Galerie Hlavního Města Prahy –
Městská knihovna, Praha . . . 009792
Galerie Hochmauren, Rottweil . . 108535
Galerie Hof 12, Amersfoort 112122
Galerie Hofberg 10, Linz 100109
Galerie Holbein, Lindau 107873
Galerie Hommage, Uelzen 108797
Galerie Hopilliart Leroux, Paris . 071395
Galerie Horizon, Hermance 116376
Galerie Horizon, Marseille 104176
Galerie Hotel Leipziger Hof,
Leipzig 019987
Galerie Hubert, Bremen 075160
Galerie Hübner & Hübner, Frankfurt
am Main 106992
Galerie Huihuaijie, Berlin 106262
Galerie Humus, Lausanne 116418
Galerie I.E.M, Antibes-Juan-les-
Pins 103258
Galerie I.S Thomas, Paris 104850
Galerie Icon, Berlin 106263
Galerie ID., Carouge 116194
Galerie IHB Spectrum, Unterseen 116677
Galerie III, Bussum 112360
Galerie il Quadro, Aachen 105934
Galerie Ilsa, Paris 071396
Galerie im 44er Haus, Leonding 002298
Galerie im Adalbert Stifter-Haus,
Oberösterreichisches Literaturhaus
im Stifterhaus, Linz 002321
Galerie im Adlon, Berlin 106264
Galerie im Ahlbecker Hof, Ahlbeck,
Seebad 105963
Galerie im alten Kloster, Köln . . 107617
Galerie im Alten Rathaus, Prien am
Chiemsee 021221
Galerie im alten Teehaus,
Havixbeck 107384
Galerie im Amtshaus, Bad
Wurzach 016916
Galerie im Andechshof,
Innsbruck 100045
Galerie im Angelburger Kunst- und
Kulturhaus, Angelburg 016581
Galerie im Arthouse, Ahrensburg 105966
Galerie im Atelier, Aalen 115952
Galerie im Behnisch-Haus, Sabine
Broekmann, Krefeld 107737
Galerie im Bettli, Dübendorf . . . 116251
Galerie im Bildhauerhaus und
Skulpturenwanderweg, Einöde 001853
Galerie im Breiten Turm,
Delitzsch 106665

Galerie im Bürger- und
Gemeindezentrum, Hofstetten-
Grünau 002093
Galerie im Bürgerhaus, Castrop-
Rauxel 106624
Galerie im Bürgerhaus,
Haigerloch 018928
Galerie im Bulgarischen Kulturinstitut,
Berlin 106265
Galerie im Einewelt-Haus,
Magdeburg 107940
Galerie im Einstein, Berlin 106266
Galerie im Elisabethenhof, Bad
Wimpfen 106099
Galerie im ersten Stock, Mödling 100121
Galerie im Finnland-Institut,
Berlin 106267
Galerie im Fontanehaus, Berlin . 017120
Galerie im Ganserhaus, Arbeitskreis
68, Wasserburg am Inn 108849
Galerie im Geiersnest,
Bollschweil 106521
Galerie im Gemeinschaftshaus,
Berlin 017121
Galerie im Gericht, Berlin 106268
Galerie im Gewölbe, Vorchdorf . 100216
Galerie im Graben, Zollikofen . . 116737
Galerie im Griesbad, Ulm 108802
Galerie im Haus Knotenbaum,
Todtnau 108770
Galerie im Helferhaus, Heimat
und Kunstverein Backnang,
Backnang 016690
Galerie im Höchhuus, Küsnacht . 116043
Galerie im Hof, Baar 116043
Galerie im Hof, Düsseldorf 106782
Galerie im Hof, Groß-Umstadt . . 076105,
107171
Galerie im Hof, Mutterstadt . . . 108232
Galerie im Hofmannschen Gut,
Dürröhrsdorf-Dittersbach 106758
Galerie im Hühnerstall, Krummin 107766
Galerie im Kapellenhof,
Obertshausen 108322
Galerie im Kehrwiederturm,
Hildesheim 019248
Galerie im Kies, Altach 099953
Galerie im Kloster, Magdeburg . 107941
Galerie im Kloster, Ribnitz-
Damgarten 108499
Galerie im Körnerpark, Kommunale
Galerie Neukölln, Berlin 017122
Galerie im Kornfeld, Zehdenick . 109003
Galerie im Kornhauskeller der Ulmer
Kunststiftung Pro Arte, Ulm . . 022259
Galerie im Krottental, Forchheim,
Oberfranken 106963
Galerie im Kulturhaus Spandau,
Berlin 017123
Galerie im Kunsthoken,
Quedlinburg 021244
Galerie im Kurpark, Bad Steben 106095
Galerie im Lend, Graz 100000
Galerie im Mänihaus, Frutigen . 116271
Galerie im Malzhaus, Plauen . . . 021148
Galerie im Maritim, Lübeck 107912
Galerie im Nachtwächterhaus,
Poysdorf 002526
Galerie im Park, Burgdorf 116182
Galerie im Polnischen Institut,
Leipzig 019988
Galerie im Prater, Berlin 017124
Galerie im Prediger, Schwäbisch
Gmünd 021748
Galerie Im Rahmen, Herten . . . 107423
Galerie Im Rahmen, Koblenz . . 107393
Galerie im Rathaus, Bad Aussee 001733
Galerie im Rathaus, Bruck an der
Mur 001802
Galerie im Rathaus Köpenick,
Berlin 017125
Galerie im Rittergut, Niederwiesa 108277
Galerie im Saalbau, Kommunale

Galerie Neukölln, Berlin 017126
Galerie im Sankt Gertrude Verlag,
Hamburg 107255
Galerie im Schauspielhaus,
Chemnitz 106634
Galerie im Schlössli,
Gontenschwil 116353
Galerie im Schloss Porcia, Spittal an
der Drau 002765
Galerie Im Schloßgarten, Leimen 107804
Galerie im Schopf, Reinach
(Aargau) 116579
Galerie im Seewinkel, Gwatt . . . 116371
Galerie im Stil, Dresden 106731
Galerie im Taschengrabais,
Dresden 106732
Galerie im Taxispalais, Innsbruck 002123
Galerie im Teisenhoferhof der
Malschule Motiv Wachau,
Weißenkirchen in der Wachau . 002906
Galerie im Theater am Saumarkt,
Feldkirch 001880
Galerie im Tor, Emmendingen . . 106880
Galerie im Traklhaus, Salzburg . 002623,
100157
Galerie im Troadkasten, Pram . . 002530
Galerie im Turm, Berlin 017127
Galerie im Unteren Tor, Bietigheim-
Bissingen 106501
Galerie Im Vest, Recklinghausen 108450
Galerie im Volkspark, Burg
Giebichstein Hochschule für Kunst
und Design Halle, Halle, Saale 018956
Galerie im Weytterturm,
Straubing 108682
Galerie im Willy-Brandt-Haus,
Berlin 017128
Galerie im Wolferlhof, Kötzting . 107708
Galerie Image, Wien 100279
Galerie Imago, Bègles 103359
Galerie Imoberdorf, Murten . . . 116523
Galerie IMP, Bordeaux 103433
Galerie Impression, Egg an der
Günz 106863
Galerie in Blau, Erlangen 106896
Galerie in C, Köln 107618
Galerie in de Kerkstraat, Arnhem 112330
Galerie in der Alten Post, Kunstverein
Eislingen, Eislingen 018147
Die Galerie in der Alten Schule,
Kommunale Galerie Treptow,
Berlin 017129
Galerie in der Brasilianischen
Botschaft, Berlin 017130
Galerie in der Burg,
Großbodungen 107172
Galerie in der Fabrik, Burgdorf . 116183
Galerie in der Feste, Karlsruhe . 107511
Galerie in der Haula, Gurmels . . 116370
Galerie in der Hauptschule,
Ulrichsberg 002851
Galerie in der Hinterstadt,
Friedrichstadt 107073
Galerie in der Klosterschule,
Glonn 018715
Galerie in der Praxis, Wien 100280
Galerie in der Scheune, Werneck 108879
Galerie in der Schmiede,
Pasching 100135
Galerie in der Ungarischen Botschaft,
Berlin 017131
Galerie in der Waldemai,
Schmallenberg 108560
Galerie In dr Sose, Sosa 108647
Galerie in Eppendorf, Hamburg . 107256
Galerie in Riedli, Belp 116114
Galerie in Touch, Zug 116883
Galerie Incontro, Eitorf 106872, 137174
Galerie Incontro, Zürich 116786
Galerie Indigo, Martigues 104629
Galerie Inge Huber, München . . 108102
Galerie Inkatt, Bremen 017586
Galerie Insel, Stuttgart 108700
Galerie Inter Tableaux, Paris . . 104851

Galerie InterArt, Tutzing 108792
Galerie International du Verre,
Biot 103391
Galerie Ippon, Vellinge 087156
Galerie Iris Steinegger, Rüdlingen 116597
Galerie Isartis, Erlangen 106897
Galerie isi, Bad Nauheim 106080
Galerie Isshorst, Hamminkeln . . 107333
Galerie J, Łódź 113675
Galerie J.L.P., Bordeaux 067332
Galerie J.M Rousseau, Ars-en-Ré 103278
Galerie Jacques Cœur,
Montpellier 104279
Galerie Jacques de la Béraudière,
Genève 116307
Galerie Jahreszeiten, Binz,
Ostseebad 106503
Galerie J'Aime, Saint-Paul-de-
Vence 105588
Galerie Jakubská, Praha 102670
Galerie Jamault, Paris 104852
Galerie Jamault, Versailles 105844,
129339
Galerie Jamborův Dům, Tišnov . 009914
Galerie Jánský Vršek, Praha . . . 102671
Galerie Jara, Lyon 104099
Galerie Jaroslava Fragnera,
Praha 009793
Galerie Jaune, Le Mans 103957
Galerie JBK, Praha 102672
Galerie Jean Charles Billy, Le Touquet-
Paris-Plage 103975
Galerie Jean Fournier, Paris . . . 104853
Galerie Jean Jaurès, Montpellier 070300
Galerie Jean Macé, Brest 103462
Galerie Jeanne, München 108103
Galerie Jeanne d'Arc, Dijon . . . 103681
Galerie Jeanne d'Arc, La Baule-
Escoublac 103865
Galerie Jeannet, Honfleur 103834
Galerie Jerome, København . . . 102882
Galerie Jérôme deNoirmont,
Paris 104854
Galerie Jiřího Trnky, Unie Výtvarných
Umělců, Plzeň 009759
Galerie Joaquin, Mandaluyong . 113396
Galerie Joaquin, Muntinlupa . . . 113408
Galerie Joaquin, San Juan 113423
Galerie Joaquin, Singapore . . . 114526
Galerie Jonas, Cortaillod 116228
Galerie Jonas, Paris . . . 071397, 104855
Galerie Josefa Sudka,
Uměleckoprůmyslové muzeum v
Praze, Praha 009794
Galerie Josephski-Neukum,
Vilgertshofen 108823
Galerie Jouvène, Marseille 104177
Galerie Joy, Zürich 116787
Galerie Jürgensen, Hoisdorf . . . 107443
Galerie Julia L, Passy-en-Valois 072028
Galerie Julien, Le Castellet 103934
Galerie Jupiter, Paris 104856
Galerie Juri, Winterthur 116722
Galerie Justitz, Unhošt 102744
Galerie K, Hannover 107363
Galerie K, Kindberg 100065
Galerie K, Landstuhl 107794
Galerie K, Zingst, Ostseebad . . 109006
Galerie k 9, Hannover 107364
Galerie K., Montreuil-sous-Bois . 104290
Galerie Kabul Art, Hamburg . . . 076225
Galerie Kaiserstraße, Dortmund . 106700,
129696
Galerie Kami, Ústí nad Labem . . 102748
Galerie Kamm, Berlin 106269
Galerie Kampl, München 108104
Galerie Kaplička, Hodonín 102592
Galerie Karenina, Wien 100281
Galerie Katharina,
Recklinghausen 108451
Galerie Kattenturm, Bremen . . . 106582
Galerie Kervoazec, Plouhinec . . 105361
Galerie Killeswerk, Stuttgart . . . 108701
Galerie Kir, Oberhausen 108316

Galerie Klaus Benden, Köln 107619
Galerie Klosterformat, Rostock . 108519
Galerie Klostermühle, Hude 107449
Galerie Königstraße, Dresden . . 106733
Galerie Konkret, Dresden 106734
Galerie Konrad Bayer, München 108105
Galerie Kontraste, Erwitte 106900
Galerie Koráb, Ústí nad Orlicí . . 102749
Galerie Kornfeld, Bern . 116133, 126714,
138030, 144047
Galerie Koruna, Hradec Králové . 102594
Galerie Krab, Praha 102673
Galerie Kriens, Kriens 116396
Galerie Křížovníků, Praha 102674
Galerie Kühn – art in boxes,
Berlin 106270
Galerie Künstlergruppe Kontra,
Hameln 058883, 107326
Galerie KuL – Kunst und Literatur,
Bruck an der Mur 099969
Galerie Kultur im Bären,
Häggenschwil 116372
Galerie Kulturwerk T66, Freiburg im
Breisgau 107057
Galerie Kulturzentrum Lerchhaus,
Eibiswald 001848
Galerie Kulurzentrum Sigristenkeller,
Bülach 116172
Galerie Kunst, Limburg an der
Lahn 107869
Galerie Kunst & Handel, Graz . . 100001,
100002
Galerie Kunst am Klimtplatz, Unterach
am Attersee 100207
Galerie Kunst am Lauffen,
Laufenburg 077033
Galerie Kunst der Gegenwart,
Salzburg 100158
Galerie Kunst im Gang, Bamberg 106115
Galerie Kunst im Heppächer,
Esslingen 106934
Galerie-Kunst ums Pferd, Vogtsburg
im Kaiserstuhl 108825
Galerie Kunstart, Ingelheim am
Rhein 107459
Galerie Kunstblick, Balingen . . . 106114
Galerie Kunstblick, Berlin 106271
Galerie Kunstbrücke Osteuropa,
Gesellschaft für Osteuropa-
Förderung.e.V, Berlin 017132
Galerie Kunstforum im Alten Haus,
Seligenstadt 108610
Galerie Kunsthandel Barbara Vogt,
Hamburg 107257
Galerie Kunstlade, Zittau 022887
Galerie Kunstpunkt Berlin, Berlin 106272
Galerie Kunststall, Kirkel 107563
Galerie Kunststück, Buxtehude . . 106622
Galerie Kunststück, Dinslaken . . 106683
Galerie Kunststück, Leipzig 107825
Galerie Kunststücke, Halle, Saale 107205
Galerie Kunstück, Oldenburg . . 108341,
137460
Galerie Kunstverein Ebersberg,
Ebersberg 018081
Galerie Kunstverein Mistelbach,
Mistelbach an der Zaya 002392
Galerie Kunstverein Wilhelm Kimmich,
Lauterbach, Schwarzwald . . . 019943
Galerie Kunstwerk, Heimbach . . 107408
Galerie L, Saint-Étienne 105510
Galerie L. J. Beaubourg, Paris . . 104857
Galerie la Belle Angèle, Pont-
Aven 105376
Galerie la Belle Epoque, Beaulieu-sur-
Mer 103349
Galerie la Bohème, Zürich 116788
Galerie la Boucherie, Noyers-sur-
Serein 104416
Galerie La Cariatide, La Martyre . 103896
Galerie La Cité, Luxembourg . . . 111851,
137811
Galerie La Couronne, La Sarraz . 116628
Galerie La Ferronnerie, Paris . . . 104858

Galerie La Ligne, Zürich 116789
Galerie la Pointe Badine, Vence . 105834,
141374
Galerie La Poutre, Marseille . . . 069936,
125638, 128767, 136835
Galerie la Santoline, Sanary-sur-
Mer 105649
Galerie La Serre, Saint-Étienne . 105511
Galerie La Spirale, Vevey 116695
Galerie La Vong, Hong Kong . . . 102079
Galerie Labor 19, Berlin 106273
Galerie Laétitia, Brie-Comte-
Robert 103468
Galerie Laforet, Martigny 116493
Galerie Laforet, Verbier 116687
Galerie Lahumière, Paris 104859, 136928
Galerie Lajarrige, La Baule-
Escoublac 068948, 103866
Galerie l'Albertine, Nice 070607
Galerie Lapislazuli, Freiburg im
Breisgau 107058
Galerie Lapislazuli, Wiesbaden . 108907
Galerie l'Aquarium, Valenciennes 016095
Galerie l'Arche, Alès 103223
Galerie Lat die Tied, Nordstrand 108287
Galerie l'Atelier, Le Castellet . . . 103935
Galerie l'Atelier, Le Mont-sur-
Lausanne 087678, 116507
Galerie Latour, Martigny 116494
Galerie Laura, Mandres-les-
Roses 104150
Galerie Laury Bailly, Paris 071398
Galerie Lazertis, Zürich 087966, 116790
Galerie le 5, Ribérac 105446
Galerie le Blason, Ruoms 105490
Galerie le Bouchon, Camaret-sur-
Mer 103495
Galerie le Breton, Port-Aven . . . 105377
Galerie Le Carré, Musée Bonnat,
Bayonne 011923
Galerie le Cèdre Rouge, Montréal 101400
Galerie le Creuset, Charleroi . . . 100678
Galerie le Deco, Tokyo 111427
Galerie Le Gué, Crevant 103646
Galerie le Patrimoine, Biot 103392
Galerie le Petit Louvre, Avignon 103311
Galerie le Point, Monaco 112003
Galerie Le Royer, Montréal 101401
Galerie le Sphinx, La Ville-Dieu-du-
Temple 103910
Galerie Lebensart, Potsdam . . . 108400
Galerie Lebenskunst,
Deisenhofen 106664
Galerie Lehmann, Die Kunsthandlung
in Harburg, Hamburg 107258
Galerie Lehner, Wien 062955
Galerie Leipziger-Schule, Leipzig 107826
Galerie Lelong, New York 122629
Galerie l'Enclume, Bôle 087330
Galerie l'Enfant, Washington . . . 097837
Galerie Leonardo da Vinci,
Mamer 082377, 111871
Galerie Léoni, Marseille 104178
Galerie les Angelots, Fontevraud-
l'Abbaye 068580
Galerie les Art du Feu, Caen . . . 103484
Galerie Les Beaux Arts, Ratingen 108435
Galerie Les Contemporains,
Bruxelles 100611
Galerie les Cyclades, Antibes-Juan-les-
Pins 103259
Galerie Les Deux-Fontaines,
Bursins 116184
Galerie les Hirondelles, Coppet . 116226
Galerie les Modernes, Montréal . 101402
Galerie les Nabis, Caluire-et-
Cuire 103493
Galerie les Tournesols,
Montbrison 104256
Galerie les Tournesols, Vichy . . 105857
Galerie Leuebrüggli, Langenthal 116404
Galerie Leviathan, Bordesholm . 106543
Galerie Liberté, Pont-Aven 105378
Galerie Librairie Australienne, Saint-

Denis (Seine-Saint-Denis) 105502
Galerie Lichtblick, Köln 107620
Galerie Lichterloh, Wien 062955
Galerie Lidiya, Potsdam 108401
Galerie Lieve Hemel, Amsterdam 112204,
137833
Galerie Lindenplatz, Gelfingen . 116274
Galerie Lions'Y, Lyon 104100
Galerie Liotard, Genève 116308
Galerie Litera, Praha 102675
Galerie Lithoart, Charlottenlund . 102785
Galerie LiTo, Detmold 106674
Galerie Litvinov, Litvínov 009678
Galerie LM Art, Trier 108775
Galerie l'Objet d'Art, Brignoles . 103470
Galerie Löwe, Hannover 076371
Galerie Loft, Paris 104860
Galerie Loïc de Pors-Mellec,
Locquirec 104043
Galerie l'Oiseau Bleu, Vertou . . 105848
Galerie Loisel Grafik, Pörtschach am
Wörther See 100141
Galerie l'Orange, Le Havre 103945
Galerie l'Orangerie, Chambrelien 116201
Galerie Louis-Pierre, Leuven . . . 063731
Galerie Lous Martin, Delft 112361
Galerie Louvre, Praha 102676
Galerie Lovers of Fine Arts,
Gstaad 116366
Galerie Ludvika Kuby a Městské
Muzeum, Březnice 009458
Galerie Ludwig, Kulmbach 107772
Galerie Ludwig Bürgel, Seekirchen am
Wallersee 002749
Galerie Luftschloß, Münster . . . 108216
Galerie Luise, Bielefeld 106492
Le Galerie Luministe, Miami . . . 121790
Galerie LWS, Paris . . . 136929, 141121
Galerie M, Altenrhein 116017
Galerie M, Berlin 017133
Galerie M & K, Jihlava 102599
Galerie M & N Uzal, Bruxelles . . 063440,
100612
Galerie M & N Uzal, Paris 071399,
104861
Galerie M + R Fricke, Berlin . . . 106274
Galerie M Bochum, Bochum . . . 106511
Galerie M.IQ, Kunsthandlung,
Mannheim 107967
Galerie M.R., Angoulême 103238
Galerie Ma, Villach 100209
Galerie Mad, Graz 100003
Galerie Maegle, im Palais Harrach,
Wien 100283, 140079
Galerie Maendel-Lawrence,
München 108106
Galerie Mainetti, Lausanne 116419
Galerie Majestic, Paris 104862
Galerie Malovaný Dům, Třebíč . 102740
Galerie Malwerk 2000, Erdweg . 106888
Galerie Mammana, Kaufbeuren . 107540
Galerie Mana, Wien 100284
Galerie Mansarte, Sankt Anton am
Arlberg 100183
Galerie Marco Edition, Bonn . . . 106528,
137114
Galerie Marengo, Angoulême . . 103239
Galerie Marenzihaus, Leibnitz . . 100092
Galerie Margot, Eindhoven 112457
Galerie Maria Lund, Paris 104863
Galerie Marie-Christine, Bruxelles 063441,
125311
Galerie Marie de Holmsky, Paris 071400,
104864
Galerie Marie-Laure de l'Ecotais,
Paris 071401, 104865
Galerie Marie Louis, La Baule-
Escoublac 103867
Galerie Marie-Louise, Grenoble . 103787
Galerie Marina, Avignon 103312
Galerie Marina, Cassis 103537
Galerie Marine, Saint-Malo (Ille-et-
Vilaine) 072918

Galerie Marion, Perpignan 072081,
105337
Galerie Maritim, Hamburg 107259
Galerie Markt Bruckmühl,
Bruckmühl 017637
Galerie Marktschlößchen, Halle,
Saale 018955
Galerie Marktstätte, Konstanz . . 107713
Galerie Marschall, Bernried 017339
Galerie Martin Janda, Wien 100285
Galerie Martin Mertens, Berlin . . 106275
Galerie Martina Detterer, Frankfurt am
Main 106993
Galerie Matignon, Paris 104866
Galerie Matignon 32, Paris 104867
Galerie Maulberger, München . . 108107
Galerie Mauritiushof, Bad
Zurzach 116045
Galerie Max 21, Iphofen 107467
Galerie Maxim, Villefranche-sur-
Mer 074149
Galerie May, Savigny-sur-Orge . 105663
Galerie Mazarine, Nice 070608
Galerie Mazarine, Paris 104868
Galerie Mazur, Kroměříž 102609
Galerie Méditerranée, Bonifacio . 103405
Galerie Méduse, Montréal 064423,
101403
Galerie Meisterschueler, Berlin . 106276
Galerie Melantrich, Polabské muzeum
Poděbrady, Rožďalovice 009867
Galerie Mémoires, Albi   066514, 103217
Galerie Mensing, Baden-Baden . 106108
Galerie Mensing, Berlin 106277
Galerie Mensing, Düsseldorf . . 106783
Galerie Mensing, Frankfurt am
Main 106994
Galerie Mensing, Hamburg 107260
Galerie Mensing, Hamm,
Westfalen 107331
Galerie Mensing, Hannover . . . 107365
Galerie Mensing, Konstanz 107714
Galerie Mensing, München 108108
Galerie Mensing, Sylt 108746
Galerie Mercure, Paris 071402
Galerie Města Blanska, Blansko 009448
Galerie Městského Muzea Čáslav,
Čáslav 009492
Galerie Meyer Kainer, Wien 100286
Galerie MG, Paris 071403
Galerie Michael, Rottach-Egern . 108528
Galerie Michael Heufelder,
München 108109
Galerie Michael Neff, Frankfurt am
Main 106995
Galerie Michael Werner, Köln . . 107621
Galerie Michael Werner, Märkisch
Wilmersdorf 107934
Galerie Michael Wiesehöfer, Köln 107622
Galerie Michel, Paris . . 104869, 141122
Galerie Michel-Angelo, Montréal 101404
Galerie Michelangelo, Las Vegas 121351
Galerie Micro, Stockholm 115855
Galerie Miejskie, Łódź . 084612, 113676
Galerie Milada, Praha 102677
Galerie Millennium, Praha 102678
Galerie Minnepand, Gent 100693
Galerie Mirages, Edinburgh 117727
Galerie Mirko Mayer, Köln 107623
Galerie Miro, Praha 102679
Galerie mit der blauen Tür,
Nürnberg 108300
Galerie Mitte, Köln 106735
Galerie Mitte, Bezirksamt Mitte von
Berlin, Berlin 017134
Galerie Mobile, Bad Wimpfen . . 106100
Galerie Modena Art, Wien 100287
Galerie Modern Picture, Leipzig . 107827
Galerie Moderne, Bad
Zwischenahn 106104, 137038
Galerie Moderne, Bruxelles 125312
Galerie Moderne, Silkeborg 102973
Galerie Moderního Umění, Hradec
Králové 102595

Galerie Moderního Umění, Hradec
Králové 009576
Galerie Mohamed El Fassi, Rabat 112048
Galerie Mohamed Kacimi, Un nouvel
espace d'art contemporain, Fès 112044
Galerie Mollwo, Riehen 116587
Galerie Mona Lisa, Paris 104870
Galerie Monaco Fine Arts,
Monaco 112004
Galerie Mondsee, Mondsee 100123
Galerie Mongeon, Portland 096407
Galerie Monique Bornstein,
Villefranche-sur-Mer 105866
Galerie Mont-Blanc, Chamonix-Mont-
Blanc 103561, 140820
Galerie Montaigne, Monaco 082510
Galerie Montana, Apeldoorn . . . 112322
Galerie Montcalm, Gatineau . . . 005868
Galerie Montgautier, Poitiers . . . 105367
Galerie Montparnasse, Genève . 116309,
144076
Galerie Moosgasse, Kempen . . 107543
Galerie Morgenland, Hamburg . . 107261
Galerie Moulay Youssef,
Casablanca 082562
Galerie Münsterland, Emsdetten 018199
Galerie Municipale, Montesquieu 014252
Galerie Municipale, Vitry-sur-
Seine 016311
Galerie Municipale d'Art Contemporain
Julio-Gonzalez, Arcueil 011712
Galerie Municipale Edouard Manet,
Gennevilliers 013090
Galerie Municipale, Salle Aliénor,
Biscarrosse 103397
Galerie Municipale, Salle Pourbal,
Biscarrosse 103398
Galerie Municipales, Sion 041991
Galerie Museal, Luxembourg . . 082371
Galerie-Musée Le Petit Chapitre,
Fosses-la-Ville 003604
Galerie Museum Rabalderhaus, Kunst
in Schwaz, Schwaz 002739
Galerie Můstek, Praha 102680
Galerie MXM, Praha 102681
Galerie MY, Jablonec nad Nisou 102597
Galerie N, Kunstvereinigung Wasgau
e.V., Dahn 017808
Galerie Nadia B, Dieulefit 103670
Galerie nächst Sankt Stephan,
Wien 100288, 136589
Galerie Naja, Le Cap-d'Agde . . 103928
Galerie Nardo, Lausanne 116420
Galerie Nationale d'Art, Dakar . . 038114
Galerie Nationale de la Tapisserie et
de l'Art Textile, Beauvais 011955
Galerie Naumann, Stuttgart 108702
Galerie Nei Liicht, Dudelange . . 111821
Galerie Nero, Wiesbaden 108908
Galerie Neu, Berlin 106278
Galerie neu & alt, Freudenberg . 075901
Galerie Neue Meister, Staatliche
Kunstsammlungen Dresden,
Dresden 017982
Galerie Neues Rathaus, Wetzlar 108892
Galerie Neumühle, Edenkoben . . 106862
Galerie Neuse Kunsthandel,
Bremen 075161
Galerie Neverland, Wien 100289
Galerie Niepel, Düsseldorf 106784
Galerie Nieves Fernandez,
Madrid 115223, 137992
Galerie Nilius, Köln 107624
Galerie No Smoking, Strasbourg 105689
Galerie No. 9, Praha 102682
Galerie Noa Noa, Papeete 105891
Galerie Noah, Augsburg 106029
Galerie Noctua, Říčany u Prahy . 140538
Galerie Nőfa, Wels 100220, 136569
Galerie Noodlebärg, Basel 116071
Galerie Noord, Groningen 112493
Galerie Noran, Lübeck 107913
Galerie Noran, Lüdinghausen . . 107927,
107928

Galerie Noran, Panker 108357
Galerie Norbert Blaeser GmbH & Co. KG,
Steffeln 108668, 137513
Galerie Nord-Est, Strasbourg . . . 105690
Galerie Nord-Kunstverein Tiergarten,
Berlin 017135
Galerie Nord-Sam, Salzburg . . . 100159
Galerie Normand, Paris 104871
Galerie Nota Bene, Genève 116310
Galerie Nothburga, Innsbruck . . 100046
Galerie November, Berlin 106279
Galerie Nový Svět, Praha 102683
Galerie Nr 7, Bad Doberan 106049
Galerie Nuances, Saint-Gilles-Croix-de-
Vie 105525
Galerie Nuances et Lumière,
Lyon 104101
Galerie Numero 16, Noordbroek 032696
Galerie O, Schaffhausen 116630
Galerie Oben, Chemnitz 106635
Galerie Oberkampf, Paris 104872
Galerie Obkirchen, Sachseln . . . 116719
Galerie of Art, Bad Bentheim . . . 106039
Galerie of Modern Art, Berlin . . 106280
Galerie Onze, La Colle-sur-Loup 103884
Galerie Opaque, Berlin 106281
Galerie Open, Berlin 106282
Galerie Orangerie, Köln 107625
Galerie Orlando, Zürich 116791
Galerie Orly, O. & P. Wille, Basel 116072
Galerie Orsel, Paris 104873
Galerie Ost-Art, Berlin 106283
Galerie Ostrov, Praha 102684
Galerie Otto, Wien . . . 062956, 100290
Galerie Otto OG, Wien . 062957, 062958
Galerie Otto, Art Haus, Wien . . . 100291
Galerie Ovadia, Nancy 104321
Galerie Oviri, Punaauia 105893
Galerie P, Carouge 116195
Galerie P, Wittingen 108952
Galerie P, Zürich 116792
Galerie Pablo Picasso,
Amsterdam 112205
Galerie Packschuppen, Baruth . . 106123
Galerie Page, Biarritz 103384
Galerie Painen, Berlin 106280
Galerie Palais Starhemberg, Wien 062959
Galerie Palais Walderdorff, Trier 022198
Galerie Palette, Oldenburg 108342
Galerie Palme, Hamburg 107262
Galerie Palü, Pontresina 116569
Galerie Pamyr, Paris 071404
Galerie Panache, Hong Kong . . . 102080
Galerie Pankow, Berlin 017136
Galerie Pappelhof, Appenzell . . 116630
Galerie Parchemine, Montréal . . 101405
Galerie Parcours, Metz 070096
Galerie Paris-Isphahan, Paris . . 071405
Galerie Parterre, Die Galerie in
Prenzlauer Berg, Berlin 017137
La Galerie Passage, Tallinn 103115
Galerie Patro, Ostrava 102634
Galerie Páv, Liberec 102614
Galerie Pazyryk, Aix-en-Provence 103196,
128427
Galerie Pendo, Zürich 116793
Galerie Peveling, Bergisch-
Gladbach 106137
Galerie Pi, København 102883
Galerie Picazur, Morges 116519
Galerie Picot-le Roy, Crozon . . . 103648
Galerie Pictures, Aachen 105935
Galerie Pidoux, Meißen 107989
Galerie Pierre, Paris 104874
Galerie Pikanta, Kunstverein Pikanta e.V.,
Leipzig 058884, 107828
Galerie Pillango, Berlin 106285
Galerie Pittoresque, Paris 071406
Galerie Place des Arts,
Montpellier 104280
Galerie Plaisance, Paris 104875
Galerie pod Zámkem, Rožmitál pod
Tremšínem 009870
Galerie Pokusa, Polnischer Kultursalon

e.V., Wiesbaden 108909
Galerie Port-Maurice, Saint-
Léonard 006919
Galerie Portalis, Aix-en-Provence 103197
Galerie Porte Avion, Marseille . . 104179
Galerie Portyč, Písek 009755
Galerie Pravěk Očima Zdeňka Buriana,
Dvůr Králové nad Labem 009544
Galerie Prisma, Bolzano 109877
Galerie Privée Hôtel les Templiers,
Collioure 012620
Galerie Privilège, Biarritz 103385
Galerie Pro Arte, Basel   087233, 116073
Galerie Pro Arte, Freiburg im
Breisgau 107059
Galerie Pro Arte, Wien 100292
Galerie Pro Arte Kasper, Morges 116520
Galerie Pro Arte, Justus & Partner,
Hamburg 107263
Galerie pro Každého, Ostrava . . 102635
Galerie pro Radost, Ostrava . . . 102636
Galerie Proarta, Zürich 116794
Galerie Proarta, Zug 116884
Galerie Profil, Weimar 108862
Galerie Prospect, Rellingen 108472
Galerie Pullach, Pullach im
Isartal 108419
Galerie Pur, Lübbecke 107903
Galerie Pygmalion, Saint-Nazaire 105560
Galerie Pythéas, Marseille 104180
Galerie Pytheas, Paris 071407
Galerie Q, Wiesbaden 108910
Galerie Quartier, Leipzig 107829
Galerie Quattro, Glattfelden 116350
Galerie Quedar, Strasbourg 105691
Galerie Quincampoix, Paris 104876
Galerie R, Eindhoven 083216
Galerie Racine, La Ferté-Milon . 069031
Galerie Rainmühle,
Emmenbrücke 041463
Galerie Raphael, Taguig 113424
Galerie Raphael 12, Frankfurt am
Main 106996
Galerie Raugraff, Nancy 104322
Galerie Rauhenstein, Wien 062960
Galerie Rebus, Hamburg 107264
Galerie Recknagel, Weiden in der
Oberpfalz 108853
Galerie Réfractaire, Paris 071408
Galerie Regard, Ajaccio 103212
Galerie Regard, Genève 116311
Galerie Regard, Le Cap-d'Agde . 103929
Galerie Reile, München 108110
Galerie Remise, Bad Nauheim . . 106081
Galerie Remise, Halle / Westfalen 018949
Galerie Remise, Rheinbach 108494
Galerie Remixx, Graz 100004
Galerie Renée-Blain, Brossard . . 005490
Galerie Repfergasse 26,
Schaffhausen 116631
Galerie Resonanz, Wien 062961, 100293
Galerie Reygers, München 108111
Galerie Rhéa, Zürich 087693
Galerie Rheinland, Solingen . . . 078495
Galerie Rhenus, Wangen im
Allgäu 108843
Galerie Riance, Courroux 116229
Galerie Rieder, München 108112
Galerie Rive Gauche, La Colle-sur-
Loup 069001, 103885
Galerie Rive Gauche, Paris 104877
Galerie Rivolta, Genève 116312
Galerie Römerapotheke, Zürich . 116795
Galerie Roessler, Ravensburg . . 108441
Galerie Roger Katwijk,
Amsterdam 112206
Galerie Rondo, Hradec Králové . 102596
Galerie Rose, Hamburg 107265
Galerie Rose, Landshut 107790
Galerie Roseg, Pontresina 116570
Galerie Rosella Junck, Venezia . 106286
Galerie Rosenfeld, Bad Kissingen 106071
Galerie Rote Rose, Regensburg . 116577
Galerie Roter Pavillon, Bad

Doberan 106050
Galerie Rothamel Erfurt, Erfurt . 106894
Galerie Rothamel Frankfurt, Frankfurt
am Main 106997
Galerie Rouge, Denver 120721
Galerie Rouge, New Orleans . . . 122169
La Galerie Rouge, Paris 071409
Galerie Roull, Brest 067470
Galerie Route de la Soie, Paris . 104878
Galerie Royale, Montréal 101406
Galerie Royale, New Orleans . . . 122170
Galerie Royale, Rochefort (Charente-
Maritime) 105453
Galerie Rozehnal, Praha 102685
Galerie Rubikon, Olomouc 102624
Galerie Rudolf, Fürstenfeldbruck 075920,
107079
Galerie Rudolfinum, Praha 009795
Galerie Rüdiger Schöttle,
München 108113
Galerie Ruf, Egling 106865
Galerie Ruhnke, Potsdam 108402
Galerie Ruine, Genève 116313
Galerie Ruländer – Altes Rathaus,
Worpswede 022770
Galerie Rupert Pfab, Düsseldorf 106785
Galerie Růžový Kopeček, Cheb . 102587
Galerie RX, Paris 104879
Galerie S, Amsterdam 112207
Galerie Saalbach, Saalbach 100152
Galerie Sahara, Rabat 112049
Galerie Saint-Antoine, Bruxelles 063442,
125313
Galerie Saint-Benoît, Chambéry . 067794
Galerie Saint Charles, Lyon 069740
Galerie Saint-Christophe, Meaux 104211
Galerie Saint Etienne, New York 122630
Galerie Saint-Georges, Lyon . . . 104102
Galerie Saint-Germain, Monaco . 082511
Galerie Saint-Germain, Segny . . 105663
Galerie Saint-Guénolé, Fouesnant 068591
Galerie Saint-Hubert, Lyon 104103
Galerie Saint-Jacques, Bruxelles 100613
Galerie Saint-Jacques, Grignan . 103794
Galerie Saint Jacques, Toulouse 073770
Galerie Saint James, McHenry . . 127312
Galerie Saint-Jean, Chalon-sur-
Saône 103555
Galerie Saint-Jean, Genève 087469,
116314
Galerie Saint-Laurent, Plérin . . . 072180,
105358, 129169
Galerie Saint-Louis, Saintes . . . 105639
Galerie Saint-Louis, Versailles . . 074056,
129340
Galerie Saint Luc, Andernos-les-
Bains 103232
Galerie Saint-Marceau des Prés,
Orléans 104428
Galerie Saint-Martial,
Châteauroux 103582
Galerie Saint-Martin, Bretenoux . 067482
Galerie Saint-Martin, Brive-la-
Gaillarde 103474
Galerie Saint-Martin – Côte Ocèan,
Arcachon 103266
Galerie Saint-Ménélée, Vichy . . . 074107
Galerie Saint-Merri Art Inuit,
Paris 104880
Galerie Saint-Michel, Bruxelles . 063443
Galerie Saint-Michel, Guérande . 103802
Galerie Saint-Michel, Nice 070609
Galerie Saint-Nicolas, Remich . . 111873
Galerie Saint-Paul, Lyon 104104
Galerie Saint-Philibert, Octeville-sur-
Mer 070775, 104421
Galerie Saint-Pierre, Bar-le-Duc 066936
Galerie Saint-Pierre, Limoges . . 104034
Galerie Saint-Roch, Paris 104881
Galerie Saint-Sauveur, Dinan . . . 103686
Galerie Saint-Valérien, Tournus . 073825,
105758
Galerie Saint-Vincent, Nevers . . 070539
Galerie Saint-Wandrille, Saint-

Wandrille-Rancon 105634
Galerie Sainte-Barbe, Nevez . . . 104355
Galerie Sainte-Catherine,
Honfleur 103835
Galerie Sainte-Catherine,
Quimper 105415
Galerie Sainte-Catherine, Rodez 105454
Galerie Sainte-Claire, Saint-Paul-de-
Vence 105589
Galerie Sainte-Marie, Montluçon 070269
Galerie Sala Terrena, Mödling . . 002399
Galerie Salvador, Paris 104882
Galerie Samarcande, Paris 071410
Galerie Sambre, Fougères (Ille-et-
Vilaine) 103752
La Galerie Samedi, Montfort-
l'Amaury 104264
Galerie Samson, Bourges 067429
Galerie Samuel Lallouz, Montréal 101407
Galerie Samy Kinge, Paris 104883
Galerie San Carlo, Paris 071411, 104884
Galerie San Jordy, Rodes 072508
Galerie Sanct Lucas, Wien 100294
Galerie Sandra Goldie,
Westmount 101848
Galerie Sankt Barbara, Hall in
Tirol 100030
Galerie Sankt Johann,
Saarbrücken 108544
Galerie Sankt Ulrich, Wien 062962
Galerie Sans Nom, Moncton . . . 006348
Galerie Sanssouci, Potsdam . . . 108403
Galerie Saphir, Dinard 103696
Galerie Saphir, Paris 104885
Galerie Saqqârah, Gstaad 116367
Galerie Sassen & Edition, Bonn . 106529,
137115
Galerie Satellite, Paris 104886
Galerie Sattelgut, Neunkirchen-
Seelscheid 108248
Galerie Sauruck, Graz 100005
La Galerie Scandinave, Paris . . . 071412
Galerie Schlichtenmaier, Grafenau, Kreis
Böblingen . . 076073, 107154, 137232
Galerie Schloß Albeck, Sirnitz . . 002758
Galerie Schloss Borbeck, Essen 018294
Galerie Schloss Damtschach,
Wernberg 100223
Galerie Schloß Mochental,
Ehingen 106866
Galerie Schloß Mondsee,
Mondsee 002412
Galerie Schloss Puchheim, Attnang-
Puchheim 099956
Galerie Schloß Rimsingen, Breisach
am Rhein 017571
Galerie Schloß Weyer, Gmunden 001951
Galerie Schmidt, Reith im
Alpbachtal 062773
Galerie Schmidtgasse 1, Wels . . 100221
Galerie Schnitt-art,
Schwarzenbach 116634
Galerie Schnorr 37, Bremen . . . 106583
Galerie Schübbe Projekt,
Düsseldorf 106786
Galerie Schüller im Bayerischen Hof,
München 108114
Galerie Schutzengel, Singen . . . 108627
Galerie Schwarz/Weiss,
Osnabrück 108345
Galerie Schwedenkate, Eutin . . . 106945
Galerie Sculptur, Bamberg 106116
Galerie Seestrasse 96, Berlingen 116115
Galerie Sehens-Wert, Krefeld . . . 076955
Galerie Sehens-Wert, Leipzig . . . 077059
Galerie Seine 51, Paris 104887
Galerie Septentrion, Marcq-en-
Barœul 104155
Galerie Serafin, Wien 100295
Galerie Severine, Bad Doberan . 106051
Galerie Siebzehn, Wädenswil . . . 116706
Galerie Siècles, Beirut 082283
Galerie Sieglinde Dietz, Bonn . . 106530
Galerie Sigfrid, Orléans 104429

Galerie Sillage, Gramat 103774
Galerie Sillage, Paimpol 104434
Galerie Simone, Paris 104888
Galerie Sionnet, Jussy 116381
Galerie SixFriedrichLisaUngar,
München 108115
Galerie Skala, Köln 107626
Galerie Skell, Schmiedeberg . . . 108568
Galerie Slim Bouchoucha, Paris . 071413
Galerie SMA, Carouge 116196
Galerie Smend, Köln 107627
Galerie SO, Solothurn 116650
Galerie Soleil, Montréal 101408
Galerie Soleilo, Hampton 099108
Galerie Sonnenfels, Wien 062963,
100296
Galerie Sonnensegel, Wien 002993
Galerie Sophien-Edition, Berlin . 106287
Galerie Spalentor, Basel 116074
Galerie Spandow, Berlin 106288
Galerie Speicher, Speicher 116657
Galerie Spektrum, Dortmund . . . 106701
Galerie Spektrum, München . . . 108116
Galerie Spektrum, Sinsheim . . . 108630
Galerie Spiekerhof, Münster . . . 108217
Galerie SR, Paris 104889
Galerie Stadtgraben, Büren an der
Aare 116177
Galerie Stadtpark, Krems 100084,
136530
Galerie Stallburg, Wien 062964
Galerie Stanislas, Paris 071414
Galerie Start, Monaco 112005
Galerie Steinek, Wien 100297
Galerie Steinreich, Düsseldorf . . 106787
Galerie Steinzeit, Aham 105959
Galerie Stephanie Hollenstein,
Lustenau 002347
Galerie Šternberk, Šternberk . . . 009890
Galerie Sternenstaub, Kraichtal . 107728
Galerie Stift Eberndorf, Eberndorf 139978
Galerie Stihl Waiblingen,
Waiblingen 022375
Galerie Stolz Berlin, Berlin 106289
Galerie Studio 10, Atelier Willi E.
Christen, Zürich 116796
Galerie Studio 38, Wien 100298
Galerie Stühler, Berlin 106290
Galerie Süd, Leipzig 107830
Galerie Sur, Wien 100299
Galerie sur Cour, Paris 104890
Galerie Symbole Art, Montréal . . 101409
Galerie T, Middelburg 112593
Galerie t' Elgee, Philadelphia . . . 123380
Galerie Taisei, Tokyo 029507
Galerie Tango, Lutry 116463
Galerie Tanit, München 108117
Galerie Tanner, Stuttgart 108703
Galerie Tannzapfenland,
Eschlikon 116259
Galerie Tansu, Montréal 101410
Galerie Taube, Berlin 106291
Galerie Témoin, Genève 116315
Galerie A Tempera, Paris 104891
Galerie Templ, Mladá Boleslav . 102620
Galerie Terminus, München 108118
Galerie Terra Viva, Saint-Quentin-la-
Poterie 015529
Galerie Teyssen, Inh. Jürgen Teyssen,
Bremerhaven 106595
Galerie Theis – Keramik, Berlin . 106292
Galerie Théma, Bordeaux 103434
Galerie Theodor von Hörmann,
Imst 002113
Galerie Thiessen, Düren 106752
Galerie Thomas, München 108119
Galerie Thomas Flor, Düsseldorf 106788
Galerie Thomas Schulte, Berlin . 106293
Galerie Thomas Zander, Köln . . . 107628
Galerie Thomire, Paris 071415
Galerie Tonart, Wien 100300
Galerie Topley, Salzburg 100160
Galerie Tourbillon, Paris 071416, 104892
Galerie Transit, Mechelen 100787

Galerie Trebchen, Marco Trebchen,
Salzburg 100161
Galerie Treff, Hamburg 107266
Galerie Triple A, Saint-Saphorin-sur-
Morges 116602
Galerie Trittligasse, Zürich 116797
Galerie Turenne, Montréal 101411
Galerie Turquoise, Sidi Bou Saïd 116985
Galerie Tvar, Praha 102686
Galerie Tvrdohlaví, Praha 102687
Galerie Twenty, Biarritz 067199
Galerie Typograph, Nürnberg . . . 108301
Galerie Uenzen 73, Süstedt 108742
Galerie Uhu, Lörrach 107886
Galerie Ulrike Hrobsky, Wien . . . 100301
Galerie Umění, Karlovy Vary . . . 102605
Galerie Umění Karlovy Vary, Karlovy
Vary 009617
Galerie un autre Regard, Rennes (Ille-
et-Vilaine) 105436
Galerie un deux trois, Genève . . 087470,
116316
Galerie Unart, Villach 100210
Galerie und Edition Artelier, Graz 100006,
136509
Galerie und Edition Wolfgang Scherer,
Miltenberg 108001
Galerie und Museum im Franck-Haus,
Marktheidenfeld 020296
Galerie UNE, Auvernier 116039
Galerie Unie Výtvarný Umělců Teplice,
Teplice 009906
Galerie unik.at, Wien 100302
Galerie Unikat, Bad Muskau . . . 106078
Galerie Unikat, Hamburg 107267
Galerie Unikat, Staufen im
Breisgau 108666
Galerie Unique, Treuenbrietzen . 078698
Galerie Urpresse, Schwerin 108596
Galerie Utopie, Leysin 116439
Galerie Utrecht, Amsterdam . . . 112208
Galerie Utrecht, Utrecht 112698, 112699
Galerie V & V, Wien 100303
Galerie v Lazarské, Praha 102688
Galerie Vallesia, Bern 116134
Galerie Van Waning, Rotterdam . 112639
Galerie Van.M, Paris 104893
Galerie ve Dvore, Litoměřice . . . 102615
Galerie ve Věži, Mělník 102619
Galerie Vendôme Rive Gauche,
Paris 104894
Galerie Venezia, Pirmasens 108381
Galerie Venise-Cadre,
Casablanca 112039
Galerie Verlag Dresden, Radebeul 108423,
137474
Galerie Verneuil, Paris 104895
Galerie von Vertes, Pfäffikon
(Schwyz) 116563
Galerie Via Art, Praha 102689
Galerie Vibrations, Paris 104896
Galerie Vice Versa, Lausanne . . 116421
Galerie Victor Hugo, Saintes . . . 073374,
105640
Galerie Victor Hugo, Toulouse . . 073771
Galerie Victoria, Aix-les-Bains . . 103207
Galerie Victoria, København 065902
Galerie Vieille du Temple, Paris . 104897
Galerie Vienna, Mödling 100122
Galerie Vier Jahreszeiten,
Hainfeld 062624
Galerie Villa Blau,
Sommerhausen 108641
Galerie Villa Bösenberg, Leipzig 107831
Galerie Villa Hühn, Gevelsberg . 107122
Galerie Villa Maria, Bad Aibling . 106037
Galerie Villa Rolandseck, Zentrum
für phantastische Künste e.V.,
Remagen 108474
Galerie Village Créole, Les Trois-
Ilets 103991
Galerie Vincent, Ottawa 101472
Galerie Viola, Aschaffenburg . . . 116009
Galerie Violetta, Wien 100304

Galerie Vita, Feldbrunnen 116261
Galerie Vitesse, Paris 104898
Galerie Vlasta, Praha 102690
Galerie Vltavín, Praha 102691
Galerie Vömel GmbH, Düsseldorf 106789,
    141717
Galerie Voltaire, Ferney-Voltaire . 103737
Galerie Voltaire, Grenoble 103788
Galerie Volubilis, Tanger 082607
Galerie vom P, Hamburg 107268, 125967
Galerie von Abercron, München 108120,
    142155
Galerie von Braunbehrens,
  München 108121
Galerie von Hobe GmbH, Reinbek 108469
Galerie von Loeper, Hamburg .. 107269
Galerie von Miller & von Miller,
  Frankfurt am Main 106998
Galerie von Miller & von Miller,
  München 108122
Galerie Vor Ort, Wien 100305
Galerie Voss, Rüdiger, Düsseldorf 106790
Galerie Vrais Rêves, Lyon 104105
Galerie Vue sur Mer, Dinard ... 103697
Galerie Výheň, Jihlava 102600
Galerie Výšehrad, Praha 009796
Galerie Výtvarného Umění, Havlíčkův
  Brod 009556
Galerie Výtvarného Umění,
  Hodonín 009562
Galerie Výtvarného Umění, Most 009708
Galerie Výtvarného Umění,
  Náchod 009710
Galerie Výtvarného Umění,
  Ostrava 009740
Galerie Výtvarného Umění, Roudnice
  nad Labem 009865
Galerie W, Paris 104899
Galerie Wagram, Paris 104900
Galerie Waldhof in Birkenwerder,
  Birkenwerder 017396
Galerie Walfischgasse, Wien ... 100306
Galerie Wall, Orléans . 104430, 136845
Galerie Wallstraße,
  Mönchengladbach 108014
Galerie Walter Heseler ek,
  München 108123
Galerie Walu, Zürich .. 116798, 126766,
    138110,   144177
Galerie Wandelbar, Gstaad 116368
Galerie We do Arts, Leingarten . 107806
Galerie Weberhaus, Weiz 002912
Galerie Weib-art, Berlin 017138
Galerie Weihergut, Salzburg ... 062792,
    100162,   136549
Galerie Weihergut Altstadtgalerie,
  Salzburg ..062793, 100163, 136550
Galerie Weissenseer Freitag,
  Berlin 106294
Galerie Weisser Elefant, Berlin . 106295
Galerie Weisses Häusle, Kunstverein
  Hechingen, Hechingen 019122
Galerie Wendelinskapelle, Marbach am
  Neckar 107977
Galerie Werner Munk, Günzburg 107182
Galerie Westend, München 108124
Galerie Westensee, Westensee .. 106396
Galerie Westphal, Berlin 074797, 106296
Galerie WFK, Maastricht 112573
Galerie white8, Villach 100211
Galerie Wildeshausen,
  Wildeshausen 108938
Galerie Willy Huybrechts, Paris . 071417
Galerie Wimmer, München 108125
Galerie Windrad, Kutzenhausen . 107776
Galerie Windrose, Heikendorf . 107403
Galerie Winter, Wiesbaden ... 108911
Galerie Wittenberg 418,
  Steckborn 116659
Galerie Wittenbrink, München . 108126,
    108127,   137403
Galerie Wittenstein, Wuppertal . 108990,
    108991,   130794
Galerie WM, Lyon 104106

Galerie Wolfshütt, Essen 106915
Galerie Xenios, Frankfurt am
  Main 106999
Galerie XVIIIème XXIème, Pont-
  Aven 072243
Galerie Ya Zamane, Rabat 082600
Galerie Yan, Quiberon 105405
Galerie Ypsilon, Gelnhausen ... 107106
Galerie Ypsilon, Zürich 087968, 133980
Galerie Z, Aigues-Mortes 103180
Galerie Z, Landau in der Pfalz . 107785
Galerie Z, Norderstedt 108283
Galerie Zabbeni, Vevey 116696
Galerie Zázvorka, Nové Město nad
  Metují 009722
Galerie Zeitkunst, Amaliendorf-
  Aalfang 127707
Galerie Zeitlos, Kamen 107498
Galerie Zeitlupe, Heidenheim an der
  Brenz 107401
Galerie Zelle e.V., Reutlingen .. 108480
Galerie Zenit, København 102884
Galerie Zero, Berlin 106297
Galerie Zeuthen, Vedbæk 066185
Galerie Žlutý Slon, Brno 102568
Galerie zum Alten Bergwerk,
  Wiesloch 108936
Galerie zum alten Ötztal, Oetz .. 002477
Galerie zum Alten Rathaus,
  Stockerau 002797
Galerie zum Burgtor, Rothenburg ob
  der Tauber 108524
Galerie zum Goldrahmen, Murg . 108229
Galerie zum Hof, Luzern 087640
Galerie zum kleinen Mann, Fulda 107089
Galerie zum kleinen Schwan,
  Baden 116048
Galerie zum Schäfli, Urnäsch .. 116680
Galerie zum Stiefel, Sankt
  Ingbert 108555
Galerie zum Strauss, Sankt
  Gallen 116611
Galerie zur grünen Tür, Uznach . 116683
Galerie zur Hammermühle, Ibach 116379
Galerie zur Heubühne,
  Oberdiessbach 116544
Galerie Zur Krone, Eschenz ... 087417
Galerie zur Stellmacherei,
  Westerstede 108887
Galerie Zur Stockeregg, Zürich . 116799
Galerie zur Uralten Uhr, Rotkreuz 087760
Die Galerie, Afrikanische Kunst,
  München 108073
Galerie.be, Bruxelles 100614
galerie.bruehl, Brühl 117642
Galeriehaus, Nürnberg 020850
Galerienhaus Stuttgart, Stuttgart 108704
Galerienverband Mannheim,
  Mannheim 058885
Les Galeries de l'Historie, Antenne
  du Musée d'Art et d'Histoire,
  Neuchâtel 041795
Galeries de Paléontologie et
  d'Anatomie Comparée, Musée
  National d'Histoire Naturelle,
  Paris 014659
Galeries des Ducs, Chatenoy-le-
  Royal 103585
Les Galeries du Fresne, Fresnay-sur-
  Sarthe 068610, 103755
Galeries du Panthéon Bouddhique
  du Japon et de la Chine, Musée
  national des Arts Asiatiques Guimet,
  Paris 014660
Galeries Nationales du Grand Palais,
  Paris 014661
Galeries Rambert, Paris 104901
Galeries Saint-Louis, Bruxelles . 063444
Galerieverein, Wendlingen am
  Neckar 058886
Galerieverein Leonberg e.V.,
  Leonberg 020044, 058887
Die Galeriewerkstatt, Sylt 108747

Galerii 2, Tallinn 103116
Galerii-G, Tallinn 103117
Galeriile de Artă Emilia Dumitrescu,
  Muzeul Brăilei, Brăila 036262
Galeriile de Artă Plastică Ştefan
  Luchian, Botoşani 036254
Galerij C.G., Mechelen 063789
Galerij De Hamer, Apeldoorn ... 082922
Galerij Goya, Mechelen 100788
Galerij in Situ, Aalst 100472
Galerij Ka, Hasselt 063679
Galerij 't Leerhuijs, Brugge ... 100534
Galerij S65, Aalst 100473
Galerij ter Gentpoort, Brugge ... 063274
Galerij Uilenspiegel, Leuven ... 063732
Galerij Zuid, Antwerpen 100489
Galerija 107, Zemun 114467
Galerija 12+, Beograd 114401
Galerija 2, Vrhnika 114750
Galerija 56, Ljubljana 114703
Galerija 73, Beograd 114402
Galerija 99, Arandjelovac 114389
Galerija AC, Žalec 114751
Galerija Ada, Beograd 114403
Galerija Akademije Likovnih Umetnosti
  Sarajevo, Sarajevo 004274
Galerija Antuna Augustinčića,
  Klanjec 008927
Galerija Arheo, Beograd 114404
Galerija Art, Smederevo 114460
Galerija Ata Raj, Sofia 101153
Galerija Avsenik, Begunje na
  Gorenjskem 114674
Galerija Ažbe, Ljubljana 114704
Galerija Balkon', Burgas 101120
Galerija Bazalt, Beograd 114405
Galerija Beograd, Beograd 114406
Galerija Bevisa, Bled 114676
Galerija Bevisa, Kranj 114689
Galerija Biblioteke grada Beograda,
  Beograd 038130
Galerija BKC, Sarajevo 100861
Galerija Božidar Jakac, Kostanjevica
  na Krki 038530
Galerija Brčko, Brčko 004260
Galerija Bukovac, Cavtat 008892
Galerija Cate Dujšin Ribar, Trogir 009007
Galerija CDA, Sarajevo 100862
Galerija Centar, Podgorica 112027
Galerija Centăr za Săvremenno
  Izkustvo, Varna 005238
Galerija Centralnog Kluba Vojske Srbije
  i Crne Gore, Beograd 038131
Galerija Contra, Koper 114684
Galerija Daut Pašin Amam,
  Skopje 030625
Galerija Denis, Arandjelovac .. 114390
Galerija Diva, Varna 101190
Galerija Doma kulture Studentski Grad,
  Beograd 038132
Galerija Doma Omladine,
  Beograd 038133
Galerija Dvor, Tolmin 114748
Galerija Edmond, Sofia 064229, 101154
Galerija Epsilon, Šumen 101180
Galerija Fabris, Beograd 114407, 114408
Galerija Fara, Škofja Loka 114744
Galerija Fresaka, Narodni Muzej,
  Beograd 038134
Galerija Fundacji FAIT, Kraków . 113597
Galerija Fundacji Sztuki Osób
  Niepełnosprawnych, Kraków . 113598
Galerija Furlan, Ljubljana 114705
Galerija Grad, Ljubljana 114706
Galerija Grada Krapine, Krapina 008938
Galerija Grafičnog Kolektiva,
  Beograd 114409
Galerija Harizma, Beograd 085338,
    114410
Galerija Irida, Sofia 101155
Galerija Izvorne Umjetnosti,
  Zlatar 009057
Galerija Juraj Plančić, Stari Grad 009004
Galerija Kamen, Radovljica 114743

Galerija Klovićevi Dvori, Zagreb . 009029
Galerija Koprivnica, Muzej Grada
  Koprivnice, Koprivnica 008930
Galerija Kos, Ljubljana 114707
Galerija Kozole, Trbovlje 114749
Galerija Kralj, Ljubljana 114708
Galerija Kristina, Sofia 101156
Galerija Kulturnog Centra Politehničke
  Akademije, Beograd 038135
Galerija Kupola, Beograd 114411
Galerija Kuršumli An, Skopje .. 030626
Galerija Lala, Ljubljana 114709
Galerija Lazar Vozarević, Sremska
  Mitrovica 038244
Galerija Legat Milice Zorić i Radoljuba
  Čolakovića, Beograd 038136
Galerija Leonardo, Sofia 101157
Galerija Likovni Susret, Subotica 038249
Galerija Likovnih Samorastnikov,
  Trebnje 038624
Galerija Likovnih Umetnosti, Slovenj
  Gradec 038612
Galerija Likovnih Umjetnosti,
  Osijek 008961
Galerija Likovnih Umjetnosti Republike
  Srpske Banja Luka, Banja Luka 004255
Galerija Lina, Ljubljana 114710
Galerija LM, Maribor 114727
Galerija Lorina, Sofia 101158
Galerija Luwigana, Ljubljana .. 114711
Galerija Majski Zatvor, Sliven .. 101154
Galerija Matice Srpske, Novi Sad 038214
Galerija Media'nox, Maribor .. 114728
Galerija Meduza, Koper 114685
Galerija Mersada Berbera,
  Sarajevo 100863
Galerija Mestne Hiše, Kranj ... 038531
Galerija Milan Konjović, Sombor 038240
Galerija Milene Pavlović Barili,
  Požarevac 038225
Galerija Miroslav Kraljević,
  Zagreb 009030
Galerija Monmartr, Beograd ... 114412
Galerija Most, Podgorica 112028
Galerija Mozaik, Celje 114678
Galerija MS, Murska Sobota .. 114737
Galerija Murska Sobota, Murska
  Sobota 038581
Galerija-Muzej Lendava, Lendavski
  grad, Lendava 038538
Galerija na Centar za Kultura,
  Mostar 004268
Galerija na Ikoni, Muzej na
  Makedonija, Skopje 030627
Galerija Naivne Umetnosti – Donacija
  Ivana Generalića, Hlebine ... 008916
Galerija Naivnih Slikara i Etnološka
  Zbirka, Kovačica 038200
Galerija Okvir, Celje 114679
Galerija Ozon, Beograd 114413
Galerija Paleta, Beograd 114414
Galerija Paleta, Sarajevo 100864
Galerija Plevna, Škofja Loka .. 114745
Galerija Portreta Tuzla, Univerzitet u
  Tuzli, Tuzla 004288
Galerija Pozorišta Mladih,
  Sarajevo 100865
Galerija Progres, Beograd 114415
Galerija Rafaela, Sofia 101159
Galerija Rampino, Beograd 114416,
    143681
Galerija Remont, Beograd 038137
Galerija Rigo, Novigrad 008957
Galerija Savremene Likovne Umetnosti,
  Niš 038211
Galerija Savremene Umetnosti Likovna
  Jesen, Sombor 038241
Galerija Sazdovski, Skopje 111883
Galerija Sebastian, Dubrovnik . 008902
Galerija Singidunum, Beograd .. 038138
Galerija Skrinja, Maribor 114729
Galerija Škuc, Ljubljana 114712
Galerija Skulptura Ivan Sabolić, Muzej
  Grada Koprivnice, Peteranec . 008967

Galerija Slika Sava Šumanović,
  Šid 038234
Galerija Slika Umetnickih Predmeta,
  Beograd 114417
Galerija Snežinka, Prebold 114742
Galerija Sodobne Umetnosti,
  Celje 038500
Galerija Spektar, Dobrič 101124
Galerija SPKD, Sarajevo 100866
Galerija Sqart, Koper 114686
Galerija Srpske Akademije Nauka i
  Umetnosti, Beograd 038139
Galerija Stara Kapetanija, Zemun 114468
Galerija Status, Arandjelovac . . . 114391
Galerija Suluj, Beograd 114418
Galerija ULUS, Udruženje Likovnih
  Umetnika Srbije, Beograd 038140
Galerija ULUS, Udruženje Likovnih
  Umetnika Srbije, Beograd 038141
Galerija Umetničkih Dela,
  Beograd 038142
Galerija Umjetnina, Split 008995
Galerija Umjetnina Narodnog Muzeja
  Zadar, Zadar 009021
Galerija Valer, Smederevo 114461
Galerija Venier, Vodnjan 102525
Galerija Vincent iz Kastva, Istarsko
  narodno kazalište, Pula 008972
Galerija Vitosha, Sofia 101160
Galerija Vjekoslav Karas, Gradski
  Muzej Karlovac, Karlovac 008924
Galerija XX, Panevėžys 111791
Galerija Zadužbine Ilije M. Kolarca,
  Beograd 038143
Galerija Zala, Ljubljana 114713
Galerija Zavjetnih Slika Brodova,
  Dubrovnik 008903
Galerija Zbona, Koper 114687
Galerija Zebra, Ljubljana 114714
Galerija Zibka, Ljubljana 114715
Galerija Žula, Ljubljana 114716
Galerija Žula 2, Maribor 114730
Galerija Zvono, Beograd 114419
Galerije Capitolum i Diana, Gradska
  palaca, Pula 008973
Galerije Sivčeva Hiša, Muzeji
  radovljiške občine, Radovljica . 038596
Galerist İstanbul, İstanbul 117043
Galeriy Dylag, Kraków 113599
Galeriy Unik, Nairobi 111560
Galery Catoire, Kinsale 109657
Gale's, Dunedin 083926, 126316
Galesburg Civic Art Center,
  Galesburg 049021
Galetas, Bozel 067443
Galetti Galleria, Milano 110318
Galetti Galleria d'Arte, Milano . 080550,
                                      110319
Galeyrand, Philippe, Saint-Jean-de-
  Monts 072852, 129236
Galfard, Eric, Paris . . 104902, 136930
Galica, Milano 110320
Galicja, Kraków 143473
Galicja Muzeum, Kraków 034873
Gáliková, Eva, Nové Zámky . . . 133056
Galilee Art School, Singapore . 056325
Galina, Jacksonville 093994
Galindos Mizraji, Jacobo,
  Montevideo 092000
Galinier, Marie-Louise, Buzet-sur-
  Tarn 067540
Galinski, Uwe, Hamburg 076226
Le Galion, Oostende 063816
Le Galion, Toulon (Var) 073710
Galion Historical Museum, Galion 049026
Galitch de Metsenaere, Piere et Nicole,
  Neuchâtel 087703
Galitch de Metsenaere, Piere et Nicole,
  Vevey 087878
Gałkowska & Gesing, Kraków . . 084577
Gall, Didier le, Aigues-Vives . . . 066439
Galla, Bucureşti 114142
Galla, M., Hattenhofen 076417

Gallagher, Denver 093290, 145126
Gallagher-Christopher, Boston . . 092471
Gallagher Gallery, Royal Hibernian
  Academy, Dublin 024881
Gallander, Cathleen, New York . 122631
Gallant, Philip, Baltimore 092327
Gallarato, Silvia, Torino 132141
Gallardo Fuentes, Victoria, Palma de
  Mallorca 086351
Gallaria Chesina, Jacqueline u. Rudolf
  Maag, Sils Maria 116639
Gallas, Dr. Klaus, Weimar 137548
Gallathe, Ulrich, Löhne 077140
Gallé, Porto 085030
Galle National Museum, Galle . . 040482
Gallé & C., Rita, Milano 110321
Gallego, Ourense . . . 086322, 086323
Gallego, Ray, Los Angeles 094341
Gallegos, Robert V., Albuquerque 092053
Gallen-Kallelan Museo, Espoo . 010772
The Gallereum, Perth 062078
Galleri +, Akureyri 109216
Galleri 1, Göteborg 115634
Galleri 111, Lomma 115730
Galleri 17, Stockholm 115856
Galleri 2000, Lillestrøm 113176
Galleri 21, Malmö . . . 040825, 115745
Galleri 25, Göteborg 115635
Galleri 27, Oslo 113203
Galleri 300m3 Art Space Göteborg,
  Göteborg 115636
Galleri 5, Middelfart 102930
Galleri 5, Reykjavík 109230
Galleri 54, Göteborg 040650
Galleri 63, Ringkøbing 102958
Galleri 67, Stockholm 115857
Galleri 68, København 065903
Galleri 7, Frederikshavn 102809
Galleri 72, Sønderborg 102982
Galleri 8, Sundsvall 115944
Galleri A, Oslo 113204
Galleri A44, Kristiansand 113168
Galleri ABCD, Malmö 115746
Galleri Adamstuen, Oslo 113205
Galleri Ängeln, Lund 115731
Galleri Ängerö, Kramfors 115713
Galleri Åkern, Kongsberg 113166
Galleri Allmenningen, Bergen . . 113125
Galleri Alstrup, Farsø . . . 065649, 102799
Galleri Amanda, Haugesund . . . 113161
Galleri Amare, Stavanger 113266
Galleri Anima, Kivik 115710
Galleri Anna H., Göteborg 115637
Galleri Arctandria, Oslo 113206
Galleri Argo, Stockholm 115858
Galleri Arnagergård, Rønne . . . 102962
Galleri Art 21, Norrköping 115781
Galleri Art So To, Aalborg 102760
Galleri Artis, Gävle 115618
Galleri Asur, Menton 104219
Galleri Asur, Oslo 113207
Galleri Athene, Drammen 113145
Galleri Aveny, Göteborg 115638
Galleri Babette, Knebel 102857
Galleri Bäckalyckan, Jönköping . 115695
Galleri Bagdad Café, Limhamn . 115723
Galleri Bagdad Café, Malmö . . . 115747
Galleri Baghuset, Aalborg 102761
Galleri Bergklinten, Orrefors . . . 115801
Galleri Bi-Z, Kristiansand 113169
Galleri Blå, Garsnäs 115621
Galleri Blå, Linköping 115724
Galleri BNK, Helsingborg 115681
Galleri Bodekull, Karlshamn . . . 115706
Galleri Bodøgaard, Bodø 113139
Galleri Borg, Reykjavík 109231
Galleri Borris, Viborg 103005
Galleri Bouhlou, Bergen 113126
Galleri Brevik, Tromsø 113283
Galleri Bryggen, Bergen 084216, 113127
Galleri By Art, Odense 102946
Galleri C, Lund 115732
Galleri C, Södertälje 115816
Galleri C J, Västervik 115985

Galleri Christian, Bergen 084217
Galleri Classic, Hobro 102841
Galleri Cobra, Stockholm 115859
Galleri Cosmopolitan, Göteborg . 115639
Galleri Cupido, Stockholm 115860
Galleri D-40, Oslo 113208, 113209
Galleri Dagmar, Fanø 102798
Galleri Dalarna, Borlänge 115592
Galleri Decó, Aalborg 102762
Galleri Dialog, Stockholm 115861
Galleri Diana, Stockholm 115862
Galleri Dianahuset, Norrtälje . . . 115784
Galleri DOBAG, Kristiansund . . . 113173
Galleri Doktor Glas, Vaxholm . . 115996
Galleri E.V., Esbjerg 102792
Galleri Elenor, Oslo 113210
Galleri Elsa, Namsos 113184
Galleri Embla, Stockholm 115863
Galleri Emilie, Fredrikstad 084238
Galleri Epoke, Vejle . . 066192, 103000
Galleri Espolin, Kabelvåg 033689
Galleri Estetica, Malmö 115748
Galleri Ett, Uppsala 115970
Galleri F 12, Forus 113152
Galleri F 15, Moss 113180
Galleri Fabian Kyrkbacken,
  Västerås 115981
Galleri Fabrikkveien 12,
  Stavanger 113267
Galleri Falkum, Skien 113259
Galleri Ferm, Göteborg 115640
Galleri Final, Malmö 115749
Galleri Flamingo, Falkenberg . . . 115607
Galleri Fløistad, Arendal 113120
Galleri Fold, Reykjavík . 109232, 109233
Galleri Format, Oslo 113211
Galleri Forsa Skola, Mönsterås . 115778
Galleri Freden, Nacka 115780
Galleri Friends of Art, Stockholm 115864
Galleri Futura, Stockholm 115865
Galleri G, Førde 113151
Galleri G, Sandnes 113255
Galleri G, Stavanger 113268
Galleri Gaalaas Øvre, Furnes . . 113154
Galleri Gallica, Malmö 115750
Galleri Gamla Badet, Falun 115611
Galleri Gamla Väster, Malmö . . 115751
Galleri Gårdinger, Kalmar 115699
Galleri Geitodden, Flekkerøy . . . 113150
Galleri Giga, Stord 113278
Galleri GKM, Malmö 115752
Galleri Gl. Lejre, Lejre 102925
Galleri God Konst, Tingsryd . . . 115951
Galleri Göteborgs Konstförening,
  Göteborg 115641
Galleri Grafik, Umeå 115963
Galleri Granlo, Sundsvall 115945
Galleri Graphique, Fredrikstad . . 113153
Galleri Grappa, Hjo 115688
Galleri Gude, Oslo 113212
Galleri Gula Rummet, Huddinge . 115692
Galleri Guldfisk, Rudkøbing . . . 102967
Galleri Gullbakke, Lillehammer . 113175
Galleri Gullmarn, Uddevalla . . . 115959
Galleri Gummeson, Stockholm . . 115866
Galleri Gunnrummet, Stockholm . 115867
Galleri Gyllene Snittet, Gnosjö . . 115622
Galleri Haaken, Oslo 113213
Galleri Halland, Halmstad 115675
Galleri Hamnen, Viken 115998
Galleri Hanne, Trondheim 113287
Galleri Helco, Hadsund 102823
Galleri Hemma hos Modins, Sala 115868
Galleri Hera, Stockholm 115868
Galleri Himmerland, Frøstrup . . . 102813
Galleri Hippo, Stockholm 115869
Galleri Hlemmur, Reykjavík 109234
Galleri Ho, Viborg 103006
Galleri Humlum, Struer 066146, 102986
Galleri Hyttestigen, Ängelholm . 115569
Galleri IF, Ängelholm 115570
Galleri Ikaros, Göteborg 115642
Galleri Ikon, Stockholm 115870
Galleri Imma, Mariestad 115774

Galleri Inez, Lund 115733
Galleri Ingrid, Fauske . 084234, 113149
Galleri Inter Art, Göteborg 115643
Galleri Inuit, Aalborg 102763
Galleri Ismene, Trondheim 113288
Galleri J2, Oslo 113214
Galleri Jägeren, Östersund 115797
Galleri Jæren, Bryne 113142
Galleri JMS, Oslo 113215
Galleri K, Oslo 113216
Galleri K, Trollhättan 115596
Galleri K 16, Ålesund 113118
Galleri Kaffeslottet, Gudhjem . . 102818
Galleri Karlslund, Örebro 115789
Galleri Karma, Stockholm 115871
Galleri Karneval, Degerhamn . . 115596
Galleri Katedralen Havnen,
  Skagen 102976
Galleri Kavaletten, Stockholm . . 115872
Galleri KC, Fjerritslev 102800
Galleri KC, Göteborg 115644
Galleri Keriō, Kópavogur 109227
Galleri KIK, Kerteminde 102854
Galleri Kimen, Molde 113178
Galleri Klettur, Hafnarfjörður . . . 109225
Galleri Knabro, København 065904,
                                      102885
Galleri København, København . 102886
Galleri Købmandsgården, Haslev 065724
Galleri Koll, Stavanger 113269
Galleri Konst och Antik, Örebro . 086900,
                                      115790
Galleri Konstgården, Tranås . . . 115954
Galleri Konstkännaren,
  Helsingborg 115682
Galleri Konstnären, Karlskrona . 115708
Galleri Kontrast, Stockholm . . . 115873
Galleri Krinkelkrogen, Herning . 102829
Galleri Kungsgården Ottenby,
  Degerhamn 115597
Galleri Kvarnen, Sundborn 115942
Galleri Langegården Tablå,
  Bergen 113128
Galleri Langeland, Rudkøbing . . 066100,
                                      102968
Galleri Lars, Karlstad 086779
Galleri Leivo, Stockholm 115874
Galleri LenArt, Gentofte 102814
Galleri Lilja, Stockholm 115875
Galleri Liljan, Värnamo 087137
Galleri Lilla Uppåkra, Staffanstorp 115823
Galleri Lily, Narvik 113185
Galleri Lista Fyr, Borhaug 113141
Galleri LNM, Oslo 113217
Galleri Lønstrup, Hjørring 102836
Galleri Lora, Stockholm 115876
Galleri Lucidor, Stockholm 115877
Galleri M, Örnsköldsvik 115794
Galleri M, Sandnes 084353
Galleri Madigan, Svendborg . . . 102988
Galleri Magnus, Hamar 113156
Galleri Magnus, Hinnerup 102883
Galleri M'am, Östersund 115798
Galleri Marie, Halden 113155
Galleri Marius, København 102887
Galleri Markveien, Oslo 113218
Galleri Marta, Frederikshavn . . . 102810
Galleri Max, Ystad 116001
Galleri MDA, Helsingborg 115683
Galleri MGM, Oslo 113219
Galleri Miró, Reykjavík 109235
Galleri Miva, Malmö 115753
Galleri Mölndal, Mölndal 115776
Galleri Moment, Ängelholm . . . 115571
Galleri Mors Mörssa, Göteborg . 115645
Galleri Mowestor, Stockholm . . 115878
Galleri Munken, Løkken 102928
Galleri Nakkehage, Nykøbing
  Sjælland 102942
Galleri Neuf, Kivik 115711
Galleri New Form i Sparreholm,
  Sparreholm 115822
Galleri Nobel, Oslo . . . 113220, 126371
Galleri Nør, Højbjerg 102843

Galleri Nørballe, Middelfart 102931
Galleri Nørby, København 102888
Galleri Nord, Örebro 115791
Galleri Nordens Bilder, Torslanda 115953
Galleri Nordenvind, Sæby 066105
Galleri Nordica, Ystad 116002
Galleri Nostalgi Antik, Roskilde . 066088
Galleri Nygaten 7, Bergen 113129
Galleri Nytorget 13, Stockholm . 115879
Galleri ÖK, Värnamo 115979
Galleri Örgryte, Göteborg 115646
Galleri Örsta, Kumla 115715
Galleri Överkikaren, Stockholm . 115880
Galleri Olab, Göteborg 115647
Galleri Opus, Bjärred 115581
Galleri Originalen, Ballerup ... 102783
Galleri Osebro, Porsgrunn 113250
Galleri Paletten, Umeå 115964
Galleri Panorama Art, Stockholm 115881
Galleri Parken, Bergen 113130
Galleri Perlen, Nykøbing Falster . 102935
Galleri Pierrot Le Fou, Valby ... 102993
Galleri Ping Pong, Malmö 115754
Galleri Pingvin, Oslo 113221
Galleri Plus, Hillerød 102831
Galleri Pop, Karlshamn 115707
Galleri PP, Varde 066182, 102997
Galleri Prinsen, Aalborg 102764
Galleri Profil, Bergen 113131
Galleri Profilen, Århus 102772
Galleri Projekt, København ... 102889
Galleri Puckelns, Stockholm .. 115882
Galleri Puls Art, København .. 102890
Galleri Python, København 102891
Galleri R, Linköping 115725
Galleri Rackargården, Kalmar .. 115700
Galleri Ravnsager, Ringsted 102959
Galleri Redningshuset, Skagen . 102977
Galleri Reiky, Reykjavík 109236
Galleri Remi, Östersund 115799
Galleri Reykjavík, Kópavogur .. 109228
Galleri Rimfaxe, København 102892,
                                 136773
Galleri Rosas Källare, Växjö ... 115992
Galleri Rose, Vejle 103001
Galleri Rose Marie, Malmö 115755
Galleri Rosengången, Lerum .. 115720
Galleri Rostrum, Malmö 115756
Galleri S, Östersund 115800
Galleri Sævars Karls, Reykjavík . 109237
Galleri Sala, Sala 115805
Galleri Salongen, Vadstena ... 115977
Galleri Sankt Erik, Stockholm .. 115883
Galleri Sankt Gertrud, København 102893
Galleri Sankt Lars, Linköping .. 115726
Galleri Scandinavia, Göteborg . 115648
Galleri SE, Falun 115612
Galleri Selle, Bjärred 115582
Galleri Sharp, Tranekær 102991
Galleri Sigma, Växjö 115993
Galleri Silkeborg Kunstnerhus,
  Silkeborg 102974
Galleri Sinklar, Sundsvall 115946,
                                 138018
Galleri Sjöhästen, Lidingö 115721
Galleri Sjöhästen, Nyköping ... 115786
Galleri Skagen, Skagen 102978
Galleri Skrautla, Akureyri 109217
Galleri Skuggi, Reykjavík 109238
Galleri Slottet, Sunne 115947
Galleri Slugen, Esbjerg 102793
Galleri Småland, Alvesta 115574
Galleri Smedbyn, Huskvarna .. 115693
Galleri Smiegården, Sandnes .. 113256
Galleri Sølvberget, Stavanger . 113270
Galleri Solsiden, Trondheim .. 113289
Galleri Specta, København 102894
Galleri Stenhallen, Borgholm . 115591
Galleri Stensgård, Torsby 115952
Galleri Sulegaarden, Assens .. 102782
Galleri Svalbard, Longyearbyen . 033756
Galleri Svaneke, Svaneke 066150,
                                 102987

Galleri Svanlunda, Malmö 115757
Galleri Svarta Soffan, Stockholm 115884
Galleri Svenshög, Lund 115734
Galleri Syd, Nykøbing Falster . 102936
Galleri Tapper-Popermajer,
  Teckomatorp 115950
Galleri Tarp, Sæby 102971
Galleri Tersæus, Stockholm ... 115885
Galleri To, Hinnerup 102834
Galleri Torp, Frändefors 086678, 115617
Galleri Torvet, Grindsted 102817
Galleri Tre, Kristiansand 113170
Galleri Trollhättan, Trollhättan .. 115957
Galleri Tvedestrand, Tvedestrand 113292
Galleri TviTvi, Trondheim 113290
Galleri Uggerby, Hjørring 102837, 102838
Galleri Úmbra, Reykjavík 109239
Galleri Urd, Bergen 113132
Galleri Utstein, Mosterøy 084282
Galleri V H, Göteborg 115649
Galleri Värend, Växjö 115994
Galleri Værket, Aalborg 102765
Galleri Vättern, Askersund 115577
Galleri Valvet, Malmö 115758
Galleri van Rijn, Stavanger 084367
Galleri Varden, Moss 113181
Galleri Veggerby, København ... 102895
Galleri Vejle, Vejle 066193, 103002
Galleri Vela, Frederiksberg 102804
Galleri Vera, Drøbak 113146
Galleri Veslemøy, Egersund ... 113148
Galleri Victoria, Halmstad 115676
Galleri Victoria, Örebro 086901
Galleri Vifladt, Dokka 113144
Galleri Viktoria, Dala-Husby ... 115595
Galleri Viktoria, Göteborg 115650
Galleri Villvin, Risør 084346
Galleri Vincent, København ... 102896
Galleri Voss, Voss 113293
Galleri Vretbacka, Gävle 115619
Galleri Wernskjöld, Kalmar ... 115701
Galleri Win-Win, Göteborg 115651
Galleri Würth, Hagan 033629
Galleri X, Rungsted Kyst 102970
Galleri X4, Linköping 115727
Galleri XIA, Sandnes 113257
La Galleria, Castell'Arquato ... 109940
La Galleria, London 118332
La Galleria, Modena 110498
Galleria, Mumbai 109435
Galleria, Pittsburgh 123544
La Galleria, Venezia . 111105, 137758
Galleria 13, Reggio Emilia 110710
Galleria 28, Timişoara 114175
Galleria 61, Palermo 110604
Galleria 70, Milano 110322
Galleria 733, Chicago 092714
Galleria 900, Parma 081113
Galleria A, Parma 081114
Galleria A+A, Centro d'Arte
  Contemporanea Slovena,
  Venezia 111106
Galleria AAA, Ascona 116035
Galleria Accademia, Torino ... 110968
Galleria Accademia Albertina,
  Torino 110969
Galleria Africa Curio, Milano .. 110323
Galleria Aion, Ascona 087200
Galleria Akka, Roma 081378
Galleria Alberoni, Piacenza ... 027293
Galleria Albert IX, Helsinki 103139
Galleria Anfitrite, Lugano 087615,
  087616,   144102,   144103
Galleria Annmari's, Tampere ... 066396
Galleria Antelami, Lemignano .. 080312
Galleria Antiquaria, Merano ... 080353
Galleria Antiquaria, Sassuolo ... 081618
Galleria Antiques, Atlanta 092137
Galleria Arabeske, Dortmund .. 106702
Galleria L'Ariete, Bologna 109825
Galleria Ars Antiqua, Bergamo . 109793
Galleria Art Alex, Helsinki 103140
Galleria Arte Borgogna, Milano . 110324
Galleria Arte Capital, Brescia .. 109901

Galleria Arte Corso, Roma 110776
Galleria Arte Oggi, Finale Ligure 110022
Galleria Arte Solferino, Milano .. 080551
Galleria Arts & Crafts, Calgary . 101239
Galleria Aurora Pallavicini, Roma 027562
Galleria Aversa, Torino 110970
Galleria Baumgartner, Mendrisio 041739
Galleria BE'19, Helsinki 103141
Galleria Bellissima, Philadelphia 123381
Galleria Bergamo, Bergamo ... 109794
Galleria Bessone, Aosta 079746
Galleria Bianconi, Milano 080552,
                                 110325
Galleria Bixio 2, Milano 110326
Galleria Blu, Milano 110327
Galleria Bo-Art, Bologna 109826
Galleria Bordone, Milano 110328
Galleria Borromeo, Padova 110565
Galleria Bosi, Roma 110577
Galleria Bosoni, Milano 080553, 110329
Galleria Brà, Verona 111181
Galleria Ca' Venier dei Leoni,
  Venezia 111107
Galleria Caldarese, Bologna 079844,
                                 109827
Galleria Cardi, Milano 110331
Galleria Cardi & Co., Milano ... 110330
Galleria CarloLivi 115, Prato ... 110690
Galleria Castiglione, Bologna ... 109828
Galleria Cavour, Firenze 110053
Galleria Ceribelli, Bergamo ... 109795
Galleria Cherubini, Padova 110566
Galleria Cilea, Napoli 110525
Galleria Ciovasso, Milano 110332
Galleria Città di Riva, Riva del
  Garda 110719
Galleria Città di Treviso, Treviso 111037
Galleria Civica, Modena 110499
Galleria Civica Anna e Luigi
  Parmiggiani, Reggio Emilia .. 027500
Galleria Civica Brunico, Brunico 025649
Galleria Civica d'Arte Contemporanea,
  Caltagirone 025697
Galleria Civica d'Arte Contemporanea,
  Suzzara 028047
Galleria Civica d'Arte Contemporanea,
  Termoli 028075
Galleria Civica d'Arte Contemporanea,
  Trento 028170
Galleria Civica d'Arte Moderna,
  Santhià 027845
Galleria Civica d'Arte Moderna,
  Spoleto 028004
Galleria Civica d'Arte Moderna e
  Contemporanea, Torino 028115
Galleria Civica, Villa Valle,
  Valdagno 028251
Galleria Civico, Modena 026848
Galleria Clas Art, Lodi . 080341, 110190
Galleria Colonna, Napoli 110526
Galleria Colonna, Roma 027563
Galleria ComoArte, Como 109969
Galleria Comunale d'Arte,
  Cagliari 025672
Galleria Comunale d'Arte Moderna e
  Contemporanea, Arezzo 025357
Galleria Comunale d'Arte, Museo
  Manzoniano, Lecco 026583
Galleria Comunale Santa Croce,
  Cattolica 025879
Galleria Cons Arc, Chiasso ... 116214
Galleria Conti, Prato 110691
Galleria Continua, Beijing 101951
Galleria Continua, San Gimignano 110894
Galleria Corsini, Firenze 026254
Galleria Credito Siciliano,
  Acireale 025233
Galleria Credito Valtellinese,
  Sondrio 027984
Galleria Cristiani, Courmayeur . 109995
Galleria Cristiani, Torino 110971
Galleria Daldiniana, Vezia 116698
Galleria d'Arte, Freiburg im

Breisgau 075865
Galleria d'Arte, Genova 110105
Galleria d'Arte, Gorizia 110145
Galleria d'Arte, Jesolo 110151
Galleria d'Arte, München 108128
Galleria d'Arte 2 di Quadri,
  Bologna 109829
Galleria d'Arte 2000, Milano ... 110333
Galleria d'Arte 37, Palermo 110605
Galleria d'Arte A. Cefaly, Catania 109951
Galleria d'Arte Alexandra, Rovigo 110887
Galleria d'Arte Ambrosiana,
  Milano 080554
Galleria D'Arte Antica, Baden-
  Baden 074634
Galleria d'Arte Antichità, Alassio 109742
Galleria d'Arte Athena, La Spezia 110155
Galleria d'Arte Bongiovanni,
  Bologna 109830
Galleria d'Arte Busellato, Asiago 109757
Galleria d'Arte Il Castello, Trento 111029
Galleria d'Arte Centro Storico,
  Firenze 110054
Galleria d'Arte Cinquantasei,
  Bologna 109831
Galleria d'Arte Colussa, Udine . 111061
Galleria d'Arte Contemporanea, Ascoli
  Piceno 025385
Galleria d'Arte Contemporanea delle
  Pro Civitate Christiane, Cittadella
  Cristiana, Assisi 025397
Galleria d'Arte Contemporanea L.
  Spazzapan, Gradisca d'Isonzo . 026458
Galleria d'Arte Contemporanea V.
  Stoppioni, Santa Sofia 027825
Galleria d'Arte Contini, Cortina
  d'Ampezzo 109982, 109983
Galleria d'Arte Contini, Venezia . 111108,
                                 111109
Galleria d'Arte del Caminetto,
  Bologna 079845
Galleria d'Arte di Palazzo Duca,
  Pavullo nel Frignano 110638
Galleria d'Arte Dusatti, Rovereto 110880
Galleria d'Arte G.S., Faenza ... 110010
Galleria d'Arte Gelmi, Sesto San
  Giovanni 110928
Galleria d'Arte Internazionale,
  Venezia 111110
Galleria d'Arte Lelii, Cagliari ... 109923
Galleria d'Arte Leonardiana,
  Alassio 109743
Galleria d'Arte Leonardo, Bolzano 109878
Galleria d'Arte L'Incontro, Verona 111182
Galleria d'Arte L'Occhio, Venezia 111111
Galleria d'Arte Maggiore,
  Bologna 109832
Galleria d'Arte Mainetti, Milano . 080555,
                                 110334
Galleria d'Arte Martinazzo,
  Montebelluna 110507
Galleria d'Arte Moderna, Firenze 026255
Galleria d'Arte Moderna, Udine . 028226
Galleria d'Arte Moderna 31,
  Napoli 110527
Galleria d'Arte Moderna A. Discovolo,
  Bonassola 025588
Galleria d'Arte Moderna Arnoldo
  Bonzagni, Cento 025895
Galleria d'Arte Moderna Carlo Rizzarda,
  Feltre 026192
Galleria d'Arte Moderna Comunale,
  San Severino Marche 027797
Galleria d'Arte Moderna e
  Contemporanea, Bergamo .. 025492
Galleria d'Arte Moderna e
  Contemporanea R. Guttuso,
  Bagheria 025432
Galleria d'Arte Moderna F. Montanari,
  Moncalvo 026882
Galleria d'Arte Moderna Giannoni,
  Novara 027058
Galleria d'Arte Moderna Moretti,
  Civitanova Marche 026016

Galleria d'Arte Moderna O. Marchesi,
Copparo 026063
Galleria d'Arte Moderna Ricci-Oddi,
Piacenza 027294
Galleria d'Arte Orler, Abano
Terme 079736, 109734, 126185
Galleria d'Arte Orler, Favaro
Veneto 080070, 110016, 126189
Galleria d'Arte Orler, Madonna di
Campiglio 080343, 110193, 126200
Galleria d'Arte Paolo Nanni,
Bologna 109833
Galleria d'Arte Paris, Treviso . . . 110938
Galleria d'Arte Plurima, Udine . . 111062
Galleria d'Arte Ponte Rosso,
Milano 110335
Galleria d'Arte Rubens, Gradisca
d'Isonzo 110147
Galleria d'Arte Sacerdoti, Milano 080556,
110336
Galleria d'Arte Sacra dei
Contemporanei, Milano 026795
Galleria d'Arte Sant' Isaia,
Bologna 109834
Galleria d'Arte Sk, Taranto 110941
Galleria d'Arte Venezia, Venezia 081838,
111112
Galleria d'Arte Vintage, Spotorno 110937
Galleria d'Arte Viola, Magenta . . 110194
Galleria Davico, Torino 110972
Galleria Davide Di Maggio,
Milano 110337
Galleria de' Foscherari, Bologna 109835
Galleria de' Fusari, Bologna . . . 109836
Galleria de las Islas, Manila . . . 113405
Galleria de Nisi, Caserta 109936
Galleria de Nisi, Roma 110778
Galleria degli Arazzi, Musei Vaticani,
Città del Vaticano 054680
Galleria degli Artisti, Roma 110779
Galleria degli ex Voto del Santuario,
Santuario di Montenero, Livorno 026603
Galleria degli Uffizi, Firenze 026256
Galleria dei Benci, Firenze 110055
Galleria dei Coronari, Roma . . . 081379
Galleria dei Cosmati, Roma . . . 081380,
126226,   126227
Galleria dei Greci-Editrice, Roma 110780,
137734
Galleria del Barcon, Milano . . . 110338
Galleria del Cortile, Roma 110781
Galleria del Costume, Firenze . . 126257
Galleria del Leone, Castelnuovo di
Farfa 109941
Galleria del Maggio, Villa
Minozzo 028388
Galleria del Manifesto, Bolzano . 109879
Galleria del Manifesto, Verona . . 111183
Galleria del Manifesto d'Arte,
Bellinzona 116113
Galleria del Mobile d'Arte Bigolin,
Treviso 111039
Galleria del Mosaico, Chiasso . . 116215
Galleria del Ponte, Torino 081702
Galleria del Quadro, San
Marzano 110897
Galleria del Restauro, Cagliari . . 109924,
131069
Galleria del Tasso Arte
Contemporanea, Bergamo . . . 109796
Galleria della Pigna, Roma 110782
Galleria della Tartaruga, Roma . 110783
Galleria dell'Accademia, Firenze 026258
Galleria dell'Accademia di Belle Arti,
Napoli 026989
Galleria dell'Accademia Nazionale di
San Luca, Roma 027564
Galleria dell'Accademia Tadini,
Lovere 026635
Galleria dell'Altabella, Bologna . 109837
Galleria delle Battaglie Arte
Contemporanea, Brescia 109902
Galleria delle Stampe Antiche,
Catania 080035

Galleria dell'Immagine, Savignano sul
Rubicone 110914
Galleria dell'Incisione, Brescia . 109903
Galleria Dello Scudo, Verona . . 111184
Galleria dello Spedale degli Innocenti,
Firenze 026259
Galleria dello Stretto, Palermo . . 110606
Galleria di Arte Moderna Palazzo Forti,
Verona 028343
Galleria di Palazzo degli Alberti,
Prato 027446
Galleria di Piazza San Marco,
Fondazione Bevilacqua la Masa,
Venezia 028296
Galleria di Sorrento, Las Vegas . 121352
Galleria Disegno, Mantova 110203
Galleria Dix, Helsinki 103143
Galleria Don Chisciotte, Roma . . 110784
Galleria Doria Pamphilj, Roma . 027565
Galleria d'Orlane, Casalmaggiore 080014
Galleria I Duchi, Trieste 111047
Galleria Duemila, Mandaluyong . 113397
Galleria Duemila, Pasay 113413
Galleria Editrice Latina, Roma . . 110785,
110786
Galleria Emy, Messina 110220
Galleria Estense, Modena 026849
Galleria Eustachi, Milano 113339
Galleria F. Russo, Roma 110787, 110788
Galleria Falchi, Levico Terme . . 110167
Galleria Farneti, Forlì 110079
Galleria Faustini, Firenze 110056
Galleria Ferrari, Maranello 026706
Galleria Fine Art and Graphics,
Oakland 123252
Galleria Stefano Forni, Bologna . 109838
Galleria Paola Forni, Bologna . . 109839
Galleria Franco Cancelliere,
Messina 110221
Galleria Fumagalli Bergamo,
Bergamo 109797
Galleria G, Helsinki 103144
Galleria G. Pedriali, Forlì 026321
Galleria Ghighini, Varese 111076
Galleria Gianna, Cleveland 120421
Galleria Gioacchini, Ancona . . . 109750
Galleria Giorgio Franchetti alla Ca'
d'Oro, Venezia 028297
Galleria Giovani Artisti, Bolzano . 109880
Galleria Giulia, Roma 110789
Galleria Goethe, Bolzano 109881
Galleria Goethe 2, Bolzano 109882
Galleria Goldoni, Livorno 110179
Galleria Grafica Tokio, Tokyo . . . 111428
Galleria Gruppo Credito Valtellinese,
Milano 026796
Galleria Gualtieri, Cesenatico . . 109960
Galleria Guicciardini, Firenze . . 080144
Galleria Hofburg, Bressanone . . 109917
Galleria II, San Antonio 124030
Galleria III Millennio, Venezia . . 111113
Galleria Il Polittico, Roma 110790
Galleria Imperiale, Genova 080261,
126196
Galleria Inter, Turku 066402
Galleria Internacional,
Washington 124907
Galleria J Antonio, New York . . . 122632
Galleria L' Immagine, Galleria d'Arte
Contemporanea, Bari 109769
Galleria La Casa, Vaglio 116684
Galleria La Colomba, Viganello . 116699
Galleria La Cornice, Norderstedt 108284
Galleria La Meridiana, Verona . . 111185
Galleria La Nevera, Gordola . . . 116354
Galleria La Rocca, Aosta 109753
Galleria La Rocca, Torino 110973
Galleria La Sirena, Tucson 124735
Galleria Lapiccirella, Firenze . . . 080145
Galleria Lapidaria, Musei Vaticani,
Città del Vaticano 054681
Galleria Le Muse, Messina 110222
Galleria Liguria, Genova 110106
Galleria Lo Spazio, Brescia 109904

Galleria Luce, Venezia 111114
Galleria M. Rizzi, Sestri Levante 027943
Galleria Magenta, Brescia 109905
Galleria Magenta, Magenta 110195
Galleria Malair, Piacenza 081150
Galleria Marabini, Bologna 109840
Galleria Marconi, Cupra
Marittima 109999
Galleria Marescalchi, Bologna . . 109841
Galleria Marescalchi, Cortina
d'Ampezzo 109984
Galleria Maria Theresia, Brunico 109918,
109919
Galleria Marieschi, Milano 110340
Galleria Marilena Bonomo, Bari . 109770
Galleria Masini, Firenze 110057
Galleria Matasci, Museo d'Arte,
Tenero 042049
Galleria Maze, Torino 110974
Galleria Menhir, La Spezia 110156
Galleria Mentana, Firenze 110058
Galleria Meria, Linköping 115728
Galleria Meridiana, Pordenone . 110674
Galleria Micati, Giulianova 110144
Galleria Michelangelo Buonarroti,
Brugge 063275
Galleria Michelangelo, Arte Antica e
Moderna, Bergamo 109798
Galleria Mignanelli, Roma 110791
Galleria Milano, Milano 110341
Galleria Massimo Minini, Brescia 109906
Galleria Modigliani, Milano 110342
Galleria Mondiale, Surry Hills . . 099746
Galleria Morone, Milano 110343
Galleria Muciaccia, Roma 110792
Galleria Municipale, Francavilla al
Mare 026343
Galleria Museo, Bolzano 109883
Galleria Mystica, Pune 079413
Galleria Narciso, Torino 110975
Galleria Nazionale, Parma 027208
Galleria Nazionale d'Arte Antica,
Trieste 028202
Galleria Nazionale d'Arte Antica
Palazzo Barberini, Roma 027566
Galleria Nazionale d'Arte Antica
Palazzo Corsini, Roma 027567
Galleria Nazionale d'Arte Moderna e
Contemporanea, Roma 027568
Galleria Nazionale delle Marche,
Urbino 028238
Galleria Nazionale dell'Umbria,
Perugia 027256
Galleria Nazionale di Arte Moderna e
Contemporanea, San Marino . 038096
Galleria Nazionale di Cosenza,
Cosenza 026086
Galleria Nazionale di Palazzo Spinola,
Genova 026394
Galleria Nella Longari, Milano . . 080557
Galleria Niccoli, Parma 110627
Galleria No Code, Bologna 109842
Galleria Nova, Ronco sopra
Ascona 116593
Galleria Novelli, Verona 111186
Galleria Nuovo, Lahti 103163
Galleria Nuovo Spazio, Udine . . 111063
Galleria Nuovo Spazio, Venezia . 111115
Galleria Oddi Baglioni di Emanuela
Lenzi, Roma 110793
Galleria on Third, New York . . . 122633
Galleria Orti Sauli, Genova 110107
Galleria Otso, Espoo 103126
Galleria Otto, Gallery Arte
Contemporanea, Bologna . . . 109843
Galleria Pace, Milano 110344
Galleria PaciArte Contemporanea,
Brescia 109907
Galleria Pack, Milano 110345
Galleria Palatina e Appartamenti Reali,
Firenze 026260
Galleria Palazzo Cini, Venezia . . 028298
Galleria Palazzo Dami, Firenze . 080146
La Galleria Pall Mall, London . . 090140

Galleria Palladio, Lugano 116456
Galleria Pananti, Firenze 080147, 110059,
126191,   137650
Galleria Perera, Roma . 081381, 110794
Galleria Perugi, Padova 110567
Galleria Petrofil Arte, Milano . . . 110346
Galleria Pianetti, Falconara
Marittima 110014
Galleria Piemonte Reale Armi Antiche,
Torino 081703
Galleria Pinacotheca,
Jyväskylän Yliopiston Museo –
Kultuurihistoriallinen Osasto,
Jyväskylä 010932
Galleria Pinciana, Roma 081382
Galleria Pinocchio, Birchington . 088468
Galleria Pollin, Treviso 111040
Galleria Previtali, Bergamo 079792
Galleria Primavera, Helsinki . . . 103145
Galleria Quadreria, San Stino di
Livenza 110901
Galleria Quadreria Den Hertog,
Bari 109771
Galleria Quadrum, Villafranca di
Verona 111208
Galleria Ravagnan, Venezia . . . 111116
Galleria Refinishers Too, Seattle 136370
Galleria Regionale d'Arte Moderna e
Contemporanea, Sassoferrato . 027874
Galleria Regionale della Sicilia,
Palermo 027176
Galleria Regionale di Palazzo Bellomo,
Siracusa 027967
Galleria Rettori Tribbio 2, Trieste 111048
Galleria Rinaldo Carnielo, Firenze 026261
Galleria Riviera, Padova 110568
Galleria Romana dell'Ottocento,
Roma 081383
Galleria Rusticana, Oslo 084311
Galleria Sabauda, Torino 028116
Galleria San Bernardo, Genova . 110108
Galleria San Carlo, Milano 110347
Galleria San Giorgio, Modena . . 080793
Galleria San Giovanni, Perugia . 081143
Galleria San Lorenzo Al Ducale,
Genova 110109
Galleria San Maurizio, Venezia . 111117
Galleria San Tomaso, Milano . . 110348
Galleria Sant'Agostino, Torino . . 110976,
126234
Galleria Sant'Antonin, Venezia . . 111118
Galleria Santo Stefano, Venezia . 111119
Galleria Schreiber, Brescia 109908
Galleria Sculptor, Helsinki 103146
La Galleria Second Classe, New
York 122634
Galleria Senato, Milano 110349
Galleria Serego, Verona 111187
Galleria Serodine, Ascona 087201
Galleria Silva, Milano 080558
Galleria Silvy Bassanese, Biella . 109809
Galleria Simonini, Modena 080794
Galleria Soave, Alessandria . . . 109748
Galleria Spada, Roma 027509
Galleria Spagnoli Arte,
Valtournenche 111073
Galleria Spazia, Bologna 109844
Galleria Spinetti, Firenze 110060
Galleria Stendhal, Parma 081115
Galleria Studio G 7, Bologna . . 109845
Galleria sul Po, Torino 110977
Galleria Tartaglia Arte, Roma . . 110795
Galleria Tecne, Padova 110569
Galleria Tega, Milano 110350
Galleria Il Tempo, Conselve . . . 109769
Galleria Terre Rare, Bologna . . . 109846
Galleria Tiffany, Bologna 079846
Galleria Titanik, Turku 103169
Galleria Tornabuoni, Firenze . . . 110061
Galleria Tornabuoni, Pietrasanta 110655
Galleria Toscani, Gradisca
d'Isonzo 110148
Galleria Traghetto, Venezia 111120
Galleria Uusitalo, Helsinki 103147

Galleria Vanna Casati, Bergamo 109799
Galleria Varotto, Susano di Castel
D'Ario 110939
Galleria Vecchiato, Padova 110570
Galleria Veneziani, Roma 081384
Galleria Verde, Roma 081385
Galleria Veronese, New Orleans 122171
Galleria Vesterinen, Helsinki . . . 103148
Galleria Vinciana, Milano 110351
Galleria Vittoria, Roma 110796
Galleria Wanda, Wolverhampton 119382
Galleria Wehrlin, Milano 080559
La Galleria, Studio Numismatico,
Pescara 081149
Gallerie 91, Charlottenlund . . 065624,
102786
Gallerie Alegria, Birmingham . . 119797
Gallerie am Markt, Kirchhain . . 107558
Gallerie am Rathausmarkt,
Viersen 108820
Gallerie Antiques, Hainault . . 089406,
134430
Gallerie Arts Graphiques, Villars-
Colmar 016225
Gallerie Australis, Adelaide . . . 098611
Gallerie Bella, Athinai 109034
Gallerie d'Arte del Design e
dell'Arredamento, Cantù . . . 109931
Gallerie dell'Accademia, Venezia 028299
Gallerie di Palazzo Leoni Montanari,
Vicenza 028366
Gallerie Dragehøj, Holmegaard . 065777
Gallerie Ethnos, Ettlingen . . . 075718
Gallerie Exclusiv, Plymouth . . 118920
Gallerie FeinSinn, Überlingen . . 108793
Gallerie Ganesha, Delhi 109317
Gallerie Lavelle, Bangalore . . . 079348
Gallerie l'Orientaliste, Fitzroy
North 099034
Gallerie Marin, Appledore, Devon 117244
Gallerie Moderne, New York . . . 122635
Gallerie Picamm, Harpstedt . . . 107378
Gallerie Q14, Oderberg 108328
Gallerie Rasmus, Kolding 102918
Gallerie Rasmus, Odense 102947
Gallerie Restigouche,
Campbellton 005545
Gallerie Rosini & C., Riccione . . 110714
Gallerie Vedbæk, Charlottenlund 065625,
102787
Gallerie Veronique, Enfield . . 089210,
134369
Gallerie Via Borromei, Milano . . 110352
Gallerie Zen, Bangalore 109257
Galleriea Graziosa Giger, Leuk
Stadt 116437
Galleries America, Las Vegas . . 121353
Galleries' Association of Korea,
Seoul 059527
The Galleries at Moore, Moore College
of Art and Design, Philadelphia 052052
Galleries at Raven's Nest, Denver 120722
Galleries of Fine Arts, Quezon
City 113419
Galleries of Justice, Nottingham 045412
The Galleries of Texas Tech, Texas
Tech University, Lubbock 050516
Galleries Primitf, Woollahra . . . 099915
Galleriet, Kolding 102919
Galleriet, Stavanger 113271
Gallerihuset, København 102897
Gallerikroken, Oslo 113222
Gallerina, Darlington 117660
Gallerix, Lidköping 115722
The Gallery, Abbotsbury 117203
Gallery, Aberdeen 088221, 134010
The Gallery, Al-Harthy 113294
The Gallery, Ardingly 117248
The Gallery, Beer 117328
The Gallery, Blandford Forum . 117389
The Gallery, Blenheim 112821
The Gallery, Burnham-on-Sea . 117485
The Gallery, Chichester 117579
The Gallery, Clare 117594

The Gallery, Dunfanaghy 109631
Gallery, El Paso 120832
The Gallery, Gateshead 044381
The Gallery, Gisborne 112898
The Gallery, Greystones 109641
The Gallery, Hamilton 100824
The Gallery, Hampton 117907
The Gallery, Kidderminster . . . 118040
The Gallery, Leek 089777
Gallery, London 118333
Gallery, Los Angeles 121489
The Gallery, Lynton 118684
The Gallery, Mount Victoria . . . 061923
Gallery, New Orleans 122172
The Gallery, New York 122636
Gallery, Oklahoma City 123281
The Gallery, Oundle 118862
Gallery, Praha 102692
The Gallery, Reigate . . 091089, 134964
The Gallery, Rhosneigr 118965
The Gallery, Richmond 123733
The Gallery, La Romana 010375
The Gallery, Saint Helens 119014
Gallery, Tbilisi 105904
Gallery, Vallauris 073951
The Gallery, Vancouver 101787
The Gallery, Virginia Beach . . 124828
The Gallery, Wymondham 119399
Gallery – University of Toronto,
Scarborough 006996
The Gallery & Beyond Words Gallery,
Marygrove College, Detroit . . 048306
Gallery & Innovations, Toronto . 101610
Gallery @ 49, New York 122637
Gallery 10, Washington 124908
Gallery 101, Houston 121042
Gallery 101, Malmö 115759
Gallery 101, Melbourne 099307
Gallery 101, University of Wisconsin-
River Falls, River Falls 052585
Gallery 11, Reykjavík 109240
Gallery 110 North, Plymouth Arts
Foundation, Plymouth 052218
Gallery 114, Portland 123623
Gallery 117, Long Beach 121400
Gallery 119, Coorparoo 098925
Gallery 12 18, Holmfirth 117979
Gallery 13, Epworth 117762
Gallery 138, New York 122638
Gallery 14, Chalkida 109061
Gallery 14, Sandwich 119060
Gallery 14, Sutton Coldfield . . 119187
Gallery 148, Holywood 117991
Gallery 16, San Francisco . . . 124285
Gallery 177, Northampton . . . 118817
The Gallery 18, Falmouth 117783
Gallery 181, Iowa State University,
Ames 046441
Gallery 185, Halifax 117901
Gallery 19, London 118334
Gallery 2, Chicago 047660
Gallery 2 on Edward, Brisbane . 098790
Gallery-2000, Aldershot 117223
Gallery 2000, Baltimore 119737
Gallery 2000, Bonn 106531
Gallery 2000, Seoul 111640
Gallery 2000, West End 099880
Gallery 203, Chicago 120132
Gallery 203, Nelson 112958
Gallery 206 and Foyer Gallery,
Missouri Western State University,
Saint Joseph 123792
Gallery 21, Bexhill-on-Sea . . . 117356
Gallery 21+Yo, Tokyo 111429
Gallery 210, University of Missouri
Saint Louis, Saint Louis 052800
Gallery 218, Walker's Point Artists
Association, Milwaukee 050913
Gallery 22, Cincinnati 120346
Gallery 23, San Diego 124123
Gallery 23 Antiques, Chalfont Saint
Giles 088818
Gallery 23 Inc., New York 122639
Gallery 25, Fresno 120896

Gallery '25, London . . 090141, 126961
Gallery 253, Bangkok 116951
Gallery 27, London 118335
Gallery 2700, Virginia Beach . . 124829
Gallery 286, London 118336
Gallery 3, North Shore 112971
Gallery 303, Statesboro 053506
Gallery 31, Nelson 112959
Gallery 33, Coleraine 117616
Gallery 33, Portland 123624
Gallery 33 Downtown, Portland . 123625
Gallery 34, Llangollen 118138
Gallery 360, Applecross 098651, 127553
Gallery 360, Joondalup 099162, 127619
Gallery 360, Minneapolis 121990
Gallery 360, Mount Lawley . . 099400,
127649
Gallery 360, Subiaco . 099725, 127680
Gallery 38, Saint Ives, Cornwall 119023
Gallery 39, Osaka 111355
Gallery 400 at the University of Illinois
at Chicago, Chicago 047661
Gallery 41, Taipei 116916
Gallery 41, Vancouver 064823
Gallery 415, Chicago 120133
Gallery 42, Delhi 109318
Gallery 42, Kirkcaldy 118055
Gallery 42, Tadcaster 119202
Gallery 43, Denver 120723
Gallery 435, Hamilton 101317
Gallery 44, Centre for Contemporary
Photography, Toronto 007174
Gallery 440, New York 122640
Gallery 444, San Francisco . . . 124286
Gallery 460 and Sculpture Park, Green
Point 099094
Gallery 47, New York 095408
Gallery 48, Lavenham 118077
Gallery 482, Fortitude Valley . . 099051
Gallery 500 Wells, Chicago . . . 120134
Gallery 515, Atlanta . . 119567, 144966
Gallery 52, Brailsford 117413
Gallery 53, Cairns 098837
Gallery 53, Hastings 117925
Gallery 532, New York 095409
Gallery 54, Penola 099558
Gallery 547, Waikato 113037
Gallery 55, Bangkok 116952
Gallery 55, Invercargill 112922
Gallery 56, Salt Lake City 123961
Gallery 56, Stratford 099723
Gallery 57, Cambridge 047360
Gallery 6+, Varna 101191
Gallery 63 Antiques, New York . 095410
Gallery 66, Twickenham 115759
Gallery 7, Mumbai . . . 109436, 130822
Gallery 7, Nelson 112960
Gallery 7–18, Hiroshima 111259
Gallery 7000, Tucson 124736
Gallery 702, Saint Paul 123907
Gallery 71, New York 122641
Gallery '72, Omaha 123313
Gallery 75, Limerick 109668
Gallery '76, Wenatchee Valley College,
Wenatchee 054278
Gallery 79, Taranaki 113015
Gallery 7916, Los Angeles . . . 121490
Gallery 8, Loganholme 099237
Gallery 8, Varna 101192
Gallery 811, New Orleans . . . 095102,
145309
Gallery 825, Los Angeles Art
Association, West Hollywood . . 054307
Gallery 83, Banff 117282
Gallery 88, Vancouver 101788
Gallery 888, Toronto 101611
Gallery 9, Los Altos 050404
Gallery 93, Derby 117673
Gallery A, New York 122642
Gallery A and Downstairs Gallery, San
Pedro 053059
Gallery Abarth, Yamanakako . . 029700
Gallery Akasaka, Fukuoka . . . 111250
Gallery AM, Seoul 111641

Gallery Americana, Houston . . 121043
Gallery Ann, Gaborone 100871
Gallery Apokalypse, Lefkosia . . 102536
The Gallery Apparao, Chennai . 109283
Gallery Arcturus, Toronto 007175
Gallery Area 53, Wien 100307
Gallery Art Center, Columbus . . 120494
The Gallery Art Centre, Celbridge 109545
Gallery Art Composition, Tokyo . 111430
Gallery Art Korner, Den Haag . 112385
Gallery Art Resource Trust,
Mumbai 109437
Gallery at 4584 Federal Boulevard,
San Diego 097089
The Gallery at Blakemere,
Blakemere 117387
Gallery at Bristol-Myers Squibb,
Princeton 052365
Gallery at Buck House, New York 095411
The Gallery at Lakeside, West
Thurrock 119327
Gallery at Lands End, San Diego 124124
Gallery at Library Square, Salt Lake
City 123962
Gallery at Market East,
Philadelphia 123382
Gallery at Midlane, Houston . . 121044
Gallery at Peachtree Hill, Atlanta 119568
Gallery at the Gildenburgh,
Peterborough 118907
Gallery at The Rocks, Sydney . . 099781
Gallery at The Wentworth,
Sydney 099782
Gallery at the Wharf, Coupeville 048038
Gallery at the Willows, Honolulu 120937
The Gallery at UTA, Arlington . . 046533
Gallery at Ward Centre, Honolulu 120938
Gallery at Werribee South,
Werribee 099877
Gallery Auctions, Houston 127250
Gallery B, Toledo 124688
Gallery B at the Market, Toledo 124689
Gallery Beaux, Tokyo 111431
The Gallery Beckenham,
Beckenham 117323
Gallery Bethlehem, Tauranga . . 113022
Gallery Bizarre, Double Bay . . 061404
Gallery Blackheath, Blackheath . 098750
The Gallery Bookstore, Chicago . 145043
Gallery Bouquet, Nagoya 111315
Gallery Bridgetown, Bridgetown 098773
Gallery Cafe, Cleveland 120422
The Gallery Café, Colombo . . . 115565
Gallery Carla, San Francisco . . 124287
Gallery Celtica, Phoenix 123493
Gallery Chavin, Wien 100308
Gallery Chine, Cairo . . 066246, 103039
Gallery Connexion, Fredericton . 005840
Gallery Dalma, Riyadh 085316
Gallery D'art, Lucknow 109406
Gallery Delaive, Amsterdam . . 112209
Gallery Departmental, Gaborone 100872
Gallery Different, London 118337
Gallery East, North Fremantle . . 099468
Gallery East, College of Eastern Utah,
Price 052358
Gallery Edition 88, Luxembourg 111852,
137812
Gallery Editions, Saint Paul . . . 123908
Gallery-ef, Tokyo 111432
Gallery Eighty, Singapore 114527
Gallery Eleven, Dorking 089082, 134334
Gallery Espace, Delhi 109319
Gallery Estitique, Portarlington . 099593
Gallery Euro, Seoul 111642
Gallery Europe, Ashford 117258
Gallery Express, Yokohama . . . 111536
Gallery F, Kyoto 111292
Gallery Felicie, Woodbury . . . 125036,
138480
Gallery Fenelon, Paris 129012
Gallery Fifty One, Philadelphia . 123383
Gallery Focus, Paju 111606
Gallery Focus, Seoul 111643

Gallery for Foreign Art Sveti Kiril i
Metodi, Sofia 101161
Gallery Four Seven Nine, Ottawa 101473
Gallery Frames, Dallas 120603
Gallery Fusion, Seoul 111644
Gallery G, Seoul 111645
Gallery Gabor Limited, Gabor Paul
Mezei Studio, Toronto 101612
Gallery Gachet, Vancouver ... 007284
Gallery Gashi, Tiranë 098524
Gallery Gazelle, Oregon State
University, Philomath 052110
Gallery GEN, New York 095412
Gallery Genesis, Mandaluyong . 113398
Gallery Georges de Bartha,
Genève 116317
Gallery Gertrude Stein, New York 122643
Gallery Glass, Portland 123626
Gallery Gold, Francistown 100870
Gallery Gondwana Australia, Alice
Springs 098631
Gallery Guide, Athinai 138766
Gallery Guide, New York 139370
Gallery HAM, Nagoya 111316
Gallery Han, Seoul 111646
Gallery Henoch, New York 122644
Gallery Hiro, Tokyo 111433
Gallery Hiyoshi, Yokohama .. 111537,
111538
Gallery Holly Snapp, Venezia .. 111121,
111122
Gallery House, Louisville 121645
Gallery Ichiban-cho, Sendai ... 111382
Gallery ID 8, Baltimore 119738
Gallery II, Bradford 043694
Gallery II, Philadelphia 123384
Gallery II, Western Michigan University,
College of Fine Arts, Kalamazoo 049857
Gallery in Cork Street, London . 118338
Gallery in The Square, Hobart . . 099136
Gallery in the Square, Pittsburgh 123545
Gallery in Wax, Austin 119672
Gallery International, Hartford . 119739
Gallery International, Cape Town 114767
Gallery IV, Los Angeles 121491
Gallery J, Bowral 098767
Gallery K, Kingston 111222
Gallery K, Lefkosia 102537
Gallery K, London 118339
Gallery K, Washington 124909
Gallery Kaleidoscope, London . 118340,
134668
Gallery Kobe, Kobe 111280
Gallery Korea, New York 051392
Gallery Koza, Rhodos 109084
Gallery Lambton, Sarnia 006966
Gallery Lasés, Paris 104903
Gallery Lavaud, Akaroa 083743
Gallery Leela, Mumbai 109438
Gallery Limited, Milwaukee ... 121919
Gallery Link International, Sydney 099783
Gallery Lolani, Windward Community
College, Kaneohe 049868
Gallery Louisiane, New Orleans . 122173
Gallery M, Baden 116049
Gallery M, Borrby 115593
Gallery M, Denver 120724
Gallery M, New York 122645
Gallery M E K, Kingston 111223
Gallery M, Marion Cultural Centre,
Oaklands Park 001379
Gallery Market East, Philadelphia 123385
Gallery Maysa, Cairo 103040
Gallery Milano, Sendai 111383
Gallery Miro, Cairo 103041
Gallery Miro, Yokohama 111539
Gallery Modern, Changwon 111577
Gallery Moos Ltd., Toronto 101613
Gallery Motomachi, Kobe 111281
Gallery Muhammed, Aşgabat .. 088154
Gallery NAGA, Boston 119868
Gallery Nature and Temptation, Kaji
Aso Studio, Boston 047113
Gallery Nii, Osaka 111356

Gallery Nine, Amsterdam 112210
Gallery Nine, Cleveland 120423
Gallery Noor, Manama 100433
Gallery North, Kansas City 121290
Gallery North, Minneapolis 121991
Gallery North, Setauket 053273
Gallery Now, Seoul 111647
Gallery NTK, Croydon 098936
Gallery Oceanic, Dunedin 112879
Gallery of Antiques and Collectibles,
Jacksonville 093995
Gallery of Art, Dickinson State
University, Dickinson 048326
Gallery of Art, Fort Valley State
University, Fort Valley 048884
Gallery of Art, Johnson County
Community College, Overland
Park 051846
Gallery of Art, University of Northern
Iowa, Cedar Falls 047476
Gallery of Art, Washington university,
Saint Louis 052801
Gallery of Arts, London 101341
Gallery of British Columbia Ceramics,
Granville Island 005919
Gallery of British Columbia Ceramics,
Vancouver 101789
Gallery of Canadian Folk Art,
Calgary 064256, 101240
The Gallery of Contemporary
Art, Sacred Heart University,
Fairfield 048667
Gallery of Contemporary Art, University
of Colorado, Colorado Springs . 047885
Gallery of Costume, Manchester City
Galleries, Manchester 045202
Gallery of Early Canadian Flight,
Wiaton 007413
Gallery of Fine and Performing Arts,
Edison Community College, Fort
Myers 057097
Gallery of Fine Art, Greenmount 099096
Gallery of Florina Artists, Florina 023079
Gallery of Graphic Arts, New
York 122646
Gallery of Henry Street Settlement,
Abrons Arts Center, New York . 051393
Gallery of History, Las Vegas . . 094145
Gallery of Holography, Fitzroy
North 099035
Gallery of Hualien County,
Hualien 042338
Gallery of Legends, Las Vegas . 094146
Gallery of Modern Art, Glasgow 044398
Gallery of Modern Art, London . 118341
Gallery of Modern Art, South
Brisbane 001502
Gallery of Modern Greek Art of
Kalamata, Kalamata 023101
Gallery of Musical Instruments,
Sangeet Natak Akademi, Delhi 024223
Gallery of National Taiwan Normal
University, Taipei 042515
Gallery of Photography, Dublin . 109585
Gallery of Prehistoric Paintings, New
York 051394
Gallery of Sculpture Art and Antiquities,
Las Vegas 094147, 121354
Gallery of Southern Art am Marienbad,
Ullrich H. Flechshar, Freiburg im
Breisgau 107060
Gallery of Sports Art, Baltimore . 119740
Gallery of Sports Art, Milwaukee 121920
Gallery of Stars, Columbus .. 093020,
145098
Gallery of Surrealism, New York 122647
Gallery of the Department of Art and
Art History, University of Nebraska,
Lincoln 050290
Gallery of the Department of Art,
Eastern New Mexico University,
Portales 052270
Gallery of the Faculty of Architecture,
Bangkok 042658

Gallery of the Municipality of Patras,
Patras 023216
Gallery of the National Institute of Art
and Disabilities, Richmond . . . 052531
Gallery of the Republic, Austin . 119673
Gallery of the Round – Multimedia Art
Magazine, Pittsburgh 139371
Gallery of the Society for Epirot
Studies, Ioannina 023090
Gallery of the Southwest, San
Antonio 124031
Gallery of the State University College
at Fredonia, State University College
at Fredonia, Fredonia 048965
Gallery of Traditional Ornamental Art,
Minsk 003247
Gallery of Visual Arts, University of
Montana, Missoula 050958
Gallery of West Indian Art, Montego
Bay 111230
Gallery Off, Nagoya 111317
Gallery Old Town, San Diego ... 124125
Gallery Oldham, Oldham 045435
Gallery on 43 Street, Pittsburgh 123546
The Gallery on Broadway, Cardiff 117529
Gallery on Broadway, San Diego 124126
Gallery On Lock, Hamilton 101318
Gallery On the Bay, Hamilton . . 101319
Gallery on the Green, London . . 118342
Gallery on the Net, Harlow 117910
Gallery on the Net, Rotherham . 118989
Gallery on the Side,
Johannesburg 114794
Gallery on the Square,
Johannesburg 114795
Gallery One, Albuquerque 119476
Gallery One, Cahirciveen 109538
Gallery One, Ellensburg 048546
Gallery One, Grayshott 117877
Gallery One, Hong Kong 065045
Gallery One, Indianapolis 121187
Gallery One, Kilkenny 109649
Gallery One, Southport 119125
Gallery One Arts, Toronto 101614
Gallery One-Fillmore Plaza,
Denver 120725
Gallery One Nine, Swansea ... 099756
Gallery One One One, University of
Manitoba, Winnipeg 007444
Gallery One-Writers Square,
Denver 120726
Gallery Ophelia, Glendale 120914
Gallery Orbit, Saint Paul 123909
Gallery Pangolin, Chalford 117553
Gallery Park, New York 122648
Gallery Paule Anglim, San
Francisco 124288
Gallery Persia, Inverness 089642, 134499
Gallery Plaza, Fresno 120897
Gallery Plus, Los Angeles 121492
Gallery Q, Tokyo 111434
Gallery Quynh, Ho Chi Minh City 125119
Gallery Rama-Art, Moskva 114257
Gallery Renee, Corvara in Badia 109989
Gallery Revel, Dallas 120604
Gallery S.P., Seoul 111648
Gallery Saigon, Ho Chi Minh City 125120
Gallery Saint James, Sutton
Coldfield 119188
Gallery Savah, Paddington, New South
Wales 099519
Gallery Scene, San Diego . . 124127,
124128
Gallery Sculpturama, Bornem . 003381
Gallery Sekiguchi, Tokyo 111435
Gallery Senda, Hiroshima 111260
Gallery Sendai, Sendai 111384
Gallery Senior, Cairo 103042
Gallery Services, Birmingham . 119798
Gallery Seventeen, Beckenham . 117324
Gallery Seventy-Nine, Boston . 119869
Gallery Shop at London Museum,
London 101342
Gallery Six, Mona Vale 099363

Gallery Space, Hyderabad 109360
Gallery Spirit, Amsterdam 112211
Gallery Stratford, Stratford ... 007084
Gallery Sumukha, Bangalore ... 109258
Gallery TE, Amsterdam 112212
Gallery Teneleven, Armadale,
Victoria 098660
Gallery Thirty Nine Seventeen,
Houston 121045
Gallery Thirty Three, Dunedin . 112880
Gallery Tresco, Tresco 045996
Gallery Umeda, Osaka 111357
The Gallery Upstairs Torquil, Henley-in-
Arden 117955
Gallery Urbis Orbis, Saint Louis . 123826
Gallery V, Columbus 120495
Gallery Vetro, San Antonio 124032
Gallery W, Sacramento 123767
Gallery West, Alexandria 046390
Gallery West, Claremont 098893
Gallery West, Denver 120727
Gallery West, Tucson 124737
Gallery WM, Amsterdam 112213
Gallery X, Seattle 124530
Gallery XII, Wichita 054386
Gallery Yacou, London 090142
Gallery Yasmeen, Sidi Bou Saïd 116986
Gallery YUME, Mitaka 111308
Gallery Z, Providence 123705
Gallery Zen, Portland 096408
The Gallery, Mercer County
Community College, Trenton . 053796
Gallerytop, Rowsley 118990
Gallesi, Carlo, Bolzano 079897
Gallher, Granada 115089
Galli Vermigli Giandotti, Firenze 131177
Galli, Francesca, Parma 081116
Galli, Luigi, Carate Brianza ... 080009
Galliani, Genova 110110
Galliata, Alassio 109744
Gallier House, New Orleans 051292
Gallier's Stripwinkel, Sint-Truiden 140260
Galligan, Providence 096505
Galligano, Reynaldo, Buenos
Aires 127522
Gallimard, Paris 136931, 141123
Gallimaufry, Broadway 088675
Gallimaufry Gallery, Blairmore . 117386
Gallina, Adriano, Napoli 137703
Gallista, San Antonio 124033
Gallo, Århus 102773
Il Gallo, Pisa 081156
Gallo Antiquariato, Padova ... 080983
Gallo Pinto, Amsterdam 112214
Gallo-Romeins Museum,
Tongeren 004074
Gallo, Alberto, Venezia 132245
Gallo, Gianni, Roma 081386
Gallo, Mario, Milano 080560
Gallop & Rivers, Crickhowell ... 089008
Gallori Turchi Antichità, Firenze . 080148
Gallot Franck, Paray-Vieille-Poste 070851
Galloway Deer Museum,
Clatteringshaws 043964
Galloway Galleries & Fine Arts Centre,
Bowen Hills 061178, 098765
Galloway House and Village-Blakely
Museum, Fond du Lac 048772
Galloway Station Museum, Edson 005771
Galloway, Ursula, Soham 135054
Gallus, Innsbruck 139998
Gallus, Ljubljana 114717
Gallus, Nendeln 142958
Gallus, E., Düsseldorf 075473
Galpão de Arte, Fortaleza 100950
Galpón de Campo, Buenos Aires 060903
Galschiøt, Jens, Odense 102948
Galston Antiques, Galston 061502
Galtebosamlingen, Spydeberg . 033967
Galtung, Oslo 113223
Galtze Mühle, Seelbach 021816
Galuchat, Bruxelles 063445
Galvan, Marta, Modena 131571
Galvani, Fabrice, Toulouse 105743

Galvani, Laura, Brescia 079940
Galveston Arts Center, Galveston 049032
Galveston County Historical Museum,
Galveston 049033
Galvic, Maria, Bochum 075033
Galway Arts Centre, Galway .. 024925
Galy, Geneviève, Saint-Girons .. 072815
Galy, Janine, Pau 072038
D'Gama, La Eliena 137975
GamaArt, Kunstbedarf, Bühl .. 075212
Gamain, Christian, Nancy 070420
Gambara, Madrid 086189
Gambarini, Malemort-sur-Corrèze 069825
Gambarini, Vigeois 074121
Gambarini, Catherine, Brive-la-
Gaillarde 067509
Gambarini, Catherine, Ussac ... 073918
Gambelli, Giancarlo, Siena ... 081625
Gambia National Museum, Banjul 016384
Gambit, Praha 102693
Gamborg, Horsens 065793
Game and Fisheries Museum,
Entebbe 043101
Gameiro, Manuel, Lisboa 084893
Gamella, Madrid 086190, 126545,
143764
Gamella Rodriguez, Luis, Madrid 086191
Gamet, Romain, Paris 129013
Gamila Gallery, Cairo 103043
Gamilarart Gallery, Tamworth .. 099804
Gamine, Philadelphia 123386
Gaminek, Jörg, Bad Karlshafen . 106070
Gamisch, W., Göppingen 076031
Gamla Bankgården, Vrigstad ... 041179
Gamla Brukets Muséer, Munkfors 040867
Gamla Bryggeriet, Laholm 115716
Gamla Carolina, Göteborg 086707
Gamla Hantverkargården, Ystad 087168
Gamla Krukmakarverkstaden,
Arvika 040543
Gamla Län, Paddington, New South
Wales 062039
Gamla Linköping, Linköping ... 040786
Gamla Saker och Ting, Linköping 086813
Gamla Saker och Ting, Skövde . 086937
Gamla Skeninge, Skänninge ... 040948
Gamla Smedby Antik, Kalmar .. 086768
Gamla Tider, Staffanstorp 086953
Gamla Tiders Antik & Kuriosa,
Göteborg 086708
Gamla Ting i Järpen, Järpen ... 086757
Gamla Ting i Rönnäs, Leksand . 086802
Gamlaklockor.se, Stockholm ... 087000
Gamle Bergen Museum, Bergen 033531
Den Gamle By, Danmarks
Købstadmuseum, Århus 010006
Det Gamle Fiskerhus og Skibshallerne,
Gilleleje 010078
Den Gamle Gård, Faaborg 010053
Gamle Hvam Museum, Årnes .. 033505
Den Gamle Krambod, Trondheim 084386
Gamle Kvernes Bygdemuseum,
Averøy 033517
Den Gamle Stue, Fanø 065646
Det Gamle Teglværk, Helsinge .. 065737
Gamle Ting, København 065905
Gamle Trehus, Oslo 084312
Gamleby Antikvariat, Gamleby . 143838
Gamlebyen Brukt og Antik,
Jessheim 084258
Gamlehaugen, Paradis 033876
Gamlesmia, Oslo 084313
Gamlestua, Kongsberg 084261
Gamli lundur, Akureyri 109218
Gamli Prestagarðurinn og Bygdarsavnið,
Nes, Eysturoy 010736
Gammalt och Antikt, Filipstad . 086675
Gammel Estrup – Herregårdsmuseet,
Auning 010017
Gammel Havn, Fur 010076
Gammel Rye Antik, Ry 066104
Gammelgården Friluftsmuseet,
Bengtsfors 040560
Gammelgårdens Antikaffär Le Village,

Göteborg 086709
Gammelgruva, Orkla Industrimuseum,
Løkken Verk 033748
Gampelen, Claude Daniel,
Gampelen 116272
Gamuza, Wrocław 084768
Gamvik Museum, Gamvik 033614
Gan County Museum, Ganxian . 007786
Gan-Song Art Museum, Seoul .. 029996
Gana Art Gallery, Seoul 111649
Gana-Beaubourg, Paris 104904
Ganada, Broome 061221
Gananoque Historical Museum,
Gananoque 005856
Ganbara, Pamplona 086389
Gandahus, Vals 042099
Gander, Michael, Hitchin 089537
Gandhi Centenary Museum,
Karimnagar 024341
Gandhi Memorial Museum,
Madurai 024390
Gandhi Museum, Bhavnagar ... 024166
Gandhi Museum, Lucknow 024379
Gandhi National Memorial, Pune 024446
Gandhi Sangrahalaya, Patna ... 024432
Gandhi Smarak Sangrahalaya,
Ahmedabad 024124
Gandhi Smarak Sangrahalaya,
Barrackpore 024158
Gandhi Smarak Sangrahalaya,
Sevagram 024472
Gandini, Jacques, Nice 136841, 140997
Gandini, Luigi, Roma 081387
Gandit, Madeleine, Grenoble ... 068728
Gandy Gallery, Praha 102694
Ganech, Dortmund 075384
Ganesa Art Gallery, Da Nang .. 125050
Ganesh, Chennai 109284
Ganesh, Mumbai 109439
Ganesh, Torino 081704
Ganesh, Troyes 073904
Ganesha, Mersch 082381
Ganesha, Rosengarten, Kreis
Harburg 108514
Ganet, Gérard, Paris 141124
Gang, Jeanne, Chicago 120135
Ganga Government Museum,
Bikaner 024179
Ganghofer-Museum, Leutasch .. 002302
Gangidori Bijutsukan, Joetsu ... 028733
Ganglberger, Gebhard, Umhausen 127888
Gangloff, Mulhouse 140973
Gangloff, Strasbourg 141350
Gangneung Municipal Museum,
Gangneung 029927
Gangol, Oberwart 062758
Gangwon Folk Museum, Cheongil 029905
Ganiro, Seoul 111650
Gannan Storezu Zone Museum,
Xiahe 008391
Gannon House, Sydney 062287
Gannuso, Ruggiero, Montevideo 092001
Ganondagan State Historic Site,
Victor 053979
Gans, Udo, Arnsberg 074425
Ganseforth, Ludwig, Düsseldorf . 141718
Gansel, Pfungstadt 078047
Gansevoort Gallery, New York . 095413
Ganske, Hamburg 137245
Ganske, M., Oldenburg 077929
Gansow, Gido, Buseck 075234
Ganss House, Landsberg am
Lech 077009
Gansu Museum, Lanzhou 007996
Gantner, Bonn 075071
Gantner, Benno Michael,
Starnberg 130652
Gantois, Cannes 103511
Ganyu Museum, Ganyu 007787
Ganz, Kate, New York 122649
Ganzhou Museum, Ganzhou ... 007788
Ganzhou Zone Museum,
Ganzhou 007789
Ganzi Storezu Museum,

Kangding 007988
Gao'an Museum, Gao'an 007790
Gaomi Museum, Gaomi 007791
Gaoming Museum, Gaoming ... 007792
Gaona, Cesar C., Buenos Aires . 060904,
125169
Gaowish Museum, Zhaoqing .. 008532
Gaozhou Museum, Gaozhou ... 007793
Gapp, Nathalie, Charleville-
Mézières 067847
Gar, Grafing bei München 076076
Garabelli, Giorgio, Torino 132142
The Garage, Genève 116318
Le Garage, Halle, Saale 107206
Le Garage Antiques, New
Orleans 095103
Garage Kunsthandel, Hamburg . 107270
Garagem de Arte, Porto Alegre . 100983
Garai, Luca, Bologna 079847
Garakuta-Do, Honolulu 093544
Garam, Seoul 111651
Garance, Monpazier 070168
Garance, Paris 071418
Garance, Saint-Germain-en-Laye 105519
Garance, Sainte-Foy-la-Grande . 105637
Garasu, Mudgeeraba 099411
Garattini Malzani, Brescia 131033
Garau, Raimondo, Milano 080561
Garaudel, Guy, Kamen 076667
Garay, Tarbes 073625, 129273
Garbarino, Genova 110111
Garbary, Wrocław 084769
Garbary 48, Poznań 113738
Garbato, Federica, Cagliari 079983
Garbato, Giovanni, Cagliari 079984
Garbett, Michael, Bristol 144280
Garbini, Paolo, Lugano 087617
Garces Carrascon, Lucia,
Zaragoza 086596
Garces, Alonso, Bogotá 102335
Garces, Maria Isabel, Bogotá .. 102336
Garcia, Grasse 068703
Garcia, Mechelen 140251
García, Palencia 115332
Garcia, Zamora 086582
Garcia Cabarcos, Carlos, Donostia-San
Sebastián 133246
Garcia Cervera, Vicente, Valencia 115502
Garcia de Vinuesa Catalan, Pilar,
Madrid 133337
Garcia Ferrero, Miguel, Palma de
Mallorca 086352
Garcia Gil, Jesus, Valladolid .. 086546
García Hijos, Barcelona 133118
García Monge, José Joaquín, San
José 102506
Garcia-Montes, Louis,
Lafrançaise 069137
Garcia Naharro, Jose Luis,
Madrid 086192, 133338
Garcia Reyes, Antonio, Las Palmas de
Gran Canaria 115377
Garcia Sayes, Fernando, Madrid 086193
Garcia Toldra, Carlos, Terrassa . 086487
Garcia Uriburu, Sara, Buenos
Aires 098570
Garcia Vargas, Agustin, Acapulco 132356
Garcia Zalvez, Diego, Girona .. 086027
Garcia, Annick, Raincourt 072358
Garcia, Cristobal, Zaragoza ... 086597
Garcia, Emile, Issoire 068881
Garcia, Gérard, Le Pin-au-Haras 069330
García, Hijos de Albaladejo García,
Torrevieja 126558
Garcia, Jean, Marseille 069937
Garcia, Jean-Pierre, Portel-des-
Corbières 072286
Garcia, José, Saint-Genis-Laval 072795
Garcia, Laurence, Lancon-
Provence 069160
García, Leon Jaime, Bogotá ... 102337
Garcia, Marie-Rose & Victorio,
Castres 067729
Garcia, Maud, Paris 071419

Garcia, Sande, Miami 121791
Garcia, Victoria, Valencia 086518
Garcin, Michel, Gap 103761
Garda Museum, Dublin 024882
Gardani, Ben, Rotsterhaule 143215
La Garde, Bordeaux 067333
Garde Rail, Seattle 124531
Gardella, Mauro, Genova 131314
Gardella, Ombretta, Genova .. 080262,
131315
Gardelle, Marcel, Pont-du-
Château 072251
Garden Art, Hungerford 089606
The Garden Art Gallery,
Carrickfergus 117537
Garden Blues, Velleron 073999, 129332
Garden County Museum,
Oshkosh 051825
Garden Court Antiques, San
Francisco 097238
Garden Critters, Albuquerque .. 119477
Garden District Book Shop, New
Orleans 145310
Garden Gallery, Apeldoorn 112323
The Garden Gallery, Southwold . 119130
Garden Gallery, Virginia Beach . 097782
Garden Gate Antiques, Oklahoma
City 095972
Garden Gate Antiques Market of
Chestnut Hill, Philadelphia .. 096108
Garden-Jardin, Paris 071420
Garden of Earthly Delights,
Philadelphia 096109
Garden of the Gulf Museum,
Montague 006354
Garden of the Zodiac, Omaha . 123314
Garden Park Antiques, Nashville 095000
Garden Park Farm Museum, Alberta
Beach 005309
Garden Spot, San Francisco ... 097239
Garden Suburb Gallery, London . 118343
Gardenghi, Stefano, Bologna .. 142631
Gardenia Galleri, Bærums Verk . 113121
Gardenstone Antiques, San Diego 097090
Gardi, Carlo Antonio, Milano .. 131450
Le Gardian du Passé, Aubagne . 066773,
128449
Gardin, Danielle, La Sentinelle . 069100
Gardiner Art Gallery, Oklahoma State
University, Stillwater 053531
Gardiner Houlgate, Corsham .. 126870
Gardiner Museum of Ceramic Art,
Toronto 007176
Gardiner, Charles, Lurgan 090654
Gardiner, John, Somerton 091360
Gardner, Kilbarchan 089682
Gardner, Tulsa 145627
Gardner & Barr, New York 122650
Gardner House Museum, Albion 046357
Gardner Museum, Gardner ... 049050
The Gardner Museum of Architecture
and Design, Quincy 052430
Gardner, Martin, Montréal 101412
Gardner, Richard & Janice,
Petworth 090998
Gárdonyi Géza Emlékmúzeum, Dobó
István Vármuzeum, Eger 023617
Gårdshuset i Stråkanäs, Kalix . 086764
La Gare de Rivière Blanche, Saint-
Ulric-de-Rivière-Blanche 006939
De Garenkoker, Haarlem 083328
Garets, Philippe des, Marolles-en-
Hurepoix 069889
Garfield, Pittsburgh 123547
Garfield Farm Museum, Lafox . 050064
Garforth Gallery, Market
Weighton 090715, 134839
Garganigo, Alice, Zürich 087969
Gargiulo, Giuseppe, Napoli ... 080882
Gargiulo, Marina, Roma 081388
Gargiulo, Michele, Napoli 080883,
110528
Gargiulo, Vincenzo, Merano ... 110212
Gargola, A Coruña 133233

Gargola, Madrid 086194
Gargrave Gallery, Gargrave 089321
Garhi Studio, Delhi 109320
Garibaldi-Meucci Museum, Staten
Island 053497
Gariban, İstanbul 088119
Garibay, San Diego 136295
De Garilhe, Lyon 128714
Garisenda, Bologna 142632
Garland, Merbein 061843
Garland, Paris 071421, 129014
Garlato, Elisabetta, Venezia 081839
Garlich, Kurt, Dreieich 075404
Garlin, S., Glun 103769
Garlogie Mill Power House Museum,
Garlogie 044379
Garmilla, Jean-Michel, La
Rochelle 069086
Garmonija, Sankt-Peterburg . . . 085242
Garmonija Kontrastov, Moskva . 114258
Garnavillo Historical Museum,
Garnavillo 049052
Garner Memorial Museum,
Uvalde 053918
Garner, John, Uppingham 091691,
135171
Garnet, Reading 138279
Garnier, A., Brugge 063276
Garnier, Alain & Brigitte, Damme 063609
Garnier, Guy, Valros 073956
Garnier, Jean Eric, Eze 103729
Garnier, Jean-Michel, Chartres . 148024
Garnier, Maurice, Paris 104905
Garnier & Cie., Maurice, Paris . . 104906
Garnier, Serge, Is-sur-Tille 068875
Garnier, Yves, Toulouse 073772, 129301
Garnisonmuseum Nürnberg,
Nürnberg 020851
Garnisonsmuseet, Boden 040568
Garnisonsmuseet i Skövde,
Skövde 040956
Garnisonsmuseum Graz-
Kanonenbastei, Stadtmuseum Graz,
Graz 001972
Garnisonsmuseum, Förderverein
Garnisonsmuseum, Zossen . . . 022895
Garo, Montréal 101413
Garo, Gaby & Christian, Knokke-
Heist 100738
Gåröströms Industrimuseum, Gnosjö
Industrimuseum, Gnosjö 040644
Garon, Paul, Chicago 145044
Garosi & Mazzoni, Firenze 131178
Garow, Ban, Osaka 111358
Garrard, Ludlow 090648
Garraud, Gilles, Carcassonne . . . 067673
Garrec, Yves le, Inzinzac-Lochrist 068872
The Garrett, Albury 098625
Garrett County Historical Museum,
Oakland 051687
Garrett Historical Museum,
Garrett 049055
Garrett, G. & N., Doncaster East 098972
Garric, Michel, Montpellier 070301
Garrido Martinez, Sergi,
Barcelona 133119
Garrignot, Jérôme, Vichy 074108
Garrigue, Denise, Balaruc-les-
Bains 103323
Garrigues, Jean, Forcalquier 068586
Garrigues, Robert, Paris 071422
Garrison Gallery Museum, Millers
Point 001295
Garritano, Filomena, Bari 142621
Garrone, Annie, Pegomas 105331
Garry, Saint-Ouen 073091
Garst Museum, Greenville 049265
Garstang Fine Art, Garstang 117816
The Garstang Museum of Archaeology,
University of Liverpool, School of
Archaeology, Classics and Egyptology,
Liverpool 044853
Garstedt, Norderstedt 126082
De Garstkamp, Overasselt 032759

Gartell, Cincinnati 092858
Gartenatelier-Galerie-Berlin,
Berlin 017139
Gartenkunst-Museum Schloss
Fantasie, Bayerische Verwaltung der
staatlichen Schlösser, Gärten und
Seen, Eckersdorf 018091
Garter Lane Arts Centre,
Waterford 024997
Garth, Delaware 127201
Garth, Harrogate 089428
Gartland Brendan, Castleblaney . 079475
Garuti, Ennio, Modena 080795
Garvens, Aerzen 074356
Garvey, Waikato 084128
Gary & Wendy, Long Beach 094202
Gary D.'s, Louisville 094551
Gary-Roche, Paris 071423, 104907
Gary Tatintsian Gallery, Moskva 114259
Gary, David, Millburn 127321
Gary's Antiques, Dallas 093129
Gary's Country Collectables,
Mildura 061852
Garyudo, Yokohama 082226
Garza, Frank, Denver 120728
Garzanti, Milano 137678
Garzon, Alvaro, Bogotá 065206, 140449
Gas Light Antique Shoppe, San
Antonio 096973
Gas Pump Heaven, Omaha 096039
Gasan Gallery, Seoul 111652
Gasca Ajuria, Julia, Donostia-San
Sebastián 133247
Gascoigne Gallery, Harrogate . . . 044516
Gascon, Rosa, Zaragoza 133499
Gascoyne Junction Museum,
Gascoyne Junction 001089
Gåsebäck, Helsingborg 086739
Gasendo, Marui, Osaka 082081
Gaseum, Essen 018295
Gąsior, Jerzy, Wrocław 132914
Gaslaternen-Freilichtmuseum Berlin,
Berlin 017140
Gaslight & Shadows, San
Francisco 097240
Gaslight Antiques, Tampa 097559
Gaslight Books, Fyshwick 139745
Gaslight Collectables and Old Books,
Burra 061238, 139672
Gaso Venus, Nagakute 082053
Gasoline Alley Antiques, Seattle . 097447
Gaspar, Donna J., Baltimore . . . 119741
Gaspar, Joan, Barcelona 114913
Gaspar, Joan, Madrid 115224
Gaspard, Marseille 128768
Gasparetto, Jean-Mario, Rolle . . 087752
Gasparini, Valentina, Venezia . . . 111123
Gasparrini, Claudio, Roma 081389,
110797
Gasparrini, Emanuele, Firenze . . 131179
Gasper, Hans, Niederzier 108278
Gaspesian British Heritage Village,
New Richmond 006469
Gass Forest Museum,
Coimbatore 024204
Gassama, New York 095414
Gasseling, M.C., Amsterdam . . . 082756
Gasser, Bolzano 079898
Gasser, Haguenau 068776
Gasser & Grunert, New York . . . 122651
Gasser Parc, Bolzano 109884
Gasser, Martin, Solothurn 133927
Gassner, Franz, Vaduz 082338
Gasson, Tadley 091538
Gasson, Herbert Gordon, Rye . . 091162,
134985
Gastaldelli, Giuseppe, Milano . . . 110353
Gastaud, Clermont-Ferrand 103604
Gaster, Anna L. & Peter, Nyborg 066009
Gasthuismuseum Sint-Dimpna,
Geel 003610
Gastl, Tübingen 142427
Gaston County Museum of Art and
History, Dallas 048107

Gastou, Yves, Paris 071424
Gastown Antique Emporium,
Vancouver 064824
Gastronomiaren Euskal Museoa,
Museo Vasco de Gastronomía,
Laudio 039548
Gasu no Kagakukan, Takaishi . . 029430
Gasu no Kagakukan, Tokyo 029508
Gasworks Arts, Albert Park 098622,
098623
Gatčinskij Filial Sankt-Peterburgskogo
Gosudarstvennogo Universiteta
Technologii i Dizajna, Gatčina . 056272
Gatčinskij Literaturno-memorialnyj
Muzej-usadba P.E. Ščerbova,
Gatčina 037012
The Gate Gallery, Alnwick 117227
The Gate Gallery & Framing Studio,
Skibbereen 109691
Gate P, Berlin 017141
Gateau, Annick, Artigues-
Foulayronnes 066749
Gatehouse Antiques, Macclesfield 090672
The Gatehouse Gallery, Glasgow 117841
Gatekeeper's Museum and Marion
Steinbach Indian Basket Museum,
Tahoe City 053653
Gatell, Marta, Bilbao 133190
Gately House, East London 038710
Gates House, Machiasport 050555
Gates Mills Historical Society Museum,
Gates Mills 049060
Gates, B.R., Fort Worth 120865
Gateway Antiques, Burford 088705
Gateway Antiques, Dublin 079557
Gateway Antiques and Art,
Denver 093291
Gateway Antiques and Auctions,
Lebanon 127281
The Gateway Gallery, Dunedin . . 112881
Gateway Pastel Artists, Saint
Louis 060492
Gateway to Science, Frances Leach
High Prairie Arts and Science
Complex, Bismarck 047021
Gateway to the Panhandle, Gate 049059
Gatewest, Winnipeg 064924
The Gathering, Cleveland 092940
Gathering, Dallas 093130
A Gathering of the Tribes, New
York 139372
Gatherum Books, Uralla 139946
Gathman, Thomas, Chicago 120136
Gathy, Hasselt 063680
Gathy, J., Liège 063758
Gatineau Valley Historical Society,
Chelsea 005607
Gatipon-Bachette, Christian,
Terraube 073637
El Gatito, Chiba 111243
Gatlyktans Antik, Lund 086833
El Gato Rojo, Bogotá 102338
Gaton Sanchez, Maria Caridad,
Valladolid 086547
Gatov Gallery, Alpert Jewish
Community Center, Long Beach 050370
Gatsch, Steffen, Dresden 129709
Gatti, San Francisco 136319
Gatti, Leonardo, Brescia 131034
Gattis, Nashville 095001
Gatzaren Ekomuseoa, Ecomuseo de la
Sal, Leintz-Gatzaga 039552
Gau, Salzgitter 078342
Gauchet, Gérard, Montbrison . . . 070218
Gauchet, Loïc, Chalain-d'Uzore . 067763
Gauchet, Louis, Nice 070610
Gaud & Pirman, Villiers 074234
Gaud, Emile, Saint-Tropez 105627
Gaudard, Reth, Villeurbanne . . . 074208
Gaudet-Blavignac, Marc, Genève 116319
Gaudete, Museum voor Volksdevotie,
Boxtel 032093
Gaudi, Madrid 115225

Gaudi Galleri, Hjørring 102839
Gaudin, Le Havre 103946
Gaudin, Madeleine, Marans 069844
Gaudnek-Museum, Altomünster . 016561
Gaudron, Jean, Champagnole . . . 103565
Gaudry, Jacques, Vallan 073950
Gauermann-Museum,
Miesenbach 002387
Gauguin, Århus 125431
Gauguin, Gentofte 125436
Gauguin, København 125450
Gauguin, Kolding 125456
Gauguin, Rio de Janeiro 101034
Gauguin, Sønderborg 125468
Gaujoux, Carla, Blois . . 067254, 103401
Gauld, Marion H., Killin 089686
Gaulin & Oestmann, Lübeck . . . 142086
Gaultier, Patrick, Quimper 105416,
141309
Gaumann, Hume 061641
Gaus & Hongler, Sankt Gallen . . 087778,
133905
Gausepohl, Detmold 106675
Gauthier, Patrick, Saint-Denis-de-
Mailloc 072730, 125798
Gautier, Granville 103779
Gautier, Grenoble 068729
Gautier, Christian, Caudebec-lès-
Elbeuf 103545
Gautier, Goxe & Belaisch, Enghien-les-
Bains 125579
Gautier, Jacques, Saint-Félix . . . 072783
Gautier, Laurent, Nice 070611
Gautier, Richard, Niort . 104405, 128873
Gautier, Sylvain, Dijon 125569
Gautron, Kostia, Saumur 073437
Gauville, Thierry, Ribaute-les-
Tavernes 141315
Gauvin, Serge, Montréal 064424
Gauzy, J.L., Agen 140758
Gavage Longrée, Liège 063759
Gavala Aboriginal Art & Culture,
Darling Harbour 061386
Gavallet, Nicole, Paris 129015
Gavar, Paris 104908
Gavar Local Tradition Museum,
Gavar 000699
Gavardie, Ivana de, Paris 104909
Gavcom, Braşov 085074
Gavello, Franco, Moncalieri 080828
Gavilan Community College Art Gallery,
Gilroy 049101
Gavin, Alain, Marnac 069882
Gavlhuset, Aalborg . . . 065581, 102766
Gavriilidis, Efthym., Thessaloniki 079210
Gawjus Fings, Hawkes Bay 083952
Gawler, Willaston 139960
Gawler Antiques, Gawler 061504
Gawler Museum, Gawler 001091
Gawlik, A., Halle, Saale 076165
Gawroński i Syn, Gdynia 084517
Gawsworth Hall, Macclesfield . . 045183
Gawthorpe Hall, Rachel Kay
Shuttleworth Textile Collections,
Padiham 045461
Gaxotte, Laurent, Porrentruy . . . 138082
Gay Nineties Button and Doll Museum,
Eureka Springs 048631
Gay, Jean-Philippe, Couchey . . . 068172
Gaya Museum, Gaya 024260
Gayan-Sourgen, Jean-Louis & Henri,
Dax 068279
Gayaud, Saint-Michel-de-Veisse . 072979
Gayck, A., Berlin 074798
Gaye, Houston 093672
Gayer Anderson Museum, Cairo . 015040
Gayer, Paul, Baccarat 066900
Gayet, Martine, Dinard 103698
Gaylord Antiques, Denver 093292
Gaylord & Co., Charles, San
Francisco 097241
Gaylords, Titchfield 091603
Gayma, Bilbao 085871, 133191
Gayndah Museum, Gayndah . . . 001092

Gayo Museum, Takengon 024662
Gayot, Madeleine, Verdun
(Meuse) 074020
Gayraud, Dominique, Béziers .. 067176
Gayraud, Jacques, Béziers 067177
Gaytal-Park, Körperich 019744
Gazado, Daniel, La Coucourde . 069011
Gazado, Daniel, Saint-Paul-Trois-
Châteaux 073247
Gazai, Nagoya 082058
Gazdaház, Cserépfalu 023584
Gaze & Son, Diss 126875
Gazebo, Fujisawa 111246
Gazebo Antiques, Los Angeles . 094342
Gazelle Fine Art, Detroit 120807
Gazendam, J.U.R., Groningen ... 083288
Gazendam, Jur, Tynaarlo 143238
Gazeta Antykwaryczna, Kraków . 138966
Gazeta Malarzy i Poetów, Poznań 138967
La Gazette de l'Hôtel Drouot,
Paris 138644
Gazette des Beaux-Arts, Paris .. 138645
Gazette Gallery, Goomalling 099089
Gazi Müzesi, Samsun 043048
Gaziantep Arkeoloji Müzesi,
Gaziantep 042903
Gaziantep Devlet Güzel Sanatlar
Galerisi, Gaziantep 042904
Gazirah Museum, Cairo 010541
Al-Gazirah, Gowhara, Riyadh ... 085317
Gázmúzeum, Budapest 023482
Gazonrouge, Athinai 109035
GB Design, Houston 093673
GBV, Avignon 103313
GCR Art, Miami 121792
Gdańska Galeria Fotografii, Muzeum
Narodowe w Gdańsku, Gdańsk 034726
Gdańska Galeria Rzeźby, Gdańsk 113473
Ge Li Dai, Hongguang, Beijing . 101953
Ge, Changle, Wuxi 102252
Ge, Lianhai, Guangzhou 102027
Ge, Qingliang, Beijing 101952
Ge, Tang-Ying, Taipei 088068
Geamia Esman Sultan, Mangalia 036557
Geary Antiques, Leeds 089768, 134534
Geazy Vase, Cairo 066247
Geba, Harald, Graz 100007
Gebauer, Berlin 129494
Gebauer, Wolfram, Dresden 075412
Gebecke, Quedlinburg 142309
Gebert, Ulf, Nürnberg 077821
Gebhardt, Margit, Zwickau 130805
Gebhardt, Wolfgang, Chemnitz . 141673
Gebirgsjägermuseum,
Traditionssammlung der
Gebirgstruppe Sonthofen,
Sonthofen 021919
Gebken, H., Dötlingen 106687
Geble, Udo, Radolfzell am
Bodensee 078118, 126101
Gebrüder-Lachner-Museum, Rain 021265
Geburtshaus des heiligen Konrad von
Parzham, Bad Griesbach 016776
Geburtshaus Levi Strauss, Museum
Jeans und Kult, Buttenheim .. 017706
Gecko Gallery, Broome 098808
Den Gecroonden Boschboom, 's-
Hertogenbosch 083389
Gedächtnisstätte Mida Huber,
Landsee 002262
Gedächtnisstätte von Scheidt-Saalfeld,
Bad Bergzabern 016720
Geddes Historic District Village,
Geddes 049063
Geddes, Graham, Armadale,
Victoria 061027
Gede Museum and National
Monument, Watamu 029871
Gedenk- u. Dokumentationsstätte KZ
Drütte, Salzgitter 021574
Gedenk- u. Dokumentationsstätte
Opfer politischer Gewaltherrschaft,
Frankfurt (Oder) 018397
Gedenkräume Schloss Hartheim, Lern-

und Gedenkort Schloss Hartheim,
Alkoven 001695
Gedenkraum 1945,
Hochwolkersdorf 002092
Gedenkstätte "Erinnern für die
Zukunft", Leipzig 019989
Gedenkstätte "Roter Ochse", Stiftung
Gedenkstätten Sachsen-Anhalt, Halle,
Saale 018958
Gedenkstätte Abraham a Sancta Clara,
Leibertingen 019957
Gedenkstätte Bergen-Belsen,
Lohheide 020105
Gedenkstätte Berlin-
Hohenschönhausen, Berlin ... 017142
Gedenkstätte Berliner Mauer,
Berlin 017143
Gedenkstätte Breitenau, Archiv und
Ausstellung der Universität Kassel,
Guxhagen 018902
Gedenkstätte Buchenwald,
Weimar 022492
Gedenkstätte der Luftfahrt,
Gollenberg 018762
Gedenkstätte Deutsche Teilung
Marienborn, Marienborn 020281
Gedenkstätte Deutscher Widerstand,
Berlin 017144
Gedenkstätte Ehemalige
Synagoge, Dorfmuseum Sennfeld,
Adelsheim 016466
Gedenkstätte ehemaliges SS-
Sonderlager / KZ Hinzert, Hinzert-
Pölert 019263
Gedenkstätte für die Opfer der
NS-"Euthanasie" Bernburg,
Bernburg 017332
Gedenkstätte für die Opfer politischer
Gewaltherrschaft 1945–1989,
Magdeburg 020205
Gedenkstätte Goldener Löwe,
Eisenach 018132
Gedenkstätte Großbeeren 1813,
Großbeeren 018838
Gedenkstätte Hadamar, Hadamar 018906
Gedenkstätte in der JVA Wolfenbüttel,
Stiftung Niedersächsische
Gedenkstätten, Wolfenbüttel .. 022731
Gedenkstätte Köpenicker Blutwoche
Juni 1933, Berlin 017145
Gedenkstätte Langenstein-Zwieberge,
Halberstadt 018937
Gedenkstätte Münchner Platz Dresden,
Dresden 017983
Gedenkstätte Museum in der "Runden
Ecke", Stasi- Macht und Banalität,
Leipzig 019990
Gedenkstätte Plötzensee für Opfer
des Nationalsozialismus, Berlin 017146
Gedenkstätte Synagoge, Dornum 017945
Gedenkstätte und Museum
Sachsenhausen, Stiftung
Brandenburgische Gedenkstätten,
Oranienburg 021000
Gedenkstätte und Museum Trutzhain,
Schwalmstadt 021760
Gedenkstätte zum Konzentrationslager
im Schloss Lichtenburg,
Gedenkstätten-Stiftung des Landes
Sachsen-Anhalt, Prettin 021216
Gedenkstätte/Museum Seelower
Höhen, Seelow 021818
Gederta Eliasa Jegavas Vēstures un
Mākslas Muzejs, Jelgava 030131
Gedexnus, Dietmar, Amsterdam 132400
Gedimino Pilies Bokštas, Lietuvos
Nacionalinis Muziejus, Vilnius . 030514
Het Gedistilleerd Museum,
Schiedam 032851
GEDOK Gemeinschaft der
Künstlerinnen und Kunstförderer e.V.,
Gruppe Sachsen, Leipzig 058888
GEDOK Gemeinschaft der
Künstlerinnen und Kunstfreunde,

Gruppe Hannover, Hannover .. 058889
GEDOK Hannover e.V., Hannover 107366
GEDOK Künstlerinnenforum,
Karlsruhe 019507
GEDOK Verband der Gemeinschaften
der Künstlerinnen und Kunstförderer
e.V., Bonn 058890
Gedser, Nykøbing Falster 102937
Geedon Gallery, Fingringhoe .. 117797
Geelen, Amsterdam 082757
Geelong Antiques, Geelong 061507
Geelong Art Society, Geelong . 058201,
099069
Geelong Gallery, Geelong 001093
Geelong Heritage Centre,
Geelong 001094
Geelong Naval and Maritime Museum,
North Geelong 001363
Geelong Street Antique Centre,
Fyshwick 061498
Geels & Co. Koffie- en Theemuseum,
Amsterdam 031941
Geels Restauratieatelier, Geel . 128045
Geerts, Henricus, Alet-les-Bains 066534
Geertsen, Tom, København 065906
Geervliet, Bas, Breda 132454
Geesmann, Franz, Himberg 062632,
127767
Geeson Brothers Motor Cycle Museum,
South Witham 045821
Geevor Tin Mining Museum,
Pendeen 045474
Geffrye Museum, London 044975
Gefle Antikvariat, Gävle 143837
Gefüge und Gefache,
Crimmitschau 129662
El Gegant de la Ciutat, Barcelona 114914
Gehain, M., Bruxelles 063446
Gehan, Cairo 066248
Géhan, Hubert, Paimpol 070841
Gehlert, Siegfried, Plauen 078056
Gehlsen, J. U., Hannover 141913
Gehrt, Karin, Berlin 074799
Geib, Renate, Ockenheim 142259
Geidai Bijutsukan, Tokyo Geijutsu
Daigaku, Tokyo 029509
Geier, Herbert, Bad Staffelstein . 125857
Geier, Josef, Sankt Gallen 133906
Geigenbau-Museum, Bubenreuth 017650
Geigenbaumuseum, Mittenwald . 020423
Geiger, Dietramszell 075357
Geiger, Konstanz 107715
Geiger, Ravensburg 078147
Geiger, Leopold, Ebenfurth ... 062551
Geiger, Ludwig, Riehen 116588
Geiger, M., Miami 094736
Geijutsu Shinbun Sha, Tokyo .. 137778
Geijutsu Shincho, Tokyo 138889
Geilenkirchen, P., Stolberg,
Rheinland 078566
Geinitz, Jürgen, Erfurt 129802
Geipel, Katharina, Berlin 129495
Geis, Helmut, Wilnsdorf 108944
Geisel, Jürgen, Würzburg 079081,
130783
Geiseler, Peter, Berlin 129496
Geiseltalmuseum, Martin-Luther-
Universität Halle-Wittenberg,
Inst.für Geowissenschaften, Halle,
Saale 018959
Geisenheyner, Winfried, Münster 142201
Geishinsha, Tokyo 111436
Geisler, Joachim, Xanten 079128
Geisler, Michael, Berlin 129497
Geisler, Ralph, Grünhainichen . 076116
Geismar, Dany, Colmar 114921
Geissbühler, Peter, Obernau .. 087719,
133879
Geissendörfer, Ernst, Rothenburg ob
der Tauber 078285
Geiter, Peter, Rottendorf 126107
Geitner, Johann, Velburg 130719
Geja, Maastricht 112574
Gekkeikan Okura Kinenkan,

Kyoto 028927
Gekko Gallery, Utrecht 083646
Gekoski, R.A., London 144607
Gelardini, M.&C., Roma 110798
Gelateria per Elisa, Gelnhausen . 107107
Gelati, Maurizio, Colorno 080054
Gelbard, Roland, Paris 071425
Das Gelbe Haus, Museum Flims, Flims
Dorf 041487
Gelbert, Gundel, Köln 142008
Gelbke, L., Delitzsch 075330
Het Geld-en Bankmuseum,
Utrecht 032946
Geldart Gallery, Henbury 117952
Gelders Geologisch Museum,
Velp 032980
Gelders Schuttersmuseum,
Didam 032220
Gelderse Smalspoor Museum,
Erlecom 032318
Geldgeschichtliches Museum,
Köln 019684
Geldmuseum der Deutschen
Bundesbank, Frankfurt am Main 018422
Geldmuseum der Oesterreichischen
Nationalbank, Wien 002994
De Gelegenheid, Den Haag 112386
Gelendžikskij Istoriko-kraevedčeskij
Muzej, Gelendžik 037016
Gelereja Iskusstva Uzbekistana,
Toshkent 054653
Geležinkelio Muziejus, Šiauliai . 030466
Geležinkelių Muziejus, Vilnius . 030515
Gelink, Annet, Amsterdam 112215
Gell, Markus, Rankweil 100146, 136542
Gellé-Lacroix, Laurent, Marseille 069938
Gellert-Museum Hainichen,
Hainichen 018933
Gelli, Giuliana & Marco, Firenze 080149
Gellmann, Leo, Mörfelden-
Walldorf 077402
Gellner, Bob, Omaha 096040
Gelmar, S., Lauterstein 077035
Geloni, Joseph, Antibes-Juan-les-
Pins 066672
Gelos-Naslelie, Moskva 085180, 114260,
126474
Gelot, Philippe, Paris 071426
Gels, Lingen 077130
Gelsenkirchen im Nationalsozialismus,
Gelsenkirchen 018625
Gély, Madeleine, Paris 071427
Gély, Monique, Perpignan 072082
Gem Antiques, Maidstone 090677
Gem Antiques, New York 095415
Gem Antiques, Sevenoaks 091269,
135017
Gem Furniture and Antiques,
Providence 096506
Gema Art, Praha 102695, 125421,
128345, 136763
Gemälde- und Skulpturensammlung,
Nürnberg 020852
Gemälde und Antik, Düsseldorf . 075434
Gemäldegalerie Alte Meister,
Museumslandschaft Hessen Kassel,
Kassel 019532
Gemäldegalerie Alte Meister,
Staatliche Kunstsammlungen
Dresden, Dresden 017984
Gemäldegalerie Dachau,
Zweckverband Dachauer Galerien
und Museen, Dachau 017796
Gemäldegalerie der Akademie der
Bildenden Künste, Wien 002995
Gemäldegalerie, Staatliche Museen
zu Berlin – Stiftung Preußischer
Kulturbesitz, Berlin 017147
Das Gemäldehaus, München ... 077580
Gemälderestaurierung im
Kunsthistorischen Museum,
Wien 127920
Gembala-Dauksza, Maria,
Düsseldorf 129742

Gemeenschap Beeldende Kunstenaars,
Arnhem 059603
Gemeentegrot Valkenburg, Valkenburg,
Limburg 032962
Gemeentelijk Archaeologisch Museum,
Grobbendonk 003660
Gemeentelijk Archeologisch Museum,
Aardenburg 031885
Gemeentelijk Heemkundig Museum,
Hoeilaart 003704
Gemeentelijk Historisch Museum,
Rijkevorsel 003966
Gemeentelijk Historisch Museum
Ouder-Amstel, Ouderkerk aan de
Amstel 032752
Gemeentelijk Museum, Melle . 003863
Gemeentelijk Museum Gevaert-Minne,
Sint-Martens-Latem 004017
Gemeentelijk Museum 't Oude Slot,
Veldhoven 032979
Gemeentelijk Museum van Evere,
Bruxelles 003444
Gemeentelijk Museum voor
Geschiedenis en Heemkunde,
Destelbergen 003541
Gemeentelijk Streekmuseum De Veste,
Sint-Stevens-Woluwe 004024
Gemeentelijke Academie voor
Beeldende Kunsten, Heusden-
Zolder 055028
Gemeentelijke Academie
voor Beeldende Kunsten,
Waasmunster 055056
Gemeentelijke Academie voor schone
Kunsten, Arendonk 054997
Gemeentelijke Academie voor
Schone Kunsten, Filiaal Kasterlee,
Kasterlee 055031
Gemeentelijke Expositieruimte
Kampen, Kampen 032533
Gemeentelijke Kunstacademie,
Maldegem 055038
Gemeentelijke Oudheidkamer,
Ermelo 032319
Gemeentelijke Oudheidkamer,
Geldrop 032336
Gemeentelijke Tekenacademie,
Wetteren 055060
Gemeentelijke Tekenschool,
Westerlo 055059
Gemeentemuseum, Maassluis . 032619
Gemeentemuseum, Temse . . 004052
Gemeentemuseum, Weesp . . 033033
Gemeentemuseum Den Haag, Den
Haag 032171
Gemeentemuseum Elburg, Elburg 032294
Gemeentemuseum Helmond,
Helmond 032460
Gemeentemuseum Het Hannemahuis,
Harlingen 032413
Gemeentemuseum Het Markiezenhof,
Bergen op Zoom 032067
Gemeentemuseum 't Behouden Huijs,
Terschelling-West 032918
Gemeentemuseum 't Sterkenhuis,
Bergen 032065
Gemeentemuseum Weert – De
Tiendschuur, Weert 033031
Gemeentemuseum Weert – Jacob van
Horne, Weert 033032
Gemeinde- und Forstmuseum,
Oftersheim 020979
Gemeinde-Heimatmuseum Harsum,
Harsum 019078
Gemeindegalerie Benzeholz,
Meggen 116496
Gemeindegalerie der Evangelischen
Brüdergemeinde Korntal, Korntal-
Münchingen 019768
Gemeindegalerie im Alten Rathaus,
Wiener Neudorf 003103
Gemeindemuseum, Bad
Krozingen 016807
Gemeindemuseum, Hittisau . . 002090

Gemeindemuseum, Krauchthal . 041622
Gemeindemuseum, Regensdorf . 041866
Gemeindemuseum, Schweiggers 002745
Gemeindemuseum Absam,
Absam 001679
Gemeindemuseum im Historischen
Rathaus, Seeheim-Jugenheim . 021815
Gemeindemuseum Rothus,
Oberriet 041815
Gemeinschaft bildender Künstler,
Wien 058254
Gemeinschaft Bildender Künstler
Straubing e.V., Straubing 058891
Gemeinschaft der Ikonenfreunde,
Hannover 058892
Gemelli, Milano 080562
Gemelli, Flavio, Milano 080563
Gemersko-Malohontské Múzeum,
Rimavská Sobota 038435
Gemgold, Amsterdam 082758
Gemi Verlagsgesellschaft,
Reichertshausen 137484
Gemica, Roma 081390, 081391
Gemini, Maastricht 083464
Gemini G.E.L., Los Angeles . . 121493,
138348
Gemini G.E.L., West Hollywood . 124994
Gemini Galerie, München 108129
Gemini Trading, Leek 089778
Gemma, Kazan 085148
Le Gemme – Il Veliero, Livorno . 080325
Gemmeren, Peter Van, Belum . . 074716
Gemmological Association of Great
Britain, London 060144
Gemmrig, Berlin 074800
Gems Antiques, Chichester 088873
Gems of Africa, Atlanta 119569
Gems Studio, London 090143
Gemst Inc., Montréal 101414
Genadendal Museum,
Genadendal 038717
Genbrugsbøger, Taastrup 140665
Genç Kültür Merkezi, Genç . . . 042907
Gence, Danièle, Coustellet . . . 068206
Genchi, Teresa, Bari 079773
Gencor Art Gallery, Auckland
Park 038648
Gendai Chokoku Bijutsukan,
Tokyo 029510
Gendai Gangu Hakubutsukan,
Mimasaka 029025
Genders, Carolyn, Danehill . . . 117659
Gendre, Michel, Proissans 072312
Gendron, Jacques, Céret 103547
Gendrot, Yvon, Languet 069179
Gene Fornby, Örnsköldsvik . . . 040899
Gene Stratton-Porter House, Rome
City 052676
Geneen, Lionel, Bournemouth . 088535
Generaal Maczek Museum,
Breda 032102
General Adam Stephen House,
Martinsburg 050725
General Crook House Museum,
Omaha 051763
General Currie Heritage School,
Richmond 006794
General Daniel Bissell House, Saint
Louis 052802
General Douglas L. McBride Museum,
New Mexico Military Institute,
Roswell 052690
General Douglas MacArthur Memorial,
Norfolk 051574
General George C. Marshall Museum,
Zwijndrecht 033123
General Grant National Memorial, New
York 051395
General Henry Knox Museum,
Thomaston 053734
General Jacob Brown Historical
Society Museum, Brownville . 047283
General John J. Pershing Boyhood
Home, Laclede 050054

General Lew Wallace Museum,
Crawfordsville 048053
General Lewis B. Hershey Museum,
Trine University, Angola 046483
General Patton Memorial Museum,
Chiriaco Summit 047727
General Patton Memorial Museum,
Ettelbruck 030583
General Restoration, New York . 135929
General Sterling Price Museum,
Keytesville 049966
General Sweeney's Museum,
Republic 052518
General Wait House, Waitsfield . 054029
Generali Foundation, Wien 002996,
136590
Generations, Cincinnati 092859
Generations-Antiques, Buffalo . 092527
Generations Antiques, Salt Lake
City 096893
Generations Furniture, Tulsa . . 097731
Genesee Country Village and Museum,
Mumford 051112
Genesini, Giorgio, Modena . . . 131572
Genesio, Ernesto, Cherasco . . 080053
Genesis, Chicago 120137
Génesis, Punta del Este 119444
Genesis Art Gallery, Kolkata . . 109400
Genesis Art Gallery, Portland . . 123627
Genesis Art Line, New York . . . 122652
Genesis Publications, London . 138170
Genette, Daniel, Versailleux . . 074073
Geneva, Archibald, Louisville . . 094552
Genghini, Maria Vittoria, Milano 131451
Genghis Khan Furniture, San
Diego 097091
Geni, Samuel, Montréal 064425, 128182
Genia Schreiber University Art Gallery,
Ramat Aviv 025171
Genicot, Pierre, Bruxelles 140176
Genina, Québec 064533
Genji, San Francisco 097242
Genkai, Ljubljana 085610
Genluna Art Gallery, Davao . . 113376
Gennaro, Domenico Pietro,
Modena 080796
Genner & Co., August, Duisburg 075548
Genner, Gerd, Duisburg 075549
Studio Di Genninger Leslie Ann, Vetro
Artistico Glassart, Venezia . . . 111124
Genoa Courthouse Museum,
Genoa 049073
Genocido Aukų Muziejus, Vilnius 030516
Genolhac, Olivier, Saint-Cristol-lès-
Alès 072720
Genootschap Kunstliefde, Utrecht 059604,
112700
Genre, Diane, New Orleans . . . 095104
Gensing Antiques, Saint Leonards-on-
Sea 091197
Gensinger, Elke, Ramstein-
Miesenbach 078124
Genson, Daniel, Bains-les-Bains 066920
Gent, Elisabeth, Bad Neuenahr-
Ahrweiler 074560
Gent, Theodor, Braunschweig . 075123
Genter, Gebhard, Kehl 076740
Genth, D., Dürmentingen 141711
Gentil-Haus, Aschaffenburg . . 016616
Gentile, M., Cleveland 120424
Gentile, Mattia, Roma 131938
Gentile, Melissa, Firenze 080150
Gentilini, Valter Emilio, Torino . 081705
Gentils, Annie, Antwerpen . . . 100490,
136610
Gentle, Rupert, Milton Lilbourne 090761
Gentry Antiques, Polperro . . . 091028
Gentse Kunstveilingen en Galerie Sint-
John, Gent 125334
Genuine Collection, Miami . . . 121793
Genzkow & Willig, Rosemarie,
Hamburg 107271
Geo, Deutsch Schützen 099972
Geo-Ethnics, Tucson 097663

Geo Inn, Oosterbeek 032722
Geo, San, Seoul 111653
Geochang Museum, Geochang . 029930
Géode, Saint-Étienne 105512
Geoffrey Kaye Museum of
Anaesthetic History, ANZCA House,
Melbourne 001270
Geoffroy, Jean-Renaud, Royan . 125795
Geographical Center Historical
Museum, Rugby 052700
Geography Museum, Faizabad . 024255
Geoje Museum, Geoje 029931
Geológiai Múzeum, Tata 023997
Geologian museo, Luonnontieteellinen
Keskusmuseo, Helsinki 010819
Geological Curators' Group,
Dublin 059377
Geological Museum, Dublin . . . 024883
Geological Museum, Ho Chi Minh
City 054803
Geological Museum,
Johannesburg 038748
Geological Museum, Khartoum . 040495
The Geological Museum, Laramie 050142
Geological Museum, Lucknow . 024380
Geological Museum, Mizpe
Ramon 025158
Geological Museum, Saskatoon . 006972
Geological Museum and Art Gallery,
University of Exeter, Redruth . 045598
Geological Museum of China,
Beijing 007619
Geological Museum of Iran,
Tehrän 024756
Geological Museum, Chulalongkorn
University, Bangkok 042659
Geological Museum, Gauhati
University, Guwahati 024275
Geological Museum, Institute of
Geology of the Armenian Academy of
Sciences, Yerevan 000730
Geological Museum, King Saud
University, Riyadh 038109
Geological Museum, Prof. Rajneth
Memorial Museum, Varanasi . 024515
Geological Survey and Mines Museum,
Department of Geological survey and
mines, Entebbe 043102
Geological Survey Department
Museum, Accra 022922
Geologial Survey Museum,
Dhaka 003221
Geological Survey Museum, Ipoh 030664
Geological Survey Museum,
Windhoek 031861
Geologičeskij i Mineralogičeskij Muzej,
Biškek 030071
Geologičeskij Muzej im. A.A. Černova,
Institut geologii Komi naučnogo
centra Rossijskoj akademii nauk,
Syktyvkar 037908
Geologičeskij Muzej im. Professora V.E.
Eršova MGGU, Moskva 037273
Geologičeskij Muzej Vserossijskogo
NII Mineralnogo Syrja im. N.M.
Fedorovskogo, Moskva 037274
Geologičeskij Muzej, Muzej OAO
Poljarnouralgeologija, Vorkuta . 038061
Geologijos ir Mineralogijos Muziejus,
Vilniaus Universiteto Gamtos mokslų
fakultetas, Vilnius 030517
Geologisch-Mineralogische Ausstellung
der Eidgenössischen Technischen
Hochschule Zürich, Zürich . . . 042195
Geologisch-Mineralogische Sammlung,
Hillesheim 019252
Geologisch Museum Hofland, Laren,
Noord-Holland 032561
Geologisch-Paläontologisches
Museum der Universität Hamburg,
Hamburg 018989
Geologisch Streek-Museum de
IJsselvallei, Olst 032715
Geologische Landessammlung von

Vorpommern, Institut für Geographie und Geologie, Greifswald 018806

Geologische Sammlung Schloß Kapfenstein, Kapfenstein 002157

Geologische Schausammlung Eselsmühle, Leinfelden-Echterdingen 019964

Geologisches Freilichtmuseum, Bad Bentheim 016715

Geologisches Freilichtmuseum, Gmünd, Niederösterreich 001947

Geologisches Freimuseum, Bettendorf 017347

Geologisches Museum, Kamp-Lintfort 019491

Geologisches Museum München, München 020527

Geologisches und Mineralogisches Museum der Christian-Albrechts-Universität, Institut für Geowissenschaften, Kiel 019602

Geologo-mineralogičeskij Muzej, Kazan 037100

Geologo-mineralogičeskij Muzej im. A.I. Kozlova, Dal'nevostočnyj gosudarstvennyj techničeskij universitet, Vladivostok 038031

Geologo-mineralogičeskij Muzej, Moskovskaja selskochozjajstvennaja akademija im. K.A. Timirjazeva, Moskva 037275

Geology and Geophysics Museum, Roorkee 024461

Geology Collections, University College London, London 044976

Geology Department Museum, Aberdeen 043356

Geology Education Museum, Durban 038700

Geology Museum, Amman .. 029767

Geology Museum, Auckland .. 033148

Geology Museum, Ha Noi ... 054794

Geology Museum, Kampala 043106

Geology Museum, Kingston .. 028441

Geology Museum, Legon 022937

Geology Museum, Lucknow .. 024381

Geology Museum of the University of Otago, Dunedin 033191

Geology Museum of the University of Stellenbosch, Stellenbosch 038834

Geology Museum, Marshall University, Huntington 049625

Geology Museum, National Cheng Kung University, Tainan 042485

GeoMuseum der Universität, Köln 019685

Geomuseum der Universität Münster, Münster 020597

Georg Elser Gedenkstätte, Königsbronn 019722

Georg-Grabe, Eike, Kassel 141977

Georg Jensen Museum, København 010175

Georg-Kolbe-Museum, Berlin ... 017148

Georg-Meistermann-Museum, Städtische Galerie für Moderne Kunst, Wittlich 022702

Georg Olms Verlag, Hildesheim . 137277

Georg Papendicks Faschings- und Karnevalsordenmuseum, Bad Reichenhall 016846

Georg Rendl-Museum, Sankt Georgen bei Salzburg 002655

Georg Trakl Forschungs- und Gedenkstätte, Salzburg 002624

Georg, Reppa, Pirmasens 078049

Georg, Sophia, Denver 120729

George & Sons, New Orleans 128187

George A. Spiva Center for the Arts, Joplin 049842

The George Adams Gallery, The Arts Centre, Melbourne 001271

George B. Dorr Museum of Natural History, Bar Harbor 046784

George Bush Presidential Library and

Museum, College Station 047872

George C. Marshall Museum, Lexington 050265

The George D. and Harriet W. Cornell Fine Arts Museum, Winter Park 054498

George Eliot Hospital Museum, Nuneaton 045425

George Fraser Gallery, University of Auckland, Auckland 033149

George Hooey Wildlife Collection, North Battleford 006505

George I. Ashby Memorial Museum, Copper Center 047979

George J. Doizaki Gallery, Japanese American Cultural & Community Center, Los Angeles 050424

George Johnston Tlingit Indian Museum, Teslin 007135

George L. Sturman Museum, University of Nevada at Las Vegas, Las Vegas 050161

George Meany Memorial Exhibition, Silver Spring 053341

George Museum, George 038718

George Paton Gallery, Victoria .. 001609

George Rogers Clark Park Museum, Vincennes 053988

George Street Antiques Centre, Hastings 089465

George Taylor House, Catasauqua 047464

George the V Antiques, Portland 096409

George W. Brown jr. Ojibwe Museum, Lac du Flambeau 050051

George W. Somerville Historical Library, Chillicothe 047715

George Walter Vincent Smith Art Museum, Springfield 053460

George Washington Birthplace National Monument, Washington's Birthplace 054194

George Washington Carver House, Diamond 048324

George Washington Carver Museum, Tuskegee Institute 053866

George Washington Masonic National Memorial, Alexandria 046391

George Washington's Ferry Farm, Stafford County 053481

George Washington's Headquarters, Cumberland 048091

George Washington's Mount Vernon, Mount Vernon 051104

George, Jill, London 118344

George, N., New York 095416

George, Paul, Spincourt 073559

Georges, Ambarès-et-Lagrave .. 066544

Georges, Guildford 127607

Georges, Alain, Lyon 104107

Georges, Annick, Chamalières .. 128538

Georges, Anselme, Marseille ... 069939

Georges, Jacques, Tintigny 082393

Georgetown Antiques Center, Washington 097838

Georgetown Art Studio, Washington 124910

Georgetown College Gallery, Georgetown 049079

Georgetown Gallery of Art, Washington 124911

Georgetown Museum, Columbus 047932

Georgetown Plating and Polishing, Washington 136456

Georgetown Refinishing, Washington 136457

Georgetown University Art Collection, Washington 054117

Georgi, Peter, Greußen 129918

Georgia Agrirama, 19th Century Living History Museum, Tifton ... 053745

Georgia Antiques, Pulborough . 091060

Georgia Association of Museums and Galleries, Atlanta 060493

Georgia Berlin, Galerie Hillmer und

Partner, Berlin 106298

Georgia Book Store, Atlanta ... 144967

Georgia Capitol Museum, Atlanta 046619

Georgia Coin and Jewelry, Atlanta 092138

Georgia Council for the Arts, Atlanta 060494

Georgia Historical Society Museum, Savannah 053176

Georgia Lawyers for the Arts, Atlanta 060495

Georgia Mountains History Museum, Gainesville 049009

Georgia Museum of Art, University of Georgia, Athens 046593

Georgia Museum of Natural History, University of Georgia, Athens . 046594

Georgia Music Hall of Fame, Macon 050570

Georgia O'Keeffe Museum, Santa Fe 053107

Georgia Salzburger Society Museum, Rincon 052574

Georgia Southern University Museum, Statesboro 053507

Georgia State Cotton Museum, Vienna 053984

Georgia State University Art Gallery, Atlanta 046620

Georgia State University School of Art, Atlanta 056682

Georgia Veterans Memorial Museum, Cordele 047989

Georgiadis Sofoklis, N., Thessaloniki 079211

The Georgian, Lancaster 050115

Georgian Antiques, Christchurch 083873

Georgian Antiques, Cork 079493

Georgian Antiques, Edinburgh . 089174, 134362

Georgian Antiques and Gallery, Christchurch 083874, 132706

Georgian Cabinets Manufacturers, London 134669

Georgian Gallery, Hasselt 063681

Georgian Gems, Swanage 091525

Georgian Group, London 060145

Georgian House, Bristol 043750

Georgian House Antiques, Chipping Norton 088888, 134262

The Georgian House, National Trust of Scotland, Edinburgh 044242

Georgian Manor Antiques, New York 095417

Georgian National Gallery, Tbilisi 016403

Georgian Painter, Tbilisi 015905

The Georgian Rooms, Bridlington 088603

Georgian Silk Museum, Tbilisi .. 016404

Georgian State Museum of Oriental Art, Tbilisi 016405

Georgian Theatre Royal Museum, Richmond, North Yorkshire .. 045609

Georgian Village, Kansas City .. 094075

Georgian Village Antiques, Sligo 079700

Georgian Village Antiques Market, London 090144

Georgia's Stone Mountain Park, Stone Mountain 053545

Georgie Lane Antiques, Maleny . 061767

Georgie's Gifts, Charlotte 092599

Georgiev, Nedialko, Sofia 140322

Georgikon Majormúzeum, Keszthely 023739

Georgina Pioneer Village, Keswick 006087

Georgiou & Kerr, London 090145

Georgiou, Chr. & A., Larnaka .. 102533

Georgiou, Iraklis, Athinai 079164

Georg's Galerie, Sylt 108748

Geosammlung der Technischen Universität Clausthal, Clausthal-Zellerfeld 017743

Geoscience Museum, Tshwane . 038847

Geoskop-Urweltmuseum,

Thallichtenberg 022139

Géothermia-Musée de la Géothermie et du Thermalisme, Chaudes-Aigues 012524

Geowissenschaftliche Sammlungen und Geologischer Garten, Martin-Luther-Universität Halle-Wittenberg, FB Geowissenschaften, Halle, Saale 018960

Geowissenschaftliche Sammlungen, TU Bergakademie Freiberg, Freiberg, Sachsen 018470

Geowissenschaftliches Museum am Geowissenschaftlichen Zentrum der Universität Göttingen, Göttingen 018748

GeoZentrum Gams, Gams bei Hieflau 001925

Geppert-Rahm, Marth, Frankfurt am Main 075783

Geppert, Dieter, Essen 075674

Geppert, Horst, Ederheim 129772

Gerätemuseum des Coburger Landes, Ahorn 016478

Gerätesammlung Georg Bauer, Stadtlauringen 021956

Gerätesammlung Koch, Königsberg in Bayern 019721

Gerald R. Ford Museum, Grand Rapids 049190

Geraldine, Geraldine 083938

Geraldine's, Portland 096410

Geraldton Regional Art Gallery, Geraldton 001097

Geralka Rural Farm, Spalding . 001511

Geraniums Antiques, Tucson ... 097664

Gerard, Bruxelles 063447

Gerard, Nijmegen 083511

Gérard, Christophe, Rumilly ... 072647

Gérard, Emond, Montfort-sur-Argens 128803

Gerard, Hubert, Aurillac 072359

Gerard, Jonas, Miami 121794

Gérard, Timo, Bremen 129632

Gérardin & Cie., G., Lyon 069741

Gerardo, Vigo 133478

Gerat, Jan, Ráztočno 133060

Geraud, Philippe, Honfleur 103836

Gerbe d'Or, Brive-la-Gaillarde .. 067510

Gerbenzon & Zoon, P., Leeuwarden 083415

Gerber, Niederhelfenschwil 087711

Gerber, Rosario 060951, 127548

Gerber, Almut, Köln ... 076869, 107629

Gerber, Anabella, Kronau 076970

Gerber, Cyril, Glasgow 117842

Gerber, Daniel, Bern ... 087282, 133678

Gerber, Daniel, Suhr .. 087834, 133935

Gerber, René, Illkirch-Graffenstaden 068866

Gerber, René, Strasbourg 073589

Gerber, Waldemar F., Basel ... 116075, 144081

Gerberhaus mit Gerbermuseum, Bretten 017620

Gerbermuseum Lohmühle, Frickingen 018513

Gerbeth, Frank, Hof 076548

Gerbeth, Frank, Schöneck / Vogtland 078385

Gerbrandt, Anja, Hamburg 076227

Gerçek, İstanbul 117044

Gercev, Andrej, Moskva 114261

Gerdau, Angela, Oldenburg ... 130489

Gerdsen, Magnus P., Hamburg . 107272

Gerðuberg Menningarmiðstöð, Reykjavík 024094

Gereedschap Museum Mensert, Delft 032154

Geremi, Trieste 081789

Gerencsér, Ildikó, Budapest ... 079258

Gerg, Magdalena, München ... 130347

Gerg, Michael, Lenggries 130238

Gerg, Michael, Ruhpolding ... 078310

Gerhard Hirsch Nachf., München 077510,
126049
Gerhard-Kurt-Müller-Stiftung,
Leipzig 019991
Gerhard-Marcks-Haus, Das
Bildhauermuseum im Norden,
Bremen 017587
Gerhard Wolf Janus press, Berlin 137070
Gerhard, Gerd, Bad Berleburg .. 106042
Gerhardsen, Atle, Berlin 106299
Gerhardt, Michael, Knau 130135
Gerhardt, Renate, Berlin 137071
Gerhart-Hauptmann-Haus, Kloster/
Hiddensee 019656
Gerhart-Hauptmann-Museum,
Erkner 018254
Gerhof, Tilburg 083615
Gerichtsmuseum Bad Fredeburg,
Schmallenberg 021681
Gering & Lopez Gallery, New
York 122653
Gering, Ulrich, Frankfurt am
Main 107000
Gerini, Stefano, Firenze 131180
Geri's Antikschöpfli, Cham 087355
Gerlach, Mespelbrunn 130301
Gerlach, A., Winsen, Luhe 079024
Gerlach, Anita, Trebbin 130706
Gerlach, Birgit,
Mönchengladbach 142128
Gerlach, Günter, Stuttgart 130673
Gerlich, Salzburg 100164
Gerling, Bettingen 076081
Gerling, A., Grassau 076081
Gerloff, Ursula, Braunschweig .. 129622
Gerly, København 102898
Germain, Elisabeth, Aubagne .. 125497
Germain, Isabelle, Paris 129016
Germain, Jacques, Montréal .. 101415
Germain, Jacques & Françoise,
Boulogne-Billancourt 067391
German Historical Books, Zossen 142532
German Occupation Museum,
Forest 044358
German Village Antiques,
Columbus 093021
German Village Furniture,
Columbus 093022
German, Hassan Yacoub,
Valencia 086519, 126559
German, John, Burton-upon-
Trent 126841
German, Michael C., London .. 090146
Germani, Rossana, Roma 081392
Germanisches Nationalmuseum,
Nürnberg 020853
Germann, Zürich 126767
Germantown Historical Society
Museum, Philadelphia 052053
Germond, G., Zürich 144178
Germond, Michel, Paris 129017
Gerner, Henrik, Moss 131182
Gernhardt, Klaus, Bad Lausick .. 129416
Gernon House and Blacksmith Shop,
Russell 052707
Gerntrup, U., Bückeburg 106610
Gerold, H., Ehrenkirchen 075579
Geroldsecker-Museum im
Storchenturm, Lahr 019853
Geroldsecker Waffenschmiede,
Seelbach 021817
Geronimo Berenguer de los Reyes jr.
Museum, Cavite City 034452
Geronimo Springs Museum, Truth or
Consequences 053817
Geronimo's Cadillac, Saint Louis 096709
Geronzi, Italo, Torino 081706
Gerringong and District Museum,
Gerringong 001099
Gerrit Rietveld Academie,
Amsterdam 056168
Gerrit Valk's Bakkerij- en Ijsmuseum,
Hellendoorn 032451
El Gerro, Valencia 086520

Gerry's, Portland 136150
Gershel, Marseille 069940
Gerson, D., Antibes-Juan-les-
Pins 066673
Gerstenberg, Hildesheim 137278
Gerstenlauer, Renate, Lindau .. 107874
Gerstenlauer, W. & R.,
Asselfingen 106020
Gerster, W., Nagold 077686
Gerstner, Heidi, Moers 108021
Gerth, U., Stadthagen 078533
Gerth, Uwe, Bad Salzuflen 074580
Gertrud Theiler Tick Museum,
Onderstepoort 038790
Gertrude Contemporary Art Spaces,
Fitzroy 001072
Gertrude Herbert Institute of Art,
Augusta 046660
Gertrude Posel Gallery,
Johannesburg 038749
Gertrude Smith House, Mount
Airy 051080
Gertrude Street Artist's Spaces,
Fitzroy 099025
Gertrudenhof, Marxzell 077305
Gertrudenkapelle, Ernst Barlach
Stiftung Güstrow, Güstrow .. 018884
Gerusel, E., Bonn 075072
Gerva Deposito Mobili Antiquariato,
Venezia 081840, 132246
Gervais Wheels Museum, Alida . 005313
Gervais, Christian, Avranches .. 066890
Gervais, Gilbert, Avranches 066891
Gervais, Gilbert, Saint-Martin-des-
Champs (Manche) 072944
Gervil, Jette, Otterup 140640
Gervis, Daniel, Paris 104910
Gervis, Richard, San Francisco . 097243
GESA Galerie, Aarau 116004
Gescheidle, Susan, Chicago .. 120138
Geschichte auf Rädern, Automuseum
Melle, Melle 020353
Geschichte(n)haus, Bildein 001783
Geschichtlich-heimatkundliche
Sammlung, Aschheim 016632
Geschichtliche Begegnungsstätte
Hachborn, Ebsdorfergrund .. 018086
Geschichtliche Sammlung der
Schule-Ausbildung für Blinde,
Landesbildungszentrum für Blinde,
Hannover 019050
Geschichtliches Museum Enns-
Donauwinkel, Sankt Valentin .. 002702
Geschichts- und Kunstverein
Aschaffenburg e.V.,
Aschaffenburg 058893
Geschichtsmuseum Lüdenscheid,
Museen der Stadt Lüdenscheid,
Lüdenscheid 020170
Geschichtswerkstatt Herrenwyk,
Lübeck 020156
Geschied- en Oudheidkundig Museum
't Freulekeshuus, Venray 032984
Geschiedkundig Museum Mesen,
Mesen 003868
Geschirr-Museum, Wilhelmsburg an
der Traisen 003117
Geschke, Bettina, Hohenkammer 130064
Gesellschaft bildender Künstler
Österreichs, Wien .. 058255, 136591
Gesellschaft der Freunde der
bildenden Künste, Wien 058256
Gesellschaft der Freunde der Neuen
Galerie Graz, Graz 058257
Gesellschaft der Freunde des Museum
Haus Cajeth, Heidelberg 058894
Gesellschaft der Freunde junger Kunst,
Baden-Baden 058895, 106109
Gesellschaft der Keramikfreunde,
Deggendorf 058896
Gesellschaft der Musikfreunde in Wien,
Archiv, Bibliothek und Sammlungen,
Wien 002997
Gesellschaft Deutscher Kunsthändler

e.V, München 058897
Gesellschaft für Aktuelle Kunst,
Bremen 017588
Gesellschaft für Bildende Kunst Trier,
Trier 058898
Gesellschaft für Bildende Kunst
und Vaterländische Altertümer,
Emden 058899
Gesellschaft für fotografische Kunst
GFFK e.V., Köln 058900
Gesellschaft für Goldschmiedekunst
e.V., Hanau 058901
Gesellschaft für Kunst und Gestaltung
e.V., Bonn 058902
Gesellschaft für Kunst und Kultur,
Sigmaringen 058903
Gesellschaft für Kunst und Volksbildung,
Wien 058258, 136592
Gesellschaft für Moderne Kunst in
Dresden e.V, Dresden 058904
Gesellschaft für Schweizerische
Kunstgeschichte, Bern 059934, 116135
Gesellschaft für vergleichende
Kunstforschung, Wien 058259
Gesellschaft Photo Archiv e.V.,
Bonn 058905
Gesellschaft zur Förderung Frankfurter
Malerei e.V., Frankfurt am Main 058906
Gesenkschmiede Lubenbach, Zella-
Mehlis 022860
Gesler, Frank & Brigitte,
Neckarsulm 077697
Geslin, Patrick, La Chapelle-
Janson 068976, 128639
Gessani & C., Paolo, Roma ... 131939
Gessinger, Wiesbaden 108912
Gessler, Robert, Friedrichshafen 137217
Gesta, New York 139373
Gestaltung und Kunst, Hochschule der
Künste Bern, Bern 056418
Gestarte, Roma 081393
Gestas & Martine Simorre-Gestas, Jean-
Pierre, Pau 072039, 125775
Gesterbyn Kartano – Kirkkonummen
Kunnan Museoalue,
Kirkkonummi 011007
Gestütsmuseum Offenhausen,
Gomadingen 018763
Gestures, Saint Kilda, Victoria .. 062180
Get Rare, Savigny-sur-Orge ... 073493
Get Stuffed, London .. 090147, 134670
Get the Picture, Los Angeles ... 121494
Getaway Gallery, Columbus 120496
Getchell & Getchell, Minneapolis 094915
Gethmann, Berlin 106300
Getler, Helen, New York 122654
Getty Images Gallery, London .. 118345
Getty Publications, Los Angeles 138349
Gettysburg National Military Park,
Gettysburg 049090
Getz, Michael, Washington 097839
Geuer, Grevenbroich 107163
Geukens & de Vil, Knokke-Heist 100739
Geursen, J. & M.I., Richmond,
Tasmania 099627
Geurten, Maria, Norwich 118830
Geus, K.H. de, Eindhoven 083217
Geusau, Pia, Wels 112743
Geuthner, Paul, Paris .. 136932, 141125
Gevaert, Yves, Bruxelles 136639
Het Geveltje, Middelburg 083493
Gevik, Toronto 101615
Gevorg Grigoryan – Giotto – Studio-
Museum, National Gallery of
Armenia, Yerevan 000731
Gevorgyan Gallery, Jerewan ... 098601
Gewehr, T., Murnau am
Staffelsee 108230
Gewehr, Vera Lucia, Porto Alegre 063972
Gewerbe-Museum, Spaichingen 021923
Gewerbemuseum, Winterthur .. 042150
Gewerbemuseum im Germanischen
Nationalmuseum, Nürnberg .. 020854
Gewölbe der Geheimnisse,

Viechtach 022323
Gewölbe Galerie, Biel 116157
Geyer, Fritz, Plauen .. 108387, 130524
Geymüller Schlössel, Aussenstelle
MAK, Wien 002998
Geymüller, Dr. Johannes von,
Essen 106916
Geyns, S., Bruxelles 063448
Gezira Art Center-Galleries of Fine
Arts, Cairo 103044
GFL Fine Art, Nedlands 061946
Ggantija, Xaghra 082449, 111926
Ggantija Prehistoric Temples,
Xaghra 030768
GH Art, Atlanta 119570
Ghafari, Sharjah 088199
Ghafary Habibullah, Philadelphia 123387
Ghalib Museum, Delhi 024224
Ghalim, Bruxelles 063449
Ghana Armed Forces Museum,
Kumasi 022933
Ghana Museums and Monuments
Board, Cape Coast 059321
Ghanassia, Danielle, Paris 104911
Ghandour, Soraya, Beirut 082284
Ghani, Abdul, Jeddah 085302
Ghar Dalam Cave and Museum,
Birzebbuga 030737
Gharb Folklore Museum, Gharb . 030738
Gharibi, M., Konstanz 107716
Ghattas, Pierre, Beirut 082285
Ghazarian, Mulhouse 070391
Ghazarian, Alain, Benfeld 067086
Ghazni Provincial Museum,
Ghazni 000001
Ghelfi, Vicenza 111204
Ghelfi, Giorgio, Verona 111188
Ghent, Norfolk 095857
Gheorghiu, A., Bucureşti 085094
Gherardi, Cristiano, Roma 081394,
081395
Ghetto Gloss, Los Angeles ... 121495
Ghezzi, Johannes, Teesdorf ... 127881
Ghiglieri, Finale Ligure 110023
Ghiglione, Biot 103393
Ghini, Aldo, Firenze 131181
Ghinolfi, Andreas, Zürich 087970
Ghio, Dario, Monaco 082512
Ghio, Hervé, Pont-Aven 105379
Ghiordian Knot, New York 095453
Ghiotti, Alain, Buno-Bonnevaux . 067535
Ghirardini, Silvana, Torino 081707
Ghiretti, Cendrine, Montauban . 104250
La Ghirlandina, Formigine 080229
Ghislain, Saint-Ouen .. 073092, 073093
Ghislaine, David, Paris 071428
Ghisoni-Giordano, Bruxelles ... 063450
Ghobadi, M., Göttingen 076044
Ghoniem, Cairo 066249
Ghoochan Museum, Ghoochan . 024693
Ghost Dog Gallery, Nashville .. 122683
Giaccherini, Riccardo, London . 134671
Giacla, Genova 080263
Giacobbo, Bruno, Ossingen ... 116570
Giacobbo, Bruno, Rheinau 116580
Giacomini, Vincenzo, Firenze .. 131182
Giacomo Lo Bue, Córdoba ... 098590,
136476
Giacomoni, Henri, Monaco ... 132366,
132367
Giaconi, Firenze 080151
Giacosa, Paola, Firenze 131183
Giada Restaurati, Firenze 131184
Giafferi, Paris 125718
Giagnoni, Patrizia Assunta,
Cagliari 079985
Giai-Miniet, Daniel, La Bazoge . 068953
Giallo, Bologna 130963
Giambattista, Sabrina, Brescia . 131035
Giamblanco, Torino .. 081708, 110978,
132143
Giammona, Giuseppe, Palermo . 131698
Gian Ferrari, Claudia, Milano ... 110354

Giancarlo Politi Editore, Milano . 137679, 142722
Giancatarina, Maria, L'Isle-sur-la-Sorgue 069569
Giancorrado Forin, Roma 081396
Giandebiagi, Gino, Noceto 080952
Gianella, Lucien, Genève 087471
Gianferrari, Giovanna, Sassuolo . 081619
Gianferro, Ascenzio, Roma 131940
GianFu Classic Art Museum, Beijing 007620
Giangiacomi, Monaco 132368
Giangiacomo Ottorino, Roma . . . 081397
Giangrossi, Rodolfo, Milano 131452
Giangualano, Renato, Brescia . 131036
Il Gianicolo, Perugia 110640
Giannetti, Elena, Firenze 131185
Giannini & Marchi, Firenze 131186
Giannone, Pisa 110664
Giannoudis, Sozos, Chalandri . 109056
Gianolla, Claudio, Venezia 081841
Giantomassi & Zari, Roma . . . 131941
Giant's Castle, Marazion 045222
Giant's Causeway and Bushmills Railway, Bushmills 043817
Giant's Causeway Visitor Centre, Bushmills 043818
Giaquinto & Turrian, Lausanne . 087571
Giard, Fr., Marcq-en-Barœul . . . 140936
Il Giardino, Trieste 081790
Giardino di Boboli, Firenze . . . 026262
Giardino Storico e Villa Garzoni, Pescia 027281
La Giarina, Verona 111189
Giarla's Coin Studio, Boston . . . 092472
Giavolucci, Loretta, San Marino . 085284
Gibault, Paris 129018
Gibbard, A. & T., Eastbourne . . 144382
Gibberd, Jane, London 144608
Gibbes Museum of Art, Charleston 047548
Gibbins, Carol, Houston 093674
Gibbins, Colin, Houston 093675
Gibbins, David, Woodbridge . . . 091911
Gibbons, Stanley, London 126962, 144609
Gibbs Farm Museum, Museum of Pioneer and Dakota Life, Falcon Heights 048684
Gibbs, Christopher, London . . . 090148
Gibbs, Paul, Conwy 088971
Gibernau, Barcelona 143717
Giberton, Marie-Claude, Vauvillers 073991
Gibier, Chantal, Chatellerault . . 067905
Gibraltar Museum, Gibraltar . . . 022941
Gibran Museum, Gibran National Committee, Bsharri 030274
Gibson Barham Gallery, Lake Charles 050072
Gibson Gallery, Fort Malden Guild of Arts and Crafts, Amherstburg . 005330
Gibson House Museum, Boston . 047114
Gibson House Museum, North York 006525
Gibson-Riley, Houston 121046
Gibson Street Gallery, Cooks Hill 098918
Gibson, Colin, Newtownards . . . 118809
Gibson, G. 514 E Pike St, Seattle 124532
Gibson, Michael, Atlanta 092139
Gibson, Michael, London 101343
Gibson, Robin, Darlinghurst . . . 098944
Gibson, Roderick, Nantwich . . . 090796
Le Gibus, La Couronne 069015
GIC, Rouen 072586
Gicquel, Pierre, Vouvray 129360
Gidalya, New York 122655
Gidding, Anton, Saint-Cyprien . 105500
Giddings, J.C., Wiveliscombe . . 091893
Giddy Bullfrog Gallery, Red Hill . 099612
Gideon, West Hollywood 124995
Gidiemme, Venezia 111125
Gidts, Amsterdam 082759
Giefarte, Lisboa 114032

Gieling, Marcel, Utrecht 083647
Giellevivi, Milano 080564
Giencke, Gisela, Graz 127746
Gieraths, Heiko, Gerwisch 129902
Gierek, Joseph, Tulsa 124803
Gierhards, Ralph, Düsseldorf . . 075475
Gierig, Timm, Frankfurt am Main 107001, 137198
Gierke, Andreas, Boostedt 129617
Gierre Antichitá, Roma   081398, 081399
Giesberts, Haarlem 083329
Gieschen, U. & B., Lindlar 107879
Giese & Schweiger, Wien 062965, 100309
Giese, Peter, Hamburg 076228
Giesicke, Dr. Barbara, Badenweiler 074646
Gießler, M., Hohberg 076557
Gietzelt, Michael, Berlin 074801
Giffard, Patrick, Dieppe 125568
Gifford, Auckland 112771
Gifford-Mead, Nicholas, London . 090149
Gift Horse, Cardiff 117530
Gifts for the Soul, Takapuna . . . 113010
Gifu-ken Bijutsukan, Gifu 028572
Gifu-ken Gendai Togei Bijutsukan, Tajimi 029426
Gifu-ken Hakubutsukan, Seki . . 029355
Gifu-ken Toji Shiryokan, Tajimi . 029427
Gig Harbor Peninsula Historical Society Museum, Gig Harbor 049095
Gigante, Fabio, Varese 081813
Gigengack, Gerardus, Hombrechtikon 087528
Gigi's, Birmingham 119799
Giglio & C., Pasquale, Milano . . 080565, 126209, 131453
I Giglioli de' Castagni, Castrocaro Terme 109944, 109945
Gijon, Francisco, Madrid 133339
Gil Fernandez, Gregorio, Madrid 133340
Gil Navares, Nemesio, Madrid . . 086195
Gil, Claudio, Rio de Janeiro . . . 101035, 101036
Gil, Matt, San Francisco 124289
Gil, Natalia, Valencia 115503
Gila Cliff Dwelling Visitor Center, Silver City 053337
Gila County Historical Museum, Globe 049131
Gilan, İstanbul 088120
Gilardoni Arte, Novara 110548
Gilbach, Gabriele, Essen 129815
Gilbert, Canberra 061271, 139685
Gilbert, Detroit 093363
Gilbert, Miami 094737, 135795
Gilbert, Philadelphia 136087
Gilbert, Turner 062352, 139938
Gilbert & Dale, Ilchester 089632
Gilbert Collection, London 044977
Gilbert House, Atlanta 046621
The Gilbert Stuart Museum, Saunderstown 053169
Gilbert White's House and the Oates Museum, Selborne 045753
Gilbert, Albert, Zürich 116800
Gilbert, David, Sanadage 091237
Gilbert, John, Robin Hood's Bay 091123, 134973
Gilbert, M., Uppingham 091692
Gilbert, Philip, Horsell . 089590, 134476
Gilbert's Bar House of Refuge, Stuart 053578
Gilbreath, Kay, Houston 093676
Gilchrist, Sacramento 123768
Gilchrist & Gilchrist, Nashville . . 095002
Gilcrease Museum, Tulsa 053848
Gild, Katariina, Tallinn 103118
Gildart, Roma 110799
Gilde-Museum, Museumshof, Oldenburg in Holstein 020992
The Gilded Cage Gallery, Calgary 101241
Gilded Griffin, Tulsa 097732
Gildehaus, Bardowick 016964

Het Gildenhuys, Blokzijl 032073
Gilden's Art, London 118346
Gilden's Art, San Rafael 138467
Gilder, Tampa 097560
Gilders Tip, Saint Louis 136200
Gildeskål Bygdesamling, Gildeskål 033617
Gildi, Roberto, Roma 142813
Gildija Masterov, Sankt-Peterburg 114341
Gilding, Market Harborough . . . 090711
Gilding Arts, New Orleans 095105, 135858
Gilding Arts and Antiques, Washington 136458
The Gilding Studio, Dublin 130856
Giles, John, Petworth 090999
Giles, Larry, Saint Louis 096710
Giles, Norman, Dublin 109586
Giles, Norman, Kinsale 109658
Giles, Sebastian, London 134672
Gilford, Patti, Chicago 120139
Gilgamesh, Portland 096411
Gilgandra Museum, Gilgandra . . 001100
Gilgen Batia, Basel 087234
Gilhofer & Ranschburg, Luzern . 144107
Gili, Gustavo, Barcelona 137959
Gilibert, Torino 142857
Gill, Heerlen 083370
Gill, Angelika, Wiesbaden 078975
Gill, David, London . . . 118347, 118348
Gillam Community Museum, Gillam 005875
Gillan, Auckland 112772
Gillard, Colette, Champniers . . . 067818
Gillard, Joseph, Plombières-lès-Dijon 072193
Gillardeau, Charles, Brive-la-Gaillarde 125538
Gillardeau, Charles, Paris 071429
Gille-Stiernet, Bruxelles 100615
Gille, Jean-Pierre, Passy-Grigny 072029
Gilleleje Museum, Gilleleje 010079
Gilles, Birgit, Bochum 075034
Gilles, Gilbert, Crest . 068236, 128573
Gilles, Thomas, Le Bourg-Saint-Léonard 069225
Gillesgården, Pohja 011259
Gillespie, Houston 093677
Gillespie Museum of Minerals, Stetson University, De Land 048184
Gillespie, Simon, London 134673
Gillet-Seurat, Anne, Nanterre . . 125653
Gillet, Michel, Paris 104912
Gilli, Joseph-Jean, Trept 073881
Gillian, Toronto 101616
Gillicienne, Charleroi . . 063601, 100679
Gillies, David, Toronto 064657
Gilligans, Leek 089779
Gillingham Museum, Gillingham, Dorset 044385
Gillio, Santa Barbara 097382
The Gillis-Browne Gallery, Harrogate 117915
Gillman, Barbara, Miami 121795
Gillman, Dale, Tulsa 097733
Gillmark, Hertford . . . 134460, 144475
Gillmeister, Peine . . . 108369, 142298
Gillo, John, Dartmouth 117666
Gillot, Seraing 063828
Gillott, Chris, Waitakere 132754
Gilman Garrison House, Exeter . 048652
Gilman Museum, Hellertown . . . 049434
Gilman Town Hall Museum, Issaquah History Museum, Issaquah . . . 049735
Gilman, Paul, Liège 063760
Gilmore Car Museum, Hickory Corners 049460
Gilmore, Charles, Holywood . . . 117992
Gilmours, New Plymouth 126322
Gilmozzi, B., München 130348
Gilot, Jean-Pierre, Paris 141126
Gilpin County Arts Association, Golden 060496
Gilpin History Museum, Central

City 047504
Gilquin, Jean-Luc, Arcis-sur-Aube 066705
Gil's Upholstery, Miami 094738, 135796
Gilsa, Bettina von, Tübingen . . . 130710
Gilson, Ernst, Aschaffenburg . . 106010
Gilstrap Heritage Centre, Newark-on-Trent 045330
Gilt, Portland 096412
Giltwood, Camberwell . 061257, 127577
Giltwood, Cheltenham . 088832, 134236
Gilty, Unley 127695
Gimaray, Arlette, Paris 104913
Gimbert, Marc, Saint-Zacharie . 073337
Gimel, Jerusalem 109717
Gimle Antikk, Oslo 084314
Gimmick, Amsterdam 082760
Gimpel Fils, London 118349
Gimsa-Dittmer, Soltau 078503
Ginac, Romain, Nice 070612
Ginac, Romain, Nîmes 070695
Ginco Gallery, Malvern, Victoria . 099269
Giner, Fernando, Madrid 115226
Gingell Littlejohn, Sliema 126260
Ginger, Bristol 117447
Ginger, Omaha 123315, 123316
Ginger, G. & D., Ludlow 090649
Gingham, Verbier 087869
Gingko Press, Corte Madera . . . 138329
Gingras Trading Post, Walhalla . 054035
Ginhart, Walter, Raubling 108437
Ginioux, Catherine, Aurillac . . . 066825
Ginja Jar, New Orleans 095106
Ginkgo, Chêne-Bourg 087365
Ginkgo-Depot, Thônex 087845
Ginkgo Leaf Books, Chicago . . . 145045
Ginkgo Museum, Weimar 022493
Ginkgo Projects, Axbridge 117263
Ginkgo, Biloba & Seiffert, Nürnberg 077822
The Ginnel, Harrogate 089429
Ginny Williams Family Foundation, Denver 048264
Gino, Sacramento 123769
Ginsberg, Nicky, Paddington, New South Wales 062040
Ginsburg, Cora, New York 095419
Ginsha, Tokyo 111437
Ginsig, Markus, Winkel 087905
Gintaro Galerija-Muziejus, Nida . 030428
Gintaro Galerija-Muziejus, Vilnius 030518
Gintzel, Annemarie, Neuburg an der Donau 108238
Ginučių Vandens Malūno Ekspozicija, Ginučiai 030337
Ginza Bargins, Palmerston North 084055
Ginza Ippodo Salon, Tokyo . . . 111438
Ginza Shibuya Gallery, Tokyo . . 111439
Ginza Shouun, Tokyo 082147
Gioacchini, G., Milano 110355
Giocanti, Hervé, Marseille 128769
Gioffreda, Mariella, Torino 132144
Gioia, Giorgio, Genova 131316
Gioia, Giorgio, Torino 132145
Gioie Ed Arti, Verona 081913
Gion Art Gallery, Kyoto 111293
Giordani, Albertino, Bologna . . . 109847
Giordano, Edmonton 101289
Giordano, Antonio, Genova . . . 131317
Giorgetti, Milano 110356
Giorgi Chitaia Ethnography Museum, Georgian National Museum, Tbilisi 016406
Giorgi Chitaia Museum of Architecture, Georgian National Museum, Tbilisi 016407
Giorgi Studio, Milano 110357
Giorgi, Franco, Firenze 080152
Giorgi, Giorgio, Roma 081400
Giorgi, M. de, Nordhorn 077801
Giorgione, Ettore, Armadale, Victoria 061028, 127555
Il Giornale dell'Arte, Torino . . . 138831

Giorni, Maria, Firenze . 080153, 142674
Giosi, Napoli 110529
Giosi, Dora, Toronto 064658
Giovagnoni, Elisabetta, Roma . . 110800
Giovanangeli, Ghislaine,
　Roquevaire 072542
Giovani d'Arte, Modena 059458
Giovannetti, Cristiana, Roma . . 131942
Giovannini, Andrea, Lumino . . . 133826
Giovannoni, Enzo, Roma 131943
Giovinazzo, Vincenzo, Torino . . 132146
Giovine & De Vero, Torino 132147
Gip Art, Postojna 114740
Gippsland Art Gallery, Sale 001479
Gippsland Auction House,
　Traralgon 125243
Gippsland Galleries, Cranbourne 098929
Gippsland Heritage Park, Moe . . 001300
Gipsformerei, Staatliche Museen
　zu Berlin – Stiftung Preußischer
　Kulturbesitz, Berlin 017149
Gipsmuseum Oberwiesen,
　Schleitheim 041954
Gipsoteca, Lucca 026637
Gipsoteca Canoviana e Casa del
　Canova, Possagno 027429
Gipsoteca del Castello Normanno
　Svevo, Bari 025453
Gipsoteca F. Jerace, Catanzaro . 025873
Gipsoteca Istituto d'Arte, Firenze 026263
Giraffe, Ottawa 101474
Giraffe Antiques, Los Angeles . . 094343
Giraffe Art Africain, Montréal . . 101416
Giraffen-Museum Dortmund,
　Dortmund 017956
Giraldi-Delmas, Philippe, Tarascon
　(Bouches-du-Rhône) 073620
Giraldi, Fabrizio, Livorno 110180
Giraldo, Treviso 111041
Giran, Christiane, Salin-de-Giraud 105643
Girand, Christine, Paris 104914
La Girandole, Lyon 069742
Girandole, Metz 070097
Girandole, Ville-d'Avray 074129
Girard, Lyon 069743
Girard, André, Paris 104915
Girard, Eric, Brest 067471
Girard, Jacques, Talence 129272
Girard, Jacques, Toulouse 105744
Girard, Jean, Cahors 067597
Girard, Jean-Jacques, Dunkerque 125577
Girard, Véronique, Paris 071430
Girard, Vincent, Villars 074126
Girardello, Bruno, Venezia 081842
Girardet, Jacqueline, Paris 071431
Girardet, Robert, Lons-le-Saunier 069627
Girardi, Pamela & Ramon,
　Chermside 098881
Girardin, Gilbert, Genève 087472
Girardot, Stéphane, Saint-Ouen . 073094
Il Girasole, Legnano 110164
Giraud, Les Ulis 069435
Giraud-Badin, Paris . . . 071432, 141127
Giraud, Dominique, Saint-Étienne 072763,
　 129226
Giraud, Patrick, Draguignan . . . 068403
Giraud, Pissarro & Segalot, New
　York 122656
Giraud, Pissarro & Ségalot, Paris 104916
Giraudo, Giorgio, Cuneo 080063
Girault, Christiane, Saint-Ouen . 073095
Giresun Müzesi, Giresun 042908
Giri, Joseph, Long Beach 121401
Girier, Dommartin (Rhône) 068375
Girndt, Uwe, Rheurdt 130588
Girne Kalesi, Girne 033459
Girnus, Andrea & Klaus, Xanten 109000
Girod, Claudia & Mike, Berlin . . 074802
Giromini, Danièle, Saint-Ouen . . 073096
Giron, Alain, Selles-sur-Cher . . . 073483
Giron, Christian, Noyers-sur-Cher 070758
Giron, Jeanne, Meymac 072510
Giron, Philippe, Saint-Aignan (Loir-et-
　Cher) 072661

Gironcoli Museum, Sankt Johann bei
　Herberstein 002667
Girondini, Francesco, Verona . . . 111190
Girotto, Dr. Maria Beatrice,
　Padova 131664
Girou, Gilles, Angers 066607
Giroud, Régine, Zürich 087971
Giroud, Robert, Grenoble 068730,
　 103789
Girouette, Crest 103644
Giroux, Frédéric, Paris 071433
Gisborne Steam Park, New
　Gisborne 001350
Gisel, Andi, London 090150
Gisinger, Ferdinand, Dornbirn . . 062547
Gislao, Robert Angelo,
　Vordingborg 066214
Gislaveds Industrimuseum,
　Gislaved 040640
Gisler, Migli, Glarus 087510
Gislöfs Smidesmuseum,
　Simrishamn 040945
Gismondi & Pastor, Monaco . . . 082513,
　 112006
Gismondi, Jean, Antibes-Juan-les-
　Pins 066674
Gismondi, Sabrina, Paris 071434
Gissler-Volm, Ernst H.,
　Badenweiler 141445
Gitanes, Los Angeles 094344
El Gitano, Cali 065250
Gitstappermolen, Vlodrop 033000
Gittel, H., Frankfurt am Main . . . 075784
Gittins & Son, G.J., Caerphilly . . 088754
Gittins, Kathy, Meifod 118731
Gitxsan Treaty Office, Hazelton . 005995
Giubboni, Prof. Paolo, Perugia . . 131751
Giubergia, Franco, Torino 081709,
　 081710
Giubilo, Trieste 081791
Lo Giudice Giampiero, Roma . . . 081401
Giuliani, Alvaro, Roma 131944
Giunta, Carlo, Roma 081402
Giunti Editore, Firenze 137651
Giuriato, Roberto, Padova 131665
Giurleo, Roberto, Firenze 131187
Giuseppe, San Gwann 111920
Giuseppe Upholstering,
　Milwaukee 135817
Giustarini, Alberto, Firenze 080154
Giusti & Bandinelli, Firenze 080155,
　 131188
Giusti, S. & Podestá, U., Milano 080567
Giusti Antichità, Formagine 080227
Giusti, Luisa Silvia, Prato 131775
Giusti, Silvia, Milano 080566
Giustina Gallery, Corvallis 048018
Giustina, S., Padova 131666
Giustiniani Palace, Chios 023046
Give-Egnens Museum, Give . . . 010080
Givingness, Austin 119674
Givon, Tel Aviv 109724
Giza Zoological Museum, Giza . . 010570
Gizard, Besançon 067123
Gizard, Antoine, Besançon 067124
Gjenreisningsmuseet,
　Hammerfest 033636
Gjesdal Bygdemuseum, Ålgård . 033487
Gjotas, Charalambos, Horw 087530,
　 133771
GJ's Consignment, Miami 094739
Gkikas, Alexandros, Athinai 079165
Glaasker, H. & G., Bokel, Kreis
　Cuxhaven 106520
Glace Bay Heritage Museum, Glace
　Bay 005879
Glacier, Zermatt 087920, 116732
Glacier National Park Museum, West
　Glacier 054300
Glade Antiques, High Wycombe . 089527
Gladis, Thomas, Stadtroda 130650
Gladius, Jaraiz de la Vera 139055
Gladstone and District Museum,
　Gladstone 005880

Gladstone Court Museum, Biggar 043596
Gladstone Gaol, Gladstone, South
　Australia 001103
Gladstone Maritime Museum,
　Gladstone, Queensland 001101
Gladstone Regional Art Gallery
　and Museum, Gladstone,
　Queensland 001102
Gladstone Working Pottery Museum,
　Stoke-on-Trent 045883
Gladstone, Barbara, New York . . 122657
Gladwell & Co, London 090151, 118350,
　 134674
Gläser, Rainer, Gottesgabe 076065
Gläsermuseum, Gerersdorf 001933
Glage, Königstein im Taunus . . . 107706
Glahn, Thomas von, Schiffdorf . . 078362
Glamis Castle, Glamis 044390
Glamour Art Gallery, New York . . 122658
Glance Back Bookshop & Gallery,
　Chepstow 134245, 144329
Glandford Shell Museum,
　Glandford 044391
Glanmore National Historic Site of
　Canada, Belleville 005414
Glans, Berith, Stockholm 143947
Glanum, Saint-Rémy-de-
　Provence 015537
Glanz, H. & I., Erftstadt 106889
Glanz, I., Erftstadt 075631
Glarner Naturmuseum, Engi . . . 041467
Glas- und Keramikmuseum,
　Großalmerode 018837
Glas- und Steinmuseum, Gmünd,
　Niederösterreich 001948
Glas Galerie Linz, Linz 100110
Glas, M.A. Ella, Hamburg 129962
Glasboden, Malmö 086851
Glasbruksmuseet i Surte, Surte . 041071
Glaser, W., Crimmitschau 075281
Glasfachschule, Kramsach 054946
Glasgalerie Eulenburg, Hamburg 107273
Glasgalerie Hittfeld, Seevetal . . . 108605
Glasgalerie Kuhler, Amsterdam . 112216
Glasgow Art Fair, Glasgow 098377
Glasgow Print Studio, Glasgow . 044399,
　 138169
The Glasgow Room, Glasgow . . 117843
Glasgow School of Art, Glasgow 056537
Glasgow School of Art – Mackintosh
　Collection, Glasgow 044400
Glasgow Society of Women Artists,
　Glasgow 060146
Glashaus-Derneburg, Holle 019318
Glashütte Annenwalde, Templin . 022571
Glashütte Gernheim, Petershagen 021107
Glashüttenmuseum des Erzgebirges,
　Neuhausen / Erzgebirge 020709
Glaskünstlervereinigung Nordrhein-
　Westfalen e.V., Düsseldorf . . . 058907
Glasmuseet Ebeltoft, Ebeltoft . . 010040
Glasmuseum, Frauenau 018459
Glasmuseum, Gehlberg 018602
Glasmuseum, Hergiswil 041592
Glasmuseum, Immenhausen . . . 019393
Glasmuseum, Warmensteinach . 022433
Glasmuseum Alter Hof Herding,
　Ernsting Stiftung Alter Hof Herding,
　Coesfeld 017755
Glasmuseum Boffzen, Boffzen . . 017466
Glasmuseum der Dorotheenhütte,
　Wolfach 022726
Glasmuseum Grünenplan,
　Grünenplan 018870
Glasmuseum Hadamar, Hadamar 018907
Glasmuseum in der Glashütte,
　Monschau 020456
Glasmuseum Passau, Passau . . 021081
Glasmuseum Rheinbach,
　Rheinbach 021395
Glasmuseum Steina, Bad Sachsa 016850
Glasmuseum Weisswasser,
　Weißwasser 022550
Glasmuseum Wertheim,

Wertheim 022587
Glasmuseum Zalto, Nagelberg . . 002427
Glasmuseum, Nederlands Museum
　voor Glas en Glastechniek,
　Schiedam 032852
Glass, Tucson 097665
The Glass & Art Gallery, Consett 117625
Glass Art, Innsbruck 100047
Glass Art Gallery, Toronto 101617
Glass Art Society, Seattle 060497
Glass Art Society Journal, Seattle 139374
Glass Artists' Gallery, Glebe . . . 099079
Glass Beveling, Indianapolis . . . 135632
The Glass Chick, Albuquerque . . 119478
Glass Creations, Portland 096413
Glass Eye Gallery, Seattle 124533
Glass Gallery and Main Gallery,
　University of Texas at El Paso,
　Department of Art, El Paso . . 048510
Glass Garage Gallery, West
　Hollywood 124996
Glass House Gallery, Truro 119258
Glass Lady, Richmond 096544
Glass Menagerie, Memphis 094634
The Glass Museum, Dunkirk . . . 048397
Glass Museum of Hsinchu City,
　Hsinchu 042330
Glass Onion Studio, Vancouver . 101790
Glass Revue, İstanbul 088121
Glass Roots, Denver 093293
Glass Roots Gallery, Richmond . 123734
The Glass Stopper, Kingsgrove . 061695
Glass, Design and Antique Repair,
　Hapert 132524
Glass, Horst, Essen 075675
Glass, Richard, Whaley Bridge . 091816
Glasschmelzofenbau-Hütte mit
　Glasausstellung, Plößberg . . . 021161
Glassdrumman Antiques, Tunbridge
　Wells 091650
Glassell School of Art, Houston . 057218
Glasshouse Studio, Seattle 124534
Glassic Art, Las Vegas 121355
Glassic Arts, Los Angeles 121496
Glasslight Gallery, New Orleans . 122174
Glassmith, Portland 136151
Glasstetter, Fritz Mario, Basel . . 087235
Glassworks, Louisville 121646
Glassworks, Seattle 097448
Glassworks 78209, San Antonio 124034
Glassworks 78258, San Antonio 124035
Glastonbury Abbey Museum,
　Glastonbury 044423
Glastonbury Art Guild,
　Glastonbury 060498
Glastonbury Gild Arts Center, New
　Canaan 051247
Glastonbury Lake Village Museum,
　Glastonbury 044424
Glastonbury Museum,
　Glastonbury 049104
Glastra, Eddie, Paddington, New South
　Wales 099520
Glatzer, Franz, Kierling 127785
Glatzl, Franz, Askim 115578
Glauberg-Museum, Glauburg . . . 018709
Glaza, Tomasz, Gdańsk 113474
Glazar, Annecy 066641
Glazer, George, New York 095420,
　 145340
Glazovskij Kraevedčeskij Muzej,
　Glazov 037017
GLB Création, Jouy (Eure-et-Loir) 068922,
　 128634
GLBT Historical Society Museum, Gay,
　Lesbian, Bisexual Transgender
　Historical Society, San
　Francisco 053001
Glebe House and Gallery, Derek Hill
　Collection, Letterkenny 024951
The Glebe House Museum,
　Woodbury 054519
Gledel, Claude, Fougères (Ille-et-
　Vilaine) 068595

Gledel, Gérard, Laval (Mayenne) 069202
Gledel, Philippe, Fougères (Ille-et-Vilaine) 068596
Gledhill, Christchurch . 083875, 132707
Gledhill, San Diego 097092
Gledswood Farm Museum and Homestead, Catherine Fields .. 000952
Gleeson, Terry, Dural 127593
Gleiche, Dr. Rudolf, Kaiserslautern 076661
Gleichweit-Strasser, Leonore, Salzburg 127848
Das Gleimhaus, Halberstadt ... 018938
Gleise, Betty, Digne-les-Bains .. 068306
Glem, København 065907
Glémeau, Raymond, Dinard 103699
Glen Antiques, Bansha 079452
Glen Eira City Council Gallery, Caulfield South 000953
Glen Ewen Community Antique Centre, Glen Ewen 005881
Glen Helen Ecology Institute Trailside Museum, Yellow Springs 054575
Glen Innes Museum, Land of the Beardies History House, Glen Innes 001105
Glen, Robert & Julia, Nairobi ... 111561
Glenara, Boorowa 098761
Glénat, Grenoble 136820
Glénat, Lyon 140899
Glenbow Museum, Calgary 005521
Glencairn Museum, Bryn Athyn . 047291
Glencoe and North Lorn Folk Museum, Glencoe 044427
Glencolmcille Folk Village Museum, Glencolmcille 024929
Glenconnor Antiques, Clonmel .. 079486
Glencorse, Kingston-upon-Thames 089706
Glendale, Halifax 117902
Glendon Gallery, Glendon College, York University, Toronto 007177
Glendower State Memorial, Lebanon 050206
Glenelg Antique Centre, Glenelg 061531
Glenelg Art Gallery, Glenelg 099083
Glenelg Fine Art Galleries, Glenelg 099084
Glenesk Folk Museum, Brechin . 043711
Glenesk Folk Museum, Glenesk 044428
Glenfarclas Distillery Museum, Ballindalloch 043473
Glenfinnan Station Museum, Fort William 044364
Glengarry, Paddington, Queensland 062052
Glengarry Pioneer Museum, Dunvegan 005732
Glengarry Sports Hall of Fame, Maxville 006279
Glenhyrst Art Gallery of Brant, Brantford 005470
Glenleigh Antiques, Armadale, Victoria 061029
Glenluce Motor Museum, Glenluce 044429
Glenlyon Gallery, Aberfeldy .. 117215
Glenn, Nashville 095003, 127330
Glenn H. Curtiss Museum, Hammondsport 049333
Glenn House, Cape Girardeau .. 047417
Glennie & Partners, Fraser, Cirencester 126860
Glenreagh Memorial Museum, Glenreagh 001107
Glenrhydding Gallery, Otley .. 118860
Glenross, Holbrook 061639
Glen's Collectibles, Honolulu .. 093545
Glensheen-Historic Congdon Estate, University of Minnesota Duluth, Duluth 048382
Glentworth Museum, Glentworth 005884
Glenview Antiques, Oakland ... 095893
Glenview Area Historical Museum,

Glenview 049125
Glenz, Erbach, Odenwaldkreis .. 129796
Glerum, Amsterdam 126272
Glesner, Poul H. & Inger, Hellerup 065732
Glessner House Museum, Chicago 047662
Gletschergarten Luzern, Stiftung Amrein-Troller, Luzern 041705
Glickman, Bruce, New York 095421
Glimmingehus, Hammenhög ... 040686
Glims Talomuseo, Espoon Kaupinginmuseo, Espoo 010773
Glineur, Patrice, Saint-Martin-de-Ré 072938
Gliorio, Samuele, Perugia 131752
Global, Groningen 083289
Global Antiques, Las Vegas 094148
Global Art Services, Bruxelles .. 100616
Global Art Venue, Seattle 124535
Global Arts, Manchester 138264
Global Arts Link, Ipswich 001167
Global Gallery, Paddington, New South Wales 099521
Global Health Odyssey, Atlanta . 046622
Global Outlet, Chicago 092715
Global Street Arts, Amsterdam . 082761
Global Village, Helensville ... 083954
Global Works, Murwillumbah ... 061932
Globalart4u, Glasgow 117844
Globe Art Gallery, Hawkes Bay . 112912
Globe Books, Seattle 145584
Globe Poster Printing Corporation, Baltimore 119742
Globe Restoration, Seattle 136371
Globenmuseum der Österreichischen Nationalbibliothek, Wien ... 002999
Il Globo, Roma 081403
Globus Antiquitäten, Essen ... 075676
Globus Galerie, Leipzig 107832
Glocken-Museum, Siegen 021852
Glockenmuseum, Apolda 016600
Glockenmuseum, Laucha an der Unstrut 019924
Glockenmuseum der Glockengießerei, Innsbruck 002124
Glockenschmiede, Hammerschmiede-Museum, Ruhpolding 021547
Gloddfa Ganol Slate Mine, Blaenau Ffestiniog 043647
Glodowski, B., Denzlingen 141683
Glöckner, Claudia, Köln 107630
Glöde, Anette, Uppsala 133597
Gloires du Passé, Paris 071435
Glomdalsmuseet, Elverum 033585
Glommen, Brent, Saint Paul ... 136220
Glommersträsks Hembygdsmuseum, Glommersträsk 040643
Gloor & Co, Jules, Aarau 116005
Glor na nGael Museum, Carrickmacross 024841
Glore Psychiatric Museum, Saint Joseph 052784
Gloria Aguirre Pagazaurtundua, Getxo 086009
Gloria Maris Schelpengalerie, Giethoorn 032350
Gloria Monasterio, Santander .. 086425
Las Glorias Antiguedades, Barcelona 085779
Gloria's Art Gallery, Omaha .. 096041, 123317
Glorious Gallery, Toronto 064659
Glory, Los Angeles 094345
Glory Days Antiques, Islington . 061652
The Glory Hole, Earl Shilton .. 089124
Gloßner, Berchtesgaden 074722
Gloßner, Anita, Berchtesgaden .. 074723
Glossop Heritage Centre, Glossop 044431
Gloth, Gerhard, Hamburg 076229
Glotin, Frédéric, Marseille 128770
Glotin, Pierre, Wolfratshausen .. 079056
Gloucester County Historical Society Museum, Woodbury 054521
Gloucester Folk Museum,

Gloucester 044433
Gloucester Historic Museum, Gloucester 001108
Gloucester Museum, Gloucester 005885
Gloucester Road Bookshop, London 144610
Gloucestershire Warwickshire Railway, Toddington 045984
Gloux, Concarneau 103623
Gloux, Josiane, Thourie 073676
Glove Museum, New York 051396
Glover-Baylee & Co, Thornbury . 125241
Glovertown Heritage Museum, Glovertown 005886
Glowienko-Reinhard, Gisela, Wolfenbüttel 130776
Glowlab, New York 122659
Gluckselig, Kurt, New York 095422
Glud Museum, Horsens 010141
Glück, E., Dachau 075286
Glücksgriff, Bad Harzburg 106059
Glustin, Saint-Ouen ... 073097, 073098
Gluza, Ryszard, Wrocław 113940
Glyde Gallery, Mosman Park ... 099391, 127647
Glydon & Guess, Kingston-upon-Thames 089707, 134515
Glyndor Gallery and Wave Hill House Gallery, Bronx 047229
The Glynn Art Association, Saint Simons Island 052870
Glynn Interiors, Knutsford 089731, 134524
Glynn Vivian Art Gallery, Swansea 045934
Glyptoteks Galleriet, København 102899
Gmeiner, Hanny, Dornbirn 062548
GML Fine Art, Saint Paul 123910
GMS Antiques, Baltimore ... 092328
Gmünder Kunst-& Schmuckauktionshaus, Schwäbisch Gmünd 126114
Gmünder Kunstverein, Schwäbisch Gmünd 058908
Gmurzynska, Sankt Moritz ... 116623
Gmurzynska, Zug 116885
Gmyrek, Wolfgang, Düsseldorf . 106791
Gnaccarini Galleria d'Arte, Bologna 109848
Gnadenhütten Historical Park and Museum, Gnadenhutten ... 049139
Gnägi, Bendicht, Lyss 126746
Gnani, P., Singapore 114528
Gneisenau Museum, Schildau, Gneisenaustadt 021646
Gnoli, Maria Elisabetta, Modena 080797, 131573
Gnoss & Horstrup, Oelde 077895
Go Antiques, Columbus 093023
GO Gallery, Amsterdam 112217
Goa Science Centre, Panaji ... 024427
Goanna Gallery, Quindalup ... 099607
Goanna Hill Gallery, Paringa .. 099551
Goarnisson-Busson, Emile, Saint-Benoît-des-Ondes 072689
Goat, Billy, Kumara 112933
Gobabis Museum, Gobabis 031837
Gobbini, Francesco, Bologna .. 130964
Gobert, Maria, Meisenthal 070041
Goble, Paul, Brighton 088622
Goclette La Recouvrance, Brest . 012205
Godalming Museum, Godalming 044439
Godang, Busan 082253
Godar, Genevieve, Lille 104009, 104010
Godard Desmarest, Paris 071436
Godard, Mira, Toronto 104101
Goddard & de Fiddes, West Perth 099882
Goddard, Keith R., London 090152
Goddess Gallery, Portland 096414
Godéas, Evelyne, Toulouse 073773
Godeau, Antoine, Paris 125719
Godeccy, Andrzej i Bożena, Łódź 084613
Godefroy, Olivier, Cambrai 067614
Godegårds Bruks- och

Porslinsmuseum, Godegård Manor Porcelain Museum, Motala .. 040857
Godehardt, F., Bochum 129599
Godel & Co., New York 122660
Godement, La Haye-Malherbe .. 069052
Godfrey & Watt, Harrogate ... 117916
Godfrey Dean Art Gallery, Yorkton 007499
Godfrey, Jemima, Newmarket .. 090831
Godfroy, Monique, Vallauris ... 105807
Godin, Liège 063761
Godin, Laurent, Paris 104917
Godo, Seoul 111654
Godøy Kystmuseum, Godøy ... 033619
Godolphin, Chagford .. 088814, 117551
Godon, Lille 140870
Godoy, George & Sandra, Tokyo 082148
Gods Artefacts, Crows Nest, New South Wales 098935
Godson & Coles, London ... 090153, 118351
Godwin-Ternbach Museum, Flushing 048765
Göbel, Hannelore, Speyer 078522
Goecke, Christoph, Berlin 074803, 129498
Goeckler, Bernd H., New York .. 095423
Göcseji Falumúzeum, Göcseji Múzeum, Zalaegerszeg 024051
Göcseji Múzeum, Zalaegerszeg . 024052
De Goede Oude Tijd, Eindhoven 083218
Goede, Gudrun, Bremen 106584
Gödecke, Kerken 076758
Gödecke, Axel, Issum 076629
Goedhuis, London 118352
Goedhuis, Michael, New York .. 122661
Goedkoper, Zutphen 083717
Gödöllői Királyi Kastélymúzeum, Gödöllő 023644
Gödöllői Városi Múzeum, Gödöllő 023645
Goeffriault, Alexandre, Montpellier 140966
Gögler, Roland, Leutkirch im Allgäu 142072
Goeij, Wim de, Antwerpen ... 140146
Göke, München 077511
Goeken, Münster 108218, 108219
Gölles, Fürstenfeld 062563
Gölles-Valda, Martina & Oswald, Riegersburg, Steiermark 062779, 127838
Göllner, Franz, Bergheim 127718
Götzschtalgalerie-Nicolaikirche, Auerbach, Vogtland 016642
Goemanszorg – Streek- en Landbouwmuseum, Dreischor . 032253
Gömöri Múzeum, Putnok 023865
Göreme Açıkhava Müzesi, Nevşehir 043033
Görg, Benzweiler 106133
Görg, Kurt, Goldbach 076058
Goering Institut e.V., München . 130349
Goeritz, Flensburg 075744
Görl, Naumburg, Saale 108233
Das Görlitzer Antiquariat, Görlitz 141832
Görög-Római Szobormásolatok Kiállítása, Tata 023998
Görög Templom Kiállítóhely, Tragor Ignác Múzeum, Vác 024024
Görögkatolikus Egyházművészeti Gyűjtemény, Nyíregyháza ... 023811
Görsan, İstanbul 117045
Goerz, Luxembourg 111853
Göschenhaus / Seume-Gedenkstätte, Grimma 018818
Gösseln, Anya von, Clonegal .. 109552
Gösta Serlachiuksen Taidemuseo, Mänttä 011138
Göta Kanalutställning, Göta Canal Exhibition, Motala 040858
Göteborgs Antik, Göteborg ... 086710
Göteborgs Antik och Konstmässa, Göteborg 098319
Göteborgs Auktionsverk, Göteborg 126587

Göteborgs Konservatorateljé,
 Landvetter 133544
Göteborgs Konstförening,
 Göteborg 059882
Göteborgs Konsthall, Göteborg . 040651
Göteborgs Konstmuseum,
 Göteborg 040652
Göteborgs Naturhistoriska Museum,
 Göteborg 040653
Göteborgs Stadsmuseum,
 Göteborg 040654
Goetgheluck, Hubert, Caestre . 067586
Göthberg, Lars, Tullinge 133592
Goethe- und Schiller-Archiv, Klassik
 Stiftung Weimar, Weimar ... 022494
Goethe-Gedenkstätte im Amtshaus,
 Museum der Stadt Ilmenau,
 Ilmenau 019388
Goethe-Gedenkstätte, Friedrich-
 Schiller-Universität Jena, Jena 019447
Goethe Institut – Galerie Condé,
 Paris 104918
Goethe-Institut New York – Exhibitions,
 New York 051397
Goethe-Institut Sydney, Woollahra 001660
Goethe-Museum Düsseldorf, Anton-
 und-Katharina-Kippenberg-Stiftung,
 Düsseldorf 018029
Goethe-Museum Stützerbach
 mit Museum zur Geschichte des
 technischen Glases, Klassik Stiftung
 Weimar, Stützerbach 022021
Goethe-Nationalmuseum, Klassik
 Stiftung Weimar, Weimar 022495
Goethes Gartenhaus 2, Bad
 Sulza 016888
Goethes Gartenhaus, Klassik Stiftung
 Weimar, Weimar 022496
Goettert, A. & A., Zirndorf 109007
Götting, H., Münster 077662
Götting, Heinrich, Münster 077663
Göttinger Antiquariat, Göttingen . 141835
Göttinger Münzhandlung,
 Göttingen 076045
Göttler, Armin, Wasserburg am
 Inn 078874, 130736
Göttler, Christine, Wasserburg am
 Inn 078875
Götz, G., Landau in der Pfalz .. 077003
Götz, Karl-Heinz & Doris,
 Weinstadt 142475
Götz, Kornelius, Oettingen in
 Bayern 130486
Götz, Kurt, Karlsruhe 141971
Götz, Rainar, Dresden . 075413, 106736,
 141696
Götz, Ulrike, Berlin ... 074804, 106301
Goetz/Fleischack Museum,
 Potchefstroom 038809
Goetzke, Andreas,
 Oberottmarshausen 130482
Goez, Los Angeles ... 121497, 121498
Goff & Rosenthal, New York .. 122662
Goffi Carboni, Roma 081404
Goffin, E., Liège 063762
Goffinet, Bernard, Lutterbach . 069670
Gofish Gallery, Maylands, Western
 Australia 099291
Gogat, Nicole, Aigues-Mortes .. 103181
Gogherman, A., Westervoort .. 132668
Goh, Karin, Bühlertal 106612
Goh, Swar Teck, Singapore ... 085410
Gohard, Robert, Paris 129019
Gohltalmuseum, Musée de la Vallée
 de la Gueule, Neu-Moresnet .. 003907
Goins, R.L., Atlanta 119571
Góis & Ferro, Lisboa 132936
Góis, Abel, Funchal 084806
Gojan, Tomasz, Warszawa ... 132879
Gojun Gallery, Gwangju 082255
Gol Bydgemuseet, Gol 033620
Golan Archaeological Museum,
 Qatzrin 025169

Gold, Atlanta 092140, 127126
Gold- und Silberschmiedemuseum,
 Mülsen 020501
Gold- und Silberschmiedemuseum,
 Wien 003000
Gold & Coin Buyer, Boston 092473
Gold & Diamond Exchange, Las
 Vegas 094149
Gold & Silver Exchange, Great
 Yarmouth 089383
Gold & Silver Exchange Action,
 Sacramento 096626
Gold and Silver Lady, Toledo .. 097605
Gold Art Gallery, Bucureşti ... 114143
Gold Cadre, Casablanca 082563
Gold Coast City Art Gallery, Surfers
 Paradise 001528
Gold Coast Fine Art Gallery,
 Chicago 120140
Gold Coast Gallery, Chicago .. 092716,
 135412
Gold Coast Prepress, Saint Louis 123827
Gold Coast Railroad Museum,
 Miami 050829
Gold Creek Curiosity Shoppe,
 Nicholls 061973
Gold Creek Gallery, Nicholls .. 099457
Gold Dragon Gallery, Tampa .. 124653
Gold Dust Coins, Chicago 092717
Gold Exchange, Las Vegas ... 094150
Gold Leaf Conservation, Chardon 135386
Gold Leaf Restoration, Portland . 136152
Gold Money, Busto Arsizio 079968
Gold Museum, Ballarat 000817
Gold 'n Grate Gallery, Pemberton 099557
Gold 'n' Kiwi Antiques,
 Otorohanga 084048
Gold 'n Silver, Omaha 096042
Gold of Africa Museum, Cape
 Town 038678
Gold Rush History Museum,
 Coloma 047880
Gold Rush Jewellers, Bendigo .. 061132
Gold Rush Numismatics, San
 Francisco 097244
Gold Street Antiques, San Jose . 097369
Gold Strike, Chicago 092718
Gold Traders, Toronto 064660
Gold Treasury Museum,
 Melbourne 001272
Gold, Fay, Atlanta 119572
Gold, Vilma, London . 118353, 144611
Goldammer, Breisach am Rhein 075137
Goldberg, West Hollywood ... 136471
Goldberg-Museum, Riesbürg ... 021431
Goldberg, David, New Orleans . 095107
Goldborgsund Museum, Nyköbing
 Falster 010243
Goldcare, Middlesborough 134852
Goldcoast Auctioneers, Miami . 127316
Golden Age Antiques, New York 095425
Golden and District Museum,
 Golden 005890
Golden Art, Brasilia 100916
Golden Art, Helsinki 066316
Golden Arts Gallery, Singapore . 114529
Golden Ball Tavern Museum,
 Weston 054349
Golden Bay Museum and Gallery,
 Golden Bay 033209
Golden Business Company,
 Antwerpen 063207
Golden Center, Damascus 088035
Golden Chair, Hampton 061591
Golden City Paddle Steamer Museum,
 Ballarat 000818
Golden Crust Gallery, Armadale,
 Victoria 061030
Golden Dragon, Regina 101529
Golden Dragon Museum,
 Bendigo 000850
Golden Dynasty Gallery, Chicago 092719
Golden Eagle Antiques,
 Allentown 127118

Golden Eagle Antiques, Tucson . 097666
Golden Era Antiques, Miami .. 094740
Golden Flower Art Gallery, Pine
 Creek 099567
Golden Gallery, Miercurea-Ciuc . 114171
Golden Gargoyles, Philadelphia . 096110
Golden Gate Antiques, San
 Francisco 097245
Golden Gate Railroad Museum, San
 Francisco 053002
Golden Goose Books, Lincoln .. 144539
Golden Harvest Antiques, Saint
 Louis 096711
Golden Hind Museum, Brixham . 043757
Golden Hinde Educational Trust,
 London 044978
Golden Horn, İstanbul 117046
Golden Isles Arts & Humanities
 Association, Brunswick 060499
Golden Lion, Los Angeles 094998
Golden Lion Antiques, Saint Paul 096825
Golden Lion Bazaar, Townsville . 062338
Golden Lotus Gallery, Ha Noi ... 125075
Golden Lotus Gallery, Ho Chi Minh
 City 125121
Golden Memories of York, York . 099177
Golden Oldies, Penkridge 090952
Golden Pacific Arts, San Diego . 124129
Golden Peart, Cairo 066250
The Golden Pendulum, Griesheim 076094
Golden Pioneer Museum, Golden 049148
Golden Ram Gallery, Cincinnati . 120347
Golden Saloon, Damascus 088036
Golden Spike National Historic Site,
 Brigham City 047209
Golden State Model Railroad Museum,
 Point Richmond 052229
Golden Ta Art Gallery, Wuxi .. 102253
Golden Tortoise, Bangkok 088078
Golden Triangle, Chicago 092720
Golden Turtle, San Francisco .. 136320
Golden Wong Enterprise,
 Singapore 085411
Golden Years Antiques, Kyoto .. 082014
Golden, Caren, New York 122663
Golden, Suzanne, New York ... 095424
Goldenflow Studios, Minneapolis 121992
Goldenlands Gallery, Saint Louis 123828
Golder Halm, Locarno 116446
Goldesberry, Nancy & Oliver,
 Houston 121047
Goldfield, West Hollywood 124997
Goldfield Gallery, Portsmouth .. 118942
Goldfields Aboriginal Art Gallery,
 Kalgoorlie 099166
Goldfields Arts Centre Gallery,
 Kalgoorlie 099167
Goldfinger, Lyon 069744
Goldfish Contemporary Fine Art,
 Penzance 118895
Goldhorn, M., Groningen 083290
Goldin, Natalia, Stockholm ... 115886
Golding Young & Co., Grantham 126906
Goldman, Judy Ann, Boston ... 119870
Goldman, Mary, Los Angeles ... 121497
Goldmark, Uppingham . 119286, 144902
Gold'n Time, Washington 097840
Goldrach, Alain, New York ... 135930
Goldsmith, Lenham 118101
Goldsmith, Oundle 127004
Goldsmith, Brion, London 101344
Goldsmiths' Hall, London 044979
Goldstein, Bielefeld 106493
Goldstein, Houston 121048
The Goldstein Museum of Design,
 University of Minnesota, Saint
 Paul 052839
Goldstream Region Museum,
 Victoria 007325
Goldthorpe, P., Passmarsh 118883
Goldwasser, Bernardo, Rio de
 Janeiro 064062
Goldwasser, Thomas A., San
 Francisco 145556

Goldwater, Auckland 112773
Golem, Denver 120730
Golemoto Staklo, Skopje 138898
Golestan Palace Museum, Tehrān 024757
Golf Museum, Far Hills 048694
Golf Museum, Newport News . 051545
Golfiana Gallery, London 144612
Golfmuseum, Regensburg 021321
Goliad State Historical Park,
 Goliad 049155
Goliard, Paul, Espinasse-Vozelle 068470,
 128595
Golkar, H., Köln 107631
Goll, Karl H., Zwingenberg, Kreis
 Bergstraße 079157
Gollasch, Werner, Berlin 074805
Goller, Ernestine, Frensdorf ... 075899
Gollong, Frédéric, Saint-Paul-de-
 Vence 105590
Golmard, Dijon 068331
Gołogórski, M., Kraków 113600, 113601
Golovanoff, Paris 071437
Het Gols-Station, Winterswijk .. 033055
Goltz, Bettina von der, Hannover 130008
Goltz, Hans, München 142156
Golubaja Gostinnaja, Sankt-
 Peterburg 114342
Golubarsch, Karl, Krefeld 076956
Goluc, Chicago 120141
Golyester, Los Angeles 094347
Goma Laca, Granada 133265
Gomart, México 082475
Gombert, André, Le Blanc-Mesnil 069222
Gombert, André, Paris 071438
Gomboc, Ron & Terrie, Middle
 Swan 099335
Gomér & Andersson, Linköping . 126623
Gomersbach, Ellen, Zeist 083711
Dias Gomes, Lisboa 084894
Gomes, Armando Fernandes,
 Lisboa 132937
Gomes, Bernardino, Paris 071439
Gomes, Hebert, Rio de Janeiro . 064063
Gomes, N.V., Porto Alegre 063973
Gomez, Baltimore 119743
Gómez, Bucaramanga 102390
Gomez, Tarragona 086483
Gómez Benito, Carmelo,
 Valladolid 086548, 143811
Gomez Costa, Begoña, Madrid . 115227
Gomez de Leon, Ernesto Nuñez,
 Barcelona 085780
Gomez Foundation for Mill House,
 Marlboro 050700
Gomez Gil, Eduardo Jose,
 Estepona 085993
Gomez Rodrigo, Jose Maria,
 Madrid 115228
Gomez Salcedo, Alejandro, Palma de
 Mallorca 115349
Gomez Turu, Barcelona 085781
Gomez Turu, Josep, Barcelona . 114915
Gómez, Cecilia, Bogotá 102339
Gómez, Clara de, Bogotá 102340
Gomez, Emilio, Donostia-San
 Sebastián 085974
Gomez, Felix, Sevilla 115450
Gómez, Isabel, Madrid 133341
Gomez, Manuel, Perpignan ... 129164
Gómez, Marco Isaac, Bogotá .. 128269
Gomez, Victor, Miami 121796
Gomiero, Diego, Padova 080984
Gomiti, Nicoló, Firenze 080156
Gommalacca, Milano 131454
Gomollon Tobajas, Manuel,
 Zaragoza 115552
Gomringer, B., Mertingen 077352
Gomshall Gallery, Gomshall .. 044444
Goncalves, Joao, Joué-lès-Tours 068917
Gonçalves, João, Porto 132971
Gonçalves, José Pereira,
 Amadora 084775
Goncalves, Lobo Joao, Joué-lès-
 Tours 068918

La Gondolina, Venezia 081843
Gone West Books, Auckland . . . 143282
Gongju Folk Drama Museum,
Gongju 029935
Gongju National Museum, Gongju 029936
Gongpyeong Art Center, Seoul . . 029997
Gongsan, Daegu 111584
Gongwangfu Art Cooperative,
Beijing 101954
Gonin-Flambois, Jean-Philippe, Sainte-
Geneviève 073352
Gonnelli, Firenze 142675
Gonnet, Christian, Paris 071440
Gonod, Jean-Bernard, Aurillac . . 066826
Gonon, Nicole, La Chapelle-de-
Guinchay 068972
Gonsior, S., Hanau 076353
Gontier, Christophe, Saint-Sébastien-
sur-Loire 073304
Gonzales Memorial Museum,
Gonzales 049157
Gonzalez, Buenos Aires 060905
González, Pamplona 086390
Gonzalez, Rio de Janeiro 101037
González Angel, Fabiola, Cali . . 102409
Gonzalez Berlanga, Rodolfo,
Madrid 133342
Gonzalez Camones, Julio, Lima . 084455,
113361
Gonzalez de la Peña, Beatriz Zamora,
Sevilla 133431
Gonzalez Garcia, Ricardo, Getxo 086010
Gonzalez Gonzalez, Miguel Angel,
Lleida 115116, 115117
Gonzalez Rial, Victor, A Coruña . 085954
González, Cecilia, Lima 113360
Gonzalez, Christopher, Runaway
Bay 111239
Gonzalez, Elvira, Madrid 115229
Gonzalez, Jacqueline, Toulouse . 073774
Gonzalez, Maria Pilar, Bilbao . . 133192
Gonzalo Aurtenechea Urbina,
Madrid 086196
Gonzalo de Jesús, Mejía Gabiria, San
José 102507
Gonzen-Museum, Sargans 041935
Goochland County Museum,
Goochland 049158
Good Day Antiques, Rowlands
Castle 091148, 134981
Good Design Shop, New York . . 095426
Good Earth Antiques, Buffalo . . 092528
Good Earth Gallery, Philadelphia 123388
Good Hope Antiques, Beaminster 088415,
134082
Good Neighbor Thrift Store, Long
Beach 094203
Good Old Days, Chicago 092721
Good Old Days Antikviteter,
København 065908
Good Old Things, New York . . . 095427
The Good Old Times, Almelo . . 082628
Good Reading, Benalla 139640
Good Stuff, Albuquerque 092054
Good Stuff, Oklahoma City . . . 095973
Good, Bad and Ugly, Sacramento 096627
Good, Daniel, Luzern 144108
Good, Martha, Brisbane 061210
Good, Robert J., Saint Louis . . . 136201
The Good, the Bad and the Ugly,
Margate 061794
The Good, the Bad and the Ugly,
Oakland 095894
Goodacre Engraving, Sutton
Bonington 091514
Goodall, Richard, Manchester . . 118707
Gooday, Debbie, Richmond,
Surrey 091095
Goodbreys, Framlingham 089303
Goodchild, Dallas 093131, 135504
Goode & Co., Thomas, London . 090154,
134675
Goodenow, Steve, Honolulu . . . 093546

Goodhue County Historical Society
Museum, Red Wing 052492
Goodman Museum, Tyler 053873
Goodman, James, New York . . . 122664
Goodman, Marc, Cleveland 092941
Goodman, Marian, Paris 104919
Goodnow Museum, Manhattan . 050636
Goods and Chattels, Mount
Gambier 061910
Goods and Chattels,
Woolloongabba 062488
Goodsoil Historical Museum,
Goodsoil 005893
Goodstuff Consignments,
Richmond 096545
Goodwater Gallery, Toronto . . . 101619
Goodwill, Seattle 145585
Goodwin, Edinburgh . . 089175, 089176
Goodwin, Loretta, Birmingham . 119800
Goodwin, Pamela, Tunbridge
Wells 091651
Goodwood House, Chichester . . 043942
Goodwood House Antiques,
Burswood 061242
Goody, Virginia Beach 097783
Goodyer, Nicholas, London . . . 144613
Goois Museum, Hilversum 032485
Goold Catholic Museum, East
Melbourne 001045
Goole Museum and Art Gallery,
Goole 044446
Goolen, Wilfried, Aurillac 125500
Goolwa National Trust Museum,
Goolwa 001109
Goomeri Antiques, Goomeri . . . 061541
Gooran, Yossef, Essen 075677
Goors Historisch Museum, Goor 032358
Goos, Spiekeroog 108656
Goose Gallery, Milton-on-Stour . 118748
Goosens, Wilfried, Mechelen . . . 063790
Goosmann, Bad Bergzabern . . . 106041
Goossens, D.&H., Bruxelles . . . 100617,
128029
Goppelt & Lell, Karlsruhe 130096
Goppert Gallery, University of Saint
Mary, Leavenworth 050204
Goralsky, Obernai 104419
Gora's House of Antiques,
Ottawa 064503
Gorcums Museum, Gorinchem . 032360
Gordin & Assoc, Victor, Toronto . 064661
Gordion Müzesi, Polatlı 043043
Gordon, Geelong 099070
Gordon Brown Collection,
Ravenshead 045584
Gordon Gallery, Tel Aviv 109725
Gordon Highlanders Museum,
Aberdeen 043357
Gordon Museum, London 044980
Gordon Snelgrove Art Gallery,
Saskatoon 006973
Gordon, Leah, New York 095428
Gordon, Robert, London 090155
Gordon, Roy, Fort Worth 120866
Gordon, Victor, Philadelphia . . . 096111,
127392
Gordon's Antiques, Waikato . . . 084129
Gordon's Discoveries, North
Perth 061994
Gore Airforce Museum, Gore . . 033211
Gore Historical Museum and Hokonui
Heritage Research Centre, Gore 033212
Gore Place, Waltham 044050
Gorenjski Muzej Kranj, Grad Khislstein,
Kranj 038532
Gorepani, Albany 098619
Gorgan Museum, Gorgan 024694
Gorgas House Museum, University of
Alabama, Tuscaloosa 053858
Gorge Road Gallery, Dunedin . . 112882
Gorgi, Mahrajène 116975
Gorgo, New Orleans 122175
La Gorgona, Trieste 132220
Gori Antichità, Padova 080985

Gori, Amanda, Prato 131776
Goris Local Tradition Museum,
Goris 000701
Goriški Muzej, Nova Gorica . . . 038584
Gorky Antiquités, Paris 071441
Gorlik, Joachim, Kempten 076751
Gorman, Manitou Springs 127305
Gormleys', Belfast 117338
Gorney, Bravin & Lee, New York 122665
Gornjesavski Muzej Jesenice,
Jesenice 038520
Górnośląski Park Etnograficzny,
Chorzów 034694
Gorny & Mosch, Giessener
Münzhandlung, München 077512,
126050
Gornyj Muzej, Sankt-Peterburg . 037733
Gorodeckij Kraevedčeskij Muzej,
Gorodec 037021
Gorodka Galeries, Galerie d'Art
Contemporain, Sarlat-la-Canéda 015688
Gorodok Chudožnikov, Art Center –
City of Artists at the Union of Artists
of the Kyrgyz Republic, Biškek 059537
Gorodskoj Istoriko-Kraevedčeskij
Muzej, Ridder 029840
Gorodskoj Kraevedčeskij Muzej,
Elec 037003
Gorodskoj Muzej Boevoj Slavy,
Kaluga 037088
Gorodskoj Muzej Iskusstvo Omska,
Omsk 037591
Gorodskoj Muzej Teatralnogo
Iskusstva, Omsk 037592
Gorodskoj Vystavočnyj Zal,
Petrozavodsk 037651
Gorostarzu, Marie-Odile de,
Noailles 070720
Gorring, Tunbridge Wells 144892
Gorringe, Bexhill-on-Sea 126818
Gorringe, Lewes 126942, 126943
Gorringe, Worthing 127102
Gorringes, Tunbridge Wells . . . 127071
Gorrotxategi Konfiteri Museoa,
Museo Gorrotxategi de Confitería,
Tolosa 040245
Gorry, James A., Dublin 109587, 130857
Gorry, P.T., Moruya 139836
Gorsium Szabadtéri Múzeum, Szent
István Király Múzeum, Tác . . . 023990
Górska, Urszula, Wrocław 113941
Gorvitz, Christian, Neuilly-sur-
Seine 070521
Gosa Lo, Chicago 135413
Goschenhoppen Folklife Museum,
Green Lane 049230
Gosford Books, Coventry 144356
Gosford Regional Gallery and Arts
Centre, Gosford 001113
Goshen Historical Society Museum,
Goshen 049165
Goslar, Michele, Bruxelles 063451
Goslarer Museum, Goslar 018765
Goslett, Roland, Richmond,
Surrey 091096, 134967
Gosling, Alison, Honiton 089562
Gosling, Jenny, Upper
Woolhampton 144899
Gospodinov, G., Montana 101128
Gosport Museum, Gosport 044451
Goss and Crested China Centre,
Horndean 139180
Gossart, Antoine, Paris 071442
Gossieaux, Martine, Paris 104920
Gossipgate Gallery, Alston 117232
Gossot, Renée, Dijon 068332
Gost, Jaime, Palma de Mallorca 086353
Goston, Sergio C. A., Rio de
Janeiro 101038
Gosudarstvennaja Kollekcija
Unikalnych Muzykalnych
Instrumentov, Moskva 037276
Gosudarstvennaja OK
Istoriko-ètnografičeskij i

Architekturnyj Muzej-zapovednik
Staraja Sarepta, Volgograd . . . 038048
Gosudarstvennaja Tretjakovskaja
Galerja, Moskva 037277
Gosudarstvennaja Tretjakovskaja
Galerja na Krymskom Valu,
Moskva 037278
Gosudarstvennoe Muzejnoe
Objedinenie Chudožestvennaja
Kultura Rossijskogo Severa,
Archangelsk 036862
Gosudarstvennyj Biologičeskij Muzej
im. K.A. Timirjazeva, Moskva . 037279
Gosudarstvennyj Borodinskij Voenno-
istoričeskij Muzej-zapovednik,
Borodino 036912
Gosudarstvennyj Centr Sovremennogo
Iskusstva, Moskva 037280
Gosudarstvennyj Centralnyj Muzej
Muzykalnoj Kultury im. M.I. Glinki,
Moskva 037281
Gosudarstvennyj Centralnyj Muzej
Sovremennoj Istorii Rossii,
Moskva 037282
Gosudarstvennyj Centralnyj Teatralnyj
Muzej im. Bachrušina, Moskva 037283
Gosudarstvennyj Chudožestvenno-
architekturnyj Dvorcovo-parkovyj
Muzej-zapovednik Oranienbaum,
Lomonosov 037201
Gosudarstvennyj Chudožestvenno-
architekturnyj Dvorcovo-parkovyj
Muzej-zapovednik Pavlovsk,
Pavlovsk 037622
Gosudarstvennyj Chudožestvennyj
Istoriko-architekturnyj i prirodno-
landšaftnyj Muzej-zapovednik
Kolomenskoe, Moskva 037284
Gosudarstvennyj Chudožestvennyj
Muzej Altajskogo Kraja, Barnaul 036883
Gosudarstvennyj Dom-muzej P.I.
Čajkovskogo, Klin 037145
Gosudarstvennyj Dom-muzej I.V.
Stalina, Gori 016389
Gosudarstvennyj Ėrmitaž, Sankt-
Peterburg 037734
Gosudarstvennyj Geologičeskij Muzej
im V.I. Vernadskogo, Moskva . 037285
Gosudarstvennyj Istoričeskij Muzej,
Moskva 037286
Gosudarstvennyj Istoričeskij Muzej
Kyrgyzskoj Respubliki, Biškek . 030072
Gosudarstvennyj
Zapovednik Gorki Leninskie, Gorki
Leninskie 037019
Gosudarstvennyj Istoriko-architekturnyj
i Chudožestvennyj Muzej-zapovednik
Drevnij Derbent, Derbent 036967
Gosudarstvennyj Istoriko-architekturnyj
i Ėtnografičeskij Muzej-zapovednik
Kiži, Kiži 037143
Gosudarstvennyj Istoriko-architekturnyj
i Prirodno-landšaftnyj Muzej
Zapovednik Izborsk, Izborsk . . 037046
Gosudarstvennyj Istoriko-architekturnyj
i Prirodnyj Muzej-Zapovednik
Monrepo, Vyborg 038075
Gosudarstvennyj Istoriko-
chudožestvennyj Dvorcovo-parkovyj
Muzej-zapovednik Gatčina,
Gatčina 037013
Gosudarstvennyj
Istoriko-ètnografičeskij Muzej-
zapovednik Šušenskoe,
Šušenskoe 037905
Gosudarstvennyj Istoriko-kulturnyj i
Prirodnyj Muzej-zapovednik
A.S. Griboedova – Chmelita,
Chmelita 036956
Gosudarstvennyj Istoriko-kulturnyj
Muzej-zapovednik "Moskovskij
Kreml", Moskva 037287
Gosudarstvennyj Istoriko-literaturnyj i
Prirodnyj Muzej-zapovednik A.A.

Bloka, Solnečnogorsk 037880
Gosudarstvennyj Istoriko-literaturnyj
Muzej-Zapovednik A.S. Puškina s
usadbami Vjazemy i Zacharovo,
Bolšie Vjazemy 036907
Gosudarstvennyj Istoriko-memorialnyj
Sankt-Peterburgskij Muzej Smolnyj,
Sankt-Peterburg 037735
Gosudarstvennyj Istoriko-memorialnyj
Zapovednik Rodina V.I. Lenina,
Uljanovsk 037991
Gosudarstvennyj Literaturno-
memorialnyj Muzej Anny Achmatovoj
v Fontannom Dome, Sankt-
Peterburg 037736
Gosudarstvennyj Literaturno-
memorialnyj Muzej N.A.
Dobroljubova, Nižnij Novgorod . 037545
Gosudarstvennyj Literaturno-
memorialnyj Muzej im. M.Ju.
Lermontova, Pjatigorsk 037658
Gosudarstvennyj Literaturno-
memorialnyj Muzej N. Ostrovskogo,
Soči 037876
Gosudarstvennyj Literaturno-
memorialnyj Muzej A.N. Radiščeva,
Radiščevo 037683
Gosudarstvennyj Literaturno-
memorialnyj Muzej-zapovednik A.P.
Čechova, Melichovo 037222
Gosudarstvennyj Literaturno-
memorialnyj Muzej-zapovednik
N.A. Nekrasova "Karabicha",
Karabicha 037095
Gosudarstvennyj Literaturno-
memorialnyj Muzejnyj Kompleks
Sabita Mukanova i Gabita Musrepova,
Almaty 029799
Gosudarstvennyj Literaturnyj Muzej,
Moskva 037288
Gosudarstvennyj Literaturnyj Muzej
A.M. Gorkogo, Nižnij Novgorod 037546
Gosudarstvennyj Literaturnyj Muzej im.
I.S. Turgeneva, Orël 037601
Gosudarstvennyj Memorialnyj
Dom-muzej Bulata Okudžavy,
Peredelkino 037635
Gosudarstvennyj Memorialnyj Dom-
muzej N.A. Rimskogo-Korsakova,
Tichvin 037935
Gosudarstvennyj Memorialnyj i
Prirodnyj Muzej-zapovednik
A.N. Ostrovskogo – Ščelykovo,
Ščelykovo 037850
Gosudarstvennyj Memorialnyj i
Prirodnyj Muzej-zapovednik
I.S. Turgeneva, Spasskoe-
Lutovinovo 037886
Gosudarstvennyj Memorialnyj
Istoriko-chudožestvennyj i Prirodnyj
Muzej-zapovednik V.D. Polenova,
Strachovo 037902
Gosudarstvennyj Memorialnyj
Muzej im. A.V. Suvorova, Sankt-
Peterburg 037737
Gosudarstvennyj Memorialnyj Muzej
Oborony i Blokady Leningrada, Sankt-
Peterburg 037738
Gosudarstvennyj Memorialnyj Muzej-
usadba V.G. Belinskogo, Literaturnyj
muzej Penza, Belinskij 036890
Gosudarstvennyj Moskovskij Muzej
Naivnogo Iskusstva, Moskva . . 037289
Gosudarstvennyj Muzej –
Gumanitarnyj Centr 'Preodolenie' im.
N.A. Ostrovskogo, Moskva . . . 037290
Gosudarstvennyj Muzej Detskich
Teatrov, Moskva 037291
Gosudarstvennyj Muzej K.A. Fedina,
Saratov 037843
Gosudarstvennyj Muzej Istorii Aviacii,
Sankt-Peterburg 037739
Gosudarstvennyj Muzej Istorii i Kultury
Belarusa, Minsk 003248

Gosudarstvennyj Muzej Istorii
Kosmonavtiki im. K.É. Ciolkovskogo,
Kaluga 037089
Gosudarstvennyj Muzej Istorii Religii,
Sankt-Peterburg 037740
Gosudarstvennyj Muzej Istorii Sankt-
Peterburga, Sankt-Peterburg . . 037741
Gosudarstvennyj Muzej Izobrazitelnych
Iskusstv Respubliki Kalmykija,
Élista 037004
Gosudarstvennyj Muzej A.I. Kuprina,
Narovčat 037539
Gosudarstvennyj Muzej V.V.
Majakovskogo, Moskva 037292
Gosudarstvennyj Muzej-masterskaja
Skulptora A.S. Golubkinoj,
Gosudarstvennaja Tretjakovskaja
Galerja, Moskva 037293
Gosudarstvennyj Muzej Oborony
Moskvy, Moskva 037294
Gosudarstvennyj Muzej Palechskogo
Iskusstva, Palech 037620
Gosudarstvennyj Muzej-panorama
Stalingradskaja Bitva, Volgograd 038049
Gosudarstvennyj Muzej Političeskoj
Istorii Rossii, Sankt-Peterburg . 037742
Gosudarstvennyj Muzej A.S. Puškina,
Moskva 037295
Gosudarstvennyj Muzej-Reservat
Pamjatniki Drevnego Taraza,
Taraz 029850
Gosudarstvennyj Muzej Vadima Sidura,
Moskva 037296
Gosudarstvennyj Muzej L.N. Tolstogo,
Moskva 037297, 037298
Gosudarstvennyj Muzej L.N. Tolstogo
"Jasnaja Poljana", Jasnaja Poljana
(Tula) 037068
Gosudarstvennyj Muzej-usadba
Archangelskoe, Archangelskoe 036866
Gosudarstvennyj Muzej-usadba P.I.
Čajkovskogo, Votkinsk 038071
Gosudarstvennyj Muzej Vyborgskij
Zamok, Vyborg 038076
Gosudarstvennyj Muzej-zapovednik
S.A. Esenina, Konstantinovo . 037149
Gosudarstvennyj Muzej-zapovednik
M.A. Šolochova, Vešenskaja . . 038019
Gosudarstvennyj Muzej Životnovodstva
im. Akademika E.F. Liskuna,
Moskovskaja selskochozjajstvennaja
akademija im. K.A. Timirjazeva,
Moskva 037299
Gosudarstvennyj Nacionalnyj Muzej
Ust-Ordynskogo Burjatskogo
Avtonomnogo Okruga, Ust-
Ordynsk 038007
Gosudarstvennyj Naučno-
issledovatelskij Institut Restavracii,
Moskva 133011
Gosudarstvennyj Naučno-
issledovatelskij Muzej Architektury
im. A.V. Ščuseva, Moskva 037300
Gosudarstvennyj Okružnoj Muzej
Prirody i Čeloveka Goroda Chanty-
Mansijsk, Chanty-Mansijsk . . . 036955
Gosudarstvennyj Russkij Muzej, Sankt-
Peterburg 037743, 037744
Gosudarstvennyj Severnyj Morskoj
Muzej, Archangelsk 036863
Gosudarstvennyj Vladimiro-
Suzdalskij Istoriko-architekturnyj i
Chudožestvennyj Muzej-zapovednik,
Vladimir 038028
Gosudarstvennyj Voenno-istoričeskij i
Prirodnyj Muzej-zapovednik Kulikovo
Pole, Tula 037960
Gosudarstvennyj Voenno-istoričeskij
Muzej-zapovednik Prochorovskoe
Pole, Prochorovka 037665
Gosudarstvennyj Vystavočnyj Zal
Beljaevo, Moskva 037301
Gosudarstvennyj Vystavočnyj Zal
Chodynka, Moskva 037302

Gosudarstvennyj Vystavočnyj Zal
Kovčeg, Moskva 037303
Gosudarstvennyj Vystavočnyj Zal-
muzej Nasledie, Moskva 037304
Gosudarstvennyj Vystavočnyj Zal Na
Kaširke, Moskva 037305
Gosudarstvennyj Vystavočnyj Zal Na
Soljanke, Moskva 037306
Gosudarstvennyj Vystavočnyj Zal
Peresvetov Pereulok, Moskva . 037307
Gosudarstvennyj Vystavočnyj Zal
Solncevo, Moskva 037308
Gosudarstvennyj Vystavočnyj Zal
Zamoskvorečie, Moskva 037309
Got Bears, Oklahoma City 095974
Got to Have It, Cleveland 092942
Got, Frédéric, Barbizon 103331
Gotay, Paul, Salt Lake City . . . 123963
Gotenyama Ippodo Gallery, Tokyo 111440
Gotham, New York 145341
Gotham Edicions, Valencia . . . 138004,
                                    143802
Gothic City Collector's Cove,
Buffalo 092529
Gothic Fine Arts Gallery,
Woodford 099901
Gothow & Motzke, Berlin 141518
Gothsland, Barcelona . 085782, 114916
Gotico Restauraciones, Cáceres 133219
Gotisches Haus, Kulturstiftung
DessauWörlitz, Wörlitz 022717
Gotlands Konstmuseum, Länsmuseet
på Gotland, Visby 041175
Gotlands Naturmuseum, Visby . 041176
Goto Sumio Bijutsukan,
Kamifurano 028761
Gotoh Bijutsukan, Tokyo 029511
Gotta Have Art, Birmingham . . 119801
Gotta Have It, New York 095429, 145342
Gottfried-August-Bürger-Museum,
Molmerswende 020452
Gottfried-Keller-Zentrum,
Glattfelden 041551
Gottfried, Walter, Schwerin . . . 142356
Gotthelf-Stube, Schriftsteller-Museum,
Lützelflüh 041692
Gottlieb Daimler-Gedächtnisstätte,
Stuttgart 022031
Gottlieb-Häußler-Heimatmuseum,
Filderstadt 018361
Gottliebin-Dittus-Haus, Bad
Liebenzell 016818
Gottlob Auwärter Museum,
Stuttgart 022032
Gottmer, Becht, Aramith,
Haarlem 137859
Gottschalch, Erik, København . 065909
Gottschalk, Klaus, Friesack . . . 129869
Gottschalk, Traute, Würzburg . . 142512
Gottscheer Stube, Bad Wildbad . 016898
Gottschick & Friedrich, Tübingen 108787
Gottsmann, Sandra, Mannheim . 130275
Gotty, Denis & Michèle, Agnetz . 066430
Goudal, Claude, Lyon 104108
Goudard, Laurent, Limoges . . . 069523
Goudemare, S., Paris 141128
De Gouden Eend, Leiden 112555
De Gouden Kroon, Alkmaar . . . 082619
De Gouden Zon, Bergen 082954
't Gouden Mijntje, Breda . 082983, 132455
Goudreau Museum of Mathematics
in Art and Science, New Hyde
Park 051274
Goudriaan, Dick, Haarlem 132516
Gougis, Emmanuelle, Bercheres-les-
Pierres 067094
Gougis, Emmanuelle, Chartres . 067859
Gougis, Philippe, Saint-Denis-en-
Val 072733
Gouin, Olivier, Saint-Palais-sur-
Mer 073237
Gouin, Pascale, Nantes 070463
Goujon, Jacques, Paris 129020
Goulandris Collection, Athinai . 022988

Goulburn Regional Art Gallery,
Goulburn 001116
Gould, Newtown, New South
Wales 139858
Gould, South Yarra 099706
Gould Street Art Gallery, Bondi
Beach 098758
Gould, Gillian, London 090156
Gould, Hal D., Denver 120731
Gould & Co, Herbert, Holywood . 089555
Gould, John, Hawthorn 099118
Gould, Richard, West Hollywood 097894
Gould, Robert, Woollahra 062469
Goulet & de Cazes, Montréal . . 064426
Goulet, Danièle, Nice 070613
Goulignac, Agnès, Bordeaux . . 067334
Gouma, Maria, Nafplion 109075
Goumain, Francis, L'Isle-
d'Espagnac 069543
Gounaropoulos Museum of Zografou,
Zografou 023324
Gounod Diffusion, Boulogne-
Billancourt 103449
Goupil, Catherine, Marseille . . . 128771
Goupil, Jacques, Alençon 103221
Gour Nitai, Singapore 085412
Gourd Sculpture Museum,
Shanlin 042447
Gourtorbe, Daniel, Miribel 070131
Gous, Julian, Johannesburg . . . 133073
Le Gousset, Durmenach 068420
Gousset, Saint-Étienne 072764, 129227
Gout, Alain, Varagéville 129329
Goutis, Athinai 079166
Gouverneur Historical Association
Museum, Gouverneur 049168
Het Gouverneurshuis, Heusden . 032480
Gouw, J.ter, Amsterdam 143032
Govary'S, Barcelona . 085783, 114917
Governale, Antonello, Palermo . 081046
Governale, Edmondo, Palermo . 081047
Government Art Collection,
London 044981
Government Arts and Commerce
College, Indore 055803
Government Arts College,
Rajahmundry 055831
Government Arts College,
Srikakulam 055836
Government Central Museum, Jaipur,
Rajasthan 024318
Government College of Art and Craft,
Kolkata 055812
Government College of Art, Punjab
University, Chandigarh 055786
Government College of Arts and Crafts,
Chennai 055787
Government College of Fine Arts,
Gwalior 055798
Government Educational Museum,
Bulandshahar 024183
Government Educational Museum,
Deoria 024241
Government Educational Museum,
Etawah 024253
Government House, Sydney . . . 001538
Government House, Winchcombe 091863
Government House Museum and
Heritage Property, Regina 006768
Government Industrial and Commercial
Museum, Kolkata 024359
Government J.T.C. Museum,
Faizabad 024256
Government Museum, Ajmer . . 024130
Government Museum, Alwar . . . 024139
Government Museum, Bharatpur 024162
Government Museum, Chennai . 024197
Government Museum, Hassan . . 024290
Government Museum, Jodhpur . 024331
Government Museum, Kodagu . 024396
Government Museum, Mathura . 024396
Government Museum, Mount
Abu 024399
Government Museum, Pudukottai 024441

Government Museum, Udaipur . 024503
Government Museum and Art Gallery,
Chandigarh 024191
Government Museum Jhalawar,
Jhalawar 024328
Government Museum Vellore,
Vellore 024517
Governmental College of Engineering
and Textile Technology Museum,
Hooghly 024292
Governmental College of Music and
Fine Arts, University of Kashmir,
Srinagar 055837
Governmental Educational Museum,
Muzaffarnagar 024413
Governor Bent Museum, Taos . . 053675
Governor General's Foot Guards
Museum, Ottawa 006589
Governor Henry Lippitt House Museum,
Providence 052383
Governor John Langdon House,
Portsmouth 052310
Governor Printz Park, Essington 048611
Governor Ross Plantation,
Seaford 053222
Governor Seay Mansion,
Kingfisher 049975
Governor Stephen Hopkins House,
Providence 052384
Governor's Gallery, Santa Fe . . 053108
Governor's Mansion, Columbia . 047909
The Governor's Mansion,
Frankfort 048924
Governor's Mansion, Sacramento 052727
Govetosa, Boranga, Padova . . . 080986
Govett-Brewster Art Gallery, New
Plymouth 033256
Govi, Alberto, Modena 142750
Govinda, Washington 124912
Govoni, Franco, Bologna 130965
Govreau, Fresno 120898
Gow Langsford, Auckland 112774
Gow, Jeremy, Forfar . . 089296, 134392
Gowan, Bud, London 064348
Gowans, Moonah 125213
Gowen, Albuquerque 119479
Gower Heritage Centre, Gower . 044457
Gower, Tom, Saint Ives, Cornwall 119024
Gowganda and Area Museum,
Gowganda 005896
Gowrie, Woollahra 099916
Goy, Renate, Chemnitz 075250
Goya, Madrid 139056
Goya Galleries, Docklands 098969
Goya Galleries, Melbourne 099308
Goya-Girl Press, Baltimore 119744,
                                        138302
Goya Sanat Galerisi, İstanbul . . 117047
Goyas, Bogotá 102341
Goyert, Köln 107632, 142009
Goyet, Amaury, Aix-en-Provence 066463
Goyet, Xavier, Marseille 069941
Góźdź, Poznań 084639, 143495
Gozo Antiques, Victoria 082446
Gozo Cathedral Museum, Victoria 030762
Gozo Heritage, Gozo 030739
GP Antiguedades, Guadalajara . 082467
GP, Galerie Perplies, Aachen . . 105936
GPM, Zlonice 065568
GR 20 – Arti Applicate, Padova . 110571
Graaf-Reinet Museum, Graaff-
Reinet 038720
Graaf, de, Nieuwkoop 143207
Graaff, R.J. de, Haarlem 132517
Graaham, Barry Joseph, Palma de
Mallorca 115350
Graal, Agen 103174
Graal, Moissac 104244
du Graal, Paris 141129
Graal, Toulouse 105745
Grabados, Madrid 086197
Grabados y Libros, Madrid 143765
Grabe, Düsseldorf 075476
Grabenheinrich, Horst, Gütersloh 107185

Graber, Minneapolis 121993
Graber, San Diego 097093
Grabherr-Häfliger, Walter,
Lüchingen 087601, 133817
Grabherr, Rosita Ina, Wien 062966
Grabinski-Waldeck, Grenoble . . 068731
Grabkapelle auf dem Württemberg,
Stuttgart 022033
Grabnegger, Karin, Klagenfurt . . 062673
Grabner, Ernst, Gmunden 062569
Gråbo Minnenas Museum,
Karlskoga 040735
Grabow, Ludwig, Berlin 074806
Grabski Galeria, Lublin 113698
Grabungsmuseum Kirchhof,
Coburg 017750
Grabungsmuseum Langenburg,
Langenburg 019905
Grace, El Paso 135550
Grace, Kyoto 082015
Grace & Co., San Francisco . . . 097246
Grace, Sendai 111385
Grace Campbell Gallery, Prince
Albert 006710
Grace Darling Museum,
Bamburgh 043483
Grace Gallery, Arezzo . 079747, 079748
Grace Hudson Museum and The Sun
House, Ukiah 053878
Grace Li Gallery, Zürich 116801
The Grace Museum, Abilene . . . 046307
Grace Place, Memphis 094635
Grace, Ross Nelson, Armadale,
Victoria 061031
Gracefield Arts Centre, Dumfries 044154
Graceful Spirit Showcase,
Portland 096415
Graceland, Memphis 050779
Graceys, South Auckland 084081
Grach, Brigitte, Aldrans 127704
Gracia de Castro, Pilar, Donostia-San
Sebastián 133248
Graciano, Joseph, Le Havre . . . 069281
Gracie, New York 095430
Gracie Mansion, New York 051398
Gracie Mansion Fine Art, New
York 122666
Gracie Mansion Gallery, New
York 122667
Gracieuse Orient, Montreuil-sous-
Bois 070326
Grácio, Rio de Janeiro 101039
Gracis, Pier Giorgio, Milano . . . 080568
Grad Bogenšperk, Šmartno pri
Litiji 038617
Grad Dobrovo, Goriški Muzej, Dobrovo
v Brdih 038508
Grad Podsreda, Podsreda 038589
Grad pri Gradu, Grad 038514
Gradl & Hinterland, Nürnberg . . 126084
Gradsign, Singapore 056326
Gradska Chudožestvena Galerija,
Berkovica 004985
Gradska Chudožestvena Galerija,
Plovdiv 005127
Gradska Chudožestvena Galerija Petko
Čurčuliev, Dimitrovgrad 005013
Gradska Galerija Bihać, Bihać . . 004257
Gradska Galerija Labin, Labin . . 008944
Gradski Archeologičeski Muzej,
Chisar 005005
Gradski Archeologičeski Muzej,
Nesebär 005096
Gradski Archeologičeski Muzej,
Sandanski 005157
Gradski Istoričeski Muzej,
Asenovgrad 004970
Gradski Istoričeski Muzej, Balčik 004975
Gradski Istoričeski Muzej,
Botevgrad 004991
Gradski Istoričeski Muzej, Čirpan 005008
Gradski Istoričeski Muzej,
Dalgopol 005010
Gradski Istoričeski Muzej,

Karnobat 005050
Gradski Istoričeski Muzej,
Kavarna 005054
Gradski Istoričeski Muzej, Klisura 005068
Gradski Istoričeski Muzej, Lom . 005083
Gradski Istoričeski Muzej, Melnik 005090
Gradski Istoričeski Muzej, Nova
Zagora 005097
Gradski Istoričeski Muzej,
Orjachovo 005101
Gradski Istoričeski Muzej,
Panagjurište 005102
Gradski Istoričeski Muzej,
Peruštica 005112
Gradski Istoričeski Muzej,
Samakov 005156
Gradski Istoričeski Muzej,
Sevlievo 005159
Gradski Istoričeski Muzej,
Teteven 005226
Gradski Istoričeski Muzej Aleko
Konstantinov, Svištov 005221
Gradski Istoričeski Muzej Iskra,
Kazanläk 005057
Gradski Muzej, Bečej 038126
Gradski Muzej, Bjelovar 008884
Gradski Muzej, Kosovska
Mitrovica 038199
Gradski Muzej, Križevci 008941
Gradski Muzej, Požega 008969
Gradski Muzej, Senj 008984
Gradski Muzej, Umag 009013
Gradski Muzej – Városi Muzeum,
Subotica 038250
Gradski Muzej Bakar, Bakar . . . 008880
Gradski muzej Čazma, Čazma . . 008894
Gradski Muzej Jastrebarsko i Gradska
Galerija, Jastrebarsko 008923
Gradski Muzej Karlovac, Karlovac 008925
Gradski Muzej Korčula, Korčula . 008933
Gradski Muzej Kruševac,
Kruševac 038206
Gradski Muzej Makarska,
Makarska 008946
Gradski muzej Nova Gradiška, Nova
Gradiška 008955
Gradski Muzej Sisak, Sisak . . . 008991
Gradski Muzej Sombor, Sombor 038242
Gradski Muzej Varaždin, Varaždin 009014
Gradski Muzej Virovitica,
Virovitica 009017
Gradski Muzej Vršac, Vršac . . . 038257
Gradski Muzej, Muzej Grada Novog
Sada, Sremski Karlovci 038246
Graduate Institute of Museology,
Tainan National University,
Kuantien 056454
Graduate Institute of Museum
Studies, Fu Jen Catholic University,
Sinjhuang 056456
Graduate Program in Art History,
Rutgers State University of New
Jersey, New Brunswick 057517
Graduate School of Architecture,
Columbia University, New York 057553
Graduate School of Figurative Art,
New York Academy of Art, New
York 057554
Graduate School of Visual Arts
Management, Yuan Ze University,
Jhongli 056453
Grady, Stewart, Philadelphia . . . 136088
Gräbner, Maria Elisabeth,
Aschaffenburg 106011
Graeco-Roman Memoirs, London 139181
Graeco-Roman Museum,
Alexandria 010515
Graef, Thomas, Saarbrücken . . 078322,
                                        108545
Gräfe, H., Hesel 076518
Gräfe, K., Dessau 075336
Gräff, A., Karlsruhe 107512
Gräff, Margot, Bad Kreuznach . . 074543
Gräfliche Sammlungen und

Afrikanisches Jagdmuseum, Erbach,
Odenwaldkreis 018230
Graeme Park/Keith Mansion,
Horsham 049562
Grängesbergsbanornas
Järnvägsmuseum, Grängesberg 040665
Gränna Antik, Gränna 086724
Grænse-Antik, Kruså 065985
Grässlin, Bärbel, Frankfurt am
Main 107002
Græsted Antik, Græsted 065707
Graetz, Frédéric, Liège 140241
Graf, Kandern 076676
Graf, La Rochelle 069087
Graf & Schelble, Basel 116076
Graf Harrach'sche Familiensammlung,
Rohrau 002603
Graf-Luxburg-Museum, Museen
Schloss Aschach, Bad Bocklet 016734
Graf-Pauli, A., Wolfratshausen . . 079057
Graf, Dr. Henriette, München . . . 077513
Graf, Jürgen, Norderstedt 077798
Graf, M., Ludwigshafen am
Rhein 077160
Graf, Peter, Pfäffikon (Zürich) . . 087736
Graf, Robert, Klagenfurt 100073
Graf, W., Minden 142127
De Grafelijke Torenmolen,
Zeddam 033089
Grafenwald, Hans, Borken,
Westfalen 075100
Grafer, Wolf-Dieter, Leutenbach . 077091
Graff, Braunschweig 141633
Graff, Montréal 101417, 136678
Graff, A., Charleroi 063602
Graffan, Roger, Ollioules 141012
Graffio, Bologna 109849
Graffiti, Verona 081914
Graffiti Edition, Wien 136593
Graffiti Gallery, Winnipeg 101879
-grafia, Kraków 138968
Grafia, Kraków 137893
Grafic Encadrement, Niort 104406
Grafica, Saint Louis 123829
Grafica Antica, Venezia 111126, 142882
Grafica e Arte, Bergamo 137620
Gráfica Latina, Asunción 084418
Grafica Manzoni, Torino 110979
Gráficas Estella, Estella 137977
Graficas Santamaria, Vitoria-
Gasteiz 138006
Grafička Zbirka Nacionalne i
Sveučilišne Knjižnice, Zagreb . 009031
Graficki Fakultet, Sveučilište u
Zabgrebu, Zagreb 055282
Grafični Muzej, Rogaska Slatina 038602
Grafični Muzej Rogoška Slatina,
Rogoška Slatina 038604
Grafiekwinkel Inkt, Den Haag . . 112387
Grafik, London 139182
Grafik-Cabinet Artis, Zell am See 063126,
                                        140138
Grafik i Väst, Göteborg 115652
Grafik Museum Stiftung Schreiner,
Bad Steben 016885
Grafik och Skulptur, Stockholm . 115887
Grafik Verlag, Frankfurt am Main 137199
Grafika, Kraków 084578, 143474
Grafika Publishing, Praha 136764
Grafikens Hus, Mariefred 040836
Grafikk Senteret, Oslo 113224
Grafikwerkstatt Dresden in den
Technischen Sammlungen,
Dresden 106737, 137148
Grafioteket, Göteborg 115653
Grafique Art Gallery, Bologna . . 109850
Grafisch Atelier Twente,
Enschede 083250
Grafisch Historisch Centrum –
Drukkerijmuseum, Etten Leur . 032321
Grafisch Musem-Atelier in den
Groenen Zonck, Wouw 033069
Grafisch Museum Groningen,
Groningen 032377

Grafisch Museum Zutphen,
Zutphen 033117
Grafische Sammlung,
Kunstsammlungen und Museen
Augsburg, Augsburg 016654
Grafische Sammlung, Museen der
Stadt Nürnberg, Nürnberg 020855
Grafisk Skole, Kongelige Danske
Kunstakademi, København .. 055306
Grafiska Sällskapet, Stockholm . 115888
Den Grafiske Højskole,
København 055307
Grafiti, Gdynia 113501
Grafix, Washington 124913
Grafo, Brescia 137631
Grafos Verlag, Vaduz 137804
Grafschafter Museum im Moerser
Schloss, Moers 020444
Grafschaftsmuseum, Wertheim . 022588
Grafstein, James, New York ... 095431
Grafström Gallery, Stockholm .. 115889
Grafton Galleries, Double Bay . 061405
Grafton Historical Society,
Grafton 049170
Grafton Regional Gallery, Grafton 001118
Graftons, Market Harborough .. 090712,
118720
Graglia, Michel, Dijon . 068333, 068334
Graha, Jakarta 109510
Graham, Burghfield Common .. 088717,
117479,   134185
Graham, Kembla Heights 099179
Graham, London 090157, 118354
Graham, Memphis 094636
Graham & Green, London 090160
Graham County Historical Museum,
Thatcher 053727
Graham-Stewart, M.J., London . 090161
Graham, Arthur, Oklahoma City . 095975
Graham, Charlton, Nicholls 099458
Graham, Gavin, London 090158, 118355
Graham & Sons, James, New
York 122668
Graham, Joss, London 090159
Graham, Neil, Brentwood 088590,
134143
Graham, Timothy, Twickenham . 119277
Grahame, Milton 099347
Grahams of Colchester,
Colchester 088950, 134288
Grahn, Minneapolis 135829
Grahn, Patrik, Espoo 066289
Grainbag, Pittsburgh 123548
Graindorge, Sées 073470
Graindorge, Jacky, Sées 073471
Grainger Museum, University of
Melbourne, Melbourne 001273
Grains Industry Museum,
Cabanatuan 034448
Grais, Valérie, Paris 129021
Gralglasmuseum, Dürnau 018023
Gramentz, Harms, Plau am See 078055,
130523
Grammatopoulou, Yianna, Athinai 109036
Grammdorf, Johanna, Aichstetten 074367
Grammelsbacher, Jochen,
Augsburg 074459
Gramophone Records Museum, Cape
Coast 022929
The Gramophone Shop, Haarlem 083330
Grampian Transport Museum, Alford,
Aberdeenshire 043391
Grampy, Omaha 096043
Grams, Brita, München 130550
Gramsch, Lahnstein 107780
Gramza, Eugeniusz, Poznań ... 084640
Gramzower Heimatstuben,
Gramzow 018794
Gran Cala, Barcelona 085784
Gran Mercato dei Mercati di
Collezionismo e Antichità,
Baganzola 098193
Gran Mercato dell'Antico
Internazionale, Modena 098194

La Gran Ocasion, Palma de
Mallorca 086354, 086355
Graña Lagoa, Francisco, Vigo . 086560,
133479
Granary Collectables, Kendal ... 089670
Granatium, Radenthein 002558
Granby Gallery, Bakewell 117267
Grance, Eloïc, Lorette 069629
Grancher, Jacques, Paris 136933
Grand Antique Mall, Cincinnati . 092860
Grand Army of the Republic Civil War
Museum, Philadelphia 052054
Grand Army of the Republic Memorial
and Veteran's Military Museum,
Aurora 046674
Grand Art, Kyiv 117135
Grand Art Gallery, Singapore .. 114530
Grand Arts, Kansas City 121291
Grand Auction Mart, New York . 127348
Grand Canyon National Park Museum
Collection, Grand Canyon 049173
Grand Central, Montréal 064427
Grand Central Auction, San
Francisco 127444
Grand Central Gallery, Boston .. 119871
Grand Central Station Antiques, San
Francisco 097247
Le Grand Cercle, Eragny-sur-
Oise 103718, 136816, 140834
Grand Coteau Heritage and Cultural
Centre, Shaunavon 007010
Grand County Museum, Hot Sulphur
Springs 049567
Grand Curtius, Liège 003793
Grand Encampment Museum,
Encampment 048582
Grand Entrances, Sans Souci ... 062191
Grand Falls Museum, Grand Falls 005900
Grand Forks Art Gallery, Grand
Forks 005905
Grand Forks County Historical Society
Museum, Grand Forks 049175
Grand Gallery, Kansas City 121292
Grand Gulf Military State Park
Museum, Port Gibson 052249
Grand Hill Gallery, Saint Paul .. 123911
Grand-Hornu, Hornu 003710
Grand Image, Seattle 124536
Grand Island Trader's Cabin,
Munising 051118
Grand Lake Area Historical Museum,
Grand Lake 049184
The Grand Lodge of Scotland Museum,
Edinburgh 044243
Le Grand Magasin, Saint-Rémy-de-
Provence 115611
Grand Manan Museum and Walter B.
McLaughlin Marine Gallery, Grand
Manan 005907
Grand Marais Art Colony, Grand
Marais 057147
Grand Marché d'Art Contemporain,
Paris 098083, 098084
Grand Monde, Paris 129022
Grand Oriental Rugs, Saint Paul 096826
Grand Portage National Monument,
Grand Portage 049187
Grand Prairie Arts Council,
Stuttgart 060500
Grand-Pré National Historic Site,
Grand-Pré 005908
Grand Prix Museum, Macau ... 008067
Grand Quartier Général Allemand,
Musée de la Resistance, Brûly-de-
Pesche 003426
Grand Rapids Art Museum, Grand
Rapids 049191
Grand Rapids Children's Museum,
Grand Rapids 049192
Grand Rapids Kendall College of Art &
Design, Ferris State University, Big
Rapids 056738
Grand River Historical Society &
Museum, Chillicothe 047716

Grand Shaft, Dover 044132
The Grand Style Gallery,
Baltimore 119745
Grand Traverse Heritage Center,
Museum and Con Foster Collection,
Traverse City 053792
Grand View Fine Art Gallery,
Cleveland 098896
Grand Village of the Natchez Indians,
Natchez 051195
La Grande Fenêtre, Lyon 140900
La Grande Ferme, Centre d'Initiation
au Ppatrimoine, Saint-Joachim 006888
La Grande Ile, Bruxelles 063452
Grande Museo, Fabriano 026165
Grande Prairie Museum, Grande
Prairie 005912
Grande Victoria Gardens, Saint
Paul 096827
Grandfather Clocks, Cincinnati . 092861,
135443
Grandfather's Clock Gallery and Clinic,
Tulsa 097734, 136430
Grandfather's Clockshop, Ballarat 061084
Grandfather's Collections,
Singapore 085413
Grandi Antichità, Ozzano Taro .. 080954
Grandi, Claudia, Bologna 130966
Grandière Frères, Paris 129023
Grandin, Christian, Paris 125720
Grandits, Johann, Stegersbach . 062847
Grandmaître, Jean-Henri, Châtillon-sur-
Marne 067922
Grandmaître, Jean-Henri, Dormans-en-
Champagne 068388
Grandma's Attic, Saint Paul ... 096828
Grandma's Corner, Miami 094741
Grandma's Place, Tampa 097561
Grandma's Things, Jacksonville 093996
Grandma's Top Drawer, Whyalla 062427
Grandmont, Detroit 120808
Grandmother Treasures, Dallas . 093132
Grandoni, Claudio, Roma 131945
Grandpa Snazzys Hardware,
Denver 093294
Grandpa's Antiques, Miami 094742
Grandpa's Antiques, Sacramento 096628
Grandpa's Shed, Fitzroy Falls .. 061472
Grandview Antiques, Columbus . 093024
Grandview Auction, Roanoke .. 127420
Grandview Mercantile Company,
Columbus 093025
Grandville & Old Nick, Basel ... 144020
Granec & Mitteldorf, Augsburg . 106030
Granelli, Livorno 110181
Grangärde Hembygdsgården,
Grangärde 040670
Grange, Cassel 067700
The Grange, Toronto 007178
La Grange, Zaltbommel 083701
La Grange Antichità, Milano ... 080569
Grange Antiques, Latrobe 061720
Grange aux Affaires, Châteauneuf-les-
Martigues 067895
La Grange Aux Arts, Peroy-les-
Gombries 072068
La Grange aux Moines, Megève 070034
La Grange aux Oies, Collonges-la-
Rouge 012623
La Grange Béranger, Musée d'Arts
Populaires, Beblenheim 011960
La Grange de Concise, Thonon-les-
Bains 073666
Grange de l'Ermitage, Armoy .. 066727
Grange de Liselotte, Vouzan ... 074282
Grange des Artisans, Sainte-
Suzanne 015632
Grange des Ongliers, Chandai .. 067823,
128543
La Grange du Cap Lande, La Teste-de-
Buch 069103
Grange Gallery, Oldtown 109681
Grange Gallery, Saint Saviour . 091211,
134996

La Grange Moulin, Les Ardillats 069399
Grange Museum of Community History,
London 044982
Grange Notre-Dame, Labastide-
d'Armagnac 069125, 128648
Grange Old Style Furniture,
Grange 061554
Grange Road Antiques, Glen
Huntly 061527
La Grange Saint-Firmin, Nempont-
Saint-Firmin 070500
Grange, Allan, Landulph 118072
Grange, Gilbert, Dijon 068335
Grangemouth Museum,
Grangemouth 044459
Granger, Audes 066804
Granger Homestead Society Museum,
Canandaigua 047397
Granger House, Marion 050689
Granger, Jean-Claude, Dreux .. 125575
Granger, Jenny, Whitstable ... 119341
Granges de Servette, Musée Paysan
et Espace d'Exposition, Chens-sur-
Léman 012548
Granier, Christian, Bielefeld .. 141597,
141598
Granier, Didier, Castelnaudary . 067711
Granier, Nadine, Albi 103218
Granier, Olivier, Valros 073957
Granini, Mesola 110216
Graniou, Christian, Monflanquin . 070164,
104246
Granisle Museum, Granisle ... 005917
Granit Sanat, İstanbul 117048
Granitabbaumuseum, Königshain 019732
Granitsa Folklore Museum,
Granitsa 023086
Granma Memorial, La Habana . 009145
Grannie Used To Have One,
Longhaven 090630
Grannie's Parlour, Kingston-upon-
Hull 089699
Grannona, Portbail 072284
Granny's Cupboard, Wisbech .. 091886
Granny's Antique Furniture Mall,
Tucson 097667
Granny's Antiques, Mildura ... 061853
Granny's Attic, Naas 079680
Granny's Attic, Ramsgate 091069
Granny's Attic Antiques,
Cootamundra 061366
Grannys Bottom Drawer, Kinsale 079650
Granny's Goodies, Oklahoma City 095976
Granny's Jumble Sale, Zaandam 083695
Granny's Market, Armadale,
Victoria 061032
Granny's Wonders, Oklahoma
City 095977
Granösunds Fiskeläge, Södra
Vallgrund 011363
Granoff, Katia, Cannes 103512
Granoff, Katia, Honfleur 103837
Granow, Hans-Dieter, Trier ... 142418
Grans Lantbruksmuseum, Piteå . 040914
Grant, Barnard Castle 117287
Grant, Droitwich 144375
Grant, Newcastle, Co. Down .. 118779
Grant & Gower, Henley 117953
Grant & Shaw, Edinburgh 144393
Grant Arts Center, San Francisco 124504
Grant County Historical Museum,
Canyon City 047415
Grant County Historical Museum,
Elbow Lake 048520
Grant County Historical Museum,
Ephrata 048592
Grant County Museum, Elgin .. 048523
Grant County Museum, Hyannis 049647
Grant County Museum, Medford 050756
Grant County Museum, Sheridan 053308
Grant County Museum, Aka Historic
Adobe Museum, Ulysses 053881
Grant-Humphreys Mansion,
Denver 048265

Grant-Kohrs Ranch, Deer Lodge 048214
Grant Medical College Museum,
  Mumbai 024403
Grant Museum of Zoology and
  Comparative Anatomy, London 044983
Grant Street Art Gallery, Ballarat 098693
Grant, Anthony, New York 122669
Grant, Denzil, Drinkstone 089108
Grant, Gary, London 118356
Grant, Jonathan, Auckland 112775,
  132690
Grantham, Cleveland 120425
Grantham Clocks, Grantham ... 089354,
  134416
Grantham Furniture Emporium,
  Grantham 089355
Grantham Museum, Grantham . 044462
Grantown Museum, Grantown-on-
  Spey 044464
Grants, Detroit 135542
Grant's Birthplace State Memorial,
  Point Pleasant 052226
Grant's Books, Prahran 139883
Grants Old Mill, Winnipeg 007445
Grants Pass Museum of Art, Grants
  Pass 049204
Granviarte, Lisboa 114033
Granville Antiques, Petworth ... 091000,
  134939
Granville Collections, Granville .. 068697
Granville County Historical Society
  Museum, Oxford 051865
Granville Gallery, Chicago 120142
Granville Gallery, Granville 103780
Granville Historical Museum,
  Granville 049210
Granvin Bygdamuseum, Granvin 033621
Grapevine, Norwich 118831
Grapevine Museum, Chiltern ... 000963
Grapheo, Roma 131946
Graphia, Sint-Niklaas .. 058377, 138550
Graphic Art Division, Milwaukee Area
  Technical College, Milwaukee . 057451
Graphic Art Publishing, Miami .. 138359
Graphic Artists Guild, New York 060501
Graphic Artists Guild's Directory of
  Illustration, Santa Barbara 139375
Graphic Arts and Visual
  Communication Faculty, Eszterházy
  Károly Főiskola, Eger 055770
Graphic Arts Gallery, Miami ... 121797
Graphic Concepts, Omaha 096044
Graphic Conservation Company,
  Chicago 135414
Graphic Design Department, Salt Lake
  Community College, Salt Lake
  City 057836
Graphic Design Department, State
  of Connecticut Tunxis Community
  College, Farmington 057076
Graphic Design: USA, New York 139376
Graphic Presentations,
  Philadelphia 123389
Graphic Station Gallery,
  Kitakyushu 111276
Graphic Studio Gallery, Dublin . 109588
Graphic World, Singapore 114531
Graphicom, Vicenza 137763
Graphics Division, Western Wisconsin
  Technical Institute, La Crosse . 057296
Graphics Gallery, San Ramon .. 127446
Graphics Home Gallery, Breda .. 082984
Graphics Plus Frames, Saint
  Albans 119004
Graphics World, Maidstone 139183
Graphicstudio, Tampa 138476
Graphigro, Paris 141130
Graphik Edition Dornick, Emmerich am
  Rhein 137177
Graphiker-Verein e.V., München 058909
Graphikmuseum Pablo Picasso
  Münster, Münster 020598
Graphique de France, Noisiel .. 136843
Graphis, Toronto 101620

Graphis, Woollahra 099917
Graphis Arte, Livorno . 110182, 137662
Graphische Kunst, Memmingen . 138698
Graphische Revue Österreichs,
  Wien 138515
Graphische Sammlung am
  Kunsthistorischen Institut der
  Universität Tübingen, Tübingen 022215
Graphische Sammlung der
  Eidgenössischen Technischen
  Hochschule Zürich, Zürich 042196
Graphische Sammlung der Universität,
  Erlangen 018260
Graphische Sammlung,
  Museumslandschaft Hessen Kassel,
  Kassel 019533
Graphische Werkstatt im Traklhaus,
  Lithowerkstatt, Salzburg 136551
Graphothek Berlin, Berlin 017150
Grapin, Joël, Thonon-les-Bains . 073667,
  129279
Gras, Eric, Romilly-sur-Seine .. 072525
Grashey, Ursula, Konstanz 107717
Grasiela, Vincennes .. 074242, 141390
Grasl, Baden bei Wien 136494
Graslitzer Gedenk- und
  Informationsraum,
  Aschaffenburg 016617
Grass, Wien 140080
Grassagården Miljömuseum,
  Strängnäs 041059
Grassi-Museum für Angewandte Kunst,
  Leipzig 019992
Grassi-Museum für Musikinstrumente
  der Universität Leipzig, Leipzig 019993
Grassi-Museum für Völkerkunde zu
  Leipzig, Staatliche Ethnographische
  Sammlungen Sachsen, Leipzig 019994
Grassi, P., Hohenberg an der
  Eger 076558
Grassin, Jean, Carnac . 103534, 136805,
  140816
Grassland Gallery, Minneapolis . 121994
Grassroots, Regina 064552
Gratan Sur S.L., Granada 133266
Grate Expectations, Washington 091756
Gratsos, Constantin, Wien 062967
Il Grattacielo, Livorno 110183
Grattepanche, Pascal, Artres ... 066753
Grau Massana, Marta, Barcelona 085785
Graue Mill and Museum, Oak
  Brook 051663
Grauer, W. jun., Sonnenbühl ... 078505
Grauer'sche Antiquariatsbuchhandlung,
  Nürnberg 142241
Grauert & Zink, Berlin 137072
Grauwert, Hamburg 107274
Grauwin, Nicolas, La Bassée ... 068942
Grava, Milwaukee 121921
Gravante & Lindemann,
  Mettmann 077365
Gravdal, Anne, Sylling 084374
Grave Creek Mound Archaeology
  Complex, Moundsville 051079
Grave Matters Mysteries,
  Cincinnati 145072
Gravelbourg and District Museum,
  Gravelbourg 005920
Gravensteen Museum voor
  Gerechtsvoorwerpen, Gent ... 003624
Graves Art Gallery, Sheffield .. 045772
Graves-Johnston, Michael,
  London 144614
Graves, Clark, Saint Louis 096712
Gravesham Museum, Gravesend,
  Kent 044467
Graviassy, Sébastien, Toulouse . 129302
Gravier, Philippe, Paris 104921
Gravigny, Josiane, Le Neubourg 069320
Gravura, Lisboa 114034, 137919
Gravura Brasileira, Rio de
  Janeiro 101040
Gravy, Sacramento ... 096629, 136189
Grawath, Peter & Silvia,

Hafenlohr 076146
Grawe, Maria, Schmallenberg .. 108567
Gray, Cincinnati 120348
Gray, Dunedin 112883
Gray, New York 095432
Gray, Toronto 064662
Gray Door, Houston 093678
Gray Gallery, Quincy 052431
Gray Matters Art Gallery, Dallas 120605
Gray Stone Press, Goodlettsville 138335
Gray, Alexandra J., Portsmouth . 091044,
  134952
Gray, George, Los Angeles ... 121500
Gray, Julia, New York 095433
Gray, Kimberly Jayne, Dallas ... 093133
Gray, Laila, Kingsthorpe 089696
Gray, Lenore, Providence 123706
Gray, Martin, New York 095434
Gray, Philip, Cobh 109553
Gray, Richard, Chicago 120143
Gray, Richard, New York 122670
Graybar, Craig L., Milwaukee .. 135818
Graybeal, Atlanta 135309
Graydon, New Farm 099425
Graylings Antiques, Andover ... 088281
Grays, London 090162
Gray's, Worcester 091944
Gray's Printing Press, Strabane . 045896
Grays School of Art Gallery and
  Museum, Aberdeen 043358
Gray's School of Art, The Robert
  Gordon University, Aberdeen .. 056505
Grayson, Houston 093679
Grayson, Donald & Barbara,
  Louisville 094553, 145245
Grayson, G.I., Ringwood 062150
Graystone, Portland 123628
Graystone, San Francisco 124291
Grazer Kunstverein, Graz 001973
Grazia Marotta, Napoli 080884
Grazioli, Kronberg im Taunus .. 107756
Gré du Van, Chalon-sur-Saône . 067777
Great Acquisitions Book Store, New
  Orleans 145311
Great Alpine Gallery, Omeo 099502
Great American Antique Mall,
  Tulsa 097735
Great American Art Works,
  Cincinnati 120349
Great American Ink, Los Angeles 121501
Great Art Gallery, Delhi 109321
Great Atlantic Church Street Gallery,
  Monmouth 118752
Great Atlantic Falmouth Gallery,
  Falmouth 117784
Great Atlantic Map Works Gallery,
  Saint Just 119042
Great Bardfield Cage, Great
  Bardfield 044470
Great Bardfield Cottage Museum,
  Great Bardfield 044471
Great Basin Historical Society &
  Museum, Delta 048236
Great Blacks in Wax Museum,
  Baltimore 046748
Great Brak River Museum, Great Brak
  River 038729
Great Brampton House Antiques,
  Hereford 089515
Great Central Railway Museum,
  Loughborough 045152
Great Cobar Heritage Centre,
  Cobar 000979
Great Escape, Louisville 094554, 145246
Great Escape, Nashville 145296
Great Escape Book Store,
  Toronto 140407
Great Exchange, Alamo Heights 144952
Great Expectations, Blackburn,
  Lancashire 088494, 134118
Great Expectations, Horncastle . 089582
Great Expectations, New Orleans 095108
Great Expectations, Poole 091033,
  134946

Great Explorations, Saint
  Petersburg 052862
Great Fish River Museum,
  Cradock 038693
Great Gatsby, Atlanta . 092141, 127127
Great Grandfathers Bookshop,
  Leyland 144536
Great Grooms of Dorking,
  Dorking 089083
Great Grooms of Hungerford,
  Hungerford .. 089607, 134490, 144491
Great Hall of the Clans Museum,
  Gaelic College of Celtic Arts and
  Crafts, English Town 005787
Great Hang-Up, Houston 121049
The Great Hanshin-Awaji Earthquake
  Memorial, Disaster Reduction
  Museum & Human Renovation
  Museum, Kobe 028840
Great Harbor Maritime Museum,
  Northeast Harbor 051629
Great Jones, Seattle 097449
Great Lakes Art Studio, Chicago 120144
Great Lakes Art Studio, Evanston 135553
Great Lakes Coin Company,
  Cleveland 092943
Great Lakes Historical Museum,
  Tuncurry 001598
Great Lakes Lighthouse Museum,
  Mackinaw City 050558
Great Lakes Naval Memorial and
  Museum, Muskegon 051134
Great Lakes Pastel Society, Grand
  Rapids 060502
Great Lakes Polishing, Cleveland 135463
Great Lakes Science Center,
  Cleveland 047815
Great Laxey Wheel and Mines Trail,
  Manx National Heritage, Laxey 044769
Great Lord Antiguidades e Objetos,
  Rio de Janeiro 064064
Great Modern Pictures, New York 122671
Great Northern Antiques,
  Minneapolis 094916
The Great Northern Architectural Antique
  Co., Tattenhall 091542, 135120
Great Northwest Bookstores,
  Portland 145452
The Great Oak Bookshop,
  Llanidloes 144547
Great Orme Tramway, Llandudno 044877
Great Ormond Street Hospital for
  Children Museum, London .. 044984
Great Overland Station, Topeka . 053771
Great Plains Art Collection in the
  Christlieb Gallery, Lincoln .. 050291
Great Plains Black Museum,
  Omaha 051764
Great Plains Transportation Museum,
  Wichita 054387
Great Sandhills Museum, Sceptre 006998
Great Sheffield, Sheffield 119083
Great South Land, Drummoyne . 098979
Great Stuff, Myrtleford 061934
Great Sydney Antiques Fair, Moore
  Park 097993
Great Things on High, Columbus 120497
Great Valley Museum of Natural
  History, Modesto 050984
Great Wall Painting and Calligraphy
  Society, Beijing 101955
Great Walsingham Gallery,
  Walsingham 119301
Great War Flying Museum,
  Cheltenham 005608
Great Warrior Aboriginal Art,
  Berry 061148
Great Wave, Buffalo 092530
Great West Warehouse,
  Saskatoon 064562, 125371
Great Western Auctions, Glasgow 126902
Great Yarmouth Museums, Great
  Yarmouth 044475
Great Zimbabwe Site Museum,

Masvingo 054856
Greater Columbus Antique Mall,
  Columbus 093026
Greater Harvard Area Historical
  Society, Harvard 049397
Greater Hazleton Historical Society
  Museum, Hazleton 049424
Greater Loveland Historical Society
  Museum, Loveland 050507
Greater Manchester Police Museum,
  Manchester 045203
Greater Philadelphia Book Bank,
  Philadelphia 145412
Greater Port Jefferson Museum,
  Historical Society of Greater Port
  Jefferson, Port Jefferson 052252
The Greater Reston Arts Center,
  Reston 052520
Greater Vernon Museum, Vernon,
  British Colombia 007313
Greca, Barcelona 085786
Greca Trading, Singapore 085414
Grecic & Assoc., Cicero 145063
El Greco, Bogotá 102342
El Greco, Bucureşti 114144
El Greco, Montevideo 092002
El Greco, Santiago de Chile 128250
Greco, Eloisa, Messina 080369
Gredler, John J., New York 095435
Gredt, Léa, Luxembourg 111854
Greef & Jansen-Greef, Bad
  Arolsen 074490, 129402
Greek Orthodox Museum,
  Jerusalem 025078
Green, Chennai 109285
Green, London 090163
Green, Toronto 064663
Green & Hughes, Los Angeles . . 094349
Green & Stone, London 090167, 134676
Green Apple Books, San
  Francisco 145557
Green Art Gallery, Jumeira 117192
Green Bay Auctions, Auckland . . 126305
Green Bay Packers Hall of Fame,
  Green Bay 049225
Green Boutique, Tampa 097562
Green Dolphin Antiques, Boston . 092474
Green Door Furniture, Denver . . 093295
Green Door Gallery, Bellville . . . 114755
Green Door Gallery, Kansas City . 121293
Green Dragon Antiques Centre,
  Wincanton 091858
Green Dragon Museum, Stockton-on-
  Tees 045878
Green Family Forge – A Blacksmith
  Museum, Trinity 007235
Green Gables House,
  Charlottetown 005595
The Green Gallery, Aberfoyle . . . 117216
Green Gallery, Auckland 112776
The Green Gallery, Dublin 109589
Green Gallery, Kram 116974
The Green Gallery, Rhosili 118964
Green Gates Gallery, Nicholson . 099460
Green Goat Art Studio, O'Connor 099498
The Green Hall Gallery,
  Craigellachie 117638
Green Hill Center for North Carolina
  Art, Greensboro 049249
Green Horse Gallery, Bardon . . . 098708
Green Horses, Kinsale 109659
Green House Gallery, Milton . . . 099348
Green Howards Regimental Museum,
  Richmond, North Yorkshire . . . 045610
Green Lane Antiques, Ormskirk . 090919,
  134910
Green Lion Books, Saint Paul . . . 145502
Green Mountain Club, Waterbury
  Center 054201
Green Palace Museum, Sa'd-Abad
  Museum, Tehrān 024758
Green Palm Gallery, Ha Noi . . . 125076
Green Park Shipbuilding Museum and
  Yeo House, Prince Edward Island

Museum and Heritage Foundation,
  Port Hill 006687
Green River Museum, Woodbury 054520
Green Square Copenhagen,
  København 065910
Green Wall Gallery, Dublin 109590
Green, Andrew, Portland 096416, 136153
Green, Arthur, Los Angeles 094348
Green, Bert W., Los Angeles . . . 121502
Green, Essie, New York 122672
Green, Frank, Liverpool 118129
Green & Son, J., Queniborough . 091062
Green, John, Glasgow 117845
Green, Michael, Harrogate 089430
Green & Sons, N., Chicago 092722
Green, Richard, London 090164, 090165,
  090166,   118357
Green, Ron, Towcester 091626, 135145
Greenan Farm Museums,
  Greenan 024931
Greenan Gallery, Ayr 117264
Greenaway Art Gallery, Kent
  Town 099182
Greenaway Gallery, Camberwell 000927
Greenbelt Museum, Greenbelt . . 049233
Greenberg & Van Doren, New
  York 122674
Greenberg & Van Doren, Saint
  Louis 123830
Greenberg, Howard, New York . 122673
Greendoor Antiques, Orange . . . 062030
Greene, Fort Worth 093452
Greene County Historical Museum,
  Waynesburg 054258
Greene County Historical Society
  Museum, Xenia 054563
Greene Memorial Museum,
  Milwaukee 050914
Greene Naftali, New York 122675
Greene Space Gallery, New York 122676
Greene Street Antiques, New
  York 095436
Greene, Michael A., Woollahra . 062470
Greeneberg Van Doren, New York 122677
Greenfield, Milwaukee 094859
Greenfield Historical Society Museum,
  Greenfield 049246
Greenfield Museum, Greenfield . 049245
Greenfield Valley Museum,
  Greenfield 044479
Greenfield & Son, H.S.,
  Canterbury 088777, 134206
Greengage Gallery, Chalfont Saint
  Giles 117552
Greengrass Antiques, Chobham . 088899
Greenhill Covenanter's House,
  Biggar 043597
Greenhill Galleries, Birmingham 117372
Greenhill Galleries, Perth 099561
Greenhill Galleries Adelaide, North
  Adelaide 099467
The Greenhouse Gallery,
  Sheffield 119084
Greenhouse Gallery of Fine Art, San
  Antonio 124036
Greenlane Gallery, Dingle 109565
Greenoaks Gallery, Launceston . 099219
Greens of Cheltenham,
  Cheltenham 088833
Green's Windmill and Science Centre,
  Nottingham 045413
Greensboro Artists' League,
  Greensboro 060503
Greensboro Artists' League Gallery,
  Greensboro 049250
Greensboro Historical Museum,
  Greensboro 049251
Greenslade, Taylor & Hunt,
  Taunton 127063
Greensleeves, Aldgate 060988
Greenspond Museum,
  Greenspond 005923
Greenstein, Saul, London 090168
Greensward Gallery, Oregon State

University, Corvallis 048019
Greentown Glass Museum,
  Greentown 049260
Greenville County Museum of Art,
  Greenville 049268
Greenville Museum of Art,
  Greenville 049263
Greenwald, June, Cleveland . . . 092944
Greenwall, Jonathan, Sandgate . 091238
Greenway, Dallas 145114
Greenway, Colin, Witney 091888
Greenwich Art Society,
  Greenwich 060504
Greenwich Heritage Centre,
  London 044985
Greenwich Local History Centre,
  London 044986
Greenwich Printmakers, London 118358,
  138215
Greenwood Antique Mall, Seattle 097450
Greenwood Books, Seattle 145586
Greenwood Centre for Living History,
  Hudson 006016
Greenwood County Historical Society,
  Eureka 048628
Greenwood Great House Museum,
  Greenwood 028438
Greenwood Military Aviation Museum,
  Greenwood, Nova Scotia 005925
Greenwood Museum, Greenwood,
  British Columbia 005924
Greenwood Press, London 138216
Greenwood, David, Ilkley 118014
Greenwood, Judy, London 090169
Greenwood, W., Burneston 088719,
  134186
Greenwood, William, Darmstadt 106655
Greer, Robin, London 144615
Grefe, Dr. Uta, Puchheim 078099
Greger, Malmö 143895
Greger, Stockholm 115890
Gregg & Co., Albuquerque 119480
Gregg County Historical Museum,
  Longview 050391
Greggie, Atlanta 119573, 144968
Gregor Rare Books, Langley . . . 145202
Gregori, Anna, Verona 132284
Gregorics, Alexander, Oslip 127830
Gregoris, Elena, Brescia 131037
Gregor's Studios, Dallas 093134
Gregory Kondos Gallery, Sacramento
  City College, Sacramento 052728
Gregory, Bottley & Lloyd, London 090171
Gregory, Henry, London 090170
Gregory, Martyn, London 118359
Gregory's Antiques and Lights,
  Fitzroy 061468
Gregory's of Greerton, Tauranga 084104
Greg's Antiques, Cincinnati 092862,
  127174
Greg's Antiques, Cleveland 092945
Greg's Upholstery, Long Beach . 135695
Gregson, Thornleigh 127688
Greifengalerie, Greifswald 107158
Greim, Dr. Gerhard, Forchheim,
  Oberfranken 075756, 141766
Greiner, Albert, Garmisch-
  Partenkirchen 075947
Greiner, Daniel, Paris 104922
Greinwald, Adelgunde & Manfred,
  Schliersee 078373
Greinwald, R., Immenstadt im
  Allgäu 076603
Greisen, Praha 125422
Greiser, Olaf, Schriesheim 108579,
  142351
Greisinger, A., Baden-Baden . . . 074635
Grekiskt Kulturcentrum,
  Stockholm 041002
El Greko, Skopje 111884
Grelet, Eric, Manosque 069838
Gremes, Augsburg 074460
Gremese, Roma 137735
Gremi, Barcelona 085787

Gremi d'Antiquaris de Catalunya,
  Barcelona 059866
Gremi d'Antiquaris i Brocanters de les
  Comarques Gironines, Girona . 059867
Gremi de Llibreters de Vell de
  Catalunya, Barcelona 143718
Gremillion & Co., Austin 119675
Gremillion & Co., Dallas 120606
Gremillion & Co., Houston 121050
Gremm-Roloff, Thomas,
  Schopfheim 078388
Gremm, T., Mannheim 077260
Grenada National Museum, Saint
  George's 023344
De Grenadier, Groningen 083291
Grenen Kunstmuseum, Skagen . 010304
De Grenen Schuur, Leiden 083430
Grenfell and District History Museum,
  Grenfell 001121
Grenfell Community Museum,
  Grenfell 005926
Grenfell House Museum, Saint
  Anthony 006854
Grenier, Ajaccio 066497
Le Grenier, Montpellier 070302
Le Grenier, Saint-Denis 085065
Le Grenier, Saint-Nicolas-de-
  Redon 072991
Le Grenier, Villers-Allerand 074176
Le Grenier, Villuis 074236
Grenier & Chocolat, New Orleans 095109
Le Grenier Animé, Sainte-Foy-de-
  Peyrolières 073348
Grenier Antiquité Bar, Blonville-sur-
  Mer 067262
Le Grenier Antiquités, Quimper . 072345
Grenier au Relais, Belfort 067069
Grenier Bovary, Ry 072650
Le Grenier Crémolan, Crémieu . 068232
Grenier d'Adèle, Porto-Vecchio . 072289
Grenier d'Alexandre, Saïx 073378
Le Grenier d'Aliénor, Castillon-la-
  Bataille 067718
Le Grenier d'Aliénor, Sainte-Terre 073371
Grenier d'Anaïs, Rennes (Ille-et-
  Vilaine) 072437
Grenier d'Annick, Gavray 068641
Le Grenier d'Argenteuil,
  Argenteuil 066714
Grenier d'Aurelianis, Orléans . . . 070815
Grenier de Bricquebec,
  Bricquebec 067490
Grenier de Chine, Larmor Plage 069188
Grenier de Françoise, La Roche-
  Bernard 069063
Le Grenier de la Chapelle,
  Bruxelles 063453
Le Grenier de la Reine Margot, Saint-
  Saturnin 073299
Le Grenier de la Vallée, Salins-les-
  Thermes 073382
Grenier de l'Autan, Caraman . . . 067663
Grenier de Margot, Saint-Germain-en-
  Laye 072809
Grenier de Marie, Thierville 073652
Grenier de Montesson,
  Montesson 070239
Grenier de Pierrot, Arpajon 066733
Grenier de Recouvrance, Brest . 067472,
  128521
Grenier de Saint-Léger, Saint-Léger-
  des-Vignes 072889
Grenier de Sucy, Sucy-en-Brie . 073603
Grenier de Thomas, Chaumont (Haute-
  Marne) 067935
Le Grenier de Valenciennes,
  Valenciennes 073944
Le Grenier de Vauban, Briançon 067487
Le Grenier de Vulaines, Vulaines 074284
Grenier de Xavier, Champs-sur-
  Yonne 067819
Grenier d'Emilie, Blere 067246
Grenier des Collectionneurs,
  Épinal 068455

Grenier d'Hélène, Versailles 074057
Le Grenier d'Olivier, Molsheim . 070150
Grenier du Cambrésis, Le Cateau-
Cambrésis 069243
Le Grenier du Cap, Lege-Cap-
Ferret 069380
Grenier du Chapeau Rouge,
Montélimar 070233
Le Grenier du Château, La
Rochefoucauld 069075
Le Grenier du Chti, Roubaix ... 072549
Le Grenier du Moulin, Argent-sur-
Sauldre 066709
Grenier du Pévèle, Orchies ... 070794,
                              070795
Le Grenier du Pilon, Sainte-
Maxime 073357
Le Grenier du Pin, Castelnau-le-
Lez 067710
Le Grenier du Pin, Lattes 069197
Grenier du Val, Le Val-Saint-
Germain 069369
Grenier Lamalgue, Toulon (Var) . 073711
Grenier Moretain, Moret-sur-
Loing 070345
Le Grenier Nivernais, Champlemy 067816
Grenier Picard, Mauregny-en-
Haye 070007
Grenier Saint Pierre, Carnac ... 067687
Grenier Savoyard, Aix-les-Bains 066481
Aux Greniers de l'Espérance,
Guise 068773
Les Greniers de Saint-Martin, Saint-
Georges-sur-Moulon 072803
Greniers de Sophie, Choisy-la-
Victoire 067981
Les Greniers d'Ici et d'Ailleurs,
Megève 070035
Les Greniers du Père Sauce, Rancourt-
sur-Ornain 072363
Greniers du Père Sauce, Vandœuvre-
lès-Nancy 073961
Grenik, Bicheno 061154
Grenna Museum, Stiftelsen
Grännamuseerna, Gränna 040668
Grenot, Jean-Claude & Marie-Agnès,
Urimenil 073915
Grenouillère, La Rochelle 069088
Grenslandmuseum, Dinxperlo . . 032228
Grenzhuus, Schlagsdorf 021660
Grenzland- und Trenckmuseum,
Waldmünchen 022400
Grenzland-Galerie, Aachen 105937
Grenzland-Museum Bad Sachsa, Bad
Sachsa 016851
Grenzlandheimatstuben des
Heimatkreises Marienbad,
Neualbenreuth 020681
Grenzlandmuseum,
Schnackenburg 021688
Grenzlandmuseum Eichsfeld,
Teistungen 022118
Grenzlandmuseum Raabs an der
Thaya, Raabs an der Thaya . . 002555
Grenzlandmuseum Swinmark,
Schnega 021692
Grenzmuseum Philippsthal (Werra),
Philippsthal 021131
Grenzmuseum Schifflersgrund, Asbach-
Sickenberg 016615
Grenzsteinmuseum Ostrach,
Ostrach 021034
Grenzwald-Destillation Museum,
Crottendorf 017786
Gresham, Crewkerne 144361
Gresham History Museum,
Gresham 049277
Grésilières, Marie-Hélène, Villefranche-
de-Rouergue 074144
Gresse, Monique, Montpellier . . 070303
Gressler, Patrick, Ars-en-Ré .. 066747
Gressler, Rainer, Frankfurt am
Main 075785
Greta Antiques, Greta 061560

Grether, Eberhard, Freiburg im
Breisgau 129860
Gretillat, Jean-Claude, Neuchâtel 087704
Grétry, Liège 063763
Greul, Aree, Frankfurt am Main . 141774
Greulich, Potsdam 078075
Greulich, Achim, Wiesbaden .. 130758
Greulich, Andreas, Frankfurt am
Main 107003
Greutert, M.H., Amsterdam ... 132401
Greuzat, Patrice, Pionsat 072165
Greuze, Jean-Baptiste, Tournus . 073826
Grev Wedels Plass Auksjoner,
Oslo 126372
Greve Museum, Greve 010085
Greve, Dr. Werner, Berlin 141519
Greve, H., Hamburg 076230
Greve, Karsten, Köln 107633
Greve, Karsten, Milano 110358
Greve, Karsten, Paris 104923
Greve, Karsten, Sankt Moritz .. 116624
Greven, Köln 137309
Greven, Karel, 's-Hertogenbosch 083390
Grevenbroek Museum, Hamont-
Achel 003670
Grévendal, J.-L., Bruxelles 063454
Grevesmühlener Mühle,
Grevesmühlen 018815
Grewe, R., Bissendorf 075008
Grewe, Theodor, Haltern am See 076176
Grey Art Gallery, New York University
Art Collection, New York ... 051399
Grey Dog Trading, Tucson ... 124738
Grey Goose Studio, Salt Lake
City 123964
Grey Goose Studio @ Trolley Square,
Salt Lake City 123965
Grey Gums Gallery, MacLean,
Queensland 099248
Grey-Harris & Co, Bristol 088657
Grey Roots Heritage, Owen
Sound 006604
Grey Suit, Video for Art & Literature,
Adamsdown 139184
Grey Wall Gallery, Tampa 124654
Greyabbey Antiques & Fine Art,
Greyabbey 089388, 117888
Greybull Museum, Greybull .. 049278
Greyer, Margita, Bad Gastein .. 062523
Greyfriars, Colchester 144348
Greyhound Hall of Fame, Abilene 046304
Greystoke, Sherborne 091303
Greystones Antiques, Greystones 079635
Greythorn Galleries, Toorak ... 099830
Greytown Books, Masterton .. 083990,
                              143331
Greytown Museum, Greytown .. 038731
GRG Editions, Paris 141131
Grian, Sankt-Peterburg 085243
Gribaudi, Simonetta, Torino ... 142858
Gribaudo, Paul, Paris 141132
Grice, Alan, Ormskirk . 090920, 134911
Grich, Chicago 092723
Grieb, Heiner, Coburg 129657
Griebel, Margit & Franz, Bad Neustadt
an der Saale 074563
Griemann, Maria, Wien 062968
Griepentrog, Detmold 075339
Grier-Musser Museum, Los
Angeles 050425
Grieshaber, Gerhard, Sankt
Gallen 087779
Griesinger, Frankfurt am Main . 141775
Griesser, Ute, Erftstadt 129798
Grießhaber, Tübingen 108788
Grietje Tump Museum,
Landsmeer 032559
Grife & Escoda, Barcelona 114918
Griff, Budapest 079259
Griffa Liberty & Deco, Torino .. 081711,
                              110980
Griffart, Jean-Louis, Bévilliers . 067159
Griffault, Brigitte, Montauban . 070196
Griffel, Michael, München 077514

Griffelkunst-Vereinigung Hamburg e.V.,
Hamburg 058910, 137246
Les Griffes, Pordenone 110675
Griffin, Dallas 093135
Griffin, Hungerford 089608
Griffin, Nashville 095004
Griffin, Pittsburgh 096290
Griffin & Mace, Lewes 089814
Griffin Museum of Photography,
Winchester 054464
Griffin, Gerald, Chicago 120145
Griffin, Gerard, Oldtown 079688, 130888
Griffin, Max, Richmond 096546
Griffin, Simon, London 090172
Griffins & Gargoyles, Chicago . 092724
Griffith, Condong 098914
Griffith & Partners, London 144616
Griffith Artworks, Queensland
College of Art, Griffith University,
Nathan 001346
Griffith Cottage Gallery, Griffith . 099100
Griffith Gallery, Griffith 099101
Griffith Pioneer Park Museum,
Griffith 001122
Griffith Regional Art Gallery,
Griffith 001123
Griffith, Bill, Houston 135604
Griffiths & Co, Worcester 127101
Griffiths Sea Shell Museum and
Marine Display, Lakes Entrance 001205
Griffiths, David, London 090173
Griffon, Cleveland 120426
Grifon, Kyïv 117136
Il Grifone, Livorno 080326
Grigoletti, Pordenone 110676
Grigoropoulos, Chicago 120146
Grijpma, Groningen 083292
Grijpma & Van Hoogen,
Groningen 083293
Grijse, de, Kortrijk 100773
Grilfeldt, Ole, Sylt 078658
Grili, Marco, Belo Horizonte ... 063903,
                              063904
Grillet, Jeanne, Juzennecourt . 068932
Grillon, Hermanville-sur-Mer .. 088812
Grillon, Paris 104924
Grillparzer-Gedenkzimmer, Wien 093001
Grima, Barcelona 133120
Grimaldi, Jorge, México 082476
Grimaldi, Marzio Alfonso, Napoli 142760
Grimaldi, Vital, Pézenas 072138
Grimaldis, Constantine, Baltimore 119746
Grimaud, Thierry, Marseille ... 069942,
                   128772,   128773
Grimberg, Felipe, Miami Beach . 121881
Grimbergen, Lisse 143192
Grimes House, Moreton-in-Marsh 118756
Grimes House Antiques, Moreton-in-
Marsh 090782
Grimes, Chris, Bristol 088658
Grimm & Rosenfeld, New York . 122678
Grimm, Daniel, Gdynia 132822
Grimm, Rolf, Schramberg 078396
Grimm, Ulrich, Magdeburg 107942,
                              137364
Grimme, Andreas, Dornstetten . 075378
Grimme, Wolfgang, Unterdießen 078762
Grimoldi Milano, Milano 080570, 131455
Grims, Wolfgang, Salzburg ... 062794
Grimsby Museum, Grimsby ... 005927
Grimsby Public Art Gallery,
Grimsby 005928
Grimstad Kunstforening,
Grimstad 059657
Grindheim Bygdemuseum,
Kollungtveit 033702
Grindsted Museum, Grindsted .. 010087
Grinnell College Art Gallery,
Grinnell 049280
Grinnell Historical Museum,
Grinnell 049281
Grinning, Byrd, Cleveland 120427
Grinter Place, Kansas City 049874

Grinter Place State Historical Site,
Fairway 048683
Gripes Modelltheatermuseum,
Nyköping 040887
Grippaldi, Filippo, Monaco 082514,
                              132369
Gripsholms Slott, Mariefred 040837
Gris, Jean, Lisboa 084895
Gris, Juan, Madrid 115230
Grisard, Francine, Marseille ... 128774
Grisoni, Andrea, Bologna 079848
Grissom Air Museum, Peru 052002
Grist Mill, Keremeos 006084
Grit, Roma 110801
Gritsch, Imst 127772
Grivot, Anne, Paris 071443
Grivot, Jean-François, Chalon-sur-
Saône 067778
Grizio, Oostende 063817
Grizolle, Colette, Tours 073848
Grizot, Cyril, Saint-Ouen 073099
Grizzly Bear Prairie Museum,
Wanham 007354
Groam House Museum,
Rosemarkie 045634
Grob, Zürich 133981
Grob, Marianne, Berlin 106302
Grobbel, E., Bad Lippspringe ... 106075
Grobe & Rastner, Wien 062969
Grobusch, Aachen 074331
Grochal, Kraków 132839
Gród Piastowski w Gieczu,
Archeologiczny Muzeum Pierwszych
Piastów na Lednicy, Giecz 034754
Grodée, Stéphane, Amiens 066573
Grodentz, Wellington 084164
Grodnenski Dziarzhauny Hystorika-
Archealagichny Muziej, Grodno . 003244
Grødaland Bygdetun, Jarmuseet,
Nærbø 033794
Groeflin & Maag, Basel 116677
Gröger, Uwe, Tutzing .. 078736, 130712
Grömling, Max, Karlstadt 076707
Gröna Paletten Galleri, Stockholm 115891
Grønboengen Auksjonsforretning,
Kongsberg 126355
De Groene Markies, Utrecht ... 083648
De Groene Schuur, Culemborg . 032140
Grönegau-Museum, Melle ... 020354
Groenenboom, J.C., Zoelen ... 132674
Groenewoud, Adriaan, Rotterdam 083570
Groeningemuseum, Musea Brugge,
Brugge 003411
Groenink Van der Veen, M.,
Leeuwarden 083416
Grønlands Kunstskole,
Eqqumiitsuliornermik Ilinniarfik,
Nuuk 055757
Grønlund's Forlag, Maribo 136778
Grønnegade Kunst, Antik og
Rammecenter, Næstved 066004
Grönsöö Slott, Enköping 040603
Gröpl, Kai, Wolfratshausen 079058
Groesser, Omaha 123318
Gröstlinger, Markus, Zell am
Pettenfirst 063125
Grözinger, Lutz, Sereetz 078457
Gróf Esterházy Károly Kastély- és
Tájmúzeum, Pápa 023831
Grofe, Egbert, Geldern 129891
Grofe, Egbert, Kevelaer 130116
Grogan & Co., Dedham 127198
Groharjeva Hiša v Sorici, Spodnja
Sorica 038620
Grolée-Virville, Alain de, Paris . 141133
Groll, Helma-Konstanze,
Magdeburg 130260
Grollemund, Michel, Biarritz ... 067200
Grommen, Michel, Liège 125342, 140242
Grona, Joan, San Antonio 124037
Gronauer Kunstverein, Gronau,
Westfalen 058911
Gronchi, Loredana, Verona ... 081915
Gronert, Traunstein 108774

Gronert, Ulrich, Berlin 074807
Grong Bygdamuseum, Grong ... 033626
Gronghe de Cabourg, Cabourg . 067545
Groninger Museum, Groningen . 032378
Groninger Schaatsmuseum,
Sappemeer 032841
Groot Constantia Manor House and
Wine Museum, Constantia 038691
Groot Constantin Homestead and Wine
Museum, Constantia 038692
Groot Kasteel van Loppem,
Loppem 003829
Groot, A.W.J. de, Utrecht 083649
Groot, Anneloes, Den Haag ... 112388
Groot, Barbara de, München 130351
Groot, de, Wiesbaden 108913
Groot, G. de, Noordeloos 132597
Groote, Zwolle 083732
Groote, de, Antwerpen 063208
Groote, de, Bruxelles 063455
Grootegoed, Hilversum 083401
Grootfontein Museum Das Alte Fort,
Grootfontein 031838
Grootmoeder's Boetiek,
Apeldoorn 082923
Grootmoeders Keukenmuseum,
Nijmegen 032685
Groovy, Pittsburgh 096291
Gropius House, Lincoln 050287
Gropler, Wolfgang, Berlin 074808
Gropp & von Loesch, Berlin ... 074809
Groppler Ulfert, Firenze 131189
Gros & Delettrez, Paris 125721
Gros, Jean-Pierre, Paris 071444
Gros, Robert, Courmangoux ... 068191
Gros, Robert, Verjon 074024
Gros, Robert, Villeurbanne 074209
Groscaux, Michel, Cricqueboeuf 068244
Grosch, Harry, Dresden 075414
Grosche, Heiko, Köln 107634
Grosdidier, Claude, Vendargues . 074007
Grosell, Peter, København 140610
Grosman, Arnhem 082938
The Grosmont Gallery, Grosmont 117893
Grosos, Michel, Villers-Farlay .. 074177
Gross, Halle, Saale 076166
Groß, Kleinfischlingen 076817
Gross McCleaf, Philadelphia ... 123390
Gross, Barbara, München 108130
Gross, Brian, San Francisco ... 124292
Groß, D., Binz, Ostseebad 074999
Gross, G., Kassel 076722
Gross, Joe, Cleveland 120428
Groß, Norbert, Meckenheim,
Pfalz 077314
Gross, Roger, New York 145343
Grosse Ile Historical Museum, Grosse
Ile 049282
Große Kunstschau im Roselius
Museum, Worpswede 022771
Großer Wappensaal im Landhaus,
Klagenfurt 002188
Grossetti Arte Contemporanea,
Milano 110359
Großheide, A., Osnabrück 077967
Großherzoglich-Hessische
Porzellansammlung, Darmstadt 017820
Grossi, Pierre-Hélen, Apt 103263
Grossi, Vincenzo, Napoli 080885
Grosslicht, Paul, Wien 062970
Großmann, Pirna 108383
Großmann, Anett, Pirna 078052
Großmann, B., Schönebeck ... 078384
Großmann, Wolfgang,
Eberswalde 075562
Großmitterberger Troadkasten,
Laussa 002276
Großschmidt, Marianne, Tulln .. 140048
Grosvenor Fine Art, Auckland .. 112777
Grosvenor Galleries, Kingston .. 111224
Grosvenor Gallery, London 118360
Grosvenor House, London 090174,
118361
Grosvenor House Interiors,

Beaconsfield 088413
Grosvenor Kunama Galleries,
Jindabyne 099159
Grosvenor Museum, Chester ... 043933
Grosvenor Prints, London 090175,
134677,   144617
Grosvenor, Brant, Cincinnati ... 092863
Grosz, Marvin L., Denver 093296
Grote, Welf-W. Freiherr von,
Worpswede 079068, 130779
Grotemeyer, Petra,
Gummersbach 129930
Groth Verlag, Bad Vilbel 137037
Groth, Heinz & Erika, Felsberg . 075733
Grothaus, Enger 075620
Grothe, Bernd, Berlin 074810
Grothe, Knud, Charlottenlund .. 102788
Groton Historical Museum,
Groton 049285
Grotowski, Trier 108776, 130707
Grotte de Bédeilhac, Musée de
Préhistoire, Tarascon-sur-Ariège 015910
Grotte de Blanot, Blanot 012063
Grotte de la Vache, Alliat 011602
Grotte de Pair non Pair, Prignac-et-
Marcamps 014998
Grotte de Rouffignac, Rouffignac-Saint-
Cernin 015175
Grotte d'Osselle, Saint-Vit 015592
Grotte et Parc Préhistoriques, Musée
Cavernicole, Foissac 012997
Grotte Préhistorique Ornée, Miers 014167
Grotte Scladina et Centre
Archéologique, Sclayn 004005
Grottes d'Arcy-sur-Cure, Arcy-sur-
Cure 011713
Grottes de Vallorbe, Vallorbe .. 042096
Grottes Préhistoriques de Sare-Lezea,
Sare 015682
Grotto, Hong Kong 102081
Groudle Glen Railway, Onchan . 045440
Ground Zero Designs, Tampa .. 124655
Groundfloor Gallery, Balmain ... 098698
Grounds for Coffee & Antiques,
Edmonton 064294
Grounds for Sculpture, Hamilton 049323
Group 2 Gallery, Bruxelles 100618
Group for Directors of Museums,
Chester 060147
Group for Education in Museums,
Gillingham, Kent 060148
Group, Philippe, Antwerpen ... 063209,
100491
Groupe des Salons Sélectionnés
d'Antiquaires, Dijon 058651
Groupement des Amis de Musées
de la Région Midi-Pyrénées,
Montauban 058652
Groupement des Amis des Musées de
la Région Centre, Orléans 058653
Groupement des Associations d'Amis
de Musées de la Région Bretagne,
Nevez 058654
Groupement des Associations d'Amis
de Musées de la Région Languedoc-
Roussillon, Montpellier 058655
Groupement des Associations d'Amis
de Musées de la Région Limousin,
Limoges 058656
Groupement des Associations des
Amis de Musées Région Provence
Alpes Côte d'Azur, Toulon (Var) 058657
Groupement National des
Entrepreneurs de Restauration de
Monuments Historiques, Paris . 058658
Groupement Rhône-Alpes des Amis de
Musée, Lyon 058659
Grout Museum of History and Science,
Waterloo 054204
Groux, Reginald, Paris 104925
Grove, Darwen 089032
The Grove Antiques Centre,
Honiton 089563

Grove Farm Gallery, Catfield ... 117546
Grove Gallery, Saint Paul 123912
Grove House Museum, Manx National
Heritage, Ramsey 025007
Grove Rare Books, Skipton ... 135050,
144848
Grove Road Antiques, Blenheim 083841
Grovebrook, Dee Why 061393
Groveland Gallery, Minneapolis . 121995
Groveland Yosemite Gateway Museum,
Groveland 049287
Grover & Thurston, Seattle 124537
Grover Cleveland Birthplace,
Caldwell 047346
Grover M. Hermann Fine Arts Center,
Marietta 050680
Grover, Ravi, Lucknow 109407
Grovewood Gallery, Asheville .. 046555
Growrich, Seoul 111655, 111656
The Grub Street Bookshop,
Fitzroy 139734
Grubb Shaft Gold and Heritage
Museum, Beaconsfield 000834
Grube Alte Hoffnung Erbstolln,
Mittweida 020426
Grubenmann-Sammlung, Teufen 042051
Gruber, München 130352
Gruber, A., Altötting 074398
Gruber, Christa, Sankt Georgen am
Steinfelde 062824
Gruber, Gerhard, Heilbronn ... 141934
Gruca, Wojciech, Gdańsk 132815
Grübel Nobre-Madeira, Álvaro,
Wiesbaden 078976
Grüger, Klaus, Pfaffenhofen an der
Ilm 078038
Gruel, Dany, Sens-de-Bretagne . 141345
Gruen, Chicago 120147
Gruen & Speer, Wiesbaden ... 108914
Grün, A. & M., Püttlingen 078101
Grün, Herbert, Heidelberg 076443
Grünauer Wassersportmuseum,
Stadtmuseum Berlin, Berlin ... 017151
Grünbichler, Johann, Bad Ischl . 062525
Gründ, Paris 136934
Gründer, Braunschweig 141634
Gründerzeitmuseum im Gutshof
Mahlsdorf, Berlin 017152
Gründler, Ulrich, Striegistal 130663
Grüne Galerie, Fränkisch-
Crumbach 106964
Grüner, Joachim, Berlin 074811
Grüner, Maria, Blaufelden 075017
Grünes Gewölbe, Staatliche
Kunstsammlungen Dresden,
Dresden 017985
Grünwald, Dr. Michael D.,
München 108131
Grünzig, Mathias, Eppendorf .. 129794
Gruet Pol, Nantes 128836
Grüttert Uhrenmuseum Bremen,
Bremen 017589
Gruetunet Museum, Kirkenær i
Solør 033696
Grützner, Ingo, München 130353
Gruia, Christine, Friesoythe ... 107075
Gruillot, Marie-Françoise,
Besançon 067125
Grulier, Courchevel 103636
Grumman Memorial Park,
Calverton 047354
Grumps, Los Angeles 094350
Grumpy, Las Vegas 094151
Grund, Andrea, Krakow am See 130188
Grund, Jürgen, Berg, Kreis
Ahrweiler 129457
Grundmann, Andreas, Münster . 142202
Grundner, Heinz, München 077515
Grundtvigs Mindestuer i Udby,
Lundby 010225
Grundy Art Gallery, Blackpool .. 043644
Grune, Haimhausen 076156
Gruner, Sittard 083606
Grunewald, Klaus, Kansas City . 145194

Grunewald, Peter, Essen 075678
Grunow, Dieter, Berlin 074812
Grunspan, Marcel, Paris 071445, 071446,
104926
Grunt, Vancouver 101791
Grunt Gallery, Vancouver 007285
Grunwald Center for the Graphic Arts,
University of California, Hammer
Museum of Art and Cultural Center,
Los Angeles 050426
Grup 62, Barcelona 137960
Grup d'Art Escolà, Torres de
Segre 059868
Grupello Verlag, Düsseldorf ... 137160
Grupo Alternativo Empresarial,
Málaga 059869, 115303
Gruppe 3zehn, Heidelberg 076444
Gruppo '78, Trieste 111049
Gruse, Andrea, Kamen 076668
Grusenmeyer & Bilquin, Bruxelles 063456
Grusenmeyer, Philippe, Boisseron 067274
Grusenmeyer, Serge, Gent 063639
Grusenmeyer, Stéphanie,
Bangkok 088079
Gruttmann, Dortmund 106703
Gruvan Antik, Norrköping 086880
Gruvik, Oskarshamn 126653
Gruvmuseet, Sala 040932
Gruyer, Paul, Lentilly 069387
Gryfton Galleries, Ryeish Green . 118999
Gryksa, Lorenz, Wien 127921
Grytøy Bygdetun, Lundenes ... 033758
Grzelachowski, Detlef, Berlin ... 074813
Grzimek, Brita, Berlin 074814
GSA Centrum voor Beeldende Kunst,
Hilversum 112538
Gschwind & Obertreis, Planegg . 130522
Gsellmanns Weltmaschine, Edelsbach
bei Feldbach 001839
Gselman, Jadranka, Maribor .. 085617,
114731
GSIS Gallery, Pasay 034563
Gslerie Crameri, Chur 116218
GSR Gallery, Beijing 101956
GSR Gallery, Shanghai 102180
GSR Gallery, Taichung 116893
Gstöttner, Friedrich, Salzburg .. 127849
Gu-Ba-Csó, Budapest 079260
Gu City Museum, Gucheng,
Hubei 007816
Gu Yue Xuan, Shanghai 065136
Guach, Juan, Palma de Mallorca 115351
Guadagno, Gabriella, Bologna . 130967
Guadalimar, Madrid 139057
Guadalupe Cultural Arts Center, San
Antonio 052938
Guagenti, Luciano, Milano 080571
Guainans, Le Castellet 069237
Guainans, Alain, Le Beausset .. 069214
Guaita, Joan, Palma de Mallorca 115352
Gualco, Caterina, Genova 110112
Gualda Hernandez, Ramon,
Granada 086037
Gualtieri, Galatée, Paris 104927, 104928
Guam Museum, Agana 023356
Guan Return to the Ancient Dian Art
Museum, Beijing 007621
Guan Xian, Beijing 101957
Guan Xian, Taipei 116917
Guanabara, Rio de Janeiro ... 101041
Guandu Museum, Guandu ... 007796
Guangdong Beginxing Museum,
Shixing 008277
Guangdong Folk Arts and Crafts
Museum, Guangzhou 007800
Guangdong Museum, Guangzhou 007801
Guangdong Museum of Art,
Guangzhou 007802
Guangdong Museum of Art Journal,
Guangzhou 138601
Guangdong Revolutionary History
Museum, Guangzhou 007803
Guangning Museum, Guangning 007798
Guangshui Museum, Xiaogan . 008416

Guangxi Academy of Fine Art, Nanning ... 055226
Guangxi Fine Arts Publishing House, Nanning ... 136734
Guangxi Natural Museum, Nanning ... 008121
Guangxi Zhuang Museum, Nanning ... 008122
Guangzhou Academy of Fine Arts, Guangzhou ... 055210
Guangzhou Ancient Works, Guangzhou ... 140440
Guangzhou Art Gallery, Guangzhou ... 007804
Guangzhou Artists Gallery, Guangzhou ... 102028
Guangzhou Fine Arts Museum, Guangzhou ... 007805
Guangzhou Huangpu War Exhibition Center, Guangzhou ... 007806
Guangzhou International Art Fair, Guangzhou ... 098048
Guangzhou Liwan District Museum, Guangzhou ... 007807
Guangzhou Lu Xun Museum, Guangzhou ... 007808
Guangzhou Museum, Guangzhou 007809
Guangzhou Photo Biennal, Guangzhou ... 098049
Guangzhou Triennal, Guangzhou 098050
Guani, Silvia, Genova ... 131318
Guard House and Soldiers Barracks, Fredericton ... 005841
Guardadero Blanco y Negro, Medellín ... 065288
Guardhouse Museum, Tattershall 045949
Guardia, Gérard, Le Castellet ... 069238
El Guardian de lo Pequeño, Madrid ... 115231
Guardian of the East, Singapore 085415
Guardini, Berlin ... 106303
The Guards Museum, London ... 044987
Guarisco, Washington ... 124914
Guarisse, Andre Caruso, Porto Alegre ... 063974
Guarnieri, Rosanna, Parma ... 110628
Guarracino, Ciro, Napoli 080886, 110530, 131622, 142761
Guayamurí Antigüedades, Caracas ... 097955
Guazzolini, Roma ... 081405, 081406
Gubala, Darius, New York ... 095437, 135931
Gubbi, Sergio, Someo ... 133928
Gucity Museum, Gucheng, Jiangsu ... 007817
Guckelberger & Co., F.A., Philadelphia ... 123391
Der Guckkasten, München ... 142157
Guckloch Galerie, Maur ... 116495
Gudagnini, Ferdinando, Carano 109933
Gudbrandsdal Krigsminnesamling, Kvam ... 033723
Gudbrandsdalen Antik, Vinstra .. 084393
Guddal, G., Rosendal ... 113251
Gude, Amsterdam ... 082762
Gudhjem Kunsthandel, Gudhjem 065714
Gudrun Spielvogel Galerie + Edition, München ... 108132, 137404
Guedel, Gildas, Paris ... 071447
Guedj, Célia B., Lyon ... 104109
Guedj, Didier, Saint-Ouen ... 073100
Guedot, Bernard, Montpellier ... 070304
Guée, Jean-Claude, Thury-Harcourt ... 073678
Gueguen, Jean-Yves, Crozon ... 068250
Gührs, Anne, Hamburg ... 076231
Güldaglar, T., Heidelberg ... 076445
Guelle, Serge, Orléans ... 070816
Guelman, Moskva ... 114262
Guelph Civic Museum, Guelph Museums, Guelph ... 005932
Le Guen, Gilles, Saint-Fargeau . 105516
Guenard-d'Abbadie, Andrée, Toulon

(Var) ... 073712
Gueneau, Jean-Michel, Paris ... 071448
Güney, Yasar, Berlin ... 074815
Güngör Dekorasyon, İstanbul ... 088122
Guennet, Ventabren ... 074015
Guénot, Jacqueline, Paris ... 071449
Günter Grass Haus, Forum für Literatur und bildende Kunst, Die Lübecker Museen, Lübeck .. 020157
Günter Raup Antiquitäten, Bremen ... 075162
Günter, Anna, Traunkirchen ... 062861
Günter, Jan, Flensburg ... 106956
Günter, Katharina & Thomas, Zürich ... 087972
Günter, Stefan, Volkach ... 078813
Günther, Mönchengladbach ... 077390
Günther, Albert, Dortmund ... 129697
Günther, Albert J., Eichstätt ... 129781
Günther, Dieter, Issum . 076630, 130080
Günther, Dr. Jörn, Hamburg ... 141865
Günther, Johannes, Offenbach am Main ... 130487
Guenther, Karin, Hamburg ... 107275
Günther, Knut, Frankfurt am Main ... 075786
Günther, M., Leipzig ... 077060
Günther, Monika, Hamburg ... 129963
Guenther, Robert, Milwaukee ... 121922
Günther, Sabine, Weißenfels ... 078917
Güntheroth, Claudia, Herzberg, Mark ... 141943
Guépin, Jean-Jacques, Ruaudin 072635
Gür, B. C., Neuhausen am Rheinfall ... 116532
Gueraiche, Ghodjah, Hydra ... 060840
Guerdener, Victor, Chantemerle-les-Blés ... 067829
Guereta, Madrid ... 115232
Guericke-Zentrum (Otto-von-Guericke-Museum in der Lukasklause), Magdeburg ... 020206
Guérin & Fils, La Baule-Escoublac ... 068949
Guérin, Guy, Pouancé ... 072293
Guerin, Jean-Claude, Paris ... 071450
Guérin, Michel, Saint-Dié-des-Vosges ... 125799
Guérin, Pierre, Nantes ... 070464
Guerlotté, Philippe, Casteljaloux 067704
Guermeur, Jeannine, Le Faou ... 069266
Guernsey, New York ... 127349
Guernsey County Museum, Cambridge ... 047375
Guernsey Museum and Art Gallery, Saint Peter Port ... 045712
Guerpillon-Bergeron, Joëlle, Quimper ... 125783
Guerra, Antonietta, Modena ... 080798
Guerra, Carmen de la, Madrid .. 115233
Guerra, Cristina, Lisboa ... 114035
Guerra, Oriana & Cecilia, Santiago de Chile ... 064968
Guerra, Paolo, Roma ... 131947
Guerre, Gérard, Avignon 066881, 128456
Guerrero, Bogotá ... 102343
Guerrero Mancilla, Eugenio, Madrid ... 133343
Guerrero, Enrique, México ... 111962
Guerreri, Gianluigi, Roma ... 081407
El Guerro Antiguedades, Valencia 086521
Guerrots, François & Chantal des, Heugleville-sur-Scie ... 068817
Gürtler, Pierre, Laufen ... 144091
Guesne, Pierre, Couptrain ... 068182
Guess, Fredrick, New Orleans .. 122176
Guest & Philips, Driffield ... 089107
Guest House Museum, Fort Bragg ... 048785
Guest, G.P. & M.J., Subiaco ... 062255
Guest, Karen, Beverley 088449, 134096
Guest, Melissa, Subiaco ... 127681
Gützlaf, Wolfgang, Berlin ... 074816
Guevara, Buenos Aires ... 060906

Guevenoux, André, Paris ... 071451
Gueyraud Transactions, Laragne-Montéglin ... 069187
Güzel Sanatlar Bölümü, Bogaziçi Universitesi, İstanbul ... 056478
Güzel Sanatlar Fakültesi, Atatürk Universitesi, Erzurum ... 056476
Güzel Sanatlar Fakültesi, Marmara Universitesi, İstanbul ... 056479
Güzel Sanatlar Fakültesi, Selçuk Universitesi, Konya ... 056480
Güzel Sanatlar Müzesi, Girne ... 033460
Güzel Sanatlar ve Musiki Bölümü, Orta Dogu Teknik Universitesi, Ankara ... 056473
Güzel Sanatlar, Tasarım ve Mimarlık Fakültesi, Bilkent Universitesi, Bilkent ... 056475
Güzelyurt Müzesi, ArKeoloji ve Doğa Müzesi, Güzelyurt ... 033462
Guffroy, François, Amiens ... 125487
Gufi, Sofia ... 064230
Gugel, Nürnberg ... 077823
Guggenberger, Ilse, Salzburg ... 062795
Guggenheim & Asher, New York 122679
Guggenheim Bilbao Museoa, Bilbao ... 039144
Guggenheim Hermitage Museum, Las Vegas ... 050162
Guggenheim Museum Soho, New York ... 051400
Guggisberg, Beat, Zimmerwald . 087921
Guglielmetti, Serge, Saint-Cyr-les-Colons ... 072725
Guglielmi, Giuseppe, München . 130354
Guicciardi, Marco, Modena ... 080799
Guichard, Chicago ... 120148
Guichard, Montbellet ... 070212
Guichard & Juillan, Saint-Brieuc 125797
Guichard, Guy, Ouzouer-sur-Loire 070831
Guichard, Henriette, Paris ... 129024
Guicheney, Philippe, Castelnaudary ... 067712
Guichoux, François, Carhaix-Plouguer ... 067683, 128529
Guidarelli, Federico, Roma ... 131948
Guidarini, Pia, Genova ... 080264
Guidat, Daniel, Cannes ... 103513
Guide Emer, Paris ... 138646
Guide to National Endowment for the Arts, Washington ... 139377
Guidi & Schoen Arte Contemporanea, Genova ... 110114
Guidi Bruscoli, Fabrizio, Firenze 080157
Guidi, Guido, Genova ... 110113
Guidi, Maya, Carouge ... 116197
Guidi, Silvia, San Marino ... 133028
Guidieri, Gina, Genova ... 080265
Guido Gezellemuseum, Musea Brugge, Brugge ... 003412
Guifei Muwen Guan (Museum), Xinping ... 008440
Guignard, Caudan ... 125548
Guignard, Paris ... 071452
Guignard, Bruno, Rochecorbon . 072493
Guignard, Gérard, Saint-Ouen .. 073101
Guignard, Jean-Louis, Vevey ... 087879
Guigon, Paris ... 104929
Guigue, Paris ... 104930
Guigue & Locca, Paris ... 129026
Guigue, Jean-François & Muriel, Paris ... 129025
Guiho, Claude, Varengeville-sur-Mer ... 073979
Guilbert, Benoît, Paris ... 141134
Guild Antiques, New York ... 095438
Guild Art Corporation, Minneapolis ... 121996
The Guild Art Gallery, Mumbai . 109440
Guild Gallery, Bristol ... 043751
Guild Gallery at El Con, Tucson . 124739
Guild Hall Museum, East Hampton ... 048432
Guild House Antiques, Vancouver 064825

Guild News, New York ... 139378
Guild of Antique Dealers and Restorers, Shrewsbury ... 060149
The Guild of Artists and Artisans, Ann Arbor ... 060505
Guild of Boston Artists, Boston . 060506
Guild of Creative Art, Shrewsbury 053327
Guild of Master Craftsmen, Lewes ... 060150, 138176
Guild of Railway Artists, Warwick 060151
Guild of Taxidermists, Glasgow . 060152
The Guild Shop, Toronto ... 101621
La Guilde Graphique, Montréal . 058446, 101418
Guilded Salvage Antiques, Minneapolis ... 094917
Guildford Cathedral Treasury, Guildford ... 044486
Guildford House Gallery, Guildford ... 044487
Guildford Museum, Guildford ... 044488
Guildhall, Exeter ... 044311
Guildhall, Leicester ... 044794
Guildhall, Thaxted ... 045966
Guildhall Art Gallery, London ... 044988
Guildhall Collection, Kingston-upon-Hull ... 044713
Guildhall Galleries, Bury Saint Edmunds ... 117491
Guildhall Gallery, Poole ... 118931
Guildhall Gallery, Winchester ... 046149
Guildhall Museum, Carlisle ... 043874
Guildhall Museum, Chichester ... 043943
Guildhall Museum, Rochester ... 045628
Guildhall Museum, Sandwich ... 045736
Guildnotes, New York ... 139379
Guilford College Art Gallery, Greensboro ... 049252
Guilford Courthouse National Military Park, Greensboro ... 049253
Guilhabert, Françoise, La Rochelle ... 069089
Guilhem, Jean-Pierre, Béziers .. 067178
Guilhemsans, Jean, Ousse ... 070829
Guilhemsans, Jean, Pau ... 072040
Guilin Museum, Guilin ... 007818
Guilin Stone Inscription Museum, Guilin ... 007819
Guillais, Hubert-Pierre, Paris ... 071453
Guillard, Patrick, Paris ... 071454
Guillaume, Evelyne, Warcq ... 074289
Guillaume, J.-M., Namur ... 063808
Guillaumot, Georges, Lyon ... 125628
Guillaumot, Pierre-Yves, Lyon .. 125629
Guillaumot, Pierre-Yves, Villefranche-sur-Saône ... 125841
Guillebaud, Georges, L'Isle-d'Espagnac ... 069544
Guillemin, Alain, Romilly-sur-Seine ... 072526
Guillemin, Jean-Jacques, Tournus ... 073827
Guillemin, Michelle, Chalon-sur-Saône ... 140818
Guillemin, Pascal, Lyon ... 069745
Guilleminot, Françoise, Villars-le-Pautel ... 074127
Guillemot, Claude, Paris ... 104931
Guillemot, Monique, Nantes ... 070465
Guiller, Medellín ... 065289
Guiller, Paris ... 071455
Guilleminet, Pierre, Bourg-en-Bresse ... 067410
Guillet, Paris ... 104932
Guillon-Laffaille, Fanny, Paris . 104933, 136935, 141135
Guillot-Renou, Cloyes-sur-le-Loir 068045
Guillou, Paris ... 141136
Guilmette, Julie, Firenze ... 131190
Guimaraes, Rio de Janeiro ... 064065
Guimarães, Alencastro, Porto Alegre ... 100984
Guimaraes, Maguered P., Belo Horizonte ... 063905

Guimiot, Philippe, Bruxelles 063457
Guinand, Anse 066660
Guindani, Gianna, Milano 131456
Guinevere Antiques, London ... 090176
Guinhut, Daniel, Montrevault ... 070330
Guinness Storehouse, Dublin ... 024884
Guinness World of Records Museum,
  Las Vegas 050163
Guinot, Philippe, Jons 103859
Guiot, Paris 104934
Guiping Museum, Guiping 007822
Guiraud, Hervé, Bordeaux 128491
Guiraud, Raoul, Paris 071456
Guirrierc, Hervé, Bordeaux 067335
Guitar Museo Ex Teresa Arte Actual,
  México 031158
Guiter, Jean-Luc, Perpignan 072083
Guitine, Evelyne, Saint-Ouen ... 073102
Guitton, Bernard, Le Lion-
  d'Angers 069294
Guiyuan Art Cooperative,
  Shanghai 102181
Guizhou Provincial Museum,
  Guiyang 007823
Guizzetti & Collet, Reims 125785
Guizzi, Pietro, Brescia 079941
GUK + Garður, Selfoss 109245
GUK + Udhus, Lejre 102926
Gukgiwon Taekwondo Museum,
  Seoul 029998
Gulbenes Vēstures un Mākslas Muzejs,
  Gulbene 030124
Gulden Galeria, Gdańsk 113475
De Gulden Roos, Antwerpen .. 100492,
                                136620
Guldklumpen, Silkeborg 066110
Gule Længe Antikviteter, Hjørring 065762
Gulessian, André, Marseille 069943
Gulf Antiques, Abu Dhabi 088158
Gulf Antiques, Manama 063144
Gulf Antiques, Sharjah 088200
Gulf Auctions, Manama 063145, 125290
Gulf Beach Art Center, Indian Rocks
  Beach 049672, 057231
Gulf Coast Artworks, Houston .. 121051
Gulf Coast Exploreum, Mobile .. 050975
Gulf Museum, Port-aux-Basque .. 006677
Gulf of Georgia Cannery,
  Richmond 006795
Gulf Shores Museum, Gulf
  Shores 049295
Gulf Station Historic Farm, Yarra
  Glen 001669
Gulino, Giovanna, Catania 080036,
                                131090
Gull Gallery, Merimbula 099328
Gullà, Carmen Rosetta, Roma .. 142814
Gulli Skoletun, Andebu 033496
Gullia, Carolyn, Atlanta 135310
Gully Antiques, Boronia 061176
Gully, Rex, Seaview 091258
Gumanitarnyj Centr-muzej Tvorčestva
  Autsajderov, Moskva 037310
Gumbez Manace Literary Ethnographic
  Museum, Taš-Aryk 030083
Gumer & Co., Louisville 094555, 094556
Gumeracha and District History Centre,
  Gumeracha 001125
Gumnut Antiques, Gosford 061546
Gumtree Museum of Art, Tupelo 053853
Gunarsa Museum, Museum Seni Lukis
  Klasik Bali, Klungkung 024605
Gundagai Historical Museum,
  Gundagai 001126
Gundelsheimer Oldtimermuseum,
  Gundelsheim, Württemberg ... 018892
Gunder, Vera, Mannstadt 129669
Gunderman, Anne, Buffalo 119957
Gunderson House, Kenyon 049936
Gundry, Houston 093680
Gunma-kenritsu Kindai Bijutsukan,
  Takasaki 029445
Gunma-kenritsu Rekishi Hakubutukan,

Takasaki 029446
Gunma-kenritsu Shizenshi
  Hakubutsukan, Tomioka 029629
Gunn, Lynton 118685
Gunn Memorial Library and Museum,
  Washington 054084
Gunn, Alastair, Balerno 134051
Gunnebo Slott och Trädgårdar,
  Mölndal 040848
Gunnels Gamla Grejor, Strängnäs 087082
Gunnersbury Park Museum,
  London 044989
Gunnes gård, Upplands Väsby .. 041108
Gunning, John, Limerick 079656, 130880
Gunnison County Pioneer and
  Historical Museum, Gunnison . 049296
Gunnison Museum of Natural History,
  Pawling 051949
Gunston Hall Plantation, Mason
  Neck 050736
Gunten, Wilfried von, Thun .. 116669,
                                133937
Gunter Callsen Antiquitäten,
  Dresden 075415, 129710
Guntermann, Düsseldorf 075477
Guntlisbergen, G., Best 082968
Guntlisbergen, John, Eindhoven . 083219
Gunyulgup Galleries, Yallingup .. 099929
Gunze Museum, Ayabe 028500
Guovdageainnu Gilisillju, Kautokeino
  Bygdetun, Kautokeino 033695
Gupta, Kavi, Chicago 120149
Gurari, Boston 092475, 145016
Gurich, Madrid 086198
Gurkha Museum, Winchester .. 046150
Gurlino, Luciano, Torino 132148
Gurriaran, Luis, Madrid 115234
Gurtner-Meier, Walter, Basel ... 087236,
                                133644
Gurtner, Christian, Wien 127922
Guruban Tajore Art Gallery,
  Chennai 109286
Guruda, Delhi 079384
Guruji Bros, Mumbai 109441
Gurusaday Museum, Museum of
  Bengal Folk Art, Kolkata 024360
Guruve, London 118362
Gusevicius, Vilnius 132322
Gush Ezyon Museum, Gush
  Ezyon 025040
Gussio, Lucia, Roma .. 081408, 131949
Gustaf-Dalman-Institut für biblische
  Landes- und Altertumskunde,
  Greifswald 018807
Gustafs Galleri i Lund, Lund ... 115735
Gustafson Gallery, Colorado State
  University, Fort Collins 048797
Gustafsson, Gudrun, Stockholm . 087001
Gustav-Adolf-Gedenkstätte,
  Lützen 020187
Gustav-Adolf-Museum, Geleitshaus,
  Museum Schloß Neu-Augustusburg,
  Weißenfels 022542
Gustav-Bauernfeind-Museum, Sulz am
  Neckar 022078
Gustav-Freytag-Archiv und Museum,
  Wangen im Allgäu 022420
Gustav III:s Antikmuseum,
  Stockholm 041003
Gustav III:s Paviljong, Solna ... 040976
Gustav Jeeninga Museum of Bible and
  Near Eastern Studies, Anderson 046475
Gustav-Lübcke-Museum, Hamm,
  Westfalen 019029
Gustav Wolf Kunstgalerie,
  Östringen 020964
Gustava Skiltera Memoriālais Muzejs,
  Rīga 030183
Gustavsbergs Porslinmuseum,
  Gustavsberg 040674
Guszahn, P. & R., Kirchlengern . 076810,
                                076811
Gut of Canso Museum, Port
  Hastings 006686

Gut Seekamp der Hans Kock-Stiftung,
  Kiel 019603
Gut Zernikow, Großwoltersdorf . 018867
Gut, Rahel, Zürich 116802
Gut, Werner & Mägi, Triengen .. 087856
Gutao Ure Museum, Beijing 007622
Das Gute Buch, Zürich 144179
Gutenberg Al Colosseo, Roma .. 110802,
                                142815
Gutenberg-Gedenkstätte, Eltville am
  Rhein 018169
Gutenberg-Museum, Mainz 020226
Guth, Tangermünde 108755
Guth-Maas & Maas, Reutlingen . 108481
Gutharc-Ballin, Paris 104935
Guthmann, Steffen,
  Langenbernsdorf 077023
Guthschmidt, Den Haag 112389
Gutian Museum, Gutian 007825
Gutierrez Carrasquilla, Enrique,
  Sevilla 133432
Gutierrez, Eric, Dallas 135505
Gutierrez, Raul, San Antonio ... 124038
Gutlin, London 090177, 134678
Gutman, Jean-Pierre, Versailles . 105845
Gutshaus Steglitz, Berlin 017153
Gutshof Pinnow, Pinnow 021133
Guttmann, Max, Helsingør 065741
Guust, den, Breda 082985
Guwahati Medical Museum, Medical
  College Guwahati, Guwahati . 024276
Guwu Xuan Art, Suzhou 102239
The Guy On Motor, Los Angeles 094351
Guy, Adam, Bruxelles 140177
Guy, Gerard, Bewdley . 088457, 134101
Guy, Jacques, Buchères 067530
Guy, Jean, Mons 070170
Guyana Museum, Georgetown .. 023389
Guyart, Brugge 100535
Guyomarc, Noël, Montréal 101419
Guyon, Alain, Besançon 067126
Guyot, Camille, Dinan 068351
Guyot, Florence, Paris 071457
Guyot, Gérard, Vanves 073975
Guyot, Marion, Atlanta 119574
Guyot, Thierry, Notre-Dame-du-
  Pre 070752
Guyuan Museum, Guyuan 007826
Guzman, Valencia 086522, 115504
Gvozdenovic, Ivica, Stuttgart ... 078606
Gwangju Municipal Folk Museum,
  Gwangju 029941
Gwangju National Museum,
  Gwangju 029942
Gwanhun Gallery, Seoul 111657
Gwen, Fort Worth 120867
Gwili Steam Railway, Bronwydd 043771
Gwilliam, Edred A.F., Cricklade . 089009
Gwinnett History Museum,
  Lawrenceville 050191
Gwydir Street Antiques,
  Cambridge 088763
Gwynby, Cleveland 092946
Gwynedd Education and Culture,
  Archives, Museums and Galleries,
  Caernarfon 043824
Gwynedd Museum and Art Gallery,
  Amgueddfa ac Oriel Gwynedd,
  Bangor, Gwynedd 043490
Gwynfair, Holyhead 089554
Gx gallery, London 118363
Gyard, Christian, Saint-Martin-en-
  Haut 072952
Gyemyung University Museum,
  Daegu 029913
Gyeongbo Museum of Paleontology,
  Namjeong 029974
Gyeongbuk National University
  Museum, Daegu 029914
Gyeonggi Provincial Museum,
  Yongin 030060
Gyeongju National Museum,
  Gyeongju 029945

Gyeongju University Museum,
  Gyeongju 029946
Gyermek- és ifjúságszervezet-történeti
  Múzeum, Zánka 024056
Gyertyás Antikvitás, Szeged ... 079324
Gyewon School of Art and Design,
  Uiwang 056105
Gyla Byggningen, Gustavsberg . 115669
Gyldenløwe, Oslo 084315
Gyllene Ängeln, Stockholm 087002
Gyllner, Stockholm 115892
Det Gylne Snitt, Oslo 084316
Gympie and District Historical and
  Gold Mining Museum, Gympie 001129
Gympie Antiques, Gympie 061569,
                                127608
Gyöngyszem Galéria, Pécs 109206
Győr Egyházmegyei Könyvtár es
  Kincstár, Győr 023661
Győrffy István Nagykun Múzeum,
  Karcag 023716
Győry & Osiris, Miami 121798
Győző, Farkas, Fertőd 130816
Gyokudo Bijutsukan, Ome 029224
Gyokuei-do Shoten, Tokyo 142922
Gyor-Moson-Sopron Megyei Tűzoltó
  Kiállítás, Mosonmagyaróvár .. 023798
Gypsy Caravan Antiques, Las
  Vegas 094152
Gypsy Savage, Houston 093681
Gysbers & Van Loon, Arnhem . 143085
It Gysbert Japicxhûs Museum,
  Bolsward 032077
Gysin, Bob, Zürich 116803
Gyűjtemények Háza, Thorma János
  Múzeum, Kiskunhalas 023747
Gyzmo Projects, Gent 100694
H & B Gallery, Minneapolis 094918
H & G Furniture Restoration,
  Atlanta 135311
H & H Fine Art, Bad Liebenstein 074547
H & L Galerie, Waiblingen 108834
H & N, Praha 128346
H & R Antiques, Memphis 094637
H & S Art, New York 122680
H D Konst, Stockholm 115893
H-Galerie, Carin Schlee, Dürnau 106757
H Gallery, Bangkok 116953
H Gallery, New York 122681
H&M Kunstwinkel, Dordrecht .. 112437
H. Earl Clack Museum, Havre .. 049409
H.a.Déco, Fecamp 068519
H.B. Antiquités, Marckolsheim .. 069857
H.C. Ørsteds Antikvariat,
  Frederiksberg 140569
H.C.Andersen-koen Kodomotachi
  Hakubutsukan, Funabashi ... 028567
H.G. Albee Memorial Museum,
  Estherville 048615
H.H. Antique Mall, Toledo 097606
H.H. Maharaja Jayaji Rao Scindia
  Museum, Gwalior 024281
H.J. van de Kamp Museum, Sint
  Oedenrode 032876
H.L.B. Antiques, Bournemouth .. 088536
H.M. Frigate Unicorn, Dundee .. 044163
H.R. MacMillan Space Centre,
  Vancouver 007286
H2 – Zentrum für Gegenwartskunst im
  Glaspalast, Kunstsammlungen und
  Museen Augsburg, Augsburg . 016655
Hå Bygdemuseum, Varhaug .. 034071
Hå Gamle Prestegard, Varhaug . 034072
Ha Noi Fine Arts Association, Ha
  Noi 060825
Ha Noi Gallery, Ha Noi 125077
Ha Noi Museum, Ha Noi 054795
Ha Noi Studio, Ha Noi 125078
Ha Noi Studio, Ho Chi Minh City 125122
Ha Noi University of Fine Arts, Ha
  Noi 058154
Haaf, Franz-Josef, Meckesheim 077316,
                                130294
Haag & Scott, Bad Kreuznach . 074544

Haag, Gérard, Zellwiller 074319
Haagen, Los Angeles 121503
Haager Heimatstuben, Haag am
   Hausruck 002032
Haagmans, Essen 129816
Haags Historisch Museum, Den
   Haag 032172
Haags Openbaar Vervoer Museum,
   Den Haag 032173
Haagse Kunstkring, Den Haag .. 059605,
                               112390
Haaken, Oslo 113225
Haaksman, D.I., Amsterdam .. 082763
Haakú Museum, Sky City Cultural
   Center, Pueblo of Acoma .. 052412
Haapasalo, Kari, Helsinki 066317
Haardplatenmuseum Klarenbeek,
   Klarenbeek 032545
Haardt, Anton, New Orleans ... 122177
Haarener Mühle, Waldfeucht ... 022390
Haarlem Antiek, Haarlem 083331
Gebr. Haas, Bedburg-Hau 141459
Haas, Budapest 109147
Haas, Cicero 135430
Haas, Zürich 116804
Haas & Fischer, Zürich 116805
Haas & Fuchs, Berlin 106304
Haas Bowmaker, Steven,
   Milwaukee 094860
Haas Gallery of Art, Bloomsburg 047055
Haas-Lilienthal House, San
   Francisco 053003
Haas, de, Amsterdam 082764
Haas, Dr. Peter, Teningen 130702
Haas, Heinrich, Schaan 082333
Haas, Johann, Wels 140052
Haas, Mariette, Ingolstadt 107461
Haas, Otto, London 144618
Haas, Raul, Cincinnati 092864
Haas, Reinhold, Randersacker . 130554
Haas, Susan P., Seattle 124538
Haas, Tom, Phoenix 123494
Haas, Wolfgang, Beichlingen .. 141460
Haas, Wolfgang, Berlin 074817
Haase, Chemnitz 075251
Haase, S., Mülheim an der Ruhr . 077434
Haase, S., Horn-Bad Meinberg . 076576
Haase, Torsten, Neukirchen/
   Erzgebirge 077735
Haasner, Brigitte, Wiesbaden .. 108915
Haastrup Folkemindesamling,
   Faaborg 010054
Hab und Gut, Aachen 076372
Hab und Gut, Sankt Pölten 062829
Habari, Wien 100310
Habashi, Alexandria 066219
Habatat, Chicago 120150
Habbekrats, Gent 063640
Habelt, Dr. Rudolf, Bonn 137116, 141616
Georg Haber & Joh.L.L. Brandner,
   Berlin 129499
Georg Haber & Joh.L.L. Brandner,
   Regensburg 130561
Haber-Feist, Osterhofen 130498
Haber, Robert, New York 095439
Haberfellner, Gerda, Wien 062971
Habermann, Ibi, Innsbruck 100048
Habernicht, Marianne, Trier ... 078701
Habernickel, K. & A., Much 077422
Habersetzer, Schortens 078394
Habert, Eric & Catherine, Le
   Passage 069322
Haberzettl, Adolf, Feuchtwangen 075734
Haberzettl, Anna-Maria, München 077516
Habitat Art and Craft, Mumbai . 109442
Habitare, Genova 080266
l'Habitat, Indianapolis 093886
Habitat Antiques, Port Macquarie 062095
Habite, San Francisco 097248
Haboldt & Co., Bob, Paris 104936
Haboldt, Bob, New York 122682
Habres & Partner, Wien 062972
Hachem, Moussa, Paris 104937

Hachette, Paris 136936
Hachidai, Hiroshima 081977
Hachiga Minzoku Bijutsukan,
   Takayama 029452
Hachinohe-shi Bijutsukan,
   Hachinohe 028576
Hachinohe-shi Hakubutsukan,
   Hachinohe 028577
Hachinohe-shiritsu Rekishi
   Hakubutsukan, Hachinohe 028578
Hachioji-shi Rekishi Hakubutsukan,
   Hachioji 028580
Hachmeister, Petershagen 142299
Hachmeister, Dr. H., Münster . 108220
Hacı Bayram Camii, Ankara .. 042821
Hacı Ömer Sabancı Kültür Merkezi,
   Adana 042792
Haçıbektaş Müzesi, Hacıbektaş . 042910
Hacienda Art Gallery, Phoenix . 123495
Hacienda Buena Vista, Ponce . 036137
La Hacienda de Soldeca,
   Zaragoza 133500
Hacienda El Paraíso, El Cerrito . 008674
Hack, Rainer, Zella-Mehlis 109004
Hacke, Axel, München 130355
Hackel, Manfred & Linda, Bernau an
   Chiemsee 106471
Hackelbury, London 118364
Hacken- und Holzmuseum,
   Köflach 002217
Hackenberger Mühle, Saarburg . 021563
Hacker Strand Art, New York .. 145344
Hacker, Roland, Hamburg 076232
Hackett & Freedman, San
   Francisco 124293
Hackettstown Historical Society
   Museum, Hackettstown 049302
Hackhofer, B., Sankt Wendel .. 142346
Hackler's, Preston 091055
Hackley & Hume Houses,
   Muskegon 051135
Hackney House Antiques, Barlow 088352
Hackney Museum, London 044990
Hadalhondi, Bilbao 085872, 133193
Hadassa, New York ... 095440, 135932
Haddo House, Methlick 045254
Haddon Galleries, Torquay 119243
Haddon Hall, Maastricht 083465
Haddon Hall, Venlo 083677
Haddonfield Museum,
   Haddonfield 049304
Haddy, Jeffery, Prahran 062111
Hadeland Bergverksmuseum,
   Grua 033627
Hadeland Folkemuseum, Jaren . 033684
Hadeland Glassverks Museum,
   Hadeland Glassverk, Jevnaker . 033686
Haden Hall House, Cradley Heath 044028
Hadernverlag, Salzburg 136552
Hadersbeck, Berlin 074818, 125878
Hadersbeck, Potsdam . 078076, 126098
Haderslev Antikvariat, Haderslev 140578
Haderslev Museum, Haderslev . 010092
Hadfield, G.K., Great Salkeld .. 089377
Hadiqat al-Asmak Museum,
   Cairo 010542
Hadjer, Reynold & Fils, Paris .. 071458,
                               129027
Hadler, Graz 062588
Hadley, New York 095441
Hadley Farm Museum, Hadley . 049307
Hadlow Down Antiques, Hadlow
   Down 089405, 134429
Hadrovský, Josef, Pardubice ... 102639
Hadsund Antikvariat, Hadsund . 140579
Hadsund Egns Museum,
   Hadsund 010096
Hadtörténeti Aero Múzeum,
   Hadtörténeti Múzeum Budapest,
   Szolnok 023978
Hadtörténeti Múzeum, Budapest 023483
Hadyna, Brno 065348
Haeberlin, Jacques, La Croix-de-
   Rozon 116235

Hädicke, B., Leipzig 142057
Häffner, P., Gaggenau 075935
Haegang Ceramics Museum,
   Icheon 029948
Hægebostad Bygdemuseum,
   Tingvatn 034014
Hägnan Friluftsmuseet i Gammelstad,
   Gammelstad 040639
Hähnel, R., Bünde 075214
Haeju Historical Museum, Haeju 029872
Häkkilän Museo, Toholampi ... 011425
Hälleforsnäs Gjuterimuseum,
   Hälleforsnäs 040676
Hälleviks Fiskemuseum,
   Sölvesborg 040972
Hällisch-Fränkisches Museum,
   Schwäbisch Hall 021751
Hällristningsmuseet, Norrköping 040877
Hälsinge Antik, Hudiksvall 086750
Hälsinge Byrån, Forsa 126582
Hälsinglands Antikvariat, Järvsö 143866
Hälsinglands Museum, Hudiksvall 040709
Häme-Galleria, Lahti 103164
Hämeen Linna, Hämeenlinna . . 010787
Hämeen Museo, Tampereen Museot,
   Tampere 011394
Hämeenkyrön Kotiseutumuseo,
   Hämeenkyrö 010783
Hämeenlinnan Kaupungin Historiallinen
   Museo, Hämeenlinna 010788
Hämeenlinnan Puhelinmuseo,
   Hämeenlinna 010789
Hämeenlinnan Taidemuseo,
   Hämeenlinna 010790
Hämmerle, Lisi, Bregenz 099966
Haendehen, Paulo Roberto, Porto
   Alegre 063975
Haenel, Cape Town 114768
Haenel-Pancera-Familienmuseum, Bad
   Ischl 001748
Hänel, Frank, Wiesbaden 108916
Haenle, Göppingen 076032
Haenle, O., Böhmenkirch 075054
Hänold, Joachim, Berlin 074819
Het Haentje te Paart, Middelburg 112594
Haering, Jürgen, Freiburg im
   Breisgau 075866
Häring, Thomas, Leipzig 130224
Härjedalens Fjällmuseum, Jämtlands
   Läns Museum, Funäsdalen .. 040635
Härkätien Museo, Renko 011311
Haertel, Karl, Kitzbühel 140002
Härtl-Kasulke, Dr. Claudia,
   Dietzenbach 106680
Härtl, Ulf D., Bamberg 074662
Härtl, Ulf D., Hemhofen 076489
Härtsfeld-Museum, Neresheim . 020662
Härtschildmuseum,
   Neresheim 020663
Hässleholms Auktionskammare,
   Hässleholm 126591
Hässleholms Museum,
   Hässleholm 040679
Häuser, Markus, Rottenbach .. 078299
Häuser, Susanne, Öhringen ... 077893
Häusler, Bischberg 075003
Häusler Contemporary, München 140813
Häusler Contemporary, Zürich . 116806
Häusser, Friedrich, Ribe 066060
Haeyer, d', Charleroi 063603
Haezelaer, Alain, Stuttgart 142391
Hafemann, Gottfried, Wiesbaden 108917
Haferland, Ulrich, Finsterwalde . 129837
Hafez, New York 095442
Haffen, Philippe, Reims 129191
Haffenreffer Museum of Anthropology,
   Brown University, Bristol 047217
Haffke, Altötting 105980
Haffmans, F.J., Utrecht 083650
Haffmuseum Ueckermünde,
   Ueckermünde 022246
Haffner, Saint Louis 096713
Haffner, Désiré, Nousseviller-Saint-
   Nabor 070755

Hafnarborg Art Gallery,
   Hafnarfjörður 024078
Hafnarhúsið, Listasafn Reykjavíkur,
   Reykjavík 024095
Hafner, Bern 087283
Hafner, Dieter, Tannheim 142409
Hafner, Paul, Sankt Gallen 116612
Haft Tappeh Museum, Haft
   Tappeh 024696
Hafter, Anatol & Natalie,
   Küsnacht 087549, 133786
Hafter, Anatol & Natalie, Zürich . 087973,
                               133982
Haga, Paris 071459
Haga Antik och Design,
   Sundsvall 087089
Hagadorn House Museum, Almond
   Historical Society, Almond .. 046419
Hagalund Antikhandel, Skänninge 086927
Hagan, Eggedal 033577
Hagedorn, M., Detmold 075340
Hagedorn, R., Paderborn 078003
Hagelandse Academie voor Beeldende
   Kunsten, Aarschot 054995
Hagelin, Stockholm 087003
Hagelmüller, Wolfgang, Hopfgarten im
   Brixental 139997
Hagelstam, Helsinki . 066318, 103149,
                               125478
Hagelstam, C., Helsinki 125479, 140701
Hagemeier, Frankfurt am Main . 107004
Hagemoser, Andreas, Berlin ... 141520
Hagen, Offenburg 077906
Hagen, Rastede 078131
Hagen, Bert, Castricum 143103
Hagen, Bert & Reiny, Amsterdam 082765,
                               143033
Hagen, Hans-J., Bayreuth 141457
Hagen, L., Selfkant 108607
Hagen, Richard, Broadway 088676,
                               117455
Hagena & Schulte, Bonn 141617, 141618
Hagenauer, Karl, Wien 100311
Hagenring e.V., Bund Bildender
   Künstler, Hagen 058912
Hager, Gaëlle, Kaltenhouse ... 068938
Haggai, Charles, Paris 071460
The Haggin Museum, Stockton . 053541
Haggith, Paul, Holgate 099144
Haghna Museum, Museum Zahal –
   Beit Eliyahu, Tel Aviv 025196
Hagi Hakubutsukan, Hagi 028583
Hagi-shi Uragami Bijutsukan,
   Hagi 028584
Hagley Hall, Hagley 044494
The Hagley Museum, Wilmington 054437
Hagley Museum and Library,
   Wilmington 054438
Hagman & Co, Stockholm 115894
Hagn, Silvia, Neumarkt am
   Wallersee 100130
Hagner, Louis, Paradiso 087729, 133885
Hagondokoff, A-C., Paris 071461
Hagstrom, Judy, Saint Paul ... 123913
Hahn, Philadelphia 123392
Hahn, Christine, Regensburg .. 078168
Hahn, Dietrich, Hamburg 107276
Hahn, Gertraud, Erlangen 075651
Hahn, Margarete, Frielendorf .. 107074
Hahn, Maria, Berlin 074820
Hahn, Udo, Haßloch 076414
Hahn, Ulrich, Erkrath 129809
Hahndorf Academy Public Art Gallery
   and Museum, Hahndorf 001130
Hahne, Bernd, Braunschweig .. 075124
Hahnenberg, Kintzheim 068941
Hahns Peak Area Historical Museum,
   Clark 047768
Hahoedong Mask Museum,
   Andong 029885
HAHS Museum, Hattiesburg Area
   Historical Society, Hattiesburg . 049403
Hai Ci Tang, Shanghai 065137
Hai-Ho Upholstery, Windsor,

Ontario 128235
Hai Wangcun Antiques Market,
Beijing 065001
Hai Yun Ming Hua, Shenzhen .. 102223
Haian Museum, Haian 007827
Haida Galleries, Lytham Saint
Annes 118687
Haida Gwaii Museum at Qay'llnagaay,
Skidegate 007045
Haider, München 108134
Haider, Harald, Purkersdorf 062772
Haiderová, Šárka, České
Budějovice 065369
Haidhausen Kunst und Antiquitäten,
München 077517
Haidhausen-Museum, München 020528
Haidinger, Elisabeth, Passau ... 078020
Haifa Artists Association, Haifa . 059385
Haifa City Museum, Haifa 025050
Haifa Museum of Art, Haifa ... 025051
Haig-Brown House, Campbell
River 005543
Haig Colliery Mining Museum,
Whitehaven 046121
Haig, Axel, Holcot 117975
Haikou Museum, Haikou 007828
Haikun Antique, Shanghai 065138
Haile, Sacramento 123770
Haileybury Heritage Museum,
Haileybury 005940
Hailsham Heritage Centre,
Hailsham 044496
Hailuodon Kotiseutumuseo Kniivilä,
Hailuoto 010795
Haim, Paul, Paris 071462
Haime, Nohra, New York 122683
Haimerl, Bad Tölz 106096
Haimerl, Sigisbert, Königsdorf .. 107705
Hain, Gabi, Deggendorf 075318
Hainan Museum, Haikou 007829
Hainan Nationality Museum,
Tongshi 008340
Hainan Photo-Arts Publishing House,
Haikou 136712
Hainanzhou Nationality Museum,
Gonghe 007795
Haindl, Anita, Mürzzuschlag ... 062743
Haindl, Josef, Langenwang .. 062708,
                              127801
Haines, Los Angeles 121504
Haines, San Francisco 124294
Haines Museum, Waukegan .. 054232
Haining Museum, Haining 007830
Haiser, Hoyerswerda 107448
Haiti Arts, Tucson 124740
Haiti Images Gallery, New York . 122684
Haitian Art Factory, Miami 121799
Haixia Art Cooperative, Tonling 102246
Haixizhou Nationality Museum,
Delingha 007736
Haiyan Museum, Haiyan 007831
Haiyang Museum, Haiyang ... 007832
Haiyun Xuan Gu Dong, Shenzhen 102224
Hajdul, Gdańsk 084504, 132816
Hajdusági Múzeum,
Hajdúböszörmény 023682
Hajózástörténeti Múzeum,
Zebegény 024057
Hak, Go Jae, Seoul ... 111658, 111659
Håkansson, Søren, Gilleleje .. 102815
HaKaren, Singapore 114532
Hakasalmen Huvila, Helsingin
Kaupunginmuseo, Helsinki .. 010820
Hakco, Houston 121052
Hake, Frankfurt am Main 107005
Hakel Fossils Museum, Byblos . 030275
Hakeney, David, Beverley 088450
Hakeney, David K., Kingston-upon-
Hull 089700
Haker, Thomas, Berlin . 141521, 141522
Hakhamaneshi Museum, Shiraz 024735
Hakka Cultural Museum of Shigang,
Shigang 042449
Hakka Culture and History Museum,

Kaohsiung 042358
Hakka Relic and Folk Custom
Museum, Meinong 042396
Hakodate Bijutsukan, Hakodate . 028587
Hakodate-shi Meijikan, Hakodate 028588
Hakodate-shiritsu Hakubutsukan,
Hakodate 028589
Hakone Bijutsukan, Hakone ... 028592
Hakone-Jinja Homotsuden,
Hakone 028593
Håkonshallen og Rosenkrantztårnet,
Bryggens Museum, Bergen .. 033532
Hakubutsukan Kenkyu, Tokyo .. 138890
Hakubutsukan Meiji-Mura,
Inuyama 028688
Hakuna Matata, Philadelphia .. 123393
Hakutosha, Nagoya 111318
Hakutsuru Fine Art Museum,
Kobe 028841
Hakutsuru Sake Brewery Museum,
Kobe 028842
Hal, Marcus, El Paso 120833
Halasi Csipke Közalapítvány –
Csipkeház, Kiskunhalas 023748
Halbach, Celle 106629
Halbach, R., Burgdorf 106618
Halbedel, Salzburg 062796
Halbert, Austin 092232
Halcyon, Richmond 096547
Halcyon, San Francisco 097249
Halcyon, Stockport 091442
Halcyon Days, London 090178
Halcyon Daze, Mittagong 061866
Halcyon Gallery, Birmingham . 117373,
                              117374
Halcyon Gallery, London 118365
Haldemann, Margit, Bern 116136
Halden Historiske Samlinger,
Halden 033630
Haldensvassdragets Kanalmuseum,
Ørje 033812
Haldimand County Museum,
Cayuga 005583
Haldy, Jacques, Montreux ... 087681,
                              087682
Hale Farm and Village, Bath .. 046811
Hale School Museum, Wembley
Downs 001632
Hale, Ivy, Tunbridge Wells .. 091652,
                              135159
Hale, Robert, London 138217
Hale, William F., Washington .. 145634
Hales, S.J., Bovey Tracey ... 126831
Halesworth and District Museum,
Halesworth 044498
Halewijn Missiaen, Gent 100695
Halewood & Sons, Preston .. 144791
Haley, Albuquerque 092055
Haley, Pat, Malvern, Victoria .. 061782,
                              125210
Half Back, Dee Why 139713
Half Moon Antiques, Prospect .. 062124
Half Price Books Records Magazines,
Kensington 145198
Half Retro, Bucureşti 085095
Halfa Museum, Wadi Halfa .. 040503
Halff, Harry, San Antonio 124039
Halfmann, Essen 075679
Halfmann, Peter, Duisburg 141728
Halfmoon Antique Mall, San
Antonio 096974
Halgene, New York 095443
Hali, Lidingö 086807, 143878
The HALI Fair, Antique Carpet and
Textile Art Fair, London 098378
Hali Publication, London 138218
Hali Publications, London 139185
Halı, Kilim ve Düz Dokuma Yaygılar
Müzesi, İstanbul 042938
Haliburton Highlands Museum,
Haliburton 005942
Haliburton House Museum, Windsor,
Nova Scotia 007422
Haliden, Bath 088388

Halifax Antiques Centre, Halifax 089413
Halifax Citadel, Halifax 005950
Halifax Historical Museum, Daytona
Beach 048179
Halifax Police Museum, Halifax . 005951
Halikkon Museo, Halikko 010796
Halin Metsäkämppämuseo, Virrat 011515
Halina, Warszawa 113851
Halk Sanatları Müzesi, Girne .. 033461
Halkjær, Eva M., Kalundborg .. 065808
Halkyone, Hamburg 141866
Hall, Chester 126856
Hall, Dallas 120607
Hall & Knight, London 090180
Hall & Knight, New York 095444
La Hall aux Arts, Angoulême ... 103240
Hall i'th' Wood Museum, Bolton 043670
Hall Museum, Meeteetse Museums,
Meeteetse 050769
Hall of Fame, Buffalo 092531
Hall of Fame, Zouk Mikayel .. 030286
Hall of Fame and Classic Car Museum,
Weedsport 054262
Hall of Fame for Great Americans,
Bronx Community College,
Bronx 047230
Hall of Flame – Museum of
Firefighting, Phoenix 052117
Hall of Health, Childen's Hospital and
Research Center, Berkeley .. 046935
Hall Place, Bexley 043588
Hall, Anthony C., Twickenham . 144896
Hall, Harry, Belfast 144240
Hall & Son, Peter, Staveley .. 135078
Hall, Robert, London 090179
Hall, Thomas, Dallas 120608
Hallands Auktionsverk, Halmstad 126593
Hallar Gallery, Kansas City .. 121294
Hallard, Nigel, Mousehole ... 118758
Hallaton Museum, Hallaton .. 044506
Die Halle, Langnau am Albis .. 116405
HALLE 14 e.V., Leipzig 107833
Halle Aux Antiquités, Buhl-
Lorraine 067532
La Halle aux Oiseaux, Marseillan 014038
Halle für Kunst Lüneburg e.V.,
Lüneburg 020179
Halle Heart Center, Tempe ... 053703
Halle Saint-Pierre Museum, Paris 014662
Hallén, Stockholm 143948
Hallen für neue Kunst,
Schaffhausen 041947
De Hallen Haarlem, Haarlem .. 032393
Haller & Evans, Kingston-upon-
Hull 126933
Haller Akademie der Künste,
Schwäbisch Hall 055706
Haller Feuerwehrmuseum, Schwäbisch
Hall 021752
Haller, H., Pforzheim 078041
Haller, Stephen, New York ... 122685
Hallerbosmuseum, Halle 003664
Hallertauer Heimat- und
Hopfenmuseum, Mainburg .. 020219
Hallertauer Hopfen- und
Heimatmuseum, Geisenfeld ... 018609
Les Halles de la Brocante, Nice 070614
Hallescher Kunstverein, Halle,
Saale 058913
Halliday, Swindon 119200
Hallidays Antiques, Dorchester-on-
Thames 089075
Hallie Ford Museum of Art, Willamette
University, Salem 052889
Halligans, Auckland 143283
Hallin, Sven, Färnäs 126578
Hallingdal Folkemuseum,
Nesbyen 033803
Halliwell's House Museum,
Selkirk 045756
Hallmark, Brighton 088623
Hallmark, San Diego 097094
Hallmark Society, Victoria ... 007326
Hallockville Museum Farm,

Riverhead 052588
Hall's Bookshop, Tunbridge Wells 144893
Hall's Croft, Shakespeare Birthplace
Trust, Stratford-upon-Avon ... 045902
Halls Fine Art Auctioneers,
Shrewsbury 127045
Hallwalls Arts Center, Buffalo . 047305
Hallward, Dublin 109591
Hallwylska Museet, Stockholm . 041004
Hallynck, Jean-Philippe, Croix .. 128575
Halm, E., Köln 076870
Halmens Hus, Halmslöjdsmuseum,
Bengtsfors 040561
Halmer, Bernhard, Wien 140081
Halmstads Auktionskammare,
Halmstad 126594
Halos, Utrecht 143244
Halosenniemi Museo, Tuusula . 011458
Haloux, Félix, Paris 104938
Halper, Susan L., New York ... 122686
Halpern, Herbert, New Orleans . 122178
Halsey Gallery, College of Charleston,
Charleston 047549
Halszka-Polonez, Poznań 113739
La Halte, Plaisance 072170
Halter & Colledani, Luzern ... 133834
Halter, K., Türkheim 078732
Halter, Yves, Rennes (Ille-et-
Vilaine) 105437
Halton Region Museum, Milton . 006317
Haluška, Pavel, Liptovský
Mikuláš 085575
Halwany, Ismat, Beirut 082286
Ham House, Richmond, Surrey . 045612
Hama National Museum, Hama . 042261
Hamacher, H., Aachen 074332
Hamacher, Joachim, Krefeld .. 130193
Hamada Gallery, Cairo 013045
Hamadryade, Paris 071463
Hamaland-Museum und
Westmünsterländische Hofanlage,
Kreismuseum Borken, Vreden . 022357
Hamamatsu-shi Gakki Hakubutukan,
Hamamatsu 028598
Hamamatsu-shi Hakubutsukan,
Hamamatsu 028599
Hamamatsu-shiritsu Bijutsukan,
Hamamatsu 028600
Hamamatsu-shiritsu Kagakukan,
Hamamatsu 028601
Hamana-ko Orugorukan,
Hamamatsu 028602
Hamana-ko Tateyama-ji Bijutsukan,
Hamamatsu 028603
Hamard, Annette, Bar-le-Duc . 066937,
             066938, 066939, 128460
Hamarøy Bygdetun, Harmarøy . 033637
Hamayu, Yokohama 082227
Hambacher Schloss, Neustadt an der
Weinstraße 020759
Hamberger, Stefan & Roswitha,
Stammham, Kreis Eichstätt .. 108682
Hambidge Center for the Creative Arts
and Sciences, Rabun Gap .. 052438
Hambleton, Vancouver 101792
Hamblin, John, Yeovil . 091969, 135267
Hamboise, Montpellier 104281
Hambros Antik og Kunst,
Ålesund 084208
Hamburg Dungeon, Hamburg . 018990
Hamburg State Park Museum,
Mitchell 050966
Hamburger Antiquariat, Hamburg 141867
Hamburger Bahnhof – Museum für
Gegenwart – Berlin, Staatliche
Museen zu Berlin – Stiftung
Preußischer Kulturbesitz, Berlin 017154
Die Hamburger Galerie, Hamburg 107277
Hamburger Kunsthalle, Hamburg 018991
Hamburger Schulmuseum,
Hamburg 018992
Hameau du Fromage, Cléron .. 012602
Hameau en Beaujolais, Romanèche-
Thorins 015135

Hameau Troglodytique de la Fosse,
Dénezé-sous-Doué 012760
Hamecher, Horst, Kassel 141978
Hamelin, Sébastien, Les Pieux . 069419
Hamer, Amsterdam 112218
Hamer, Großhansdorf 107175
Hamers & Penz, Halle, Saale .. 107207
Hames, Peter, Bideford 144246
Hametner, Roland, Stoob 100203
Hamhung Historical Museum,
Hamhung 029873
Hami Gang, Seoul 111660
Hamich-Rafensteiner, Ilse,
Nördlingen 108280
Hamidèche, Dominique, Lens .. 069385
Hamill, Boston 119872
Hamilton, Knokke-Heist 063703
Hamilton, Phoenix 136115
Hamilton, Woodbridge 091912
Hamilton & Miller, Camperdown 125185
Hamilton & Selway, West
Hollywood 124998
Hamilton Art Gallery, Hamilton,
Victoria 001135
Hamilton Arteffects, Hamilton . 099105
Hamilton Artists, Hamilton 005969
Hamilton Billiards & Games,
Knebworth 089726, 134523
The Hamilton Children's Museum,
Hamilton 005970
Hamilton Grange National Memorial,
New York 051401
Hamilton Heritage Museum, Hamilton,
Tasmania 001134
Hamilton House, South Berwick 053398
Hamilton Hume Museum, Yass . 001673
Hamilton Kerr Institute,
Cambridge 134199
Hamilton Library and Two Mile House,
Carlisle 047427
Hamilton Low Parks Museum,
Hamilton 044509
Hamilton Military Museum,
Hamilton 005971
Hamilton Museum of Steam and
Technology, Hamilton 005972
Hamilton Osborne King, Dublin . 126169
Hamilton Psychiatric Hospital Museum,
Hamilton 005973
Hamilton van Wogener Museum,
Clifton 047832
Hamilton & Co, A.D., Glasgow .. 089330
Hamilton, Anne, Burnham Market 088722
Hamilton, Beth, Fairlight 099013
Hamilton, Jane, Tucson 124741
Hamilton, Rosemary, London .. 134679
Hamilton, Ross, London 090181
Hamiltons, London 118366
Hamilton's, Tacoma 127470
Hamilton's Gallery, London ... 044991
Haminan Kaupunginmuseo,
Hamina 010797
Hamiota Pioneer Club Museum,
Hamiota 005980
Hamish, Clark, Woollahra 062471
Hamizgaga Museum, Kibbutz
Nachsholim 025141
Hamlin Garland Homestead, West
Salem 054323
Hamline University Galleries, Saint
Paul 052840
Hamlyn Lodge, Ollerton 090914, 134907
Hamm, Lampertheim 076999
Hamm, Gitta, Darmstadt 129670
Hamm, Pierre, Benfeld 067087
Hammarlund, Stockholm 143949
Hammarö Auktionsverk, Skoghall 126663
Hammars Antik, Örebro 086902
Hammaslääketieteen Museo,
Helsinki 010821
Hammelehle & Ahrens, Köln .. 107635
Hammelrath, Birgit, Köln 107636
Hammer, Augsburg 106031
Hammer, New York 122687

Hammer- und Waffenschmiede-
Museum Hexenagger,
Altmannstein 016551
Hammer Museum, Frankfurt am
Main 018423
Hammer Museum, University
of California Los Angeles, Los
Angeles 050427
Hammer, Carl, Chicago 120151
Hammer, Ingo R., Köfering .. 076841,
130139
Hammer, Klaus, Olching 077921
Hammer, Ludwig, Regensburg . 108462
Hammer, Udo, Werbach 078925
Hammerberg, Bob, Seattle 097451
Hammerblow, Tucson 097668
Hammered Design, Denver 120732
Hammerichs Hus, Ærøskøbing . 010002
Hammermuseum zu Hannover,
Hannover 019051
Hammerschmiede Gröningen,
Satteldorf 021621
Hammerschmiede und Heimatmuseum
Lahr, Lahr 019854
Hammerschmiede und Stockerhof,
Neuburg an der Kammel 020693
Hammersite, Ramat Aviv 126181
Hammerstein, Brigitte von,
Hamburg 129964
Hammerstein, Hans, München . 142158
Hammill, Janet, Thornbury ... 127687
Hamminoff, Chris, Caen 067569, 140810
Hammond, Saint Louis 145496
Hammond Castle Museum,
Gloucester 049134
Hammond Hall Antiques, Wagga
Wagga 062397
Hammond-Harwood House Museum,
Annapolis 046501
Hammond Historical Museum,
Hammond 049331
Hammond Museum, North Salem 051617
Hammond, Jeffery, Leominster . 089806
Hammond, John, Fürth 107084
Hammond, Thomas, Tampa 124656
Hamnladans Antikt, Norrtälje .. 086887
Hamnmuseet, Varberg 041153
Hamon, Claude, Lampaul-
Guimiliau 069158
Hamon, Jean-Pierre, Le Havre . 103947
Hampden-Booth Theatre Library at the
Players, New York 051402
Hampden Junque, Baltimore ... 092329
Hampden Street Antique, Denver 093297
Hampe, Richard, Klosterneuburg 062679
Hampel Kunstauktionen München,
München 126051
Hampel, A., Friedenweiler 075912
Hampel, Dr. Frithjof, Schönwalde-
Glien 078386
Hampshire Antiques, Vancouver 064826
Hampshire Contemporary Artists,
Lymington 118680
Hampshire Farm Museum, Botley,
Hampshire 043678
Hampson Archeological Museum,
Wilson 054452
Hampstead Antique and Craft
Emporium, London 090182
Hampstead Museum, London .. 044992
Hampton, Bentleigh 061136
Hampton & Littlewood,
Alphington 126796
Hampton County Historical Museum,
Hampton 049338
Hampton Court Antiques,
Hampton 061592
Hampton Court Emporium, East
Molesey 089135
Hampton Court Palace Antiques, East
Molesey 089136, 134349
Hampton Court Palace, Historic Royal
Palaces, East Molesey 044210
Hampton History Museum,

Hampton 049341
Hampton House, Richmond 096548
Hampton National Historic Site
Museum, Towson 053787
Hampton Plantation,
McClellanville 050546
Hampton-Preston Mansion,
Columbia 047910
Hampton Roads Antique Connection,
Norfolk 095858
Hampton Roads Naval Museum,
Norfolk 051575
Hampton University Museum,
Hampton 049342
Hampton, Briggs & Chine, New
York 095445
Hamptonne Country Life Museum,
Saint Lawrence 045701
Hamptons, Christchurch 088906
Hamptons, Great Malvern ... 126909
Hamptons, Marlborough 126986
Hamura-shiritsu Hakubutsukan,
Hamura 028604
Hamza Gyűjtemény és Jász Galéria,
Jászberény 023700
Hamza Hakim-Jade Niazy Muzeyi,
Farg'ona 054632
Hamza Hakim-zade Niazy Yodgorlik
Muzeyi, Qo'qon Viloyat Ülkashunoslik
Muzeyi, Qo'qon 054639
Hamzavi, Mohamed, Strasbourg 073590
Han Fong Galerie, Paris 071464
The Han Horse, New York 095446
Han Ji Yun Contemporary Space,
Beijing 101958
Han Ji Yun Contemporary Space,
Seoul 111662
Han Kuk Art Museum, Yongin .. 030061
Han Military Forces Yong Museum,
Xuzhou 008456
Han Mo Xuan, Hong Kong 102082
Han Palace, San Francisco ... 097250
Han Tomb Cultural Relics Exhibition
Hall, Mawangdui 008077
Han, Jiwon, New York 122688
Han, Kee Sook, Daegu 111585
Han, Kwang, Singapore 085416
Han, Sun, Seoul 111661
Hana, Seoul 111663
Hana Cultural Center, Hana, Maui 049434
Hana Gallery, Vancouver 064827
Hana Senesh House, Kibbutz Sdot
Yam 025147
Hanaire, Madeleine, Paris 129028
Hanak, Los Angeles 094352
Hanall, Buchon 082252
Hanamaki-shiritsu Hakubutsukan,
Hanamaki 028605
Hanappe, Els, Athinai 109037
Hanart Gallery, Hong Kong ... 102083
Hanart Gallery, Taipei 116918
Hanart TZ Gallery, Hong Kong . 102084
Hanauer Museum, Kehl 019572
Hanbat Education Museum,
Daejeon 029919
Hanbisuan Mikado Pheasant Ecology
Museum, Su-Ao 042468
Hanbury Hall, Droitwich 044142
Hanby House, Westerville Historical
Society, Westerville 054334
Hanch Hall, Lichfield 044823
Hancheng Museum, Hancheng . 007833
Hanchuan Museum, Hanchuan . 007834
Hancock, Moss Vale 061903
Hancock & Harwell, Atlanta ... 092142
Hancock County Historical Museum,
Carthage 047451
Hancock Historical Museum,
Findlay 048733
Hancock Historical Society Museum,
Hancock 049347
Hancock House, Hancock's
Bridge 049348
Hancock Museum, Newcastle-upon-

Tyne 045347
Hancock Shaker Village, Pittsfield 052183
Hancock, Peter, Chichester 088874,
144338
Hancocks & Co., London 090183
Hand Art Galerie, Rapperswil . 116576
Hand in Hand, Coldstream 088956
Hand Made Studio, Vientiane .. 111749
Hand Workshop Art Center,
Richmond 052550, 123735
Hand Woven Rugs, Los Angeles 094353
Handa-shiritsu Hakubutsukan & Niimi
Nankichi Kinenkan, Handa .. 028610
Handan Museum, Handan 007835
Handan Painting Gallery, Handan 102034
Handarbeitsmuseum,
Traunkirchen 002840
Handboekinderij De Pers,
Antwerpen 127994
Handdancer's Art, Louisville ... 121647
Handel House Museum, London 044993
Handels- og Søfartsmuseet paa
Kronborg, Helsingør 010107
Handelsbua, Skjeberg 084359
Handelsgesellschaft für historische
Wertpapiere, Wien 125273
Handelskompaniet i Gävle, Gävle 086681
Handjani-Leleu, Paris 071465
Handley-Hicks, Fort Worth 120868
Handmark Galleries, Battery
Point 098714
Hands Across, Phoenix 096213
Hands on Children's Museum,
Olympia 051757
Hands on History, Kingston-upon-
Hull 044714
Hands-on House, Children's Museum
of Lancaster, Lancaster 050117
Hands On! Regional Museum, Johnson
City 049825
Handschriftenabteilung der Landes-,
Universitäts- und Murhardschen
Bibliothek, Kassel 019537
Handsworth Saint Mary's Museum,
Handsworth 044510
Handweaving Museum and Arts
Center, Clayton 047785
Handweberei Henni Jaensch-Zeymer,
Schwielowsee 021805
Handwebmuseum Rupperath, Bad
Münstereifel 016827
Handwerk & Gewerbe Museum,
Krauschwitz 019781
Handwerker- und Heimatstube
Gernsbach, Gernsbach 018660
Handwerker Gallery of Art, Ithaca
College, Ithaca 049738
Handwerker, Helmut, Karlsfeld . 141968
Handwerkermuseum, Freinsheim 018496
Handwerkermuseum,
Schönecken 021706
Handwerkermuseum Laufenbach,
Taufkirchen an der Pram 002811
Handwerkerstuben,
Feuchtwangen 018638
Handwerkskunst-Museum, Rattenberg,
Inn 002575
Handwerksmuseum, Deggendorf 017839
Handwerksmuseum, Horneburg . 019344
Handwerksmuseum, Sankt Leonhard
am Hornerwald 002674
Handwerksmuseum am Mühlenberg,
Suhlendorf 022077
Handwerksmuseum Groß-Gerau, Groß-
Gerau 018833
Handwerksmuseum Kork, Kehl . 019570
Handwerksmuseum Ovelgönne,
Ovelgönne 021056
Handwert, Jena 107479
Haney House Museum, Thomas Haney
House, Maple Ridge 006261
Hanf-Museum, Berlin 017155
Hanford Mills Museum, East
Meredith 048444

Hang, Hong Kong 065046
Hang Art, Mount Lawley 099401
Hang Art, Paris 104939
Hang Art, San Francisco 124295, 124296
Hang It, Minneapolis .. 121997, 121998
Hang Ups, Fyshwick 127601
Hang-Ups, Red Hill 099613
Hang Ups of Greytown, Greytown 112901
Hanga Ten, Contemporary Japanese
  Prints, London 118367
Hangar-7, Salzburg 100165
Hangar Bicocca – spazio d'arte
  contemporanea, Milano 026797
Le Hangar des Chineurs, Perros-
  Guirec 072102
Hangar du Chineur, Dardilly ... 068275
Hangguk Minsok-chon, Kihung . 029964
Hanging Around, Enfield 117753
Hanging Tree Gallery,
  Albuquerque 092056
Hangleiter, Hans Michael,
  Otzberg 130502
Hangups Gallery, Omaha 123319
Hangzhou Painting Gallery,
  Hangzhou 102039
Hanhai, Beijing 125398
Hanham, William, London ... 090184,
  118368
Haniel Museum, Duisburg 018055
Hanif, Rüsselsheim 108536
Hanila Muuseum, Hanila 010612
Hanina, London 118369
Hanita Museum, Kibbutz Hanita 025133
Hank Snow Country Music Centre,
  Liverpool 006202
Hankard, Jean-Jacques,
  Bruxelles 140178
Hankasalmen Kotiseutumuseo,
  Hankasalmi 010799
Hanko Front Museum, Hanko ... 010800
Hanko Museo, Hanko 010801
Hanks, Buffalo 092532
Hanks, Steve, Albuquerque ... 092057
Hanks, W., Austin 092233
Hanley House, Clayton 047783
Hanley, Jack, San Francisco ... 124297
Hanlin, Hong Kong 065047
Hanlon, Matthew, New York ... 135933
Hanlon, Terry, Dallas 093136
Hanmo Art Center, Beijing 101959
Hanna, Birmingham .. 092409, 145001
Hanna Pioneer Museum, Hanna 005982
Hannah & Sisters, Charlotte ... 092600
Hannah Lindahl Children's Museum,
  Mishawaka 050953
Hannah, Catherine, Rotorua ... 084070
Hannah, Eva, Sydney 099784
Hannah's Art Studio, Miami 121800
Hannak, Gudrun, Murnau am
  Staffelsee 142216
Hannas, Dublin 142602
Hanna's Place, Louisville 094557
Hanne's Antik og Genbrug, Valby 066173
Hannes Hus og Fonden Gamle
  Sønderho, Fanø 010059
Hannoversches Straßenbahn-Museum,
  Sehnde 021826
Hanns Schell Collection,
  Österreichisches Museum für
  Schloss, Schlüssel, Kästchen,
  Kassetten und Eisenkunstguss,
  Graz 001974
Hannseli Spycher, Rickenbach bei
  Winterthur 041876
Hanover, Scarborough 091251
Hanover Gallery, Hannover ... 107367
Hanover Historical Society Museum,
  Hanover 049354
Hanover House, Clemson University,
  Clemson 047799
Hanover Museum, Lucea,
  Hanover 028452
Hanover Square Gallery, Brooklyn 119941
Hanrieder-Gedenkraum,

Putzleinsdorf 002552
Hans-Böckler-Geburtshaus,
  Trautskirchen 022177
Hans Breinlinger Museum, Sammlung
  Doris und Frieder Knittel,
  Konstanz 019759
Hans Christian Andersen Museum,
  Odense 010259
Hans Christian Andersens
  Barndomshjem, Odense 010260
Hans Erni Museum, Hans Erni-Stiftung,
  Luzern 041706
Hans-Fallada-Haus, Feldberger
  Seenlandschaft 018345
Hans-Grade-Museum, Borkheide 017515
Hans Grohmann Kunstverlag –
  Gemälde – Großhandel,
  München 108135
Hans Gross Kriminalmuseum,
  Graz 001975
Hans Herr House and Museum, Willow
  Street 054429
Hans Mauracher-Museum,
  Landesmuseum Joanneum,
  Graz 001976
Hans Nielsens Hauges Minne,
  Rolvsøy 033906
Hans Schatzdorfer Haus, Pramet 002534
Hans-Thoma-Geburtshaus, Bernau im
  Schwarzwald 017328
Hans-Thoma-Gedächtnisstätte und
  Vortaunusmuseum, Oberursel . 020935
Hans-Thoma-Museum, Bernau im
  Schwarzwald 017329
Hans Trudel-Haus Stiftung,
  Baden 041251
Hans van Riessen Museum,
  Vledderveen 032996
Hans, Marcus, Düsseldorf 141719
Hansági Múzeum,
  Mosonmagyaróvár 023799
Hansberg, Peter, Groß Sarau .. 076103,
  129922
Hanse Galerie, Stralsund 108679
Hanse, Norbert, Paris 104940
Hanseatic Booth or The Pier House,
  Symbister 045942
Hanseatische Auktionatoren,
  Ratekau 126102
Hanseatische Kunst- und
  Antiquitätentage, Hamburg ... 098142
Hanseatische Münzenhandlung,
  Bremen 075163
Hanseatisches Auktionshaus für
  Historica, Bad Idesloe 125856
Hanseatisches Buch Kontor,
  Bremen 141651
Det Hanseatiske Museum og
  Schøtstuene, Bergen 033533
Hansen, Bonn 106532
Hansen, Honolulu 093547
Hansen, Perpignan 072084
Hansen Chard Antiques, Pershore 090972,
  134933
Hansen-Gerret, Perpignan 072085
Hansen Gerret, Reynès 072448
Hansen, A. & M., Karlsruhe 076694
Hansen, B., Mannheim 077261
Hansen, Børge Chr., Horsens .. 065794
Hansen, Christian, Berlin 074821
Hansen & Co, Cornelius,
  Roskilde 066089
Hansen, Egon, Oslo 084317
Hansen, Ingrid, Washington ... 124915
Hansen, Jan, Frederiksberg ... 140570
Hansen, Jesper, København ... 065911
Hansen, Jørn, Middelfart 140625
Hansen, Kjær, Herning 065750
Hansen, Torben Bo, Odder 066023
Hansens, Canberra 061272
Hanshan Tang Books, London .. 144619
Hanshin, Arb, Osaka 082082
Hansjakobmuseum im Freihof, Haslach
  im Kinzigtal 019086

Hansmann, C., Bühl 075213
Hanson, Houston 121053, 121054,
  121055
Hanson, New Orleans 122179
Hanson, San Francisco 124298
Susan Hanson & Paul Weber,
  Berlin 141523
Hansong Womens College Museum,
  Seoul 029999
Hansord & Son, Lincoln 089829, 134560
Hansord at No 2, Lincoln 089830
Hansson, Höör 143862
Hansson, Malmö 115760
Hansson, Stockholm 087004
Hanstein, Marlies, Saarbrücken . 108546
Hansure, Hong Kong 065048
Hantho Galleri, Haugesund ... 084248
Hantise, La Ciotat .. 068992, 128640
Hantke, Jan, Valencia 133464
Hantverks- och Sjörfartsmuseét
  på Norra Berget, Medelpads
  Fornminnesförening, Sundsvall 041065
Hantverksloftet, Simrishamn ... 040946
Hantvkmuseet, Ängelholm 040513
Hanul cu Tei, Bucureşti 114145
Hanul Domnesc, Suceava 036736
Hanuš & Aufricht, Cheb 065376
Hanusek, Heinz, Nürnberg 077824
Hanwon Art Museum, Seoul ... 030000
Hanza-Kruger, Szeged 079325
Haos Galerija Beograd, Beograd 114420
Haour, Marie, Paris 129029
Hap Moore, York 127501
Hapart, Philippe, Paris 071466
Ha'Penny Bridge Galleries, Dublin 079558
Hapetian, J.Z., Lisboa . 084896, 132938
Happach, Hans-P., Starnberg .. 130653
Happening, Frankfurt am Main . 075787
Happy Angel Antiques, Los
  Angeles 094354
Happy Art, New Orleans 122180
Happy Days For Old Toys,
  Amsterdam 082766
Happy Thought's, Milwaukee .. 121923
Hapu Galerie, Wien 100312
Har-Ber Village, Grove 049286
Har-El, Jaffa 137611
Har., Sjøfn, Reykjavík 109241
Hara Gendai Bijutsukan, Tokyo . 029512
Hara Museum ARC, Shibukawa . 029371
Hara-shobo Japanese Prints,
  Tokyo 082149
Harajuku, Piga, Tokyo 111441
Harák, Praha 065497
Haralambous, Martin, Hamburg . 076233
Harald, H., Vienenburg 078796
Harare School of Art, Harare .. 058163
Harbeck, Carindale 139886
Harbin Museum, Harbin 007846, 007847
Harbor Branch Oceanographic
  Institution, Fort Pierce 048856
Harbor Defense Museum of New York
  City, Brooklyn 047259
Harbor Gallery, Norfolk 123226
Harborough Museum, Market
  Harborough 045229
Harbottle, Patricia, Gillingham,
  Dorset 089325
Harbour Antiques, Tewantin ... 062315
Harbour Cottage Gallery,
  Kirkcudbright 118057
Harbour Gallery, Dundee 117693
Harbour Gallery, Isles-of-Scilly . 118029
Harbour Gallery, Whitstable ... 119342
Harbour Life Exhibition, Bridport 043728
Harbour Lights Gallery, Porthgain 118937
Harbour Museum, Derry 044090
Harbour View Gallery, Port-of-
  Ness 118936
Harboyan, Krikor, Los Angeles . 094355
Harcourt Antiques, London ... 090185
Harcourt House Antiques,
  Harcourt 061596
Hard 2 Find Toyz, Las Vegas ... 094153

Hard To Find, Whangarei 143378
Hard To Find Books, Auckland . 143284,
  143285
Hard To Find Bookshop,
  Auckland 143286
Hardanger Antikk, Norheimsund 084286
Hardanger Folkemuseum, Utne . 034058,
  034059
Hardeman, G., Leersum 132556
Hardenburg, Wuppertal 142524
v. Harder, Martinsried 142118
Harder-Kuhner, Jakob, Benken (Sankt
  Gallen) 087269
Harder-Kuhner, Jakob, Kaltbrunn 087540
Harder, Paul, Utrecht 083651
Hardie, Perth 090977
Hardie, Rebecca, London 144620
Hardin County Historical Museum,
  Kenton 049935
Harding Home and Museum,
  Marion 050693
Harding Museum, Franklin 048938
Harding, Keith, Northleach 090860,
  134888
Harding, Yiri, Lisarow 061739
Hardner, Gerd, Dresden 075416, 141697
Hardt, Peter, Radevormwald ... 078116,
  108426
Hardtmuseum, Durmersheim ... 018068
Hardware Fine Art, Saint
  Leonards 099671
Hardwick Hall, Chesterfield ... 043936
Hardwick Historical Museum,
  Hardwick 049357
Hardy Country, Melbury Osmond 090738
Hardy, Barbara, Trier 130708
Hardy & Co., James, London .. 090186,
  134680
Hardy, Joyce, Hacheston 089399
Hardy's Cottage, Higher
  Bockhampton 044584
Haremere Hall, Etchingham ... 044305
Harenberg, Dortmund . 106704, 137144
Hareng Saur, Decazeville 068286,
  103661
Hares, Cirencester .. 088918, 134273
Hare's, Northallerton 118815
Harestanes Countryside Visitor Centre,
  Jedburgh 044663
Haret, Serge le, Huisseau-en-
  Beauce 068846
Harewood House, Harewood ... 044511
Harford, Camperdown 061267
Frank Hargesheimer & Maik Günther,
  Düsseldorf 075478, 125929
Hari, Jeju 111602
Hari Om, Mumbai 109443
Haring, A., Amsterdam 143034
Hariphunchai National Museum,
  Lampun 042696
Haris, Warszawa 084724, 084725
Harishin, Kobe 081994
Hariya Razna and Ahmed Urabi
  Museum, Zaqaziq 010637
Harjamäen Sairaalamuseo,
  Siilinjärvi 011355
Harjukosken Vesimylly, Mikkeli . 011157
Harjumaa Muuseum, Keila 010627
Harkányi Galéria, Pécs 109207
Harkins, Brian, London 090187
Harkins, Hammond, Columbus . 120498
Harkort, U., Wiesbaden 078977
Harlan & Weaver, New York ... 122689
Harlan-Lincoln House, Mount
  Pleasant 051093
Harland, Tom, North Ferriby ... 118813
Harlech Castle, Harlech 044512
Harlekin-Antik, Steyr 062850
Harlem Art, New York 122690
Harlem Art Gallery, New York . 122691
Harlequin, Budapest 109148
Harlequin Antiques, Armadale,
  Victoria 061033

Harlequin Antiques, Edinburgh . 089177
Harlequin Antiques, Grantham . . 089356
Harlequin Antiques, Nottingham 090893,
134898
Harlequin Antiques, Porthcawl . 091038,
144790
Harlequin Books, Bristol 144281
Harlequin Books, London 144621
Harlequin Gallery, Lincoln 144540
Harlequin Gallery, London . . 118370
Harlequin Horse Antiques,
Kalamunda 061663
Harleston, Harleston 117908
Harley, Christian Malford 088908
Harley Gallery, Worksop 046197, 119396
Harlinger Aardewerk Museum,
Harlingen 032414
Harlinghausen, Barbara,
Lengerich 077080
Harlinghausen, Klaus, Osnabrück 142276
Harlow Carr Museum of Gardening,
Royal Horticultural Society,
Harrogate 044517
Harlow Ceramics Collection & Kleven
Print Collection, Bemidji State
University, College of Arts and
Letters, Bemidji 046912
Harlow Gallery, Kennebec Valley Art
Association, Hallowell 049319
Harmakhis, Bruxelles 063458
Harman, Paul & Nicholas,
Dorking 089084, 134335
Harmar Station, Marietta 050681
Harmattan, Fouesnant 068592
Harmatys, Maria, Kraków 143475
Harmel, Berlin 074822
Harmer, H.R., New York 127350
Harmer, Johnson, New York . . . 145345
Harmers, London 126963
Harmeyer, M., Bremen 075164
Harmonies, Beir-Mery 137801
Harmonikamuseum Zwota, Sammlung
historischer und neuzeitlicher
Zungeninstrumente, Zwota . . . 022921
Harmonikazentrum, Graz 001977
Harmonise Fine Art Prints,
London 118371
Harmonium-Art-Museum, Schelle 004000
Harmonium-Museum, Liestal . . . 041675
Harmony, Dol-de-Bretagne 068363
Harmony, Roaring Spring 127421
Harmony Arts, Ahmedabad 109247
Harmsen, B., Erlangen 075652
Harn Homestead and 1889er Museum,
Oklahoma City 051727
Harness Racing Museum and Hall of
Fame, Goshen 049166
Harnett, William, Newcastle West 079685,
130887
Harney County Historical Museum,
Burns 047330
Harnisch, Michael, Stuttgart . . . 078607
Harold, Sacramento . . 096630, 136190
Harold M. Freund American Museum
of Baking, Manhattan 050637
Harold Warp Pioneer Village
Foundation, Minden 050928
Haroonian, Los Angeles 094356
Harouni, New Orleans 122181
Harper, Poynton 091051, 134954
Harper, York 091978
Harper City Historical Museum,
Harper 049363
Harper Collins, New York 138396
Harper-James, Montrose 090770,
134856
Harper, Jeff, Wellington 113067
Harper, Martin & Dorothy,
Bakewell 088335, 134049
Harpers, New Orleans 095110
Harpers Ferry National Historical Park,
Harpers Ferry 049365
Harpley, Peter, London 118372
Harput Müzesi, Elazığ 042889

Harraca-Roehl, Paris 104941
Harrassowitz, Otto, Wiesbaden . . 142487
Harrewyn, Frédéric, Cucq 068254
Harrewyn, Frédéric, Le Touquet-Paris-
Plage 069363
Harriann, Minneapolis 135830
Harries, Hamburg 076234
Harriet Beecher Stowe Center,
Hartford 049385
Harriman, John, Lima 113362
Harringer, Rutzenmoos 062782
Harringer, Alexander, Gmunden 062570
Harrington, Albion 060979
Harrington College of Design,
Chicago 056886
Harrington Street Gallery,
Chippendale 098886
Harrington, Adrian, London . . . 134681,
144792
Harrington, Peter, London 134682,
144623
Harris, Fort Worth 093453
Harris, Houston 121056
Harris, Richmond 136169
Harris, Toronto 101622
Harris & Barr, Baltimore 092330, 127133
Harris & Holt, Chester 088860
Harris Art Gallery, James M. Lykes
Maritime Gallery and Hutchings
Gallery, Galveston 049034
Harris Auction Galleries,
Baltimore 127134
Harris Courtin, Paddington, New South
Wales 099522
Harris Exhibit Hall, Oxford 051866
Harris Heritage and Museum,
Harris 005989
Harris Lindsay, London 090189
Harris Museum and Art Gallery,
Preston 045565
Harris, Colin, Eversley 089225
Harris, Elizabeth, New York . . . 122692
Harris, James, Seattle 124539
Harris, Jonathan, London 090188
Harris, Lisa, Seattle 124540
Harris, Lorraine, Colac 061357
Harris, Marlene, Pittsburgh . . . 096292
Harris, Michael, Toronto 064664
Harris, Michael, Winnipeg 064925
Harris, O.K., New York 122693
Harris, Richard, Forster 125193
Harris, Rupert, London 134683
Harris, Thelma, Oakland 123253
Harris, Tom, Marshalltown 127308
Harrison, Richmond . . 096549, 136170
Harrison, Tauranga 113023
Harrison, Vancouver . . 101793, 101794
Harrison County Historical Museum,
Marshall 050715
Harrison Galleries, Calgary 101242
Harrison Galleries, Vancouver . . 101795
Harrison House, Branford 047179
Harrison, Anna, Gosforth, Tyne and
Wear 089346, 134411
Harrison, F., Plymouth 144788
Harrison, John, Aston Tirrold . . . 088317
Harrison, Steven, Stockholm . . . 087005
The Harris's Arcade, London . . . 090190
Harrods, London 090191
Harrogate Antique and Fine Art Fair,
Harrogate 098379
Harrogate Antique Fair, Harrogate 098380
Harrogate House, Singapore . . . 085417
Harrogate Pavilions Antiques and Fine
Art Fair, Harrogate 098381
Harrop Fold Clocks, Bolton-by-
Bowland 088515, 134123
The Harrow Decorative and Fine Arts
Society, Pinner 060153
Harrow Museum, Harrow 044521
Harrströmin Myllymäkimuseo,
Harrström 010803
Harry, Los Angeles 094357
Harry D. Hendren Gallery, Saint

Charles 052756
Harry J. Music Museum,
Kingston 028442
Harry Oppenheimer Diamond Museum,
Ramat Gan 025173
Harry Ransom Humanities Research
Center, Austin 046692
Harry S. Truman Birthplace,
Lamar 050105
Harry S. Truman Home,
Independence 049664
Harry S. Truman Little White House
Museum, Key West 049954
Harry S. Truman Office and Courtroom,
Independence 049665
Harry S. Truman Presidential Library
and Museum, Independence . . 049666
Harry S. Washbrook Museum,
Edam 005741
Harry's Coin Center, Philadelphia 096112
Harry's Hamburger Hafenbasar,
Hamburg 076235
Harry's Rare Coins, San Antonio 096975
Harry's Truman Farm Home,
Grandview 049200
Harsefelder Museumszug,
Harsefeld 019076
Harstad Antikvitet, Harstad . . . 084245
Hart, Houston . 093682, 093683, 127251
Hart, Linby 118114
Hart, London 118373
Hart & Rosenberg, London 090192
Hart-Cam Museum, Hartney . . . 005992
H'Art Gallery, Bucureşti 114146
Hart House, Toronto 007179
Hart-Witzen, Charlotte 119989
Hart, Abigail, Kew 127624
Hart, Bob, Jacksonville 093997
Hart, Catherine, Rouen 072588
Hart, David, Mooloolaba 099369
Hart, Edna, Los Angeles 094358
Hart, Grace, Houston 093684
Hart, Julie, Broken Hill 098797
Hart, Kym, Main Beach 099254
Hart, Leonard & Tracy, Houston . 121057
Hart, M., Amsterdam 082767
Hart, Penine, New York 095447
Hartan, Thomas, Ahorn 129369
Harté, Florent, Bordères-et-
Lamensans 067367
Hartenberger, Klaus, Ratzeburg . 078140
Hartenbos Museum, Hartenbos . 038733
Harter, Eliane, Nice 070615
Harter, Heinz, Nürnberg 142242
Harter, Jamil, Montbrison 070219
Harter, Jean, Montbrison 070220
Harter, Sybille, Schaanwald . . . 111778
Harteveld, Fribourg 144068
Hartford Art School, University of
Hartford, West Hartford 058084
Hartford House Museum,
Manhattan 050638
Hartgen, Stadtbergen 108661
Harthan, Angelika, Stuttgart . . . 108705
Hartig, Hans-Christian & Ilse,
Goch 076017
Hartig, Peter, Charlottenlund . . . 065626
Hartistas, Lisboa 132939
Hartl, Ammerbuch 105987
Hartl, Wien 100313
Hartl-Prager, Helmut, Graz 062589
Hartl, Georg L., Bernried, Kreis
Deggendorf 074963
Hartl, Georg L., München 077518
Hartl, Hans & Sylvia, Poing 108393
Hartland Art Council, Hartland . . 060507
Hartland Quay Museum, Bideford 043593
Hartland Quay Museum, Hartland 044522
Hartleib, Michael, Lahr 076995
Hartleitner, Walter, Hofheim in
Unterfranken 130057
Hartlepool Art Gallery, Hartlepool 044523
Hartley, J., Ripley, Surrey 091112
Hartley, John, Petersfield 134936

Hartleys, Ilkley 126929
Hartlich, Petra, Steinberg 078548
Hartman, Columbus 093027
Hartman, New York 095448
Hartman, Charles A., Portland . . 123629,
145453
Hartman, Zsu Zsi, Subiaco 099726
Hartmanin Mökki, Oripää 011211
Hartmann, Barcelona 114919
Hartmann, Karlsruhe 076695
Hartmann, København 065912
Hartmann, Saint Louis 096714
Hartmann, Beatrix, Wiesbaden . 078978
Hartmann, Dr. Roland, Sankt
Gallen 144127
Hartmann, E., Homberg (Efze) . . 076569
Hartmann, Frédéric, Toulouse . . 141364
Hartmann, H., Aschendorf 074443
Hartmann, J., Augsburg 074461
Hartmann, N., Berlin 129500
Hartmann, P., Bielefeld 074977
Hartmann, Robert, Dessau 129679
Hartmann, V., Gräfenhainichen . 076072
Hartmann, Werner W., Düsseldorf 106792
Hartnell's Art Gallery, Salinas . . 052899
Hartnoll, Julian, London 090193, 118374
Den Hartogh Ford Automobielmuseum,
Hillegom 032481
Hartogsohn, København 065913
Hartskeerl, R.E., Haarlem 132518
Hartsville Museum, Hartsville . . 049394
Hartung & Hartung, München . . 126052,
142159
Hartung's Auto and License Plate
Museum, Glenview 049126
Hartware MedienKunstVerein,
Dortmund 058914
Hartwich, Sellin, Ostseebad . . . 108612
Hartwig, Kalchreuth 076665
Hartwig, Robert, Berlin 141524
Hartworks, Dartmouth 117667
Harty, Lela, San Diego 124130
Hartz, Geeveston 099076
Hartz, Saint Louis 123831
Hartz, Alejandra von, Miami . . . 121801
Hartz, Jean-Claude, Obernai . . . 070770
Harumi, Toronto 101623
Harvard Gallery, Sporting and Wildlife
Art, Books, New York 122694
Harvard Historical Society Collection,
Harvard 049399
Harvard Historical Society Museum,
Still River 053528
Harvard House, Shakespeare
Birthplace Trust, Stratford-upon-
Avon 045903
Harvard Museum of Natural History,
Cambridge 047361
Harvard University Art Museums, Fogg
Art Museum, Arthur M. Sackler
Museum, Busch-Reisinger Museum,
Cambridge 047362
Harvard University Semitic Museum,
Cambridge 047363
Harvery Art Gallery, Harvey . . . 001137
Harvest Home, Bowraville 098768
Harvest Moon, Saint Louis 096715
Harvest, Rose, Irymple 061651
Harvey & Co., New York 095449, 135934
Harvey & Gore, London 090195, 134684
Harvey County Historical Society,
Newton 051552
Harvey Grant Heritage Centre,
Springdale 007066
Harvey House Museum, Florence 048758
Harvey-Lee, Elizabeth, North
Aston 090845
Harvey Studio, Nashville 122084
Harvey, Emily, New York 122695
Harvey, James, Clovelly 098899
Harvey, Patricia, London 090194
Harvey, Trevor Victor, Seaforth . 099683
Harvey & Co., W.R., Witney . . . 091889,
135246

Harvey, Y. Philippe, Montréal . . . 064428
Harwath, Arno, Donaueschingen 129692
Harwich Antiques Centre,
  Harwich 089455
Harwich Maritime Museum and
  Harwich Lifeboat Museum,
  Harwich 044527
Harwich Redoubt Fort, Harwich . 044528
Harwood, Albuquerque 119481
Harwood Museum of Art of the
  University of New Mexico, Taos 053676
Haryana Prantiya Puratatva
  Sangrahalaya, Jhajjar 024327
Harz, Syke 078655
Harzenetter, Ludwig, Westerheim,
  Kreis Unterallgäu 078947
Harzer Roller-Kanarien-Museum, Sankt
  Andreasberg 021585
Harzir, Everard de, Bruxelles . . . 063459
Harzmuseum, Wernigerode 022580
Has, Zürich 087974, 144180
Al-Hasam, Abu, Sharjah 088201
Hasan Süzer Etnografya Müzesi,
  Gaziantep 042905
Hasbach, A.L., Wien 140082
Haschemi Edition Cologne, Köln 137310
Hase, Grażyna, Warszawa 113852
Hasebe-ya, Tokyo 082150
Hasegawa, Nagoya 111319
Hasegawa Machiko Kinen Bijutsukan,
  Tokyo 029513
Haselünner Heimathäuser,
  Haselünne 019085
Hasenauer, Peter, Saalfelden am
  Steinernen Meer 127841
Hasenclever, Michael, München 108136
Hasenmuseum, Bubikon 041372
Hasenstab, Aschaffenburg 074432
The Hash Marihuana Hemp Museum,
  Amsterdam 031942
Hashimoto Kansetsu Kinenkan,
  Kyoto 028928
Hashimoto, Kuniya, Kyoto 082016
Hasičské Muzeum, Fire-fighting
  Museum, Bechyně 009437
Hasikova, Bratislava 085551
The Haskell County Historical Museum,
  Sublette 053587
Haslam & Whiteway, London . . . 090196
Haslauer, Oberalm 062753
Haslem, Jane, Washington 124916
Haslemere Antiques Market,
  Haslemere 089456
Haslemere Educational Museum,
  Haslemere 044530
Haslev Museum, Haslev 010098
Haslinger, Quirin, Linz 140011
Haspengouwse Academie voor
  Beeldende Kunsten, Sint-
  Truiden 055049
Hassalls Gallery, Eltham 061444
Hassan, Dubai 088177
Hassan & Hussein, Cairo 066251
Hassan, Gilles, Montpellier 140967
Hassan, Pascal, Saint-Ouen . . . 073103
Hassan, Roger, Paris 071467
Hassanbey Zardabi Museum of
  Natural History, Bakı 003162
Haßdenteufel, Christa, Sankt
  Wendel 078355
Hassel, Barbara, Frankfurt am
  Main 129847
Hasselberg, Klaus, Potsdam . . . 130530
Hasselmann, Klaus, Bremen . . . 106585
Hasselmann, Martin, Höslwang . 130053
Hasselt Art Gallery, Hasselt 100715
Hassett's Uralla Military Museum,
  Uralla 001604
Hassfurther, Wolfdietrich, Wien . 062973,
      100314,   125274,   140083
Hassler, Christian, Bourron-
  Marlotte 067436
Hassler, Christian, Moret-sur-
  Loing 070346

Hassloch, Bad Kissingen 074535
Hassold, Peter & Robert,
  Augsburg 141414
Haßold, Robert, Königsbrunn . . . 076923
Hassoldt, J.C, Amsterdam 132402
Haste, Ipswich 089646, 118025, 134502
Hasted Hunt, New York 122696
Hastert, Thomas, Trier 078702
Hasting Valley Fine Art Gallery,
  Wauchope 099872
Hastings, Roma 081409
Hastings and Rother Decorative
  and Fine Arts Society, Bexhill-on-
  Sea 060154
Hastings Antique Centre, Saint
  Leonards-on-Sea 091198
Hastings Arts, Hastings 117926
Hastings Community Arts Centre,
  Hastings 033217
Hastings Museum, Hastings 049402
Hastings Museum and Art Gallery,
  Hastings 044532
Hastings-Western Port Historical
  Museum, Hastings 001138
Hat Sanatları Müzesi, İstanbul . . 042939
Hat Works, Museum of Hatting,
  Stockport 045875
Hata, Nagasaki 111309
Hatakeyama Memorial Museum of
  Fine Art, Tokyo 029514
Határőrség Központi Múzeuma,
  Körmend 023753
Hatay Devlet Güzel Sanatlar Galerisi,
  Hatay 042911
Hatay Müzesi, Hatay 042912
Hatch-Billops Collections, New
  York 051403
Hatch, Gideon, London 090197, 134685
Hatchards, London 144624
Hatchuel, Paris 141137
Hatfield House Antiques, Fort
  Worth 093454
Hatfield & Sons, H.J., London . . 134686
Hatfield, Ian, Haugesund 084249
Hathaway Ranch Museum, Santa Fe
  Springs 053126
Hatherleigh Antiques, Hatherleigh 089470
Hatje Cantz Verlag, Ostfildern . . 137463
Hatjigeorgiou, Vassili, Quillan . . 072338
Hatle, Praha 140519
Hatry, Heidelberg 141926
Hatten, Richard de, Poids-de-
  Fiole 072213
Hattendorf, Wiebke, Hamburg . . 129965
Hatter, Mad, Auckland 143287
Les Hattes, L'Isle-sur-la-Sorgue 069570
Hattis, Phyllis, New York 122697
Hatton-Eielson Museum, Hatton 049405
Hatton Gallery, Newcastle-upon-
  Tyne 045348
Hatton Gallery, Colorado State
  University, Fort Collins 048798
Hattowska, Grażyna, Kraków . . . 132840
Hatzig, Hermann-Josef,
  Rauschenberg 078146
Haub, Bianca, Würzburg 108974
Haubensack, Hanna, Zürich 144181
Haubrich, Bernd, Trier 078703
Haubs, Markus, München 130356
Hauchs Physiske Cabinet, Sorø
  Akademis, Sorø 010324
Hauck & Aufhäuser, München . . 077519,
                      126053
Hauck, Anne, West Hollywood . . 097895
Hauck, Bernd, Rodewald 108505,
                      137487
Haude & Spener, Berlin 137073
Haudrecy, d', Knokke-Heist 100740
Haufe & Lutz, Karlsruhe 141972
Hauff, Prof. Gottfried, Potsdam . 130531
Hauff, Reinhard, Stuttgart 108706
Hauffs Märchen-Museum,
  Baiersbronn 016934
Haugar Vestfold Kunstmuseum,

Tønsberg 034017
Haugesund Billedgalleri,
  Haugesund 033642
Haugesund Kunstforening,
  Haugesund 059658
Haugg, Buxheim 129648
Haughey, Kirkby Stephen 089714
Haughton, Brian, London 090198
Haughton, Stephen, Hawthorn . . 061603,
                      127613
Haugwitz, Ralph Rüdiger, Berlin 074823
Hauhon Esinemuseo, Hauho . . . 010805
Hauke, Udo, Lützen 130258
Haukiputaan Koulumuseo,
  Haukipudas 010806
Haunch of Venison, London 118375
Haunch of Venison, Zürich 116807
Haunsperger, Ingolstadt 076612
Haupt Verlag, Bern 138031
Haupt, Horst, Berlin 074824
Haupt, Rudolf, Augsburg 074462
Hauptmann & Kampa, Zürich . . . 116808
Hauptmann, Burkhard, Bamberg 074663
Hauptmann, U., Bielefeld 074978
Hauptstaatsarchiv Stuttgart,
  Landesarchiv Baden-Württemberg,
  Stuttgart 022034
Haurie Vigne, Magdalena, Sevilla 115451
Haus am Kleistpark, Kunstamt
  Tempelhof-Schöneberg, Berlin . 017156
Haus am Lützowplatz, Berlin . . . 017157
Haus am Waldsee, Berlin 017158
Haus der Auktionen, Frankfurt am
  Main 125947
Haus der badisch-pfälzischen
  Fasnacht, Speyer 021930
Haus der Brandenburgisch-
  Preußischen Geschichte,
  Potsdam 021184
Haus der Donauschwaben,
  Sammlungen heimatlichen Erbes,
  Sindelfingen 021872
Haus der Dorfgeschichte Bliesmengen-
  Bolchen, Mandelbachtal 020243
Haus der erzgebirgischen Tradition,
  Deutschneudorf 017881
Haus der Fotografie – Dr. Robert-
  Gerlich-Museum, Burghausen . 017689
Haus der Gemälde, Nürnberg . . 108302,
                      130465
Haus der Geschichte, Bad
  Harzburg 016781
Haus der Geschichte Baden-
  Württemberg, Stuttgart 022035
Haus der Geschichte der
  Bundesrepublik Deutschland,
  Bonn 017490
Haus der Geschichte Wittenberg,
  Lutherstadt Wittenberg 020197
Haus der Heimat, Olbernhau . . . 020982
Haus der Keramik, Wilhelmsburg an
  der Traisen 100402
Haus der Kulturen der Welt,
  Berlin 017159
Haus der Kunst, Graz 100008
Haus der Kunst, München 020529
Haus der Kunst, Nürnbrecht 020841
Haus der Kunst, Remshalden . . . 078202,
          078203,   130581
Haus der Kunst Sankt Josef,
  Solothurn 116651
Haus der Kunsteditionen,
  Freilassing 075895
Haus der Natur, Eichendorf 018115
Haus der Natur, Museum für Natur
  und Technik, Salzburg 002625
Haus der Naturfreunde,
  Naturkundemuseum der
  Naturfreunde Deutschlands OG
  Duisburg e.V., Duisburg 018056
Haus der Ortsgeschichte, Schwäbisch
  Hall 021753
Haus der Ost- und Westpreußen,
  Oberschleißheim 020920

Haus der Seidenkultur
  Paramentenweberei Hubert Gotzes
  e.V., Krefeld 019785
Haus der Stadtgeschichte, Offenbach
  am Main 020971
Haus der Stadtgeschichte, Städtische
  Museen, Donauwörth 017932
Haus der tausend Bilder, Kiel . . 076775,
                      107553
Haus der Völker, Museum für Kunst
  und Ethnographie, Schwaz . . . 002740
Haus der Wannsee-Konferenz,
  Berlin 017160
Haus des Apfels, Puch bei Weiz 002543
Haus des Gastes mit heimatkundlicher
  Sammlung, Gößweinstein 018745
Haus des Moores,
  Heidenreichstein 002065
Haus des Waldes, Köln 019686
Haus Fischerzunft, Bad
  Säckingen 016852
Haus für Kunst Uri, Altdorf 041209
Haus Kickelhain, Stadtmuseum
  Mosbach, Mosbach 020470
Haus Konstruktiv, Zürich 042197
Haus Mährisch-Schönberg, Bad
  Hersfeld 016783
Haus of Prints & Multiples,
  Antwerpen 136621
Haus Opherdicke, Holzwickede . 019330
Haus Rottels, Stadthistorische
  Abteilung des Clemens-Sels-
  Museums Neuss, Neuss 020744
Haus Ruhrnatur, Mülheim an der
  Ruhr 020494
Haus Sandfeld, Wuppertal 079104
Haus Stöcken, Haan 018904
Haus Tannenbusch,
  Schutzgemeinschaft Deutscher Wald
  Dormagen, Dormagen 017940
Haus Völker und Kulturen –
  Ethnologisches Museum, Sankt
  Augustin 021588
Haus Wittgenstein, Bulgarisches
  Kulturinstitut, Wien 003002
Haus zum Kirschgarten, Historisches
  Museum Basel, Basel 041267
Haus zum Palmbaum mit
  Bachausstellung Arnstadt,
  Arnstadt 016609
Haus zum Stockfisch – Stadtmuseum,
  Erfurt 018243
Haus zum Torggel, Berneck 041329
Haus zum Wasen, Wagenhausen 116709
Haus zur Geschichte der IBM
  Datenverarbeitung, Sindelfingen 021873
Hausammann, Cortina
  d'Ampezzo 109985
Hauschild, H.M., Bremen 137126
Hausen, Luisa, Euskirchen 075720,
                      106942
Hausenhäusl, Reit im Winkl 021354
Hauser, München 142160
Hauser, Schwarzenburg 087808, 133918
Hauser & Wirth, London 118376
Hauser & Wirth, Zürich 116809
Hauser, Walter, Lyon 069746
Hausgalerie, Müllendorf 100125
Hausjärven Kotiseutumuseo,
  Hausjärvi 010807
Hausmuseum Kloster Kreuzlingen,
  Kreuzlingen 041624
Hausmuseum Sankt Katharinental,
  Diessenhofen 041440
Hausser, M-P, Soufflenheim . . . 073547
Haußmann, B., Mössingen 130312
Haußmann, Martin, Besigheim . 129585
Hausstätter, Leopold, Altötting . 105981
Hauswedell & Nolte, Hamburg . 125968,
                      141868
Hauswedell & Nolte, New York . 127351
Hauswedell & Co., Dr. Ernst,
  Stuttgart 137520
Haut-Pavé, Paris 104942

Haut Quartier, Pézenas 141305
Haute Antiques 207, Bruxelles . 063460
Haute Ecole d'Art Appliqués HES de
  Genève, Genève 056427
Haute Ecole d'Art de Perpignan ESAP,
  Perpignan 055408
Haute école d'art et de design,
  Genève 056428
Haute Epoque, Flin 068546
Hautecoeur, Paris 104943
Les Hautes-Mynes, Le Thillot .. 013702
Hauteville House, Musées de la Ville
  de Paris, Saint Peter Port 045713
Hautin, Jean-Louis, Cosne-Cours-sur-
  Loire 125562
Havacılık Müzesi, İstanbul 042940
Havana Galerie, Zürich 116810
Havana Gallery, Chicago 120152
Havana Gallery, Vancouver ... 101796
Havant Museum, Havant 044537
Havard Père & Fils, Paris ... 129030,
                    129031,   129032
Havard & Havard, Cowbridge .. 088987
Havas, Christophe, Rouen ... 072589
Havas, Frédéric, Caen 067570
Havayeda Museum, Holon 025066
Havelková, Elena, Mariánské
  Lázně 140485
Havelock Museum, Havelock ... 033218
Havelocks, Harrogate . 089431, 134436
Havemann, Minke, Heinbockel . 107409
Haven Antiques, Wainfleet 091706
Haven Art, Sacramento 123771
Het Havenmuseum, Rotterdam . 032802
Havenplan's Architectural Emporium,
  Killamarsh 089684
Haver, Peter, Blaricum 132441
Haverbeque, Patrick, Marcq-en-
  Barœul 069863
Haverford Township Historical Society
  Museum, Havertown 049408
Haverhill and District Local History
  Centre, Haverhill 044540
Havers, Hamilton, Caversham . 134224
Havesdonckhoeve, Museum van Oud
  Landbouwalaam en Volkshuisraad,
  Bornem 003382
Havets Hus, Lysekil 040817
De Havilland Aircraft Heritage Centre,
  Mosquito Aircraft Museum, Saint
  Albans 045663
Havilland, Adele de, London ... 090199
Havilland, de, New York 095450
Havillands, de, Clifton Hill 139700
Havlíček, Znojmo 128373
Havlik, Jan, London 090200
Havlik, Rudolf, Wien .. 100315, 127923
Havråtunet, Lonevåg 033754
Havre de Grace Decoy Museum,
  Havre de Grace 049410
Havstad, Drøbak 113147
Havza Atatürk Evi, Havza 042913
Hawaii All-Collectors Show,
  Honolulu 098449
Hawaii Antique Center, Honolulu 093548
Hawaii Children's Discovery Center,
  Honolulu 049530
Hawaii Maritime Center, Honolulu 049531
Hawaii Museums Association,
  Honolulu 060508
Hawaii Pacific University Gallery,
  Honolulu 049532
Hawaii State Foundation on Culture
  and the Arts, Honolulu 049533
Hawaiian Antiquities, Honolulu . 093549
Hawaiian Art, Indianapolis 121188
Hawaiian Islands Coin, Honolulu 093550
Hawaii's Plantation Village,
  Waipahu 054028
Hawari, S. & H., München ... 077520,
                    130357
Hawick Museum, Hawick 044543
Hawk, Columbus 120499
Hawk, Louisville 121648

Hawk, Pittsburgh 123549
Hawke, Kansas City 121295
Hawke House, Bordertown 000868
The Hawker Gallery, Amersham 117238
Hawker Museum, Hawker, South
  Australia 001139
Hawke's Bay Museum, Napier . 033250
Hawkesbury Creative Art Centre,
  Windsor, New South Wales .. 099890
Hawkesbury Regional Museum,
  Windsor, New South Wales .. 001647
Hawkesbury Valley Auctions, Windsor,
  New South Wales 125245
Hawkeye Log Cabin, Burlington 047317
Hawkins, John, Moss Vale 061904
Hawkins, Philip, Shaftesbury .. 135020
Hawks Hill Gallery, Kingsley ... 099198
Hawks Inn, Delafield 048227
Hawks Nest State Park, Ansted . 046506
Hawkstrega, Indianapolis 093887
Hawley Antiques, Beverley 088451,
                    134097
Hawley Library, Los Angeles ... 138350
Hawley, Nathan, San Francisco . 097251
Hawn, W.H., Oklahoma City ... 095978
Haworth, Binghamton 138305
Haworth Art Gallery, Accrington 043376
Haworth, Peter, Beetham 088430,
                    134089
Haws & Co., Denver 120733
Hawthorn, Birmingham 119802
Hawthorn, Galway 109636
Hawthorn, London 134687
Hawthorn, Gerard, London ... 090201,
                    134688
Hawthorne, Hull 134486
Hawthorne, New York 122698
Hawthorne Boulevard Books,
  Portland 145454
Hawthorne Cottage, National Historic
  Site, Brigus 005480
Hawthorns Urban Wildlife Centre,
  Southampton 045825
Hay, Bunbury 098817
Hay Gaol Museum, Hay 001141
Hay Hill Gallery, London 118377
Hay House Museum, Macon ... 050571
Hay, Ashley, Clayfield 061333
Hayart Cultural Center, Yerevan . 000732
Hayashibara Bijutsukan,
  Okayama 029212
Hayashibara Shizenkagaku
  Hakubutsukan, Okayama ... 029213
The Haybarn Antiques, Upper
  Moutere 084116
Haycraft, Roland, Wallingford . 135183
Haydar, Beirut 082287
Hayde, Georg, Feldafing 129834
Haydn-Geburtshaus, Rohrau ... 002604
Haydn-Gedenkstätte mit Brahms-
  Gedenkraum, Wien Museum,
  Wien 003003
Haydn-Haus Eisenstadt,
  Eisenstadt 001860
Haydon Gallery, Nebraska Art
  Association, Lincoln 050292
Haye, Danielle, Paris 071468
Al-Hayek, Bassam, Damascus .. 088037
Hayem, François, Paris 071469
Hayes, Burswood 125184
Hayes, Penn Yan 127386
Hayes, Jamie, New Orleans ... 122182
Hayes, Joseph M., Columbus .. 093028
Hayes, M., Baltimore 135343
Haygate Gallery, Telford 091554
Haygood, Don & Pola, Glyfada . 109064
Hayhurst, Jeanette, London ... 090202
Hayko Restoration, New York . 135935
Hayles, Derek, Maidenhead ... 144724
Haylett, A.P. & M.A., Outwell . 090928
Hayloft Gallery, Palace Stable Heritage
  Centre, Armagh 043432
The Haymakers Gallery, Hay-on-
  Wye 117937

Haymar, Busselton 061243
Haymes & Sons, Pokolbin ... 139874
Haymon Verlag, Innsbruck 136520
Haynes, Broadway ... 117456, 134169
Haynes Motor Museum,
  Sparkford 045846
Haynes, Frank, Great Bowden .. 117878
Haynes, Roger, Leek 089780
Hayo, Châtenois 067910
Hayot, Monelle, Saint-Rémy-en-
  l'Eau 137005
Hayot, Pascal, Mesigny 104220
Hays, Louisville ... 094558, 127297
Hay's Gallery, London 118378
Hays, Daniel, Valdivienne 073936
Haystack Mountain School of Crafts,
  Deer Isle 056994
Haystack Mountain School of Crafts
  Gallery, Deer Isle 048213
Hayward, Dunedin ... 083927, 126317
Hayward Area Museum, Hayward 049419
Hayward China Museum, Saint
  John 006891
Hayward Gallery, London 044994
Hayward Gallery Bookshop,
  London 144625
Hayworth Art Gallery, Los
  Angeles 121505
Hazaël-Massieux, Philippe,
  Avignon 128457
Hazama Inosuke Bijutsukan,
  Kaga 028739
Hazard Press, Christchurch ... 137881
Hazeborg, Bernhard ter,
  Hamburg 076236
Hazel of Brecon, Brecon 088588
Hazelbrook Cottage Antique Centre,
  Hazelbrook 061610
Hazelhoff, Janna, Groningen .. 143149
Hazelhurst Regional Gallery and Arts
  Centre, Gymea 001128
Hazelton Pioneer Museum,
  Hazelton 005996
Hazledine, Anthony, Fairford .. 089245,
                    134380
Hazlitt, Gooden & Fox, London . 090203
Hazlitt, Gooden & Fox, New York 095451
Hazor Museum, Ayelet-Hashahar 025023
Hazou, Elie, San Diego 124131
Hazrati Museum, Semnan 024732
HBT Art Gallery, Wuppertal ... 108992
HD Antiques, New York 095452
HDM Galerie, Moers 108022
He County Museum, Hexian ... 007862
He Xiang Ning Art Museum,
  Shenzhen 008262
He Xuan, Tian, Singapore 114533
Head House Antiques and Uniques,
  Louisville 094559
Head-Smashed-In Buffalo Jump, Fort
  Macleod 005823
Head Street Gallery, Halstead .. 117905
Head, D.R., Weybridge 119331
Head, Peter, Fukuoka 081965
Headfooters, Cleveland 120429
Headland Historical Museum,
  Nambucca Heads 001336
Headlands Center for the Arts,
  Sausalito 053171
Headley & Menzies, Memphis .. 094638
Headley-Whitney Museum,
  Lexington 050243
Headrow Antiques Centre, Leeds 089769
Headrow Gallery, Leeds 118088
Heads 'n' Tails, Wiveliscombe .. 091894,
                    135248
Headwaters Heritage Museum, Three
  Forks 053740
Healdton Oil Museum, Oklahoma
  Historical Society, Healdton ... 049426
Healesville Art Gallery,
  Healesville 099121
Healing & Arts Studio, Boston .. 119873

Healry Gallery, Singapore 085418
Health Adventure, Asheville ... 046556
Health Museum, Hyderabad ... 024297
Health Museum, Yangon 031833
Healthspace Cleveland, Cleveland 047816
Healy House and Dexter Cabin,
  Leadville 050200
Healy, Justin, Malvern 139818
Heaman's Antique Autorama,
  Carman 005569
Heanor Antiques Centre, Ilkeston 089633
Heape's Antiques, Burnham-on-
  Sea 088727
Heard Museum, Phoenix 052118
Heard Natural Science Museum,
  McKinney 050564
Hearst, New York 138397
Hearst Art Gallery, Saint Mary's
  College, Moraga 051049
Hearst Castle, San Simeon ... 053064
Heart, Paris 104944
Heart Field Gallery, Nagoya ... 111320
Heart Of Art, Christchurch 112843
Heart of Diamonds, Miami 094743
Heart of Dixie Railroad Museum,
  Calera 047350
Heart of Glass 2003, Greytown . 112902
Heart of West Texas Museum,
  Colorado City 047881
Heart Spoken, Portland 123630
Heart Strings, Trafalgar 062342
Heart Wood, Kilkenny 130875
Heart, Lion, New York 145346
Hearthstone Historic House Museum,
  Appleton 046512
Heartland, Maitland 061761
Heartland -California Museum of the
  Heart, Rancho Mirage 052460
Heartland Antiques, Jacksonville 093998,
                    135655
Heartland Gallery, Austin 119676
Heartland Mall, Oklahoma City . 095979
Heart's Content Cable Station, Heart's
  Content 006000
Heath-Bullock, Godalming 089340,
                    134409
Heath Woodworks, Salt Lake City 136239
Heath, A.R., Bristol 144282
Heath, Bob, Parkhurst 125222
Heath, Claudia, Charlotte 119990
Heath, Mike, Newport, Isle of
  Wight 090835, 134882
Heathcote Antiques, Cross Hills . 089014
Heather, J.C., Woolpit 091937
The Heatherley School of Fine Art,
  London 056560
Heatherslaw Mill, Cornhill-on-
  Tweed 044010
Heathfield Antiques & Country Pine,
  Holt, Norfolk 089548
Heaton Antiques, Bradford 088558
Heaton Hall, Manchester City Galleries,
  Manchester 045204
Heaven & Earth, Sydney 099785
Heaven & Earth Antiques,
  Edmonton 064295
Heaven and Earth, Sydney 099786
Heaven Farm Museum, Uckfield 046015
Heaven Hi Originals, Saint Paul . 123914
Heaven, Barrie, Adelaide 060968
Heavenly Handbags, Charlotte .. 092601
The Heavenly One, Singapore . 085419
Heaven's Art Gallery, Montego
  Bay 111231
Hebbel-Museum, Wesselburen . 022595
Hebecker, Michel, Weimar 108863
Hebei Art School Shijiangzhuang
  Fenxiao, Shijiazhuang 055240
Hebei Fine Arts Publishing House,
  Shijiazhuang 136740
Hebei Museum, Shijiazhuang .. 008272
Hebei Painting Gallery,
  Shijiazhuang 102237
Hebei Provincial Museum,

Shijiazhuang 008273
Hebei Provincial Museum,
Wuchang 008367
Heberlé, Jean-Jacques,
Guebwiller 068754
Heberri, Donostia-San Sebastián 133249
Hébert, Serge, Yébleron 074304
Hebezeug-Museum, Witten 022696
Hebi Museum, Hebi 007850
Hebler, Bochum 106512
Hebling, R., Gochsheim 076024, 129907
Hebraica Judaica, Paris 141138
Hécey, Erna, Luxembourg 111855
Hecher, Gert, Wien 127924
Hecht-Lang, Dipl. Restauratorin
Brigitte, Bad Waldsee 129427
Hecht, A., München 077521
Hecht, Richard, Dallas 093137
Hechter-Schulz Afro-Culture Museum,
Boksburg 038665
Heck, Thomas Leon, Reutlingen 078212,
142330
Heck, Thomas Leon, Tübingen . 078722,
126129,   142428
Heck, V.A., Wien 140084
Heckel & Börner, Winsen, Luhe . 130770
Heckenhauer, J.J., Tübingen . . 108789,
137534,   142429
Hecker, Maik David, Düsseldorf . 129743
Heckers, Martin,
Mönchengladbach 077391
Heckington Windmill, Heckington 044552
Heckl, Heinrich, Wien 127925
Heckmann, Stuttgart 130674
Heckmann, Jean-Pierre, Paris . 129033
Heckmann, Manfred, Berlin . . . 074825
Heckmann, Peter-Carsten,
Neuhardenberg 077730
Heckner, W., Iphofen . . 076616, 107468
Heckscher Museum of Art,
Huntington 049621
Heckt, Sönke, Laboe 076988
Hecq, J.-M., Charleroi . 063604, 125330
Hector Exhibit Centre, Pictou . . 006648
Hector Pieterson Museum and
Memorial, Johannesburg 038750
Heddal, Notodden 084287
Heddal Bygdetun, Notodden . . . 033810
Hedderich-Hoffmann, Thomas,
Solingen 078496
Heddon Museum, Dowagiac . . . 048359
Hedel, Ansbach 105995
Hedenborg, Roskilde 140652
Hedenius, Stockholm 115895
Hédevin, Jean-Pierre, Marck . . 069854
Hedge, Penny, Whitby 119335
Hedgecoe & Freeland, Chobham 134268
Hediger, Thomas, Sankt Gallen . 087780
Hedingham Antiques, Sible
Hedingham 091336, 135048
Hedley, Peter, Wareham 119304, 135165
Hedleys Humpers, Saint-Ouen . 073104
Hedley's of Hexham, Hexham . 089523,
134462
Hedmark Auksjoner, Nes på
Hedmark 126363
Hedmarksmuseet og Domkirkeodden,
Hamar 033633
Hedon Museum, Hedon 044554
Hedrich, Ruth, Ehringshausen . . 106868
Hedrick, Natalie, Charlotte . . . 092602
Hedrum Bygdetun, Kvelde 033724
Hee Loppe- og Antikmarked,
Ringkøbing 066069
Heeb, Iris, Schaan 132319
Heede & Moestrup, København . 065914,
102900,   136774
Heemann, Annette, Münster . . . 077664
Heemerf De Schutsboom, Goirle 032357
Heemhuis en Sigarenmuseum,
Arendonk 003320
Heemkamer Barthold Van Heessel,
Aarle Rixtel 031886
Heemken, Frank, Karlsruhe . . . 130097

Heemkundemuseum Beek, Beek 032050
Heemkundemuseum Paulus Van
Daesdonck, Ulvenhout 032939
Heemkundig Hagelands Museum, Tielt-
Winge 004067
Heemkundig Museum, Asse . . . 003325
Heemkundig Museum,
Borgerhout 003380
Heemkundig Museum, Bree . . . 003399
Heemkundig Museum, Moen . . . 003872
Heemkundig Museum, Ommeren 032718
Heemkundig Museum, Tervuren 004056
Heemkundig Museum,
Waasmunster 004121
Heemkundig Museum, Wilrijk . . 004153
Heemkundig Museum, Zolder . . 004163
Heemkundig Museum Boekhoute,
Boekhoute 003375
Heemkundig Museum De Drie Rozen,
Gravenwezel 003657
Heemkundig Museum Gerard
Meeusen, Essen 003583
Heemkundig Museum Sincfala,
Knokke 003750
Heemkundig Museum Slag van Lafelt,
Vlijtingen 004120
Heemkundig Museum 't Smiske,
Rijmenam 003967
Heemkundig Museum Tempelhof,
Beerse 003352
Heemkundig Museum Wetteren,
Wetteren 004148
Heemkundig Museum Wissekerke,
Kruibeke 003760
Heemkundig Streekmuseum Jan Uten
Houte, Etten Leur 032322
Heemmuseum Bystervelt, Schelle 004001
Heemmuseum De Kaeck,
Wommelgem 004155
Heemmuseum De Pastorij,
Edegem 003570
Heemmuseum de Zuiderkempen,
Westerlo 004147
Heemmuseum Die Swane, Heist-op-
den-Berg 003688
Heemmuseum Eeklo, Eeklo . . . 003572
Heemschuur Maasland, Heesch 032437
Heemskerk, Gerrit, Amsterdam . 112219
Heer, Oslo 113226
Heeramaneck & Son, Mumbai . . 109444
Heerd, Tucson 097669
Heeren, Monika von, Rottenbuch 130605
Héeren, Viviann, Lund 133548
Heeresgeschichtliches Museum,
Wien 003004
Heerkens Thijssen, Aerdenhout . 132380
De Heerlijckheid, Middelburg . . 083494
Hees, Daniel, Mühlacker 130317
Heesberg Museum, Gevensleben 016882
Heeseler, Georg, Krefeld 076957
Heeswyk, J., Ediger-Eller 075571
Hefaistos, Bucureşti 132997
Hefaistos, Děčín 102590
Hefeng Museum, Hefeng 007854
Heffel, Montréal 125370
Heffel, Ottawa 125372
Heffel, Toronto 125383
Heffel, Vancouver 125389
Heffernan, John, Douglas 079513,
130840
Heffner, Bratislava 114653
Heffner, Tegernsee 078671
Hefner Zoology Museum, Miami
University, Oxford 051867
Hegarda-Verlag, Linz am Rhein . 137354
Hegartys, Bandon 130828
Hegau-Bodensee-Galerie, Singen 108628
Hegau-Museum, Singen 021876
Hegdehaugen Antikvariat, Oslo . 143402
Hegelhaus, Stuttgart 022036
Hegemann, Hagen 107196
Hegemann, A & G, Hamburg . . 076237
Hegemann, H., Hannover 076373

Hegenbart, Günther, Wien 062974,
127926
Hegenbart jun., Werner, Wien . . 062975
Hegenberg, Hans sen., Essen . . 075680
Hegetschweiler, Marlène,
Ottenbach 116558
Hegewisch-von Perfall, Katharina,
München 108137
Heggeli, Oslo 084318
Hegglin, Lotti, Unterkulm 133946
Hegmataneh Museum, Samedan 024730
Hegnauer & Schwarzenbach,
Bern 144048, 144049
Hegyvidék Galéria, Budapest . . 023484
Hegyvidéki Helytörténeti Gyűjtemény,
Budapest 023485
Heian, Houston 093685
Heian, Kyoto 082017, 111294
Heiando, Kyoto 142898
De Heibergske Samlinger – Sogn
Folkemuseum, Kaupanger 033694
Heick, W., Oldenburg 077930
Heickmann, Chemnitz 075252
Heidbüchel, B., Düren 106753
HeiDe, Velen 078787
Heide Museum of Modern Art,
Bulleen 000905
Heidefeld & Partner, Krefeld . . . 107738
Heidegalerie, Babenhausen,
Hessen 106036
Heidelberger Kunstverein,
Heidelberg 019134, 058915
Heidema, Groningen 083294
Heidemuseum Dat ole Huus,
Bispingen 017400
Heidemuseum Rischmannshof
Walsrode, Freilichtmuseum,
Walsrode 022412
Heidenberg, Lillian, New York . . 122699
Heidenreich, Manfred, Hemhofen 076490
Heidenreich, Manfred,
Heroldsbach 076506
Het Heidenrijck, Nijmegen 083512
Heidi Weber-Museum, Zürich . . 042198
Heidi's Attic Wares, Taree 062308
Heidkamp, Ludwig, Düsseldorf . 075479,
129744
Heidrich, Buenos Aires 127523
Heidrichs Kunsthandlung, Berlin 106305
Heights Antiques on Yale,
Houston 093686
Heights Gallery, Houston 121058
Heights Station Antiques,
Houston 093687
Heights Theater Antiques,
Houston 093688
Heigl, Traunreut 108773
Heihe Museum, Heihe 007855
Heijboer, Anton, Amsterdam . . . 112220
Heijokyuseki Shiryokan, Nara
Bunkazaikenkyusho, Nara 029143
Heikintupa, Pohjaslahti 011260
Heikkilä, Helsinki 066319
Heilers, W., Isterberg 076633
Heilig, A., Bruchsal 106601
Heilig, Andreas B., Frankfurt am
Main 075788
Heiligenstedten, Tornesch 078688
Heiliggeistkirche, Museen der Stadt
Landshut, Landshut 019883
Heilongjiang Nationality Museum,
Harbin 007848
Heilsarmee – Museum und Archiv,
Bern 041314
Heim, Miami 121802
Heim & Kollmann, Waldsolms . . 078847
Heim, Chanin, New York 122700
Heim, Jean-François, Paris 104945
Heim, Philippe, Paris 104946
Heim, Reinhold, Chemnitz 129652
Heimat- und Bauernmuseum Bruck,
Bruck in der Oberpfalz 017636
Heimat- und Bauernmuseum Sulz,
Sulz am Neckar 022079

Heimat- und Bergbaumuseum Lugau,
Lugau 020189
Heimat- und Bergbaumuseum
Nentershausen, Nentershausen 020661
Heimat- und Bergbaumuseum
Reinsdorf, Reinsdorf bei
Zwickau 021353
Heimat- und Brauereimuseum
Pleinfeld, Pleinfeld 021153
Heimat- und Braunkohlemuseum
Steinberg, Steinberg, Kreis
Schwandorf 021980
Heimat- und Buddelmuseum Osten,
Osten 021022
Heimat- und Bürstenbindermuseum,
Ramberg 021270
Heimat- und Burgmuseum,
Kirkel 019633
Heimat- und Feuerwehrmuseum,
Schauenstein 021632
Heimat- und Flößermuseum Calmbach,
Bad Wildbad 016899
Heimat- und Grimmelshausenmuseum,
Oberkirch 020900
Heimat- und Hafnermuseum,
Heidenheim 019151
Heimat- und Handfeuerwaffenmuseum,
Kemnath 019581
Heimat- und Heringsfängermuseum
Heimsen, Petershagen 021108
Heimat- und Hugenotten-Museum,
Stutensee 022023
Heimat- und Humboldtmuseum,
Eibau 018110
Heimat- und Industriemuseum
Kolbermoor, Kolbermoor 019755
Heimat- und Industriemuseum
Wackersdorf, Wackersdorf 022360
Heimat- und Keramikmuseum
Kandern 019494
Heimat- und Kleinbrennermuseum,
Steinach, Ortenaukreis 021973
Heimat- und Krippenmuseum,
Zirl 003140
Heimat- und Landlermuseum Bad
Goisern 001744
Heimat- und Miedermuseum Heubach,
Heubach 019239
Heimat- und Montanmuseum in
Schloß Oberkinberg, Kindberg 002168
Heimat- und Naturkunde-Museum
Wanne-Eickel, Emschertal-Museum,
Herne 019217
Heimat- und Ortsmuseum,
Zürich 042199
Heimat- und Palitzsch-Museum
Prohlis, Dresden 017986
Heimat- und Pfarrmuseum, Museum
Wildalpen, Wildalpen 003111
Heimat- und Rebbaumuseum,
Spiez 042012
Heimat- und Schiffahrtsmuseum,
Heinsen 019176
Heimat- und Schmiedemuseum,
Widdern 022619
Heimat- und Ski-Museum,
Braunlage 017551
Heimat- und Ständemuseum,
Tragöß 002829
Heimat- und Torfmuseum,
Gröbenzell 018825
Heimat- und Uhrenmuseum, Villingen-
Schwenningen 022335
Heimat- und Waldmuseum,
Eppenbrunn 018220
Heimat- und Weinbaumuseum, Graach
an der Mosel 018778
Heimat- und Zunftmuseum
Adenau 016469
Heimat-Haus Großderschau,
Großderschau 018843
Heimat-Museum Maintal, Maintal 020223
Heimat- und Bildhauer Kern Museum,
Forchtenberg 018383

Heimat-und Handwerkermuseum,
  Leutershausen-Wiedersbach .. 020052
Heimatarchiv Zwittau, Esslingen 018317
Heimatdiele Kutenholz, Kutenholz 019835
Heimatecke am Seifenbach,
  Beierfeld 017017
Heimatgeschichtliche Sammlung des
  Kirchspiels, Schwabstedt ... 021746
Heimathaus, Königswiesen 002219
Heimathaus, Litzelsdorf 002335
Heimathaus, Neufelden 002434
Heimathaus, Obernberg am Inn . 002461
Heimathaus, Pregarten 002538
Heimathaus, Raab 002553
Heimathaus, Sankt Georgen an der
  Gusen 002653
Heimathaus, Sankt Peter im
  Sulmtal 002692
Heimathaus, Schörfling am
  Attersee 002726
Heimathaus, Steinbach am
  Attersee 002778
Heimathaus, Ulrichsberg 002852
Heimathaus Alte Mühle,
  Schladen 021659
Heimathaus Aying, Aying 016684
Heimathaus Beandhaus, Sankt Johann
  am Walde 002665
Heimathaus Bevergern, Hörstel . 019283
Heimathaus De Theeshof,
  Schneverdingen 021693
Heimathaus der Stadt Lauingen,
  Lauingen 019937
Heimathaus der Stadt Warendorf,
  Warendorf 022431
Heimathaus des Rupertiwinkels und
  Heimatstube der Sudetendeutschen,
  Tittmoning 022162
Heimathaus Dingden,
  Hamminkeln 019034
Heimathaus Ebensee, Ebensee . 001835
Heimathaus Gallneukirchen,
  Gallneukirchen 001921
Heimathaus Greven Worth,
  Selsingen 021840
Heimathaus-Handwerkermuseum,
  Scheidegg 021637
Heimathaus im Alten Turm,
  Haslach 002059
Heimathaus Julbach, Julbach . 002152
Heimathaus Mehedorf,
  Bremervörde 017615
Heimathaus mit Photographiemuseum,
  Mariazell 002363
Heimathaus Mörbisch, Mörbisch 002408
Heimathaus Nesselwang,
  Nesselwang 020668
Heimathaus Neuchl-Anwesen,
  Hohenschäftlarn 019309
Heimathaus Neudorf – Seoski muzej,
  Neudorf bei Parndorf 002433
Heimathaus Ollershof, Munster . 020621
Heimathaus Pfarrkirchen,
  Pfarrkirchen 021115
Heimathaus Plettenberg,
  Plettenberg 021155
Heimathaus Richard Eichinger und
  Steinmuseum, Enzenkirchen . 001872
Heimathaus Sittensen, Sittensen 021886
Heimathaus Sonthofen,
  Sonthofen 021920
Heimathaus Spitalskirche/Bürgerspital,
  Bad Leonfelden 001755
Heimathaus-Stadtmuseum
  Vöcklabruck, Vöcklabruck ... 002867
Heimathaus und Schimuseum,
  Saalbach 002615
Heimathaus und Stadtmuseum,
  Perg 002498
Heimathaus Wartberg an der Krems,
  Wartberg an der Krems 002896
Heimathaus Welver, Welver ... 022555
Heimathaus Wenigzell, Wenigzell 002923
Heimathausanlage Schafstall,

Bremervörde 017616
Heimathof Emsbüren, Emsbüren 018197
Heimathof Hüll, Drochtersen ... 018017
Heimatkabinett Westerholt,
  Herten 019231
Heimatkundehaus und Münzkabinett,
  Sankt Marien 002677
Heimatkundesammlung Inzersdorf,
  Herzogenburg 002080
Heimatkundliche Sammlung,
  Uznach 042090
Heimatkundliche Sammlung
  Bergendorf, Holzheim 019326
Heimatkundliche Sammlung Blaustein,
  Blaustein 017417
Heimatkundliche Sammlung Burg
  Rheinfels, Sankt Goar 021595
Heimatkundliche Sammlung Elzach,
  Elzach 018171
Heimatkundliche Sammlung Heideck,
  Heideck 019129
Heimatkundliche Sammlung
  Heroldsbach, Heroldsbach 019223
Heimatkundliche Sammlung Isen,
  Isen 019424
Heimatkundliche Sammlung Pfronten,
  Pfronten 021124
Heimatkundliche Sammlung Strick,
  Bad Mitterndorf 001757
Heimatkundliche Sammlung und
  Ausstellung, Hermannsburg ... 019209
Heimatkundliche Sammlung, Haus
  zum Bären, Richterswil 041873
Heimatkundliche Sammlung, Mühle
  Oberkleinich, Kleinich 019640
Heimatkundliches Museum,
  Friedeburg 018519
Heimatkundliches Museum Medaria,
  Matrei in Osttirol 002369
Heimatkundliches Museum Wetzlhäusl,
  Sankt Gilgen 002659
Heimatmuseum, Abstatt 016458
Heimatmuseum, Adelsdorf 016464
Heimatmuseum, Aesch 041196
Heimatmuseum, Affalterbach . 016472
Heimatmuseum, Ahlen 016475
Heimatmuseum, Aken 016488
Heimatmuseum, Alfdorf 016501
Heimatmuseum, Algermissen .. 016507
Heimatmuseum, Allendorf 016510
Heimatmuseum, Allensbach ... 016511
Heimatmuseum, Altenkirchen, Kreis
  Kusel 016538
Heimatmuseum, Althütte 016547
Heimatmuseum, Altlichtenwarth 001708
Heimatmuseum, Amöneburg .. 016576
Heimatmuseum, Angermünde . 016584
Heimatmuseum, Annaburg ... 016592
Heimatmuseum, Arendsee 016603
Heimatmuseum, Arneburg 016605
Heimatmuseum, Arzberg 001723
Heimatmuseum, Aßlar 016637
Heimatmuseum, Attiswil 041238
Heimatmuseum, Auetal 016644
Heimatmuseum, Bad Aibling .. 016699
Heimatmuseum, Bad Dürrheim . 016754
Heimatmuseum, Bad Eilsen ... 016756
Heimatmuseum, Bad Endbach . 016761
Heimatmuseum, Bad König ... 016798
Heimatmuseum, Bad Laer 016810
Heimatmuseum, Bad
  Langensalza 016812
Heimatmuseum, Bad Lauterberg 016814
Heimatmuseum, Bad Oldesloe . 016840
Heimatmuseum, Bad Rodach .. 016848
Heimatmuseum, Bad Schandau 016864
Heimatmuseum, Bad Teinach-
  Zavelstein 016890
Heimatmuseum, Bad Überkingen 016892
Heimatmuseum, Bad Vilbel 016896
Heimatmuseum, Balingen 016938
Heimatmuseum, Bammental ... 016960
Heimatmuseum, Baunach 016973
Heimatmuseum, Beelitz 017013

Heimatmuseum, Belgern 017021
Heimatmuseum, Bergneustadt . 017061
Heimatmuseum, Berkatal 017063
Heimatmuseum, Bermel 017323
Heimatmuseum, Bernau bei
  Berlin 017325
Heimatmuseum, Biebesheim am
  Rhein 017371
Heimatmuseum, Blaubeuren ... 017412
Heimatmuseum, Bleicherode ... 017422
Heimatmuseum, Blindheim ... 017424
Heimatmuseum, Bobenheim-
  Roxheim 017427
Heimatmuseum, Bodenfelde ... 017445
Heimatmuseum, Brackenheim .. 017529
Heimatmuseum, Breckerfeld ... 017569
Heimatmuseum, Brigachtal ... 017626
Heimatmuseum, Brüssow 017646
Heimatmuseum, Brunsbüttel ... 017648
Heimatmuseum, Buchenberg .. 017653
Heimatmuseum, Buchloe 017656
Heimatmuseum, Büddenstedt .. 017661
Heimatmuseum, Burg-Reuland . 003503
Heimatmuseum, Burg Stargard . 017682
Heimatmuseum, Calau 017710
Heimatmuseum, Camburg 017717
Heimatmuseum, Crivitz 017785
Heimatmuseum, Dachwig 017800
Heimatmuseum, Dahlen 017804
Heimatmuseum, Dahme 017806
Heimatmuseum, Darlingerode . 017817
Heimatmuseum, Denkendorf .. 017854
Heimatmuseum, Dienheim ... 017884
Heimatmuseum, Dietenhofen .. 017890
Heimatmuseum, Dietzenbach .. 017894
Heimatmuseum, Dirlewang 017910
Heimatmuseum, Dömitz 017920
Heimatmuseum, Dohna 017924
Heimatmuseum, Domersleben . 017925
Heimatmuseum, Donzhausen .. 041445
Heimatmuseum, Dossenheim .. 017969
Heimatmuseum, Duderstadt ... 018020
Heimatmuseum, Dunningen ... 018066
Heimatmuseum, Ebermannstadt 018076
Heimatmuseum, Ebern 018077
Heimatmuseum, Echzell 018088
Heimatmuseum, Eggenstein-
  Leopoldshafen 018101
Heimatmuseum, Eibelstadt ... 018111
Heimatmuseum, Eich 018113
Heimatmuseum, Eichenzell ... 018116
Heimatmuseum, Eisenbach ... 018138
Heimatmuseum, Ellrich 018157
Heimatmuseum, Elsenfeld 018166
Heimatmuseum, Eltmann 018168
Heimatmuseum, Elze 018172
Heimatmuseum, Emskirchen .. 018201
Heimatmuseum, Epfenbach ... 018216
Heimatmuseum, Eppelborn 018218
Heimatmuseum, Ergoldsbach .. 018250
Heimatmuseum, Eriskirch 018251
Heimatmuseum, Erlach 001873
Heimatmuseum, Eschenburg .. 018275
Heimatmuseum, Eschwege 018278
Heimatmuseum, Falkensee 018335
Heimatmuseum, Falkenstein,
  Vogtland 018340
Heimatmuseum, Finsterbergen . 018364
Heimatmuseum, Föhren 018377
Heimatmuseum, Forchheim,
  Breisgau 018379
Heimatmuseum, Freiamt 018467
Heimatmuseum, Freigericht ... 018493
Heimatmuseum, Freudenstadt .. 018505
Heimatmuseum, Friedewald ... 018520
Heimatmuseum, Friedland 018521
Heimatmuseum, Friedrichsdorf . 018524
Heimatmuseum, Friesenheim .. 018531
Heimatmuseum, Fußgönheim .. 018574
Heimatmuseum, Gadebusch ... 018577
Heimatmuseum, Gaildorf 018584
Heimatmuseum, Garbsen 018586
Heimatmuseum, Geisa 018606
Heimatmuseum, Geislingen an der

Steige 018613
Heimatmuseum, Geretsried 018654
Heimatmuseum, Gersfeld 018667
Heimatmuseum, Gerstungen ... 018673
Heimatmuseum, Geversdorf ... 018683
Heimatmuseum, Giesen 018687
Heimatmuseum, Gimbsheim ... 018698
Heimatmuseum, Ginsheim-
  Gustavsburg 018701
Heimatmuseum, Glonn 018716
Heimatmuseum, Görwihl 018740
Heimatmuseum, Gößnitz 018742
Heimatmuseum, Graben-Neudorf 018780
Heimatmuseum, Grafing bei
  München 018792
Heimatmuseum, Gransee 018795
Heimatmuseum, Greußen 018813
Heimatmuseum, Grindelwald .. 041567
Heimatmuseum, Großenlüder .. 018848
Heimatmuseum, Großröhrsdorf . 018859
Heimatmuseum, Grüningen ... 041568
Heimatmuseum, Grünstadt 018873
Heimatmuseum, Gschwend 018875
Heimatmuseum, Günzburg 018881
Heimatmuseum, Gundheim 018884
Heimatmuseum, Haiger 018925
Heimatmuseum, Haimhausen . 018929
Heimatmuseum, Hallau 041580
Heimatmuseum, Hallstadt 018972
Heimatmuseum, Haßloch 019092
Heimatmuseum, Haynsburg ... 019119
Heimatmuseum, Hebertsfelden . 019120
Heimatmuseum, Heiligenhafen . 019167
Heimatmuseum, Herbrechtingen 019198
Heimatmuseum, Hergensweiler . 019205
Heimatmuseum, Hettenleidelheim 019237
Heimatmuseum, Heuchelheim bei
  Frankenthal 019242
Heimatmuseum, Heusenstamm . 019243
Heimatmuseum, Hillscheid 019253
Heimatmuseum, Himmelpforten 019259
Heimatmuseum, Hohberg 019296
Heimatmuseum, Hohenroda .. 019308
Heimatmuseum, Hohenwestedt . 019314
Heimatmuseum, Holle 019319
Heimatmuseum, Holzgerlingen . 019324
Heimatmuseum, Hornburg 019343
Heimatmuseum, Hoya / Weser . 019347
Heimatmuseum, Igel 019380
Heimatmuseum, Ihringen 019381
Heimatmuseum, Ilvesheim 019391
Heimatmuseum, Immendingen . 019392
Heimatmuseum, Immenstaad am
  Bodensee 019394
Heimatmuseum, Kaprun 002158
Heimatmuseum, Karlstein am
  Main 019525
Heimatmuseum, Kastl, Kreis Amberg-
  Sulzbach 019554
Heimatmuseum, Kefenrod 019567
Heimatmuseum, Kehl 019571
Heimatmuseum, Keltern 019580
Heimatmuseum, Kenn 019591
Heimatmuseum, Kirchen 019621
Heimatmuseum, Kirchhain 019623
Heimatmuseum, Klipphausen .. 019650
Heimatmuseum, Kölleda 019671
Heimatmuseum, Königs
  Wusterhausen 019717
Heimatmuseum, Königsdorf ... 019727
Heimatmuseum, Kraiburg am Inn 019775
Heimatmuseum, Kroppenstedt . 019809
Heimatmuseum, Küssnacht am
  Rigi 041631
Heimatmuseum, Kuhardt 019824
Heimatmuseum, Ladbergen ... 019833
Heimatmuseum, Landau an der
  Isar 019868
Heimatmuseum, Langenzenn .. 019916
Heimatmuseum, Leimersheim .. 019960
Heimatmuseum, Leipheim 019971
Heimatmuseum, Lengenfeld ... 020035
Heimatmuseum, Lenggries 020036
Heimatmuseum, Leutenbach ... 020050

Heimatmuseum, Lilienthal 020070
Heimatmuseum, Lommatzsch . . 020112
Heimatmuseum, Lorch,
Württemberg 020116
Heimatmuseum, Luckenwalde . . 020123
Heimatmuseum, Lügde 020175
Heimatmuseum, Magstadt . . . 020215
Heimatmuseum, Mainz-Kostheim 020236
Heimatmuseum, March 020278
Heimatmuseum, Markgröningen 020286
Heimatmuseum, Markranstädt . 020292
Heimatmuseum, Marktl 020297
Heimatmuseum, Maßbach . . . 020314
Heimatmuseum, Mauthausen . . 002378
Heimatmuseum, Meerane . . . 020325
Heimatmuseum, Mehrstetten . . 020339
Heimatmuseum, Meinhard . . . 020340
Heimatmuseum, Mellrichstadt . . 020356
Heimatmuseum, Melsungen . . 020358
Heimatmuseum, Merchweiler . . 020370
Heimatmuseum, Mering 020371
Heimatmuseum, Merkendorf . . . 020372
Heimatmuseum, Merzenich . . . 020378
Heimatmuseum, Miesbach . . 020404
Heimatmuseum, Millstatt 002388
Heimatmuseum, Mindelheim . . 020409
Heimatmuseum, Mitterfels . . . 020424
Heimatmuseum, Möckmühl . . 020432
Heimatmuseum, Montlingen . . 041751
Heimatmuseum, Moosburg an der
Isar 020462
Heimatmuseum, Moringen . . . 020466
Heimatmuseum, Mügeln 020477
Heimatmuseum, Müllrose . . . 020500
Heimatmuseum, Münsingen . . 020589
Heimatmuseum, Murg 020623
Heimatmuseum, Nauheim . . . 020643
Heimatmuseum, Naumburg,
Hessen 020644
Heimatmuseum, Neckargerach . 020655
Heimatmuseum, Neidenstein . . 020658
Heimatmuseum, Nellingen . . . 020659
Heimatmuseum, Neu-Ulm 020679
Heimatmuseum, Neukirch/Lausitz 020713
Heimatmuseum, Neukirchen,
Schwalm-Eder-Kreis 020716
Heimatmuseum, Neumagen-
Dhron 020721
Heimatmuseum, Neunkirchen, Neckar-
Odenwald-Kreis 020730
Heimatmuseum, Neupotz 020733
Heimatmuseum, Neureichenau . 020735
Heimatmuseum, Neuried,
Ortenaukreis 020736
Heimatmuseum, Neustadt in
Sachsen 020771
Heimatmuseum, Neustadt, Wied 020772
Heimatmuseum, Nidda 020781
Heimatmuseum, Nideggen . . . 020785
Heimatmuseum, Nieder-Olm . . 020789
Heimatmuseum, Niederaichbach 020790
Heimatmuseum, Niederhall . . . 020797
Heimatmuseum, Niederstetten . 020799
Heimatmuseum, Northeim 020836
Heimatmuseum, Nüdlingen . . . 020840
Heimatmuseum, Ober-Flörsheim 020885
Heimatmuseum, Oberderdingen 020891
Heimatmuseum, Oberndorf am
Lech 020906
Heimatmuseum, Oberndorf am
Neckar 020907
Heimatmuseum, Obernfeld . . . 020909
Heimatmuseum, Oberstdorf . . . 020930
Heimatmuseum, Ockenheim . . 020945
Heimatmuseum, Oederan 020948
Heimatmuseum, Oettingen in
Bayern 020967
Heimatmuseum, Osterhofen . . 021026
Heimatmuseum, Osterwieck . . 021030
Heimatmuseum, Ostrach 021035
Heimatmuseum, Otterberg . . . 021041
Heimatmuseum, Papenburg . . 021072
Heimatmuseum, Perl 021103
Heimatmuseum, Perleberg . . . 021105

Heimatmuseum, Pfaffenhausen . 021109
Heimatmuseum, Pfullendorf . . . 021125
Heimatmuseum, Philippsburg . . 021130
Heimatmuseum, Plankstadt . . . 021144
Heimatmuseum, Polch 021167
Heimatmuseum, Polling 021169
Heimatmuseum, Postbauer-Heng 021172
Heimatmuseum, Preetz 021211
Heimatmuseum, Pressath 021215
Heimatmuseum, Pretzsch 021218
Heimatmuseum, Prien am
Chiemsee 021222
Heimatmuseum, Pritzwalk 021224
Heimatmuseum, Queidersbach . 021250
Heimatmuseum, Quierschied . . 021253
Heimatmuseum, Radeburg . . . 021259
Heimatmuseum, Radevormwald 021261
Heimatmuseum, Radstadt 002560
Heimatmuseum, Rattenberg . . 021291
Heimatmuseum, Raunheim . . . 021297
Heimatmuseum, Rauschenberg . 021299
Heimatmuseum, Rehlingen-
Siersburg 021339
Heimatmuseum, Reilingen 021348
Heimatmuseum, Reinach (Basel-
Land) 041870
Heimatmuseum, Reinheim 021352
Heimatmuseum, Renningen . . 021376
Heimatmuseum, Rheda-
Wiedenbrück 021390
Heimatmuseum, Rieneck 021427
Heimatmuseum, Rietz 002599
Heimatmuseum, Rodenberg . . 021444
Heimatmuseum, Rodewald . . . 021445
Heimatmuseum, Rötha 021458
Heimatmuseum, Rötz 021460
Heimatmuseum, Ronnenberg . . 021470
Heimatmuseum, Roßwein 021485
Heimatmuseum, Rotenburg,
Wümme 021500
Heimatmuseum, Rothrist 041893
Heimatmuseum, Sandhausen . . 021582
Heimatmuseum, Sankt Aegyd am
Neuwalde 002644
Heimatmuseum, Sankt Egidien . 021590
Heimatmuseum, Sankt Martin am
Wöllmißberg 002680
Heimatmuseum, Schalkau 021628
Heimatmuseum, Scheeßel 021634
Heimatmuseum, Schermbeck . 021642
Heimatmuseum, Schifferstadt . . 021645
Heimatmuseum, Schliersee . . . 021672
Heimatmuseum, Schlitz 021673
Heimatmuseum, Schnaittach . . 021689
Heimatmuseum, Schöneck /
Vogtland 021705
Heimatmuseum, Schöneiche bei
Berlin 021707
Heimatmuseum, Schöningen . . 021712
Heimatmuseum, Schwalmtal . . 021762
Heimatmuseum, Schwanenstadt 002732
Heimatmuseum, Schwanfeld . . 021765
Heimatmuseum, Seeg 021814
Heimatmuseum, Seelze 021819
Heimatmuseum, Sersheim . . . 021844
Heimatmuseum, Seßlach 021847
Heimatmuseum, Siegbach 021848
Heimatmuseum, Simbach am Inn 021864
Heimatmuseum, Sinzig 021883
Heimatmuseum, Sohland an der
Spree 021896
Heimatmuseum, Spangenberg . 021926
Heimatmuseum, Speicher 021928
Heimatmuseum, Sprendlingen . 021941
Heimatmuseum, Stade 021946
Heimatmuseum, Stadt Wehlen . 021951
Heimatmuseum, Stadtilm 021954
Heimatmuseum, Stadtsteinach . 021959
Heimatmuseum, Staufenberg . 021966
Heimatmuseum, Steinbach,
Taunus 021979
Heimatmuseum, Sternberg . . . 021997
Heimatmuseum, Stockstadt am
Main 022001

Heimatmuseum, Strasburg 022012
Heimatmuseum, Strausberg . . 022015
Heimatmuseum, Stromberg . . 022018
Heimatmuseum, Talheim 022096
Heimatmuseum, Tambach-
Dietharz 022097
Heimatmuseum, Todtmoos . . . 022163
Heimatmuseum, Trebur 022181
Heimatmuseum,
Trechtingshausen 022183
Heimatmuseum, Treffurt 022185
Heimatmuseum, Triptis 022207
Heimatmuseum, Undenheim . . 022271
Heimatmuseum,
Unterschneidheim 022282
Heimatmuseum, Vallendar 022299
Heimatmuseum, Varel 022302
Heimatmuseum, Velburg 022311
Heimatmuseum, Venningen . . . 022316
Heimatmuseum, Viernheim . . . 022329
Heimatmuseum, Wadern 022361
Heimatmuseum, Wächtersbach . 022365
Heimatmuseum, Wald 042119
Heimatmuseum, Waldaschaff . 022379
Heimatmuseum, Waldfischbach-
Burgalben 022392
Heimatmuseum, Waldheim . . . 022393
Heimatmuseum, Waldsee 022402
Heimatmuseum, Waldthurn . . . 022403
Heimatmuseum, Wanna 022425
Heimatmuseum, Wedel 022448
Heimatmuseum, Weener 022450
Heimatmuseum, Weisendorf . . 022530
Heimatmuseum, Weissach im Tal 022533
Heimatmuseum, Weißenborn . . 022535
Heimatmuseum, Wemding 022558
Heimatmuseum, Wennigsen . . 022565
Heimatmuseum, Wertach 022586
Heimatmuseum, Wertingen . . . 022589
Heimatmuseum, Westerhausen . 022598
Heimatmuseum, Westheim . . . 022600
Heimatmuseum, Wilfersdorf . . . 003115
Heimatmuseum, Windeck 022676
Heimatmuseum, Winden 022677
Heimatmuseum, Wippra 022690
Heimatmuseum, Wolfratshausen 022736
Heimatmuseum, Wolfstein 022750
Heimatmuseum, Wusterhausen . 022830
Heimatmuseum, Zarrentin am
Schaalsee 022842
Heimatmuseum, Zemmer 022863
Heimatmuseum, Zerf 022865
Heimatmuseum, Zingst,
Ostseebad 022883
Heimatmuseum, Zorge 022894
Heimatmuseum, Zusmarshausen 022904
Heimatmuseum – Traktoren-
und Landmaschinenmuseum,
Großengottern 018846
Heimatmuseum "Grafschaft Molbach",
Kreuzau 019794
Heimatmuseum Aarburg, Aarburg 041192
Heimatmuseum Achental,
Achenkirch 001683
Heimatmuseum Adelboden,
Adelboden 041194
Heimatmuseum Adlhoch-Haus, Altdorf,
Kreis Landshut 016523
Heimatmuseum Adolphshöhe,
Wallerfangen 022409
Heimatmuseum Ahlem, Hannover 019052
Heimatmuseum Ahrbergen,
Giesen 018688
Heimatmuseum Allschwil,
Allschwil 041208
Heimatmuseum Alte Weberstube,
Niedercunnersdorf 020792
Heimatmuseum Alte Windmühle Groß
Reken, Reken 021355
Heimatmuseum Alter Pfarrhof,
Raisting 021268
Heimatmuseum Altes Rathaus, Böhl-
Iggelheim 017457
Heimatmuseum Altes Schulhaus, Böhl-

Iggelheim 017458
Heimatmuseum am
Kobernaußerwald, Lohnsburg am
Kobernaußerwald 002340
Heimatmuseum am Pfäffikersee,
Pfäffikon (Zürich) 041838
Heimatmuseum Amt Blankenstein,
Gladenbach 018705
Heimatmuseum Arnold Bärtschi,
Dulliken 041450
Heimatmuseum Aschen, Diepholz 017885
Heimatmuseum Assel,
Drochtersen 018018
Heimatmuseum auf dem Klosterhof,
Steinheim am Albuch 021986
Heimatmuseum Bad Großpertholz, Bad
Großpertholz 001746
Heimatmuseum Barbaramühle,
Longkamp 020113
Heimatmuseum Berching mit Ritter-
v.-Gluck-Archiv, Berching . . . 017036
Heimatmuseum Bergen-Enkheim,
Frankfurt am Main 018424
Heimatmuseum Bernhardsthal,
Bernhardsthal 001779
Heimatmuseum Beuel, Bonn . . 017491
Heimatmuseum Beutelsbach,
Weinstadt 022525
Heimatmuseum Bezau, Bezau . 001782
Heimatmuseum Biebrich,
Wiesbaden 022632
Heimatmuseum Bierstadt,
Wiesbaden 022633
Heimatmuseum Bislich, Wesel . 022590
Heimatmuseum Blankenau,
Hosenfeld 019346
Heimatmuseum Blankenloch,
Stutensee 022024
Heimatmuseum Bockwindmühle,
Lebusa 019947
Heimatmuseum Börgerende,
Börgerende-Rethwisch 017464
Heimatmuseum Boizenburg,
Boizenburg 017474
Heimatmuseum Boizenburg,
Außenstelle am Elbberg,
Boizenburg 017475
Heimatmuseum Borgloh, Hilter am
Teutoburger Wald 019256
Heimatmuseum Boxberg,
Boxberg 017526
Heimatmuseum Brueckenhof,
Königswinter 019738
Heimatmuseum Bucheggberg,
Buchegg 041376
Heimatmuseum Bühl, Baden,
Bühl 017666
Heimatmuseum Carl Swoboda,
Schirgiswalde 021656
Heimatmuseum Charlottenburg-
Wilmersdorf, Berlin 017161
Heimatmuseum Crumstadt,
Riedstadt 021422
Heimatmuseum Dachtel,
Aidlingen 016485
Heimatmuseum Davos, Davos
Dorf 041432
Heimatmuseum Debstedt, Langen,
Kreis Cuxhaven 019894
Heimatmuseum Delkenheim,
Wiesbaden 022634
Heimatmuseum der Albert-Edelmann-
Stiftung, Ebnat-Kappel 041452
Heimatmuseum der Deutschen aus
Bessarabien, Stuttgart 022037
Heimatmuseum der Gemeinde
Weißenkirchen, Perschling . . . 002499
Heimatmuseum der Hauptschule
Taufkirchen, Taufkirchen an der
Pram 002812
Heimatmuseum der Insel Hiddensee,
Kloster/ Hiddensee 019657
Heimatmuseum der Parabutscher
Donauschwaben, Bad

Schönborn 016867
Heimatmuseum der Stadt Bad Tölz,
Bad Tölz 016891
Heimatmuseum der Stadt Gernsheim,
Gernsheim 018663
Heimatmuseum der Stadt Herrnhut,
Herrnhut 019227
Heimatmuseum der Stadt Ketzin,
Ketzin 019596
Heimatmuseum der Stadt Rerik, Rerik,
Ostseebad 021378
Heimatmuseum der Stadt Rheinau in
Freistett, Rheinau 021394
Heimatmuseum der Stadt Wilsdruff,
Wilsdruff 022672
Heimatmuseum der Talschaft
Lauterbrunnen, Lauterbrunnen 041664
Heimatmuseum der
Verbandsgemeinde Wallhalben,
Schmitshausen 021687
Heimatmuseum des Daadener Landes,
Daaden 017793
Heimatmuseum des Heimatbundes
Jauernig, Vaihingen an der Enz 022295
Heimatmuseum des Landkreises
Regensburg, Altenthann 016542
Heimatmuseum des Nordböhmischen
Niederlandes, Böblingen 017453
Heimatmuseum Dietingen,
Dietingen 017893
Heimatmuseum Dykhus, Borkum 017516
Heimatmuseum Ebenhausen,
Oerlenbach 020962
Heimatmuseum Ebersbach in der
Humboldt-Baude, Ebersbach,
Sachsen 018080
Heimatmuseum Egg, Egg 001842
Heimatmuseum Eisenbach, Obernburg
am Main 020904
Heimatmuseum Elgg, Elgg 041461
Heimatmuseum Elsbethen,
Elsbethen 001866
Heimatmuseum Erbenheim,
Wiesbaden 022635
Heimatmuseum Esterau,
Holzappel 019323
Heimatmuseum Eversberg,
Meschede 020383
Heimatmuseum Flacht, Weissach 022532
Heimatmuseum Frankenberg,
Frankenberg, Sachsen 018391
Heimatmuseum Freudenthal,
Memmingen 020361
Heimatmuseum Friedrichshain,
Berlin 017162
Heimatmuseum Fronfeste, Limbach-
Oberfrohna 020074
Heimatmuseum Fügen, Fügen .. 001911
Heimatmuseum für Stadt- und
Landkreis, Kirchheimbolanden . 019629
Heimatmuseum für Stadt und Kreis
Gelnhausen, Gelnhausen 018620
Heimatmuseum Gaaden, Gaaden 001917
Heimatmuseum Gablitz, Gablitz . 001919
Heimatmuseum Gadernheim,
Lautertal 019945
Heimatmuseum Garbenheim,
Wetzlar 022608
Heimatmuseum Gersau, Gersau 041543
Heimatmuseum Gnas, Gnas ... 001956
Heimatmuseum Grabow, Grabow 018782
Heimatmuseum Grafenwörth,
Grafenwörth 001963
Heimatmuseum Grafling,
Oberschneiding 020927
Heimatmuseum Grahhof, Ramsau am
Dachstein 002567
Heimatmuseum Greene,
Kreiensen 019790
Heimatmuseum Grimmen "Im
Mühlentor", Grimmen 018824
Heimatmuseum Gröbming,
Gröbming 001995
Heimatmuseum Groß-Enzersdorf, Groß-

Enzersdorf 001998
Heimatmuseum Großengersdorf,
Großengersdorf 002003
Heimatmuseum Großes Walsertal,
Sonntag 002761
Heimatmuseum Großschönau,
Großschönau 002015
Heimatmuseum Großseelheim,
Kirchhain 019624
Heimatmuseum Großzschepa,
Hohburg 019298
Heimatmuseum Guntramsdorf,
Guntramsdorf 002022
Heimatmuseum Hadersdorf,
Hadersdorf 002035
Heimatmuseum Harthausen,
Winterlingen 022688
Heimatmuseum Haueneberstein,
Baden-Baden 016921
Heimatmuseum Haus der
Südostdeutschen, Böchingen . 017456
Heimatmuseum Haus Horn,
Niederzier 020805
Heimatmuseum Heidelsheim,
Bruchsal 017632
Heimatmuseum Heinrich Zoller,
Wittelshofen 022695
Heimatmuseum Helfs Hof,
Bochum 017435
Heimatmuseum Herzogenweiler,
Villingen-Schwenningen 022336
Heimatmuseum Heuchelheim-
Kinzenbach, Heuchelheim ... 019241
Heimatmuseum Hochzeitshaus,
Homberg (Efze) 019331
Heimatmuseum Höchstädt, Höchstädt
an der Donau 019276
Heimatmuseum Hoheneggelsen,
Söhlde 021888
Heimatmuseum Holzgau, Holzgau 002103
Heimatmuseum Hopfenhaus,
Aidlingen 016486
Heimatmuseum Hossingen,
Meßstetten 020389
Heimatmuseum Hus tu Löwenberg,
Löwenberger Land 020099
Heimatmuseum Ilz, Ilz 002111
Heimatmuseum Im Ackerbürgerhaus,
Vohenstrauß 022353
Heimatmuseum im alten Bruderhaus
und in der Dechanathofscheune,
Altenmarkt im Pongau 001705
Heimatmuseum im Alten Kloster,
Erbendorf 018231
Heimatmuseum im Alten Leuchtturm,
Wangerooge, Nordseebad ... 022424
Heimatmuseum im alten Marktturm,
Fischamend 001891
Heimatmuseum im alten Pfarrhaus,
Schwieberdingen 021804
Heimatmuseum im Alten Rathaus,
Bruchköbel 017630
Heimatmuseum im alten Zollhaus,
Kobersdorf 002215
Heimatmuseum im Amonhaus,
Lunz 002346
Heimatmuseum im Berchtoldshof,
Uhingen 022255
Heimatmuseum im Einlegerhaus,
Obertrum am See 002470
Heimatmuseum im Fünfeckigen Turm,
Neckarbischofsheim 020651
Heimatmuseum im Hartmannhaus,
Marktoberdorf 020298
Heimatmuseum im Haus Catoir, Bad
Dürkheim 016749
Heimatmuseum im Herrenhaus,
Burghaun 017686
Heimatmuseum im Hintermeierhaus,
Donauwörth 017933
Heimatmuseum im Kolb'schen Haus,
Bickenbach 017365
Heimatmuseum im Oberen Torturm,
Lauchheim 019927

Heimatmuseum im Schelfenhaus,
Volkach 022355
Heimatmuseum im Schillerhaus,
Oberkochen 020902
Heimatmuseum im Schlösschen,
Ochsenfurt 020941
Heimatmuseum im Schloß,
Zörbig 022892
Heimatmuseum im Studierstüberl,
Krieglach 002243
Heimatmuseum im Tuchmacherhaus,
Thannhausen 022143
Heimatmuseum im Unteren Turm,
Leutershausen-Wiedersbach . 020053
Heimatmuseum im Vogteigebäude,
Hüttlingen 019364
Heimatmuseum im Wachtturm,
Geyer 018684
Heimatmuseum im Wiggensbacher
Informationszentrum,
Wiggensbach 022649
Heimatmuseum im Wolfschneider-Hof,
Taufkirchen, Kreis München . 022112
Heimatmuseum in der Mühle Gettorf,
Gettorf 018681
Heimatmuseum in der Scheune,
Mössingen 020447
Heimatmuseum Ingersleben/
Neudietendorf, Ingersleben ... 019404
Heimatmuseum Ittersbach,
Karlsbad 019500
Heimatmuseum Kalchofengut,
Unken 002854
Heimatmuseum Kals, Kals am
Großglockner 002155
Heimatmuseum Kappe,
Zehdenick 022843
Heimatmuseum Karlsdorf, Karlsdorf-
Neuthard 019501
Heimatmuseum Katsdorf,
Katsdorf 002161
Heimatmuseum Kaumberg,
Kaumberg 002163
Heimatmuseum Kaunertal,
Kaunertal 002164
Heimatmuseum Kautzen,
Kautzen 002165
Heimatmuseum Kematen, Kematen an
der Ybbs 002166
Heimatmuseum Kiechlinsbergen,
Endingen am Kaiserstuhl 018203
Heimatmuseum Kirberg,
Hünfelden 019359
Heimatmuseum Kirchdorf, Insel
Poel 019416
Heimatmuseum Kirchham,
Kirchham 002176
Heimatmuseum Kirchheim,
Heidelberg 019135
Heimatmuseum Kirchschlag,
Kirchschlag 002177
Heimatmuseum Klösterle, Schönau im
Schwarzwald 021697
Heimatmuseum Klosterlangheim,
Lichtenfels 020061
Heimatmuseum Klus, Klus 041618
Heimatmuseum Köpenick, Berlin 017163
Heimatmuseum Kohbauernhaus,
Köstendorf 002220
Heimatmuseum Kornburg,
Nürnberg 020856
Heimatmuseum Kuchl, Kuchl ... 002249
Heimatmuseum Lahnau-Waldgirmes,
Lahnau 019847
Heimatmuseum Laisa,
Battenberg 016970
Heimatmuseum Lamspringe,
Lamspringe 019865
Heimatmuseum Langelsheim,
Langelsheim 019890
Heimatmuseum Langenau,
Langenau 019899
Heimatmuseum Langenlois,
Langenlois 002263

Heimatmuseum Langenseifen, Bad
Schwalbach 016874
Heimatmuseum Langenselbold,
Langenselbold 019915
Heimatmuseum Langenzersdorf,
Langenzersdorf 002268
Heimatmuseum Laubach,
Fridericianum, Laubach 019923
Heimatmuseum Lauchhammer,
Lauchhammer 019925
Heimatmuseum Lauda-Königshofen,
Lauda-Königshofen 019928
Heimatmuseum Lauenau,
Lauenau 019929
Heimatmuseum Leeheim,
Riedstadt 021423
Heimatmuseum Leer, Leer 019950
Heimatmuseum Lembach,
Lembach 020288
Heimatmuseum Lengsdorf, Bonn 017492
Heimatmuseum Lette, Coesfeld . 017756
Heimatmuseum Leuk Stadt, Leuk
Stadt 041669
Heimatmuseum Lich, Lich 020059
Heimatmuseum Lichtenberg,
Berlin 017164
Heimatmuseum Liebenwalde im
historischen Stadtgefängnis,
Liebenwalde 020066
Heimatmuseum Litschau,
Litschau 002334
Heimatmuseum Losenstein,
Losenstein 002343
Heimatmuseum Lübbecke und
Kunstgalerie im Speicher,
Lübbecke 020149
Heimatmuseum Lütgendortmund,
Dortmund 017957
Heimatmuseum Mank, Mank .. 002352
Heimatmuseum Marchegg,
Marchegg 002355
Heimatmuseum Markt Piesting, Markt
Piesting 002367
Heimatmuseum Marner Skatklub von
1873, Marne 020307
Heimatmuseum Mautern, Mautern in
Steiermark 002376
Heimatmuseum Medenbach,
Wiesbaden 022636
Heimatmuseum Meerholz,
Gelnhausen 018621
Heimatmuseum-Menton,
Teningen 022129
Heimatmuseum Metzgerhaus,
Kirchdorf 002175
Heimatmuseum Michelbacher
Schlößchen, Alzenau in
Unterfranken 016567
Heimatmuseum Michelhausen,
Michelhausen 002386
Heimatmuseum mit Folterkammer,
Pöggstall 002516
Heimatmuseum mit Franz Jonas-
Gedenkraum, Teesdorf 002813
Heimatmuseum mit ostdeutscher
Heimatstube, Hanerau-
Hademarschen 019045
Heimatmuseum mit Raimund
von Montecuccoli-Gedenkraum,
Hafnerbach 002040
Heimatmuseum Mittelbuchen,
Hanau 019036
Heimatmuseum Möhringen,
Stuttgart 022038
Heimatmuseum Münchingen, Korntal-
Münchingen 019769
Heimatmuseum Mureck, Mureck 002426
Heimatmuseum Naunheim,
Wetzlar 022609
Heimatmuseum Naurod,
Wiesbaden 022637
Heimatmuseum Neckarau,
Mannheim 020247
Heimatmuseum Neuenwalde, Langen,

Kreis Cuxhaven 019895
Heimatmuseum Neunkirchen,
Neunkirchen 002447
Heimatmuseum Neustadtgödens,
Sande 021581
Heimatmuseum Nied, Frankfurt am
Main 018425
Heimatmuseum Niederleis,
Niederleis 002452
Heimatmuseum Niederndorf,
Niederndorf 002454
Heimatmuseum Niemes und
Prachatitz, Ingolstadt 019409
Heimatmuseum Nürnberg,
Nürnberg 020857
Heimatmuseum Nutli-Hüschi,
Klosters 041616
Heimatmuseum Obbornhofen,
Hungen 019365
Heimatmuseum Oberbichl,
Prägraten 002529
Heimatmuseum Obere Mühle,
Loßburg 020121
Heimatmuseum Obergünzburg
mit Südseesammlung,
Obergünzburg 020894
Heimatmuseum Oberhofen, Oberhofen
im Inntal 002460
Heimatmuseum Obermühle, Ach 001682
Heimatmuseum Oberwölz und
Österreichisches Blasmusikmuseum,
Oberwölz 002473
Heimatmuseum Ole Schüne,
Drakenburg 017971
Heimatmuseum Oltingen-Wenslingen-
Anwil, Oltingen 041827
Heimatmuseum Pabneukirchen,
Pabneukirchen 002486
Heimatmuseum Peldemühle,
Wittmund 022705
Heimatmuseum Persenbeug-Gottsdorf,
Persenbeug 002500
Heimatmuseum Pfarrscheuer,
Römerstein 021455
Heimatmuseum Pfunds, Pfunds . 002506
Heimatmuseum Plieningen,
Stuttgart 022039
Heimatmuseum Prättigau,
Grüsch 041571
Heimatmuseum Pressbaum,
Pressbaum 002540
Heimatmuseum Prestegg,
Altstätten 041214
Heimatmuseum Prieros,
Heidesee 019156
Heimatmuseum Pulkau, Pulkau . 002545
Heimatmuseum Purkersdorf,
Purkersdorf 002551
Heimatmuseum Rabenstein,
Rabenstein an der Pielach . . . 002557
Heimatmuseum Rain am Lech,
Rain 021266
Heimatmuseum Rapperswil,
Rapperswil 041859
Heimatmuseum Ratzenried,
Argenbühl 016604
Heimatmuseum Rauchkate, Zetel 022866
Heimatmuseum Rauchstubenhaus
Edelschachen, Anger 001713
Heimatmuseum Reinickendorf,
Berlin 017165
Heimatmuseum Reischenau,
Dinkelscherben 017906
Heimatmuseum Reiterhaus, Neusalza-
Spremberg 020742
Heimatmuseum Reith, Reith im
Alpbachtal 002587
Heimatmuseum Reutlingen,
Reutlingen 021381
Heimatmuseum Rheiderland,
Weener 022451
Heimatmuseum Rheinwald,
Splügen 042015
Heimatmuseum Rhüden, Seesen 021821

Heimatmuseum Rodenstein, Fränkisch-
Crumbach 018386
Heimatmuseum Römstedthaus,
Bergen, Kreis Celle 017046
Heimatmuseum Rohrbach,
Heidelberg 019136
Heimatmuseum Rohrhirschmühle,
Bühl 017667
Heimatmuseum Rother Hof,
Pottendorf 002522
Heimatmuseum Runder Turm,
Sigmaringen 021862
Heimatmuseum Sammlung Fuchs,
Kirchbach in Steiermark 002169
Heimatmuseum Sandweier, Baden-
Baden 016922
Heimatmuseum Sanhofen,
Mannheim 020248
Heimatmuseum Sankt Laurentius,
Dahlenburg 017805
Heimatmuseum Schanfigg, Arosa 041232
Heimatmuseum Schierstein,
Wiesbaden 022638
Heimatmuseum Schinznach-Dorf,
Schinznach Dorf 041953
Heimatmuseum Schloß Brenz,
Sontheim an der Brenz 021918
Heimatmuseum Schloß Fels, Fels am
Wagram 001886
Heimatmuseum Schloß Hochhaus,
Vorchdorf 002877
Heimatmuseum Schloß Schönebeck,
Bremen 017590
Heimatmuseum Schloß Wolkenstein,
Wolkenstein 022754
Heimatmuseum Schönau,
Günselsdorf 002018
Heimatmuseum Schrems,
Schrems 002729
Heimatmuseum Schulte-Wessels,
Neuenhaus 020699
Heimatmuseum Schwanheim,
Frankfurt am Main 018426
Heimatmuseum Schwarzbubenland,
Dornach 041446
Heimatmuseum Schwarzes Tor, Sankt
Georgener Museen, Sankt Georgen
im Schwarzwald 021592
Heimatmuseum Seehof, Buonas 041383
Heimatmuseum Sissach, Sissach 041998
Heimatmuseum Spittel, Büren an der
Aare 041379
Heimatmuseum Sprucker Mühle,
Guben 018876
Heimatmuseum Spycher, Rickenbach
(Luzern) 041875
Heimatmuseum Stadt Bernstadt,
Bernstadt auf dem Eigen 017344
Heimatmuseum Stadt Nagold,
Nagold 020636
Heimatmuseum Stadt Teltow,
Teltow 022123
Heimatmuseum Stadt und
Landkreis Neudek im Erzgebirge,
Augsburg 016656
Heimatmuseum Stammheim,
Stuttgart 022040
Heimatmuseum Stein, Stein am
Rhein 042032
Heimatmuseum Steinbach,
Ottweiler 021051
Heimatmuseum Steinfischbach,
Waldems 022384
Heimatmuseum Steyregg,
Steyregg 002791
Heimatmuseum Strümpfelbach,
Weinstadt 022526
Heimatmuseum Suhr, Suhr 042039
Heimatmuseum Tannheimer Tal,
Tannheim 002808
Heimatmuseum Tarrenz, Tarrenz 002809
Heimatmuseum Thaya, Thaya . . 002824
Heimatmuseum Tieringen,
Meßstetten 020390

Heimatmuseum Traun, Traun . . 002838
Heimatmuseum Treptow, Berlin . 017166
Heimatmuseum Trimmis, Trimmis 042069
Heimatmuseum Trubschachen,
Trubschachen 042071
Heimatmuseum Uelzen mit
Gläsersammlung Röver, Uelzen 022247
Heimatmuseum Uhl'sches Haus,
Göllheim 018726
Heimatmuseum und
Dokumentationszentrum zur
Deutschen Nachkriegsgeschichte,
Wanfried 022417
Heimatmuseum und Galerie Neue
Diele, Jork 019469
Heimatmuseum und Gedenkstätte
Bützow, Bützow 017677
Heimatmuseum und Hedwig Courths-
Mahler Archiv, Nebra 020650
Heimatmuseum und Karl-
August-Forster-Bienenmuseum,
Illertissen 019384
Heimatmuseum und Keltenhaus,
Ligist 002310
Heimatmuseum und Klöpferhaus,
Eibiswald 001849
Heimatmuseum und
landwirtschaftliche Sammlung,
Alberschwende 001694
Heimatmuseum und
Naturalienkabinett, Waldenburg, Kreis
Chemnitzer Land 022388
Heimatmuseum und Schiele Museum,
Neulengbach 002441
Heimatmuseum und Textil-
Schauwerkstatt, Greiz 018810
Heimatmuseum Untergrombach,
Bruchsal 017633
Heimatmuseum Usseln, Willingen 022668
Heimatmuseum Vietze, Höhbeck 019279
Heimatmuseum Vilsburg – Kröninger
Hafnermuseum, Vilsburg 022343
Heimatmuseum Vollmers Mühle,
Seebach 021810
Heimatmuseum Walldorf, Mörfelden-
Walldorf 020442
Heimatmuseum Waltersweier,
Offenburg 020975
Heimatmuseum Wandsbek,
Hamburg 018993
Heimatmuseum Warnemünde,
Rostock 021492
Heimatmuseum Wartberg, Wartberg
im Mürztal 002897
Heimatmuseum Weesen, Weesen 042126
Heimatmuseum Weiler,
Schorndorf 021726
Heimatmuseum Weitenung, Bühl 017668
Heimatmuseum Weng, Weng im
Innkreis 002922
Heimatmuseum Wiesenbach,
Wiesenbach 022643
Heimatmuseum Windischgarsten,
Windischgarsten 003125
Heimatmuseum Winterberg im
Böhmerwald, Freyung 018509
Heimatmuseum Wißmar,
Wettenberg 022604
Heimatmuseum Wörgl, Wörgl . . 003126
Heimatmuseum Wolfskehlen,
Riedstadt 021424
Heimatmuseum Wommelshausen, Bad
Endbach 016762
Heimatmuseum Worben, Worben 042169
Heimatmuseum Worms Abenheim,
Worms 022760
Heimatmuseum Worms Horchheim,
Worms 022761
Heimatmuseum Worms-Weinsheim,
Worms 022762
Heimatmuseum Zehlendorf,
Berlin 017167
Heimatmuseum Zwischen Venn und
Schneifel, Sankt Vith 003998

Heimatmuseum, Arcun da Tradiziun,
Waltensburg 042123
Heimatmuseum, Das lebendige
Volkskundemuseum,
Dornstetten 017943
Heimatmuseum, Goethehaus
Volpertshausen, Hüttenberg . . . 019363
Heimatmuseum, Schloss Dätzingen,
Grafenau, Kreis Böblingen . . . 018784
Heimatmuseum, Stäblistübli,
Brugg 041369
Heimatmuseum, Städtisches Museum,
Mülheim an der Ruhr 020495
Heimatmuseum, Zürcher Unterländer
Museumsverein, Oberwangen 041817
Heimatscheune, Zschepplin . . . 022900
Heimatstätte der Kuhländer,
Ludwigsburg 020126
Heimatstube, Altenbrak 016533
Heimatstube, Altenstadt an der
Waldnaab 016540
Heimatstube, Benneckenstein . . 017030
Heimatstube, Brüel 017640
Heimatstube, Calbe, Saale 017712
Heimatstube, Dabel 017794
Heimatstube, Derenburg 017860
Heimatstube, Ditfurt 017914
Heimatstube, Eggesin 018102
Heimatstube, Elsteraue 018167
Heimatstube, Enkirch 018215
Heimatstube, Esslingen 018318
Heimatstube, Fehrbellin 018344
Heimatstube, Gittelde 018702
Heimatstube, Goslar 018766
Heimatstube, Großraming 002011
Heimatstube, Hasselfelde 019088,
019089
Heimatstube, Hinternah 019260
Heimatstube, Hohenzieritz 019315
Heimatstube, Lahstedt 019856
Heimatstube, Lenzkirch 020040
Heimatstube, Liepgarten 020069
Heimatstube, Marlow 020306
Heimatstube, Matzlow-Garwitz . 020317
Heimatstube, Neulußheim 020719
Heimatstube, Neuruppin 020738
Heimatstube, Quellendorf 021251
Heimatstube, Remseck am
Neckar 021366
Heimatstube, Schäplitz 021626
Heimatstube, Schochwitz 021695
Heimatstube, Schönwalde,
Altmark 021715
Heimatstube, Schonach im
Schwarzwald 021719
Heimatstube, Schwarzenberg am
Böhmerwald 002735
Heimatstube, Schwedt / Oder . . 021776
Heimatstube, Tanne 022105
Heimatstube, Zörbig 022893
Heimatstube "Der Große Hof",
Wienrode 022627
Heimatstube 1813 Möckern,
Möckern 020431
Heimatstube Abbenrode,
Abbenrode 016453
Heimatstube Adlergebirge,
Waldkraiburg 022397
Heimatstube Altenau, Altenau . . 016527
Heimatstube Ansprung, Zöblitz . 022890
Heimatstube Arenborn,
Oberweser 020938
Heimatstube Batzsch-Brestowatz,
Königsbach-Stein 019720
Heimatstube Bienrode,
Braunschweig 017556
Heimatstube Bliedersdorf,
Bliedersdorf 017423
Heimatstube Börnecke,
Blankenburg 017406
Heimatstube Bredenbeck,
Wennigsen 022566
Heimatstube Brunkensen, Alfeld 016503
Heimatstube Bühlau,

Großharthau 018850
Heimatstube der Stadt Saaz,
Roth 021502
Heimatstube des Heimatkreises
Reichenberg, Augsburg 016657
Heimatstube Dessau-Alten,
Dessau 017865
Heimatstube Dreetz, Dreetz ... 017972
Heimatstube Eime, Eime 018128
Heimatstube Elbingerode, Elbingerode,
Harz 018150
Heimatstube Eltze, Uetze 022251
Heimatstube Endersbach,
Weinstadt 022527
Heimatstube Feldberg, Feldberger
Seenlandschaft 018346
Heimatstube Fischbach,
Niedereschach 020794
Heimatstube Frauenwald und
Bunkermuseum, Frauenwald .. 018463
Heimatstube Freest, Kröslin ... 019799
Heimatstube Freiwaldau-Bieletal,
Städtisches Museum im Kornhaus,
Kirchheim unter Teck 019625
Heimatstube für Volkskunst,
Zschadraß 022899
Heimatstube Gadenstedt,
Lahstedt 019857
Heimatstube Gräfenhain/ Nauendorf,
Georgenthal 018643
Heimatstube Groß Denkte,
Denkte 017858
Heimatstube Groß Gleidingen,
Vechelde 022304
Heimatstube Groß Lafferde,
Lahstedt 019858
Heimatstube Groß Nemerow, Groß
Nemerow 018835
Heimatstube Großbodungen,
Großbodungen 018839
Heimatstube Gutenberg,
Oberostendorf 020914
Heimatstube Hänigsen, Uetze .. 022252
Heimatstube Hetzdorf, Uckerland 022240
Heimatstube Hohenhameln,
Hohenhameln 019304
Heimatstube im "Spieker Anno 1754",
Tarmstedt 022106
Heimatstube im alten Rathaus,
Hessigheim 019234
Heimatstube im Burgturm, Plau am
See 021147
Heimatstube im Elbtor, Werben . 022569
Heimatstube in der Wasserburg,
Gerswalde 018674
Heimatstube Kappel, Lenzkirch . 020041
Heimatstube Lebus, Lebus 019946
Heimatstube Lichtenstadt,
Zirndorf 022885
Heimatstube Mederitsch,
Pflückuff 021118
Heimatstube Mühlhausen, Villingen-
Schwenningen 022337
Heimatstube Münstedt, Lahstedt 019859
Heimatstube Neckarrems, Remseck
am Neckar 021367
Heimatstube Nendingen,
Tuttlingen 022234
Heimatstube Neuhaus-Schierschnitz,
Neuhaus-Schierschnitz 020708
Heimatstube Oberg, Lahstedt .. 019860
Heimatstube Petznick, Templin . 022126
Heimatstube Podersam-Jechnitz,
Kronach 019803
Heimatstube Radegast, Radegast 021260
Heimatstube Rautheim,
Braunschweig 017557
Heimatstube Röckwitz, Röckwitz 021450
Heimatstube Röhrnbach-Kaltenbach,
Röhrnbach 021454
Heimatstube Rothenklempenow,
Rothenklempenow 021516
Heimatstube Ruhland, Ruhland . 021544
Heimatstube Schandelah,

Cremlingen 017780
Heimatstube Schlesien,
Kaufbeuren 019560
Heimatstube Seeheilbad Graal-Müritz,
Graal-Müritz, Seeheilbad 018779
Heimatstube Sitzenroda, Schildau,
Gneisenaustadt 021647
Heimatstube Sperenberg, Am
Mellensee 016569
Heimatstube Stadt und Landkreis
Saaz, Georgensgmünd 018640
Heimatstube Tribsees, Tribsees . 022192
Heimatstube Trochtelfingen,
Bopfingen 017506
Heimatstube Untere Baranya, Gingen
an der Fils 018699
Heimatstube Usedom, Usedom . 022288
Heimatstube Vietmannsdorf,
Templin 022127
Heimatstube Wahrenbrück, Uebigau-
Wahrenbrück 022245
Heimatstube Warthe, Boitzenburger
Land 017470
Heimatstube Wendhausen,
Schellerten 021640
Heimatstube Wiesental,
Waghäusel 022368
Heimatstube Wilthen, Wilthen .. 022674
Heimatstube, Heimat- und
Förderverein e.V Stapelburg,
Stapelburg 021960
Heimatstuben, Langelsheim 019891
Heimatstuben, Waldbronn 022381
Heimatstuben der Stadt, Titisee-
Neustadt 022160
Heimatstuben im Arrestturm,
Spalt 021925
Heimatstuben im Oberen Tor,
Scheinfeld 021638
Heimatstuben Weipert und
Erzgebirgsschau, Gunzenhausen 018896
Heimatstube der Gemeinde und des
Harzklub Zweigvereins Bad Suderode,
Bad Suderode 016886
Heimatstube Klaushagen,
Boitzenburger Land 017471
Heimatstube Staupitz, Pflückuff . 021119
Heimattreff Gersweiler,
Saarbrücken 021556
Heimatverein Räntzelstecher Güsten
Osmarsleben e.V, Güsten ... 018882
Heimatvertriebenen-Stuben,
Traun 002839
Heimerdinger, Debra, Brevard . 119936
Heimethüs, Denzlingen 017859
Heimilisidnardarsafnid Halldórustofa,
Blönduós 024068
Heimroth, Alain, Paris 071470
Heimschneidermuseum,
Großwallstadt 018865
Heimtali Koduloomuuseum,
Heimtali 010613
Heimweberei-Museum,
Schalkenmehren 021629
Hein, Bitburg 075013
Hein, Christoph, Dresden 129711
Hein, Jörg, Berlin 129501
Heinänen, Helsinki 066320
Heindl, Günter, Heidelberg 130028
Heindl, Karin, Oldenburg 142267
Heine-Haus, Altonaer Museum in
Hamburg, Hamburg 018994
Heine, Thorsten, Dannenberg . 075297
Heineck, Ralph, Heidelberg ... 076446
Heinecke Experience,
Amsterdam 031943
Heinemann, Paul, Starnberg .. 142374
Heinemann, William, London .. 138219
Heinicke, Hans-Jürgen, Potsdam 078077
Heinicke, Thomas, Oberwiera . 130484
Heinisch, Wieland, Knetzgau .. 076828
Heinlein, Christian, Windheim . 079020,
                    130769

Heinley, Boston 119874
Heino Lubja Kaalumuuseum,
Mustvee 010640
Heinolan Kaupunginmuseo,
Heinola 010809
Heinonen, Helsinki 128413
Heinrich, Wien 100316
Heinrich-Blickle-Museum, Sammlung
Gußeiserner Ofenplatten,
Rosenfeld 021472
Heinrich-Büssing-Haus,
Wolfsburg 022741
Heinrich-Ehmsen-Stiftung, Stadtgalerie
Kiel, Kiel 019604
Heinrich-Hansjakob-Gedenkstätte,
Freiburg im Breisgau 018480
Heinrich Harrer-Museum,
Hüttenberg 002108
Heinrich Heine Antiquariat,
Düsseldorf 141720
Heinrich-Heine-Institut-Museum,
Düsseldorf 018030
Heinrich Hugendubel Verlag,
München 137405
Heinrich-Schliemann-Gedenkstätte,
Neubukow 020684
Heinrich-Schliemann-Museum,
Ankershagen 016585
Heinrich-Schütz-Haus Weißenfels,
Weißenfels 022543
Heinrich-Schütz-Haus, Forschungs-
und Gedenkstätte im Geburtshaus
des Komponisten, Bad Köstritz 016803
Heinrich-Sohnrey-Archiv und
Gedächtnisstätte, Mautturm, Jühnder
Schloß, Jühnde 019471
Heinrich-Vogeler Museum,
Barkenhoff Stiftung Worpswede,
Worpswede 022772
Heinrich Vogeler Stiftung Haus im
Schluh, Worpswede 022773
Heinrich-von-Zügel-Gedächtnisgalerie,
Wörth am Rhein 022720
Heinrich Zille Museum, Berlin .. 017168
Heinrich, G., Bielefeld 074979
Heinrich, Herbert, Rottenburg an der
Laaber 130608
Heinrich, J., Olching 077922
Heinrich, Marie-Luce, Köln ... 076871
Heinrich, Michael, Braunschweig 075125
Heinrich, Nick, Boppard 075098
Heinrich, Peter, Hamburg 076238
Heinrich, Peter, Schorndorf ... 130629
Heinrichs, Bernd, Celle 075246
Hein's Mühle, Bendorf 017024
Heins, Marlene, Niederhausen . 108276
Heinsch, Hans, Denkte 075333
Heinsch, Hans, Wolfenbüttel .. 079052
Heintzelman, Albert, Los Angeles 135720
Heinz & Heinz, Hannover 107368
Heinz & Heinz, Lüneburg 107930
Heinz Nixdorf MuseumsForum,
Paderborn 021061
Heinz, Anneli, Hofheim am
Taunus 076553
Heinz, Marcel, Carcès 103531
Heinze, Friedrich, Salzburg ... 100166
Heinzel, Aberdeen 117209
Heinzel, B., Kassel 107533
Heinzerling & Achenbach-Heinzerling,
Hungen 107452
Heinzl, Luise, Viechtach 078793
Heinzl, Walter, Wien 062976, 127927
Heinzmann, Villingen-
Schwenningen 078801
Heirler, Dresden 106738
Heirloom, Harrogate 117917
Heirloom, Honolulu 093551
Heirloom & Howard, West Yatton 091791
Heirloom Antiques, Winnipeg .. 064926
Heirloom Furniture, Denver ... 093298,
                    135532
Heirloom Jewels, Baltimore ... 092331
Heirlooms, Coconut Grove 061348

Heirlooms, Dún Laoghaire 079605
Heirlooms, Durban 085642
Heirlooms, Sandton 085664
Heirlooms, Wareham .. 091725, 135187
Heirlooms, Worcester 091945
Heirlooms International, Tulsa .. 097736
Heisando, Tokyo 082151, 132310
Heise, T., München 077522
Heisey Museum, Lock Haven .. 050350
Heisey News, Newark 139380
Heiss, Günther, München 077523
Heiss, Julia, Mainleus 077229
Heiß, Margot, Zapfendorf 079132
Heissler, Willi F. G., Bobenheim am
Berg 125897
Heitmeyer-Löns, Sabine,
Havixbeck 130022
Heitsch, Françoise, München .. 108138
Heitsch, Jörg, München 108139
Heitz et Darmancier, Bourges . 125535
Heitz, Anthony & Daniel, Basel . 087237
Heitzmann, Alain, Beaune 067023
Heitzmann, Jérôme, Chevigny-Saint-
Sauveur 067967
Heitzmann, Monique, Chagny .. 067761
Heitzmann, Patrick, Arc-sur-Tille 066669
Heitzmann, Thierry, Mormant .. 070351
Heiwajima Antique Fair, Tokyo .. 098229
Heizungs- und Schulmuseum Wien,
Wien 003005
Hejtmanek, Regina, Vitry-sur-
Seine 129358
Hekkers & Van Spijker, Wilp .. 132670
De Heksenbal, Breda 082986
De Heksenwaag, Oudewater .. 032755
Helander, M., Helsinki 125480
Helander, M., Viitasaari 125483
Helbling-Schatt, Max & Margrit,
Benken (Sankt Gallen) 133666
Held, Stuttgart 142392
Held, Tönisvorst 130705
Held Joan Museum, Carlton
South 000946
Held-Poage Memorial Home &
Research Library, Ukiah 053879
Helderse Kunstliga, Den Helder . 059606,
                    112425
Hélen Antiquités, Guecelard .. 068756
Helen Day Art Center, Stowe ... 053560
Helen E. Copeland Gallery, Montana
State University, Bozeman 047165
Helen King Kendall Memorial Art
Gallery, San Angelo 052925
Helen Louise Allen Textile Collection,
Madison 050595
Helen Pitt Gallery, Vancouver .. 101797
Helen, Mary, Tucson 124742
Helena & Dieb, Lisboa 114036
Helena Gallery, Helena 049431
Helena Modjeska Historic House,
Modjeska Canyon 050983
Helena Thompson Museum,
Workington 046696
Helenas Antik och Prylbod,
Eskilstuna 086668
Helena's Collectables, Shere .. 091313
Helenbeck, Nice 104371
Helene Antiques, Armadale,
Victoria 061034
Helen's Depression, San Antonio 096976
Helen's Jewelry, Saint Louis ... 136202
Helen's on Discovery, Helensvale 061509
Helensville and District Pioneer
Museum, Helensville 033221
Helfertová, Praha 128347
Helfrich, Rödermark 108507
Helga's Antiques, Cork 079494
Helgenæs Præstegård, Knebel .. 010164
Helgers, Georg, Xanten 079129
Helgheim, Terje Johan, Bergen . 126339
Heli, Mannheim 142106
Hélian, Martin, Montpellier ... 128809
L'Hélice Terrestre, Saint-Georges-des-
Sept-Voies 015332

The Helicopter Museum, Weston-
super-Mare 046100
Helier, Jean-Luc, Mormant 070352
Heliez, Stephan, Chamalières . . 067791
Heliloojate Kappide Majamuuseum,
Viljandi Muuseumi, Suure-Jaani 010661
Helin, Hauho 066291, 066292
Helinä Rautavaaran museo,
Espoo 010774
Héliodore, Saint-Étienne 072765
Helios, Nashville 122085
Helios & Co, Weedon 091763
Helios Arts, Pittsburgh 123550
Helios at the Spinney, Tanworth-in-
Arden 119204
Hell, Christian, Wolfratshausen . 130777
Hell, Claudia, Großgmain 062614,
127758
Hell, Doris, Großgmain   062615, 127759
Hell, Dr. Bernhard, München . . 142161
Hellbrügge, Dr. Christoph,
Ascheberg 129390
Hellenic American Union Gallery,
Athinai 109039
Hellenic Antiquities Museum,
Melbourne 001274
Hellenic Children's Museum-The
Station, Athinai 022989
Hellenic Folklore Research Centre,
Athinai 022990
Hellenic Museum and Cultural Center,
Chicago 047663
Hellenic Theatre Museum and Study
Center, Athinai 022991
Hellens Manor, Much Marcle . . . 045304
Hellenthal, Berlin 106306
Heller, Omaha 123320
Heller-Irmscher, B., Zwingenberg,
Kreis Bergstraße 109011
Heller, Douglas, New York 122701
Heller, Joshua, Washington . . . 145635
Heller, Petra, Bayreuth 074694
Heller, Thomas, Gefrees 075958
Heller, Thomas, Passau 078021
Hellfire Gallery, Merivale 099330
Hellhof, Kronberg im Taunus . . 107757
Helling, Oslo 084319
Hellmann, Meiningen 077328
Hellmann, Kurt, Wien 062977
Hello-Five Antiques, Edmonton . 064296
Hellweg-Museum Unna, Unna . . 022273
Hellwig, Wuppertal 079105
Hellwig, Wilhelm, Wuppertal . . . 079106
Helm, Mellrichstadt 077340
Helm, Witten 079031
Helman, Warszawa . . . 084726, 084727,
132880
Helmcken House, Royal British
Columbia Museum, Victoria . . . 007327
Helme Kihelkonnamuuseum,
Helme 010614
Helmeke, Jörg, Stendal 108672
Helmering, P., Berlin 074826
Helmeringhausen Museum,
Helmeringhausen 031839
Helmersen, Hélène d', Paris . . . 071471
Helmert, Dr. Gunter, Chemnitz . 075253
Helmhaus Zürich, Zürich 042200
Helmich, Bernd M., Berlin 129502
Hellmig, Bielefeld 074980
Hellmig, Dissen am Teutoburger
Wald 075365
Helming, L. & R., Rheine 078225
Helminger, Hof bei Salzburg . . . 127768
Helmrich, Barbara, Isernhagen . 130077
Helms-Museum, Hamburger Museum
für Archäologie und die Geschichte
Harburgs, Hamburg 018995
Helmshore Mills Textile Museum,
Helmshore 044558
Helmsley Bookshop, Helmsley . . 144465
Helmstedter Freundeskreis für
Paramentik und christliche Kunst,
Helmstedt 019184

Helmstetter, Bruno, Schöntal . . . 130628
Helmstetter, Markus, Mörschbach 077410
Helmut Newton Stiftung im Museum
für Fotografie, Staatliche Museen
Berlin-Stiftung Preußischer
Kulturbesitz, Berlin 017169
Helo, Adriana Carlota, São Paulo 101096
Helsingborgs Antik och Konstmässa,
Helsingborg 098320
Helsingborgs Auktionskammare,
Helsingborg 126599
Helsingborgs Auktionsverk,
Helsingborg 126600
Helsingborgs Idrottsmuseum,
Helsingborg 040692
Helsingborgs Skolmuseum,
Helsingborg 040693
Helsinge Hembygdsmuseum,
Vantaa 011490
Helsinge Hestevogns Museum,
Helsinge 010103
Helsingin Antiikki, Järvenpää . . 066375
Helsingin Antikvariaatti, Helsinki 140702
Helsingin Automuseo, Helsinki . 010822
Helsingin Diakonissalaitoksen Museo,
Helsinki 010823
Helsingin Huutokauppakamari,
Helsinki 125481
Helsingin Kaupungin Taidemuseo,
Meilahti, Helsinki 010824
Helsingin Kaupungin Taidemuseo,
Tennispalatsi, Helsinki 010825
Helsingin Kaupunginmuseo,
Helsinki 010826
Helsingin Taidehalli, Helsinki . . 010827
Helsingin Väestönsuojelumuseo,
Helsinki 010828
Helsingin Yliopiston Taiteiden
Tutkimuksen Laitos, Helsinki . . 055342
Helsingin Yliopistomuseo Arppeanum,
Helsinki 010829
Helsingin Yliopiston Maatalousmuseo,
Helsinki 010830
Helsingør Bymuseum, Helsingør 010108
Helsted, Klaus, Haslev 065725
Helston Folk Museum, Helston . 044561
Heltborg Museum, Sydthy Kunst- og
Kulturcenter, Hurup Thy 010148
Helvetia Museum, Helvetia 049435
Helvetisches Goldmuseum,
Burgdorf 041384
Helweg, Bitten, Holbæk 065774
Helwing, Anna, Los Angeles . . . 121506
Helytörténeti Emléktár,
Dunaharaszti 023608
Helytörténeti Gyűjtemény és Tájház,
Várpalota 024032
Helytörténeti Gyűjtemény, Derkovits
Gyula Művelődési Központ,
Tiszaújváros 024013
Helytörténeti Kiállítás és J. Haydn
Emlékszoba, Fertőd 023636
Helytörténeti Múzeum,
Csempeszkopács 023582
Hemani, Abu Dhabi 088159
Hembygdsgården, Bjurholm . . . 040565
Hembygdsgården i Kristvalla,
Nybro 040885
Hembygdsgården Labbas, Storby 011374
Hembygdsgården Rots Skans,
Älvdalen 040511
Hembygdsmuseet i Korsberga,
Korsberga 040757
Hemeleers, Frantz, Bruxelles . . 063461
Hémery, Francois, Paris 129034
Hémery, François-Paul, Ho Chi Minh
City 136475
Hémery, François-Paul, Le
Faouët 128657
Hemet Museum, Hemet 049436
Hemingway Museum, Oak Park   051666
Hemken, Helgard, Siegen 078462
Hemlängtan Antikt, Trelleborg . . 087101
Det Hemmelige Galleri, Herning 102830

Hemmerle, München 077524
Hemne Bygdemuseum,
Kyrksæterøra 033727
Hemnes Museum, Bjerka 033544
Hempe, R., Vierkirchen 078797
Hempel, Christian, Dresden . . . 129712
Hempel, F.H. Jürgen, Monschau 108029
Hempenius, R., Almelo 082629
Hemphill, Washington 124917
Hemphill, George, Wellington . . 132760
Hemsedal Bygdatun, Hemsedal . 033648
Hemstra, Faaborg 065644
Hemswell Antique Centres, Hemswell
Cliff 089499, 134452, 144466
Hemudu Ruins Museum, Yuyao . 008517
Henan Fine Arts Publishing House,
Zhengzhou 136747
Henan Museum, Zhengzhou . . . 008537
Henåns Antikgård, Henån 086743
Henau, Antwerpen 063210
Henaut, Paris 104947
Hendel, Renate, Taufkirchen, Vils 108758
Henderson, Louisville . 094560, 135758
Henderson, Perth 090978
Henderson Art & Framing, King's
Lynn 118046
Henderson Phillips, Washington . 124918
Henderson, Carol, Fort Worth . . 120869
Hendlinger, N., Berlin 129503
Hendricks Gallery & Framing Ltd,
Edmonton 101290
Hendricks, Jörg, Burgwedel . . . 075232,
141669
Hendricks, Marc F.J., Meinier . . 087669
Hendrickson House Museum and Old
Swedes Church, Wilmington . . 054439
Hendrickx-Gourdin, Namur 063809
Hendrik's Welt, Erftstadt 075632
Hendrik's Welt, Hürth 076586
Hendrik's Welt, Bad Münstereifel 074556
Hendrik's Welt, Nettersheim . . . 077703
Hendriks, E.W., Amsterdam . . . 082768
Hendriks, Holger, Hattingen . . . 076418
Hendriks, Twan, Tilburg 112679
Hendrix, Dallas 120609
Hendrycks, Andre, Gent 063641
Heneage, Thomas, London 144626
Héneaux, Marie-Emmanuel, Saint-
Brieuc 072697
Heney, Nice 104372
Henfield Museum, Henfield 044565
Heng, Singapore 114534
Hengchuen Peninsula Aboriginal Arts
and Crafts Museum, Mudan . . . 042402
Hengesbach, Rolf, Köln 107637
Hengevoss-Duerkop, Hamburg . 107278
Hengfeng Museum, Hengfeng . . 007857
Hengistburgmuseum & Volkskundliche
Sammlung, Hengsberg 002075
Hengyang Museum, Hengyang . . 007858
Henham, Martin, London 090204,
134689
Henhouse, Birmingham 092410
Henie Onstad Kunstsenter,
Høvikodden 033658
Henin & Planchar, Bruxelles . . . 063462
Henk, Budapest 079261
Henke, B., Bonn 075073
Henke, Heinrich, Passau 142294
Henke, Manfred, Berlin 141525
Henke, Piet, Alkmaar 082620
Henken, Los Angeles 121507
Henkerhaus, Bernau bei Berlin . 017326
Henki, Ajan, Helsinki 066321
Henley & Grange Arts Society, Henley
Beach 058202, 099126
Henley Antiques Centre, Henley-in-
Arden 089505
Henley Art and Crafts Guild,
Reading 060155
Henley Auctions, Mile End 125212
Henley House Antiques, Rumford 091152
Henley-in-Arden Auction Sales, Henley-
in-Arden 126921

Henn, Maastricht 112575
Henn, Andreas, Stuttgart 108707
Henn, U., Peine 078031
Henneberg-Museum,
Münnerstadt 020586
Hennebergisches Kloster
Veßra, Kloster Veßra 019654
Hennemann, Marianne, Bonn . . 106533
Hennen, Christian, Leverkusen . 107861
Hennen, Hardy, Mettlach 107998
Hennepin History Museum,
Minneapolis 050937
Hennessee, Rob, Tampa 127472
Hennessy, Crewkerne 089005
Hennessy, Clive, Vevey 087880
Hennig, Birgit, Zerbst 079139
Hennig, Dr. Karl, Hamburg 076239
Hennig, H., Löffingen 077139
Hennig, Marie-Louise, Köln 076872
Hennige, Renate, Heilbronn . . . 076473
Henning, Enno, Wittmund 079039
Hennings, Paul, Hamburg 141869
Henningsen, Karen, Ringkøbing . 066070
Hennwack, Berlin 141526
Hennwack, Kyritz 142040
Henocq, Sprinkange 082389
Henocq, Robert, Sprinkange . . . 082390
Henocque, Marie-Sylvie, Paris . . 129035
Henot, Marc, Enghien-les-Bains   103714
Henri-Arnaud-Haus –
Waldensermuseum, Ötisheim . 020966
Henri Godts Libraire – Expert –
Antiquaar, Bruxelles 140179
Henri, Stephan, Condé-sur-
Marne 068123
Henrich, München 077525
Henrich, Hella, Schillingen 078363
Henrichshütte Hattingen,
Hattingen 019098
Henrietta's, Jacksonville 121255
Henrik Antiquités, Cannes 067640
Henrik Ibsen Museum, Skien . . 033943
Henrike Höhn Kunstprojekte,
Berlin 106307
Henrique & Eduardo, Lisboa . . . 084897
Henriques de Carvalho, Manuel,
Lisboa 084898
Henro, New York 095453
Henry, Pau 105329
Henry Art Gallery, University of
Washington, Seattle 053235
Henry B. Plant Museum, Tampa   053668
Henry Blogg Lifeboat Museum,
Cromer 044047
Henry County Historical Museum, New
Castle 051251
Henry County Museum and Cultural
Arts Center, Clinton 047840
Henry-Dumant-Museum, Heiden 041587
Henry Ford Estate, Universty of
Michigan-Dearborn, Dearborn . 048194
Henry Ford Museum, Dearborn . 048195
Henry Guest House, New
Brunswick 051243
Henry H. Blommel Historic Automotive
Data Collection, Connersville . . 047964
Henry Kendall Cottage and Museum,
Gosford 001114
Henry Lawson Centre, Gulgong . 001124
Henry Moore Foundation, Much
Hadham 045303
Henry Moore Institute, Leeds . . 044780
Henry Radford Hope School of Fine
Art, Indiana University Bloomington,
Bloomington 056753
Henry S. Reuss Ice Age Visitor Center,
Campbellsport 047393
The Henry Sheldon Museum of
Vermont Histoy, Middlebury . . . 050857
Henry-Stuart House, Québec . . . 006729
Henry van de Velde-Gesellschaft e.V.,
Hagen 058916
Henry van de Velde Museum,
Kunstsammlungen Chemnitz,

Chemnitz 017732
Henry Whitfield State Historical
Museum, Guilford 049291
Henry, Gene L ., Seattle 097452
Henry, Hubert, Saint-Julien-du-
Sault 072863
Henry, Karel & Willis, Marshfield 127309
Henry, Ph., Quimper 072346
Henry's, Ash 088293
Henry's Antiques, Seattle 097453
Henry's Auktionshaus,
Mutterstadt 126076
Hens, Susanne, Inning am
Ammersee 130075
Henschel Verlag, Leipzig 137343
Henschel, Edgardo, Buenos Aires 139572
Henschke, Hannover 076374
Henschke, Iserlohn 076618
Hense, Thomas, Kiel 076776
Henseleit, Frank, Köln 107638
Henseler, Luzern 087641
Henshelwood Gallery, Newcastle-upon-
Tyne 118787
Hensing & Döring, Bremen 075165
Hensoldt, Rainer M., Wetzlar ... 078959
Hentrich, Berlin 141527
Hentrich, L., Mayen 107984
Hentschel & Visser, Bocholt ... 075028
Henty Old Wares, Henty 061618
Henty on the Hill, Olinda 062026
Henz-Malzach, Johann, Erschwil 087415,
133734
Henze & Butkus, Alheim 074376, 129376
Henze & Ketterer, Wichtrach ... 116715,
138098,
144150
Henze, Ernst-Dieter, Hamburg .. 076240
Henze, M., Halle, Saale 076167
Hepner, Guy, London 118379
Hepplewhite, Singapore 085420
Heptonstall Museum, Heptonstall 044569
Hepu Museum, Hepu 077860
Hepworth, Anthony, Bath 117306,
117307,
144231
Hepworth, Gordon, Newton Saint
Cyres 090840, 118806
Her House, London 118380
Her Majesty Customs and Excise
National Museum, Liverpool .. 044854
Her Majesty Tower of London, Historic
Royal Palaces, London 044995
Hera Gallery, Hera Educational
Foundation, Wakefield 054033
Herail, B., Leeds 134535
Herakles, Pullach im Isartal 078103
Herakles Numismatics, Charlotte 092603
Herald, Marcelline, Eton 089216
Heraldic Museum, Dublin 024885
Heraldry Today, Ramsbury 144795
Heran, Jeanne, Paris 071472
Heras Institute of Indian History and
Culture, Mumbai 024404
Heras Tibaita, Maria Luisa de las,
Madrid 086199
Herat Provincial Military Museum,
Herat 000002
Herat Provincial Museum, Herat 000003
De Heraut, Rotterdam 143220
Herbart, Wolfgang, Mogán 133406
Herbelet, Denis, Saint-Memmie . 072977
Herbelin, Christophe, Chinon ... 125558
Herbener, Thomas, Neugersdorf 077729
Herbert, Paris 071473
Herbert Art Gallery and Museum,
Coventry 044019
Herbert CPR Train Station Museum,
Herbert 006001
Herbert E. Clark Collection of Near
Eastern Antiquities, Jerusalem 025079
Herbert F. Johnson Museum of Art,
Ithaca 049739
Herbert-Gerisch-Stiftung,
Neumünster 020725
Herbert Hoover National Historic Site,
West Branch 054290

Herbert Hoover Presidential Library-
Museum, West Branch 054291
Herbert Press, London 138220
Herbert, Ingeborg, Einhausen .. 075589
Herbert, Matthias, Nürnberg .. 077825,
130466
Herbette, Danièle, Beauval 067056
Herbette, Denis, Doullens 125573
Herbich, Karl, Cebu .. 084474, 132812
Herbich, Karl, Maisach 130272
L'Herbier Déco, Saint-Denis ... 085066
Herbinet, Jacques, Paris 141139
Herbita, Palermo 137710
Herbst, Christiane, Lüneburg ... 077210
Herbst, Hagen & Holger,
Greifswald 076085
Herbst, Kristina, Berlin 129504
Herbst, Thomas, Detmold 125916
Herburger, Heinz, Saarbrücken . 108547
Hercher, Hans W., Umkirch 078760
Herckenrode, Hasselt 063682
Hercouët, Catherine, Loheac ... 069609
Herda, Waltershausen 077858
Herdeiros, José Luís, Lisboa ... 084899
Herdenkingsbordenmuseum,
Lochem 032601
Herdla Museum, Herdla 033649
Here in the Past, San Diego ... 097095
Hereford Antique Centre,
Hereford 089516, 134458
Hereford Inlet Lighthouse, North
Wildwood 051621
Hereford Museum and Art Gallery,
Hereford 044571
Herefordshire Regiment Museum,
Hereford 044572
Herendi Porcelánművészeti, Múzeum
Alapítvány, Herend 023684
Hereward, Ely 144408
Hereward Gallery, Bourne 117398
Herfatz, Osuji, Wangen im Allgäu 078862
Herforder Kunstverein im Daniel-
Pöppelmann-Haus, Herford ... 058917
Herfurthsches Haus, Hainichen . 018934
Hergat, Danièle, Bergerac 067103
Herget, George, New Orleans .. 145312
Herholdt Jensen, Nærum 125459
Heriard & Cimino, New Orleans . 122183
Héricé, Christophe, Montcaret .. 070222
Herinek, Gerhard, Wien 062978
Hering, Dr. M. H. Bernd, Fürth . 129871
Hering, Jens-Uwe, Marienberg . 077292
Herinneringscentrum Kamp
Westerbork, Hooghalen 032503
Heriot-Watt University Museum,
Edinburgh 044244
Héritage, Boulogne-sur-Mer ... 067401
Heritage Antique, Baltimore ... 092332
Heritage Antique Jewellery & Art,
Toronto 064665
Héritage Antique Métropolitain,
Montréal 064429
Heritage Antiques, Calgary 064257
Heritage Antiques, Charlotte ... 092604
Heritage Antiques, Chichester .. 088875
Heritage Antiques, Hamilton ... 064319
Heritage Antiques, London 064349
Heritage Antiques, Mosta 082417
Heritage Antiques, Wem 135201
L'Héritage Antiquité, Québec ... 064534
Heritage Art and Antiques,
Mumbai 079402
Heritage Art Center, Quezon City 034581
Heritage Auctions, Swaffham .. 127061
Heritage Book Shop, West
Hollywood 145643
Heritage Books, Ryde 144805
Heritage Center Museum,
Berryville 046949
Heritage Center of Clark County,
Clark County Historical Society,
Springfield 053470
Heritage Center of Lancaster County,
Lancaster 050118

Heritage Center, Harrisonburg-
Rockingham Historical Society,
Dayton 048177
The Heritage Center, Red Cloud Indian
School, Pine Ridge 052139
Heritage Centre Canada – HVACR,
Mississauga 006341
Heritage Centre of Saint Louis
Convent, Monaghan 024967
Heritage Collectables and Antiques,
Houston 093689
Heritage Collection, Dallas 093138
Heritage Collection Catherina Brand,
Ladybrand 038776
Heritage Conservancy,
Doylestown 048365
Heritage Custom Furniture,
Houston 135605
Heritage Farm & Village, Western
Development Museum, North
Battleford 006506
Heritage Farmstead Museum,
Plano 052191
Heritage for the Blind, Miami .. 094744
Heritage Frame & Picture,
Tarrytown 124679
Heritage Furniture Restoration,
Buffalo 135372
Heritage Galleries, Calgary 064258
Heritage Gallery, Ha Noi 097981
Heritage Gallery, Nelson Bay .. 061948
Heritage Gallery, Singapore ... 085421
Heritage Gallery, Tetbury 091571, 119220
Heritage Gallery, Waikato 084130
Heritage Gift Shop, Tucson ... 097670
Heritage Hall Museum and Archives,
Freeman 048968
Héritage Harwood, Montréal ... 064430
Heritage Hazenmore Museum,
Hazenmore 005998
Heritage Hill Museum,
Dandenong 001014
Heritage Hill State Park, Green
Bay 049226
Heritage Hjemkomst Interpretive
Center, Moorhead 051047
Heritage House, Hamilton 100825
Heritage House Antiques,
Auckland 083790
Heritage House Museum, Smiths
Falls 007049
Heritage House Museum of
Montgomery County, Mount Ida 051091
Heritage House of Orange County
Museum, Orange 051790
Heritage Lane Furniture Restoration,
Concord 128160
Heritage Library Museum,
Anchorage 046471
Heritage Models Centre, River
Herbert 006808
Heritage Motor Centre, Gaydon . 044383
Heritage Museum, Baltimore .. 046749
Heritage Museum, Deer Lake .. 092809
Heritage Museum, Idaho Springs 049655
Heritage Museum, Kuah 030677
Heritage Museum and Potton House,
Big Spring 046980
The Heritage Museum at Falfurrias,
Falfurrias 048685
Heritage Museum Foundation of Tate
County, Senatobia 053268
Heritage Museum of Fine Arts for
Youth, Detroit 048307
Heritage Museum of Northwest Florida,
Valparaiso 053937
The Heritage Museum, Clatsop
Historical Society, Astoria 046585
Heritage Museums, Sandwich .. 053073
Heritage Music Repair, Kansas
City 135679
Heritage North Museum,
Thompson 007140
Heritage of the Americas Museum, El

Cajon 048494
Heritage Park, Fort McMurray .. 005824
Heritage Park – Historical Village,
Calgary 005522
Heritage Park Folk Museum, Barbados
Museum, Saint Phillip 003241
Heritage Park Museum, Terrace 007132
Heritage Railway Association,
Tywyn 060156
Heritage Restorations, Llanfair
Caereinion 089870, 134574
The Heritage Shop, Singapore .. 085422
The Heritage Society Museum,
Houston 049577
Heritage Square Museum, Los
Angeles 050428
Heritage Square Museum,
Ontario 051782
Heritage Sutton, Sutton, Québec 007106
Heritage Trading, Dubai 088178
Heritage Trading Company,
Auckland 083791
Heritage Vietnamese Arts and Crafts,
Ho Chi Minh City 097983
Heritage Village, Machiko Lake 051105
Heritage Village, Sharonville .. 053287
Heritage Village – Pinellas County
Historical Museum, Largo 050148
Heritage Village Museum,
Woodville 054538
Heritage Walk Museum,
Escondido 048606
Héritier, Carmen, Mirmande ... 104240
Héritier, Maurice, Autun 066835
Héritier, Michelle, Fauconnières . 068507
Héritier, Pierre-Antoine, Carouge 133713
Heriz, Cincinnati 092865
Herkimer County Historical Society
Museum, Herkimer 049450
Herkimer Home, Little Falls ... 050332
Herkomer-Museum am Mutterturm,
Neues Stadtmuseum Landsberg am
Lech, Landsberg am Lech ... 019877
Herkules Villa, Budapesti Történeti
Múzeum, Aquincumi Múzeum,
Budapest 023486
Herles, Michael, Oberursel 077885
Herlitze, Bodo, Berlin 074827
Herlyn, Freiburg im Breisgau .. 141803
Herman & Wilkinson, Dublin .. 126170
Herman Antiquités, Ukkel 063852,
128091
Herman Ottó Múzeum, Miskolc . 023783
Herman Ottó Múzeum – Herman Ottó
Emlékház, Miskolc 023784
Herman Ottó Múzeum – Kiállítóhelye,
Miskolc 023785
Herman Ottó Múzeum Ásványtára,
Miskolc 023786
Herman Teirlinckmuseum,
Beersel 003354
Herman Son, J., New Orleans . 095111
Hermann, Paris 136997
Hermann, Wollombi 127698
Hermann-Allmers-Haus,
Sandstedt 021583
Hermann Broch-Museum,
Teesdorf 002814
Hermann-Grima House, New
Orleans 051293
Hermann-Grochtmann-Museum
Datteln, Datteln 017834
Hermann-Hesse-Höri-Museum,
Gaienhofen 018582
Hermann-Hesse-Museum, Calw 017714
Hermann Hesse Museum,
Hiroshima 028646
Hermann Historica, München .. 077526,
126054
Hermann-Köhl-Museum, Pfaffenhofen
an der Roth 021113
Hermann Krause Kunsthandel GmbH,
Köln 076873, 142010
Hermann-Oberth-Raumfahrt-Museum,

Feucht 018354
Hermann, Arthur, Hamburg 107279
Hermann, Rudi, Altenmarkt im
Pongau 062517
Hermanns, Peter, Geilenkirchen . 075960,
129888
Hermano Blodés Muziejus, Nida 030429
Hermano Zudermano Memorialinis
Muziejus, Šilutės Muziejus,
Macikai 030413
Hermanos Cebader, Madrid . . . 133344
Hermanos Garcia, Zaragoza . . . 086598
Hermans, Elisabeth, Bruxelles . 140180
Hermantier, Dany, Florac 068550
Hermay & Bernar, Hautot-sur-
Mer 068800
Hermeling, E., Arnhem 132426
Hermeling, Tobias, Wien 100317
Hermeneutics of Art, New York . 139381
Hermenier, Marcel, Saint-
Berthevin 072691
Hermenier, Pascal, La Croixille . 069020
Herment, Daniel, Zürich 087975
Herment, Paule et Edouard, Nogent-le-
Rotrou 070726
Hermerén, Karin, Landskrona . . 133543
Hermès, Annemasse 140766
Hermes, La Plata 139603
Hermes, Salzkotten 078345
Hermes, Sarajevo 063867
Hermes Restauro, Pisa 131765
Hermes, Dipl.-Restauratorin Sabine,
Köln 130148
Hermes, Elke, Berlin 141528
Hermesdorf, K., Trier 078704
Hermesvilla, Wien Museum,
Wien 003006
Hermet, Paule, Brou 067517
Hermille, Mireille, L'Isle-sur-la-
Sorgue 069571
Herminette, Mornant 070353
L'Herminette, Paris 071474
The Hermitage, Ho-Ho-Kus . . . 049493
The Hermitage – Home of President
Andrew Jackson, Hermitage . . 049453
Hermitage Amsterdam,
Amsterdam 112221
Hermitage Antiquarian Bookshop,
Denver 145127
Hermitage Antiques, Dallas . . . 093139
Hermitage Antiques, Geelong . . 061508
Hermitage Antiques, Honiton . . 089564
Hermitage Foundation Museum,
Norfolk 051576
Hermitage Gallery, Whitby 119336
Hermitage Gatehouse Museum,
Ancaster 005335
Hermitage Rooms, London 044996
Hernandez, Albuquerque 119482
Hernández, Buenos Aires 060907
Hernández, Cali 102410
Hernani, Buenos Aires 060908
Herne Bay Museum and Gallery,
Herne Bay 044578
Herning Auktionshus, Herning . . 125441
Herning Kunstmuseum, Herning 010114
Herning Museum, Herning 010115
Hernitz, Lilo, München 077527
Herode, Avon 066889
Herodek, Peter, Konstanz 076936
Herodote, Lyon 069747
Heroes Aren't Hard to Find,
Charlotte 092605
Heroes Collectibles, Houston . . 093690
Herøy Bygdesamling, Herøy . . . 033650
Herøy Kystmuseum, Herøy 033651
Herold, Helga, Berlin 141529
Herold, O., Leipzig 107834
Herold, Rainer, Hamburg 107280
Herold, Rainer, Kampen 107502
Herold, Thessa, Paris 104948
Heron, Cottbus 141675
Heron, Johannesburg 114796
Heron Corn Mill and Museum of

Papermaking, Beetham 043554
Heron & Son, H.W., Yoxall 091984
Herons Gallery, Castlemaine . . 098876
Herons Reef Historic Gold Diggings,
Fryerstown 001088
Heros, Vilseck 078806
Herovits, Johann, Goldach 133762
Herpin, Hubert, Cannes 067641
Herr der Münzen, Landau in der
Pfalz 077004
Herr, Arnold, Los Angeles 145228
Herr, Holger, Potsdam 108405
Herr, Werner G., Köln . 076874, 126005
Herrbach, Bernard, Châtenois . . 067911,
128552
Herrchen, Erika & Andreas,
Wiesbaden 078979
Herregården Museum, Larvikmuséene,
Larvik 033733
Herrenhaus Altenhof, Altenhof . . 016537
Herrenschmidt, M. & D.,
Prichsenstadt 078090
Herrera, Dallas 135506
Herréra-Gutierrez, Jean-Claude,
Thiers 073651
Herrera de Reyes, Inés, Bogotá . 065207
Herrera, Carlos, Los Angeles . . . 094359,
135721
Herrera, Manuel, Caracas 097956
Herrero Zurdo, Antonio, Madrid . 115235
Herreshoff Marine Museum/America's
Cup Hall of Fame, Bristol 047218
Herrett Center for Arts and Science,
Twin Falls 053867
Herriz, Venezia 081844
Herrle, Christine, Erding 106887
Herrlein, Andreas, Frankfurt am
Main 129848
Herrliberger-Sammlung, Museen Maur,
Maur 041731
Herrmann & Wagner, Berlin . . . 106309
Herrmann, Antonimus, Mertloch 130298
Herrmann, Bärbel, Dresden . . . 075417,
129713
Herrmann, Gabriele, Gräfelfing . 141838
Herrmann, Gunter, Berlin 129505
Herrmann, Hans, Drachselsried . 075403
Herrmann, Peter, Berlin 106308
Herrmann, R., Dachsberg 075290
Herrmann, Thomas, Neumarkt in der
Oberpfalz . . . 077737, 108245, 142224
Herrnhuter Künstlergilde e.V.,
Herrnhut 058918
Herron & Hill, Memphis 145254
Herron Gallery, Indiana University –
Purdue University Indianapolis,
Indianapolis 049684
Herron School of Art, Indiana
University – Purdue University
Indianapolis, Indianapolis 057234
L'Herrou, Vincent, Paris 071475, 071476
Herry, Daniel, Beaune 125508
Hers, Rotterdam 112640
Hersche, Urs, Winterthur 087908, 133968
The Herschel Museum of Astronomy,
Bath 043518
Herschell Carrousel Factory Museum,
North Tonawanda 051618
Herscher, Paris 136938
Hershey Museum, Hershey 049454
Herstein & Co., L.A., Baltimore . 092333,
135344
Hertel Antiquitäten, Niederzier . . 077780,
130460
Herter, Stuttgart 078608
Herter Art Gallery, University of
Massachusetts, Amherst 046450
Hertford Antiques, Hertford 089518
Hertford Museum, Hertford 044579
Hertfordshire Fine Art and Antiques
Fair, Hatfield 098382
Hertling, Edmund, Mudershausen 142132
Hertling, Michael, Berlin 129506
Hertoghe, Denis, Antwerpen . . . 063211

Herts and Essex Antique Centre,
Sawbridgeworth . . . 091249, 135010
Herttoniemen Museo, Helsinki . . 010831
Hertz, Louisville 121649
Hertz, C., Bremen 075166
Hertzmann, Paul M., San
Francisco 124299
Hervas, Granada 115090
Hervas, Paris 136939
Hervas Paya, Virginia, Murcia . . 115314
Hervas Rios, Francisco, Alicante 133082
Hervé & Micheloud, Sion 133923
Hervé, Aline, Saint-Ouen 073105
Hervé, Hugot, Saint-Julien-les-
Villas 072869
Hervé, Iseli, Lausanne 133798
Hervé, Jean François, Nancy . . . 128822
Hervé, Jean-Luc, Geneston 068648
Hervé, Jeannine, Paris 104949, 129036
Hervey Bay Historical Village and
Museum, Scarness 001486
Hervey Bay Regional Art Gallery,
Pialba 001404
Hervieu, Paul, Nice 104373
Hervieux & Motard, Rouen 072590
Hervieux, Jean-Paul, Charmoy . . 067855
Hervieux, Jean-Paul, Saint-Aubin-sur-
Yonne 072681
Hervold, Hamar 113157
Hervy, Eric, Villedieu-les-Poêles 074137
Hervy, Fernand, Villedieu-les-
Poêles 074138
Hervy, Jacqueline, Courthezon . . 068199
Hervy August Bibliothek,
Herzlinger 074132
Herwald, Susan, Cleveland 120430
Herwald, Thomas, Karlsruhe . . . 076696
Herweg, Ursula, Leverkusen . . . 077093
Herwick, Egon, Berlin 074828
Herwig, Aalst 100474
Héry, Georges & Jacques, Chalon-sur-
Saône 067779
Herz, Wilfried, Essen 075681
Herzberg-Rebel, Corina, Ebersbach an
der Fils 129769
Herzenstreu, Düsseldorf 075480
Wolfgang Herzer & Gabriele Hammer,
Weiden in der Oberpfalz 108854
Herzliya Museum of Contemporary Art,
Herzliya 025062
Herzog, Bamberg 106117
Herzog, Küssnacht am Rigi 133787
Herzog Anton Ulrich-Museum,
Kunstmuseum des Landes
Niedersachsen, Niedersächsische
Landesmuseen Braunschweig,
Braunschweig 017558
Herzog August Bibliothek,
Wolfenbüttel 022732
Herzog, Bernhard, Salzburg . . . 127850
Herzog, Christiane, Saint-Ouen . 073106
Herzog, Enrico, Bahretal 141446
Herzog, Georges, Büren zum Hof 116178
Herzog, Helena, Lübeck 077174
Herzog, Lucette, Paris . 104950, 136940
Herzog, Peter, Basel 087238
Herzogkasten, Stadtmuseum
Abensberg, Abensberg 016457
Herzoglicher Kunstbesitz, Schloss
Callenberg, Stiftung der Herzog von
Sachsen-Coburg und Gotha'schen
Familie, Coburg 077751
Herz'sche Heimatstiftung Hilgerhof,
Pittenhart 021142
Heshan Museum, Heshan 007861
Heske, A., Krummendeich 107763
Heskens, Eindhoven 112458
Hesketh Hubbard Art Society,
London 060157
Heskin Hall Antiques, Chorley . . 088902
Hespe Gallery, San Francisco . . 124300
Hesperia, Princeton 139382
Hesperus, Hannover 141914
Hess, New York 135936
Hess-Divo, Zürich 087976
Hess, Dietrich, Karlsruhe 107513

Hess, Hanspeter, Hochdorf 130051
Hess, John M., Manheim 127304
Hess, John M., Mannheim 127306
Hessaby Museum, Tehrän 024759
Hesse, Otego 127383
Hesse, Claire, Lyon 128715
Hesse, Isa, Potsdam 130532
Hessel, Alan, Moreton-in-Marsh 134859
Hesselbach, Lauenförde 107800
Hessing, Manfred, Münster 077665
Hessische Landes- und
Hochschulbibliothek, Darmstadt 017829
Hessische Landesbibliothek Fulda,
Fulda 018561
Hessischer Museumsverband e.V.,
Kassel 058919
Hessisches Braunkohle
Bergbaumuseum, Borken,
Hessen 017513
Hessisches Forst-Kulturhistorisches
Museum, Biebergemünd 017368
Hessisches Forstmuseum Alte
Fasanerie, Hanau 019037
Hessisches Kutschen- und
Wagenmuseum, Lohfelden . . . 020104
Hessisches Landesmuseum Darmstadt,
Darmstadt 017822
Hessisches Puppenmuseum,
Hanau 019038
Hest, Ljubljana 114718, 114719
Hest, Maribor 114732
Hester Rupert Art Museum, Graaff-
Reinet 038721
Hestié Antichità, Taranto 081634
Hesz, Jürgen, Wels 062874
Het Zotte Kunstkabinet,
Mechelen 003854
Hétéroclite, Bordeaux 103435
Hetjens-Museum, Deutsches
Keramikmuseum, Düsseldorf . . 018031
Hétroy, Hugues, Béhen 128465
Hettinger County Historical Society
Museum, Regent 052504
Hettmer, V., Burgthann 075228
Hetzel, Herbert, Goch 076018
Hetzler, Gerhard, Würzburg 079082
Hetzler, Max, Berlin 106310
Hetzner, Ingeborg, Ansbach . . . 105996
Heubeck, Nürnberg 142243
Heubel, L., Bergisch Gladbach . 074731
Heubel, Lothar, Düsseldorf 075481,
075482
Heubel, Lothar, Frankfurt am
Main 075789
Heubel, Lothar, Köln 076875
Heuberger, Roman, Köln 142011
Heuchel, Ute, Stuttgart 130675
Heugas, Michèle, Pau 072041
Heuneburgmuseum, Herbertingen 019196
Heuner, Jürgen, Neu-Anspach . . 130442
Heupgen, Th. J., Haag in
Oberbayern 076140
L'Heure Bleue, Paris 071477
L'Heure Bleue, Rouen 072591
Heureka – The Finnish Science Centre,
Vantaa 011491
L'Heureux, Claude, Montréal . . . 064431
Heurich House Museum,
Washington 054118
Heurtebise, Dijon 140832
Heusburg, Brachtenbach 132326
Heuschneider, Ronny, Potsdam . 130533
Heusel, Monika, Frankfurt (Oder) 075763
Heuser Art Center, Peoria 051986
Heuser-Mantell, Margret,
Düsseldorf 106793
Heuser, Christa, Karlsdorf-
Neuthard 076680
Heuson-Museum im Rathaus,
Büdingen 017513
Heuss, Didier, Le Thillot 069353
Heussenstamm Stiftung, Frankfurt am
Main 018427
Heuts, M., Köln 076876

Register der Institutionen und Firmen

Heutz, J., Gangelt – Hillsboro Museums, Hillsboro

Heutz, J., Gangelt 075940
Hever Castle and Gardens, Edenbridge 044229
Heves Megyei Sportmúzeum, Eger 023618
Hevesi Múzeumi Kiállítóhely, Heves 023685
Hewitt, Charles, Paddington, New South Wales 099523, 127656, 139865
Hewitt, Muir, Halifax 089414
Hewitt, Sue, Mosman 099384
Hexagone, Aachen 105938
Hexehäusl, Schwetzingen 078435
Hexenbürgermeisterhaus, Städtisches Museum, Lemgo 020029
Hexenmuseum, Riegersburg, Steiermark 002598
Hexenmuseum, Ringelai 021436
Hexham Old Gaol, Hexham 044580
Hexter, London 064350
Hey Betty, Pittsburgh 096293
Heybutzki, Gerd, Köln 142012
Heydar Aliyev Museum, Naxçivan 003187
Heydenbluth, E., Barsinghausen 074681
Heydenryk jr., New York 095454, 135937
Heyder, Johannes, Wildenfels . . 079011
Heyduck, Manfred, Düsseldorf . 075483
Heye, Friedrich.W, Unterhaching 137542
Heyer, François, Pfetterhouse . . 072148
Heykel Atölye/Galeri, İstanbul . . 117049
Heym-Oliver House, Russell . . . 052708
Heyman, H., Zaltbommel 112720
Heymann, Darmstadt 075302
Heymann, F., Hannover 076375
Heymann, Henrique, Lisboa . . . 114037
Heymans, P.B.W., Den Haag . . . 132474
Heyn, Johannes, Klagenfurt . . . 140003
Heyne, Wilhelm, München 137406
Heytesbury, Farnham 089263
Heyuan Museum, Heyuan 007863
Heyvaert, M., Gent 063642
Heyward-Washington House, Charleston 047550
Heywood, Minneapolis 121999
Heywood Hill, G., London 144627
Heywood, W.O., Whangarei 132769
Heze Museum, Heze 007864
Heze Zone Museum, Heze 007865
Hezlett House, Coleraine 043996
HfG-Archiv, Ulmer Museum, Ulm 022260
HGST Antik, Göteborg 086711
Hi-Desert Nature Museum, Yucca Valley 054609
Hi Qo, Kingston 111225
Hibberd, H., Philadelphia 145413
Hibbing Historical Museum, Hibbing 049456
Hibel Museum of Art, Jupiter . . 049852
Hibernia-Antiques.com, Dublin . 079559
Hibiscus, Modena 080800
Hick, David, Saint Helier 091183
Hick, David, Saint Lawrence . . . 091193
Hickl, Reinhard, Koppl 062683
Hickl, Reinhard, Salzburg 062797
Hickman, Miami 135797
Hickmet, David, London 090205
Hickories Museum, Elyria 048573
Hickory Grove Rural School Museum, Ogden 051712
Hickory Museum of Art, Hickory 049458
Hicks, London 118381
Hicks Art Center, Newtown 051557
Hicks, Purdy, London 118382
Hickson, Lewis E., Gilberdyke . . 089324
Hicksville Gregory Museum, Hicksville 049461
Hið Íslenzka Reðasafn, Húsavík . 024083
Hida Kokusei Kougei Gakuen, Takayama 056035
Hida Minzoku Kokokan, Takayama 029453
Hida Minzoku-mura, Takayama . 029454
Hida Takayama Bijutsukan, Takayama 029455
Hida Takayama Shunkei Kaikan, Takayama 029456
Hidalgo Pumphouse Heritage, Hidalgo 049462
Hidalgo, Victoria, Madrid 115236
Hidde Nijland Museum Hindeloopen, Hindeloopen 032488
Hidden House, Whangarei 113105, 132770
Hidden Place, Miami 094745
Hidden Splendor, Salt Lake City 123966, 123967
Hidden Treasures, Los Angeles . 094360
Hidden Treasures, Minneapolis . 094919
Hidden Treasures, New York . . . 095455
Hidden Treasures, Philadelphia . 096113, 136089
Hidden Treasures, Richmond . . . 096550
Hidden Treasures of Tulsa, Tulsa 097737
Hiddenite Center, Hiddenite 049463
Hideaway, West Hollywood 097896
Hideaway Antiques, Toronto . . . 064666
Hidell Brooks, Charlotte 119991
Hideout, Hawkes Bay 112913
Hidra, Eivissa 115059
Hidson, Pat, Milwaukee 121924
Hiekan Taidemuseo, Tampere . . 011395
Hieke, Dr. Ursula, Wien 062979, 100318
Hien, Minh, Ho Chi Minh City . . 097984
Hienert, Wolfgang, Wien 127928
Hier & Ailleurs, Paris 071478
Hier Aujourd'hui Demain, Paris . 071479
Hier Comme Aujourd'hui, Calais 067602
Hier et Ailleurs, Montceau-les-Mines 070225
Hier et Aujourd'hui, Avallon 066864
Hier et Aujourd'hui, Rivière-Salée 082456
Hier et Avant Hier Antiguidades, Rio de Janeiro 064066
Hier le Der, Montier-en-Der 070250
Hierapolis Arkeoloji Müzesi, Denizli 042877
Hiermeier, Karl-Heinz, Eggenfelden 075575
Hieronymus, Dresden 106739
Hieronymus, Ludwigsburg 142083
Hierse, K., Neuenkirchen bei Neubrandenburg 077722
Higashi-Hiroshima-shiritsu Bijutsukan, Higashi-Hiroshima 028617
Higashi Kiyo Gallery, West Hollywood 124999
Higashi-Osaka Art Center, Higashi-Osaka 028620
Higashi-Osaka-shiritsu Kyodo Hakubutsukan, Higashi-Osaka . 028621
Higashi-yamate District Historic Preservation Center, Nagasaki . 029091
Higashimurayama Mingeikan, Higashimurayama 028622
Higashiyama Kaii Gallery, Nagano 029071
Higdon &Sons, Jo, Tucson 136420
Higgins & Maxwell, Louisville . . 121650
Higgins Armory Museum, Worcester 054544
Higgins Art Gallery, Cape Cod Community College, West Barnstable 054287
Higgins Museum, Okoboji 051742
Higgins Press, Lostwithiel 090634
Higgins, Barbara, South Yarra . . 062226
Higgins & Sons, Michael D., Tucson 097671
Higgins, Richard, Longnor, Shropshire 134815
Higgins, Sandra, London 118383
High Cliff General Store Museum, Sherwood 053311
High Desert Museum, Bend . . . 046914
High Museum of Art, Atlanta . . . 046623
High Noon, Los Angeles 094361
High on Art, Armadale, Victoria . 098661
High Plains Heritage Center, Cascade County Historical Museum & Society, Great Falls 049218
High Plains Museum, Goodland . 049159
High Plains Museum, McCook . . 050548
High Point Museum, High Point . 049466
High Prairie and District Museum, High Prairie 006003
High Street Antiques, Alcester . . 088239
High Street Antiques & Jewellery, Bantry 079454
High Street Book Shop, Hastings 144455
High Street Books, Honiton 144485
High Touch, Manama 100434
High Wire Gallery, Philadelphia . 123394
Highbrow, Winnipeg 140434
Highfields Fine Art, Highfields . . 099128
Highgate Antiques, Highgate, South Australia 061622
Highgate Fine Art, London 118384
Highgate Gallery, London 118385
Highland Antiques, Cleveland . . 092947
Highland Cultural Arts Gallery, Highland Cultural Center, Highland 049469
Highland House Museum, Hillsboro 049478
Highland Landscapes, Gairloch . 117815
Highland Maple Museum, Monterey 051019
Highland Museum of Childhood, Strathpeffer 045910
Highland Park Antiques, Dallas . 093140
Highland Park Antiques, New York 095456
Highland Park Gallery, Dallas . . 120610
Highland Park Historical Museum, Highland Park 049472
Highland Pioneers Museum, Great Hall of the Clans, Baddeck 005368
Highland Trading Company, Edmonton 064297
Highland Village Museum, An Clachan Gàidhealach, Iona 006039
Highlanders of Strathalbyn, Strathalbyn 062247
Highlands, Fort Washington 048890
Highlands Museum and Discovery Center, Ashland 046562
Highlands Museum of the Arts, Sebring 053256
Highlands Vault, Birmingham . . 092411
Highpoint Center for Printmaking, Minneapolis 122000, 138363
Highsmith, David, San Francisco 145558
Hightower, Fred, Philadelphia . . 136090
Highway, København 065915
Highway Antique Mall, Miami . . 094746
Highway Antiques, Coffs Harbour 061351
Highway Galleria, Williams 099886
Highway Gallery, Mount Waverley 099409
The Highway Gallery, Upton-upon-Severn 091698
Highwic, Auckland 033150
Highwire Gallery, Philadelphia . . 123395
Higuiner, Bruno, Châteauneuf-la-Forêt 067894
Hiihtomuseo, Lahden Kaupunginmuseo, Lahti 011071
Hiiumaa Muuseum, Kärdla 010622
Hijman, D.J., Utrecht 132646
Hikarinotani Metaru Bijutsukan, Imba 028685
Hikmet, Pinar, İstanbul 088123
Hiko-Antik, Salzburg 062798
Hiko-Antik, Seewalchen am Attersee 062842
Hikobae, Kobe 111282
Hikone Castle Museum, Hikone . 028624
Al-Hilal, Manama 100435
Hilary Chapman Fine Prints, Richmond, Surrey 118971
Hilary, Vincent, Igoville 068864
Hilber, Basel 116078
Hilbert, Ferdinand, Mettenheim . 077361
Hilbert, Franz & Ute, Aachen . . . 074333
Hilbrandt, Katja, Haiger 141846
Hild, Bernd, Weilburg 142467
Hildebrand, Kashya, New York . 122702
Hildebrand, Kashya, Zürich 116811
Hildebrandt, Harald, Neuss 142225
Hildebrandt, Petra, Hamburg . . . 141870
Hildene, Manchester 050632
Hilding, Frank, Svendborg 066158
Hildt, Jeremy, Chicago 092725
Hiles, Kansas City 135680
Hilgemann, Kai, Berlin 106311
Hilgenberg, U., Knüllwald 076829
Hilger, Ernst, Paris 104951
Hilger, Prof. Ernst, Wien 100319, 136594
Hill, Gdańsk 084505, 143451
Hill, Oklahoma City 136050
Hill & Wang, New York 138398
Hill Aerospace Museum, Hill Air Force Base 049473
Hill Country Museum, Kerrville . 049942
Hill Cumorah Visitors Center, Palmyra 051899
Hill Farm Antiques, Leckhampstead 089763, 134531
The Hill Gallery, Tetbury 119221
Hill-Hold Museum, Montgomery 051028
Hill House, Helensburgh 044555
The Hill House, Portsmouth Historical Association, Portsmouth 052323
Hill of Tarvit Mansion House, Cupar 044059
Hill-Smith, Sam, Adelaide 098612
Hill-Stead Museum, Farmington 048701
Hill Top, Ambleside 043412
Hill Tribes Museum, Bangkok . . 042660
Hill, Alan, Sheffield 144836
Hill, David, Kirkby Stephen 089715
Hill, Davis, Omaha 136061
Hill, Erin, Mosman 099385
Hill, G.A., Brasted 088580
Hill, J., San Francisco 097252
Hill, Jonathan A., New York . . . 145347
Hill, Laurel, Rokeby 062165
Hill, Marsh, Milwaukee 094861
Hill, Michael, Auckland 083792
Hill, Michael, Queenstown 084064
Hill, Michael, Wellington 084165
Hill, Michael, Whangarei 084203
Hill, Mike, Hamilton . . 064320, 125367
Hill, Petra & Uwe, Merxheim . . . 107997
Hill, T.S., Dorking 144373
Hillcrest Gallery, Hampton 099109
Hillcrest Museum, Souris 007058
Hillcroft Furniture, Houston 093691, 135606
Hille, Henning, Bad Bevensen . . 074494, 125854, 129404
Hillebrand, Wolfgang & Renate, Gauting 075953
Hillel Jewish Student Center Gallery, Cincinnati 047745
Hilleman, Johan Th., Arnhem . . 082939
Hiller, Erlenbach 116257
Hiller Aviation Institute, San Carlos 052959
Hillert, G., Sinzheim 130642
Hilleshuis Museum, Rotterdam . 032803
Hillestad, Dølemo 113143
Hillforest House Museum, Aurora 046677
Hilliard, Kansas City 121296
The Hilliard Society of Miniaturists, Wells 060158
Hilliard, Jane, Tralee 109699
Hilligoss, Tom, Chicago 120153
Hillje, S., Oldenburg 077931
Hills Antiques, Macclesfield 090673
Hills Native Art, Vancouver 101798
Hills Studio, Kalamunda 099163
Hills, J.K., Portland 096417
Hillsboro Area Historical Society Museum, Hillsboro 049481
Hillsboro Fine Art, Dublin 109592
Hillsboro Museums, Hillsboro . . 049475

Hillsborough Railway Museum,
Hillsborough 006005
Hilltop, Hamilton 064321
Hilltop, Stefani, Fish Creek 099018
Hillview Artspace, Wedmore . . . 119316
Hillview Museum, Bagenalstown 024821
Hillwood Art Museum, Long Island
University, Brookville 047275
Hillwood Museum, Hillwood Museum
Foundation, Washington 054119
Hilmarton Manor Press, Calne . . 138144,
144293
Hilprecht-Sammlung Vorderasiatischer
Altertümer der Friedrich-Schiller-
Universität Jena, Jena 019448
Hilsenbeck, Pia Maria, Baunatal 129448
Hilt, Isabelle, Cannes 067642
Hilton, Roch 118983
Hilton Hordern, Cremorne 098932
Hilton, P.J., London 144628
Himachal State Museum, Shimla 024477
Himalaija, Freiburg im Breisgau 075867
Himalayan Vision II, New York . 095457
Himalayan Vision IV, New York . 095458
Himangan Kotiseutumuseo,
Himanka 010885
Himeji-jo, Himeji 028625
Himeji-kenritsu Budokan, Himeji 028626
Himeji-shiritsu Bijutsukan, Himeji 028627
Himeji-shirutsu Bungaku
Hakubutsukan, Himeji 028628
Himezaki Lighthouse Museum,
Ryotsu 029287
Himiko Oriental Gallery,
Stockholm 087006
Himley Hall, Himley 044587
Himmelkeller, Hermann Bauch-
Museum, Kronberg im
Weinviertel 002246
Himmelreich, Beat, Bern 087284, 133679
Himmelsberga Museum – Ölands
Museum för Konst och Kulturhistoria,
Borgholm 040577
Himovitz, Michael, Sacramento . 123772
Hince, Kenneth, Prahran 125226, 139884
Hinchingbrooke House,
Huntingdon 044617
Hinchman House Natural Science
Museum, Maria Mitchell Association,
Nantucket 051155
Hinck, W., Varel 108809
Hinckley and District Museum,
Hinckley 044588
Hinckley Fire Museum, Hinckley 049487
Hinckley Foundation Museum,
Ithaca 049740
Hincz Gyula Állandó Gyűjtemény,
Tragor Ignác Múzeum, Vác . . 024025
Hind, Paddington, Queensland . . 062053
Hinde, Phillippa, Navan 130886
Hinderickx & Winderickx, Utrecht 143245
Hindevik, Jan, Orrefors 126651
Hindhede, Jens Jakob, Græsted 065708
Hindi Sangrahalaya, Allahabad . 024137
Hindley Museum, Hindley 044589
Hindley, Marilyn, Townsville . . . 062339
Hindman, Leslie, Chicago 127167
Hindsbo, Sys, Ringe 066066
Hindsight Books, London 144629
Hindsight Books, Waikato 143354
Hines, Judy, Holt, Norfolk 089549,
117985, 134467
Hingerl, Josef, Bodenkirchen . . 075050
Hinge's Antikvariat, Århus 140553
Hingley, Richmond 136171
Hings, Toronto 064667
Hingstons, Wilton, Salisbury . . . 091854
Hinkel, Flonheim 075753
Hinkes, Gerda, Nürnberg 130467
Hinkler House Museum,
Bundaberg 000911
Hinna Antik, Hinna . . . 084252, 084253
Hinner, Klaus, Pfalzfeld 078039
Hino Gallery, Tokyo 111442

Hino-shiritsu Shinsengumino Furusato
Hakubutsukan, Hino 028631
Hinrichsmeyer, Johannes, Berlin 129507
Hinsdale Historical Society Museum,
Hinsdale 049490
Hinstorff Verlag, Rostock 137490
Hinterglasmuseum, Sandl 002643
Hinterland Art Centre,
Landsborough 099217
Hinterlandmuseum Schloss
Biedenkopf, Biedenkopf 017372
Hintermayer, Sankt Pölten 140035
Hintermayer, Wien . . . 062980, 140085,
140086, 140087, 140088
Hintermayer, Kurt, Salzburg . . . 140024
Hinterreiter, Manfred,
Schwertberg 062839
Hinterschweiger, Friedrich,
Murau 140018
Hintgen, Joël, Mamer 082378
Hintz, H.-Dieter, Edesheim 075569
Hinz, Jutta, Fredersdorf 129856
Hiper Ocasion, Palma de
Mallorca 086356
Hipfl, Eva, Zell am See 140139
Hipologické Múzeum, Múzeum Koni,
Topoľčianky 038468
Hipp-Halle, Gmunden 001952
Hipp, Ed, Dallas 093141
Hippe, Horst, Rednitzhembach . . 078158
L'Hippocampe, Aigues-Mortes . . 103182
Hippologické Muzeum, Slatiňany 009884
Hirabayashi, New Orleans 135859
Hirado Kanko Shiryokan, Hirado 028632
Hiraide Gatokan, Shiojiri 029390
Hiraizumi Historical Artifacts Museum,
Hiraizumi 028636
Hirakata Municipal Gotenyama Art
Center, Hirakata 028639
Hiram Blauvelt Art Museum,
Oradell 051789
Hiram Smith Octagon House,
Neenah 051209
Hirano Image Library, Hirano
Hakabutsukan, Osaka 029243
Hirano Soundscape Museum, Yamato-
koriyama 029703
Hirata Kinenkan, Takayama 029457
Hiratsuka Bijutsukan, Hiratsuka . 028642
Hiratsuka-shi Hakubutsukan,
Hiratsuka 028643
Hiraya Gallery, Manila 113406
Hird, Kalgoorlie 099168
Hiret & Nugues, Laval (Mayenne) 125609
Hirlam, Gérard, Nice 140998
Hirmer Verlag, München 137407
Hirnickel, Elisabeth, Bad
Kissingen 106072
Hirnsperger, Rupert, Koppl 127792
Hirosaki-shiritsu Hakubutsukan,
Hirosaki 028644
Hirose-machi Rekishi oyo Minzoku
Hakubutsukan, Hirose 028645
Hiroshima Art Club, Hiroshima . 059493
Hiroshima Bijutsukan, Hiroshima 028647
Hiroshima Castle, Hiroshima . . . 028648
Hiroshima Children's Museum,
Hiroshima 028649
Hiroshima City Museum of History and
Traditional Crafts, Hiroshima . 028650
Hiroshima Heiwa Kinen Shiryokan,
Hiroshima 028651
Hiroshima-kenritsu Bijutsukan,
Hiroshima 028652
Hiroshima-shi Gendai Bijutsukan,
Hiroshima 028653
Hirota, Tatsuo, Tokyo 111443
Hirsaver, Miloš, Hradec Králové . 065390
Hirsch, Miami 121803
Hirsch, Catherine, Paris 071480
Hirsch, H., Bedburg 074706
Hirsch, Siegfried-H., Bamberg . . 141449
Hirsch, W., Richmond 096551
Hirschberger Heimatstuben,

Alfeld 016504
Hirschheydt, Harro von,
Wedemark 142465
Hirschi, Peter B., Oberhofen am
Thunersee 087718, 144118
Hirschl & Adler, New York 122703
Hirschl & Adler Galleries, New
York 122704
Den Hirschsprungske Samling,
København 010176
Hirschwirtscheuer-Museum für
die Künstlerfamilie Sommer,
Künzelsau 019816
Hirsel Homestead Museum,
Coldstream 043995
Hirsh, London 090206, 090207
Hirshhorn Museum and Sculpture
Garden, Smithsonian Institution,
Washington 054120
Hirsmäen Museoalue, Valkeala . 011484
Hirst, London 090208
Hirt, Andreas, Pöllau 127835
Hirt, Patrice, La Neuveville 087709,
133874
Hirt, Robert & Margrit,
Richterswil 116582, 144122
His & Hers Collectibles, Houston 093692
His Galleri, His 113163
His Place Too, San Antonio 096977
Hischier, V., Meggen 133855
Hiscock House, Trinity 007236
Hispa-Bel, Bruxelles 063463
Hispanic American Arts, New
York 139383
Hispanic Society of America, New
York 060509
The Hispanic Society of America, New
York 051404
Hispantics, Antwerpen 063212
Histo-Rail Limousin du Chemin de Fer,
Saint-Léonard-de-Noblat 015419
Histoire Ancienne, Bruxelles . . . 063464
Histoire d'A, Bruxelles 063465
Histoire d'un Siècle, Vitré 074256
Historama – Museum für Technik und
Verkehr, Nostalgiebahnen in Kärnten,
Ferlach 001889
Historama Kloster Hornbach,
Hornbach 019341
Historia Auktionshaus,
Ganderkesee 125955
Historia Elstal, Wustermark 022831
Historia Hamburg, Hamburg . . . 076241
Historia Verlag, Basel 144021
Historia Verlag, Freiburg im
Breisgau 075868, 137211
Historia Verlag, Wolfurt 063123
Historial de la Grande Guerre,
Péronne 014835
Historial du Gouffre, Padirac . . . 014602
Historial/Musée de Cire, La Chaise-
Dieu 013375
Historiallinen Museo, K.H. Renlundin
Museo, Kokkola 011016
Historic Alexandria Antiques Show,
Alexandria 098450
Historic Allaire Village, Allaire . . 046404
Historic and Ethnic Instrument
Collection, University of Western
Australia, Nedlands 001347
Historic Annapolis Antiques Show,
Annapolis 098451
Historic Annapolis Foundation,
Annapolis 046502
Historic Babcock Mill, Odessa . . 006540
Historic Bath, Bath 046810
Historic Bethabara Park, Winston-
Salem 054489
Historic Bethany, Bethany 046953
Historic Bowens Mills and Pioneer
Park, Middleville 050869
Historic Brattonsville, McConnells 050547
Historic Burlington County Prison
Museum, Mount Holly 051089

Historic Camden Revolutionary War
Site, Camden 047385
Historic Carson House, Marion . 050692
Historic Charleston Foundation,
Charleston 047551
Historic Charlton Park Museum,
Hastings 049401
Historic Cherry Hill, Albany 046341
Historic Columbia Foundation,
Columbia 047911
Historic Columbus Foundation,
Columbus 060510
Historic Costume & Textiles Collection,
Ohio State University, College of
Human Ecology, Columbus . . . 047938
Historic Crab Orchard Museum and
Pioneer Park, Tazewell 053692
Historic Cragfont, Castalian
Springs 047460
Historic Daniel Boone Home and
Boonesfield Village, Lindenwood
University, Defiance 048221
Historic Deepwood Estate, Salem 052890
Historic Deerfield, Deerfield 048218
Historic District Antique Mall,
Phoenix 096214
Historic Dockyard Chatham,
Chatham 043910
Historic Edenton, Edenton 048471
Historic Fallsington, Fallsington . 048692
Historic Ferryland Museum,
Ferryland 005805
Historic Fort Steilacoom, Tacoma 053643
Historic Fort Stockton, Fort
Stockton 048878
Historic Fort York, Toronto 007180
Historic Gardner's Basin, Atlantic
City 046643
Historic General Dodge House, Council
Bluffs 048034
Historic Georgetown, Georgetown 049076
Historic Governors' Mansion,
Cheyenne 047629
Historic Halifax, Halifax 049318
Historic Hat Creek Ranch, Cache
Creek 005511
Historic Heights Antiques,
Houston 093693
Historic Hermann Museum,
Hermann 049642
Historic Hope Plantation, Windsor 054471
Historic House of Fine Art,
Yandina 099934
Historic Houses of Odessa, Delaware,
Odessa 051708
Historic Hudson Valley, Tarrytown 053685
The Historic Indian Agency House,
Portage 052268
Historic Jefferson College,
Washington 054183
Historic Kenmore, Fredericksburg 048954
Historic Landmarks Foundation of
Indiana, Indianapolis 049685
Historic Langhorne Museum,
Langhorne 050131
Historic Latta Plantation,
Huntersville 049611
Historic Liberty Jail Visitors Center,
Liberty 050273
Historic Lincoln, Lincoln 050304
Historic London Town, Edgewater 048475
Historic Lyme Village, Bellevue . 046885
Historic Madison House, Madison 050583
Historic Mann House, Concord . 047956
Historic Marine, Bruxelles 063466
Historic Markerville Creamery,
Markerville 006264
Historic Mill Creek, Mackinaw
City 050559
Historic Mobile Preservation Society
Museum, Mobile 050976
Historic Morven, New Jersey State
Museum, Princeton 052366
Historic Museum of Arkansas, Little

Rock 050335
Historic New England, Harrison Gray
 Otis House, Boston 047115
Historic New Harmony, New
 Harmony 051259
The Historic New Orleans Collection,
 New Orleans 051294
Historic Newton Home, Decatur 048205
Historic Northampton,
 Northampton 051625
Historic Occoquan, Occoquan . . 051698
Historic O'Keefe Ranch, Vernon,
 British Colombia 007314
Historic Palmyra, Palmyra 051900
Historic Pensacola Village,
 Pensacola 051980
Historic Pithole City, Drake Well
 Museum, Pleasantville 052207
Historic Pittsford, Pittsford 052184
Historic Preservation Association of
 Bourbon County, Fort Scott . . . 048870
Historic Resources Centre, Annan 043419
Historic Richmond Town, Staten
 Island 053498
Historic Rock Ford, Lancaster . . 050119
Historic Rosedale, Charlotte . . 047573
Historic Rugby, Rugby 052701
Historic Saint Mary's City, Saint Mary's
 City 052832
Historic Sauder Village, Archbold 046522
Historic Schaefferstown,
 Schaefferstown 053191
Historic Scotland, Edinburgh . . 044245
Historic Shepherdstown Museum,
 Shepherdstown 053307
Historic Smithfield, Blacksburg . 047027
Historic Spanish Point, Osprey . 051833
Historic Speedwell, Morristown . 051063
Historic Stagville, Bahama 046711
Historic Stagville, Durham 048408
Historic Waco Foundation, Waco 054016
Historic Ward-Meade Park Museum,
 Topeka 053772
Historic White Pine Village,
 Ludington 050521
Historic Yale Museum, Yale . . . 007493
Historic Zion Schoolhouse,
 Toronto 007181
Historica Antiquariat, Dresden . 141698
Historical and Ethnological Museum of
 Cappadocia, Kavala 023124
Historical and Ethnological Museum of
 Patras, Patras 023217
Historical and Ethnological Museum of
 the Mani, Kranae 023145
Historical and Folk Art Museum of
 Gavalohori, Chania 023040
Historical and Folklore Museum of
 Rethymnon, Rethymnon 023244
Historical and Genealogical Society of
 Indiana County, Indiana 049673
Historical and Natural History Museum
 of Natick, Natick 051200
Historical Arms Series, Alexandria
 Bay 139384
Historical Center for Southeast New
 Mexico, Roswell 052691
Historical Collection Electronical
 Engineering, Technische Universiteit
 Delft, Faculteit ITS, Delft 032155
Historical Design Collection, New
 York 095459
Historical Electronics Museum,
 Linthicum 050319
Historical Heritage Complex in Old
 Harry – CAMI, Old Harry 006547
Historical Investments, Buffalo . 092533
Historical Living Village and Pioneer
 Museum at Shandro, Shandro . 007008
Historical Museum, Jerusalem . 025080
Historical Museum, Mahébourg . 030787
Historical Museum and Pioneer Village,
 Sundre 007100
Historical Museum at Fort Missoula,

Missoula 050959
The Historical Museum at Saint
 Gertrude, Cottonwood 048027
Historical Museum Chokokan,
 Saga 029293
Historical Museum of Anthiros,
 Karditsa 023112
Historical Museum of Arta, Arta 022970
Historical Museum of Azogire,
 Azogire 023034
Historical Museum of Bay County, Bay
 City 046831
Historical Museum of Cecil County,
 Elkton 048542
Historical Museum of Crete,
 Iráklion 023096
Historical Museum of Frederick County,
 Frederick 048945
Historical Museum of Old Randolph,
 Randolph 052463
Historical Museum of Palm Beach
 County, West Palm Beach 054316
Historical Museum of Portitsa,
 Karditsa 023113
Historical Museum of Private
 Schools at the Foreign Settlement,
 Nagasaki 029092
Historical Museum of Saint James-
 Assiniboia, Winnipeg 007446
Historical Museum of Sansia Town,
 Sansia 042435
Historical Museum of Southern Florida,
 Miami 050830
Historical Museum of Talbot County,
 Easton 048455
Historical Museum of the National
 Resistance of Rentina, Agrapha,
 Karditsa 023114
Historical Portraits, London . . . 118386
Historical Society Museum,
 Hudson 049603
Historical Society Museum of Carroll
 County, Westminster 054342
Historical Society Museum of
 Ocean Grove, New Jersey, Ocean
 Grove 051701
Historical Society Museum of Saint
 Kilda, Saint Kilda, Victoria 001469
Historical Society Museum of the
 Town of Greenwich, Greenwich 049271
Historical Society of Berks County
 Museum, Reading 052482
Historical Society of Bloomfield
 Museum, Bloomfield 047034
Historical Society of Brookfield,
 Brookfield 047240
Historical Society of Kent County
 Museum, Chestertown 047623
Historical Society of Middletown
 and the Wallkill Precinct,
 Middletown 050866
Historical Society of Oak Park and
 River Forest, Oak Park 051667
Historical Society of Old Yarmouth
 Museum, Yarmouth Port 054571
Historical Society of Pennsylvania,
 Philadelphia 052055
Historical Society of Quaker Hill and
 Pawling, Pawling 051950
Historical Society of Wells and
 Ogunquit Museum, Wells 054272
Historical Woolscour Blackall,
 Blackall 000862
Historical, Folk Art and Natural History
 Museum of Kozani, Kozani 023144
Historicana, Burlingame 145025
Historické a Národopisné Múzeum
 Želovce, Želovce 038484
Historické Múzeum ve Slavkově u
 Brna, Slavkov u Brna 009885
Historické Múzeum, Národní Muzeum,
 Praha 009797
Historické Múzeum, Slovenské
 Národné Múzeum, Bratislava . 038308

Historicum, Sargans 041936
Historicum – Sammlung für
 Bergbaukunde, Modellsammlung,
 Winkler-Gedenkstätte und Karzer, TU
 Bergakademie Freiberg, Institut für
 Wissenschaft- und Technikgeschichte,
 Freiberg, Sachsen 018471
Historie, Emmen 083242
Historiekamer Hardenberg,
 Hardenberg 032408
Historiengewölbe, Rothenburg ob der
 Tauber 021506
Historijski Muzej Bosne i Hercegovine,
 Sarajevo 004275
Historik Ambiente, Dietfurt an der
 Atlmühl 075354
Historisch-Archäologisches Museum,
 Österreichischer Archäologie Bund,
 Markgrafneusiedl 002366
Historisch Documentatiecentrum voor
 het Nederlands Protestantisme 1800-
 Heden, Amsterdam 031944
Historisch Museum, Capelle aan den
 IJssel 032134
Historisch Museum Arnhem,
 Arnhem 032021
Historisch Museum De Bevelanden,
 Goes 032355
Historisch Museum de Scheper,
 Eibergen 032280
Historisch Museum De Tien Malen,
 Putten 032770
Historisch Museum Den Briel,
 Brielle 032109
Historisch Museum Deventer,
 Deventer 032217
Historisch Museum Ede, Ede . . . 032265
Historisch Museum Grenadiers
 en Jagers, Oranje Kazerne,
 Arnhem 032022
Historisch Museum Haarlemmermeer,
 Hoofddorp 032501
Historisch Museum Hedel, Hedel 032424
Historisch Museum Hengelo, Hengelo,
 Overijssel 032464
Historisch Museum Het Kleine Veenlo,
 Veenendaal 032973
Historisch Museum Het Palthe Huis,
 Oldenzaal 032712
Historisch Museum Piet Dorenbosch,
 Boxtel 032094
Historisch Museum Rotterdam Het
 Schielandshuis, Rotterdam . . . 032804
Historisch Museum Ter Aar, Ter
 Aar 031884
Historisch Museum Tweestromenland,
 Beneden Leeuwen 032059
Historisch Museum Warsenhoeck,
 Nieuwegein 032674
Historisch Museum Wolfheze,
 Wolfheze 033059
Historisch Museum Zuid-
 Kennemerland, Haarlem 032394
Historisch Muzeum Hazerswoude,
 Hazerswoude Dorp 032423
Historisch Openluchtmuseum
 Eindhoven, Eindhoven 032284
Historisch Pijp en Tabaksmuseum,
 Sint-Niklaas 004018
Historisch-Technisches
 Informationszentrum,
 Peenemünde 021088
Historische Arztpraxis, Burgstädt 017699
Historische Ausstellung "Fragen an die
 deutsche Geschichte", Berlin . 017268
Historische Ausstellung Krupp,
 Essen 018296
Historische Bibliothek, Städtische
 Museen Quedlinburg,
 Quedlinburg 021245
Historische Collectie UMCG,
 Universiteitsmuseum, Groningen 032379
Historische Dauerausstellung Schloss
 Borbeck, Essen 018297

Historische Edition, Vellmar 142450
Historische Einungsmeisterstube,
 Dogern 017923
Historische Fahrzeugsammlung,
 Simmelsdorf 021865
Historische Fraunhofer-Glashütte,
 Benediktbeuern 017028
Historische Genie Verzameling,
 Vught 033013
Historische Gerberei, Burg bei
 Magdeburg 017679
Historische Gesellenherberge,
 Herbergsmuseum, Blankenburg 017407
Historische Hammerschmiede,
 Blaubeuren 017413
Historische Handwerkerstuben,
 Gingst 018700
Historische Höhler, Stadtmuseum Gera,
 Gera 018646
Historische Käsküche,
 Wiggensbach 022650
Historische Kuranlagen und Goethe-
 Theater Bad Lauchstädt, Bad
 Lauchstädt 016813
Historische Lehrsammlung der
 Feuerwehr Köln, Köln 019687
Historische Lochgefängnisse im Alten
 Rathaus, Museen der Stadt Nürnberg,
 Nürnberg 020858
Historische Mühle, Stiftung Preußische
 Schlösser und Gärten Berlin-
 Brandenburg, Potsdam 021185
Historische Ochsentretanlage
 im Brunnenhausmuseum,
 Schillingsfürst 021650
Historische Ölmühle, Simonswald 021870
Historische Ortssammlung,
 Reigoldswil 041868
Historische Räume des Stadtschlosses,
 Fulda 018560
Historische Sammlung der von
 Bodelschwingh'schen Anstalten,
 Bielefeld 017375
Historische Sammlung Eltville, Eltville
 am Rhein 018170
Historische Sammlungen im
 Gesundheitspark, Bad Gottleuba 016775
Historische Schlosserei Pelzer,
 Städtische Museen Jena, Jena 019449
Historische Senfmühle,
 Monschau 020457
Historische Spinnerei Gartetal,
 Gleichen 018713
Historische Turmstuben, Städtisches
 Museum, Winnenden 022683
Historische Verzameling Aan- en
 Afvoertroepen, Stroe 032910
Historische Verzameling Intendance,
 Bussum 032129
Historische Verzameling Korps
 Nationale Reserve, Harskamp . 032415
Historische Verzameling
 Luchtdoelartillerie, Ede 032266
Historische Verzameling Militair
 Geneeskundige Dienst,
 Loosdrecht 032602
Historische Verzameling Regiment van
 Heutsz, Arnhem 032023
Historische Wassermühle Dickendorfer
 Mühle, Dickendorf 017882
Historische Weberei, Braunsdorf 017568
Historische Wehranlage, Mühlhäuser
 Museen, Mühlhausen/Thüringen 020484
Historische Werkstatt Nordenstadt,
 Wiesbaden 022639
Historischen Fabrikanlage Maste-
 Barendorf mit Nadelmuseum und
 Gelbgießerei, Iserlohn 019425
Historischer Ausstellungsraum im
 Heizkraftwerk Moabit, Berlin . . 017171
Historischer Eisenhammer, Roth 021503
Historischer Hafen Berlin, Berlin 017172
Historischer Kunstbunker, Museen der
 Stadt Nürnberg, Nürnberg 020859

Historischer Saal der Fischerzunft,
Würzburg 022790
Historischer Schieferbergbau Lehesten,
Lehesten 019954
Historischer Straßenbahnhof Leipzig-
Möckern, Leipzig 019975
Historisches Archiv der Stadt Köln,
Köln 019688
Historisches Bergamt Bad Wildungen-
Bergfreiheit, Bad Wildungen .. 016900
Historisches Binz Museum, Binz,
Ostseebad 017392
Historisches Centrum Hagen,
Hagen 018914
Historisches Erzbergwerk im
Silbergbergwerk Bodenmais,
Bodenmais 017447
Historisches Feuerwehrmuseum
Lüchow-Dannenberg,
Dannenberg 017811
Historisches Glasapparate-Museum,
Cursdorf 017788
Historisches Heimatmuseum Östringen,
Östringen 020965
Historisches Kabinett Burgruine
Eckartsburg, Eckartsberga ... 018089
Historisches Käsereimuseum,
Altusried 016565
Historisches Museum,
Bischofszell 041348
Historisches Museum am Hohen Ufer,
Hannover 019053
Historisches Museum am Strom –
Hildegard von Bingen, Bingen am
Rhein 017389
Historisches Museum Arbon,
Arbon 041227
Historisches Museum Aurich,
Aurich 016681
Historisches Museum Bamberg,
Museen der Stadt Bamberg,
Bamberg 016948
Historisches Museum Basel,
Basel 041268
Historisches Museum Bayreuth,
Bayreuth 016989
Historisches Museum Bern, Musée
Historique de Berne, Bern 041315
Historisches Museum Bielefeld,
Bielefeld 017376
Historisches Museum Blumenstein,
Historische Museum Solothurn,
Solothurn 042001
Historisches Museum Bremerhaven,
Bremerhaven 017607
Historisches Museum der Pfalz Speyer,
Speyer 021931
Historisches Museum der Stadt Baden,
Baden 041252
Historisches Museum Frankfurt am
Main, Frankfurt am Main 018428
Historisches Museum für Mittelanhalt
und Bach-Gedenkstätte im Schloss
Köthen, Köthen 019747
Historisches Museum Hanau – Schloß
Philipsruhe, Hanau 019039
Historisches Museum Heiden,
Heiden 041588
Historisches Museum im Marstall,
Paderborn 021062
Historisches Museum im Spital,
Dinkelsbühl 017903
Historisches Museum Luzern,
Luzern 041707
Historisches Museum Obwalden,
Sarnen 041939
Historisches Museum Olten,
Olten 041822
Historisches Museum Regensburg,
Museen der Stadt Regensburg,
Regensburg 021322
Historisches Museum Saar,
Saarbrücken 021557
Historisches Museum Schloß Gifhorn,

Museen des Landkreises Gifhorn,
Gifhorn 018695
Historisches Museum Schloß Urach,
Bad Urach 016893
Historisches Museum Thurgau, Schloß
Frauenfeld, Frauenfeld 041490
Historisches Museum Uri, Altdorf 041210
Historisches Museum Verden,
Domherrenhaus, Verden ... 022318
Historisches Museum Wiedlisbach,
Wiedlisbach 042136
Historisches Portfolio, Völklingen 078812,
142454
Historisches Schmucksteinbergwerk
Silberschacht, Bach an der
Donau 016687
Historisches Schuhmuseum,
Landsberg am Lech 019878
Historisches Silberbergwerk "Hella-
Glückstollen", Neubulach 020685
Historisches Silbererzbergwerk Grube
Samson und Heimatmuseum
Sankt Andreasberg, Sankt
Andreasberg 021586
Historisches Straßenbahn-Depot Sankt-
Peter, Nürnberg 020860
Historisches Straßenbahndepot, Halle,
Saale 018961
Historisches und Völkerkundemuseum,
Stiftung St.Galler Museen, Sankt
Gallen 041921
Historisches Zentrum Engelshaus,
Museum für Frühindustrialisierung,
Wuppertal 022820
Historisches Zentrum, Museum
für Frühindustrialisierung,
Wuppertal 022819
Historisk Museum, Slangerup .. 010321
Historisk Museum, Kulturhistorisk
Museum, Universitetet i Oslo,
Oslo 033835
Historiska Museet, Stockholm . 041005
Historiska Museet, Lunds Universitet,
Lund 040807
Historium de Sedan, Sedan 015748
History, Pittsburgh 096294
History, Trieste 081792
History and Legendary Museum,
Kuah 030678
History and Natural History Museum,
Alger 000038
History and Traditions Museum,
Lackland Air Force Base 050053
History Banque, Houston 093694
History Center of the United Methodist
Church, Madison 050589
History Merchant, Dallas 093142, 145115
History Museum, Ahmednagar . 024129
History Museum & Historical Society
of Western Virginia, Roanoke . 052603
History Museum and Nature Museum,
Los Gatos 050478
History Museum for Springfield-
Greene County, Springfield ... 053467
History Museum of Armenia,
Yerevan 000733
History Museum of the Taiping
Heavenly Kingdom in Nanjing,
Nanjing 008106
History Museum, Rhodes University,
Grahamstown 038724
History of Art, School of Arts, Culture
and Environment, Edinburgh . 056528
History of Medicine Collections, Duke
University, Durham 048409
History of Transportation, Western
Development Museum, Moose
Jaw 006420
The History Place, Morehead City 051053
History San José, San Jose ... 053038
History Shop, Wigan 046134
Historyczny Oddział Muzeum w
Grudziądzu, Grudziądz 034774
Hita Salamanca, Horacio Perez,

Barcelona 085788
Hitachi Kyodo Hakubutsukan,
Hitachi 028658
Hitchcocks, Limpley Stoke 118112
Hitchcox, Rupert, Chalgrove ... 088821
Hitchin Museum and Art Gallery,
Hitchin 044591
Hitit, Ankara 088093
Hitomi Bijutsukan, Eigenji 028530
Hittite, Toronto 101624
Hittleman, Michael, West
Hollywood 125000
Hitzeroth, Dr. Wolfram, Marburg 137374
Hitzler, Dr., Bochum 141606
The Hive, Cambridge 088764
Hiwa Kagaku Hakubutsukan,
Hiwa 028659
Hiwan Homestead Museum,
Evergreen 048648
Hiwwelhaus, Marpingen 020308
Hixon House, La Crosse 050028
Hiyake, Grimaud 103796
Hiyas ng Bulakan Museum,
Malolos 034513
Hjärne, C., Helsingborg 115684
Hjallese Bogantikvar, Odense . 140638
Hjedding Mejerimuseum, Ølgod . 010263
Hjemmevaernsmuseet, Padborg 010272
Hjemsted Oldtidspark, Skærbæk 010302
Hjerl Hedes Frilandsmuseum,
Vinderup 010362
Hjertholm, Bergen 113133
Hjo Auktionstjänst, Hjo 126602
Hjorteds Antikvariat, Hjorted .. 143860
HK Antikt, Västerås 087141
HK International, Paris 071481
Hlubuček, Ivan, Hradec Králové . 065391
Hlucháň, Petr & Chaloupková,
Brno 065349
HMAS Castlemaine – Museum Ship,
Williamstown 001642
HMCS Haida, Hamilton 005974
HMCS Sackville, Halifax 005952
HMS Belfast, London 044997
HMS Ganges Museum, Ipswich . 044645
HMS Trincomalee, Hartlepool . 044524
HMS Victory, Portsmouth 045553
HMS Warrior 1860, Portsmouth . 045554
Ho, Hong Kong 102085
Ho-Am Art Museum, Yongin .. 030062
Ho Art Antiques, Singapore ... 085423
Ho Chi Minh City Fine Arts Museum,
Ho Chi Minh City 054804
Ho Chi Minh City University of Fine Art,
Ho Chi Minh City 058158
Ho Chi Minh Memorial House, Thua
Thien Hue 054818
Ho Chi Minh Museum, Ha Noi . 054796
Ho Chi Minh Museum, Ho Chi Minh
City 054805
Ho Chi Minh Museum, Thua Thien
Hue 054819
Ho Ho, Taichung 116894
Ho Phrakèo, Vientiane 030088
Ho Sam Choi, Hong Kong 065050
Ho, Cheung, Hong Kong 065049
Ho, Doris, Hong Kong 102086
Hoa et B Art, Sélestat 105667
Hoa Sanh, Vancouver 064828
Hoa, Sen, Ha Noi 125079
Hoalu, Neumarkt in der
Oberpfalz 077738
Hoam, Seoul 111664
Hoang, Hac, Ho Chi Minh City . 125123
Hoang, Kim, Ho Chi Minh City . 125124
Hoard Historical Museum, Fort
Atkinson 048779
Hoarders, Auckland 083793
Hoare, Appley, London 090209
Hoare, Bob, Ringmer 091104
Hoare, Peter, Southborough .. 091375
Hoban, Hindmarsh 061624
Hobart, San Francisco 124301
Hobart Antiques, Hobart 061631

Hobart Heritage Museum, Hobart 001144
Hobart Historical Society Museum,
Hobart 049494
Hobbies Museum of Fretwork,
Dereham 044089
The Hobbit Bookstore, Tanilba
Bay 139927
Hobbit Hill Antiques, Vancouver . 064829
Hobbits Antiques, Ballydehob .. 079451
Hobbs & Bishops, Hong Kong . 065051
Hobbs Parker, Ashford 126802
Hobbs, Stuart, Horley 134473
Hobbs, Susan, Toronto 101625
Hobby, Gdynia 084518
Hobby- och Leksaksmuseum,
Stockholm 041006
Hobby Lobby, Phoenix 096215
Hobden, Michael, Battery Point . 061107
Hobe, John & George T.,
Philadelphia 096114, 136091
Hobel, Graz 062590
Hobelmuseum Langenfeld,
Langenfeld 019913
Hoberg, Hermann, Düsseldorf . 075484
Hobi Sanat Galerisi, İstanbul ... 117050
Hobin & Blunt, Los Angeles ... 094263
Hobro Museum, Sydhimmerlands
Museum, Hobro 010125
Hobrovejens Bogantikvariat,
Aalborg 140545
Hobson, Ursula, Philadelphia ... 096115
Hobsons, Wellington 084166
Hobstar, Ramona 139385
Hoch + Partner, Galerie und Werkstatt
für Holzschnitt und Hochdruck,
Leipzig 107835
Hochburg-Museum,
Emmendingen 018182
Hochhauser, Mainburg 062727
Hochheimer Kunstsammlung,
Hochheim am Main 019270
Hochhuth, Walter D., Hamburg . 141871
Hochman, Irena, New York 122705
Hochofen-Museum Bundschuh,
Thomatal 002826
Hochofenmuseum Radwerk IV,
Vordernberg 002880
Hochreuther, Max S, Zürich ... 116812
Hochrheinmuseum, Bad
Säckingen 016853
Hochschul-und Landesbibliothek,
Fulda 018561
Hochschule der Bildenden Künste Saar,
Saarbrücken 055702
Hochschule der Künste Bern, Berner
Fachhochschule, Bern 056419
Hochschule für Bildende Künste
Braunschweig, Braunschweig . 055491
Hochschule für Bildende Künste
Dresden, Dresden 055516
Hochschule für bildende Künste
Hamburg, Hamburg 055571
Hochschule für Gestaltung Offenbach
am Main, Kunsthochschule des
Landes Hessen, Offenbach am
Main 055676
Hochschule für Gestaltung und Kunst
HGK FHNW, Basel 056407
Hochschule für Gestaltung und Kunst
Luzern, Luzern 056434
Hochschule für Gestaltung
Fachhochschule Schwäbisch-Hall,
Schwäbisch Hall 055707
Hochschule für Grafik und Buchkunst,
Leipzig 055625
Hochschule für Grafik und Buchkunst
Leipzig, Leipzig 055626, 137344
Hochschwabmuseum, Sankt
Ilgen 002663
Hochstattffl, Michael, Klafstein .. 062703
Hochstift Meissen, Meißen 020346
Hochwaldmuseum, Hermeskeil . 019215
Hock, B., Miami 121804
Hock, Hong, Singapore 114535

Hockaday Museum of Arts,
Kalispell 049864
Hockey Hall of Fame, Toronto . . 007182
Hockey, Garry, Hamilton, Victoria 061585
Hockin, Keith, Stow-on-the-Wold 091471
Hocking Valley Museum of Theatrical
History, Nelsonville 051213
Hockley, William, Petworth 091001
Hodesh & Assoc., Cincinnati . . . 127175
Hodge, Sarah, Worcester 091946
Hodgeville Community Museum,
Hodgeville 006009
Hodgins, Calgary 125363
Hodgkins & Co, Ian, Slad 144849
Hodgkinson, Kenthurst 099184
Hodgson, Alan, Great Shefford . 089378,
134424
Hodiny & Hubeny, Praha 065498
Hodkins, Ian, Stroud 144866
Hodnet Antiques, Hodnet 089541
Hodsoll, Christopher, London . . . 090210
Höbarth- und Madermuseum der Stadt
Horn, Horn 002104
Hoëbeke, Paris 136941
Höch, Janine, Villach 100212
Höche, Gabriele, Berlin 074829, 141530
Höchsmann, Sven,
Neckarwestheim 077698
Höchstetter, Renate, Bamberg . . 074664
Höchtberger, Hans, München . . 142162
Hoeckhuys, Breda . . . 082987, 132456
Höckner, Walter, Salzburg 062799
Höderath, Ulla & Peter, Bergisch
Gladbach 074732
Hoedt, W., Frankfurt am Main . . 075790
van der Hoeff, Korst, 's-
Hertogenbosch 126289
Höfler, Joachim, Darmstadt 129671
Höganäs Museum och Konsthall,
Höganäs 040699
Högarps Bymuseum, Vetlanda . . 041161
Höglinger, Barbara, Gmunden . . 062571
Höglinger, Wolfgang, Laakirchen 062705
Höglund, Linköping 086814
Höh, Helmut von der, Hamburg . 141872
Höher, Dolores, Feistritz an der
Drau 139979
Höhere Abteilung für Kunst und
Design, HTBLVA Graz-Ortweinschule,
Graz 054930
Höhere Bundeslehranstalt für
künstlerische Gestaltung, Linz . 054949
Höhere Grafik und
Kommunikationsdesign, HTL Bau und
Kunst – Innsbruck, Innsbruck . . 054942
Höhere Graphische Bundes-Lehr- und
Versuchsanstalt, Wien 054965
Höhlenkundemuseum Dechenhöhle,
Iserlohn 019426
Höhlenmuseum, Frasdorf 018458
Höhlenmuseum, Obertraun 002469
Höhlenmuseum, Sundlauenen . . 042041
Höhlenmuseum Eisensteinhöhle, Bad
Fischau-Brunn 001740
Höhlenmuseum in der Lurgrotte,
Landesmuseum Joanneum,
Peggau 002490
Höhlenmuseum Kubacher Kristallhöhle
und Freilicht-Steinemuseum,
Weilburg 022479
Höhn, Marie, Wien 062981
Höhn, Ulrich, Flonheim 129844
Høiland, Thomas, Egå 125433
Høiland, Thomas, København . . 125451
Højbo Antik, Århus 065600
Højer Mølle- og Marskmuseum,

Højer 010129
Højgaard Nielsen, Lis, Kongens
Lyngby 065983
Højgaard, Poul, Ringe 066067
Højriis, Susanne, Vejby 066188, 102999
Højrup Antik, Hjørring 065763
Hoek, Groningen 083295, 143150
't Hoekje, Den Haag 083085
Hoekstra, Jubbega 132550
Höland, Karl-Günter, Erlbach . . . 129811
Hölder, Doris, Ravensburg 108442
Hölderlinturm, Tübingen 022216
Höllbacher, Erentrud, Salzburg . 062800
Hoelle, Christian, Bordeaux 074152
Hoelle, Christian, Villegouge . . . 074152
Höller, Alfred N., Wuppertal . . . 079107
Höller, Hanne, Bielefeld 074981
Hølonda Skimuseum of Bygdasamling,
Gåsbakken 033615
Höls, Maria, Düren . . . 075445, 075446,
129732
Hölscher, Detlev, Enger 075621
Höltje, Jochen, Tübingen 108790
Hölty-Stube, Göttingen 141836
Hölz, Christine, Düsseldorf 106794
Hölzl, Carl, Wien 140089
Hölzl, Cora, Düsseldorf 106795
Hölzle, Jörg, Ladenburg 130204
Hölzli & Weiss, Bad Waldsee . . 074609
Hömberg, Klaus, Arnsberg 106000
Hönig, Silke, Halle, Saale 129938
Hönigl, Reit im Winkl 108470
Höpfner, Martin, Großostheim . . 129810
Hoepli, Ulrico, Milano 137680
Höpner, Gudrun, Herne 130036
Hoeppe, Brigitte, Witzenhausen . 137557
Höpperger, Innsbruck 062648
Hoeppner, Hans & Marc, Essen . 106917,
137184
Hörby Museum, Hörby 040703
Höribach Verein, Sankt Lorenz . 058260,
100187
Hoermann, Gaby, Deidesheim . . 075321
Hörner, Hans-Joachim,
Wiesbaden 108918
Hörrle, Oskar, Heidelberg 076447,
125987
Hörselbergmuseum, Wutha-
Farnroda 022835
Hørsholm Egns Museum,
Hørsholm 010131
Hörstemeier, Melle 077336
Hörter, Aschaffenburg 074433
Hösch, Wolfgang, Elsenfeld . . . 075603
Hoeschler, Harsewinkel 107379
Hösel, Magdeburg 077219
Høvåg Museum, Høvåg 033657
Hoeve, Victoria, Sittard 112676
Hövelmann, Michael, Bedburg-
Hau 074708
Hövels, Wendelin, Dreieich 075405
Hövermann, G., Verden 078788
Høy, Frantz, Horsens 065795
Hof & Hof, Groningen 112494
Hof & Huyser, Amsterdam 112222
Hof ter Leiffebrughe, Gent 063643
Hof van Edens, Zaltbommel . . . 083702
Hof van Hessen, Huissen 032517
Hof, Wolfgang, Augsburg 125851
Hofbauer, Ken, Las Vegas 121356
Hofer, London 118387
Hofer, Brigitte, Freistadt 062562
Hofer, Christine, Luzern 133835
Hofer, Dr. Tiina, Wien 062983
Hofer, Herbert, Aschach an der
Donau 127709
Hofer, Ingrid, Graz 062591
Hofer, Jürgen, Saarbrücken 078323
Hofer & Sohn, Karl, Wien 062982
Hoferick, Frank, Radebeul 130546
Hoferick, Ines, Radebeul 130547
Hoff, Ahaus 105960

Hoff, Bernard, Wasselonne 074291,
074292
Hoff, Markus, Nordhausen 077799
Hoffeld & Co., Jeffrey, New York 122707
Hoffer, Jean-Jacques, Paris 071482
Hoffinger, Wieland, Wien 127929
Hoffman, Columbus 145099
Hoffman Gallery, Oregon College of Art
and Craft, Portland 052294
Hoffman, A., Stockholm 087007
Hoffman, Nancy, New York 122708
Hoffman, Rhona, Chicago 120154
Hoffman, Ronald, New York 095460
Hoffmann, Bad Griesbach 106058
Hoffmann, Dortmund 075385
Hoffmann, Nordhorn 142236
Hoffmann, Radeberg 108422
Hoffmann, Sønderborg 066138
Hoffmann & Schneider,
Hückelhoven 076579
Hoffmann Galerie, Zwolle 112732
Hoffmann-Schimpf, Bernd,
Mannheim 130276
Hoffmann-von-Fallersleben-Museum,
Museum zur Geschichte deutscher
Dichtung und Demokratie im 19. Jh.,
Wolfsburg 022742
Hoffmann, A., Lienen 077103
Hoffmann, A., Rheine 108498
Hoffmann, Adelheid, Friedberg,
Hessen 107071, 137214
Hoffmann, Anne, Bremen 129633
Hoffmann, Claudia, Erlangen . . . 129810
Hoffmann, D., Freiburg im
Breisgau 141804
Hoffmann, F., Troisdorf 078713
Hoffmann, Judith, Lebach 077037
Hoffmann, Lothar, Ahrensburg . . 129370
Hoffmann, M., Paderborn 108352
Hoffmann, Michael, Rott 078292
Hoffmann, Natascha, Gattikon . . 087439
Hoffmann, Roland, Paris 071483
Hoffmann, Roland, Saint-Ouen . 073107
Hoffmann, Rosa F., Koblenz . . . 076834
Hoffmann, U. & M., Schwäbisch
Gmünd 078403
Hoffmann, Ulrich, Stuttgart 108708
Hoffmann, Utz, Usedom 078771
Hoffmann, V., Ludwigshafen am
Rhein 077161
Hoffmeister, Wolfenbüttel 142503
Hoffmeister, Eva, Heidelberg . . . 130029
Hoffmeister, Ulrike, Lüdenscheid 107923
Hoffschult, Heidelberg 141927
Hoffstead II, Negril 111234
Hofft, Joanna, Maastricht 132575
Hofft, Joanna, Melick-Roerdalen . 132581
Hofgalerie, Füssen 100087
Hofgalerie, Wien . . . 062984, 100320
Hofgalerie am See & Kreativ Werkstatt,
Karin Weiß, Buchheim 106751
Hofgarten an der Residenz, Bayerische
Verwaltung der staatlichen Schlösser,
Gärten und Seen, München . . 020530
Hofinger, Andreas, Steinerkirchen an
der Traun 127879
Hofjagd- und Rüstkammer,
Kunsthistorisches Museum, Wien 003007
Hofje Van Gratie, Delft 032156
Hofland, Rogier, Amsterdam . . . 132403
Hofmann & Co., Fürth 129872
Hofmann, Luzern 087642
Hofmann, Oberursel 077886
Hofmann & von Sell, Berlin 106312
Hofmann, Angelika, Pforzheim . . 142303
Hofmann, B., München 142163
Hofmann, Dieter, München 142164
Hofmann, Erich, Innsbruck 062649
Hofmann, Gerd, Olbernhau 077919
Hofmann, Gerhard, Neustadt an der
Weinstraße 108258
Hofmann, Gunther, Glashütten . . 076006
Hofmann, Heinrich & Petra,
Itzgrund 076635

Hofmann, Jürgen, Heidelberg . . 076448
Hofmann, Petra, Frankfurt am
Main 137200
Hofmann, R., Offenburg 077907
Hofmann, Rupert, Oberkirch . . . 077867
Hofmann, Ursula, Nürnberg 130468
Hofmarkmuseum, Eggenfelden . 018098
Hofmarkmuseum Schloß Eggersberg,
Riedenburg 021414
Hofmatt Galerie, Sarnen 116627
Hofmeister, Hans & Emmi,
Colmberg 075269
Hofmeister, Margot, Massing . . . 107982
Hofmeister, Sabine, Stuttgart . . 130676
Hofmobiliendepot, Möbelmuseum
Wien, Wien 003008
Hofner, Gerhard, Nürnberg 142244
Hofstätter, Ingrid, Wien 100321
Hofstätter, Reinhold, Wien 062985
Hofstede, Deventer 083157
Hofsteenge, C., Groningen 083296
Hofstetter, Jean-Jacques,
Fribourg 116270
Hofstra Museum, Hempstead . . 049437
Hofu Tenmangu History Museum,
Hofu 028660
Hofwyl-Broadfield Plantation,
Brunswick 047284
Hog Hollow Art Gallery, Saint
Louis 123832
Hog Wild Collectible Toys, Austin 092235
Hoga, Benjamin Alexander, San
Diego 124132
Hogan, Athlone 079440
Hogan, Collingwood 098913
Hogarte Deco, Caracas 097957
Hogarth, Paddington, New South
Wales 099524
Hogarth's House, London 044998
Hoge, Jürgen, Wuppertal 142525
Hogervorst, M., Dordrecht 083182
Hogger, Hannelore, Uchte 130713
Hoglund, Armadale, Victoria . . . 098662
Hoglund, Auckland 112778
Hoglund, Sydney 099787
Hoglund Art, Nelson . 112961, 112962
The Hoglund Art, Wellington . . . 113068
Hogue Gallery Waikiki, Honolulu 120939
Hoguet, Anne, Paris 071484
Hoguet, Yves, Amiens 066574
Hohberger Bienenmuseum,
Hohberg 019297
Hohemer, Miami 121805
Hohenbalken, Agnes, Wien 127930
Hohenegg, G., Sonthofen 078508
Hohenhof, die Jugendstil-Villa
des Osthaus Museums Hagen,
Hagen 018915
Hohenlohe, Wien 100322
Hohenlohe-Museum, Neuenstein,
Württemberg 020704
Hohenloher Freilandmuseum,
Schwäbisch Hall 021754
Hohenloher Kunstverein e.V.,
Künzelsau 019817, 058920
Hohenloher Kunstverein e.V.,
Langenau 019906, 058921
Hohenloher Urweltmuseum,
Waldenburg, Hohenlohekreis . . 022387
Hohensinner, Lukas, Grein an der
Donau 062609
Hohenthal & Bergen, Berlin 106313,
106314
Hohenzollerisches Landesmuseum,
Hechingen 019123
Hohhaus-Museum, Lauterbach,
Hessen 019942
Hohhot City Museum, Hohhot . . 007186
Hohl, Marcel, Zürich 133983
Hohmann, Walsrode 108841
Hohmann, Waltraud, Bleckede . . 141604
Hohmann, Wilhelm, Stuttgart . . 142393
Hoiles-Davis Museum, Greenville 049261
Hojing Cultural Museum,

Hokkaido Daigaku Nogakubu Hakubutsukan, Sapporo – Homes, Pubs and Clubs, Coleraine

Index of Institutions and Companies

Kaohsiung 042359
Hokkaido Daigaku Nogakubu Hakubutsukan, Sapporo 029330
Hokkaido Home-Making School Museum, Tsugaru 029656
Hokkaido Kaitaku Kinenkan, Sapporo 029331
Hokkaido-kenritsu Kita-hitobito Hakubutsukan, Abashiri 028460
Hokkaido-ritsu Bungakukan, Sapporo 029332
Hokkaido Sogo Bijutsu Senmon Gakko, Sapporo 056032
Hokkaidoritsu Kindai Bijutsukan, Sapporo 029333
Hokkaidoritsu Migishi Kotaro Bijutsukan, Sapporo 029334
Hokkaidoritsu Obihiro Bijutsukan, Obihiro 029199
Hokken, Gouda 083264
Hokuetsu Bijutsukan, Nakajo . . . 029133
Hol Bygdemuseum, Hol i Hallingdal 033662
Holarium – 3D-Museum und Digital Art Gallery, Esens 018281
Holasek, Bernd F., Graz 062592, 100009
Holasek, Erich, Graz 062593
Holbak, Gitte & Niels, Valby . . . 066174
Holbein, Rijssen 126295
Holburne Museum of Art, Bath . 043519
Holden Gallery, Manchester 045205
Holden Historical Society Museum, Holden 006010
Holden Wood Antiques Centre, Haslingden 089461
Holden, Ernest, Baltimore 135345
Holden, Hélène, Montréal 064432
Holden, Robert, Craven Arms . . 117643
Holden, Robert, London 118388
Holdenhurst Books, Bournemouth 144255
Holder, Harald, Kempten 141984
Holderried, Bernd, Pfaffenhofen an der Ilm 130512
Holdhus Skulemuseum, Eikelandsosen 033584
Holdich, Raymond D., London . . 090211
Holding, Carole, Saint George's . 100841
Holdredge, Long Beach 121402
The Hole-in-the-Wall, Armagh . . 088286
Holé, Janine, Miribel 070132
Holgado de Latorre, Ricardo, Córdoba 133229
Holgården, Mora, Färnäs 086671
Holiday Antique Gift Show, Miami Beach 098452
Holiday Treasures, Birmingham . 092412
Holistico, Irene, Caracas 097958
Holje, Olofström 143911
Holladay Art and Antique, Salt Lake City 096894
Die Holländer, Berlin 129508
Holländer-Windmühle, Boiensdorf 017469
Holland, Warendorf 078867
Holland & Holland, London 090212, 144630
Holland & More, Osnabrück . . . 077968
Holland Area Arts Council, Holland 049507
Holland Art Fair-Den Haag, Den Haag 098257
Holland Art Fair-Utrecht, Utrecht 098258
Holland Art Gallery, Amsterdam . 112223
Holland Art Gallery, Den Haag . . 112391
Holland Art Gallery, Eindhoven . 112459
Holland Art Gallery, Rotterdam . 112641
Holland Art Group, Eindhoven . . 059607
Holland Experience, Amsterdam 031945
Holland-Hibbert, Hazlitt, London 118389
Holland Historical Society Museum, Holland 049510
Holland House, Louisville 094561
Holland Land Office Museum, Batavia 046807
Holland Museum, Holland 049508
Holland Toys, Amsterdam 082769
Holland Tunnel Art, Brooklyn . . . 047260
Holland, Herbert, Wilhelmsburg an der Traisen 063115, 063116
Holland, Lothar, Essen 075682, 125939
Holland, S., Portland 145455
Hollander York Gallery, Toronto . 101626
Hollander, E., Dorking 134336
Hollands & Assoc., Baltimore . . 092334
Het Hollands Kaasmuseum, Alkmaar 031891
De Hollandsche Schouwburg, Amsterdam 031946
Hollandsworth, Martin, Saint Louis 096716
Hollar, Praha 102696
Hollard, Olivier, Damville 068270
Hollenberg Pony Express Station, Hanover 049351
Hollendergaten Antikvariat, Bergen 143383
Holler, Eckart, Chemnitz 075254, 129653
Hollett & Son, R.F.G., Sedbergh . 091259, 144829
Holleville, Jean-Pierre, Pont-Remy 072260
Hollevout, Lille 104011
Hollex, Bratislava 085552, 133045
Holliday & Sons , Christchurch . 083876, 132708
Holliday, G.P., Christchurch 083877, 132709
Hollingshead, Chris, Teddington . 144878
Hollingsworth, Norfolk . 095859, 136031
Hollingsworth, Andrew, Chicago 092726
Hollis, Memphis 121695
Hollis, Susanne, West Hollywood 097897
Hollmann, Isolde & Wolfgang, Schwäbisch Gmünd 078404
Hollming & Co., V., Helsinki 066322
Holló László Emlékmúzeum, Déri Múzeum, Debrecen 023594
Holló László Galéria, Putnok . . . 023866
Hollow, Peter, Coburg 127588
Holloway, Edward & Diana, Suckley 091501
Holloway, Elizabeth, Caulfield North 061303
Holloways, Banbury 126807
Holloways of Ludlow, Ludlow . . 090650
Hollville, Gérard, Daours 068273
Holly Farm Antiques, Rotherham 091145
Hollycombe Steam Collection, Liphook 044843
The Hollycroft Foundation, Ivoryton 049743
Hollyhock, West Hollywood 097898
Hollyhock Antiques, Toowoomba 062329
Hollys Book Rack, Mansfield . . . 145249
Hollytrees Museum, Colchester . 043990
Hollywood Antique Showcase, Portland 096418
Hollywood Book City, Los Angeles 145229
Hollywood Bowl Museum, Los Angeles 050429
Hollywood Classics, Saint Louis 096717
Hollywood Coins, Nashville 095005
Hollywood Dream Factory, Toledo 097607
Hollywood Entertainment Museum, Los Angeles 050430
Hollywood Galleries, Hong Kong 065052
Hollywood Guinness World of Records Museum, Los Angeles 050431
Hollywood Heritage Museum, Los Angeles 050432
Hollywood Movie Posters, Los Angeles 121508
Hollywood Poster Company, Los Angeles 094363
Hollywood Poster Exchange, West Hollywood 125001
Hollywood Posters, Denver 120734
Hollywood Road Gallery, London 118390, 134690
Hollywood Square Antiques, Milwaukee 094862
Hollywood Wax Museum, Branson 047180
Hollywood Wax Museum, Los Angeles 050433
Hollywoodland, Los Angeles . . . 094364
Holm, Malmö 115761
Holmängens Auktionshall, Vänersborg 126695
Holman Museum, Holman 006011
Holmasto, Helsinki 066323
Holmes, San Jose 124449
Holmes à Court Gallery, East Perth 001046
Holmes County Historical Society Museum, Millersburg 050897
Holmes-Samsel, San Francisco . 097253
Holmes, Andy, Nottingham 144757
Holmes, D., London . . 090213, 134691
Holmestova, Frekhaug 033611
Holmestrand Museum, Holmestrand 033663
Holmgrens Volkswagenmuseum, Pålsboda 040912
Holmöns Båtmuseum, Holmön . 040705
Holmsbu Billedgalleri, Holmsbu . 033665
Holmwood Antiques, Dorking . . 089085
Holocaust Documentation and Education Center, Miami 050831
Holocaust Memorial Center of Central Florida, Maitland 050606
Holocaust Museum Houston, Houston 049578
Hologrammen 3-D, Amsterdam . 112224
Holokausto Ekspozicija, Valstybinis Vilniaus Gaono Žydų Muziejus, Vilnius 030519
Holokauszt Emlékközpont, Budapest 023487
Holowood – Holographiemuseum Bamberg, Bamberg 016949
Holroyd & Co., F.W., North Ballachulish 118811
Holseybrook, Tampa 136387
Holst Birthplace Museum, Cheltenham 043926
Holst, Hans-Jürgen, Groß Rönnau 076102
Holstebro Auktioner, Holstebro . 125447
Holstebro Kunstmuseum, Holstebro 010135
Holstebro Museum, Holstebro . . 010136
Holstein, H., Bremen 106586
Holsworthy Museum, Holsworthy 044595
Holt Antique Centre, Holt, Norfolk 089550
Holt Skolemuseum, Tvedestrand 034046
Holt, Harris, Charlotte 119993
Holt, John, Houston 093695
Holt & Co., R., London 090214
Holter Museum of Art, Helena . . 049432
Holtmann, Heinz, Köln . 107639, 137311
Holtmann, Heinz, Schüller 108581
Holtreman, João Lopes, Algés . . 143546
Holtschulte, Recklinghausen . . . 108452
Holtz, Albert, Thonon-les-Bains . 125820
The Holtzman Art Gallery, Towson 053788
Holtzman, Karen, Washington . . 124919
Holub, Reinhard, Lüdenscheid . . 107924
Holunder Hof, Kelkheim 076743, 130110
Holy Defense Museum, Kerman 024706
Holy Oriental Art Gallery, Beijing 101960
Holyhead Maritime Museum, Holyhead 044598
Holyland Arts Museum, Bethlehem 034159
Holyland Exhibition, Los Angeles 050434
Holyland Tapestries, Miami 094747
Holyrood Architectural Salvage, Edinburgh 089178
Holyrood Galleries, Newport, Isle of Wight 118800
Holz Hoch 3, Singen 078478
Holz-Vonderbeck, Frankfurt am Main 075791
Holz-Vonderbeck, Offenbach am Main 077902
Holz, Dr. Dagmar, Königswinter . 076928
Holz, Françoise, Arles 125496
Holzapfel, Paul, La Ferté-Macé . 069029
Holzapfel, Wiliam-Philippe, Bagnoles-de-l'Orne 066913
Holzapfel, William-Philippe, La Ferté-Macé 069030
Holzemer, Shelly, Minneapolis . . 122001
Holzen, Urs & Pia von, Bubendorf 087335
Holzer, Thomas, Emmendingen . 075609
Holzer, Werner, Wien . . 062986, 127931
Holzgerätemuseum, Schlüchtern 021675
Holzhauer, J., Kamp-Lintfort . . . 076669
Holzhausen, S., Glücksburg 076010
Holzinger, Thomasroith 062857
Holzkanumuseum, Stuttgart . . . 022041
Holzknechtmuseum im Salzkammergut, Lumberjack Museum, Bad Goisern 001745
Holzknechtmuseum Ruhpolding, Ruhpolding 021548
Holzschnitt-Museum Klaus Herzer, Mössingen 020448
Holztechnisches Museum, Rosenheim 021475
Holztreiger, Bela, Rio de Janeiro 064067
Die Holzwerkstatt, Henstedt-Ulzburg 130035
Der Holzwurm, Dossenheim . . . 075401
Holzwurm-Museum, Quedlinburg 021246
Homan, Cincinnati 135444
Homan-Do, Tokyo 082152
Homan, Grace, Austin 092236
Homan, Jim, Albuquerque 092058, 135284
Homburger, Hildegard, Berlin . . 129509
Home Again, Denver 093299
Home Again, Epping 061450
Home Again, Phoenix 096216
Home and Colonial, Berkhamsted 088444, 144243
Home and Garden, Toronto 064668
Home Antiquariato, Roma 081410
Home Antiques, Mönchengladbach 077392
Home Art, İstanbul 088124
Home Etc., Portland 096419
Home Galéria, Budapest 109149
Home Hook & Ladder Antiques, New Orleans 095112
Home-Museum of Azim Azimzadeh, Bakı 003163
Home Objet, Paris 071485
Home of 100 Happiness, Singapore 085424
Home of Chales Darwin – Down House, Downe 044135
Home of Franklin D. Roosevelt, Hyde Park 049650
Home of History, Rotterdam . . . 032805
Home of Stone – a Ford County Museum, Dodge City 048335
The Home Place, Kansas City . . 121297
Home-Sweet-Home, Cologny . . . 087378
Home Sweet Home, Providence 096507
Home Sweet Home Museum, East Hampton 048433
Homebase Collectibles, Indianapolis 093888
Homenco, São Paulo 064186
Homeopathy Works, Berkeley Springs 046942
Homer & Zachary, Virginia Beach 136444
Homer Watson Gallery, Kitchener 101326
Homer Watson House and Gallery, Kitchener 006123
Homes, Pubs and Clubs,

Coleraine 088958, 134293
Homespun Furniture, Wyandotte 136474
Homestead Antique Museum,
Drumheller 005718
Homestead Clock Repair, Dallas 135507
Homestead Manor, Oklahoma
City 095980
Homestead Museum, Biggar ... 005426
Homestead National Monument of
America, Beatrice 046845
Homestead Park Pioneer Museum,
Port Augusta 001419
Homesteader, Oakland 095895
Homestyle, Sacramento 096631
Hometown Connection, Phoenix 096217
Homeville Antique Fire Department,
Homer 049522
Homeville Museum, Homer 049523
Homewood Art Gallery,
Birmingham 119803
Homewood Hill Gallery, Kaikoura 112928
Homewood House Museum, Johns
Hopkins University, Baltimore . 046750
Homewood Museum, Prescott .. 006706
Homey, Chicago 120155
Homme Debout, Brive-la-
Gaillarde 067511
Hommes et Matières, Paris 071486
Homolka, Martina, Berlin 129510
Homolovi Ruins, Winslow 054484
Homs Museum, Homs 042262
Homs, Llucia, Barcelona 114920
Honan, Tony, Ennis 079617
Honan's Antiques, Gort 079633
Honarhaye Moaser Tehran,
Tehrän 024760
Honarhaye Tazini – Rakibkhaneh,
Isfahan 024701
Honda, New York 122709
Honda Collection Hall, Motegi .. 029058
Honda Museum, Kanazawa 028770
Hondenmuseum, Eindhoven ... 032285
Hondo-shi Amakusa Kirishitankan,
Hondo 028662
Hondo-shiritsu Mingeikan, Hondo 028663
Hondt, Y. de, Brugge 063277
Hone & Son, A.W., Birmingham . 088480,
134110
Honess, Lymington 118681
Honest, Jon, Salt Lake City ... 096895
Honey & Noney, Hamburg 076242
Honey Hive Emporium,
Semaphore 062201
Honeybee Science Center, Tamagawa
University, Tokyo 029515
Honeychurch, Hong Kong 065053
Honeychurch, Seattle 097454
Honeycombe, K., Framlingham . 089304
Honeywood, Hennebont 068806
Honeywood Heritage Centre,
London 044999
Hong, Guangzhou 102029
Hong-Gah Museum, Taipei 042516
Hong-ik University Museum,
Seoul 030001
Hong Kong Academy of the Fine Arts,
Hong Kong 055216, 136722
Hong Kong Art School, Hong
Kong 055217
Hong Kong Gallery Centre, Hong
Kong 102087
Hong Kong Heritage Museum, Hong
Kong 007872
Hong Kong Institute of Architects,
Hong Kong 055218
Hong Kong International Contemporary
Art Fair, Hong Kong 098051
Hong Kong Museum of Art, Hong
Kong 007873
Hong Kong Museum of History, Hong
Kong 007874
The Hong Kong Racing Museum, Hong
Kong 007875
Hong Kong Railway Museum, Hong

Kong 007876
Hong Kong Science Museum, Hong
Kong 007877
Hong Kong Space Museum, Hong
Kong 007878
Hong Kong Visual Arts Centre, Hong
Kong 007879
Hong, Choi, Las Vegas 121357
Hong, Hac, Ho Chi Minh City . 125125
Hong, Kuiung, Chicago 145046
Hong, Kwong Tai, Hong Kong . 065054
Hongdong Museum, Hongdong . 007894
Honghu Revolution History Museum,
Honghu 007895
Hongkong Craft, Singapore ... 085425
Hongli Art Guan, Shenzhen .. 102225
Hongsan Museum, Seoul 030002
Honingen, Gouda 112483
Honiton Antique Centre, Honiton 089565
Honiton Antique Toys, Honiton . 089566
Honiton Clock Clinic, Honiton . 089567,
134470
Honiton Fine Art, Honiton 089568,
117993, 134471
Honiton Lace Shop, Honiton ... 089569
Honiton Old Bookshop, Honiton . 144486
Honiton Pottery, Honiton 089570
Honjin Hakubutsukan Futagawa Syuku,
Toyohashi 029644
Honkala, Erkki, Helsinki 066324
Honma Bijutsukan, Sakata 029319
Honnelaitre, Claude, Paris 129037
Honoart, Solingen 078497
Honolulu Academy of Arts,
Honolulu 057209
Honolulu Academy of Arts Museum,
Honolulu 049534
Honolulu Furniture Antiques and Gallery,
Honolulu 093552, 135577
Honolulu House Museum,
Marshall 050709
Honolulu Printmakers, Honolulu . 060511
Honor Oak Gallery, London ... 118391
Honourable Silver Objects Antiquairs,
Antwerpen 063213
Honsbeek, P. & J., Den Haag . 083086
Honscha, S., Isernhagen 076624
Honscha, Siegward, Hannover . 076376
Honterus, Budapest . 126151, 142558
Honts, Jacqui von, San Antonio 124040
Hontza Natur Zientzien Museoa,
Museo de Ciencias Naturales del
Durangoaldea, Mañaria 039714
Hoo-Hoo International Forestry
Museum, Gurdon 049297
Hood, San Diego 097096
Hood Museum of Art, Hanover . 049353
Hood River County Historical Museum,
Hood River 049550
Hoodoo, Portland 096420
Hoodoo Gallery, Tamworth ... 099805
Hooge Crater Museum, Zillebeke 004160
Hooge, G. de, Middelburg 083495,
132584
Hoogendam, W., Den Haag ... 112392
Hoogendijk, Pieter, Baarn 082951
Hoogeveen, Bart, Ter Aar 132378
Hoogkamp, Amsterdam 143035, 143036,
143037
Hoogstede, G., Ravenstein ... 132611
Hoogsteder & Hoogsteder, Den
Haag 083087
Hook, Jo Anne, Main Beach .. 099255
Hooke, Houston 093696
Hooked on Glass, Denver 093300
Hooks-Epstein, Houston 121059
Hooky's Books, Auckland 143288
Hoon, Bangkok 088080
Hooper, David & Karol, Leamington
Spa 089754
Hoopers, London 118392
Hoopman, Amsterdam 112225
Hoorn-Ashby, New York 122710
Hooshmand, Los Angeles 094365

Hoosier Salon Gallery,
Indianapolis 049686
Hoosier Salon Patrons Association,
Indianapolis 060512
Hooter, O.J., New Orleans 095113
Hooton Park Exhibition Centre,
Ellesmere Port 044284
Hoover Dam Museum, Boulder
City 047146
Hoover Historical Center, North
Canton 051598
Hooy Kaye Museum, Bruxelles . 003445
Hopareboden, Stockholm 087008
Hope, Indianapolis 093889
Hope and Glory, London 090215
Hope Cottage, Kingscote 001193
The Hope Gallery, Hope Valley . 117997
Hope Historical Society Museum,
Hope 049551
Hope Lodge and Mather Mill, Fort
Washington 048891
Hope Museum, Hope 006012
Hope Sawmill Museum,
Peterborough 006637
Hope, R.J., Calgary 101243
Hopefulmonster Editore, Torino . 137748
Hopetoun House, Sendai 082112
Hopetoun House, South
Queensferry 045817
Hopewell Culture National Historic
Park, Chillicothe 047718
Hopewell Furnace Hopewell Furnace
National Historic Site, Elverson 048570
Hopewell Museum, Hopewell .. 049553
Hopewell Museum, Paris 051918
Hopf, M., Grünkraut 076117
Hopfen Erlebnis Hof,
Altmannstein 016552
Hopfenmuseum, Tettnang 022134
Hopfinger, Wolfgang, Hagenberg 139993
Hopi Cultural Center Museum, Second
Mesa 053257
The Hopi Gallery, Albuquerque . 119483
Hôpital Notre-Dame de la Rose,
Lessines 003783
Hopke, K., Liebenwalde 077101
Hopkin & Sons, H., Dublin 079560
Hopkins, Caloundra 098848
Hopkins-Custot, Paris . 071487, 104952
Hopkins House Gallery, Haddon
Township 049303
Hopmuseum De Stadsschaal,
Poperinge 003952
Hopp Ferenc Kelet-Ázsiai Művészeti
Múzeum, Iparművészeti Múzeum,
Budapest 023488
Hopp-Gantner, Maria-Luise,
Starnberg 078538
Hoppe, R., Hoyerswerda 076578
Hoppen, Michael, London 118393
Hoppen, Stephanie, London ... 090216,
118394
Hoppenot, Hubert, Lyon 140901
Hoppenstock, Detlef, Pattensen . 142296
Hopper, Pegge, Honolulu 120940
Hoppmann, Andreas, Köln 130149
Hoppstädter, Willi, Mandelbachtal 077248
Hopsewee Plantation,
Georgetown 049081
Hopwell, Paul, West Haddon ... 091787,
135206
Hopwood, Manela & Stacey,
Houston 121060, 121061
Hor, Ivan, Saint-Tropez 105628
Hora, Jiří, Náchod 140487
Horace Greeley House, New Castle
Historical Society, Chappaqua . 047530
Horace Young, Cleveland 135464
Horácka Galerie, Nové Město na
Moravě 009720
Horácké Muzeum, Nové Město na
Moravě 009721
Horajji-san Shizenkshi Hakubutsukan,
Hourai 028665

Horaiya, Kobe 081995
Horan, Vivian, New York 122711
Horatio Colony House Museum,
Keene 049912
Horay, Pierre, Paris 136942
Hordago, Saint-Jean-de-Luz .. 105533
Hordaland Kunstsenter, Bergen . 113134
Hordamuseet, Fana 033593
Hordern House, Potts Point ... 139881
Hordern, David, Aylesbury 134047
Horeau Reynaud, Pierrette, Les
Alleuds 069394
Horehronské Múzeum, Brezno . 038333
Horhor Sanat Galerisi, İstanbul . 117051
Horie Orugorukan, Nishinomiya . 029187
Horikawa, Kobe 111283
Horikawa, Michael D., Honolulu 093553
Horim Museum, Seoul 030003
Horiuchi, Noriyoshi, Tokyo 111444
Horizon, Utrecht 126301
Horizon Antiques, Tulsa 097738
Horizon Books, Seattle 145587, 145588
Horizon Chimérique, Bordeaux . 067337
Horizon Galleries, Broken Hill . 098798
Horizon Galleries, New Orleans . 122184
Horizon Gallery, Silverton 099687
Horizons Unlimited Supplementary
Educational Center, Salisbury . 052904
Horizont 2000, Cairo 103046
Hork, Billy, Chicago 120156
Horka, Chlumec u Ústi nad
Labem 102588
Horlbeck, Nürnberg 077826
L'Horloge, Touvois 073868
L'Horloger, Lausanne . 087572, 133799
Horlogeriemuseum, Museum
Woning voor antieke Horlogerie,
Mechelen 003855
Horn, Poznań 084641
Horn, Albert, Gotha 137230
Horn, Jerry W., Pittsburgh 123551
Horn, Joyce, Houston 093697
Horn, Linda, New York 095461
Horn, Manfred, Hannover 107369
Horn, Penelope, Adelaide 139614
Horn, Simon, London . 090217, 134692
Horn, Traudel, Nürnberg 077827
Hornbostel, Kassel 107534
Hornby Books, Christchurch .. 143304
Hornby Museum, Hornby 049559
Horncastle Antique Centre,
Horncastle 089583, 134475
Horncastle Antiques, Motueka .. 084001
Hornchurch Fine Art Gallery,
Hornchurch 118000
Horne, Salt Lake City 123968
Horne, Jonathan, London 090218
Horner, Magherafelt 118694
Horner, Jack, Saint Peters 062182
Horners, Acle 126795
Horngacher, François, Genève . 087473
Hornické Muzeum Krásno,
Krásno 009648
Hornické Muzeum OKD, Ostrava 009741
Hornické Muzeum Příbram,
Příbram 009852
Hornický Skanzen důl Mayrau,
Hornické Muzeum OKD Ostrava,
Vinařice u Kladna 009952
Horniman Museum, London ... 045000
Hornington, Woburn Sands ... 119373
Hornsbury Mill, Chard 043907
Hornsby, London 134693
Hornsea Museum of Village Life,
Hornsea 044601
Hornsey, London 126964
Hornsey, Ripon ... 091118, 144801
Hornung, Anton, Dresden 075418
Horny, Ferdinand, Kirchdorf an der
Krems 127787
Horological Workshops, Guildford 089395
Hororata Historic Museum,
Darfield 033183
Horowicz, Nathan, New York ... 095462

Horowitz, Seattle 097455
Horowitz, Glenn, New York ... 145348
Horrach Moyá, Palma de
    Mallorca 115353
Horry County Museum, Conway 047968
Hors du Temps, Nîmes 070696
Hors du Temps, Quimperlé ... 072352
Hors-Jeu, Genève 116320
Hors-Série, Bruxelles 140181
Horsch, Bob, Chicago 120157
The Horse and Buggy, Hamilton 143319
Horse Feathers, Indianapolis .. 093890
Horsefeathers Antiques, Toronto 064669
Horseman, New York . 095463, 135938
Horsens Kunstgalleri, Horsens . 065796,
                                  102849
Horsens Kunstmuseum, Horsens 010142
Horsens Museum, Horsens 010143
Horseshoe Antiques and Gallery,
    Burford 088706
Horseshoe Bend National Military Park,
    Daviston 048165
Horseshoe Creek Country Relics,
    Everton 061457
Horseshoe Gallery, Kelso 118035
Horseshoe Lake Road Antiques,
    Christchurch 083878
Horsforth Village Museum,
    Horsforth 044603
Horsham Arts Centre, Horsham . 044605
Horsham Bookshop, Horsham . 144489
Horsham Museum, Horsham ... 001153,
                                  044606
Horsham Regional Art Gallery,
    Horsham 001154
Horst-Boskamp, ter, Zutphen ... 083718
horst interior.galerie.art, Berlin . 106315
horst interior.galerie.art,
    München 108140
Horst-Janssen-Museum,
    Oldenburg 020985
Horsten, København 065916
Horster Motorrad-Museum,
    Gelsenkirchen 018626
Horstmann & Godfrey, Hong
    Kong 065055
Horswell Art Gallery, Plymouth . 118921
Horswill, Helen, West Kirby ... 091788
Horszowski, Henryk, Wrocław .. 084770,
                  126430,   143542
Horta, Bruxelles 125314
Horta e Costa, Frederico, Lisboa 084900
Horta, Miguel E., Belo Horizonte 100888
Horte, Max, Goslar 076060
Horten Bilmuseum, Horten 033670
Hortensia Förlag och Galleri,
    Göteborg 115654, 138007
Hortobágyi Pásztormúzeum,
    Hortobágy 023697
Horton, Richmond, Surrey 091097
Horton, D., London 090219
Horton, Joe, Nashville 122086
Horton, Robert, Hertford 089519
Horton, Tony, Dallas 120611
Horus, New York 122712
Horus Gallery, Cairo 103047
Horváth, Budapest 142559
Horváth József Festőművész Állandó
    Kiállítása, Sopron 023891
Horwarth, Reutlingen 108482
Horwich, Joy, Chicago 120158
Horyu-ji Homotsukan, Tokyo National
    Museum, Tokyo 029516
Horyuji Kondo Hekiga Mosha Tenjikan,
    Aichi-kenritsu Geijutsu Daigaku,
    Nagakute 029068
Hos Linda, Aalborg 128376
Hosains, London 090220, 144631
Hosan, Busan 111566
Hosane, Beijing 125399
Hosane, Garbsen 125956
Hosane, Shanghai 125409
Hosane, Taipei 126781
Hosanna Museum, Chai-Yi 042279

Hosegood, Alan, Bristol 134160
Hosen Gakuen Tanki Daigaku,
    Tokyo 056044
Hosfelt, New York 122713
Hosfelt, San Francisco 124302
Hoshino, Kyoto 111295
Hoskin, John, Toronto 064670
Hosking, Ballarat 061085
Hosking, Susanne, Leura 061736
Hoskins House, Burlington 047324
Hosko, Saint Paul 123915
Hosokawa, Chizuko, Paris 104953
Hosomi Bijutsukan, Kyoto 028929
Hosotte, Georges, Irancy 103854
Hosp, Edith, Sankt Gilgen 100184
Hospitaalmuseum Onze-Lieve-
    Vrouw ter Potterie, Musea Brugge,
    Brugge 003413
Hospitaalmuseum Sint-Janshospitaal,
    Musea Brugge, Brugge 003414
Hospital de los Venerables,
    Sevilla 040166
Hospital de San Juan Bautista,
    Hospital Tavera, Toledo 040233
Hospitals and Clinics Medical Museum,
    University of Iowa City 049708
Hospitalsmuseet, Vadstena 041125
Hoss, Tulsa 097739
Hoss, Annette, Kesseling 130115
Hoss, Martino, Seattle 124542
Hoss, Marwan, Paris 104954
Hossari, Samir, Randers 140643
Hostettler, Christian, Belp 144043
L'Hostis, Guy, Lannion 103920
L'Hostis, Guy, Paimpol 104435
Hostis, Guy, Paimpol 070842
Hostrup-Pedersen & Johansen,
    Valby 136784
Hosyokan Byodoin Museum, Uji,
    Kyoto 029675
Hot Art Framed Prints, Sutton
    Coldfield 119189
Hot Artz Gallery, Taihape 113009
Hot House Art Gallery,
    Indianapolis 121189
Hot Poppentheatermuseum,
    Gouda 032362
Hot Property, Toronto 064671
Hot Shops Art Center, Omaha . 123321
Hot Springs County Museum,
    Thermopolis 053729
Hotbath Gallery, Bath 117308
Hotcakes Gallery, Milwaukee ... 121925
Hotch Potch, Dannenberg 075298
Hotei, Leiden 083431
Hotel, London 118395
Hotel de Paris Museum,
    Georgetown 049077
Hôtel de Ventes des Chartrons,
    Bordeaux 125528
Hôtel de Ventes Flagey, Bruxelles 125315
Hôtel de Ventes Saint-Georges,
    Toulouse 125825
Hôtel des Ventes, Autun 066836, 125501
Hôtel des Ventes, Chatou 125555
Hôtel des Ventes, Entzheim .. 125580
Hôtel des Ventes, Nantes 125656
Hôtel des Ventes, Tours 125827
Hôtel des Ventes – Demade Encheres,
    Bernay 125513
Hôtel des Ventes Bordeaux Rive Droite,
    Bordeaux 125529
Hôtel des Ventes Couton et Veyrac,
    Nantes 125657
Hôtel des Ventes de la Seine, Elbeuf-
    sur-Seine 125578
Hôtel des Ventes du Boischaut, Saint-
    Amand-Montrond 125796
Hôtel des Ventes Elysée, Liège . 125343
Hôtel des Ventes Mobilières,
    Bayonne 125507
Hôtel des Ventes Mosan,
    Bruxelles 125316
Hôtel des Ventes Mosan, Liège . 125344

Hôtel des Ventes Périgord,
    Bergerac 125512
Hôtel des Ventes Saint-Georges,
    Liège 125345
Hôtel des Ventes Universelles,
    Québec 125375
Hôtel Dieu – Musée Greuze,
    Tournus 016010
Hotel Drouot Antiguidades, Rio de
    Janeiro 064068
Hôtel Groslot, Orléans 014583
Hotel-Museum Arthur Merghelynck,
    Ieper 003724
Hotelli- ja Ravintolamuseo,
    Helsinki 010832
Hotelmuseum, Flims Waldhaus . 041488
Hotspur, London 090221
Hott, Armin, Kandel 107504
Hottelmann-Schmidt, Irene,
    Bamberg 074665
Hottowitz, Antje, Bad Soden am
    Taunus 074599
Hotvedt, Arne, Oslo 132786
Hotz-Behofsits, Rudolf,
    Oberpullendorf 140020
Hotz, Manfred, Lindau 077118
Hotze, Thomas, Oberursel 077887
Hou Shi, Tiang, Singapore ... 114536
Hou Wang Chinese Temple and
    Museum, Atherton 000807
Houbraken, J., Veldhoven 132654
Houck, Portland 136154
Houdebert, Jack, Evian-les-Bains 068491
Houdek, J., Praha 128348
Houdini Museum, Scranton ... 053216
Houf, Thomas, Köln . 076877, 130150
Houg, Olivier, Lyon 104110
Houghton Antiques, Houghton,
    Cambridgeshire 089592
Houghton County Historical Museum,
    Lake Linden 050082
Houghton Hall Antiques, Market
    Weighton 090716, 134840
Houghton Hall Soldier Museum, King's
    Lynn 044703
Houis, Lionel, Paimboeuf 070839
Houk, Edwynn, New York 122714
Houlberg, Laila & Claus,
    Gudhjem 102819
Houldsworth, London 118396
Houle, George J., Los Angeles . 094366,
                                 145230
Houllier, Marseille 104181
Hound Dog Antiques, Vancouver 064830
Hourcadette, Pierre, Saint-Jean-de-
    Luz 072845
Hourian, San Francisco 097254, 136321
Hourrière, Jacques-Denis, Paris 129038
R. Hours & Hugues de Valaurie, Aix-
    en-Provence 125485
Housatonic Museum of Art,
    Bridgeport 047202
House & Son, Bournemouth .. 126829
The House, Miami 121806
The House 1860–1925,
    Monmouth 090769
House Antiques, Birmingham . 092413
House Gallery, London 118397
House Gallery, München 108141
House Gallery, Oklahoma City . 123282
House in the Horseshoe, Sanford 053082
House Mill, London 045001
House-Museum of Uzeyir Hajibeyov,
    Baki 003164
House of a Thousand Dolls,
    Loma 050359
House of Antiquity, Nether
    Stowey 144738
The House of Art, Beirut 111772
House of Art, Varna 101193
House of Art Books, Mandeville 145248
House of Arts and Letters,
    Lemesos 102549

House of Books, Auckland 083794,
                                 143289
House of Books, Lismore 139806
House of Bottles, Kinglake 001192
House of Bottles and Bottle Museum,
    Tewantin 001569
House of Cachet, Pittsburgh .. 123552
House of Castlegate,
    Knaresborough 118061
House of Cedar, Springwood .. 062238
House of Christian, Ash Vale ... 088294
House of Clocks, Ampthill 088276,
                                 134031
House of Color Designs, Las
    Vegas 121358
House of Commons, Ottawa ... 006590
House of David Auctions, Haynes
    City 127241
House of Dun, Montrose 045291
House of Glass, Houston 093698
House of Hadjigeorgakis Kornesios,
    Lefkosia 009417
House of Happy Walls, Jack London
    State Historic Park, Glen Ellen 049107
The House of Harriette & John,
    Austin 092237
House of Heirlooms, San Diego . 097097
House of History, Franklin County
    Historical and Museum Society,
    Malone 050613
House of Huanghuali, Singapore 085426
The House of Ikat, Auckland ... 083795
House of Lamps, Austin 092238
The House of Lords, Dublin 024886
House of Manannan, Manx National
    Heritage, Peel 045471
House of Memories, Latchford . 006168
House of Memories Museum,
    Wilson 054453
House of Mirrors, London 090222
House of Music Parma, Parma . 027209
House of Noble Wood, Singapore 085427
House of Our Own, Philadelphia 145414
House of Oz, Alice Springs 098632
House of Persian Rugs, Calgary 064259
House of Riches, Cottesloe ... 061368,
                                 127590
House of Roses, Senator Wilson Home,
    Deadwood 048190
House of Seconds, Louisville ... 094562
House of Siam Ltd., Bangkok .. 088081
House of Silver, Pittsburgh ... 096295
House of Silver, Winnipeg 064927
House of Stapleton Collection, San
    Diego 097098
House of the Anchors, Sea of Galilee
    Fishing Museum, Kibbutz Ein
    Gev 025127
House of the Binns, Linlithgow . 044838
House of the Redtails Gallery,
    Tucson 053830
House of the Scribe, Almog ... 025015
The House of the Seven Gables,
    Salem 052879
House of the Temple,
    Washington 054121
House of Wickersham, Juneau . 049849
House of Windsor, Mooloolaba . 061875
House on Cary, Richmond 096553
House on Crutches Museum, Bishop's
    Castle 043637
House on the Hill Museums Adventure,
    Stansted Mountfitchet 045860
The House on the Rock, Spring
    Green 053444
The House that Moved, Exeter . 089231
House Things Antiques, Hinckley 089530
House, Bernard G., Wells 091771,
                                 135200
House, Howard, Seattle 124543
Household Cavalry Museum,
    Windsor 046163
Housesteads Roman Fort and
    Museum, Housesteads 044608

Houston, Old Noarlunga 099500
Houston Antique Dealers,
  Houston 093699
Houston Antique Dealers Association,
  Houston 060513
Houston Art Dealers Association,
  Houston 060514
Houston Center for Photography,
  Houston 049579
Houston Fine Art Press, Houston 138340
Houston Fire Museum, Houston 049580
Houston Glass Company,
  Houston 093700
Houston Harte University Center, San
  Angelo 052926
Houston Museum of Decorative Arts,
  Chattanooga 047594
Houston Museum of Natural Science,
  Houston 049581
Houston Numismatic Exchange,
  Houston 093701
Houston Police Museum,
  Houston 049582
Houston Precious Metals,
  Houston 093702
Houston Women's Caucus for Art,
  Houston 060515, 121062
Houston, William P., Frankfurt am
  Main 075792
Hout Bay Gallery, Hout Bay 114786
De Houtkrul, Alkmaar 132382
Houvet, Christophe, Fontaine-
  Simon 068563
Hovanec, Jozef, Spišská Nová
  Ves 133061
Hovden Jernvinnemuseum, Bykle 033565
Hove Museum and Art Gallery,
  Hove 044610
Hovedvejens Antik, Ringe 066068
Hovercraft Museum, Gosport .. 044452
Hoves, Eberhard, Krefeld 107739
Hovhannes Hovhanisyan House-
  Museum, Echmiadzin 000694
Hovhannes Tumanyan Museum,
  Yerevan 000734
Hovingh & Zoon, C., Haarlem .. 143160
How we lived then – Museum of
  Shops, Eastbourne 044215
Howald's Erben, Max, Bern 087285
Howard, Stratford-upon-Avon . 091493,
                          135101
Howard, Taipei 116919
Howard & Co, Toronto 064672
Howard Antiques & Fine Art,
  Southport 091379
Howard B. Owens Science Center,
  Lanham-Seabrook 050134
Howard County Arts Council, Ellicott
  City 060516
Howard County Center of African
  American Culture, Columbia . 047897
Howard County Historical Museum,
  Kokomo 050019
Howard County Historical Society
  Museum, Ellicott City 048554
Howard-Dickinson House Museum,
  Henderson 049443
The Howard Gallery, Dorking .. 089086,
                          134337
Howard Gardens Gallery, University of
  Wales Institute, Cardiff 043865
Howard House Antiques, Silvan . 062210
Howard House of Artifacts, Old
  Perlican 006549
Howard Steamboat Museum,
  Jeffersonville 049807
Howard University Gallery of Art,
  Washington 054122
Howard University Museum,
  Washington 054123
Howard W. Cannon Aviation Museum,
  Henderson 049441
Howard, Austin, Miami 121807
Howard, Joe, Los Angeles 121509

Howard, John, Woodstock 091931
Howard, Jonathan, Chipping
  Norton 088889, 134263
Howard, Patrick, Dublin 079561
Howards, Baldock 088340, 134050
Howards, Broadway .. 088677, 134170
Howarth Gallery, Burnley 088730,
                          117486
Howe, Christopher, London 090223
Howe, Kate, London 118398
Howeg, Thomas, Zürich 116813, 138111,
                          144182
Howell & Howell, Woollahra ... 062472
Howell Harris Museum, Brecon . 043713
Howell Historical Society & Committee
  Museum, Howell 049598
Howell Living History Farm,
  Titusville 053751
Howell, Charles, Oldham 090911, 134905
Howell & Assoc., Linda C., Oklahoma
  City 123283
Howell, Patti, Austin 092239
Howlett, Ruth, Smythesdale .. 062215
Howes About Antiques, Toronto . 064673
Howes Bookshop, Hastings 144456
Howick Contemporary Art,
  London 118399
Howick Historical Village,
  Pakuranga 033273
Howick Museum, Howick 038737
Howisey Brothers, Seattle 136372
Howkins, Peter, Great Yarmouth 089384,
                          134426
Howland House, Plymouth 052209
Howorth, J., Leeds 089770
Howth Harbour Gallery, Howth . 109642
Hoxie Gallery, Westerly Public Library,
  Westerly 054331
Hoxton Books, London 144632
Hoy en el Arte, Buenos Aires .. 098571
Hoy, Renee, Singapore 085428
Hoyer, A. & O., Jettingen 076652
Hoypoloi, Chicago 120159
Hoyt Antiques, Toronto 064674
Hoyt Institute of Fine Arts, New
  Castle 051253
Hoyt Sherman Place, Des Moines 048283
Hozat Kültür Merkezi, Hozat ... 042914
Hozhoni, Seattle 124544
HP Antiquités, Paris 071488
HP Antiquités, Saint-Ouen 073108
HPH Siebdruck.Galerie, Leinfelden-
  Echterdingen 107805
HR Antiquités, Bellac 067074
Hrách, Jiří, Praha 140520
Hrad a Muzeum Hodin,
  Šternberk 009891
Hrad Devín, Mestské Múzeum v
  Bratislave, Bratislava 038309
Hrad Helfštýn, Týn nad Bečvou . 009927
Hrad Krásna Hôrka, Slovenské
  Národné Múzeum – Múzeum Betliar,
  Krásnohorské Podhradie 038373
Hrad Roštejn, Muzeum Vysočiny
  Jihlava, Telč 009904
Hradčany, Praha 065499
Hradecký, Pavel, Liberec 065413
Hrdlicka, Jaroslav, München ... 077528
Hrdličkovo Muzeum Člověka,
  Praha 009798
Hřebíček, Brno 102569
Hriesik, M., Worpswede 108968
Hron, Otto, Traunreut 078690
Hrozen, České Budějovice 102577
Hruschka, Weil im Schönbuch .. 108857
Hruschka, Frank, Radebeul 108424
Hruška & Hrušková, Písek 065446,
                          140496
Hrvatska Akademija Znanosti i
  Umjetnosti, Dubrovnik 055280
Hrvatska Akademija Znanosti i
  Umjetnosti, Zagreb 055283
Hrvatska Akademija Znanosti i
  Umjetnosti – Gliptoteka, Zagreb 009032

Hrvatski Muzej Arhitekture, Hrvatska
  Akademija Znanosti i Umjetnosti,
  Zagreb 009033
Hrvatski Muzej Naivne Umjetnosti,
  Zagreb 009034
Hrvatski Nacionalni Komitet ICOM,
  Zagreb 058521
Hrvatski Poštansko-telekomunikacijski
  Muzej, Zagreb 009035
Hrvatski Povijesni Muzej, Zagreb 009036
Hrvatski Prirodoslovni Muzej,
  Zagreb 009037
Hrvatski Školski Muzej, Zagreb . 009038
Hrvatski Sportski Muzej, Zagreb 009039
Hrvatsko Društvo Likovnih Umjetnika
  Split, Split 058522
Hrvatsko Društvo Likovnih Umjetnika
  Varaždin, Varaždin 058523
Hrvatsko Muzejsko Društvo,
  Zagreb 058524
Hrvatsko Numizmatičko Društvo,
  Croation Numismatic Society,
  Zagreb 058525
HRW Antiques, London 090224
Hryciow, Stefan, Darmstadt 075303
HS Galerie, Heidelberg 076449
HS-Styl, Wrocław 132915
Hsin Wang Ceramics Memorial Gallery,
  Yingge 042595
Hsinchu City Fire Museum,
  Hsinchu 042331
HSR Antiquitäten, Bonn 075074
Hsu, Jeff, Taipei 088069, 116920
HT, Madrid 115237
HTL1 Bau und Design, Höhere
  technische Bundeslehranstalt,
  Linz 054950
HTS-Museum, Rødovre 010283
The HTV, Amsterdam 138929
Hu Shih Memorial Hall, Academia
  Sinica, Taipei 042517
Hua Feng, Beijing 065002
Hua Mei, Seattle 124545
Hua Ren, Beijing 101961
Hua-Ströfer, Hai-Yen, Mannheim 130277
Hua Zhai, Qing, Singapore 114537
Huacheng, Shenzhen 102226
Huadu, Beijing 101962
Huafan Cultural Gallery, Huafan
  University, Shihtin 042450
Huahan, Beijing 101963
Huai Hong, Paris 071489
Huai'an Museum, Huai'an 007896
Huaibei Museum, Huaibei 007899
Huaihua Zone Museum, Huaihua 007900
Huaiji Museum, Huaiji 007901
Huainan Museum, Huainan 007902
Huaiyang Museum, Huaiyang .. 007903
Huaiyin Museum, Huai'an 007897
Hualien County Stone Sculpture
  Museum, Hualien 042339
Hualien Railway Museum,
  Hualien 042340
Huaming, Qingdao 065127
Huan, Les Mureaux 069417
Huan Shi Wen Bao, Beijing 101964
Huanaki Cultural Centre, Alofi .. 033448
Huanduoci Curiosity Museum,
  Hengshan 042326
Huang, Indianapolis 093891
Huang Guei-li Memorial Hall,
  Banciao 042270
Huang Mun Art Center, Tongsiao 042577
Huanggang Museum, Huanggang 007905
Huanghe Museum, Zhengzhou . 008538
Huangmei Museum, Huangmei . 007906
Huangping Nationality Holiday
  Museum, Huangping 007907
Huangpu Hua Yuan, Shenzhen . 102227
Huangshan Beauty Spot Museum,
  Huangshan 007908
Huangshan Museum, Huangshan 007909
Huangshi Museum, Huangshi .. 007913
Huangsi Art Exhibition, Changhua 042281

Huangyan Museum, Huangyan . 007914
Huangzhong Museum,
  Huangzhong 007915
Huangzhou Museum, Huangzhou 007916
Huapai Collectables, Huapai ... 083957
Huaqiao, Shanghai 065139
Huashan Literature and Fine Arts
  Press, Shijiazhuang 136741
Huat & Co., Aik, Singapore 085429
Huating Museum, Huating 007917
Huaxia Art Cooperative, Beijing . 101965
Huazhou, Manzhouli 065126
Huazhou Museum, Huazhou ... 007918
Hub City Collectables, Saskatoon 064563
Hub Local History Collection, Aberfoyle
  Park 000753
HUB Robeson Galleries, University
  Park 053898
Huba, Pavol, Dolný Kubín 085560
Hubbard, Ipswich 089647, 134503
Hubbard, Surrey Hills, Victoria .. 062261
Hubbard Museum of the American
  West, Ruidoso Downs 052702
Hubbard, Andre C., Flatts 128094
Hubbard, John, Blakedown 088504,
                          134119
Hubbard, Rick, Romsey 091137
Hubbardton Battlefield Museum,
  Hubbardton 049600
Hubbell Trading Post, Ganado .. 049041
Hubbers Emporium, Invercargill . 083967
Hubeaux, Liège 063764
Hubei Academy of Fine Art,
  Wuhan 055247
Hubei Fine Arts Press, Wuhan .. 136745
Huber, Einsiedeln 087405
Huber, Landshut 107791
Huber & Treff, Jena 107480
Huber House Gallery, New
  Orleans 122185
Huber, Andreas, Wien 100323
Huber, Bruno & Susanne,
  Lommis 087600
Huber, Camillo-Carlo, Soulce .. 116656
Huber, Franz, Offenburg 137457
Huber, Heidi, Bern 116137
Huber, Laurent, Paris 129039
Huber, Manfred, Manching 077247,
                          130273
Huber, P., Norderney 077795
Huber, Rudger, Mörnsheim 077404
Huber, Ursula, Basel 116079
Huber, Volker, Offenbach am
  Main 108330, 137455
Huberhus Museum, Lech 002283
Hubers & Hawkins, Chandler .. 098880
Huber, Andreas, Wien 100323
Hubert, Boynes 067442
Hubert, Montréal 064433
Hubert, New York 122715
Hubert, Worpswede 108969
Hubert, Armin, Mannheim 107968
Hubert, Christian, Chevaigne .. 067965
Hubert, Marie, Caen .. 067571, 128526
Hubert, Marie, Dreux 068407
Hubert's Antiques, Burford 088707
Hubertus, Horst, Seligenstadt . 078453
Huble Homestead, Prince George 006715
Hubley & Co., F.B., Cambridge . 127153
Hubschraubermuseum,
  Hubschrauberzentrum Bückeburg,
  Bückeburg 017659
Huc, Marie-Claude, Toulouse ... 073775
Hucher, Claude, Tocane-Saint-
  Apre 073685
Huchet, Patrick, Singapore 085430
La Huchette, Genève 087474
Huckleberry & Chin, Edmonton . 064298
Hudcová Dita, Praha 065500
Huddersfield Art Gallery,
  Huddersfield 044613
Huddersfield Picture Framing Company,
  Huddersfield 089598, 134482
Huddleston Farmhouse Inn Museum,
  Cambridge City 047376

Hudební Antikvariat, Ostrava . . . 140495
Hudel, Antoine, Bordeaux 128492
Hudetz-Turm, Wiesent 022645
Hudiksvalls Antikvariat,
Hudiksvall 143864
Hudler, Emil, Dießen am
Ammersee 129684
Hudobná Expozícia – Rodný Dom
J.N. Hummela, Mestské Múzeum v
Bratislave, Bratislava 038310
Hudobné Múzeum, Slovenské Národné
Múzeum, Bratislava 038311
Hudson, Jacksonville 145188
Hudson, Oakland 095896
Hudson Bay Gallery, Virginia
Beach 124830
Hudson Bay Museum, Hudson
Bay 006017
Hudson City Antiques, New York 095464
Hudson Hills Press, Manchester 138355
Hudson Museum, University of Maine,
Orono 051816
Hudson River Maritime Museum,
Kingston 049983
The Hudson River Museum,
Yonkers 054577
Hudson Valley Art Association,
Stamford 060517
Hudson Valley Children's Museum,
Nanuet 051160
Hudson, Franklin, New York . . . 122716
Hudson, J.N., Deniliquin 061398
Hudson's Hope Historical Society
Museum, Hudson's Hope 006018
Hübens, Theo, Maastricht 112576
Hübert, Jelena, Belzig 074717
Huebmer-Gedächtnisstätte,
Naßwald 002429
Hülsmeier, Th., Osnabrück 077969
Hümmer, Johannes & Margit,
Nürnberg 126085
Huemul, Buenos Aires 139573
Hünersdorff, London 144633
Hünteler, Hans-Joachim,
Hamburg 141874
Hüntelmann, Werlte 108875
Hüren, Hans-Joachim, Tönisvorst 078685
Hürland, Raymund, München . . 077529
Huerta de San Vicente, Casa-Museo
Federico García Lorca, Granada 039463
Hürten-Heimatmuseum, Bad
Münstereifel 016828
Huertos, Tomas Hernandez,
Burgos 085911
Hüseyn Cavidin Ev-Muzeyi, Bakı 003165
Hüseyn Cavidin Ev-Muzeyi,
Naxçivan 003188
Hüsing-Hoge, Bodo-U., Entre-os-
Rios 084797
Hüsken, André, Hamburg 076244
Hüsken, Norbert, Würzburg . . . 079083
Hüsstege, 's-Hertogenbosch . . 112528
Hüsstege Iconen, Haaren 112506,
132512
Huet, Jean Pierre, Fougères (Ille-et-
Vilaine) 068597
Hütten- und Technikmuseum
Ilsenburg, Ilsenburg 019390
Hüttenberger Heimatmuseum,
Linden 020085
Hüttenmuseum Thale, Thale
(Harz) 022137
Hütter, Dr. Harald, Trofaiach . . 140047
Hüttner, München 077530

Huf – und Hackenschmiede
Lindermayr, Spital am Pyhrn . . 002763
Huff, I., Passau 078022
Hufkens, Xavier, Bruxelles 100619
Hufnagel, E. & W., Lennestadt . 077086
Hufschmid, Esther, Zürich 116814
Hufschmiede Museum Frehrking,
Neustadt am Rübenberge 020751
Hufschmiedemuseum,
Engelhartszell 001867
Hufthamartunet, Storebø 033995
Reinhard Hug & Birgit Norpoth-Hug,
Rheinberg 078224
Hug, Beat, Colombier (Neuchâtel) 133723
Hug, Daniel, Los Angeles 121510
Hugendubel, Kreuzlingen 138058
Hugendubel, München 142165
Hugenotten-Museum, Berlin . . . 017173
Huggele, Esslingen 106935
Hugh, Fullarton 099065
Hugh Cash Antiques, Maynooth . 079673
Hugh Hardinge Book's,
Cambridge 144297
The Hugh Lane Municipal Gallery of
Modern Art, Dublin 024887
Hugh Mercer Apothecary Shop,
Fredericksburg 048955
Hugh Miller's Cottage, Cromarty 044045
Hughenden Manor, High
Wycombe 044582
Hughes, Saint Louis 123833
Hughes Day, Judith, New York . 122717
Hughes Fine Arts Center, Grand
Forks 049176
The Hughes Gallery, Belfast . . . 117339
Hughes, Christopher, Worcester . 119394
Hughes, E., Clayton 145079
Hughes, Harriet, Liverpool 134567
Hughes, John, Chippendale . . . 127586
Hughes, Michael, London 090225
Hughes, Michael C., New York . 095465
Hughes, Paul, Charlotte 119994
Hughes, Ray, Surry Hills 099747
Hughlin, San Francisco 097255
Hugo, New York 095466
Hugo Wolf-Haus, Perchtoldsdorf 002495
Huguan Museum, Huguan 007919
Huguenot Historical Society Museum,
New Paltz 051308
Huguenot Memorial Museum,
Franschhoek 038715
Hugues Pénot, La Baule-
Escoublac 103868
Huguet Michelle, Fleurance 068545
Huguette, Jean Luc de,
Montpellier 070305
Huhn, E., Köln 076878
Hui Hua Arts, Chengdu 102013
Hui No'eau Visual Arts Center,
Makawao 050609
Hui, Yang, Singapore 085431
Hui'an Museum, Hui'an 007920
Huichang Museum, Huichang . . . 007921
Huidong Museum, Huidong 007922
Huihua Group of the Xue Qin Art
Cooperative, Beijing 101966
Huilai Museum, Huilai 007923
Huis de Lalaing, Oudenaarde . . 003939
Huis Doorn, Doorn 032240
Huis Hellemans, Edegem 003571
Huis Marseille, Museum voor
Fotografie, Amsterdam 031947
Huis met de Kogel, Haarlem . . . 083332
Huis Singraven, Denekamp 032211
Huis Ten Bosch Museums,
Sasebo 029349
Het Huis van Alijn, Gent 003625
Huis Verwolde, Laren, Gelderland 032560
Huis voor Beeldende Kunst,
Utrecht 112701
Huis Zypendaal, Arnhem 032024
Huisgen, Manfred, Neuss 130452
Huisman, Hetty, Amsterdam . . . 137834
Huittisten Museo, Huittinen Museum,

Huittinen 010887
Huiyou, Shanghai 102182
Huize Betje Wolff,
Middenbeemster 032654
Huize Keizer, Denekamp 032212
Huize Nijenstede, Hardenberg . . 032409
Huizenga, M.L., Amsterdam . . . 143038
Huizer Museum Het Schoutenhuis,
Huizen 032519
Huizhou Museum, Huizhou 007924
Hujer, Anna, Erlangen 141745
Hula, Honolulu 120941
Huldigungssaal und Rathaus,
Goslar 018767
Huldreheimen, Bykle 033566
Huleh Valley Regional Prehistoric
Museum, Kibbutz Maayan
Baruch 025140
Hulihee Palace, Kailua-Kona . . . 049854
Hull and East Riding Museum,
Kingston-upon-Hull 044715
Hull Antique Centre, Kingston-upon-
Hull 089701
Hull Lifesaving Museum, The Museum
of Boston Harbor Heritage, Hull 049608
Hull Maritime Museum, Kingston-upon-
Hull 044716
Hull University Art Collection, Kingston-
upon-Hull 044717
Hulme, J. Alan, Chester 144331
Hulsch, Carlos, Berlin 106316
Hulsey Gallery, Oklahoma City
University, Oklahoma City 051728
Huludao Museum, Huludao 007925
Humačka Arheološko Zbirka,
Franjevački Samostan, Ljubuški 004267
Human & Rousseau, Roggebaai 137947
Human Horizons Art School,
Miami 057432
Humanities Division, Herkimer County
Community College, Herkimer . 057196
Humbert Collection, Münster . . . 020599
Humbert, Bruno, Oyonnax 070832
Humbert, Dominique, Courtavon 068196
Humberts, Bourton-on-the-Water 126830
Humble, Fyshwick 061499
Humble, Owen, Newcastle-upon-
Tyne 090825
Humbleyard Fine Art, London . . 090226
Humboldt and District Museum and
Gallery, Humboldt 006020
Humboldt County Historical
Association Museum, Dakota
City 048105
Humboldt County Historical
Association Museum, Humboldt 049609
Humboldt County Old Settler's
Museum, Livermore 050341
Humboldt Historical Museum,
Humboldt 049610
Humboldt-Museum Schloß Tegel,
Berlin 017174
Humboldt State University Natural
History Museum, Arcata 046520
Hume, Dudley, Brighton 088624
Humeurs, Paris 071490
Hummel, Alois, Freiburg im
Breisgau 075869
Hummel, Eberhard, Schlüchtern 130625
Hummel, Emil, Hofheim in
Unterfranken 130058
Hummel, I., Fridolfing 075907
Hummel, Jörg, Spangenberg . . . 078514
Hummel, Julius, Wien . . 062987, 100324
Hummel, Klaus & Mona, Salem,
Baden 078340
Hummel, Richard, Frankfurt am
Main 075793
Hummel, Richard, Wiesbaden . . 078980
Hummeluhre, Viby 103008
Hummer, Edith, Freistadt 127734
Humpal Holcombe, Portland . . . 123631
Humpback Rocks Mountain Farm
Visitor Center, Waynesboro . . 054257

Humphrey-Carrasco, London . . . 090227
Humphrey Forum, Minneapolis . 050938
Humphrey, Gil, Memphis 094639
Hun, New York 122718
Hunan Fine Arts Publishing House,
Changsha 136705
Hunan Literature and Fine Arts
Publishing House, Changsha . . 136706
Hunan Provincial Museum,
Changsha 007674
Hunan Revolutionary Cemetery,
Changsha 007675
Hund, Isabella, München 108142
Hundertmark, Armin, Las Palmas de
Gran Canaria 115378, 137999
Hundörat, Stockholm 143950
Hundred & One Antiques,
Toronto 064675
Hundsmarktmühle, Thalgau 002822
Hundt, Wesel 078939
Hunebedcentrum, Borger 032085
Huněk, Jan, Praha 065501
Hunerkada, Islamabad 113300
Hung-Yen Museum, Hung-Yen . . 054814
Hung, Hing Hong, Hong Kong . . 065056
Hung, Long, Ho Chi Minh City . . 125126
Hungarian Folk-Art Museum, Port
Orange 052255
Hungerbühler, Urs, Wald 087891
Hungerford Arcade, Hungerford . 089609
Hungry Ghost, Bourton-on-the-
Water 088547
Hunold, H. G., Greven 107161
Hunsche, Licie Fayet, Porto
Alegre 063976
Hunsrück-Museum, Simmern . . 021868
Hunsrücker Holzmuseum,
Morbach 020464
Hunsu Nationality Museum,
Xingyi 008434
Hunt, San Antonio 124041
Hunt, San Francisco 097256
Hunt, Sidney 127459
Hunt & Kastner, Praha 102697
Hunt & Lomas, Cirencester 134274
The Hunt Gallery, Sandwich . . . 119061
Hunt Gallery, Mary Baldwin College,
Staunton 053511
Hunt Institute for Botanical
Documentation, Pittsburgh . . . 052165
Hunt-Morgan House, Lexington . 050244
The Hunt Museum, Limerick . . . 024954
Hunt Valley Antiques Show,
Timonium 098453
Hunt, Adele, Dallas 093143
Hunt, Daniel, London 118400
Hunt, Linda, Wexford 079723
Hunt, Stanley, Gainsborough . . 089316,
134398
Hunt, Stanley, Retford 091092
Hunter, Memphis 135778
Hunter College Art Galleries, New
York 051405
Hunter-Dawson State Historic Site,
New Madrid 051282
Hunter do all, Louisville 135759
Hunter Fine Art, Glasgow 117846
The Hunter Gallery, Long Melford 118655
Hunter House, East Kilbride . . . 044205
Hunter Museum of American Art,
Chattanooga 047595
Hunter Street Gallery, Parramatta 099553
Hunter Valley Antiques, Branxton 061191
Hunter Valley Museum of Rural Life,
Scone 001488
Hunter, Donald, Cheltenham . . . 134237
Hunter, Jill, Stamford . 091405, 135072
Hunter, Moa, Auckland 143290
Hunter, Pat, Fresno 120899
Hunterdon Museum of Art,
Clinton 047841
Hunterian Art Gallery, Glasgow . 044401
Hunterian Museum, Glasgow . . . 044402
Hunterian Museum, London . . . 045002

Hunterian Museum, Zoology Section, Glasgow 044403
Hunters and Collectors Antiques, Mosman 061894
Hunters and Selectors, North Shore 084033
Hunter's Antiques, Seattle 097456
Hunter's Find, Tampa 097563
Hunters Hill Historical Museum, Hunters Hill 001156
Hunter's of Hampton, Hampton . 089418
Huntingdon County Museum, Huntingdon 049612
Huntingdon Trading Post, Huntingdon 089620
Huntington, Stow-on-the-Wold . 091472
Huntington Art Collections, San Marino 053053
Huntington Beach Art Center, Huntington Beach 049629
Huntington Beach International Surfing Museum, Huntington Beach . 049630
Huntington County Historical Society Museum, Huntington .. 049617
Huntington Frontiers, San Marino 139386
Huntington Library Press, San Marino 138466
Huntington Museum of Art, Huntington 049626
Huntington Railroad Museum, Huntington 049627
Huntington Trade School, Huntington Historical Society, Huntington . 049622
Huntington, Archer M., Austin .. 119677
Huntley, Diana, London 090228
Huntly, Campbell 098856
Huntly and Districts Historical Museum, Huntly 001157
Huntsman Marine Science Centre Aquarium & Museum, Atlantic Reference Centre, Saint Andrews 006848
Huntsville Art League, Huntsville 060518
Huntsville Museum of Art, Huntsville 049635
Hunziker, Ulrich, Rümlang 116598
Huon Valley Apple and Heritage Museum, Huonville 001158
Huppert, Bruno, Forcalquier ... 068587
Hurda Talumuuseum, Hurda ... 010615
Hurdal Bygdetun, Hurdal 033676
Hurds, Seattle 097457
Huret, Jean Etienne, Paris 141140
Hurl, Rosemarie & Anthony, Tusmore 062356
Hurley Patentee Manor, Hurley . 049638
Hurnall, Marvin, Prahran 062112
Huron City Museums, Port Austin 052246
Huron County Museum, Goderich 005889
Huronia Museum, Midland 006304
Hurrah's Nest, Austin 092240
Hurricane Valley Heritage Park Museum, Hurricane 049642
Hurst Ranch, West Covina 054297
Hurst, Anthony, Woodbridge .. 091913, 135253
Hurst, D., London 134694
Hurstmere Antiques, North Shore 084034
Hurston, A.J.V., Los Angeles .. 094367
Hurt & Assoc., Norfolk 123227
Hurta, Houston 093703
Hurtebize, Cannes 067643, 103514
Hurthe, Dieter, Berlin 074830
Hurtigrutemuseet, Stokmarknes 033992
Hurtownia Antyków, Katowice .. 084532
Hurworth Cottage, New Plymouth 033257
Hus-Museum, Konstanz 019760
Hus Var Fine Art, Buffalo 119958
Huschka, Josef Alois, Kirchanschöring 076800
Huseby Bruk, Grimslöv 040672
Huser, Robert, Flawil 087425
Huser, Rose-Marie, Billom 067221
Huset Med Det Rare I, Tromsø . 084381

Husi, Heinz, Chippis 087369
Húsipari Múzeum, Budapest ... 023489
Húsipari Üzemtörténeti Gyűjtemény, Gyula 023674
Husitské Muzeum v Táboře, Tábor 009902
Huskisson Trading Post, Huskisson 061643
Huskvarna Antik, Huskvarna ... 086752
Huskvarna Stadsmuseum, Huskvarna 040712
Husqvarna Fabriksmuseum, Huskvarna 040713
Huss, Gottfried, Olching 077923
Huss, Sally, San Diego 124133
Huss, Sally, Tucson 124743
Le Hussard, Metz 070098
Hussenot-Desenonges, Bruno, Vierzon 074118
Hussenot Desenonges, Marie-Claude, Nançay 070408
Hussenot, Ghislaine, Paris 104955
Hussey, Paul, Dublin 130858
Hussian School of Art, Philadelphia 057672
Huste, Wolfgang, Bad Honnef .. 074529, 125855, 141424
Husted, Seattle 124546
Huster, L., Detmold 141686
Huster, V., Baden-Baden 074636
Husumhus Antik, Brønshøj 065619
Huszármúzeum, Csillagvár – Középkori életpanoptikum, Balatonszentgyörgy 023429
Huszita-Ház – Helytörténeti Gyűjtemény, Gönc 023649
Hut, Manama 100436
The Hut Gallery, Ferntree Gully . 099015
Hutchinson Art Association, Hutchinson 060519
Hutchinson Art Center, Hutchinson Art Association, Hutchinson .. 049644
Hutchinson County Museum, Borger 047097
Hutchinson House Museum, Waupaca 054234
Hutchinson-Scott, Ripon 127021
Hutchinson, Martha, Philadelphia 096116, 136092
Hutchinson Antiques, Chelmsford 088828, 134229
Hutchison House Museum, Peterborough 006638
Huth, U., Prisannewitz 078096
Huther, Koblenz 076835
Hutin, Emmanuel, Paris 141141
Hutin, Renaud, Sarlat-la-Canéda 073411
Hutnícke Múzeum Železiarne Podbrezová, Podbrezová 038424
Hutong a Roma, Roma 081411
Hutong Gallery, Milano 080572
Hutt Art Society, Wellington ... 059635
Hutt Street Books, Adelaide .. 139615
Hutt, Andrew, Kirkcaldy 118056
Hutter, Ira, New York .. 095467, 127352
Hutterer, Jutta, Wiener Neustadt 100401
Hutterloh, Herbert, Arnsberg .. 106001
Huttermuseum, Erdweg 018234
Hutto, Houston 135607
Hutton, Leonard, New York ... 122719
Huvé, Philippe, Saint-Denis-sur-Sarthon 072734
Huvé, Philippe, Sées 073472
Huygensmuseum Hofwijck, Voorburg 033004
Huyghe-Cariou, Henri, Tréogat . 073880
Huyse Molensloot, Antwerpen . 100493
Huysser, d', Bruxelles 100620
Huzette, Jocelyne, Épernay 068450
Huzhou Museum, Huzhou 007926
HV Galleri, Stockholm 115896
HVB Kunst Palais, München ... 020531
Hvitträsk, Kirkkonummi 011008
HW Kunst, Vejle 066194

Hwa Kang Museum, Chinese Culture University, Taipei 042518
Hwajeong Museum, Seoul 030004
Hwang, Jason, Portland 096421
Hwas Gallery, Shanghai 102183
Hwataoyao – Kiln Exhibition Gallery, Yuanli 042600
HWL-Galerie und Edition, Düsseldorf 106796, 137161
Hyacinths, Boston 092476
Hyart & Fils, Compiègne 128569
Hyatt One-Room Schoolhouse, Milby 006309
Hybrid, Seattle 097458
Hyde & Seek Antiques, San Francisco 097257
Hyde Art Gallery, Grossmont College, El Cajon 048495
The Hyde Collection, Glens Falls 049123
Hyde Hall, Cooperstown 047974
Hyde Hall, East Springfield 048447
Hyde Park Antiques, New York . 095469
Hyde Park Art Center, Chicago . 047664
Hyde Park Barracks Museum, Sydney 001539
Hyde Park Jewellery, Sydney .. 062288
Hyde Park Refinishing, Tampa . 136388
Hyde Road Antiques, Paignton . 090941, 134922
Hyde, Liza, New York 095468
Hyder, Jane, Wellington 113069
Hyderabad Museum, Hyderabad 034113
Hydo, Petersham 099564
Hydra Antiques, South Yarra ... 062227
Hydro Industripark-Museum, Porsgrunn 033878
Hydrobiological Museum, Alexandria 010516
Hyeonchungsa Shrine, Asan ... 029888
Hyland, Melbourne 139827
Hyland House, Guilford 049292
Hyldig, Ane-Mette & Finn, Nykøbing Mors 066017
Hylleveien, Arendal 084210
Hylteberga Galeri, Skurup 115815
Hylténs Industrimuseum, Gnosjö Industrimuseum, Gnosjö 040645
Hyman and Sons General Store, Magasin Hyman, Gaspé 005862
Hyman, James, London 118401
Hyman, Linda, New York 122720
Hymers Museum, Hymers 006023
Hymne, Parlow Valley 114807
Hyna, U., Rottach-Egern 108529
Hyogo-ken Togeikan, Sasayamashi 029348
Hyogo-kenritsu Bijutsukan, Kobe 028843
Hyogo kenritsu Hito to Shizen no Hakubutsukan, Sanda 029327
Hyogo-kenritsu Rekishi Hakubutsukan, Himeji 028629
Hyperion Antique Centre, Saint Ives, Cambridgeshire . 091189, 127028
Hypocaust, Saint Albans 045664
Hyrax Hill Museum, Hyrax Hill .. 029858
Hyšpler, Lubomír, Příbram 102732, 140537
Hythe, Kingston 099199
Hythe Local History Room, Hythe, Kent 044623
Hytteballe Antikviteter, Odense . 066036, 066037
Hyundai, Seoul 111665
Hyundai Art Gallery, Seoul 111666, 111667

I-20 Gallery, New York 122722
I 20 Gallery, New York 122721
I-35 North Antiques, Kansas City 094076
I-40 Antique Mall, Albuquerque . 092225
I-44 Antique Mall, Tulsa 097740
I-5 Gallery, Los Angeles 050435
I Dialoghi, Pavia 137715
I Galleriet Magne Augestad, Sandnes 113258

I Lanzi, Firenze 080158
I Leonarte, Valencia 115505
I-Punkt, Sigmaringen 078470
I Saw Designs, Milwaukee 094863
I Saw It First, Minneapolis 094920
I Segni, Torino 081712
I Space, University of Illinois at Urbana-Champaign, Champaign 047518
I Thought You Knew, San Francisco 097258
I Tid och Rum, Nacka 086875
I. Nikoladze House Museum, Tbilisi 016408
I.C. Editions, New York 138399
I.D., International Design, New York 139387
I.DE.A. – Internationales Designcenter Austria, Schrems ... 100195, 136565
I.E.P. Inventaires Estimations & Partages, Paris 071491
I.G.A. Old Masters, Saint Lawrence 091194
I.P. Stanback Museum, South Carolina State University, Orangeburg .. 051795
I.Q. Gallery, Dublin 109593
I.V.K. Rajwade Sanshodhan Mandal Museum, Dhule 024249
i8, Reykjavík 109242
Iaccarino, Francesco, Napoli ... 080887
Iacoletti, Mariano, Napoli 080888
Ian Kenny, Ian, Hamilton 117906
The Ian Potter Museum of Art, Melbourne 001275
Ianira Antique Shop, Lemesos . 065328
Ian's Antiques, Oakland 095897
Iarca Gallery, Chicago 120160
Iavarone, Antonio, Napoli 080889
Iavarone, Luigi, Napoli 080890, 142762
Ibala, Bussum 083018
Ibañez Redondo, Angel, Madrid . 133345
Ibara Kyodo Denchu Bijutsukan, Ibara 028666
Ibaraki Daigaku Izura Bijutsu Bunka Kenkyujo, Kitaibaraki 028822
Ibaraki-ken Kindai Bijutsukan, Mito 029034
Ibaraki-ken Tenshin Kinen Izura Bijutsukan, Kitaibaraki 028823
Ibaraki-kenritsu Rekishikan, Mito 029035
Ibaraki-kenritsu Shizenshi Hakubutsukan, Iwai 028719
Ibaraki-kenritsu Tougeikan, Kasama 028788
Ibarbia, Elena, Donostia-San Sebastián 133250
Ibars Pallas, Silvia, Barcelona .. 133121
Ibbett & Mosely, Sevenoaks ... 127036
Ibel, Rebecca, Columbus 120500
Ibenfeldt, Holte 065785, 102847
Iberia, Bern 144050
Iberoamericana International, North Bay Village 139388
Ibex Gallery, Unterseen 116678
IBRA Collection, International Bee Research Association, Cardiff . 043866
Ibrahim Hussein Foundation Museum, Pulau Langkawi ... 030722
Ibresinskij Étnografičeskij Muzej pod Otkryrym Nebom, Čuvašskij Nacionalnyj Muzej, Ibresi 037022
Ibsen-museet, Norsk Folkemuseum, Oslo 033836
Ibsenhuset og Grimstad Bymuseet, Grimstad 033624
Ibu, Paris 104956, 136943
Ibunsai, Nara 082069
ICA Monthly Bulletin, London .. 139186
ICAMT International Committee for Architecture and Museums Techniques, Amsterdam 059608
Icaro Gallery, Long Beach 121403
ICAS Gallery, Letchworth 118105
ICDAD International Committee for Decorative Arts and Design

Museums, Wien 058261
Ice Museum, Kamikawa 028762
Icebox Gallery, Minneapolis .. 122002
Icelandic Museums Association,
Patreksfjörður 059348
Iceni Village and Museums, Cockley
Cley 043986
Ichabod, Denver 145128
Ichan Qala Davlat Muzej Qòriqxonasi,
Xiva 054673
Ichikawa-shiritsu Koko Hakubutsukan,
Ichikawa 028669
Ichikawa-shiritsu Shizenshi
Hakubutsukan, Ichikawa ... 028670
Ichimai-no-e, Sapporo 082099
Ichinomiya-shiritsu Hakubutsukan,
Ichinomiya 028672
Ichinoseki-shiritsu Hakubutsukan,
Ichinoseki 028673
Ici, Warszawa 084728
Ickleton Antiques, Saffron
Walden 091168
Icko, Los Angeles 121511
Icks, Horst, Bramsche 129620
Ickworth House, Horringer 044602
ICLM International Committee for
Literary Museums, Paradis .. 059659
ICM, Haarlem 083333
ICMAH International Committee for
Archaeology and History Museums,
Luxembourg 059555
ICME – International Committee
for Museums and Collections of
Ethnography, Miami Beach ... 060520
ICOFOM International Committee for
Museology, Barnaul 059772
Icograda Message Board,
Montréal 138580
ICOM – Belgian National Comittee,
Hornu 058378
ICOM – Comité National Monégasque
du Conseil International des Musées,
Monaco 059580
ICOM Afghanistan, Kabul ... 058164
ICOM Andorra, Aixovall 058168
ICOM Angola, Luanda 058169
ICOM Arab, Giza 058566
ICOM Argentina, Buenos Aires . 058176
ICOM Asia-Pacific, Eumseong . 059528
ICOM Australia, South Brisbane . 058203
ICOM Azerbaijan, Bakı 058341
ICOM Bangladesh, Dhaka 058345
ICOM Barbados, Saint Michael . 058347
ICOM Bénin, Porto-Novo 058393
ICOM Bolivia, La Paz 058395
ICOM Bosnia-Hercegovina,
Sarajevo 058396
ICOM Botswana, Gaborone ... 058397
ICOM Bulgaria, Sofia 058405
ICOM Burkina Faso,
Ouagadougou 058408
ICOM Burundi, Gitega 058410
ICOM Cambodia, Phnom Penh . 058411
ICOM Cameroon, Yaoundé 058413
ICOM Canada, Ottawa 058447
ICOM Chile, Santiago de Chile . 058485
ICOM Chinese National Committee,
Beijing 058497
ICOM Comitato Nazionale Italiano,
Milano 059459
ICOM Comité Haïtien, Port-au-
Prince 059337
ICOM Comité National du Niger,
Niamey 059647
ICOM Comité Sénégalais, Saint-
Louis 059784
ICOM Committee for University
Museums and Collections,
Berlin 058922
ICOM Comores, Moroni 058513
ICOM Costa Rica, San José ... 058517
ICOM Côté d'Ivoire, Grand
Bassam 058519
ICOM Crne Gore, Cetinje 059582

ICOM Cuba, La Habana 058530
ICOM Cyprus, Lefkosia 058535
ICOM du Tchad, N'Djamena ... 058481
ICOM Ecuador, Guayaquil 058563
ICOM Egypt, Cairo 058567
ICOM Estonia, Tallinn 058571
ICOM F.Y.R. Macedonia, Skopje . 059559
ICOM Finland, Helsinki 058575
ICOM Ghana, Cape Coast 059322
ICOM Grèce, Athinai 059329
ICOM Guatemala, Guatemala .. 059336
ICOM Iceland, Reykjavík 059349
ICOM India, Kolkata 059362
ICOM Indonesia, Jakarta 059373
ICOM Iran, Tehrān 059374
ICOM Ireland, Ballina 059378
ICOM Jamaica, Kingston 059487
ICOM Jordan, Amman 059518
ICOM Kazakhstan, Almaty 059520
ICOM Kenya, Nairobi 059521
ICOM Kyrgyzstan, Biškek 059538
ICOM Lao, Vientiane 059540
ICOM Latin America and the
Caribbean Regional Organization,
San Jose 058518
ICOM Latvijas Nacionāla Komiteja,
Rīga 059542
ICOM Liban, Beirut 059546
ICOM Libyan Arab Jamahiriya,
Tripoli 059547
ICOM Lietuva, Vilnius 059552
ICOM Luxembourg, Luxembourg . 059556
ICOM Madagascar, Isoraka 059563
ICOM Magyar Nemzeti Bizottság,
Budapest 059339
ICOM Malaysia, Kuala Lumpur . 059564
ICOM Mali, Bamako 059565
ICOM Maroc, Rabat 059585
ICOM Mauritanie, Nouakchott .. 059568
ICOM Mauritius, Port Louis 059569
ICOM México, México 059573
ICOM Moldova, Chişinău 059576
ICOM Mongolia, Ulaanbaatar ... 059581
ICOM Nacionalni Komitet Serbia,
Beograd 059787
ICOM Namibia, Windhoek 059586
ICOM National Commitee of Belarus,
Minsk 058350
ICOM Nederland, Amsterdam .. 059609
ICOM Nepal, Lalitpur 059588
ICOM New Zealand, Rotorua ... 059636
ICOM News, Paris 138647
ICOM Nigeria, Calabar 059649
ICOM Northern Ireland, Belfast . 060159
ICOM-Österreichisches
Nationalkomitee, Wien 058262
ICOM Oman, Muscat 059678
ICOM Panamá, Ciudad de
Panamá 059683
ICOM Paraguay, Asunción 059686
ICOM Philippines, Manila 059688
ICOM Poland, Poznań 059695
ICOM Portugal, Lisboa 059756
ICOM Qatar, Doha 059763
ICOM Regional Organization for West
Africa, Bamako 058540
ICOM República Dominicaine, Santo
Domingo 058562
ICOM Republique Congo,
Brazzaville 058516
ICOM République Democratique du
Congo, Kinshasa 058514
ICOM Romania, Ploieşti 059765
ICOM Schweiz, Internationaler
Museumsrat, Zürich 059935
ICOM Seychelles, Victoria 059795
ICOM Singapore, Singapore ... 059800
ICOM Slovakia, Bratislava 059804
ICOM Slovenia, Ljubljana 059807
ICOM South Africa, Sunnyside .. 059817
ICOM Sri Lanka, Colombo 059877
ICOM Swaziland, Lobamba 059878
ICOM Sweden, Stockholm 059883
ICOM Tanzania, Dar-es-Salaam . 060000

ICOM Thailand, Pathumthani ... 060001
ICOM Togo, Lomé 060003
ICOM Tunisia, Tunis 060004
ICOM Turkish National Committee,
Ulus 060008
ICOM Ukraine, Lviv 060011
ICOM Uruguay, Punta del Este . 060280
ICOM Uzbekistan, Toshkent ... 060821
ICOM Venezuela, Caracas 060823
ICOM Viet Nam, Ha Noi 060826
ICOM Zambia, Livingstone 060829
ICOM Zimbabwe, Ascot 060830
ICOMAC Regional Organization for
Central Africa, Kinshasa 058515
ICOMAM International Committee of
Museums and Collections of Arms
and Military History, Delft ... 059610
Icon Art Gallery, Picton 112989
Icon Books, Byron Bay 139675
Icon Boulevard 409 First Av S,
Seattle 124547
Icon Gallery, Wilmslow 119353
Icon News, London 139187
Icon World Studio, Leederville .. 061730
Iconastas, London 090229
Iconografisch Bureau, Den Haag . 032174
IConomix Journal, Paris 138648
ICR International Committee for
Regional Museums, Heraklion . 059330
Icram, Barcelona 133122
ICU Hachiro Yuasa Memorial Museum,
Mitaka 029032
ID Collection, Portland 123632
ID Galerie, Helmond 032461
Ida Aalbergin Lapsuudenkoti,
Leppäkoski 011108
Ida Green Gallery, Austin College,
Sherman 053310
IDAG-International Dynamic Art Gallery,
Antwerpen 100494
Idaho Association of Museums,
Boise 060521
Idaho Museum of Mining and Geology,
Boise 047075
Idaho Museum of Natural History,
Pocatello 052222
Idaho State Historical Museum,
Boise 047076
Ida's Idea, Washington 124920
Idea Art and Design Center,
Hsinchu 042332
Idea Books, Amsterdam 137835
Idea Books, Viareggio 137762
Idea-Castellum, Budapest 109150
Idea Studio, Milano 080573
Ideal Baseball Cards, Cincinnati . 145073
Ideal Book Store, Yonkers 145647
Idéal Passion, Marseille 104182
Ideal Upholstery and Furniture, San
Francisco 136322
Ideart, Barcelona 133123
Idearte Antichità, Assago 079753
Ideas and More, Miami 094748
Ideas Galería de Arte, Cali ... 102411
Idéco, Fontenay-le-Comte 068573
I'Idee, Cincinnati 120350
Idee Arcobaleno, Milano 080574
Idee nel Tempo, Padova 080987
Idee nel Tempo, Verona 081916
Ideja, Pleven 101130
Idemitsu Bijutsukan, Tokyo ... 029517
Identity Gallery, Kolkata 109401
identity of art Ltd., Kunst für
Unternehmen, Aumühle 074474
IDF Gallery, Nagoya 111321
Idiom Studio, Wellington 113070
Idle Hour Book Inn, Rotorua ... 143341
Idle Time Books, Washington .. 145636
Idojiri Kokokan, Fujimi 028537
Idrottsantikvariátet, Vällingby . 143997
Idrottsmuseet, Malmö 040826
Idstedt-Gedächtnishalle, Idstedt . 019378
Idube.Co.Uk, Chobham 117586
Idvies Antiques, Letham 089810

Idyllwild Arts Academy, Idyllwild . 057229
Iégor, Montréal 125371
Iemmolo, Angelo, Padova ... 131667
Ien, Amersfoort 112123
Ieri e Oggi, Milano 080575
Ieri e Oggi, Venezia 081845
Iermano Antiquités, Napoli ... 080891,
                                080892
Iesu, Trieste 081793
Ievos Simonaitytès Memorialinis
Muziejus, Priekulė 030448
If Museum Inner Spaces, Wyższa
Szkola Nauk Humanistycznych i
Dziennikarstwa, Poznań ... 035116
IF Museum Innerspaces, Poznań 113740
ifa-Galerie Berlin, Institut für
Auslandsbeziehungen e.V.,
Berlin 017175
ifa-Galerie Stuttgart, Institut
für Auslandsbeziehungen e.V.,
Stuttgart 022042
IFAR Journal, New York 139389
Ifield Watermill, Crawley 044034
Ifj. Lele J. Néprajzi Gyűjteménye,
Szeged 023925
Iföverkens Industrimuseum,
Bromölla 040584
IFPDA Print Fair, New York 098454
IG Bildende Kunst, Wien 100325
IG Halle, Rapperswil 041860
Iga and Shigaraki Pottery Museum,
Ueno 029672
Iga-ryu Ninja Yashiki, Ueno ... 029673
Igarskij Kraevedčeskij Kompleks Muzej
Večnoj Merzloty, Igarka 037023
Igartzako Museoa, Beasain ... 039112
Igavel, New York 127353
Igel, Manya, London .. 090230, 118402
Iggy's, San Antonio 096978
Iglesia Caro, Jose de la, Madrid 086200
Iglesia ni Cristo Museum and Gallery,
New Era University, Quezon City 034582
Iglińska, Anna, Kraków 113602
Iglinski, Grzegorz, Hamburg ... 076245
Iglou Art Esquimau, Douai ... 103703
Isabel Ignacio, Galería de Arte,
Sevilla 115452
Ignaszewski, Georg, Berlin ... 129511
Ignatowicz-Woźniakowska, Dorota,
Warszawa 132881
Ignaz Günther Museum,
Altmannstein 016553
Ignaz J. Pleyel Museum,
Großweikersdorf 002016
Ignotus Technika, Budapest ... 142560
Iguana Gallery, Stroud 119173
Ihle-Wirth, Franziska & Geroon,
Fulda 129882
Ihler, Christian, La Garde 069046
Ihn Gallery, Seoul 111668
Ihring, Rolf, Berlin 141531
Ihrke, Christine, Berlin 106317
II. Yesemek Heykel Atölyesi ve
Açıkhava Müzesi, İslahiye 042916
Iida-shi Bijutsu Hakubutsukan,
Iida 028674
Iida-shiritsu Kokokan, Iida 028675
III Wash Design, Houston ... 121063
Iisakin Jussin Tupa, Kauhava . 010973
Iisaku Muuseum, Iisaku 010617
Iittalan Iasimuseo, Iittala 010894
Ijff en van Velzen, Maastricht . 132576
IK Galerie, Maastricht 112577
Ikaros, Salzy 065556, 107235
Ike Taiga Bijutsukan, Kyoto ... 028930
Ikebana Arts, San Jose 124450
Ikebana Bijutsukan, Kyoto 028931
Ikeda & Lokker, Tokyo 111445
Ikeda 20-seiki Bijutsukan, Ito . 028708
Ikeder, Bilbao 114993
Ikegami & Takebi, Kyoto 082018
Ikegami Bijutsu Mingeikan,
Takatomachi Inashi 029450
Ikels, Rena, Rellingen 108473

Ikkan, New York 122723
Iklektik Designs, Houston 121064
IKO Gemeentelijk Instituut
voor Kreatieve Opvoeding,
Hoogstraten 055029
Ikodespotiko Paleopolio, Lefkosia 065323
Ikodinovic & Co., Nenad,
Waterloo 063855
Ikon- och Konstföreningen,
Stockholm 059884
Ikon Gallery, Birmingham 043620
Ikon Show Room, Haderslev . . 102820
Ikona Photo Gallery, Venezia . 111127
Ikonen Galerie, Berlin 074831
Ikonen-Galerie, Bern 116138
Ikonen-Galerie Dritsoulas,
München 077531, 130358
Ikonen Galerie Sophia, Zürich . . 087977,
133984
Ikonen-Museum der Stadt Frankfurt
am Main, Stiftung Dr. Schmidt-Voigt,
Frankfurt am Main 018429
Ikonen-Museum, Museen
der Stadt Recklinghausen,
Recklinghausen 021306
Ikonengalerie Helmut Brenske,
Hannover . . . 076377, 125984, 137267
Ikonenmuseum, Autenried 016683
Ikonenmuseum, Bad Breisig . . . 016738
Ikonenmuseum, Potzneusiedl . . 002524
Ikonenmuseum Schloß Autenried,
Ichenhausen 019375
Ikonenzentrum Alexej Saweljew,
Traben-Trarbach 022170
Ikonos, Burgos 133211
Ikor, Sapporo 111376
Ikuno Kobutsukan no Kogyo
Hakubutsukan, Ikuno 028681
Ikuo Hirayama Bijutsukan,
Onomichi 029233
Il Etait Une Fois, Beausoleil . . 067054
Il était une fois, Plaisance-du-
Touch 072171
Il Etait Une Fois, Remiremont . . 072413
Il-Kartell, Tarxien 082437
Il Mauriziano, Reggio Emilia . . 027501
Ilana Goor Museum, Tel Aviv . . 025197
Ilchester Museum, Ilchester . . . 044624
Ilchmann, Dagmar, Templin . . . 108762
Ídánach, Dublin 109594
Ildiko, Malvern, Victoria 061783
Île-a-la-Crosse Heritage Museum, Île-
a-la-Crosse 006025
Ile au Trésor, Biarritz 067201
Ile aux Antiquités, Victot-Pontfol 074114
Ile aux Trésors, Grenoble 068732,
140847
Ile aux Trésors, Neufchâtel-en-
Bray 070511
Ile du Temps, Libourne 069460
Ilene & Wayne, Honolulu 093554
Iles, Francis, Rochester 091127, 118985,
134974
ILevel, New York 122724
Ilfracombe Arts Society,
Ilfracombe 118012, 118013
Ilfracombe Machinery and Heritage
Museum, Ilfracombe 001162
Ilfracombe Museum, Ilfracombe 044626
Ilg, Hansruedi, Zürich 133985
Ilgner, Jürgen, Melle 077337
İlhami Atalay Sanat Galerisi,
İstanbul 117052
Iliad, New York 095470
Ilias Lalaounis Jewelry Museum,
Athinai 032992
Ilias, Antonis, Angers . 066608, 128443
Ilić, Boza, Prokuplje 114459
Iliev, Rumen, Köln 107640, 137312
Iligan Museum, Iligan 034479
Iliopoulou, Anna, Glyfada 079193
Ilisso, Nuoro 137707
Ilja Chavchavadze House Museum,
Tbilisi 016409

Ilja Chavchavadze Literature-Memorial
Museum, Tbilisi 016410
Ilko, C., Salzburg 062801
Illawarra Museum, Wollongong . 001653
Illegaard, Tartu 103121
Illg, Karin, Balingen 074654
Illi, Raymond & Alexandre,
Genève 087475, 116321, 144077
Illies, Mainz 107950
Illingworth Kerr Gallery, Calgary 005523
Illinois and Michigan Canal Museum,
Lockport 050352
Illinois Art Gallery, Chicago . . . 047665
Illinois Artisans and Visitors Centers,
Whittington 054382
Illinois Arts Alliance, Chicago . 060522
Illinois Association of Museums,
Springfield 060523
Illinois Citizen Soldier Museum,
Galesburg 049022
Illinois Institute of Art, Chicago . 056887
Illinois Railway Museum, Union . 053884
Illinois State Museum,
Springfield 053451
Illmer Kunst + Design,
Osnabrück 108346
Illmert, Sylvia, Saarbrücken . . . 108548
Illo Tempore, Bari 130915
Illos, Buffalo 135373
Illulian, Milano 080576
Illum, Miami 094749
Illuminary Art, Toronto 101627
Illuminata Gallery, Houston . . . 121065
Illuminati, Milano 080577
Het Illuseum, Gent 003626
Illusoria-Land, Museum und Galerie
für optische Täuschungen und
Holographien, Ittigen 041608
Illustrated Gallery, Philadelphia . 123396
Illustration House, New York . . . 095471,
127354
Illustration Magazine, Saint Louis 139390
Illustrations and Graphic Design
Department, Harcum College, Bryn
Mawr 056811
Illustrator, Minneapolis 139391
Illustratoren Organisation e.V.,
Frankfurt am Main 058923
Illy, Günther, Timelkam 062859
Ilmajoen Museo, Ilmajoki 010895
Ilmatorjuntamuseo, Tuusula . . . 111459
Ilmavoimien Viestikoulun
Perinnehuone, Tikkakoski 011420
Ilmin Museum of Art, Seoul . . . 030005
Ilo, Tallinn 136788, 140680
Ilocaniana Cultural Museum,
Magsingal 034501
Iloniemen Pajamuseo, Raatala . 011294
Îlot des Palais, Québec 006730
L'Ilotrésor, Chavannes-près-
Renens 087362
Iloulian, John, Los Angeles . . . 094368,
135722
Iloulian, John, West Hollywood . 097899
Ils Sont Fous ces Romains, Ville-
d'Avray 074130
Ilse B., Paris 071492
ILTE- Industria Libraria Tipografica
Editrice, Moncalieri 137701
Ilulissani Katersugaasivik,
Ilulissat 023329
Ilumina-Art, Bochum 106513
Ilwaco Heritage Museum, Ilwaco 049659
Im Autorenregister, Köln 142013
IM Gallery, Singapore 114538
Im Kinsky, Wien 125275
Im Kinsky Kunst Auktionen, Wien 062988,
125276
Im Obersteg, Sankt Gallen 087781
Im Trend, Amberg 105985
Imabari-shiritsu Kono Bijutsukan,
Imabari 028683
Imabayashi, Fukuoka 081966
Imafronte, Murcia 139058

Image, Łomianki 113683
Image 54 Gallery, Calgary 101244
L'image amusée, Céret 103548
Image Art, Gillingham . 117824, 138164
Image Arts Etc ., Omaha 123322
Image Arts Etc., San Diego . . . 124134
Image Business, Christchurch . . 112844,
137882
Image Complete, Denver 120735
Image Direct, Amsterdam 112226
Image Document, Châtillon (Hauts-de-
Seine) 067916
Image Document, Paris 071493
Image en Plus, Nîmes 104402
Image et le Livre, Strasbourg . . 141351
Image Factory, Indianapolis . . . 121190
Image Fine Art & Photography,
York 119410
Image Fotografisk Galleri, Århus 102774
Image Graphics, Oakland 123254
Image in, Gérardmer 068652
Image Museum of Hsinchu City,
Hsinchu 042333
Image of Art, Hyderabad 109361
Image One International, Houston 121066
Image Shed, Halifax 117903
Image Showcase, Albuquerque . 119484
Image State, New York 095472
Image Visions Gallery, South
Yarra 099707
Imagem Galeria, Fortaleza 100951
Imagen, Ciudad de Panamá . . . 113318
Imagerie, Den Haag 143112
Imagerie, Lannion 103921
Imagerie d'Epinal, Épinal 136815
Images, London 090231
Les Images, Newtownabbey . . . 118807
Images a Gallery, San Francisco 124303
Images and Frames, Boston . . . 119875
Images Art, Toronto 101628
Images Art & Life, Modena 138832
Images Art Gallery, Bangalore . 109259
Images Art Gallery, Kansas City 121298
Images By Woolf, Wellington . . 113071
Images de la Mer, La Trinité-sur-
Mer 069109
Images du Faubourg, Annecy . . 066642
Images du Passé, Lignieres . . . 069464
Images Festival, Toronto 058448
Images Fine Art, Pittsburgh . . . 123553
Images Gallery, Toledo 124690
Images International, Honolulu . 120942
Images of Austin and the Southwest,
Austin 119678
Images of Nature, Denver 120736
Images of Nature, Las Vegas . . 121359
Images of Nature, Omaha 123323
Images of the North, San
Francisco 124304
Images Under Glass, Minneapolis 122003
Images Under Glass, Saint Paul 123916,
123917
Imagina, Venezia 111128
Imagina-Museo Interactivo Puebla,
Puebla 031374
Imaginaciones Arte, Cali 102412
L'Imaginaire, Bruxelles 140182
The Imaginarium, Anchorage . . 046472
Imaginarium Hands-On Museum, Fort
Myers 048852
Imaginary Places, Philadelphia . 123397
Imagination Station Science Museum,
Wilson 054456
Imagination Unlimited, Miami . . 094750
Imagine, Fuchu 081962
Imagine, Luxembourg 111856
Imagine, Maastricht 112578
Imagine Arts Espaces Matières,
Bordeaux 103436
Imagine Interiors, Memphis . . . 094640
Imagine Venezia London, Venezia 111129
Imaging Solutions, Richmond . . 123736
Imago, Wernigerode 078935
imago fotokunst, Berlin 106318

Imago Grafica, Perugia 110641
Imago Lignea, Torino . 081713, 110981,
132149
Imago Mundi, Buenos Aires . . . 139574
Imago-Verlag, Aichwald 137024
Imaian, Las Rozas 115104
Imaian, Madrid 115238
Imaizumi Hakubutsukan,
Shiozawa 029392
Imari, Porto Alegre 063977
Imari-shi Rekishi Minzoku Shiryokan,
Imari 028684
Imasterpiece, Miami 121808
Imatran Kulttuurihistoriallinen
Kaupunginmuseo, Imatra 010898
Imatran Taidemuseo, Imatra . . . 010899
Imavision Gallery, Taipei 116921
Imbard, Patrick, Paris 071494
Imbert, Didier, Paris 104957
Imelda, Knokke-Heist 100741
Imeson Collection, Windsor,
Ontario 064903
IMF Center Exhibit, Washington . 054124
Imhoff Art Gallery, Saint Walburg 006944
Imker- und Heimatmuseum
Wassertrüdingen,
Wassertrüdingen 022443
Imkereigeschichtliche und
bienenkundliche Sammlungen,
Celle 017723
Imkereimuseum, Grüningen . . . 041569
Imkereimuseum, Pöggstall 002517
Imkereimuseum, Pramet 002535
Imkereimuseum Grüningen,
Wetzikon 042133
Imkerijmuseum Poppendamme,
Grijpskerke 032370
Imkermuseum, Sitzenberg-
Reidling 002759
IMM Münz Institut, München . . . 077532
Immagine, Milano 110360
L'Immagine, Napoli 110531
L'Immagine, Padova 110572
L'Immagine Ritrovata, Bologna . 130968
Immagini Spazio Arte, Cremona 109997
Immendorf, San Francisco 124305
Immers, H.W.J., Delft 083032
Immig, Anna, Göppingen 076033,
107136
Immigrant City Museum,
Lawrence 050190
Immigrant-institutets Museum,
Borås 040575
Immigration Museum, Melbourne 001276
Immingham Museum,
Immingham 044630
Immler, Georg, Röthenbach,
Allgäu 078268
IMMOCARD, Essen 141748
Immortal Treasures, West
Hollywood 097900
Imogene Antique and Fine Jewellery,
Fitzroy North 061473
Imogene Herbert Historical Museum,
Manistique 050645
IMP Jewellery, Toorak 062324
Impact Fine Art Gallery, Buffalo . 119959
Impact Livres, Paris . . 141142, 141143
Impala, Armadale, Victoria 061035,
127556
l'Impasse, Bruxelles 063467
Impasse Saint-Jacques Antiquités,
Bruxelles 063468
Les Impatients, Montréal 006373
Imperato Collection of West
African Artifacts, Safari Museum,
Chanute 047524
Imperial, Fuengirola 085999
Imperial, Indianapolis . 127261, 135633
L'Impérial, Nice 104374
Imperial and District Museum,
Imperial 006029
Imperial Antiques, Armadale,
Victoria 061036

Imperial Antiques, Kingston-upon-
Hull 089702
Imperial Antiques, Stockport ... 091443
Imperial Bookshop, Hobart ... 139775
Imperial Calcasieu Museum, Lake
Charles 050073
Imperial Coins, Richmond 096554
Imperial Dream, Honolulu 093555
Imperial Fine Books, New York . 145349
Imperial Gallery, Virginia Beach . 124831
Imperial Garden, Hong Kong ... 065057
Imperial Glass and Plywood House,
Jaipur, Rajasthan 109380
Imperial Half Bushel, Baltimore . 092335
Imperial Oriental Art, New York . 095473
Imperial War Museum, London . 045003
Imperial War Museum Duxford,
Duxford 044192
Imperial War Museum North,
Manchester 045206
Imperio Antiguedades, Granada . 086038
Império dos Sentidos, Porto .. 085031
Imported Books, Dallas 145116
Imprésario, Paris 104958
Impresija, Trjavna 101182
Impression, Dunajská Streda ... 114659
Impression 5 Science Center,
Lansing 050136
Impressions, New York 139392
Impressions, Taipei 116922
Impressions Gallery of Photography,
Bradford 043695
Impressions in Leather by Georgie,
Milwaukee 121926
Impressions of Daylesford,
Daylesford 098956
Imprévu, Chaville 067949
L'Imprévu, Paris 071495
Imprimer, Houston 093704
Imprimerie Monsieur le Prince,
Paris 136944
Imprimerie Nationale, Paris ... 136945
Imprint, Fairfield 145145
Impulse Art, Houston 121067
Impulsion B., Paris 071496
Impulsion's, Cugnaux 103650
Imran, Singapore 085432
Imrie, Bridge-of-Earn 088594
Imschoot, Gent 136654
IMT Restaurering, Odense 128401
Imura Art Museum, Kyoto 028932
In Art We Trust, Miami 121809
In Buri National Museum, Sing
Buri 042718
In de Bolle Baaf, Middelburg ... 083496
In de Gecroonde Duijvekater,
Zaandam 033075
In de Gouden Schmouck Sceer,
Maastricht 083466
In de Zevende Hemel, Apeldoorn 032010
In Depot 68, Middelburg 112595
In Dreams, Oakland 095898
In Druk, 's-Hertogenbosch 112529
In Fine, Paris 136946
In Flanders Fields Museum, Ieper 003725
In focus Galerie, Köln 107641
In-Frames, Dunfermline 117695
In Graphic Detail, Vancouver .. 101799
In Her Image Gallery, Portland . 123633
In Illo Tempore, Miami 094751
In Period, Winchcombe 091864
IN QUARTO, Marseille 140943
In Site Art, Indianapolis 121191
In-Situ Manchester, Manchester 090694,
134831
In 't Paradijs, Maastricht 083467
In The Frame, Blackrock 109530
In the Groove, Baltimore 092336
In the Loupe, Boston 139393
In Vogue, Waihi 084118
Inamori, Tenri 082127
Inard, Montesquieu-Volvestre . 104262
Inart, Vordingborg 103010

Inarteks, Moskva 085181
Inasia Art, Singapore 114539
Inasta, Dogana 085280, 126480
Inatura – Erlebnis Naturschau
Dornbirn, Dornbirn 001824
Inauen, Hannes, Sankt Gallen .. 116613
Inauen, Hans & Carmela,
Appenzell 116026
Inax Tile Museum, Tokoname .. 029481
Inazawa City Oguiss Memorial Art
Museum, Inazawa 028687
Het Inboedelhuis, Gouda 083265, 083266
Het Inboedeltje Antiekhal, Breda 082988
Incado, Esbjerg 102794
Incarnat, Grignan 103795
L'Incartade, Lille 104012, 128678
İnce Minare Müzesi, Konya 043003
Inchata, Den Haag 083088
Incheon Metropolitan City Museum,
Incheon 029952
Inchmartine Fine Art, Inchture .. 089637,
134498
Inch's Books, Pickering 144787
INCIBA-Comisión Nacional de Artes
Plásticas, Montevideo 060281
Incontrart, Genova 110115
L'Incontro, Chiari 109962
L'Incontro, Lugano 116457
Incontro d'Arte, Roma 110803
Incorporate World Arts, New York 122725
Incredible Deals, Miami 094752
Incrocci, Marco, Genova 080267
Incubator, London 118403
L'Incunabolo, Firenze 131191
Incurable Collector, Toronto ... 101629
Incurável Colecionador, Porto
Alegre 063978
Incutti, Lia, Milano 110361
Incutti, Lia, Napoli 110532
Indeco, Seoul 111669
Independence Coin and Stamp,
Charlotte 092606
Independence Hall, Tel Aviv ... 025198
Independence Hall of Korea,
Cheonan 029903
Independence Historical Museum,
Independence 049661
Independence National Historical Park,
Philadelphia 052056
Independence Seaport Museum,
Philadelphia 052057
Independence Visitors Center,
Independence 049667
Independent Artisans, Pittsburgh 123554
Independent Gallery, London .. 118404
Independenz Galerie, Aachen .. 105940
Inderøy Museums- og Historielag,
Inderøy 033679
Index, New York 139394
Index Bookcentre, London 144634
India Arts Museum, Chennai .. 024198
India International Centre, Delhi 109322
India Street Antiques, San Diego 097099
Indian Academy of Fine Arts,
Amritsar 055776
Indian Academy of Fine Arts Gallery,
Amritsar 024143
Indian Art Gallery, Wädenswil . 116707,
144145
Indian Artists Network, Mumbai 059363,
109445
Indian Arts and Crafts Association,
Albuquerque 060524
Indian Arts and Crafts Board,
Washington 054125
Indian Arts Research Center, Santa
Fe 053109
Indian Association for the Study of
Conservation of Cultural Property,
Delhi 059364
Indian Ceramics, Kolkata 138776
Indian City U.S.A., Anadarko .. 046463
Indian College of Fine Arts,
Lashkar 055815

Indian Head Museum, Indian
Head 006031
Indian Hill Historical Society Museum,
Cincinnati 047746
Indian Institute of Art in Industry,
Kolkata 055813
Indian King Tavern Museum,
Haddonfield 049305
Indian Land Museum, Gossau
(Zürich) 041557
Indian Mill Museum State Memorial,
Upper Sandusky 053906
Indian Museum, Kolkata 024361
Indian Museum, Lemmer 032594
Indian Museum of Lake County-Ohio,
Painesville 051890
Indian Museum of North America,
Crazy Horse 048057
Indian Museum of the Carolinas,
Laurinburg 050183
Indian Pueblo Cultural Center,
Albuquerque 046361
Indian Society of Oriental Art,
Kolkata 059365
Indian Springs State Park Museum,
Flovilla 048762
Indian Style, West Hollywood .. 125002
Indian Sun, Houston 121068
Indian Temple Mound Museum, Fort
Walton Beach 048885
Indian War Memorial Museum,
Delhi 024225
Indiana Basketball Hall of Fame, New
Castle 051252
Indiana Medical History Museum,
Indianapolis 049687
Indiana State Museum,
Indianapolis 049688
Indiana Territory Capitol,
Vincennes 053989
Indiana Transportation Museum,
Noblesville 051569
Indiana University Art Museum,
Bloomington 047047
Indiana University Press,
Bloomington 138307
Indiana War Memorials Museum,
Indianapolis 049689
Indianapolis Art Center, Churchman-
Fehsenfeld-Gallery, Indianapolis 049690
Indianapolis Motor Speedway Hall of
Fame Museum, Indianapolis . 049691
Indianapolis Museum of Art,
Indianapolis 049692
Indianapolis Museum of Art –
Columbus Gallery, Columbus .. 047926
Indianas Antiques, Atlanta ... 092143,
135312
Indianermuseum, Zürich 042223
Indians on Columbus, New York 122726
Indica, Delhi 109323
L'Indicibile, Firenze 080159
Indicible, Rouen 072592, 072593
Indidjinart Aboriginal Art Gallery,
Dubbo 098984
Indien Cinq Nations, Québec ... 101500
Les Indiennes, Paris 071497
L'Indifférent, Lyon 069748
Indigenart, Fremantle 061489
Indigenart, Subiaco 062256
Indigenous Art, Christchurch .. 083879
Indigo, Kathmandu 112075
Indigo, London 090232
Indigo, Manningford Bruce ... 090697
Indigo, Minneapolis 094921
Indigo, Philadelphia ... 096117, 123398
Indigo Blue Art, Singapore ... 114540
Indigo for Contemporary Art, Saint
Andrews 119007
Indigo Uptown, Minneapolis .. 094922
Indik, Adriana, Buenos Aires .. 098572
Indira, Cáceres 085921
Indira Gandhi National Centre for the
Arts, Delhi 024226

Individual Artists of Oklahoma, Oklahoma
City 051729, 060525
Indo American Arts Council, New
York 060526
Indo-Cuban Bani Museum, Banes 009068
Indonesian Art, Richmond,
Victoria 099631
Indonésie Tropicale, Montgeron . 070245
Indre Sør-Troms Distriktsmuseum,
Tennevoll 034012
Indruk, Zwolle 143274
Indulgence Antiques,
Templestowe 062313
Indunabor, Bogotá 102344
Induni, Antonia Maria, Torino . 132150
Industrial and Agricultural Museum,
York 054587
Industrial and Commercial Museum,
Lahore 034129
Industrial Art, Ha Noi 125080
Industrial Design Institute, Hochschule
Magdeburg-Stendal (FH),
Magdeburg 055632
Industrial Designers Society of
America, Dulles 060527
Industrial Heritage Complex –
Merrickville Lockstation,
Merrickville 006294
Industrial Museum, Nottingham . 045414
Industrie- und Filmmuseum Wolfen,
Wolfen 022729
Industrieel Smalspoor Museum,
Erica 032316
Industriemuseen in der Euregio Maas-
Rhein e.V., Euskirchen 058924
Industriemuseum, Limbach-
Oberfrohna 020075
Industriemuseum Brandenburg,
Brandenburg an der Havel ... 017543
Industriemuseum Elmshorn,
Elmshorn 018164
Industriemuseum Hartha, Hartha 019080
Industriemuseum Lauf, Lauf an der
Pegnitz 019931
Industriemuseum Lohne, Lohne . 020108
Industrieviertel-Museum, Museum
für Industrie und Arbeitsgeschichte,
Wiener Neustadt 003105
Industriemuseen, Horsens 010144
Industrimuseet, Kopperå 033711
Industrimuséet C.W. Thorstensons
Mekaniska Verkstad, Åmål ... 040524
Industrimuseet Tollereds Öfvre
Kraftstation, Tollered 041079
Industrion Museum for Industry and
Society, Kerkrade 032543
Inecos, Roma 081412
Ineichen, Zürich 126768
Ines Art Gallery, Singapore ... 114541
Inetllibre, Barcelona 143719
Infanterie Museum, Harskamp . 032416
Infantry and Small Arms School Corps
Weapons Collection, Warminster 046054
Infinito, Sapporo 082100
Infinity Antique Furniture, Hong
Kong 065058
Infinity Fine Arts, New York ... 122727
Info-Musées, Hornu 138551
Informations- und
Dokumentationszentrum Berlin,
Berlin 017801
Informationsdienst KUNST,
Regensburg 138699
Informationsstelle des
Naturparks Fichtelgebirge, Zell,
Oberfranken 022858
Informationszentrum am
Wasserstraßenkreuz Minden,
Minden 020416
Informationszentrum Donnersberghaus,
Dannenfels 017813
Informationszentrum Schloss
Paffendorf, RWE Rheinbraun AG,
Bergheim (Erft) 017049

l'Informatore Europeo d'Arte e di
 Antiquariato, Bologna 138833
Infusion Art Studio, Tucson 124744
Infusion Design Art Glass Gallery,
 Tucson 124745
Ingatestone Hall, Ingatestone . . 044631
Ingbar, Michael F., New York . . 122728
Ingeborgmuseet Fredheim,
 Tangen 034010
Ingeborrarps Friluftsmuseum,
 Örkelljunga Hembygdsförening,
 Örkelljunga 040898
Ingebrigt Vik's Museum, Øystese 033818
Ingelbrecht, Helge Velo, Oslo . . 113227
Ingelmark Antik och Design,
 Ängelholm 126562
Ingelmark Antik och Design,
 Visby 126700
Ingels-Humblot, Véronique,
 Roanne 125788
Ingels, F., Gent 128049
Inger & Dan, Stockholm 087009
Ingerslev, København 065917
Ingersol, Tonya, Baltimore 119747
Ingersoll Cheese Factory Museum and
 Sports Hall of Fame, Ingersoll . 006032
Ingiulla, Roma 081413
Ingle, Virginia Beach 097784
Ingle, D.D. & A., Nottingham . . 090894
Ingleby Gallery, Edinburgh 117728
Ingledale, Stoney Creek 007078
Inglenook, Harpole 089420
Inglenook, Ramsbury . 091066, 134957
Inglett, Susan, New York 122729
Inglewood District Historical Museum,
 Inglewood 001164
Ingrained, Rockingham 062164
Ingram, Toronto 064676
Ingram Gallery, Toronto 011630
Ingrao, New York 095474
Ingrids Art, Hawkes Bay 112914
Ingrow Loco Museum, Keighley . 044669
Ingstrup Antik, Løkken 065992
Ingurugiro Etxea, Museo Medio
 Ambiental, Azpeitia 039031
Ingušskij Gosudarstvennyj Muzej
 Kraevedenija im. T.Ch. Malsagova,
 Nazran 037542
Inherited Arts, Singapore 114542
Inhof, Ryde 127672
Inhofer, Edith, Regensburg 078169
Inhouse, Toronto 064677
Iniesta Gloria, Moissac 070143
Inishie, Nara 082070
Initial Tree Gallery, San Antonio . 124042
Initiative, Sankt-Peterburg 114343
Initiative von Münchner Galerien
 zeitgenössischer Kunst,
 München 058925
Injalak Arts & Crafts, Oenpelli . . 062024
Inje, Den Haag 112393
Injuka, München 077533
Ink Factory Auction, Saint Marys 127431
Ink Miami, Miami Beach 098455
Ink Tree, Küsnacht 138059
Inkjet Galerie, Spaichingen 078509
Inkoart, Kyïv 117137
Inkspecta, Mitcham 099354
Inkt Delft, Delft 112362
Inkworks, Los Angeles 145231
Inky Parrot Press, Oxford 138273
Inland Seas Maritime Museum,
 Vermilion 053959
Inland-Water Ecological Laboratory
 Collection, Jerusalem 025081
Inlibris, Wien 140090
Inlight Art Glass, Buffalo 092534, 135374
Inliquid, Philadelphia 123399
Inmaculada Concepcion, Bogotá 065208
Inman, Kerry F., Houston 121069
Inman, Raymond P., Hove 126927
Inn-Museum, Wasserbau- und
 Schiffahrtstechnische Sammlung,
 Rosenheim 021476

The Inner Bookshop, Oxford . . . 144767
Inner City Art Gallery, New York 122730
Inner Eye Vision, Leven 118108
Inner Galeria Nacisku Brow-art,
 Wrocław 113942
Inner Mongolia Museum, Hohhot 007867
Inner Vision Antiques, Los
 Angeles 094369
Innervisions Posters, Seattle . . . 124548
Inniscara Art & Framing,
 Rathcoole 109684
Innisfail and District Historical
 Museum, Innisfail 001165
Innisfail and District Historical Village,
 Innisfail 006033
Inniskeen Folk Museum,
 Inniskeen 024934
Innisville and District Museum,
 Innisville 006034
Innocent Fine Art, Clifton 117600
Innocenti, Ferdinando & Filippo,
 Firenze 080160, 142676
Innocenti, Roberto, Firenze . . . 080161
Innovanet, Auckland 112779
Innovatum Kunskapens Hus,
 Trollhättan 041088
Inns of Court and City Yeomanry
 Museum, London 045004
Innuit Gallery, London 101345
Innviertler Freilichtmuseum
 Brunnbauerhof, Andorf 001712
Innviertler Künstlergilde, Ried im
 Innkreis 058263
Inoue, Tokyo 142923
Inouye & Son, T. Edo, Tokyo . . . 082153
Inpa Coleções Zoológicas,
 Manaus 004551
Inprint, Stroud 144867
Inprints, Seattle 124549
Inquadrature Arte e Cornici,
 Trieste 111050
Inquadro, Piobesi Torinese 110660
A l'Inquet, Carbonne 067664
Inquisitive, Toronto 064678
Inquisitor's Palace, National
 Museum of Ethnography of Malta,
 Vittoriosa 030765
Inro Bijitsukan, Takayama 029458
Insa Gallery, Seoul 111670
Insam, Josef & Angelika,
 Regensburg 078170
Insam, Peter, Regensburg 078171,
 130562
Insect Collections at the Entomology
 Department, National Taiwan
 University, Taipei 042519
Insectarium de Montréal, Muséums
 Nature de Montréal, Montréal . 006374
Insects Museum, Amman 029768
Insects World, Wunstorf 022812
Insel Verlag, Frankfurt am Main 137201
Inselgalerie, Berlin 116319
Inselgalerie, Frauenchiemsee . . 107049
Inselgalerie, Wittdün 108949
Inselmuseum, Pellworm 021098
Inselmuseum Spiekeroog,
 Spiekeroog 021936
Insensatez, Lisboa 114038
Inshallah Egyptian Imports,
 Paraparaumu 084059
Inside, Cleveland 120431
Inside Out, Penkridge 090953
Inside Out, San Antonio 096979
Inside Outside, Houston 093705
Inside Secrets, Portland 096422
Inside Space, London 090233
Inside, Magazine di Arte e Cultura
 Contemporanea, Brescia 138834
Insight Art Gallery, Springbrook . 099713
Insight Sculpture Gallery,
 Norwich 118832
Insights – El Paso Science Center, El
 Paso 048511

InSite-Museum Exhibition, San
 Diego 052964
Insolite, Bordeaux 067338
Insolite, Saint-Ouen 073109
Insolite, Troyes 073905
Inspiration Gallery, Memphis . . . 121696
Inspirations, London 118405
Inspirations Gallery, Mildura . . . 099339
Inspire, Listowel 079665
Inspire – Hands-on Science Centre,
 Norwich 045400
Inspire Fine Art, Chicago 120161
Inspires Art Gallery, Oxford 118864
Installation Gallery, San Diego . . 124135
Instanbul to Samarkand Gallery,
 Austin 092241
L'Instant Durable, Clermont-
 Ferrand 136809
Institut Art et Archéologie de Byzance,
 Strasbourg 055420
Institut Culturel Avataq,
 Westmount 007386
Institut d'archéologie et des sciences
 de l'antiquité, Université de
 Lausanne, Lausanne 056431
Institut d'Art Conservation et Couleur,
 Paris 129040
Institut d'Art Contemporain,
 Villeurbanne 016289
Institut d'Art Médiéval, Université Paul
 Valery, Montpellier 055384
Institut d'Arts Visuels, Orléans . 055391
Institut de France, Saint-Jean-Cap-
 Ferrat 015367
Institut d'Egypte, Cairo 055334
Institut der Künste, Pädagogische
 Hochschule Freiburg, Freiburg im
 Breisgau 055544
Institut der Künste, Pädagogische
 Hochschule Schwäbisch Gmünd,
 Schwäbisch Gmünd 055705
Institut des Beaux-Arts Section IV,
 Université Libanaise, Chouf . . . 056116
Institut des Beaux-Arts, Section I,
 Université Libanaise, Beirut . . . 056115
Institut des Beaux-Arts, Section
 II, Université Libanaise, Furn El-
 Chebbak 056118
Institut des Beaux-Arts, Section III,
 Université Libanaise, Tripoli . . 056120
Institut des Communications
 Graphiques, Montréal 055144
Institut Design- und Kunstforschung,
 Hochschule für Gestaltung und Kunst
 HGK FHNW, Basel 056408
Institut d'Histoire de l'Art et de
 Muséologie, Université de Neuchâtel,
 Neuchâtel 056438
Institut et Musée Voltaire,
 Genève 041525
Institut Finlandais, Paris 104959
Institut for Kunsthistorie, Københavns
 Universitet, København 055308
Institut für Ästhetisch-Kulturelle
 Bildung, Universität Flensburg,
 Flensburg 055535
Institut für Architektur und Medien,
 Technische Universität Graz,
 Graz 054931
Institut für Architektur und Städtebau,
 Fakultät Gestaltung, Berlin . . . 055465
Institut für Architektur, Fachgebiet Bau-
 und Kunstgeschichte, Technische
 Universität Berlin, Berlin 055464
Institut für Architekturgeschichte,
 Universität Stuttgart, Stuttgart . 055719
Institut für Architekturtheorie, Kunst-
 und Kulturwissenschaft, Technische
 Universität Graz, Graz 054932
Institut für Architekturzeichnen
 und Raumgestaltung, Technische
 Universität Carolo-Wilhelmina zu
 Braunschweig, Braunschweig . 055492
Institut für Ausbildung in bildender

Kunst und Kunsttherapie,
 Bochum 055487
Institut für Auslandsbeziehungen e.V. –
 Abt. Kunst, Stuttgart 022043
Institut für Auslandsbeziehungen e.V.-
 Büro Berlin, Berlin 017177
Institut für Baugeschichte,
 Architekturtheorie und
 Denkmalpflege, Technische
 Universität Dresden, Dresden . 055517
Institut für Baugestaltung, Universität
 Karlsruhe (TH), Karlsruhe 055599
Institut für Bildende Künste,
 Universität Karlsruhe (TH),
 Karlsruhe 055600
Institut für Bildende Kunst
 und Kulturwissenschaften,
 Kunstuniversität Linz, Linz . . . 054951
Institut für Bildende Kunst und
 Kunstwissenschaft, Universität
 Hildesheim, Hildesheim 055589
Institut für Bildende Kunst, Akademie
 der bildenden Künste, Wien . . 054966
Institut für Bildende Kunst, Kunst-
 und Werkpädagogik, Universität
 "Mozarteum" Salzburg,
 Salzburg 054961
Institut für Bildende und Mediale
 Kunst, Universität für angewandte
 Kunst Wien, Wien 054967
Institut für Buch- und
 Handschriftenrestaurierung,
 Bayerische Staatsbibliothek,
 München 055650
Institut für Buchgestaltung und
 Medienentwicklung, Staatliche
 Akademie der Bildenden Künste,
 Stuttgart 055720
Institut für Bühnengestaltung,
 Universität für Musik und
 darstellende Kunst Graz, Graz . 054933
Institut für Byzantinistik, Byzantinische
 Kunstgeschichte und Neogräzistik,
 Ludwig-Maximilians-Universität
 München, München 055651
Institut für Darstellen und Gestalten,
 Universität Stuttgart, Fakultät
 1-Architektur+Stadtplanung,
 Stuttgart 055721
Institut für das künstlerische Lehramt,
 Akademie der bildenden Künste,
 Wien 054968
Institut für Design, Hochschule für
 Gestaltung und Kunst Luzern,
 Luzern 056435
Institut für Design, Universität für
 angewandte Kunst Wien, Wien 054969
Institut für Designforschung,
 Oldenburg 055678
Institut für Elementares Formen/
 Institut für Bildende Kunst,
 Technische Universität Carolo-
 Wilhelmina zu Braunschweig,
 Braunschweig 055493
Institut für Entwerfen und
 Gebäudelehre, Fakultät für
 Architektur und Landschaft,
 Hannover 055578
Institut für Entwurf, Medien und
 Darstellung EMD, Technische
 Universität Carolo-Wilhelmina zu
 Braunschweig, Braunschweig . 055494
Institut für Ergonomie und
 Designforschung (IED), Universität
 Duisburg-Essen, Essen 055533
Institut für Europäische
 Kunstgeschichte (IEK), Ruprecht-
 Karls-Universität Heidelberg,
 Heidelberg 055584
Institut für Experimentelles
 Bekleidungs- und Textildesign,
 Fakultät Gestaltung, Berlin . . . 055466
Institut für Gebäudelehre und
 Entwerfen, Technische Universität

Carolo-Wilhelmina zu Braunschweig,
Braunschweig 055495
Institut für Geschichte der Medizin der
Robert Bosch Stiftung, Stuttgart 022044
Institut für Geschichte und
Kunstgeschichte, Technische
Universität Berlin, Berlin 055467
Institut für Geschichte und Theorie der
Architektur, Gottfried Wilhelm Leibniz
Universität Hannover, Hannover 055579
Institut für Geschichte und Theorie der
Gestaltung GTG, Fakultät Gestaltung,
Berlin 055468
Institut für Gestaltung und Darstellung,
Fakultät für Architektur und
Landschaft, Hannover 055580
Institut für Gestaltung, Leopold-
Franzens-Universität Innsbruck,
Innsbruck 054943
Institut für Grundlagen der Gestaltung,
Universität Karlsruhe (TH),
Karlsruhe 055601
Institut für Grundlagen der Planung in
der Architektur, Universität Stuttgart,
Stuttgart 055722
Institut für Grundlagen moderner
Architektur und Enwerfen, Universität
Stuttgart, Stuttgart 055723
Institut für Hermeneutik und
Religionsphilosophie (IHR), Universität
Zürich, Zürich 056445
Institut für Kirchenbau und kirchliche
Kunst der Gegenwart, Marburg 055640
Institut für Kirchengeschichte,
Christliche Archäologie und
Kirchliche Kunst, Universität Wien,
Wien 054970
Institut für Konservierung und
Restaurierung, Akademie der
bildenden Künste, Wien 054971
Institut für Konservierung und
Restaurierung, Universität für
angewandte Kunst Wien, Wien 054972
Institut für konstruktive Kunst und
konkrete Poesie, Rehau 055696
Institut für Künste und Medien,
Philosophische Fakultät,
Potsdam 055692
Institut für Künstlerische Keramik und
Glas (IKKG), Fachhochschule Koblenz,
Höhr-Grenzhausen 055590
Institut für Kultur- und
Kunstwissenschaft, Humboldt-
Universität zu Berlin, Berlin . . . 055469
Institut für Kulturtheorie,
Kulturforschung und Kunst, Leuphana
Universität, Lüneburg 055631
Institut für Kunst- und
Designwissenschaften (IKUD),
Universität Duisburg-Essen,
Essen 055534
Institut für Kunst- und
Kulturwissenschaft, Akademie der
bildenden Künste, Wien 054973
Institut für Kunst- und
Musikwissenschaft, Technische
Universität Dresden, Dresden . 055518
Institut für Kunst und Architektur,
Akademie der bildenden Künste,
Wien 054974
Institut für Kunst und Gestaltung,
Kunstuniversität Linz, Linz . . . 054952
Institut für Kunst und Gestaltung,
Technische Universität Wien,
Wien 054975, 136595
Institut für Kunst und Handwerk, Kloster
Lehnin 055610, 107568
Institut für Kunst und Kunsttheorie,
Universität zu Köln, Köln 055614
Institut für Kunst und Matielle
Kultur, Universität Dortmund,
Dortmund 055513
Institut für Kunst und Musik,
Pädagogische Hochschule Karlsruhe,

Karlsruhe 055602
Institut für Kunst und Technologie,
Universität für angewandte Kunst
Wien, Wien 054976
Institut für Kunst, Hochschule für
Gestaltung und Kunst Luzern,
Luzern 056436
Institut für Kunst, Musik und
Sport, Pädagogische Hochschule
Ludwigsburg, Ludwigsburg . . . 055630
Institut für Kunsterziehung, Universität
Regensburg, Regensburg 055694
Institut für Kunstgeschichte mit
Arbeitsbereich Christliche Archäologie
und Byzantinische Kunstgeschichte,
Johannes Gutenberg-Universität
Mainz, Mainz 055635
Institut für Kunstgeschichte und
Archäologie, Rheinische Friedrich-
Wilhelms-Universität Bonn,
Bonn 055490
Institut für Kunstgeschichte und
Archäologien Europas, Martin-Luther-
Universität Halle-Wittenberg, Halle,
Saale 055552
Institut für Kunstgeschichte,
Abt. Ostasien, Zentrum für
Ostasienwissenschaft (ZO),
Heidelberg 055585
Institut für Kunstgeschichte,
Bayerische Julius-Maximilians-
Universität Würzburg, Würzburg 055744
Institut für Kunstgeschichte,
Denkmalpflege und
Industriearchäologie, Technische
Universität Wien, Wien 054978
Institut für Kunstgeschichte, Humboldt-
Universität zu Berlin, Berlin . . . 055470
Institut für Kunstgeschichte,
Justus-Liebig-Universität Gießen,
Gießen 055547
Institut für Kunstgeschichte, Karl-
Franzens-Universität Graz, Graz 054934
Institut für Kunstgeschichte, Leopold-
Franzens-Universität Innsbruck,
Innsbruck 054944
Institut für Kunstgeschichte, Ludwig-
Maximilians-Universität München,
München 055652
Institut für Kunstgeschichte, Paris-
Lodron-Universität Salzburg,
Salzburg 054962
Institut für Kunstgeschichte, Rheinisch-
Westfälische Technische Hochschule,
Aachen 055440
Institut für Kunstgeschichte,
Universität Bern, Bern 056420
Institut für Kunstgeschichte,
Universität Karlsruhe (TH),
Karlsruhe 055603
Institut für Kunstgeschichte,
Universität Leipzig, Leipzig . . . 055627
Institut für Kunstgeschichte,
Universität Regensburg,
Regensburg 055695
Institut für Kunstgeschichte,
Universität Stuttgart, Stuttgart . 055724
Institut für Kunstgeschichte,
Universität Wien, Wien 054977
Institut für Kunstgeschichte,
Westfälische Wilhelms-Universität
Münster, Münster 055665
Institut für Kunstpädagogik, Johann
Wolfgang Goethe-Universität,
Frankfurt am Main 055539
Institut für Kunstpädagogik, Justus-
Liebig-Universität Gießen,
Gießen 055548
Institut für Kunstpädagogik, Ludwig-
Maximilians-Universität München,
München 055653
Institut für Kunstpädagogik, Universität
Leipzig, Leipzig 055628
Institut für Kunsttherapie und

Forschung, Universität Ottersberg,
Ottersberg 055684
Institut für Kunstwissenschaft und
Bildende Kunst, Universität Koblenz/
Landau, Campus Landau, Landau in
der Pfalz 055623
Institut für Kunstwissenschaft und
Kunstpädagogik, Universität Bremen,
Fachbereich 09 Kulturwissenschaften,
Bremen 055499
Institut für Kunstwissenschaft und
Philosophie, Katholisch-Theologische
Hochschule Linz, Linz 054953
Institut für Kunstwissenschaft,
Universität Koblenz-Landau,
Koblenz 055612
Institut für Kunstwissenschaften,
Kunstpädagogik und
Kunstvermittlung, Universität für
angewandte Kunst Wien, Wien 054979
Institut für Liturgiewissenschaft,
Christliche Kunst und Hymnologie,
Karl-Franzens-Universität Graz,
Graz 054935
Institut für Medien, Kunstuniversität
Linz, Linz 054954
Institut für moderne Kunst,
Nürnberg 020861
Institut für Museumsforschung,
Staatliche Museen zu Berlin –
Stiftung Preußischer Kulturbesitz,
Berlin 017178
Institut für Museumskunde, Staatliche
Akademie der Bildenden Künste,
Stuttgart 055725
Institut für Naturwissenschaften und
Technologie in der Kunst, Akademie
der bildenden Künste, Wien . . . 054980
Institut für Papierrestaurierung,
Wien 054981
Institut für Produkt- und
Prozessgestaltung, Fakultät
Gestaltung, Berlin 055471
Institut für Raum und Design,
Kunstuniversität Linz, Linz . . . 054955
Institut für Raumgestaltung,
Technische Universität Graz,
Graz 054936
Institut für Realienkunde des
Mittelalters und der Frühen Neuzeit,
Österreichische Akademie der
Wissenschaften, Krems 054948
Institut für Restaurierungs- und
Konservierungswissenschaften,
Fachhochschule Köln, Fakultät für
Kulturwissenschaften, Köln . . . 055615
Institut für Transmediale Gestaltung,
Fakultät Gestaltung, Berlin . . . 055472
Institut für Vogelforschung –
Vogelwarte Helgoland, Heinrich-
Gätke-Halle, Wilhelmshaven . . 022659
Institut für Weiterbildung, Kunst im
Kontext, Fakultät Bildende Kunst,
Berlin 055473
Institut für Wissenschaft und Kunst,
Wien 054982
Institut für Zeitgenössische Kunst,
Technische Universität Graz,
Graz 054937
Institut Heidersbergen, Wolfsburg 058926
Institut HyperWerk, Hochschule für
Gestaltung und Kunst HGK FHNW,
Basel 056409
Institut Industrial Design, Hochschule
für Gestaltung und Kunst HGK FHNW,
Aarau 056404
Institut Innenarchitektur und
Szenografie, Hochschule für
Gestaltung und Kunst HGK FHNW,
Basel 056410
Institut Iskusstv i Dizajna, Udmurtskij
Gosudarstvennyj Universitet,
Iževsk 056273
Institut Kunst, Fakultät Bildende Kunst,

Berlin 055474
Institut Kunst, Hochschule für
Gestaltung und Kunst HGK FHNW,
Basel 056411
Institut Kunstdidaktik und Ästhetische
Erziehung, Fakultät Bildende Kunst,
Berlin 055475
Institut Kunstwissenschaft und
Ästhetik, Fakultät Bildende Kunst,
Berlin 055476
Institut Lehrberufe für Gestaltung und
Kunst, Hochschule für Gestaltung
und Kunst HGK FHNW, Basel . . 056412
Institut l'Homme et le Temps,
Centre de Restauration, Musée
International d'Horlogerie, La Chaux-
de-Fonds 133717
Institut Mathildenhöhe, Museum
Künstlerkolonie, Darmstadt . . . 017823
Institut Mode-Design, Hochschule für
Gestaltung und Kunst HGK FHNW,
Basel 056413
Institut National d'Archéologie et des
Arts, Tunis 056471
Institut National des Arts,
Kinshasa 055273
Institut National des Arts de Bamako,
Bamako 056133
Institut Nationale des Beaux-Arts,
Tétouan 056163
Institut Néerlandais, Paris 104960
Institut Provincial des Arts et Metiers,
Nivelles 055044
Institut Régional d'Art Visuel, Fort-de-
France 056135
Institut Reine Astrid, Sections
Artistiques, Mons 055042
Institut Royal du Patrimoine Artistique,
Koninklijk Instituut voor het
Kunstparimonium, Bruxelles . . 136640
Institut Seni Indonesia Yogyakarta,
Yogjakarta 055848
Institut Suisse pour l'Étude de l'Art
(ISEA), Université de Lausanne,
Lausanne 056432
Institut Supérieur des Beaux Arts de
Tunis, Université de Tunis, Tunis 056472
Institut Supérieur pour l'Étude du
Langage Plastique, Bruxelles . 058379,
100621, 136641, 140183
Institut Technique des Arts Plastiques,
Lubumbashi 055274
Institut Tessin, Paris 014663
Institut und Museum für Geologie und
Paläontologie, Tübingen 022217
Institut Universitaire de formation des
maitres IUFM, Maxéville 055380
Institut za Oblikovanje, Akademija za
Likovno Umetnost, Ljubljana . . 056335
The Institute For American Indian
Studies, Washington Green . . . 054193
Institute for Art Research and
Documentation, Madison 057386
Institute for Arts and Humanities
Education, New Brunswick . . . 057397
Institute for Jewish Studies Gallery,
Jerusalem 025082
Institute for the Arts, Rice University,
Houston 057219
Institute Menezes Braganza,
Panaji 024428
Institute of Aesthetics and Art History,
Graduate School of Letters, Kyoto
University, Kyoto 056012
Institute of American Indian Arts,
Santa Fe 057873
Institute of American Indian Arts
Museum, Santa Fe 053110
Institute of Archaeology and Museum
Studies, Jos 033049
Institute of Architecture and Art,
Baki 054988
Institute of Architecture and Islamic
Arts, Al-Albayt University, Al-

Mafraq 056067
Institute of Arts and Crafts,
Dhaka 054992
Institute of Arts and Crafts Gallery,
Dhaka 003222
Institute of Conservation, London 060160,
134695
Institute of Contemporary Art,
Boston 047116
Institute of Contemporary Art, Maine
College of Art, Portland 052280
Institute of Contemporary Art,
University of Pennsylvania,
Philadelphia 052058
Institute of Contemporary Arts,
London 045005
Institute of Contemporary Plastic Art
Kitakanto, Maebashi 056018
Institute of Design, Illinois Institute of
Technology, Chicago 056888
Institute of Fine Arts and Design,
Philippine Womens University,
Manila 056228
Institute of Fine Arts, New York
University, New York 057555
Institute of Geological and Nuclear
Sciences, Hutt City 033224
Institute of Humanities and Creative
Arts, University of Worcester,
Worcester 056615
Institute of Jamaica, Kingston . . 028443
Institute of Meteoritics Meteorite
Museum, Albuquerque 046362
Institute Of Modern Art, Judith Wright
Centre of Contemporary Arts,
Fortitude Valley 001078
Institute of Paper Conservation, Upton-
upon-Severn 060161
Institute of Texan Cultures, San
Antonio 052939
Institute of Visual Arts, University of
Wisconsin-Milwaukee,
Milwaukee 050915, 057452
Instituti za Kiparstvo, Slikarstvo i
Restavratorstvo, Akademija za
Likovno Umetnost, Ljubljana . . 056336
Institutionen för Konstvetenskap,
Stockholms Universitet,
Stockholm 056399
Instituto Amatller de Arte Hispánico,
Barcelona 039066
Instituto Cultural Cabañas,
Guadalajara 030999
Instituto de Arte Americano y Regional,
Universidad Nacional de Tucumán,
Tucumán 054880
Instituto de Arte e Comunicação
Social, Universidade Federal
Fluminense, Rio de Janeiro . . . 055094
Instituto de Arte Visual y
Comunicaciones, Universidad
Nacional de Tucumán, Tucumán 054881
Instituto de Arte y Educación,
Universidad Nacional de Tucumán,
Tucumán 054882
Instituto de Artes e Ofícios,
Lisboa 056250
Instituto de Artes Gráficas de Oaxaca,
Oaxaca 031330
Instituto de Artes Visuais, Universidade
Estadual Paulista, Ipiranga . . . 055079
Instituto de Artes, Universidade de
Brasília, Brasília 055069
Instituto de Bellas Artes Magdalena,
Universidad del Magdalena, Santa
Marta 055270
Instituto de Diseño, Universidad
Nacional de Tucumán, Tucumán 054883
Instituto de Investigaciones
Arqueológicas y Museo Prof. Mariano
Gambier, Albardón 000118
Instituto de Investigaciones de
Arquitectura Urbanismo y Artes,
Universidad Mayor de San Andrés,

La Paz 055062
Instituto de Investigaciones de
Arte y Cultura de Oriente Antiguo,
Universidad Nacional de Tucumán,
Tucumán 054884
Instituto de Investigaciones Históricas
de la Manzana de las Luces, Buenos
Aires 000162
Instituto de Letras e Artes,
Universidade Federal de Pelotas,
Pelotas 055087
Instituto de Mexique, Paris 104961
Instituto de Valencia de Don Juan,
Don Juan Institute of Valencia,
Madrid 039625
Instituto Departamental de Bellas
Artes, Cali 055266
Instituto Gómez-Moreno de la
Fundación Rodríguez-Acosta,
Granada 039464
Instituto Hondureño de Cultura
Interamericana, Comayagüela . 055763
Instituto Mexicano Norteamericano de
Jalisco, Guadalajara 031000
Instituto Moreira Sales, São
Paulo 004860
Instituto Moreira Salles, Belo
Horizonte 004342
Instituto Moreira Salles, Poços de
Caldas 004635
Instituto Moreira Salles, Centro
Cultural IMS – Rio de Janeiro, Rio
de Janeiro 004695
Instituto Nacional de Bellas Artes,
México 056146
Instituto Nacional Indigenista,
México 056147
Instituto Paulista de Restauro, São
Paulo 055105, 136669
Instituto Português de Museus,
Lisboa 059757
Instituto Potósino de Bellas Artes, San
Luis Potosí 056156
Instituto Regional de Bellas Artes de
Orizaba, Orizaba 056152
Instituto Superior de Arte, La
Habana 055286
Instituto Superior de Artes, San
José 055278
Instituto Zacatecano de Bellas Artes,
Zacatecas 056160
Instituts- und Studiensammlung des
Instituts für Ur- und Frühgeschichte,
Wien 003009
Institutt for Arkeologie, Konservering
og Historiske Studier, Universitet og
Oslo, Oslo 056211
Institutt for Kunsthistorie, Universitetet
i Bergen, Bergen 056209
Institutul de Arheologie şi Istoria Artei,
Cluj-Napoca 056262, 137923
Institutul de Arhitectură Ion Mincu,
Bucureşti 056258
Institutul de Istoria Artei George
Oprescu, Bucureşti 056259
Instituut voor Sociale Geschiedenis,
AMSAB, Gent 003627
Instytut Artystyczny, Uniwersytet
Mikołaja Kopernika, Wydział Sztuk
Pięknych, Toruń 056239
Instytut Historii Architektury i
Konserwacji Zabytkow, Politechnika
Krakowska im. Tadeusza Kosciuszki,
Kraków 056233
Instytut Historii Sztuki, Kraków . 056234
Instytut Historii Sztuki, Warszawa 056243
Instytut Historii Sztuki, Wrocław 056246
Instytut Sztuk Pięknych, Uniwersytetu
Marii Curie-Sklodowskiej, Wydzial
Artystyczny, Lublin 056236
Instytut Sztuki PAN, Warszawa . 056244,
137906
Instytut Sztuki Wyspa, Gdańsk . 034727
Instytut Sztuki, Wydzial Artystyczny UŚ

w Cieszynie, Uniwersytet Śląski w
Katowicach, Cieszyn 056230
Instytut Zabytkoznawstwa i
Konserwatorstwa, Uniwersytet
Mikołaja Kopernika, Wydział Sztuk
Pięknych, Toruń 056240
Insulae, Palermo 131699
In't Profijtelijk Boeksken – Winkel
Antwerpen, Antwerpen 140147
In't Profijtelijk Boeksken – Winkel
Leuven, Leuven 140235
In't Profijtelijk Boeksken – Winkel
Westerlo, Westerlo 140264
Intaglio Gallery, Saint Ives 099664
IntAkt – Internationale
Aktionsgemeinschaft bildender
Künstlerinnen, Wien 058264
Intaria, Madrid 086201
Intarsija, Sankt-Peterburg 133019
Intarta, Knokke-Heist 100742
Intel Museum, Santa Clara 053095
Intemporel, Paris 104962, 141144
Intemporel, Saint-Gaudens 072788
Intemporelle, Metz 104226
Intependent Art Space, Phoenix 123496
Inter Achats Dieuze, Dieuze . . . 068304
Inter African, Silverton 085666
Inter-American Art Consultants, New
York 122731
Inter Antiquariaat, Amsterdam . 143039
Inter-Art, Helsinki 103150
Inter Art – Sonsbeek Art and Design,
Arnhem 032025
Inter Art Actuel, Québec 138581
Inter Art Center, Hasselt 100716
Inter Art Galerie Reich, Köln . . . 107642
Inter Art Gallery, New York 122732
Inter Art Ring, Moskva 059773, 114263
Inter Art, Agentur für Kunst GmbH,
Berlin 106320, 137074
Inter Art, Union Freischaffender Künstler
und Kunstfreunde Stuttgart,
Stuttgart 058927, 108709
Inter-Atrium, Porto 114098
Inter-Continental, Vancouver . . 101800
Inter-Fun, Lyon 140902
Inter Galereja Sovremennogo
Iskusstva, Moskva 114264
Inter-Galerie Nikolaisaal Potsdam,
Potsdam 108406
Inter-Livres, Paris 136947
Inter Militaria en Collectibles,
Oss 083527
Inter Prints, Houston 121070
Interaction Designs, West Valley
City 136472
Interart, Genève 116322
Interart, Rotterdam 083571
Het Interbellum, Amsterdam . . . 082770
Interbook, Trier 142419
Interbook International, Schiedam 143226
Intercisa Múzeum, Dunaújváros . 023610
Intercoin, Newcastle-upon-Tyne 090826
Intercol London, London 090234, 134696,
144635
INTERCOM International Committee
for Museums Management,
Castries 059783
Intercraft Gallery, Sharjah 117199
Intercultural Museum and Art Gallery,
Baltimore 046751
Interdisziplinäres Forschungszentrum
für Kunstwissenschaften und
Ästhetik, Freie Universität Berlin,
Berlin 055477
Interessengemeinschaft
niederösterreichische Museen und
Sammlungen, Krems-Stein . . . 058265
Interfennica, Helsinki 140703
Intergalerias, Acapulco 111931
Interieur, Budapest 079262
Interieur 13, Gent 063644
Interieur 2000, Judenburg 062659,
127782

Intérieur Design, Marly 133853
Interieurs, New York 095475
Interior Art Australia, Richmond,
Victoria 099632
Interior Details, Louisville 121651
Interior Elements, Nashville 095006
Interior Visions, San Francisco . 097259
Intérior's, Le Havre 069282
Interiors Market, Atlanta 092144
Interiors Market, Birmingham . . 092414
Interiors Marketplace, Charlotte . 092607
Interiors Society, Milano 080578
Interkult, Wien 100326
Interlaken Historical Society Museum,
Interlaken 049703
Intermedia Arts Minnesota-Galleries,
Minneapolis 050939
Intermezzo, Cogolin 068066
Intermezzo, Dordrecht 112438
Intermobilia, Lisboa 084901
Intermuseum Conservation Association,
Oberlin 060528
Internacional Antique, Madrid . . 086202
Internasjonale Barnekunstmuseet,
Oslo 033837
Internasjonalt Kultursenter og Museum,
Oslo 033867
Internationaal Exlibriscentrum, Sint-
Niklaas 004019
International Academy of Media Arts
and Sience, Ogaki 056023
International Altex, Madrid 115239
International Antique Books & Coins,
Singapore 085433, 143685
International Antique Buyers, New
York 095476
International Antique Fair, Nagoya,
Aichi 098230
International Antiques, Denver . . 093301
International Antiques, Norfolk . 095860,
136032
International Antiques, Tulsa . . . 097741
International Antiquités, Saint-
Ouen 073110
International Art, Eskilstuna 115603
International Art and Design Fair, New
York 098456
International Art and Frame Gallery,
Tampa 124657
International Art Center, Miami . 050832
International Art Centre,
Auckland 083796, 126306
International Art Designs, Windsor,
Ontario 101855
The International Art Fair of The
Netherlands, Rotterdam 098259
International Art Gallery,
Nashville 122087
International Art Gallery, San
Francisco 124306
International Art Gallery,
Washington 124921
International Art Gallery Saint-Anne,
Bruxelles 100622
International Art Products, Egmond aan
Zee 083204, 112451
International Art Space,
Kellerberrin 099178
International Artists, Phoenix . . . 123497
International Asian Art Fair, New
York 098457
International Association for the
Fantastic in the Arts Newsletter,
Boca Raton 139395
International Association of
Agricultural Museums, Rožnov pod
Radhoštěm 058540
International Association of Arts
Critics, Skopje 059560
International Association of
Astronomical Artists, Lyme
Regis 060162
International Association of
Astronomical Artists, Orlando . 060529

International Association of Custom
Museums, Antwerpen 058380
International Association of Dealers in
Ancient Art, London 060163
International Association of Libraries
and Museums of Performing Arts
(SIMBAS), Roma 059460
International Association of Museum
Facility Administrators, New
York 060530
International Association of Pastel
Societies, Falls Church 060531
International Association of Transport
and Communication Museums,
Ängelholm 059885
International Auctioneers, Genève 059936,
139093
International Bluegrass Music
Museum, Owensboro 051852
International Bowling Museum and
Hall of Fame, Saint Louis 052803
International Center of Medieval Art.
Newsletter, Columbia 139396
International Center of Photography,
New York 051406
International Centre for the Study of
the Preservation and the Restoration
of Cultural Property (ICCROM),
Roma 059461
International Ceramics Fair & Seminar,
London 098383
International Child Art,
Washington 124922
International Clown Hall of Fame and
Research Center, West Allis .. 054285
International Coins Unlimited,
Philadelphia 096118
International Committee for
Conservation, Roma 059462
International Committee for Egyptology
in ICOM – CIPEG, Baltimore . 060532
International Committee for Exhibition
Exchange, ICOM-ICEE, Ottawa . 058449
International Committee for Marketing
and Public Relations, ICOM-MPR,
Oslo 059660
International Committee for Monetary
Museums, Utrecht 059611
The International Committee for
Museums and Collections of
Costume, London 060164
International Committee for Museums
and Collections of Glass, Toledo 060533
International Committee for Museums
and Collections of Modern Art,
CIMAM, Barcelona 059870
International Committee for Museums
and Collections of Natural History,
Frankfurt am Main 058928
International Committee for Museums
Security, ICMS, Berlin 058929
International Committee on the
Training of Personnel, ICTOP,
Berlin 058930
International Committee on
Management, ICOM-INTERCOM,
Ottawa 058450
International Company, Chofu . 081960
International Confederation
of Architectural Museums,
Rotterdam 059612
International Congress of Maritime
Museums, Essex 060534
International Conservation Services,
Chatswood 127585
International Contemporary Art Fair
Association, Paris 058660
International Contemporary Art Fair for
Photo-Based Art, Video and New
Media, Miami 098458
International Corporate Art, Oslo 084320
International Council of African
Museums (AFRICOM), Nairobi . 059522
International Council of Graphic Design

Associations, Montréal 058451, 136679
International Council of Museums
ICOM, Paris 058661
International Council on Monuments
and Sites, Paris 058662
International Crane Foundation
Museum, Baraboo 046786
International Cultural Center for Youth
in Jerusalem, Jerusalem .. 025083
International Custom Upolstery,
Philadelphia 136093
The International Design Magazine,
Cincinnati 139397
International Exposition of Sculpture
Objects and Functional Art, Chicago,
Chicago 098459
International Exposition of Sculpture
Objects and Functional Art, New York,
New York 098460
International Fair Poster Competition,
Plovdiv 098029
International Festival of Photography
and Digital Imaging, Delray
Beach 098461
International Fine Art and Antique
Dealers Show, New York 098462
International Fine Art Biennial
Exhibition, Beijing 098052
International Fine Art Club, Bel-Air
Fine Art, Genève 116323
International Fine Art Conservation
Studios, Bristol 134161
International Fine Art Fair, New
York 098463
International Fine Art Print Dealers
Association, New York 060535
International Fox Museum,
Summerside 007098
International Furniture, Phoenix . 136116
International Furniture Imports,
Indianapolis 093892
International Galleries, New York 122733
International Gallery of African Art,
Washington 124923
International Gospel Music Hall of
Fame and Museum, Detroit ... 048308
International Graphics, Eggenstein-
Leopoldshafen 137170
International Guide to Art Fairs and
Antiques Shows, New York ... 139398
The International Guild of Artists,
Ilkley 060165
International Hippocratic Foundation of
Kos, Kos 023141
International Ice Hockey Federation
Museum, Kingston 006100
International Images, Sewickley 053279
International Importers,
Sacramento 096632
International Institute for Conservation
of Historic and Artistic Works,
London 060166
International Institute of Metropolitan
Detroit, Detroit 048309
International Jewish Sports Hall of
Fame, Netanya 025163
International Journal of Art and Design
Education, Oxford 139188
The International Journal of Arts
Management, Montréal 138582
International Los Angeles Photographic
Art Exposition, Santa Monica . 098464
International Map Collectors' Society,
Newton Abbot 060167
International Motor Racing Research
Center, Watkins Glen 054224
International Movement for a New
Museology, Zuni 060536
International Museum of Art, El
Paso 048512
International Museum of Cartoon Art,
Boca Raton 047068
International Museum of Cartoon Art,
Stamford 053482

International Museum of Cultures,
Dallas 048118
International Museum of Peace and
Solidarity, Samarqand 054645
International Museum of Photography
and Film, George Eastman House,
Rochester 052617
International Museum of Surgical
Science, Chicago 047666
International Museum of the Horse,
Lexington 050245
International Museum of Women, San
Francisco 053004
International Museum Theatre Alliance,
Washington 060537
International New York Photographic
Art Exposition, New York 098465
International Poster Gallery,
Boston 119876
International Print Center New York,
New York 051407
International Printing Museum,
Carson 047441
International Ravissant, New York 095477
International Review of African
American Art, Hampton 139399
International Sculpture Center,
Hamilton 049324
International Sculpture Park,
The Chicago Athenaeum,
Schaumburg 053192
International Sewing Machine
Collectors' Society, London ... 060168
International Society for Jewish Art,
Mohegan Lake 060538
International Society Of Antique Scale
Collectors, Los Angeles 060539
International Society of Appraisers,
Renton 060540
International Society of Copier
and Computergraphic Artists,
Brooklyn 060541
International Society of Marine
Painters, Eagle 060542
International Sport Aviation Museum,
Lakeland 050095
International Spy Museum,
Washington 054126
International Swimming Hall of Fame,
Fort Lauderdale 048826
International Tennis Hall of Fame
Museum, Newport 051533
International Trading Post,
Houston 093706
International Visions Gallery,
Washington 124924
International Watch Exchange, New
York 095478
International Wildlife Museum,
Tucson 053831
International Women's Air and Space
Museum, Cleveland 047817
International World Of Art Gallery,
Dubai 117180
Internationale Antikmesse Aabenraa,
Rødekro 098066
Internationale Arbeitsgemeinschaft
der Archiv-, Bibliotheks- und
Graphikrestauratoren – IADA,
Göttingen 058931
Internationale Biënnale Antiek, Actuele
Kunst en Levensstijl, Kortrijk .. 098020
Internationale Fachmesse
für Museumswesen,
Sammlungen, Restaurierung und
Ausstellungstechnik, Leipzig .. 098143
Internationale Gesellschaft der
Bildenden Künste e.V., Berlin . 058932
Internationale Gesellschaft für Kunst,
Gestaltung und Therapie, Basel 059937
Internationale Gutenberg-Gesellschaft
in Mainz e.V., Mainz 058933
Internationale Jugendbibliothek,
München 020532

Internationale Kunstmesse Basel,
Kunst des 20. und 21. Jahrhunderts,
Basel 098329
Internationale Kunstmesse Zürich,
Zürich 098330
Internationale Messe für
zeitgenössische Kunst mit Fokus
Zentral- und Osteuropa, Wien . 098002
Internationale Sommerakademie für
Bildende Kunst, Salzburg ... 100167,
136553
Internationale Spitzensammlung,
Nordhalben 020825
Internationale Tage – Boehringer
Ingelheim, Ingelheim am Rhein 019402
Internationale Verband der
KunstkritikerInnen (AICA), Sektion
Österreich, Wien 058266
Internationalen Antiquariatsmesse,
Berlin 098144
Internationaler Kunstkritiker-Verband
(AICA), Sektion Schweiz, Zürich 059938
Internationaler Kunstkritikerverband
e.V. (AICA), Sektion der
Bundesrepublik Deutschland,
Köln 058934
Internationales Artistenmuseum in
Deutschland, "Die Rumeshalle der
Artisten und des Fahrenden Volkes",
Klosterfelde 019658
Internationales Ausstellungszentrum
NÖ, Schallaburg 002711
Internationales Design Zentrum Berlin
e.V., Berlin 058935
Internationales Keramik-Museum, Die
Neue Sammlung München, Weiden
in der Oberpfalz 022459
Internationales Luftfahrtmuseum,
Villingen-Schwenningen 022338
Internationales Maritimes Museum
Hamburg, Hamburg 018996
Internationales Muschelmuseum,
Wangerland 022423
Internationales Radiomuseum, Bad
Laasphe 016809
Internationales Wind- und
Wassermühlenmuseum, Gifhorn 018696
Internationales Zeitungsmuseum der
Stadt Aachen, Aachen 016437
Internationella Barnkonstmuseet,
Vårby 041155
Internazionale, Milano 110362
Internazionale d'Arte Contemporanea a
Torino, Torino 098195
Internet Auction Group,
Pittsburgh 127403
Interni, Milano 080579
Interno & Dum Dum, Bologna . 109851
Internos Books, London 138221
Interntional Chinese Snuff Bottle
Society. Journal, Baltimore ... 139400
Intersection for the Arts, San
Francisco 060543
Intersection for the Arts Gallery, San
Francisco 053005
L'Intersigne Livres Anciens, Paris 141145
Intertrus Antiques, New York ... 095479
Interventi Classensi, Ravenna .. 138835
Interventia, Den Haag 083089
Interview, New York 139401
Intervista – Multiculture Nomadismo
Trasgressioni, Milano 138836
Intipunku-Art, Amstelveen 112131
Intrepid Sea-Air-Space Museum, New
York 051408
L'Introuvable, Paris 141146
Intrude Gallery, Fitzroy 099026
Inttueri, Rio de Janeiro 101042
Inuit Art Foundation, Ottawa . 136680
Inuit Art Quarterly, Ottawa 138583
Inuit Gallery of Vancouver,
Vancouver 101801
Inuit Heritage Center, Baker Lake 005374
Inuyama Kenbunkan, Inuyama . 028689

Inventa, Münster 137435
L'Inventaire, Montauban 104251
The Invention Factory, Trenton . 053797
Inveraray Castle, Inveraray 044635
Inveraray Jail, Inveraray 044636
Inveraray Maritime Museum,
  Inveraray 044637
Inverbeg Galleries, Alexandria . 117225
Inverell Art Society, Inverell 058204,
  099153
Inverell Books, Inverell 139782
Inverkeithing Museum,
  Inverkeithing 044638
Inverness Miners Museum,
  Inverness 006037
Inverness Museum and Art Gallery,
  Inverness 044641
Invernizzi, Osvaldo, Zürich ... 087978
Inversiones en Arte y Naturaleza,
  Bilbao 114994
Inversiones en Arte y Naturaleza,
  Zaragoza 115553
Inversiones i Reservats,
  Barcelona 126512
Inversora Aranga, Caracas 097959
Investart, Genova 110116, 110117
Investitionsbank Berlin, Berlin .. 017179
Invictus Arc, Providence 096508
Invisible Culture. An Electronic Journal
  for Visual Culture, University of
  Rochester, Rochester 139402
L'Invitation au Voyage, Perpignan 072086
The Inya Gallery of Art, Yangon . 112055
Inyaka African Art, Paddington, New
  South Wales 099525
Inženernyj (Michailovskij) Zamok,
  Gosudarstvennyj Russkij Muzej,
  Sankt-Peterburg 037745
Ioannidis, Nik., Piraeus 109082
Ioannidou, Parthena, Thessaloniki 079212
Ioannina Museum, Ioannina 023091
Iodice, Salvatore, Napoli 131623
Iolani Palace, Honolulu 049535
Ion Graphix, Chicago 120162
Iona Abbey and Nunnery, Isle-of-
  Iona 044655
Iona Antiques, London 090235
Iona Heritage Centre, Isle-of-Iona 044656
Iona House Gallery, Woodstock . 119388
Ionad Arann Heritage Centre, Aran
  Islands 024817
Ionescu, M., Heinsberg 076481
Ionnina Byzantine Museum,
  Ioannina 023092
Iook, Dunedin 083928
Iorietti, Carlo, Torino 132151
Iorio, Antonio, Napoli 080893
Iosco County Historical Museum, East
  Tawas 048448
Ioseb Grishashvili Tbilisi History
  Museum, Georgian National Museum,
  Tbilisi 016411
Iotti, Gianfranco, Reggio Emilia . 081184
Iowa Aviation Museum,
  Greenfield 049244
Iowa Great Lakes Maritime Museum,
  Arnolds Park 046545
Iowa Masonic Library and Museum,
  Cedar Rapids 047488
Iowa Museum Association,
  Waterloo 060544
Iowa Railroad Historical Museum,
  Boone 047087
Iparművészeti Múzeum,
  Budapest 023490
Ipogeo dei Volumni, Ponte San
  Giovanni 027399
Ippedico, Guiseppe, Saint-Ouen . 073111
Ippogrifo, Milano 080580
Ippuku Bijutsukan, Asahi 028482
Ipreciation, Singapore 114543
Ipsos Galerie, Soual 073541
Ipswich Art Society, Ipswich ... 060169
Ipswich Historical Society, Heard

House Museum & Whipple House
  Museum, Ipswich 049715
Ipswich Museum, Ipswich 044646
Ipswich Transport Museum,
  Ipswich 044647
Iqbal Museum, Javed Manzil,
  Lahore 034130
Ir-Art, Moskva 114265
Iraan Museum, Iraan 049717
Iram, Saint-Ouen 073112
Iran Bastan Museum, Tehrän .. 024761
Iran Bazar, Jerusalem 109718
Iranian National Museum of Water,
  Sa'd-Abad Museum, Tehrän .. 024762
Iraq Natural History Research Centre
  and Museum, Baghdad 024803
Iraqi National Museum, Baghdad 024804
Irazu Mendizabal, J.A., Vitoria-
  Gasteiz 115543
Irbid Archaeological Museum, Yarmouk
  University, Irbid 029780
Irchel Antiquitäten, Buch am
  Irchel 087336
Irchester Narrow Gauge Railway
  Museum, Wellingborough ... 046076
Iredell Museum of Arts & Heritage,
  Statesville 053509
Ireland, Bad Dürkheim 074505
Ireland House at Oakridge
  Farm, Burlington's Museums,
  Burlington 005500
Ireland's Historic Science Centre,
  Birr 024834
Irene Cullis Gallery, Greensboro
  College, Greensboro 049254
Irg, F., Schwendi 078426
Irgens, Bror, Stockholm 087010
Irina & Sons, Sofia 101162
Irina's Antiques, San Francisco . 136323
L'Iris, Milano 080581
Iris & B. Gerald Cantor Center for
  Visual Arts, Stanford University,
  Stanford 053488
Iris and B. Gerald Cantor Art
  Gallery, College of the Holy Cross,
  Worcester 054545
Iris Antique Globes and Maps,
  Almen 082638
Iris Photography Gallery, Parafield
  Gardens 099549
Irish Academic Press, Dublin ... 137604
Irish Agricultural Museum,
  Wexford 025002
Irish American Heritage Museum, East
  Durham 048426
Irish American Museum, Albany 046342
Irish Antique Dealers Association,
  Adare 059379
Irish Architectural Archive, Dublin 024888
The Irish Art Group, Ballymoney 117274
Irish Arts Review, Dublin 138781
Irish Auctioneers and Valuers Institute,
  Dublin 059380
Irish Historical Picture Company,
  Dublin 109595
Irish Horse Museum, Kildare ... 024936
Irish International Antique and Fine Art
  Fair, Dublin 098179
Irish Jewish Museum, Dublin ... 024889
Irish Labour History Society Museum,
  Dublin 024890
Irish Linen Centre and Lisburn
  Museum, Lisburn 044844
Irish Moss Interpretive Centre and
  Museum, Miminegash 006320
Irish Museum of Modern Art,
  Dublin 024891
Irish Museums Association,
  Dublin 059381
Irish Music Hall of Fame, Dublin 024892
Irish Palatine Association Museum,
  Rathkeale 024981
Irish Professional Conservators' and
  Restorers' Association, Dublin . 059382

Irkutskij Architekturno-ètnografičeskij
  Muzej Talcy, Irkutsk 037024
Irkutskij Oblastnoj Chudožestvennyj
  Muzej im. V.P. Sukačeva,
  Irkutsk 037025
Irkutskij Oblastnoj Kraevedčeskij
  Muzej, Irkutsk 037026
Irlen, Medellín 128302
Irma Dahn-Anderssons Nävermuseum,
  Bollnäs 040570
Irma Stern Museum of the University
  of Cape Town, Rosebank 038826
Irmão, Juazeiro, Rio de Janeiro . 064069
Irmer, Herten 076513
Irodalom Háza – Mészöly Miklós
  Múzeum, Wosinsky Mór Megyei
  Múseum, Szekszárd 023948
Iroise Collections, Brest 067473
Iron and Steel Museum of Alabama,
  McCalla 050545
Iron Art, Birmingham 092415
Iron Bridge Historical Museum, Iron
  Bridge 006041
Iron County Museum, Caspian . 047458
Iron Creek Museum, Lougheed . 006222
Iron Garden, Bridgetown 003236
Iron Gate, Nashville 095007
Iron Horse Design, Albuquerque 119485
Iron Mission Museum, Cedar City 047474
Iron Mountain Iron Mine Museum,
  Vulcan 054010
Ironbridge Gorge Museum Trust,
  Coalbrookdale 060170
Ironbridge Tollhouse, Ironbridge 044649
Ironchurch Antiques Centre,
  Bolton 088513, 134122
Ironcraft of Omaha, Omaha 136062
Ironeyes Gallery, Indianapolis .. 121192
Ironside Market, San Antonio .. 096980
Ironton, Denver 120737
Ironwood Area Historical Museum,
  Ironwood 049719
Ironworld Discovery Center,
  Chisholm 047728
Iroquois County Historical Society
  Museum, Watseka 054228
Iroquois Falls Pioneer Museum,
  Iroquois Falls 006043
Iroquois Indian Museum, Howes
  Cave 049599
Iroquois Indian Museum Notes, Howes
  Cave 139403
Irresistibles, Houston 093707
Irrigaray, Bertrand, Jurancon .. 068927
Irrigaray, Bertrand, Pau 072042
Irrseer Heimathaus, Sammlung Hans
  Mairhofer-Irrsee, Zell am Moos 003135
IRSA, Kraków 137894
Iruma-shi Hakubutsukan Alit,
  Iruma 028694
Irvin Goodon International Wildlife
  Museum, Boissevain 005434
Irvine, Washington 124925
Irvine Burns Club Museum, Irvine 044651
Irvine Fine Arts Center, Irvine .. 049721
Irvine Museum, Irvine 049722
Irvine Arts Center, Irving 049724
Irving House Historic Centre and
  New Westminster Museum, New
  Westminster 006473
Irving, Mark & Dorien, Berwick-upon-
  Tweed 117348
Irving, Ruth, One Tree Hill 062028
Irwin, Anne, Atlanta 119575
Irwin, Rex, Woollahra 062473
Is a Brocante, Nogent-sur-Marne 070728
Is Art, Katoomba 099175
Isaac, Verona 111191
Isaac & Ede, London 090236
Isaac Farrar Mansion, Bangor .. 046778
Isaac, John, Albuquerque 092060
Isaac, Sylvia, Heidelberg Heights 127614
Isaacs, Louisville 094563
Isaacs & Isaacs, Louisville 094564

Isaacs-Innuit Gallery, Toronto .. 101631
Isaacs, Segovia, Barcelona 114921
Isaak Kaplan Old Yishuv Coourt
  Museum, Jerusalem 025084
Isaakevskij Sobor – Memorialnyj
  Muzej-pamjatnik, Sankt-
  Peterburg 037746
Isabel Anderson Comer Museum and
  Arts Center, Sylacauga 053628
Isabel Miller Museum, Sitka 053360
Isabella Antiques, Warminster .. 091734
Isabella Brant Galerij, Antwerpen 100495,
  127995
Isabella M., Modena 080801
Isabella Stewart Gardner Museum,
  Boston 047117
Isabella's Jewellery, Noosa
  Heads 061978
Isabelle, Dijon 068336
Isacsson, Gamleby 086685
ISAD – Istituto Superiore di
  Architettura e Design, Milano . 055924
Isadora, Beirut 082288
Isadora, Seattle 097459
Isahakyan House Museum,
  Gyumri 000703
Isamu Noguchi Garden Museum, Long
  Island City 050379
Isanti County Museums,
  Cambridge 047371
Isao, Nara 111335
Isarte, Milano 080582
Isaza, Martha Isabel, Medellín . 128303
Isberian, Oscar, Chicago 092727
Isca Gallery, Budleigh Salterton . 117471
ISCA Quarterly, Brooklyn 139404
Ischua Valley Historical Society,
  Franklinville 048942
ISCO, Camberley 138145
ISE Art Foundation, New York .. 051409
Iseema, Nagoya 082059
Isele, Klaus, Eggingen 137171
Isenbaye Art Gallery and Cultural
  Troupe, Oshogbo 033439
Isenmann, Heike, Potsdam 108407
Isenring, Ersigen ... 087416, 133735
Isergebirgs Museum Neugablonz,
  Kaufbeuren 019561
Iserhardt, Volker, Bad Münster am
  Stein-Ebernburg 129422
Iseum Romkert, Savaria Múzeum,
  Szombathely 023981
Isfahan, Essen 075683
Isfahan, Alexander Joles, Cape
  Town 085633, 133070
Isgro & Co., San Francisco 097260
ishak Paşa Sarayi, Ağri 042797
Ishavsmuseet Aarvak, Brandal . 033557
Ishibashi Museum of Art and
  Asian Gallery, Ishibashi Zaidan,
  Kurume 028911
Ishii, Tokyo 082154
Ishii Chawan Bijutsukan, Hagi .. 028585
Ishii, David, Seattle 145589
Ishii, Taka, Tokyo 111446, 137779
Ishikawa -kenritsu Shizenshi
  Shiryokan, Kanazawa 028771
Ishikawa-kenritsu Bijutsukan,
  Kanazawa 028772
Ishikawa-kenritsu Dento Sangyo
  Kogeikan, Kanazawa 028773
Ishikawa-kenritsu Rekishi
  Hakubutsukan, Kanazawa ... 028774
Ishikawa Takuboku Memorial Hall,
  Tamayama 029467
Ishimaru, Sapporo 142911
Ishimitsu Art Club, Osaka 059494
Ishinomaki Bunkakan, Ishinomaki 028702
ISIA Faenza – Istituto Superiore per le
  Industrie Artistiche, Faenza .. 055888
ISIA Firenze – Industrial and
  Communication Design, Firenze 055894
ISIA Roma – Istituto Superiore per le
  Industrie Artistiche, Roma 055968

ISIA Urbino – Istituto Superiore per le
 Industrie Artistiche, Urbino . . . 055985
Isidro, Eduardo Nobre, São João Das
 Lampas 085059
Ising, Dr., Heubach 076522
Isinger Dorfmuseum – Alte Kelter,
 Tübingen 022218
Isis District Historical Complex,
 Childers 000959
İskele İon Müzesi, İskele 033464
Iskitimskij Kraevedčeskij Muzej,
 Novosibirskij oblastnoj kraevedčeskij
 muzej, Iskitim 037034
Iskolamúzeum, Békés 023437
Iskolatörténeti és Egyházművészeti
 Múzeum, Cegléd 023579
Iskolatörténeti Kiállítás, Katona József
 Múzeum, Kecskemét 023724
Iskusstvo, Moskva 143640
Iskusstvo, Sankt-Peterburg 143667
Iskusstvo – The Russian Art Magazine,
 Moskva 139010
Isla, Burgos 085912
Isla Center for the Arts at the
 University of Guam, Mangilao . 023358
Islamabad Museum, Islamabad . 034116
Islamia College Museum,
 Peshawar 034143
Islamic Arts and Architecture
 Organization, Columbus 060545
Islamic Museum, Abbasiya 010509
Islamic Museum, Amman 029769
Islamic Museum, Jerusalem . . . 025085,
 034165
Islamic Museum, Tehrān 024763
Islamic Museum, Department of
 Antiquities, Tripoli 030296
Island Antique Mall, Jacksonville 093999
Island Art Galleries, Honolulu . . 120943,
 120944
Island Artbiz, Newhaven 099438
Island Auction Rooms, Shanklin 127038
Island County Historical Society
 Museum, Coupeville 048039
Island Fine Arts, Bembridge . . . 117347
Island Mountain Arts Public Gallery,
 Wells 007374
Island Nantiques, Kingscote . . . 061694
Island Treasures Antique Mall,
 Honolulu 093556
Island Treasures at the Marina,
 Honolulu 120945
Islands Art Center, California State
 University, Channel Islands,
 Camarillo 047355
Islands Museum, Tiverton 007151
Isle d'Avray, Ville-d'Avray 074131
Isle La Motte Historical Society, Isle La
 Motte 049731
Isle of Wight Courthouse,
 Smithfield 053368
Isle of Wight Steam Railway and Isle
 of Wight Railway Heritage Museum,
 Havenstreet 044538
Isle of Wight Wax Works,
 Brading 043702
Isleford Historical Museum,
 Isleford 049732
Isler, Theodor, Zürich 144183
Isles of Scilly Museum, Saint
 Mary's 045705
Isles, Andrew, Prahran 139885
Islesboro Historical Museum,
 Islesboro 049733
Islevbro, Rødovre 140648
Islington Antiques, Islington . . . 061653
Islington Education Artefacts Library,
 London 045006
Islington Museum, London 045007
Islip Art Museum, East Islip . . . 048438
Ismailia Museum, Ismailia 010579
Isman-Fänder, Anat, Hamburg . 076246
Ismer, Angela, Amelinghausen . 074406
L'Isola, Trento 111030

Isonkyrön Kotiseutumuseo,
 Isokyrö 010907
Isopo, Funchal 132927
Isoz, Daniel, Domdidier 087396
İsparta Devlet Güzel Sanatlar Galerisi,
 İsparta 042917
İsparta Müzesi, İsparta 042918
Israel Artists' Association,
 Jerusalem 059386
Israel Bible Museum, Zefat . . . 025223
Israel Children's Museum, Holon 025067
Israel Defense Forces History Museum,
 Tel Aviv 025199
Israel Goor Theater Archive and
 Museum, Jerusalem 025086
Israel Museum Products,
 Jerusalem 137613
Israel Museum, Jerusalem,
 Jerusalem 025087
Israel National Committee of ICOM,
 Ramat-Gan 059387
Israel Oil Industry Museum, Haifa 025052
Israel Railway Museum, Haifa . . 025053
Israel, B.M., Amsterdam 143040, 143041
Israel, B.M., Egmond aan den
 Hoef 143133
Israel, Barbara, New York 095480
Israeli Air Force Museum,
 Hatzerim 025061
Israeli Museum of Photography, Tel
 Hai 025215
Iss, Dominique, Illkirch-
 Graffenstaden 068867, 128626
Issaly & Pichon, Cannes 125545
Issaly, Jean-Pierre, Saint-
 Chamarand 072712
Issaquah Depot Museum, Issaquah
 History Museums, Issaquah . . 049736
Isseido, Tokyo 142924
Isselbacher, New York 122734
Issert, Catherine, Saint-Paul-de-
 Vence 105591
Issyk-Kulskij Gosudarstvennyj Istoriko-
 Kulturnyj Muzej-Zapovednik, Čolpan-
 Ata 030079
Istana Batu, Kota Bharu 030673
İstanbul Arkeoloji Müzeleri,
 İstanbul 042941
İstanbul Büyükşehir Belediyesi
 Karikatür ve Mizah Müzesi,
 İstanbul 042942
İstanbul Devlet Güzel Sanatlar Galerisi,
 İstanbul 042943
İstanbul Kültür ve Sanat Vakfı,
 İstanbul 060009
İstanbul Menkul Kıymetler Borsası
 Sanat Galerisi, İstanbul 117053
İstanbul Modern Sanat Müzesi,
 İstanbul 042944
İstanbul Sanat Fuarı, İstanbul . . 098342
İstanbul Sanat Müzesi, İstanbul . 042945
İstanbul Şişli'deki Ataürk Evi, İnkılap
 Müzesi, İstanbul 042946
IsTinto, Milano 110363
Istituti Culturali ed Artistici, Forli 055905
Istituto Centrale per il Restauro,
 Roma 055969
Istituto Centrale per il Restauro, Sede
 Distaccata, Roma 055970
Istituto d'Arte del Figurino,
 Milano 055925
Istituto d'Arte F. Palazzi, Napoli . 055935
Istituto della Cultura Arbereshe
 Giuseppe Gangale, Caraffa . . 025765
Istituto di Archeologia, Università
 Cattolica del Sacro Cuore,
 Milano 055926
Istituto di Cultura Berlin, Berlin . 106321
Istituto di Fotografia Alpina Vittorio
 Sella, Biella 025509
Istituto di Storia dell'Arte e di Estetica,
 Urbino 055986
Istituto di Storia dell'Arte, Fondazione
 Giorgio Cini, Venezia 055991

Istituto di Storia dell'Arte, Universita
 degli Studi di Parma, Parma . . 055946
Istituto di Storia e di Arte del Lazio
 Meridionale, Anagni 059463
Istituto di Studi Romani, Roma . 055971
Istituto e Museo di Storia della
 Scienza, Firenze 026264
Istituto Europeo d'Arte e Cultura,
 Milano 055927
Istituto Europeo di Design,
 Madrid 056373
Istituto Europeo di Design,
 Milano 055928
Istituto Geografico de Agostini,
 Novara 137706
Istituto Grafico Italiano, Milano . 137681
Istituto Italiano dei Castelli,
 Milano 059464
Istituto Italiano di Numismatica,
 Palazzo Barberini, Roma 055972
Istituto Nazionale di Archeologia e
 Storia dell'Arte, Roma 055973
Istituto per il Restauro Maria Teresa
 Caiazzo, Salerno 055976
Istituto per la Storia della Resistenza e
 della Societa Contemporanea
 nelle Province di Biella e Vercelli,
 Varallo 028270
Istituto per la Storia dell'Arte Lombarda,
 Palazzo Aresa Jacini, Cesano
 Maderno 059465, 137635
Istituto Poligrafico e Zecca dello Stato,
 Roma 137736
Istituto Statale d'Arte, Faenza . . 055889
Istituto Statale d'Arte, Grottaglie 055910
Istituto Statale d'Arte, Lecce . . . 055910
Istituto Statale d'Arte, Urbino . . 055987
Istituto Statale d'Arte Adolfo Venturi,
 Modena 055932
Istituto Statale d'Arte di Gubbio con
 Scuola Media Annessa, Gubbio 055911
Istituto Statale d'Arte e Liceo Statale
 d'Arte, Riccione 055957
Istituto Statale d'Arte E.Giannelli,
 Parabita 055944
Istituto Statale d'Arte Franco Russoli
 di Pisa, Pisa 055951
Istituto Statale d'Arte Nardò,
 Nardò 055938
Istituto Statale d'Arte Osvaldo Licini,
 Ascoli Piceno 055871
Istituto Statale d'Arte Poggiardo Nino
 Della Notte, Poggiardo 055952
Istituto Statale d'Arte U. Boccioni,
 Napoli 055936
Istituto Statale d'Arte Venezia,
 Venezia 055992
Istituto Statale d'Arte Vincenzo
 Bellisario, Avezzano 055872
Istituto Superiore per le Industrie
 Artistiche-Ceramica, Faenza . . 055890
Istituto Universitario di Magistero,
 Fisciano 055902
Istituto Universitario Olandese di Storia
 dell'Arte, Firenze 055895
Istituto Veneto di Scienze, Lettere ed
 Arti, Venezia 055993
Istoki, Belozerskij Istoriko-
 chudožestvennyj Muzej,
 Belozersk 036893
Istoria, Beirut 082289
Istoričeskaja Usadba Botik Petra I,
 Pereslavl-Zalesskij gosudarstvennyj
 istoriko- architekturnyj i
 chudožestvennyj muzej-zapovednik,
 Veskovo 038020
Istoričeski Muzej, Bansko 004977
Istoričeski Muzej, Blagoevgrad . 004990
Istoričeski Muzej, Bracigovo . . . 004993
Istoričeski Muzej, Burgas 004994
Istoričeski Muzej, Ciprovci 005006
Istoričeski Muzej, Dimitrovgrad . 005014
Istoričeski Muzej, Dobrič 005018
Istoričeski Muzej, Drjanovo 005023

Istoričeski Muzej, Gabrovo 005031
Istoričeski Muzej, Goce Delčev . 005034
Istoričeski Muzej, Gorna
 Orjachovica 005035
Istoričeski Muzej, Ichtiman 005038
Istoričeski Muzej, Isperich 005039
Istoričeski Muzej, Karlovo 005048
Istoričeski Muzej, Loveč 005085
Istoričeski Muzej, Omurtag 005099
Istoričeski Muzej, Petrič 005113
Istoričeski Muzej, Pomorie 005133
Istoričeski Muzej, Popovo 005134
Istoričeski Muzej, Pravec 005137
Istoričeski Muzej, Provadia 005140
Istoričeski Muzej, Razgrad 005148
Istoričeski Muzej, Razlog 005149
Istoričeski Muzej, Svištov 005222
Istoričeski Muzej, Velingrad . . . 005260
Istoričeski Muzej – Preminavane
 na Ruskite Vojski pri Svištov,
 Svištov 005223
Istoričeski Muzej Batak, Batak . 004980
Istoričeski Muzej Belogradčik,
 Belogradčik 004981
Istoričeski Muzej Pavlikeni,
 Pavlikeni 005104
Istoričeski Muzej Silistra, Silistra 005163
Istoričeski Muzej Targovište,
 Targovište 005225
Istoričeski Muzej, Tutrakan 005234
Istoričeskij Muzej Dom na Naberežnoj,
 Moskva 037311
Istoričnyj Muzej, Dnipropetrovskyj
 deržavnij istoričnij muzej im
 Javornic'kogo, Dnipropetrovs'k 043148
Istorijski Muzej Crne Gore, Narodni
 Muzej Crne Gore, Cetinje 031751
Istorijski Muzej Srbije, Beograd . 038144
Istoriko-Architekturnyj i
 Chudožestvennyj Muzej
 Ivangorodskaja Krepost,
 Ivangorod 037035
Istoriko-architekturnyj Muzej
 Kuzneckaja Krepost',
 Novokuzneck 037573
Istoriko-Architekturnyj,
 Chudožestvennyj i Archeologičeskij
 Muzej "Zarajskij Kreml",
 Zarajsk 038080
Istoriko-chudožestvennyj Muzej,
 Uglič 037985
Istoriko-étnografičeskij Muzej Usadba
 Galskich, Čerepovec 036942
Istoriko-Kraevedčeskij Muzej,
 Ivanovo 037040
Istoriko-Kraevedčeskij Muzej,
 Merke 029827
Istoriko-Kraevedčeskij Muzej,
 Rudnyj 029841
Istoriko-Kraevedčeskij Muzej,
 Tokmok 030084
Istoriko-Kraevedčeskij Muzej g.
 Kronštadta, Kronštadt 037177
Istoriko-kulturnyj Centr Krasnye Palaty
 XVII Veka, Moskva 037312
Istoriko-Kulturnyj Muzej-zapovednik
 Azret Sultan, Turkestan 029852
Istoriko-kulturnyj Muzej-zapovednik
 Idnakar, Glazov 037018
Istoriko-literaturnyj Muzej I.A.
 Gončarova, Uljanovskij oblastnoj
 kraevedčeskij muzej, Uljanovsk . 037992
Istoriko-memorialnyj i Architekturno-
 chudožestvennyj Muzej, Tichvin 037936
Istoriko-memorialnyj Kompleks
 Gerojam Stalingradskoj Bitvy na
 Mamaevom Kurgane, Volgograd 038050
Istoriko-memorialnyj Muzej M.V.
 Lomonosova, Lomonosovo . . . 037205
Istoriko-memorialnyj Muzej
 Polityčnych V'jazniv, Ternopil's'kyj
 oblasnyj krajeznavčyj muzej,
 Ternopil' 043307
Istoriko-memorialnyj Muzej Semji

Fadeevyh-Chan-Blavackij,
Dnipropetrovskyj deržavnij
istoričnij muzej im Javornic'kogo,
Dnipropetrovs'k 043149
Istoriko-memorialnyj Muzejnyj
Kompleks Bobriki, Donskij 036971
Istoriko-sportivnyj Muzej, Rossijskaja
Gosud. Akademija fizičeskoj kultury,
Moskva 037313
Istoriski Muzej, Kruševo 030615
Istoryko-architekturna Pam'jatka-
muzej Kyïvska Fortecja, Kyïv . . 043194
d'Istria, Michel, Paris 071498
István Borpince és Nemzeti
Bormúzeum, Eger 023619
Iszatt, Casimir, Heidelberg 076450
Ita, Egidio, Innsbruck 127775
Itabashi Kuritsu Bijutsukan,
Tokyo 029518
Itaca, Venezia 111130
Itä-Hämeen Museo, Hartola . . . 010804
Itä-Helsingin Antikvariaatti,
Helsinki 140704
Itai Pan, Rio de Janeiro 064070
Ital Reding-Haus, Schwyz 041970
Italarte, Milano 110364
Italarte, Roma 110804
Italarte, Galleria Indicatore, Roma 110805
Italarte, Galleria Portuense, Roma 110806
Italarte's, Bogotá 065209, 128270
Italcoinvest, Ferrara 080071
Italia. Bollettino d'Arte, Roma . . 138837
Italiaander, Amsterdam 082771
Italian American Museum, New
York 051410
Italiana Antiquariato, Firenze . . . 080162,
142677
Italienisches Kulturinstitut, Berlin 017180
Italo Camattini & C., Parma . . . 110629
Itami-shiritsu Bijutsukan, Itami . 028705
Itami-shiritsu Hakubutsukan,
Itami 028706
Itasca Heritage Center Museum, Grand
Rapids 049197
Itaya Hazan Kinenkan,
Shimodate 029381
Itayanagi-machi Mingeikan,
Itayanagi 028707
Itel, Martin, Olten 116553
Itfaiye Müzesi, İstanbul 042947
Ithemba, Paris 104963
Itimuseum, Museo Itinerante de Arte
Contemporaneo del Mercosur,
Buenos Aires 000163
Itinérante, Château-Gontier 067872
Itinere, A Coruña 115034
Ito, Kyoto 111296
ITO Antiques, Miami 094753
Itojen, Singapore 085434
It's a Blast, Tucson 124746
It's About Time, Westcliff-on-Sea 091794
It's Antiques, Miami 094754
It's New to You, Philadelphia . . . 096119
It's Random, New York 122735
Itsukaichi Folklore Museum,
Akiruno 028467
Itsukushima-jinja Homotsukan,
Hatsukaichi 028614
Itsuo Bijutsukan, Ikeda 028677
Itsutsubashikan, Sendai 029357
Ittah, Anath & Jacques, Genève 087476,
116324
Ittinger Museum, Warth 042124
Ituna Cultural and Historical Museum,
Ituna 006047
Iturralde, Los Angeles 121512
Iturriate, C., Lima 084456
Iturrioz Leza, Roberto, Logroño . 086074
Iuliintik, Timişoara 085134
IUP Denis Diderot, Université de
Bourgogne, Dijon 055369
Iupui Cultural Arts Gallery,
Indianapolis 049693
Iuric, Emilia, Milano 080583

Iv Cantons, Girona 086028
IVAM Centre Julio González,
Valencia 040305
IVAM, Institut Valencià d'Art Modern,
Valencia 115506
Ivan Dougherty Gallery, University of
New South Wales, Paddington, New
South Wales 001391
Ivan Franko Museum, Winnipeg 007447
Ivan Meštrović Gallery, Ivan Mestrovič
Foundation, Split 008996
Ivana, Dimitrie, Paris 071499
Ivane Javakhishvili Samtskhe-Javaxeti
History Museum, Akhaltsikhe . 016385
Ivánffy & Uhler, Dallas 120612
Ivanhoe Collectibles Corner,
Ivanhoe 061654
Ivano-Frankivs'kyj Chudožnij Muzej,
Ivano-Frankivs'k 043156
Ivano-Frankivs'kyj Krajeznavčyj Muzej,
Ivano-Frankivs'k 043157
Ivanovskij Oblastnoj Chudožestvennyj
Muzej, Ivanovo 037041
Ivanyi, Cherokee 098883
Ivar Aasen-tunet, Hovdebygda . . 033675
Ivar Lo-Museet, Stockholm 041007
Ivarsson Auktioner, Tibro 126686
I've Been Framed, Miami 121810
I've Been Framed, Tulsa 124804
Iveland og Vegusdal Bygdemuseum,
Vatnestrom 034075
Ives, John, Twickenham 144897
Ivey-Selkirk, Saint Louis 127427
Ivorra Segura, Facundo, Alcoy . . 133078
Ivory Art, Miami 121811
Ivory Press, London 138222
L'Ivre Livre, Foix 140841
Ivy, Lewes 138177
Ivy Antiques, Wakefield 084141
Ivy Cottage, San Antonio 096981
Ivy Green, Birthplace of Helen Keller,
Tuscumbia 053864
Ivy Hall Antiques, Birr 079458
Ivy House Antiques, Acle 088236
Ivy's Antiques, Memphis 094641
Iwaki-shi Bijutsukan, Iwaki 028721
Iwakuni Historical Museum & Kikkawa
Family Museum, Iwakuni 028722
Iwakuni-shi Bijutsukan no Kashiwara
Shuu, Iwakuni 028723
Iwalewa Haus, Afrikazentrum der
Universität Bayreuth, Bayreuth 016990
Iwami Adachi Bijutsukan,
Hamada 028597
Iwami Bijutsukan, Masuda 029992
Iwan de vent nouveau, Jal-el-Dib 082316
Iwand-Kussin, Heidrun, Berlin . . 074832
Iwantja Arts & Crafts, Indulkana
Everard 061649
Iwasaki Bijutsukan, Ibusuki 028667
Iwasaki Kinen Kenbunkan, Nisshin-city,
Aichi Prefecture 029192
Iwasaki Museum, Yokohama . . . 029712
Iwata-gun Kaori Hakubutsukan,
Toyoda 029643
Iwata Senshinkan, Inuyama . . . 028690
Iwata-shi Kyu Mitsukegakko, Iwata,
Shizuoka 028725
Iwate-kenritsu Bijutsukan,
Morioka 029051
Iwate-kenritsu Hakubutsukan,
Morioka 029052
Iwate-kenritsu Meisho Hakubutsukan,
Morioka 029053
Iwayama Urushi Bijutsukan,
Morioka 029054
IWK – Mitteilungen des Instituts für
Wissenschaft und Kunst, Wien 138516
Iynedjian Art, Lausanne 116422
Iynedjian Art, Verbier 116688
Iyoboya Salmon Museum,
Murakami 029061
Izaak Walton's Cottage,
Shallowford 045767

Izarn, Pascal, Paris 071500
Izaro, Donostia-San Sebastián . . 143743
Izba Muzealna im. Antoniego
Krajewskiego, Lanckorona 034933
Izba Pamięci Żydów Lublina,
Lublin 034991
Izcandé, San José 102508
Izdatelstvo na Bălgarskata akademija
na naukite, Sofia 136671
Iziko Museums of Cape Town, Cape
Town 059818
Izkustvo i Spomeni, Stara Zagora 064241,
101179
Izložba na Stari Kotlenski Kilimi, Muzej
na Kotlenskite Vazrožđenci i Panteon
na Georgi S. Rakovski, Kotel . . 005075
Izložben Kompleks, Veliko
Tărnovo 005253
Izložbena Zala Sirak Skitnik,
Sliven 005165
Izmaïl's'ka Kartynna Galereja,
Izmaïl 043161
Izmaïl's'kyj Krajeznavčyj Muzej
Pridunav'ja, Izmaïl 043162
Izmaïl's'kyj Muzej O.V. Suvorova,
Izmaïl 043163
İzmir Arkeoloji Müzesi, İzmir . . 042973
İzmir Etnografya Müzesi, İzmir . 042974
İznik Müzeleri, Iznik Müzesi ve Iznik
Ayasofya Müzesi, İznik 042978
Izolda International, Gießen 107127
İzstāžu Zāle, Madonas
Novadpētniecības un Mākslas Muzejs,
Madona 030159
Izu Orugorukan, Ito 028709
Izukogen Kirie Bijutsukan, Ito . . 028710
Izumi, Sendai 082113
Izumi-mura Rekishi no Sato, Aoba no
Fue Museum, Anauma Minzoku
Museum & Kyodoshiryo Museum,
Izumi, Fukui 028727
Izumi, J., Kyoto 082019
Izumo Mingeikan, Izumo 028731
Izumo Taisha Homutsuden,
Izumo 028732
Izzard, Sebastian, New York . . . 122736
Izzo, Giovanna, Napoli 131624
Izzo, Matthew, Philadelphia 123400
J & A Brass Polishing,
Sacramontosco 136194
J & D Collectibles, Norfolk 095861
J & E Antiques, Saint Paul 096829
J & G Gallery, Chichester 117580
J & H Antiques, Minneapolis . . . 094923
J & H China Repair, Vancouver . 128226
J & J Antiques, Fresno 093503, 135564
J & J Sportscards & Collectables,
Albuquerque 092061
J & K Furniture, Phoenix 096218
J & L New Art, Houston 121071
J & L Refinishing, San Francisco 136324
J & M Gallery, Hong Kong 102088
J & P Fine Art, Zürich 116815
J & P Gallery, Seoul 111671
J & P Timepieces, New York . . . 095481,
135939
J & R Gallery, Tucson 124747
J & V International, New York . . . 135940
J 70, Donostia-San Sebastián . . 085975
J Å B Art, Falun 115613
J B M Art Consultancy, South
Yarra 062228
J&C Furniture Specialist, New
Orleans 135860
J&G Art, Milano 110365
J. & A. Antiques, Bridgend 088595
J. A. Stein / Dürer Verlag,
Nürnberg 142245
J. Evetts Haley History Center,
Midland 050875
J. F. Schreiber-Museum,
Esslingen 018319
J. Nikoladze Art College, Tbilisi . 055435
J. Paul Getty Museum, Los

Angeles 050436
J. V. Pianos and Cambridge Pianola,
Cambridge 134200
J. Wayne Stark University Center
Galleries, Texas A&M University,
College Station 047873
J.A., Castres 067730
J.A.M, Saint-Ouen 073113
J.A.N. Fine Art, London 090237
J.A.V. David Museum, Killarney . 006088
J.B. Antiques, Wimborne Minster 091857,
135231
J.B. Wallis Museum of Entomology,
Winnipeg 007448
J.B.F. Editions d'Art, Paris 104964,
136948
J.B.'s Hole in the Wall, Charlotte 092608
J.E. Reeves Home and Museum,
Dover 048354
J.F. Willumsens Museum,
Frederikssund 010071
J.F.G, Villeurbanne 074210
J.F.H., Aix-en-Provence 066464
J.G.M., Paris 104965
J.J. Jackson Memorial Museum,
Weaverville 054260
J.J.College of Architecture, Mumbai
University, Mumbai 055819
J.J.'s Collectables, Oklahoma
City 095981
J.K. MacCarthy Museum, Goroka 034190
J.L. Runebergin Koti, Porvoo . . . 011272
J.L. Shellshear Museum of Physical
Anthropology and Comparative
Anatomy, Sydney 001540
J.L.F., Saint-Ouen 073114
J.M. Antiquités, Paris 071501
J.M. Davis Arms and Historical
Museum, Claremore 047762
J.M.W. Turner House, London . . 045008
J.N. Antiques, Redbourn 091078
J.P. Art Centre, Kaohsiung 116891
J.P.C., Lyon 136830, 140903
J.P.L.D., Paris 071502
J.R. Toys and Model Car Gallery,
Dallas 093144
J.T. Wilson Museum of Human
Anatomy, Sydney 001541
J.U. Polimski Muzej, Berane . . . 031747
J.V. Snellmanin Kotimuseo,
Kuopio 011051
J.W. Parmely Historical Home Museum,
Ipswich 049716
J.Z. Antik, København 065918
JA Antik, Århus 065601, 128377
Jaafar, André, Paris 129041
Jaap's Antik, Hisings Backa . . . 086746
Jabatan Muzium Sabah, Kota
Kinabalu 030675
Al-Jaber, Dubai 088179
Jaberg, Miriam M., Minneapolis . 122004
Jabiru Australia, Saint Marys . . . 062181
Jabiru Gallery of Fine Art, Sanctuary
Cove 099674
Jablonka, Berlin 106322
Jablonka, Köln 107643
Jablonski, Anröchte 074415
Jabloński, Adam, Warszawa . . . 113853,
113854
Jabonga Parish Museum,
Jabonga 034482
Jabotinsky Museum, Tel Aviv . . 025029
Jaboulet, Jean-Paul, Brando . . . 067447
Jacalone, Jerome, New York . . . 095482
Jacazzi, C., Saint-Hilaire-Saint-
Florent 072830
Jaccaz, Saint-Julien-en-Genevois 072864
Jacek, Gdańsk 084506, 143452
Jachthuis Sint-Hubertus,
Hoenderloo 032494
Jacinto Courthouse, Corinth . . . 047991
Jack Gallery, Chicago 120163
Jack Gallery, Las Vegas 121360, 121361
Jack Gallery, Los Angeles 121513

Jack Gallery, New Orleans 122186
Jack Heath Art Gallery, University of
  Kwazulu-Natal, Pietermaritzburg 038799
Jack Lynn Memorial Museum,
  Horsefly 006015
Jack Miner Museum, Kingsville . 006115
Jack S. Blanton Museum of Art,
  Austin 046693
Jack the Stripper, Edmonton ... 128164
Jackfield Tile Museum, Jackfield 044661
Jackie Antiquités, La Celle-Saint-
  Cloud 068966
Jackman, Saint Kilda, Victoria .. 099667
Jackpot, Milwaukee 094864
Jack's, Montréal 064434
Jack's Antiques, Sandringham . 062189
Jack's Art, Arnhem 112331
Jack's Art Gallery, New York ... 122737
Jackson, Atlanta 119576
Jackson, Braidwood 061189
Jackson, Dallas 120613
Jackson, Seattle 097460
Jackson & Wickliff, Carmel 127154
Jackson Art Center, Jackson School,
  Washington 054127
Jackson Artworks, Omaha 123324
Jackson County Historical Museum,
  Lakefield 050094
Jackson County Historical Museum,
  Maquoketa 050662
Jackson County Historical Society
  Museum, Black River Falls ... 047024
Jackson Design, Berlin 074833
Jackson Design, Stockholm ... 087011
Jackson-Grant, D.M., Teynham . 091592
Jackson Hall Gallery, Frankfurt . 048925
Jackson Hole Historical Museum,
  Jackson 049762
Jackson Siegel, Aaron, New York 095483
Jackson Street Gallery, Seattle . 097461
Jackson-Washabaugh Historical
  Museum, Kadoka 049853
Jackson, Allan K.L., Edinburgh . 089179
Jackson, B., Atlanta 119577
Jackson, D.T., Saint Louis 096718
Jackson, Lee, London 144636
Jackson, Peter, Nottingham ... 118841
Jackson's, Oakland 136042
Jackson's Mill Historic Area,
  Weston 054353
Jacksonville Fire Museum,
  Jacksonville 049770
Jacksonville Museum of Modern Art,
  Jacksonville 049771
Jacmaire & Buteux, Saint-Ouen
  (Somme) 129241
Jacob, Callian 103492
Jacob, Paris 104966
Jacob Breda Bullmuseet,
  Rendalen 033887
Jacob-Kalähne, Syke 108743
Jacob Kelley House Museum,
  Hartsville 049395
Jacob Walden House, Historical
  Society of Walden and Wallkill Valley,
  Walden 054034
Jacob, Ann, Atlanta 119578
Jacob, Dietmar, Reichenbach im
  Vogtland 130574
Jacob, G., Waldmohr 078843
Jacob, John, Woodbridge 144935
Jacob, René, Grenoble 068733
Jacob, Samuel, Santa Monica .. 138469
Jacob, Stéphane, Paris 104967
Jacobi, Tel Aviv 079734
Jacobs, Buffalo 092535
Jacobs, Cardiff 088786, 134213
Jacobs & Bouwknegt, Paris ... 104968
Jacobs & Hunt, Petersfield ... 127014
Jacobs & Uys, Tshwane 114820
Jacobs, Erika & Iris, Hamburg .. 076247
Jacobs, F., Bruxelles 063469
Jacobs, Frans, Amsterdam 112227

Jacobs, H., Bremen 075167
Jacobs, Lionel, Richmond, Surrey 091098
Jacobsa, Wolfgang, Nürnberg .. 077828,
                                108303
Jacobsen, Salt Lake City 136240
Jacobsen, Britta, København ... 065919
Jacobsen, Dorthe & Morten,
  Esbjerg 065638
Jacobsen, Margo, Portland 123634
Jacobson & Howard, New York . 122738
Jacobson, Bernard, London 118406
Jacobsson, K., Svedala 115948
Jacomart, Madrid 115240
Jacomet, Nice 128863
Jacomo, Pascale, Paris 129042
Jacono-Muntean, Domenico & Raluca,
  Wien 140091
Jacotte, Paris 104969
Jacotti, Giovanna, Brescia 131038
Jacquart, Danielle, Brugge 063278
Jacquelin, Yvette & Pierre,
  Blanzy 067239
Jacqueline Casey Hudgens Center for
  the Arts, Duluth 048378
Jacqueline's Room, Böblingen .. 075053
Jacquemart, Christiane,
  Montpellier 070306
Jacquemin, Gérard, Allonzier-la-
  Caille 066538
Jacquemin, Serge, Presly 072308
Jacques, Houston 093708
Jacques, Oslo 084321
Jacques Marchais Museum of Tibetan
  Art, Staten Island 053499
Jacques, Jason, New York 095484
Jacqueson, Jean-Guy, Gevrey-
  Chambertin 068656
Jacquet, Gérard, Lyon . 069749, 104111
Jacquet, Gérard, Rugles 072643
Jacquey, Catherine, Paris 071503
Jacquey, Philippe, Besançon .. 067127
Jacquier, Paul-Francis, Kehl ... 076741
Jacquin, Christiane, Saint-Ouen 073115
Jacquin, Marie-Claude, Grande-
  Rivière 068695
Jacqulene, Singapore 114544
Jacquot, Patrick, Conflans-en-
  Jarnisy 068132
Jade, Amsterdam 112228
Jade Antique House, Singapore . 085435
The Jade Dragon, Ann Arbor ... 092095
Jade Union Gallery, Hong Kong . 102089
Jadestone Gallery, Portland 096423
Jadis, Bath 088389
Jadis, Brive-la-Gaillarde 067512
Jadis, Cerny 067753
Jadis, Dommartin-aux-Bois 068378
Jadis, Lamorlaye 069152
Jadis, Québec 064535
Jadis, Saint-André-de-Sangonis 072670
Jadis, Toulon (Var) 073713
Jadis, Vichy 074109
Jadis Antiquités, Lausanne ... 087573,
                                126736
Jadis Antiquités, Pouillé 072297
Jadis et Naguère, Aubenas ... 066779
Jadis et Naguère, Bruxelles ... 063470
Jadis et Naguère, Casablanca . 082564
Jadis et Naguère, Deuil-la-Barre 068291
Jadis et Naguère, Laissey 069145
Jadis et Naguère, Paris 104970, 141147
Jadite, New York 122739
J'Adore, Kansas City 094077
Jaeckin, Anne & Just, Paris ... 104971
Jaeger, Dr. Roland, Hamburg .. 141875
Jäger, Dr. Siegfr., Kelkheim ... 130111
Jaeger, F. de, Gent 128050
Jäger, Julius, Wiesbaden 126139
Jäger, R., Wächtersbach 078820
Jaeger, Winand, Mainz 077235
Jägerndorfer Heimatstuben,
  Ansbach 016594
Jægerspris Slot, Jægerspris ... 010152
Jäggi, Aeschi 087182

Jael Stichting Mystic Museum,
  Dronten 032255
Jämtlands Flyg och Lottamuseum,
  Östersund 040901
Jämtlands Läns Museum,
  Östersund 040902
Jäneda Muuseum, Jäneda 010619
Jaep, Anne-Marie, Chissay-en-
  Touraine 067980
Jærmuseet, Nærbø 033795
Järnerot, Eva, Uppsala 087121, 133598
Järntorgsboden, Stockholm 087012
Järnvägs- och Industrimuseum,
  Hagfors 040680
Järnvägsmuseet, Regionmuseet i
  Skåne, Kristianstad 040759
Järva-Jaani Tultörjeseltsi Muuseum,
  Järva-Jaani 010620
Järvamaa Muuseum, Paide 010649
Järvids Antik och Möbelmåleri,
  Arvika 086632
Järvsö Antik och Kuriosa, Järvsö 086758
Jäschke, Jens, Görlitz 076036
Jaeschke, Olaf, Braunschweig . 106562,
                          129623, 141635
Jaeschke, Paweł, Poznań 084642
Jafar Jabarly State Museum of
  Theatre, Bakı 003166
Jafe, San Francisco 136325
Jaffa Antigüedades, Lima 084457
Jaffna National Museum, Jaffna 040484
Jaffray, Alan, Melksham 090740
Jaffre, Michel, La Flèche 069034
Jaffrey Civic Center, Jaffrey ... 049779
JAG, Paris 104972
Jagamarra Collection, Potts Point 099597
Jagd- und Falknereimuseum,
  Riedenburg 021415
Jagd- und Fischereimuseum,
  Adelsdorf 016465
Jagd- und Fischereimuseum Schloss
  Tambach, Weitramsdorf 022553
Jagd- und Fischereimuseum Schloß
  Wolfstein und Galerie Wolfstein,
  Freyung 018510
Jagd- und Naturkunde Museum,
  Niederstetten 020800
Jagd- und Schloßmuseum
  Spangenberg, Spangenberg .. 021927
Jagd und Naturkundemuseum, Hahn
  am See 018923
Jagdhaus Gabelbach, Klassik Stiftung
  Weimar, Ilmenau 019389
Jagdmuseum, Ellwangen 018161
Jagdmuseum Gußwerk, Gußwerk 002025
Jagdmuseum Schloss Stainz,
  Landesmuseum Joanneum,
  Stainz 002773
Jagdmuseum Wulff, Dedelstorf . 017838
Jagdschau im Jagdschloss Springe,
  Springe 021942
Jagdschloß, Gelbensande 018617
Jagdschloß Granitz, Binz,
  Ostseebad 017393
Jagdschloss Grunewald, Stiftung
  Preußische Schlösser und Gärten
  Berlin-Brandenburg, Berlin ... 017181
Jagdschloß Paulinzella, Museum zur
  Geschichte des Klosters, des Forstes
  und der Jagd, Rottenbach ... 021519
Jagdschloss Stern, Stiftung
  Preußische Schlösser und Gärten
  Berlin-Brandenburg, Potsdam . 021186
Jagdschloss Letzlingen, Stiftung Dome
  und Schlösser in Sachsen-Anhalt,
  Letzlingen 020048
Jager, Hélène, Metz 070099
Jager, Ronald de, Ouderkerk aan de
  Amstel 132609
Het Jagershuis, Zaandam 083696
Jaggard, Sydney 062289
Jaggedart, London 118407
Jagro, Espoo 066290
Jagua, Tampa 136389

Jaguar Daimler Heritage Trust,
  Coventry 044020
Jahangir Art Gallery, Mumbai .. 109446
Jahn, Diethelm, Vellmar 130723
Jahn, Fred, München . 108143, 137408
Jahn, Karin, Wangerland 108844
Jahn, Lothar, Garbsen 141820
Jahn, Manfred, Berlin 074834
Jahnke Road Antique Center,
  Richmond 096555
Jahns, Wolfgang, Köln 076879, 107644
Jahnsen, Bill, Smørum 066135
Jahr, Ulrich, Hebertsfelden ... 130025
Jahre, Eckehard, Senftenberg . 078456
Jail Training School Museum,
  Lucknow 024382
Jailhouse Galleries, Burke Arts Council,
  Morganton 051054
Jaime's Studio, San Francisco . 136326
Jain, Delhi 109324
Jaipur, Los Angeles 094370
Jairos, Bogotá 102345
Jaja, Amsterdam 112229
Jajuta, Singapore 114545
Jakisch, L., Hofgeismar 076552
Jakob, Norrtälje 143908
Jakob-Grünenwald-Gedächtnisstätte,
  Ebersbach an der Fils 018078
Jakob-Kocher, Jeannette, Môtier,
  Vully 133863
Jakob, Ernst, Basel 116080
Jakob, Gerhard, Frankfurt am
  Main 075794
Jakob, Günther, Lichtenfels ... 077100,
                                130241
Jakob, Karlheinz, Prutting 130543
Jakobi, Witten 108951
Jakobi, Kurt, Florstadt 075754
Jakobs, Duisburg 075550
Jakobstads Museum, Jakobstad 010910
Jakopičeva Galerija, Arhitekturni Muzej
  Ljubljana, Ljubljana 038544
Jákováli Hasszán Múzeum, Pécs 023840
Jakovljev, Gordana, March ... 130285
Jaktvårdsmuseum, Eksjö 040600
Jakubowski & de Weydenthal,
  Paris 129043
Jakubowski, Pawel, Warszawa . 132882
Jakutskij Gosudarstvennyj Muzej
  Istorii i Kultury Narodov Severa im.
  Jaroslavskogo, Jakutsk 037055
Jala, Cirfontaines-en-Ornois ... 067991
Jalan, Toronto 064680
Jalan-Jalan, Nashville 095008
Jalasjärven Museo, Jalasjärvi .. 010914
Jalkaväkimuseo, Mikkeli 011158
Jallay, Camille, Kleinbettingen . 082356
Jallet, Gérard, Varces-Allieres-et-
  Risset 073978
Jallet, Marie-Christine, Tours ... 073849
Jallier, Alain, Magny-en-Vexin . 069803
Jalna, Little Haywood . 089845, 134564
Jalokivi Galleria, Kemi 010983
Jamaat Art Gallery, Mumbai ... 109447
Jamahiriya Museum, Tripoli ... 030297
Jamaica Center for Arts,
  Jamaica 049781
Jamaica Folk Museum, Kingston 028444
Jamaican People's Museum of Craft
  and Technology, Spanish Town 028456
Jamalo-Neneckij Okružnoj
  Kraevedčeskij Muzej im. I.S.
  Šemanovskogo, Salechard ... 037702
Jamal's Corner Shop, Chennai . 079373
Jamandic, Chester 088861
Jamar, Antwerpen 100496
Jamar, Tracy E., New York 135941
Jamault, Laurent, La Flotte ... 103888
Jamault, Philippe, Brest 067474
Jamberoo Antiques, Jamberoo . 061655
James, Bangor 117286
James, Dalkey 109367
James, Dublin 109596
James, Pittsburgh 123555

James A. Garfield National Historic
Site, Mentor 050798
James A. McFaul Environmental
Center of Bergen County,
Wyckoff 054559
James A. Michener Art Museum,
Doylestown 048366
James and Meryl Hearst Center for
the Arts, Cedar Falls 047477
James Art Studio, Saskatoon . . 101546
James Bond 007 Museum,
Nybro 040886
James Buchanan Foundation for
the Preservation of Wheatland,
Lancaster 050120
James Cook Museum, Cooktown 000989
James Dunklin House, Laurens . 050182
James E. Lewis Museum of
Art, Morgan State University,
Baltimore 046752
James Ford Bell Museum of Natural
History, University of Minnesota,
Minneapolis 050940
James Hall Museum of Transport,
Johannesburg 038751
James Hogg Exhibition, Selkirk . 045757
James House Museum,
Bridgetown 005474
James Howe Gallery, Kean University,
Union 053888
James J. Hill House, Saint Paul . 052841
James J. O'Mara Pharmacy Museum,
Saint John's 006903
James Joyce Centre, Dublin . . . 024893
James Joyce Museum, Dublin . . 024894
James K. Polk Ancestral Home,
Columbia 047919
James K. Polk Memorial,
Pineville 052141
James-Krüss-Turm, München . . 020533
The James Madison Museum,
Orange 051793
James Madison's Montpelier,
Montpelier Station 051038
James Mitchell Museum, Galway 024926
James Mitchell Varnum House and
Museum, East Greenwich 048427
James Monroe Museum and Memorial
Library, Fredericksburg 048956
James of Saint Albans, Saint
Albans 091174, 144810
James Paterson Museum,
Moniaive 045286
James R. Slater Museum of Natural
History, Tacoma 053644
James W. Dillon House Museum,
Dillon 048331
The James Watrous Gallery, Wisconsin
Academy, Madison 050596
James Whitcomb Riley Birthplace and
Museum, Greenfield 049242
James Whitcomb Riley Museum,
Indianapolis 049694
James, Angela, Vancouver 064831
James & Son, Anthony, London 090238
James, Anthony, Whimple 091822
James, Brian, Telford . 091555, 135123
James, Dominique, Saint-Ouen . 073116
James, Joe, Ocho Rios 111238
James, Jonathan, London 090239
James, Joseph, Penrith 090957, 134927
James, Murray & Thomson,
Woolloongabba 062489
James, R.A., Sileby 091344
James, R.A., Tutbury . 091668, 135166
James, Renée, Saint-Ouen 073117
Jamestown Museum, Jamestown 001168,
049790
Jamestown Settlement,
Jamestown-Yorktown Foundation,
Williamsburg 054413
Jamestown Visitor Center & Museum,
Jamestown 049792
Jamieson Museum, Moosomin . 006424

Jamin, Reims 072392
Jamin, Laurent, Enghien-les-
Bains 068444
Jamin, Laurent, Saint-Gratien . 072816
Jammes, Paul, Paris 141148
Jammu and Kashmir Academy of Art,
Srinagar 055838
Jamshid, Cincinnati 092866
Jamtli, Jämtlands Läns Museum,
Östersund 040903
Jamy Antiquités, Toulon (Var) . 073714
Jan Becher Muzeum, Karlovy
Vary 009618
Jan Bouman Haus, Potsdam . . 021187
Jan Kochanowski Muzeum, Jacek
Malczewski Muzeum Radom,
Czarnolas 034702
Jan Steen Galerie, Amsterdam . 112230
Jan Van Eyck Academie,
Postacademic Institute for Research
and Production in Fine Arts, Design
and Theory, Maastricht 056179
Jan Vissermuseum, Museum voor
Landbouw, Helmond 032462
Jāṇa Akuratera Muzejs, Rīga . . 030184
Jāṇa Jaunsudrabiņa Muzejs Riekstiņi,
Nereta 030165
Jaṇa Rozentāla Saldus Vēstures un
Mākslas Muzejs, Saldus 030231
Jaṇa Rozentāla un Rūdolfa Blaumaṇa
Muzejs, Rīga 030185
Jāṇa Zabera Memoriālais Muzejs,
Meirani 030163
Janalan, Roma 099650
Janani, Bangalore 079349
Janbroers, J.W., Den Haag . . . 143113
Jančík, Jaromír, Opava 102629
Janco-Dada Museum, Ein Hod . 025036
Jandl, Wien 127932
Jandolo, Antonio, Roma 081414
Jandovo Muzeum, Budyně nad
Ohři 009490
Jane Addams' Hull-House Museum,
University of Illinois Chicago,
Chicago 047667
Jane Austen's House, Chawton . 043914
Jane L. and Robert H. Weiner Judaic
Museum, Rockville 052651
Jane Neville-Rolfe Gallery, Alpha 000795
Jane Voorhees Zimmerli Art Museum,
Rutgers State University of New
Jersey, New Brunswick 051244
Jane Welsh Carlyle's House,
Haddington 044492
Janeček, Václav, Jablonec nad
Nisou 065395
Janeiro, Noémia, Lisboa 084902
Janek & Synowie, Marek,
Kobylnica 084543, 132828
Jane's Art & Craft, Singapore . . 085436
Jane's Books, Wellington 143366
Janet Turner Print Museum, California
State University, Chico, Chico . 047711
Janetschek, Christian,
Heidenreichstein 139994
Janfrey, Milwaukee 121927
Janfusun Museum, Gukeng . . . 042322
Janggigot Lighthouse Museum,
Pohang 029977
Jangva, Helsinki 103151
Janin & Ruggeri, Bonnieux . . . 103407
Janine Arnold Art Consulting,
Wien 062989
Janinos Monkutės-Marks Muziejus-
Galerija, Kėdainiai 030379
Janis Rozentals Upper Secondary Riga
Art School, Rīga 056109
Janke, Egon, Köln 130151
Janker, B., Beratzhausen 129454
Jankuv, Dávid, Bardejov 085538
Jannin, Jean-Marc, L'Isle-sur-la-
Sorgue 069572
Jannis Spyropoulos Museum,
Ekali 023067

Le Jannou, Gildas, Perros-Guirec 072103
Jannowitz, Juist 076656
Janoray, Charles, New York . . . 122740
Janos Gat Gallery, New York . . 122741
Jan's & Co, Los Angeles 094371, 127293
Jan's Antik, Valby 066175
Jan's Antiques, Chicago 092728
Jan's Antiques Mall, Long Beach 094204
Jan's Art Shop, Teignmouth . . . 119217
Jans, Chris, Amsterdam 132404
Jansem, Flora, Paris 104973
Jansen, Fredenbeck 075840
Jansen, Krefeld 076958
Jansen, Philadelphia 096120
Jansen, Alexa, Köln 107645
Jansen, H., Viersen 142451
Jansen, M.F.A., Breda 082989
Jansen, Ronald, Weesp 143265
Janses Antik, Hedemora 086735
Jansonius, Lugano 116458
Jansonius, Morcote 116514
Janssen, Rotterdam 083572
Janßen, Thedinghausen 078677, 130703
Janssen, Antje, Bergisch
Gladbach 129459
Janssen, Heinz, Kevelaer 107549,
130117, 141988, 141989
Janssen, Ina, Hamburg 129966
Janssen, Jörg, Geldern 107104
Janssen, L.-A., Paderborn 108353
Janssen, Michael, Köln 107646
Janssen, Rodolphe, Bruxelles . . 100623
Janssen, Stephan, Lent 132566
Janssens, Breda 082990
Janssens van der Maelen, Francis,
Bruxelles 063471
Janssens, Eric, Lenggries 077085
Janssens, Jean-Louis, Visé 128092
Jansson, Enköping 143828
Jansson, Sven, Kalmar 126611
Jantar, Stralsund 108680
Jantti, Matti, Bodø 113140
Jantyik Mátyás Múzeum, Békés 023438
Jantyik Mátyás Múzeum – Békési
Galéria, Békés 023439
Jantyik Mátyás Múzeum – Békési
Tájház, Békés 023440
Jantzen, Jean-Claude, Nancy . . 070421
Jantzen, Klaus, Hamburg 076248
Jantzen, Laurence, Paris 071504
Jantzen, Laurence, Ville-d'Avray 074132
Janus-Antik, Frankfurt am Main 075795
Janus Antiques & Interiors,
Kinsale 079651
Janus Pannonius Múzeum, Pécs 023841
Janus, Peter, Wuppertal 079108
Janzen, Gevelsberg 107123
Janzen-Köppen, L., Bad
Bellingen 074491
Janzen, Axel, Altensteig 074396
Janzen, Frank, Lilienthal 130242
Janzen, Gisela, Gescher 107120
Japan Amateur Art Museum, Ogi 029204
Japan Art, Frankfurt am Main . . 107006
Japan Art Academy, Tokyo . . . 056045
Japan Chic, Kobe 081996
Japan Clay Doll Museum,
Nakano 029137
Japan Design Foundation, Osaka 059495
Japan Football Mueum, Tokyo . 029519
Japan Footwear Museum & Japan
Folk Toy and Doll Museum,
Fukuyama 028562
Japan Gallery, New York 122742
Japan Graphic Arts, Tokyo 137780
Japan Graphic Designers Association,
Tokyo 059496
Japán Múzeum, Füzesgyarmat . 023641
Japan Print Gallery, London . . . 118408
Japan Publications, Tokyo 137781
Japan Society Gallery, New York 051411
Japan Sword, Tokyo 082155
Japan Traditional Craft Center,
Tokyo 082156

Japan Ukiyo-e Club, Nagoya . . . 059497
Japanache, West Hollywood . . . 097901
Japanese American National Museum,
Los Angeles 050437
Japanese Antiques, Los Angeles 094372
Japanese Antiques and Art, San
Antonio 096982
Japanese Art Crafts Association,
Tokyo 059498
Japanese Art Dealers Association of
New York, New York 060546
Japanese Art Society of America, New
York 060547
Japanese Association of Art Critics,
Tokyo 059499
Japanese Canadian National Museum,
Burnaby 005504
Japanese Council of Art Museums,
Kanazawa 059500
Japanese Cultural Center of Hawaii,
Honolulu 049536
Japanese Gallery, Hiroshima . . 081978
Japanese Gallery, London 090240
Japanese National Committee for
ICOM, Tokyo 059501
Japanese Ski Memorial Museum,
Joetsu 028734
Japanese Sword and Oriental
Antiques, Minneapolis 094924
Japanese Sword Society, San
Francisco 060548
Japanese Sword Society of the US,
Breckenridge 060549
Japanese Works Of Art, Milano . 110366
Japaneseantiques, Dublin 079562
Japanesque, Washington 124926
Japanisches Kulturinstitut, Köln . 019689
Japansalon, München 077534
Japantown Art and Media Collection,
San Francisco 053006
Japanz Gallery, Christchurch . . 112845
Japhet, Christiane & Stéphane,
Nice 125663
Japon Antique, Paris 071505
Japonesque Gallery, San
Francisco 124307
Japour, A. J., Miami Beach . . . 121882
Jaquenoud, Paula, Marseille . . . 069944
Jaques, Jean, Philadelphia 123401
Jaquiery, Michel, Genève 126724
Jara, Victor, Salamanca 086407
Al-Jarah, Manama 063146
Jarash Archaeological Museum,
Jarash 029782
Járdányi Paulovics István Romkert,
Savaria Múzeum, Szombathely 023982
Jarden, Kraków 113603, 113604
Jardim-Museu Agrícola Tropical,
Lisboa 035759
Le Jardin Antique, Sacramento . 096633
Le Jardin aux Papillons, Musée Vivant,
Vannes (Morbihan) 016118
Jardin de l'Estérel, Castillon-du-
Gard 067716
Le Jardin-de-Saint-Julien, Pierrefitte-
sur-Sauldre 072156
Jardin de Sculptures et d'Agrumes du
Palais, Menton 014131
Jardin d'Ébène, Cliousclat 068041
Jardin des Arts, Bergen 052841
Le Jardin des Arts, Ris-Orangis 105449
Le Jardin des Arts, Saint-Denis . 114118
Le Jardin des Arts et Encadrements,
Barbizon 103332
Jardin des Lettres, Bourges . . . 140804
Jardin des Muses, Saint-Pierre-du-
Val 141339
Le Jardin des Papillons, Paris . . 014664
Jardin du Temps, Lucciana 069655
Le Jardin Ferroviaire, Chatte . . 012523
Jardin Géologique d'Obourg,
Obourg 003916
Jardinières et Intérieurs, Saint-
Ouen 073118

Jardinique, Beech 088427
Jardins, São Paulo 140304
Jardins del Príncep, Museu
d'Escultures a l'Aire Iliure de
Santiago, Tortosa 040267
Jared's, Tulsa 097742
De Jaren, Breda 082991
Jareteg, Borås 115588
Jarman, David, Lewes 144534
Jarmarka Knig po Iskusstvu, Dizajnu i
Architekture, Moskva 098281
Jarmuschek & Partner, Berlin .. 106323
Jarnac, Louis, Yssingeaux 074313
Jarndyce, London 144637
Jarneau Venisse Maryvonne,
Granville 068698
Jaroslavskij Chudožestvennyj Muzej –
Gubernatorskij Dom, Jaroslavl . 037060
Jaroslavskij Istoriko-architekturnyj i
Chudožestvennyj Muzej-zapovednik,
Jaroslavl 037061
Jaroslavskij Muzej Boevoj Slavy,
Jaroslavskij gosudarstvennyj istoriko-
architekturnyj muzej-zapovednik,
Jaroslavl 037062
Jarovit, Lviv 117155
Jarrell Plantation Georgia,
Juliette 049845
Jarrett, Billy, Dallas 135508
Jarry, Dominique, Saint-Remy-du-
Val 073293
Jarsbo, Århus 102775
Jartunet, Åsgreina 033509
Jarver, Peter, Cairns .. 098838, 098839,
098840
Jarver, Peter, Kuranda 099209
Jarvis, Matlock 144731
Jas, Utrecht 112702
Jas, M., Breda 082992
Jasañ, Igor, Košice 133050
Jasha's Bohemian Bazaar,
Memphis 094642
Jaski Art Gallery, Amsterdam ... 082772
Jaskmanický, Dr. Jiří, Praha ... 102698
Jaskula, Jens, Berlin 129512
Jasmine Fine Arts, Singapore .. 114546
Jason, Auckland 143291
Jason, Mark, London 118409
Jasper, Federal 099014
Jasper Cultural and Historical Centre,
Maple Creek 006259
Jasper Rand Art Museum,
Westfield 054336
Jasper-Yellowhead Museum and
Archives, Jasper 006048
Jasper's Antique Radio Museum, Saint
Louis 052804
Jaspers, Michael, München 108144
Jassamine House, Celbridge ... 079478
Jasu Gallery, Margate 099282
Jász Múzeum, Jászberény 023701
Jászai Mari Emlékház, Ászár ... 023408
Jászai Mari Emlékszoba,
Rábatamási 023867
Jatkodd, Los Angeles 121514
Jauberthie, Alain, Baladou 066923
Jaud, Georg, Garmisch-
Partenkirchen 075948
Jaudel, Marie-Pierre, Paris 071506
Jaudel, Marie-Pierre, Saint-Ouen 073119
Jauert, Ove, Rødovre 066078
Jauffroy, Bernard, Royan 072629
Jauker, Hans, Wien 140092
Jaunmoku Pils Muzejs, Tukums 030239
Jaunpils Muzejs, Jaunpils 030126
Java Antiques, Washington 097841
Java Art, Virginia Beach 124832
Java Nook, Toronto 101633
Javaherat, Tehrān 024764
Javaheri, Mohammad Bagher,
Firenze 131192
Javier Antigüedades, Lima 084458
Javier Navascues Aoiz, Donostia-San
Sebastián 133251

Javogue & Ingall, Miami 121812
Javor, Burgos 085913
Javouhey, Patrick, Beaune 067024
Jawa, Torino 081714
Jawaharlal Nehru State Museum,
Itanagar 024316
Jaworski, Sven & Ingram,
Werbach 130746
Jaws Antiques, Toronto 064681
JAX Art Gallery, New Orleans .. 122187
Jaxquart Antiques, Holywood ... 089556
Gwenda Jay & Dan Addington,
Chicago 120164
Jay Books, Edinburgh 144394
Jay I. Kislak Foundation, Miami
Lakes 050850
Jay, Bruno, La Tronche 069110
Sri Jayachamarajendra Art Gallery,
Mysore 024416
Jayalakshmi & Co., Bangalore .. 079350
Jayer, Sanilhac 073405
Jaylac, Renaud, Toulouse 073776
Jayroe, Dallas 093145
Jay's Antiques, North Petherton 090850
Jay's Place, San Antonio 096983
Jāzepa Vītola Memoriālais Muzejs,
Gaujiena 030123
Jazmin Asian Arts, Singapore .. 085437
jazzART Galerie, Gelsenkirchen . 107110
Jazzens Museum, Kolbäck 040755
Jazzothèque, Nice 014477
JBL International Antiques,
Miami 094755
JBPA, New York 122743
JC Antikhörna, Karlstad 086780
JC Antiquités, Lacroix-Falgarde . 069132
JC Diffusion, Villeurbanne 074211
Jean Art Gallery, Seoul 111672
Jean Christophe, Mandelieu-La
Napoule 069831
Jean Lafitte National Historical Park
and Preserve, New Orleans .. 051295
Jean-Louis Antiquités, Caen 067572
Jean-Louis Pressensé Libraire – La
Fourbithèque, Nantes 140984
Jean P. Haydon Museum, Pago
Pago 000084
Jean-Pascal, Bordeaux 128493
Jean-Paul-Museum der Stadt
Bayreuth, Bayreuth 016991
Jean Paul Slusser and Warren Robbins
Galleries, University of Michigan,
School of Art and Design, Ann
Arbor 046489
Jean-Paul-Zimmer in der Rollwenzelei,
Bayreuth 016992
Jean-Pierre Antiquités, Lyon ... 069750
Jean-Pierre Pescatore Collection Villa
Vauban, Luxembourg 030597
Jean, Isabelle & Thierry, Aix-en-
Provence 066465
Jean, René, Maransin 069845
Jeane Dixon Museum and Library,
Strasburg 053566
Jeanette's Collectibles, Saint
Louis 096719
Jeanmarie Gallery, New York ... 122744
Jeanne B. Créations, Casablanca 082565
Jeanne et Jeremy, Paris 071507
Jeanneau, Christian, Angers ... 066609
Jeannelle, Patrice, Paris 071508
Jeannerat, Christian, Neuchâtel . 087705
Jeannet, Jean-François, Bourges 067430
Jean's Accents, Los Angeles ... 094373
Jean's Military Memories, Great
Harwood, Lancashire 089368
Jean's Silversmiths, New York .. 095485
Jeantet, Jean-Jacques, Balsièges 066927
Jeck, Eleanor, Tucson 124748
Jedburgh Castle Jail Museum,
Jedburgh 044664
Jedlik Ányos Emlékkiállítás, Győr 023662
Jeff Matthews Memorial Museum,
Galax 049013

Jefferds Tavern, York 054580
Jeffers Petroglyphs Historic Site,
Comfrey 047946
Jefferson & Co., Los Angeles ... 135723
Jefferson Barracks, Saint Louis . 052805
Jefferson Coin Rarities, New
Orleans 095114
Jefferson County Historical Association
Museum, Steubenville 053523
Jefferson County Historical Museum,
Madison 050584
Jefferson County Historical Museum,
Pine Bluff 052136
Jefferson County Historical Society
Museum, Port Townsend 052259
Jefferson County Historical Society
Museum, Watertown 054215
Jefferson Historical Museum,
Jefferson 049800
Jefferson Landing State Historic Site,
Jefferson City 049804
Jefferson National Expansion
Memorial, Saint Louis 052806
Jefferson Patterson Museum, Saint
Leonard 052789
Jefferson Telephone Museum,
Jefferson 049797
Jefferys, Lostwithiel 126978
Jeffrey & McConnell, San Diego 097100
Jeffrey Charles & Henry Peacock,
London 118410
Jeffrey Ruesch Gallery, New York 122745
Jeffries, Catriona, Vancouver ... 101802
Jeff's 3-D Poster Art, Miami ... 121813
Jeff's Antiques, Los Angeles ... 094374
Jegge, Martin, Biel ... 087318, 133694
Jegun, Antoine, Sénas 073488
Jehangir Art Gallery, Mumbai .. 024405
Jehl, Philadelphia 123402
Jehle, Mathäus, Eppingen 129795
Jehu, Ron, San Francisco 124308
Jeju-do Folklore and Natural History
Museum, Jeju 029955
Jeju Education Museum, Jeju ... 029956
Jeju Folklore Museum, Jeju ... 029957
Jeju National University Museum,
Jeju 029958
Jēkaba Priņa Anatomijas Muzejs,
Rīga 030186
Jēkabpils Vēstures Muzeja Brīvdabas
Nodaļa, Jēkabpils 030127
Jēkabpils Vēstures Muzejs,
Jēkabpils 030128
Jekyll Island Museum, Jekyll
Island 049808
Jelanik, Olomouc 065431
Jelínek, Praha 065502
Jellicoe Arts, Nelson 112963
Jellicoe, Roderick, London 090241
Jellinek, Tobias, Twickenham ... 091674
Jelling, Vejle 066195
Jelly van den Bosch Museum,
Roden 032791
Jeltsch, T., Bönnigheim 075055
Jem, Kelmscott 061674
JEM – Journal of Education in
Museums, Gillingham 139189
Jem Coins, Fresno 093504
Jemez State Monument, Jemez
Springs 049809
Jemo, Villeurbanne 074212
Jems, El Paso 093404
Jena, Tokyo 142925
Jenack, William J., Chester 127161
Jenaer Kunstverein e.V., Jena .. 058936
Jenbacher Museum, Jenbach .. 002146
Jenco, Chicago 120165
Jenei Béla Betű, Budapest 142561
Jeneverhuis Van Hoorebeke,
Eeklo 003573
Jenin, Claude, Lille 128679
Jenior, Winfried, Kassel 076723, 137292
Jenisch Haus, Museum für Kunst und
Kultur an der Elbe, Hamburg . 018997

Jenischek, O., Bergisch Gladbach 141464
Jenkins Johnson Gallery, San
Francisco 124309
Jenko's, Charlotte 092609
Jenkova Kasarna, Zgornje
Jezersko 038637
Jenle Museum, Nanna og Jeppe
Aakjærs Hjem, Roslev 010295
Jennemann, Dr. S., Isny im
Allgäu 107474
Jenner, Le Havre 069283
Jenner Museum, Berkeley 043570
Jennes, M., Antwerpen 140148
Jennings-Brown House Female
Academy, Bennettsville 046918
Jennings, Paul, Angarrack 088282
Jennrich, Merzenich 077355
Jennrich, E. & U., Düren 075447
Jennywell Hall Antiques, Crosby
Ravensworth 089013
Jens Nielsens & Olivia Holm-Møller
Museet, Holstebro 010137
Jens Søndergaard Museet,
Lemvig 010218
Jensen, Auckland 112780
Jensen, Helsingør 102827
Jensen Arctic Museum, Western
Oregon University, Monmouth . 050994
Jensen, Birgit, Rønne 066082
Jensen, Dean, Milwaukee 121928
Jensen, Else, København 065920
Jensen, Frank, Otterup 066048
Jensen, Georg, København 065921
Jensen, Gert, Farum 065650
Jensen, Helge, Valby 140667
Jensen, Joan, Svaneke 066151
Jensen, Karin, Gilleleje 065696
Jensen, Leif, Valby .. 066176, 102994
Jensen, Niels, Nykøbing Falster 140629
Jensen, Ove & Nina, Grindsted . 065712
Jensen, Soren, New York 095486
Jensen, Tommy, Sorø 066143
Jente, Saint Louis 136203
Jentsch, Remplin 078197
Jentsch, Detlef, Gütersloh 076123,
125962
Jentzsch, Carl-Rainer, Adelberg . 105958
Jeonju Municipal Museum,
Jeonju 029960
Jeonju National Museum, Jeonju 029961
Jeoufen Art Gallery, Rueifang .. 042431
Jeph's Stuff, Portland 096424
Jéquel, Nathalie, Lyon 128716
Jered, San Diego 124136
Jeremiah Lee Mansion,
Marblehead 050665
Jeremy, London 090242
Jerewan State Art Academy Gallery,
Jerewan 098602
Jerewan State Humanitarian College,
Jerewan 054886
Jerewan State Institute of Fine Arts,
Jerewan 054887
Jérez, Jean-Luc, Saint-Julien-de-
Lampon 072861
Jeri, Jean, Fontaine-de-Vaucluse 068558
Jericho, Oxford 144768
Jericho Historical Society, Jericho
Corners 049813
Jerilderie Doll World, Jerilderie .. 001170
Jermy & Westerman, Nottingham 144758
Jermy, Pete, Ulverstone 139942
Jernander, A., Bruxelles 063472
Jerner, Stockholm 087013
Jernigan & Wicker, San
Francisco 124310
Jerome County Historical Society,
Jerome 049815
Jerome K. Jerome Birthplace Museum,
Walsall 046040
Jerome State Historic Park,
Jerome 049814
Jérome, Jean-Luc, Cravant 068224

Jérôme, Jean-Luc, Meung-sur-
Loire 070105
Jerrari, Youssef, Casablanca ... 082566
Jerrye, Ann, Indianapolis 093893
Jerry's Fine Art, Miami 121814
Jerry's Refinishing, Fresno ... 135565
Jerschewski, Volker, Flensburg . 106957
Jersey Antiques, Osborne Park . 062037
Jersey Battle of Flowers Museum,
Saint Ouen 045709
Jersey City Museum, Jersey City 049819
Jersey Heritage Trust, Saint Dominick,
Jersey 045685
Jersey Museum, Saint Helier .. 045692
Jersey Photographic Museum, Saint
Helier 045693
Jersey War Tunnels, Saint
Lawrence 045702
Jersin, Nikki, Denver 093302
Jerstedt, Goslar 076061
Jerusalem Art's, Houston 121072
Jerusalemhaus mit Verwaltung der
Städtischen Sammlungen Wetzlar,
Städtische Sammlungen Wetzlar,
Wetzlar 022610
Jerxheimer Kunstverein,
Jerxheim 058937
Jeschek, Angelika, Augsburg .. 074463
Jeschke, Hauff & Auvermann,
Berlin 125879
Jeschke, Hauff & Van Vliet,
Berlin 125880
Jesenski & Turk, Zagreb 065314
Jeske, Norbert, Stollberg,
Erzgebirge 078570
Jeske, O., Bevern, Holstein 074969
Jeske, Wolfgang, Berlin 141532
Jessa, Denise, Montereau-Fault-
Yonne 070237, 128801
Jesse Besser Museum, Alpena . 046420
Jesse James Bank Museum,
Liberty 050274
Jesse James Farm and Museum,
Kearney 049906
Jesse James Home Museum, Saint
Joseph 052785
Jesse Peter Museum, Santa Rosa
Junior College, Santa Rosa ... 053139
Jesse, Dr. Jürgen & Claudia,
Bielefeld 106494, 137109
Jessie's Art Work, Los Angeles . 094375
Jessop, London 090243
Jessy & Raymond's Gallery, Ciudad de
Panamá 113319
Jesteburger Mühle, Jesteburg .. 076647
Jester Antiques, Tetbury 091572
Jesuitenkolleg, Mindelheim 020410
Jesus Jones and Justice Museum of
Art, Los Angeles 050438
Jeta, Singapore 114547
Jetter, Edgar, Vreden 130729
Jeu de l'Oie, Biarritz 067202
Jeu de Paume, Site Concorde,
Paris 014665
Jeu de Paume, Site Sully, Paris 014666
Jeudy, I., Tournai 063845
Jeunesse, E.R., Paris 141149
Jeunink, Brigitte, Gleize 068674
Jevons, Francis, London 090244, 134697
Jevrejski Istorijski Muzej,
Beograd 038145
Jewel Antiques, Leek 089781
Jewel Box, Dallas 093146
The Jewel Casket, Dublin 079563
Jewel of Tibel, Toronto 064682
Jewel of Tibet, San Francisco .. 097261
Jewel Tower, London 045009
The Jewel Tree, Cape Town ... 085634
Jewell County Historical Museum,
Mankato 050650
Jewellery Exchange, Auckland . 083797
Jewelry and Coin Center, Buffalo 092536
Jewelry Box Antiques, Kansas
City 094078

Jewelry by Mimi, Philadelphia .. 096121
Jewelry Palace, Las Vegas 094154
Jewett Hall Gallery, Augusta .. 046666
Jewish Battalions Museum,
Ahivil 025012
Jewish Holocaust Centre,
Elsternwick 001058
Jewish Institute for the Arts, Boca
Raton 047069
Jewish Museum, London 045010
The Jewish Museum, New York 051412
Jewish Museum Finchley,
London 045011
Jewish Museum of Australia, Gandel
Centre of Judaica, Saint Kilda,
Victoria 001470
Jewish Museum of Florida, Miami
Beach 050847
Jewish Museum of Maryland,
Baltimore 046753
Jewish Museum of Thessaloniki,
Thessaloniki 023291
Jewish National Fund Museum,
of Provisional Peoples Council and
Administration, Tel Aviv 025201
Jewitt &Co., J.B., Cleveland ... 135465
Jewry Wall Museum, Leicester
Museums, Leicester 044795
Jeypore Branch Museum, Jaipur,
Orissa 024317
Jezebel, New Orleans 095115
Jeżewski, Wiesław, Warszawa . 132883
JF Galerie, Zürich 116816
JFF Fire Brigade & Military
Collectables, Bedhampton ... 088425
JFK Special Warfare Museum, Fort
Bragg 048788
JGalerie, Paris 104974
Jhalawar Archaeology Museum,
Jhalawar 024329
Jharokha, Delhi 109325
Jheng Cheng-Gong Memorial Hall,
Sikou 042454
Jhenlangong Mazu Exhibition Hall,
Dajia 042297
JHS Antiques, Ashbourne 088299
Ji Ce's Antikt, Nyköping 086890
JI Gallery, Los Angeles 094376
Ji Ya, Shenzhen 102228
Ji, Zhon, Shenyang 102212
Jia-Art, Taipei 116923
Jia Friendship Union, Beijing ... 101967
Jia Society Yiyuan, Beijing ... 101968
Jiading District Museum,
Shanghai 008227
Jian City Museum, Jian, Jilin .. 007929
Jian Museum, Jian 042351
Jian Museum, Jian, Jiangxi ... 007928
Jiangdu Museum, Jiangdu 007932
Jiangmen Museum, Jiangmen . 007935
Jiangnan, Shanghai 102184
Jiangnan Gongyuan History Display
Center, Nanjing 008107
Jiangning Museum, Jiangning .. 007936
Jiangshan Museum, Jiangshan . 007937
Jiangsu, Singapore 085438
Jiangsu Chinese Painting Gallery,
Nanjing 102149
Jiangsu Fine Arts Publishing House,
Nanjing 136732
Jiangsu Literature and Fine Arts
Publishing House, Nanjing 136733
Jiangxi Fine Arts Publishing House,
Nanchang 136731
Jiangxi Provincial Museum,
Nanchang 008095
Jiangxian Museum, Jiangxian .. 007938
Jiangyin Calligraphy and Painting
Gallery, Jiangyin 102136
Jiangyin Museum, Jiangyin 007939
Jian'ou Museum, Jian'ou 007941
Jianyang Museum, Jianyang ... 007942
Jianzhou, Singapore 114548
Jiaoling Museum, Jiaoling 007943

Jiaonan Museum, Jiaonan 007944
Jiaoshan Beike Stone Museum,
Zhenjiang 008540
Jiaosi Museum, Jiaosi 042352
Jiaozhou Museum, Jiaozhou ... 007945
Jiaozhuang Hu Didao Station,
Beijing 007623
Jiaozuo Museum, Jiaozuo 007946
Jiashan Museum, Jiashan 007947
Jiaxing Museum, Jiaxing 007948
Jiayuguan Changcity Museum,
Jiayuguan 007949
Jibac, Tampa 124658
Jibei Cultural Relics Exhibition Hall,
Baisha 042267
Jibreen, Muscat 084400, 113298
Jidaiya, Kobe 081997
Jie Jie Art, Shanghai 102185
Jierjia Sweet Potato Museum,
Taichung 042473
Jiexi Museum, Jiexi 007950
Jiexiu Museum, Jiexiu 007951
Jieyang Museum, Jieyang 007952
Jigsaw Gallery and Bridgetown
Heritage Museum, Bridgetown 000884
Jihočeské Motocyklové Muzeum,
České Budějovice 009505
Jihočeské Muzeum v Českých
Budějovicích, České Budějovice 009506
Jihomoravské Muzeum ve Znojmě,
Znojmo 009990
Jiji Railway Museum, Jiji 042354
Jika Kram, Hjørring 065764
Jikjiseongbo Museum, Gimcheon 029932
Jikyu-an Zen Art, Kyoto 082020
Jilin City Museum, Jilin 007953
Jilin Provincial Museum,
Changchun 007662
Jilin Provincial Museum of Natural
History, Changchun 007663
Jilin Provincial Museum of Revolution,
Changchun 007664
Jilin University Museum,
Changchun 007665
Jilin Wenmiao Museum, Jilin .. 007954
Jillin Fine Arts Publishing House,
Nanjing 136703
Jillings, Doro & John, Newent .. 090829,
134878
Jim & Shirley's Antiques, Long
Beach 094205
Jim Clark Room, Duns 044182
Jim Crow Museum, Ferris State
University, Big Rapids 046978
Jim Gatchell Memorial Museum,
Buffalo 047311
Jim Kempner Fine Art, New York 122746
Jim Savage Art Gallery and Museum,
Sioux Falls 053355
The Jim Thompson House,
Bangkok 042661
Jim Thorpe Home, Oklahoma
Historical Society, Yale 054567
Jiménez Alvárez, Carmen Maria, La
Laguna 133279
Jimenez Garcia, Mario, Madrid . 143766
Jimenez, Christine, Pougues-les-
Eaux 072296
Jiménez, Javier, Madrid 086203
Jimmie Rodgers Museum,
Meridian 050805
The Jimmy Stewart Museum,
Indiana 049674
Jimnette, Tshwane 085676
Jimo Museum, Jimo 007955
Jimpy's Antiques, Nelson 084016
Jim's Antiques, Phoenix 096219, 136117
Jim's Shop, Fresno 135566
Jin, Kitakyushu 111277
Jin, Tokyo 082157
Jin Ancestral Hall Museum,
Taiyuan 008312
Jin Hing, Los Angeles 094377
Jin-Zhi Gallery, Yuanli 116943

Jin, Rong Zhai, Taiyuan 102242
Jinan Museum, Jinan 007956
Jinan Painting Gallery, Jinan .. 102138
Jinan Revolutionary Mausoleum of
Fallen Heroes, Jinan 007957
Jincheng-gu Construction Art Museum,
Jincheng 007962
Jindu, Suzhou 102240
Jing Niao, Shenyang 102213
Jingchuan Museum, Jingchuan . 007963
Jingdezhen Ceramic History Museum,
Jingdezhen 007964
Jingdu Art Cooperative, Beijing . 101969
Jinggangshan Revolution Museum,
Jinggangshan 007965
Jingmen Museum, Jingmen ... 007966
Jingsha Art Museum, Jingsha .. 007967
Jingsha Jingzhou Museum,
Jingsha 007968
Jingtong Mining Museum, Pingsi 042412
Jingu Bijutsukan, Ise 028695
Jingu Hakubutsukan, Ise 028696
Jingu Nogyokan, Ise 028697
Jingxi Chuang Museum, Jingxi . 007969
Jingyang Museum, Jingyang ... 007970
Jingyi, Beijing 101970
Jining Museum, Jining 007971
Jinjiang Museum, Jinjiang 007972
Jinjiuhuang, Shanghai 065140
Jinju National Museum, Jinju .. 029963
Jinlong, Taiyuan 102243
Jinpra, Tokyo 082158
Jinsenko, K.B., Kobe 081998
Jinshan Museum, Shanghai ... 008228
Jinshi Art Guan, Handan 102035
Jinsun, Seoul 111673
Jinta Desert Art Gallery, Sydney 099788
Jintai Art Museum, Beijing 007624
Jinyu Art, Shanghai 102186
Jinze Xuan Art Center, Shenzhen 008263,
102229
Jinzhou District Museum, Dalian 007717
Jinzhou Museum, Jinzhou 007974
Jioufen Kite Museum, Rueifang . 042432
Jiro Osaragi Kinenkan, Yokohama 029713
Jishan Museum, Jishan 007975
Jishui Museum, Jishui 007977
Jitensha Bunka, Tokyo 029520
Jitensha Hakubutsukan Saikuru Senta,
Sakai 029311
Jitter, Bug, Salt Lake City 096896
Jitter, Magazin für Bildgestaltung,
Berlin 138700
Jiujiang Museum, Jiujiang 007978
Jiulio's Furniture Restoration,
Memphis 135779
Jiuquan Museum, Jiuquan 007979
Jivoult, Boris, Vesoul 125838
Jixian Display Center, Jixian ... 007980
Jiyang Museum, Jiyang 007981
Jízdárna Pražského Hradu, Praha 009799
Jizzax Viloyat Ülkashunoslik Muzeyi,
Jizzax 054632
JJ Antiek, Zaandam 083697
JJNT Antiques, Portland 096425
JLG, Veigne 073995
Jllien, Andy, Zürich 116817
JLM Antiquités, Paris 071509
JLM Négoce, Lyon 069751
JLP-Antik, Bratislava .. 085553, 133046
JLR Arts Graphiques, Cachan .. 136804
JLT, Le Bouscat 069227
JM Arts, Paris 104975
JM Galleries, Pittsburgh 123556
JM Modern, San Francisco 097262
JMA Gallery, Wien 100327
JMB Diffusion, Nice 070616
JMJ, Créteil 068240
JMK Artwork, Mount Gravatt .. 061912
JMP Sonorisation, Venelles ... 105836
JMS & Eva, New York 095487
JMS Gallery, Philadelphia 123403
JMW Gallery, Boston 092477

Jo, Seoul 111674
Jo An, 's-Hertogenbosch ... 083391
Jo-An Pictures, New York 122747
Jo Ann's Antiques, Houston 093709
Jo, Hyun, Busan 111567
Joachim Ringelnatz Museum,
Cuxhaven 017789
Joaillerie, Lille 069491
Joaillerie, Francine, Marseille .. 069945
Joalheria, Giam & Carlos, Rio de
Janeiro 064071
Joan Guaita Art, Palma de
Mallorca 115354
Joana Village, Mandelieu-La
Napoule 069832
Joanin, Pau 072043
Joanin, Patricia, Biarritz 103386
Joanny, Jacques, Albi 125486
Joanny, Jacques, Castres 125547
Joan's Antiques, Littlehampton . 089849
Joan's on Miller, North Sydney . 099476
Joan's Ridge Street Art Gallery, North
Sydney 099477
Joantik, Kornwestheim 076942
Joao, Manuel, La Primaube ... 069058
Joao, Manuel, Rodez 072511
Joas, K., Korb 076939
Jobst, Alois, Weilheim in
Oberbayern 108858
Jocham, Anna, Sankt Gallen ... 087782
Jochum, Hans-Peter, Berlin 106324
Jodaline, Austin 119679
Jodie's Gallery, Charlotte 119995
Jodrell Bank Visitor Centre,
Macclesfield 045184
Joe and Emily Lowe Art Gallery,
Syracuse 053634
Joe Kubert School of Cartoon and
Graphic Art, Dover 057020
Joe Upholstry, New York 135942
Joe Weatherly Museum,
Darlington 048152
Jöchler, J., Landeck 140005
Jødisk Museum, Trondheim ... 034033
Jöhle, H., Vöhrenbach 078809
Joel and Lila Harnett Museum
of Art, University of Richmond,
Richmond 052551
Joel and Lila Harnett Print Study
Center, University of Richmond,
Richmond 052552
Joel, Leonard, South Yarra 125234
Joella's House of Antiques,
Omaha 096045, 127379
Jöllenbeck, Köln 107647
Jølstramuseet og Kunstgalleriet,
Vassenden 034074
Jönköpings Antik- och Konsthandel,
Jönköping 086762
Jönköpings Läns Museum,
Jönköping 040719
Jönköpings Skiv och Bokantikvariat,
Jönköping 143868
Jöns-Halvars Antik, Bollnäs ... 086646
Jöns, Karen, Wohlde 079048
Jöns, M. & M., Laatzen 076987, 107777
Joensen, Helgi, Stavanger 084368
Joensuun Taidemuseo, Joensuu 010917
Jörg-Metzler-Stube, Ravenstein . 021304
Jørgens Antik, Ribe 066061
Jørgensen, Birthe, Nykøbing
Sjælland 066019
Jørgensen, Niels Åge, Randers . 066054
Jørgensen, Tamira Mariann,
Silkeborg 066111
Jørgensen, Verner, Helsingør ... 065742
Jörger, Claus-Peter, Berlin 074835
Joe's Coin Shop, Long Beach . 094206
Joe's Cooperative Antiques,
Omaha 096046
Joe's Furniture, Los Angeles ... 094378
Joetsu-shiritsu Hakubutsukan,
Joetsu 028735
Jõevere Talumuuseum, Lindora . 010636

Jövo Háza Központ, Budapest .. 023491
Jofa, Phoenix 123498
Joffo, Roby, Paris 071510
Joffr'Antiq', Aix-en-Provence ... 066466
Jogglejoint Gallery, Regina 064553,
                                 101530
Johan Nygaardsvold Museum,
Hommelvik 033667
Johann-Adams-Mühle, Tholey .. 022150
Johann-Baptist-Graser-Schulmuseum,
Bayreuth 016993
Johann-David-Köhler-Haus, Colditz
Museum, Colditz 017758
Johann Friedrich Böttger und die
Geschichte der Porzellanmanufaktur
Meissen, Schleiz 021661
Johann-Friedrich-Danneil-Museum,
Museum des Altmarkkreises
Salzwedel, Salzwedel 021580
Johann Jacobs Museum, Sammlung
zur Kulturgeschichte des Kaffees,
Zürich 042202
Johann Köleri Muuseum,
Vastemõisa 010712
Johann Michael Haydn-Museum,
Salzburg 002626
Johann-Peter-Frank Museum,
Rodalben 021443
Johann Strauß Wohnung, Wien
Museum, Wien 003010
Johann Ulrich Steiger-Freilichtmuseum,
Bad Zurzach 041249
Johanna Museet, Skurup 040961
Johanna-Spyri-Museum im Alten
Schulhaus, Hirzel 041596
Johannes, Helsinki 140705
Johannes a Lasco Bibliothek,
Emden 018175
Johannes Brahms-Museum,
Mürzzuschlag 002419
Johannes-Brahms-Museum, Johannes
Brahms-Gesellschaft Internationale
Vereinigung, Hamburg 018998
Johannes-Buchhandlung,
München 142166
Johannes Jørgensens Mindestuer,
Svendborg 010333
Johannes Larsen Museet,
Kerteminde 010159
Johannes Linnankosken
Nuoruudenkoti, Askola 010760
Johannes Lohlammen Museo,
Sammatti 011342
Johannes-Molzahn-Centrum für
Documentation und Publication,
Kassel 019535
Johannes Stegmann Art Gallery,
Bloemfontein 038657
Johannesburg Art Gallery,
Johannesburg 038752
Johannessen, Fredrikstad 132773
Johanniter- und Maltesermuseum,
Heitersheim 019177
Johanniter-Museum, Krautheim . 019783
Johanniterhalle, Schwäbisch Hall 021755
Johannitermuseum, Bubikon ... 041374
Johannsen, Chr., Husum 076589
Johannsen, Harry J., Saint Paul 096830
Johanson-Thor, I., Stockholm .. 143951
Johansson, Borås 086647
Johansson, Bo, Gamleby 086686
Johansson, Lars Y., Stockholm . 087014
Johansson, Per-Anders, Aplared 126566
Johansson, Ulla & Arne, Ystad . 087169
Johen, Singapore 085439
Johenneken, U., Bonn 075075
John A. Hermann jr. Memorial Art
Museum, Bellevue 046886
John A. Logan College Museum,
Carterville 047450
John Abbott II House, Hamilton . 049325
John Adams and John Quincy Adams
Birthplaces, Quincy 052435
The John and Mable Ringling Museum

of Art, State Art Museum of Florida,
Sarasota 053149
John B. Davis Gallery of Fine Art,
Pocatello 052223
John Berggruen Gallery, San
Francisco 124311
John Brown Farm, Lake Placid . 050085
John Brown House, Osawatomie 051823
John Brown House, Providence . 052385
John Brown Wax Museum, Harpers
Ferry 049366
John Buchan Centre, Biggar ... 043598
John Bunyan Museum, Bedford 043553
John Burroughs Memorial,
Roxbury 052698
John Creasey Museum, Salisbury 045724
John Curtin Gallery, Curtin University
of Technology, Bentley 000852
John Day Fossil Beds National
Monument, Kimberly 049969
John Deere House, Dixon 048332
John Dony Field Centre,
Bushmead 043815
John E. and Walter D. Webb Museum
of Vintage Fashion, Island Falls 049730
John E. Conner Museum,
Kingsville 049989
John E. Weatherhead Gallery,
University of Saint Francis, Fort
Wayne 048895
John Elliott Classics Museum,
University of Tasmania, Sandy
Bay 001485
John F. Kennedy Presidential Library-
Museum, Boston 047118
John Fitzgerald Kennedy House,
Brookline 047247
John G. Diefenbaker Replica Law
Office, Wakaw 007350
John G. Johnson Collection,
Philadelphia Museum of Art,
Philadelphia 052059
John G. Neihardt Center, Nebraska
State Historical Society,
Bancroft 046773
John G. Voigt House, Plymouth . 052219
John Gorrie Museum,
Apalachicola 046509
John Greenleaf Whittier Home,
Amesbury 046444
John H. Vanderpoel Art Association,
Chicago 060550
John H. Vanderpoel Art Gallery,
Chicago 047668
John Hancock Warehouse, York 054581
John Hansard Gallery, University of
Southampton, Southampton .. 045826
John Harris/Simon Cameron Mansion,
Harrisburg 049371
John Hastie Museum, Strathaven 045909
John Humphrey House,
Swampscott 053622
John Hutchings Museum of Natural
History, Lehi 050216
John J. Harvey Fireboat Collection,
New York 051413
John J. McDonough Museum of Art,
Youngstown 054600
John James Audubon Museum,
Henderson 049438
John Jarrold Printing Museum,
Norwich 045401
John Jay French House,
Beaumont 046858
John Jay Homestead, Katonah . 049901
John King Workshop Museum,
Pinxton 045504
John Knox's House, Edinburgh . 044246
John Lennon Museum, Saitama 029302
John McDouall Stuart Museum,
Kirkcaldy 044727
John Mariani Art Gallery, Greeley 049221
John Marshall House, Richmond 052553
John Michael Kohler Arts Center,

Sheboygan 053296
John Moore Countryside Museum,
Tewkesbury 045963
John Muir's Birthplace, Dunbar . 044159
John N. Leamon Museum,
Brigus 005481
John Nicholas Brown Center for the
Study of American Civilization,
Brown University, Providence . 052386
John P. Barclay Memorial Gallery,
Pittsburgh 052166
John P. Fisher Memorial Museum,
Kingston 006101
John P. McGovern Museum of Health
and Medical Science, Houston 049583
John Paul Jones Birthplace Museum,
Arbigland 043423
John Paul Jones House Museum,
Portsmouth 052311
The John Q. Adams Center for the
History of Otolaryngology – Head
and Neck Surgery, Alexandria . 046392
John R. Jackson House, Winlock 054477
John R. Park Homestead, Harrow 005991
John Rains House, Rancho
Cucamonga 052457
John Ralston Museum,
Mendham 050791
John Rivers Communications Museum,
Charleston 047552
John Rylands University Library-
Special Collections, Manchester 045207
John S. Barry Historical Society
Museum, Constantine 047965
John Sinclair Railway Collection,
Killingworth 044695
John Slade Ely House, New
Haven 051262
John Smart House, Medina County
Historical Society, Medina 050763
The John Southern Gallery,
Dobwalls 044105
John Stark Edwards House,
Warren 054065
John Strong Mansion, Addison . 046320
John Sydnor's 1847 Powhatan House,
Galveston 049035
John Szoke Editions, New York . 138400
John Taylor Bellfoundry Museum,
Loughborough 045153
John-Wagener-Haus Sievern, Langen,
Kreis Cuxhaven 019896
John Walter Museum, Edmonton 005754
John Weaver Sculpture Museum,
Hope 006013
John Wesley Powell Memorial
Museum, Page 051888
John Whitehead Works of Art,
London 090246
John Woolman Memorial, Mount
Holly 051090
John Wornall House Museum,
Independence 049668
John, C., London 090245, 134698
Johnathan Novak Contemporary Art,
Los Angeles 121515
Johndandy, Toronto 101634
Johnen, Berlin 106325
Johnen & Schöttle, Köln 107648
John's Antiques, Saint Louis ... 096720
John's Antiques, Saint Paul ... 096831
John's Antiques Studio, Chicago 092729,
                                 135415

John's Art and Antiques,
Baltimore 092337
John's Book Shop, Mosman .. 139838
The Johns Hopkins University
Archaeological Collection,
Baltimore 046754
John's Krambu, Stavanger 084369
Johns, Derek, London . 090247, 118417
Johnson, Chicago 092730
Johnson, Edmonton .. 101291, 101292
Johnson, Long Beach 135696

Johnson, Memphis ... 094643, 135780
Johnson, West Hollywood 125003
Johnson & Johnson, Seattle ... 097463
Johnson Atelier, Mercerville ... 050803
Johnson Collection of Photographs,
Movies and Memorabilia, Safari
Museum, Chanute 047525
Johnson County Historical Society
Museum, Coralville 047988
Johnson County Historical Society
Museum, Tecumseh 053694
Johnson County Historical Society
Museum, Warrensburg 054071
Johnson County History Museum,
Franklin 048935
Johnson County Museums,
Shawnee 053292
Johnson Ferry House Museum,
Titusville 053752
Johnson-Feugère, Dun-le-
Palestel 068411, 103707
Johnson Hall, Johnstown 049830
Johnson Heritage Post Art Gallery,
Grand Marais 049186
Johnson-Humrickhouse Museum,
Coshocton 048022
Johnson Willard, E., Oklahoma
City 123284
Johnson & Sons, Arthur,
Nottingham 126999
Johnson, David, San Antonio . 096984
Johnson, Ernest, Ottawa 064504
Johnson, Eron, Denver 093303
Johnson, Howard, Saint Louis . 136204
Johnson, Jan, Montréal 140363
Johnson, Lucy, London 090248
Johnson, Margot, New York .. 095488
Johnson, Matthew R., Seattle .. 097462
Johnson, Pat, Houston 093710
Johnson, Peter, Penzance 090965,
134931
Johnson, R. Stanley, Chicago . 120166
Johnson & Sons, Rex,
Birmingham 088481
Johnson, Robert, Brasília 063934
Johnson, Robert R., San
Francisco 097263
Johnson, Sam, Los Angeles .. 145232
Johnson, Samuel, Armadale,
Victoria 098663
Johnson, Sarah, San Francisco . 136327
Johnson, Steve, Newcastle-upon-
Tyne 090827
Johnson, Walker & Tolhurst,
London 090249, 134699
Johnson, William E., Dallas ... 120614
Johnsonese, Chicago 120167
Johnson's, Leek 089782
Johnssen, Carol, München 108145
Johnston, Dublin 079564
Johnston, Hunters Hill 099147
Johnston, Mayen 077310
Johnston, Pittsburgh 123557
Johnston & Shine, Tampa 097564
Johnstone, Saint Paul 096832
Johnstown Area Heritage Association,
Johnstown 049833
Johnstown Historical Society Museum,
Johnstown 049831
Joigny, Renaud de, Nantes 128837
Joint Jacky, Saint-Quentin
(Aisne) 073277
Joint Publishing, Hong Kong ... 136723
Joint Venture Antiques, Saint
Louis 096721
Jojo's Place, Vancouver 064832
Jojo's Treasures, Fort Worth ... 093455
Jojuma! Electronic Art Gallery, El
Paso 120834
Jókai Mór Emlékmúzeum, Petőfi
Irodalmi Múzeum, Balatonfüred 023424
Jókai Mór Emlékszoba, Budapest 023492
Jókai Mór Emlékszoba és Tájház,
Tardona 023996

Jokioisten Museorautatie ja
Kapearaidemuseo, Jokioinen .. 010919
Jokioisten Naulamuseo,
Jokioinen 010920
Jokioisten Pappilamuseo,
Jokioinen 010921
Jokkmokks Stencenter,
Jokkmokk 040724
Joli, Memphis 121697
Joliat, André, Dübendorf 087397
Joliat, André, Zürich 087979
Joliet Area Historical Museum,
Joliet 049835
Jolly Farm Antiques, Villers-sur-
Mer 074180
Jolly, Thomas, Chicago 092731
Joly Antiquités, Namur 063810
Joly, Alain, La Frette 069038
Joly, Claude, Ligueil 069466
Joly, Marc, Belfort 067070
Joly, Patrick, Thezy-Glimont ... 073648
Jolyon Warwick, James,
Woollahra 062474
Jomain, Marie-Claire, Orange .. 070789
Jomamo, Palma de Mallorca ... 086357
Jomier, Daniel, Meaulne 070024
Jomon Bijutsukan, Nakaniida .. 029135
Jomon-Kan, Kizukuri 028837
Jon, New Orleans 095116
Jonard, Robert, Paris 141150
Jonas Hörna, Huskvarna 086753
Jonas Verlag, Marburg 137375
Jonathan Hager House and Museum,
Hagerstown 049310
Jonathan Swift Gallery,
Carrickfergus 117538
Jonavos Krašto Muziejus, Jonava 030339
Jonckers, Sittard 083607
Jones, Gräfelfing 076068
Jones, London 090250, 134700
Jones, Stockholm 143952
Jones & Jacob, Watlington ... 127079
Jones & Vicente, Punta del Este 119445
Jones Art Gallery, Dungarvan . 109632
The Jones Museum of Glass and
Ceramics, Sebago 053253
Jones, Alan & Margaret,
Auckland 083798
Jones, C.V., Ballarat .. 061086, 098694
Jones, Cameron, New Orleans . 122188
Jones, Casey, Louisville 094565
Jones, Christopher, Flore 089286
Jones, Christopher, London ... 090251
Jones & Co., Clark, New York . 095489
Jones, Dave & Pamela,
Yarrahapinni 099938
Jones, Elfed, Colwyn Bay 144352
Jones, Helen, Sacramento 123773
Jones, Herb, Norfolk 123228
Jones, Howard, London 090252
Jones, Jane E., Dallas 120615
Jones, Kevin, Dublin 079565
Jones, Louis & Susan, Norfolk . 123229,
123230
Jones, Louis & Susan, Virginia
Beach 124833
Jones, Michael, Northampton .. 090855
Jones, Paul, Clayfield 061334
Jones, Peter, London 090253
Jones, Solomon, Indianapolis .. 093894
Jones, Solway, Los Angeles ... 094379
Jones, Stella, New Orleans 122189
Jones, Steven, Beaumaris 117320
Jones, Tiffany, Buderim 061227
Jonesborough-Washington County
History Museum, Jonesborough 049840
de Jong, Eindhoven 083220
Jong, A. de, Utrecht 132647
Jong, Anne de, Berlicum 132438
Jong, de, Amsterdam 082773
Jong, Harry, Amsterdam 082774
Jong, Leo de, Rotterdam 112642
Jong, R.F.G.J. de, Den Haag ... 083090
Jong, Ria, Amsterdam 082775

Jong, Sophie, Alkmaar 112107
Jongbloed, A., Leiden 143183
Jonge, Fabery de, Middelburg .. 083497
Jonge, Jon de, Mount Nelson .. 099403
Jonge, Maartje de, Amsterdam . 082776
Jongeneel, Pieter-Pierre, Bergen 112341
Jongert & Jongert, Amsterdam . 112231
Jongh, Matthys de, Zutphen ... 143272
Jongma, Juliette, Amsterdam .. 112232
Jonina, Maastricht 083468
Joniskeit, Reiner D., Stuttgart .. 108710
Joniškio Istorijos ir Kultūros Muziejus,
Joniškis 030340
Jonker, Bert, Zwolle 132676
Jonker, Henk, Egmond ad Hoef . 132498
Jonkers, Christian, Henley-on-
Thames 144468
Jonny's Antiques, Toronto 064683
Jono Biliūno Memorialinis Muziejus,
Anykščiai 030319
Jono Bugailiškio, Vilnius 111797
Jonquet, Etienne, Paris 125722
Jonson Gallery, University of
New Mexico Art Museum,
Albuquerque 046363
Jonsson, K.A., Stockholm 133573
Jonsson, Stig, Leksand 086803
Joods Historisch Museum,
Amsterdam 031948
Joods Museum van Deportatie en
Verzet, Mechelen 003856
Joodse Schooltje, Leek 032563
Joongang, Daegu 111586
Joongang University Museum,
Seoul 030006
Joop, Den Haag 112394
Joos, Ursula, Würzburg 130784
Joosten, Yolanda, Arnhem 132427
Jopie Huisman Museum,
Workum 033063
Jordan, Denver 093304
Jordan Archaeological Museum,
Amman 029770
Jordan-Fahrbach, Eva,
Braunschweig 129624
Jordan Folklore Museum,
Amman 029771
Jordan Historical Museum of the
Twenty, Jordan 006054
Jordan Museum of Popular Tradition,
Amman 029772
Jordan National Gallery of Fine Arts,
Amman 029773
Jordan Natural History Museum,
Irbid 029781
Jordan Schnitzer Museum of
Art at the University of Oregon,
Eugene 048618
Jordan-Witte, Meike, Todendorf . 108769
Jordan, Helmut, Stockelsdorf .. 130661
Jordan, James A., Lichfield 089822,
134556
Jordan, Stefan, München 077535
Jordan, T.B. & R., Stockton-on-
Tees 091448, 119148, 135090
Jorde, Ursula, München 077536
Jordi Ramos, Barcelona 133124
Jordi, Peter & Ursula, Wabern .. 133958
Jordi, Rosa, La Bisbal d'Empordà 085903
Jorge B. Vargas Museum and Filipiniana
Research Center, University of
Philippines, Quezon City 034583
Jorge Barlin National Monument,
Baao 034426
Jorgensen, Dublin 109597
Joris, E.E., Breda 082993
Jorns, Markus, Unterseen 087865
Joroisten Kotiseutumuseo,
Joroinen 010922
Joron-Derem, Christophe, Paris . 125723
Jorvik Museum, York 046224
Jos Art, Amsterdam 112233
Jo's Gallery, Detroit 120809
Jos National Museum and Museum of

Traditional Nigerian Architecture,
Jos 033430
Jósa András Múzeum,
Nyíregyháza 023812
Jose Antonio, Sevilla 086451
Jose Drudis-Biada Art Gallery, Los
Angeles 050439
Jose P. Laurel Memorial Museum,
Manila 034531
José P. Laurel Monument Museum,
Santo Tomas 034593
José Rizal Shrine Museum,
Calamba 034450
José Rizal Shrine Museum,
Dapitan 034473
José Rizal Shrine Museum,
Manila 034532
Josée Bienvenu Gallery, New
York 122748
Josef Albers Museum Quadrat Bottrop,
Moderne Galerie, Museum für Ur-
und Ortsgeschichte, Bottrop .. 017523
Josef Weinheber-Museum,
Kirchstetten 002178
Josefine-Weihrauch-Heimatmuseum
Neudenau, Neudenau 020694
Joseloff Gallery, University of
Hartford, Hartford Art School, West
Hartford 054301
Joselsson, Delft 083033
Joseph, West Hollywood 097902
Joseph & Pearce, London 090254
Joseph A. Cain Memorial Art Gallery,
Del Mar College, Corpus Christi 048005
The Joseph A. Callaway
Archaeological Museum, Southern
Baptist Theological Seminary,
Louisville 050491
Joseph A. Tallman Museum,
Cherokee 047604
Joseph Brant Museum, Burlington's
Museums, Burlington 005501
Joseph Gross Gallery, University of
Arizona, Tucson 053832
Joseph Manigault House,
Charleston 047553
Joseph Misson-Gedenkstätte,
Mühlbach am Mannhartsberg . 002416
Joseph Moore Museum,
Richmond 052534
Joseph Parry's Ironworkers Cottage,
Merthyr Tydfil 045251
Joseph Priestley House,
Northumberland 051639
Joseph Schneider Haus,
Kitchener 006124
Joseph Smith Historic Center,
Nauvoo 051202
Joseph, Paul, Las Vegas 121362
Joseph, R.N., New York 095490
Joseph, Robert, Cincinnati ... 120351,
145074
Josephraj, Austin 119082
Joseph's Antique Repair, Los
Angeles 135724
Joseph's Antiques, Kansas City . 094079
Joseph's Oriental Rug,
Indianapolis 093895
Jose's Rare Coins, Cleveland .. 092948
Josette's Antiques, Las Vegas .. 094155
Josh, Birmingham 119804
Josh, Debenham 089044
Joshi Bijutsu Daigaku-Sagamihara,
Sagamihara 056029
Joshi Bijutsu Daigaku-Suginami,
Tokyo 056046
Joshibi Bijutsukan, Joshi Bijutsu
Daigaku, Sagamihara 029296
Josiah Day House, West
Springfield 054280
Josiah Quincy House, Quincy .. 052584
Josie at Place, New York 095491
Joslowitzer Heimatstube,
Zwingendorf 003148

Joslyn Art Museum, Omaha ... 051765
Joslyn Members Magazine,
Omaha 139405
Josnin, Paris 129044
Jossaume, Jack, Nice 070617
Josseaud, A.P., Haarlem 132519
Josselin-Vecten, Alain, Belloy .. 067081
Jost House Museum, Sydney .. 007115
Jost, Bernhard, Weiningen ... 087894
Jost, Leonhard, Backnang 129401
Jotenkaku Hakubutsukan, Kyoto 028933
Jottier, M., Sint-Niklaas 063832
Jouan, Paris 129045
Joubert, Paris 071511
Joubert, Anne, Hallivillers 128618
Joubert, Jean-Pierre, Paris 104976,
                                 136949
Joubier, Gérard, Quimperlé 072353
Jouëts d'Autrefois Rêves d'Aujourd'hui,
Azay-le-Rideau 011856
Jouets Passion, Liège 063765
Jouffroy, Concarneau 068118
Jouglar, William, Saint-Sauveur . 105615
Jouissance, Pittsburgh 096296
Joumana, Dubai 117181
Jourdain, Mennecy 070055
Jourdan, New York 095492
Jourdan-Bachman Pioneer Farm,
Austin 046694
Jourdan, Franck, Gaillon-sur-
Montcient 068623
Le Journal des Arts, Paris 138649
Journal für Kunstgeschichte,
Karlsruhe 138701
Journal of Aesthetics and Art Criticism,
Cincinnati 139406
Journal of Canadian Art History,
Annales d'histoire de l'art canadien,
Montréal 138584
Journal of Cultural Research in Art
Education, Greenville 139407
Journal of Decorative and Propaganda
Arts, Miami Beach 139408
Journal of Design History, Oxford 139190
Journal of Early Southern Decorative
Arts, Winston-Salem 139409
Journal of Egyptian Archaeology,
London 139191
Journal of Museum Education,
Washington 139410
The Journal of Stained Glass,
London 139192
Journal of the History of Collections,
Oxford 139193
Journal of the Warburg and Courtauld
Institutes, London 139194
The Journal of William Morris Studies,
London 139195
Journée Professionelles Internationales
de l'Antiquités et de la Brocante,
Douai 098086
Journée Professionelles Internationales
de l'Antiquités et de la Brocante,
Lyon 098087
Journée Professionelles Internationales
de l'Antiquités et de la Brocante,
Montpellier 098085
Journées des Antiquaires,
Bourges 098088
Journet, Jean-Claude, Chambéry 067795
Journey East, Singapore 085440
Journey Museum, Rapid City .. 052470
Journeyman Antiques Centre, East
Molesey 089137
Journey's End, Toronto 064684
Journey's End Cottage and Laishley
House, Onehunga Fencible,
Auckland 033151
Journiac, Gilles, La Chapelle-sur-
Oudon 068982
Journiac, Luc, Segré 073476
Jours Anciens, Paris 141151
Jousse Entreprise, Paris 104977
Jousseaume, François, Paris .. 141152

Joutsan Kotiseutumuseo, Joutsa 010923
Joutsen, Taina, Helsinki 140706
Jouvent, Madeleine, Toulouse .. 073777
Jouvent, Mady, Alvignac 066542
Jouvent, Michel, Rignac 072465
Jovanović, Zeljko, Beograd 114421
Jovenart, Madrid 115241
Joverhoff, Freiburg im Breisgau 075870
Jovinas, San Francisco 136328
Jovis-Verlag, Berlin 137075
Joy, Lillebonne 069509
Joy Pratt Markham Gallery and
Cynthia H. Coughlin Gallery, Walton
Arts Center, Fayetteville 048715
Joyas de Ocasion, Madrid 086204
Joyce, Paris 104978
Joyce & Co., Genova 110118
Joyce & Co., Roma 137737
Joyce-Royce, Antwerpen 140149
Joyce, John, Ballina 109527
Joyce, Lincoln, Leatherhead ... 118082
Joyce & Co., Thomas J., Chicago 145047
Joyerot, Gagny 128605
The Joyo Geibun Center, Mito .. 029036
Joy's Antique, Pittsburgh 096297
Joys Antiques, Wellington 084167
Joysmith, Memphis 121698
Jozef Gregor Tajovský Pamätný Dom,
Tajov 038463
Jozefien, Antwerpen 063214
Jozefowski, Maciej,
Korschenbroich 076945, 107727
József Attila Emlékház, Öcsöd .. 023814
József Attila Emlékmúzeum,
Attila József Memorial Museum,
Balatonszárszó 023426
József Attila Emlékszoba,
Budapest 023493
József Attila Múzeum, Makó ... 023769
JP Antik, Audun-le-Roman 066807
JP Antik, Beuvillers 067154
JP Antik, Söderbärke 086941
JP Art, Cannes 103515
JP Art Gallery, Beijing 101971
JPC, Clamart 068001
JRB Art at the Elms, Oklahoma
City 123285
JRM International, Singapore .. 114549
JRS Fine Art, Providence 123707
JS Design, Bergen op Zoom .. 082960
JSC, Paris 104979
JTT-Just the Thing, Westbury . 091793
Ju, Xinjia, Shenzhen 102230
Juan Alberto, Rosario 060952
Juan Luna Shrine Museum,
Badoc 034430
Juanas Cerrada, Fernando,
Madrid 115242
Juanes, Juan de, Alicante 114830
Juaniquet, Jean, Toulouse 073778,
                                 141365
Juanita Studio Gallery, Austinmer 098676
Juaristi Larragon, Maria Pilar,
Barcelona 133125
Jubb, Margaret R., Bristol 088659
Jubilee, Houston 093711
Jubilee Antiques, Tottenhill 091624
Jubilee Hall Antiques Centre,
Lechlade 089760
Jubilee Photographica, London . 090255
Jubilee Terrace Antiques, Bardon 061100
Jucileia, Patricio, Porto Alegre .. 063979
Jud & Co., Felix, Hamburg 141876
Juda, Annely, London 118412
Judah L. Magnes Museum,
Berkeley 046936
The Judah P. Benjamin Confederate
Memorial at Gamble Plantation,
Ellenton 048549
Judaica, Amstelveen 082666
Judaica Classics by Doina, New
York 095493
Judaica Museum of Central
Synagogue, Irvington 049727

The Judaica Museum, The Hebrew
Home, Bronx 047231
Judd Street Books, London 144638
Judenbad, Wetterau-Museum,
Friedberg, Hessen 018516
La Juderia, Sevilla 086452
Judge Roy Bean Visitor Center,
Langtry 050133
Judge's Lodging, Presteigne ... 045564
Judges' Lodgings, Lancaster ... 044750
Judiska Museet i Stockholm,
Stockholm 041008
Judith, Barcelona 085789
Judith Basin County Museum,
Stanford 053490
Judy Garland Museum, Grand
Rapids 049198
Judy's Antiques, Las Vegas ... 094156
Judy's Antiques, Philadelphia .. 096122
Jüdische Galerie Berlin, Berlin . 017182
Jüdische Gedenkstätte und ehemalige
Synagoge, Wallhausen 022411
Jüdisches Kulturmuseum und
Synagoge, Veitshöchheim ... 022306
Jüdisches Kulturzentrum, Fulda . 018562
Jüdisches Museum, Creglingen . 017776
Jüdisches Museum, Frankfurt am
Main 018430
Jüdisches Museum, Göppingen . 018728
Jüdisches Museum, München .. 020534
Jüdisches Museum, Steinbach am
Glan 021977
Jüdisches Museum Berlin, Berlin 017183
Jüdisches Museum der Schweiz,
Basel 041269
Jüdisches Museum der Stadt Wien,
Wien 003011
Jüdisches Museum Emmendingen,
Emmendingen 018183
Jüdisches Museum Franken in Fürth,
Fürth 018458
Jüdisches Museum Franken in
Schnaittach, Schnaittach 021690
Jüdisches Museum Georgensgmünd,
Georgensgmünd 018641
Jüdisches Museum Hohenems,
Hohenems 002097
Jüdisches Museum im Raschi-Haus,
Worms 022763
Jüdisches Museum Rendsburg
und Dr.-Bamberger-Haus,
Stiftung Schleswig-Holsteinische
Landesmuseen Schloß Gottorf,
Rendsburg 021372
Jüdisches Museum Westfalen, Jewish
Museum of Westphalia, Dorsten 017947
Jülich, Alfred, Speyer . 078523, 142368
Jülich, Klaus, Kirchhain 076803
Jülicher, Niederkrüchten 108275
Juemo, Südbrookmerland 108740
Juenemann, Robert, Porto Alegre 063980
Jünger Haus, Langenenslingen . 019909
Jünger, Andrea, Baden bei Wien 099959
Jürgen Valdeig Kunsthandlung und
Verlag, Erfurt 075639, 137178
Jürgens, G., Schloß Holte-
Stukenbrock 108565
Jürgens, U., Quedlinburg 142310
Juers, Dieter, Sylt 078659
Juette, Christian, Saint-Pern ... 073253
Jüttner, Dr. Angelika, Wien ... 100328
Jüttner, Paul, Wernigerode 142478
Juffa, Bad Staffelstein 114944
Juffermans, Utrecht . 083652, 112703,
                                 143246
Jugal Kishore, Chennai 079374
Jugend Museum Schöneberg,
Berlin 017184
Jugendhuset, Varde Museum,
Varde 010350
Jugendkunstschule Pankow,
Berlin 017185
Jugendstil, Beograd 085339
Jugglers Gallery, Foster 099055

Jugoslovenska Galerija Umetnickih
Dela, Beograd 038146
Juhani Ahon Museo, Iisalmi ... 010892
Juhás, Banská Štiavnica 085537
Juhasz, Axel, München 142167
Juhls Silver Gallery, Bergen ... 113135
Juhokusti, Lahti 066381
Juilien, Bernard, Camiers 067620
Juillard, Marcel, Mortagne-au-
Perche 070356
Juillard, Pascal, Saint-Ouen ... 073120
Juillerat, Claude, Hyèvre-Paroisse 068861
Juillet, Michel, Sévrier 073517
Juju's Gallery, Atlanta 119579
Juke Box Classic, Malakoff ... 069820
Jukkasjärvi Museet, Jukkasjärvi 040725
Jukurrpa Artists, Alice Springs . 060991,
                                 098633
Julémont, Paul, Leuven 063733
Jules, Den Haag 083091
Julfa Regional History Museum,
Julfa 003183
Juli, Christiane, Perpignan 072087
Julia A. Purnell Museum, Snow
Hill 053373
Julia C. Butridge Gallery, Austin 046695
Julia Gallery, Taipei 116924
Julia, James D., Fairfield 127219
Julian Antiques, Hurstpierpoint . 089627
Julian Antiques, New York 095494,
                                 135943
Julian Ashton Art School, Sydney 099789
Julian H. Sleeper House, Saint
Paul 052842
Juliana Gallery, Seoul 111675
Julian's Contemporary Art,
Cleveland 120432
JuliArt, Toronto 101635
Julia's Art Studio, Windsor,
Ontario 101856
Julien-Laferrière, Eric, Paris ... 071512
Julien, Anne, Paris 104980
Julienne, Patrick, Saint-Grégoire 072819
Juliet Art Magazine, Trieste ... 138838
Juliette Gordon Low Birthplace,
Savannah 053177
Julio, Jean, Prades 072303
Julistemuseo, Lahden
Kaupunginmuseo, Lahti 011072
Julita Sveriges Lantbruksmuseums,
Julita 040726
Julius, Helsinki 138620
Julius & Ørenberg, Oslo 132787
Julius Gordon Africana Centre,
Riversdale 038822
Jullié, Pierre, Mazamet 070016
Jullien, Bernard, Paris 071513
Jumalon Museum and Art Gallery,
Cebu 034462
Jumbo's, Timaru 084109
Jumeau Yves, Tarascon (Bouches-du-
Rhône) 073621
Juming Museum, Chin-shan .. 042288
Jumoke, Surry Hills 062263
Jumpoff, Madrid 115243
Jun, Philadelphia 123404
Jun Pottery Shu House,
Tongxiang 008341
Junagadh Museum, Junagadh . 024334
Junck, Rossella, Venezia 081846
Junction Antiques, Guildford ... 061566
Junction City Historical Society
Museum, Junction City 049847
Junction Cottage, Pontypool ... 045522
Jundt Art Museum, Gonzaga
University, Spokane 053441
June, West Hollywood 125004
Juneau-Douglas City Museum,
Juneau 049850
Junee Historical Museum, Junee 001171
Juneja Art Gallery, Jaipur,
Rajasthan 109381
Jung, Kaiserslautern 107494
Jung, Albert, Hameln 107327

Jung, Cornelia & Joachim,
  Stuttgart 108711
Jung, H., Herrenberg 076509
Jung, Heinrich, Ginsheim-
  Gustavsburg 076002
Jung, K., Aachen 141396
Jung, Marlise, München 077537
Jung, Peter, Weimar 130738
Jungara Gallery, Port Douglas . . 099580
Jungbauer, Dr. Rainer, Straubing 078575
Die Junge Galerie, Gladbeck . . . 107134
Junge Kunst, Frechen 138702
Junge Kunst, Wolfsburg 022743, 058938
Junge Kunsthalle, Staatliche
  Kunsthalle Karlsruhe, Karlsruhe 019508
das junge museum, Bottrop . . 017524
Junger, Karl-Heinz, Küssaberg . . 076986
Jungersted & Brostrøm,
  Frederiksberg 102805
Junghans, Karl, Berlin . 074836, 137076,
                                    141533
Jungi, Rudolph, Zürich 087980
Jungle Bookshop, Port Douglas . 139875
Jungmann, Eric, Sarrebourg . . . 073416
Jungmun Folklore Museum,
  Seogwipo 029980
Jungschläger, Dordrecht 083183, 132489
Jungvirtová, Jiřina, České
  Budějovice 065370
Jungwirth-Müller, Hildegard, Tulln 140049
Junho Ontwerpen, Den Haag . . 132475
Juniata College Museum of Art,
  Huntingdon 049613
Junion, Moskva 114266
Het Junior, Nieuw Buinen 143204
The Junior Museum, Troy 053812
The Junior Museum of Bay County,
  Panama City 051907
Junior Museum, Newark Museum,
  Newark 051497
JuniorMuseum im Ethnologischen
  Museum, Staatliche Museen
  zu Berlin – Stiftung Preußischer
  Kulturbesitz, Berlin 017186
Juniper Arts, Christchurch 112846
Juniset-Art, Moskva 085182
Junk, Amsterdam 143042
The Junk Box, London 090256
Junk 'n Treasures, Chicago . . . 092732
The Junk Shop, London 090257
Junken-Warnecke, A., Wedel . . 130737
Junker & Partner, Wien 062990
Junkers Bolighus, Haderslev . . . 065719
Junketeria, Fort Worth 093456
Junktion, New Bolingbroke . . . 090809
Junktiques, Calgary . 064260, 128151
Junle, Singapore 114550
Junma, Shanghai 102187
Juno, Brugge 063279
Juno, Gilleleje 065697
Juno, New York 122749
Juno Design Gallery, Dunoon . . 117697
Juno Kunstverlag GmbH, Berlin . 137077
Junod, Danielle, Nyon 116541
Junqua-Lamarque, Alain, Pau . . 072044
Junque, Houston 093712
The Junque Cellar, Edmonton . . 064299
Junque Shop, Detroit 093364
Junqueira 220, Lisboa 132940
Junta Nacional de Arqueología y
  Etnología, La Habana 058531
Junten Gakuen Design Academy,
  Tokyo 056047
Juntiques Fifth Avenue,
  Jacksonville 094000
Juodkrantės Parodų Salėa, Lietuvos
  Dailės Muziejus, Juodkrantė . . 030341
Juozo Gruodžio Memorialinis Muziejus,
  Kaunas 030352
Juozo Grušo Memorialinis Muziejus,
  Maironio Lietuvių Literatūros
  Muziejus, Kaunas 030353
Juozo Naujalio Memorialinis Muziejus,
  Raudondvaris 030452

Juozo Paukštelio Memorialinis
  Muziejus, Kėdainių Krašto Muziejus,
  Kėdainiai 030380
Juozo Tumo-Vaižganto Memorialinis
  Muziejus, Maironio Lietuvių
  Literatūros Muziejus, Kaunas . 030354
Juozo Zikaro Memorialinis Muziejus,
  Nacionalinis M.K. Čiurlionio Dailės
  Muziejus, Kaunas 030355
Juozo Zikaro Memorialinis Muziejus,
  Panevėžio kraštotyros muziejus,
  Paliukai 030439
Jupiter, La Chapelle-la-Reine . . 103880
Jura-Bauernhof-Museum,
  Hitzhofen 019269
Jura-Museum Eichstätt, Eichstätt 018120
Jura-Troc, Porrentruy 087739
Juran & Co., Alexander, London 090258
Juraschek, Winsen, Luhe 108946
Juraschek, Helmut, Lauben . . . 130217
Jurbarko Krašto Istorijos Muziejus,
  Jurbarkas 030344
Jurgeleit, H., Wegberg 078883
Jurica Nature Museum, Benedictine
  University, Lisle 050322
Jurídica Yny, Buenos Aires . . . 139575
Jurinskij Istoriko-chudožestvennyj
  Muzej, Jurino 037075
Juris & Perl, Venezia 111131
Jurjeveckij Kraevedčeskij
  Muzej, Muzej goroda Jurjevca,
  Jurjevec 037076
Jurka, Rob, Amsterdam 112234
Jūrmalas Pilsētas Muzejs,
  Jūrmala 030137
Juro Antiques, Worcester 091947
Jury Moreau, Geneviève, Digoin 068308
Jus Afrika Imports, Tampa 124659
Juschka, Priska C., New York . . 122750
Jussi Björlingmuseet, Borlänge . 040581
Just, Faaborg 102797
Just a Second, London 090259, 134701
Just a Second Variety,
  Philadelphia 096123
Just Antiques, Midland 061851
Just Art, Fochabers 117802
Just Bearly New Orleans,
  Cincinnati 092867
Just Because, Denver 093305
Just Books, Milwaukee 145267
Just Books, Tampa 145615
Just Books, Truro 144890
Just Browsing, Blenheim . . . 083842,
                                    143297
Just Clicked Photo Gallery,
  Hamilton 100826
Just Dandy, Seattle 097464
Just Fine Art, Coleraine 117617
Just Frames and Art, Cleveland 120433
Just Glass, Alston 088264
Just Imagine, Russell 113004
Just Like Grandma's, Tulsa . . . 097743
Just Painting, Meopham 118740
Just Pictures, Saint Louis 123834
Just Plain Positive, Philadelphia 123405
Just Plane, Auckland 083799
Just Scottish Gallery, Edinburgh 117729
Just Stuff, Oakleigh 062021
Just Us, Göteborg 143847
Just, Andreas, Mommenheim . . 130313
Just, F., Niederkassel 108274
Juste à Côté, Lyon 104112
Juste Autrement, Saint-Julien-en-
  Genevois 072865
Justeau, La Boisse 068954
Justice and Police Museum,
  Sydney 001542
Justice Brothers Racing Museum,
  Duarte 048371
Justice, George, Lewes 134551
Justin & Burks, Portland 096426
Justin Smith Morrill Homestead,
  Strafford 053561
Justina M. Barnicke Gallery, Hart

House, University of Toronto,
  Toronto 007183
Justinus-Kerner-Haus, Weinsberg 022521
Justis, Martha, Jacksonville . . 127266
Justuff, Louisville 094566
Justum, Nelly, Cucuron 068256
Justus, Solna 086946
Justus Brinckmann Gesellschaft e.V. –
  Freundeskreis, Hamburg 058939
Jusupov, Sankt-Peterburg 114344
Jusupovskij Dvorec, Sankt-
  Peterburg 037747
Jusupovskij Sad, Sankt-
  Peterburg 085244
Juszczak, Janusz, Poznań 132859
Jute Museum, Kolkata 024362
De Jutter, Delft 083034
't Jutterke, Breda 082994
Jutulheimen Bygdemuseum, Museet i
  Nord-Gudbrandsdalen, Vaagaa 034061
Jutz, Susanne, Celle 106630
Jutzi, Volker, Rohrmoos 130593
Juuan Pitäjänmuseo, Juuka . . . 010926
Juvan Museo, Juva 010929
Juvelir Torg-A, Moskva 085183
Juxian Museum, Juxian 007983
Juxtapoz, San Francisco 139411
Južno-Kazachstanskij Istoriko-
  Kraevedčeskij Muzej, Čimkent . 029817
Jydsk Antikvariat, Århus 140554
Jydsk Boghandel og Antikvariat,
  Skive 140656
Jysk Automobilmuseum, Gjern . 010081
Jyväskylä Lyseon Museo,
  Jyväskylä 010933
Jyväskylän Näkövammaisten Koulun
  Museo, Jyväskylä 010934
Jyväskylän Taidemuseo,
  Jyväskylä 010935
Jyväskylän Vanha Antikvariaatti,
  Jyväskylä 140726
Jyväskylän Yliopiston Museo –
  Keski-Suomen Luontomuseo,
  Jyväskylä 010936
Jyväskylän Yliopiston Museo –
  Kulttuurihistoriallinen Osasto,
  Jyväskylä 010937
JZ Antiguidades, Porto Alegre . . 063981
K & C Agencies, Greenacre . . . 061555
K & G – Keshıshıan, İstanbul . . 088125
K & G Old Orchard Antique Mall,
  Toledo 097608
K & K-Museum, Türnitz 002843
K & K Restoration, San Antonio 136263
K & K Schmuckhandel, Wien . . 062991
K & P Collectibles, Indianapolis 093896
K-B Sales, Fort Worth 093457
K D Antiques, Chester 088862
K-D Did It, Portland 096427
K Galerie, München 108146
K Gallery, Bratislava 114654
K Gallery, Bruxelles 100624
K-Konservator, Göteborg 133528
K Livres, Aix-en-Provence 140759
K Museum, Shimizu 029376
K-Pro Galerie, Brno . . 102570, 125414
K&K Keramikmuseum, Bad
  Radkersburg 001760
K. Banio Etnografinė Žvejo Sodyba-
  Muziejus, Rusne 030458
K. Hajdarov Yodgorlik Muzeyi, Qo'qon
  Viloyat Ülkashunoslik Muzeyi,
  Qo'qon 054640
K.B.M. Antiquités, Paris 071514
K.C. Antiques, Darwen 089033
K.C.S. Paniker Gallery,
  Thiruvananthapuram 024492
K.E. Kivirikon Lintu- ja
  Nisäkäskokoelma, Helsingin Yliopisto,
  Helsinki 010833
K.F.C., Bristol 117448
K.G.O., Zagreb 102526
K.H. Renlundin Museo ja
  Kamerakokoelma, Kokkola . . . 011017

K.K. Gallery, Pickering 118912
K.u.k. Hausmuseum, Weyregg am
  Attersee 002930
k/haus News, Wien 138517
K20 Kunstsammlung am Grabbeplatz,
  K20K21 Kunstsammlung Nordrhein-
  Westfalen, Düsseldorf 018032
K21 Kunstsammlung im Ständehaus,
  K20K21 Kunstsammlung Nordrhein-
  Westfalen, Düsseldorf 018033
K4, Lisboa 132941
K4 Galerie, Saarbrücken 108549
KA, Burgas 101121
Ka-Do-Ha Indian Village Museum,
  Murfreesboro 051122
Ka Lui, Puerto Princesa 113415
Kaaber, København 140611
Kaan Zamaan, Kerikeri 112932
Kaanal, Wijnegem 063856
Kaandy Gallery, South Fremantle 099695
Kaarinan Koulumuseo, Littoinen 011121
Kaarlelan Kotiseutumuseo,
  Kokkola 011018
Kaarlo Kangasniemen Urheilumuseo,
  Kullaa 011048
Kaas, Peter & Rigmor, Viborg . . 066207
Kaasboerderijmuseum de Weistaar,
  Maarsbergen 032612
Kaasmuseum Bodegraven,
  Bodegraven 032074
Kaaswaag Gouda, Gouda 032363
Kaatsmuseum, Franeker 032328
Kaatza Station Museum, Lake
  Cowichan 006154
Kaban ng Hiyas ng Lungsod ng
  Mandaluyong, Mandaluyong . . 034516
Kabanadze, Boris, Berlin 129513
Kabar, Biškek 111748
Kabardino-Balkarskij Gosudarstvennyj
  Muzej Izobrazitelnych Iskusstv,
  Nalčik 037532
Kabarnet Museum, Kabarnet . . 029859
Kabayan National Museum,
  Kabayan 034486
Kabel & Lang, Darmstadt 075304
Kabella, Wolfgang, Paderborn . 078004
Kabinet, Sankt-Peterburg 139011
Kabinet 18, Almere 112116
Kabinet České Grafiky, Praha . . 122699
Kabinet Grafike, Hrvatska Akademija
  Znanosti i Umjetnosti, Zagreb . 009040
Kabinett, Bern 116139
Kabinett für aktuelle Kunst,
  Bremerhaven 106596
De Kabinetten van de Vleeshal,
  Middelburg 032648
Kaboom, New Orleans 145313
Kaboom, Virginia Beach 097785
Al-Kabra, Al-Khobar 085293
Kabul Antique Center, Manila . . 084479
Kabusa Konsthall, Köpingebro . 040754
Kabuth, Markus, Gelsenkirchen . 107111
Kachch Museum, Bhuj 024177
Kachelmuseum De Drie Kronen,
  Boekel 032075
Kachinas, Navaho, Albuquerque 119486
Kachler, Walldorf 078849
Kaczmarczyk, Roman, Kraków . 084579
Kaczmarek, Triberg im
  Schwarzwald 078699
Kaczorowski, Philippe, Nantes . 125658
Kadel, Iris, Karlsruhe 107514
Kadi, Bruxelles 063473
Kadıköy Belediyesi Kültür Merkezi,
  İstanbul 117054
Kadın Eserleri Kütüphanesi ve Bilgi
  Merkezi, İstanbul 060010
Kadina Antiques, Kadina 061662
Kadlec, David, Indianapolis . . . 121193
Kadman Numismatic Pavilion, Eretz-
  Israel Museum, Tel Aviv 025202
Kadrinoff, Léré 069391
Kadrinoff, Alain, Saint-Ouen . . 073121
Kadrioru Loss Väliskunsti Muuseum,

Eesti Kunstimuuseum, Tallinn . 010676
Kaede, Kamakura 111268
Käfer, Hubert, Kronberg im
  Taunus 107758
Käfer, Wilfried, Marburg 077284
Käferstein, Martin,
  Kleinmachnow 130131
Kähler, Cordula, Frankfurt am
  Main 129849
Kähler, Ingrid, Hamburg 076249
Kämmer, Norbert & Hedwig,
  Deidesheim 075322
Kämpf, Waltraud, Bad
  Windsheim 074617
Kaempfer, Friedel & Horst,
  Erftstadt 106890
Kärnan, Helsingborg Museum,
  Helsingborg 040694
Kärntner Druck- und
  Verlagsgesellschaft, Klagenfurt 136526
Kärntner Fischereimuseum,
  Seeboden 002748
Kärntner Freilichtmuseum mit
  Bauernmöbelsammlung, Maria
  Saal 002358
Kärntner Landesgalerie,
  Klagenfurt 002196
Kärsämäen Kotiseutumuseo,
  Kärsämäki 010947
Kaes, Antonia, Holler 130065
Käschtle, Meitingen 077334
Käseglocke, Museum für Worpsweder
  Möbel und Kunsthandwerk,
  Worpswede 022774
Käser, Hans, Metzingen 130302
Käserei-Museum, Endingen am
  Kaiserstuhl 018204
Käsereimuseum, Lochen 002336
Kaess-Weiss, Gudrun, Stuttgart . 108712
Kaesser, Monika, Wien 062992
Kätelhön Druckgrafik, Möhnesee 108012,
                                137381
Käthe-Kollwitz-Museum Berlin,
  Berlin 017187
Käthe Kollwitz Museum Köln,
  Köln 019690
Käthe Kruse-Poppenmuseum, Den
  Helder 032206
Käthe-Kruse-Puppen-
  Museum, Städtische Museen,
  Donauwörth 017934
Käufeler, M., Sursee 116665
Kaffeekannenmuseum,
  Schöppenstedt 021717
Kafkas, Miami Beach 145260
Kafsack, Margareta, Büren 075218
De Kaft, Gent 140223
Kaga-shi Bijutsukan, Kaga 028740
Kagaku Gijutsukan, Nihon Kagaku
  Zaidan, Tokyo 029521
Kagan, Alona, New York 122751
Kagan, Beverlee, Miami 094756
Kaganczuk, Berlin 074837
Kagaribi, Osaka 082083
Kagawa-ken Bunka Kaikan,
  Takamatsu 029431
Kagawa-kenritsu Higashiyama Kaii
  Setouchi Bijutsukan, Sakaide . 029316
Kagawa-kenritsu Rekishi
  Hakubutsukan, Takamatsu ... 029432
Kagedo, Seattle 124550
Kagoshima City Modern Literature
  Museum & Marchen Fairy Tale
  Museum, Kagoshima 028741
Kagoshima-kenritsu Hakubutsukan,
  Kagoshima 028742
Kagoshima-shiritsu Bijutsukan,
  Kagoshima 028743
Kagoshima-shiritsu Kokokan,
  Kagoshima 028744
Kagu no Hakubutsukan, Tokyo . 029522
Kahan, Honolulu 093557
Kahan, Alexander, New York .. 122752
Kahan, Jane, New York 122753

Kahei Hakubutsukan, Nippon Ginko
  Kin'yu Kenkyusho, Tokyo 029523
Kahitsukan – Kyoto Museum of
  Contemporary Art, Kyoto 028934
Kahl, D., Naumburg, Saale 077689
Kahl, Dieter, Berlin 106326, 141534
Kahl, Holger, Oberoppurg 077870
Kahler, Rudolf & Ingeborg, Selb 078450
Kahlhammer, Maria Saal 062730
Kahmen, Dr. Hartmut, Rheinbach 078223
Kahn-Dumousset, Paris 125724
Kahn, Helen R., Montréal 140364
Kahnweilerhaus, Rockenhausen 021439
Kahoku, Seattle 111386
Kahramanmaraş Devlet Güzel Sanatlar
  Galerisi, Kahramanmaraş 042979
Kahramanmaraş Müzesi,
  Kahramanmaraş 042980
Kahutara Canoes and Taxidermy
  Museum, Featherston 033204
Kaiblinger, Siegfried, Sankt Leonhard
  am Forst 100186
Kaieji Buddhist Temple Site Museum,
  Sennan 029364
Kaien, Sandnes 084354
Kaifeng Museum, Kaifeng 007984
Kaigado, Tokyo 111447
Kaigan Bijutsukan, Chikura 028514
Kaikodo, New York 095495, 122754
Kaikohe Pioneer Village and Museum,
  Kaikohe 033232
Kaikoura District Museum,
  Kaikoura 033233
Kaim, A., Köln 076880
Kainuun Museo, Kajaani 010948
Kaiping Huaqiao Museum,
  Kaiping 007987
Kairė-Dešin, Vilnius 111798
Kaisendo Museum, Kaminoyama 028764
Kaiser Franz-Josef-Museum für
  Handwerk und Volkskunst, Baden bei
  Wien 001772
Kaiser Maximilian von Mexiko
  Museum, Hardegg 002056
Kaiser, Anne, Chur 087371, 116219
Kaiser, G., Friesenheim 075917
Kaiser, Rudolf, Weimar 078897
Kaiser, Rüdiger, Frankfurt am
  Main 075796
Kaiserappartements, Sisi Museum-
  Silberkammer, Wien 003012
Kaiserburg Nürnberg, Germanisches
  Nationalmuseum, Nürnberg ... 020862
Kaisergruft, Wien 003013
Kaiserliche Hofburg zu Innsbruck,
  Innsbruck 002125
Kaiserpfalz mit Sankt Ulrichskapelle,
  Goslar 018768
Kaiserpfalzruine, Verwaltung der
  Staatlichen Schlösser und Gärten
  Außenstelle Steinau/Gelnhausen,
  Gelnhausen 018622
Kaisersaal Schloß Schwarzburg,
  Thüringer Landesmuseum
  Heidecksburg Rudolstadt,
  Schwarzburg 021771
Kaiserstühler Heimatmuseum,
  Endingen am Kaiserstuhl 018205
Kaiserstühler Weinbaumuseum,
  Vogtsburg im Kaiserstuhl 022351
Kaiservilla, Bad Ischl 001749
Kaishu Ichizo Museum,
  Aizuwakamatsu 028466
Kaišiadorių Muziejus, Kaišiadorys 030345
Kaitsevää Ühendatud Õppeasutuste
  Muuseum, Tartu 010693
Kaivolan Museo, Laitila 011083
Kaivopuiston Antiikki, Helsinki . 066325
Kaizen Gallery, Broadstairs ... 117453
Kaizen International, Rochester . 091128,
                                134975
Kaja, Jan & Jacek Soliński,
  Bydgoszcz 113447
Kajaanin Taidemuseo, Kajaani . 010949

Kajan, W., Troisdorf 078714
Kaji, Kyoto 082021
Kakabdze Art Gallery, Kutaisi . 016390
Kakamigahara-shiritsu Uchuu
  Kagakukan, Kakamigahara ... 028751
Kake Bijutsukan, Kurashiki 028898
Kake Tribal Heritage Foundation,
  Kake 049855
Kakegawa-shi Ninomaru Bijutsukan,
  Kakegawa 028752
Kakh-e-Rejat va Ebrat & Mellat,
  Tehrän 024765
Kakiemon, Tokyo 082159
Kaklamanou, Chalandri 109057
Kakunodate Museum of Traditional
  Crafts and Industry, Kakunodate 028754
Kakurin-ji Treasure House,
  Katsuura 028795
Kala, Santeny 073408
Kala Bharati Art Foundation,
  Mumbai ...059366, 109448, 130823
Kala Bharati Art School, Mumbai 055820
Kala Bhavana College of Fine Arts and
  Crafts, Santiniketan 055833
Kala-Bhavana Department, Visva
  Bharati University, Birbhum ... 055782
Kala Gallery, Kala Art Institute,
  Berkeley 046937
Kala Kendra, Dehradun 055788
Kala Mandir Art Gallery,
  Hyderabad 109362
Kala Niketan, Kolhapur 055808
Kaláb, Dr. Bedřich, Kroměříž .. 065410
Kalae, Montréal 064435
Kalahari-Oranje Museum,
  Upington 038872
Kalajoen Kalastusmuseo, Kalajoki 010950
Kalajoen Kotiseutumuseo,
  Kalajoki 010951
Kalakriti, Hyderabad 109363
Kalakruti, Chennai 109287
Kalamandir School of the Arts,
  Hanumanthanagar 055800
Kalamazoo Aviation History Museum,
  Kalamazoo 049858
Kalamazoo Institute of Arts,
  Kalamazoo 049859
Kalamazoo Valley Museum, Kalmazoo
  Valley Community College,
  Kalamazoo 049860
Kalamunda History Village,
  Kalamunda 001174
Kalan Torppa, Backnang 010974
Kalanag-Museum, Backnang .. 016691
Kalannin Kotiseutumuseo, Kalanti 010952
Kalantiaw Shrine Museum, Batan 034435
Kalao, La Coucourde 069012
Kalart, San Francisco 124312
Kalash, Lemesos 065329
Kalaupapa Historical Park,
  Kalaupapa 049861
Kalbininko Kazimiero Būgos
  Memorialinis Muziejus, Zarasų Krašto
  Muziejus, Paziegė 030445
Kalbininko, Kunigo Kazimiero Jauniaus
  Kletele-muziejus, Lembo 030412
Kalcher, Margarita, Salzburg ... 062802
Kaldas, Johanna, Frankfurt am
  Main 107007
Kaldewey, Vera & Olaf, Hamburg 129960
Kaleden, Miami 094757
Kaleidoscope, Kansas City 049881
Kaleidoscope, Phoenix 123499
Kaleidoscope, Richmond 096556
Kaleidoscópio, Rio Grande do Sul 140299
Kaleidoskop, Zürich 144184
Kaleidoskop Galerie, Pfungstadt 108379
Kalejdoskop, Åhus 139080
Kaleko Mølle, Faaborg 010055
Kalenbach, Philipp, Albstadt ... 141403
Der Kalenderverlag Mannheim GmbH –
  Weingarten, Unterhaching ... 137543
Kalesse, B., Haynsburg 076423
Kalfas, Barbara, Kraków 132841

Kalfayan Galleries, Athinai 109040
Kalfayan, Arsen & Roupen,
  Thessaloniki 109088
Kalfon, Guy, Paris 071515
Kalfon, Patrick, Saint-Ouen ... 073122
Kalgoorlie Galleries, Kalgoorlie . 099169
Kalgsbrun, Nicole, New York ... 122755
Kali-Bergbaumuseum Volpriehausen,
  Uslar 022290
Kalidoscope Gallery, Calgary .. 101245
Kaligos & Co., N., Thessaloniki . 109089,
               137577,   142544
Kalihi-Palama Culture and Arts Society,
  Honolulu 060551
Kalil & Lauar, Belo Horizonte .. 100889
Kaliman, Paddington, New South
  Wales 099526
Kalimuseum, Buggingen 017678
Kalina, Regen 108454
Kalinčeva Kášta, Trjavna 005227
Kaliningradskaja Chudožestvennaja
  Galereja, Kaliningrad 037081
Kaliningradskij Oblastnoj Istoriko-
  chudožestvennyj Muzej,
  Kaliningrad 037082
Kaliningradskij Oblastnoj Kraevedčeskij
  Muzej, Kaliningrad 037083
Kalinka Antique Gallery, Bondi .. 061170
Kalinowska, Uta, Kraków 084580,
                                143476
Kalinowski, Moorabbin 099372
Kalipong Art Gallery, Mumbai .. 109449
Kalix Flottningsmuseum, Kalix .. 040727
Kalkaska County Historical Museum,
  Kalkaska 049865
Kalkmann, Hans-Werner & Jens, Bad
  Salzdetfurth 106086
Kalkofenmuseum, Steinbach am
  Ziehberg 002780
Kalkofenmuseum, Untermarchtal 022279
Kalkuhl, Hans, Sternberg 108674
Kállay Gyűjtemény, Nyíregyháza 023813
Kallenaution Kievari ja Kotiseutumuseo,
  Juupajoki 010928
Kallensee & Sohn, Kurt, Potsdam 130534
Kaller-Kimche, R., New York ... 122756
Kalligramm, Berlin 141535
Kalligraphie Edition, Hardheim . 137270
Kallinger, Hofkirchen 130060
Kallinger, Dr. Winfried, Wien .. 062993
Kallinger, Johannes, Regensburg 130563
Kallion, Helsinki 066326, 140707
Kallioniemi, Taivalkoski 011388
Kallisté 21, Paris 104981
Kallmann-Museum, Ismaning ... 019432
Kálmán Imre Emlékmúzeum,
  Siófok 023883
Kalmar Antikvariat, Kalmar 143873
Kalmar Antiques, Sydney 062290
Kalmar Konst och Ram, Kalmar 115702
Kalmar Konstmuseum, Kalmar . 040728
Kalmar Läns Museum, Kalmar . 040729
Kalmar Sjöfartsmuseum, Kalmar 040730
Kalmar Slott, Kalmar 040731
Kalmyckaja Gosudarstvennaja
  Kartinnaja Galereja, Ėlista ... 037005
Kalmyckij Respublikanskij
  Kraevedčeskij Muzej im. Prof. N.N.
  Palmova, Ėlista 037006
Kalncempju Pagasta Novadpētniecības
  Muzejs Ates Dzirnavās,
  Kalncempji 030143
Kalnjali, Hyderabad 079390
Kalocsai Érseki Kincstár – Városi
  Képzőművészeti Gyűjtemény,
  Kalocsa 023702
Kalocsai Foegyházmegyei
  Gyűjtemények – Foszékesegyházi
  Könyvtár, Kalocsa 023703
Kalos, Barcelona 114922
Kalos, Taipei 116925
Kalo's Antiquariato, Messina ... 080370,
                                142705
Kalousek, Praha 102700

Kalpa de Boer, Amsterdam 112235
Kalpers, Marie-Josette, Liège .. 063766
Kals, Burkhard, Kirchberg in Tirol 062662
Kalt-Zehnder, Zug 138126
Kalt, F., Riedlingen 078235
Kalteis, Walter, Krems 062688
Kaltenstadler, Hann. Münden ... 076356
Kalthoff, Jürgen, Essen 106918
Kalundborg og Omegns Museum,
  Kalundborg 010156
Kalusche, Dr. Bernd, Frankfurt am
  Main 141776
Kalusky, K.-J., Verden 078789
Kalusteneuvos, Helsinki 066327
Kałużny, Antoni, Kraków 132842
Kalužskij Filial Moskovskogo
  Gosudarstvennogo Universiteta
  Kultury i Iskusstv, Kaluga .. 056275
Kalužskij Oblastnoj Chudožestvennyj
  Muzej, Kaluga 037090
Kalužskij Oblastnoj Kraevedčeskij
  Muzej, Kaluga 037091
Kalvystės Muziejus, Mažosios Lietuvos
  Istorijos Muzieus, Klaipėda .. 030387
Kalwaryjski, Erazm & Polak-
  Kalwaryjska, Alina, Szczecin .. 113778
Kam-Ho, Edmonton 064300
Kamaaina, Honolulu 135578
Kamakura Gallery, Kamakura .. 111269
Kamakura Kindai Bijutsukan,
  Kamakura 028755
Kamakura Kokuhokan, Kamakura 028756
Kamakura-shi Bungaku,
  Kamakura 028757
Kamakuragu Homotsukan,
  Kamakura 028758
Kamali, New York 095496
Kaman, Tampa 124660
Kamčatskij Oblastnoj Kraevedčeskij
  Muzej, Petropavlovsk-
  Kamčatskij 037647
Kamčatskij Voenno-istoričeskij Muzej,
  Petropavlovsk-Kamčatskij 037648
Kameja, Samara 085225
Kameleon, Warszawa 113855
Kamelot, Philadelphia 127393
Kamen, Leo, Toronto 101636
Kamena Gallery, Edmonton ... 101293
Kamenný Dům, České Muzeum
  Stříbra, Kutná Hora 009657
Kameo, Tokyo 111448
De Kamer, Breda 082995
Kamera, Kraków 113605
Kamera- und Fotomuseum Leipzig,
  Leipzig 019996
Kamera-lehti, Helsinki 138621
Kamernyj Šaljapinskij Zal, Nacionalnyj
  Kulturnyj Centr Kazan, Kazan . 037101
Kames, Brodick and Whiting Bay 088683
Kami no Hakubutsukan, Tokyo . 029524
Kamienica Celejowska, Muzeum
  Nadwiślańskie, Kazimierz Dolny 034823
Kamienica Hipolitów, Muzeum
  Historycznego Miasta Krakowa,
  Kraków 034874
Kamienica pod Gwiazdą, Muzeum
  Okręgowe w Toruniu, Toruń .. 035283
Kamigata Ukiyo-e Bijutsukan,
  Osaka 029244
Kaminokuni Mingeikan,
  Kaminokuni 028763
Kamińska-Samek, Krystyna,
  Kraków 143477
Kaminski House Museum,
  Georgetown 049082
Kaminski, Frank C., Beverly ... 127136
Kamiss, Houston 093713
Kam'jans'ka Kartynna Galereja,
Kam'jans'kyj deržavnyj
  istoriko-kul'turnyj zapovidnyk,
  Kam'janka 043168
Kam'jans'kyj Deržavnyj
  Istoriko-kul'turnyj Zapovidnyk,
  Kam'janka 043169

Kam'jans'kyj Istoryčnyj Muzej,
Kam'jans'kyj deržavnyj
  istoriko-kul'turnyj zapovidnyk,
  Kam'janka 043170
Kaml, Isabella, Kammern im
  Liesingtal 127784
Kaml, Isabella, Wien 127933
Kamloops Art Gallery, Kamloops 006056
Kamloops Museum & Archives,
  Kamloops 006057
Kamm, Bettina, München 130359
Kamm, Erika, Ennenda 087409
Kammari, Alexandran, Helsinki . 066328
Kammer, Renate, Hamburg 107282
Kammerer, Christina, Erlangen . 075653
Kammerer, Jutta, Graz 100010
Kammerer, W., Heidelberg 076451
Kammerhof – Baníctvo na Slovensku,
  Slovenské Banské Múzeum, Banská
  Štiavnica 038286
Kammerhof Museen Gmunden,
  Gmunden 001953
Kammerhofgalerie der Stadt Gmunden,
  Gmunden 001954
Kammerhofmuseum, Bad Aussee 001734
Kammerlander, Walter, Dornbirn 062549
Kammermann & Amrein-Kammermann,
  Kriens 087546, 133783
Kamniški Muzej, Kamnik 038522
Kamogawa-shiritsu Tougeikan,
  Kamogawa 028766
Kamoliddin Bekhzad National Insitute
  of Arts and Design, Academy of Arts
  of Uzbekistan, Toshkent 058147
Kamolz, Karin, Braunschweig .. 106563
Kamon Antiques, Tokyo 082160
Kamp, Chicago 120168
Kamp-Lintforter Münzfreunde, Kamp-
  Lintfort 058940
Kamper Tabaksmuseum, Kampen 032534
Kampf, Heinz-Hubertus, Muri bei
  Bern 087693, 144114
Kampffmeyersche Holzlöffelsammlung,
  Bad Dürkheim 016750
Kamphaeng Phet National Museum,
  Kamphaeng Phet 042690
Kamphans, T., Beelen 074710
Kamphuis, J., Zwolle 112733
Kampintorin Antikvariaatti,
  Helsinki 140708
Kampmann, Henrik, København 102901
Kampschulte, Ingrid, Friedrichsthal,
  Saar 129868
Kamratowski, Bernhard, Berlin . 074838
Kamsack and District Museum,
  Kamsack 006063
Kamthieng House, Bangkok ... 042662
Kan, Sendai 082114
Kan Kan Kyo Art Gallery, Tokyo . 111449
Kan Kikuchi Kinenkan,
  Takamatsu 029433
Kan, Toyo, Tokyo 082161
Kana Art, Beograd 085340
Kanabec History Center, Mora .. 051048
Kanagawa-kenritsu Kanazawabunko
  Museum, Yokohama 029714
Kanagawa Kenritsu Kindai Bijutsukan,
  Hayama 028615
Kanagawa-kenritsu Rekishi
  Hakabutsukan, Yokohama 029715
Kanagawa-kenritsu Seimei-no-Hoshi
  Chikyu Hakubutsukan, Odawara 029201
Kanagawa Kindai Bungakukan,
  Yokohama 029716
Kanagawa Prefectural Gallery,
  Yokohama 029717
Kanai-machi Hakabutsukan,
  Kanai 028767
Kanalgalleriet, Notodden 113187
Kanalmuseet, Trollhättan 041089
Kanalmuseum Skantzen,
  Hallstahammar 040681
Al-Kananah, Dubai 088180
Kanatzer, Fort Worth 135557

Kanavamuseo, Varkaus 011497
Kanazawa Bijutsu Kogei Daigaku,
  Kanazawa 056005
Kanazawa Daigaku Hakubutsukan,
  Kanazawa 028775
Kanda Kotetsudo, Tokyo 082162
Kandahar Provincial Museum,
  Kandahar 000006
Kandavas Novada Muzejs,
  Kandava 030144
Kandeil, Cairo 066252
Kandeloos Museum, Kojoor ... 024715
Kandelous Museum, Kandelous . 024703
Kandinsky, San José 102509
Kandiyohi County Historical Society
  Museum, Willmar 054428
Kandler, Toulouse 105746
Kandos Bicentennial Industrial
  Museum, Kandos 001177
Kandy National Museum, Kandy 040485
Kandyan Antiques, Colombo .. 086612,
                            086613
Kane, Boston 119877
Kane, Marie, Virginia Beach .. 124834
Kane, Paul, Dublin 109598
Kaneesha Art Gallery,
  Ahmedabad 109248
Kaneesha Boutique, Ahmedabad 109249
Kanegendo, Sendai 082115
Kanekichi, Nagoya 082060
Kaneko, Yokohama 142949
Kanel, Gilbert, Savigneux 073450
Kanel, Marie-José, Saint-Étienne 072766
Kanelhylden, Gilleleje 065698
Kanemori Museum with Baccarat
  Collection, Hakodate 028590
Kaner & Kaner, Freiburg im
  Breisgau 141805
Kanev, Den Haag 112395
Kanfer, Larry, Minneapolis ... 122005
Kang, New York 095497
Kang Byung, New York 122757
Kang Da, Shanghai ... 065141, 065142
Kang Xicao Museum, Beijing ... 007625
Kang, Philip, Seoul 111676
Kanga Books, Chifley 139696
Kangaroo House, Hurstbridge . 099149
Kangaroo Island Gallery,
  Kingscote 099197
Kangasalan Kotiseutumuseo,
  Kangasala 010953
Kangaslammin Kotiseutumuseo,
  Kangaslampi 010956
Kangasniemen Kotiseutumuseo,
  Kangasniemi 010957
Kangomar, Kraków 084581
Kangourou, Pulnoy 072325
Kania, Bad Füssing 106057
Kaniv College of Culture and Arts,
  Kaniv 056489
Kaniv's'kyj Istoryčnyj Muzej,
  Kaniv 043173
Kanivs'kyj Muzej Narodnogo
  Dekoratyvnogo Mystectva,
  Kaniv 043174
Kanizsai Dorottya Múzeum,
  Mohács 023795
Kankaanpään Kaupunginmuseo,
  Niinisalo 011186
Kankakee County Historical Society
  Museum, Kankakee 049869
Kanken Kanji Hakubutsukan,
  Kyoto 028935
Kanlungan ng Sining, Manila ... 113407
Kannada Research Institute Museum,
  Karnataka University, Dharwad 024246
Kanne & Kruike, Rotterdam .. 083573
Kannenberg-Rentschler, Manfred,
  Berlin 141536
Kannengießer, Ute, Wien 127934
Kannonkosken Kotiseutukokoelma,
  Kannonkoski 010959
Kannuksen Museo, Kannus .. 010960

Kraków 113606
Kanra Ancient Art Museum,
  Kanra 028780
Kansai, Osaka 111359
Kansainvälinen Kahvikuppimuseo,
  Posio 011278
Kansainvälinen Pullomuseo,
  Sonkajärvi 011370
Kansallispukukeskus, Suomen
  Käsityönmuseo, Jyväskylä 010938
Kansas African American Museum,
  Wichita 054388
Kansas Aviation Museum,
  Wichita 054389
Kansas City Art Institute, Kansas
  City 057279
Kansas City Art Institute Gallery,
  Kansas City 049882
Kansas City Artists Coalition, Kansas
  City 060552, 121299, 138343
Kansas City Collectibles, Kansas
  City 094080
The Kansas City Jewish Museum,
  Overland Park 051847
Kansas City Legends, Kansas
  City 094081
Kansas City Museum, Kansas
  City 049883
Kansas Cosmosphere and Space
  Center, Hutchinson 049645
Kansas Health Museum, Kansas
  Learning Center for Health,
  Halstead 049320
Kansas Museum of History,
  Topeka 053773
Kansas Museums Association,
  Topeka 060553
The Kansas Teachers' Hall of Fame,
  Dodge City 048336
Kansas Watercolor Society,
  Hutchinson 060554
Kansetsu Hashimoto Bijutsukan,
  Hakusasonso Villa, Kyoto 028936
Kant- en Textielcentrum, Diest .. 003548
Kant, Renate, Singapore 133037
Kanta Museum Argungu,
  Argungu 033415
Kantcentrum, Brugge 003415
Kante, Mariana, Bruxelles 063474
Kantelberg, Jill, Toronto 064685
Kantipur Art Gallery, Kathmandu 112076
De Kantlijn, Bredevoort 143098
Kantmuseum, Uithuizen 032935
Kantmuseum Olsene, Olsene ... 003919
Kantner, Hermann, Wien 140093
Kantonales Fischereimuseum Mettlen,
  Netstal 041790
Kantonales Museum Altes Zeughaus,
  Wehrhistorisches Museum,
  Solothurn 042002
Kantonsmuseum Baselland,
  Liestal 041678
Kantor, Los Angeles 121516
Kantor & Feuer, Los Angeles .. 121517
Kantor Kupiecki Antyk, Gdańsk . 084507
Kántor Sándor Fazekasház,
  Györffy István Nagykun Múzeum,
  Karcag 023717
Kantor, Jacques, Paris 071516
Kantz Aranyozó, Budapest ... 130811
Kanuma Municipal Art Museum of
  Kawakami Sumio, Kanuma ... 028781
Kanumis, Lyon 069752
Kanzeon-ji Temple Treasure House,
  Dazaifu 028523
Kanzler, Wilhelm, Hof bei
  Salzburg 127769
Kao Chong, Trang 042726
Kao Luang, Nakhon Si
  Thammarat 042702
Kao Ta Phet, Sukhothai 042720
Kao Yai, Saraburi 042717
Kaohsiung Astronomical Museum,
  Kaohsiung 042360

Kaohsiung County Museum of Natural
History, Kao Yuan University,
Lujhu 042387
Kaohsiung Museum of Fine Arts,
Kaohsiung 042361
Kaohsiung Museum of History,
Kaohsiung 042362
Kapan Local Tradition Museum,
Kapan 000709
Kapan Museum of Contemporary Art,
Kapan 000710
Kapenguria Museum, Kapenguria 029860
Kaperdalen Samemuseum, Midt-Troms
Museum, Senja 033934
Kapfer, Niederwaldkirchen 062749
Kapfhamer, Ingrid, Aichach 141402
Kaphammel, Thomas,
Braunschweig 106564
Kaphammel, Thomas, Goslar . . 107150
Kapilavastu Museum, Lumbani . 031881
Kapiolani Community College,
University of Hawaii, Honolulu . 057210
Kapisa Provincial Museum and
Archeological Site, Mahmud-
i-Raqi 000007
Kapitän Tadsen Museum,
Langeneß 019912
Kapiteinshuis Pekela, Nieuwe
Pekela 032673
Kapiti Coast Museum Waikanae,
Wellington 033347
Kaplan, Austin 092242, 135326
Kaplan, Stockholm 126673
Kaplan & Son, Alec, Sydenham . 085668
Kaplan, Howard, New York 095498
Kaplan, Leo, New York 095499, 095500,
122758
Kaplanstöckl, Hohenzell 002099
Káplár Miklós Gyűjtemény, Hajdúsági
Múzeum, Hajdúböszörmény . . 023683
Kaple Sv. Kříže, Praha 009800
Kapoor, New York 122759
Kapopoulos, Eyag., Athinai 079167
Kaposvar Historic Site, Our Lady of
Assumption Church Museum,
Esterhazy 005791
Kappa Pi International Honorary Art
Fraternity, Cleveland 060555
Kappan Hakubutsukan, Tokyo . . 029525
Kappatos Gallery, Athinai 109041
Kappel jun., Günther, Bad
Mitterndorf 062529
Kappel, Manfred, Salzburg 062803,
127851
Kappelhoff, Herbert, Emmerich am
Rhein 075612
Kappler, Chris, Seattle 097465
Kappler, Doris, Darmstadt 106656
Kapucijnenmuseum, 's-
Hertogenbosch 032468
Kapula, B., Hamburg 107283
Kapunda Museum, Kapunda . . . 001179
Kapusta, Matej, Kremnica 133052
Kara, Paris 071517
Kara Art, Jussy 116382, 138057
Karabela, Katowice 084533
Karačaevo-Čerkesskij Istoriko-kulturnyj
i Prirodnyj Muzej-Zapovednik,
Čerkessk 036947
Karacolskij Istoričeskij Kraevedčeskij
Muzej, Karacol 030080
Karagandinskij Oblastnoj
Istoriko-Kraevedčeskij Muzej,
Qaraghandy 029838
Karagandinskij Oblastnoj Muzej
Izobrazitelnogo Iskusstva,
Qaraghandy 029839
Karaimų Etnografinė Paroda, Trakų
Istorijos Muziejus, Trakai 030490
Karajian, Jean, New York 095501
Karakter, Amsterdam 082777
Karam, Jorge Alberto, Porto
Alegre 100985
Karaman Müzesi, Karaman 042982

Karamanlidis, Theodoros, Berlin 141537
Karambamba & Attalidou,
Lefkosia 102538
Karan Art Gallery, Jaipur,
Rajasthan 109382
Karascho GbR, Siebke & Könitz,
Essen 106919
Karasek-Museum,
Seifhennersdorf 021830
Karatay Çini Eserler Müzesi,
Konya 043004
Karatepe-Aslantaş Açıkhava Müzesi,
Osmaniye 043042
Karausz, Jorge, Montevideo . . . 092003,
119431
Karava, Maria, Athinai 079168
Karavanserai, Maastricht 112579
Karbstein, Peter, Düsseldorf . . . 125930
Kard Bar Prints & Posters, Newcastle-
upon-Tyne 118788
Kardemir Demir Çelik Müzesi,
Karabük 042981
Kardinal-Bea-Museum, Blumberg 017426
Kardinolo Vincento Sladkevičiaus
Memorialinis Butas-Muziejus,
Kauno Arkivyskupijos Muziejus,
Kaunas 030356
Kardos, Péter, Zürich 144185
Kare Sanat Galerisi, İstanbul . . . 117055
Kareem, Atlanta 119580
Karelskij Gosudarstvennyj
Kraevedčeskij Muzej,
Petrozavodsk 037652
Karen Blixen Museet, Rungsted
Kyst 010297
Karen Blixen Museum, Nairobi . 029866
Karen's Collectible Dolls, Miami 094758
Karen's Creative Centre,
Singapore 114551
Karen's Gifts, Louisville 094567
Karepa Kalame Talumuuseum,
Vihula 010716
Kareum – Gamla Bilsalongen,
Sparreholm 040983
Karfunkel, Groß Kummerfeld . . . 107170
Karg, Natalie, New York 122760
Karges, William A., West
Hollywood 125005
Kargl, Georg, Wien 100329
Kargopolskij Gosudarstvennyj Istoriko-
Architekturnyj i Chudožestvennyj
Muzej, Kargopol 037096
Kariandusi Prehistoric Site Museum,
Gilgil 029857
Karibil Cottage, Griffith 061562
Karich, S., Üxheim 078750
Karijoen Kotiseutumuseo, Karijoki 010961
Karikatür ve Mizah Müzesi,
İstanbul 042948
Karikatürcüler Derneği Sanat Galerisi,
İstanbul 117056
Karikatur und Cartoon Museum Basel,
Basel 041270
Karikaturmuseum Krems, Krems 002233
Karins Antik Stube, Kerpen 076761
Kariya-shiritsu Bijutsukan, Kariya 028782
Kariye Müzesi, Ayasofya Müzesi,
İstanbul 042949
Karjalainen Kotitalo, Imatra 010900
Karjalan Kotiseutumuseo,
Mynämäki 011175
Karjalan Portti, Ky, Rauha 066393
Karkaralinsky Rajonnyj
Istoriko-Kraevedčeskij Muzej,
Karkaralinsk 029821
Karkkilan Ruukkimuseo, Karkkila 010963
Karkmann, Heidi, Hamburg 076250
Karl & Faber Kunstauktionen,
München . . . 077538, 126055, 142168
Karl Bacher-Stüberl, Zellerndorf 003138
Karl Drerup Fine Arts Gallery, Dept of
Fine Arts, Plymouth State College,
Plymouth 052216
Karl-Ernst-Osthaus-Bund, Hagen

Kunstverein, Hagen 058941
Karl Ernst von Baeri Muuseum,
Tartu 010694
Karl Heinrich Waggerl-Haus,
Wagrain 002884
Karl-Hofer-Gesellschaft, Freundeskreis
der Universität der Künste Berlin e.V.,
Berlin 058942
Karl-Marx-Buchhandlung, Berlin 141538
Karl-May-Haus, Hohenstein-
Ernstthal 019311
Karl-May-Museum, Karl-May-Stiftung,
Radebeul 021257
Karl-Otto-Braun-Museum,
Ludwigshafen am Rhein 020134
Karl Ristikivi Majamuuseum,
Tartu 010695
Karl the Twelfth Sweedish Antiques,
West Hollywood 097903
Karl und Magdalene Haberstock
Stiftung, Kunstsammlungen und
Museen Augsburg, Augsburg . 016658
Karl-Wagenplast-Museum,
Schwaigern 021757
Karl-Wörn-Haus, Schwetzingen . 021801
Karl, Helmut, Stephanskirchen . 130659
Karla Antik, Stockholm 087015
Kārļa Skalbes Memoriālais Muzejs
Saulrieti, Vecpiebalga 030250
Kārļa Ulmaņa Piemiņas Muzejs Pikšas,
Bērze 030105
Karle, M., Staufen im Breisgau . 108667
Karlebo Museum, Kokkedal 010204
Karlenström, Borås 115589
Karlenström, Göteborg 115655
Karlie, Philadelphia 123406
Karlinger-Schmiede,
Unterweissenbach 002858
Karlmann, Philipp, Absam 062512
Karlovarské Muzeum, Karlovy
Vary 009619
Karlovské Muzeum, Velké
Karlovice 009944
Karlshamns Museum, Karlshamn 040723
Karlskoga Konsthall, Karlskoga . 040736
Karlsson, Trelleborg 087102
Karlsson, Leo, Frankfurt am Main 075797
Karlsson, Magnus, Stockholm . . 115897
Karlucci, Saint Paul . . . 096833, 136221
Karma, Cairo 103048
Karma, Den Haag 112396
Karmal-Skillington, Nashville . . . 095009
Karmazyn, Kraków 084582
Karmøy Fiskerimuseum,
Vedavågen 034076
Karmsund Folkemuseum,
Haugesund 033643
Karnatak Fine Arts College, Karnataka
University, Dharwad 055794
Karnataka Chitrakala Parishath,
Bangalore 055779, 109260
Karnataka Government Museum
and Venkatappa Art Gallery,
Bangalore 024150
H. Karner & M Kosco, Wien . . . 127935
Karnevalmuseum,
Mönchengladbach 020438
Karnevalsmuseum, Blankenheim 017411
Karnowsky, Merry, Los Angeles . 121518
Karo, Heidelberg 076452
Karol Plicka Múzeum, Slovenské
Národné Múzeum, Blatnica . . . 038295
Károli Gáspár Múzeum és
Bibliakiállítás, Gönc 023650
Karolina, Akureyri 109219
Karoline Antikt, Oslo 084322
Karoly, Harry, Kritzendorf 100087,
127797
Karoly, Harry, Wien . . . 100330, 127936
Károlyi-Kastély, Fehérvárcsurgó . 023634
Károlyi-Kastély, Füzérradvány . . 023640
Karos, Milwaukee 094865
Karottenmuseum-Musée de la Carotte,
Raeren 003958

Karoutsos, Sotirios, Bologna . . . 130969
Karowski, M., Rosbach vor der
Höhe 108512
Karp, J. Louis & Phyllis T.,
Cincinnati 092868, 127176
Karp, Jules, New York 095502
Karpa, L., Roßla 078278
Karpatendeutsches Museum,
Karlsruhe 019509
Karpeles Manuscript Museum,
Buffalo 047306
Karpeles Manuscript Museum,
Charleston 047554
Karpeles Manuscript Museum,
Duluth 048383
Karpeles Manuscript Museum,
Jacksonville 049772
Karpeles Manuscript Museum,
Tacoma 053645
Karpf, Helmut, Wiener Neustadt 063114
Karpio & Facchini, Miami Beach 121883
Karpisek, Yves, Ceyras 067758
Karrasburg Museum Coswig,
Coswig 017762
Karrebæk, Anne, København . . . 102902
Karrenmuseum, Essen 003584
Karrer, Jürgen, Berlin 074839
Kars Devlet Güzel Sanatlar Galerisi,
Kars 042984
Kars Müzesi, Kars 042985
Karsenty, Jean-Louis, Paris 104982
Karsh-Masson Gallery, Ottawa . 006591
Karst- und Höhlenkundliche Abteilung,
Naturhistorisches Museum Wien,
Wien 003014
Karst, Jean, Schaan 132320
Karst, Michael, Heide 076429
Karsten, L., Bielefeld 141599
Karstens, Mike, Münster 108221
Karstulan Kotiseutumuseo,
Karstula 010965
Kartäusermuseum Tückelhausen,
Ochsenfurt 020942
Kartause Aggsbach, Aggsbach
Dorf 001689
Kartause Gaming, Gaming 001923
Karthaus-Prüll Museen des
Bezirksklinikums Regensburg,
Regensburg 021323
Kartinnaja Galereja A.A. Plastova,
Uljanovskij oblastnoj chudožestvennyj
muzej, Uljanovsk 037993
Kartinnaja Galereja K. Vasiljeva,
Kazan 037102
Kartinnaja Galereja Ministerstva
Kultury Respubliki Adygeja,
Majkop 037217
Kartinnaja Galereja Pejzažej P.M.
Grečiškina, Stavropol 037897
Kartinnaja Galerija im. P.I. Šolochova,
Borisoglebsk 036911
Kartiny-Baget-Antikvariat,
Moskva 085184, 085185
Kartnaller, Anton, Innsbruck . . . 062650
Das Kartoffelmuseum, Stiftung Otto
Eckart, München 020535
Karttulan Kotiseutumuseo,
Karttula 010966
Kartuli Khelovneba, Tbilisi 105906
Kartynna Galereja,
Kam'janec'-Podil's'kyj 043167
Kartynna Galereja, Volyns'kyj
Krajeznavčyj Muzej, Luc'k . . . 043226
Karuizawa Wild Flower Museum of Art,
Karuizawa 028784
Karvian Museo ja Ulkomuseoalue,
Karvia 010968
Karwah, Toronto 101637
Kary, Al, Fortaleza 100952
Karyes-Protaton, Agion Oros . . . 022945
Kaş Galerisi, İstanbul 117057
Kasagi, Kamakura 111270
Kasama Inari Bijutsukan, Kasama 028789
Kasama Nichido Bijutsukan,

Kasama 028790
Kaschenbach, Peter, Trier 108777
Kaschner-Wendland, Werdohl . . 078928
Kasden, Los Angeles 094380
Kashba, Amsterdam 082778
Kasher, Steven, New York 122761
Kashi Zone Silu Museum, Kashi 007989
Kashima-jingu Homotsukan, Kashima,
Ibaraki 028793
Kashishian, Long Beach 094207
Kashma House, Nelson 084017
Kashmir Art Emporium, Manama 100437
Kashmir Art Palace, Chennai . . . 079375
Kashmir Carpets, Hyderabad . . . 079391
Kasia, Kay, Chicago 120169
Kasimir, Wien 062994
Kasimir's, Brugge 063280
Kaskade Infohaus Wasser, Jade 019439
Kaskarots, Ciboure 067990
Kaskö Hembygdsmuseum,
Kaskinen 010970
Kasmin, Paul, New York 122762, 122763
Kasori Kaizuka Hakubutsukan,
Chiba 028510
Kaspar, Toronto 101638
Kaspar, Hanne, Voerde 142456
Kasper, Dr. Alfons, Bad
Schussenried 137036
Kasper, Ralf, Kiel 076777
Kaspers & Co., Den Haag 083092
Kasr Tin Hinan, El Madania 060837
Kass & Meridian, Chicago 120170
Kass Galéria, Móra Ferenc Múzeum,
Szeged 023926
Kass, Bertrand, Innsbruck 100049
Kassai, R. & P., New York 095503,
135944
Kassák Múzeum, Petőfi Irodalmi
Múzeum, Budapest 023494
Kasseler Kunstverein e.V., Kassel 058943
Kaßler, Winfried, Langenhagen . 077025
Kast, Thomas, Röfingen 078260
Käšta-muzej Akademik Dončo Kostov,
Nacionalen Selskostopanski Muzej
Sofia, Lokorsko 005082
Käšta-muzej Aleksandär Mutafov,
Sozopol 005208
Käšta-muzej Angel Känčev,
Trjavna 005228
Käšta-muzej Angel Karalijčev, Dom na
Detskata Kniga, Sofia 005173
Käšta-muzej Asen Razcvetnikov,
Draganovo 005020
Käšta-muzej Asen Zlatarov,
Chaskovo 005003
Käšta-muzej Canko Cerkovski, Bjala
Čerkva 004988
Käšta-muzej Chadži Dimitär, Okräžen
Istoričeski Muzej, Sliven 005166
Käšta-muzej Christo Botev,
Kalofer 005044
Käšta-muzej Christo G. Danov,
Istoričeski Muzej, Plovdiv 005128
Käšta-muzej Christo Smirnenski,
Nacionalen Muzej na Bälgarskata
Literatura, Sofia 005174
Käšta-muzej Dimcho Debeljanov,
Koprivštica 005069
Kašta-muzej Dimitär Blagoev,
Bankja 004976
Käšta-muzej Dimitär Blagoev,
Sofia 005175
Käšta-muzej Dimitär Dimov,
Nacionalen Muzej na Bälgarskata
Literatura, Sofia 005176
Käšta-muzej Dimitär Pešev, Istoričeski
Muzej, Kjustendil 005063
Käšta-muzej Dimitär Poljanov,
Karnobat 005051
Käšta-muzej Dobri Voinikov,
Šumen 005214
Käšta-muzej Emilijan Stanev, Veliko
Tärnovo 005254
Käšta-muzej Geo Milev, Stara

Zagora 005209
Käšta-muzej Georgi Benkovski,
Koprivštica 005070
Käšta-muzej Georgi Dimitrov,
Kovačevci 005078
Käšta-muzej Iljo Vojvoda, Istoričeski
Muzej, Kjustendil 005064
Käšta-muzej Ivan Lazarov, Nacionalna
chudožestvena galerija, Sofia . 005177
Käšta-muzej Ivan Vazov,
Berkovica 004986
Käšta-muzej Ivan Vazov, Nacionalen
muzej na bälgarskata literatura,
Sofia 005178
Käšta-muzej Jordan, Jovkov,
Žeravna 005269
Käšta-muzej Lamartine, Plovdiv 005129
Käšta-muzej Ljuben Karavelov,
Koprivštica 005071
Käšta-muzej Neofit Rilski,
Bansko 004978
Käšta-muzej Nikola Ěnkov Vapcarov,
Bansko 004979
Käšta-muzej Nikola Vapcarov,
Nacionalen Muzej na Bälgarskata
Literatura, Sofia 005179
Käšta-muzej Panaët Chitov,
Sliven 005167
Käšta-muzej Panaët Volov,
Šumen 005215
Käšta-muzej Pančo Vladigerov,
Šumen 005216
Käšta-muzej Pejo Javorov, Nacionalen
Muzej na Bälgarskata Literatura,
Sofia 005180
Käšta-muzej Pejo K. Javorov,
Čirpan 005009
Käšta-muzej Petko i Pencho
Slavejkovi, Nacionalen Muzej na
Bälgarskata Literatura, Sofia . . 005181
Käšta-muzej Rajna Knjaginja,
Panagjurište 005103
Käšta-muzej Rusi Čorbadži,
Žeravna 005270
Käšta-muzej s Dom-pametnik Jordan
Jovkov, Dobrič 005019
Käšta-muzej Sava Filaretov,
Žeravna 005271
Käšta-muzej Slavejkovi, Trjavna 005229
Käšta-muzej Stanislav Dospevski,
Pazardžik 005108
Käšta-muzej Stojan i Vladimir Zaimovi,
Pleven 005117
Käšta-muzej Svetoslav Obretenov,
Provadia 005141
Käšta-muzej Todor Kableškov,
Koprivštica 005072
Käšta-muzej Zachari Stojanov,
Medven 005089
Käšta-muzej Zachari Stojanov,
Ruse 005152
Kastamonu Arkeoloji Müzesi,
Kastamonu 042986
Kastamonu Devlet Güzel Sanatlar
Galerisi, Kastamonu 042987
Kastaniealle 1a, Sønderborg . . . 066139
Kasteel Achtendries, Gent 003628
Kasteel Amerongen, Amerongen 031906
Kasteel Ammersoyen,
Ammerzoden 031917
Kasteel Beauvoorde,
Wulveringem 004157
Kasteel Cannenburch, Vaassen . 032960
Kasteel De Doornenburg,
Doornenburg 032242
Kasteel de Haar, Collectie Th. Baron
Van Zuylen Van Nijevelt Van de Haar,
Haarzuilens 032401
Kasteel Doorwerth, Doorwerth . . 032243
Kasteel Duivenvoorde,
Voorschoten 033006
Kasteel Groeneveld, Baarn 032044
Kasteel Heeswijk, Heeswijk-
Dinther 032439

Kasteel Heeze – Collectie H.N.C.
Baron Van Tuyll Van Serooskerken
Van Heeze en Leende, Heeze . 032441
Kasteel Hernen, Hernen 032467
Kasteel Het Nijenhuis, Museum de
Fundatie, Heino 032449
Kasteel Hoensbroek, Hoensbroek 032495
Kasteel Middachten, De Steeg . 032149
Kasteel-Museum Sypesteyn,
Loosdrecht 032603
Kasteel Ooidonk, Deinze 003535
Kasteel Radboud, Medemblik . . 032639
Kasteel Rosendael, Rozendaal . . 032834
Kasteel van Beersel, Beersel . . . 003355
Kasteel van Bornem, Bornem . . 003383
Kasteel van Gaasbeek, Gaasbeek 003608
Kasteel van Horst, Sint-Pieters-
Rode 004023
Kasteel van Leeuwergem,
Zottegem 004166
Kasteel van Wijnendale, Torhout 004077
Kasteelmuseum Slot van Laarne,
Laarne 003769
Kasteelruïne Valkenburg, Valkenburg,
Limburg 032963
Kasteeltoren IJsselstein,
IJsselstein 032524
Kaštela Gradski Muzej, Kaštel
Lukšić 008926
Kaštelet Crikvine, Ivan Mestrović
Foundation, Split 008997
Kastelholms Slott, Sund 011378
Kasten & Kisten, Dordrecht 083184
Kasten, Dr. Friedrich W.,
Mannheim 107969
Kastern, Hannover 076378, 125985,
141915
Kaštiel a Kysuce Galéria,
Oščadnica 038418
Kaštiel' Imre Madácha a Literárne
osobnosti, SNM-Múzeum Kultúry
Mad'arov na Slovensku, Dolná
Strehová 038344
Kastner, Honolulu 120946
Kastoria Byzantine Museum,
Kastoria 023120
Kastrupgårdsamlingen, Kastrup . 010157
Kasuga Taisha Homostuden,
Nara 029144
Kasumisou, Tokyo 082163
Kasuo, Iwamura Pierre, Valognes 105814
Kasvimuseo, Turku 011436
Kasvimuseo, Luonnontieteellinen
Keskusmuseo, Helsinki 010834
Kat, Gilles, Beaulieu-sur-Mer . . 066998
Katachi, Hamburg 076251
Katamatite Museum, Katamatite 001180
Katanning Historical Museum,
Katanning 001181
Katarina-Antikvariatet, Stockholm 143953
Kate, Hemswell Cliff 089500
The Kate Chopin House and Bayou
Folk Museum, Cloutierville . . . 047849
Kate Shelley Railroad Museum,
Boone 047088
Kate, Joyce, Seattle 124551
Katebi, G., Köln 076881
Katedra Vytvarné Vychovy, Jihočeska
univerzita, Pedagogická fakulta,
České Budějovice 055297
Katedralna Riznica, Split 008998
Katelman, Omaha 096047, 136063
Kateri Galleries, National Shrine
of the North American Martyrs,
Auriesville 046669
Kates, Mary, Auckland 083800
Kathareze, Tilburg 112680
Katharina-Luther-Stube, Stadt- und
Kulturgeschichtliches Museum
Torgau, Torgau 022167
Katharinas Puppenhaus,
Pleystein 021156
Katharine House Gallery,

Marlborough . 090722, 118722, 144729
Katherine Art Gallery, Katherine 099173
Katherine Mansfield Birthplace,
Wellington 033348
Katherine Museum, Katherine . . 001182
Katherine Nash Gallery, University of
Minnesota, Minneapolis 050941
Katherine's Collection, Dallas . . 093147
Kathleen Cullen Fine Arts, New
York 122764
Kathmandu – Hot Spring in the Art
Museum, Yuzawa 029752
Kathryn's Antiques, Portland . . . 096428
Kathy's Antiques, Long Beach . . 094208
Katia, Genova 110119
Katia, Paris 071518
Katina Paxinou Museum, Athinai 022993
Katiny Muziejus, Šiauliai 030467
Katir, Maria, Rosenheim 078273
Katlin, Auckland 083801
Kato, Kyoto 082022
Kato, Tokyo 111450
Katona József Emlékház,
Kecskemét 023725
Katona József Múzeum,
Kecskemét 023726
Katonah Museum of Art, Katonah 049902
Katong, Singapore 085441
Katoomba Antiques, Katoomba . 061672
Katoomba Fine Art, Katoomba . . 099176
Katori-jingu Homotsukan, Sawara 029351
Katowicki Dom Sztuki, Katowice 084534,
113513
Katowicki Salon Antyków,
Katowice 084535
Katowickie Stowarzyszenie Artystyczne,
Katowice 059696, 137891
Katrineholms Hembygdsmuseet,
Kristineholm 040762
Kats, Rotterdam 083574
Katschtaler Heimatmuseum,
Rennweg 002588
Katsura, Paris 071519, 104983
Katsushika Hokusai Bijutsukan,
Tsuwano 029663
Katsushika-ku Kyoudo to Tenmon no
Hakubutsukan, Tokyo 029526
Katsuyama Kenbunkan,
Katsuyama 028797
Katten Kabinet, Amsterdam . . . 031949
Kattenburg, Rob, Amsterdam . . 082779,
143043
Kattlunds Museigård, Havdhem . 040688
Katumuseo, Helsingin
Kaupunginmuseo, Helsinki . . . 010835
Katwijks Museum, Katwijk aan
Zee 032540
Katxibatxes, Donostia-San
Sebastián 085976
Katy's Art Gallery, Indianapolis . 121194
Katz, Atlanta 119581
Katz, Boston 119878
Katz, Allie, Tampa 097565
Katz, Daniel, London 090260
Katz, David, Basel 116081
Katz, Karla, New Orleans 095117
Katzbichler, Bernd, Wiesenfelden 142497
Katzen, C.Y., Washington 124927
Katzenmeyer, Günther, Feldkirch 062560
Katzenmuseum, Traunstein 022173
Katzer, Wolfgang, Augsburg . . . 074464
Kauai Children's Discovery Museum,
Kapaa 049899
Kaua'i Museum, Lihu'e 050278
Kauba, Indianapolis 121195
Kauffman Gallery, Shippensburg 053315
Kauffman Museum, North
Newton 051609
Kauffmann, Gudrun, Möttingen . 108026
Kaufman, Tel Aviv 079735
Kaufman, Inna, Amsterdam 112236
Kaufman, Sam, Los Angeles . . . 094381
Kaufmann, Toulouse 073779
Kaufmann, Elisabeth, Zürich . . . 116818

Kaufmann, Francesca, Milano – Kellogg Historical Society Museum, Kellogg

Index of Institutions and Companies

Kaufmann, Francesca, Milano .. 110367
Kaufmann, Stefan, Grünstadt .. 076118
Kaufmannsmuseum, Heimatverein Haslach, Haslach 002060
Kauhajoen Museo, Kauhajoki .. 010972
Kauhavanmuseo, Kauhava 010975
Kauhl, H., Hückelhoven 076580
Kauka, Giesenhausen 107125
Elisabeth Kaul & Leokadia Streibel, Gerabronn 075979
Kaulbach-Haus, Museum Bad Arolsen, Bad Arolsen 016703
Kaunislehdon Talomuseo, Hyrynsalmi 010890
Kauno Arkivyskupijos Muziejus, Kaunas 030357
Kauno Galerija, Kaunas 111784
Kauno IX Forto Muziejus, Kaunas 030358
Kauno Langas, Kaunas 111785, 111786
Kauno Lėlių Muziejus, Kaunas .. 030359
Kauno Paveikslų Galerija, Nacionalinis M.K. Čiurlionio Dailės Muziejus, Kaunas 030360
Kauno Tado Ivanausko Zoologijos Muziejus, Kaunas 030361
Kaupel, Jürgen, Recklinghausen 078155
Kaupp, Karlheinz, Sulzburg 126127
Kauppamakasiini, Helsinki 140709
Kauppilan perinnetalo, Humppila 010888
Kauppilan Umpipiha, Laitila 011084
Kauppilanmäen Museo, Valkeakoski 011481
Kauri Books, Birkenhead 143296
The Kauri Museum, Matakohe . 033244
Kaurilan Koulumuseo, Tohmajärvi 011423
Kaustisen Kotiseutumuseo, Kaustinen 010977
Kausynoé, Bruxelles 063475
Kautsch, Veronica, Michelstadt . 107999
Kauttuan Tehtaan Museo, Kauttua 010979
Kautzsch & Sohn, Schöppenstedt 078387
Kavalerfløjens Antikviteter og Kunst, Kerteminde 065812
Kavalet, Varna 101194
Kavalierhaus Gifhorn – Museum für bürgerliche Wohnkultur, Museen des Landkreises Gifhorn, Gifhorn .. 018697
Kavalkad Antikt och Kuriosa, Örebro 086903
Kavanagh Antiques, Montréal .. 064436
Kavanagh, E., South Crossmalina 130892
Kavanagh, Kevin, Dublin 109599
Kaw Mission, Council Grove ... 048036
Kawagoe-shi Bijutsukan, Kawagoe 028798
Kawagoe-shiritsu Hakubutsukan, Kawagoe 028799
Kawaguchi-ko Motor Museum, Narusawa 029157
Kawaguchi-shiritsu Kagakukan, Kawaguchi 028800
Kawaguchiko Bijutsukan, Kawaguchiko 028803
Kawai Kanjiro's House, Kyoto .. 028937
Kawakami, Gasendo, Kyoto 111297
Kawamura Bijutsukan, Yuda ... 029748
Kawamura Memorial Museum of Art, Sakura 029324
Kawanabe Kyosai Memorial Museum, Warabi 029694
Kawasaki, Kyoto 082023
Kawasaki Juvenile Science Museum, Kawasaki 028806
Kawasaki Peace Museum, Kawasaki 028807
Kawasaki-shi Hakubutsukan, Kawasaki 028808
Kawasaki-shiritsu Nihon Minkaen, Kawasaki 028809
Kawase, Yukiko, Paris 104984
Kawhia Regional Museum Gallery, Kawhia 033235
Kawilihan Art Gallery, Mandaluyong 113399
Kawsara Fall, Amsterdam 082780
Kay Kee, Hong Kong 065059
Kay, Frances, San Diego 124137
Kay, Gary, Armadale, Victoria .. 061037
Kay & Assoc., Hanna, San Francisco 124313
Kaya Steam Locomotive Square, Kaya 028813
Kayafas, Boston 119879
Kaye, Anne, Providence 096509
Kaye, Lita, Lyndhurst 090665
Kayes, Tshwane 085677
Kayes of Chester, Chester 088863, 134252
Kayliash Lacroix, Alain, Lutry .. 087626
Kayo, San Francisco 145559
Kay's Antiques, Claremont 085638
Kay's Antiques, Tulsa 097744
Kayser, Enrico, Marburg 077285
Kayseri Arkeoloji Müzesi, Kayseri 042992
Kayseri Devlet Güzel Sanatlar Galerisi, Kayseri 042993
Kayseri Etnografya Müzesi, Kayseri 042994
Kayu, Hohenlinden 076559
Kazakstan Paleolit Muzeji, Al-Farabi Atyndagy Kazak Ylttyk Universiteti, Almaty 029800
Ķazaķstan Respublikasynyŋ Memlekettik Ävilchan Ķasteev Atyndaghy Öner Muzeji, Almaty 029801
Kazama, Pizançon 072167
Kazanlăška Roza, Gradski istoričeski muzej, Kazanlăk 005058
Kazanskaja Gosudarstvennaja Akademija Iskusstv i Kultury, Kazan 056276
Kazanskaja Starina, Kazan 085149
Kazantzakis Museum, Iráklion .. 023097
Kazari, Prahran 062113
Kazari, Richmond, Victoria 062146
Kaze, Osaka 111360
Kazemattenmuseum, Oosterend 032723
Kazemattenmuseum Kornwerderzand, Kornwerderzand 032554
Kazerooni, Manama 063147
Kazinczy Ferenc Emlékcsarnok, Sátoraljaújhely 023877
Kazinczy Ferenc Múzeum, Sátoraljaújhely 023878
Kazio Varnelio Namai-Muziejus, Lietuvos Nacionalinis Muziejus, Vilnius 030520
Kazmine, Michelle, Montceau-les-Mines 104259
Kazuaki Iwasaki Space Art Gallery, Ito 028711
Kazuno Cultural Property Preservation Center, Kazuno 028815
Kazuo Kotera, Kyoto 082024
Kazuo Museum, Chaoyang 007686
KB Art, Cosenza 109990
KB Auksjoner, Hamar 126351
KCM Auctions, Seattle 127453
KD Coins, Cincinnati 092869
KdF-Museum in der Kulturkunststatt Prora, Prora 021226
Ke-Shan, Taipei 116926
Ke, Chua, Beijing 101972
Kealley's Gemstone Museum, Nannup 001337
Keane Antiques, Dunmanway .. 079612
Keane on Ceramics, Kinsale ... 109660
Kearney Area Children's Museum, Kearney 049909
Kearney County Historical Museum, Minden 050929
Kearney Mansion Museum, Fresno 048980
Kearney & Sons, T.H., Belfast .. 088433, 134091
Kearny Cottage, Perth Amboy .. 051999
Keating, J.J., Kennebunk 127274
Keating, Michael, San Francisco 124314
Keating, Nevill, London 118413
Keats Gallery, Shanklin 119073
Keats House, London 045012
Keats-Shelley House, Roma ... 027570
De Kebof, Zwolle 083733
Kebon Binatang Bandung, Bandung 024529
Kechichian, Claude, Paris 104985
Keck, Dieter, Deggingen 129673
Keck, Siegfried, Dillingen an der Donau 106681
Keckler & Donovan, Seattle ... 124552
Kedah Royal Museum, Alor Setar 030660
Kedah State Art Gallery, Alor Setar 030661
Kėdainių Krašto Muziejus, Kėdainiai 030381
Kedzie Coins, Chicago 092733
Kędzierski, Ryszard, Wrocław .. 143543
Kee, Chan Shing, Hong Kong .. 065060, 065061
Kee & Co., Cheong, Singapore . 085442
Kee, Meng Cheng, Singapore .. 114552
Keeble, Langport 089747, 144513
Keeble, Ottawa 064505
Keegan, Peter, Dublin 130859
Keel, Birmingham 135354
Keel Row, North Shields 144745
Keel, Magdolna, Killwangen ... 116385, 116386
Keeler Tavern Museum, Ridgefield 052570
Keelung Mid-Summer Ghost Festival Museum, Keelung 042371
Keene, San Antonio 124043
Keene, Barry M., Henley-on-Thames 089508, 117961, 134455
Keens, Surbiton 119182
Keepsakes, Innerleithen 089640
Keepsakes, Regina 101531
Keepsakes, Saint Louis 096722
Keepsakes & Kollectibles, Houston 093714
De Keersmaeker, Sint-Niklaas .. 063833
Keetmanshoop Museum, Museum Association of Namibia, Keetmanshoop 031840
Keevil, Frances, Woollahra 099918
Keewatin Maritime Museum, Douglas 048344
Kegel-Konietzko, Boris, Hamburg 076252
Kegel, Gerhard, Buchholz in der Nordheide 106608
Kegelmann, P. Michael, Frankfurt am Main 075798, 125948
Kegworth Museum, Kegworth .. 044666
Kegyeleti Szakgyűjtemény, Budapesti Temetkezési Intézet, Budapest 023495
Kehdinger Küstenschiffahrts-Museum, Wischhafen 022691
Kehoe, Lesley, Melbourne 061819
Kehrein, Peter, Neuwied 142230
Kehrer, Klaus, Heidelberg 137273
Kehwa, Amsterdam 082781
Kehwa Art, Amsterdam 112237, 137836
Kehys ja Kultaus, Turku 066403
Kehysaitta, Turku 066404
Kei, Amersfoort 112124
Kei Fujiwara Bijutsukan, Bizen . 028504
Kei Mino, Tokyo 082164
Kei, Wing, Hong Kong 065062
Keiflin, Roger, Saint-Raphaël ... 105605
Keigado, Fukuoka 081967
Keighery, R.J., Waterford 079719
Keighley and Worth Valley Railway Museum, Haworth 044548
Keijzer, Jan, Sint Maartensvlotbrug 132629
Keijzer, T. de, Amersfoort 112125
Keiko, Boston 119880
Keiko Bijutsukan, Hirara 028641
Keikyän Kotiseutumuseo, Keikyä 010980
Keil, New Orleans 095118
Keil, Dr. Robert, Wien . 062995, 100331, 125277
Keil, H.W., Broadway .. 088678, 134171
Keil, John, London 090261
Keilhacker, Ludwig P., Taufkirchen, Vils 130699
Keillor House Museum, Dorchester 005713
Keim, Stuttgart 108713
Keinath & Decurtins, Zürich ... 087981
Keindl, Landshut 107792
Keio Daigaku Bijutsu Senta, Tokyo 029527
Keio Gallery, Tokyo 111451
Keip & von Delft, Stockstadt am Main 142377
Keir Memorial Museum, Malpeque 006250
Keis og Bichel, Frederiksværk .. 140575
Keishodo, Kamiya, Osaka 082084
Keitele-Museo, Suolahti 011382
Keitelman, Maurice, Bruxelles .. 100625
Keith Harding's World of Mechanical Music, Northleach 045392
Keith Museum, Keith 001183
De Keizerin, Eindhoven 083221
Kekavas Novadpētniecības Muzejs, Kekava 030147
Kékfestő Múzeum, Pápa 023832
Keladi Museum, Keladi 024343
Kelantan Malay Arts and Crafts, Kelantan 082398
Kelderman, Lith 143195
Kelet, Bruxelles 100626
Kelham Island Museum, Sheffield 045773
Kelifa, Gilles, Marseille 128775
Kell House Museum, Wichita Falls 054397
Kelleher, Oliver, Blarney 130830
Kellenberger, Eric, Blonay 087328, 116163, 144056
Keller, Bautzen 141455
Keller, Dinkelsbühl 075361
Keller Galerie, Donauwörth 106689
Keller Galerie, Zürich 116819
Keller-Pfyl, Sonja, Basel 087239
Keller Restauro, Birmenstorf ... 133700
Keller, Dieter, Luzern .. 087643, 133836
Keller, Fritz, Stuttgart . 078609, 142394
Keller & Assoc., Kelsey, Tampa . 124661
Keller, Robert, Kandern 107505
Keller, Simson, Güglingen 076120
Keller, Tracy, New Farm 099426
Keller, Ursula, Mannheim 107970
Keller, W., Soden 130643
Keller, Wolfgang, Sulzbach am Main 078647
Kellergalerie – Büchergilde Wiesbaden, Wiesbaden 108919, 142488
Kellergedenkstätte Krippen, Stadtverwaltung Bad Schandau, Bad Schandau 016865
Kellermann, Essen 106920
Kellermann, Detlef, Aachen 074334
Kellermuseum Preßhaus, Großkrut 002010
Kellerviertel Heiligenbrunn, Freilichtmuseum, Heiligenbrunn 002072
Kelley, Atlanta 092145
Kelley, Cincinnati 120352
Kelley, San Antonio 124044
Kelley House Museum, Mendocino 050792
Kelley, Anthony, New Orleans .. 122190
Kelley, Dale, Chicago 092734
Kellie Castle, Pittenweem 045509
Kellie's Antiques, Brisbane 061211
Kelliher and District Heritage Museum, Kelliher 006069
Kellner, Gottfried, Theresienfeld . 127883
Kellner, Hans, München 130360
Kellner, Michael, Hamburg 137247
Kello, Charles, Norfolk 123231
Kellogg Historical Society Museum,

Kellogg 049914
Kellokosken Sairaalan Museo,
Kellokoski 010982
Kellross Heritage Museum,
Leross 006183
Kellum, København 102903
Kelly, Albuquerque 135285
Kelly, Saint Louis 096723
Kelly Antiques, Rodley 091131
Kelly-Griggs House Museum, Red
Bluff 052489
Kelly-Randall, Cleveland 120434
Kelly, Alison, Armadale, Victoria 098664
Kelly, Frank R., Sheffield 091285
Kelly, June, New York 122765
Kelly, Sean, New York 122766
Kelly's Refinishing, Jacksonville 135656
Kellyville Antiques, Kellyville . . . 061673
Kelmès Krašto Muziejus, Kelmé 030382
Kelmscott, Baltimore 144985
Kelmscott House Antiques,
Waikato 084131
Kelmscott Manor, Kelmscott . . . 044671
Kelnhof-Museum, Bräunlingen . . 017532
Kelowna Art Gallery, Kelowna . . 006074
Kelowna Museum, Kelowna 006075
Kelsey Museum of Archaeology, Ann
Arbor 046490
Kelsey, Ken, Auckland 132691
The Kelso House Museum, Kent 049931
Kelsterbacher Museum für Porzellan,
Kelsterbach 019578
Kelston Fine Arts, Bath 117309
Kelten Römer Museum Manching,
Manching 020242
Keltendorf Mitterkirchen,
Mitterkirchen 002395
Keltenmuseum, Hallein 002050
Keltenmuseum Gracarca, Sankt
Kanzian 002670
Keltenmuseum Hochdorf/Enz,
Eberdingen 018072
Keltenpark Schwarzenbach,
Schwarzenbach 002734
Kelter- und Weinbaumuseum,
Niederstetten 020801
Die Kelter-Galerie, Notzingen . . 108289
Kelterborn, Etta & Klaus,
Schortens 078395
Kelton House Museum,
Columbus 047939
Kelvedon Hatch Secret Nuclear Bunker,
Kelvedon Hatch 044673
Kelvingrove Art Gallery and Museum,
Glasgow 044404
Kemajuan Kraftangan Malaysia,
Rawang 030725
Kembery Antique Clocks, Bath . 088390,
134077
Kemble, Rodney F., Glusburn . 134408
Kemeny, Ulrike, Wangen im
Allgäu 078863
Kemerer Museum of Decorative Arts,
Bethlehem 046960
Kemerovskij Gosudarstvennyj
Universitet Iskusstv i Kultury,
Kemerovo 056277
Kemerovskij Istoriko-architekturnyi
Muzej Krasnaja Gorka,
Kemerovo 037123
Kemerovskij Oblastnoj Kraevedčeskij
Muzej, Kemerovo 037124
Kemerovskij Oblastnoj Muzej
Izobrazitelnych Iskusstv,
Kemerovo 037125
Kemijärven Kotiseutumuseo,
Kemijärvi 010988
Kemin Kirjadivari, Kemi 140729
Kemin Kultuurihistoriallinen Museo,
Kemi 010984
Kemin Taidemuseo, Kemi 010985
Kemin Työläismuseo ja
Kähertäjämuseo, Kemi 010986

Keminmaan Kotiseutumuseo,
Keminmaa 010989
Kemle, H., Heidelberg 076453
Kemmer, Eberhard, Hamburg . . 141877
Kemmerich, M., Lindlar 077124
Kemp-McCarthy Memorial Museum,
Rowe 052696
Kemp & Assoc., Karl, New York 095504,
095505
Kemp, Melville, Nottingham . . . 090895,
134899
Kempe-Wolter, Freia, Gräfelfing . 076069,
107153
Kempeleen Kotiseutumuseo,
Kempele 010990
Kempenaar, Haarlem 083334
Kemper Museum of Contemporary Art,
Kansas City 049884
Kempf, Stephan J., Mollis 087677,
116506, 144112
Kempin, Hermann, Hannover . . . 107370
Kempisch Museum, Brecht . . . 003397
Kempkens, Siegfried, Düsseldorf 075485
Kemps, Bristol 088660
Kempsey Antiques, Kempsey . . 061676
Kempsey Museum, Kempsey . . 001184
Kempton Old Books, Kempton . . 013789
Ken Domon Museum of Photography,
Sakata 029320
Ken et Dy, Salon-de-Provence . . 105646
Kenal, Roman, Edmonton 064301
Kendal Museum, Kendal 044675
Kendal Studios Antiques, Kendal 089671
Kendall Campus Art Gallery,
Miami-Dade Community College,
Miami 050833
The Kendall College Mitchell
Museum of the American Indian,
Evanston 048639
Kendall College of Art and Design,
Ferris State University, Grand
Rapids 057149
Kendall Gallery, Kendall College of Art
and Design, Ferris State University,
Grand Rapids 049193
Kendall Whaling Museum,
Sharon 053285
Kendall, Beryl, London 090262, 118414
Kendall's, Cowes 117632
Kende, Tübingen 078723
Kenderdine Art Gallery,
Saskatoon 006974
Kendle Green Antiques, Albury . 060983
Kendons, Ingatestone 089639
Kendzia, Klaus D., Hamburg . . . 125969
Kenilworth Historical Society Museum,
Kenilworth 049918
Kenji Igarashi Memorial Museum of
Laundry, Tokyo 029528
Kenji Taki Gallery, Nagoya 111322
Kenji Taki Gallery, Tokyo 111452
Kenkeleba, New York 122767
Kenkoukanri Hakubutsukan,
Hiroshima 028654
Kenlzo, Lima 084459, 132809
Kenmare Antiques & Collectables,
Kenmare 079639
Kenmare Art Gallery, Kenmare . 109647
Kenmore Gallery, Kenmore 099181
Kennard, Salt Lake City 096897
Kennebunkport Historical Museum,
Kennebunkport 049921
Kennedy, Boston 119881
Kennedy, Dallas 093148
Kennedy, London . . . 090263, 134702
Kennedy, Miami 121815
Kennedy, New York 122768
Kennedy, Sacramento 096634
Kennedy, Tehrän 079423
Kennedy-Douglas Center for the Arts,
Florence 048752
Kennedy House, Selkirk 007002
Kennedy Museum of Art, Ohio
University, Athens 046599

Kennedy Wolfenden, Portrush . . 127016
Kennedy, Fred, Atlanta 092146
Kennedy, Michael, Sligo 109693
Kennedy, Peter, Woking 144928
Kennedy, Robin, Canterbury . . . 061279
Kennedy, Robin, Richmond,
Surrey 091099
The Kennedys, Berlin 017188
Kenner, T. & B., Steißlingen . . . 078555
Kennerknecht, E., Immenstadt im
Allgäu 076604
Kennesaw Civil War Museum,
Kennesaw 049923
Kennesaw Mountain National
Battlefield Park, Kennesaw . . 049924
Kennet and Avon Canal Trust Museum,
Devizes 044095
Kenneth, Haymarket 099119
Kenneth G. Fiske Museum of the
Claremont Colleges, Claremont 047756
Kenneths Konst och Antik,
Hillerstorp 086745
Kennis, Dirk, Antwerpen 063215
The Kenny Gallery, Galway 109637
Kenny, Ian, Strathaven 119168
Kenny, Paul, Botany 061177
Kenny's Antiques, Hamlyn
Heights 061586
Kennys Bookshop, Galway 142608
Kenny's Coins, Oklahoma City . . 095982
Kenny's Collectibles, Austin 092243
Keno City Mining Museum, Keno
City 006077
Keno, Leigh, New York 095506
Kenosha County Museum,
Kenosha 049926
Kenosha Public Museum,
Kenosha 049927
Kenoz, Cairo 103049
Kenpu Hakubutsukan, Komoro . 028883
Kenpu Hakubutsukan, Kokusai Boeki
Kanko Kaikan, Yokohama . . . 029718
Kenrick, Saint Louis 096724
Kenritsu Kagakukan Joetsu,
Joetsu 028736
Ken's Antique and Fine Art,
Singapore 085443
Ken's Antiques, Pittsburgh 096298
Ken's Art Gallery, Firenze 110062
Ken's Gallery, Fort Worth 120870
Kensei Kinenkan, Niigata 029168
Kensington, Barcelona 085790
Kensington, Sevilla 086453
Kensington, Tokyo 082165
Kensington Antiques, Ascot Vale 061063
Kensington Art Gallery, Norwood 099489
Kensington Church Street Antique
Centre, London 090264
Kensington Gallery, Calgary . . . 101246
Kensington House Collections,
Richmond 096557
Kensington Palace State Apartments,
Historic Royal Palaces, London 045013
Kensington Road Antiques,
Norwood 062014
Kensoon, Singapore 085444
Kent, New York 122769
Kent & Kent, Hobart 061632
Kent & Sussex Gold Refiners, Tunbridge
Wells 091653, 135160
Kent and East Sussex Railway,
Tenterden 045960
Kent and Medway Museum, West
Malling 046091
Kent and Sharpshooters Yeomanry
Museum, Edenbridge 044230
Kent and Vicki Logan Galleries
and Capp Street Project, Wattis
Institute for Contemporary Arts, San
Francisco 053007
Kent Antiques, Kent Town 061678
Kent Art Association, Kent 049929,
060556
Kent Auction Galleries,

Folkestone 126897
Kent Battle of Britain Museum,
Hawkinge 044545
Kent Campus Museum and Gallery,
Jacksonville 049773
Kent Collector, Plattsburgh 139412
Kent-Delord House Museum,
Plattsburgh 052198
Kent House Antiques,
Haverfordwest 089473
Kent Police Museum, Chatham . 043911
Kent Potters Gallery, Maidstone . 118697
Kent State University Art Galleries,
Kent 049932
Kent State University Museum,
Kent 049933
Kent-Stone, Dallas 093149
Kent, Cheslava, Mordialloc 127645
Kent, J., Wallasey 127075
Kentdale Antiques, Tunbridge
Wells 091654, 135161
Kenthurst Galleries, Kenthurst . . 099185
Kentmere House, York 119411
Kenton, Portland 096429
Kenton, J., Brooklyn 127150
Kentshire, New York 095507
Kentuck Museum and Art Center,
Northport 051636
Kentuckiana Pastel Society,
Louisville 060557
Kentucky Association of Museums,
Murray 060558
Kentucky Derby Museum,
Louisville 050492
Kentucky Folk Art Center,
Morehead 051052
Kentucky Guild of Artists and
Craftsmen, Berea 060559
Kentucky Historical Society,
Frankfort 048926
Kentucky Military History Museum,
Frankfort 048927
Kentucky Museum of Arts and Design,
Louisville 050493
The Kentucky Museum, Western
Kentucky University, Bowling
Green 047157
Kentucky Railway Museum, New
Haven 051269
Kentucky State Capitol, Frankfort 048928
Kentucky Woodworks, Louisville 135760
Kenulf Fine Arts, Stow-on-the-
Wold 091473, 119160, 135096
Kenwood, Cincinnati 120353
Kenwood House, London 045014
Kenworthy, Latrobe 061721
Keny, Columbus 120501
Kenya Museum Society, Nairobi 059523
Kenyon, Gerald, Dublin 079566
Kenz, Casablanca 082567, 112040,
132373
Kenzen, G., Rothenburg ob der
Tauber 078282
Keokuk Art Center, Keokuk . . . 049937
Keokuk River Museum, Keokuk . 049938
Keown, John T., Tulsa 124805
Képcsarnok, Budapest 109151, 109152
Képcsarnok – Gulácsy Lajos Terem,
Szeged 023927
Le Képi Rouge, Paris 071520
Kepler, Weil der Stadt 142466
Kepler-Gedächtnishaus, Museen der
Stadt Regensburg, Regensburg 021324
Kepler-Museum, Weil der Stadt . 022473
Kepler, Herwig, Hollabrunn 062640
Keppel, Haarlem 083335
't Keppeltje, Doetinchem 083170
Képzőművészek Gulácsy Lajos,
Budapest 023469
Ker, Iliz de, Perros-Guirec 072104,
105343
Ker, Maria, Guérande 068762
Keräily-, antiikki- ja taide-esineiden
messut, Tampere 098068

Keram, Kraków 084583
Keramaikos, Arnhem 112332
Kerameiki Techni, Nea Smyrni . 138767
Kerámevo, Amersfoort 112126
Keramiek Studio, Enschede 112477
Keramiekcentrum Tiendschuur,
Tegelen 032914
Keramik Art, Atelier für experimentelle
Keramik Jürg C. Bächtold,
Wilihof 116719
Keramik Kabinett Dießen, Dießen am
Ammersee 137140
Keramik Magazin/Ceramics Magazine,
Frechen 138703
Keramik-Museum, Triengen .. 042068
Keramik-Museum Berlin, Berlin . 017189
Keramik-Museum Bürgel, Bürgel 017674
Keramik Museum Mettlach,
Mettlach 020393
Keramikatelier, Hesperange ... 111829
Keramikgalerie Terra Rossa,
Leipzig 107836
Keramikkunst Bayreuth, Bayreuth 058944
Keramikmuseum, Rheinsberg . . 021401
Keramikmuseum Ludwigsburg,
Landesmuseum Württemberg,
Ludwigsburg 020127
Keramikmuseum Scheibbs,
Scheibbs 002716
Keramikmuseum Schloss Obernzell,
Zweigmuseum des Bayerischen
Nationalmuseums, Obernzell . . 020912
Keramikmuseum Staufen, Badisches
Landesmuseum Karlsruhe, Staufen
im Breisgau 021964
Keramikmuseum und Maria-Felchin-
Sammlung, Matzendorf 041730
Keramikmuseum Westerwald,
Deutsche Sammlung für Historische
und Zeitgenössische Keramik, Höhr-
Grenzhausen 019280
Keramikos Muziejus, Nacionalinis
M.K. Čiurlionio Dailės Muziejus,
Kaunas 030362
Keramiksammlung der Virgilkapelle,
Wien Museum, Wien 003015
Keramisch Atelier, Utrecht 112704
Keramisch Museum Goedewaagen,
Nieuw Buinen 032668
Keramos, Mayen 138704
Keramos, Paris 129046, 129047
Keramos, Pune 079414
Kerang Antiques, Kerang 061681
Keranroue, Hyacinthe, Paris ... 129048
Keraudren, Marie, Trégastel ... 073874
Keravan Museo, Kerava 010991
Keravan Taidemuseo, Kerava . . 010992
Kerber Verlag, Bielefeld 137110
Kerber Verlag, Leipzig 137345
Kerbiriou, Eugène-André, Pont
Aven 105380
Kerbyville Museum, Kerby 049941
Kerčs'kyj Deržavnyj Istoriko-kul'turnyj
Zapovidnyk, Kerč 043177
Kerdour, Michel de, Québec ... 101501
Kerdrain, Lionel, Meudon 070104
Kerebel, Alain, Temae, Moorea . 105894
Keresztény Múzeum, Esztergom 023631
Kerhir, Laurent, Auray 066817
Kerkhof, W., Dordrecht 083185, 132490
Kerkmuseum Janum, Janum . . 032527
Kerl, J., Weißenburg in Bayern . 078916
Kerler, G., Lorch, Württemberg . 077150
Kerler, Heinrich, Ulm 142443
Kerley, Geelong 125196
Kerlikowsky, Dr. Horst, München 077539
Kerlin, Dublin 109600
Kerlin Gallery, Belfast 117340
Kerls, Dr. Ingo, Bremen 075168
Kermani, Deira 088165, 117172
Kermes, Madrid 133346
Kermes, Roma 131950
Kermis- en Circusmuseum Steenwijk,
Steenwijk 032902

Kern County Museum,
Bakersfield 046719
Kern, Anton, New York 122770
Kern, Jens, Chemnitz 075255
Kern, M., Neubrandenburg 077713
Kern, Rosemarie, Bad Wimpfen . 074615
Kern, Theo, Ransbach-Baumbach 108429
Kernaghan, Bryan, Southport . . 144858
Kernavės Archeologijos ir Istorijos
Muziejus, Kernavė 030383
Kerner, Cecile, Bruxelles 063476
Kerner, Claude, Elbeuf-sur-Seine 068439
Kerner, G., Wendlingen am
Neckar 078923
Kerngast, Heinrich, Lichendorf . 127805
Kernstock-Museum, Bruck an der
Lafnitz 001800
Kerpaci, Tiranë 098525
Kerr, Jacksonville 094001
Kerr Place, Onancock 051770
Kerr, Albertina, Portland 096430
Kerr, Jody, Oklahoma City 095983
Kerr, Louella, Glebe 139752
Kerr, Louella, Petersham 139873
Kerr, Norman, Cartmel 144318
Kerr, Thomas, London 090265
Kerrobert and District Museum,
Kerrobert 006086
Kerry Anne's Fine Arts, Kew . . 061683
Kerry Anne's Fine Arts, Toorak . 062325
The Kerry Bog Village Museum,
Glenbeigh 024928
Kerry County Museum, Tralee . . 024994
Kers, Apeldoorn 082924
Kersa, Kerava 140730
Kerseboom, Willem, Bergen .. 082956
Kershaw County Historical Society
Museum, Camden 047386
Keršovani, Otokar, Rijeka 136751
Kerssemakers, Emile, Heerlen . 143169
Kersten, Kraków 113607
El Kerubin, Lima 084460
Kesälahden Museo, Kesälähti . . 010994
Keshett, Melbourne 061820
Keshike Tengqi Museum, Chifeng 007704
Keshishian, London 090266
Keshishian, New York 095508
Keski-Suomen Ilmailumuseo,
Tikkakoski 011421
Keski-Suomen Museo, Jyväskylä 010939
Keski-Suomen Sairaanhoitopiirin
Sairaalamuseo, Jyväskylä . . . 010940
Keski-Suomen Tieliikennemuseo,
Kintaus 011006
Keskin, B., Eutingen im Gäu ... 075724
Keskiner Antika, İstanbul 088127
Keskiner, Mehmet Kamil, İstanbul 088126
Keskisuomalaisen Museo,
Jyväskylä 010941
Kesner, Jan, Los Angeles 121519
Kesselaar & Klar, Buxtehude . . 075238
Kesselmark, Christiane, Cancale 103497
Kessler, Kreuzlingen 126734
Kessler Ford, Ashburton 088307
Kessler, Annette, Stuttgart 130677
Kessler, Marina, Miami 121816
Kesteloot, Bruxelles 063477
Kestler, Gerhard, Korneuburg . . 140004
Kestnergesellschaft, Hannover . . 019054,
058945
Kestrel, Vancouver 140424
Keswick Bookshop, Keswick ... 144501
Keswick Mining Museum,
Keswick 044680
Keswick Museum and Art Gallery,
Keswick 044681
Keta's Antiques, Toledo 097609
Ketelbuters, R., Bruxelles 100627
Ketley, Carol, London 090267
Kéttemplom, Kecskemét 109199
Kettenschmiedemuseum,
Fröndenberg 018536
Ketter, Hamburg 076253, 125970,
141878

Ketterer Kunst, München 077540,
126056, 137409, 142169
Ketterer-Haus-Museum, Biberach,
Ortenaukreis 017364
Kettle River Museum, Midway . . 006308
Kettle's Yard, Cambridge 043843
Kettumäki, Aaro, Nokia 066389
Ketz, Hans-Jürgen, Münster .. 142203
Keuck, Bad Harzburg 074516
Keuken, Lehrte 077048
Keul & Sohn, Wiesbaden 077981,
108920
Keulen, Köln 130152
Keulenaer, Fanny de, Reims ... 072393
Keulenaer, G. de, Gent 100696
Keumho Museum of Art, Seoul . 030007
Keumsan Gallery, Paju 111607
Keune, Sabine, Duisburg 141729
Keup, Regensburg 126103
Keuruun Kotiseutumuseo, Keuruu 010996
Keuruun Museo, Keuruu 010997
Keussen, E., Ohlsbach 077917
Keussen, E., Ortenberg, Baden . 077954
De Keuster, Antwerpen 125296
Kever, Hermann, Varel 078780
Keverne, Roger, London 090268, 134703
Kevin's Book Nook, Springwood 139913
Kevin's Collectibles, Charlotte . 092610
Kevorkian, Paris 071521
Kew & Cie, Hong Kong 065063
Kew Auctions and Antiques, Richmond,
Surrey 127019
Kew Bridge Steam Museum,
Kew 044686
Kew Gallery, Kew 099188
Kew Gallery, New York 122771
Kew Historical Museum, Kew . . 001186
Kew Palace Museum & Queen
Charlotte's Cottage, Historic Royal
Palaces, Kew 044687
Kewaunee County Historical Museum,
Kewaunee 049948
Kewenig Galerie, Köln 107649
Kewenig, Jule, Palma de
Mallorca 115355
Kewriosity, Kew 061684
Key, Houston 135608
Key Antiques, Chipping Norton . 088890
Key Antiques, Tugun 062350
The Key to It All, Phoenix 096220
Key West Company, Tokyo 082166
Key West Lighthouse Museum, Key
West 049955
Key West Museum of Art and History,
Key West 049956
Keya Gallery, New York 095509, 145350
Keymer Son & Co, Brasted 088581
Keys, Aylsham 126803
Keys & Sons, Indianapolis ... 093897,
135634
Keyser, Annette de, Antwerpen . 100497
Keyser, Wilfried de, Kapellen .. 063691
Keystone, Coalville 088937
Keystone, Singapore 085445
Keystone Area Museum,
Keystone 049963
Keystone Furniture, Indianapolis 093898
Keystone Pioneer Museum,
Roblin 006816
Kforumvienna, Vienna 058267
KG Books, New Farm 139853
KGB Kongide Muuseum, Tartu
Linnamuuseum, Tartu 010696
KGM-Atelje, Zaprešić 102528
Kgosi Bathoen II (Segopotso) Museum,
Kanye 004297
Khachatour Abovyan House-Museum,
Yerevan 000735
Khai, Singapore 085446
Khairalla, Camille, Beirut 082290
Khairallah, Elias Antoine, Torino 132152
Khairallah, Ghida, Baabda 111769
Khajana Buildings Museum,
Hyderabad 024298

Al-Khaleej, Zahrat, Manama ... 063148,
100438
Khalid, Marrakech 082590
Khalid, Singapore 085447
Khalifa's House Museum,
Omdurman 040501
Khalif's House Museum Abdallah at-
Ta'ayushi, Khartoum 040496
Khalkhal Museum, Khalkhal ... 024711
Khallouf, Faddoui, Joué-lès-Tours 128633
Khama III Memorial Museum,
Serowe 004300
Khami Ruins Site Museum,
Bulawayo 054844
Khan El-Khalily, Alexandria 066220
Khan el-Maghraby Gallery, Cairo 103050
The Khan Museum, Hadera 025041
Khan, Shaker Bin Murad,
Muttrah 084403
Khanate, Portland 096431
Khanh Hoa Provincial Museum, Nha
Trang 054815
Khariton Akhvlediani State Museum,
Batumi 016386
Khatchikian, Jacques, Croignon 068245
R. Khawam & Cie, Khepri Culture Arts
& Design, Paris 071522
Khayam, Souffelweyersheim .. 073545
Khayyer, Mohammad, Wiesbaden 108921
Khazana, Minneapolis 122006
Khedr, Krikor, Damascus 088038
Khelovani, Tbilisi 105907
KHG-Galerie, Graz 100011
Khiching Museum, Khiching ... 024346
Khim Art, Sartrouville 073422
Khoja, Manama 063149
Khong, Linda, Luzern 087644
Khorassan Natural History Museum,
Shahid Beheshti Collage,
Mashad 024721
Khoren Ter-Harutyan Museum,
Echmiadzin 000695
Khot-la-Cha, Vancouver 101803
Khotan, Milano 080584
Khoy Museum, Khoy 024714
Khris, Saint Louis 123835
Khushbu Art Gallery, Pune ... 109497
Khutorsky, Gregory, New York . 095510
Khyber Digital Media Centre,
Halifax 005953
Kiama Art Society, Kiama 058205,
099190
Kiang, Marilyn, Atlanta 119582
Kiangan National Museum,
Kiangan 034490
Kiasma Nykytaiteen Museo, Valtion
Taidemuseo, Helsinki 010836
Kiato Archaeological Museum,
Kiato 023127
Kibbutz Art Gallery, Tel Aviv .. 025203
Kibbutz Negba Museum, Kibbutz
Negba 025142
Kibi Kokoshizou, Yamateson .. 029702
Kibiji Kyodo Hakubutsukan, Soja 029409
Kibler, Alain, Haguenau 068777
Kibo Art Gallery, Marangu 042634
KIBO Gallery, Seattle 124553
Kiboko Projects, New York 051414
KiC – Kunst in der Carlshütte,
Büdelsdorf 106611
Kichline, H., Atlanta 092617
Kick Gallery, Northcote 099485
KickArts, Cairns 098841
Kicken Berlin, Berlin 106327
Kicki's Antiques, Kungsgården . 086796
Kidder & Smith, Boston 119882
Kidderminster Railway Museum,
Kidderminster 044689
Kidding Awound, Toronto 064686
Kidner, George, Lymington ... 126981
Kido, Tokyo 111453, 137782
Kidogo Arthouse, Fremantle ... 099062
Kids Can Press, Toronto 136684
Kids' Science Museum of Photons,

Kizu 028836
Kidspace Children's Museum,
Pasadena 051934
Kidsquest Children's Museum,
Bellevue 046889
Kidwelly Industrial Museum,
Kidwelly 044691
Kiebler, Kurt, Bad Wurzach ... 129430
Kiecken Duthoit, Dunkerque ... 068415
Kiefer, Hans, Basel 133645
Kiefer, Mariele, Bad Kreuznach . 074545
Kiefer, Peter, Pforzheim 126094, 142304
Kiefer, T., Mannheim 077262
Kiehle Gallery, Saint Cloud State
University, Saint Cloud 052762
Kiek es Antikes, Borken,
Westfalen 075101
Kiek in beim Turm, Lemgo 077075
Kiek in de Kök, Tallinna
Linnamuuseum, Tallinn 010677
Kiekeboe, Gent 140224
Kiekebusch, Jörg, Wunstorf ... 108981,
137564
't Kiekenkot, Gent 063645
't Kiekhuus, Wolvega 033060
Kiel, Albert, Darmstadt 075305
Kieler Stadt- und Schiffahrtsmuseum,
Kiel 019605
Kiener, Paris 071523
Kiener, Chantal, Paris 104986
Kiener, Martin, Zürich 087982
Kienhorst, I.Jzendijke 083411
Kienzle & Gmeiner, Berlin 106328
Kienzle, Andr. Claus, Augsburg . 129396
Kienzler-Schmiede, Kirchzarten . 019631
Kiepert, Berlin 141539
Kieppi-Luontomuseo ja
Mineraalikokoelma, Kokkola .. 011019
Kierat 1 – Galeria Sztuki Zwiazek
Polskich Artystów Plastyków,
Szczecin 113779
Kierat 2 – Galeria Sztuki Zwiazek
Polskich Artystów Plastyków,
Szczecin 113780
Kiernan, Mosman Park 061899
Kierondino, Groningen 083297
Kiesel, G., Ismaning 130078
Kieselbach, Budapest . 079263, 109153
Kievartcentr, Kyïv 117138
Kievskaja Galereja, Kyïv 117139
Kihniön Museo, Kihniö 011000
Kii-Fudoki-no-oka Museum of History,
Wakayama 029690
Kiido, Tokyo 082167
Kiihtelysvaaran Museo,
Heinävaara 010808
Kiikalan Kotiseutumuseo, Kiikala 011001
Kiikoisten Kotiseutumuseo,
Torpparimuseum/Esinemuseo,
Kiikoinen 011002
Kiimingin Museo, Kiiminki 011003
Kiippa, Tapani, Helsinki 103152
Kijarat Kiadó, Budapest 137587
Kijk- en Luistermuseum, Bant . . 032046
Kijk en Luister Museum,
Bennekom 032060
KiK – Kultur im Kloster,
Klostermarienberg 002204
Kikava Sanat Evi, Samsun 088153
Kikel, A., Bodman-Ludwigshafen 075051
Kikkermuseum, Den Haag 032175
Kiko, Wilhelm, Soest 078487
Kikori, Nishitokyo 082079
Kiko's Fossil and Mineral Museum,
Banciao 042271
Kiku, San Francisco 097264
Kikusui Handicraft Museum,
Kawanishi 028805
Kikuta, Sendai 111387
Kikuya, Kyoto 082025
Kilátó, Budapest 109154
Kilback Bros, Vancouver 128227
Kilby Store and Farm, Harrison
Mills 005990

Kilchmann & Martín, México . . . 111963
Kilchmann, Peter, Zürich 116820
Kilchoer, Olivier, Fribourg 133746
Kilcock Art Gallery, Kilcock ... 109648
Kildahl, Wenche, Tromsø 084382
Kildevang, Næstved 066005
Kildonan Museum, Museum Chill
Donnan, Kildonan, Isle of South
Uist 044693
Kildonan Roofing, Winnipeg ... 128241
Kile Sanat Galerisi, İstanbul ... 117058
Kilen Galleri, Hvasser 113164
Kilian, Françis, Strasbourg 105692
Kilian, Hans, Marburg 077286
Kiliansboden, Lund 086834
Kilim Arte & Antichità, Verona . . 081917
Kilin, København 065922, 102904
Kilin, Singapore 114553
Kiliński, Jacek, Warszawa 113856
Kilkivan Historical Museum,
Kilkivan 001188
Kill City Crime Books, Prahran . 139886
Killarney Centennial Museum,
Killarney 006089
Killen, David, New York 095511, 122772
Killhope, the North of England Lead
Mining Museum, Weardale ... 046071
Killin Gallery, Killin 089687
Kilmacolm Antiques, Kilmacolm 089688,
134512
Kilmainham Gaol and Museum,
Dublin 024895
Kilmallock Museum, Kilmallock . 024943
Kilmister, Masterton 132730
Kilmorack Gallery, Beauly 117319
Kilmore Old Wares, Kilmore ... 061691
Kilmore Quay Maritime Museum,
Kilmore Quay 024944
Kilmore Quay Maritime Museum,
Wexford 025003
Kilsyth's Heritage, Kilsyth 044700
The Kilvert Gallery, Clyro 117604
Kilwinning Abbey Tower,
Kilwinning 044702
Kim, Hong Kong 065064
Kim, Murcia 086314, 115315
Kim 3, Los Angeles 094382
Kim 3, San Francisco 097265
Kim Turner Vaadia Fine Art, New
York 095512, 122775
Kim, Hyon Joo, Seoul 111677
Kim, Jae Sun, Busan 111568
Kim, Jin Hye, Seoul 111678
Kim, Leong, Singapore 085448
Kim, Long Seng, Singapore 085449
Kim, Phuong, Ho Chi Minh City . 125127
Kim, San, Singapore 133038
Kim, Thye Lee, Singapore 085450
Kim, Tina, New York 122773
Kim, Young Seob, Seoul 111679
Kim, Young Hee, New York 122774
Kimba and Gawler Ranges Historical
Society Museum, Kimba 001189
Kimball Art Center, Park City ... 051921
Kimball House Museum, Battle
Creek 046825
Kimbell Art Museum, Fort Worth 048904
Kimbell Art Museum, Publications
Department, Fort Worth 138333
Kimball, Patsy, Buderim 061228
Kimber & Son, Malvern Link ... 090685,
134828
Kimberley, Bilbao 085873
Kimberley, Melbourne 099309
Kimberley District Heritage Museum,
Kimberley 006090
Kimberley Mine Museum,
Kimberley 038764
Kimberly Crest House, Redlands . 052494
Kimberly's Fine Art, Dallas ... 120616
Kimble County Historical Museum,
Junction 049846
Kimcherova, New York 095513
Kimchi Field Museum, Pulmuone

Gimchi Museum, Seoul 030008
Kime, Robert, Marlborough 090723
Kimga, Yokohama 111540
Kimin, Singapore 085451
Kimjack, Calgary 064261
Kimmerich, Dennis, Düsseldorf . 106797
Kimmerich, Horst, Düsseldorf . . 075486
Kimmich, Klaus, Köln 107650
Kimo Bruks Museum and Gallery,
Oravais 011207
Kimpfbeck, J., Nersingen 108234
Kimpton, K., San Francisco ... 124315
KiMuncho, Seoul 111680
KiMuncho, Singapore 114554
Kimura, Kyoto 082026
Kimura, San Francisco 124316
Kimyon, Adnen, Frankfurt am
Main 075799
Kimzey & Miller, Seattle 124554
Kin Fat, Hong Kong 065065
Kin Liou, Paris 071524
Kina NZ, Taranaki 113016
Kina Slott, Drottningholm 040594
Kincaid Museum, Kincaid 006091
Kincová, Zlín 102751
Kincsem Múzeum,
Tápiószentmárton 023994
Kind, Parisa, Frankfurt am Main 107008
Kindbom & Co, Olof, Karlsborg . 126612
Kinder- und Jugendgalerie
Sonnensegel, Brandenburg an der
Havel 017544
Kinder- und Jugendmuseum,
Nürnberg 020863
Kinder- und Jugendmuseum im
G.-Lübcke-Museum, Hamm,
Westfalen 019030
Kinder- und Jugendmuseum München,
München 020536
Kinder- und Jugendmuseum
Prenzlauer Berg, Berlin 017217
Kinder-Akademie Fulda
Werkraummuseum, Fulda 018563
Kinder Fine Arts, Christchurch . 112847
Kindergarten Museum, Potsdam 021188
Kindergartenmuseum, Bruchsal . 017634
Kindermann, Thomas,
Wiebelsdorf 078962
Kindermuseum, Aflenz 001688
Kindermuseum, Frankfurt am
Main 018431
Kindermuseum, Karlsruhe 019508
Kindermuseum im FEZ-Berlin,
Berlin 017191
Kindermuseum, Schaufenster, Schule
und Kinderkunst e.V., Wuppertal 022821
Kindersley and District Plains Museum,
Kindersley 006092
Kinderwagenmuseum, Nieuwolda 032680
Kinderwagens van toen,
Dwingeloo 032258
Kinderwelt Walchen, Vöcklamarkt 002869
Kindheitsmuseum, Marburg ... 020272
Kindheitsmuseum, Schönberg,
Holstein 021698
Kindler, Matthias Johannes Andreas,
Graz 062594, 125255
Kinenkan Mikasa, Yokosuka ... 029738
Kineone, Istres 103856
Kinet, J., Bruxelles 063478
Kineya, Tokyo 082168
King, Cleveland 120435
King, Fort Worth 093458
King, Houston 093715
King, Salt Lake City 123969
King, San Antonio 096985
King & Co., San Francisco 124317
King, Wymondham, Norfolk 091962
King & Chasemore, Midhurst ... 126988
King County Arts Commission Gallery,
Seattle 053236
King Feng Arts, Hong Kong ... 065066
King Griffin, San Diego 124138
King Hooper Mansion,

Marblehead 050666
King Island Historical Museum,
Currie 001010
King John's Castle, Limerick ... 024955
King John's Hunting Lodge,
Axbridge 043454
King Kamehameha V – Judiciary
History Center, Honolulu 049537
King Mac Craw, Trappes 073871
King Manor, Jamaica 049782
King-Prime Gallery, Bangalow .. 098704
King Richard's, Atlanta 092148
King Saud University Museum,
Riyadh 038110
King Sejong The Great Memorial
Exhibition, Seoul 030009
King-Smith & Nutter, Stow-on-the-
Wold 091474
King Street Antiques, Toledo ... 097610
King Street Curios, Melksham .. 090741
King Street Curios, Tavistock .. 091548
King Street Gallery, Darlinghurst 098945
King Street Gallery, Newcastle,
Staffordshire 118782
King Street Gallery, Newtown, New
South Wales 099450
King Township Historical Museum,
King City 006093
King Tut, San Diego 124139
King Tut Egyptian Imports,
Tampa 097566
King, A., San Antonio 127437
King, Ann, Bath 088391
King, Anthony, Stuttgart 078610
King, David H., Dallas 093150
King, Dennis, Enfield 079616
King, Donald, New York 095514, 135945
King, Erika, Miami 121817
King, Henry, New York 122776
King, John, London 090269
King, John K., Detroit 145138
King, John K., Ferndale 145147
King, Kevin, Dunedin 132718
King, Laurence, London 138223
King, Peter, Ti Tree 062319
King, Philip, Auckland 132692
King, Roger, Dallas 093151
King, Roger, Hungerford 089610
Kingaroy Shire Art Gallery,
Kingaroy 001191
Kingdom Antiques, Tralee 079712
Kingfisher Gallery, Cowbridge . . 117630
Kinglake Gallery, Kinglake 099193
Kingman County Historical Museum,
Kingman 049977
Kingman Museum of Natural History,
Battle Creek 046826
Kingman Tavern Historical Museum,
Cummington 048092
Kings Art Gallery, Vancouver ... 101804
King's Cottage Antiques, Leamington
Spa 089755
Kings County Museum, Hampton 005981
Kings County Museum, Kentville 006083
Kings Court Antiques,
Ngaruawahia 084025
King's Court Galleries, Dorking . 089087,
144374
King's Court Galleries, London . 090270,
144659
Kings Fireplaces, Antiques & Interiors,
Cardiff 088787, 134214
Kings Gallery, Canterbury 117518
King's House Antiques,
Birmingham 092416
Kings Landing Historical Settlement,
Prince William 006721
King's Lynn Arts Centre, King's
Lynn 044704
King's Lynn Museum, King's
Lynn 044705
King's Mill Antique Centre,
Burnley 088731, 134187
King's Mill Visitor Centre,

Wrexham 046204
Kings Mountain National Military Park,
Blacksburg 047025
King's Own Royal Regiment Museum,
Lancaster 044751
King's Own Scottish Borderers
Regimental Museum, Berwick-upon-
Tweed 043577
King's Own Yorkshire Light Infantry
Regimental Museum, Doncaster 044110
King's Point Museum, King's
Point 006094
Kings Quair, Aberdeen 144212
Kings Road Galleries, London ... 118415
The King's Royal Hussars Museum,
Winchester 046151
Kings Weston Roman Villa,
Bristol 043752
Kingsbridge Auction Sales,
Kingsbridge 126931
Kingsbury Watermill Museum, Saint
Albans 045665
Kingsclere Old Bookshop,
Kingsclere 144504
Kingsgate Gallery, London 045015
Kingsland Homestead, Queens
Historical Society, Flushing .. 048766
Kingsley & Co, Wirral 127091
Kingsley Plantation, Jacksonville 049774
Kingston, Boston 119883
Kingston, Moorabbin 099373
Kingston Antiques, Fitzroy North 061474
Kingston Antiques Market, Kingston-
upon-Thames 089708, 134516, 144505
Kingston Archaeological Centre,
Kingston 006102
Kingston Galleries, Kingston-upon-
Hull 118049
Kingston Gallery, Kingston 118048
Kingston Mills Blockhouse, Elgin 005776
Kingston Museum, Kingston-upon-
Thames 044721
Kingston Pioneer Museum,
Kingston 001195
Kingston, Richard J., Henley-on-
Thames 089509, 144469
Kingstone, Honolulu 093558
Kingsway Auction Rooms, Lytham
Saint Anne's 126982
Kingussie Museum, Highland Folk
Museum, Kingussie 044724
Kinistino and District Pioneer Museum,
Kinistino 006116
Kinizsi Pál Vármúzeum,
Nagyvázsony 023807
Kinkade, Fort Worth 120871
Kinkade, Thomas, Atlanta 119583
Kinkade, Thomas, Bangsar Baru 111901
Kinkade, Thomas, Broadway ... 117457
Kinkade, Thomas, Dallas 120617
Kinkade, Thomas, Dartford ... 117661
Kinkade, Thomas, Edinburgh ... 117730
Kinkade, Thomas, Fresno 120900
Kinkade, Thomas, Glasgow ... 117847
Kinkade, Thomas, Jacksonville . 121256
Kinkade, Thomas, Kansas City . 121300
Kinkade, Thomas, London 118416
Kinkade, Thomas, Louisville ... 121652
Kinkade, Thomas, Minneapolis . 122007
Kinkade, Thomas, New Orleans 122191
Kinkade, Thomas, Oklahoma City 123286,
145398
Kinkade, Thomas, Portland 123635
Kinkade, Thomas, San Diego .. 124140
Kinkade, Thomas, Solihull 119111
Kinkade, Thomas, Toronto 101639
Kinkead, James W., Memphis .. 135781
Kinkel, Ludwigshafen am Rhein 077162
Kinkelin, Severin, Näfels 087695, 133868
Kinky Beaux Arts, Stuttgart 138705
Kinlough Folk Museum, Kinlough 024945
Kinnear, Glenreagh 061535
Kinneil Museum and Roman Fortlet,
Bo'ness 043674

Kinney Pioneer Museum, Clear
Lake 047789
Kinnulan Kotiseutumuseo,
Kinnula 011005
Kinokuniya, Tokyo ... 111454, 142926
Kinpo-cho Rekishi Akusha, Kinpo 028817
Kinross Museum, Kinross 044726
Kinsale Art Gallery, Kinsale 109661
Kinsale Regional Museum,
Kinsale 024947
Kinsella, Edge Hill 061437
Kinski, Heidenheim an der Brenz 076465
Kinsman, Robinson, Toronto ... 101640
t'Kint de Roodenbeke, Harold,
Bruxelles 063479, 100628
Kinter, Prof. Klaus & Jutta,
Remshalden 108478
Kintolai Gallery, Adelaide 098613
Kinz, Tillou + Feigen, New York 122777
Kinzel & Bode, Braunschweig .. 075126
Kinzelt, Karina, Peitz 130508
Kioskshop Berlin, Berlin 071192
Kiosques, Toulon (Var) . 141356, 141357
Kipling, Houston 121073
Kipling District Museum, Kipling 006117
Kipp Gallery, Indiana University of
Pennsylvania, Indiana 049675
Kipp, Manfred, Lengerich 077081
Kipp, P.J., Utrecht 132648
Kippert, Bruno, Erlangen 075654
Kipping, Venlo 083678
Kipton, New York 122778
Király, Gabriella, Aszód 109123
Kirbižu Meža Muzejs, Vitrupe .. 030257
Kirby, R., Acrise 088237
Kirche am Hohenzollernplatz,
Berlin 017193
Kirchengeschichtliches Museum in der
Pfalzkapelle, Bad Wimpfen ... 016907
Kirchenhäusl, Sankt Oswald bei
Freistadt 002685
Kirchenmuseum, Ernen 041471
Kirchenmuseum, Marienhafe ... 020282
Kirchenschatz, Bremgarten
(Aargau) 041361
Kirchenschatz, Glarus 041548
Kirchenschatz, Sursee 042042
Kirchenschatz des Chorherrenstiftes,
Beromünster 041331
Kirchenschatz im Sankt
Fridolinsmünster, Bad
Säckingen 016854
Kirchenschatz-Museum Sankt Martin,
Altdorf 041211
Kirchenschatz-Museum, Katholische
Stadtpfarrkirche, Baden 041253
Kirchenschatz Sankt Pelagius Kirche,
Bischofszell 041349
Kirchhof, Bad Berleburg 074492
Kirchhoff, J., Kiel 076778
Kirchliches Heimatmuseum Bärenthal,
Bärenthal 016932
Kirchmeyer, Bitschwiller-les-
Thann 067233
Kirchner Museum Davos, Davos
Platz 041434
Kirchner, Ilse, Baiersdorf 074650
Kirchner, Rainer, Essen 075684
Kirchwehm, Gudrun, Berlin ... 106329
Kirikiño, Bilbao 085874
Kirikkale Kültür Merkezi,
Kırıkkale 042996
Kirillo-Belozerskij Istoriko-
Architekturnyj i Chudožestvennyj
Muzej-Zapovednik, Kirillov .. 037127
Kirishitan Shiryokan, Hirado .. 028633
Kirisits, Werner, Stubenberg ... 062856
Kiriyama Gallery, Tokyo 111455
Kirja- ja Lehtilinna, Kuopio .. 140734
Kirja- ja Paperikonservointi,
Mikkeli 128421
Kirjakaruselli, Kauhava 140728
Kirjapaja, Helsinki 136791
The Kirk, Surry Hills 099748

Kirk Ports Gallery, North Berwick 090846
Kirkaldy Testing Museum,
London 045016
Kirkcaldy Museum and Art Gallery,
Kirkcaldy 044728
Kirkgate, Thirsk 135135
Kirkham House, Paignton 045464
Kirkhoff, Birgitte, København ... 102905
Kirkkomuseo, Nokia 011191
Kirkland, Leumeah 099227
Kirkland Art Center, Clinton ... 047844
Kirkland Fine Arts Center-Perkinson
Gallery, Decatur 048201
Kirkland Museum of Fine and
Decorative Art, Denver 048266
Kirkland, David, Wellington ... 084168,
132761
Kirkland, Graham, London 090271
Kırklareli Müzesi, Kırklareli ... 042997
Kirkleatham Museum, Redcar ... 045594
Kirkpatrick Mostert, Jeanie,
Houston 093717
Kirkpatrick Science and Air Space
Museum at Omniplex, Oklahoma
City 051730
Kirkpatrick, Jim R., Houston ... 093716
Kirk's Kards, Cleveland 092949
Kirkuk Museum, Kirkuk 024808
Kirms-Krackow-Haus, Klassik Stiftung
Weimar, Weimar 022497
Kiro Studio International, New
York 122779
Kiron, Paris 104987
Kirovograd'skyj Chudožno-
memorial'nyj Muzej O. Osmorkina,
Kirovograd 043178
Kirovograd'skyj Literaturno-
memorial'nyj Muzej M.L.
Kropyvnyc'kogo, Kirovograd'skyj
oblasnyj krajeznavčyj muzej,
Kirovograd 043179
Kirovograd'skyj Literaturnyj Muzej,
Kirovograd 043180
Kirovograd'skyj Muzej Musyčnoï
Kul'tury im. Karolja Šymanovs'kogo,
Kirovograd 043181
Kirovograd'skyj Oblasnyj Chudožnij
Muzej, Kirovograd 043182
Kirovograd'skyj Oblasnyj Krajeznavčyj
Muzej, Kirovograd 043183
Kirovskij Oblastnoj Chudožestvennyj
Muzej im. Viktora i Apollinarija
Vasnecovych, Kirov 037131
Kirovskij Oblastnoj Kraevedčeskij
Muzej, Kirov 037132
Kirriemuir Gateway to the Glens
Museum, Kirriemuir 044737
Kirsch, Wien 062996
Kirsch, Alfred, München 077541
Kirschenmuseum, Endingen am
Kaiserstuhl 018206
Kırşehir Devlet Güzel Sanatlar Galerisi,
Kırşehir 042998
Kırşehir Müzesi, Kırşehir 042999
Kirst, Siegfried, Wunstorf 079097
Kirsten, Beedenbostel 074709
Kirsten Flagstad Museum,
Hedmarksmuseet og Domkirkeodden,
Hamar 033634
Kirsten Kjærs Museum, Frøstrup 031004
Kirsten, Nicholas, Seattle 124555
Kirsti Rauman Museo, Rauma .. 011302
Kirtland Temple Historic Center,
Kirtland 049995
Kirton Antiques, Kirton 089719
Kirwan, Roger, Clonmel 079487
KIS Galerie, Amsterdam 112238
Kis Géza Ormánság Múzeum és
Skanzen, Sellye 023879
Kisa, Tokyo 082169
Kisbey Museum, Kisbey 006121
Kiscelli Múzeum, Budapesti Történeti
Múzeum, Budapest 023497

Kisek, Båstad 086640
Kisfaludy Károly Emlékkiállítás,
Tét 024004
Kisfaludy Sándor Emlékház,
Sümeg 023906
Kisgaléria, Győr 109192
Kish, New York 122780
Kishon, Sara, Tel Aviv 109726
Kishu-furitsu Hakubutsukan,
Shirahama 029395
Kisjankó Bori Emlékház,
Mezőkövesd 023776
Kiskastély – Művészetek Háza,
Nagykanizsa 023803
Kiskon Kotiseutumuseo, Kisko . 011009
Kiskun Múzeum,
Kiskunfélegyháza 023744
Kisler Rosenvald, Paris 071525
Kislovodskij Istoriko-kraevedčeskij
Muzej Kreposť, Kislovodsk ... 037139
Kismet, San Jose 124451
Kismit, Toledo 124691
Kiso-ji Bijutsukan, Agematsu .. 028462
Kispesti Helytörténeti Gyűjtemény,
Budapest 023498
Kiss, Szeged 109231
KISS – Temporäres Museum
zeitgenössischer Kunst, Kunst
im Schloss Untergröningen,
Abtsgmünd 016459
Kiss Pál Múzeum, Tiszafüred . 024011
Kiss, Ilona, Budapest 109155
Kiss, Marlène & Andreas, Cham 075249
Kissi Gallery, Tiranë 098526
Kißler, Sabine, Köln 130153
Kistefos-museet, Jevnaker 033687
Kistemann, Wolfgang, Berlin .. 141540
Kistler, Lyss 087654, 116486
Kistner, E. & R., Nürnberg 142246
Kisumu Museum, Kisumu 029861
Kit & Caboodle, Toronto 064687
Kit Carson Historical Society Museum,
Kit Carson 049997
Kit Carson Museum, Las Animas 050151
KIT Tropenmuseum, Amsterdam 031950
Kita, Kyoto 082027
Kita, Yamato-koriyama 111531
Kita-kamakura Museum,
Kamakura 028759
Kitai Hakubutsukan, Kodaira .. 028866
Kitajskij Dvorec-muzej, Muzej-
zapovednik Oranienbaum,
Lomonosov 037202
Kitakami-shiritsu Hakubutsukan,
Kitakami 028824
Kitakyushu-shiritsu Bijutsukan,
Kitakyushu 028827
Kitakyushu-shiritsu Shizenshi
Hakubutsukan oyo Ningenteki
Rekishi, Kitakyushu 028828
Kitale Museum, Kitale 029862
Kitamae-senshu-no-yakata Ukon-ke,
Echizen 028529
Kitami-Abashiri-furitsu Bunkakan,
Kitami 028831
Kitami Peppermint Museum,
Kitami 028832
Kitamura Bijutsukan, Kyoto ... 028938
Kitano Bijutsukan, Nagano ... 029072
Kitano Bunkakan, Nagano 029073
Kitano Temman-gu Treasure House,
Kyoto 028939
Kitayama Cedar Museum, Kyoto 028940
Kitazawa, Tokyo 142927
Kitazawa Bijutsukan, Suwa ... 029416
Kitchen Witch Country Furnishings,
Toronto 064688
Kitchenaid, Columbus 135480
Kitchener, Hamilton 064226
Kitchener-Waterloo Art Gallery,
Kitchener 006125
Kiteen Kotiseutumuseo, Kitee Local
History and Crafts Museum,
Kitee 011010

Kithur Rani Channamma Memorial
Museum, Kittur 024347
Kitikmeot Heritage Society, Cambridge
Bay 005539
Kitimat Centennial Museum,
Kitimat 006127
Kitna Antik, Viborg 066208
Kitovras, Moskva 114267
Kitsap County Historical Museum,
Bremerton 047189
Kitsch, Montréal 064437
Kitsch und Kunst, Heilbronn ... 076474
Kitsos Makris Folk Art Center,
Volos 023315
Kitte no Hakubutsukan, Tokyo ... 029529
Kittel, Michael, Rosengarten,
Württemberg 130598
Kittelberger, Pittsburgh 096299
Kittelsenhuset, Berg-Kragerø Museum,
Kragerø 033713
Kittelty's Auction Rooms, Ballarat 125178
Kittery Historical and Naval Museum,
Kittery 049998
Kittilän Kotiseutumuseo, Kittilä . 011011
Kittilbu Utmarksmuseum, Østre
Gausdal 033815
Kittiwake Gallery, Eyemouth ... 117780
Kittredge Art Gallery, University of
Puget Sound, Tacoma 053646
Kittrell & Riffkind, Dallas 120618
Kittson County History Center Museum,
Lake Bronson 050069
Kitzig, Erdmann, Hemmingen ... 130033
Kitzig, Erdmann, Wennigsen ... 130744
Kitzinger, J., München 142170
Kiuchi Gallery, Ichikawa 028671
Kiuruveden Kotiseutumuseo,
Kiuruvesi 011012
Kivikauden Kylä, Saarijärvi 011330
Kivimuseo, Tampereen Museot,
Tampere 011396
Kivirannan Kotimuseo, Säkylä ... 011334
Kivisakasti, Mikkeli 011159
Kiwanis Van Slyke Museum
Foundation, Caldwell 047344
Kiya, Yokohama 082228
Kiyoharu Shirakaba Museum,
Nagasaka 029084
Kiyomizu Sannenzaka Museum,
Kyoto 028941
Kiyosato Kitazawa Bijutsukan,
Takane 029439
Kiyosato Museum of Photographic
Arts, Takane 029440
Kiyose-shi Kyodo Hakubutsukan,
Kiyose 028834
Kız Teknik Öğretim Müzesi,
Ankara 042822
Kiži – Gosudarstvennyj Istoriko-
architekturnyj i Ētnografičeskij Muzej-
zapovednik, Petrozavodsk 037653
Kizilcelik, Yusuf, Ravensburg ... 108443
Kızıltoprak Sanat Galerisi,
İstanbul 117059
Kizljarskij Kraevedčeskij Muzej
im. P.I. Bagrationa, Dagestanskij
gosudarstvennyj objedinennyj istoriko-
architekturnyj muzej, Kizljar . . 037144
Kjær, Viborg 140673
Kjær, Flemming, København ... 065923
Kjærsgaard, Randers 066055
Kjærsgaard, Lisbeth, København 065924
Kjærulff, Nykøbing Sjælland ... 066020
Kjærulff, Taastrup 102989
Kjarvalsstaðir, Listasafn Reykjavíkur,
Reykjavík 024096
Kjeldaas Kunsthandel, Oslo 113228
Kjellman, Uppsala 143992
Kjelsmark, Connie, Holstebro ... 065781
Kjerringøy Gamle Handelssted,
Nordland Fylkesmuseet, Bodø . 033548
KJT Art Factory, Houston 121074
Kjubh Kunstverein, Köln 058946
Klaavola Museo, Tuusula 011460

Klaber & Klaber, London 090272, 134704
Klabisz, Marian, Warszawa 132884
Klabund, Hannover 141916
Klabund, Elke, Hannover 141917
Klæbu Bygdemuseum, Klæbu .. 033700
Klæmmensen, Henning, Odense 066038
Klag, A., Duchroth 075436
Klagsbrun, Nicole, New York ... 122781
Klaiber, Daniela, Seeheim-
Jugenheim 108603
Klaipėdos Galerija, Klaipėda ... 111788
Klaipėdos Paveikslų Galerija, Lietuvos
Dailės Muziejus, Klaipėda 030388
Klamath County Museum, Klamath
Falls 050000
Klamath Falls Art Association, Klamath
Falls 060560
Klamath National Forest Interpretive
Museum, Yreka 054604
Klampfl, Wien 127937
Klange, Osaka 111361
Klapka György Múzeum,
Komáron 023756
Klapka, Georg, Budapest 126152
Klappers, Wolfgang, Rodgau ... 078254
Klappzahlenwecker Museum,
Leipzig 019997
Klaproth, Bernd, Hagen 076148
Klar, Manfred, Berlin 074840
Klara Wallner Gallerie, Berlin ... 106330
De Klare Lijn, Leiden . 083432, 083433
Klaricová, Kateřina, Praha 102701
Klaschewski, Jürgen, Bleicherode 075020
Klasema, Bergen op Zoom 082961
Klasing, U., Lemgo 077076
Klassen, M., Marienheide 107979
Das Klassenzimmer,
Tauberbischofsheim 022107
Klassik Stiftung Weimar, Weimar 022498
Klášter Sv. Anežky České, Národní
Galerie v Praze, Praha 009801
Klašter Sv. Jiří, Národní galerie v
Praze, Praha 009802
Klasztorna, Muzeum w Grudziądzu,
Grudziądz 034775
Klau, S., Uelzen 142439
Klaus, Bucureşti 085096
Klaus-Groth-Museum und Neue
Museuminsel Lüttenheid,
Heide 019127
Klaus, Dietmar, Marienberg ... 077293
Klaus, Kurt, Güstrow 129928
Klaus, Kurt, Lübeck 130255
Klausberger, Roman, Wien 062997
Klausner, Louisville 121653
Klausner Stadtgalerie, Chiusa .. 025974
Klaussner, Martin, Fürth 141817
Klavdianos, Angelique, Kifissia . 109071
Klaver, H.A.M., Amsterdam 082782
't Klavertje, Antwerpen 063216
Klax-Galerie für Kinderkunst,
Berlin 106331
Klayman, Leon, New York 138401
KLB, Paris 125725
Kleanthous, London 090273
Ruth Klebel & Johannes Weiss,
Bregenz 127719
Kléber, Béziers 067179
Kléber Broc Antic, Dunkerque .. 068416
Kleber, H. & B., Jestetten 141965
Kleber, Horst, Bad Bramstedt ... 074496
Kleboniškiu Kaimo Buities
Ekspozicija, Daugyvenės kultūros
istorijos muziejus-draustinis,
Kleboniškes 030395
Klederdracht en Visserijmuseum,
Bunschoten Spakenburg 032118
Kledo, Tiranë 098527
Klee, Brigitte, Darmstadt 106657
Klee, Erwin, Wiesbaden 078982
Kleff, Jacques, Erbach,
Odenwaldkreis 075627
Klefing, Nordhorn 077802
Kleidt, Stephanie, München 130361

Kleihues, Sigrid, Berlin 074841
Kleijn, Harderwijk 132526
Kleijn, Alexander, Zutphen 083719
Kleijn & Zoon, W., Rijswijk, Zuid-
Holland 132615
Kleim, Obernai 070771
Klein, Essen 106921, 129817
Klein, Leverkusen 077094
Klein, Sceaux 105664
Klein, West Hollywood 125006
Klein & Scholer, Merzig 077357
Klein & Zwickmann, Peter, Bonn 075076
Klein 14 Stichting, Rotterdam .. 112643
Klein Beekman, Zwolle 083734
Klein Chrétien, Lemberg 069383
Klein Kras, Arnhem 082940
Klein Museum, Mobridge 050982
Klein-Reesink, Hamminkeln ... 141904
Klein Saab Museum d'Oude Bolneus,
Woerden 033057
Klein-st-e Galerie, Arneburg ... 105999
Klein Westphal, C., Amsterdam . 082783
Klein, André, Sainte-Colombe-sur-
Seine 073344
Klein, Bertrand, Mulhouse 070392
Klein, Dominique, Bayonne 066379
Klein, Erhard, Bad Münstereifel . 106077
Klein, Eric, Nice 070618
Klein, Erich & Anneliese, Neu-
Ulm 077710
Klein, François, Paris 129049
Klein, Fritz, Hannover 076379
Klein, G. & E., Achim 105957
Klein, H., Köln 076882
Klein, Isabelle, Paris 071526
Klein, Isabelle, Toulouse 073780
Klein, J. Alex, Adenau 074353
Klein, K.-H., Nidda 142231
Klein, Katharina, Pirmasens ... 108382
Klein, Lawrence & Tan, North
Shore 084035, 132733
Klein, Lutz, Stuttgart 078611
Klein, Lysiane, Pézenas 072139
Klein, Martina, Wörrstadt 079044
Klein, Michel, Saint-Ouen 073123
Klein, Peter, Wörrstadt 079045
Klein, Prof. Dr. Peter, Hamburg . 129968
Klein, Robert, Boston 119884
Klein, Stefan, Augsburg 074465
Klein, Stéfan, Beblenheim 067060
Klein, Toan, Toronto 101641
Klein, Werner, Köln 107651
Kleinbauer, Dieter, Merseburg . 077350
Kleindienst, Matthias, Leipzig .. 107837,
137346
Die kleine Galerie, Bad
Brückenau 106046
Die kleine Galerie, Braunschweig 106565
Kleine Galerie, Den Haag 117257
Kleine Galerie, Deutschlandsberg 099973
Kleine Galerie, Ebersbach an der
Fils 106852
Kleine Galerie, Eberswalde 106853
Die Kleine Galerie, Neunkirchen,
Siegerland 108249
Kleine Galerie, Paderborn 078005
Kleine Galerie, Regensburg ... 108463
Kleine Galerie, Venezia 081847
Kleine Galerie, Waiblingen 108835
Kleine Galerie, Wien 003016
Kleine Galerie Bad Waldsee, Bad
Waldsee 106097
Kleine Galerie Eckes, Nieder-Olm 108273
Kleine Humboldt-Galerie, Humboldt-
Universität, Berlin 017194
De Kleine Kunsthandel, Heerlen 083371
Der Kleine Kunstladen,
Regensburg 078172
Das Kleine Museum,
Gelsenkirchen 018627
Das Kleine Museum, Hagnau am
Bodensee 018922
Kleine Schloßgalerie Weiher,
Ahorntal 105964

Das kleine Spielzeugmuseum, Baden-
Baden 016923
Das kleine Wäschemuseum,
Otter 021039
Kleine Welt, Puppen- und
Puppenstubenmuseum, Freiburg im
Breisgau 018481
Kleine, J., Gummersbach 076128
Kleinehelleforth-Kaunitz, Ina, Verl 078791
Kleiner Prinz, Aschau im
Chiemgau 074440
Kleiner, Gottfried, Ober-Ramstadt 130475
Kleiner & Co., Robert, London .. 090274
Kleinert-Nicolas, Isabelle, Paris . 129050
Kleinert, D., Berlin 074842
Kleines Antiquariat Jona,
Andernach 141409
Kleines Bücherhaus, Zürich 144186
Kleines Heimatmuseum Windmühle,
Werdum 022574
Kleines Kirchengeschichtsmuseum
Sankt Gumbertus, Evang.-Luth.
Pfarramt Illesheim, Illesheim . . 019385
Kleines Militärmuseum Burg
Laurenburg, Laurenburg 019940
Kleines Museum – Kinder der Welt,
Frankfurt (Oder) 018398
Kleines Museum am Hafen,
Romanshorn 041888
Kleines Plakatmuseum, Bayreuth 016994
Kleingärtner, Lothar, Natendorf . 126077
Kleinhaarhuis, Almelo 082630
Kleinheinrich, Dr. Josef, Münster 137436
Kleinjan, Willem, Utrecht 083653
Kleinman, Dick, Cleveland 120436
Kleinöder, R., Düsseldorf 075487
Kleinsasser, Graz 127747
Kleinschmidt, I., Bonn 075077
Kleinschmidt, Stefan, Hamburg . 076254
Kleinsorge, R., Köln 076883
Kleinwächter, Helma, Rosbach vor der
Höhe 108513
Kleio, Haarlem 083336
Kleisen, Jacqueline, Amsterdam 082784
Ingrid Kleiss & Rudolf Polzer,
Riedlingen 108500
Kleist, Idstein 076595, 107457
Kleist-Museum, Kleist-Gedenk- und
Forschungsstätte e.V., Frankfurt
(Oder) 018399
Klemenčič-Maj, Dore, Žalec ... 114752
Klement, Anna, Neulengbach ... 140019
Klement, Manfred, Schwerin ... 137505
Klemettimuseo, Kuortane 011057
Klemke, Hilmar M., Düsseldorf . 106798
Klemm Gallery, Siena Heights
University, Adrian 046321
Klemm, Holger, Hochheim am
Main 130052
Klemm, Kristin, Sankt Ingbert . 130617
Klemme, Bad Salzuflen 106087
Klemmer, Gerd, Dortmund 075386,
129866
Klemm's, Berlin 106332
Klenoden, Karlskrona 086775
Klenoden, Piteå 086915
Klenoden Antik, Sundsvall 087090
Kleplisch, Raphaël, Brunoy ... 067524
Klepner, Melbourne 061821
Klepp Bygdemuseum, Kleppe .. 033701
Klepper Museum, Rosenheim .. 021477
Klerksdorp Museum, Klerksdorp 038772
Klesel, Gerhard, Bleicherode .. 075021
Klesper, Karin, Jesteburg 107481
Kless, Elvira, Dresden 129714
Klever Schwanenturm, Kleve ... 019642
Klevering, Amsterdam . 082785, 082786
Klevfos Industrimuseum,
Ådalsbruk 033462
Kley, Werner, Hamm, Westfalen 107332,
137261
Kleyboldt, Alice, Olfen 130492
KLH Gallery, Newtown, New South
Wales 099451

Klick Kindermuseum Hamburg,
Hamburg 018999
Klickitat County Historical Society
Museum, Goldendale 049150
Kliemchen, Franz & Doris,
Wuppertal 079109
Klier, Alois, Frankfurt am Main . 107009
Klier, Heinz, Rosenheim 078274
Klijn, Johan, Den Haag 132476
Klim, Toronto 136685
Klimas, H.-D., Baden-Baden .. 074637
Klimas, Oleksander, Neu-Ulm . 077711
Klimaty, Warszawa 113857
Klimbacher, Dietmar, Innsbruck . 100050
Klimp, de, Venlo 083679
Klincker, D., Kastorf 076734
Klincksieck & Cie., Paris 136950, 141153
Kline Creek Farm, Winfield .. 054475
Kline, Eric, Santa Monica 145572
Kling & Bang, Reykjavik 109243
Kling, Hendrik, Stuttgart 078612
Klinger & Co., Harwich 127239
Klinger, Dietmar, Passau 137469
Klinger, Eberhard, Görlitz 107141
Klinger, Johannes, Aschau im
Chiemgau 106017
Klinger, Klaus, Frechen 075838
Klinger, Udo, Lützelbach 077216
Klinghammer, Esther, Köln .. 107652,
130154
Klinghoffer, Andrea, Kreuzlingen 087545
Klinghoffer, Charlotte Daniela,
Salenstein 126749
Klingner, Bruxelles 063480
Klingspor-Museum, Offenbach am
Main 020972
Klinker, Leslie John, Frespech .. 068614
Klinkhamer, Groningen 083298
Klinkhammer, Ruth, Bornheim, Rhein-
Sieg-Kreis 129619
Klinkhoff, Walter, Montréal 101420
Klinskoe Muzejnoe Obedinenie,
Klin 037146
Klintehuset Antik, Brørup 065621
Klintehuset Antik, Esbjerg 065639
Klio 1, Pleven 064225, 140318
Klio Buchhandlung und Antiquariat,
Zürich 144187
Klismos, Parma 131726
Klis'sche Buchhandlung, Hohenstein-
Ernstthal 141951
Klit Christensen, Bo, Skørping .. 066128
Klittich-Pfankuch, Adelheid,
Braunschweig 106566, 125904, 141636
Klix, Ulrich, Berlin 074843
Kljuev, Nikolaj, Moskva 085186
KLM Enterprises, Dallas 093152
Kloane Galerie im Gwölb, Zwettl 100406
Klocke, Alois, Viersen 078800
Klocker, Dr. Hubert, Wien 100332
Klöble, Orestes, Bensheim 129453
Klöckner, Stefan, Biebergemünd 129587
Klöden, Kurt, Neuss 108252
Klöppelmuseum, Abenberg 016454
Klöpping, Marion, Witten 079032
Klös, Bodo, Lich 107864
Kloess, Werner, Wien 062998
Klöter, Peter, Grafenau, Kreis
Böblingen 076074, 125960
Kloidt-Haus, VG Buch- und
Pressehandel, Essen 141749
Klokkengieterij Museum,
Heiligerlee 032443
Klokkengieterijmuseum, Vries .. 033008
De Klokkenkamer, Groningen .. 083299
Klokkenmuseum, Frederiksoord . 032330
Klokkenmuseum, Hechtel 003687
Klokkenmuseum van de Sint-
Martinusbasiliek, Halle 003665
Klokkergården Bygdetun Lyngdal
Misjonsmuseum, Lyngdal 033760
Klokowski, Magdeburg 077220
Klomp, Nijmegen 083513
Klompenmuseum De Platijn, Best 032068

Klompenmuseum Gebr. Wietzes,
Eelde 032270
Klompenmuseum 't Oale Ambacht,
Goor 032359
Klompenmuseum 't Schöpke Enter,
Enter 032312
De Klomphoek, Beilen 032054
Klondike Gold Rush National Historical
Park, Seattle 053237
Klondike National Historic Sites,
Dawson City 005687
Klondyke, Almere 143003
Klonk & Hartmann, Wetter 130752
Klooster Ter Apel, Ter Apel 032007
De Kloosteresch, Almelo 112111
Klopfer, Herbert, Sankt Gallen . 087783,
133907
Klopstock Museum, Städtische
Museen Quedlinburg,
Quedlinburg 021247
Klose, Bielefeld 106495
Klose, D., Bad Wünnenberg .. 074620
Klose, Gustav, Hamburg 076255, 129969
Klose, Ilka, Würzburg 108975
Kloster Alpirsbach, Klostermuseum,
Alpirsbach 016515
Kloster Chorin, Chorin 017742
Kloster Denkendorf, Denkendorf 017855
Kloster Ebstorf, Ebstorf 018087
Kloster Frauenchiemsee,
Frauenchiemsee 018460
Kloster Hadmersleben,
Hadmersleben 018909
Kloster Irsee, Irsee 019422
Kloster Isenhagen, Hankensbüttel 019046
Kloster Lüne mit Textilmuseum,
Lüneburg 020180
Kloster Mariawald, Heimbach .. 019172
Kloster Mariensee, Neustadt am
Rübenberge 020752
Kloster Medingen, Bad Bevensen 016730
Kloster-Museum, Sankt Märgen 021604
Kloster Sankt Marienberg,
Helmstedt 019185
Kloster Schöntal, Schöntal 021714
Kloster Seeon, Seeon-Seebruck 021820
Kloster Walsrode, Walsrode ... 022413
Kloster Wiblingen, Museum im
Konventbau und Bibliothekssaal,
Ulm 022261
Kloster Wienhausen, Wienhausen 022626
Kloster zum Heiligen Kreuz,
Kulturhistorisches Museum Rostock,
Rostock 021493
Klosteranlage mit Klostermuseum,
Maulbronn 020319
Klosterbräu Brauereimuseum,
Irsee 019423
Klosterfelde, Berlin .. 106333, 106334
Klostergalerie, Fürstenfeldbruck 107080
Klostergalerie, Zehdenick 022844
Klostergårdens Antik, Skara ... 086931
Klosterkælderen Antik og Porcelæn,
Mariager 065996
Klosterkirche Blaubeuren,
Blaubeuren 017414
Klosterkirche Himmelpfort,
Fürstenberg, Havel 018539
Klosterlund Museum og Naturcenter,
Engesvang 010041
Klostermann, K., Brand-Erbisdorf 075115
Klostermann, Rudolf, Havixbeck 130023
Klostermühle, Boitzenburger
Land 017472
Klostermühlenmuseum,
Thierhaupten 022148
Klostermuseum, Roggenburg .. 021462
Klostermuseum der Benediktiner-Abtei,
Ottobeuren 021047
Klostermuseum der Sankt
Josefskongregation, Ursberg .. 022286
Klostermuseum Disentis, Disentis 041444
Klostermuseum Heggbach,
Maselheim 020313

Klostermuseum Hirsau, Calw .. 017715
Klostermuseum in der
Pfarrkirche Sankt Magnus, Bad
Schussenried 016870
Klostermuseum Jerichow,
Jerichow 019461
Klostermuseum Müstair, Müstair 041780
Klostermuseum Sankt Georgen, Stein
am Rhein 042033
Klostermuseum und Caspar Wolf-
Kabinett, Kulturstiftung St. Martin,
Muri 041782
Klostermuseum Wittichen,
Schenkenzell 021641
Klosterruine Memleben,
Memleben 020359
Klostersammlung der
Benediktinerabtei,
Michaelbeuern 002384
Klostertal Museum, Dalaas 001808
Klostertunet, Sunnhordland
Folkemuseum, Halsnøy Kloster 033631
Klostervelvet Antik, Stockholm . 087016
Klostervilla Adelberg, Adelberg . 016463
Klotz, Alan, New York 122782
Klotz, Carol, Atlanta 092149
Klotz, Christa, Passau 108362
Klotz, Ferdinand, Landau in der
Pfalz 077005
Klotz, François, Obernai 070772
Klotz, Martine, Paris 129051
Klotzbücher, Bad Reichenhall . 074572
Klotzer, Mira & Dieter, New York 135946
Kluane Museum of Natural History,
Burwash Landing 005508
Klub Artystyczny Rio, Warszawa 113858
Klub Galeria Kisielice, Poznań .. 035117
Klubbslaget, Mariestad 126641
Kluckert, Torsten H., Berlin 129514
Kluckow, Alan, Sunningdale ... 119180
Klucznik Galeria, Gdańsk 113476
Klüber, Weinheim 078907
Klügel, Walter, Wien 140094
Klümann, Hans-Gerd, Rheine . 078226
Klün, Anja, Ludwigsburg 103240
Kluge, von, Zürich ... 087983, 133986
Klughammer, Markt Indersdorf . 130289
Kluizenaar, M. de, Leiden 083434
Klujian & Sons, Mathew, Chicago 092735
Klun, Ernesto, Dallas 093153
Klute, Felix, Schmallenberg ... 078376
Kluuvin Galleria – Glogalleriet,
Helsingin Kaupungin Taidemuseo,
Helsinki 010837
Klyne Esopus Museum, Ulster
Park 053880
KM Antiques of London, New
York 095515
KM Art, Milwaukee 121929
Kmetty János Múzeum and Kerényi
Jenő Gyűjtemény, Ferenczy Múzeum,
Szentendre 023956
Kmieć, Piotr, Toruń 084682
KN Gallery, Chicago 120171
Knaap, Nancy, Oosterlittens .. 132606
Knäpper & Baumgarten,
Stockholm 115898
Knall, Stockholm 115899
Knapp, E.L., Louisville 094568
Knapp, Reinhard & Karola,
Reichenbach, Kreis Birkenfeld . 078183
Knapp, Siegfried, Isny im Allgäu 076627
Knappe, Silkeborg 140654
Knappogue Castle, Quin 024978
Knapstein, Sundern 078653
Knaresborough Castle and
Old Courthouse Museum,
Knaresborough 044738
Knauf-Museum, Iphofen 019419
Knauf, D., Dahlem 075291
Knaus, Albrecht, München 137410
Knaus, Bernhard, Frankfurt am

Main 107010, 137202, 141777
Knauß, Maren, Helgoland .. 076487,
141938
Knaut Rhuland House Museum,
Lunenburg 006229
Knauth, Jens, Chemnitz 129654
Knebworth House, Knebworth .. 044740
Knecht & Wrzal, Neuchâtel 133871
Knecht, Alfred, Karlsruhe 107516
Knee Hill Historical Society Museum,
Three Hills 007141
Kneer, Monika, Munderkingen .. 130435
Kneidinger, Thomas, Linz 062717
Kneipp-Museum, Bad Endbach . 016763
Kneisel, Bernd, Aschaffenburg . 074434
Knepper, G.& K.-H.,
Güntersberge 076121
Kner Nyomdaipari Múzeum,
Gyomaendrőd 023656
Knick and Knack, Bangalore ... 079351
Knickknack, Phoenix 096221
Knicks Knacks Emporium, Sutton-on-
Sea 091522
Kniebel, Eva, Nürnberg 077829
Knieß, Uwe, Berlin 074844
Knife River Indian Villages National
Historic Site, Stanton 053491
Knight, Houston 093718
Knight, Singapore 114555
Knight Museum, Alliance 046413
Knight & Sons, B.R., Saint Ives,
Cambridgeshire 091190
Knight, Darren, Waterloo 099866
Knights Antiques, Henley-on-
Thames 089510
Knights of Columbus Museum, New
Haven 051263
Knightsbridge, Tampa 097567
Knightsbridge Coins, London ... 090275
Knightsbridge Gallery, Newent .. 118794
Knigočej, Moskva 143641
Knijff, Zeist 083712
Kninski Muzej, Knin 008928
De Knipoog, Gouda 083267
Knippschild, Hans-Jürgen,
Meschede 077359
Knižnaja Lavka, Sankt-Peterburg 085245,
143668
Knižnaja Lavka Architektora,
Moskva 143642
Knižnaja Lavka Pisatelej, Moskva 143643
Knižnaja Lavka Vostokoveda,
Moskva 143644
Knižny Magazin N60, Sankt-
Peterburg 143669
Knižnyj Aukcion v Centralnom Dome
Literatorov, Moskva 126475
Knjižara i Antikvarijat Srpske Književne
Zadruge, Beograd 143682
Knjižna Starinarna, Radovljica .. 143700
Knobloch, Stephan, Frankfurt am
Main 129850
Knock Gallery, Crathie 117642
Knock Museum, Knock 024950
Knock on Wood, Calgary 128152
Knockengorroch, Castle Douglas 117542
Knödgen, Koblenz 076836
Knoedler & Co., New York 122783
Knödler, A., Blumberg 075026
Knöll, Friederike, Wien 062999
Knöll, Niklaus, Basel . 087240, 133646
Knöll, Niklaus, Bern .. 116140, 133680
Knöll, Thomas, Basel . 087241, 133647
Knöll, Thomas, Berlin . 074845, 129515
Knörle-Jahn, Cornelia, München 130362
Knoester, Den Haag 083093
Knol, Reinier, Leeuwarden ... 112550
Knole, Sevenoaks 045762
Knoll, Hans, Dillingen, Saar ... 129686
Knoll, Hans, Wien .. 100333, 136596
Knollmayer, Othmar, Tulln 062863,
127887
Knollwood, New York 095516
Knoor, Rheurdt 078232

Knoor, Heiner, Issum .. 076631, 130081
Knop, Klaus, Görlitz 076037
Knopek, Nadja, Köln .. 076884, 126007
Knopp, Klaus, Ohlsdorf 062760
Knops, C.J.M., Munstergeleen .. 132590
Knops, Michel, Sittard 112677
Knospe, Herrstein 076512
Knott House Museum,
Tallahassee 053658
Knotty Pine Auction Gallery, West
Swanzey 127494
Knous, Bill & Sue, Denver 093306
Knowles, Beverley, London 118417
Known Space Books, Adelaide . 139616
Knox, Denver 120738
Knox County Historical Society
Museum, Edina 048476
Knox County Museum, Benjamin 046916
Knox County Museum, Knoxville 050002
Knox's Headquarters State Historic
Site, Vails Gate 053922
Knoxville Museum of Art,
Knoxville 050012
Knud Rasmussens Hus,
Hundested 010147
Knudaheio – Garborgheimen,
Undheim 034054
Knudsen, Erhardt, Odense 066039
Knudsen, Helle, Stockholm 115900
Knudsen, Lars, Millfield 099342
Knüppel, Helga, Hamburg 107284
Knüttel, Kurt & Maria, Königstein im
Taunus 130179
Knuf, Frits, Vendôme 141377
Knuht, A., Berlin 074846
Knust, Sabine, München 108147
Knuth, Potsdam 142308
Knuts, Bruxelles 063481
Knutsford Antiques Centre,
Knutsford 089732, 134525
Knutsford Heritage Centre,
Knutsford 044742
Knutsson, Stockholm 087017
Knutsson, Bo, Vänersborg 087134
Knyrim, Brigitte, Regensburg .. 108464
Knysna Museum, Knysna 038773
Ko-Ra, Skopje 111885
Kô-Zen, Montréal 101421
KOA Art Gallery, University of Hawaii,
Kapiolani Community College,
Honolulu 049538
Koala Country Commodities, Port
Macquarie 062096
Kôbányai Helytörténeti Gyüjtemény,
Budapest 023499
Kobariški Muzej, Kobarid 038526
Kobe Doll Museum, Kobe 028844
Kobe Lamp Museum, Kobe 028845
Kobe Lampwork Glass Museum,
Kobe 028846
Kobe Maritime Museum, Kobe . 028847
Kobe-shi Fukushoku Hakabutsukan,
Kobe 028848
Kobe-shi Kagakukan, Kobe 028849
Kobe-shiritsu Hakubutsukan,
Kobe 028850
Kobelius, E., Lassan 130216
Kober, Köln 076885
Kober, Wertheim 108882
Koberstein, E., Kleinostheim .. 076820
Kobie – Serie Bellas Artes,
Bilbao 139059
Kobijutsu Kaito, Kyoto 082028
Kobisenmühle, Sankt Georgener
Museen, Sankt Georgen im
Schwarzwald 021593
Kobo, Seattle 124556
Kobo Aya, Yokohama 082229
Kobold, Josef, Vilsbiburg 078805
Kobrine, Danièle, Paris 071527
Kobrossi, Mouna, Jounieh 111775,
142953
Kobs, Detlef, Ausacker 074478
Kobsik, Meersburg 077321

Kobsik, Oberstaufen 077877
Kocábek, Praha 065503
Kocaeli Devlet Güzel Sanatlar Galerisi,
İzmit 042976
Kocaeli Müzeleri ve Örenyerleri,
İzmit 042977
Kocal, A., Sandhausen 078347
Koch, Den Haag 132477
Koch, Wolfenbüttel 079053
Koch & Englert, Gießen 107128
Koch & Kesslau, Berlin 106335
Koch-McErlain, New York 095517
Koch-Westenhoff, Margot,
Lübeck 107914
Koch, Anette von, Stockholm .. 087018
Koch, Carla, Amsterdam 112239
Koch, Horst, Idar-Oberstein ... 076594
Koch, Ingrid, München 142171
Koch, Jean-Marie, Schlierbach . 073462
Koch, Jochen, Reichartshausen . 130572
Koch, Jürgen, Berlin 141541
Koch, Jürgen, Hannover 107371
Koch, Matthias, Duderstadt ... 075437
Koch, Michaela, Minden 130307
Koch, O., Braunschweig 075127
Koch, Peter, Suhl 130698
Koch, Richard, Münster 077666
Koch, Robert, San Francisco ... 124318
Koch, Silvia, Berlin 129516
Koch, T., Nieheim 108279
Koch, Wilfried, Neubrandenburg 077714
Kochan, Pat, Dallas 120619
Kochan, W., Bremen 141652
Kochanek, Frank, Gelsenkirchen 075967
Kocher & Allemann, Selzach .. 116636
Kochi Bungakukan, Kochi 028856
Kochi-ken Kaitokukan, Kochi .. 028857
Kochi-kenritsu Bijutsukan, Kochi 028858
Kochi Liberty and People's Rights
Museum, Kochi 028859
Kochinian Ovanes, San Francisco 097266,
124319
Kochukyo, Tokyo 082170
Kocioł Artystyczny, Kraków ... 113608
Kocken, Willi, Kevelaer 076768, 107550
Kockler, Paul, Le Luc 069297
Kocsimúzeum, Bábolna Nemzeti
Ménesbirtok, Bábolna 023411
Kocsis-Seger, Ilse, Stuttgart ... 130678
Kodai Orient Hakubutsukan,
Tokyo 029530
Kodai Yuzen-en Bijutsukan, Kyoto 028942
Kodály Zoltán Emlékmúzeum és
Archívum, Budapest 023500
Kodama Museum, Kagoshima .. 028745
Kodansha International, Tokyo .. 137783
Kodisjoen Kotiseutumuseo,
Kodisjoki 011013
Koditek, Ursula, Emmerich am
Rhein 106881
Kodner, Saint Louis ... 096725, 123830
Kodolányi Emlékmúzeum, Vajszló 024029
Kodric, Friederike, Salzburg ... 062804
Kodymý, Miroslav, Praha 128349
Kö-Galerie am Kaisergarten,
Siegen 108619
Koe, Lilian, Petaling Jaya 137814
Köbelin, Rainer, München 142172
København Museum for Moderne
Kunst, København 010177
Københavns Antik, København . 065925
Københavns Antikcenter,
København 065926
Københavns Auktioner,
København 125452
Københavns Bymuseum,
København 010178
Köberle, Ulrich, Überlingen ... 078746,
078747
Købmands Gårdens Antik, Skive 066119
Koechlin, Dr., Basel 144022
Köchling, Marie-Theres, Münster 077667
Köck, Rückersdorf 062781
Köck, Günter Frank, München .. 130363

Köckert, Anne Marie, Bad Homburg
v.d.Höhe 074520
Køge Museum, Køge 010202
Köhl, Peter H., Saarbrücken ... 142343
Köhler, Ottobrunn 108347
Köhler, Wittlich 108954
Köhler, Christian,
Mönchengladbach 108015
Köhler, Günther, Wien . 063000, 127938
Köhler, Heiko, Frankfurt (Oder) . 075764
Köhler, Josef, Erwitte 075660
Köhler, M., Naumburg, Saale ... 077690
Köhlereimuseum, Hasselfelde .. 019090
Köhn, J., Eching, Kreis Freising . 075563
Köhn, Joachim, Neufahrn bei
Freising 077727
Köhnlein, Werner, Reutlingen ... 108483
Köhrmann, G., Oldenburg 077932
Koek, T.H., Beverwijk 132440
Kölcsey Ferenc Emlékszoba,
Szatmárcseke 023918
Köll, Karl-Heinz, Seefeld in Tirol 127872
Koeller, Frédéric, Beure 067153
Kölling, Torsten, Burg bei
Magdeburg 075222
Köllisch, Nürnberg 077830
Köllner, Isabel, Weimar 142472
Köln International School of Design
(KISD), Fachhochschule Köln,
Köln 055616
Kölner Graphikwerkstatt, Köln .. 107653
Kölner Münzkabinett, Köln 076886
Kölner Münzstube, Köln 076887
Kölner MuseumsBulletin, Berichte,
Forschungen und Aktuelles aus den
Museen der Stadt Köln, Köln .. 138706
Kölner Weinmuseum, Köln 019691
Kölnischer Kunstverein, Köln .. 019692,
058947
Kölnisches Stadtmuseum, Köln . 019693
Koelsch, Houston 121075
Kölsch, Hans-Werner, Wuppertal 108993
Koeltz, Sven, Königstein im
Taunus 142025
Kölzer, Arnold, Mastershausen . 077307
Koenders, M.C.J., Voorthuizen .. 132662
Köndgen, Heinrich, Wuppertal .. 142526
Koendo, Iida, Tokyo 082171
Koenig, Bruxelles 063482
König, Darmstadt 075306
König, Großostheim 076112
Koenig, Pézenas 072140
König, Soest 078488
König, Andreas, Vetschau 130724
König, B & R., Ebersdorf bei
Coburg 075561
König, Harry, Deckenpfronn ... 075317
König, Hartmut, Nauen 142218
Koenig, Hermann, Stuhr 078586
König, Johann, Berlin 106336
König, Johann, Waldfischbach-
Burgalben 078839
Koenig, Leo, New York 122784
König, M., Leichlingen 077051
König, Marianne,
Rednitzhembach 142321
König, O., Köln 076888
König, Ursula, Hanau 107337
König, Walther, Berlin . 141542, 141543,
141544
König, Walther, Bonn 141619
König, Walther, Dresden 141699, 141700,
141701
König, Walther, Düsseldorf ... 141721
König, Walther, Frankfurt am
Main 141778, 141779
König, Walther, Köln . 137313, 142014
Koenig, Walther, London 144640
König, Walther, München 142173
König, Walther, Münster 142204
König, Walther, Nürnberg 142247
König, Walther, Stuttgart 142315
König, Wolfgang, Rhodt unter
Rietburg 078233

Königliches Schloss Berchtesgaden
und Rehmuseum, Sammlung
von Herzog Albrecht von Bayern,
Berchtesgaden 017038
Königshaus am Schachen, Bayerische
Verwaltung der staatlichen Schlösser,
Gärten und Seen, Garmisch-
Partenkirchen 018591
Königshofer, Wien 063001
Königspesel, Hooge 019336
Königspfalz Freilichtmuseum,
Tilleda 022157
Koenitz, Leipzig 107838, 142058
Koenitz, Martin, Dresden 141702
Könner, Neustadt am
Rübenberge 077753
Koenraad Bosman Museum,
Rees 021310
Könyvesbolt & Antikvárium,
Szolnok 142583
Könyvtár és Múzeum – Borsodi Tájház,
Muvelodési Központ, Edelény . 023611
Köpings Museum, Köping 040752
Koepp, Jeanne, Berlin 106337
Köppen, Bernd, Berlin 129517
Koerner, Ann, New Orleans ... 095119
Körner, Ralf, Olsberg 130494
Körnig, Uwe, Frankfurt am Main 141780
Körönd, Budapest 109156
't Körrefke, Maastricht 083469
Körs, Wilhelm, Düsseldorf 106799
Kössl, Ried im Innkreis 062774, 136544
Kößler, Johannes, Babenhausen, Kreis
Unterallgäu 074484
Köster, D.C.H., Kiel 076779
Köster, H., Hilden 076529
Köster, Hubert, Braunschweig . 075128
Kösters, Wilhelm, Münster 077668
Koether, Mario, Falkenstein,
Vogtland 075727
Koetser, David, Zürich 116821
Koetsudo, Tokyo 111456
Köyhän Maalarin Kehys- ja
Taidekauppa, Tampere 066397
Köyliön Torpparimuseo, Tuiskula 011431
Középkori O-Zsinagóga, Soproni
Múzeum, Sopron 023892
Középkori Romkert – Nemzeti
Emlékhely, Szent István Király
Múzeum, Székesfehérvár ... 023936
Középkori Zsidó Imaház, Budapesti
Történeti Múzeum, Budapest . 023501
Közlekedési Múzeum, Budapest 023502
Közlekedési Múzeum – Kiskörösi
Közúti Múzeum, Kiskörös 023741
Központi Bányászati Múzeum,
Sopron 023893
Központi Kohászati Múzeum, Országos
Müszaki Múzeum, Miskolc ... 023787
Kofalk, L., Solingen 078498
Koffie- en Theemuseum,
Amsterdam 031951
Koffie- en Winkelmuseum,
Pieterburen 032766
Koffie Verkeert, Rotterdam ... 112644
Koffiebanderij De Gulden Tas,
Bree 003400
Koffler Centre School of Visual Arts,
Toronto 055160
The Koffler Gallery, Toronto ... 007184
Kofukuji Kokuhokan, Nara ... 029145
Koga Masao Museum of Music,
Tokyo 029530
Kogado, Tokyo 082172
Kogal, Bern 116141
Kogando, Osaka 082085
Koganei City Hakenomori Art Museum,
Koganei 028873
Kogei Murata, Tokyo 111457
Kogire-Kai, Kyoto 126224
Koglin, Matthias, Hamburg ... 076529
Kogoj, Oskar, Miren 114736
Kogoj, Oskar, Žalec 114753
Kogsbøll, København 065927

Kogsbøll, Erik, Hellerup 065733
Kogure, Tokyo 111458
Koh Art Gallery, New York 122785
Koh-i-Noor Art Galery, Karlsruhe 107517
Kohán György Múzeum, Gyula . 023675
Kohl Children's Museum of Greater
  Chicago, Glenview 049127
Kohl, Dr. Bernhard A., Stuttgart . 142396
Kohl, Frank, Limburg an der
  Lahn 107870
Kohl, Georg, Rosenheim 078275
Kohl, Marie-Luise, Aachen ... 105941
Kohlbacher, Peter, Saarbrücken . 078324
Kohler, Rorschach 116594
Kohler, Villingen-Schwenningen . 078802,
                            130725
Kohler, Walldorf 078850
Kohler, Douglas B., Columbus . 093029
Kohlhammer, W., Stuttgart 137521
Kohli, Emil, Saanen 087765
Kohlmann, Dres. Matthias & Thomas,
  Gräfelfing 076070
Kohlmann, Linda Maria, Berlin . 074847
Kohlpoth, Offenbach am Main .. 077903
Kohlross, A., Haiger 076155
Kohlweyer, Wolfgang,
  Braunschweig 141637
Kohn, Francine, Saint-Ouen ... 073124
Kohn, Marc-Arthur, Paris 125726
Kohn, Michael, Los Angeles ... 121520
Kohok, Tain-l'Hermitage 105700
Kohtla Kaevanduspark-muuseum,
  Ida 010616
Kohut, Maciej Jan, Torino 132153
Koichi Yanagi, New York 122786
Koine, Napoli 110533
Koine, Roma 131951
Koinz, Pittsburgh 096300
Koishibara-mura Yakimono Bijutsukan,
  Koishibara 028875
Koita, Mory, Milano 080585
Koizumi Yakumo Kinenkan,
  Matsue 029004
Koji Pottery Museum, Sindian . 042458
Kojima Torajiro Kinenkan, Ohara
  Bijutsukan, Kurashiki 028899
Kok & Zoon, A., Amsterdam .. 143044
Koke'e Natural History Museum,
  Kauai 049903
Kokemäen Maatalousmuseo ja
  Sotilaspoikamuseo, Kokemäki . 011014
Kokemäen Ulkomuseo, Kokemäki 011015
Kokerbeck, Bielefeld 074982
Kokerei Zollverein, Essen 018298
Kokh-Wosnitza, Tatiana A.,
  Dienheim 129682
Kokka-sha, Tokyo 138891
Kokkie, Haarlem 083337
Koko, New York 095518
Kokodo, Sendai 082116, 142916
Kokon, New York 122787
Kokon, Tilburg 112681
Kokonozi, Tiranë 098528
Kokopelli, Mönchengladbach .. 108016
Kokorian & Co., Wien 063002
Kokoro, Paris 071528
Kokoschka-Haus Pöchlarn,
  Pöchlarn 002513
Koktanek, Radevormwald 078117
Kokubunji-shi Bunkazai Hozonkan,
  Kokubunji 028876
Kokugado, Tokyo 082173, 132311
Kokugakuin University Archaeology
  Museum, Tokyo 029532
Kokuritsu Kagaku Hakubutsukan,
  Tokyo 029533
Kokuritsu Kagaku Hakubutsukan ,
  Shinju-ku, Tokyo 029534
Kokuritsu Kokusai Bijutsukan,
  Osaka 029245
Kokuritsu Minzokugaku Hakubutsukan,
  Minpaku, Suita 029410
Kokuritsu Miraikan, Tokyo 029535
Kokuritsu Rekishi Minzoku

Hakubutsukan, Sakura 029325
Kokuritsu Seiyo Bijutsukan,
  Tokyo 029536
Kokusai Bijutsu Renmei Nihon Iinkai,
  Tokyo 059502
Kola, Anja L., Boston 119885
Kolaitis, Giannis, Patras 079199
Kolans, Belo Horizonte 100890
Kolar Gallery, Slovenj Gradec .. 114747
Kolarik, Gudrun M, Wien 140095
Kolarin Kunnan Kotiseutumuseo,
  Sieppijärvi 011352
Kolart, Gdynia 113502
Kolb Indonesische Bilderkunst,
  München 077542
Kolb & Son Ltd., Andrew, Bronx 119937
Kolb, Anne-Jacqueline, Bern .. 133681
Kolb, Erwin, Wetzikon 144149
Kolb, Gerhard, Lauterach 127803
Kolb, Hartwig W., Lohr am Main 107891
Kolb, Martin Arnulf, Wien 063003
Kolb, Wolfgang, Ulm 078755
Kolb, Wolfgang & Sylvia,
  Stiefenhofen 078560
Kolbäcks Konst och Antikt,
  Kolbäck 086786
Kolbe, Hildegard, Urbar 078768
Kolbenheyer-Archiv und Gedenkstätte,
  Geretsried 018655
Kolbien, R. & M., Garbsen 107099,
                            137219
Kolbinger, Alfred, Regensburg . 130564
Kolcze, Jozsef, Løkken 065993
Koldehoff, G., Badbergen 074625
Kolding Antik- og Kræmmermarked,
  Kolding 065976
Koldo Mitxelena Kulturunea
  Erakustaret, Donostia-San
  Sebastián 039350
Kolekcija Historyczno-regionalna
  Rodziny Makowskich, Muzeum
  Historyczno-Etnograficzne,
  Chojnice 034688
Kolekcjoner, Poznań 084643
Kolen, I., Montevideo 092004
Kolhammer, Alfred,
  Klosterneuburg 062680
Kolhammer, Alfred, Wien 063004
Kolhapur Museum, Kolhapur .. 024351
Koliaiy, Göttingen 076046
Kolinda Gallery, Alstonville ... 098644
Kolisch, Heinz, Wien 140096
Kolk Raamatuantikvariaat, Tallinn 140681
Kolkmann, Stephan, Coesfeld .. 075266
Kollar, A.J., Seattle 124557
Kollar, Ondrej, Mödling 127821
Kollectible Kingdom, Las Vegas . 094157
Kollector's Kastie, Tucson 097672
Kolledž Iskusstv, Meždunarodnyj
  Vostočno-Evropejskij Universitet,
  Iževsk 056274
Kollegger, Andreas, Kesswil .. 087541
Kollegium der Restauratoren,
  Berlin 129518
Kollekcija, Sankt-Peterburg .. 085246
Kollekcija Drevnostej, Moskva .. 085187
Kollekcioner, Sankt-Peterburg . 085247
Kollekcioner-M, Sankt-Peterburg 085248
Koller, Braine-l'Alleud 125300
Koller, Düsseldorf 125931
Koller, Genève 126725
Koller, London 126965
Koller, München 126057
Koller, Paris 125727
Koller, Shanghai 125410
Koller, Sydney 125237
Koller, Žukovka 126479
Koller Auktionen AG, Zürich .. 116822,
                       126769, 144188
Koller Galéria, Budapest 109157
Koller-Piskernik, Dr. Elke, Sankt
  Gallen 087784
Koller, Andrea, Iserlohn 076619
Kollmann, Kurt, Breddorf 129627

Kollschegg, Uwe, Ahrensburg .. 074361
Kolmanová, Eva, Běrunice u
  Poděbrad 128317
Kolmanskop Museum,
  Kolmanskop 031841
Kolmorgen, København 065928
Koło – Galeria Sztukij Współczesnej
  Związku Polskich Artystów Plastyków,
  Gdańsk 113477
Koło Naukowe Studentów Historii
  Sztuki UKSW, Warszawa ... 059697
Kolodziejski-Zieba, Wieslaw-M.,
  Roding 130591
Kolomoki Mounds State Park Museum,
  Blakely 047032
Kolonaki, Antwerpen 063217
Koloniaal, Antwerpen 063218
Kolonija, Moskva 114268
Kolouch, Vlastimil, Ústí nad
  Labem 065565
Kolpings- und Handwerksmuseum im
  Faltertor, Dettelbach 017878
Kolsva Auktioner, Kolsva 126615
Kolta, Kyïv 117140
Kolthek, F., Amsterdam 082787
Kolumba, Kunstmuseum des
  Erzbistums Köln, Köln 019694
Komagane Kogen Bijutsukan,
  Komagane 028878
Komagata Jukichi Bijutsu Kinenkan,
  Nagaoka 029075
Komárek, Pavel, Praha 128350
Komatsu Hitoshi Art Museum,
  Kyoto 028943
Komatsu Shiritsu Hakubutsukan,
  Komatsu 028880
Komatzki, Günter, Feldmeilen .. 116262,
                            144066
Komecki, Jan, Szczecin 084676
Komet Berlin, Berlin 106338
Komi-Permjackij Okružnoj
  Kraevedčeskij Muzej im. Subbotina-
  Permjaka, Kudymkar 037181
Komi Respublikanskij Istoriko-
  Kraevedčeskij Muzej, Syktyvkar 037909
Komis Meblowy, Warszawa ... 084729,
                            132885
Komiszenkér Kiállítás, Komáron 023757
Komiyama, Tokyo 142928
Komlói Természettudományi
  Gyűjtemény, Komló 023759
Kommandør Chr. Christensen's
  Hvalfanger Museum, Sandefjord 033921
Kommissiehuis Kanapee,
  Amersfoort 082652
Kommissionshaus Sankt Andrä,
  Salzburg 062805
Kommodeskuffen, Aalborg ... 065582
Kommunale Galerie, Darmstadt . 017824
Kommunale Galerie Berlin, Kulturamt
  Charlottenburg-Wilmersdorf von
  Berlin, Berlin 017195
Kommunale Galerie Gelsenkirchen,
  Gelsenkirchen 018628
Kommunale Galerie im Leinwandhaus,
  Frankfurt am Main 018432
Komorowska-Sławiec, Anna,
  Warszawa 132886
Komperon Kotiseutumuseo,
  Viinijärvi 011510
Kompiuterijos Muziejus, Vilnius . 030521
Kompleksi Monumental i Lidhjes
  Shqiptare të Prizrenit, Prizren . 038228
Komponierstube Gustav Mahlers,
  Steinbach am Attersee 002779
Komposch, Helmut, Altach 062516
Kompozitorimus Antano Budriūno
  Kambarys, Vilnius 030522
Kon-Tiki Museum, Oslo 033839
Konak, İstanbul 088128
Konak Knegine Ljubice, Beograd 038147
Konak Kneza Miloša, Beograd .. 038148
Konan-ji Homotsukan, Setoda .. 029367
Konark, Kolkata 079307

Konceptkonstmuseum,
  Rydboholm 040926
Koncz & Társa, Balatonfüred ... 079227
Kondo-Tchouko, Eyoutekebiye, Les
  Abymes 079217
Kondo Yuzo Kinenkan, Kyoto ... 028944
Kondor Gallery, Silverwater ... 099689
Kondras, John, New York 135947
Koneczny, P., London 134705
Konfeksjonsmuseet, Romsdalsmuseet,
  Molde 033782
Kong Shu Ming, Hong Kong ... 065067
Det Kongelige Akademi for de skønne
  Kunster, København 055309
Den Kongelige Mønt- og
  Medaillesamling, Nationalmuseet,
  København 010179
Den Kongelige Mynts Museum,
  Kongsberg 033704
Det Kongelige Nordiske
  Oldskriftselskab, København . 058554
Kongenhus Mindeparks Museum,
  Viborg 010359
Konginkankaan Kotiseutumuseo,
  Konginkangas 011022
Kongkan, Busan 111569, 111570
Kongosho-ji Treasure House, Ise 028698
Kongsberg Antik, Kongsberg ... 084262
Kongsberg Ski Museum, Department
  of Norwegian Mining Museum,
  Kongsberg 033705
Kongsberg Våpenfabrikks Museum,
  Kongsberg 033706
Kongshantang, Beijing 101973
Kongsvinger Festningsmuseum,
  Kongsvinger 033709
Konigorski, Gabriele, Kleve 076823
De Koningshoek, Amsterdam ... 082788
't Koningshof, Deventer 083158
Koninklijk Militair Historisch Museum,
  Armamentarium, Delft 032157
Koninklijk Museum voor Midden-Afrika,
  Tervuren 004057
Koninklijk Museum voor Schone
  Kunsten Antwerpen, Antwerpen 003289
Koninklijk Oudheidkundig Genootschap,
  Amsterdam 031953
Koninklijk Paleis te Amsterdam,
  Amsterdam 031953
Koninklijk Tehuis voor Oud-Militairen
  en Museum Bronbeek, Arnhem 032026
Koninklijke Academie van Beeldende
  Kunsten, Den Haag 056171
Koninklijke Academie voor Beeldende
  Kunsten, Mechelen 055039
Koninklijke Academie voor Schone
  Kunsten, Antwerpen 054996
Koninklijke Academie voor Schone
  Kunsten, Dendermonde ... 055017
Koninklijke Academie voor Schone
  Kunsten Kortrijk, Kortrijk 055032
Koninklijke Hoofdgilde Sint-Sebastiaan,
  Brugge 003416
Koninklijke Kring voor Oudheidkunde,
  Letteren en Kunst van Mechelen,
  Mechelen 058381
Koninklijke Kunstkring Heikracht,
  Lommel 058382
Koninklijke Vlaamse Academie van
  België voor Wetenschappen en
  Kunsten, Bruxelles 058383
Konishi, Los Angeles 094383
Konjušenny Muzej-Muzej
  Avtomobilnogo Transporta, Sankt-
  Peterburg 037748
Konkol, Steffen, Rathmannsdorf 078152
Konkursbuch, Tübingen 137535
Konneveden Kotiseutumuseo,
  Konnevesi 011023
Konoha, Tokyo 111459
Konoki one Artist Museum, Nasu 029161
Konouz, Manama 063150
Konovalenko, Wien 063005
Konrad, Dinslaken 075362

Konrad-Falz, Wolfgang,
Hildesheim 130045
Konrad Theiss Verlag, Stuttgart . 137522
Konrad, Anton H., Weißenhorn . 137550
Konrad, Nils, Ellhofen 075594
Konrad's Furniture Finishing,
Winnipeg 128242
Konservator Gregor, Solna 133562
Konservierung und Restaurierung,
Hochschule der Künste Bern,
Bern 056421
Konservointikeskus, Jyväskylä .. 128420
Konst Akuten, Halmstad 115677
Konst i Fik, Bjärred 115583
Konst i Lund, Lund 115736
Konst och Kultur Konservering,
Falun 133522
Konstantinas Sankenis Memorialinis
Muziejus, Veleniskiai 030501
Konstantino Glinskio Sodyba-Muziejus,
Jurbarka 030342
Konstantinovskij Dvorec, Sankt-
Peterburg 037749
Konstepidemin, Göteborg 115656
Konstfack, Hägersten 056396
Konstföreningen Aura Krognoshuset,
Lund 059886, 115737
Konstfrämjandet, Örebro 059887
Konstfrämjandet i Västerbotten,
Umeå 059888, 115965
Konstfrämjandet i Västmanland,
Västerås 059889, 115982
Konstfrämjandet Skåne, Malmö . 059890,
115762
Konstfrämjandet Stockholms Län
Galleri, Stockholm 115901
Konstfrämjandet Uppsala Län,
Uppsala 059891, 115971
Konstgalleri Vättern, Motala ... 115779
Konstgalleriet i Åhus, Åhus 115572
Konstgrafiska Tryckeriet,
Norrköping 115782
Konsthallen i Hishult, Hishult .. 133538
Konsthallen Köpmanbrinken 2,
Stockholm 115902
Konsthallen Sandviken,
Sandviken 040938
Konsthallen Trollhättan,
Trollhättan 041090
Konsthallen, Himmelsberga Museum,
Borgholm 040578
Konsthistorisk Tidskrift,
Stockholm 139081
Konsthistoriska Sällskapet,
Konstvetenskapliga institutionen,
Stockholm 059892
Konstkällaren, Kalmar 115703
Konstkatalogen, Stockholm 139082
Konstkungen, Malmö 115763
Konstmuseum, Lund 040810
Konstnärernas Riksorganisation,
Stockholm 059893
KonstnärsGården, Romakloster . 115802
Konstnärshuset, Stockholm 059894
Konstperspektiv, Malmö 139083
Konstruktiv Tendens, Stockholm 115903
Konstrum Verkligheten, Umeå .. 115966
Konstsalongen, Helsinki 066329
Konstvärlden & Disajn,
Stockholm 139084
Konstvetenskapliga Institutionen,
Uppsala Universitet, Uppsala .. 056403
Kontainer, Los Angeles 121521
Konteksty. Polska Sztuka Ludowa,
Warszawa 138969
Kontextverlag, Berlin 137078
Kontiolahden Museo, Kontiolahti 011024
Kontny, Roswitha, Gerabronn .. 075980
Kontrast Galerie, Wohlen 116725
Konventmuseum der Barmherzigen
Brüder, Wien 003017
Konvolut, Roma 142816
Konya Devlet Güzel Sanatlar Galerisi,
Konya 043005

Konzeptgalerie Buepa, Berlin ... 106339
Konzett, Philipp, Graz 100012
Konzett, Philipp, Wien 100334
Koo, New York 095519
Koo de Kir Art, Boston 119886
Koo, Alfred, Singapore 085452
Koochiching County Historical Museum
& Bronko Nagurski Museum,
International Falls 049704
Kooiman, P., Ottoland 132607
kooio, Innsbruck 138518
Kooistra, Rotterdam 112645
Kook Museum, Jerusalem 025088
Kook, Petra, Werl 108874
Kookaburra Antiques, Cairns ... 061247
Kookabura Books, Battery Point 139635
Kookaburras Gallery, Bonny Hills 098759
Kookmin College Museum, Seoul 030010
Kool Braun, Negril 111235
Kool Kitchens, Pittsburgh 096301
Kool Kollectables, Las Vegas ... 094158
Kool Toiz Collectibles, El Paso .. 093405
Koolau, Honolulu 120947
Koolen, Françoise, Venterol 105837
Kools, C., Amsterdam 132405
Koompartoo, San Gwann 082428
Koop, Oldenzaal 132602
Koop, Heinz, Papenburg 078015
Koop, Klaus Dieter, Helmut .. 107442
Koop, Wolfgang, Düsseldorf ... 141722
Kooperative Restaurierungen,
Graz 127748
Koopjes Bazar, Almelo 082631
Koopkunst, Rotterdam 112646
Koopman, London 090276
Koopmans-De Wet House Museum,
Iziko Museums of Cape Town, Cape
Town 038679
Koorda Museum, Koorda 001196
Koorenman, Leiden 083435
Koori Kullas, Goulburn 061551
Koorie Heritage Trust Cultural Centre-
Galleries, Melbourne 001277
Koorn, Pieter, Alkmaar 132383
Koornmarktspoort, Kampen ... 032535
Kootenay Gallery of Art, History and
Science, Castlegar 005577
Kootenay School of Art, Nelson . 055146
Kooymans, Den Haag 083094
Kop, Arnhem 082941, 132428
Kop van Zuid, Rotterdam 112647
Kopala, Tbilisi 105908
Kopáncsi Tanyamúzeum,
Tornyai János Múzeum,
Hódmezővásárhely 023690
Kopatz-Baumann, Sigrun,
Homburg 141952
Kopecká, Hradec Králové 128324
Kopeikin, Paul, Los Angeles ... 121522
De Koperen Bol, Enschede 083251
Kopermolen, Vaals 032959
Koperslagersmuseum Van de Beele,
Horst 032512
Kopetz, Hanny, Bayreuth 074695, 141458
Kopiergerätemuseum, Wien 003018
Kopitzki, Michael, Bernau bei
Berlin 074956, 129581
Koplin del Rio, West Hollywood . 115007
Kopp, Bad Homburg v.d.Höhe .. 074521
Kopp, Saarbrücken 078325
Kopp, Rainer, Berlin 074848
Koppány, László, Budapest 109158
Koppe, Hofheim am Taunus 076554
Koppel, Alan, Chicago 120172
Koppelwiser, Schwerin 078429
Koppen, Bussum 083019
Koppreck jun., H., Pattensen ... 078026
Koprikorn, Varna 101195
Kopriva, Manfred, Krems 100085,
136531
Kopsendal, Mook 132589
Koptisches Museum,
Borgentreich 017511
KOPTY Auktione, Vejle 125472

Koraalberg, Antwerpen 100498
Korablinkskij Kraevedčeskij Muzej,
Korablino 037150
Koralewski, Paris 104988
Korallen- und Heimatmuseum,
Nattheim 020641
Korb, Heinrich, Linz 140012
Korber, Ralph, Berlin 074849
Korbinian-Brodmann-Museum,
Hohenfels 019303
Korbmacher-Museum,
Beverungen 017352
Korbmacher-Museum, Malsfeld . 020241
Korbmacher, C., Burg bei
Magdeburg 075223
Kordansky, David, Los Angeles . 121523
Kordegarda, Zachęta – Narodowa
Galeria Sztuki, Warszawa 035323
Kordeuter, Felix, Zürich 087984
Korea Antique Company, Seoul . 082260
Korea Art Gallery, Busan 111571
Korea Art Gallery, Seoul 111681
Korea Craft Artist Association,
Seoul 059529
Korea Culture and Art Foundation,
Seoul 059530
Korea Maritime University Maritime
Museum, Busan 029898
Korea University Museum, Seoul 030011
Korea War Memorial, Seoul 030012
Korean Agricultural Museum,
Seoul 030013
Korean American Museum, Los
Angeles 050440
Korean Art Gallery, Pyongyang . 029875
Korean Central Ethnographic Museum,
Pyongyang 029876
Korean Central Historical Museum,
Pyongyang 029877
Korean Christian Museum, Soongsil
University, Seoul 030014
Korean Culture And Arts Foundation,
Venezia 111132
Korean Fine Arts Association,
Seoul 059531
Korean Furniture Museum, Seoul 030015
Korean International Art Fair,
Seoul 098236
Korean Magazine Museum, Seoul 030016
Korean Museum Association,
Seoul 059532
Korean Museum of Contemporary
Clothing, Seoul 030017
Korean National Committee for the
International Council of Museums
(ICOM), Seoul 059533
Korean National University of Arts,
Seoul 056102
Korean Professional Artists Association,
Seoul 059534
Korean Revolutionary Museum,
Pyongyang 029878
Korean Ski Museum, Ganseong . 029929
Korean Union of Architects,
Pyongyang 059525
Korean War Veterans National
Museum, Rantoul 052467
Korean War Veterans National
Museum, Tuscola 053862
Koreana Art and Antiques, New
York 095520
Koreana Folkcraft, Seoul 082261
Korecký, Dr. Miroslav, Praha ... 065504,
102702
Korekawa Archaeological Museum,
Hachinohe 028579
Koreman, Frans, Amsterdam ... 132406
Koren en Pelmolen De Noordstar,
Noordbroek 032697
Korenke, H., Frankfurt am Main 141781
Korenmolen de Hoop,
Hellendoorn 032452
Korenmolen de Hoop, Klarenbeek 032546
Korenmolen de Phenix, Nes ... 032665

Korenmolen de Zandhaas,
Santpoort 032839
Koreshan Historic Site, Estero .. 048612
Koridon, J.G.M., Almelo 082632
Koriyama City Museum of Art,
Koriyama 028885
Koriyama-shi Fureai Kagakukan,
Koriyama 028886
Korkea Vuoren Antiikki, Helsinki 066330
Korkenzieher Museum Kaiserstuhl,
Vogtsburg im Kaiserstuhl 022352
Korkos, Hong Kong 102090
Korn, Harry, Handschuheim 068786
Korn, M.E., London 144641
Kornbrennerei-Museum, Telgte .. 122119
Kornby, Cairns 061248, 139677
Kornelius-Galerie, Aachen 105942
Galerie Kornfeld & Cie., Zürich . 116823
Kornhamnstorgs Antikvariat,
Stockholm 143954
Kornhaus Georgenthal-Klosterruinen,
Georgenthal 018644
Kornhaus-Museum, Weiler-
Simmerberg 022481
Kornhausforum, Bern 041316
Kornye, Dallas 093154, 120620
Kornye, Fort Worth 120872
Koroit Antiques, Koroit 061703
Korom Gyula Fodrászati Eszközök
Gyűjtemény, Budapest 023503
Koroneou, Eleni, Athinai 109042
Koronowicz, Mark, Prahran 062114
Koropp, Elionore Maria, München 108148
Koroški Muzej Ravne na Koroškem,
Ravne na Koroškem 038598
Koroški Pokrajinski Muzej, Slovenj
Gradec 038613
Korper, Olga, Toronto 101642
Korpilahden Kotiseutumuseo,
Korpilahti 011025
Korppoon Kotiseutumuseo,
Korppoo 011026
Korsch, A., Gilching 137225
Korsnäs Hembygdsmuseum,
Korsnäs 011027
Korsør By- og Overfartsmuseum,
Korsør 010211
Kortárs Művészeti Intézet,
Dunaújváros 055769
Kortás Galéria, Debrecen 079301,
109189
Kortás Galéria, Kecskemét 109200
Kortedala Museum, Göteborg .. 040655
Kortesjärven Luontomuseo,
Kortesjärvi 011028
Kortforlaget, Oslo 137886
Korth, A., Krummhörn 076981
Korth, Benedikt, Treis-Karden .. 078696
Kortjakje, Nijmegen 083514
Kortländer, Karin, München 077543
Kortmann, U. & E., Nordkirchen 108286
Kortright Centre Museum,
Woodbridge 007487
Kortrijk 1302, Kortrijk 003757
Kortum, Leipzig 077061
Koru, Hong Kong 102091
Koru-Kivi Antiques, Helsinki ... 066331
Koru NZ Art, Dunedin 112884
Korvenkylän Nuorisoseuran Museo,
Muhniemi 011165
Korvensuun Voimalaitos- ja
Konepajamuseo, Mynämäki .. 011176
Koryo Museum of Art, Kyoto .. 028945
Koryo Taehakyo Pakmulgwan,
Seoul 030018
Koryu-ji Reihoden, Kyoto 028946
Kosanji Temple Museum, Setoda 029368
Koschatmuseum, Klagenfurt ... 002190
Koschel, Anja, Bergisch Gladbach 129460
Koscielak, Gosia Malgorzata,
Chicago 120173
Kosciusko-Society of Solothurn,
Solothurn 042004

Kosciuszko-Museum, Kosciuszko-
Gesellschaft Solothurn,
Solothurn 042004
Kosetsu Bijutsukan, Kobe 028851
Koshare Indian Museum, La
Junta 050044
Kosian, Silvia, Wien 100335
Kosido, Ruse 101133
Kosken Kotiseutumuseo, Koski . 011029
Koskenniskan Mylly- ja Kievarimuseo,
Tuupovaara 011456
Kosman, Louise, Edinburgh 117731
Kośmicki, Michał, Poznań 132860
Kosnowski, Andrzej J., Warszawa 143523
Kosova Graščina, Gornjesavski Muzej
Jesenice, Jesenice 038521
Kosowski, Włodzimierz,
Warszawa 132887
Kossann, Roger, Bremen 129634
Kossert, Oschatz 077955
Kossuth Lajos Emlékmúzeum, Magyar
Nemzeti Múzeum, Monok . . . 023796
Kossuth Múzei, Kútahya 043012
Kossuth Múzeum, Cegléd 023580
Kossuth Múzeumhajó, Közlekedési
Múzeum, Budapest 023504
Kost, Gabriela, Hamburg 076257
Kostanajskij Memorialnyj Muzej im. I.
Altynsarina, Kostanaj 029823
Kostanajskij Oblastnoj Istoriko-
Kraevedčeskij Muzej, Kostanaj 029824
Kostede Museum für Überlebenskunst,
Kamen 019486
Kostens hus, Luleå 040802
Kostercraft, North Richmond . . . 061997
Kostia, Jacques, Lyon 128717
Kostorek, Radoslav, Olomouc . . 065432
Kostoulakos, Peter, Lowell 135771
Kostromskoj Gosudarstvennyj Istoriko-
architekturnyj Muzej-zapovednik,
Kostroma 037154
Kostromskoj Gosudarstvennyj
Obedinennyj Chudožestvennyj Muzej,
Kostroma 037155
Kostrzewa, Bernard, Blénod-les-
Toul 067244, 128472
Kostuummuseum De Gouden Leeuw,
Noordhorn 032699
Koszta József Múzeum, Szentes 023965
Kotalla, Ralf, Haigerloch 129936
Koterba, Edward, Los Angeles . . 135725
Kotikadun Mainio, Helsinki 066332
Kotiseutumuseo, Jurva 010925
Kotiseutumuseo Lukkarin Puustelli,
Loppi 011126
Kotiseutumuseo Muina, Rauma . 011303
Kotiseutumuseo Pellava ja Sahti,
Alfred Kordelinin Ajokalumuseo,
Lammi 011087
Kotiseututalo Kahari, Noormarkku 011193
Kotkaniemi – Presidentti P.E.
Svinhufvudin koti, Luumäki . . . 011131
Koto-ku Fukagawa Edo Siryokan,
Tokyo 029537
Koto-ku Nakagawa Funabansho
Museum, Koto Community
Foundation, Tokyo 029538
Kotobuki, Olching 108333, 126089
Kotohira-gu Museum, Kotohira . . 028888
Kotomiz, Tokyo 137784
Kotsu Kagaku Hakubutsukan,
Osaka 029246
Kottmann, Renate, Oberstaufen . 108317
Kottmeier, Hans-Jörg, Trier 078705,
. . . . 142420
Kotuku Gallery, Whataroa 113117
Kotulova Dřevěnka, Muzeum Těšínska,
Havířov 009553
Kotzebue Museum, Kotzebue . . . 050020
Kotziok, Roman, München 108149,
. . . . 137411
Kougea, Eleni, Athinai 109043
Kougeas, Ioannis, Athinai 142538
Koukos, Nikol., Thessaloniki . . . 109090

Koulumuseo, Somero 011365
Koulumuseo, Helsingin
Kaupunginmuseo, Helsinki . . . 010838
Kourion Museum, Episkopi 009402
Kouros, New York 122788
Kourtidis, Ioan E., Athinai 079169
Kousbroek, Rudy, Arnhem 112333
Kouseiroudoushou Sangyo Anzen
Gijutsukan, Tokyo 029539
Koušek, Zdeněk, Praha 128351
Koussam, Cannes 103516
Kouvolan Lukusoppi, Ay, Kouvola 140732
Kouvolan Taidemuseo, Kouvola . 011035
Kova, Witka & Dr. László,
Hamburg 137248
Kovacek & Zetter, Wien 063006, 100337
Kovacek, Regine & Michael,
Wien 100336
Kovacic, Wien 063007
Kovács Margit Állandó Kerámia
Kiállítás, Városi Muvészeti Múzeum,
Győr 023663
Kovács Margit Kerámiagyűjtemény,
Ferenczy Múzeum, Szentendre 023957
Kovács, Endréné, Budapest 079264
Kovacs, Patrick, Wien 063008
Kovář, Pavel, Praha 065505
Kovaski Muzej Kropa, Muzeji
radovljiške občine, Kropa 038536
Kovčeg, Moskva 114269, 137925
Kovčeg, Odesa 117159
Koveks 91, Sofia 101163
Kovels on Antiques and Collectibles,
Pepper Pike 139413
Kovjoki Museo, Kovjoki 011038
Kowal, Jean-Michel, Roanne . . . 072491
Kowalkowski, Michał, Sopot . . . 143502
Kowall, Heinrich, Mödling 062740
Kowallik, B., Neuss 142226
Kowalska, Krystyna, Wrocław . . . 113943
Kowalski, Andrzej, Warszawa . . . 132888
Kowalski, Stefan, Potsdam 078078
Kowasa Gallery, Barcelona 114923,
. . . . 143720
Kown Young Ok, Roma 081415
Koxinga Museum, Tainan 042486
Koyama Keizo Bijutsukan,
Komoro 028884
Koyama, Tomio, Tokyo 111460
Koyanagi, Tokyo 111461
Koyasan Reihokan, Koya 028889
Koyman, Toronto 101643
Koyoen Pastel Society,
Nishinomiya 059503
Koyunoğlu Şehir Müzesi, Konya . 043006
Koz, Mandurah 099273
Kozak, Philippe, Valenciennes . . 073945
Kozáková, Eva, Praha 140521
Kozan, Chicago 120174
Kozin, Chiba 111244
Kozloff & Meaders, Pittsburgh . . 096302
Kozminsky, Melbourne 061822, 099310
Kozu Kobunka Museum, Kyoto . . 028947
KPM Welt, Berlin 017196
De Kraal, Venlo 083680, 083681
Kraay, Tonnie, Aalst 132377
Krabb, Herbert, Kuchl 062699
De Krabbedans, Eindhoven 112460
Kracer, Miami 121818
Krächan, Klaus, Quierschied . . . 078112
Kraeckel-Hansum, Thomas,
Hildesheim 130046
Kraemer & Cie., Paris 071529
Kraemer & Hansen, Osnabrück . 142277
Krämer, Anton, Mitterteich 077377
Krämer, Karl, Stuttgart 137523
Krämer, Klaus, Eisenheim 075593
Krämer, Matthias, Henneberg . . . 076492
Krämer, Michael, Meiningen 077329
Krœmmerhjørnet, Græsted 065709
Kränz, Regine, Hamburg 119970
Kränzl, Göppingen 107137
Kränzl, Trude, Gaienhofen 107096
Krätzer, Hermann, Hannover . . . 076380

Krätzler, W. Mona, München . . . 077544
Kräutler, Anton, Koblach 062681, 127789
Kraevaja Detskaja Kartinnaja Galereja,
Vladivostok 038032
Kraevedčeskij Muzej Evrejskoj
Avtonomnoj Oblasti, Birobidžan 036900
Kraevedčeskij Muzej g. Lomonosova,
Lomonosov 037203
Kraevedčeskij Muzej g. Puškina,
Puškin 037674
Kraevedčeskij Muzej g. Puškino,
Puškino 037678
Kraevedčeskij Muzej Novgorodskogo
Rajona, Velikij Novgorod 038013
Kraevoj Muzej Istorii Literatury,
Iskusstva i Kultury Altaja,
Barnaul 036884
Krafft & Partner, Gebenstorf . . . 116273
Krafft, Maria-Teresa, Pully 138083
Kraft, Åhus 115573
Kraft, Salt Lake City 096898
Kraft, Alfred, Spiegelau 108655
Kraft, J.L., Chicago 120175
Kraft, Konrad, Bockhorn 129601
Kraft, Stuart, Dallas 120621
Kraft, Thomas, Perg 127822
Kraft, Vera von, Knokke-Heist . . 100743,
. . . . 128066
Kraftwerkmuseum Löntsch,
Netstal 041791
Kragh & Halling, Charlottenlund 102789
Krahforst, J., Datteln 075314
Krahnstöver, Matthias, Kohren-
Sahlis 130181
Krahuletz-Museum, Eggenburg . 001843
Kraichgau, Sinsheim 078483
Kraichgaubooks, Mosbach 142131
Krajanské Múzeum MS Bratislava,
Matica Slovenska, Bratislava . 038312
Krajčírik, Ľubomír, Levoča 133053
Krajnja Chata, Luc'k 117152
Krajská Galerie Výtvarného Umeni ve
Zlíně, Zlín 009983
Krajské Múzeum, Prešov 038430
Krajské Muzeum Cheb, Cheb . . 009515
Krajské Muzeum Sokolov,
Sokolov 009888
Kråkö Båtbyggarmuseum, Kråkö 011039
Krakow, Barbara, Boston 119887
Krakowska, Kraków 113609
Krakowski Antykwariat Naukowy,
Kraków 143478
Kråksundet notnaust Sjøbruksmuseum,
Tustna 034045
Kral, Oakland 123255
Kral, Markus Johann, Hörsching 062637
Kral, Werner, Baden bei Wien . . 062531
Kralik, Wien 127939
Kralingsmuseum, Rotterdam . . . 032806
Královská Mincovna v Jáchymove,
Jáchymov 009592
Královský Letohrádek, Praha . . . 009983
Kramboden, Espergærde 065642
Kramboden, Holbæk 065775
Krambuvika Bygdemuseum,
Tennevoll 034013
Kramer, Minneapolis 122008
The Kramer House, Baltimore . . 119748
Kramer, D., Offenburg 142261
Kramer, E., Offenburg 142262
Kramer, Edith, Wackersberg . . . 108831
Kramer, Karin, Berlin 137079
Kramer, Waldemar, Frankfurt am
Main 137203
Kramer, Wilhelm, Volkmarsen . . 078815
Kramm, Hans, Düsseldorf 129745
Kramm, Uwe-Karsten, Stuttgart . 142397
Krammig & Pepper, Berlin 106340
Kramp & Kramp, Lemgo 130236
Krampitz, Hans-Jochen, Bremen 075169
Krampitz, Peter, Flensburg 075745,
. . . . 075746
Kramps, Moers 077407
Krampus, Trieste 081794

Kramski, M., Burgdorf 106619
Krane Galleri, Tromsø 113284
Kranenburg, A., Borken,
Westfalen 106548
Kranendonk, Amsterdam 082789
Das Kranichhaus, Museum des alten
Landes Hadeln, Otterndorf . . . 021042
Krankenhaus-Museum am Klinikum
Bremen-Ost, Bremen 017591
Krankenhausgalerie, Castrop-
Rauxel 106625
Krankenhausmuseum, Nürnberg 020864
Krannert Art Museum, University of
Illinois, Champaign 047519
Kranz, Edith, Hennef 107415
Kranz & Sohn, Kristina,
Braunschweig 075129, 125905
Kranz, Manfred, Frankfurt am
Main 075800
Kranzbühler, Gero, Osnabrück . . 077970
Krarup & Ingwersen,
Charlottenlund 065627
Kras, Varsselder Veldh 132653
Krasko, Pavol, Nitra 085580
Krasl Art Center, Saint Joseph . 052781
Krāslavas Vēstures un Mākslas
Muzejs, Krāslava 030148
Krasnodarskij Chudožestvennyj Muzej
im. F.A. Kovalenko, Krasnodar . 037158
Krasnodarskij Gosudarstvennyj Istoriko-
archeologičeskij Muzej-zapovednik,
Krasnodar 037159
Krasnodarskij Gosudarstvennyj
Universitet Kultury i Iskusstv,
Krasnodar 056280
Krasnodarskij Vystavočnyj Zal,
Krasnodar 037160
Krasnojarskij Chudožestvennyj Muzej
im. V.I. Surikova, Krasnojarsk . 037162
Krasnojarskij Gorodskoj Vystavočnyj
Zal, Krasnojarsk 037163
Krasnojarskij Gosudarstvennyj
Chudožestvennyj Institut,
Krasnojarsk 056281
Krasnojarskij Istoričeskij Muzej
Memorial Pobedy, Krasnojarsk 037164
Krasnojarskij Kraevoj Kraevedčeskij
Muzej, Krasnojarsk 037165
Krasnojarskij Kulturno-istoričeskij
Muzejnyj Kompleks, Krasnojarsk 037166
Krasnokamskaja Kartinnaja Galereja,
Krasnokamsk 037171
Krasnokamskij Kraevedčeskij Muzej,
Krasnokamsk 037172
Krasnoturinskij Kraevedčeskij Muzej,
Krasnoturinsk 037173
Krasnye Palaty XVII Veka,
Moskva 085188, 114270
Krass, Rosario 098591
Kraszewski-Museum, Stadtmuseum
Dresden, Dresden 017987
Kratochvil, Josef, Klatovy 065404,
. . . . 128328
Kratochwill, Karl, Wien 127940
Krattiger, Roland, Basel 133648
Kratz, Dr. Annette-Isabell,
Hamburg 076258
Kratzer, Gerald, Marchtrenk 062728
Kraus, Nashville 122088
Kraus & Silbernagel, Berlin 125881
Kraus, Barbara, Øyslebø 113189
Kraus, E., Overath 078000
Kraus, Gérard, Versailles 074058
Kraus, Jörg-Rudolf, Hörsching . . 062638
Kraus, Sylvia, Santiago de Chile 064969
Kraus, Werner, München 077545
Krausbauer, Günther, Neu-
Anspach 108236
Krause, Almere 082644
Krause, Iola 138341
Krause, Helmut, Mörfelden-
Walldorf 077403, 130311
Krause, Joachim, Mannheim . . . 142107
Krause, Jochen, Bielefeld 074983

Krause, Oliver, Berlin 129519
Krause, Oskar, Pfäffikon (Zürich) 116564, 138081
Krause, Peter-Jörg, Berlin 106341
Kraushaar Galleries Inc, New York 122789
Krausmann, Walter, Ried im Innkreis 100149
Krauß, Dennis, Perleberg 078036
Krauß, Rudolf, Oelsnitz 077897
Krausz, Ron, München 077546
Kraut, František, Praha 128352
Kraut, Michael, Bleiburg 062538, 099963
Krautgartner, Karin, Ried im Innkreis 062775
Kravets & Wehby, New York ... 122790
Krawarik, Gebr., Wien 063009
Krawczyk, Edmund, Lublin 084628
Kraxenberger, R., Erharting 075647
Krazie Gal's Kloset, Jacksonville 094002
Kražių Kraštotyros Muziejus, Kelmės krašto muziejus, Kražiai ... 030396
Krbušek, Jaroslav, Praha ... 102703
Krchňavý, Telč 065560
Kreations Gallery, Rosebud .. 099653
Kreativhofladen, Estedt 075714
Kreativo, Cancún 111936
Kreatum Interieur, Genk 063618
Sabine Krebber & B. Michael Schmid, Werne 108878
Krebs, Bianka, Fürth 107085
Krebs, Dieter, Frankfurt am Main 075801
Krebs, Elisabeth, Wien 127941
Krebs & Co., Karl Heinz, Bremen 137127
Krebs, Martin, Bern 116142
Krebs, R., Meppen 107994
Krebs, Roswitha, Witten 079033, 079034
Krebs, Thekla, Hamburg 129971
Krebsen, København 102906
Krebsmühle, Oberursel 142256
Kreckovic, L. & E., London 090277, 134706
The Kreeger Museum, Washington 054128
Krefelder Kunstverein e.V., Krefeld 058948
Dieter Kreft & Lilo Kreft-Hirschinger, Dahn 106649
Kreft, I., Spenge 108649
Kregl, Renate, Kirchberg-Thening 062663
Kreil, München 108150
Kreinberg, Sidney, Cleveland ... 092950, 135466
Kreis- und Heimatmuseum, Wittstock/Dosse 022707
Kreis- und Heimatmuseum Bogen, Bogen 017467
Kreis- und Stadtmuseum Dieburg, Dieburg 017883
Kreis- und Universitätsmuseum Helmstedt, Helmstedt 019186
Kreis Feuerwehrmuseum Lohmen, Lohmen, Kreis Güstrow 020106
Kreisagrarmuseum, Dorf Mecklenburg 017938
Kreisbergbaumuseum Sassenroth, Herdorf 019202
Kreischer, Jürgen, Ubstadt-Weiher 078743
Kreisgalerie, Dahn 017809
Kreisgalerie, Mellrichstadt ... 020357
Kreisheimatmuseum, Grimma .. 018819
Kreisheimatmuseum Demmin, Demmin 017853
Kreisheimatmuseum Frankenberg, Frankenberg, Eder 018387
Kreisheimatmuseum Striediecks Hof, Museum Bünde, Bünde 017672
Kreisheimatmuseum Weißes Haus, Rotenburg an der Fulda 021499
Kreisheimatstube, Ellzee 018163
Kreisler, Juan F., Madrid 115244
Kreisler, Lesley, Taranaki 113017

Kreismuseum, Osterburg 021023
Kreismuseum Bad Liebenwerda, Bad Liebenwerda 016817
Kreismuseum Bitburg-Prüm, Bitburg 017403
Kreismuseum Bitterfeld, Bitterfeld 017405
Kreismuseum der Heimatvertriebenen, Seligenstadt 021836
Kreismuseum Finsterwalde, Finsterwalde 018365
Kreismuseum Geilenkirchen, Geilenkirchen 018605
Kreismuseum Heinsberg, Heinsberg 019175
Kreismuseum Herzogtum Lauenburg, Ratzeburg 021294
Kreismuseum Jerichower Land, Genthin 018639
Kreismuseum Lodron-Haus, Mühldorf am Inn 020481
Kreismuseum Oberhavel, Oranienburg 021001
Kreismuseum Peine, Peine 021093
Kreismuseum Prinzeßhof, Itzehoe 019437
Kreismuseum Schönebeck, Schönebeck 021702
Kreismuseum Syke, Syke 022090
Kreismuseum Walderbach, Walderbach 022389
Kreismuseum Wewelsburg, Büren 017673
Kreismuseum Zons, Dormagen . 017941
Kreitz-Matthies, B., Karlsruhe . 076698, 141973
Kreitz, J., Bad Homburg v.d.Höhe 074522
Krekel, Elisabeth, Mengerskirchen 107993
Krell'sche Schmiede, Wernigerode 022581
Kreloufi, Gilbert, La Varenne-Saint-Hilaire 069118
Kremer, Ostbevern 077985
Kremer, Hugo, Ankum 074412
Kremer, Karl, Osnabrück 077971
Kremer, Margret, Spangdahlem . 078513
Kremer, U., Paderborn 078006
Kremmer'n Antikk, Helleland ... 084251
Kremser, Karl-Heinz, Wien ... 063010
Kremstaler, Bangalore 062734
Křenek, Karel, Dobřichovice .. 140469
Křenek, Karel, Praha .. 102704, 140522
Krenzel, Haan 107193
Kréo, Paris 104989
Kreps, Andrew, New York 122791
Kresge Art Museum, East Lansing 048440
Kress, Michael, Birstein 129594
Kress, Willi, Brugg 133706
Kressibucher, Beat, Aadorf 087170, 133606
Kressmann, Vincent, Saint-Sauveur-le-Vicomte 105617
Kresz Géza Mentomúzeum, Budapest 023505
Kretingos Muziejus, Kretinga ... 030397
Kretschmann, Buchholz in der Nordheide 129643
Kretz, Pascal et Carine, Beaugency 066993
Kretzer, Marc Daniel, Kirchhain . 141909
Kreussel, Lutz-Peter, Kulmbach . 142038
Kreuter, Björn, Stockholm 087019
Kreutz, B., Ediger-Eller 075572
Kreutz, Inge, Aschaffenburg ... 074435
Kreutzberger, W.-D., Peine 078032
Kreutzer, Joachim, Saarbrücken 078524
Kreutzer, Johannes, Langgöns .. 130215
Kreutziger, A. Caroline, Worpswede 108970
Kreuz-Galerie, Bülach 116173
Kreuzberg Museum, Bezirksmuseum Friedrichshain-Kreuzberg, Berlin 017197
Kreuzer, Maria, Amorbach 105989

Kreuzhof, Melanie M., Mondsee 100124
Kreuzstadelmuseum, Türkenschlacht-Gedenkraum, Mogersdorf ... 002409
Krevsky, George, San Francisco 124320
Krichel, Mönchengladbach ... 077393
Krichel, B., Zornheim 079148
Krick, Eibelstadt 137172
Krickmeyer, Edgar, Göttingen . 076047
Kriebel, Michael, Stuttgart 130679
Krieg, Oberwil (Basel-Land) 087721, 116550
Krieger & Stahlmann, Düsseldorf 129746
Krieger & Stahlmann, Essen ... 129818
Krieger Lassen, Per, Holte 065786
Krieger, Stephanie, New York .. 095521
Krienke, Christine, Wiesbaden . 130759
Kriivari, Rovaniemi 140743
Krijgsman & Minderman, Den Haag 083095
Krikl, Andreas, Wien 140097
Krikorian Gallery, Worcester Center for Crafts, Worcester 054546
Kriminalmuseum der Kantonspolizei Zürich, Zürich 042203
Krimmel, Karin, Köln 130155
Krings-Ernst Galerie, Köln 107654
Krings, Antonio, Köln 076889
Krins, Werner, Sieverstedt 078469
Krinzinger, Dr. Ursula, Wien ... 100338, 136597
Krippana – Exposition Internationale de Crèches d'Église, Büllingen 003501
Krippen- und Spielzeugmuseum, Bubendorf 041371
A. Krippendorf & D. Lanz, Emmendingen 129792
Krippenmuseum, Dornbirn 001825
Krippenmuseum, Fulpmes 001913
Krippenmuseum, Glattbach 018708
Krippenmuseum, Telgte 022120
Krippenmuseum, Vösendorf ... 002872
Krippensammlung, Großraming . 002012
Krippenstube-Heimatstube, Plößberg 021162
Krips, Maximilian, Köln 107655
Krisal, Carouge 087351
Krischke, Dragan, Berlin 074850
Krishna Menon Museum and Art Gallery, Kozhikode 024371
Krishtie, Bangalore 079352
Krišjāņa Barona Muzejs, Rīga .. 030187
Krispen, Houston 093719
Krißler, Ursula, Kirchheim unter Teck 130127
Krist, Martin, Báhoň 133044
Kristall-Museum, Obergesteln .. 041809
Kristallgrotte, Guttannen 041577
Kristallmuseum, Guttannen 041578
Kristallmuseum, Viechtach 022324
Kristallmuseum Riedenburg, Riedenburg 021416
Kristalmuseum 't Los Hoes, Borculo 032082
Kristen, Dr. Lorenz, Berlin 141545
Kristensen, Anne Marie & Leif, Skive 066120
Kristensen, Frede, Struer 066147
Kristensen, Jette, Brøndby Strand 065616
Kristensen, Mogens, Helsinge .. 065738
Kristianhus Motormuseum, Vik i Sogn 034087
Kristiansand Kanonmuseum, Kristiansand 033715
Kristiansands Auksjonsforretning, Kristiansand 126356
Kristina, Sofia 101164
Kristinehamns Auktionskammare, Kristinehamn 126616
Kristinehamns Konstmuseum, Kristinehamn 040761
Kristiregi Galleries, Las Piñas . 113414
Kristi's Selections, Antwerpen .. 063219
Kristjan Raua Majamuuseum, Eesti

Kunstimuuseum, Tallinn 010678
Kriszio, Olaf, Frankfurt am Main 141782
Kriterion, Società di Restauro Scientifico e Manutenzione, Bologna .. 059466, 130970
Kritische Berichte, Marburg ... 138707
Krivoriz'kyj Krajeznavčyj Muzej, Kryvyj Rig 043190
Křížek, Miroslav, Praha 128353
Krkonošské Muzeum, Jilemnice 009606
Krkonošské Muzeum – Augustiniánský Klášter, Vrchlabí 009958
Krkonošské Muzeum – Tři historické domy, Vrchlabí 009959
Kro Art Gallery, Wien 100339
Krobath & Wimmer, Wien 100340
Kröber, Mirko, Soest 078489
Krøderbanen, Krøderen 033722
Kröger, Brissago 116168
Kröger, Christian, Löhne 130244
Kröger, Edzard, Horst 076577
Kröger, Hinrich, Berlin 106342
Kröger, Roland, Herford 076501
Kröll, Rafaela, Tutzing 078737
Kröller-Müller Museum, Otterlo . 032739
Krönungssaal des Aachener Rathauses, Aachen 016438
Krogsgaard, Egon, Kjellerup .. 065814
Krohn, Katharina, Basel 116082
Krohn, Richard M., Süderbrarup 078641
Kroker, Torsten, Straubing 078576
Krokin Gallery, Moskva 114271
Kroma, Saint Louis 123837
Kronacher Kunstverein, Kronach 058949, 107752
Kronberg, Esbjörn, Stockholm . 087020
Kronberger Antiquariat, Kronberg im Taunus 142037
Kronberger Haus, Frankfurt am Main 018445
Kronbloms Kök, Örebro 040892
Kronborg Slot, Helsingør 010109
Krone, København 065929
Krone, Zürich 087985
Krone, Hans-Dieter, Lünen 137363, 142095
Krone, U., Euskirchen 141758
Kronen Galerie, Zürich 116824
Kronenberg, Basel .. 126708, 144023
Kronenbühl, Rehetobel 116578
Kroning, Ringkøbing 140646
Kronis, Dimitri, Paris 141154
Kronobergs Lantbruksmuseum, VÄXJÖ 041144
Kronwitter, Straubing 078577
Krook, Ulf, Mariefred 133558
Krook, Ulf, Stockholm 133574
Krookilan Kotiseutukeskus, Raisio 011296
Kroos, Brenda, Cleveland 120437
Kropáček, Petr, Písek 065447
Kropka J.W. Śliwczyńscy, Września 137914
Krosna, Moskva 114272
Kroumova Trulli, Gallia, Roma .. 110807
Kroupa, Ernst K., Wien 063011
Kroyer, Helmut, Hamburg 076259
Kru-Kü-Ku-Krei Kleine Galerie, Krumpendorf 100088
Krudttaarnsmuseet, Frederikshavn 010070
Krudtværksmuseet, Frederiksværk 010073
Krüge-Museum, Creußen 017781
Krüger, Korbach 076940
Krueger, Saint Louis 123838
Krüger Espantoso, Germán, Lima 113363
Krüger, Christiane, Rödermark . 108508
Krüger, D., Vechelde 078781
Krüger, Frank, Rheinsberg 078231
Krüger, O. & E., Emsbüren 075616
Krüger, Patrick Urs, Saarbrücken 078327, 130614
Krüger, Peter, Berlin 074851
Krüger, Stefan, Essen 141750

Krüger, Stefan, Köln 142015
Krüger, Wolfgang, Leipzig 077062
Krügersches Haus, Museum Gesthacht,
    Geesthacht 018601
Krug, Manfred, Betzdorf 074967
Krug, R., Schönaich 078383
Kruger, Manfred, Tshwane 038848
Kruger y Tejada, Madrid 086205
Krugier & Cie., Jan, Genève ... 116325
Kruidentuin de Groene Kruidhof,
    Elburg 032295
Kruiersmuseum, Balen 003337
Kruimel, Nico, Bussum 083020
't Kruithuis, Alkmaar 082621
Kruj, Andor, Budapest 079265
Kruk, Gerry, Calgary 064262
Kruk, W., Poznań 084644
Krul, René, Nieuw Vennep 143206
Krulik, Betty, New York 122792
Krumbiegel, Karin, Kaarst 076658
Krumhorn, Adalbert, Hamburg . 076260
Kruml, Pierluigi, Trieste 142877
Krumpendorfer Künstler Kultur Kreis,
    Kru-Kü-Ku-Krei, Krumpendorf . 058268
Krumpeter, Christian, Lübeck . 077175
Krumschmid, Theresia,
    Winzendorf 063119
Kruna, Beograd 114422
Krunk, Sankt-Peterburg 114345
Krupinski, B., Waldkirch 078841
Krupka, Baltmannsweiler 141447
krupp stadt museum, Berndorf . 001777
Krupp, H. Gerhard, Weinbach . 108865
Krupp, Nicolas, Basel 116083
Kruppa, S., Auhagen 074472
Kruså Handelshus, Kruså 065986
Kruschel, Martina, Radebeul .. 108425
Kruse, Auburn 127128
Kruse, Flensburg 106958
Kruse, Hagen 076149
Kruse House Museum, West
    Chicago 054294
Kruse-Klefisch, Herford 076502
Kruse, D., Dipperz 075364
Kruse, Ed, Roswell 127422
Kruse, Eva, Weede 078882
Krusemark-Camin, Elke, Mühlheim am
    Main 077429
Krusenbaum, Carl, Bochum ... 075035
Kruszone, Poznań 137901
Kruydt, La Baule-Escoublac .. 068950
Kruzina, Antoine, Sin-le-Noble .. 073522
Kryms'ke Chudožne Učylyšče im. M.
    Samokiša, Simferopol' 056500
Krymski Respublikanskij Kraevedčeskij
    Muzej, Simferopol' 043300
Krypte en Schatkamer van de O.L.V.
    Basiliek, Halle 003666
Kryptemuseum onder de Sint-
    Hermeskerk, Ronse 003984
Krysztyna Gallery, Dundee ... 117694
Krzisnik, Milan, Werbach 078926
Krzywa Galeria, Toruń 113790
Krzyżanowski, Kraków 143479
Krzyzanowski, U. & M.,
    Waldhufen 142459
KS Art, New York 122793
K's Selection, Chiba 081959
Ksan Historical Village and Museum,
    Hazelton 005997
KSH, Siegen 078463
Ksiegarnia Naukowa, Gdańsk . 143453
Ktema, L'Isle-sur-la-Sorgue ... 069573
Kuala Lumpur Craft Museum, Kuala
    Lumpur 030685
Kuandu Museum of Fine Arts, Taipei
    National University of the Arts,
    Taipei 042520
Kuballe, Reinhard, Osnabrück . 142278
Kubalu Skolas Muzejs, Dundaga 030119
Kubasch, Karl, Heiligenstadt,
    Heilbad 076478

Kube, Jan K., München 077547, 126058
Kube, Jan K., Sugenheim 078646,
                                126126
Kubera, Nice 070619
Kubeš, Dobruška 065380
Kubicki, Józef, Warszawa 143524
Kubin-Haus, Oberösterreichische
    Landesmuseen, Wernstein am
    Inn 002927
Kubiner Heimatstube, Weinsberg 022522
Kubinkabinett im Badhaus,
    Leogang 002296
Kubinyi Ferenc Múzeum, Nógrádi
    Megyei Múzeum Szervezet,
    Szécsény 023921
Kubo, Itsuo, São Paulo 101097
KuboShow, Kunstmesse im Revier,
    Herne 098145
Kuboso Memorial Museum of Arts,
    Izumi, Izumi, Osaka 028728
Kubowicz, Maria i Piotr,
    Warszawa 113859
Kubus Hannover, Städtische Galerie,
    Hannover 019055
Kuc, Berlin 074852, 129520
Kuća Cveća, Muzej Istorije Jugoslavije,
    Beograd 038149
Kucera, Christian, Duisburg 075551
Kucera, Greg, Seattle 124558
Kucharczyk-Łojan, Jadwiga,
    Katowice 084536
Kuckei + Kuckei, Berlin 106343
Kuckó, Eger 079304
Kuczynski, Paulo, São Paulo ... 064187
Kuddnäs Museum, Nykarleby ... 011200
Kudełko, Robert, Zabierzów ... 132920
Kuderski, Piotr, Kraków 084584
Kudlek & Van der Grinten, Köln . 107656
Kudos Gallery, Paddington, New South
    Wales 099527
Kübelbeck, Sylvia, Schaan 082334
Kübler, Armin, Esslingen 075710
Kübli, Astrid, Basel ... 087242, 133649
Küchler, Bernhard, Bad
    Schandau 074588
Küfer-Museum, Langenau 019900
Küferei-Museum, Eberbach 018069
Kügelnhaus, Museum der Dresdner
    Romantik, Dresden 017988
Kühling, Wilhelm, Augsburg ... 074466
Kühn, Hans, Osterode am Harz .. 077989
Kühn, Harald, Alheim 105974
Kuehn, Sara, Bruxelles 063483
Kühn, V. & E., Lilienthal 117867
Kühne, D., Hünxe 076584
Kühne, Frank, Dresden 075419, 106740
Kühne, Franz, Hedingen 144085
Kühnemuth, Reinhard, Berlin ... 129521
Kühner, W., Rümmingen 078308
Kuehnert, Houston 127252
Kühnert, Ruth, Sebnitz 078437, 108602
Kühr, Wolfgang, Jagsthausen ... 107477
Küker, Bernd & Karin, Stadt
    Wehlen 108660
Kümmel, Duderstadt 075438
Kümmel, Lüdenscheid 077199
Kümmerle, Hans & Dagmar,
    Göppingen 141831
Kuenburg Kunststätte, Payerbach 002488
Künker, München 077548
Künker, Fritz Rudolf, Osnabrück 077972,
                                126090
Künkler, Mainz 077236
Künstler – Kritisches Lexikon der
    Gegenwartskunst, München .. 138708
Künstler der Filder, Filderstadt .. 058950
Künstleratelier Gamsburgschule,
    Annaberg 001715
Künstlerbahnhof Ebernburg, Bad
    Münster am Stein-Ebernburg . 016825
Künstlerbund Dresden e.V.,
    Dresden 058951
Künstlerbund Graz, Graz 058269
Künstlerbund Klosterneuburg,

Klosterneuburg 058270
Künstlerbund Mecklenburg und
    Vorpommern e.V. im BBK,
    Schwerin 058952
Künstlerbund Steinburg e.V.,
    Itzehoe 058953
Künstlerbund Tübingen e.V.,
    Tübingen 058954
Künstlerbunker, Leverkusen ... 020057
Künstlerei im Sudhaus, Tübingen 108791
Künstlerforum Bonn, Bonn ... 017493
Künstlergarten Alfred Kurz,
    Lichtenau 002303
Die Künstlergilde e.V., Esslingen 058955,
                     106936,   138709
Künstlergilde Buslat e.V.,
    Pforzheim 058956, 108376
Künstlergilde Salzkammergut,
    Traunkirchen 058271
Künstlergilde Ulm, Ulm 058957
Künstlergilde Wels, Wels 058272
Künstlergruppe Experiment Stuttgart
    e.V., Stuttgart 058958
Künstlerhaus 188 e.V., Halle,
    Saale 107208
Künstlerhaus Berlin, Berlin 106344
Künstlerhaus Bethanien, Berlin . 017198,
                                137080
Künstlerhaus Bremen, Bremen . 106587
Künstlerhaus Büchsenhausen,
    Innsbruck 002126
Künstlerhaus des Berufsverbandes
    Bildender Künstler Niederbayern,
    Passau 108363
Künstlerhaus Dortmund,
    Dortmund 058959, 106705
Künstlerhaus Estis, Hamburg ... 107285
Künstlerhaus Exter, Schloss- und
    Gartenverwaltung Herrenchiemsee,
    Übersee 022244
Künstlerhaus Graz am Landesmuseum
    Joanneum, Graz 001978
Künstlerhaus Hohenossig, Krostitz 107762,
                                137331
Künstlerhaus Hooksiel, Hooksiel . 019337,
                                058960
Künstlerhaus Lenz, Gladenbach . 018706
Künstlerhaus Marktoberdorf, Kunst-
    und Kulturstiftung Dr. Geiger-Haus,
    Marktoberdorf 020299
Künstlerhaus mit Galerie,
    Göttingen 018749
Künstlerhaus Palais Thurn und Taxis,
    Bregenz 099967
Künstlerhaus S11, Solothurn ... 116652
Künstlerhaus Schloß Balmoral, Stiftung
    Rheinland-Pfalz für Kultur, Bad
    Ems 016759
Künstlerhaus Stuttgart, Stuttgart 058961,
                                108714
Künstlerhaus Walter Helm,
    Aschaffenburg 016618
Künstlerhaus Wien, Wien 003019
Künstlerhausgalerie, Karlsruhe . 107518
Künstlerinnenhof Die Höge e.V,
    Bassum 106124
Künstlerkeller, Annweiler am
    Trifels 105993
Künstlerkreis Ortenau, Offenburg 058962
Künstlerverband-Nachrichten
    österreichischer Bildhauer, Wien 138519
Künstlerverband österreichischer
    Bildhauer, Wien 058273
Künstlerverein Malkasten,
    Düsseldorf 058963
Künstlervereinigung Maerz, Linz 058274
Künstlerzeche Unser Fritz 2/3,
    Herne 019218
Künstner, Norbert, Wien 063012
Künzle, Fredy, Lichtensteig ... 133806
Küpper, Eugen, Münster 142205
Küppers, Illerkirchberg 076599
Küppers, C., Mönchengladbach . 077394
Kürbismühlenmuseum, Preding . 002537

Kürkman, Garo, İstanbul 138133
Küstenmuseum, Juist 019476
Küstenmuseum am Bontekai,
    Wilhelmshaven 022660
Küster, Limburg an der Lahn ... 077112
Kütahya Devlet Güzel Sanatlar Galerisi,
    Kütahya 043013
Kütahya Müzeleri ve Örenyerleri,
    Kütahya 043014
Kufa Gallery, London 045017
Kugel, Irmhild, Duisburg 075552
Kugel, Nicolas, Paris . 071530, 104990
Kugler, Bernd, Innsbruck 100051
Kuhl-Linscomb, Houston 093720
Kuhlberg, Brigitte von, Berlin .. 074853
Kuhlemann, Katharina, Bad Homburg
    v.d.Höhe 129411
Kuhlmann, Brandenburg an der
    Havel 075116
Kuhlmann King Historical House and
    Museum, Boerne 047072
Kuhlmann, Dieter & Margot,
    Oberhausen 142252
Kuhlmann, Heinrich, Münster .. 077669
Kuhlmann, L., Bielefeld 141600
Kuhmalahden Kotiseutumuseo,
    Kuhmalahti 011043
Kuhmoisten Kotiseutumuseo,
    Kuhmoinen 011046
Kuhn, Anne-Grit, Gera 129897
Kuhn, Karl-Josef, Düsseldorf ... 106800
Kuhn, Nick, Bath 088392
Kuhnert, Dieter, San Francisco . 136329
Kuhnke, Siegfried, Pähl 078013
Kuijper, Breda 082996
Kuiper, M., Meerssen 132579
Kuivaniemen Kotiseutumuseo,
    Kuivaniemi 011047
Kujumzian, Dicki, München ... 077549
Kuk, Moskva 114273
Kuka-Muzej na Narodniot Cheroj
    Christijan Todorovski Karpoš,
    Naroden Muzej Kumanovo,
    Kumanovo 030616
Kukje Gallery, Seoul 111682
Kukkola Fiskemuseum,
    Haparanda 040687
Kuknip Minsok Pakmulgwan,
    Seoul 030019
Kukolnaja Galereja Vachtanov,
    Moskva 037314
Kukutsch, Elfriede, Spittal an der
    Drau 062844
Kulay-Diwa Art Galleries,
    Parañaque 113411
Kulbach, Gregor, Döttlingen ... 129691
Kulbach, Richart, Heidelberg .. 141928
Kulcsar, Peter, Wien 063013
Kuldīgas Novada Muzejs, Kuldīga 030104
Kulik Art Inuit, Québec 101502
KULIMU, Zeitschrift für Kunst &
    Literatur & Musik, Lappersdorf 138710
Kulíšek, Dr. Ladislav, Znojmo .. 065571
Kulkukirja, Tampere 140746
Kulla Gunnarstorps Mölla,
    Helsingborg 040695
Kullaan Kotiseutumuseo, Kullaa . 011049
Kullängsstugan, Askersund 040693
Kullagårdens Auktioner, Hökerum 126603
Kulli, Susanna, Zürich . 116825, 138112
Kulma-Antikvari, Mikkeli 140736
Kulø, Erry Marie, Oslo 113229
Kulstrunk, Gabriele, Baden 116050
Kulta Galleria Ailio, Jyväskylä . 103162
Kultamuseo Tankavaara,
    Tankavaara 011415
Kultaraami, Lahti 066382
Kulthaus, Haigerloch 107200
Kulttuurien Museo, Helsinki ... 010839
Kultur- und Heimattreff, Tholey . 022151
Kultur- und Kommunikationszentrum
    Lagerhalle Osnabrück,
    Osnabrück 021014

Kultur- und Militärmuseum,
Grafenwöhr 018791
Kultur- und Museumszentrum Schloss
Glatt, Sulz am Neckar 022080
Kultur- und Stadthistorisches Museum
Duisburg, Duisburg 018057
Kultur & Kommunikation, Bern . 116143
Kultur Antiquariat, Hamburg ... 141879
Kultur Bahnhof Eller, Düsseldorf 018034
Kultur-Gut-Museum, Sammlung
Lehner, Aigen im Mühlkreis .. 001691
Kultur-Historisches Museum Grenchen,
Grenchen 041564
Kultura-Extra, Reutlingen 138711
Kultura Weekly, Sofia 138566
Kulturama, Museum des Menschen,
Zürich 042204
Kulturamt Mitte, Berlin 017199
Kulturamt Neukölln, Berlin 017200
Kulturas Books, Washington ... 145637
Kulturbryggarna, Malmö 143896
Kulturbunker Mülheim, Köln .. 019695
Kulturcentrum i Ronneby,
Ronneby 040921
Kulturelles Forum im Freiherr-vom-
Stein-Haus, Langenfeld 019914
Kulturen Zivot, Skopje 138899
Kulturfabrik Hainburg, Hainburg an der
Donau 002042
Kulturforum Alte Post, Neuss .. 020745
Kulturforum Burgkloster und Museum
für Archäologie, Die Lübecker
Museen, Lübeck 020158
Kulturforum Würth, Chur 041416
Kulturforum Würth, Kolding ... 010205
Kulturgården på Lassfolk, Lillby 011116
Kulturgeschichtliches Museum
Bügeleisenhaus, Hattingen ... 019099
Kulturgeschichtliches Museum
Osnabrück, Osnabrück 021015
Kulturgut, Essen 075685
Kulturgut Höribach, Sankt Lorenz 002675
Kulturhalle Tübingen, Tübingen . 022219
Kulturhaus der Otto-Hellmeier-Stiftung,
Raisting 021269
Kulturhistorie Sønderborg
Slot, Museum Sønderjylland,
Sønderborg 010323
KulturhistorieTønder, Museum
Sønderjylland, Tønder 010342
Kulturhistorische Sammlung,
Frankenmarkt 001901
Kulturhistorische Sammlung,
Iphofen 019420
Kulturhistorische Sammlung am
Landesmuseum Joanneum,
Graz 001979
Kulturhistorisches Karnevalsmuseum
des Regionalverband Düren,
Niederzier 020806
Kulturhistorisches Museum
Barockhaus, Städtische Sammlungen
für Geschichte und Kultur Görlitz,
Görlitz 018735
Kulturhistorisches Museum der
Hansestadt Stralsund, Stralsund 022010
Kulturhistorisches Museum
Franziskanerkloster, Städtische
Museen Zittau, Zittau 022888
Kulturhistorisches Museum
Kaisertrutz/Reichenbacher Turm/
Barockhaus Neißtor 30, Städtische
Sammlungen für Geschichte und
Kultur Görlitz, Görlitz 018736
Kulturhistorisches Museum Prenzlau
im Dominikanerkloster, Prenzlau 021213
Kulturhistorisches Museum Schloß
Merseburg, Merseburg 020374
Kulturhistorisches Museum,
Magdeburger Museen,
Magdeburg 020207
Kulturhistorisk Museum, Randers 010273
Kulturhistoriska Museet i Bunge,
Fårösund 040622

Kulturhistoriska Museet 'Kulturen i
Lund', Lund 040809
De Kulturhistoriske Samlinger,
Bergen 033534
Kulturhuset, Stockholm 041009
Kulturkunststatt Prora, Prora ... 021227
Kulturmühle Lützelflüh, Lützelflüh 041693
Kulturno-obrazovatelnyj centr Škola
L.N. Tolstogo, Gosudarstvennyj Muzej
L.N. Tolstogo, Moskva 037315
Kulturno Povijesni Muzej,
Dubrovnik 008904
Kulturno-vystavočnyj Centr Raduga,
Samara 037709
Kulturno-vystavočnyj Centr Raduga,
Uljanovsk 037994
Kulturno-Vystavočnyj Kompleks
Sokolniki, Moskva 037316
Kulturnyj Centr Kazan, Kazan .. 037103
Kulturnyj Centr-muzej V.S. Vysockogo,
Moskva 037317
Kulturos Barai, Vilnius 142965
Kulturpfad Root, Root 041890
Kulturreich, Hamburg 107286
Kulturschloß Reichenau, Reichenau an
der Rax 002579
Kulturspeicher im Schamuhn Museum,
Uelzen 022248
Kulturverein Landstrich,
Brunnenthal 138520
Kulturverkstedet, Mosjøen 113179
Kulturwerk des Berufsverband
Bildender Künstlerinnen und
Künstler Südbaden e.V., Freiburg im
Breisgau 058964
Kulturwerk des Berufsverbandes
Bildender Künstler Berlins,
Berlin 058965
Kulturwerk des Berufsverbandes
Bildender Künstler Niederbayern e.V,
Fürstenzell 058966
Kulturwissenschaftliche Fakultät,
Universität Bayreuth, Bayreuth 055456
Kulturzentrum der Aktion Lebensqualität,
München 108151, 142174
Kulturzentrum Ehrenhof, NRW-Forum,
Düsseldorf 018035
Kulturzentrum Hans-Reiffenstuel-Haus,
Pfarrkirchen 021116
Kulturzentrum Kolvenburg,
Billerbeck 017388
Kulturzentrum Ostpreußen,
Ellingen 018154
Kulturzentrum Schloß Bonndorf,
Bonndorf im Schwarzwald ... 017505
Kulturzentrum Sinsteden
des Rhein-Kreises Neuss,
Landwirtschaftsmuseum und
Skulpturenhallen Ulrich Rückriem,
Rommerskirchen 021467
Kulunkian-Joumier, Marie-Geneviève,
Saint-Cybardeaux 072722
Kulunkian, Charles, Saint-Yrieix-la-
Montagne 073336
Kulunkian, Stéphane, Garat 068632
Kulunqi Religion Museum, Kailu 007986
Kum, Hing, Hong Kong 065068
Kuma Bijutsukan, Kuma 028890
Kumagai Morikazu Museum,
Tokyo 029540
Kumaki, Tokyo 082174
Kumamoto Castle, Kumamoto .. 028891
Kumamoto-kenritsu Bijutsukan,
Kumamoto 028892
Kumamoto-kenritsu Gigei Bijutsukan,
Kumamoto 028893
Kumamoto Kokusai Mingeikan,
Kumamoto 028894
Kumamoto-shiritsu Gendai Bijutsukan,
Kumamoto 028895
Kumamoto-shiritsu Hakubutsukan,
Kumamoto 028896
Kumar Gallery, Delhi ... 079386, 109327,
137593

Kumar, Jain-Virendra, Delhi 079385,
109326
Kumar, Prakash, Bangalore 079353
Kumar, Sushil, Hyderabad 079392
Kumaya Bijutsukan, Hagi 028586
Kumbez, Almaty 137798, 138895
Kumbu Kumbu, Nairobi 111562
Kume Bijutsukan, Tokyo 029541
Kumer, Navin, New York 122794
Kumlehn, Inge, Sandhausen .. 130616
Kummer, Beat, Bern 087286
Kummer, Wolfram, Pattensen . 130507
Kumon, Los Angeles 121524
Kumpf, Daniela, Wiesbaden .. 078983
Kumps, L., Antwerpen 063220
Kumquat, San Francisco 124321
Kumra & Co., Singapore 085453
Kun-Shan University Art Center,
Yongkang 042597
K'Un Sun, Bogotá 065210
Kun Zsigmond Lakásmúzeuma,
Óbudai Múzeum, Budapest .. 023506
Kunaicho Sannomaru Shozokan,
Tokyo 029542
Kundencenter der Stadtwerke
Strausberg, Strausberg 022016
Kunffy Lajos Emlékmúzeum,
Somogytúr 023886
Kung Long Mineral Museum,
Sincheng 042455
Kungajaktmuseet, Vargön 041159
Kungang Mei Ye Art Galley,
Shanghai 102188
Kungl. Akademien för de fria
Konsterna, Stockholm 056400
Kungl. Konsthögskolan,
Stockholm 056401
Kungl. Myntkabinettet, Sveriges
Ekonomiska Museum,
Stockholm 041010
Kungl. Vitterhets Historie och Antikvitets
Akademien, Stockholm 056402, 059895
Kungliga Akademien för de Fria
Konsterna, Stockholm 041011
Kungliga Slottet i Stockholm,
Stockholm 041012
Kungsantik, Stockholm 087021
Kungsbacka, Lübeck 077176
Kungsbutiken, Gävle 086682
Kungsholmens Antik och Kuriosa,
Stockholm 087022
Kungsstugan och Cajsa Wargs Hus,
Örebro 040893
Kungsudden Local Museum,
Kungsör 040766
Kungurskij Kraevedčeskij Muzej,
Kungur 037182
Kunigo Stanislovo Rimkaus Muziejus,
Laukuva 030406
Kunisaki-machi Rekishi Minzoku
Shiryokan, Kunisaki 028897
Kunitachi Bijutsu Kenkyujo,
Kokubunji 056008
Kunizakari Sake Museum, Handa 028611
Kunj Art Gallery, Karachi 113305
Kunozan-Toshogu-ji Homotsukan,
Shizuoka 029399
Kunsch, Berlin 129522
Kunshan Painting Gallery,
Kunshan 102144
Die Kunskamer, Waterfront 114823
Kunst & Co., Amstelveen 112132
Kunst, Christchurch 112848
Kunst & Ko., Eindhoven 112461
Kunst, Tallinn 136789
Kunst – Werk, Utrecht 112705
Kunst- og Antikvitetshandler Ringen,
Farum 058555
Kunst- und Antiquitäten-Messe Bonn,
Bonn 098146
Kunst- und Antiquitäten-Messe
München, München 098147
Kunst- und Antiquitäten-Messe Sylt,
Wenningstedt 098148

Kunst- und Antiquitäten-Tage,
Münster 098149
Kunst- und Antiquitätenhandel in der
Orangerie, Berlin 098150
Kunst- und Antiquitätentage,
Braunschweig 098151
Kunst- und Auktionsgalerie am
Baldeneysee, Essen . 075686, 125940
Kunst- und Auktionshaus Schloß
Hagenburg, Hagenburg 125963
Kunst- und Auktionshaus Wiesbaden,
Wiesbaden 126140
Kunst- und Auktionshaus Wilhelm M.
Döbritz, Frankfurt am Main ... 075802,
125949
Kunst- und Ausstellungshalle der
Bundesrepublik Deutschland,
Bonn 017494
Kunst- und Gewerbeverein Regensburg,
Regensburg 021325, 058967
Kunst- und Heimathaus,
Paderborn 021063
Kunst- und Kulturkontakt
Perchtoldsdorf, Perchtoldsdorf . 058275
Kunst- und Kulturkreis Rastede,
Rastede 058968
Kunst- und Kulturpädagogisches
Zentrum der Museen in Nürnberg,
Nürnberg 058969
Kunst- und Kulturverein Angeln,
Stoltebüll 058970
Kunst- und Museumsbibliothek,
Museen der Stadt Köln, Köln .. 019696
Kunst- und Museumskreis, Bad
Essen 058971
Kunst- und Museumsverein
Starnberger See e.V., Feldafing 058972
Kunst- und Museumsverein Wuppertal
e.V., Wuppertal 022822, 058973
Kunst- und Skulpturenmuseum,
Städtische Museen Heilbronn,
Heilbronn 019160
Kunst & Antiek Journaal,
Schiedam 138930
Kunst & Antiekrevue, Den Haag 138931
Kunst & Beterschap, Amsterdam 112240
Kunst & Brezen, Nürnberg 108304
Kunst & Cotto, Sattledt 062833
Kunst & Cultuur, Bruxelles 138552
Kunst & Forum, Bad Wimsbach-
Neydharting 127714
Kunst & Kultur, Kulturpolitische
Zeitschrift der Ver.di, Tübingen 138712
Kunst & Kunst, Eindhoven 112462
Kunst & Kunst, Groningen 083300
Kunst & Kuriosa, Heidelberg .. 125988
Kunst & Rahmen, Burg,
Dithmarschen 129645
Kunst & Raum, Düsseldorf 106801
Kunst & Vliegwerk, Hilversum .. 083402
Kunst & Vorm, Dordrecht 083186
Kunst & Werk, Ingelheim am
Rhein 107460
Kunst 7, Zürich 116826
Kunst ab Werk, Miltach 108000
Kunst Adelt, Maastricht 137865
Kunst-aktuell, Bad Schmiedeberg 138713
Kunst am Schützenhof, Bad Neuenahr-
Ahrweiler 106083
Kunst am Wall, Bremen 106588
Kunst Archiv Darmstadt,
Darmstadt 017825
Kunst aus NRW, Aachen 016439
Kunst-Buch, Leipzig 142059
Kunst-Design & Antiquitäten,
Hermagor 062631
Kunst en Vliegwerk, Voorburg .. 083691
Kunst-Event Noord Nederland,
Groningen 098260
Kunst-Event Oost Nederland,
Arnhem 098261
Kunst Galerie Am Rathaus,
Emden 106878
Kunst Genuss, Hamburg 076261

Kunst Haus Wien, Wien 003020
Kunst i Skolen, Oslo 084323
Kunst in den Alten Schützenhaus,
   Zofingen 042184
Kunst in Burggrafiat, Alzey 105983
Kunst in Keller, Halle, Saale . . . 107209
Kunst in Rahmen, München 108152
Kunst in Raum, Coesfeld 106641
Kunst in Regenbogenstadl,
   Polling 108394
Kunst in Rohnerhaus, Lauterach 002277
Kunst in Tal, Gudrun Höritzsch +
   Konrad Hunger, Wilischthal . . 108941
Kunst in Turm, Sömmerda 021889
Kunst-in-Ahoy, Rotterdam 098262
Kunst in der Schranne,
   Feuchtwangen 106950
Kunst in der Zehntscheune, Bad
   Homburg v.d.Höhe 106065
Kunst in Frankfurt e.V., Frankfurt am
   Main 058974
Kunst in Huis, Antwerpen 100499
Kunst in Huis, Bruxelles 100629
Kunst in Opdracht, Almere . . . 082645
Kunst-Kabinett Usedom, Benz, Kreis
   Ostvorpommern 106132
Kunst Klubben, Oslo . . 113230, 126373
Kunst Kreis Jura Neumarkt, Neumarkt
   in der Oberpfalz 058975
De Kunst Kringloop, Bussum . . . 083021
Kunst Kurier, Köln 138714
Kunst Mammern, Mammern . . . 116491
Kunst og Antikk-Senteret,
   Drammen 084230
Kunst og Antikviteter, Aalborg . 065583
Kunst og Antikviteter,
   Kristiansand 084266
Kunst og Håndverksmessen,
   Lillestrøm 098267
Kunst og Kultur, Oslo 138945
Kunst Pavillon, München 020537
Kunst Raum Goethestraße, Linz 022322
Kunst-Raum Haerten,
   Kusterdingen 107775
Kunst Raum Riehen, Riehen . . . 041880
Kunst-Schaefer, Wiesbaden . . 108922,
                   130760
Kunst und Antiquitäten,
   Hinterbrühl 062633
Kunst und Antiquitäten an der Lahn,
   Limburg an der Lahn 077113
Kunst und Antiquitäten Ausstellung,
   Baddeckenstedt 098152
Kunst und Antiquitäten Ausstellung,
   Wotersen 098153
Kunst und Antiquitäten Galerie, Zell
   am See 063127
Kunst und Antiquitäten München,
   München 098154
Kunst und Auktionen, München . 138715
Kunst und Bild, Berlin 137081
Kunst und Kirche, Darmstadt . . 138517
Kunst und Kirche, Kremsmünster 138521
Kunst und Kledage, Wendorf . . 108870
Kunst und Kommunikation e.V.,
   München 058976
Kunst und Literatur-Verein Lahn-Dill
   e.V., Sinn 058977
Kunst und Me(e)r, Krummhörn . 107764
Kunst und Museum, Hollfeld . . 019320
Kunst und Rahmen, Artgroup, Frankfurt
   am Main 107011, 129851
Kunst und Recht, Berlin 138717
Kunst und Stein, Bern 139094
Kunst und Unterricht, Seelze . . 138718
Kunst Werk, Herford 107419
Kunst Werk Berlin, Berlin 074854
Kunst Werk-Kunstverein Ladbergen,
   Ladbergen 058978
Kunst-Werkstatt, Rheda-
   Wiedenbrück 108488
Kunst(Zeug)Haus, Sammlung Bosshard,
   Rapperswil 041861
kunst:art, Köln 138719

Kunst:raum Sylt Quelle, Rantum 021278
Kunstacademie De Lei, Stedelijke
   Academie voor Beeldende Kunst,
   Leuven 055034
Kunstakademie Bad Reichenhall, Bad
   Reichenhall 055451
Kunstakademie Düsseldorf,
   Düsseldorf 055524
Kunstakademie Münster, Hochschule
   für Bildende Künste, Münster . 055666
Kunstakademiet i Trondheim, Norges
   Teknisk-Naturvitenskapelige
   Universitet, Trondheim 056215
Kunstamt Tempelhof-Schöneberg,
   Berlin 017201
Kunstantikvariat Pama, Oslo . . . 113231,
                  143403
Kunstantikvariat, Herne 107421
Kunstantiquariat R. Brehmer,
   München 142175
KunStart – Vineart, International
   Modern and Contemporary Art Fair of
   Bolzano, Bolzano 098196
Kunstatelier Carte Blanche,
   Hasselt 100717
Kunstatelier, Thommes Nentwig,
   Vechta 108813
Kunstauktionshaus an der
   Frauenkirche, Dresden 125922
Kunstauktionshaus Leipzig,
   Leipzig 126024
Kunstauktionshaus Schloss Ahlden,
   Ahlden 125847
Kunstauktionshaus Schlosser,
   Bamberg 125861
Kunstausstellung des Rothenburger
   Künstlerbundes, Rothenburg ob der
   Tauber 021507
Kunstausstellung Stift Herzogenburg,
   Herzogenburg 002081
Kunstbeeld, Diemen 138932
Kunstbehandlung, München . . . 108153
Kunstberatung Eva Müller,
   Grünwald 107179
Kunstbibliothek, Staatliche Museen
   zu Berlin – Stiftung Preußischer
   Kulturbesitz, Berlin 017202
Kunstblick-Galerie im Internet,
   Adnet 099951
Der Kunstblitz, Wuppertal 138720,
                  138721
Kunstbuchanzeiger, Königstein im
   Taunus 138722
Kunstbüro Berlin, Berlin 106345
Kunstbulletin, Zürich 139095
Kunstbunker Tumulka, München 020538,
                  020539
De Kunstcaemer van Blijkshof,
   Deventer 112428
Kunstcentralen, Århus 065602
Kunstcentralen, Oslo 113232
Kunstcentrum De Waagh,
   Oldenzaal 032713
Kunstcentrum II Ventuno, Hasselt 100718
Kunstcentrum Pand Paulus,
   Schiedam 032853
Kunstcentrum Zaanstad,
   Zaandam 112718, 137877
Kunstcentrum Zaanstad – Filiat
   Uitgeest, Uitgeest 112689
Kunstchronik, Nürnberg 138723
Kunstdepot 24, Berlin 106346
Kunstenaarscentrum De Fabriek,
   Eindhoven 032286
Kunstentreprenørene, Bergen . . 084218
Kunstetagen-Galerie, Remscheid 108477
Kunstfabrik am Flutgraben,
   Berlin 017203
Kunstfabrikken, Aalborg 065584, 102767,
                  136770
Kunstfleck am Eck, Wolnzach . . 079063
Kunstförderverein Schöningen e.V.,
   Schöningen 058979, 108573
Kunstförderverein Weinheim e.V.,

Weinheim 058980
Kunstforeningen GL Strand,
   København 010180, 058556
Kunstforum Arabellapark,
   München 020540
Kunstforum Bâloise, Basel 041271
Kunstforum der Berliner Volksbank,
   Berlin 017204
Kunstforum Ebendorf Raumacht,
   Wien 003021
Kunstforum Gummersbach, Bergisch
   Gladbach 058981
Kunstforum Montafon, Schruns . 058276,
                  100196
Kunstforum Ostdeutsche Galerie,
   Regensburg 021326
Kunstforum Palastweiher,
   Königswinter 076929
Kunstforum Rheinhessen, Essenheimer
   Kunstverein e.V, Essenheim . . 058982
Kunstforum Seligenstadt e.V.,
   Seligenstadt 058983
Kunstforum Waldviertel e.V,
   Schrems 058277
Kunstforum Weil der Stadt, Weil der
   Stadt 058984
Kunstforum Wien, Wien 002944
Kunstforum Würth, Turnhout . . 004097
Kunstforum, International,
   Ruppichteroth 138724
Kunstgalerie Vreden, W. Schröder,
   Vreden 108827
Kunstgalleri R.B., Genk 100684
Kunstgalleriet, Odense 066040, 102949
Kunstgalleriet, Slagelse 066131, 102979
Kunstgalleriet, Stavanger 113272
Kunstgarten, Aarwangen 116009
Kunstgeschichte aktuell, Wien . . 138522
Kunstgeschichte I, insbes.
   Mittelalterliche Kunstgeschichte,
   Otto-Friedrich-Universität Bamberg,
   Bamberg 055452
Kunstgeschichte II, insbes. für Neuere
   und Neueste Kunstgeschichte, Otto-
   Friedrich-Universität Bamberg,
   Bamberg 055453
Kunstgeschichtliche Gesellschaft zu
   Berlin, Berlin 058985
Kunstgeschichtliches Institut, Albert-
   Ludwigs-Universität Freiburg,
   Freiburg im Breisgau 055545
Kunstgeschichtliches Institut, Johann
   Wolfgang Goethe-Universität,
   Frankfurt am Main 055540
Kunstgeschichtliches Institut, Philipps-
   Universität Marburg, Marburg . 055641
Kunstgeschichtliches Institut, Ruhr-
   Universität Bochum, Bochum . 055488
Kunstgeschichtliches Seminar und
   Kunstsammlung, Georg-August-
   Universität Göttingen, Göttingen 055549
Kunstgeschichtliches Seminar,
   Universität Hamburg, Hamburg 055572
Kunstgesellschaft Grenchen,
   Grenchen 059939
Kunstgesellschaft Luzern, Luzern 059940
Kunstgewerbemuseum, Staatliche
   Kunstsammlungen Dresden,
   Dresden 017989
Kunstgewerbemuseum, Staatliche
   Museen zu Berlin – Stiftung
   Preußischer Kulturbesitz, Berlin 017205
Kunstgewerbesammlung, Museum
   Bellerive, Zürich 042205
Kunstgriff, Leipzig 107839
Kunstgußmuseum Lauchhammer,
   Lauchhammer 019926
Kunsthal Rotterdam, Rotterdam . 032807
Kunsthal Sint-Pietersabdij, Gent 003629
Kunsthalle Arbon, Arbon 041228
Kunsthalle Barmen, Von der Heydt-
   Museum, Wuppertal 022823
Kunsthalle Basel, Basler Kunstverein,
   Basel 041272

Kunsthalle Bern, Bern 041317
Kunsthalle Bielefeld, Bielefeld . . 017377
Kunsthalle Bremen, Der Kunstverein in
   Bremen, Bremen 017592
Kunsthalle Darmstadt, Darmstadt 017826
Kunsthalle der Europäischen
   Kunstakademie, Trier 022199
Kunsthalle der Hypo-Kulturstiftung,
   München 020541
Kunsthalle der Sparkasse Leipzig,
   Leipzig 019998
Kunsthalle Dominikanerkirche,
   Osnabrück 021016
Kunsthalle Dresden im Art'otel
   Dresden, Dresden 017990
Kunsthalle Düsseldorf, Düsseldorf 018036
Kunsthalle Emden, Stiftung Henri und
   Eske Nannen und Schenkung Otto
   van de Loo, Emden 018176
Kunsthalle Erfurt im Haus zum Roten
   Ochsen, Erfurt 018244
Kunsthalle Exnergasse, Wien . . 003023
Kunsthalle Faust, Hannover . . . 019056
Kunsthalle Fridericianum, Kassel 019536
Kunsthalle Gießen, Gießen 018690
Kunsthalle Göppingen, Göppingen 018729
Kunsthalle Harry-Graf-Kessler Weimar,
   Weimar 022499
Kunsthalle im Schloss, Isny im
   Allgäu 019434
Kunsthalle Jesuitenkirche, Museen
   der Stadt Aschaffenburg,
   Aschaffenburg 016619
Kunsthalle Koblenz, Koblenz . . . 107572
Kunsthalle Krems, Krems 002234
Kunsthalle Leoben, Leoben 002291
Kunsthalle Lingen, Kunstverein Lingen,
   Lingen 020091
Kunsthalle Loppen, Loppem . . . 003830
Kunsthalle Mainz, Mainz 020227
Kunsthalle Mannheim, Mannheim 020249
Kunsthalle Nürnberg, Nürnberg . 020865
Kunsthalle Palazzo, Liestal 041677
Kunsthalle Rostock, Rostock . . . 021494
Kunsthalle Sankt-Annen und Sankt-
   Annen-Museum, Die Lübecker
   Museen, Lübeck 020159
Kunsthalle Sankt Gallen, Sankt
   Gallen 041922
Kunsthalle Tübingen, Tübingen . 022220
Kunsthalle Vierseithof,
   Luckenwalde 020124
Kunsthalle Villa Kobe, Halle,
   Saale 018962
Kunsthalle Weishaupt, Ulm 022262
Kunsthalle Wien, Standort 1,
   Wien 003024
Kunsthalle Wien, Standort 2,
   Wien 003025
Kunsthalle Wil, Wil 042139
Kunsthalle Wilhelmshaven,
   Wilhelmshaven 022661
Kunsthalle Winterthur, Winterthur 042151
Kunsthalle Würth, Schwäbisch
   Hall 021756
Kunsthalle Ziegelhütte, Appenzell 041223
Kunsthalle zu Kiel, Christian-Albrechts-
   Universität, Gemäldegalerie und
   Grafische Sammlung, Kiel . . . 019606
Kunsthalle Zürich, Zürich 042206
Kunsthalle, Kunstverein Bremerhaven
   von 1866, Bremerhaven 017608
Kunsthallen Brandts, Odense . . 010261
Kunsthandel, København 065930
Der Kunsthandel, Neu-Isenburg . 138725
Kunsthandel am Fasanenplatz,
   Berlin 074855
Kunsthandel Anne Soppe, Essen 075687
Kunsthandel De Kool, Dordrecht 083187
Kunsthandel Esthetica,
   Rotterdam 083575
Kunsthandel Faethe, Berlin 074856
Kunsthandel Klefisch, Köln 076890,
                  126008

Kunsthandel Michael Draheim,
Wiesbaden 078984
Kunsthandel P. de Boer BV,
Amsterdam 082790
Kunsthandel Venlo, Venlo 083682
Der Kunsthandel Verlag, Neu-
Isenburg 137441
Kunsthandel Wolfgang Werner,
Berlin 074857, 106347
Kunsthandel Wolfgang Werner,
Bremen 075171, 106589
Kunsthandelsverband der Schweiz,
Association des Commerçants d'Art
de la Suisse, Zürich 059941
Kunsthandlerforeningen af 1943,
Slagelse 058557
Kunsthandlung Ahrenshoop,
Ahrenshoop, Ostseebad 074365
Kunsthandlung Alstertal,
Hamburg 076262
Kunsthandlung Elke, Bottrop ... 075109
Kunsthandlung in der Marktstraße,
Erfurt 075640
Kunsthandlung Mag.art Vera Weber,
Galerie am Roten Hof, Wien .. 100341
Kunsthandlung Sankt Jürgen,
Flensburg 075747
Kunsthandwerk & Design,
Frechen 138726
Kunsthandwerk am Markt,
Sebnitz 078438
Kunsthaus, Ratingen 078136
Kunsthaus Adlerstraße, Stuttgart 078613
Kunsthaus am Friesenpark,
Delmenhorst 106667
Kunsthaus Am Markt, Schmalkalden,
Kurort 078375
Kunsthaus am Museum, Trier .. 142421
Kunsthaus am Niederrhein,
Mönchengladbach 077395
Kunsthaus Apolda Avantgarde,
Apolda 016601
Kunsthaus Baselland, Muttenz . 041786
Kunsthaus Benrath, Düsseldorf . 075488
Kunsthaus Bregenz, Bregenz ... 001793
Kunsthaus Bregenz, Verlagsabteilung,
Bregenz 136497
Kunsthaus Bühler, Stuttgart ... 078614,
                               108715
Kunsthaus Centre PasquArt, Biel 041337
Kunsthaus Christine, Bamberg . 074666
Kunsthaus City, Hamburg 076263
Kunsthaus Dresden, Städtische Galerie
für Gegenwartskunst, Dresden 017991
Kunsthaus Eckernförde,
Eckernförde 106855, 141731
Kunsthaus Erfurt, Erfurt 018245
Kunsthaus Erfurt e.V., Erfurt .. 058986
Kunsthaus Essen, Essen 018299
Kunsthaus Freudenstadt,
Freudenstadt 075903
Kunsthaus Gahlberg-Strodehne e.V.,
Potsdam 058987
Kunsthaus Glarus, Glarus 041549,
                               059942
Kunsthaus Graz am Landesmuseum
Joanneum, Graz 001980
Kunsthaus Grenchen, Grenchen . 041565
Kunsthaus Hamburg, Hamburg . 019000
Kunsthaus Hannover, Hannover . 107372
Kunsthaus Hinter den Zäunen,
Schöneck 137499
Kunsthaus in Stade, Stade 021947
Kunsthaus Kannen, Museum für Art
Brut + Outsider Art, Münster . 020600
Kunsthaus Kaufbeuren,
Kaufbeuren 019562
Kunsthaus Kloster Gravenhorst,
Hörstel 019284
Kunsthaus Köflach, Köflach 002218
Kunsthaus Langenthal,
Langenthal 041637
Kunsthaus Lübeck, Lübeck 107915
Kunsthaus Magdeburg,

Magdeburg 077221
Kunsthaus Meyenburg,
Nordhausen 020827
Kunsthaus Müllers, Rendsburg . 078205
Kunsthaus Mürzzuschlag,
Mürzzuschlag 002420
Kunsthaus Mürzzuschlag,
Verlagsabteilung, Mürzzuschlag 100126,
                               136538
Kunsthaus Nordwalde,
Nordwalde 108288
Kunsthaus Nova, München 077550
Kunsthaus Nürnberg, Nürnberg . 020866
Kunsthaus Oberkassel,
Düsseldorf 075489
Kunsthaus Oggersheim, Ludwigshafen
am Rhein 077163, 107898
Kunsthaus Raskolnikow, Dresden 106741
Kunsthaus Rehau und Archiv Eugen
Gomringer, Rehau 021335
Kunsthaus Sankt Alban, Basel .. 116084
Kunsthaus Santa Fé, San Miguel de
Allende 111992
Kunsthaus Schill, Stuttgart 108716
Kunsthaus Schwanheide,
Schwanheide 108590
Kunsthaus Stadt Bocholt, Bocholt 017428
Kunsthaus Strohdehne, Potsdam 058988,
                               108408
Kunsthaus Tacheles, Berlin 106348
Kunsthaus Zürich, Zürich 042207
Kunsthaus Zug, Zug 042250
Kunsthaus zur Linde, Naters ... 087696,
                               116524
KunstHausUelzen, Uelzen 022249
Kunsthistoriker aktuell, Wien ... 138523
Kunsthistorisch Instituut – Afdeling
Kunstgeschiedenis, Radboud
Universiteit, Nijmegen 056181
Kunsthistorische Arbeitsblätter,
Köln 138727
Kunsthistorische Gesellschaft Wien,
Wien 058278
Kunsthistorisches Institut in Florenz,
Max-Planck-Institut, Firenze .. 055896
Kunsthistorisches Institut, Christian-
Albrechts-Universität zu Kiel,
Kiel 055608
Kunsthistorisches Institut, Eberhard-
Karls-Universität Tübingen,
Tübingen 055734
Kunsthistorisches Institut, Freie
Universität Berlin, Berlin 055478
Kunsthistorisches Institut, Friedrich-
Alexander-Universität Erlangen-
Nürnberg, Erlangen 055530
Kunsthistorisches Institut, Universität
zu Köln, Köln 055617
Kunsthistorisches Institut, Universität
Zürich, Zürich 056446
Kunsthistorisches Museum,
Admont 001685
Kunsthistorisches Museum Wien,
Wien 003026
Kunsthistorisches Seminar und
Kustodie, Friedrich-Schiller-
Universität Jena, Jena 055594
Kunsthistorisches Seminar, Universität
Basel, Basel 056414
Kunsthjørnet, Sønderborg 066140
Kunsthochschule Berlin-Weißensee,
Hochschule für Gestaltung,
Berlin 055479
Kunsthochschule für Medien Köln
(KMH), Köln 055618
Kunsthochschule Kassel, Fachbereich
20 der Universität Kassel,
Kassel 055607
Kunsthøgskolen i Oslo, Fakultet for
Design, Oslo 056212
Kunsthøgskolen i Oslo, Fakultet for
Visuell Kunst, Oslo 056213
Kunsthøgskolen in Bergen,
Bergen 056210

Kunsthof Bad Salzelmen,
Schönebeck 021703
Kunsthof e.V., Halberstadt 058989
Kunsthof Zürich, Hochschule für
Gestaltung und Kunst, Zürich . 042208
Het Kunsthuis, Almere 112117
Kunsthuis 1927, Emmen 112472
Kunsthuis Marc Chagall,
Amsterdam 112241
Kunsthuis Secretarie, Meppel .. 032647
Kunstuset, Frederiksberg 065666
Kunstuset, Odense 102950
Kunstuset, Oslo 113233
Kunstigalerii, Narva 010641
Kunstindustrimuseet, Nasjonalmuseet
for Kunst, Arkitektur og Design,
Oslo 033840
kunstinform, Dresden 138728
Kunstjahr, Regensburg 138729
Kunstkabinett, Regensburg 108465
Kunstkabinett am Rheintor, Rees 108453
De Kunstkamer, Gouda 112484
Kunstkammer der Pfarrkirche Sankt
Georg, Bocholt 017429
Kunstkammer des Innsbrucker
Servitenklosters, Innsbruck ... 002127
Kunstkanzlei, Wien 100342
Kunstkaten Ahrenshoop, Ahrenshoop,
Ostseebad 016480
Kunstkeller, Liestal 116441
Kunstkeller Annaberg, Annaberg-
Buchholz 016590
Kunstkeller Bern, Bern 116144
Kunstklause, Bern 087287, 133682
Kunstkontakt, Dornbirn 099977, 136500
Kunstkontakt, Oberursel 058990
Kunstkontor Hartmut Rampoldt,
Berlin 106349
Kunstkontor Westnerwacht,
Regensburg 108466
Kunstkreis-Galerie, Lauda-
Königshofen 107799
Kunstkreis Hameln, Gesellschaft zur
Förderung der bildenden Künste e.V.,
Hameln 058991
Kunstkreis Lauda-Königshofen e.V.,
Lauda-Königshofen 058992
Kunstkreis Niederrhein, Emmerich am
Rhein 058993
Kunstkreis Schenefeld e.V., Schenefeld,
Bezirk Hamburg 058994
Kunstkreis Südliche Bergstraße,
Wiesloch 058995
Kunstkreis Verlag, Korntal-
Münchingen 137327
Kunstkring Jacques Gorus,
Antwerpen 058384, 100500
KunstLANDing, Aschaffenburg .. 016620
Kunstleben im Hamburger
Gängeviertel, Hamburg 107287
Kunstlocatie Würth, 's-
Hertogenbosch 032469
Het Kunstmagazijn, Dordrecht .. 112439
Kunstmanifest – Kunstbüro,
Pinkafeld 136541
Kunstmann, Dr. Joanna, Palma de
Mallorca 115356
Kunstmann, Dr. Joanna, Santanyi 115436
Kunstmann, G., Halle, Saale ... 076168
Kunstmarkt.com, Würzburg 138730
Kunstmesse Salzburg,
Publikumsmesse für zeitgenössische
bildende Kunst, Bergheim ... 098003
Kunstmesse Schwetzingen,
Schwetzingen 098155
Kunstmühle Schwaan, Schwaan 021743
Kunstmuseen Krefeld, Kaiser Wilhelm
Museum, Museum Haus Lange,
Museum Haus Esters, Krefeld . 019786
Kunstmuseet Brundlund Slot, Museum
Sønderjylland, Aabenraa 009997
Kunstmuseet i Tønder, Museum
Sønderjylland, Tønder 010343
Kunstmuseet Køge Skitsesamling,

Køge 010203
Kunstmuseum, Sankt Gallen ... 041923
Kunstmuseum Ahlen, Theodor F.
Leifeld-Stiftung, Ahlen 016476
Kunstmuseum Alte Post, Mülheim an
der Ruhr 020496
Kunstmuseum Basel, Basel 041273
Kunstmuseum Bayreuth mit
Tabakhistorischer Sammlung,
Bayreuth 016995
Kunstmuseum Bern, Bern 041318
Kunstmuseum Bochum, Bochum 017436
Kunstmuseum Bonn, Bonn 017495
Kunstmuseum Celle mit Sammlung
Robert Simon, Celle 017724
Kunstmuseum des Kantons Thurgau,
Warth 042125
Kunstmuseum Dieselkraftwerk Cottbus,
Brandenburgische Kulturstiftung
Cottbus, Cottbus 017766
Kunstmuseum Dr. Krupp,
Privatmuseum für Kunst und
Geschichte, Weinbach 022513
Kunstmuseum Gelsenkirchen,
Gelsenkirchen 018629
Kunstmuseum Heidenheim,
Heidenheim an der Brenz ... 019152
Kunstmuseum Hohenkarpfen, Hausen
ob Verena 019111
Kunstmuseum Kloster Unser Lieben
Frauen, Magdeburger Museen,
Magdeburg 020208
Kunstmuseum Liechtenstein,
Vaduz 030305
Kunstmuseum Luzern, Luzern .. 041708
Kunstmuseum Nörvenich, Museum
Arno Breker, Sammlung Europäische
Kunst, Nörvenich 020817
Kunstmuseum Olten, Olten 041823
Kunstmuseum Poschiavo, Fondazione
Ernesto Conrad, Poschiavo ... 041846
Kunstmuseum Solingen, Solingen 021898
Kunstmuseum Solothurn,
Solothurn 042005
Kunstmuseum Stuttgart, Stuttgart 022045
Kunstmuseum Thun, Thun 042055
Kunstmuseum Thun, Thunerhof,
Thun 042056
Kunstmuseum Walter, Augsburg 016659
Kunstmuseum Winterthur,
Winterthur 042152
Kunstmuseum Wolfsburg,
Wolfsburg 022744
KunstNaturLandschaft, Heideblick 019128
Kunstnerforbundet, Oslo 059661, 113234
Kunstnerforeningen af 18. November,
København 058558
Kunstnernes Informasjonskontor,
Oslo 113235
Kunstogenblik, Gent 063646
Kunstpanorama Luzern, Luzern . 041709
Kunstpark Rost, Grasbrunn 107155
Kunstpassage, Wiesbaden 078985
Kunstpaviljoen, Nieuw Roden .. 032669
Kunstpaviljongen, Høvik 084256
De Kunstproeverij, Groningen .. 112495
Kunstpunkt, Darmstadt 106658
Kunstpunkt, Emden 106879
Kunstpunkt Berlin, Berlin 106350
Kunstpunkt-Edewecht, Edewecht 075570
kunstquartal im hatje cantz verlag,
Ostfildern 138731
Kunstquartier Hagen, Osthaus
Museum Hagen und Emil
Schumacher Museum Hagen,
Hagen 018916
Kunsträume, Zermatt 116733
Kunstransmitter, München 108154
Kunstrat im Deutschen Kulturrat,
Frankfurt am Main 058960
Kunstraum, Essen 106922
KunsTraum, Tannhausen 108756
KunstRaum, Trier 108778
Kunstraum Aarau, Aarau 041368

Kunstraum am Cospudener See,
Markkleeberg 107980
Kunstraum Baden, Baden . . 041254
Kunstraum Bruxelles, Bruxelles . 003446
Kunstraum Buchberg, Gars am
Kamp 058279, 136503
Kunstraum Dornbirn, Dornbirn . . 001826
Kunstraum Düsseldorf,
Düsseldorf 018037
Kunstraum Eching am Ameersee-
La Frascetta, Eching am
Ammersee 106854
Kunstraum Falkenstein, Hamburg 107288
Kunstraum Farmsen, Galerie der
Hamburger Volkshochschule,
Hamburg 019001
Kunstraum Hebel_121, Basel . . 116665
Kunstraum Innsbruck, Innsbruck 002128
Kunstraum Klosterkirche,
Traunstein 022174
Kunstraum Kreuzberg/Bethanien,
Berlin 017206
Kunstraum Leuphana, Universität
Lüneburg, Lüneburg 020181
Kunstraum Mühlenstraße, Lübeck 107916
kunstraum muenchen, München 058997,
108155, 137412
Kunstraum Notkirche, Essen . . . 018300
Kunstraum Pettneu, Pettneu am
Arlberg 100138
Kunstraum Sankt Virgil, Salzburg 002627
Kunstraum Schwifting,
Schwifting 108601
Kunstraum Stein am Rhein, Stein am
Rhein 116661
Kunstraum Wohnraum, Akureyri 109220
Kunstraum, Thurgauische
Kunstgesellschaft, Kreuzlingen 116393
Kunstraum69, Hanau 107338
kunstraumno.10,
Mönchengladbach 108017
Kunstreich, Bern 116145
Kunstreich am Pergamon-Museum,
Berlin 074858
Kunstring Folkwang e.V., Essen 058998
Kunstsalon e.V., Köln 107657
Kunstsalon Pullach, Pullach im
Isartal 108420
Kunstsalon Wolfsberg, Zürich . . 116827
Kunstsammlung der Universität
Göttingen, Göttingen 018750
Kunstsammlung des Herzoglichen
Georgianums, München 020542
Kunstsammlung Hans und Marlis
Suter, Oberhofen am Thunersee 041811
Kunstsammlung im Schloß Rheinstein,
Trechtingshausen 022184
Kunstsammlung im Stadtmuseum,
Städtische Museen Jena, Jena 019450
Kunstsammlung Lorenzkapelle
Rottweil, Rottweil 021526
Kunstsammlung Neubrandenburg,
Neubrandenburg 020682
Kunstsammlung Robert Spreng,
Reiden 041867
Kunstsammlung und
Kupferstichkabinett der Martin-Luther-
Universität Halle-Wittenberg, Halle,
Saale 018963
Kunstsammlung Villa Schüpbach,
Steffisburg 042029
Kunstsammlungen Böttcherstraße,
Paula Modersohn-Becker Museum
& Museum im Roselius-Haus,
Bremen 017593
Kunstsammlungen Chemnitz,
Chemnitz 017733
Kunstsammlungen Chemnitz –
Museum Gunzenhauser,
Chemnitz 017734
Kunstsammlungen der Ruhr-
Universität Bochum, Bochum . 017437
Kunstsammlungen der Stadt Limburg,
Limburg an der Lahn 020077

Kunstsammlungen der Veste Coburg,
Coburg 017752
Kunstsammlungen des Stiftes
Kremsmünster, Kremsmünster 002240
Kunstsammlungen und Graphische
Sammlung, Benediktinerstift
Göttweig, Furth bei Göttweig . . 001915
Kunstsammlungen und Museen
Augsburg, Augsburg 058999
Kunstsammlungen Zwickau,
Städtische Museen Zwickau,
Zwickau 022909
Kunstsammlungen, Staatliches
Museum Schwerin, Schwerin . 021792
KunstSandra, Sünching 078644
Kunstschacht Katernberg, Essen 018301
Kunstscheune Barnstorf, Wustrow,
Ostseebad 108999
Kunstscheune Zernikow,
Plattenburg 108386
Kunstschiff Arte Noah, Kunstverein
Würzburg, Würzburg 108976
KunstSchnitt, Passau 108364
Kunstschule Bad Essen, Bad
Essen 055450
Kunstschule Rostock e.V.,
Rostock 055699
Kunstschule Wandsbek, Staatl.
anerkannte Berufsfachschule
für Kommunikationsdesign,
Hamburg 055573
Kunstseminar-Galerie, Luzern . . 116477
Kunstsjop, Bergen op Zoom . . . 082962
Kunststätte Bossard, Jesteburg . 019462
Der Kunststall, Dahlem 106648
Kunststation Kleinsassen,
Hofbieber 107439
Kunststiftung Baden-Württemberg,
Stuttgart 022046
Kunststiftung Erich Hauser,
Rottweil 021527
Kunststiftung Hohenkarpfen,
Kunstverein Schwarzwald-Baar-
Heuberg, Hausen ob Verena . 059000
Kunststiftung Poll, Berlin 017207
Kunststoff, Leipzig 138732
Kunststoff-Museums-Verein e.V.,
Düsseldorf 018038
Die Kunststube, Giengen an der
Brenz 107124
Kunststube, Wien 063014
Kunststübchen am Teich, Neustadt,
Hessen 077767
Kunststücke, Linz 062718, 127808
Kunstsupermarkt, Sylt 078660
Kunsttempel, Kassel 107535
Kunsttermine, Neu-Isenburg . . . 138733
Kunsttijdschrift Vlaanderen, Tielt, West-
Vlaanderen 138553
Kunstuniversität Linz, Universität für
künstlerische und industrielle
Gestaltung, Linz 054956
't Kunstuus, Heinkenszand 032447
Kunstværket, Bærums Verk 113122
Kunstverein "Talstrasse" e.V, Halle,
Saale 018964, 059001
Kunstverein AA-galerie, Wels . . 058280,
100222
Kunstverein Aalen e.V., Aalen . . 059002
Kunstverein ACUD, Berlin 059003
KunstVerein Ahlen e.V., Ahlen . . 059004
Kunstverein Aichach e.V.,
Aichach 059005
Kunstverein Amstetten, Sankt Peter in
der Au 058281
Kunstverein Arnsberg e.V.,
Arnsberg 059006
Kunstverein Augsburg, Augsburg 059007
Kunstverein Aurich, Aurich 059008,
106035
Kunstverein Bad Aibling, Bad
Aibling 059009, 106038
Kunstverein Bad Homburg Artlantis
e.V., Bad Homburg v.d.Höhe . 059010

Kunstverein Bad Salzdetfurth, Bad
Salzdetfurth 016858, 059011
Kunstverein Baden, Baden bei
Wien 058282, 099960
Kunstverein Bamberg, Bamberg 059012
Kunstverein Basellland, Muttenz . 059943
Kunstverein Bayreuth, Bayreuth . 059013
Kunstverein Bergisches Land,
Wipperfürth 059014
Kunstverein Bergneustadt,
Bergneustadt . . . 059015, 106139
Kunstverein Biberach e.V.,
Ummendorf 059016
Kunstverein Biel, Biel 059945
Kunstverein Binningen, Binningen 059945
Kunstverein Bochum, Hattingen . 059017
Kunstverein Braunschweig e.V.,
Braunschweig . . . 017559, 059018
Kunstverein Bremerhaven von 1886,
Bremerhaven 059019
Kunstverein Bretten e.V., Bretten 059020,
106598
Kunstverein Buchholz/Nordheide e.V.,
Buchholz in der Nordheide . . . 059021,
106609
Kunstverein Buxtehude e.V,
Buxtehude 059022
Kunstverein Celle e.V., Celle . . . 059023
Kunstverein Coburg e.V., Coburg 059024,
106639
Kunstverein Darmstadt,
Darmstadt 059025
Kunstverein Das Damianstor Bruchsal
e.V., Bruchsal 059026, 106602
Kunstverein Deggendorf,
Deggendorf 059027
Kunstverein der Diözese Rottenburg-
Stuttgart, Bad Buchau 059028
Kunstverein der Stadt Glauchau,
Glauchau 059029
Kunstverein Dillingen im Alten Schloß,
Dillingen, Saar 059030
Kunstverein Donauwörth,
Donauwörth 059031
Kunstverein Ebersberg,
Ebersberg 059032
Kunstverein Eisenturm Mainz e.V,
Mainz 059033
Kunstverein Eislingen, Eislingen 059034
Kunstverein Ellwangen e.V.,
Ellwangen 059035
Kunstverein Elmshorn, Elmshorn 059036,
106876
Kunstverein Emmerich, Emmerich am
Rhein 059037
Kunstverein Emsdetten,
Emsdetten 059038
Kunstverein Engen e.V.,
Stubengesellschaft 1599, Engen 059039
Kunstverein Erlangen, Erlangen . 059040
Kunstverein Feldberger Land e.V,
Feldberger Seenlandschaft . . 059041
Kunstverein Flensburg, Flensburg 059042
Kunstverein Freiburg, Freiburg im
Breisgau 059043
Kunstverein Friedberg, Friedberg,
Hessen 059044
Kunstverein Friedrichshafen,
Friedrichshafen 059045
Kunstverein Friedrichstadt, Berlin 017208,
059046
Kunstverein für die Rheinlande und
Westfalen, Düsseldorf 018039
Kunstverein für die Rheinlande und
Westfalen e.V., Düsseldorf . . . 059047
Kunstverein Galerie-Werkstatt Bayer
Dormagen e.V., Dormagen . . 059048,
106690
Kunstverein Ganderkesee,
Ganderkesee 059049
Kunstverein Gauting, Gauting . . 059050
Kunstverein Gelderland, Geldern 018618,
059051
Kunstverein Gelsenkirchen e.V,

Gelsenkirchen 059052
Kunstverein Germersheim,
Germersheim 059053
Kunstverein Giannozzo, Verein zur
Förderung der aktuellen plastischen
Kunst, Berlin 059054
Kunstverein Göppingen,
Göppingen 059055
Kunstverein Göttingen e.V,
Göttingen 059056
Kunstverein Grafschaft Bentheim,
Neuenhaus 059057, 108241
Kunstverein Greven e.V., Greven 059058
Kunstverein Gundelfingen,
Gundelfingen, Breisgau 059059
Kunstverein Hannover, Hannover 019057,
059060
Kunstverein Harburger Bahnhof,
Hamburg 059061
Kunstverein Hattingen, Hattingen 059062
Kunstverein Hechingen,
Hechingen 059063
Kunstverein Heidenheim e.V,
Heidenheim an der Brenz 059064
Kunstverein Heilbronn, Heilbronn 019161,
059065
Kunstverein Hildesheim,
Hildesheim 059066
Kunstverein Hochrhein, Bad
Säckingen 059067
Kunstverein Hockenheim,
Hockenheim 059068
Kunstverein Hof, Hof . . 019286, 059069
Kunstverein Horn, Horn 058283
Kunstverein Hürth, Hürth 059070
Kunstverein Husum und Umgebung
e.V., Schobüll 059071
Kunstverein Ibbenbüren,
Ibbenbüren 059072
Kunstverein im Reuchlinhaus e.V.,
Pforzheim 059073
Der Kunstverein in Bremen,
Bremen 059074
Kunstverein in Hamburg e.V,
Hamburg 019002, 059075
Kunstverein Ingolstadt, Ingolstadt 019410,
059076
Kunstverein Jülich, Jülich 059077
Kunstverein Kärnten, Klagenfurt 002191
Kunstverein Kaponier Vechta e.V,
Vechta 059078
Kunstverein Kehdingen, Freiburg,
Elbe 059079
Kunstverein Kirchheim unter Teck,
Kirchheim unter Teck 059080
Kunstverein Kirchzarten e.V.,
Kirchzarten 059081
Kunstverein KISS-Kunst im Schloss
Untergröningen e.V., Abtsgmünd 059082
Kunstverein Köln rechtsrheinisch e.V.,
Köln 019697, 059083
Kunstverein Konstanz e.V.,
Konstanz 019761, 059084
Kunstverein Kornwestheim,
Kornwestheim 059085
Kunstverein Kreis Gütersloh e.V.,
Gütersloh 059086
Kunstverein Langenhagen,
Langenhagen 059087
Kunstverein Leimen '81 e.V.,
Leimen 059088
Kunstverein Leipzig, Leipzig . . . 059089
Kunstverein Leverkusen Schloß
Morsbroich e.V., Leverkusen . 059090
Kunstverein Lippstadt e.V.,
Lippstadt 059091
Kunstverein Ludwigsburg e.V.,
Ludwigsburg 059092
Kunstverein Ludwigshafen am Rhein
e.V., Ludwigshafen am Rhein . 059093
Kunstverein Lüneburg,
Lüneburg 059094
Kunstverein March e.V., March . 059095
Kunstverein Mittleres Kinzigtal,

Haslach im Kinzigtal 059096
Kunstverein München, München 059097
Kunstverein Münsterland,
Coesfeld 059098
Kunstverein Neckar-Odenwald e.V.,
Mosbach 059099
Kunstverein Neuhausen e.V.,
Neuhausen auf den Fildern . . . 059100
Kunstverein Neustadt am Rübenberge
e.V., Neustadt am Rübenberge 059101
Kunstverein Neustadt an der
Weinstraße e.V., Neustadt an der
Weinstraße 059102
Kunstverein Nordenham e.V.,
Nordenham 059103
Kunstverein Oberaargau,
Langenthal 059946
Kunstverein Oberhausen,
Oberhausen 059104
Kunstverein Oberwallis, Brig . . . 059947
Kunstverein Oerlinghausen e.V.,
Oerlinghausen 059105
Kunstverein Osterholz, Osterholz-
Scharmbeck 059106
Kunstverein Ostseebad Kühlungsborn,
Kühlungsborn, Ostseebad . . . 059107,
107769
Kunstverein Ottobrunn e.V.,
Ottobrunn 059108
Kunstverein Paderborn e.V.,
Paderborn 059109
Kunstverein Panitzsch, Panitzsch 059110,
108356
Kunstverein Passau – Sankt-Anna-
Kapelle, Passau 021082
Kunstverein Passau e.V., Passau 059111
Kunstverein Plauen-Vogtland e.V,
Plauen 059112
Kunstverein Radolfzell e.V., Radolfzell
am Bodensee 059113
Kunstverein Rastatt e.V., Rastatt 059114
Kunstverein Region Heinsberg e.V.,
Heinsberg 059115
Kunstverein Reutlingen,
Reutlingen 021382, 059116
Kunstverein Rhein-Ruhr e.V., Köln 059117
Kunstverein Ribnitz-Damgarten e.V.,
Ribnitz-Damgarten 059118
Kunstverein Röderhof, Huy 059119
Kunstverein Rosenheim,
Rosenheim 059120, 137488
Kunstverein Rügen e.V., Putbus . 059121
Kunstverein Ruhr e.V., Essen . . 018302,
059122
Kunstverein Salzgitter e.V.,
Salzgitter 059123
Kunstverein Sankt Gallen, Sankt
Gallen 059948
Kunstverein Schaffhausen,
Schaffhausen 059949
Kunstverein Schieder-Schwalenberg,
Schieder-Schwalenberg 059124
Kunstverein Schopfheim,
Schopfheim 059125
Kunstverein Schorndorf,
Schorndorf 059126
Kunstverein Schwäbisch Hall,
Schwäbisch Hall 059127
Kunstverein Schwedt e.V., Schwedt /
Oder 059128
Kunstverein Schwerte, Schwerte 021798,
059129
Kunstverein Schwetzingen,
Schwetzingen 059130
Kunstverein Siegen, Siegen 059131
Kunstverein Speyer, Speyer . . 021932
Kunstverein Speyer e.V., Speyer 059132
Kunstverein Springhornhof e.V.,
Neuenkirchen, Kreis Soltau-
Fallingbostel 020700, 059133
Kunstverein Südsauerland, Olpe 059134
Kunstverein Tauberbischofsheim,
Tauberbischofsheim 059135
Kunstverein Teterow e.V.,

Teterow 059136
Kunstverein Trier Junge Kunst e.V.,
Trier 059137
Kunstverein Uelzen, Uelzen 059138
Kunstverein Ulm, Ulm . 022263, 059139
Kunstverein Unna, Unna 059140
Kunstverein Uri, Altdorf 059950
Kunstverein Viernheim e.V.,
Viernheim 059141
Kunstverein Villa Streccius Landau e.V.,
Landau in der Pfalz 059142
Kunstverein Weiden, Weiden in der
Oberpfalz 022460
Kunstverein Wesseling,
Wesseling 059143
Kunstverein Wilhelm Kimmich e.V.,
Lauterbach, Schwarzwald 059144
Kunstverein Wilhelmshöhe Ettlingen,
Ettlingen 059145
Kunstverein Wiligrad, Lübstorf . . 059146
Kunstverein Winterthur,
Winterthur 059951
Kunstverein Wolfenbüttel,
Wolfenbüttel 059147
Kunstverein Wolfsburg, Wolfsburg 059148
Kunstverein Würzburg e.V.,
Würzburg 059149
Kunstverein Xanten, Xanten . . 059150
Kunstverein zu Frechen e.V.,
Frechen 018464, 059151
Kunstverein zu Rostock, Rostock 059152
Kunstvereinigung Wasgau e.V.,
Dahn 059153
Kunstvereniging Diepenheim,
Diepenheim 032221, 059613
KunstVerket Galleri, Oslo 084324, 113236
Kunstverlag Heselmann, Krefeld 137328
Kunstverlag Hofstetter, Ried im
Innkreis 136545
Kunstverlag R. Schumacher GmbH,
Edition Fils, Düsseldorf 106802, 137162
Kunstverlag Ver Sacrum, Rottenburg
am Neckar 137492
Kunstverlag Würzburg, Würzburg 137562
Kunstweb, Rotterdam 083576
Kunstwelt Rochus von Probst,
Graz 062595, 100013
Kunstwerk, Köln . . . 019698, 059154
Kunstwerk, Mannheim 107971
Das Kunstwerk, Wien . 063015, 063016,
125278, 125279
KunstWerk artifex, Braunschweig 106567
Kunstwerkplaats, Amsterdam . . 059614
Die Kunstwerkstatt, Stuttgart . . 130680
KunstWerkStatt Lake,
Cloppenburg 106638
Kunstwerkstatt, Felix Striebeck,
Hagen 076150, 107197
Kunstwinkel Leo Steinhauzer,
Eindhoven 112463
Kunstzaken, Genk 100685
Kunstzeitung, Regensburg 138734
Kunstzentrum Koppelschleuse,
Meppen 107995
Kuntsin Taidekokoelma, Vaasa . 011468
Kuny Domokos Megyei Múzeum,
Tata 023999
Kunying Art and Culture Museum,
Beimen 042272
Kunz-Werle, A., Häusern 076145
Kunz, Alain, Basel 144024
Kunz, Gunther, Sonnefeld 078504
Kunz, Vreny, Uster 133947
Kunze-Kania, I., Karlsbad 076679
Kunze, Claus, Dresden 141703
Kunze, Gerhard, Berlin 129523
Kunze, Lilli, Düsseldorf 075490
Kuo Chang-Hsi Knife and Sword
Museum, Jiading 042350
Kuo Yuan Ye Museum of Cake and
Pastry, Taipei 042521
Kuo, Georg, Berlin 074859
Kuo, Veronika, Fürstenwalde . . . 075923
Kuokkalan Museoraitti, Lempäälä 011104

Kuopion Korttelimuseo, Kuopio . 011052
Kuopion Luonnontieteellinen Museo,
Kuopio Kulttuurihistoriallinen Museo,
Kuopio 011053
Kuopion taidemuseo, Kuopio . . . 011054
Kuparas, Vilnius 111799
Kuperion Gallery, Merano 110213
Kupf – Kulturplattform Oberösterreich,
Linz 058284, 136533
Kupferberg-Museum, Mainz . . . 020228
Kupferbergwerk, Beckingen . . . 017008
Kupfergraben Verlag, Berlin 137082
Kupferschmiede-Museum,
Tangermünde 022100
Kupferstich-Kabinett, Staatliche
Kunstsammlungen Dresden,
Dresden 017992
Kupferstichkabinett – Sammlung der
Zeichnungen und Druckgraphik,
Staatliche Museen zu Berlin –
Stiftung Preußischer Kulturbesitz,
Berlin 017209
Kupferstichkabinett, Akademie der
bildenden Künste, Wien 003027
Kupferstichkabinett, Kunstmuseum
Basel, Basel 041274
Kupina, Moskva 085189, 085190
Kupiškio Etnografijos Muziejus,
Kupiškis 030401
Kuppelwieser, Henndorf am
Wallersee 100034
Kupper, Linda, Neu-Isenburg . . 108237
Kupperman, Amsterdam 082791
Kupris, A., Westerkappeln 078950
Kupšina, Jiří, Kolín 065406
Kur Gallery, Donostia-San
Sebastián 115053
Kur-und Stadtmuseum Bad Ems, Bad
Ems 016760
Kura, Auckland 112781
Kura, Waikato 113038
Kura Gallery, Wellington 033349
Kuralan Kylämäki, Turun
Maakuntamuseo, Turku 011437
Kuramayama Museum, Kyoto . . 028948
Kuranda Arts Cooperative,
Kuranda 099210
Kurashiki Geijutsu Kagaku Daigaku,
Kurashiki 056011
Kurashiki Mingeikan, Kurashiki . 028900
Kurashiki-shi Bijutsukan,
Kurashiki 028901
Kurashiki-shi Nihon Token Bijutsukan,
Kurashiki 028902
Kurashiki-shiritsu Kokokan,
Kurashiki 028903
Kuraya, Sakai 082097
Kuraya, San Francisco 097267
Kurayoshi Hakubutsukan,
Kurayoshi 028908
Kurdi, Talal, Jeddah 085303
The Kurdish Museum, Brooklyn . 047261
Kure – Hakubutsukan, Izumi . . 028726
Kure-shiritsu Bijutsukan, Kure . 028909
Kurer, Bernadette, Sankt Gallen . 133987
Kurfner, Michael, München 077551
Kurfürst, K., Offenburg 077908
Kurfürstliches Freihaus, Stadt- und
Kulturgeschichtliches Museum
Torgau, Torgau 022168
Kurfürstliches Gärtnerhaus, Bonn 017496
Kurganskij Oblastnoj Chudožestvennyj
Muzej, Kurgan 037184
Kurganskij Oblastnoj Kraevedčeskij
Muzej, Kurgan 037185
Kurikan Museo, Kurikka 011059
Kurimanzutto, México 111964
Kuriosaboden, Arboga 086630
Kurioses Muschel-Museum,
Nordholz 020831
Kurioses Muschel-Museum,
Spiekeroog 021937
Kuriositet & Antikk, Stavanger . 084370
Kurita Hakubutsukan, Ashikaga . 028488

Kurmann, André, Biberist 087313
Kurmann, André & Paulette,
Solothurn 087821
Kurmuseum, Bad Tatzmannsdorf 001764
Kurmuseum Bad Wildungen, Bad
Wildungen 016904
Kurnikova Hiša, Tržiški Muzej,
Tržič 038625
Kuroda Kinenshitsu, Tokyo . . . 029543
Kurofune, Tokyo 082175
Kurpfälzische Münzhandlung,
Mannheim 077263, 126033
Kurpfälzisches Museum der Stadt
Heidelberg, Heidelberg 019137, 019138
Kurrajong Antique Centre,
Kurrajong 061708
Kurrer, Mölln 077380
Kuršėnų Kalendorių Muziejus,
Kuršėnai 030404
Kuršių Nerijos Nacionalino Parko
Gamtos Muziejus, Klaipėda . . 030389
Kursiv – eine Kunstzeitschrift,
Graz 138524
Kurskaja Kartinnaja Galereja im. A.A.
Dejneki, Kursk 037186
Kurskij Gosudarstvennyj Oblastnoj
Kraevedčeskij Muzej, Kursk . . 037187
Kurskij Gosudarstvennyj Oblastnoj
Muzej Archeologii, Kursk 037188
Kurt Tucholsky Literaturmuseum,
Rheinsberg 021402
Kurt, Andreas, Murnau am
Staffelsee 077683
Kurta, Axel, Berlin 141546
Kurtej, Miloš, Kutná Hora 065411,
102610
Kurth, Hanspeter, Laupen 087557
Kurt's Furniture Stripping, Buffalo 135375
Kurts, Lisa, Memphis 121699
Merve Kurttepeli & Barbara Ther,
İstanbul 088129
Kurukshetra Panorama & Science
Centre, Kurukshetra 024373
Kurun Ulkomuseo, Kuru 011060
Kurun Warun, Pomona 062087
Kurz, Wien 063017
Kurz-Immetsberger, Schopfheim 078389
Kurz, Lothar, Lenting 107857
Kurz, Rainer, Oberaudorf 142250
Kurza Stopa – Galeria Współczesnej
Sztuki Polskiej, Muzeum Historyczno-
Etnograficzne, Chojnice 034689
Kurzemes Cietokšņa Muzejs,
Zante 030258
Kurzemes Hercogu Kapenes, Rundāles
Pils Muzejs, Jelgava 030132
Kurzreuther, Christa, Monheim am
Rhein 077413
Kusabue, Nagoya 111323
Kusber, Jochen, Rastede 108431
Kuser Farm Mansion, Hamilton . 049326
Kush, New York 122795
Kushiro Art Museum, Hokkaido,
Kushiro 028914
Kushiro City Museum, Kushiro . 028915
Kushnir & Taylor, New York . . . 122796
Kusmikim, Amsterdam 082792
Kusseneers, Paul, Antwerpen . . 100501,
136622
Kustannus Oy Taide Art Publishers,
Helsinki 136622
Kustavin Kotiseutumuseo, Kustavi 011061
Kuster, Hugo, Sankt Gallen . . . 087785,
133909
Kustom Korners, Pittsburgh . . . 123558
Kuszelevic, Henri, Saint-Ouen . . 073125
Kuszelewic, Henri, Paris 071531
Kutaisskij Gosudarstvennyj Muzej
Istorii i Etnografii, Kutaisi . . . 016837
Kutjarra Arts, Alice Springs . . . 098634
Kutschen- und Heimatmuseum,
Obertilliach 002468
Kutschen- und Schlittenmuseum,

Großraming 002013
Kutschen-, Schlitten- und
 Wagenmuseum Rottach-Egern,
 Rottach-Egern 021517
Kutschenkabinett, Landau in der
 Pfalz 019872
Kutschenmuseum, Augustusburg 016673
Kutschenmuseum, Laa an der
 Thaya 002255
Kutschenmuseum, Ludwigslust . 020142
Kutschenmuseum, Neustadt,
 Dosse 020765
Kutschenmuseum Hessisches
 Landgestüt, Dillenburg 017897
Kutschenmuseum, Historisches
 Museum Basel, Basel 041255
Kutschensammlung, Eigeltingen 018126
Kutschensammlung Robert Sallmann,
 Amriswil 041217
Kuttig, Peter, Falkenstein,
 Vogtland 141762
Kuttner & Siebert, Berlin 106351
Kutynski, Piet, Tilburg 083616
Kuurojen Museo, Helsinki 010840
Kuurstra, J.J., Amsterdam 137837
Kuusamon Kotiseutumuseo,
 Kuusamo 011062
Kuusiston Kartano, Kuusisto . . . 011064
Kuva-yi Milliye Müzesi, Balıkesir 042842
Kuvataideakatemia, Helsinki . . . 055343
Kuwait National Museum, Kuwait 030068
Kuwana-shiritsu Hakubutsukan,
 Kuwana 028916
Kuwayama Bijutsukan, Nagoya . 029117
Kuyper, Bubb, Haarlem 126287, 143161
Kuypers, S., Aachen 074335
Kuzbik, Joëlle, Montcenis 070228
Kuzuu-machi Yoshizawa Kinen
 Bijutsukan, Kuzuu 028917
Kvadrat, Omsk 114327
Kvadrat, Sankt-Peterburg 114346
Kvæfjord Bygdetun, Sør-Troms
 Museum, Harstad 033638
Kvällsantikvariatet, Sundsvall . . 143975
Kvæven Bygdetun, Tjørhom 034016
Kvam Bygdemuseum,
 Norheimsund 033808
Kvariaatti Antikvaarinen Kiirjakauppa,
 Turku 140749
Kvarkens Båtmuseum, Malax . . 011141
Kvarnbacken Prylmuseum,
 Sandviken 040939
Kvevlax Hembygdsmuseum,
 Kvevlax 011065
Kvikmyndasafn Islands,
 Hafnarfjörður 024079
Kvindemuseet i Danmark, Århus 010007
Kvist, Horsens 102850
Kviteseid Bygdetun, Kviteseid . . 033726
KW Institute for Contemporary Art,
 Kunst-Werke Berlin e.V., Berlin 017210
Kwa Muhle Museum, Durban . . 038701
Kwade, Alexander, Hannover . . 076381,
                                  130009
Kwai Fung Hin, Hong Kong . . . 102092
Kwak, Los Angeles . . . 094384, 135726
Kwame Nkrumah University of
 Science and Technology Museum,
 Kumasi 022934
Kwan, Hua, Singapore 114556
Kwang-Ya Art Center, Yu-da
 Institute of Business Technology,
 Zaociao 042609
Kwangdong College Museum,
 Gangneung 029928
Kwartalnik Fotografia, Września 138970
Kwartler, Alice, New York 095522
KwaZulu Natal Society of Arts –
 KZNSA, Durban 059819
Kwidzinski, Kasimir J.,
 Spechbach 130648
Kwinitsa Station Railway Museum,
 Prince Rupert 006718
Kwok, Singapore 085454

Kyaratei, Narita 082078
Kyburz, Rudolf, Schindellegi . . . 087806
Kyffhäuser-Denkmal, Bad
 Frankenhausen 016764
Kyhnau, Sorø 066144
Kyïvs'kyj Literaturno-memorial'nyj
 Budinok-muzej M. Ryl's'kogo,
 Kyïv 043195
Kyïvs'kyj Literaturno-memorial'nyj
 Budinok-muzej Tarasa Ševčenka,
 National Taras Shevchenko Museum,
 Kyïv 043196
Kyïvs'kyj Literaturno-memorial'nyj
 Muzej-kvartyra P. Tyčyny, Kyïv 043197
Kyïvs'kyj Memorial'nyj Budinok-muzej
 M. Zan'kovec'koï, Kyïv 043198
Kyïvskyj Muzej Puškina, Kyïv . . 043199
Kyïvs'kyj Muzej Rosijs'koho Mystectva,
 Kyïv 043200
Kyïvs'kyj Nacional'nyj Universitet
 Kul'tury i Mystectv, Kyïv 056491
Kyle and Carrick District Library and
 Museum, Ayr 043460
Kyle, Francis, London 118418
Kylia Arte, Bucaramanga 102391
Kymenlaakson maakuntamuseo,
 Kotka 011031
Kymmel, Peter von, Oderwitz . . 077891
Kyneton Museum, Kyneton 001201
Kyngston House Museum, Saint
 Albans 045666
Kynigalakis Dimitris, S., Galatsi . 079192
Kynologisches Museum,
 Klostermarienberg 002203
Kyocera Museum of Fine Ceramics,
 Kyoto 028949
Kyösti- ja Kalervo Kallion Museo,
 Nivala 011189
Kyoko chirathivat, Bangkok . . . 116954
Kyomo, Nueva Andalucía 126554
Kyongju University Museum,
 Kyongju 029968
Kyongnam College Museum,
 Masan 029969
Kyoto Antique Center, Kyoto . . 082029
Kyoto Arashiyama Orugoru
 Hakubutsukan oyo Guido Reuge
 Hakubutsukan, Kyoto 028950
Kyoto Asahi Beer Oyamazaki Villa
 Museum, Oyamazaki 029284
Kyoto Bazaar, Kyoto 082030
Kyoto Bisho, Kyoto 082031
Kyoto Chopstick Culture Museum,
 Kyoto 028951
Kyoto City Dohda Senior High School
 of Arts, Kyoto 056013
Kyoto Daigaku Bungakukan,
 Kyoto 028952
Kyoto Daigaku Hakubutsukan,
 Kyoto 028953
Kyoto-fu Kyoto Bunka Hakubutsukan,
 Kyoto 028954
Kyoto Gigei Bijutsukan, Kyoto . . 028955
Kyoto Gion Oil Lamp Museum,
 Kyoto 028956
Kyoto-kenritsu Insho-Domoto
 Bijutsukan, Kyoto 028957
Kyoto-kenritsu Tango Minzoku
 Shiryokan, Miyau 029042
Kyoto-kenritsu Yamashiro Minzoku
 Shiryokan, Yamashiro 029701
Kyoto Kokuritsu Hakubutsukan,
 Kyoto 028958
Kyoto Kokuritsu Kindai Bijutsukan,
 Kyoto 028959
Kyoto Museum for World Peace,
 Kyoto 028960
Kyoto-shi Bijutsukan, Kyoto . . . 028961
Kyoto-shi Seishonen Kagaku,
 Kyoto 028962
Kyoto-shiritsu Bijutsu Daigaku, Bijutsu
 Gakuin, Kyoto 056014
Kyoto-shiritsu Gakkou Rekishi
 Hakubutsukan, Kyoto 028963

Kyoto-shiritsu Geijutsu Daigaku,
 Kyoto 056015
Kyoto-shiritsu Kokokan, Kyoto . . 028964
Kyoto Shoin, Kyoto 142899
Kyoto Yuzen Bunka Kaikan,
 Kyoto 028965
Kyoto Zokei Geijutsu Daigaku,
 Kyoto 056016
Kypriotis & Sia, E., Glyfada 079194
Kyrgyzskij Gosudarstvennyj Muzej
 Literatury i Iskusstva im. Toktogula
 Satylganova, Biškek 030073
Kyrgyzskij Nacionalnyj Muzej
 Izobrazitelnych Iskusstv im. Gapara
 Ajtijeva, Biškek 030074
Kyrios Books, Nottingham 144759
Kyser-Hollingsworth, Washington 097842
Kystmuseet Hvaler, Vesterøy . . 034083
Kystmuseet i Nord-Trøndelag,
 Rørvik 033901
Kystmuseet i Øygarden, Rong . . 033907
Kystmuseet i Sør-Trøndelag,
 Hitra 033653
Kystmuseet i Sogn og Fjordane,
 Florø 033604
Kysucka Galéria, Čadca 038337
Kysucke Múzeum, Čadca 038338
Kyu Ikegamike, Takatomachi
 Inashi 029450
Kyukodo, Sendai 082117
Kyunghee University Museum,
 Seoul 030020
Kyungin Museum of Fine Art,
 Seoul 030021
Kyungsung Museum, Busan . . . 029899
Kyushu Rekishi Shiryokan,
 Dazaifu 028524
Kyushu Sangyo Daigaku
 Hakubutsukan, Fukuoka 028550
Kyushu Zokei Art College, Kyushu
 Sangyo Daigaku, Fukuoka . . . 056000
Kyyjärven Kotiseutumuseo,
 Kyyjärvi 011066
Kyzylordynskij Istoriko-Kraevedčeskij
 Muzej, Kyzylorda 029825
KZ-Gedenk- und Begegnungsstätte,
 Ehemalige Außenstelle des
 Konzentrationslagers Neuengamme,
 Ladelund 019842
KZ-Gedenkstätte Dachau, Stiftung
 Bayerische Gedenkstätten,
 Dachau 017797
KZ-Gedenkstätte Flossenbürg,
 Flossenbürg 018376
KZ-Gedenkstätte Husum-Schwesing,
 Schwesing 021800
KZ-Gedenkstätte Mauthausen,
 Mauthausen 002379
KZ-Gedenkstätte Mittelbau-Dora,
 Nordhausen 020828
KZ Gedenkstätte Moringen,
 Moringen 020467
KZ-Gedenkstätte Neckarelz,
 Mosbach 020471
KZ-Gedenkstätte Neuengamme,
 Hamburg 019003
KZ-Gedenkstätte Sandhofen,
 Mannheim 020250
KZ-Gedenkstätte und Zeitgeschichte-
 Museum Ebensee, Ebensee . . 001836
KZNSA Gallery, Durban 038702
L & B Kunstauktionen, Essen . . 125941,
                                  125942
L & B Viewing Room, Portland . . 123586
L & L Antiklädchen, Merchweiler 077347
L & L Coins, Denver 093307
L & M Antiques, Tampa 097568, 136304
L & M Arts, New York 122797
L & N Productions, Philadelphia 123407
L & R Antiques, Houston 093721
L & S Coins, Jacksonville 094003
L Art, Kyïv 117141
L. & Co., Saint-Cyr-en-Pail 072723
L. Lange's Ovnmuseum,

Svendborg 010334
L.A. Galerie, Frankfurt am Main 107012
L.A. Mayer Museum for Islamic Art,
 Jerusalem 025089
L.A.C. Lieu d'Art Contemporain,
 Sigean 015808
L.A.R.A., Genova 131319
L.B. Contemporary Art, Bologna 109852
L.C. Bates Museum, Hinckley . . 049486
L.D. Brinkman Art Foundation,
 Kerrville 049943
L.D. Upholstery and Refinishing,
 Edmonton 128165
L.E.D.A., Paris 071532
L.G. Antique Restoration, Los
 Angeles 135727
L.G.N. Antik, Storvik 087080
L.M., Torino 081715
L.MA France, Paris 071533
L.O.G.G. Lights, Tynemouth . . . 091683
L.P. Antiques, Walsall 091717
L.S. Raheja School of Art,
 Mumbai 055821
L.V.S. Antiquités, Paris 071524
La Boheme Gallery, Macclesfield 118690
La Boutique del Piano, A Coruña 085955,
                                  133234

La Brea Antique Collection, Los
 Angeles 094385
La Butte aux Cailles, Paris 071535
La Casa del Barrio Art Gallery, San
 José 102510
La Casa della Nonna, Roma . . . 081416
La Casa della Nonna,
 Valmontone 081812
La Centrale Powerhouse Gallery,
 Montréal 101422
La Chaumière de Granièr,
 Thiézac 015935
La Criée Centre d'Art Contemporain,
 Rennes (Ille-et-Vilaine) 015065
La Ferme de Gribelle, Gedinne . 082354
La Ferme Musée de Marcq,
 Enghien 003577
La Galeria, Sant Cugat del Vallès 115411
La Galerie d'Art, Bowness-on-
 Windermere 117409
La Hougue Bie Museum,
 Grouville 044485
La Hune Brenner, Paris 104991
La Loie Fuller, Paris 071536
La Maison Bleue, Craon 103640
La Maison Bleue, Paris Italici 105909, 105910
La Malle des Indes, Metz 104227
La Part de l'Oeil, Bruxelles 138554
La Petite Galerie, Tampa 097569
La Piscine – Musée d'Art et
 d'Industrie André Diligent,
 Roubaix 015162
La Porte Étroite, Paris 141155
La Provence par Marc Ferrero,
 Eze 103730
La Raza-Galeria Posada,
 Sacramento 052729
La Remise Musée de la Vie Rurale,
 Offaing 003917
La Rocca, Leonardo, Palermo . . 110607
La Salle University Art Museum,
 Philadelphia 052060
La Serre aux Papillons, La Queue-lèz-
 Yvelines 013452
La Tibetaine, Paris 071537
La Tine, Troistorrents . 087858, 133942
La Trobe Street Gallery,
 Melbourne 099311
La Trobe University Art Museum,
 Bundoora 000912
La Trouvaille, Grône 087515
La vie des musées, Hornu 138555
Laakes, W., München 077552
Laarman, Goes 143145
Laat, H. de, 's-Hertogenbosch . 083392
Laatzen, Hermann, Hamburg . . 141880
The Lab, San Francisco 053008

Labadie, Charles, Saint-Jean-de-
Luz 105534
Labaëre, Dourges 128586
Laban, Jean-Michel, Oloron-Sainte-
Marie 070782
Laband Art Gallery, Loyola Marymount
University, Los Angeles 050441
Labarre A, Saint-Germain-
Lembron 072811
Lábasház, Soproni Múzeum,
Sopron 023894
L'Abat Jour de Charme, Trouville-sur-
Mer 073891
Labatut, Paris 104992
Labbé, Armel, La Chartre-sur-le-
Loir 068986
Labet, Michel, Étiolles 068483
Labeyrie, Alain, Bordeaux 128494
Labeyrie, Pascal, Bordeaux 128495
Labiche, Nicole, Montaure 070204
Labiche, Nicole, Rouen 072594
Il Labirinto, Milano 142723
Labirynt, Kraków 113610
Labirynt II, Kraków 113611
Labisch, Hubert, Unterpleichfeld 130715
Labonne, Albert-Claude, La Réole 069061
Labor Fruit, Los Angeles 121525
Laboratório de Ensino de Ciências,
Faculdade de Filosofia, Ciências e
Letras de Ribeirão Preto, Ribeirão
Preto 004674
Il Laboratorio, Firenze 142678
Il Laboratorio, Roma 131952
Laboratorio Antiquariato, Milano 131457
Laboratório de Anatomia e
Identificação de Madeiras,
Manaus 004552
Laboratorio de Conservacion y
Restauracion, Barcelona 133126
Laboratório de Demonstração do
Instituto de Fisica, Intituto de Fisica /
IF, São Paulo 004861
Laboratorio de Restauracion y
Enmarcacion, Barcelona 133127
Laboratorio del Legno, Verona . 132285
Laboratorio di Restauro, Milano 131458
Laboratorio di Restauro e
Manutenzione d'Arte Tessile,
Firenze 131193
Il Laboratorio di Restauro San
Prospero, Milano 131459
Laboratorio di Restauro Santo Stefano,
Bologna 130971
Laboratorio Lucidatura Mobili Antichi,
Milano 131460
Laboratorio per l'Arte Contemporanea,
Ospedaletto 110552
Laboratorio Restauri d'Arte,
Trieste 132221
Laboratorio Restauro Dipinti,
Parma 131727
Laboratorio Restauro e Conservazione,
Padova 131668
Laboratorios Dopla, Sevilla 133433,
133434
Labour of Love, Jacksonville . . 135657
Labrada, Cali 102413
Labrador Interpretation Centre, North
West River 006521
Labrador Straits Museum, L'Anse-au-
Loup 006163
Labråten, Asker Museum, Valstads
samlinger, Hvalstad 033678
Labre, Michael, Sassey 073425
L'Abri Pataud, Gisement Préhistorique
et Musée, Les Eyzies-de-Tayac-
Sireuil 013748
Labriffe, M., Marcq-en-Barœul . 069864
Labrouche, Montendre 070235
Labrouche, Line & Michel,
Messac 070080
Labry, Dominique-Mathieu,
Montpeyroux 070319

Laburnum, Poole 091034, 134947
Laburnum Antiques, Richmond . 096558
Laburnum Cottage Antiques, Eye 089242
Labyrint, Praha 136765, 138613
Labyrinth, Eggenstein-
Leopoldshafen 075577
Labyrinth, Valletta 082441
Labyrinth Kindermuseum Berlin,
Berlin 017211
Labyrinthos, Firenze 138839
Lac Cardinal Regional Pioneer Village
Museum, Grimshaw 005929
Lac-Du-Bonnet and District Historical
Society Museum, Lac-Du-
Bonnet 006142
Lac-La-Hache Museum, Lac-La-
Hache 006143
Lac Lagorio Arte Contemporanea,
Brescia 109909
Lac Qui Parle County Historical
Museum, Madison 050587
Lac Sainte-Anne Historical Society
Pioneer Museum, Rochfort
Bridge 006818
Lac, Hong, Ho Chi Minh City . . 125128
Lac, Isabelle du, Toulouse 073781
Lacam, Nice 070620
Lacarin, Madeleine, Montluçon . 070270
Lacaze, Alain, Québec 101503
L'Accent Sud, Saint-Tropez 073314
Laccroix, Jean-Claude, Lorient . 069638
Lace Antiguidades, Rio de
Janeiro 064072
Lace House Museum, Black
Hawk 047023
The Lace Museum, Sunnyvale . 053607
Lacerba, Ferrara 110020
Lacerte, Madeleine, Québec . . . 101504
Lacewing Fine Art, Winchester . 119359
Lacey Historical Society Museum,
Forked River 048775
Lach, Grzegorz, Kraków 132843
Lachaert, Sofie, Tielrode 100817
Lachaux, Alain, Beaune 067025
Lachaux, Jean-Christophe, Paris 071538
Lachaux, Jean-Pierre, Beaune . . 067026
Lachenaud, Jean-Luc, Tours . . . 105763
Lacher-Biene, Silvia, Müllheim . 077438
Lachin, Venezia 081848, 081849
Lachin Antichità, Esposizione Vendita
Restauro, Venezia 081850
Lachman, Dušan, Praha 128354
Lachmayr, Hannes, Adnet 062514
Lachmund, U., Abbenrode 074348
Lachmuth, Wolfgang,
Schopfheim 078390
Lacidogna, Michele, Torino 081716
Låckereds Gård Antik,
Vänersborg 087135
Lackey, Oklahoma City 095984
Lackey, David, Houston 093722
Lackey, Hugh, San Antonio . . . 096986
Lackham Museum of Agriculture and
Rural Life, Lacock 044747
Lackmann-Wisskirchen, L.,
München 077553
Lackmann, Sabine, Essen 075688
Lackner, G., Salzburg 062806
Lackus, H., Kraichtal 076947
Lacombe, Claudine, Auneau . . . 066814
Lacombe, Eric, Bordeaux 125530
Lacombe, Jean-Pierre, Rodez . . 072512
Lacoste, Jacques, Paris 104993
Lacote, Philippe, Beaune 067027
Lacour, Guy, Goven 068691
Lacour, Jean-Marc, Bruz 067527, 128525
Lacour, Jean-Marie, L'Isle-sur-la-
Sorgue 069574
Lacour, Josette, Genève 087477, 116326
Lacour, Pierre, Fourdrain 068601
Lacourbas, Hélène et Romain,
Bellenaves 067077
Lacourt, Maurice, Charmes . . . 067853
The Lacquer Chest, London . . . 090278

Lacquerware Museum, Pagan . . 031824
Lacrimosa, Osaka 082086
Lacroix, Québec 101505
Lacroix, Alexandre, Paris 104994
Lacroix, Serge, Saint-Julien-en-
Genevois 072866
The Lacrosse Museum and National
Hall of Fame, Baltimore 046755
Lacruz, Valencia 133465
Lacueva, Eduardo, Buenos Aires 139576
Lacy, London 090279, 118419
Lacy, West Hollywood 125008
Lacy Scott & Knight, Bury Saint
Edmunds 126842
Lacy Scott & Knight, Stowmarket 127058
Lacydon, Marseille 104183, 136836
Laczkó Dezső Múzeum,
Veszprém 024040
Ladakh, Delhi 109328
Ladbyskibsmuseet, Kerteminde . 010160
Laden für Nichts, Leipzig 107840
Ladenburger Spielzeugauktion,
Ladenburg 126021
Ladengalerie Berlin, Berlin 106352
Laderer, Pittsburgh 123559
Ladeuille, Roger, Lamorlaye . . . 069153
Ladevèze, Michel, Cormeilles . . 068153
Ladew Manor House, Monkton . 050990
Ladics-ház, Corvin János Múzeum,
Gyula 023676
Ladner, Hubert & Helma, Urbar . 078769
Ladoucette, Marc de, Paris 104995
Ladrière, Sandrine, Paris 071539
Lads, Osaka 111362
Ladson, Malvern, Victoria 127634
Ladstätter, Markus, Wien 127942
Ladue Antique Galleries, Saint
Louis 096726
Lady and the Tramp Antiques Mall,
Saint Paul 096834
Lady Cant Bis, Antwerpen 063221
Lady Denman Maritime Museum,
Huskisson 001161
Lady Di Régiségbolt, Budapest . 079266
Lady Franklin Gallery, Lenah
Valley 001218
Lady Jane, Rotterdam 083577
Lady Lever Art Gallery, Port
Sunlight 045536
Lady Matilda Gallery, Nana Glen 099419
Lady of the Lake, Austin 092244
Lady of the Sea Gallery, Norfolk 095862
Lady Piú, Torino 081717
Lady Steptoe, Orange 062031
Lady Waterford Hall, Berwick-upon-
Tweed 043578
Lady Wilson Museum,
Dharampur 024244
Ladybug, San Diego 124141
Ladyship Oils, Killurin 109655
Ladysmith Railway Museum,
Ladysmith 006151
Läänemaa Muuseum, Haapsalu 010608
Lääninkivalteri Aschanin Talo,
Heinola 010810
Läcktymsuseo, Helsinki 010841
Lährm, Gerhard, Salzburg 062807
Lækkert, Bærre, Oslo 084325
Lämmle, B., Vaihingen an der
Enz 108808
Lände, Kressbronn am Bodensee 019792
Ländliche Bildergalerie,
Dischingen 017912
Längelmäen Kirkkomuseo,
Längelmäki 011067
Längelmäen Korsumuseo,
Länkipohja 011068
Längelmäen Pellavamuseo,
Länkipohja 011069
Länsmuseet Gävleborg, Gävle . . 040637
Länsmuseet Halmstad, Halmstad 040683
Länsmuseet Västernorrland,
Härnösand 040677

Länsmuseet Varberg, Varberg . . 041154
Lærken, Nykøbing Falster 102938
Læsehesten, Aalborg 140546
Læsehesten, Frederikshavn 140574
Læsehesten, Silkeborg 140655
Læsehesten Kolding Antikvariat,
Kolding 140622
Læsø Museum, Læsø 010214
Läzer, Weimar 108864, 137549
Lafarge, Jean-Pierre, Cognac . . 068059
Lafaurie, Bénédicte de, Paris . . 071540
Lafay, Benoît, La Bégude-de-
Mazenc 128637
Lafayette, Cleethorpes 144342
Lafayette, Saint-Ouen 073126
Lafayette Antics, Paris 071541
Lafayette College Art Gallery,
Easton 048456
Lafayette Miners Museum,
Lafayette 050058
Lafayette Museum – Alexandre
Mouton House, Lafayette 050061
Lafayette Museum Association,
Lafayette 060561
Lafayette Natural History Museum,
Lafayette 050062
Lafayette Ragsdale, Memphis . . 121700
Lafeara, Manama 100439
Lafferty, Tampa 097570
Laffitte, Jeanne, Marseille 140944
Lafite, Céline, Paris 129052
Lafitte, Christchurch 117590
Laflamme Galleries, Calgary . . . 101247
Lafleur, Amiens 066575
Lafon, Didier, Brunoy 067525
Lafon, François, Paris 125728
Lafon, Josée, Saint-Pierre-
d'Oléron 073259
Lafond, Pittsburgh . . . 123560, 123561
Lafond, Bernard, Régnié-Durette 072372
Laforet Hakuba Bijutsukan,
Hakuba 028596
Laforge, Pierre, Saint-Maximin-la-Sainte-
Baume 072972, 129240
Lafosse, Jean, Pontaubert 072268
Lafourcade, Antonio y Eduardo Roman,
Madrid 133347
Lafourcade, Yannick, Tours 073850
Lafouresse, Jean-Bruno,
Bayonne 066975, 066976, 128462,
128463
Lafreniere & Pai, Ottawa 101475
Lagaborg Antik, Laholm 086798
Lagalerie, Luxembourg 111857
Lagan Lookout Centre, Belfast . 043557
Laganda, Johann Josef, Unterach am
Attersee 062864
Laganland Bilmuseum, Lagan . . 040769
Lagard, Buenos Aires 098573
Lagarde, Fabrègues 068502
Lagarde, Hugues, Brin-sur-Seille 067500
Lagazio, Valparaíso 064992
Lågdalsmuseet, Kongsberg 033707
Lage, Madrid 086206
L'Age d'Homme, Paris 141156
L'Age d'Or, Saint-Gaudens 072789
L'Age d'Or, Valenciennes 073946
Lagénie, Françoise, Paris 129053
Das Lager, Freiburg im Breisgau 075871
Lagerbauer, Ferdinand, Hamburg 076265
Lagerquist, Evelyn & Kay, Atlanta 119584
Laget, Guy, Ambronay . 066558, 066559
Laget, Léonce, Paris 141157
Laget, Yves, Valence (Drôme) . . 105801
Laggner, Matthias, Wien 127943
Laghzaoui, Mohamed, Genève . 116327
Lagier, Annie, L'Isle-sur-la-
Sorgue 104039
Lagier, Gilbert, Orange 070790
Lagière, Jacques, Issoire 128628
Lagios, Kostas, Agia Paraskeri
Attikis 109012
Lagnel-Tastemain, Caen 067573
Lagrain, P., Gent 128051

Lagrand, C., Bruxelles 063484
Lagrand, Marc, Paris 071542
Lagrange-Choux, Monique,
  Marseille 069946
Lagrange, André, Beauzelle ... 067059
Lagravère, Nicole, Bracieux ... 067444
Lagravinese, Leonardo, Roma .. 081417,
  131953
Lagrille, Marc, La Guerche-de-
  Bretagne 069051
LaGuardia and Wagner Archives,
  LaGuardia Community College, Long
  Island City 050380
Laguna, Seattle 097466
Laguna Art Museum, Laguna
  Beach 050066
Lahaina, San Francisco 124322
Lahaina Arts Society, Lahaina . 060562
Lahaina Restoration Foundation,
  Lahaina 050067
Lahall, Örebro 115792
Laham, Rio de Janeiro 064073
LaHave Islands Marine Museum,
  LaHave 006153
Lahcen, Eva, Köln 107658
Lahden Historiallinen Museo, Lahden
  Kaupunginmuseo, Lahti 011073
Lahden Kaupunginmuseo, Lahti 011074
Lahden Kuvastin ja Taidesoppi,
  Lahti 066383
Lahden Taidemuseo, Lahden
  Kaupunginmuseo, Lahti 011075
Lahmi, Claude, L'Isle-sur-la-
  Sorgue 069575
Lahn-Marmor-Museum, Villmar . 022342
Lahnsteiner Fastnachtsmuseum,
  Lahnstein 019848
Laholms Målerikonservering,
  Laholm 133542
Lahore Art Gallery, Lahore 113309
Lahore Arts Council, Lahore ... 059680
Lahore Fort Museum, Lahore ... 034131
Lahore Museum, Lahore 034132
Lahoreau, Patrick, Pezou 072145
Lahovari, Bucureşti 114147
Lahrichi, Rabat 082601, 112050
Lai, Vientiane 111750
Lai, C.C., Toronto 064689, 101644
Lai, Sandro, Cagliari 079986
Laibstain, Harry, Norfolk 095863
Laick, Jean & Marcelle, Dol-de-
  Bretagne 068364
Laidet, Gérard, Nice 104375
Laidhay Croft Museum,
  Dunbeath 044160
Laidley District Historical Village,
  Laidley 001202
Laigle, Elisabeth, Roquebilliere . 072535
Laihian Kirkkomuseo, Laihia ... 011081
Laihian Museo, Laihia 011082
Laiho Museum, Changhua 042282
Laïk, Jean-Marc, Koblenz 107573,
  137299
Laikrodžių Muziejus, Lietuvos Dailés
  Muziejus, Klaipéda 030390
Laila's Antik, Fredericia 065653
Laimbacher, Karl, Appenzell ... 087195,
  133621
Laimer Schlößl, München 077554
Laine, Henri, Nantes 128838
Lainé, Pascal, Ménerbes 104217
Laing Art Gallery, Newcastle-upon-
  Tyne 045349
Laing Museum, Newburgh 045338
Laing, Shayna, Montréal 101423
Laipa – Galerija Valmiera,
  Valmiera 030247
Laiptai, Šiauliai 111793
L'Air d'Antan, Neuville-sur-Sarthe 070529
Lair, Christian, Falaise 068504
Lairaud, Jean-Pierre, Bordeaux . 067339
L'Aire, Saint-Pierre-d'Eyraud ... 073258
Lairgate Galleries, Beverley ... 117352
Lair's Gallery of Fine Arts, Vevey 087881

Laist, Christine, Seeheim-
  Jugenheim 142359
Laisvés Kovų Muziejus Krauji
  Takais, Lazdijų krašto muziejus,
  Lazdijai 030408
Laiyang Museum, Laiyang 007993
Laizhou Museum, Laizhou 007994
Lajba, John E., Omaha 123325
Lajoix, Dr. Anne, Paris 071543
Lajos Ferenc és Kortársai Grafikai
  Gyűjtemény, Corvin János Múzeum,
  Gyula 023677
Lajos, Dargay, Kalocsa 109196
Lakdiva Gallery, Palmerston ... 099546
Lake Antiques, Lake 089734
Lake Austin Fine Arts, Austin .. 119681
Lake Bathurst Cottage Collectables,
  Braidwood 061190
Lake Broadwater Natural History
  Museum, Dalby 001012
The Lake Champlain Maritime
  Museum, Vergennes 053958
Lake Country Museum,
  Okanagan 006544
Lake Country Pastel Society, White
  Bear Lake 060563
Lake County Discovery Museum,
  Wauconda 054230
Lake County Historic Courthouse
  Museum, Lakeport 050098
Lake County Historical Museum,
  Tavares 053689
Lake County Historical Society
  Museum, Kirtland Hills 049996
Lake County Museum, Two
  Harbors 053870
Lake Daylesford Book Barn,
  Daylesford 139712
Lake Erie Nature and Science Center,
  Bay Village 046835
Lake Forest-Lake Bluff Historical
  Society Museum, Lake Forest . 050077
The Lake Gallery, Waikato 113039
The Lake Gallery, Wicklow 109709
Lake George Historical Association
  Museum, Old Country Courthouse,
  Lake George 050079
Lake Guernsey Museum,
  Guernsey 049290
Lake Hopatcong Historical Museum,
  Landing 050129
Lake House Arts Centre,
  Auckland 033152
Lake House Bookshop, Colombo 143818
Lake Jackson Historical Museum,
  Lake Jackson 050080
Lake Jackson Mounds Archaeological
  Park, Tallahassee 053659
Lake Macquarie City Art Gallery,
  Booragul 000867
Lake Malawi Museum, Mangochi 030655
Lake Meredith Aquatic and Wildlife
  Museum, Fritch 048987
Lake of the Red Cedars Museum,
  Cedar Lake 047482
Lake of the Woods County Museum,
  Baudette 046829
Lake of the Woods Museum,
  Kenora 006078
Lake Perris Regional Indian Museum,
  Perris 051991
Lake Placid-North Elba Historical
  Society Museum, Lake Placid . 050086
Lake Russell Gallery, Emerald
  Beach 099001
Lake Shore Railway Museum, North
  East 051603
Lake State Rare Coin, Saint Paul 096835
Lake Superior Maritime Visitors Center,
  Duluth 048384
Lake Superior Railroad Museum,
  Duluth 048385
Lake Tabourie Museum, Lake
  Tabourie 001204

Lake Tahoe Museum, South Lake
  Tahoe 053410
Lake Taupo Museum and Art Gallery,
  Taupo 033304
Lake Tuggeranong College,
  Tuggeranong 054923
Lake Waccamaw Depot Museum,
  Lake Waccamaw 050089
Lake Wales Depot Museum, Lake
  Wales 050090
Lake, D & E, Toronto 101645
Lake, Randall, Salt Lake City .. 123970
Lakeer, Mumbai 109450
Lakeeren, Mumbai 109451
Lakeland Antiques Art Gallery,
  Ambleside 088269
Lakeland Antiques Art Gallery,
  Grasmere 089359, 117874
Lakeland College, Sheboygan .. 057907
Lakeland Motor Museum, Grange-over-
  Sands 044458
Lakeman, Oscar, Miami 121819
Lakes Antiques Gallery,
  Grasmere 089360, 144441
Lakes District Museum,
  Arrowtown 033134
Lakes District Museum Society, Burns
  Lake 005507
Lakes Entrance Art and Print Gallery,
  Lakes Entrance 099215
Lakeshore Gallery, Milwaukee .. 121930
Lakeside Interiors Classic, Berlin 074860
Lakeside Interiors Country, Berlin 074861
Lakeside Joondalup Bookshop,
  Joondalup 139785
Lakeside Museum, Bulyea 005494
Lakeside Treasures, Chicago ... 092736
Lakeview Furniture Refinishing, Stoney
  Creek 128204
Lakeview Museum of Arts and
  Sciences, Peoria 051987
Lakewood, Dallas 093155
Lakewood Antiques Market,
  Atlanta 092150
Lakewood Lighting, Dallas 135509
Lakewood's Heritage Culture and the
  Arts Galleries, Lakewood ... 050100
Laks, New York 095523, 135948
Lakshana, Bangalore 109261
Lakshana, Chennai 109264
Lakshana, Hyderabad . 109364, 109365
Lal Antik, Ankara 088094
Lalanne, Jean-Claude, Paris ... 071544
Lalbagh Fort Museum, Dhaka .. 003223
Lalbaltry, Gilbert, Saint-Ouen .. 073127
Laleham Antiques, Laleham ... 089735
L'Alfiere, Torino 110982
Laligant, Patricia, New York ... 095524
Lalit Kala Akademi, Delhi 055791,
  109329
Lalla Wool Shed Gallery,
  Launceston 099220
Lallemand, Honfleur 103838
Lallier, Fontenay-le-Fleury 068575
Lally & Co., J.J., New York ... 095525
Lalmi, Hachemi, Herzeele 068816
Laluque, Claude, Saint-Ouen ... 073128
Lam & Co., Hong Kong 065069
Lam & Carly, Hong Kong 065072
Lâm Dông Museum, Dà Lat ... 054784
LAM-Librairie, Art et Musée,
  Ouagadougou 140331
Lam, Maria, Hong Kong 065070
Lam, Yuet Wah, Hong Kong ... 065071
Lama Arts, Kathmandu 112077
Lamai, Hamburg 076266
Lamanai Archaeological Site,
  Lamanai 004177
Lamantia jr., James R., New
  Orleans 122192
Lamar Dodd Art Center, La
  Grange 050034
Lamarche, Patrick, Benfeld ... 067088
Lamare, Frédéric, Fontaine-

Heudebourg 068559
Lamare, Pierre, Saint-Aquilin-de-
  Pacy 072676
Lamarkeez, Manama 100440
Lamarque, Hélène, Paris 104996
Lamasat, Jdeidet El-Metn 082318
Lamasat, Manama 063151
Lamazza, Francesco, Hannover . 130010
Lamb Arcade, Wallingford 091711,
  135184
Lamb, Alexander, Vancouver ... 064833
Lamb, Jerry, Portland 096432
Lamb, Roger, Stow-on-the-Wold 091475
Lambaa Art House, Silkeborg .. 102975
Lambach, Arno, Essen 106923
Lambay House, Howth 109643
Lambda Rising Bookstore,
  Baltimore 144986
Lambert & Son, Woodbridge ... 091914
Lambert & Foster, Tenterden ... 127066
Lambert & Lambert, Paris 071545
Lambert-Barnett, G., Paris 129054
Lambert-Pelco, Paul E.,
  Columbus 093030
Lambert Rouland, Paris 104998
Lambert, Dorrian, Lincoln 089831,
  134561
Lambert, J.-F., Bruxelles 063485
Lambert, James, La Ferté-
  Gaucher 069027
Lambert, Michel, Marseille 140945
Lambert, Philippe, Saint-Maur-des-
  Fossés 072965
Lambert, William M., Oklahoma
  City 095985
Lambert, Yvon, New York 122798
Lambert, Yvon, Paris 104997
Lambertini, Luigi, Bologna 130972
Lambertini, Leandro, Bologna .. 130973
Lambertsmühle, Verein zur Förderung
  der Lambertsmühle, Burscheid 017703
Lamberty, London 090280
Lambertz-Abraham, Karin,
  Düsseldorf 075491, 106803
Lambiek, Amsterdam 112242
Lambing Flat Museum, Young . 001677
Le Lambrequin, Rosiers-
  d'Egletons 072546
Lambrich, Christiane, Mainz ... 077237
Lambropoulos Archaeological
  Collection, Polygyros 023231
Lamb's Ears, Birmingham 092417
Lambton Heritage Museum, Grand
  Bend 005899
Lame 92, Bologna 109853
L'Ame des Objets, Beaujeu 066994
Lamelli Galeria Śródmiejskiego
  Ośrodka Kultury, Kraków ... 113612
Lamere, Houston 121076
Lameris, Gouda 083268
Lamers, Hannover 076382
Lamers, A. & W., Düsseldorf .. 106804
Lamers, Günther, Hannover ... 076383
Lamers, Heinrich, Dortmund ... 106706,
  137145
Lamers, W.M., Winterswijk 132671
Lamine, Guy, Hirson 068822
Lamm, Bad Wildungen 074612
Lamme, Claes, Vetlanda 087158
Lammek, Oldenburg 142268
Lammel, Manfred, Nettersheim . 108235
Lammertse, Matthijs, Amsterdam 132402
Lammi, J., Helsinki 066333
Lamon, Cathy, Castres 103543
Lamont, Evesham 117768
Lamont, London 090281
The Lamont Gallery, Phillips Exeter
  Academy, Exeter 048653
Lamontage, New York 122799
Lamort, Anne, Paris 141158
Lamot, Bruxelles 063486, 100630
Lamouille, Bernard, Ville-la-
  Grand 074134
Lamouret, Patrick, Nantes 070466

Register der Institutionen und Firmen

Lamoureux Ritzenhoff, Montréal – Lang, Helmut R., Wiesbaden

Lamoureux Ritzenhoff, Montréal 101424
The Lamp Gallery, Neustadt am Rübenberge 077754
The Lamp Gallery, Ripley, Surrey 091113
Lamp och Möbelboden, Örebro . 086904
Lamp Post Antiques, Jacksonville 094004
Lamp Post Gallery, Portland . . . 123637
Lamp Repair 'n Shade Studio, Tampa 136391
Lamp Restorations, Botany 127566
Lampa, Daniel F., Plaidt 108384
La Lampada di Aladino, Genova 131320
Lampade, Roma 081418
Lamparas Europeas, Lima 084461
Lamparas Pergamino, Lima 084462
El Lamparin, Lima 084463
Lampe à Huile, Flers 068541
La Lampe d'Aladin, Marrakech . 082591
Lampekælderen, Kjellerup 065815
Lampenmuseum, Geel 003611
Lampenmuseum, Wezemaal . . . 004150
Lampi-Museo, Liminka 011118
Lamplighter, Columbus 135481
Lamplighters, Portland 136155
Lampronti – Lebole, Milano . . . 080586
Lampronti, Carlo, Roma 081419
Lampronti, Cesare, Roma 081420
Lampronti, Giulio, Roma 081421
Lamps & Lighting, San Antonio . 096987
Lamsa, Riyadh 085318
Lamson, Ho Chi Minh City 125129
Lamu Museum, Lamu 029863
Lamur & Hijos, Santiago de Chile 064970
Lamus, Bydgoszcz . . . 084496, 143439
Lamus, Katowice 084537
Lamus, Kraków 084585, 084586, 132844
Lamus, Poznań 143496
Lamus, Warszawa . . . 084730, 084731, 126422, 143525, 143526
L'Amusant Musée, Juigne-sur-Sarthe 013338
L'Amusée, La Gacilly 069041
Lamutov Likovni Salon, Kostanjevica na Krki 114688
Lamy, Véronique, Le Titre 069357
Lamyantic, Vouillé 074276
L'an 1767, Thenon 073646
L'An 1920, Annecy 066643
Lan Sang, Tak 042725
Lanai, México 082477, 111965
Lanan, Berry 061149
Lanark and District Museum, Lanark 006156
Lanark Gallery, Lanark 118068
Lanark Museum, Lanark 044748
Lancaster, Auckland 083802
Lancaster, Oklahoma City 095986
Lancaster, Toowoomba 062330
Lancaster Antique Toy and Train Collection, Portsmouth Museums, Portsmouth 052324
Lancaster City Museum, Lancaster 044752
Lancaster County Art Association, Strasburg 060564
Lancaster County Museum, Lancaster 050121
Lancaster Historical Society Museum, Lancaster 050113
Lancaster Leisure Park Antiques, Lancaster 089741
Lancaster Maritime Museum, Lancaster 044753
Lancaster Museum of Art, Lancaster 050122
Lancaster Museum/ Art Gallery, Lancaster 050110
Lancastrian Antiques & Co, Lancaster 089742
Lancefield Courthouse Museum, Lancefield 001206
Lancefield, David M., Sandgate . 091239, 135005
Lancelot Hill Antiques, Bowral . . 061182
Lancer Centennial Museum, Lancer 006158
Lancioni, Francesco, Bologna . . 079849
Lancz, Patrick, Bruxelles 100631
Land- en Tuinbouwmuseum, Etten Leur 032323
Land-Art, Dannstadt-Schauernheim 075299
Land Fossils and Minerals Museum, Yongkang 042598
Land of the Yankee Fork Historical Museum, Nampa 051152
Land of Was, San Antonio 096988
Land Reform Museum, Taipei . . 042522
Het Land van Strijen, Strijen . . . 032909
Het Land van Thorn/ Panorama Thorn, Thorn 032920
Land, Janne, Wellington 084169, 113072
Landau, Montréal 101425
Landberg, Ivar, Hägersten 115670
Landbouw- en Juttersmuseum Swartwoude, Buren, Friesland . 032121
Landbouwmuseum Erve Niehof, Diepenheim 032222
Landbouwmuseum Leiedal, Bissegem 003368
Landbrugs- og Interiørmuseet, Farsø 010061
Landbrugsmuseet Melstedgård, Gudhjem 010090
Landbrugsmuseet Skarregaard, Nykøbing Mors 010246
Landbruksmuseet for Møre og Romsdal, Vikebukt 034088
Landcommanderij Alden Biesen, Bilzen 003365
Landelle, Françoise, Paris 071546
Landells, Keith, New Orleans . . 135861
Landen, Rupert, Reading 091075
Landenberger, Ulrich, Bodelshausen 129602
Landenberger, Ulrich, Hechingen 130026
Landenberger, Viola, Esslingen . 129824
Lander, Truro 119259
Landers, Ross, El Paso 093406
Landes-Feuerwehrmuseum, Stendal 021994
Landesbergbaumuseum Baden-Württemberg, Sulzburg 022087
Landesbibliothek Oldenburg, Oldenburg 020986
Landesbibliothek und Murhadsche Bibliothek der Stadt Kassel, Universitätsbibliothek Kassel, Kassel 019537
Landesfeuerwehrmuseum Meetzen Mecklenburg-Vorpommern, Meetzen 020336
Landesgalerie Linz, Oberösterreichische Landesmuseen, Linz 002323
Landesgeschichtliche Sammlung der Schleswig-Holsteinischen Landesbibliothek, Kiel 019607
Landesgremium Wien des Handels mit alter und moderner Kunst, Antiquitäten sowie Briefmarken und Numismatika, Wien 058285
Landesmuseum Burgenland, Eisenstadt 001861
Landesmuseum für Kunst und Kulturgeschichte Oldenburg, Schloss, Augusteum und Prinzenpalais, Oldenburg 020987
Landesmuseum für Kunst und Kulturgeschichte, Stiftung Schleswig-Holsteinische Landesmuseen Schloß Gottorf, Schleswig 021664
Landesmuseum für Natur und Mensch, Oldenburg 020988
Landesmuseum für schaumburg-lippische Geschichte, Landes- und Volkskunde, Bückeburg 017660
Landesmuseum für Technik und Arbeit in Mannheim, Mannheim 020251
Landesmuseum für Vorgeschichte Sachsen-Anhalt, Halle, Saale . 018965
Landesmuseum für Vorgeschichte, Landesamt für Archäologie, Dresden 017993
Landesmuseum Kärnten, Klagenfurt 002192
Landesmuseum Koblenz, Staatliche Sammlung technischer Kulturdenkmäler, Koblenz 019662
Landesmuseum Mainz, Mainz . . 020229
Landesmuseum Württemberg, Stuttgart 022047
Landesrabbiner Dr. I.E. Lichtigfeld-Museum, Michelstadt 020400
Landesstelle für die nichtstaatlichen Museen in Bayern, München . . 059155
Landesverband Berliner Galerien e.V., Berlin 059156
Landesverband der Museen zu Berlin e.V., Berlin 059157
Landesverband der Niederösterreichischen Kunstvereine, Niederösterreichisches Dokumentationszentrum für Moderne Kunst, Sankt Pölten 058286
Landesverband Galerien in Baden Württemberg e.V, Grafenau, Kreis Böblingen 059158
Landesverband Salzburger Volkskultur, Salzburg 058287
Landeszeughaus am Landesmuseum Joanneum, Graz 001981
Landfall Press, Santa Fe 138468
Landgate Books, Rye 144808
Landhausgalerie Ausstellungsbrücke, Sankt Pölten 002695
Landherr, Detlef, Berlin 129524
Landik & Měchura, Praha 128355
Landings Gallery, Edinburgh . . . 044247
Landini, Renato, Milano 080587
Landis Valley Museum, Lancaster 050123
Landis, Alan, Sydney 062291
Landjunk, Arthur, Hamburg 125971
Landlmuseum Sulzbürg, Mühlhausen, Oberpfalz 020482
Landlord's Manor House Museum, Dayi 007733
Landman, Santiago, Santiago de Chile 064971
Landmark Art, Mumbai 109452
Landmark Forest Heritage Park, Carrbridge 043880
Landmark Gallery, Cairns 098842
Landmark Gallery, Winnipeg . . . 101880
Landmaschinenmuseum – Sammlung Speer, Rimbach 021432
Landré, G.N., Amsterdam 143045
Landrieux, Pierre, Paris 071547
Landrot, Paris 104999
Landry, Montréal 064438
Landry, L.A., Essex 127216
Lands Beyond, New York 095526, 122800
Lands Museum, Dokka 033570
Landsborough Shire Historical Museum, Landsborough 001207
Landscapes of the Holy Land Park, Eretz-Israel Museum, Tel Aviv . 025204
Landschaftsinformationszentrum (LIZ) Hessisches Kegelspiel Rasdorf, Rasdorf 021279
Landschaftsmuseum, Seligenstadt 021837
Landschaftsmuseum Angeln, Langballig 019888
Landschaftsmuseum der Dübener Heide, Bad Düben 016748
Landschaftsmuseum der Kulmregion, Pischelsdorf 002509
Landschaftsmuseum im Schloß Trautenfels, Landesmuseum Joanneum, Trautenfels 002841
Landschaftsmuseum Obermain, Kulmbach 019830
Landschaftsmuseum Schönhengstgau, Göppingen 018730
Landschaftsmuseum Westerwald, Hachenburg 018905
Landschloss Pirna-Zuschendorf – Botanische Sammlungen der TU Dresden, TU Dresden, Pirna . . 021138
Landschulmuseum Göldenitz, Dummerstorf 018065
Landskron, Gabi, Regensburg . . 130565
Landskrona Konsthall, Landskrona 040773
Landskrona Museum, Landskrona 040774
Landt, Frederiksberg 140571
Landtechnik-Museum Braunschweig Gut Steinhof, Braunschweig . . 017560
Landtechnikmuseum, Leiben . . . 002285
Landtechnisches Museum Burgenland, Sankt Michael im Burgenland . 002682
Landwehr, Hans-Jürgen, Kassel 141979
Landwehr, Heinrich, Langenargen 130210
Landwehr, Stephan, Berlin 106353, 129525
Landwirtschaftliches Museum und Pfarrer Jungblut-Museum, Prinzendorf 002541
Landwirtschaftliches Museum Wetzlar, Wetzlar 022611
Landwirtschafts- und Heimatmuseum, Karben 019499
Landwirtschaftsmuseum, Rhede 021393
Landwirtschaftsmuseum, Weil am Rhein 022468
Landwirtschaftsmuseum Lüneburger Heide, Museumsdorf Hösseringen, Suderburg 022071
Landwirtschaftsmuseum Reitscheid, Freisen 018497
Landwirtschaftsmuseum Schloss Ehrental, Klagenfurt 002193
Landwirtschaftsmuseum Schloss Stainz, Landesmuseum Joanneum, Stainz 002774
Landy, New York 122801
Lane, Stockbridge 091433
Lane Community College Art Gallery, Eugene 048619
Lane County Historical Museum, Dighton 048327
Lane County Historical Museum, Eugene 048620
Lane Fine Art, London 090282
The Lane Gallery, Auckland 112782
Lane Gallery, Manchester 118708
Lane House, Roseburg 052680
The Lane Place, Crawfordsville . 048054
Lane, André, Châteauroux 125553
Lane, Eileen, New York 095527
Lane, Neil, Los Angeles 094386
Lane, Peter, Woollahra 062475
Lane, Russell, Warwick 091745
Lane & Son, W.H., Penzance . . . 127010
Laneri, Lucia, Roma 081422
Lanes Armoury, Brighton 088625
Lanesfield School, Edgerton . . . 048473
The Laneside Gallery, Coleraine 117618
Laney College Art Gallery, Oakland 051677
Laney, Philip, Great Malvern . . . 126910
Lang, Shanghai 102189
Lang, Wien 100343
Lang-Levin, Chicago 092737
Lang Pioneer Village, Keene . . . 006068
Lang Sommer, Rudkøbing 102969
Lang Water Powered Grist Mill, Peterborough 006639
Lang, Dominique, Dudelange . . 111822
Lang, Franz, Aschenberg 127710
Lang, Helmut R., Rennerod 142328
Lang, Helmut R., Wiesbaden . . . 108923, 142489

Lang, Manfred, Grafenwörth ... 099996
Lang, Marion E., Indianapolis ... 093899
Lang, Max, New York . 122802, 138402
Lang, Paul, Zürich 087986
Lang, Siegfried & Ernestine,
  Traunstein 078694
Langage Plus, Alma 005318
Langauer, Elisabeth, Wien ... 063018
Långbans Gruvby, Värmlands Museum,
  Filipstad 040624
Langbeinmuseum, Hirschhorn . 019265
Lange, Bayonne 066977
Lange, Bremen 075172
Lange, Halver 076177
Lange, Kierspe 076794
Lange, Siegburg 108614
Lange & Lange, Gebr., Hamm,
  Westfalen 076349
De Lange IJzeren Brug,
  Dordrecht 083188
Lange, de, Emmen 083243
Lange, Dietmar, Paderborn ... 078007
Lange, Dr. Petra, Berlin 106355
Lange, Gerd & Hilde, Greding .. 107157
Lange, Gerhard, Hamburg 076267
Lange, H., Tarnow 078667
Lange, H.-J., Twist 142437
Lange, Juliane, Karlsruhe 130098
Lange, Ludwig, Berlin 106354
Lange, Ursula, Berlin 074862
Langeberg, Zutphen 083720
Långedrags Konst, Västra
  Frölunda 115987
Långelanda Tingshus, Årjäng . 040534
Langelands Museum, Rudkøbing 010296
Langeloh, Elfriede, Weinheim .. 078908
Langemann, M., München 077555
Langen Foundation, Neuss 020746
Langen, Patricia, Bonn 129608
Langenbach, Kingston 127275
Langenbach, Dr. H., Freiburg im
  Breisgau 075872
Langenbacher, Urs, Füssen 129879
Langenberg, Amsterdam 112243
Langenfeld, D., Burgwald 075229
Langenhof, de, Zwolle 112734
Langenholt, Thomas, München . 130364
Langenströer, Thomas, Ispringen 076628
Langer, Udo, Pogeez 078063
Langer, Wolfgang, Zürich 144189
Långeruds Auktionskammare,
  Kristinehamn 126617
Langeveer, Haarlem 112513
Langeweg & Langeweg, Tilburg 083617,
                                 132641
Langewiesch, Karl Robert, Königstein
  im Taunus 137324
Langewiesche-Brandt,
  Ebenhausen 137168
Langford, J. & R., Llangollen .. 089871
Langgaard, Kaj, Hinnerup 128390
Langgadå 62, Sandnes 084355
Langgeng Gallery, Magelang .. 109517
Langhagen & Harnisch, Hamburg 076268
Langham and District Heritage Village
  Museum, Langham 006159
Langham Cultural Centre Galleries,
  Kaslo 006066
Langhans Galerie, Praha 102705, 128356
Langheinz, F., Darmstadt 106659
Långholmens Fängelsemuseum,
  Stockholm 041013
Langhorne Place Antiques, Oklahoma
  City 095987
Het Langhuis, Zwolle 112735
Langi Morgala Museum, Ararat . 000800
Langianni, Letizia, Prato 131777
Langingkoski Imperial Fishing Lodge,
  Kotka 011032
Langlade County Historical Society
  Museum, Antigo 046507
Langlet, Jean-Marie, Cambrai .. 067615
Langley, Rochester . 091129, 134976
Langley Centennial Museum and

National Exhibition Centre, Fort
  Langley 005821
Langley Park Spring Antiques Fair,
  Loddon 098384
Langlois Antiquités, Louhans ... 069643
Langlois Tapisseries, Blois 067255
Langlois, Danièle, Amiens 140763
Langlois, Marie-Pierre, Saint-
  Ouen 073129
Langloy, Brigit, Lyon 104113
Langmann Uno, Vancouver 101805
Langner, K.D., Bayreuth 074696
Langobardenwerkstatt Zethlingen,
  Zethlingen 022868
Langohr, J.-F., Liège 063767
Langolo, Las Vegas 121363
Langsdorfer Heimatmuseum,
  Lich 020060
Langstaff, Edward, Bruxelles ... 128030
Langston, B., Jacksonville 094005,
                                 127267
Langston, Tom, San Jose 097370
Langton, Sheffield 091286, 091287
Langton Matravers Museum, Langton
  Matravers 044759
Lanham, Louisville 135761
Lanhydrock, Bodmin 043664
Lanier Mansion, Madison 050585
Lanier Museum of Natural History,
  Buford 047312
Lanigan and District Heritage Centre,
  Lanigan 006161
Lanik, Atlanta 119585
L'ànima – abast de Tothom, Sant
  Cugat del Vallès 115412
Lanino – Il Balon, Torino 081718
Lani's Antiques, Tampa 097571
Lankershim Art Gallery, Los
  Angeles 121526
Lankes, Döhlau 125919
Lankes & Spaan, Eilsbrunn ... 137173
Lankes, Heinz-Dieter, Hof 076549
Lankester, Saffron Walden 091169,
                                 144809
Lanman Museum, Framlingham 044368
Lann, Henri le, Poitiers 072224
Lanna Museum, Lampang 042693
Lannan, Boston 092478
Lannaud, T., Portbail 072285
Lanneau, Léon, La Chapelle-Basse-
  Mer 103879
Lannel, Brionne 067502
Lannier, Monique, Tours 073851
Lannino, Giovanni, Palermo ... 131700
Lanno Antichità, Napoli 080894
Lannody, São Paulo 101098
Lannoo, Tielt, West-Vlaanderen . 136662
Lansdown Antiques, Tetbury ... 091573
Lansdowne Auction Galleries,
  Lansdowne 127278
Lansing Art Gallery, Lansing ... 050137
Lansing Manor House Museum, North
  Blenheim 051596
Lansing Veterans Memorial Museum,
  Lansing 050135
Lansquenet, Lille 069492
De Lantaarn, Den Haag 083096
Lantbruksmuseum, Ingatorp ... 040714
Lanteiron, Marie-Agnès,
  Monflanquin 070165
Lantelme, Jean-Claude, Paris .. 071548
Lantern, Dublin 079567
Lantern Bryn Mawr Bookshop,
  Washington 145638
Lantern Gallery of Fine Art, Elgin 117747
Lanterna Magica, Genova 080268
La Lanterna Magica, Torino ... 081719
La Lanterne, Bayeux 066969
Lanterne de Hans, Grasse 068704
Lanting, Shanghai 102190
Lantriac, Lyon 069753
Lantsoght, Alphonse, Paris ... 105000
Lanxi Museum, Lanxi 007995
Lanyang Museum of Yilan County,

Yilan 042590
Lanz, Dr., Graz 062596
Lanz, Dr. Hubert, München ... 077556,
                                 126059
Lanza, San Francisco 136330
Lanzani, Gaëtan, Paris 129055
Lanzenberg, Fred, Bruxelles .. 100632
Lanzhou Museum, Lanzhou ... 007997
Lanzone 16, Milano 080588
Lao Gallery, Vientiane 111751
Laoag Museum, Vigan 034613
Laocheng Xian Art Academy,
  Shanghai 055231
Laocheng Xian Painting and
  Calligraphy Association,
  Shanghai 058498
Laocoonte, Madrid 133348
Laohekou Museum, Laohekou .. 007998
Laorden Garzon, Maria Dolores,
  Madrid 133349
The LAPADA Autumn Antiques and
  Fine Art Fair, Cheltenham 098385
Lapalus, Jean-Pierre,
  Meursanges 070108
Lapchin, Marc, Lyon 128718
LaPelle, Rodger, Philadelphia ... 123408
Lapeyre, Françoise, Brive-la-
  Gaillarde 067513
Lapeyrière, Catherine,
  Champcevinel 103566
Lapham-Patterson House,
  Thomasville 053736
Lapidarij Muzeja Grada Trogira,
  Trogir 009008
Lapidario Civico, Gradisca
  d'Isonzo 026459
Lapidario Comunale, Osimo .. 027120
Lapidarium, Neustadt am Main . 020750
Lapidarium der Stiftsbibliothek St.
  Gallen, Sankt Gallen 041924
Lapidarium, Berlinische Galerie,
  Berlin 017212
Lapierre, Thierry, Challes les
  Eaux 067767
Lapin Maakuntamuseo,
  Rovaniemi 011321
Lapin Metsämuseo, Rovaniemi . 011322
Lapin Tl. Kotiseutumuseo, Lappi 011096
Lapinlahden Taidemuseo ja Eemil
  Halosen Museo, Halosten
  Museosäätiö, Lapinlahti 011089
Lapis Box, Vancouver 064834
Lapis Fine Art, London 118420
Lápiz, Revista Internacional de Arte,
  Madrid 139060
Laplace, Bertrand, Rouen 072595
Lapointe, Edith, Vigoulet-Auzil . 074124
Laporta del Rio, Antonio, Madrid 133350
Laporte, Nicole, Soyaux 073558
Lappa Valley Steam Railway,
  Newquay 045369
Lappan, Oldenburg 137461
Lappsdaten, Arvidsjaur 040540
Laprugne, Jean-Pierre, Paris ... 071549
Lapu-Lapu City Science Museum,
  Lapu-Lapu 034492
Lapuente, Carlos de, New York . 095528
Lapworth Museum of Geology,
  Birmingham 043621
Laqua, Carsten, Berlin 106356, 141547
Lar, Donostia-San Sebastián ... 085977
Lara Arte, Roma 110808
Lara, Elida, Barranquilla 102265
Lara, Isabel, Madrid 086207
Laramie Plains Museum, Laramie 050143
Larasati, Jakarta 126163
Larasati, Jakarta 126485
L'Arcadia, Livorno 110184
Larchmere Antiques, Cleveland . 092951
Larchmont Historical Society Museum,
  Mamaroneck 050617
Lårdal Bygdemuseum,
  Høydalsmo 033659
Lardanchet, Paris 141159

Lardanchet, Henri, Lyon 140904
Lardenois, Pierre, Forest 068588
Lardeux, Joël, Crosmieres 068248
Laredo Children's Museum,
  Laredo 050147
Lanza, San Francisco 136330
LaReido, Kerava 066378
The Large Art Company,
  Baltimore 119749
Large Print Book, Englewood . 145144
Largeaud, André & Jacqueline,
  Paris 105001
Largeault, Didier, Paris 071550
L'Argile, Sancerre 073400
Largs Museum, Largs 044761
Lari, Rossella, Firenze 131194
Larici Numismatica, Erba 080066
Larimer Exchange, Denver 093308
Lark, Asheville 138300
Lark Lane Motor Museum,
  Liverpool 044855
Larking-Coste, Marc, Paris ... 129056
Larks, Ballater 117270
L'Armada, Loupian 069646
Les Larmes d'Éros, Paris 105002,
                            136951, 141160
Larnaca District Archaeological
  Museum, Larnaka 009406
Larnaca District Museum,
  Larnaka 009407
Larne Museum, Larne 044762
Laroche & Denis, Québec 101506
Laroche, Gabrielle, Paris 071551, 071552
Laroche, Yves, Montréal 101426
Larock-Granoff, Paris 105003
Laroy, Houston 093723, 135609
Larraulgo Museoa, Museo de Larraulh,
  Larraul 039545
Larrea, Buenos Aires 127524
Larrea, Martínez 060949, 127546
Larrea Fernandez-Valdes, Julian,
  Madrid 086208
Larrieu, André, Martigues 069986
Larrison, Indianapolis 093900
Larrivet, Jean-Philippe, Camblanes-et-
  Meynac 067612
Larrouilh, Ginou, Biarritz 103387
Larruy, Ramon, Vitoria-Gasteiz . 086575
Larrybane and Carrick-A-Rede,
  Ballycastle 043475
Larry's Antiques, Buffalo 092537
Larry's Antiques, Virginia Beach 097786
Larry's Place, Houston 093724
Lars Noak Blacksmith Shop, Larson/
  Ostlund Log Home & One-Room
  Capitol School, New Sweden . 051315
Lars, Cultura y Ciudad, Valencia 139061
Larsberger, Linköping 086815
Larsen, Exeter 139728
Larsen, Anna, Odder 066024
Larsen, Ejler, Odense 140639
Larsen, Jens E., København ... 065931
Larsen, Jesper, Frederiksberg . 065667,
                                 128382
Larsen, Poul, Thisted 066162
Larsen, Thormod, Helsingborg . 115685
Larsmo Hembygdsmuseum,
  Larsmo 011099
Larson Gallery, Yakima Valley
  Community College, Yakima . 054565
Larson, Sophia, Tiefenfucha ... 100205
Larsson, Halmstad 086733
Larsson, Regina, Siselen 133924, 138090
L'Arte, Molinella 110505
L'Arte Antica di Silverio Salamon,
  Torino 081720, 110983, 142859
Larue, Chantal, Colombes 068105
Larunda, Timişoara 085135
Larvik Sjøfartsmuseum,
  Larvikmuséene, Larvik 033734
Larwood, Roderick, Larling ... 134528
Larz Anderson Auto Museum,
  Brookline 047248
Las Casas Gil, Manuel Enrique de,
  Zaragoza 133501

Las Cruces Museum of Natural History,
  Las Cruces 050153
Las Luces, Buenos Aires 139577
Las Vegas Antique Jewelry and Watch
  Show, Las Vegas 098466
Las Vegas Antique Slot Machine, Las
  Vegas 094159
Las Vegas Art Museum, Las
  Vegas 050164
Las Vegas International Scouting
  Museum, Las Vegas 050165
Las Vegas Natural History Museum,
  Las Vegas 050166
Lasalle College of the Arts, Institute
  of Contemporary Arts Singapore,
  Singapore 056327
LaSalle County Historical Society
  Museum, Utica 053913
Lasarettimuseo, Turku 011438
Lasartis, Paris 071553
Lasbleiz, Roger, Melesse 070042
Lascaux II, Fac-similé de Lascaux,
  Montignac 014264
Lascelles, Roger, London 090283
Lasch, Mark, Mainz 077238
Laschütza, Rödinghausen 108510
Laserow, Malmö 086852
Lashburn Centennial Museum,
  Lashburn 006167
Lashenden Air Warfare Museum,
  Headcorn 044549
Laski, Stanislas, Saint-Ouen . . . 073130
Lasnier, Jean-Pascal, Latresne . . 069196
Lasry, Joëlle, Paris 071554
Lassa & Scalese, Genova 080269
Lassaletta Delclos, Ignacio,
  Barcelona 114924, 114925
Lassalle, Judith, London 090284, 134707
Lassandro, Maria Vita, Cagliari . 079987
Lassaner Mühle, Lassan 019922
Lassaussois, Chartres 067860
Lassco St. Michael's, London . . 090285
Lassen, Honolulu 120948
Lassen, Las Vegas 121364
Lassen Historical Museum,
  Susanville 053619
Lassen Museum Waikiki,
  Honolulu 049539
Lasserre-du-Rozel, Gérard,
  Bénodet 067093
Lassetters, Fyshwick 099066
Lassiter, Joie, Charlotte 119996
Lassnig, Alexander, Oberalm . . . 127824
Lasso de la Vega, Alvaro, Málaga 086293
Lassotta, Gabi, Balingen 129437
Lassus, Catherine, Bordeaux . . . 067340
Last, Zürich 116828
Last Century Antiques,
  Innerleithen 089641
Last Century Shoppe, Midhirst . 083995
The Last Chance Saloon, London 118421
The Last Hurrah, Bibra Lake . . . 139644
Last Indian Raid Museum,
  Oberlin 051690
The Last Post Book Shop, Wagga
  Wagga 139948
Last Roundup, Phoenix 096222
Last Touch, Gharghur 111913
Lasta, Moskva 114274
Lastenmuseo, Helsingin
  Kaupunginmuseo, Helsinki . . . 010842
Lastentarhamuseo, Helsinki 010843
Lasting Impressions, San Antonio 096989
Lasting Impressions Gallery,
  Kenilworth 099180
Lasting, A., Artarmon 061059
Lastrucci, Francesco, Prato 081170
Lata Dwudzieste Lata Trzydzieste,
  Kraków 113613
Latcham Gallery, Stouffville 007081
Latchford, K. & R., Cheltenham . 088834
Late Bloomers, New York 095529
Late Bloomers Treasures, Tampa 097572
The Late Joy's, Houston 093725

L'Atelier, Thoiry 073661
L'Atelier Brocante, Cannes 067644
Latemse Verkoopzaal, Sint-Martens-
  Latem 125356
Laténium, Parc et Musée
  d'Archéologie de Neuchâtel,
  Hauterive 041586
Laterna Magica, Helsinki 103153,
  140710
Die Laterne, Ovelgönne 077999
Latgales Kultūrvēstures Muzejs,
  Rēzekne 030174
Latham, Lionel, Genève 087478
Latham, Polly, Boston 092479
Latham, R.H., Tarleton 091539
Lathière, André, Le Blanc 069220
Latin American Art Museum,
  Miami 050834
Latin American Cultural Center
  Museum, Koyang 029966
Latin Art Broker, Los Angeles . . 121527
Latin Inspiration, Northampton . . 118818
Latin Sun, Los Angeles 121528
Latino Art Gallery, Philadelphia . 123409
Latinovits Zoltán Emlékmúzeum,
  Balatonszemes 023427
Latitude 53 Contemporary Visual
  Culture, Edmonton 005755
Latitudes, Houston 121077
Latona, Saltsjö-Duvnas 115806
Latorre, Fernando, Madrid 115245
Latorre, Fernando, Zaragoza . . . 115554
Latragna, Franco, Aschaffenburg 106012
Latrobe Bicycle Race Club Museum,
  Latrobe 001209
Latrobe Court House Museum,
  Latrobe 001210
Latrobe Regional Gallery,
  Morwell 001316
Latrobe Street College of Art and New
  Media, Melbourne 054908
Latroc-Ant, Ascoux 066757
Lattelekom Telekomunikāciju Muzejs,
  Rīga 030188
Lattemann, Reinhard, Mühltal . . 108037
Lattimer Gallery, Vancouver . . . 101806
De Lattre, Paris 141161
Lattre, Dominique de, Saint-
  Ouen 105567, 141336
Latude, Hugues de, Gardouch . . 140843
Latvian Centre for Contemporary Art,
  Rīga 030189
Latvian History Museum, Toronto 007185
Latvian Museum, Rockville 052652
Latvian Museum, Wayville 001630
Latvian Museum Council, Rīga . 059543
Latvijas Arhitektūras Muzejs,
  Rīga 030190
Latvijas Dabas Muzejs, Rīga . . . 030191
Latvijas Dzelzceļa Vēstures Muzeja
  Jelgavas Nodaļa, Jelgava 030133
Latvijas Dzelzceļa Vēstures Muzeja
  Rīgas Nodaļa, Rīga 030192
Latvijas Etnogrāfiskais Brīvdabas
  Muzejs, Rīga 030193
Latvijas Fotogrāfijas Muzejs, Rīga 030194
Latvijas Kara Muzejs, Rīga 030195
Latvijas Ķīmijas Vēstures Muzejs,
  Rīga 030196
Latvijas Kultūras Muzejs Dauderi,
  Rīga 030197
Latvijas Lauksaimniecības Muzejs,
  Talsi 030235
Latvijas Lauksaimniecības
  Universitātes Muzejs, Jelgava . 030134
Latvijas Mākslas Akadēmija, Rīga 056110
Latvijas Mākslas Akadēmija –
  Restaurācijas Nodaļas, Rīga . . 056111
Latvijas Muzeju Asociācija, Rīga 059544
Latvijas Okupācijas Muzejs, Rīga 030198
Latvijas Republikas Meliorācijas un
  Zemkopības Muzejs, Mālpils . . 030162
Latvijas Sporta Muzejs, Rīga . . . 030199
Latvijas Ugunsdzēsības Muzejs,

Rīga 030200
Latvijas Universitātes Botānikas
  Muzejs, Rīga 030201
Latvijas Universitātes Geologijas
  Muzejs, Rīga 030202
Latvijas Universitātes Pedagogijas
  Muzejs, Rīga 030203
Latvijas Universitātes Skaitļošanas
  Tehnikas un Informatikās Muzejs,
  Museum of the History of Science
  and Technologies of the University of
  Latvia, Rīga 030204
Latvijas Universitātes Vēstures Muzejs,
  Rīga 030205
Latvijas Universitātes Zoologijas
  Muzejs, Rīga 030206
Latvijas Vēstures Muzejs, Rīga . 030207
Latzer, Josef, Kreuzlingen 116394
Latzko, H., Marxzell 077306
Lau, Helga, Aidlingen 105970
Lauaxeta, Bilbao 085875
Laub, Claudia, Los Angeles 121529
Laube, August, Zürich 144190
Lauber, Wolfgang, Bad Endorf . . 129407
Laubinger, K.-D., Delbrück 075325
Laubinger, Rudolf, Osnabrück . . 077973
Laubry, Jean-Paul, Le Châtelet-en-
  Brie 069248
Laucirica, Carlos, Bilbao 085876
Laucournet, Dominique, Paris . . 141162
Laudal Museum, Øyslebø 033816
Laudenbach, Helmuth, Berg, Kreis
  Starnberg 074726
Lauder & Howard, East
  Freemantle 061430
Lauder Museum, Amitville Historical
  Society, Amityville 046457
Lauderdale House, London 045018
Laudoueneix, Pierre, L'Isle-sur-la-
  Sorgue 069576
Laue, Dr. Nathalia, Frankfurt am
  Main 107013
Laue, Georg, München 077557
Laue, J.G., München 077558
Lauenburgischer Kunstverein e.V.,
  Mölln 059159
Lauer, Boris, Wiesbaden 108924
Lauer, H.-Dieter, Freisen 075896
Lauer, Klaus, Zweibrücken 079151
Laufener Stiftsschatz, Laufen . . . 019934
Laug, Frederiksberg 128383
Laug, Rüdiger, Herford 141940
Lauge Falkersby, Paris,
  København 065932
Laughing at the Sun Art Gallery,
  Austin 119682
The Laughing Hippo, Townsville 099838
Laughing Horse Gallery, Byron
  Bay 098831
Laughing Ogre, Columbus 093031
Laughlin, Seattle 124559
Laughlin, Kenneth, Philadelphia . 096124
Laugier, Jacques, Beaulieu-sur-
  Mer 066999
Laugø Smedie Museum, Helsinge 010104
Laukmiškių Kaimo Muziejus,
  Kupiškio Etnografijos Muziejus,
  Laukminiškes 030405
Laumeier Sculpture Park, Saint
  Louis 052807
Launceston Steam Railway,
  Launceston 044766
Launderette Gallery, Cork 109557
Lauper, Allschwil 116016
Laura and Charles Ferguson Museum,
  Mentone 001284
Laura and Paul Mesaros Galleries,
  Morgantown 051056
Laura Ann Antiques, Indianapolis 093901
Laura Antiques, Vancouver 064835
Laura Brocante, Yutz 074314
Laura Ingalls Wilder Museum, Walnut
  Grove 054045
Laura Ingalls Wilder-Rose Wilder

Lane Historic Home and Museum,
  Mansfield 050655
Laura Secord Homestead, Niagara
  Falls 006478
Laura, Luigi Anton, Monaco . . . 082515
Lauraine Designs, Miami 135798
Laurans, Joël, Maumusson 070005,
  128790
Laurantica, Toulon (Var) 073715
Laura's Collectibles, Milwaukee . 094866
Laure A. Sprague Art Gallery,
  Joliet 049836
Laure, Edouard, Saint-Ouen . . . 073131
Laurel and Hardy Museum,
  Ulverston 046019
Laurel Antiguedades, Sevilla . . . 086454
Laurel Hill Antiques, Providence 096510
The Laurel Museum, Laurel . . . 050177
Laurel und Hardy-Museum,
  Solingen 021899
Laurella, Hyères 068857
Laurelwood, Memphis 121701
Laurelwood, San Jose 097371
Lauren Rogers Museum of Art,
  Laurel 050180
Laurence C. Jones Museum, Piney
  Woods 052143
Laurencier, Bordeaux 140794
Laurencier, Patrick, Bordeaux . . 140795
Laurencin, Lyon 104114
Laurens, Albuquerque 092062
Laurens, Whitstable . . 091836, 135225
Lauren's Galerie, Bourges 103453
Laurens Van der Post Museum,
  Philippolis 038797
Laurens, Marie Helene, Venezia 081851
Laurent Hubert Musée, Orbey . . 014572
Laurent-Lasson, Eric, Paris 129057
Laurent, Brigitte, Pazayac 072053,
  129163
Laurent, Brigitte, Terrasson-
  Lavilledieu 073636, 129275
Laurent, Corentin, Pleuven 072188
Laurent, Corentin, Quimper 072347
Laurent, Danièle, Niort 070712
Laurent, François, Vic-le-Comte 074095
Laurent, Guy, Vichy . . . 074110, 125839
Laurent, Jacques, L'Isle-sur-la-
  Sorgue 069577
Laurent, Jean-Claude, Ennordres 068445,
  128590
Laurent, Patrick, Mougins 070362
Laurent, Serge, Paris 105004
Laurentin, Antoine, Paris 105005
Laures, Georges, Luxembourg . . 111858
Lauretta, Michel, Mandelieu-La
  Napoule 069833, 128736
Lauri Viidan Museo, Tampere . . 011397
Lauri, Bruno, Villeparisis 074172
Laurie, James & Mary,
  Minneapolis 145278
Laurie, John, London . 090286, 134708
Laurie, Peter, London 090287
Laurier House, Ottawa 006592
Laurie's, Pittsburgh 096303
Laurin, Zürich 116829
Laurin, Claire M., Aix-en-
  Provence 103198
Laurin, Didier, Lannion 069183
Laurin, Guilloux, Buffetaud & Tailleur,
  Paris 125729
Laurinmäen Torpparimuseo,
  Janakkala 010915
Laurino, Giusi, Napoli 110534
Lauriol, Simone, Saint-Hippolyte-du-
  Fort 072831
Lauriston Castle, Edinburgh . . . 044248
Lauritsen, N.J., Aabenraa 128375
Lauritz.com, Aalborg 125429
Lauritz.com, Askim 126569
Lauritz.com, Düsseldorf 125932
Lauritz.com, Egå 125434
Lauritz.com, Esbjerg 125435
Lauritz.com, Hamburg 125972

Lauritz.com, Helsingborg 126601
Lauritz.com, Helsingør 125439
Lauritz.com, Herlev 125440
Lauritz.com, Herning 125442
Lauritz.com, Kolding 125457
Lauritz.com, Malmö . 126638, 133556
Lauritz.com, Næstved 125460
Lauritz.com, Odense 125462
Lauritz.com, Oslo 126374
Lauritz.com, Roskilde 125466
Laursen, Roskilde 066090
Laursen, Birthe, Paris 115006
Laursen, Per Michael, Humlebæk 128392
Laursen, Sonja, Frederikssund . 065686
Lausanne, Rio de Janeiro 101043
Lausberg, Toronto 101646
Lausberg, Bernd A., Düsseldorf . 106805
Lausecker, Claude, Champigny-sur-
Marne 067815
Lausitzbuch, Werpeloh 142479
Lausitzer Bergbaumuseum
Knappenrode, Hoyerswerda .. 019349
Lauta, Varna 101196
Lauterbach, Hans, Berlin 074863,
106357
Lauterbach, Rolf-P., Göttingen .. 076048
Lauterhahn, Ludwigshafen am
Rhein 107899
Lauth, Werner, Ludwigshafen am
Rhein 107900, 130251
Lauth, Werner, Mannheim 077264,
130278
L'Authie, Tigny-Noyelle 073680
Lautier, Claude, Vetheuil 074080
Lauvring, Middelfart 065999
Lava Beds National Monument,
Tulelake 053845
Laval, Alain, Souillac 073549
Laval, Didier, Levroux 069450
Lavalier, New York 095530
Lavanchy, Evian-les-Bains 068492
Lavanchy, Thonon-les-Bains ... 073668
Lavanchy, Henri, Evian-les-Bains 068493
Lavande, Saint-Paul-de-Vence .. 105592
Lavanttaler Heimatmuseum,
Wolfsberg 003127
Lavanttaler Obstbaumuseum, Sankt
Paul 002689
Lavda, Anastasia, Athinai 109044
Laveen, Sydney 062292
Lavelle Art Gallery, Clifden 109547
Lavelli, Guéret 068763, 140855
Låven Antik og Auksjon,
Porsgrunn 084344, 126383
Lavender, Boston 092480
Lavender, Maldon 061765
Lavender & Roses, Den Haag .. 083097
Lavender Bay Gallery, North
Sydney 099478
Lavender Blue, Dartmouth 117668
Lavender Hill, Philadelphia 096125
Lavender Passion Gallery,
Longwood 099238
Lavender, D.S., London 090288
Lavender, Lida, London 090289, 134709
Lavenham Contemporary,
Lavenham 118078
Lavergne, Bertrand de, Paris ... 071555
Laverroux, Jean-Pierre, Clermont-
Ferrand 068029
Lavest, Georges, Clermont-
Ferrand 068030
Lavezzo, New York 095531
Lavian Kotiseutumuseo, Lavia .. 011101
Lavian, Joseph, London 090290, 134710
Lavian, Max, New York 095532
Lavignes-Bastille, Paris 115007
Lavigueur, Marc-André, Québec 128195
Lavilgrand, François, Monthléry . 070262
Lavín, Bilbao 085877
Lavirotte, Philippe, Chalon-sur-
Saône 067780
Lavka Antikvara, Sankt-Peterburg 085249,
085250

Lavka Chudožnikov, Sankt-
Peterburg 114347
Lavka Drevnostej, Almaty 082237
Lavka knigoljuba, Moskva 143645
Lavka Starevščika, Sankt-
Peterburg 085251
Lavocat Insignes Militaires, Prades-le-
Lez 072304
Le Lavoir, Mougins 014346
Lavoissière & Gueilhers, La
Rochelle 125605
Lavorazione Antica del Cuoio,
Torino 132154
Lavori Artigiani Romani, Roma . 081423
Lavradio de Radios, Rio de
Janeiro 064074
Lavrille, J., Gent 132052
Lavy, Bernard, Bons-en-Chablais 067297
Lavy, Corinne, La Roche-sur-
Foron 069066
Law, Hong Kong 102093
Law Enforcement Museum,
Roselle 052681
Law Uk Folk Museum, Hong
Kong 007880
Lawen, Ferdinand, Briedel 129640
Lawes, Anne, Neuenstadt am
Kocher 077725
Lawford, Suzanne, Rushcutters
Bay 062175
Lawler-Taylor, Birmingham ... 119805
Lawn Heritage Museum, Lawn . 006174
Lawndale, Philadelphia 096126
Lawndale Art Center, Houston . 049584
Lawniczak, Lüdinghausen 077203
Lawrence, Kansas City 121301
Lawrence, Vancouver 140425
Lawrence & Scott, Seattle 097467
Lawrence County Gray House
Museum, Ironton 049718
Lawrence County Historical Museum,
Bedford 046868
Lawrence Gallery, Portland ... 123638
Lawrence Hall of Science, University
of California, Berkeley, Berkeley 046938
Lawrence House Museum,
Launceston 044767
Lawrence House Museum,
Maitland 006246
Lawrence L. Lee Scouting Museum,
Manchester 050627
Lawrence Wilson Art Gallery,
University of Western Australia,
Crawley 001001
Lawrence, Bob, London 090291, 134711
Lawrence, Brad J., Erskineville . 061452
Lawrence, David, Virginia Beach 097787
Lawrence & Sons, F.G., Redhill . 091082
Lawrence, Hermann,
Schwanewede 078414
Lawrence, Martin, San Francisco 124323
Lawrences, Bletchingley 126824
Lawrences, Crewkerne 126871
Lawrens & Cie., Paris 071556
Lawry, John, Malvern East 099267
Law's Artland, Hong Kong 065073
Laws Railroad Museum, Bishop 047714
Lawson, Albuquerque . 092063, 135286
Lawson, Cambridge 144298
Lawson, Dallas 120622
Lawson, Hurst Green 089625
Lawson, San Francisco 124324
Lawson & Co, E.M., East
Hagbourne 144379
Lawson, F. & T., Richmond,
Surrey 091100
Lawson, Keith, Scratby 091252
Lawson, Peter, Townsville 099839
Lawsons, Sydney 125238
Lawton, Hughes 139780
Lawton Gallery, University of
Wisconsin-Green Bay, Green
Bay 049227
Laxa, Tilo, Rüdenhausen 078306

Laxfield and District Museum,
Laxfield 044771
Lay-Bonnot, Madeleine, Nice .. 070621
Lay, Kenneth, Prahran 062115
Le Lay, Philippe, Châteaulin .. 067884
Layan, Roselyne, Montauban .. 140958
Layard, Siegrid, Eutin 075723
Layland Museum, Cleburne ... 047796
Laymance, Houston 121078
Layne, A.C., Carlisle 134219
Layr Wuestenhagen Contemporary,
Wien 100344
Layug, Budji, Makati 113388
Laywood, Anthony W., Knipton . 144510
Lazar, Beograd 114423
Lazareth, Annick, Mesigny 070077
Lazaro Montelongo, Maria Magdalena,
Santa Cruz de Tenerife 115422
Lazarovici, Bernard, Saint-Ouen 073132
Laz'art Déco, Lyon 069754
Lazartigues, Josiane, Nérac ... 070502
Lazarus, David, Hartley Wintney 089453
Lazarus, Julie, Fort Worth 120873
Lazarus, Steven, London 090292
Lazdijų Krašto Muziejus, Lazdijai 030409
Lazdynų Pelėdos Muziejus,
Paragiai 030443
Lazer, Buffalo 092538
Lazerhead, Louisville 121654
Lazic, Rubens, Saint-Ouen 073133
Lazoukine, Deauville 103659
Lazrak, Khadija, Casablanca .. 082568
Lazulum, Marseille 128776
Lazy Daisies, Hazel Grove 117942
Lazy Dog Antiques, Chicago .. 092738
Lazzarato, Giovanni, Padova .. 080988
Lazzareschi, Riccardo, Livorno . 131346
Lazzarini, Torino 081721, 132155
Lazzarini, Alain, Le Ban-Saint-
Martin 069213
Lazzaro by Corsi, Milano 110368
Lazzeri, Filippo, Firenze 131195
Lazzoni, Chantal, Saint-Étienne . 072767
LB Antike, Hechingen 076424
LBD Fine Art, Chicago 120176
LB's Antiques, Dallas 093156
LBW Gallery, Indianapolis 121196
LCE School of Arts and Design,
Singapore 056328
Le Bihan, Gildas, Carnoët 067691
Le Bihan, Jacques, Paris 071557
Le Bouffo, Magdeleine, Loudéac 104048
Le Bric a Brac, Negril 111236
Le Broisin, Champéry 116202
Le Brun, Guy, Livarot 104041
Le Calonnec, Pierre & René,
Pleurtuit 072186
Le Cam, Françoise & Philippe
Belhache, Nantes 070467
Le Carmel, Tarbes 015913
Le Chanjour-Marais, Paris 071558,
105008
Le Clerc, Bruxelles 063487
Le Couvent des Ursulines, Liège 063768,
128076
Le Coz, Gilbert, Pommerit-le-
Vicomte 072228
Le Creuset, Centre Culturel Privé,
Waterloo 004134
Le Dean, David-Pierre, Nantes . 070468
Le Domaine Perdu, Meyrals ... 104235
Le Fell, Jean-Marie, Paris 105009,
141163
Le Floch, Ludovic, Tousson ... 073867
Le Fuel de l'Espée, Paris 071559
Le Gaillard & Associés, Alain,
Paris 105010
Le Gatti, Jean-Pierre, Bruxelles . 125317
Le Goff, François-Xavier & Sylvie,
Pléneuf-Val-André 072178
Le Gouil, Dominique, Asnières-sur-
Seine 066760
Le Grand Filon, Saint-Georges-
d'Hurtieres 015333

Le Havre Enchéres, Le Havre .. 125610
Le Houelleur, Guy, Deauville .. 125567
Le-Lay, Philippe, Locronan 069607
Le Mans, Wendover 119321
Le mevel antiquités, Arts de la table,
Lille 069493
Le Mouël, Jean-Marie, Paris ... 125730
Le Musée Archéologique de Montaigle,
Falaën 003593
Le Nost, Alain, Paimpol 104436
Le Paon Artistic Circle, Singapore 114557
Le Passé Simple, Morges 087685
Le Petit Musée, Wavre 004140
Le Petit Musée du Pain, Chievres 003518
Le Point Fortifié Hillmann, Colleville-
Montgomery 012617
Le Puits-Marie, Maison de la
Métallurgie et de l'Industrie
Liégeoise, Blegny 003371
Le Puy Collection, Le Puy-en-
Velay 069343
Le Quai, École d'Art et Design,
Mulhouse 055385
Le Roy House and Jell-o Gallery, Le
Roy 050197
Le Roy, Bianca, La Varenne-Saint-
Hilaire 069119
Le Roy, Sabine, Lisieux 104038
Le Salon du Cadeau Culturel et
Touristique, Museum Expressions,
Paris 098089
Le Sault de Sainte Marie Historical
Sites, Sault Sainte Marie 053166
Le Strange Emporium,
Horncastle 089584
Le Sueur County Historical Museum,
Elysian 048575
Le Vesconte, Peter, Saint Helier 091184
Le Village des Automates, Musée,
Saint-Cannat 015247
Le, Xuan, Ho Chi Minh City ... 125130
Lea County Cowboy Hall of Fame and
Western Heritage Center, Hobbs 049495
Lea, Klaus, München 108156
Lea, Rocky, Bairnsdale 098683
Leach, Derek, Cincinnati 092870
Leach, Elizabeth, Portland 123639
Leader, Tampa 097573
Leadhills Miners' Library Museum,
Leadhills 044772
Leading Tickles Heritage Centre,
Leading Tickles 006176
Leadon, Eric, Las Vegas 121365
Léadouze, Alexandre, Cannes . 103517
Leadouze, Alexandre, Paris 105011
Leadshan Museum, Qianshan,
Jiangxi 008169
Leaf Galleries, Maidstone 118698
Leaf Rapids National Exhibition Centre,
Leaf Rapids 006177
Léage, François, Paris 071560
Leah's Gallery, New York 095533
Leakey, Charles, Inverness 144499
Leal Antiquario, Rio de Janeiro . 064075
Léal, Eduardo, Blois 067256
Leali, Alberto, Roma 131954
Leali, Alessio, Roma 131955
Leali, Andrea, Roma 131956
Leali, Angelo, Roma 131957
Leaming's Run Garden and Colonial
Farm, Cape May Court House . 047421
Leamington Art Gallery,
Leamington 006178
Leamington Spa Art Gallery and
Museum, Leamington Spa .. 044773
Le'an Museum, Le'an 007999
Leanders Antikvariat, Västerås . 144000
Leanin' Tree Museum of Western Art,
Boulder 047143
Leapfrog Studio, Mooloolaba .. 099370
Leaping Rabbit, Ottawa 064506
Learmonth Museum, Learmonth 001217
Learsy, Miami 094759
Leather Bucket, Philadelphia ... 096127

Leather Museum, Chicago 047669
Leather Museum, Haram 010577
Leather Museum, Gerberei Museum,
Swakopmund 031854
Leatherhead Museum of Local History,
Leatherhead 044774
Leatherstocking Brush & Palette Club,
Cooperstown 060565
Leavin, Margo, West Hollywood 125009
Lebaillif, Daniel, Le Breuil-en-
Auge 069229
LeBaron, Sacramento 123774
Lebas, François, Issoudun 068882
Lebas, François, Saint-Georges-sur-
Arnon 072801
Lebbing, F.G., Amsterdam 082793
Lebeer Hossmann, Bruxelles . 136642
Lebel, Antoine, Bruxelles 063488
Lebell Residence,
Kristiinankaupunki 011041
Lebendes Textilmuseum, Groß-
Siegharts 002001
Lebendiges Museum, Bad
Wildungen 016902
Lebert, San Francisco 097268
Lebert, Jean-Claude, Gondrecourt-le-
Château 068678
Lebesgue, Vincent, Laroque-des-
Arcs 069191
Leblanc, Christian, Lyon 128719
Leblanc, Danièle, Wiege-Faty . 074296
Leblay, Bernard, Clermont-
Ferrand 128562
Leblond, Les Vans 069438
Lebon, Baudoin, Paris 105012
Lebourg, A., Le Havre 128662
Lebourg, Marie-Nicole, Dinard . 068356
Lebouteiller, Jacky, Vire 074248
Lebouteiller, Joël, Mesnil-
Clinchamps 070079
Lebouteiller, Joël, Saint-Sever-
Calvados 073309
Lebovic, Josef, Paddington, New
South Wales 099528
Lebra, François, Nice 070622
Lebraly, Guy, Cannes 128528
Lebran, Monica, Roma 131958
Lebreton, Alain, Bonneville-sur-
Touques 067292
Lebreton, Alain, Deauville 068281
Lebriz Sanat Galerisi, İstanbul .. 117060
Lebrun, Paris 071561, 129058
Lebrun, Adrienne, Sartilly 073421
Lebrun, Marie-Thérèse, Saint-
Honoré 072832
Lebsack, Linda, Denver 145129
Lebzelterei- und
Wachszieherereimuseum, Pfaffenhofen
an der Ilm 021111
Leccese, Pasquale, Milano 110369
Lecchi, Jeannine, Paris 129059
Lecerf, Acquigny 066418
Lechang Museum, Lechang ... 008000
Lechevallier, Coutances 068208
Lechfeldmuseum, Königsbrunn . 019725
Lechflößermuseum, Lechbruck am
See 019948
Lechlade Arcade, Lechlade 089761
Lechleitner, Herwig, Braz 062540
Lechmere, Great Malvern 144443
Lechner Museum, Ingolstadt ... 019411
Lechner, Gérard, Bourgoin-Jallieu 067434
Lechner, Rene,
Mönchengladbach 077396
Lechthaler, Herbert, Ampaß ... 062519
Leckie, Maldon 099259
Leclerc, Prahran 062116
Leclerc Franck, Le Translay ... 069365
Leclerc, Alain, Villeneuve-sur-Lot 129351
Leclercq de Lescure, Bernard,
Lille 128680
Leclercq, Jamy, Martainneville . 069978
Leclercq, Jean-Louis, Amiens .. 066576
Leclercq, Philippe, Prouzel 072314

Lecoeur, Jean, Saint-Michel-Chef-
Chef 072978
Lecointre-Drouet, Paris 141164
Lecomte-Durouil, Marc, Toulon
(Var) 105714
Lecomte, Alain, Paris 105013
Lecomte, Bernard, Paris 105014, 141165
Lecomte, Daniel, Héric 068810
Lecomte, Max, Château-Gontier 067873
Leconey, Elizabeth, Dallas 093157
Leconte, Jean-Marc, Paris 071562
Leconte, Pierre, Veyreau 074082
Lecourt, Gilbert, Pleine-Fougerès 105357
Lecourt, Gilbert, Saint-Malo (Ille-et-
Vilaine) 105550
Lecourtois, Jean-Jacques,
Carentan 067681
Lectori Salutem, Weesp 143266
Les Lectures d'Ysengrin,
Bordeaux 140796
Lécuyer, Vincent, Paris 105015
Leda Décors, Gambais 068625
Leda Décors, Saint-Ouen 073134
Ledaal Museum, Stavanger 033975
Ledbury Heritage Centre,
Ledbury 044776
Leder- und Gerbermuseum, Mülheim
an der Ruhr 020497
Leder, Hugo, Geltendorf 129893
Leder, Jutta, Wolfratshausen ... 079059
Ledermann, Burggen 075225
Ledermann, Charles, Laronxe .. 069190
Lederwerkstätte, Laa an der
Thaya 127799
Ledesma, Pablo, Buenos Aires . 127525
Ledeur, Jean-Paul, Paris 129060
Ledford & Ledford, Oklahoma
City 095988, 136051
Ledoux, M., Miami 145257
Ledra, Lefkosia 065324
Ledru, Guy, Livarot 069600
Ledru, Servane, Nantes 070469
Leduc, Sylvia, Saint-Ouen 073135
Lee, Los Angeles 094387
Lee Chapel and Museum,
Lexington 050266
Lee County Historical Society Museum,
Loachapoka 050349
Lee-Fendall House Museum,
Alexandria 046393
Lee Gallery, Taipei 116927
Lee Tze-Fan Memorial Art Gallery,
Hsinchu 042334
Lee Verre, Elena, Montréal ... 101427
Lee, Diane, Oklahoma City 095989
Lee, Harvey, Jacksonville 121257
Lee, Hwaik, Seoul 111683
Lee, Hyun, Daegu 111587
Lee, Hyun Ryung, Seoul 111684
Lee, Jacques, Paris 129061
Lee, Jain, Buffalo 092539
Lee, James K., King's Lynn ... 089691
Lee, Janie C., Houston 121079
Lee, Jian Qiu, Hong Kong 065074
Lee, Kwong Tin, Hong Kong ... 065075
Lee, Lori, Ivybridge 118031
Lee, Lothar & Marianne, Berlin . 074864,
074865
Lee, Ron, Las Vegas 121366
Lee, Sabina, Los Angeles 121530
Lee, Tom, Memphis 094644
Lee, Wing Hong, Hong Kong .. 065076
Lee, Y., Los Angeles 121531
Leeber Corgan, Sharon, Dallas . 120623
Leech, John, Auckland 112783
Leedel Moorehead-Graham Fine Arts
Gallery, Pine Bluff 052137
Leeds Industrial Museum, Leeds
Museums and Galleries, Leeds 044781
Leeds Metropolitan University Gallery,
Leeds 044782
Leeds Museum, Leeds 044783
Leeds Museums and Galleries Review,
Leeds 139196

Leedy-Voulke's Art Center, Kansas
City 049885
Leedy, Sherry, Kansas City 121302
Leeflang Oude Dingen, Den Haag 083098
Leegenhoek, J. O. & Henriette,
Paris 105016, 129062
Leegenhoek, Jacques, Paris ... 105017
Leek Antiques Centre, Leek .. 089783,
134540
Leek Art Gallery, Leek 044790
Leek Old Books, Leek 144521
Leelanau Historical Museum,
Leland 050217
Leemark, Christchurch 083880
Leepa-Rattner Museum, Tarpon
Springs 053683
Lee's Comm. Art Institute, Kuala
Lumpur 056129
Lee's Horsey Fine Art Studio,
Kaohsiung 042363
Lees & Sons, M., Worcester ... 091948
Lees, Michael Scott, Bungendore 098823
Lees, Michael Scott, Jindabyne . 099160
Leest, Den Haag 143114
Leeuw, Jef Theo de, Ekeren .. 100681,
136652
Lefakis, Chalandri 109058
Lefakis, Néa Kifissia 109077
Lefebvre & Petrenko, Paris 105019
Lefebvre, Arnaud, Paris 105018
Lefebvre, Eric, Saint-Pryvé Saint-
Mesmin 137004
Lefèbvre & Fils, Georges, Paris . 071563
Lefèbvre, Jacques, Montebourg 070230
Lefèbvre, Marc, Paris . 071564, 141166
Lefèbvre, Véronique, Beaucamps-
Ligny 066991
Lefèvre, Saint-Maur-des-Fossés 072966
Lefèvre, Blandine, Paris 071565
Lefèvre, Blandine, Saint-Ouen .. 073136
Lefevre, François, Bazoches-les-
Hautes 066989
Lefevre, Jean-Claude, Mayenne 070013
Lefèvre, Nicolas, Granville 068699
Lefèvre, Pierre-Yves, Paris 125731
Lefèvre, Romain, Gisors 068672
Lefèvre, Romain, Vesly 074076
Lefèvre, Valérie, Conty 068140
Lefèvre, W. & V., Kortrijk 063721
Lefferts Historic House, Brooklyn 047262
Leffingwell House Museum,
Norwich 051648
Leffler, Toledo 097611
Lefkowicz, Edward J.,
Providence 145467
Lefor-Openo, Paris 105020
Leforestier, Alain, Forges-les-
Eaux 068590
Lefranc, Frédéric, Auxerre 125502
Lefranc, Monique-Valérie,
Marmande 069881
Lefrançois, F.J., Bordeaux 067341
Left Bank Antiques, Athlone ... 079441
Left Bank Art Consultants, Surry
Hills 062264
Left Bank Art Gallery, Greymouth 033213
Left Bank Artists Co-Op,
Beechworth 098721
Left Bank Books, Saint Louis .. 145497
Left Bank Books, Seattle 145590
Left Bank Gallery, Norfolk 123232
Left Gallery, Toronto 101647
Lefteris Kanakakis Gallery,
Rethymnon 023245
Leg Laki, Edit, Pécs 109208
Legabilleux, Jean-Paul, Parville . 072027
Legacies Antiques, Vancouver .. 064836
Legacies Limited, Milwaukee .. 094867
Legacy, Ciudad de Panamá .. 113300
Legacy, Shannon 084076
Legacy Antiques, Athinai 079170
Legacy Antiques, Cleveland ... 092952
Legacy Antiques, Dallas 093158
Legacy Art, Saint Paul 123918

Legacy Artworks, Boston 119888
Legacy Furniture, Milwaukee .. 094868
Legacy Gold, Dallas 093159
Legacy Home Collections,
Singapore 085455
Legacy Limited, Seattle 124560
Legacy's Antiques, San Diego .. 097101
Legado Valeriano Salas, Béjar . 039114
Legal Book Store, Los Angeles . 145233
Legart, Perugia 131753
Legaspi City Museum, Legaspi . 034493
Legat Nedeljka Gvozdenovića,
Beograd 038150
Legat-Verlag, Tübingen 137536
Legend Antiques, Oakland 095899
Legend Press, Artarmon 136478
Legenda, Budapest 142562
La Légende, Nice 070623
Légende des Siècles, Besançon 067128
Legends Concert Posters,
Portland 123640
Legends in Time, Toledo 097612
Legends of Art, Las Vegas 121367
Legends of Hollywood, Omaha . 123326
Legends of New Mexico Museum,
Algodones 046400
Legends of the Game Baseball
Museum, Arlington 046534
Leger, Malmö 115764
Leger, Helmut, München 108157, 137413
Léger, Joël, Le Havre 069284
Legg & Son, E.C., Cirencester . 134275
Legg, Jerry, Dorchester 089072, 134331
Legg, Michael, Dorchester ... 089073
Legge, Dublin 109601
Legge, Redfern 099621
Legge, Christopher, Oxford ... 090932,
134917
Legge, Elisabeth, Toronto 064690,
140408
Legge, Joan, Salt Lake City ... 096899
Legge, Joy, Subiaco 062257
Leggett, Arthur, Toronto 101648
Légier, Nathalie, L'Isle-sur-la-
Sorgue 069578
Legiewicz, Dinslaken 075363
Legion of Honor, Fine Arts Museum of
San Francisco, San Francisco . 053009
Legislative Building Art Galleries,
Regina 006769
Legnaghi, Mattea, Verona 132286
Legni e Restauro, Milano 131461
Legot-Thisse, Bruxelles 063489, 128031
Legović, Pula 102522
Legra Gallery, Leigh-on-Sea .. 118100
Legrado Garcia, Jose Maria,
Madrid 133351
Legrand, Alain, Paris 129063
Legrand, Claudine, Paris 105021
Legrand, Jean-Luc, Cormainville 068151
Legrand, Nadine, Armentières .. 103273
Legros, Martine, Saint-Martin-de-
Ré 072939
Legroux, Pascale, Fontenay-le-
Comte 128601
Legue, Jacques, Bondy 067286
Lehalle, Paris 105022
Lehár-Schikaneder-Schlößl, Wien 003028
Lehár-Villa, Bad Ischl 001750
Lehembre, Philippe, Marcilly-la-
Campagne 069852
Lehembre, Philippe, Verneuil-sur-
Avre 074029
Leher, Alfred, Landshut 077014
Lehigh County Historical Society
Museum, Allentown 046410
Lehigh University Art Galleries/
Museum, Zoellner Arts Center,
Bethlehem 046961
Lehle, Agathe & Pierre,
Kolbermoor 107709
Lehman College Art Gallery,
Bronx 047232
Lehman, Morgan, New York .. 122803

Lehmann, Bad Füssing 074513
Gebr. Lehmann, Berlin 106358
Gebr. Lehmann, Dresden 106742
Lehmann-Dronke, Clemens,
  Kerpen 141987
Lehmann-Haupt, Roxana, New
  York 135949
Lehmann Maupin, New York ... 122804
Lehmann Maupin, (Bowery), New
  York 122805
Lehmann-Özyurt, Stephan, Varel 142447
Lehmann, Bernd, Dresden 075420
Lehmann, D., Köln 142016
Lehmann, Heinz, Stans 087831
Lehmann, J., Chemnitz 075256, 125913
Lehmann, Jens, Hannover 130011
Lehmann, Leskiw & Schedler,
  Toronto 101649
Lehmann, Marc, Lausanne 087574
Lehmann, Uwe, Potsdam 078079
Lehmeier, Andrea & Roland,
  Pyrbaum 078105
Lehmkuhl, Thomas, Steinfurt .. 130657
Lehmmuseum, Buchberg 017651
Lehn, Joseph, Baltimore 092338
Lehnemann, Wilfried, Bochum .. 075036
Lehner, Maria, Wolfsberg im
  Schwarzautal 100404
Lehner, Nikolaus, Altdorf, Kreis
  Landshut 129381
Lehnert & Landrock, Cairo ... 103051,
                            136787
Lehr- und Schauberwerk Frisch
  Glück, Johanngeorgenstadt ... 019467
Lehr- und Schauberwerk Herkules
  Frisch Glück, Beierfeld 017018
Lehrerhaus-Friemersheim,
  Duisburg 018058
Lehrstuhl für Bildnerische Gestaltung,
  Rheinisch-Westfälische Technische
  Hochschule Aachen, Aachen .. 055441
Lehrstuhl für Bildnerisches Gestalten,
  Technische Universität München,
  München 055654, 055655
Lehrstuhl für Entwerfen und
  Denkmalpflege, Technische
  Universität München, München 055656
Lehrstuhl für Grundlagen der
  Gestaltung und Darstellung,
  Technische Universität München,
  München 055657
Lehrstuhl für Industrial Design,
  Technische Universität München,
  München 055658
Lehrstuhl für Islamische
  Kunstgeschichte und Archäologie,
  Otto-Friedrich-Universität Bamberg,
  Bamberg 055454
Lehrstuhl für Kunst und Kunsttheorie,
  Kulturwissenschaftliche Fakultät,
  Frankfurt (Oder) 055536
Lehrstuhl für Kunsterziehung,
  Universität Passau, Passau ... 055686
Lehrstuhl für Kunstgeschichte
  und Christliche Archäologie,
  Philosophische Fakultät, Universität
  Passau, Passau 055687
Lehrstuhl für Kunstgeschichte und
  Professur für Kunsterziehung,
  Katholische Universität,
  Eichstätt 055526
Lehrstuhl für Kunstgeschichte,
  Universität Augsburg, Augsburg 055448
Lehrstuhl für Kunstpädagogik und
  Kunstdidaktik, Otto-Friedrich-
  Universität Bamberg, Bamberg 055455
Lehrstuhl für Kunstpädagogik,
  Department Fachdidaktiken,
  Nürnberg 055671
Lehrstuhl für Kunstpädagogik,
  Universität Augsburg, Augsburg 055449
Lehrstuhl für Restaurierung,
  Kunsttechnologie und
  Konservierungswissenschaften,

Technische Universität München,
  München 055659
Lehrstuhl für Theorie und Geschichte
  von Architektur, Kunst und Design,
  Technische Universität München,
  München 055660
Lehrstuhl Kunst, Pädagogische
  Hochschule Weingarten/ Fakultät II,
  Weingarten, Kreis Ravensburg . 055738
Lehtimäen Kotiseutumuseo,
  Lehtimäki 011102
Lehtse Koduloomuuseum, Lehtse 010634
Lehuard, Jacques, Jonquières-Saint-
  Vincent 068908
Lehuard, Ludovic, Nîmes 070697
Lei Cheng Uk Han Tomb Museum,
  Hong Kong Museum of History, Hong
  Kong 007881
Leia Livraria, São Paulo 140305
Leiber, Steven, San Francisco . 124325
Leibfried, Fr., Sindelfingen 078475
Leib'sches Haus, Abteilung
  Stadtgeschichte und Volkskunde,
  Oberhessisches Museum und
  Gail'sche Sammlungen, Gießen 018691
Leibzig-Biland, Marly 133854
Leica Gallery, New York 122806
Leica Gallery Prague, Praha ... 009804
Leica Stammbaum und Museum,
  Solms 021909
Leicester Antiques Warehouse,
  Leicester 089791
The Leicester Galleries, London 090293,
                    118422,   138224
Leicestershire CCC Museum,
  Leicester 044796
Leicestershire County Council
  Museums Service, Glenfield .. 060171
Leicestershire Pastel Society, Leicester
  LE4 4LD 060172
Leichter, Danica, Frankfurt am
  Main 075803
Leida, Wageningen 137876, 143261
Leidel, Herbert, München 108158
Leidelmeijer, Amsterdam 082794
Leidl, Monika, München 142176
De Leidse Wagenmaker, Leiden 132559
Leigh, Chicago 120177
Leigh Gallery Books, Southend-on-
  Sea 144856
Leigh Heritage Centre, Leigh-on-
  Sea 044805
Leigh Jewellery, Leigh,
  Lancashire 089794
Leigh Yawkey Woodson Art Museum,
  Wausau 054236
Leigh, Jamie, Oldbury West
  Midlands 118849
Leigh, Laurie, Oxford 090933
Leighton Art Centre, Calgary ... 005524
Leighton Art Gallery, Weymouth 119333
Leighton Buzzard Railway, Leighton
  Buzzard 044807
Leighton Hall, Carnforth 043878
Leighton House Museum, London 045019
Leighton-Morris, Parkhurst ... 085659
Leijon, Ulf, Uppsala 133599
Leijser, Margot, Arnhem 082942
Leilivende, Amadora 126431
A Leiloeira, Lisboa 126445
A Leiloeira, Porto 126459
Leiloeira Corte Real, Porto ... 085032,
                            126460
A Leiloeira Invicta do Norte,
  Porto 126461
Leilões União, Porto 126462
Leiloës, Cabral Monca da, Lisboa 126446
Leimarel Museum and Research
  Centre, Imphal 024309
Leimberger, Hanns, Bodenmais . 106517
Leinen- und Spitzenmuseum
  Haigerseelbach, Haiger 018926
Leinen und Lavendel, Lübeck .. 077177
Leiner, Beckum 074705

Leinfelder Haus, Leinfelden-
  Echterdingen 019965
Leingruber, Rolf, Basel 133650
Leinorte, Porto 126463
Leinster, Dublin 109602
Leinweber, Dr. J., Marburg 142114
Leinz, Gabriele, Bonn 129609
Leipold, Günther, Fulda 075929
Leipziger Antiquariat, Leipzig .. 142060
Leipziger Antiquariatsmesse,
  Leipzig 098156
Leipziger Grafikbörse e.V.,
  Leipzig 059160, 107841
Leipziger Jahresausstellung e.V.,
  Leipzig 059161
Leipziger Münzhandlung und Auktion,
  Leipzig 077063, 126025
Leirfjord Bygdesamlinger,
  Leirfjord 033735
Leiria & Nascimento, Lisboa .. 084903,
            084904,   114039,   126447
Leiria & Nascimento, Porto ... 085033,
                            126464
Leiris, Louise, Paris .. 105023, 136952
Leis, Karl, Karlsruhe 076699
Leis, Ulla, Mömbris 077384
Leiser-Hofmann & Conen, Karlstein
  am Main 076708
Leismann, Hedwig, Sankt
  Wendel 108558
Leiss, Hilde, Hamburg 107289
Leist, F., Osnabrück 077974
Leiston Trading Post, Leiston .. 089800
Leisure Art Gallery, Singapore . 114558
Leisure Painter, Tenterden 139197
Leitch Collieries, Crowsnest Pass 005667
Leith Gallery, Edinburgh 117732
Leith Hall, Huntly 044620
Leith, Nan, Canterbury 088778
Leitheimer Schloß-Museum,
  Kaisheim 019484
Leithenmühle, Windhaag bei
  Freistadt 003120
Leitner-Hofherr, Lermoos 100095
Lejko, Július, Vel'ké Kapušany . 085596
Lejre Forsøgscenter, Lejre 010217
Leka Bygdemuseum, Leka 033736
Lekon, Essen 106924
Leks, Paris 141167
Leksands Antik och Kuriosa,
  Leksand 086804
Leksands Museum, Leksand ... 040775
Leksbergs Skolmuseum, Vadsbo
  Museum, Mariestad 040842
Leksvik Bygdesamling, Leksvik . 033737
Leland, Kippie L. & Ted,
  Nashville 095010
Leland, Perry H., San Francisco 097269
Lelandais, Marcel, Bourseul ... 067437
Leleu, Gérald, Leubringhen ... 069441
Lelievre, Jean-Marie, Athies-sous-
  Laon 066765
Lellis, Keith de, New York 122807
Lellouche, Abraham Guy, Paris . 071566
Lelong, Argenteuil 103271
Lelong, New York 122808
Lelong, Paris 105024, 136953
Lelong, Zürich 116830, 138113
Lelouch, Jean-Marc, Paris 071567
Lemaire, Amsterdam 082795
Lemaire SA, Ancienne Maison L.
  Michiels, Bruxelles 063490
Lemaire, André, Paris . 105025, 105026
Lemaire, Bernard, Tours 129320
Lemaire, Chantal, Alençon 066523
Lemaire, Dominique, Templeuve . 073694
Lemaire, Régis, Moulins (Allier) . 070372
Lemaitre d'Estève de Bosch, Marie-
  Hélène, Paris 071568
Lemaître d'Estève de Bosch,
  Marie-Hélène, Saint-Ouen-de-
  Thouberville 073233
Lemand, Claude, Paris 105027
Lemarchand, Philippe, Saint-Sauveur-

le-Vicomte 073303
Lemarié, Jean, Notre-Dame-du-
  Touchet 070753
Lemarie, Sylvie, Brive-la-
  Gaillarde 140808
Lemasson, Gaetan, Roquebilliere 072536
Lemaster, Austin 092245, 135327
Lemay, Los Angeles 121532
Lemazo, Jean-Claude, Dole 068367
Lembaga Muzium Negeri, Pulau
  Pinang 030723
Lembcke, Wolfgang, Ammersbek 074408
Lemberg, Venezia 111133
Lemberg-Richtermeier, Ilona
  Marieluise, Hamburg 129972
Lembessi, Antoinette, Paris ... 071569
Lemburg, Hamburg 076269
Lemee House, Natchitoches ... 051199
Lemeni, Valentin, Vandœuvre-lès-
  Nancy 073962
Lemercier, Philippe, Laval
  (Mayenne) 069203
Lemhi County Historical Museum,
  Salmon 052908
Lemieux, André, Québec 101507
Lemin Kotiseutumuseo, Lemi Heritage
  Museum, Lemi 011103
Lemke-Kellner, Brigitte, Nürnberg 077831
Lemke, M., Berlin 074866
Lemke, Olaf, Berlin 074867
Lemkow, Sara & Rachel, Ely ... 089200
Lemmer, Eva, Peißenberg 108372
Lemmers, S.C., Lisse 143193
Lemmons Contemporary, New
  York 122809
Lemoine, Arras 066738
Lemoine, Catherine, Villegouge . 129348
Lemoine, J., Paris 105028
Lemoine, Stanislas, Caen 067574
Lemoine, Yves, Dolus-le-Sec ... 068368
Lemon Grass Arts, Singapore .. 085456
Lemon Hill Mansion, Philadelphia 052061
Lemon Pie, Virginia Beach 097788
Lemon Sky, Los Angeles 121533
Lemon Sky, Miami 121820
Lemon Sky, Miami Beach 121884
Lemon Street Gallery, Dublin .. 109603
Lemon Street Gallery, Truro ... 119260
Lemon Tree Antiques, Art & Books, Fort
  Worth 093459, 145150
Lemonnier, Bruxelles .. 063491, 125318
Lemonnier Frères, Versailles ... 074059
Lemousse, Michel, Guignen ... 068768
LeMoyne Art Foundation,
  Tallahassee 053660
Lemp, Helmut, Dresden 141704
Lempart, Renate, Hamburg 141881
Lempertz, Berlin 106359, 125882
Lempertz, Bruxelles 125887
Lempertz, Köln 076891, 126009
Lempertz, München 126060
Lempertz, Tokyo 126247
Lempertz, Versailles 125833
Lempertz, Matthias, Königswinter 142028
Lempio & Ern, Wiehl 078965
Lemun Kotiseutumuseo, Lemu . 011107
Lemuria Gallery, Philadelphia .. 123410
Lemvig Museum Vesterhus,
  Lemvig 010219
Leña Gomariz, Córdoba 085943
Lenaerts, Dirk, Antwerpen 127996
LenArt, Sankt-Peterburg 114348
Lenbachmuseum,
  Schrobenhausen 021738
Lencastre, Isabel, Porto 085034
Lencquesaing, Hughes de, Boulogne-
  Billancourt 067392
Lendemar, Pia, Trelleborg 133591
Lendl, Eugen, Graz ... 100014, 139988
Lendo & Pintando, Brasília ... 140274
Lenduce, Basel 116086
Lenfant, Fernand, Launois-sur-
  Vence 069199
Lengertz, Lund 143889

Lenggenhager, Hans, Neuhausen am Rheinfall 087707
Lengil, Porto Alegre 063982
Lenglet, Christine, Paris 071570
Lenhart, San Jose 124452
Lenin-Museo, Tampere 011398
Lenk, Michael, Wels 140053
Lenky Meno Galerija, Vilnius ... 111800
Lennard Antiques, Sible Hedingham 091337
Lennartz, M., Kaarst 076659
Lennon & Weinberg, New York . 122810
Lennon House, Liverpool 044856
Lennon School of Art and Design, Lemesos 055293
Lennox, London 118423
Lennox, Parramatta ... 099554, 127657
Lennox and Addington County Museum, Napanee 006454
Lennox, Karen, Chicago 120178
Lennoxlove House, Haddington . 044493
Lennoxville Ascot Historical and Museum Society, Lennoxville . 006182
Lenoir & Assoc., San Antonio . 124045
Lenoir Forge Museum, Arichat . 005349
Lenoir-Retaud, Bernadette, Cruzilles-Mepillat 068253
Lenoir, Jean-Marc, Chécy 067951
Lenorman, Mathieu, Paris 071571
Lenormand, Jacques, Paris 071572, 125732
Lenormand, Xavier, Paris 071573
L'Enrajolada Casa-Museu Santacana, Martorell 039729
Lensfire, Toronto 101650
Lente, Jean-Marc, Cormeilles-en-Parisis 068155
Lenth, Andreas, Alsfeld 074384
Lentini, Vincenzo, Palermo 131701
Lentjou, Christina, Kifissia 109072
Lenton, A., Leicester 144525
Lentos Kunstmuseum Linz, Linz . 002324
The Lentz Center for Asian Culture, Lincoln 050293
Lenvik Bygdemuseet, Finnsnes . 033597
Lenz, Milwaukee 121931
Lenz, Murnau am Staffelsee ... 077684
Lenz, Andreas, Coburg 075263, 129658
Lenz, Horst, Wennigsen 130745
Lenze, Renate, Waldkirch 142460
Lenzen, Michael & Stefan, Düsseldorf 141723
Lenzi, Raffaela, Palermo 081048
Lenzy, Houston 093726
Leo, New York 095534, 095535
Leo & Coppi, Berlin 106360
Leo Baeck Institute New York, Jewish Museum Berlin, Berlin 017213
Leo, Thomas, Rødovre 066079
Leo, Vincenzo, Palermo 131702
Leobon, Frédéric, Cholet 067984
Leodina Antiguidades, Belo Horizonte 063906
Leoframes, Brighton . 134151, 144266
Leominster Antiques Market, Leominster 089807
Leominster Folk Museum, Leominster 044810
Leon, Atlanta 119586
León, Bilbao 133194
Leon, Miami 094760
Leon, Washington 136459
Léon & Claverie, Paris 071575
Leon Apacible Historical Landmark Museum, Taal 034598
León Campos, Guatemala 109121
Léon, Albert de, Paris 071574
Leon, F., Los Angeles 135728
Léon, Philippe, Plouigneau ... 072208
Leona, Madrid 115246
Leona Paegles Memoriālais Muzejs, Vidriži 030256
Leonard, Burrows 135383
Leonard, Nashville 135847

Leonard & Bina Ellen Art Gallery, Montréal 006375
Leonard, C.M., New York 095536
Leonard, M., Charleville-Mézières 067848
Leonard, Marjorie, Enniskillen . 117756
Leonardis, David, Chicago 120179, 120180
Leonardo, Cambridge 139414
Leonardo, Lausanne 116423
Leonardo, Omaha 123327
Léonardo, Paris 105029
Leonardo, Skopje 111886
Leonardo, Tiranë 098529
Leonardo da Vinci Society Newsletter, London 139198
Leonardof, Bogotá 065211
Léondar, Knokke-Heist 100744
Leone, Rio de Janeiro 101044
Il Leone, Roma 110809
Leone & C., Antonino, Messina . 080371
Leone, Giacinto, Taranto 081635
Leone, Pasquale, Napoli 080895
Leong Thin Lien, Penang 082405
Leong & Sons, Seow Kok, Singapore 085457
Leonhard, Graz ... 062597, 100015
Leonhard, Hans Jørgen, Aabenraa 065574
Leonhardi-Museum, Dresden .. 017994
Leonhardi-Museum Aigen, Bad Füssing 016773
Leonhardt, Hans-Ulrich, Görlitz . 076038, 125959, 129909
Leonhardt, Jochen, Ibbenbüren . 125993
Leonhardt, Monika, Rehlingen-Siersburg 078180
Léoni, Saint-Galmier 105517
Leoni, Lilia, Roma 081424
Léoni, Patrice, Marseille 104184
Leonora Gwalia Historical Museum, Leonora 001220
Leopaldi, Donato, Torino 132156
Leopard Books, Bristol 144283
Leopold, Kansas City 121303
Léopold, Stockholm .. 087023, 143955
Léopold d'Or, Agen 066428
Leopold Figl-Museum, Rust im Tullnerfeld 002613
Leopold-Hoesch-Museum Düren, Düren 018021
Leopold Museum, Wien 003029
Léopoldi-Ansaldi, Bastia 066960
Leorbide Contratas, Donostia-San Sebastián 133252
Leo's Antiques, Houston 093727
Leo's Tenne, Schweisweiler .. 021787
Leow, Singapore 085458
Lepage, Claudine, Châteaudun . 075873
Lepage, Raymond, Paris 129064
Lepanto, Dr. Wassili, Heidelberg 107394
Lepape, Claude, Nancy 070422
Lepavec, Alain, Paris 129065
Leperre, Carole, Reims 072394
Lepert & Scheler, Paris 141168
Lepic, Paris .. 105030, 136954, 141169
Lepic, Roland, Paris 071576
L'Epicerie, Tourrettes-sur-Loup . 073830
Lepikon Torppa, Pielavesi 011249
Lepina, Juan, Vigo 086561
Leping Museum, Leping 008001
Leporatti, Alessandro, Pisa ... 081157
Lepoultel, Philippe, Heugueville-sur-Sienne 068818
Lepoutre et Fils, Roubaix 072550
Lepoutre, André, Hardelot-Plage 068789
Leppävirran Kotiseutumuseo, Leppävirta 011109
Leppen, Schmidt & Pirec, Ahaus 105961
Lepramuseet Sant Jørgens Hospital, Bergen 033535
Lepre, Aurelio, Palermo 110608
Leprévost, Claude, Isneauville . 068879
Leprince, Christian, Paris 071577, 105031

Leptis Magna Museum, Department of Antiquities, Leptis Magna ... 030291
Lerch, Lauf an der Pegnitz 077030
Lerch, Dr. Tom, Berlin 129526
Lerch, Robert M., New York .. 095537
Lerchenfeld, Ingrid Freifrau von, München 077559
Lerchl, Klaus, Lippstadt 130243
Lereec, Elizabeth, Saint-Malo (Ille-et-Vilaine) 072919
Lerendu, Alain, Marcey-les-Greves 069850
Lermer, Michael J., München .. 130365
Lermontovskij Gosudarstvennyj Muzej-zapovednik Tarchany, Lermontovo 037193
Lerner, Leonardo, Buenos Aires . 139578
Lernon, Rouen 105476
Leroux, Philippe, Nice 070624
Leroux, Régis, Thiberville 073649
Leroy, Nancy 125652
Leroy and District Heritage Museum, Leroy 006184
Leroy, Dominique, Sens 137008
Leroy, Olivier, Bondues 067284
Lerverk Galleri, Göteborg 115657
Lervoire, Thénac 073645
Lervold, Lena 126357
Les Antiquaires en Céramique Ancienne, Bruxelles 058385
Les Antiquitees, Roma 081425
Les Chances de L'Art, Bolzano . 109885
Les Ex-Voto, Monthermé 014263
Les Filles du Calvaire, Bruxelles 100633
Les Filles du Calvaire, Paris ... 105032
Les Grottes de Gargas, Aventignan 011830
Les Jacobins, Musée de Morlaix, Morlaix 014326
Les Liles, Barcelona 085791
Les Murs ont la Parole, Aigues-Mortes 103183
Les Peintres Témoin de leur Temps, Paris 105033
Les Roues de L'Areuse, Ecomusée de la Haute-Areuse, Saint-Sulpice (Neuchâtel) 041911
Les Territoires de la Mémoire, Liège 003794
Les Vieilles Choses, Saint-Hilaire-du-Harcouët 072827, 129234
Lesage, Montpellier 104282
LesArt e.V., Überregionaler Verein lesbischer Künstlerinnen und für kunstinteressierte Lesben, Köln 059162
Lesbian Herstory Educational Foundation, Brooklyn 047263
Lesch, I., Freiburg im Breisgau . 075873
Leschevin, Christophe, Lille ... 069494
Leschevin, Sabine, Roubaix ... 141317
Lescomère, François, Yvrac ... 074317
Lescop de Moy, Antoine, Paris . 071578
Lescoules, Christian, Albi 066515
Lescouzères, Rousseau, Bazas . 066986
Lescroart, Marcel, Toucy 073692
Lescure Collections, Mazamet . 070017
Lescureux, Christian, Amiens .. 066577, 066578
Lesekauz, Augsburg 141415
Leseratte, Braunschweig 141638
Lesigne, Michel, Orval 070825
Lesja Bygdatun, Museet i Nord-Gudbrandsdalen, Lesja 033738
Lesjöforsmuseum, Lesjöfors .. 040776
Leske, W., Einbeck 075587
Leskowsky Hangszergyüjtemény Közalapítvány, Kecskemét 023727
Leslie & Leslie, Haddington 089401, 126914
Leslie Feely Fine Art, New York . 122811
Leslie Hill Open Farm, Ballymoney 043481
Leslie-Lohman Gay Art Foundation, New York 051415

Leslie M. Frost Natural Resources Centre, Minden 006322
Leslie Sacks Fine Art, Los Angeles 121534
Leslie's, Seattle 097468
Leslie's Collectibles, Long Beach 094209
Lesnícke a Drevárske Múzeum, Zvolen 038489
Lesnické, Myslivecké a Rybářské Muzeum, Národní Zemědělské Muzeum Praha, Hluboká nad Vltavou 009561
Lesotho National Museum, Maseru 030287
Lesourd, Lucien, Talmont-Saint-Hilaire 073616
Lespinasse, Bry-sur-Marne 067528
Lespinasse, François, Rouen ... 105477
L'Esprit du Cap, Lege-Cap-Ferret 069381
Lessing-Museum, Kamenz 019488
Lessirard, Marie-Louise, Chandai 067824
L'Estanco, Tassin-la-Demi-Lune . 073629
Lester-Garland Premises, Trinity 007237
Lester, M., Albuquerque 119487
Lestijärven Kotiseutumuseo, Lestijärvi 011110
Lestrade, Michel, Toulouse 141366
L'Estrange, Christchurch 112849
Lestranger, Paris 105034
Lestranger, Saint-Rémy-de-Provence 105612
Leszczyłowski, Jan, Sopot 084663
Leszko, J., Weinheim 078909
L'Etang d'Art, Bages (Aude) ... 011863
Letchworth Museum and Art Gallery, Letchworth 044813
Letecké Muzeum VHÚ Praha, Historický Ústav Armády České Republiky, Praha 009805
Letellier, Jean, Paris 071579
Leterreau, Briec, Rennes (Ille-et-Vilaine) 105438
Leth Jensen, Jakob, Frederiksberg 065668
Letham, Ponsonby 112992
Lethert, Christian, Köln 107659
Lethert, P. & S., Bad Münstereifel 074557
Lethgau, Lübeck 107917
Letif Kerimov ad. Azerbaycan Xalçasi ve Xalq Tetbiqi Seneti Dövlet Muzeyi, Bakı 003167
Letnisko, Miejski Ośrodek Kultury, Milanówek 035014
Letohrádek Ostrov, Galerie Umění Karlovy Vary, Ostrov nad Ohří . 009746
Letostruj, Sofia 064231, 140323
Let's Collect Devon Pottery Magazine, Saint Albans 139199
Let's Dish, Austin 092246
Let's Go Antiquing!, New Orleans 095120
Let's Talk Antiques, Daytona Beach 127197
Letschert, Peter, Ransbach-Baumbach 078125
Letschiner Heimatstuben, Letschin 020047
Lett, Michael, Auckland 112784
Letter a Gallery, Whittlesey 119346
Letteratura Tattile, Rimini 142793
Le Lettere, Firenze 137652
Letterkundig Museum en Kinderboekenmuseum, Den Haag 032176
Letters Bookshop, Toronto 140409
Lettl-Atrium, Museum für surreale Kunst, Augsburg 116660
La Lettre Écarlate, Paris 141170
La Lettre volée, Bruxelles 100634, 136643
Lettrée, Anne, Paris 105035
Letu, Bernard, Genève 116328, 138050, 144078
Lëtzebuerger Artisten Center – Salon

de Printemps, Luxemburg 098241
Letzelter, D., Baltmannsweiler .. 074657
Letzner, Georg, Heimertingen .. 076480
Leu, Kerstin & Holger, München 108159
Leu, Michael, San Francisco .. 124326
Leuchten-Hammes,
　Mönchengladbach 077397
Leuchter, Ruth, Düsseldorf 106806
Leudet, Christian, Le Bec-
　Hellouin 069216
Il Leudo, Genova 110120
Leuenberger, Zürich .. 087987, 116831,
　　　　　　　　　　　　　133987
Leupin, Brigitta, Basel 116087
Leupolt, Fred, Templin 078674
Leura Fine Woodwork Gallery,
　Leura 099230
Leuro-Arte, Málaga 115304
Lev, London 090294, 134712
Levanger Museum, Levanger .. 033740
Levasseur, Noyon 070761
Levasseur, Paris 105036
Levé, François, Paris 071580
Levecher, Michel, Juziers 068936
Leveillé, Jean-Claude, Pacy-sur-
　Eure 070836, 128881
Léveilley, Jean-Pierre, Angers .. 066610
Leven Antiques, Ulverstone 062372
Levene, M.P., London 090295
Levent, Alain, Athies-sous-Laon 066766
Leventis Municipal Museum of Nicosia,
　Lefkosia 009418
Lévêque, André, Montlouis-sur-
　Loire 128804
Leveque, Claude, Fresnes 068611
Lever, Vancouver 064837
Levere Memorial Temple,
　Evanston 048640
Leverenz & Ritter, Hamburg .. 076270
Leverett Historical Museum,
　Leverett 050222
Levers, Wollongong ... 099894, 127699
Levesque, Jeanne, Paris 071581
Levesque, Valérie, Paris 071582
Levi Coffin House, Fountain City 048914
Lévi, Liana, Paris 136955
Levi, Sara, Tel Aviv 109727
Levin Fine Arts Gallery, Levin
　Manawatu 112934
Levin, Betty, New York 122812
Levina-Swannet, Toulon (Var) .. 073716
Levine Museum of the New South,
　Charlotte 047574
Levine, Diana, Boston 119889
Levine & Assoc., Elizabeth, New
　York 095538
Levine, Jonathan, New York ... 122813
Levine, Leona, Norwich 090876, 134894
Levionnais, Marie-Jeannne, Blainville-
　sur-Orne 067236
Levis, Calgary 125364
Levison, Don, San Francisco .. 097270
Levitske Brothers, Brentwood .. 145019
Levoir, Rose, Paris 129066
Lévy-Alban, Sylvain, Paris 071585
Levy Galerie, Hamburg 107290, 137249
Levy Gallery, Moore College of Art and
　Design, Philadelphia 052062
Levy, Bernard & S. Dean, New
　York 095539
Levy, Claudia, Santiago de Chile 064972
Lévy, Etienne, Paris .. 071583, 129067
Lévy, Gérard, Paris 071584
Lévy, Jacques, Paris 105037
Lévy, Johann, Paris 105038
Lévy, Luc, Montpellier 128810
Lévy, Michael, Long Beach 121404
Lévy, Pierre, Paris 105039
Lévy, Richard, Albuquerque 119488
Levy, Stan, New Orleans 095121
Levy, Wendy J., Manchester .. 118709
Lewandowski, Jean-Pierre,
　Blanzy 067240
Lewcock, John, Ely 144409

Lewendon, Nedlands 061947
Lewerentz, Sigrid, Kamen 107499
Lewes Antiques Centre, Lewes . 089815,
　　　　　　　　　　　　　134552
Lewes Book Centre, Lewes 144535
Lewes Castle and Barbican House
　Museum, Lewes 044817
Lewes Clock Shop, Lewes 089816,
　　　　　　　　　　　　　134553
Lewes Historical Society Museum,
　Lewes 050223
Lewin, Harriett, London 090296
Lewin, Henri, Paris 071586
Lewin, Henri, Saint-Ouen 073137
Lewis, Miami 121821
Lewis, Omaha 123328
Lewis, Toronto 064691
Lewis & Lloyd, Redhill 091083
Lewis and Clark Boat House and
　Nature Center, Saint Charles . 052757
Lewis Collection, Cambridge ... 043844
Lewis County Historical Museum,
　Chehalis 047601
Lewis County Historical Society
　Museum, Lowville 050514
Lewis County Historical Society
　Museum, Lyons Falls 050542
Lewis Elton Gallery, Guildford .. 044489
Lewis Textile Museum, Blackburn,
　Lancashire 043642
Lewis, Doris, Los Angeles 094388
Lewis, Frank, Killarney 109653
Lewis, H., Baltimore 119750
Lewis, Ivor & Patricia,
　Peterborough 090983
Lewis, James and Nellie Stratton
　Gallery, Stephens College,
　Columbia 047900
Lewis, Julianne, Kew 061685
Lewis, M. & D., London 090297
Lewis, Michael, Bruton 088692, 117469,
　　　　　　　　　　　　　144287
Lewis, Michael & Amanda,
　Wellington 091770, 135198
Lewis, Richard, New Orleans .. 122193
Lewis, Sandra, Kasendorf 076710
Lewisham Local Studies Centre,
　London 045020
Lewistown Art Center, Lewistown 050237
Lewitzka, Victor Harbor 099852
Lewitzki, Robert, Celle 075247
Lewry, David C.E., Gosport 134414
Lex, Hans, Bad Endorf 106055
Lex, Horst & Hinrika, Weyarn .. 108893
Lexa a Syn, Litoměřice 065417
Lexica Örebro Antikvariat, Örebro 143910
Lexington Art, New York 122814
Lexington Art League, Lexington 060566
Lexington County Museum,
　Lexington 050262
Lexington Historical Society, Buckman
　Tavern, Munroe Tavern, Hancock-
　Clarke House, Antique Fire
　Equipment Museum, Lexington 050255
Lexington History Museum,
　Lexington 050246
Ley, Joe & Sheila, Louisville .. 094569,
　　　　　　　　　　　　　127298
Leye Museum, Leye 008002
Leye, Bernard de, Bruxelles ... 063492
Leyendecker, Santa Cruz de
　Tenerife 115423
Leyer, R., Kaarst 076660
Leyh, Michael, Rothenburg ob der
　Tauber 108525
Leyrolles, D., Liège 128077
Leyrolles, Jean-Marie, Rodez .. 072513
Leyshon, M., Columbus 145100
Leyson, Kay, London 090298
Lez-Arts-Cachés, Bruxelles ... 063493
Lézards de la Table, Saint-Ouen 073138
Lézarts, Cap-d'Ail 128817
LG Objetos de Arte, São Paulo . 064188
LGC Arte Hoje, Rio de Janeiro . 101045

Lgovskij Kraevedčeskij Muzej,
　Lgov 037194
Lgovskij Literaturno-memorialnyj
　Muzej N.N. Aseeva, Lgov 037195
Lgovskij Literaturno-memorialnyj
　Muzej A.P. Gajdara, Lgov ... 037196
Lhalia, Rio de Janeiro 101046
L'Harmattan, Paris 141171
Lhasa Museum, Lhasa 008003
L'Héritier, Pont-Aven 105381
Lhermitte, Paris 141172
Lhomme, Liège 125346, 140243
L'Homme-de-Plume, Saint-Ouen 073139
Lhoste, Eric, Pau 072045
Lhotte, Nicole, Nancy 070423
LHU Arts Center, Lunghwa University
　of Science and Technology,
　Gueishan 042320
L'Huillier-Coburg-Kastély, Edelény 023612
L'Huillier, Thierry, Houilles 068842
Li, New York 135950
Li Bros., Shanghai ... 065143, 065144
Li County Museum, Lixian 008042
Li Mei-Shu Memorial Gallery,
　Sansia 042436
Li Tien-lu Hand Puppet Historical
　Museum, Sanjhi 042434
Li Vigni, Pietro, Palermo 081049
Li Xian Jun Former Residence,
　Nanjing 008108
Li Yun Ge, Shanghai 102191
Li Zong Ren Display Center,
　Guilin 007820
Li, An, Singapore 114559
Li, Bab, Singapore 114560
Li, Bai, Singapore 085459
Li, H.K., Vancouver 101807
Li, Hing, Los Angeles 094389
Li, L., Albuquerque 092064
Li.Arte Giotto, Palermo 110609
Liagre, Hugues, L'Isle-Jourdain . 069546,
　　　　　　　　　　　　　069547
Liaigre, Jacqueline, Gençay .. 068645
Liancourt, Paris 129068
Liane and Danny Taran Gallery, Saidye
　Bronfman Centre for the Arts,
　Montréal 006376
Liangshan Yizu Thrall Society Museum,
　Xichang 008421
Liangshanzhou Guan, Xichang .. 008422
Liangzhu Culture Museum,
　Hangzhou 007840
Lianjiang Museum, Lianjiang,
　Fujian 008005
Lianjiang Museum, Lianjiang,
　Guangdong 008006
Liannan Nationality Museum,
　Qingyuan 008182
Lianping Museum, Lianping 008007
Lianshan Chuang Yaozu Self-Rule
　Nationality Museum, Qingyuan 008183
Lianyungang Museum,
　Lianyungang 008008
Lianzhou International Photo Festival,
　Lianzhou 098053
Lianzhou Museum, Qingyuan .. 008184
Liao, Philadelphia 096128
Liaocheng Zone Museum,
　Liaocheng 008009
LiaoJin Cheng Yuan, Beijing ... 007626
Liaoning Art Gallery, Shenyang . 008256
Liaoning Fine Arts Publishing House,
　Shenyang 136739
Liaoning Jin Modern City Museum,
　Shenyang 008257
Liaoning Painting Gallery,
　Shenyang 102214
Liaoning Provincial Museum,
　Shenyang 008258
Liaoyang Museum, Liaoyang .. 008010
Lib Lab, Lisboa 084905
Libbrecht, A., Gent 063647
Libby, Cincinnati 092871
Libby Museum, Wolfeboro 054513

Libby, John A., Toronto 101651
Libelle, Basel 144025
Liber Antiquus, Washington ... 145639
Liberace Museum, Las Vegas .. 050167
Liberal Art Program, Laurinburg 057326
Liberal Arts Department, Gloucester
　County College, Sewell 057904
Liberal Arts Division, Arts and
　Humanities Dept., Saint Louis
　Community College at Florissant
　Valley, Ferguson 057085
Liberal Arts Division, Elgin Community
　College, Elgin 057043
Liberal Arts Division, Northeast
　Community College, Norfolk . 057582
Liberal Arts Gallery, Detroit 120810
Librería Anticuaria Astarloa,
　Bilbao 143735
Liberman, Anne-Marie, Mazères
　(Ariège) 070019
Libermann, Bruxelles 063494
Libert, Damien, Paris 125733
La Liberté, Clermont-Ferrand .. 068031
Liberté, Straubing 078578
Liberti, Antonio Pietro, Roma .. 131959
Libertino, Francesco, Martigny . 087662
Liberty, Kortrijk 063722
Liberty, London 090299
Liberty Books, Christchurch .. 143305,
　　　　　　　　　　　　　143307
Liberty Coins, Minneapolis ... 094925
Liberty County Museum, Chester 047614
Liberty Hall, Frankfort 048929
Liberty Hall Historic Center,
　Lamoni 050108
Liberty Hall Museum, Union ... 053889
Liberty Investments, Gent 100697
Liberty Kunsthandel, Rotterdam . 083578,
　　　　　　　　　　　　　132621
Liberty Memorial Museum of World
　War One, Kansas City 049886
Liberty Village Arts Center and Gallery,
　Chester 047615
Liberty's, Deauville 068282
Libertyville-Mundelein Historical
　Society Museum, Libertyville . 050275
Libos, Borås 086648
Libra Antiques, Bournemouth .. 088537
Libra Antiques, London 090300
Libra Auctioneers, Bangalore .. 079354,
　　　　　　　　　　　　　126160
Il Libraio, Vicenza 142888
Libraire Antiquaire, Montréal .. 064439
Les Libraires Entre les Lignes,
　Paris 141173
Librairie A la Recherche du Passé,
　Sarzeau 141343
Librairie à Micita, Bordeaux ... 140797
Librairie Ancienne, Metz 070100, 140954
Librairie Ancienne, Paris 141174
Librairie Andante, Paris 141175
Librairie Arcade, Osaka 142906
Librairie Ardennaise, Redu ... 140258
Librairie Art des Jardins, Sauliac-sur-
　Célé 141344
Librairie Arte, Bruxelles 140184
Librairie Auguste Blaizot, Paris . 141176
Librairie aux Belles Images,
　Rabat 142998
Librairie Bonaparte, Paris 141177
Librairie Book Shop, New
　Orleans 145314
Librairie Brocante du Palais,
　Nantes 140985
Librairie Chevalier Paul, Toulon
　(Var) 141358
Librairie Chrétien, Paris 141178
Librairie d'Argences, Paris 141179
Librairie de la Seine, Paris 141180
Librairie de la Tour, Vichy 141385
Librairie de l'Abbaye-Pinault,
　Paris 129069, 141181
Librairie de l'Amateur,
　Strasbourg 141352

Librairie de l'Ameublement et de la Décoration, Paris 141182
Librairie de l'Avenue, Saint-Ouen 141337
Librairie de l'Escurial, Nice 140999
Librairie de Montbel, Paris 141183
Librairie de Sèvres, Paris 141184
Librairie Denis, Tours 141369
Librairie des Alpes, Grenoble .. 140848
Librairie des Alpes, Paris 141185
Librairie des Arts et Métiers, Lormaye 140875
Librairie des Canuts, Lyon 140905
Librairie des Carrés, Versailles . 141381
Librairie des Eléphants, Bruxelles 125320, 140185
Librairie des Galeries, Bruxelles 140186
Librairie des Hautebises, Grenoble 140849
Librairie des Nouveautés, Lyon . 140906
Librairie des Pyrénées, Bayonne 140778
Librairie des Terreaux, Lyon ... 140907
Librairie Diamond, Montréal ... 140365
Librairie Diogène, Lyon 140908
Librairie du Bât d'Argent, Lyon . 140909
Librairie du Camée, Paris 141186
Librairie du Change, Lyon 140910
Librairie du Château, Neuchâtel 144115
Librairie du Château de Villeneuve, Vence 137012, 141375
Librairie du Cygne, Paris 141187
Librairie du Métro, Lyon 140911
Librairie du Musée d'Art Moderne de la Ville de Paris, Paris 141188
Librairie du Musée des Beaux-Arts, Nantes 140986
Librairie du Musée des Beaux-Arts, Rouen 140987
Librairie du Musée d'Orsay, Paris 141189
Librairie du Musée du Grand Palais, Paris 141190
Librairie du Musée du Louvre, Paris 141191
Librairie du Musée du Moyen-Age, Paris 141192
Librairie du Musée Matisse, Nice 141000
Librairie du Nouveau Monde, Québec 140383
Librairie du Sénéchal, Villefranche-de-Rouergue 141387
Librairie du Spectacle Garnier Arnoul, Paris 141193
Librairie Edition Guénégaud, Paris 136956, 141194
Librairie Epigrammes, Paris ... 141195
Librairie Epsilon, Paris 141196
Librairie et Documentation de l'Auto Moto-Avion-Train, Paris 141197
Librairie Expérience, Lyon 140912
Librairie Farfouille, Paris 141198
Librairie Fischbacher, Paris ... 141199
Librairie Fl. Tulkens, Bruxelles .. 140367
Librairie Footnotes, Montréal ... 140366
Librairie Godot de Mauroy, Paris 141200
Librairie Gourmande, Paris 141201
Librairie Henner, Paris 141202
Librairie Henri-Julien, Montréal . 140367
Librairie Impériale, Arras 140767
Librairie Jakin, Bayonne 103344, 140779
Librairie Jules Verne, Paris 141203
Librairie La Bourse, Lyon 140913
Librairie La Gryffe, Lyon 140914
Librairie la Procure-Léo, Lyon .. 140915
Librairie L'Arnaqueur, Paris 141204
Librairie Laumière, Paris 141205
Librairie le Moniteur, Lyon ... 140916
Librairie Le Petit Prince, Paris . 141206
Librairie Les Arcades, Paris ... 141207
Librairie Les Beaux Livres, Monaco 142996
Librairie Lyonnaise, Lyon 140917
Librairie Maitre Albert, Paris ... 141208
Librairie Mazarine, Paris 140767
Librairie Mécanique, Lyon 140918
Librairie Méridiés, Lyon 140919

Librairie Monte Cristo, Paris ... 141210
Librairie Nicaise, Paris 141211
Librairie Notre-Dame, Grenoble . 140850
Librairie Nouvel Age, Montréal . 140368
Librairie Occitania, Toulouse ... 141367
Librairie Oriens, Paris 141212
Librairie Orientale Samuelian, Paris 141213
Librairie Page 17, Charleville-Mézières 140822
Librairie Pages Volantes, Paris . 141214
Librairie Paris-Musées, Paris ... 141215
Librairie Philippe Lucas, Lyon . 069755, 140920
Librairie Philoscience, Malicorne-sur-Sarthe 140935
Librairie Philosphique J. Vrin, Paris 141216
Librairie Picasso, Paris 141217
Librairie Pont-Neuf, Paris 141218
Librairie Réunion des Musées Nationaux, Lyon 140921
Librairie Roger J. Ségalat, Lausanne 144093
Librairie Romain, Paris 141219
Librairie Thomas-Scheler, Paris . 141220
Librairie Ulysse, pays et voyages, Hendaye 140856
Librairie Ulysse, pays et voyages, Paris 141221
Librairie Vers et Prose, Paris ... 141222
Librantea, Stuttgart 142398
Librantica, Libramont 098021
Libraria Kontshantverk, Stockholm 133575
Library & Gallery, Cambridge .. 005537
The Library Cafe, Seattle 145591
Library of Congress, Washington 054129
Library of the Greek Orthodox Patriarchate of Alexandria and All Africa, Alexandria 010517
Libre Boek, Warffum 143263
Libre Cours, Bruxelles 100635
Le Libre Regard, Agde 103171
Libreria Albertina, Torino 142860
Libreria Anticuaria, Valencia ... 143803
Librería Anticuaria de México, Puebla 142993
Librería Anticuaria de México, Toluca 142994
Librería Anticuaria Studio, Barcelona 143721, 143722
Libreria Antiquaria, Roma 142817
Libreria Antiquaria Giulio Cesare, Roma 142818
Libreria Antiquaria Mirabilia, Torino 142861
Libreria Antiquaria Palatina, Firenze 142679
Libreria Antiquaria Piemontese, Torino 142862
Libreria Antiquaria Regina, Napoli 142763
Libreria Antiquaria Sant'Agostino, Roma 142819
Libreria Antiquaria Umberto Saba, Trieste 142878
Libreria Antiquaria Xodo, Torino 142863
Libreria Aradia, Roma 142820
Libreria Arrigucci, Firenze 142680
Libreria Balague, Barcelona ... 143723
Librería Bonaire, Valencia 143804
Libreria Boulandier, Bilbao 143736
Librería Castalia, México 142989
Libreria Dante & Descartes, Napoli 142764
Libreria Dante Alighieri, Torino . 142864
Librería de Avila, Buenos Aires . 139579
Librería de la Escalinata, Madrid 143767
Libreria degli Studenti, Roma .. 142821
Libreria dei Passi Perduti, Fratticiola Selvatica 142690
Librería del Ayer, Buenos Aires . 139580
Libreria del Mosaico, Chiasso .. 144060
Libreria del Porcellino, Firenze . 142681

Libreria Docet, Bologna 142633
Libreria Ed Oggettistica Avalon, Genova 080270
Libreria Editrice Goriziana, Gorizia 142697
Librería el Foro, Buenos Aires .. 139581
Librería el Jagüel, Buenos Aires 139582
Librería el Túnel, Buenos Aires . 139583
Librería Errata, Bogotá 065212, 140450
Librería Fabre, Barcelona 143724
Librería Facultades, Valencia ... 143805
Librería El Filobiblión, Madrid .. 143768
Librería Fuad Audi, Modena 142751
Libreria Goya, Barcelona 143725
Libreria Guida, Napoli 142765
Librería Gulliver, Madrid 143769
Librería Helena de Buenos Aires, Buenos Aires 139584
Libreria Incunable, A Coruña ... 143740
Libreria La Fenice, Brescia 142656
Libreria Laia, Valencia 143806
Libreria L'Archivolto, Milano ... 142724
Libreria L'Atalante, Milano 142725
Librería Lux, Alicante 143707
Libreria Maxtor, Valladolid 143812
Libreria Millenium, Milano 142726
Libreria Morgana, Bologna 142634
LIbreria Musicale Gallini, Milano 142727
Librería Nacional, La Plata 139604
Libreria Naturalistica, Bologna .. 142635
Libreria Ogarde, Madrid 143770
Libreria Oreste Gozzini, Firenze . 142682
Libreria Roges, Mataró 143786
Libreria Rotondi, Roma 142822
Libreria Sapienza, Padova 142775
Librería Selvaggio, Barcelona .. 143726
Librería Solar del Bruto, Puente Tocinos-Murcia 143792
Libreria Tara, Roma .. 081426, 142823
Libreria Toletta, Venezia 142883
Libreria Universal, Guatemala .. 140451
Libreria Vitorio, Madrid 143771
Libreria Vobiscum, Alicante 143708
Libresso, Bargstedt 141453
Libresso – Antiquariat in der Universität, Hamburg 141882
Libretto, Langnau im Emmental 144090
Librex, Berlin 141548
Libri Antichi di Lorenzo Pellielo, Biella 142623
Libri di Ieri, Roma 142824
De Librije, Vlaardingen 143257
Libris, Tampere 140747
Libro, Dornach 144064
Libropiel, Madrid 143772
Libros el Glyptodon, Buenos Aires 139585
Libros la Cueva, Buenos Aires . 139586
Libros Lenzi, La Plata 139605
Libros Lord Byron, Buenos Aires 139587
Libros Madrid, Madrid 143773
Libros Romo, Madrid 143774
Licandro, Gianfranco, Wien 100345
Licastro, Francesco, Roma 110810
Liceo Artistico Statale, Cassino . 055882
Liceo Artistico Statale A. Martini, Schio 055979
Liceo Artistico Statale Anton Giulio Bragaglia, Frosinone 055907
Liceo Artistico Statale Busto Arsizio, Busto Arsizio 055878
Liceo Artistico Statale Mattia Preti, Reggio Calabria 055956
Liceo Artistico Statale Medardo Rosso, Lecce 055915
Liceo Artistico Statale Verona, Verona 055996
Lices Antiquités, Rennes (Ille-et-Vilaine) 072438
Licheng District Museum, Licheng 008011
Lichfield County Antiques Show, Kent 098467
Lichfield Heritage Centre,

Lichfield 044824
Licht, Blairsville 127143
Lichtdruck-Kunst Leipzig e.V., Leipzig 019999, 059163
Lichtdruck-Werkstatt-Museum, Dresden 017995
Lichtenberg, Vibeke H. de, Greve 128387
Lichtenburg Museums and Art Gallery, Kultuurhistories, Diamant Delwers en Landboumuseum, Lichtenburg 038777
Lichtentaler Pfarrmuseum, Wien 003030
Lichter, Dr. Walter, Metterich ... 077362
Lichterfelder Antiquariat, Berlin . 141549
Lichtfuß, J., Köln 142017
Lichtlstall, Kitzbühel 062664
Lichtungen, Graz 138525
Lichtzone, Groningen 112496
Licking County Art Association Gallery, Newark 051501
Licking County Historical Society, Newark 051502
Licorne, Antony 066684
La Licorne, Beirut 082291
La Licorne, Clermont-Ferrand .. 068032
La Licorne, Landas 069161
La Licorne, Neaufles-Saint-Martin 070498
Licorne, West Hollywood 097904
La Licorne Bleue, Flavigny-sur-Ozerain 103743
Licorne Saint-Antoine, Castanet-Tolosan 067703
Licorne Saint-Antoine, Egletons . 068430
Lidia, Cluj-Napoca 085116
Lidköpings Konstförening, Lidköping 059896
Lidköpings Konsthall, Lidköping 040780
Lidman, Malmö 086853
Lidon, Patrick, Orléans 070817
Lids, Saint Louis 096727
Lidtke, Kurt, Seattle 124561
Lie, K., Nienhagen 077785
Liebaert, Thierry, Roncq 072532
Liebel, Detlef, Rheinfelden ... 133890
Liebenberger Geschichtsmuseum, Löwenberger Land 020100
Liebens, Piet, Mechelen 128081
Die Liebensteiner Schule, Liebenau 100096
Lieber, Andreas, Frankfurt am Main 075804
Lieber & Sohn, Dr., Radeburg . 078115, 130549
Lieber, Gebr., Bannewitz 074677
Liebermann House, Nahariya Heritage Museum, Nahariya 025159
Liebermann-Villa am Wannsee, Berlin 017214
Liebertee, Berlin 106361
Lieberuks, U., Naumburg, Saale 077691
Liebetrau, Katharina, Bonn 129610
Liebhaveren Antik, Espergærde . 065643
Liebig-Museum, Gießen 018692
Lieblinger Heimatstube, Willstätt 022617
Liebscher, Gerd, Panketal 078014
Liebsohn, Daniel, México 082478
Liechtenstein Museum, Die Fürstlichen Sammlungen, Wien 003031
Liechtenstein Schloss Wilfersdorf, Wilfersdorf 003116
Liechtensteinische Kunstgesellschaft, Vaduz 059548
Liechtensteinische Kunstschule, Nendeln 056122
Liechtensteinische Landesbibliothek, Vaduz 030306
Liechtensteinische Staatliche Kunstsammlung, Vaduz 030305
Liechtensteinisches Landesmuseum, Vaduz 030308
Lied Discovery Children's Museum, Las Vegas 050168
Liedenpohjan Kylämuseo, Liedenpohja 011111

Lieder, Roland, Lutherstadt
 Wittenberg 077217
Liedermann, Krems ... 062689, 127795
Liedermann, Erich, Rohrendorf bei
 Krems 062780
Liedigk, Thomas, Hamburg 107291
Liedon Vanhalinna, Vanhalinna . 011489
Lief, West Hollywood 097905
Lieh, Hsin Hang, Singapore ... 085460
Liehrmann, Liège 100783
Liel, Saarbrücken 108550
Liemers Museum, Zevenaar ... 033098
Lien, Ha Noi 125081, 125082
Lien, Minneapolis 145279
Liepāja Okupaciju Rezimos, Liepājas
 Muzejs, Liepāja 030153
Liepājas Muzeja Izstāžu Zāles,
 Liepāja 030154
Liepājas Muzejs, Liepāja 030155
Lier Bygdetun, Tranby 034023
Lierheimer, H., Niedereschach .. 077777
Lierne museer, Nordli 033806
Liers, Dr.-Ing. D., Berlin 129527
Lierz, Manfred, Hennef 130034
Liesch, Jens, Langenleuba-
 Niederhain 130214
Liesegang, Karl, Schleswig ... 142348
Liesenberg, W., Kirchzarten ... 076814
Lietti, Roberta, Como 109970
Lietuvos Aido, Vilnius 111801
Lietuvos Aviacijos Muziejus,
 Kaunas 030363
Lietuvos Banko Muziejus, Vilnius 030523
Lietuvos Dailės Muziejus, Vilnius 030524
Lietuvos Energetikos Muziejus,
 Vilnius 030525
Lietuvos Etnokosmologijos Muziejus,
 Kulionių 030400
Lietuvos Geologijos Muziejus,
 Vievis 030506
Lietuvos Jūrų Muziejus, Klaipėda 030391
Lietuvos Laisvės Kovų ir Kančių
 Muziejus, Merkinės Kraštotyros
 Muziejus, Merkinė 030421
Lietuvos Liaudis Buities Muziejus,
 Rumšiškes 030456
Lietuvos Matematikų Muziejus,
 Vilniaus Universiteto Matematikos ir
 Informatikos fakultetas, Vilnius 030526
Lietuvos Medicinos ir Farmacijos
 Istorijos Muziejus, Kauno medicinos
 universitetas, Kaunas 030364
Lietuvos Nacionaline Dailės Galerija,
 Vilnius 030527
Lietuvos Nacionalinis Muziejus,
 Vilnius 030528
Lietuvos Prezidento K. Griniaus
 Memorialinis Muziejus,
 Marijampolė 030416
Lietuvos Radijo ir Televizijos Muziejus,
 Vilnius 030529
Lietuvos Šaulių Sąjungos Muziejus,
 Kaunas 030365
Lietuvos Savanoriu Muziejus,
 Vilnius 030530
Lietuvos Sporto Istorijos Ekspozicija,
 Vytauto Didžiojo Karo Muziejus,
 Kaunas 030366
Lietuvos Teatro, Muzikos ir Kino
 Muziejus, Vilnius 030531
Le Lieu, Lorient 104045
Lieu d'Art Contemporain,
 Colomiers 012634
Lieu Historique National de Côteau-du-
 Lac, Côteau-du-Lac 005653
Lieu Historique National des Forges du
 Saint-Maurice, Trois-Rivières . 007245
Lieu Historique National du Canada
 de la Bataille-de-la-Châteauguay,
 Ormstown 006558
Lieu Historique National du Canada de
 la Bataille-de-la-Ristigouche, Pointe-
 à-la-Croix 006664
Lieu Historique National du Manoir

Papineau, Montebello 006355
Lieu Historique National Sir Wilfrid
 Laurier, Ville-des-Laurentides . 007340
Le Lieu, Centre en Art Actuel,
 Québec 101508
Lieutaud, James, Toulon (Var) .. 073717
Lieutenant General Ashton Armoury
 Museum, Victoria 007328
Lieve Hemel, Amsterdam 112244
Life, Colchester 117612
Life Art, Zürich 116832
Life Gallery, Manama 100441
Life Sciences Museum,
 Johannesburg 038753
Lifeboat House, Margate 045225
Lifeboat Museum, Eastbourne .. 044216
Lifeline, Cairns 139678
Lifeline, Lutwyche 139811
Lifeline, Toowoomba 139932
Lifetime Gallery, West Hollywood 097906
Lifton, Judy, Cleveland 120438
Lifu Museum of Chinese Medicine,
 Taichung 042474
Liga Nieuw Beelden, Leiden .. 112556
Ligeard, Catherine, Goult 068687
Ligems, Kuala Lumpur 111906
Liger, Jean-Philippe, Orléans ... 070818
Liger, Jean-Philippe, Yèvre-le-
 Chatel 074309
Liget, Budapest 109159
Liggett, Tulsa 124806
Liggieri, Roberto, Padova 080989
Lighezzolo, Robert, Le Lardin-Saint-
 Lazare 069292, 128665
Light and Greenery Art Museum,
 Sagamihara 029297
The Light Factory, Charlotte ... 047575
Light Factory, Charlotte 119997
Light Flight, Seattle 124562
The Light Gallery, Leamington
 Spa 089756
Light Infantry Museum,
 Winchester 046152
Light Opera Gallery, San
 Francisco 124327
Light Power, Boston .. 092481, 135365
Light Square Gallery, Adelaide .. 000766
Light Street Gallery, Baltimore . 119751
The Lightbox, Woking 046176
Lightgallery, London 118424
Lighthouse Antiques, Miami .. 094761
Lighthouse Gallery, Mamble .. 118701
Lighthouse Gallery, Penzance .. 118896
Lighthouse Gallery, Tequesta .. 053714
Lighthouse Gallery, Truro ... 119261
Lighthouse Gallery, Poole Centre for
 the Arts, Poole 045526
Lighthouse Glass, New Orleans . 122194
Lighthouse Images, Saint Paul . 123919
Lighthouse Keepers Cottage,
 Carnarvon 000950
Lighthouse Rockart Gallery,
 Tewantin 099817
Lightner Museum, Saint
 Augustine 052745
Lights, Camera, Action United
 Kingdom, Nottingham 144760
Lightship Portsmouth Museum,
 Portsmouth Museums,
 Portsmouth 052325
Lightstorm Photo Gallery, Berry . 098743
Lightview Gallery, Baltimore .. 092339
Lightworks, Minneapolis 094926, 135831
Lightworks Magazine,
 Birmingham 139415
Ligija, Kaunas 085150
Lignan Art Publishing House,
 Guangzhou 136710
Ligne Art Sud, Montolieu 070285,
 104268
Ligne Claire, Mons 140253
Ligne Treize, Carouge 116198
Lignier, Guy, Boulogne-sur-Mer . 103450
Lignier, Guy, Hardelot-Plage ... 103807

Lignorama – Holz- und
 Werkzeugmuseum, Riedau ... 002596
Lignum, Prato 131778
Ligozat, André, Gap 068630
Ligüerre, Pablo, Zaragoza 133502
Liguori Editore, Napoli 137704
Ligures, Bordighera 138840
Lihula Muuseum, Lihula 010635
Liicht, Nei, Dudelange 111823
Liinelot, Helsinki 066334
Liisa's Antiikki, Helsinki 066335
Lijiang Museum, Lijiang 008012
Lijiang Publishing House, Guilin 136711
Lijiashan Bronze Museum,
 Jiangchuan 007931
Lijn 3, Den Haag 112398
Lik, Peter, Aspen 119528
Lik, Peter, La Jolla 121325
Lik, Peter, Noosa Heads 099461
Like Before, Cagliari 079988
Likeio Ton Ellinidon Collection,
 Rethymnon 023246
Likha Antiques and Crafts Inc.,
 Manila 084480
Liki Rossii, Sankt-Peterburg .. 137931
Likin, Sandanski 101135
Likovni Klub Scena Crnjanski,
 Beograd 114424
Lil-Tob-Lou, Los Angeles 094390
Lila Chrif, Casablanca 082569
Lila Laden, Bonn 075078
Lile & Lav, Beograd 114425
L'Ile au Trésor, Annecy 066644
L'Ile au Trésor, Saint-Dizier 072740
Lilfords, New Milton 118777
Lilian, Andrée, Riehen 116589
Lilian's Fine Collectables, Toronto 064692
Lilias Lane, Nelson 084018
Lilienhof, Lilienthal 020071
Liliesleaf Gallery, Rosebank 114811
Liling Museum, Liling 008013
Liliput, Köln 142018
Liljendal Hembygdsmuseum,
 Liljendal 011115
Liljevalchs Konsthall, Stockholm 041014
Lilla Galleriet, Sölvesborg 115819
Lilla Galleriet, Umeå 115967
Lilla Galleriet Helsingborg,
 Helsingborg 115686
Lilla Hönsahuset, Lund 086835
Lilla Huset Antikt, Skövde 086938
Lilla Konstgalleriet, Kalmar ... 115704
Det Lille Antikvariat, Århus ... 140555
Den Lille Butik, Fanø 065647
Det Lille Galleri, Haderslev ... 102821
Det Lille Galleri, Holstebro 102846
Det Lille Galleri, Kristiansand .. 113171
Det Lille Galleri, Vejle 066196
Lille Heddinge Rytterskole, Rødvig
 Stevns 010284
Lille Kunterbunt, Mitmach Museum
 Minden, Minden 020417
Lillebonne, Nancy .. 104323, 136840
Lillehammer Kunstmuseum,
 Lillehammer 033741
Lillesand By-og Sjøfartsmuseum,
 Lillesand 033745
Lillian and Albert Small Jewish
 Museum, Washington 054130
Lillian and Coleman Taube Museum of
 Art, Minot 050950
Lillian Stevenson Nursing Archives
 Museum, Saint John's 006904
Lillie Art Gallery, East Dunbartonshire
 Museums, Milngavie 045270
Lilliput, Pittsburgh 096304, 136132
Lilliput Antique Doll and Toy Museum,
 Brading 043703
Lillooet District Museum, Lillooet 006194
Lillstreet Art Center, Chicago ... 120181
Lilly's Art, Boston 119890
Lil's Treasure, Mulgoa 061927
Liltorp, Elin, Aalborg 065585
Lily White, Portland 096433

Lilydale Antique Centre, Lilydale 061737
Lily's Antiques, Maryborough,
 Queensland 061797
Lily's Antiques, Pittsburgh 096305
Lim Antiqua, Lucca ... 137663, 142701
Lim, Dae, Seoul 111685
Lim, Raywin, Whitianga 084205
Lim, Woo, Seoul 111686
Lima Hembygdsgård, Lima ... 040782
L'Image-en-Provence, Marseille .. 104185
Limardo, Buenos Aires 127526
Limassol District Museum,
 Lemesos 009426
Limassol Municipal Art Gallery,
 Lemesos 009427
Limb, R.R., Newark-on-Trent ... 090811
Limbažu Muzejs, Limbaži 030157
Limberlost State Historic Site,
 Geneva 049067
Limbrock, Gerd, Hemer 076488
Limbrock, J., Rhede 108491
Limburg, Den Haag 083099
Limburgs Miniatuurmuseum, Sint
 Geertruid 032872
Limburgs Museum, Venlo 032982
Limburgs Openluchtmuseum
 Eynderhoof, Nederweert Eind . 032663
Lime Tree Antiques, Hadlow ... 089404
Lime Tree Gallery, Stanhope ... 119140
Limelight Gallery, Surry Hills ... 099749
Limerick City Gallery of Art,
 Limerick 024956
Limerick Museum, Limerick ... 024957
Limesmuseum Aalen, Zweigmuseum
 des Archäologischen
 Landesmuseums Baden-
 Württemberg, Aalen 016449
Limfjordsmuseet, Løgstør 010222
Limhamns Auktionsbyrå,
 Trelleborg 126690
Limhamns Museum, Malmö ... 040827
Liming Fine Arts Gallery, Qingdao 102159
Limingan Kotiseutumuseo,
 Liminka 011119
Limited, London 138225
Limited Edition Contemporary
 Graphics, Pershore 138276
Limited Edition Graphics, London 118425
Limited Edition Mint, Philadelphia 096129
Limited Editions, Ashburton ... 143276
Limited Editions, Mobberley ... 090764,
 134854
LIMN Gallery, San Francisco .. 124328
Limon Heritage Museum, Limon 050281
Limongelli, Michelangelo, Torino 132157
Limpsfield Watercolours,
 Limpsfield .. 089826, 118113, 134557
Lin & Keng, Beijing 101974
Lin & Keng, Taipei 116928
Lin An-Tai Historical Home,
 Taipei 042523
Lin Art, Košice 085568
Lin Liu-Hsin Puppet Theatre Museum,
 Toa-Thiu-Thia Puppet Centre,
 Taipei 042524
Lin San Zhi Art Display Center,
 Zhujiang 008553
Lin Yuan Simplicity Art Museum,
 Yuchi 042605
Lin Yutang House, Taipei 042525
Lin, Anthony, Hong Kong 065077
Lin, Peter, Los Angeles 135729
Lin, Polly, Los Angeles 094391
Linan Museum, Hangzhou 007841
Linard, Eric, La Garde-Adhemar 103889
Linardi & Risso, Montevideo .. 144947
Linares, Palma de Mallorca ... 086358
Lina's Antiques, Fresno 093505
Linc Real Art, San Francisco .. 124329
Linchuan Museum, Linchuan ... 008014
Linckersdorff, München 077560
Linckersdorff, von, Berlin 074868
Linckh, Andrea, Bamberg 129440
Lincoln, Buffalo 092540

Lincoln Antique Mall, Chicago . . 092739
Lincoln Arts Council, Lincoln . . . 060567
Lincoln Cathedral Treasury,
 Lincoln 044831
Lincoln Children's Museum,
 Lincoln 050294
Lincoln County Historical Museum,
 Davenport 048157
Lincoln County Historical Society,
 Lincoln 050283
Lincoln County Historical Society
 Museum, North Platte 051614
Lincoln County Historical Society
 Museum of Pioneer History,
 Chandler 047523
Lincoln County Pioneer Museum,
 Hendricks 049445
Lincoln Home, Springfield 053452
Lincoln Homestead, Springfield . 053458
Lincoln Log Cabin, Lerna 050220
Lincoln Memorial Shrine,
 Redlands 052495
The Lincoln Museum, Fort
 Wayne 048896
Lincoln Parish Museum, Ruston 052712
Lincoln Park Historical Museum,
 Lincoln Park 050309
Lincoln School of Art and Design,
 University of Lincoln, Lincoln . 056548
The Lincoln-Tallman Restorations,
 Janesville 049793
Lincoln Tomb, Springfield 053453
Lincoln Train Museum,
 Gettysburg 049091
Lincoln, Ebrahimzadeh,
 Saarbrücken 078328
Lincolns House, Los Angeles . . . 094392
Lincoln's New Salem Historic Site,
 Petersburg 052008
Lincolnshire Fine Arts, Grimsby . 117891
Lincolnshire Road Transport Museum,
 Lincoln 044832
Lind, David, Louisville 121655
Lind, Gregory, San Francisco . . . 124330
Linda Durham Contemporary Art,
 Santa Fe 124466
Linda Gallery, Beijing 101975
Linda Gallery, Jakarta 109511
Linda Gallery, Shanghai 102192
Linda Gallery, Singapore 114561
Linda Loves Lace, San Antonio . 096990
Lindale Art Gallery, Paraparaumu 112987
Lindart, Birmingham 119806
Lindas, Mallow 079671
Linda's Antiques, Kinsale 079652,
 142609
Lindas Gamle Skap, Fauske . . . 084235
Lindås Skulemuseum, Isdalstø . 033682
Linda's Treasures, Dallas 093160
Lindberg & Persson, Björkö 143820
Lindberg & Persson, Täby 143979
Lindberg, B. & M., Helsinki 066336
Linde, Lübeck 107918
Linde Auktionstjänst, Lindesberg 126621
Linde-Schöpfli, Winterthur 087909
Linde, de, Aalst 063176
Lindemann, Dülmen 075442
Lindemann, Dorothea,
 Hildesheim 130047
Lindemann, Felix, Altenrhein . . . 116018
Lindemann, Oliver, Hamburg . . . 107292
Lindemann, Winfried,
 Schermbeck 078361
Lindemanns, H., Stuttgart 142399
Linden, Bilambil 098747
Linden, Tampa 124662
Linden – Saint Kilda Centre for
 Contemporary Arts, Saint Kilda,
 Victoria 001471
Linden Art Center, Birmingham . 119807
Linden House Antiques,
 Leominster 089808, 134548
Linden House Antiques, Stansted
 Mountfitchet 091414

Linden-Museum Stuttgart, Staatliches
 Museum für Völkerkunde,
 Stuttgart 022048
Linden Railroad Museum, Linden 050313
Linden, Peter, Blackrock 079459
Lindenau-Museum, Altenburg . . 016534
Lindenauer, Isak, San Francisco 097271
Lindenberg, Rotterdam 143221
Lindenfelser Museum, Lindenfels 020088
Linder, Fritz, Gars am Kamp . . . 062566
Linder, Fritz, Horn 062641
Linder, Gisèle, Basel 116088
Linder, Lars, Hedesunda 126597
Linder, Reinhard, Gummersbach 076129
Linderholm, Lena, Stockholm . . 115904
Lindesbergs Museum,
 Lindesberg 040784
Lindesnes Bygdemuseum, Sør-
 Audnedal 033957
Lindgaard Thomsen, Jimmy,
 Skagen 066116
Lindgren, Lisa, Dragør 065633
Lindholm, Göteborg 133529
Lindholm, Stelle 078556
Lindholm Høje Museet,
 Nørresundby 010239
Lindigtmühle am Lindenvorwerk,
 Kohren-Sahlis 019752
Lindinger & Schmid, Regensburg 137480
Lindisfarne Priory, Holy Island . . 144597
The Lindisfarne Room,
 Gateshead 117819
Lindkvist & Sjöberg, Stockholm . 087024
Lindleinturm-Museum,
 Creglingen 017777
Lindner, Stuttgart 078615
Lindner, Wien 100346
Lindner, Andreas, Wien 063019
Lindner, H., Eichstätt 141734
Lindner, Hans, Mainburg 142099
Lindner, Johannes, Radebeul . . 130548
Lindner, Konrad, Bernried, Kreis
 Deggendorf 129584
Lindner, Ramon, Essen 075689
Lindner, U., Callenberg 075242
Lindorf, Hans-Karl, Kiel 076780
Lindquist, Solveig, Helsinki 128414
Lindsay, Den Haag 143115
Lindsay, London 090301
Lindsay Gallery, Lindsay 006196
Lindsay Wildlife Museum, Walnut
 Creek 054044
Lindsay, Burns & Co., Perth . . . 127012
Lindsay, Duff, Columbus 120502
Lindsay, Lillian, Oberon 062023
Lindsay, Robert, Melbourne 099312
Lindsay's Antiques, Invercargill . 083968
Lindsell Art Gallery, Lindsell . . . 118120
Lindsey, Nashville 095011
Lindsey Antiques, North Berwick 090847
Lindsey Court Architectural,
 Horncastle 089585
Lindskoug, Mats, Lund 133549
Lindstrand-Bengtsson, Ginger,
 Mjölby 126642
Lindströms, Kenth, Stockholm . . 087025
The Line Gallery, Linlithgow . . . 118121
La Linea d'Ombra, Roma 142825
Linea, Marina, Wien 063020
LINEART, the Art Fair, Gent 098022
Lineaverde, Vomero 138841
Linecker, Leopold, Schalchen . . 127867
Lines, D. & J., Wychbold 091960,
 135265
Linfen Museum, Linfen 008016
Linfen Zone Museum, Linfen . . . 008017
Linga Longa, Tirau 113033
Lingao Museum, Lingao 008018
Lingard, Ann, Rye 091163
Lingbeek & Van Daalen, Haarlem 132520
Lingelsheim, Eckhard von,
 Lübeck 142087
Lingen, Axel, Recklinghausen . . 078156
Lingenauber, Eckart, Monaco . . 112007

Lingenauber, Eckart, Düsseldorf 075492
Lingier-Sermeus, Gent 063648, 100698
Lingnan Fine Arts Museum, Academia
 Sinica, Taipei 042526
Lingnan Hua Pai Memorial Hall,
 Guangzhou 007810
Lingner, K., Osterode am Harz . 142285
Lingshan Museum, Qinzhou . . . 008189
Lingtai Museum, Lingtai 008019
Lingtong Exhibition, Changzhou . 007684
Lingyuan Museum, Lingyuan . . . 008021
Lingyun Museum, Lingyun 008022
Linhai Museum, Linhai 008023
Linhart, Henrich, Košice 085569
Lini, Cristiano, Milano 131462
Linija-Art, Moskva 114275
Link, B., Buchen 075207
Link, E., Fridingen an der Donau 075906
Linka, Israël, Amsterdam 112245
Linke, Friedr.-A., Goslar 129915
Linke, Günter, Berlin . . 074869, 141550
Linke, Stefan, Düsseldorf 075493,
 129747
Linköpings Antikvariat, Linköping 143880
Linköpings Konsthall, Linköping 040787
Linköpings Slotts- och
 Domkyrkomuseum, Linköping . 040788
Linkowski, Giorgio A., Zürich . . . 116833
Linley Sambourne House,
 London 045021
Linli Museum, Linli 008024
Linlithgow Palace, Linlithgow . . 044839
Linlithgow Story, Linlithgow 044840
Linlithgow Union Canal Museum,
 Linlithgow 044841
Linn County Historical Museum and
 Moyer House, Brownsville 047279
Linn County Historical Society
 Museum, Cedar Rapids 047489
Linn County Museum, Pleasanton 052203
Linn-Sui, Torino 110984
Linnan Museo ja Ortodoksinen
 Kirkkomuseo, Olavinlinna 011205
Linnanmäen Museo, Helsinki . . 010844
Linnartz, H., Köln 076892
Linné, Uppsala 115972
Linnéas Antik, Motala 086874
Linneborn, Berlin 106362
Linnémuseet, Uppsala 041113
Linnenschmidt'sche Vormals
 Landesherrliche Mühle zu Venne,
 Ostercappeln 021025
Linnés Hammarby, Uppsala 041114
Linnés Råshult, Diö 040592
Linnet, Lotte, Århus . . 065603, 102776
Linnhoff, Birgit, Kiel 130119
Linnoitus- ja Kotiseutumuseo,
 Luumäki 011132
Lino, Surrey Hills, Sydney 136489,
 138496
Liño – Revista de Arte, Oviedo . 139062
Linossier, Gilles, Lyon 069756
Linossier, Gilles, Paris 071587
Linotype Museum, Schopfheim . 021721
Linqing Museum, Linqing 008025
Linquan Museum, Linquan 008026
Lins, Ursula, Schleswig 130621
Linschoten, J., Utrecht 083654
Linscott, King River 099192
Linsolas-Chaux, Bernard & Véronique,
 Bellevesvre 067079
Linsolite, Liège 063769, 140244
Linssen, Venlo 083683
Linssen, H.J.C., Venlo 083684
Linssen, P.H.A., Venlo . 083685, 083686
Lint, Stadl-Paura 127877
Lintel, Thomas von, New York . . 122815,
 138403
Lintner, Fritz, Spit Junction 062235
Lintner, Joachim M., Salzburg . . 062808
Linton & Shaw, Subiaco 099727
Linton and District Museum,
 Linton 001222
Linton Court Gallery, Settle 119067

Linton & Assoc., Richard,
 Frankston 099057
Lintong County Museum, Lintong 008027
Lintow, Joachim, Trebbin 078695
Linwood Community Arts Centre,
 Christchurch 112850
Linxiazhou Museum, Tiyuchang . 008335
Linyi Museum, Linyi, Shandong 008030
Linyi Museum, Linyi, Shanxi . . . 008032
Linyou Museum, Linyou 008033
Linz, Peer, Amsterdam 112246, 143046
Linzen, H.J., Heinsberg 076482
LinzGenesis, Linz 002325
Liohuo Aboriginal Stone-roofed Hut
 Museum, Neipu 042409
Lion and Eagle, San Antonio . . . 096991
Lion Antiques, Pershore 090973
Lion Bazaar, Wellington 084170
Lion Books, Bewdley 144245
Le Lion d'Argent, Lyon 069757
Lion d'Or, Rotterdam 083579
Lion Fine Arts, Hay-on-Wye 089479
Lion Gallery, Leominster 118103
Lion Gallery, Penrith 118888
The Lion Gallery and Bookshop,
 Knutsford . . . 089733, 118065, 134526,
 144311
Le Lion Qui Nous Regarde,
 Fontvieille 103750
Lion Saint Book's, Hereford 144473
Lion Salt Works, Northwich 045393
Lion, Roland, Gignac 068664
Lionel Wendt Art Gallery,
 Colombo 040477
Lionel, François, Nœux-les-Mines 070723
Liongoren Art Gallery, Quezon
 City 113420
Lionheart Gallery, Louisville 121656
Lion's Kingdom Museum,
 Toucheng 042578
The Lions Mark, Vancouver 064838
Lions-Trend, Bern 087288
Liova, Paris 071588, 071589
Lipan, Ljubljana 114720
LipanjePuntin, Roma 110811
LipanjePuntin, Trieste 111051
Lipeckij Oblastnoj Kraevedčeskij Muzej,
 Lipeck 037197
Lipicai Múzeum, Szilvásvárad . 023975
Lipitch, Michael, Barnet 088360
Lipitch, Peter, London 090302
Lipp, Karl M., München 137414
Lippen, Heike, Gladbeck 076004
Lippertsche Buchhandlung &
 Hallesches Antiquariat, Halle,
 Saale 141852
Lippische Gesellschaft für Kunst e.V.,
 Detmold 059164
Lippische Landesbibliothek,
 Detmold 017875
Lippischer Künstlerbund e.V,
 Detmold 059165
Lippisches Landesmuseum,
 Detmold 017876
Lippl, Alois, Viechtach 078794
Lippold, K., Roßlau 078279
LIPS Slotenmuseum, Dordrecht . 032247
Lipschitz, Cape Town 114769
LIPSIKUS Kinder- und Jugendmuseum,
 Stadtgeschichtliches Museum Leipzig,
 Leipzig 020000
Lipton, Honolulu 127247
Lipton, William, New York 095551
Liptovské Múzeum, Ružomberok 038438
Liquid Paper Correction Fluid Museum,
 Santa Fe 053111
Liquid Paper Museum, Nambe . . 051151
La Lira, Torino 081722
Lire Entre les Vignes, Paris 105040
Lire et Relire, Chomedey 140341
Liron Pean Reynald, Saint-Julien-sur-
 Calonne 072870
Lis Tes Ratures, Nantes 140987
Lis, Claire, Plaisance-du-Touch . 072172

Lis, Gilles, Perthes 072111
Lisa J, Kingswinford 118052
Lisabeth, P., Brugge 063281
Lisakovskij Muzej Istorii i Kultury
Verchnego Pritobolja, Lisakovsk 029826
Lisboa 20, Lisboa 114040
Lisbon Historical Society Museum,
Lisbon 050321
Lischa Line, Nice 070625
Lischka-Elfenspeich, Berlin 074870
Liselund Gamle Slot, Borre 010024
Lisht, Rio de Janeiro 064076
Lishui Museum, Lishui, Jiangsu 008037
Lishui Museum, Lishui, Zhejiang 008038
Lisio Arte della Séta, Firenze .. 055897
Liška, Praha 065506
Lisle Station Park, Lisle 050323
Lismed, Lisboa 126448
Lismore Regional Gallery,
Lismore 001223
Liss, Toronto 101652
Liss Ohayon, Josephine, Toronto 064693
Liss, L., Montreuil-sous-Bois ... 070327
Lissauer, H.M., Elwood 061445
Lisse, Heinsberg 107410
Lisson Gallery, London 118426
The List, Edinburgh 139200
List & Francke, Meersburg 142121
List Gallery, Swarthmore College,
Swarthmore 053623
List-Petersen, Husum 076590
List Visual Arts Center, Massachusetts
Institute of Technology,
Cambridge 047364
List, Monika, Stuttgart 130681
List, Thomas, Berlin 129528
Lista Museum, Vanse 034068
Listafelag Føroya, Tórshavn 024063
Listaháskóli Íslands, Reykjavík . 055773
Listasafn Ásí, Ásmundarsalur,
Reykjavík 024097
Listasafn Einars Jónssonar,
Reykjavík 024098
Listasafn Íslands, Reykjavík ... 024099
Listasafn Kópavogs – Gerðasafn,
Kópavogur 024087
Listasafn Sigurjóns Ólafssonar,
Reykjavík 024100
Listasafnið á Akureyri, Akureyri . 024063
Listasavn Føroya, Tórshavn 010743
LISTE-The Young Art Fair, Basel 098331
Lister, West Perth 062422
Lister Calder Gallery, Subíaco .. 099728
Liszt Ferenc Emlékmúzeum és
Kutatóközpont, Budapest 023507
Liszt-Museum, Klassik Stiftung
Weimar, Weimar 022500
Lit, Oklahoma City 123287
Lit Sing Kwang Chinese Temple,
Innisfail 001166
Lit Verlag, Münster 137437
Litarartury Muzej Janki Kupaly,
Minsk 003249
Litcham Village Museum,
Litcham 044846
Litchfield, Saratoga 099678
Litchfield County Auctions,
Litchfield 127286
Litchfield Historical Society Museum,
Litchfield 050324
Literair Museum, Hasselt 003680
Literarisches Museum,
Badenweiler 016930
Literarisches Museum, Städtisches
Museum im Kornhaus, Kirchheim
unter Teck 019626
Literárna Expozícia – Múzeum Janka
Jesenského, Mestské Múzeum v
Bratislave, Bratislava 038313
Literárne a Hudobné Múzeum, State
Scientific Library Banská Bystrica,
Banská Bystrica 038278
Literárne a Vlastivedné Múzeum,
Svätý Jur 038460

Literárne Múzeum Pavla Horova,
Bánovce nad Ondavou 038277
Literární Oddělení, Moravské zemské
muzeum, Brno 009470
Literary Documentation Centre H.B.
Yasin, Jakarta 024564
Literary History Museum, Minsk 003250
Literate Veteran, Cleveland 145086
Literaturarchiv Sulzbach-Rosenberg,
Sulzbach-Rosenberg 022083
Literaturhaus, Magdeburg 020209
Literaturhaus München,
München 020543
Literatūrinis Aleksandro Puškino
Muziejus, Vilnius 030532
Literaturladen, Anbach 141408
Literaturmuseum Altaussee,
Altaussee 001700
Literaturmuseum Brecht-Weigel-Haus,
Buckow 017657
Literaturmuseum der Moderne,
Deutsches Literaturarchiv Marbach,
Marbach am Neckar 020260
Literaturmuseum im Baumbachhaus,
Meininger Museen, Meiningen 020341
Literaturmuseum im Gartenhaus der
Stadtbibliothek Baden-Baden, Baden-
Baden 016924
Literaturmuseum Theodor Storm,
Heiligenstadt, Heilbad 019171
Literaturno-chudožestvennyj Muzej
M. Cvetaevoj, Nacionalnyj muzej
respubliki Baškortostan, Usen-
Ivanovskoe 038005
Literaturno-chudožestvennyj Muzej-
usadba Prijutino, Vsevoložsk .. 038073
Literaturno-Memorial'nyj Muzej N.A
Ostrovskogo, Šepetovka 043289
Literaturno-Memorialny Muzej
A.A. Achmatovoï, Slobidka
Šelechivs'ka 043302
Literaturno-memorial'nyj Budinok-
muzej M. Bulgakova, Kyïv 043201
Literaturno-memorial'nyj Dom-Muzej
D.N. Mamina-Sibirjaka, Municipalnyj
objedinennyj muzej pisatelej Urala,
Ekaterinburg 036980
Literaturno-memorialnyj Dom-Muzej
F.M. Rešetnikova, Municipalnyj
objedinennyj muzej pisatelej Urala,
Ekaterinburg 036981
Literaturno-memorialnyj Dom-muzej
Sulejmana Stalskogo, Ašaga-
Stalsk 036869
Literaturno-memorialnyj Muzej
O.S. Puškina ta P.I. Čajkovskogo,
Kam'jans'kyj deržavnyj
istoriko-kul'turnyj zapovidnyk,
Kam'janka 043171
Literaturno-memorialnyj Muzej F.A.
Abramova, Verkola 038018
Literaturno-memorialnyj Muzej F.M.
Dostoevskogo, Novokuzneck .. 037574
Literaturno-memorialnyj Muzej F.M.
Dostoevskogo, Sankt-Peterburg 037750
Literaturno-memorialnyj Muzej F.M.
Dostoevskogo, Semey 029843
Literaturno-memorialnyj Muzej
Vasilija Erošenko, Starooskolskij
kraevedčeskij muzej, Staryj
Oskol 037894
Literaturno-memorialnyj Muzej
A.M. Gorkogo, Samara 037710
Literaturno-memorialnyj Muzej
A.M. Gorkogo, Nacionalnyj Muzej
Respubliki Tatarstan, Kazan ... 037104
Literaturno-memorialnyj Muzej Ivana
Franko, Lviv 043238
Literaturno-Memorial'nyj Muzej Lesi
Ukrainky, Volynskoj Krajeznavčyj
Muzej, Kolodjažne 043185
Literaturno-Memorial'nyj Muzej M.
Eminesku, Černivec'kyj krajeznavčyj
muzej, Černivci 043130

Literaturno-memorial'nyj Muzej
Puškina, Odes'kyj literaturnyj muzej,
Odesa 043269
Literaturno-memorialnyj Muzej
Stancionnyj Smotritel, Vyra ... 038077
Literaturno-memorial'nyj Muzej T.G.
Ševčenka, Ševčenkove 043298
Literaturno-memorialnyj Muzej M.M.
Zoščenko, Sankt-Peterburg ... 037751
Literaturno-muzykalnyj Muzej Dača
Šaljapina, Kislovodsk 037140
Literaturno-mystec'kyj Muzej,
Chmel'nyc'kyj 043143
Literaturno-teatralnyj Muzej N.M.
Djakonova, Syktyvkar 037910
Literaturnyj Memorialnyj Dom-muzej
Abaja, Semey 029844
Literaturnyj Muzej Abakumcevo,
Gosudarstvennyj literaturno-
memorialnyj muzej-zapovednik
N.A. Nekrasova "Karabicha",
Abakumcevo 036839
Literaturnyj Muzej Goroda Nalčik,
Nacionalnyj Muzej Kabardino-
Balkarskoj Respubliki, Nalčik . 037533
Literaturnyj Muzej im. K.V. Ivanova,
Čeboksary 036931
Literaturnyj Muzej Instituta Russkoj
Literatury, Rossijskaja Akademija
Nauk, Sankt-Peterburg 037752
Literaturnyj Muzej Literaturna
Chersonščina, Oblasnyyj krajeznavčyj
muzej, Cherson 043138
Literaturnyj Muzej Maksima
Bogdanoviča, Minsk 003251
Literaturnyj Muzej F.I. Tjutčeva,
Brjanskij gosudarstvennyj
obedinennyj kraevedčeskij muzej,
Ovstug 037616
Litho Galerie, Amsterdam 112247
Lithographie, Paris 105041
Lithographiesteinarchiv und
druckhistorische Werkstätte,
München 020544
Lithos, Saint-Restitut 105614
Lithos, Venezia 132247
Lithuanian Artists' Association,
Vilnius 059553
The Lithuanian Museum, Chicago 047670
Lithuanian Museum-Archives of
Canada, Mississauga 006342
Lithuanian Museums Association,
Vilnius 059554
Litografia Tucano, Rio de Janeiro 101047,
101048
Litra, Paris 141223
Litt, David, Bedford 088424
Litterer, R.& H., Wiesenbach .. 079001
Little Africa, New York 095541
Little Antique Shop, Nashville .. 095012
Little Antique Shop, New York .. 095542
Little Bighorn Battlefield Museum,
Crow Agency 048076
Little Bits, Gisborne 083941
The Little Brick House, Vandalia 053949
Little Claybelt Homesteaders Museum,
New Liskeard 006468
Little Collectables, Congleton . 088967
Little Compton Historical Society
Museum, Little Compton 050328
Little Cowboy Bar and Museum,
Fromberg 048988
Little English, Memphis 094645
Little Falls Historical Museum, Little
Falls 050333
The Little Gallery, Hamilton ... 100827
Little Gallery, Hawkes Bay 112915
The Little Gallery, Johannesburg 114797
The Little Gallery, Pittenweem . 091017
The Little Gallery, Saint-Émilion . 105508

Little Gallery, Singapore 114562
The Little Gallery, Tetbury 119222
Little Gem, Shrewsbury 091330, 135045
Little Holland House, London ... 045022
Little House Heirloom Gallery,
Birmingham 092418
Little House on the Parking Lot,
Toronto 064694
Little Houses, Dunkeld 044177
Little Jem's, Penzance 090966
Little Known Gallery, Chicago .. 120182
Little Ladies, Chicago 092740
Little Lane Books, Camden 139682
Little London Gallery, Holloway . 117976
Little Lu, San Francisco 097272
Little Malop Gallery, Geelong .. 099071
Little Moreton Hall, Congleton .. 044000
Little Museum, Tewkesbury 045964
Little Nemo, Torino ... 110985, 142865
Little Norway, Blue Mounds ... 047058
Little Paddocks Art Gallery,
Timaru 113029
Little Paris Antiques, Los Angeles 094393
Little Picture Gallery.Com,
Greenford 117885
Little Prairie Heritage Society,
Chetwynd 005613
Little Red School House Museum,
Weyauwega 054359
Little Red Schoolhouse-Living Library,
Beloit 046903
Little River Gallery, Christchurch 112851
Little School Museum, Lockeport 006206
Little Shop of Crafts, New York . 095543
Little Stone School, Saskatoon . 006975
The Little Strawberry Gallery,
Bristol 117449
Little Studio, San Antonio 124046
Little Summerhouse, Miami 135799
Little Sweets Museum, Hirano
Hakabutsukan, Osaka 029247
Little Theatre Antiques Centre,
Jesmond 089659, 134507
Little Things, Las Vegas 094160
Little Traverse Historical Museum,
Petoskey 052021
Little Treasures, Caulfield North 061304
Little Treasures, Dornoch 089099
Little Treasures, Tshwane 085678
Little van Gogh, Bad Honnef ... 074530
Little Van Gogh, Bruxelles 100636
Little White House, Warm
Springs 054057
Little White Schoolhouse, Ripon 052579
Little White Schoolhouse, Truro . 007253
Little World Museum of Man,
Inuyama 028691
Littlebury Antiques, Saffron
Walden 091170, 134986
Littlehampton Museum,
Littlehampton 044848
Littlejohn Contemporary, New
York 122816
Littleton & Hennessy, London .. 090303
Littleton & Hennessy, New York 095544
Littleton Historical Museum,
Littleton 050340
Littleton, Maurine, Washington . 124929
Littman Gallery, Portland State
University, Portland 052295
Littmann, Klaus, Basel 116089
Litvich, Judith, San Francisco .. 124331
Litvin, Salt Lake City 096900
Litwin, Fred, Washington 097843
Litwinschuh, Manfred & Renate,
Weiskirchen 078914
Liudo Truikio ir Marijonos
Rakauskaitès Memorialinis Muziejus,
Nacionalinis M.K. Čiurlionio Dailès
Muziejus, Kaunas 030367
Liuhe County Display Center,
Liucheng 008039
Liuksialan Maatalousmuseo,
Kangasala 010954

Liuligongfang, Taipei 116929
Liuzhou Museum, Liuzhou 008041
Liv An Amzer, Ploudalmézeau .. 072196
Livadijskij Dvorec, Jalta 043165
Livani, Athinai 137574
Livapaşa Konağı Etnografya Müzesi,
Kastamonu 042988
Livarot Troc, Livarot 069601
Live-Art, Lippstadt 077133
Livellini, Luigi, Verona 132287
Liven, Isai, Köln 076893, 130156
Livengood, Los Angeles 094394
Liverpool Academy of Art,
Liverpool 118130, 118131
Liverpool School of Art and
Design, Art and Design Academy,
Liverpool 056550
Liverpool Scottish Regimental
Museum, Liverpool 044857
Liverpool Street Gallery, Surry
Hills 099750
Liverpool University Press,
Liverpool 138179
Livesey Museum for Children,
London 045023
Livesey, Scott, Armadale, Victoria 061038
Livia de Alencar Chaucs, Barcellos,
Porto Alegre 100986
Livinec, Gauducheau & Jézéquel,
Rennes (Ille-et-Vilaine) 125786
Livinec, Gauducheau et Jezequel,
Rennes (Ille-et-Vilaine) 125787
Living, Waldenburg,
Hohenlohekreis 078836
Living Art Gallery, Chippendale . 098887
Living Art Gallery, Portland 123641,
127410
The Living Arts and Science Center,
Lexington 050247
Living Arts of Tulsa, Tulsa 124807
Living Desert Art Gallery, Broken
Hill 098799
Living History Farms Museum,
Urbandale 053911
Living In The Past, Ballarat 061087
The Living Museum, Springfield 139416
Living Prairie Museum, Winnipeg 007449
The living room, Amsterdam 112248
The Living Room, New Orleans . 122195
The Living Store, Salzburg 062809
Living with Art, Amsterdam 138933
Living Word National Bible Museum,
Aledo 046377
Livingroom Galerie, Besigheim . 106476
Livingston County Historical Society
Museum, Geneseo 049065
Livingston Depot Center,
Livingston 050344
Livingston, Thomas, San
Francisco 097273
Livingstone, Den Haag 112399
Livingstone and Stanley Memorial,
Tabora 042639
Livingstone Museum, National
Museum of Zambia, Livingstone 054836
Livingstone, Neil, Dundee 089115
Livonijas Ordeņa Pils Tornis,
Tukums 030240
Livraghi, Pietro, Milano 131463
Livraria Académica, Porto 143604
Livraria Alfarrabista, Coimbra . 143550
Livraria Alfarrabista, Lisboa 143571
Livraria Amadeu, Belo Horizonte 140267,
140268
Livraria Antiquária do Calhariz,
Lisboa 143572
Livraria Antiquario, Fortaleza .. 140280
Livraria Antiquario, Rio de
Janeiro 140294
Livraria Artes e Letras, Lisboa . 143573
Livraria AZ do Livro, Lisboa 143574
Livraria Barateira, Lisboa 143575
Livraria Bizantina, Lisboa 143576
Livraria Brasileira, Rio de Janeiro 140295

A Livraria do Messias, São Paulo 140306
Livraria Dom Pedro V, Lisboa .. 143577
Livraria e Sebo Horizonte, Belo
Horizonte 140269
Livraria Elcla Boavista, Porto ... 143605
Livraria Histórica e Ultramarina,
Lisboa 143578
Livraria João do Rio, Rio de
Janeiro 140296
Livraria Leonardo, Rio de Janeiro 140297
Livraria Maria de Historias, Rio de
Janeiro 140298
Livraria Mosaico, Interior de Minas
Gerais 140285
Livraria Olisipo, Lisboa 143579
Livraria Pindorama, Brasilia ... 140275
Livraria Santo Piamarta,
Fortaleza 140281
Le Livre, Salleboeuf 141342
Livre à Lili, Lyon 140922
Livre Classique, Saint-Pierre-de-
Clages 144124
Livre Comme l'Air, Nantes 140988
Livre et Passion, Redu 140259
Livre Mon Ami, Paris 141224
Le Livre Penseur, Auray 140769
Livres d'Antan, Paris 141225
Livres d'Autrefois, Angoulême .. 140765
Livres Sans Frontières, Paris .. 141226
Livresse, Feytiat 140836
Livrustkammaren, Stockholm .. 041015
Liwan, Casablanca 082570
Lixia Qidong Art, Jinan 102139
Lixin, Hang, Singapore 085461
Lizabeth Oliveria Gallery, Los
Angeles 121535
Lizan Tops, New York 122817
Lizana, A., Santiago de Chile .. 064973
Lizardi-Harp, West Hollywood .. 125010
Lizauli Cultural Village, Lizauli .. 031842
Lizène, Jacques, Liège 136661
Lizhor Gallery, Bogotá 102346
Liznjekova Hiša, Gornjesavski Muzej
Jesenice, Kranjska Gora 038534
Lizzadro Museum of Lapidary Art,
Elmhurst 048563
Lizzani, Saverio, Roma 110812
Lizzi, Enrico, Merano 080354
Lizzie's Place, South Brisbane .. 062221
Ljósmyndasafn Reykjavíkur Grófarhús,
Reykjavík 024101
Ljudmilova, Cvetanka, Gorna
Orjachovica 101126
Ljungbergmuseet, Ljungby 040791
Ljungby Museum, Ljungby 040792
Ljungsbro Antik, Ljungsbro 086821
Ljusdalsbygdens Museum,
Ljusdal 040793
Ljusterö Hembygdsmuseum,
Ljusterö 040795
LKAB Gruvmuseum, Kiruna 040748
LL Worldwide Art Industry,
Amsterdam 112249
LLAC Sala d'Exposicions,
Tarragona 086484, 143798
Lladó Collectors Society, New
York 060568
Lladro Museum, New York 051416
Llanberis Lake Railway, Llanberis 044872
Llancaiach Fawr Living History
Museum, Nelson 045319
Llandudno Museum and Art Gallery,
Llandudno 044878
Llangollen Motor Museum,
Llangollen 044889
Llangollen Railway, Llangollen .. 044890
Llanidloes Museum, Llanidloes .. 044892
Llanishen Antiques, Cardiff 088788
Llarenas, Las Palmas de Gran
Canaria 086375
Llechwedd Slate Caverns, Blaenau
Ffestiniog 043648
Lleo i Gall, Céret 103549
Lleonart Claraso, Jaime,

Barcelona 114926
Llewellyn, Alexander, London .. 118427
Lleyn Historical and Maritime Museum,
Nefyn 045317
Llibrería Al Tossal, Valencia 143807
Llibreria Antiquària Catedral,
Tarragona 143799
Llibreria Antiquària Els Gnoms,
Sedó 143795
Llibreria Gabernet, Barcelona .. 143727
Lloan, Janine, Montréjeau 070321
Llobet Germans, Barcelona 133128
La Llonja, Palma de Mallorca .. 039887
Llorenc Coll Gelado, Girona 115081
Llorente, Madrid 086209
La Llorona Art Gallery, Chicago . 120183
Lloyd, Eastport 127212
Lloyd-Davies, Sue, Carmarthen . 144316
Lloyd George Museum, Cricciert 044042
Lloyd Jones, Walter, Barmouth . 126808
Lloyd Jones, Walter, Dolgellau . 126876
Lloyd, John, Ditchling Common . 134325
Lloyd, Michael, Llandysul 089867
Lloyd, Robin, Ross-on-Wye 091140,
134980
Lloyd's of Kew, Kew 144502
Lluch Fluxa Rossello, Maria de, Palma
de Mallorca 115357
Llwyngwril Gallery, Llwyngwril . 118140
Llywernog Silver-Lead Mine Museum,
Ponterwyd 045520
LM, Lyon 104115
LM Escritório de Arte, Fortaleza 100953
LMAK Projects, New York 122818
LMAN, Los Angeles 121536
LMB Arts, London 118428
LMS Galerija, Rīga 111759
Lo, Napoli 110535
Lo, Saint Louis 096728
Lo Biondo, Salvatore, Palermo . 081050
Lo Scrigno, Trieste 081795
Lo, Fung, Hong Kong 102094
Loango, Strasbourg 105693
Loas, Nice 104376
Lobasso, Giuseppe, Milano ... 131464
Lobato Ranger, Oscar, Barcelona 085792
Lobaumuseum, Wien 003032
Lobdengau-Museum, Ladenburg 019845
Lobel, New York 095545
Lobel, Christian, Nice 104377
Lober, Valette Thierry, Limoges . 069524
Lobera Amann, Jose, Bilbao 133195
Lobis, Klaipėda 142962
L'Objet, Noirmoutier-en-l'Ile .. 104414
Lobkovicky Palác, National Museum,
Praha 009806
Lobo, Albuquerque 092065
Local and Maritime Museum,
Newhaven 045356
Local Antiquities Museum,
Chitradurga 024202
Local Art Museum, Pingtung ... 042410
Local Color Gallery, Nashville .. 122089
Local Colors Artworks, Salt Lake
City 123971
Local Cultural Museum, Lujhu .. 042388
Local Cultural Museum, Keelung
Municipal Cultural Center,
Keelung 042372
Local Images Art Gallery, Penola 099559
Local Museum, Ashkelon 025020
Local Museum, Bhanpura 024161
Local Museum, Kibbutz Maabarot 025139
Local Museum, Kibbutz Sasa .. 025145
Local Studies Collection @ Library,
Marrickville 001256
Local_30, Warszawa 113860
Localbahnmuseum, Bayerisch
Eisenstein 016980
Localbahnmuseum, Tiroler Museums
Bahnen, Innsbruck 002129
Locatelli, Alain, Lyon 069758
Locci, Luciano, Roma 131960
Loch & Mayberry, Winnipeg ... 101881

The Loch Ewe Gallery, Bridport . 117429
Loch Fyne Gallery, Tarbert 119205
Loch Gallery, Toronto 101653
Loch Ness 2000 Exhibition,
Drumnadrochit 044143
Loch, Silvia, Eschwege 141747
Lochaber, Fort William 117806
Locher, René, Horw 133772
Lochhead, Ken, Haddington .. 117899
Lochmann, Hans-Rudolf, Neustadt an
der Weinstraße 077760
Lochtenberg, L.M., Hilversum .. 083403
Lociuro, Alberto, Roma 081427
Lock Museum of America,
Terryville 053722
Lock-up Museum and John Paynter
Gallery, Newcastle Historic Reserve
Trust, Newcastle 001353
Lock jr., John H., Philadelphia .. 096130
Lock, Wilma, Sankt Gallen ... 116614
Locke & England, Leamington
Spa 126937
Locke Street Antiques, Hamilton 064323
Locke & Assoc., Suzy, Oakland . 123256
Lockehaven Schoolhouse Museum,
Enfield 048586
Locke's Distillery Museum,
Kilbeggan 024935
Lockhouse-Friends of the Delaware
Canal, New Hope 051272
Lockmaster's House Museum,
Elgin 005777
Lockner, Hermann P., Würzburg 079084
Lockport Gallery, Illinois State
Museum, Lockport 050353
Locks, Marian, Philadelphia ... 123411
Locksmith's House, Willenhall Lock
Museum, Willenhall 046136
Lockwood-Mathews Mansion Museum,
Norwalk 051644
Loco Chingo Studio, Dallas 120624
Lococo Fine Art, Saint Louis ... 138449
Locomotion – The National Railway
Museum at Shildon, Shildon .. 045787
Locus Gallery, Saint Louis 123839
Locus Solus, Napoli 110536
Locus Solus, Paris 105042
Locust Grove, Louisville 050494
Locust Grove, Samuel Morse Historic
Site, Poughkeepsie 052339
Locust Lawn and Terwilliger House,
Gardiner 049047
Loder, York 062509
Lodewijk, Eindhoven 083222
Lodewijk, M.J., Den Haag 083100
Lodge Antiques, Robe 062154
Lodge Antiques, Rosedale 062168
Lodge, Fenny, Bletchley 117391
Lodi, Roberto, Modena 080802
Lodi jr., Silvano, Milano 080589
Lodiana, Nice 070626
Lodrini, Rosa, Brescia 079942
Łódzkie Towarzystwo Fotograficzne,
Łódź 059698
Loe-Van, Singapore .. 114563, 114564
Loeb-Larocque, Béatrice, Paris . 141227
Loeb, Albert, Paris 105043
Löb, Uwe, Worms 142506
Löber, Alois, Faulbach 075729
Löbig, Richard, Dieburg 129681
Löblein, Peter, Bamberg 074667
Loecher, Roger P., Denver 093309
Loeckx, Gent 063649, 125335
Loeding, Dominique, Hamburg . 129973
Lödöse Museum, Lödöse 040796
Loeffel, Tobias, Basel . 116090, 133651
Löffelsend, Rolf & Angelika,
Düsseldorf 129748
Löffler, Maria, Achern 074352
Löfstad Slott Museum, Östergötlands
Museum, Norrköping 040878
Löhe-Zeit-Museum,
Neuendettelsau 020698
Löhle, Fritz, Uhldingen-

Mühlhofen 078752
Loehner, Gauting 141824
Löhr, H., Alfter 074375
Löhrl, Christa & Dietmar,
Mönchengladbach 108018
Loek Kok, Haarlem 083338
Løken Truls Restaureringatelier,
Hamar 132777
Løkken 6, Stavanger 084371
Løkken Antik, Løkken 065994
Løkken Museum, Løkken 010224
Lönnströmin Taidemuseo, The Teresia
and Rafael Lönnström Foundation,
Rauma 011304
Løonberg Høpfner, Torben & Bente,
Roskilde 066091
Loeper, Gabriele von, Hamburg . 107293
Loerakker, Amsterdam 112250
Lörcher, Margot, Aystetten 141422
Lørenskog Bygdemuseum,
Skårer 033938
Loertscher & Partner, Thomas,
Spiegel 087824
Loes & Reinier, International Ceramics,
Deventer 112429
Loës, Diane de, Genève 087479
Lösch, Erich, Monsheim 126038
Löschenkohl, Helmut, Gschwendt 062619
Loest, Robert, Schwerin 142357
Løten Bankmuseum, Løten 033750
Lötschentaler Museum, Kippel . 041615
Loetscher, Marc-André & Serge,
Biel 087319
Lövångers Sockenmuseum,
Lövånger 040797
Lövei, Jánosné, Miskolc 142579
Loevenbruck, Hervé, Paris 105044
Löw, Seltisberg 087810
Löwe, Edmund, München 077561
Loewe, Oskar, Recklinghausen . 142319
Löwel, Stockach 078562
Löwen-Drogerie und Museum,
Oelsnitz 020960
Löwenburg, Museumslandschaft
Hessen Kassel, Kassel 019538
Löwenpalais, Stiftung Starke,
Berlin 017215
Löwenstein, Dr., Göllersdorf . . 139983
Loewinstein, Frankfurt am Main 075805,
075806
Lofanta, Kaunas 142961
Lofoten Krigsminnemuseum,
Svolvær 034005
Lofotmuseet, Kabelvåg Fiskerimuseum,
Kabelvåg 033690
Loft, Cincinnati 120354
Loft Antiques, Ballyboughhal . . 079448
Loft Antiques, Minneapolis 094927
Loft Arte, Valdagno 111070
The Loft Frames & Gallery,
Clonakilty 109550
Loft Gallery, Hillsborough 117970
Loftet Antikviteter, Oslo 084326
Loftus-Potter, Nicole, Deal 089036
Log Cabin Antiques, Austin . . . 092247
Log Cabin Museum, Murray
Harbour 006439
Log Cabin Museum, Saanich Pioneer
Society, Saanichton 006836
Log Cabin Train Shop, Pittsburgh 096306
Log Cabin Village, Fort Worth . . 048905
The Log Farm, Nepean 006461
Log House Museum, Columbiana 047922
Log of Mystic Seaport, Mystic . . 139417
Logan, Ottawa 046507
Logan, Woodquay 109710
Logan Art Gallery, Logan 001228
Logan County Historical Society
Museum, Bellefontaine 046878
Logan County Museum, Paris . . 051915
Logan Hall, Natural Resources
Canada's Geological Survey of
Canada, Ottawa 006593
Logan Museum of Anthropology,

Beloit 046905
Logan, Martin & Mo, Ashburton 088308
Loganberry Books, Cleveland . . 145087
Logans, Washington 097844
Logello, Allemand, Montfavet . . 070242
Logge dell'Orafo, San Marino . . 114382
La Loggia, Brescia 079943
La Loggia, Lugano . . 087618, 133824
La Loggia, Sansepolcro 087797
La Loggia, Udine 111064
Logging Museum, Silver Lake Forestry
Centre, Peachland 066619
Logie Steading Art Gallery,
Dunphail 117699
Logisoft, Oslo 126375
Loglio, Ivano, Torino 132158
Logos, Warszawa 143527
Loh, Lüneburg 142093
Lohaus, Stella, Antwerpen 100502
Lohéac, Landerneau 069163
Lohgerber-, Stadt- und Kreismuseum,
Dippoldiswalde 017908
Lohjan Museo, Lohja 011122
Lohl, Dieter, Göttingen 107146
Lohmann, A., Emsdetten 129793
Lohmann, Günther, Halle /
Westfalen 141847
Lohmann, Michael, Timmendorfer
Strand 078681, 078682
Lohner, A., Inchenhofen 076606
Thomas Lohr & Martina Hahn,
Thomas, Mainz 077239
Lohrbacher Heimatstuben,
Mosbach 020472
Lohri, Andreas, Zug 134006
Lohse, Henri, Berlin 129529
Lohtajan Kotiseutumuseo,
Lohtaja 011124
Loi, Cristina, Cagliari 079989
Loïc, Monnier, Chinon 067979
Loidl, Matthias, Unterreit 137544,
142445
Loimaan Kotiseutumuseo,
Loimaa 011125
Lois E. Woods Museum, Norfolk 051577
Loiseau, Pornichet 072281
Loiseau & Schmitz, Saint-Germain-en-
Laye 125803
L'Oiseau Bleu, Pont-Aven 105382
Loisium, Langenlois 002264
Loison, Savasse 073447
Loizillon, Dominique, Compiègne 125560
Łój, Bogusław, Warszawa 132889
Loja da Calçada, Lisboa 084906
Loja das Colecções, Lisboa 143580
Loja do Atalaia, Lisboa 084907, 084908
Loja do Colecionador, Rio de
Janeiro 064077
Lojas de Antiguidades Vila Real, São
Paulo 064189
Lojkásková, Věra, Český Těšín . 065375,
140466
Lokaal 01, Antwerpen 100503
Lokaal 01, Breda 112355
Lokalhistorisches Pulkautalmuseum,
Hadres 002038
De Lokeend, Breda 082997
De Lokeend, Venlo 083687
Lokpark Ampflwang,
Oberösterreichsiches Eisenbahn- und
Bergbaumuseum, Ampflwang im
Hausruckwald 001710
Lokpark Brosigstraße,
Braunschweig 017561
Lokstallet Strömstad Konsthall,
Grebbestad 040671
Lokwelt Freilassing, Freilassing . 018494
Lola Gallery, San Francisco . . . 124332
Loldia, Constanţa 085118
Loleo, Mauro, Genova 131321
Loliée, Bernard, Paris 141228
L'Olifant Jeannoutot et Fils, Paris 129070
Lolita's Art Garden, London . . . 118429
Lolland-Falsters Stiftsmuseum,

Maribo 010228
Lolland-Falsters Traktor- og
Motormuseum, Eskildstrup . . . 010047
Lolli, Dino, Prato 081171
Lollichon, François, Fay-de-
Bretagne 068515
Lollipop, San Francisco 097274
Lom Bygdamuseum, Museet i Nord-
Gudbrandsdalen, Lom 033752
Lomakovskij Muzej Starinnych
Avtomobilej i Motociklov,
Moskva 037318
Lombard, Stepney 099719
Lombard Antiques, Newark-on-
Trent 090812
Lombard Historical Museum,
Lombard 050360
Lombard Korona, Saratov 085278
Lombardi, Napoli 142766
Lombardi, Alessandro, Roma . . 081428
Lombardi, Enrico, Roma 110813
Lombardo & Partners Antiques,
Rivoli 081187
Lombart, Marcel, Le Havre . . . 069285,
069286
Lombrail-Teucquam, Paris 125734
Lometsch, Kassel 141980
Lomillo, Florentino, Burgos . . . 133212
Lomma Antik och Kuriosa,
Lomma 086823
Lommedalsbanen, Rykkinn . . . 033911
Lomonds Gallery, Gateside . . . 117821
Lomonosov, Deventer 143127
Lompoc Museum, Lompoc 050362
Lon C. Hill Home, Rio Grande Valley
Museum, Harlingen 049360
London Antique Gallery, London 090304,
134713
London Antique Shop, Auckland 083803
London Antiquités, La Baule-
Escoublac 068951
London Antiquités, Saint-Nazaire
(Loire-Atlantique) 072990
London Art Fair, London 098386
London Borough of Bromley Museum,
Orpington 045442
London Brass Rubbing Centre,
London 045024
The London Brass Rubbing Centre in
Washington D.C., Gaithersburg 049012
London Canal Museum, London 045025
The London Cigarette Card,
Somerton 127048
London Coin, London 064351
London College of Communication,
University of the Arts London,
London 056561
London College of Fashion, University
of the Arts London, London . . 056562
London Federation of Museums and
Art Galleries, London 060173
London Film School, London . . 056563
London Fine Arts, Singapore . . 114565
London Fire Brigade Museum,
London 045026
London Frameworks, Dubai . . . 117182
London Gallery, Bruxelles 063495
London Gallery, Detroit 093365
London Gallery, Tokyo 111462
London House, Harrogate 089432,
134437
London House Antique Centre,
Moreton-in-Marsh 090783
London House Antiques,
Westerham 091799, 135211
London House Oriental Rugs and Carpets,
Boston Spa 088522, 134127
London Map Fair, London 098387
London Market, Lyon 069759
London Militaria Market, London 090305
London Motorcycle Museum,
Greenford 044480
London Museum of Archaeology,

London 006212
London Museums and Galleries
Marketing Association, London 060174
London Original Print Fair,
London 098388
London Picture Centre, London . 118430
London Regional Children's Museum,
London 006213
London Scottish Regimental Museum,
London 045027
London Sewing Machine Museum,
London 045028
London Shop, Royan 072630
The London Silver Vaults, London 090306
London Street Projects, Los
Angeles 121537
Londonderry Mines Museum,
Londonderry 006217
Londoño Villegas, Mario,
Medellín 065290
Londoño, María Teresa, Medellín 128304
London's Transport Museum,
London 045029
Lone Star Coins, San Antonio . . 096992
Lone Star Flight Museum/
Texas Aviation Hall of Fame,
Galveston 049036
Lone Star Pastel Society,
Amarillo 060569
Lone Star Toys, Dallas 093161
Lone Star Trains, San Antonio . . 096993
Long & Ryle, London 118431
Long Beach Antique and Collectible
Mall, Long Beach 094210
Long Beach Arts, Long Beach . . 121405,
121406
Long Beach Museum of Art, Long
Beach 050371
Long Branch Historical Museum, Long
Branch 050378
The Long Curve Gallery, Chipping
Campden 117584
Long Eaton Town Hall, Long
Eaton 045144
Long Fang Ge, Nanjing 102150
Long Island Children's Museum,
Garden City 049045
Long Island Culture History Museum,
Commack 047947
Long Island Maritime Museum, West
Sayville 054325
Long Island Museum, Long
Island 003200
Long Island Museum of American
Art, History and Carriages, Stony
Brook 053548
Long Island Museum of Science and
Technology, Garden City 049046
Long Life for Art, Christoph Waller,
Eichstetten am Kaiserstuhl . . 129782
Long Lost Friends, Houston . . . 145171
Long Melford Antique Centre, Long
Melford 090619
Long-Nan Museum of Natural Lacquer
Ware, Puli 042418
Long Shop Museum, Leiston . . 044808
Long Street Antiques, Tetbury . 091574
Long Warehouse, Coalbrookdale 043975
Long, Bao, New York 095546
Long, Meredith, Houston 121080
Long, Michael D., Nottingham . 090896
Long, Stephen, London 090307
Long, Timothy, Sevenoaks 135018
Long, Van, Hue 125161
Longaberger, Saint Louis 096729,
096730
Longanesi, Milano 137682
Longboat Key Center for the Arts,
Longboat Key 050388
Longchamp, Olivier, Villars-sur-
Glâne 133954
Longchuan Museum, Longchuan 008043
Longdale Craft Centre and Museum,
Ravenshead 045585

Longden, George, Macclesfield . 144721
Longelin, Michel, Saint-Laurent-
Blangy 072880
Longfellow, Portland 145456
Longfellow-Evangeline State Historic
Site, Saint Martinville 052827
Longfellow House, Cambridge . . 047365
Longfellow's Wayside Inn Museum,
Sudbury 053588
Longford Antiques, Longford . . 061750
Longgang Ke Jia Folklorish Museum,
Shenzhen 008264
Longhorn, San Antonio 096994
Longhorn Museum, Pleasanton . 052205
Longhu Painting and Calligraphy
Association, Beijing 058499
Longhua Mausoleum of Fallen Heroes,
Shanghai 008229
Longhua Museum, Longhua 008044
Longjing Koreazu Glaswegiansu Guan,
Longjing 008045
Longkou Museum, Longkou 008046
Longleat House, Warminster . . . 046055
Longmire, London 090308
Longmont Museum, Longmont . 050390
Angelo Longo Editore, Ravenna . 137725
Longo, Christine, Grézet-
Cavagnan 068742
Longo, Franco, Milano 131465
Longo, L. E. Manuel, Milano . . 080590
Longo, Sergio, São Paulo 101099
Longo, Vittorio, Milano 080591
Longquan Museum, Longquan . 008047
Longquanyi District Museum,
Chengdu 007694
Longquanyi Mingshu Wangling
Museum, Chengdu 007695
Longreach Power House Museum,
Longreach 001230
Longridge Gallery, Longridge . . 118660
Longrock Museum, Longyan . . . 008049
Longstreet Farm, Holmdel 049515
Longue Vue House, New Orleans 051296
Longuestre Franck, Brest 067475
Longuet, Bernard, Saint-Vincent 073330
Longueville, Daniel, Maisons-
Laffitte 069815
Longview Museum of Fine Art,
Longview 050392
Longwood, Salt Lake City 096901
Longwood Center for the Visual Arts,
Farmville 048711
Longyear Museum, Chestnut Hill 047624
Longyou Museum, Longyou 008050
La Lonja, Zaragoza 040444
Lonnes, Hannelore, Bremen . . . 075173
Lonnie's Art, El Paso 120835
Lonsdale, Toronto 101654
Lonsdale Antiques, Southend-on-
Sea 091377
Lonsdale Antiques, Vancouver . 064839
Lonsdale Quay Native Arts,
Vancouver 101808
Lonsdale, Elaine, Hemswell Cliff 144467
Lonsky, Alexander Nikolaus,
Wien 127944
Lónyai-Kastély, Tuzsér 024021
De Looier Antiek Centrum,
Amsterdam 082796
Look, Almaty 111551
Look Art Gallery, Helmsley 117947
Look Galerie und Edition, Krefeld 107740,
137329
Look Gallery, Minneapolis 122009
Look Out Discovery Centre,
Bracknell 043688
Look, John, Boston 092482
Looking Glass 1950's and Carnival
Memorabilia, Saint Louis 096731
Looking Glass of Bath, Bath . . . 088393,
134078
Loomes, Brian, Pateley Bridge . 090946
Loomes, Robert B., Stamford . . 135073
Les Looms Gallery, New York . . 095547

Loong, Singapore 085462, 133039
Loong, Thng, Singapore 085463
Loop, Melbourne 099313
Loop, Toronto 101655
Loop Museum and Art Center,
Seagraves 053224
Loos Koboltgruva, Loos 040798
Loosduins Museum De Korenschuur,
Den Haag 032177
Loose, R., Den Haag 143116
Loosveldt, Oostende 100802
Loot, Oakland 095900
Looze, A. de, Rotterdam 083580
Lopard-Dessolin, Mâcon 069794, 125632
Lopburi, Bath 088394
Lopburi, Singapore 085464
Lopdell House Gallery, Auckland 112785
Lopdell House Gallery, Waitakere 033327
Lopera, Verona 132288
Lopes, Anabela Oliveira, Lisboa . 084909
Lopes, Benedito Gilberto Almeida, São
Paulo 140307
Lopes, Luís, Lisboa 084910
Lopes, Marciano, Fortaleza . . . 063944
Lopes, Tania Maria Weckerle, Porto
Alegre 100987
López, Madrid 133352
López, Montevideo 092005
López, New York 135951
Lopez Island Historical Museum,
Lopez Island 050395
Lopez-Linares, Madrid . 086210, 086211
Lopez Memorial Museum, Pasig 034569
Lopez Morales, Enrique,
Zaragoza 133503
López Quiroga, Ramón, México . 111966
Lopez Rey, Francisco Javier, A
Coruña 115035
López Salgado, Fred, Bogotá . . 128271
López, Florence, Paris 071590
López, Javier, Madrid 115247
López, Jean-François, Nîmes . . 070698
López, Joaquin & Tony, Miami . 121822
López, José, Lima 084464, 132810
López, Ricardo, Medellín 065291
López, Roberto, Montevideo . . . 092006
López, Yvette, Chêne-Bourg . . 087366,
126716
Łopieńska-Lipczyk, Anna,
Warszawa 132890
Lopinski, Michel, Argelès-sur-
Mer 128446
Lopreito, Buenos Aires 060909
Loquat Museum of Sinshe,
Sinshe 042464
Loquens Gallery, Cheltenham . . 088835,
134238
Loqui Delfa, Torino 081723
L'Or du Temps, Pamiers 104438, 128883
Lora D Art Gallery, Chicago . . . 120184
Lora Robins Gallery of Design from
Nature, University of Richmond,
Richmond 052554
Lora, Luciano, Valdagno 111071
Lorang, Robert, Bamberg 141450
L'Orange Bleue, Coutances . . . 068209
L'Orangerie-de-la Motte, Saint-Quentin-
la-Motte-Croix-au-Bailly 073281
Lorant, Eva, Vallauris 115808
Lorcan O' Neill, Roma 110814
Daniela Lorch & Jürgen Schulz-Lorch,
Sigmaringen 130639
Lorch + Seidel Gallery, Berlin . . 106363
Lord Frontenac, Napier 084006
Lord Howe Island Museum, Lord
Howe Island 001232
Lord Ponsonby Antiques,
Auckland 083804
Lord Strathcona's Horse Museum,
Calgary 055225
Lord, Alan, Folkestone 089289
Lord, Michael H., Milwaukee . . 121932
Lordereau, Serge, La Chaise-
Dieu 068967

Lore Degenstein Gallery, Susquehanna
University, Selinsgrove 053265
Loreau, Pierre, Riec-sur-Belon . 072461
Loreck-Zimmermann, Sylvia,
Garmisch-Partenkirchen 075949
Lorelei, Amsterdam 143047
Lorelei, Bruxelles 100637
Loreley Galerie, Oberwesel . . . 108325
Loreley-Museum im historischen
Stadtturm, Sankt Goarshausen 021596
Lorene's Antiques, Tampa 097574
Lorentz, Thomas, Gröbenzell . . 076096,
129919
Lorentzen, Kirsten, Skagen . . . 066117
Lorentzon, Stockholm 115905
Lorenz, Bolzano 131008
Lorenz, Toronto 064695, 064696, 101656
Dr. Lorenz & Meyer, Cloppenburg 125914
Lorenz Books, London 138226
Lorenz, Dieter, Düsseldorf 075494
Lorenz, Hans, Jüchen 076655
Lorenz, Ingrid, Frankfurt am
Main 141783
Lorenz, Lutz, Chemnitz 129655
Lorenz, Nancy, New York 135952
Lorenz, Ute, Erfurt 129803
Lorenzelli Arte, Milano 110370
Lorenzelli, Jacopo, Bergamo . . 079793
Lorenzen, Brigitte, Gelting 107115
Lorenzi & Meier, Zürich 133988
Lorenzi, Flávio José de, Porto
Alegre 063983
Lorenzo State Historic Site,
Cazenovia 047471
Lorette, Jean-Paul, Nancy 070424
Lorgère, Hervé, Paris 071591
Lori Local Tradition Museum,
Vanadzor 000715
Lori, Marco, Milano 110371
Lorient Collections, Lorient . . . 069639
Lorin & Kristy, Singapore 114566
Loriot, Jean-Claude, Fitilieu . . . 068535
Lorn Museum, Ledaig 044775
Lorne Antiques, Lorne 061751
Lorne Scots Regimental Museum,
Brampton 005456
Lornoy, A., Bruxelles 063496
Lornyon, Budapest 079267
Lorrai, Eric, Marseille 069947
Lorraine, Genk 063619
Lorraine's Antiques and Collectables,
Yarragon 062504
Lorraine's Collectibles, Saint
Louis 096732
Lorre, Zwolle 083735
Lorrie, Albuquerque 092066
Lortal, Didier, Revel 105445
Lortal, Jean, Saint-Mandé 072924
Lorthois, Catherine, Confolens . 068135
Los Alamos County Historical Museum,
Los Alamos 050402
Los Altos History Museum, Los
Altos 050405
Los Angeles Academy of Figurative
Art, Van Nuys 058040
Los Angeles Antiques, Los
Angeles 094395
Los Angeles Antiques Show, Santa
Monica 098468
Los Angeles Art Association, West
Hollywood 060570
Los Angeles Art Exchange, Los
Angeles 121538
Los Angeles Artcore, Los
Angeles 121539
Los Angeles Artcore Center, Los
Angeles 121540
Los Angeles Arts Show, Santa
Monica 098469
Los Angeles Asian and Tribal Art
Show, Santa Monica 098470
Los Angeles Center for Photographic
Studies, Los Angeles 050442
Los Angeles Contemporary Exhibitions,

Los Angeles 050443
Los Angeles County Museum of Art,
Los Angeles 050444
Los Angeles Craft and Folk Art
Museum, Los Angeles 050445
Los Angeles Fine Art Gallery, West
Hollywood 125011
Los Angeles Maritime Museum, San
Pedro 053060
Los Angeles Modernism Show, Santa
Monica 098471
Los Angeles Municipal Art Gallery,
Barnsdall Art Park, Los Angeles 050446
Los Angeles Museum of the Holocaust,
Los Angeles 050447
Los Angeles Photo Group, Los
Angeles 060571, 121541
Los Angeles Police Museum, Los
Angeles 050448
Los Angeles Printmaking Society,
South Pasadena 060572
Los Angeles Rugs and Antiques, Los
Angeles 094396
Los Angeles Stripping and Finishing
Center, Los Angeles 135730
Los Angeles Valley College Art Gallery,
Valley Glen 053936
Los Nogales Museum, Seguin . . 053264
Los Tintoretto's, Bucaramanga . 102392
Loschwitzer Antiquariat, Dresden 141705
Loscorpione, Bari 130916
Loseley House, Guildford 044490
Losenscky, Monika, Westersteede 078951
Losert, Günter, Marktredwitz . . 077301
Loshi Stone Art Museum, Ershuei 042310
Loški Muzej, Škofja Loka 038609
Lossiemouth Fisheries and Community
Museum, Lossiemouth 045149
Lost & Found Antiques, Oakland 095901
Lost & Foundry Antiques, New
Orleans 095122
Lost Art, Cincinnati 135445
Lost City Art, Washington 124930,
124931
Lost City Arts, New York 095548
Lost City Museum, Overton . . . 051848
Lost Dutchman, Indianapolis . . 093902
The Lost Gallery, Strathdon . . . 119169
Lost Nation Casting, New
Orleans 122196
Lost River State Park Museum,
Mathias 050740
Lost World of Wonders,
Milwaukee 094869
Lostalot-Dutel, de, Valence
(Drôme) 125828
Lostwithiel Museum, Lostwithiel 045150
Lószerszám Múzeum, Sümeg . . 023907
Lot Antique, Cahors 067598
Lotempio Antiques, Buffalo . . . 092541,
135376
Lotherton Hall, Leeds Museums and
Galleries, Aberford 043365
Lothon, Mireille, Coulombs . . . 068179
Lothringer 13, Städtische Kunsthalle
München, München 020545
Lots of Furniture Antiques, Dallas 093162
Lots Road Auctions, London . . . 126966
Lotsabooks, Waikato 143355
Lotschnauer Heimatstube,
Eislingen 018148
Lotte Gallery, Seoul 111687
Lotte Glob, Altnaharra 117233
Lotte Reimers-Stiftung, Kunstzentrum,
Deidesheim 017842
Lotte Reiniger Museum, Stadtmuseum
Tübingen, Tübingen 022221
Lottehaus, Städtische Sammlungen
Wetzlar, Wetzlar 022612
Lottermann & Söhne, Mannheim 077265
Lotti, Dallas 120625
Lottin, Henriette, Ommeél 070784
Lotton, Charles, Chicago 120185
Lotus, Bangkok 088082

Lotus Antiquities, Los Angeles . 094397
Lotus Art Gallery, Ho Chi Minh
City 125131
Lotus Collection, San Francisco . 097275
Le Lotus d'Eze, Eze 103731
Lotus Explorer Museum, Baihe . 042266
Lotus Gallery, Austin 092248
The Lotus Gallery, Bath 117310
Lotus Gallery, Salt Lake City . . . 096902
Lotus Gallery, Surfers Paradise . 099737
Lotus House, Los Angeles 121542
Lotus Image Museum, Luodong 042389
Lou Carreyrou, Villefranche-du-
Périgord 074146
Lou Holtz and Upper Ohio Valley Hall
of Fame, East Liverpool 048442
Lou Solei, Lavey-Village 116431
Lou, Meggie, Long Beach 094211
Loubatières, Francis, Portet-sur-
Garonne 136998
Loudean, Memphis 094646
Louden-Henritze Archaeology Museum,
Trinidad State Junior College,
Trinidad 053806
Loudon & Thys, Würzburg 079085
Loudon Heritage Farm Museums,
Sterling 053522
Loudoun Mansion, Philadelphia . 052063
Loudoun Museum, Leesburg . . . 050213
Lougheed House Conservation Society,
Calgary 005526
Loughnan, Nicholas, Youghal . . . 079726
Louhisaaren Kartanolinna,
Askainen 010759
Louie, Harry F., San Francisco . 136331
Louie's Rock-n-Reels,
Albuquerque 092067
Les Louis, Peymeinade 072123
Louis' Antiques, Wembley 062418
Louis Bernal Gallery, Pima College-
Center for the Arts, Tucson . . . 053833
Louis Couperus Museum, Den
Haag 032178
Louis d'Or, Monaco 082516
Le Louis d'Or, Monaco 082517
Louis E. May Museum, Fremont 048974
Louis H. and Lena Firn Grover
Museum, Shelbyville 053303
Louis-S.-Saint-Laurent, Compton 005643
Louis Spohr-Gedenk- und
Forschungsstätte, Museum der
Geschichte des Violinspiels,
Kassel 019539
Louis Tussaud Wax Museum,
København 010181
Louis Tussaud's Waxworks Museum,
Niagara Falls 006479
Louis, Françoise, Cogolin 068067
Louis, James, Domme 068380
Louis, René, Caen 067575
Louisa County Historical Society,
Louisa 050482
Louisa's Antiques, Hobart 061633
Louisbourg Marine Museum,
Louisbourg 006224
Louisburg College Art Gallery,
Louisburg 050483
Louise-Henriette-Stift, Kloster
Lehnin 019651
Louise Wells Cameron Art Museum,
Wilmington 054446
Louisiana Arts and Science Center,
Baton Rouge 046814
Louisiana Association of Museums,
Baton Rouge 060573
Louisiana Children's Museum, New
Orleans 051297
Louisiana Country Music Museum,
Marthaville 050721
Louisiana Museum of Modern Art,
Humlebæk 010146
Louisiana Naval War Memorial U.S.S.
Kidd, Baton Rouge 046815
Louisiana Old State Capitol, Center for

Political and Governmental History,
Baton Rouge 046816
Louisiana State Exhibit Museum,
Shreveport 053319
Louisiana State Museum, New
Orleans 051298
Louisiana State Museum,
Patterson 051947
Louisiana State University Museum of
Art, Baton Rouge 046817
Louisiana Tech Museum, Louisiana
Tech University, Ruston 052713
Louistine, Ottawa 128191
Louisville Antique Mall, Louisville 094570
Louisville Science Center,
Louisville 050495
Louisville Slugger Museum,
Louisville 050496
Louisville Visual Art Association,
Louisville 060574
Louisville Visual Art Museum,
Louisville 050497
Louly, Cairo 066253
Lounais-Hämeen Museo, Forssa 010781
Lounge Gallery, London 118432
Loup de Sable, Rouen 072596
Loupret, Patrick, Saint-Sever
(Landes) 073308, 129246
Loupỳ, Miroslav, Jilemnice 065400
Lourdes Carcedo, Burgos 115000
Le Lourec-Weber, Marie-Paule,
Luxembourg 111859
Loureiro, Bilbao 085878
Loureiro, Leonardo, Porto Alegre 100988
Louro, Porto 085035
Louro, Maria Sousa, Braga 084779
L'Ours Sam, Montapas 070182
Lous, Johan, Frederiksberg 065669
Louth Museum, Louth 045157
Louvencourt, Anaury de, Paris . . 105045
Louvre, Montevideo 092007
Louvre, Tehrãn 079424
Louvre Antiq'Bijoux, Paris 071592
Le Louvre French Antiques,
Dallas 093163
Louvre Gallery, Los Angeles . . . 094398,
135731
Louvre's Gallery, Picnic Bay . . . 099565
Louwman Collection – Het
National Automobilmuseum,
Raamsdonksveer 032774
Loux Bazar, Cairo 066254
Lovački Muzej, Zagreb 009041
Lovart – International Noir d'Ivoire,
Montreuil-sous-Bois 104291
Lovasmúzeum, Bábolna Nemzeti
Ménesbirtok, Bábolna 023412
Love, Fresno 135567
Love Field Antique Mall, Dallas . 093164,
145117
Love Lane Antiques, Nantwich . 090797
Love Saves the Day, New York . 095549
Love, David, Harrogate 089433
Love, Richard H., Chicago 120198
Lovecko Lesnicke Muzeum, Úsov 099935
Lovegrove, Karyn, Los Angeles . 121543
Loveland Museum and Gallery,
Loveland 050506
Lovely Lane Museum, Baltimore 046756
Lovers Lane Antique Market,
Dallas 093165
Loves, Perth 127013
Loves Auctioneers, Richmond . . 125376
Lovett, Houston 093728
Lovett, Palm Beach 099544
Lovett, Tulsa 124808
Loviisan Kaupunginmuseo,
Loviisa 011127
Lovisan Merenkulkumuseo,
Loviisa 011128
Lovriner Stube, Donauwörth . . . 017935
Low, Michael, Forres 089297
Low, Rebecca, Fort Worth 120874
Lowak, J., Austin 119683

Lowe, Atlanta 119587
Lowe & Sons, Chester 088864
Lowe, Los Angeles 121544
Lowe, Loughborough . 090636, 134817
Lowe Art Museum, University of
Miami, Coral Gables 047986
The Lowe Gallery at Hudson Guild,
New York 051417
Lowe, James, New York 145351
Lowe, Lynn Rae, Tucson 124749
Lowe, Michael, Cincinnati 120355
Lowell Damon House,
Wauwatosa 054239
Lowenstein, Diana, Miami 121823
Lower Cape Fear Historical Society
Museum, Wilmington 054447
Lower Columbia College, The Arts
Gallery, Longview 050393
Lower East Side Tenement Museum,
New York 051418
Lower Fort Garry, Saint Andrews 006849
Lower House Fine Antiques,
Redditch 091081, 134960
Lower Methil Heritage Centre, Lower
Methil 045159
Lower Nupend Gallery, Great
Malvern 117881
Lower White River Museum, Prairie
County Museum, Des Arc 048281
Lower, Graham, Flimwell 089284
Lowes Court Gallery, Egremont . 117746
Lowestoft and East Suffolk Maritime
Museum, Lowestoft 045161
Lowestoft Museum, Lowestoft . . 045162
Lowewood Museum, Hoddesdon 044593
Lowndes County Museum,
Valdosta 053926
Lowood Antiques, Lowood 061754
Lowrence, West Hollywood 097907
The Lowry Gallery, Salford 045720
Lowy, Julius, New York 135953
Loxton Historical Village, Loxton 001235
Loy, Rastede 108432
Loy, Versmold 078792
Loy, D., Bad Rothenfelde 074576
Loyal Edmonton Regiment Military
Museum, Edmonton 005756
Loyal Heights Upholstery & Design,
Seattle 136373
Loyal Upholstery Shop,
Philadelphia 136094
Loyalist Cultural Centre, Bath . 005395
Loyalist House Museum, Saint
John 006892
Loyer & Fils, Poissy 072217
Loyola, Salt Lake City 096903
Lozano, Málaga 086294
Lozano, Oscar, México 082479
Lozicarte, San Marino 114383
Lozinguez, Marie-José, Ambert . 066549
Ložinský, Aleš, Praha 128357
LP Décoration Inter, Tournai . . . 063846
LR Galeria de Arte, Brasília 063935
LRB Antiquities, Brahma Lodge . 061187
LS Antiquariaat, Amsterdam . . . 143048
LT Antiques, Baltimore 092340
LTG Gallery, Delhi 109330
Lu Ju Troupe Xiyuan Art Gallery,
Jinan 102140
Lu Xun and Qiu Jin Museum,
Shaoxing 008253
Lu Xun House, Shanghai 008230
Lu Xun Museum, Beijing 007627
Lu, Le Ge, Singapore 085465
Luanping Museum, Luanping . . . 008052
Lubbe, E., San Francisco 145560
Lubbock Art Association,
Lubbock 060575
Lube, San Antonio . . . 124047, 124048
Lubelskie Towarzystwo Zachęty Sztuk
Pięknych, Lublin 034992, 059699
Luber Galerias, Santander 115431
Lubespére, Pierrette, Le Castera 069240
Lubeznik Center for the Arts, Michigan

City 050851
Lubicki, Estelle, Paris 071593
Lubie, Castres 067731
Lubin, Jocelyne, Moissac 070144
Lublin, Mary, New York 122820
Lubovnianske Múzeum – Hrad, Stará
Ľubovňa 038456
Lubovniansky Skanzen, Stará
Ľubovňa 038457
Lubrano, Gaetano, Roma 081429
Lubrano, Michel, Paris . 071594, 129071
Lubuskie Muzeum Wojskowe w
Zielonej Górze z/s w Drzonowie,
Letnica 034944
Luca, Zaltbommel 112721
Luca Verlag, Lingen 137353
Lucan Antique Show, Lucan . . . 098037
Lucarelli, Irma, Roma 131961
La Lucarne, Lausanne 087575, 133800
Lucas &Son, Las Vegas 135685
Lucas, Noordhorn 132598
Lucas, Dany, La Baule-Escoublac 103869
Lucas, Fay, London 090309
Lucas, G.C., Indianapolis 121197
Lucas, Gilles, Pierrelaye 072159
Lucas, Joël, Beauvoir-sur-Mer . 067057,
128464
Lucas, Phyllis, New York 095550
Lucca, Barcelona 085793
Lucca, Lucien, Saint-Ouen . . . 073140
Lucchetti, Mario, Traversetolo . . 111023
Lucchini, Anna Maria, Milano . . 131466
Lucchini, Aristodemo, Roma . . . 081430
I Lucci di Negro, Torino 132159
Lucciola, Tucson 097673
Luce, Brett, Paris 071595
Luce, Jean-Marc, Crest 068237, 128574
Lucente, Antonio, Torino 132160
Le Lucernaire, Paimpol 070843
Lucero, Eduardo, Los Angeles . . 121545
Lucey, Con P., Ballintemple . . . 079447,
130826
Luchetta, Daniele, Venezia 111134
Luchetti, Buenos Aires 060910, 125170
Luchmann, Roland, Berlin 129530
Luchsinger, Hans, Riehen 087748,
133895
Luchtvaart Museum Texel, De
Cocksdorp 032144
Luchtvaart Museum Twenthe,
Enschede 032307
Lucia, Gino, Roma 131962
Luciani, Lidia, Roma 081431
Luciano Inga-Pin, Milano 110372
Lucien-Paris, Nogent-sur-Marne 125667
Lucien-Paris, Paris 125735
Luciferlag im Kunsthaus Lübeck,
Lübeck 137360
Lucifersmuseum, Geraardsbergse
Musea, Geraardsbergen 003647
Lucinda, Baltimore 119752
Lucioli, Laura, Firenze 131196
Lucioni, Irela & Mario, Lima . . . 084465
Luck, J., Ladbergen 076989
Luckhard Museum, The Indian
Mission, Sebewaing 053254
Luckhurst, J., Sandgate 091240, 135006
Luckman, Coventry . . . 088984, 144357
Luckman Gallery, California State
University Los Angeles, Los
Angeles 050449
Luckow, Kai H., Düsseldorf 129749
Lucks, München 077562
Lucky Dog Antiques, Portland . . 096434
Lucky Draw, Fort Worth 120875
Lucky Eddie, Oslo 143404
Lucky Find, London 064352
Lucky Find, Saint Paul 096836
Lucky Seven Antiques,
Providence 096511
Lucky Strike Oldwares, Dunolly . 061417
Lucky's World, Lubbock 127300
Luckystar, Milwaukee 121933
Lucoff, K., Los Angeles 121546

Lucrèce, Paris 105046
Lucretia Vaile Museum, Palmer
   Lake 051897
Lucullus, New Orleans 095123
Lucy Craft Laney Museum of Black
   History, Augusta 046661
Lucy Hayes Heritage Center,
   Chillicothe 047719
Lucy Maud Montgomery Birthplace,
   Charlottetown 005596
Lucy's Attic, Milwaukee 094870
Luda Bērziņa Memoriālais Muzejs,
   Jūrmala 030138
Ludescher, Gabriele, Krefeld . . 076959
Ludi-Bretscher, Esther, Neuchâtel 144116
Ludion, Gent 136655
Le Ludion, Paris 071596
Ludlow 38, New York 122821
Ludlow Museum, Shropshire County
   Museum, Ludlow 045165
Ludopassion, Toulon (Var) 073718
Ludorff, Düsseldorf 106807
L'udovit Fulla Galéria, Slovenská
   Národná Galéria, Ružomberok . 038439
Ludvig, Budapest 109160
Ludvika Gammelgård och
   Gruvmuseum, Ludvika 040801
Ludwig, Fredericia 102801
Ludwig, Rio de Janeiro 064078
Ludwig-Dörler-Gemäldemuseum,
   Schillingsfürst 021651
Ludwig Forum für Internationale Kunst,
   Aachen 016440
Ludwig Galerie Schloss Oberhausen,
   Oberhausen 020896
Ludwig-Gebhard-Museum, Tiefenbach,
   Oberpfalz 022155
Ludwig-Harms-Haus, Ev.-luth.
   Missionswerk in Niedersachsen,
   Hermannsburg 019210
Ludwig Museum im
   Deutschherrenhaus, Koblenz . . 019663
Ludwig Múzeum Budapest – Kortárs
   Művészeti Múzeum, Budapest . 023508
Ludwig-Thoma-Haus, Tegernsee 022114
Ludwig Wittgenstein – Gedenkraum,
   Trattenbach 002836
Ludwig Wittgenstein-Dauerausstellung,
   Kirchberg am Wechsel 002172
Ludwig, Eberhard, München . . . 130366
Ludwig, Franz, Holzminden . . . 076567
Ludwig, Günter, Oberasbach . . 077855
Ludwig, Jürg, Zizers 087922
Ludwig, Martin, Kronach 107753
Ludwig, Martin, Kulmbach 107773
Ludwika, Sopot 084664
Ludzas Novadpētniecības Muzejs,
   Ludza 030158
Lübbe, Bergisch Gladbach 137045
Lübeck, Goch 076020
Die Lübecker Museen, Kulturstiftung
   Hansestadt Lübeck, Lübeck . . 059166
Lübking, Dorothea, Minden . . . 077373,
                                        108006
Lüchinger, Sankt Gallen 144128
Lück, Mannheim 130279
Lücke, Schweinfurt 142355
Lücker, M., Aachen 074336
Lücker, R., Kinheim 076797
Lüddecke, Wilhelm, Chemnitz . . 075257
Lüdemann, I., Rotenburg,
   Wümme 078283
Lüderitz Museum, Lüderitz 031843
Lüders, Axel, Hamburg 111883
Lüdin, Paul, Basel . . . 087243, 116091
Lüdtke, Monika, Bad Münder am
   Deister 129419
Lüers, Günter, Lauterecken . . . 077034
Lügenmuseum, Kyritz 019836
Luegger, Waltraud, Wien 127945
Lühikese Jala Galerii, Tallinn . . 103119
Lühn, Linn, Köln 107660
Lühr, Ernst, Oldenburg 077933

Lührs, Joachim, Hamburg 076271,
                                        141884
Lülf-Steffen, A., Herford 076503
Luen Chai, Hong Kong . 065078, 065079
Lüpke, Herbert, Rodenbach . . . 108503
Lüpnitz, Berlin 141551
Lütge, Cape Town 114770
Lüth, H.-H., Schobüll 108571
Lüthard, Marlise, Murgenthal . . 087691
Lüthi, Beat, Basel 087244, 133652
Lüthi, Beat & Alexandra, Bern . . 087289,
                  087290,   133683
Lüthi, Monika, Bern 133684
Lüthi, Ursula, Bern . . . 087291, 133685
Lüthy, Solothurn 144139
Lütke, Manfred, Leverkusen . . . 077095
Lütt Galerie, Fahrdorf 106946
Lüttgens, Hubert, Aachen 074337
Lüttgens, K., Aachen 074338
Lüttgens, Klaus, Aachen 074339, 129362
Lüttje Koophus, Emden 075606
Lüttmerding, László, München . . 077563
Luez, Richard, Beauvais 125509
Lüzinyan Evi, Lefkoşa 033473
Lufeng Saurian Museum, Lufeng 008053
Luffman, J., Haslington 089463
Luft, Achim, Bergen, Kreis
   Traunstein 074727
Luftfahrt- und Technik- Museumspark
   Merseburg, Merseburg 020375
Luftfahrt-Museum Laatzen-Hannover,
   Laatzen 019837
Luftfahrtausstellung im Alten
   Straßenbahndepot, Brandenburg an
   der Havel 017545
Luftfahrthistorische Sammlung
   Flugplatz Finow, Schorfheide . . 021723
Lufttechnisches Museum,
   Rechlin 021305
Lufttechnisches Museum,
   Rothenburg/Oberlausitz 021514
Luftwaffenmuseum der Bundeswehr,
   Förderverein des Luftwaffenmuseums
   der Bundeswehr e.V, Berlin . . 017216
Lugan, Xavier, Lafitte-sur-Lot . . 069135
Lugang Folk Arts Museum,
   Lugang 042381
Lugani, Dr., Berlin 074871
Lugans'kyj Chudožnij Muzej,
   Lugans'k 043229
Lugans'kyj Deržavnyj Institut Kul'tury i
   Mystectv, Lugans'k 056495
Lugans'kyj Oblasnyj Krajeznavčyj
   Muzej, Lugans'k 043230
Lugauer, Carina, München 142177
Lugdunumis, Lyon 069760
Lugert, Peter, Frankfurt am Main 137204
Luggage Store, San Francisco . . 053010
Lugger, Hans, Treglwang 127885
Lughien, Amsterdam 112251
Lugley Street Antiques, Newport, Isle of
   Wight 090836, 134883
Lugmair, Hans, Wien 140098
Lugo & Poiesi, Verona 081918
Lugouqiao Display Center, Beijing 007628
Lugu Museum, Lugu 042384
Lugu Tea Museum, Lugu 042385
Luguet, Didier, Montdidier 070229
Luh, W., Trossingen 078715
Luhr, K., Köln 076894
Luhring, Augustine, New York . . 122822
Luiéve Galeria Arte, Funchal . . 113974
Luigi, San-Pawl-Il-Bahar 082431, 132340
Luigi Antiquités, Voiron 074273
Luigi Franco Arte Contemporanea,
   Torino 110986
Luis-B, Zürich 126770
Luis Muñoz Rivera Museum,
   Barranquitas 036119
Luíscius, 's-Hertogenbosch . . . 143173
Luisenhütte, Balve 016943
Lujain, Damascus 088039
Luk, Kee, Hong Kong 065080
Luka, Praha 102706

Lukacs & Donath, Roma 081432
Lukas Verlag, Berlin 137083
Lukas, S., Brno 128319
Lukaszewska, Maria, Poznań . . 084645
Luke's Antiques, Birmingham . . 092419
Lukoschek, W., Lichtenau, Kreis
   Paderborn 107866
Luksic Castex, Buenos Aires . . 098574
Lula B.'s Antique Mall, Dallas . . 093166
Luleå Auktionsverk, Luleå 126628
Lullaby, Indianapolis 121198
Lullingstone Roman Villa,
   Eynsford 044318
Lulu & Lavigne, Toronto 064697
Lum and Abner Museum and Jot 'Em
   Down Store, Pine Ridge 052138
Luma, Luxembourg 111860
Lumas, Berlin . 106364, 106365, 106366
Lumas, Frankfurt am Main 107014
Lumas, Hamburg 107294
Lumas, München 108160
Lumas, Stuttgart 108717
Lumas Editionsgalerie,
   Düsseldorf 106808
Lumb & Sons, Charles,
   Harrogate 089434
Lumberman's Museum, Patten . 051946
Lumen, Biot 103394
Lumen, Venlo 083688
Lumen Travo, Amsterdam 112252
Lumgair, Irene, San Diego 124142
Lumière, Küsnacht 116400
Lumière, Tokyo 082176
Lumière de l'Oeil, Paris 071597, 129072
Lumina Gallery, Malvern, Victoria 099270
Luminaires Anciens, Vannes
   (Morbihan) 073972
Luminaires de Prestige, Fontenay-le-
   Comte 103748, 128602
Lumma, N., Ronnenberg 078270
Lummis Home 'El Alisal', Los
   Angeles 050450
Luna, Amsterdam 082797
Luna, Milano 131467
Luna, Nottingham 090897
La Luna Casa y Arte, San
   Salvador 103101
Luna Collectibles, Jacksonville . 094006
Luna Creazioni, Venezia 081852
La Luna di Glifada, Martina
   Franca 110207
Luna Hernandez, Francisco,
   Málaga 086295
Luna Mimbres Museum, Deming 048239
La Luna nel Pozzo, Venezia . . . 081853
Luna, Josue, São Paulo 064190
Lunacy, Philadelphia 096131
Lunadei, Franco, Roma 081433
Lund Bygdemuseum, Moi 033780
Lund University Art Collection,
   Lund 040810
Lund, Charlotte, Stockholm . . . 115906
Lund, Mogens, Thisted 066163
Lundar Museum, Lundar 006227
Lundgren, Stockholm 087026
Lundgren, Kurt, Nyborg 066010
Lundin, A., København 065933, 125453
Landmark, Stockholm 115907
Lundquist, Göteborg 143848
Lunds, Victoria 125392
Lunds Konsthall, Lund 040811
Lunds Konsthandel, Lund 086836
Lunds Universitets Antikmuseum,
   Lund 040812
Lundsbjerg Kræmmer- og Auktionshaller,
   Aabenraa 065575, 125428
Lundströmska Gården, Sigtuna . 040940
Lundvall, Regina, Stockholm . . 133576
Lundy's Lane Historical Museum,
   Niagara Falls 006480
Lune, Pierre de, Liège 063770
Lunel, Annie, Beaumont-sur-
   Sarthe 067008
Lunel, Patrick, Douains 068394

Lunenburg Art Gallery,
   Lunenburg 006230
Lung, Robert, Reichenau 130573
Lungauer Heimatmuseum Tamsweg,
   Tamsweg 002807
Lungauer Landschaftsmuseum,
   Mauterndorf 002377
Luniddi, Ettore, Roma 131963
Luniver & Co, Montréal 101428
Luniverre Gallery, Paris 105047
L'Univers, Le Mans 069307
Lunn Antiques, London 090310, 134714
Lunsford, Stephen, Vancouver . . 140426
Lunt Roman Fort, Baginton 043464
Lunwerg, Barcelona 137961
Lunwerg, Madrid 137993
Lunwerg, México 137817
Luochuan Museum, Luochuan . . 008054
Luoding Museum, Luoding 008055
Luohan, Paris 071598
Luohe Wieldyang Museum,
   Wuyang 008384
Luonan Museum, Luonan 008056
Luonnonhistoriallinen Museo,
   Porvoo 011273
Luonnontieteellinen museo,
   Luonnontieteellinen keskusmuseo,
   Helsinki 010845
Luontotalo Arkki, Satakunta Museo,
   Pori 011264
Luopioisten Museo, Luopioinen . 011129
Luopioisten Torpparimuseo,
   Luopioinen 011130
Luostarinmäen Käsityöläismuseo,
   Turun Maakuntamuseo, Turku . 011439
Luotsitupa, Uusikaupunki 011464
Luoyang Ducheng Guan,
   Luoyang 008058
Luoyang Glaswegiansu Guan,
   Luoyang 008059
Luoyang Museum, Luoyang . . . 008060
Luoyang Tumulus Museum,
   Luoyang 008061
Lupe Gallery, London 118433
Luperini, Beatrice, Verona 081919
Lupi, Monica, Firenze . 080163, 080164
Lupin, Arsène, Montréal 064440
Lupo'Art, Palermo 110610
Luppo, Antwerpen 063222
Lupton Square Gallery, Honley . 117995
Lupu, Jean, Paris 071599
Luquet, Roger, Saint-Maurice-
   l'Exil 072969
Lura Watkins Museum,
   Middleton 050861
Lure Me In, Minneapolis 094928
Lurie, Gertrude, Basel 087245
Lusa, Patricia, Buenos Aires . . 060911
Lusaka National Museum,
   Lusaka 054839
Luscious Conceptions, Windsor . 119365
Luseland and Districts Museum,
   Luseland 006231
Lusetti, Luca, Brescia 079944
Lush, Auckland 112786
Lush, Jacksonville 094007
Lushan Museum, Lushan 008062
Lusher, Nicholas, Hamilton . . . 100828
Lushun Museum, Dalian 007718
Lushun Snakes Dao Natural Museum,
   Dalian 007719
Lushun Su Jun Mausoleum of Fallen
   Heroes, Dalian 007720
L'Usine, Menton 070061
Lusk, David, Memphis 121702
Lussan, Régine, Paris 105048
Lust, B.N.J., Amsterdam 082798
Lustinger, H., München 077564
Lusto-Suomen Metsämuseo,
   Punkaharju 011283
Lustr'Antic, Saint-Quentin (Aisne) 073278
Lustre, Auckland 083805
Lustyk, Louisville 121657
Łuszczyk, Antoni, Pyrzyce 084660

Lutèce, Narbonne 070493
Lutèce, Paris 071600, 141229
Lutèce, Saint-Pierre-Eglise .. 073261
Lutece, San Francisco 124333
Lutet Toti, Saint-Maur-des-
  Fossés 129239
Luther Burbank Home, Santa
  Rosa 053140
Luther College Fine Arts Collection,
  Decorah 048207
Luther-Stube Mansfeld, Mansfeld 020258
Luther W. Brady Art Gallery,
  Washington 054131
Luther, H.M., New York 095551, 095552
Luther, Heidrun, Dinslaken 106684,
                          137141
Luther, Reinhard, Duisburg 106847
Luther, Tracy, North Saint Paul . 127373
Lutheran Brotherhood Gallery,
  Minneapolis 050942
Lutherer, Nettetal 077706
Lutherer, Andreas, Nettetal .. 137439
Lutherhaus mit Bibel-Café,
  Eisenach 018133
Lutherhaus Wittenberg, Stiftung
  Luthergedenkstätten in Sachsen-
  Anhalt, Lutherstadt Wittenberg 020198
Lutherkirche und Lutherzimmer
  im Dorfgemeinschaftshaus,
  Moorgrund 020461
Luttár, Budapest 109161
The Lutterworth Press,
  Cambridge 138147, 144299
Lutz, Delft 112363
Lutz, Stuttgart 078616, 108718
Lutz Children's Museum,
  Manchester 050622
Lutz Custom Upholstery, Fresno 135568
Lutz Mountain Meeting House,
  Moncton 006349
Lutz, Edmond, Zürich ... 087988, 133989
Lutz, Franz Johann, Gallbrunn .. 127736
Lutz, Hans-Rudolf, Zürich 138114
Lutz, Jutta, Mainz 107951
Lutz jun., Martin,
  Schrobenhausen 130630
Lutz, Yvan, Trouville-sur-Mer .. 073892
Lutze, Bernd, Friedrichshafen .. 107072
Luusuan Kylämuseo, Luusua .. 011133
Luvane, Walsall 119300
Luvian Kotiseutumuseo, Luvia .. 011134
Luweizhao Painting and Calligraphy
  House, Pinghu 008150
Lux, Koblenz 126002
Lux, Münster 077670
Lux Art, Köln 107661
Lux Art Institute, Encinitas .. 057054
Lux, Kenneth, New York 122823
Luxe Art Museum, Singapore .. 038270
Luxe Gallery, New York 122824
Luxford, Rosedale 062169
Luxl, Peter, Wien 063021
Luxor Gallery, Caracas 097960
Luxor Museum of Ancient Art, Al-
  Uqsur 010511
Luxor Trading, Houston 093729
Luxun Academy of Fine Arts,
  Shenyang 055238
Luz, Denver 120739
La Luz de Jesus, Los Angeles .. 121547
Luz Gallery, Makati ... 084475, 113389
Luz, Paula, Braga 084780
Luza Acht, Lübben ... 020150, 107906
Luzán, Bremen 106590
Luzerne County Museum, Wilkes-
  Barre 054406
Luzernische Glasmalerei, Luzern 133837
Luzhou Bowu Guan, Luzhou ... 008063
Lvivska Akademia Mystetstv, Lviv 043240
Lvivs'ka Galereja Mystectv, Lviv 043241
Lvivs'kyj Istorýčnyj Muzej, Lviv . 043241
Lvivs'kyj Muzej Istorii Religii,
  Lviv 043242
LVR-Archäologischer Park Xanten,

Xanten 022837
LVR-Industriemuseum, Schauplatz
  Bergisch Gladbach, Bergisch
  Gladbach 017052
LVR-Industriemuseum, Schauplatz
  Engelskirchen, Engelskirchen . 018210
LWL-Freilichtmuseum Detmold,
  Westfälisches Landesmuseum für
  Volkskunde, Detmold 017877
LWL-Freilichtmuseum Hagen,
  Westfälisches Landesmuseum für
  Handwerk und Technik, Hagen 018917
LWL-Industriemuseum, Westfälisches
  Landesmuseum für Industriekultur,
  Dortmund 017958
LWL-Landesmuseum für Kunst und
  Kulturgeschichte, Münster 020601
LWL-Museum für Archäologie,
  Herne 019219
LWL-Museum für Naturkunde,
  Westfälisches Landesmuseum mit
  Planetarium, Münster 020602
Lyall, Alexander, Long Melford . 090620,
                          134809
Lycée Technique des Arts et Métiers,
  Luxembourg 056124
The Lyceum – Alexandria's History
  Museum, Alexandria 046394
Lyceum Antique Store, Shanghai 065145
Łyczak, Barbara, Toruń 113791
Łyczak, Barbara i Ryszard, Toruń 113792
Lydd Town Museum, Lydd 045171
Lydia Koidula Memoriaalmuuseum,
  Pärnu 010647
Lydiard Antiques, Ballarat 061088
Lydiard House, Swindon 045938
Lydia's Jewelry, San Francisco . 097276
Lydon, Chicago 120187
Lye Antique Furnishings, Lye .. 090657
Lyet, J.P, Villeurbanne 074213
Lyktan, Arboga 086631
Lyktan, Borrby 143824
Lyle Sopel Studio, Vancouver .. 101809
Lylyn Viestivarikon Museo, Lyly . 011135
Lyman Allyn Museum of Art, New
  London 051277
Lyman House Memorial Museum,
  Hilo 049484
Lyme Academy College of Fine Arts,
  Old Lyme 057614
Lyme Art Association, Old Lyme 060576
Lyme-Bay Auction Galleries,
  Seaton 127034
Lyme Hall, Disley 044100
Lyme Regis Philpot Museum, Lyme
  Regis 045174
Lymington Antiques Centre,
  Lymington 090660, 144720
Lyn and Exmoor Museum, Lynton 045177
Lynch, Tuam 079715
Lynch, Florence, New York ... 122825
Lynch, Mary, Colchester 134289
Lynch, Merrill, Dallas 093167
Lynchburg Fine Arts Center,
  Lynchburg 050531
Lynchburg Museum System,
  Lynchburg 050532
Lyndale House Antiques,
  Minneapolis 094929
Lynden, San Francisco 124334
Lynden Pioneer Museum, Lynden 050536
Lyndhurst, Tarrytown 053686
Lyndhurst Antiques Centre,
  Lyndhurst 090666
Lyndon, Louisville 094571
Lyndon B. Johnson National Historical
  Park, Johnson City 049829
Lyndon Baines Johnson Museum,
  Austin 046696
Lyndon House Art Center, Athens 046695
Lynes & Lynes, Cork 079495
Lynette, Long Beach 094212
Lyngby Antik, Kongens Lyngby . 065984
Lyngby Antikvariat, Kongens

Lyngby 140624
Lynge & Søn, København 140612
Lyngerup Antik, Jægerspris ... 065806
Lynn Canyon Ecology Centre, North
  Vancouver 006516
Lynn Lake Mining Town Museum,
  Lynn Lake 006232
Lynn Museum, Lynn 050538
Lynnhaven House, Association for the
  Preservation of Virginia Antiquities,
  Virginia Beach 053998
Lynven, New York 122826
Lyon, Kansas City 121304
Lyon County Historical Society and
  Museum, Emporia 048579
Lyon County Historical Society
  Museum, Marshall 050711
Lyon, Oliver, Malmesbury 090681
Lyon, Richard, Hurstpierpoint . 144496
Lyonel-Feininger-Galerie,
  Quedlinburg 021248
Lyonnet, Aline, Saint-Didier-au-Mont-
  d'Or 072736
Lyons Corner House, Northwood 118823
The Lyons Redstone Museum,
  Lyons 050539
Lyons Wier Ortt Contemporary Art,
  New York 122827
Lyra, Amsterdam 082799
La Lyre, Rabat 082602
Lyrum, Århus 102777
Lys Antique, Antibes-Juan-les-
  Pins 066675
Le Lys Antique, Louhans 069644
Lys d'Or, Cogolin 068068
Le Lys d'Or, Grenoble 068734
Lysand, Eze 103732
Lyse Antic, Saint-Lye 072904
Lysekils Auktionsbyrå, Lysekil . 126631
Lysoňkovi, Stanislav a Marta,
  Brno 065350
Łyszczarz, Wiesława, Wrocław . 084771
Lyth Arts Centre, Lyth 045178
Lytham Hall, Lytham Saint
  Anne's 045179
Lytham Heritage Centre, Lytham Saint
  Anne's 045180
Lytham Lifeboat Museum, Lytham
  Saint Anne's 045181
Lytham Saint Annes Art Society,
  Lytham Saint Anne's 060175
Lytham Windmill Museum, Lytham
  Saint Anne's 045182
Lythe Cottage Antiques, Lythe . 090671
Lyttelton Historical Museum,
  Christchurch 033173
Lyttleton Gallery, North
  Melbourne 099471
Lytton Museum, Lytton 006233
Lyytikäinen, Lahti 066384
Lyzon, Nashville 122090
M – M Antiques, Kansas City . 107295
M – The New York Art World, New
  York 139418
M & B Antiques, Kansas City . 094082
M & G Upholstery, Houston ... 135610
M & H Art, New York 122828
M & K, Praha 102707
M & M, Barcelona 133129
M & M, Edmonton 064302
M & M Antiques, Blackheath .. 061161
M & M Coins, Buffalo 092542
M & S Antiquitäten, Ahrensburg 074362,
                          129371
M & S Curio Shop, Houston .. 093730
M & S Sales, Miami 094762
M & W Graphics, New York .. 135954
M-Ars, Kranj 114690
M-Art, Issy l'Eveque 068885
M-Art Arnhem, Arnhem 112334
M-Art Prestige, Maribor 085618, 114733,
                          133068
M de G, Paris 071601
M Gallery, Tbilisi 105911

M K Art Foundation, Ahmedabad 109250
M-Oriental Art, Los Angeles ... 121548
M% Gallery, Cleveland 120439
M&M, Vrhnika 085625
M&S Rare Books, Providence . 145468
M. & V. Anticuario, Medellín .. 065292
m. a. x. museo, Chiasso 041411
M. Christina Geis Gallery, Georgian
  Court College, Lakewood 050101
M. Javakhishvili Museum, Tbilisi 016412
M. Kostava Memorial House Museum,
  Tbilisi 016413
M.A.C, Paris 071602
M.A.R.E. – Movimento Artistico-
  Culturale Relazioni Europee,
  Riccione 059467
M.Arts Graphiques, Paris 136957
M.C. Mehta Gallery, Ahmedabad 024125
M.C.M. Antiquités, Bourg-en-
  Bresse 067411
M.C.N. Antiques, London 090311
M.D.C.A., Paris 105049
M.E.C. Art Gallery, Delhi 109331, 109332
M.E.'s Gallery, Houston 121081
M.F. Husain's Museum of Art and
  Cinema, Hyderabad 024299
M.G. Publicité, Paris 141230
M.G.R. Exports, Bruton 088693
M.I.C., Faenza 110011
M.I.T. Press, Cambridge 138314
M.J. Decoarts, Philadelphia ... 123412
M.J. Photos, Philadelphia 123413
M.J.'s Bookshop, Boronia 139650
M.L.H.A. Shop, New York 095553
M.M.B., Avignon 103314
M.M.C. – Midwest Museums
  Conference, Saint Louis 139419
M.O.M.A, New York 139420
M.S.M., Cagliari 079990
M.T. Oibek nomidagi Özbekiston
  Khalqlari Tarikhi Muzyi, Özbekiston
  Respublikasi Fanlar Akademiyasi,
  Toshkent 054654
M.T. Oibek Yodgorlik Muzeyi,
  Toshkent 054655
M.V.D. Antichità, Milano 080592
M.V.M., Bucureşti 114148
M.V.T, Greenford 117886
M.W. Williams Geological Museum,
  Vancouver 007287
M16 Studios, Fyshwick 099067
M2, Milano 110373
Ma Bohème, Orcet 070793
Ma Boutique, Toulon (Var) ... 073719
Ma City Revolution Museum,
  Jiangling 007934
Ma Maison, Torino 132161
Ma Maison de Papier, Bruxelles 140188
Ma Tasse de Thé, Vitry-le-
  François 074259
Ma, Feng-Chun, Amsterdam .. 082800
Ma.Ni Restauri, Parma 131728
Ma.R.C. Art Laboratorio Restauro,
  Verona 132289
Maack, Salt Lake City 123972
Maagdenhuismuseum, Openbaar
  Centrum voor Maatschappelijk
  Welzijn, Antwerpen 003290
Maali, Mia, Henley Brook 099127
Maanmittausmuseo, Helsinki .. 010846
Ma'anshan Museum, Ma'anshan 008064
Maarjamäe Loss, Eesti
  Ajaloomuuseum, Tallinn 010679
Maarten van Rossum Museum,
  Zaltbommel 033086
Al-Maas, Sharjah 088202
Maas- en Scheepvaartmuseum,
  Maasbracht 032616
Maas en Waal, Druten 132620
Maas en Waals Museum 1939–1945,
  Winssen 033053
Maas Gallery, London . 090312, 118434
Maaser, Klaus, Wuppertal ... 108994
Maasland, Maastricht 143199

Maaß, Jörg, Berlin 106367
Maaßen, Wassenberg 142464
Maassen, Ralf-Dieter, Blieskastel 129597
Mabee-Gerrer Museum of Art, Saint
  Gregory's University, Shawnee 053294
Mabel and Victor D'Amico Studio and
  Archive, Amagansett 046431
Mabel Hartzell Historical Home, Mount
  Union College, Alliance 046414
Mabel Larson Fine Arts Gallery,
  Austin Peay State University,
  Clarksville 047777
Mabela, Sarajevo 063868
Mabel's, Cleveland 135467
Mabillon, Aix-en-Provence ... 103199
Måbødalen Kulturlandskapsmuseum,
  Eidfjord 033579
Mabry-Hazen House, Knoxville . 050013
Mc Buffers, Cleveland 135468
The Mac Gallery, Dallas 120626
Mac Humble, Bradford-on-Avon 088562
MAC, Midlands Art Centre,
  Birmingham 043622
MAC/VAL, Musée d'art contemporain
  du Val-de-Marne, Vitry-sur-
  Seine 016312
McAdams, Alice, New York 095554
McAdams, Christopher, Oakland 136043
Macalester College Art Gallery, Saint
  Paul 052843
McAllen International Museum,
  McAllen 050544
McAllister House Museum, Colorado
  Springs 047886
The McAllister Institute, New
  York 057556
McAnthony, Fort Worth 120876
Macari, Ernesto, Roma 131964
MacArthur, Ballarat 098695
Macassar, Dijon 068337
Macau Artes Antigas, Rio de
  Janeiro 064079
Macau Museum of Art, Macau . 008068
Macau Wine Museum, Macau .. 008069
Macaulay Heritage Park, Picton . 066646
Macaulay Museum of Dental History,
  Charleston 047555
McAuley, Currajong 098938
Macaux, Yves, Bruxelles 063497
McAvenue, Francis J., Belturbet 079457
McBains, Exeter 089232
MacBride Museum, Whitehorse . 007405
McBride, J. Michael, Omaha ... 123329
McBurnie & Cutler, Toronto ... 140410
McCabe, Gerard, Adelaide 060969
Maccacaro, Egidio, Verona ... 081920
McCall, Betsy, Denver . 093310, 135533
McCalls, Aberdeen 088222
MacCallum More Museum, Chase
  City 047587
McCallum Museum, Sibley 053329
McCanse, Stephen, Toronto ... 101657
MacCarney, Frank, Nenagh ... 079684
Maccarone, New York 122829
McCarron, Paul, New York ... 122830
McCarthy Dermot Studio & Art Gallery,
  Killarney 109654
McCarthy, Ann, Cork 079496
McCarthy, Ian, Clutton . 088936, 134282
McCarthy, Patrick, Los Angeles . 121549
McCartney House, Liverpool ... 044858
McCartneys, Ludlow 126980
McCarty, Philadelphia 096132
McCaughen & Burr, Saint Louis 123840
Macchi, Pisa 110665
Macchiaroli, Gaetano, Napoli .. 137705
Macchioni, Marino, Camposanto 080007
McClain, David, Chicago 092741
McClain & Co., Robert, Houston 121082
McClellan, Judith, Houston .. 093731,
                         127253
McClelland, Camberwell 098851
McClelland & Stewart, Toronto . 136686

McClelland Gallery, Langwarrin . 001208
McClelland, Joshua, Melbourne . 061823
Macclesfield Silk Museum,
  Macclesfield 045185
McClure, Tulsa 136431
McClurg Mansion, Chautaqua County
  Historical Society, Westfield . 054338
McColl Shiraz, Charlotte 119998
McCollum-Chidester House Museum,
  Camden 047377
McConaghy House, Hayward ... 049420
McCone County Museum, Circle 047752
McConn, Castlerea 079476
MacConnal-Mason, London ... 090313,
              118435,   134715
McConnell Mansion, Moscow .. 051075
McConnell, Nick, Deal 144367
McCool, Richmond 123737
McCord and District Museum,
  McCord 006235
McCormack, R. & C., Dublin ... 079568
McCormick & McDonnell, Potts
  Point 062103
McCormick Gallery, Midland College,
  Midland 050876
McCormick, Thomas, Chicago .. 120188
McCormick, Tim, Woollahra ... 062476,
                         139965
McCoy House, Lewistown 050238
McCoy, Jason, New York 122831
McCoy, Mary Helen, Birmingham 092420
McCrae Gallery, McCrae 099242
McCrae Homestead Museum,
  McCrae 001236
McCrae House, Guelph Museums,
  Guelph 005933
McCrossin's Mill Museum, Uralla 001605
McCulloch Campbell, Eden 098995
Macculloch Hall Historical Museum,
  Morristown 051064
McCulloch House Museum and Hector
  Exhibit Centre, Pictou 006649
McCulloch, John, Felixstowe ... 089269
McCullough, Theresa, London .. 090314
McCully, Honolulu 093559
McCune, Stephen, San Antonio . 136264
McCutchen Overland Inn,
  McCutchenville 050550
McCutcheon, John, Columbus .. 120503
McDade Museum, McDade 050551
MacDonald, Gosforth, Tyne and
  Wear 089347, 134412
Macdonald, Newcastle-upon-
  Tyne 118789
MacDonald House Museum, East Lake
  Ainslie 005738
Macdonald Stewart Art Centre,
  Guelph 005934
Macdonald, Donald, Glasgow ... 144435
McDonald, Joy, London 090315, 134716
Macdonald, Magdalene, Rüthen 108537
McDonald, Mary, Albuquerque .. 119489
McDonald, Rosemary, Surry Hills 127682
McDonnell, Bridget, Carlton
  South 098868
McDonnell, Nadia, Hampton ... 061593
McDonnell, Stephen, Dún
  Laoghaire 079606, 130868
McDonnel's Antique Furniture,
  Portarlington 079690
McDonough, Pittsburgh 096307
McDougal, A., New York 122832
McDougall Mill Museum,
  Renfrew 006783
Macdougalls, Breakish 126833
McDowell House Museum,
  Danville 048146
McDowell, Rosalie, Denver 093311
Macé, Andrée, Paris 071603
Macé, Christophe, Biarritz 067203
Macé, Roland, Verneuil-sur-Avre 074030
Mace, Victor, Milton 099349
Macedo, Manoel, Belo Horizonte 100891
Macedon Historical Society Museum,

Macedon 050552
Macedonian Museum of Contemporary
  Art, Thessaloniki 023292
McElwain, Saint Louis 123841
McEvoy & Assoc., John, San
  Francisco 124335
McEwan, Ballater 088343, 134052,
                         144224
McEwan, Garry, Saint Kilda,
  Victoria 099668
McFadden, Tulsa 097745
McFaddin-Ward House,
  Beaumont 046859
McFarland Historical Society Museum,
  McFarland 050553
McFarland House, Niagara Falls 006481
Macfarlane, Salt Lake City 096904
McFee, Indianapolis 121199
McGahern, Patrick, Ottawa 140378
McGaw, Bruce, West Nyack 138479
McGeary, Bruxelles 100638
McGill-Duncan, Castle Douglas . 117543
McGillis Pioneer Home, Saint
  Victor 006941
McGilloway Gallery, Londonderry 118654
McGilvery, Laurence, La Jolla . 138345
McGinnis, Cheryl, New York ... 122833
McGlinn, Jack, Philadelphia ... 096133
MacGowan, Saskatoon 064564
McGowan, Vincent, Bundoran .. 079465
McGrath & Braun, Denver 120740
McGrath, B.J., Bathurst 061105
McGrath, Earl, Los Angeles ... 121550
McGrath, Earl, New York 122834
McGraw-Hill, New York 138404
Macgregor, Saint Andrews 127026
Macgregor Museum, Harare 054849
McGregor Museum and Duggan
  Cronin Gallery, Kimberley ... 038765
McGregor Wright, Wellington .. 113073,
                         132762
MacGregor, Nicola Ann, Firenze . 131197
McGriff, Dallas 093168
McGroarty Cultural Art Center,
  Tujunga 053844
McGuire, Houston 121083
McGuire, San Francisco 097277
Macha, Mary, Subiaco 062258
Machado, Rio de Janeiro 064080,
                         064081
Machado de Jesus, Lisboa 126449
Machado, Avelar, Lisboa 143581
McHale, Gaines, Baltimore 092341
Machan Museum, Lagrange 050065
MacHardy Coombs, London 118436
MacHenry, Newtownabbey 090843
McHenry County Historical Society
  Museum, Union 053885
McHenry Museum, Modesto 050985
Macheret-Allée, Paris 071604
Macheret, Michel, Langres 069177
Macherez, Catherine, Nice 070627
Machida-shiritsu Hakubutsukan,
  Machida 028990
Machida-shiritsu Kokusai Hanga
  Bijutsukan, Machida 028991
Machida-shiritsu Shashin Bijutsukan,
  Machida 028992
Machin Chouette, Québec 064536
Machine-Age Corporation, Boston 092483
Machine Age Gallery, Geelong .. 061509
Machine Age Modern, Toronto . 064698
Machine Age Wonders,
  Vancouver 064840
MACHmit! Museum für Kinder,
  Berlin 017217
Machoi, Monika, Bernau bei
  Berlin 074957
Machoïr-Bailly, Versailles 115249
Machold, Bremen 075174, 129635
Machon, Antonio, Madrid 115248
McHugh, Richmond 136172
Macia's Photography, Los
  Angeles 121551

Macias Zambrano, Manuel,
  Madrid 115249
Macias, Christina, Honolulu 120949,
                         120950
Macik, Paula, Charlotte 092611
Maciky Lagerio Karceris-muziejus,
  Šilutės Muziejus, Macikai ... 030414
McIlreavy, Alan & Fiona, Buderim 139670
MacInnes, Don, Wellington ... 084171
McIntire Department of Art, University
  of Virginia, Charlottesville .. 056861
McIntosh County Historical Society
  Museum, Ashley 046574
McIntosh Gallery, Kingussie ... 118053
McIntosh Gallery, London 006214
Macintosh, Patrick & Elona,
  Sherborne 091304
McIntyre Gallery, Regina 101532
Macio, Bordeaux 103437
McIvor, John & Fran, New
  Lambton 061952
Mack, Eglfing 129773
Mack, Houston 135611
Mack, Seattle 124563
Mack, Christian, Gordes 068683
Mack, D.J. & S.A., Carlton, New South
  Wales 061283
Mack, David, Branksome 088575,
                         134139
Mack, E.A., Baltimore 092342
Maçka Mezat, İstanbul 117061, 126790
Maçka Sanat Galerisi, İstanbul . 117062
McKay, Unley 062376
McKay, Hamish, Wellington .. 113074
McKechnie Institute, Girvan .. 044387
McKee, Casino 061292
Mackenbach, J.P., Den Haag .. 083101
McKenna & Co., London 090316
McKenna, North Adelaide 061982
McKenna Children's Museum, New
  Braunfels 051233
McKenzie, Claremont 098894
Mackenzie & Smith, Ludlow .. 090651,
                         134822
Mackenzie and District Museum,
  Mackenzie 006236
MacKenzie Art Gallery, Regina . 006770
Mackenzie Crossroads Museum, High
  Level 006002
McKenzie Fine Art, New York . 122835
MacKenzie Heritage Printery Museum,
  Queenston 006748
Mackenzie House, Toronto 007186
Mackenzie Thorpe Arthaus, Richmond,
  North Yorkshire 118966
Mackenzie, Ian, London 118437
McKenzie, J.W., Ewell 144416
McKenzie, Robert, Cannon Hill . 061275
Mackey, Denver 120741
Mackey, Houston 121084
McKinn, Austin 092250
McKinney Avenue Antique Market,
  Dallas 093169
McKinney Avenue Contemporary,
  Dallas 120790
McKinnon & Gibbons, Deloraine 061395
McKinnon, Karen, Charlotte ... 092612
McKissick Museum, University of
  South Carolina, Columbia ... 047912
Macklin and District Museum,
  Macklin 006237
Macklin, Terence, Whitstable .. 119343
Macklowe, New York 095555
McKnight, Michael, Tampa 097575
Mackowitz, Felix, Innsbruck ... 127776
Macktown Living History Site and
  Whitman Trading Post, Rockton 052650
MacLachlan Woodworking Museum,
  Kingston 006103
McLain & Brash, Pittsburgh ... 136133
Mclaine, Mark, Toronto 064699
McLane Center at Silk Farm, Audubon
  Society of New Hampshire,
  Concord 047958

Mclaren, Houston 093732
MacLaren Art Centre, Barrie . . 005386
Maclaren Roberts, Mosman . . . 099386
McLarty Treasure Museum, Vero
  Beach 053968
Maclaurin Art Gallery, Ayr 043461
MacLean, London 118438
McLean-Alanreed Area Museum,
  McLean 050565
McLean County Arts Center,
  Bloomington 047043
McLean County Historical Society
  Museum, Washburn 054081
McLean County Museum of History,
  Bloomington 047044
McLean Mill, Port Alberni . . . 006673
McLean Museum and Art Gallery,
  Greenock 044482
McLeavey, Peter, Wellington . . 084172
Macleay Museum, University of
  Sydney, Sydney 001543
McLellan Galleries, Glasgow . . 044405
McLeod, Braidwood 098770
McLeod, Paddington, New South
  Wales 062041
MacLeod, Vancouver 140427
McLoughlin, Paul & Linda,
  Strathalbyn 062248
McLurg Museum, Wilkie 007415
McMahon Peadar,
  Carrickmacross 130833
McMahon, Gene, Seattle 124564
McMahons Antiques, Mallow . . 079672
McMahons, Keith, Seaford . . . 062198
MacMaison, New Orleans . . . 095124
McManus Galleries, Dundee . . 044164
McMaster Museum of Art,
  Hamilton 005975
McMichael Canadian Art Collection,
  Kleinburg 006129
Mcmillan, Omaha 096048
McMinn County Living Heritage
  Museum, Athens 046602
McMorland, J&J, Gourock . . . 117870
McMullen Museum of Art, Boston
  College, Chestnut Hill 047625
McMullens, Los Angeles 094399
McMurray, Avoca 098679
McMurtrey, Houston 121085
McNamara House Museum,
  Victoria 053981
McNaught, Hamilton 101320
McNaughtan, Edinburgh 144395
McNay Art Museum, San Antonio 052940
McNee, Tracey, Glasgow 117848
McNeill, Radlett 118952
McNoon, Los Angeles 094400
Macon County Museum Complex,
  Decatur 048202
Macon, Jacques, Ferrières
  (Loiret) 068526
Macon, Jacques, Montargis . . 070185
Macondo, Pittsburgh 123562
Macoun, Dietrich, Hamburg . . 107296,
                                129974
MacPheadris/Warner House,
  Portsmouth 052312
McPhee, G.D., Prahran 062117
McPheeters, Richard, Tulsa . . . 097746
McPherson & Co., Kingston . . 138344
McPherson College Gallery,
  McPherson 050576
McPherson County Old Mill Museum,
  Lindsborg 050317
McPherson Library Special Collections,
  University of Victoria, Victoria . 007329
McPherson Museum, McPherson 050577
McPherson, R. & G., London . . 090317,
                                134717
MacPherson's Mill and Farm
  Homestead, New Glasgow . . 006467
McQuade, Kilbarchan 089683
McQueen, Eddi, Tralee 079713, 130894
McRae, Neil, Auchterarder . . . 117264

Macroom Museum, Macroom . . 024962
Macrorie House Museum,
  Pietermaritzburg 038800
Macrorie Museum, Macrorie . . 006238
Macrosun, Saint Louis 123842, 123843
MAC's – Musée des Arts
  Contemporains, Hornu 003711
Mac's Antique Car Museum,
  Tulsa 053849
McSpadden & Haskell, Chicago . 092742
McSwain, Charlotte . . 092613, 135392
McSweeney, Gerald, Midleton . . 142610
McTague, Harrogate 089435
McTague of Harrogate, Harrogate 117918
McTear, Glasgow 126903
Macula, Paris 136958
Macula, Santa Cruz de Tenerife 115424
McVeigh & Charpentier, London 090318,
                                134718
McWane Center, Birmingham . . 047006
McWhirter, London 090319
Maczulat, E., Braunschweig . . 141639
Mad Art, Saint Louis 123844
Mad Booking, Nashville 145297
Mad Dog and the Pilgrim Booksellers,
  Lander 145201
Mad-Elf-Art, Stockholm 115908
MAD Gallery, Lancefield 099216
Mad Musée, Liège 003795
Madaba, Paris 105050
Madaba Archaeology Museum,
  Ajloun 029758
Madách Imre Emlékmúzeum, Nógrádi
  Történeti Múzeum, Csesztve . . 023587
Madaire, Thierry, Sainte-Agathe-la-
  Bouteresse 073339
Madalyn, Norfolk 095864
Madam Blå Antik, Hadsund . . . 065723
Madam Blå Brukt og Antikk,
  Drammen 084231
The Madam Brett Homestead,
  Beacon 046842
Madam Olsen, Bergen 084219
Madame Tussaud's, London . . 045030
Madame Tussaud's Amsterdam,
  Amsterdam 031954
Madame Tussaud's Wax Museum,
  New York 051419
Madames, Mice & More,
  Indianapolis 093903
Madang Museum, Yomba 034191
Madara Nacionales Istoriko-
  archeologičeski Rezervat,
  Madara 005088
Madcap Antiques, Toronto . . . 064700
Maddaluno, Napoli 131625
Maddermarket Antiques, Norwich 090877
Maddich, Chicago 120189
Maddi's Gallery, Charlotte . . . 119999
Made By Hand, Ashbourne . . . 117254
Made From Australia Galleries,
  Deakin 098960
Made Gallery, Carlton North . . 098866
Made in France, Houston 093733
Made in France, Nashville 095013
Made in Japan, Hyde Park . . . 061645
Made in Nippon, Auckland . . . 083806
Made in the Murray Mallee, Swan
  Hill 062271
Madec, Eliane, Fécamp 125586
Madel, Pierre, Paris 071605
La Madeleine, Bouin 067386
La Madeleine, Cabourg 067546
Madeline & Laupin, Toulon-sur-
  Arroux 105719
Madeline Island Historical Museum, La
  Pointe 050046
Mader, Milwaukee 121934
Madera County Museum, Madera 050579
Maderas Gallery, Tucson 124750
Madert, Mehring 107988
Madhavan Nair Foundation,
  Edapally 024250
Madhouse, Rotorua 113000

Madhouse Fun Furnishings,
  Buffalo 092543
Madhya Pradesh Tribal Research and
  Development Institute, Bhopal . 024170
Madison, Atlanta 119588
Madison Avenue Antique Mall,
  Memphis 094647
Madison Avenue Galleries,
  Toledo 097613
Madison Children's Museum,
  Madison 050597
Madison County Historical Museum,
  Edwardsville 048487
Madison County Historical Society
  Museum, Winterset 054500
Madison Fine Art, New York . . . 095556
Madison Historical Society Museum,
  Madison 050580
Madison-Morgan Cultural Center
  Collection, Madison 050582
Madison Museum of Contemporary
  Art, Madison 050598
Madison Press Books, Toronto . 136687
Madoi, Alessandro, Parma . . . 131729
Madonas Novadpētniecības un
  Mākslas Muzejs, Madona . . . 030160
Madonna House Museum,
  Combermere 005640
Madonna Massimo, Palermo . . 081051
Madoura, Vallauris 105809
Madras Christian College Museum,
  Chennai 024199
MADRE – Museo d'Arte Donnaregina,
  Napoli 026990
Madres Dominicanas – Monasterio
  Sancti Spiritus, Toro 040256
Madrières, Olivier, Laguepie . . 069142
Madrigal Center, Muntinlupa . . 113409
La Madrileña, Madrid 086212
Madron, Chicago 120190
Madsen, Lisbeth, Herning . . . 065751
Madureira, Maria, Porto Alegre . 100989
Madurodam, Den Haag 032179
Mäder, Bern 087292
Mäder, Franz, Basel . . 116092, 138024
Mädi's Antiquitätenlädeli, Aeschlen ob
  Gunten 087183
Maeght, Barcelona 114927
Maeght, Paris 105051, 136959
Maeght Musée, Paris 014667
Mährisch-Schlesisches
  Heimatmuseum, Klosterneuburg 002208
Maejima Memorial Museum,
  Joetsu 028737
Maek, Hyang, Daegu 111588
Maekawa, Luana, Firenze 131198
Mäkelänkadun Antikvariaatti,
  Helsinki 140711
Maels.L, Floriane, Eze 103733
Mälsåkers Slott, Stallarholmen . 040985
Die Männer von Brettheim, Rot am
  See 021497
Mäntsälän Kotiseutumuseo,
  Mäntsälä 011136
Mäntyharjun Museo, Mäntyharju 011139
Mäntyviidan Kijakauppa, Helsinki 140712
Märchen- und Puppenmuseum,
  Schloß Rosenau 002722
Märchenhain Gröhl, Gröhl 001934
Märchenhaus und Heimatmuseum
  Ottenschlag, Ottenschlag . . . 002484
Märchenofen, Fachmuseum
  Gußeiserne Öfen/
  Feuerstättenmuseum, Neu-Ulm 020680
Maere, Jan de, Bruxelles 063498,
                                100639
Märkisches Glasmuseum,
  Stechlin 021968
Märkisches Museum, Kulturforum
  Witten, Witten 022697
Märkisches Museum, Stiftung
  Stadtmuseum Berlin, Berlin . . 017218
Märkisches Ziegelei-Museum Glindow,
  Werder 022571

Märklin-Museum, Göppingen . . 018731
Maertzdorf, Doetinchem 083171
März, Ladenburg 107778
März, Mannheim 107972
Maerzgalerie, Berlin 106368
Maerzgalerie, Leipzig 107842
Maes, Knokke-Heist . . 100745, 128067
Maese, Lewis, Houston 093734
Mäsinge Antik och Konsthantverk,
  Båstad 086641
Maessen, Bad Kissingen 074536, 074537
Maessen, Petra, Bad Kissingen . 074538
Mästerby Antik, Stockholm . . . 087027
Maestro Gallery, Taipei 116930
Mätzelt, R., Hörstel 076545
Mafe, São Paulo 101100
Maffart, Guy, Rennes (Ille-et-
  Vilaine) 072439, 129203
Maffei, São Paulo 101101
Maffei, Giorgio, Torino 142866
Maffeischächte der Grube
  Auerbach-Nitzlbuch, Bergbau-
  und Industriemuseum Ostbayern,
  Auerbach, Kreis Deggendorf . 016641
Maffra Sugarbeet Museum,
  Maffra 001240
Maffucci, Manuel, Lugano . . . 133825
Mafikeng Museum, Mafikeng . . 038780
Mafra & Mafra, Lisboa 132942
Mag, Miami 139421
Mag Presse, Brignoles 140807
Maga Magò, Trieste 142879
Magadanskij Oblastnoj Kraevedčeskij
  Muzej, Magadan 037212
Magan Sangrahalaya Samiti,
  Wardha 024520
Magana Gallery, Windsor, Ontario 101857
Magapa, Granada 133267
Magasin, Duisburg 075553
Le Magasin, Haarlem 083339
Magasin – Centre National d'Art
  Contemporain, Grenoble . . . 013177
Magasin 3 Stockholm Konsthall,
  Stockholm 041016
La Magasin de Cadres, Lille . . 069495,
                                128681
Magasin de Collection Privée,
  Genève 087480
Magasin Eclair, Angers 066611
Le Magasin Pittoresque, Nice . . 141001
Magatani, Tokyo 111463
El Magatzem, La Bisbal
  d'Empordá 085904
Het Magazijn, Amsterdam 082801
Het Magazijn, Gent 063650
Magazin 4, Bregenz 099968
Magazin Etalon, Bucureşti . . . 085097
The Magazine Antiques, New
  York 139422
Magazine Arcade Antiques, New
  Orleans 095125
Magazine Street Gallery, New
  Orleans 122197
MAGAZINPRESSE Verlag GmbH
  Zeitschriften & Co. KG,
  München 137415
Magazinul Colecţiilor, Bucureşti . 085098
Magazyn MM, Warszawa 084732
Magazyn Sztuki + Obieg, Gdańsk 138971
Magazzino d'Arte Moderna,
  Roma 110815
Il Magazzino di Aristide, Milano 080593
Magd.factor, Sellingen 137873
Magdalen College Museum, Wainfleet
  All Saints 046028
Magdalene Gallery, Taunton . . 119212
Magdalenen-Verlag, Holzkirchen 137281
Magdeburger Antiquariat,
  Magdeburg 142097
Magdeburger Museen, Direktion,
  Magdeburg 020210
Magellan Trading, Tucson 097674
Magenta, Florida 138488
Magenta, Nijmegen 112608

Magenta, Roermond 112622
Magenta Fine Arts, West Kirby . 119325
Magers & Quinn, Minneapolis . . 145280
Magersfontein Battlefields Museum,
  McGregor Museum, Kimberley 038766
Mages, Christian de, Nice 070628
Magevney House, Memphis . . . 050780
Maggi, Gabriella, Trieste 132222
Maggia, Giovanni, Verona 081921
Maggie L. Walker National Historic
  Site, Richmond 052555
Maggie May's, North Shields . . . 090851,
                                    134886
Maggie's Book Bazaar, Highgate
  Hill 139771
Maggies Place, Columbus 093032
Il Maggiolino, Bari 079774
Maggs Bros., London 144642
Magg's Shipping, Liverpool 089857
Maggs, John, Falmouth 134382, 144422
Al-Maghraby, Khan, Cairo 103052
Magi Arte, Milano 110374
Magi, Cezar, New Orleans 122198
Magia Imagem, Lisboa 114041
Magic Art Corporation, Tshwane 085679
Magic Changes, Schaumburg . . 139423
Magic Dragon, Los Angeles . . . 121552
Magic Faces, Buffalo 119960
Magic Flute Artworks, Lower
  Swanwick 118668
Magic House-Saint Louis Children's
  Museum, Saint Louis 052808
Magic Lantern Castle Museum, San
  Antonio 052941
Magic Lanterns, Saint Albans . . 091175
Magic Moments, Albuquerque . . 092068
Magic of Quimper, Littlehampton 089850
Magic Oriental Antiques, Abu
  Dhabi 088160
Magic Shop, Columbus 093033
Magica, Beograd 114426
Magical Artroom, Tokyo 111464
Le Magicien, Orchies 070796
La Magie du Verre, La Tour-de-
  Peilz 116672
Magil Fine Art, Dublin 109604
Magill Antiques and Art Gallery,
  Magill 061757
Magin, Liège 140245
Magis, J.-J., Paris 141231
Magistrati, Sandro, Milano 131468
Magistri, Roma 081434, 131965
Magli, Ettore, Bologna 130974
Maglio, Davide, Torino 132162
Magliocchetti, Roma 131966
Magloire, Philippe & Claude,
  Paris 071606
Magna, Coimbra 113965
Magna Gallery, Oxford 090934, 144769
Magnan, Alberto, New York 122836
La Magnanerie du Coudray, Maison
  de la Soie d'Anjou, Le Coudray-
  Macquard 013621
Magnanerie Muséalisée de la Roque,
  Molezon 014192
Magne, Michel, Rouen 141322
Magne, René, Saint-Julien-de-
  Jordanne 072860
Magnesitmuseum, Radenthein . 002559
Magnet, Wilfried, Völkermarkt . . 100214,
                                    136568
Magnet, Wilfried, Wien 100347, 140099
Magnetawan Historical Museum,
  Magnetawan 006240
Magnificat, Imola 080306
Magnifico, Albuquerque 119490
Magnin, François, Paris 141232
Magnino, Mondeville 070158
Magnitogorskaja Kartinnaja Galereja,
  Magnitogorsk 037214
Magnitogorskij Kraevedčeskij Muzej,
  Magnitogorsk 037215
Magnolia Editions, Oakland 138437
Magnolia Grove-Historic House

Museum, Greensboro 049248
Magnolia Manor Museum, Cairo 047343
Magnolia Mound Plantation, Baton
  Rouge 046818
Magnolia Plantation, Charleston 047556
Magnolias, Houston 121086
Des Magnolias comme Autrefois,
  Paris 071607
Magnoni, Massimo, Roma 131967
Magnum Ars, Moskva . 085191, 114276,
                        133012,   143646
Magnus Antikt, Alingsås 086627
Magnus Dagestad Museet, Voss
  Folkemuseum, Voss 034094
Magnus Edizioni, Fagagna 137640
Magnussen, Bergen 084220
Magnussen, Birte, København . . 065934
Magoffin Home, El Paso 048513
Magome Studio, Tokyo 132312
Magpie, Hinckley 089531
Magpie, Windsor, Ontario 064904
Magpie Antiques, Long Melford . 090621
Magpie Antiques, Swansea 091529,
                                    135111
Magpie Auctions, Largs Bay . . . 125205
Magpie Books, Fremantle 139741
Magpie Gallery, Uppingham . . . 119287
Magpie Jewellers and Antiques,
  Evesham 089226
Magpies, Rushden 091154
Magpie's Nest, Salt Lake City . . 096905,
                                    123973
Magrutsch, Verena, Wien 127946
Magsingal National Museum,
  Magsingal 034502
Maguire, Martin, Ballina 079444
Magus, Frankfurt am Main 075807
Magus, Seattle 145592
Maguy, Laurent, Font-Romeu-Odeillo-
  Via 068557
Magyar Állami Operaház Múzeuma,
  Budapest 023509
Magyar Alumíniumipari Múzeuma,
  Székesfehérvár 023937
Magyar Cserkészszövetség
  Gyűjtemény, Budapest 023510
Magyar Elektrotechnikai Múzeum,
  Budapest 023511
Magyar Építészeti Múzeum,
  Kulturális Örökségvédelmi Hivatal,
  Budapest 023512
Magyar Építőipari Múzeum,
  Veszprém 024041
Magyar Ferences Könyvtár Gyöngyösi
  Muemlék Könyvtára, Gyöngyös 023652
Magyar Filmtörténeti Gyűjtemény,
  Budapest 023513
Magyar Földrajzi Gyűjtemény, Érd 023626
Magyar Fotográfiai Múzeum,
  Kecskemét 023728
Magyar Fotográfusok Háza,
  Budapest 109162
Magyar-Francia Galéria,
  Budapest 109163
Magyar Fuszerpaprika Múzeum,
  Népművészeti Tájház, Kalocsa . 023704
Magyar Iparművészeti Egyetema,
  Budapest 055765
Magyar Képzőművészek és
  Iparművészek Szövetsége,
  Budapest 023514, 059340
Magyar Kepzőművészeti Egyetem,
  Budapest 055766
Magyar Kereskedelmi és
  Vendéglátóipari Múzeum,
  Budapest 023515
Magyar Mezőgazdasági Múzeum,
  Budapest 023516
Magyar Mezőgazdasági Múzeum
  Tanyamúzeuma, Magyar
  Mezőgazdasági Múzeum,
  Lajosmizse – Bene 023767
Magyar Motorok Múzeuma,
  Bodrogkeresztúr 023448

Magyar Múzeumok, Budapest . . 138772
Magyar Naiv Művészek Múzeuma,
  Katona József Múzeum,
  Kecskemét 023729
Magyar Nemzeti Bank
  Látogatóközpont, Budapest . . . 023517
Magyar Nemzeti Galéria,
  Budapest 023518
Magyar Nemzeti Múzeum,
  Budapest 023519
Magyar Numizmatikai Társulat,
  Budapest 059341
Magyar Olajipari Múzeum,
  Zalaegerszeg 024053
Magyar Ortodox Egyházművészeti
  Múzeum, Miskolc 023788
Magyar Régészeti és
  Művészettörténeti Társulat,
  Budapest 059342
Magyar Technikai Múzeum Alapítvány,
  Budapest 023520
Magyar Térképészet Szakmatörténeti
  Múzeum, Hadtörténeti Múzeum,
  Budapest 023521
Magyar Természettudományi Múzeum,
  Budapest 023522
Magyar Tudományos Akadémia
  Művészettörténeti Kutatóintézet,
  Budapest 055767
Magyar Vasúttörténeti Park,
  Budapest 023523
Magyar Vegyészeti Múzeum,
  Várpalota 024033
Magyar Zsidó Múzeum és Levéltár,
  Budapest 023524
Magyarországi Kalcitkristályok Színes
  Világa, Fertőrákos 023637
Maha Al Saqqa Center – Palestinian
  Folk Museum, Bethlehem 034160
Maha Wirawong National Museum,
  Nakhon Ratchasima 042701
Mahaboudha Art Centre,
  Kathmandu 112078
Mahaffie Stagecoach Stop and Farm,
  Olathe 051743
Mahagonny, Berlin 074872
Mahaki, Kawakawa 112930
Mahakoshal Art Gallery, Raipur . 024454
Mahal, Martina, Hünstetten 141954
Mahana Club Galerie, Dole 103702
Mahana Gallery, Gisborne 112899
Mahant Ghasidas Memorial Museum,
  Raipur 024455
Mahara Gallery, Paraparaumu . . 112988
Maharaja Banaras Vidya Mandir
  Museum, Varanasi 024516
Maharaja Fate Singh Museum,
  Vadodara 024508
Maharaja Sawai Man Singh II
  Museum, Jaipur, Rajasthan . . 024319
Maharaja's Emporium, Montréal 064441
Mahatma Gandhi Hindi Sangrahalaya,
  Kalpi 024336
Mahatma Gandhi Museum,
  Mangalore 024394
Mahatma Gandhi Museum,
  Verulam 038874
Mahatma Phule Vastu Sangrahalaya,
  Pune 024447
Mahaux, Pierre, Bruxelles 063499
Mahboubian, London 090320
Mahé, Gérard, Saint-Ouen 073141
Mahé, Joël, Châteaudun 067882
Mahe, Ros, Moustiers-Sainte-
  Marie 070377, 104309
Mahendra Museum, Kathmandu 031872
Mahérault, Pascal, Lagny-sur-
  Marne 069140
Maherault, Pascal, Vaires-sur-
  Marne 073929
Mahgoub, Mostafa Hassan,
  Alexandria 066221
Mahieu, Jouarre 140859
Mahieu, Paris 129073

Mahl, Lesley, Schwelm 078424
Mahla & Co., Pittsburgh 096308
Mahler, Launceston 061726
Mahler, Irene, Seattle 124565
Mahler, Jakob, Badenweiler . . . 074647
Mahler, K., Hemmoor 076491
Mahler, Norbert, Schwarzach am
  Main 078415
Al-Mahmood, Manama 063152, 100442
Mahn- und Gedenkstätte Düsseldorf,
  Düsseldorf 018040
Mahn- und Gedenkstätte
  Isenschnibber-Feldscheune,
  Gardelegen 018588
Mahn- und Gedenkstätte Ravensbrück,
  Stiftung Brandenburgische
  Gedenkstätten, Fürstenberg,
  Havel 018540
Mahn- und Gedenkstätte
  Steinwache, Stadtarchiv Dortmund,
  Dortmund 017959
Mahn- und Gedenkstätte Wernigerode,
  Wernigerode 022582
Mahn- und Gedenkstätten Wöbbelin,
  Wöbbelin 022716
Mahogany Shop, San Diego . . . 136396
Mahone Bay Settlers Museum,
  Mahone Bay 006242
Mahoney, Geelong 099072
Mahonia, Helsinki 066337
Mahr, Thomas, Sonnberg 127875
Mahr, Thomas, Wien 127947
Mahtra Talurahvamuuseum,
  Juuru 010621
Mahymobiles – Musée Automobile,
  Leuze-en-Hainaut 003788
Mai, Ha Noi 125083, 125084
Mai, Ho Chi Minh City 125132
Mai, Sint-Niklaas 100816
Mai Hien & Ahn Khanh, Ha Noi . 125085
Mai 36 Galerie, Zürich 111384
Mai, Phuong, Ho Chi Minh City . 125133,
                                    125134
Maia, Alice R. Silva, Funchal . . 084807
Maia, Ana Paula Silva, Lisboa . . 084911
Maia, Paulo Silva, Porto 085036
Maia, Philip, Dallas 093170
Maibaum, Walter, New York . . . 122837
Maici Nonoi, Philadelphia 096134
Maidens & Foster, Hawkes Bay . 126318
Maidhof Bros., San Diego 097102
Maidstone and District Museum,
  Maidstone 006243
Maidstone Museum and Bentlif Art
  Gallery, Maidstone 045193
Maier & Stumvoll, Lengenfeld bei
  Krems 062709
Maier-Hahn, Kiki, Düsseldorf . . . 106809
Maier-Mangold, N., Hüttisheim . 076587
Maier Museum of Art, Randolph-
  Macon Woman's College,
  Lynchburg 050533
Maier jun., A., Ellwangen 075599
Maier, B., Kirchheim am Neckar 076805
Maier, Bernhard, Ellwangen . . . 106873
Maier, Frieder, Nürtingen 142249
Maier, Ingrid, Neuburg an der
  Donau 077719
Maier, Josef, Innsbruck 100052
Maier, Konrad, Schwäbisch
  Gmünd 108586
Maier, L., Bruchsal 075597
Maier, M., Neresheim 077700
Maier, Manfred, Heidenheim an der
  Brenz 076466
Maier, Manfred, Wetzikon 087900
Maier, René, Saint-Rémy-de-
  Provence 073290
Maier, Thomas, Olbernhau 077920
Maier & Co., Thomas, Stuttgart . 078617
Maier, Werner, Vandans 062866, 127891
Maierhofer, Krems 062690
Maigne Arsene, Gramat 013140
Maigret, Thierry de, Paris 135736

Maihaugen, De Sandvigske Samlinger,
  Lillehammer 033742
Maillard, Julie, Paris 105052
Maillard, Olivier, Villard-de-Lans 074125
Maillard, Valérie, Chatenoy-le-
  Royal 067914
Maille, Bernard, Paris 141233
Maillon Manquant, Rouen . . . 072597
Mailly, Chantal, Versailles . . . 074060
Mailors Flat Antiques, Mailors
  Flat 061758
Mailout, Chesterfield 139201
Mailund Toft, Aage, Kolding . . 065977
Main Access Gallery, Winnipeg . 101882
Main Art Gallery, California State
  University, Fullerton, Fullerton . 048994
Main Art Tree, Pacific Paradise . 099507
Main Centre Heritage Museum, Main
  Centre 006244
Main d'Or, Arzon 103280
Main Fine Art, Glasgow 117849
Main Furniture, Hamilton 064324
Main Gallery of Henry Street
  Settlement, New York 051420
Main Line Art Center, Haverford 049406
Main Place for Antiques, San
  Antonio 096995
Main Point Books, Edinburgh . 144396
Main Street Collectors Mall,
  Portland 096435
Main Street Gallery, Hahndorf . 099103
Main Street Gallery, Minneapolis 122010
Main Street Gallery, Montville . 099364
Main Trend, Taipei 116931
Main, C. & K.M., Green
  Hammerton 089386
Mainardi, Livorno 131347
Mainart, Philadelphia 136095
Maine Antique Digest, Waldoboro 139424
Maine Art Gallery, Wiscasset . . 054506
Maine Association of Museums,
  Augusta 060577
Maine College of Art, Portland . 057706
Maine Durieu, Paris 105053
Maine Historical Society Museum,
  Portland 052281
Maine Maritime Museum, Bath . 046809
Maine Photographic Workshops,
  Rockport College, Rockport . . 057789
Maine State Museum, Augusta . 046667
Mainfränkische Münzhandlung,
  Würzburg 079086
Mainfränkisches Museum Würzburg,
  Würzburg 022791
Mainhill Gallery, Ancrum 117240
Mainhill Gallery, Jedburgh . . . 089655
Maini & Veli, Helsinki 066338
Mainichi Art Collection, Nagoya . 029118
Mainieri, Giuseppe, Milano . . . 080594
Mainly Books, Perth 139871
Mainstream Publishing Company,
  Edinburgh 138156
Maio, Michel de, Ippling 068873
Maior, Palma de Mallorca 115358
Maior, Pollença 115394
Maiquez, Tampa 097576
Mair, Prof. Ursula, Salzburg . . 127852
Mairangi Arts Centre, Auckland . 033153
Le Maire, Liège 063771
Maironio Lietuvių Literatūros Muziejus,
  Kaunas 030368
Mairose, Rösrath 078263
De Mais, Getxo 086011
Maisa Gallery, Cairo 103053
Maisel, Albuquerque 119491
Maisel's Brauerei- und Büttnerei-
  Museum, Privatbrauerei Gebr. Maisel,
  Bayreuth 016996
Maisenbacher, Christoph, Trier . 108779,
                                       137533
La Maison, Bourne End 088529, 134128
Maison, Limerick 079657
La Maison, London . . . 090321, 134719
La Maison, Memphis 094648

Maison & Jardin, Paris 071608
Maison & Maison, Houston . . . 093735
Maison 14, Washington 105052
Maison à Pondalez, Musée de Morlaix,
  Morlaix 014327
Maison A. Ravier, Morestel . . . 014318
Maison Adolphe Hardy, Dison . 003557
Maison André-Benjamin-Papineau,
  Laval 006171
Maison André Derain,
  Chambourcy 012426
Maison Antique, Catania 080037
Maison Arthur, Clermont-Ferrand 103605
Maison-Atelier Foujita, Villiers-le-
  Bacle 016295
Maison Bakerdijan, Paris 071609
Maison Beys, Saint-Ouen . . . 073142
Maison Bibelot, Firenze 080165, 126192
Maison Blanche, Singapore . . 085466
La Maison Bleue, Chinon 103593
Maison Buttin-de-Loës,
  Grandvaux 041563
Maison Centrale des Filles de la
  Charité du Sacré-Coeur de Jésus,
  Sherbrooke, Québec 007023
Maison Champenoise, Oeuilly . 014546
Maison Chaussavoine, Brece . 067453
Maison Chevalier, Québec . . . 006731
La Maison Coloniale, Verona . . 081922
Maison Communale Gérard Philippe,
  La Garde 013420
Maison Creole Eureka, Moka . 030789
Maison Cyr – Musée de Chapelle,
  Saint-Basile 006856
Maison d'Abraham Mazel à Falguières,
  Mialet 014165
La Maison d'Amandine, Bruxelles 063500
Maison d'Art, Monaco 082518
Maison D'Art, Toowoomba . . . 062331
La Maison d'Art, Us 105782
Maison d'Art Alsacienne,
  Strasbourg 015694
Maison d'Art Bourgogne du Sud,
  Marcigny 104154
Maison d'Art Montegrappa,
  Bologna 079850
Maison d'Artiste de la Grande Vigne,
  Dinan 012793
Maison d'Auguste Comte, Paris . 014668
Maison Daumm, Saint-Ouen . . 073143
La Maison d'Aure, Musée Paysan –
  Arts et Traditions Populaires,
  Sailhan 015201
Maison d'Autrefois, Chasseneuil-du-
  Poitou 012475
Maison de Balzac, Paris 014669
Maison de Boua Kang Bung, Luang
  Prabang 030086
Maison de Brian, Simiane-La-
  Rotonde 105674
La Maison de Charrier, Valbonne 073934
Maison de Famille, Montargis . 070186
Maison de Géologie et de
  l'Environnement de Haute-Alsace,
  Sentheim 015773
Maison de George Sand, Nohant-
  Vic 014517
Maison de Gergovie, La Roche-
  Blanche 013461
Maison de Jean Monnet, Bazoches-
  sur-Guyonne 011930
La Maison de Jules, Gassin . . 103763
Maison de Jules Verne, Amiens 011631
Maison de Kerner, Riantec . . . 015084
Maison de la Baie, Courtils . . 012705
Maison de la Baie du Mont Saint-
  Michel, Vains 016082
Maison de la Beauce, Orgeres-en-
  Beauce 014576
Maison de la Bible, Lyon 140923
Maison de la Broderie, Villers-
  Outréaux 016285
Maison de la Canne, Les Trois-
  Ilets 030779

Maison de la Chataigne, Mourjou 014362
Maison de la Châtaigne et du
  Châtaignier, Saint-Martin-de-
  Boubaux 015447
La Maison de la Chicorée,
  Orchies 014574
Maison de la Culture Ahuntsic-
  Cartierville, Montréal 006377
Maison de la Culture Côte-des-Neiges,
  Montréal 006378
Maison de la Culture Mercier,
  Montréal 006379
Maison de la Culture Plateau-Mont-
  Royal, Montréal 006380
Maison de la Culture Pointe-aux-
  Trembles, Montréal 006381
Maison de la Culture Rosemont-Petite-
  Patrie, Montréal 006382
Maison de la Dentelle, Urçay . 016071
Maison de la Dernière Cartouche,
  Bazeilles 011928
La Maison de la Dordogne
  Quercynoise, Souillac 015846
Maison de la Faïence,
  Badonviller 011861
Maison de la Faïence, Musée de la
  Céramique, Desvres 012763
Maison de la Faune, Murat
  (Cantal) 014384
Maison de la Faune, Vaujany . 016138
Maison de la Flore,
  Longchaumois 013857
Maison de la Fonderie, Vrigne-aux-
  Bois 016330
Maison de la Forêt, Raismes . . 015030
La Maison de la Forêt et du Bois,
  Ecomusée de la Bresse, Saint-Martin-
  en-Bresse 015455
Maison de la Fourdonne, Saint-Cirq-
  Lapopie 015261
Maison de la Lentille, Chouday . 012572
Maison de la Loire, Montlouis-sur-
  Loire 014270
Maison de la Magie Robert Houdin,
  Blois 012077
Maison de la Manufacture d'Armes
  Blanches, Klingenthal 013352
Maison de la Mariée, Saint-
  Joachim 015392
Maison de la Mer, Courseulles-sur-
  Mer 012700
Maison de la Metallurgie et de
  l'Industrie de Liège, Liège . . 003796
Maison de la Meunerie, Nieul-sur-
  l'Autise 014495
Maison de la Musique Mécanique,
  Mirecourt 014183
Maison de la Nature, Eze 012944
Maison de la Nature, Paris . . . 014670
Maison de la Nature et de l'Oiseau,
  Musée Henri Boileau, Neuville-sur-
  Ailette 014461
Maison de la Négritude et des Droits
  de l'Homme, Champagney . . 012432
Maison de la Pataphonie, Dinant 003554
Maison de la Pétanque, Vallauris 016101
Maison de la Pierre de Volvic,
  Volvic 016322
Maison de la Pierre et du Ciment,
  Montalieu-Vercieu 014217
Maison de la Pisciculture, Sizun 015821
Maison de la Poesie, Montréal 101429
Maison de la Pomme et de la Poire,
  Barenton 011897
Maison de la Pomme et de la Poire,
  Carrouges 012347
Maison de la Pomme et Forge, Sainte-
  Opportune-la-Mare 015629
Maison de la Poterie, Bouffioulx 003386
Maison de la Réserve Naturelle du
  Lac de Remoray, Labergement-
  Sainte-Marie 013512
Maison de la Résistance, Chasseneuil-
  sur-Bonnieure 012476

Maison de la Rivière, Châteauneuf-sur-
  Sarthe 012497
Maison de la Rivière, Chinon . . 012560
Maison de la Rivière et du Pêcheur,
  Saint-Georges-de-Montaigu . . 015330
Maison de la Rivière, de l'Eau et de la
  Pêche, Sizun 015822
Maison de la Ruralité, Thuillies . 004062
Maison de la Sainte-Victoire, Saint-
  Antonin-sur-Bayon 015224
Maison de la Science, Liège . . 003797
Maison de la Sirène, Collonges-la-
  Rouge 012624
Maison de la Thiérache, Liart . 013790
Maison de la Treille, Paris . . . 116645
Maison de la Vallée d'Ossau,
  Laruns 013580
Maison de la Vie Paysanne, Han-sur-
  Lesse 003671
Maison de la Vie Rurale, Musée
  d'Archéologie et Minéralogie,
  Massiac 014078
Maison de la Vigne, Essoyes . 012913
Maison de l'Abbé Grégoire,
  Embermenil 012866
Maison de l'Abeille et du Miel,
  Riez 015090
Maison de l'Agoût, Fraisse-sur-
  Agout 013041
Maison de l'Air, Paris 014671
Maison de l'Amandier, Saint-Rémy-de-
  Provence 015538
Maison de l'Amérique Latine,
  Paris 105054
Maison de l'Amiral, Pouldreuzic 014980
Maison de l'Âne, Beauvoir-sur-
  Mer 011959
Maison de l'Antiquité, Moncé-en-
  Belin 070153
Maison de l'Araire, Yzeron . . . 016366
Maison de l'Archéologie des Vosges
  du Nord, Niederbronn-les-Bains 014494
Maison de l'Ardoise, Rimogne . 015094
Maison de l'Argile, Ateliers Thérèse
  Neveu, Aubagne 011764
La Maison-de-L'Autographe,
  Paris 141234
Maison de l'Automate, Lunas . 069659
Maison de l'Automate,
  Tournemire 016008
Maison de l'Eau, Auxerre 011821
La Maison de l'Eau, Mallièvre . 013984
Maison de l'Eau, Neuvy-sur-
  Barangeon 014464
Maison de l'Eau et de la Pêche,
  Besse-en-Chandesse 012007
Maison de l'Eau et de la Pêche,
  Neuvic (Corrèze) 014455
Maison de l'Eclusier, Saint-Malo-de-
  Guersac 015442
Maison de l'Émail, Morez 014321
Maison de l'Espace Vert, Nancy 014402
Maison de l'Etain, Villedieu-les-
  Poêles 016235
Maison de l'Éventail, Bèze . . . 012019
Maison de l'Histoire, Larra . . . 013576
Maison de l'Histoire de la Terre de
  Gorze, Gorze 013134
Maison de l'Industrie, Saint-
  Berthevin 015235
Maison de Loire, Belleville-sur-
  Loire 011978
Maison de l'Oiseau et du Poisson,
  Outines 014598
Maison de l'Oiseau, Musée des
  Oiseaux, Lancheres 013545
Maison de l'Olivier, Le Val . . . 013708
Maison de Louis Pasteur, Arbois 011703
Maison de Louise, Les Quatre-Routes-
  du-Lot 069421
La Maison de l'Ours, Saint-Lary-
  Soulan 015406
Maison de l'Outil et de la Pensée
  Ouvrière, Troyes 016049

Maison de Magayant, Sainte-Marie-de-Re 015619
Maison de Marie-Louise, Samoens 073393
Maison de Nos Aïeux, Sainte-Famille 006949
Maison de Nostradamus, Salon-de-Provence 015661
Maison de Pauline, Gréoux-les-Bains 013188
Maison de Pays, Bage-le-Chatel 011862
Maison de Pays, Hede 013246
Maison de Pays, Melle 014116
Maison de Pays, Mornant 014333
Maison de Pays, Pont-sur-Sambre 014947
Maison de Pays, Sainte-Marie-aux-Mines 015617
Maison de Pays, Seuilly 015788
Maison de Pays, Tourzel-Ronzieres 016027
La Maison de Poupée, Paris ... 071610
Maison de Provence, Atlanta ... 092151
Maison de Provence, New Orleans 095126
Maison de Robert Schumann, Scy-Chazelles 015745
La Maison de Russie, Aix-les-Bains 066482
Maison de Sainte Marie-Madeleine-Postel, Barfleur 011898
Maison de Sainte-Thérèse, Lisieux 013827
Maison de Troglodytique Seigneuriale, Turquant 016064
La Maison de Van Gogh, Auvers-sur-Oise 011815
Maison de Victor Hugo, Paris .. 014672
Maison de Victory, Le Lavandou 069293
La Maison dell'Antico Arredo, Modena 080803
Maison d'Elsa Triolet et Louis Aragon, Saint-Arnoult-en-Yvelines ... 015227
Maison d'Émile Zola, Musée de l'Affaire Dreyfus, Medan 014108
Maison Départementale de l'Innovation, Centre de culture scientifique et technique, Clermont-Ferrand 012594
Maison d'Erasme, Bruxelles ... 003447
Maison des Abeilles, Brignoles . 012225
Maison des Alpages, Besse-en-Oisans 012009
Maison des Amis du Blanc, Le Blanc 013607
Maison des Ancêtres de Bossuet, Seurre 015791
Maison des Archers, Musée de la Villemarque et des Traditions Bretonnes, Quimperlé 015023
Maison des Artisans, Angers .. 011651
Maison des Artisans & Créateurs, Espace Saint-Eloi, Séez 015751
Maison des Artisans d'Art, Sarzeau 105657
La Maison des Arts, Bages (Aude) 011864
Maison des Arts, Malakoff ... 013979
Maison des Arts, Saint-Martin . 109112
Maison des Arts du Belvédère Dar El-Founoun, Tunis 042782
Maison des Arts et du Vin, La Réole 013453
Maison des Arts et Tradition Populaires de la Landes Médocaine, Carcans 012326
Maison des Arts et traditions Populaires, Couvent des Cordeliers, Châteauroux 012504
Maison des Arts Ruraux, Saint-Germain-de-Marencennes ... 015338
Maison des Chapais, Saint-Denis 006865
Maison des Chapais, Saint-Denis-de-la-Bouteillerie 006866

Maison des Châteaux Forts, Obersteinbach 014545
Maison des Cinq Sens, Paris ... 014673
Maison des Compagnons du Devoir du Tour de France, Strasbourg ... 015871
Maison des Deux Marines, Briare 012217
Maison des Eaux Minérales, Vic-sur-Cère 016196
Maison des Esclaves, Dakar ... 038115
Maison des Etangs, Saint-Viâtre 015587
Maison des Fortifications, Maubeuge 014082
Maison des Fromages, Les Moussières 013757
Maison des Fromages d'Auvergne, Egliseneuve-d'Entraigues ... 012858
Maison des Géants, Ath 003327
Maison des Insectes, Charlesbourg 005589
Maison des Jardies-Musée Léon Gambetta, Centre des Monuments Nationaux, Sèvres 015794
Maison des Jésuites de Sillery, Sillery 007038
Maison des Johnnies, Roscoff .. 015153
Maison des Liqueurs de l'Auzonnet, Saint-Florent-sur-Auzonnet ... 015319
Maison des Marais, Marchesieux 014002
Maison des Marais Mouillés, Coulon 012692
Maison des Métiers du Cuir, Graulhet 013167
Maison des Métiers du Monde Rural, Tamnay-en-Bazois 015901
Maison des Minéraux, Crozon .. 012726
Maison des Paludiers, Saille ... 015202
Maison des Papillons, Saint-Tropez 015578
Maison des Rochers, Graufthal . 013166
Maison des Sirènes et des Siréniens, Réserve Géologique de Haute-Provence, Castellane 012353
Maison des Sports Bretons, Berrien 011997
Maison des Traditions, Chassignolles 012477
Maison des Traditions, La Chapelle-des-Marais 013381
Maison des Tresses et des Lacets, La Terrasse-sur-Dorlay 013494
Maison des Ventes, Saint-Gaudens 072790
Maison Design, Salt Lake City . 096906
Maison d'Histoire Locale, Cocheren 012610
Maison Dieu, Faversham 044339
Maison Dieu, Ospringe 045443
Maison dite de François 1er, Aubigny-sur-Nère 011775
Maison dite de la Duchesse Anne, Morlaix 014328
Maison Docteur Joseph-Frenette, Causapscal 005581
Maison d'Offwiller, Offwiller ... 014549
Maison d'Ossau, Musée du Parc National, Arudy 011753
Maison du 15ème Antiquités, Nogent-l'Artaud 070724
La Maison du Bal, Los Angeles . 094401
Maison du Bâti, Sansac-de-Marmiesse 015677
Maison du Bison d'Europe, Sainte-Eulalie-en-Margeride 015610
La Maison du Blé et du Pain, Echallens 041453
Maison du Boutis Les Cordelles, Calvisson 012295
Maison du Buronnier, Laveissière 013598
La Maison du Cadre, Paris ... 105055
La Maison du Cerf, Ligny-le-Ribault 013800
Maison du Cheval, Corlay 012672
Maison du Cidre et du Terroir, Le Hezo 013648

Maison du Collectionneur, Bordeaux 067342
La Maison du Collectionneur, Paris 071611
Maison du Colon, Ville-Marie .. 007342
Maison du Comté, Poligny 015617
Maison du Fier, Les Portes-en-Ré 013759
Maison du Fleuve Rhône, Givors 013119
Maison du Forgeron et Maréchal Ferrand, Carnusse 012339
Maison du Fromages et de la Vache Salers, Salers 015646
Maison du Gabardan, Gabarret . 013070
Maison du Granit, Lac-Drolet .. 006141
Maison du Haut-Rhône, Centre Culturel, Seyssel 015804
Maison du Jazz de Liège & de la Communauté Française, Liège 003798
Maison du Journalier, Pleaux .. 014886
Maison du Kochersberg, Truchtersheim 016057
La Maison du Lac, Les Salles-sur-Verdon 013767
Maison du Lauzeron, Champclause 012435
Maison du Lin, Routot 015182
Maison du Littoral, Perros-Guirec 014848
La Maison du Livre, Leura 139805
Maison du Livre et de l'Affiche, Chaumont (Haute-Marne) 012527
Maison du Livre, de l'Image et du Son, Artothèque, Villeurbanne . 016290
Maison du Luthier-Musée, Jenzat 013317
Maison du Marbre, Rinxent ... 015095
Maison du Mont Aiguille Vercors-Trièves, Chichilianne 012556
Maison du Notaire, Trois-Pistoles 007241
Maison du Pain d'Alsace, Sélestat 015755
Maison du Parc, Sorèze 015837
Maison du Parc et des 4 Montagnes, Autrans 011808
Maison du Parc National des Ecrins, Vallouise 016109
La Maison du Parc Naturel des Plaines de l'Escaut, Bon-Secours 003376
Maison du Parc Naturel Régional de Camargue, Saintes-Maries-de-la-Mer 015639
Maison du Parc Naturel Régional de Lorraine, Bisping 012049
Maison du Parc Naturel Régional des Ballons des Vosges, Munster . 014383
Maison du Parc Naturel Régional des Vosges du Nord, La Petite-Pierre 013449
Maison du Parc Naturel Régional du Luberon, Musée de Paléontologie, Apt 011699
Maison du Parc naturel régional Normandie-Maine, Carrouges . 012348
Maison du Patrimoine, Argentat 011719
Maison du Patrimoine, Beauchastel 011935
Maison du Patrimoine, Blanquefort 012064
Maison du Patrimoine, Bourguignon-les-Morey 012173
Maison du Patrimoine, Grande-Synthe 013147
Maison du Patrimoine, Lampaul-Guimiliau 013543
Maison du Patrimoine, Le Crès . 013622
Maison du Patrimoine, Lesterps 013776
Maison du Patrimoine, Marsannay-la-Côte 014037
Maison du Patrimoine, Pleyben . 014894
Maison du Patrimoine, Sarcelles 015680
Maison du Patrimoine, Villard-de-Lans 016220
Maison du Patrimoine Archéologique, Voingt 016319

Maison du Patrimoine Culturel, Grand-Bassam 008874
Maison du Patrimoine de Saint-Chef, Saint-Chef 015252
Maison du Patrimoine des Kerhorres, Le Relecq-Kerhuon 013694
Maison du Patrimoine d'Oloron et du Haut Béarn, Oloron-Sainte-Marie 014562
Maison du Patrimoine et Ilsle, Crémieu 012718
Maison du Patrimoine Historique du Donezan, Rouze 015183
Maison du Patrimoine Maritime, Camaret-sur-Mer 012297
Maison du Patrimoine, Musée de Matour, Matour 014080
Maison du Pays des Étangs, Tarquimpol 015918
Maison du Pays Roussillonnais, Salle de l'Edit, Roussillon (Isère) ... 015179
La Maison du Père Mousset, Vrigny 016331
Maison du Platin, Flottille en Pertuis, La Flotte 013418
Maison du Portal, Levens 013779
Maison du Protestantisme, Nîmes 014499
Maison du Protestantisme Poitevin Beaussais, La Couarde 013401
Maison du Recteur et de la Paroisse, Loqueffret 013869
Maison du Recteur et des Pilhaoueriens, Loqueffret 013870
Maison du Roi-Musée de la Ville, Broodhuis-Museum van de Stad, Bruxelles 003448
Maison du Roy, Saint-Antonin-Noble-Val 072674
Maison du Rugby, Avignonet-Lauragais 011845
Maison du Sabot Neufchâtelois, Neufchâtel-en-Saosnois 014450
Maison du Sabotier, Ecomusée de la Brière, La Chapelle-des-Marais 013382
Maison du Saint-Curé d'Ars, Ars-sur-Formans 011749
Maison du Saint-Nectaire, Saint-Nectaire 015475
Maison du Sauvetage, Grand-Fort-Philippe 013145
Maison du Seigle, Menessaire . 014124
Maison du Sucre, Frasnes-lez-Anvaing 003606
Maison du Terril, Rieulay 015088
Maison du Terroir, Saint-Martin-Lacaussade 015456
Maison du Tisserand, Musée Atelier des Toisles, Quintin 015025
Maison du Val, Abondance 011533
Maison du Val d'Azun, Arrens-Marsous 011746
Maison du Val de Villé, Albé ... 011576
Maison du Verre, Puy-Guillaume 015011
Maison du Vieill Alby, Albi 011582
Maison du Viet Nam, Paris 105056
Maison du Vieux-Zinal, Zinal ... 042183
Maison du Village de Val-David, Val-David 007264
Maison Dumulon, Rouyn-Noranda 006833
Maison et Atelier de Jean-François Millet, Barbizon 011892
Maison et la Tour de Mélusine, Vouvant 016328
Maison et Sculptures de Jean Cacoud, Fransèches 013045
Maison Européenne de la Photo, Paris 014674
La Maison Fassier, Rémalard .. 072407
Maison Française, West Hollywood 097908
Maison Francis Jammes, Orthez 014593
Maison Gallet, Paris 071612
Maison Gerard, New York 095557

Maison Guillaume de Rubrouck,
Rubrouck 015186
Maison Hamel-Bruneau, Sainte-
Foy 006951
Maison Henri IV, Cahors 012287
Maison Historique de Léon et Jeanne
Blum, Jouy-en-Josas 013333
Maison Ideale, San Francisco . . 097278
Maison J.A. Vachon, Sainte-Marie-de-
Beauce 006956
Maison Jacques Prévert, Omonville-la-
Petite 014564
Maison Jean-François Millet, Gréville-
Hague 013192
Maison Jean Giono, Association des
Amis de Jean Giono, Manosque 013993
Maison Jean-Vilar, Bibliothèque
Nationale de France, Avignon . 011833
Maison Jeanne d'Arc, Orléans . . 014584
Maison Jonathas, Musée Communal
de la Tapisserie, Enghien 003578
Maison la Laurière, Québec 101509
Maison Lamontagne, Rimouski . 006805
Maison Lansyer, Loches-en-
Touraine 013845
Maison Léon-Provancher, Cap-
Rouge 005552
Maison LePailleur, Châteauguay 005603
Maison Levanneur, Chatou 012521
Maison Littéraire de Victor Hugo,
Bièvres 012035
Maison Lorraine de la Polyculture,
Musée Vivant, Lucey 013915
Maison Louis Bertrand, L'Isle-
Verte 006199
Maison Louis-Hippolyte Lafontaine,
Boucherville 005445
La Maison Lune de Miel, Antony 066685
Maison Marc, Saint-Ouen 073144
Maison Marie Henry, Clohars-
Carnoët 012607
Maison Médard-Bourgault, Saint-Jean-
Port-Joli 006881
Maison Michel, Coppet 041427
Maison-Musée de Jeanne Jugan,
Cancale 012307
Maison-Musée de Kerlard, Ile-de-
Groix 013286
Maison-Musée des Charcot, Neuilly-
sur-Seine 014453
Maison-Musée des Sciences, Lettres
et Arts de Cholet et de sa Région,
Cholet 012567
Maison-Musée du Parc Naturel
Régional du Haut-Jura, Lajoux 013535
Maison Musée René Descartes,
Descartes 012762
Maison-Musée Ted Jacobs, Les
Cerqueux-sous-Passavent 013746
Maison Natale de Claude Gellée,
Chamagne 012413
Maison Natale de Jean Giraudoux,
Bellac 011971
Maison Natale de Jeanne d'Arc,
Domremy-la-Pucelle 012811
Maison Natale de Jeanne d'Arc et
Centre d'Interprétation, Domremy-la-
Pucelle 012812
Maison Natale de la Religieuse Sophie
Barat, Joigny (Yonne) 013319
Maison Natale de Saint-Vincent-de-
Paul, Saint-Vincent-de-Paul . . 015589
Maison Natale du Brave Crillon, Fidèle
Compagnon d'Henri IV, Murs . . 014390
Maison Natale du Maréchal
Foch, Musée du Maréchal Foch,
Tarbes 015914
Maison Natale du Maréchal Jourdan,
Limoges 013815
Maison Natale Sainte Therese de
l'Enfant-Jésus, Alençon 011589
Maison Nationale Bonaparte,
Musée de la Maison Bonaparte,
Ajaccio 011569

Maison Nationale de la Pêche et de
l'Eau, Ornans 014589
Maison Nationale des Patriotes, Saint-
Denis-sur-Richelieu 006867
Maison Neumann, Metz 104228
La Maison Papon, Blonay 087329
Maison Peyne, Laurède 013584
Maison Philippe, Paris 071613
Maison Philippe, Pézenas 072141
Maison Philippe T, La Queue-en-
Brie 069059
Maison Pierre-Lacroix, Sainte-Marie-
de-Beauce 006957
Maison Platt, Paris 071614
Maison Renaud, Troyes 073906
Maison Res Rara, Den Haag . . . 083102
Maison Rodolphe-Duguay, Nicolet 006494
Maison Ronsard, Paris 071615, 129074
Maison Rouge Galerie, Chicago . 120191
Maison Rouge, Fondation Antoine de
Galbert, Paris 014675
Maison Rurale de l'Outre-Forêt,
Kutzenhausen 013354
Maison Saint-Dominique,
Fanjeaux 012950
Maison Saint-Vincent, Villepreux 105875
Maison Satie, Honfleur 013268
Maison Sergeur, Faumont 068508
Maison Stendhal, Grenoble 013178
Maison Tavel, Genève 041526
La Maison Tournaisienne – Musée de
Folklore, Tournai 004081
Maison Traditionnelle de la Boucherie,
Limoges 013816
Maison Trestler, Vaudreuil-Dorion 007308
Maison Tsawenhohi, Wendake . 007378
Maison V Collections, Vancouver 064841
Maison Vincent Van Gogh,
Cuesmes 003528
Maison Vinette, Malmedy 003839
Maison Visinand, Centre Culturel,
Montreux 041754
Maison Zina, Chéraga 060836
Maison, M.A., Saint Paul 096837
Maison, Michel, Seattle 097469
Maisonneuve et Larose, Paris . . 136962
Maisonneuve, Adrien, Paris 136960
Maisonneuve, Grégoire, Paris . . 105057
Maisonneuve, Jean, Paris 136961,
. . . . 141235
Maisons-Ateliers de Jean Arp et
Sophie Taeuber, Fondation Arp,
Clamart 012583
Maisternia Sens, Kyïv 117142
Maitland Art Center, Maitland . . 050607
Maitland City Art Gallery, Maitland,
New South Wales 001241
Maitland Historical Museums, Maitland
Historical Museum, Telephone
Museum, Historic Waterhouse
Residence Museum, Maitland . 050608
Maitland Museum, Maitland, South
Australia 001242
Maitre, Jean-Jacques, Bignoux . 067219
Les Maîtres Anciens, Québec . . 064537
Les Maîtres Argenteurs, Paris . . 129075
Maîtres de Forme Contemporains,
Bruxelles 100640, 136644
La Maîtrise, Hasselt 063683
Maiwald, Klaus, Langen, Hessen 077021,
. . . . 142046
Maizuru Museum of Salvage, Maizuru
Repatriation Memorial Museum,
Maizuru 028994
Maizuru-shiritsu Sekei Renga
Hakubutsukan, Maizuru 028995
Maj, Z., Wolfenbüttel 079054
Maja, Gdańsk 113478
La Maja Antigüedades, Buenos
Aires 060912
Majesteit, Rotterdam 083581
Majestic Coins, Salt Lake City . 096907
Majesty Promotional Products, San
Diego 097103

Majewska, Lucyna, Lublin 113699
Majewski, Zdzisław, Wołomin . . 113914
Majlath, Judith, Wien 063022
Al-Majlis, Dubai 117183
Majmua Art Gallery, Karachi . . . 113306
Majnabbe, Göteborg 115658
Majo, Denia 085963
La Majolica, New York 095558
Majolika Galerie, Karlsruhe 019511
The Major General Frederick Funston
Boyhood Home and Museum,
Iola 049707
Major John Bradford House,
Kingston 049980
Majorelle Museum, Marrakech . 031781
Majori, Hasselt 063684
Majorstuen, Oslo 143405
Majuscules, Marseille 069948
MAK, Sarajevo 100867
MAK – Österreichisches Museum
für angewandte Kunst /
Gegenwartskunst, Wien 003033
Mak-Art Salon, Toruń 084683
MAK Art Society, Wien 058288
MAK Center for Art and Architecture,
Pearl M. Mackey House, Los
Angeles 050451
MAK Center for Art and Architecture,
Schindler House, West
Hollywood 054308
MAK Verlag, Wien 136598
Mak, A., Dordrecht 126285
Makassar, Bochum 075037
Makassar, France, Paris 071616
Makatka, Warszawa 084733
Make It Big, Philadelphia 123414
Makedonsko Archeološko Društvo –
Zavod i Muzej Prilep, Prilep . . 030623
Makerere Art Gallery, Makerere
University, Kampala 043107
Makers, Milano 080595
Makers on Mair, Ballarat 098696
Maket, Michel, Paris 071617
Makhoul, Claudine Varan Keutnedjian,
São Paulo 064191
Makina Erremintaren Museoa,
Fundación Museo Máquina
Herramienta, Elgoibar 039374
Makk, Honolulu 120951, 138338
Makler & Hope, Philadelphia . . . 123415
Makonde African Arts, Wembley 119319
Makonde Art Museum,
Bammental 016961
Makonde Bijutsukan, Futami . . . 028569
Makoschey, G., Bad Essen 074509
Makoveć, J., Wien 127948
Makowski, Moers 077408
Makowski, Prichsenstadt 078091,
. . . . 078092
Makowski, Ingrid, Mülheim an der
Ruhr 077435
Maksymiuk, Janusz, Biała
Podlaska 113426
Makta, Sofia 101165
Maktabi & Sons, Ali, Beirut 082292
Maktabi & Sons, Hassan, Beirut 082293
Maktabi, Iwan, Beirut 082294
Makuhari Kitazawa Bijutsukan,
Chiba 028511
Mała Galeria, Warszawa 113861
Mała Galeria Białostockiego Ośrodka
Kultury w Famie, Białystok . . . 034639
Mała Galeria Sztuki, Gdańsk . . . 084508,
. . . . 113479
Mala Galerija Udruženja Likovnih
Umetnika Primenjenih Umetnosti i
Dizajnera Srbije, Beograd 038151
Mala Galerija, Moderna Galerija,
Ljubljana 038545
Malá Sova, Praha 140523
Malabar Farm State Park, Lucas 050520
Malacañan Palace Presidential
Museum, Manila 034533
Malafaia, Rio de Janeiro 064082

Malagash Area Heritage Museum,
Malagash 006248
Malagoli, Rosa, Modena 080804
Malahide Antiques, Malahide . . 079668
Malahide Castle, Malahide 024964
Malaika Art & Design, Almere . . 112118
Malaise, V., Bruxelles 063501
Malakademie, Per von Spee,
Köln 055619
Malakološki Muzej, Makarska . . 008947
Malanda Antiques, Malanda . . . 061763
Målaratéljén, Eskilstuna 115604
Malatesta, Brescia 079945
Malatya Devlet Güzel Sanatlar Galerisi,
Malatya 043015
Malatya Müzesi, Malatya 043016
Malatya Sabancı Kültür Merkezi,
Malatya 043017
Malausséna, Simone, Nice 070629
Malavasi, Maurizio, Milano 142728
Malavolti, Giliana, Modena 131574
Malavoy, Brigitte, Paris 129076
Malay Art Gallery, Singapore . . . 114567
Malay Ethnographic Museum,
University of Malaya, Kuala
Lumpur 030686
Malaysian Youth Museum,
Melaka 030705
Malazgirt Kültür Merkezi,
Malazgirt 043018
Malba – Fundación Costantini, Museo
de Arte Latinoamericano de Buenos
Aires, Buenos Aires 000164
Malbasa, Strasbourg 073591
Malbert, Chicago 092743, 120192
Malbork, Montréal 101430
Malbos, Charles, Avignon 066882
Malca, New York 122838
Malchers, Anne, Bergisch
Gladbach 106138
Malco, Helsinki 066339
Malcolm, Franklin, Chicago . . . 092744
Malcolm's No. 1, Sherburn-in-
Elmet 127043
Malda Museum, Malda 024392
Malden Public Library Art Collection,
Malden 050610
Maldis, Vilnius 111802, 142966
Maldon and District Agricultural and
Domestic Museum, Goldhanger 044441
Maldon Antiques and Collectors
Market, Maldon 090680
Maldon District Museum, Maldon 045196
Maldon Museum, Maldon 001243
Maldon's Old Grain Store Antiques,
Maldon 061766
Maldoror, Toulouse 105747
Malé Máslovické Muzeum Másla, The
Little Butter Museum of Maslovice,
Vodochody 009956
Malek Museum, Tehrän 024766
Maleny Antiques, Maleny 061768
Maleny Art Gallery, Maleny 099263
Malerifabrikken, Vejle 066197
Målerikonservator, Gimo 133526
Malerisalongen, Hafrsfjord 042042
Måleriyrkets Museum, Stockholm 041017
Malermuseum Wetzikon,
Wetzikon 042134
Malerschulen Baden-Leesdorf, HTL
Baden, Malerschule Leesdorf, Baden
bei Wien 054924
Malerskole 1, Kongelige Danske
Kunstakademi, København . . . 055310
Malerskole 2, Kongelige Danske
Kunstakademi, København . . . 055311
Malerskole 3, Kongelige Danske
Kunstakademi, København . . . 055312
Malerstübchen Willingshausen,
Willingshausen 022670
Malesa, Jeffery, Sorrento 099692
Malesherbes Antiquités, Paris . . 071618
Malet, Montmorency 070278
Malgrati, Elena, Milano 131469

Mali Neboder, Rijeka .. 065313, 140453
Malibu Galleries, Omaha 123330, 123331
Malibu Gallery, Minneapolis ... 122011
Malibu Lagoon Museum, Malibu 050612
Malichin, Hildegard, Baden-
  Baden 106110
Malignon, Josiane, Paris 071619
Malik & Son, David, London ... 090322
Mȧlilla Hembygdspark, Mȧlilla .. 040819
Mȧlilla Motormuseum, Mȧlilla .. 040820
Mȧlilla Sanatoriums Museum,
  Mȧlilla 040821
Maling Road Antique Market,
  Canterbury 061280
Málinga- og Bátasavnið, Leirvik 010735
Malingarte, Amadora 084776
Malingue, Daniel, Paris 105058
Malinka, Köln 076895, 130157
Malinowski, Gerhard, Hamburg .. 141885
Malinowski, Zdzisław, Gdynia .. 084519
Malits, Ewald, Guntramsdorf ... 062621
Malki Museum, Banning 046780
Malkia, Tucson 124751, 124752
Malkin, Jeanne, Chicago 120193
The Mall Antiques Arcade,
  London 090323
Mall Galleries, London 045031
Mall of Saint Paul, Saint Paul .. 096838
Mallaig Heritage Centre, Mallaig 045197
Mallala and Districts Historical
  Museum, Mallala 001244
Mallalieu Motor Museum, Christ
  Church 003237
Mallams, Abingdon 126794
Mallams, Cheltenham 126854
Mallams, Oxford 127006
Mallawy Museum, Mallawy 010580
La Malle à Bart, Bart 066955
La Malle Anglaise, Cormeilles .. 068154
La Malle Anglaise, Pont-d'Ouilly 072250
La Malle Anglaise, Rennes (Ille-et-
  Vilaine) 072440
La Malle au Trésor, Lausanne .. 087576,
                              133801
La Malle aux Chimères,
  Châteauneuf 067888
Malle aux Idées, Niort 070713
Malle aux Trésors, Blois 067257
La Malle aux Trésors, Bruxelles 063502
La Malle aux Trésors, Genève .. 087481
Malle Cauchoise, Gonneville-la-
  Mallet 068681
Malle du Martroi, Orléans 141013
Malle Pietje, Zutphen 083721
Malle Poste, Annecy 066645
Mallee, Billabong 098748
Mallek, Kay, Tucson 097675
Mallet, Edith, Montrond-les-Bains 070332
Mallet, Marco Antonio Serveira, São
  Paulo 064192
Mallett, London 090324
Mallett, Mosman 063534
Mallett, New York 095559
Mallette & Assoc., Cincinnati .. 092872,
                              127177
Malleville, Étienne, Rouen 072598
Malliaris, A., Thessaloniki 142545
Malling Beck, Hillerød 102832
Mallon, J.M., Indianapolis 121200
Mallory & Sons, San Diego 097104
Mallory-Neely House, Memphis .. 050781
Mallory & Sons, E.P., Bath 088395
Mally, Wolfgang, Adendorf 074354
La Malmaison, Cannes 012313
Malmgran, Göteborg 115659
Malmin Taide ja Kehys, Helsinki 066340
Malmö Antik och Konstmässa,
  Malmö 098321
Malmö Auktionsbyrȧ, Åkarp ... 126564
Malmö Konsthall, Malmö 040828
Malmö Konstmuseum, Malmö ... 040829
Malmö Museer, Malmö 040830
Malmska Gȧrden, Jakobstads
  Museum, Jakobstad 010911

Malo Deux, Leiden 083436
Maloberti-Miermont, Marie-Claude,
  Courpière 068195
Malojaroslaveckaja Gorodskaja
  Kartinnaja Galereja,
  Malojaroslavec 037218
Malojaroslaveckij Istoriko-
  kraevedčeskij Muzej,
  Malojaroslavec 037219
Malojaroslaveckij Muzej Voennoj Istorii
  1812 Goda, Malojaroslavec ... 037220
Malojaroslaveckij Muzejno-vystavočnyj
  Centr, Malojaroslavec 037221
Malokarpatské Múzeum v Pezinku,
  Pezinok 038419
Malombra, Vicenza 142889
Malomipari Múzeum, Budapest . 023525
Malone, Begoña, Madrid 115250
Les Malouines, Charleroi 100680
Malouinière-Musée, Saint-Malo (Ille-et-
  Vilaine) 015437
Malouvier, Le Havre 103948
Malouvier, Rouen 072599
Malovat, Chicago ... 092745, 120194
Malow, P., Nossentiner Hütte .. 077806
Mȧlselv Bygdemuseum, Midt-Troms
  Museum, Moen 033776
Malt House Antiques, Narbeth .. 090800
Malt House Antiques, Stow-on-the-
  Wold 091476
The Malta at War Museum,
  Vittoriosa 030766
Malta Aviation Museum, Ta'Qali 030749
Malta Classic Car Collection,
  Qawra 030745
Malta Maritime Museum,
  Vittoriosa 030767
Malta National Committee of ICOM,
  Valletta 059567
Maltby, Winchester 119360
Maltepe, İstanbul 117063
Maltese Cross Antiques,
  Christchurch 083881
Malteser-Museum, Mailberg ... 002349
Malthouse Antiques, Bridgnorth . 088599,
                              134145
Malthouse Antiques, Dorking ... 089088
Malthouse Antiques Centre,
  Alcester 088240
Malthouse Arcade, Hythe, Kent . 089628
Maltinti, Bruno, Milano 080596, 131470
Malton, Cincinnati 120356
Malton Antique Market, Malton . 090603
Malton Museum, Malton 045200
Maltwood Art Museum and Gallery,
  Victoria 007330
Malu, Maastricht 083470
Malucelli, Forlì 110080
Malungs Hembygdsgȧrd, Malung 040835
Malura Museum, Unterdießen .. 022276
Malurt, Frederikshavn 102811
Malvaux, Bertrand, Nantes 070470
Malvern Antiques, Malvern, South
  Australia 061773
Malvern Antiques and Collectors Fair,
  Malvern 098389
Malvern Artists' Society, Malvern,
  Victoria 099271
Malvern Bookshop, Great
  Malvern 144444
Malvern Museum, Great Malvern 044473
Malvern Studios, Great Malvern 089372,
                              134422
Malvisi, Stefania, Parma 131730
Maly, Severin, Paris 105059
Malynowsky, Daniel J, Montréal 064442
MAM, Wien 100348
MAM Mario Mauroner Contemporary Art
  Salzburg, Salzburg . 100168, 136554
MAM Mario Mauroner Contemporary
  Art Sculpture Garden, Salzburg 100169
MAM Mario Mauroner Contemporary
  Art Vienna, Wien 100349
MAMA, Showroom for Media and

Moving Art, Rotterdam 112648
Mamco, Musée d'Art Moderne et
  Contemporain, Genève 041527
Mamie Doud Eisenhower Birthplace,
  Boone 047089
Mamma Kunsthuis, Brugge 100536
Mammad Seid Ordubad Home
  Museum, Ordubad 003194
Mammoth Site of Hot Springs, Hot
  Springs 049566
Mammutmuseum,
  Niederweningen 041803
Mamnu"gohi Bostonshinosii Panjakent,
  Pandžakent 042622
Mamoa de Lamas, Lamas 035745
Man and His Work Center in Memory
  of S. Avitsur, Eretz-Israel Museum,
  Tel Aviv 025205
Man and the Living World Museum,
  Ramat Gan 025174
Man at Arms Magazine,
  Woonsocket 139425
Man From Snowy River Folk Museum,
  Corryong 000995
Man in the Galilee Museum, Kibbutz
  Ginosar 025132
MAN-Museum, Augsburg 016661
Man of Ross Gallery, Ross-on-
  Wye 118987
Man Po, Hong Kong 065081
Mana'art, Vendôme 105835
Manahil Art Works, Manama ... 100443
Manaia Gallery, Waikato 113040
Manakova Kuća, Etnografski Muzej
  Srbije, Beograd 038152
Manama Auction Center,
  Manama 125291
Manana's Art Salon, Tbilisi 105912
Mañanas Campestres, Buenos
  Aires 127527
Manao, Argentan 066710
Manassas Museum System,
  Manassas 050619
Manassas National Battlefield Park,
  Manassas 050620
Manasse, Pierugo, Roma 081435
Manatee Village Historical Park
  Museum, Bradenton 047170
Manawatu Upholstery, Manawatu 132726
Manca, Mérignac (Gironde) 070072
Manceaux, Jean-Denis, Reims . 129192
Mancebo, São Paulo 064193
Manchester Antique Company,
  Stockport 091444
Manchester Art Gallery,
  Manchester 045208
Manchester Art Show,
  Manchester 098390
Manchester College of Arts and
  Technology, Manchester ... 056580
Manchester Craftsmen's Guild,
  Pittsburgh 052167
Manchester Historic Association
  Museum, Manchester 050628
Manchester Historical Society
  Museum, Manchester 050624
Manchester House Antiques Centre,
  Chipping Norton 088891
Manchester Jewish Museum,
  Manchester 045209
The Manchester Museum, The
  University of Manchester,
  Manchester 045210
Manchester United Museum and Tour
  Centre, Manchester 045211
Manchester University Press,
  Manchester 138265
Mancina, Saverio, Autun 066837
Mancini, Perugia 131754
Mancini & C., Leonardo, Perugia 131755
Mancini, Letizia & Alessandra,
  Roma 110816
Mancini, Simone, Perugia 131756
Mancino, Ivana, Palermo 131703

Mancuso Casa d'Aste, Salsomaggiore
  Terme 126230
Mancuso, Gianpaolo, Trieste ... 132223
Mancy, Jean-Luc, Meaulne 070025
Mandal Bymuseum, Mandal ... 033763
Mandal Kunstforening, Mandal . 059662
Mandal og Opplands Folkemuseum,
  Øyslebø 033817
Mandala Fine Arts, Hong Kong . 065082
Mandalá, Giuseppa, Palermo ... 081052
Mandallaz, Michel, Le Vivier-sur-
  Mer 069377
Mandarin, Hong Kong 065083
Mandarin, Seattle 097470
Mandarin Antique, Atlanta 092152
Mandarin Gallery, Otford 090924, 134912
Mandel, Ludwig, Landshut 077015
Mandelli, Sergio & Thao,
  Seregno 110924
Mandells Gallery, Norwich 090878,
                              118833
Mander and Mitchenson Theatre
  Collection, London 045032
Manderley Manor, Toronto 064701
Mandeville Gallery, Union College,
  Schenectady 053195
La Mandigotte-Galerie Renaissance,
  Royan 072631
Mandi's Gallery, Sacramento ... 123775
Mandl, Edgar, Wien 127949
Mandl, Helmut, München 130367
Mandos-Feldmann, Westerstede 108888,
                              137551
Mandos, Ron, Rotterdam 112649
Mandra, Rodolphe, Presly 072309
La Mandragola, Milano 080597
Mandragora Verlag, Münster ... 137438
Mandragore, Prades 129178
Mandrake, Seattle ... 097471, 136374
Mandurah Community Museum,
  Mandurah 001246
Manduras, John, Jerrabomberra 099158
Mandy, Dresdner Produzentengalerie
  Leipzig, Leipzig 107843
Mané-Katz Museum, Haifa 025054
Al-Maneei, Mohammed Abdulrahman,
  Riyadh 085319
Maneesi Sotamuseo, Helsinki .. 010847
Le Manége, Paris 071620
Manège des Hogues, Saint-Léonard
  (Seine-Maritime) 072890
Manège du Temps, Orgeval ... 070799
Manelli, Saint-Maur-des-Fossés 105558,
                              137003
De Manenstegell, Bathmen 132435
Manenstegell, Deventer 132485
Manera, East Freemantle 061431
Manes, Francesco, Napoli 131626
Manescau, Raymonde, Avignon . 066883
Manetstötter, Karlheinz, Maisach 107959
Manetti & Masini, Firenze 131199
Manevy, Madeleine, Paris 071621
Manfra, Tordella & Brookes, New
  York 095560
Manfredini, Luca, Pallanza ... 081085
Manfreds Haus- und Hofmuseum,
  Weisweil 022551
Mang, Wolfgang, Saarbrücken . 142344
Mangalam, Paris 071622, 105060
Mangapps Railway Museum, Burnham-
  on-Crouch 043791
Mangel, Bad Wildungen 074613
Mangel, Benjamin, Philadelphia . 123416
Mangel, Hans-Joachim, Bad
  Wildungen 074614
Manggha Centrum Sztuki i Techniki
  Japońskiej, Muzeum Narodowe w
  Krakowie, Kraków 034500
Mangiacasale, Claudio, Milano . 142729
Mangiardi, Massimo, Torino ... 081724,
                              132163
Mangili Bartolacci, Carlo, Brescia 079946
Mangin, Charlotte, Mirepoix ... 070127
Mangin, Stéphane, Paris 131702

Mangionello, Silvia, Firenze 131200
Mångkulturellt Centrum, Tumba 041093
Manglam Arts, Jaipur, Rajasthan 109383
Manglano Gil, Matilde, Madrid . 133353, 133354
Mango Explorer Museum, Yujing 042606
Mangold, Leipzig 107844
Mango's Fine Art Gallery, Saint Peter 100466
Mangystauskij Oblastnoj Istoriko-Kraevedčeskij Muzej, Aktau .. 029788
Manhattan Arts and Antiques Center, New York 095561
Manhattan Arts Center, Manhattan 050639
Manhattan Arts International, New York 139426
Manhattan Books, New York ... 145352
Manhattan Galeria, Łódź 113677
Manhattan Gallery, Nottingham . 118842
Manhattan Shoppes, Tampa ... 097577
Manhattan West Timeless Jewelry, Denver 093312
Manheim, Ida, New Orleans ... 095127
Manhyia Palace Museum, Kumasi 022935
Mania, Kay, Hessisch Oldendorf 076519, 130042
Manial Palace Museum, Cairo .. 010543
Manic, Paris 071624, 129077
Manicó, Mercato, Catania 080038
Manido Chiman, Golden Lake .. 005892
Maniero, Roma 059468, 110817
Manifesta, Torino 081725
Manigand Mobilier, Saint-Cyr-sur-Menton 072727
Maniitsup Katersugaasivia, Sukkertoppen Museum, Maniitsoq 023331
Manila Galería, Medellín 102452
Manila Police Department Museum and Gallery, Manila 034534
Manilla Historical Royce Cottage Museum, Manilla 001247
Maningrida Arts & Culture, Winnellie 062441
Manion, V.J., Ashbourne 088300
Maniototo Early Settlers Museum, Naseby 033251
Manipur State Museum, Imphal 024310
Manipur University Museum, Canchipur 024188
Manisa Devlet Güzel Sanatlar Galerisi, Manisa 043020
Manisa Müzesi, Manisa 043021
Maniscalco, Santo, Catania 131091
Manistee County Historical Museum, Manistee 050644
Manitoba Agricultural Hall of Fame, Brandon 005465
Manitoba Agricultural Museum, Austin 005362
Manitoba Amateur Radio Museum, Brandon 005466
Manitoba Automobile Museum, Elkhorn 005778
Manitoba Baseball Hall of Fame & Museum, Morden 006426
Manitoba Children's Museum, Winnipeg 007450
Manitoba Crafts Museum and Library, Winnipeg 007451
Manitoba Electrical Museum, Winnipeg 007452
Manitoba Museum, Winnipeg . 007453
Manitoba Printmakers Association, Winnipeg 058452, 136692
Manitoba Sports Hall of Fame and Museum, Winnipeg 007454
Manitowoc County History Museum, Manitowoc 050647
Manjimup Timber Museum, Manjimup 001248
Manjiya, Osaka 142907

Manjusha Museum, Dharmasthala 024245
Manke, W., Bremen 075175
Manko, Hans-Werner, Wiesbaden 078986
Mankota Museum, Mankota 006253
Manley, Ashleigh, Surfers Paradise 099738
Manley, Brett, Armadale, Victoria 061039, 127557
Manley, J., Eton 089217, 134372
Manly Art Gallery and Museum, Manly 001249
Manly Collectables, Manly 061786
Manly Gallery, Manly 099274
Gebr. Mann, Berlin 137084
Mann Museum, Opelika 051785
Mann-Simons Cottage, Columbia 047913
Mann & Sons, David, Cranleigh 134303
Mann, Gunilla, Simrishamn 115810
Mann, John, Canonbie 088769, 134202
Mann, Mina, New York 095562
Mann, Robert, New York 122839
Mann, Thomas, New Orleans .. 122199
Mann, Vincent, New Orleans ... 095128
Manna, Den Haag 083103
Manna Gum Gallery, Dromana . 098978
Manna, Emilia, Napoli 080896
Mannaminne, Nordingrå 040872
Manneken-Pis-Museum, Geraardsbergen 003648
Mannenzaal van het Sint Pieters en Bloklands Gasthuis, Amersfoort 031910
Mannerheim museo, Helsinki .. 010848
Mannerheimintien Antiikki, Helsinki 066341
Manners, E. & H., London 090325
Manneschi, Rovigo, Firenze ... 131201
Johannes Mannhardt & Matthias Ruhdorfer, Rottach-Egern 078294
Mannhart, Buchs (Sankt Gallen) 087337
Mannheimer Kunstverein e.V., Mannheim 059167
Mannies Arts, Kathmandu 112079
Manning Park Visitor Centre, Manning Park 006254
Manning Regional Art Gallery, Taree 001564
Manning, K., San Francisco .. 145561
Manning, Oliver, Canterbury ... 134207
Manning, Shedd, Atlanta 135313
Manningham Gallery, Doncaster 098971
Mannsfeld, Colmar 013620
Mannum Old Wares and Curios, Mannum 061789
Manny, Chicago 092746, 135416
Del Mano, Los Angeles 121553
Mano, Seoul 115668
Mano, Jose M. de la, Madrid .. 086213, 115251
Manoel Theatre and Museum, Valletta 030751
Le Manoir, Rabat 082603, 112051
Manoir Automobile, Lohéac 013853
Manoir de Folie Vallée, Pont-de-l'Arche 072248
Manoir de Folie Vallée, Rouen .. 072600
Manoir de Kernault, Mellac ... 014115
Manoir de la Ville de Martigny, Martigny 041725
Manoir de Veygoux, Scénomusée des Combrailles, Charbonnières-les-Varennes 012453
Le Manoir de Villedoin, Velles .. 105821
Manoir de Villers, Saint-Pierre-de-Manneville 115511
Manoir du Huis Bois-Maison P.N.R. du Boulonnais, Le Wast 013717
Manoir Leboutillier, Anse-au-Griffon 005344
Manoir Mauvide-Jenest, Saint-Jean 006880
Manolaki, Thessaloniki 109091
Manoni, Goffredo, Milano 080598
Manor Antiques, Adare 079433

Manor Antiques, Gravesend, Kent 089362
Manor Antiques, Wilstead 091851, 135230
Manor Barn, Skipton 091348
The Manor Bindery, Fawley ... 134388
Manor Cottage Heritage Centre, Southwick 045841
Manor Farm, Bursledon 043799
Manor Farm Antiques, Standlake 091412
Manor Fine Arts, Banbridge .. 117278
Manor Gallery, Gravesend ... 117876
Manor House, Cardiff 134215
Manor House, Donington-le-Heath 044113
Manor House, Sandford Orcas . 045731
Manor House Antique Mall, Indianapolis 093904
Manor House Antiques, Blackheath 061162
The Manor House Antiques, Brasted 088582
Manor House Antiques, Fishguard 089279
Manor House Antiques, Sacramento 096635, 136191
Manor House Art Gallery and Museum, Ilkley 044628
Manor House Gallery, Cheltenham . 088836, 117563, 134239
Manor House Museum, Kettering 044585
Manor Museum, Manor 006255
Manorism, Atlanta 092153
Manos, Denver 120742
Manos Faltaits Museum, Skyros 023269
Las Manos Gallery, Chicago ... 120195
Las Manos Magicas, Houston .. 121087
Manou, Ronchamp .. 072529, 105457
Manoukian Noseda, Giacomo, Milano 080599, 131471
Manrique, Martin, Los Angeles . 135732
Manro-Classic – Auto & Musik Museum, Koppl 002223
Le Mans Collections, Le Mans . 069308
La Mansarda, Medellín 065293
La Mansarde, Dampierre-en-Yvelines 068267
Mansart Rémy, Bruxelles 063503
Mansell, G., Hoylake 089595
Mansell, William C., London .. 134720
Mansers Antiques, Shrewsbury . 091331, 135046
Mansfeld-Museum Hettstedt, Hettstedt 019238
Mansfelder Bergwerksbahn e.V., Benndorf 017029
Mansfeldgalerie im Kulturhaus, Lutherstadt Eisleben 020192
Mansfield, Deloraine 061396
Mansfield Art and Antiques, Christchurch 083882
Mansfield Art Center, Ohio State University, Mansfield 050656
Mansfield Fine Arts Guild, Mansfield 060578
Mansfield Gallery, Mansfield ... 099278
Mansfield Historical Society Museum, Storrs 053554
Mansfield Museum and Art Gallery, Mansfield 045221
Mansfield Park Gallery, Glasgow 117850
Mansfield State Historc Site, Mansfield 050654
Mansfield, Louise, Stillorgan ... 109696
Manship House Museum, Jackson 049749
Manshodo, Hiroshima 081979
Mansion Antique House, Singapore 085467
La Mansión del Fraile, Bucaramanga 065235
Mansion House, Warkworth ... 033338
Mansions, Lincoln 089832
Manske, D., Bernburg 074959
Mansky, Claudia, Rotenburg an der

Fulda 078282
Manson, Edward, Woodbridge .. 091915, 135254
Manson, Yves, La Flèche 125603
Mansour, Cairo 103054
Mansour Bros., Damascus 088040
Mansour, London 090326, 118439
Al-Mansour Gallery, Cairo 103056
Mansour, Ahmed, Cairo 103055
Manten, W., Vreeland 132663
Manthey, Berlin 074873
Manthey, Cornelia, Hamburg .. 141886
Manthey, Gerhard, Hamburg .. 129975
Manthey, Gerhard, Kiel 076781
Mantler, Wolfgang, Wien 140100
Manton, F.T., Uxbridge 135176
Mantra, Torino 110987
Mantu, N., Galaţi 114168
Manu, Breda 082998
Manualidades, La Paz 100845
Manuel L. Quezon Memorial Shrine Museum, Quezon City 034584
Manuelian, Georges, Nice 070630
Manuella, Nice 070631
Manufacture Robert Four, Aubusson 066795, 128451
Manufakturgalerie Eva Brosché, Berlin 106369
Manujlivs'kyj Literaturno-memorialnyj Muzej O.M. Gor'kogo, Verchnja Manujlivka 043317
Manuka, Fisher 061465
Manumedia, Graz 136510
Manurewa Community Arts Centre, Auckland 033154
Manus, Männedorf 138074
Manus Presse, Stuttgart 108719
Manus, Magnus, Amsterdam .. 112253
Manusardi, Marina, Milano ... 080600
Manuscribe, Tucson 145623
Manuscript, Rotterdam 126296, 143222
Manuscript Museum, Bibliotheca Alexandrina, Alexandria ... 010518
O Manuscrito Histórico, Lisboa . 143582
Manwarning, M.G, Bristol 144284
Manx Museum, Manx National Heritage, Douglas 025006
Manyara Museum, Manyara .. 042633
Man'yo Bunkakan, Wakayama . 029691
Manyung Gallery, Mount Eliza . 099395
Manzanal, Donostia-San Sebastián 133253
Manzano, Juan B., Sevilla 086455
Manzione, Punta del Este 119446
Manzke, Dipl. Gemälderestaurator Holger, Potsdam 130535
Manzo, Oklahoma City 095990, 136052
Manzoni Villa, Musei Civici, Lecco 026584
Mao Dun Former Residence, Beijing 007629
Maoling Museum, Xingping ... 008433
Maoming Museum, Maoming .. 008076
The Map House of London, London 144643
Map World, London 144644
Mapes, David W., Vestal 127484
Mapes, Shirley K., Omaha 145401
Maple Antiques, Surbiton 091510
Maple Europe, Paris 071625
Maple Leaf Furniture Refinishing, Toronto 128212
Maple Ridge Art Gallery, Maple Ridge 006262
Maple Ridge Museum, Maple Ridge 006263
Maple Street Book Shop, New Orleans 145315
Maple Syrup Museum, Saint Jacobs 006877
Maple Valley Historical Museum, Maple Valley 050661
Maples, Mosta 082418
Maplewood Antique Mall, Saint

Louis 096733
Maplewood Farm, North
Vancouver 006517
Mappa Mundi and Chained Library,
Hereford 044573
Il Mappamondo, Milano . . . 110375
Mappamundi, Knokke-Heist . . 100746,
140234
Mapungubwe Museum, Pretoria 038818
Maqdisi, Damascus 088041
Maqi, Tbilisi 105913, 105914
Maqueda Abreu, Maria Luisa,
Granada 115091
La Mar, Denia 085964
Mar Pau, Viña del Mar 101926
Mar, Grup d'Art Escolà,
Barcelona 114928
Mára! SunArt Verlag, Meppen . . 137380
Marabini, Bologna 109854
Maragall, Barcelona . . 085794, 114929,
126513, 137962
Maragheh Museum, Maragheh . 024718
Marais Des Cygnes Memorial Historic
Site, Pleasanton 052204
Marais, Jean-Claude, Pezou . . 072146
Maraldi, Walter, Roma 142826
Maralt, Kyra, Berlin 106370
Maramec Museum, The James
Foundation, Saint James 052775
Marando, Bochum 141607
Marani, Ferny Creek 061464
Marani, Roma 081436
Maraninchi, Marseille 128777
Maraş Ikon Müzesi, Gazimağusa 033452
Marat Art, Bratislava 114655
Maratha History Museum, Deccan
College, Pune 024448
Marathon County Historical Society
Museum, Wausau 054237
Maraut, Michèle & Didier, Saint-Jean-
Cap-Ferrat 072834
Maraux, Patrick, Nevers 070540
Marazzani Visconti Terzi Arte,
Piacenza 110649
Marazzi, Loris, Venezia 111135
Marbach, Jean, Mulhouse 070393,
104314
Mårbacka Manor, Home of Selma
Lagerlöf, Östra Ämtervik 040905
Marbeau, Paris 105061
Marbella, New York 122840
Marbella, David, New Orleans . . 095129
Marble Arts, Richmond 096559
Marble Hill Gallery, Twickenham 091675
Marble Hill House, Twickenham 046008
Marble Historical Museum,
Marble 050664
Marble Moon, Glenelg 099085
Marble Mountain Library and Museum,
West Bay 007380
Marble Palace, Houston 093736
Marble Palace Art Gallery and Zoo,
Kolkata 024363
Marble Springs State Historic
Farmstead Governor John Sevier
Memorial, Knoxville 050014
Marblehead Arts Association,
Marblehead 060579
Marblehead Museum and J.O.J. Frost
Folk Art Gallery, Marblehead . . 050667
Marbles Kids Museum, Raleigh . . 052448
La Marbrerie, Lille 069496
Marbrier, Thierry, Mauregny-en-
Haye 070008
Marbs, Uwe, Baden-Baden . . 074638
Marburger Kunstverein, Marburg 059168
Marburger Kunstverein e.V.,
Marburg 059169
Marburger Universitätsmuseum
für Kunst und Kulturgeschichte,
Marburg 020273
Marc, Fougères (Ille-et-Vilaine) . 068598
Marc Chagall Galeria de Arte, Rio de
Janeiro 101049

Marc Chagall Kunsthuis,
Amsterdam 031955
Marc Jancou Contemporary, New
York 122841
Marc, Jean-Louis, Paris 105062
Marc, Philip, Chantenay Saint-
Imbert 067830
Marca Puig, Maria Teresa,
Barcelona 085795
Marcar'art, Albi 066516
Marcazzan, Fabien, Monferran-
Saves 070162
Marce, Jean Philippe, Montpellier 128811
Marceau, Paris 071626
Marceau, Frédéric, Saint-Genis-les-
Ollières 072796
Marcel-Ordas, Lyon 140924
Marcela M. Agoncillo Historical
Museum, Taal 034599
Marcelin, Franck, Aix-en-
Provence 066467
Marcella Sembrich Opera Museum,
Bolton Landing 047080
Marcellus Historical Society Museum,
Marcellus 050670
Marcelo H. del Pilar Museum,
Malolos 034514
Marcelpoil, Alain, Paris 071627
Marcelpoil, Alain, Rillieux-la-Pape 072468
Marcelpoil, Jeanne, Saint-Just . . 072874
Marcel's Antik Shop, Basel 087246
Marcet Books, London 144645
March and District Museum,
March 045223
March Ballester, Antonio Maria,
Barcelona 085796, 114930
March Field Museum, Riverside . 052593
March-Museum, Vorderthal 042116
March, Brigitte, Stuttgart 108720
March, Danielle, Paris 105063
March, David & Sally, Abbots
Leigh 088217
March, Tomas, Valencia 115507
Marchal, Étienne, Paris 071628
Marchand d'Art, Belo Horizonte . 100892
Marchand d'Oubli, Paris 071630
Marchand d'Oublis, Arnay-le-Duc 066730
Marchand, Fréderic, Paris 071629
Marchand, Joséphine,
Villeurbanne 074214
Marchand, Nathalie, Beaune . . . 067028
Marchand, P.J., Rotterdam 132622
Marchandet, Pascale, Paris 125737
Les Marchands du Design,
Bordeaux 067343
Marchant, Mark, Coggeshall . . . 088946,
134286
Le Marche, Norfolk 095865
Le Marché, Tokyo 082177
Marché aux Puces du Canal,
Villeurbanne 074215
Le Marche Biron, Saint-Ouen . . 073145
Marché d'Art, Fortaleza 063945
Marché Dauphine, Saint-Ouen . . 073146
Marché des Antiquaires, Vallauris 073952
Marché des Antiquités, Saint-
Étienne 072768
Marché Jules Vallès, Saint-Ouen 073147
Marché Paul Bert, Walincourt-
Selvigny 074286
Marche, Eric, La Croix-Saint-
Leufroy 069019
Marcheguet, Alain, Le Mesnil-sur-
Blangy 103959
Marches à Suivre, Paris 071631, 071632
Marches du Palais, Aix-en-
Provence 066468
Marchese, Antonio, Prato 110692
Marchese, Paola, Bologna 130975
Marchesi Gino & Figlio, Bologna 079851
Marchesi, Maria Luisa, Modena . 080805
Marchesini, Agnese, Ravenna . . 110701
Marchetti, Roma 110818
Marchetti, San Diego 124143

Marchetti & Fontanini, Brescia . 131039
Marchetti, Paolo, Udine 111065
Marchewka, Edmond, Mairé-
Levescault 069810
Marchfeldmuseum, Weikendorf . 002901
Marchgraber, Erwin, Marchtrenk 062729
Marchi, Roberto, Roma 110819
Marchin, Simonetta, Roma 110820
Marchina Arte Contemporanea,
Brescia 109910
Marchió, Fabrizio, Merano 080355
Marchis-Mouren, Jean-François,
Frankfurt am Main 075808
Marchisio, Felice, Torino 132164
Marchpane, London 144646
Márcia & Molduras, Belo
Horizonte 100893
Marciana, Venezia 081854, 081855
Marciana SRL, Oggetti D'Arte
Antichita', Venezia 081856
Marciano, Gilbert, Montréal . . . 064443
Marciano, Mickael, Paris 105064
Marcilhac, Félix, Paris . 071633, 105065
Marckle Myers, New York 095563
Marco, San Diego 124144
Marco Canepa Gallery, Genova . 110377
Marco Canepa Gallery, Milano . . 110376
Marco Noire Contemporary Art, San
Sebastiano Po 110900
Marco Polo, Harstad 084246
Marco Polo, Torino 081726
Marco Polo, Valencia 086523
Marco Polo Antiquités,
Marrakech 082592
Marco Polo Precious Metals, New
York 095564
Marco Rinaldi Numismatica,
Verona 081923
Marcoe, Leonard, Winnipeg . . . 101883
Marcoff, Hasselt 063685
Marcoleo, New York . . 135955, 135956
Marcolino, Arnaud, Rio de
Janeiro 128115
Marcombo, Barcelona 137963
Marcon, Bucureşti 085099
Marcon, Charles, Polignac 072226
Marcon, Paul, Cussac-sur-Loire . 068258
Marcon, Robert, Le Puy-en-Velay 069344
Marcon, Robert, Saint-Paulien . . 073250
Marconcini, Romano, Pisa 081158
Marconi, Gió, Milano 110377
Marcoola Antiques, Marcoola . . 061793
Marcorama, Ciudad de Panamá . 113321
Marcos de Matos Bicalho, Antônio,
Belo Horizonte 063907
Marcos Museum, Batac 034434
Marcos y Cuadros, Santiago de
Chile 064974
Marcos, Miguel, Barcelona 114931
Marcos, Miguel, Zaragoza 115555
Marcou, Marcel, Lyon 069761
Marcoux, Sandra, Mably 069789
Marcoz, Boston 092484
Marcq Antiquités, Marcq-en-
Barœul 069865
Marcum, Louisville 121658
Marcum, Paris 071634
Marcus, Birmingham 117375
Marcus, Curt, New York 122842
Marcus, Johannes, Amsterdam . . 082802,
132408, 143049
Marcy's Things, Milwaukee 094871
Mardfeld, Samuel, Franconville . 068604
Mardfeld, Samuel, Saint-Ouen . . 073148,
073149
Mardi Gras Arts, Saint Louis . . . 096734
Mardi Gras Gallery, New Orleans 122200
Mardi Gras Museum, Biloxi 046991
Mardin Müzesi, Mardin 043022
Mardom-shenasi-ye Kavir, Nayin 024725
De Mare, Leiden 112557
Marec, Yves, Le Folgoët 069268
Marechal, Oupeye 140256
Maréchal, Alexis, Bordeaux 067344

Maréchal, B., Saint-Martin-du-
Mont 072949
Maréchal, Jacques, Rouen 072601
Maréchaux, Paris 071635
Mareeba District Rodeo Museum,
Mareeba 001253
Mareeba Heritage Museum,
Mareeba 001254
Marek, Warszawa 113862
Marek, West Hollywood 125012
Marek i Art, Warszawa 084734
Marel, Karel, Friedberg, Hessen 141815
Marella Gallery, Beijing 101976
Marella Gallery, Milano 110378
Marella Project Space 1, Como . 109971
Marelle, Bruno, Rouen 072602
Marelli, Omaha 123332
Marelli, Roberto Gaetano, Milano 080601
Maren Gallery, México 111967
Marengo, Marco, Genova 131322
Marescalchi, Monaco 112008
Maresch, Wien 127950
Maretti, Monaco 112009
Marève, Collobrières . . 068077, 068078
Marfé Gioielli, Napoli 080897
Marg, Mumbai 138777
Marg Publications, Mumbai . . . 137568
Margam County Park, Port Talbot 045538
Margaree Salmon Museum, North East
Margaree 006512
Margaret Fort Trahern Gallery,
Clarksville 047778
Margaret Harvey Gallery, Saint
Albans 045667
Margaret Harwell Art Museum, Poplar
Bluff 052240
Margaret Hutchinson Compton Gallery,
Cambridge 047366
Margaret Laurence Home,
Neepawa 006458
Margaret Lokey Gallery, Murray State
College Campus Center Gallery,
Tishomingo 053750
Margaret Mitchell House and Museum,
Atlanta 046624
Margaret River Galleries, Margaret
River 099281
Margaret Trowell School of Industrial
and Fine Art, Kampala 056482
Margarete Blank Gedenkstätte,
Borsdorf 017522
Margarete-Steiff-Museum, Giengen an
der Brenz 018685
Margaretha Museum, Laag
Keppel 032558
Margaret's Antiques, Wellington 084173
Margaret's Cupboard, Denver . . 093313
Margaron, Alain, Nice 141002
Margaron, Alain, Paris 105066
Margate Old Town Hall Local History
Museum, Margate 045226
La Marge, L'Isle-sur-la-Sorgue . 140874
Margetić, Beograd 114427
Margev, Bratislava . . . 085554, 114656
Margherita, Taranto 081636
Margherita, Trento 111031
Margie, L., Basel 087247
Margie's Menagerie, Indianapolis 093905
Marginessi, Kyïv 117143
Marginiai, Vilnius 111803
Margo Veillon Gallery, for
Contemporary Egyptian Art,
Cairo 103057
Margo, Adair, El Paso 120836
Margo, Thierry, Paris 141236
Margolet, Baltimore 092343
Margolis, Nancy, New York 122843
Margot's Galerie, Westerstede . . 108609
Margret-Knoop-Schellbach-Museum,
Körle 019743
Margrove – South Cleveland Heritage
Centre, Boosbeck 043676
Marguerat, Bernard & Marialyne,
Genève 087482

Margutta 102, Roma 110821
Margy's Antiquetates Azure,
  Berwick 061152
Mari Arte Contemporanea,
  Imbersago 110149
Mari-jo, Volkach 108826
Mari Valles, Maria Isabel,
  Barcelona 114932
Mari, Sylvie, Mougins 104303
Maria Antik, Stockholm 087028
Maria de Lisitzin Fine Art and
  Historical Museum, Porvoo .. 011274
Maria del Valle Quintana Dominguez,
  Madrid 115252
Maria Martinez Museum, Santa
  Fe 053112
Mária-ruhák Múzeum, Andocs .. 023406
Mariadevadassan, Joseph, Pantin 070848
Mariager Antik, Mariager 065997
Mariager Museum, Mariager ... 010226
Mariah's Antiques, Vancouver .. 064842
Marialust, Apeldoorn 112324
Marian Goodman Gallery, New
  York 122844, 138405
Marian Koshland Museum,
  Washington 054132
Marianelli Antiquari, Firenze .. 080166
Mariani, Claudio, San Francisco 097279,
  097280, 136332, 136333
Mariani, Luigi, Milano 131472
Marianini, Cinzia, Firenze 131202
Marianini, Foresto, Firenze ... 131203
Marianist, Saint Louis 123845
Mariann Galéria, Győr 109193
Marianna Kistler Beach Museum
  of Art, Kansas State University,
  Manhattan 050640
Marianna-Lee County Museum,
  Marianna 050672
Marianna's Fine Arts, Saint Louis 123846
Marianne, Châtenois 067912
Mariano Rodríguez Gallery, La
  Habana 009146
Mariano, João, Lisboa 084912
Marian's Collection, Surry Hills . 062265
Mariantiquités, Sainte-Savine .. 073369
Mariaplans Antikvariat, Göteborg 143849
Maria's Antik Boutique, Mainz .. 077240
Marias Antikvariat, Hanko 140687
Maria's Antiques, Virginia Beach 097789
Marias Museum of History and Art,
  Shelby 053301
Mariaschin, Shirley, New York . 095565
Mariaux, Marcel, Le Châble ... 087354
Maricampo Antichtà, Milano ... 080602
Maridadi, C.C, Tshwane 085680
Marie-Andrée, Olivet 070778
Marie Antoinette, Toronto 064702
Marie-Elisabeth-Lüders-Haus,
  Berlin 017219
Marie-Hager-Haus, Burg
  Stargard 017683
Marie-Hélène, Rouen 072603
Marie Hull Gallery, Hinds Community
  College, Raymond 052478
Marie-Pierre Antiquités, Beaune 067029
Marie Rawdon Museum,
  Matjiesfontein 038782
Marie Sekei Tokei oyo Orugoru
  Hakubutsukan, Inuyama 028692
Marie Sophie D, Wimereux 105883
Marie Tak van Poortvliet Museum,
  Domburg 032238
Marie-Tortue, Compiègne 068114
Marie Walsh Sharpe Art Foundation,
  Space-Program, New York .. 051421
Marie, Daniel, Rémalard 072408
Marie, Marc, Nice 070632
Mariefreds Hembygdsmuseum,
  Mariefred 040838
Mariehamns Glas & Ramaffär,
  Mariehamn 066386
Mariehamns Konsthandel,
  Mariehamn 066387

Mariella Antichitá, Bologna 079852
Mariemont, Cincinnati 092873
Marien-Antiquariat, Berlin 141552
Marienglashöhle Friedrichroda,
  Friedrichroda 018523
Maries Antiques & Pianos,
  Barberstown Straffan 079455
Maries Nostalgi, Kongsberg 084263
Maries Pianos, Staffan 079701
Marie's Treasures, San Antonio . 096996
Mariestads Industrimuseum, Vadsbo
  Museum, Mariestad 040843
Marietta House Museum, Glenn
  Dale 049121
Marietta Museum of History,
  Marietta 050675
Marigorri, Donostia-San
  Sebastián 085978
Marijampolės Kraštotyros Muziejus,
  Marijampolė 030417
Marije, Tilburg 083618
Marijos ir Jurgio Šlapelių Namas-
  muziejus, Vilnius 030533
Marijos Savickienės Gamtos Išdaigų
  Muziejus, Tytuvėnai 030493
Marika Kotopouli Museum,
  Zografou 023325
Marikas Antique, Enskede 086661
Marika's Antique Shop, Boston . 092485
Mariko, Denver 120743
Marília de Dirceu Antiguidades, Belo
  Horizonte 063908
Marilla Historical Society Museum,
  Marilla 050683
Marimeks, Veliko Târnovo 101201
Marimon, Marisa, Ourense 115320
Marín, Medellín 065294
Marin, Umag 102524
Marin County Watercolor Society,
  Corte Madera 060580
Marin History Museum, San
  Rafael 053062
Marin Museum of the American Indian,
  Novato 051658
Marin Orts, Elena, Alicante ... 085698
Marin Society of Artists, Ross .. 060581
Marín, F., Burgos 085914, 133213
Marin, Gérard, Saint-Maurice .. 072968
Marina, Los Angeles 094402
Marina Gallery, Port Douglas .. 099581
Marine & Wildlife, Lincoln 118119
Marine Art Gallery, Padstow ... 118873
Marine Arts Gallery, Toronto ... 101658
Marine Corps Art Collection,
  Washington 054133
Marine d'Autrefois, Paris 071636, 129078
Marine-Ehrenmal und Technisches
  Museum U 995, Laboe 019838
Marine Fine Arts, Terrey Hills .. 099816
Marine Gallery, Busan 111572
The Marine Gallery, Cowes 117633
Marine House, Beer 117329
Marine Museum, Toronto 007187
Marine Museum at Fall River, Fall
  River 047233
Marine-Museum Limburg, Limburg an
  der Lahn 020078
Marine Museum of Manitoba,
  Selkirk 007003
Marine Museum of the Great Lakes,
  Kingston 006104
Marine Science Museum, Tokai
  University, Shizuoka 029400
Marine Sciences Museum,
  Mayagüez 036135
Marine Transportation Museum,
  Ogi 029205
Marineland Ocean Museum,
  Marineland 050684
Marinelli, Lorenzo, Torino 132165
Marinemuseet, Horten 033671
Marinemuseum, Den Helder ... 032207
Marinemuseum Stralsund,
  Stralsund 022011

The Mariners' Museum, Newport
  News 051546
Mariners' Park Museum, Milford 006311
Marinesse, Henri, Groslay 068753
Marinette County Historical Museum,
  Marinette 050685
Maringalleriet, Stockholm 115909
Maringer, Dr. Karl-Heinz, Sankt
  Pölten 100189
Marini, Alessio, Roma 081437
Marini, Giovanna, Valenzano ... 142880
Mariniersmuseum, Rotterdam .. 032808
Marinmuseum, Karlskrona 040740
Marino, Dublin 079569
Marino, Firenze 131204
Marino Affaitati, Gabriele, Milano 131473
Marino Meubles, Valette-du-Var 129327
Marino, Emanuela, Roma 131968
Marino, Massimiliano, Genova .. 131323
Marinov, Emil, Kjustendil 064224
Marinov, Emil, Šumen 140326
Mario Lanza Museum,
  Philadelphia 052064
Marion, Perpignan 105338
Marion Art Center, Marion 050691
Marion County Historical Society
  Museum, Marion 050694
Marion County Historical Society
  Museum, Salem 052891
Marion Heritage Center, Marion . 050690
Marion Public Library Museum,
  Marion 050687
Marion Scott Gallery, Vancouver 007288,
  101810
Marion, Josiane, La Bresse ... 068957
Marionettmuseet, Stockholm .. 041018
Marion's, Iron Acton 118027
Mario's Antique Shop,
  Philadelphia 096135
Mariotti & Materassi, Firenze .. 131205
Mariotti, Franco, Firenze 080167
Mariposa Gallery, Albuquerque . 119492
Marischael, Nicolas, Paris 129079
Marischal Museum, Marischal
  College, University of Aberdeen,
  Aberdeen 043359
Marit, Antwerpen 063223
Maritiem en Jutters Museum,
  Oudeschild 032754
Maritiem Museum Rotterdam,
  Rotterdam 032809
Maritiem Shop, Antwerpen ... 063224
Maritim Antik, Hamburg 076272
Maritiman, Göteborgs Maritima
  Upplevelsecentrum, Göteborg . 040656
Maritime and Seafood Industry
  Museum, Biloxi 046992
Maritime Antiques, København . 065935
Maritime Archaeology Museum,
  Melaka 030706
Maritime Centre Newcastle,
  Newcastle 001354
Maritime Centre, Iziko Museums of
  Cape Town, Cape Town 038680
Maritime Command Museum,
  Halifax 005954
Maritime Heritage Prints, Boston 092486
Maritime Heritage Society of
  Vancouver, Britannia Beach .. 055484
Maritime Industry Museum at Fort
  Schuyler, SUNY Maritime College,
  Bronx 047233
Maritime Museum, Cohasset ... 047860
The Maritime Museum, Dublin . 024896
Maritime Museum, Low Head .. 001233
Maritime Museum, Nadur, Gozo 030743
Maritime Museum, Singapore .. 038271
Maritime Museum of Andros,
  Andros 022959
Maritime Museum of British Columbia,
  Victoria 007331
Maritime Museum of Crete,
  Chania 023041
Maritime Museum of Macau,

Macau 008070
Maritime Museum of Monterey,
  Monterey 051011
Maritime Museum of Tasmania,
  Hobart 001145
Maritime Museum of the Atlantic,
  Halifax 005955
Maritime Museum of Thera (Santorini),
  Oia 023203
Maritime Museum of Townsville,
  Townsville 001588
Maritime Museum Zierikzee,
  Zierikzee 033102
Maritime Prints, Crewkerne 117646
Maritime Replicas America,
  Miami 094763
Maritime Self-Defense Force Sasebo
  Museum, Sasebo 029350
Maritime Tradition Museum,
  Piraeus 023221
Mariu s Comercio de Antiguidades,
  Rio de Janeiro 064083
Mariupol's'kij Krajeznavčyj Muzej,
  Mariupol' 043259
Marjan, Vancouver 064843
Marjankov, Bohumil, Litoměřice 140482
La Marjolaine, Breteuil-sur-Iton . 067484
Marjolaine, Saint-Martin-sur-Oust 072956
Marjorie and Herman Platt Gallery,
  University of Judaism, Los
  Angeles 050452
Marjorie Kinnan Rawlings Historic
  State Park, Cross Creek 048074
Mark, London . 090327, 118440, 134721
Mark Pasek Gallery, New York . 122845
Mark Twain Birthplace,
  Stoutsville 053558
Mark Twain Home and Museum,
  Hannibal 049350
Mark Twain House, Hartford ... 049386
Mark, I., Chicago 120196
Mark, Uwe, Oederan 077892
Marka Mesat Antika, İstanbul .. 088130
Markarian, Cincinnati 092874
Markarydsortens Hembygdsmuseum,
  Markaryd 040845
Marke & Münze, Graz 098004
Markedshallen, Nykøbing Falster 066014
Markeim Art Center, Haddonfield 049306
Markel, Kathryn, New York ... 122846
Marker Museum, Marken 032637
Markert, Julia, Firenze 131206
Market-Antiques, Dallas 093171
Market-Antiques, Fort Worth .. 093460
Market-Antiques, Houston 093737
The Market Antiques, San
  Antonio 096997
Market Art, Houston 121088
Market Books, Whangarei 143379
Market Central, Memphis 094649
Market Coins, Seattle 097472
Market Collectibles, Seattle ... 097473
Market Cross Gallery, Debenham 117672
The Market Cross Gallery, Sturminster
  Newton 119177
Market Deeping Antiques, Market
  Deeping 090706
Market Five Gallery, Washington 124932
Market Gallery, Toronto 007188
Market Graphics, Seattle 124566
Market House, Burnham Market 088372
The Market House, Monaghan . 109676
Market House Gallery, Marazion 118716
Market Lane Antiques, Tralee .. 079714
Market Lavington Village Museum,
  Market Lavington 045231
Market Street Mews, Penryn .. 118890
Market the Champion Village,
  Houston 093738
Marketplace Antiques Gallery,
  Detroit 093366
Markey, Lawrence, San Antonio 124049
Markfield Beam Engine and Museum,
  London 045033

Register der Institutionen und Firmen

Markgräfler Museum Müllheim, Müllheim
– Martin Luther King jr. Center and Preservation District,

Markgräfler Museum Müllheim, Müllheim 020499
Markgräfliches Opernhaus, Bayerische Verwaltung der staatlichen Schlösser, Gärten und Seen, Bayreuth 016997
Markgraf, Gunzenhausen 107189
Markgrafenmuseum Ansbach, Ansbach 016595
Markham Museum and Historic Village, Markham 006265
Markham & Son, E.J., Colchester 088951, 134290
Marki, Brian, Portland 123642
De Markies, Breda 082999
Markies, J., Forest Row 089295
Markiza, Beograd 085341
Markoff, Youri, Sablonnières .. 072653
Markov, W., Bonn 141620
Marks & Marks, Denver 120744
Marks Antikt och Antikvariat, Fritsla 086680, 143836
Mark's Antique Tubs, New Orleans 095130
Mark's Book Barn, Turrella 139939
Mark's China Pottery, Fresno .. 135569
Marks-Hirshfeld Museum of Medical History, University of Queensland, Herston 001142
Marks, Anthony, London 090328
Marks, B. J. & S., Oldham 090912
Marks, Barrie, London 144647
Marks, Matthew, New York 122847, 122848, 122849
Marks, Percy, Sydney 062293
Markt- und Schaustellermuseum, Essen 018303
Markt Einersheimer Heimatstuben, Markt Einersheim 020293
't Marktje, Den Haag 083104
Markus, Michael, Lingen 117881
Markwalder-Hoch, Beatrice, Würenlos 087916
Markworth, Nidda 142232
Marland's Grand Home, Ponca City 052233
Marlboro County Historical Museum, Bennettsville 046919
Marlboro Gold Arts, Cork 079497
Marlboro Historical Society, Marlboro 050701
Marlborough, London . 118441, 144648
Marlborough, Milton . 061857, 127640
Marlborough, Monaco 112010
Marlborough, New York 122850, 122851
Marlborough, Santiago de Chile . 101912
Marlborough Gallery, Blenheim . 112822
Marlborough Parade Antique Centre, Marlborough 090724, 134844
Marlborough Provincial Museum, Blenheim 033162
Marle Place Gallery, Brenchley . 117419
Marletta, Pino, Firenze 131207
Marlière, Hélène, Saint-Tropez . 073315
Mårlind, Göteborg 115660
Marlipins Museum, Shoreham-by-Sea 045790
Marlo, Philadelphia 145415
Marlow Antique Centre, Marlow . 090728
Marlowe Books, Canterbury ... 144307
Marly, Biel 087320
Marma Gully Antiques, Murtoa . 061931
Marmar, Madrid 115253
Marmaras, Timotheos, Athinai . 079171
Marmaris Kalesi ve Arkeoloji Müzesi, Marmaris 043023
Marmet, Peter, Igis 087535
Marmoles Carrigarc, Barcelona . 133130
Marmorbad, Museumslandschaft Hessen Kassel, Kassel 019540
Marmori, Elisabetta, Roma 131969
Marmormuseum, Adnet 001687
Marmorpalais, Stiftung Preußische Schlösser und Gärten Berlin-Brandenburg, Potsdam 021189

Marnardal Museum, Marnardal . 033766
Marney, Patrick, Long Melford . 090622, 134810
Marnier, Edward, Shepton Mallet 091300
Marninart, Washington 124933
Marokunst, Rotterdam 083582
Maronie, Kyoto 111298
Maronne, Elisabeth, Bordeaux . 140798
Maros Antikvárium, Budapest .. 142563
Marotte, Nancy 070425
Die Marotte, Schweinfurt 078422
Marotte de Foucade, Boulogne-sur-Mer 067402
Marotte, Pascal, Passenans ... 129162
Marotti, Silvio, Montevideo ... 092008, 144948
Marouchian, Kevork, München . 077565
Maroux, Simone, Corbigny 068146
Marquage, Marseille 104186
Marquand, Seattle 138471
Marquard, Rotterdam 083583
Marquardsen, H., Breklum 075142
Marquardt, Curt, Ottersberg ... 077996
Marquardt, Dieter, Vechelde .. 108811
Marquardt, Martin, Freiberg am Neckar 129857
Marquardt, Nikki Diana, Paris .. 105067
Marque, Blakehurst 139649
Marque, Edouard de la, Paris .. 071637
Marqueria Esart, La Paz 100846
O Marquês, Lisboa 084913
Marquès, Magin, Bedoin 103358
Marques, Maria E., Amadora .. 113954
Marques, Maxime, Aubagne ... 140768
Marques, Paulo Tavares, Lisboa 084914
Marques, Silvia, Bilbao 085879
Marquet, Karine, Paris 105068
Marquetabla, Bucaramanga ... 102393
La Marqueteria, Asunción 113332
Marquetería y Galería Acrópolis, Bucaramanga 102394
Marqueterie d'Art, Saverne 073449
Marquette, Cincinnati 120357
Marquette County History Museum, Marquette 050705
Marquette County Museum, Marquette County Historical Society, Westfield 054339
Marquette, Frédéric, Abbeville . 066410
Marquette, Frédéric, Cocquerel . 068055
Marquez, Lima 113364
Marquez, Laura, Madrid 115254
Marquis, Los Angeles 121554
Marquis' Auctions, Earlville 127209
Le Marquis de Carabas, Boissy-la-Rivière 067277
Marquis, Olivier, Épinal 125582
Marquisat de Pont-Croix, Musée du Patrimoine, Pont-Croix 014933
Marquit, Andrea, Boston 119891
Marquiz, Eze 103734
Marquois, Jacky, Niort 070714, 128874
Marr, Iain, Beauly 088416
Marra & Carbone, Roma 081438, 131970
Marrari, Enrico, Roma 081439
Marrassé, Philippe, Saint-Hilaire-de-Lusigna 072822
Marrero, Las Palmas de Gran Canaria 115379
Marrett House, Standish 053486
Marri, Tito, Prato 131779
Marrin & Sons, G. & D.I., Folkestone . 089290, 134391, 144428
Marringhurst Pioneer Park Museum, Pilot Mound 006652
Marriott, Corinda 139704
Marriott, Stafford Hits 139914
Marriott, Strathpine 139915
Marriott, Michael, London 099320, 134722
Marro, Ferdinando, Tübingen .. 078724
Marrone, München 077566
Marry, Lyon 128720

Marryat, Richmond, Surrey 091101, 134968
M'Ars, Moskva 114277
Mars Gallery, Chicago 120197
Mars, Laura, Berlin 106371
Marsa, Kobarid 114682
Marsal, Asunción 113333
Marsaleix, Christian, Bordeaux . 128496
Marsaleix, Pierre, Paris 129080
Marsam, Rabat 112052
Marsault, Biarritz 067204
Marschner, Berlin 074874
Marschollek, Nicolle, Groningen 083301
Marschpoort Kunst en Antiek, Zutphen 083722
Marsden Galleries, Bathurst ... 098712
Marsden Hall Antiques, Taupo .. 084095
Marsden, Barrett, London 118442
Marsden, Franck, Hardinghen .. 068791
Marsden, Jane J., Atlanta 092154
Marse Taltavull, Maria Amor, Barcelona 114933
Marseille Collections, Marseille . 069949, 140946
Marsel, Beograd 114428
Marselis, Århus 065604
Marsh Collection Society, Amherstburg 005331
Marsh, Gerald E., Winchester . 091870, 135238
Marshall, London 090330, 134723
Marshall, San Antonio 096998
Marshall Arts, Memphis 121703
Marshall County Historical Museum, Holly Springs 049512
Marshall County Historical Museum, Plymouth 052208
Marshall County Historical Society Museum, Warren . 054063, 054064
Marshall County Museum, Marshalltown 050718
Marshall M. Fredericks Sculpture Museum, Saginaw Valley State University, University Center .. 053895
Marshall Phillips, London 090331, 134724
Marshall Seifert, Dunedin 112885
Marshall & Co., Frank R., Knutsford 126935
Marshall, Kevin, Kingston-upon-Hull 089703
Marshall, Lyne, Tallegalla 099802
Marshall, Nancy, New York 095566
Marshalls of Wells, Wells 091772
Marshs, Cork 079498
Marsilio, Venezia 137759
Marsilius, Speyer 142369
La Marsillargue, Avignon 066884
Marsjandisen, Oslo 084327
Marsk-Antik, København 140613
Marskandiserbutikken, Roskilde 066092
Marskandiseren, Middelfart ... 066000
Marskandiseren, Varde 066183
Marskandiseren i Gilleleje, Gilleleje 065699
Marskantik, Roskilde 066093
Marsol, Milano 080603
Marson, Lecrico Terme 110168
Marstal Søfartsmuseum, Marstal 010230
Marstallmuseum, Bayerische Verwaltung der staatlichen Schlösser, Gärten und Seen, München .. 020546
Marstallmuseum, Fürst Thurn und Taxis Museen, Regensburg .. 021327
Marsteen, Geelong 061510
Marston, Abingdon 138135
Marston House, San Diego 052965
Marston, Luce, Washington ... 097846
Marszalec-Milanowska, Katarzyna, Gdańsk 113480
Marszand, Łódź 113678
Marszand, Warszawa 113863
Mart Saare Majamuuseum, Viljandi Muuseumi, Suure-Jaani 010662

Marta Herford, Museum für Kunst, Design und Architektur, Herford 019203
Martagon, Malaucene 104147
Martano, Torino 081727
Martayan Lan, New York 145353
M'Arte, Milano 110379
Marteau, Thierry, Bonneuil Matours 067290
Martel Greiner, Paris 105069
Martel Maides, Saint Peter Port . 127030
Martel, Thierry, Goult 128612
Martella, Francesco, Roma 081440
Martelli Castaldi, Napoli 131627
Martelli, Giorgio, Roma 131971
Martello d'Arte, Palermo 081053
Martelot, Gisèle, Sainte-Marie-du-Mont 073355
Marten River Logging Museum, Marten River 006267
Martens, Gent 100699
Martens, Monica, Aardenburg . 132379
Mårtenson & Persson, Båstad .. 115579
Martha Berry Museum, Rome .. 052671
Martha Street Studio, Winnipeg . 101884
Martha's Mixture, Richmond .. 096560, 136173
Martha's Vineyard Exhibition, Edgartown 048472
Marti, Barcelona 133131
Martial, Philippe, Paris 129081
Martin & Co, Cheltenham 088837
Martin, Los Angeles 121555
Martin, Montpellier 070307
Martin & Co., Norfolk 123233
Martin, Reims 129193
Martin, Saint Louis 096735
Martin & Chausselat, Versailles . 125835
Martin & G. Chausselat, J., Versailles 125836
Martin & Lawrence, Boston ... 119892
Martin & Lawrence, New Orleans 122202, 127335
Martin & Lawrence, New York . 122854
Martin & Lawrence, San Francisco 124337
Martin & Parke, Farnborough, Hampshire 089257, 144424
Martin & Pole, Wokingham 127097
Martin & Stein, Sydney 062294
Martin & Zambito, Seattle 124567
Martin and Osa Johnson Safari Museum, Chanute 047526
Martin-Anselme, Moncontour-de-Bretagne 070156
Martin Art Gallery, Allentown .. 046411
Martin Benka Museum, Martin . 038390
Martin-Caille Matignon, Paris .. 105072
Martin-Caille, Bruno, Paris 105071
Martin County Historical Museum, Stanton 053492
Martin County Pioneer Museum, Fairmont 048677
The Martin D'Arcy Museum of Art, Loyola University Museum of Medieval, Renaissance and Baroque Art, Chicago 047671
Martin du Daffoy, Jean-Luc, Paris 071639, 071640
Martin du Louvre, Paris 105073
Martin du Nord- de Bouvet, Sainte-Geneviève-des-Bois 125810
Martin Filatelia e Numismatica, Venezia 081857
Martin-Gropius-Bau, Berlin ... 017220
Martin-Heidegger-Museum, Meßkirch 020386
Martin Hernandez, Jose Manuel, Salamanca 086408
Martin-Lauterburg-Stiftung, Bern 041319
Martin Ledesma, Miguel Juan, Madrid 086214
Martin Luther King jr. Center and Preservation District, Birth Home Museum, Ebenezer Baptist Church

Museum, Atlanta 046625
Martin Luthers Geburtshaus, Stiftung
Luthergedenkstätten in Sachsen-
Anhalt, Lutherstadt Eisleben . . 020193
Martin Luthers Sterbehaus, Stiftung
Luthergedenkstätten in Sachsen-
Anhalt, Lutherstadt Eisleben . . 020194
Martin-Malburet, Marc, Paris . . . 105074,
141238
Martin Museum of Art, Waco . . 054017
Martin of London, Los Angeles . 094403
Martin of Melbourne, Armadale,
Victoria 061040
Martin of Melbourne, East
Prahan 127595
Martin-Quick, Wolverhampton . . 091904
Martin Rodriguez, Jose Luis,
Salamanca 086409
Martin Saenz, Francisco, Málaga 086296
Martin Sanchez, Ricardo Antonio,
Madrid 115255
Martin Street Antiques,
Gardenvale 061503
Martin-Taylor, David, London . . 090334
Martin Van Buren Historic Site,
Kinderhook 049971
Martín Villena, Ignacio, Granada 143746
Martin-von-Wagner-Museum
der Universität Würzburg,
Antikensammlung und
Gemäldegalerie mit Graphischer
Sammlung, Würzburg 022792
Martin, Alain Georges, Roma . . . 081441
Martin, Ambrosio, Miami 121824
Martin, Andreas, Dresden 075421
Martin, C.J., London 090332
Martin, Chantal, Toulouse 129303
Martin, Christophe & Sylvain,
Chauffailles 067930
Martin, Christopher H., Dallas . . 120628
Martin, Clorinde, Paris 105070
Martin, Daniel, Chatellerault 067906
Martin, Didier, Aucamville 128452
Martin, E., Bonn 075079
Martin, Florence, Paris 071638
Martin, Gail, New York 095567, 122852
Martin, Georges, Vence 074004
Martin, Günter, Bürgstadt 075219
Martin, Guy, Paris 141237
Martin, Heidrun & Rolf, Edertal . 075567
Martin, J., Indianapolis 121201
Martin, Jean, Paris 129082
Martin, Jean Claude, Chavanoz . 128554
Martin, Jean-Pierre, Bourg-Saint-
Maurice 067418
Martin, John, London . 118443, 118444
Martín, Juan, México 111968
Martin, Maria, Bram-le-Lauragais 067446
Martin, Mary-Anne, New York . . 122853
Martin, Michael, Heinsberg . . . 076483
Martin, Michael, San Francisco . 124336
Martin, Michel, Quiberon 105406
Martin, Olivier, Montréal 101431
Martin, Patrice, Saint-Étienne-de-
Chigny 072779
Martin, Patrice, Souppes-sur-
Loing 073551
Martin, Paul, Marlborough 090725
Martin, Peter, Clayfield 061335
Martin, Peter J., Eton . 089218, 134373
Martin, Richard, Gosport 089349, 134415,
144440
Martin, Richard, Paddington, New
South Wales 099529
Martin, Robin, London 090333
Martin, Steve, New Orleans . . . 122201
Martin, Ted, Toronto 101659
Martin, Thierry, Nantes 070471
Martin, Tony, Looe 090632
Martin, Vance, Kensington 121321
Martin, Yvona, Berlin 074875
Martinarte, Torino 110988
Martinborough Gallery,
Martinborough 112945

Martinčoková-Husárová, Katarína,
Košice 133051
Martine, Dudelange 082352
Martine Encadrement, Viry-
Chatillon 105882
Martinek, Düsseldorf 075495
Martinelli, Castaldi & C., Napoli . 131628
Martinelli, Elisa, Modena 110500
La Martinerie, Amboise 103226
Martines, Luca, Torino 132166
El Martinete, Málaga 086297
Martinetti, Gerard, Toulon (Var) . 129289
Martinez, Madrid 086215
Martinez, Troy 124703
Martinez Antiquité, Saint-Gratien 127607
Martinez-Barbeito Manovel, Beatriz, A
Coruña 133235
Martínez Charry, Octavio, Bogotá 128272
Martinez-Fleurot, Paris 071641, 141240
Martínez García, Antonio, Cádiz . 133222
Martínez Gómez, José María,
Montevideo 092009
Martinez Museum, Martinez . . . 050723
Martinez Romero, M., Valencia . 115508
Martinez Rueda, Barcelona . . . 133133
Martinez Sanchez, M.L., Madrid 115256
Martinez, Alfredo, Miami 121825
Martinez, Antonio Gil, Sevilla . . 086456
Martinez, Didier, Paris 141239
Martínez, Eugenio, Barcelona . . 133132
Martinez, Félix, Valencia 133466
Martinez, Guy, Amboise 066555
Martinez, J., Edinburgh 089180
Martinez, Jean-Claude, Paris . . 115075
Martinez, José, Lyon 104116
Martinez, Lorenzo, Madrid 086216
Martinez, Luis Fernando,
Barranquilla 102266
Martínez, M., Liège . . . 063772, 125347
Martinho, Ricardo P. Ferreira,
Lisboa 143583
Martini, Phuket 116968
Martini, Roma 081442
Martini & Ronchetti, Genova . . . 110122
Martini Girl Art Gallery, Atlanta . 119589
Martini, Brigitte, Landsberg am
Lech 077010
Martini, Fabrizio, Livorno 080327
Martini, Giovanni, Padova 080990
Martini, Laurent, Paris 071642
Martini, Mario, Roma 131972
Martini, Roland, Lafox 069136
Editions de la Martinière, Paris . 136963
Martinis, Marika de, Flims
Waldhaus 116267
Martino, Mostar 100856
Martinoja, Angelo, Roma 081443
Martinos, Eleni, Athinai 079172
Martinot, Alain, Paris 071643
Martins, Fortaleza 100954
Martin's Auction, Purvis 127415
Martins, Henri, Venette 129334
Martins, João, Setúbal 114113
Martins, João Moura, Porto . . . 085037
Martins, Manuel L. Murteira,
Lisboa 084916, 084917
Martins, Manuel Murteira, Lisboa 084915
Martinsen, Bo, Esbjerg 065640
Martinus, Werner, Gräfelfing . . . 129916
Martinustoren, Losser 032607
Martlesham Antiques,
Martlesham 090730
Martock Gallery, Martock 118726
Marton Historic Village Museum,
Marton 033239
Martorano, Luigi, Taranto 132081
Martorano, Paolo M., Milano . . . 080604
Martre, Caroline, Pont-l'Abbé . . 072257
Marttilan Kirkkomuseo, Marttila . 011147
Marttilan Kotiseutumuseo,
Marttila 011148
Martuni Art Gallery, Martuni . . . 000711

Marty & Pat's Art Gallery, Fort
Worth 120877
Marty-Parrot, Nathalie, Zürich . . 133990
Marty, Claude, Désandans 068287
Marty, Laure, Toulouse 073782
Marty, Wilfrid-Emmanuel,
Toulouse 105748, 141368
Martyn Ferenc Múzeum, Janus
Pannonius Múzeum, Pécs . . . 023842
Martyno Jankaus Muziejus,
Bitenai 030326
Martyr's Memorial at Sports City,
Amman 029774
Martyrs' Shrine, Midland 006305
Marue Ohno Shoten, Tokyo . . . 082178
Marugame Bijutsukan,
Marugame 028996
Marugame Gallery, Marugame . 111306
Marugame Genichiro-Inokuma Gendai
Bijutsukan, Marugame 028997
Marugame Hirai Museum,
Marugame 028998
Marugan de Santos Santi, Saint-
Ouen 073150
Marugan Perez, Miguel Angel,
Palencia 086332
Marui, Hiratsuka 081972
Marukiya, Kobe 081999
Maruku Arts, Ayers Rock 061072
Marull Dalmau, Jesus Maria,
Barcelona 133134
Marulla, Giuseppe, Roma 131973
Marumo, Claude, Paris 105076
Marunaka, Hiroshima 081980
Marunouchi, A. Jain, New York . 122855
Marusho, Kadoma 111266
Marusu Museum, Ishinomaki . . 028703
Maruyo, Nagoya 082061
Maruzen, Tokyo 142929
Marvelous Arts & Antiques of China,
Vancouver 064844
Marvin, San Marino 085285
Marvin Newton House, Brookfield 047241
Marvin, Alexander, New York . . . 095568
Marvol Museum, Kuilsrivier 038774
Marv's Collectibles Plus, Portland 096436
Marway, Winnipeg 064928
Marwen Foundation, Chicago . . 047672
Marx, Düsseldorf 075496
Marx & Saunders, Chicago . . . 120198
Marx, Etienne, Compiègne 068115
Marx, Heather, San Francisco . . 124338
Marx, Joachim, Mühlenbecker
Land 077424
Marx, Klaus-Günter, Düsseldorf . 075497
Marx, M., Lengerich 077082
Marxhausen Art Gallery, Seward 053278
Mary and Leigh Block Museum
of Art, Northwestern University,
Evanston 048641
Mary Ann's Antiques, Denver . . 093314
Mary Arden's House and the
Countryside Museum, Shakespeare
Birthplace Trust, Stratford-upon-
Avon 045904
Mary Baker Eddy Library for the
Betterment of Humanity, Boston 047119
Mary Ball Washington Museum,
Lancaster 050127
Mary Brogan Museum of Art and
Science, Tallahassee 053661
Mary Catherine's Antiques,
Columbus 093034
Mary E. Black Gallery, Nova Scotia
Centre for Craft and Design,
Halifax 005956
Mary Leu Fine Carving Gallery and
Meili Museum, Sansing 042438
Mary Mackillop Place Museum, North
Sydney 001365
Mary McLeod Bethune House,
Washington 054134
Mary March Regional Museum and
Logging Exhibition, Grand Falls-

Windsor 005903
Mary Mary, Glasgow 117851
Mary Moffat Museum,
Griquatown 038732
Mary Place Gallery, Paddington, New
South Wales 099530
Mary Porter Sesnon Art Gallery,
University of California at Santa Cruz,
Porter College, Santa Cruz . . 053100
Mary Queen of Scots' House,
Jedburgh 044665
Mary Rose Museum, Portsmouth 045555
Mary Todd Lincoln House,
Lexington 050248
Mary Washington College Galleries,
Fredericksburg 048957
Mary Washington House,
Fredericksburg 048958
Maryborough Art Society, Maryborough,
Queensland 058206, 099284
Maryborough Station Antique
Emporium, Maryborough,
Victoria 061799
Mayfield Museum, Mayfield . . . 006269
Marygrove College, Dept. of Art,
Detroit 057011
Maryhill Museum of Art,
Goldendale 049151
Maryland Arms Collectors Association,
Baltimore 060582
Maryland Art Place, Baltimore . 046757,
119753
Maryland Association of History
Museums, Annapolis 060583
Maryland College of Art and Design,
Silver Spring 057916
Maryland Federation of Art,
Annapolis 060584
Maryland Historical Society Museum,
Baltimore 046758
Maryland Institute College of Art,
Baltimore 056706
Maryland Institute Museum, College of
Art Exhibitions, Baltimore . . . 046759
Maryland Pastel Society,
Riderwood 060585
Maryland Science Center,
Baltimore 046760
Marymont Plantation Antique Shop,
Nashville 095014
Marymoor Museum of Eastside History,
Redmond 052498
Marymount Manhattan College, New
York 057557
Marymount Manhattan College
Museum, New York 051422
Maryol, Johannesburg 114798
Maryol, Rosebank 114812
Maryport Maritime Museum,
Maryport 045232
Mary's Attic, Virginia Beach . . . 097790
Mary's Imports, Jacksonville . . . 094008
Mary's Stuff, Chicago 092747
Mary's Swap Meet, Oklahoma
City 095991
Mary's Upholstery, Memphis . . . 094650,
135782
Marysa, Paris 071644
Maryse, Trouville-sur-Mer 073893
Marystown Heritage Museum,
Marystown 006270
Marywood University Art Galleries,
Scranton 053217
Marzano, Roberto, Roma 081444
Marzano, Stefano, Verona 132290
Marzee, Nijmegen 112609
Marzellus-Buchhandlung, Köln . 142019
Marziart, Hamburg 107297
Marzocchi, Luciano, Cordenons . 109977
MAS, Amsterdam 082803
Mas d'Antic, Valencia 086524
Mas de Bony, Marsillargues . . . 069975
Mas de Curebourg, L'Isle-sur-la-
Sorgue 069579

Mas de la Pyramide, Saint-Rémy-de-Provence 015539
Le Mas des Arnaud, Saint-Martin-de-Crau 105555
Mas des Chevaliers, Les Baux-de-Provence 069403
Mas Maury, André, Pescadoires 072116
Mas Muntaner, Palma de Mallorca 086359
Mas Oliveras, Santiago, Barcelona 143728
Mas Suárez, Elvira, A Coruña .. 085956, 143741
Mas, Christian, Saint-Barthélemy 109110
Mas, Colette, Paris 105077, 141241
Masa Art, Miami 121826
Masago, Osaka 111363
Masaki Art Museum, Osaka ... 029248
Masamune, Saron, Osaka 082087
Masan Art Center, Masan 111605
Masar, Vero de, Wien 100350
Masarykovo Muzeum, Hodonín . 009563
Masbanaji, Pierre, Lyon 104117
Mascalchi, Mariapia, Prato 081172
Maschenmuseum, Albstadt 016492
Mascherino, Roma 110822
Il Mascherone, Palermo 081054
Maschinen-u.Heimatmuseum, Eslohe 018284
Maschinenmuseum-Musée de la Machine-Centre Müller, Biel .. 041338
Maschio, Fernand, Vercia 129335
Masciantonio, Gabriele, Verona . 132291
Masclef, Amiens 066579
Mascot, New York 122856
Maser, Pittsburgh 123563
Masfrand, Denis, Lomme 069612
Mashantucket Pequot Museum, Mashantucket 050733
Mashiko Sankokan, Mashiko .. 029001
Mashrabia, Cairo 103058
Al-Mashrabia, Manama 063153
Masi, Alessandra, Firenze 131208
Masi, Antonio, Milano 080605
Masi, Lamberto, Palermo 081055
Masi, Ninetta, San Marino 085286
Masi, Rosario, Torino 081728
Masia Museu Can Magarola, Alella 038938
Masiero, Pietro, Milano 131474
Masini, Amel, Paris 129083
Masini, Francesco, Firenze ... 131209
Masini, Stefania, Firenze 080168
Masip Pascual, Jose, Barcelona 085797
Masis-Metna, Genève 087483
Maskaronen, Stockholm 115910
Maske, Siegfried, Rietz-Neuendorf 108502
Maskell & Co., Don, San Francisco 124339
Maskenmuseum, Denkendorf .. 017856
Maskun Museo, Masku 011149
Maslak McLeod, Toronto 101660, 101661
Maslan, Michael, Seattle 097474, 145593
Maslewski, K. & R., Kirchheim unter Teck 076808
Masliah, Guy, Saint-Ouen ... 073151
Maslin, Betty, Murchison 061929
Maslo, Rüdiger, Winsen, Luhe .. 079025
La Masmédula Galería, México . 116169
Masnata, Genova 110123
Masnica, Karol, Kysucké Nové Mesto 085573
Måsøy Museum, Havøysund .. 033645
Mason, Cincinnati 092875
Mason, Darwin 098950
Mason, Dublin 079570
Mason, Fresno 120901
Mason, Kings Park 099195
Mason, New Orleans 122203
Mason County Museum, Mason 050734
Mason Gray Strange, Kilkenny . 125203
Mason Gross Art Galleries, RutgersUniversity, New

Brunswick 051245
Mason, David, Toronto 140411
Mason, Harry, Brighton 088626, 134152
Mason, Ian, Ballarat 139632
Mason, Jane, North Shore 112972
Masonic Grand Lodge Library and Museum, Waco 054018
The Masonic Library and Museum of Pennsylvania, Philadelphia ... 052065
Al-Masri, Mohammed Bassam, Damascus 088042
Mass Gallery, Fitzroy North 099036
Mass Moca, Massachusetts Museum of Contemporary Art, North Adams 051590
Massa Múzeum, Országos Műszaki Múzeum, Miskolc 023789
Massachusetts College of Art, Boston 056770
Massachusetts Historical Society, Boston 047120
Massacre Rocks State Park Museum, American Falls 046438
Massada, London 090335
Massalia, Marseille 104187
Massalme, Dr. Elisabeth, Sankt Michaelisdonn 108556
Massarella, Saltaire 119055
Massarelli, Pedro H., São Paulo 101102
Massaroni, Rosina, Roma 081445
Masse, Vincent, Troyes 073907
Masseguin Lou Brocantou, Mende 070047
Massenet, Aliette, Paris 071645
Massey, Saint Louis 127428
Massey Area Museum, Massey . 006272
Massey & Son, D.J., Macclesfield 090674
Massias, Jacques, Nantes 070472
Massieu, Martine, Champeaux-et-la-Chapelle-Pommier 067808
Massillon Museum, Massillon .. 050737
Massimiliani, Marco, Roma 110823
Massin, Paris 136964
Massingham, Roy, Brasted 088583
Massnes Villmarksmuseum, Bjordal 033546
Massol, Jacques, Paris 105078
Masson, Paris 071646
Le Masson Brunot, Paris 129084
Masson, Andrée, Villeurbanne . 074216
Masson, Jean, Cucq 068255
Masson, Michèle, Bordeaux ... 140799
Masson, Philippe, Jaunay-Clan . 068898
Masson, Pictet & Boissonnas, Zürich 133991
Massone, Giuseppe, Torino ... 081729
Massot, Régis, Changé 067825
Massoud, Mervat, Cairo 103059
Massow, Dagmar von, Berlin .. 074876
Massua-Educational Museum on the Holocaust, Tel Yitzhak 025217
Mastelinck, Jean-Pierre, Avesnes-le-Comte 066867
Mastellaro, Luciano, Padova ... 131669
Master, Amsterdam 082804
Master Drawings, New York ... 139427
Master Drawings Association, New York 060586, 138406
Master Framers, Saint Paul ... 136222
Master International Art, Venezia 111136, 111137
Master Mirror, Modena 080806
Master Pieces Antiques, Cincinnati 092876, 127178
Master Rug Gallery, Los Angeles 094404
Master, Thomas, Chicago 120199
Masterpeace Fine Art Studio, Gabriola Island 101309
Masterpeach, Kenmare 114805
Masterpiece, Hobart 061634
Masterpiece Editions Limited, Chicago 120200
Masterpiece Gallery, Cincinnati . 120358

Masterpiece Gallery, Toronto ... 101662
Masterpiece Gallery, Vancouver . 101811
Masters Art, Barcelona 114934
Masters Art, Palma de Mallorca 115359, 115360
Masters Gallery, Woodend 099899
Masters Gallery Ltd, Calgary ... 101248
Masterworks, Oakland 123257
The Masterworks Foundation Collection, Paget 004194
MasterWorks Gallery, Auckland . 112787, 112788
Masterworks Gallery, Tampa .. 124663
Masterworks of New Mexico, Albuquerque 119493
Mastny & Dietrichstein, Wien .. 063023
Mastracci, Pittsburgh 096309
I Mastri Paoli, Bologna 130976
Mastrigt & Verheul, Den Haag .. 112400
Mastro, Beppe, Cagliari 131070
Mastruzzi, Umberto, Roma 131974
Masuhr, Ch., Bernkastel-Kues .. 074962
Masur Museum of Art, Monroe . 050998
Masure, Arnaud, Châteaulin ... 067885
Masuya, Hiroshima 081981
Masuzawa, Sapporo 082101
Mata-Hari Antiques, Singapore . 085468
Mata Saralegui, Jose Manuel, Donostia-San Sebastián 085979
Mata Valles, Teresa, Barcelona . 085798
Mataatua Gallery, Whakatane .. 113103
Matagorda County Museum, Bay City 046832
Matahari, Napier 084007
Matanza, Miami 094764
Matarazzo, Gesualdo, Catania .. 080039
Matchbox Road Museum, Newfield 051525
Matchboxmuseum Latent, Prinsenbeek 032768
Matchimawas National Museum, Songkhla 042719
Matei, Soltau 108637
Matei, Elvira, Bucureşti 085100
Matenadaran Manuscript Museum, Yerevan 000736
Mateo e Hijo, Rafael, Sevilla ... 133435
Mateos Osorio, Antonio, Málaga 143785
Materazzo, Michele, Bologna .. 130977
Materia, Québec 101510
Material Culture, Philadelphia .. 096136
Materialis Verlag, Biberach an der Riß 137108
Materiały Starożytne, Warszawa 138972
Matériaux d'Antan, Aix-en-Provence 066469
Matern, Christiana, Salzburg ... 062810, 100170, 140025
Materna, Daniel, Vancouver ... 101812
Materne, Hans-Jürgen, Dortmund 075387
Materne, Hans-Jürgen, Herdecke 076498
Maters & Begeer, Franeker 132504
Mateu, Montserrat, Barcelona .. 137964
Mateus, Joaquim Cunha, Porto . 085038
Mateyka, Marsha, Washington . 124934
Mathaf Gallery, London 090336, 118445
Mathé, Villeurbanne 074217
Mathé, Frédéric, Lyon 069762
Mathé, Lucile, Nice 070633
Mathé, Monique & Jean, Sens . 073499
Mathematikum, Gießen 018693
Mathematisch-Physikalischer Salon, Staatliche Kunstsammlungen Dresden, Dresden 017996
Mather Homestead, Wellsville .. 054275
Mather Post Office Museum, Cartwright 005574
Matheson Museum, Gainesville . 049002
Mathews, Gloria, Omaha 096049
Mathewson, Tracy, Warkworth . 113053
Mathias Ham House, Dubuque . 048375
Mathias Schmid-Museum, Ischgl 002144
Mathias, Jean-Jacques, Paris .. 125738
Mathias, Ph., Daisendorf 075293

Mathies, Bad Driburg 129405
Mathieson, Tom, Ramsgate 062136
Mathieu, Besançon 067129
Mathieu, Lyon 104118
Mathieu, Alain, Champeix 067809
Mathieu, Jean-Claude, Dommartin-aux-Bois 068379
Mathieu, Marie-France, Etival-Clairefontaine 068484
Mathieu, Patrick, Châtillon-sur-Loire 067921
Mathilda Traditional Lab & Gallery, Malmö 115765
Mathis, Jean-Louis, Bléré 140784
Mathisen, Bjørn, Oslo 084328
Matilda Antiques, Long Jetty .. 061748
Matilda Art Rentals, Sydney ... 062295
Matilda Roslin-Kalliolan Kirjailijakoti, Merikarvia 011153
Matilda's Antique Centre, North Fremantle 061988
Matinais, Herbignac 068807
Mativet, Georges, Paris 129085
Matiz, Belo Horizonte 063909
Matiz, Miguel Hidalgo 138914
Matiz Arte Galeria, Belo Horizonte 100894
Matlock Antiques, Matlock 090737
Matlosz & Co., W., Phoenix ... 096223, 136118
Matombo – Schönes aus Afrika, Salzburg 100171
Matong Memories, Matong 061801
Matopos National Park Site Museums, Bulawayo 054845
Matos, Maria Oliveira, Setúbal . 114114
Matos, Rafael, México 126263
Matousek, Peter, Oberteuringen 077882
Mátra Múzeum, Gyöngyös 023653
Matre & Penley, Atlanta 119590
Matriart, Toronto 138585
Matrica Múzeum, Százhalombatta 023920
Matrixarts, Sacramento 052730
Mats Kuriosa, Malmö 086854
Matschgerermuseum, Absam .. 001680
Matschnig, Wien 063024
Matsell, Brian, Derby 089055
Matsqui Sumas Abbotsford Museum, Abbotsford 005301
Matsudo Museum, Matsodo ... 029003
Matsui, Nara 111336
Matsukyu, Kyoto 082032
Matsumae-cho Bunkakan, Matsumae 029008
Matsumori Art, Tokyo 082179
Matsumoto Museum of Folk and Arts, Matsumoto 029009
Matsumoto-shiritsu Bijutsukan, Matsumoto 029010
Matsumoto-shiritsu Hakubutsukan, Matsumoto 029011
Matsumoto, K ., San Jose 124453
Matsumura, Tokyo 142930
Matsuo Museum, Takigawa ... 029464
Matsuoka Museum of Art, Tokyo 029544
Matsura Shiryo Hakubutsukan, Hirado 028634
Matsusaka Oroshi Kinenkan, Matsusaka 029013
Matsushima, Sendai 082118
Matsushima Historical Museum, Matsushima 029017
Matsushima Orugoru Hakubutsukan, Matsushima 029018
Matsushita Bijutsukan, Fukuyama, Aira-gun 028566
Matsushita Hakubutsukan, Aira . 028464
Matsuyama, Sapporo 111339
Matsuyama Municipal Shiki-Kinen Museum, Matsuyama 029022
Matsuzaki, Yokohama 082230
Matt, Dr. Hansjakob von, Zürich 087989, 144191

Matt, Frederic, Dommartin
(Rhône) 068376
Matt, Martin von, Stans 144140
Matt, Serge, San Francisco .. 097281,
.... 136334
Matt, Sibylle von, Stans 133932
Matta, John, New Orleans 135862
Mattancherry Palace Museum,
Kochi 024348
Mattapoisett Museum and Carriage
House, Mattapoisett 050741
The Mattatuck Museum of the
Mattatuck Historical Society,
Waterbury 054199
Mattawa and District Museum,
Mattawa 006276
Matte, K.-Dieter, Berlin 074877
Matteoli, Daniele, Firenze .. 131210
Matter, Willy, Weissenburg .. 087895
Mattern, Charles, Saint-Ouen . 073152
Matthei, Santiago de Chile .. 101913
Matthes, Ursel, Lübben 077168
Matthews, Tampa 124664
Matthews Museum of Maine Heritage,
Union 053887
Matthews, Richard, Auckland . 083807
Matthews, William, Denver .. 120745
Matthias, J., Houston 121089
Matthias, Zapfe, Weimar 078898
Matthies, Joachim D., Hamburg 076273
Matthiesen, London .. 090337, 118446
Matthijsse, Thijs, Middelburg . 132585
Matthys, Bruxelles 063504
Matti, Stuttgart 078618
Mattiello, Corinna, Venezia .. 132248
Mattioli, Gérard, Guebwiller . 068755
Mattioli, Manuela, Bologna .. 130978
Mattisrud Småbruksmuseum,
Løten 033751
Mattke, Jörg, Lübeck 077178
Mattox, Wayne, Woodbury 127499
Mattress Factory Museum,
Pittsburgh 052168
Matt's Gallery, London . 045034, 138227
Matt's Museum, Lawtell 050194
Mattsson, Per-Johan, Gävle .. 133525
Mattye, Houston 093739
Mattye Reed African Heritage Center,
Greensboro 049255
Matuizos Manto, Vilnius 142967
Maturango Museum of the Indian
Wells Valley, Ridgecrest .. 052568
Maturen, Antonio Esteban,
Zaragoza 086599, 133504
Matusalen, Salamanca 086410
Matuschek, B., Taunusstein .. 108759
Matuschowitz, Bochum 075038
Mátyás Király Palota Múzeum, Magyar
Nemzeti Múzeum, Visegrád .. 024047
Matyó Múzeum, Herman Ottó
Múzeum, Mezőkövesd 023777
Matzke, Helga, Grünwald 076119
Mau, Otto, Koblenz 076837
Mauad Seba, Michel, Montevideo 127112
Maubach, Magdalene & Susanne,
Goch 129906
Maubert, Jean-Marie, Hiers-
Brouage 068819
Maubrie, Gabrielle, Paris 105079
Mauch, Brigitte, Göppingen .. 107138
Mauchard, François, Tendon .. 073633
Maud Railway Museum, Maud . 045239
Maude I. Kerns Art Center,
Eugene 048621
Maude Kerns Art Center School,
Eugene 057061
Maudet, Bruno, Tours . 105764, 129321
Maudieu, Valérie, Issoudun .. 125597
Maudrich, Wilhelm, Wien 140101
Maudrux-Miniconi, Marie-France,
Alata 066500
Maud's Attic, Ipswich 089648
Maud's Place – The Putnam House,
Brookfield 005488

Mauer-Design, Ettlingen 137187
Mauermann, Essen 106925
Mauermuseum – Museum Haus am
Checkpoint Charlie, Berlin . 017221
Mauersberg, R., Nienburg 077783
Mauersberger-Museum,
Großrückerswalde 018861
Maui Gallery, Westport 113101
Mauilotus Asian Arts Gallery,
Minneapolis 122012
Maulbecker, Heiko & Marianne,
Köln 107662
Mauléon, Cordes-sur-Ciel 103630
Maulet, Didier, Paris 129086
Maune, J., Saint Louis 096736
Maunier & Noudel-Denian, Toulon
(Var) 125822
Maunoury Dufour, Jacqueline,
Chablis 067760
Mauny, Philippe, Mortagne-sur-
Gironde 070358
Maupas, Catherine, Montoire-sur-le-
Loir 070284
Maupilier, Belz 067084
Maupilier et Fils, Les Clayes-Sous-
Bois 069406
Maupin, Constance, Paris 071647
Maupin, Frédéric, Boulogne-sur-
Mer 067403
Maupin, Olivier, Tigy 129281
Maurel Franck, Montescot 070238
Maurel, Guy, Orange 070791
Mauren, Gisela, Köln 076896
Maurer, Frankfurt am Main 107015
Maurer, Küsnacht 116401
Maurer, A. & F., Buchs (Zürich) 116171
Maurer, Annette M., Madrid .. 133355
Maurer, Eduard, Alchenflüh .. 087186
Maurer, H.-J. & R., Bretzfeld .. 075194
Maurer, Horst & Waltraud,
Dietzenbach 141688
Maurer, R., Düsseldorf 075498
Maurer, Ruth, Zürich 116835
Maurer, Wolfgang, Gmund am
Tegernsee 076013
Maurer, Wolfgang, München ... 108161
Maurermuseum und
Dachdeckermuseum,
Langenlois 002265
Mauri Antik, Rudkøbing 066101
Mauri Ora Gallery, Kaikoura .. 112929
Mauriac, Rony, Larressore 069193
Maurice Antiquités, Saint-Étienne 072769
Maurienne, La Chambre 068968
Maurier, Frédérique, Paris 129087
Mauries, Denis, Coulomby 068180
Maurin, Paul, Paris 071648
Maurin, Raymond, Arles 066723
Mauritianum, Naturkundliches
Museum Altenburg, Altenburg . 016535
Mauritius Postal Museum, Port
Louis 030792
Maurits Galerie, Utrecht 112706
Mauritshuis, Koninklijk Kabinet van
Schilderijen, Den Haag 032180
Mauro Bi, Torino 081730
Mauron, Pierre, La Teste-de-
Buch 069104, 128647
Mauroner, Mario, Wien 100351, 136599
Maury, Bernard, Crouttes-sur-
Marne 068249
Maury, P.C., Paris 129088
Maus Park Antiques, Toronto .. 064703
Maus, Phillip, Groß Nordende . 076101,
.... 129921
Maus, Phillip, Kampen 076671
Mausefallen- und Kuriositätenmuseum
mit Galerie der Stillen Örtchen,
Güntersberge 018878
Mausefallen-Museum, Neroth . 020664
Mauseum, Bergheim (Erft) 017050
Mausoleum im Schlossgarten
Charlottenburg, Stiftung Preußische
Schlösser und Gärten Berlin-

Brandenburg, Berlin 017222
Mauthe, Markus, Langenargen . 107796
Mauz, Elisabeth, Albstadt 105971
Mava, Antibes-Juan-les-Pins .. 103260
Mava, Singapore 114568
Mavec & Co., Janet, New York . 095569
Mavelle, Thierry, Reims 129194
Maverick, Atlanta 119591
Maverick, Cincinnati 145075
Mave's Gallery, Stow-on-the-
Wold 119161
Mavie, Aubagne 066774
Mawer & Son, Thomas, Lincoln 126948
Mawr-Vassar, Bryn, Pittsburgh . 145436
Maws, Telford 119219
Maws Gallery, Bridgnorth 117423
Maws Gallery, Ironbridge 118028
Max Beckmann Gesellschaft e.V.,
München 059170
Max Böhlen Museum, Kandern . 019495
Max Ernst Museum Brühl, Brühl . 017643
Max et Isa, Megève 070036
Max Euwe Centrum, Amsterdam 031956
Max Liebermann Haus, Stiftung
"Brandenburger Tor", Berlin . 017223
Max-Reger-Gedächtniszimmer,
Brand 017538
Max-Slevogt-Galerie im Schloß Villa
Ludwigshöhe, Edenkoben .. 018092
Max Studio, Venezia 111138
Max, Erich, Döhlau 075367
Maxa, Martin, Banská Bystrica . 085536
Maxfield House Antiques,
Warminster 091735
Maxhütten-Museum, Bergen, Kreis
Traunstein 017047
Maxi Livres, Lyon 140925
Maxies Antiikki, Helsinki 066342
Maxim, Prato 081173
Maxim-Gorki-Gedächtnisstätte,
Heringsdorf, Seebad 019208
Maxime Brocante, Bressuire .. 067460
Maxime, Marie, Paris 071649
Maximilian-Gesellschaft e.V.,
Stuttgart 059171
Maximilianmuseum, Kunstsammlungen
und Museen Augsburg,
Augsburg 016662
Maximo Art, Albuquerque 119494
Maxine's, Saint-Ouen 073153
Maxoly, Miami 121827
Maxwell, Roleystone 099648
Maxwell & Dockrees, Woodford . 091920,
.... 127100
Maxwell Museum of Anthropology,
Albuquerque 046364
Maxwell, Helen, Braddon 098769
Maxwood, Cape Town . 085635, 133071
Maxxi – Museo delle Arti del XXI
secolo, Roma 027571
May, Aschaffenburg 137031
May, Megève 104215
May, Seoul 111689
May & Son, Shipton Bellinger . 127044
May and Herbert Gibbs Art Collection,
Heritage House Cultural Centre,
South Perth 001507
May Antique Center Mall, Oklahoma
City 095992
May-Lüllemann, Christa,
Wiesbaden 078987
May Moré, Madrid 115257
May Museum and Park,
Farmville 048710
May Natural History Museum and
Museum of Space Exploration,
Colorado Springs 047887
May Street Gallery, Saint Peters 099672
May, Ilona, Benshausen 129452
May, J.E., Amsterdam 082805
May, K.H., Amsterdam 082806
May, Maggie, Denver 093315
May, Marita, Daun 106661
May, Monika, Marburg 130283

May, Thierry, Fontainebleau ... 068567
Maya, Maastricht 083471
Maya Plastik Sanat Galerisi,
İstanbul 117064
Maya Polsky Gallery, Chicago . 120201
Mayang Miaozu Self-Rule Museum,
Mayang 008078
Mayart, Enschede 112478
Mayborn Museum Complex,
Waco 054019
Mayence, Mainz 107952
Mayer, Wien 140102
Nadja Mayer & Martin Indys, Simbach
am Inn 078471
Mayer & Mayer, Stuttgart 078619
Mayer-Jung, Paul G,
Jagsthausen 076640
Mayer, B.H., Buffalo 092544
Mayer, Bernward, Berlin 106372
Mayer, Bridgette, Philadelphia . 123417
Mayer, Christine, München 108162
Mayer, Dr. Gunter, Gunskirchen . 136514
Mayer, Elizabeth, New York .. 122857
Mayer, Hans, Düsseldorf 106810
Mayer jun., Harald, Tübingen .. 078725
Mayer, Harald, Wien 063025
Mayer, Helga, Prerow, Ostseebad 108411
Mayer, Joachim, Kampen 076672
Mayer, Ludwig, Jerusalem 142614
Mayer, Norbert, Neustadt an der
Weinstraße 108259
Mayer, Oswald, Kreimbach-
Kaulbach 107746
Mayer, Peter, Oberndorf bei
Salzburg 127828
Mayer, Richard H., Bamberg .. 074668,
.... 106118
Mayer, Susanne, Freiberg,
Sachsen 129858
Mayer, Trace, Louisville 094572
Mayer, U. & S., Göttingen 076049
Mayer, Willi, Langenenslingen . 130212
Mayer, Y. & D., New York 095570
Mayerfeld, Hamburg 107298
Mayerling, Paris 071650
Mayer'sche Hofkunstanstalt,
München 130368
Mayes, Hilary, Saint Ives,
Cornwall 119025
Mayfair & Co., New York 095571
Mayfair Antiques, Dorking 089089
Mayfair Antiques, Otley 090926, 134914
Mayfair Antiques, Tampa 097578
Mayfair Galleries, Moorabbin . 125214
Mayfair Gallery, London 090338
Mayfair Gallery, West Hollywood 097909
Mayfair Pawnbrokers, London . 090339
The Mayfield Gallery,
Bournemouth 117404
Mayflower Antiques at Admiral Vernon,
London 090340
Mayflower Gold Mill, Silverton . 053343
Mayflower Society Museum,
Plymouth 052210
Mayhem, New York 095572
Mayibuye Centre Collection, University
of the Western Cape, Bellville . 038652
Maylands Antiques, Maylands, South
Australia 061803
Maylis, Tours 073852
Maymont, Richmond 052556
Maynard, Mark, London 090341
Mayne Gallery, Kingsbridge 118047
Mayne Island Museum, Mayne
Island 006280
Mayo County Museum, Ballina . 024822
Mayor Gallery, London 118447
Mayor, Gudrun, Carouge 144059
Mayoral, Andrés, Madrid 133356
Mayoral, Manuel, Barcelona .. 085799,
.... 114935
Mayorga, Saint-Jean Pied-de-
Port 105539
The Mayor's Residence Arts Salon,

Taipei 042527
Maypole, Charlotte 092614
Mayr, Hans, Kreuth 130199
Mayr, Rudolf, München 077567
Mayr, Stephan, Sistrans 127873
Mayr, Thomas, Singen 108629
Mayrberger, Ute, Neusiedl am
  See 062748
Mayrhans, Roland, Tübingen ... 142430
Mayrhofer, Dagmar, Pürbach ... 140022,
                              140023
Mayrhofer, G., Wolfern 063120
Mayrhofer, Max, Obernberg am
  Inn 062757
Mayrische Musikalienhandlung,
  Salzburg 140026
Mayrl, Kelkheim 076744
Mays, Maggie, Buxton 088748
Maysbury, Elsternwick 061442
Maysnow, Boston 119893
Mayuyama & Co., Tokyo 082180
Mayville Area Museum of History and
  Genealogy, Mayville 050746
Mayville Historical Society Museum,
  Mayville 050748
Mazaleyrat, Daniel, Pont-sur-
  Yonne 072266
Mazanec, Martin, Litoměřice .. 065418
Mazar Islamic Museum, Al-
  Mazar 029760
Mazarin, Albi 066517
Mazarin, Toulon (Var) 105715
Mazarine, Montréal 101432
Mazarini, Lyon 104119
Mazda Creative, Nagpur 109488
Mazda Motor Museum,
  Hiroshima 028655
Maze, Jonathan, Auckland 132693
Mazeau, Edith, Périgueux 072059
Mažeikių Muziejus, Mažeikiai .. 030420
Mazenaud, Fabien, Bû 067529
Mazerand, Joseph, Massaguel . 069990
La Mazière, Cordes-sur-Ciel ... 068149
Mazières, Serge, Nice 070634
Mazo Lebouc, Paris . 105080, 141242
Mazo, Alain, Paris 136965
Mazomanie Historical Society Museum,
  Mazomanie 050749
Mažosios Lietuvos Istorijos Muzieus,
  Klaipéda 030392
Mazot, Annecy 066646
Mazot Musée, Troistorrents 042070
Mazot Savoyard, Combloux ... 068109
Mazu Folklore Museum of Tianhou
  Temple, Lugang 042382
Mazuel, Jacques, Savigny-sur-
  Orge 073454
Mazur, Elliot, New York 095573
Mazure, Florence, Bordeaux ... 067345
Mazurier, Myriam, Jonzac 068910
Mazza, Erminio, Torino 132167
Mazzella Di Bosca, Paul,
  Marseille 128778
Mazzi, Francesca, Firenze 131211
Mazzi, Marco, Lodi 110191
Mazzini-Gedenkstätte, Grenchen 041566
Mazzitelli, Riccardo, Roma 081446
Mazzocchi, Giorgio, Parma ... 110630
Mazzola, Gustavo, Milano 110380
Mazzoleni, Torino 110989
Mazzoleni Arte, Milano 080606
Mazzoli, Emilio, Modena 110501
Mazzon Bruna, Venezia 081858
Mazzone, Enrico, Torino 081731
Mazzoni, Piacenza 110650
Mazzoni, Fabrizio, Firenze 080169
Mazzoni, Gianni, Bologna 079853
Mazzoni, Pietro, Roma 081447
Mazzotta, Gabriele, Milano ... 137683
Mazzù, Francesco, Messina ... 131356
Mazzucchi, Venezia 111139
MB Antigüedades, Buenos Aires 060913
MB-Art, Stuttgart 108721
MB Galerie, Kaufbeuren 107541

Mbantua Gallery, Alice Springs . 098635
MBD, Waikato 132752
MBG Antiques, Newark-on-Trent 090813,
                              134871
MBRC, San Diego 097105
MC Art, London 118448
MC Collectables, Indianapolis .. 093906
MC-Museum Gyllene Hjulet,
  Surahammar 041069
MCAD Gallery, Minneapolis College of
  Art and Design, Minneapolis .. 050943
MCC Museum, London 045035
McCaffrey, Victoria, South Yarra 127678
McClelland, Whangarei 113106, 113107
McClenaghan, North Shore 084036
McCormack, Wellington 113075
McCrostie & Co, Christchurch .. 083883
MCD Art-Gallery, Hilpoltstein .. 107434
McDonnell, Bridget, Carlton,
  Victoria 098870
MCF Antiques, Waikato 084132
McKearney, Hawkes Bay 083953
McKenzie, Kenneth, Malvern East 061771
MCM Antiquités, Montauroux .. 070206
McPherson, Auckland 112789
MCS, Jacksonville 135658
MDA, Museum Documentation
  Association, Cambridge 060176
MDG Fine Arts, London 118449
MDI Gallery, Philadelphia 123418
Mdina, Valletta 082442
Me Magazine, White Plains ... 139428
Meacci Freisleben, Firenze 080170
Mead, Mount Martha 061918
Mead Art Museum, Amherst ... 046451
Mead Gallery, Coventry 044021
Meade County Historical Society
  Museum, Meade 050750
Meadow Brook Art Gallery, Oakland
  University, Rochester 052607
Meadow Brook Hall, Rochester . 052608
Meadow Creek Galleries, Saint
  Paul 123920
Meadow Farm Museum, Glen
  Allen 049105
Meadow Garden Museum,
  Augusta 046662
Meadow Lake Museum, Meadow
  Lake 006281
Meadow, M., San Francisco ... 097282
Meadowcroft Museum of Rural
  Life, Historical Society of Western
  Pennsylvania, Avella 046707
Meadowgate Antiques, Dallas .. 093172
Meadowlands Gallery, Banff .. 117283
Meadowlands Museum,
  Rutherford 052714
Meadows & Passmore, Brighton 134153
Meadows Lamp Gallery,
  Edinburgh 089181
Meadows Museum of Art, Centenary
  College, Shreveport 053320
Meadows Museum, Southern
  Methodist University, Dallas .. 048119
Meadowtide, Bartestree 117291
Meaford Museum, Meaford 006282
Meagan, Miami 094765
Meagher County Historical Association
  Castle Museum, White Sulphur
  Springs 054374
Meakk, Roosendaal 083543
Meandar, Zagreb 136755
Meandr, Moskva 085192
Meanskinisht Museum, Kitwanga 006128
Measham Museum, Measham . 045241
MEB Sanat Galerisi, İstanbul ... 117065
Mebane Antique Auction Gallery,
  Mebane 127313
Mebel, Singapore 085469
Meca,Angi & Detrás del Rollo,
  Murcia 115316
Mecano Catalunya, L'Hospitalet de
  Llobregat 133275
Meccoci, Ilaria, Pisa 131766

La Mecedora, Estepona 085994
Il Mecenate, Milano 080607
Mécène, Paris 141243
Mecha Arte y Porcelana, La Paz 100847
Mechanical Music and Doll Collection,
  Chichester 043944
Mechanical Music Museum and
  Bygones, Cotton 044014
Mechanische Schau-Werkstätte,
  Museum am Maxplatz, Rehau . 021336
Mechanisches Musikmuseum,
  Sursee 042043
Méchiche, Frédéric, Paris 105081
Méchiche, Jean-Luc, Paris 071651,
                              105082
Meckenstock, Hans Holger,
  Düsseldorf 129750
Mecki's Basar, Berlin 074878
Mecklenburger Antik-Diele,
  Neustrelitz 077770
Mecklenburgisches Orgelmuseum,
  Malchow 020238
Mecklenburgisches
  Volkskundemuseum Schwerin,
  Freilichtmuseum Schwerin-Mueß,
  Schwerin 021793
Mecoen, Moissac 070145
Mecosta County Historical Museum,
  Big Rapids 046979
Mecrom & Holey, Feldkirchen, Kreis
  Straubing-Bogen 075731
Mecseki Bányászati Múzeum –
  Uránérc Bányászati Kiállítás,
  Központi Bányászati Múzeum,
  Kővágószőlős 023766
Mecseki Bányászati Múzeum,
  Bányászattörténeti Kiállítás és
  Könyvtár, Pécs 023843
Mecseki Bányászati Múzeum,
  Földalatti Bányászati Kiállítás,
  Pécs 023844
Meda Riquier Rare Books,
  London 144649
Meda, Massimo, Milano 110381
Meda, Massimo, Santa Margherita
  Ligure 110907
Médaillier Municipal, Bordeaux . 012110
Medalhão Galeria de Arte,
  Brasília 100917
Medalių, Vilnius 111804
Meddowes, Alexander, Edinburgh 089182
Mede, Robert, Toronto 101663
Medea, Winnipeg 100467
Medeja, Minsk 100467
Medelhavsmuseet, Stockholm . 041019
Medelpads Auktionsservice,
  Sundsvall 126679
Medem, Otterndorf 142287
Meden en Perzen, Dordrecht ... 083189
Medgyessy Ferenc Emlékmúzeum,
  Déri Múzeum, Debrecen ... 023595
Média, Neuchâtel 138078
Media & Communications Department,
  Holland College, Charlottetown 055121
Média 3000, Paris 105083
Media And Moving Art,
  Rotterdam 112650
Media Art Farm, Tbilisi 058700, 105915
Media Xchange 123, Saint Paul 096839
Mediadesign, Hochschule für Design
  und Informatik (FH), Berlin ... 055480
Medialia Gallery, New York 122858
Médiart, Paris 105084
Medical Art Society, London ... 060177
Medical Artists Association of Great
  Britain, Stotfold 060178
Medical History Museum, University of
  Melbourne, Melbourne 001278
Medical Museum, Amman 029775
Medicea, Dogana 085281
Medicea, Rivista interdisciplinare di
  studi medici, Firenze 138842
Medici, Solothurn 116653
Medici Gallery, London 118450

Medici Gallery, Sarasota 124475
Medici Scuola di Lingua Arte e
  Cultura, Firenze 055898
The Medici Society, London ... 138228
Medici, Giovanni, Roma 081448
Medicine Hat Clay Industries National
  Historic District, Medicine Hat . 006283
Medicine Hat Museum and Art Gallery,
  Medicine Hat 006284
Medicine Man Gallery, Tucson .. 097676,
                              124753
Medicinhistoriska Museet,
  Göteborg 040657
Medicinhistoriska Museet, Lund 040813
Medicinhistoriska Museet,
  Vänersborg 041130
Medicinhistoriska Museet Eugenia,
  Stockholm 041020
Medicinsk-Historisk Museum,
  København 010182
Medicinski Muzej, Beograd 038153
Medicinskij Muzej, Russkaja
  Akademija medicinskih nauk,
  Moskva 037319
Médicis, Besançon 103369
Medienhaus Hannover, Hannover 059172,
                              137268
Medieval Art, Antwerpen 063225
Medieval Art, London 090342
Medika Gallery, Blyth 098755
Medina, Tunis 116990
Medina-Bi, Donostia-San
  Sebastián 133254
Medina Cobo, Carmen, Donostia-San
  Sebastián 085980
Medina Gallery, Barnstaple 088362
Medina Railroad Museum,
  Medina 050761
Medinabeitia, Isabel, Bilbao ... 085880
Mediodía Vidrio y Restauración,
  Madrid 133357
Mediolanum, Milano 142730
Medisch Farmaceutisch Museum de
  Griffioen, Delft 032158
Meditativ Art Galerie, Krefeld .. 107741
La Mediterranea Arte, Napoli .. 110537
Mediterranean Art, Palma de
  Mallorca 115361
Medium Modernes Antiquariat,
  Münster 142206
Medizin- und Pharmaziehistorische
  Sammlung Kiel der Christian-
  Albrechts-Universität, Kiel 019608
Medizinhistorische Sammlung
  der Ruhr-Universität Bochum,
  Institut für Geschichte der Medizin,
  Bochum 017438
Medizinhistorische Sammlungen,
  Klinikum Chemnitz, Chemnitz . 017735
Medizinhistorisches Museum, Institut
  für Geschichte der Medizin der
  Medizinischen Universität Wien,
  Wien 003034
Medizinhistorisches Museum,
  Universität Zürich, Zürich 042209
Medjnarodni Grafični Likovni Centar,
  Ljubljana 038546
Medlar, San Antonio 096999
Medola, Mauro, Bologna 130979
Medranda Abenza, Justo, Madrid 133358
Medrese Müzesi, Akşehir 042802
Medu Book, Atlanta 144969
Meduna, Kryrstyna, Wien 100352
Medunarodna Izložba Umetnosti, Novi
  Sad 098290
Medusa, Hamburg 076274
La Medusa, Palermo 081056
Medusa Antik, Stockholm 087029
Medway Heritage Centre,
  Chatham 043912
Medway Towns Gallery,
  Rochester 045629
Medzinárodná Výstava Výtvarného
  Umenia, Bratislava 098292

Mee, Seoul 111690
Mee-Mee's & Makin's, Saint
Louis 096737
Mee, Victor, Cloverhill 079490
Meed, Kansas City 135681
Meehan, M.E., New York 095574
Meeker County Museum and G.A.R.
Hall, Litchfield 050326
Meeker Home Museum, Greeley 049222
Meekins & Co., Fort Worth 093461
Meeks & Co, F., Birmingham . . 088482,
134111
Meeky, Jennifer, Nhill 061971
Meemelink, Jan, Den Haag . . . 143117
Sri Meenakshi Sundaresvara Temple
Museum, Madurai 024391
Meeniyan Art Gallery, Meeniyan 099294
Meent Art, Hilversum 112539
Meeraner Kunstverein, Meerane 059173
Meerbuscher Kunstauktionshaus,
Meerbusch 077318, 126034
Meeresmuseum, Öhringen 020953
Meerrettich, Berlin 106373
Meerrettichmuseum, Baiersdorf . 016935
Meert & Rihoux, Bruxelles 100641
Meert, Jef, Knokke 140233
Meesenburg, Kruså . . . 065987, 065988
Meesters, Hans, Amsterdam . . 112254
Meeting, Venezia 111140
The Meeting House, Ilminster . . 118015
Meezkeret-Batia Museum, Meezkeret-
Batia 025153
Mefisto, Praha 128358
Méga Troc 63, Aigueperse 066435
Megalo Chorio Folklore Museum,
Megalo Chorio 023162
Megan Fox, Kelly, New York . . . 122859
Megaride Antichità, Napoli 080898
Megarry's and Forever Summer,
Blackmore 088496
Megart, Kraków 132845
Megaw & Hogg Antiques, Kent
Town 061679
Megerian, New York 095575
Megginch Castle, Errol 044303
Meghalaya State Museum,
Shillong 024474
Megisti Museum, Kastellorizo . . 023118
Megna, Fabio Massimo, Roma . 081449
Meguro Gajoen Bijutsukan, Tokyo 029545
Meguro Kiseichukan, Tokyo . . . 029546
Meguro Museum of Art, Tokyo . . 029547
Megye- és Várostörténeti Kiállítás,
Eger 023620
Méhészeti Gyűjtemény, Gödöllő . 023646
Mehler-Koch, M., Juist 107488
Mehlhorn, Andra, Hamburg 129976
Mehlig, Axel, Bonn 141621
Mehlis, Jens, Plauen 126096
Mehlkopf, O., Aachen 074340
Mehlman, Justine, Washington . 097847
Mehmet Akif Ersoy Evi, Ankara . 042823
Mehner, Holger, Marienberg . . . 077294
Mehr am Meer, Wangerland . . . 078865
Mehran, West Hollywood 097910
Mehrangarh Museum, Jodhpur . 024332
Mehrhoff, Cincinnati . . 092877, 135446
Mehringer, Ernst, Spitz 062845
Mehta, Mumbai 079403
Mehu, New York 122860
Mei Lian, Roma 081450
Mei Ling Arts Museum, Puzi . . . 042424
Mei Museum, Meixian 088080
Mei Museum, Shueili 042451
Mei-Nong Country Gallery,
Qishan 042425
Meichle, Sankt Gilgen 062825
Meiderle, Martina, Wien 063026
Meidiyas Museum of Art and
Papertole, Linkou 042377
Meier, Aerzen 074357
Meier, Münster 077671
Meier-Za, E., Oftringen 087723, 133881
Meier-Za, E., Zofingen 087923

Meier, Anthony, San Francisco . 124340
Meier, Anton, Genève 116329
Meier, Astrid, Castrop-Rauxel . . 106626
Meier, Bernard, Kapellen 063692, 128064
Meier, Christophe, Bruyères le
Châtel 128524
Meier, Corina, Kammerstein . . . 130089
Meier, David, San Antonio 136265
Meier, Dr. Helmut, Arth 116031
Meier, Erich-Franz, Hannover . . 130012
Meier, Gisela, München 077568, 108163
Meier, H., Gutenzell-Hürbel . . . 076137
Meier, Horst Dieter, Freiburg im
Breisgau 107061, 141806
Meier, Jack, Houston 121090
Meier, Karin, Mantel 077282
Meier, Michael Th., Bad Münder am
Deister 129420
Meierhofer, Arnulf, Salzburg . . . 127853
Meierhoff, Kansas City 094083
Meierimuseet, Institutt for
Næringsmiddelfag, Ås 033506
Meierische Museumboerderij,
Heeswijk-Dinther 032440
Meigle Gallery, Meigle 118732
Meigle Museum of Sculptured Stones,
Meigle 045242
Meignen, Christian, Gent 128053
Meigs County Museum, Pomeroy 052232
Meijer, Amsterdam 132409
Meijer, Utrecht 112707
Meijerimuseo, Saukkola 011345
Meijerink & Zoon, G.A.,
Enschede 083252, 132502
Meijerink, T., Visvliet 132657
Meiji, Tokyo 111465
Meiji-Daigaku Hakubutsukan,
Tokyo 029548
Meiji-Daigaku Kouko Hakubutsukan,
Tokyo 029549
Meiji-Daigaku Shouhin Hakubutsukan,
Tokyo 029550
Meiji Jingu Homotsuden, Tokyo . 029551
Meiji Memorial Picture Gallery,
Tokyo 029552
Meiji Shobo, Tokyo 142931
Meile, Urs, Luzern 116478
Meili, Herbert, Horgen 116378
Meincke Danielsen, Lone,
Gentofte 065690
Meindl & Sulzmann, Wien 140103,
140104
Meindl, P., Hofheim in
Unterfranken 076556, 130059
Meineche, Flemming, Ringkøbing 066071
Meinema, A., Enschede 083253
Meiners, Enschede 083254
Meininger Museen, Schloss
Elisabethenburg, Meiningen . . 020342
Meinong Hakka Museum,
Meinong 042397
Meintzinger, Frickenhausen am
Main 075904
Meir Dizengoff Museum, Tel Aviv 025206
Meire, Belo Horizonte 063910
Meire, Jean-Marie & Dominique,
Brugge 063282
Meirelles, Maurício, Belo
Horizonte 063911, 128101
Meirelles, Miguel, Malvern,
Victoria 127635
Meise, Brigitta, Hamburg 141887
Meisel, Louis K. & Susan P., New
York 122861
Meishan Farming Village Museum,
Meishan 042398
Meisinger, Helmut, Stadl-Paura . 062846,
127878
Meisnitzer, Heinz, Wien 127951
Meiso, Sendai 111388
Meissen Galerie, Berlin 106374
Meißener-Porzellan- und Christofle
Tafelsilber-Ausstellung- Schloß Weyer,
Gmunden 001955

Meissener Porzellan-Sammlung Sigrid
Pless, Bad Soden am Taunus . 016880
Meißener Porzellan-Sammlung Stiftung
Ernst Schneider, Zweigmuseum des
Bayerischen Nationalmuseums,
Oberschleißheim 020921
Meissl, Gerhard, Bad Sauerbrunn 127713
Meissl, Gerhard, Rohrbach bei
Mattersburg 127839
Meissner, Praha 065507, 140524
Meissner & Neumann, Praha . . 065508,
125423
Meissner, Hans, Meilen 116497
Meissner, Rose, Söhrewald . . . 108632
Meißner, W., Tännesberg 078665
Meissner's Auction Service, New
Lebanon 127333
Gebr. Meister, Koblenz 076838
Meister-Götz, Brigitte, Hettstadt 107427
Meister, Michael, Ludwigsburg . 077156
Le Meitour, Portland . . 096437, 123643
Meiwald, Ingrid & Uwe, München 108164
Meixner, Klaus, Würzburg 142513
Meja, Renate, Wenden 078922
Méjanès, Claude, Auzits 066859
Mejean, Christine & Gilles,
Perpignan 072088
Mejerimuseet vid Korsholms
Lantbruksskolor, Vaasa 011469
Mejía Vallejo, Roxana, Medellín . 065297
Mejía, José Manuel, Medellín . . 065295
Mejia, Roxana, Medellín 065296
Mekaanisen Musiikin Museo,
Varkaus 011498
Mekanik, Antwerpen . . 100504, 140150
Mekong Gallery, Singapore . . . 114569
Mel Fisher Maritime Heritage Museum,
Key West 049957
La Mela e il Serpente, Padova . 080991
La Melagrana, Massa 110209
Melaleuca Gallery, Anglesea . . 098647
Melanchthonhaus Bretten,
Bretten 017621
Melanchthonhaus Wittenberg, Stiftung
Luthergedenkstätten in Sachsen-
Anhalt, Lutherstadt Wittenberg 020199
Melanide, Palermo 081057
Mélanson, Chantal, Annecy . . . 103249
Melberg, Jerald, Charlotte 120000
Melbourne, London . . 090343, 134725
Melbourne Affordable Art Show,
Carlton South 097994
Melbourne Antique Prints and Maps,
Camberwell 061258
Melbourne Art Fair, Melbourne . 097995
Melbourne Clocks Museum,
Hampton 001136
Melbourne Contemporary Arts,
Melbourne 118733
Melbourne Fine Art Gallery,
Melbourne 061824, 099314
Melbourne Hall, Melbourne . . . 045243
Melbourne Maritime Museum, Home
of Polly Woodside, Southbank . 001509
Melbourne Museum of Printing,
Footscray 001075
Melbourne Museum, Museum Victoria,
Carlton South 000947
Melbourne Studio of Painting and
Drawing, Fitzroy 099027
Melbourne Tank Museum, Narre
Warren North 001343
Melbourne's Living Museum of the
West, Maribyrnong 001255
Melcher, Dirk, Monheim am
Rhein 077414
Melcher, Gernot, München 077569
Melchior, F., Merseburg 077351
Melchior, Karin, Kassel 107536
Melchiori, Anna, Genova 131324
Meldal Bygdemuseum, Trøndelag
Folkemuseum, Meldal 033770
Meldrum House, Whangarei . . . 084204

Melelli, Fernando, Trieste 081796
Melesi, Lecco 110163
Melet, Laurent, Macornay 069798
Meletta, Eric, München 077570
Melford Antique Warehouse, Long
Melford 090623
Melford Fine Arts, Long Melford . 118656
Melford Hall, The National Trust, Long
Melford, Sudbury, Suffolk . . . 045147
Melfort and District Museum,
Melfort 006287
Melgarejo Peñuela, Alonso,
Bucaramanga 102395
Melgosa Sanchez, Martin Sergio,
México 111970
Méli-Mélo, Minneapolis 094930
Meli-Melo, Montréal 101433
Méli-Mélo, Saint-Macaire 072905
Meli-Melo Artist's Alliance, Hong
Kong 058500
Melich-Mautner, Daniela, Wien . 063027
Melicharovo Městské Muzeum,
Unhošť 009933
Melin, Malmköping 086840
Melindre, São Paulo 064194
Melissa Publishing House,
Athinai 137575
Melk, Frans, Hilversum 143174
Melkowska, Grażyna, Szczecin . 113781
Mellado, San Lorenzo de El
Escorial 133419
Mellado Lopez, Jose Antonio,
Badalona 133088
Mellat Museum, Sa'd-Abad Museum,
Tehrān 024767
Melle & Sowada, Twistringen . . 078740
Mellemværftet, Nordmøre Museum,
Kristiansund 033719
Mellendorf, Peter, Lübeck 077179
Mellerstain House, Gordon 044448
Mellette House, Watertown 054218
Melli Museum, Tehrān 024768
Mellier, Jean-Louis, Castillon-la-
Bataille 067719
Mellilän Museoriihi, Mellilä 011150
Mellilän Radio- ja Puhelinmuseo,
Mellilä 011151
Melling, K., Burnley 117487
Mellinger, J. Ch., Stuttgart 137524
Mellon Financial Corporation's
Collection, Pittsburgh 052169
Mellors & Kirk, Nottingham . . . 127000
Mellow, Honolulu 093560
Melngalvju nams, Rīga 111760
Melnick House Gallery, Ascot . . 117252
Melnikow, M., Heidelberg 076454
Meloche, Monique, Chicago . . . 120202
Melodia, Vincenzo, Firenze . . . 131212
Mélodie du Temps, Paris 071652
Melody Art, Alresford 117231
Melody Art Studio, Tugun 099842
Melody's Antique, Chester 088865
Meløy Bygdemuseum, Ørnes . . 033813
Meloni, Mauro, Cagliari 131071
Melori & Rosenberg, Venezia . . 111141,
111142
La Meloria, Livorno 080328
Meloun, Karel, Klášterec nad
Ohří 102608
Melrose Abbey Museum, Melrose 045245
Melrose Courthouse Museum,
Melrose 001283
Melrose House, Tshwane 038849
Melrose Lightspace, Los Angeles 121556
Melrose Plantation Home Complex,
Melrose 050771
Melrose Vintage, Phoenix 096224
Mel's Antiques, Baltimore 092344
Mel's Antiques, Kitchener 064338
Melse & Rietema, Groningen . . 083302
Melsheimer, Hubertus, Köln . . . 076897
Melson, R., Kent 076782
Melting Art, Limoges 069525
Melton Antiques, Woodbridge . . 091916

Melton Art Reference Library Museum, Oklahoma City 051731
Melton Carnegie Museum, Melton Mowbray 045246
Melton Park Gallery, Oklahoma City 123288
Meltons, London 090344
Meltzer, Neal, New York 122862
Melun Encheres, Melun 125642
Melusine, Asnières-sur-Seine .. 066761
Melville Discovery Centre, Applecross 000797
Melville Heritage Museum, Melville 006290
Melville Railway Museum, Melville 006291
Melville, Donald, Takapuna 084088
Melvin Art Gallery, Florida Southern College, Dept. of Art, Lakeland 050096
Melvin B. Tolson Black Heritage Center, Langston 050132
Melwood Photography Gallery, Pittsburgh 123564
Melzer, Michael, Lüdenscheid .. 142091
Mem-Erie Historical Museum, Erie 048595
Members Calender, New York .. 139429
Membrado, Buenos Aires 060914
Meme et Moi, Los Angeles 094405
Memi Pasa Fürdoje, Pécs 023845
Mémin, Gilles, Le Mans 069309
Memme-Taadi Kamber, Karksi .. 010626
Memmed Seid Ordubadinin Xatire Muzeyi, Memorial Museum of Mammad Said Ordubady, Bakı 003168
Memmel, Fritz, Speyer 078524
Mémoire de la Nationale 7, Piolenc 014872
Mémoire de la Photo, Donzère . 012814
Mémoire de Saumon, Petite Camargue Alsacienne, Saint-Louis-la-Chaussée 015426
Mémoire d'Enfance, Cannes 067645
Mémoire du Livre, Paris 141244
Mémoire du Rhin, Petite Camargue Alsacienne, Saint-Louis-la-Chaussée 015427
La Mémoire du Temps, Bruxelles 063505
Mémoire du Temps, Byblos 082314
Mémoires, Rouen 141323
Memoires Antiguidade, Rio de Janeiro 064084
Mémoires de Grenier, Vincennes 074244
Mémoires de Greniers, Fargues-Saint-Hilaire 068506
Les Mémoires d'Obersoultzbach, Obersoultzbach 014544
Memoirs of the American Academy in Rome, New York 139430
Memoirs of the American Academy in Rome, Roma 138843
Memola, Angela, Bologna 109855
Memoli Numismatica e Antiquariato, Milano 080608
Memorabilia di Andrea Balloni, Milano 080609
Memorable Treasures, Atlanta . 119592
Memorear Antiguidades, Rio de Janeiro 064085
Mémorial 1939–1945, Fort de la Cité d'Alet, Saint-Malo (Ille-et-Vilaine) 015438
Memorial Art Gallery of the University of Rochester, Rochester 052618
Memorial Day Museum, Waterloo 054209
Le Mémorial de Caen, Caen 012275
Mémorial de la Shoah, Musée, Centre de Documentation Juire Contemporaine, Paris 014676
Memorial de Sergipe, Aracaju .. 004311
Mémorial de Vendée, Musée du Logis de la Chabottière, Saint-Sulpice-le-Verdon 015570
Mémorial des Guerres en Indochine,

Fréjus 013047
Mémorial des Soldats de La Nouvelle France, Le Château-d'Oléron .. 013618
Le Mémorial d'Essigny le Grand, Musée Privé de la Grande Guerre 14–18, Essigny-le-Grand 012911
Mémorial d'Évocation Vaudois et Huguenot, La Roque-d'Anthéron 013486
Memorial di Imigração Ucraniana, Curitiba 004426
Memorial do Imigrante, Departamento de Museus e Arquivos, São Paulo 004862
Mémorial du Maréchal Leclerc de Hauteclocque et de la Libération de Paris et Musée Jean Moulin, Paris 014677
Mémorial Fort Montbarey, Brest 012206
Memorial Galleries, Houston 121091
Memorial Gallery of Tako Child Atelier, Tako 029465
Memorial Geroičnoï Oboroni Odesi – 411 Beregova Linija, Odes'kyj istoriko-krajeznavčyj muzej, Odesa 043270
Memorial Hall Museum, Deerfield 048219
Memorial Hall of the Zunyi Conference, Zunyi 008560
Memorial JK, Brasília 004357
Memorial Library and Art Gallery, Saskatoon 006976
Mémorial Musée du Régiment, Fondation Roi Baudouin, Spa . 004039
Memorial Museum, Naxçivan .. 003189
Memorial Museum & Pioneer Museum, Ada 046317
Memorial Museum B.P. Hasdeu, Câmpina 036342
Memorial Museum of the Kanto Earthquake Disaster, Tokyo .. 029553
Memorial Museum of the War of Liberation, Pyongyang 029879
Memorial Museum Passchendaele 1917, Zonnebeke 004165
Memorial Pengisytiharan Kemerdekaan, Melaka 030707
Memorial Picture Gallery, Tokyo 111466
Memorial Pontes de Miranda da Justiça do Trabalho em Alagoas, Maceió 004540
Memorial Prins Karel, Het Domein Raversijde, Oostende 003929
Memorial Sultan Tidore Museum, Tidore 010411
Memorial Tancredo Neves, São João Del Rei 004832
Memorial Union Art Collection, Arizona State University, Tempe 053704
Memorial Union Art Gallery, Corvallis 048020
Memorial Union Art Gallery, University of California, Davis 048161
Memorial Union Gallery, North Dakota State University, Fargo 048698
Memorialinė Tremtines, Gydytojos Dalios Grinkevičiūtės Ekspozicija, Laukuva 030407
Memorialmuseum Friedrich Fröbel, Oberweißbach 020937
Memorialnaja Gvardejskaja Krasnoznamënnaja Podvodnaja Lodka S-56 1939 g., Voenno-istoričeskij Muzej Tichookeanskogo Flota, Vladivostok 038033
Memorialnaja Kvartira Andreja Belogo, Gosudarstvennyj muzej A.S. Puškina, Moskva 037320
Memorialnaja Kvartira G.M. Kržižanovskogo, Gosudarstvennyj centralnyj muzej sovremennoj istorii Rossii, Moskva 037321
Memorialno-chudožestvennyj Dom-muzej Vladimira Aleksandroviča Serova, Tverskaja oblastnaja

kartinnaja galereja, Ėmmaus .. 037007
Memorialno-literaturnyj Muzej A.P. Bondina, Ekaterinburg 036982
Memorial'nyj Budinok-muzej V. Dalja, Muzej istoriï ta kul'tury mista Lugans'ka, Lugans'k 043231
Memorial'nyj Budinok-muzej A.A. Kocky, Užgorod 043311
Memorial'nyj Budinok-muzej F.F. Manajla, Užgorod 043312
Memorial'nyj Budinok-muzej Tarasa Ševčenka, Kyïv 043202
Memorialnyj Chudožestvennyj Muzej Valentina Aleksandroviča Serova v Domotkanovo, Tverskaja oblastnaja kartinnaja galereja, Krasnaja Nov 037157
Memorialnyj Dom-Muzej Aaly Tokombaeva, Biškek 030075
Memorialnyj Dom-muzej Akademika S.P. Koroleva, Moskva 037322
Memorialnyj Dom-muzej Akademika I.V. Kurčatova, Moskva 037323
Memorialnyj Dom-muzej Akademika A.A. Uchtomskogo, Rybinskij gosudarstvennyj istoriko-architekturnyj i chudožestvennyj muzej-zapovednik, Rybinsk .. 037698
Memorialnyj Dom-muzej Akademika I.M. Vinogradova, Velikie Luki . 038011
Memorialnyj Dom-muzej S.T. Aksakova, Nacionalnyj muzej Respubliki Baškortostan, Ufa .. 037978
Memorialnyj Dom-Muzej P.P. Bažova, Municipalnyj objedinennyj muzej pisatelej Urala, Ekaterinburg .. 036983
Memorialnyj Dom-muzej Gercena, Gosudarstvennyj Literaturnyj muzej, Moskva 037324
Memorialnyj Dom-muzej P.D. Korina, Gosudarstvennaja Tretjakovskaja galerja, Moskva 037325
Memorialnyj Dom-muzej G.D. Krasilnikova, Nacionalnyj muzej Udmurtskoj Respubliki im. K. Gerda, Alnaši 036848
Memorialnyj Dom-muzej I.I. Levitana, Plesskij gosudarstvennyj istoriko-architekturnyj i chudožestvennyj muzej-zapovednik, Ples 037660
Memorialnyj Dom-muzej Mažita Gafuri, Ufa 037979
Memorialnyj Dom-muzej G.V. Plechanova, Lipeck 037198
Memorialnyj Dom-muzej Podpolnaja Tipografija Permskogo Komiteta RSDRP 1906 g., Permskij oblastnoj kraevedčeskij muzej, Perm .. 037639
Memorialnyj Dom-muzej I.P. Požalostina, Rjazanskij Oblastnoj Chudožestvennyj Muzej im. I.P. Požalostina, Solotča 037881
Memorialnyj Dom-Muzej I.I. Šiškina, Gosudarstvennyj muzej izobrazitelnych iskusstv Respubliki Tatarstan, Elabuga 036997
Memorialnyj Dom-muzej im. N.G. Slavjanova, Permskij oblastnoj kraevedčeskij muzej, Perm 037640
Memorialnyj Dom-muzej L.V. Sobinova, Jaroslavskij gosudarstvennyj istoriko-architekturnyj i chudožestvennyj muzej-zapovednik, Jaroslavl . 037063
Memorialnyj Dom-muzej Uljanovych, Astrachanskij Muzej-zapovednik, Astrachan 036875
Memorialnyj Dom-muzej A.A. Vaneeva, Gosudarstvennyj istoriko-ėtnografičeskij muzej-zapovednik Šušenskoe, Ermakovskoe 037009
Memorialnyj Dom-muzej Viktora i Apolinarija Vasnecovych, Kirovskij Oblastnoj Chudožestvennyj Muzej

im. V.M. i A.M. Vasnecovych, Rjabovo 037686
Memorialnyj Dom-muzej Marko Vovčok, Nalčik 037534
Memorialnyj Dom-muzej N.E. Žukovskogo, Orechovo 037596
Memorialnyj Istoriko-biografičeskij Dom-muzej Akademika N.V. Melnikova, Muzej istorii i kultury Srednego Prikamja, Sarapul .. 037839
Memorialnyj Kabinet A.V. Lunačarskogo, Rossijskaja Akademija muzyki im. Gnesinych, Moskva 037326
Memorialnyj Kabinet-muzej Akademika A.P. Vinogradova, GEOChI Rossijskoj Akademii Nauk, Moskva 037327
Memorialnyj Kompleks G. Tukaja, Nacionalnyj muzej Respubliki Tatarstan, Novyj Kyrlaj 037587
Memorialnyj Kompleks 'Vorošilovskaja Batareja', Voenno-istoričeskij Muzej Tichookeanskogo Flota, Vladivostok 038034
Memorialnyj Muzej Akademika K.I. Satpaeva, Almaty 029802
Memorialnyj Muzej Akademika V.I. Vernadskogo, GEOChI Rossijskoj Akademii Nauk, Moskva 037328
Memorialnyj Muzej Akademikov A.E. i B.A. Arbuzovych, Kazan 037105
Memorialnyj Muzej N.I. Beloborodova, Tula 037961
Memorialnyj Muzej A.N. Burdenko, Moskva 037329
Memorialnyj Muzej-dača A.S. Puškina, Puškin 037675
Memorialnyj Muzej Jurija Gagarina, Gagarin 037011
Memorial'nyj Muzej V. Ivasjuka, Černivci 043131
Memorialnyj Muzej-kabinet Akademika N.V. Belova, Moskva 037330
Memorialnyj Muzej-kabinet Akademika P.L. Kapicy, Moskva 037331
Memorialnyj Muzej-kabinet Akademika M.V. Keldyša, Rossijskaja Akademija nauk, Moskva 037332
Memorialnyj Muzej-kabinet Akademika A.N. Nesmejanova, Rossijskaja Akademija nauk, Moskva 037333
Memorialnyj Muzej-kabinet Akademika N.D. Zelinskogo, Moskva 037334
Memorialnyj Muzej-kabinet M.V. Frunze, Ivanovo 037042
Memorialnyj Muzej Karima Chakimova, Nacionalnyj muzej Respubliki Baškortostan, Ufa 037980
Memorialnyj Muzej I.S. Ključnikova-Palantaja, Joškar-Ola 037070
Memorialnyj Muzej Kosmonavtiki, Moskva 037335
Memorialnyj Muzej-kvartira Aktёrov Samojlovych, Sankt-Peterburgskij Muzej Teatralnogo i Muzykalnogo Iskusstva, Sankt-Peterburg ... 037753
Memorialnyj Muzej-kvartira E.F. Gnesinoj, Rossijskaja Akademija muzyki im. Gnesinych, Moskva 037336
Memorialnyj Muzej-kvartira S.M. Kirova, Vladikavkaz 038024
Memorialnyj Muzej-kvartira Vadima Alekseeviča Kozina, Magadan . 037213
Memorialnyj Muzej-kvartira P.K. Kozlova, Institut Estestvoznanija i Techniki Rossijskoj Akademii Nauk, Sankt-Peterburg 037754
Memorialnyj Muzej-kvartira G.M. Kržižanovskogo i V.V. Starkova, Minusinskij Regionalnyj Kraevedčeskij Muzej im. N.M. Martjanova, Minusinsk 037227
Memorialnyj Muzej-kvartira N.A. Nekrasova, Sankt-Peterburg . 037755
Memorialnyj Muzej-kvartira I.P.

Pavlova, Institut Fiziologii im. I.P.
Pavlova Rossijskoj Akademii Nauk,
Sankt-Peterburg 037756
Memorialnyj Muzej-kvartira A.S.
Puškina, Gosudarstvennyj Muzej A.S.
Puškina, Moskva 037337
Memorialnyj Muzej-kvartira Rimskogo-
Korsakova, Sankt-Peterburg . . 037757
Memorialnyj Muzej-kvartira A.A.
Šogencukova, Nalčik 037535
Memorialnyj Muzej-kvartira M.S.
Spiridonova, Čeboksary 036932
Memorialnyj Muzej-kvartira A.M.
Vasnecova, Gusadarstvennaja
Tretjakovskaja galereja, Moskva 037338
Memorialnyj Muzej-laboratorija A.S.
Popova, Kronštadt 037178
Memorialnyj Muzej Licej, Nacionalnyj
Muzej Puškina, Puškin 037676
Memorial'nyj Muzej-majsternja Teodizïï
Briž, Lvivs'ka Galereja Mystectv,
Lviv 043243
Memorialnyj Muzej A.G. Malyškina,
Mokšan 037231
Memorialnyj Muzej-Masterskaja
Skulptora S.T. Konenkova,
Moskva 037339
Memorialnyj Muzej Alii Moldagulovoj,
Aktöbe 029790
Memorial'nyj Muzej Mychajla
Hruševskoho u Lvovi, Lviv . . . 043244
Memorial'nyj Muzej V.I. Nemiroviča-
Dančenka, Donec'kij oblasnyj
krajeznavčyj muzej, Neskučne 043267
Memorialnyj Muzej B.L. Pasternaka,
Nacionalnyj muzej Respubliki
Tatarstan, Čistopol 036959
Memorialnyj Muzej A.S. Popova, Sankt-
Peterburg 037758
Memorialnyj Muzej A.S.
Popova – Izobretatelja Radio,
Krasnoturinskij kraevedčeskij muzej,
Krasnoturinsk 037174
Memorialnyj Muzej A.N. Skrjabina,
Moskva 037340
Memorialnyj Muzej-usadba Akademika
I.P. Pavlova, Rjazan 037688
Memorialnyj Muzej-usadba
Chudožnika N.A. Jarošenko,
Kislovodsk 037141
Memorialnyj Muzej-usadba M.P.
Musorgskogo, Naumovo 037541
Memorialnyj Muzej Konstantina
Vasiljeva, Nacionalnyj Muzej
Respubliki Tatarstan, Vasiljevo 038010
Memorialnyj Muzej N.I. Vavilova,
Institut obščej genetiki im. N.I.
Vavilova Rossijskoj Akademii Nauk,
Moskva 037341
Memorialnyj Muzej-zapovednik N.A.
Rimskogo-Korsakova Ljubensk-
Večaša, Pljussa 037663
Memorialul Ipoteşti-Centrul Naţional
de Studii Mihai Eminescu,
Ipoteşti 036522
Memorialul Victimelor Comunismului
şi al Rezistenţei, Sighetu
Marmaţiei 036707
Memories, Bramley 088567
Memories, Rochester 091130
Memories, Sheffield 144837
Memories, Vancouver 064845
Memories & More, Passau 078023
Memories Antiques, San Diego . 097106
Memories from Yvonne, Regina 064554
Memories in Glass, Tulsa 097747
Memories of Old Antiques,
Houston 093740
Memories of Vail, Denver 120746
Memories of Yesteryear, Torquay 062336
Memories Old & New,
Jacksonville 094009
Memorijalna Galerija Aleksandar
Rukavina, Brtonigla 008889

Memorijalna Zbirka Eugena Kumičića,
Brseč 008888
Memorijalni Muzej Džuro Jakšić,
Srpska Crnja 038248
Memorijalnyj Budinok-muzej D.
Javornic'kogo, Dnipropetrovskyj
deržavnyj istoričnij muzej im
Javornic'kogo, Dnipropetrovs'k 043150
Memory House Antiques, San
Francisco 097283
Memory Lane, Sowerby Bridge . 091393
Memory Lane, Wellington 084174
Memory Lane Antique, Los
Angeles 094406
Memory Lane Antiques, Ascot
Vale 061064, 127560
Memory Lane Antiques, Great
Bookham 089365
Memory Lane Antiques, London 064353
Memory Lane Antiques, Lower
Stondon 090640
Memory Lane Antiques, Pomona 062088
Memory Lane Antiques, Regina . 064555
Memory Lane Antiques, San
Jose 097372
Memory Lane Antiques & Collectibles,
Malahide 079669
Memory Lane Antiques Centre, Ripley,
Derbyshire 091110
Memory Lane Collection,
Goodwood 061538
Memory Lane Museum at Seattle
Goodwill, Seattle 053238
Memory Lane Photographics,
Hastings 099110
Memory Lane Upholstery,
Calgary 128153
Memory Market, Oklahoma City 095993
Memoryville USA, Rolla 052668
Mempel, Dagmar, Vieselbach . . 108822
Memphis, Cairo 066255
Memphis Antique Market,
Memphis 094651
Memphis Belle B17 Flying Fortress,
Memphis 050782
Memphis Brooks Museum of Art,
Memphis 050783
Memphis Coin Gallery, Memphis 094652
Memphis College of Art,
Memphis 057423
Memphis College of Art Gallery,
Memphis 050784
Memphis Pink Palace Museum,
Memphis 050785
Memphis Plating Works,
Memphis 135783
Menage A Trois, Christchurch . . 083884
Ménager, Benoît, Mûrs-Erigné . 070402
La Ménandière, Alpe-d'Huez . . . 066541
Menard Art Museum, Komaki . . 028879
Menard, Henri, Marseille 069950
Ménard, Hervé, La Chapelle-Saint-
Mesmin 068980
Menardi, Uta, Leiden 083437
Menarini, Giampiero, Bologna . . 142636
Menart, Aude, Monpazier 104247
Menat, Roma 081451
Mencendorfa Nams – Rīdzinieku Mäja-
Muzejs, Rīga 030208
Menczer Museum of Medicine and
Dentistry, Hartford 049387
Mende, Karlsruhe 141974
Mende-Kohnen, Susanne,
Wiesbaden 130761
Mende, B. & F., Havixbeck 107385
Mende, P., Binzen 075000
Mendel-Antiquariat, Berlin 141553
Mendel Art Gallery, Saskatoon . 006977
Mendelianum – Památník Gregora
Mendela, Moravské zemské muzeum,
Brno 009471
Mendelssohn-Remise,
Geschichtsforum Jägerstraße e.V.,
Berlin 017224

Mendes, São Paulo . . . 140308, 140309
Mendès, Martial, Toulouse 073783
Mendez, Vezin-le-Coquet 129343
Méndez, Marvia, Caracas 097961
Mendez, Oscar, Barcelona 133135
Mendl, Reinhard, Bad
Reichenhall 074573
Mendo, Amsterdam 112255
Mendocino Art Center,
Mendocino 057424
Mendocino Art Center Gallery,
Mendocino 050793
Mendocino County Museum,
Willits 054427
Mendoza, Caracas 125048
Mendoza Rodriguez, Hector,
Bogotá 065213, 065214
Menec, Helsinki 140713
Meneghel, Susanne, Wien 063028
Menem, Alex, Paris 105085
Ménéndez, Claude, Fontiers-
Cabardès 068581
Ménéndez, Pierre, Fontiers-
Cabardès 068582
Menez, Marie-Françoise, Aurillac 066827
Menezes, Gessy Geyer De, Porto
Alegre 100990
Menezes, Roberto, Miami 094766,
                                    135800
Meng, Singapore 133040
Mengelmoes, Arnhem 082943
Menges, A., Oldenburg 077934
Menges & Söhne, Aloys,
Hannover 076384
Mengès, Christophe, Monbalen . 070151
Mengès, Gérard, Monbalen 070152
Mengzhou Museum, Mengzhou . 008081
Menhir, Bern 144051
Menier Gallery, London 118451
Menikos, Menikou, Pafos 065335
The Menil Collection, Houston . 049585
Menim Restorations, Langport . 134527
Las Meninas, Barcelona 085800, 085801
Menine, Millau 070121
Měnínská Brána, Muzeum města Brna,
Brno 009472
Menka-s Antiques, Clifton Hill . 061343
Menke, Chr.-Peter, Dortmund . . 075388,
                                    129699
Menkhaus, Antonius,
Georgsmarienhütte 107116
Mennello Museum of American Folk
Art, Orlando 051805
Mennickheim, Kassel 130104
Mennink, L.J., Amsterdam 082807
Mennitti, Guido, Firenze 131213
Mennonite Heritage Center,
Harleysville 049359
Mennonite Heritage Museum,
Goessel 049141
Mennonite Heritage Museum,
Rosthern 006830
Mennonite Heritage Village,
Steinbach 007072
Mennonite Historical Society of British
Columbia, Abbotsford 005302
Mennonite Library and Archives
Museum, North Newton 051610
Mennour, Kamel, Paris 105086
Menocal, Nina, México 111971
Menominee County Heritage Museum,
Menominee 050795
Menomonee Falls Antiques,
Milwaukee 094872
Menories Antiguidades, Rio de
Janeiro 064686
Menotti, Baden bei Wien 099961
Menotti, Croix 068246
Menozzi, Dr. Valeria, Modena . . 131575
Menrad, W., Ladenburg 076990, 107779
Menssen, Norderstedt 108285
Menßen, U., Esens 075666
Mentalvårdsmuseet Säter, Säter 040929
Mentjes, Meike, Berlin 129531

Mentlhof-Bergbauernmuseum,
Heiligenblut 002071
Mentor Graham Museum, Blunt 047062
Mentougou Huayang Antique,
Beijing 065003
MenTouGou Museum, Beijing . . 007630
Menu, Jean, Saint-Calais 072701
Menuiserie au Fil du Temps, Evaux et
Menil 068489
Menuiserie Claude, Giromagny . 128611
Menuiserie-Ebénisterie d'Art
Retrouvailles, Stadtbredimus . 082391,
                                    132332
Menyea, Los Angeles 094407
Menzel, Wien 063029
Menzel, Christa & Werner,
Fürstenwalde 107082
Menzel, Hermann, Hamburg . . . 076275
Menzel, Manfred, Bad Honnef . . 074531
Menzer Heimatstube, NaturParkHaus
Stechlin, Stechlin 021969
Menzerath, A., Simmerath 078472
Menzoyan, Marc, Saint-Andeol-le-
Château 072667
Meopham Windmill, Meopham . 045242
Mephisto Antiquariat, Paderborn 142290
Meppener Kunstkreis e.V.,
Meppen 059174
Mer et Découvertes, Bonifacio . 067288
La Mer les Bateaux, Noirmoutier-en-
l'Ile 070736
Mer Terre et Soleil, Sète 073515
Mera, Jaime, Santiago de Chile 101914
Meråker Bygdemuseum, Meråker 033771
Meramec Antiques, Saint Louis . 096738
Merano Arte – Kunst Meran, Galleria
d'Arte Contemporanea, Merano 026762
Merc & Cia, Sevilla 086457
Mercadante, Antonio, Torvilliers . 073690
Mercadé, Richard, Céret 067750
Mercader, Houston 093741
El Mercader de Venecia,
Badalona 085711
El Mercadillo de Salado, Sevilla 086458
Mercado, Brasilia 100918
Mercado das Pulgas Antiguidades, Rio
de Janeiro 064087
Mercado de Tesouros de Objetos Antigos,
Rio de Janeiro 064088, 128116
El Mercado del Mueble Usado, La
Plata 060947
Mercado Negro Antiguidades, Porto
Alegre 063984
Mercado Puerta de Toledo,
Madrid 115258
Mercador da Sé, Lisboa 084918
Il Mercante, Napoli 080899
Il Mercante d' Oriente, Verona . 081924
Il Mercante di Stampe, Milano . 142731
Mercantic, Sant Cugat del Vallès 126556
Mercantile Arts Centre Kintyre,
Tarbert 045946
Al Mercatino, Milano 080610
Mercatino dell'800 e del '900,
Napoli 080900
Mercatino dell'Usato, Bologna . 079854
Mercato dell'Antiquariato,
Brescia 079947, 142657
Mercato d'Epoca, Miami 094767
Mercato, Roberto, Claro 087376
Mercato, Utrecht 143247
Mercator Graventoren-Watermolen,
Rupelmonde 003990
Mercator Museum, Sint-Niklaas 004020
Mercatorfonds, Bruxelles 136645
Mercatu Arte, Belo Horizonte . . 100895
La Merced, Bilbao . . . 085881, 133196
Merced County Courthouse Museum,
Merced 050801
Mercedes-Benz Museum,
Stuttgart 022049
Mercer Art Gallery, Harrogate . . 044518
Mercer County Genealogical and
Historical Society, Princeton . . 052364

Mercer County Historical Museum, The Riley Home, Celina 047494
Mercer County Museum, Mercer 050802
Mercer Gallery, Monroe Community College, Rochester 052619
Mercer Museum of the Bucks County Historical Society, Doylestown . 048367
Mercer Street Books and Records, New York 145354
Mercer Union – Centre for Contemporary Visual Art, Toronto 007189
Mercer, Noel, Long Melford 090624
Mercer, Thomas, London 090345
The Merchandise Mart International Antiques Fair, Chicago 098472
Merchant Adventurers' Hall, York 046225
Merchant House Antiques, Honiton 089571
Merchant House Antiques, Ipswich 089649
The Merchant of Welby, Welby . 062416
Merchant's House Museum, New York 051423
Merchant's House Museum, Plymouth 045512
Merchat, Gérard, Besançon 067130
Merchionne, Loano 110189
Merchionne, Patrizia, Pietra Ligure 110654
Mercier & Cie., Lille 125614
Mercier & Kalck, Paris 125739
Mercier Press, Blackrock, Cork . 137602
Mercier, Chantal, Clermont-Ferrand 068033
Mercier, Gaëtan, Aubigny-sur-Nère 066792
Mercier, Jean-Marie, Biarritz 067205
Mercier, Jean-Yves, Cholet 067985
Mercier, Josiane, Saint-Geniez-d'Olt 072793
Mercier, Michel, Marseille 069951
Merciér, Thierry, Paris 105087
Merck hoe Sterck, Rotterdam . 132623
Mercure, Toulon (Var) 073720
Mercuri & Talarico, Roma 131977
Mercuri, Giuseppe, Roma 131976
Mercurio Arte, Milano 110382
Mercurio Arte, Palermo 110611
Mercurio Arte, Rapallo 110696
Mercurio Arte Contemporanea, Viareggio 111200
Mercurio, Th., Levallois-Perret . 069446
Mercurius, Almelo 082633
Mercury, London 090346
Mercury Bay and District Museum, Whitianga 033367
Mercury Gallery, Boston 119894
Mercury Signs, Houston 121092
Merdøgård, Arendal 033502
Mere Antiques, Topsham 091613
Mere Museum, Mere 045248
Meredith, Baltimore 119755
Meredith Gallery, Baltimore ... 046761
Meredith Palmer Gallery, New York 122863
Merello, Massimo, Genova 080271
Merenwijk, Leiden 112558
Mereside Books, Macclesfield . 134825, 144722
Merge, Auckland 112790
Mergemeier, Düsseldorf 106811, 129751
Mergen, G., Steinfurt 078550
Mergenthaler, Gerd, Wiesloch . 079002
Mergenthaler, Walter, Arnstein .. 141411
Mergozzi, Ursula, Bad Wurzach . 129431
Mergus Verlag, Melle 137378
Merhaba, Ankara 088095
Meri-Valentino, Sofia 140324
Meriden Historical Society Museum, Meriden 050804
Meridian Galerie, Zürich 087990, 133992
Meridian Gallery, San Francisco 124341

Meridian Gallery at Artstore, Boston 117397
Meridian International Center – Cafritz Galleries, Washington 054135
Meridian Museum of Art, Meridian 050806
Meridian Street Antiques, Indianapolis 093907
La Meridiana, Messina 110223
Meridiana, Milano 080611
La Meridiana, Piacenza 110651
Meridiana Restauri, Firenze 131214
Meridiani, Torino 081732
La Meridienne, Haarlem 083340
La Méridienne Antique, Rabat . 082604, 132374
Meriel-Bussy, Yves, Ploudalmézeau 105360
Merighi, Diano Marina 110003
Merighi, Varazze 111074
Merijärven Kotiseutumuseo, Merijärvi 011152
Merikarvian Kalastusmuseo, Merikarvia 011154
Merimiehen Koti, Uusikaupunki . 011465
Merimiehenkotimuseo, Pohjois-Pohjanmaan Museo, Oulu .. 011215
De Merindol, Tshwane 085681
Merino, Monsheim 108030
Merino, Edouard 93506581, Monaco 082519
Meritalon Museo, Salo 011341
Merivale Fine Art, Christchurch . 112852
Meriwether Lewis Dredge Museum, Brownville 047282
Meriwether Lewis National Monument, Hohenwald 049500
Merkator 1, Varna 140329
Merkinės Kraštotyros Muziejus, Merkinė 030422
Merkl, Thomas, Bretten 129639
Merklappenmuseum, Dieteren .. 032225
Merkle, Den Haag 112401
Merkle, Stuttgart 108722, 137525
Merkle, Thierry, Strasbourg 129269
Merkz, Peter, Bad Homburg v.d.Höhe 129412
Merle, Bernadette, Beaumont-sur-Grosne 067006
Merlet, Alain-Claude, Saint-Martin-Petit 072955
Merlet, Odette, Barbezieux-Saint-Hilaire 066944
Merlin, Singapore 114570
Merlin, Tetbury 091575
Merlin Gallery, Swindon 119201
Merlin-Lochet, Aix-en-Provence 066470
Merlin Verlag, Vastorf 137547
Merlin, Jean-Michel, Paris 071653
Merlins Antiques, Carmarthen . 088798
Merlo, Norfolk 095866, 136033
Mermal Books, Bangkok 144204
Mermier, Jean-Eric, Vésenaz .. 087872, 133951
Mermod, Patrick, Les Arcs 069398
Mermoz-Santo Micali, Paris ... 071654, 071655, 105088, 105089
Meroni, Daniele Mario, Milano . 131475
Meroogal House, Nowra 001374
Merowe Museum, Merowe 040500
Merredin Military Museum, Merredin 001287
Merredin Museum, Merredin .. 001288
Merrepen Arts, Daly River 061385
The Merrick Art Gallery, New Brighton 051238
Merrifield, Tom, London 118452
Merrill, Detroit 120811
Merrill & Johnson, Denver 120747
Merrill Historical Museum, Merrill 050808
Merrill Museum, Carrollton ... 047439
Merrill & Co., Duane, Milton ... 127322
Merrill & Assoc., Todd, New York 095576
Merrin, New York 122864

The Merriscourt Gallery, Sarsden 119062
Merritt Museum of Anthropology, Oakland 051678
Merriwa Colonial Museum, Merriwa 001289
Merry Old England, Garmisch-Partenkirchen 075950, 125957
Merry Old England, Kelkheim .. 076745, 130112
Merry, Robert, Saint Louis 127429
Merry's, Northampton 126997
Merryvale Antiques, Pittsburgh . 096310
Merschbrock, B., Bad Salzuflen 074581
Mersea Island Museum, West Mersea 046092
Merseny, Thomas, Starnberg .. 078539
Mersey Lane Antiques, Gore ... 083942
Merseyside Maritime Museum, Liverpool 044859
Mersin Müzesi, Mersin 043025
Merson, Paris 071656
Mertens, Merzen 077353
Mertens & Pomplun, Berlin 141554
Mertes-Schön, Eva, Bonn 129611
Mertig, Oliver & Gardana, Essen 075690
Mertinat, Harald, Wuppertal ... 079110
Mertins, Jürgen, Berlin 137085
Mertins, Till, Berlin 129532
Merton Abbey Mills, London ... 126967
Meru Museum, Meru 029864
Merv, Dinmore 061401
Les Merveilles d'Alice, Saint-Ouen 073154
Les Merveilles de la Mer, Jdeidet El-Metn 030278
Merville Antiques & Fine Art, Newtownabbey 090844
Mervyn Quinlan Museum, Benenden 043567
Merwin and Wakeley Galleries, Illinois Weslayn University, Bloomington 047045
Merwin House, Stockbridge ... 053534
Meryem Ana Evi, Selçuk 043055
Merz & Solitude, Stuttgart 137526
Merz Akademie, Hochschule für Gestaltung Stuttgart, Stuttgart . 053519
Merz, Ulrike, München 130369
Merzbach, Rainer, Koblenz 076839
Mesa Historical Museum, Mesa 050812
Mesa Southwest Museum, Mesa 050813
Mesa Street Antique Mall, El Paso 093407
Mesa Verde National Park Museum, Mesa Verde 050814
La Mésangère, Liège 063773
Mesarici, Alfred, Hartberg 062629
Meschik, Bernd, Treffen 127884
Mese-Fischer, Del, Meisterschwanden 116501
Meseguer Jordan, Madrid 115259
Meseház, Békéscsaba 023442
Meseke, A., Salzhemmendorf .. 078344
Meshkinshahr Museum, Meshkinshahr 024723
Mesiti, Barbara, Torino 132168
Mesmer, Hanspeter, Basel 087248, 133653
Mesna Antikvariat, Lillehammer 143389
Mesnier, Jacques, La Cluse-et-Mijoux 069000
Mesnil aux Abeille, Beautheil .. 011953
Mesnil mon Temps, Le Mesnil-Esnard 069317
Mesnil, Lucien, Saint-Ursanne . 087771
Mesolithic Museum, Abinger Common 043375
Mesquite Heritage Museum, Mesquite 050816
Messageot, Charles-Henri, Versailles 074061
Messageot, Charles-Henri, Voisins-le-Bretonneux 074275
Messager, Michele, Saint-Brieuc 072698
Messages, Washington 139431

Méssaggio, Célestin, Les Andelys 069396, 103981
Messe für Kunst und Antiquitäten Hofburg Wien, Wien 098005
Messe für Kunst und Antiquitäten Residenz Salzburg, Salzburg . 098008
Messel & Wildhagens, Oslo 143406
Messer, Stephen, Windsor 135242
Messerermuseum, Steinbach an der Steyr 002781
Der Messerkönig, Wien 127952
Messerli, Marcel, Bern 087293
Messerschmidt, Thomas, Flensburg 075748
Messiez-Poche, Thierry, Toulouse 129304
Messina Museum, Messina 038783
Messina, Giuseppe, Palermo ... 131704
Messina, Orazio, Milano 131476
Messineo, New York 122865
Messink, Apeldoorn 132424
Messiqua, Marcia, New York ... 095577
Messner Mountain Museum Dolomites, Cibiana di Cadore 025980
Messner Mountain Museum Juval, Castelbello 025816
Messori, Roncaglia Giancarlo, Modena 080807
Messum, David, London 090347, 118453, 134726
Messum, David, Marlow 118723
Mest 50-tal, Farsta 086674
Mestdach, Knokke-Heist 063704
Mestdagh, Patrick, Bruxelles ... 063506, 140189
Mestna Galerija, Soštanj 038618
Mestna Galerija Ljubljana, Ljubljana 038547
Mestni Muzej Idrija, Grad Gewerkenegg, Idrija 038516
Mestni Muzej Ljubljana, Ljubljana 038548
Mestni Muzej Radovljica, Muzeji radovljiške občine, Radovljica . 038597
Mesto Pechati, Moskva 139012
Městská Galéria, Rimavská Sobota 038436
Městská Galerie Art Club, Týn nad Vltavou 009928
Městská Galerie Beroun, Beroun 009444
Městská Galerie Josefa Lieslerova, Kadaň 009611
Městská Galerie Karla Havlíčka, Kadaň 009612
Městská Galerie Špejchar, Chomutov 009520
Městská Galerie v Zámeckém Parku, Veselí nad Moravou 009949
Městská Galerie, Městské Muzeum a Galerie, Polička 009770
Městská Výstavná Sieň Barónka Kežmarok, Kežmarok 038361
Městské Muzeum, Aš 009436
Městské Muzeum, Bechyně ... 009438
Městské Muzeum, Bystřice nad Pernštejnem 009493
Městské Muzeum, Čáslav 009493
Městské Muzeum, Čelákovice .. 009497
Městské Muzeum, Česká Třebová 009503
Městské Muzeum, Chotěboř ... 009522
Městské Muzeum, Chrast u Chrudimě 009523
Městské Muzeum, Chrastava .. 009525
Městské Muzeum, Dobruška .. 009537
Městské Muzeum, Františkovy Lázně 009547
Městské Muzeum, Frenštát pod Radhoštěm 009548
Městské Muzeum, Horažd'ovice 009567
Městské Múzeum, Ilava 038358
Městské Muzeum, Jesenice ... 009599
Městské Muzeum, Klobouky u Brna 009632
Městské Muzeum, Králíky 009646
Městské Muzeum, Lanškroun . 009659

Městské Muzeum, Ledeč nad
Sázavou 009662
Městské Muzeum, Letohrad ... 009664
Městské Muzeum, Mnichovo
Hradiště 009702
Městské Muzeum, Moravská
Třebová 009704
Městské Muzeum, Nepomuk ... 009715
Městské Muzeum, Netvořice ... 009717
Městské Muzeum, Nové Město nad
Metují 009723
Městské Muzeum, Nové Strašecí 009724
Městské Muzeum, Nový Bydžov 009726
Městské Muzeum, Počátky ... 009764
Městské Muzeum, Přelouč ... 009849
Městské Muzeum, Přibyslav ... 009854
Městské Muzeum, Protivín ... 009857
Městské Muzeum, Radnice u
Rokycan 009859
Městské Muzeum, Rtyně v
Podkrkonoší 009873
Městské Muzeum, Rýmařov ... 009877
Městské Múzeum, Sabinov ... 038441
Městské Muzeum, Sadská ... 009878
Městské Muzeum, Sedlčany ... 009879
Městské Muzeum, Skuteč ... 009881
Městské Muzeum, Slapanice ... 009883
Městské Muzeum, Strážnice ... 009895
Městské Muzeum, Stříbro ... 009896
Městské Muzeum, Valašské
Klobouky 009939
Městské Muzeum, Velká Bíteš . 009943
Městské Muzeum, Veselí nad
Moravou 009950
Městské Muzeum, Vimperk ... 009951
Městské Muzeum, Volyně ... 009957
Městské Muzeum, Žamberk ... 009969
Městské Muzeum, Zbiroh ... 009972
Městské Múzeum, Zlaté Moravce 038487
Městské Múzeum a Galéria, Vel'ký
Šariš 038481
Městské Muzeum a Galerie,
Břeclav 009457
Městské Muzeum a Galerie,
Dačice 009532
Městské Muzeum a Galerie, Dvůr
Králové nad Labem 009545
Městské Muzeum a Galerie, Hlinsko v
Čechách 009559
Městské Muzeum a Galerie, Hořice v
Pokrkonoší 009568
Městské Muzeum a Galerie, Lomnice
nad Popelkou 009679
Městské Muzeum a Galerie,
Svitavy 009901
Městské Muzeum a Galerie,
Vodňany 009955
Městské Muzeum a Galerie J. Karse,
Velvary 009947
Městské Muzeum a Galerie Julie W.
Mezerové, Úpice 009934
Městské Muzeum a Galerie Otakara
Spaniela a Josefa Wágnera,
Jaroměř 009593
Městské Muzeum a Galerie, Okresní
vlastivědné muzeum Přerov,
Hranice 009579
Městské Muzeum Antonína Sovy,
Pacov 009747
Městské Muzeum Dr. E. Beneše,
Kožlany 009644
Městské Múzeum Fil'akovo,
Fil'akovo 038349
Městské Muzeum Klenotnice, Nová
Paka 009718
Městské Múzeum Michala Tillnera,
Malacky 038386
Městské Múzeum Poltár, Poltár . 038425
Městské Múzeum Rajec, Rajec . 038434
Mestské Múzeum v Bratislave,
Bratislava 038314
Městské Muzeum v Kralupech nad
Vltavou, Kralupy nad Vltavou ... 009647
Městské Muzeum v Mariánských

Lázních, Mariánské Lázně 009691
Mestské Múzeum v Pezinku,
Pezinok 038420
Městské Muzeum, Jihomoravské
Muzeum Znojmo, Moravský
Krumlov 009706
Městské Muzeum, Městské Muzeum a
Galerie, Polička 009771
Městské Muzeum, Muzeum v Bruntále,
Krnov 009652
Městské Muzeum, Oddělení Oblastní
Muzeum Jihovýchodní Moravy,
Napajedla 009712
Městské Muzeum, Okresní muzeum
Kroměřížska, Holešov 009565
Městské Muzeum, Okresní Vlastivědné
Muzeum Nový Jičín, Bilovec . 009447
Městské Muzeum, Okresní Vlastivědné
Muzeum Nový Jičín, Příbor . 009851
Městské Vlastivědné Mudruňkovo
Muzeum, Říčany u Prahy 009862
Městské Vrbasovo Muzeum,
Ždánice 009975
Mestský Vlastivedný Dom,
Žitnoostrovné Múzeum Dunajská
Streda, Šamorín 038443
Mesuret, Claude, Lesparre-
Médoc 069440
Mesureur, Karine, L'Isle-sur-la-
Sorgue 069580
Met Room, Barcelona 114936
Meta-Aesthetix, Port Melbourne 099591
La Métairie, Paris 071657
Metais, Claude, Compiègne 128570
Métais, Henri, Rouen . 072604, 141324
Metal, Warszawa 113864
Metal Art Company, San
Francisco 136335
Metal Art Studio, New York 095578,
135957
Metal Arts, Vestavia Hills ... 136440
The Metal Arts Guild, Toronto . 101664
The Metal Gallery, London 118454
Metal Maddness Art Gallery, Salt Lake
City 123974
Metallhandwerksmuseum, Steinbach-
Hallenberg 021978
Metallmuseum, Tullnerbach-
Lawies 002849
Metalocus, Madrid 139063
Metamora Courthouse, Metamora 050818
Metamorforse, Amsterdam 143050
Metamorphis, Ashburton 117257
Metamorphosis Church Museum,
Meteora 023169
Metaphores, Nijmegen 083515
Metaphysical Art Gallery, Taipei . 116932
Metaphysical Book Store Center,
Denver 145130
Metára, Rio de Janeiro 101050
Metaux Précieux, Aix-en-
Provence 066471
Métaux Précieux, Carpentras ... 067695
Métayer, Jean-Jacques, Mérignac
(Charente) 070070
Métayer, Marie-Antoinette, Paris 071658
Metcalfe-Hofmann, Deal 117670
Metchosin School Museum,
Victoria 007332
Meteor Crater, Museum of
Astrogeology, Winslow ... 054485
Meteorite Museum of the Center for
Meteorite Studies, Tempe ... 053705
Meteorkrater-Museum, Steinheim am
Albuch 021987
Methe, Reiner, Ringgau 078241
Methil Heritage Centre, Methil . 045253
Methmann & Co., Hamburg ... 076276
Metiers d'Art Sant Roch, Céret . 012385
Les Métiers du Champagne, Musée
de la Cave des Champagnes de
Castellane, Épernay 012876
Metin Sanat Galerisi, İstanbul .. 117066
Metis, Amsterdam 112256

Metivier, Nicholas, Toronto 101665
Metlewicz, Dr. Thomasz, Wien . 063030
Métra, Raoul, Autun 066838
Metras, Rene, Barcelona 114937
Métraux, Raynald, Lausanne ... 116424
Metro 5 Gallery, Armadale,
Victoria 098665
Metro Antiques, New York ... 095579
Metro Antiques, Philadelphia ... 096137
Metro Art, New York 122866
Metro Arts, Brisbane 000887
Metro Furniture, Mumbai ... 079404
Metro Gallery, New York 095580
Metro Pictures, New York 122867
Metro Retro, Dallas 093173
Metrocoins, Pirmasens 078050
Metroform, Tucson 124754
Metrolina Antiques, Charlotte . 092615
Metrònom, Barcelona 114938
Metrònom, Fundació Rafael Tous d'Art
Contemporani, Barcelona ... 039067
Metropolart, Paris 105090
Metropole Gallery, Folkestone . 044355
Metropolis, Bordeaux 067346
Métropolis, Lyon 104120
Métropolis, Paris . 071659, 105091
Metropolis M, Utrecht 138934
Metropolitan Antique Gallery,
Denver 093316
Metropolitan Antiques, New York 095581
Metropolitan Arts Institute,
Phoenix 057678
Metropolitan Gallery, Austin .. 119684
Metropolitan Garage Door,
Milwaukee 094873
Metropolitan Graphic Art Gallery, New
York 122868
Metropolitan Home, Vancouver . 064846
The Metropolitan Museum of Art, New
York 051424
Metropolitan Museum of Art Bulletin,
Middle Village 139432
Metropolitan Museum of Art Journal,
New York 139433
Metropolitan Museum of Art, Special
Publications, New York 138407
Metropolitan Museum of Manila,
Manila 034535
Metropolitan Pavilion, New York 095582
Metropolitan Police Historical
Collection, London 045036
Metropolitana Leiloeira, Porto . 126465
Métrot, André, Paris . 071660, 071661
Metsers, G.L., Den Haag ... 112402
Metta, Madrid 115260
Mettchen, Dieter, Darmstadt ... 075307
Mettlach Steinzeugsammler e.V.,
Rehlingen-Siersburg 059175
Metz Collection, Pierrevillers ... 072160
Metz de Benito, München 077571,
077572
Metz-Noblat, Bertrand de, Nancy 070426
Metz von Tessin'sche
Kutschensammlung, Starzach . 021962
Metz, Hilde, Antwerpen 100505
Metz, J. & M., Heidelberg 076455,
125989
Metze, Manfred, Blankenrath . 129596
Metzen, T., Tübingen 078726
Metzgalerie, Koblenz 107574
Metzger, A., Kielce 143462
Metzger, Jean-Pierre, Prunières 072320
Metzger, Marcel, Pouancé ... 072294
Metzgermuseum Zunft zum Widder,
Zürich 042210
Metzner, Bernd, Seevetal ... 108606
Metzner, Eberhard, Mainz ... 077241
Meubelmakerij G. de Reus,
Amsterdam 132410
Au Meuble d'Antan, Pont-
à-Celles 128086
Meubles & Cie, Orvault 128880
Les Meubles d'Antan, Mehun-sur-
Yèvre 070039

Meubles d'Art, Pully 087744
Meubles d'Art Chartres, Ribérac 072455
Meubles d'Autrefois, Augy ... 066810
Meubles d'Autrefois, Besançon . 067131
Meubles Décennies 30'a 70',
Québec 064538
Meubles d'Hier et d'Aujourd'hui,
Poilley 072214
Meubles et Objets en Périgord, Piegut-
Pluviers 072154
Meubles la Merisière, Fréjus ... 066860
Meubles Oubliés, Vanier 064899, 128233
Meubles Ray, Québec 064539
Meubles Style Restauration,
Bonne 128476
Meubl'Occase, Montpellier ... 070308
Meugnier, Jean-Louis, Cordes-sur-
Ciel 068150
Meunier, Franck, Ferney-Voltaire 103738
Meunier, Jacqueline, Saint-Martin-de-
Gurson 072932
Meunier, Jean-Jacques, Jarnac . 068894
Meunier, Marie-Yolande, Toulouse 129305
Meunier, Patrice, Bordeaux 128497
Le Meur, Dijon 140833
Meurier, Jocelyne, Nîmes 070699
Meuron, Anita, Honolulu 120952
Meurs, I.J.A., Leiden 143184
Meusburger, Tina, München .. 077573
Meuschel, Konrad, Bad Honnef . 141425
Meux Home Museum, Fresno .. 048981
Mevagissey Fine Art, Mevagissey 118741
Mével, Jacques, Plouelech ... 072209
Mével, Jacques, Trébeurden ... 073872
Mevel, Jean-Pierre, Toulon (Var) 073721
Mevlânâ Müzesi, Konya 043007
Mevlevi Tekke Müzesi, Lefkoşa . 033474
Mewes, Lübeck 077180
Mews, Dallas 093174, 093175
Mews Antique Emporium, Holt,
Norfolk 089551
Mexic-Arte Museum, Austin .. 046697
Mexican Fine Arts Center Museum,
Chicago 047673
Mexican Gallery, Dallas 120629
The Mexican Museum, San
Francisco 053011
México en el Tiempo, Miguel
Hidalgo 138915
Mey, Jürgen, Geldern 107105
Mey, Jürgen an, Kerken 107545, 137295
Meyer, Aesch 144008
Meyer, Ahaus 074359
Meyer, Marseille 104188
Meyer, Paris 071662, 105092
Meyer, Recklinghausen 078157
Meyer & Deubert, Neustadt an der
Weinstraße 108260
Meyer & Riegger, Karlsruhe ... 107519
Meyer Boswell, San Francisco . 145562
Meyer-Buhr, Ute, Nürnberg ... 130469
Meyer-Dommert, M., Berlin ... 129533
Meyer-Ellinger, Köln 077663
Meyer-Lindenberg, Reinhard,
Lehrte 077049
Meyer-Oceanic Art, Paris ... 071666
Meyer, Alexander, Sankt Martin . 078353
Meyer, Antje & Andreas,
Artlenburg 074426
Meyer, Bernard, Strasbourg ... 129270
Meyer, Claude, Paris 071663
Meyer, Don, Dallas 093176
Meyer, Dr. Claus-Peter, Aachen . 074341
Meyer, Françoise, Hamburg ... 076277
Meyer, H. Jürgen, Lüneburg ... 107931
Meyer, Harald, Passau 142295
Meyer, Harry, Hamburg 076278
Meyer, Harry, Reinbek 078186
Meyer, Irin von, Hamburg ... 129977
Meyer, Jean-Claude, Paris ... 071664
Meyer, Joe, Fulenbach 087437, 133750
Meyer, Jutta, Lüneburg 107932
Meyer, Marion, Paris 105093
Meyer, Martin, Fürth 129873

Meyer, Michel, Nice 141003
Meyer, Michel, Paris 071665
Meyer, Philippe, Vevey 144144
Meyer, Rainer F., Berlin 141555
Meyer, Rolf, Marienberg 077295
Meyer, Thomas, Regensburg ... 142323
Meyer, Thomas V., San Francisco 097284
Meyerdiercks, Peter, Hamburg . 076279
Meyerhof Scheeßel, Scheeßel .. 021635
Meyerovich, San Francisco 124342
Meyerowitz, Ariel, Long Island
City 121417
Meyniel, Georges, Laval
(Mayenne) 069204
Meynier, Maria-Hélène, Toulouse 073784,
129306
Meysenburg, Ulrich F., Essen .. 141751
Meysenburg, Ulrich-Ferdinand,
Essen 137185
Meysey-Thompson, Sarah,
Woodbridge 091917
Meystre, Catherine Noëlle, Zürich 116836
Meža Muzejs Boju Pilī, Aizpute . 030097
Mézaute, Michel, Lalinde 069146
Meždunarodna Restavracija,
Ruse 101134
Meždunarodnaja Chudožestvennaja
Jarmarka, Moskva 098282
Meždunarodnaja Vystavka Architektury
i Sredovovo Disajna, Moskva . 098283
Meždunarodnaja Vystavka-Prodaža
Živopisi, Grafiki, Skulptury i Izdelij
Dekorativno-Prikladnogo Iskusstva,
Perm 098284
Meždunarodnyj Centr-Muzej im. N.K.
Rericha, Moskva 037342
Meždunarodnyj Muzejno-vystavočnyj
Centr, Primorskij gosudarstvennyj
muzej im. V.K. Arseneva,
Vladivostok 038035
Mèze, Yoland, Tavel 073630
Mézédes emlékeink – Mézeskalácsos,
Cukorka- és Gyertyakészíto Múzeum,
Szekszárd 023949
Mézerac, Arnaud de, Caen 103485
Mezinárodní Muzeum Keramiky,
Bechyně 009439
Mezinárodní Veletrh Poštovních
Známek, Mincí, Telefonních Karet,
Minerálů a Sběratelství, Praha 098059
Mezőgazdasági Tájmúzeum – Holland-
ház, Dég 023600
Mezőföldi Tájház, Dég 023601
Mezogazdasági és Élelmiszeripari
Szakképzo Intézet,
Szabadkigyós 023911
Mezogazdasági Eszköz- és
Gépfejlodéstörténeti Szakmúzeum,
Szent István Egyetem, Gödöllő 023647
Mezőgazdasági Gépmúzeum,
Mezőkövesd 023778
Mezogazdaságtudományi Kar Történeti
Múzeum, Nyugat-Magyarországi
Egyetem, Mosonmagyaróvár .. 023800
Mezohegyesi Állami Ménes
Kocsimúzeuma, Mezőhegyes .. 023775
Mezquita Catedral de Córdoba,
Córdoba 039304
Mezzacqui, Costante e Tino,
Modena 080808, 142752
Mezzani, Giuseppe, Verona 081925
Mezzanin, Wien 100353
Mezzanine, Cleveland 092953
La Mezzanine, Montréal 101434
The Mezzanine Gallery, The
Metropolitan Museum of Art, New
York 051425
Mezzomonaco, Pascal, Lyon ... 128721
MFA Circle Gallery, Maryland
Federation of Art, Annapolis . 119526
MFA City Gallery, Maryland Federation
of Art, Baltimore 119756
MFF, Saint-Paul-de-Vence 105593
MFK Mund- und Fußmalende Künstler

Verlag GmbH, Stuttgart 137527
MG Artes, Fortaleza 063946
MGJ Jewellers, Wallingford 091712
MGM Galeria, Warszawa 113865
MGM Münzgalerie München,
München 077574
MGM Studio, Roma 110824
MGS & NSI, Minneapolis 094931
MH, Châteaurenard (Bouches-du-
Rhône) 067898
Mher Abeghyan Museum,
Echmiadzin 000696
Mher Mkrtchyan House Museum,
Gyumri 000704
Mhiripiri, Minneapolis 122013
Mi Mi Gallerie, Chicago 122003
Mi, Alma, Los Angeles 094408
Mia, Taman Melawati 111909
Mia Lou, Philadelphia 123419
Mia Mia Gallery, Templestowe .. 099813
MIAAO – Museo Internazionale delle
Arti Applicate oggi, Torino 028117
Mialkas, G., Magdeburg 077222
Miami Art Central, Miami 050835
Miami Art Museum, Miami 050836
Miami Beach Antique Jewelry and
Watch Show, Miami Beach ... 098473
Miami Beach Fantasy, Miami .. 094768
Miami Books, North Miami 145383
Miami Children's Museum, Miami 050837
Miami Contemporary Art Fair, Coconut
Groove 098474
Miami County Museum, Peru ... 052003
Miami Gold Coins, Miami 094769
Miami International Map Fair,
Miami 098475
Miami Mid-Century, Miami 094770
Miami Museum, Miami 006298
Miami Museum of Science,
Miami 050838
Miami Shores Antiques, Miami . 094771
Miami Station Museum, Miami . 006299
Miami Twice, Miami 094772
Miami University Art Museum,
Oxford 051868
Miami University Hiestand Galleries,
Oxford 051869
Miami Watercolor Society, Miami 060587
Mian County Museum, Mianxian 008082
Miandoab Museum, Urmieh ... 024789
Miani, Adriano, Venezia 132249
Miano, Guido, Milano 137684
Mianposhteh Palace Museum, Bandar-
Anzali 024688
Mianyang Museum, Mianyang .. 008083
Miaoli Ceramics Museum,
Gongguan 042316
Miaoli Railway Museum, Miaoli . 042399
Miaoli Wood Sculpture Museum,
Sanyi 042441
Miasskij Kraevedčeskij Muzej,
Miass 037223
Mibedo, Paris 071667
Micawber Antiques, Bridgnorth . 088600
Miccinesi, Livio, Firenze 131215
Micello, Luigi, Rolle 087753
Michael, Portland 123644
Michael, San Francisco 097285, 124343
Michael Bruce Cottage Museum,
Kinnesswood 044725
Michael C. Carlos Museum, Emory
University, Atlanta 044626
Michael C. Rockefeller Arts Center
Gallery, Fredonia 048966
Michael Davitt National Memorial
Museum, Straide 024991
Michael-Ende-Museum, München 020547
Michael Faraday's Laboratory and
Museum, London 045037
Michael G. Schüll, Büro für Kunst,
Obertrubach 108322
Michael Imhof Verlag GmbH & Co KG,
Petersberg, Kreis Fulda 137470
Michael og Anna Anchers Hus og

Saxilds Gaard, Skagen 010305
Michael Pacher-Haus, Congress-
Wolfgangsee, Sankt Wolfgang im
Salzkammergut 002706
Michael Richardson Contemporary Art,
London 138229
Michael Werner Gallery, New
York 122869
Michael, Alan, Virginia Beach .. 124835
Michael, Dr., North Tamborine . 062002
Michael, Heike, Betzdorf 074968
Michael, Judith, Corbridge ... 088977
Michael, S., Larnaka 065318
Michaelis Collection, Iziko Museums of
Cape Town, Cape Town 038681
Michaelis School of Fine Art,
University of Cape Town, Cape
Town 056341
Michaels, México 082480
Michael's Antique Clocks, Oklahoma
City 095994, 136053
Michael's Antiques, Birmingham 092421
Michael's Antiques, Neu-Isenburg 077708
Michael's Antiques, New York .. 095583
Michael's Auction Gallery,
Cedartown 127155
Michaels Bookshop, Ramsgate . 144796
Michael's China Porcelain,
Seattle 136375
Michael's Old Books, Oklahoma
City 145399
Michaels Privatmuseum,
Nellingen 020660
Michael's Rug Gallery, Baltimore 092345
Michaels, J., Minneapolis 135832
Michaels, Lawrence, Milwaukee 121935
Michaelsen, Anke-Friedericke,
Berlin 074879
Michaelsen, Prof. Hans, Berlin . 129534
Michail & Lombardo, New York . 122870
Michail, Louise, Milano 080612
Michajlovskij Istoričeskij Muzej,
Rjazanskij istoriko-architekturnyj
muzej-zapovednik, Michajlov .. 037224
Michajlovskoe – Gosudarstvennyj
Memorialnyj Istoriko-literaturnyj
i Prirodno-landšaftnyj Muzej-
Zapovednik A.S. Puškina, Puškinskie
Gory 037681
Michalak, Goch 076021
Michalak, T., Gdynia 132823
Michalak, Władysław, Łódź ... 084614
Michalakakos, Michalis, Athinai . 079173
Michalko, Praha 128359
Michallek, Alain, Reugny 072446
Michallon, René, Le Paradou .. 103961
Michalopoulos, James, New
Orleans 122204
Michalovský, Tibo, Košice 085570
Michałowska, Marzena,
Warszawa 132891
Michałowski, Bartłomiej, Lublin . 113700
Michalská Antik, Praha 065509
Michalski, Passau 078024
Michaud, Nevers 125661
Michaud, Henry-François, Reims 072395
Micheau, Jean-Pierre, Paris ... 071668
Michel Antiquités, Meylan 070112,
128798
Michel Brouillet House and Museum,
Vincennes 053990
Michel-Buck-Gedenkstätte im
Kaplaneihaus, Ertingen 018272
Michel Moder, Santiago de Chile 064975
Michel-Natal-Sparwood Heritage
Museum, Sparwood 007064
Michel, Albin, Paris 136966
Michel, Claude, Lausanne 133802
Michel, Hélène et Eric, Saint-Saturnin-
les-Avignon 073300
Michel, Jean, La Chaux-de-
Fonds 087359, 133718
Michel, Klaus, Gießen 075994

Michel, Leo, Milano .. 080613, 080614
Michel, Lotti, Biel 116158
Michel, Luc, Paris 129089
Michel, Marie-Claude, Bordeaux 067347
Michel, Perle, Saint-Amand-
Montrond 072664
Michel, Raymond, Champs-sur-
Marne 103570
Michel, Robert, Nice 128864
Michel, Romuald, Ellhofen 075595
Michelaine, Aigues-Mortes 103184
Michelangeli, Sergio, Roma ... 131978
Michelangelo, Alassio 109745
Michelangelo, La Paz . 063863, 128096
Michelangelo, Mostar 100857
Michelangelo Refinishing Company,
New York 135958
Michelanngelo, Rimini 110716
Michele, C., Los Angeles 094409
Michelet, Antibes-Juan-les-Pins 066676
Michelet-Huot, Jacques, Reims . 072396
Micheletti, Colette, Arles 066724
Michelham Priory, Hailsham ... 044497
Michelis, U., Gifhorn 076000
Michelle Rosenfeld Gallery, New
York 122871
Michelotti, Marseille .. 104189, 128779
Michelotti, Claudio, Parma 081117
Michelsen, Christian, Wald ... 087892
Michelson Museum of Art,
Marshall 050716
Michelstettner Schule, Asparn an der
Zaya 001728
Micheluzzi, Venezia 081859
Michener House Museum,
Lacombe 006149
Michi, Houston 093742
Michie Tavern Ca. 1784,
Charlottesville 047582
Michiels, J., Bruxelles 063507
Michielssen, P., Knokke-Heist .. 063705
Michif Historical and Cultural
Preservation Society Museum,
Surrey 007101
Michigan Avenue Antiques,
Detroit 093367, 127207
Michigan Chapter Gallery, National
Conference of Artists, Detroit . 048310
Michigan Historical Museum,
Lansing 050138
Michigan Hot Glass Workshop Studio,
Detroit 120812
Michigan Maritime Museum, South
Haven 053407
Michigan Museums Association,
Lansing 060588
Michigan Sports Hall of Fame,
Detroit 048311
Michigan State University Museum,
East Lansing 048441
Michigan Street Antique Center,
Indianapolis 093908
Michigan Transit Museum, Mount
Clemens 051084
Michitsch, Elisabeth, Wien ... 063031,
100354
Michot, M., Brugge 063283
Michoutouchkine Pilioko Foundation,
Port Vila 054675
Mickelson, Washington 124935
Mickey, Hixson 127245
Mickey's Autograph Arena,
Indianapolis 145180
Mick's Place, Gent 063651
Mickyz Curios, Opotiki 084046
Mico INAFCA Museum, Kingston 028445
Mičovna, Praha 009807
Micro-Musée, La Trinité 030777
Micro-Musée de la Lessive, Seyne-les-
Alpes 015799
Micro-Musée de l'École, Seyne-les-
Alpes 015800
Micro-Musée de l'Ex-Colonie
Pénitentiaire, Mettray 014149

Micro-Musée du Tailleur, Seyne-les-
Alpes 015801
Micro Rolled Thread, Chicago . . 092748
Micromusée de la Légion Etrangère,
Douzillac 012831
Micromusée de la Maison de Retraite
Jeanne Pierretée Carnot, Nolay 014523
Micus, Herbert, Siegburg 078460
Mid-America Air Museum,
Liberal 050271
Mid-America All-Indian Center
Museum, Wichita 054390
Mid America Antique Center,
Milwaukee 094874
Mid America Print Conseil, New
Albany 060589
Mid-America Science Museum, Hot
Springs 049564
Mid Atlantic Air Museum,
Reading 052483
Mid-Atlantic Association of Museums,
Baltimore 060590
Mid-Atlantic Center for the Arts, Cape
May 047419
Mid Century Design, Auckland . 083808
Mid-Century Design, San Diego 097107
Mid Century Modern, Atlanta . . 092155
Mid-Century Modern, Toronto . . 064704
Mid-City Coins, Saint Louis 096739
Mid-Continent Railway Museum, North
Freedom 051605
Mid Cornwall Galleries, Saint Blazey
Gate 119010
Mid Hants Railway 'Watercress Line',
Alresford 043403
Mid-Hudson Children's Museum,
Poughkeepsie 052340
Mid-Richmond Historical Museum,
Coraki 000991
Mid-South Coin Company,
Memphis 094653
Mid-States Coin, Tulsa 097748
Mid-Town Specialties,
Sacramento 096636
Mid-Valley Art League, Temple
City 060591
MidAmerica Pastel Society, Overland
Park 060592
Mida's Touch, Memphis 094654
Middelfart Antikvitetshandel,
Middelfart 066001
Middelfart Museum, Middelfart . . 010232
Middelheimmuseum, Antwerpen 003291
Middendorpshuis, Den Ham . . . 032204
Middione, Angelo, Torino 132169
Middle Border Museum of
American Indian and Pioneer Life,
Mitchell 050970
Middle East Museum, Yerevan . 000737
Middle East Public Auction,
Riyadh 126483
Middle Eastern Culture Centre in
Japan, Mitaka 029033
Middle Kingdom, Ottawa 064508
Middle Lake Museum, Middle
Lake 006300
Middle Park Fine Art Gallery, Middle
Park 099334
Middleborough Historical Museum,
Middleborough 050853
The Middlebury College Museum of
Art, Middlebury 050858
Middlebury Historical Society Museum,
Wyoming 054560
Middleham Antiques, Middleham 090755
Middlesbrough Institute of Modern Art,
Middlesbrough 045261
Middlesex Canal Collection, Center for
Lowell History, Lowell 050510
Middlesex County Historical Society
Museum, Middletown 050864
Middlesex University, Art and Design
Learning Resources, London . . 056564
Middleton Place Foundation,

Charleston 047557
Middleton Top Engine House,
Middleton-by-Wirksworth 045263
Middleton, Arthur, London 090348
Middleton, Bobbie, Tetbury 091576
Middletown Antiques, Louisville . 094573
Middletown Fine Arts Center,
Middletown 050867
Middletown Valley Historical Society
Museum, Middletown 050865
Middleville and District Museum,
Lanark 006157
Midgaard Miljø, København 065936
Midhurst, Matfen 090735, 118728
Midiri, Francesco, Milano 080615
Midland Air Museum, Baginton . 043465
Midland Arts, Indianapolis 121202
Midland Arts and Antiques Market,
Indianapolis 093909
Midland Center for the Arts School,
Midland 057442
Midland County Historical Museum,
Midland 050872, 050877
Midland Motor Museum,
Bridgnorth 043720
Midland Pictures, Birmingham . . 117376
Midland Railway Butterley, Ripley,
Derbyshire 045618
Midlands Federation of Museums and
Art Galleries, Telford 060179
Midland's Fine Furniture, Omaha 096050,
136064
Midmarch Arts Press, New York 138408
Midnight Pickle, Jacksonville . . . 094010
Midpoint Gallery, Omaha 123333
Midt-Telemark Auksjonsforretning, Bø
i Telemark 126341
Midtown Antiques, New York . . . 095584
Midtown Art Center, Houston . . 049586
Midtown Artist Market, Memphis 122704
Midtown Collectibles, Tulsa 097749
Midtown Galleries and Auction,
Memphis . . . 094655, 127315, 135784
Midtown Gallery, Nashville 122091
Midtsønderjyllands Museum,
Gram 010083
Midttunet, Skei i Jølster 033942
Midway Museum, Midway 050882
Midway Village and Museum,
Rockford 052637
Midwest Coin Company, Saint
Louis 096740
Midwest Money, Saint Louis . . . 096741
Midwest Museum Newsletter,
Elkhart 139434
Midwest Museum of American Art,
Elkhart 048533
Midwest Old Settlers and Threshers
Association Museum, Mount
Pleasant 051094
Midwinter, Richard, Market
Drayton 090710
Midwood, Charlotte 092616
Mie-kenritsu Bijutsukan, Tsu . . . 029653
Mie-kenritsu Hakubutsukan, Tsu 029654
Miedler-Ludi, Ursula, Gstaad . . . 087519
Miedler, Hans, Wien 063032
Międzynarodowe Centrum Sztuki w
Łódźi, International Center of Art in
Łódź, Łódź 034964
Międzynarodowe Forum Sztuki,
Łódź 098270
Miegunyah Pioneer Women's Memorial
House Museum, Bowen Hills . . 000872
Miehseler, F., Zülpich 079150
Miejsce Pamięci i Muzeum Auschwitz-
Birkenau, Oświęcim 035088
Miejsce Sztuki Officyna, Szczecin 113782
Miejska, Wrocław 113944
Miejska Galeria Arsenał, Poznań 035118
Miejska Galeria Sztuki,
Galeria Malarstwa Polskiego,
Częstochowa 034705
Mielcarek-Kulak, Warszawa . . . 132892

Miele-Museum, Gütersloh 018888
Miele, Frank J., New York 122872
Mielenz, Gerhard, Hannover . . . 076385
Mieli, David, Roma 081452
Mieli, Ruben, Roma 081453
Mielke & Mielke, Hannover 141918
Miellerie des Gorges de la Loire,
Roche-la-Molière 015116
Miellerie, Le Monde des Abeilles,
Villeneuve-de-Berg 016262
Miers Home Museum, Sutton county
Historical Society, Sonora 053390
Miers, J. & P., Mogo 099359
Miervalža Ķemera Memoriālais
Muzejs, Miervaldis Kemers Memorial
Museum, Jūrmala 030139
Mies-Pilsner-Heimatmuseum,
Dinkelsbühl 017904
Mies van der Rohe Haus, Berlin 017225
Mieszkanie 23, Kraków 113614
Mieszkanie-Pracownia Kazimiery
Iłłakowiczówny, Biblioteka
Raczyńskich, Poznań 035119
Miethe, H., Borken, Westfalen . . 075102
Le Mieux Galleries, New Orleans 122205
Mifflinburg Buggy Museum,
Mifflinburg 050884
Migale, Nice 104378
Migard, Riquewihr . . . 072473, 105448
Migart, Praha 102708
Migdał, Joanna, Wrocław 113945
Mighty Eighth Air Force Museum,
Pooler 052239
Migliarini, Alberto, Firenze 131216
Migliorini & Co, Claudio, Firenze 080171,
131217
Migliozzi, Alessandro, Napoli . . . 131629
Migneco, Franzina Cinzia,
Palermo 081058
Mignon, Haßloch 107381
Mignot, Saint-Sébastien-sur-
Loire 073305
Migration Museum, Adelaide . . . 000767
Migros Museum für Gegenwartskunst,
Zürich 042211
Miguaise, Jean-Paul, Provins . . 072318
Miguaise, Jean-Paul, Voulton . . 074278
Miguel Malvar Museum, Santo
Tomas 034594
Miguelenea, Pamplona 086391
Miguet Frères, Paris 071669
Miguet, Maité, Saint Anatoly . . . 129216
Mihálka, Budapest 079268
Mihalyi, U., Bexbach 074970
Miharu-machi Ningyo Hakubutsukan,
Miharu 029024
Mihkli Talumuuseum, Saaremaa
Museum, Kihelkonna 010628
Miho Museum, Shigaraki 029373
Mihrab, Vancouver 064847
Mijabi, Warszawa 084735, 113866
Mijangos, Alberto, San Antonio . 124050
Mijn Winkeltje, Haarlem 083341
Mijndert, Oegstgeest 132600
Mik, Wang, Busan 111573
Mika Gallery, New York 122873
Mikado, Claude, Toulon (Var) . . 073722
Mikael, Dan, Paris 071670
Mikaélian, Edouard, Grenoble . . 068735
Mikaeloff, Robert, Monaco 082520
Mikaeloff, Robert, Paris 071671, 071672
Mikaeloff, Yves, Paris 071673
Mikalojaus Konstantino Čiurlionio
Memorialinis Muziejus, Nacionalinis
M.K. Čiurlionio Dailės Muziejus,
Druskininkai 030333
Mike & Társa, Budapest 142564
Mike's Antiques, Detroit 093368
Mike's Antiques, Tulsa 097750
Mike's Art, Louisville 121659
Mikes Polishing, Austin 135328
Mikes World of Books, Hamilton 140347
Mikeska, J., Bad Homburg
v.d.Höhe 074523

Miki Gallery, Tokyo 111467
Mikkeli, Tartu 103122
Mikkeli Muuseum, Eesti
Kunstimuuseum, Tallinn 010680
Mikkelin Taidemuseo, Mikkeli . . 011160
Mikki, F., Knokke-Heist 100747
Miklautz, Fischertratten 127733
Miklautz, Gmünd, Kärnten 062568
Miklin-Kniefacz, Silvia, Wien . . . 127953
Miklova Hiša, Ribniški Muzej in
Galerija, Ribnica 038600
Miko ir Kipro Petrauskų Lietuvių
Muzikos Muziejus, Kaunas . . . 030369
Mikołajczyk, Kraków 113615
Mikołajczyki, Kraków 113616
Mikschik, Ingolstadt 076613
MIKST, Kyïv 117144
Mikszáth Kálmán Emlékház, Nógrádi
Történeti Múzeum, Horpács . . 023696
Mikumi Museum, Mikumi 042635
Mil, Warszawa 113867
Mil Objetos, Fortaleza . 063947, 063948
Mila, Buenos Aires 060915
Milagros 1530 Post Alley, Seattle 124568
Milam County Museum and Annex,
Cameron 047387
Milan, Fresno 135570
Milan, Houston 093743
Milan Gallery, Fort Worth 120878
Milan Historical Museum, Milan . 050885
Milan Pedersen, Arne, Græsted . 065710
Milanese & Son, San Francisco . 136336
Milanese, Angelo, Torino 132170
Milanese, Gennaro, Milano 131477
Milano, Warszawa 113868
Milano Arte Antica e Moderna,
Milano 098197
Milano Flash Art Fair, Milano . . . 098198
Milano Fumetto, Milano 142732
Milano International Design and Art
Show, Milano 098199
Milano International Fine Art and
Antique Show, Milano 098200
Milano International Modern Arts
Show, Milano 098201
Milano Internazionale Antiquariato,
Milano 098202
Milano Model and Toy Museum,
Elmhurst 048564
Milano, Daniel, Paris 071674
Milanoarte, Milano 080616
Mil'Art, Poznań 113741
Milarte, Milano 110383
Milas Antiques, Chicago 092749
Milas Kültür Merkezi, Milas 043026
Milas Müzesi, Milas 043027
Milazzo, Orazio, Catania 131092
Milazzotto, Alberto, Firenze 131218
MilbenkäseMuseum Würchwitz,
Würchwitz 022784
Milbrado, Timişoara 085136
Milchpintli, Rubigen 041895
Milchwirtschaftliches Museum,
Kiesen 041613
Milden Community Museum,
Milden 006310
Mildenhall, Brighton, Victoria . . . 061203
Mildenhall and District Museum,
Mildenhall, Suffolk 045264
Mildenhall Antiques, Brighton,
Victoria 139657
Mildred Lane Kemper Art Museum,
Saint Louis 052809
Mildred M. Mahoney Silver Jubilee
Dolls House Gallery, Fort Erie . 005816
Mildred's, Minneapolis 094932
Mildred's Antiques, Edmonton . . 064303
Mildura Arts Centre, Mildura . . . 001291
Milefsky, Lee, Baltimore 092346, 119757
Mileham, Tom, Buffalo 092545
Milenario, Logroño 086075
Milenium, Beograd 114429
Milenium Unic Art, Bucureşti . . . 085101
Milenković, Dragan, Beograd . . . 114430

Miler, Sonia di, Lugano 087619
Miles, Kinross 089711, 134519
Miles, New York 122874
Miles, Round Corner 099656
Miles and District Historical Village
  Museum, Miles 001293
Miles B. Carpenter Museum,
  Waverly 054241
Miles Mineral Museum, Eastern New
  Mexico University, Portales ... 052271
Miles, Anna, Auckland 112791
Miles, Archie, Gosforth, Cumbria 144439
Miles, Roy, London 118455
Milesi, Bruno, Le Vaudioux ... 069371
Milestone Antiques, Easingwold 089126
Milestone Antiques, Lichfield ... 089823
Milestones: Hampshire's Living History
  Museum, Basingstoke 043509
Milet Müzesi, Söke 043066
Miletitscher Heimatstube,
  Schönaich 021696
Milevské Muzeum, Milevsko ... 009697
Milford, Dunedin 112886, 112887
Milford Galleries, Auckland ... 112792
Milford Haven Antiques, Milford
  Haven 090758
Milford Haven Museum, Milford
  Haven 045265
Milford Historical Museum,
  Milford 050891
Milford Historical Society Museum,
  Milford 050889
Milford Museum, Milford 050890
Milhes, Bruno, Toulouse 129307
Mili Weber-Haus, Sankt Moritz . 041931
Milic, Zoran, Ljubljana 133067
Le Milieu du Monde, Saint-Paul-de-
  Vence 105594
Milieu Educatie Centrum,
  Eindhoven 032287
Milieucentrum De Grote Rivieren,
  Heerewaarden 032431
Il Milione, Milano 110384
Militär- und Jagdgeschichtliche
  Sammlung, Museumslandschaft
  Hessen Kassel, Bad Wildungen 016903
Militär-Museum Kossa, Kossa .. 019774
Militärgeschichtliche Ortssammlung,
  Schnifis 002725
Militärhistorisches Museum der
  Bundeswehr, Dresden 017997
Militärmuseum Luzern, Kriens .. 041628
Militärmuseum Rosenau,
  Sonntagberg 002762
Militärmuseum St. Luzisteig,
  Maienfeld 041718
Militair Historisch Museum De
  Veteraan, Eefde 032269
Militaire Luchtvaart Museum,
  Soesterberg 032894
Militaria, Kraków 084587, 084588
Militaria Barcelona, Barcelona .. 085802
Militaria Historica, Hamburg ... 076280
Militarial Musée Mémorial pour la Paix,
  Boissezon 012094
Military Academy Museum, Seoul 030022
Military Arms and Collectibles, San
  Diego 097108
Military Aviation Museum,
  Tangmere 045945
Military Best, Jacksonville 094011
Military Bookman, New York ... 145355
Military Bookroom, Malvern,
  Victoria 139820
Military Collectables, San Antonio 097000
Military Collectible Items Wanted,
  Detroit 093369
Military Communications and
  Electronics Museum, Kingston 006105
Military Dependants Village Museum,
  Hsinchu 042335
Military Heritage Museum,
  Lewes 044818
Military Heritage Tours, Longford 024959

Military Historians Museum,
  Westbrook 054329
Military History Bookshop, Hythe,
  Kent 144497
Military Medal Museum, San
  Jose 053039
Military Museum, Aden 054823
Military Museum, Bodmin 043665
Military Museum, Cairo 010544
Military Museum, El-Alamein ... 010568
Military Museum, Macau 008071
Military Museum, Sanaa 054830
Military Museum, Tehrän 024769
Military Museum, Ulaanbaatar .. 031730
Military Museum Fort,
  Bloemfontein 038658
Military Museum of Devon and Dorset,
  Dorchester 044118
Military Museum of Southern New
  England, Danbury 048137
Military Museum, Jamaica Defence
  Force Museum, Kingston 028446
Military Parade Bookshop,
  Marlborough 144730
Military Vehicle Museum, Newcastle-
  upon-Tyne 045350
Military Veterans Museum,
  Oshkosh 051828
Militos, Xalandri 137578, 142546
Militz, Ernst, Kamen 130088
Miljømuseet Dokken, Karmsund
  Folkemuseum, Haugesund ... 033644
Milk Creek Crossing, Portland .. 096438
Milk Gallery & Spotty Dog Design,
  Wellington 113076
Milkyuan Yaozu Self-Rule Nationality
  Museum, Ruyuan 008213
The Mill, Haarlem 083342, 083343
Mill, Philadelphia 113420
Mill Antiques, Sleaford 091350, 135051
Mill Antiques of Bibury, Bibury . 088458
Mill Art, New York 135959
Mill Cove Gallery, Castletownbere 109543
Mill Creek Antiques, Nashville .. 095015
Mill Farm Antiques, Disley 089060,
                                134323
The Mill Gallery, Coverack 117627
Mill Gallery, Ermington 089212, 134370
Mill Green Museum and Mill,
  Hatfield 044536
Mill House Antiques, Goudhurst 089351
Mill House Cider Museum and
  Dorset Collection of Clocks,
  Owermoigne 045447
Mill House Gallery, Parbold 118882
Mill Lane Antiques, Woodford
  Green 091921
Mill of Kintail Conservation Area,
  Robert Tait McKenzie Memorial
  Museum, Almonte 005320
Mill Race Historical Village,
  Northville 051640
Mill Street Studio, Clare 098890
Mill Studio, New York 118751
Mill Trail Visitor Centre, Alva ... 043409
Millan, Cádiz 085923
Millard, Omaha 096051
Millard Fillmore House, Aurora
  Historical Society, East Aurora . 048423
Millard, Jon, Louisville 121660
Mill'Broc, Guignes 068769
Millbrook, Baltimore 092347
Millbrook Antiques, Las Vegas . 094161
Millchell, Fort Worth 093462
Millcourt Antiques, Donegal ... 079512
Millcreek Antiques, Cincinnati .. 092878
Mille et Une Feuilles, Beaune .. 140780
Mille Facettes, Paris 071675
Millefiori, Villeneuve-sur-Lot ... 074168
Millénaires, Cogolin 068069
Millenium-Antikvarnica, Herceg-
  Novi 082536
Millenium Fine Arts, Roma 110825
Millennic Glass, Kansas City .. 121305

Millennium, Dallas 093177
Millennium Art Gallery, New York 122875
Millennium Art Gallery,
  Philadelphia 123421
Millennium Arts Center International,
  Washington 054136
Millennium Decorative Arts,
  Washington 097848
Millennium Galleries, Sheffield . 045774
Millenniumi Földalatti Vasúti
  Múzeum, Budapesti Közlekedési
  Részvénytársaság, Budapest .. 023526
Miller, Cincinnati 120359
Miller, Dallas 093178
Miller, Kansas City 121306
Miller, Paris 071676
Miller, Tulsa 097751
Miller, Virginia Beach 124836
Miller & Arney, Washington ... 097849
Miller & Block, Boston 119895
Miller Art Center, Springfield .. 053475
Miller Art Museum, Sturgeon Bay 053584
Miller Bakehouse Museum,
  Palmyra 001393
Miller-Cory House Museum,
  Westfield 054337
Miller House Museum,
  Hagerstown 049311
Miller Museum of Geology,
  Kingston 006106
Miller, Austin T., Columbus 093035
Miller, D., New York 135960
Miller, Duncan R., London 090349,
  090350, 118456, 118457, 134727,
                                134728
Miller, James, Wooler 091935
Miller, Janie, North Sydney ... 062000
Miller, Jason S., Miami 121828
Miller, Johannes von, Bad
  Wiessee 129428
Miller, John, Newcastle 099432
Miller, Laurence, New York ... 122876
Miller, Maggie, Bovey Tracey .. 117408
Miller, Marc, Tucson 124755
Miller, Marlene, Sorrento 062220
Miller, Mary Frances, Charlotte . 092617
Miller, Max, Houston 093744
Miller, Nick, Cullompton 089024
Miller, Paulina, Phoenix 123500
Miller, Peter, Chicago 120204
Miller, Robert, New York 122877
Miller, Ron, Los Angeles 094410
Miller, Steve, New York 122878
Miller, Susan, San Francisco ... 124344
Miller, Thomas N., Newcastle-upon-
  Tyne 126995
Miller, Virginia, Miami 121829
Millers, Ringwood 091106
Millers, Frantz, Tournon-sur-
  Rhône 073821, 129319
Millertown Museum, Millertown . 006313
Milles, Hans-Jürgen, Namborn . 077687
Millescamps, Henri-Pierre,
  Périgueux 141302
Millesgården, Lidingö 040778
Millet, Saint-Martin-du-Mont ... 072950
Millet, San Isidro 060959
Millet and District Museum,
  Millet 006314
Millet, Jean-Claude, La Charité-sur-
  Loire 068985
Millet, René, Paris 105094
Millet, Roger, Torce-en-Vallee .. 073688
Millgate Pine and Antiques,
  Thirsk 091597
Millhouse Art Society Gallery,
  Milton 099350
Milliard, Jacques, Montréal 064444
Milliarede Besch, Lyon 125630
Milliat, Robert, Rive-de-Gier ... 072479
Millicent, Los Angeles 121557
Millicent, Pittsburgh 096311
The Millicent Gallery, Millicent . 099343
Millicent National Trust Museum,

Millicent 001296
Millicent Rogers Museum of Northern
  New Mexico, Taos 053677
Millich, Wien 063033
Milliken Gallery, Converse College,
  Spartanburg 053428
Milliken, Mark, New York 122879
The Millinery Works, London ... 090351
Millington, K.L. & J., Lepton ... 089809
Milliyet Sanat Dergisi, İstanbul . 139114
Millmount Museum and Tower,
  Drogheda 024866
Millner, Arthur, London 090352
De Millo Fine Art, Monaco ... 082521
Millon & Associés, Mathias, Paris 125740
Millon-Garnier, Christelle, Chalon-sur-
  Saône 067781
Millord, Crécy-la-Chapelle 068228
Millpetersch Haus Güdesweiler,
  Oberthal 020933
Millpool Gallery, Mousehole ... 118759
Mills, Boston 119896
Mills, Cork 079499, 079500
Mills, Honolulu 093561
Mills Antiques, Cullompton ... 089026
Mills College Art Museum,
  Oakland 051679
Mills County Museum, Glenwood 049129
Mills Observatory Museum,
  Dundee 044165
Mills, Robert, Bristol 088661
Mills, Robert H., Philadelphia .. 096138
Mills, Rocky, Indianapolis ... 121203
Millstreet Local Museum,
  Millstreet 024966
Milltown-Head Bay d'Espoir Museum,
  Milltown 006315
Millwall FC Museum, London .. 045038
Millwheel Gallery, Minehead .. 118749
Millwood Gallery, Wellington .. 113077
Millyard Gallery, Uppermill ... 119284
Milmo-Penny, Dominic, Dublin . 109605
Milne Bay Military Museum,
  Toowoomba 001584
Milne, John, Aberdeen 126793
Milne, Nigel, London 090353
Milne, T., London 144650
Milnthorpe & Daughters, Mary,
  Settle 091264
Milo, Seattle 124569
Milone, Giovanni, Torino 081733, 137749
Milos Crnjanski Emlékszoba,
  Csongrád 023588
Milou, Atlanta 092156
Milou, Perry, Philadelphia 123422
Milovanović, Kraljevo 133033
Milpitas, San Jose 097373
Milrose, I., Mettmann 077366
Milstain, Marc, Bruxelles 100642
Milton, Jacksonville .. 094012, 135659
Milton Antiques, Blandford Forum 088505,
                                134120
The Milton Art & Craft Gallery,
  Banchory 117260
Milton Blacksmith Shop Museum,
  Milton Heritage Society, Milton 006318
Milton Historical Museum, Milton 050907
Milton Holgate, Ripon 091119
Milton House Museum, Milton .. 050908
Milton J. Rubenstein Museum
  of Science and Technology,
  Syracuse 053635
Milton Keynes Gallery, Milton
  Keynes 045273
Milton Keynes Museum, Milton
  Keynes 045274
Milton, Lionel, New Orleans ... 122206
Milton's Cottage, Chalfont Saint
  Giles 043902
Milwaukee Art Museum,
  Milwaukee 050916
Milwaukee Auction Galleries,
  Milwaukee 127324
Milwaukee County Historical Society

Museum, Milwaukee 050917
Milwaukee Institute of Art and Design,
Milwaukee 057453
Milwaukee Public Museum,
Milwaukee 050918
Milwaukee Underground Ink,
Milwaukee 121936
Milwaukee Very Special Art Center,
Milwaukee 050919
Milwaukee's Finest Antique Store,
Milwaukee 094875
Milwaukie Antique Mall, Portland 096439
Milwe, Maria-Helene von der,
Aachen 105943
Milwee, Frank, Washington . . 097850
Mimarlık Fakültesi, Orta Dogu Teknik
Universitesi, Ankara 056474
Mimax-Inn Gallery, Singapore . . 114571
Mimbridge Antiques Centre,
Chobham 088900
Mimer, Stockholm 143956
Mimi, Chicago 120205
Mimi Ferzt Gallery, New York . . 122880
Mimi Fifi, London 090354
Mimi Restauro Ligneo Dorture,
Venezia 132250
Mimi's African Art Gallery,
Minneapolis 122014
Mimi's Memories, Jacksonville . 094013
Mina Andreina, Torino 132171
Minaaz Art Gallery, Hyderabad . 109366
Minaret, Philippe, Paris 105095
Miñarte, Málaga 086298, 133404
Minas Avetisyan Museum,
Yerevan 000738
Minassian, Dominique, Marennes 069872
Minatoya, Tokyo 111468
Minault, Rene, Ayl 129400
Mind Ahead and Art, New York . 122881
Mind and Matter Gallery, South
Surrey 007060
Mindata, Bath 138137
Minden Hills Museum, Minden . . 006323
Mindener Museum für Geschichte,
Landes- und Volkskunde,
Minden 020418
Minder, Daniel, Zürich 133993
Mindermann, Rolf, Juist 076657, 107489
Minds Eye, Wellington 143367
Mindszenty Emlékhely, Keresztény
Múzeum, Esztergom 023632
Mine d'Argent Saint-Barthélemy,
Sainte-Marie-aux-Mines 015618
La Mine-Image, Musée de la Mine, La
Motte-d'Aveillans 013443
Mine Sanat Galerisi, İstanbul . . 117067
Mine Témoin d'Alès, Alès 011593
Minelli, Giuseppe, Genova 131325
Miner, A., Seattle 124570
Minera Lead Mines, Minera . . 045277
Mineral- och Gruvmuseet,
Holsbybrunn 040707
Mineral and Gem Geological Museum,
Parrsboro 006613
Mineral and Lapidary Museum
of Henderson County,
Hendersonville 049444
Mineral County Museum,
Superior 053610
Mineral Museum, Butte 047336
Mineral Resources Divisional Museum,
Dodoma 042632
Mineralien- und Heimatmuseum
Pleystein, Pleystein 021157
Mineralien-Fossilien-Bergbau, Privat-
Museum Helmut und Monika
Steininger, Langenlois 002266
Mineralien-Kabinett des Mineralogisch-
Petrografischen Museums der TU
Braunschweig, Braunschweig . . 017562
Mineralien Museum, Seedorf . . . 041974
Mineralien-Museum Andreas Gabrys,
Lam 019863
Mineralien-Museum Zirbenhof, Ebene

Reichenau 001833
Mineralienmuseum, Bad
Säckingen 016855
Mineralienmuseum, Einsiedeln . 041457
Mineralienmuseum, Freisen 018498
Mineralienmuseum, Neubulach . 020686
Mineralienmuseum Essen, Museen der
Stadt Essen, Essen 018304
Mineralienmuseum Pforzheim-
Dillweißenstein, Pforzheim . . . 021120
Mineraliensammlung Lauingen,
Lauingen 019938
Mineraliensammlung Steger, Wiener
Neustadt 003106
Mineralienschau, Vohenstrauß . . 022354
Mineralienzentrum Steinstadel,
Aggsbach Dorf 001690
Mineralogical Museum of Harvard
University, Cambridge 047367
Mineralogical Museum of Kamarzia,
Attica 023032
Mineralogical Museum of Lavrion,
Lavrion 023152
Mineralogičeskij Muzej im. A.E.
Fersmana Rossijskoj Akademii Nauk,
Moskva 037343
Mineralogičeskij Muzej im. A.V.
Sidorova, Irkutsk 037027
Mineralogičeskij Muzej Moskovskoj
Geologorazvedyvatelnoj Akademii,
Moskva 037344
Mineralogičeskij Muzej, Geologičeskij
institut Kol'skogo naučnogo centra
RAN, Apatity 036856
Mineralogické muzeum – PřF-UK,
Přírodovědecká fakulta Univerzity
Karlovy v Praze, Praha 009808
Mineralogisch-Geologisch Museum,
Delft 032159
Mineralogisch-Geologische Gesteins-
Sammlung, Schwarzenfeld 021775
Mineralogisch Museum, Grou . . 032386
Mineralogisch Museum, Academy for
Mineralogy, Merksem 003866
Mineralogisch-petrographische
Sammlung der Universität Leipzig,
Leipzig 020001
Mineralogische Sammlung,
Oberthal 020934
Mineralogische Schau- und
Lehrsammlung, Tübingen 022222
Mineralogisches Museum (Kristalle
und Gesteine), Universität Münster,
Münster 020603
Mineralogisches Museum der Philipps-
Universität Marburg, Marburg . 020274
Mineralogisches Museum der
Universität, Würzburg 022793
Mineralogisches Museum der
Universität Bonn, Bonn 017497
Mineralogisches Museum der
Universität Hamburg, Hamburg 019004
Mineralogisches Museum Oberkirchen,
Freisen 018499
Mineralogy Museum, Bangkok . 042663
Mineralogy, Petrography and
Volcanology Museum, Catania . 025867
Mineralovodskij Kraevedčeskij Muzej,
Mineralnye Vody 037225
Mineralų Muziejus, Geologijos ir
geografijos institutas, Vilnius . 030534
Minerva, Trieste 111052
Minerva Antik, Malmö 086855
Minerva Antiques, London 090355
Minerva Editrice, Assisi 137617
Minerva Quilt Centre, Llanidloes 118139
Minerva Restauri, Milano 131478
Minerva, The International Review
of Ancient Art and Archaeology,
London 139202
Minerva's Antiques, Birmingham 092422
Minerve, Toulon (Var) 073723
Minervini, Giorgio, Roma 081454
Mines d'Asphalte de la Presta,

Travers 042066
Mines de Sel de Bex, Bex 041334
Minet Frères, Bruxelles 140190
Ming, Cincinnati 120360
Ming, Honolulu 093562
Ming, Shenzhen 102231
Ming, Taichung 116895
Ming Art, Bad Zwischenahn . . . 074623
Ming Art Gallery, Beijing 101977
Ming Art Gallery, Taipei 116933
Ming-Ching, Singapore 085470
Ming Direct, Waikato 084133
Ming Direct, Weikato 084149
Ming Furniture, New York 095585
Ming Gallery, Louisville 094574
Ming-K'I, Waardamme 063854
Ming Mai, Santa Monica 097384
Ming og Qing, Charlottenlund . 065628
Ming Qing Antiques, Shanghai . 065146
Ming Things, Ottawa 064509
Ming Tong, Lille 104013
Mingardi, Gianfranco, Brescia . . 131040
Mingazzini, Michela, Roma 081455
Mingboy, Singapore 085471
Mingcheng Yuan Shi Museum,
Nanjing 008109
Mingei, Atlanta 092157
Mingei International Museum, San
Diego 052966
Mingei International Museum, North
County Satellite, Escondido . . 048607
Mingenback Art Center Gallery,
Lindsborg 050318
Mingenew Museum, Mingenew . 001297
Ming's Gallery, Hong Kong 102095
Mingshan Chashi Museum,
Mingshan 008085
Mingus, Praha 140525
Mingus, Umeå 143986
Mingxi Museum, Mingxi 008086
Minh, Chau, Ha Noi 125086
Minhauzena Muzejs, Dunte 030120
Minhe Hui Tuzu Self-Rule Museum,
Minhe 008087
Minhwa, Paris 071677
Mini-Château, Amboise 011622
Mini-Mart of Tucson, Tucson . . 097677
Mini-Musée, Habsheim 013222
Mini-Musée Architectural,
Charnat 012461
Mini Musée des Miniatures,
Sospel 015839
Mini Print International de Cadaqués,
Cadaqués 098313
Mini Rastro Casavella, Vigo 086562
Miniaci Art Gallery, Milano 110385
Miniatur Wunderland Hamburg,
Hamburg 019005
Miniaturbuchverlag Leipzig, Neufahrn
bei Freising 137442
Miniature Books Museum, Baki . 003169
Miniature Magic, Austin 092251
Miniatures, Lyon 128722
Miniatures Museum of Taiwan,
Taipei 042528
Miniaturmuseum Arikalex, Berlin 017226
Miniaturpark Klein-Erzgebirge,
Oederan 020949
Miniatuur Walcheren, Middelburg 032649
Miniatuurstad, Antwerpen 003292
Minigallery.Co.Uk, Thornton
Heath 119229
Minilista, F., Miami 121830
Minimotor Muzejs, Riga 030209
Minimundus, Klagenfurt 002194
Minimuseum, Groningen 032380
The Mining Museum, Platteville . 052195
Minion, Kyoto 082033
Miniota Municipal Museum,
Miniota 006325
Minischeepvaartmuseum,
Ouwerkerk 032757
Minisink Valley Historical Society
Museum, Port Jervis 052253

Minjasafn Austurlands,
Egilsstadir 024073
Minjasafn Egils Ólafssonar,
Patreksfjörður 024090
Minjasafnid Burstarfelli,
Vopnafjörður 024120
Minjasafnida á Akureyri, Akureyri 024064
Minkler, Thomas, New York . . . 095586,
145356
Minlaton National Trust Museum,
Minlaton 001298
Minle Museum, Minle 008088
Minna Street Gallery, San
Francisco 124345
Minnamoora, Wanniassa 062405
Minneapolis College of Art and Design,
Minneapolis 057459
Minneapolis Institute of Arts,
Minneapolis 050944
Minneapolis Institute of Arts. Annual
Report, Minneapolis 139435
Minneapolis Uptown Antiques,
Minneapolis 094933
Minnedosa and District Museum,
Minnedosa 006326
Minnenas Bod, Göteborg 086712
Minnesota Air National Guard Museum,
Saint Paul 052844
Minnesota Association of Museums,
Minneapolis 060593
Minnesota Children's Museum, Saint
Paul 052845
Minnesota Historical Society Museum,
Saint Paul 052846
Minnesota Museum of American Art,
Saint Paul 052847
Minnesota Pioneer Park Museum,
Annandale 046496
Minnesota State Capitol Historic Site,
Saint Paul 052848
Minnesota Transportation Museum,
Saint Paul 052849
Minnette, Los Angeles 094411
Minnick, Honolulu 135579
Minnie's Thirteen Forty One,
Jacksonville 094014
Minnilusa Pioneer Museum, Rapid
City 052471
Mino, Alain, Beinheim . 067064, 128466
Mino, Alvaro, Roma 081456
Minokamo-shiritsu Hakubutsukan,
Minokamo 029027
Minoletti, Gianfranco, Genova . . 080272,
131326
Minoletti, Marco, Genova 131327
Minoriten Galerien, Graz 100016
Minority Arts Resource Council Studio
Art Museum, Philadelphia 052066
Minot, Paris 071678
Minot Art Association, Minot . . 060594
Le Minotaure, Alger 060834
Le Minotaure, Paris . . . 071679, 105096
Minotauro, La Spezia 110157
Minotauro, Roma 081457, 131979
Minouche, Marin-Epagnier 116492
Minqin Museum, Minqin 008089
Minqing Museum, Minqing 008090
Minski, H., Gummersbach 076130
Minsky, Paris 105097
Minssen, Ile-Tudy 103851
Minster Abbey Gate House Museum,
Minster-in_Sheppey 045278
Minster Abbey Gatehouse Museum,
Sheerness 045768
Minster Books, Wimborne
Minster 135232, 144925
Minster Gate Bookshop, York . . 135270,
144941
Mint Museum, Osaka 029029
Mint Museum of Art, Charlotte . 047576
Mint Museum of Craft and Design,
Charlotte 047577
Mint, Franklin, Las Vegas 094162
Mintage, Mumbai 109453

Mintai Tieshi Museum, Quanzhou 008196
Minten, Kees, Beers bij Cuyk . . 132436
Minter Field Air Museum, Shafter 053282
Mintmaster, Atlanta 092158
Minto Museum, Minto 006327
Minton, Damien, Newcastle . . . 099433
Minton, Joseph, Dallas 093179
Mintsas, Theodoros, Athinai . . 079174
Les Minuscules, Marseille 069952
Minusinskaja Chudožestvennaja
Kartinnaja Galereja, Minusinsk 037228
Minusinskij Regionalnyj Kraevedčeskij
Muzej im. N.M. Martjanova,
Minusinsk 037229
Minute Man, National Historical Park,
Concord 047952
Minx, Dr. Rainer, Wünsdorf 142508
Minxi Revolution History Museum,
Minxi 008091
Minymaku Arts, Amata 060996
Minzoku Shiryokan, Miyama . . . 029040
Miqueleiz, Pamplona . . 086392, 143791
Mir, Sankt-Peterburg 143670
MIR Appraisal Services, Chicago 092750
Mir Dizajna, World of Design, Sankt-
Peterburg 139013
Mir-Emad Calligraphy Museum, Sa'd-
Abad Museum, Tehrän 024770
Mir i Muzej, Tula 139014
Mir Iskusstv, Moskva 085193
Mir Iskusstv, Vinnycja 117164
Mir Russki, Linlithgow 089840
Mira Antique, Bucureşti 085102
Mira Fine Art Gallery, Melbourne 061825
Mirabeau, Houston 093745
Mirabeau, Paris 105098, 129090
Mirabella, Giacomo, New York . . 135961
Mirabella, Patrizia, Napoli 080901
Mirabilia, Roma 110826
Mirabilia, Venezia 111143
Miracle Metal, Cleveland 135469
Miracle of America Museum,
Polson 052231
Mirage, Douglas 139436
Mirage, Kraków 113617
Mirage Art Gallery, Delhi 109333
Mirage Studio, Cleveland 120440
Miraglia, Salvador, Lyon 140926
Mirakhore, Esther, Les Acacias . 087181
Mirakov, Jurij, Moskva 114278
Miralles Guillen, Vicente H.,
Valencia 133467
Miramar Zeemuseum, Vledder . 032992
Miramichi Natural History Museum,
Miramichi 006330
Miramón Kutxa Espacio de la Ciencia,
Donostia-San Sebastián 039351
Miramont Castle Museum, Manitou
Springs 050646
Miramont, Céline, Les Ternes . . 069433
Miranda Libreria, Madrid 143776
Miranda Sáchez, Santiago de
Chile 064976, 128251
Miranda, Dino, Roma 081458
Miranda, Jorge, Paris 129091
Miranda, Miguel, Madrid 143775
Miranda, Roberto, San Isidro . . 060960
Mirandas, Carmen, Salt Lake
City 096908
Mirande, Bernard, Fortaleza . . . 100955
Mirani, Bucureşti 085103
Mirari, Windsor, Ontario 101858
Miras, Moskva 114279
Miras, Ufa 114375
Mira's Antik, Ingelheim am Rhein 076607
Mirasaka Peace Museum of Art,
Mirasaka 029028
Mirate, Francesco, Venezia 081860
Mire, New Orleans 095131
Mirehouse, Keswick 044682
Mirer, R., Obersaxen Affeier . . . 116549
Mirer, Rudolf, Zernez . 116734, 138101
Mireuksa Temple Museum, Iksan 029949
Mireya, Buenos Aires 139588

MIRI-Antiquitäten, Berlin 074880
Miriam and Ira D. Wallach Art Gallery,
New York 051426
Miriam Switching Post, Saint
Louis 096742
Mirishet, Bergen 082957
Mirisola, Giuseppe, Palermo . . . 142779
Mirkovich, N., Houston 121093
Miro, Ljubljana 085611, 114721
Miro 464, Sofia 064232
Miro Art, Kortrijk 063723
Miro, Isabel, Madrid 086217
Miro, Victoria, London 118458
Miroir de Fief, Orgeval 070800
Miroir des Maîtres, Genève 116330
Miroirs & Meubles, Hilversum . . 083404
Miroiterie, Villefranche-sur-Saône 074150
Miroiterie Pays, Mâcon 069795
Miromesnil Fine Art, Paris 105099
Miron, Paris 071680
Miron Antic, Nice 070635
Miroo, Namu, Yeongchen 111746
Mirou, Lille 069497
Mirouge Museum, Mizhi 008092
Mirror and District Museum,
Mirror 006333
Mirror Fair, New York . . 095587, 135962
Mirror Image, Dallas 120630
Mirror Image, Dún Laoghaire . . . 079607
Mirror Mirror, New Orleans 095132
Mirski, Boris, Boston 119897
Mirso, Podgorica 112029
Mirus, Wiesbaden 078988
Mirvish, David, Toronto 140412
Mirviss, Joan B., New York 122882
Miryang Municipal Museum,
Miryang 029971
Mirza Fatali Akhundov House Museum,
Tbilisi 016414
Miscellanea, Mumbai 109454
Miscellanea, San Diego 097109
Miscellany Antiques, Great
Malvern 089373
Mischell, Hans H., Köln 076898
Mischkind, Raphaël, Lille 104014,
128682
Miscio, Giorgio, Roma 131980
Mise en Page, Lyon 140927
Mise en Scène, Roanne 105451
Mis'en Scène 40, Tours 073853
Miser, Saint Louis 096743
Mishima Taisha Treasure House,
Mishima 029029
Mishkan Le'Omanut, Ein Harod . 025035
Misia Arte, Bari 079775
Misijné Múzeum Nitra, Nitra . . . 038408
Misirloglou, Christos, Frankfurt am
Main 141784
Misis Mozaik Müzesi, Yakapınar 043082
Misjonsmuseet på Fjellhaug,
Oslo 033841
Misjonsmuseet, Misjonshogskolen,
Stavanger 033976
Mis'ka Chudožna Galereja im. S.I.
Vasyl'kivs'kogo, Charkiv 043135
Miskolci Egyetemtörténeti
Gyüjtemények, Miskolc 023790
Miskolci Galéria – Városi Művészeti
Múzeum – Rákóczi-ház,
Miskolc 023791
Misora Hibari Kinenkan, Kyoto . 028966
Misrachi, México 111972
Misrachi, Alberto, México 111973,
111974
Miss Baltimore Agency, New
York 095588
Miss Edna's Antiques, New
Orleans 095133
Miss Elany, Long Eaton 090612
Miss-Luig, Werdohl 108872
Missa, Yongin 111747
Missa Atelier, Bourges 103454
Missal, Birgit, Hamburg 076281
Missal, Gisela, Wenningstedt-

Braderup 078924
Misseri, Andrea & Palombi,
Alessandro, Roma 081459
Missiemuseum Steijl, Steijl 032904
Missillier, Philippe, Lyon 069763
The Missing Link, New York 095589
Mission Antiques, Portland 096440
Mission Art Gallery, Pécs 109209
Mission Bell Museum, Coweta . 048046
Mission Cultural Center for Latino Arts,
San Francisco 053012
Mission Gallery, San Diego 097110
Mission Gallery, Swansea 045935
The Mission House, Stockbridge 053535
Mission Houses Museum,
Honolulu 049540
Mission Inn Museum, Riverside . 052594
Mission Mill Museum, Salem . . . 052892
Mission Museum, Mission 006335
Mission Oak Shop, San
Francisco 097286
Mission Oak Shop, Seattle 097475
Mission San Carlos Borromeo del Rio
Carmelo, Carmel 047435
Mission San Juan Capistrano Museum,
San Juan Capistrano 053047
Mission San Luis Obispo de Tolosa,
San Luis Obispo 053048
Mission San Luis Rey Museum,
Oceanside 051704
Mission San Miguel, San Miguel 053057
Mission, Cliff, San Diego 127440
Missionart, Miskolc 109201
Missionary Museum of the N.G. Kerk,
Stellenbosch 038835
Missionary Museum, Amathole
Museum, King William's Town . 038771
Missions-Ethnographisches Museum
St. Gabriel, Mödling 002400
Missionsmuséet Nårunga,
Vårgårda 041156
Missionsmuseum, Sankt Ottilien 021607
Missionsmuseum, Schwarzach am
Main 021769
Missionsmuseum Bug, Bamberg 016950
Missionsmuseum Dachsberg,
Prambachkirchen 002533
Missionsmuseum der Liebenzeller
Mission, Bad Liebenzell 016819
Missionsmuseum der Mariannhiller
Missionare, Würzburg 022794
Missionsmuseum der Pallottiner
Limburg, Limburg an der Lahn 020079
Missionsmuseum des päpstlichen
Missionswerks der Kinder in
Deutschland, Aachen 016441
Missirlakis, Madrid 133359
Missisquoi Museum, Stanbridge
East 007069
Missisquoi Valley Historical Society,
North Troy 051620
Mississauga Heritage Foundation,
Mississauga 006343
Mississippi Art Colony, Stoneville 060595
Mississippi Baptist Historical
Commission, Clinton 047839
Mississippi County Historical Society,
Charleston 047539
Mississippi Crafts Center,
Ridgeland 052571
Mississippi Governor's Mansion,
Jackson 049750
Mississippi Museum of Art,
Jackson 049751
Mississippi Museum of Natural
Science, Jackson 049752
Mississippi Museums Association,
Natchez 060596
Mississippi Petrified Forest, Flora 048750
Mississippi River Museum,
Dubuque 048376
Mississippi River Museum at Mud
Island River Park, Memphis . . 050786
Mississippi University for Women

Museum, Columbus 047931
Mississippi Valley Textile Museum,
Almonte 005321
Mississippi Agriculture and Forestry/
National Agricultural Aviation
Museum, Jackson 049753
La Missive, Besançon 067132
Mißler, Herbert, Peiting 078035
Missoula Art Museum, Missoula 050960
Missouri Coin, Saint Louis 096744
Missouri Historical Society, Saint
Louis 052810
Missouri Mouse, Saint Paul 096840
Missouri Museums Association,
Kansas City 060597
Missouri State Museum, Jefferson
City 049805
Missouri Town 1855, Lees
Summit 050212
Mistasinihk Place Interpretive Centre,
LaRonge 006164
Misteckij Kurin, Kyïv 117145
Misteckij Ljoch, Kyïv 117146
Mister Daves, Winnipeg 064929
Mister Poster, Indianapolis 121204
Mister R's, San Francisco 136337
Mister Thrifty Book Shop, Sydney 139925
Misterek, Uwe, Jever . 076653, 130084
The Mistletoe Gallery, Masham . 118727
Místní Muzeum, Mladá Vožice . . 009701
Místní Muzeum, Muzeum Dr. Otakara
Kudrny, Netolice 009716
Místní Vlastivědné Muzeum,
Železnice 009979
Místní Vlastividní Muzeum,
Kopidlno 009637
Mistral, La Bégude-de-Mazenc . 103875
Mistrali, Emilio, Parma 081118
Mistretta, Diana, Palermo 081059
Misty's Antiques, Portland 096441
Misura Su Misura, Messina 080372,
131357
Misuraca, Prof. Donatello, Roma 131981
MIT Museum, Cambridge 047368
The MIT Press, London 138230
Mita Arts, Tokyo 111469
Mitcham Village Art and Crafts
Association, Mitcham . 058207, 099355
Mitchell, Cockermouth 126866
Mitchell, Jacksonville 121258
Mitchell Algus, New York 122883
Mitchell County Historical Museum,
Osage 051822
Mitchell Gallery of Flight,
Milwaukee 050920
Mitchell House, Maria Mitchell
Association, Nantucket 051156
Mitchell-Innes & Nash, New York 122884,
122885
Mitchell Library Exhibition,
Glasgow 044406
Mitchell Museum at Cedarhurst,
Mount Vernon 051102
Mitchell Museum of the American
Indian, Evanston 048642
Mitchell School of Fine Arts,
Baltimore 056707
Mitchell, Anthony, Nottingham . . 090898,
134900
Mitchell, Brett, Cleveland 120441
Mitchell, Christina, Milton 099351
Mitchell, Duncan, Aldridge 117224
Mitchell & Son, John, London . . 090356,
134729
Mitchell, L.K., Richmond 136174
Mitchell, Mike, Washington 124936
Mitchell, Paul, London 090357
Mitchell, Sally, Tuxford 091671, 119275
Mitchell, T.J., Dublin . . 097571, 130860
Mitchell's, Blackburn, Lancashire 088495
Mitchell's Refinishing, Buffalo . . 135377
Miter Box, Cincinnati 120361
Mitered Corner, Seattle 124571
Mithrasheiligtum, Mundelsheim . 020616

Mithrász-szentély, Soproni Múzeum,
Fertőrákos 023638
Mitmannsgruber, Sankt Pölten .. 062830
Mitnitzky, Joaquim, Rio de
Janeiro 064089
Mito, Warszawa 113869
Mito Geijutsu-kan, Gendai Bijutsu
Gallery, Mito 029037
Mitochu, Tokyo 082181
Mitofskys, Dublin 079572
Mito's Antigüedades, Caracas .. 097962
Mitre Box, Baltimore 119758
Mitre Box, San Francisco 124346
Mitre House Antiques, Ludlow .. 090652
Mitro-Henseler, Ludmila, Rastede 130555
Mitropoličji Palaty – Muzej
Drevnerusskogo Iskusstva,
Jaroslavskij chudožestvennyj muzej,
Jaroslavl 037064
Mitschkowski, Henny, Bremen .. 075176
Mitspielzeugmuseum, Leipzig .. 020002
Mitsubishi Auto Gallery, Okazaki 029218
Mitsubishi Minatomirai Indasutoriaru
Hakubutsukan, Yokohama ... 029719
Mitsui Kinen Bijutsukan, Tokyo . 029554
Mitsukoshi Gallery, Tokyo 111470
Mitsumura, Tokyo 137785
Mitte Museum am Festungsgraben,
Berlin 017227
Mitte Museum am Gesundbrunnen,
Berlin 017228
Mitteilungen der Österreichischen
vergleichende Kunstforschung,
Wien 138526
Mitteilungen der Österreichischen
Numismatischen Gesellschaft,
Wien 138527
Mittelalterliches Foltermuseum,
Rüdesheim am Rhein 021538
Mittelalterliches Kriminalmuseum,
Rothenburg ob der Tauber ... 021508
Mittelalterliches Museum, Wien . 003052
Mittelmoselmuseum im Barockhaus
Böcking, Traben-Trarbach 022171
Mittelrhein-Museum Koblenz,
Koblenz 019664
Mittelschwäbisches Heimatmuseum
Krumbach, Krumbach 019810
Mittendorf, Karl-Heinz, Goslar .. 076062
Mittentzweg, M., Mannheim 077266
Mittermüller, A., Heufeld 130043
Mitterrand, Edward, Genève ... 116331
Mitton, Michelle, Clonakilty 115065
Mitzlaff, Otto von, Wächtersbach 078821
Mitzpe Revivim Museum, Kibbutz
Revivim 025144
Miu, Singapore 114572
Miura Arts, Tokyo 111471
Mix, Buffalo 092546
Le Mix, Pittsburgh 096312
Mix, Toronto 138586
Mix Media Gallery, Jacksonville . 121259
Mixed Company, Philadelphia . 096139
Mixed Greens, New York 122886
Mixed Media, Las Vegas 121368
Mixed Media Art Gallery, San
Diego 124145
Mixed Media Gallery, Columbus 120504
Mixografía Workshop & Gallery, Los
Angeles 121558, 138351
Mixture Contemporary Art,
Houston 121094
Miyagi-ken Bijutsukan, Sendai . 029358
Miyagi Michio Kinenkan, Tokyo . 029957
MIYAI Furoshiki and Fukusa Gallery,
Kyoto 028967
Miyake, Tokyo 111472
Miyake Bijutsukan, Kagoshima . 028746
Miyakonojo-shi Bijutsukan,
Miyakonojo 029039
Miyakoya, Kyoto 082034
Miyama-cho Kayabuki Bijutsukan oyo
Kyodo Hakubutsukan, Miyama 029041
Miyamota, New York 095590

Miyandoab Museum, Miyandoab 024724
Miyashita, Sapporo 111378
Miyawaki, Ichiro, Kyoto 111299
Miyazaki-ken Sogo Hakubutsukan,
Miyazaki 029043
Miyazaki-kenritsu Bijutsukan,
Miyazaki 029044
Miyazawa Kenji Kinenkan,
Hanamaki 028606
Miyoshi, Kobe 082000
Miyuki, Tokyo 111473
Mizel Museum, Denver 048267
Mizener, Hamilton 064325
Mizener, Toronto 064705
Mizon, Maree, Paddington, New South
Wales 099531
Mizu-to-Midori-no Fureaikan,
Okutama 029222
Mizuhomachi Kyodo Hakubutsukan,
Mizuho 029048
Mizuma, Tokyo 111474
Mizunami-shi Kaseki Hakubutsukan,
Mizunami 029049
Mizunami Toji Shiryokan,
Mizunami 029050
Mizuno, West Hollywood 125013
Mizzi, Marsa 082415
Mjellby Konstmuseum, Halmstad 040684
MJM Antiques, Hungerford 089611
MJM Arte Antigo, São Paulo ... 128126
MJM Global Arts, Bruxelles 063508
Mjölby Hembygdsgård, Mjölby . 040846
Mjølstad, Gjøvik 126349
Mjs Antiques, Tauranga 084105
MK Fine Art, New York 122887
MK Galerie, Rotterdam 112651
MK Kunst Galerie, Mülheim an der
Ruhr 108039
MKG Art Management, Houston 121095
MKgalerie, Berlin 106375
MKJ Art Gallery, New York 122888
Mkrtchyan, A.H., Jerewan 098603
MLA North East Annual Review,
Newcastle-upon-Tyne 139203
MLA West Midland, Regional Council
for Museums, Libraries and Archives,
Birmingham 060180
Mladinska Knjiga, Ljubljana ... 137944,
143697
MLD, Saint-Ouen 073155
Mleczko, Andrzej, Kraków 113618
MM Galerie, Grebenstein 107156
MME Fine Art, New York 095591
MMST Galerie, Thessaloniki ... 109092
MNAC Publications, Barcelona . 137965
MO-Artgallery, Amsterdam 082808
Mo Lan, Beijing 101978
Mo Tu Uv – Museum of Mechanical
Toys, Ulaanbaatar 031731
Mo Yuan Ge, Beijing 101979
MOA Bijutsukan, Atami 028497
Moanalua Gardens Foundation,
Honolulu 049541
Moat Park Heritage Centre,
Biggar 043599
Moatti, Alain, Paris ... 071681, 105100
Moatti, Emmanuel, Paris 105101
Moaven Al- Molk Museum,
Kermanshah 024710
Moba, Bruxelles 100643
Mobach Keramiek Museum,
Utrecht 032947
Mobbs, Annette, Ashford 091241
Mobèche, Pablo, La Tine 087849
Moberg, Kristinehamn 086792
Moberg, Bert, Borås 143822
Mobilart C & R, Montréal 128183
Mobile Books, Lomma 143883
Mobile Home, London 111474
Mobile Medical Museum, Mobile 050977
Mobile Museum of Art, Mobile . 050978
Mobile Städtische Galerie im Museum
Folkwang, Essen 018305
Mobiles Kindermuseum im

Freizeitheim Vahrenwald,
Hannover 019058
Mobili & Oggetti della Nonna,
Roma 081460
Mobilia-Auto- ja Tiemuseo,
Kangasala 010955
Mobilia International, Verona ... 081926
Mobilier des Vallées de l'Himalaya,
Aussois 066830
Mobilus, Strasbourg 129271
Mobius, Boston 119898
Mobius Gallery, Boston 047121
Mobles Monthy, Sabadell 115399
Mobydick Antiquités, Saint-Malo (Ille-
et-Vilaine) 072920
Mocenni, M., Den Haag 083105
Mockbeggar Plantation,
Bonavista 005439
Mockler, Stephen, Berchtesgaden 129455
Mocko, H., Esslingen 075711
Mocny, L., Lindlar 107880
Mocza, Renate, Wien 127954
Moda, Buffalo 092547
Moda, San Francisco 097287
Modalen Skulemuseum, Modalen 033774
ModAnOn.de – Modernes Antiquariat
Online, Elze 137176, 141739
Mode Moderne, Philadelphia ... 096140
Mode Museum Feigel, Ehingen . 018105
Model House Gallery, Llantrisant 044893
Modelbouwmuseum, Wassenaar 033026
Modell Äldre, Stockholm 087030
Modellbahnmuseum-Muggendorf,
Wiesenttal 022646
Modello, A., Paris 105102
Modellpark Mönichkirchen,
Mönichkirchen 002407
Modell's Restoration, New
Orleans 135863
Modemuseum im Schloss
Ludwigsburg, Landesmuseum
Württemberg, Ludwigsburg .. 020128
Modemuseum Provincie Antwerpen,
Antwerpen 003293
Modemuseum Schloss Meyenburg,
Meyenburg 020397
Moder Nena, Santiago de Chile . 064977
Modern & Antique Clock Repair,
Dallas 135510
Modern Art, Argelès-sur-Mer .. 103270
Modern Art, Chicago 092751
Modern Art, Mechelen 100789
Modern Art, Tokyo 111475
Modern Art and Showbusiness
Academy, Tbilisi 055436
Modern Art Gallery, Karlsruhe . 107520
Modern Art Gallery, Los Angeles 121559
Modern Art Gallery, Singapore . 114573
Modern Art Gallery, Taichung . 116896
Modern Art Gallery, Trondheim . 113291
Modern Art Museum of Fort Worth,
Fort Worth 048906
Modern Art Museum of Yerevan,
Yerevan 000739
Modern Art Oxford, Oxford 045453
Modern Art Oxford Bookshop,
Oxford 144770
Modern Art Products, Hong Kong 102096
Modern Art Research Institute, Academy
of Arts of the Ukraine, Kyïv ... 056492,
138134
Modern Artifacts, San Francisco 097288
Modern Artists Gallery, Whitchurch-on-
Thames 119339
Modern at Sundance Square, Fort
Worth 120879
Modern British Artists, London . 118461
Modern Day, Miami 135801
Modern Fuel Artist – Run Centre,
Kingston 006107
Modern Gallery, Hyderabad ... 109367
Modern House, Cleveland 092954,
120442
Modern Hut, Houston 093746

Modern Magyar Képtár I, Janus
Pannonius Múzeum, Pécs 023846
Modern Magyar Képtár II, Janus
Pannonius Múzeum, Pécs 023847
Modern Masters Gallery, San
Francisco 124347
Modern Mural Arts Institute,
Atami 111240
Modern Painters, London 139204
Modern Painting & Framing,
Ottawa 101476
Modern Primitive Gallery, Atlanta 119593
Modern Sculpture.com, San
Francisco 124348
Modern Studio, Bruxelles 063509
Modern Time Antique, Vancouver 064848
Modern Times, Chicago 092752
Modern Woodworking, Louisville 135762
Moderna Galerija, Ljubljana 038549
Moderna Galerija, Podgorica .. 031765
Moderna Galerija, Valjevo 038253
Moderna Galerija, Zagreb 009042
Moderna Möbelklassiker, Malmö 086856
Moderna Museet, Stockholm .. 041021
Moderna Museet Butik,
Stockholm 138015, 143957
Modernaire, Oakland 095902
Modernariato, Milano 110386
Moderne, Amsterdam 112257
Moderne Antiques, Los Angeles 094412
Moderne Galerie im Weinstadtmuseum
Krems, Krems 002235
Moderne Gallery, Philadelphia . 096141,
123423, 145416
Modernism, Saint-Ouen 073156
Modernism, San Francisco ... 124349
Modernism Gallery, Miami 094773
Modernism, A Century of Art and
Design, New York 098476
Il Modernista, Bologna 130980
Modernista, Praha 065510
Modernity, New York 122889
Modernpast, San Francisco ... 097289
Modernt och Antikt, Järfälla ... 086755
Modes & Dugardyn, Bruxelles . 063510
Modestie, Deauville 068283
Modified Art Gallery, Phoenix .. 052119
Modique Antiques, Vancouver . 064849
Modjopahit, Amsterdam 143051
Modoc County Historical Museum,
Alturas 046427
Modschiedler, Hamburg 076282
Modulo, Lisboa 114042
Modulo, Porto 114099
Modulverlag, Wien 136600
Modus, Paris 105103
Modus, Rotterdam 112652
Modus Vivendi, Baden-Baden . 074639
Modus Vivendi, Carentan 067682
Modus Vivendi Gallery, Lemesos 102550
Moe Historical Museum, Moe .. 001301
Moebel Angèle, Bartenheim .. 066956
Möbelberget Antik Storvik,
Storvik 087081
Möbelflicker, Baden bei Wien .. 127715
Möbelgruppen, Hägersten 086727
Der Möbelhof, Lehrberg 077046
Möbelkällaren, Enskede 086662
Møbelklassikeren Randers,
Randers 066056
Møbelverkstedet Restaurering,
Oslo 132788
Moebli, Nürnberg 077832
Möcking, Ingo, Bielefeld 129589
Mödlinger Künstlerbund, Wien . 058289
Mödlinger Stadtverkehrsmuseum,
Mödling 002401
Möhkön Ruukki, Ilomantsin
Museosäätiö, Möhkö 011164
Möhring, Dortmund 075389
Møllenberg, Trondheim 084387
Möllenhoff, Irmgard, Porta
Westfalica 078067
Møller, Lemvig 102927

Möller, Ostercappeln 077987
Möller, Waibstadt 078830
Møller Andersen, Aksel & Eva,
Herning 065752
Moeller Fine Art Ltd., New York 122890,
138409
Møller Witt, Århus 102778
Moeller, Dr. Martin, Hamburg .. 076284
Möller, Frank-C., Hamburg 076283
Møller, Hanne, Svaneke 066152
Möller, Herbert, Hagen 076151
Möller, J., Dortmund .. 075390, 106707
Möller, Lothar, Neufra 108242
Möller, Rico, Michendorf 130305
Möller, Robert & Co., Kiel 076783
Möller, Sibylle Karola, Bad Soden am
Taunus 106094
Møller, Susanne, Haderslev ... 065720
Möller, Thomas, Berlin 129535
Möller, Ulrike-S., Rostock 108520
Möllering, Beate, München 108165
Möllers, Dr. Doris, Münster 108222
Möllers, K., Birkenau 105066
Möllner Museum Historisches Rathaus,
Mölln 020434
Mölndals Antik, Mölndal 086870
Mölndals museum, Mölndal 040849
Mølsteds Museum, Dragør 010033
Moelv Auksjonsforretning, Moelv 126358
Mönchehaus Museum für Moderne
Kunst Goslar, Goslar 018769
Mönchehaus Museum Goslar, Verein zur
Förderung moderner Kunst e.V. Goslar,
Goslar 059176, 137227
Mönchguter Museen im Ostseebad
Göhren, Göhren, Ostseebad ... 018725
Mönchhof Galerie, Kilchberg
(Zürich) 116384
Mönikes jun., F., Steinheim,
Westfalen 078554
Möniste Muuseum, Möniste ... 010639
Møns Museum, Stege 010328
Mønsted Antik og Ting, Viborg . 066209
Mørch Jensen, Palle,
Frederiksberg 065670
Moerchen, Anne, Hamburg 107299
Møre Auksjonsforretning, Tornes I
Romsdal 126393
Mörfelden Museum, Mörfelden-
Walldorf 020443
Mörgeli Bilderrahmen, Zürich . 087991,
116837,   133994
Mörike Museum im alten Schulhaus,
Neuenstadt am Kocher 020702
Mörike Zimmer Wermutshausen,
Niederstetten 020802
Mörikehaus Ochsenwang, Bissingen
an der Teck 017401
Mörmann, Elmar, Rothenburg ob der
Tauber 078287
Mörner, Axel, Stockholm 115911
Mörsbacher Museum, Mörsbach 020446
Mörth, Wien 140105
Moervarststede, Wachtebeke . 004123
Moes, Anny, Bar-le-Duc 066940
Kuno Mösch & Marianne Baumann,
Balgach 116051
Moesgaard, Faaborg 065645
Moesgaard, Jytte Elise, Kampen 076673,
076674
Moesgård Museum, Højbjerg .. 010127
Mösl, Kristina, Berlin 129536
Möwe Bay Museum, Möwe Bay 031845
Moffarts, Benoît de, Bruxelles .. 063511
Moffat Museum, Moffat 045281
Moffatt-Ladd House, Portsmouth 052313
Mogadishni, Valby 102995
Mogador, Aubagne 103281
Mogador, Bruxelles 063512
Mogens & Angelo, København .. 065937
Moggia, Roberta, Genova 131328
Moghul, Adelaide 060970
Mogliani, Lamanon 103916
Mogliasso Antichità, Torino 081734

Mohan, Kathmandu 112080
Mohapp, Emmerich, Graz 127749
Mohave Museum of History and Arts,
Kingman 049976
Mohawk Arms, Bouckville 127147
Mohawk Trail School Museum,
Hamilton 005976
Moheda Skolmuseum, Moheda . 040851
Mohler, Basel 087249, 144026
Mohler, Philipp, Liestal 116442, 133808
Mohn, Johan Albert, Bergen 084221
Mohnmuseum, Ottenschlag 002485
Moholdt, Hønefoss ... 084254, 126352
Mohomed, Carimo, Lisboa 143584
Mohonk Preserve, Gardiner 049048
Mohr, Helsingør 065743
Bernhard Mohr & Gabriele Kübler-
Mohr, Herdwangen-Schönach . 107417
Mohr & Suckel, Lindau 077119
Mohr, Arno, Offenbach an der
Queich 077905
Mohr, Heinz, Wiesbaden 142490
Mohr, I., Hirschberg an der
Bergstraße 076541
Mohr, I., Schriesheim 078400
Mohr, M. & A., Göppingen 107139
Mohri Hakubutsukan, Hofu 028661
Mohrmann, Edgar, Hamburg ... 125973
Mohtashem, Milano 080617
Mohtashemi, Amir, London 090358
Mohzee, London 090359
Moin, Seoul 111691, 111692
Moinard, Gérard, La Rochelle .. 069090
Moinat, Genève 087484
Moinat, Charles-Emile, Rolle ... 087754
Moinat, H.C., Amsterdam 143052
Moineau, Nadine, Nantes 104341
Moinet, Civray 067992
Moir Street Books, Hawthorn . 139767
Moira, London 090360
Moira Furnace, Moira 045282
Moirandat, Alain François, Basel 144027
Mõisaküla Muuseum, Mõisaküla 010638
Moisan, Martine, Paris 105104
Moisan, Richard, Cannes 013518
Moisan, Richard, Vallauris 105810
Moisant Weyl, Joni, Los Angeles 121560
Moisburger Mühlenmuseum,
Außenstelle des Freilichtmuseums
am Kiekeberg, Moisburg 020449
Moisel, W., Allmersbach im Tal . 074380,
129377
Moissinac, Alain, Aurillac 066828
Moisson, Dominique & Valentin, Doué-la-
Fontaine 068398, 128585
Moissonnier, Laurent, Clessé ... 068039
Moissonnier, Laurent, Pommiers 072229
Al-Moiz, Sharjah 088203
Mojahid, Casablanca .. 082571, 082572
Moji-ko Orugorukan, Kitakyushu 028829
Moji-kuritsu Idemitsu Bijutsukan,
Kitakyushu 028830
Mojraghi, Mauro & Music, Steven,
Parma 110631
Mok Am Museum of Art, Koyang 029967
Mok, Cheung, Hong Kong 065084
Moka Buddhist Museum,
Gangcheon 029923
Mokichi Saito Museum,
Kaminoyama 028765
Mokka-Antiquitäten, Mettmann . 077367
Mokkumto, Seoul 111693
Mokrzycki, Stanisław, Gdańsk . 132817
Mokum, Amsterdam 112258
Mol, Conny, Amsterdam 082809
Mola- het Provinciaal Molenmuseum,
Wachtebeke 004124
Mola e Serradora de Cal Pal,
Ordino 000099
Molard, Aline, Brégnier-Cordon . 067456
Molarum, Stockholm 087031
Molayem, Albert, Roma 081461
Mold Museum, Mold 045283
Moldai Studios, Keighley 118033

Moldenhauer, A., Prerow,
Ostseebad 108412
Molding Art, Pittsburgh 096313
Molducci, San Lazzaro di Savena 110896
Moldurarte, Zaragoza . 115556, 115557
Molduras & Pardini, Belo
Horizonte 100896
Molduseg, Bogotá 102347, 128273
Mole Hall Antiques, Aldeburgh . 088241
Mole Hole, Birmingham 092423
Moleiro, Barcelona 137966
Molen-, Bakkerij- en Stoommuseum,
Essen 003585
Molen De Vriendschap, Weesp . 033034
Molen van Frans Expositie, Mander
(Tubbergen) 032634
Molen van Sloten, Amsterdam . 031957
Molenaars, Breda 112356
Molènes, Quentin, Genève 144079
Molenmuseum, Koog aan de
Zaan 032552
Molenmuseum, Lommel 003825
Molenmuseum de Assumburg, Nieuw
Vossemeer 032671
Molenmuseum De Valk, Leiden . 032580
Molenmuseum De Wachter,
Zuidlaren 033112
Molenmuseum, Centrum voor
Molinologie, Sint-Amands 004013
Molerhiisli, Hausach 019108
Molermuseet, Nykøbing Mors .. 010247
Molesworth Gallery, Dublin 119606
Moléty Krašto Muziejus, Molėtai 030423
Molevitten, Skive 066121
Molhant, V., Herent 128062
Moli d'en Gaspar, Lluchmajor .. 039592
Molica, Giuseppe, Roma 110827
Molin, Inger, Stockholm 115912
Molina Art, Valladolid . 115526, 115527
Molina, Gabriel, Madrid 143777
Molina, Juan, Arévalo 133085
Molina, Pedro, Bogotá 065215
Molina, Modena 080809
Molinari, Cristina, Genova 131329
Molinari, Prof. Paola, Parma ... 131731
Molinaro, Livia, Versailles 074062
Moline, Florian, Nîmes 070700
Molínek, Jan, Rychnov nad
Kněžnou 065554, 140539
Molkerei-Museum, Bernbeuren . 017331
Moll, Lima 113365
Moll, Evert, Amsterdam 132411
Molland, Leek 089784
Mollbrink, Linköping 086816
Mollbrink, Uppsala 115973
Molle, Gilbert, Lyon 069764
De Mollegangen, Bergen op
Zoom 082963
Mollenhoff, Rolf, Aubenas 066780
Mollet, Grigny 068745
Mollnar, Wien 063034
Molloy, Dallas 135511
Molloy, Southport 091380
Molly Brown House Museum,
Denver 048268
Molmar, Pamplona 115389
Molnar, Bergen op Zoom 082964
Molnár C. Pál Műterem-múzeum,
Budapest 023527
Molnar, Attila, Stuttgart 078620
Molnár, László, Eger 079305
Molong Historical Museum,
Molong 001302
Mols Galleriet, Ebeltoft 102790
Moltke-Gedächtnisstätte, Museum der
Stadt Parchim, Parchim 021074
Moltkei Werkstatt, Köln 107664
Moluks Historisch Museum,
Utrecht 032948
Molwitz-Stube, Mitterteich 020425
Mom & Pop Antiques,
Washington 097851
Le Moma, Casablanca 082573
MOMA Wales, Machynlleth 045189

MoMA, Museum of Modern Art, New
York 051427
MoMA, Museum of Modern Art,
Publications Department, New
York 138410
Momenta Art, Brooklyn 119942
Momentoes, Chennai 109289
Momentous Arts, Singapore .. 114574
Mompart, Michel, Donzenac .. 068385
Mompesson House, Salisbury .. 045725
Mom's Ole Stuff, Kansas City . 094084
Momu Galleries, Houston 121096
Momus, Atlanta 119594
Mon D'Art, Galați 114169
Mon Grenier, Alès 066530
Mon-Jam Arts, Deal 117671
Mon State Museum,
Mawlamyine 031820
Mon Style à Moi, Bruxelles ... 063513
Mona and Don's Antiques, Austin 092252,
135329
Mona Incha Workshops, Roscrea 079694
Mona Lisa, Chamonix-Mont-
Blanc 103562
Mona Lisa, Viersen 108821
Monacensia, Literaturarchiv und
Bibliothek, München 020548
Monaco Modern Art, Monaco .. 112011
Monaco Modern Art Museum,
Monaco 031716
Monaghan County Museum,
Monaghan 024968
El Monaguillo, Pamplona 086393
Monahan, Thomas, Chicago .. 120109
Monarch Antiques, Saint Leonards-on-
Sea 091199, 134992
Monarch Books, Oxford 138274
Monarch Coin Corporation, Salt Lake
City 096909
Monarch Collectibles, Houston . 093747
Monarch Collectibles, San
Antonio 097001
Monarch Gallery, Greenock ... 117887
Monarch Historical Museum,
Williamtown 001645
Monaro, Lydia, Montréal 101435
Monart, Dwellingup 098991
Monart Saint Vital, Winnipeg .. 101886
Monas Antiques, Cork 079501
Monash Gallery of Art, Wheelers
Hill 001638
Monash University Museum of Art,
Clayton 000971
Monastère de Mauriac, Mauriac 014089
Monastère Royal de Brou, Bourg-en-
Bresse 012153
Monasterio, Madrid ... 086218, 086219
Monasterio Cisterciense de Santa
María, Sobrado de los Monjes 040190
Monasterio de la Cartuja, Sevilla 040167
Monasterio de la Encarnación,
Madrid 039626
Monasterio de la Encarnación,
Osuna 039854
Monasterio de las Descalzas Reales,
Madrid 039627
Monasterio de San Juan de los Reyes,
Toledo 040234
Monasterio de San Lorenzo de
El Escorial, San Lorenzo de El
Escorial 040064
Monasterio de San Pedro de Cardeña,
Castrillo del Val 039253
Monasterio de Santa Catalina,
Arequipa 034211
Monasterio de Santa Clara de
Tordesillas, Tordesillas 040250
Monasterio de Santa María de la
Rábida, Palos de la Frontera .. 039907
Monasterio de Santa María San
Salvador, Cañas 039216
Monasterio de San Juan de Duero,
Sección Medieval del Museo
Numantino, Soria 040195

Monasterio Santa María la Real de
  las Huelgas, Museo de Ricas Telas,
  Burgos 039167
Monasterium De Wijngaard,
  Brugge 003417
Monastery of Aghiou Tessarakonta,
  Spárti 023273
Monastery of Evangelistria Daponte,
  Skopelos 023267
Monastery of Leimon Collection,
  Kalloni 023107
Monastery of Petra, Kataphygi . 023122
Monastery of Saint Catherine, Gebel
  Katherîna 010569
Monastery of the Archangel Michael
  at Panormitis, Panormitis 023209
Monastery of the Taxiarchs, Megalo
  Livadi 023163
Monastery of the Zoodochos Pighi,
  Poros 023234
Monberg, Morten, Vedbæk 066186
Monbrison, Alain de, Paris ... 071682
Monbrison, Naïla de, Paris 105105
Moncada, Fernando, Lisboa 084919
Moncada, Francisco Cabral,
  Lisboa 084920
Moncada, Francisco Xavier,
  Lisboa 084921, 084922
Moncada, Isabel Cabral, Lisboa . 084923
Moncada, Valentina, Roma 110828
Monchy, de, Zutphen 112728
Monclus Rodriguez, Enrique,
  Madrid 086220
Moncton Museum, Moncton ... 006350
Moncur Gallery, Boissevain 005435
Mondadori, Verona 137761
Mondadori Electa, Milano 137685
Mondak Heritage Center, Sidney 053331
Mondazzi, Torino 132172
Monde à l'Envers, Chevreuse ... 067972
Le Monde Antiques, Dallas 093180
Le Monde de Belibaste, Épopée
  Cathare, Villerouge-Termenès . 016280
Le Monde de Gonor, Plan-
  d'Orgon 072173
Le Monde du Voyage, Saint-
  Ouen 073157
Mondial Art, Bougival 067378
Mondial Art, Paris 105106
Mondial Art Exchange, Saint-
  Ouen 073158
Mondial International, New York 095592
Mondial Phil, Legball 079991
Il Mondo dell'Arte, Roma 110829,
  110830,   110831
Il Mondo Stijl, Den Haag 083106
Mondriaanhuis – Museum voor
  Constructieve en Concrete Kunst,
  Amersfoort 031911
Monemvasia Archaeological Collection,
  Monemvasia 023177
Monesterio de Santes Creus, Museu
  d'Història de Catalunya, Santes
  Creus 040117
Monestir de Cura, Randa 039993
Monet, Skopje 111887
Monet Exhibition Centre, Shepton
  Mallet 119089
Moneta, Warszawa 084736
Moneta Nova, Bremen 075177
Monetti, Antonio, Napoli 131630
Monetto, Grimaud 103797
Monety i Antyki, Kalisz 084528
Money Museum, Kingston 028447
Money Museum, Tehrän 024771
Money Museum of the UFJ Bank,
  Nagoya 029119
The Money Museum, Federal Reserve
  Bank of Kansas City, Kansas
  City 049887
money trend – Das Münzenmagazin,
  Wien 138528
Monfregola, Emilio & Doris,
  Riehen 116590

Mongers, Hingham 089535
Mongerson, Chicago 120207
Mongin, Robert, Paris . 071683, 105107
Mongolian and Tibetan Cultural Center,
  Taipei 042529
Mongolian Art Museum,
  Ulaanbaatar 031732
Mongolian National Modern Art Gallery,
  Ulaanbaatar 031733
Mongolian Theatre Museum,
  Ulaanbaatar 031734
The Monhegan Museum,
  Monhegan 050989
Monica, Leuven 063734
Monier, Patrice, Nice 128865
Monier, Trevoux 073882
Monin, Villeurbanne 074218
Monin, Anne Marie, Paris 071684,
                              105108
Monin, Jacques, Lyon 069765
Monini, Massimo, Toulouse ... 073785
Monis alte Stube, Südlohn ... 078643
Möniste Muuseum, Kuutsi 010632
Monitor, Roma 110832
Monjuan Bijutsu-ten, Nara 082071
Monkdogz Urban Art, New York 122891
Monks' Dormitory Durham Cathedral,
  Durham 044189
Monkwearmouth Station Museum,
  Sunderland 045920
Monleon, Lenore, New York ... 095593
Monluc, Gérard, Paris 071685
Monmouth County Historical Museum,
  Freehold 048967
Monmouth Museum, Lincroft .. 050311
Monna Lisa, Vechta 108814
La Monna Lisa Galería de Arte,
  Cali 102414
Monnaie d'Or, Paris 071686
Les Monnaies de Lyon, Paris .. 071687
Les Monnaies de Notre-Dame,
  Lyon 069766
Monnaies Médailles 17, La
  Rochelle 069091
Monnais, Montréal 064445
Monnet, François, Clermont-
  Ferrand 128563
Monney, Daniel, Fribourg 144069
Monnier-Debroux, Jacqueline,
  Quingey 072355
Monnier, Bernard, La Valette-du-
  Var 069112
Monnier, Corinne le, Le Havre .. 103949
Monnier, Frédéric, Busy 067538
Le Monnier, Philippe, Strasbourg 073592,
                              073593
Monnier, Raymond, Lyon 069767
Monnier, Virginie, Besançon .. 067133
Monnier, Wilhelmina, Le Pré-Saint-
  Gervais 069339
Monnin, Patrick, Chambéry ... 067796
Monnot, Nolay 070742
Monocacy National Battlefield,
  Frederick 048946
Monogramma, Roma 110833
Monogrammit, Houston 093748
Monopol, Magazin für Kunst und
  Leben, Berlin 138735
Monopteros Fine Art, Chesham . 117568
Monostori Erőd, Hadtörténeti Múzeum
  Budapest, Komárom 023758
Monotype Guild of New England,
  Sharon 060598
Monreal & Poppen, Bonn 106534
Monro, London 090361, 134730
Monroe, Tampa 136392
Monroe County Heritage Museum,
  Monroeville 051004
Monroe County Historical Museum,
  Monroe 051000
Monroe County Historical Society,
  Bloomington 047048
Monroe County Local History Room
  and Library, Sparta 053427

Monroe Historical Society Museum,
  Monroe 050995
Mons Breidvik Galleries,
  Norheimsund 033809
Mons-Montmartre, Mons 128083
Mons, Sandrine, Nice 104379
Monsalve, Madrid 133360
Monsantic.com, Mons 125350
Monseau, Claude, Bordeaux .. 067348
Monsenhor, Braga 084781, 132923
Monserrato Arte, Roma 081462
Monsieurs Restaurateur M.C.E.,
  Ees 132497
Monsignor Hawes Priest House
  Museum, Mullewa 001323
Monsignor Yatco Ecclesiastical
  Museum, Batangas 034436
Monsoon, Düsseldorf 075499
Monsoon Traders, Woodside .. 062454
Monsoonasia, Singapore 114575
Mont De Lancey Historical Museum,
  Wandin 001617
Mont-Martre, Eindhoven 083223
Mont Orgueil Castle, Saint Martin 045704
Mont Vert, Apeldoorn 082925
Montador, Dieppe 103666
Montafoner Bergbaumuseum Silbertal,
  Silbertal 002757
Montafoner Heimatmuseum Schruns,
  Schruns 002731
Montafoner Tourismusmuseum
  Gaschurn, Gaschurn 001929
Montage, Castleton 117545
Montage, Nelson 118775
Montage, Wellington 084175
Montage Gallery, Baltimore ... 119759
Montagnac, Marion, Arpajon .. 066734
Montagnon, Jeanney Sylvie,
  Montbéliard 104252
Montagu Museum, Montagu .. 038785
Montague Museum, Montague . 051005
Montaigne, Chantilly 067832
Montaigne, Yves, Lamorlaye .. 069154
Montalto, Attilio, Catania 080040
Montalto, Elena, Venezia 081861
Montalvo Center for the Arts,
  Saratoga 053153
Montana Auto Museum, Deer
  Lodge 048215
Montana Berlin, Berlin 106376
Montana Historical Society,
  Helena 049433
Montana Museum of Art and Culture,
  University of Montana, Missoula 050961
Montanari, Aurora, Modena ... 080810
Montanari, Roberto, Vicenza .. 111205
Montanaro-Visconteum, C & A,
  Roma 081463
Montanini, Gian Paolo, Parma .. 142782
Montanmuseum, Gußwerk 002026
Montanmuseum, Hieflau 002082
Montansier, Versailles 074063
Montaperto, Angelo L., New York 135963
Montariol, Jacqueline, Condom . 068129
Montarras, Jean Francois, Raon-
  l'Etape 072367
Montauban Numismatique,
  Montauban 070197
Montauk, Clermont 047801
Montauk Point Lighthouse Museum,
  Montauk 051006
Montaut, Bordeaux 103438, 128498
Montaut, Saint-Sulpice-et-
  Cameyrac 137007
Montauti, Siegfried, Nice 104380,
                              128866
Montavoci, Marco, Milano 131479
Montázs, Miskolc 109202
Montcada, Barcelona 114939
Montcada Taller, Barcelona ... 114940
Montcades Dos, Palma de
  Mallorca 086360
Montchaud, André, Provins ... 072319

Montclair, Oakland ... 095903, 127375
Montclair Art Museum, Montclair 051007
Montclair Art Museum. Bulletin-
  Newsletter, Montclair 139437
Montclair Art School, Montclair . 057473
Montclair Historical Society Museum,
  Montclair 051008
Montclair State University Art Galleries,
  Upper Montclair 053905
Monte-Carlo Antiquités, Monaco 082522
Monte-Carlo Antiquités, Salon
  International des Antiquaires,
  Monaco 098250
Monte-Carlo International Fine Art and
  Antiques Fair, Monaco 098251
Monte Clark, Vancouver 101813
Monte Clark Gallery, Toronto .. 101666
Monte della Farina, Roma 142827
Monte L. Bean Life Science Museum,
  Provo 052404
Monte Titano Arte, San Marino . 114384
Monte Wade, San Antonio ... 124051
Monteblanco Galería, Medellín . 102453
Montecatini, Concepción 007518
Monteiro & Fernandes, Sesimbra 114110
Montemaggi, Linda K., New York 145357
Montemayor, Barcelona 085803
Montemayor Fortea, Julian,
  Barcelona 085804, 114941
Montenegro Antiguedades,
  Madrid 086221
Montepío Antigüedades, Buenos
  Aires 060916
Monter, Robert, Saint-Amand-
  Montrond 072665
Monterey Art Gallery, South
  Auckland 113007
Monterey History and Art Association,
  Monterey 051012
Monterey Museum of Art,
  Monterey 051013
Monterey Museum of Art at La Mirada,
  Monterey 051014
Monterey Museum of Art News,
  Monterey 139438
Montero Gomez, Juan, Las Palmas de
  Gran Canaria 115380
Montes, Luisa, Donostia-San
  Sebastián 085981
Montesi, Grazia, Milano 080618
Montet, Champagne-au-Mont
  d'Or 067806
Monteverde & Cia., Montevideo 144949
Monteverdi Car Collection, Schweizer
  Automuseum, Binningen 041343
Montezuma, Erftstadt 106891
Montezuma Galerie, Köln 107665
Montfort-Museum, Tettnang ... 022135
Montgomery, San Francisco ... 124350
Montgomery County Historical Society
  Center, Dayton 048172
Montgomery County Historical Society
  Museum, Rockville 052653
Montgomery House Museum,
  Frankville 005835
Montgomery Museum and Lewis
  Miller Regional Art Center,
  Christiansburg 047731
Montgomery Museum of Fine Arts,
  Montgomery 051025
Montgomery Street Antique Mall, Fort
  Worth 093463
Montgomery, W.H.C., Greyabbey 117889
Montheillet, Pierre, Lyon 104121
Monthly Art Magazine, Tokyo .. 138892
Monti, Isabella, Salzburg 100172
Monti, Miriam, Bologna 079855
Monti, Piero, Roma ... 131982, 131983
Monti, Roberto, Modena 110502
Montialoux, Jacques, Saint-
  Pargoire 073238
Montibeller, Wien 127955
Monticello, Norfolk ... 095867, 136034
Monticello, Portland 096442

Monticello Gallery, Dallas 120631
Monticello Railway Museum,
  Monticello 051030
Monticello, Home of Thomas Jefferson,
  Charlottesville 047583
Monticone Gioielli, Torino 081735
Montleau, Gérald de, La
  Rochefoucauld 069076
Montleau, Gérald de, Paris 105109
Montmartre, Den Haag 083107
Montmartre, Parma 110632
Montmeat, Renaud, Paris 071688
Montoli, Claudio, Verona 132292
Montoriol, Esther, Barcelona ... 114942
Montoya Antiguedades, Madrid . 086222
Montoya Jimenez, Enrique,
  Madrid 086223
Montparnasse, Casalecchio di
  Reno 109935
Montpelier Cultur Arts Center,
  Laurel 050178
Montpellier Clocks, Cheltenham 088838,
                                134240
Montpellier Mansion, Laurel ... 050179
Montpellier Mews, Harrogate .. 089436
Montpellier Numismatique,
  Montpellier 070309
Montrasio Arte, Milano 110387
Montre, Alicante 085699
Montreal Science Centre,
  Montréal 006383
Montrose, Albury 060984
Montrose Air Station Museum,
  Montrose 045292
Montrose County Historical Museum,
  Montrose 051040
Montrose Museum and Art Gallery,
  Montrose 045293
Montserrat, New York . 122892, 122893
Montserrat College of Art,
  Beverly 056736
Montserrat, Jean-Pierre, Limoux 069537
Montshire Museum of Science,
  Norwich 051654
Montville Antiques, Montville ... 061874
Montville Township Historical Museum,
  Montville 051043
Monty's, Fortitude Valley 061483
Montys, Pantin 104439
Montys, Paris 105110
Monty's Moorabbin Auctions,
  Moorabbin 125215
Monum, Paris 136967
Monumen Nasional, Jakarta ... 024565
Monument Lefebvre,
  Memramcook 006292
Monumental Brass Society, Stratford
  Saint Mary 060181
Monumental Sculptures,
  Washington 124937
Monumenti Musei e Gallerie Pontificie,
  Città del Vaticano 054682
Monumento a Cristo Rey,
  Belalcázar 008580
Monveneur, Francis, Saint-Georges-de-
  Reneins 072798
Monvoisin, Lille 128683
Monzani, Massimiliano, Modena 080811
Monzon, Uria, Ciboure 103594
Mood Indigo, New York 095594
Moods, Mechelen 100790
Moody & Assoc., Birmingham .. 119808
Moody, Houston 121097
Moody County Museum,
  Flandreau 048743
The Moody Mansion Museum,
  Galveston 049037
Moog jr., Henry B., Atlanta 092159
Mooi Per Stuk, Amsterdam 082810
Mooij, Utrecht 083655
Mook, Seoul 111694
Sree Moolam Shastyabdapurti
  Memorial Institute,
  Thiruvananthapuram 024493

Moon, Seoul 111695
Moon & Star Antiques, Cleveland 092955
Moon Gate, Singapore 085472
Moon River Fine Art, Taipei 116934
Moon-Stone Gallery, Paris 071689
Moon, Michael, Whitehaven ... 144920
Moonah Arts Centre, Moonah .. 001304
Moonee Ponds Antiques, Moonee
  Ponds 061878
Mooney, J.R., San Antonio 124052
Mooney, Riro D., Duxford 089123,
                                134345
Moongate Oriental Antiques and
  Gallery, Armadale, Victoria ... 061041
Moonlight Antiques, Miami 094774
Moonraker Gallery, Fowey 117810
Moonta Antiques, Moonta 061880
Moonta Mines Museum, Moonta 001305
Moonta National Trust Museum,
  Moonta 001306
Moor- und Fehnmuseum, Barßel 016967
Moor Antiques, Ashburton 088309
Moor-Galerie, Friedhelm, Hille .. 107433
Moor Hall Antiques, Chester ... 088866,
                                134253
Moor Street Art Gallery, Fitzroy . 099028
Moor und Bauer-Museum,
  Wardenburg 022427
Moor, Alain de, Bischoffsheim .. 067230
Moor, Christian, Vichy 074111
Moorabbin Air Museum,
  Cheltenham 000957
Moorabool Antique Galleries,
  Geelong 061511
Moore, Birmingham 119809
Moore College of Art and Design,
  Philadelphia 057673
Moore County Historical Museum,
  Dumas 048389
Moore Gallery Ltd, Toronto 101667
Moore Home, Lerna 050221
Moore Museum, Mooretown ... 006418
Moore, Allen & Innocent,
  Cirencester 126861
Moore, Carl, Houston 093749
Moore, D.C., New York 122894
Moore, E., Tulsa 136432
Moore, Elizabeth, New York 122895
Moore, Eric T., Hitchin 144478
Moore, Jack, Trawden 091627, 135146
Moore, Linda, San Diego 124146
Moore, Marcus, Stanton upon Hine
  Heath 091416, 135076
Moore, Mark, Santa Monica ... 124471
Moore, Sigrid, Stuttgart 078621
Moore, T., Ottawa 064510
Moore, Widney, Portland 123645
Mooreichensammlung Johann Weber,
  Göppingen 018732
Mooren, Francine, Liège 063774
Moores Creek National Battlefield,
  Currie 048097
Moores & Son, Walter, Market
  Harborough 090713
Moorfoot Contemporary Art,
  Amisfield 117239
Moorhead, Patrick, Brighton ... 088627
Moorland, Uppermill 144900
Moorman, Memphis .. 094656, 135785
Moormuseum Moordorf,
  Südbrookmerland 022073
Mooroopna Hospital and District
  Museum, Mooroopna 001308
Moorside Books, Clitheroe 144345
Moorthamers, Paris 141245
Moosavi, New York 095595
Moose Factory Centennial Museum,
  Kirkland Lake 006119
Moose Jaw Museum and Art Gallery,
  Moose Jaw 006421
Moose River Gold Mines Museum,
  Middle Musquodoboit 006301
Moose, Jeffrey, Seattle 124572
Moosehorn Heritage Museum,

Moosehorn 006423
Moosim, Chongju 111579
Moot Hall Museum, Aldeburgh . 043381
Mope, Sevilla 086459, 133436
Mopsi, Den Haag 083108
Mor, Giovanna & C., Brescia ... 079948
Mora & Upham, London 090362, 134731
Móra Ferenc Emlékház,
  Kiskunfélegyháza 023745
Móra Ferenc Múzeum, Szeged . 023928
Mora Rodriguez, Lluis, Barcelona 085805
Mora, Marie, Salzburg 140027
Mora, Migliore, Palermo 081060
Mora, William, Richmond,
  Victoria 099633
Morabito, Catania 109952
A Morada, Lisboa 084924
Moraes, Eurema Fatima Gallo, Porto
  Alegre 100991
Moraes, Julio, São Paulo 128127
Moraes, Luciano Diniz de,
  Brasília 100919
Moraes, Oliveira & Oliveira,
  Brasília 100920
Morain, Gilbert, Callian 067605
Moran, Miami 135802
Moran, San Antonio 097002
Moran Open-air Museum,
  Namyangju 029976
Moran, Joe, Regina 101533
Moran, John, Altadena 127119
Moran, John, Pasadena 127385
Morand, Chantal, Genève 087485
Morand, Corinne, Lalinde 069147
Morand, Richard & Ludovic,
  Paris 125741
Morandell, A. & T., Bolzano 109886
Morandi, Enrico, Firenze 131219
Morandi, Rita, Porto Alegre 100992
Morange Candas, Arras 066739
Morans Furniture, Goresbridge . 079632
Morant, Richard, London 090363,
                                134732
Morantes, Bucaramanga 065236
Morard, Claude, Nyon . 087714, 116542
Morassutti Amistani, Padova ... 131670
Morat-Institut für Kunst und
  Kunstwissenschaft, Freiburg im
  Breisgau 018482
Morate, Cuenca 085960
Morate Tornero, Manuel,
  Valladolid ... 086549, 086550, 133476,
                                133477
Morató Esmaltador, Barcelona . 133136
Moravec, Vít, Brno 128320
Moravian Historical Society Museum,
  Nazareth 051204
Moravian Museum, Pudsey 045571
Moravian Museum of Bethlehem,
  Bethlehem 046962
Moravská Galerie Brno –
  Místodržitelský Palác, Brno ... 009473
Moravská Galerie Brno – Pražákův
  Palác, Brno 009474
Moravská Galerie Brno –
  Uměleckoprůmyslové Muzeum,
  Brno 009475
Moravské Zemské Muzeum,
  Brno 009476
Morawa, Wien 100355, 140106, 140107,
                                140108
Morawa Old Police Station Museum,
  Morawa 001309
Moray, Dunedin 112888
Moray Firth Wildlife Centre, Spey
  Bay 045848
Morchard Bishop Antiques, Morchard
  Bishop 090771
Morcos, Patrick, Paris 071690
Mord + Totschlag, Braunschweig 141640
Mordecai Historic Park, Raleigh . 052449
Morden and District Museum,
  Morden 006427
Mordialloc Historical Society Museum,

Mentone 001285
Mordoch, Lélia, Paris 105111
Mordovskij Muzej Izobrazitelnych
  Isskustv im. S.D. Érzi, Saransk 037833
Mordovskij Respublikanskij
  Kraevedčeskij Muzej, Saransk . 037834
More & More Antiques, New
  York 095596
More Gallery, Philadelphia 123424
More Nucci, Milano ... 080619, 131480
More Stars, Portland 096443
Moreau, Saint-Ouen 073159
Moreau, Tours 073854
Moreau-Billy, Danielle, La Baule-
  Escoublac 103870, 103871
Moreau-Guionneau, Ginette,
  Bergerac 067104
Moreau, Annette, Saint-Hilaire-de-
  Riez 072824
Moreau, Aurélie, Lyon 104122
Moreau, B. Freiherr von, Bad
  Tölz 129424
Moreau, Catherine Philomène,
  Paris 071691
Moreau, Jean-Luc, Paris 071692
Moreau, P. C. E., Jemeppe-sur-
  Sambre 136840
Moreaux-Jouannet, Isabelle, Aix-les-
  Bains 128429
Moree Gallery, Moree 099374
Moree Plains Gallery, Moree ... 001310
Morefield, Port Macquarie 099589
Moreira & Galveias, Lisboa 114043,
                                114044
Morel, Viriat 074251
Morel de Westgaver, Alain & Evelyne,
  Bruxelles 125321, 140191
Morel, Christophe, Paris 125742
Morel, Dury, Lyon 069768
Morel, Henri, Champigny-en-
  Beauce 067813
Morel, Laurent, Neuchâtel 133872
Morel, Olivier, Paris 129092
Morel, Philippe, Arras 103276
Moreland, Brunswick 125182
Morell, Jim, Cleveland 092956
Morelli, Jean, Orange 070792
Morelli, Sandro, Firenze 080172
Morena-Kalopisi, Chalandri ... 079190
Moreno, San Antonio 136266
Moreno Alaix, Carlos, Bogotá .. 102348
Moreno Herrera, Carmen Rosa,
  Roma 110834
Moreno Sanchez, Antonio Jose,
  Córdoba 085944
Moreno Sanz, Juan Salvador,
  Badalona 085712
Moreno, Antonio – Hijo,
  Salamanca 086411
Moreno, Leticia, Cali 065251
Morera Alvarez, Isabel, Toledo . 115470
Morera Soler, Francisco, Sabadell 086404,
                                133414
Morero, Jean-Claude, Cruis ... 068251
Moreschi, Martin, Ohlsbach ... 077918
Moréteau, Paule, Lyon 128723
Moreton, C.S., Inchture 089638
Moreton's Harbour Museum, Moreton's
  Harbour 006428
Moretti, Alfredo, Prato 081174
Moretti, Attilio, Roma 081464
Moretti, Fabrizio, Firenze 080173
Moretti, Ferdinando, Roma ... 131984,
                                131985
Morettini, Ortensia, Roma 131986
Moretto, Sante, Monticello Conte
  Otto 110510
Moreux, Laurent, Paris 071693
Morfi, Lemesos 102551
Morgama, Bilbao 114995
Morgan, Chicago 120208
Morgan, Kansas City 121307
Morgan, Norfolk 095868
Morgan, Omaha 096052

Morgan, Pittsburgh 123565
Morgan, Stockholm 115913
The Morgan Library, New York . 051428
Morgan Roberts, Jean, Conwy . 117626
Morgan Row Museum,
Harrodsburg 049379
Morgan Street Gallery, Auckland 112793
Morgan-West, New Orleans ... 122207
Morgan, Jeff, Hawker, South
Australia 099112
Morgan, Linda, New York 095597
Morgan, R., Ringwood 091107
Morgana, Cairo 103060
Morgand, Evelyne, Mereville . 070069
Morgans Auctions, Belfast 126816
Morge, Jean-Jacques,
Chevrainvilliers 067969
De Morgenster, Tilburg 083619
Morgenstern, Miami 094775
Morgenstern, Stralsund 142380
Morgenstern, Claude, Saint-
Mandé 072925
Morgenstern, Günter, Reinbek . 078187
Morgondagens Minnen,
Stockholm 143958
Morgulis, Michael, Buffalo 119961
Mori Bijutsukan, Tokyo 029556
Mori Gallery, Sydney 099790
Mori-no-Naka no Chiisana Ongakukan,
Nikko 029181
Mori-Ōgai-Gedenkstätte, Berlin . 017229
Mori, Gianni, Parma 081119
Mori, Gino, Parola 081135
Mori, Giorgio, Parma 131732
Mori, Hanae, Tokyo 082182
Mori, Ryozo, Sapporo 142912
De Moriaan Antiek, Leiden ... 083438
Moriah, New York 095598, 095599
Moriani, Milena, Pisa 110666
Moriarty, Madrid 115261
Moriarty, Salt Lake City 096910
Moriarty Historical Museum,
Moriarty 051058
Móricz Zsigmond Emlékház,
Prügy 023864
Móricz Zsigmond Emlékház, Jósa
András Általános, Tiszacsécse . 024010
Morigi & Figlio, G., Bologna ... 130981
Morija Museum, Morija 030288
The Morikami Museum and Japanese
Gardens, Delray Beach 048232
Morikuma, Saitama 082096
Morillas, Donostia-San Sebastián 133255
Morin, Claude, Saint-Paul-de-
Vence 105595
Morin, Jacques, Saint-Brieuc . 072699
Morin, Jean, Plougonven 072200
Morin, Jean-Yves, Sainte-Pience 073363
Morin, Michel André, London .. 090364
Morin, Thierry, Lyon 069769
Morini, Lucio, Brescia 079949
Morioka City Local Hall, Morioka 029055
Morion, Helsinki 066343
Morisco, Federico, Roma 131987
Morisco, Gianfranco, Roma ... 081465
Morishita Bijutsukan, Hinase ... 028630
Morisot, Patrick, Colombey-les-
Belles 068106
Morita, Tokyo 082183
Morita-mura Rekishi Mingeikan,
Morita 029056
Moritz, Dipl.-Restauratorin Martina,
Bergisch Gladbach 129461
Moritz, E., Goslar 076063
Morlan Gallery, Lexington 050249
Morley College, London 056565
Morley Gallery, London . 045039, 118462
Morley, David, Twickenham ... 091676
Morley, Patrick & Gillian,
Warwick 091746
Morley & Co, Robert, London .. 090365,
134733
Morlock, Barbara, Schenefeld . 108561
Mormon Station Museum, Genoa 049074

Mormon Visitors Center,
Independence 049669
Mormors Antik och Kuriosa,
Ljusdal 086822
Mormors Källare, Vadstena 087132
Mormors Prylar, Norrtälje 086888
Mormors Skattkista, Stockholm . 087032
Mormors Spegel Antik,
Stockholm 087033
Mormorsgrufwans Antik,
Åtvidaberg 086635
Mornet-Tristan, Colette, Oloron-Sainte-
Marie 070783
Morning Glory Antiques, Kansas
City 094085
Morning Mountain Arts, Berry . 061150
Morning Star Antiques, Tucson . 097678
Morning Star Traders, Tucson . 124756
Morningside Antiques,
Albuquerque 092069
Morningside Antiques, Miami .. 094776
Morningside Art, Houston 121098
Morningside Gallery, Edinburgh . 117733
Morningside Nature Center,
Gainesville 049003
Mornington Peninsula Regional Gallery,
Mornington 001313
Moro de Venecia, Valencia ... 086525
Morohashi Kindai Bijutsukan,
Kitashiobara 028833
Moromi Mingeikan, Okinawa .. 029220
Moroni, Renzo, Roma 081466
Moroni, Umberto, Roma 081467
Morosini, Padova 080992
Morpeth Art Studio, Morpeth ... 099380
Morpeth Chantry Bagpipe Museum,
Morpeth 045297
Morpeth Gallery, Morpeth 099381
Morpeth Trading Post, Morpeth . 061890
Morph, Denver 120748
Morphets, Harrogate 126918
Morphett's Enginehouse Museum,
Burra 000916
Morpho Gallery, Chicago 092753, 120209
Morpurgo, J.M., Amsterdam ... 082811
Morra Arte Studio, Napoli 110538
Morrab, Truro 119262
Morrab Studio, Penzance 118897
Morrin, Dublin 079573
Morrin Centre, Literary and Historical
Society of Quebec, Québec ... 006732
Morrin Museum, Morrinsville ... 033247
Morring, Robert, Atlanta 092160
Morrinsville Antiques,
Morrinsville 083997
Morris, Södertälje 143921
Morris & Bricknell, Ross-on-Wye 127022
Morris & Shirley, Southampton . 091372,
127049, 135064
Morris and District Centennial
Museum, Morris 006430
Morris and Helen Belkin Art Gallery,
University of British Columbia,
Vancouver 007289
Morris Belknap Gallery, Dario
Covi Gallery and SAL Gallery,
Louisville 050498
Morris-Butler House Museum,
Indianapolis 049695
Morris Graves Museum of Art,
Eureka 048627
Morris-Jumel Mansion, New York 051429
Morris Marshall & Poole,
Llanidloes 126952
Morris Marshall & Poole,
Montgomery 126991
The Morris Museum, Morristown 051065
Morris Museum of Art, Augusta 046663
Morris, Chris, Sassafras 062193
Morris, Darlene, Tucson 097679
Morris, Donald, New York 122896
Morris, Frances, Enniskillen ... 117757
Morris, M.A.J., Burton-upon-

Trent 088734, 144291
Morris, Peter, Bromley 088685
Morris, Robert Lee, New York .. 122897
Morrison, Chicago 120210
Morrison & West, Lyttelton 083980
Morrison Museum of the Country
School, Islay 006046
Morrison, David, Portland 145457
Morrison, David J., Christchurch 132710
Morrison, Guy, London 090366
Morrison, Jan, Bristol 088662
Morrison & Son, Robert, York .. 091979
Morrison, Thomas, Dallas 093181,
135512
Morristown National Historical Park,
Morristown 051066
Morro Bay State Park Museum of
Natural History, Morro Bay ... 051072
Morrow Mountain State Park Museum,
Albemarle 046351
Mors Doré, Paris 129093
Morsak Verlag, Grafenau, Kreis
Freyung-Grafenau 137233
Morse Museum, Morse 006432
Morse Road Antique Center,
Columbus 093036
Morseburg, West Hollywood .. 125014
Morsink, Jan, Amsterdam 082812
Morsink, Saskia, Enschede ... 083255
Morskoj Muzej-Akvarium,
Tichookeanskij naučno-
issledovatelskij centr rybolovstva,
Vladivostok 038036
Morskoj Muzej, Karelskij morskoj
istoriko-kulturnyj centr i
morskoj klub Poljarnyj Odissej,
Petrozavodsk 037654
Morslands Historiske Museum,
Nykøbing Mors 010248
Morsø Kunsthandel, Thisted .. 066164
Morten, Eric J., Manchester .. 144726
Mortensen & Markman,
København 102907
Mortensen, Anna, Toreby 102990
Mortgage, Portland 123646
Morthland, Sarah, New York ... 122898
Mortier-Valat, Joëlle, Paris 105112
Mortimer, Brian, Exeter 089233
Morton, Fort Worth 093464
Morton & Co, Timaru 084110
Morton Grove Historical Museum,
Morton Grove 051073
Morton Homestead, Prospect
Park 052375
Morton House Museum, Benton
Harbor 046924
Morton J. May Foundation Gallery,
Maryville University, Saint Louis 052811
Morton Lee, J., Hayling Island . 089483,
117940
Morton Museum of Cooke County,
Gainesville 049011
Morton Square Antiques,
Nashville 095016
Morton, Betty, London 118463
Morton, John, Dublin 079574
Morton, Louis C., Guadalajara . 082468
Morton, Louis C., México 082481
Morton, Louis C., Monterrey .. 082493
Morueco Rodriguez, Luis, Madrid 086224
Morus-Verlag, Berlin 137086
Moruzzi Numismatica, Roma .. 081468
Morvan, Alain, München 077575
Morven, Timaru 084111
Morven Gallery, Barvas 117294
Morven Historical Museum,
Morven 001315
Morven Park Museum, Leesburg 050214
Morwellham Quay Historic Port &
Copper Mine, Morwellham ... 045298
Mory-Wydeven, Ann M.,
Milwaukee 121937
Mo's Gallery, Tucson 124757
Mosaic (St. Petersburg), Saint

Petersburg 139439
Mosaic Gallery, Amman 029776
Mosaic Planet, Blackheath 061163
Il Mosaico, Udine 111066
Mosaico Gal Arte, Porto Alegre . 100993
Mosaicon, Locarno 133812
Mosaïque, Briançon 103467
Mosaique Gallery, Chicago ... 120211
Mosaïque Gallo-Romaine, Amou 011638
Mosarm, Liège 063775
Moscatelli, Claudio, London ... 134734
Moschakeion, International Museum of
Marine Research, Moschato . 023178
Moscow Art Magazine, Moskva . 139015
Moscow Business Plaza Gallery,
Moskva 114280
Mose, Odense 066041, 102951
Mose Toidze House Museum,
Tbilisi 016415
Mosebach, Rainer, Halle, Saale . 076169
Mosel & Tschechow, Galerie und Verlag,
München 108166, 137416
Mosel-Weinmuseum, Bernkastel-
Kues 017335
Moseley Emporium, Birmingham 088483,
134112
Moser, İstanbul 088131
Moser & Klang, Stockholm ... 087034
Eva Moser-Seiberl & Wilfried Moser,
Bad Aussee 127711
Moser, Adolf, Salzburg 127854
Moser, Andreas, Wien 140109
Moser, Charlotte, Genève 116332
Moser, Michael, Osnabrück ... 077975
Moser, Paul, Mittelbiberach ... 077376
Moser, Roland, Landshut 142045
Moser, Theo, Oberau 127827
Moser, Walter, Wien 063035
Moses, Dallas 120632
Moses Lake Museum and Art Center,
Moses Lake 051078
Moses Myers House, Norfolk .. 051578
Moses Saidian, Tehrän 079425
Moses, Dexter, Croydon Park . 061381
Moses, Manuel, München 130370,
130371
Moses, Nancy, Philadelphia ... 123425
Mosie & Midge, Baltimore 092348
Moskat, Walter, Wolfurt 063124
Moskovskaja Gosudarstvennaja
Kartinnaja Galereja A. Šilova,
Moskva 037345
Moskovskaja Gosudarstvennaja
Specializirovannaja Škola Akvareli
Sergeja Andrijaki, Moskva 056285
Moskovskij Dom Knigi, Moskva . 143647
Moskovskij Akademičeskij
Chudožestvennyj Licej, Rossijskaja
Akademija Chudožestv, Moskva 056286
Moskovskij Centr Iskusstva,
Moskva 037346
Moskovskij Dom Fotografii,
Moskva 037347
Moskovskij Dom Skulptorov,
Moskva 114281
Moskovskij Dvor Art-Agentstvo,
Moskva 114282
Moskovskij Gosudarstvennyj Muzej
S.A. Esenina, Moskva 037348
Moskovskij Gosudarstvennyj
Universitet Kultury i Isskustv,
Chimki 056269
Moskovskij Gosudarstvennyj
Vystavočnyj Zal Novyi Manež,
Moskva 037349
Moskovskij Literaturnyj Muzej-centr
K.G. Paustovskogo, K.G. Paustovsky
Museum-Centre of Literature,
Moskva 037350
Moskovskij Meždunarodnyj Salon
Isjaščnych Iskusstv, Moskva . 098285
Moskovskij Muzej Sovremennogo
Iskusstva, Moskva 037351
Moskovskij Muzej-usad'ba Ostankino,

Register der Institutionen und Firmen
Mosman Art Gallery, Mosman
– Mount Desert Island Historical Museum, Somesville

Moskva 037352
Mosman Art Gallery, Mosman . . 099387
Mosnier, Laure, Montpellier 128812
Mosquera, Ciudad de Panamá . 113322
Moss, Cheltenham 144326
Moss, New Orleans 095134
Moss & Co., Washington 097852
Moss Antikvariat, Moss 143390
Moss Antikvitets- og Auksjonsforretning, Moss 084281, 126360
Moss End Antique Centre, Warfield 091727, 135188
Moss Kunstgalleri, Moss 113183
Moss Mansion Museum, Billings 046985
Moss-Thorns Gallery of Arts, Hays 049415
Moss, Alan, New York 095600
Moss, Brendan M., Vancouver . . 064850
Moss, Ralph & Bruce, Baldock . 088341
Moss, Sydney L., London 090367
Moss, Tobey C., Los Angeles . . 121561
Mossakowski, Marek, Warszawa 084737, 084738, 143528, 143529
Mossbank and District Museum, Mossbank 006433
Mossette, Fresno 093506
Mossini, Massimo, Mantova . . . 080346
Mossman Gallery, Mossman . . . 099393
Most čerez Stiks – Pons per Styx, Muzej Nonkonformistskogo Iskusstva, Sankt-Peterburg 037759
Most, Peter, Berlin 129537
MostBirnHaus Stift Ardagger, Ardagger 001718
Mosteiro de S. Martinho de Tibães, Mire de Tibães 035857
Mosteiro de Santa Clara-a-Velha, Instituto Português do Património Arquitectónico, Coimbra 035618
Mosteiro de São João de Tarouca, São João de Tarouca 036013
Mostert, Dorothy, Houston 093750
Mostly Bali, New York 095601
Mostly Bears, Tucson 097680
Mostly Boxes, Eton . . . 089219, 134374
Mostly Movables, Toronto 064706
Mostly Posters, Boston 119899
Mostmuseum, Neumarkt im Mühlkreis 002444
Mostmuseum, Sankt Leonhard am Forst 002673
Mostmuseum und Heimathaus, Sankt Marienkirchen an der Polsenz . 002678
Mostra Archeologica G. Venturini, San Felice sul Panaro 027746
Mostra Cabriniana, Codogno . . . 026027
Mostra Cartografica, Mendatica . 026758
Mostra d'Antiquariato, Brescia . 098203
Mostra d'Arte ed Antiquariato, Kunst- und Antiquitätenausstellung, Bolzano 098204
Mostra degli Antichi Mestieri di Calabria, Tropea 028219
Mostra dei Pupi Siciliani, Teatro Stabile dell'Opera, Caltagirone 025698
Mostra della Civiltà Contadina, Lavello 026577
Mostra della Civiltà Contadina, Massa Marittima 026732
Mostra della Civiltà Contadina e Pastorale, Avezzano 025426
Mostra di Antiquariato e d'Arte, Forlì 098205
Mostra di Cimeli del Risorgimento, Convento delle Clarisse, Salemi 027706
Mostra di Palazzo Farnese, Museo delle Carrozze, Piacenza 027295
Mostra Etnografica Museo Contadino della Piana, Capannori 025747
Mostra Europea d'Arte Antica, Genova 098206
La Mostra Gallery, Cardiff 117531
Mostra Internazionale di Modernariato, Antichità e Collezionismo, Baganzola 098207
Mostra Mercato d'Arte Contemporanea Cremona, Cremona 098209
Mostra Mercato d'Arte Moderna e Contemporanea, Vicenza 098210
Mostra Mercato d'Arte, Artisti ed Associazioni Culturali, Forlì . . . 098208
Mostra Mercato dell'Antiquariato, Piacenza 098211
Mostra Mercato di Alto Antiquariato + Mostra Mercato di Antiquariato per Parchi, Giardini e Ristrutturazioni, Modena 098212
Mostra Mercato di Arte e Antiquariato, Padova 098213
Mostra Mineraria Permanente, Sutri 028043
Mostra Nazionale di Antichità, Baganzola 098214
Mostra Nazionale di Pittura Contemporanea, Marsala 026716
Mostra Permanente del Costume Arbereshe, Vaccarizzo Albanese 028245
Mostra Permanente del Presepio, Muggia 026982
Mostra Permanente della Biblioteca Estense, Modena 026850
Mostra Permanente della Ceramica, San Lorenzello 027769
Mostra Permanente della Civiltà Contadina, Montefoscoli 026942
Mostra Permanente della Cultura Materiale, Levanto 026596
Mostra Permanente della Giudaica e Raccolta di Minerali, Laino Borgo 026544
Mostra Permanente della Resistenza, Lugo 026651
Mostra Permanente della Resistenzo, Sala Consiliare, Massa Marittima 026733
Mostra Permanente della Tradizione Mineraria, Tarvisio 028058
Mostra Permanente di Archeologia, Rivello 027536
Mostra Permanente di Paleontologia, Terni 028076
Mostra Permanente di Xilografie di Pietro Parigi, Firenze 026265
Mostra Permanente Le Carrozze d'Epoca, Roma 027572
Mostra Permanente P. Mariani, Istituto Internazionale di Studi Liguri, Bordighera 025592
Mostviertelmuseum Haag, Haag, Niederösterreich 002033
Mostviertler Bauernmuseum, Amstetten 001711
Mosul Museum, Mosul 024809
Mosvik Museum, Mosvik 033791
Mot, J., Bruxelles 100644
Motala Brandförsvarsmuseum, Motala 040859
Motala Industrimuseum, Motala 040860
Motala Motormuseum, Motala . . 040861
Motala Museum – Charlottenborgs Slott, Motala 040862
Motel Court Antiques, Seattle . . 097476
Mother Armenia Military Museum, Yerevan 000740
Mother Hubbard's Cupboard, Manawatu 132727
Mother Wouldn't Like It?, Sandy Bay 062190
Mother's Tankstation, Dublin . . . 109607
Motherwell Heritage Centre, Motherwell 045299
Motherwell Homestead, Abernethy 005304
Moti, Delhi 109334
Moti Hasson Gallery, New York . 122899
Motiejaus Gustaičio Memorialinis Muziejus, Lazdijų krašto muziejus, Lazdijai 030410
Motion Picture Arts Gallery, New York 122900
Motivarte, Escuela de Fotografía, Buenos Aires 054866
Motive Gallery, Amsterdam 112259
Motley's, Richmond 127419
Moto, Tokyo 111476
Moto Moto Museum, Mbala . . . 054842
Motobu-cho Hakubutsukan, Motobu 029059
Motoki, Mitaka 082051, 142903
Motoori Norinaga Kinenkan, Matsusaka 029014
Motor & Nostalgimuséet, Grängesberg 040666
Motor Books World, Camberwell 139680
Motor Museum, Filching 044343
Motor-Sport-Museum Hockenheimring, Hockenheim 019272
Motorbåtmuseet Museihuset, Vreta Kloster 041178
The Motorboat Museum, Basildon 043506
Motorcycle Hall of Fame Museum, Pickerington 052127
Motormuseet, Strømmen 033997
Motorola Museum, Schaumburg 053193
Motorrad-Museum, Augustusburg 016674
Motorrad Museum, Ibbenbüren . 019371
Motorrad-Museum, Otterbach . . 021040
Motorrad-Museum Krems-Egelsee, Krems 002236
Motorrad-Träume, Zschopau . . . 022901
Motorrad-Veteranen- und Technikmuseum, Großschönau 018863
Motorradmuseum, Gossau (Sankt Gallen) 041556
Motorradmuseum, Neunkirchen . 002448
Motorradmuseum, Sammlung Waldmann, Sulz 002805
Motorradwelt der DDR, Kulturkunststatt Prora, Prora . . 021228
Motown Historical Museum, Detroit 048312
Motsuji Homotsukan, Hiraizumi . 028637
Motta, Federico, Milano 137686
Motta, Luigi, Milano 080620
Motte-Barrois, Marie-Christine, Marcq-en-Barœul 069866
Motte, Alexandre, Dinard 103700
Motte, Dominique, Dinard 068357
Motte, Rémy, Villeneuve-d'Ascq 074160
Motte,Jean-Pierre y Otro, Valencia 115509
Mottier, Alexandre, Genève 116333
Mottin, Jean-Marie, Caligny . . . 067604
Mottola, Dario, Milano 080621
Motts Military Museum, Groveport 049288
Mottys, Broke 061216
Motueka District Museum, Motueka 033248
Motyw, Warszawa 113870
Moucheron, Joël, Châlons-en-Champagne 067786
Mouchez, Paris 129094
Moudgalya Antiques, Bangalore 079355
Mouël-Chouffot, Lucienne, Anglet 125490
Le Mouel Enchères, Anglet 066622, 125491
Moufflet, Oslo 084329
Moufflet, Jane, Los Angeles . . . 121562
Moufflet, Jane, Saint-Ouen 073160, 141338
Moufflet, Pascal, Nice 070636
Mougenot, Patricia, Le Thillot . . 069354
Mougin, Paris 105113
Mougin-Berthet, Paris 071694
Mougins Antiquités d'Oriano, Mougins 070363
Mougins Art Prestige, Mougins . 104304
Mougins, Pierre de, Paris 105114
Mouhtar Ashrafi Yodgorlik Muzeyi, Toshkent 054656
Moulagensammlung des Universitätsspitals und der Universität Zürich, Zürich 042212
Moulart, E., Bruxelles 063514
Moulay, Ismaïl, Rabat 112053
Mould, Anthony, London 118464
Mouldings, C.C., Albuquerque . . 092070
Mouldsworth Motor Museum, Mouldsworth 045301
Mouliet, Olivier, Jarret 068895
Le Moulin, Champagné 103563
Le Moulin, Plan-les-Ouates 087737
Moulin à Eau Maître Marcel, Sainte-Agathe-en-Donzy 015597
Moulin à Musique Mécanique, Mormoiron 014330
Le Moulin à Papier de Brousses, Brousses-et-Villaret 012238
Moulin à Papier Vallis-Clausa, Fontaine-de-Vaucluse 013003
Moulin à Vent Gaillardin, Chapelon 012449
Moulin de Beaumont, Beaumont 005404
Moulin de Boschepe, Boeschepe 012088
Moulin de la Brocante, Juziers . 068937
Moulin de la Chevrotière, Deschambault 005703
Moulin de la Herpinière, Musée de l'Outil, Turquant 016065
Le Moulin de la Pleugère, Saint-Ouen-de-Sécherouvr 073232
Moulin de l'Epinette, Angles . . . 066617
Moulin de Malet, Clermont-Soubiran 012601
Moulin de Mérouvel, L'Aigle . . . 069144
Moulin de Nouvet, Courville-sur-Eure 068204
Moulin de Pierre, Artenay 011751
Moulin de Pierre, Hauville 013244
Moulin de Rotrou et son Musée, Vaas 016079
Moulin de Traou-Meur-Écomusée, Pleudaniel 014889
Moulin des Grands Vignes, Musée de la Vigne et du Vin, Périssac . . 014832
Moulin des Jésuites, Charlesbourg 005590
Moulin d'Eschviller, Eschviller . . 012897
Moulin du Bois-Landon, Beaumont-sur-Sarthe 011947
Moulin du Vieux Guingamp, Bégard 067062
Moulin Fleming, LaSalle 006165
Moulin Légaré, Saint-Eustache . 006868
Moulin Michel de Gentilly, Bécancour 005409
Moulin-Musée de la Brosserie, Saint-Félix 015316
Moulin Pomper, Baden 066905
Moulin Rouge, Portland 096444
Les Moulins, Sagy 015199
Moulins, Toulouse 105749
Les Moulins de l'Îsle-aux-Coudres, Saint-Louis-de-Île-aux-Coudres 006921
Moulins Souterrains, Le Locle . . 041685
Moullec, Gabriel, Rosnoen 072547
Moulton, Mosman 099388
Moundarren, Millemont 136839
Moundbuilders State Memorial and Museum, Newark 051503
The Mount, Woore . . . 091938, 144937
Mount Airy Museum of Regional History, Mount Airy 051081
The Mount Antiques Centre, Carmarthen 088799, 134220
Mount Barker Antiques, Mount Barker 061905
Mount Bruce Pioneer Museum, Masterton 033241
Mount Clare Museum House, Baltimore 046762
Mount Dandenong Antiques Centre, Olinda 062027, 127655
Mount Desert Island Historical

Museum, Somesville 053388
Mount Dora Center for the Arts, Mount
Dora 051087
Mount Edgcumbe House, Cremyll 044038
Mount Ephraim Prints, Tunbridge
Wells 119271, 144894
Mount Holly Community Historical
Museum, Belmont 046902
Mount Holyoke College Art Museum,
South Hadley 053405
Mount Kembla Collectables and Old
Wares, Mount Kembla 061914
Mount Kenya Art Galleries,
Nanyuki 111565
Mount Laura Homestead Museum,
Whyalla 001639
Mount Manaia Art Gallery,
Whangarei 113108
Mount Mary College Costume
Museum, Milwaukee 050921
Mount Morgan Historical Museum,
Mount Morgan 001320
Mount Pleasant Antiques Centre, Market
Weighton 090717, 134841
Mount Pleasant Furniture,
Vancouver 064851
Mount Pleasant Historical Society
Museum, Mount Pleasant 051097
Mount Pleasant, Philadelphia Museum
of Art, Philadelphia 052067
Mount Prospect Historical Society
Museums, Mount Prospect . . . 051100
Mount Pulaski Courthouse, Mount
Pulaski 051101
Mount Rainier National Park Museum,
Ashford 046560
Mount Royal College Gallery,
Calgary 005527
Mount Rushmore National Memorial,
Keystone 049964
Mount Saint Vincent University Art
Gallery, Halifax 005957
Mount Stewart House,
Newtownards 045382
Mount Street Gallery, Brecon . 117418,
144261
Mount Vernon Hotel Museum, New
York 051430
Mount Vernon Museum of
Incandescent Lighting,
Baltimore 046763
Mount Victoria and District Historical
Museum, Mount Victoria 001321
Mount Victoria Antiques and Book Shop,
Mount Victoria 061924, 139844
Mount Washington Museum, North
Conway 051601
Mountain Art, Llanberis 118134
Mountain Bookstore, Hamilton . . 140348
Mountain Farm Museum,
Cherokee 047606
Mountain Gateway Museum, Old
Fort 051748
Mountain Heritage Center, Western
Carolina University, Cullowhee . 048087
Mountain Life Museum, London 050364
Mountain Mills Museum, Saint
Catharines 006859
Mountain Peaks Antiques, Salt Lake
City 096911
Mountain-Plains Museums Association,
Littleton 060599
Mountain Resort Museum,
Chengde 007689
Mountain View Doukhobor Museum,
Grand Forks 005906
Mountain View Furniture Centre,
Stratford 084087
Mountain View Gallery, Highfields 099129
Mountain View Museum, Olds . . 006550
Mountaineering Museum,
Darjeeling 024210
Mountaineering Museum,
Hanamaki 028607

Mountbatten Exhibition, Romsey 045633
Mounted Branch Museum, East
Molesey 044211
Mountford, Michael, Boyle 079462,
130831
Mourad, Cairo 103061
Mouradian, Hélène, Paris 105115
Mouret, Gilbert, Marseille 069953,
128780
Mourlot, J.L, Marseille 104190
Mourmans, Knokke-Heist 100748
Mourne, Letterkenny 079653
Mouseio Akropoleos, Athinai . . . 022994
Mouseio Ellinikis Laïkis Technis,
Athinai 022995
Mouseio Ellinikis Paidikis Tehnis,
Athinai 022996
Mouseio Ethnikon Archaiologikon,
Athinai 022997
Mouseio Isrorias tis Ellinikis
Endymasias, Athinai 022998
Mouseio Istorias Laografias Aharnon,
Athinai 022999
Mouseio Kerameikos, Athinai . . . 023000
Mouseio Kompologiou, Nafplion . 023193
Mouseio Laikon Organon, Athinai 023001
Mouseio Mpenaki, Athinai 023002
Mouseio Trenon, Athinai 023003
Mouseio Vorre, Paiania 023206
Mouseio Zakynthou, Zakynthos . 023321
Mouseion Goulandri Fysikis Istorias,
Kifissia 023128
Mousley Museum of Natural History,
Yucaipa 054607
Moussa Castle Museum,
Beiteddine 030271
Moussé, Claude, Toulouse 129308
Mousseau, Séverine, Chef-
Boutonne 067952
Mousset, Daniel, Lyon 128724
Moussion, Paris 105116
Moust, J., Brugge 063284
Moustache & Trottinette,
Montpellier 140968
Mout- en Brouwhuis De Snoek,
Alveringem 003266
Moutain Lodge Pottery and Art School,
Newport 055865
Mouthon, Sylvain, Faverges . . . 068511
Mouton, Michel, Lanester 069165
Mouvances, Paris 105117
Mouvement Socio-Culturel de la
Tuque, La Tuque 006140
Mouvements Modernes, Paris . . 071695
Mouysset, Francine, Rodez 072514
Mouzannar, Pierre, Beirut 082295
Moveis, Alvinus, Rio de Janeiro 064090
Movelanjos, Rio de Janeiro 064091
Movements in the Arts, Westport 139440
Movie Arts Posters, San Jose . . 124454
The Movie Museum, Owosso . . 051857
Movie Shop, Norwich . 090879, 144752
Movieland Wax Museum of the Stars,
Niagara Falls 006482
Movilla, Leichhardt 127629
Movimiento Artistico del Rio Salado,
Phoenix 060600
Moving Image Centre, Auckland 112794
Mowbray Gallery, Port Douglas . 099582
Mowhay Gallery, Trebetherick . . 119253
Moxhams Antiques, Bradford-on-
Avon 088563
Moy Antique Pine, Moy 090788
Moy Antiques, Moy 090789
Moy, J., London 134735
Moy, Jakob Graf von, Salzburg . 062811
Moyano, Barcelona 085806
Moyano 21, Madrid 086225
Moyard, Morges 087686, 133862
Moycullen Village Antiques,
Moycullen 079677
Moylurg, Castletownsend 079477
Moynihan & Riehlman, New York 122901
Moyse's Hall Museum, Bury Saint

Edmunds 043810
Mozaik Müzesi, İstanbul 042950
Mozart Antiquités, Paris 071696
Mozart Wohnhaus, Internationale
Stiftung Mozarteum, Salzburg . 002628
Mozarthaus St.Gilgen, Sankt
Gilgen 002660
Mozarthaus Vienna, Wien Museum,
Wien 003036
Mozarthaus, Kunstsammlungen und
Museen Augsburg, Augsburg . 016663
Mozarts Geburtshaus, Internationale
Stiftung Mozarteum, Salzburg . 002629
MPA Studio, Winnipeg 101887
Mpellos, Agios Dimitrios 019013
MPF Gallery, Philadelphia 123426
MPG Contemporary, Boston . . . 119900
MPG Town Gallery, Odiongan . . 034559
MPLS Numisatics, Minneapolis . 094934
MPM Dizajn, Ohrid 111877
MPP 83, Pleven 128138
MPR International Committee
for Museums Marketing and PR,
Malmö 059897
Mr Antiguedades, Estepona 085995
Mr Pickwicks Trading Company,
Christchurch 083885, 143308
Mr. George, Luzern 087645
Mr. Lucky, San Francisco 097290
Mr. Paint Stripper, Los Angeles . 135733
Mr. Yeste, Barcelona . 085807, 133137
Mramornyj Dvorec, Gosudarstvennyj
Russkij Muzej, Sankt-Peterburg 037760
Mronz, Ute, Köln 107666
Mrs Mills Antiques, Ely 089201
Mrugalla, Peter, Regensburg . . . 078173
MRW Antiguidades, Rio de
Janeiro 064092
MS Art, Praha 102709
MS Libros, Buenos Aires 139589
Ms. B Art Gallery, Oklahoma City 123289
MSA Limited, Saint Louis 096745
MSCC Forsyth Center Galleries,
Texas A & M University, College
Station 047874
Msvar, Tampa 097579
MTA Briefings, Washington 139441
MTA Tabiat Tarihi Müzesi, Ankara 042824
Mtengatenga Postal Museum,
Namaka 030657
MU – De Witte Dame, Eindhoven 112464
Mu-Zee-Um, Oostende 003930
Mu.ZEE – Kunstmuseum aan Zee,
Oostende 003931
muba – Museum für Baukultur,
Neutal 002451
Mucciaccia, Lucio, Roma 110835
Muccioli, Detroit 120813
Much Wenlock Museum, Much
Wenlock 045305
Much, Ado, Atlanta 092161
Mucha, La Rochelle 069092
The Muchnic Gallery, Atchison . 046590
Muchovo muzeum, Praha 009809
Muci, Caracas 125049
Muck, Stefan, Windach 130768
Muckel, Ratingen 078137
Muckenhofer, Robert, Trieben . . 127886
Muckenthaler Gallery and Mansion
Museum, Fullerton 048995
Muckleburgh Collection, Holt . . . 044596
Muckross House Gardens and
Traditional Farms, Killarney . . 024940
Mücsarnok, Budapest 023528
mudac – Musée de Design et
d'Arts Appliqués Contemporains,
Lausanne 041652
Mudanya Mutareke Evi Müzesi,
Mudanya 043028
Mudde, A.M., Haarlem 083344
Muddi's Antik, Horsens 065797
MudFire Pottery Center, Atlanta . 119595
Mudfish Pottery, Albuquerque . . 119495
Mudgee Colonial Inn Museum,

Mudgee 001322
Mudhead, Denver 120749, 120750
Mudrány Kúria, Jósa András Múzeum,
Szabolcs 023913
Mü-Terem, Budapest . 079269, 109164,
126153
La Mueblería, Medellín 065298
Muebles Baque, Barcelona 085808
Muebles la Ganga, Fuengirola . . 086000
Muebles Usados el Callejón,
Medellín 065299
Muebles y Ceramicas Colombia,
Barranquilla 128256
Mück, Brombachtal 075198
Mühl, Regensburg 078174
Mühl, Dieter, Kempten 141985
Mühlan, Adelheid & Michael, Landau
in der Pfalz 142042
Mühlbauer & Partner,
Regensburg 130566
Mühlbauer, Franz, Rosenheim . . 130599
Mühlbauer, Peter, Pocking 078061
Mühlbauer, Stephan, Regensburg 130567
Mühlberger, Gudrun, Pichl bei
Wels 127834
Mühlberger, Michael, Zapfendorf 079133
Mühlberger, Wolfgang, Wien . . . 140110
Mühlburger, Werner, Lienz 062711
Die Mühle, Eberswalde 018083
Mühle Laar, Zierenberg 022876
Mühle Oberneuland, Bremer
Landesmuseum für Kunst-und
Kulturgeschichte, Bremen . . . 017594
Mühlen- und Landwirtschaftsmuseum,
Fehmarn 018342
Mühlen- und Landwirtschaftsmuseum,
Jever 019463
Mühlen- und Mennoitenmuseum
Bettinger Mühle, Schmelz . . . 021686
Mühlenbeck, Holger, Wuppertal . 079111
Mühlenbein-Schelkle, Berlin . . . 129538
Das Mühlendorf, Ottenhöfen im
Schwarzwald 021038
Mühlenfachmuseum, Aurich . . . 016682
Mühlengalerie, Großefehn 107173
Mühlenhof Altkalen, Altkalen . . . 016548
Mühlenhof-Freilichtmuseum Münster,
Münster 020604
Mühlenmuseum, Haren (Ems) . . 019073
Mühlenmuseum, Oberkrämer . . 020903
Mühlenmuseum, Pfullingen 021126
Mühlenmuseum, Rheinsberg . . . 021403
Mühlenmuseum Brüglingen,
Münchenstein 041774
Mühlenmuseum Katzbrui,
Apfeltrach 016599
Mühlenmuseum Mitling-Mark
mit Sammlung Omas Küche,
Westoverledingen 022601
Mühlenmuseum Moisburg,
Moisburg 020450
Mühlenmuseum Teichmühle,
Steinwiesen 021991
Mühlenmuseum und Mühlenweg,
Maria Luggau 002357
Mühlerama, Museum in der Mühle
Tiefenbrunnen, Zürich 042213
Mühlfeld + Stohrer, Frankfurt am
Main 107016
Mühlhaus, Alena, Wiesbaden . . 108925
Mühlich & Neumann, Berlin . . . 129539
Mühlviertler Keramikwerkstätte
Hafnerhaus, Leopoldschlag . . 002300
Mühlviertler Kulturgütermuseum,
Sankt Johann am Wimberg . . 002666
Mühlviertler Schlossmuseum Freistadt,
Oberösterreichische Landesmuseen,
Freistadt 001906
Mühlviertler Vogelkundeweg,
Gutau 002028
Mühlviertler Waldhaus, Windhaag bei
Freistadt 003121
Mülheimer Kunstverein e.V., Mülheim
an der Ruhr 059177

Mülhens-Molderings, Barbara,
Köln 130158
Müllauer, Thomas Franz, Krems 062691
Müller, Aschaffenburg . 106013, 129388
Müller, Chemnitz 075258
Müller, Neuss 077748
Müller, Offenburg 142263
Müller, Wurzbach 079123
Müller & Gräff, Stuttgart 142400
Müller & Schindler, Simbach am
Inn 137507
Mueller Bhavan, Max, Bangalore 109262
Müller-Feldmann, Annemarie, Bad
Neuenahr-Ahrweiler 141432
Müller-Kögler, Klaus, Mainaschaff 137365
Müller-Linow, Seeheim-
Jugenheim 108604
Müller Miklós Kiállítóterem,
Orosháza 023818
Müller-Pröls, Herbert, Pettstadt . 130511
Müller-Radloff, Christine, Leipzig 130225
Müller-Rössler, R., Gernsbach .. 107119
Mueller-Roth, Stuttgart 108723
Müller-Schneevoigt, Friedrich,
Neckargerach 142219
Müller, Alfred Otto, Köln 107667, 130159
Müller, Andreas, Sinn . 078480, 130640
Müller, Barbara, Chemnitz 129656
Müller, C.-V. & M., Furtwangen im
Schwarzwald 107095
Müller, Christian, Großpösna .. 076113
Müller, Dorothea, Xanten 142530
Mueller, Dr. Ulrich, Köln 076899
Müller, E., Oberkirch 077868
Müller, E.W., Kiel 076784
Müller, Elfriede, Salzwedel 078346
Mueller, Evelyn, Biebergemünd . 106481
Müller's Nachf., Ferdinand Rudolf,
Wien 140111
Müller Franz Xaver, Würzburg .. 142514
Müller, Gabriele, Horw 133773
Müller, Gerhard, Essen 129819
Müller, Günter, Augsburg 074467
Müller, H.-J., Leonberg 077089
Müller, Harald, Nürnberg 077833
Müller, Hartmuth & Christa,
Mengen 077345
Müller, Henry, Winterthur 144153
Müller, Herman & Frederik, Sint-
Niklaas 063834
Müller, Hubertus, Zwickau 079152
Müller, Jakob, Esslingen 075712
Müller, Johannes, Bern 087294
Müller, Johannes, Salzburg 140028,
140029
Müller, Josef, Bad Tölz 141437
Müller, Karl, München 130372
Müller, Karlheinz, Villmar 078804,
126135, 130726, 142452
Müller, Karsten, Berlin . 074881, 141556
Müller, Ken, Koblenz 107575
Müller, Klaus, Göttingen 076050, 129911
Müller, Klaus, Landau in der
Pfalz 130207, 142043
Müller, Klaus-Dieter, Deutsch
Evern 075349
Müller, Lars, Baden 138021
Müller, Libuse, København 102908
Müller, Magnus, Berlin 106377
Müller, Manfred, Kempten 076752
Müller, Manfred, Morbach 077419
Müller, Manfred, Tengen 078675
Müller, Marion, Gräfelfing 076071
Müller, Mark, Zürich 116838
Müller, Max, Chemnitz 141674
Müller, Meike, Ratingen 078138
Müller, Monika, Viechtach 078795
Müller, Monika & Dietmar,
Velbert 108816
Müller, Otto, Salzburg 136555
Müller, Reinhardt, Bad
Langensalza 129415
Müller, Robert, Hamburg 129978

Müller, Rudolf, Saarbrücken ... 078329
Müller, Stefan, Düsseldorf 141724
Müller, Stefan, Mönchengladbach 142129
Müller, T., Schaffhausen 144134
Müller, Thomas, Kelkheim 130113
Müller, W., Bonn 075080
Müller, Walter, Genève 087486
Müller, Walter, Neusäß 077745
Müller, Werner, Flums Hochwiese 116268
Müller, Willi, Hamburg 129979
Müllertz, Birte, Charlottenlund .. 065629
Müllmuseum, Bad Säckingen .. 016856
Mülstroh, Klaus, Heinsberg 107411,
141937
Münch, Mainz-Kastel 130270
Münch, Rainer, Gutenstetten ... 076136
Münchau, Melanie, Potsdam ... 130536
München Kereskedohaz, Miskolc 079314
Münchhausen-Museum,
Bodenwerder 017448
Münchner Feuerwehrmuseum,
München 020549
Münchner Puppenmuseum,
München 020550
Münchner Stadtmuseum,
München 020551
Münchner Tiermuseum,
München 020552
Muenchow, Oliver, Zürich 116839
Münkel & Brettschneider,
Eiterfeld 129785
Münkwitz, Michael, Rostock ... 130602
das münster, Regensburg 138736
Münstersches Kunst- und
Auktionshaus, Eric M.A. Wilbois,
Münster 126075
Münter-Haus, Murnau am
Staffelsee 020624
Münz Zentrum Rheinland,
Solingen 078499, 126117
Die Münze, Greiz 076086
Münze Hall und Münzerturm, Hall in
Tirol 002047
Münze und Antik, Bad Kissingen 075977
Der Münzen- und Medaillensammler,
Freiburg im Breisgau 138737
Münzen am Zoo, Berlin 074882
Münzen im Europa-Center, Berlin 074883
Münzen Revue, Regenstauf ... 138738
Münzen und Medaillenkunst,
Ettenheim 075716
Münzenlocher, Dorfen .. 075375, 129693
Münzenstube, Jena 076644
Münzer, Basel 087250
Münzhandelsgesellschaft,
Braunschweig 075130
Münzkabinett am Landesmuseum
Joanneum, Graz 001982
Münzkabinett der Preussag AG,
Hannover 019059
Münzkabinett Frankfurt am Main,
Frankfurt am Main 075809
Münzkabinett Gera, Gera 075978
Münzkabinett im Stadtarchiv Mainz,
Mainz 020230
Münzkabinett und Antikensammlung
der Stadt Winterthur, Winterthur 042153
Münzkabinett, Staatliche
Kunstsammlungen Dresden,
Dresden 017998
Münzmuseum, Bad Iburg 016793
Müri, Matthias, Zürich 087992
Müritzeum, Das Naturerlebniszentrum,
Waren (Müritz) 022429
Müser, Monika, Düsseldorf 075500
Müßler, Thomas, Limbach-
Oberfrohna 077108
Muestra Arqueológica, Armenia . 008571
Muestra Arqueológica,
Montenegro 008730
Muestra Arqueológica, Quimbaya 008774
Muestra Arqueológica Colección
Rosmary, Colosó 008666
Muestrario Arqueológico de Amazonas,

Chachapoyas 034246
Muestra's, Valencia 115510
Müthing, Bernhard, Billerbeck .. 129591
Müvezettörteneti Ertesitö,
Budapest 138773
Muga, Nagoya 111324
Mugal India, Jaipur, Rajasthan . 109384
Mugart, Donostia-San Sebastián . 115054
Mugele, Andre, Fichtenau 129836
Muggenthaler, Margit, München 077576
Mughal Art Corner, Mumbai ... 079405
Mughal Museum, Lahore 034133
Muğla Devlet Güzel Sanatlar Galerisi,
Muğla 043030
Muğla Kültür Merkezi, Muğla .. 043031
Muğla Müzesi, Muğla 043032
Mugler, Nicole, Paris 071697
Mugnaini & Tinacci, Firenze ... 131220
Muguet-Lhermine, Daniele, Paris 129095
Muh Sheng Museum of Entomology,
Puli 042419
Muhammad Ali Center, Louisville 050499
Muhammadi, London 144651
Muhl, Gerd, Gessertshausen .. 129903
Muhoksen Kotiseutumuseo,
Muhos 011166
Muhr, Claus-Maria, Wels 062875
Mui, Hiroshima 081982
Muiderslot, Muiden 032659
Muijsers, P.R., Tilburg 112682
Muir, Patrick, Kingston 081955
Muir, Robert, Nedlands 139850
De Muizon & Le Coënt, Senlis . 125816
Mújica, Santiago de Chile 064978,
101915
Muka Gallery, Auckland 112795
Muka Studio, Auckland 112796, 137808
Mukai, Kobe 111284
Mukai Junkichi Fuzoku Bijutsukan,
Tokyo 029557
Mukashi & Mukashi, Sapporo .. 082102
Mukhtar Museum, Giza 010571
Mukimi- Hudjra Muzeyi, Qo'qon
Viloyat Ülkahunoslik Muzeyi,
Qo'qon 054641
Mukti Juddha Museum, Dhaka . 003224
Mulberry, Pittsburgh 096314
Mulberry Heights Antiques,
Birmingham 092424
Mulberry House, Vancouver ... 064852
Mulberry Phospate Museum,
Mulberry 051110
Mulcahy, Dallas 120633
Mulder, Frederick, London 090368,
118465, 144652
Mulet, Jean-François, Hyères .. 068858
Mulet, Pilar, Madrid 115262
Mulford House and Farm, East
Hampton Historical Society, East
Hampton 048434
Mulgara Gallery, Yulara 099949
Mulgi Külamuuseum, Halliste .. 010610
Mulgrave Art Gallery, Mulgrave . 099412
Mulgrave Settlers Museum,
Gordonvale 001112
Mulhall, Susan, Naas 130885
Mulherin, Katharine, Toronto .. 101668
Mulholland, Susan, Tucson 124758
Mulier & Mulier, Knokke-Heist . 100749
Mulin, Shanghai 102193
Il Mulino, Gorizia 110146
Il Mulino, Savona 110916
Mull Museum, Tobermory 045983
Mullach Ban Folk Museum, Mullach
Ban 045306
Mullan Gallery, Belfast 117341
Mullee, Kolding 102920
Mullen, Robin, Guildford 117895
Mullenders, Georges, Liège 063776
Mullendorff, Antwerpen 063226
Muller, Almelo 082634
Muller, Forbach 068585
Muller von Rohrbach, Jean,
Lausanne 133803

Muller, Claude, Esch-sur-Alzette 132329
Muller, Frederik, Bergum 143090
Muller, Jean-David, Plan-les-
Ouates 087738
Muller, Jean-François, Appoigny 066689,
066690
Muller, Nicolas, Nice 070637
Muller, Roland, Strasbourg 073594
Muller, Rudolph H., El Paso ... 093408
Mulligan, Richard, West
Hollywood 097911
Mullingar Military Museum,
Mullingar 024971
Mullion Gallery, Mullion 118764
Mullis, Robert, Swindon 135113
Mulorz, Schwerte 078433
Mulot, Saint-Gilles-Croix-de-Vie . 105526
Mulsant, Yves, Brie-Comte-
Robert 067492
Multatuli Museum, Amsterdam . 031958
Multi Art i Göteborg, Västra
Frölunda 115988
Multi Art International, Bonn ... 075081
multi-druck Verlag, Neusäß 137445
Multi Dutch, Amsterdam 082813
Multi Studio, Brasília 100921
Multian Kotiseutumuseo, Multia . 011167
Multiart, Verona 132293
Multiarte, Asunción 113334
Multicollection, Angers 066612
Multicolor, Madrid 086226, 143778
Multicultural Arts, Ottawa 101477
Multicultural Heritage Centre, Stony
Plain 007079
Multigestion Puerto Mont, Getxo 115066
Multigestion Puerto Mont,
Hondarribia 115098
Multigraphic, Venezia 111144
Multimedia Arte Contemporanea,
Brescia 109911
Multipel 13, Knokke-Heist 100750
Multipla de Arte, São Paulo ... 101103
Multiple Box, Waterloo 099867
Multiple Box Hamburg, Hamburg 107300,
137250, 141888
Multiple Choice, Amsterdam ... 112260
Multiple Design, Huizen 112542, 137860
Multiple Impressions, New York 122902
Multiples, Brian, Santa Monica . 138470
Multiplicidad, Madrid 115263
Multivision "Worpswede und das
Teufelsmoor", Worpswede 022775
Multnomah, Portland 096445
Mulvane Art Museum, Topeka .. 053774
Mulvany Bros., Dublin 109608
Mulzer, Georg, Lautrach 077036, 130219
MuMa – Musei del Mare e della
Navigazione, Castello d'Albertis,
Genova 026395
Mummery, Andrew, London ... 118466
Munakata Prints Museum,
Kamakura 028760
Munakata Shiko Kinenkan,
Aomori 028476
Munakata Shikou Art Station,
Muika 029060
Munakata Taisha Shinpōkan,
Genkai 028571
Munari, René, Grenoble 140851
Muncaster Castle, Ravenglass . 045581
Muncaster Watermill, Ravenglass 045582
Munce Art Center, Zionsville .. 054617
Munch-museet, Oslo 033842
Munchforlaget, Oslo 113237
Munchs Hus, Åsgårdstrand ... 033508
Muncie Children's Museum,
Muncie 051115
Mund, A., Gütersloh 107186
Mundaneum – Centre d'Archives de la
Communauté Française, Mons 003879
Mundaring Arts Centre,
Mundaring 099413
The Mundesley Gallery,
Mundesley 118766

El Mundo de Carmen, Caracas . 097963
El Mundo de la Ceramica,
  Baranquilla 128255
El Mundo de la Moneda, México 082482,
  126264
El Mundo de Muñecas, Icod de los
  Vinos 039508
O Mundo do Livro, Lisboa 114045,
  143585
Mundy, C.W., Indianapolis 121205
Mundy, Ewan, Glasgow 089331, 117852,
  134405
Muné, Jean-Philippe, Paris 071698
Muné, Mercédès, Saint-Quentin
  (Aisne) 125808
Muñeca, Barniz de, Alcalá de
  Henares 085689, 133077
Muñecas, Madrid 126546
Munerotto, Gianfranco, Venezia . 132251
Munganbana Aboriginal Art Gallery,
  Cairns 098843
Municipal Art Gallery, Molyvos . 023176
Municipal Art Gallery,
  Thessaloniki 023293
Municipal Art Gallery of Xanthi,
  Xanthi 023320
Municipal Art Society, New York 051431
Municipal Art Society of New York,
  New York 060601
Municipal d'Arts Plastiques, La
  Menuiserie, Choisy-le-Roi . . . 012566
Municipal Gallery of Athens,
  Athinai 023004
Municipal Gallery of Corfu, Corfu 023054
Municipal Gallery of Herakleion,
  Iráklion 023098
Municipal Gallery of Lamia,
  Lamia 023148
Municipal Gallery of Livadia,
  Livadia 023156
Municipal Gallery of Piraeus,
  Piraeus 023222
Municipal Gallery of Rhodes,
  Rhodos 023249
Municipal Gallery of Samos,
  Samos 023253
Municipal Gallery of Volos, Volos 023316
Municipal Museum, Dimona . . . 025032
Municipal Museum Gwalior,
  Gwalior 024282
Municipality Museum, Hebron . . 034162
Municipalnaja Chudožestvennaja
  Galereja g. Kostromy, Kostroma 037156
Municipalnyj Objedinennyj Muzej
  Pisatelej Urala, Ekaterinburg . . 036984
Municipalnyj Muzej Anna
  Achmatova – Serebrjannyj vek,
  Sankt-Peterburg 037761
Municypal'nyj Muzej Pryvatnych
  Kollekcij im. O. Bleščunova,
  Odesa 043271
Munier, Rodolphe, Paris 129096
Munkácsy Mihály Emlékház,
  Békéscsaba 023443
Munkácsy Mihály Múzeum,
  Békéscsaba 023444
Munkeborg, Saltsjö-Boo 086922, 133561
Munkkiniemen Antiikkipuuseppä,
  Helsinki 066344
Muñoz Antiguedades, Málaga . . 086299
Muñoz Jimenez, Alfredo, Estella 126523
Muñoz Turégano, Valencia 086526
Muñoz, Carlos, San José 102511
Muñoz, Jose, Estepona 085996
Munro, Vera, Hamburg 107301
Munroe, Victoria, Boston 119901
Munro's Mill Antique Centre,
  Tamworth 062303
Munsala Hembygdsmuseum,
  Munsala 011170
Munsin Museum, Masan 029970
Munson-Williams-Proctor Arts Institute

Museum of Art, Utica 053916
Munson-Williams-Proctor Arts Institute.
  Bulletin, Utica 139442
Munster Literature Centre, Cork 024859
Munt- en Penningkabinet van
  de Spaar- en Voorschotbank,
  Surhuisterveen 032911
Munting Museo ng Namayan,
  Mandaluyong 034517
De Muntkoerier, Apeldoorn . . . 082926,
  138935
Muntwisselkantoor, Apeldoorn . 082927
Munzi, Gabriella, Roma 081469
Muonion Kotiseutumuseo,
  Muonio 011171
Mur Art, Interlaken 133777
Le Mur Vivant, London 090369, 118467
Mura Clay Gallery, Newtown, New
  South Wales 099452
Mura-Todesco, Béatrice, Dijon . 068338
Murad, Mohammed Isam,
  Damascus 088043
Murakami Folk Museum,
  Murakami 029062
Murakami Hojido, Tokyo 082184
Muraki, Ebetsu 081961
Mural da História, Lisboa 132943
Muramooto, Akiko, Hong Kong . 065085
Murari, Max, Villeneuve-sur-Lot 074169
Murarka's Art Gallery, Kolkata . . 109402
Murat, Pamplona 086394
Murat-David, Philippe, Paris . . . 071699
Murata & Friends, Berlin 106378
Murator, Warszawa 137907
Murauchi Bijutsukan, Hachioji . . 028581
Al-Murbati, Manama 063154
Murch, Jill, Moonta 099371
Murchison Falls National Park
  Museum, Murchison Falls 043111
Murchison Museum, Murchison 033249
Murchison Settlement Museum,
  Mullewa 001324
Murdaca, Natale, Milano 131481
Murder by the Book, Cranston . 145104
Murder by the Book, Denver . . . 145131
Murder by the Book, Houston . 145172
Murder by the Book, Portland . 145458
Murder Ink, New York 145358
Murdocca, Vincenzo, Roma . . . 131988
Murdoch & Barclay, Moorabbin . 127643
Murdoch McLennan, Auckland . 083809
Muré, Robert, Toulouse 073786, 129309
Mureau, Dreux 128588
Murer, Milano 080622
Murgtal-Museum, Forbach 018378
Murias Numismatica, A Coruña . 048597
Murie Museum, Kelly 049915
Muriel, Boris, West Hollywood . 125015
Murillo, Edmonton 128166
Murillo, Lisboa 114046
Murino, Mario, Milano 131482
Murmanskij Oblastnoj
  Chudožestvennyj Muzej,
  Murmansk 037527
Murmanskij Oblastnoj Kraevedčeskij
  Muzej, Murmansk 037528
Les Murmures du Temps,
  Crémieu 068233
Murney Tower Museum,
  Kingston 006108
Il Muro di Tessa, Milano 142733
Muroran-shi Seishonen Kagakukan,
  Muroran 029064
Muroran-shi Tontenkan, Muroran 029065
Muros Vivos, Bucaramanga . . . 102396
Murphy, Los Angeles 135734
Murphy, Jan, Fortitude Valley . . 099052
Murphy, Jennifer, Charlotte . . . 120001
Murphy, John,
  Newtownshandrum 079687
Murphy, Michael, Tampa 124665
Murphy, Roger, Kingston 061696
Murphy's Furniture, Cork 079502
Murphy's Loft, Cleveland 120443

Murr, Gabriele, Bamberg 074669
Murr, Renate, Esslingen 141755
Murray Hill Art, Cleveland 120444
Murray-Lindsay Mansion,
  Lindsay 050315
Murray, Guy, New York 122903
Murray, Kevin, Armadale, Victoria 061042
Murray's Motorcycle Museum,
  Laxey 044770
Murrell Home, Oklahoma Historical
  Society, Park Hill 051923
Murrer, Werner, München 077577,
  130373
Murrough, H. O'Brien, Portland . 145445
Murrumburrah-Harden Historical
  Museum, Murrumburrah 001328
Murrurundi Museum, Murrurundi 001329
Mursia, Ugo, Milano 137687
Murston House, Geelong 099073
Murtada, Abdul Khaliq Najwani,
  Muttrah 084404
Murtoa Water Tower Museum,
  Murtoa 001330
Murtovaaran Talomuseo, Valtimo 011486
Murtuz, Moskva 114283
Mus-Olles Museum, Nälden . . . 040870
Musa, Newcastle-upon-Tyne . . 118790
Musa Heritage Gallery, Kumbo-
  Nso 005294
MUSA Museum auf Abruf, Wien 003037
MUSAC, Museo de Arte
  Contemporáneo de Castilla y León,
  León 039555
Musaem Chontae Dhún na nGall,
  Letterkenny 024952
Músaem Corca Dhuibhne, Corca
  Dhuibhne Regional Museum,
  Ballyferriter 024827
Musafir-Khana Palace, Cairo . . . 010545
Musandi, Chipo, Berlin 106379
La Musardiére, Plauzat 105356
Musardière, Renescure 072416
Musashi-murayama-shi Rekishi
  Minzoku Shiryokan, Musashi-
  murayama 029066
Musashino Art University Museum and
  Library, Kodaira 028867
Musashino Bijutsu Daigaku,
  Tokyo 056048
Al-Musbahi, Jeddah 085304
Muscarelle Museum of Art, College of
  William and Mary, Williamsburg 054414
Muscariello, Giuseppe, Napoli . 131631
Muscariello, Mario, Napoli 080902
Muscariello, Pasquale, Napoli . . 131632
Muscariello, Vincenzo, Napoli . . 080903
Muscat Gate Museum, Muscat . 034098
Muscatine Art Center, Muscatine 051132
Muschel- und Schneckenmuseum,
  Norden 020818
Muschelkalkmuseum Hagdorn,
  Ingelfingen 019401
Muschi & Licheni, Trieste 081797
Muschietti, Walter, Udine 081811
Muschik, Aschaffenburg 106014
Muschter, A., Berlin 074884
Le Muse, Milano 080623
Muse Art Gallery, Philadelphia . 052068
Muse Gallery, Jackson 049763
Muse Gallery, Philadelphia 123427
The Muse Gallery, Milawa 099338
MUSE-O, Stuttgart 022050
Muse Studio, Point Lonsdale . . 099568
Museactron, Maaseik 003834
Muséal Adolphe Pégoud,
  Montferrat 014257
Musée, Aigues-Mortes 011548
Musée, Bages (Pyrénées-
  Orientales) 011865
Musée, Bischwiller 012048
Musée, Brignon 012227
Musée, Chaillac 012399
Musée, Chazé-sur-Argos 012539
Musée, Cransac 012712

Musée, Cruzy 012729
Musée, Evolène 041481
Musée, Frontignan-la-Peyrade . 013063
Musée, Gramat 013141
Musée, Joinville (Haute-Marne) . 013322
Musée, Kampong Thom 005284
Musée, Le Landeron 041634
Musée, Laragne-Montéglin 013572
Musée, Lisieux 013828
Musée, Livry 013840
Musée, Lupiac 013918
Musée, Moyen 014371
Musée, Orainville 014567
Musée, Plouay 014899
Musée, Quiberon 015016
Musée, Saint-Hilaire-en-Lignière 015360
Musée, Tauriers 015920
Musée 1900, Arpaillargues-et-
  Aureillac 011738
Musée 1900, Cauterets 012373
Musée 1900 Arts et Techniques,
  Champlitte 012437
Musée 2ème Guerre Mondiale,
  Ambleteuse 011617
Musée 39–45, Roscoff 015154
Musée 39–45 du Grand Blockhaus,
  Batz-sur-Mer 011907
Musée 40–44, Monceau-
  Imbrechies 003875
Musée à Flot de l'Escorteur d'Escadre,
  Musée Naval Maillé Brézé,
  Nantes 014412
Musée à Flot du Galion,
  Plouharnel 014909
Musée à la Ferme de Rome, Bézu-la-
  Forêt 012026
Musée à la Mémoire des Combattants
  et de Victimes de Guerre,
  Mâcon 013957
Musée a la Recherche du Temps
  Perdu, Bricquebec 012219
Musée A la Rencontre des Vieux
  Métiers, Bouray-sur-Juine 012142
Musée Aal Braaneburg,
  Brandenbourg 030567
Musée Abbaye d'Airvault,
  Airvault 011557
Musée-Abbaye de Charroux,
  Charroux 012465
Musée-Abbaye Saint-Germain,
  Auxerre 011822
Musée Abbé Deletoille, Fondation
  Boudenoot, Fruges 013064
Musée Abbé Jules Lemire, Vieux-
  Berquin 016211
Musée Acadien, Archigny 011710
Le Musée Acadien, Caraquet . . 005556
Musée Acadien, Cheticamp . . . 005610
Musée Acadien de Pubnico-Ouest,
  West Pubnico 007381
Musée Acadien du Québec,
  Bonaventure 005436
Musée Acadien, Prince Edward Island
  Museum and Heritage Foundation,
  Miscouche 006334
Musée Acadien, Université de
  Moncton, Moncton 006351
Musée Accous-Fermiers Basco-
  Béarnais, Accous 011535
Musée Adam Mickiewicz, Paris . 014678
Musée Adja Swa, Yamoussoukro 008879
Musée Adjaman Mihiin Raphaël,
  Vavoua 008878
Musée Adrien Mentienne, Bry-sur-
  Marne 012249
Musée Adzak, Paris 014679
Musée Aéronautique, La Baule-
  Escoublac 013363
Musée Aérorétro, Albon 011586
Musée Africain, Ile-d'Aix 013283
Musée Africain, Lyon 013933
Musée Africain de Namur, Namur 003895
Musée Agathois, Agde 011536
Musée Agricole, Coffrane 041420

Register der Institutionen und Firmen
Musée Agricole, Villy-le-Maréchal
– Musée Arts et Traditions Populaires, Firmi

Musée Agricole, Villy-le-Maréchal 016299
Musée Agricole Bras de Brosne, Marles-sur-Canche 014022
Musée Agricole de Brullioles ou la Cadole, Brullioles 012241
Musée Agricole de la Haute Hesbaye, Musée La Vie rurale, Liernu . . 003819
Musée Agricole Départemental Marcel-Mouilleseaux, Botans 012126
Musée Agricole des Ruralies, Prahecq 014990
Musée Agricole du Château de Didonne, Semussac 015762
Musée Agricole du Parlement, Gap 013077
Musée Agricole Vivant, Bissey-la-Pierre 012050
Musée Agricole, Association La Gerbe, Sainte-Geneviève (Meurthe-et-Moselle) 015613
Musée Agro-Pastoral d'Aussois, Aussois 011805
Musée Airborne, Sainte-Mère-Eglise 015626
Musée Al Mathaf El Lubnani, Musée du Liban Féodal, Beiteddine . . 030272
Musée Alain Fournier et Jacques Rivière, La Chapelle-d'Angillon 013380
Musée Albert André, Bagnols-sur-Cèze 011871
Musée Albert Schweitzer, Gunsbach Village 013219
Musée Albert Schweitzer, Kaysersberg 013348
Musée Alésia, Alise-Sainte-Reine 011599
Musée Alexandre Dumas, Villers-Cotterêts 016282
Musée Alexandre-Louis Martin, Carnières 003506
Musée Alexis Forel, Morges . . . 041759
Musée Alfred Bonnot, Chelles . . 012543
Musée Alfred de Vigny, Champagne-Vigny 012431
Musée Alice Taverne, Ambierle . 011616
Musée Alphonse Daudet, Fontvieille 013024
Musée Alphonse Daudet, Saint-Alban-Auriolles 015206
Musée Alpin, Chamonix-Mont-Blanc 012427
Musée Alsacien, Haguenau 013223
Musée Alsacien, Strasbourg . . . 015872
Musée America-Gold Beach, Ver-sur-Mer 016162
Musée Amérindien de Mashteuiatsh, Mashteuiatsh 006271
Musée Amérindien et Inuit de Godbout, Godbout 005888
Musée Amora, Musée de la Moutarde Amora, Dijon 012781
Musée Amphoralis, Musée des Potiers Gallo-Romains, Sallèles-d'Aude 015653
Musée André Abbal, Carbonne . 012325
Musée André Dunoyer-de-Ségonzac, Boussy-Saint-Antoine 012177
Musée André-Marie Ampère, Musée de l'Électricité, Poleymieux-au-Mont-d'Or 014920
Musée Angladon, Avignon 011834
Musée Animalier, Ville-sous-Anjou 016229
Musée Animalier à Jougne, Jougne 013330
Musée Animé des Arts et Traditions Populaires, Serralongue 015779
Musée-Animé du Jouet et des Petits Trains, Colmar 012627
Musée Animé du Vin et de la Tonnellerie, Chinon 012561
Musée Anne de Beaujeu, Moulins (Allier) 014354
Musée Antiquités Gallo-Romain, Aoste 011694
Musée Antoine Brun, Sainte-Consorce 015606
Musée Antoine Lécuyer, Saint-Quentin (Aisne) 015525
Musée Antoine Vivenel, Compiègne 012643
Musée Antoine Wiertz, Musées Royaux des Beaux-Arts de Belgique, Bruxelles 003449
Musée Août 1944, La Bataille de la Poche de Falaise, Falaise 012949
Musée Apicole, Nedde 014440
Musée-Aquariophile, Dunkerque 012842
Musée Aquarium, Bergerac 011988
Musée-Aquarium d'Arcachon, Arcachon 011708
Musée Aquarium de la Rivière Dordogne, Creysse 012724
Musée-Aquarium du Laboratoire Arago, Banyuls-sur-Mer 011888
Musée Archéologique, Antigny . 011691
Musée Archéologique, Arlon . . . 003321
Musée Archéologique, Bagnols-en-Forêt 011870
Musée Archéologique, Banassac 011881
Musée Archéologique, Bezouce . 012025
Musée Archéologique, Brumath . 012243
Musée Archéologique, Champagnole 012433
Musée Archéologique, Charleroi 003511
Musée Archéologique, Corseul . 012679
Musée Archéologique, Cutry . . . 012740
Musée Archéologique, Delme . . 012752
Musée Archéologique, Dijon . . . 012782
Musée Archéologique, Ensisheim 012867
Musée Archéologique, Entrains-sur-Nohain 012870
Musée Archéologique, Escolives-Sainte-Camille 012898
Musée Archéologique, Gafsa . . . 042743
Musée Archéologique, Gouzeaucourt 013137
Musée Archeologique, L'Aigle . . 013530
Musée Archéologique, Laissac . . 013534
Musée Archéologique, Larache . 031777
Musée Archéologique, Lectoure 013721
Musée Archéologique, Magnac-Laval 013961
Musée Archéologique, Martizay . 014072
Musée Archéologique, Marvejols 014074
Musée Archéologique, Monségur 014203
Musée Archéologique, Mormoiron 014331
Musée Archéologique, Murol . . . 014387
Musée Archéologique, Namur . . 003896
Musée Archéologique, Nîmes . . 014500
Musée Archéologique, Peyriac-de-Mer 014857
Musée Archéologique, Pithiviers-le-Vieil 014877
Musée Archéologique, Rabat . . . 031788
Musée Archéologique, Saint-Gilles-du-Gard 015351
Musée Archéologique, Saint-Martin-de-Bromes 015448
Musée Archéologique, Saint-Pal-de-Mons 015489
Musée Archéologique, Saint-Paulien 015503
Musée Archéologique, Sainte-Agnès 015598
Musée Archéologique, Saverne . 015726
Musée Archéologique, Simiane-Collongue 015810
Musée Archéologique, Sollières-Sardières 015828
Musée Archéologique, Soulosse-sous-Saint-Élophe 015853
Musée Archéologique, Soyons . . 015858
Musée Archéologique, Strasbourg 015873
Musée Archéologique, Suippes . 015891
Musée Archéologique, Tétouan . 031799
Musée Archéologique, Thérouanne 015930
Musée Archéologique, Toulouse 015986
Musée Archéologique, Vachères 016080
Musée Archéologique, Vaison-la-Romaine 016084
Musée Archéologique, Vienne-en-Val 016207
Musée Archéologique, Villemagne-l'Argentière 016252
Musée Archéologique, Villeneuve-d'Ascq 016255
Musée Archéologique – Eglise Saint-Laurent, Grenoble 013179
Musée Archéologique – Hôtel de Sade, Saint-Rémy-de-Provence 015540
Musée Archéologique Armand Viré, Luzech 013928
Musée Archéologique Blasimon, Musée Municipal de Blasimon, Blasimon 012066
Musée Archéologique d'Argentomagus, Saint-Marcel (Indre) 015444
Musée Archéologique d'Autelbas, Bastogne 003344
Musée Archéologique de Chimtou, Jendouba 042747
Musée Archéologique de Djemila, Djemila 000059
Musée Archéologique de la Basse-Meuse, Oupeye 003943
Musée Archéologique de la Porte du Croux, Nevers 014467
Musée Archéologique de la Préhistoire à l'Époque Médiévale, Nyons . 014539
Musée Archéologique de la Région de Breteuil, Breteuil 012212
Musée Archéologique de Lamta, Lamta 042753
Musée Archéologique de l'Université Americaine, Beirut 030265
Musée Archéologique de Makthar, Makthar 042756
Musée Archéologique de Rom-Sainte-Soline, Rom 015133
Musée Archéologique de Saint-Pierre, Musées de Vienne, Vienne . . . 016203
Musée Archéologique de Sbeïtla, Sbeïtla 042767
Musée Archéologique de Sfax, Sfax 042768
Musée Archéologique de Site, Liffol-le-Grand 013794
Musée Archéologique de Sousse, Sousse 042774
Musée Archéologique de Thésée-la-Romaine, Thésée-la-Romaine . 015931
Musée Archéologique de Touraine, Tours 016016
Musée Archéologique de Viuz-Faverges, Faverges 012954
Musée Archéologique Départemental, Jublains 013336
Musée Archéologique Départemental, Saint-Bertrand-de-Comminges 015236
Musée Archéologique Départemental du Val-d'Oise, Guiry-en-Vexin . 013214
Musée Archéologique des Vestiges et Musée Agricole, Jonvelle 013323
Musée Archéologique du Cap Bon, Nabeul 042762
Musée Archéologique du Château Féodal, Bressieux 012200
Musée Archéologique du Gâtinais, Montargis 014221
Musée Archéologique du Minervois, Olonzac 014561
Musée Archéologique du Théâtre de Guelma, Guelma 000065
Musée Archéologique d'Uxellodunun, Vayrac 016143
Musée Archéologique et de Paléontologie, Minerve 014175
Musée Archéologique et d'Histoire Locale, Saint-Prix 015523
Musée Archéologique et Historique Cantonal de Gimont, Gimont . . 013108
Musée Archéologique et Historique du Comte de Logne, Vieuxville . . . 004112
Musée Archéologique et Historique Municipal, Bibliothèque André Malraux, Ermont 012890
Musée Archéologique et Lapidaire, Blois 012078
Musée Archéologique et Lapidaire de Montauban, Musées Gaumais, Montauban-sous-Buzenol 003887
Musée Archéologique et Missionnaire, Saxon-Sion 015739
Musée Archéologique Gallo-Romain, Revel-Tourdan 015080
Musée Archéologique Henri Prades, Lattes 013582
Musée Archéologique Hôtel Dieu, Cavaillon 012375
Musée Archéologique Jean Régnier, Mont-Saint-Vincent 014213
Musée Archéologique Le Cloître, Elne 012862
Musée Archéologique Léon Alègre, Bagnols-sur-Cèze 011872
Musée Archéologique Ludna, Saint-Georges-de-Reneins 015331
Musée Archéologique Municipal, Martres-Tolosane 014073
Musée Archéologique Municipal, Pélissanne 014820
Musée Archéologique Municipal de Fréjus, Fréjus 013048
Musée Archéologique Régional d'Orp-le-Grand, Orp-le-Grand 003938
Musée Archéologique Régional d'Oupeye-(M.A.R.O.), Vivegnis . 004119
Musée Archeologique René Galloux, Musées du Donjon, Montrichard 014309
Musée Archeologique, Grottes d'Azé, Azé 011858
Musée Archéologique, Le Trésor d'Eauze, Eauze 012850
Musée Archéologique, Musée de la Boulangerie, Blaye 012068
Musée Archéologique, Musée de Saint-Bazeille, Sainte-Bazeille 015603
Musée Archéologique, Musées Municipaux, Saintes 015634
Musée Archéologique Géologique et Ethnologique des Vans, Les Vans 013770
Musée Archipélitude, Saint-Pierre 038091
Musée-Archives de Saint-Pierre, Saint-Pierre 038092
Musée Ardeche d'Autrefois, Thueyts 015957
Musée Ardoisier, Saint-Julien-Mont-Denis 015402
Musée Ariana, Musée Suisse de la Céramique et du Verre, Genève 041528
Musée Arménien de France, Nourhan Fringhian Foundation, Paris . . . 014680
Musée Art du Chocolat, Lisle-sur-Tarn 013836
Musée Art et Culture, Lescar . . 013772
Musée Art Paysan, Villeneuve-de-Berg 016263
Musée Arteum, Châteauneuf-le-Rouge 012494
Musée Arthur Batut, Labruguiere 013515
Musée Arthur Le-Duc, Musée du Château, Torigni-sur-Vire 015971
Musée Arthur Rimbaud, Charleville-Mézières 012455
Musée Artisanal et Rural, Clion-sur-Seugne 012606
Musée Artisanal et Rural d'Arts, Ars 011748
Musée Arts d'Afrique et d'Asie, Vichy 016199
Musée Arts et Histoire, Bormes-les-Mimosas 012123
Musée Arts et Traditions Populaires, Firmi 012983

Musée Atelier d'Automates, Lusigny-sur-Barse 013920
Musée Atelier de Poterie, Neuvic-sur-l'Isle 014458
Musée-Atelier de Poupées Francépoque, Nances-Lac d'Aiguebelette 014399
Musée-Atelier du Cuivre, Villedieu-les-Poêles 016236
Musée-Atelier du Cuivre et de l'Argent, Paris 014681
Musée-Atelier du Trompe l'Oeil et du Décor Peint, Périgueux 014828
Musée-Atelier Werner Lichtner-Aix, Sérignan-du-Comtat 015776
Musée Atger, Montpellier 014288
Musée Athois, Ath 003328
Musée au Fil de la Pierre, Antoing 003276
Musée Au fil du papier, Musée Municipal de Pont-à-Mousson, Pont-à-Mousson 014928
Musée Aubois d'Histoire de l'Éducation, Troyes 016050
Musée Audemars Piguet, Le Brassus 041359
Musée Auguste Jacquet de Beaucaire, Beaucaire 011933
Musée Auguste Rodin, Paris 014682
Musée Augustin Bernard, Musée de Bourbon-l'Archambault, Bourbon-l'Archambault 012147
Musée Auto Moto Vélo, Châtellerault 012508
Musée Auto-Retro, Trooz 004095
Musée Automobile, La Clayette . 013394
Musée Automobile, Saint-Germain-Laval 015343
Musée Automobile, Saint-Jacques 006879
Musée Automobile Bellenavois, Bellenaves 011975
Musée Automobile de la Sarthe, Le Mans 013654
Musée Automobile des Voitures des Chefs d'État, Sauvigny-Le-Bois 015724
Musée Automobile Espera Sbarro, Pontarlier 014948
Musée aux Dominicaines, Espace Culturel des Dominicaines, Pont-l'Évêque 014941
Musée Baccarat, Paris 014683
Musée Baillet-Latour, Latour ... 003776
Musée Bajén-Vega, Monesties . 014197
Musée Balzac, Saché 015196
Musée Barbey d'Aurévilly, Saint-Sauveur-le-Vicomte 015559
Musée Barbier-Mueller, Genève . 041529
Musée Baron Gérard, Bayeux ... 011919
Musée Baron Martin, Gray 013173
Musée Baroncelli, Saintes-Maries-de-la-Mer 015640
Musée Barrois, Bar-le-Duc 011890
Musée Barthélemy Thimonnier de la Machine à Coudre et du Cycle, Amplepuis 011639
Musée Bartholdi, Colmar 012628
Musée-Base V 3 de Mimoyecques, Landrethun-le-Nord 013549
Musée Basque et de l'Histoire de Bayonne, Bayonne 011924
Musée Bastien Lepage, Montmédy 014275
Musée Baud, L'Auberson 041240
Musée Baudy, Giverny 013113
Musée Bayard et de la Chevalerie, Pontcharra 014952
Musée Beaubois, Carcassonne . 012328
Musée Beauceron du Grand Breau, Tivernon 015969
Musée Beaulne, Coaticook 005630
Musée Belge de Radiologie-Belgisch Museum voor Radiologie, Bruxelles 003450

Musée Benoist, Le Gavre 013634
Musée Benoît de Puydt, Bailleul 011875
Musée Benoît XIII, Châteaurenard (Bouches-du-Rhône) 012500
Musée Bernadette, Lourdes ... 013889
Musée Bernadotte, Pau 014806
Musée Bernard Buffet, Paris ... 014684
Musée Bernard Buffet, Shizuoka 029401
Musée Bernard d'Agesci, Niort . 014506
Musée Bernard Palissy, Lacapelle-Biron 013516
Musée Bert Flint, Agadir 031770
Musée Bert Flint, Marrakech .. 031779
Musée Bibliothèque, Anost 011678
Musée-Bibliothèque de l'Histoire du Protestantisme, Joigny (Yonne) 013320
Musée-Bibliothèque François Pétrarque, Fontaine-de-Vaucluse 013004
Musée-Bibliothèque Humaniste, Sélestat 015756
Musée-Bibliothèque Pierre-André Benoît, Alès 011594
Musée Bigouden, Pont-l'Abbé . 014939
Musée Bijoux de Braque, Saint-Dié-des-Vosges 015291
Musée Binger de Zaranou, Abengourou 008866
Musée Biochet-Brechot, Association des Amis du Vieux Caudebec, Caudebec-en-Caux 012368
Musée Bizarre und Mäusemuseum, Rieden bei Baden 041877
Musée Boilhet-Christofle, Saint-Denis (Seine-Saint-Denis) 015284
Musée Boleslaw Biegas, Paris . 014685
Musée Bon-Pasteur, Québec .. 006733
Musée Bonaparte, Auxonne ... 011826
Musée Bonaventure Fieullien, Regniowez 015045
Musée Bonnat, Bayonne 011925
Musée Bonnevallais, Bonneval . 012101
Musée Bossuet, Meaux 014107
Musée Botanique, Angers 011652
Musée Bouchard, Paris 014686
Musée Boucher-de-Perthes, Abbeville 011531
Musée Bourbonnais, Bellerive-sur-Allier 011976
Musée Bourbonnais, Moulins (Allier) 014355
Musée Bourdelle, Paris 014687
Musée Bourgeon-Jallieu, Bourgoin-Jallieu 012169
Musée Bourguiba de Art Islamique, Monastir 042761
Musée Bourguignon Perrin de Puycousin, Tournus 016011
Musée Boutdeville, Langueux . 013559
Musée Brasserie des Fagnes, Mariembourg 003851
Musée Bres Marcel, Saint-Léger-près-Troyes 015417
Musée Breton, Quimper 015018
Musée Briard et de la Crypte Mérovingienne, Jouarre 013328
Musée Bruno Danvin, Saint-Pol-sur-Ternoise 015516
Musée Bruxellois de la Gueuze, Museum van de Geuze, Bruxelles 003451
Musée Bruxellois de l'Industrie et du Travail, Bruxelles 003452
Musée Buffon, Buffon 012250
Musée C.C. Olsommer, Veyras . 042111
Musée Cadou, Louisfert 013887
Musée Calbet, Grisolles 013198
Musée Calvet, Avignon 011835
Musée Calvin, Noyon 014534
Musée Camille Lemonnier, Camille Lemonniermuseum, Bruxelles . 003453
Musée Camille Pissarro, Pontoise 014959
Musée Campagnard Agricole, Fillières 012982

Musée Campagnard du Haut-Beaujolais, Saint-Christophe . 015256
Musée Campanaire, L'Isle-Jourdain 013833
Musée Campanaire Bollée, Saint-Jean-de-Braye 015374
Musée Campredon, Maison René Char, L'Isle-sur-la-Sorgue 013834
Musée Canadien de la Nature, Gatineau 005869
Musée Canadien de la Poste, Gatineau 005870
Musée Canadien des Civilisations, Gatineau 005871
Musée Cantini, Marseille 014044
Musée Cantonal d'Archéologie et d'Histoire, Lausanne 041653
Musée Cantonal d'Archéologie, Musées Cantonaux du Valais, Sion 041992
Musée Cantonal de Géologie, Lausanne 041654
Musée Cantonal de Zoologie, Lausanne 041655
Musée Cantonal des Beaux-Arts, Lausanne 041656
Musée Cantonal d'Histoire Naturelle, Musées Cantonaux du Valais, Sion 041994
Musée Cantonal d'Histoire, Musées Cantonaux du Valais, Sion . 041993
Musée Cappon, Marans 013996
Musée Cappus-Fähndrich, Develier 041438
Musée Carnavalet – Histoire de Paris, Paris 014688
Musée Casa De Les Alberes, Argelès-sur-Mer 011716
Musée Casimir Delavigne, Pressagny-l'Orgueilleux 014996
Musée Casino de Sa, Saxon ... 041944
Musée Castral, Wintzenheim .. 016351
Musée Catalan des Arts et Traditions Populaires, Casa Pairal Joseph Deloncle, Perpignan 014839
Musée Caverne du Dragon, Oulches-la-Vallée-Foulon 014597
Musée Cavernicole de la Grotte de Balme, La Balme-les-Grottes . 013356
Musée Cavernicole de la Grotte de Thaïs, Saint-Nazaire-en-Royans 015474
Musée Célestin Freinet, Bonnut . 012105
Musée Celte – Gallo Romain, Musée Krumacker, Seltz 015758
Musée-Centre Charles Péguy, Orléans 014585
Musée Centre Culturel, Vichy . 016200
Musée Centre d'Art et de Littérature, L'Echelle 013719
Musée-Centre d'Arts Plastiques Chanot, Clamart 012584
Musée Centre de Documentation Alfred Desmasures, Hirson . 013266
Musée Centre de l'Alchimie, Tiffauges 015962
Musée-Centre de l'Imaginaire Arthurien, Concoret 012651
Musée Centre Historique des Sapeurs-Pompiers, Firminy 012986
Musée Centre National Jean Moulin, Bordeaux 012111
Musée Centre Nature de Botrange, Robertville 003970
Musée Cernuschi, Musée des Arts de l'Asie de la Ville de Paris, Paris 014689
Musée Césaire Phisalix, Mouthier-Haute-Pierre 014366
Musée César Filhol, Annonay . 011676
Musée Cévenol, Le Vigan (Gard) 013715
Musée Chalutier Le Kifanlo, OCEAM, Les Sables-d'Olonne 013763
Musée Champenois de l'Imprimerie, Musée de la Cave des Champagnes de Castellane, Épernay 012877

Musée Champollion, Figeac ... 012979
Musée-Chapelle Bellini, Cannes 012314
Musée-Chapelle Saint-Blaise-des-Simples, Milly-la-Forêt 014172
Musée Charbonneau-Lassay, Loudun 013881
Musée Charles Cros, Fabrezan . 012946
Musée Charles de Bruyères, Remiremont 015059
Musée Charles de Gaulle, Lille . 013803
Musée Charles et Marguerithe Bieth, Abengourou 008867
Musée Charles Léandre, Salle d'Exposition, Domfront 012804
Musée Charles-Louis Philippe, Cérilly 012388
Musée Charles Milcendeau-Jean Yole, Soullans 015852
Musée Charles-Portal, Cordes-sur-Ciel 012669
Musée Charles VII, Pôle de la Porcelaine, Mehun-sur-Yèvre . 014112
Musée Charlier, Charliermuseum, Bruxelles 003454
Musée Charolais du Machinisme Agricole, Neuvy-Grandchamp . 014463
Musée-Château, Grignan 013196
Musée-Château, La Ferté-Saint-Aubin 013412
Musée-Château, Montreuil-Bellay 014303
Musée-Château, Senonches ... 015770
Musée Château Abbaye de Cassan, Roujan 015177
Musée Château d'Aigremont, Awirs 003334
Musée Château d'Annecy et Observatoire Régional des Lacs Alpins, Annecy 011671
Musée-Château de Camponac, Pessac 014853
Musée-Château de Cheverny, Contres 012664
Musée-Château de Fontaine-Henry, Fontaine-Henry 013009
Musée-Château de Goulaine, Haute-Goulaine 013236
Musée Château de La Barben, La Barben 013357
Musée-Château de la Garnache, La Garnache 013423
Musée-Château de La Verrerie, Oizon 014556
Musée-Château de Pommard, Pommard 014924
Musée-Château de Pyrénée, Lourdes 013890
Musée-Château de Rochebrune, Etagnac 012916
Musée-Château de Saint-Germain-de-Livet, Saint-Germain-de-Livet . 015337
Musée-Château de Saint-Ouen, Saint-Ouen 015484
Musée-Château des Adhémar, Montélimar 014247
Musée Château des Princes de Ligne et Musée Lapidaire, Antoing . 003277
Musée-Château des Roures, La Bastide-de-Virac 013360
Musée-Château d'Espeyran, Saint-Gilles-du-Gard 015352
Musée-Château d'Etoges, Etoges 012927
Musée Château du Bucheneck, Soultz-Haut-Rhin 015854
Musée-Château du Moine Sacristain, Aurec-sur-Loire 011796
Musée-Château-Ferme de Falaën, Falaën 003594
Musée-Château Saint-Brisson, Saint-Brisson-sur-Loire 015245
Musée-Château, Château de Ville-d'Avray, Ville-d'Avray 016228
Musée Chez Manuel, Migne-Auxances 014168
Musée Chintreuil, Pont-de-Vaux 014934

Musée Christhi, Lourdes 013891
Musée Christian-Dior, Granville . 013150
Musée Ciotaden, La Ciotat . . . 013392
Musée Claude Bernard, Fondation
  Marcel Merieux, Saint-Julien-en-
  Beaujolais 015399
Musée Claude Chappe, Maison de la
  Culture, Brûlon 012242
Musée Claude Debussy, Saint-
  Germain-en-Laye 015340
Musée Claude Monet, Giverny . 013114
Musée Claude Nicolas Ledoux Saline
  Royale, Arc-et-Senans 011707
Musée Clément Ader, Muret . . . 014386
Musée Cloître Roman Saint-André-le-
  Bas, Musées de Vienne, Vienne 016204
Musée Cognacq-Jay, Paris 014690
Musée Cointreau, Saint-Barthélemy-
  d'Anjou 015232
Musée Colby-Curtis, Stanstead
  Historical Society, Stanstead . 007070
Musée Colette, Saint-Sauveur-en-
  Puisaye 015558
Musée Collection Eugène Pesch,
  Differdange 030575
Musée Collection Ferroviaire,
  Miramas 014179
Musée Colombophile du Mont-Valérien,
  Suresnes 015893
Musée Communal, Clérac 012590
Musée Communal, Pierre-
  Perthuis 014864
Musée Communal, Xhoris 004158
Musée Communal d'Archéologie et
  d'Art Religieux, Amay 003268
Musée Communal d'Archéologie
  Hesbignonne, Saint-Georges-sur-
  Meuse 003991
Musée Communal d'Archéologie, d'Art
  et d'Histoire, Nivelles 003914
Musée Communal de Braine-l'Alleud,
  Braine-l'Alleud 003392
Musée Communal de Folklore Léon
  Maes, Mouscron 003982
Musée Communal de Huy, Huy . 003720
Musée Communal de la Pierre de
  Sprimont, Sprimont 004044
Musée Communal de la Résistance et
  de la Déportation, Vénissieux . 016157
Musée Communal de la Ville
  de Braine-le-Comte, Braine-le-
  Comte 003396
Musée Communal de Woluwe-Saint-
  Lambert, Bruxelles 003455
Musée Communal du Comte de Jette,
  Bruxelles 003456
Musée Communal Georges Mulpas,
  Elouges 003574
Musée Communal Herstalien
  d'Archéologie et de Folklore,
  Herstal 003698
Musée Communal Robert Tatin, Cossé-
  le-Vivien 012683
Musée Communautaire des Soeurs de
  Sainte-Croix, Montréal 006384
Musée Comtadin, Carpentras . . 012341
Musée Comtadin-Duplessis,
  Carpentras 012342
Musée Comtois, Besançon 012001
Musée Condé, Chantilly 012445
Musée Condorcet, Ribemont . . . 015085
Musée Conrad-et-Marcel-
  Schlumberger, Crévecœur-en-
  Auge 012723
Musée Conservation, Saint-Paul-Trois-
  Châteaux 015499
Musée-Conservatoire de la Dentelle,
  Luxeuil-les-Bains 013922
Musée Conservatoire de la Vie
  Agricole et Rurale d'Autrefois,
  Gimont 013109
Musée Conservatoire de la Vie
  Agricole et Rurale, Ecomusée des
  Pays de l'Oise, Hétomesnil . . . 013260

Musée Conservatoire de l'Agriculture,
  Chartres 012468
Musée-Conservatoire de l'Outil,
  Brières 012198
Musée Conservatoire des Cépages,
  Saint-Ilpize 015365
Musée Conservatoire des Productions
  Fruitières, Saint-Jean-d'Arvey . 015371
Musée-Conservatoire des Traditions
  Rurales, Mauriac 014090
Musée-Conservatoire du Patrimoine,
  La Garde-Freinet 013422
Musée Conservatoire Ethnologique de
  Haute-Provence, Mane 013987
Musée Conservatoire Maritime et
  Rural, Narbonne 014427
Musée-Conservatoire Nationale de la
  Typographie, Paris 014691
Musée-Conservatoire Rural
  d'Arts et Traditions Populaires,
  Villefagnan 016240
Musée Constantin Meunier, Musées
  Royaux des Beaux-Arts de Belgique,
  Bruxelles 003457
Musée Costi, Bergerac 011989
Musée Crozatier, Le Puy-en-
  Velay 013692
Musée Curie, Paris 014692
Musée da Guine Portugesa,
  Bissau 023387
Musée da Silva des Arts et de la
  Culture, Porto-Novo 004186
Musée d'Abéché, Abéché 007508
Musée d'Allard, Montbrison . . . 014238
Musée d'Allauch, Allauch 011600
Musée Damien Bec, Laval-
  Roquecezière 013592
Musée d'Anatomie Delmas-Orfila-
  Rouvière, Paris 014693
Musée d'Anatomie Lesbres-Tagand de
  l'Ecole Vétérinaire de Lyon, Marcy-
  l'Étoile 014012
Musée dans la Mairie Vitrine,
  Panossas 014612
Musée d'Ansembourg, Musées
  d'Archéologie et d'Arts Décoratifs,
  Liège 003799
Musée d'Antan, Aigremont 011542
Musée d'Antan, La Léchère . . . 013437
Musée d'Anthropologie, Kinshasa 008834
Musée d'Anthropologie, Sainte-
  Foy 006952
Musée d'Anthropologie Préhistorique,
  Monaco 031717
Musée Dapper, Paris 014694
Musée d'Aquitaine, Bordeaux . . 012112
Musée Dar Cheraït, Tozeur 042779
Musée Dar-El-Annabi, Sidi Bou
  Saïd 042771
Musée Dar Hammamet,
  Hammamet 042745
Musée Dar Sghir, Mahdia 042754
Musée Dar-Si-Said, Marrakech . 031780
Musée Dar Taïeb, Nabeul 042763
Musée d'Archéologie, Antibes-Juan-
  les-Pins 011685
Musée d'Archéologie,
  Battambang 005282
Musée d'Archéologie, Capdenac-le-
  Haut 012321
Musée d'Archéologie, Izernore . 013305
Musée d'Archéologie, Le Pègue . 013678
Musée d'Archéologie, Séverac-le-
  Château 015793
Musée d'Archéologie, Tournai . . 004082
Musée d'Archéologie Bargoin,
  Clermont-Ferrand 012595
Musée d'Archéologie du Jura, Lons-le-
  Saunier 013866
Musée d'Archéologie et de Sciences
  Naturelles, Douai 012820
Musée d'Archéologie et d'Ethnologie,
  Villeneuve (Alpes-de-Haute-
  Provence) 016254

Musée d'Archéologie et d'Histoire
  Locale, Denain 012754
Musée d'Archéologie et Folklore,
  Musée Communal, Verviers . . . 004106
Musée d'Archéologie Industrielle,
  Vaulx 004102
Musée d'Archéologie Méditerranéenne,
  Marseille 014045
Musée d'Archéologie Précolombienne
  et de Préhistoire de la Martinique,
  Fort-de-France 030775
Musée d'Archéologie Préhistorique de
  l'Université de Liège, Liège . . . 003800
Musée d'Archéologie Sublacustre,
  Musée des Lacs, Sanguinet . . 015675
Musée d'Archéologie Tricastin, Saint-
  Paul-Trois-Châteaux 015500
Musée d'Archéologie, du Costume,
  d'Art et Traditions Populaires, Saint-
  Jean-de-Maurienne 015380
Musée d'Archéologieque, Saint-Ciers-
  sur-Gironde 015259
Musée d'Archéologique, Saint-
  Raphaël 015531
Musée d'Architecture – La Loge,
  Bruxelles 003458
Musée d'Arenberg, Rebecq 003961
Musée d'Argonne, Varennes-en-
  Argonne 016128
Musée d'Armes Anciennes, La Cluse-
  et-Mijoux 013395
Musée d'Armes-Citadelle de Dinant,
  Dinant 003555
Musée d'Armes de Liège, Liège . 003801
Musée d'Armes du Bordj Nord,
  Fèz 031774
Musée d'Armes et d'Historie Militaire,
  Tournai 004083
Musée d'Arromanches, Musée du
  Débarquement, Arromanches-les-
  Bains 011747
Musée d'Art, Toulon (Var) 015976
Musée d'Art Africain, Piré-sur-
  Seiche 014875
Musée d'Art Africain Religieux,
  Gentinnes 003644
Musée d'Art Africain, Musée de
  l'Institut Fondamental d'Afrique Noire,
  Dakar 038116
Musée d'Art Américain, Giverny . 013115
Musée d'Art Ancien, Musées Royaux
  des Beaux-Arts de Belgique,
  Bruxelles 003459
Musée d'Art Breton, Saint-
  Vougay 015593
Musée d'Art Contemporain,
  Dunkerque 012843
Musée d'Art Contemporain de Lyon,
  Lyon 013934
Musée d'Art Contemporain de
  Montréal, Montréal 006385
Musée d'Art Contemporain de Tanger,
  Tanger 031796
Musée d'Art Contemporain en Plein
  Air le Cyclop, Milly-la-Forêt . . . 014173
Musée d'Art Contemporain
  Fernet Branca, Saint-Louis-la-
  Chaussée 015428
Musée d'Art Contemporain, La Croix
  Blanche, Noyers-sur-Serein . . . 014532
Musée d'Art Contemporain, MAC,
  Marseille 014046
Musée d'Art Contemporarin Africain à
  Cotonou, Cotonou 004182
Musée d'Art Contemporarin des
  Laurentides, Saint-Jérôme . . . 006885
Musée d'Art de Joliette, Joliette 006051
Musée d'Art de Mont-Saint-Hilaire,
  Mont-Saint-Hilaire 006353
Musée d'Art du Valais, Sion . . . 041995
Musée d'Art et d'Archéologie,
  Aurillac 011800
Musée d'Art et d'Archéologie,
  Laon 013569

Musée d'Art et d'Archéologie,
  Senlis 015767
Musée d'Art et d'Archéologie, Soulac-
  sur-Mer 015850
Musée d'Art et d'Archéologie
  de l'Université d'Antananarivo,
  Antananarivo 030644
Musée d'Art et d'Archéologie de
  Périgord, Périgueux 014829
Musée d'Art et d'Archéologie, Musée
  Municipal, Hyères 013281
Musée d'Art et d'Artisanat,
  Corsavy 012678
Musée d'Art et de Folklore Régional,
  Fessy 012976
Musée d'Art et de Tradition Populaire,
  Bazoges-en-Pareds 011932
Musée d'Art et de Traditions
  Populaires André Abet, Saint-Laurent-
  de-Cerdans 015407
Musée d'Art et des Traditions
  Populaires, Sizun 015823
Musée d'Art et d'Histoire, Belfort 011968
Musée d'Art et d'Histoire,
  Bulgnéville 012252
Musée d'Art et d'Histoire, Chaumont
  (Haute-Marne) 012528
Musée d'Art et d'Histoire, Cholet 012568
Musée d'Art et d'histoire,
  Fribourg 041500
Musée d'Art et d'Histoire,
  Genève 041530
Musée d'Art et d'Histoire,
  Kaysersberg 013349
Musée d'Art et d'Histoire, Livry-
  Gargan 013841
Musée d'Art et d'Histoire,
  Narbonne 014428
Musée d'Art et d'Histoire, Nice . 014478
Musée d'Art et d'Histoire, Rueil-
  Malmaison 015189
Musée d'Art et d'Histoire, Saint-Denis
  (Seine-Saint-Denis) 015285
Musée d'Art et d'Histoire, Saint-Julien-
  du-Sault 015398
Musée d'Art et d'Histoire, Thorens-
  Glières 015951
Musée d'Art et d'Histoire Alfred Douët,
  Saint-Flour (Cantal) 015321
Musée d'Art et d'Histoire de Baugé,
  Baugé 011909
Musée d'Art et d'Histoire de Cognac,
  Cognac 012611
Musée d'Art et d'Histoire de la
  Chapelle Notre-Dame, Luz-Saint-
  Sauveur 013927
Musée d'Art et d'Histoire de Langres,
  Langres 013557
Musée d'Art et d'Histoire de Lisieux,
  Lisieux 013829
Musée d'Art et d'Histoire de Provence,
  Grasse 013156
Musée d'Art et d'Histoire de Puisaye,
  Villiers-Saint-Benoît 016297
Musée d'Art et d'Histoire de Toul,
  Toul 015975
Musée d'Art et d'Histoire des Côtes
  d'Armor, Saint-Brieuc 015242
Musée d'Art et d'Histoire du Judaïsme,
  Paris 014695
Musée d'Art et d'Histoire du Pays
  Blayais, Blaye 012069
Musée d'Art et d'Histoire Local,
  Pezens 014860
Musée d'Art et d'Histoire Louis
  Senlecq, L'Isle-Adam 013832
Musée d'Art et d'Histoire Marcel
  Dessal, Dreux 012838
Musée d'Art et d'Histoire Neuchâtel,
  Neuchâtel 041796
Musée d'Art et d'Histoire Roger
  Rodière, Montreuil-sur-Mer . . . 014306
Musée d'Art et d'Histoire Romain
  Rolland, Clamecy 012585

Musée d'Art et d'Histoire, Maison d'Armande Béjart, Meudon –
Musée de la Batterie de Merville, Merville-Franceville-Plage

Index of Institutions and Companies

Musée d'Art et d'Histoire, Maison d'Armande Béjart, Meudon . . . 014153
Musée d'Art et d'Histoire, Musées Municipaux, Rochefort (Charente-Maritime) 015122
Musée d'Art et d'Industrie, Saint-Étienne 015302
Musée d'Art et Historique, Montbéliard 014234
Musée d'Art et Tradition Populaire, Lamballe 013539
Musée d'Art et Traditions Populaires, Binic 012041
Musée d'Art et Traditions Populaires, Tunis 042783
Musée d'Art et Traditions Populaires, Musée Lorrain, Nancy 014403
Musée d'Art Funeraire de Marville, Marville 014075
Musée d'Art Haïtien, Port-au-Prince 023392
Musée d'Art Inuit Brousseau, Québec 006734
Musée d'Art Islamique, Marrakech 031781
Musée d'Art Islamique de Rakkada, Rakkada 042764
Musée d'Art Islamique Raqqada, Kairouan 042749
Musée d'Art Juif, Paris 014696
Musée d'Art Juif J.C. Katz, Colmar 012629
Musée d'Art Local, Oisemont . . 014555
Musée d'Art Mercian Karuizawa, Miyota 029047
Musée d'Art Militaire, Vincey . . 016302
Musée d'Art Moderne, Céret . . 012386
Musée d'Art Moderne, Troyes . . 016051
Musée d'Art Moderne de la Ville de Paris, Paris 014697
Musée d'Art Moderne de Saint-Étienne Métropol, Saint-Étienne 015303
Musée d'Art Moderne et Contemporain, Strasbourg . . 015874
Musée d'Art Moderne et Contemporain, Les Abattoirs, Toulouse 015987
Musée d'Art Moderne et d'Art Contemporain, Liège 003802
Musée d'Art Moderne et d'Art Contemporain, Nice 014479
Musée d'Art Moderne Grand-Duc Jean, Luxembourg 030588
Musée d'Art Moderne Jean Peské, Collioure 012621
Musée d'Art Moderne Lille Métropole, Villeneuve-d'Ascq 016256
Musée d'Art Moderne-Museum voor Moderne Kunst, Musées Royaux des Beaux-Arts de Belgique, Bruxelles 003460
Musée d'Art Moderne Richard Anacréon, Granville 013151
Musée d'Art Moderne, Donation Maurice Jardot, Belfort . . 011969
Musée d'Art Naïf, Lasne-Chapelle-Saint-Lambert 003774
Musée d'Art Naïf, Nice 014480
Musée d'Art Nègre, Yaoundé . . 005297
Musée d'Art Populaire, Le Monastier-sur-Gazeille 013665
Musée d'Art Populaire du Hauran, Muséd d'Antiquites de Bosra, Bosra 042255
Musée d'Art Populaire Religieux, Saint-Julien-Molhesabate 015401
Musée d'Art Religieux, Blois . . 012079
Musée d'Art Religieux, Castelnau-Magnoac 012356
Musée d'Art Religieux, Fréland . 013052
Musée d'Art Religieux, Marsal . 014035
Musée d'Art Religieux, Vermand 016171
Musée d'Art Religieux et d'Art Mosan, Liège 003803

Musée d'Art Roger Quilliot, Clermont-Ferrand 012596
Musée d'Art Rustique, Musée de la Poupée et du Joeut Edouard Wolff, Vianden 030603
Musée d'Art Sacré, Dijon . . . 012783
Musée d'Art Sacré, Hambye . . 013225
Musée d'Art Sacré, Le Monêtier-les-Bains 013667
Musée d'Art Sacré, Le Val . . . 013709
Musée d'Art Sacré, Lectoure . . 013722
Musée d'Art Sacré, Lorgues . . 013871
Musée d'Art Sacré, Saint-Béat . 015233
Musée d'Art Sacré, Saint-Cirq-Lapopie 015262
Musée d'Art Sacré, Senez 015765
Musée d'Art Sacré, Viviers-en-Ardèche 016315
Musée d'Art Sacré – Francis Poulenc, Rocamadour 015112
Musée d'Art Sacré Copte et Byzantin Scete, Le Revest-les-Eaux . . . 013696
Musée d'Art Sacré de Fourvière, Lyon 013935
Musée d'Art Sacré de Rostilas Loukine, Arsonval 011750
Musée d'Art Sacré du Gard, Pont-Saint-Esprit 014944
Musée d'Art Sacré Occidental, Le Revest-les-Eaux 013697
Musée d'Art Sacré, Trésor de la Cathédrale Notre-Dame, Paris . 014698
Musée d'Art Singulièr Raymond Reynaud, Sénas 015764
Musée d'Art Urbain, Montréal . . 006386
Musée d'Artisanat Monuments en Allumettes et Sciences Naturelles, Fontet 013023
Musée d'Arts Africains, Langonnet 013556
Musée d'Arts Décoratifs, Saint-Géoire-en-Valdaine 015329
Musée d'Arts et de Traditions Populaires, Amélie-les-Bains-Palalda 011626
Musée d'Arts et de Traditions Populaires, Cervione 012396
Musée d'Arts et de Traditions Populaires, Palinges 014609
Musée d'Arts et de Traditions Populaires, Poncé-sur-le-Loir . 014926
Musée d'Arts et de Traditions Populaires, F. Arnaudies, Saint-Paul-de-Fenouillet 015492
Musée d'Arts et de Traditions Populaires, Musée André Voulgre, Mussidan 014392
Musée d'Arts et d'Histoire du Donjon, Niort 014507
Musée d'Arts et Traditions Populaires, Cancale 012308
Musée d'Arts et Traditions Populaires, La Guérinière 013431
Musée d'Arts et Traditions Populaires, Marmoutier 014026
Musée d'Arts et Traditions Populaires, Yssingeaux 016361
Musée d'Arts et Traditions Populaires de Castellane et du Moyen Verdon, Castellane 012354
Musée d'Arts et Traditions Rurales, Saint-Alban-Auriolles 015207
Musée d'Arts Populaires, Laroque-des-Albères 013575
Musée d'Arts Populaires de Lille-Sud, Lille 013804
Musée d'Arts Religieux, Verdelais 016163
Musée d'Assier, Feurs 012977
Musée Daubigny, Auvers-sur-Oise 011816
Musée Dauphinois, Grenoble . . 013180
Musée David, Angers 011653
Musée David et Alice Van Buuren, Bruxelles 003461

Musée d'Avions Montélimar, Montélimar 014248
Musée de Baalbeck, Baalbeck . 030263
Musée de Bagnes, Le Châble . . 041397
Musée de Basse-Navarre et des Chemins de Saint-Jacques, Institut Culturel Basque, Saint-Palais (Pyrénées-Atlantiques) 015490
Musée de Bastia, Bastia 011905
Musée de Bataille, Mutzig 014394
Musée de Béjaïa, Béjaïa 000052
Musée de Beni Abbès, Bechar . . 000051
Musée de Bible et Terre Sainte, Paris 014699
Musée de Bibracte, Saint-Léger-sous-Beuvray 015418
Musée de Biot, Musée d'Histoire et de Céramique Biotoise, Biot . . 012043
Musée de Boinchoux, Besse-et-Saint-Anastaise 012010
Musée de Boîtes à Musique et Automates, Sainte-Croix 041913
Musée de Bonoua, Bonoua . . . 008871
Musée de Borda, Dax 012747
Musée de Bordj Moussa, Béjaïa 000053
Musée de Botanique, Lausanne . 041657
Musée de Bouxwiller et du Pays de Hanau, Bouxwiller 012180
Musée de Bretagne, Rennes (Ille-et-Vilaine) 015066
Musée de Bulla Regia, Jendouba 042748
Musée de Cahors Henri-Martin, Cahors 012288
Musée de Calès, Lamanon 013538
Musée de Calligraphie et Epigraphie Arabe, Damascus 042257
Musée de Carouge, Carouge . . . 041391
Musée de Castel-Merle, Sergeac . 015717
Musée de Centre Social, Luzy . . 013929
Musée de Cerdagne, Sainte-Léocadie 015614
Musée de Cerdagne d'Eyne, Eyne 012942
Musée de Cerisiers La Grange a Janou, Cerisiers 012389
Musée de Cervières, Cervières . 012395
Musée de Charlevoix, La Malbaie 006134
Musée de Charroux et de son Canton, Charroux-d'Allier 012466
Musée de Château, Durtal 012848
Musée de Château Guillaume, Lignac 013795
Musée de Châtillon-Coligny, Châtillon-Coligny 012514
Musée de Chauny, Chauny 012533
Musée de Cherchell, Cherchell . 000055
Musée de Chine, China Museum Scheut – Anderlecht, Bruxelles 003462
Musée de Circonscription Archéologique, Sétif 000073
Musée de Cire de l'Historial, Sainte-Anne-d'Auray 015601
Musée de Cire et Salle de Documents Historiques, Perros-Guirec 014849
Musée de Cire-Historial de la Haute-Auvergne, Aurillac 011801
Musée de Cire Historique, Byblos 030276
Musée de Cires, Historial des Princes de Monaco, Monaco 031718
Musée de Citadelle, Sisteron . . . 015814
Musée de Clichy et Société Historique, Clichy 012605
Musée de Cohennoz, Cohennoz . 012611
Musée de Cox, Cox 012711
Musée de Dar Jamaï, Meknès . . 031784
Musée de Diamare, Maroua . . . 005295
Musée de Djelfa, Djelfa 000058
Musée de Dol, Dol-de-Bretagne . 012799
Musée de Domaine de Lacroix-Laval, Marcy-l'Étoile 014013
Musée de Drummond Castle, Ile-de-Molène 013287
Musée de Enchanters, Antrain-sur-Couesnon 011692

Musée de Faïences Christine Viennet, Béziers 012021
Musée de Faykod, Aups 011794
Musée de Ferroviaire à Orient-Express, Sospel 015840
Musée de Figurines Végétales, Basse-Pointe 030774
Musée de Flandre, Bailleul 011876
Musée de Folklore, Nismes 003910
Musée de Folklore de la Chée, Heiltz-le-Maurupt 013247
Musée de Folklore et d'Histoire Armand Pellegrin, Ancienne Ecole Communale des Garçons, Opheylissem 003936
Musée de Folklore Flamand Jeanne Devos, Wormhout 016355
Musée de France de Berck-sur-Mer, Berck-sur-Mer 011987
Musée de Fresques, Blain 012053
Musée de Gajac, Villeneuve-sur-Lot 016274
Musée de Géologie, Lozanne . . 013912
Musée de Géologie, Sainte-Foy . 006953
Musée de Géologie Botanique – Espace Geologique, La Bourboule 013368
Musée de Germigny et Grange Germignonne, Germigny-des-Prés 013096
Musée de Glozel, Ferrières-sur-Sichon 012973
Musée de Grenoble, Grenoble . 013181
Musée de Groesbeeck de Croix, Namur 003897
Musée de Guérin, Guérin 005935
Musée de Hottemme, Barvaux . 003339
Musée de Jean Chouan, Saint-Ouen-des-Toits 015485
Musée de Jouet et da la Poupée Ancienne, L'Isle-sur-la-Sorgue . 013835
Musée de Jouques, Musée d'Art et d'Histoire, Jouques 013331
Musée de Jules Gounon-Loubens, Loubens-Lauragais 013879
Musée de Kabgayi, Kabgayi . . . 038087
Musée de Kamouraska, Kamouraska 006062
Musée de Kent, Bouctouche . . . 005446
Musée de Kerhinet, Saint-Lyphard 015431
Musée de la Baleine, Port-la-Nouvelle 014968
Musée de la Bande Dessinée et Musée Municipal, Angoulême . 011663
Musée de la Banderette, Travers 042067
Musée de la Banque Nationale de Belgique, Bruxelles 003463
Musée de la Basilique et des Thermes Romains, Arlon 003322
Musée de la Bataille, Fontenoy-en-Puisaye 013020
Musée de la Bataille, Rocroi . . . 015128
Musée de la Bataille de Tilly sur Seulles 1944, Tilly-sur-Seulles 015966
Musée de la Bataille des Ardennes, La Roche-en-Ardenne 003766
Musée de la Bataille des Ardennes, Wiltz 030607
Musée de la Bataille du 6 Août 1870, Woerth 016354
Musée de la Bataille Mai-Juin 1940, Semuy 015763
Musée de la Batellerie, Briennon 012224
Musée de la Batellerie, Conflans-Sainte-Honorine 012658
Musée de la Batellerie de l'Ouest, Redon 015042
Musée de la Batellerie d'Offendorf, Offendorf 014547
Musée de la Batellerie et du Plan Incliné, Saint-Louis-Arzviller . 015425
Musée de la Batterie de Merville, Merville-Franceville-Plage 014145

Musée de la Bécane à Grand-Père, Sault-lès-Rethel 015711
Musée de la Bendrologie, Manéga 005275
Musée de la Bière, Armentières 011735
Musée de la Bière, Sankt Vith .. 013999
Musée de la Bière, Stenay 015861
Musée de la Blanchisserie Artisanale Joseph Gladel, Craponne (Rhône) 012714
Musée de la Bohème, Yviers .. 016364
Musée de la Boissellerie, Bois-d'Amont 012092
Musée de la Boite à Biscuits, Opheylissem 003937
Musée de la Boîte en Fer Blanc Lithographiée, Grand-Hallet .. 003656
Musée de la Bonneterie et du Négoce de la Toile, Quevaucamps 003956
Musée de la Bonneterie, Musée de Vauluisant, Troyes 016052
Musée de la Boulangerie, Bonnieux 012104
Musée de la Boulangerie Rurale, La Haye-de-Routot 013432
Musée de la Bourrellerie et de la Ferronnerie, Lahamaide 003771
Musée de La Bouteille, Saint-Émilion 015297
Musée de la Brasserie-Museum van de Belgische Brouwers, C.B.B. Museum, Bruxelles 003464
Musée de la Bresse – Domaine des Planons, Saint-Cyr-sur-Menton 015279
Musée de la Broderie, Fontenoy-le-Château 013022
Musée de la Cabane des Bangards, Thann 015926
Musée de la Cadillac, Saint-Michel-sur-Loire 015471
Musée de la Calligraphie et de l'Imagerie Ancienne, Musée Christhi, Capvern-les-Bains 012323
Musée de la Camargue, Arles .. 011729
Musée de la Caricature, La Cassagne 013372
Musée de la Carte Postale, Mortagne-sur-Gironde 014339
Musée de la Casse, Orcières .. 014575
Musée de la Castre, Cannes ... 012315
Musée de la Cavalerie, Saumur 015714
Musée de la Cave, Damery 012743
Musée de la Cave des Champagnes de Castellane, Épernay 012878
Musée de la Céramique, Digoin 012775
Musée de la Céramique, Lezoux 013787
Musée de la Céramique, Rouen 015164
Musée de la Céramique d'Andenne, Andenne 003272
Musée de la Céramique et de l'Ivoire, Musée Municipal de Commercy, Commercy 012641
Musée de la Céramique Saint-Jean-l'Aigle, Longwy 013864
Musée de la Cervoise, du Gruyt et des Bières Mediévales, Musée de la Bière et du Peket, Anthisnes .. 003274
Musée de la Chalosse, Montfort-en-Chalosse 014258
Musée de la Chapellerie, Espéraza 012907
Musée de la Chapellerie, Le Somail 013700
Musée de la Charronnerie, Brienne-la-Vieille 012222
Musée de la Chartreuse et Fondation Bugatti, Molsheim 014194
Musée de la Chasse et de la Nature, Baurech 011915
Musée de la Chasse et de la Nature, Chambord 012425
Musée de la Chasse et de la Nature, Foix 012998
Musée de la Chasse et de la Nature,

Paris 014700
Musée de la Chasse et de la Venerie, Villiers-le-Duc 016296
Musée de la Chasse, de la Vénerie et de la Protection de la Nature, Lavaux-Sainte-Anne 003777
Musée de la Chasse, Musée de la Faune, Boutenac 012178
Musée de la Châtaigneraie, Joyeuse 013335
Musée de la Chemiserie et de l'Elégance Masculine, Argenton-sur-Creuse 011722
Musée de la Chevalerie, Armes et Archeries, Carcassonne 012329
Musée de la Chimie, Jarrie 013309
Musée de la Chouannerie et des Guerres de l'Ouest, Plouharnel 014910
Musée de la Citadelle Vauban, Belle-Île-en-Mer 011972
Musée de la Civilisation, Québec 006735
Musée de la Cloche, Sévrier ... 015796
Musée de la Cloche et du Carillon, Tellin 004051
Musée de la Cloches et de Sonnailles, Espace Campanaire André Malraux, Hérépian 013251
Musée de la Cochonnaille, Tourzel-Ronzieres 016028
Musée de la Cohue, Vannes (Morbihan) 016119
Musée de la Coiffe, Blesle 012072
Musée de la Coiffe et du Costume Oléronnais, Le Grand-Village-Plage 013636
Musée de la Coiffe et du Pays Pelebois le Prieuré, Souvigne (Deux-Sèvres) 015856
Musée de la Colombophilie, Bouvignies 012179
Musée de la Comédie, Preignac 014994
Musée de la Commanderie, Viaprès-le-Petit 016193
Musée de la Communication en Alsace, Poste-Diligences-Télécoms, Riquewihr 015102
Musée de la Compagnie des Indes, Port-Louis 014969
Musée de la Compagnie Royale des Anciens Arquebusiers de Visé, Visé 004116
Musée de la Comtesse de Ségur, Aube 011767
Musée de la Conciergerie, Paris 014701
Musée de la Confiture, Pau 014807
Musée de la Confrérie des Vignerons, Vevey 042106
Musée de la Conscription, Strasbourg 015875
Musée de la Conserverie Le Gall, Loctudy 013849
Musée de la Construction Navale, Noirmoutier-en-l'Ile 014518
Musée de la Contrefaçon, Paris 014702
Musée de la Corbillière, Mer 014135
Musée de la Cordonnerie, Alby-sur-Chéran 011587
Musée de la Correrie, Saint-Pierre-de-Chartreuse 015507
Musée de la Corse, Corte 012681
Musée de la Course Landaise, Bascons 011902
Musée de la Coutellerie, Nogent-en-Bassigny 014513
Musée de la Coutellerie, Thiers . 015934
Musée de la Création Franche, Bègles 011965
Musée de la Crèche, Chaumont (Haute-Marne) 012529, 103586
Musée de la Crèche Provençale, Cavaillon 012376
Musée de la Cuisine, Saint-Charles-de-Drummond 006862
Musée de la Curiosité et de la Magie,

Paris 014703
Musée de la Dame aux Camélias, Gacé 013071
Musée de la Défense Aérienne, Alouette 005323
Musée de la Dentelle, Caudry .. 012370
Musée de la Dentelle, Chamalières-sur-Loire 012416
Musée de la Dentelle, Marche-en-Famenne 003843
Musée de la Dentelle, Mirecourt 014184
Musée de la Dentelle à la Main, Arlanc-en-Livradois 011726
Musée de la Dentelle du Puy-la-Galerie, Le Puy-en-Velay ... 013693
Musée de la Déportation et de la Résistance, Tarbes 015915
Musée de la Distillerie, Pontarlier 014949
Musée de la Distillerie Combier, Saumur 015715
Musée de la Douane et des Frontières, Hestrud 013259
Musée de la Draperie, Vienne .. 016205
Musée de la Fabuloserie, Dicy . 012765
Musée de la Faculté de Médecine, Strasbourg 015876
Musée de la Faïence, Forges-les-Eaux 013026
Musée de la Faïence, Marseille . 014047
Musée de la Faïence, Montereau-Fault-Yonne 014251
Musée de la Faïence de la Manufacture Royal Boch, Centre de la Faïence Royal Boch, La Louvière 003764
Musée de la Faïence et de la Poterie, Ferriére-la-Petite 012970
Musée de la Faïence et des Arts de la Table, Samadet 015667
Musée de la Faïence Jules Verlingue, Quimper 015019
Musée de la Faïencerie de Gien, Gien 013101
Musée de la Famille, Chasselay 012474
Musée de la Faucillonnaie, Vitré 016306
Musée de la Faune, Bellevaux . 011977
Musée de la Faune, Lanhélin .. 013560
Musée de la Faune Alpine, Servoz 015783
Musée de la Faune et de la Nature, Séez 015752
Musée de la Faune et la Flore, Etaples 012919
Musée de la Fausse Monnaie, Maison Farinet, Saillon 041905
Musée de la Faux, Pont-Salomon 014946
Musée de la Fédération Nationale de Spéléologie, Revel 015079
Musée de la Femme Henriette Bathily, Gorée 038118
Musée de La Ferblanterie, La Tour-Blanche 013496
Musée de la Ferme d'Antan, Ploudiry 014901
Musée de la Ferme et des Vieux Métiers, Bosquentin 012125
Musée de la Ferme, Musée Municipal, Blangy-sur-Bresle 012058
Musée de la Ferronnerie, Francheville (Eure) 013043
Musée de la Fête Foraine, Plancher-les-Mines 014882
Musée de la Fève et de la Crèche, Musée des Arts et Traditions Populaires, Blain 012054
Musée de la Figurine, La Grande-Fosse 013428
Musée de la Figurine, Toulon (Var) 015977
Musée de la Figurine, Tulette .. 016058
Musée de la Figurine Historique, Compiègne 012644
Musée de la Figurine Historique et du Jouet Ancien, Le Val 013710

Musée de la Figurine Historique, Château de Mongaston, Charre 012464
Musée de la Filature, Angoustrine-Villeneuve 011669
Musée de la Flore et de la Faune du Haut Bugey, Hotonnes 013273
Musée de la Foire et du Théâtre Itinérant, Saint-Ghislain 003992
Musée de la Folie Marco, Barr . 011900
Musée de la Fontaine et de l'Eau, Genval 003645
Musée de la Forêt, Saugues-en-Gévaudan 015703
Musée de la Forêt Ardennaise, Renwez 015070
Musée de la Forêt, Centre d'Interpretation de la Nature Pierre Noe, Spa 004040
Musée de la Forge, Denazé ... 012757
Musée de la Forge, Romedenne 003981
Musée de la Forteresse, Festungsmuseum, Luxembourg 030589
Musée de la Fortification Cuirassée, Villey-le-Sec 016294
Musée de la Foudre, Marcenat . 013999
Musée de la Fourche et de la Vie Rurale, Mortier 003891
Musée de la Fourme et du Sabot, Sauvain 015721
Musée de la Fraise et du Patrimoine, Plougastel-Daoulas 014903
Musée de la Fraise et du Terror Wépionnais, Wépion 004143
Musée de la Franc-Maçonnerie, Musée du Grand Orient de France, Paris 014704
Musée de la France Protestante de l'Ouest, Monsireigne 014204
Musée de la Gare, Temiscaming 007131
Musée de la Gaspésie, Gaspé .. 005863
Musée de la Gendarmerie, Melun 014119
Musée de la Gendarmerie, Toamasina 030651
Musée de la Géologie et de l'Archéologie, Musée Municipal, Blangy-sur-Bresle 012059
Musée de la Glace, Mazaugues 014098
Musée de La Glacerie, La Glacerie 012569
Musée de la Gourmandise, Hermalle-sous-Huy 003697
Musée de la Grenouillère, Maison Joséphine, Croissy-sur-Seine . 012725
Musée de la Grosse Forge, Aube 011768
Musée de la Grotte de Cougnac, Payrignac 014818
Musée de la Grotte des Huguenots, Vallon-Pont-d'Arc 016108
Musée de la Guerre, Calais ... 012293
Musée de la Guerre, Clervaux . 030568
Musée de la Guerre, Fromelles . 013062
Musée de la Guerre 1939–45, Pourrain 014982
Musée de la Guerre au Moyen Âge, Castelnaud-la-Chapelle 012359
Musée de la Guerre de 1870, Gravelotte 013170
Musée de la Halte du Pèlerin, Larressingle 013577
Musée de la Hardt, Fessenheim 012974
Musée de la Haute-Auvergne, Saint-Flour (Cantal) 015322
Musée de la Haute Haine, Carnières 003507
Musée de la Haute Sûre, Martelange 003852
Musée de la Hesbaye, Remicourt 003964
Musée de la Houille Blanche et de ses Industries, Lancey 013564
Musée de la la Faune des Pyrénées, Nébias 014439
Musée de la Laine, du Vêtement de Laine et du Tissage, Louhossoa 013886

Musée de la Lauze, Annoisin-
Châtelans 011675
Musée de la Lavande, Cabrières-
d'Avignon 012267
Musée de la Lavande, Saint-
Remèze 015532
Musée de la Légion Étrangère,
Aubagne 011765
Musée de la Lessive, Spa 004041
Musée de la Liberation, Aywaille 003335
Musée de la Libération, Cherbourg-
Octeville 012549
Musée de la Libération 15 Août 1944,
Le Muy 013674
Musée de la Libération de la
Normandie, Surrain 015896
Musée de la Ligne Koningshooikt-
Wavre, Chaumont 003517
Musée de la Ligne Maginot,
Bambiderstroff 011880
Musée de la Ligne Maginot,
Hatten 013232
Musée de la Ligne Maginot,
Hunspach 013276
Musée de la Ligne Maginot,
Lembach 013728
Musée de la Ligne Maginot Alpine,
Sainte-Agnès 015599
Musée de la Ligne Maginot Immerhof,
Association du Petit Ouvrage,
Hettange-Grande 013261
Musée de la Ligne Maginot, Fort de
Fermont, Longuyon 013862
Musée de la Locomotion et de
l'Attelage de l'Île de France,
Coignières 012616
Musée de la Loire, Cosne-Cours-sur-
Loire 012682
Musée de la Lunette, Morez .. 014322
Musée de la Lunette, Musées des
Techniques et Cultures Comtoises,
Morez 014323
Musée de la Lutherie et de
l'Archèterie Françaises,
Mirecourt 014185
Musée de la Machine à Vapeur, Arc-
en-Barrois 011706
Musée de la Machine à Vapeur,
Oignies 014552
Musée de la Machine Agricole,
Vroncourt 016332
Musée de la Machine Agricole
Ancienne, Saint-Loup 015430
Musée de la Machine Agricole et à
Vapeur, Ambert 011613
Musée de la Machine de Guerre,
Espalion 012900
Musée de la Maille, Riorges .. 015098
Musée de la Main, Fondation Claude
Verdan, Lausanne 041658
Musée de la Mairie de Crécy, Crécy-la-
Chapelle 012716
Musée de la Maison Carrée, Nay 014437
Musée de la Maison de Tartarin,
Tarascon (Bouches-du-Rhône) . 015908
Musée de la Maison s'Kleenderfel,
Weyersheim 016348
Musée de la Maquette, Clervaux 030569
Musée de la Marine, Etaples .. 012920
Musée de la Marine, Grasse ... 013257
Musée de la Marine, Saint-Brévin-les-
Pins 015241
Musée de la Marine de la Seine,
Caudebec-en-Caux 012369
Musée de la Marine de Loire,
Châteauneuf-sur-Loire 012496
Musée de la Marine et de l'Économie
de Marseille, Marseille 014048
Musée de la Marine et de la Marne
d'Ethnographie et d'Art Populaire,
Honfleur 013269
Musée de la Médaille Miraculeuse, La
Vie de Sainte Catherine Labouré,
Lourdes 013892

Musée de la Médecine, Bruxelles 003465
Musée de la Mémoire de la
Vigne et du Vin, Saint-Pierre-de-
Mézoargues 015512
Musée de la Mémoire des Murs et
d'Archéologie Locale, Verneuil-en-
Halatte 016175
Musée de la Mémoire du Royans,
Rochechinard 015117
Musée de la Mémoire et de la Paix,
Clerval 012604
Musée de la Mémoire Paysanne,
Estissac 012914
Musée de la Mer, Biarritz 012028
Musée de la Mer, Cannes 012316
Musée de la Mer, Gorée 038119
Musée de la Mer, Havre-Aubert 005993
Musée de la Mer, Îles-de-la-
Madeleine 006028
Musée de la Mer, Le Grau-du-
Roi 013638
Musée de la Mer, Paimpol 014604
Musée de la Mer de Pointe-au-Père,
Rimouski 006806
Musée de la Mer des Faluns, Doué-la-
Fontaine 012825
Musée de la Mer OCEAM, Les Sables-
d'Olonne 013764
Musée de la Mer, Centre National de
Recherches Océanographiques, Nosy-
Bé 030650
Musée de la Mer, Musée de Ua Huka,
Hane 016373
Musée de la Métallurgie, Bogny-sur-
Meuse 012089
Musée de la Métallurgie, Saint-Chély-
d'Apcher 015253
Musée de la Météorite, L'Aigle . 013531
Musée de la Météorologie,
Valleraugues 016105
Musée de la Meunerie,
Bannegon 011886
Musée de la Meunerie, Challain-la-
Potherie 012403
Musée de la Meunerie, Le Breuil 013613
Musée de la Meunerie et de la
Boulangerie, Harzé 003678
Musée de la Meunerie-Moulin de
la Petite Strument, La Roche-en-
Ardenne 003767
Musée de la Meunerie, Moulin à Eau,
Courtelevant 012704
Musée de la Meunerie, Moulin Banal,
Braine-le-Château 003395
Musée de la Mine, Brassac-les-
Mines 012192
Musée de la Mine, Bruay-la-
Buissière 012239
Musée de la Mine, Champagnac-les-
Mines 012430
Musée de la Mine, La Machine . 013439
Musée de la Mine, La Ricamaire 013456
Musée de la Mine, Nœux-les-
Mines 014512
Musée de la Mine, Saint-Étienne 015304
Musée de la Mine, Sainte-Maime 015615
Musée de la Mine de Bois-du-Luc,
Houdeng-Aimeries 003714
Musée de la Mine de Cap Garonne,
Le Pradet 013691
Musée de la Mine de Cuivre et de
Soufre, Saint-Pierre-de-la-Palud 015509
Musée de la Mine et de l'École,
Harnes 013228
Musée de la Mine et de Meunerie-
Moulin de Marcy, Le Molay-
Littry 013664
Musée de la Mine et des Hommes,
Blanzy 012065
Musée de la Mine et des Techniques
Minières, Giromagny 013112
Musée de la Mine Lucien-Mazars,
Aubin 011778
Musée de la Mine Marcel Maulini,

Ronchamp 015144
Musée de la Mine-Mémoire Ouvrière,
Harchies 003673
Musée de la Mine, Centre d'Exposition
Langevin, Grande-Synthe ... 013148
Musée de la Mine, de la Verrerie et
du Chemin de Fer, Epinac 012884
Musée de la Mine, Musée de la
Clouterie, Musée du Gazomètre,
Fontaine-l'Evêque 003603
Musée de la Mine, Puits de Mine Hély
d'Oissel, Gréasque 013176
Musée de la Miniature,
Montélimar 014249
Musée de la Miniature, Saint-Sauves-
d'Auvergne 015556
Musée de la Mode, Marseille .. 014049
Musée de la Mode, Yverdon-les-
Bains 042171
Musée de la Mode et du Textile, Les
Arts Decoratifs, Paris 014705
Musée de la Mode Retrouvée,
Digoin 012776
Musée de la Moineaudière, Xonrupt-
Longemer 016358
Musée de la Moisson, Sagy (Magny-
en-Vexin) 015200
Musée de la Monnaie de Paris,
Paris 014706
Musée de la Montagne, Le Haut-du-
Them 013640
Musée de la Montagne Saint-Pierre,
Lanaye 003772
Musée de la Montre Droz-Flores,
Villers-le-Lac 016283
Musée de la Mosaïque et des Émaux,
Briare 012218
Musée de la Mosaïque Gallo-Romaine,
Sorde-l'Abbaye 015834
Musée de la Moto, Marseille ... 014050
Musée de la Moto, Savigny-les-
Beaune 015734
Musée de la Motocyclette,
Entrevaux 012873
Musée de la Musique, Anduze . 011645
Musée de la Musique de
Ouagadougou, Ouagadougou .. 005278
Musée de la Musique et des
Instruments, Musée Municipal,
Blangy-sur-Bresle 012060
Musée de la Musique Mécanique,
Haubourdin 013234
Musée de la Musique Mécanique, Les
Gets 013750
Musée de la Musique Mécanique,
Mormoiron 014332
Musée de la Musique, Cité de la
Musique, Paris 014707
Musée de la Mytiliculture,
Esnandes 012899
Musée de la Mytiliculture,
Pénestin 014822
Musée de la Nacre et de la Tabletterie,
Méru 014142
Musée de la Naissance de l'Aviation
Suisse, Payerne 041835
Musée de la Natation, Chissey-les-
Mâcon 012565
Musée de la Nativité, Lourdes . 013893
Musée de la Nature, Ettelbruck . 030584
Musée de la Nature, Sivry 004031
Musée de la Nature et de la Chasse,
Thuir 015959
Musée de la Nature et des Sciences,
Sherbrooke, Québec 007024
Musée de la Navigation Fluviale/
Ascenseur à Bateaux, Arques (Pas-
de-Calais) 011742
Musée de la Pailletterie, Amilly . 011635
Musée de la Paléontologie et de
l'Évolution, Montréal 006387
Musée de la Papeterie-Canson et
Montgolfier, Davézieux 012746
Musée de la Parfumerie, Paris . 014708

Musée de la Parfumerie Cyrnarom,
Bastia 011906
Musée de la Parfumerie Fragonard,
Grasse 013158
Musée de la Parfumerie Fragonard,
Paris 014709
Musée de la Parfumerie Galimard,
Grasse 013159
Musée de la Parfumerie Molinard,
Grasse 013160
Musée de la Parfumerie Molinard,
Nice 014481
Musée de la Paysannerie, Baguer-
Morvan 011873
Musée de la Paysannerie et des Vieux
Métiers, Valentigney 016098
Musée de la Pêche, Concarneau 012648
Musée de la Pêche et de la Nature,
APNLE, Levallois-Perret 013778
Musée de la Pêche et des Poissons,
Amberloup 003269
Musée de la Pelote Basque, Sare 015683
Musée de la Perle, Papeete ... 016377
Musée de la Petite Gare, Olonne-sur-
Mer 014559
Musée de la Pétrification,
Savonnières 015738
Musée de la Pharmacie, Tence . 015922
Musée de la Pharmacie Albert Ciurana,
Montpellier 014289
Musée de la Photo, Gracay 013138
Musée de la Photographie,
Charleroi 003512
Musée de la Photographie,
Mougins 014347
Musée de la Photographie, Port
Louis 030793
Musée de la Pierre, Trambly .. 016030
Musée de la Pierre à Fusil, Luçay-le-
Male 013914
Musée de la Pierre à Fusil,
Meusnes 014157
Musée de la Pierre Écrite,
Vézelay 016190
Musée de la Pierre et du Marbre,
Basècles 003340
Musée de la Pierre et Site des
Carrières, Maffle 003837
Musée de la Pierre Philosophale,
Issoire 013298
Musée de la Pince,
Montécheroux 014246
Musée de la Pointe Saint-Mathieu,
Plougonvelin 014904
Musée de la Poire Tapée,
Rivarennes 015106
Musée de la Police de l'Île de
Montréal et de l'Île Bizard,
Montréal 006388
Musée de la Police et de la
Gendarmerie, Floreffe 003600
Musee de la Pomme, Bourg-Saint-
Maurice 012155
Musée de la Pomme et du Cidre,
Pleudihen-sur-Rance 014890
Musée de la Pomme et du Cidre,
Vaudeurs 016136
Musée de la Pomme et du Cidre
et des Métiers Traditionnels,
Rosay 015152
Musée de la Porcelaine,
Couleuvre 012689
Musée de la Porcelaine, Saint-Yrieix-
la-Perche 015595
Musée de la Porte, Tubize 004096
Musée de la Porte de Hal-
Hallepoortmuseum, Bruxelles . 003466
Musée de la Porte de Thann,
Cernay 012392
Musée de la Poste, Amélie-les-Bains-
Palalda 011627
Musée de la Poste, Nantes ... 014413
Musée de la Poste, Paris 014710
Musée de la Poste, Prunelli-di-

Fium'Orbu 015005
Musée de la Poste, Saint-Flour
(Cantal) 015323
Musée de la Poste et des Techniques
de Communication de Basse-
Normandie, Caen 012276
Musée de la Poterie, La Borne . 013365
Musée de la Poterie, Vallauris . . 016102
Musée de la Poterie de Grès,
Betschdorf 012014
Musée de la Poterie du Sartre, Saint-
Alban-Auriolles 015208
Musée de la Poterie et de la
Céramique, Sadirac 015197
Musée de la Poterie et Musée du
Jambon d'Ardenne, La Roche-en-
Ardenne 003768
Musée de la Poterie, Töpfereimuseum,
Nospelt 030598
Musée de la Poterne, Preuilly-sur-
Claise 014997
Musée de la Poudrerie Nationale,
Association des Amis du Parc
Forestier National de Sevran,
Vaujours 016140
Musée de la Poupée, Aubeterre-sur-
Dronne 011771
Musée de la Poupée, La Côte-
d'Aime 013397
Musée de la Poupée, Le
Beausset 013605
Musée de la Poupée, Montbrison 014239
Musée de la Poupée, Paris 014711
Musée de la Poupée et des Jouets
Anciens, Guérande 013203
Musée de la Poupée et du Jouet
Ancien, Wambrechies 016336
Musée de la Préfecture de Police,
Paris 014712
Musée de la Préhistoire, Jemeppe-sur-
Sambre 003742
Musée de la Préhistoire, Le Mas-
d'Azil 013659
Musée de la Prehistoire, Rânes . 015035
Musée de la Préhistoire, Saint-
Flovier 015324
Musée de la Préhistoire, Saint-
Porchaire 015518
Musée de la Préhistoire,
Solesmes 015827
Musée de la Préhistoire Charentaise,
La Rochefoucauld 013469
Musée de la Préhistoire du Vercors,
Vassieux-en-Vercors 016132
Musée de la Préhistoire en
Wallonie, Préhistosite de Ramioul,
Flémalle 003599
Musée de la Préhistoire-Le Cairn,
Centre Archéologique de Recherche
sur le Néolithique, Saint-Hilaire-la-
Forêt 015361
Musée de la Préhistoire, Goyet la
Préhistorique, Goyet-Gesves . 003655
Musée de la Préhistoire/Paléontologie,
Peyzac-le-Moustier 014858
Musée de la Préhistorique,
Mozet 003893
Musée de la Première Guerre
Mondiale, Villeneuve-les-
Convers 016269
Musée de la Princerie, Verdun
(Meuse) 016166
Musée de la Protohistoire, Beynac-et-
Cazenac 012018
Musée de la Prune et du Pruneau,
Lafitte-sur-Lot 013523
Musée de la Publicité, Les Arts
Decoratifs, Paris 014713
Musée de la Radio, Mons 003880
Musée de la Radio, Villers-le-Lac 016284
Musée de la Radio et du Phonographe,
Lanobre 013562
Musée de la Radio Galletti, Saint-
Maurice-de-Rotherens 015461

Musée de la Radio Les Années 30,
Saint-Victor-sur-Arlanc 015588
Musée de la Reddition, Reims . 015049
Musée de la Régence, Ensisheim 012868
Musée de la Reine Bérengère, Le
Mans 013655
Musée de la Reliure, Beaumesnil 011944
Musée de La Remise, Gex 013099
Musée de la Réparation Navale,
Marseille 014051
Musée de la Reproduction du Son,
Saint-Fargeau 015314
Musée de la Résistance, Cabanès
(Aveyron) 012262
Musée de la Résistance,
Clavières 012586
Musée de la Résistance, Fussy . 013067
Musée de la Résistance, La Balme-de-
Thuy 013355
Musée de la Résistance,
Limoges 013817
Musée de la Résistance, Mussy-sur-
Seine 014393
Musée de la Résistance, Sospel 015841
Musée de la Résistance, Varennes-
Vauzelles 016129
Musée de la Résistance Bretonne,
Saint-Marcel (Morbihan) 015445
Musée de la Résistance d'Anterrieux
de la 7éme Cie et des Maquis du
Réduit de la Truyère, Anterrieux 011683
Musée de la Résistance du Mont
Mouchet, Auvers (Haute-Loire) 011813
Musée de la Résistance en Morvan,
Saint-Brisson (Nièvre) 015244
Musée de la Résistance en Zone
Interdite, Denain 012755
Musée de la Résistance et de la
Déportation, Agen 011538
Musée de la Résistance et de la
Déportation, Angoulême 011664
Musée de la Résistance et de la
Déportation, Besançon 012002
Musée de la Résistance et de la
Déportation, Grenoble 013182
Musée de la Résistance et de la
Déportation, Montauban 014224
Musée de la Résistance et de
la Déportation en Picardie,
Fargniers 012951
Musée de la Résistance et de la
Déportation, Centre Régional
d'Histoire de la Résistance,
Castelnau-le-Lez 012355
Musée de la Résistance et
Déportation, Blois 012080
Musée de la Résistance et des
Anciens Combattants, Floing . 012994
Musée de la Résistance Henri
Queuille, Neuvic (Corrèze) . . . 014456
Musée de la Résistance Nationale,
Champigny-sur-Marne 012436
Musée de la Résistance, de
l'Internement et de la Déportation,
Chamalières 012415
Musée de la Résistance, Fort de
Bondues, Bondues 012097
Musée de la Révolution Française,
Vizille 016317
Musée de la Rose et des Arts, Les
Rosiers-sur-Loire 013761
Musée de la Roue Tourne, La Mothe-
Achard 013442
Musée de la Route, Henri-
Chapelle 003692
Musée de la Route, Mons 003881
Musée de la Rubanerie Cominoise,
Comines 003522
Musée de la Ruralité, Donzac . . 012813
Musée de la Sablière,
Châteaubriant 012488
Musée de la Sahla, Audun-le-
Tiche 011788
Musée de la Schlitte et des Métiers

du Bois, Muhlbach-sur-Munster 014373
Musée de la Science-Fiction,
de l'Utopie et des Voyages
Extraordinaires, Maison d'Ailleurs,
Yverdon-les-Bains 042172
Musée de la Scierie Ariel-Vincent,
Sainte-Croix-aux-Mines 015607
Musée de la Seconde Guerre
Mondiale, Digne-les-Bains . . 012770
Musée de la Seconde Guerre
Mondiale, Le Val-Saint-Père . . 013712
Musée de la Seconde Guerre
Mondiale Roger Bellon, Conlie . 012659
Musée de la Seconde Guerre
Mondiale, Message Verlaine,
Tourcoing 016005
Musée de la Sénatorie, Guéret . 013205
Musée de la SNSM, Ile-de-
Molène 013288
Musée de la Société Archéologique de
Corseul, Corseul 012680
Musée de la Société Archéologique
et Historique de la Charente,
Angoulême 011665
Musée de la Société de l'Histoire du
Protestantisme Français, Paris 014714
Musée de la Société d'Histoire et
d'Archéologie, Musée de Briord,
Briord 012228
Musée de la Soie, Monoblet . . . 014199
Musée de la Soierie, Charlieu . . 012457
Musée de la Sorcellerie,
Blancafort 012056
Musée de La Source, Valignat . . 016099
Musée de la Station Biologique,
Roscoff 015155
Musée de la Statue Religieuse,
Lourdes 013894
Musée de la Straize, Gy-en-
Sologne 013221
Musée de la Taillanderie, Nans-sous-
Sainte-Anne 014410
Musée de la Tannerie Granval,
Maringues 014020
Musée de la Tapiserie et des Arts du
Tissu, Tournai 004084
Musée de la Tapisserie
Contemporaine, Angers 011654
Musée de la Tapisserie de Bayeux,
Bayeux 011920
Musée de la Targette, Neuville-Saint-
Vaast 014460
Musée de la Terre, Rambervillers 015031
Musée de la Terre et du Bois, Saint-
George 041906
Musée de la Thiérache, Vervins . 016185
Musée de la Toinette, Murat-le-
Quaire 014385
Musée de la Tonnellerie, Cadillac 012271
Musée de la Tonnellerie, Chécy . 012541
Musée de la Tonnellerie, Generac 013086
Musée de la Tonnellerie Animée,
Vergèze 016169
Musée de la Tonnellerie et du Vin,
Aloxe-Corton 011605
Musée de la Tonnellerie, Collection
d'Objets Chai, Tonnellerie et vieux
Millésimes, Listrac-Médoc 013838
Musée de la Tour, Coucy-le-Château-
Auffrique 012685
Musée de la Tour aux Puces,
Thionville 015937
Musée de la Tour Carrée, Rosans 015151
Musée de la Tour de l'Abbaye,
Jouarre 013329
Musée de la Tour de l'Orle d'Or,
Semur-en-Auxois 015759
Musée de la Tour des Echevins,
Luxeuil-les-Bains 013923
Musée de la Tour d'Honneur, Lesparre-
Médoc 013774
Musée de la Tour du Moulin,
Marcigny 014008
Musée de la Tour Romaine,

Arlon 003323
Musée de la Tour Salamandre,
Beaumont 003347
Musée de la Tour, Histoire et
Traditions Locales, Antibes-Juan-les-
Pins 011686
Musée de la Tournerie, Lavans-lès-
Saint-Claude 013593
Musée de la Tradition Champenoise,
Musée de la Cave des Champagnes
de Castellane, Épernay 012879
Musée de la Tradition de la Brigade
des Sapeurs-Pompiers de Paris,
Paris 014715
Musée de la Tuile, Blajan 012055
Musée de la Tuilerie, Malbrans . 013981
Musée de la Vache et de Alpages,
Frangy 013044
Musée de la Vallée à Barcelonnette,
Barcelonnette 011894
Musée de la Vallée à Jausiers,
Jausiers 013314
Musée de la Vallée à Lauzet, La
Lauzet 013436
Musée de la Vallée à Meyronnes,
Meyronnes 014161
Musée de la Vallée à Pontis,
Pontis 014956
Musée de la Vallée à Saint-Paul-sur-
Ubaye, Saint-Paul-sur-Ubaye . . 015498
Musée de la Vallée d'Aure,
Ancizan 011640
Musée de la Vallée de la Creuse,
Eguzon 012860
Musée de la Vannerie, Cadenet . 012269
Musée de la Vannerie, Remilly-sur-
Lozon 015058
Musée de la Vannerie des
Compagnons du Terroir, Villaines-les-
Rochers 016215
Musée de la Vannerie et de l'Artisant,
Vallabrègues 016100
Musée de la Vannerie L'Osarium, Le
Mayet-en-Montagne 013660
Musée de la Vénerie et des Spahis,
Senlis 015768
Musée de la Verrerie, Blangy-sur-
Bresle 012061
Musée de la Verrerie, Veauche . 016144
Musée de la Verrerie, Atelier
du Souffleur de Verre, Saint-
Galmier 015326
Musée de la Vie, Bidon 012031
Musée de la Vie Artisanale,
Ambachtenmuseum, Ham-sur-
Heure 003668
Musée de la Vie Bourguignonne Perrin-
de-Puycousin, Dijon 012784
Musée de la Vie d'Autrefois,
Marchin 003846
Musée de la Vie d'Autrefois,
Vivoin 016316
Musée de la Vie Lensoise, Lens 003761
Musée de la Vie Locale de Walhain,
Walhain 004128
Musée de la Vie Locale et de la
Coutellerie, Gembloux 003612
Musée de la Vie Monastique, Centre
d'Art Sacré, Saint-Hilaire-du-
Harcouët 015359
Musée de la Vie Paysanne, Musées
Gaumais, Montquintin 003889
Musée de la Vie Populaire,
Soumagne 004038
Musée de la Vie Régionale,
Cerfontaine 003509
Musée de la Vie Régionale des Rièzes
et des Sarts, Cul-des-Sarts . . 003529
Musée de la Vie Romantique, Hôtel
Scheffer-Renan, Paris 014716
Musée de la Vie Rurale, Mont-Saint-
Jean 014211
Musée de La Vie Rurale, Saint-
Quentin-Fallavier 015527

Musée de la Vie Rurale,
Steenwerck 015860
Musée de la Vie Rurale, Tusson 016066
Musée de la Vie Rurale Condruze,
Fondation Christian Blavier,
Nandrin 003905
Musée de la Vie Rurale d'Huissignies,
Huissignies 003719
Musée de la Vie Rurale en Wallonie,
Saint-Hubert 003994
Musée de la Vie Rurale et Forestière,
Saint-Michel-en-Thiérache ... 015470
Musée de la Vie Rurale Li Vîle Grègne,
Musée de la Vie Paysanne d'Antan,
Erezée 003581
Musée de la Vie Rurale, Musée
Municipal, Desaignes 012761
Musée de la Vie Tihangeoise,
Tihange 004070
Musée de la Vie Wallonne, Liège 003804
Musée de La Vieille École, Les
Valagans 013769
Musée de la Vieille Prune, Distillerie
Louis Roque, Souillac 015847
Musée de la Vigne, La Roche-
Vineuse 013468
Musée de la Vigne, Verzenay .. 016186
Musée de la Vigne Benoit-Raclet,
Musée Départemental du
Compagnonnage, Romanèche-
Thorins 015136
Musée de la Vigne et de la Vie Rurale,
Plauzat 014885
Musée de la Vigne et du Patrimoine,
Brioude 012230
Musée de la Vigne et du Terroir, Saint-
Pourçain-sur-Sioule 015519
Musée de la Vigne et du Vin,
Aigle 041200
Musée de la Vigne et du Vin,
Ansouis 011681
Musée de la Vigne et du Vin,
Aubière 011774
Musée de la Vigne et du Vin, Boën-
sur-Lignon 012084
Musée de la Vigne et du Vin,
Boudry 041356
Musée de la Vigne et du Vin, Buzet-
sur-Baise 012261
Musée de la Vigne et du Vin,
Cabara 012263
Musée de la Vigne et du Vin,
Cérons 012394
Musée de la Vigne et du Vin,
Gabarnac 013069
Musée de la Vigne et du Vin,
Gigondas 013104
Musée de la Vigne et du Vin,
Gruissan 013200
Musée de la Vigne et du Vin, Le
Mesnil-sur-Oger 013663
Musée de la Vigne et du Vin,
Lézignan-Corbières 013786
Musée de la Vigne et du Vin,
Lhomme 013789
Musée de la Vigne et du Vin, Mareau-
aux-Près 014015
Musée de la Vigne et du Vin,
Onzain 014565
Musée de la Vigne et du Vin,
Pierreclos 014865
Musée de la Vigne et du Vin, Vaux-en-
Beaujolais 016142
Musée de la Vigne et du Vin d'Anjou,
Saint-Lambert-du-Lattay 015405
Musée de la Vigne et du Vin de
Franche-Comté, Arbois 011704
Musée de la Vigne et du Vin-Galerie
d'Art, Lods 013852
Musée de la Vigne et du Vin, Château-
Musée de Mongenan, Portets . 014974
Musée de la Vigne-Moulin du Haut-
Benauge, Gornac 013133
Musée de la Vilaine Maritime, La

Roche-Bernard 013460
Musée de la Villa Romaine de Pully,
Pully 041854
Musée de la Ville de Saint-
Quentin-en-Yvelines, Montigny-le-
Bretonneux 014267
Musée de la Ville de Saverne,
Saverne 015727
Musée de la Ville d'Eaux, Spa . 004042
Musée de la Viscose, Échirolles 012853
Musée de la Viticulture,
Santenay 015678
Musée de la Voie Sacrée, Souilly 015849
Musée de la Voiture de Course,
Savigny-lès-Beaune 015735
Musée de la Voiture Hippomobile et
de la Citadelle de Bourg, Bourg-sur-
Gironde 012158
Musée de l'Abbaye, Echternach 030578
Musée de l'Abbaye, Gaillac ... 013072
Musée de l'Abbaye de Bourgueil,
Bourgueil 012170
Musée de l'Abbaye de Clairvaux, Ville-
sous-la-Ferté 016230
Musée de l'Abbaye de Font-Caude,
Cazedarnes 012379
Musée de l'Abbaye de Port-Royal,
Magny-lès-Hameaux 013965
Musée de l'Abbaye de Saint-Mathieu,
Plougonvelin 014905
Musée de l'Abbaye d'Orval, Villers-
devant-Orval 004113
Musée de l'Abbaye Romane, La
Sauve-Majeure 013489
Musée de l'Abbaye Saint-Vigor, Cérisy-
la-Forêt 012390
Musée de l'Abbaye Sainte-Croix, Les
Sables-d'Olonne 013765
Musée de l'Abeille, Château
Richer 005602
Musée de l'Abeille, Le Tronchet 013707
Musée de l'Abeille, Tilff 004071
Musée de l'Abeille et du Braconnage,
Salbris 015645
Musée de l'Abeille Vivante, Vitré 016307
Musée de l'Abri, Hatten 013233
Musée de l'Absinthe, Auvers-sur-
Oise 011817
Musée de l'Académie Saint-Anselme,
Aosta 025338
Musée de l'Accordéon,
Montmagny 006356
Musée de l'Accordéon, Siran . 015813
Musée de Lachine, Lachine ... 006146
Musée de l'Affiche de la Carte Postale
et de l'Art Graphique, Toulouse 015988
Musée de Laghouat, Laghouat . 000068
Musée de l'Agriculture Auvergnate,
Coltines 012635
Musée de l'Agriculture Catalane, Saint-
Michel-de-Llotes 015466
Musée de l'Agriculture et de la Vie
Rurale, Labrousse 013514
Musée de l'Aile Delta, Athis-
Mons 011762
Musée de l'Air et de l'Espace, Le
Bourget (Seine-Saint-Denis) . 013612
Musée de l'Alambic, Quimper . 015020
Musée de l'Alambic, Saint-
Desirat 015288
Musée de l'Albanais, Rumilly .. 015191
Musée de l'Algérie Française,
Perpignan 014840
Musée de l'Alta Rocca, Lévie . 013782
Musée de l'Amande et du Macaron,
Montmorillon 014281
Musée de l'Amérique Française,
Québec 006736
Musée de l'Amour et des ses
Traditions, Oger 014550
Musée de l'Ancien Collège des
Jésuites, Reims 015050
Musée de l'Ancien Évêché,
Grenoble 013183

Musée de l'Ancien Havre, Le
Havre 013643
Musée de l'Ancienne Abbaye de
Landévennec, Landévennec .. 013546
Musée de l'Ancienne Chartreuse de la
Verne, Collobrières 012622
Musée de L'Ancienne Forge,
Pesmes 014851
Musée de l'Ancienne Orangerie-Site
Buffon, Montbard 014229
Musée de l'Âne, Venosc 016158
Musée de l'Arc de Triomphe,
Paris 014717
Musée de l'Archéologie, Céret . 012387
Musée de l'Archerie Traditionnelle,
Ayssenes 011853
Musée de l'Architecture Traditionnelle,
Sfax 042769
Musée de l'Ardenne, Charleville-
Mézières 012456
Musée de l'Ardoise, Renazé .. 015061
Musée de l'Ardoise, Trélazé .. 016040
Musée de l'Ardoise, Couvent des
Carmélites, Fumay 013066
Musée de l'Areuse, Boudry ... 041357
Musée de l'Argonne, Sainte-
Menehould 015624
Musée de l'Arles et Provence
Antiques, Arles 011730
Musée de l'Armagnac, Condom 012655
Musée de l'Armée, Paris 014718
Musée de l'Armée des Alpes,
Sospel 015842
Musée de l'Armée Nationale Populaire,
Alger 000039
Musée de l'Art Campanaire,
Labergement-Sainte-Marie ... 013513
Musée de l'Art Culinaire, Fondation
Auguste Escoffier, Villeneuve-
Loubet 016271
Musée de l'Art du Sucre, Cordes-sur-
Ciel 012670
Musée de l'Art Forain et de la
Musique Mécanique, Conflans-en-
Jarnisy 012657
Musée de l'Art Wallon, Salle Saint-
Georges, Liège 003805
Musée de l'Artillerie, Draguignan 012834
Musée de l'Artisanat et des Traditions
Rurales, Brebotte 012196
Musée de l'Artisanat Rural Ancien,
Tigy 015964
Musée de l'Assistance Publique,
Bruxelles 003467
Musée de l'Assistance Publique-
Hôpitaux de Paris, Paris 014719
Musée de l'Association des Marches
Folkloriques de l'Entre-Sambre-et-
Meuse, Gerpinnes 003652
Musée de l'Atelier de Rosa Bonheur,
Thomery 015941
Musée de l'Atelier de Xaintonge, Pont-
l'Abbé-d'Arnoult 014940
Musée de l'Atlantique, La
Rochelle 013471
Musée de l'Attelage, Marquillies 014032
Musée de l'Attelage et du Cheval,
Sérignan (Hérault) 015775
Musée de l'Auberge Symmes,
Gatineau 005872
Musée de l'Audiovisuel,
Montarcher 014219
Musée de l'Auditoire, Sainte-
Suzanne 015633
Musée de l'Auld Alliance Franco-
Écossaise, Aubigny-sur-Nère . 011776
Musée de l'Auto, Moto et Cycle Jean
Tua, Genève 041531
Musée de l'Automate et de la
Robotique, Souillac 015848
Musée de l'Automobile, Martigny 041726
Musée de l'Automobile, Orgon . 014578
Musée de l'Automobile, Plan-les-
Ouates 041840

Musée de l'Automobile, Valençay 016089
Musée de l'Automobile, Velaine-en-
Haye 016146
Musée de l'Automobile de Vendée,
Talmont-Saint-Hilaire 015899
Musée de l'Automobile Française,
Collection Philippe Charbonneaux,
Reims 015051
Musée de l'Automobile La Belle
Époque, Pont-l'Évêque 014942
Musée de l'Automobile Miniature,
Fontenay-sur-Mer 013018
Musée de l'Automobile Sportive,
Sanary-sur-Mer 015672
Musée de l'Automobiliste,
Mougins 014348
Musée de l'Avallonnais, Avallon 011829
Musée de l'Aventure Industrielle du
Pays d'Apt, Apt 011700
Musée de l'Aviation, Perpignan . 014841
Musée de l'Aviation, Savigny-les-
Beaune 015736
Musée de l'Aviation, Warluis .. 016339
Musée de l'Aviation Légère de
l'Armée de Terre, Dax 012438
Musée de l'Eau, Pont-en-Royans 014937
Musée de l'Eau de Vie et des Vieux
Métiers, Valognes 016110
Musée de l'Échevinage, Saintes 015635
Musée de l'Éclairage, Dinant . 003556
Musée de l'École, Bourges ... 012161
Musée de l'École, Bruxelles ... 003468
Musée de l'École, Carcassonne . 012330
Musée de l'École, Chartres .. 012469
Musée de l'École, Châtelus
(Allier) 012510
Musée de l'Ecole, Echery ... 012852
Musée de l'École, Saint-Étienne-de-
Tinée 015306
Musée de l'École, Saint-Remy . 015535
Musée de l'Ecole, Villeneuve-
d'Ascq 016257
Musée de l'Ecole 1900, Saint-Martin-
des-Olmes 015453
Musée de l'Ecole Biblique et
Archéologique Française,
Jerusalem 034166
Musée de l'Ecole de Nancy,
Nancy 014404
Musée de l'École Jules Ferry,
Fraize 013042
Musée de l'Ecole Navale, Lanvéoc-
Poulmic 013568
Musée de l'Ecole Rurale,
Trégarvan 016034
Musée de l'École Rurale d'Auvergne,
Messeix 014148
Musée de l'Ecorché d'Anatomie,
Musée Charron Forgeron, Le
Neubourg 013675
Musée de Leers, Leers 013724
Musée de l'Eglise Abbatiale de
Floreffe, Floreffe 003601
Musée de l'Église de Lorris,
Lorris 013876
Musée de l'Eglise Notre-Dame,
Montréal 006389
Musée de l'Électricité, Dijon . 012785
Musée de l'Électricité, Saint-
Aignan 015205
Musée de l'Elysée, Lausanne . 041659
Musée de l'Empéri, Salon-de-
Provence 015662
Musée de l'Empéri de Salon et de la
Crau, Salon-de-Provence ... 015663
Musée de l'Enfance, Alger ... 000040
Musée de l'Entité de Walcourt,
Walcourt 004125
Musée de l'Éphèbe, Agde .. 011537
Musée de l'Epicerie et des
Commerces, Lignerolles ... 013797
Musée de l'Erotisme, Paris ... 014720
Musée de l'Escrime Charles Debeur,
Bruxelles 003469

Musée de l'Espace, Kourou 016371
Musée de l'Estérel, Les Adrets-de-
l'Estérel 013731
Musée de l'Etang de Thau,
Bouziques 012182
Musée de l'Etiquette, Langon .. 013555
Musée de l'Evéché, Sion 041996
Musée de l'Eventail, Paris 014721
Musée de l'Habitat Ancien,
Bessans 012006
Musée de l'Habitat Rural en Pays
d'Aigues, La Tour-d'Aigues .. 013498
Musée de l'Habitat Rural Passé et des
Traditions, La Garnache 013424
Musée de l'Heure et du Feu,
Villerest 016279
Musée de l'Histoire, Sartrouville 015700
Musée de l'Histoire de France à
travers son Armée, Montet et
Bouxal 014254
Musée de l'Histoire de France,
Archives nationales, Paris ... 014722
Musée de l'Histoire de la Médecine,
Vandœuvre-les-Nancy 016117
Musée de l'Histoire de la Médecine de
Toulouse, Toulouse 015989
Musée de l'Histoire de la Poterie, La
Chapelle-des-Pots 013383
Musée de l'Histoire de l'Abbaye, Saint-
Sauveur-le-Vicomte 015560
Musée de l'Histoire de l'Emigration
Percheronne au Canada,
Tourouvre 016012
Musée de l'Histoire des Bastides
et de l'Evolution de l'Habitat,
Bassoues 011904
Musée de l'Histoire du Costume, Scey-
sur-Saône-et-Saint-Albin .. 015742
Musée de l'Histoire du Fer,
Jarville 013310
Musée de l'Histoire du Feu,
Musée des Sapeurs-Pompiers,
Campuzan 012305
Musée de l'Histoire et de la Vie
Salmiennes, Vielsalm 004110
Musée de l'Histoire Paysanne, Maraye-
en-Othe 013997
Musée de l'Histoire Vivante Parc
Montreau, Montreuil-sous-Bois 014305
Musée de l'Historie de Montpellier,
Montpellier 014290
Musée de l'Historire de Rosny, Rosny-
sous-Bois 015160
Musée de l'Holographie, Paris . 014723
Musée de l'Homme de Néandertal, La
Chapelle-aux-Saints 013379
Musée de l'Homme et de l'Industrie,
Le Creusot 013625
Musée de l'Hôpital Général de Dijon,
Dijon 012786
Musée de l'Hôpital Psychiatrique,
Brumath 012244
Musée de l'Hôpital, Musée Saint-Roch,
Yssingeaux 016362
Musée de l'Horlogerie, Pesmes . 014852
Musée de l'Horlogerie, Saint-Nicolas-
d'Aliermont 015477
Musée de l'Horlogerie Ancienne,
Fougères (Ille-et-Vilaine) 013032
Musée de l'Horlogerie du Haut-Doubs,
Morteau 014340
Musée de l'Horlogerie et de
l'Emaillerie, Genève 041532
Musée de l'Horlogerie et du
Décolletage, Cluses 012609
Musée de l'Hospice Comtesse,
Lille 013805
Musée de l'Hospice Saint-Roch,
Issoudun 013299
Musée de l'Hôpital, Apothicairerie, La
Rochefoucauld 013470
Musée de l'Hôtel de Barral,
Soissons 015825
Musée de l'hôtel de Berny,

Amiens 011632
Musée de l'Hôtel de Vermandois,
Senlis 015769
Musée de l'Hôtel de Ville,
Amboise 011623
Musée de l'Hôtel de Ville,
Hondschoote 013267
Musée de l'Hôtel-Dieu, Beaune . 011948
Musée de l'Hôtel-Dieu, Mantes-La
Jolie 013995
Musée de l'Hôtel-Dieu,
Porrentruy 041844
Musée de l'Hôtel Sandelin, Saint-
Omer 015482
Musée de l'Hydraviation,
Biscarrosse 012045
Musée de Lignan et du Canton de
Créon, Lignan-de-Bordeaux ... 013796
Musée de l'Iguanodon, Bernissart 003362
Musée de l'Île de France,
Sceaux 015740
Musée de l'Île d'Oleron Aliénor
d'Aquitaine, Saint-Pierre-
d'Oléron 015513
Musée de l'Image, Épinal 012885
Musée de l'Image Populaire,
Pfaffenhoffen 014861
Musée de l'Impression sur Etoffes,
Mulhouse 014375
Musée de l'Imprimerie, Bordeaux 012113
Musée de l'Imprimerie de Lyon,
Lyon 013936
Musée de l'Imprimerie-Museum van
de Boekdrukkunst, Cabinet des
Estampes, Bruxelles 003470
Musée de l'Imprimerie, Association
Pro Arte Graphica, Nantes ... 014414
Musée de l'Industrie Régionale de
la Chaussure, Saint-André-de-la-
Marche 015217
Musée de l'Infanterie,
Montpellier 014291
Musée de l'Insolite, Cabrerets . 012264
Musée de l'Insolite, Manent Max,
Loriol-sur-Drôme 013873
Musée de l'Institut du Monde Arabe,
Paris 014724
Musée de l'Institut Pasteur de Lille,
Lille 013806
Musée de l'Invasion Aéroportée,
Musée du Café Gondrée,
Bénouville 011985
Musée de l'Océanie, Saint-
Symphorien-sur-Coise 015572
Musée de Lodève, Lodève 013851
Musée de l'Œuf et de l'Icône,
Banne 011885
Musée de l'Œuvre, Avignon ... 011836
Musée de l'Œuvre Notre-Dame,
Strasbourg 015877
Musée de l'Œuvre Viollet-le-Duc,
Vézelay 016191
Musée de l'Olivette, Sorède .. 015836
Musée de l'Olivier, Les Baux-de-
Provence 013738
Musée de l'Olivier, Nyons 014540
Musée de l'Olivier des Mées,
Mées 014109
Musée de l'Opéra, Vichy 016201
Musée de l'Opinel, Saint-Jean-de-
Maurienne 015381
Musée de l'Optique, Biesheim . 012033
Musée de l'Optométrie, Institut et
Centre d'Optométrie, Bures-sur-
Yvette 012254
Musée de l'Or, Jumilhac-le-
Grand 013341
Musée de l'Orangerie, Paris ... 014725
Musée de l'Oratoire Saint-Joseph,
Montréal 006390
Musée de l'Orce, Saint-Georges-sur-la-
Prée 015334
Musée de l'Ordre de la Libération,
Paris 014726

Musée de l'Ordre de Malte,
Bardonnex 041259
Musée de l'Orfevrerie de la
Communaute Française, Château de
Seneffe, Seneffe 004007
Musée de l'Ostréiculture,
Marennes 014017
Musée de l'Ostrévant, Bouchain 012127
Musée de Lourdes, Lourdes .. 013895
Musée de l'Ours des Cavernes, Musée
Communal, Entremont-le-Vieux 012872
Musée de l'Outil, Magny-en-
Vexin 013964
Musée de l'Outil des Métiers du Bois,
Montjoie-Saint-Martin 014269
Musée de l'Outil et de la Vie au
Village, Saint-Privat-des-Prés . 015521
Musée de l'Outil et des Métiers,
Tinténiac 015968
Musée de l'Outil et des Vieux Métiers,
La Celle-Guenand 013373
Musée de l'Outil, la Forge Médiévale,
Wy-Dit-Joli-Village 016357
Musée de l'Outillage Agricole et
Viticole du Château Aney, Cussac-
Fort-Médoc 012736
Musée de l'Outillage Ancien et des
Traditions, Briançonnet 012216
Musée de l'Outillage Artisanal Rural
et Bourguignon, Cruzille-en-
Mâconnais 012728
Musée de l'Outillage et des
Techniques Traditionnelles de la
Taille de Pierre, Coucouron ... 012684
Musée de Louvain-la-Neuve, Louvain-
la-Neuve 003831
Musée de l'Uniforme Légionnaire,
Musée de la Légion Étrangère,
Puyloubier 015012
Musée de l'Union Compagnonnique
des Devoirs Unis, Compagnons du
Tour de France, Nantes ... 014415
Musée de l'Université de Tuléar,
Tuléar 030653
Musée de l'Uranium, Lachaux . 013519
Musée de Machines Agricoles
Miniatures, Guerlesquin 013208
Musée de Maison-Alfort, Maisons-
Alfort 013974
Musée de Maquettes Animées, Vallon-
en-Sully 016106
Musée de Marine, Faranui ... 016372
Musée de Maroc Telecom, Rabat 031789
Musée de Marrakech, Fondation Omar
Ben Jelloun, Marrakech 031782
Musée de Matériel Agricole, Saint-
Laurent-des-Autels 015410
Musée de Matériel Viticole, Saint-
Laurent-des-Autels 015411
Musée de Matière Médicale,
Paris 014727
Musée de Mayombe, Kinshasa . 008835
Musée de Mbandaka, Mbandaka 008841
Musée de Megève, Megève ... 014110
Musée de Melun, Melun 014120
Musée de Miellerie, Miellerie des
Butineuses, Saint-Saturnin-les-
Avignon 015555
Musée de Minéralogie,
Strasbourg 015878
Musée de Minéralogie de l'Ecole des
Mines de Paris, Paris 014728
Musée de Minéralogie et de
Géologie, Musées de l'Université,
Bruxelles 003471
Musée de Minéralogie et du
Pétrographie, Ambazac 011610
Musée de Miquelon, Miquelon . 038090
Musée de Modèles Réduits, La
Rochelle 013472
Musée de Modélisme, Savigny-les-
Beaune 015737
Musée de Montaut-Centre Culturel,
Musée Préhistorique et de Peinture,

Montaut 014228
Musée de Montmartre, Musee de
France, Paris 014729
Musée de Moulages, Strasbourg 015879
Musée de Musiques Méchanique,
L'Auberson 041241
Musée de Natitingou, Natitingou 004183
Musée de Nevers Magny-Cours,
Magny-Cours 013963
Musée de Nogent, Nogent-sur-
Marne 014515
Musée de Normandie, Caen .. 012277
Musée de Notre-Dame-de-Paris,
Paris 014730
Musée de Ouadane, Nouakchott 030784
Musée de Paléontologie et
Préhistoire, Palaja 014607
Musée de Paléontologie, La Voulte-sur-
Rhône 013505
Musée de Paléontologie, Menat 014121
Musée de Paléontologie, Villers-sur-
Mer 016286
Musée de Paléontologie Christian Guth,
Chilhac 012557
Musée de Paléontologie Humaine de
Terra Amata, Nice 014482
Musée de Paléontologie, Sciences de
la Terre, Villeurbanne 016291
Musée de Palmyra, Palmyra ... 042263
Musée de Payerne et Abbatiale,
Payerne 041836
Musée de Peinture J. Drevon, Saint-
Jean-de-Bournay 015373
Musée de Peinture Louis Français,
Plombières-les-Bains 014896
Musée de Peinture Mario Prassinos,
Saint-Rémy-de-Provence 015541
Musée de Peintures Contemporaines
Hors du Temps, Pierrefeu (Alpes-
Maritimes) 014867
Musée de Personnages de Cire,
Waterloo 004135
Musée de Picardie, Aniebs Metropole
Museums, Amiens 011633
Musée de Pierre Corneille, Le Petit-
Couronne 013679
Musée de Plein Air, Parakou .. 004185
Musée de Plein Air, Seurre .. 015792
Musee de Plein-Air de Cherchell,
Cherchell 000056
Musée de Plein Air des Maisons
Comtoises, Nancray 014400
Musée de Plein Air du Quercy de
Cuzals, Sauliac-sur-Célé ... 015707
Musée de Plein Air, Maison pour Tous,
Gueugnon 013210
Musée de Plein Air, Parc Préhistorique,
Saint-Vallier-de-Thiey 015584
Musée de Pont-Aven, Pont-Aven 014931
Musée de Pont-Saint-Esprit, Saint-
Paulet-de-Caisson 015501
Musée de Pontarlier, Pontarlier . 014950
Musée de Poteries de Terre Vernissée,
Saint-Jean-de-Fos 015375
Musée de Préhistoire, Aurignac . 011799
Musée de Préhistoire, Coulmiers 012690
Musée de Préhistoire,
Echternach 030579
Musée de Préhistoire, Kinshasa 008836
Musée de Préhistoire Amédée Lemozi,
Cabrerets 012265
Musée de Préhistoire des Gorges du
Verdon, Quinson 015024
Musée de Préhistoire d'Île-de-France,
Nemours 014442
Musée de Préhistoire James Miln-
Zacharie le Rouzic, Carnac ... 012338
Musée de Préhistoire Libanaise,
Beirut 030266
Musée de Préhistoire Mésolitique
L. Coulonges, Sauveterre-la-
Lémance 015723
Musée de Préhistoire Régionale,
Menton 014132

Musée de Provins et du Provinois,
Provins 015002
Musée de Pully, Pully 041855
Musée de Racour, Musée d'Histoire et
de la Vie d'Autrefois, Racour . . 003957
Musée de Radio France, Paris . . 014731
Musée de Rauranum, Rom . . . 015134
Musée de Région Auguste Chabaud,
Graveson 013171
Musée de Rome, Université de Caen
Basse-Normandie, Caen 012278
Musée de Rouget de Lisle, Donation
André Lançon, Lons-le-Saunier 013867
Musée de Royan, Royan 015184
Musée de Saint-Boniface, Saint-
Boniface 006857
Musée de Saint-Eustache et de ses
Patriotes, Saint-Eustache . . . 006869
Musée de Saint-Gengoux-le-Royal,
Saint-Gengoux-le-National 015328
Musée de Saint-Georges, Saint-
Georges, Manitoba 006871
Musée de Saint-Germain-Laval, Saint-
Germain-Laval 015344
Musée de Saint-Gervais, Saint-Gervais-
sur-Mare 015347
Musée de Saint-Martin, Saint-
Martin 023352
Musée de Saint-Maurin, Saint-
Maurin 015463
Musée de Saint-Paul, Saint-Paul-de-
Vence 015496
Musée de Saint-Pierre-Jolys, Saint-
Pierre-Jolys 006933
Musée de Sainte-Anne-de-Beaupré,
Sainte-Anne-de-Beaupré 006946
Musée de Sainte-Sigolène, Écomusée
La Fabrique, Sainte-Sigolène . . 015631
Musée de Salers, Maison des
Templiers, Salers 015647
Musée de Sarreguemines,
Sarreguemines 015696
Musée de Schirmeck, Schirmeck 015743
Musée de Sétif, Sétif 000074
Musée de Sismologie et Magnétisme
Terrestre, Strasbourg 015880
Musée de Site, Sare 015684
Musée de Site Galloromain Villa
Loupian, Musée Archéologique et
Historique, Loupian 013888
Musée de Site Le Bazacle,
Toulouse 015990
Musée de Skikda, Skikda 000076
Musée de Sologne, Cheverny . . 012554
Musée de Sologne, Moulin du
Chapitre, Romorantin-Lanthenay 015142
Musée de Sonis, Bataille du 2
Décembre 1870, Loigny-la-
Bataille 013854
Musée de Sonvilier, Sonvilier . . 042010
Musée de Soueida, Soueida . . . 042264
Musée de Souvigny, Souvigny . . 015857
Musée de Spéléologie, Ecomusée du
Gouffre, Fontaine-de-Vaucluse . 013005
Musée de Statues d'Enfants, Musée
Schoenenberger, Les Mages . . 013754
Musée de Sucy, Sucy-en-Brie . . 015889
Musée de Suresnes René Sordes,
Suresnes 015894
Musée de Tabarka, Tabarka . . . 042776
Musée de Tahiti et des Îles – Te Fare
Manaha, Punaauia 016380
Musée de Tauroentum, Saint-Cyr-les-
Lecques 015278
Musée de Tautavel-Centre Européen
de la Préhistoire, Tautavel 015921
Musée de Tazoult, Tazoult 000078
Musée de Téboursouk,
Téboursouk 042778
Musée de Temouchent, Ain
Temouchent 000037
Musée de Tessé, Le Mans 013656
Musée de Tigran, Champvoux . . 012441
Musée de Timgad, Timgad-Batna 000080

Musée de Tipasa, Tipasa 000081
Musée de Tissage, Chauffailles . 012525
Musée de Tlemcen, Tlemcen . . 000082,
000083
Musée de Tourisme et d'Artisanat, Lys-
Saint-Georges 013956
Musée de Traditions Populaires et
d'Archéologie, Chauvigny 012534
Musée de Ua Huka, Musée
Archeologique, Vaipaee 016381
Musée de Ventabren, Musée
Archéological La Calade,
Ventabren 016160
Musée de Verrières-le-Buisson,
Verrières-le-Buisson 016179
Musée de Vieux Cormeilles,
Cormeilles-en-Parisis 012675
Musée de Vieux-la-Romaine, Vieux-la-
Romaine 016212
Musée de Vieux-Lavaux, Cully . . 041430
Musée de Vieux Logis, Sainte-
Enimie 015609
Musée de Vieux Sallèles, Sallèles-
d'Aude 015654
Musée de Voitures Anciennes et
Sellerie, Menetou-Salon 014126
Musée de Voitures Hippomobiles
Anciennes, Tour-en-Sologne . . 016002
Musée de Vulliod Saint-Germain,
Pézenas 014859
Musée de Wanne, Trois-Ponts . 004094
Musée de Warcoing, Warcoing . 004082
Musée de Waterloo, Waterloo . . 004136
Musée de Zarzis, Zarzis 042787
Musée d'Eben, Eben-Emael . . . 003565
Musée d'Egyptologie, Villeneuve-
d'Ascq 016258
Musée d'El-Oued, El-Oued 000062
Musée Delphinal, Beauvoir-en-
Royans 011958
Musée d'Enfidha, Enfidaville . . . 042742
Musée d'Ennery, Paris 014732
Musée Dentaire de Lyon, Lyon . 013937
Musée d'Entomologique,
Chaveyriat 012537
Musée d'Entrecasteaux,
Entrecasteaux 012871
Musée Départemantal des Sapeurs-
Pompiers de l'Orne, Bagnoles-de-
l'Orne 011869
Musée Départemental, Baume-les-
Messieurs 011912
Musée Départemental Albert Kahn,
Boulogne-Billancourt 012135
Musée Départemental Alexandre
Franconie, Cayenne 016369
Musée Départemental Breton,
Quimper 015021
Musée Départemental d'Archéologie,
Hastingues 013231
Musée Départemental d'Archéologie,
Montrozier 014313
Musée Départemental d'Art Ancien et
Contemporain, Épinal 012886
Musée Départemental d'Art
Contemporain de Rochechouart,
Rochechouart 015118
Musée Départemental d'Art Religieux,
Digne-les-Bains 012771
Musée Départemental d'Art Religieux,
Sées 015749
Musée Départemental d'Art Sacré
Contemporain, Saint-Pierre-de-
Chartreuse 015508
Musée Départemental de Flandre,
Musée d'Art et d'Histoire,
Cassel 012350
Musée Départemental de la
Préhistoire, Le Grand-Pressigny 013635
Musée Départemental de la
Résistance, Le Teil 013701
Musée Départemental de la
Résistance en Vercors, Vassieux-en-
Vercors 016133

Musée Departemental de la
Résistance et de la Déportation,
Forges-les-Eaux 013027
Musée Départemental de la
Résistance et de la Déportation,
Lorris 013877
Musée Départemental de la
Résistance et de la Déportation,
Manneville-sur-Risle 013989
Musée Départemental de la
Résistance et de la Déportation,
Tulle 016059
Musée Départemental de la
Résistance et de la Déportation Jean
Philippe, Toulouse 015991
Musée Départemental de la Tapisserie,
Aubusson 011780
Musée Départemental de la Vigne et
des Pressoirs, Champlitte 012438
Musée Départemental de l'Abbaye de
Saint-Riquier, Saint-Riquier . . . 015549
Musée Départemental de l'École
Publique, Chevregny 012555
Musée Départemental de
l'École Publique, Saint-Clar-de-
Lomagne 015265
Musée Départemental de l'Ecole
Publique, Vergne 016170
Musée Départemental de l'Éducation,
Saint-Ouen-l'Aumône 015486
Musée Départemental de l'Oise,
Beauvais 011956
Musée Départemental de Préhistoire,
Solutré-Pouilly 015831
Musée Départemental de Préhistoire
Corse, Sartène 015699
Musée Départemental de Saint-
Antoine, Saint-Antoine-l'Abbaye 015221
Musée Départemental del'Ariège,
Château des Comtes de Foix,
Foix 012999
Musée Départemental des Antiquités
de la Seine-Maritime, Rouen . . 015165
Musée Départemental des Arts et
Traditions Populaires de la Haute-
Loire et du Massif Central, Saint-
Didier-en-Velay 015290
Musée Départemental des Arts et
Traditions Populaires du Perche,
Saint-Cyr-la-Rosière 015276
Musée Départemental des Hautes-
Alpes, Gap 013078
Musée Départemental des Pays de
Seine-et-Marne, Saint-Cyr-sur-
Morin 015282
Musée Départemental des Sapeurs-
Pompiers, Sciez 015744
Musée Départemental des Sapeurs-
Pompiers du Val-d'Oise, Osny . 014594
Musée Départemental du Sel,
Marsal 014036
Musée Departemental du Textile,
Lavelanet 013599
Musée Départemental Georges de La
Tour, Vic-sur-Seille 016197
Musée Départemental Hébert, La
Tronche 013501
Musée Départemental Ignon-Fabre,
Mende 014122
Musée Départemental Jérôme-
Carcopino, Aléria 011591
Musée Départemental Maurice Denis,
Saint-Germain-en-Laye 015341
Musée Départemental Pascal Paoli,
Morosaglia 014336
Musée Départemental Stéphane
Mallarmé, Vulaines-sur-Seine . 016333
Musée des 18 Jours-Mémoires de
Guerre, Pipaix 003950
Musée des 4 Saisons,
Generauges 013087
Musée des Abénakis, Odanak . . 006539
Musée des Affiches Léonetto-
Cappiello, Épernay 012880

Musée des Alambics, Saint-Romain-
de-Benet 015553
Musee des Allues, Maison Bonnevie,
Meribel 014137
Musée des Alpilles, Saint-Rémy-de-
Provence 015542
Musée des Amis de Castrum Vetus,
Châteauneuf-les-Martigues . . . 012495
Musée des Amis de Montmélian,
Montmélian 014276
Musée des Amis de Thann,
Thann 015927
Musée des Amis du Vieux Corbie,
Corbie 012667
Musée des Amis du Vieux Donzère,
Donzère 012815
Musée des Amis du Vieux Guérigny,
Guérigny 013207
Musée des Amis du Vieux Lormont,
Lormont 013874
Musée des Amis du Vieux
Montrichard, Musees du Donjon,
Montrichard 014310
Musée des Amis du Vieux Nolay,
Nolay 014524
Musée des Amis du Vieux Saint-
Étienne, Saint-Étienne 015305
Musée des Amoureux,
Werentzhouse 016346
Musée des Anciens Enfants de Troupe,
Autun 011809
Musée des Anciens Outils Agricoles,
Saint-Mathurin 015459
Musée des Années 30, Boulogne-
Billancourt 012136
Musée des Antiquités Nationales,
Saint-Germain-en-Laye 015342
Musée des Appareils Photographiques
et de Cinéma d'Amateurs,
Marsilly 014065
Musée des Archives, Saint-
Hubert 003995
Musée des Archives Départementales
de l'Aveyron, Conseil Général de
l'Aveyron, Rodez 015129
Musée des Armes, Tulle 016060
Musée des Arômes de Provence,
Saint-Rémy-de-Provence 015543
Musée des Arômes et du Parfum,
Graveson 013172
Musée des Art et Tradition, Reuille-
Vergy 015078
Musée des Arts Africains, Océaniens,
Amérindiens, Marseille 014052
Musée des Arts Anciens du Namurois,
Namur 003898
Musée des Arts Asiatiques, Nice 014483
Musée des Arts Décoratifs,
Bordeaux 012114
Musée des Arts Décoratifs, Lyon 013938
Musée des Arts Décoratifs,
Strasbourg 015881
Musée des Arts Décoratifs et de
la Modernité, Gourdon (Alpes-
Maritimes) 013135
Musée des Arts Décoratifs Sabatier
d'Espeyran, Montpellier 014292
Musée des Arts Decoratifs, Les Arts
Decoratifs, Paris 014733
Musée des Arts du Cognac,
Cognac 012612
Musée des Arts et de l'Enfance,
Fécamp 012959
Musée des Arts et des Meubles,
Poilley 014916
Musée des Arts et des Sciences,
Sainte-Croix 041914
Musée des Arts et d'Histoire,
Locronan 013848
Musée des Arts et Metiers, Paris 014734
Musée des Arts et Métiers, Saint-
Gildas-de-Rhuys 015349
Musée des Arts et Métiers de la Vigne
et du Vin, Moulis-en-Médoc . . 014360

Musée des Arts et Tradition Apicoles, Fontan 013014
Musée des Arts et Traditions Populaires, Djerba 042740
Musée des Arts et Traditions Populaires, Ermont 012891
Musée des Arts et Traditions Populaires, Esse 012909
Musée des Arts et Traditions Populaires, Le Sel-de-Bretagne 013699
Musée des Arts et Traditions Populaires, Saint-Arnoult-en-Yvelines 015228
Musée des Arts et Traditions Populaires, Saint-Esprit 030780
Musée des Arts et Traditions Populaires, Treignac 016037
Musée des Arts et Traditions Populaires, Wattrelos 016344
Musée des Arts et Traditions Populaires 'Dar Ben Abdallah' de Tunis, Tunis 042784
Musée des Arts et Traditions Populaires de la Haute-Loire, Lavaudieu 013596
Musée des Arts et Traditions Populaires de la Vallée de la Cère, Arpajon-sur-Cère 011740
Musée des Arts et Traditions Populaires de Moyenne Provence, Draguignan 012835
Musée des Arts et Traditions Populaires, Abbaye de Valbonne, Valbonne 016087
Musée des Arts et Traditions Populaires, Musée Municipal, Bressuire 012201
Musée des Arts et Traditions Populaires, Positifs de Moules à Gâteaux, La Petite-Pierre 013450
Musée des Arts Forains, Paris . . 014735
Musée des Arts Patrimoine et des Traditions, Nanteuil-en-Vallée . 014424
Musée des Arts Populaires, Noyant 014531
Musée des Arts Populaires, Reillanne 015048
Musée des Arts Populaires et Traditions, Alger 000041
Musée des Arts Populaires et Traditions Dar Jallouli, Sfax . . . 042770
Musée des Arts Sahariens de Laâyoune, Laâyoune 031776
Musée des Arts Traditionnels, Cussac-Fort-Médoc 012737
Musée des Arts Traditionnels et de la Culture, Musée du Palais des Rois Bamoun, Foumban 005293
Musée des Attelages de la Belle Epoque, Saint-Martin-du-Lac . 015454
Musée des Attelages, de la Carrosserie et du Charronnage, Vonnas 016325
Musée des Augustines de l'Hôpital Général de Québec, Québec . . 006737
Musée des Augustines de l'Hôtel Dieu de Québec, Québec 006738
Musée des Augustins, Toulouse 015992
Musée des Automates, La Rochelle 013473
Musée des Automates, Neuilly-sur-Seine 014454
Musée des Automates, Sauclières 015701
Musée des Automates de Grenoble, Grenoble 013184
Musée des Automates Éma, Lyon 013939
Musée des Ballons, Musée des Montgolfières, Balleroy 011878
Musée des Bastides, Espace Bastides, Monflanquin 014198
Musée des Batailles, Montesquieu-des-Albères 014253
Musée des Bateliers de la Dordogne, Port-Sainte-Foy 014971

Musée des Beaux-Arts, Agen . . 011539
Musée des Beaux-Arts, Angers . 011655
Musée des Beaux-Arts, Arras . . 011744
Musée des Beaux-Arts, Beaune 011949
Musée des Beaux-Arts, Beirut . . 030267
Musée des Beaux-Arts, Béziers . 012022
Musée des Beaux-Arts, Blois . . 012081
Musée des Beaux-Arts, Brest . . 012207
Musée des Beaux-Arts, Caen . . 012279
Musée des Beaux-Arts, Carcassonne 012331
Musée des Beaux-Arts, Chambéry 012418
Musée des Beaux-Arts, La Chaux-de-Fonds 041406
Musée des Beaux-Arts, Dijon . 012787
Musée des Beaux-Arts, Dole . . 012800
Musée des Beaux-Arts, Gaillac . 013073
Musée des Beaux-Arts, Kinshasa 008837
Musée des Beaux-Arts, La Ferté-Macé 013407
Musée des Beaux-Arts, La Rochelle 013474
Musée des Beaux-Arts, Libourne 013791
Musée des Beaux-Arts, Lille . . 013807
Musée des Beaux-Arts, Le Locle 041686
Musée des Beaux Arts, Lons-le-Saunier 013868
Musée des Beaux-Arts, Lyon . . 013940
Musée des Beaux-Arts, Marseille 014053
Musée des Beaux-Arts, Menton 014133
Musée des Beaux-Arts, Mirande 014180
Musée des Beaux-Arts, Montbard 014230
Musée des Beaux-Arts, Nancy . 014405
Musée des Beaux-Arts, Nantes . 014416
Musée des Beaux-Arts, Nîmes . 014501
Musée des Beaux-Arts, Orléans 014586
Musée des Beaux-Arts, Pau . . 014808
Musée des Beaux-Arts, Quimper 015022
Musée des Beaux-Arts, Reims . 015052
Musée des Beaux-Arts, Rennes (Ille-et-Vilaine) 015067
Musée des Beaux-Arts, Rouen . 015166
Musée des Beaux-Arts, Strasbourg 015882
Musée des Beaux-Arts, Tourcoing 016006
Musée des Beaux-Arts, Tournai . 004085
Musée des Beaux-Arts, Tours . . 016017
Musée des Beaux-Arts, Valenciennes 016096
Musée des Beaux-Arts, Vannes (Morbihan) 016120
Musée des Beaux-Arts, Villeneuve-sur-Yonne 016276
Musée des Beaux-Arts Charleroi, Charleroi 003513
Musée des Beaux-Arts de Bordeaux, Bordeaux 012115
Musée des Beaux-Arts de Dunkerque, Dunkerque 012844
Musée des Beaux-Arts de Montréal, Montréal 006391
Musée des Beaux-Arts de Sherbrooke, Sherbrooke, Québec 007025
Musée des Beaux-Arts Denys Puech, Rodez 015130
Musée des Beaux-Arts et Arts Décoratifs Raymond Ritter, Morlanne 014329
Musée des Beaux-Arts et Céramique, Verviers 004107
Musée des Beaux-Arts et d'Archéologie, Besançon 012003
Musée des Beaux-Arts et d'Archeologie Joseph Déchelette, Roanne 015110
Musée des Beaux-Arts et d'Archéologie, Musées de Vienne, Vienne 016206
Musée des Beaux-Arts et de la Dentelle, Alençon 011590
Musée des Beaux-Arts et de la

Dentelle, Calais 012294
Musée des Beaux-Arts et d'Histoire, Saint-Lô 015423
Musée des Beaux-Arts Hyacinthe Rigaud, Perpignan 014842
Musée des Beaux-Arts Jules Chéret, Nice 014484
Musée des Beaux-Arts, Musée Municipal, Angoulême 011666
Musée des Beaux-Arts, Musées Municipaux, Saintes 015636
Musée des Beaux-Arts, Villa Steinbach, Mulhouse 014376
Musée des Bêtises de Cambrai, Cambrai 012301
Musée des Bières Belges, Lustin 003833
Musée des Blasons, Saint-Jean-de-Veleriscle 015384
Musée des Blindés, Saumur . . . 015716
Musée des Bois Jolis, Ecomusée de la Région de Fourmies, Felleries . 012964
Musée des Bois Sculptes, Cancale 012309
Musée des Bories, Gordes 013129
Musée des Bujoliers, Saint-Cezaire 015249
Musée des Calèches, Apremont-sur-Allier 011697
Musée des Camions de Brasserie, Romedenne 003982
Musée des Canadiens Nord-Bourguignons, Sainte-Colombe-sur-Seine 015605
Musée des Canonniers Sédentaires de Lille, Lille 013808
Musée des Carrosses, Versailles 016180
Musée des Caves de la Cantrie, Saint-Fiacre-sur-Maine 015317
Musée des Celtes, Libramont . . 003789
Musée des Champs, Saint-Ségal 015561
Musée des Chantiers de Jeunesse, Châtel-Guyon 012506
Musée des Charmettes, Chambéry 012419
Musée des Chartrons, Bordeaux 012116
Musée des Chasseurs à Pied, Charleroi 003514
Musée des Chasseurs Ardennais, Marche-en-Famenne 003844
Musée des Chemins de Fer Belges-Museum van de Belgische Spoorwegen, Bruxelles 003472
Musée des Chemins de Fer Militaires et Industriels, La Neuville-les-Bray 013446
Musée des Civilisations, Saint-Just-Saint-Rambert 015404
Musée des Civilisations de Côte d'Ivoire, Abidjan 008869
Musée des Civilisations de l'Europe et de la Méditerranée, Marseille . 014054
Musée des Coiffes, Fresnay-sur-Sarthe 013055
Musée des Coiffes et des Traditions, Les Ponts-de-Cé 013758
Musée des Commandos, Flawinne 003597
Musée des Commerces Anciens, Doué-la-Fontaine 012826
Musée des Communications et d'Histoire de Sutton, Sutton, Québec 007107
Musée des Contenants Vinaires, Pouzols-Minérvois 014985
Musée des Corbières, Sigean . . 015809
Musée des Cordeliers, Saint-Jean-d'Angély 015369
Musée des Cultures Guyanaises, Cayenne 016370
Musée des Cultures Légumières et ses Réserves, La Courneuve . . 013403
Musée des Cultures Taurines, Nîmes 014502
Musée des Débuts de l'Aviation,

Douzy 012832
Musée des Distilleries Limougeaudes, Limoges 013818
Musée des Docks Romains, Marseille 014055
Musée des Eaux de Vie, Lapoutroie 013571
Musée des Egouts de Paris, Paris 014736
Le Musée des Enfants-Het Kindermuseum, Bruxelles 003473
Musée des Engins de Travaux Publics, Centre de Formation des Travaux Publics Emile Pico, Mallemort . 013983
Musée des Épaves Sous-Marines, Commes 012642
Musée des Équipages, Maincy . 013972
Musée des Évolutions, Maison de Pays, Bousies 012175
Musée des Faïences Anciennes de la Tour d'Aigues, La Tour-d'Aigues 013499
Musée des Faïences de Thiviers, Thiviers 015939
Musée des Faïences de Varages, Varages 016124
Musée des Feuillardiers, Chalus 012412
Musée des Filles de Jésus, Trois-Rivières 007246
Musée des Flûtes du Monde, Château la Croix-Chabrière, Bollène . . . 012096
Musée des Fontaines Salées, Saint-Père-sous-Vezelay 015505
Musée des Forges, Brocas-les-Forges 012236
Musée des Fossiles, Montceau-les-Mines 014241
Musée des Francs et de la Famenne, Marche-en-Famenne 003845
Musée des Fusiliers Mont-Royal, Montréal 006392
Musée des Galvachers, Anost . . 011679
Musée des Glaciers, Lourtier . . 041690
Musée des Goemoniers, Plouguerneau 014907
Musée des Grenouilles, Estavayer-le-Lac 041476
Musée des Guerres de Vendée, Le Pin-en-Mauges 013682
Musée des Guerres de Vendée, Saint-Florent-le-Vieil 015318
Musée des Hospices Civils de Lyon, Lyon 013941
Musée des Hospitalières de l'Hôtel-Dieu de Montréal, Montréal . . . 006393
Musée des Imprimés en Luxembourg, Redu 003962
Musée des Industries du Viemeu, Friville-Escarbotin 013060
Musée des Instruments de Musique, L'Aigle 013532
Musée des Instruments de Musique à Vent, La Couture-Boussey 013404
Musée des Instruments de Musique, Musées Royaux d'Art et d'Histoire, Bruxelles 003474
Musée des Ivoires, Yvetot 016363
Musée des Jacobins, Saint-Sever (Landes) 015566
Musée des Jacobins, Musée d'Auch, Auch 011781
Musée des Jasseries des Monts du Forez, Saint-Anthème 015220
Musée des Lettres et Manuscrits, Paris 014737
Musée des Liqueurs Cherry-Rocher, La Côte-Saint-André 013398
Musée des Maîtres et Artisans, Montréal 006394
Musée des Majorettes, Saint-Quentin-Fallavier 015528
Musée des Maladies de la Peau, Paris 014738
Musée des Manufactures de Dentelles,

Retournac 015074
Musée des Maquettes, Forges-les-
Eaux 013028
Musée des Maquettes, Saint-Privat-
des-Près 015522
Musée des Marais Salants, Batz-sur-
Mer 011908
Musée des Mariniers du Rhône,
Serrières 015781
Musée des Mariniers et de la
Batellerie, Chouzé-sur-Loire . . 012573
Musée des Marionnettes, Aubeterre-
sur-Dronne 011772
Musée des Médailles et Vieux
Courseulles, Courseulles-sur-
Mer 012701
Musée des Mégalithes et Centre
d'Exposition de Wéris, Wéris . . 004144
Musée des Meilleurs Ouvriers de
France, Bourges 012162
Musée des Merveilles, Tende . . 015923
Musée des Métiers Anciens et de la
Tradition Rurale, Estinnes-au-
Mont 003587
Musée des Métiers d'Autrefois,
Pierrefort 014870
Musée des Métiers de France, Argent-
sur-Sauldre 011717
Musée des Métiers de la Pierre, Saint-
Beauzély 015234
Musée des Métiers du Bois,
Labaroche 013507
Musée des Métiers et des Légendes
de la Forêt d'Orléans, Loury . . 013907
Musée des Métiers et Traditions
Paysannes, Argy 011725
Musée des Métiers Ruraux,
Eternoz 012923
Musée des Mille et une Racines,
Cornimont 012676
Musée de la Minéralogie, Université
Louis Pasteur, Strasbourg . . . 015883
Musée des Minéraux et de la Faune
des Alpes, Le Bourg-d'Oisans . 013609
Musée des Minéraux et Fossiles,
Rougemont 041894
Musée des Minéraux et Fossiles des
Ardennes, Bogny-sur-Meuse . . 012090
Musée des Mines d'Argent des
Gorges du Fournel, L'Argentière-la-
Bessée 013574
Musée des Mines d'Aumetz,
Aumetz 011791
Musée des Mines de Fer,
Aumetz 011792
Musée des Mines de Fer de Lorraine,
Neufchef 014451
Musée des Missions,
Plouguerneau 014908
Musée des Monnaies et Médailles J.
Puig, Perpignan 014843
Musée des Moulages, Lyon 013942
Musée des Moulages,
Montpellier 014293
Musée des Moulins, Aram, Villeneuve-
d'Ascq 016259
Musée des Musiques Arabes et
Méditerranéennes Ennejma Ezzahra,
Palais du Baron Rodolphe d'Erlanger,
Sidi Bou Saïd 042772
Musée des Musiques Populaires,
Montluçon 014271
Musée des Oiseaux, Trignac . . . 016043
Musée des Ondes Emile Berliner,
Montréal 006395
Musée des Ostensions et de la
Religon Populaire, Esse . . 012910
Musée des Outil d'Hier, Chauvirey-le-
Vieil 012535
Musée des Outils Agraires,
Châteaurenard (Bouches-du-
Rhône) 012501, 012502
Musée des Outils du Bois et de la Vie
Paysanne, Bligny-sur-Ouche . . 012074

Musée des Outils et Appareils
Ménagers, Merceuil 014136
Musée des Papes, Grande-Anse 005909
Musée des Papeteries Canson et
Montgolfier, Annonay 011677
Musée des Papillons, Fuveau . . 013068
Musée des Papillons, Saint-Chély-
d'Apcher 015254
Musée des Papillons de Liège,
Liège 003806
Musée des Parachutistes, Pau . . 014809
Musée des Patenôtriers, Brain-sur-
Allonnes 012186
Musée des Peintres de l'Ecole de
Murol, Murol 014388
Musée des Peintres du Pays, Saint-
Cyprien 015274
Musée des Pénitents Blancs, Marsac-
en-Livradois 014034
Musée des Petits Hommes en
Pyrénées, Arreau 011745
Musée des Petits Métiers de la Ferme,
Wervicq-Sud 016347
Musée des Petits Meubles Bretons,
Plozevet 014914
Musée des Phares, La Martre . . 006136
Musée des Pioneers, Saint-André-
Avellin 006845
Musée des Pioniers et de
Chanoinesses, Notre-Dame-de-
Lourdes 006529
Musée des Pionniers, Grande-
Digue 005910
Musée des Plans-Reliefs, Paris . . 014739
Musée des Plantes Médicinales et de
la Pharmacie, Musées de l'Université
Libre de Bruxelles, Bruxelles . . 003475
Musée des Pompiers, Wasquehal 016340
Musée des Pompiers Auxiliaires de
Mont-Royal, Montréal 006396
Musée des Pompiers, Centre
de Secours Principal,
Fontainebleau 013010
Musée des Postes et
Telecommunications,
Luxembourg 030590
Musée des Poupées d'Antan,
Malicorne-sur-Sarthe 013982
Musée des Poupées d'Antan et
de la Tonnellerie, Maisons-lès-
Chaource 013977
Musée des Poupées, Collection Rohan,
Josselin 013327
Musée des Processions, Hyon-
Ciply 003723
Musée des Rangers, Grandcamp-
Maisy 013146
Musée des Records, La Tour-
Blanche 013497
Musée des Religions, Nicolet . . 006495
Musée des Rochers Sévigné,
Vitré 016308
Musée des Santons Animés,
Maussane-les-Alpilles 014094
Musée des Santons des Baux, Les
Baux-de-Provence 013739
Musée des Santons Di-Landro,
Aubagne 011766
Musée des Sapeurs Pompiers,
Bièvres 012036
Musée des Sapeurs-Pompiers,
Fessenheim 012975
Musée des Sapeurs-Pompiers,
Thônes 015945
Musée des Sapeurs-Pompiers de
France, Montville 014317
Musée des Sapeurs-Pompiers du
Grand Lyon, Lyon 013943
Musée des Sapeurs Pompiers-
L'Univers des Pompiers, Athus . . 003331
Musée des Sciences et Techniques,
Laval (Mayenne) 013588
Musée des Sciences Naturelles,
Orléans 014587

Musée des Soeurs de Miséricorde,
Montréal 006397
Musée des Soeurs Grises de Montréal,
Montréal 006398
Musée des Soldats du Feu,
Wambrechies 016337
Musée des Sous-Stations R.A.T.P.,
Paris 014740
Musée des Souvenirs Napoléoniens et
des Archives du Palais, Monaco 031719
Musée des Spahis, La Horgne . . 013434
Musée des Spahis, Valence
(Drôme) 016092
Musée des Suisses dans le Monde,
Museum der Schweizer im Ausland,
Pregny-Chambésy 041851
Musée des Tanneurs, Montargis 014222
Musée des Tapisseries, Aix-en-
Provence 011560
Musée des Techniques Faïencières,
Sarreguemines 015697
Musée des Techniques Fromagères,
Saint-Pierre-sur-Dives 015514
Musée des Technologies Ferroviaires,
La Seyne-sur-Mer 013491
Musée des Télécommunications,
Perros-Guirec 014850
Musée des Télécoms, Pleumeur-
Bodou 014893
Musée des Téléphones de Genève,
Plan-les-Ouates 041841
Musée des Temps Barbares,
Marle 014021
Musée des Terre-Neuvas et de la
Pêche, Fécamp 012960
Musée des Thoniers, Etel 012922
Musée des Timbres et des Monnaies,
Monaco 031720
Musée des Tisserands Mayennais,
Ambrières-les-Vallées 011625
Musée des Tissus, Lyon 013944
Musée des Toiles Peintes, Colombier
(Neuchâtel) 041423
Musée des Tourneurs, Aiguines . 011552
Musée des Tradition de l'Ile,
Noirmoutier-en-l'Ile 014519
Musée des Traditions, Les Avanchers-
Valmorel 013736
Musée des Traditions Agricoles du
Ségala, Pradinas 014989
Musée des Traditions Béarnaises,
Salies-de-Béarn 015648
Musée des Traditions Bugistes, Saint-
Rambert-en-Bugey 015530
Musée des Traditions Comtadines,
Pernes-les-Fontaines 014833
Musée des Traditions du Ban-de-la-
Roche, Neuviller-la-Roche . . . 014462
Musée des Traditions du Vieux Dirinon,
Dirinon 012798
Musée des Tradition et Arts
Normands, Martainville-Epreville 014068
Musée des Traditions et de l'Histoire
de Biscarrosse, Biscarrosse . . 012046
Musée des Traditions et des Barques
du Léman, Saint-Gingolph . . . 041907
Musée des Traditions Locales,
Plounéour-Trez 014912
Musée des Traditions Locales, Sainte-
Maxime 015622
Musée des Traditions Namuroises,
Namur 003899
Musée des Traditions Populaires,
Escaudain 012895
Musée des Traditions Populaires,
Saint-Gilles-du-Gard 015353
Musée des Traditions Populaires Cap
Al Campestre, Lherm 013788
Musée des Traditions Populaires
Marius Audin, Beaujeu 011940
Musée des Traditions Populaires,
Mémoire des Olonnes, Olonne-sur-
Mer 014560
Musée des Traditions Rurales et

Artisanales, Revel-Tourdan . . . 015081
Musée des Traditions Stéphanoises,
Saint-Étienne-de-Tinée 015307
Musée des Traditions Vésubiennes,
Saint-Martin-Vésubie 015458
Musée des Traditions Vigneronnes, Le
Caveau Bugiste, Vongnes 016324
Musée des Tramways à Vapeur et
des Chemins de Fer Secondaires
Français, Butry-sur-Oise 012259
Musée des Tramways et Autobus,
Luxembourg 030591
Musée des Transmissions,
Suresnes 015895
Musée des Transports – Amitram,
Marquette-Lez-Lille 014031
Musée des Transports de Pithiviers,
Bouzonville-en-Beauce 012184
Musée des Transports en Commun du
Pays de Liège, Liège 003807
Musée des Transports, du Tourisme et
des Communications, Toulouse 015993
Musée des Trophées de Chasse, La
Chapelle-Glain 013384
Musée des Troupes de Marine,
Fréjus 013049
Musée des Troupes de Montagne,
Grenoble 013185
Musée des Tumulus de Bougon,
Bougon 012132
Musée des Ursulines, Mâcon . . 013958
Musée des Ursulines, Québec . . 006739
Musée des Ursulines, Trois-
Rivières 007247
Musée des Ustensiles de Cuisine
Anciens, Saint-Denis-la-
Chevasse 015287
Musée des Vallées Cévenoles,
Acropolis Museum Montpellier, Saint-
Jean-du-Gard 015388
Musée des Vieilles Automobiles,
Canet-en-Roussillon 012310
Musée des Vieilles Landes, Lit-et-
Mixe 013839
Musée des Vieux Métiers, Ardes-sur-
Couze 011714
Musée des Vieux Métiers,
Saissac 015643
Musée des Vieux Métiers et des Eaux-
de-Vie, Dabo 012741
Musée des Vieux Métiers et des
Traditions Populaires, Sainte-Croix-
Volvestre 015608
Musée des Vieux Outils, Auterive 011807
Musée des Vieux Outils, Fourcès 013038
Musée des Vieux Outils Noël-Morard,
Caromb 012340
Musée des Vieux Outils Vignerons,
Aragon 011702
Musée des Vieux Tacots, Les
Riceys 013760
Musée des Vins de Touraine,
Tours 016018
Musée des XVes Jeux Olympiques
d'Hiver, Albertville 011579
Musée Despiau Wlérick, Mont-de-
Marsan 014208
Musée d'Estavayer, Estavayer-le-
Lac 041477
Musée d'Ethnographie, Genève . 041533
Musée d'Ethnographie,
Neuchâtel 041533
Musée d'Ethnographie de Genève –
Annexe de Conches, Conches . 041425
Musée d'Ethnographie de l'Université
de Bordeaux II, Bordeaux 012117
Musée d'Ethnographie et d'Art
Populaire, Honfleur 013270
Musée d'Ethnographie et d'Histoire de
Conflans, Albertville 011580
Musée d'Evreux, Evreux 012935
Musée d'Exposition des Fouilles
Archéologiques, Javols 013315
Musée d'Eysses, Villeneuve-sur-

Lot 016275
Musée d'Héritage de Notukeu,
Ponteix 006669
Musée d'Hippone, Annaba . 000050
Musée d'Histoire, La Chaux-de-
Fonds 041407
Musée d'Histoire, Elne 012863
Musée d'Histoire, La Neuveville 041800
Musée d'Histoire, Obernai 014543
Musée d'Histoire, Ouidah 004184
Musée d'Histoire Contemporaine,
Bibliothèque de Documentation
Internationale Contemporaine,
Paris 014741
Musée d'Histoire de la Brasserie de
Diekirch, Diekirch 030572
Musée d'Histoire de la Médecine,
Paris 014742
Musée d'Histoire de la Résistance et
de la Déportation de l'Ain et du Haut-
Jura, Nantua 014425
Musée d'Histoire de la Ville de
Luxembourg, Luxembourg . . 030592
Musée d'Histoire de la Ville et
d'Ethnographie du Pays Malouin,
Saint-Malo (Ille-et-Vilaine) 015439
Musée d'Histoire de Marseille,
Marseille 014056
Musée d'Histoire des Sciences,
Genève 041534
Musée d'Histoire du 20e Siècle,
Estivareilles 012915
Musée d'Histoire et d'Archéologie, Les
Baux-de-Provence 013740
Musée d'Histoire et d'Archéologie,
Longjumeau 013860
Musée d'Histoire et d'Archéologie,
Marcinelle 003847
Musée d'Histoire et d'Archéologie,
Vannes (Morbihan) 016121
Musée d'Histoire et d'Archéologie de
Die et du Diois, Musée de France,
Die 012766
Musée d'Histoire et d'Archéologie de
la Chartreuse, Molsheim 014195
Musée d'Histoire et d'Archéologie du
Pays d'Apt, Apt 011701
Musée d'Histoire et d'Archéologie
René Bauberot, Châteauponsac 012498
Musée d'Histoire et d'Archéologie,
Academie de la Val d'Isère, Moutiers
(Savoie) 014367
Musée d'Histoire et d'Archéologie,
Société d'Histoire de Comines-
Warneton, Warneton 004132
Musée d'Histoire d'Art,
Mirande 014181
Musée d'Histoire et d'Art, Villeneuve-
Loubet 016272
Musée d'Histoire et d'Arts Populaires,
Pellafol-le-Vieux 014821
Musée d'Histoire et de l'Éclairage,
Liège 003808
Musée d'Histoire et de Tambow,
Amnéville-les-Thermes 011636
Musée d'Histoire et de Vie Locale,
Genappe 003614
Musée d'Histoire et des Arts
Décoratifs, Tournai 004086
Musée d'Histoire et d'Ethnographie,
Aneho 042728
Musée d'Histoire et du Patrimoine de
Dorval, Montréal 006399
Musée d'Histoire et Traditions
Populaires, L'Oudon-Notre-Dame-de-
Fresnay 013880
Musée d'Histoire Jean Garcin
1939–1945, L'Appel de la Liberté,
Fontaine-de-Vaucluse 013006
Musée d'Histoire Local, Billom . 012040
Musée d'Histoire Local,
Maubourguet 014084
Musée d'Histoire Locale, Belin-
Beliet 011970

Musée d'Histoire Locale, Cagnes-sur-
Mer 012283
Musée d'Histoire Locale, Deuil-la-
Barre 012764
Musée d'Histoire Locale, Ecaussinnes-
Lalaing 003568
Musée d'Histoire Locale,
Mougins 014349
Musée d'Histoire Locale, Saint-Paul-
de-Vence 015497
Musée d'Histoire Locale, Saint-
Renan 015548
Musée d'Histoire Locale et des
Traditions, Fréjus 013050
Musée d'Histoire Locale et des Vieux
Mètiers, Saint-Ciers-sur-Gironde 015260
Musée d'Histoire Locale et Traditions
Populaires, Arts et Traditions
Populaires, Olargues 014557
Musée d'Histoire Locale Pierre Salvi,
Viarmes 016194
Musée d'Histoire Médiévale, Châtel-
sur-Moselle 012507
Musée d'Histoire Militaire à Bandera,
Ajaccio 011570
Musée d'Histoire Naturelle, Aix-en-
Provence 011561
Musée d'Histoire Naturelle,
Amiens 011634
Musée d'Histoire Naturelle,
Angers 011656
Musée d'Histoire Naturelle, Baume-les-
Messieurs 011913
Musée d'Histoire Naturelle,
Besançon 012004
Musée d'Histoire Naturelle,
Chambéry 012420
Musée d'Histoire Naturelle, La Chaux-
de-Fonds 041408
Musée d'Histoire Naturelle,
Eloyes 012865
Musée d'Histoire Naturelle,
Fribourg 041501
Musée d'Histoire Naturelle, Gray 013174
Musée d'Histoire Naturelle, La
Rochelle 013475
Musée d'Histoire Naturelle,
Nantes 014417
Musée d'Histoire Naturelle de Mons,
Mons 003882
Musée d'Histoire Naturelle de Toulon
et du Var, Toulon (Var) 015978
Musée d'Histoire Naturelle du Parc de
Miguasha, Nouvelle 006532
Musée d'Histoire Naturelle et de
Géologie, Lille 013809
Musée d'Histoire Naturelle et de la
Préhistoire, Nîmes 014503
Musée d'Histoire Naturelle et
d'Ethnographie, Colmar 012630
Musée d'Histoire Naturelle et
d'Histoire Locale, Musée Municipal,
Elbeuf-sur-Seine 012861
Musée d'Histoire Naturelle et
Régionale, Gruffy 013199
Musée d'Histoire Naturelle et Vivarium,
Tournai 004087
Musée d'Histoire Naturelle Gabriel
Foucher, Bourges 012163
Musée d'Histoire Naturelle Philadelphe
Thomas, Gaillac 013074
Musée d'Histoire Naturelle, Aquarium,
Amnéville-les-Thermes 011637
Musée d'Histoire Urbaine de Bergerac,
Bergerac 011990
Musée d'Histoire Naturelles,
Boulogne-sur-Mer 012139
Musée d'Histore Naturelle Victor-Brun,
Montauban 014225
Musée d'Homme Palais Royaux de
Porto-Novo, Porto-Novo 004187
Musée d'Horlogerie du Locle, Le
Locle 041687
Musée d'Huez et de l'Oisans, L'Alpe-

d'Huez 013537
Musée Dicks, Vianden 030604
Musée d'Iconographie Historique,
Onoz 003920
Musee d'Ifri, Ifri 000066
Musée d'Initiation à la Nature de
Normandie, Caen 012280
Musée d'Initiation à la Préhistoire,
Tamnies 015902
Musée d'Institut National de
Archéologique d'El-Jem, El-Jem 042741
Musée d'Intérêt National du Tabac,
Bergerac 011991
Musée Diocésain d'Art Sacré, Moulins
(Allier) 014356
Musée Diocésain et Trésor de la
Cathédrale Saint-Aubain, Namur 003900
Musée d'Istres, Istres 013302
Musée d'Ixelles-Museum van Elsene,
Bruxelles 003476
Musée d'Izieu Mémorial des Enfants
Juifs Extermineis, Izieu 013306
Musée d'Métiers et Traditions du Pays
d'Olliergues, Olliergues 014558
Musée Dobrée, Nantes 014418
Musée Docteur Joseph Fau,
Conques 012660
Musée Documentaire Dar Ayed, Ksar
Hellal 042751
Musée Domaine de Gourjade,
Castres 012365
Musée Domaine de Samara, La
Chaussée-Tirancourt 013389
Musée Don Bosco, Doékoué . . 008873
Musée d'Orainville, Orainville . . 014568
Musée d'Orange, Culture Espace,
Orange 014569
Musée d'Orbe, Orbe 041829
Musée d'Orbigny-Bernon, La
Rochelle 013476
Musée d'Orgues de Kermesse,
Bruxelles 003477
Musée Dorothée Salhab Kazemi,
Brumana 030273
Musée d'Orsay, Paris 014743
Musée Doubrère, Marciac 014005
Musée d'Ourthe-Amblève, Musée
Communal, Comblain-au-Pont . 003521
Musée d'Sacré, Mours-Saint-
Eusèbe 014363
Musée du 11 Septembre 1709,
Musée de la Bataille de Malplaquet,
Bavay 011916
Musée du 3è Lanciers, Namur . 003901
Musée du Bailles, Nismes 003911
Musée du Bain Rituel Juif,
Bischheim 012047
Musée du Barreau de Paris,
Paris 014744
Musée du Bas Saint-Laurent, Rivière-
du-Loup 006813
Musée du Bassin Houiller Lorrain,
C.C.S.T.I., Petite-Rossele 014856
Musée du Bateau-Lavoir Le Saint-
Julien, Laval (Mayenne) 013589
Musée du Batha, Fèz 031775
Musée du Batiment, Moulins
(Allier) 014357
Musée du Béret, Nay 014438
Musée du Berry, Bourges 012164
Musée du Biterrois, Béziers . . . 012023
Musée du Blé et du Pain, Ecomusée
de la Bresse Bourguignonne, Verdun-
sur-le-Doubs 016167
Musée du Bocage Normand, Saint-
Lô 015424
Musée du Bois, Aubonne 041242
Musée du Bois, Seyssel 015805
Musée du Bois de Maredret-Galerie A
la Fontaine, Maredret 003850
Musée du Bois et de la Forêt, La
Chaise-Dieu 013376
Musée du Bois et de l'Outil,
Montgobert 014262

Musée du Bois, Musée de Ua Huka,
Papuakeikaha 016379
Musée du Bonbon Haribo, Uzès 016077
Musée du Bonsaï, Châtenay-
Malabry 012512
Musée du Bouchardais, L'Ile-
Bouchard 013801
Musée du Bouton-Château
d'Ambleville, Ambleville 011619
Musée du Breuil, Bourbon-Lancy 012143
Musée du Bronze d'Inverness,
Inverness 006038
Musée du Bugey-Valmorey,
Lochieu 013846
Musée du Buis, Parc des Topiaires,
Durbuy 003562
Musée du Café Gondrée,
Bénouville 011986
Musée du Calvados et des Métiers
Anciens de Pont-l'Évêque, Pont-
l'Évêque 014943
Musée du Calvaire de Pontchateau,
Pontchâteau 014953
Musée du Camembert,
Vimoutiers 016300
Musée du Camp d'Internement
du Vernet d'Ariège, Le Vernet
d'Ariège 013714
Musée du Camp Romain, Saint-Jean-
le-Vieux 015389
Musée du Canal, Cleden-Poher . 012589
Musée du Canal, Ecuisses 012857
Musée du Canal de Berry, Audes-
Reugny 011785
Musée du Canal d'Orléans, Fay-aux-
Loges 012956
Musée du Canal du Duc de Berry,
Dun-sur-Auron 012841
Musée du Capitellu, Ajaccio . . . 011571
Musée du Cardinal Verdier, Lacroix-
Barrez 013521
Musée du Cartonnage et de
l'Imprimerie, Valréas 016116
Musée du Casque de Combat, Saint-
Vincent-sur-Jabron 015590
Musée du Centre Acadien de
l'Université Sainte-Anne, Pointe de
l'Eglise 006666
Musée du Centre de Recherches et de
Documentation du Sénégal, Saint-
Louis 038121
Musée du Centre de Recherches
sur les Monuments Historiques,
Paris 014745
Musée du Centre Élisabeth Bergeron,
Saint-Hyacinthe 006875
Musée du Cerf-Volant, Perrigny-lès-
Dijon 014847
Musée du Chablais, Bex 041335
Musée du Chablais, Thonon-les-
Bains 015949
Musée du Champignon, Saumur 015717
Musée du Champignon, Tillet-Saint-
Ode 004072
Musée du Chanvre, Bligny-sur-
Ouche 012075
Musée du Chapitre, Soignies . . 004034
Musée du Charroi Rural,
Salmiech 015660
Musée du Châtaignier, Marrons et
Champignons, Villefranche-du-
Périgord 016244
Musée du Château, Baugé 011910
Musée du Château, Flers 012991
Musée du Château, Fougères (Ille-et-
Vilaine) 013033
Musée du Château, Lunéville . . 013916
Musée du Château, Margaux . . 014037
Musée du Château, Montbard . 014231
Musée du Château, Montmaur . 014273
Musée du Château, Noirmoutier-en-
l'Ile 014520
Musée du Château, Pauillac . . . 014813
Musée du Château, Roquebrune-Cap-

Martin 015146
Musée du Château, Siorac-en-
Périgord 015812
Musée du Château Comtal,
Rochefort 003972
Musèe du Château de Bourbilly, Vic-
de-Chassenay 016195
Musée du Château de Dourdan,
Dourdan 012829
Musée du Château de Fagnolles,
Fagnolles 003591
Musée du Château de Flaugergues,
Montpellier 014294
Musée du Château de Franchimont,
Theux 004059
Musée du Château de la Guignardière,
Avrillé 011851
Musée du Château de la Louviere,
Montluçon 014272
Musée du Château de Mascarras,
Garlin 013083
Musée du Château de Mesnières-en-
Bray, Neufchâtel-en-Bray 014448
Musée du Château de Monbazillac,
Monbazillac 014196
Musée du Château de Montrond-les-
Bains, Montrond-les-Bains . . . 014312
Musée du Château de Pontécoulant,
Pontécoulant 014954
Musée du Château de Ravel,
Ravel 015040
Musée du Château de Vincennes,
Vincennes 016301
Musée du Château de Vitré, Vitré 016309
Musée du Château des Baux, Les
Baux-de-Provence 013741
Musée du Château des Comtes de
Marchin, Modave 003871
Musée du Château des Ducs de
Bretagne, Château des Ducs de
Bretagne, Nantes 014419
Musée du Château des Ducs de
Wurtemberg, Montbéliard 014235
Musée du Château d'Henri IV,
Nérac 014443
Musée du Château d'Oigny-en-Valois,
Oigny-en-Valois 014553
Musée du Château du Marais, Val
Saint-Germain 016086
Musée du Château Dufresne,
Montréal 006400
Musée du Château-Fort, Ecaussinnes-
Lalaing 003569
Musée du Château Fort-Liberia,
Villefranche-de-Conflent 016241
Musée du Château Paul Pastre,
Marsillargues 014064
Musée du Château Ramezay,
Montréal 006401
Musée du Château Saint-Jean,
Nogent-le-Rotrou 014514
Musée du Châtillonnais, Châtillon-sur-
Seine 012520
Musée du Chemin de Fer, Apremont-
la-Forêt 011696
Musée du Chemin de Fer Miniature,
Clécy 012587
Musée du Cheminot, Ambérieu-en-
Bugey 011612
Musée du Cheval, La Sarraz . . 041942
Musée du Cheval en 1900, Cussac-
Fort-Médoc 012738
Musée du Chien, Villeneuve-les-
Convers 016270
Musée du Chocolat,
Barcelonnette 011895
Musée du Cidre, Domagné . . . 012802
Musée du Cidre, Montaure . . . 014227
Musée du Cidre, Saint-Jean-des-
Champs 015385
Musée du Cidre Breton, Alain Bourvon,
Argol 011723
Musée du Cidre du Pays d'Othe, Eaux-
Puiseaux 012849

Musée du Cimetière Américain Henri-
Chapelle, Hombourg 003706
Musée du Cinéma, Bruxelles . . 003478
Musée du Cinéma, Cine-Chaplin,
Nismes 003912
Musée du Cinquantenaire-
Jubelparkmuseum, Musées Royaux
d'Art et d'Histoire, Bruxelles . . 003479
Musée du Circuit de Spa-
Francorchamps, Stavelot 004047
Musée du Cirque, Vatan 016134
Musée du Cirque, Wasquehal . . 016341
Musée du Clocher, Haute-Rivoire 013237
Musée du Cloître, Tulle 016061
Musée du Cloître Notre-Dame en Vaux,
Châlons-en-Champagne 012407
Musée du Clou, Saint-Cornier-des-
Landes 015272
Musée du Collège de Lévis,
Lévis 006192
Musée du Colombier, Alès . . . 011595
Musée du Colombier, Sarniguet . 015689
Musée du Commissariat de l'Armée
de Terre, Montpellier 014295
Musée du Compagnonnage,
Paris 014746
Musée du Compagnonnage,
Tours 016019
Musée du Coquillage, Le
Lavandou 013649
Musée du Coquillage, Papara . 016375
Musée du Corp des Magasins
Militaires, Montréal 006402
Musée du Corps de la Cavalerie
Française 1940, Jandrain-
Jandrenouille 003739
Musée du Corps Franc Pommiès,
Castelnau-Magnoac 012357
Musée du Costume, Château-
Chinon 012478
Musée du Costume Breton,
Paimpol 014605
Musée du Costume Breton, Sainte-
Anne-d'Auray 015602
Musée du Costume Comtadin, Pernes-
les-Fontaines 014834
Musée du Costume Comtois,
Ornans 014590
Musée du Costume et de la Dentelle,
Museum voor het Kostuum en de
Kant, Bruxelles 003480
Musée du Costume Militaire,
Manoir de Sainte-Mesme, Sainte-
Mesme 015627
Musée du Costume Provençal, Solliès-
Ville 015829
Musée du Costume Savogard,
Sévrier 015797
Musée du Costume Savoyard, Bourg-
Saint-Maurice 012156
Musée du Coticule, Vielsalm . . 004111
Musée du Couteau de Laguiole,
Laguiole 013528
Musée du Crest-Cherel, Ugine . 016068
Musée du Cristal, Bayel 011918
Musée du Crozet, Le Crozet . . 013627
Musée du Cuir et de la Tannerie, Bort-
les-Orgues 012124
Musée du Cuir et de la Tannerie,
Château-Renault 012484
Musée du Cuir et des Industries du
Péruwelzis, Péruwelz 003948
Musée du Cuir et du Parchemin,
Levroux 013784
Musée du Cuivre à Stolzembourg,
Stolzembourg 030602
Musée du Cwarmê, Malmedy . 003840
Musée du Cycle, Huccorgne . . 003718
Musée du Cycle et de la Moto,
Lunéville 013917
Musée du Cyclisme, Lavaux-Sainte-
Anne 003778
Musée du Débarquement à Utah
Beach, Sainte-Marie-du-Mont . 015620

Musée du Désert, Mialet 014166
Musée du Dessin et de l'Estampe
Originale, Gravelines 013169
Musée du Diamant, Durbuy . . . 003563
Musée du Docteur Faure, Musée
Municipal, Aix-les-Bains 011567
Musée du Dolder, Riquewihr . . 015103
Musée du Dolon, Chanas 012442
Musée du Donjon de la Toque,
Huriel 013279
Musée du Faouët, Le Faouët . . 013630
Musée du Felicien, Argentine . . 011721
Musée du Fer, Arles-sur-Tech . . 011733
Musée du Fer, Reichshoffen . . . 015047
Musée du Fer à Repasser,
Longwy 013865
Musée du Fer Blanc Ancien et
Moderne, Saint-Arcons-d'Allier . 015226
Musée du Fer et de la Métallurgie
Ancienne, Saint-Hubert 003996
Musée du Fer et du Chemin de Fer,
Vallorbe 042097
Musée du Fer et du Fil, Dompierre
(Orne) 012808
Musée du Feu Adjudant-Chef Gérard,
Châlons-en-Champagne 012408
Musée du Feutre, Mouzon 014370
Musée du Fjord, La Baie 006131
Musée du Flacon à Parfum, La
Rochelle 013477
Musée du Florentinois, Saint-
Florentin 015320
Musée du Florival, Guebwiller . 013201
Musée du Foie Gras, Frespech . 013058
Musée du Foie Gras, Roquettes 015150
Musée du Foie Gras et des Traditions
Populaires, Samatan 015668
Musée du Foie Gras, Maison de l'Oie
et du Canard, Thiviers 015940
Musée du Folklore, Ittre 003732
Musée du Folklore et de la Vie
Montoise, Mons 003883
Musée du Folklore, Musée du
Boubonnais, Moulins (Allier) . . 014358
Musée du Football en France – Musée
du Sport, Paris 014747
Musée du Fort de Battice,
Battice 003346
Musée du Fort de Huy, Huy . . . 003721
Musée du Fort de la Pompelle,
Puisieulx 015008
Musée du Fort de la Pompelle,
Reims 015053
Musée du Fort de Vaux, Vaux-devant-
Damloup 016141
Musée du Fort d'Embourg, Musée
sur les Deux Guerres Mondiales,
Embourg 003576
Musée du Fort Saint-Jean, Saint-Jean-
sur-Richelieu 006883
Musée du Fort Vauban,
Ambleteuse 011618
Musée du Fossile, Maison du Fossile,
Lion-sur-Mer 013825
Musée du Frère Mutien-Marie,
Malonne 003842
Musée du Fromage Cheddar, Saint-
Prime 006934
Musée du Fromage de la Chèvrerie,
Noirmoutier-en-l'Ile 014521
Musée du Fromage et de la Vache
Salers, Saint-Paul-de-Salers . . 015493
Musée du Fusil Rouillé, Bioul . . 003367
Musée du Gabarier, Saint-Simon
(Charente) 015567
Musée du Gâteau Basque, Sare 015685
Musée du Gemmail, Lourdes . . 013896
Musée du Gemmail, Art de Lumière,
Tours 016020
Musée du Général Thys, Dalhem 003530
Musée du Grand-Lens, Lens . . . 041665
Musée du Grand Saint-Bernard,
Grand-Saint-Bernard 041561
Musée du Granit, Bécon-les-

Granits 011962
Musée du Granit, Saint-Michel-de-
Montjoie 015467
Musée du Grenier du Roy Louis XIII,
Villemur-sur-Tarn 016253
Musée du Grès, Saint-Amand-en-
Puisaye 015210
Musée du Groupe Jovinien de
Résistance Bayard, Musée de Joigny,
Joigny (Yonne) 013321
Musée du Guesclin, Châteauneuf-de-
Randon 012492
Musée du Hard Rock Café, Paris 014748
Musée du Haut-Poitou, Mirebeau 014182
Musée du Haut-Richelieu, Musée
d'Histoire Régionale et de Céramique
Québécoise, Saint-Jean-sur-
Richelieu 006884
Musée du Haut-Rouergue,
Laguiole 013529
Musée du Haut-Val-d'Arly,
Megève 014111
Musée du Huchier, Poucharramet 014977
Musée du Jamais Vu, Puymaurin 015013
Musée du Jeu Ancien, Saint-Hilaire-
de-Brethmas 015356
Musée du Jeu de l'Oie, Collection
Dietsch, Rambouillet 015032
Musée du Jeu de Paume,
Versailles 016181
Musée du Jouet, Bruxelles . . . 003481
Musée du Jouet, Clervaux 030570
Musée du Jouet, Moirans-en-
Montagne 014188
Musée du Jouet, Poissy 014917
Musée du Jouet Ancien "Atlantrain",
Saint-Just-Luzac 015403
Musée du Jouet Ancien Automobile,
Rocamadour 015113
Musée du Jouet et de l'Enfant,
Ferrières 003596
Musée du Jus de Fruit, Espace
Muséographique les Pressoirs de
Provence, Saint-Andiol 015216
Musée du Lac, Nernier 014445
Musée du Lac de Paladru, Maison
d'Art et d'Histoire des Trois Vals,
Charavines 012451
Musée du Lait, Belvedere 011981
Musée du Lait, Montebourg . . . 014245
Musée du Lait et des Traditions, Saint-
Étienne-de-Tinée 015308
Musée du Lapidaire, Lamoura . . 013542
Musée du Lavage et du Repassage,
Verneuil-en-Bourbonnais 016174
Musée du Lavandin, Musée Aromes,
Carennac 012335
Musée du Léman, Nyon 041806
Musée du Léon, Lesneven 013773
Musée du Libertaire, Fontenay-les-
Briis 013017
Musée du Liège, Maureillas-las-
Illas 014088
Musée du Liège et du Bouchon,
Mézin 014164
Musée du Littoral et de des Fours à
Chaux, Regnéville-sur-Mer . . . 015043
Musée du Livre, Beaune 140781
Musée du Livre, Lyon 013558
Musée du Livre et Cabinets de
Donations de la Bibliothèque Royale
de Belgique, Bruxelles 003482
Musée du Livre, de la Pierre et du
Verre, Conches-en-Ouche 012649
Musée du Loup, Le Cloître-Saint-
Thégonnec 013620
Musée du Louvre, Paris 014749
Musée du Louvre, Paris 014750
Musée du Lycée Militaire de Saint-Cyr,
Saint-Cyr-l'École 015277
Musée du Machinisme Agricole,
Neuvic (Corrèze) 014457
Musée du Machinisme Agricole,
Prissac 015000

Musée du Machinisme Agricole,
Revest-Saint-Martin 015082
Musée du Machinisme Agricole,
Tullins-Fures 016062
Musée du Machinisme Agricole et du
Cheval de Trait de 1830 à nos Jours,
Juvigny-sous-Andaine 013346
Musée du Machinisme Agricole,
Musée des Métiers du Bois, Bourbon-
Lancy 012144
Musée du Maïs, Morlaàs 014324
Musée de la Maison Forte,
Rarécourt 015037
Musée du Malgré-Tout, Treignes 004091
Musée du Manoir, Saussey ... 015720
Musée du Manoir d'Argentelles,
Villebadin 016233
Musée du Maquettisme, de la
Fonderie et de la Moulerie, Musée
Municipal, Blangy-sur-Bresle . 012062
Musée du Marbre, Rance ... 003960
Musée du Marbre et de la Pierre
Bleue, Bellignies 011979
Musée du Maréchal Murat, Labastide-
Murat 013510
Musée du Matériel Agricole, La
Villeneuve-les-Convers 013504
Musée du Memoire de la Terre,
Tataouine 042777
Musée du Mémorial de Sidi Ferruch,
Port-Vendres 014972
Musée du Meuble Normand, Villedieu-
les-Poêles 016237
Musée du Miel et de l'Abeille,
Gramont 013142
Musée du Miel, Espace de Découverte
du Miel et de l'Abeille, Lobbes 013823
Musée du Mineur, Ensisheim . 012869
Musée du Mineur, La Grand-
Combe 013427
Musée du Modèle Réduit, Varces 016127
Musée du Monde Rural, Rettel . 015075
Musée du Monde Souterrain, Han-sur-
Lesse 003672
Musée du Monde Souterrain et
Historique de la Spéléologie
Française, Saint-Remèze 015533
Musée du Mont Corbier, Saint-Jean-
de-Maurienne 015382
Musée du Mont-de-Piété et de la
Condition des Soies, Avignon . 011837
Musée du Mont-Repais,
Montmelon 041752
Musée du Mont Riant, Alger ... 000042
Musée du Montreux, Montreux . 041755
Musée du Moteur, Saumur ... 015718
Musée du Moteur et de l'Outil à la
Ferme, Juvigné 013345
Musée du Moudjahid, Musée de la
Révolution, Skikda 000077
Musée du Moulin, Saint-Jean-Saint-
Nicolas 015391
Musée du Moulin, Wissant 016352
Musée du Moulin à Huile,
Rustrel 015192
Musée du Moulin à Marée,
Trégastel 016035
Musée du Moulin de Boly,
Lourdes 013897
Musée du Moulins des Bouillons,
Gordes 013130
Musée du Mouton et de la Bergerie
Nationale, Rambouillet 015033
Musée du Mur de l'Atlantique,
Audinghen 011787
Musée du Mur de l'Atlantique,
Ouistreham 014595
Musée du Nord de la Vendée,
Montaigu (Vendée) 014216
Musée du Nougat Arnaud Soubeyran,
Montélimar 014250
Musée du Nouveau-Brunswick, Saint
John 006893
Musée du Nouveau Monde, La

Rochelle 013478
Musée du Noyonnais, Noyon ... 014535
Musée du Pain, Uhlwiller 016069
Musée du Pain Bouilli et du Four
Communal, Villar-d'Arène .. 016218
Musée du Pain d'Épices d'Alsace et
des Douceurs, Gertwiller 013097
Musée du Palais de la Reine,
Antananarivo 030645
Musée du Palais des Evêques, Saint-
Lizier 015422
Musée du Palais du Dey, Casbah,
Alger 000043
Musée du Palais Royal, Phnom
Penh 005285
Musée du Panache Gascon, Termes-
d'Armagnac 015924
Musée du Panorama de la Vie de
Sainte-Bernadette, Lourdes ... 013898
Musée du Panthéon National Haïtien,
Port-au-Prince 023393
Musée du Papier Chiffon,
Muzillac 014396
Musée du Papier Le Nil, Musée
Municipal, Angoulême 011667
Musée du Papier Peint, Museum of
Wall Paper, Rixheim 015108
Musée du Papillon et des Arts
Africains, Aubeterre-sur-Dronne 011773
Musée du Papogay, Rieux-
Volvestre 015089
Musée du Parc M'Ploussoue,
Bonoua 008872
Musée du Parchemin et de
l'Enluminure, Duras 012847
Musée du Passé Agricole de
Beauvechain, Beauvechain ... 003350
Musée du Pastel, Magrin 013967
Musée du Patrimoine, Guellala . 042744
Musée du Patrimoine, Néris-les-
Bains 014444
Musée du Patrimoine, Roquebrune-sur-
Argens 015147
Musée du Patrimoine Berlugan André
Cane, Beaulieu-sur-Mer 011941
Musée du Patrimoine de l'Isle
Crémieu, Hières-sur-Amby ... 013263
Musée du Patrimoine des Pères de
Bétharram, Lestelle-Bétharram 013775
Musée du Patrimoine et de la Dentelle,
Chantilly 012446
Musée du Patrimoine Montagnard,
Nages 014397
Musée du Patrimoine Religieux et des
Croyances Populaires, Moustey 014364
Musée du Patrimoine Rural,
Dompierre (Orne) 012809
Musée du Patrimoine, Maison
du Patrimoine, Six-Fours-les-
Plages 015820
Musée du Paye de Guérande,
Guérande 013204
Musée du Pays, Ussel 016073
Musée du Pays Brignolais,
Brignoles 012226
Musée du Pays d'Albe, Sarralbe 015690
Musée du Pays de Luchon, Bagnères-
de-Luchon 011868
Musée du Pays de Retz, Bourgneuf-
en-Retz 012168
Musée du Pays de Sarrebourg,
Sarrebourg 015695
Musée du Pays de Thônes,
Thônes 015946
Musée du Pays du Der, Sainte-Rémy-
en-Bouzemont 015630
Musée du Pays et Val de Charmey,
Charmey 041402
Musée du Pays Rabastinois,
Rabastens 015027
Musée du Pays Vaurais, Lavaur 013597
Musée du Pays Welche, Fréland 013053
Musée de Peigne et de la Plasturgie,
Oyonnax 014601

Musée du Père Gèdèon, Centre
d'interprétation des Pionniers
de l'Aviation, Sainte-Marie-de-
Beauce 006958
Musée du Père Jean-Marie de la
Mennais, Ploërmen 014895
Musée du Père Noël, Chailly-en-
Bière 012402
Musée du Petit Bateau et des
Aventures de la Mer, Canet-en-
Roussillon 012311
Musée du Petit Format, Nismes 003913
Musée du Petit Lourdes, Lourdes 013989
Musée du Petit-Palais, Avignon . 011838
Musée du Pétrogliphe, Musée de Ua
Huka, Hokatu 016374
Musée du Pétrole, Merkwiller-
Pechelbronn 014141
Musée du Peuple Haïtien,
Fermath 023390
Musée du Phare de Cordouan,
Cordouan's Lighthouse Museum, Le
Verdon-sur-Mer 013713
Musée du Phonographe et de la
Musique Mécanique, Sainte-
Maxime 015623
Musée du Phonographe et de Musique
Mécanique, Limoges 013819
Musée du Pilori, Niort 014508
Musée du Plan Relief, Ancienne
Chapelle de Citadelle, Bitche . 012052
Musée du Poète Ferrailleur, Lizio 013843
Musée du Premier Empire, Musée
Noisot, Fixin 012988
Musée du Préservatif, Condom . 012656
Musée du Président Jacques Chirac,
Sarran 015692
Musée du Présidial Louis Lacrocq,
Guéret 013206
Musée du Présidial, Musée
et Moulin de Castelnaudary,
Castelnaudary 012361
Musée du Prieuré, Charolles .. 012462
Musée du Prieuré, Harfleur .. 013227
Musée du Prieuré d'Amilhac,
Servian 015782
Musée du Prieuré de Graville, Le
Havre 013644
Musée du Prieuré, Ecomusée du
Creusot-Montceau, Perrecy-les-
Forges 014846
Musée du Protestantisme Dauphinois,
Le Poët-Laval 013686
Musée du Protestantisme en Haut
Languedoc, Société de l'Histoire du
Protestantisme Français, Ferrières
(Tarn) 012971
Musée du Quai Branly, Arts et
Civilisations d'Afrique, d'Asie,
d'Océanie et des Amériques,
Paris 014751
Musée du Quercorb, Puivert .. 015009
Musée du Radar Allemand, Douvres-la-
Délivrande 012830
Musée du Rail, Cajarc 012291
Musée du Rail, Dinan 012794
Musée du Ranquet, Clermont-
Ferrand 012597
Le Musée du Rêve, Paris 105118
Musée du Revermont, Treffort . 016032
Musée du Rhum, Sainte-Marie 030783
Musée du Rhum, Sainte-Rose . 023353
Musée du Rocher des Fourches,
Cornus 012677
Musée du Roi de Cuba,
Mushenge 008842
Musée du Rouergue, Espalion . 012901
Musée du Rouergue Arts Métiers,
Salles-la-Source 015659
Musée du Roy Soleil, Mont-Louis 014210
Musée du Royal 22e Régiment,
Québec 006740
Musée du Sable, Collège Jean-
Monnet, Le Château-d'Olonne . 013619

Musée du Sabot, Buironfosse . 012251
Musée du Sabot, Etang-sur-
Arroux 012918
Musée du Sabot, La Haye-de-
Routot 013433
Musée du Sabot, Porcheresse . 003954
Musée du Sabotier, Soucht ... 015845
Musée du Safran, Boynes 012185
Musée du Santon, Generagues . 013088
Musée du Santon, Le Val 013711
Musée du Santon, Marseille .. 014057
Musée du Santon, Pierrefeu-du-
Var 014868
Musée du Santon et des Traditions de
Provence, Fontaine-de-Vaucluse 013007
Musée du Santons, Gréoux-les-
Bains 013189
Musée du Sapeur-Pompier,
Bouzigues 012183
Musée du Sapeur Pompier, Espace
Marti-Ducros, Saint-Hippolyte-du-
Fort 015362
Musée du Sapeurs-Pompiers, Ille-sur-
Têt 013293
Musée du Saut du Tarn, Saint-
Juéry 015393
Musée du Savignéen, Savigné-sur-
Lathan 015729
Musée du Savon de Marseille Marius
Fabre, Salon-de-Provence 015664
Musée du Scaphandre, Espalion 012902
Musée du Sceau Alsacien, La Petite-
Pierre 013451
Musée du Scribe, Saint-Christol-lès-
Alès 015255
Musée du Sel, Salies-de-Béarn . 015649
Musée du Séminaire de Saint-
Hyacinthe, Saint-Hyacinthe ... 006876
Musée du Septennat, Château-
Chinon 012479
Musée du Service de Santé des
Armées, Paris 014752
Musée du Site Balnéaire, Dinard 012797
Musée du Site de l'Outil au Moutier-
d'Ahun, Ahun 011541
Musée du Ski, Besse-en-
Chandesse 012008
Musée du Ski des Laurentides, Saint-
Sauveur 006935
Musée du Ski et de la Tradition
Rousselande, Les Rousses ... 013762
Musée du Ski, Exposition Parc des
Ecrins, Briançon 012215
Musée du Sous-Officier, Saint-Maixent-
l'École 015436
Musée du Souvenier, Seillans . 015754
Musée du Souvenir, Guer 013202
Musée du Souvenir, Soulac-sur-
Mer 015851
Musée du Souvenir 1914–1918,
Ablain-Saint-Nazaire 011532
Musée du Souvenir des Guerres
1914–18 et 1940–45, Ecaussinnes-
d'Enghien 003567
Musée du Souvenir du Combattant,
Chalon-sur-Saône 012404
Musée du Souvenir du Génie Militaire,
Angers 011657
Musée du Souvenir Militaire de Lyon,
Lyon 013945
Musée du Souvenir Militaire de
Thiérache, Martigny 014070
Musée du Squelette, Île-Verte .. 006027
Musée du Suaire, Cadouin ... 012273
Musée du Sucre d'Orge, Moret-sur-
Loing 014319
Musée du Tacot, Cléron 012603
Musée du Tapis et des Arts Textiles,
Clermont-Ferrand 012598
Musée du Téléphone, Warneton 004133
Musée du Téléphone Narbonne,
Narbonne 014429
Musée du Téléphone, Maison de la
Communication, Nancy 014406

Musée du Témiscouata, Notre-Dame-du-Lac 006530
Musée du Temple de Minerve, Tebessa 000079
Musée du Temps, Besançon ... 012005
Musée du Temps, Ceillac 012380
Musée du Temps des Baigneurs, Cordemais 012668
Musée du Temps Passé, Sainte-Agnès 015600
Musée du Terrain d'Aviation de Conde-Vraux, Vraux 016329
Musée du Terroir, Sisteron 015815
Musée du Terroir, Villeneuve-d'Ascq 016260
Musée du Terroir de Peille, Peille 014819
Musée du Terroir des Clots, Mizoën-en-Oisans 014187
Musée du Terroir et Musée de la Volaille, Romenay 015140
Musée du Terroir Hérissonnais, Hérisson 013253
Musée du Terroir Marseillais, Marseille 014058
Musée du Terroir Normand, Caugé 012371
Musée du Terroir Saint-Dalmas, Valdeblore 016088
Musée du Terroire, Conches-en-Ouche 012650
Musée du Textile, Cholet ... 012570
Musée du Textile des Vosges, Ventron 016161
Musée du Textile et de la Vie Sociale, Ecomusée de la Région Fourmies-Trélon, Fourmies 013039
Musée du Textilen, Labastide-Rouairoux 013511
Musée du Thé, Mariage Frères, Paris 014753
Musée du Théâtre, Couilly-Pont-aux-Dames 012687
Musée du Théâtre Forain et Musée Archéologique, Artenay ... 011752
Musée du Théâtre Royal de Toone, Bruxelles 003483
Musée du Tire-Bouchon, Ménerbes 014123
Musée du Tissage et de la Soierie, Bussières 012256
Le Musée du Tisserand Dauphinois, La Batie-Montgascon 013361
Musée du Touage et Canal de Saint-Quentin, Riqueval 015100
Musée du Touquet, Le Touquet-Paris-Plage 013704
Musée du Tour Automatique et d'Histoire de Moutier, Langdrehautomaten- und Geschichtsmuseum, Moutier .. 041768
Musée du Tour de France, Arvieux-en-Queyras 011755
Musée du Tourneur, Cinquétral . 012575
Musée du Train, Clairac 012580
Musée du Train à Vapeur, Sainte-Foy-l'Argentière 015612
Musée du Train et de la Miniature, Musée de la Vie Rurale, Saint-Sauves-d'Auvergne 015557
Musée du Train et des Equipages Militaires, Tours 016021
Musée du Train ou du Jouet, Arpaillargues-et-Aureillac ... 011739
Musée du Transmanche, Escalles 012894
Musée du Transport Urbain Bruxellois, Bruxelles 003484
Musée du Travail, Montfermeil . 014255
Musée du Travail Charles Peyre, Montfermeil 014256
Musée du Trégor, Guimaec 013212
Musée du Tribunal Révolutionnaire, Tinchebray 015967
Musée du Trièves, Mens 014127
Musée du Trophée des Alpes, La

Turbie 013502
Musée du Val-de-Cher, Selles-sur-Cher 015757
Musée du Valois, Vez 016189
Musée du Valois et de l'Archerie, Crépy-en-Valois 012719
Musée du Veinazès, Lacapelle-del-Fraisse 013517
Musée du Vélo, Cormatin 012674
Musée du Vélo, Fenneviller ... 012967
Musée du Vélo, Periers 014826
Musée du Vélo et de la Moto, Domazan 012803
Musée du Vélo Le Belle Échappée, La Fresnaye-sur-Chédouet 013419
Musée du Vélocipède, Cadouin . 012274
Musée du Veneur, Céré-la-Ronde 012384
Musée du Verre, Blaye-les-Mines 012070
Musée du Verre, Marcinelle ... 003848
Musée du Verre, Sars-Poteries . 015698
Musée du Verre, Musées d'Archéologie et d'Arts Décoratifs, Liège 003809
Musée du Verrier, Saint-Prex .. 041910
Musée du Vieaux Château, Bricquebec 012220
Musée du Vieil Aix, Aix-en-Provence 011562
Musée du Vieil Argenteuil, Argenteuil 011720
Musée du Vieil Auvillar, Auvillar 011818
Musée du Vieil Eygalières, Eygalières 012937
Musée du Vieil Outil Hastière, Hastière 003685
Musée du Vieux-Baulmes, Baulmes 041294
Musée du Vieux Biarritz, Musée Historique, Biarritz 012029
Musée du Vieux Bourg, Mimizan 014174
Musée du Vieux Brest, Brest ... 012208
Musée du Vieux-Château, Collections d'Art Naïf, Laval (Mayenne) ... 013590
Musée du Vieux-Cimetière, Soignies 004035
Musée du Vieux Clairac, Clairac 012581
Musée du Vieux Courseulles, Courseulles-sur-Mer 012702
Musée du Vieux Figeac, Figeac . 012980
Musée du Vieux Granville, Granville 013152
Musée du Vieux Lambesc, Lambesc 013541
Musée du Vieux L'Arbresle, L'Arbresle 013573
Musée du Vieux Marseille, Marseille 014059
Musée du Vieux Mazères, Hotel d'Ardouin, Mazères (Ariège) .. 014101
Musée du Vieux Merville, Merville 014144
Musée du Vieux-Monthey, Monthey 041750
Musée du Vieux Montpellier, Montpellier 014296
Musée du Vieux-Moudon, Château du Rochefort, Moudon 041766
Musée du Vieux Moulin, Villar-Loubière 016219
Musée du Vieux Moulin à Vent, Vensac 016159
Musée du Vieux Nîmes, Nîmes . 014504
Musée du Vieux Nimy, Mons ... 003884
Musée du Vieux Pays, Le Châble 041398
Musée du Vieux Pays, Dieulouard 012769
Musée du Vieux-Pays d'Enhaut, Château-d'Oex 041404
Musée du Vieux Pérouges, Pérouges 014837
Musée du Vieux-Phare, Matane 006274
Musée du Vieux Plassac, Plassac 014844
Musée du Vieux Pommiers, Pommiers-en-Forez 014925
Musée du Vieux Pressoir et de la

Vigne, Coulanges-la-Vineuse .. 012688
Musée du Vieux-Saxon, Saxon . 041945
Musée du Vieux Sisteron, Musée Municipal, Sisteron 015816
Musée du Vieux Toulon, Toulon (Var) 015979
Musée du Vieux Toulouse, Toulouse 015994
Musée du Vieux-Tréport, Le Tréport 013706
Musée du Vieux Warcq, Warcq . 016338
Musée du Vigneron, Buxy 012260
Musée du Vigneron, Chaumont-le-Bois 012531
Musée du Vigneron, Les Marches 013755
Musée du Vigneron, Verdigny .. 016165
Musée du Vigneron Paul Coulon et Fils, Rasteau 015038
Musée du Vigneron, Cave Vinicole du Vieil Armand, Wuenheim 016356
Musée du Vignoble et des Vins d'Alsace, Kientzheim 013350
Musée du Vignoble Nantais, Le Pallet 013676
Musée du Village, Moknine ... 042760
Musée du Village, Saint-Sorlin-d'Arves 015568
Musée du Vin, Cheignieu-la-Balme 012542
Musée du Vin, Cissac-Médoc .. 012577
Musée du Vin, Ehnen 030580
Musée du Vin, Nissan-lez-Ensérune 014510
Musée du Vin, Paris 014754
Musée du Vin, Saint-Hilaire-de-Loulay 015357
Musée du Vin dans l'Art, Pauillac 014814
Musée du Vin de Bourgogne, Beaune 011950
Musée du Vin de Cahors et de la Table Lotoise, Cahors 012289
Musée du Vin de Champagne Piper-Heidsieck, Reims 015054
Musée du Vin de la Batellerie et de la Tonnellerie, Bergerac 011992
Musée du Vin et de Saint-Jacques de Compostelle, Pomerol 014923
Musée du Vin-Mimoria di Vino, Luri 013919
Musée du Vin, Cave Brotte, Châteauneuf-du-Pape 012493
Musée du Vin, Gourmet Quercynois, Saint-Cirq-Lapopie 015263
Musée du Vitrail, Curzac-sur-Vonne 012734
Musée du Vitrail et du Verre, Gordes 013131
Musée du Vivarais Protestant, Pranles 014991
Musée du Zinc, Montpellier ... 014297
Musée Ducal, Bouillon 003390
Musée Duhamel du Monceau, Pithiviers-le-Vieil 014878
Musée d'Unterlinden, Colmar .. 012631
Musée Dupleix et Amas, Landrecies 013548
Musée Dupuytren, Université Paris VI Pierre et Marie Curie, Paris ... 014755
Musée d'Uxellodunum et Musée de Martel, Martel 014069
Musée d'Yverdon-les-Bains et sa Région, Yverdon-les-Bains .. 042173
Musée E. Chenon, Château-Meillant 012482
Musée Eclaté d'Arts et de Traditions Populaires, Cardaillac 012334
Musée-Ecole de Broderie, Cilaos 036182
Musée École de la Perrine, Laval (Mayenne) 013591
Musée-École du Grand Meaulnes, Épineuil-le-Fleuriel 012887
Musée Ecole Publique en Charente, Saint-Fraigne 015325

Musée-École, Centre d'Etudes et des Patrimoines, Saint-Christophe-en-Brionnais 015258
Musée Écologique Vanier, Laval 006172
Musée EDF Electropolis, L'Aventure de l'Electricité, Mulhouse 014377
Musée EDF Hydrelée, Vaujany .. 016139
Musée Edgar Melik, Cabriès ... 012268
Musée Edgard Clerc, Musée Précolombien, Le Moule 023347
Musée Edith Piaf, Paris 014756
Musée Edmond Rostand, Cambo-les-Bains 012300
Musée Edouard Branly, Paris .. 014757
Musée Edouard Mahé, Retiers . 015073
Musée Éducatif de Préhistoire, Villeneuve-la-Comtesse 016265
Musée Éducatif d'Entomologie de Levens, Levens 013780
Musée Église Catholique de l'Assomption, Rosenwiller ... 015156
Musée El Kobba des Arts et Traditions, Sousse 042775
Musée Elise Rieuf, Massiac ... 014079
Musée Emile Aubry, Béjaïa ... 000054
Musée Emile Chénon, Châteaumeillant 012491
Musée Émile Guillaumin, Musée du Folklore, Ygrande 016360
Musée Émile Jean, Villiers-sur-Marne 016298
Musée Émile Verhaeren, Honnelles 003707
Musée Emir Abdelkader, Miliana 000069
Musée Emmanuel de la Villéon, Fougères (Ille-et-Vilaine) 013034
Musée en Herbe, Paris 014758
Musée en Piconrue, Bastogne .. 003345
Musée en Plein Air de Sculptures Contemporaines, Étiolles 012925
Musée en Plein Air du Sart-Tilman, Centre d'Animation et d'Intégration des Arts Plastiques de la Communauté française de Belgique, Liège 003810
Musée Entomologique, Musée des Papillons, Saint-Quentin (Aisne) 015526
Musée Ermitage John Io Scorpione, Saint-Ame 015214
Musée Ernest Cognacq, Saint-Martin-de-Ré 015451
Musée Ernest Renan, Tréguier . 016036
Musée Espace Bérenger Saunière, Rennes-le-Château 015069
Musée Espéranto a Gray, Gray . 013175
Musée Estève, Bourges 012165
Musée Estrine, Saint-Rémy-de-Provence 015544
Musée et Atelier d'Art Animalier, La Chapelle-Thouarault 013386
Musée et Centre de Transmission de la Culture Daniel Weetaluktuk, Inukjuak 006035
Musée et du Vieux Chinon, Chinon 012562
Musée et Maison des Vins, Bandol 011884
Musée et Parc Archéologiques de Montauban, Musées Gaumais, Buzenol 003505
Musée et Plaisir du Chocolat, Biarritz 012030
Musée et Salon Permanent de l'Automobile de Sport Prestige et Collection, Chauffailles 012526
Musée et Site Archéologiques de Nice-Cimiez, Nice 014485
Musée et Sites Archéologiques de Saint-Romain-en-Gal-Vienne, Saint-Romain-en-Gal 015554
Musée et Temple des Bastides, Labastide-d'Armagnac 013509
Musée et Trésor de l'Hospice, Bourg-Saint-Pierre 041358

Musée Ethnographique, Chefchaouen 031772
Musée Ethnographique, Mouans-Sartoux 014344
Musée Ethnographique, Tétouan 031800
Musée Ethnographique Alexandre Senou Adande, Porto-Novo .. 004188
Musée Ethnographique de l'Olivier, Cagnes-sur-Mer 012284
Musée Ethnographique des Oudaïa, Rabat 031790
Musée Ethnographique du Donjon, Niort 014509
Musée Ethnographique du Peuple Basque, Isturitz 013303
Musée Ethnographique Regional, Bouar 007507
Musée Ethnologique Intercommunal du Vermandois, Vermand 016172
Musée Ethnolopique Provincial, Tshikappa 008843
Musée Etienne-Jules Marey, Beaune 011951
Musée Étienne Martin, Noyal-sur-Vilaine 014530
Musée Etival dans le Temps, Ferme-Musée, Etival-Clairefontaine . 012926
Musée Eudore-Dubeau, Montréal 006403
Musée Eugène Boudin, Honfleur 013271
Musée Eugène Burnand, Moudon 041767
Musée Eugène Farcot, Sainville . 015642
Musée Eugène le Roy et Vieux Métiers, Montignac 014265
Musée Européen de Sculptures Contemporaines en Plein Air, Musée à Ciel, Launstroff 013583
Musée-Expo de la Grange aux Abeilles, Giffaumont-Champaubert 013103
Musée Expo Forêt, Gérardmer .. 013093
Musée Extraordinaire Georges-Mazoyer, Ansouis 011682
Musée Fabre, Montpellier 014298
Musée Faniahy, Musée de l'Université de Fianarantsoa, Fianarantsoa . 030649
Musée Fantastique de la Bête du Gévaudan, Saugues-en-Gévaudan 015704
Musée Fayet, Béziers 012024
Musée Fenaille, Rodez 015131
Musée-Ferme Conservatoire Bigourdane, Péré 014825
Musée Ferme d'Autrefois et Matériel Artisanal, Musée Hunault Joseph, Thourie 015956
Musée Ferme de la Patte d'Oie, Le Plessis-Patte-d'Oie 013684
Musée-Ferme des Castors du Chili, Musée de la Faune et Traditions du Béarn, Arzacq-Arraziguet 011758
Musée Ferroviaire, Longueville . 013861
Musée Ferroviaire, Saint-Sulpice-de-Pommeray 015569
Musée Ferroviaire de Saint-Géry, Saint-Géry 015348
Musée Ferroviaire de Treignes, Treignes 004092
Musée Ferroviaire et des Vieilles Gares, Guîtres 013217
Musée Fesch, Ajaccio 011572
Musée Flaubert et d'Histoire de la Médecine, Rouen 015167
Musée Flottant-Architecture Navale, Audierne 011786
Musée Folklorique de Ghardaïa, Ghardaïa 000063
Musée Folklorique et Archéologique, Châtelus-le-Marcheix 012511
Musée Folklorique et Viticole à Possen, Bech-Kleinmacher 030563
Musée-Fondation Alexandra David Néel, Digne-les-Bains 012772
Musée-Fondation Bemberg, Toulouse 015995

Musée Fondation Deutsch, Belmont-sur-Lausanne 041303
Musée-Fondation Louis Jou, Les Baux-de-Provence 041502
Musée-Fort de Leveau, Feignies 012963
Musée Fort de Sucy, Sucy-en-Brie 015890
Musée Fort Lagarde, Prats-de-Mollo-la-Preste 014992
Musée Fougou, Montpellier 014299
Musée Fournaise, Chatou 012522
Musée Fragonard de l'Ecole Nationale Vétérinaire d'Alfort, Maisons-Alfort 013975
Musée Français – 1e Armée Française Mai 1940, Cortil-Noirmont 003524
Musée Français de la Brasserie, Saint-Nicolas-de-Port 015479
Musée Français de la Carte à Jouer, Issy-les-Moulineaux 013300
Musée Français de la Photographie, Bièvres 012037
Musée Français de la Spéléologie, Courniou-les-Grottes 012698
Musée Français des Phares et Balises, Ile-d'Ouessant 013291
Musée Français du Chemin de Fer, Mulhouse 014378
Musée Francisque Mandet, Riom 015096
Musée Franco-Australien, Villers-Bretonneux 016281
Musée Franco-Tchécoslovaque, Darney 012744
Musée François Desnoyer, Saint-Cyprien 015275
Musée François Duesberg des Arts Décoratifs, Mons 003885
Musée François Mauriac, Saint-Maixant (Gironde) 015435
Musée François-Mauriac, Vemars 016148
Musée François Mitterrand, Espace Culturel, Jarnac 013308
Musée François Pilote, La Pocatière 006137
Musée François Pompon, Saulieu 015708
Musée Frédéric Japy, Beaucourt 013168
Musée Frédéric Mistral, Maillane 013968
Musée Friry, Remiremont 015060
Musée Fromages et Patrimoine, Ambert 011614
Musée Frugès – Le Corbusier, Pessac 014854
Musée Gadagne, Musée Historique de Lyon et Musée International de la Marionnette, Lyon 013946
Musée Galerie Honoré Camos, Bargemon 011899
Musée Gallé-Juillet, Musée Municipal, Creil 012717
Musée Galliera, Musée de la Mode de la Ville de Paris, Paris 014759
Musée Gallo-Romain, Berneau . 003361
Musée Gallo-Romain, Biesheim . 012034
Musée Gallo-Romain, Blicquy .. 003372
Musée Gallo-Romain, Périgueux 014830
Musée Gallo-Romain, Petit-Bersac 014855
Musée Gallo-Romain, Rumes ... 003988
Musée Gallo-Romain d'Aoste, Aoste 011695
Musée Gallo-Romain de Lyon-Fourvière, Lyon 013947
Musée Gallo-Romain de Tauroentum, Saint-Cyr-sur-Mer 015281
Musée Gallo-Romain de Waudrez, Waudrez 004138
Musée Gallo-Romain d'Octodure, Martigny 041727
Musée Gallo-Romain, Musée des Potiers de Portout, Chanaz . 012443
Musée Gantner, La Chapelle-sous-Chaux 013385
Musée Gardanne Autrefois, Gardanne 013080

Musée Gaspar, Arlon 003324
Musée Gassendi, Digne-les-Bains 012773
Musée Gaston Fébus, Associaton Escole Gastou Fébus, Mauvezin (Hautes-Pyrénées) 014095
Musée Gaston Grégor, Musée d'Arts et d'Histoire, Salles-d'Angles 015658
Musée Gatien-Bonnet, Lagny-sur-Marne 013525
Musée Gaumais, Musées Gaumais, Virton 004115
Musée Gay-Lussac, Saint-Léonard-de-Noblat 015420
Musée Général Estienne, Berry-au-Bac 011998
Musée Géo Charles, Échirolles . 012854
Musée Géologique, Conakry ... 023380
Musée Géologique, Dakar 038117
Musée Géologique, La Charguia 042752
Musée Geologique, Vernet-les-Bains 016173
Musée Géologique du Rwanda, Kigali 038088
Musée Géologique et Paléontologique, Langé 013551
Musée Géologique et Salle Archéologique, Saint-Avit-Senieur 015231
Musée Géologique Fossiles Préhistoire, Neuil 014452
Musée Geologique Sengier-Cousin, Jadotville 008832
Musée George Sand et de la Vallée Noire, La Châtre 013388
Musée Georges Borias, Uzès .. 016078
Musée Georges Clemenceau, Paris 014760
Musée Georges Garret, Vesoul . 016187
Musée Georges Labit, Toulouse . 015996
Musée Georges Perraudin-Résistance en Morvan, Musée de la Résistance, Saint-Honoré-les-Bains 015364
Musée Georges Pompidou, Montboudif 014237
Musée Georgette Lemaire, Angoulême 011668
Musée Gilles-Villeneuve, Berthierville 005420
Musée Girodet, Montargis 014223
Musée Girouxville, Girouxville .. 005877
Musée Goethe, Sessenheim ... 015784
Musée Goetz-Boumeester, Villefranche-sur-Mer 016247
Musée Gorsline, Bussy-le-Grand 012258
Musée Goupil, Bordeaux 012118
Musée Goya, Castres 012366
Musée Granet, Aix-en-Provence 011563
Musée Grétry, Liège 003811
Musée Grévin, La Rochelle 013479
Musée Grévin, Le Mont-Saint-Michel 013669
Musée Grévin, Paris 014761
Musée Grévin, Saint-Jean-de-Luz 015378
Musée Grévin de la Provence, Salon-de-Provence 015665
Musée Grévin de Lourdes, Lourdes 013900
Musée Grobet-Labadié, Marseille 014060
Musée Grotte de Limousis, Limousis 013823
Musée Gruérien, Bulle 041382
Musée Guerre et Paix en Ardennes, Novion-Porcien 014529
Musée Guillaume Apollinaire, Stavelot 004048
Musée Guillion, Musée Départemental du Compagnonnage Pierre-François Guillon, Romanèche-Thorins . 015137
Musée Gustave Moreau, Musée National, Paris 014762
Musée Gustave Stoskopf, Brumath 012245
Le Musée Gutenberg des Arts

Graphiques et de la Communication, Fribourg 041502
Musée Gyger, Vallorbe 042098
Musée Hansi, Riquewihr 015104
Musée Hardy, Clécy 012588
Musée Harmas Jean-Henri Fabre, Sérignan-du-Comtat 015777
Musée Haut Savoyard de la Résistance, Bonneville 012103
Musée Haviland, Limoges 013820
Musée Hector Berlioz, La Côte-Saint-André 013399
Musée Helmut Warzecha, Saint-Philbert-de-Grand-Lieu 015506
Musée Henri Barbusse, Aumont-en-Halatte 011793
Musée Henri-Barre, Thouars ... 015955
Musée Henri Boez, Maubeuge .. 014083
Musée Henri Chapu, Le Mée-sur-Seine 013661
Musée Henri Dupuis, Saint-Omer 015483
Musée Henri Giron, Le Vigan (Lot) 013716
Musée Henri Malartre, Rochetaillée-sur-Saône 015126
Musée Henri Mathieu, Bruyères 012248
Musée Henry Clews, Mandelieu-La Napoule 013985
Musée Henry de Monfreid, Ingrandes (Indre) 013295
Musée Héritage, Saint-Albert .. 006842
Musée Héritage, Saint-Pierre .. 038093
Musée Hermès, Paris 014763
Musée Hippomobile, Saint-Michel-des-Andaines 015468
Musée Historial, Grézolles 013194
Musée Historial, Mazères (Gironde) 014102
Musée Historie Locale, Musée des Amis de Marchiennes, Marchiennes 014004
Musée Historique, Antananarivo 030646
Musée Historique, Basse-Rentgen 011903
Musée Historique, Bourail 033130
Musée Historique, Carhaix-Plouguer 012336
Musée Historique, Cérisy-la-Forêt 012391
Musée Historique, Châtillon-sur-Saône 012519
Musée Historique, Clervaux 030571
Musée Historique, Fénetrange . 012966
Musée Historique, Gourdon (Alpes-Maritimes) 013136
Musée Historique, Haguenau .. 013224
Musée Historique, La Couvertoirade 013405
Musée Historique, Mont-Saint-Michel 014212
Musée Historique, Mulhouse ... 014379
Musée Historique, Strasbourg .. 015884
Musée Historique d'Abomey, Abomey 004181
Musée Historique de Gorée, Gorée 038120
Musée Historique de Graffiti Anciens et d'Archéologie, Marsilly .. 014066
Musée Historique de la Faïence, Moustiers-Sainte-Marie 014365
Musée Historique de la Ligne Maginot Aquatique, Sarralbe 015691
Musée Historique de la Principauté de Stavelot-Malmedy, Stavelot ... 004049
Musée Historique de la Tour de la Chaîne, La Rochelle 013480
Musée Historique de Lausanne, Lausanne 041660
Musée Historique de l'Ile-de-Sein, Ile-de-Sein ....
Musée Historique de Saint-Gilles-les-Hauts, Saint-Paul 036188
Musée Historique de Saint-Paul, Saint Paul 006928
Musée Historique de Saint-Pierre,

Saint-Pierre 030781
Le Musée Historique de Tracadie,
Tracadie-Sheila 007230
Musée Historique de Troyes et de la
Champagne, Musée de Vauluisant,
Troyes 016053
Musée Historique de Verfeil et des
Petites Filles Modèles, Verfeil . 016168
Musée Historique de Vevey,
Vevey 042107
Musée Historique de Villèle, Saint-
Gilles-les-Hauts 036186
Musée Historique des Communications
Eberdt, Sutton, Québec 007108
Musée Historique des Soeurs de
l'Assomption, Nicolet 006496
Musée Historique du Bailliage de
Rouffach, Rouffach 015174
Musée Historique du Centre-Var, Le
Luc 013650
Musée Historique du Château,
Langoiran 013554
Musée Historique du Cidre, La
Duretière, Melleray-la-Vallée . . 014118
Musée Historique du Domaine
National, Saint-Cloud 015270
Musée Historique du Madawaska,
Edmundston 005770
Musée Historique du Papier,
Ambert 011615
Musée Historique du Papier, Couze-et-
Saint-Front 012710
Musée Historique du Port, La
Rochelle 013481
Musée Historique et Archéologique,
Civaux 012578
Musée Historique et Archéologique,
Montségur 014315
Musée Historique et Archéologique de
l'Orléanais, Orléans 014588
Musée Historique et de la Pêche,
Talmont-sur-Gironde 015900
Musée Historique et des Porcelaines,
Nyon 041807
Musée Historique et Diorama
1914–1918, Souchez 015844
Musée Historique et Erckmann-
Chatrian, Phalsbourg 014862
Musée Historique et Lapidaire, Isle-
Aumont 013296
Musée Historique et Logis Thiphaine,
Le Mont-Saint-Michel 013670
Musée Historique et Militaire,
Freyming-Merlebach 013059
Musée Historique et Militaire,
Huningue 013275
Musée Historique et Militaire,
Mémorial du Finistère, Brest . . 012209
Musée Historique et Préhistorique Van
den Steen, Jehay-Amay 003740
Musée Historique Valdec-de-Lessart,
Portets 014975
Musée Historique, Institut National des
Jeunes Aveugle, Paris 014764
Musée Hofer-Bury, Lavérune . . 013600
Musée Honoré Mercier,
Sabrevois 006838
Musée Horlogerie Automates Yves
Cupillard, Morteau 014341
Musée Horreum Romain,
Narbonne 014430
Musée Horta, Hortamuseum,
Bruxelles 003485
Musée Hospitalier, Charlieu . . . 012458
Musée Hospitalier – Apothicairerie,
Hôtel-Dieu, Bourg-en-Bresse . 012154
Musée Hospitalier et Apothicairerie,
Louhans 013884
Musée-Hôtel Bertrand,
Châteauroux 012505
Musée-Hôtel de Beaumont,
Valognes 016111
Musée Hôtel de la Dentelle,
Brioude 012231

Musée Hôtel de Ville, Thann . . . 015928
Musée-Hôtel Granval-Caligny,
Valognes 016112
Musée Hôtel Lallemand, Bourges 012166
Musée-Hôtel Le Vergeur, Reims 015055
Musée Humoristique Albert Dubout,
Palavas-les-Flots 014608
Musée Huperel de l'Histoire Cathare,
Minerve 014176
Musée Huppy Autrefois, Huppy . 013277
Musée Hydrogéologique,
Chaudfontaine 003516
Musée Ianchelevici, La Louvière 003765
Musée Impérial de la Miniature,
Verneuil-sur-Avre 016176
Musée In Situ des Bains Romains, Aix-
les-Bains 011568
Musée Industriel Corderie Vallois,
Notre-Dame-de-Bondeville . . . 014526
Musée Industriel du Val-de-Travers,
Fleurier 041486
Musée Industriel et d'Ethnologie,
Lille 013810
Musée Ingres, Montauban 014226
Musée Ingres: Bulletin,
Montauban 138650
Musée Intercommunal d'Histoire et
d'Archéologie, Louvres 013910
Musée International d'Art Naïf,
Vicq 016202
Musée International d'Art Naïf Yvon-M.
Daigle, Magog 006241
Musée International d'Arts Modestes,
Sète 015786
Musée International de Akum,
Bamenda 005289
Musée International de la Base
Aérienne de Chièvres, Chievres 003519
Musée International de la Chasse,
Gien 013102
Musée International de la Chaussure,
Romans-sur-Isère 015139
Musée International de la Croix-Rouge
et du Croissant-Rouge, Genève 041535
Musée International de la Faune,
Argelès-Gazost 011715
Musée International de la Faune,
Québriac 015015
Musée International de la Marionnette,
Internationaal Marionettenmuseum
Peruchet, Bruxelles 003486
Musée International de la Miniature,
Lyon 013948
Musée International de la Parfumerie,
Grasse 013161
Musèe International de la Reforme,
Genève 041536
Musée International de l'Automobile,
Le Grand-Saconnex 041560
Musée International de l'Etiquette,
Aigle 041201
Musée International de Mobilier
Miniature, Vendeuvre 016153
Musée International des Hussards,
Tarbes 015916
Musée International des Polices
et des Gendarmeries, Charvieu-
Chavagneux 012473
Musée International d'Horlogerie,
Institut l'Homme et le Temps, La
Chaux-de-Fonds 041409
Musée International du Carnaval et du
Masque, Binche 003366
Musée International du Coquillage,
Saint-Barthélemy 023350
Musée International du Long Cours
Cap Hornier, Saint-Malo (Ille-et-
Vilaine) 015440
Musée International, Pétanque
et Boules, Saint-Bonnet-le-
Château 015238
Musée Isérables, Fondation Pro
Aserablos, Isérables 041605
Musée Itinérant d'Art Byzantin,

Montréal 006404
Musée Ivan Tourguéniev,
Bougival 012131
Musée Ivanoff, La Borne 013366
Musée J. Armand Bombardier,
Valcourt 007268
Musée J.-E. Blanche, Offranville 014548
Musée J.-F. Leyraud, La Roche-sur-le-
Buis 013465
Musée J.B. Mathon et André Durand,
Neufchâtel-en-Bray 014449
Musée Jacquemart-André, Fontaine-
Chaalis 013002
Musée Jacquemart-André, Paris 014765
Musée Jacques Cartier à Saint-Malo,
Manoir de Limoelou, Saint-Malo (Ille-
et-Vilaine) 015441
Musée Jacques Pellan, Giverny . 013116
Musée Jadis Allevard, Allevard-les-
Bains 011601
Musée-Jardin Paul Landowski,
Boulogne-Billancourt 012137
Musée Jean Aicard, Solliès-Ville 015830
Musée Jean Aicard – Paulin Bertrand,
La Garde 013421
Musée Jean-Charles Cazin,
Samer 015669
Musée Jean Cocteau, Menton . . 014134
Musée Jean de La Fontaine, Château-
Thierry 012486
Musée Jean Gabin, Mériel 014138
Musée Jean-Henri Fabre, Saint-
Léons 015421
Musée Jean-Honoré Fragonard,
Grasse 013162
Musée Jean-Jacques Rousseau,
Genève 041537
Musée Jean-Jacques Rousseau,
Montmorency 014279
Musée Jean-Jacques Rousseau,
Môtiers 041763
Musée Jean-Lurçat, Angers . . . 011658
Musée Jean Lurçat, Saint-Laurent-les-
Tours 015414
Musée Jean Macé, Beblenheim 011961
Musée Jean-Marie Somet, Villars
(Loire) 016224
Musée Jean-Marie Souplet, Musée
Minéralogique et Paléontologique,
Engis 003579
Musée Jean-Mermoz, Aubenton 011770
Musée Jean-Moulin, Paris 014766
Musée Jean Racine, La Ferté-
Milon 013410
Musée Jeanne d'Aboville, La
Fère 013406
Musée Jeanne d'Arc, Chinon . 012563
Musée Jeanne d'Arc, Rouen . . . 015168
Musée Jeanne d'Arc, Sainte-Catherine-
de-Fierbois 015604
Musée Jenisch, Musée des Beaux-
Arts de Vevey et Cabinet Cantonal
des Estampes, Vevey 042108
Musée Joachim du Bellay, Liré . 013826
Musée Johannique, Vaucouleurs 016135
Musée Joseph Abeilhe, Marciac 014006
Musée Joseph Denais, Beaufort-en-
Vallée 011938
Musée Joseph Durand, Saint-Clément-
des-Baleines 015267
Musée Joseph Filion, Sainte-
Thérèse 006960
Musée Joseph Lhoménède, Frugières-
le-Pin 013065
Musée Joseph Vaylet, Espalion . 012903
Musée Jouenne, Saint-Rémy-de-
Provence 015545
Musée Judéo-Alsacien,
Bouxwiller 012181
Musée Juif Comtadin, Cavaillon 012377
Musée Juif de Belgique-Joods
Museum van België, Bruxelles 003487
Musée Juin 1944, L'Aigle 013533
Musée Jules Desbois, Parçay-les-

Pins 014614
Musée Jules Destrée, Charleroi . 003515
Musée Jules Garinet, Châlons-en-
Champagne 012409
Musée Jules Lamy de Lac-à-la-Croix,
Métabetchouan-Lac-à-la-Croix 006297
Musée Jules Romains, Saint-Julien-
Chapteuil 015396
Musée Jules Verne, Nantes 014420
Musée Jurassien d'Art et d'Histoire,
Delémont 041437
Musée Jurassien des Arts,
Moutier 041769
Musée Jurassien des Sciences
Naturelles, Porrentruy 041845
Musée Juste pour Rire, Montréal 006405
Musée Kasba, Tanger 031797
Musée Kateri Tekakwitha,
Kahnawake 006055
Musée Kaysone Phomvihane,
Vientiane 030089
Musée Kio-Warini, Loretteville . . 006221
Musée Kotama, Jijel 000067
Musée Kwok-On, Paris 014767
Musée La Batterie d'Azeville,
Azeville 011859
Musée La Diana, Montbrison . . . 014240
Musée la Faïencerie, Ancy-le-
Franc 011642
Musée La Ferme Forézienne, Saint-
Bonnet-les-Oules 015239
Musée la Magie des Automates, Lans-
en-Vercors 013563
Musée la Main et l'Outil, Pannes 014611
Musée La Maison du Bucheron,
Pourcy 014981
Musée La Maison du Passé, Royat-les-
Bains 015185
Musée la Maison du Pêcheur et "Le
Hope", Saint-Gilles-Croix-de-Vie 015350
Musée la Muse, Mérindol 014140
Musée la Percée du Bocage 1944,
Saint-Martin-des-Besaces 015452
Musée La Pouponnière, Bourg-
Charente 012151
Musée la Taille de Pierre, Saint-
Antoine-l'Abbaye 015222
Musée la Vie des Jouets,
Mauleon 014086
Musée la Villa Aurélienne, Fréjus 013051
Musée Labasso, Bangassou . . . 007505
Musée l'Abeille Vivante, Le
Faouët 013631
Musée Labenche d'Art et d'Histoire,
Brive-la-Gaillarde 012235
Musée Lacoune, Lacaune 013518
Musée Laennec, Nantes 014421
Musée Lamartine, Mâcon 013959
Musée Lamartine, Pierreclos . . . 014866
Musée Lambinet, Versailles 016182
Musée Lamothe-Cadillac, Saint-
Nicolas-de-la-Grave 015478
Musée Langlois, Beaumont-en-
Auge 011946
Musée Languédocien,
Montpellier 014300
Musée Lapérouse, Albi 011583
Musée Lapidaire, Aire-sur-
l'Adour 011556
Musée Lapidaire, Avignon 011839
Musée Lapidaire, Baume-les-
Messieurs 011914
Musée Lapidaire, Carpentras . . . 012343
Musée Lapidaire, Morlaàs 014325
Musée Lapidaire, Saint-Germain-
Laval 015345
Musée Lapidaire, Saint-Guilhem-le-
Désert 015355
Musée Lapidaire, Saint-Ursanne 041912
Musée Lapidaire, Villeveyrac . . . 016292
Musée Lapidaire de l'Abbaye, Savigny
(Rhône) 015730
Musée Lapidaire de l'Abbaye Saint-
Pierre, Maillezais 013971

Musée Lapidaire de Mozac, Mozac 014372
Musée Lapidaire du Cloître de l'Ancienne Cathédrale, Vaison-la-Romaine 016085
Musée Lapidaire et d'Art Religieux, Société des Amis des Arts, Charlieu 012459
Musée Lapidaire et Exposition Permanente sur la Restauration, Carcassonne 012332
Musée Lapidaire, Palais des Archevêques, Narbonne 014431
Musée Larrey, Beaudean 011937
Musée L'Art en Marche Hauterives Art Brut, Neuve Invention, Hauterives 013241
Musée L'Art en Marche Lapalisse Art Brut, Lapalisse 013570
Musée Laumonier de la Locomotive à Vapeur, Vierzon 016209
Musée Laurier, Victoriaville 007338
Musée Lautrecois, Lautrec 013587
Musée Le Grenier de l'Histoire, Lèves 013781
Musée Le Jacquemart, Langeac 013552
Musée Le Masque en Liberté, Nibelle 014472
Musée le Mémoire d'Oÿé, Oÿé . 014600
Musée Le Minerve, Yzeures-sur-Creuse 016367
Musée Le Mont-Saint-Michel, Le Mont-Saint-Michel 013671
Musée Le Patrimoine du Sapeur-Pompier, Beaune 011952
Musée Le Petit Louvre, La Pacaudière 013447
Musée le Petit Mont-Saint-Michel, Saint-Marcan 015443
Musée Le Pionnier, Saint-Malo . 006923
Musée Le Secq des Tournelles – Musée de la Ferronnerie, Rouen 015169
Musée le Soum, Saint-Véran .. 015585
Musée le Temple du Soleil, Saint-Michel-en-l'Herm 015469
Musée Le Vieux Cormenier, Champniers-en-Vienne 012440
Musée Leblanc-Duvernoy, Auxerre 011823
Musée Leclerc Briant, Épernay . 012881
Musée Lella Hadria, Midoun ... 042759
Musée Lénine, Paris 014768
Musée Leon de Smet, Deurle .. 003542
Musée Léon Dierx, Saint-Denis . 036184
Musée Léon Marès, Lovagny .. 013911
Musée Léon Perrin, Môtiers ... 041764
Musée Les 3 Salles, Lectoure . 013723
Musée Les Amis de l'Outil, Bièvres 012038
Musée Les Ateliers du Bois, Vadencourt 016081
Musée Les Flocons, Ittre 003733
Musée Les Sanglots Longs, Reguiny 015046
Musée Les Sources d'Hercule, Musée Archéologique de Site Gallo-Romain, Deneuvre 012758
Musée les Thermes d'Antonin, Carthage 042737
Musée Les Trésors de Lisette, Archingeay 011711
Musée les Trois Pignons, La Société Saint-Pierre, Cheticamp ... 005611
Musée Les Voltigeurs de Québec Manège Militaire, Québec 006741
Musée Libertador du Général San Martin, Boulogne-sur-Mer 012140
Musée Ligier Richier, Saint-Mihiel 015472
Musée Lillet, Podensac 014915
Musée Littéraire Victor Hugo, Vianden 030605
Musée Local, Boën-sur-Lignon . 012085
Musée Local de Préhistoire, Sceaux-du-Gatinais 015741
Musée Local des Arts et Traditions Populaires, Auxi-le-Château .. 011825
Musée Loire Métiers, Musée Tradition Populaires et Métiers, Saint-Clement-des-Levées 015269
Musée Lombard, Doullens 012828
Musée Longines, Saint-Imier .. 041908
Musée Lorrain, Nancy 014407
Musée Lorrain des Cheminots, Rettel 015076
Musée Lorrain du Cheval, Gondrecourt-le-Château 013126
Musée Lou Férouil, Gilette ... 013105
Musée Louis Braille, Coupvray . 012694
Musée Louis Cornu, Wassigny . 016342
Musée Louis Hémon, Péribonka 006631
Musée Louis Jourdan, Saint-Paul-de-Varax 015494
Musée Louis Lutaud, Ferrière-la-Grande 012969
Musée Louis Pasteur, Strasbourg 015885
Musée Louis Pergaud, Belmont . 011980
Musée Louis Philippe, Eu ... 012931
Musée Louis-Vouland, Avignon . 011840
Musée Louise Cottin, Vigneulles-lès-Hattonchâtel 016213
Musée Louise Weiss, Saverne .. 015728
Musée Louison Bobet, Saint-Méen-le-Grande 015464
Musée Lucien Roy, Beure 012015
Musée McCord, Montréal 006406
Musée Maçonnique de la Grande Loge de France, Paris 014769
Musée Mado Robin, Yzeures-sur-Creuse 016368
Musée Magic de la Musique Mécanique et de la Machine Parlante, Mesnil-Raoul 014147
Musée Magie des Abeilles, Villechauve 016234
Musée Maginot de Rohrbach-Fort Casso, Rohrbach-lès-Bitche . 015132
Musée Magnanerie, Lagorce ... 013526
Musée Magnelli, Musée de la Céramique, Vallauris 016103
Musée Magnin, Dijon 012788
Musée Mai Juin 1940, Margut . 014019
Musée Maillol, Fondation Dina Vierny, Banyuls-sur-Mer 011889
Musée Maillol, Fondation Dina Vierny, Paris 014770
Musée Mainssieux, Voiron ... 016320
Musée Mairie, Le Pont-de-Beauvoisin 013687
Musée-Maison d'Art Contemporain Chaillioux, Fresnes (Val-de-Marne) 013057
Musée-Maison de Chateaubriand, Châtenay-Malabry 012513
Musée-Maison de Clemenceau, Saint-Vincent-sur-Jard 015591
Musée Maison de George Sand, Gargilesse-Dampierre 013082
Musée Maison de la Baleine, Luc-sur-Mer 013913
Musée Maison de la Brique, Saint-Martin-d'Aubigny 015446
Musée Maison de la Dame de Brassempouy, Brassempouy .. 012195
Musée-Maison de la Nature, Samer 015670
Musée-Maison de la Nature et de l'Estuaire de l'Orne, Sallenelles 015655
Musée Maison de la Radio TSF, Saint-Appolinard 015225
Musée Maison de la Soie, Saint-Hippolyte-du-Fort 015363
Musée-Maison de la Vie Rurale, La Flocellière 013416
Musée Maison de l'Armateur, Le Havre 013645
Musée-Maison de l'Aspre, Arts et Traditions Populaires, Thuir ... 015960
Musée-Maison de l'Erdre, Nantes 014422
Musée Maison de l'Huître, Gujan-Mestras 013218
Musée-Maison d'Ecole, Montceau-les-Mines 014242
Musée Maison Déodat Roché, Maison du Catharisme, Arques (Aude) . 011741
Musée-Maison des Amis de la Forêt, Mervent 014143
Musée-Maison des Arts des Hauts-Cantons, Bédarieux 011963
Musée-Maison des Canuts, Lyon 013949
Musée-Maison des Dentelles et du Point d'Argentan, Argentan ... 011718
Musée-Maison des Milelli, Propriété des Bonaparte, Ajaccio 011573
Musée-Maison des Pierreux, Massangis 014077
Musée-Maison des Vanniers, Feytiat 012978
Musée-Maison du Café, Vieux-Habitants 023354
Musée-Maison du Canal, Fégréac 012962
Musée-Maison du Jambon de Bayonne, Arzacq-Arraziguet .. 011759
Musée-Maison du Jardinier, Gétigné 013098
Musée-Maison du Pain, Commeny 012640
Musée Maison du Pays, Moisdon-la-Rivière 014189
Musée-Maison du Paysan, Touvois 016029
Musée-Maison du Pêcheur, La Chevrolière 013390
Musée-Maison du Petit Poitou, Chaillé-les-Marais 012400
Musée Maison du Potier, Le Fuilet 013633
Musée-Maison du Saumon et de la Rivière, Brioude 012232
Musée-Maison du Terroir, Châteauponsac 012499
Musée-Maison du Verre, Meisenthal 014114
Musée-Maison Lohobiague dite Maison Louis XIV, Saint-Jean-de-Luz 015379
Musée-Maison Marcel Proust, Maison de Tante Léonie, Illiers-Combray 013294
Musée-Maison Natale de Jean de Lattre de Tassigny, Mouilleron-en-Pareds 014351
Musée-Maison Natale Gustave Courbet, Ornans 014591
Musée-Maison Paternelle de Bernadette, Lourdes 013901
Musée Maison Picassiette, Chartres 012470
Musée Maison Robert Doisneau, Gentilly 013091
Musée Maison Saint-Gabriel, Montréal 006407
Musée Malraux, Le Havre 013646
Musée M'Am Jeanne, Fontenoy-en-Puisaye 013021
Musée Manoir de Kerazan, Loctudy 013850
Musée-Manoir de la Touche, Nantes 014423
Musée Manoir de Vains, Vains . 016083
Musée-Manoir des Sciences, Réaumur 015041
Musée Marc-Aurèle Fortin, Montréal 006408
Musée Marc Deydier, Cucuron . 012730
Musée Marc Leclerc, La Ménitré 013440
Musée Marc Mureau, Restigné . 015072
Musée Marcel Collet, Musée d'Histoire et d'Archéologie, Loverval ... 003832
Musée Marcel de Marcheville, Romorantin-Lanthenay 015143
Musée Marcel Lenoir, Montricoux 014311
Musée Marcel Sahut, Volvic ... 016323
Musée Maréchal Joffre, Rivesaltes 015107
Musée Marguerite Audoux, Aubigny-sur-Nère 011777
Musée Marguerite Bourgeoys, Montréal 006409
Musée Marguerite de Bourgogne, Tonnerre 015970
Musée Marguerite Vidal des Arts et Traditions Populaires, Moissac . 014190
Musée Marguerite Yourcenar, Saint-Jans-Cappel 015366
Musée Marial de Beauraing, Beauraing 003348
Musée Marie Baz, Deir-el-Qamar 030277
Musée Marie Laurencin, Chino . 028515
Musée Marie-Rose Durocher, Lieux Historiques des Soeurs des Saints-Noms de Jésus et de Marie, Longueuil 006219
Musée-Marinarium du Haut Lavédan, Pierrefitte-Nestalas 014869
Musée Marine et Montagne, Mons (Var) 014202
Musée Maritime, Camaret-sur-Mer 012298
Musée Maritime, Carantec ... 012324
Musée Maritime, Le Mont-Saint-Michel 013672
Musée Maritime Cameroun, Douala 005291
Musée Maritime Chantereyne, Cherbourg-Octeville 012550
Musée Maritime de Charlevoix, Saint-Joseph-de-la-Rive 006916
Musée Maritime de l'Île de Tatihou, Saint-Vaast-la-Hougue 015580
Musée Maritime de l'Ile de Tatihou, Saint-Vaast-la-Hougue 015579
Musée Maritime de Regnéville, Les Fours à Chaux, Regnéville-sur-Mer 015044
Musée Maritime de Tadoussac, Tadoussac 007123
Musée Maritime du Québec, L'Islet 006200
Musée Maritime Islais, Port-Joinville 014967
Musée Maritime La Rochelle, La Rochelle 013482
Musée Maritime l'Esvale et Ostréicole, La Tremblade 013500
Musée Maritime, Fluvial et Portuaire, Rouen 015170
Musée Marius Barbeau, Saint-Joseph-de-Beauce 006915
Musée Marmottan Monet, Paris 014771
Musée Marsa El-Kharez, El-Kala 000060
Musée Marsil – Musée du Costume et du Textile du Québec, Saint-Lambert 006917
Musée Martin Duby, Auriol ... 011803
Musée Massey, Tarbes 015917
Musée Matheysin, La Mure-d'Isère 013445
Musée Mathurin Méheut, Lamballe 013540
Musée Matisse, Le Cateau-Cambrésis 013617
Musée Matisse, Nice 014486
Musée Maurice Beaumont, Musée du Cobattant de Haute-Soane, Luxeuil-les-Bains 013924
Musée Maurice Carême, Bruxelles 003488
Musée Maurice Dufresne, Azay-le-Rideau 011857
Musée Maurice et Eugénie de Guérin, Andillac 011643
Musée Maurice Genevoix, Saint-Denis-de-l'Hôtel 015286
Musée Maurice Gottlob, Espace

Culturel, Mougins 014350
Musée Maurice Poignant, Bourg-sur-
Gironde 012159
Musée Maurice Ravel, Montfort-
l'Amaury 014259
Musée Maxime Mabilleau,
Saumur 015719
Musée Mazot de Plan-Cerisier,
Martigny 041728
Musée Médiéval, Brain-sur-
Allonnes 012187
Musée Medieval, Serralongue . 015780
Musée Médiumnique Fernand
Desmoulin, Brantôme 012189
Musée Mémoires de la Terre et
Horizons Cathares, Mazamet . . 014096
Musée Mémoires du Moyen Âge,
Carcassonne 012333
Musée Mémorial 1914–1918,
Mémorial du Vieil Armand,
Wattwiller 016345
Musée Memorial Alain, Mortagne-au-
Perche 014337
Musée-Mémorial de la Bataille de
l'Atlantique, Camaret-sur-Mer . 012299
Musée Mémorial de la Bataille de
Normandie, Bayeux 011921
Musée Mémorial de la Bataille
de Verdun, Fleury-devant-
Douaumont 012992
Musée Mémorial de la Ligne Maginot
du Rhin, Marckolsheim 014009
Musée Mémorial des Combat de la
Poche de Colmar, Turckheim . 016063
Musée Mémorial du Front d'Orient
1915–1918, Troissereux 016044
Musée Mémorial du Général de Gaulle,
Bayeux 011922
Musée-Mémorial du Linge, Orbey 014573
Musée Mémorial Maginot de Haute-
Alsace, Mémorial de Haute-Alsace,
Uffheim 016067
Musée-Mémorial National du
Débarquement en Provence (1944),
Toulon (Var) 015980
Musée Memorial Omaha Beach 6 Juin
1944, Saint-Laurent-sur-Mer . 015415
Musée Mémorial Pegasus,
Ranville 015036
Musée Mer et Désert, Maison de
l'Eau, Villerville 016288
Musée Metlili, Ghardaïa 000064
Musée Michel Aboujaoudé, Zalka 030284
Musée Michel Braibant, Conservatoire
des Arts et Métiers du Livre,
Montolieu 014285
Musée Militaire, Cambrai 012302
Musée Militaire, Colombier
(Neuchâtel) 041424
Musée Militaire, Veckring 016145
Musée Militaire Baie-des-Chaleurs,
New Richmond 006470
Musée Militaire de Bure, Bure-
Tressange 012253
Musée Militaire de la Guerre
1914 – 1918, Musée de la Bataille
du Saillant de Saint-Michel,
Thiaucourt 015932
Musée Militaire des Gloires
et Souvenirs du Périgord,
Périgueux 014831
Musée Militaire du 12e Régiment
Blindé du Canada, Trois-
Rivières 007248
Musée Militaire Faller, Mars-la-
Tour 014033
Musée Militaire Genevois,
Cologny 041422
Musée Militaire Ligne Mareth,
Mareth 042758
Musée Militaire National,
Manouba 042757
Musée Militaire Vaudois, Morges 041760
Musée Militaire, Musée d'Armes et

Objets Anciens, Sonvilier 042011
Musée Mine, Cagnac-les-Mines 012281
Musée Mineralexpo, La
Bourboule 013369
Musée Minéralogique, Nîmes . . 014505
Musée Minéralogique d'Asbestos,
Asbestos 005353
Musée Minéralogique de la Société
Industrielle, Mulhouse 014380
Musée Minéralogique de l'École des
Mines d'Alès, Alès 011596
Musée Minéralogique et Minier de
Thetford Mines, Thetford Mines 007139
Musée Minéralogique et
Paléontologique, Saint-Pierre-de-la-
Palud 015510
Musée Mini-Trains-Lyonnais,
Lyon 013950
Musée Miniatures et Poupées, Gréoux-
les-Bains 013190
Musée Minier d'Halanzy, Halanzy 003662
Musée Missionnaire, Pontmain . 014958
Musée Mistorique, Jaulny 013313
Musée Mnemosina, Prunelli-di-
Fium'Orbu 015006
Musée Moissan, Paris 014772
Musée Monétaire Cantonal,
Lausanne 041661
Musée Monopoli, Barsy-Flostoy . 003338
Musée Monsieur Pigneau de Behaine,
Origny-en-Thiérache 014580
Musée Montagnard, Les Houches 013751
Musée Montagnard du Lavedan,
Aucun 011784
Musée Montagnard et Pictural, Les
Haudères 041583
Musée Montebello, Galerie du Musée,
Trouville-sur-Mer 016046
Musée Montenot Agricole et
Archéologique, Genay 013085
Musée Moralès, Port-de-Bouc . . 014965
Musée Morice Lipsi, Rosey 015157
Musée Moto du Douhet, Le
Douhet 013628
Musée-Moulin à Vent de Rairé,
Sallertaine 015657
Musée Moulin Cézanne, Le
Tholonet 013703
Musée-Moulin de Guerin, Saint-Julien-
Chapteuil 015397
Musée-Moulin de l'Espine, Fougax-et-
Barrineuf 013031
Musée Moulin de Maupertuis,
Donzy 012817
Musée-Moulin du Papé Malaval,
Tornac 015972
Musée Moulin Faber, Hotton . . . 003712
Musée Municipal, Barentin 011896
Musée Municipal, Bernay 011994
Musée Municipal, Bourbonne-les-
Bains 012148
Musée Municipal, Châlons-en-
Champagne 012410
Musée Municipal, Chartres 012471
Musée Municipal, Châtellerault . 012509
Musée Municipal, Coutances . . . 012707
Musée Municipal, Cusset 012739
Musée Municipal, Dudelange . . . 030576
Musée Municipal, Épernay 012882
Musée Municipal, Forcalquier . . 013025
Musée Municipal, Gerberoy 013094
Musée Municipal, Hazebrouck . . 013245
Musée Municipal, Hennezel-
Clairey 013250
Musée Municipal, La Charité-sur-
Loire 013387
Musée Municipal, La Réole 013454
Musée Municipal, La Roche-sur-
Yon 013467
Musée Municipal, Le Monastier-sur-
Gazeille 013666
Musée Municipal, Louhans 013885
Musée Municipal, Mazan 014097
Musée Municipal, Nuits-Saint-

Georges 014537
Musée Municipal, Saint-Amand-les-
Eaux 015211
Musée Municipal, Saint-Antonin-Noble-
Val 015223
Musée Municipal, Saint-Calais . 015246
Musée Municipal, Saint-Dizier . . 015295
Musée Municipal, Saint-Jean-
d'Angély 015370
Musée Municipal, Sainte-Maure-de-
Touraine 015621
Musée Municipal, Semur-en-
Auxois 015760
Musée Municipal, Villandraut . . . 016216
Musée Municipal, Vire 016304
Musée Municipal A. Danicourt,
Péronne 014836
Musée Municipal A.G. Poulain,
Vernon 016177
Musée Municipal Albert Marzelles,
Marmande 014023
Musée Municipal Alfred Canel, Pont-
Audemer 014930
Musée Municipal Ancienne Abbaye
Saint-Léger, Soissons 015826
Musée Municipal Auguste Grasset,
Varzy 016130
Musée Municipal Bourbon-Lancy,
Bourbon-Lancy 012145
Musée Municipal d'Archéologie,
Roquefort-sur-Soulzon 015149
Musée Municipal d'Archéologie,
Murviel-les-Montpellier 014391
Musée Municipal d'Art et d'Histoire,
Colombes 012632
Musée Municipal d'Art Naïf, Noyers-
sur-Serein 014533
Musée Municipal d'Arts Asiatiques,
Toulon (Var) 015981
Musée Municipal d'Avranches,
Avranches 011848
Musée Municipal de Cambrai, Musée
des Beaux-Arts, Cambrai 012303
Musée Municipal de Civray,
Civray 012579
Musée Municipal de Draguignan,
Musée de France, Draguignan . 012836
Musée Municipal de la Chartreuse,
Douai 012821
Musée Municipal de la Poterie, La
Chapelle-aux-Pots 013378
Musée Municipal de la Toile de Jouy,
Jouy-en-Josas 013334
Musée Municipal de la Vie Sauvage,
Le Bugue 013614
Musée Municipal de l'Évêché,
Limoges 013821
Musée Municipal de Louviers,
Louviers 013909
Musée Municipal de Millau,
Millau 014170
Musée Municipal de Mokolo,
Mokolo 005296
Musée Municipal de Montmorillon,
Montmorillon 014282
Musée Municipal de Moret, Moret-sur-
Loing 014320
Musée Municipal de Pithiviers,
Pithiviers-le-Vieil 014879
Musée Municipal de Préhistoire,
Fismes 012987
Musée Municipal de Préhistoire, Les
Matelles 013756
Musée Municipal de Préhistoire des
Matelles, Viols-le-Fort 016303
Musée Municipal de Préhistoire
Régionale, Saint-Pons-de-
Thomières 015517
Musée Municipal de Saint-Barthélemy,
Saint-Barthélemy 023351
Musée Municipal de Saint-Cloud,
Musée des Avelines, Saint-
Cloud 015271
Musée Municipal de Saint-Gaudens,

Saint-Gaudens 015327
Musée Municipal de Saint-Maur, La
Varenne-Saint-Hilaire 013503
Musée Municipal de Sault, Sault
(Vaucluse) 015710
Musée Municipal de Vendôme,
Vendôme 016154
Musée Municipal Département Jouet,
Les Bains Douches, La Ferté-
Macé 013408
Musée Municipal d'Ernée, Ernée 012892
Musée Municipal des Beaux-Arts et
d'Archéologie, Valence (Drôme) 016093
Musée Municipal des Beaux
Arts et d'Histoire Naturelle,
Châteaudun 012490
Musée Municipal des Capucins,
Coulommiers 012691
Musée Municipal des Fossiles,
Tourtour 016025
Musée Municipal des Fours Banaux,
Vierzon 016210
Musée Municipal des Mosaïques
Romaines, Diekirch 030573
Musée Municipal d'Étampes,
Etampes 012917
Musée Municipal d'Histoire et
d'Archéologie, Harnes 013229
Musée Municipal d'Orbec, Orbec 014571
Musée Municipal du Château, Musée
Delarue, Gannat 013075
Musée Municipal du Peigne, Art et
Traditions, Ezy-sur-Eure 012945
Musée Municipal du Présidial,
Bazas 011927
Musée Municipal Emile Villez,
Guînes 013213
Musée Municipal François Guiguet,
Corbelin 012665
Musée Municipal Frédéric Blandin,
Nevers 014468
Musée Municipal Gautron du Coudray,
Marzy 014076
Musée Municipal Georges Turpin,
Musée d'Art et d'Histoire,
Parthenay 014804
Musée Municipal Joseph Jacquiot,
Montgeron 014261
Musée Municipal Max Claudet, Salins-
les-Bains 015650
Musée Municipal Méditerranéen de
Cassis, Cassis 012351
Musée Municipal-Musée d'Art et
d'Archéologie, Hôtel Fouquet,
Château-Gontier 012481
Musée Municipal Nicolas Poussin, Les
Andelys 013732
Musée Municipal Paul Lafran, Saint-
Chamas 015251
Musée Municipal Raymond Touchard,
Lussac-les-Châteaux 013921
Musée Municipal Robert Dubois-
Corneau, Brunoy 012247
Musée Municipal Saint-John-Perse,
Pointe-à-Pitre 023348
Musée Municipal Saraleguinea,
Guéthary 013209
Musée Municipal Urbain-Cabrol,
Villefranche-de-Rouergue 016243
Musée Municipal Vendôme,
Naveil 014436
Musée Municipal Yvon Gueret,
Pierrelatte 014871
Musée Municipal, Musée Arts et
Traditions, Mauriac 014091
Musée Municipal, Musée d'Art et
d'Histoire, Bergues 011993
Musée Municipal, Musée d'Arts et
d'Histoire, Hesdin 013258
Musée Municipal, Musée de l'Hôtel de
Ville, Richelieu 015086
Musée Municipal, Musée du Jardin
Jean Rostand, Lillebonne 013812
Musée Municipale, Lay-Saint-

Register der Institutionen und Firmen
Musée Municipale, Mont-de-Lans
– Musée Paysan et Artisanal, La Chaux-de-Fonds

Christophe . . . 013602
Musée Municipale, Mont-de-Lans 014206
Musée Municipale, Saint-Laurent-d'Olt . . . 015412
Musée Municipale, Villefranche-de-Lonchat . . . 016242
Musée Nader, Port-au-Prince . . 023394
Musée Namesokanjic, Lac-Mégantic . . . 006144
Musée Napoléon, Ligny . . . 003820
Musée Napoléon Ier et Tresors des Eglises, Brienne-le-Château . . 012223
Musée Napoléonien, Antibes-Juan-les-Pins . . . 011687
Musée Napoléonien, Ile-d'Aix . . 013284
Musée Napoléonien d'Art et d'Histoire Militaire, Fontainebleau . . . 013011
Musée Napoléonien du Château de Grosbois, Boissy-Saint-Léger . . 012095
Musée Napoléonien-Mémoire 1814, Essises . . . 012912
Musée Nasréddine Dinet de Bou-Sâala, M'Sila . . . 000070
Musée Natale de Pierre Corneille, Rouen . . . 015171
Musée National, Beirut . . . 030268
Musée National, Moroni . . . 008831
Musée National, N'Djamena . . . 007509
Musée National Adrien Dubouché, Limoges . . . 013822
Musée National Ahmed Zabana – Demaeght Museum, Oran . . . 000071
Musée National Auguste Rodin, Villa des Brillants, Meudon . . . 014154
Musée National Barthélémy Boganda, Bangui . . . 007506
Musée National Cirta de Constantine, Constantine . . . 000057
Musée National d'Archéologie de Sétif, Sétif . . . 000075
Musée National d'Art Brassicole et Musée de la Tannerie, Wiltz . . 030608
Musée National d'Art Moderne, Paris . . . 014773
Musée National de Burkina Faso, Ouagadougou . . . 005279
Musée National de Carthage, Carthage . . . 042738
Musée National de Céramique, Sèvres . . . 015795
Musée National de Conakry, Conakry . . . 023381
Musée National de Géologie, Antananarivo . . . 030647
Musée National de Gitega, Gitega . . . 005281
Musée National de Guinée, Conakry . . . 023382
Musée National de Kananga, Kananga . . . 008833
Musée National de la Céramique, Safi . . . 031793
Musée National de la Coopération Franco-Américaine, Blérancourt 012071
Musée National de la Légion d'Honneur et des Ordres de Chevalerie, Paris . . . 014774
Musée National de la Marine, Brest . . . 012210
Musée National de la Marine, Paris . . . 014775
Musée National de la Marine, Port-Louis . . . 014970
Musée National de la Marine, Rochefort (Charente-Maritime) . 015123
Musée National de la Marine, Toulon (Var) . . . 015982
Musée National de la Renaissance, Ecouen . . . 012855
Musée National de la Résistance, Bruxelles . . . 003489
Musée National de la Résistance, Esch-sur-Alzette . . . 030582
Musée National de la Voiture et du Tourisme, Compiègne . . . 012645
Musée National de l'Assurance Maladie, Lormont . . . 013875
Musée National de l'Éducation, Montréal . . . 006410
Musée National de l'Éducation, Rouen . . . 015172
Musée National de Lubumbashi, Lumumbashi . . . 008840
Musée National de Monaco, Collection de Galéa, Monaco . . . 031721
Musée National de Nouakchott, Nouakchott . . . 030785
Musée National de Phnom Penh, Phnom Penh . . . 005286
Musée National de Préhistoire, Les Eyzies-de-Tayac-Sireuil . . . 013749
Musée National de Préhistoire et d'Ethnographie du Bardo, Alger 000044
Musée National de Sarh, Sarh . 007511
Musée National de Yaoundé, Yaoundé . . . 005298
Musée National d'Enéerune, Site et Archéologique, Nissan-lez-Ensérune . . . 014511
Musée National des Antiquités, Alger . . . 000045
Musée National des Arts Asiatiques Guimet, Paris . . . 014776
Musée National des Arts d'Afrique et d'Océanie, Paris . . . 014777
Musée National des Arts et Traditions, Libreville . . . 016382
Musée National des Arts et Traditions Populaires, Paris . . . 014778
Musée National des Beaux-Arts d'Alger, Alger . . . 000046
Musée National des Beaux-Arts du Québec, Québec . . . 006742
Musée National des Châteaux de Malmaison et de Bois-Préau, Rueil-Malmaison . . . 015190
Musée National des Châteaux de Versailles et de Trianon, Versailles . . . 016183
Musée National des Deux Victoires, Mouilleron-en-Pareds . . . 014352
Musée National des Douanes, Bordeaux . . . 012119
Musée National des Granges de Port Royal, Magny-lès-Hameaux . . . 013966
Musée National des Jeux de Paume, Ath . . . 003329
Musée National des Mines de Fer Luxembourgeoises, Rumelange 030601
Musée National des Monuments Français, Paris . . . 014779
Musée National des Prisons, Fontainebleau . . . 013012
Musée National d'Histoire et d'Art (MNHA), Luxembourg . . . 030593
Musée National d'Histoire Militaire, Diekirch . . . 030574
Musée National d'Histoire Naturelle, Nationalmuseum für Naturgeschichte, Luxembourg . . . 030594
Musée National du Bardo, Alger 000047
Musée National du Bardo, Le Bardo . . . 042734
Musée National du Château de Compiègne – Musée du Second Empire, Musée de la Voiture et du Tourisme, Compiègne . . . 012646
Musée National du Château de Fontainebleau, Fontainebleau . 013013
Musée National du Château de Pau, Musée Henri IV, Pau . . . 014810
Musée National du Congo, Brazzaville . . . 008844
Musée National du Costume, Grand-Bassam . . . 008875
Musée National du Gabon, Libreville . . . 016383
Musée National du Jihad, Alger 000048
Musée National du Mali, Bamako 030734
Musée National du Moudjahid, Alger . . . 000049
Musée National du Moudjahid, El Madania . . . 000061
Musée National du Moyen Age, Paris . . . 014780
Musée National du Niger, Niamey . . . 033408
Musée National du Papier, Malmedy . . . 003841
Musée National du Rwanda, Butare . . . 038086
Musée National du Scoutisme en France, Thorey-Lyautey . . . 015953
Musée National du Sport, Paris . 014781
Musée National du Togo, Lomé . 042730
Musée National d'Utique, Utique 042786
Musée National Ernest Hébert, Paris . . . 014782
Musée National Eugène Delacroix, Paris . . . 014783
Musée National Fernand Léger, Biot . . . 012044
Musée National Jean-Jacques Henner, Paris . . . 014784
Musée National Lao, Vientiane . 030090
Musée National Message Biblique Marc Chagall, Nice . . . 014487
Musée National N'Djamena, N'Djamena . . . 007510
Musée National Picasso La Guerre et la Paix, Vallauris . . . 016104
Musée National Suisse – Château de Prangins, Prangins . . . 041848
Musée National Suisse de l'Audiovisuel – Audiorama, Fondation Audiorama, Montreux . . . 041756
Musée Nature en Provence, Riez 015091
Musée Naturel du Sud, Chenini Gabes . . . 042739
Musée Naval, Monaco . . . 031722
Musée Naval de Nice, Nice . . . 014488
Musée Naval de Sarcelles, Sarcelles . . . 015681
Musée Naval Fort Balaguier, La Seyne-sur-Mer . . . 013492
Musée Navale du Québec, Québec . . . 006743
Musée Nicéphore Niépce, Chalon-sur-Saône . . . 012405
Musée Nicholas Ibrahim Sursock, Beirut . . . 030269
Musée Nissim de Camondo, Les Arts Décoratifs, Paris . . . 014785
Musée Nivernais de l'Education, Nevers . . . 014469
Musée No.4 Commando, Ouistreham . . . 014596
Musée Normandie-Niemen, Les Andelys . . . 013733
Musée Notre-Dame, Le Folgoët . 013632
Musée Notre Dame de la Pierre, Ecomusée de la Vallee d'Aspe, Sarrance . . . 015693
Musée Notre-Dame, Musée du Gemmail, Lourdes . . . 013902
Musée Nungesser et Coli, Etretat 012929
Musée Oberlin, Waldersbach . . . 016334
Musée-Observatoire de la Vallée d'Anjou, Saint-Mathurin-sur-Loire . . . 015460
Musée Océanien de Cuet, Montrevel-en-Bresse . . . 014308
Musée Océanographique, Bonifacio . . . 012098
Musée Océanographique, Kingersheim . . . 013351
Musée Océanographique, L'Ile-Rousse . . . 013802
Musée Océanographique Dar-El-Hout de Salammbô, Salammbô . . . 042765
Musée Océanographique de l'Odet, Ergué-Gabéric . . . 012889
Musée Océanographique de Monaco, Monaco . . . 031723
Musée Océanographique, Aquariophilie, Saint-Macaire . . 015432
Musée Olivier-de-Serres, Mirabel (Ardèche) . . . 014177
Musée Ollier, Les Vans . . . 013771
Musée Olympique Lausanne, Lausanne . . . 041662
Musée Omega, Biel . . . 041339
Musée Organistrum et des Vielles à Roues du Périgord Noir, Belvès 011982
Musée Ornithologique Charles Payraudeau, La Chaize-le-Vicomte . . . 013377
Musée Oscar Roty, Jargeau . . . 013307
Musée Ostréicole, Bourcefranc-le-Chapus . . . 012149
Musée Ostréicole, Saint-Trojan-les-Bains . . . 015576
Musée P. Dubois-A. Boucher, Nogent-sur-Seine . . . 014516
Musée Pablo Casals, Prades . . . 014987
Musée Paderewski, Morges . . . 041761
Musée Palais el Bahia, Marrakech . . . 031783
Musée Palais Lascaris, Musée des Arts et Traditions Populaires du Pays Niçois, Nice . . . 014489
Musée Palaisien du Hurepoix, Palaiseau . . . 014606
Musée Paléo-Écologique, Marchamp . . . 014000
Musée Paléontologique, Aguessac . . . 011540
Musée Paléontologique et Minéraux, Rougemont . . . 015176
Musée-Palombière, Près de l'Aire de Repos de l'Autoroute, Le Mas-d'Agenais . . . 013658
Musée Parc de la Droséra, Jeansagnière . . . 013316
Musée Passé Simple, Pleudihen-sur-Rance . . . 014891
Musée Pasteur, Dole . . . 012801
Musée Pasteur, Paris . . . 014786
Musée Pastoral le Monde Paysan d'Autrefois, Saint-Pé-de-Bigorre 015504
Musée Patrimoine et Traditions, Mirepoix . . . 014186
Musée Paul Arbaud, Aix-en-Provence . . . 011564
Musée Paul Charnoz, Musée de la Faïence, Paray-le-Monial . . . 014613
Musée Paul Dini, Villefranche-sur-Saône . . . 016250
Musée Paul Dupuy, Toulouse . . . 015997
Musée Paul-Emile Victor, Centre Polaire, Prémanon . . . 014995
Musée Paul Gauguin, Le Carbet 030778
Musée Paul Gauguin, Papeete . . 016378
Musée Paul Gauguin, Pont-Aven 014932
Musée Paul Géradin, Jodoigne . 003743
Musée Paul-José Gosselin, Saint-Vaast-la-Hougue . . . 015581
Musée Paul Raymond, Pont-Saint-Esprit . . . 014945
Musée Paul Surtel, Reuilly (Indre) 015077
Musée Paul Valéry, Sète . . . 015787
Musée Paul Voivenel, Capoulet-Junac . . . 012322
Musée Paysan, Espezel . . . 012908
Musée Paysan, Le Grand-Village-Plage . . . 013637
Musée Paysan, Viuz-en-Sallaz . . 016313
Musée Paysan de Bourgogne Nivernaise, La Celle-sur-Loire . 013374
Musée Paysan de la Save, Espaon . . . 012904
Musée Paysan du Moulin Neuf, Saint-Diéry . . . 015294
Musée Paysan du Sundgau, Oltingue . . . 014563
Musée Paysan et Artisanal, La Chaux-

de-Fonds 041410
Musée Peleforo Gbon Coulibay,
Korhogo 008876
Musée Percheron, Mortagne-au-
Perche 014338
Musée Petiet, Limoux 013824
Musée Peugeot, Sochaux 015824
Musée Peynet et du Dessin
Humoristique, Antibes-Juan-les-
Pins 011688
Musée Picasso, Antibes-Juan-les-
Pins 011689
Musée Picasso, Paris 014787
Musée Pierre Bayle, Carla-Bayle . 012337
Musée Pierre Borrione, Aime-en-
Tarentaise 011554
Musée Pierre Boucher, Trois-
Rivières 007249
Musée Pierre de Luxembourg,
Villeneuve-lès-Avignon 016268
Musée Pierre Gaudin, Puteaux . . 015010
Musée Pierre-Loti, Musées
Municipaux, Rochefort (Charente-
Maritime) 015124
Musée Pierre-Mondanel, Musée de la
Batellerie, Pont-du-Château . . 014936
Musée Pierre Noël, Saint-Dié-des-
Vosges 015292
Musée Pierre Vigne, Boucieu-le-
Roi 012128
Musée Pillon, Chaumont-en-
Vexin 012530
Musée Pincé, Angers 011659
Musée Piperie Léonard,
Andenelle 003271
Musée-Placard d'Erik Satie, Paris 014788
Musée Plain Air Régional, Guérin 005936
Musée Pobe Mengao, Mengao . . 005276
Musée Poche de Royan, Le Gua 013639
Musée Pointe-des-Chênes, Sainte-
Anne-des-Chênes 006948
Musée Pol Mara, Gordes 013132
Musée Pontois, Le Pont-de-
Beauvoisin 013688
Musée Portuaire, Dunkerque . . . 012845
Musée Postal, Rabat 031791
Musée Postal d'Aquitaine, Saint-
Macaire 015433
Musée Postal de Tunis, Tunis . . . 042785
Musée Poveal, Battambang 005283
Musée-Préau des Accoules,
Marseille 014061
Musée Préfectoral de Boké, Boké 023379
Musée Préfectoral de Kissidougou,
Kissidougou 023384
Musée Préfectoral de Koundara,
Koundara 023385
Musée Préfectoral de N'Zerekore,
N'Zerekore 023386
Musée Préhistorique Charles Nardin,
Sainte-Foy-la-Grande 015611
Musée Préhistorique du Tumulus de
Barnenez, Plouézoch 014902
Musée Préhistorique Finistérien,
Penmarch 014824
Musée Préhistorique, Grotte,
Marsoulas 014067
Musée Préhistorique, Musée d'Art et
d'Histoire, Eymet 012939
Le Musée Privé, Paris 105119
Musée-Privé de Cire du Vieil,
Albi 011584
Musée Privé de Documentation
Ferroviaire-Mupdofer, Bruxelles 003490
Musée Privé Saint-Jean l'Aigle,
Herserange 013256
Musée Privee de N'Douci,
N'Douci 008877
Musée-Promenade Marly-le-Roi,
Louveciennes 013908
Musée Protestant de la Grange de
Wassy, Wassy 016343
Musée Provençal des Arts et
Traditions Poplaires, Donzère . 012816

Musée Provençal des Transports
Urbains et Régionaux, La
Barque 013358
Musée Provençal du Costume et du
Bijou, Grasse 013163
Musée Provençal et Folklorique
Taradeau, Les Arcs 013734
Musée Provincial de Douala,
Douala 005292
Musée Provincial de la Forêt,
Namur 003902
Musée Provincial du Houet, Bobo-
Dioulasso 005272
Musée Provincial du Poni, Gaoua 005273
Musée Provincial du Sanmatenga,
Kaya 005274
Musée Provincial Félicien Rops,
Namur 003903
Musée Prytanée Militaire, La
Flèche 013415
Musée Pyrénéen, Lourdes 013903
Musée Pyrénéen de Niaux, Niaux 014471
Musée Québécois de Culture
Populaire, Trois-Rivières 007250
Musée Quentovic, Etaples 012921
Musée Rabelais, Seuilly 015789
Musée Raimu, Cogolin 012613
Musée Rambolitrain, Rambouillet 015034
Musée Raoul Dastrac, Collection
d'Arts et de Traditions Populaires,
Aiguillon 011551
Musée Rath, Genève 041538
Musée Raymond Lafage, Lisle-sur-
Tarn 013837
Musée Raymond Peynet, Brassac-les-
Mines 012193
Musée Raymond Poincaré,
Sampigny 015671
Musée Raymond Rochette, Saint-
Sernin-du-Bois 015562
Musée Réattu, Arles 011731
Musée Reflet Brionnais,
Iguerande 013282
Musée Regards sur le Passé,
Hourtin 013274
Musée Régimentaire les Fusiliers
de Sherbrooke, Sherbrooke,
Québec 007026
Musée Regional, Kisangani 008839
Musée Régional André Grenard
Matsoua, Kinkala 008845
Musée Régional d'Ethnologie,
Béthune 012013
Musée Regional d'Argenteuil, Saint-
André-d'Argenteuil 006847
Musée Régional d'Armes, Mutzig 014395
Musée Régional d'Arts et de Traditions
Populaires, Grimaud 013197
Musée Régional d'Auvergne,
Riom 015097
Musée Régional de Fouras,
Fouras 013037
Musée Régional de Géologie Pierre
Vetter, Decazeville 012750
Musée Régional de la Côte-Nord,
Sept-
Îles 007005
Musée Régional de la Poterie bas
Normande, Ger 013092
Musée Régional de la Résistance et
de la Déportation, Thionville . . 015938
Musée Régional de la Sagne, La
Sagne 041904
Musée Régional de la Vigne et du Vin,
Montmélian 014277
Musée Régional de l'Air, Groupement
Préservation Patrimoine Aéronautique,
Marcé 013998
Musée Régional de l'Alsace Bossue,
Sarre-Union 015694
Musée Régional de l'Orléanais,
Beaugency 011939
Musée Regional de Mahdia,
Mahdia 042755
Musée Régional de Préhistoire,

Orgnac-l'Aven 014577
Musée Régional de Rimouski,
Rimouski 006807
Musée Régional de Sikasso,
Sikasso 030736
Musée Régional de Sokadé,
Sokadé 042731
Musée Régional de Vaudreuil-
Soulanges, Vaudreuil-Dorion . . 007309
Musée Régional de Zinder,
Zinder 033409
Musée Régional des Arts de la Table,
Ancien Hospice Saint-Pierre du XVIIe
s, Arnay-le-Duc 011736
Musée Régional des Enrôles de Force,
Dudelange 030577
Musée Régional des Métiers, La
Cheze 013391
Musée Régional des Mines de
Malartic, Malartic 006249
Musée Régional des Savanes,
Dapaong 042729
Musée Régional des
Télécommunications en Flandres,
Marcq-en-Barœul 014011
Musée Régional d'Histoire et
d'Archéologie, Musée de Visé,
Visé 004117
Musée Régional d'Histoire et
d'Artisanat du Val-de-Travers,
Môtiers 041765
Musée Régional d'Histoire et
d'Ethnographie, Fort-de-France 030776
Musée Régional du Cerel,
Toamasina 030652
Musée Régional du Chemin de Fer,
Rosny-sous-Bois 015161
Musée Régional du Chemin de Fer
Nord-Pas-de-Calais, Denain . . 012756
Musée Régional du Cidre et du
Calvados, Valognes 016113
Musée Régional du Machinisme
Agricole, La Ferté-Milon 013411
Musée Régional du Sahel, Gao . . 030735
Musée Régional du Timbre et de la
Philatélie, Le Luc 013651
Musée Régional Dupuy-Méstreau,
Saintes 015637
Musée Régional Ex-Dixième Riaom,
Thiès 038122
Musée Régional Ma-Loango Diosso,
Pointe-Noire 008846
Musée Relais du Cheval de Trait
Comtois, Levier 013783
Musée Religieux d'Art Franciscain,
Nice 014490
Musée Religieux Franco-Canadien,
Tours 016022
Musée Remember 39–45, Léhon 013726
Musée Remise aux Outils Champenois,
Dormans-en-Champagne 012818
Musée Renaudot, Loudun 013882
Musée René Davoine, Charolles 012463
Musée Renoir, Cagnes-sur-Mer . 012285
Musée Requien, Avignon 011841
Musée Rêve Auto-Jeunesse,
Mosnac 014342
Musée Rêve et Miniatures,
Brantôme 012190
Musée Revoir le Passé, Les Loges-
Marchis 013753
Musée Ribauri, Lasne-Chapelle-Saint-
Lambert 003775
Musée Rignault, Saint-Cirq-
Lapopie 015264
Musée Rizla Croix, Mazères-sur-
Salat 014104
Musée Rochelais d'Histoire
Protestante, La Rochelle 013483
Musée Rodava, Cercle d'Histoire et
d'Archéologie, Rèves 003965
Musée Rodriguez Huguette, Saint-
Macaire 015434
Musée Roger Thières, Payrac . . . 014817

Musée Rolin, Musées Municipaux,
Autun 011810
Musée Romain, Avenches 041245
Musée Romain, Nyon 041808
Musée Romain, Vallon 042094
Musée Romain de Lausanne-Vidy,
Lausanne 041663
Musée Romand de la Machine
Agricole, Gingins 041544
Musée Ronsard, La Riche 013459
Musée Rosine Perrier,
Villargondran 016221
Musée Royal de l'Armée et d'Histoire
Militaire, Koninklijk Museum van
het Leger en de Krijgsgeschiedenis,
Bruxelles 003491
Musée Royal de Mariemont,
Morlanwelz-Mariemont 003890
Musée Royaume de l'Horloge,
Villedieu-les-Poëles 016238
Musée Roybet Fould, Courbevoie 012696
Musée Rude, Dijon 012789
Musée Rue Etienne Lenoir, Mussy-la-
Ville 003894
Musée Rupert de Chièvres,
Poitiers 014918
Musée Rural, Foucherans 013030
Musée Rural, Saint-André-
d'Embrun 015218
Musée Rural Berry-Marche,
Boussac 012176
Musée Rural d'Arts et Traditions
Populaires La Combà, Autrafé, La
Combe-de-Lancey 013396
Musée Rural d'Auris-en-Oisans
des Choses d'Autrefois, Auris-en-
Oisans 011804
Musée Rural de la Sologne
Bourbonnaise, Beaulon 011943
Musée Rural de l'Educations dans les
Côtes d'Armor, Saint-Nicolas-du-
Pelem 015480
Musée Rural des Art Populaires,
Laduz 013522
Musée Rural du Bois, Coupiac . 012693
Musée Rural du Champsaur, Bénévent-
et-Charbillac 011984
Musée Rural du Porzay,
Plomodiern 014897
Musée Rural du Porzay, Saint-Nic-
Pontrez 015476
Musée Rural Jurassien, Les
Genevez 041542
Musée Rural, Écomusée des Monts du
Forez, Usson-en-Forez 016074
Musée Ruzo, Montreux 041757
Musée Sabatier, Saint-Bonnet-de-
Joux 015237
Musée Sabatier d'Eyspéran,
Montpellier 014301
Musée Saharien de Ouargla,
Ouargla 000072
Musée Sahut, Simiane-Collongue 015811
Musée Saint-Brieux, Saint Brieux
Museum, Saint-Brieux 006858
Musée Saint-Étienne, Honfleur . 013272
Musée Saint-François-Régis, La
Louvesc 013438
Musée Saint-Jacques, Moissac . 014191
Musée Saint-Jean, Anneyron . . 011674
Musée Saint-Jean, La Rochette
(Savoie) 013485
Musée Saint-Joachim, La
Broquerie 006132
Musée Saint-Joseph, Saint-
Joseph 006914
Musée Saint-Loup, Troyes 016054
Musée Saint-Martin, Tours 016023
Musée Saint-Nicolas, Vitré 016310
Musée Saint-Pierre Fourier,
Mattaincourt 014081
Musée Saint-Raymond, Musée des
Antiques de Toulouse, Toulouse 015998
Musée Saint-Remi, Reims 015056

Musée Saint-Sauveur Nibelle Autrefois, Nibelle 014473
Musée Saint-Simon, La Ferté-Vidame 013413
Musée Saint-Vic, Saint-Amand-Montrond 015212
Musée Sainte-Anne, Lachine 006147
Musée Sainte-Croix, Poitiers 014919
Musée Sainte-Jeanne-Antide-Thouret, Sancey-le-Long 015673
Musée Sainte-Marie, Church Point 005617
Musée Sainte-Martine, Sainte-Martine 006959
Musée Salies, Musée des Beaux-Arts, Bagnères-de-Bigorre 011866
Musée-Salle Communale, Yvoire 016365
Musée-Salle Documentaire, Coucy-le-Château-Auffrique 012686
Musée Santons du Monde, Huilerie, Clermont-l'Hérault 012600
Musée Sarret de Grozon, Arbois 011705
Musée Saturnin Garimond, Fons 013001
Musée Savoisien, Chambéry ... 012421
Musée Schiller et Goethe, Châlons-en-Champagne 012411
Musée Schœlcher, Pointe-à-Pitre 023349
Musée Scout Baden-Powell, Sisteron 015817
Musée Sculpture sur Bois, Lans-en-Vercors 013564
Musée Sellier, Cogolin 012614
Musée Serret, Saint-Amarin ... 015213
Musée Servie de Santé des Armées au Val-de-Grâce, Paris 014789
Musée Sherlock Holmes, Lucens 041691
Musée Sidi Ben-Aissa, Le Kef .. 042750
Musée Sidi Bou Saïd, Sidi Bou Saïd 042773
Musée Sidi Mohamed Ben Abdellah, Essaouira 031773
Musée Sidi Zitouni, Houmt Souk 042746
Musée Simon-Segal, Aups 011795
Musée Site Départemental d'Archéologie de Bavay, Bavay 011917
Musée Skolig al Louarn, Skolig al Louarn, Plouvien 014913
Musée Sobirats, Carpentras ... 012344
Musée Somme 1916, Albert ... 011578
Musée Souleïado Charles Demery, Musée du Tissu Provençal, Tarascon (Bouches-du-Rhône) 015909
Musée Sous-Glaciaire, La Grave 013429
Musée Sous Marin Pays Lorient, Lorient 013872
Musée souterrain de la Poterie, Hospices de la Madeleine, Saint-Émilion 015298
Musée Souterrain de l'Outil, Chacé 012398
Musée Souvenir du Dirigeable R101, Beauvais 011957
Musée Spadois du Cheval, Spa . 004043
Musée Spectacle de Musiques du Monde, Montoire-sur-le-Loir .. 014283
Musée Spitfire, Florennes 003602
Musée Sterna de l'Armée et des Trois Guerres, Seynod 015803
Musée Stewart au Fort de l'Île Sainte-Hélène, Montréal 006411
Musée Suffren et du Vieux Saint-Cannat, Saint-Cannat 015248
Musée Suisse de la Figurine Historique, Morges 041762
Musée Suisse de la Machine à Coudre, Fribourg 041503
Musée Suisse de la Marionnette, Fribourg 041504
Musée Suisse de l'Appareil Photographique, Vevey 042109
Musée Suisse de l'Orgue, Schweizer Orgelmuseum, Roche 041885
Musée Suisse de Spéléologie, Chamoson 041400

Musée Suisse du Jeu, La Tour-de-Peilz 042064
Musée Suisse du Sel, Aigle ... 041202
Musée Sundgauvien, Altkirch .. 011607
Musée sur l'Artillerie, la Cavalerie et l'Infanterie, Seclin 015747
Musée Tancrède, Hauteville-la-Guicharde 013242
Musée Tauromachique, Pomarez 014922
Musée Tavet-Delacour, Pontoise 014960
Musée Tchantchès – Musée de la République Libre d'Outre-Meuse, Liège 003812
Musée Terbol, Ecomusée de la Békaa, Zahlé 030283
Musée Terrasse, Viverols 016314
Musée Territorial de Nouvelle-Calédonie, Nouméa 033131
Musée Terrus, Elne 012864
Musée Testut-Latarjet de Médecine et d'Anatomie, Lyon 013951
Musée Textile de Haute-Alsace, Husseren-Wesserling 013280
Musée Théâtre Ki-Yi, Abidjan .. 008870
Musée Théodore Rousseau, Musée Departemental de l'École de Barbizon, Barbizon 011893
Musée Théophile Jouglet, Anzin 011693
Musée Thimonnier Deyon, Saint-Germain-au-Mont-d'Or 015336
Musée Thomas-Henry, Cherbourg-Octeville 012551
Musée Tomi Ungerer – Centre International de l'illustration, Musées de Strasbourg, Strasbourg .. 015886
Musée Toulouse-Lautrec, Albi .. 011585
Musée Toulouse-Lautrec, Naucelle 014435
Musée Tour de la Lanterne, La Rochelle 013484
Musée Tour d'Eben-Ezer, Musée du Silex, Eben-Emael 003566
Musée Tradition et Modernité, Verbier 042101
Musée Traditions et Vie, Châtillon-sur-Chalaronne 012516
Musée Traditions Verrieres, Eu . 012932
Musée Trésor d'Église, Art Sacré de la Cathédrale Saint-Étienne, Metz 014151
Musée Troglodytique, Louresse . 013905
Musée Truschbaum, Elsenborn . 003575
Musée Ukraina, Saskatoon 006978
Musée Une Halte sur le Chemin de Saint-Jacques, Ecomusée de la Vallée d'Aspe, Borce 012106
Musée Urbain Tony Garnier, Lyon 013952
Musée Utrillo-Valadon, Sannois . 015676
Musée Valaisan de la Vigne et du Vin, Walliser Reb- und Weinmuseum, Sierre 041986
Musée Valentin Haüy, Paris 014790
Musée Valery Stuyver, Seraing-le-Château 004010
Musée Van Œuvren, Bourgueil . 012717
Musée Vauban, Neuf-Brisach .. 014447
Musée Vauban et de la Fortification dans les Alpes, Mont-Dauphin 014205
Musée Vendéen, Fontenay-le-Comte 013016
Musée Verger Tarin, Autun 011811
Musée Verlaine, Juniville 013342
Musée Vert, Naturall History Museum of Le Mans, Le Mans 013657
Musée Victor Aubert, Maule ... 014085
Musée Victor Hugo, Villequier .. 016278
Musée Vie et Métiers d'Autrefois dans le Haut-Jura, La Pesse 013448
Musée Vieux Presbytère de Batiscan, Batiscan 005397
Musée Vigneron, Vendres 016155
Musée Villa Tamaris Pacha, La Seyne-sur-Mer 013493
Musée Villageois d'Orgon, Orgon 014579
Musée Villeneuvien, Villeneuve-sur-

Yonne 016277
Musée Villien, Musée de la Société Archéologique, Avesnes-sur-Helpe 011831
Musée Vini-Viticole Gallo-Romain, Beaucaire 011934
Musée Vissoie, Vissoie 042113
Musée Viticole, Saint-Yzans-de-Médoc 015596
Musée Viticole Charles Bernard, Fixin 012989
Musée Vivant d'Archéologie Industrielle Ferroviaire, Sprimont 004045
Musée Vivant de Bujumbura, Bujumbura 005280
Musée Vivant de la Brasserie à Vapeur, Pipaix 003951
Musée Vivant de la Faisanderie, Milly 014171
Musée Vivant de la Forge, Seyne-les-Alpes 015802
Musée Vivant de la Laine et du Mouton, Saint-Pierreville .. 015515
Musée Vivant de la Métairie de Crampet, Mazères (Gironde) .. 014103
Musée Vivant de la Mine, Auchel 011782
Musée Vivant de la Passementerie, Jonzieux 013326
Musée Vivant de la Plante Aquatique, Limal 003821
Musée Vivant de la Pomme de Terre, Genappe 003615
Musée Vivant de la Porcelaine, Foëcy 012996
Musée Vivant de l'Abeille, Corbeny 012666
Musée Vivant de l'Abeille, Valensole 016097
Musée Vivant de l'Apiculture Gâtinaise, Châteaurenard (Loiret) ... 012503
Musée Vivant de l'Égypte pour Enfants, Jambes 003738
Musée Vivant Denon, Chalon-sur-Saône 012406
Musée Vivant des Vieux Métiers, Argol 011724
Musée Vivant du Chemin de Fer Vicinal, Thuin 004065
Musée Vivant du Cheval, Chantilly 012447
Musée Vivant du Cheval de Trait, Waret-l'Evêque 004131
Musée Vivant du Cheval de Trait et de l'Attelage, Pradelles 014986
Musée Vivant du Cinéma, Lyon . 013953
Musée Vivant du Cochon, Chambonas 012423
Musée Vivant du Jambon – La Ferme des Sanglochons, Verlaine ... 004104
Musée Vivant du Roman d'Aventures, Jassans-Riottier 013311
Musée Vivant et Historique du Train Historique du Lac de Rillé, Rillé 015093
Musée Vivant Européen du Cirque, Piolenc 014873
Musée Vivant Médiéval, Murol . 014389
Musée Voland, Cerny 012393
Musée Vosgien de la Brasserie, Ville-sur-Illon 016231
Musée Vulcanologique Franck-Arnold Perret, Saint-Pierre 030782
Musée Wagon de l'Armistice, Compiègne 012647
Musée Wellington, Waterloo ... 004137
Musée Westercamp, Wissembourg 016353
Musée Würth, Erstein 012893
Musée Yan Dagent, Saint-Servais 015563
Musée Yani Faux, Orbais-l'Abbaye 014570
Musée Yves Brayer, Cordes-sur-Ciel 012671
Musée Yves Brayer, Les Baux-de-

Provence 013743
Musée Zadkine, Les Arques ... 013735
Musée Zadkine, Paris 014791
Musée Zervos, Préfiguration de la Fondation Zervos, Vézelay 016192
Musée Ziem, Martigues 014071
Musée Zoologique de l'Université Louis Pasteur et de la Ville de Strasbourg, Strasbourg 015887
Musée, Champagne J.M. Gobillard, Hautvillers 013243
Musée, Maison du Suc, Saint-Julien-Molhesabate 015400
Musée, Société des Amis du Musée, Ville-d'Avray 016227
Museen Alte Bischofsburg, Wittstock/Dosse 022708
Museen der Stadt Bad Kösen, Romanisches Haus, Kunsthalle, Käthe-Kruse-Puppenmuseum, Bad Kösen 016801
Museen für Kunst und Kulturgeschichte der Hansestadt Lübeck – Verwaltung, Lübeck . 020160
Museen im Alten Rathaus, Bürkel-Galerie, Scherenschnitt-Kabinett, Schuhmuseum, Pirmasens .. 021134
Museen im Alten Schloss, Stadtmuseum und Aischgründer Karpfen-Museum, Neustadt an der Aisch 020754
Museen im Kulturzentrum, Historisches Museum Rendsburg/Druckmuseum, Rendsburg ... 021373
Museen im Markgrafen-Schloss, Stadtgeschichtliche Sammlung und Fotomuseum Hirsmüller, Emmendingen 018184
Museen im Rathaus, Osmanenmuseum, Deutschmeistermuseum, Weinbaumuseum, Perchtoldsdorf 002496
Museen im Wehrturm, Perchtoldsdorf 002497
Museerne i Fredericia, Fredericia 010064
Musées d'Archéologie et d'Arts Décoratifs, Liège 003813
Musées de la Citadelle, Direction, Villefranche-sur-Mer 016248
Musées de la Cour d'Or, Metz .. 014152
Musées de La Poèslerie et de la Dentelle, Villedieu-les-Poêles . 016239
Musées de la Réole, Musée Militaire, Automobile, Ferroviaire et d'Agricole, La Réole 013455
Musées de l'Horlogerie Pater, Longueville 003828
Musées de Sens, Sens 015771
Musées de Vieille Forge, de la Traversée du Siècle et des Automates, Musées Barlier, Fréland 013054
Musées Départementaux Albert Demard, Champlitte 012439
Musées Départementaux de la Guadeloupe, Pointe-à-Pitre .. 059334
Musées des Colombiers Cauchois, Oherville 014551
Musées des Technique et Cultures Comtoises, Salins-les-Bains . 015651
Musées et Collections Publiques de France, La Lettre des Musées, Paris 138651
Musées et Société en Wallonie, Namur 058386
Musées Galerie, Nancy 104324
Musées Royaux d'Art et d'Histoire, Bruxelles 003492
Musées Universitaire, Kinshasa . 008838
Museet Ett Hem, Turku 011440
Museet Færgegården, Jægerspris 010153
Museet for Dansk Bladtegning, København 010183

Museet for Fotokunst, Odense . 010262
Museet for Holbæk og Omegn,
  Holbæk 010132
Museet for Religiøs Kunst,
  Lemvig 010220
Museet for Samtidskunst,
  Roskilde 010288
Museet for Samtidskunst,
  Nasjonalmuseet for Kunst, Arkitektur
  og Design, Oslo 033843
Museet for Thy og Vester Hanherred,
  Thisted 010340
Museet for Varde By og Omegn,
  Varde 010351
Museet Holmen, Løgumkloster . 010223
Museet i Mælandsgården,
  Skudeneshavn 033950
Museet Kvarnen, Konstgalleri och
  Ferlinmuseum, Filipstad 040625
Museet Kystens Arv, Stadsbygd 033969
Museet Leksaksland, Malmö . . 040831
Museet Näktergalen, Vimmerby . 041170
Museet på Gudhjem Station,
  Gudhjem 010091
Museet på Koldinghus, Kolding . 010206
Museet på Sønderskov, Brørup . 010028
Museet Psykiatrisk Hospital i Århus,
  Risskov 010282
Museet Tikøb Frysehus, Tikøb . . 010341
Museet ved Sct. Ansgar Kirke,
  København 010184
Museet ved Trelleborg, Slagelse 010319
Musehome, Genova 110124
Musei Anatomici, Modena 026851
Musei Capitolini, Roma 027573
Musei Civici, Lecco 026585
Musei Civici, Reggio Emilia . . . 027502
Musei Civici di Padova, Padova . 027144
Musei Civici di Palazzo d'Avalos,
  Vasto 028284
Musei Civici di Pesaro, Pesaro . 027264
Musei Civici e Pinacoteca L. Sturzo,
  Caltagirone 025699
Musei Civici Veneziani, Venezia . 059469
Musei Civici, Castello Visconteo,
  Pavia 027231
Musei Comunali di Roma. Bollettino,
  Roma 138844
Musei del Castello Scaligero,
  Malcesine 026680
Musei della Civiltà Contadina del Friuli
  Imperiale, Aiello del Friuli 025257
Musei della Cultura Popolare Radana,
  San Benedetto Po 027727
Musei e Gallerie d'Italia, Roma . 138845
O Musei Moveis Antigos, São
  Paulo 064195
Musei Provinciali di Gorizia,
  Gorizia 026453
Musei Sistini del Piceno, San
  Benedetto del Tronto 027722
Museifartyget Pommern,
  Mariehamn 011145
Museion – Museo d'Arte Moderna e
  Contemporanea, Bolzano 025580
Museion Versand, Berlin 074885
Museisällskapet Jädraås-Tallås
  Järnväg, Jädraås 040715
Museispårvägen Malmköping,
  Malmköping 040822
El Museo, Bogotá 102349, 136749,
  138607
Museo, Helsinki 138622
Museo – Pinacoteca San Francesco,
  San Marino 038097
Museo 13 de Marzo, Pinar del
  Río 009285
Museo 9 de Abril, Santa Clara . 009332
Museo a la Bandera y Santuario
  de la Patria, Iguala de la
  Independencia 031065
Museo A. Binda, Cittiglio 026009
Museo A. Cassioli, Asciano . . . 025382
Museo A. Colocci, Jesi 026529

Museo A. e C. Gaffoglio, Rapallo 027479
Museo A. Lia, La Spezia 026537
Museo A. Murer, Falcade 026180
Museo A. Ricci dei Minerali Elbani,
  Rio nell'Elba 027527
Museo A. Versace, Bagnara
  Calabra 025435
Museo A.R. Giorgi, Reggiolo . . . 027507
Museo Abarth, Marietta 050676
Museo Abel Santamaría, La
  Habana 009147
Museo Abel Santamaria, Santa
  Clara 009333
Museo Abraham González, Vicente
  Guerrero 031650
Museo Acacia, La Habana 009148
Museo Academia de Historia Leonardo
  Tascón, Buga 008641
Museo Activo del Pago de los Arroyos,
  Rosario 000542
Museo Ada e Giuseppe Marchetti,
  Fumone 026357
Museo ADIAC, Museo Indígena de
  Sutiaba, León 033383
Museo Adolfo López Mateos, Atizapán
  de Zaragoza 030835
Museo Adolfo Lozano Sidro, Priego de
  Córdoba 039973
Museo Aeronáutico, Baradero . . 000141
Museo Aeronáutico, Montevideo 046254
Museo Aeronautico Caproni di Taliedo,
  Roma 027574
Museo Aeronautico Caproni-Taliedo,
  Vizzola Ticino 028417
Museo Aeronáutico Colonel Luis
  Hernan Paredes, Maracay 054764
Museo Aeronáutico del Perú,
  Lima 034319
Museo Aeronáutico Municipalo, Los
  Alcázares 038922
Museo Aeronáutico Torreón de Gando,
  Museo del Aire, Telde 040217
Museo Aeronautico y del Espacio,
  Quito 010470
Museo Aerospaziale Monte di Apollo,
  Perugia 027257
Museo Africano, Roma 027575
Museo Africano, Verona 028344
Museo Africano Mundo Negro,
  Madrid 039628
Museo Afro Antillano de Panamá,
  Ciudad de Panamá 034173
Museo Agrario Tropicale, Istituto
  Agronomico per l'Oltremare,
  Firenze 026266
Museo Agrarista de Tzurúmutaro,
  Pátzcuaro 031364
Museo Agrícola El Patio, Teguise 040213
Museo Agrícolo, Ronco Briantino 027668
Museo Agricolo Brunnenburg,
  Landwirtschaftsmuseum, Tirolo di
  Merano 028089
Museo Agricolo della Civiltà Contadina
  L. Carminati, San Giuliano
  Milanese 027763
Museo Agricolo e del Vino Ricci
  Curbastro, Capriolo 025762
Museo Agricolo e Museo dell'Arte
  Conciaria, Castelfranco Veneto 025825
Museo Agro-Forestale San Matteo,
  Erice 026158
Museo Agro-Pastorale dell'Area
  Ellenofona, Bova Marina 025615
Museo Agrumario, Reggio
  Calabria 027494
Museo Aguilar y Eslava, Cabra . 039178
Museo Alberto Arvelo Torrealba,
  Barinas 054704
Museo Alcamese d'Arte Sacra,
  Alcamo 025280
Museo Alcázar de Colón, Santo
  Domingo 010385
Museo Aldrovandiano, Musei di
  Palazzo Poggi, Bologna 025534

Museo Alejandro Otero, Caracas 054714
Museo Alejo Carpentier,
  Camagüey 009087
Museo Alessi, Enna 026152
Museo Alfar, Santa Elena de
  Jamuz 040102
Museo Alfieriano, Centro Nazionale di
  Studi Alfieriani, Asti 025405
Museo Alfonsiano, Pagani 027165
Museo Alfredo Ramírez de Arellano &
  Rosello, San Germán 036148
Museo All'Aperto di Storia
  dell'Agricultura, Facoltà di Agraria,
  Sassari 027866
Museo Almacén El Recreo,
  Chivilcoy 000304
Museo Almafuerte, La Plata . . . 000407
Museo Alpino Duca degli Abruzzi,
  Courmayeur 026094
Museo Alpujarreño de Costumbre
  Populares, Pedro Antonio de Alarcón,
  Capileira 039226
Museo Alta Valle Scrivia, Sezione
  Etnologica, Valbrevenna 028250
Museo Altar de la Patria, Ixcateopan
  de Cuauhtémoc 031068
Museo Alternativo Remo Brindisi,
  Comacchio 026041
Museo Alto Bierzo, Bembibre . . 039123
Museo Amano, Lima 034320
Museo Amazigh, Melilla 039744
Museo Ambientale del W.W.F.,
  Centro di Educazione Ambientale,
  Orbetello 027095
Museo Ambientalistico Madonita,
  Polizzi Generoso 027386
Museo Ambiente Cubano, Gibara 009123
Museo Amedeo Bocchi, Parma . 027210
Museo Americanista, Lomas de
  Zamora 000430
Museo Americanistico Federico
  Lunardi, Genova 026396
Museo Amparo, Puebla 031375
Museo Anatomia Veterinaria,
  Napoli 026991
Museo Anatomico, Siena 027950
Museo Anatómico, Valladolid . . 040335
Museo Anatomico G. Tumiati,
  Ferrara 026205
Museo Anatómico Pedro Ara,
  Córdoba 000319
Museo Anatomico Veterinario,
  Pisa 027353
Museo Anconetani del Acordeón,
  Buenos Aires 000165
Museo Andrés Avelino Cáceres,
  Lima 034321
Museo Andrés Cué, Amancio . . 009063
Museo Anenecuilco, Ayala 030839
Museo Anestesiológico y Biblioteca
  Histórica de la Asociación de
  Anestesiología de Buenos Aires,
  Buenos Aires 000166
Museo Angel Orensanz y Artes de
  Serralup, Instituto de Cultura y Arte
  Contemporáneo, Sabiñánigo . . 040032
Museo Antártico, Montevideo . . 046255
Museo Antica Miniera di Talco
  Brunetta, Cantoira 025743
Museo Antiguo Colegio de San
  Ildefonso, México 031159
Museo Antiguo Cuartel de Caballería
  del Ejército Español, Camagüey 009088
Museo Antiguo Palacio de la Medicina
  Mexicana, México 031160
Museo Antimperialista, Nueva
  Gerona 009270
Museo Antiquarium Archeologico,
  Falerone 026181
Museo Antón, Candás 039217
Museo Antoniano, Padova 027145
Museo Antonio Lussich, Punta
  Ballena 046288
Museo Antonio Maceo, San

Pedro 009321
Museo Antonio Manuel Campoy,
  Cuevas del Almanzora 039338
Museo Antonio Raimondi, La
  Merced 034300
Museo Antonio Raimondi, Lima . 034322
Museo Antonio Ricaurte, Villa de
  Leyva 008821
Museo Antrológico, Cereté 008662
Museo Antropologia e Historia Valle de
  Sula, San Pedro Sula 023401
Museo Antropología e Historia,
  Universidad Autónoma de
  Tamaulipas, Victoria 031653
Museo Antropológico de Antofagasta,
  Antofagasta 007513
Museo Antropológico de Iquique,
  Iquique 007523
Museo Antropológico de Tunja,
  Tunja 008812
Museo Antropológico del Banco
  Central, Guayaquil 010444
Museo Antropológico Montané, La
  Habana 009149
Museo Antropológico Reina Torres de
  Araúz, Ciudad de Panamá 034174
Museo Antropológico Shuar,
  Quito 010471
Museo Antropológico Tairona, Santa
  Marta 008784
Museo Antropológico, Universidad del
  Tolimo, Ibagué 008697
Museo Antropológico, Universidad
  Mayor, Real y Pontifica de San
  Francisco Xavier, Sucre 004242
Museo Antzetik Ta Jtelum,
  Zinacantán 031708
Museo Anzoátegui, Pamplona . . 008742
Museo Aperos de Labranza,
  Lerdo 031111
Museo Apistico, Bologna 025535
Museo Apistico Didattico,
  Bregnano 025622
Museo Archaeologico di Santa
  Restituta, Lacco Ameno 026542
Museo Archeologica, Savignone 027884
Museo Archeologico, Acqualagna 025235
Museo Archeologico, Amelia . . . 025313
Museo Archeologico, Arborea . . 025349
Museo Archeologico, Artena . . . 025380
Museo Archeologico, Atri 025413
Museo Archeologico, Bonorva . . 025590
Museo Archeologico,
  Caltanissetta 025701
Museo Archeologico, Camaiore . 025706
Museo Archeologico, Castello di
  Godego 025836
Museo Archeologico, Cesena . . 025920
Museo Archeologico, Dorgali . . 026141
Museo Archeologico, Enna 026153
Museo Archeologico, Faenza . . 026171
Museo Archeologico, Fermo . . . 026197
Museo Archeologico, Frosinone . 026353
Museo Archeologico, Isola della
  Scala 026522
Museo Archeologico, Lentini . . . 026591
Museo Archeologico, Marianopoli 026712
Museo Archeologico, Molfetta . . 026872
Museo Archeologico, Ostiglia . . 027129
Museo Archeologico, Palazzuolo sul
  Senio 027172
Museo Archeologico, Povegliano
  Veronese 027433
Museo Archeologico, Priverno . . 027458
Museo Archeologico, Pula 027460
Museo Archeologico, Rocca di
  Mezzo 027541
Museo Archeologico, San Miniato 027785
Museo Archeologico, Santadi . . 027827
Museo Archeologico, Santorso . 027848
Museo Archeologico, Scansano . 027898
Museo Archeologico, Siena . . . 027951
Museo Archeologico, Teramo . . 028069
Museo Archeologico, Venafro . . 028290

Museo Archeologico, Ventotene . 028333
Museo Archeologico, Veroli 028340
Museo Archeologico, Vetulonia . 028356
Museo Archeologico, Villasimius 028396
Museo Archeologico al Teatro Romano,
Verona 028345
Museo Archeologico Antiquarium
Arborense, Oristano 027105
Museo Archeologico Antonio Santarelli,
Forlì 026322
Museo Archeologico Aquaria,
Soncino 027981
Museo Archeologico Attrezzato e
Cattedrale di San Zeno, Pistoia 027360
Museo Archeologico B. Antonucci,
Pietrasanta 027318
Museo Archeologico Baglio Anselmi,
Marsala 026717
Museo Archeologico Civico,
Carassai 025767
Museo Archeologico Civico,
Forlimpopoli 026331
Museo Archeologico Civico-Diocesano
G. B. Leopardi, Penne 027245
Museo Archeologico Comprensoriale,
Teti 028084
Museo Archeologico Comunale,
Vicchio 028363
Museo Archeologico Comunale di
Artimino, Carmignano 025774
Museo Archeologico dei Campi Flegrei,
Bacoli 025430
Museo Archeologico del Finale, Finale
Ligure 026237
Museo Archeologico del Territorio,
Cupra Marittima 026119
Museo Archeologico del Territorio di
Populonia, Piombino 027341
Museo Archeologico del Territorio,
Antiquarium Suasanum, San Lorenzo
in Campo 027770
Museo Archeologico della Badia,
Badia di Licata, Licata 026597
Museo Archeologico della Lucania
Occidentale, Padula 027164
Museo Archeologico della Valle Sabbia,
Gavardo 026376
Museo Archeologico dell'Abruzzo,
Crecchio 026095
Museo Archeologico dell'Agro
Nocerino, Nocera Inferiore ... 027041
Museo Archeologico dell'Alto Adige,
Bolzano 025581
Museo Archeologico dell'Alto
Canavese, Cuorgnè 026118
Museo Archeologico dell'Antica,
Alife 025294
Museo Archeologico dell'Antica Capua,
Santa Maria Capua Vetere ... 027819
Museo Archeologico dell'Antica
Corfinio, Corfinio 026070
Museo Archeologico dell'Arte della
Lana, Consorzio Sistena Museale
Valle di Liri, Arpino 025375
Museo Archeologico di Calatia,
Maddaloni 026668
Museo Archeologico di Casteggio,
Casteggio 025809
Museo Archeologico di Morgantina,
Aidone 025256
Museo Archeologico di Naxos, Giardini
Naxos 026436
Museo Archeologico di Sepino,
Sepino 027922
Museo Archeologico di Vallecamonica,
Cividate Camuno 026013
Museo Archeologico di Villa Sulcis,
Carbonia 025769
Museo Archeologico e Antiquarium,
Sala Consilina 027705
Museo Archeologico e Antiquarium
Nazionale, Formia 026333
Museo Archeologico e d'Arte della
Maremma, Grosseto 026473

Museo Archeologico e della Ceramica,
Montelupo Fiorentino 026947
Museo Archeologico e della Collegiata,
Casole d'Elsa 025803
Museo Archeologico e della via
Flaminia, Cagli 025670
Museo Archeologico e Paleontologico,
Asti 025406
Museo Archeologico e Paleontologico
di Farneta, Cortona 026083
Museo Archeologico e Pinacoteca
Comunale, Foligno 026313
Museo Archeologico e Pinacoteca,
Musei di Palazzo Farnese,
Piacenza 027296
Museo Archeologico e Storico Aristico,
Palazzo Traversa, Bra 025618
Museo Archeologico ed Etnografico,
Ittireddu 026526
Museo Archeologico Eoliano,
Lipari 026600
Museo Archeologico Etneo,
Adrano 025243
Museo Archeologico Etnografico,
Paulilatino 027229
Museo Archeologico F. Milizia,
Oria 027100
Museo Archeologico G. Allevi,
Offida 027082
Museo Archeologico G. Cilnio
Mecenate, Arezzo 025358
Museo Archeologico G. Moretti, San
Severino Marche 027798
Museo Archeologico G. Rambotti,
Desenzano del Garda 026131
Museo Archeologico Girolamo
Rossi, Forte dell'Annunziata,
Ventimiglia 028331
Museo Archeologico Ibleo,
Ragusa 027472
Museo Archeologico Iulium Carnicum,
Palazzo Leschiutta, Zuglio 028435
Museo Archeologico Lametino,
Lamezia Terme 026548
Museo Archeologico Liutprando, San
Giovanni in Persiceto 027759
Museo Archeologico Lomellino,
Gambolò 026369
Museo Archeologico Lucus Feroniae,
Capena 025750
Museo Archeologico Luigi Donini, San
Lazzaro di Savena 027764
Museo Archeologico Medievale,
Attimis 025417
Museo Archeologico Michele Janora,
Irsina 026514
Museo Archeologico Nazionale,
Adria 025244
Museo Archeologico Nazionale,
Aquileia 025345
Museo Archeologico Nazionale,
Capaccio 025746
Museo Archeologico Nazionale,
Cassino 025806
Museo Archeologico Nazionale,
Cividale del Friuli 026011
Museo Archeologico Nazionale,
Crotone 026109
Museo Archeologico Nazionale,
Eboli 026147
Museo Archeologico Nazionale,
Firenze 026267
Museo Archeologico Nazionale, Gioia
del Colle 026443
Museo Archeologico Nazionale,
Manfredonia 026693
Museo Archeologico Nazionale,
Mantova 026696
Museo Archeologico Nazionale,
Napoli 026992
Museo Archeologico Nazionale,
Parma 027211
Museo Archeologico Nazionale, Quarto
d'Altino 027463

Museo Archeologico Nazionale,
Sperlonga 028000
Museo Archeologico Nazionale,
Spoleto 028005
Museo Archeologico Nazionale,
Taranto 028055
Museo Archeologico Nazionale,
Venezia 028300
Museo Archeologico Nazionale,
Viterbo 028406
Museo Archeologico Nazionale
d'Abruzzo, Campi 025725
Museo Archeologico Nazionale del
Melfese, Melfi 026757
Museo Archeologico Nazionale della
Sibaritide, Cassano allo Jonio . 025804
Museo Archeologico Nazionale
dell'Abruzzo, Sede la Civitella,
Chieti 025966
Museo Archeologico Nazionale
dell'Abruzzo, Sede Villa Frigerij,
Chieti 025967
Museo Archeologico Nazionale delle
Marche, Ancona 025319
Museo Archeologico Nazionale
dell'Umbria, Perugia 027258
Museo Archeologico Nazionale di
Cagliari, Cagliari 025673
Museo Archeologico Nazionale di
Cosa, Orbetello 027096
Museo Archeologico Nazionale di Locri
Epizefiri, Locri 026612
Museo Archeologico Nazionale di Luni,
Ortonovo 027115
Museo Archeologico Nazionale di
Metaponto, Bernalda 025499
Museo Archeologico Nazionale di
Palestrina, Palestrina 027192
Museo Archeologico Nazionale di
Venosa, Venosa 028329
Museo Archeologico Nazionale Etrusco,
Cerveteri 025914
Museo Archeologico Nazionale Etrusco,
Tuscania 028224
Museo Archeologico Oliveriano,
Pesaro 027265
Museo Archeologico per la Preistoria e
Protostoria del Tigullio, Chiavari 025953
Museo Archeologico Provinciale,
Potenza 027430
Museo Archeologico Provinciale,
Salerno 027708
Museo Archeologico Provinciale F.
Ribezzo, Brindisi 025640
Museo Archeologico Provinciale,
Palazzo Ateneo, Bari 025454
Museo Archeologico Regionale,
Agrigento 025252
Museo Archeologico Regionale,
Aosta 025339
Museo Archeologico Regionale,
Gela 026382
Museo Archeologico Regionale A.
Salinas, Palermo 027177
Museo Archeologico Regionale Paolo
Orsi, Siracusa 027968
Museo Archeologico Remo Fumagalli,
Broletto di Novara 025645
Museo Archeologico San Giovanni
in Compito, Savignano sul
Rubicone 027883
Museo Archeologico Sentinate,
Sassoferrato 027875
Museo Archeologico Statale,
Arcevia 025351
Museo Archeologico Statale, Ascoli
Piceno 025386
Museo Archeologico Statale,
Cingoli 025986
Museo Archeologico Statale,
Teano 028061
Museo Archeologico Statale,
Urbisaglia 028243
Museo Archeologico Statale V.

Capialbi, Vibo Valentia 028361
Museo Archeologico Statale V. Laviola,
Amendolara 025314
Museo Archeologico Tifernum
Mataurense, Sant'Angelo in
Vado 027832
Museo Archeologico Villa Abbas,
Sardara 027854
Museo Archeologico, Centro
Documentazione e Studi sui Castelli
e le Fortificazioni in Calabria, Santa
Severina 027823
Museo Archeologico, Galleria d'Arte
Moderne e Contemporanea, San
Gimignano 027750
Museo Archeologico, Gruppo
Archeologico Cavriana, Cavriana 025886
Museo Archeologico, Palazzi La
Cinquantina, Cecina e Marina . 025887
Museo Archeologico, Palazzo Pretorio,
Colle di Val d'Elsa 026031
Museo Archeologico, Villa Soranzo,
Varallo Pombia 028271
Museo Archiconfradía de la Sangre,
Museo Coloraos, Murcia 039794
Museo-Archivio di Fotografia Storica,
Roma 027576
Museo-Archivo de Falset y Comarca,
Falset 039394
Museo Archivo Rubén Darío,
León 033384
Museo Arcivescovile, Ravenna . 027483
Museo Argentino de Caza Mayor,
Buenos Aires 000167
Museo Argentino de Ciencias
Naturales Bernardino Rivadavia,
Buenos Aires 000168
Museo Argentino de Motos Antiguas,
Mendoza 000451
Museo Argentino del Títere, Buenos
Aires 000169
Museo Armando Reverón,
Caracas 054715
Museo Armando Reverón,
Macuto 054761
Museo Armería 9 de Abril, La
Habana 009150
Museo Arqueología Regional, La
Romana 010376
Museo Arqueológico, Gandía .. 039425
Museo Arqueològic, Llíria 039587
Museo Arqueològic Municipal,
Vallada 040331
Museo Arqueológica de Mongua,
Mongua 008728
Museo Arqueológical Rafael Larco
Herrera, Lima 034323
Museo Arqueológico, Aguilares . 000117
Museo Arqueológico, Aranzazu . 008569
Museo Arqueológico, Cacabelos 039182
Museo Arqueológico, La Carolina 008569
Museo Arqueológico, Cartago .. 008660
Museo Arqueológico, Curumaní . 008671
Museo Arqueológico, Gata de
Gorgos 039429
Museo Arqueológico,
Gualeguaychú 000382
Museo Arqueológico, Inzá 008700
Museo Arqueológico, Manizales 008706
Museo Arqueológico, Novelda . 039821
Museo Arqueológico, Oiba 008739
Museo Arqueológico, Paéz 008740
Museo Arqueológico, Pasca ... 008746
Museo Arqueológico, Puerto
Berrío 008770
Museo Arqueológico, Puerto de la
Cruz 039979
Museo Arqueológico, Sagunto . 040035
Museo Arqueológico, Salamina . 008781
Museo Arqueológico, Sogamoso 008799
Museo Arqueológico, Tafí del
Valle 000638
Museo Arqueológico, La Tebaida 008806
Museo Arqueológico, Trinidad .. 004252

Museo Arqueológico Adàn Quiroga,
San Fernando del Valle de
Catamarca 000571
Museo Arqueológico Alejandro Pezzia
Aseretto, La Angostura 034297
Museo Arqueológico Alto Maestrazgo,
Benasal 039127
Museo Arqueológico Atahualpa,
Ibarra 010455
Museo Arqueológico Calima,
Calima 008655
Museo Arqueológico Carlos Aiiercado,
Esmeraldas 010439
Museo Arqueológico Casa de Madame
Augustín, Santa Marta 008785
Museo Arqueologico Casa del
Marqués de San Jorge, Bogotá 008592
Museo Arqueológico Chiametlán,
Rosario 031417
Museo Arqueológico Chimica,
Istituto de Cultura de El Copey, El
Copey 008677
Museo Arqueológico Comarcal, Palacio
de Rubalcaba, Orihuela 039844
Museo Arqueológico Comunal,
Carquín 034240
Museo Arqueológico Cuiculco,
México 031161
Museo Arqueológico de Ancash,
Huaraz 034280
Museo Arqueológico de Apaxco,
Apaxco 030829
Museo Arqueológico de Asturias,
Oviedo 039861
Museo Arqueológico de Benalmádena,
Benalmádena 039126
Museo Arqueológico de Cajamarca y
Museo Médico, Cajamarca . . . 034228
Museo Arqueológico de Campeche,
Campeche 030849
Museo Arqueológico de Cancún,
Cancún 030857
Museo Arqueológico de Chasicó,
Chasicó 000301
Museo Arqueológico de Chusís,
Sechura 034399
Museo Arqueológico de Ciudad
Guzmán, Guzmán 031035
Museo Arqueológico de Comayagua,
Comayagua 023396
Museo Arqueológico de Comitán,
Comitán de Domínguez 030902
Museo Arqueológico de Esteli, San
Juan de Limay 033404
Museo Arqueológico de Guacarí,
Guacarí 008687
Museo Arqueológico de La Caleta, La
Caleta 010369
Museo Arqueológico de la Ciudad,
Emiliano Zapata 030961
Museo Arqueológico de la Costa
Grande, Zihuatanejo 031706
Museo Arqueológico de La Serena, La
Serena 007525
Museo Arqueológico de la Universidad
Nacional del Altiplano, Pucara . 034390
Museo Arqueológico de Lagos de
Moreno, Lagos de Moreno . . . 031101
Museo Arqueológico de Lumbrales,
Lumbrales 039603
Museo Arqueológico de Martins
Sarmento, Guimarães 035728
Museo Arqueológico de Mazatlán,
Mazatlán 031130
Museo Arqueológico de Muzo,
Muzo 008732
Museo Arqueológico de Nextlalpan,
Nextlalpan 031324
Museo Arqueológico de Puerto
Esperanza, Puerto Esperanza . 000506
Museo Arqueológico de Saipina, Santa
Cruz 000452
Museo Arqueológico de Santa Trega, A
Guardia 039481

Museo Arqueológico de Solentiname,
Solentiname 033407
Museo Arqueológico de Tepeapulco,
Tepeapulco 031542
Museo Arqueológico de Tilcara,
Tilcara 000647
Museo Arqueológico de Tizatlán,
Tizatlán 031572
Museo Arqueológico de Tonalá, Tonalá,
Chiapas 031605
Museo Arqueológico de Tula Jorge R.
Acosta, Tula de Allende 031618
Museo Arqueológico de Úbeda, Casa
Mudéjar, Úbeda 040282
Museo Arqueológico de Venustiano
Carranza, Venustiano Carranza 031644
Museo Arqueológico de Xochimilco,
México 031162
Museo Arqueológico d'Eivissa i
Formentera, Eivissa 039360
Museo Arqueológico del Banco Central
de Bahia, Bahia de Carácluez . 010416
Museo Arqueológico del Centro de
Estudios Histórico Sociales Julio
Espejo Nuñez, Jauja 034295
Museo Arqueológico del Cerro de la
Estrella, México 031163
Museo Arqueológico del Cesar,
Valledupar 008818
Museo Arqueológico del Colegio Juan
XXIII, Lima 034324
Museo Arqueológico del Huila,
Neiva 008734
Museo Arqueológico del Instituto
de Arqueología y Etnología,
Mendoza 000452
Museo Arqueológico del Sinú,
Tierralta 008808
Museo Arqueológico del Soconusco,
Tapachula 031511
Museo Arqueológico Dr. Román Piña
Chan, Tenango del Valle 031532
Museo Arqueológico e Histórico de
Andahuaylas, Andahuaylas . . . 034210
Museo Arqueológico Etnográfico, San
Pedro del Pinatar 040071
Museo Arqueológico-Etnológico
Gratiniano Baches, Pilar de la
Horadada 039943
Museo Arqueológico Etnológico
Municipal, Guardamar de
Segura 039480
Museo Arqueológico Frederico Galvez
Durand, Huancayo 034275
Museo Arqueológico Gregorio Aguilar
Barea, Juigalpa 033380
Museo Arqueológico Guamuhaya,
Trinidad 009373
Museo Arqueológico Horacio Urteaga,
Universidad Nacional de Cajamarca,
Cajamarca 034229
Museo Arqueológico José Casinelli,
Santa Inés 034398
Museo Arqueológico José María
Morante de la Universidad Nacional
de San Augustín de Arequipa,
Arequipa 034212
Museo Arqueológico Juan Sepúlveda y
Museo Eduardo Ruiz, Uruapán 031630
Museo Arqueológico Julio César
Cubillos, Cali 008645
Museo Arqueológico Julio César
Cubillos, San Agustín 008782
Museo Arqueológico La Bagatela, Villa
del Rosario 008824
Museo Arqueológico La Encomienda,
Fundación Emilio Pérez Piñero,
Calasparra 039197
Museo Arqueológico La Merced,
Cali 008646
Museo Arqueológico Luis G. Urbina,
México 031164
Museo Arqueológico Marco Fidel
Micolta, Guamo 008692

Museo Arqueológico Mesoamericano,
Tijuana 031565
Museo Arqueológico Municipal,
Alcúdia (Baleares) 038933
Museo Arqueológico Municipal,
Burriana 039176
Museo Arqueológico Municipal,
Cartagena 039237
Museo Arqueológico Municipal,
Cehegín 039259
Museo Arqueológico Municipal,
Crevillente 039327
Museo Arqueológico Municipal,
Elda 039372
Museo Arqueológico Municipal, Jerez
de la Frontera 039528
Museo Arqueológico Municipal,
Lorca 039595
Museo Arqueológico Municipal,
Marchena 039724
Museo Arqueológico Municipal,
Montoro 039784
Museo Arqueológico Municipal, San
Fulgencio 040059
Museo Arqueológico Municipal
Alejandro Ramos Folqués, Palacio de
Altamira, Elche 039365
Museo Arqueológico Municipal Antonio
Ballester, Callosa de Segura . . 039204
Museo Arqueológico Municipal
Camilo Visedo Molto, Museu d'Alcoi,
Alcoy 038928
Museo Arqueológico Municipal
Cayetano Mergelina, Yecla . . . 040435
Museo Arqueológico Municipal de
Alhambra, Alhambra 038943
Museo Arqueológico Municipal de
Obulco, Porcuna 039962
Museo Arqueológico Municipal José
María Soler, Edificio de Ayuntamiento,
Villena 040416
Museo Arqueológico Municipal y
Centro de Interpretación del Mar,
Águilas 038894
Museo Arqueológico Municipal y
Museo Municipal Miguel Justino
Ramírez, Piura 034385
Museo Arqueológico Municipal, Casa
de la Cultura, Santaella 040111
Museo Arqueológico Municipal, Iglesia
de la Soledad, Caravaca de la
Cruz 039227
Museo Arqueológico Nacional,
Madrid 039629
Museo Arqueológico Nacional Brüning,
Lambayeque 034301
Museo Arqueológico Padre Martín
Recio, Estepa 039388
Museo Arqueológico-Paleontológico
Municipal de Rojales, Rojales . 040013
Museo Arqueológico Pío P. Díaz,
Cachi 000285
Museo Arqueológico Provincial,
Alicante 038944
Museo Arqueológico Provincial,
Badajoz 039034
Museo Arqueológico Provincial,
Huesca 039499
Museo Arqueológico Provincial,
Ourense 039857
Museo Arqueológico Provincial, San
Salvador de Jujuy 000607
Museo Arqueológico Provincial,
Sevilla 040168
Museo Arqueológico Provincial
Andalgalá, Andalgalá 000123
Museo Arqueológico Provincial Condor
Huasi (interim), Belén 000146
Museo Arqueológico R.P. Gustavo
Le Paige S.J., San Pedro de
Atacama 007534
Museo Arqueológico Regional, Alcalá
de Henares 038913
Museo Arqueológico Regional Anibal

Montes, Río Segundo 000538
Museo Arqueológico Regional Guane,
Floridablanca 008682
Museo Arqueológico Regional Inca
Huasi, La Rioja 000416
Museo Arqueológico San Miguel de
Azapa, Arica 007515
Museo Arqueológico-Torre del Agua,
Osuna 039855
Museo Arqueológico y Colonial, Fuerte
Quemado 000374
Museo Arqueológico y Etnográfico,
Betancuria 039132
Museo Arqueológico y Etnológico
Córdoba, Consejería de Cultura Junta
de Andalucia, Córdoba 039305
Museo Arqueológico y Etnológico de
Granada, Granada 039465
Museo Arqueológico y Etnológico de
Lucena, Palacio Castillo del Moral,
Lucena 039599
Museo Arqueológico y Etnológico
Municipal Dámaso Navarro,
Petrer 039942
Museo Arqueológico y Paleontológico
de Guane, Barichara 008575
Museo Arqueológico, Cueva de Siete
Palacio Barrio de San Miguel,
Almuñécar 038965
Museo Arqueológico, Fundación Cueva
de Nerja, Nerja 039817
Museo Arqueológico, Museo de la
Naturaleza y el Hombre-Antiguo
Hospital Civil, Santa Cruz de
Tenerife 040098
Museo Arqueológico, Universidad
Estatal de Cuenca, Cuenca . . . 010425
Museo Arqueológico, Universidad
Mayor de San Simón,
Cochabamba 004205
Museo Arqueolóogico y Etnográfico,
Ibarra 010456
Museo Arqueolóxico e Histórico, A
Coruña 039318
Museo Arquidiocesano, Santiago de
Cuba 009344
Museo Arquidiocesano de Arte
Religioso, Pamplona 008743
Museo Arquidiocesano de Arte
Religioso de Popayán, Popayán 008762
Museo Arquitectura de Trinidad,
Trinidad 009374
Museo Arrese, Corella 039312
Museo Art Nouveau-Art Déco,
Fundación Manuel Ramos Andrade,
Salamanca 040039
Museo Artequin, Santiago de
Chile 007535
Museo Artes Decorativo, Gibara 009124
Museo Artes Plásticas, Pinar del
Río 009286
Museo Artes y Costumbres Populares
Alto Guadalquivir, Cazorla . . . 039258
Museo Artesanal de Gualaceo,
Gualaceo 010442
Museo Arti e Mestieri di un Tempo,
Cisterna d'Asti 025996
Museo Arti e Tradizioni Popolari del
Gargano Giovanni Tancredi, Monte
Sant'Angelo 026916
Museo Artístico de Roncesvalles,
Roncesvalles 040017
Museo Artistico della Bambola,
Suverato 028045
Museo Artistico Industriale F. Palizzi,
Istituto d'Arte di Napoli, Napoli 026993
Museo Arturo Michelena,
Caracas 054716
Museo Arxiu Capitular, Tortosa . 040268
Museo Astronomico Copernicano,
Roma 027577
Museo Astronomico e Geofisico,
Modena 026852
Museo Atelier Antonio Ortiz Echagüe,

Carro Quemado 000295
Museo Audiovisual, Caracas . . . 054717
Museo Augustín Landívar,
   Cuenca 010426
Museo Augusto Righi, Montese . 026960
Museo Aurelio Castelli, Siena . . 027952
Museo Aurelio Espinosa Polit,
   Cotocollao 010423
Museo Aurelio Espinoza Pólit,
   Quito 010472
Museo Aurelio Marena, Bitonto . 025521
Museo Austral de Pintura Primitiva
   Moderna Naif, Esquel 000369
Museo Auto d'Ecopa, San Marino 038098
Museo Azuayo del Folklore,
   Cuenca 010427
Museo B. Gigli, Recanati 027489
Museo Badia di San Gemolo,
   Valganna 028256
Museo Bagatti Valsecchi, Milano 026798
Museo Bahía de los Angeles,
   Ensenada 030965
Museo Balaa Xte Guech Gulal,
   Teotitlán del Valle 031539
Museo Balear Ciencies Naturals,
   Sóller 040191
Museo Balseros del Mar del Sur,
   Salango 010502
Museo Baluarte de Santiago,
   Veracruz 031645
Museo Banco Central de Reserva del
   Perú, Lima 034325
Museo Banco del Pacifico,
   Guayaquil 010445
Museo Bandini, Fiesole 026225
Museo Bardini, Firenze 026268
Museo Barjola, Gijón 039435
Museo Baroffio e del Santuario
   del Sacro Monte sopra Varese,
   Varese 028277
Museo Barovier e Toso, Venezia 028301
Museo Barracco, Roma 027578
Museo Beata Camilla Battista Varano,
   Camerino 025710
Museo Beata P. Morosini, Albino 025272
Museo Beato Angelico, Vicchio . 028364
Museo Beato Angelo, Acri 025241
Museo Belisario Porras, Las
   Tablas 034184
Museo Bellapart, Santo Domingo 010386
Museo Bello Piñeio, Centro Cultural
   Antigo Hospital da Caridade,
   Ferrol 039397
Museo Benalauría, Benalauría . . 039125
Museo Benedettino Nonantolano
   e Diocesano di Arte Sacra,
   Nonantola 027049
Museo Benigno Malo, Cuenca . . 010428
Museo Benito Hortiz, Sancti-
   Spíritus 009322
Museo Benito Juárez, José
   Azueta 031088
Museo Benito Juárez, Casa del
   Benemérito de las Américas, La
   Habana 009151
Museo Berenziano, Seminario
   Vescovile, Cremona 026099
Museo Bernabé de las Casas,
   Mina 031276
Museo Bernardino Caballero,
   Asunción 034195
Museo Bernardo de Muro, Tempio
   Pausania 028066
Museo Bernardo Martínez Villegas,
   Bácum 030843
Museo Bernardo Samper Sordo,
   Bogotá 008593
Museo Berrocal, Vilanueva de
   Algaidas 040393
Museo Bersano delle Contadinerie
   e Stampe antiche del Vino, Nizza
   Monferrato 027038
Museo Biblioteca Comunale,
   Etroubles 026164

Museo Biblioteca Municipal Cesar
   Fernández Navarro, Santa Fé . 000613
Museo Biblioteca Nueva Cádiz y Casa
   Natal de Juan Bautista Arismendi,
   Nueva Cádiz 054770
Museo Biblioteca Palacio Postal,
   México 031165
Museo Biblioteca Palafoxiana,
   Puebla 031376
Museo Biblioteca Pape,
   Monclova 031282
Museo Biblioteca Rosauro Rosa
   Acosta, Pampatar 054771
Museo Bicknell, Bordighera . . . 025593
Museo Bicocchi, Pomarance . . . 027389
Museo Bodoniano, Parma 027212
Museo Bodoniano. Bolletino,
   Parma 138846
Museo Bolivar, Ciudad de
   Panamá 034175
Museo Bolívar Periodista, Piura . 034386
Museo Bolívariano, Caracas . . . 054718
Museo Bolivariano, Pativilca . . . 034380
Museo Bolivariano de Arte
   Contemporáneo, Fundación Museo
   Bolivariano, Santa Marta 008786
Museo Bologna-Buonsignori,
   Siena 027953
Museo Boncompagni Ludovisi,
   Roma 027579
Museo Bonsor, Mairena del Alcor 039696
Museo Borromeo, Palazzo Borromeo,
   Isola Bella 026520
Museo Botánico, Córdoba 000320
Museo Botánico, Montevideo . . 046256
Museo Botánico, Tuxtla Gutiérrez 031621
Museo Botanico Cesare Bicchi,
   Lucca 026638
Museo Botanico della Lessinia,
   Fumane 026356
Museo Bottacin, Musei Civici di
   Padova, Padova 027146
Museo Bottega della Tarsia Lignea,
   Sorrento 027993
Museo Brandolini e Giol, Oderzo 027077
Museo Brembano di Scienze Naturali,
   Villa Speranza, San Pellegrino
   Terme 027791
Museo C. Nivola, Orani 027094
Museo Ca' la Ghironda, Bologna 025536
Museo Cabrera, Ica 034286
Museo Cabriniano, Sant'Angelo
   Lodigiano 027833
Museo Cacique Balata, La
   Cumbre 000400
Museo Calabozo de Hidalgo,
   Chihuahua 030875
Museo Calabrese di Etnografia e
   Folklore Raffaele Corso, Palmi . 027196
Museo Camillo Leone, Vercelli . 028335
Museo Camilo Egas del Banco Central,
   Quito 010473
Museo Camilo Ortega de la Ciudad
   Masaya, Masaya 033393
Museo Camón Aznar de Ibercaja,
   Zaragoza 040445
Museo Campagna G. Bruno,
   Campagna 025722
Museo Campamento Irritila, San
   Pedro 031475
Museo Camuno e Biblioteca Civica,
   Breno 025623
Museo Canario, Las Palmas de Gran
   Canaria 039901
Museo Cándido Bido, Bonao . . 010368
Museo Canonicale, Verona 028346
Museo Cantonale d'Arte, Lugano 041695
Museo Cantonale di Storia Naturale,
   Lugano 041696
Museo Capilla Alfonsina, México 031166
Museo Capitán Juan de Zevallos, Valle
   Hermoso 000664
Museo Capitolare, Atri 025414
Museo Capitolare, Cagliari 025674

Museo Capitolare, Ortona 027112
Museo Capitolare di Arte Sacra,
   Gravina di Puglia 026468
Museo Capitolare, Cattedrale di San
   Rufino, Assisi 025398
Museo Cappella San Severo,
   Napoli 026994
Museo Cappuccini Emiliani, Reggio
   Emilia 027503
Museo Capuaniano, Mineo 026841
Museo Carducci – Biblioteca e Casa,
   Bologna 025537
Museo Carlo Venturini, Massa
   Lombarda 026729
Museo Carlo Zauli, Faenza 026171
Museo Carlos Emilio Grijalva,
   Tulcán 010505
Museo Carlos Torres Barraza, Valentín
   Gómez Farías 031632
Museo Carnico delle Arti Popolari
   Michele Gortani, Tolmezzo . . . 028106
Museo Carozzi, Seriate 027924
Museo Casa, Valverde del
   Camino 040358
Museo Casa Alejandro de Humboldt,
   La Habana 009152
Museo Casa Blanca, San Juan . 036158
Museo Casa Busoni, Empoli . . . 026149
Museo Casa Carlos Gardel, Buenos
   Aires 000170
Museo Casa Cautiño, Guayama . 036130
Museo Casa Clautana E. Borsatti,
   Claut 026022
Museo Casa Colonial, Pamplona 008744
Museo Casa Cultural "Alfonso López
   Pumarejo", Honda 008695
Museo Casa de Africa, La
   Habana 009153
Museo Casa de Agustín Arrieta,
   Puebla 031377
Museo Casa de Asia, La Habana 009154
Museo Casa de Benalcazar,
   Quito 010474
Museo Casa de Bernardo A. Houssay,
   Buenos Aires 000171
Museo Casa de Bolívar, Academia
   de Historia de Santander,
   Bucaramanga 008637
Museo Casa de Carmen Montilla, La
   Habana 009155
Museo Casa de Carranza,
   Cuatrocienegas 030921
Museo Casa de Carranza,
   México 031167
Museo Casa de Colón, Las Palmas de
   Gran Canaria 039902
Museo Casa de Colón, San Sebastián
   de La Gomera 040075
Museo Casa de Cultura de Ixtlán,
   Ixtlán del Río 031073
Museo Casa de Dulcinea, Museo del
   Santa Cruz, El Toboso 040229
Museo Casa de Hidalgo la Francia
   Chiquita, San Felipe 031440
Museo Casa de Isla Negra, San
   Felipe 007533
Museo Casa de la Bola, México 031168
Museo Casa de la Cultura,
   Montemorelos 031286
Museo Casa de la Cultura "Luis
   Camacho Rueda", Socorro . . . 008798
Museo Casa de la Cultura de
   Barichara, Barichara 008576
Museo Casa de la Cultura de
   Cuacutecuta, Instituto de Cultura de
   Norte Santander, Cuacutecuta . 008668
Museo Casa de la Cultura de
   Pensilvania, Pensilvania 008756
Museo Casa de la Cultura Simón
   Patiño, Universidad Técnica de Oruro,
   Oruro 004229
Museo Casa de la Libertad,
   Sucre 004243
Museo Casa de la Masacre de Ponce,

Ponce 036138
Museo Casa de la Moneda,
   Mariquita 008711
Museo Casa de la Obrapía, La
   Habana 009156
Museo Casa de la Orfebrería, La
   Habana 009157
Museo Casa de la Plomada,
   Vallecillo 031638
Museo Casa de la Poesía, La
   Habana 009158
Museo Casa de la Tecnología,
   Monterrey 031287
Museo Casa de la Troya, Santiago de
   Compostela 040121
Museo Casa de León Trotsky,
   México 031169
Museo Casa de les Rocas,
   Valencia 040306
Museo Casa de los Mártires, La
   Habana 009159
Museo Casa de Morelos, Ecatepec de
   Morelos 030955
Museo Casa de Morelos, Morelia 031301
Museo Casa de Policarpa Salavarrieta,
   Guaduas 008597
Museo Casa de Puerto Rico y Sala del
   Tabaco, La Habana 009160
Museo Casa de Ramón López Velarde,
   Jerez de García Salinas 031083
Museo Casa de Ricardo Rojas, Buenos
   Aires 000172
Museo Casa de Sais Montes de Oca
   Hermanos, Pinar del Río 009287
Museo Casa de Sucre, Quito . . 010475
Museo Casa de Yrurtia, Buenos
   Aires 000173
Museo Casa de Zela, Tacna . . . 034403
Museo Casa del Agua, Tehuacán 031527
Museo Casa del Alfeñique,
   Puebla 031378
Museo Casa del Deán, Puebla . 031379
Museo Casa del Diego, Celaya . 030862
Museo Casa del Dr. Mora,
   Comonfort 030905
Museo Casa del Duca, Venezia . 028302
Museo Casa del Fundador Ponzalo
   Suárez Rendón, Tunja 008813
Museo Casa del General Alvaro
   Obregón, Huatabampo 031052
Museo Casa del Libertador Simón
   Bolívar, La Habana 009161
Museo Casa del Maestro José
   Márquez Figueroa, Puebla 031380
Museo Casa del Marques de Aguayo
   y Museo Municipal, Mazapil
   Zacatecas 031129
Museo Casa del Padre, Orselina 041832
Museo Casa del Teatro, Buenos
   Aires 000174
Museo Casa di Dante, Firenze . 026269
Museo Casa Diego Rivera,
   Guanajuato 031022
Museo Casa Dorada-Maison d'Or,
   Universidad Autonoma Juan Misael
   Saracho, Tarija 004249
Museo Casa Dr. Belisario Domínguez,
   Comitán de Domínguez 030903
Museo Casa Estudio Diego Rivera y
   Frida Kahlo, México 031170
Museo Casa Francisco Villa, Hidalgo
   del Parral 031045
Museo Casa-Fuerte Juan de Ponce de
   León, Higuëy 010372
Museo Casa Fundación Oswaldo
   Guayasamín, La Habana 009162
Museo Casa Galimberti, Como . 026115
Museo Casa Generalísimo Máximo
   Gómez, Montecristi 010374
Museo Casa Hacienda San Jacinto,
   San Jacinto 033403
Museo Casa Ibáñez, Olula del
   Río 039833
Museo Casa José Lezama Lima, La

Habana 009163	
Museo Casa Mata, Matamoros . 031127	
Museo Casa Maya, Cancún 030858	
Museo Casa Mosquera, Popayán 008763	
Museo Casa Murillo, Sevilla . . 040169	
Museo Casa Natal de Antonio Maceo,	
Santiago de Cuba 009345	
Museo Casa Natal de Calixto García,	
Holguín 009217	
Museo Casa Natal de Camilo	
Cienfuegos, La Habana 009164	
Museo Casa Natal de Carlos Manuel	
de Céspedes, Bayamo 009072	
Museo Casa Natal de Celia Sanchez	
Manduley, Media Luna 009260	
Museo Casa Natal de Cervantes,	
Alcalá de Henares 038914	
Museo Casa Natal de Frank País,	
Santiago de Cuba 009346	
Museo Casa Natal de Hermanos	
Ameijeiras, Pueblo Viejo 009297	
Museo Casa Natal de Ignacio	
Agramonte, Camagüey 009089	
Museo Casa Natal de Isabel la	
Católica, Palacio de Juan II, Madrigal	
de las Altas Torres 039692	
Museo Casa Natal de José María	
Heredia, Santiago de Cuba . . . 009347	
Museo Casa Natal de José Martí, La	
Habana 009165	
Museo Casa Natal de Jovellanos,	
Gijón 039436	
Museo Casa Natal de Morelos,	
Morelia 031302	
Museo Casa Natal de Pedro Martínez	
Brito, Ciego de Avila 009100	
Museo Casa Natal de Serafín Sánchez,	
Sancti-Spíritus 009323	
Museo Casa Natal General Santander,	
Villa del Rosario 008825	
Museo Casa Natal Juan Manuel	
Márquez, La Habana 009166	
Museo Casa Natal Rubén Darío,	
Darío 033375	
Museo Casa Natal y Museo de Niceto	
Alcalá-Zamora y Torres, Priego de	
Córdoba 039974	
Museo Casa Natale di Giotto,	
Vicchio 028365	
Museo Casa Natale di San Pio X,	
Riese Pio X 027515	
Museo Casa Natale di Tiziano Vecellio,	
Pieve di Cadore 027320	
Museo Casa Natale G. D'Annunzio,	
Pescara 027273	
Museo Casa Natale Giacomo Puccini,	
Lucca 026639	
Museo Casa Negret, Popayán . . 008764	
Museo Casa Paoli, Ponce 036139	
Museo Casa Pascoli, Barga . . . 025451	
Museo Casa Rafael Núñez,	
Cartagena 008656	
Museo Casa Roig, Humacao . . . 036133	
Museo Casa Rosada Abelardo Díaz	
Alfaro, Caguas 036122	
Museo Casa Rossini, Pesaro . . 027266	
Museo Casa Ruth Lechuga,	
México 031171	
Museo Casa Studio F. Melani,	
Pistoia 027361	
Museo Casa Taller Celia Chevalier,	
Buenos Aires 000175	
Museo Casa Walser, Macugnaga 026665	
Museo Casabianca, Malo 026683	
Museo Casita Blanca de Agustín Lara,	
Boca del Río 030847	
Museo Castellano-Manchego del	
Juguete, Mohedas de la Jara . 039758	
Museo Castello Masegra, Sondrio 027985	
Museo Castillo de la Fortaleza, Santa	
Lucía 040105	
Museo Castillo del Macho,	
Peñíscola 039936	
Museo Castillo Serrallés, Ponce 036140	

Museo Catedral, Palma de	
Mallorca 039888	
Museo Catedral de La Plata, La	
Plata 000408	
Museo Catedralicio, Almería . . . 038961	
Museo Catedralicio, Badajoz . . 039035	
Museo Catedralicio, Baeza . . . 039041	
Museo Catedralicio, Burgo de	
Osma 039165	
Museo Catedralicio, Burgos . . . 039168	
Museo Catedralicio, Cádiz 039189	
Museo Catedralicio, Ciudad Real 039287	
Museo Catedralicio, Cuenca . . 039331	
Museo Catedralicio, Guadix . . . 039479	
Museo Catedralicio, Jaén 039522	
Museo Catedralicio, Málaga . . . 039700	
Museo Catedralicio, Ourense . . 039858	
Museo Catedralicio, Santo Domingo	
de la Calzada 040144	
Museo Catedralicio, Segovia . . 040155	
Museo Catedralicio, Sigüenza . . 040182	
Museo Catedralicio, Zamora . . . 040437	
Museo Catedralicio de La Seo de	
Zaragoza, Zaragoza 040446	
Museo Catedralicio de Tui, Tui . 040279	
Museo Catedralicio Diocesano, Ciudad	
Rodrigo 039293	
Museo Catedralicio-Diocesano,	
León 039556	
Museo Catedralicio Diocesano,	
Pamplona 039909	
Museo Catedralicio-Diocesano,	
Valencia 040307	
Museo Catedralicio, Iglesia Catedral,	
Palencia 039872	
Museo Catedralico y Giralda,	
Sevilla 040170	
Museo Católica de Santa María,	
Cercado 034243	
Museo Católico Peninsular,	
Mérida 031133	
Museo Cavour, Santena 027844	
Museo Cemento Rezola, Donostia-San	
Sebastián 039352	
Museo Cemí, Jayuya 036134	
Museo Centrale del Risorgimento,	
Roma 027580	
Museo Centrale dell'Istituto Storico	
della Resistenza, Imperia 026509	
Museo Centrale Montemartini, Musei	
Capitolini, Roma 027581	
Museo Centro Cultural Pachamama,	
Amaicha del Valle 000122	
Museo Centro de Cultura Mayo de el	
Júpare Blas Mazo, Huatabampo 031053	
Museo Centro de Interpretacion de	
Monesterio, Monesterio 039765	
Museo Centro de la Imagen,	
México 031172	
Museo Centro Studi A.G. Barrili,	
Carcare 025771	
Museo Centro Taurino Potosino, San	
Luis Potosí 031454	
Museo Ceramoteca, Salamanca 031420	
Museo Cerezo Moreno,	
Villatorres 040414	
Museo Cerlogne, Saint Nicolas . 027701	
Museo Cerralbo, Madrid 039630	
Museo Cerro de la Campana, Santiago	
Suchilquitongo 031495	
Museo Cerro de los Huizaches, San	
Pablo Huixtepec 031474	
Museo Cervantino, El Toboso . . 040230	
Museo Cervi, Gattatico 026375	
Museo Chairama, Santa Marta . 008787	
Museo Chiaramonti e Braccio	
Nuovo, Musei Vaticani, Città del	
Vaticano 054683	
Museo Chihuahuense de Arte	
Contemporáneo, Chihuahua . . 030876	
Museo Chileno de Arte Precolombino,	
Santiago de Chile 007536	
Museo Chillida Leku, Hernani . . 039488	
Museo Chorotega-Nicarao llamado	

"Enrique Berio Mántica Deshon",	
Chinandega 033373	
Museo Chorro de Maita,	
Yaguajay 009392	
Museo Christiano, Cividale del	
Friuli 026012	
Museo Ciénaga de San Pedro,	
Madera 031119	
Museo Ciencias Naturales, Nueva	
Gerona 009271	
Museo Ciencias Naturales Carlos de la	
Torre, Sancti-Spíritus 009324	
Museo Ciudad de Baeza, Baeza 039042	
Museo Ciudades del Mundo, La	
Habana 009167	
Museo Civico, Aci Castello . . . 025232	
Museo Civico, Agrigento 025253	
Museo Civico, Airola 025258	
Museo Civico, Alatri 025263	
Museo Civico, Amalfi 025306	
Museo Civico, Apice 025342	
Museo Civico, Argenta 025363	
Museo Civico, Ariano Irpino . . . 025367	
Museo Civico, Asolo 025394	
Museo Civico, Atina 025411	
Museo Civico, Avola 025429	
Museo Civico, Baranello 025445	
Museo Civico, Bardonecchia . . 025449	
Museo Civico, Barga 025452	
Museo Civico, Bassano del	
Grappa 025470	
Museo Civico, Belluno 025481	
Museo Civico, Bisacquino 025515	
Museo Civico, Bolzano 025582	
Museo Civico, Borgo Velino . . . 025601	
Museo Civico, Bormio 025605	
Museo Civico, Cabras 025668	
Museo Civico, Campagnano di	
Roma 025723	
Museo Civico, Campochiaro . . . 025732	
Museo Civico, Caprino Veronese 025751	
Museo Civico, Casale Monferrato 025789	
Museo Civico, Casamicciola	
Terme 025795	
Museo Civico, Castel Bolognese 025810	
Museo Civico, Castelleone 025833	
Museo Civico, Castelnovo	
Bariano 025839	
Museo Civico, Castelnuovo	
Scrivia 025845	
Museo Civico, Castelraimondo . 025846	
Museo Civico, Castroreale 025861	
Museo Civico, Centuripe 025898	
Museo Civico, Champdepraz . . 025940	
Museo Civico, Cittadella 026006	
Museo Civico, Collelongo 026036	
Museo Civico, Conversano 026061	
Museo Civico, Cori 026071	
Museo Civico, Crema 026096	
Museo Civico, Crotone 026110	
Museo Civico, Cuneo 026116	
Museo Civico, Erba 026155	
Museo Civico, Feltre 026193	
Museo Civico, Ferentino 026196	
Museo Civico, Foggia 026308	
Museo Civico, Fondi 026318	
Museo Civico, Fornovo San	
Giovanni 026336	
Museo Civico, Fratta Polesine . 026349	
Museo Civico, Fucecchio 026355	
Museo Civico, Gibellina 026439	
Museo Civico, Guardiagrele . . . 026487	
Museo Civico, Iesi 026497	
Museo Civico, Larciano 026565	
Museo Civico, Larino 026567	
Museo Civico, Lodi 026614	
Museo Civico, Lucignano 026646	
Museo Civico, Luino 026653	
Museo Civico, Maddaloni 026669	
Museo Civico, Magliano Sabina . 026672	
Museo Civico, Manerbio 026692	
Museo Civico, Mattinata 026748	
Museo Civico, Mazara del Vallo . 026749	
Museo Civico, Medicina 026753	

Museo Civico, Mirandola 026845	
Museo Civico, Monsampolo del	
Tronto 026896	
Museo Civico, Montelparo 026946	
Museo Civico, Noto 027054	
Museo Civico, Osimo 027121	
Museo Civico, Patrica 027224	
Museo Civico, Pegognaga 027241	
Museo Civico, Pescia 027282	
Museo Civico, Piedimonte	
Matese 027312	
Museo Civico, Pieve di Cento . . 027325	
Museo Civico, Pistoia 027362	
Museo Civico, Pizzighettone . . 027373	
Museo Civico, Pontecorvo 027402	
Museo Civico, Putignano 027461	
Museo Civico, Ragogna 027470	
Museo Civico, Rende 027509	
Museo Civico, Rieti 027516	
Museo Civico, Riva del Garda . . 027535	
Museo Civico, Rovereto 027683	
Museo Civico, San Bonifacio . . 027730	
Museo Civico, San Cesario di	
Lecce 027736	
Museo Civico, San Ferdinando di	
Puglia 027747	
Museo Civico, San Gimignano . 027751	
Museo Civico, San Severo 027801	
Museo Civico, Sanremo 027810	
Museo Civico, Sansepolcro . . . 027812	
Museo Civico, Santa Giusta . . . 027815	
Museo Civico, Sesto Calende . . 027937	
Museo Civico, Siena 027954	
Museo Civico, Sulmona 028030	
Museo Civico, Susa 028038	
Museo Civico, Taverna 028060	
Museo Civico, Terrasini 028082	
Museo Civico, Tolfa 028105	
Museo Civico, Trevignano	
Romano 028182	
Museo Civico, Trinitapoli 028213	
Museo Civico, Troia 028217	
Museo Civico, Vasanello 028283	
Museo Civico, Viterbo 028407	
Museo Civico A. Giacomelli, Castel	
San Zeno, Montagnana 026903	
Museo Civico A. Klitsche de la Grange,	
Allumiere 025295	
Museo Civico A. Parazzi, Viadana 028357	
Museo Civico A. Tubino, Masone 026724	
Museo Civico A.E. Baruffaldi, Badia	
Polesine 025431	
Museo Civico Ala Ponzone,	
Cremona 026100	
Museo Civico Albano, Albano	
Laziale 025266	
Museo Civico Alta Val Brembana,	
Valtorta 028268	
Museo Civico Antiquarium,	
Nettuno 027030	
Museo Civico Antiquarium Platina,	
Piadena 027301	
Museo Civico Antonino Olmo e	
Gipsoteca, Savigliano 027880	
Museo Civico Antonio Cordici,	
Erice 026159	
Museo Civico Archeologica e	
Quadreria, Monterubbiano . . . 026958	
Museo Civico Archeologico,	
Angera 025327	
Museo Civico Archeologico,	
Anzio 025336	
Museo Civico Archeologico,	
Arona 025373	
Museo Civico Archeologico, Barbarano	
Romano 025446	
Museo Civico Archeologico, Bene	
Vagienna 025485	
Museo Civico Archeologico,	
Bergamo 025493	
Museo Civico Archeologico,	
Bologna 025538	
Museo Civico Archeologico,	
Camerino 025711	

Museo Civico Archeologico, Canosa di Puglia 025741
Museo Civico Archeologico, Castelfranco Emilia 025822
Museo Civico Archeologico, Castelleone di Suasa 025834
Museo Civico Archeologico, Castro dei Volsci 025860
Museo Civico Archeologico, Castrovillari 025863
Museo Civico Archeologico, Cavaion Veronese 025882
Museo Civico Archeologico, Chiomonte 025972
Museo Civico Archeologico, Ciró Marina 025994
Museo Civico Archeologico, Cologna Veneta 026039
Museo Civico Archeologico, Concordia Sagittaria 026056
Museo Civico Archeologico, Cosenza 026087
Museo Civico Archeologico, Fiesole 026226
Museo Civico Archeologico, Gazzo Veronese 026381
Museo Civico Archeologico, Gottolengo 026456
Museo Civico Archeologico, Grotte di Castro 026480
Museo Civico Archeologico, Latronico 026575
Museo Civico Archeologico, Marciana 026708
Museo Civico Archeologico, Massa Marittima 026734
Museo Civico Archeologico, Mel 026756
Museo Civico Archeologico, Mondragone 026888
Museo Civico Archeologico, Montalto Marche 026907
Museo Civico Archeologico, Nepi 027028
Museo Civico Archeologico, Nicotera 027034
Museo Civico Archeologico, Norma 027051
Museo Civico Archeologico, Padria 027163
Museo Civico Archeologico, Pitigliano 027368
Museo Civico Archeologico, Ramacca 027474
Museo Civico Archeologico, Remedello 027508
Museo Civico Archeologico, Roccagloriosa 027543
Museo Civico Archeologico, Rosignano Marittimo 027675
Museo Civico Archeologico, Rutigliano 027695
Museo Civico Archeologico, Santa Marinella 027822
Museo Civico Archeologico, Sarteano 027863
Museo Civico Archeologico, Tolentino 028098
Museo Civico Archeologico, Treia 028167
Museo Civico Archeologico, Verucchio 028353
Museo Civico Archeologico Arsenio Crespellani, Bazzano 025477
Museo Civico Archeologico C. Cellini, Ripatransone 027528
Museo Civico Archeologico del Castello San Giorgio U. Formentini, La Spezia 026538
Museo Civico Archeologico della Val Tenesi, Manerba del Garda . . 026691
Museo Civico Archeologico di Fregellae, Ceprano 025900
Museo Civico Archeologico e Paleobotanico, Perfugas 027248
Museo Civico Archeologico Eno Bellis, Oderzo 027078

Museo Civico Archeologico Etnologico, Modena 026853
Museo Civico Archeologico F. Saverio Majellaro, Bisceglie 025516
Museo Civico Archeologico Genna Maria, Villanovaforru 028394
Museo Civico Archeologico L. Fantini, Monterenzio 026955
Museo Civico Archeologico M. Petrone, Vieste 028378
Museo Civico Archeologico O. Nardini, Velletri 028286
Museo Civico Archeologico P. Giovio, Como 026047
Museo Civico Archeologico Paleoambientale, Budrio 025653
Museo Civico Archeologico 'Romualdi', Notaresco 027053
Museo Civico Archeologico Sa Domu Nosta, Senorbì 027921
Museo Civico Archeologico Trebula Mutesca, Monteleone Sabino . 026944
Museo Civico Archeologico U. Granafei, Mesagne 026772
Museo Civico Archeologico, Isola d'Elba, Portoferraio 027419
Museo Civico Archeologico, Palazzo Comunale, Salò 027713
Museo Civico Arecheologico delle Acque, Chianciano Terme 025945
Museo Civico Aufidente, Castel di Sangro 025812
Museo Civico Aufidenate De Nino, Alfedena 025287
Museo Civico B. Romano, Termini Imerese 028074
Museo Civico Belliniano, Catania 025868
Museo Civico C. G. Nicastro, Bovino 025617
Museo Civico C. Verri, Biassono 025508
Museo Civico Casa Cavassa, Saluzzo 027716
Museo Civico Castel Nuovo, Napoli 026995
Museo Civico Castello Ursino, Catania 025869
Museo Civico Chiusa, Chiusa . . 025975
Museo Civico Craveri di Storia Naturale, Bra 025619
Museo Civico D. dal Lago, Valdagno 028252
Museo Civico d'Arte, Modena . . 026854
Museo Civico d'Arte, Pordenone 027410
Museo Civico d'Arte Antica e Palazzo Madama, Torino 028118
Museo Civico d'Arte Contemporanea, Albissola Marina 025277
Museo Civico d'Arte Contemporanea U. Apollonio, San Martino di Lupari 027776
Museo Civico d'Arte Industriale e Galleria Davia Bargellini, Bologna 025539
Museo Civico dei Fossili di Besano, Besano 025501
Museo Civico dei Villaggi Scomparsi, Villa Estense 028387
Museo Civico del Castello, Conegliano 026058
Museo Civico del Marmo, Carrara 025783
Museo Civico del Risorgimento, Modena 026855
Museo Civico del Risorgimento, Museum of the Risorgimento, Bologna 025540
Museo Civico del Sigillo, La Spezia 026539
Museo Civico del Territorio, San Daniele del Friuli 027744
Museo Civico della Ceramica, Nove 027067
Museo Civico della Cultura Materiale, Roseto degli Abruzzi 027674
Museo Civico della Laguna Sud

San Francesco Fuori le Mura, Chioggia 025971
Museo Civico della Media Valle del Liri, Sora 027989
Museo Civico della Navigazione Fluviale, Battaglia Terme 025476
Museo Civico della Torre, Treviglio 028181
Museo Civico della Val Fiorentina Vittorino Cazzetta, Selva di Cadore 027912
Museo Civico delle Armi L. Marzoli, Brescia 025629
Museo Civico delle Arti e Tradizioni Popolari, Micigliano 026777
Museo Civico delle Centuriazioni Romane Padovana, Borgoricco 025602
Museo Civico delle Ceramiche, Santo Stefano di Camastra 027847
Museo Civico delle Scienze, Pordenone 027411
Museo Civico di Archeologia Ligure, Genova 026397
Museo Civico di Arte Contadina, Romagnese 027664
Museo Civico di Arte Moderna, Mombercelli 026880
Museo Civico di Belle Arti, Lugano 041700
Museo Civico di Belriguardo, Castello di Belriguardo, Voghiera 028420
Museo Civico di Merano – Stadtmuseum Meran, Merano . 026763
Museo Civico di Palazzo Te, Mantova 026697
Museo Civico di Paleontologia, Empoli 026150
Museo Civico di Paleontologia e Archeologia, Ugento 028230
Museo Civico di Paleontologia e Paletnologia, Borgosesia 025603
Museo Civico di San Francesco, Montefalco 026931
Museo Civico di San Francesco, Montone 026970
Museo Civico di Santa Caterina, Treviso 028184
Museo Civico di Santa Maria della Consolazione, Ex Convento dei Domenicani, Altomonte 025300
Museo Civico di Schifanoia, Ferrara 026206
Museo Civico di Scienze Naturali, Faenza 026172
Museo Civico di Scienze Naturali, Lovere 026636
Museo Civico di Scienze Naturali, Voghera 028418
Museo Civico di Scienze Naturali E. Caffi, Bergamo 025494
Museo Civico di Scienze Naturali ed Archeologia della Valdinievole, Pescia 027283
Museo Civico di Storia Naturale, Carmagnola 025773
Museo Civico di Storia Naturale, Casalnuovo Monterotaro 025792
Museo Civico di Storia Naturale, Chies d'Alpago 025964
Museo Civico di Storia Naturale, Cittanova 026008
Museo Civico di Storia Naturale, Crocetta del Montello 026106
Museo Civico di Storia Naturale, Ferrara 026207
Museo Civico di Storia Naturale, Milano 026799
Museo Civico di Storia Naturale, Morbegno 026976
Museo Civico di Storia Naturale, Trieste 028203
Museo Civico di Storia Naturale, Verona 028347
Museo Civico di Storia Naturale del

Liceo Glassico N. Machiavelli, Lucca 026640
Museo Civico di Storia Naturale del Salento, Calimera 025696
Museo Civico di Storia Naturale Giacomo Doria, Natural History Museum, Genova 026398
Museo Civico di Vignola, Vignola 028384
Museo Civico di Villa Rathgeb, Abano Terme 025225
Museo Civico di Zoologia, Roma 027582
Museo Civico Didattico di Scienze Naturali, Pinerolo 027333
Museo Civico Didattico Gian Andrea Irico, Trino 028214
Museo Civico Diocesano, Tropea 028220
Museo Civico Diocesano La Castellina, Norcia 027050
Museo Civico Domenico Mambrini, Galeata 026363
Museo Civico Don Queirolo, Vado Ligure 028246
Museo Civico e Antiquarium, Latina 026572
Museo Civico e Archeologico, Castello Visconteo, Locarno 041683
Museo Civico e d'Arte Sacra, Palazzo dei Priori, Colle di Val d'Elsa . . 026032
Museo Civico e della Ceramica, Cerreto Sannita 025908
Museo Civico e Diocesano d' Arte Sacra, Montalcino 026905
Museo Civico e Pinacoteca, Asti 025407
Museo Civico e Pinacoteca, Fano 026183
Museo Civico e Pinacoteca, Sassoferrato 027876
Museo Civico e Pinacoteca, Urbania 028234
Museo Civico e Pinacoteca A. Vernarecci, Palazzo Ducale Corte Alta, Fossombrone 026341
Museo Civico e Pinacoteca E. Lancerotto, Noale 027039
Museo Civico e Pinacoteca P.F. Crociani, Montepulciano 026952
Museo Civico e Risorgimentale Don Giovanni Verità, Modigliana . . 026869
Museo Civico e Teatro e Serbatoio Romano, Falerone 026182
Museo Civico E. Barba, Gallipoli 026368
Museo Civico E. Nardi, Poggio Mirteto 027382
Museo Civico Etno-Antropologico, Petralia Sottana 027289
Museo Civico Etnografico, Atri . . 025415
Museo Civico Etnografico, Casale di Scodosia 025788
Museo Civico Etnografico, Stanghella 028011
Museo Civico Etnografico C.G. Franchini, Oleggio 027087
Museo Civico Etnografico La Lessinia, Bosco Chiesanuova 025609
Museo Civico Etnografico M.G. Fontana, Sappada 027851
Museo Civico F. Bodini, Gemonio 026385
Museo Civico F. de Rocco, San Vito al Tagliamento 027802
Museo Civico F. L. Belgiorno, Modica 026867
Museo Civico F. Rittatore Von Willer, Palazzo Comunale, Farnese . . 026186
Museo Civico Fedrigo, Sona . . . 027980
Museo Civico Filippo Meli, Ciminna 025984
Museo Civico G. Sutermeister, Legnano 026589
Museo Civico G. Ugonia, Brisighella 025642
Museo Civico G. Zannato, Montecchio Maggiore 026926
Museo Civico G.B. Adriani, Palazzo Gotti di Salerano, Cherasco . . 025943
Museo Civico Gaetano Filangieri,

Napoli 026996
Museo Civico-Galleria di Palazzo
   Pretorio, Prato 027447
Museo Civico Geo-Speleologico,
   Garessio 026373
Museo Civico Geologia e Etnografia,
   Predazzo 027454
Museo Civico Giacomo Rodolfo,
   Carignano 025772
Museo Civico Giovanni Fattori,
   Livorno 026604
Museo Civico Giulio Ferrari, Carpi 025779
Museo Civico Goffredo Bellini,
   Asola 025393
Museo Civico Gonzaga, Novellara 027068
Museo Civico Guiseppe Fiorelli,
   Lucera 026645
Museo Civico Il Correggio, Palazzo dei
   Principi, Correggio 026076
Museo Civico L. Mallé, Dronero 026145
Museo Civico Luigi Bailo, Treviso 028185
Museo Civico Luigi Dalla Laita,
   Ala 025259
Museo Civico M. Trucco, Albisola
   Superiore 025275
Museo Civico Marinaro Gio-Bono
   Ferrari, Camogli 025721
Museo Civico Marinaro U. Mursia,
   Milano 026800
Museo Civico Medievale, Bologna 025541
Museo Civico Messapico, Alezio 025286
Museo Civico Naturalistico, Maserada
   sul Piave 026723
Museo Civico Naturalistico dei
   Monti Prenestini, Capranica
   Prenestina 025756
Museo Civico Naturalistico Severo Sini,
   Villa d'Almè 028386
Museo Civico Orientale Santa Cardu,
   Cagliari 025675
Museo Civico Ornitologico,
   Lonato 026619
Museo Civico Ornitologico e di
   Scienze Naturali Luigi Scanagatta,
   Varenna 028275
Museo Civico P. A. Garda, Ivrea 026527
Museo Civico P. Rosario, Ascoli
   Satriano 025391
Museo Civico Padre Michele Jacobelli,
   Casalvieri 025793
Museo Civico Paleontologico,
   Macerata Feltria 026664
Museo Civico per la Preistoria del
   Monte Cetona, Cetona 025937
Museo Civico Pietro Lotti, Ischia di
   Castro 026515
Museo Civico-Pinacoteca,
   Vicenza 028367
Museo Civico Pio Capponi,
   Terracina 028081
Museo Civico Polivalente,
   Mistretta 026846
Museo Civico Preistorico, Pofi . . 027377
Museo Civico Rocca Flea, Gualdo
   Tadino 026483
Museo Civico Rogadeo, Bitonto 025522
Museo Civico Speleo-Archeologico,
   Nuoro 027073
Museo Civico Storico Archeologico,
   Leno 026590
Museo Civico Storico Artistico,
   Antrodoco 025334
Museo Civico Storico ed Etnografico,
   Primaluna 027457
Museo Civico Storico-Etnografico,
   Ripatransone 027529
Museo Civico, Convento Gesuitico,
   Salemi 027707
Museo Civico, Ex-Convento degli
   Agostinianoi Scalzi, Cerchio . . 025901
Museo Civico, Museo dell'Artigianato e
   del Lavoro Contadino, Andreis . 025325
Museo Civico, Palazzo Sgadari,
   Gangi 026371

Museo Civico, Pinacoteca Comunale,
   Nocera Umbra 027042
Museo Civico, Pinacoteca della
   Resistenza, Caldarola 025692
Museo Civico, Raccolta del Giocattolo
   G. Superti Furga et Oglio-Chiese,
   Canneto sull'Oglio 025740
Museo Civico, Sezione Archeologia e
   Storia Naturale, Finale Emilia . 026235
Museo Civico, Sezione Archeologica,
   Bojano 025527
Museo Civico, Sezione Paleontologica,
   Bojano 025528
Museo Civico, Villa Colloredo Mels,
   Recanati 027490
Museo Civico, Villa Pallavicino,
   Busseto 025663
Museo Civicodi Scienze Naturali,
   Randazzo 027476
Museo Civicodi Storia Naturale, Villa
   Gardella, Stazzano 028012
Museo Clarisse Coulombie de Goyaud,
   Ituzaingó 000391
Museo Clemente Rospigliosi,
   Pistoia 027363
Museo Coahuila y Texas,
   Monclova 031283
Museo Colección de Veterinaria Militar,
   Madrid 039631
Museo-Colección del Observatorio
   Astronómico Nacional, Madrid . 039632
Museo-Colección Específico del
   Suboficial, Tremp 040273
Museo-Colección Específico Militar El
   Desnarigado, Ceuta 039271
Museo Colecciones ICO, Instituto de
   Crédito Oficial, Madrid 039633
Museo Colombiano de Informática,
   Bogotá 008594
Museo Colonial, El Tocuyo 054750
Museo Colonial Charcas, Universidad
   Mayor de San Francisco Xavier,
   Sucre 004244
Museo Colonial de Acolman,
   Acolman 030805
Museo Colonial del Conde de Sierra
   Gorda, Santander Jiménez . . . 031492
Museo Colonial e Histórico, Complejo
   Museográfico Enrique Udaondo,
   Luján 000433
Museo Coltzin, Culiacán 030934
Museo Comandancia La Plata, Santo
   Domingo 009366
Museo Comarcal, Zalduondo . . . 040436
Museo Comarcal da Fonsagrada, A
   Fonsagrada 039403
Museo Comarcal de Arte Sacro de
   Peñafiel, Iglesia de Santa María,
   Peñafiel 039929
Museo Comarcal Velezano Miguel
   Guirao, Hospital Real, Vélez
   Rubio 040362
Museo Come Eravamo,
   Borgamanero 025594
Museo Comercial, Hellín 039486
Museo Comico Verscio, Verscio . 042102
Museo Comité Olímpico, México 031173
Museo Communale Antiquarium,
   Atena Lucana 025410
Museo Comparativo de Biología
   Marina, Viña del Mar 007559
Museo Complesso della Collegiata,
   Castiglione Olona 025858
Museo Comunal de Chañar Ladeado,
   Chañar Ladeado 000299
Museo Comunal Pedro Bargero, Emilio
   Bunge 000364
Museo Comunal Regional de San
   Guillermo, San Guillermo 000579
Museo Comunale, Montaione . . . 026904
Museo Comunale, Nova Ponente 027055
Museo Comunale, Praia a Mare 027436
Museo Comunale, Roggiano
   Gravina 027550

Museo Comunale A. Mendola,
   Favara 026191
Museo Comunale Archeologico,
   Viddalba 028377
Museo Comunale d'Arte
   Contemporanea, Polignano a
   Mare 027384
Museo Comunale d'Arte Moderna,
   Ascona 041235
Museo Comunale della Ceramica,
   Cutrofiano 026125
Museo Comunale della Manifattura
   Chini, Borgo San Lorenzo 025599
Museo Comunale della Valle dei
   Nuraghi, Torralba 028147
Museo Comunale della Valle del Sarno,
   Sarno 027858
Museo Comunale E. Durio,
   Civiasco 026010
Museo Comunale La Malaria e la sua
   Storia, Pontinia 027405
Museo Comunale Storico, Acquaviva
   delle Fonti 025238
Museo Comunale Trappeto Maratea,
   Vico del Gargano 028374
Museo Comunale, Cassero, Monte San
   Savino 026915
Museo Comunidad Las Terrazas, San
   Cristobal 009315
Museo Comunitario, Mineral del
   Monte 031278
Museo Comunitario, Puente de
   Ixtla 031403
Museo Comunitario Altagracia de
   Arauz, Ensenada 030966
Museo Comunitario Amuzgo de
   Xochistlahuaca, Xochistlahuaca 031677
Museo Comunitario Colonia
   Vicente Guerrero, Colonia Vicente
   Guerrero 030897
Museo Comunitario Concuemitl,
   Cuencamé 030925
Museo Comunitario Coronela Amelia
   La Güera Robles, Xochipala . . 031676
Museo Comunitario Crónica y Arte
   Agropecuario de la Sierra Gorda,
   Landa de Matamoros 031104
Museo Comunitario Cuarenta Casas,
   Valentín Gómez Farías 031633
Museo Comunitario de Antropología e
   Historia, Apatzingán 030828
Museo Comunitario de Atoyac,
   Atoyac 030837
Museo Comunitario de Carrillo Puerto,
   Alvaro Obregón 030823
Museo Comunitario de Ciudad Lerdo,
   Lerdo 031112
Museo Comunitario de Estación
   Coahuila, Mexicali 031142
Museo Comunitario de Historia de
   San José de Gracia, San José de
   Gracia 031443
Museo Comunitario de Huitzapula,
   Huitzapula 031061
Museo Comunitario de Ixtlahuacán,
   Ixtlahuacán 031072
Museo Comunitario de la Ballena, La
   Paz 031096
Museo Comunitario de la Sal,
   Armería 030834
Museo Comunitario de las Varas,
   Compostela 030906
Museo Comunitario de Pejelagartero,
   Huimanguillo 031059
Museo Comunitario de Pipillola,
   Españita 030974
Museo Comunitario de San Andrés
   Mixquic, México 031174
Museo Comunitario de San Nicolás
   Zoyatlán, San Nicolás Zoyatlán 031472
Museo Comunitario de San Rafael del
   Sur, San Rafael del Sur 033406
Museo Comunitario de Santa Eulalia,
   Santa Eulalia 031489

Museo Comunitario de Santa Martha,
   Mulegé 031315
Museo Comunitario de Tototepec,
   Totototepec 031617
Museo Comunitario de Villa El
   Salvador, Villa El Salvador . . . 034417
Museo Comunitario de Villa Regina,
   Villa Regina 000683
Museo Comunitario de Xalisco,
   Xalisco 031671
Museo Comunitario de Xolalpan,
   Ixtacamaxtitlán 031070
Museo Comunitario del Valle de
   Guadalupe, Ensenada 030967
Museo Comunitario el Asalto a las
   Tierras, Mexicali 031143
Museo Comunitario El Rosario, El
   Rosario 030960
Museo Comunitario Felipe Carrillo
   Puerto, Motul 031313
Museo Comunitario Francisco I.
   Madero, Francisco I. Madero,
   Coahuila 030977
Museo Comunitario General Francisco
   Villa, Durango 030947
Museo Comunitario Hicupa,
   Tehuacán 031528
Museo Comunitario Hitalulu, San
   Martín Huamelulpan 031464
Museo Comunitario Huitza Chilin, San
   Pedro 031476
Museo Comunitario Iluikatlachiyalistli,
   Yahualica 031680
Museo Comunitario Inti Raymi,
   Lima 034326
Museo Comunitario Ismael Girón
   González, Pánuco, Zacatecas . 031360
Museo Comunitario Itzmal Kauil,
   Izamal 031074
Museo Comunitario Ji, Ocosingo 031342
Museo Comunitario Jna Niingui, San
   Miguel Tequixtepec 031469
Museo Comunitario Joyonaque, Tuxtla
   Gutiérrez 031622
Museo Comunitario Juan García
   Aldama de la Comunidad Cucapá, El
   Mayor Indígena 030958
Museo Comunitario Kumkuy Ys Untzi,
   Copainalá 030909
Museo Comunitario Maika,
   Poanas 031373
Museo Comunitario Note Ujía, Santa
   María Asunción Tlaxiaco 031490
Museo Comunitario Nu-kuiñe,
   Cuquila 030940
Museo Comunitario Raramuri,
   Guachochi 030992
Museo Comunitario Rubén Jaramillo,
   Tlaquiltenango 031585
Museo Comunitario Serafín Olarte,
   Papantla 031361
Museo Comunitario Shan-Dany, Santa
   Ana del Valle 031482
Museo Comunitario sin Paredes
   Colonia Orizaba, Mexicali 031144
Museo Comunitario Snuuvico, San
   Juan Mixtepec Juxtlahuaca . . . 031452
Museo Comunitario Tejamen, Nuevo
   Ideal 031328
Museo Comunitario Tzakualli,
   Zacualpan 031697
Museo Comunitario Unidad Indígena
   Emiliano Zapata, Hueyapan de
   Ocampo 031057
Museo Comunitario Unión y Progreso,
   Cuatrocienegas 030922
Museo Comunitario Xiximes, Gómez
   Palacio 030988
Museo Comunitario Xolalpan Calli,
   Hueyapan 031056
Museo Ya Nfädi Yu Nohño
   Los Conocimientos de los Otomíes,
   Tolimán 031593
Museo Comunitario Yucu-Iti, Santa

Register der Institutionen und Firmen
Museo Comunitario Ze Acatl Topiltzin Quetzalcoatl, Tepoztlán – Museo de Arte Colonial de Palacio Valle Iznaga,

María Yucu-Iti 031491
Museo Comunitario Ze Acatl Topiltzin Quetzalcoatl, Tepoztlán 031556
Museo Concha Ferrant, La Habana 009168
Museo Conchiliologico e della Carta Moneta, Bellaria Igea Marina . 025480
Museo Contadino, Montalto Pavese 026909
Museo Contadino, Varese Ligure 028281
Museo Contadino della Bassa Pavese, San Cristina e Bissone 027743
Museo Contadino Foiano – Bauernmuseum Völlan, Lana . . 026550
Museo Contalpa de Nombre de Dios, Nombre de Dios 031327
Museo Contisuyo, Moquegua . . 034374
Museo Convento de San Diego, Quito 010476
Museo Convento de Santa Clara, Salamanca 040040
Museo Convento del Desierto de la Candelaria, Ráquira 008775
Museo Convento Santo Ecce Homo, Sutamarchán 008805
Museo Conventual de las Carmelitas Descalzas de Antequera, Monasterio de San José de Carmelitas Descalzas, Antequera 038975
Museo Conventual de Santa Paula, Sevilla 040171
Museo Cooperativo Eldorado, Eldorado 000361
Museo Coronel Félix Luque Plata, Guayaquil 010446
Museo Correale di Terranova, Sorrento 027994
Museo Correr e Biblioteca d'Arte e Storia Veneziana, Musei Civici Veneziani, Venezia 028303
Museo Costumbrista de Sonora, Alamos 030819
Museo Costumbristá Juan de Vargas, La Paz 004211
Museo Coyug-Curá, Pigüé 000492
Museo Criollo de los Corrales, Buenos Aires 000176
Museo Cristero Ing. Efrén Quezada, Encarnación de Díaz, Jalisco . . 030964
Museo Cristero Señor Cura Cristóbal Magallanes, Totatiche 031614
Museo Cristóbal Mendoza, Trujillo 054776
Museo Cruz Herrera, La Línea de la Concepción 039567
Museo Cuadra de Bolívar, Caracas 054719
Museo Cuartel de Emiliano Zapata, Tlaltizapán 031580
Museo Cuartel Zapatista, México 031175
Museo Cuevas de Las Maravillas, La Romana 010377
Museo Cultura Antigua Ribera del Bernesga, Cuadros 039330
Museo Cultura e Musica Popolare dei Peloritani, Messina 026773
Museo Cultura y Tradición de Jalcomulco, Jalcomulco 031076
Museo Cultural del Instituto Geografico Militar, Quito 010477
Museo Cultural Iijdio Guaranga, Guaranga 010443
Museo d' Arte e Scienza, Fondazione Gottfried Matthaes, Milano . . . 026801
Museo D. Agostinelli, Roma . . . 027583
Museo D. Chalonge, Erice 026160
Museo da Terra de Melide, Melide 039743
Museo Dantesco, Ravenna 027484
Museo d'Art Contemporani, Elche 039366
Museo d'Arte Antica, Milano . . . 026802
Museo d'Arte Contemporanea D. Formaggio, Teolo 028068
Museo d'Arte Contemporanea di Roma – MACRO, Roma 027584
Museo d'Arte Contemporanea di Villa Croce, Genova 026399
Museo d'Arte Costantino Barbella, Chieti 025968
Museo d'Arte d'Ammobigliamento, Palazzina di Caccia, Stupinigi . 028028
Museo d'Arte della Città, Ravenna 027485
Museo d'Arte dello Splendore, Giulianova 026448
Museo d'Arte e Ammobiliamento, Nichelino 027033
Museo d'Arte e Archeologia I. Mormino, Fondazione Banco di Sicilia, Palermo 027178
Museo d'Arte e Storia Antica Ebraica, Casale Monferrato 025790
Museo d'Arte G. Bargellini, Pieve di Cento 027326
Museo d'Arte Mendrisio, Mendrisio 041740
Museo d'Arte Moderna, Brescia 025630
Museo d'Arte Moderna, Tertenia 028083
Museo d'Arte Moderna di Bologna – MAMbo, Istituzione Galleria d'Arte Moderna die Bologna, Bologna 025542
Museo d'Arte Moderna e Contemporanea, Ferrara 026208
Museo d'Arte Moderna e Contemporanea – MAM, Gazoldo degli Ippoliti 026379
Museo d'Arte Moderna e Contemporanea di Trento e Rovereto, Rovereto 027684
Museo d'Arte Moderna e Contemporanea di Trento e Rovereto, Trento 028171
Museo d'Arte Moderna Mario Rimoldi, Musei delle Regole d'Ampezzo, Cortina d'Ampezzo 026080
Museo d'Arte Moderna Pagani, Castellanza 025829
Museo d'Arte Moderna, Villa Malpensata, Lugano 041698
Museo d'Arte Naturale, Gioiosa Ionica 026444
Museo d'Arte Nuoro, Nuoro 027074
Museo d'Arte Orientale, Venezia 028304
Museo d'Arte Orientale Edoardo Chiossone, Genova 026400
Museo d'Arte P. Pini, Milano . . . 026803
Museo d'Arte Pietro Cavoti, Galatina 026362
Museo d'Arte Preistorica, Pinerolo 027334
Museo d'Arte S. Pertini, Savona 027890
Museo d'Arte Sacra, Baiardo . . . 025442
Museo d'Arte Sacra, Camaiore . 025707
Museo d'Arte Sacra, Chiaramonte Gulfi 025947
Museo d'Arte Sacra, Montespertoli 026963
Museo d'Arte Sacra, Ponzone . . 027407
Museo d'Arte Sacra, San Casciano in Val di Pesa 027735
Museo d'Arte Sacra, San Gimignano 027752
Museo d'Arte Sacra, San Giovanni in Persiceto 027760
Museo d'Arte Sacra, San Leo . . 027767
Museo d'Arte Sacra, Tavarnelle Val di Pesa 028059
Museo d'Arte Sacra Cardinal Agnifili, Rocca di Mezzo 027542
Museo d'Arte Sacra della Insigne Collegiata di San Lorenzo, Montevarchi 026964
Museo d'Arte Sacra della Marsica, Celano 025892
Museo d'Arte Sacra della Santa Maria Assunta, Gallarate 026365
Museo d'Arte Sacra della Val d'Arbia, Buonconvento 025658
Museo d'Arte Sacra L. Acquarone, Lucinasco 026647
Museo d'Arte Sacra San Marco, Mombaroccio 026877
Museo d'Arte Sacra Silvestro Frangipane, Zagarise 028429
Museo d'Arte Sacra, Chiesa di Santa Maria Maddalena, Gradoli 026462
Museo d'Arte Sacra, Palazzo Corboli, Asciano 025383
Museo d'Arte Sacra, Palazzo dei Papi, Viterbo 028408
Museo d'Arte Sacra, Palazzo del Vescovado, Faenza 026173
Museo d'Arte Sacro San Martino, Alzano Lombardo 025305
Museo d'Arte Siamese Stefano Gardu, Cagliari 025676
Museo d'Arti e Mestieri Antichi, Montelupone 026948
Museo das Mariñas, Betanzos . 039134
Museo das Peregrinacións, Santiago de Compostela 040122
Museo David Paltán, San Gabriel 010504
Museo David Paltan, Tulcán . . . 010506
Museo David Ramírez Lavoignet, Misantla 031279
Museo de Abancay, Abancay . . 034209
Museo de Academia de San Carlos, Universidad Nacional Autónoma de México, México 031176
Museo de Acaponeta, Acaponeta 030800
Museo de Aduanas y Puerto, Buenos Aires 000177
Museo de Aeronáutica y Astronáutica, Museo del Aire, Madrid 039634
Museo de Aguascalientes, Aguascalientes 030808
Museo de Ahumada, Villa Ahumada y Anexas 031655
Museo de Akil Uyotoch Cah, Akil 030818
Museo de Albacete, Albacete . . 038904
Museo de Albarracín, Albarracín 038908
Museo de Alcalá la Real, Alcalá la Real 038917
Museo de Alfarería Tradicional Vasca, Elosu 039375
Museo de Almería, Almería 038962
Museo de Ámbar Dominicano, San Felipe de Puerto Plata 010378
Museo de Ambiente Histórico Cubano, Santiago de Cuba 009348
Museo de Ameca, Vetagrande . . 031648
Museo de América, Madrid 039635
Museo de Anatomía, México . . . 031177
Museo de Animales Venenosos, Buenos Aires 000178
Museo de Antioquia, Medellín . . 008714
Museo de Antofagasta, Antofagasta 007514
Museo de Antropolocía Forense, Paleopatología y Criminalística Profesor Reverte, Facultad de Medicina, Madrid 039636
Museo de Antropología, Ibarra . 010457
Museo de Antropología, Tecolotlán 031523
Museo de Antropología David J. Guzmán, San Salvador 010593
Museo de Antropología de Tenerife, La Laguna de Tenerife 039540
Museo de Antropología de Xalapa, Universidad Veracruzana, Xalapa 031667
Museo de Antropología e Historia de Allende, Allende 030821
Museo de Antropología e Historia de Rivas, Rivas 033401
Museo de Antropología e Historia de Santiago Papasquiaro, Santiago Papasquiaro 031494
Museo de Antropología e Historia Natural Los Desmochados, Casilda 000296
Museo de Antropología e Historia, Fundación Lisandro Alvarado, Maracay 054765
Museo de Antropología y Ciencias Naturales, Concordia 000315
Museo de Antropología y Municipal de Boaco, Juigalpa 033381
Museo de Antropología, Universidad del Atlántico-Facultad de Ciencias Humanas- Departamento de Historia, Barranquilla 008578
Museo de Antropología, Universidad Nacional de Córdoba, Córdoba 000321
Museo de Anzoátegui, Barcelona 054702
Museo de Armas de la Nación, Buenos Aires 000179
Museo de Armería de Álava, Vitoria-Gasteiz 040425
Museo de Arqueología, Acapulco 030801
Museo de Arqueología, La Habana 009169
Museo de Arqueología, Ixtepec . 031071
Museo de Arqueología Colonial Rodolfo I. Bravo, Cafayate 000286
Museo de Arqueología de Álava, Vitoria-Gasteiz 040426
Museo de Arqueología de Córdoba, Córdoba 030910
Museo de Arqueología de la Universidad Nacional de Trujillo, Trujillo 034410
Museo de Arqueología de la Universidad Nacional de Tucumán, San Miguel de Tucumán 000595
Museo de Arqueología de León, León 031107
Museo de Arqueología de Occidente de México, Universidad de Guadalajara, Guadalajara 031001
Museo de Arqueología de Salta, Salta 000554
Museo de Arqueología e Historia Natural de Yungay, Yungay . . . 034421
Museo de Arqueología el Chamizal, Juárez 031090
Museo de Arqueología Josefina Ramos de Cox, Instituto Riva-Agüero, Universidad Católica del Perú, Lima 034327
Museo de Arqueología Samuel Humberto Espinoza Lozano, Huaytará 034284
Museo de Arqueología y Antropología, Universidad Nacional Mayor de San Marcos, Lima 034328
Museo de Arqueometalurgia de Cerro Muriano, Obejo 039824
Museo de Arquitectura, Caracas 054720
Museo de Arquitectura Leopoldo Rother, Universidad Nacional de Colombia, Bogotá 008595
Museo de Arte Abstracto del Sur, Fundación Adrián Dorado, Buenos Aires 000180
Museo de Arte Abstracto Español, Fundación Juan March, Cuenca 039332
Museo de Arte Abstracto Manuel Felguérez, Zacatecas 031687
Museo de Arte Alberto Mena Caamaño, Quito 010478
Museo de Arte Americano de Maldonado, Maldonado 046240
Museo de Arte Carrillo Gil, México 031178
Museo de Arte Colonial, Antigua 023360
Museo de Arte Colonial, Bogotá 008596
Museo de Arte Colonial, Caracas 054721
Museo de Arte Colonial, Mérida 054767
Museo de Arte Colonial, Morelia 031303
Museo de Arte Colonial, Quito . . 010479
Museo de Arte Colonial de la Habana, Casa Conde Bayona, La Habana 009170
Museo de Arte Colonial de Palacio Valle Iznaga, Sancti-Spíritus . . 009325

Museo de Arte Colonial de San Francisco, Santiago de Chile –
Museo de Artes y Tradiciones Populares, Universidad Popular,

Index of Institutions and Companies

Museo de Arte Colonial de San
Francisco, Santiago de Chile . . 007537
Museo de Arte Colonial Pedro de
Osma, Lima 034329
Museo de Arte Contemporánea de
Vigo, Marco, Vigo 040379
Museo de Arte Contemporáneo, Bahía
Blanca 000133
Museo de Arte Contemporáneo,
Bogotá 008597
Museo de Arte Contemporáneo,
Ciudad de Panamá 034176
Museo de Arte Contemporáneo,
Córdoba 000322
Museo de Arte Contemporáneo,
Cusco 034258
Museo de Arte Contemporáneo,
Montevideo 046257
Museo de Arte Contemporáneo, San
Salvador 010594
Museo de Arte Contemporáneo,
Santiago de Chile 007538
Museo de Arte Contemporáneo,
Valdivia 007553
Museo de Arte Contemporáneo al Aire
Libre, Aracena 038978
Museo de Arte Contemporáneo Alfredo
Zalce, Morelia 031304
Museo de Arte Contemporáneo Ángel
Miguel de Arce, Sasamón . . . 040147
Museo de Arte Contemporáneo Ángel
Zárraga, Durango 030948
Museo de Arte Contemporáneo Ateneo
de Yucatán, Mérida 031134
Museo de Arte Contemporáneo Beato
Angélico, La Plata 000409
Museo de Arte Contemporáneo de
Caracas Sofía Imber, Caracas . 054722
Museo de Arte Contemporáneo de
Chihuahua, Chihuahua 030877
Museo de Arte Contemporáneo de
Monterrey, Monterrey 031288
Museo de Arte Contemporáneo de
Oaxaca, Oaxaca 031331
Museo de Arte Contemporáneo de
Puerto Rico, San Juan 036159
Museo de Arte Contemporáneo
Esteban Vicente, Segovia . . . 040156
Museo de Arte Contemporáneo
Francisco Narváez, Porlamar . 054774
Museo de Arte Contemporáneo José
María Moreno Galván, La Puebla de
Cazalla 039977
Museo de Arte Contemporáneo
Moderno de Ibagué, Ibagué . . 008698
Museo de Arte Contemporáneo
Número 8, Aguascalientes . . . 030809
Museo de Arte Contemporáneo Pajar
de Agustín, Valle de Hecho . . 040352
Museo de Arte Contemporáneo Union
Fenosa – MACUF, A Coruña . . 039319
Museo de Arte Contemporáneo,
Museo del Museo de Santa Cruz,
Toledo 040235
Museo de Arte Coro y Museo Alberto
Henríquez, Falcón 054751
Museo de Arte Costarricense, San
José 008854
Museo de Arte de Caguas,
Caguas 036123
Museo de Arte de Lima, Lima . . 034330
Museo de Arte de Matanzas,
Matanzas 009252
Museo de Arte de Mazatlán,
Mazatlán 031131
Museo de Arte de Orizaba y del
Estado de Veracruz, Orizaba . . 031348
Museo de Arte de Pereira,
Pereira 008758
Museo de Arte de Ponce, The Luis A.
Ferre Foundation, Ponce 036141
Museo de Arte de Puerto Rico, San
Juan 036160
Museo de Arte de Querétaro,

Querétaro 031409
Museo de Arte de Sinaloa,
Culiacán 030935
Museo de Arte e Historia,
Durango 039357
Museo de Arte e Historia Alberto
Mena Caamaño, Quito 010480
Museo de Arte e Historía Casa de Los
Celis, Valencia 054779
Museo de Arte e Historia de la
Ciudad "Jorge y Elisa", Instituto
de Cultura de Norte de Santander,
Cuacutecuta 008669
Museo de Arte e Historía de Los
Toldos, Los Toldos 000431
Museo de Arte e Historia, Universidad
Nacional Mayor de San Marcos,
Lima 034331
Museo de Arte E. Minnicelli, Río
Gallegos 000530
Museo de Arte en Ciudad Juárez,
Instituto Nacional de Bellas Artes,
Juárez 031091
Museo de Arte Español Enrique
Larreta, Buenos Aires 000181
Museo de Arte Francisco Oller,
Bayamón 036121
Museo de Arte Gauchesco La Recova,
San Antonio de Areco 000564
Museo de Arte Gráficas Juan Pablos,
México 031179
Museo de Arte Guillermo Ceniceros,
Durango 030949
Museo de Arte Habana, La
Habana 009171
Museo de Arte Hispanoamericano
Isaac Fernández Blanco, Buenos
Aires 000182
Museo de Arte Italiano, Lima . . 034332
Museo de Arte José Luis Bello y
González, Puebla 031381
Museo de Arte José Luis Bello y
Zetina, Puebla 031382
Museo de Arte Moderno, Cuenca 010429
Museo de Arte Moderno, Mérida 054768
Museo de Arte Moderno, México 031180
Museo de Arte Moderno, Santiago de
Chile 007539
Museo de Arte Moderno, Santo
Domingo 010387
Museo de Arte Moderno de
Barranquilla, Barranquilla 008579
Museo de Arte Moderno de Bogotá,
Bogotá 008598
Museo de Arte Moderno de Bogotá II,
Bogotá 008599
Museo de Arte Moderno de
Bucaramanga, Bucaramanga . . 008638
Museo de Arte Moderno de Buenos
Aires, Buenos Aires 000183
Museo de Arte Moderno de Cartagena,
Cartagena 008657
Museo de Arte Moderno de Durango,
Gómez Palacio 030989
Museo de Arte Moderno de la Casa
de la Cultura Ecuatoriana, Quito 010481
Museo de Arte Moderno de Medellín,
Medellín 008715
Museo de Arte Moderno del Estado de
México, Toluca 031594
Museo de Arte Moderno Jesús Soto,
Ciudad Bolívar 054744
Museo de Arte Moderno La Tertulia,
Cali 008647
Museo de Arte Moderno Ramírez
Villamizar, Pamplona 008745
Museo de Arte Olga Costa,
Guanajuato 031023
Museo de Arte Oriental, Ávila . . 039020
Museo de Arte Polular, México . 031181
Museo de Arte Popular Americano,
Santiago de Chile 007540
Museo de Arte Popular de la
Universidad, Puebla 031383

Museo de Arte Popular de la
Universidad Nacional San Cristóbal
de Huamanga, Ayacucho 034221
Museo de Arte Popular de Quinua,
Quinua 034393
Museo de Arte Popular de San
Bartolo de Coyotepec, San Baltazar
Chichicapan 031432
Museo de Arte Popular José
Hernández, Buenos Aires 000184
Museo de Arte Popular Petare,
Caracas 054723
Museo de Arte Popular Poblano y
Museo de Artesanías, Puebla . 031384
Museo de Arte Popular Tlajomulco,
Tlajomulco de Zúñiga 031577
Museo de Arte Popular Venezolano,
Juangriego 054757
Museo de Arte Prehispánico,
Mascota 031125
Museo de Arte Prehispánico Carlos
Pellicer, Tepoztlán 031557
Museo de Arte Prehispánico de
México Rufino Tamayo, Oaxaca 031332
Museo de Arte Religioso, La
Habana 009172
Museo de Arte Religioso,
Huaquechula 031051
Museo de Arte Religioso Cardenal
Luque, Tenjo 008807
Museo de Arte Religioso Colonial,
Ciudad de Panamá 034177
Museo de Arte Religioso Colonial,
Comayagua 023397
Museo de Arte Religioso Concepción,
Riobamba 010498
Museo de Arte Religioso Convento de
San Francisco, Cajamarca . . . 034230
Museo de Arte Religioso de Girón,
Girón 008686
Museo de Arte Religioso de la
Catedral, Lima 034333
Museo de Arte Religioso de
la Catedral de San Nicolás,
Rionegro 008777
Museo de Arte Religioso de Monguí,
Monguí 008729
Museo de Arte Religioso de Piura,
Piura 034387
Museo de Arte Religioso de San Gil,
Atienza 039016
Museo de Arte Religioso de Santa
Mónica, Puebla 031385
Museo de Arte Religioso Ex Convento
Franciscano, Huejotzingo 031054
Museo de Arte Religioso i
Palenteológico de San Bartolomé,
Atienza 039017
Museo de Arte Religioso Juan de
Tejeda, Córdoba 000323
Museo de Arte Religioso Julio Franco
Arango, Duitama 008672
Museo de Arte Religioso la
Concepción, Baños 010418
Museo de Arte Religioso Presbítero
Antonio María Franco, Supía . . 008804
Museo de Arte Religioso Santo
Domingo de Porta Coeli, San
Germán 036149
Museo de Arte Religioso Tiberio de J.
Salazar y Herrera, Parroquia de la
Catedral de Sonsón, Sonsón . 008801
Museo de Arte Religioso y Colonial,
Cali 008648
Museo de Arte Sacra das Clarisas,
Convento das Clarisas, Monforte de
Lemos 039767
Museo de Arte Sacra de Iria Flavia,
Casa Rectoral de Iria, Padrón . 039868
Museo de Arte Sacra de Santa Clara,
Allariz 038949
Museo de Arte Sacra de Santa María
do Campo, Museo de Arte Sacro de
la Colegiata, A Coruña 039320

Museo de Arte Sacra de Vilanova,
Lourenzá 039598
Museo de Arte Sacro, Betancuria 039133
Museo de Arte Sacro, Castrojeriz 039255
Museo de Arte Sacro, Chihuahua 030878
Museo de Arte Sacro, León . . . 033385
Museo de Arte Sacro de La Rioja, La
Rioja 000417
Museo de Arte Sacro de Tineo,
Tineo 040227
Museo de Arte Sacro San Francisco,
Salta 000555
Museo de Arte Sacro San Miguel,
Oruro 004230
Museo de Arte Sacro Santiago de los
Caballeros, Gáldar 039421
Museo de Arte Sacro, Iglesia Colegial,
Osuna 039856
Museo de Arte Sacro, Iglesia
Parroquial de San Esteban,
Huete 039503
Museo de Arte Sacro, Monasterio de
San Pelayo de Antealtares, Santiago
de Compostela 040123
Museo de Arte San Miguel Arcángel,
Mula 039792
Museo de Arte Taino, San Felipe de
Puerto Plata 010379
Museo de Arte Universidad Nacional
de Colombia, Bogotá 008600
Museo de Arte Virreinal de Taxco,
Taxco de Alarcón 031513
Museo de Arte y Artesanía de Linares,
Linares 007526
Museo de Arte y Diseño
Contemporáneo, San José . . . 008855
Museo de Arte y Tradiciones
Populares 008601
Museo de Arte, Historia y Cultura
Casa Alonso, Vega Baja 036173
Museo de Arte, Universidad del
Magdalena, Santa Marta 008788
Museo de Artes Decorativas de
Palacio Salcines, Guantánamo . 009130
Museo de Artes e Industrias Populares,
Pátzcuaro 031365
Museo de Artes Gráficas, Bogotá 008602
Museo de Artes Plásticas,
Madonado 046239
Museo de Artes Plásticas Eduardo
Sívori, Buenos Aires 000185
Museo de Artes Populares, Tepic 031551
Museo de Artes Visuales Aramara,
Tepic 031552
Museo de Artes y Artesanías Enrique
Estrada Bello, Santo Tomé . . . 000632
Museo de Artes y Costumbres
Populares, Sevilla 040172
Museo de Artes y Costumbres
Populares e internacional de Arte
Naif, Jaén 039523
Museo de Artes y Costumbres
Populares, Fundacón Unicaja,
Málaga 039701
Museo de Artes y Costumbres
Populares, Museo Monográfico del
Castro de Viladonga, Pedrafita do
Cebrero 039922
Museo de Artes y Oficios,
Monóvar 039771
Museo de Artes y Tradiciones
Populares, Espinal 008678
Museo de Artes y Tradiciones
Populares, Madrid 039637
Museo de Artes y Tradiciones
Populares de Ráquira, Ráquira 008776
Museo de Artes y Tradiciones
Populares de Tlaxcala, Tlaxcala 031586
Museo de Artes y Tradiciones
Populares, Instituto Riva-Agüero,
Universidad Católica del Perú,
Lima 034334
Museo de Artes y Tradiciones
Populares, Universidad Popular, San

Sebastián de los Reyes 040077
Museo de Artesanía Iberoamericana,
La Orotava 039852
Museo de Artesanías Cefa, Quito 010482
Museo de Artesanías del Mundo,
Silvia 008794
Museo de Artesanías Tradicionales
Folklóricas, Corrientes 000346
Museo de Astronomía y Geodesia,
Universidad Complutense de Madrid,
Madrid 039638
Museo de Atoyac, Atoyac de
Alvarez 030838
Museo de Autos y del Transporte
Humberto Lobo Villarreal,
Monterrey 031289
Museo de Ávila, Ávila 039021
Museo de Aviones Históricos en Vuelo,
Fundación Infante de Orleans,
Madrid 039639
Museo de Ayutla, Ayutla de los
Libres 030840
Museo de Bahía de Cochinos, Playa
Girón 009295
Museo de Baltasar Lobo, Iglesia de
San Esteban, Zamora 040438
Museo de Baracoa, Baracoa . . . 009069
Museo de Bardados del Paso Azul,
Lorca 039596
Museo de Barquisimeto,
Barquisimeto 054705
Museo de Batopilas, Francisco I.
Madero, Coahuila 030978
Museo de Belas Artes da Coruña, A
Coruña 039321
Museo de Bellas Artes, Badajoz 039036
Museo de Bellas Artes,
Comayagüela 023398
Museo de Bellas Artes, Córdoba 039306
Museo de Bellas Artes, Valencia 040308
Museo de Bellas Artes, Viña del
Mar 007560
Museo de Bellas Artes Agustin Araujo,
Treinta y Tres 046297
Museo de Bellas Artes al Aire Libre
Caminito, Buenos Aires 000186
Museo de Bellas Artes Bonaerense, La
Plata 000410
Museo de Bellas Artes Ceferino
Carnacini, Villa Ballester 000670
Museo de Bellas Artes Claudio León
Sempere, Burzaco 000284
Museo de Bellas Artes de Álava,
Palacio Augustin, Vitoria-Gasteiz 040427
Museo de Bellas Artes de Asturias,
Palacio Verlarde, Oviedo 039862
Museo de Bellas Artes de Caracas,
Caracas 054724
Museo de Bellas Artes de Castellón,
Castellón de la Plana 039249
Museo de Bellas Artes de Granada,
Granada 039466
Museo de Bellas Artes de Gravina,
Alicante 038945
Museo de Bellas Artes de la Boca
Quinquela Martín, Buenos Aires 000187
Museo de Bellas Artes de Murcia,
Murcia 039795
Museo de Bellas Artes de Santander,
Santander 040113
Museo de Bellas Artes de Sevilla,
Sevilla 040173
Museo de Bellas Artes de Toluca,
Toluca 031595
Museo de Bellas Artes Departamental
de San José, San José de
Mayo 046295
Museo de Bellas Artes Domingo
Faustino Sarmiento, Mercedes 000460
Museo de Bellas Artes Franklin
Rawson, San Juan 000586
Museo de Bellas Artes Juan Yapari,
Posadas 000497
Museo de Bellas Artes Laureano

Brizuela, San Fernando del Valle de
Catamarca 000572
Museo de Bellas Artes Octavio de la
Colina, La Rioja 000418
Museo de Bermejillo, Bermejillo 030846
Museo de Bomberos, Madrid . . 039640
Museo de Bordados Paso Blanco,
Lorca 039597
Museo de Buenos Aires,
Tacotalpa 031504
Museo de Burgos, Casa Miranda,
Burgos 039169
Museo de Cáceres, Cáceres . . . 039183
Museo de Cádiz, Cádiz 039190
Museo de Calcos y Esculturas
Comparda Ernesto de La Cárcova,
Buenos Aires 000188
Museo de Calzado, Arnedo 038993
Museo de Campeche, Campeche 030850
Museo de Caracas, Caracas . . . 054725
Museo de Carmen Conde y Antonio
Oliver, Cartagena 039238
Museo de Carruajes de Sevilla,
Instituto Hispano-Cubano,
Sevilla 040174
Museo de Casa Popenoe,
Antigua 023361
Museo de Caza Benito Albarrán,
Zapopan 031699
Museo de Centro Cultural Recoleta,
Buenos Aires 000189
Museo de Cera, Guadalajara . . . 031002
Museo de Cera, Madrid 039641
Museo de Cera de la Ciudad de
México, México 031182
Museo de Cera de la Villa,
México 031183
Museo de Cera de Tijuana,
Tijuana 031566
Museo de Cerámica de la Cartuja-
Pickman, Salteras 040053
Museo de Cerámica La Peñuela,
Chinchilla de Monte-Aragón . . . 039277
Museo de Cerámica Ruíz de Luna,
Museo del Santa Cruz, Talavera de
la Reina 040199
Museo de Cerámica y Bellas Artes
Julián de la Herrería, Asunción 034196
Museo de Cerámica y Loza Popular,
Falcón 054752
Museo de Chilecito, Chilecito . . 000302
Museo de Ciencia, Cuernavaca . 030926
Museo de Ciencia y Juego,
Universidad del Magdalena, Santa
Marta 008789
Museo de Ciencia y Técnica, Buenos
Aires 000190
Museo de Ciencia y Tecnología del
Estado de Veracruz, Xalapa . . . 031668
Museo de Ciencias, Caracas . . . 054726
Museo de Ciencias, Huánuco . . 034278
Museo de Ciencias Antropológicas y
Naturales, Santiago del Estero 000628
Museo de Ciencias de Ensenada,
Ensenada 030968
Museo de Ciencias Explora, León 031108
Museo de Ciencias Morfológicas,
Rosario 000543
Museo de Ciencias Naturales,
Arnedo 038994
Museo de Ciencias Naturales,
Asunción 034197
Museo de Ciencias Naturales, Coronel
Pringles 000343
Museo de Ciencias Naturales,
Salta 000556
Museo de Ciencias Naturales,
Toluca 031596
Museo de Ciencias Naturales,
Villafranca del Bierzo 040405
Museo de Ciencias Naturales Augusto
G. Schulz, Resistencia 000522
Museo de Ciencias Naturales
Bartolomé Mitre, Córdoba 000324

Museo de Ciencias Naturales Carlos
de la Torre, Holguín 009218
Museo de Ciencias Naturales de Álava,
Vitoria-Gasteiz 040428
Museo de Ciencias Naturales de
Atarfe, Atarfe 039014
Museo de Ciencias Naturales de
Especialidad Zoológica Médica,
Buenos Aires 000191
Museo de Ciencias Naturales de
Oberá, Oberá 000479
Museo de Ciencias Naturales del
Colegio de San José de La Salle,
Medellín 008716
Museo de Ciencias Naturales del
Departamento San Cristóbal, San
Cristóbal 000570
Museo de Ciencias Naturales Dr.
Amado Bonpland, Corrientes . . 000347
Museo de Ciencias Naturales e
Historia, Instituto Superior del
Profesorado Antonio Ruiz de
Montoya, Posadas 000498
Museo de Ciencias Naturales Federico
Carlos Lehmann, Cali 008649
Museo de Ciencias Naturales
Florentino Ameghino, Las Flores 000424
Museo de Ciencias Naturales Luma,
Tanti 000643
Museo de Ciencias Naturales Rvdo. P.
Antonio Scasso, Colegio Don Bosco,
San Nicolás de los Arroyos . . . 000603
Museo de Ciencias Naturales Tomás
Romay, Santiago de Cuba 009349
Museo de Ciencias Naturales y
Antropológicas Juan Cornelio
Moyano, Mendoza 000453
Museo de Ciencias Naturales y
Antropológicas Prof. Antonio Serrano,
Paraná 000484
Museo de Ciencias Naturales y
Misional del Colegio San José,
Esperanza 000366
Museo de Ciencias Naturales,
Convento Rr. Pp. Dominicos, Valverde
de la Virgen 040357
Museo de Ciencias Naturales, Museo
de la Naturaleza y el Hombre, Santa
Cruz de Tenerife 040099
Museo de Ciencias Naturales,
Universidad Centroamericana,
Managua 033388
Museo de Ciencias Naturales,
Universidad de Navarra,
Pamplona 039910
Museo de Ciencias Naturales,
Universidad Nacional San Antonio
Abad, Cusco 034259
Museo de Ciencias y Antropología,
Cartago 008661
Museo de Ciudad Bolívar, Ciudad
Bolívar 054745
Museo de Ciudad Real, Ciudad
Real 039288
Museo de Ciudades Hermanas,
Guadalajara 031003
Museo de Coamo, Coamo 036128
Museo de Cochoapa, Ometepec 031347
Museo de Cocula es el Mariachi,
Cocula 030888
Museo de Colexiata do Sar, Santiago
de Compostela 040124
Museo de Colón, Colón 030895
Museo de Condega, Condega . . 033374
Museo de Conkal, Conkal 030908
Museo de Cosaltlán de Carvajal,
Cosaltlán de Carvajal 030914
Museo de Coscomatepec,
Coscomatepec de Bravo 030915
Museo de Costumbres y Artes
Populares de los Montes,
Guadamur 039478
Museo de Covadonga, Santuario de
Covadonga 040146

Museo de Criminalística de la Policía
Nacional del Perú, Lima 034335
Museo de Cuajinicuilapa,
Cuajinicuilapa 030919
Museo de Cuenca, Cuenca 039333
Museo de Cultura Popular Angel Gil
Hermidas, Villahermosa 031660
Museo de Cultura Popular Costa Rica,
Heredia 008848
Museo de Cultura Populares del
Estado de México, Toluca 031597
Museo de Culturas Populares,
Durango 030950
Museo de Culturas Populares de
Chiapas, San Cristóbal de las
Casas 031434
Museo de Culturas Populares de San
Luis Potosí, San Luis Potosí . . . 031455
Museo de Culturas Populares e
Indígenas, Obregón 031339
Museo de Culturas Populares
e Indígenas de Sonora,
Hermosillo 031039
Museo de Descubrimiento, Museo
Histórico Nacional, Montevideo 046258
Museo de Dibujo "Julio Gavín-Castillo
de Larrés", Larrés (Huesca) . . . 039546
Museo de Dinosaurios, Salas de los
Infantes 040048
Museo de Duarte, Santo
Domingo 010388
Museo de Ejército Oriental,
Holguín 009219
Museo de Emiliano Zapata, Emiliano
Zapata 030962
Museo de Escritores Bolivianos,
Universidad Mayor de San Andres,
La Paz 004212
Museo de Escuela de Educación
Técnica Número 1, Dolores . . . 000354
Museo de Escultura al Aire Libre,
Madrid 039642
Museo de Escultura al Aire Libre de
Leganés, Leganés 039550
Museo de Escultura Jacinto Higueras,
Santisteban del Puerto 040143
Museo de Escultura Novarro Santafé,
Villena 040417
Museo de Esculturas Luis Perlotti,
Buenos Aires 000192
Museo de Etnografía, Quito . . . 010483
Museo de Etnohistoria de Suyo,
Suyo 034402
Museo de Farmacia, Burgos . . . 039170
Museo de Farmacia, Llivia 039588
Museo de Farmacia, Peñaranda de
Duero 039935
Museo de Farmacia Militar,
Madrid 039643
Museo de Ferrocarriles en Miniatura
Werner Dura, La Falda 000401
Museo de Filatelia de Oaxaca,
Oaxaca 031333
Museo de Filigrana de Cobre, La
Esperanza 034298
Museo de Firmat, Firmat 000372
Museo de Física de la Universidad
Nacional de La Plata, La Plata 000411
Museo de Física Recreativa,
Puebla 031386
Museo de Fósiles, Igea 039509
Museo de Frank País, Holguín . . 009220
Museo de Galera, Galera 039423
Museo de Gendarmería Nacional,
Buenos Aires 000193
Museo de General Zaragoza, General
Zaragoza 030986
Museo de Geología de la UNAM,
México 031184
Museo de Geología del IPN,
México 031185
Museo de Geología Dr. Jenaro
González Reyna, Universidad
Michoacana, Morelia 031305

Museo de Geología y Paleontología,
  Neuquén 000477
Museo de Geología y Suelos,
  Armenia 008572
Museo de Geologia, Paleontologia i
  Prehistória del Bruc, El Bruc . . 039159
Museo de Geología, Universidad de
  Caldas, Manizales 008707
Museo de Geología, Universidad
  Nacional Autónoma de Nicaragua,
  Managua 033389
Museo de Gomellano, Gumiel de
  Hizán 039484
Museo de Granjita Siboney, Santiago
  de Cuba 009350
Museo de Guadalajara,
  Guadalajara 039476
Museo de Guadalupe, Guadalupe 031016
Museo de Guamúchil de Salvador
  Alvarado, Guamúchil . . 031020
Museo de Heráldica Alavesa,
  Mendoza 039747
Museo de Herrera, Chitré-
  Herrera 034171
Museo de Historia de Agüimes,
  Agüimes 038890
Museo de Historia de Ciencias
  Naturales, Chuquimarca 034254
Museo de Historia de Ensenada,
  Ensenada 030969
Museo de Historia de Higueras,
  Higueras 031046
Museo de Historia de la Automoción,
  Salamanca 040041
Museo de Historia de la Medicina,
  Bogotá 008603
Museo de Historia de la Medicina
  y de la Cirugía Vicente A. Risolía,
  Universidad de Buenos Aires, Buenos
  Aires 000194
Museo de Historia de la Minería de
  Santa Rosalía, Mulegé 031316
Museo de Historia de Lampazos de
  Naranjo y las Armas Nacionales,
  Lampazos de Naranjo 031103
Museo de Historia de Linares,
  Linares 031116
Museo de Historia de Panamá, Ciudad
  de Panamá 034178
Museo de Historia de Tabasco,
  Villahermosa 031661
Museo de Historia de Tenerife, La
  Laguna de Tenerife 039541
Museo de Historia del Deporte,
  Sala Polivalente Ramón Font, La
  Habana 009173
Museo de Historia del Transporte,
  Guadalupe 031017
Museo de Historia Económica,
  Bogotá 008604
Museo de Historia Mexicana,
  Monterrey 031290
Museo de Historia Natural, Alta
  Suiza 008566
Museo de Historia Natural,
  Bogotá 008605
Museo de Historia Natural, Ciudad de
  Panamá 034179
Museo de Historia Natural,
  Guatemala 023369
Museo de Historia Natural,
  Manizales 008708
Museo de Historia Natural,
  Mérida 031135
Museo de Historia Natural,
  México 031186
Museo de Historia Natural,
  Pereira 008759
Museo de Historia Natural,
  Popayán 008765
Museo de História Natural, Tunja 008814
Museo de Historia Natural,
  Villahermosa 031662
Museo de Historia Natural, Zapotlán

del Rey 031703
Museo de Historia Natural Alcide
  D'Orbigni, Cochabamba 004206
Museo de Historia Natural Alfredo
  Dugès, Universidad de Guanajuato,
  Guanajuato 031024
Museo de Historia Natural de
  Concepción, Concepción 007519
Museo de Historia Natural de El
  Salvador, San Salvador 010595
Museo de Historia Natural de la
  Universidad Antenor Orrego,
  Trujillo 034411
Museo de Historia Natural de
  la Universidad Ricardo Palma,
  Lima 034336
Museo de Historia Natural de San
  Pedro Nolasco, Santiago de
  Chile 007541
Museo de Historia Natural de
  Valparaíso, Valparaíso 007556
Museo de Historia Natural del Ateneo
  Fuente, Universidad Autónoma de
  Coahuila, Saltillo 031422
Museo de Historia Natural del
  Paraguay, Asunción 034198
Museo de Historia Natural Dr. Manuel
  M. Villada, Universidad Autónoma del
  Estado de México, Toluca . . . 031598
Museo de Historia Natural Dr. Manuel
  Martínez Solórzano, Universidad
  Michoacana, Morelia 031306
Museo de Historia Natural Dr. Ricardo
  S. Vadell, Suipacha 000636
Museo de Historia Natural Felipe Poey,
  La Habana 009174
Museo de Historia Natural Javier
  Prado, Universidad Nacional Mayor
  de San Marcos, Lima 034337
Museo de Historia Natural José Vilet
  Brullet, Mexquitic 031272
Museo de Historia Natural Noel Kempf
  Mercado, Universidad Autonoma
  Gabriel Rene Moreno, Santa
  Cruz 004239
Museo de Historia Natural Pedro
  Villalobos, Cobija 004204
Museo de Historia Natural Regional,
  Universidad Juárez del Estado de
  Durango, Lerdo 031113
Museo de Historia Natural Valle de la
  Prehistoria, Santiago de Cuba . 009351
Museo de Historia Natural Víctor Vaca
  Aguinaga, Lambayeque 034302
Museo de Historia Natural,
  Universidad Autónoma de Baja
  California Sur, La Paz 031097
Museo de Historia Natural,
  Universidad de Santiago de
  Compostela, Santiago de
  Compostela 040125
Museo de Historia Natural,
  Universidad Industrial de Santander,
  Bucaramanga 008639
Museo de Historia Natural,
  Universidad Mayor, Real y Pontificia
  de San Francisco Xavier, Sucre 004245
Museo de Historia Reginal San
  Lorenzo y Galería del Libertador
  General San Martín, San
  Lorenzo 000590
Museo de Historia Regional de Pánuco,
  Pánuco, Veracruz 031359
Museo de Historia Regional
  de Sabinas Hidalgo, Sabinas
  Hidalgo 031419
Museo de Historia Regional de Tristán
  Suárez, Trisán Suárez 000657
Museo de Historia Republicana,
  Tegucigalpa 023403
Museo de História y Artesanía de la
  Colonial Tovar, Colonia Tovar . 054746
Museo de História y Ciencias
  Naturales, Lobería 000429

Museo de Historia y Cultura Casa
  Pedrilla, Cáceres 039184
Museo de Historia y Geografía, Santo
  Domingo 010389
Museo de Histórico, Tunja 008815
Museo de Hostotipaquillo,
  Hostotipaquillo 031047
Museo de Hualpén, Concepción 007520
Museo de Huelva, Huelva 039497
Museo de Iconos – Casa Grande,
  Torrejón de Ardoz 040260
Museo de Imáges de la Semana
  Santa en Popayán, Popayán . . 008766
Museo de Inmigrantes de Alemania,
  Mataglpa 033395
Museo de Insectos, Universidad de
  Costa Rica, San José 008856
Museo de Instituto Magnasco,
  Gualeguaychú 000383
Museo de Instituto Nacional de
  Estudios de Teatro, Buenos
  Aires 000195
Museo de Instrumentos Musicales,
  Buenos Aires 000196
Museo de Instrumentos Musicales de
  Bolivia, La Paz 004213
Museo de Instrumentos Musicales
  Pedro Pablo Traversari, Casa de la
  Cultura Ecuatoriana, Quito . . . 010484
Museo de Intendencia, Madrid . 039644
Museo de Investigaciones de Zonas
  Aridas, Universidad Nacional Agraria,
  Lima 034338
Museo de Jaén, Jaén 039524
Museo de Jala, Jala 031075
Museo de Jamapa, Jamapa . . . 031080
Museo de Jérica, Jérica 039531
Museo de la Abadía del Sacro-Monte,
  Granada 039467
Museo de la Academia de Infantería,
  Toledo 040236
Museo de la Acuarela del Estado de
  México, Toluca 031599
Museo de la Administración Federal
  de Ingresos Públicos, Buenos
  Aires 000197
Museo de la Agricultura, Torremocha
  de Jarama 040262
Museo de la Alhambra, Granada 039468
Museo de la Amistad México-Cuba,
  Tuxpan de Rodríguez Cano . . 031619
Museo de la Antigua Estación
  Ferroviaria, Teocelo 031535
Museo de la Arriería Maragata,
  Santiago Millas 040134
Museo de la Artesanía Tradicional
  Juan Alfonso Carrizo, La Plata 000412
Museo de la Aseguradora, Colección
  Arte Siglo XX, Alicante 038946
Museo de la Aviación Naval Argentina,
  Bahía Blanca 000134
Museo de la Balanza, Buenos
  Aires 000198
Museo de la Basílica de Guadalupe,
  México 031187
Museo de la Basílica de Santa María,
  Elche 039367
Museo de la Basílica Nuestra Señora
  de la Candelaria, Convento de los
  Dominicos, Candelaria 039218
Museo de la Biblia, Lima 034339
Museo de la Biblioteca Nacional,
  Madrid 039645
Museo de la Cabrera, Encinedo 039376
Museo de la Caña de Azúcar, El
  Cerrito 008675
Museo de la Canción Yucateca,
  Mérida 031136
Museo de la Capilla Real,
  Granada 039469
Museo de la Caricatura, México 031188
Museo de la Caricatura Severo
  Vaccaro, Buenos Aires 000199
Museo de la Casa de Acuerdo de

San Nicolás, San Nicolás de los
  Arroyos 000604
Museo de la Casa de Cultura de
  Tepeapulco, Tepeapulco Hidalgo 031544
Museo de la Casa de Cultura de
  Zacapú, Zacapu 031685
Museo de la Casa de Gobierno, San
  Miguel de Tucumán 000596
Museo de la Casa de la Cultura,
  Pinchote 008761
Museo de la Casa de la Cultura, San
  Luis Potosí 031456
Museo de la Casa de la
  Independencia, Asunción 034199
Museo de la Casa de la Música
  Mexicana, México 031189
Museo de la Casa de las Artesanías
  del Estado de Chihuahua,
  Bocoyna 030848
Museo de la Casa del Risco,
  México 031190
Museo de la Casa Histórica de la
  Independencia Nacional, San Miguel
  de Tucumán 000597
Museo de la Casa Rosada, Buenos
  Aires 000200
Museo de la Catedral, Astorga . 039009
Museo de la Catedral, Cusco . . 034260
Museo de la Catedral, Granada . 039470
Museo de la Catedral, Murcia . . 039796
Museo de la Catedral, Trujillo . . 034412
Museo de la Catedral de Santiago
  de Compostela, Santiago de
  Compostela 040126
Museo de la Cerámica
  Contemporánea, Santo
  Domingo 010390
Museo de La Cerería,
  Tlayacapan 031590
Museo de la Charrería, México . 031191
Museo de la Ciencia de Valladolid,
  Valladolid 040336
Museo de la Ciencia y del Agua,
  Murcia 039797
Museo de la Ciencia y el Cosmos, La
  Laguna de Tenerife 039542
Museo de la Ciencia y el Juego,
  Bogotá 008606
Museo de la Ciudad, Buenos
  Aires 000201
Museo de la Ciudad, Campeche 030851
Museo de la Ciudad, Guadalajara 031004
Museo de la Ciudad, Madrid . . . 039646
Museo de la Ciudad, Melilla . . . 039745
Museo de la Ciudad, Mérida . . . 031137
Museo de la Ciudad, Murcia . . . 039798
Museo de la Ciudad, Paraná . . . 000485
Museo de la Ciudad, Quito 010485
Museo de la Ciudad, Salta 000557
Museo de la Ciudad, Tecate . . . 031522
Museo de la Ciudad, Victoria . . 000667
Museo de la Ciudad, Vinarós . . 040420
Museo de la Ciudad de Chetumal,
  Chetumal 030870
Museo de la Ciudad de Colón,
  Colón 000308
Museo de la Ciudad de Córdoba,
  Córdoba 030911
Museo de la Ciudad de Irapuato,
  Irapuato 031067
Museo de la Ciudad de La Habana, La
  Habana 009175
Museo de la Ciudad de León,
  León 031109
Museo de la Ciudad de México,
  México 031192
Museo de la Ciudad de Píritu,
  Píritu 054772
Museo de la Ciudad de Querétaro,
  Querétaro 031410
Museo de la Ciudad de San Francisco,
  San Francisco 000577
Museo de la Ciudad de Sombrerete,
  Sombrerete 031502

Register der Institutionen und Firmen

Museo de la Ciudad de Tepatitlan de Morelos, Tepatitlán de – Museo de la Solidaridad Salvador Allende, Santiago de Chile

Museo de la Ciudad de Tepatitlán de Morelos, Tepatitlán de Morelos 031540
Museo de la Ciudad de Veracruz, Veracruz 031646
Museo de la Ciudad Tuxtlán, Tuxtla Gutiérrez 031623
Museo de la Clandestinidad, Nueva Gerona 009272
Museo de la Coca, La Paz 004214
Museo de la Cofradía del Espíritu Santo de los Congos, Villa Mella 010406
Museo de la Colegiata, Medinaceli 039742
Museo de la Colegiata de Santa Juliana y Claustro, Santillana del Mar 040137
Museo de la Colegiata San Luis, Villagarcía de Campos 040406
Museo de la Colegiata, Colegiata-Parroquia de San Bartolomé Apóstol, Belmonte 039121
Museo de la Colonia Siete de Marzo, Juan Aldama 031089
Museo de la Colonización, Esperanza 000367
Museo de la Comisión Provincial de Monumentos, León 039557
Museo de la Compañía de Bomberos Italia No. 5, Bellavista 034225
Museo de la Comunidad de Agua Blanca, Puerto López 010468
Museo de la Comunidad de Chordeleg, Chordeleg 010421
Museo de la Concatedral de San Pedro, Soria 040196
Museo de la Constitución, Guáimaro 009126
Museo de la Coria, Convento, Trujillo 040275
Museo de la Cueva del Señor del Calvario, México 031193
Museo de la Cultura de Albán, Albán 008564
Museo de la Cultura Huasteca, Tampico 031509
Museo de la Cultura José Arens Berg, Sullana 034401
Museo de la Cultura Martínez de Compañón, Tambogrande 034408
Museo de la Cultura Maya, Chetumal 030871
Museo de la Cultura Popular, Cañete 039221
Museo de la Cultura Potosina, San Luis Potosí 031457
Museo de la Diócesis de Matagalpa, Matagalpa 033396
Museo de la Dirección Nacional del Antártico, Buenos Aires 000202
Museo de la Diversidad Biológica y Cultural del Pacífico, Quibdó . . 008773
Museo de la E.E.M. Número 3, Colegio Nacional, Dolores 000355
Museo de la Electricidad, Caracas 054727
Museo de la Electricidad, Lima . 034340
Museo de la Encarnación, Corella 039313
Museo de la Energía, Embalse . 000363
Museo de la Escuela Militar de Montaña y Operaciones Especiales, Huesca 039500
Museo de la Escultura Mexica Eusebio Dávalos, Acatitlán 030803
Museo de la Estampa, Cuitzeo . 030933
Museo de la Estampa del Estado de México, Toluca 031600
Museo de la Estampa y del Diseño Carlos Cruz Diez, Caracas . . . 054728
Museo de la Evangelización, Huejotzingo 031055
Museo de la Fabrica Nacional de Moneda y Timbre, Madrid 039647

Museo de la Facultad de Ciencias de la Salud, Universidad Mayor de San Francisco Xavier, Sucre 004246
Museo de la Facultad de Ciencias Naturales y del Instituto Miguel Lillo, San Miguel de Tucumán 000598
Museo de la Familia Dominicana del Siglo XIX, Museo Casa del Tostado, Santo Domingo 010391
Museo de la Farmacia Dra. D'Alessio Bonino, Buenos Aires 000203
Museo de la Farmacia Hispana, Madrid 039648
Museo de la Fauna, Aguascalientes 030810
Museo de la Fauna de Veracruz, Xalapa 031669
Museo de la Fauna y Ciencias Naturales, Monterrey 031291
Museo de la Ferrería de La Pradera, Subachoque 008803
Museo de la Fiesta, Caravaca de la Cruz 039228
Museo de la Forja, El Franco . . 039410
Museo de la Fotografía, Rafaela 000513
Museo de la Fotografía, Fototeca del INAH, Pachuca 031351
Museo de la Fuerza de Aviación Naval, Callao 034233
Museo de la Fuerza de Submarinos, Callao 034234
Museo de la Fundación Guayasamin, Quito 010486
Museo de la Fundación John Boulton, Caracas 054729
Museo de la Gaita, Museo del Pueblo de Asturias, Gijón 039437
Museo de la Gráfica de la Historia Social de Taxco Siglo XX, Taxco de Alarcón 031514
Museo de la Guardia Civil, Madrid 039649
Museo de la Guerra de Castas, Felipe Carrillo Puerto 030976
Museo de la Guerra Hispano-Cubano-Norteamericana, Santiago de Cuba 009352
Museo de la Historia de Ponce, Ponce 036142
Museo de la Hoya, Laguardia . 039538
Museo de la Huaxteca, Amatlán de los Reyes 030824
Museo de la Huerta, Almoradi . 038964
Museo de la Iglesia de la Compañía de Jesús, Ica 034287
Museo de la Iglesia de San Pedro, Chucuito 034250
Museo de la Iglesia de Santa María la Real, Sasamón 040148
Museo de la Iglesia Parroquial San Pedro González Telmo y su Claustro Histórico, Buenos Aires 000204
Museo de la Iglesia, Claustro alto de la Catedral, Oviedo 039863
Museo de La Imagen, Vista Alegre 009391
Museo de la Independencia, Dolores Hidalgo 030945
Museo de la Indumentaria Mexicana Luis Márquez Romay, México . 031194
Museo de la Industria Azucarera, Abel Santamaría 009059
Museo de la Industria Azucarera Obispo Colombres, San Miguel de Tucumán 000599
Museo de la Inmaculada Concepción, Cuenca 010430
Museo de la Inmigración Japonesa, Lima 034341
Museo de la Insurgencia, Rincón de Romos 031415
Museo de la Isla de Cozumel, Cozumel 030918
Museo de la Labranza, Francisco I.

Madero, Hidalgo 030981
Museo de la Lealtad Republicana, Chihuahua 030879
Museo de la Lucha Clandestina, Nueva Gerona 009273
Museo de la Lucha Clandestina, Santiago de Cuba 009353
Museo de la Lucha Estudiantil Jesús Suárez Gayol, Camagüey 009090
Museo de la Lucha Obrera, Cananea 030855
Museo de la Luz, México 031195
Museo de la Máquina Agrícola, Esperanza 000368
Museo de la Máscara, Morelia . 031307
Museo de la Medicine, Cuenca . 010431
Museo de la Memoria, Cuacutecuta 008670
Museo de la Memoria, Tlaxcala 031587
Museo de la Miel, Peñalver . . . 039933
Museo de la Minería, San Juan de Guadalupe 031447
Museo de la Minería del Estado de México, El Oro 030959
Museo de la Minería del País Vasco, Euskal Herriko Meatzaritzaren Museoa, Gallarta 039424
Museo de la Minería José de la Borda, Taxco de Alarcón 031515
Museo de la Minería y de la Industria de Asturias, San Martín del Rey 040066
Museo de la Moneda, Caracas . 054730
Museo de la Moto Clásica de Hervás, Hervás 039490
Museo de la Muerte, San Juan del Río 031448
Museo de la Música Alejandro García Caturla, Santa Clara 009334
Museo de la Música Grabada, Zapotlanejo 031704
Museo de la Música Puertorriqueña, Ponce 036143
Museo de la Música Rodrigo Prats, Santa Clara 009335
Museo de la Nación, Lima . . . 034342
Museo de la Nacionalidad, La Villa de Los Santos 034183
Museo de la Naturaleza de Cantabria, Carrejo-Cantabria 039235
Museo de la No Intervención Fuerte de Loreto, Puebla 031387
Museo de la Paeria, Lleida . . . 039578
Museo de la Palabra, Villa Adelina 000669
Museo de la Palabra y La Imagen, San Salvador 010596
Museo de la Parroquia de Gámbita, Gámbita 008684
Museo de la Partitura Historica, Rosario 000544
Museo de la Pasión Boquense, Buenos Aires 000205
Museo de la Patagonia Dr. Francisco P. Moreno, Bariloche 000144
Museo de la Paz de Gernika, Gernika-Lumo 039433
Museo de la Pintura Mural y Museo Manuel Gamio, Teotihuacán . . 031536
Museo de la Piratería, Santiago de Cuba 009354
Museo de La Plata, La Plata . . 000413
Museo de la Platería, Taxco de Alarcón 031516
Museo de la Pluma, México . . 031509
Museo de la Policía del Chaco, Resistencia 000523
Museo de la Policía Federal Argentina, Buenos Aires 000206
Museo de la Policía Preventiva de la Ciudad de México, México . . 031197
Museo de la Porcelana, Santo Domingo 010392
Museo de la Provincia de Granma, Bayamo 009073

Museo de la Provincia El Dorado, El Dorado 034271
Museo de la Psicología Experimental Argentina Horacio G. Piñero, Buenos Aires 000207
Museo de la Purísima Concepción, Celaya 030863
Museo de la Quemada, Guadalupe 031018
Museo de la Radio y TV – Lulio Moscoso, Santo Domingo 010404
Museo de la Radiodifusión, Puebla 031388
Museo de la Real Academia de Bellas Artes de la Purísima Concepción, Valladolid 040337
Museo de la Real Academia de Bellas Artes de San Fernando, Madrid 039650
Museo de la Real Academia Nacional de Farmacia, Madrid 039651
Museo de la Real Capilla, Arenas de San Pedro 038986
Museo de la Real Colegiata, Roncesvalles 040018
Museo de la Real Maestranza de Caballería de Ronda, Ronda . 040019
Museo de la Real Plaza de Toros de Sevilla, Museo Taurino de la Real Maestranza de Caballería de Sevilla, Sevilla 040175
Museo de la Recoleta, Arequipa 034213
Museo de la Recoleta, Sucre . 004247
Museo de la Reconquista, Orihuela 039845
Museo de la Reconquista, Tigre 000644
Museo de la Región de Ansenuza Aníbal Montes, Miramar 000467
Museo de la Reincorporación, Tacna 034404
Museo de la Resistencia Indígena, Ixcateopan de Cuauhtémoc . . 031069
Museo de la Revolución, La Habana 009176
Museo de la Revolución, Nueva Gerona 009274
Museo de la Revolución de San Pedro, San Pedro 031477
Museo de la Revolución Salvadoreña, Perquín 010590
Museo de la Revolución, Casa Hermanos Aquiles Serdán, Puebla 031389
Museo de la Rioja, Logroño . . 039594
Museo de la Sal, Zipaquirá . . . 008830
Museo de la Sal y el Azúcar, Aguadulce 034168
Museo de la Salle, Dasmariñas . 034475
Museo de la Santa Casa de Ejercicios Espirituales Sor María Antonia de la Paz y Figueroa, Buenos Aires . 000208
Museo de la Santa Iglesia Catedral, Ávila 039022
Museo de la Secretaria de Hacienda y Credito Publico Antiguo Palacio del Arzobispado, México 031198
Museo de la Selva Juan Forster, Miramar 000468
Museo de la Semana Santa de Cuenca, Cuenca 039334
Museo de la Semana Santa Marinera de Valencia Salvador Caurín Alarcón, Valencia 040309
Museo de la Sep, México 031199
Museo de la Shoá- Fundación Memoria del Holocausto, Buenos Aires 000209
Museo de la Sidra Copa de Oro, San Andrés Cholula 031431
Museo de la Sidra de Asturias, Nava 039814
Museo de la Sociedad Amigos de Laguardia, Laguardia 039539
Museo de la Soledad, Oaxaca . 031334
Museo de la Solidaridad Salvador

Allende, Santiago de Chile ... 007542
Museo de la Tortuga, Ecomuseo de
Colola, Aquila 030832
Museo de la Universidad,
Salamanca 040042
Museo de la Universidad de Valladolid,
Valladolid 040338
Museo de la Universidad Nacional
Hermilio Valdizán, Huánuco ... 034279
Museo de la Vera Cruz, Caravaca de
la Cruz 039229
Museo de la Vida Animal Herpetario,
Puebla 031390
Museo de la Vida Rural, L'Espluga de
Francolí 039383
Museo de la Villa de Ahome,
Ahome 030814
Museo de la Villa de Almonte,
Almonte 038963
Museo de Labranza de la Torre Vieja,
San Clemente 040055
Museo de Laca, Chiapa de Corzo 030872
Museo de Lagos de Moreno, Lagos de
Moreno 031102
Museo de lan deuda externa, Buenos
Aires 000210
Museo de Las Americas, Denver 048269
Museo de Las Américas, San
Juan 036161
Museo de las Artes Populares Cidap,
Cuenca 010432
Museo de las Artes Populares de
Jalisco, Guadalajara 031006
Museo de las Artes, Universidad de
Guadalajara, Guadalajara 031005
Museo de las Atarazanas Reales,
Santo Domingo 010393
Museo de las Aves de México,
Saltillo 031423
Museo de las Californias, Los
Cabos 031118
Museo de las Californias, Tijuana 031567
Museo de las Campanas, Mina
Clavero 000466
Museo de las Carmelitas, Peñaranda
de Bracamonte 039934
Museo de las Casas Reales, Santo
Domingo 010394
Museo de las Ciencias de Castilla La
Mancha, Cuenca 039335
Museo Las Conceptas,
Cuenca 010433
Museo de las Cortes de Cádiz,
Cádiz 039191
Museo de las Culturas, Cotacahi 010422
Museo de las Culturas, Saltillo . 031424
Museo de las Culturas Aborígenes,
Cuenca 010434
Museo de las Culturas Afromestizas,
Cuajinicuilapa 030920
Museo de las Culturas de la Huasteca
Potosina Tamuantzán, Valles . 031639
Museo de las Culturas de Oaxaca,
Oaxaca 031335
Museo de las Culturas de Occidente,
Colima 030890
Museo de las Culturas del Norte,
Monte Verde 031285
Museo de las Culturas Prehispánicas,
Puerto Vallarta 031405
Museo de las Culturas Prehispánicas
de Durango, Durango 030951
Museo de las Encartaciones,
Sopuerta 040194
Museo de las Escuelas, Buenos
Aires 000211
Museo de las Esculturas de Copán,
Copán 023399
Museo de las Estelas Mayas,
Campeche 030852
Museo de las Leyendas,
Guanajuato 031025
Museo de las Luchas Obreras,
Manzanillo 009246

Museo de las Máscaras Ceremoniales,
Chichicastenango 023364
Museo de las Miniaturas Poblanas,
Palmar de Bravo 031357
Museo de las Misiones Jesuitas,
Loreto 031117
Museo de las Momías,
Guanajuato 031026
Museo de las Muñecas, Píritu . 054773
Museo de las Parrandas de Remedios,
Remedios 009304
Museo de las Pinturas Rupestres
de San Ignacio, Pueblo de San
Ignacio 031402
Museo de las Revoluciones Mariano
Jiménez, San Luís Potosí ... 031458
Museo de las Telecomunicaciones,
Madrid 039652
Museo de las Telecomunicationes,
México 031200
Museo de las Termas Públicas de
Caesaraugusta, Zaragoza 040447
Museo de las Tradiciones Potosinas,
San Luís Potosí 031459
Museo de las Villas Pasiegas, Vega de
Pas 040359
Museo de l'Enginyeria de Catalunya,
Barcelona 039068
Museo de León, León 039558
Museo de León – Edificio Pallares,
León 039559
Museo de Lluc, Escorca 039380
Museo de los Altos de Chiapas, San
Cristóbal de las Casas 031435
Museo de los Andes, Socha ... 008797
Museo de los Asentamientos,
Federación 000371
Museo de los Bomberos,
Guayaquil 010447
Museo de los Bomberos,
Matanzas 009253
Museo de los Caminos, Astorga 039010
Museo de los Combates contra los
Bandidos, Trinidad 009375
Museo de los Combatientes del Morro
de Arica, Lima 034343
Museo de los Concilios y de la Cultura
Visigoda, Museo del Santa Cruz,
Iglesia de San Román, Toledo . 040237
Museo de los Condestables de
Castilla, Medina de Pomar .. 039737
Museo de los Corporales, Museo
Diocesano, Daroca 039341
Museo de los Corrales Viejos de
Parque de los Patricios, Buenos
Aires 000212
Museo de Los Ferrocarrileros,
México 031201
Museo de los Héroes de la
Restauración, Santiago de los
Caballeros 010383
Museo de los Instrumentos del Teatro
Colón, Buenos Aires 000213
Museo de los Llanos, Guanare . 054754
Museo de los Metales, Cuenca . 010435
Museo de los Minerales y Rocas,
Porriño 039964
Museo de los Molinos de Gregorio
Prieto, Valdepeñas 040293
Museo de los Molinos de Viento,
Campo de Criptana 039213
Museo de los Niños, Bogotá .. 008607
Museo de los Niños Abasto, Buenos
Aires 000214
Museo de los Niños Barrilete,
Córdoba 000325
Museo de los Niños de Caracas,
Caracas 054731
Museo de los Niños Tin Marin, San
Salvador 010597
Museo de los Orígenes, Madrid 039653
Museo de los Orishas, Asociacion
Cultural Yoruba de Cuba, La
Habana 009177

Museo de los Patricios de Buenos
Ayres, Buenos Aires 000215
Museo de los Pioneros, Río
Gallegos 000531
Museo de Málaga, Sección de
Arqueología, Málaga 039702
Museo de Málaga, Sección de Bellas
Artes, Málaga 039703
Museo de Maní y Museo del Convento
San Miguel Arcángel, Maní ... 031121
Museo de Marionetas, Bogotá . 008608
Museo de Mas de las Matas, Mas de
las Matas 039731
Museo de Medicina Legal y Ciencas
Forenses, Bogotá 008609
Museo de Medicina Maya, San
Cristóbal de las Casas 031436
Museo de Medicina Tradicional y
Herbolaria, Cuernavaca 030927
Museo de Medicina Tropical Carlos J.
Finlay, La Habana 009178
Museo de Medios de Comunicación,
Resistencia 000524
Museo de Menorca, Mahón ... 039694
Museo de Metales Preciosos
Precolombinos o Museo del Oro, La
Paz 004215
Museo de Metrología – Museo de
Pesas y Medidas, Centro Español de
Metrología, Tres Cantos 040274
Museo de Mexticacan,
Mexticacán 031273
Museo de Mier y Noriega, Mier y
Noriega 031275
Museo de Minerales del Instituto
Geofísico, Universidad Javeriana,
Bogotá 008610
Museo de Minerales y Fósiles,
Zumárraga 040469
Museo de Mineralogía Dr Miguel
Romero Sánchez, Tehuacán . 031529
Museo de Mineralogía y Geología Dr.
Alfredo Stelzner, Córdoba 000326
Museo de Mineralogía y Geología
Ing. Luis Silva Ruelas, Universidad
Michoacana, Morelia 031308
Museo de Mineralogía, Universidad
Autónoma de Hidalgo, Pachuca 031352
Museo de Mineralogía, Universidad de
Guanajuato, Guanajuato 031027
Museo de Minería de la Compañía
Real del Monte, Pachuca 031353
Museo de Minería e Historia de
Cosalá, Cosalá 030913
Museo de Miniaturas, Huaraz . 034281
Museo de Miniaturas Militares
Ciudadela de Jaca, Jaca 039518
Museo de Mitos y Leyendas, Tafí del
Valle 000639
Museo de Momias de Celaya,
Celaya 030864
Museo de Monseñor Ramón Ibarra y
González, Puebla 031391
Museo de Moyogalpa,
Moyogalpa 033398
Museo de Municipal, Bielsa .. 039138
Museo de Murcia, Sección de
Arqueología, Murcia 039799
Museo de Museos Colsubsidio,
Bogotá 008611
Museo de Navajas y Curiosidades del
Mundo, Puebla 031392
Museo de Navarra, Pamplona . 039911
Museo de Nosa Señora da Antiga,
Monforte de Lemos 039768
Museo de Nuestra Raíz Africana,
Instituto de Cultura Puertorriqueña,
San Juan 036162
Museo de Nuestra Señora del Rosario,
Talpa de Allende 031508
Museo de Numismática del Estado de
México, Toluca 031601
Museo de Numismática, Museos del
Banco Central de Costa Rica, San

José 008857
Museo de Odontología, Madrid . 039654
Museo de Odontología, México . 031202
Museo de Oftalmología Doctor
Saracíbar, Valladolid 040339
Museo de Oro Precolombino, Museos
del Banco Central de Costa Rica,
San José 008858
Museo de Orozko, Orozko 033853
Museo de Oxolotán, Tacotalpa . 031505
Museo de Palencia, Palencia .. 039873
Museo de Paleontología, Delicias 030942
Museo de Paleontología de
Guadalajara, Guadalajara 031007
Museo de Paleontología de
la Preparatoria, Apaxtla de
Castrejón 030830
Museo de Paleontología de Rincón
Gallardo, General Cepeda ... 030985
Museo de Paleontología Eliseo
Palacios Aguilera, Tuxtla
Gutiérrez 031624
Museo de Paracho, Paracho .. 031363
Museo de Patología, Buenos
Aires 000216
Museo de Patones, Patones .. 039920
Museo de Penonomé, Penonomé 034187
Museo de Pericos, Culiacán .. 030936
Museo de Piedra y Artesanía Canaria,
Ingenio 039513
Museo de Piria, Piriápolis 046287
Museo de Pontevedra,
Pontevedra 039960
Museo de Prehistoria e Arqueoloxía de
Vilalba, Vilalba 040387
Museo de Prehistoria y Arqueología de
Cantabria, Santander 040114
Museo de Prehistoria y Paleontología,
Orce 039842
Museo de Prehistoria y Servicio
de Investigación Prehistórica,
Valencia 040310
Museo de Pungarabato,
Pungarabato 031408
Museo de Quejana, Monasterio de
Dominicas, Valdegovía 040290
Museo de Relojes, Colección
de Relojes, Palacio Municipal, A
Coruña 039322
Museo de Reproducciones Artísticas,
Bilbao 039145
Museo de Rocas, Parque Nacional de
Timanfaya, Yaiza 040433
Museo de Roda de Isábena, Roda de
Isábena 040010
Museo de Ruiz, Ruiz 031418
Museo de SADAIC Vincente López y
Planes, Buenos Aires 000217
Museo de Salamanca,
Salamanca 040043
Museo de San Antolín,
Tordesillas 040251
Museo de San Esteban,
Salamanca 040044
Museo de San Esteban Tetelpa,
Zacatepec de Hidalgo 031695
Museo de San Francisco, Potosí 004235
Museo de San Jerónimo, Casa 034261
Museo de San José María de Yermo y
Parres, Puebla 031393
Museo de San Juan, San Juan . 036163
Museo de San Juan Raya, Zapotitlán
Salinas 031702
Museo de San Juan Tlacotenco, San
Juan Tlacotenco 031453
Museo de San Lorenzo Tenochtitlán,
Texistepec 031564
Museo de San Miguel, Tlaxiaco 031589
Museo de San Pablo, San Pablo
Huitzo 031473
Museo de San Pedro, Navolato . 031323
Museo de San Pedro, San Pedro
Lagunillas 031479
Museo de San Salvador, Toro . 040257

Museo de Sanidad e Higiene
Pública, Instituto de Salud Carlos III,
Madrid 039655
Museo de Santa Clara, Bogotá . 008612
Museo de Santa Cruz, Toledo .. 040238
Museo de Santa Monica, Panay 034560
Museo de Santiago de los Caballeros,
Antigua 023362
Museo de Santuario di Montevergine,
Santuario di Montevergine ... 027850
Museo de Sayula Juan Rulfo,
Sayula 031499
Museo de Segóbriga, Museo de
Cuenca, Saelices 040033
Museo de Segovia, Casa del Sol,
Segovia 040157
Museo de Semana Santa,
Orihuela 039846
Museo de Semana Santa,
Zamora 040439
Museo de Semana Santa, Iglesia de
Santa Cruz, Medina de Rioseco 039738
Museo de Sitio Arqueologico el
Mongote Real Alto, Guayaquil . 010448
Museo de Sitio Arqueológico Kuntur
Wasí, San Pablo 034397
Museo de Sitio Boca de Potrerillos,
Mina 031277
Museo de Sitio Casa de Juárez,
Oaxaca 031336
Museo de Sitio Castillo de Teayo,
Castillo de Teayo 030861
Museo de Sitio Complejo Arqueológico
de Sillustani, Atuncolla 034219
Museo de Sitio de Cacaxtla,
Nativitas 031320
Museo de Sitio de Cempoala,
Cempoala 030865
Museo de Sitio de Chalcatzingo,
Chalcatzingo 030867
Museo de Sitio de Chan Chan,
Trujillo 034413
Museo de Sitio de Chichen Itza,
Timun 031569
Museo de Sitio de Chinchero,
Urubamba 034416
Museo de Sitio de Cholula, San Pedro
Cholula 031478
Museo de Sitio de Coatetelco,
Coatetelco 030887
Museo de Sitio de Comalcalco,
Comalcalco 030901
Museo de Sitio de el Tajín,
Papantla 031362
Museo de Sitio de el Zapotal, Ignacio
de la Llave 031064
Museo de Sitio de Higueras, Vega de
Alatorre 031642
Museo de Sitio de Huallamarca,
Lima 034344
Museo de Sitio de Ingapirca,
Azogues 010415
Museo de Sitio de La Isabela,
Luperón 010373
Museo de Sitio de la Pampa Galeras,
Quinua 034394
Museo de Sitio de la Vega Vieja, La
Vega Vieja 010405
Museo de Sitio de La Venta,
Huimanguillo 031060
Museo de Sitio de la Zona
Arqueológica de Tzintzuntzan,
Tzintzuntzan 031629
Museo de Sitio de Monte Albán,
Oaxaca 031337
Museo de Sitio de Narihualá,
Catacaos 034242
Museo de Sitio de Ocuilan,
Ocuilan 031346
Museo de Sitio de Pachacamac,
Lima 034345
Museo de Sitio de Palenque Alberto
Ruz Lhuillier, Palenque 031356
Museo de Sitio de Paracas Julio C.

Tello, Pisco 034382
Museo de Sitio de Pomona, Tenosique
de Pino Suárez 031534
Museo de Sitio de Puruchuco,
Lima 034346
Museo de Sitio de Querétaro,
Querétaro 031411
Museo de Sitio de San Isidro,
Tembladera 034409
Museo de Sitio de San Lorenzo
Tenochtitlán, Tenochtitlán 031533
Museo de Sitio de San Miguel Ixtapan,
San Miguel Ixtapan 031468
Museo de Sitio de Sipán, Zaña . 034422
Museo de Sitio de Tenayuca,
Tlalnepantla 031578
Museo de Sitio de Teotihuacán,
Teotihuacán 031537
Museo de Sitio de Toniná,
Ocosingo 031343
Museo de Sitio de Tres Zapotes,
Santiago Tuxtla 031496
Museo de Sitio de Túcume,
Campo 034238
Museo de Sitio de Uxmal, Santa
Elena 031488
Museo de Sitio de Wari, Wari .. 034418
Museo de Sitio de Xochicalco,
Xochicalco 031675
Museo de Sitio de Xochitecatl,
Nativitas 031321
Museo de Sitio Degollado Jalisco,
Degollado 030941
Museo de Sitio del Alto de la Alianza,
Tacna 034405
Museo de Sitio del Claustro de Sor
Juana, México 031203
Museo de Sitio del Mirador del Cerro
San Cristóbal, Lima 034347
Museo de Sitio del Rumicucho, San
Antonio 010503
Museo de Sitio El Algarrobal, El
Algarrobal 034270
Museo de Sitio El Torreoncito,
Torreón 031608
Museo de Sitio Huaca El Dragón, La
Esperanza 034299
Museo de Sitio Huaca Pucllana,
Lima 034348
Museo de Sitio Ocotelulco,
Totolac 031615
Museo de Sitio Parque Reducto 2,
Lima 034349
Museo de Sitio Polyforum Siqueiros,
México 031204
Museo de Sitio San Agustín,
Juárez 031092
Museo de Sitio Talavera, México 031205
Museo de Sitio Tecpan, México . 031206
Museo de Sitio Wari, Ayacucho . 034222
Museo de Sitio Wari Willca,
Huancayo 034276
Museo de Siyasa, Cieza 039281
Museo de Sor de María Romero,
Granada 033378
Museo de Suelos de Colombia,
Instituto Geográfico Agustín Codazzi,
Bogotá 008613
Museo de Suelos, Universidad
Nacional Amazónica, Iquitos .. 034291
Museo de Susticacán,
Susticacán 031503
Museo de Tabar, Navarra 039815
Museo de Tabar, Tabar 040198
Museo de Tacuichamona,
Culiacán 030937
Museo de Tahdziú, Tahdziú 031506
Museo de Tapalpa, Tapalpa 031512
Museo de Tapices, Iglesia de San
Millán, Oncala 039835
Museo de Tarlac, Tarlac 034607
Museo de Taximaroa, Hidalgo .. 031044
Museo de Tecnología Aeronáutica y
Espacial, Córdoba 000327

Museo de Telecomunicaciones,
Buenos Aires 000218
Museo de Telefonía, Delicias ... 030943
Museo de Tepeticpa, Tepeticpa . 031546
Museo de Terra Santa, Santiago de
Compostela 040127
Museo de Texcalyacac,
Texcalyacac 031562
Museo de Textiles Andinos Bolivianos,
La Paz 004216
Museo de Tlahualilo, Tlahualilo de
Zaragoza 031575
Museo de Tlalancaleca, San Matías
Tlalancaleca 031465
Museo de Tlapa, Tlapa de
Comonfort 031581
Museo de Tlayacapan,
Tlayacapan 031591
Museo de Todos los Santos, La
Paz 031098
Museo de Totolapan, Totolapan . 031616
Museo de Tradiciones Populares de
Aguadas, Aguadas 008562
Museo de Trajes Regionales de
Colombia, Bogotá 008614
Museo de Transportes, Complejo
Museográfico Enrique Udaondo,
Luján 000434
Museo de Tudela, Palacio Decanal,
Tudela 040277
Museo de Ulía, Iglesia Parroquial
Nuestra Señora de la Asunción,
Montemayor 039778
Museo de Unidades Acorazadas,
Madrid 039656
Museo de Urología de Buenos Aires,
Buenos Aires 000219
Museo de Usos y Costumbres
Etnológicas, Paracuellos de
Jarama 039913
Museo de Valladolid, Palacio de Fabio
Nelli, Valladolid 040340
Museo de Velardeña, Velardeña 031643
Museo de Xiutetelco, Xiutetelco 031674
Museo de Xochitepec,
Xochitepec 031678
Museo de Xolalpancalli,
Zacapoaxtla 031684
Museo de Yautepec, Yautepec . 031681
Museo de Zamora, Palacio del Cordón,
Zamora 040440
Museo de Zaragoza, Secciones
de Arqueología y Bellas Artes,
Zaragoza 040448
Museo de Zaragoza, Secciónes de
Etnología y Cerámica, Zaragoza 040449
Museo de Zoología Alfonso L. Herrera,
México 031207
Museo de Zoología Comparada de
Vertebrados, Facultad de Ciencias
Biológicas, Madrid 039657
Museo de Zoología Juan Ormea,
Universidad Nacional de La Libertad,
Trujillo 034414
Museo de Zoología, Universidad de
Córdoba, Córdoba 000328
Museo de Zurbarán, Marchena . 039725
Museo degli Affreschi G.B.
Cavalcaselle, Verona 028348
Museo degli Alpini, Bassano del
Grappa 025471
Museo degli Alpini, Palazzo Comunale,
Savignone 027885
Museo degli Arazzi delle Madrice,
Marsala 026718
Museo degli Argenti, Ariano
Irpino 025368
Museo degli Argenti, Firenze ... 026270
Museo degli Argenti Contemporanei,
Sartirana Lomellina 027864
Museo degli Arredi Sacri, Castel
Sant'Elia 025814
Museo degli Attrezzi Agricoli,
Suno 028037

Museo degli Attrezzi Agricoli Ferretti
Florindo, Sefro 027910
Museo degli Ex Voto, Santuario della
Madonna delle Quercia, Viterbo 028409
Museo degli Indios dell'Amazonia,
Assisi 025399
Museo degli Insetti, Bisegna ... 025518
Museo degli Scavi, Piuro 027370
Museo degli Strumenti della
Riproduzione del Suono, Palazzo
Mattei, Roma 027585
Museo degli Strumenti Musica
Populare Sardi, Tadasuni 028048
Museo degli Strumenti Musicali
Meccanici, Savio 027886
Museo degli Strumenti Musicali
Meccanici, Sestola 027941
Museo degli Strumenti Musicali,
Istituto Musicale Arcangelo Corelli,
Cesena 025921
Museo degli Strumenti per il Calcolo,
Fondazione Galileo Galilei, Pisa 027354
Museo degli Usi e Costumi della
Gente di Romagna, Santarcangelo di
Romagna 027839
Museo degli Usi e Costumi della
Gente Trentina, San Michele
all'Adige 027783
Museo degli Usi e Costumi della Valle
di Goima, Zoldo Alto 028434
Museo degli Usi e Costumi delle Genti
dell'Etna, Giarre 026437
Museo dei Bambini Explora,
Roma 027586
Museo dei Benedettini, Pontida . 027404
Museo dei Bozzetti, Pietrasanta 027319
Museo dei Bronzi Dorati, Pergola 027250
Museo dei Burattini, Mantova .. 026698
Museo dei Burattini e delle Figure,
Cervia 025915
Museo dei Campionissimi, Novi
Ligure 027070
Museo dei Cappuccini, Bassano del
Grappa 025472
Museo dei Combattenti, Soncino 027982
Museo dei Costumi della Partita
a Scacchi, Castello Inferiore,
Marostica 026715
Museo dei Cuchi, Roana 027539
Museo dei Damaschi, Lorsica .. 026633
Museo dei Ferri Chirurgici,
Pistoia 027364
Museo dei Ferri Taglienti,
Scarperia 027901
Museo dei Fossili, Meride 041744
Museo dei Fossili, Vestenanova . 028355
Museo dei Fossili, Zogno 028431
Museo dei Fossili Don Giuseppe
Mattiacci, Serra San Quirico .. 027926
Museo dei Fossili e dei Minerali,
Smerillo 027973
Museo dei Fossili e di Storia Naturale,
Montefalcone Appennino 026932
Museo dei Fossili e Minerali del
Monte Nerone, Apecchio 025341
Museo dei Fossili e Mostra dei Funghi,
Pioraco 027345
Museo dei Grandi Fiumi, Rovigo 027688
Museo dei Legni Processionali,
Santuario Madonna della
Misericordia, Petriolo 027291
Museo dei Magli, Ponte Nossa . 027398
Museo dei Minerali e della Miniera,
Oltre il Colle 027090
Museo dei Minerali Elbani, Rio
Marina 027526
Museo dei Navigli, Milano 026804
Museo dei Parati Sacri,
Montemarano 026950
Museo dei Reperti Bellici e
Storici della Guerra 1915–1918,
Capovalle 025755
Museo dei Reperti della Civiltà
Contadina, Alberobello 025271

Museo dei Soldati del Montello,
Nervesa della Battaglia 027029
Museo dei Soldatino e della Figurina
Storica, Calenzano 025693
Museo dei Tasso e della Storia
Postale, Camerata Cornello . . . 025709
Museo dei Trasporti, Ranco 027475
Museo dei Trombini di Lessinia, San
Bortolo 027731
Museo del Aceite de Oliva, Cabra 039179
Museo del Agua y del Suelo,
Viedma 000668
Museo del Alcázar, Sevilla 040176
Museo del Alcázar, Toledo 040239
Museo del Almirez, Les Coves de
Vinromà 039326
Museo del Almudín, Xàtiva 040432
Museo del Ámbar de Chiapas, San
Cristóbal de las Casas 031437
Museo del Ambiente Histórico Cubano,
Santiago de Cuba 009355
Museo del Anís de Rute, Rute . . 040026
Museo del Antiguo Recinto del
Congreso Nacional, Buenos
Aires 000220
Museo del Antiquo Parque de Bombas,
Ponce 036144
Museo del Arbol El Genízaro,
Nagarote 033399
Museo del Archivo Histórico de
Granada, Granada 033379
Museo del Area Fundacional,
Mendoza 000454
Museo del Arroz, Valencia 040311
Museo del audifono, Buenos
Aires 000221
Museo del Automóvil, Buenos
Aires 000222
Museo del Automóvil, México . . . 031208
Museo del Automóvil,
Montevideo 046259
Museo del Automóvil Club Argentino,
Buenos Aires 000223
Museo del Automóvil de Santa
Teresita, Santa Teresita 000627
Museo del Automóvil Puebla,
Puebla 031394
Museo del Automóvil, Complejo
Museográfico Enrique Udaondo,
Luján 000435
Museo del Automovilismo Juan
Manuel Fangio, Balcarce 000138
Museo del Azulejo, Montevideo . 046260
Museo del Azulejo, Onda 039836
Museo del Azulejo Francés, Punta
Ballena 046289
Museo del Baile Flamenco,
Sevilla 040177
Museo del Banco Central,
Cuenca 010436
Museo del Banco Central, Loja . 010461
Museo del Banco Central,
Riobamba 010499
Museo del Banco Central de Manta,
Manta 010464
Museo del Banco Provincia de
Córdoba, Córdoba 000329
Museo del Bandolero, Ronda . . . 040020
El Museo del Barrio, New York . . 051432
Museo del Barro, Asunción 034200
Museo del Battistero, Albenga . 025268
Museo del Beisbol, Caracas . . . 054732
Museo del Belén, Museo de Historia
Local, San Javier 040062
Museo del Bigallo, Firenze 026271
Museo del Bijou, Casalmaggiore 025791
Museo del Biroccio Marchigiano e
Museo Beltrami, Filottrano . . 026233
Museo del Borgo, Amato 025312
Museo del Bosco, Sovicille 027997
Museo del Botijo, Villena 040418
Museo del Botijo, Palacio de Toral
de los Guzmanes, Toral de los
Guzmanes 040248

Museo del Brigante, Sonnino . . 027987
Museo del Café, Matagilpa 033397
Museo del Café, San Cristóbal de las
Casas 031438
Museo del Calzado, Elda 039373
Museo del Calzado El Borceguí,
México 031209
Museo del Camino Real de
Hecelchakan, Hecelchakán . . . 031037
Museo del Camoscio, Opi 027093
Museo del Campesino,
Xochitepec 031679
Museo del Camposanto Vecchio e
dell'Opera, Pisa 027355
Museo del Canal Interoceanico de
Panamá, Ciudad de Panamá . . 034180
Museo del Cañamo, Callosa de
Segura 039205
Museo del Canasto Cafetero y
Muestra Arqueológica de la Cultura
Quimbaya, Filandia 008679
Museo del Capitolo di San Lorenzo,
Perugia 027259
Museo del Capoluogo,
Castelbellino 025815
Museo del Cappello, Alessandria 025283
Museo del Cappello, Montappone 026910
Museo del Capriolo, Bisegna . . . 025519
Museo del Carajo, Cali 008650
Museo del Carbón, Lenguazaque 008704
Museo del Carmen, México 031210
Museo del Carmen, Villa de
Leyva 008822
Museo del Carmen de Maipú,
Santiago de Chile 007543
Museo del Carnaval, Santiago de
Cuba 009356
Museo del Carro, Buendía 039160
Museo del Carro y Aperos de
Labranza, Tomelloso 040246
Museo del Carruaje "El Tacú", Villa
Ciudad Parque 000673
Museo del Cartel José P.H. Hernández,
Río Grande 036146
Museo del Casin, Bagni di Lucca 025437
Museo del Castagno, Pescaglia . 027272
Museo del Castello, Fumone . . . 026358
Museo del Castello Colleoni Porto
Thiene, Thiene 028085
Museo del Castello di Porciano,
Stia 028015
Museo del Castello di San Giusto e
Lapidario Tergestino, Trieste . . 028204
Museo del Castello di Sarre,
Sarre 027859
Museo del Castello e della Città,
Piombino 027342
Museo del Castello Scaligero, Torri del
Benaco 028155
Museo del Castillo de Peralada,
Peralada 039939
Museo del Cattedrale, Ferrara . . 026209
Museo del Cavallo Giocattolo,
Grandate 026464
Museo del Cenacolo di Andrea del
Sarto a San Salvi, Firenze . . . 026272
Museo del Cenedese, Vittorio
Veneto 028413
Museo del Centro Caprense Ignazio
Cerio, Capri 025759
Museo del Centro Cultural Joaquín
Arcadio Pagaza, Valle de Bravo 031635
Museo del Centro Cultural La Cantera,
Jonacatepec 031086
Museo del Centro Cultural Patio de los
Ángeles, Guadalajara 031008
Museo del Centro Cultural Sor Juana
Inés de la Cruz, Tepetlixpa . . . 031547
Museo del Centro Cultural
Universitario Quinta Gameros,
Universidad Autónoma de Chihuahua,
Chihuahua 030880
Museo del Centro Cultural Zacualpan,
Comala 030898

Museo del Centro de Estudios de la
Revolución Mexicana, Jiquilpan 031085
Museo del Centro de Historia de Tuluá,
Tuluá 008809
Museo del Centro de Investigación
Arqueológica de Ancón, Lima . 034350
Museo del Centro Histórico, San Juan
del Río 031449
Museo del Centro Studi, Albissola
Marina 025278
Museo del Cervo Sardo, Caserma
Umberto Noci, Sinnai 027965
Museo del Chiodo, Forno di
Zoldo 026335
Museo del Chocolate, Astorga . . 039011
Museo del Cielo, Villa Carlos Paz 000672
Museo del Cine, Durango 030952
Museo del Cine Pablo C. Ducrós
Hicken, Buenos Aires 000224
Museo del Cine y la Fotografía,
Puebla 031395
Museo del Cinema, Milano 026805
Museo del Cioccolato, Norma . . 027052
Museo del Círculo de Bellas Artes,
Pozoblanco 039966
Museo del Círculo de Estudios
Ferroviarios del Uruguay,
Montevideo 046261
Museo del Cobre, Bogotá 008615
Museo del Colegio De La Salle de
Buenos Aires Hermano Esteban,
Buenos Aires 000225
Museo del Colegio Médico, Santiago
de Compostela 040128
Museo del Colegio Nacional Amauta,
Chupaca 034252
Museo del Colegio Nacional Pedro
Vicente Maldonado, Riobamba 010500
Museo del Colegio Padres
Franciscanos, Ontinyent 039840
Museo del Colegio San José, Buenos
Aires 000226
Museo del Colegio Stella Maris,
Atacames 010414
Museo del Coltello Sardo, Arbus 025350
Museo del Combattente, Modena 026856
Museo del Conservatorio die Napoli,
Napoli 026997
Museo del Convento de La Merced,
Mantas 034373
Museo del Convento de los Descalzos,
Lima 034351
Museo del Convento de San Francisco
de Asís, Lima 034352
Museo del Convento de Santa
Catalina, Cusco 034262
Museo del Convento de Santa Clara,
Carrión de Los Condes 039236
Museo del Convento de Santa Úrsula,
Salamanca 040045
Museo del Convento de Santo
Domingo, Cusco 034263
Museo del Convento di San Francesco,
Susa 028039
Museo del Corallo, Ravello 027481
Museo del Corallo, Istituto Statale
d'Arte, Torre del Greco 028150
Museo del CosmoCaixa Barcelona,
Barcelona 039069
Museo del Costume Arbereshe,
Frascineto 026347
Museo del Costume e della Tradizione
Nostra Gente, Guardiagrele . . . 026488
Museo del Costume Farnesiano,
Gradoli 026463
Museo del Costume R. Piraino,
Palermo 027072
Museo del Costume Tradizionale,
Museo Villa Visconti Venosta,
Grosio 026472
Museo del Cristallo, Colle di Val
d'Elsa 026033
Museo del Cristo Negro de Portobelo,
Portobelo 034188

Museo del Cuale, Puerto Vallarta 031406
Museo del Danza, La Habana . . 009179
Museo del Deporte, Cienfuegos . 009104
Museo del Deporte, Holguín . . . 009221
Museo del Deporte, Pabellón de
Arteaga 031350
Museo del Deporte Pierre de
Coubertin, La Falda 000402
Museo del Desierto, Saltillo . . . 031425
Museo del Disco, Zipacón 008828
Museo del Duomo, Città di
Castello 026001
Museo del Duomo, Fidenza 026221
Museo del Duomo, Guardiagrele 026489
Museo del Duomo, Massa 026726
Museo del Duomo, Milano 026806
Museo del Duomo, Ravello 027482
Museo del Duomo, Udine 028227
Museo del Duomo, Valenza 028254
Museo del Ejército, Callao 034235
Museo del Ejército, Madrid 039658
Museo del Ejército y Fuerza Aérea
Bethlemitas, México 031211
Museo del Ejército y Fuerza Aérea
Mexicanos Cuartel Colorado,
Guadalajara 031009
Museo del Ejidatario, Guasave . 031033
Museo del Emigrante Canario,
Teguise 040214
Museo del Esparto, Cieza 039282
Museo del Estado, Morelia 031309
Museo del Estado de Mérida,
Mérida 054769
Museo del Este, Monte Caseros 000471
Museo del Ex Convento de Culhuacán,
México 031212
Museo del Falegname T. Sana-Onlus,
Almenno San Bartolomeo 025296
Museo del Ferro e della Ghisa,
Ex Officine Meccaniche Ilva,
Follonica 026314
Museo del Ferrocarril,
Aguascalientes 030811
Museo del Ferrocarril, Águilas . . 038895
Museo del Ferrocarril, Castejón . 039243
Museo del Ferrocarril,
Guadalajara 031010
Museo del Ferrocarril, Mérida . . 039748
Museo del Ferrocarril, Monterrey 031292
Museo del Ferrocarril, Ponferrada 039955
Museo del Ferrocarril, Torreón . 031609
Museo del Ferrocarril de Asturias,
Gijón 039438
Museo del Ferrocarril de Madrid,
Madrid 039659
Museo del Ferrocarril, Villena . . 040419
Museo del Fin del Mundo,
Ushuaia 000660
Museo del Fiore, Acquapendente 025236
Museo del Folklore e delle Tradizioni
Popolari, Caltanissetta 025702
Museo del Foro de Caesaraugusta,
Zaragoza 040450
Museo del Fossile del Monte Baldo,
Brentonico 025624
Museo del Foto – Gabinetto di Fisica,
Università Urbino, Urbino 028239
Museo del Frer, Chiaverano 025960
Museo del Fuerte de San Felipe,
Bacalar 030842
Museo del Fuerte de San Juan de
Ulúa, Veracruz 031647
Museo del Fungo e di Scienze
Naturali, Boves 025624
Museo del Fútbol Aguileño,
Águilas 038896
Museo del Gaucho y de la Moneda,
Montevideo 046262
Museo del General Toribio Ortega,
Coyame 030916
Museo del Giocattolo, Palermo . 027180
Museo del Giocattolo, Ludoteca
Comunale, Sezze 027946
Museo del Gioco del Pallone

a Bracciale e Tamburello,
Santarcangelo di Romagna . . . 027840
Museo del Grabado a la Estampa
Digital Artes, Ribeira 040003
Museo del Grabado Español
Contemporáneo, Marbella 039722
Museo del Grano, Cerignola 025904
Museo del Hombre, Tegucigalpa 023404
Museo del Hombre Dominicano, Santo
Domingo 010395
Museo del Hombre en el Universo,
Sopó 008802
Museo del Hombre y su Entorno,
Caleta Olivia 000287
Museo del Hospital Santo Toribio de
Mogrovejo, Lima 034353
Museo del Humor, San Antonio de los
Baños 009312
Museo del Indio, Los Toldos . . . 000432
Museo del Indio y del Gaucho,
Tacuarembó 046296
Museo del Instituto Americano del
Arte, Cusco 034264
Museo del Instituto de Investigaciones
Antropológicas, México 031213
Museo del Instituto Nacional de
Cultura, Arequipa 034214
Museo del Int. Presepio in Miniatura
Guido Colitti, Campobasso . . . 025729
Museo del Istmo, La Línea de la
Concepción 039568
Museo del Jade, San José 008859
Museo del Jazz G. Dagnino,
Genova 026401
Museo del Juguete Tradicional
Mexicano, San Francisco de los
Romo 031441
Museo del Labrador, Humanes de
Madrid 039505
Museo del Lago Gutierrez Dr. Rosendo
Pascual, Bariloche 000145
Museo del Larimar, Santo
Domingo 010396
Museo de las Comunicaciones de
Mar del Plata, Mar del Plata . . 000444
Museo del Lavoro, Castelmagno 025837
Museo del Lavoro Contadino,
Brisighella 025643
Museo del Lavoro Contadino,
Campobello di Mazara 025731
Museo del Lavoro e della Civiltà
Rurale, Palaia 027167
Museo del Lavoro e della Tradizioni
Popolari della Versilia Storica,
Seravezza 027923
Museo del Lavoro Povero e della
Civiltà Contadina, Livraga . . . 026609
Museo del Lazio Meridionale,
Anagni 025317
Museo del Legno, Cantù 025744
Museo del Libro, San Javier . . . 000582
Museo del Libro Antiguo, Antigua 023363
Museo del Limari, Ovalle 007529
Museo del Lino, Pescarolo 027278
Museo del Litoral Boliviano, La
Paz 004217
Museo del Lupo Appenninico, Civitella
Alfedena 026020
Museo del Maglio, Breganze . . . 025621
Museo del Malcantone, Curio . . 041431
Museo del Mantecado, Estepa . 039389
Museo del Mar, Cádiz 039192
Museo del Mar, La Barra 046238
Museo del Mar, Mazatlán 031132
Museo del Mar, Santa Pola 040110
Museo del Mar El Rodadero, Santa
Marta 008790
Museo del Mar Mundo Marino, Santa
Marta 008791
Museo del Mar y de la Sal,
Torrevieja 040264
Museo del Mar, Cofradía de
Pescadores, San Pedro del
Pinatar 040072

Museo del Mar, Universidad de
Oriente, Boca de Río 054706
Museo del Marco, Madrid 039660
Museo del Mare, Furci Siculo . . 026359
Museo del Mare, Nardò 027023
Museo del Mare, Pesaro 027267
Museo del Mare, Piombino 027343
Museo del Mare, Pizzo 027374
Museo del Mare e della Costa M. Zei,
Sabaudia 027697
Museo del Mare, Palazzo dell'Orologio,
Ischia Ponte 026516
Museo del Mare, Palazzo Vinciprova,
Pollica 027388
Museo del Médico, Santa Fé . . . 000614
Museo del Medio Ambiente,
Valencia 040329
Museo del Medioevo e del
Rinascimento, Sorano 027992
Museo del Merletto, Offida 027083
Museo del Merletto a Tombolo,
Pescocostanzo 027287
Museo del Merletto, Musei Civici
Veneziani, Burano 025659
Museo del Milagroso, Buga 008642
Museo del Modellismo Storico,
Voghiera 028421
Museo del Molino de Papel,
México 031214
Museo del Monasterio de San Joaquín
y Santa Ana, Valladolid 040341
Museo del Monasterio de Santa Isabel,
Valladolid 040342
Museo del Monasterio de Santa María
de la Vid, La Vid 040376
Museo del Monasterio de Tulebras,
Real Monasterio de Nuestra Señora
de la Caridad, Tulebras 040281
Museo del Monastero della Beata
Filippa Mareri, Petrella Salto . 027290
Museo del Mondo Contadino,
Savoca 027887
Museo del Motociclo, Rimini . . . 027520
Museo del Motociclo Moto Guzzi,
Mondello del Lario 026887
Museo del Mundo, Buenos Aires 000227
Museo del Niño, Albacete 038905
Museo del Niño, San Juan 036164
Museo del Niño, Santa Ana del
Valle 031483
Museo del Niño – Trampolín, Santo
Domingo 010397
Museo del Niño Globo mi Ciudad
Pequeña, Guadalajara 031011
Museo del Niño Papalote, México 031215
Museo del Olivar y del Aceite,
Baena 039039
Museo del Opera Omnia R. Musa e
Pinacoteca Parmigiani, Bedonia 025478
Museo del Oppidum de Iruña, Iruña
de Oca 039516
Museo del Origen, Santiago
Ixcuintla 031493
Museo del Oro, Ipiales 008702
Museo del Oro Calima, Cali . . . 008651
Museo del Oro Nariño, Pasto . . 008747
Museo del Oro Quimbaya,
Armenia 008573
Museo del Oro Quimbaya,
Manizales 008709
Museo del Oro, Banco de la República,
Bogotá 008616
Museo del Paesaggio, Castelnuovo
Berardenga 025841
Museo del Paesaggio e Museo Storico
Artistico, Verbania Pallanza . . 028334
Museo del Paesaggio Storico
dell'Appennino, Firenzuola . . . 026303
Museo del Palacio de Bellas Artes,
México 031216
Museo del Palacio Nacional,
Guatemala 023370
Museo del Palazzo del Podestà
e Gipsoteca Libero Andreotti,

Pescia 027284
Museo del Pane, Sant'Angelo
Lodigiano 027834
Museo del Papiro, Siracusa 027969
Museo del Parco Nazionale Gran
Paradiso, Noasca 027040
Museo del Parque Abel Santamaría
Cuadrado, Santiago de Cuba . . 009357
Museo del Parque Arqueológico El
Caño, Natá 034186
Museo del Parque Nacional Los
Alerces, Villa Futalaufquen . . . 000678
Museo del Pasado Cuyano,
Mendoza 000455
Museo del Pastor, Los Barrios de
Luna 039110
Museo del Pastor, Anitgua Fragua de
Masegoso, Masegoso de Tajuña 039732
Museo del Patriarca, Valencia . . 040313
Museo del Patrimonio Histórico –
Palacio de Aguas Corrientes, Buenos
Aires 000228
Museo del Patrimonio Industriale,
Bologna 025543
Museo del Patrimonium, Sutri . . 028044
Museo del Peperoncino, Maierà 026674
Museo del Periodismo Bonaerense,
Capilla del Señor 000290
Museo del Periodismo y las Artes
Gráficas, Guadalajara 031012
Museo del Pianoforte Antico,
Accademia internazionale di
interpretazione musicale su stumenti
d'epoca, Ala 025260
Museo del Po e Young Museum,
Revere 027511
Museo del Po, Museo Civiltà
Contadina e Artigiana, Monticelli
d'Ongina 026967
Museo del Policía, Monterrey . . 031293
Museo del Presepe, Modena . . . 026857
Museo del Presepio, Dalmine . . 026128
Museo del Presepio Tipologico
Internazionale, Roma 027587
Museo del Primer Supremo Tribunal
de Justicia de la Nación, Ario de
Rosales 030833
Museo del Profumo, Milano 026807
Museo del Pueblo, Guanajuato . 031208
Museo del Pueblo, Pirané 000494
Museo del Pueblo de Asturias, Muséu
del Pueblu d'Asturies, Gijón . . 039439
Museo del Pueblo-Hospital,
Quiroga 031413
Museo del Pueblo Maya Dzibilchaltún,
Chablekal 030866
Museo del Pueblo y Museo
Pre-Colombino de Chagüitillo,
Chagüitillo 033372
Museo del Puerto de Ingeniero White,
Ingeniero White 000388
Museo del Queso, Villarejo-
Periesteban 040412
Museo del Quijote, Ciudad Real 039289
Museo del Real Monasterio de San
Juan de la Peña, Jaca 039519
Museo del Real Seminario de Bergara,
Bergara 039130
Museo del Recuerdo, Academia
Mexicana de la Lengua, México 031217
Museo del Reggimento "Giovani
Fascisti", Ponti sul Mincio 027403
Museo del Regio Esercito Italiano,
Orta 027109
Museo del Retablo, Burgos 039171
Museo del Retrato Hablado,
México 031218
Museo del Río Magdalena,
Honda 008696
Museo del Risorgimento, Brescia 025631
Museo del Risorgimento,
Castelfidardo 025818
Museo del Risorgimento,
Catanzaro 025874

Museo del Risorgimento, Genova 026402
Museo del Risorgimento, Imola . 026502
Museo del Risorgimento, Lucca 026641
Museo del Risorgimento, Massa
Marittima 026735
Museo del Risorgimento,
Solferino 027978
Museo del Risorgimento, Villafranca di
Verona 028391
Museo del Risorgimento Aurelio Saffi,
Forlì 026323
Museo del Risorgimento Duca
d'Aosta, Museo delle Ceroplastiche
e Quartiere Feudale-Castello E.
d'Arborea, Sanluri 027808
Museo del Risorgimento e della
Resistenza, Ferrara 026210
Museo del Risorgimento e della
Resistenza, Vicenza 028368
Museo del Risorgimento e Raccolte
Storiche, Milano 026808
Museo del Risorgimento F. Tanara,
Langhirano 026554
Museo del Risorgimento L. Musini,
Fidenza 026222
Museo del Risorgimento Vittorio
Emanuele Orlando, Palermo . . 027181
Museo del Risorgimento, Musei di
Palazzo Farnese, Piacenza . . . 027297
Museo del Risorgmento e Resistenza
R. Giusti, Mantova 026699
Museo del Románico, Aguilar de
Campoo 038892
Museo del Ron, Santiago de
Cuba 009358
Museo del Ron Havana Club, La
Habana 009180
Museo del Rubinetto e della
sua Tecnologia, San Maurizio
d'Opaglio 027780
Museo del Sacrario, Asiago . . . 025392
Museo del Sacro dell'Alta Val Trebbia,
Montebruno 026920
Museo del Sale, Paceco 027139
Museo del Samán, Guacarí 008688
Museo del Sannio, Benevento . . 025486
Museo del Santo Rosario, Aroche 038995
Museo del Santuario de la Virgen de
Agua Santa, Baños 010419
Museo del Santuario di Crea,
Serralunga di Crea 027927
Museo del Santuario di Montevergine,
Mercogliano 026770
Museo del Santuario di Santa Maria
Santissima della Fontenuova,
Monsummano Terme 026899
Museo del Seminario Conciliar de
Nueva Caceres, Naga 034557
Museo del Sidecar, Cingoli 025987
Museo del Siglo XIX, Bogotá . . . 008617
Museo del Sottosuolo P. Parenzan,
Instituto Sperimentale di Biologia,
Latiano 026570
Museo del Sud-Ovest Americano,
Cuveglio 026127
Museo del Tabaco Herminio Torres
Grillo, Caguas 036124
Museo del Tamburo Parlante,
Montone 026971
Museo del Tango Roberto Firpo,
Salto 000561
Museo del Tasso, Roma 027588
Museo del Teatro, Cesena 025922
Museo del Teatro, Faenza 026174
Museo del Teatro Argentina,
Roma 027589
Museo del Teatro dell'Aquila,
Fermo 026198
Museo del Teatro G. Borgatti,
Cento 025896
Museo del Teatro Municipale,
Piacenza 027298
Museo del Teatro Romagnolo,
Forlì 026324

Museo del Teatro, Teatro Nuovo,
Spoleto 028006
Museo del Teclado, Caracas . . . 054733
Museo del Tejido Sedero Valenciano,
Art Major de la Seda, Valencia 040314
Museo del Templo Mayor,
México 031219
Museo del Tequila, Tequila . . . 031559
Museo del Territorio, Foggia . 026309
Museo del Territorio, Longiano . 026623
Museo del Territorio, Riccione . . 027514
Museo del Territorio, San Severino
Marche 027799
Museo del Territorio Biellese,
Biella 025510
Museo del Territorio dell'Alta Valle
dell'Aulella, Casola in Lunigiana 025800
Museo del Territorio F.R. Fazio,
Roccapalumba 027547
Museo del Territorio, Museo Civico,
Finale Emilia 026236
Museo del Tesoro, Lodi 026615
Museo del Tesoro del Duomo,
Berceto 025489
Museo del Tesoro del Duomo,
Monza 026972
Museo del Tesoro del Duomo,
Rieti 027517
Museo del Tesoro del Duomo,
Veroli 028341
Museo del Tesoro della Cattedrale,
Anagni 025318
Museo del Tesoro della Cattedrale,
Aosta 025340
Museo del Tesoro della Cattedrale,
Savona 027891
Museo del Tesoro della Cattedrale di
San Lorenzo, Genova 026403
Museo del Tesoro Santuario della
Misericordia, Savona 027892
Museo del Tessile, Chieri 025961
Museo del Tessile e della Tradizione
Industriale, Busto Arsizio 025667
Museo del Tessuto, Prato 027448
Museo del Traje, La Rioja 000419
Museo del Traje en Tabasco,
Villahermosa 031663
Museo del Traje, Centro de
Investigación del Patrimonio
Etnológico, Madrid 039661
Museo del Transporte, Garza
García 030982
Museo del Transporte Guillermo José
Schael, Caracas 054734
Museo del Transporte Terrestre,
Santiago de Cuba 009359
Museo del Trattore Agricolo, Azienda
Tosi, Senago 027913
Museo del Tren, Murcia 039800
Museo del Tricolore, Reggio
Emilia 027504
Museo del Turismo dell'Alta Pusteria,
Villabassa 028389
Museo del Vajont, Longarone . . 026621
Museo del Valle de Tehuacán,
Tehuacán 031530
Museo del Valle de Xico, Valle de
Chalco 031636
Museo del Vetro, Musei Civici
Veneziani, Murano 026985
Museo del Vicino Oriente, Università
Roma, Roma 027590
Museo del Vidrio, Monterrey . . 031294
Museo del Vidrio, Real Fábrica Cristal
San Ildefonso, San Ildefonso . . 040060
Museo del Viento, Museu del Vent,
L'Aldea 038936
Museo del Villaggio, Fraz Gudon,
Chiusa 025976
Museo del Vino, Caldaro 025689
Museo del Vino, Villafamés 040403
Museo del Vino de la D. O. Campo de
Barja, Vera de Moncayo 040369
Museo del Vino de Montilla,

Montilla 039781
Museo del Vino e della Donna,
Ciliverghe 025983
Museo del Vino Juan Carcelén,
Jumilla 039535
Museo del Vino Valdepeñas,
Valdepeñas 040294
Museo del Vino, Fondazione
Lungarotti, Torgiano 028107
Museo del Vino, Fundación Provincial
del Vino de Valladolid, Peñafiel 039930
Museo del Virreinato, Lima 034354
Museo del Virrey Ezpeleta,
Guaduas 008690
Museo del Volcán Masaya,
Masaya 033394
Museo Deleddiano, Nuoro 027075
Museo dell' Antica Civiltà Locale,
Bronte 025646
Museo dell'Accademia Ligustica
di Belle Arti, Biblioteca di Storia
dell'Arte, Genova 026404
Museo della Badia Benedettina, Cava
dei Tirreni 025881
Museo della Badia di San Salvatore,
Vaiano 028248
Museo della Barca Lariana, Pianello
del Lario 027304
Museo della Basilica, Gandino . 026370
Museo della Basilica, Grotte di
Castro 026481
Museo della Basilica di San Michele
Arcangelo, Monte Sant'Angelo . 026917
Museo della Basilica di San Nicola,
Bari 025455
Museo della Basilica di San Pancrazio,
Roma 027591
Museo della Basilica di San Venanzio,
Camerino 025712
Museo della Basilica di Santa Maria
della Passione, Milano 026809
Museo della Basilica Patriarcale
di Santa Maria degli Angeli,
Porziuncola 027428
Museo della Basilica Patriarcale Santa
Maria degli Angeli, Assisi 025400
Museo della Battaglia, San Martino
della Battaglia 027775
Museo della Battaglia, Vittorio
Veneto 028414
Museo della Battaglia del Senio,
Alfonsine 025289
Museo della Battaglia di Marengo,
Alessandria 025284
Museo della Battaglia di San Martino,
Desenzano del Garda 026132
Museo della Biblioteca Maldotti,
Guastalla 026491
Museo della Bilancia,
Campogalliano 025733
Museo della Bonifica, Argenta . 025364
Museo della Bonifica, San Donà di
Piave 027745
Museo della Calzatura, Sant'Elpidio a
Mare 027843
Museo della Calzatura d'Autore di
Rossimoda, Strá 028021
Museo della Calzatura P. Bertolini,
Vigevano 028381
Museo della Canzone e
della Riproduzione Sonora,
Vallecrosia 028262
Museo della Carrozza, Macerata 026659
Museo della Carta, Amalfi 025307
Museo della Carta e della Filigrana,
Fabriano 026166
Museo della Carta e della Filigrana,
Pioraco 027346
Museo della Cartolina S. Nuvoli,
Isera 026518
Museo della Casa Carnica, Prato
Carnico 027453
Museo della Casa Carsica,
Monrupino 026895

Museo della Casa Contadina,
Bolognetta 025576
Museo della Casa Contadina,
Corciano 026066
Museo della Casa Trevigiana, Casa da
Noal, Treviso 028186
Museo della Casina delle Civette, Villa
Torlonia, Roma 027592
Museo della Cattedrale, Bari . . . 025456
Museo della Cattedrale, Barletta 025465
Museo della Cattedrale, Chiusi
Città 025978
Museo della Cattedrale, Fossano 026339
Museo della Centuriazione,
Cesena 025923
Museo della Ceramica, Bassano del
Grappa 025473
Museo della Ceramica,
Castellamonte 025827
Museo della Ceramica, Fiorano
Modenese 026239
Museo della Ceramica, Patti . . . 027227
Museo della Ceramica del Castello,
Castello Arechi, Salerno 027709
Museo della Ceramica di Castelli,
Palazzo Comunale, Castelli . . . 025835
Museo della Ceramica, Villa Guariglia,
Vietri 028379
Museo della Certosa, Certosa di
Pavia 025911
Museo della Certosa, Serra San
Bruno 027925
Museo della Chiesa di San Francesco,
Mercatello sul Metauro 026768
Museo della Chiesa di San Lorenzo,
Portovenere 027427
Museo della Città, Ancona 025320
Museo della Città, Aquino 025348
Museo della Città e del Territorio,
Monsummano Terme 026900
Museo della Città, Musei Comunali,
Rimini 027521
Museo della Civiltà Contadina,
Acri 025242
Museo della Civiltà Contadina,
Aliano 025292
Museo della Civiltà Contadina,
Amandola 025309
Museo della Civiltà Contadina,
Andrate 025324
Museo della Civiltà Contadina,
Ateleta 025409
Museo della Civiltà Contadina,
Barolo 025467
Museo della Civiltà Contadina,
Bastiglia 025475
Museo della Civiltà Contadina,
Borghi 025595
Museo della Civiltà Contadina, Borgo
San Lorenzo 025600
Museo della Civiltà Contadina,
Calvisano 025705
Museo della Civiltà Contadina,
Castelfranco Emilia 025823
Museo della Civiltà Contadina,
Cavenago d'Adda 025885
Museo della Civiltà Contadina,
Cervia 025916
Museo della Civiltà Contadina,
Copparo 026064
Museo della Civiltà Contadina,
Curtatone 026123
Museo della Civiltà Contadina,
Fabriano 026167
Museo della Civiltà Contadina, Figline
Valdarno 026230
Museo della Civiltà Contadina,
Filottrano 026234
Museo della Civiltà Contadina, Guardia
Piemontese 026485
Museo della Civiltà Contadina, Loreto
Aprutino 026629
Museo della Civiltà Contadina,
Maierato 026675

Museo della Civiltà Contadina,
Mairano 026678
Museo della Civiltà Contadina, Moio
della Civitella 026871
Museo della Civiltà Contadina,
Mombaroccio 026878
Museo della Civiltà Contadina,
Montechiaro d'Acqui 026927
Museo della Civiltà Contadina,
Montefiore dell'Aso 026937
Museo della Civiltà Contadina,
Novellara 027069
Museo della Civiltà Contadina,
Roscigno 027671
Museo della Civiltà Contadina, San
Nicola La Strada 027788
Museo della Civiltà Contadina,
Sappada 027852
Museo della Civiltà Contadina,
Sassocorvaro 027872
Museo della Civiltà Contadina,
Serrastretta 027930
Museo della Civiltà Contadina, Sesto
Fiorentino 027938
Museo della Civiltà Contadina,
Spinetoli 028003
Museo della Civiltà Contadina,
Teggiano 028062
Museo della Civiltà Contadina,
Todi 028094
Museo della Civiltà Contadina, Torre
Pallavicina 028152
Museo della Civiltà Contadina,
Verdello 028337
Museo della Civiltà Contadina,
Zibello 028430
Museo della Civiltà Contadina A.
Marsiano, Niscemi 027036
Museo della Civiltà Contadina D.
Bianco, Sammichele di Bari . . 027720
Museo della Civiltà Contadina del
Cilento, Montecorice 026930
Museo della Civiltà Contadina del
Mendrisiotto, Stabio 042017
Museo della Civiltà Contadina dell'Alto
Monferrato e della Bassa Langa,
Castagnole delle Lanze 025808
Museo della Civiltà Contadina e
Artigianale d'Enza, Montecchio
Emilia 026925
Museo della Civiltà Contadina e della
Canapa G. Romagnoli, Argelato 025362
Museo della Civiltà Contadina e
Tradizioni Popolari, Tuglie . . . 028222
Museo della Civiltà Contadina ed
Artigiana, Monterosso Calabro . 026956
Museo della Civiltà Contadina ed
Artigiana, Cripta Chiesa San Filippo,
Ripatransone 027530
Museo della Civilta Contadina Friulana,
Farra d'Isonzo 026187
Museo della Civiltà Contadina in Val
Vibrata, Controguerra 026060
Museo della Civiltà Contadina L.
Rubat, Piscina 027359
Museo della Civiltà Contadina
Masseria Lupoli, Crispiano . . . 026105
Museo della Civiltà Contadina
nell'Area, Montefalcone di Val
Fortore 026933
Museo della Civiltà Contadina Valle
dell'Aniene, Roviano 027687
Museo della Civiltà Contadina, Castello
De Vargas, Perdifumo 027247
Museo della Civiltà Contadina, Istituto
Agrario A. Motti, Castelnovo no'
Monti 025840
Museo della Civiltà Contadina,
Pastorale, Artigianale e della Miniera,
Seui 027247
Museo della Civiltà Contadina, Villa
Smeraldi, Bentivoglio 025488
Museo della Civiltà del Lavoro,
Campiglia Marittima 025724

Register der Institutionen und Firmen
Museo della Civiltà Locale, Buseto Palizzolo
– Museo dell'Artigianato Tessile e della Seta, Reggio Calabria

Museo della Civiltà Locale, Buseto Palizzolo 025661
Museo della Civiltà Locale – Raccolta d'Arte di San Francesco, Complesso Museale, Trevi 028179
Museo della Civiltà Messapica, Poggiardo 027378
Museo della Civiltà Montanara, Sestola 027942
Museo della Civiltà Romana, Roma 027593
Museo della Civiltà Rupestre, Zungri 028436
Museo della Civiltà Rurale, Altamura 025297
Museo della Civiltà Rurale del Vicentino, Cantina Sociale Val Leogra, Malo 026684
Museo della Civiltà Salinara, Cervia 025917
Museo della Civiltà Solandra, Malé 026681
Museo della Civiltà Valligiana, Fortezza di Bardi, Bardi 025448
Museo della Civitella, Chieti . . . 025969
Museo della Collegiata, Castell'Arquato 025830
Museo della Collegiata, Chianciano Terme 025946
Museo della Collegiata di Santa Maria a Mare, Maiori 026677
Museo della Collegiata di Sant'Andrea, Empoli 026151
Museo della Collegiata, Stiftsmuseum Innichen, San Candido 027733
Museo della Collezione Gambarotta, Piozzo 027347
Museo della Comunità Ebraica, Venezia 028305
Museo della Comunità Ebraica Carlo e Vera Wagner, Trieste 028205
Museo della Confraternita della Misericordia, Anghiari 025329
Museo della Congregazione Mechitarista dei Padri Armeni, Venezia 028306
Museo della Corte, Porto Viro . . 027418
Museo della Croce Rossa Italiana, Campomorone 025734
Museo della Cultura Alpina del Comelico, Comelico Superiore . 026045
Museo della Cultura Arbereshe, San Paolo Albanese 027789
Museo della Cultura Marinara, Tortoreto Lido 028159
Museo della Cultura Mezzadrile, Morro d'Alba 026980
Museo della Cultura Popolare, Grosseto 026474
Museo della Cultura Popolare e Contadina, Carrega Ligure 025786
Museo della Donna, Angrogna . 025331
Museo della Donna E. Ortner, Merano 026764
Museo della Esposizione Permanente, Milano 026810
Museo della Farmacia Picciola, Trieste 028206
Museo della Figurina, Modena . 026858
Museo della Figurina di Gesso e dell'Emigrazione, Coreglia Antelminelli 026069
Museo della Filigrana, Campo Ligure 025726
Museo della Flora, Fauna e Mineralogia, Auronzo di Cadore 025421
Museo della Fondazione Giovanni Scaramangà di Altomonte, Trieste 028207
Museo della Fondazione Horne, Firenze 026273
Museo della Fondazione Mandralisca, Cefalù 025889
Museo della Fondazione Querini Stampalia, Venezia 028307
Museo della Fondazione Ugo da Como, Lonato 026620
Museo della Fotografia Storica e Contemporanea, Fondazione Italiana per la Fotografia, Torino 028119
Museo della Frutticoltura Adofo Bonvicini, Massa Lombarda . . 026730
Museo della Gente dell'Appennino Pistoiese, Cutigliano 026124
Museo della Gente Senza Storia, Altavilla Irpina 025299
Museo della Geotermia, Pomarance 027390
Museo della Grancia, Serre di Rapolano 027933
Museo della Guerra, Castel del Rio 025811
Museo della Guerra Adamellina, Spiazzo 028001
Museo della Guerra Bianca, Temù 028067
Museo della Guerra Bianca, Vermiglio 028338
Museo della Guerra, Il Vittoriale degli Italiani, Gardone Riviera 026372
Museo della Linea Gotica, Montefiore Conca 026935
Museo della Liquirizia, Rossano 027678
Museo della Liuteria, Arpino . . . 025376
Museo della Macchina da Scrivere P. Mitterhofer, Schreibmaschinenmuseum P. Mitterhofer, Parcines 027202
Museo della Magnifica Comunità di Cadore, Pieve di Cadore 027321
Museo della Mail Art, Montecarotto 026922
Museo della Marineria dell'Alto e Medio Adriatico, Cesenatico . . 025934
Museo della Marionetta, Teatro Gianduia, Torino 028120
Museo della Memoria, Sant'Andrea Apostolo 027831
Museo della Miniera, Cesena . . 025924
Museo della Miniera, Massa Marittima 026736
Museo della Miniera, Prata di Pordenone 027441
Museo della Miniera, Schilpario 027904
Museo della Miniera Aurifera della Guia, Macugnaga 026666
Museo della Miniera di Zolfo di Cabernardi, Sassoferrato 027877
Museo della Misericordia, Collezione dell'Arciconfraternita della Misericordia-Palazzo Roffia, San Miniato 027786
Museo della Montagna, Macugnaga 026667
Museo della Nave Romana, Comacchio 026042
Museo della Nostra Terra, Pieve Torina 027329
Museo della Paglia e dell'Intreccio D. Michalacci, Signa 027964
Museo della Pesca, Caslano . . . 041392
Museo della Pesca, Magione . . . 026671
Museo della Pesca e delle Civiltà Marinara, Museo delle Anfore – Antiquarium Truentinum, San Benedetto del Tronto 027723
Museo della Pietà Popolare, Molfetta 026873
Museo della Pietra Serena, Firenzuola 026304
Museo della Pievania, Corciano . 026067
Museo della Pieve, Sant'Ambrogio di Valpolicella 027830
Museo della Pieve di San Pietro, Prato 027449
Museo della Pieve di Staggia, Poggibonsi 027380
Museo della Pipa, Gavirate 026378
Museo della Plastica, Pont Canavese 027394
Museo della Preistoria, Celano . 025893
Museo della Preistoria della Tuscia, Valentano 028253
Museo della Propositura di San Pietro, Montecatini Val di Nievole 026924
Museo della Regina, Cattolica . . 025880
Museo della Repubblica Partigiana, Montefiorino 026939
Museo della Resistanza e del '900, Imola 026503
Museo della Resistenza e della Civiltà Contadina, Pietrabruna 027315
Museo della Resistenza Cà Malanca, Brisighella 025644
Museo della Resistenza e del Folclore Valsabbino, Pertica Bassa 027252
Museo della Risorgimento, Genova 026405
Museo della Rocca, Cesena . . . 025925
Museo della Rocca, Dozza 026143
Museo della Sanità e dell'Assistenza, Bologna 025544
Museo della Santa, Bologna . . . 025545
Museo della SAT, Palazzo Pedrotti, Trento 028172
Museo della Scuola, Bolzano . . 025583
Museo della Scuola, Pramollo . . 027439
Museo della Scuola Grande dei Carmini, Venezia 028308
Museo della Seta Abegg, Garlate 026374
Museo della Sindone, Torino . . . 028121
Museo della Società di Esecutori di Pie Disposizioni, Siena 027955
Museo della Società di Studi Patri, Gallarate 026366
Museo della Società Geografica Italiana, Roma 027594
Museo della Specola, Bologna . 025546
Museo della Stampa, Jesi 026530
Museo della Stampa, Soncino . . 027983
Museo della Tarsia, Rolo 027552
Museo della Tecnica e del Lavoro MV Agusta, Samarate 027718
Museo della Tecnica e del Lavoro MV Augusta, Trequanda 028178
Museo della Tecnologia Contadina, Santu Lussurgiu 027849
Museo della Terza Armata, Padova 027147
Museo della Tipografia, Città di Castello 026002
Museo della Tonnara, Milazzo . . 026835
Museo della Tonnara, Stintino . . 028020
Museo della Transumanza e del Costume Abruzzese Molisano, Sulmona 028031
Museo della Val Codera, Novate Mezzola 027066
Museo della Val Gardena, Union di Ladins, Ortisei 027111
Museo della Valchiavenna, Chiavenna 025958
Museo della Valle, Cavargna . . . 025884
Museo della Valle Cannobina, Gurro 026496
Museo della Valle Intelvi, Lanzo d'Intelvi 026557
Museo della Via Ostiense, Roma 027595
Museo della Vicaria di San Lorenzo, Zogno 028432
Museo della Villa Bettoni, Bogliaco 025526
Museo della Villa San Carlo Borromeo, Senago 027914
Museo della Vita, Castelmagno . 025838
Museo della Vita Contadina, Castelnuovo Don Bosco 025843
Museo della Vita Contadina in Romagnolo, Russi 027693
Museo della Vita e del Lavoro della Maremma Settentrionale, Cecina e Marina 025888
Museo della Vita e delle Tradizioni Popolari Sarde, Nuoro 027076
Museo della Vite e del Vino, Carmignano 025775
Museo della Vite e del Vino, Villa Poggio Reale, Rufina 027692
Museo della Viticoltura, Prarostino 027440
Museo della Xilografia, Carpi . . . 025780
Museo della Zampogna, Mostra Pernamente di Cornamuse Italiane e Stranicre, Scapoli 027899
Museo della Zisa, Palermo 027182
Museo dell'Abazzia di San Colombano, Bobbio 025525
Museo dell'Abbazia, Casamari-Veroli 025794
Museo dell'Abbazia, Castiglione a Casauria 025854
Museo dell'Abbazia, San Benedetto Po 027728
Museo dell'Abbazia di Farfa, Fara Sabina 026185
Museo dell'Abbazia di San Nilo, Grottaferrata 026477
Museo dell'Abbazia di Santa Maria in Sylvis, Sesto al Reghena 027936
Museo dell'Accademia d'Arte C. Scalabrino, Montecatini Terme 026923
Museo dell'Accademia di Belle Arti, Perugia 027260
Museo dell'Accademia Etrusca, Cortona 026084
Museo dell'Accademia Ligustica di Belle Arti, Genova 026406
Museo dell'Accademia Nazionale Virgiliana, Mantova 026700
Museo dell'Aeronautica G. Caproni, Trento 028173
Museo dell'Agricoltura e del Mondo Rurale, San Martino in Rio . . . 027777
Museo dell'Agro Falisco, Civita Castellana 026015
Museo dell'Agro Veientano, Formello 026332
Museo dell'Alta Val Venosta, Curon Venosta 026122
Museo dell'Alto Medioevo, Roma 027596
Museo dell'Apicoltura, Renon . . 027510
Museo dell'Apicultura, Gruppo la Cappelletta, Abbiategrasso . . . 025229
Museo dell'Arciconfraternita, Tolentino 028099
Museo dell'Arciconfraternita dei Genovesi, Cagliari 025677
Museo dell'Ardesia, Cicagna . . . 025981
Museo dell'Aria dello Spazio, Due Carrare 026146
Museo dell'Arredo Contemporaneo, Russi 027694
Museo dell'Arte Classica, Sezione di Archeologia, Roma 027597
Museo dell'Arte Contadina, Marciana 026709
Museo dell'Arte Contemporanea e dei Pittori dell'Emigrazione, Recanati 027491
Museo dell'Arte del Cappello, Ghiffa 026434
Museo dell'Arte del Vino, Staffolo 028010
Museo dell'Arte della Seta, Catanzaro 025875
Museo dell'Arte della Tornitura del Legno, Pettenasco 027292
Museo dell'Arte e della Tecnologia Confetteria, Sulmona 028032
Museo dell'Arte Mineraria, Iglesias 026499
Museo dell'Arte Sacra, Aldino . . 025282
Museo dell'Arte Serica e Laterizia, Malo 026685
Museo dell'Arte Votiva, Cesena . 025926
Museo dell'Artigianato Tessile e della Seta, Reggio Calabria 027495

Museo dell'Attore Napoletano,
Napoli 026998
Museo dell'Automobile, San Martino in
Rio 027778
Museo dell'Automobile Carlo Biscaretti
di Ruffia, Torino 028122
Museo dell'Automobile e della Tecnica,
Farra d'Isonzo 026188
Museo dell'Automobile L. Bonfanti,
Romano d'Ezzelino 027665
Museo dell'Aviazione, Rimini . . . 027522
Museo dell'Avifauna Appenninica,
Sarnano 027855
Museo delle 3 Religioni Promoteiste,
Bertinoro 025500
Museo delle Anime dei Defunti,
Roma 027598
Museo delle Antiche Vie, Urbana 028233
Museo delle Armi, Terni 028077
Museo delle Armi Antiche,
Martinsicuro 026720
Museo delle Armi Antiche, San
Marino 038099
Museo delle Armi Antiche e Fortezza
Medievale, Acquaviva Picena . 025239
Museo delle Armi Antiche, Rocca di
Offagna, Offagna 027080
Museo delle Armi e del Martello,
Sarnano 027856
Museo delle Armi e delle Uniformi,
Urbisaglia 028244
Museo delle Armi Moderne, San
Marino 038100
Museo delle Armi, Armature e
Strumenti di Tortura, Castello
Aragonese, Ischia Ponte 026517
Museo delle Arti e delle Tradizioni,
Treviso 028187
Museo delle Arti e delle Tradizioni di
Puglia, Latiano 026571
Museo delle Arti e delle Tradizioni
Popolari, Canepina 025738
Museo delle Arti e delle Tradizioni
Popolari, Palazzo Montanari,
Sassoferrato 027878
Museo delle Arti e Tradizioni Popolari,
Civitanova Marche 026017
Museo delle Arti e Tradizioni Popolari,
Correggio 026077
Museo delle Bilance e dei Pesi,
Monterchi 026954
Museo delle Carrozze, Catanzaro 025876
Museo delle Carrozze, Napoli . . 026999
Museo delle Carrozze, Trani . . . 028162
Museo delle Carrozze dell'Ottocento,
Verona 028349
Museo delle Carrozze, Palazzo Pitti,
Firenze 026274
Museo delle Centuriazioni,
Granze 026466
Museo delle Ceramiche, Forlì . . 026325
Museo delle Ceramiche,
Tolentino 028100
Museo delle Ceramiche Abruzzesi,
Loreto Aprutino 026630
Museo delle Cere, Roma 027599
Museo delle Cere – Strumenti di
Tortura, San Marino 038101
Museo delle Cere Anatomiche Luigi
Cattaneo, Università di Bologna, Dep.
Anatomia Umana e Fisiopatologia,
Bologna 025547
Museo delle Cripta, Santa Maria degli
Angeli, Poggiardo 027379
Museo delle Culture, Castagnola 041394
Museo delle Culture del Mondo,
Castello d'Albertis, Genova . . . 026407
Museo delle Culture Extraeuropee Dinz
Rialto, Rimini 027523
Museo delle Curiosità, San
Marino 038102
Museo delle Due Guerre Mondiali,
Loro Piceno 026632
Museo delle Erbe, Teggiano . . . 028063

Museo delle Erbe, Veroli 028342
Museo delle Ferrovie, Monserrato 026898
Museo delle Ferrovie in Sardegna,
Cagliari 025678
Museo delle Genti d'Abruzzo,
Pescara 027274
Museo delle Genti delle Valli di Lanzo,
Ceres 025902
Museo delle Genti di Montagna,
Palazzuolo sul Senio 027173
Museo delle Grigne, Esino Lario 026162
Museo delle Guerre, Grassobbio 026467
Museo delle Icone, Peccioli . . . 027240
Museo delle Icone e della Tradizione
Bizantina, Frascineto 026348
Museo delle Marionette,
Campomorone 025735
Museo delle Miniere, Valle
Aurina 028259
Museo delle Mummie, Chiesa di San
Stefano, Ferentillo 026195
Museo delle Mura, Roma 027600
Museo delle Musiche dei Popoli,
Castello d'Albertis, Genova . . . 026408
Museo delle Navi e Antiche Carte
Geografiche, Università Bologna,
Bologna 025548
Museo delle Navi Romane,
Fiumicino 026305
Museo delle Navi Romane, Nemi 027027
Museo delle Origini, Roma 027601
Museo delle Palafitte, Molina di
Ledro 026876
Museo delle Pentole nella Storia, ACM
Italia, Rozzano 027690
Museo delle Porcellane, Firenze 026275
Museo delle Rimembranze,
Cosenza 026088
Museo delle Scienze della Terra e del
Lavoro Contadino, Piandimeleto 027302
Museo delle Sinopie, Opera della
Primaziale Pisana, Pisa 027356
Museo delle Statue – Stele
Lunigianesi, Pontremoli 027406
Museo delle Statue Menhir,
Laconi 026543
Museo delle Tavolette di Biccherna,
Siena 027956
Museo delle Terme, Bagni di
Lucca 025438
Museo delle Terre Marchigiane, San
Lorenzo in Campo 027771
Museo delle Tradizioni Agroalimentari
della Sardegna, Casa Steri,
Siddi 027948
Museo delle Tradizioni ed Arti
Contadine, Picciano 027310
Museo delle Tradizioni Popolari, Città
di Castello 026003
Museo delle Tradizioni Popolari,
Offida 027084
Museo delle Tradizioni Popolari, San
Vito di Cadore 027804
Museo delle Trame Mediterranee,
Gibellina 026440
Museo delle Truppe Alpine Mario
Balocco, Biella 025511
Museo delle Valli, Comacchio . . 026043
Museo delle Valli d'Argenta,
Argenta 025365
Museo delle Valli Valdesi, Torre
Pellice 028154
Museo dell'Emigrante Casa Giannini,
Favale di Malvaro 026190
Museo dell'Emigrazione G.B.
Scalabrini, Francavilla Angitola 026344
Museo dell'Energia Elettrica,
Brescia 025632
Museo dell'Enoteca Regionale
Piemontese, Grinzane Cavour . 026471
Museo dell'Homo Salvadego, Cosio
Valtellino 026090
Museo dell'Illustrazione, Ferrara 026211
Museo dell'Immagine, Cesena . 025927

Museo dell'Informazione,
Senigallia 027916
Museo dell'Ingegno e della Tecnologia
Preindustriale, Colorno 026040
Museo dell'Intreccio Mediterraneo,
Castelsardo 025848
Museo dell'Istituto di Anatomia
Patologica, Padova 027148
Museo dell'Istituto di Patologia del
Libro, Roma 027602
Museo dello Sbarco, Anzio 025337
Museo dello Sbarco Alleato, Forte
Sangallo, Nettuno 027031
Museo dello Scalpellino Povese, Pove
del Grappa 027432
Museo dello Scarpone e della Scarpa
Sportiva, Montebelluna 026918
Museo dello Sfilato Siciliano,
Chiaramonte Gulfi 025948
Museo dello Spazzacamino, Santa
Maria Maggiore 027820
Museo dello Strumento Musicale,
Reggio Calabria 027496
Museo dello Studio, Bologna . . . 025549
Museo dell'Occhiale, Pieve di
Cadore 027322
Museo dell'Olio, Cisano 025995
Museo dell'Olio in Sabina, Castelnuovo
di Farfa 025842
Museo dell'Olivo, Imperia 026510
Museo dell'Olivo e dell'Olio,
Torgiano 028108
Museo dell'Ombrello e del Parasole,
Gignese 026442
Museo dell'Opera del Duomo,
Orvieto 027117
Museo dell'Opera del Duomo,
Prato 027450
Museo dell'Opera del Santuario,
Tolentino 028101
Museo dell'Opera di Santa Croce,
Firenze 026276
Museo dell'Opera di Santa Maria del
Fiore, Firenze 026277
Museo dell'Opera di Santa
Niccolao, Fraz Buggiano Castello,
Buggiano 025656
Museo dell'Opera e del Territorio,
Caserta 025798
Museo dell'Opera Metropolitana,
Siena 027957
Museo dell'Opificio delle Pietre Dure,
Firenze 026278
Museo dell'Orologio da Torre G.B.
Bergallo, Tovo San Giacomo . 028161
Museo dell'Orso, Villavallelonga 028397
Museo dell'Osservatorio, Monte Porzio
Catone 026911
Museo dell'Osservatorio Astronomico
di Capodimonte, Napoli 027000
Museo dell'Uomo, Serrapetrona . 027929
Museo dell'Uomo, Susegana . . . 028041
Museo dell'Uomo e dell'Ambiente,
Terra del Sole 028080
Museo dell'Utensileria, Lanzo
Torinese 026558
Museo Delso, L'Alfàs del Pi . . . 038939
Museo Demantropologico, Sez. di Arte
Orafa, Cervaro 025913
Museo Demoantropologico,
Pastena 027223
Museo Demologico dell'Economia, del
Lavoro e della Storia Sociale Silana,
San Giovanni in Fiore 027758
Museo Départemental A. Demard, Le
Haut-du-Them 013641
Museo des Orígenes, La Habana 009181
Museo di Anatomia Comparata
Battista Grassi, Università di Roma,
La Sapienza, Roma 027603
Museo di Anatomia Patologica C.
Taruffi, Bologna 025550
Museo di Anatomia Patologica
Veterinaria e Teratologia, Ozzano

dell'Emilia 027136
Museo di Anatomia Patologica,
Università La Sapienza, Roma . 027604
Museo di Anatomia Umana,
Torino 028123
Museo di Anatomia Umana Normale,
Istituto di Anatomia Umana,
Parma 027213
Museo di Anatomia Umana Normale,
Università Bologna, Bologna . . 025551
Museo di Anatomia Umana, 2
Università Napoli, Napoli 027001
Museo di Anatomia Veterinaria,
Napoli 027002
Museo di Antichità Etrusche e Italiche,
Facoltà di Lettere, Roma 027605
Museo di Antichita, Collezioni
Archeologiche, Torino 028124
Museo di Antonio da Sangallo il
Giovane, Montefiascone 026934
Museo di Antropologia Criminale,
Torino 028125
Museo di Antropologia ed Etnografia,
Padova 027149
Museo di Antropologia ed Etnografia,
Torino 028126
Museo di Antropologia ed Etnologia,
Firenze 026279
Museo di Antropologia G. Sergi,
Università Roma, Roma 027606
Museo di Antropologia, Università
Bologna, Bologna 025552
Museo di Antropologia, Università di
Napoli, Napoli 027003
Museo di Apicoltura Guido Fregonese,
Piavon di Oderzo 027305
Museo di Archeologia Sacra e Profana,
Roma 027607
Museo di Archeologia Urbana e
dei Commerc Antichi in Abruzzo,
Lanciano 026552
Museo di Architettura Militare,
Bologna 025553
Museo di Architettura, Casa di Biagio
Rossetti, Ferrara 026212
Museo di Arte Cinese ed Etnografico,
Parma 027214
Museo di Arte Contadina,
Montegallo 026943
Museo di Arte Contemporanea
e del Novecento, Monsummano
Terme 026901
Museo di Arte Contemporanea,
Palagio Fiorentino, Stia 028016
Museo di Arte e Giacimenti Minerari,
Facoltà di Ingegneria, Latina . . 026573
Museo di Arte e Storia delle Miniere,
Massa Marittima 026737
Museo di Arte e Tradizione Contadina,
Olevano di Lomellina 027089
Museo di Arte Medioevale e Moderna,
Arezzo 025359
Museo di Arte Religiosa, Oleggio 027088
Museo di Arte Sacra, Agnone . . 025247
Museo di Arte Sacra, Bientina . . 025513
Museo di Arte Sacra, Castrovillari 025864
Museo di Arte Sacra, Longiano . 026624
Museo di Arte Sacra, Sabbioneta 027698
Museo di Arte Sacra, Suvereto . 028046
Museo di Arte Sacra, Vibo
Valentia 028362
Museo di Arte Sacra del Castello,
Genga 026387
Museo di Arte Sacra San Giuseppe,
Rogliano 027551
Museo di Arte Sacra, Chiesa Santa
Maria del Gamio, Saracena . . . 027853
Museo di Arti Decorative della
Fondazione Pietro Accorsi,
Torino 028127
Museo di Auto e Moto d'Epoca,
Vigolzone 028385
Museo di Biologia Marina, Porto
Cesareo 027415

Museo di Biologia Marina, Instituto Talassografico, Taranto 028056
Museo di Biologia Marina, Università Bologna, Fano 026184
Museo di Blenio, Lottigna 041689
Museo di Botanica, Museo di Storia Naturale, Firenze 026280
Museo di Capodimonte, Napoli . 027004
Museo di Casa Vasari, Arezzo .. 025360
Museo di Castelgrande, Museo Storico Archeologico e Storico Artistico, Bellinzona 041300
Museo di Castelvecchio, Verona 028350
Museo di Chimica, Roma 027608
Museo di Cimeli Storico Militari, Chiaramonte Gulfi 025949
Museo di Civiltà Preclassiche della Murgia Meridionale, Ostuni ... 027133
Museo di Criminologia Medievale, San Gimignano 027753
Museo di Criminologico, Roma . 027609
Museo di Cultura Contadina, Oriolo 027103
Museo di Cultura Contadina dell'Alta Val Trebbia, Montebruno 026921
Museo di Cultura Popolare, Egna 026148
Museo di Dipinti Sacri Ponte dei Greci, Venezia 028309
Museo di Ecologia e Storia Naturale, Marano sul Panaro 026707
Museo di Etnografico e Antropologia Giovanni Podenzana, La Spezia 026540
Museo di Etnomedicina A. Scarpa, Sala delle Conchiglie, Piazzola sul Brenta 027309
Museo di Etnomedicina A. Scarpa, Università Genova, Genova ... 026409
Museo di Etnopreistoria, Castel dell'Ovo, Napoli 027005
Museo di Fisica, Bologna 025554
Museo di Fisica, Università Degli Studi di Napoli Federico II, Napoli ... 027006
Museo di Fossili, Ronca 027667
Museo di Fotografia Contemporanea, Cinisello Balsamo 025991
Museo di Gallese e Centro Culturale Marco Scacchi, Convento di Santa Chiara, Gallese 026367
Museo di Geologia, Camerino .. 025713
Museo di Geologia, Napoli 027007
Museo di Geologia e Paleontologia, Firenze 026281
Museo di Geologia e Paleontologia, Torino 028128
Museo di Geologia e Paleontologia, Università Padova, Padova ... 027150
Museo di Geologia e Paleontologia, Università Pavia, Pavia 027232
Museo di Geologia, Università Roma, Roma 027610
Museo di Informatica e Storia del Calcolo, Pennabilli 027242
Museo di Leventina, Giornico .. 041545
Museo di Merceologia, Roma .. 027611
Museo di Milano, Milano 026811
Museo di Minerali e Fossili, Montefiore Conca 026936
Museo di Mineralogia, Massa Marittima 026738
Museo di Mineralogia, Palermo . 027183
Museo di Mineralogia e Paleontologia, Saint Vincent 027703
Museo di Mineralogia e Petrografia L. Bombicci, Università Bologna, Bologna 025555
Museo di Mineralogia e Petrografia, Università Parma, Parma 027215
Museo di Mineralogia e Petrografia, Università Pavia, Pavia 027233
Museo di Mineralogia e Petrologia, Università Padova, Padova ... 027151
Museo di Mineralogia L. De Prunner, Università Cagliari, Cagliari ... 025679
Museo di Mineralogia, di Petrologia

e Geologia, Università Modena, Modena 026859
Museo di Mineralogia, Università Roma, Roma 027612
Museo di Muggia e del Territorio, Muggia 026983
Museo di Nostra Signora di Bonaria, Cagliari 025680
Museo di Palazzo Altieri, Oriolo Romano 027104
Museo di Palazzo d'Arco, Mantova 026701
Museo di Palazzo Davanzati, Firenze 026282
Museo di Palazzo De Vio, Gaeta 026360
Museo di Palazzo della Penna, Perugia 027261
Museo di Palazzo Ducale, Mantova 026702
Museo di Palazzo Gamilli, Nocera Umbra 027043
Museo di Palazzo Mirto, Palermo 027184
Museo di Palazzo Mocenigo, Musei Civici Veneziani, Venezia 028310
Museo di Palazzo Pepoli Campogrande, Bologna 025556
Museo di Palazzo Piccolomini, Pienza 027313
Museo di Palazzo Pretorio, Certaldo 025910
Museo di Palazzo Reale, Napoli 027008
Museo di Palazzo Reale Genova, Genova 026410
Museo di Palazzo Strozzi, Firenze 026283
Museo di Palazzo Vecchio, Firenze 026284
Museo di Paleobotanica ed Etnobotanica dell'Orto Botanico, Napoli 027009
Museo di Paleontologia, Milano . 026812
Museo di Paleontologia, Modena 026860
Museo di Paleontologia, Napoli . 027010
Museo di Paleontologia, Roma . 027613
Museo di Paleontologia dell'Accademia Federiciana, Catania 025870
Museo di Paleontologia e Mineralogia, Campomorone 025736
Museo di Paleontologia e Paletnologia, Maglie 026673
Museo di Paleontologia e Speleologia E.A. Martel, Carbonia 025770
Museo di Paleontologia G. Buriani, San Benedetto del Tronto 027724
Museo di Paletnologia, Polignano a Mare 027385
Museo di Pantelleria, Pantelleria 027199
Museo di Peppone e Don Camillo, Brescello 025627
Museo di Pittura Murale, Prato . 027451
Museo di Prali e della Val Germanasca, Prali 027437
Museo di Preistoria e del Mare, Trapani 028164
Museo di Preistoria e Protostoria della Valle del Fiume Fiora, Manciano 026688
Museo di Preistorico e Paleontologico, Sant'Anna d'Alfaedo 027837
Museo di Rievocazione Storica, Mondavio 026885
Museo di Rocca Fregoso, Sant'Agata Feltria 027829
Museo di Rodoretto, Prali 027438
Museo di Roma, Roma 027614
Museo di Roma in Trastevere, Roma 027615
Museo di San Domenico, Bologna 025557
Museo di San Francesco, Aversa 025425
Museo di San Francesco, Montella 026945
Museo di San Giuseppe, Bologna 025558
Museo di San Marco, Firenze . 026285
Museo di San Marco, Venezia . 028311
Museo di San Martino, Olivone . 041821

Museo di San Michele, Sagrado 027700
Museo di San Petronio, Presso Basilica San Petronio, Bologna 025559
Museo di San Pietro, Assisi ... 025401
Museo di Santa Cecilia, San Lazzaro di Savena 027765
Museo di Santa Maria degli Angioli, Chiesa Santa Maria degli Angioli, Lugano 041699
Museo di Santa Maria delle Grazie, San Giovanni Valdarno 027762
Museo di Santa Maria di Castello, Genova 026411
Museo di Sant'Agostino, Genova 026412
Museo di Sant'Antonino, Piacenza 027299
Museo di Santo Stefano, Bologna 025560
Museo di Scienza dela Terra, San Gemini 027748
Museo di Scienza della Terra, Università Bari, Bari 025457
Museo di Scienze Archeologiche e d'Arte, Università di Padova, Padova 027152
Museo di Scienze della Terra U. Baroli, Crodo 026107
Museo di Scienze Naturali, Belvi 025484
Museo di Scienze Naturali, Bolzano 025584
Museo di Scienze Naturali, Brescia 025633
Museo di Scienze Naturali, Cesena 025928
Museo di Scienze Naturali, Città della Pieve 025997
Museo di Scienze Naturali, Malnate 026682
Museo di Scienze Naturali, Rosignano Marittimo 027676
Museo di Scienze Naturali del Collegio, Lodi 026616
Museo di Scienze Naturali Don Bosco, Alassio 025262
Museo di Scienze Naturali ed Umane, Convento di San Giuliano, L'Aquila 026560
Museo di Scienze Naturali L. Paolucci, Offagna 027081
Museo di Scienze Naturali Tommaso Salvadori, Fermo 026199
Museo di Scienze Naturali, Musei Universitari, Pavia 027234
Museo di Scienze Naturali, Università Camerino, Camerino 025714
Museo di Sculture Iperspaziali, Bomarzo 025586
Museo di Speleologia, Borgo Grotta Gigante 025598
Museo di Speleologia e Carsismo A. Parolini, Valstagna 028266
Museo di Speleologia Vincenzo Rivera, L'Aquila 026561
Museo di Stato, San Marino ... 038103
Museo di Storia Contadina, Pagnacco 027166
Museo di Storia Contemporanea, Milano 026813
Museo di Storia della Fisica, Padova 027153
Museo di Storia della Medicina, Roma 027616
Museo di Storia dell'Agricoltura, Cesena 025929
Museo di Storia dell'Agricoltura, Urbania 028235
Museo di Storia dell'Agricoltura e della Pastorizia, Morano Calabro 026975
Museo di Storia delle Mezzadria, Senigallia 027917
Museo di Storia e Arte, Musei Provinciali di Gorizia, Gorizia . 026454
Museo di Storia e Civiltà, Medesano 026752

Museo di Storia e Cultura Contadina, Genova 026413
Museo di Storia e Cultura della Val del Biois, Vallada Agordina 028258
Museo di Storia Locale, Velturno 028289
Museo di Storia Naturale, Cremona 026101
Museo di Storia Naturale, Follonica 026315
Museo di Storia Naturale, Macerata 026660
Museo di Storia Naturale, Merate 026767
Museo di Storia Naturale, Perugia 027262
Museo di Storia Naturale, Piacenza 027300
Museo di Storia Naturale, Senna Lodigiana 027920
Museo di Storia Naturale, Stroncone 028027
Museo di Storia Naturale, Sulmona 028033
Museo di Storia Naturale – Sezione Mineralogia, Firenze 026286
Museo di Storia Naturale – Sezione Zoologica La Specola, Università Firenze, Firenze 026287
Museo di Storia Naturale A. Orsini, Ascoli Piceno 025387
Museo di Storia Naturale A. Stoppani, Venegono Inferiore 028292
Museo di Storia Naturale Aquileglia, Assemini 025395
Museo di Storia Naturale del Cilento, Laureana Cilento 026576
Museo di Storia Naturale della Lunigiana, Aulla 025420
Museo di Storia Naturale della Maremma, Grosseto 026475
Museo di Storia Naturale dell'Accademia dei Fisocritici, Siena 027958
Museo di Storia Naturale Don Bosco, Torino 028129
Museo di Storia Naturale e del Territorio, Calci 025687
Museo di Storia Naturale e dell'Uomo, Carrara 025784
Museo di Storia Naturale Faraggiana Ferrandi, Palazzo Faraggiana, Novara 027059
Museo di Storia Naturale S. Ferrari, Bedonia 025479
Museo di Storia Naturale, Dipartimento per lo studio del Territorio e le sue Risorse, Genova 026414
Museo di Storia Naturale, Musei Civici Veneziani, Venezia 028312
Museo di Storia Naturale, Università di Parma, Parma 027216, 027217
Museo di Storia Naturali e Archeologia, Montebelluna 026919
Museo di Storia Quarnese, Quarna Sotto 027462
Museo di Storia, d'Arte e d'Antichità Don Florindo Piolo, Serravalle Sesia 027932
Museo di Strumenti del Conservatorio Statale di Musica Giuseppe Verdi, Torino 028130
Museo di Torcello, Venezia 028313
Museo di Tossicia, Tossicia 028160
Museo di Tradizioni Popolari e del Costume d'Epoca, San Pietro Avellana 027793
Museo di Val Verzasca, Sonogno 042009
Museo di Vallemaggia, Giumaglio 041547
Museo di Valmaggia, Cevio 041396
Museo di Villa Beatice d'Este, Cinto Euganeo 025992
Museo di Villa Cagnola, Gazzada 026380
Museo di Villa Faraggiana, Albissola Marina 025279

Museo di Vita Artigiana e Contadina,
Scarperia 027902
Museo di Vita Cappuccina,
Genova 026415
Museo di Vita Quotidiana e Lavoro in
Villa, Orgiano 027097
Museo di Zoologia, Bari 025458
Museo di Zoologia, Napoli 027011
Museo di Zoologia e Mineralogia,
Siddi 027949
Museo di Zoologia Pietro Doderlein,
Università Palermo, Palermo . . 027185
Museo di Zoologia, Università Bologna,
Bologna 025561
Museo di Zoologia, Università Roma,
Roma 027617
Museo di Zoologico e di Anatomia,
Università Torino, Torino 028131
Museo di Zoologico, Università
Catania, Catania 025871
Museo Diacesano, Enna 026154
Museo Didáctico Antonini, Nasca 034376
Museo Didáctico Artiguista,
Maldonado 046241
Museo Didáctico de Prehistoria, Alba
de Tormes 038902
Museo Didattico, Morimondo . . 026978
Museo Didattico, Niscemi 027037
Museo Didattico Archeologico,
Laterza 026569
Museo Didattico d'Arte e Vita
Preistorica, Capo di Ponte . . . 025752
Museo Didattico del Mare, Napoli 027012
Museo Didattico della Riserva
Regionale Incisioni Rupestri,
Ceto 025936
Museo Didattico della Scuola Medica
Salernitana, Salerno 027710
Museo Didattico della Seta,
Como 026048
Museo Didattico delle Maioliche,
Grottaglie 026478
Museo Didattico Giuseppe Pellizza,
Istituto Giuseppe Pellizza,
Volpedo 028422
Museo Didattico Zoologico, Oria 027101
Museo Didattico, Villa Adriana,
Tivoli 028092
Museo Diefenbach, Capri 025760
Museo Diego Aragona Pignatelli
Cortes, Napoli 027013
Museo Diego Rivera Anahuacalli,
México 031220
Museo Diffuso, I Luoghi dell'Anima,
Pennabilli 027243
Museo Dillmann S. Bullock,
Angol 007512
Museo d'Interesse Locale,
Crespadoro 026104
Museo Diocesà, Tarragona 040204
Museo Diocesà i Comarcal, Museu de
Lleida, Lleida 039579
Museo Diocesano, Acerenza . . . 025230
Museo Diocesano, Agrigento . . 025254
Museo Diocesano, Albenga . . . 025269
Museo Diocesano, Andria 025326
Museo Diocesano, Ariano Irpino 025369
Museo Diocesano, Ascoli Piceno 025388
Museo Diocesano, Bari 025459
Museo Diocesano, Benevento . . 025487
Museo Diocesano, Bisceglie . . . 025517
Museo Diocesano, Brugnato . . . 025648
Museo Diocesano, Capua 025763
Museo Diocesano, Cassano allo
Jonio 025805
Museo Diocesano, Castelvecchio
Subequo 025850
Museo Diocesano, Catanzaro . . 025877
Museo Diocesano, Ciudad Real . 039290
Museo Diocesano, Córdoba . . . 039307
Museo Diocesano, Cortona 026085
Museo Diocesano, Cuenca 039336
Museo Diocesano, Fermo 026200
Museo Diocesano, Gaeta 026361

Museo Diocesano, Gerace 026430
Museo Diocesano, Gubbio 026492
Museo Diocesano, Imola 026504
Museo Diocesano, Jesi 026531
Museo Diocesano, Lamezia
Terme 026549
Museo Diocesano, Lanciano . . . 026553
Museo Diocesano, Lanusei 026555
Museo Diocesano, Mantova . . . 026703
Museo Diocesano, Mazara del
Vallo 026750
Museo Diocesano, Milano 026814
Museo Diocesano, Monopoli . . . 026890
Museo Diocesano, Nola 027045
Museo Diocesano, Osimo 027122
Museo Diocesano, Piazza
Armerina 027306
Museo Diocesano, Pienza 027314
Museo Diocesano, Pistoia 027365
Museo Diocesano, Pitigliano . . . 027369
Museo Diocesano, Recanati 027492
Museo Diocesano, Salerno . . . 027711
Museo Diocesano, San Miniato . 027787
Museo Diocesano, Santa Lucia del
Mela 027816
Museo Diocesano, Sant'Agata de
Goti 027828
Museo Diocesano, Sassari 027867
Museo Diocesano, Spoleto 028007
Museo Diocesano, Squillace . . . 028009
Museo Diocesano, Sulmona . . . 028034
Museo Diocesano, Trani 028163
Museo Diocesano, Urbania 028236
Museo Diocesano, Valle del Espíritu
Santo 054780
Museo Diocesano, Vallo della
Lucania 028263
Museo Diocesano, Velletri 028287
Museo Diocesano A. Bernareggi,
Bergamo 025495
Museo Diocesano Albani, Urbino 028240
Museo Diocesano Antonio
Bergamaschi, Pennabilli 027244
Museo Diocesano-Catedralicio,
Segorbe 040150
Museo Diocesano-Catedralicio,
Valladolid 040343
Museo Diocesano Catedralicio de Arte
Sacro, Orihuela 039847
Museo Diocesano-Catedralicio,
Catedral y Antiguo Palacio Episcopal,
Barbastro 039051
Museo Diocesano d'Arte Sacra,
Alghero 025290
Museo Diocesano d'Arte Sacra,
Brescia 025634
Museo Diocesano d'Arte Sacra,
Caltanissetta 025703
Museo Diocesano d'Arte Sacra,
Grosseto 026476
Museo Diocesano d'Arte Sacra,
Nicotera 027035
Museo Diocesano d'Arte Sacra,
Orte 027110
Museo Diocesano d'Arte Sacra,
Pinerolo 027335
Museo Diocesano d'Arte Sacra,
Piteglio 027367
Museo Diocesano d'Arte Sacra,
Pordenone 027412
Museo Diocesano d'Arte Sacra,
Rossano 027679
Museo Diocesano d'Arte Sacra, Scaria
di Lonzo 027900
Museo Diocesano d'Arte Sacra,
Treviso 028188
Museo Diocesano d'Arte Sacra A.
Luciani, Vittorio Veneto 028415
Museo Diocesano d'Arte Sacra Santa
Apollonia, Venezia 028314
Museo Diocesano d'Arte Sacra, Trinità,
Trivento 028216
Museo Diocesano de Arte Antiguo,
Sigüenza 040183

Museo Diocesano de Arte Religioso
Nuestra Señora del Cisne, Loja 010462
Museo Diocesano de Arte Sacro,
Bilbao 039146
Museo Diocesano de Arte Sacro, Las
Palmas de Gran Canaria 039903
Museo Diocesano de Arte
Sacro, Convento de Santa Clara,
Moguer 039757
Museo Diocesano de Arte Sacro,
Palacio Episcopal, Teruel 040225
Museo Diocesano de Huesca,
Huesca 039501
Museo Diocesano de Jaca, Jaca . 039520
Museo Diocesano de San Sebastián,
Donostia-San Sebastián 039353
Museo Diocesano de Tui, Tui . . . 040280
Museo Diocesano della Basilica del
Crocefisso, Amalfi 025308
Museo Diocesano di Arte Sacra,
Chiavari 025954
Museo Diocesano di Arte Sacra,
Genova 026416
Museo Diocesano di Arte Sacra,
Lodi 026617
Museo Diocesano di Arte Sacra,
Palermo 027186
Museo Diocesano di Arte Sacra, Santa
Severina 027824
Museo Diocesano di Arte Sacra,
Sarsina 027861
Museo Diocesano di Arte Sacra,
Susa 028040
Museo Diocesano di Arte Sacra del
Duomo, Arezzo 025361
Museo Diocesano di Arte Sacra San
Gregorio Barbarigo, Padova . . 027154
Museo Diocesano di Arte Sacra,
Palazzo Vescovile, Volterra . . . 028425
Museo Diocesano di Santo Stefano,
Firenze 026288
Museo Diocesano e Catedralicio de
Lugo, Lugo 039601
Museo Diocesano e Catedralicio
Santos Sancristobal,
Mondoñedo 039764
Museo Diocesano e Collezione
dei Presepi, Museo Diocesano
Bressanone – Diözesanmuseum
Hofburg Brixen, Bressanone . . 025639
Museo Diocesano e Gallerie del
Tiepolo, Udine 028228
Museo Diocesano e Oratorio di San
Bernardino, Siena 027959
Museo Diocesano e Tesoro della
Cattedrale, Troia 028218
Museo Diocesano Intercomunale di
Arte Sacra, Castignano 025859
Museo Diocesano Intercomunale di
Arte Sacra, Comunanza 026055
Museo Diocesano Intercomunale di
Arte Sacra, Grottammare 026479
Museo Diocesano Intercomunale di
Arte Sacra, Monteprandone . . . 026951
Museo Diocesano Intercomunale di
Arte Sacra, Ripatransone 027531
Museo Diocesano Intercomunale di
Arte Sacra, Rotella 027680
Museo Diocesano Lucas Guillermo
Castillo, Falcón 054753
Museo Diocesano Regina Coeli,
Monasterio de Regina Coeli,
Santillana del Mar 040138
Museo Diocesano San Pietro,
Teggiano 028064
Museo Diocesano Teatino, Chieti 025970
Museo Diocesano Tridentino,
Trento 028174
Museo Diocesano y Catedralicio,
Salamanca 040046
Museo Diocesano y Cedralicio,
Calahorra 039195
Museo Diocesano, Antiguo Palacio
Episcopal, Palencia 039874

Museo Diocesano, Catedral,
Tudela 040278
Museo Diocesano, Oratorio della
Misericordia, Sarzana 027865
Museo Diocesano, Vecchio Episcopio,
Ancona 025321
Museo División del Norte,
Canutillo 030859
Museo División del Norte, Lerdo 031114
Museo do Humor, Fene 039396
Museo do Mosteiro de San Paio
de Antealtares, Santiago de
Compostela 040129
Museo do Moucho, Cerceda . . . 039263
Museo do Pobo Galego, Santiago de
Compostela 040130
Museo Doctor Francia, Yaguarón 034208
Museo Documentario della Città,
Gradisca d'Isonzo 026460
Museo Doganale Svizzero,
Schweizerisches Landesmuseum,
Gandria 041509
Museo Dolores Olmedo, México 031221
Museo Domenicani, Taggia 028049
Museo Don Pio Palla, Leonessa 026592
Museo Doña Rosario, Santiago de
Cuba 009360
Museo Donazione Luxoro,
Genova 026417
Museo Donazione U. Mastroianni,
Arpino 025377
Museo Donizettiano, Bergamo . 025496
Museo Dr. Alfredo Gutierrez
Valenzuela, Sucre 004248
Museo Dr. Atl, Pihuamo 031369
Museo Dr. José Gómez Panaco,
Balancán 030844
Museo Duca di Martina, Villa
Floridiana, Napoli 027014
Museo Ducati, Bologna 025562
Museo Dupré, Fiesole 026227
Museo e Biblioteca della Resistenza,
Sansepolcro 027813
Museo e Certosa di San Martino,
Napoli 027015
Museo e Chiostri Monumentali di
Santa Maria Novella, Firenze . . 026289
Museo e Pinacoteca Basilio Cascella,
Pescara 027275
Museo e Pinacoteca Civica,
Alessandria 025285
Museo e Pinacoteca Comunali,
Gubbio 026493
Museo e Pinacoteca della Basilica di
San Paolo, Roma 027618
Museo e Pinacoteca della Città,
Todi 028095
Museo e Pinacoteca della Comunità di
Fiemme, Cavalese 025883
Museo e Pinacoteca Diocesana,
Palazzo Arcivescovile, Camerino 025715
Museo e Pinacoteca Nazionale di
Palazzo Mansi, Lucca 026642
Museo e Pinacoteca S. Gentili, San
Ginesio 027756
Museo e Villaggio Africano, Basella di
Urgnano 025469
Museo E. Bernardi, Padova 027155
Museo E. Galessi, Montevarchi . 026965
Museo E. Leandro, San Cesario di
Lecce 027737
Museo E. Renzi, Coriano 026072
Museo E. Treccani, Milano 026815
Museo Ebraico, Asti 025408
Museo Ebraico, Bologna 025563
Museo Ebraico, Ferrara 026213
Museo Ebraico, Firenze 026290
Museo Ebraico, Livorno 026605
Museo Ebraico, Soragna 027990
Museo Ebraico di Roma, Roma . 027619
Museo Ebraico-Jüdisches Museum,
Merano 026765
Museo Ecclesiastico, Caorle . . . 025745
Museo Ecuatoriano de Ciencias

Register der Institutionen und Firmen
Museo Eduardo Barreiros, Museo de Automoción, Finca – Museo Etnografico Vallivo, Val Masino

Naturales, Quito 010487
Museo Eduardo Barreiros, Museo de Automoción, Finca Valmayor, Valdemorillo 040291
Museo Eduardo Carranza, Villavicencio 008827
Museo Eerola, Tuulos 011455
Museo Efrain Martínez Zambrano, Popayán 008767
Museo Egipcio, Montevideo 046263
Museo Egizio, Torino 028132
Museo Egnazia, Monopoli 026891
Museo El Cañón, Puerto Boniato 009298
Museo El Castillo Diego Echavarría, Medellín 008717
Museo el Centenario, Garza García 030983
Museo El Chapulín, Saltillo 031426
Museo El Hombre y la Naturaleza, San Juan 000587
Museo El Hurón Azul, La Habana 009182
Museo El Jonotal, Playa Vicente, Veracruz 031371
Museo El Minero, Santa Bárbara 031485
Museo El Pariancito, San Luis Potosí 031460
Museo el Pausílipo, Las Tablas . 034185
Museo El Piñero, Nueva Gerona 009275
Museo El Polvorín, Monclova ... 031284
Museo El Reencuentro, Darregueira 000352
Museo El Remate, Comala 030899
Museo El Sol de las Botellas, Barcelona 054703
Museo Elder, Las Palmas de Gran Canaria 039904
Museo Elías Nandino, Cocula .. 030889
Museo Elisarion, Minusio 041745
Museo elle Arti e Tradizioni Contadine, Roccasecca dei Volsci 027548
Museo Emidiano, Chiesa di Sant'Emidio, Agnone 025248
Museo Emilia Ortiz, Tepic 031553
Museo Emilio Greco, Catania .. 025872
Museo Emilio Greco, Orvieto ... 027118
Museo Enologico Grasso, Milazzo 026836
Museo Entomológco Mariposas del Mundo, San Miguel 000594
Museo Entomológico, León 033386
Museo Entomológico Luis E. Peña Guzman, Santiago de Chile ... 007544
Museo Entomológico, Universidad Agraria de La Molina, Lima ... 034355
Museo Epper, Ascona 041236
Museo Erbario, Roma 027620
Museo Eritreo Bottego, Parma .. 027218
Museo Ermita de Santa Elena, Necrópolis Romana Siglos I Y II, Irún 039514
Museo Ernest Hemingway, San Francisco de Paula 009317
Museo Escolar de Pusol, Elche . 039368
Museo Escolar de Sitio de la Matamba, Jamapa 031081
Museo Escolar María Goretti, Pasto 008748
Museo Escolar y Comunitario de Acamalín, Xico 031672
Museo Escultórico de Geles Cabrera, México 031222
Museo Específico de la Academia de Artillería, Segovia 040158
Museo Específico de la Academia de Caballería, Academia de Caballería, Valladolid 040344
Museo Especifico de la Academia de Ingenieros, Hoyo de Manzanares 039496
Museo Específico de la Academia General Militar, Zaragoza 040451
Museo Específico de la Brigada Paracaidista, Alcalá de Henares 038915
Museo Específico de la Legión, Ceuta 039272
Museo Estatal de Culturas Populares, Monterrey 031295
Museo Estatal de Patología, Puebla 031396
Museo Estudio Diego Rivera, México 031223
Museo Ethnográfico, Florencia . 008680
Museo Ethnográfico de las Obras Misionales Pontificias, Buenos Aires 000229
Museo Ethnografico Don L. Pellegrini delle Civiltà Contadina, San Pellegrino in Alpe 027790
Museo Ethnohistórico, Soatá ... 008796
Museo Etiopico G. Massaia, Frascati 026345
Museo Etnico-Agricolo-Pastorale ed Artigiano, Alcamo 025281
Museo Etnico Arbereshe, Civita . 026014
Museo Étnico de los Seris, Hermosillo 031040
Museo Étnico de los Yaquis, Obregón 031340
Museo Etnico della Civiltà Salentina Agrilandi Museum, Brindisi ... 025641
Museo Etno-Antropologico, Calatafimi 025686
Museo Etno-Antropologico, San Cipirello 027739
Museo Etno-Antropologico, San Mauro Castelverde 027781
Museo Etno-Antropologico, Scaletta Zanclea 027896
Museo Etno-Antropologico della Terra di Zabut, Sambuca di Sicilia .. 027719
Museo Etno-Antropologico della Valle del Belice, Casa Gucciardi, Gibellina 026441
Museo Etno-Antropologico delle Madonie, Geraci Siculo 026432
Museo Etnoarqueológico del Beni Kenneth Lee, Trinidad 004253
Museo Etnobotánico, Tuluá 008810
Museo Etnográfico, Albán 008565
Museo Etnográfico, Alcubilla del Marqués 038931
Museo Etnografico, Aprica 025344
Museo Etnografico, Aritzo 025371
Museo Etnográfico, Atauta 039015
Museo Etnografico, Bomba 025587
Museo Etnográfico, Cajamarca . 034231
Museo Etnográfico, Castrojeriz . 039256
Museo Etnográfico, Cilleros 039284
Museo Etnografico, Collinas ... 026037
Museo Etnografico, Faeto 026179
Museo Etnografico, Fossalta di Portogruaro 026337
Museo Etnografico, Fratta Polesine 026350
Museo Etnográfico, Grado 039454
Museo Etnográfico, Grandas de Salime 039473
Museo Etnográfico, Inzá 008701
Museo Etnografico, Lusevera .. 026656
Museo Etnográfico, Madruga ... 009238
Museo Etnografico, Malborghetto Valbruna 026679
Museo Etnografico, Montodine . 026969
Museo Etnografico, Novalesa .. 027057
Museo Etnografico, Novara di Sicilia 027065
Museo Etnografico, Oneta 027092
Museo Etnografico, Ortonovo .. 027116
Museo Etnografico, Ossimo Superiore 027124
Museo Etnografico, Parma 027219
Museo Etnografico, Ponte in Valtellina 027396
Museo Etnográfico, Quintana Redonda 039989
Museo Etnografico, Roccalbegna 027545
Museo Etnográfico, Romanillos de Medinaceli 040015
Museo Etnografico, Roseto Capo Spulico 027672
Museo Etnográfico, Santa Lucía 009341
Museo Etnografico, Schilpario .. 027905
Museo Etnográfico, Tuxtla Gutiérrez 031625
Museo Etnografico, Vione 028401
Museo Etnografico Africo-Mozambico, Bari 025460
Museo Etnográfico Andrés Barbero, Asunción 034201
Museo Etnografico Antico Mulino ad Acqua Licheri, Fluminimaggiore 026306
Museo Etnografico Archeologico, Palau 027168
Museo Etnográfico Benigno Eiriz, Alija del Infantado 038948
Museo Etnografico Cerginolano, Cerignola 025905
Museo Etnografico Civiltà Contadina, Forlì 026326
Museo Etnográfico Comarcal de Azuaga, Azuaga 039033
Museo Etnografico Comunale, Premana 027456
Museo Etnografico Coumboscuro della Civiltà Provenzale in Italia, Sancto Lucia de Coumboscuro 027806
Museo Etnográfico da Limia, Vilar de Santos 040394
Museo Etnográfico de Autilla, Autilla del Pino 039018
Museo Etnográfico de Cantabria, Casa de Velarde, Muriedas 039808
Museo Etnográfico de Colombia, Bogotá 008618
Museo Etnográfico de Don Benito, Don Benito 039347
Museo Etnográfico de el Cerrp de Andèvalo, Antigua Casa de la Sal, El Cerro de Andévalo 039264
Museo Etnográfico de Oyón-Oion, Oyón 039865
Museo Etnográfico de Quirós y Comarca, Quirós 039990
Museo Etnográfico de Valderredible, Valderredible 040298
Museo Etnografico del Cansiglio Servadei, Tambre 028052
Museo Etnografico del Cologio Nacional Mejia, Quito 010488
Museo Etnografico del Coumboscuro, Monterosso Grana 026957
Museo Etnografico del Ferro, Bienno 025512
Museo Etnográfico del Oriente de Asturias, Llanes 039574
Museo Etnografico del Pinerolese e Museo del Legno, Pinerolo ... 027336
Museo Etnografico del Ponente Ligure, Castello Medievale, Cervo 025918
Museo Etnográfico del Reino de Pamplona, Casa Fanticorena, Ollo 039828
Museo Etnografico della Civiltà Contadina, Marianopoli 026713
Museo Etnografico della Cultura Contadina, Morigerati 026977
Museo Etnografico della Lunigiana, Villafranca in Lunigiana 028392
Museo Etnografico della Piana del Dragone, Volturara Irpina 028428
Museo Etnografico della Provincia die Belluno, Cesiomaggiore 025935
Museo Etnografico della Torre, Albino 025273
Museo Etnografico della Trinità, Botticino 025613
Museo Etnografico della Val Varatella, Toirano 028096
Museo Etnografico della Valle, Ultimo 028232
Museo Etnografico della Valle Brembana, Zogno 028433
Museo etnografico della Valle di Muggio, Cabbio 041390
Museo Etnografico dell'Alta Valle Seriana, Ardesio 025354
Museo Etnografico dell'Etna, Linguaglossa 026599
Museo Etnografico di Servola, Trieste 028208
Museo Etnografico di Sorrentini, Chiesa Madre, Patti 027228
Museo Etnografico Don Luigi Pellegrini, Castiglione di Garfagnana 025857
Museo Etnográfico e da Historia de San Paio de Narla, Friol 039411
Museo Etnografico e del Folklore Valsesiano, Borgosesia 025604
Museo Etnografico e della Cultura Materiale, Aquilonia 025347
Museo Etnografico e della Stregoneria, Triora 028215
Museo Etnografico e di Scienze Naturali, Torino 028133
Museo Etnográfico e do Viño de Cambados, Cambados 039210
Museo Etnografico e Tradizioni Contadine Borgo Antico, Torricella Sicura 028157
Museo Etnografico Extremeño González Santana, Olivenza .. 039827
Museo Etnografico Francesco Bande, Sassari 027868
Museo Etnografico G. Carpani, Lizzano in Belvedere 026610
Museo Etnografico Galluras, Luras 026655
Museo Etnográfico Huichol Wixarica, Zapopan 031700
Museo Etnografico Il Cicle della Vita, Quartu Sant'Elena 027464
Museo Etnográfico Juán B. Ambrosetti, Buenos Aires 000230
Museo Etnografico La Steiva, Piverone 027372
Museo Etnográfico Leonés Ildefonso Fierro, León 039560
Museo Etnográfico Madre Laura, Medellín 008718
Museo Etnográfico Miguel Angel Builes, Medellín 008719
Museo Etnográfico Minero, Oruro 004231
Museo Etnográfico Municipal Dámaso Arce, Olavarría 000480
Museo Etnográfico Olimio Liste, San Cristovo de Cea 040056
Museo Etnográfico Piedad Isla, Cervera de Pisuerga 039268
Museo Etnografico Romagnolo Benedetto Pergoli, Forlì 026327
Museo Etnográfico Rural, Fuentepinilla 039419
Museo Etnográfico Rural y Comarcal, Centro Cultural, Prioro 039976
Museo Etnografico San Domu de is Ainas, Armungia 025372
Museo Etnográfico Santa Rosa de Ocopa, Concepción 034255
Museo Etnografico Siciliano Giuseppe Pitrè, Palermo 027187
Museo Etnográfico Sotelo Blanco, Santiago de Compostela 040131
Museo Etnografico sulla Lavorazione del Legno, San Vito di Leguzzano 027805
Museo Etnográfico Tanit, San Bartolomé de Lanzarote 040054
Museo Etnográfico Textil Pérez Enciso, Plasencia 039947
Museo Etnografico Tiranese, Madonna di Tirano 026670
Museo Etnografico Tiranese, Tirano 028087
Museo Etnografico U. Ferrandi, Chiuso al Pubblico, Novara 027060
Museo Etnografico Vallivo, Val Masino 028249

Museo Etnográfico y Archivo Histórico Enrique Squirru, Azul 000132
Museo Etnográfico y Colonial Juan de Garay, Santa Fé 000615
Museo Etnográfico y de Arqueología Industrial, Chillarón de Cuenca 039275
Museo Etnográfico, Musei delle Regole d'Ampezzo, Cortina d'Ampezzo 026081
Museo Etnográfico, Universidad del Magdalena, Santa Marta .. 008792
Museo Etnográfico Casa do Patrón, Lalín 039543
Museo Etnohistórico de Ayabacara, Ayabacara 034220
Museo Etnológica de Casa Mazo, Valle de Hecho 040353
Museo Etnológica del Valle Medio del Ebro, Belchite 039118
Museo Etnológico, Barca 039052
Museo Etnológico, Cehegín 039260
Museo Etnológico, San Juan de Plan 040063
Museo Etnológico Aljibes de Gasparito, Rojales 040014
Museo Etnológico de Amazonas, Puerto Ayacucho 054775
Museo Etnológico de Canet d'En Berenguer, Canet d'En Berenguer 039220
Museo Etnológico de Horcajuleo de la Sierra, Horcajuelo de la Sierra 039493
Museo Etnológico de la Diputación, Castellón de la Plana 039250
Museo Etnológico de la Huerta de Murcia, Alcantarilla 038918
Museo Etnológico de la Laguna Grande, Beteta 039135
Museo Etnológico de Morella y del Meastrazgo, Morella 039788
Museo Etnológico de Navarra Julio Caro Baroja, Ayegui 039026
Museo Etnológico de Pavías, Pavías 039921
Museo Etnológico de Ribadavia, Casa da la Fundación Martínez Vázquez, Ribadavia 040000
Museo Etnológico de Sechura, Sechura 034400
Museo Etnologico della Apuane, Massa 026727
Museo Etnológico la Posada, Chinchón 039279
Museo Etnológico Misional C.M.F., Santa Isabel 010601
Museo Etnológico Missionario, Castelnuovo Don Bosco 025844
Museo Etnológico Popular, Fuensanta 039415
Museo Etnológico y Arqueológico Municipal, Manzanares el Real 039721
Museo Etnólólogico Anselmo Buil, Blecua y Torres 039153
Museo Etnomusicale I Gigli di Nola, Nola 027046
Museo Etrusco, Asciano 025384
Museo Etrusco Gasparri, Piombino 027344
Museo Etrusco M. Guarnacci, Palazzo Desideri-Tangassi, Volterra 028426
Museo Ettore Pomarici Santomasi, Gravina di Puglia 026469
Museo Europeo degli Studenti, Bologna 025564
Museo Euskal Herria, Gernika-Lumo 039434
Museo Evita, Buenos Aires ... 000231
Museo Evocativo y de Bellas Artes Osvaldo Gasparini, San Antonio de Areco 000565
Museo Ex Convento Agustino de San Pablo, Yuriria 031683
Museo Ex Convento Agustino Siglo XVI,

Museo Etnográfico y Archivo Histórico Enrique Squirru, Azul 031224
Museo Ex-Convento de San Juan Bautista, Tlayacapan 031592
Museo Ex Convento del Carmen, Guadalajara 031013
Museo Ex Hacienda San Gabriel de Barrera, Guanajuato 031029
Museo Ex Templo de San Agustín, Zacatecas 031688
Museo Ex Voto di Santa Maria della Portella, Cervara di Roma ... 025912
Museo Ex Votos Sr. de la Misericordia, Tepatitlán de Morelos 031541
Museo-Exposición de Creencias y Religiosidad Popular, Abizanda 038883
Museo-Exposición de Oficios y Artes Tradicionales, Aínsa 038899
Museo Exposición El Hombre, Bogotá 008619
Museo-Exposición Marín Bosqued, Aguarón 038889
Museo-Exposición Permanente de Geológico, Laspaúles 039547
Museo-Exposición Permanente de Juegos Tradicionales, Campo . 039212
Museo Exposición Permanente de Juguetes, Albarracín 038909
Museo Exposición Permanente de la Electricidad, Murillo de Gállego 039809
Museo Exposición Permanente de Pedagogía, Centro de Profesores y Recursos de Huesca, Huesca . 039502
Museo-Exposición Permanente Diocesano, Albarracín 038910
Museo Extremeño e Iberoamericano de Arte Contemporáneo, Badajoz 039037
Museo F. Assetto, Frontino ... 026351
Museo F. Renzi, Borghi 025596
Museo F.A. Marcucci, Ascoli Piceno 025389
Museo Fallero, Valencia 040315
Museo Farallón Dillon, Ballenita 010417
Museo Farmacéutico, Matanzas 009254
Museo Farmacia Cinquecentesca, Roccavaldina 027549
Museo Farmacia Conventuale, Convento Cappuccini, Venezia . 028315
Museo Farmacobotánica Juan A. Domínguez de la Universidad de Buenos Aires, Buenos Aires . 000232
Museo Faro a Colón, Santo Domingo 010398
Museo Faryluk, Aristóbulo del Valle 000127
Museo Felisa Rincón de Gautier, San Juan 036165
Museo Félix de Jesús, México . 031225
Museo Félix Murga, Amurrio .. 038973
Museo Fernán Félix de Amador, Luján 000436
Museo Fernando García, Canelones 046237
Museo Ferrocarrilero, Empalme . 030963
Museo Ferroviario, Stazione di Trieste Campo Marzio, Trieste 028209
Museo Ferroviario, Alcázar de San Juan 038919
Museo Ferroviario, Tacna 034406
Museo Ferroviario de Gualeguaychú, Gualeguaychú 000384
Museo Ferroviario del Verbano, Luino 026654
Museo Ferroviario Piemontese, Savigliano 027881
Museo Ferroviario, Stazione Ferroviaria, Cuneo 026117
Museo Ferroviaro, Bogotá 008620
Museo Ferrucciano, Gavinana . 026377
Museo Ferrucciano, San Marcello Pistoiese 027772
Museo Filatélico, San José ... 008860
Museo Filatélico del Banco de la República, Medellín 008720

Museo Filemón Gutiérrez Ramírez, Ameca 030825
Museo Filipe Santiago Gutiérrez, Toluca 031602
Museo Fin de la Tierra, Francisco I. Madero, Coahuila 030979
Museo Finca El Abra, Siguanea . 009371
Museo Fiorentino di Preistoria, Firenze 026291
Museo Fioroni, Legnago 026587
Museo Fittile, Ripatransone ... 027532
Museo Flesia, Coronel Pringles . 000344
Museo Florencio de la Fuente, Huete 039504
Museo Florencio Molina Campos, Moreno 000473
Museo Folklórico, La Rioja 000420
Museo Folklórico Araucano de Cañete Juan A. Ríos M., Cañete 007516
Museo Folklórico Regional de Humahuaca, Humahuaca 000387
Museo Fomento Cultural Banamex, México 031226
Museo Fondazione C. Faina e Museo Civico Palazzo Faina, Orvieto . 027119
Museo Fondazione Famiglia Piccolo di Calanovella, Capo d'Orlando . 025754
Museo-Fondazione P. Conti, Fiesole 026228
Museo Forense de la Morgue Judicial, Buenos Aires 000233
Museo Forestal de la Policía Ecológica, Jesús María 034296
Museo Forestale C. Siemoni, Poppi 027408
Museo Fortaleza de San Carlos de la Cabaña, La Habana 009183
Museo Fortaleza de San Felipe, San Felipe de Puerto Plata 010380
Museo Fortuny, Musei Civici Veneziani, Venezia 028316
Museo Fotográfico, Calarcá ... 008644
Museo Fotográfico de Nanchital, Nanchital de Lázaro Cárdenas del Río 031319
Museo Fournier de Naipes de Álava, Palacio de Bendaña, Vitoria-Gasteiz 040429
Museo Fragua Martiana, La Habana 009184
Museo Francescano – Raccolta d'Arte, Gubbio 026494
Museo Francescano Padre Aurelio Menin, Chiampo 025944
Museo Francescano, Istituto Storico dei Cappuccini, Roma 027621
Museo Francesco Borgogna, Vercelli 028336
Museo Francesco Cilea e Nichola Antonio Manfroce, Palmi 027197
Museo Francisco Campos, Guayaquil 010449
Museo Francisco Goitia, Zacatecas 031689
Museo Francisco Huerta Rendón en la Casona Universitaria, Guayaquil 010450
Museo Francisco Javier Sauza, Tequila 031560
Museo Francisco José de Caldas, Bogotá 008621
El Museo Francisco Oller y Diego Rivera, Buffalo 047307
Museo Francisco Sarabia, Lerdo 031115
Museo Franco Azzinari, Altomonte 025301, 025302
Museo Franco Lázaro, Chiapa de Corzo 030873
Museo Franz Mayer, México ... 031227
Museo Fray Bernardo Padilla, Acámbaro 030797
Museo Fray Luís de León, Belmonte 039122
Museo Frey Pedro Bedón, Ibarra 010458
Museo Frida Kahlo, Casa Azul,

México 031228
Museo Frissel de Arte Zapoteca, Universidad de las Américas, Santa Ana del Valle 031484
Museo Friulano di Storia Naturale, Udine 028229
Museo Fuentes de Loreto y Guadelupe, Palmar de Bravo 031358
Museo Fuerte Conde de Mirasol, Vieques 036175
Museo Fuerte de la Loma, Victoria de las Tunas 009384
Museo Fuerte Independencia, Tandil 000641
Museo Fundación Antonio Maura, Madrid 039662
Museo Fundación Casa de Alba, Palacio de Liria, Madrid 039663
Museo Fundación Díaz Caneja de Arte Contemporaneo, Palencia . 039875
Museo Fundación Estero Beach, Ensenada 030970
Museo Fundación Eugenio Granell, Santiago de Compostela ... 040132
Museo Fundación Jesús Otero, Santillana del Mar 040139
Museo Fundación Liste, Vigo . 040380
Museo Fundación Naum Komp, Buenos Aires 000234
Museo G. Arcucci, Ariano Irpino 025370
Museo G. Filangieri, San Potito Sannitico 027794
Museo G. Marconi, Pontecchio Marconi 027401
Museo G. Spontini, Maiolati Spontini 026676
Museo G. Whitaker, Marsala . 026719
Museo G.B. Filippa, Rimella .. 027519
Museo G.D. Cassini, Perinaldo . 027251
Museo Gabriel García Márquez, Aracatauca 008568
Museo Gabriel Mateus, Sabanagrande 008780
Museo Gabriela Mistral de Vicuña, Vicuña 007558
Museo Galego de Arte Contemporánea Carlos Maside, Arès 038989
Museo Galego do Xoguete, Allariz 038950
Museo Galería Manuel Lepe, Puerto Vallarta 031407
Museo Galleria di Villa Borghese, Roma 027622
Museo Garibaldino, Caprera .. 025757
Museo Garibaldino, Genova ... 026418
Museo Garibaldino della Campagna dell'Agro Romano per la Liberazione di Roma 1867, Mentana ... 026760
Museo Garnelo, Montilla 039782
Museo Gauchesco Ricardo Güiraldes, San Antonio de Areco 000566
Museo Gemológico, Bogotá ... 008622
Museo General Augusto César Sandino, San Rafael del Norte 033405
Museo General Belgrano, Buenos Aires 000235
Museo General de Arequito, Arequito 000126
Museo General Francisco Villa, San Juan del Río 031450
Museo General Ignacio Zaragoza, Puebla 031397
Museo General Porfirio Rubio, Landa de Matamoros 031105
Museo Geo-Paleontologico, Ferrara 026214
Museo Geologico, Castell'Arquato 025831
Museo Geológico de la Universidad Nacional de Ingeniería del Perú, Lima 034356
Museo Geológico del Seminari, Barcelona 039070
Museo Geológico del Uruguay,

Montevideo 046264
Museo Geologico e Paleontologico G.
  Capellini, Bologna 025565
Museo Geologico G.G. Gemmellaro,
  Palermo 027188
Museo Geológico Georg Petersen,
  Pontificia Universidad Católica del
  Perú, Lima 034357
Museo Geológico José Royo y Gómez,
  Bogotá 008623
Museo Geológico Marino Arce Herrera,
  Bucaramanga 008640
Museo Geológico Petrolero,
  Universidad Surcolombiana,
  Neiva 008735
Museo Geominero, Madrid 039664
Museo Geopaleontologico Alto
  Aventino, Palena 027174
Museo Geopaleontologico Cava
  Bomba, Cinto Euganeo 025993
Museo Geopaleontologico dei Fossili
  della Lessinia, Velo Veronese . 028288
Museo Geopaleontologico del Castello,
  Lerici 026593
Museo Geopaleontologico Naturalistico
  Antropico e Ornitologico Brancaleoni,
  Piobbico 027340
Museo Germán Guglielmetti, Benito
  Juárez 000147
Museo Gianmaria da Tusa,
  Cefalù 025890
Museo Giannettino Luxoro,
  Genova 026419
Museo Gianni Visentin, Carpanè di
  San Nazario 025777
Museo Giovanni Antonio Sanna,
  Sassari 027869
Museo Giovanni Boldini, Palazzo
  Massari, Ferrara 026215
Museo Giuseppe Scarabelli,
  Imola 026505
Museo Gli Orsanti, Compiano . 026053
Museo Gonzalo Carrasco,
  Otumba 031349
Museo Gráfico de la Mujer
  Guerrenese, Chilpancingo 030883
Museo Gráfico El Peruano,
  Cercado 034244
Museo Gran Casino, Bayamo . 009074
Museo Gran Mariscal de Ayacucho,
  Cumaná 054749
Museo Gregoriano Egizio, Musei
  Vaticani, Città del Vaticano . 054684
Museo Gregoriano Etrusco, Musei
  Vaticani, Città del Vaticano . 054685
Museo Gregoriano Profano, Musei
  Vaticani, Città del Vaticano ... 054686
Museo Gregorio Prieto,
  Valdepeñas 040295
Museo Guaraní, Santa Cruz .. 004240
Museo Guerrero Malagón, Urda . 040286
Museo Guillermo Perez Chiriboga del
  Banco Central, Quito 010489
Museo Guillermo Spratling, Taxco de
  Alarcón 031517
Museo Gustavo de Maeztu, Palacio de
  los Reyes de Navarra, Estella . 039387
Museo Hacienda San Pedro,
  Universidad Autónoma de Nuevo
  León, General Zuazua 030987
Museo Hans Multscher e Museo
  Civico, Vipiteno 028402
Museo Hendrik Christian Andersen,
  Roma 027623
Museo Herbario de la Facultad de
  Ciencias, México 031229
Museo Hermanas Giralt,
  Cienfuegos 009105
Museo Hermanas Mirabal,
  Conuco 010370
Museo Hermann Hesse, Fondazione
  Hermann Hesse, Montagnola . 041748
Museo Hermanos López Rayón,
  Tlalpujahua 031579

Museo Hermanos Nacif Weiss,
  Rivadavia 000539
Museo Hermenegildo Galeana, Tecpan
  de Galeana 031525
Museo Hermila Domínguez
  de Castellanos, Comitán de
  Domínguez 030904
Museo Heroes y Martires Revolución
  Sandinista en Estelí, Estelí ... 033377
Museo Hidalgo, Salamanca ... 031421
Museo Hidráulico del Sistema de
  Drenaje Profundo, México 031230
Museo Hidráulico los Molinos del Río
  Segura, Murcia 039801
Museo História de la Medicina,
  Facultad de Medicina, Valencia 040316
Museo Historia de las Ciencias Carlos
  J. Finlay, La Habana 009185
Museo Historia del Ferrocarril de
  Antioquia, Medellín 008721
Museo Historia Municipal, Antilla 009065
Museo Historiador Hernández Galiano,
  Tostado 000650
Museo Histórica Ex Convento de
  Tepoztlán, Tepoztlán 031558
Museo Histórico, El Carmen de
  Bolívar 008673
Museo Histórico, Pinar del Río . 009288
Museo Historico, Riobamba ... 010501
Museo Historico, San Antonio de los
  Baños 009313
Museo Histórico, Zapala 000686
Museo Histórico "Juan Bautista
  Ambrosetti", Gualeguay 000381
Museo Histórico Andrés Avelino
  Cáceres, Ayacucho 034223
Museo Histórico Andrés Avelino
  Cáceres, Huamanga 034272
Museo Histórico Anton García de
  Bonilla, Ocaña 008738
Museo Histórico Arqueológico, Doña
  Mencia 039348
Museo Histórico Arqueológico
  Guillermo Magrassi, Mar del
  Plata 000445
Museo Histórico-Artístico, Cuacos de
  Yuste 039329
Museo Historico Bae Calderon,
  Guayaquil 010451
Museo Histórico BBVA, Bilbao . 039147
Museo Histórico Brigadier General
  Cornelio de Saavedra, Buenos
  Aires 000236
Museo Histórico Casa de la
  Convención de Rionegro,
  Rionegro 008778
Museo Historico Casa de los Tratados,
  Cuenca 010437
Museo Histórico Comunal, Villa
  Trinidad 000684
Museo Histórico Comunal de
  San Genaro Norte, San Genaro
  Norte 000578
Museo Histórico Comunal y de la
  Colonización Judía Rabino, Moisés
  Ville 000470
Museo Histórico Conventual San
  Carlos, San Lorenzo 000591
Museo Histórico Cultural Juan
  Santamaría, Alajuela 008847
Museo Histórico de 26 de Julio,
  Santiago de Cuba 009361
Museo Histórico de Acapulco, Fuerte
  de San Diego, Acapulco 030802
Museo Histórico de Almedinilla,
  Ecomuseo del Río Caicena,
  Almedinilla 038956
Museo Histórico de Antioquia, Palacio
  de la Cultura "Rafael Uribe Uribe",
  Medellín 008722
Museo Histórico de Arenaza,
  Arenaza 000125
Museo Histórico de Arrecifes,
  Arrecifes 000129

Museo Histórico de Ayotzinapa, Tixtla
  de Guerrero 031571
Museo Histórico de Baní, Bani . 010367
Museo Histórico de Belmez y del
  Territorio Minero, Belmez ... 039120
Museo Histórico de Cañuelas,
  Cañuelas 000289
Museo Histórico de Cartagena Casa
  de la Inquisición, Academia de la
  Historia de Cartagena de Indias,
  Cartagena 008658
Museo Histórico de Cera, Buenos
  Aires 000237
Museo Histórico de Chacamarka,
  Chacamarka 034245
Museo Histórico de Ciego de Avila,
  Ciego de Avila 009101
Museo Histórico de Corrientes
  Manuel Cabral de Mayo y Alpoín,
  Corrientes 000348
Museo Histórico de Entre Ríos
  Martiniano Leguizamón, Paraná 000486
Museo Histórico de Guachinango,
  Guachinango 030991
Museo Histórico de Guatraché,
  Guatraché 000385
Museo Histórico de la Casa de la
  Cultura, Ambalema 008567
Museo Histórico de la Ciudad,
  Torreón 031610
Museo Histórico de la Ciudad,
  Palacio del Marqués de Campo,
  Valencia 040317
Museo Histórico de la Colonia San
  Carlos, San Carlos Centro ... 000569
Museo Histórico de la Educación
  Tecnológica en México,
  Chilpancingo 030884
Museo Histórico de la Honorable
  Cámara de Diputados de la Nación,
  Buenos Aires 000238
Museo Histórico de la Policía de la
  Provincia de Misiones, Posadas 000499
Museo Histórico de la Policía Nacional,
  La Paz 004218
Museo Histórico de la Policía National,
  Bogotá 008624
Museo Histórico de la Prefectura
  Naval Argentina, Tigre 000645
Museo Histórico de la Provincia de
  Catamarca, San Fernando del Valle
  de Catamarca 000573
Museo Histórico de la Provincia Dr.
  Orestes di Lullo, Santiago del
  Estero 000629
Museo Histórico de la Revolución
  Mexicana, Chihuahua 030881
Museo Histórico de la Sierra Gorda,
  Jalpan 031079
Museo Histórico de Laborde,
  Laborde 000421
Museo Histórico de Múzquiz,
  Múzquiz 031317
Museo Histórico de Navarro,
  Navarro 000476
Museo Histórico de Osorno,
  Osorno 007528
Museo Histórico de Palma Soriano,
  Palma Soriano 009279
Museo Histórico de Ranchos,
  Ranchos 000517
Museo Histórico de Reynosa,
  Reynosa 031414
Museo Histórico de San Francisco,
  Santa Fé 000616
Museo Histórico de San Miguel de
  Allende, San Miguel de Allende 031466
Museo Histórico de San Nicolás de
  los Garza, San Nicolás de los
  Garza 031471
Museo Histórico de San Roque, San
  Roque 040073
Museo Histórico de Sargadelos,
  Cervo 039269

Museo Histórico del Banco Provincia
  Dr. Arturo Jauretche, Buenos
  Aires 000239
Museo Histórico del Bierzo,
  Ponferrada 039956
Museo Historico del Norte, Salta 000558
Museo Histórico del Oriente de
  Morelos, Cuautla 030923
Museo Histórico del Regimiento de
  Granaderos a Caballo General San
  Martín, Buenos Aires 000240
Museo Histórico del Transporte Carlos
  Hillner Decoud, Quilmes 000510
Museo Historico Dr. Rafael Angel
  Calderón Guardia, San José . 008861
Museo Histórico el Dique, Astilleros de
  Puerto Real, Puerto Real 039984
Museo Histórico Etnológico de la Villa
  de Mijas, Mijas 039751
Museo Histórico Ex Aduana de Ciudad
  Juárez, Juárez 031093
Museo Histórico Ferroviario Escribano
  Alfredo Rueda, Rueda 000551
Museo Histórico Franciscano, San
  Salvador de Jujuy 000608
Museo Histórico Fuerte Barragan,
  Ensenada 000365
Museo Histórico Fuerte de San José
  el Alto, Campeche 030853
Museo Histórico General Julio de
  Vedia, 9 de Julio 000115
Museo Histórico Hipólito Unanue,
  Ayacucho 034224
Museo Histórico José Hernández-
  Chacra Pueyrredón, Villa
  Ballester 000671
Museo Histórico Judío y del
  Holocausto Tuvie Maizel,
  México 031231
Museo Histórico La Casona de Santa
  Rosa, Liberia 008849
Museo Histórico La Gallareta Forestal,
  La Gallareta 000403
Museo Histórico Local, Puente
  Genil 039978
Museo Histórico Local de Monturque,
  Monturque 039786
Museo Histórico Manuel A. Moreira,
  Laboulaye 000422
Museo Histórico Mexicano de
  Cananea, Cananea 030856
Museo Histórico Militar, Asunción 034202
Museo Histórico Militar, La Paz . 004219
Museo Histórico Militar de Melilla,
  Melilla 039746
Museo Histórico Militar de Valencia,
  Valencia 040318
Museo Histórico Minero Don Felipe de
  Borbón y Grecia, Madrid ... 039665
Museo Histórico Minero Francisco
  Pablo Holgado, Almadén ... 038952
Museo Histórico Municipal,
  Baena 039040
Museo Historico Municipal, Bahía
  Blanca 000135
Museo Histórico Municipal, Fuente
  Tójar 039417
Museo Histórico Municipal,
  Montilla 039783
Museo Histórico Municipal,
  Ocotlán 031344
Museo Histórico Municipal, San
  Fernando 040057
Museo Histórico Municipal, Taxco de
  Alarcón 031518
Museo Histórico Municipal Alfredo E.
  Múlgura, General Belgrano ... 000377
Museo Histórico Municipal Andrés A.
  Roverano, Santo Tomé 000633
Museo Histórico Municipal Brigadier
  General Juan Martín de Pueyrredón,
  San Isidro 000581
Museo Histórico Municipal de
  Armstrong, Armstrong 000128

Museo Histórico Municipal de Cañada
  de Gómez, Cañada de Gómez . 000288
Museo Histórico Municipal de Écija,
  Palacio de Benamejí, Écija . . . 039358
Museo Histórico Municipal de
  Guanabacoa, Guanabaoa 009127
Museo Histórico Municipal de Lincoln,
  Lincoln 000428
Museo Histórico Municipal de Luque,
  Luque 000439
Museo Histórico Municipal de
  Pergamino, Pergamino 000490
Museo Histórico Municipal de Priego
  de Córdoba, Priego de Córdoba 039975
Museo Histórico Municipal de Regla,
  Regla 009303
Museo Histórico Municipal de Teba,
  Teba 040211
Museo Histórico Municipal de Villa del
  Río, Villa del Río 040400
Museo Histórico Municipal de Villa
  Gesell, Villa Gesell 000679
Museo Histórico Municipal de
  Villamartín, Villamartín 040409
Museo Histórico Municipal General
  Levalle, General Levalle 000378
Museo Histórico Municipal Guillermo
  Zegarra Meneses, Arequipa . . 034215
Museo Histórico Municipal Hércules J.
  Rabagliati, Ramallo 000516
Museo Histórico Municipal Juan
  Lavalle, Baradero 000142
Museo Histórico Municipal La Para, La
  Para 000404
Museo Histórico Municipal Monte
  Grande, Monte Grande 000472
Museo Histórico Municipal Víctor E.
  Míguez, Mercedes 000462
Museo Histórico Municipal Villa Clara,
  Villa Clara 000674
Museo Histórico Municipal Villa del
  Rosario, Villa del Rosario 000677
Museo Histórico Municipal y Colección
  de Arte Precolombino Arminio Weiss,
  Rafaela 000514
Museo Histórico Municipal, Casa de
  Cultura, Cañete de las Torres . 039222
Museo Histórico Nacional, Buenos
  Aires 000241
Museo Histórico Nacional, Santiago de
  Chile 007545
Museo Histórico Nacional – Casa
  Rivera, Montevideo 046265
Museo Histórico Nacional Casa del
  Virrey Liniers, Alta Gracia 000120
Museo Histórico Nacional del Cabildo
  de Buenos Aires y de la Revolucion
  de Mayo, Buenos Aires 000242
Museo Histórico Naval,
  Cienfuegos 009106
Museo Histórico Naval Central,
  México 031232
Museo Historico Okinawa Bolivia,
  Santa Cruz 004241
Museo Histórico Policial de La Plata,
  La Plata 000414
Museo Histórico Provincial,
  Cienfuegos 009107
Museo Histórico Provincial, Sancti-
  Spíritus 009326
Museo Histórico Provincial Agustín V.
  Gnecco, San Juan 000588
Museo Histórico Provincial Brig.
  Estanislao López, Santa Fé . . . 000617
Museo Histórico Provincial Colonel
  Manuel José Olascoaga, Chos
  Malal 000306
Museo Histórico Provincial de
  Guantánamo, Guantánamo . . . 009131
Museo Histórico Provincial de Rosario
  Dr. Julio Marc, Rosario 000545
Museo Histórico Provincial Marqués
  de Sobremonte, Córdoba 000330
Museo Histórico Regional,

Alpachiri 000119
Museo Histórico Regional – Casa
  Garcilaso Cusco, Cusco 034265
Museo Histórico Regional de
  Ayacucho, Ayacucho 000131
Museo Histórico Regional de Gaiman,
  Gaiman 000375
Museo Histórico Regional de la
  Colonia San José, San José . . 000583
Museo Histórico Regional De la Isla,
  Isla del Cerrito 000390
Museo Histórico Regional de
  Magallanes, Punta Arenas 007531
Museo Histórico Regional de Río
  Cuarto, Río Cuarto 000528
Museo Histórico Regional de Tacna,
  Tacna 034407
Museo Histórico Regional de
  Urdinarrain, Urdinarrain 000659
Museo Histórico Regional de Villa La
  Angostura, Villa La Angostura . 000680
Museo Histórico Regional Ex Cuartel
  de la Compañía Fija, Ensenada 030971
Museo Histórico Regional Gabriel
  Campomar Cervera, Salliqueló 000553
Museo Histórico Regional Ichoalay,
  Resistencia 000525
Museo Histórico Regional Municipal de
  Magdalena, Magdalena 000440
Museo Histórico Regional Pablo
  Argilaga, Santo Tomé 000634
Museo Histórico Regional Padre
  Francisco Cremasco, Villa
  Concepción del Tío 000675
Museo Histórico San Ignacio de
  Loyola, San José 000584
Museo Histórico Sanmartíniano,
  Mendoza 000456
Museo Histórico Sarmiento, Buenos
  Aires 000243
Museo Histórico y Antropológico de
  Nindirí, Nindirí 033400
Museo Histórico y Antropológico
  Mauricio Van de Maele, Valdivia 007554
Museo Histórico y Arqueológico, Belén
  de Umbría 008581
Museo Histórico y Arqueológico,
  Charalá 008663
Museo Histórico y Arqueológico Dr.
  Estrada Torres, Azua 010366
Museo Histórico y de la Tradiciones
  Populares de la Boca, Buenos
  Aires 000244
Museo Histórico y Natural de Lavalle,
  Lavalle Villa Tulumaya 000427
Museo Histórico y Numismático del
  Banco Central de la República
  Argentina "Dr. José Evaristo Uriburu",
  Buenos Aires 000245
Museo Histórico y Numismático del
  Banco de la Nación Argentina,
  Buenos Aires 000246
Museo Histórico y Regional Pico
  Truncado, Pico Truncado 000491
Museo Histórico y Tradicional de
  Barracas al Sud, Avellaneda . . 000130
Museo Histórico, Arqueológico y de
  Arte Pedro Balduín, San Pedro de
  Jujuy 000605
Museo Histórico, Colonial y de
  Bellas Artes y Museo de Ciencias
  Naturales Regional Mesopotámico,
  Mercedes 000461
Museo Histório de Aspe, Aspe . 039008
Museo Hotel Marquetá, Mariquita 008712
Museo Houssay de Historia de la
  Ciencia y la Técnología, Universidad
  de Buenos Aires, Buenos Aires 000247
Museo Huehuetla-tolli, Comala . 030900
Museo Hunucmá de Sisal,
  Hunucmá 031062
Museo I Luoghi del Lavoro Contadino,
  Buscemi 025660
Museo I. Nievo, Fossalta di

Portogruaro 026338
Museo I. Salvini, Cocquio
  Trevisago 026025
Museo Iberoamericano de Arte
  Moderno de Popayán, Popayán 008768
Museo Ibleo delle Arti e Tradizioni
  Popolari – Guastella, Modica . . 026868
Museo Iconográfico, Itagua . . . 034207
Museo Iconográfico del Quijote,
  Guanajuato 031030
Museo Icticola del Beni, Universidad
  Técnica del Beni, Trinidad 004254
Museo Ideale Leonardo da Vinci,
  Vinci 028399
Museo Iglesia del Salvador, Oña 039834
Museo Iglesia San Juan de Dios,
  Murcia 039802
Museo Ignacio Zuloaga, Pedraza de la
  Sierra 039924
Museo Igneológico del Cuerpo Ofical
  de Bomberos, Bogotá 008625
Museo Illichiano, Castell'Arquato 025832
Museo Iloilo, Iloilo 034480
Museo Imágenes de la Frontera,
  Dajabón 010371
Museo in Erba, Bellinzona 041301
Museo Inca de la Universidad
  Nacional San Antonio Abad,
  Cusco 034266
Museo Independencia, Córdoba . 000331
Museo Indiano, Università Bologna,
  Bologna 025566
Museo Indígena de Colombia Ethnia,
  Bogotá 008626
Museo Indígena de Guatavita,
  Guatavita 008693
Museo Indígena Regional Enin,
  Mitú 008726
Museo Indigenista, Esquel 000370
Museo Indigeno, San José 008862
Museo Industrial de Santa Catarina El
  Blanqueo, Santa Catarina 031486
Museo Inés Mercedes Gómez Álvarez,
  Universidad Ezequiel Zamora,
  Guanare 054755
Museo Instituto Cultural México Israel,
  México 031233
Museo Insular de la Palma, Santa
  Cruz de la Palma 040096
Museo Inter-étnico de la Amazonía,
  Florencia 008681
Museo Interactivo de Ciencia,
  Tecnología y Medio Ambiente Sol del
  Niño, Mexicali 031145
Museo Interactivo de Ciencias
  Naturales, Ituzaingó 000392
Museo Interactivo de Ciencias-Puerto
  Ciencia, Universidad Nacional de
  Entre Ríos, Paraná 000487
Museo Interactivo de la Ciencia y
  Tecnología – Principia, Málaga 039704
Museo Interactivo de la Música,
  Málaga 039705
Museo Interactivo del Centro de
  Ciencias de Sinaloa, Culiacán . 030938
Museo Interactivo del Medio Ambiente,
  México 031234
Museo Interactivo El Rehilete,
  Pachuca 031354
Museo Interactivo La Avispa,
  Chilpancingo 030885
Museo Interattivo delle Scienze,
  Foggia 026310
Museo Internacional de Arte
  Contemporáneo, Arrecife 038997
Museo Internacional de Arte Naïf,
  Jaén 039525
Museo Internacional de Electrografía,
  Cuenca 039337
Museo Internazionale del Folklore e
  Civiltà Contadina, Atina 025412
Museo Internazionale del Presepio,
  Morrovalle 026981

Museo Internazionale della Campana,
  Agnone 025249
Museo Internazionale della Caricatura,
  Tolentino 028102
Museo Internazionale della Croce
  Rossa, Castiglione delle Stiviere 025855
Museo Internazionale della
  Fisarmonica, Castelfidardo . . . 025819
Museo Internazionale delle Ceramiche,
  Fondazione, Faenza 026175
Museo Internazionale delle Marionette
  A. Pasqualino, Palazzo Fatta,
  Palermo 027189
Museo Internazionale dell'Etichetta,
  Cupramontana 026121
Museo Internazionale dell'Immagine
  Postale, Belvedere Ostrense . . 025483
Museo Interparroquial de Alaejos,
  Alaejos 038901
Museo Interparroquial de Arte Sacro,
  Iglesia de Santa María de Mediavilla,
  Medina de Rioseco 039739
Museo Irineo Germán, Azoyú . . 030841
Museo Irpino, Avellino 025423
Museo Irureta, Tilcara 000648
Museo Isabel Rubio, Isabel Rubio 009224
Museo Isidro Fabela,
  Atlacomulco 030836
Museo It Akean, Kalibo 034487
Museo Italgas, Torino 028134
Museo Italiano della Ghisa,
  Longiano 026625
Museo Italoamericano, San
  Francisco 053013
Museo Itinerante de Arte
  Contemporáneo del Mercosur,
  Buenos Aires 000248
Museo Ittico, Pescara 027276
Museo Ittico Augusto Capriotti, San
  Benedetto del Tronto 027725
Museo Ixchel del Traje Indigena,
  Guatemala 023371
Museo J.M. Cruxent, Los Teques 054760
Museo Jacinto Guerrero, Ajofrín 038900
Museo Jacinto Jijón y Caamaño,
  Quito 010490
Museo Jacobo Borges, Caracas 054735
Museo Jarocho Salvador Ferrando,
  Tlacotalpan 031574
Museo Jesuítico de Sitio, Yapeyú 000685
Museo Jesuítico Nacional de Jesus
  María, Córdoba 000332
Museo Jesútico Nacional de Jesús
  María, Jesús María 000393
Museo Joaquín Peinado de Ronda,
  Ronda 040021
Museo Jorge Eliécer Gaitán, Puerto
  Gaitán 008771
Museo Jorge Pasquini López, San
  Salvador de Jujuy 000609
Museo José Antonio Páez,
  Acarigua 054696
Museo José de Jesús Almaza,
  Xicoténcatl 031673
Museo José Domingo de Obaldía,
  David 034181
Museo José Fernando Ramírez,
  Durango 030953
Museo José Guadalupe Posada,
  Aguascalientes 030812
Museo José Luis Cuevas, Colima 030891
Museo José Luis Cuevas, México 031235
Museo José Manuel Estrada,
  Suipacha 000637
Museo José María Luis Mora,
  Ocoyoacac 031345
Museo José María Morelos y Pavón,
  Cuautla 030924
Museo José María Velasco,
  Toluca 031603
Museo José Martí, La Habana . 009186
Museo José Natividad Correa Toca
  Teapa, Teapa 031520
Museo José Ortega, Arroba de los

Montes 038998
Museo José y Tomás Chávez Morado,
Silao 031501
Museo Juan Cabré, Calaceite . 039194
Museo Juan Carlos Iramaín, San
Miguel de Tucumán 000600
Museo Juan Lorenzo Lucero,
Pasto 008749
Museo Juan Manuel Blanes,
Montevideo 046266
Museo Juárez del Jardín Borda,
Cuernavaca 030928
Museo Juárez e Historia de Mapimí y
Ojuela, Mapimí 031124
Museo Juarista de Congregación
Hidalgo, Entronque Congregación
Hidalgo 030972
Museo Judío de Buenos Aires Dr.
Salvador Kibrick, Buenos Aires 000249
Museo Julio Romero de Torres,
Córdoba 039308
Museo Kan Pepen, Teabo 031519
Museo Kartell, Noviglio 027071
Museo Ken Damy di Fotografia
Contemporanea, Brescia 025635
Museo Kusillo, Fundación Cultural
Quipus, La Paz 004220
Museo L. Garaventa, Chiavari .. 025955
Museo L. Minguzzi, Milano 026816
Museo la Antigua Casa del Agua,
Hunucmá 031063
Museo La Casa del Libro, San
Juan 036166
Museo La Chole, Petatlán 031367
Museo La Cinacina, San Antonio de
Areco 000567
Museo La Flor de Jimulco,
Torreón 031611
Museo La Fudina, Comelico
Superiore 026046
Museo La Isabelica, Santiago de
Cuba 009362
Museo La Pesca, Benito Juárez 030845
Museo La Vieja Estación, María
Teresa 000449
Museo La Vigía, Matanzas 009255
Museo-Laboratorio, Vernio 028339
Museo Laboratorio del Tessile, Soveria
Mannelli 027996
Museo Laboratorio di Archeologia,
Monsampolo del Tronto 026897
Museo Laboratorio di Arte
Contemporanea, Università La
Sapienza, Roma 027624
Museo Laboratorio Santa Barbara,
Mammola 026687
Museo Ladino di Fassa, Pozza di
Fassa 027434
Museo Ladino-Fodom, Livinallongo del
Col di Lana 026602
Museo Laguna del Caimán, Tlahualilo
de Zaragoza 031576
Museo Lapidario, Ferrara 026216
Museo Lapidario, Urbino 028241
Museo Lapidario del Duomo di
Sant'Eufemia, Grado 026461
Museo Lapidario della Canonica della
Cattedrale, Chiostro della Canonica
del Duomo, Novara 027061
Museo Lapidario e del Tesoro del
Duomo, Modena 026861
Museo Lapidario Estense,
Modena 026862
Museo Lapidario Maffeiano,
Verona 028351
Museo Lapidario Marsicano,
Avezzano 025427
Museo Lara, Ronda 040022
El Museo Latino, Omaha 051766
Museo Lázaro Galdiano, Madrid 039666
Museo Legado Hermanos Álvarez
Quintero, Utrera 040288
Museo Legislativo los Sentimientos de
la Nación, México 031236

Museo Leonardiano, Palazzina Uzielli
& Castello dei Conti Guidi, Vinci 028400
Museo Leoncavallo, Brissago .. 041366
Museo Leoniano, Carpineto
Romano 025782
Museo L'Iber, Museo de los Soldatitos
de Plomo, Valencia 040319
Museo Libisosa, Lezuza 039563
Museo Lítico de Pukara, Pukara 034391
Museo Local de Acámbaro, Dolores
Hidalgo 030946
Museo Local de Antropología
e Historia de Compostela,
Compostela 030907
Museo Local de Valle de Santiago,
Valle de Santiago 031637
Museo Local Tuxteco, Santiago
Tuxtla 031497
Museo Locale, Termeno 028072
Museo Lodovico Pogliaghi, Santa
Maria del Monte, Varese 028278
Museo Lombardo di Storia
dell'Agricoltura, Sant'Angelo
Lodigiano 027835
Museo López Claro, Santa Fé .. 000618
Museo Lorenzo Coullaut-Valera,
Marchena 039726
Museo Loringiano, Málaga 039706
Museo Los Amantes de Sumpa,
Guayaquil 010452
Museo Luciano Rosero, Pasto . 008750
Museo Lucio Balderas Márquez,
Landa de Matamoros 031106
Museo Luigi Ceselli, Subiaco .. 028029
Museo Luigi Varoli, Cotignola .. 026093
Museo Luis A. Calvo, Agua de
Dios 008561
Museo Luis Alberto Acuña, Villa de
Leyva 008823
Museo Luis Donaldo Colosio,
Francisco I. Madero, Coahuila . 030980
Museo M. Antonioni, Ferrara ... 026217
Museo Madre Caridad Brader,
Pasto 008751
Museo Madres Benedictinas,
Sahagún 040036
Museo Maestro Alfonso Zambrano,
Pasto 008752
Museo Malacologico, Erice 026161
Museo Malacologico delle Argille,
Cutrofiano 026126
Museo Malacologico Piceno, Cupra
Marittima 026120
Museo Malvinas Argentinas, Río
Gallegos 000532
Museo Manoblanca, Buenos
Aires 000250
Museo Manuel de Falla, Alta
Gracia 000121
Museo Manuel F. Zárate,
Guarraré 034182
Museo Manuel López Villaseñor,
Ciudad Real 039291
Museo Manuel Ojinaga, Manuel
Ojinaga 031122
Museo Manuela Sáenz, Quito .. 010491
Museo Manzoniano, Lesa 026594
Museo Manzoniano, Milano ... 026817
Museo Marcelo López del Instituto
Hellen Keller, Córdoba 000333
Museo Marcha del Pueblo
Combatiente, La Habana 009187
Museo Marchigiano del Risorgimento
G. e D. Spadoni, Macerata ... 026661
Museo Marcos Redondo,
Pozoblanco 039967
Museo Mariano, La Habana .. 009188
Museo Mariano Acosta, Ibarra . 010459
Museo Mariano Matamoros,
Jantetelco 031082
Museo Marina Núñez del Prado,
Lima 034358
Museo Marinaro Gio Bono Gerrari,
Cairo Montenotte 025685

Museo Marinaro Tommasino-Andreatta,
San Colombano Certenoli 027741
Museo Marino d'Arte Sacra
Contemporanea, Comacchio .. 026044
Museo Marino de Tecolutla,
Tecolutla 031524
Museo Marino Marini, Firenze .. 026292
Museo Marino Marini, Pistoia .. 027366
Museo Marino Marini, Galleria d'Arte
Moderna, Milano 026818
Museo Mario Abreu, Maracay . 054766
Museo Mario Praz, Roma 027625
Museo Marítimo "Seno de Corcubión",
Corcubión 039302
Museo Marítimo de Asturias,
Gozón 039453
Museo Marítimo de Ushuaia y Presidio,
MUSEO de Arte Marino Ushuaia –
Museo Antartico Jose Maria Sobral,
Ushuaia 000661
Museo Marítimo del Cantábrico,
Santander 040115
Museo Marítimo Torre del Oro,
Sevilla 040178
Museo Marítimo y Naval de la
Patagonia Austral, Río Gallegos 000533
Museo Marsiliano, Biblioteca
Universitaria di Bologna,
Bologna 025567
Museo Martín Cárdenas,
Cochabamba 004207
Museo Martín Gusinde, Puerto
Williams 007530
Museo Martín Perez, Montevideo 046267
Museo Martini di Storia dell'Enologia,
Pessione 027288
Museo Mascagno, Livorno ... 026606
Museo Maschera, Folklore e Civiltà
Contadina, Acerra 025231
Museo Massò, Bueu 039161
Museo Maurice Minkowski, Buenos
Aires 000251
Museo Máximo Gómez, La
Habana 009189
Museo Medardo Burbano, Pasto 008753
Museo Medardo Rosso, Barzio . 025468
Museo Medíceo della Petraia e
Giardino, Firenze 026293
Museo Medievale Borbonico,
Salle 027712
Museo Melezet, Bardonecchia . 025450
Museo Melia Cayo Coco y Museo Tryp
Colonial, Cayo Coco 009098
Museo Memoria Coronel Leoncio
Prado, Callao 034236
Museo Memorial 12 de Septiembre,
La Habana 009190
Museo Memorial 26 de Julio, Victoria
de las Tunas 009385
Museo Memorial Antonio Guiteras
Holmes, Pinar del Río 009289
Museo Memorial El Morrillo,
Matanzas 009256
Museo Memorial Ernesto Che Guevara,
Santa Clara 009336
Museo Memorial José Martí, La
Habana 009191
Museo Memorial La Demajagua,
Manzanillo 009247
Museo Memorial Los Malagones,
Moncada 009265
Museo Memorial Mártires de
Barbados, Victoria de las Tunas 009386
Museo Memorial Vicente García,
Victoria de las Tunas 009387
Museo Memorias de Yucundé, San
Pedro y San Pablo Teposcolula 031480
Museo Memorie Pollentine,
Pollenza 027387
Museo Meo-Evoli, Monopoli .. 026892
Museo Mercantile, Bolzano ... 025585
Museo Mercedes Sierra de Pérez El
Chico, Bogotá 008627
Museo Meteorológico Nacional Dr.

Benjamín A. Gould, Córdoba . . 000334
Museo Metropolitano, Buenos
Aires 000252
Museo Metropolitano de Monterrey,
Monterrey 031296
Museo Michelangelo, Caprese . 025758
Museo Michoacano de las Artesanías,
Morelia 031310
Museo Miguel Álvarez del Toro, Tuxtla
Gutiérrez 031626
Museo Militar de Cartagena, Museo
Militar Regional de Sevilla,
Cartagena 039239
Museo Militar de Colombia,
Bogotá 008628
Museo Militar de Menorca, Es
Castell 039244
Museo Militar Regional, A
Coruña 039323
Museo Militar Regional de Burgos,
Burgos 039172
Museo Militar Regional de Canarias,
Centro de Historia y Cultura Militar
del Mando de Canarias, Santa Cruz
de Tenerife 040100
Museo Militar Regional de Sevilla,
Museo Naval de San Fernando,
Sevilla 040179
Museo Militar Regional, Castell de San
Carlos, Palma de Mallorca .. 039889
Museo Mille Voci e Mille Suoni,
Bologna 025568
Museo Mineralogico, Bortigiadas 025607
Museo Mineralogico, Valle Aurina 028260
Museo Mineralogico Campano, Vico
Equense 028375
Museo Mineralogico e Naturalistico,
Bormio 025606
Museo Mineralogico e Paleontologico
delle Zolfare, Caltanissetta ... 025704
Museo Mineralogico Permanente,
Carro 025787
Museo Mineralogico Prof. Manuel
Tellechea, Mendoza 000457
Museo Mineralógico Salón Tulio
Ospina, National University of
Colombia, Medellín 008723
Museo Mineralogico Sardo,
Iglesias 026500
Museo Mineralógico y Antropológico,
Ibagué 008699
Museo Mineralógico y Geológico,
Universidad Técnica de Oruro,
Oruro 004232
Museo Minerario Alpino, Cogne . 026030
Museo Minero de Riotinto, Minas de
Riotinto 039753
Museo Minero Ferrería de San Blas,
Sabero 040031
Museo Minero, Centro Cultural
Asensio Sáez, La Unión 040285
Museo Mínimo, Corigliano
Calabro 026073
Museo Miniscalchi Erizzo, Verona 028552
Museo Mirador de Quistococha,
Iquitos 034292
Museo Miscellaneo Galbiati,
Brugherio 025647
Museo Misional de Nuestra Señora de
Regla, Chipiona 039280
Museo Missionario, Padova ... 027156
Museo Missionario Cinese e di Storia
Naturale, Lecce 026579
Museo Missionario delle Grazie,
Rimini 027524
Museo Missionario Etnologico, Musei
Vaticani, Città del Vaticano .. 054687
Museo Missionario Francescano,
Fiesole 026229
Museo Mitre, Buenos Aires ... 000253
Museo Mohicca, Mohicca 031281
Museo Molino de Papel La Villa,
Buenos Aires 000254
Museo Molino de Teodoro, Cieza 039283

Museo Molino Nant Fach,
  Trevelin 000655
Museo Monástico Medival y de
  Farmacia, Santo Domingo de
  Silos 040145
Museo Monográfico de Cástulo,
  Linares 039566
Museo Monográfico de Clunia,
  Peñalba de Castro 039931
Museo Monográfico de El Cigarralejo,
  Palacio de Menahermosa, Mula 039931
Museo Monográfico de la Alcudia,
  Elche 039369
Museo Monográfico de la Colonia
  Celsa, Museo de Zaragoza, Velilla de
  Ebro 040363
Museo Monográfico de la Villa
  Romana de la Olmeda, Saldaña 040049
Museo Monográfico de Tiermes,
  Museo Numantino de Soria, Montejo
  de Tiermes 039777
Museo Monográfico del Azafrán,
  Casa de la Cultura, Monreal del
  Campo 039772
Museo Monográfico del Castro de
  Viladonga, Castro de Rei .. 039254
Museo Monográfico y Necrópolis
  Púnica de Puig des Molins,
  Eivissa 039361
Museo Monseñor Agustín Blujaki,
  Asunción 034203
Museo Monseñor José Fagnano, Río
  Grande 000536
Museo Monseñor Juan Sinforiano
  Bogarin, Asunción 034204
Museo Monte de Piedad, Buenos
  Aires 000255
Museo Monumental de la
  Municipalidad Distrital de Huaura,
  Huaura 034283
Museo Monumento al Deportato,
  Carpi 025781
Museo Morandi – Collezioni Comunali
  d'Arte, Bologna 025569
Museo Morando Bolognini,
  Sant'Angelo Lodigiano 027836
Museo Morelos, Carácuaro .. 030860
Museo Multidisciplinario del Colegio
  La Salle, Lima 034359
Museo Mundial del Tango, Buenos
  Aires 000256
Museo Mundo de Ambar, Santo
  Domingo 010399
Museo Mundocaña, Cali 008652
Museo Mundocaña, Palmira .. 008741
Museo Municipal, Alcázar de San
  Juan 038920
Museo Municipal, Algeciras . 038942
Museo Municipal, Alzira ... 038968
Museo Municipal, Aroche ... 038996
Museo Municipal, Ayllón ... 039027
Museo Municipal, Cabra 039180
Museo Municipal, Calatayud . 039199
Museo Municipal, Ciego de Avila 009102
Museo Municipal, Consuegra . 039301
Museo Municipal, Holguín .. 009222
Museo Municipal, Iquitos .. 034293
Museo Municipal, Mier 031274
Museo Municipal, Nerva 039819
Museo Municipal, Ourense .. 039859
Museo Municipal, Palma del Río 039898
Museo Municipal, El Puerto de Santa
  María 039983
Museo Municipal, Puertollano 039985
Museo Municipal, Requena .. 039995
Museo Municipal, Sancti-Spíritus 009327
Museo Municipal, Valdepeñas 040296
Museo Municipal, Zuheros .. 040465
Museo Municipal Alex Urquiola, El
  Salvador 009309
Museo Municipal Arco Ajújar, Medina
  de Rioseco 039740
Museo Municipal Barão de Santo
  Angelo, Rio Pardo 004775

Museo Municipal C.I.P.A.S., Salto 000562
Museo Municipal Cabildo de
  Montevideo, Montevideo ... 046268
Museo Municipal Carmen Funes, Plaza
  Huincul 000495
Museo Municipal Casa de Alfaro,
  Montecristi 010465
Museo Municipal de 10 de Octubre,
  La Habana 009192
Museo Municipal de Abreus,
  Abreus 009060
Museo Municipal de Aguada de
  Pasajeros, Aguada de Pasajeros 009061
Museo Municipal de Alto Songo y La
  Maya, Alto Songo 009062
Museo Municipal de Amancio,
  Amancio 009064
Museo Municipal de Arqueología y
  Etnología, Segorbe 040151
Museo Municipal de Arqueología y
  Etnología, Villajoyosa ... 040408
Museo Municipal de Arqueología y
  Etnología, Bejís 039117
Museo Municipal de Arte Angel María
  de Rosa, Junín 000395
Museo Municipal de Arte
  Contemporáneo de Madrid,
  Madrid 039667
Museo Municipal de Arte Decorativo
  Firma y Odilo Estévez, Rosario 000546
Museo Municipal de Arte en Vidrio de
  Alcorcón, Alcorcón 038926
Museo Municipal de Arte Juan Carlos
  Castagnino, Mar del Plata . 000446
Museo Municipal de Arte Moderno,
  Cuenca 010438
Museo Municipal de Arte Moderno de
  Mendoza, Mendoza 000458
Museo Municipal de Artes Plásticas,
  Olavarría 000481
Museo Municipal de Artes Plásticas,
  Rivera 046291
Museo Municipal de Artes Plásticas
  Faustino Jorge Bonadeo, Rauch 000518
Museo Municipal de Artes Plásticas
  Pompeo Boggio, Chivilcoy . 000305
Museo Municipal de Artes Visuales,
  Concordia 000316
Museo Municipal de Artes Visuales,
  Quilmes 000511
Museo Municipal de Artes Visuales
  Sor Josefa Díaz y Clucellas, Santa
  Fé 000619
Museo Municipal de Arucas,
  Arucas 039006
Museo Municipal de Báguano,
  Báguano 009066
Museo Municipal de Bahía Honda,
  Bahía Honda 009067
Museo Municipal de Baraguá,
  Baraguá 009070
Museo Municipal de Bartolomé Masó,
  Bartolomé Masó 009071
Museo Municipal de Bayamo,
  Bayamo 009075
Museo Municipal de Baza, Baza 039111
Museo Municipal de Bellas Artes,
  Bahía Blanca 000136
Museo Municipal de Bellas Artes, La
  Calera 000397
Museo Municipal de Bellas Artes, La
  Paz 000405
Museo Municipal de Bellas Artes,
  Tandil 000642
Museo Municipal de Bellas Artes de
  Valparaíso, Valparaíso ... 007557
Museo Municipal de Bellas Artes Dr.
  Genaro Perez, Córdoba 000335
Museo Municipal de Bellas Artes Dr.
  Genaro Pérez, Río Cuarto . 000529
Museo Municipal de Bellas Artes Dr.
  Urbano Poggi, Rafaela 000515
Museo Municipal de Bellas Artes Juan
  B. Castagnino, Rosario ... 000547

Museo Municipal de Bellas Artes Juan
  Manuel Blanes, Montevideo . 046269
Museo Municipal de Bellas Artes
  Lucas Braulio Areco, Posadas . 000500
Museo Municipal de Bellas Artes Prof.
  Amelio Ronco Ceruti, Metán . 000464
Museo Municipal de Bellas Artes,
  Antiguo Convento de San Francisco,
  Santa Cruz de Tenerife ... 040101
Museo Municipal de Bolivia,
  Bolivia 009079
Museo Municipal de Boyeros, La
  Habana 009193
Museo Municipal de Buey Arriba, Buey
  Arriba 009080
Museo Municipal de Cabaiguán,
  Cabaiguán 009081
Museo Municipal de Cabana,
  Cabana 034226
Museo Municipal de Cacomun,
  Cacomun 009082
Museo Municipal de Caibarién,
  Caibarién 009083
Museo Municipal de Caimanera,
  Caimanera 009084
Museo Municipal de Calahorra,
  Calahorra 039196
Museo Municipal de Calimete,
  Calimete 009086
Museo Municipal de Camajuaní
  Hermanos Vidal Caro,
  Camajuaní 009092
Museo Municipal de Campechuela,
  Campechuela 009093
Museo Municipal de Candelaria,
  Candelaria 009094
Museo Municipal de Carlos Manuel
  de Céspedes, Carlos Manuel de
  Céspedes 009096
Museo Municipal de Cauto Cristo,
  Cauto Cristo 009097
Museo Municipal de Cerámica,
  Paterna 039919
Museo Municipal de Chambas,
  Chambas 009099
Museo Municipal de Chancay,
  Chancay 034247
Museo Municipal de Chincha, Chincha
  Alta 034249
Museo Municipal de Ciencias, Bahía
  Blanca 000137
Museo Municipal de Ciencias
  Naturales, Valencia 040320
Museo Municipal de Ciencias
  Naturales Carlos Ameghino,
  Mercedes 000463
Museo Municipal de Ciencias
  Naturales Carlos Darwin, Punta
  Alta 000509
Museo Municipal de Ciencias
  Naturales Lorenzo Scaglia, Mar del
  Plata 000447
Museo Municipal de Cifuentes,
  Cifuentes 009108
Museo Municipal de Ciro Redondo,
  Ciro Redondo 009109
Museo Municipal de Colón José R.
  Zulueta, Colón 009111
Museo Municipal de Consolación del
  Sur, Consolación del Sur . 009112
Museo Municipal de Contramaestre,
  Contramaestre 009113
Museo Municipal de Corralillo,
  Corralillo 009114
Museo Municipal de Cruces,
  Cruces 009115
Museo Municipal de Cueto,
  Cueto 009117
Museo Municipal de Cumanayagua,
  Cumanayagua 009118
Museo Municipal de Daimiel,
  Daimiel 039340
Museo Municipal de El Trébol, El
  Trébol 000360

Museo Municipal de Eldorado,
  Eldorado 000362
Museo Municipal de Esmeralda,
  Esmeralda 009119
Museo Municipal de Estepona,
  Estepona 039390
Museo Municipal de Etnografía,
  Villadiego 040401
Museo Municipal de Etnología,
  Castellón de la Plana 039251
Museo Municipal de Florencia,
  Florencia 009120
Museo Municipal de Florida,
  Florida 009121
Museo Municipal de Fomento,
  Fomento 009122
Museo Municipal de Gibara,
  Gibara 009125
Museo Municipal de Guamá, La
  Plata 009234
Museo Municipal de Guane,
  Guane 009128
Museo Municipal de Guantánamo –
  Antigua Cárcel, Guantánamo . 009132
Museo Municipal de Guayaquil,
  Guayaquil 010453
Museo Municipal de Guisa, Guisa 009133
Museo Municipal de Habana del Este,
  La Habana 009194
Museo Municipal de Historia,
  Necolí 008733
Museo Municipal de Historia del Arte,
  Montevideo 046270
Museo Municipal de Historia Malitzín
  de Huiloapan, Huiloapan de
  Cuauhtémoc 031058
Museo Municipal de Historia Natural,
  General Alvear 000376
Museo Municipal de Historia Natural,
  San Rafael 000606
Museo Municipal de Historia y
  Arqueología, González 030990
Museo Municipal de Historia y
  Arqueología, Rivera 046292
Museo Municipal de Huaytará,
  Huaytará 034285
Museo Municipal de Imías, Playitas y
  Cajobabo, Cajobabo 009085
Museo Municipal de Jagüey Grande,
  Jagüey Grande 009229
Museo Municipal de Jesús Menéndez,
  Jesús Menéndez 009229
Museo Municipal de Jiguaní,
  Jiguaní 009230
Museo Municipal de Jovellanos
  Domingo Mujica, Jovellanos . 009233
Museo Municipal de la Ciudad de
  Carcarañá, Carcarañá 000291
Museo Municipal de la Ciudad de
  Rosario, Rosario 000548
Museo Municipal de la Lisa, La
  Habana 009195
Museo Municipal de La Sierpe, La
  Sierpe 009370
Museo Municipal de La Yaya, La
  Yaya 009396
Museo Municipal de Lajas, Lajas 009235
Museo Municipal de Los Arabos
  Clotilde García, Los Arabos . 009236
Museo Municipal de Los Palacios, Los
  Palacios 009237
Museo Municipal de Madrid,
  Madrid 039668
Museo Municipal de Maisí, Maisí 009239
Museo Municipal de Majagua,
  Majagua 009240
Museo Municipal de Majibacoa,
  Majibacoa 009241
Museo Municipal de Manatí Jesús
  Suárez Gayol, Manatí 009242
Museo Municipal de Manicaragua,
  Manicaragua 009243
Museo Municipal de Mantua,
  Mantua 009244

Register der Institutionen und Firmen
Museo Municipal de Manuel Tames, Manuel Tames – Museo Nacional de la Acuarela, México

Museo Municipal de Manuel Tames, Manuel Tames . . . 009245
Museo Municipal de Manzanillo, Manzanillo . . . 009248
Museo Municipal de Marianao, La Habana . . . 009196
Museo Municipal de Martí, Martí 009249
Museo Municipal de Mayarí, Mayarí . . . 009259
Museo Municipal de Media Luna, Media Luna . . . 009261
Museo Municipal de Minas, Minas . . . 009262
Museo Municipal de Minas de Matahambre, Minas de Matahambre . . . 009263
Museo Municipal de Moa, Moa . 009264
Museo Municipal de Morón, Morón . . . 009266
Museo Municipal de Najasa, Sibanicú . . . 009368
Museo Municipal de Nazca y Casa Museo María Reiche, Nazca . . 034377
Museo Municipal de Niceto Pérez, Niceto Pérez . . . 009267
Museo Municipal de Niquero, Niquero . . . 009268
Museo Municipal de Nuevitas, Nuevitas . . . 009278
Museo Municipal de Palma Soriano, Palma Soriano . . . 009280
Museo Municipal de Pedro Betancourt, Pedro Betancourt . . . 009282
Museo Municipal de Perico, Perico . . . 009283
Museo Municipal de Pilón, Pilón 009284
Museo Municipal de Pintura, Villadiego . . . 040402
Museo Municipal de Placetas, Placetas . . . 009294
Museo Municipal de Poblado de Jimaguayú, Jimaguayú . . . 009231
Museo Municipal de Ponteareas, Ponteareas . . . 039959
Museo Municipal de Porteña, Porteña . . . 000496
Museo Municipal de Primero de Enero, Primero de Enero . . . 009296
Museo Municipal de Quemado de Güines, Quemado de Güines . . 009300
Museo Municipal de Rafael Freyre, Rafael Freyre . . . 009301
Museo Municipal de Ranchuelo, Ranchuelo . . . 009302
Museo Municipal de Río Cauto, Río Cauto . . . 009305
Museo Municipal de Rodas, Rodas . . . 009306
Museo Municipal de Ronda, Palacio de Mondragón, Ronda . . . 040023
Museo Municipal de Sagua de Tánamo, Sagua de Tánamo . . . 009307
Museo Municipal de Sagua la Grande, Sagua la Grande . . . 009308
Museo Municipal de San Cristobal, San Cristobal . . . 009316
Museo Municipal de San Juan y Martínez, San Juan y Martínez 009318
Museo Municipal de San Luis 29 de Abril, San Luis . . . 009319
Museo Municipal de San Miguel del Padrón, La Habana . . . 009197
Museo Municipal de San Telmo, Donostia-San Sebastián . . . 039354
Museo Municipal de Sandino, Sandino . . . 009330
Museo Municipal de Santa Cruz del Sur, Santa Cruz del Sur . . . 009340
Museo Municipal de Santiponce, Antiguo Matadero, Santiponce . 040142
Museo Municipal de Santo Domingo, Santo Domingo . . . 009367
Museo Municipal de Sibanicú, Sibanicú . . . 009369
Museo Municipal de Sierra de Cubitas, Cubitas . . . 009116
Museo Municipal de Taguasco, Taguasco . . . 009372
Museo Municipal de Trinidad, Trinidad . . . 009376
Museo Municipal de Unión de Reyes Juan G. Gómez, Unión de Reyes . . . 009378
Museo Municipal de Urbano Noris, Urbano Noris . . . 009379
Museo Municipal de Varadero, Varadero . . . 009380
Museo Municipal de Vedado, La Habana . . . 009198
Museo Municipal de Venezuela, Venezuela . . . 009381
Museo Municipal de Vertientes, Vertientes . . . 009382
Museo Municipal de Vigo Quiñones de León, Vigo . . . 040381
Museo Municipal de Viñales Adela Azcuy Labrador, Pinar del Río . 009290
Museo Municipal de Yaguajay, Yaguajay . . . 009393
Museo Municipal de Yara, Yara . 009394
Museo Municipal de Yateras, Yateras . . . 009395
Museo Municipal del Cerro, La Habana . . . 009199
Museo Municipal del Deporte, San Fernando del Valle de Catamarca . . . 000574
Museo Municipal del III Frente, Santiago de Cuba . . . 009363
Museo Municipal del Teatro, Lima . . . 034360
Museo Municipal Dr. Rodolfo Doval Fermi, Sastre . . . 000635
Museo Municipal Dreyer, Dreyer 034269
Museo Municipal Eduard Camps Cava, Guissona . . . 039483
Museo Municipal Elisa Cendrero, Ciudad Real . . . 039292
Museo Municipal Emilio Daudinot, San Antonio del Sur . . . 009314
Museo Municipal Estados del Duque, Malagón . . . 039711
Museo Municipal Fernando García Grave de Peralta, Puerto Padre 009299
Museo Municipal Forte de San Damián, Ribadeo . . . 040001
Museo Municipal Francisco Javier Balmaseda, Santa Clara . . . 009337
Museo Municipal Histórico Fotográfico de Quilmes, Quilmes . . . 000512
Museo Municipal Histórico Regional Almirante Brown, Bernal . . . 000148
Museo Municipal Histórico Regional Santiago Lischetti, Villa Constitución . . . 000676
Museo Municipal Ignacio Balvidares, Puan . . . 000503
Museo Municipal IV Centenario, Pinos . . . 031370
Museo Municipal Jatibonico, Jatibonico . . . 009228
Museo Municipal Jerónimo Molina, Jumilla . . . 039536
Museo Municipal Jesús González Herrera, Matamoros . . . 031128
Museo Municipal Jesús H. Salgado, Teloloapan . . . 031531
Museo Municipal José A. Mulazzi, Tres Arroyos . . . 000653
Museo Municipal José Manuel Maciel, Coronda . . . 000341
Museo Municipal José Reyes Meza, Altamira . . . 030822
Museo Municipal Juan M. Ameijeiras, El Salvador . . . 009310
Museo Municipal Lino Enea Spilimbergo, Unquillo . . . 000658
Museo Municipal López Torres, Tomelloso . . . 040247
Museo Municipal Los Guachimontones y Museo de Arquitectura Prehispánica, Teuchitlán . . . 031561
Museo Municipal Manuel Torres, Marín . . . 039728
Museo Municipal Marceliano Santa María, Burgos . . . 039173
Museo Municipal Mariano Benlliure, Crevillente . . . 039328
Museo Municipal Mateo Hernández, Béjar . . . 039115
Museo Municipal Nueva Gerona, Nueva Gerona . . . 009276
Museo Municipal Precolombino y Colonial, Montevideo . . . 046271
Museo Municipal Primeros Pobladores, San Martín de los Andes . . . 000593
Museo Municipal Punta Hermengo, Miramar . . . 000469
Museo Municipal Quetzalpapalotl, Teotihuacán . . . 031538
Museo Municipal Ramón María Aller Ulloa, Lalín . . . 039544
Museo Municipal Roberto Rojas Tamayo, Colombia . . . 009110
Museo Municipal Rosendo Arteaga, Jobabo Dos . . . 009232
Museo Municipal San Juan de Iris, San Juan de Iris . . . 034396
Museo Municipal Ulpiano Checa, Colmenar de Oreja . . . 039300
Museo Municipal Victorio Macho, Palencia . . . 039876
Museo Municipal, Cortijo de Casablanca, Pizarra . . . 039946
Museo Municipal, Museo de la Ciudad, Carmona . . . 039233
Museo Municipal, Palacio Nájera, Antequera . . . 038976
Museo Municipal Dr. Santos F. Tosticarelli, Casilda . . . 000297
Museo Mural Diego Rivera, México . . . 031237
Museo Mural Homenaje a Benito Juárez, México . . . 031238
Museo Muratoriano, Modena . . . 026863
Museo Musicale d'Abruzzo, Archivio F.P. Tosti, Ortona . . . 027113
Museo Na-Bolom, San Cristóbal de las Casas . . . 031439
Museo Nacional Antropológico Eduardo López Rivas, Oruro . . 004233
Museo Nacional Casa Guillermo Valencia, Popayán . . . 008769
Museo Nacional Centro de Arte Reina Sofía, Madrid . . . 039669
Museo Nacional de Aeronáutica, Morón . . . 000475
Museo Nacional de Aeropuertos y Transporte Aéreo, Aeroplaza, Málaga . . . 039707
Museo Nacional de Agricultura, Universidad Autónoma Chapingo, Texcoco de Mora . . . 031563
Museo Nacional de Antropología, Bogotá . . . 008629
Museo Nacional de Antropología, Madrid . . . 039670
Museo Nacional de Antropología, México . . . 031239
Museo Nacional de Antropología, Arqueología e Historia del Perú, Lima . . . 034361
Museo Nacional de Arqueología, La Paz . . . 004221
Museo Nacional de Arqueología Marítima, Centro Nacional de Investigaciones Arqueológicas Submarinas, Cartagena . . . 039240
Museo Nacional de Arqueología y Etnología, Guatemala . . . 023372
Museo Nacional de Arquitectura, Madrid . . . 039671
Museo Nacional de Arquitectura, México . . . 031240
Museo Nacional de Arte, La Paz 004222
Museo Nacional de Arte, México 031241
Museo Nacional de Arte Decorativo, Buenos Aires . . . 000257
Museo Nacional de Arte Fantástico, Córdoba . . . 030912
Museo Nacional de Arte Moderno, Guatemala . . . 023373
Museo Nacional de Arte Oriental, Buenos Aires . . . 000258
Museo Nacional de Arte Popular, Mérida . . . 031138
Museo Nacional de Arte Romano, Mérida . . . 039749
Museo Nacional de Artes Decorativas, La Habana . . . 009200
Museo Nacional de Artes Decorativas, Madrid . . . 039672
Museo Nacional de Artes Decorativas de Santa Clara, Santa Clara . . 009338
Museo Nacional de Artes e Industrias Populares, México . . . 031242
Museo Nacional de Artes Visuales, Montevideo . . . 046272
Museo Nacional de Bellas Artes, Asunción . . . 034205
Museo Nacional de Bellas Artes, Buenos Aires . . . 000259
Museo Nacional de Bellas Artes, La Habana . . . 009201
Museo Nacional de Bellas Artes, Montevideo . . . 046273
Museo Nacional de Bellas Artes de Santiago de Chile, Santiago de Chile . . . 007546
Museo Nacional de Camilo Cienfuegos, Sancti-Spíritus . . . 009328
Museo Nacional de Caza, Palacio Real, Riofrío . . . 040005
Museo Nacional de Cerámica y de las Artes Suntuarias González Martí, Valencia . . . 040321
Museo Nacional de Ciencia y Tecnología, Madrid . . . 039673
Museo Nacional de Ciencias Naturales, Madrid . . . 039674
Museo Nacional de Colombia, Bogotá . . . 008630
Museo Nacional de Costa Rica, San José . . . 008863
Museo Nacional de Culturas Populares, México . . . 031243
Museo Nacional de Escultura, Valladolid . . . 040345
Museo Nacional de Etnografía y Folklore, La Paz . . . 004223
Museo Nacional de Ferrocarriles Mexicanos, Puebla . . . 031398
Museo Nacional de Historia, Guatemala . . . 023374
Museo Nacional de Historia Colonial, Omoa . . . 023400
Museo Nacional de Historia de la Medicina Eduardo Estrella, Quito . . . 010492
Museo Nacional de Historia Natural, La Habana . . . 009202
Museo Nacional de Historia Natural, Santiago de Chile . . . 007547
Museo Nacional de Historia Natural, Santo Domingo . . . 010400
Museo Nacional de Historia Natural y Antropología, Montevideo . . . 046274
Museo Nacional de Historia Natural, Universidad Mayor de San Andres, La Paz . . . 004224
Museo Nacional de Historia, Castillo de Chapultepec, México . . . 031244
Museo Nacional de Informática, Lima . . . 034362
Museo Nacional de la Acuarela, México . . . 031245

Museo Nacional de la Alfabetización,
La Habana 009203
Museo Nacional de la Cerámica, La
Habana 009204
Museo Nacional de la Cerámica,
Tonalá, Jalisco 031606
Museo Nacional de la Cultura Peruana,
Lima 034363
Museo Nacional de la Dirección
Regional del Banco Central del
Ecuador, Quito 010493
Museo Nacional de la Estampa,
México 031246
Museo Nacional de la Historia del
Traje, Buenos Aires 000260
Museo Nacional de la Inmigración,
Ex Hotel de Inmigrantes, Buenos
Aires 000261
Museo Nacional de la Máscara, San
Luis Potosí 031461
Museo Nacional de la Música, La
Habana 009205
Museo Nacional de la Revolución,
México 031247
Museo Nacional de las Culturas,
México 031248
Museo Nacional de las Intervenciones,
México 031249
Museo Nacional de las
Telecomunicaciones, Bogotá . . 008631
Museo Nacional de los Ferrocarriles
Mexicanos, México 031250
Museo Nacional de Nicaragua,
Managua 033390
Museo Nacional de Reproducciones
Artísticas, Madrid 039675
Museo Nacional de San Carlos,
México 031251
Museo Nacional de Virreinato,
Tepotzotlán 031555
Museo Nacional del Cobre, Salvador
Escalante 031430
Museo Nacional del Grabado, Buenos
Aires 000262
Museo Nacional del Hombre del
Instituto Nacional de Antropología
y Pensamiento Latinoamericano,
Buenos Aires 000263
Museo Nacional del Ministerio del
Interior, La Habana 009206
Museo Nacional del Petróleo,
Comodoro Rivadavia 000310
Museo Nacional del Petróleo
Samuel Schneider Uribe,
Barrancabermeja 008577
Museo Nacional del Prado,
Madrid 039676
Museo Nacional del Sombrero,
Aguadas 008563
Museo Nacional del Teatro,
Almagro 038953
Museo Nacional del Teatro, National
Institute for Theatre Studies, Buenos
Aires 000264
Museo Nacional del Títere,
Huamantla 031048
Museo Nacional Ferroviario, Buenos
Aires 000265
Museo Nacional Justo José
de Urquiza, Concepción del
Uruguay 000313
Museo Nacional Mariano,
Chiquinquira 008664
Museo Nacional Paleontológico y
Arqueológico, Universidad Autonoma
Juan Misael Saracho, Tarija . . 004250
Museo Nacional y Centro de
Investigación de Altamira, Santillana
del Mar 040140
Museo Nacional y Zoológico La Aurora,
Guatemala 023375
Museo Nahim Isaias, Guayaquil 010454
Museo Nájerillense, Nájera . . . 039813
Museo Napoleónico, La Habana 009207

Museo Napoleonico, Rivoli
Veronese 027538
Museo Napoleonico, Roma 027626
Museo Napoleonico, Tolentino . 028103
Museo Napoleonico G. Antonelli,
Arcole 025352
Museo Natural, Pensilvania 008757
Museo Natural Dr. Carlos A. Marelli,
Colón 000309
Museo Naturali e Appennino, Revì nel
Lazio 027512
Museo Naturalistica-Archeologico,
Vicenza 028369
Museo Naturalistico,
Champdepraz 025941
Museo Naturalistico, Lesina . . . 026595
Museo Naturalistico, Morino . . . 026979
Museo Naturalistico, Picinisco . 027311
Museo Naturalistico, Portoferraio 027420
Museo Naturalistico Archeologico M.
Locati, Lama dei Peligni 026546
Museo Naturalistico Centro Natura,
Pescasseroli 027279
Museo Naturalistico del Frignano,
Pavullo nel Frignano 027239
Museo Naturalistico del Parco, San
Romano di Garfagnana 027795
Museo Naturalistico della Riserva
Naturale Orientata di Onferno,
Gemmano 026383
Museo Naturalistico della Vallata del
Bidente, Bagno di Romagna . . 025440
Museo Naturalistico e della Civiltà
Contadina, Berceto 025490
Museo Naturalistico ed Antropologico
Paolo Barrasso, Caramanico
Terme 025766
Museo Naturalistico F. Minà Palumbo,
Convento Badia, Castelbuono . 025817
Museo Naturalistico G. Zanardo,
Tambre 028053
Museo Naturalistico Libero Gatti,
Copanello 026062
Museo Naturalistico Missionario,
Chiusa Pesio 025977
Museo Naturalistico N. De Leone,
Penne 027246
Museo Naturalistico O. Perini, Borca di
Cadore 025591
Museo Naturalistico, Instituto Sacra
Famiglia, Chieri 025962
Museo Naucalpan de la Cultura
Tlatilca, Naucalpan de Juárez . 031322
Museo Naval, Catia la Mar 054742
Museo Naval, Madrid 039677
Museo Naval, Montevideo 046275
Museo Naval de Cartagena,
Cartagena 039241
Museo Naval de Ferrol, Ferrol . . 039398
Museo Naval de Ilo, Ilo 034289
Museo Naval de la Nación, Tigre 000646
Museo Naval de San Fernando, San
Fernando 040058
Museo Naval del Perú, Callao . . 034237
Museo Naval del Puerto de la Cruz,
Casa Iriarte, Puerto de la Cruz 039980
Museo Naval Puerto Belgrano, Puerto
Belgrano 000504
Museo Navale, Genova 026420
Museo Navale, Napoli 027016
Museo Navale Archeologico N.
Lamboglia, La Maddalena 026533
Museo Navale Didattico, Milano 026819
Museo Navale Internazionale del
Ponente Ligure, Imperia 026511
Museo Navale O. Zibetti,
Caravaggio 025768
Museo Navale Romano, Albenga 025270
Museo Navale, Ravenna 027486
Museo Nazionale Alinari della
Fotografia – MNAF, Firenze . . . 026294
Museo Nazionale Archeologico,
Altamura 025298
Museo Nazionale Archeologico,

Civitavecchia 026019
Museo Nazionale Archeologico,
Ferrara 026218
Museo Nazionale Archeologico,
Sarsina 027862
Museo Nazionale Archeologico,
Tarquinia 028057
Museo Nazionale Archeologico di
Egnazia, Fasano-Savelletri . . . 026189
Museo Nazionale Archeologico di Vulci,
Canino 025739
Museo Nazionale Atestino, Este 026163
Museo Nazionale Concordiese,
Portogruaro 027424
Museo Nazionale d'Abruzzo,
L'Aquila 026562
Museo Nazionale d'Arte Medievale e
Moderna, Matera 026746
Museo Nazionale d'Arte Orientale,
Roma 027627
Museo Nazionale degli Strumenti
Musicali, Roma 027628
Museo Nazionale del Bargello,
Firenze 026295
Museo Nazionale del Cinema,
Torino 028135
Museo Nazionale del Palazzo di
Venezia, Roma 027629
Museo Nazionale del Risorgimento
Italiano, Torino 028136
Museo Nazionale del San Gottardo,
Airolo 041204
Museo Nazionale della Attività
Subacquee, Marina di Ravenna 026714
Museo Nazionale della Certosa,
Calci 025688
Museo Nazionale della Fotografia,
Brescia 025636
Museo Nazionale della Montagna
Duca degli Abruzzi, Torino . . . 028137
Museo Nazionale della Residenza
Napoleonica dei Mulini,
Portoferraio 027421
Museo Nazionale della Scienza e
della Tecnica Leonardo da Vinci,
Milano 026820
Museo Nazionale della Siritide,
Policoro 027383
Museo Nazionale dell'Agro Picentino,
Pontecagnano 027400
Museo Nazionale dell'Alta Val d'Agri,
Grumento Nova 026482
Museo Nazionale dell'Antartide F.
Ippolito, Palazzina Millo, Genova 026421
Museo Nazionale dell'Arma della
Cavalleria, Pinerolo 027337
Museo Nazionale delle Arti e Tradizioni
Popolari, Roma 027630
Museo Nazionale delle Arti Naïves C.
Zavattini, Luzzara 026657
Museo Nazionale delle Paste
Alimentari, Roma 027631
Museo Nazionale dell'Età Neoclassica
in Romagna, Faenza 026176
Museo Nazionale di Canossa N.
Campanini, Ciano d'Enza 025979
Museo Nazionale di Casa Giusti,
Monsummano Terme 026902
Museo Nazionale di Castel
Sant'Angelo, Roma 027632
Museo Nazionale di Palazzo Reale,
Pisa 027357
Museo Nazionale di Reggio Calabria,
Reggio Calabria 027497
Museo Nazionale di San Matteo,
Lungarno Mediceo, Pisa 027358
Museo Nazionale di Santa Maria delle
Monache e Paleolitico, Isernia 026519
Museo Nazionale di Villa Guinigi,
Lucca 026643
Museo Nazionale di Villa San Martino,
Portoferraio 027422
Museo Nazionale Domenico Ridola,
Matera 026747

Museo Nazionale Etrusco di Villa
Giulia, Roma 027633
Museo Nazionale Etrusco Pompeo Aria,
Marzabotto 026721
Museo Nazionale Ferroviario, Stazione
Pietrarsa, Napoli 027017
Museo Nazionale Garibaldino, La
Maddalena 026534
Museo Nazionale Jatta, Ruvo di
Puglia 027696
Museo Nazionale Preistorico ed
Etnografico Luigi Pigorini, Roma 027634
Museo Nazionale Romano-Terme di
Diocleziano, Roma 027635
Museo Néstor, Las Palmas de Gran
Canaria 039905
Museo Nezahualcóyotl Acomiztli,
Nezahualcóyotl 031325
Museo ng Bangko Sentral ng Pilipinas,
Manila 034536
Museo ng Batangas, Lipa 034495
Museo ng Buhay Pilipino, Nayong
Pilipino, Pasay 034564
Museo ng Kalinangang Pilipino,
Cultural Center of the Philippines,
Pasay 034565
Museo ng Katipunan, Lipa 034496
Museo ng Makati, Makati 034506
Museo ng Maynila, Manila 034507
Museo ng Rebolusyon, San Juan 034589
Museo Ñiace, Tepelmeme de
Morelos 031545
Museo Nicanor Piñole, Gijón . . . 039441
Museo Ñico Lopez, Bayamo . . . 009076
Museo Niño Minero, Vetagrande 031649
Museo Norace, Rosignano
Marittimo 027677
Museo Norawa, Guachochi 030993
Museo Normanno-Bizantino, San
Marco d'Alunzio 027773
Museo Norte de Potosí, Potosí . 004236
Museo Notarial Argentino, Buenos
Aires 000266, 058177
Museo Notario Roberto Mario Arata,
Buenos Aires 000267
Museo Novarese d'Arte e Storia R.
Fumagalli, Novara 027062
Museo Nuestra Señora de Carrodilla,
Luján de Cuyo 000437
Museo Nuestro Pueblo y Museo
Etnográfico de Tecpatán,
Tecpatán 031526
Museo Nueva Cádiz, La Asunción 054758
Museo Numantino, Soria 040197
Museo Numismático, La Habana 009208
Museo Numismático del Banco
Continental, Arequipa 034216
Museo Numismático del Banco Wiese,
Lima 034364
Museo Numismatico della Zecca,
Roma 027636
Museo Numismático y Filatélico del
Banco Central, Santo Domingo 010401
Museo Nuovo – Museo del
Conservatori, Roma 027637
Museo Obispo Fray José Antonio de
San Alberto, Córdoba 000336
Museo Obispo Vellosillo, Ayllón . 039028
Museo Ocarina e degli Strumenti in
Terracotta, Budrio 025654
Museo O'Higginiano y de Bellas Artes
de Talca, Talca 007551
Museo Omar Huerta Escalante,
Jonuta 031087
Museo Omero, Ancona 025322
Museo Oncativo, Oncativo 000482
Museo Onsernonese, Casa Geo de
Giorgi, Loco 041688
Museo Ora Ton, Chamula 030868
Museo Orazano, Licenza 026598
Museo Orazio Comes, Portici . . 027414
Museo Organologico Didattico,
Cremona 026102
Museo Organológico Folklórico

Musical, Bogotá 008632
Museo Oriental, Valladolid 040346
Museo Orientale, Tagliacozzo . 028050
Museo Orlando Binaghi, Círculo Criollo
El Rodeo, Moreno 000474
Museo Ornitológico, Aragua . . 054697
Museo Ornitológico, Poppi 027409
Museo Ornitologico, San
Gimignano 027754
Museo Ornitologico C. Beni, Stia 028017
Museo Ornitologico e di Scienze
Naturali, Ravenna 027487
Museo Ornitologico Naturalistico S.
Bambini, Pietralunga 027316
Museo Ornitológico Patagónico, El
Bolsón 000358
Museo Ornitologico U. Foschi,
Forlì 026328
Museo Orocoveño de Familia Avilés,
Orocovis 036136
Museo Orve, Chaupimascos . . 034248
Museo Oscar Fernandez Morera,
Sancti-Spíritus 009329
Museo Oscar María de Rojas,
Cárdenas 009095
Museo Ossenvanza, Museo
Missionario d'Arte Cinese,
Bologna 025570
Museo Ostetrico, Bologna 025571
Museo Ostetrico, Padova 027157
Museo Ostiense, Ostia Antica . . 027127
Museo Othoniano, San Luis
Potosí 031462
Museo P. Calderini, Palazzo dei Musei,
Varallo Sesia 028272
Museo P. Vagliasindi, Randazzo . 027477
Museo Pablo Bush Romero, Puerto
Aventura 031404
Museo Pablo Casals, San Juan . 036167
Museo Pablo Gargallo, Zaragoza 040452
Museo Pablo Serrano, Zaragoza 040453
Museo Padiglione delle Carozza,
Museo Storico, Città del
Vaticano 054688
Museo Padre Coll, Smithsonian
Institution, Buenos Aires 000268
Museo Paisa, Caicedonia 008643
Museo Palacio de Don Pedro I,
Astudillo 039013
Museo Palacio de la Canal, San
Miguel de Allende 031467
Museo-Palacio Spínola, Teguise 040215
Museo Palatino, Roma 027638
Museo Palazzina Storica, Peschiera
del Garda 027280
Museo Palenteológico de la Laguna,
Torreón 031612
Museo Paleocristiano, Aquileia . 025346
Museo Paleomarino, Museo di Biologia
Marina & Paleontologia, Reggio
Calabria 027498
Museo Paleontologico, Amandola 025310
Museo Paleontologico, Bova . . 025614
Museo Paleontologico,
Fluminimaggiore 026307
Museo Paleontologico, L'Aquila . 026563
Museo Paleontológico, Mariquita 008713
Museo Paleontologico,
Monfalcone 026889
Museo Paleontologico,
Montevarchi 026966
Museo Paleontológico, Murcia . . 039803
Museo Paleontológico, Pinar del
Río 009291
Museo Paleontológico – Sala Longinos
Navás, Zaragoza 040454
Museo Paleontológico – Sala Lucas
Mallada, Facultad de Ciencias,
Zaragoza 040455
Museo Paleontológico Archeologico
V. Caccia, San Colombano al
Lambro 027740
Museo Paleontológico de Elche,
Gruppo cultural Paleontológico,

Elche 039370
Museo Paleontológico Egidio Feruglio,
Trelew 000651
Museo Paleontologico M. Gortani,
Portogruaro 027425
Museo Paleontologico Parmense,
Parma 027220
Museo Paleontologico Rinaldo Zardini,
Musei delle Regole d'Ampezzo,
Cortina d'Ampezzo 026082
Museo Paleontologico S. Lai,
Ceriale 025903
Museo Paleontológico y Petrolero
Astra, Comodoro Rivadavia . . 000311
Museo Paleontológico, Arqueológico
e Histórico de la Ciudad de Deán
Funes, Deán Funes 000353
Museo Paleontologico, Malatestiana,
Mondaino 026883
Museo Paleoveneto, Pieve di
Cadore 027323
Museo Palladio, Vicenza 028370
Museo Pambata, Manila 034538
Museo Pampeano de Chascomús,
Chascomús 000300
Museo Pancho Coímbre, Ponce . 036145
Museo Pantaleón Panduro,
Tlaquepaque 031583
Museo Panteón San Fernando,
México 031252
Museo Parigino a Roma, Roma . 027639
Museo Parma Gemma, Casarza
Ligure 025796
Museo Parque de Esculturas,
Montevideo 046276
Museo Parque de la Cristianía,
Chapala 030869
Museo Parque Los Tres Ojos, Santo
Domingo 010402
Museo Parrocchiale, Bionaz . . 025514
Museo Parrocchiale, Brescello . 025628
Museo Parrocchiale, Campo
Tures 025728
Museo Parrocchiale, Challand Saint
Victor 025938
Museo Parrocchiale, Chambave 025939
Museo Parrocchiale,
Champorcher 025942
Museo Parrocchiale, Fié 026224
Museo Parrocchiale, Gressoney Saint
Jean 026470
Museo Parrocchiale, La Salle . . 026536
Museo Parrocchiale, Ornavasso . 027107
Museo Parrocchiale, Ponte
Buggianese 027395
Museo Parrocchiale, Saint
Vincent 027704
Museo Parrocchiale, San Marco
d'Alunzio 027774
Museo Parrocchiale, Sarre . . . 027860
Museo Parrocchiale, Valgrisenche 028257
Museo Parrocchiale,
Valtournenche 028269
Museo Parrocchiale d'Arte Sacra,
Dozza 026144
Museo Parrocchiale del Santuario del
Santissimo Crocifisso, Buggiano 025657
Museo Parrocchiale della Chiesa
Collegiata, Mercatello sul
Metauro 026769
Museo Parrocchiale della Propositura
di San Martino, Lastra a Signa 026568
Museo Parrocchiale di Arte Sacra,
Valstrona 028267
Museo Parrocchiale di Arte Sacra,
Chiesa di San Maurizio, Ponte in
Valtellina 027397
Museo Parrocchiale e Museo P.
Mascagni, Bagnara di Romagna 025436
Museo Parrocchiale Madonna dei
Raccomandati, Gessopalena . . 026433
Museo Parrocchiale San Giacomo,
Massignano 026743
Museo Parrocchiale San Pietro Martire,

Venezia 028317
Museo Parroquial, Alquezar . . . 038966
Museo Parroquial, El Barco de
Ávila 039109
Museo Parroquial, Bocairent . . 039155
Museo Parroquial, Celanova . . 039262
Museo Parroquial, Chillón 039276
Museo Parroquial, Espejo 039382
Museo Parroquial, Ezcaray . . . 039393
Museo Parroquial, Lucena del
Cid 039600
Museo Parroquial, San Sebastián del
Oeste 031481
Museo Parroquial, Santa María del
Campo 040107
Museo Parroquial, Valdemoro . . 040292
Museo Parroquial Colegiata de San
Miguel, Aguilar de Campoo . . 038893
Museo Parroquial de Arte Religioso,
Turmequé 008817
Museo Parroquial de Arte Sacro y
Etnología, Anso 038974
Museo Parroquial de Calatañazor,
Calatañazor 039198
Museo Parroquial de Frómista, Iglesia
de San Pedro, Frómista 039412
Museo Parroquial de la Asunción
de Nuestra Señora de Navarrés,
Navarrés 039816
Museo Parroquial de Monterroso,
Monterroso 039779
Museo Parroquial de San Cosme y
San Damián, Covarrubias . . . 039325
Museo Parroquial de San Martiño de
Mondoñedo, Foz 039409
Museo Parroquial de San Martiño de
Mondoñedo, San Martiño-Foz . 040067
Museo Parroquial de Santa Cruz,
Iglesia de Santa Cruz, Madrid . 039678
Museo Parroquial de Santa Eulalia,
Paredes de Nava 039916
Museo Parroquial de Santa María,
Becerril de Campos 039113
Museo Parroquial de Valderas,
Valderas 040297
Museo Parroquial de Yanguas,
Yanguas 040434
Museo Parroquial de Zurbarán,
Marchena 039727
Museo Parroquial Etnográfico, La
Ercina 039378
Museo Parroquial Nuestra Sacra de la
Asunción, Museo de Arte Sacro,
Palomares del Campo 039906
Museo Parroquial Santa María de
Salvador, Chinchilla de Monte-
Aragón 039278
Museo Parroquial, Iglesia de
Nuestra Señora de la Asunción,
Pastrana 039918
Museo Parroquial, Iglesia de Nuestra
Señora de la Asunción y del
Manzano, Hondarribia 039492
Museo Parroquial, Iglesia de San
Eutropio, Paradas 039914
Museo Parroquial, Museo de Arte
Sacro Parroquial de Santa Cruz,
Écija 039359
Museo Participativo de Ciencias,
Buenos Aires 000269
Museo Particular Los Abuelos,
Franck 000373
Museo Partigiano, Carpasio . . . 025778
Museo Partigiano Alfredo Di Dio,
Ornavasso 027108
Museo Pasos de Semana Santa,
Bilbao 039148
Museo Patio del Moro, Guaduas . 008691
Museo Patio Herreriano, Museo de
Arte Contemporáneo Español,
Valladolid 040347
Museo Patria Chica de Mohamed Diaz,
Realicó 000520
Museo Pavese di Scienze Naturali,

Pavia 027235
Museo Pecharromán, Pasarón de la
Vera 039917
Museo Pedagógico Ciencias
Naturales Jesús María Hernando,
Valladolid 040348
Museo Pedagógico de Arte Infantil,
MUPAI, Madrid 039679
Museo Pedagógico de Chile, Santiago
de Chile 007548
Museo Pedagógico de Historia del
Arte, Caracas 054736
Museo Pedagógico José Pedro Varela,
Montevideo 046277
Museo Pedagógico Textil, Madrid 039680
Museo Pedagóxico de Galicia,
Santiago de Compostela 040133
Museo Pedrin Troya, San Nicolás de
Bari 009320
Museo Pedro Coronel, Zacatecas 031690
Museo Pedro Nel Gomez,
Medellín 008724
Museo Penitenciario Argentino Antonio
Ballvé, Buenos Aires 000270
Museo Peñón Blanco, Peñón
Blanco 031366
Museo per la Matematica – Giardino
di Archimede, Priverno 027459
Museo per la Storia dell'Università di
Pavia, Pavia 027236
Museo per l'Arte e l'Archeologia del
Vastese, Gissi 026446
Museo Pérez Comendador-Leroux,
Hervás 039491
Museo Perfumeria Gal, Alcalá de
Henares 038916
Museo Perrando, Sassello 027871
Museo Petatlán, Petatlán 031368
Museo Peten Ak, Valladolid . . . 031634
Museo Petrográfico del Servicio
Nacional de Geología y Minería,
Quito 010494
Museo Picasso Málaga, Málaga 039708
Museo Picasso, Colección Eugenio
Arias, Buitrago del Lozoya . . . 039162
Museo Piersanti, Matelica 026744
Museo Pietro Canonica, Roma . 027640
Museo Pietro della Vedova, Rima San
Giuseppe 027518
Museo Pietro Micca e dell'Assedio,
Torino 028138
Museo Pinacoteca A. Salvucci,
Seminario Vescovile, Molfetta . 026874
Museo Pinacoteca Comunale G. De
Nittisu, Barletta 025466
Museo-Pinacoteca de Arte
Contemporáneo, Zarzuela del
Monte 040461
Museo Pinacoteca della Santa Casa,
Loreto 026626
Museo Pinacoteca dell'Arte dei
Bambini, Smerillo 027974
Museo Pinacoteca Parrocchiale di
Santa Maria, Mogliano Marche 026874
Museo-Pinacoteca, Chiesa di
Sant'Agostino, Visso 028405
Museo Pintor Antonio Rodríguez Luna,
Montoro 039785
Museo Pio Clementino, Musei Vaticani,
Città del Vaticano 054689
Museo Pio Collivadino, Banfield . 000140
Museo Pio Cristiano, Musei Vaticani,
Città del Vaticano 054690
Museo Pio IX, Senigallia 027918
Museo Pio X, Salzano 027717
Museo Planetario, Garza García 030984
Museo Plaza de Revolución General
Antonio Maceo, Santiago de
Cuba 009364
Museo Plebano, Agno 041198
Museo Polare Etnografico Silvio
Zavatti, Fermo 026201
Museo Poldi Pezzoli, Milano . . 026821
Museo Policial, Centro de Formación

de la Policía, Ávila 039023
Museo Polifacético Regional Vottero,
Laguna Larga 000423
Museo Polivalente, Chiaramonte
Gulfi 025950
Museo Polivalente, Vittoria 028410
Museo Polivalente de Palmira,
Palmira 009281
Museo Pomposiano, Codigoro . . 026026
Museo Ponchielliano, Paderno
Ponchieli 027141
Museo Popol Vuh, Guatemala . . 023376
Museo Popoli e Culture, Milano . 026822
Museo Popular de Arte
Contemporáneo, Palacio del Bayle,
Villafamés 040404
Museo Porfirio Corona Covarrubias, El
Grullo, Jalisco 030957
Museo Poschiavino, Palazzo de
Bassus-Mengotti, Casa Tomé,
Poschiavo 041847
Museo Posta de Yatasto, Metán 000465
Museo Postal Cubano, La
Habana 009209
Museo Postal Eduardo Santos,
Bogotá 008633
Museo Postal y Filatélico del Perú,
Lima 034365
Museo Postal y Telegráfico,
Madrid 039681
Museo Postal y Telegráfico Dr. Ramón
J. Cárcano, Buenos Aires . . 000271
Museo Postale e Telegrafico della
Mitteleuropa, Trieste 028210
Museo Postale, Filatelico-Numismatico,
San Marino 038104
Museo Prehistórico de Tepexpan,
Tepexpan 031548
Museo Preindustrial de la Caña de
Azúcar, Motril 039790
Museo Preistorico, Paceco . . . 027140
Museo Preistorico Balzi Rossi,
Ventimiglia 028332
Museo Preistorico dell'Isolino, Museo
Civico di Varese, Biandronno . . 025507
Museo Preistorico e Archeologico A.C.
Blanc, Villa Paolina, Viareggio . 028359
Museo Preistorico N. Lamboglia e Val
Varatello, Toirano 028097
Museo Presidente José Evaristo
Uriburu, Salta 000559
Museo Presidio Modelo, Nueva
Gerona 009277
Museo Pringles de Arqueología e
Historia, Coronel Pringles . . . 000345
Museo Privado de Arqueología
Regional Julio Gironde, Carmen de
Patagones 000294
Museo Prof. Arturo Reyes Viramontes,
Jalpa 031077
Museo Prof. Dr. Juan Augusto
Olsacher, Zapala 000687
Museo Prof. Moisés Sáenz Garza,
Apodaca 030831
Museo Prof. Ninfa María Báez,
Acoyapa 033371
Museo Prof. Ricardo Vega Noriega,
San Ignacio 031442
Museo Profano, Città del
Vaticano 054691
Museo Provincial, Ciego de Avila 009103
Museo Provincial, Lugo 039602
Museo Provincial Arqueológico
Eric Boman, Santa María de
Catamarca 000623
Museo Provincial de Arqueología
Wagner, Santiago del Estero . . 000630
Museo Provincial de Artes de La
Pampa, Santa Rosa 000624
Museo Provincial de Bellas Artes,
Salta 000560
Museo Provincial de Bellas Artes Dr.
Juan Ramon Vidal, Corrientes . 000349
Museo Provincial de Bellas Artes Dr.

Pedro E. Martínez, Paraná 000488
Museo Provincial de Bellas Artes
Emiliano Guiñazu – Casa de Fader,
Luján de Cuyo 000438
Museo Provincial de Bellas Artes
Emilio A. Caraffa, Córdoba . . . 000337
Museo Provincial de Bellas Artes
Ramón Gómez Cornet, Santiago del
Estero 000631
Museo Provincial de Bellas Artes Rene
Brusau, Resistencia 000526
Museo Provincial de Bellas Artes Rosa
Galisteo de Rodríguez, Santa Fé 000620
Museo Provincial de Bellas Artes
Timoteo Eduardo Navarro, San
Miguel de Tucumán 000601
Museo Provincial de Ciencias
Naturales Dr. Angelo Gallardo,
Rosario 000549
Museo Provincial de Ciencias
Naturales Florentino Ameghino,
Santa Fé 000621
Museo Provincial de Ciencias
Naturales y Oceanográfico, Puerto
Madryn 000507
Museo Provincial de Historia Natural
de La Pampa, Santa Rosa 000625
Museo Provincial de Holguín La
Periquera, Holguín 009223
Museo Provincial de la Historia de
Pinar del Río, Pinar del Río . . . 009292
Museo Provincial de Matanzas,
Matanzas 009257
Museo Provincial de Villa Clara y
Museo Municipal, Santa Clara . 009339
Museo Provincial do Mar, Cervo 039270
Museo Provincial Don Emilio Bacardí
Moreau, Santiago de Cuba . . . 009365
Museo Provincial Dora Ochoa de
Masramón, San Luis 000592
Museo Provincial General Vicente
García, Victoria de las Tunas . . 009388
Museo Provincial Ignacio Agramonte,
La Vigía 009390
Museo Provincial José María Morante,
Arequipa 034217
Museo Provincial Mario Brozoski,
Puerto Deseado 000505
Museo Provincial y Parroquial,
Iglesia de San Miguel, Piña de
Campos 039944
Museo Provincial, Museo de Teruel,
Teruel 040226
Museo Provinciale, Catanzaro . . 025878
Museo Provinciale Campano,
Capua 025764
Museo Provinciale d'Arte
Contempranea, Crotone 026111
Museo Provinciale degli Usi e Costumi,
Südtiroler Landesmuseum für
Volkskunde, Brunico 025650
Museo Provinciale della Caccia e della
Pesca, Racines 027468
Museo Provinciale della Vita
Contadina, San Vito al
Tagliamento 027803
Museo Provinciale delle Miniere,
Racines 027469
Museo Provinciale delle Miniere,
Vipiteno 028403
Museo Provinciale delle Tradizioni
Popolari Abbazia Cerrate, Lecce 026580
Museo Provinciale di Storia Naturale,
Foggia 026311
Museo Provinciale di Storia Naturale,
Livorno 026607
Museo Provinciale Sigismondo
Castromediano, Lecce 026581
Museo Pucciniano, Torre del
Lago 028151
Museo Pueblo de Luis, Trelew . 000652
Museo Puertorriqueño del Deportes,
Salinas 036147
Museo Quattroruote, Rozzano . . 027691

Museo Quezaltenango,
Quezaltenango 023378
Museo Quinta Amalia Simoni,
Camagüey 009091
Museo Quinta Casa de Juan Montalvo,
Ambato 010412
Museo Quinta de San Pedro
Alejandrino, Fundación Museo
Bolivariano, Santa Marta 008793
Museo Quinta Juan Leon Mera,
Ambato 010413
Museo R. Campelli,
Pievebovigliana 027331
Museo R. Cassolo, Mede 026751
Museo R. Cuneo, Savona 027893
Museo R. De Cesare, Spinazzola 028002
Museo Rabelo, La Habana 009210
Museo-Raccolta di Fisica, Bari . 025461
Museo Radiofónico de la Ciudad de
Boulogne, Boulogne 000149
Museo Rafael Coronel, Zacatecas 031691
Museo Rafael Escriña, Santa Clara de
Saguier 000612
Museo Rafael Zabaleta, Quesada 039988
Museo Raíces, Jesús María . . . 008703
Museo Raíces de Satevó, Satevó 031498
Museo Rainis e Aspazija, Casa Carlo
Cattaneo, Castagnola 041395
Museo Ralli, Marbella 039723
Museo Ralli, Punta del Este . . . 046290
Museo Ralli, Santiago de Chile . 007549
Museo Ramón Gaya, Casa Palarea,
Murcia 039804
Museo Ramón López Velarde,
México 031253
Museo Ramón Pérez Fernández,
Rivadavia 000540
Museo Ramos Generales, Villa
María 000681
Museo Ratti dei Vini d'Alba, La
Morra 026535
Museo Rauzi, Palmira 000483
Museo Rayo, Roldanillo 008779
Museo Real Colegiata de San Isidoro,
Panteón, Biblioteca, Museo, Claustro,
León 039561
Museo Recoleto, Quezon City . . 034585
Museo Regional, San Juan Bautista
Tuxtepec 031445
Museo Regional Amazónico,
Iquitos 034294
Museo Regional Anibal Cambas,
Posadas 000501
Museo Regional Arqueológico
Rossbach, Chichicastenango . . 023365
Museo Regional Carlos Funes Derieul,
Coronel Dorrego 000342
Museo Regional Castelli, Castelli 000298
Museo Regional Cayetano Alberto
Silva, Venado Tuerto 000665
Museo Regional Comunitario
Altepepialcalli, México 031254
Museo Regional Cuauhnáhuac,
Cuernavaca 030929
Museo Regional Daniel Hernández
Morillo, Huancavelica 034274
Museo Regional de Acambay,
Acambay 030798
Museo Regional de Actopán,
Actopán 030806
Museo Regional de Aldama,
Aldama 030820
Museo Regional de Antropología,
Resistencia 000527
Museo Regional de Antropología
Carlos Pellicer Cámara,
Villahermosa 031664
Museo Regional de Antropología e
Historia de Baja California Sur, La
Paz 031099
Museo Regional de Antropología J.
Jesús Figueroa Torres, Sayula . 031500
Museo Regional de Arqueología,
Huamanga 034273

Museo Regional de Arqueología de
Junín, Chupaca 034253
Museo Regional de Arqueología de
Tuxpan, Tuxpan de Rodríguez
Cano 031620
Museo Regional de Arte e Historia,
Jerez de García Salinas 031084
Museo Regional de Arte Popular La
Huatápera, Uruapan 031631
Museo Regional de Atacama,
Copiapó 007522
Museo Regional de Bolillo,
Cartagena 039242
Museo Regional de Campeche,
Campeche 030854
Museo Regional de Casma Max Uhle,
Casma 034241
Museo Regional de Chiapas, Tuxtla
Gutiérrez 031627
Museo Regional de Cinco Saltos,
Cinco Saltos 000307
Museo Regional de Claromecó,
Balneario Claromecó 000139
Museo Regional de Cuetzalan,
Cuetzalán 030932
Museo Regional de Durango y
Museo del Niño, Universidad Juárez,
Durango 030954
Museo Regional de Guadalajara,
Guadalajara 031014
Museo Regional de Guanajuato
Alhóndiga de Granaditas,
Guanajuato 031031
Museo Regional de Guerrero,
Chilpancingo 030886
Museo Regional de Hidalgo,
Pachuca 031355
Museo Regional de Historia de
Aguascalientes, Aguascalientes 030813
Museo Regional de Historia de Colima,
Colima 030892
Museo Regional de Historia de
Tamaulipas, Victoria 031654
Museo Regional de Historia e Historia,
Universidad de Sonora,
Hermosillo 031041
Museo Regional de Ica María Reiche
Gross Newman, Ica 034288
Museo Regional de Iguala, Iguala de
la Independencia 031066
Museo Regional de Iquique,
Iquique 007524
Museo Regional de la Araucania,
Temuco 007552
Museo Regional de la Arqueológico de
Tiwanaku, Tiwanaku 004251
Museo Regional de la Cerámica,
Tlaquepaque 031584
Museo Regional de la Ciudad de La
Paz, La Paz 000406
Museo Regional de la Laguna,
Torreón 031613
Museo Regional de las Culturas
Occidente, Guzmán 031036
Museo Regional de Mascota,
Mascota 031126
Museo Regional de Nayarit,
Tepic 031554
Museo Regional de Nuevo León,
Monterrey 031297
Museo Regional de Pigüé, Pigüé 000493
Museo Regional de Pintura José
Antonio Terry, Tilcara 000649
Museo Regional de Puebla,
Puebla 031399
Museo Regional de Querétaro,
Querétaro 031412
Museo Regional de Río Verde, Río
Verde 031416
Museo Regional de Sacanta,
Sacanta 000552
Museo Regional de San Carlos, San
Carlos 046293
Museo Regional de San Martín,

Moyobamba 034375
Museo Regional de Sinaloa,
Culiacán 030939
Museo Regional de Sonora,
Hermosillo 031042
Museo Regional de
Telecomunicaciones Tomás Guzmán
Cantú, La Paz 031100
Museo Regional de Tierra Caliente,
Coyuca de Catalán 030917
Museo Regional de Tlaxcala,
Tlaxcala 031588
Museo Regional de Ucayali,
Pucallpa 034389
Museo Regional de Yucatán,
Mérida 031139
Museo Regional del Valle del Fuerte,
Ahome 030815
Museo Regional Doctor Leonardo
Oliva, Ahualulco de Mercado . . 030816
Museo Regional Dr. Adolfo Alsina,
Carhué 000292
Museo Regional Dr. Federico Escalada,
Río Mayo 000537
Museo Regional Dr. José Luro, Pedro
Luro 000489
Museo Regional El Fuerte, El
Fuerte 030956
Museo Regional Francisco Mazzoni,
Maldonado 046242
Museo Regional Guayupe de Puerto
Santander, Puerto Santander . . 008772
Museo Regional Huasteco, Valles 031640
Museo Regional José G. Brochero,
Santa Rosa de Río Primero . . . 000626
Museo Regional Las Varillas, Las
Varillas 000426
Museo Regional Lázaro Cárdenas,
Tlapehuala 031582
Museo Regional Luciano Márquez,
Zacatlán 031696
Museo Regional Malargüe,
Malargüe 000442
Museo Regional Maracó, General
Pico 000379
Museo Regional Michoacano,
Morelia 031311
Museo Regional Municipal de El
Calafate, El Calafate 000359
Museo Regional Particular Epifanio
Saravia, Santa Catalina 000611
Museo Regional Patagónico,
Comodoro Rivadavia 000312
Museo Regional Potosino, San Luis
Potosí 031463
Museo Regional Provincial Padre
Manuel Jesús Molina, Río
Gallegos 000534
Museo Regional Rescate de Raíces,
Ahualulco de Mercado 030817
Museo Regional Rincón de Vida, Villa
Mirasol 000682
Museo Regional Rufino Muñiz Torres,
Ocampo 031341
Museo Regional Rumi Mayu (Río de
Piedra), Sanagasta 000610
Museo Regional Salesiano de Rawson,
Rawson 000519
Museo Regional Salesiano Maggiorino
Borgatello, Punta Arenas 007532
Museo Regional Trevelin, Trevelin 000656
Museo Regional Valcheta,
Valcheta 000663
Museo Regional Valparaíso,
Valparaíso 031641
Museo Regional y de Arte Marino,
Puerto San Julián 000508
Museo Regionale, Messina 026774
Museo Regionale A. Pepoli,
Trapani 028165
Museo Regionale della Centovalli e del
Pedemonte, Intragna 041604
Museo Regionale della Ceramica,
Caltagirone 025700

Museo Regionale della Ceramica,
Deruta 026129
Museo Regionale di Camarina,
Ragusa 027473
Museo Regionale di Scienze Naturali,
Saint Pierre 027702
Museo Regionale di Scienze Naturali,
Affiliation Regione Piemonte,
Torino 028139
Museo Religioso de Guatavita,
Guatavita 008694
Museo Religioso Santa Domingo,
Bolivar 010420
Museo Religioso y Colonial de San
Francisco, Cali 008653
Museo Renacimiento Indígena,
Huamuxtitlán 031050
Museo Renato Brozzi,
Traversetolo 028166
Museo Ricardo J. Alfaro, Bella
Vista 034169
Museo Richard Ginori della
Manifattura di Doccia, Sesto
Fiorentino 027939
Museo Ripley, México 031255
Museo Risorgimentale,
Montichiari 026968
Museo Risorgimentale, Pieve di
Cadore 027324
Museo Rivarossi dei Treni in Miniatura,
Como 026049
Museo Roca, Buenos Aires . . . 000272
Museo Rocsen, Nono 000478
Museo Rodera Robles, Casa del
Hidalgo, Segovia 040159
Museo Rodolfo Valentino,
Castellaneta 025828
Museo Rognoni Salvaneschi,
Novara 027063
Museo Romano, Tortona 028158
Museo Romano de Astorga,
Astorga 039012
Museo Romano Oiasso, Irún . . . 039515
Museo Romántico, Madrid 039682
Museo Romántico, Trinidad 009377
Museo Romántico – Casa de
Montero, Museo Histórico Nacional,
Montevideo 046278
Museo Rómulo Raggio, Vicente
López 000666
Museo Ron Arehucas, Arucas . . 039007
Museo Rosendo Mejica, Iloilo . . 034481
Museo Rosminiano, Stresa 028024
Museo Rubén Herrera, Saltillo . . 031427
Museo Rudolf Stolz, Sesto in
Pusteria 027940
Museo Runa de Antropología, Tafí del
Valle 000640
Museo Rural, Pozorrubio de
Santiago 039968
Museo Rural de la Posta de Sinsacate,
Jesús María 000394
Museo S. Fancello, Dorgali 026142
Museo Sa Domu De Donna Nassia,
Lotzorai 026634
Museo Sac-be, Yaxcabá 031682
Museo Sacrario delle Bandiere,
Roma 027641
Museo Sacrario Galileo Ferraris,
Livorno Ferraris 026608
Museo Sacro, Città del Vaticano 054692
Museo Sacro – Tesoro di
Sant'Ambrogio, Milano 026823
Museo Sacro de Caracas,
Caracas 054737
Museo Sacro de la Catedral, San
Miguel de Tucumán 000602
Museo Sacro, Santuario della
Madonna del Rosario, Pompei . 027391
Museo Sala, San Juan Bautista
Tuxtepec 031446
Museo Sala de Arte Moderno,
Juchitán de Zaragoza 031094
Museo Sala de Arte Público David

Alfaro Siqueiros, México 031256
Museo Sala de Arte, Universidad
Autónoma de Baja California,
Tijuana 031568
Museo Sala de Exposición
Banco Central de Esmeraldas,
Esmeraldas 010440
Museo-Sala de los Símbolos Patrios,
Hermosillo 031043
Museo Salesiano, Huancayo . . . 034277
Museo Salvatore Ferragamo,
Firenze 026296
Museo Salzillo, Murcia 039805
Museo San Antonio de Padua,
Córdoba 000338
Museo San Francisco, La Paz . . 004225
Museo San Giuseppe da Copertino,
Osimo 027123
Museo San Gregorio de Polanco, San
Gregorio de Polanco 046294
Museo San José el Mogote, San José
el Mogote 031444
Museo San Juan de la Cruz,
Úbeda 040283
Museo San Juan de Letrán,
Chucuito 034251
Museo San Juan en la Historia, San
Juan del Río 031451
Museo San Miguel del Padrón, La
Habana 009211
Museo San Miguel Teotongo,
México 031257
Museo San Nicolò, Militello in Val di
Catania 026839
Museo Santa Clara de Murcia,
Murcia 039806
Museo Santa Clara la Real, Tunja 008816
Museo Santa Iglesia Catedral,
Plasencia 039948
Museo Sant'Andrea, Clusone . . . 026023
Museo-Santuario de San Pedro Claver,
Cartagena 008659
Museo Santuario Nuestra Señora
Natividad del Guayco, La
Magdalena 010463
Museo Santuarios Andinos de la
Universidad Católica de Santa María,
Arequipa 034218
Museo Sardo di Antropologia ed
Etnografia, Università di Monserrato,
Cagliari 025681
Museo Sardo di Geologia e
Paleontologia D. Lovisato, Università
Cagliari, Cagliari 025682
Museo Schaferrer, Oxapampa . . 034379
Museo Scientifico L. Guidi,
Osservatopio Vauepio, Pesaro . 027268
Museo Scout-Guía, Madrid 039683
Museo Scuola Odin-Bertot,
Angrogna 025332
Museo Sefardí, Sinagoga de Samuel
Ha Leví, Toledo 040240
Museo Selinuntino, Castelvetrano 025852
Museo Sella e Mosca, Alghero . 025291
Museo Senador Domingo Faustino
Sarmiento, Buenos Aires 000273
Museo Senen Mexic, Acatlán de
Osorio 030804
Museo Señorita Elupina Cordero,
Santa Teresa 010381
Museo Serfín de Indumentaria
Indígena, México 031258
Museo Servando Cabrera Moreno, La
Habana 009212
Museo Siciliano delle Tradizioni
Religiose, San Salvatore di
Fitalia 027796
Museo Siderúrgico de Colombia,
Nobsa 008736
Museo Silvanus G. Morley y de las
Estelas, El Petén 023366
Museo Silvestre Vargas,
Tecatitlán 031521
Museo Simón Bolívar, Ziortza-

Bolibar 040463
Museo Sinagoga de Santa María la
Blanca, Toledo 040241
Museo Sistino Vescovile, Montalto
Marche 026908
Museo Sitio Arqueológico de San
Andrés, Valle de Zapotitán . . . 010600
Museo Sitio Huellas de Acahualinca,
Acahualinca 033370
Museo Sito Arqueológico Guayabo de
Turrialba, Turrialba 008865
Museo Social Argentino de
Universidad, Buenos Aires 000274
Museo Sol y Mar, Matanzas . . . 009258
Museo Soldati della Battaglia, San
Biagio di Callalta 027729
Museo Sor María de Jesús de Agreda,
Agreda 038888
Museo Sorolla, Madrid 039684
Museo Soumaya, México 031259
Museo Speleopaleontologico,
Genga 026388
Museo Speologico, Sgonico . . . 027947
Museo Sperimentale d'Arte
Contemporanea Muspac,
L'Aquila 026564
Museo Statale, Anghiari 025330
Museo Statale, Mileto 026838
Museo Stauros di Arte Sacra
Contemporanea, Isola del Gran
Sasso 026521
Museo Stibbert, Firenze 026297
Museo Storia Memoria, Sossano 027995
Museo Storico, Montese 026961
Museo Storico, Voghera 028419
Museo Storico-Aeronautica Militare,
Bracciano 025620
Museo Storico Alfa Romeo,
Arese 025355
Museo Storico Aloisiano, Castiglione
delle Stiviere 025856
Museo Storico Alpino, Antrodoco 025335
Museo Storico Archeologico,
Cicagna 025982
Museo Storico Archeologico,
Segni 027911
Museo Storico Archeologico del
Territorio, Bagni di Lucca 025439
Museo Storico Archeologico dell'Antica
Nola, Convento delle Canossiane,
Nola 027047
Museo Storico Archeologico G. Gabetti,
Dogliani 026136
Museo Storico Artistico, Cassino 025807
Museo Storico Artistico, Tesoro, Città
del Vaticano 054693
Museo Storico Badogliano, Bari . 025462
Museo Storico C. Musazzi,
Parabiago 027200
Museo Storico Cappuccino,
Camerino 025716
Museo Storico Carlo Levi, Aliano 025293
Museo Storico Civico Cuggionese,
Cuggiono 026113
Museo Storico D. Foschi, Forlì . . 026329
Museo Storico degli Truppe Alpini,
Trento 028175
Museo Storico dei Bersaglieri,
Roma 027642
Museo Storico dei Granatieri di
Sardegna, Roma 027643
Museo Storico dei Vigili del Fuoco,
Mantova 026704
Museo Storico del Castello di
Miramare, Trieste 028211
Museo Storico del Combattente,
Palazzolo Milanese 027171
Museo Storico del Mutuo Soccorso,
Pinerolo 027338
Museo Storico del Nastro Azzurro,
Salò 027714
Museo Storico del Risorgimento della
Società Economica, Chiavari . . 025956
Museo Storico del Soldatino M.

Massacesi, Bologna 025572
Museo Storico del Trotto, Civitanova
 Marche 026018
Museo Storico del Vetro e
 della Bottiglia J.F. Mariani,
 Montalcino 026906
Museo Storico dela Caccia e del
 Territorio, Cerreto Guidi ... 025907
Museo Storico della Brigata Sassari,
 Caserma La Marmora, Sassari 027870
Museo Storico della Città,
 Bergamo 025497
Museo Storico della Didattica Mauro
 Laeng, Università degli Studi Roma
 Tre, Dto. di Scienze dell'Educazione,
 Roma 027644
Museo Storico della Fanteria,
 Roma 027645
Museo Storico della Filarmonica
 Sestrese, Genova 026422
Museo Storico della Guardia di
 Finanza, Roma 027646
Museo Storico della Guerra
 1915–1918, Roana 027540
Museo Storico della Liberazione,
 Roma 027647
Museo Storico della Linea Gotica,
 Auditore 025418
Museo Storico della Motorizzazione
 Militare, Roma 027648
Museo Storico della Resistanza S.
 Maneschi, Neviano degli Arduini 027032
Museo Storico della Resistenza,
 Stazzema 028013
Museo Storico della Tappezzeria,
 Bologna 025573
Museo Storico dell'Arma dei
 Carabinieri, Roma 027649
Museo Storico dell'Arma del Genio,
 Roma 027650
Museo Storico dell'Arte Sanitaria,
 Roma 027651
Museo Storico delle Armi, Civitella del
 Tronto 026021
Museo Storico delle Poste e
 Telecomunicazioni, Roma 027652
Museo Storico delle Saline, Margherita
 di Savoia 026710
Museo Storico dell'Oro Italiano,
 Predosa 027455
Museo Storico di Architettura Militare,
 Roma 027653
Museo Storico Dittico Alto Livenza,
 Sacile 027699
Museo Storico e Topografico, Firenze
 Com'era, Firenze 026298
Museo Storico Etnografico,
 Bruscoli 025651
Museo Storico Etnografico,
 Sampeyre 027721
Museo Storico-Etnografico,
 Sanluri 027809
Museo Storico Etnografico della Bassa
 Valsesia, Romagnano Sesia ... 027663
Museo Storico Etnografico delle Alpi
 Liguri, Mendatica 026759
Museo Storico Etnografico
 Naturalistico, Chiesa in
 Valmalenco 025965
Museo Storico G. Garibaldi,
 Como 026050
Museo Storico in Trento Onlus,
 Trento 028176
Museo Storico Italiano della Guerra,
 Rovereto 027685
Museo Storico Italo-Ungherese,
 Vittoria 028411
Museo Storico Michele Ghislieri,
 Vicoforte 028376
Museo Storico Militare, Comun
 Nuovo 026054
Museo Storico Militare e Sacrario di
 Redipuglia, Fogliano Redipuglia 026312
Museo Storico Militare Umberto I,

Turate 028223
Museo Storico Minerario,
 Perticara 027253
Museo Storico Minerario di Perticara,
 Novafeltria 027056
Museo Storico Multimediale B.
 d'Alviano, Alviano 025304
Museo Storico Navale, Venezia .. 028318
Museo Storico Nazionale di Artiglieria,
 Torino 028140
Museo Storico Novarese Aldo Rossini,
 Novara 027064
Museo Storico Perugina, Perugia 027263
Museo Storico Vaticano, Città del
 Vaticano 054694
Museo Storico Villa Ciani,
 Lugano 041700
Museo Stradivariano, Museo Civico
 Ala Ponzone, Cremona 026103
Museo Sudtirolese della Frutticoltura,
 Lana 026551
Museo Sveviano, Trieste 028212
Museo T. Nuvolari, Mantova 026705
Museo Taller Arte Africano, San
 Juan 036168
Museo Taller de Escultura, Victoria de
 las Tunas 009389
Museo Taller del Moro, Toledo . 040242
Museo Taller Litográfico, Cádiz . 039193
Museo Taller Luis Nishizawa,
 Toluca 031604
Museo Tamayo Arte Contemporáneo,
 México 031260
Museo Taminango de Artes y
 Tradiciones Populares de Nariño,
 Pasto 008754
Museo Tarike, Chihuahua 030882
Museo Taurino, Cali 008654
Museo Taurino, Córdoba 039309
Museo Taurino, Hellín 039487
Museo Taurino, Huamantla ... 031049
Museo Taurino, La Línea de la
 Concepción 039569
Museo Taurino, Madrid 039685
Museo Taurino, Murcia 039807
Museo Taurino, Salamanca .. 040047
Museo Taurino de Béjar, Béjar . 039116
Museo Taurino de Bogotá,
 Bogotá 008634
Museo Taurino de la Plaza de Acho,
 Lima 034366
Museo Taurino de Mijas, Mijas . 039752
Museo Taurino Municipal,
 Antequera 038977
Museo Taurino, County Council of
 Valencia, Valencia 040322
Museo Teatrale alla Scala,
 Milano 026824
Museo Teatrale del Burcardo,
 Roma 027654
Museo Teatro La Casa de los Títeres,
 Monterrey 031298
Museo Tecnico Navale, La
 Spezia 026541
Museo Técnico Pedagógico de la
 Región de Pasco, San Juan . 034395
Museo Tecnologia, Cultura e Civiltà
 Contadina, Guardia Lombardi . 026484
Museo Tecnológico Aeroespacial, Las
 Higueras 000425
Museo Tecnológico de la Comisión
 Federal de Electricidad, México 031261
Museo Tecnológico Ingeniero Eduardo
 Latzina, Buenos Aires 000275
Museo Telefónico Victoria,
 México 031262
Museo Templete de los Heros
 Nacionales, Colegio Militar Eloy
 Alfaro, Quito 010495
Museo Teniente José Azuela,
 Zihuatanejo 031707
Museo Teodoro Ramos Blanco, La
 Habana 009213
Museo Tepeapulco, Tepeapulco . 031543

Museo Teresiano, Alba de
 Tormes 038903
Museo Teresiano, Ávila 039024
Museo Teresiano Monasterio de la
 Encarnación, Ávila 039025
Museo Territoriale del Lago di Bolsena,
 Bolsena 025577
Museo Territoriale dell'Agro
 Foronovano, Torri in Sabina . 028156
Museo Tesoro Basilica di San
 Francesco, Assisi 025402
Museo-Tesoro Catedralicio,
 Toledo 040243
Museo-Tesoro de la Cofradía de la
 Expiración, Málaga 039709
Museo Tesoro del Tempio della Beata
 Vergine, Reggio Emilia 027505
Museo Tesoro della Collegiata di San
 Lorenzo, Parrocchia di San Lorenzo,
 Chiavenna 025959
Museo Tesoro Parroquial, Lebrija 039549
Museo Tetztzontlilco, Tizayuca . 031573
Museo Textil La Trinidad, Santa Cruz
 Tlaxcala 031487
Museo Thermalia, Museu de Caldes
 de Montbui, Caldes de Montbui 039200
Museo Thyssen-Bornemisza,
 Madrid 039686
Museo Tiflológico, Madrid ... 039687
Museo Timoteo L. Hernández,
 Villaldama 031666
Museo Tipologico del Presepe,
 Macerata 026662
Museo Tlallán, Tala 031507
Museo Tohue, Vicente Guerrero,
 Durango 031652
Museo Tolone Azzariti, Girifalco . 026445
Museo Toma de Zacatecas o Cerro de
 la Bufa, Zacatecas 031692
Museo Tomás Morel, Santiago de Los
 Caballeros 010384
Museo Tomaskitla, Epazoyucan . 030973
Museo Tomochic, Vicente
 Guerrero 031651
Museo Tonallán, Tonalá, Jalisco 031607
Museo Tonatiuh, Zempoala ... 031705
Museo Torlonia, Roma 027655
Museo Tornambe de la Universidad
 Nacional de San Juan, San
 Juan 000589
Museo Torre di Porta Romana,
 Gubbio 026495
Museo Torres García, Montevideo 046279
Museo Torricelliano, Faenza .. 026177
Museo Toti dal Monte, Pieve di
 Soligo 027328
Museo Towi, Guachochi 030994
Museo Tradicional Ik Al Ojov,
 Zinacantán 031709
Museo Tradiciones y Leyendas del
 Coronel Arechavala, León 033387
Museo Tribunal de la Santa Inquisición
 y del Congreso, Lima 034367
Museo Tridentino di Scienze Naturali,
 Trento 028177
Museo Túenl Casa Parker,
 Angangueo 030826
Museo Tuscolano, Frascati ... 026346
Museo U. Gumbu, Cervo 025919
Museo Ucraniano, Apóstoles .. 000124
Museo Ugo Carà, Muggia 026984
Museo Universitario, Medellín .. 008725
Museo Universitario, Pasto ... 008755
Museo Universitario Alejandro Rangel
 Hidalgo, Nogueras 031326
Museo Universitario Antigua Casa de
 los Muñecos, Puebla 031400
Museo Universitario Contemporáneo
 de Arte, México 031263
Museo Universitario de Arqueología de
 Manzanillo, Manzanillo 031123
Museo Universitario de Artes,
 Mendoza 000459
Museo Universitario de Artes

Populares, Colima 030893
Museo Universitario de Ciencias y
 Artes, México 031264
Museo Universitario de Ciencias,
 Universidad Autónoma de Zacatecas,
 Zacatecas 031693
Museo Universitario de la Universidad
 Nacional Federico Villarreal,
 Lima 034368
Museo Universitario de Paleontología,
 México 031265
Museo Universitario de San Carlos,
 Guatemala 023377
Museo Universitario del Chopo,
 México 031266
Museo Universitario di Storia Naturale
 e della Strumentazione Scientifica,
 Modena 026864
Museo Universitario Dr. Ricardo
 Bohorquez R., Universidad Autonoma
 Tomas Frias, Potosí 004237
Museo Universitario Florentino y
 Carlos Ameghino, Rosario 000550
Museo Universitario La Memoria
 Colectiva, Universidad Nacional Siglo
 XX, Llallagia 004227
Museo Universitario Lariab de Historia
 Antigua de la Huaxteca, Tamuín 031510
Museo Universitario Pio López
 Martínez, Cayey 036127
Museo Universitario y Museo de
 Anatomía General, Universidad
 Nacional Daniel Alcides Carrión,
 Yanacancha 034419
Museo Univérsitas, El Cerrito .. 008676
Museo Urdaneta Histórico Militar,
 Maracaibo 054763
Museo Usina Vieja, San Antonio de
 Areco 000568
Museo V. Venturi, Loro Ciuffenna 026631
Museo V.G. Rossi, Villa Durazzo, Santa
 Margherita Ligure 027817
Museo V.S. Breda, Padova 027158
Museo Valdese, Rorà 027669
Museo Valdese della Balziglia,
 Massello 026742
Museo Valenciano de la Ilustración
 y la Modernidad, MUVIM,
 Valencia 040323
Museo Valenzuela, Valenzuela . 034611
Museo Vallerano Bregagliotto,
 Stampa 042021
Museo Vallivo, Oratorio dei Disciplini,
 Valfurva 028255
Museo Valtellinese di Storia e Arte,
 Sondrio 027986
Museo Vasariano di San Croce, Bosco
 Marengo 025611
Museo Vasco de Historia de las
 Ciencias y de la Medicina,
 Leioa 039553
Museo Vasco de Quiroga, Madrigal de
 las Altas Torres 039693
Museo Vasco del Ferrocarril,
 Azpeitia 039032
Museo Vásquez Fuller, Otavalo . 010466
Museo Verbum – Casa das Palabras,
 Vigo 040382
Museo Vesuviano G. B. Alfano,
 Pompei 027392
El Museo Viajero, Buenos Aires . 000276
Museo Víctor Alejandro Jaramillo,
 Ibarra 010460
Museo Víctor Manuel, La Habana 009214
Museo Victorio Macho, Victorio Macho
 Museum, Toledo 040244
Museo Vicús, Piura 034388
Museo Villa Colloredo Mels,
 Recanati 027493
Museo Villa dei Cedri, Bellinzona 041302
Museo Villa Medicea, Poggio a
 Caiano 027381
Museo Villa Romana de Navatejera,
 Villaquilambre 040410

Register der Institutionen und Firmen

**Museo Villa Urania-Antiche Maioliche di Castelli, Fondazione
– Museu Casa do Sertão, Universidade Estadual de Feira de**

Museo Villa Urania-Antiche Maioliche
di Castelli, Fondazione Paparella
Treccia-Devlet, Pescara 027277
Museo Villa Urpo, Siivikkala . . . 011356
Museo Vincenzo Vela, Ligornetto 041680
Museo Virgen de Zapopan,
Zapopan 031701
Museo Virgiliano, Virgilio 028404
Museo Virreinal de Zinacantepec, San
Miguel Zinacantepec 031470
Museo Virtual de Artes el Pais,
Montevideo 046280
Museo Vitaliano Zuccardi,
Sincelejo 008795
Museo Vivo de al-Andalus, Fundación
Roger Garaudy, Córdoba 039310
Museo Vivo della Scienza, Napoli 027018
Museo Vostell-Malpartida de Cáceres,
Malpartida de Cáceres 039712
Museo Vulcanologico dell'Osservatorio
Vesuviano, Ercolano 026156
Museo Wakuatay, Playas de
Rosarito 031372
Museo Weilbauer, Quito 010496
Museo Wifredo Lam, La Habana 009215
Museo Wilhelm Schmid, Brè sopra
Lugano 041360
Museo Würth La Rioja, Agoncillo 038884
Museo Xochiltepec, Magdalena . 031120
Museo Xólotl, San Bartolo
Tenayuca 031433
Museo Xul Solar, Buenos Aires . 000277
Museo y Archivo Dardo Rocha, La
Plata 000415
Museo y Archivo Histórico de la
Facultad de Derecho, Universidad de
Buenos Aires, Buenos Aires . . 000278
Museo y Archivo Histórico Localista
Almirante Guillermo Brown,
Adrogué 000116
Museo y Artesanía el Molino,
Agüimes 038891
Museo y Biblioteca de Ciencias
Stephen Hawking, San Salvador 010598
Museo y Centro Cultural AGADU,
Montevideo 046281
Museo y Centro Cultural Victoria
Ocampo, Mar del Plata 000448
Museo y Centro de Estudios
Humanísticos, Universidad del
Turabo, Gurabo 036132
Museo y Centro Didáctico del Encaje
de Castilla y León, Tordesillas . 040252
Museo y Centro Estudios Históricos de
la Facultad de Odontología, Buenos
Aires 000279
Museo y Jardin Botánico Profesor
Atilio Lombardo, Montevideo . . 046282
Museo y Monumento Histórico Las
Bóvedas, Uspallata 000662
Museo y Parque Arqueológico de
Obando, Obando 008737
Museo y Parque Libres del Sur,
Dolores 000356
Museo y Parque Nacional,
Montenegro 008731
Museo Yoloxóchitl, Cuernavaca . 030930
Museo Zacatecano, Zacatecas . 031694
Museo Zambiano, Castelfidardo 025820
Museo Zoloxúchitl, Zacualpan . . 031698
Museo Zoologico Cambria,
Messina 026775
Museo Zoologico César Dominguez
Flores, Tuxtla Gutiérrez 031628
Museo Zoológico Dámaso A.
Larrañaga, Montevideo 046283
Museo Zoologico degli Vertebrati,
Avellino 025424
Museo Zoologico dell'Istituto Nazionale
per la Fauna Selvatica, Ozzano
dell'Emilia 027137
Museo Zoologico G. Scarpa,
Treviso 028189
Museo Zoologico, Università Padova,

Padova 027159
Museo Zuloaga, Iglesia de San Juan,
Segovia 040160
Museo Zumalacárregui,
Ormaiztegi 039850
Museoalus Sigyn, Forum Marinum,
Turku 011441
Muséobus, Naninne 003906
Muséobus, Musée de la Famille, de
la jeunesse et de la Découverte,
Otterburn Park 006600
Muséocabinet, Saint-Jérôme . . . 006886
Museográfica Antonio Martínez, La
Roda 040009
Museogràfica de la Parròquiad
l'Assumpció, Montesa 039780
Museogràfica de la Villa,
Camporrobles 039214
Museográfica de los Campamentos
Romanos et Petavonium, Santibáñez
de Vidriales 040135
Museogràfica Permanent de Forcall,
Forcall 039406
Museokeskus Vapriikki, Tampere 011399
Museokylä Kalluntalo, Laukaa . . 011100
Museologia Scientifica, Firenze . 138847
Museological Review, Leicester . 139205
Museon, Den Haag 032181
Museon Arlaten, Arles 011732
Museonder, Otterlo 032740
Muséoparc de la Préhistoire, Saint-
Vrain 015594
Museos Arqueológic, Portoviejo 010467
Museos de la Atalaya, Palacio del
Tiempo y El Misterio de Jerez, Jerez
de la Frontera 039529
Museos de la Universidad Nacional de
Cajamarca, Museo Herramientas
Andinas, Museo de Zoología, Museo
Botánico, Museo Entomológico,
Museo de Geología, Museo de
Fitología, Cajamarca 034232
Museos del Gaucho y de la Moneda,
Montevideo 046284
Museos Integrales de Antofagasta
de la Sierra y Laguna Blanca,
San Fernando del Valle de
Catamarca 000575
Museos Oro del Perú – Armas del
Mundo, Lima 034369
Museos Universitarios de Bolivia,
Universidades Estatales Bolivianas,
La Paz 004226
Muséoscope du Lac, Rousset . . 015178
Muséosite Cavernicole de la Grotte de
la Madeleine, Saint-Remèze . . 015534
Muséosite de l'Amphithéâtre Gallo-
Romain, Lillebonne 013813
Muséosite Gallo-Romain, Pithiviers-le-
Vieil 014880
Museotalo Warpunen,
Anjalankoski 010755
Muses Galantes, Paris 141246
Museu, Pallejà 039877
Museu a Casa do Povo, Ferro . . 035695
Museu a Cidade do Açúcar,
Funchal 035701
Museu Académico de Coimbra,
Coimbra 035619
Museu Aeroespacial, Rio de
Janeiro 004696
Museu Afro-Brasileiro, Salvador 004787
Museu Afro Brasileiro de Sergipe,
Laranjeiras 004532
Museu Agrícola da Escola Profissional
Agricultura de Cister, Alcobaça 035483
Museu Agrícola de Entre Douro e
Minho, Vila do Conde 036080
Museu Agrícola de Riachos,
Riachos 035977
Museu Agrícola e Museu de Arte
Sacra, Ribeira Chã 035979
Museu Alfredo Andersen,
Curitiba 004427

Museu Alfredo Varela, Jaguarão 004506
Museu Amsterdam Sauer de Pedras
Preciosas, Rio de Janeiro 004697
Museu Anchieta Ciencias Naturais,
Porto Alegre 004638
Museu Angelo Agostini / Casa de
Cultura Laura Alvim, Rio de
Janeiro 004698
Museu Aníbal Ribeiro Filho,
Paranaguá 004609
Museu Anjos Teixeira, Sintra . . . 036037
Ao Museu Antiguidade, São
Paulo 128128
Museu Antoniano, Aljezur 035498
Museu Antoniano, Lisboa 035760
Museu Antonino, Faro 035686
Museu Antonio Granemann de Souza,
Curitibanos 004442
Museu Antonio Lago, Rio de
Janeiro 004699
Museu Antonio Parreiras, Niterói 004589
Museu Antropológico, Coimbra . 035620
Museu Antropológico, Goiânia . . 004471
Museu Antropológico Caldas Júnior,
Santo Antônio da Patrulha . . . 004817
Museu Antropológico Diretor Pestana,
Ijuí 004490
Museu Antropológico do Instituto
Superior de Cultura Brasileira, Rio de
Janeiro 004700
Museu Antropológico do Rio Grande
do Sul, Porto Alegre 004639
Museu Arquediocesano Dom Joaquim,
Brusque 004373
Museu Arqueològic Comarcal,
Banyoles 039048
Museu Arqueològic de la Ciutat de
Dénia, Castell de Dénia, Dénia 039345
Museu Arqueològic de l'Esquerda,
Roda de Ter 040011
Museu Arqueològic del Vendrell, El
Vendrell 040366
Museu Arqueològic d'Ontinyent i de la
Vall d'Albaida, Ontinyent 039841
Museu Arqueològic i Etnogràfic
Municipal Soler Blasco, Xàbia . 040430
Museu Arqueològic i Paleontológic,
Baldomar 039044
Museu Arqueològic i Paleontológic,
Casa Natal de Rafael de Casanova,
Moià 039759
Museu Arqueològic Municipal,
Banyeres de Mariola 039046
Museu Arqueològic Municipal,
Bocairent 039156
Museu Arqueològic Municipal, La Vall
d'Uixo 040330
Museu Arqueológic Municipal, Josep
Castelló i Real, Els Prats de Rei 039971
Museu Arqueológico, Lagoa
Santa 004527
Museu Arqueológico, Oliva 039826
Museu Arqueológico da Citânia de
Sanfins, Sanfins de Ferreira . . 035988
Museu Arqueológico de Barcelos,
Barcelos 035546
Museu Arqueológico de Sambaqui
Joinville, Joinville 004513
Museu Arqueológico de São Miguel de
Odrinhas, Sintra 036038
Museu Arqueológico de Serpa,
Serpa 036020
Museu Arqueológico do Carmo,
Lisboa 035761
Museu Arqueológico e Etnográfico de
Setúbal, Setúbal 036027
Museu Arqueológico e Etnológico
António Tomás Pires, Elvas . . . 035647
Museu Arqueológico e Lapidar do
Infante D. Henrique, Faro 035687
Museu Arqueológico Municipal,
Gouveia 035722
Museu Arqueológico Regional de
Castro Marim, Castro Marim . . 035608

Museu Arquidiocesano de Mariana,
Mariana 004559
Museu Arquivo Histórico da Santa
Casa, Belém 004331
Museu Arquivo Histórico Florense,
Flores da Cunha 004450
Museu Arxiduc Lluís Salvador
d'Austria de son Marroig, Deià 039343
Museu Arxiu Comarcal, Museu
Municipal de Moià – Coves del Toll,
Moià 039760
Museu Arxiu de l'Arboç, L'Arboç 038982
Museu-Arxiu de Santa María de
Mataró, Mataró 039735
Museu Arxiu de Vilassar de Dalt,
Vilassar de Dalt 040396
Museu-Arxiu Municipal de Calella,
Calella 039203
Museu-Arxiu Parroquial, Canet de
Mar 039219
Museu-Arxiu Tomàs Balvey i Bas,
Cardedeu 039230
Museu Banespa, São Paulo . . . 004863
Museu Barão de Mauá, Mauá . . 004564
Museu Barrau, Santa Eulària del
Río 040103
Museu Bernardino Machado, Vila Nova
de Famalicão 036096
Museu Bi Moreira, Universidade
Federal de Lavras, Lavras 004533
Museu Biblioteca Anselmo
Braamcamp Freire, Santarém . 036000
Museu-Biblioteca da Casa de
Bragança, Fundação de Casa de
Bragança, Vila Viçosa 036108
Museu Biblioteca Fundació Mauri, La
Garriga 039428
Museu Bocage da Zoológico e
Antropológico, Universidade de
Lisboa, Lisboa 035762
Museu Botânico da Escola Superior
Agrária de Beja, Beja 035553
Museu Botânico Dr. João Barbosa
Rodrigues, São Paulo 004864
Museu Botânico, Museu de História
Natural, Coimbra 035621
Museu Calmon Barreto, Araxá . . 004317
Museu Calouste Gulbenkian,
Lisboa 035763
Museu Câmara Cascudo, Universidade
Federal de Rio Grande do Norte,
Natal 004581
Museu Capela Santa Luzia,
Vitória 004955
Museu Capitular de la Seu Nova,
Lleida 039580
Museu Carlos Costa Pinto,
Salvador 004788
Museu Carmen Miranda, Rio de
Janeiro 004701
Museu Carpológico do Jardim
Botânico do Rio de Janeiro, Rio de
Janeiro 004702
Museu Casa Alphonsus de
Guimaraens, Mariana 004560
Museu Casa Cel. Joaquim Lacerda,
Lapa 004529
Museu Casa da Hera, Vassouras 004947
Museu Casa d'Areny-Plandolit,
Ordino 000100
Museu Casa de Benjamin Constant,
Rio de Janeiro 004703
Museu Casa de Portinari,
Brodowski 004372
Museu Casa de Rui Barbosa, Rio de
Janeiro 004704
Museu Casa de Rui Barbosa,
Salvador 004789
Museu Casa do Anhanguera, Santana
de Parnaíba 004813
Museu Casa do Assento, Friande 035699
Museu Casa do Pontal, Rio de
Janeiro 004705
Museu Casa do Sertão, Universidade

Estadual de Feira de Santana, Feira
de Santana 004448
Museu Casa dos Contos, Ouro
Preto 004598
Museu Casa Fritz Alt, Joinville . 004514
Museu Casa Guignard, Ouro
Preto 004599
Museu Casa Guimarães Rosa,
Cordisburgo 004416
Museu Casa Pedro Américo,
Areia 004321
Museu Casa Rull, La Massana . 000097
Museu-Casa Verdaguer, Museu
d'Història de la Ciutat,
Vallvidrera 040356
Museu Castro Alves, Salvador . . 004790
Museu Cau Ferrat, Sitges 040187
Museu Central das Forças Armadas,
Luanda 000110
Museu Cerdà, Puigcerdà 039987
Museu Ceroplástico Augusto Esteves,
Faculdade de Medicina / FM, São
Paulo 004865
Museu Cívico-Religioso Padre Cicero,
Juazeiro do Norte 004517
Museu Clara Nunes,
Caetanópolis 004378
Museu Claretià, Vic 040373
Museu Comajoan Fauna del Montseny,
Santa María de Palautordera . . 040106
Museu Comarcal de Berga,
Berga 039129
Museu Comarcal de la Conca de
Barberà, Montblanc 039773
Museu Comarcal de la Garrotxa,
Olot 039830
Museu Comarcal de l'Anoia-Museu de
la Pell, Igualada 039510
Museu Comarcal de l'Horta Sud,
Torrente 040263
Museu Comarcal de l'Urgell,
Tárrega 040210
Museu Comarcal de Manresa,
Manresa 039717
Museu Comendador Ananias Arruda,
Baturit 004330
Museu Comendador José Dias de
Andrade, Arceburgo 004320
Museu Comercial, Belo Horizonte 063912
Museu Comercial e Arqueológico
do Sindicato dos Empregados,
Maceió 004541
Museu Comunitário, Ipumirim . . 004491
Museu Comunitário, Itapiranga . 004498
Museu Condes de Castro Guimarães,
Cascais 035603
Museu Convento dos Lóios, Santa
Maria da Feira 035994
Museu Corina Novelino,
Sacramento 004783
Museu Coritiba Football Club,
Curitiba 004428
Museu Cosme Bauçá, Felanitx . 039395
Museu CTT/TLP, Lisboa 035764
Museu D. João VI, Rio de
Janeiro 004706
Museu da Abolição, Centro de
Referência da Cultura Afro-Brasileira,
Recife 004660
Museu da Agricultura de Fermentões,
Guimarães 035729
Museu da Agua da Epal, Lisboa 035765
Museu da Alfaia Agrícola,
Estremoz 035667
Museu da Alfaia Agrícola, Mafra 035840
Museu da Alfaia Agricola, Montemor-
o-Novo 035865
Museu da Alfaia Agrícola, Santo
Amaro 036005
Museu da Arte Brasileira, Maceió 004542
Museu da Arte Pierre Chalita,
Maceió 004543
Museu da Associação Nacional dos
Veteranos, São João Del Rei . . 004833

Museu da Bacia do Paraná,
Maringá 004562
Museu da Baleia, Caniçal 035596
Museu da Biblioteca Nacional,
Lisboa 035766
Museu da Borracha, Rio Branco 004676
Museu da Brigada Militar, Porto
Alegre 004640
Museu da Caça, Fundação de Casa de
Bragança, Vila Viçosa 036109
Museu da Cachaça, Paty do
Alferes 004615
Museu da Capela de Sã Frutuoso
de Montélios, São Jerónimo de
Real 036010
Museu da Carris, Lisboa 035767
Museu da Casa Brasileira, São
Paulo 004866
Museu da Casa Cadaval, Évora . 035675
Museu da Casa da Cultura da Orada,
Orada 035885
Museu da Casa da Cultura de
Teresina, Teresina 004928
Museu da Casa do Povo do Raminho,
Raminho 035972
Museu da Casa Grande, Freixo de
Numão 035698
Museu da Cera, Fátima 035692
Museu da Cerâmica, Caldas da
Raínha 035587
Museu da Cerâmica de Sacavém,
Sacavém 035985
Museu da Cidade, Almada 035505
Museu da Cidade, Campinas . . . 004385
Museu da Cidade, Coimbra . . . 035622
Museu da Cidade, São Paulo . . 004867
Museu da Cidade de Lisboa,
Lisboa 035768
Museu da Cidade de Recife,
Recife 004661
Museu da Ciência e Indústria,
Porto 035935
Museu da Companhia das Lezírias,
Salvaterra de Magos 035987
Museu da Companhia Independente
do Palácio Guanabara, Rio de
Janeiro 004707
Museu da Companhia Paulista,
Jundiá 004524
Museu da Cortiça da Fábrica do
Inglês, Silves 036031
Museu da Cultura e Arte Popular, João
Pessoa 004511
Museu da Curia Metropolitana, Porto
Alegre 004641
Museu da Educação e do Brinquedo,
São Paulo 004868
Museu da Electricidade, Lisboa . 035769
Museu da Eletricidade, Rio de
Janeiro 004708
Museu da Emigração, das
Comunidades e da Luso-
Descendência, Fafe 035682
Museu da Energia, Itu 004501
Museu da Escola de Engenharia, Rio
de Janeiro 004709
Museu da Escola de Minas Gerais,
Belo Horizonte 004343
Museu da Escola Naval, Rio de
Janeiro 004710
Museu da Escola Regional de
Artesanato de Santo Amaro, Santo
Amaro 036006
Museu da Escravatura, Museum of
Slavery, Luanda 000111
Museu da Estrada de Ferro Madeira-
Mamoré, Porto Velho 004656
Museu da Estrada de Ferro
Sorocabana, Sorocaba 004918
Museu da Fábrica de Cimentos Lis,
Maceira 035837
Museu da Faculdade de Direito,
Faculdade de Direito da Universidade
de São Paulo, São Paulo 004869

Museu da Faculdade de Odontologia,
Araraquara 004315
Museu da Familia Colonial,
Blumenau 004352
Museu da Farmácia, Lisboa . . . 035770
Museu da Farmáciada Santa Casa da
Misericórdia do Rio de Janeiro, Rio
de Janeiro 004711
Museu da Fauna, Rio de Janeiro 004712
Museu da Fazenda Federal, Rio de
Janeiro 004713
Museu da Força Expedicionária
Brasileira, Rio de Janeiro 004714
Museu da Freguesia de Estômbar,
Estômbar 035664
Museu da Fundação Castro Alves, Vila
Nova de Famalicão 036097
Museu da Fundação Cupertino
de Miranda, Vila Nova de
Famalicão 036098
Museu da Fundação da Casa de
Mateus, Vila Real 036103
Museu da Fundação Nacional de
Saúde, Brasília 004358
Museu da Graciosa, Santa Cruz da
Graciosa 035989
Museu da Gravura, Curitiba . . . 004429
Museu da Gravura Brasileira,
Universidade da Região da
Campanha, Bagé 004324
Museu da Guarda, Guarda 035727
Museu da História da Medicina
Maximiano Lemos, Faculdade de
Medicina da Universidade do Porto,
Porto 035936
Museu da Igreja Matriz,
Gondomar 035721
Museu da Imagem, Campinas . . 004386
Museu da Imagem, Franca 004466
Museu da Imagem e do Som, Campo
Grande 004392
Museu da Imagem e do Som, Rio de
Janeiro 004715
Museu da Imagem e do Som de
Alagoas, Maceió 004544
Museu da Imagem e do Som de
Campinas, Campinas 004387
Museu da Imagem e do Som
de Campos, Campos dos
Goitacazes 004398
Museu da Imagem e do Som de
Cascavel, Cascavel 004406
Museu da Imagem e do Som de
Curitiba, Curitiba 004430
Museu da Imagem e do Som de e da
Imprensa da Bahia, Salvador . 004791
Museu da Imagem e do Som de
Goiás, Goiás 004477
Museu da Imagem e do Som de
Piracicaba, São Paulo 004870
Museu da Imagem e do Som de
Resende, Resende 004672
Museu da Imagem e do Som de
Santa Catarina, Florianópolis . . 004452
Museu da Imagem e do Som de São
Paulo, São Paulo 004871
Museu da Imagem e do Som de
Taubaté, Taubaté 004922
Museu da Imagem e do Som do
Ceará, Fortaleza 004461
Museu da Imagem e do Som do Pará,
Belém 004332
Museu da Imagem e do Som do
Paraná, Curitiba 004431
Museu da Imagem em Movimento,
Leiria 035749
Museu da Imigração, Santa Bárbara
d'Oeste 004805
Museu da Imprensa, Brasília . . . 004359
Museu da Imprensa de Fafe,
Fafe 035683
Museu da Inconfidência, Ouro
Preto 004600
Museu da Indústria Baleeira, São

Roque do Pico 036017
Museu da Indústria da Cortiça, Santa
Maria da Feira 035995
Museu da Indústria Têxtil,
Calendário 035589
Museu da Latoaria, Tomar 036055
Museu da Liga dos Combatentes da
Grande Guerra, Lisboa 035771
Museu da Limpeza Urbana – Casa de
Banhos de Dom João VI, Rio de
Janeiro 004716
Museu da Literatura, São Paulo 004872
Museu da Loucura, Barbacena . 004327
Museu da Margalha, Gavião . . . 035716
Museu da Marioneta, Lisboa . . . 035773
Museu da Matriz, Ponta Delgada 035918
Museu da Mina de Pirites,
Aljustrel 035502
Museu da Música, Lisboa 035773
Museu da Música Portuguesa,
Estoril 035666
Museu da Paróquia, Redondo . . 035973
Museu da Paróquia da Benedita,
Alcobaça 035484
Museu da Pessoa, São Paulo . . 004873
Museu da Polícia Civil, Polícia Civil do
Estado do Rio de Janeiro, Rio de
Janeiro 004717
Museu da Polícia Judiciária,
Barro 035550
Museu da Polícia Militar de
Pernambuco, Recife 004662
Museu da Porcelana Senhor Adelino
dos Santos Gouveia, Pedreira . 004618
Museu da Quinta das Cruzes,
Funchal 035702
Museu da Quinta de Santiago, Centro
de Arte de Matosinhos, Leça da
Palmeira 035748
Museu da Rádio, Lisboa 035774
Museu da Rainha Dona Leonor,
Beja 035554
Museu da Regiao Flåviense,
Chaves 035614
Museu da República, Rio de
Janeiro 004718
Museu da República e Resistência,
Lisboa 035775
Museu da Revolução, Maputo . . 031804
Museu da Santa Casa da Misericórdia
da Vila da Ericeira, Ericeira . . . 035651
Museu da Santa Casa da Misericórdia
de Alvor, Alvor 035519
Museu da Sé de Angra do Heroísmo,
Angra do Heroísmo 035529
Museu da Secção de Tecnologia
do Serviço Florestal, Rio de
Janeiro 004719
Museu da Sociedade Imparcial 15 de
Janeiro de 1898, Alcochete . . 035487
Museu da Televisão, Lisboa . . . 035776
Museu da Terra de Miranda, Miranda
do Douro 035855
Museu da Terra e da Vida, Mafra 004549
Museu da UFRGS, Universidade
Federal do Rio Grande do Sul, Porto
Alegre 004642
Museu da Universidade Federal do
Pará, Belém 004333
Museu da Venrável Ordem Terceira de
São Francisco da Penitência, Rio de
Janeiro 004720
Museu da Vida Submarina e da
História Submersa, Lisboa . . . 035777
Museu da Vinha e do Vinho de
Reguengos, Reguengos de
Monsaraz 035976
Museu Darder, Espai d'Interpretació
de l'Estany, Banyoles 039049
Museu d'Arqueologia de Catalunya,
Barcelona 039071
Museu d'Arqueologia de Catalunya-
Empúries, L'Escala 039379
Museu d'Arqueologia de Catalunya

Girona, Girona 039444
Museu d'Arqueología de Catalunya
  Ullastret, Ullastret 040284
Museu d'Arqueología de Catalunya,
  Castell d'Olèrdola, Olèrdola . . . 039825
Museu d'Arqueología Salvador
  Vilaseca, Reus 039997
Museu d'Arqueològia Submarina,
  Pals 039908
Museu d'Arquitectura de la Real
  Càtedra Gaudí, Barcelona 039072
Museu d'Art, Girona 039445
Museu d'Art Casa Turull,
  Sabadell 040027
Museu d'Art Contemporani, Pego 039926
Museu d'Art Contemporani de
  Barcelona, Barcelona 039073
Museu d'Art Contemporani de
  Mallorca, Sa Pobla 039949
Museu d'Art Contemporani d'Eivissa,
  Eivissa 039362
Museu d'Art Espanyol Contemporani,
  Fundación Juan March, Palma de
  Mallorca 039890
Museu d'Art Frederic Marès,
  Montblanc 039774
Museu d'Art i Història de Reus,
  Institut Municipal de Museus de
  Reus, Reus 039998
Museu d'Art Modern, Tarragona 040205
Museu d'Art Modern Jaume Morera,
  Lleida 039581
Museu d'Art Municipal,
  Cadaqués 039187
Museu d'Art Popular, Horta de Sant
  Joan 039494
Museu d'Art Sacre de Santa Eulàlia,
  Encamp 000088
Museu das Alfaias, Cachoeira . 004375
Museu das Alfândegas, Lisboa . 035778
Museu das Armas, Lapa 004530
Museu das Bandeiras, Goiás . . 004478
Museu das Ciências da Terra,
  Estremoz 035668
Museu das Comunicações,
  Lisboa 035779
Museu das Crianças, Lisboa . . 035780
Museu das Flores, Santa Cruz das
  Flores 035990
Museu das Monções, Porto Feliz 004654
Museu das Oferendas ao Soldado
  Desconhecido, Batalha 035551
Museu das Reduções, Ouro
  Preto 004601
Museu das Rendas de Bilros, Vila do
  Conde 036081
Museu das Técnicas Rurais e Museu
  Municipal de Oliveira de Frades,
  Oliveira de Frades 035884
Museu das Tropas Aerotransportadas,
  Praia do Ribatejo 035968
Museu de Alberto Sampaio,
  Guimarães 035730
Museu de Alhandra, Alhandra . . 035475
Museu de Anatomia Humana Prof. Dr.
  Alfonso Bovera, Instituto de Ciências
  Biomédicas / ICB, São Paulo . . 004874
Museu de Anatomia Veterinária
  Prof. Dr. Plínio Pinto e Silva, São
  Paulo 004875
Museu de Angra do Heroísmo, Angra
  do Heroísmo 035530
Museu de Antropologia do Vale do
  Paraíba, Jacareí 004505
Museu de Armas Ferreira da Cunha,
  Petrópolis 004627
Museu de Armas Major Lara Ribas,
  Florianópolis 004453
Museu de Arqueologia, Montemor-
  o-Novo 035866
Museu de Arqueologia, Sesimbra 036023
Museu de Arqueologia de Mértola,
  Mértola 035851

Museu de Arqueologia e Artes
  Populares, Paranaguá 004610
Museu de Arqueologia e Etnologia,
  Salvador 004792
Museu de Arqueologia e Etnologia da
  Universidade de São Paulo, São
  Paulo 004876
Museu de Arqueologia, Fundação de
  Casa de Bragança, Vila Viçosa 036110
Museu de Arte, Belém 004334
Museu de Arte, Goiânia 004472
Museu de Arte, Joinville 004515
Museu de Arte, Londrina 004537
Museu de Arte Africana da Azambuja,
  Azambuja 035541
Museu de Arte Assis Chateaubriand,
  Campina Grande 004382
Museu de Arte Brasileira, São
  Paulo 004877
Museu de Arte Contemporâea,
  Jataí 004508
Museu de Arte Contemporânea, Feira
  de Santana 004449
Museu de Arte Contemporânea,
  Funchal 035703
Museu de Arte Contemporânea da
  Universidade de São Paulo, São
  Paulo 004878
Museu de Arte Contemporânea de
  Niterói, Niterói 004590
Museu de Arte Contemporânea de
  Pernambuco, Olinda 004595
Museu de Arte Contempôrnea,
  Curitiba 004432
Museu de Arte Contemprânea da
  Bienal de Cerveira, Vila Nova de
  Cerveira 036094
Museu de Arte da Bahia,
  Salvador 004793
Museu de Arte da Pampulha, Belo
  Horizonte 004344
Museu de Arte da Universidade Fedral
  do Ceará, Fortaleza 004462
Museu de Arte de Santa Catarina,
  Florianópolis 004454
Museu de Arte de São Paulo Assis
  Chateaubriand, São Paulo . . . 004879
Museu de Arte do Paraná,
  Curitiba 004433
Museu de Arte do Rio Grande do Sul
  Aldo Malagoli, Porto Alegre . . 004643
Museu de Arte e Cultura Canjira,
  Itaobim 004497
Museu de Arte e Cultura Popular da
  UFMT, Universidade Federal de Mato
  Grosso, Cuiabá 004421
Museu de Arte e Etnografia da Ribeira
  Chã, Ribeira Chã 035980
Museu de Arte e História, Natal 004582
Museu de Arte e Histórico,
  Jaboticabal 004504
Museu de Arte Mágica e Ilusionismo
  João Peixoto dos Santos, São
  Paulo 004880
Museu de Arte Moderna,
  Resende 004673
Museu de Arte Moderna da Bahia,
  Salvador 004794
Museu de Arte Moderna de São Paulo,
  São Paulo 004881
Museu de Arte Moderna do Rio de
  Janeiro, Rio de Janeiro 004721
Museu de Arte Popular, Lisboa . 035781
Museu de Arte Popular da Fortaleza
  dos Reis Magos, Natal 004583
Museu de Arte Primitiva Moderna,
  Guimarães 035731
Museu de Arte Religiosa, Vitória 004956
Museu de Arte Sacra, Aljezur . . 035499
Museu de Arte Sacra, Amarante 035523
Museu de Arte Sacra, Castelo
  Branco 035605
Museu de Arte Sacra, Esposende 035659
Museu de Arte Sacra, Évora . . . 035676

Museu de Arte Sacra, Funchal . 035704
Museu de Arte Sacra, Gouveia . 035723
Museu de Arte Sacra, Mértola . 035852
Museu de Arte Sacra, Pedrógão
  Grande 035904
Museu de Arte Sacra, Rio de
  Janeiro 004722
Museu de Arte Sacra, Salvador . 004795
Museu de Arte Sacra, São
  Cristóvão 004825
Museu de Arte Sacra, São João Del
  Rei 004834
Museu de Arte Sacra, São Luís . 004842
Museu de Arte Sacra, São Paulo 004882
Museu de Arte Sacra, São
  Sebastião 004912
Museu de Arte Sacra, Taubaté . 004923
Museu de Arte Sacra, Vila do
  Conde 036082
Museu de Arte Sacra, Vila Viçosa 036111
Museu de Arte Sacra da Catedral-
  Basílica, Salvador 004796
Museu de Arte Sacra da Misericórdia
  do Fundão, Fundão 035712
Museu de Arte Sacra da Sé de
  Portalegre, Portalegre 035924
Museu de Arte Sacra da Universidade,
  Coimbra 035623
Museu de Arte Sacra da
  Universidade Lusíada, Vila Nova de
  Famalicão 036099
Museu de Arte Sacra de Alcoutim,
  Alcoutim 035492
Museu de Arte Sacra de Angra dos
  Reis, Angra dos Reis 004306
Museu de Arte Sacra de Arouca,
  Arouca 035536
Museu de Arte Sacra de Boa Morte,
  Goiás 004479
Museu de Arte Sacra de Iguape,
  Iguape 004489
Museu de Arte Sacra de Monsaraz,
  Monsaraz 035862
Museu de Arte Sacra de Santos,
  Santos 004819
Museu de Arte Sacra do Carmo /
  Paróquia do Pilar, Ouro Preto . 004602
Museu de Arte Sacra do Estado de
  Alagoas, Marechal Deodoro . . 004558
Museu de Arte Sacra do Pará,
  Belém 004335
Museu de Arte Sacra Dom Ranulfo,
  Maceió 004545
Museu de Arte Sacra e Arqueologia,
  Porto 035937
Museu de Arte Sacra e Arqueologia
  de Vila Alva, Vila Alva 036077
Museu de Arte Sacra e Etnografia
  Religiosa, Horta 035734
Museu de Arte Sacra e Etnologia de
  Fátima, Fátima 035693
Museu de Arte Sacra Padre Roberto
  Barbalho, Recife 004663
Museu de Arte do Mosteiro do Convento
  dos Lóios, Santa Maria da Feira 035996
Museu de Artes Decorativas,
  Évora 035677
Museu de Artes e Ofícios, Belo
  Horizonte 004345
Museu de Artes e Ofícios Francisco
  Távora, Escalhão 035655
Museu de Artes Plásticas,
  Taubaté 004924
Museu de Artes Plásticas Quirino da
  Silva, Mococa 004567
Museu de Artes Sacras Iria Josepha
  da Silva, Mococa 004568
Museu de Artes Visuais, São
  Luís 004843
Museu de Artes Visuais Ruth
  Schneider, Passo Fundo 004611
Museu de Astronomia e Ciências Afins,
  Rio de Janeiro 004723
Museu de Aveiro, Aveiro 035538

Museu de Badalona, Badalona . 039038
Museu de Barroso, Montalegre . 035863
Museu de Belas-Artes,
  Cataguases 004408
Museu de Biodiversidade do Cerrado,
  Uberlândia 004940
Museu de Biologia Prof. Mello Leitão,
  Santa Teresa 004812
Museu de Bonecos e Ventrílocos, São
  Paulo 004883
Museu de Cabrera, Centre de Visitants
  des Celler, Palma de Mallorca . 039891
Museu de Carrosses Fúnebres,
  Barcelona 039074
Museu de Cartão, São Paulo . . 004884
Museu de Cera, Barcelona 039075
Museu de Cera, Foro de Mallorca,
  Binissalem 039151
Museu de Cerámica, Barcelona . 039076
Museu de Cerâmica, Borba . . . 035562
Museu de Ceràmica de l'Alcora,
  Alcora 038924
Museu de Ceràmica de Manises,
  Manises 039715
Museu de Ciência e Técnica da Escola
  de Minas, Ouro Preto 004603
Museu de Ciência e Tecnologia,
  Salvador 004797
Museu de Ciências da Terra – DNPM,
  Rio de Janeiro 004724
Museu de Ciências Morfológicas,
  Universidade Federal de Minas
  Gerais, Pampulha 004607
Museu de Ciências Naturais,
  Guarapuava 004483
Museu de Ciências Naturais, Porto
  Alegre 004644
Museu de Ciências Naturais,
  Recife 004664
Museu de Ciências Naturais,
  Triunfo 004934
Museu de Ciências Naturais PUC
  Minas, Belo Horizonte 004346
Museu de Ciències Naturals,
  Costitx 039324
Museu de Ciències Naturals de
  Barcelona, Institut de Cultura
  Barcelona, Barcelona 039077
Museu de Ciències Naturals, Convent
  del Carme, Onda 039837
Museu de Comunicação Social
  Hipólito José da Costa, Porto
  Alegre 004645
Museu de Corrente, Corrente . . 004418
Museu de Coses del Poble,
  Collbató 039299
Museu de Curiositats Marineres,
  Vilanova i la Geltrú 040389
Museu de D. Lopo de Almeida,
  Abrantes 035474
Museu de Design e Artes Decorativas,
  Lisboa 035782
Museu de Documentos Especiais,
  Praia 007504
Museu de Electricidade, Fornelos 035697
Museu de Escola Prática de Artilharia,
  Vendas Novas 036073
Museu de Escultura Comparada,
  Mafra 035841
Museu de Esportes Mané Garrincha,
  Rio de Janeiro 004725
Museu de Etnografia e Arqueologia
  Industrial, Torres Novas 036060
Museu de Etnografia e História,
  Porto 035938
Museu de Etnografia e História
  de Póvoa de Varzim, Póvoa de
  Varzim 035965
Museu de Évora, Évora 035678
Museu de Ex-votos do Senhor do
  Bonfim, Salvador 004798
Museu de Farmácia, São Paulo . 004885
Museu de Física, Coimbra 035645
Museu de Folclore Édison Carneiro,

Rio de Janeiro 004726
Museu de Folclore Saul Martins,
Vespasiano 004949
Museu de Formentera,
Formentera 039407
Museu de Francisco Tavares Proença
Júnior, Castelo Branco 035606
Museu de Gavà, Gavà 039430
Museu de Geociências, São
Paulo 004886
Museu de Geociências da
Universidade de Brasília,
Brasília 004360
Museu de Geologia, Barcelona . 039078
Museu de Geologia, Maringá . . 004563
Museu de Geologia, Porto Alegre 004646
Museu de Geologia, Sant Celoni 040080
Museu de Geologia da Universitat
de València, Burjassot 039175
Museu de Geologia Valentí Masachs,
Manresa 039718
Museu de Geologia, Universidade de
Trás-os-Montes e Alto Douro, Vila
Real 036104
Museu de Granollers, Granollers 039474
Museu de Granollers-Ciències
Naturals, Granollers 039475
Museu de Grão Vasco, Viseu . . . 036115
Museu de História e Ciências Naturais,
Além Paraíba 004302
Museu de História Natural, Belo
Horizonte 004347
Museu de História Natural,
Campinas 004388
Museu de História Natural,
Coimbra 035625
Museu de História Natural, Crato 004420
Museu de História Natural,
Maputo 031805
Museu de História Natural de Sines,
Sines 036033
Museu de História Natural Prof.
Antonio Pergola, Atibaia 004323
Museu de História Natural, Faculdade
de Ciências da Universidade do
Porto, Porto 035939
Museu de Homem do Norte,
Manaus 004553
Museu de Homem Sergipano,
Aracaju 004312
Museu de Imagem e do Som de
Pernambuco, Recife 004665
Museu de Images do Inconsciente,
Rio de Janeiro 004727
Museu de Índio, Uberlândia . . . 004941
Museu de Instrumentos de Cálculo
Numérico, São Carlos 004824
Museu de Instrumentos Musicais do
Conservatório Nacional, Lisboa 035783
Museu de Janes e Malveira,
Alcabideche 035478
Museu de Jazigos Minerais
Portugueses, Instituto Geológico
e Mineiro, São Mamede de
Infesta 036015
Museu de José Malhoa, Caldas da
Rainha 035588
Museu de l#Aigua, Lleida 039582
Museu de la Casa Codorniú, Sant
Sadurní d'Anoia 040094
Museu de la Catedral, Barcelona 039079
Museu de la Ciència i de la Tècnica
de Catalunya, Terrassa 040222
Museu de la Ciència Pare Eduard
Vitòria, Alcoy 038929
Museu de la Ciutat Casa de Polo,
Villarreal 040413
Museu de la Ciutat, Museu de
Valls, Casa Municipal de Cultura,
Valls 040355
Museu de la Conca Dellà, Isona 039517
Museu de la Farmàcia Catalana,
Universitat de Barcelona,
Barcelona 039080

Museu de la Festa, Casal de Sant
Jordi, Alcoy 038930
Museu de la Fusta, Areu 038990
Museu de la Història de Fontcoberta,
Fontcoberta 039405
Museu de la Jugueta de Mallorca, Sa
Pobla 039950
Museu de la Mar, Peñíscola . . . 039937
Museu de la Marina de Vilassar,
Vilassar de Mar 040398
Museu de la Música, Barcelona 039081
Museu de la Noguera, Balaguer 039043
Museu de la Pesca, Palamós . . 039871
Museu de la Rajoleria de Paiporta,
Paiporta 039869
Museu de la Taronja, Burriana . 039177
Museu de la Universitat d'Alacant,
Museo de la Universidad de Alicante,
Alicante 038947
Museu de la Vall de Lord, Sant Llorenç
de Morunys 040092
Museu de la Valltorta, Tírig 040228
Museu de la Vila, Vilar-Rodona . 040395
Museu de l'Acordió, Arsèguel . . 038999
Museu de l'Adet, Torelló 040253
Museu de l'Aigua Collecció Gavarró,
Vilanova del Camí 040388
Museu de Lamego, Lamego . . . 035747
Museu de Lanifícios, Universidade da
Beira Interior, Covilhã 035644
Museu de l'Aquarela, Fundació J.
Martínez Lozano, Llança 039573
Museu de l'Art de la Pell, Vic . . 040374
Museu de l'Artiste Faller,
Valencia 040324
Museu de l'Arxiprestal, Basílica de
Santa María la Mayor, Morella 039789
Museu de l'Automòbil, Barcelona 039082
Museu De L'automoció Roda Roda,
Lleida 039583
Museu de l'Ebre, Tortosa 040269
Museu de l'Empordà, Figueres . 039399
Museu de les Arts Decoratives,
Barcelona 039083
Museu de les Arts Escèniques,
Barcelona 039084
Museu de les Ciències Príncipe Felipe,
Ciutat de les Arts i les Ciències,
Valencia 040325
Museu de les dues rodes,
Canillo 000087
Museu de l'Escola de Gràmatica
Ramon Llull, Algaida 038940
Museu de l'Estable i Troballes, Museu
Municipal, Guimerà 039482
Museu de l'Estampació de Premià de
Mar, Premià de Mar 039972
Museu de l'Horta d'Almàssera,
Antiguo Matadero Municipal,
Almàssera 038954
Museu de l'Institut de Criminologia,
Facultat de Dret, Barcelona . . . 039085
Museu de l'Institut Paleontològic
Miquel Crusafont, Sabadell . . . 040028
Museu de Llavaneres, Sant Andreu de
Llavaneres 040079
Museu de Lloret de Mar, Lloret de
Mar 039591
Museu de Malacología Cau del Cargol,
Vilassar de Dalt 040397
Museu de Mallorca, Palma de
Mallorca 039892
Museu de Maquetes d'Art Romànic,
Escaldes-Engordany 000091
Museu de Marinha, Lisboa 035784
Museu de Mataró, Mataró 039736
Museu de Medallística Enrique Giner,
Nules 039822
Museu de Medicina, Curitiba . . . 004434
Museu de Minerais e Rochas,
Uberlândia 004942
Museu de Mineralogia,
Congonhas 004412
Museu de Mineralogia Luiz Englert,

Porto Alegre 004647
Museu de Miniaturas de Arte e
Ofícios, Cartaxo 035601
Museu de Moixent, Mogente . . . 039755
Museu de Monforte, Monforte . . 035861
Museu de Monte Redondo, Monte
Redondo 035864
Museu de Montgrí i del Baix Ter,
Torroella de Montgrí 040266
Museu de Montserrat, Monistrol de
Montserrat 039769
Museu de Nampula, Nampula . . 031812
Museu de Nines Russes, Escaldes-
Engordany 000092
Museu de Nossa Senhora da Glória do
Outeiro, Rio de Janeiro 004728
Museu de Odontologia Dr. Solon de
Miranda Galvão, Natal 004584
Museu de Odontologia Professor
Salles Cunha, Rio de Janeiro . 004729
Museu de Olaria, Barcelos 035547
Museu de Ornitologia, Goiânia . 004473
Museu de Ovar, Ovar 035890
Museu de Paleontologia, Monte
Alto 004571
Museu de Patologia do Instituto
Oswaldo Cruz, Rio de Janeiro . 004730
Museu de Pedras Ramis Bucair,
Cuiabá 004422
Museu de Pesca, Santos 004820
Museu de Pintura, Lourinhã . . . 035830
Museu de Pintura e Escultura Martins
Correia, Golegã 035719
Museu de Poblet, Monistir de
Poblet 039766
Museu de Polícia Militar, São
Paulo 004887
Museu de Pré-História Professora
Márcia Angelina Alves, Perdizes 004624
Museu de Relleu, Relleu 039994
Museu de Relógio Prof. Dimas de
Melo Pimenta, São Paulo 004888
Museu de Rocafort, El Pont de
Vilomara i Rocafort 039958
Museu de Roque Gameiro,
Minde 035854
Museu de Rua Alto Paraná Ontem e
Hoje, Alto Paran 004303
Museu de Rubí, Rubí 040025
Museu de Sant Cugat, El Monestir,
Sant Cugat del Vallès 040083
Museu de Santa Maria, Vila do
Porto 036085
Museu de Santa Maria Adelaide,
Arcozelo 035534
Museu de Santa Maria de Lamas,
Santa Maria de Lamas 035999
Museu de Santo André, Santo
André 004816
Museu de São Jorge, Calheta de São
Jorge 035590
Museu de São Miguel de Entre-
Ambos-os-Rios, Ponte da Barca 035921
Museu de São Roque, Lisboa . . 035785
Museu de Saúde Pública Emílio Ribas,
São Paulo 004889
Museu de Serralves, Museu de Arte
Contemporânea, Porto 035940
Museu de Sertão, Petrolina . . . 004626
Museu de Setúbal – Convento de
Jesus – Galeria de Arte Quinhentista,
Setúbal 036028
Museu de Telecomunicações, Rio de
Janeiro 004731
Museu de Terrassa, Terrassa . . 040223
Museu de Tradições do Nordeste,
Brasília 004361
Museu de Valores, Rio de
Janeiro 004732
Museu de Valores do Banco Central
do Brasil, Brasília 004362
Museu de Venâncio Aires, Venâncio
Aires 004948
Museu de Vila Real, Vila Real . . 036105

Museu de Vila Verde de Ficalho, Vila
Verde de Ficalho 036107
Museu de Vilafamés Vicente Aguilera
Cerni, Vilafamés 040384
Museu de Vilafranca, Vilafranca del
Penedès 040385
Museu de Zoologia, São Paulo . 004890
Museu d'Eines Antigues,
Torregrossa 040259
Museu d'Eines del Camp Masia de
Can Deu, Sabadell 040029
Museu d'Eines del Pagès, Sant
Climent de Llobregat 040081
Museu del Blat i la Pagesia,
Cervera 039266
Museu del Blat, Museus d'Osona,
Taradell 040201
Museu del Camp, Vilabella 040383
Museu del Càntir, Argentona . . 038991
Museu del Castell de Onda,
Onda 039838
Museu del Centre d'Estudis
Contestans, Cocentaina 039298
Museu del Cinema, Colecció Tomàs
Mallol, Girona 039446
Museu del Convent de la Beata sor
Fracinaina de la Mare de Déu dels
Dolors, Sencelles 040161
Museu del Dolmen, Llagostera . 039571
Museu del Ferrocarril, Vilanova i la
Geltrú 040390
Museu del Joguet de Catalunya,
Figueres 039400
Museu del Monestir de Sant
Bartomeu, Inca 039511
Museu del Monestir de Santa Maria
de L'Estany, L'Estany 039386
Museu del Montsec, Artesa de
Segre 039003
Museu del Montsià, Amposta . . 038971
Museu del Pagès, Castelldans . 039247
Museu del Pagès, Linyola 039570
Museu del Pagès, Torroella de
Fluvià 040265
Museu del Perfum, Barcelona . . 039086
Museu del Perfum, Fundació Júlia
Bonet, Escaldes-Engordany . . 000093
Museu del Pin, Ordino 000101
Museu del Port, Tarragona 040206
Museu del Prat, Centre d'Interpretació
del Delta de Llobregat, El Prat de
Llobregat 039969
Museu del Suro, Museu de la Ciècia,
Palafrugell 039870
Museu del Tabac – Antiga Fàbrica
Reig, Sant Julià de Lòria 000105
Museu del Taulell Manolo Safont,
Onda 039839
Museu del Temple Expiatori de la
Sagrada Familia, Barcelona . . . 039087
Museu del Vi, Museu de Vilafranca,
Vilafranca del Penedès 040386
Museu del Vidre, Algaida 038941
Museu del Volcans, Olot 039831
Museu dels Autòmats, Barcelona 039088
Museu dels Raiers, El Pont de
Claverol 039957
Musèu d'Era Val d'Aran, Vielha e
Mijaran 040377
Museu des Comunicações Marechal
Rondon, Ji-Paran 004510
Musèu d'Estris Antics de l'Agricultura,
Bellvís 039119
Museu d'Etnografia d'Eivissa, Santa
Eulària del Río 040104
Museu d'Etnologia, Pego 039927
Museu d'Etnologia i Arqueologia,
Callosa d'En Sarrià 039206
Museu Deu, El Vendrell 040367
Museu d'Història, Sabadell 040030
Museu d'Història de Calp, Calpe 039208
Museu d'Història de Cambrils,
Cambrils 039211
Museu d'Història de Catalunya,

Barcelona 039089
Museu d'Història de la Ciudad, Sant
Feliu de Guíxols 040086
Museu d'Història de la Ciutat,
Girona 039447
Museu d'Història de la Ciutat de
Barcelona, Barcelona 039090
Museu d'Història de la Ciutat de
Palma, Palma de Mallorca ... 039893
Museu d'Història de la Joguina, Sant
Feliu de Guíxols 040087
Museu d'Història de l'Hospitalet
de Llobregat, L'Hospitalet de
Llobregat 039495
Museu d'Història de Manacor, Torre
dels Enagistes, Manacor 039713
Museu d'Historia de Nules, Nules 039823
Museu d'Història de Tarragona, Casa
Castellarnau, Tarragona 040207
Museu d'Història del Calçat,
Barcelona 039091
Museu d'Història dels Jueui,
Girona 039448
Museu di Arte Contemporânea,
Americana 004304
Museu Didáctico do Instituto de
Arqueologia, Instituto da Faculdade
de Letras, Coimbra 035626
Museu Diocesà de Barcelona,
Barcelona 039092
Museu Diocesà de Girona,
Girona 039449
Museu Diocesa de Mallorca, Palma de
Mallorca 039894
Museu Diocesa de Menorca,
Ciutadella de Menorca 039294
Museu Diocesà d'Urgell, La Seu
d'Urgell 040163
Museu Diocesà i Comarcal, Palacio
Episcopal, Solsona 040192
Museu Diocesà Santa Maria,
Eivissa 039363
Museu Dionís Bennàssar,
Pollença 039953
Museu do Ar, Air Museum, Alverca do
Ribatejo 035517
Museu do Arroz, Comporta 035637
Museu do Automóvel, Fafe 035684
Museu do Automóvel, Vila Nova de
Famalicão 036100
Museu do Automóvel Antigo, Paço
d'Arcos 035892
Museu do Automóvel, Arte & História,
São Francisco de Paula 004827
Museu do Banco de Crédito Real de
Minas Gerais, Juiz de Fora .. 004519
Museu do Banco de Portugal,
Lisboa 035786
Museu do Barreiro e Museu Industrial
da Quimiparque, Barreiro 035548
Museu do Barro, Redondo 035974
Museu do Batalhão de Sapadores
Bombeiros, Porto 035941
Museu do Bombeiro, Lisboa 035787
Museu do Bombeiro Manuel Valdés
Sobral, Valença 036070
Museu do Bonde, Rio de Janeiro 004733
Museu do Brinquedo, Évora ... 035679
Museu do Brinquedo de Sintra,
Sintra 036039
Museu do Brinquedo, Faculdade de
Educação / FE, São Paulo 004891
Museu do Café Cel. Francisco
Schmidt, Ribeirão Preto 004675
Museu do Café de Novadelta, Campo
Maior 035593
Museu do Campo da Lacam,
Carapinheira 035598
Museu do Gangaço, Triunfo ... 004593
Museu do Caramulo, Caramulo . 035597
Museu do Carnaval, Rio de
Janeiro 004734
Museu do Carnaval Hilton Silva,
Florianópolis 004455

Museu do Carro Eléctrico, Porto 035942
Museu do Castelo, Ourém 035887
Museu do Castelo de Castro Marim,
Castro Marim 035609
Museu do Ceará, Fortaleza ... 004463
Museu do Centro Científico e Cultural
de Macau, Lisboa 035788
Museu do Chiado, Museu Nacional de
Arte Contemporânea, Lisboa . 035789
Museu do Chocalho, Alcáçovas . 035480
Museu do Cinema, Lisboa 035790
Museu do Círio de Nazaré,
Betém 004351
Museu do Colégio Mauá, Santa Cruz
do Sul 004806
Museu do Colono, Santa
Leopoldina 004807
Museu do Convento de São Domingos,
Montemor-o-Novo 035867
Museu do Crime, Academia de Polícia
Dr. Coriolano Nogueira Cobra, São
Paulo 004892
Museu do Diamante, Diamantina 004443
Museu do Douro, Peso da Régua 035912
Museu do Estado da Bahia,
Salvador 004799
Museu do Estado de Pernambuco,
Recife 004666
Museu do Estado do Pará,
Belém 004336
Museu do Eucalipto, Rio Claro . 004677
Museu do Expedicionário,
Curitiba 004435
Museu do Fado e da Guitarra
Portuguesa, Lisboa 035791
Museu do Ferro e da Região de
Moncorvo, Torre de Moncorvo . 036059
Museu do Forte de Santiago,
Funchal 035705
Museu do Fuzileiro, Palhais .. 035896
Museu do Gás, Rio de Janeiro . 004735
Museu do Grupo Folclórico, Santa
Eulália de Barrosas 035993
Museu do Homem, Curitiba 004436
Museu do Homem do Nordeste,
Recife 004667
Museu do Índio, Florianópolis . 004456
Museu do Índio, Lagoa Seca .. 004528
Museu do Índio, Manaus 004554
Museu do Indio, Rio de Janeiro 004736
Museu do Instituto Arqueológico,
Histórico e Geográfico Pernam
Bucano, Recife 004668
Museu do Instituto Butantan, São
Paulo 004893
Museu do Instituto de Anatomia
Patológica, Coimbra 035627
Museu do Instituto de Anatomia,
Faculdade de Medicina da
Universidade do Porto, Porto .. 035943
Museu do Instituto de Antropologia
Cámara Cascudo, Natal 004585
Museu do Instituto de Botânica
Gonçalo Sampaio, Faculdade de
Ciências da Univsidade do Porto,
Porto 035944
Museu do Instituto de Medicina Legal,
Lisboa 035792
Museu do Instituto Geográfico e
Historico da Bahia, Salvador . 004800
Museu do Instituto Geográfico
Português, Lisboa 035793
Museu do Instituto Histórico,
Carpina 004404
Museu do Instituto Histórico e
Geográfico Brasileiro, Rio de
Janeiro 004737
Museu do Instituto Histórico e
Geográfico de Alagoas, Maceió 004546
Museu do Instituto Histórico e
Geográfico do Rio Grande do Norte,
Natal 004586
Museu do Instituto Oceanográfico,
Instituto Oceanográfico / IO, São

Paulo 004894
Museu do Jogo, Oeiras 035879
Museu do Lar dos Veteranos Militares,
Torres Vedras 036063
Museu do Mamulengo, Olinda . 004596
Museu do Mar, Sesimbra 036024
Museu do Mar, Vila do Conde . 036083
Museu do Mar – Rei D. Carlos,
Cascais 035604
Museu do Minstério do Trabalho,
Brasília 004363
Museu do Monte da Guerreira,
Tavira 036051
Museu do Monumento Nacional aos
Mortos da Segunda Guerra Mundial,
Rio de Janeiro 004738
Museu do Mosteiro de Alcobaça,
Alcobaça 035485
Museu do Mosteiro de São Martinho
de Tibães, Braga 035569
Museu do Mosteiro dos Jerónimos,
Lisboa 035794
Museu do Neo-Realismo, Vila Franca
de Xira 036088
Museu do Nordeste, Nordeste . 035874
Museu do Ouro, Sabará 004782
Museu do Ouro de Travassos,
Travassos – Póvoa de Lanhoso 036066
Museu do Papel Moeda, Porto . 035945
Museu do Papel Terras de Santa
Maria, Paços de Brandão ... 035893
Museu do Pescador, Montijo ... 035868
Museu do Piauí, Teresina 004929
Museu do Pio XII, Novo
Hamburgo 004594
Museu do Porco, Pampilhosa . 035898
Museu do Porto de Belém,
Belém 004337
Museu do Porto do Rio, Rio de
Janeiro 004739
Museu do Primeiro Reinado, Rio de
Janeiro 004740
Museu do Rádio, Juiz de Fora . 004520
Museu do Rancho Folclórico das
Papoilas do Campo, Cela ... 035611
Museu do Rancho Folclórico de
Alenquer, Alenquer 035494
Museu do Rancho Folclórico de São
Miguel do Milharado, Milharado 035853
Museu do Rancho Folclórico do
Calvário, Estômbar 035665
Museu do Regimento de Infantaria 14,
Viseu 036116
Museu do Rio, Alcoutim 035493
Museu do Santuário de Nossa
Senhora da Abadia, Bouro ... 035568
Museu do Scrimshaw, Horta .. 035735
Museu do Seridó, Caicó 004379
Museu do Seridó, Universidade
Federal de Rio Grande do Norte,
Natal 004587
Museu do Sol, Penápolis 004621
Museu do STF, Brasília 004364
Museu do Superior Tribunal de
Justiça, Brasília 004365
Museu do Teatro Nacional de São
Carlos, Lisboa 035795
Museu do Teatro Romano,
Lisboa 035796
Museu do Telephone, Rio de
Janeiro 004741
Museu do Theatro Municipal, São
Paulo 004895
Museu do Trabalho Michel Giacometti,
Setúbal 036029
Museu do Traje de Viana do Castelo,
Viana do Castelo 036075
Museu do Trajo, São Brás de
Alportel 036009
Museu do Trem, Rio de Janeiro 004742
Museu do Trem, São Leopoldo . 004840
Museu do Trem da Cidade de Recife,
Recife 004669
Museu do Tribunal de Justiça do

Estado de São Paulo, São Paulo 004896
Museu do Tropeiro, Castro 004407
Museu do Vaqueiro, Morada
Nova 004576
Museu do Vidro, Marinha Grande 035847
Museu do Vinho, Funchal 035706
Museu do Vinho, Pico da Pedra 035914
Museu do Vinho, Videira 004953
Museu do Vinho dos Biscoitos,
Biscoitos 035559
Museu Dom Avelar Brandão Vilela,
Teresina 004930
Museu Dom Diogo de Souza,
Universidade da Região da
Campanha, Bagé 004325
Museu Dom João VI, Rio de
Janeiro 004743
Museu Dom José, Sobral 004917
Museu Dom Vital, Recife 004670
Museu dos Baleeiros, Lajes do
Pico 035744
Museu dos Biscaínhos, Braga . 035570
Museu dos Bombeiros, Vila do
Conde 036084
Museu dos Bombeiros Voluntários de
Bucelas, Bucelas 035580
Museu dos Caminhos de Ferro de
Lousado, Lousado 035834
Museu dos Capelinhos, Horta . 035736
Museu dos Cordofones, Tebosa 036053
Museu dos Cristos, Borba 035563
Museu dos Cristos, Sousel ... 036049
Museu dos Fósforos, Tomar ... 036056
Museu dos Inhamuns, Tauá ... 004920
Museu dos Relógios Antigos,
Serpa 036021
Museu dos Rios e das Artes
Marítimas, Constância 035641
Museu dos Teatros do Rio de Janeiro,
Rio de Janeiro 004744
Museu dos Terceiros, Ponte de
Lima 035922
Museu dos Transportes e
Comunicações, Porto 035946
Museu dos Transportes Públicos
Gaetano Ferolla, São Paulo ... 004897
Museu dos Transportes Urbanos,
Museu da Cidade, Coimbra .. 035628
Museu Dr. José Faibes Lubianca,
Porto Alegre 004648
Museu Dr. Marcelino Moules, Cinco
Ribeiras 035616
Museu Durán i Sanpere, Cervera 039267
Museu e Arquivo Histórico, Itu . 004502
Museu e Arquivo Histórico do Centro
Cultural Banco do Brasil, Rio de
Janeiro 004745
Museu e Arquivo Histórico Municipal,
Guapor 004482
Museu e Arquivo Municipal, Bom
Jesus 004354
Museu e Arquivo Municipal,
Lagoa 035740
Museu e Arquivo Público Municipal de
Campo Belo, Campo Belo ... 004390
Museu e Biblioteca do Instituto
Geográfico e Histórico do Amazonas,
Manaus 004555
Museu e Biblioteca Pública Pelotense,
Pelotas 004620
Museu Ecologia, Nanuque 004578
Museu Educativo Gama d'Eça, Santa
Maria 004809
Museu Egipci de Barcelona,
Barcelona 039093
Museu Els Cups, Montbrió del
Camp 039775
Museu Eng. António de Almeida,
Porto 035947
Museu Enric Monjo, Vilassar de
Mar 040399
Museu Epigráfico Egitanense, Idanha-
a-Nova 035737
Museu Episcopal de Vic, Vic ... 040375

Museu-Escola de Artes Decorativas
  Portuguesas, Fundação Ricardo do
  Espírito Santo Silva, Lisboa . . . 035797
Museu-Escola Esacro do Estado da
  Paraíba, João Pessoa 004512
Museu Escolar de Marrazes,
  Leiria 035750
Museu Estadual do Carvão, Arroio dos
  Ratos 004322
Museu Etnográfico e Arqueológico de
  Ourém, Ourém 035888
Museu Etnogràfic Andino-Amazònic
  dels Caputxins, Barcelona 039094
Museu Etnogràfic Casa Cristo,
  Encamp 000089
Museu Etnogràfic de Ripoll, Antic
  Museu Folklòric Parroquial,
  Ripoll 040007
Museu Etnográfico, Santa Cruz das
  Flores 035991
Museu Etnográfico, Santa Maria da
  Feira 035997
Museu Etnográfico da Arada,
  Arada 035532
Museu Etnográfico da Casa do Povo,
  Almeirim 035510
Museu Etnográfico da Freguesia de
  Almoster, Almoster 035513
Museu Etnográfico da Lousã,
  Lousã 035832
Museu Etnográfico da Madeira,
  Ribeira Brava 035978
Museu Etnográfico da Pedreira,
  Pedreira 035902
Museu Etnográfico da Povoação,
  Povoação 035966
Museu Etnográfico da Praia de Mira,
  Praia de Mira 035967
Museu Etnográfico da Região do
  Vouga, Trofa 036067
Museu Etnográfico da Sociedade de
  Geografia de Lisboa, Lisboa . . 035798
Museu Etnográfico de Campelos,
  Campelos 035592
Museu Etnográfico de Carregueiros,
  Carregueiros 035600
Museu Etnográfico de Espiche,
  Luz 035835
Museu Etnográfico de Serpa,
  Serpa 036022
Museu Etnográfico de Vila Franca,
  Lisboa 035799
Museu Etnográfico de Vila Franca de
  Xira, Vila Franca de Xira 036089
Museu Etnográfico de Vilarinho da
  Furna, Terras de Bouro 036054
Museu Etnográfico do Cabouco,
  Cabouco 035581
Museu Etnográfico do Espinheiro,
  Espinheiro 035656
Museu Etnográfico do Espírito Santo,
  Vila do Porto 036086
Museu Etnográfico do Grupo Típico de
  Ançã, Ançã 035527
Museu Etnográfico e Arqueológico do
  Dr. Joaquim Manso, Nazaré . . 035871
Museu Etnográfico e Arte Sacra de
  São Torcato, São Torcato 036018
Museu Etnográfico e Museu Casa
  Tradicional, Glória do Ribatejo . 035717
Museu Etnografico Nacional,
  Bissau 023388
Museu Etnográfico Regional do
  Algarve, Faro 035688
Museu Etnogràficu de la Llechería,
  Morcín 039787
Museu Etnològic, Benaguasil . . 039124
Museu Etnològic de Barcelona,
  Barcelona 039095
Museu Etnologic de Dénia, Dénia 039346
Museu Etnologic de Guadalest, Casa
  Típica del Segle XVIII, Guadalest 039477
Museu Etnològic de l'Alcúdia,
  L'Alcúdia (Valencia) 038935

Museu Etnològic de Muro, Secció
  Etnología Museu de Mallorca,
  Muro 039810
Museu Etnològic del Montseny La
  Gabella, Arbúcies 038983
Museu Etnològic i Arqueològic,
  Juneda 039537
Museu Etnològic Local, Betxí . . 039136
Museu Etnologic Molí de Dalt, Sant
  Lluis 040091
Museu Etnològic Municipal,
  Torrella 040261
Museu Eugènio Teixeira Leal,
  Salvador 004801
Museu Farroupilha, Triunfo 004936
Museu Fazenda, Nanuque 004579
Museu Ferreira de Almeida,
  Nampula 031813
Museu Ferreira de Castro, Sintra 036040
Museu Ferroviário de Arco de Baúlhe,
  Arco de Baúlhe 035533
Museu Ferroviário de Chaves,
  Chaves 035615
Museu Ferroviário de Pires do Rio,
  Pires do Rio 004634
Museu Florestal Octávio Vecchi, São
  Paulo 004898
Museu Folclorico da Divisão de
  Discoteca e Biblioteca de Música de
  São Paulo, São Paulo 004899
Museu Frederic Marès,
  Barcelona 039096
Museu Frei Galvão,
  Guaratinguetá 004485
Museu Fundação Amalia Rodrigues,
  Lisboa 035800
Museu Fundação Arquivo Paes Telles,
  Ervedal 035652
Museu Fundação Laura Santos,
  Gouveia 035724
Museu Fundació Emili Vilá,
  Llagostera 039572
Museu Futbol Club Barcelona
  President Núñez, Barcelona . . 039097
Museu Galeria Manuel Cabanas, Vila
  Real de Santo António 036106
Museu GEAL, Grupo de Etnologia
  e Arqueología da Lourinhã,
  Lourinhã 035831
Museu Geológico da Bahia,
  Salvador 004802
Museu Geológico Valdemar Lefèvre,
  São Paulo 004900
Museu Geológico, Instituto Nacional
  de Engenharia, Tecnologia e
  Inovação, Lisboa 035801
Museu Getúlio Vargas, São Borja 004822
Museu Goiano Zoroastro Artiaga,
  Goiânia 004474
Museu Guido Straube, Curitiba . 004437
Museu Gustavo Barroso,
  Fortaleza 004464
Museu H. Stern, Rio de Janeiro 004746
Museu Henrique e Francisco Franco,
  Funchal 035707
Museu Henriqueta Catharino, Instituto
  Feminino da Bahia, Salvador . . 004803
Museu Hernández Mora-Hernández
  Mora, Mahón 039695
Museu Històric de la Seu,
  Manresa 039719
Museu Històric Municipal,
  Riudoms 040008
Museu Historic Municipal de Polinyà,
  Polinyà 039952
Museu Histórico, Cambé 004380
Museu Histórico, Divinópolis . . . 004444
Museu Histórico, Igarassu 004487
Museu Histórico, Lavras 004534
Museu Histórico, Londrina 004538
Museu Histórico, Nova Ponte . . 004592
Museu Histórico, Pará de Minas . 004608
Museu Histórico, Rio Pomba . . . 004776
Museu Histórico, São Francisco do

Sul 004828
Museu Histórico, São Paulo 004901
Museu Histórico, Sertanópolis . . 004915
Museu Histórico, Siqueira
  Campos 004916
Museu Histórico Abílio Barreto, Belo
  Horizonte 004348
Museu Histórico Aurélio Dolabella,
  Santa Luzia 004808
Museu Histórico Bárbara Heliodora,
  São Gonçalo do Sapucaí 004831
Museu Histórico Beato José de
  Anchieta, Pedreira 004619
Museu Histórico Casa do Imigrante,
  Bento Gonçalves 004350
Museu Histórico Corália Venites Maluf,
  Sacramento 004784
Museu Histórico Cultural, Rio do
  Sul 004773
Museu Histórico da Cidade,
  Brasília 004366
Museu Histórico da Cidade do Rio de
  Janeiro, Rio de Janeiro 004747
Museu Histórico da Imigração
  Japonesa no Brasil, São Paulo 004902
Museu Histórico da Universidade
  Federal, Viçosa 004951
Museu Histórico da Vista Alegre,
  Ílhavo 035738
Museu Histórico de Araxá – Don Beja,
  Araxá 004318
Museu Histórico de Brejo Madre de
  Deus, Brejo Madre de Deus . . 004371
Museu Histórico de Jataí Francisco
  Honório de Campos, Jataí 004509
Museu Histórico de Sergipe, São
  Cristóvão 004826
Museu Histórico de Witmarsum,
  Palmeira 004606
Museu Histórico Desembargador
  Edmundo Mercer Júnior, Tibagi 004933
Museu Histórico do Corpo de
  Bombeiros Militar do Estado do Rio
  de Janeiro, Rio de Janeiro 004748
Museu Histórico do Exército e Forte
  de Copacabana, Rio de Janeiro 004749
Museu Histórico do Piauí,
  Teresina 004931
Museu Histórico do Senado Federal,
  Brasília 004367
Museu Histórico Dona Mariana
  Joaquina da Costa, Vespasiano 004950
Museu Histórico Dr. Sellmann
  Nazareth, São José do Rio
  Preto 004837
Museu Histórico e Artístico,
  Quitandinha 004658
Museu Histórico e Artístico do
  Maranhão, São Luís 004844
Museu Histórico e de Artes, Santana
  do Ipanema 004814
Museu Histórico e de Ordem Geral
  Plínio Travassos dos Santos,
  Rebeirão Preto 004659
Museu Histórico e Diplomático do
  Itamaraty, Rio de Janeiro 004750
Museu Histórico e Geográfico de
  Monte Sião, Monte Sião 004574
Museu Histórico e Geográfico
  de Poços de Caldas, Poços de
  Caldas 004636
Museu Histórico e Geral da Cidade de
  São Vicente, São Vicente 004914
Museu Histórico e Pedagógico,
  Mococa 004569
Museu Histórico e Pedagógico
  Conselheiro Rodrigues Alves,
  Guaratinguetá 004486
Museu Histórico e Pedagógico Dr.
  Washington Luís, Batatais 004329
Museu Histórico e Pedagógico Fernão
  Dias Paes, Penápolis 004622
Museu Histórico e Pedagógico
  Marechal Cândido Rondon,

Araçatuba 004314
Museu Histórico e Pedagógico
  Monteiro Lobato, Taubaté 004925
Museu Histórico e Pedagógico
  Pe. Manoel da Nóbrega, São
  Manuel 004845
Museu Histórico e Pedagógico Prof.
  Lourenço Filho, Porto Ferreira . 004655
Museu Histórico e Pedagógico
  Zequinha de Abreu, Santa Rita do
  Passa Quatro 004811
Museu Histórico Farroupilha,
  Piratini 004633
Museu Histórico Kiyoshi Takano,
  Ura 004945
Museu Histórico Militar, Maputo 031806
Museu Histórico Muncipal José
  Chiachiri, Franca 004467
Museu Histórico Municipal,
  Carangola 004401
Museu Histórico Municipal, Dois
  Irmãos 004445
Museu Histórico Municipal,
  Montenegro 004575
Museu Histórico Municipal, Telêmaco
  Borba 004927
Museu Histórico Municipal e da
  Imigração Italiana Oswaldo Samuel
  Massei, São Caetano do Sul . . 004823
Museu Histórico Municipal Tuany
  Toledo, Pouso Alegre 004657
Museu Histórico Nacional, Rio de
  Janeiro 004751
Museu Histórico Pedagógico
  Conselheiro Dr. João da Silva Carrão,
  Americana 004305
Museu Histórico Pedagógico Índia
  Vanuíre, Tupã 004937
Museu Histórico Prof. Carlos de Silva
  Lacaz, Faculdade de Medicina / FM,
  São Paulo 004903
Museu Histórico Professor José
  Alexandre Vieira, Palmas 004605
Museu Histórico Regional,
  Apucarana 004309
Museu Histórico Regional,
  Itambacuri 004495
Museu Histórico Regional, Passo
  Fundo 004612
Museu Histórico Tenente Coronel
  PM Augusto de Almeida Garrett,
  Curitiba 004438
Museu Histórico Visconde de São
  Leopoldo, São Leopoldo 004841
Museu Homo Sapiens, Maó 004580
Museu i Centre d'Estudis de l'Esport
  Dr. Melcior Colet, Barcelona . . 039098
Museu i Centre d'Estudis Juniperians,
  Petra 039941
Museu i Fons Artistic, Porreres . 039963
Museu i Necròpolis Paleocristians,
  Museu Nacional Arqueològic de
  Tarragona, Tarragona 040208
Museu Iconogràfic, Ordino 000102
Museu Imperial, Petrópolis 004628
Museu Inaldo de Lyra Neves-Manta,
  Academia Nacional de Medicina, Rio
  de Janeiro 004752
Museu Integrado de Roraima, Boa
  Vista 004353
Museu Internacional de Arte Naïf do
  Brasil, Rio de Janeiro 004753
Museu Internacional de Titelles
  d'Albaida, Albaida 038907
Museu Jaguaribano, Aracati . . . 004313
Museu João de Deus, Lisboa . . 035802
Museu João Pedro Nunes, São
  Gabriel 004830
Museu Joaquim José Felizardo, Porto
  Alegre 004649
Museu José Bonifácio, São Paulo 004904
Museu José Luciano de Castro,
  Anadia 035525
Museu Julio de Castilhos, Porto

Register der Institutionen und Firmen
Museu Júlio Dinis, Ovar
– Museu Municipal Miquel Soldevila, Prats de Lluçanès

Alegre 004650
Museu Júlio Dinis, Ovar 035891
Museu Krekovic, Polígono del Levante, Palma de Mallorca 039895
Museu la Granja, Espai Cultural, Santa Perpètua de Mogoda 040109
Museu Laboratório Mineralógico e Geológico, Porto 035948
Museu Lasar Segall, São Paulo . 004905
Museu Leprológico, Curitiba . . . 004439
Museu Local Arqueològic, Artesa del Lleida 039004
Museu Local do Pico da Pedra, Pico da Pedra 035915
Museu Luso-Hebraico Abraham Zacut, Tomar 036057
Museu Maçônico Mário Verçosa, Manaus 004556
Museu Maçonico Português, Lisboa 035803
Museu Malinverni Filho, Lages . . 004525
Museu Manolo Hugué, Mas Manolo, Caldes de Montbui 039201
Museu Marès de la Punta, Museu d'Arenys de Mar, Arenys de Mar 038987
Museu Mariano Procópio, Juiz de Fora 004521
Museu Maricel de Mar, Sitges . . 040188
Museu Marítim de Barcelona, Barcelona 039099
Museu Marítima Almirante Ramalho Ortigão, Faro 035689
Museu Marítimo, Funchal 035708
Museu Marítimo e Regional de Ílhavo, Ílhavo 035739
Museu Marítimo Joaquín Saludes, Valencia 040326
Museu Mazzaropi, Taubaté 004926
Museu Medina, Braga 035571
Museu Memória do Bixiga, São Paulo 004906
Museu Meridional, Porto Alegre 004651
Museu Metropolitano de Arte, Curitiba 004440
Museu Militar, Barcelona 039100
Museu Militar, Ponta Delgada . . 035919
Museu Militar de Bragança, Bragança 035577
Museu Militar de Coimbra, Coimbra 035629
Museu Militar de Lisboa, Lisboa 035804
Museu Militar do Baixo Alentejo, Beja 035555
Museu Militar do Buçaco, Buçaco 035579
Museu Militar do Forte do Brum, Recife 004671
Museu Militar do Porto, Porto . . 035949
Museu Mineiro, Belo Horizonte . 004349
Museu Mineiro, São Pedro da Cova 036016
Museu Mineralógico da Escola de Minas, Ouro Preto 004604
Museu Mineralógico e Geológico, Museu de História Natural, Coimbra 035630
Museu Mineralógico e Geológico, Museu Nacional de História Natural, Lisboa 035805
Museu Moacir Andrade, Manaus 004557
Museu Molí Paperer de Banyeres de Mariola, Banyeres de Mariola . 039047
Museu Molí Paperer de Capellades, Capellades 039225
Museu Mollfulleda de Mineralogía, Museu d'Arenys de Mar, Arenys de Mar 038988
Museu Monàstic, Vallbona de les Monges 040349
Museu-Monestir de Pedralbes, Museu d'Història de la Ciutat, Barcelona 039101
Museu Monografic de Pol Lèntia,
Museo Monográfico de Pllentia, Alcúdia (Baleares) 038934
Museu Monogràfic del Castell de Llinars, Llinars del Vallés 039585
Museu Monográfico de Conimbriga, Condeixa-a-Nova 035639
Museu Monográfico do Bárrio e Campo Arqueológico, Bárrio . . 035549
Museu Monográfico do Fundão, Fundão 035713
Museu Monstruário de Manica, Manica 031802
Museu Muncipal de Victor Graeff, Victor Graeff 004952
Museu Municipal, Alcover 038927
Museu Municipal, Amadora 035521
Museu Municipal, Barbacena . . 004328
Museu Municipal, Cachoeira do Sul 004377
Museu Municipal, Caxias do Sul 004410
Museu Municipal, Conquista . . . 004413
Museu Municipal, Ferreira do Alentejo 035694
Museu Municipal, Garibaldi 004470
Museu Municipal, Itambaracá . . 004496
Museu Municipal, Llívia 039589
Museu Municipal, Manlleu 039716
Museu Municipal, Missal 004566
Museu Municipal, Paulínia 004616
Museu Municipal, Pedras Grandes 004617
Museu Municipal, Rafelcofer . . . 039991
Museu Municipal, Rolândia 004778
Museu Municipal, Rosário de Oeste 004779
Museu Municipal, São João Del Rei 004835
Museu Municipal, Sobralinho . . 036048
Museu Municipal, Tossa de Mar 040271
Museu Municipal, Uberlândia . . 004943
Museu Municipal, Visconde do Rio Branco 004954
Museu Municipal Abade de Pedrosa, Santo Tirso 036008
Museu Municipal Adão Wolski, Contenda 004415
Museu Municipal Adolfo Eurich, Turvo 004938
Museu Municipal Alcácer do Sal, Alcácer do Sal 035479
Museu Municipal Alípio Vaz, Cataguases 004409
Museu Municipal Amadeo de Souza-Cardoso, Amarante 035524
Museu Municipal Armindo Teixeira Lopes, Mirandela 035856
Museu Municipal Atílio Rocco, São José dos Pinhais 004839
Museu Municipal Can Xifreda, Sant Feliu de Codines 040085
Museu Municipal Capitão Henrique José Barbosa, Canguçu 004399
Museu Municipal Carlos Reis de Torres Novas, Torres Novas . . . 036061
Museu Municipal Carmen Miranda, Marco de Canaveses 035846
Museu Municipal da Azambuja, Museo Etnográfico Sebastião Mateus Arenque, Azambuja 035542
Museu Municipal da Chamusca, Chamusca 035613
Museu Municipal da Fotografia João Carpinteiro, Elvas 035648
Museu Municipal da Moita, Moita 035859
Museu Municipal da Ribeira Grande, Ribeira Grande 035982
Museu Municipal d'Agramunt, Agramunt 038887
Museu Municipal d'Almassora, Almazora 038955
Museu Municipal David Canabarro, Sant'Anna do Livramento 004815
Museu Municipal de Alcochete – Núcleo Sede, Alcochete 035488
Museu Municipal de Alcochete-Núcleo de Arte Sacra, Alcochete 035489
Museu Municipal de Alcochete-Núcleo do Sal, Alcochete 035490
Museu Municipal de Alvaiázere, Alvaiázere 035516
Museu Municipal de Antônio Prado, Antônio Prado 004307
Museu Municipal de Arqueologia, Albufeira 035477
Museu Municipal de Arqueologia, Aljezur 035500
Museu Municipal de Arqueologia, Amadora 035522
Museu Municipal de Arqueologia, Monte Alto 004572
Museu Municipal de Arqueologia de Silves, Silves 036032
Museu Municipal de Arqueologia e Etnologia, Sines 036034
Museu Municipal de Arte e História, Nova Era 004591
Museu Municipal de Arte Moderna Abel Manta, Gouveia 035725
Museu Municipal de Avis, Avis . 035540
Museu Municipal de Baião, Baião 035545
Museu Municipal de Bobadela, Bobadela 035560
Museu Municipal de Bombarral, Bombarral 035561
Museu Municipal de Ca l'Arrà, Cabrils 039181
Museu Municipal de Caminha, Caminha 035591
Museu Municipal de Campo Maior, Campo Maior 035594
Museu Municipal de Castro Daire, Castro Daire 035607
Museu Municipal de Ciutadella, Ciutadella de Menorca 039295
Museu Municipal de Esposende, Esposende 035660
Museu Municipal de Estremoz, Estremoz 035669
Museu Municipal de Fotografia Carlos Relvas, Golegã 035720
Museu Municipal de Grândola, Grândola 035726
Museu Municipal de la Festa d'Elx, Elche 039371
Museu Municipal de la Pagesia, Castellbisbal 039246
Museu Municipal de Loulé, Loulé 035828
Museu Municipal de Loures, Loures 035829
Museu Municipal de Mafra, Mafra 035842
Museu Municipal de Marvão, Marvão 035849
Museu Municipal de Molins de Rei, Molins de Rei 039761
Museu Municipal de Montijo, Casa Mora, Moinho de Vento do Esteval, Quinta Nova da Atalaia & Moinho de Maré do Cais, Montijo 035869
Museu Municipal de Moura, Moura 035870
Museu Municipal de Nàutica, El Masnou 039733
Museu Municipal de Óbidos, Óbidos 035875
Museu Municipal de Olhão, Olhão 035881
Museu Municipal de Paços de Ferreira, Paços de Ferreira 035894
Museu Municipal de Palmela, Palmela 035897
Museu Municipal de Paredes, Paredes 035899
Museu Municipal de Paredes de Coura, Paredes de Coura 035900
Museu Municipal de Penafiel, Penafiel 035908
Museu Municipal de Penamacor, Penamacor 035910
Museu Municipal de Peniche, Peniche 035911
Museu Municipal de Pinhel, Pinhel 035916
Museu Municipal de Pollença, Pollença 039954
Museu Municipal de Portalegre, Portalegre 035925
Museu Municipal de Portimão, Portimão 035926
Museu Municipal de Porto de Mós, Porto de Mós 035961
Museu Municipal de Sant Pol de Mar, Sant Pol de Mar 040093
Museu Municipal de Santarém, Santarém 036001
Museu Municipal de São José dos Campos, São José dos Campos 004838
Museu Municipal de Torres Vedras, Torres Vedras 036064
Museu Municipal de Vale de Cambra, Vale de Cambra 036069
Museu Municipal de Viana do Castelo, Viana do Castelo 036076
Museu Municipal de Vila Flor, Vila Flor 036087
Museu Municipal de Vila Franca de Xira, Vila Franca de Xira 036090
Museu Municipal de Vila Franca do Campo, Vila Franca do Campo 036091
Museu Municipal de Vouzela, Vouzela 036118
Museu Municipal Deolindo Mendes Pereira, Campo Mourão 004395
Museu Municipal d'Història i Arqueologia, Cullera 039339
Museu Municipal Dias de Oliveira, Valongo 036072
Museu Municipal do Cadaval, Cadaval 035583
Museu Municipal do Crato, Crato 035645
Museu Municipal do Folclore, Penápolis 004623
Museu Municipal do Funchal – História Natural, Funchal 035709
Museu Municipal do Fundão, Fundão 035714
Museu Municipal do Sabugal, Sabugal 035984
Museu Municipal do Seixal, Seixal 036019
Museu Municipal Domingos Battistel, Nova Prata 004593
Museu Municipal Dr. António Gabriel Ferreira Lourenço, Benavente . 035557
Museu Municipal Dr. João Calado Rodrigues, Mação 035836
Museu Municipal Dr. José Formosinho, Lagos 035742
Museu Municipal Dr. Santos Rocha, Figueira da Foz 035696
Museu Municipal Elisabeth Aytai, Monte Mor 004573
Museu Municipal Embaixador Hélio A. Scarabôtolo, Marília 004561
Museu Municipal Francisco Manoel Franco, Itaúna 004500
Museu Municipal Francisco Veloso, Cunha 004424
Museu Municipal Guilleries, Sant Hilari Sacalm 040088
Museu Municipal Hipólito Cabaço, Alenquer 035495
Museu Municipal Joan Pla i Gras, Llinars del Vallés 039586
Museu Municipal Joaquim de Bastos Bandeira, Perdões 004625
Museu Municipal Josep Aragay, Breda 039158
Museu Municipal Karl Ramminger, Mondaí 004570
Museu Municipal Miquel Soldevila, Prats de Lluçanès 039970

Museu Municipal Montcada, Montcada
i Reixac 039776
Museu Municipal Napoleão Joele, São
Sebastião de Paraíso 004913
Museu Municipal Oswaldo Russomano,
Bragança Paulista 004355
Museu Municipal Padre Antônio
Ribeiro Pinto, Urucânia 004946
Museu Municipal Pedro Laragnoit,
Miracatu 004565
Museu Municipal Polinucleado de
Tomar, Tomar 036058
Museu Municipal Prof. Álvaro Viana de
Lemos, Lousã 035833
Museu Municipal Prof. Hugo Daros,
Gramado 004481
Museu Municipal Severo Portela,
Almodôvar 035512
Museu Municipal Sobral Mendes,
Penafiel 035909
Museu Municipal Vicenç Ros,
Martorell 039730
Museu Municipal Visconde de
Guarapuava, Guarapuava 004484
Museu Municipal Wenceslau Braz,
Itajubá 004493
Museu Municipal Wülson
Jehovah Lütz Farias, Frederico
Westphalen 004469
Museu Municipal, Església de Sant
Domènec, Tortosa 040270
Museu Nacional, Rio de Janeiro 004754
Museu Nacional, São Tomé . . 038107
Museu Nacional Arqueològic de
Tarragona, Tarragona 040209
Museu Nacional da Ciência e da
Técnica, Coimbra 035631
Museu Nacional da Imprensa, Jornais
e Artes Gráficas, Porto 035950
Museu Nacional da Literatura,
Porto 035951
Museu Nacional da Moeda,
Maputo 031807
Museu Nacional d'Art de Catalunya
(MNAC), Barcelona 039102
Museu Nacional de Agricultura,
Escola Superior Agrária de Coimbra,
Coimbra 035632
Museu Nacional de Antropologia,
Luanda 000112
Museu Nacional de Arqueologia,
Benguela 000106
Museu Nacional de Arqueologia,
Lisboa 035806
Museu Nacional de Arte, Maputo 031808
Museu Nacional de Arte Antiga,
Lisboa 035807
Museu Nacional de Belas Artes, Rio
de Janeiro 004755
Museu Nacional de Etnologia,
Lisboa 035808
Museu Nacional de Geologia,
Maputo 031809
Museu Nacional de História Natural,
Luanda 000113
Museu Nacional de História Natural e
Museu de Ciência, Lisboa 035809
Museu Nacional de Imigração e
Colonização, Joinville 004516
Museu Nacional de l'Automòbil,
Encamp 000090
Museu Nacional de Machado de
Castro, Coimbra 035633
Museu Nacional de Soares dos Reis,
Porto 035952
Museu Nacional do Azulejo,
Lisboa 035810
Museu Nacional do Desporto,
Lisboa 035811
Museu Nacional do Mar –
Embarcações Brasileiras, São
Francisco do Sul 004829
Museu Nacional do Teatro,
Lisboa 035812

Museu Nacional do Traje, Lisboa 035813
Museu Nacional do Vinho,
Alcobaça 035486
Museu Nacional dos Coches,
Lisboa 035814
Museu Nacional Ferroviário,
Entroncamento 035649
Museu Naval de Amazônia,
Belém 004338
Museu Naval e Oceanográfico, Rio de
Janeiro 004756
Museu Navio Comandante Bauru, Rio
de Janeiro 004757
Museu Nikolaï Siadristy, Ordino . 000103
Museu Nogueira da Silva, Braga 035572
Museu Nossa Senhora Aparecida,
Aparecida 004308
Museu Nossa Senhora d'Aires, Viana
do Alentejo 036074
Museu Numismático e Filatélico, Rio
de Janeiro 004758
Museu Numismático Português,
Lisboa 035815
Museu O Mundo Ovo de Eli Heil,
Florianópolis 004457
Museu obert de geologia 'Roques al
Carrer', Andorra la Vella 000086
Museu Oceanográfico do Valo do
Itajaí, Universidade do Valo do Itajaí,
Itajaí 004492
Museu Oceanográfico Prof. Eliézer de
C. Rios, Rio Grande 004774
Museu Oceanográfico Prof. Luiz
Saldanha, Azeitão 035544
Museu Oceanográfico, Instituto
da Conservação da Natureza,
Setúbal 036030
Museu Oswaldo Aranha, Alegrete 004303
Museu Padre Joseph Cornellius Maria
Demam, Coronel Fabriciano . . 004417
Museu Padre Júlio Maria, Russas 004781
Museu Palácio Nacional da Ajuda,
Lisboa 035816
Museu Paleontológic, Museo de
Ciencias Naturales, Valencia . . 040327
Museu Paleontologica Juna Cano
Forner, San Mateo 040069
Museu Paraense Emílio Goeldi,
Belém 004339
Museu Paranaense, Curitiba . . 004441
Museu Parc Cultural de la Muntanya
de Sal, Cardona 039231
Museu Paroquial, Moncaparracho 035860
Museu Paroquial, Queiriga . . . 035970
Museu Paroquial de Óbidos,
Óbidos 035876
Museu Paroquial de Oliveira de
Azeméis, Oliveira de Azeméis . 035882
Museu Paroquial de São Sebastião,
Guimarães 035732
Museu Paroquial de Vila Praia de
Âncora, Vila Praia de Âncora . . 036102
Museu Parroquial, Castellò
d'Empúries 039248
Museu Parroquial, Santa Pau . 040108
Museu Parroquial de Sant Julià,
Campos 039215
Museu Parroquial de Sant Llorenç
des Cardassar, Sant Llorenç des
Cardassar 040090
Museu Parroquial de Santa Joan
Baptista, Deià 039344
Museu Parroquial de Santa Maria de
Sineu, Sineu 040186
Museu Parroquial de Santa Maria la
Major d'Inca, Inca 039512
Museu Parroquial dels Sants Joans,
Estivella 039392
Museu Parroquial, Parròquia Sant
Martí, Altafulla 038967
Museu Particular de Armas de Arlindo
Pedro Zatti, Porto Alegre 004652

Museu Particular del Pastor Joaquim
Gassó, Fornells de la Muntanya 039408
Museu Patríco Corrêa da Câmara,
Universidade da Região da
Campanha, Bagé 004326
Museu Pau Casals, El Vendrell . 040368
Museu Paulista da Universidade de
São Paulo, São Paulo 004907
Museu Paulo Firpo, Dom Pedrito 004446
Museu Paulo Machado de Carvalho,
São Paulo 004908
Museu Pedro Cruz, Pedrógão
Grande 035905
Museu Pedro Ludovico, Goiânia 004745
Museu Picasso, Barcelona . . . 039103
Museu Pinacoteca, Igarassu . . 004488
Museu Pio XII, Braga 035573
Museu-Poblat Ibèric, Tornabous 040254
Museu Popular Beatriz Costa,
Malveira 035845
Museu Postal, Ordino 000104
Museu Postal e Telegráfico da ECT,
Brasília 004368
Museu Professor Hugo Machiado da
Silveira, Patrocínio 004614
Museu Rafael Bordalo Pinheiro,
Lisboa 035817
Museu Regional, Vitória da
Conquista 004958
Museu Regional Abade de Baçal,
Bragança 035578
Museu Regional Carlos Machado,
Ponta Delgada 035920
Museu Regional D. Bosco, Campo
Grande 004393
Museu Regional da Fauna e Flora do
Itatiaia, Itatiaia 004499
Museu Regional da Huila,
Lumbango 000114
Museu Regional d'Artá, Artá . . 039000
Museu Regional de Arqueologia D.
Diogo de Sousa, Braga 035574
Museu Regional de Arqueologia e
Etnografia de Arganil, Arganil . 035535
Museu Regional de Cabinda,
Cabinda 000107
Museu Regional de Etnografia do
Lobito, Lobito 000109
Museu Regional de Leiria, Leiria 035751
Museu Regional de Olinda,
Olinda 004597
Museu Regional de Oliveira de
Azeméis, Oliveira de Azeméis . 035883
Museu Regional de Paredes de Coura,
Paredes de Coura 035901
Museu Regional de São João Del Rei,
São João Del Rei 004836
Museu Regional de Sintra, Sintra 036041
Museu Regional do Alto Uruguai,
Erechim 004447
Museu Regional do Huambo,
Huambo 000108
Museu Regional do Sul de Minas,
Campanha 004381
Museu Regional Dr. Alípio de Araújo
Silva, Rio Preto 004777
Museu Regional Manuel Nobre,
Azambujeira 035543
Museu Regional Olívio Otto,
Carazinho 004403
Museu Renato Lobo Garcia,
Colares 035636
Museu Republicano Convenção de Itu,
Universidade de São Paulo, Itu 004503
Museu Restauraçao, Lisboa . . 035818
Museu Rocaguinarda de Terrissa dels
Països Catalans, Oristà 039849
Museu Rodrigues Lapa, Anadia . 035526
Museu Romàntic Can Llopis,
Sitges 040189
Museu Romantic Can Papiol, Vilanova
i la Geltrú 040391
Museu Romàntic Delger, Caldes de
Montbui 039202

Museu Romântico da Quinta da
Macieirinha, Porto 035953
Museu Rondon, Universidade Federal
de Mato Grosso, Cuiabá 004423
Museu Rural, Alcáçovas 035481
Museu Rural – Centro de Arte e
Cultura, Reguengo Grande . . . 035975
Museu Rural da Casa do Povo de
Santa Maria, Estremoz 035670
Museu Rural de Beringel,
Beringel 035558
Museu Rural de Boticas, Boticas 035567
Museu Rural de Vila Franca do Campo,
Vila Franca do Campo 036092
Museu Rural de Villa-Lobos,
Borba 035564
Museu Rural do Grupo dos Amigos de
Estremoz, Estremoz 035671
Museu Rural e do Vinho do Concelho,
Cartaxo 035602
Museu Rural e Etnográfico, São João
da Ribeira 036012
Museu S. Pedro da Palhaça,
Palhaça 035895
Museu Sacro da Igreja de São
Sebastião, Araxá 004319
Museu Sacro São José de Ribamar,
Aquiraz 004310
Museu Sala de Armas, Almeida 035508
Museu Salles Cunha, Rio de
Janeiro 004759
Museu Salvador, Instituto Botánico de
Barcelona, Barcelona 039104
Museu Santa Casa da Misericórdia,
Coimbra 035634
Museu Santa Maria Rosa Molas,
Jesús 039533
Museu Santos Barosa da Fábrica do
Vidro, Marinha Grande 035848
Museu São João Batista,
Caratinga 004402
Museu São Norberto, Pirapora do Bom
Jesus 004632
Museu Sargento Edésio de Carvalho,
Sousa 004919
Museu Sebastião da Gama, Vila
Nogueira de Azeitão 036093
Museu Sítio Arqueológico Casa dos
Pilões, Rio de Janeiro 004760
Museu Sobre a Pesca do Atum,
Tavira 036052
Museu Solar Monjardim, Vitória 004957
Museu Sub-Tenente António
Bernardino Apolónio Piteira,
Arraiolos 035537
Museu Taurí de la Monumental,
Barcelona 039105
Museu Taurino, Alcochete 035491
Museu Teatro, Rio de Janeiro . 004761
Museu Técnico-Científico do
Instituto Oscar Freire, Faculdade de
Medicina / FM, São Paulo 004909
Museu Temático do Vinho,
Almeirim 035511
Museu Tempostal, Salvador . . 004804
Museu Teresa Bernardes Adami
de Carvalho, Santo Antônio do
Monte 004818
Museu Territorial do Amapá,
Macapá 004539
Museu Textil i d'Indumentària, Palau
del Marques de Llio, Barcelona 039106
Museu Théo Brandão de Antropologia
de Folclore, Maceió 004547
Museu Theodormiro Carneiro Santiago,
Itajubá 004494
Museu Tinguí-Cuera, Araucária . 004316
Museu Torre Balldovina, Santa Coloma
de Gramenet 040095
Museu-Tresor Parroquial, Esglesia de
Sant Esteve, Olot 039832
Museu Universitário, Campinas . 004389
Museu Universitário de Arte,
Uberlândia 004944

Museu Universitário Gama Filho, Rio
de Janeiro 004762
Museu Universitário Professor
Oswaldo Rodrigues Cabral,
Florianópolis 004458
Museu Valencia del Joguet, Ibi . 039506
Museu Valencià d'Etnologia,
Valencia 040328
Museu Valencià d'Història Natural,
Valencia 040329
Museu Víctor Balaguer, Vilanova i la
Geltrú 040392
Museu Victor Bersani, Santa
Maria 004810
Museu Victor Meirelles,
Florianópolis 004459
Museu Viladomat d'Escultura,
Escaldes-Engordany 000094
Museu Villa-Lobos, Rio de
Janeiro 004763
Museu Vivo da Comutação Manual,
Vilar 036112
Museu Vivo da Memória Candanga,
Brasília 004369
Museu Vivo de Vilar do Pinheiro, Vilar
do Pinheiro 036113
Museu von Martius, Teresópolis 004932
Museu Vulcano-Espeleológico
Machado Fagundes, Angra do
Heroísmo 035531
Museu Zé Didor, Campo Maior . 004394
Museu Zoobotânico Augusto Ruschi,
Passo Fundo 004613
Museu Zoológico, Coimbra 035635
Museu Zoológico, Pombal 035917
O Museu, Faculdade de Belas Artes
da Universidade do Porto, Porto 035934
Museum, Bagheria 025433
Museum, Coswig 017763
The Museum, Dunwich 044185
The Museum, Greenwood 049276
Museum, Großpösna 018858
The Museum, Newton Stewart . 045373
Museum, Zofingen 042185
Museum – Begegnung der Kulturen,
Krenglbach 002242
Museum – Herberge de Ar,
Westerbork 033038
Museum – Treasures from the
National Archives, Kew 044688
Museum "Kaffeemühle", Wolgast 022752
Museum "Währungsreform 1948"
und Konklave in Rothwesten,
Fuldatal 018565
Museum "Werkstatt Plaumer",
Öhringen 020954
Museum 1806, Institut zur
militärgeschichtlichen Forschung
Jena 1806 e.V., Jena 019451
Museum 1915–1918, Kötschach-
Mauthen 002222
Museum 1940–1945, Dordrecht 032248
Museum 3. Dimension,
Dinkelsbühl 017905
Museum '40-'45, Slochteren . . 032885
Museum Aargau, Lenzburg 041666
Museum Abdij Mariënlof, Kerniel 003749
Museum Abdij van Egmond, Egmond
Binnen 032279
Museum Abdijkerk, Thorn 032921
Museum Abodiacum, Denklingen 017857
Museum Abri Satriamandala,
Jakarta 024566
Museum Abtei Liesborn, Museum des
Kreises Warendorf, Wadersloh . 022362
Museum Abteiberg,
Mönchengladbach 020439
Museum Abtsküche,
Heiligenhaus 019168
Museum Adam Malik, Jakarta . 024567
Museum Adlerturm, Museum für
Kunst und Kulturgeschichte,
Dortmund 017960

Museum Adorf, Adorf 016471
Museum Affandi – Museum Seni,
Yogyakarta 024672
Museum Africa, Johannesburg . 038754
Muséum Agricole et Industriel de
Stella Matutina, Piton-Saint-Leu 036183
Museum Aguntum, Verein Curatorium
pro Agunto, Dölsach 001822
Museum Albert van Dyck,
Schilde 004004
Museum Ald Slot, Wergea 033035
Museum Alex Mylona, Athinai . 023005
Museum Alexandrowka, Potsdam 021190
Museum Alin, Pontresina 041842
Museum Alte Brettersäge,
Rappottenstein 002571
Museum Alte Kirche, Reken . . . 021356
Museum Alte Lateinschule Großenhain,
Großenhain 018847
Museum Alte Münze, Stolberg (Harz)
Kurort 022003
Museum Alte Pfefferküchlerei,
Weißenberg 022534
Museum Alte Post, Pirmasens . . 021135
Museum Alte Textilfabrik, Weitra 002909
Museum alte Ziegelei Westerholt,
Wardenburg 022428
Museum Altes Bad Pfäfers, Bad
Ragaz 041246
Museum Altes Bürgerhaus, Stolberg
(Harz) Kurort 022004
Museum Altes Gymnasium, Städtische
Sammlungen Schweinfurt,
Schweinfurt 021782
Museum Altes Land, Jork 019470
Museum Altes Rathaus,
Leingarten 019967
Museum Altes Rathaus, Villingen-
Schwenningen 022339
Museum Altmünster,
Altmünster 016562
Museum Alzey, Alzey 016568
Museum am Burghof, Lörrach . . 020098
Museum am Dom, Würzburg . . . 022795
Museum am Lindenbühl, Mühlhäuser
Museen, Mühlhausen/Thüringen 020485
Museum am Lindenplatz, Weil am
Rhein 022469
Museum am Markt, Schiltach . . 021654
Museum am Maxplatz, Rehau . . 021337
Museum am Meer, Büsum 017676
Museum am Modersohn-Haus,
Worpswede 022776
Museum am Odenwaldlimes,
Elztal 018173
Museum am Ostwall, Dortmund 017961
Museum am Schölerberg, Natur und
Umwelt-Planetarium, Osnabrück 021017
Museum am Thie, Göttingen . . . 018751
Museum am Widumshof, Urbach 022284
Museum Amandinhuis, Herk-de-
Stad 003696
Museum Amden, Amden 041216
Museum Amelinghausen
mit Zinnfigurenausstellung,
Museumsverein für das Fürstentum
Lüneburg, Amelinghausen . . . 016572
Museum Amtspforte Stadthagen,
Stadthagen 021953
Museum an der Krukenburg, Bad
Karlshafen 016796
Museum Anatomicum, Marburg 020275
Museum Anatomy FK UI, Jakarta 024568
Museum and Archives of Georgia
Education, Georgia College and State
University, Milledgeville 050896
Museum and Art Gallery,
Burdwan 024184
Museum and Art Gallery, Forfar 044359
Museum and Art Gallery,
Hooghly 024293
Museum and Art Gallery, Newport,
Gwent 045361
Museum and Art Gallery,

Nuneaton 045426
Museum and Art Gallery of the
Northern Territory, Darwin . . . 001021
Museum and Art Gallery, Universiti
Teknologi Mara, Shah Alam . . 030728
Museum and Arts Center in
the Sequim Dungeness Valley,
Sequim 053272
Museum and Exhibition Design Group,
London 060182
Museum and Gallery of Art and
Archaeology, Karnataka University,
Dharwad 024247
Museum and Study Collection, Central
Saint Martins College of Art and
Design, London 045040
Museum Anna Nordlander,
Skellefteå 040954
Museum Anna Paulowna,
Breezand 032106
Museum-Appartment of Abdulla Shaig,
Bakı 003170
Museum-Appartment of Niyazi,
Bakı 003171
Museum Appeleshof, Gechingen 018599
Museum Appenzell, Appenzell . . 041224
Muséum-Aquarium de Nancy,
Nancy 014408
Museum Arbeit am Rhein, Eichesheim-
Illingen 018153
Museum Arbeitswelt Steyr, Steyr 002783
Museum Arnold Vander Haeghen,
Gent 003630
Museum Arnoldstein, Arnoldstein 001721
Museum-Art Collection of the National
Bank, Athinai 023006
Museum Artha Suaka, Jakarta . 024569
Museum Aschenbrenner, Stiftung
Aschenbrenner, Garmisch-
Partenkirchen 018592
Museum Asia Africa, Bandung . 024530
Museum Asimbugo, Bima 024543
Museum Asmat, Jakarta 024570
Museum Aspern-Essling 1809,
Wien 003038
Museum Asserajah El Hasyimiah
Palace, Bengkalis 024540
Museum Association of Arizona,
Phoenix 060602
Museum Association of Israel, Rishon
Le-Zion 059388
Museum Association of Montana,
West Glacier 060603
Museum Association of New York,
Troy 060604
Museum Association of Newfoundland
and Labrador, Saint John's . . . 058453
The Museum at Drexel University,
Philadelphia 052069
Museum at Mountain Home, Johnson
City 049826
Museum at Prophetstown, Battle
Ground 046827
Museum at Southwestern Michigan
College, Dowagiac 048360
Museum at the Fashion Institute of
Technology, New York 051433
The Museum at Warm Springs, Warm
Springs 054058
Museum Atelier Franz Zeh,
Heidenreichstein 002066
Museum Auerbach, Auerbach,
Vogtland 016643
Museum auf dem Burghof,
Springe 021943
Museum auf dem Sachsenberg,
Schwerin 021794
Museum auf der Burg, Raron . . 041863
Museum auf der Osterburg,
Weida 022458
Museum August Kestner,
Hannover 019060
Museum Autovision, Altlußheim . 016550
Museum Bad Abbach, Bad

Abbach 016698
Museum Bad Arolsen, Bad
Arolsen 016704
Museum Bad Orb, Bad Orb 016841
Museum Baden, Solingen 021898
Museum Bärengasse, Zürich . . . 042214
Museum Bahari, Jakarta 024571
Museum Bahnhof Bexbach,
Bexbach 017354
Museum Bakkerij Mendels,
Middelstum 032652
Museum Balanga, Palangkaraya 024624
Museum Bangga & Lore, Poso . 024637
Museum Bargteheide,
Bargteheide 016965
Museum Barockschloß Delitzsch,
Delitzsch 017846
Museum Barockschloß Rammenau,
Rammenau 021272
Museum Batara Guru, Palopo . . 024628
Museum Batara Guru, Pare-Pare 024630
Museum Batik, Pekalongan 024631
Museum Batik, Yogyakarta 024673
Museum Bautzen, Bautzen 016977
Museum Bayerische Metallwarenfabrik
(BMF), Nürnberg 020867
Museum Bayerisches Vogtland,
Hof 019287
Museum Bebinghehoes, Exloo . . 032324
Museum Beeckestijn, Velsen
Zuid 032981
Museum Beelden aan Zee, Den
Haag 032182
Museum bei der Kaiserpfalz,
Ingelheim am Rhein 019403
Museum beim Markt, Badisches
Landesmuseum Karlsruhe,
Karlsruhe 019512
Museum beim Solenhofer Aktien-
Verein, Gunzenhausen 018895
Museum Bellerive, Museum für
Gestaltung Zürich, Zürich 042215
Museum Bergér, Eichstätt 018121
Museum Berggruen – Picasso und
seine Zeit, Staatliche Museen
zu Berlin – Stiftung Preußischer
Kulturbesitz, Berlin 017230
Museum Bergsdorf, Zehdenick . 022845
Museum Berlin-Karlshorst, Berlin 017231
Museum Bernhard Brühl,
Landsberg 019876
Museum Bevrijdende Vleugels,
Best 032069
Museum Biologi UGM,
Yogyakarta 024674
Museum Bischofsheim,
Bischofsheim 017397
Museum BL, Liestal 041678
Museum Blaafarveværket, Åmot i
Modum 033493
Museum Blauer Anger, Bergen, Kreis
Traunstein 017048
Museum Blindenwerkstatt Otto Weidt,
Stiftung Gedenkstätte Deutscher
Widerstand, Berlin 017232
Museum Bockwindmühle,
Falkenhain 018334
Museum Bodmann, Bodman-
Ludwigshafen 017449
Museum Boerhaave, Leiden . . . 032581
Museum Bookshop, London . . . 144653
Museum Bramberg "Wilhelmgut",
Bramberg 001791
Museum Brandhorst, Bayerische
Staatsgemäldesammlungen –
Pinakothek der Moderne,
München 020553
Museum Brawijaya, Malang . . . 024610
Museum Bredius, Den Haag . . . 032183
Museum Breitenfurt, Breitenfurt 001798
Museum Brigade en Garde Prinses
Irene, Oirschot 032708
Museum Briner und Kern,
Winterthur 042154

Museum Broeker Veiling, Broek op
Langedijk 032110
Museum Bruder Klaus, Sachseln 041902
Museum Bürgstadt, Bürgstadt . . 017675
Museum Bundo Kandung,
Bukittinggi 024551
Museum Buren en Oranje, Buren,
Gelderland 032123
Museum Burg Abenberg, Haus
fränkischer Geschichte
Klöppelmuseum, Abenberg . . . 016455
Museum Burg Bederkesa, Bad
Bederkesa 016709
Museum Burg Brome, Brome . . 017628
Museum Burg Creuzburg,
Creuzburg 017782
Museum Burg Eisenhardt, Belzig 017022
Museum Burg Falkenstein, Stiftung
Dome und Schlösser in Sachsen-
Anhalt, Falkenstein, Harz 018336
Museum Burg Frankenberg,
Aachen 016442
Museum Burg Golling, Golling an der
Salzach 001962
Museum Burg Heidenreichstein,
Heidenreichstein 002067
Museum Burg Hohnstein,
Hohnstein 019316
Museum Burg Hornberg,
Neckarzimmern 020657
Museum Burg Kriebstein,
Kriebstein 019795
Museum Burg Linn, Krefeld . . . 019787
Museum Burg Mylau, Mylau . . . 020630
Museum Burg Posterstein,
Posterstein 021173
Museum Burg Pottenstein,
Pottenstein 021209
Museum Burg Querfurt, Querfurt 021252
Museum Burg Ramsdorf, Velen . 022312
Museum Burg Ranis, Ranis 021276
Museum Burg Ronneburg, Ronneburg,
Hessen 021468
Museum Burg Scharfenstein,
Scharfenstein 021631
Museum Burg Stein, Hartenstein 019079
Museum Burg Stickhausen,
Detern 017873
Museum Burg Wendelstein,
Vacha 022293
Museum Burghalde, Lenzburg . . 041667
Museum Buurt Spoorweg,
Haaksbergen 032388
Museum by the Bay, Lewisporte 006193
Museum Caledon River,
Smithfield 038831
Museum Cards, Amersfoort 137823
Museum de Casteelse Poort,
Wageningen 033018
Museum Castellum, Mainz-Kastel 020235
Museum Catharijneconvent,
Utrecht 032949
Museum Center, Maysville 050745
Museum Ceuclum, Cuijk 032139
Museum Charles Delporte,
Damme 003531
Museum Charlotte Zander,
Bönnigheim 017460
Museum Chasa Jaura, Valchava 042093
Museum Cheann A'Loch, Laxay 044768
Museum Chornhuus, Gränichen . 041559
Museum Chris Van Niekerk,
Boshof 038666
Museum Collection and Old Pharmacy
of the Friars Minor, Dubrovnik . 008905
Museum Computer Network, National
Office, Laguna Beach 060605
Museum Connection Die, Cape
Town 038682
Museum Constant Permeke,
Jabbeke 003737
Museum Council, Moskva 059774
Museum Daerah Bangkalan,
Bangkalan 024535

Museum Daerah Blambangan,
Banyuwangi 024539
Museum Daerah Sumenep,
Sumenep 024652
Museum Dara Yuanti, Sintang . . 024649
Museum de 5000 Morgen,
Hoogeveen 032502
Museum De Bres, Ruisbroek,
Antwerpen 003987
Museum de Burghse Schoole, Burgh-
Haamstede 032125
Museum De Cateraars,
Groningen 032381
Museum De Cruquius, Cruquius 032137
Museum De Doornboom,
Hilvarenbeek 032482
Museum De Eendracht, Anjum . 032006
Museum de Géologie Provençale, La
Roque-d'Anthéron 013487
Museum de Gevangenpoort, Den
Haag 032184
Museum De Ghulden Roos,
Roosendaal 032797
Museum De Goudse Glazen – Sint
Janskerk, Gouda 032364
Museum De Grutterswinkel,
Leeuwarden 032572
Museum De Helpoort, Maastricht 032622
Museum de Historische Drukkerij,
Maastricht 032623
Museum De Kijkuit, Merksem . . 003867
Museum De Kluis, Boekel 032076
Museum De Koloniehof,
Frederiksoord 032331
Museum de Koperen Knop,
Hardinxveld-Giessendam 032411
Museum de Maurits 1940–1945,
Doesburg 032230
Museum De Moriaan, Gouda . . 032365
Museum De Paviljoens, Almere . 031901
Museum De Ronde Venen,
Vinkeveen 032988
Museum De Roode Tooren, Museum
voor Stad en Ambt Doesburg,
Doesburg 032231
Museum De Schilpen, Maasland 032618
Museum de Speeltoren,
Monnickendam 032656
Museum de Stratemakerstoren,
Nijmegen 032686
Museum de Striid tsjin it Wetter,
Wommels 033062
Museum de Timmerwerf, De Lier 032146
Museum de Trije Grietenijen, Grou 032387
Museum de Valkhof, Hellendoorn 032453
Museum de Verzamelaar,
Zuidhorn 033111
Museum De Vier Quartieren, Museum
van het Brabantse Volksleven
1800–1950, Oirschot 032709
Museum De Weem,
Westeremden 033040
Museum De Wieger, Deurne . . . 032214
Muséum de Zoologie et
d'Anthropologie de l'ULB,
Bruxelles 003493
Museum de Zwarte Tulp, Lisse . 032600
Museum Delfzijl, Delfzijl 032165
Museum der 50er Jahre, Bad
Bevensen 016731
Museum der 50er und 60er Jahre,
Vaihingen an der Enz 022296
Museum der Alltagskultur – Schloss
Waldenbuch, Außenstelle des
Landesmuseums Württemberg,
Waldenbuch 022385
Museum der Anthropologie, Universität
Zürich-Irchel, Zürich 042216
Museum der Arbeit, Hamburg . . 019006
Museum der Arbeit, Hann.
Münden 019047
Museum der Barbarossastadt
Gelnhausen, Gelnhausen 018623
Museum der Belgischen Streitkräfte in

Deutschland, Museum van de BSD,
Soest 021892
Museum der Berliner Bürgerbräu
Brauerei, Berlin 017233
Museum der bildenden Künste Leipzig,
Leipzig 020003
Museum der Brotkultur, Ulm . . . 022264
Museum der Cistercienserinnen-Abtei
Lichtenthal, Baden-Baden 016925
Museum der Deutschen
Binnenschifffahrt Duisburg-Ruhrort,
Duisburg 018059
Museum der Deutschen
Spielzeugindustrie, Neustadt bei
Coburg 020763
Museum der deutschen
Sprachinselorte bei Brünn, Erbach,
Alb-Donau-Kreis 018226
Museum der Donauschwaben,
Herbrechtingen 019199
Museum der Freiwilligen Feuerwehr,
Heiligenhaus 019169
Museum der Freiwilligen Feuerwehr,
Laxenburg 002280
Museum der Göttinger Chemie,
Göttingen 018752
Museum der Grafschaft Mark,
Altena 016525
Museum der Grafschaft Rantzau,
Barmstedt 016966
Museum der Granitindustrie,
Haselbachtal 019083
Museum der Heimatvertriebenen,
Vöcklabruck 002868
Museum der Heimatvertriebenen,
Stadtmuseum Wels-Burg, Wels 002918
Museum der heimischen Tierwelt,
Nabburg 020634
Museum der historischen
Waschtechnik, Ostbevern 021021
Museum der Katakombenstichting,
Valkenburg, Limburg 032964
Museum der Klosterruine mit
Lapidarium, Rüeggisberg 041896
Museum der Kommunikations- und
Bürogeschichte, Bamberg 016951
Museum der Kulturen Basel,
Basel 041276
Museum der Landschaft, Saanen 041900
Museum der Landschaft Hasli,
Meiringen 041734
Museum der Landsmannschaft
Ostpreußen, Gelsenkirchen . . . 018630
Museum der Landtechnik, Land- und
Forstwirtschaftliche Fachschule,
Knittelfeld 002214
Museum der Marktgemeinde Sankt
Johann in Tirol, Sankt Johann in
Tirol 002669
Museum der Mechitharisten-
Congregation, Wien 003039
Museum der Moderne Salzburg
Mönchsberg, Salzburg 002630
Museum der Moderne Salzburg
Rupertinum, Salzburg 002631
Museum der Natur und Umwelt
Cottbus, Cottbus 017767
Museum der Natur, Stiftung Schloss
Friedenstein Gotha, Gotha 018775
Museum der Oberzent,
Beerfelden 017014
Museum der Photographie,
Fischamend 001892
Museum der Porzellan-Manufaktur
Meissen, Meißen 020347
Museum der Puppengeschichte –
Joyce Merlet, Köln 019699
Museum der Schildbürger, Schildau,
Gneisenaustadt 021648
Museum der Schwalm,
Schwalmstadt 021761
Museum der Schweizer
Kapuzinerprovinz, Sursee 042044
Museum der schweizerischen

Luftwaffe, Dübendorf 041448
Museum der Seefahrt,
Geiselhöring 018608
Museum der Staatlichen Münze Berlin,
Berlin 017234
Museum der Stadt Alfeld (Leine),
Alfeld 016505
Museum der Stadt Aue, Aue . . . 016640
Museum der Stadt Bad Berleburg, Bad
Berleburg 016726
Museum der Stadt Bad Gandersheim,
Bad Gandersheim 016774
Museum der Stadt Bad Ischl, Bad
Ischl 001751
Museum der Stadt Bad Neuenahr-
Ahrweiler, Bad Neuenahr-
Ahrweiler 016834
Museum der Stadt Bad Staffelstein,
Bad Staffelstein 016883
Museum der Stadt Bensheim,
Bensheim 017034
Museum der Stadt Bludenz,
Bludenz 001788
Museum der Stadt Boppard,
Boppard 017510
Museum der Stadt Borna, Borna 017520
Museum der Stadt Butzbach
im Solms-Braunfelser Hof,
Butzbach 017707
Museum der Stadt Calw, Calw . 017716
Museum der Stadt Dommitzsch,
Dommitzsch 017926
Museum der Stadt Eberbach,
Eberbach 018070
Museum der Stadt Ehingen,
Ehingen 018106
Museum der Stadt Eschborn,
Eschborn 018274
Museum der Stadt Ettlingen,
Ettlingen 018324
Museum der Stadt Füssen,
Füssen 018552
Museum der Stadt Gladbeck,
Stadtgeschichte und Computerkunst,
Gladbeck 018703
Museum der Stadt Gmünd, Gmünd,
Kärnten 001942
Museum der Stadt Gronau (Leine),
Gronau, Leine 018830
Museum der Stadt Güstrow,
Güstrow 018885
Museum der Stadt Hagenow,
Hagenow 018921
Museum der Stadt Kapfenberg,
Kapfenberg 002156
Museum der Stadt Korneuburg,
Korneuburg 002225
Museum der Stadt Kraichtal,
Kraichtal 019777
Museum der Stadt Lahnstein im
Hexenturm, Lahnstein 019849
Museum der Stadt Lahr, Villa Jamm
im Stadtpark, Lahr 019855
Museum der Stadt Langen, Langen,
Hessen 019892
Museum der Stadt Lauffen am Neckar,
Lauffen am Neckar 019936
Museum der Stadt Lennestadt,
Lennestadt 020038
Museum der Stadt Lienz, Lienz . 002307
Museum der Stadt Ludwigsfelde,
Ludwigsfelde 020132
Museum der Stadt Lünen,
Gesindehaus Schloß Schwansbell,
Lünen 020185
Museum der Stadt Miltenberg und
Porzellansammlung Kreis Dux,
Miltenberg 020408
Museum der Stadt Mittweida "Alte
Pfarrhäuser", Mittweida 020427
Museum der Stadt Nauen, Nauen 020642
Museum der Stadt Neu-Isenburg Haus
zum Löwen, Neu-Isenburg . . . 020673
Museum der Stadt Neustadt an

Register der Institutionen und Firmen

Museum der Stadt Neustadt in Holstein, Neustadt in Holstein
– Museum für Film und Fernsehen, Stiftung Deutsche Kinemathek,

der Weinstraße, Neustadt an der
Weinstraße 020760
Museum der Stadt Neustadt in
Holstein, Neustadt in Holstein . 020769
Museum der Stadt Neustrelitz,
Neustrelitz 020773
Museum der Stadt Neutraubling,
Ortsgeschichtliche Dokumentation,
Neutraubling 020774
Museum der Stadt Parchim,
Parchim 021075
Museum der Stadt Pasewalk und
Künstlergedenkstätte Paul Holz,
Pasewalk 021079
Museum der Stadt Pegau, Pegau 021091
Museum der Stadt Ratingen,
Ratingen 021288
Museum der Stadt Schkeuditz,
Schkeuditz 021657
Museum der Stadt Schopfheim,
Schopfheim 021722
Museum der Stadt Schwartau, Bad
Schwartau 016875
Museum der Stadt Schwaz auf Burg
Freundsberg, Schwaz . . . 002741
Museum der Stadt Steyr, Steyr . 002784
Museum der Stadt Troisdorf,
Bilderbuchmuseum, Troisdorf . 022208
Museum der Stadt Villach,
Villach 002862
Museum der Stadt Vils, Vils . . 002865
Museum der Stadt Waiblingen,
Waiblingen 022376
Museum der Stadt Weinheim,
Weinheim 022520
Museum der Stadt Worms,
Worms 022764
Museum der Stadt Zerbst mit
Sammlung Katharina II., Zerbst 022864
Museum der Stadtpfarrkirche
St. Xaver – museum sacrum,
Leoben 002292
Museum der Stadtwerke,
Pirmasens 021136
Museum der Trauerarbeitsplätze,
Institut für Philosophie und
Gruppendynamik, Klagenfurt . 002195
Museum der unerhörten Dinge,
Berlin 017235
Museum der Votivkirche, Wien . 003040
Museum der Weltkulturen mit Galerie
37, Frankfurt am Main 018434
Museum der Westlausitz Kamenz,
Kamenz 019489
Museum der Winterthur-
Versicherungen, Winterthur . . 042155
Museum Dermbach, Dermbach . 017861
Museum des Augustiner
Chorherrenstifts, Reichersberg 002582
Museum des Blindenwesens,
Wien 003041
Museum des Chorherrenstifts
Klosterneuburg, Klosterneuburg 002209
Museum des ehemaligen Kreises
Stuhm/Westpr., Bremervörde . 017617
Museum des Geschichtsvereins
Keltisches Noreia, Mühlen . . . 002417
Museum des Haager Landes, Haag in
Oberbayern 018903
Museum des Handwerks, Bad
Bederkesa 016710
Museum des Heimatkundlichen
Vereins, Scheßlitz 021643
Museum des Historischen Vereins
Freising, Freising 018501
Museum des Kreises Plön mit
norddeutscher Glassammlung,
Plön 021160
Museum des Landes Glarus,
Näfels 041788
Museum des Nötscher Kreises, Nötsch
im Gailtal 002458
Museum des Ohrekreises,
Wolmirstedt 022755

Muséum des Sciences Naturelles de
Belgique, Bruxelles 003494
Museum des Stiftes Admont,
Admont 001685
Museum des Teltow, Zossen . . . 022896
Museum des Todesmarsches im
Belower Wald, Wittstock/Dosse 022709
Museum des Vereins für Heimatkunde
im Land Birkenfeld, Birkenfeld 017394
Muséum des Volcans, Aurillac . . 011802
Museum des Waldes, Akademie der
Sächsischen Landesstiftung Natur
und Umwelt, Tharandt 022146
Museum des Wiener Männergesang-
Vereins, Wien 003042
Museum des Zisterzienserstifts
mit Margret Bilger-Galerie,
Schlierbach 002721
Museum Dewantara Kirti Griya,
Yogyakarta 024675
Museum Dharma Wiratama,
Yogyakarta 024676
Muséum d'Histoire Naturell,
Auxerre 011824
Muséum d'Histoire Naturelle,
Autun 011812
Muséum d'Histoire Naturelle,
Bagnères-de-Bigorre 011867
Muséum d'Histoire Naturelle,
Blois 012082
Muséum d'Histoire Naturelle,
Chartres 012472
Muséum d'Histoire Naturelle,
Genève 041539
Muséum d'Histoire Naturelle,
Grenoble 013186
Muséum d'Histoire Naturelle, Le
Havre 013647
Muséum d'Histoire Naturelle,
Lyon 013954
Muséum d'Histoire Naturelle,
Marseille 014062
Muséum d'Histoire Naturelle,
Neuchâtel 041798
Muséum d'Histoire Naturelle,
Nice 014491
Muséum d'Histoire Naturelle,
Perpignan 014844
Muséum d'Histoire Naturelle,
Rouen 015173
Muséum d'Histoire Naturelle, Saint-
Denis 036185
Muséum d'Histoire Naturelle,
Troyes 016055
Muséum d'Histoire Naturelle de
Bordeaux, Bordeaux 012120
Muséum d'Histoire Naturelle de la
Ville de Bayonne, Bayonne . . . 011926
Muséum d'Histoire Naturelle de
Toulouse, Toulouse 015999
Muséum d'Histoire Naturelle de Tours,
Tours 016024
Muséum d'Histoire Naturelle,
Fauniscope, Gras 013155
Muséum d'Histoire-Naturelle, Museum
Parc Emanuel Liais, Cherbourg-
Octeville 012552
Muséum d'Histoire Naturelle, Vivarium,
Dijon 012790
Museum Dhondt-Dhaenens,
Deurle 003543
Museum Dingolfing-Herzogsburg,
Dingolfing 017902
Museum Diponegoro, Magelang . 024607
Museum Dirgantara Mandala,
Yogyakarta 024677
Museum Donkmeer, Klim naar
Democratie, Berlare 003360
Museum Dorestad, Wijk bij
Duurstede 033047
Museum Dorpsbehoud Papendrecht,
Papendrecht 032763
Museum Dr. Guislain, Gent 003631
Museum du Coquillage, Les Sables-

d'Olonne 013766
Museum du Parc Tsimbazaza,
Antananarivo 030648
Museum Echt, Echt 032262
Museum Eckernförde,
Eckernförde 018090
Museum Education Roundtable,
Washington 060606
Museum ehemalige Klöppelschule
Tiefenbach, Tiefenbach,
Oberpfalz 022156
Museum Eisenhüttenwerk Peitz,
Peitz 021097
Museum Elbinsel Wilhelmsburg,
Hamburg 019007
Museum Elisabeth Weeshuis,
Culemborg 032141
Museum en oud Kempisch Dorp,
Heist-op-den-Berg 003689
Museum Engiadina Bassa, Scuol 041971
Museum Enschedé, Haarlem . . . 032395
Museum Ephraim-Palais, Stiftung
Stadtmuseum Berlin, Berlin . . . 017236
Museum Erding, Erding 018233
Museum Erzgebirgische Volkskunst
"Buntes Holz" im Postgut am
Altmarkt, Hohenstein-Ernstthal 019312
Museum Eschenz, Eschenz 041474
Museum Ethnographers Group,
Exeter 060183
Museum Europäischer Kulturen,
Staatliche Museen zu Berlin –
Stiftung Preußischer Kulturbesitz,
Berlin 017237
Museum Expressiver Realismus,
Kißlegg 019636
Museum Fahrzeug – Technik –
Luftfahrt, Lauffen 002274
Museum Fahrzeug-Technik-Luftfahrt,
Bad Ischl 001752
Museum Falak-ol-Aflak, Khorram-
Abaad 024712
Museum Falsters Minder, Nyköbing
Falster 010243
Museum Fatahillah, Museum Sejarah
Jakarta atau Museum Batavia,
Jakarta 024572
Museum Félix de Boeck,
Drogenbos 003560
Museum Festung Dresden-Kasematten,
Dresden 017999
Museum Feuerbachhaus Speyer,
Speyer 021933
Museum Fischerhaus, Öhningen 020951
Museum 't Fiskershúske,
Moddergat 032655
Museum Flederwisch, Furth im
Wald 018567
Museum Flehite Amersfoort,
Amersfoort 031912
Museum Fluxus+, Potsdam 021191
Museum Folkwang Essen, Essen 018306
Museum for African Art, Long Island
City 050381
Museum for Beduin Culture, Kibbutz
Lahav 025138
Museum for Local Culture, Beipu 042273
Museum for the Victims of Political
Repression, Ulaanbaatar 031735
Museum for Wood Culture,
Muraoka 029063
Museum Fort Kijkduin, Den
Helder 032208
Museum Fort Fectio, Puurs . . . 003955
Museum Forum Arenacum Rindern,
Kleve 019643
Museum der Völker,
Völkerkundemuseum der
Franziskaner, Werl 022575
Museum Frankomarchia,
Frankenmarkt 001901
museum franz gertsch, Burgdorf 041385
Museum Franz Schubert und sein
Freundeskreis, Atzenbrugg . . . 001731

Museum Freriks, Winterswijk . . 033056
Museum Frerikshuus, Aalten . . 031883
Museum Freudenthaler Glassammlung,
Weißenkirchen im Attergau . . . 002905
Museum-Freundschaften, Prora . 021229
Museum Frieder Burda, Baden-
Baden 016926
Museum Friedhof Ohlsdorf,
Hamburg 019008
Museum Friedrichshagener
Dichterkreis, Berlin 017238
Museum Friesenheim, Ludwigshafen
am Rhein 020135
Museum Fritz von Graevenitz,
Stuttgart 022051
Museum für Abgüsse Klassischer
Bildwerke, München 020554
Museum für Angewandte Kunst,
Frankfurt am Main 018435
Museum für Angewandte Kunst Gera,
Gera 018647
Museum für Angewandte Kunst,
Museen der Stadt Köln, Köln . . 019700
Museum für Antike Schiffahrt
des Römisch-Germanischen
Zentralmuseums, Mainz 020231
Museum für Archäologie,
Frauenfeld 041491
Museum für Archäologie und
Volkskunde, Kösching 019745
Museum für Archäologie Westfälisches
Außenstelle, Olpe 020996
Museum für Asiatische Kunst,
Staatliche Museen zu Berlin –
Stiftung Preußischer Kulturbesitz,
Berlin 017239
Museum für bäuerliche Arbeitsgeräte,
Imst 002114
Museum für bäuerliche Arbeitsgeräte
des Bezirks Oberfranken,
Bayreuth 016998
Museum für bäuerliche und sakrale
Kunst, Ruhpolding 021549
Museum für bäuerliches Handwerk
und Kultur, Wilhelmsdorf, Kreis
Ravensburg 022656
Museum für bergmännische
Volkskunst Schneeberg,
Schneeberg 021691
Museum für Beschirrung und
Besattelung, Hufbeschlag und
Veterinär-Orthopädie, Wien . . . 003043
Museum für bildende Kunst im
Landkreis Neu-Ulm, Nersingen 020665
Museum für Billard- und
Kaffeehauskultur, Wien 003044
Museum für Bodenschätzung,
Eickendorf 018125
Museum für brandenburgische
Kirchen- und Kulturgeschichte,
Bischofsresidenz Burg Ziesar,
Ziesar 022878
Museum für das Fürstentum Lüneburg,
Lüneburg 020182
Museum für Design der 60er & 70er
Jahre im PopDom, Köln 019701
Museum für deutsche
Fernsehgeschichte, Wiesbaden 022640
Museum für die Archäologie des
Eiszeitalters, Neuwied 020776
Museum für dörfliche Alltagskultur
Rubenheim, Gersheim 018668
Museum für Druck, Mosbach . . 020473
Museum für Druckgrafik,
Rankweil 002569
Museum für Druckkunst Leipzig,
Stiftung Werkstattmuseum für
Druckkunst Leipzig, Leipzig . . 020004
Museum für Eisenkunstguß,
Hirzenhain 019266
Museum für Energiegeschichte(n),
E.ON Avacon AG, Hannover . . . 019061
Museum für Film und Fernsehen,
Stiftung Deutsche Kinemathek,

Berlin 017240
Museum für Fotografie – Sammlung
Fotografie, Kunstbibliothek, Staatliche
Museen Berlin – Stiftung Preußischer
Kulturbesitz, Berlin 017241
Museum für Frühgeschichte
des Landes Niederösterreich,
Traismauer 002832
Museum für Frühislamische
Kunst, Bummiler Kunststiftung,
Bamberg 016952
Museum für Gegenwartskunst Siegen,
Siegen 021853
Museum für Gegenwartskunst,
Öffentliche Kunstsammlung Basel,
Basel 041277
Museum für Gerberei- und
Stadtgeschichte, Hirschberg, Saale-
Orla-Kreis 019264
Museum für Gestaltung Basel,
Basel 041278
Museum für Gestaltung Zürich,
Zürich 042217
Museum für Gestaltung Zürich –
Plakatraum, Zürich 042218
Museum für Gestaltung Zürich –
Sammlungen für Design und Grafik,
Zürich 042219
Museum für Glaskunst, Lauscha 019941
Museum für Hamburgische
Geschichte, Stiftung Historische
Museen Hamburg, Hamburg . . 019009
Museum für Haustierkunde Julius
Kühn, Halle, Saale 018966
Museum für Heimatgeschichte und
Nordpfalzgalerie, Alsenz 016518
Museum für historische Schreib- und
Rechenmaschinen, Schaan . . . 030301
Museum für historische Wehrtechnik,
Röthenbach an der Pegnitz . . . 021459
Museum für Höhlenkunde,
Laichingen 019861
Museum für Holzhandwerk,
Sinzig 021884
Museum für Islamische Kunst,
Staatliche Museen zu Berlin –
Stiftung Preußischer Kulturbesitz,
Berlin 017242
Museum für Jagdtier- und Vogelkunde
des Erzgebirges, Augustusburg 016675
Museum für Kaffeetechnik, Emmerich
am Rhein 018188
Museum für Kinematographie,
Ribbesbüttel 021407
Museum für klösterliche Kultur,
Weingarten, Kreis Ravensburg . 022516,
022517
Museum für Kloster- und
Heimatgeschichte, Harsefeld . . 019077
Museum für Kommunikation,
Bern 041320
Museum für Kommunikation,
Hamburg 019010
Museum für Kommunikation Berlin,
Museumsstiftung Post- und
Telekommunikation, Berlin . . . 017243
Museum für Kommunikation Frankfurt,
Frankfurt am Main 018436
Museum für Kommunikation Nürnberg
(Verkehrsmuseum), Nürnberg . 020868
Museum für Konkrete Kunst,
Ingolstadt 019412
Museum für Krippen, Eß- und
Trinkkultur und Brauereigeschichte,
Burghausen 017690
Museum für Kunst in Steatit, Frankfurt
am Main 018437
Museum für Kunst und Gewerbe
Hamburg, Hamburg 019011
Museum für Kunst und
Kulturgeschichte Dortmund,
Dortmund 017962
Museum für Kunsthandwerk,
Leipzig 019992

Museum für Kutschen, Chaisen,
Karren, Zweigmuseum des
Landesmuseums Württemberg,
Heidenheim an der Brenz 019153
Museum für Lackkunst, Münster 020605
Museum für ländliches Kulturgut,
Landwirtschaft, Forsten und Jagd,
Ulrichstein 022269
Museum für Landtechnik und
Landarbeit, Emmerthal 018193
Museum für landwirtschaftliche
Maschinen, Erbach, Alb-Donau-
Kreis 018227
Museum für Licht und Beleuchtung,
Arnsberg 016607
Museum für Literatur am Oberrhein,
Karlsruhe 019513
Museum für Luftfahrt und Technik,
Wernigerode 022583
Museum für Magie und
Hexenverfolgung, Penzlin 021101
Museum für Medizin-Meteorologie,
Museum Dürnhof, Zwettl 003144
Museum für Militär- und
Zeitgeschichte, Kolitzheim 019756
Museum für Mission- und
Völkerkunde, Sankt Wendel . . 021612
Museum für Modelleisenbahnen und
Modellautos, Steißlingen 021992
Museum für Moderne Kunst Frankfurt
am Main, Frankfurt am Main . 018438
Museum für modernes Glas,
Öhringen 020955
Museum für Mühlenbautechnik,
Syrau 022095
Museum für Musikautomaten,
Sammlung Dr. H. Weiss-Stauffacher,
Seewen 041976
Museum für Musikinstrumente,
Musikakdemie Kürnbach,
Kürnbach 019822
Museum für Nachrichten- und
Hochfrequenztechnik, Aachen . 016443
Museum für Naive Kunst,
Bönnigheim 017460
Museum für Natur und Umwelt,
Lübeck 020161
Museum für Naturkunde,
Chemnitz 017736
Museum für Naturkunde,
Dortmund 017963
Museum für Naturkunde der Stadt
Gera, Gera 018648
Museum für Naturkunde und
Vorgeschichte, Dessau 017866
Museum für Naturkunde, Leibniz-
Institut für Evolutions- und
Biodiversitätsforschung an der
Humboldt-Universität zu Berlin,
Berlin 017244
Museum für Naturkunde,
Magdeburger Museen,
Magdeburg 020211
Museum für Neue Kunst, Städtische
Museum Freiburg, Freiburg im
Breisgau 018483
Museum für Niederrheinische
Sakralkunst, Kempen 019582
Museum für Nummernschilder,
Verkehrs- und Zulassungsgeschichte,
Großolbersdorf 018856
Museum für Österreichische
Zeitgeschichte, Scharnstein . . . 002713
Museum für Optik, Meßtechnik
und mathematische Geräte,
Nothweiler 020839
Museum für Ortsgeschichte,
Mutterstadt 020628
Museum für Ortsgeschichte und
Weinbau, Zellertal 022862
Museum für Ostasiatische Kunst,
Museen der Stadt Köln, Köln . . 019702
Museum für Outsiderkunst,
Schleswig 021665

Museum für Photographie,
Braunschweig 017563
Museum für Puppen, Bären und
Spielzeug, Solingen 021901
Museum für Rechtsgeschichte,
Pöggstall 002518
Museum für Regionalgeschichte
und Volkskunde, Stiftung Schloss
Friedenstein Gotha, Gotha 018776
Museum für Sächsische Fahrzeuge,
Chemnitz 017737
Museum für Sächsische Volkskunst
mit Puppentheatersammlung,
Staatliche Kunstsammlungen
Dresden, Dresden 018000
Museum für Sakrale Kunst und
Liturgie, Heidelberg 019139
Museum für schlesische
Landeskunde im Haus Schlesien,
Königswinter 019739
Museum für Schnitzkunst
und Kulturgeschichte,
Oberammergau 020888
Museum für Seenotrettungsgeräte,
Neuharlingersiel 020707
Museum für Sepulkralkultur,
Kassel 019541
Museum für Stadtgeschichte,
Alpirsbach 016516
Museum für Stadtgeschichte, Breisach
am Rhein 017572
Museum für Stadtgeschichte, Höhr-
Grenzhausen 019281
Museum für Stadtgeschichte,
Neuenburg am Rhein 020696
Museum für Stadtgeschichte Dessau,
Dessau 017867
Museum für Stadtgeschichte im Adam-
und-Eva-Haus, Paderborn 021064
Museum für Stadtgeschichte und
Volkskunde, Heppenheim 019193
Museum für Stadtgeschichte,
Naturkunde und Völkerkunde Julius
Riemer, Städtische Sammlungen,
Lutherstadt Wittenberg 020200
Museum für Thüringer Volkskunde
Erfurt, Erfurt 018246
Museum für Uhren und mechanische
Musikinstrumente, Oberhofen am
Thunersee 041812
Museum für Uhren und Schmuck,
Frankfurt am Main 018439
Museum für Unterhaltungskunst,
Österreichisches Circus- und Clown-
Museum, Wien 003045
Museum für Unterwasserarchäologie,
Landesamt für Kultur und
Denkmalpflege-Archäologie und
Denkmalpflege, Sassnitz 021618
Museum für Ur- und Frühgeschichte,
Freiburg im Breisgau 018477
Museum für Ur- und Frühgeschichte,
Stillfried 002794
Museum für Ur- und Frühgeschichte,
Wieselburg an der Erlauf 003110
Museum für Ur- und Frühgeschichte
Thüringens, Weimar 022501
Museum für Ur- und Frühgeschichte,
Willibaldsburg, Eichstätt 018122
Museum für Urgeschichte, Asparn an
der Zaya 001729
Museum für Urgeschichte,
Koblach 002216
Museum für Urgeschichte(n), Zug 042251
Museum für verfolgte Kunst-Israel,
Ashdod 025019
Museum für visuelle Kommunikation,
Köln 019703
Museum für Völkerkunde,
Burgdorf 041386
Museum für Völkerkunde, Wien 003046
Museum für Völkerkunde der
Universität Kiel, Kiel 019609
Museum für Völkerkunde Dresden,

Staatliche Ethnographische
Sammlungen Sachsen, Dresden 018001
Museum für Völkerkunde Hamburg,
Hamburg 019012
Museum für Volkskultur, Spittal an der
Drau 002766
Museum für Volkskunst, Haslach 002061
Museum für Vor- und Frühgeschichte,
Gunzenhausen 018898
Museum für Vor- und Frühgeschichte,
Langenau 019901
Museum für Vor- und
Frühgeschichte – Archäologisches
Museum, Frankfurt am Main . 018405
Museum für Vor- und Frühgeschichte
sowie Stadtgeschichte, Egeln . 018096
Museum für Vor- und Frühgeschichte,
Staatliche Museen zu Berlin –
Stiftung Preußischer Kulturbesitz,
Berlin 017245
Museum für Vor- und Frühgeschichte,
Stiftung Saarländischer Kulturbesitz,
Saarbrücken 021558
Museum für Waage und Gewicht,
Balingen 016939
Museum für Waffentechnik,
Ausrüstung, Auszeichnungswesen,
Vilshofen 022345
Museum für Wattenfischerei,
Wremen 022781
Museum für Weinbau und
Stadtgeschichte, Edenkoben . . 018093
Museum für Weinkultur,
Deidesheim 017843
Museum für Westfälische Literatur,
Oelde 020958
Museum für Wohnkultur des
Historismus und des Jugendstils,
Hilterfingen 041594
Museum für Zeit – Pfälzisches
Turmuhrenmuseum,
Rockenhausen 021440
Museum für Zeitgenössische
Glasmalerei, Langen, Hessen . 019893
Museum Furthmühle, Pram 002531
Muséum Gallery, Albi 103219
Museum Garching an der Alz,
Garching an der Alz 018587
Museum Gayo, Aceh Tengah . . . 024521
Museum Gedong Kirtya,
Singaraja 024648
Museum Gedung Arca, Gianyar . 024559
Museum Gedung Joang '45,
Jakarta 024573
Museum Geelvinck Hinlopen Huis,
Amsterdam 031959
Museum gegenstandsfreier Kunst,
Otterndorf 021043
Museum Geiserschmiede,
Bühlertal 017669
Museum-Gemaal de Hoogte,
Nieuwolda 032681
Museum Gensungen, Felsberg . 018353
Museum Geologi, Bandung 024531
Museum Geologie/Paläontologie,
Heidelberg 019140
Museum Georg Schäfer,
Schweinfurt 021783
Museum Gesigt van't Dok,
Hellevoetsluis 032457
Museum GGZ-Drenthe, Assen . . 032036
Museum Giersch, Frankfurt am
Main 018441
Museum Goa Bala Lompoa,
Sungguminasa 024653
Museum Goch, Goch 018721
Museum Godshuis Belle, Ieper . 003726
Museum Göltzsch, Rodewisch . . 021446
Museum Goldener Steig,
Waldkirchen 022396
Museum Goldenes Dachl,
Innsbruck 002130
Museum Gottfried Silbermann,
Frauenstein 018462

Museum Graphia, Urbino 028242
Museum Griesheim, Griesheim . 018817
Museum Grootseminarie, Brugge 003418
Museum Grossauheim, Hanau .. 019040
Museum Großklein, Hallstattzeitliches
  Museum, Großklein 002007
Museum Großkrotzenburg,
  Großkrotzenburg ... 018853
Museum Gülden Creutz, Leinefelde-
  Worbis 019961
Museum Gugging, NÖ Museum
  Betriebs GmbH, Maria Gugging 002356
Museum Gula, Klaten 024604
Museum Gunnar-Wester-Haus,
  Städtische Sammlungen Schweinfurt,
  Schweinfurt 021784
Museum Gust de Smet, Deurle . 003544
Museum Gustavianum, Uppsala 041115
Museum Gut Grottenhof, Leibnitz 002286
Museum H.R. Giger, Gruyères .. 041573
Museum Haldensleben,
  Haldensleben 018943
Museum Halle, Halle / Westfalen 018950
Museum Hameln, Hameln 019027
Museum Haus Esters, Kunstmuseen
  Krefeld, Krefeld 019788
Museum Haus Kast, Gaggenau . 018578
Museum Haus Kasuya, Yokosuka 029739
Museum Haus Lange, Kunstmuseen
  Krefeld, Krefeld 019789
Museum Haus Löwenberg,
  Gengenbach 018636
Museum Haus Ludwig für
  Kunstausstellungen, Saarlouis . 021564
Museum Haus Martfeld,
  Schwelm 021788
Museum Havezate Mensinge,
  Roden 032792
Museum Hedendaagse Glaskunst,
  Museums Vledder, Vledder .. 032993
Museum Hedendaagse Grafiek,
  Museums Vledder, Vledder . 032994
Museum Heimathaus Irmintraut,
  Ottersberg 021045
Museum Heimathaus Münsterland,
  Telgte 022121
Museum Heineanum, Museum für
  Vogelkunde, Halberstadt 018939
Museum Henriette Polak,
  Zutphen 033118
Museum Heraldiek Benelux,
  Temse 004053
Museum Herbarium Bogoriensis,
  Bogor 024545
Museum Herisau, Herisau 041593
Museum Herman De Cuyper,
  Blaasveld 003369
Museum Hermann der Lahme,
  Altshausen 016564
Museum Herzogsburg, Braunau am
  Inn 001792
Museum Hessenstube, Baunatal 016974
Museum Het Alaam, Genk 003617
Museum Het Gebied van Staden,
  Staden 004046
Museum Het Groot Graffel,
  Warnsveld 033022
Museum Het Oude Raadhuis,
  Megen 032644
Museum Het Petershuis, Gennep 032343
Museum Het Prinsenhof, Delft . 032160
Museum Het Rembrandthuis,
  Amsterdam 031960
Museum Het Schip, Amsterdam 031961
Museum het Tramstation,
  Schipluiden 032861
Museum Het Valkhof, Nijmegen 032687
Museum Het Verscholen Dorp,
  Vierhouten 032987
Museum Heylshof, Stiftung Kunsthaus
  Heylshof, Worms 022765
Museum Highlights, Wickenburg 139443
Museum Historische Verzameling

Nederlandse Politie, Zaandam . 033076
Museum Historischer Bierkrüge,
  Hainfeld 002044
Museum Höxter-Corvey, Höxter . 019285
Museum Hoffmann'sche Sammlung,
  Kohren-Sahlis 019753
Museum Hofmühle, Immenstadt im
  Allgäu 019396
Museum Hohenau an der March,
  Hohenau an der March 002095
Museum Hohenzollern in Franken,
  Bayerische Verwaltung der
  staatlichen Schlösser, Gärten und
  Seen, Kulmbach 019831
Museum Holstentor, Die Lübecker
  Museen, Lübeck 020162
Museum Huelsmann, Kunst + Design,
  Bielefeld 017378
Museum Hüsli-Sammlung
  Schwarzwälder Volkskunst,
  Grafenhausen 018788
Museum Humanum, Waldkirchen an
  der Thaya 002893
Museum Huta Bolon Simanindo,
  Pematangsiantar 024633
Museum Huta Bolon Simanindo,
  Samosir 024640
Museum Huthaus Einigkeit, Brand-
  Erbisdorf 017539
Museum Huthaus Zinnwald, Galerie
  Ewald Schönberg, Altenberg . . 016532
Museum Ianchelevici, Goudriaan 032369
Museum Idar-Oberstein, Idar-
  Oberstein 019377
Museum Illingen, Illingen, Kreis
  Neunkirchen 019386
Museum im Adler, Benningen am
  Neckar 017033
Museum im alten Brauhaus, Homberg
  (Ohm) 019332
Museum im alten Pfarrhaus,
  Wiedensahl 022622
Museum im Alten Rathaus, Bad
  Brückenau 016740
Museum im Alten Rathaus,
  Mönchberg im Spessart 020437
Museum im Alten Rathaus,
  Neckargemünd 020653
Museum im Alten Rathaus,
  Waghäusel 022369
Museum im Alten Schloß,
  Altensteig 016541
Museum im Alten Schloß,
  Neckarbischofsheim 020652
Museum im alten Zeughaus, Bad
  Radkersburg 001761
Museum im Ammes Haus,
  Hatzfeld 019105
Museum im Amtshaus, Grünsfeld 018872
Museum im Amtshausschüpfla,
  Erlangen 018261
Museum im Astorhaus, Walldorf 022405
Museum im Ballhaus, Imst 002115
Museum im Bellpark, Kriens ... 041629
Museum im Bierlinghaus, Bad
  Bayersoien 016708
Museum im Blaahaus am Unteren
  Römerweg, Kiefersfelden ... 019600
Museum im Bock, Leutkirch im
  Allgäu 020055
Museum im Boyneburgischen Schloß,
  Sontra 021921
Museum im Bürgerspital, Städtisches
  Kießling-Museum, Drosendorf an der
  Thaya 001829
Museum Im Dorf, Wittingen ... 022701
Museum Im Dorf, Außenstelle des
  Heimatmuseums Reutlingen,
  Reutlingen 021383
Museum im ehemaligen Augustiner-
  Chorherrenstift, Schloss- und
  Gartenverwaltung Herrenchiemsee,
  Herrenchiemsee 019224
Museum im 'Fressenden Haus' –

Burgkasten Weißenstein, Regen 021311
Museum im Frey-Haus, Brandenburg
  an der Havel 017546
Museum im Fürstenstöckl,
  Ebenau 001832
Museum im Goetz-Haus, Leipzig 020006
Museum im Goldschmiedehaus,
  Ahlen 016477
Museum im Gotischen Haus, Bad
  Homburg v.d.Höhe 016788
Museum im Graulturm, Gau-
  Algesheim 018597
Museum im Greuterhof, Islikon . 041606
Museum im Grünen Haus, Reutte 002594
Museum im Heimathaus,
  Traunstein 022175
Museum im Herrenhaus,
  Hausach 019109
Museum im Hexenturm,
  Markdorf 020284
Museum im Hirsch, Remshalden 021369
Museum im Hollerhaus, Dietfurt an
  der Altmühl 017892
Museum im Jost-Sigristen-Haus,
  Ernen 041472
Museum im Kastenturm,
  Bischofshofen 001784
Museum im Kinzigtorturm,
  Gengenbach 018637
Museum im Kleihues-Bau, Museen
  der Stadt Kornwestheim,
  Kornwestheim 019771
Museum im Klösterle, Peiting . 021096
Museum im Kloster – Museum
  des Landkreises Osnabrück,
  Bersenbrück 017345
Museum im Kloster Grafschaft,
  Schmallenberg 021682
Museum im Koffer, Kinder- und
  Jugendmuseum Nürnberg,
  Nürnberg 020869
Museum im Kornhaus, Altheim . 016543
Museum im Kornhaus,
  Rorschach 041892
Museum im Kräuterkasten,
  Albstadt 016493
Museum im Kreuzgang, Museen der
  Stadt Landshut, Landshut 019884
Museum im Kulturspeicher,
  Würzburg 022796
Museum im Lagerhaus, Sankt
  Gallen 041925
Museum im Ledererhaus, Purgstall an
  der Erlauf 002548
Museum im Leuchtturm,
  Putgarten 021240
Museum im Malhaus, Wasserburg,
  Bodensee 022441
Museum im Marstall, Winsen,
  Luhe 022686
Museum im Mendelssohn-Haus,
  Leipzig 020007
Museum im Mesnerhaus, Pfaffenhofen
  an der Ilm 021112
Museum im Mönchenkloster,
  Jüterbog 019475
Museum im Pflegschloss,
  Schrobenhausen 021739
Museum im Porsche Werk Leipzig,
  Leipzig 020008
Museum im Prediger, Schwäbisch
  Gmünd 021749
Museum im Rathaus, Gleisdorf . 001936
Museum im Rathaus Möhringen,
  Museen der Stadt Tuttlingen,
  Tuttlingen 022235
Museum im Ritterhaus,
  Offenburg 020976
Museum im Ritterhaus, Osterode am
  Harz 021029
Museum im Römerbad, Heidenheim
  an der Brenz 019154
Museum im Schäferhaus,
  Wahlsburg 022373

Museum im Schafstall, Neuenstadt am
  Kocher 020703
Museum im Schlößle, Freiberg am
  Neckar 018468
Museum im Schlössle, Kressbronn am
  Bodensee 019793
Museum im Schloss, Bernstadt, Alb-
  Donau-Kreis 017343
Museum im Schloß, Frohburg .. 018537
Museum im Schloß Bad Pyrmont, Bad
  Pyrmont 016843
Museum im Schloß Lützen,
  Lützen 020188
Museum im Schloss Wolfenbüttel,
  Stadtmuseum, Wolfenbüttel .. 022733
Museum im Schloss,
  Porzellanmanufaktur Fürstenberg,
  Fürstenberg, Kreis Holzminden 018541
Museum im Schmidt-Haus,
  Nabburg 020632
Museum im Schottenstift, Wien . 003047
Museum im Schulhaus,
  Niederrohrdorf 041802
Museum im Schweizer Hof,
  Bretten 017622
Museum im Seelhaus, Bopfingen 017507
Museum im Stadtturm, Groitzsch 018829
Museum im Stadtturm, Waidhofen an
  der Ybbs 002890
Museum im Stasi-Bunker, Museum in
  der "Runden Ecke", Machern . 020202
Museum im Steinhaus, Bad
  Wimpfen 016908
Museum im Steintor, Anklam .. 016586
Museum im Steintorturm,
  Brandenburg an der Havel ... 017547
Museum im Stern, Warburg ... 022426
Museum im Stockalperschloß,
  Brig 041365
Museum im Strumpfhaus,
  Oberstaufen 020929
Museum im Tabor, Feldbach ... 001878
Museum im Trafohaus, Leipzig . 020009
Museum im Troadkostn zu Giem,
  Feldbach 001879
Museum im Turm, Neuleiningen 020718
Museum im Turmhof, Steckborn 042207
Museum im Umweltschutz-
  Informationszentrum Oberfranken
  mit Kinder-Erlebnis-Museum,
  Bayreuth 016999
Museum im Vorderen Schloss,
  Mühlheim an der Donau 020490
Museum im Wächterhaus,
  Nürnberg 020870
Museum im Wasserschloß
  Kottingbrunn, Kottingbrunn ... 002227
Museum im Wasserwerk, Berliner
  Wasserbetriebe, Berlin 017246
Museum im Wehener Schloss,
  Taunusstein 022113
Museum im Westrich, Ramstein-
  Miesenbach 021273
Museum im Wetterbergschen Adelshof,
  Bad Münder am Deister 016824
Museum im Wittelsbacher Schloss
  Friedberg, Friedberg, Kreis Aichach-
  Friedberg 018518
Museum im Zendenrathaus,
  Ernen 041473
Museum im Zeughaus,
  Schaffhausen 041948
Museum im Zeughaus, Vechta . 022305
Museum im Zeughaus, Tiroler
  Landesmuseum Ferdinandeum,
  Innsbruck 002131
Museum im Zisterzienserkloster
  Lehnin, Kloster Lehnin 019652
Museum im Zwinger, Weida ... 018770
Museum In Black, Los Angeles . 050453
Museum in de Stiftskerk, Thorn 032922
Museum in den Berglen, Berglen 017060
Museum in den Halven Maen,
  Rockanje 032789

Museum in der Adler-Apotheke,
Eberswalde 018084
Museum in der Alten Schule, Efringen-
Kirchen 018095
Museum in der Bruderkirche,
Altheim 016544
Museum in der Burg, Bad
Driburg 016747
Museum in der Burg,
Coppenbrügge 017760
Museum in der Fronfeste, Neumarkt
am Wallersee 002443
Museum in der Kaiserpfalz,
Paderborn 021065
Museum in der Majolika, Badisches
Landesmuseum Karlsruhe,
Karlsruhe 019514
Museum in der Präparandenschule,
Höchberg 019274
Museum in der Remise, Bad
Harzburger und Schreiberhauer
Heimatstube, Burgbergmuseum, Bad
Harzburg 016781
Museum in der Scheuergasse,
Zwingenberg, Kreis Bergstraße 022917
Museum in der Stadthalle Stern
(Interim), Riesa 021428
Museum in Docklands, London . 045041
Museum in Progress, Wien 003048
Museum In 't Houten Huis, De
Rijp 032147
Museum in the Community,
Hurricane 049643
The Museum in the Park, Stroud 045916
Museum Indonesia, Jakarta ... 024574
Museum Industriekultur,
Osnabrück 021018
Museum Industriekultur mit
Motorradmuseum, Museen der Stadt
Nürnberg, Nürnberg 020871
Museum Innviertler Volkskundehaus
und Galerie der Stadt Ried, Ried im
Innkreis 002595
Museum Insel Hombroich, Stiftung
Insel Hombroich, Neuss 020747
Museum-Institute of the Armenian
Genocide, Armenian National
Academy of Sciences, Yerevan 000741
Museum International, Oxford . . 139206
Museum Internationales Baum-Archiv,
Winterthur 042156
Museum Istana Mangkunegaran,
Surakarta 024657
Museum Istana Sultan Ternate, North
Malaku 024621
Museum Istant Siak Sri Indapura,
Bengkalis 024541
Museum It Kokelhûs van Jan en Sjut,
Earnewald 032261
Muséum Jacobs van den Hof,
Amersfoort 031913
Museum Jacques van Mourik,
Mook 032658
Museum Jagd und Wild auf Burg
Falkenstein, Falkenstein, Kreis
Cham 018339
Museum Jagdschloß Kranichstein,
Stiftung Hessischer Jägerhof,
Darmstadt 017827
Museum Jakob Smits, Mol 003873
Museum Jamu Nyonya Menir,
Semarang 024644
Museum Jan Boon, De Rijp .. 032148
Museum Jan Cunen, Oss 032737
Museum Jan Heestershuis,
Schijndel 032857
Museum Jan Karlsgården och Vita
Björn, Sund 011379
Museum Jan Lont,
Hippolytushoef 032490
Museum Jan Vaerten, Beerse . 003353
Museum Jan van der Togt,
Amstelveen 031920
Museum Jannink, Enschede ... 032308

Museum Jerusalem Panorama
Kreuzigung Christi, Altötting . . 016556
Museum Joang '45, Medan ... 024614
Museum Joang '45, Surabaya . 024654
Museum Josephine Baker, Castelnaud-
la-Chapelle 012360
Museum Joure, Joure 032529
Museum Judengasse, Jüdisches
Museum, Frankfurt am Main . 018442
Museum Judenplatz Wien, Wien 003049
Museum jüdischer Geschichte und
Kultur, Aschaffenburg 016621
Museum Junge Kunst, Frankfurt
(Oder) 018400
Museum Junkerhaus, Lemgo .. 020030
Museum Kalimantan Tengah,
Palangkaraya 024625
Museum Kalt-Heiß,
Deckenpfronn 017837
Museum Kameleon Dorp en Museum
Hans Brinker, Terherne 032915
Museum Kampa, Nadace Jana a Medy
Mládkových, Praha 009810
Museum Karl-Marx-Haus, Trier . 022200
Museum Kartause Astheim,
Volkach 022356
Museum Kartini, Jepara 024600
Museum Kartini Rembang,
Rembang 024639
Museum Kasper, Norrtälje 040881
Museum Kasteel Wijchen,
Wijchen 033045
Museum Katendrecht, Rotterdam 032810
Museum Katharinenhof mit
Mühlenturm und Stadtscheune,
Kranenburg 019780
Museum Kebangkitan Nasional,
Jakarta 024575
Museum Kebudayaan dan Kemajuan
Asmat, Irian Jaya 024562
Museum Kebun Raya Bogor,
Bogor 024546
Museum Kedokteran, Jakarta . 024576
Museum Kellinghusen,
Kellinghusen 019576
Museum Kempenland, Eindhoven 032288
Museum Kempenland, Lommel . 003826
Museum Kendil Riau,
Tanjungpandan 024664
Museum Kennemerland,
Beverwijk 032070
Museum Kepolisian Negara Republik
Indonesia, Museum Kriminal-Mabak,
Jakarta 024577
Museum Kereta, Ambarawa . 024524
Museum Kerkelijke Kunst,
Workum 033064
Museum Kesselhaus Herzberge,
Berlin 017247
Museum Kierling, Verein für Kunst,
Kultur und Erwachsenenbildung,
Heimatkunde, Heimatforschung,
Denkmalschutz sowie Wissenschaft
und Forschung, Kierling 002167
Museum Kinderwereld, Roden . 032793
Museum Kindheit und Jugend,
Stiftung Stadtmuseum Berlin,
Berlin 017248
Museum Kirche zum Heiligen Kreuz,
Zittau 022889
Museum Kitzbühel, Kitzbühel .. 002182
Museum Kleines Klingental,
Sammlung Basler Münsterskulpturen,
Kloster Klingental, Basel 041279
Museum Kleines Schloß,
Blankenburg 017408
Museum Klösterli, Ettiswil .. 041480
Museum de Klompenmaker,
Keijenborg 032541
Museum Kloster Asbach, Bayerisches
Nationalmuseum München,
Rotthalmünster 021524
Museum Kloster Bentlage, Städtische
Museen Rheine, Rheine 021396

Museum Kloster Hude, Sammlung zur
Klostergeschichte, Hude 019352
Museum Kloster Kamp, Kamp-
Lintfort 019492
Museum Kloster Stift zum
Heiligengrabe, Heiligengrabe . . 019166
Museum Kloster Zeven, Zeven . 022873
Museum Kloster Zinna, Kloster
Zinna 019655
Museum Knoblauchhaus, Stiftung
Stadtmuseum Berlin, Berlin .. 017249
Museum Knochenstampfe,
Zwönitz 022918
Museum Kölner Karnevalsorden,
Köln 019704
Museum Komodo, Jakarta ... 024578
Museum Kraftwerk Plessa,
Plessa 021154
Museum Kranenburgh, Bergen . 032066
Museum Kraton Kasepuhan,
Cirebon 024554
Museum Kraton Yogyakarta,
Yogyakarta 024678
Museum Kruijsenhuis, Oirschot . 032710
Museum Kuenburggewölbe,
Werfen 002925
Museum Künstlerhaus Gasteiger,
Bayerische Verwaltung der
staatlichen Schlösser, Gärten und
Seen, Utting am Ammersee .. 022292
Museum Küppersmühle für Moderne
Kunst, Duisburg 018060
Museum Kulturerbe Hallstatt,
Hallstatt 002052
Museum Kulturgeschichte der Hand,
Wolnzach 022757
Museum Kunst der Westküste,
Alkersum 016509
Museum Kunstraum Neuss,
Neuss 020748
Museum Der Kunststall, Dahlem 017801
Museum Kurá Hulanda,
Willemstad 009398
Museum Kurhaus Kleve, Ewald
Mataré-Sammlung, Kleve 019644
Museum Kutschen-Wagen-Karren,
Museen der Stadt Gescher,
Gescher 018675
Museum La Congiunta, Haus für
Relief und Halbfiguren von Hans
Josephsohn, Giornico 041546
Museum La Diligence, Heerlen . 032434
Museum La Truaisch, Sedrun .. 041973
Museum Lambert Van Meerten,
Delft 032161
Museum Lammert Boerma,
Borgercompagnie 032086
Museum Langenargen/ Bodensee,
Langenargen 019897
Museum Langenthal, Langenthal 041638
Museum Langes Tannen,
Uetersen 022250
Museum Langmatt, Stiftung
Langmatt Sidney und Jenny Brown,
Baden 041255
Museum Laufental, Laufen 041642
Museum Lauriacum, Enns 001870
Museum Laxenburg, Laxenburg 002281
Museum Le Mayeur, Sanur ... 024641
Museum Leben am Meer, Esens 018282
Museum Lettergieten 1983,
Westzaan 033041
Museum Leuchtenburg,
Seitenroda 021831
Museum Lichtenberg im Stadthaus,
Berlin 017250
Museum Lindengut, Winterthur . 042157
Museum Lindwurm, Stein am
Rhein 042034
Museum Liner Appenzell,
Appenzell 041225
Museum Lionardo da Vinci Ideale,
Karlskrona 040741
Museum Löffingen, Löffingen . . 020097

Museum Löwenburg, Christoph Merian
Stiftung, Ederswiler 041454
Museum Loka Budaya, Jayapura 024598
Museum London, London 006215
Museum Ludwig, Museen der Stadt
Köln, Köln 019705
Museum Maarssen, Maarssen .. 032614
Museum Magic Jukebox, Menen 003864
Museum Malerwinkelhaus,
Marktbreit 020295
Museum Malikusaleh, Aceh
Utara 024523
Museum Management and
Curatorship, Kidlington 139207
Museum Mandala Bakti,
Semarang 024645
Museum Mandala Wangsit Siliwangi,
Bandung 024532
Museum Manggala Wanabakti,
Jakarta 024579
Museum Mannersdorf,
Mannersdorf 002354
Museum Mariastein, Kirchbichl . 002174
Museum Mayer van den Bergh,
Antwerpen 003294
Muséum Mécanique Naturelle,
Giverny 013117
Museum mechanische Klangfabrik,
Haslach 002062
Museum Meerhem, Gent 003632
Museum Meermanno-Westreenianum,
Den Haag 032185
Museum Meinhardt, Mainhardt . 020220
Museum Memorial Sultan Ternate,
Ternate 024667
Museum Mengeringhausen, Bad
Arolsen 016705
Museum Menkemaborg,
Uithuizen 032936
Museum Mensch und Natur,
München 020555
Museum Mesdag, Van Gogh Museum
Amsterdam, Den Haag 032186
Museum Military Intelligence,
Intelligence Corps Museum
and Medmenham Collection,
Chicksands 043948
Museum Militer, Jakarta 024580
Museum Mini Koram, Sintang .. 024650
Museum Mistra, Mystras 023185
Museum Mobil, Bogor 024547
Museum Modern Art, Hünfeld .. 019357
Museum Moderner Kunst – Wörlen,
Gemeinnützige Gesellschaft,
Passau 021083
Museum Moderner Kunst Kärnten,
Klagenfurt 002196
Museum Moderner Kunst Stiftung
Ludwig Wien, Wien 003050
Museum Modest Huys, Zulte . . 004169
Museum Mödling, Thonetschlössl,
Mödling 002402
Muséum Møhlman, Appingedam 032016
Museum Moorseer Mühle,
Nordenham 020820
Museum Morsbroich, Leverkusen 020058
Museum MPU Tantular, Surabaya 024655
Museum Münster, Münster 041778
Museum Municipal, Beira 031801
Muséum Municipal d'Histoire Naturelle
Henri-Lecoq, Clermont-Ferrand 012599
Museum Murten, Murten 041784
Museum Musica, Stadskanaal .. 032898
Museum Nagele, Nagele 032662
Museum Nan Eilean, Lionacleit . 044842
Museum Nan Eilean, Steornabhagh,
Stornoway 045889
Museum Nanning Hendrik Bulthuis,
Museum Kerkelijke Kunst Workum,
Leeuwarden 032573
Museum Nasional, Jakarta ... 024581
Muséum National d'Histoire
Naturel Arboretum de Chèvreloup,
Rocquencourt 015127

Muséum National d'Histoire Naturelle, Paris 014792

Museum National d'Histoire Naturelle, Rabat 031792

Museum Natur und Mensch, Greding 018800

Museum Necca and New England Center for Contemporary Art, Brooklyn 047249

Museum Nederlandse Cavalerie, Amersfoort 031914

Museum Negeri Bali, Denpasar . 024555

Museum Negeri Jambi, Jambi . 024597

Museum Negeri Java Tengah, Semarang 024646

Museum Negeri Jawa Barat, Bandung 024533

Museum Negeri Kalimantan Barat, Pontianak 024636

Museum Negeri Kalimantan Timur Mulawarman, Tenggarong 024666

Museum Negeri Lampung, Bandarlampung 024528

Museum Negeri Nusa Tenggara Barat, Mataram 024613

Museum Negeri of Aceh, Banda Aceh 024527

Museum Negeri of Bengkulu, Bengkulu 024542

Museum Negeri of Irian Jaya, Jayapura 024599

Museum Negeri of La Galigo, Ujungpanang 024670

Museum Negeri of Lambung Mangkurat, Banjarbaru ... 024537

Museum Negeri Propinsi Nusa Tenggara Timur, Kupang 024606

Museum Negeri Propinsi Sumatera Selatan Balaputra Dewa, Palembang 024626

Museum Negeri Sulawesi Tenggara, Kendari 024603

Museum Negeri Sulawesi Utara, Manado 024611

Museum Negeri Sumatera Barat, Padang 024622

Museum Negeri Sumatera Utara, Medan 024615

Museum Negeri Timor-Timur, Dili 010409

Museum Nekara, Selayar 024642

Museum Neue Mühle, Erfurt ... 018247

Museum Neuhaus, Biel 041340

Museum Neuhaus – Sammlung Liaunig, Neuhaus 002435

Museum Neukirchen-Vluyn, Neukirchen-Vluyn 020717

Museum Neukölln, Museum für Stadtkultur und Regionalgeschichte, Berlin 017251

Museum Neuruppin, Neuruppin . 020739

Museum Neustadt an der Orla, Neustadt an der Orla 020755

Museum News, Washington ... 139444

Museum Nienburg, Nienburg ... 020807

Museum Niesky, Niesky 020811

Museum Nikolaikirche, Stiftung Stadtmuseum Berlin, Berlin ... 017252

Museum No 1, Richmond, Surrey 045613

Museum Nordenham, Nordenham 020821

Museum NordJura, Weismain .. 022531

Museum Nostalgie der 50er Jahre, Burgpreppach 017697

Museum Notes (Providence), Providence 139445

Museum Nusantara, Museum voor het Indonesisch Cultuurgebied, Delft 032162

Museum Nymphenburger Porzellan – Sammlung Bäuml, Bayerische Verwaltung der staatlichen Schlösser, Gärten und Seen, München .. 020556

Museum Ober-Ramstadt, Ober-Ramstadt 020886

Museum Oberes Donautal, Heimatmuseum Fridingen, Fridingen an der Donau 018515

Museum Oberschützen, Oberschützen 002466

Museum Oculorum, Buenos Aires 000280

Museum of Aboriginal Affairs, Gombak 030663

Museum of Acre, Acre 025009

Museum of African American Art, Los Angeles 050454

The Museum of African Tribal Art, Portland 052282

Museum of Afro-American History, Boston 047122

Museum of Agios Stephanos, Meteora 023170

Museum of Alaska Transportation and Industry, Wasilla 054197

Museum of Amana History, Amana 046432

Museum of Ambleside, Ambleside 043413

Museum of American Architecture and Decorative Arts, Houston 049587

Museum of American Finance, New York 051434

Museum of American Frontier Culture, Staunton 053512

Museum of American Glass at WheatonArts, Millville 050901

Museum of American Historical Society of Germans from Russia, Lincoln 050295

The Museum of American Illustration at the Society of Illustrators, New York 051435

Museum of American Political Life, University of Hartford, West Hartford 054302

Museum of American Presidents, Strasburg 053567

Museum of American West, Los Angeles 050455

Museum of Anatomy, Glasgow . 044407

Museum of Anatomy of the Department of Anatomy, Athinai 023007

Museum of Ancient and Modern Art, Penn Valley 051976

Museum of Ancient Artifacts, Chicago 047674

Museum of Ancient Artifacts, Las Vegas 050169

Museum of Ancient Culture, Kilmartin 044698

Museum of Animal Husbandary and Veterinary, Assam Veterinary College, Guwahati 024277

Museum of Anthracite Mining, Ashland 046572

Museum of Anthropology, Denver 048270

Museum of Anthropology, Goudi . 023085

Museum of Anthropology, Lawrence 050185

Museum of Anthropology, Pullman 052415

Museum of Anthropology, Vancouver 007290

Museum of Anthropology, Winston-Salem 054490

Museum of Anthropology and Archaeology, Tshwane 038850

Museum of Anthropology, California State University, Chico, Chico . 047712

Museum of Anthropology, California State University, Fullerton, Fullerton 048996

Museum of Anthropology, National Taiwan University, Taipei 042530

Museum of Anthropology, Northern Kentucky University, Highland Heights 049470

Museum of Anthropology, University of Michigan, Ann Arbor 046491

Museum of Anthropology, University of Missouri, Columbia 047901

Museum of Anthropology, Wayne State University, Detroit 048313

Museum of Antique Armour Helmets and Swords, Kyoto 028968

Museum of Antiquities, Alnwick . 043401

Museum of Antiquities, Jamnagar 024322

Museum of Antiquities, Newcastle-upon-Tyne 045351

Museum of Antiquities, Saskatoon 006979

Museum of Antropology, History and Art, University of Puerto Rico, Rio Piedras 052577

Museum of Appalachia, Norris . 051589

Museum of Aqaba Antiquities, Aqaba 029779

Museum of Archaeological Site Moenjodaro, Dokri 034109

Museum of Archaeology, Batumi 016387

Museum of Archaeology, Durham 044190

Museum of Archaeology, Legon . 022938

Museum of Archaeology, Vadodara 024509

Museum of Archaeology and Ethnology, Burnaby 005505

Museum of Archaeology and History, National Museum of Ireland Ard-Mhúsaem na hÉireann, Dublin 024897

Museum of Armenian-Russian Friendship, Abovyan 000688

Museum of Army Flying, Middle Wallop 045258

Museum of Art, Fort Lauderdale 048827

Museum of Art (MoA), Seoul National University, Seoul 030023

Museum of Art and Archaeology, University of Missouri-Columbia, Columbia 047902

Museum of Art and History, Santa Cruz 053101

Museum of Art and History, Weston 054346

Museum of Art at Brigham Young University, Provo 052405

Museum of Art, Archaeology and Folklore, Mysore 024417

Museum of Art, Rhode Island School of Design, Providence 052387

Museum of Art, Saint Louis University, Saint Louis 052812

Museum of Art, University of Maine, Orono 051817

Museum of Art, University of Michigan, Ann Arbor 046492

Museum of Art, University of New Hampshire, Durham 048403

Museum of Art, Washington State University, Pullman 052416

Museum of Arthropoda, Pune .. 024449

Museum of Arts and Crafts, Lucknow 024383

Museum of Arts and Design, New York 051436

The Museum of Arts and Sciences, Daytona Beach 048180

Museum of Arts and Sciences, Macon 050572

Museum of Asian Art, Corfu ... 023055

Museum of Audio Visual Technology, Wellington 033350

Museum of Automobile Art and Design, Syracuse 053636

Museum of Automobiles, Morrilton 051059

Museum of Aviation, Paraparaumu 033278

Museum of Aviation at Robins Air Force Base, Warner Robins ... 054060

Museum of Bad Art, Dedham .. 048211

Museum of Bamboo Art, Nantou 042404

Museum of Banknotes of the Ionian Bank, Corfu 023056

Museum of Barnstaple and North Devon, Barnstaple 043501

Museum of Bath at Work, Bath . 043520

Museum of Belize, Belize City .. 004175

Museum of Berkshire Aviation, Woodley 046186

Museum of Biblical Art, New York 051437

Museum of Biloxi, Biloxi 046992

Museum of Brewing, Burton-upon-Trent 043803

The Museum of Broadcast Communications, Chicago 047675

Museum of Bronx History, Bronx 047234

Museum of Bus Transportation, Hershey 049455

Museum of Byzantine Culture, Thessaloniki 023294

Museum of Cannock Chase, Hednesford 044553

Museum of Cape Breton Heritage, North East Margaree 006513

Museum of Carousel Art and History, Sandusky 053071

Museum of Casts and Archaeological Collection, Thessaloniki 023295

Museum of Central Australia, Alice Springs 000789

Museum of Central Connecticut State University, New Britain 051239

Museum of Ceramics, East Liverpool 048443

Museum of Changhua, Changhua 042283

Museum of Chia-Yi City, Chai-Yi 042280

Museum of Chicot County Arkansas, Lake Village 050088

Museum of Childhood, Edinburgh 044249

Museum of Childhood, Masterton 033242

Museum of Childhood, Sudbury, Derbyshire 045918

Museum of Childhood, Toronto . 007190

Museum of Childhood Memories, Beaumaris 043543

Museum of Childhood, Edith Cowan University, Perth 001402

Museum of Children's Art, Oakland 051680

Museum of Chinese Chive Farming, Cingshuei 042292

Museum of Chinese in the Americas, New York 051438

Museum of Chungju National University, Chungju 029910

Museum of Church History and Art, Salt Lake City 052915

The Museum of Clallam County Historical Society, Port Angeles 052242

Museum of Classical Archaeology, Cambridge 043845

Museum of Classical Archaeology, University of Adelaide, Adelaide 000768

Museum of Coast and Anti-Aircraft Artillery, Green Point 038730

Museum of Communication, Burntisland 043797

Museum of Comparative Zoology, Harvard University, Cambridge 047369

Museum of Connecticut History, Hartford 049388

The Museum of Contemporary Art, Atlanta 046627

Museum of Contemporary Art, Chicago 047676

Museum of Contemporary Art, Cleveland 047818

Museum of Contemporary Art, Fort Collins 048799

Museum of Contemporary Art, Miami 050839

Museum of Contemporary Art, Washington 054137

Museum of Contemporary Art Denver, Denver 048271

Museum of Contemporary Art London, London 045042
Museum of Contemporary Art Los Angeles, Los Angeles 050456
Museum of Contemporary Art San Diego – Downtown, San Diego 052967
Museum of Contemporary Art San Diego – La Jolla, La Jolla 050040
Museum of Contemporary Art Sapporo, Sapporo 029335
Museum of Contemporary Art Sydney, Sydney 001544
Museum of Contemporary Art, Basil and Elise Goulandris Foundation, Andros 022960
Museum of Contemporary Art, Taipei, Taipei 042531
Museum of Contemporary Articrafts, Tehrān 024772
Museum of Contemporary Arts, Luther Burbank Center for the Arts, Santa Rosa 053141
Museum of Contemporary Canadian Art, Toronto 007191
Museum of Contemporary Craft, Portland 052296
Museum of Contemporary Japanese Poetry, Tanka and Haiku, Kitakami 028825
The Museum of Contemporary Photography, Columbia College Chicago, Chicago 047677
Museum of Contemporary Religious Art, Saint Louis 052813
Museum of Cosmetology, Chicago 047678
Museum of Costume, Bath 043521
Museum of Costume and Textiles, Nottingham 045415
Museum of Country Life, National Museum of Ireland Ard-Mhúsaem na hÉireann, Castlebar 024845
Museum of Craft and Folk Art, San Francisco 053014
Museum of Cretan Ethnology, Vori 023317
Museum of Cultural and Natural History, Central Michigan University, Mount Pleasant 051095
Museum of Cycladic and Ancient Greek Art, Athinai 023008
Museum of Daejeon History, Daejeon 029920
Museum of Dartmoor Life, Okehampton 045433
Museum of Death, Los Angeles 050457
Museum of Decorative Art, Chicago 047679
Museum of Decorative Arts and History, National Museum of Ireland Ard-Mhúsaem na hÉireann, Dublin 024898
Museum of Delphic Celebrations of Angelos and Eva Sikellianou, Athinai 023009
Museum of Dentistry, Liverpool . 044860
Museum of Dionysios Solomos and Eminent Zakynthians, Zakynthos 023322
Museum of Discovery and Science, Fort Lauderdale 048828
Museum of Discovery, Arkansas Museum of Science and History, Little Rock 050336
Museum of Domestic Design and Architecture, Middlesex University, Barnet 043496
Museum of Dr. Sun Yat-Sen, Zhongshan 008545
Museum of Drinking Water, Taipei 042532
Museum of Early Southern Decorative Arts, Winston-Salem 054491
Museum of Early Trades and Crafts,

Madison 050590
Museum of East Alabama, Opelika 051786
Museum of East Anglian Life, Stowmarket 045895
Museum of East Asian Art, Bath 043522
The Museum of East Texas, Lufkin 050523
Museum of Ecclesiastical Art, Samos 023254
Museum of Ecclesiastical Art of Serrai and Nigrita, Serres 023257
Museum of Ecclesiastical Art of the Metropolis of Trikki and Stagoi, Trikala 023306
The Museum of Edinburgh, Edinburgh 044250
Museum of Education and Documents Library, Cairo 010546
Museum of Electricity, Christchurch 043956
Museum of English Rural Life, Reading 045589
Museum of Engravings and Graphic Arts, Athinai 023010
Museum of Entertainment, Whaplode 046112
Museum of Epirot Folk Art, Baron Tositsa Foundation, Metsovo .. 023172
Museum of Ethnography, Accra . 022923
Museum of Ethnography and Folklore, Eretz-Israel Museum, Tel Aviv . 025207
Museum of Ethnology and Folk Art, Ratnapura 040486
Museum of Eton Life, Windsor .. 046164
Museum of Farming Life, Ellon . 044286
Museum of Farnham, Farnham . 044335
Museum of Fife and Drum, Ivoryton 049744
The Museum of Films for Children, Weston Woods Museum, Weston 054347
Museum of Fine Art, Las Cruces 050154
Museum of Fine Arts, Beijing .. 007631
Museum of Fine Arts, Chandigarh 024192
Museum of Fine Arts, Hobart .. 001146
The Museum of Fine Arts, Houston 049588
Museum of Fine Arts, Missoula . 050962
Museum of Fine Arts, Saint Petersburg 052863
Museum of Fine Arts, Santa Fe . 053113
Museum of Fine Arts, Springfield 053461
Museum of Fine Arts and Cultural Center, Alexandria 010519
Museum of Fine Arts, Boston, Boston 047123
Museum of Fire, Edinburgh 044512
Museum of Flight, Seattle 053239
Museum of Florida History, Tallahassee 053662
Museum of Flying, Santa Monica 053133
Museum of Folklore and Popular Art, Sifnos 023263
Museum of Folklore Arts and Culture, Salamina 023251
Museum of Freemasonry, London 045043
Museum of Fulham Palace, London 045044
Museum of Garden History, London 045045
Museum of Gems and Minerals, Ratnapura 040487
Museum of Geology, Edmonton . 005757
Museum of Geology, South Dakota School of Mines and Technology, Rapid City 052472
Museum of Geometric and Madi Art, Dallas 048120
Museum of Giannoulis Halepas, Pyrgos 023239
Museum of Glass and Ceramics, South Portland 053414

Museum of Glass, International Center for Contemporary Art, Tacoma 053647
Museum of Greek Folk Art, Athinai 023011
Museum of Greek Folk Art-Pottery Collection, Athinai 023012
Museum of Greek Musical Instruments, Thessaloniki 023296
Museum of Gujarat Vidyasabha, Ahmedabad 024126
Museum of Harlow, Harlow 044513
Museum of Hartlepool, Hartlepool 044525
Museum of Health Care, Kingston 006109
Museum of Historical Archaeology, Kingston 028448
Museum of Historical Geology and Geological Survey of Pakistan, Quetta 034148
Museum of History and Art, Ontario 051781
Museum of History and Industry, Seattle 053240
Museum of History, Anthropology and Art, San Juan 036169
Museum of Holography, Chicago 047680
Museum of Human Sciences, Harare 054850
Museum of Hygiene, Cairo 010547
Museum of Idaho, Idaho Falls .. 049653
Museum of Imaging Technology, Chulalongkorn University, Bangkok 042664
Museum of Independent Telephony, Abilene 046305
Museum of Indian Arts and Culture, Santa Fe 053114
Museum of Indian Culture, Allentown 046412
Museum of Industry, Stellarton . 007073
Museum of Inhaca Island, Maputo 031810
Museum of Installation, London 045046
Museum of Instruments, Royal College of Music, London 045047
Museum of International Folk Art, Santa Fe 053115
Museum of International Friendship, Beijing 007632
Museum of Irish Transport, Killarney 024941
Museum of Iron and Darby Furnace, Coalbrookdale 043976
Museum of Islamic Art, Cairo .. 010548
Museum of Islamic Art (MIA), Doha 036180
Museum of Islamic Ceramic, Gezira Art Center, Cairo 010549
Museum of Island History, Newport, Isle of Wight 045365
Museum of Islay Life, Port Charlotte 045532
The Museum of Israeli Art, Ramat Gan 025175
Museum of Jewish Heritage – A Living Memorial to the Holocaust, New York 051439
Museum of Jurassic Technology, Culver City 048088
Museum of Kaigarayama Shell Mound, Kiyosu 028835
Museum of Kalymnos, Kalymnos 023108
Museum of Kent Life, Sandling . 045733
Museum of Kerala History, Edapally 024251
Museum of Komotini, Komotini . 023137
Museum of Korean Buddhist Art, Seoul 030024
Museum of Korean Commercial History, Yongin 030063
Museum of Korean Embroidery, Seoul 030025
Museum of Korean Traditional Music, Seoul 030026

Museum of Lakeland Life, Kendal 044676
Museum of Lancashire, Preston 045566
Museum of Latin American Art, Long Beach 050372
Museum of Lead Mining, Wanlockhead 046048
Museum of Lighting, Edinburgh 044252
Museum of Lillydale, Lillydale .. 001221
Museum of Lincolnshire Life, Lincoln 044833
Museum of Liverpool Life, New Museum of Liverpool, Liverpool 044861
Museum of Local History, Burnley 043792
Museum of Local History, Hastings 044533
Museum of Local History and Industry, Llandrindod Wells 044874
Museum of London, London ... 045048
Museum of Long Island Natural Sciences, Stony Brook 053549
Museum of Macao, Macau ... 008072
Museum of Making Music, Carlsbad 047431
Museum of Malawi, Blantyre .. 030654
Museum of Man in the Sea, Panama City 051908
Museum of Marine Life of Cyprus, Aghia Napa 009399
Museum of Maroccan Arts, Tanger 031798
Museum of Maryland History, Maryland Historical Society, Baltimore 046764
Museum of Medical Equipment, Bangkok 042665
Museum of Medical History, Shanghai 008231
Museum of Medical Missionary in Remote Areas, Puli 042420
Museum of Medical Sciences, Hong Kong 007882
Museum of Medieval Torture Instruments, Praha 009811
Museum of Meteorology, Tainan 042487
Museum of Methodism, London 045049
Museum of Migrating People, Harry S. Truman High School, Bronx ... 047235
Museum of Mineralogy, Petrology and Geology, Zografou 023326
Museum of Mobile, Mobile 050979
Museum of Modern Art, Cairo .. 010550
Museum of Mohamed Mahmoud Khalil and his wife, Giza 010572
Museum of Mongolian Ethnography, Ulaanbaatar 031736
The Museum of Museums, Waregem 004130
Museum of NaJu Pears, Naju .. 029973
Museum of National Architecture and Urban Life of Gyumri, Gyumri . 000705
Museum of Natural and Cultural History, University of Oregon, Eugene 048622
Museum of Natural History, Jerusalem 025090
Museum of Natural History, Nanaimo 006448
Museum of Natural History, Providence 052388
The Museum of Natural History, Stevens Point 053524
Museum of Natural History, Tshwane 038860
Museum of Natural History, Ulaanbaatar 031737
Museum of Natural History and Historical Photography, Polygyros 023232
Museum of Natural History and Primitive Technology, Somers . 053379
Museum of Natural History and Technology, Shiraz University,

Shiraz 024736
Museum of Natural History, National
Museum of Ireland Ard-Mhúsaem na
hÉireann, Dublin 024899
Museum of Natural History, Southwest
State University, Marshall 050712
Museum of Natural History, University
of Massachusetts, Amherst . . . 046452
Museum of Natural History, University
of the Philippines Los Baños, College
Los Baños 034471
Museum of Natural History, Wayne
State University, Detroit . . . 048314
Museum of Natural Resources
in Xiaoxinganling, Yichun,
Heilongjiang 008481
Museum of Natural Science, Baton
Rouge 046819
Museum of Natural Sciences,
Saskatoon 006980
Museum of Nautical Art, Symi . 023278
Museum of Nebraska Art,
Kearney 049910
Museum of Nebraska History,
Lincoln 050296
Museum of Neon Art, Los
Angeles 050458
Museum of Net Manufacture,
Bridport 043729
Museum of New Art, Pontiac . . 052237
Museum of New Hampshire History,
Concord 047959
Museum of New Mexico, Santa
Fe 053116
The Museum of Newport History,
Newport 051534
Museum of North Carolina Minerals,
Spruce Pine 053479
Museum of North Craven Life,
Settle 045761
Museum of North Idaho, Coeur
d'Alene 047856
Museum of Northern Arizona,
Flagstaff 048740
Museum of Northern British Columbia,
Prince Rupert 006719
Museum of Northern History at the
Sir Harry Oakes Château, Kirkland
Lake 006120
Museum of Northwest Art, La
Conner 050024
Museum of Northwest Colorado,
Craig 048049
The Museum of Nursing,
Camperdown 000934
The Museum of Nursing History,
Philadelphia 052070
Museum of Old Photographs and
Museum of Unearthed Artifacts,
Nagasaki 029093
Museum of Old Western Time,
Singdian 042461
Museum of Orient and East Marion
History, Oysterponds Historical
Society, Orient 051800
The Museum of Outdoor Arts,
Englewood 048587
Museum of Oxford, Oxford . . . 045454
Museum of Paleontology, University of
California, Berkeley, Berkeley . 046939
Museum of Palestinian Popular
Heritage, Al-Bireh 034155
Museum of Pathology, University of
Sydney, Sydney 001545
Museum of Pediatric Surgery,
Chulalongkorn University,
Bangkok 042666
Museum of Peoples and Cultures,
Provo 052406
Museum of Philippine Art, Manila 034539
Museum of Philippine Political History,
Manila 034540
Museum of Photographic Arts, San
Diego 052968

Museum of Pioneer Settlement,
Kibbutz Yifat 025149
Museum of Piping, National Piping
Centre, Glasgow 044408
Museum of Porn in Art, Zürich . 042220
Museum of Post and Telecomunication
and Mokhaberat Museum,
Tehrän 024773
Museum of Postindependence Athens,
Athinai 023013
Museum of Prehistory,
Jerusalem 025091
Museum of Prehistory, Shaar
Hagolan 025183
Museum of Prehistory and
History, Arkansas Tech University,
Russellville 052711
Museum of Primitive Art and Culture,
Peace Dale 051956
Museum of Primitive Methodism,
Englesea Brook, Crewe 044039
Museum of Printing History,
Houston 049589
Museum of Prints and Printmaking,
Albany 046343
Museum of Psalms, Jerusalem . 025092
Museum of Radio and Technology,
Huntington 049628
Museum of Rail Travel, Ingrow . 044632
Museum of Reading, Reading . . 045590
Museum of Regional and
Mediterranean Archaeology, Nir-
David 025164
Museum of Religious Art,
Summerside 007099
Museum of Religious Art, Saint Louis
University, Saint Louis 052814
Museum of Richmond, Richmond,
Surrey 045614
Museum of Robotics, Orinda . . . 051801
Museum of Rugby, Twickenham 046009
Museum of Rural Life, Denton . 048245
Museum of Russian Art,
Minneapolis 050945
Museum of Russian Arts,
Yerevan 000742
Museum of Russian Culture, San
Francisco 053015
Museum of Sacred Art, Macau . 008073
Museum of Sacred Icons, Athinai 023014
Museum of Saint Albans, Saint
Albans 045668
Museum of Samoa, Apia 038094
Museum of Science, Boston . . . 047124
Museum of Science and History of
Jacksonville, Jacksonville 049775
Museum of Science and Industry,
Chicago 047681
Museum of Science and Industry,
Tampa 053669
Museum of Science and Industry in
Manchester, Manchester 045212
Museum of Science and Technology,
Accra 022924
Museum of Science and Technology,
Tshwane 038852
Museum of Science at the Toledo Zoo,
Toledo 053759
Museum of Scotland, National
Museums of Scotland,
Edinburgh 044253
Museum of Scottish Country Life,
National Museums of Scotland, East
Kilbride 044206
Museum of Scottish Lighthouses,
Fraserburgh 044371
Museum of Sculpture and Figurines of
Loukia Georganti, Athinai 023015
Museum of Seminole County History,
Sanford 053080
Museum of Sex, New York . . . 051440
Museum of Shadow Puppet,
Gangshan 042315
Museum of Shanxi Province,

Taiyuan 008313
Museum of Shells of the Greek Seas,
Néa Moudiana 023201
Museum of Sisiyat Culture,
Nanjhuang 042403
Museum of Site, Most I, Hong
Kong 007883
Museum of Site, Most II, Hong
Kong 007884
Museum of South Somerset,
Yeovil 046217
Museum of South Texas History,
Edinburg 048478
Museum of South Yorkshire Life,
Doncaster 044111
Museum of Southern African Rock Art,
Johannesburg 038755
Museum of Southern History,
Jacksonville 049776
The Museum of Southern History,
Sugarland 053592
Museum of Southwestern Biology,
Albuquerque 046365
Museum of Soviet Repression,
Tbilisi 016416
Museum of Spanish Colonial Art,
Santa Fe 053117
Museum of Speed, Pendine . . . 045476
Museum of Surveying, Lansing . 050139
Museum of Sussex Archaeology,
Lewes 044819
Museum of Sydney, Sydney . . . 001546
Museum of Taxes, Jerusalem . . 025093
Museum of Teenage Art,
Housatonic 049570
Museum of Telecommunications, Néa
Kifissia 023199
Museum of Television and Radio,
Beverly Hills 046973
The Museum of Television and Radio,
New York 051441
Museum of Texas Tech University,
Lubbock 050517
Museum of Thai Herbal Medicine
and Thai Pharmaceutical History,
Chulalongkorn University,
Bangkok 042667
Museum of the 40's, Point Clear
Bay 045516
Museum of the Agricultural
College and Research Institute,
Coimbatore 024205
Museum of the Albemarle, Elizabeth
City 048525
Museum of the Aleutians,
Unalaska 053882
Museum of the Alphabet, Jaars
Summer Institute of Linguistics,
Waxhaw 054243
Museum of the American Cowboy,
Colorado Springs 047888
Museum of the American Hungarian
Foundation, New Brunswick . 051246
Museum of the American Numismatic
Association, Colorado Springs . 047889
Museum of the American Piano, New
York 051442
Museum of the American Quilter's
Society, Paducah 051884
Museum of the Arkansas Grand
Prairie, Stuttgart 053586
Museum of the Bedford Historical
Society, Bedford 046869
Museum of the Berkeley Springs,
Berkeley Springs 046943
Museum of the Big Bend, Alpine 046421
Museum of the Cape Fear Historical
Complex, Fayetteville 048721
Museum of the Cariboo Chilcotin,
Williams Lake 007417
Museum of the Central Marine
Fisheries Research Station,
Mandapam Camp 024393
Museum of the Cherokee Indian,

Cherokee 047607
Museum of the Cherokee Strip,
Oklahoma Historical Society,
Enid 048588
Museum of the City of Athens, Vouros-
Eutaxias, Athinai 023016
Museum of the City of Lake Worth,
Lake Worth 050091
Museum of the City of New York, New
York 051443
Museum of the College of Engineering,
Chennai 024200
The Museum of the Confederacy,
Richmond 052557
Museum of the Convent of
Agios Andreas, Agios Andreas
Arkadias 022946
Museum of the Cultural Palace of
National Minorities, Beijing . . . 007633
Museum of the Cumbraes,
Millport 045268
Museum of the Democratic
Movements, Machida 028993
Museum of the Dog, Saint Louis 052815
Museum of The Duke of Wellington's
Regiment, Halifax 044502
Museum of the Egyptian People's
Assembly, Cairo 010551
Museum of the Faculty of Arts,
Alexandria 010520
Museum of the Franciscan Convent,
Jerusalem 025094
Museum of the Fur Trade,
Chadron 047511
Museum of the Geological Sciences,
Blacksburg 047028
Museum of the George and Nefeli
Giabra Pierides Collection, Bank
of Cyprus Cultural Foundation,
Lefkosia 009419
Museum of the Gonia Monastery,
Kissamos 023133
Museum of the Great Plains,
Lawton 050195
Museum of the Grove of the Village
Shrine, Hirano Hakabutsukan,
Osaka 029250
Museum of the Gulf Coast, Port
Arthur 052245
Museum of the Highwood, High
River 006004
Museum of the Historical Society
of Trappe, Collegeville, Perkiomen
Valley, Trappe 053789
Museum of the History and Art
of the Holy City of Missolonghi,
Mesolongion 023167
Museum of the History of Children's
Games and Books, Myrina . . . 023183
Museum of the History of Cypriot
Coinage, Lefkosia 009420
Museum of the History of Medicine,
Baki 003172
Museum of the History of Science,
Oxford 045455
Museum of the History of Science,
Sydney 001547
Museum of the History of Tiberias and
the Lake Kinnereth, Tiberias . . 025219
Museum of the Holmesdale Natural
History Club, Reigate 045602
Museum of the Holy Trinity Church of
Ethiopia, Addis Ababa 010723
Museum of the Hudson Highlands,
Cornwall-on-Hudson 047999
Museum of the Imperial Palace of
Manchu State, Chiaoxchun . . . 007666
Museum of the Institute of African
Studies, Ibadan 033422
Museum of the Institute of African
Studies, Oshogbo 033440
Museum of the Institute of Ethnology,
Academia Sinica, Taipei 042533
Museum of the Institute of History

and Philology, Academia Sinica,
Taipei 042534
Museum of the Iron Age,
Andover 043417
Museum of the Isles, Isle-of-
Skeye 044658
Museum of the Jewellery Quarter,
Birmingham 043623
Museum of the Jimmy Carter Library,
Atlanta 046628
Museum of the Korean Cultural Center,
Los Angeles 050459
Museum of the Llano Estacado,
Plainview 052190
Museum of the Manchester Regiment,
Ashton-under-Lyne 043446
Museum of the Mausoleum of King
Nanyue, Guangzhou 007811
Museum of the Medical College,
Mysore 024418
Museum of the Meiji Restoration,
Kagoshima 028747
Museum of the Merchant Marine,
Piraeus 023223
Museum of the Metropolis Square,
Naxos 023197
Museum of the Mid-Suffolk Light
Railway, Wetheringsett 046104
Museum of the Mitsubishi Heavy
Industries, Nagasaki 029094
Museum of the Mountain Man,
Pinedale 052140
Museum of the National Center of
Afro-American Artists, Boston . 047125
Museum of the Northamptonshire
Regiment, Northampton 045390
Museum of the Northern Great Plains,
Fort Benton 048782
Museum of the Olive and Greek Olive
Oil, Spárti 023274
Museum of the Order of Saint John,
London 045050
Museum of the Percy Tenantry
Volunteers 1798–1814, Alnwick 043402
Museum of the Plains Indian,
Browning 047276
Museum of the Puerto Rican Family of
the 19th Century, Instituto de Cultura
Puertorriqueña, San Juan 036170
Museum of the Red River, Idabel 049652
Museum of the Regiments,
Calgary 005528
Museum of the Rockies,
Bozeman 047166
Museum of the Royal College of
Physicians, London 045051
Museum of the Royal College of
Surgeons of England, London . 045052
Museum of the Royal Leicestershire
Regiment, Leicester 044797
Museum of the Royal Military School
of Music, Twickenham 046010
Museum of the Royal Pharmaceutical
Society of Great Britain, London 045053
Museum of the Royal Thai Air Force,
Bangkok 042668
Museum of The Royal Westminster
Regiment Historical Society, New
Westminster 006474
Museum of the Rural Municipality of
Roland, Roland 006823
Museum of the Sacrifice of the People
of Kalavryta, Kalavryta 023105
Museum of the Sarvajanik
Wachanalaya, Nasik 024423
Museum of the Shwenadwar Pagoda,
Bago 031814
The Museum of the South Dakota
State Historical Society, Pierre 052129
Museum of the Southern Jewish
Experience, Jackson 049754
Museum of the Southwest,
Midland 050878
Museum of the Statesman Eleftherios

Venizelos and the Corresponding
Historical Period, Leshi Fileleytheron,
Athinai 023017
Museum of the Taipei Municipal
Cheng Kung Senior High School,
Taipei 042535
Museum of the Underground
Prisoners-Acre, Acre 025010
Museum of the University of
Connecticut Health Center,
Farmington 048702
Museum of the Waxhaws and Andrew
Jackson Memorial, Waxhaw . . 054244
Museum of the Weathersfield
Historical Society, Reverend Dan
Forster House, Weathersfield . 054259
Museum of the Western Jesuit
Missions, Hazelwood 049423
Museum of the Western Prairie,
Altus 046428
Museum of the Weston Historical
Society, Weston 054350
Museum of the White Mountains of
Samaria, Chania 023042
Museum of the Worcestershire
Regiment, Worcester 046192
Museum of Tinian Artists, Pyrgos 023240
Museum of Tolerance, A Simon
Wiesenthal Center Museum, Los
Angeles 050460
Museum of Tolerance, Simon
Wiesenthal Center, Los Angeles 050461
Museum of Traditional Costumes, Alor
Gajah 030658
Museum of Traditional Nan Bel Music
and Theater, Changhua 042284
Museum of Transport, Glasgow . 044409
Museum of Transport – Greater
Manchester, Manchester 045213
Museum of Transport and Technology,
Auckland 033155
Museum of Transportation, Saint
Louis 052816
Museum of Transportation and Rural
Industries, Boyanup 000876
Museum of Tropical Queensland,
Townsville 001589
Museum of Vertebrate Zoology,
University of California, Berkeley,
Berkeley 046940
Museum of Veterinary, Agricultural
University, Guwahati 024278
Museum of Victorian Reed Organs and
Harmoniums, Shipley 045788
Museum of Victorian Whitby,
Whitby 046115
Museum of Vintage Fashion,
Lafayette 050057
Museum of Vintage Fashion, Lincoln
Hills 050308
Museum of Vision Science and
Optometry, Waterloo 007365
Museum of Vision, Foundation of the
American Academy of Ophthalmology,
San Francisco 053016
Museum of Waldensian History,
Valdese 053924
Museum of Weapons, Gjirokastër 000017
Museum of Wellington, City and Sea,
Wellington 033351
Museum of Welsh Life, Saint
Fagans 045686
The Museum of Western Art,
Kerrville 049944
Museum of Western Australian Sport,
Claremont, Western Australia . 000968
Museum of Western Colorado, Grand
Junction 049182
Museum of Witchcraft, Cornwall 044011
Museum of Wonders, Imbil 001163
Museum of Woodcarving, Shell
Lake 053304
Museum of Woodwork and
Ornamental Turning, Ashburton 033137

Museum of Worcester Porcelain,
Worcester 046193
Museum of World Religions,
Yungho 042607
Museum of Yachting, Newport . 051535
Museum of Yarmouth History,
Yarmouth 054570
Museum of Yarmukian, Kibbutz Shaar
Hagolan 025148
Museum of Yilan Government History,
Yilan 042591
Museum of Yokohama Urban History,
Yokohama 029720
Museum of York County, Rock
Hill 052625
Museum of Yorkshire Dales Lead
Mining, Earby 044195
Museum of Zoology, University of
Michigan, Ann Arbor 046493
Museum on Main Street,
Pleasanton 052202
Museum on the Boyne, Alliston . 005316
Museum on the Mound,
Edinburgh 044254
Museum on the Seam,
Jerusalem 025095
Museum onder de N.H. Kerk,
Elst 032300
Museum Ons' Lieve Heer op Solder,
Amsterdam 031962
Museum Ons Mijnverleden,
Houthalen 003716
Museum Oorlog en Vrede, Breda 032103
Museum Orth, Orth an der
Donau 002481
Museum Orvelte, Orvelte 032733
Museum Oskar Reinhart am
Stadtgarten, Winterthur 042158
Museum Osterzgebirgsgalerie im
Schloß, Dippoldiswalde 017909
Museum Osthusschule, Bielefeld 017379
Museum Otto Ludwig, Eisfeld . . 018146
Museum Otto Schäfer, Buchkunst,
Graphik, Kunsthandwerk,
Schweinfurt 021785
Museum Otzberg, Sammlung zur
Volkskunst in Hessen, Otzberg 021055
Museum Oud-Asperen, Asperen 032031
Museum Oud Overschie,
Rotterdam 032811
Museum Oud-Rijnsburg,
Rijnsburg 032784
Museum Oud Soest, Soest 032893
Museum Oud Vriezenveen,
Vriezenveen 033010
Museum Oud Westdorpe,
Westdorpe 033037
Museum Oude Boekdrukkunst,
Almen 031900
Museum Pachen, Rockenhausen 021441
Museum Pachten, Dillingen, Saar 017901
Museum Palagan, Ambarawa . . 024525
Museum Palthehof, Nieuwleusen 032679
Museum Pancasila Sakti,
Jakarta 024582
Museum Pangeran Diponegoro,
Yogyakarta 024679
Museum Papageorgiou,
Limenaria 023154
Museum Papiermühle Homburg,
Triefenstein 022193
Museum Passeier – Andreas Hofer,
San Leonardo in Passiria 027768
Museum Pater Valentinus Paquay-
Heilig Paterke, Hasselt 003681
Museum Paul Delvaux, Sint-
Idesbald 004015
Museum Paul Tétar Van Elven,
Delft 032163
Museum Pemda Balige, Balige . 024526
Museum Pemirintah Daerah Grobogan,
Purwodadi Grobogan 024638
Museum Pendidikan Islam,
Yogyakarta 024680

Museum Perjuangan, Bogor . . . 024548
Museum Perjuangan, Yogyakarta 024681
Museum Perjuangan Bukit Barisan,
Medan 024616
Museum Pers, Surakarta 024658
Museum Pers Antara, Jakarta . . 024583
Museum Petersberg, Petersberg bei
Halle, Saale 021106
Museum Pfalzgalerie,
Kaiserslautern 019480
Museum Pfleggerichtshaus,
Anras 001716
Museum Plagiarius, Solingen . . 021902
Museum Plantin-Moretus /
Prentenkabinet, Antwerpen . . . 003295
Museum Pos & Giro, Bandung . 024534
Museum Prabu Geusan Ulun,
Yayasan Pangeran Sumedang,
Sumedang 024651
Museum Practice, London 139208
Museum Presidency,
Bloemfontein 038659
Museum Priesterhäuser, Städtische
Museen Zwickau, Zwickau . . . 022910
Museum Professionals Group,
Duxford 060184
Museum Prüm, Prüm 021233
Museum Pugung Ulago Sembah,
Donggala 024558
Museum Pura Mangkunegara,
Surakarta 024659
Museum Purbakala Mojokerto,
Mojokerto 024618
Museum Purbakala Trowulan,
Mojokerto 024619
Museum Puri Lukisan, Foundation
Ratna Wartha, Ubud 024669
Museum Quality Finishes, New
York 135964
Museum Quintana – Archäologie in
Künzing, Künzing 019820
Museum Raabe-Haus,
Eschershausen 018276
Museum Rade am Schloß Reinbek,
Sammlung Rolf Italiaander, Hans
Spegg, Reinbek 021349
Museum Radio-Wereld, Diever . 032226
Museum Radya Pustaka,
Surakarta 024660
Museum Raesfeld, Raesfeld . . . 021263
Museum Ratna Wartha, Gianyar 024560
Museum Region Neulengbach,
Neulengbach 002442
Museum Regiunal, Savognin . . 041943
Museum Regiunal Surselva Casa
Carniec, Ilanz 041603
Museum Reich der Kristalle,
Mineralogische Staatssammlung
München, München 020557
Museum Reichenau, Reichenau 021342
Museum Reichenfels,
Hohenleuben 019305
Museum Rekor Indonesia Jamu Jago,
Semarang 024647
Museum Reksa Artha, Jakarta . 024584
Museum René de Clercq,
Deerlijk 003534
Museum Retz, Stadtsammlung und
Südmährische Galerie, Retz . . 002589
Museum Rietberg Zürich, Zürich 042221
Museum Rijn- en Binnenvaart,
Antwerpen 003296
Museum Rijswijk Het Tollenshuis,
Rijswijk, Zuid-Holland 032787
Museum Ritter, Sammlung Marli
Hoppe-Ritter, Waldenbuch . . . 022386
Museum Roemervilla, Bad Neuenahr-
Ahrweiler 016835
Museum Römervilla, Grenzach-
Wyhlen 018812
Museum Rolf Werner, Bansin,
Seebad 016963
Museum Rosenegg, Kreuzlingen 041625
Museum Rudolf Wachter, Neues

Schloß Kißleg, Kißlegg 019636
Museum Rufferdinge,
Bezoekerscentrum Rufferdinge,
Landen 003773
Museum Ruman Bolon,
Pematang 024632
Museum Rupelklei, Terhagen . 004055
Museum Sacrum, Tiel 032924
Museum Sächsisch-Böhmisches
Erzgebirge, Marienberg 020280
Museum Saet en Cruyt, Andijk . 032004
Museum Saigerhütte Olbernhau-
Kupferhammer, Olbernhau ... 020983
Museum Salz und Moor, Grassau 018798
Museum Salzbütte, Huttwil 041602
Museum Sankt Blasien, Sankt
Blasien 021589
Museum Sankt Dionys, Esslingen 018320
Museum Sankt Ingbert, Sankt
Ingbert 021599
Museum Sankt Leon-Rot, Sankt Leon-
Rot 021603
Museum Sankt Urbanhof Sursee,
Sursee 042045
Museum Sankt Veit, Sankt Veit an der
Glan 002703
Museum Sankt Wendel, Mia-Münster-
Haus, Sankt Wendel 021613
Museum Saoraja Lapawowoi,
Watampone 024671
Museum Sarganserland, Sargans 041937
Museum Sasmita Loka,
Yogyakarta 024682
Museum Sasmita Loka A. Yani,
Jakarta 024585
Museum Satria Mandala, Jakarta 024586
Museum Scheuerle Waldrems,
Backnang 016692
Museum Scheveningen, Den
Haag 032187
Museum Schietkamp-Harskamp,
Harskamp 032417
Museum SchillerHaus, Mannheim 020252
Museum Schloß Adelsheim,
Heimatmuseum, Berchtesgaden 017039
Museum Schloß Bernburg,
Bernburg 017333
Museum Schloß Bruck, Museum der
Stadt Lienz, Lienz 002308
Museum Schloß Burgk, Burgk .. 017694
Museum Schloss Colditz, Colditz 017759
Museum Schloß Ehrenstein,
Ohrdruf 020980
Museum Schloss Fasanerie, Hessische
Hausstiftung, Eichenzell 018117
Museum Schloss Fellenberg,
Kulturstiftung für den Landkreis
Uerzig-Wadeln, Merzig 020382
Museum Schloss Friedrichsfelde,
Stiftung Stadtmuseum Berlin,
Berlin 017253
Museum Schloß Glücksburg,
Römhild 021456
Museum Schloss Hardenberg,
Velbert 022309
Museum Schloss Hellenstein,
Heidenheim an der Brenz 019155
Museum Schloss Herzberg, Herzberg
am Harz 019232
Museum Schloß Hohenlimburg,
Hagen 018918
Museum Schloss Homburg,
Nümbrecht 020842
Museum Schloss Hundshaupten,
Egloffstein 018103
Museum Schloß Klippenstein,
Radeberg 021255
Museum Schloß Kuckuckstein,
Liebstadt 020068
Museum Schloss Kyburg, Kyburg 041632
Museum Schloß Lackenbach,
Lackenbach 002257
Museum Schloss Lembeck,
Dorsten 017948

Museum Schloss Lichtenberg,
Fischbachtal 018366
Museum Schloß Lichtenburg,
Prettin 021217
Museum Schloß Molsdorf, Erfurt 018248
Museum Schloß Moritzburg,
Zeitz 022852
Museum Schloss Münsingen,
Münsingen 041776
Museum Schloss Netzschkau,
Netzschkau 020671
Museum Schloss Neuenburg, Freyburg
(Unstrut) 018508
Museum Schloss Oberschwappach,
Knetzgau 019659
Museum Schloss Ottersbach,
Großklein 002008
Museum Schloss Peigarten,
Dobersberg 001819
Museum Schloß Ratibor, Roth . 021504
Museum Schloß Rochlitz,
Rochlitz 021438
Museum Schloß Rochsburg,
Lunzenau 020191
Museum Schloß Schwarzenberg,
Schwarzenberg 021774
Museum Schloss Seefeld, Staatliches
Museum für Ägyptische Kunst,
Seefeld 021813
Museum Schloss Steinau an
der Straße mit Brüder-Grimm-
Gedenkstätte, Steinau an der
Straße 021975
Museum Schloß Steinheim,
Hanau 019041
Museum Schloss und Gut Liebenberg,
Löwenberger Land 020101
Museum Schloss Waldegg,
Feldbrunnen 041483
Museum Schloß Weesenstein,
Müglitztal 020478
Museum Schloß Wilhelmsburg,
Schmalkalden, Kurort 021678
Museum Schmelzra, Scuol 041972
Museum Schmiede am Burgplatz,
Backnang 016693
Museum Schneggli, Reinach
(Aargau) 041869
Museum Schnütgen, Museen der
Stadt Köln, Köln 019706
Museum Schöne Stiege,
Riedlingen 021418
Museum Schöngrabern,
Hollabrunn 002101
Museum Schokland, Ens 032306
Museum Schorfheide,
Schorfheide 021724
Museum Schwab, Museum für
Urgeschichte und Archäologie,
Biel 041341
Museum Schwarzes Ross,
Hilpoltstein 019254
Museum Science and Technology,
Jakarta 024587
Museum Seitingen-Oberflacht,
Seitingen-Oberflacht 021832
Museum Sejarah Mesjid Banten,
Bogor 024549
Museum Sender Dobl, Dobl ... 001821
Museum Seni, Denpasar 024556
Museum Seni, Jakarta 024588
Museum Seni, Medan 024617
Museum Seni, Padang 024623
Museum Seni Rupa dan Keramik,
Jakarta 024589
Museum Sensenwerk,
Deutschfeistritz 001811
Museum Sepakat Segenap, Aceh
Tenggara 024522
Museum Services Australia,
Pymble 127665
Museum Shalom Europa,
Würzburg 022797
Museum Ship Norgoma, Saint Mary's

River Marine Centre, Sault Sainte
Marie 006990
Museum Simalungun,
Pematangsiantar 024634
Museum Sint Bernardushof,
Aduard 031888
Museum Sint Sellian, Sellingen . 032868
Museum Siwa Lima, Ambon ... 010407
Museum Slag bij Heiligerlee,
Heiligerlee 032444
Museum Slag der Zilveren Helmen,
Halen 003663
Museum Slager, 's-
Hertogenbosch 032470
Museum Smallingerland,
Drachten 032252
Museum Smidt van Gelder,
Antwerpen 003297
Museum Soltau, Soltau 021911
Museum Sono Budoyo,
Yogyakarta 024683
Museum Sonyine Malige, Halmahera
Tengah 010410
Museum Sophie La Roche,
Bönnigheim 017462
Museum Spaans Gouvernement,
Maastricht 032624
Museum Specken, Bad
Zwischenahn 016918
Museum SPUR Cham, Cham ... 017726
Museum Stad Appingedam,
Appingedam 032017
Museum Stadt Bad Hersfeld, Bad
Hersfeld 016784
Museum Stadt Königsberg,
Duisburg 018061
Museum Stammertal,
Unterstammheim 042084
Museum Stapelen, Boxtel 032095
Museum Starnberger See,
Starnberg 021961
Museum Stedhûs Sleat, Sloten . 032887
Museum Stein, Krems 002237
Museum Steinarbeiterhaus,
Hohburg 019299
Museum Steinau... das Museum
an der Straße, Steinau an der
Straße 021976
Museum Steinmetzhaus,
Eggenburg 001844
Museum Stellingwerff-Waerdenhof,
Hasselt 003682
Museum Stemmler,
Schaffhausen 041949
Museum Stichting van het
Rotterdamsche Tramweg
Maatschappij, Ouddorp 032746
Museum Stift Börstel, Berge ... 017043
Museum Stift Melk, Melk 002381
Museum Stoffels Säge-Mühle,
Hohenems 002098
Museum Stoomdepot, Rotterdam 032812
Museum Stoomgemaal Winschoten,
Winschoten 033051
Museum Strauhof,
Literaturausstellungen, Zürich . 042222
Museum Strom und Leben,
Umspannwerk Recklinghausen,
Recklinghausen 021307
Museum Studies, Chicago 139446
The Museum Studio and Arts Centre,
Nairobi 029867
Museum Suaka Budaya,
Surakarta 024661
Museum Subak Bali, Kediri 024602
Museum Sudirman, Magelang .. 024608
Museum Sulawesi Tengah, Palu 024629
Museum Sumpah Pemuda,
Jakarta 024590
Museum Suriname, Amsterdam . 031963
Museum Sursilvan, Trun 042072
Museum Swaensteyn, Voorburg 033005
Museum Synagoge Gröbzig,
Gröbzig 018826

Museum Synthese, Hallbergmoos 018947
Museum 't Coopmanshûs,
Franeker 032329
Museum 't Gevang, Doetinchem 032233
Museum 't Kniphof, Essen 003586
Museum 't Nieuwhuys Hoegaarden,
Hoegaarden 003702
Museum 't Schilderhuis, Driebergen-
Rijsenburg 032254
Museum Taf Wallet, Koksijde .. 003751
Museum Takstil, Jakarta 024591
Museum Taman Laut Ancol,
Jakarta 024592
Museum Taman Prasati, Jakarta 024593
Museum Taruna Akbari Udarat,
Magelang 024609
Museum Tauernbahn,
Schwarzach 002733
Museum Tegernseer Tal,
Tegernsee 022115
Museum Telefonzentrale Magden,
Museum für Kommunikation, CH-
3000 Bern 6, Magden 041717
Museum Theo Jans, Klarenbeek 032547
Museum Theo Kerg, Museum der
Stadt Schriesheim, Schriesheim 021736
Museum Theulegium, Tholey ... 022152
Museum Thonet, Stuhlmuseum,
Frankenberg, Eder 018388
Museum Thurnfels, Völs 002871
Museum Tierpark Sababurg,
Hofgeismar 019291
Museum Tinguely, Basel 041280
Museum Tni A.L. Loka Jala Carana,
Surabaya 024656
Museum Ton Schulten,
Ootmarsum 032729
Museum Tre Kronor, Stockholm 041022
Museum Trustee Association,
Washington 060607
Museum Tugu Nasional, Jakarta 024594
Museum Tysmans, Houthalen .. 003717
Museum U.P.T. Balitung,
Tanjungpandan 024665
Museum und Begegnungsstätte Alter
Krug, Zossen 022897
Museum und Besucherbergwerk,
Goslar 018771
Museum und Besucherbergwerk
der Graphit Kropfmühl AG,
Hauzenberg 019114
Museum und Galerie der Stadt,
Schwabmünchen 021745
Museum und Galerie in Uigendorf,
Unlingen 022272
Museum und Kunstsammlung Schloß
Hinterglauchau, Glauchau ... 018711
Museum und Kupferzeche Hüttau,
Hüttau 002105
Museum und Stiftung Anton
Geiselhart, Münsingen 020590
Museum unter der Y-Burg, Kernen im
Remstal 019593
Museum Uranbergbau, Bad
Schlema 016866
Museum Uslar, Uslar 022291
Museum Valkenheide,
Maarsbergen 032613
Museum Valse Kunst, Museums
Vledder, Vledder 032995
Museum Van Bommel Van Dam,
Venlo 032983
Museum van de Abdij van Roosenberg,
Waasmunster 004122
Museum van de Kanselarij der
Nederlandse Orden, Apeldoorn 032011
Museum van de Koninklijke
Marechaussee, Buren,
Gelderland 032124
Museum van de Kruisbooggilde Sint-
Joris, Brugge 003419
Museum van de Onze-Lieve-
Vrouwkerk, Brugge 003420
Museum van de Sint-Jakobskerk,

Antwerpen 003298
Museum van de Stichting Bisdom van
Vliet, Haastrecht 032402
Museum van de Twintigste Eeuw,
Hoorn 032506
Museum van de Vrouw, Echt . . 032263
Museum van de Willebroekse Vaart,
Willebroek 004151
Museum van Deinze en de Leiestreek,
Deinze 003536
Museum van Egmond, Egmond aan
Zee 032277
Museum Van Gerwen-Lemmens,
Valkenswaard 032969
Museum van Hedendaagse
Kantwerken, Sint-Truiden 004027
Museum van Hedendaagse Kunst
Antwerpen, Antwerpen 003299
Museum Van Hemessen,
Woubrugge 033067
Museum van het Abtenhuis,
Geraardsbergse Musea,
Geraardsbergen 003649
Museum van het Ambacht, Den
Haag 032188
Museum van het Heilig Bloed,
Baseliek van het Heilig Bloed,
Brugge 003421
Museum van het Kamp van Beverlo,
Leopoldsburg 003782
Museum van het Nederlandse
Uurwerk, Zaandam 033077
Museum van het Vlaams
Studentenleven, Leuven 003784
Museum van Knipkunst,
Westerbork 033039
Museum van Lien, Fijnaart 032327
Museum Van Loon, Amsterdam . 031964
Museum van Zwerfstenen in de
Rosmolen, Zeddam 033090
Museum Veluwezoom, Doorwerth 032244
Museum Verbindingsdienst, Ede . 032267
Museum Viadrina, Frankfurt
(Oder) 018401
Museum Victoria, Melbourne Museum,
Carlton South 000948
Museum Viechtach, Viechtach . . 022325
Museum Villa Erckens,
Grevenbroich 018814
Museum Villa Haiss, Zell am
Harmersbach 022854, 022855
Museum Villa Rot – Kunstsammlung
Hoenes-Stiftung, Burgrieden . . 017698
Museum Villa Rustica,
Möckenlohe 020430
Museum Villa Sarabodis,
Gerolstein 018664
Museum Villa Stahmer,
Georgsmarienhütte 018645
Museum Villa Stuck, München . . 020558
Museum Village, Monroe 051001
Museum Vitkovických Železáren,
Ostrava 009742
Museum Vitznau-Rigi, Vitznau . . 042114
Museum Vlaamse Minderbroeders,
Sint-Truiden 004028
Museum Vleeshuis, Antwerpen . 003300
Museum Vliegbasis Deelen,
Deelen 032152
Museum Volunteers of the Philippines,
Dasmariñas 059689
Museum voor Anesthesie Verantare,
Antwerpen 003301
Museum voor Communicatie, Den
Haag 032189
Museum voor de Geschiedenis van de
Wetenschappen, Gent 033633
Museum voor de Oudere Technieken,
Grimbergen 003658
Museum voor Figuratieve Kunst De
Buitenplaats, Eelde 032271
Museum voor Folklore, Zottegem 004167
Museum voor Geschiedenis van
de Geneeskunde, Universiteit

Gent, Faculteit Geneeskunde en
Gezondheidswetenschappen,
Gent 003634
Museum voor Heem- en
Oudheidkunde, Kontich 003754
Museum voor Heem- en Volkskunde,
Wenduine 004142
Museum voor Heemkunde Almelo,
Almelo 031897
Museum voor het Kruideniersbedrijf,
Utrecht 032950
Museum voor het
Radiozendamateurisme Jan Corver,
Budel 032115
Museum voor Industriele Archeologie
en Textiel, Gent 003635
Museum voor Keramiek Pablo Rueda
Lara, Rotterdam 032813
Museum voor Kerkelijke Kunst Sint-
Carolus, Antwerpen 003302
Museum voor Klederdracht en Oud
Speelgoed, Warnsveld 033023
Museum voor Moderne Kunst,
Arnhem 032027
Museum voor Moderne Kunst
Oostende (PMMK), Oostende . . 003931
Museum voor Naaldkunst,
Wedde 033030
Museum voor Natuur- en Wildbeheer,
Nederlands Jachtmuseum,
Doorwerth 032245
Museum voor Religieuze Kunst,
Uden 032933
Museum voor Schone Kunsten,
Oostende 003933
Museum voor Schone Kunsten Gent,
Gent 003636
Museum voor Sleutel en Slot,
Oostduinkerke 003922
Museum voor Torhouts Aardewerk,
Torhout 004078
Museum voor Volkskunde,
Dendermonde 003538
Museum voor Volkskunde, Musea
Brugge, Brugge 003422
Museum voor Vrede en
Geweldloosheid, Amsterdam . . 031965
Museum Voswinckelshof,
Dinslaken 017907
Museum Vrolik, Amsterdam . . . 031966
Museum Waemena, Irian Jaya . . 024563
Museum Wäschefabrik, Bielefeld 017380
Museum Wald und Umwelt,
Ebersberg 018082
Museum Waldhof, Bielefelder
Kunstverein e.V., Bielefeld . . . 017381
Museum Walzengravieranstalt
Guntramsdorf, Guntramsdorf . . 002023
Museum Wanua Paksinanta,
Manado 024612
Museum Wasseralfingen, Aalen . 016450
Museum Wasseramt, Halten . . . 041582
Museum Wasserburg Anholt,
Isselburg 019436
Museum Wasserburg, Kunst- und
kulturgeschichtliche Sammlungen der
Stadt, Wasserburg am Inn . . . 022439
Museum Wateringhuis, Lommel . 003827
Museum Wattens – Industrie &
Vorgeschichte, Wattens 002898
Museum Wayang, Jakarta 024595
Museum Weiler Textilgeschichte, Weil
am Rhein 022470
Museum Weißenfels Schloß Neu-
Augustusburg, Weißenfels . . . 022544
Museum Weltkulturen,
Mannheim 020253
Museum Wendener Hütte,
Wenden 022561
Museum West, San Jose 124455
Museum Wierdenland, Ezinge . . 032326
Museum Wiesbaden, Wiesbaden 022641
Museum Wildberg, Wildberg . . . 022651
Museum Willem van Haren,

Heerenveen 032430
Museum Willet-Holthuysen,
Amsterdam 031967
Museum Wirtles Haus,
Rottenacker 021518
Museum Wolfhalden, Wolfhalden 042168
Museum Wolfram von Eschenbach,
Wolframs-Eschenbach 022735
Museum Works Galleries, New
York 122904
Museum Works Galleries, West
Hollywood 125016
Museum Workshop and Stores,
Grangemouth 044460
Museum Würth, Künzelsau 019818
Museum Wurzen mit
Ringelnatzsammlung, Wurzen . 022828
Museum Wustrow, Wustrow,
Wendland 022834
Museum Wuyts-Van Campen en Baron
Caroly, Lier 003816
Museum Yadnya, Tangawi 024663
The Museum Yamato Bunkakan,
Nara 029146
Museum Yerseke, Yerseke 033072
Museum Yogya Kembali,
Yogyakarta 024684
Museum Zahal, Hakirya 025060
Museum Zeitreise Mensch,
Cortaccia 026079
Museum Zella-Mehlis, Zella-
Mehlis 022861
Museum Zinkenbacher Malerkolonie,
Sankt Gilgen 002661
Museum Zitadelle Jülich,
Stadtgeschichtliches Museum Jülich,
Jülich 019472
Museum zu Allerheiligen,
Schaffhausen 041950
Museum Zum Arabischen Coffe Baum,
Stadtgeschichtliches Museum Leipzig,
Leipzig 020010
Museum zum Schiff, Laufenburg 041644
Museum Zunftturm, Bacharach . 016688
Museum zur brandenburg-preußischen
Geschichte – Zinnfigurenmuseum,
Gusow-Platkow 018900
Museum zur Dresdner Frühromantik
im Kügelgen-Haus, Dresden . . 017988
Museum zur Geschichte der
Herrnhuter Brüdergemeine,
Königsfeld im Schwarzwald . . 019730
Museum zur Geschichte der Juden im
Ostalbkreis, Bopfingen 017508
Museum zur Geschichte der Juden
in Kreis und Stadt Heilbronn,
Obersulm 020931
Museum zur Geschichte Hohenheims,
Stuttgart 022052
Museum zur Geschichte von Christen
und Juden, Laupheim 019939
Museum zur Kloster- und
Stadtgeschichte, Steinheim an der
Murr 021988
Museum zur Ronmühle, Schötz . 041962
Museum zur Stadtgeschichte,
Königslutter am Elm 019734
Museum Zwaantje Hans Stockman's
Hof, Schoonebeek 032863
Museum Zwijgershoek, Stedelijke
Musea, Sint-Niklaas 004021
Museum, Ancient Synagogue and
Necropolis, Beit Shearim 025028
museumbickel, Walenstadt 042120
Museumboerderij, Leens 032566
Museumboerderij, Paasloo 032762
Museumboerderij, Staphorst . . . 032901
Museumboerderij de Karstenhoeve,
Ruinerwold 032836
Museumboerderij de Tip, Herveld 032478
Museumboerderij de Zwemkolk,
Markelo 032635
Museumboerderij Erve Hofman,
Hellendoorn 032454

Museumboerderij Eungs Schöppe,
Markelo 032636
Museumboerderij Gilde Koat,
Zeddam 033091
Museumboerderij New Greenwich
Village, Zuidwolde, Drenthe . . 033114
Museumboerderij Scholten, Den
Ham 032205
Museumboerderij 't Olde Maat Uus,
Giethoorn 032351
Museumboerderij Tante Jaantje,
Callantsoog 032132
Museumboerderij Vreeburg,
Schagen 032844
Museumboerderij Wendezoele, Ambt-
Delden 031903
Museumboerderij Westfrisia,
Hoogwoud 032504
Museumbrief, Antwerpen 138556
Museumbrouwerij de Hemel,
Nijmegen 032688
Museumbrouwerij De Roos,
Hilvarenbeek 032483
Museumcenter Leoben, Leoben 002293
Museumgemaal Caners, 's-
Hertogenbosch 032471
MuseumgoudA, Gouda 032366
Museumhaven Zeeland, Zierikzee 033103
'T Museumke, Handel 032406
Museumkelder Derlon,
Maastricht 032625
Museumkerk Sint Pieter, Rekem 003963
Museumleven, Antwerpen 138557
Museummolen, Schermerhorn . . 032848
Museummolen de Nieuwe Palmboom,
Schiedam 032854
Museummolen Jan Pol, Dalen . . 032143
Museums- und Kunstverein Osnabrück
e.V., Osnabrück 059178
Museums Alaska, Anchorage . . . 060608
Museums and Galleries Disability
Association, Hove 060187
Museums and Relevant Products and
Technologies Exposition, Beijing 098054
Museums Aotearoa, Museum
Association of New Zealand,
Wellington 059637
Museums Association, London . . 060188
Museums Association of India,
Delhi 059367
Museums Association of Namibia,
MAN, Windhoek 059587
Museums Association of Nigeria,
Lagos 059650
The Museums Association of Pakistan,
Karachi 059681
Museums Association of
Saskatchewan, Regina 058454
Museums Association of the
Caribbean, Bridgetown 058348
Museums Australia, Civic Square 058208
Museums Australia Magazine,
Parkes 138497
Museums-Bergwerk Schauinsland,
Freiburg im Breisgau 018485
Museums-Gutshof Sonnekalb, Bad
Kösen 016802
Museums in Essex Committee,
Chelmsford 060189
Museums Journal, London 139209
Museums-Journal, Kulturprojekte
Berlin, Berlin 138739
Museums Magazine, New York . 139447
Museums North, Northern Federation
of Museums and Art Galleries,
Hartlepool 060190
Museums of Filipinana and Rizaliana,
Sampaloc 034587
The Museums of Oglebay Institute,
Wheeling 054082
Museums of South Lanarkshire,
Ferniegair 044341
Museums Weapons Group, Leeds 060191
Museums Yearbook, London . . . 139210

Register der Institutionen und Firmen

Museums, Libraries and Archives Council, London
– Mutter. Museum für Kunst. Sammlung Berger, Amorbach

Museums, Libraries and Archives
Council, London 060185
Museums, Libraries and Archives
North East, Newcastle-upon-
Tyne 060186
Museumsanlage Kulturstiftung
Landkreis Osterholz, Osterholz-
Scharmbeck 021027
Museumsbäckerei Imsweiler,
Imsweiler 019399
Museumsbahn Bremerhaven-
Bederkesa, Bad Bederkesa ... 016711
Museumsbauernhof Wennerstorf,
Wennerstorf 022564
Museumsberg Flensburg, Städtische
Museen und Sammlungen für den
Landesteil Schleswig, Flensburg 018372
Museumsboerderij Oud Noordwijk,
Noordwijk, Zuid-Holland 032700
Museumsbund Österreich,
Leonding 058290
Museumsbygningen Kunstauktioner,
København 125454
Museumscenter Hanstholm,
Hanstholm 010097
Museumschiff Elbe 1, Cuxhaven 017790
Museumschip Mercuur, Den
Haag 032190
Museumsdampfzug Rebenbummler,
Freiburg im Breisgau 018486
Museumsdorf Baruther Glashütte,
Baruth 016969
Museumsdorf Bayerischer Wald,
Tittling 022161
Museumsdorf Cloppenburg,
Niedersächsisches Freilichtmuseum,
Cloppenburg 017747
Museumsdorf Düppel, Stiftung
Stadtmuseum Berlin, Berlin ... 017254
Museumsdorf Krumbach,
Krumbach 002248
Museumsdorf Trattenbach,
Ternberg 002817
Museumsdorf Volksdorf mit
Spiekerhus, Hamburg 019013
Museumseisenbahn, Hamm,
Westfalen 019031
Museumseisenbahn Payerbach-
Hirschwang, Hirschwang 002086
Museumseisenbahn Schwalm-Knüll,
Fritzlar 018534
Museumsfarm Tomita,
Nakafurano 029132
Museumsfeuerschiff Amrumbank/
Deutsche Bucht, Emden 018177
Museumsfriedhof Tirol, Kramsach 002229
Museumsgalerie, Altomünster ... 016563
Museumsgalerie Allerheiligenkirche,
Mühlhäuser Museen, Mühlhausen/
Thüringen 020486
Museumsgården, Læsø 010215
Museumshafen Oevelgönne,
Hamburg 019014
Museumshof, Bad Oeynhausen .. 016838
Museumshof, Neuruppin 020740
Museumshof, Rahden 021264
Museumshof, Roßtal 021484
Museumshof am Sonnenluch,
Heimatmuseum Erkner, Erkner 018255
Museumshof auf den Braem, Museen
der Stadt Gescher, Gescher ... 018676
Museumshof Emmerstedt,
Helmstedt 019187
Museumshof Ernst Koch,
Wernigerode 022584
Museumshof-Galerie, Oldenburg in
Holstein 020993
Museumshof Historischer Moorhof
Augustendorf, Gnarrenburg ... 018720
Museumshof-Senne, Bielefeld .. 017382
Museumshof und Heimatmuseum,
Falkenstein, Harz 018337
Museumshof Zirkow, Zirkow ... 022884
Museumskirche Sankt-Katharinen, Die

Lübecker Museen, Lübeck ... 020163
Museumskunde, Berlin 138740
Museumsland Donauland Strudengau,
Waldhausen 058291
Museumslandschaft Deilbachtal,
Essen 018307
Museumslandschaft Hessen Kassel,
Direktion und Verwaltung,
Kassel 019542, 059179
Museumslogger AE7 Stadt Emden,
Emden 018178
Museumsmühle, Sankt Julian .. 021602
Museumsmühle, Woldegk 022723
Museumsmühle Hasbergen,
Delmenhorst 017850
Museumsmühle mit heimatkundlicher
Sammlung, Varel 022303
Museumsnytt, Oslo 138946
Museumspädagogische Beratung,
Museumslandschaft Hessen Kassel,
Kassel 019543
Museumspädagogische Gesellschaft
e.V., Köln 059180
Museumspark, Rüdersdorf bei
Berlin 021534
Museumspavillon Flavia Solva,
Landesmuseum Joanneum,
Wagna 002882
Museumspoorlijn Star,
Stadskanaal 032899
Museumsscheune, Brest 017619
Museumsscheune, Helmenzen .. 019183
Museumsscheune Fränkische Schweiz,
Hollfeld 019321
Museumsscheune Kremmen,
Kremmen 019791
Museumsschiff Cap San Diego,
Hamburg 019015
Museumsschiff FMS Gera,
Bremerhaven 017609
Museumsschiff Mannheim des
Landesmuseums für Technik und
Arbeit, Mannheim 020254
Museumsspinnerei Neuthal,
Bäretswil 041257
Museumsstadl, Bernried, Kreis
Deggendorf 017340
Museumstjenesten, Viborg 058559
Museumstoomtram Hoorn-Medemblik,
Hoorn 032507
Museumsverband Baden-Württemberg
e.V., Villingen-Schwenningen . 059181
Museumsverband des Landes
Brandenburg e.V., Potsdam ... 059182
Museumsverband für Niedersachsen
und Bremen e.V., Hannover ... 059183
Museumsverband für Niedersachsen
und Bremen e.V. – Mitteilungsblatt,
Hannover 138741
Museumsverband Hamburg,
Hamburg 059184
Museumsverband Mecklenburg-
Vorpommern e.V., Güstrow ... 059185
Museumsverband Rheinland-Pfalz e.V.,
Ludwigshafen am Rhein 059186
Museumsverband Sachsen-Anhalt e.V.,
Bernburg 059187
Museumsverband Schleswig-Holstein
e.V., Rendsburg 059188
Museumsverbund Thüringen e.V.,
Erfurt 059189
Museumsverbund Pankow –
Heynstraße, Berlin 017255
Museumsverbund Pankow –
Prenzlauer Allee, Berlin 017256
Museumsverbund Südniedersachsen
e.V., Göttingen 059190
Museumsverein Aachen e.V.
Aachen 059191
Museumsverein Bahnhofmuseum e.V,
Benneckenstein 017031
Museumsverein Dokumentation
2. Weltkrieg Hürtgenwald e.V,
Hürtgenwald 019362

Museumsverein Düren e.V.,
Düren 059192
Museumsverein Feuerwehr
Obererbach, Obererbach 020893
Museumsverein Hameln e.V.,
Hameln 059193
Museumsverein Kassel e.V.,
Museumslandschaft Hessen Kassel,
Kassel 059194
Museumsverein Mönchengladbach
e.V, Städtisches Museum Abteiberg,
Mönchengladbach 059195
Museumswelten, Saarbrücken . 098157
Museumswohnung WBS 70,
Berlin 017257
Museumszentrum Lorsch, Lorsch 020117
Museumszug Hespertalbahn,
Essen 018308
Museumvisie, Amsterdam 138936
Museumwerf 't Kromhout,
Amsterdam 031968
Museumwinkel Albert Heijn,
Zaandam 033078
Museu de História Natural e
Etnologia Indígena, Centro de Ensino
Superior, Academia SVD, Juiz de
Fora 004522
Museu Municipal de Almada, Núcleo
Museológico de Arqueologia e
História, Núcleo Museológico da
Água e Núcleo Naval, Almada . 035506
Musexpo, Clairfayts 012582
Muséye Chay-e Lahijan, Lahijan 024716
Muséye Honarha-ye Melli,
Tehrän 024774
Muséye Kerman, Kerman 024707
Muséye Mardom Shenassi,
Tehrän 024775
Musgrave Bickford, Crediton .. 089000
Mushakoji Saneatsu Memorial Hall,
Chofu 028521
Musiani, Nicoletta, Bologna 079856
Music Box of Zami Museum,
Metulla 025155
The Music House Museum,
Acme 046315
Music Information Museum,
Kaohsiung 042364
Musica Kremsmünster,
Kremsmünster 002241
Musical Box Society International
Museum, Houston 049590
Musical Boxes Museum, Hall of Halls
Rokko, Kobe 028852
Musical Instruments Museum,
Jerusalem 025096
Musical Museum, London 045054
Musical Treasures, Los Angeles 094413
Musical Wonder House,
Wiscasset 054507
Musik- und Wintersportmuseum,
Klingenthal 019649
Musik-Antiquariat, Zentralantiquariat
Leipzig, Leipzig 142061
Musikantiquariat zum grossen C,
Oberried 142255
Musikhistorische Sammlung Jehle,
Albstadt 016494
Musikhistorisk Museum og Carl
Claudius' Samling, København 010185
Musikinstrumente-Museum der Völker,
Sankt Gilgen 002662
Musikinstrumenten- und
Puppenmuseum, Goslar 018772
Musikinstrumenten-Ausstellung des
Händel-Hauses, Musikmuseum der
Stadt Halle, Halle, Saale 018967
Musikinstrumenten-Museum,
Markneukirchen 020289
Musikinstrumenten-Museum,
Staatliches Institut für
Musikforschung – Stiftung
Preußischer Kulturbesitz, Berlin 017258

Musikinstrumenten-Restaurator
Wolfgang Wenke, Eisenach ... 129783
Musikinstrumentenmuseum Lißberg,
Ortenberg, Hessen 021007
Musikinstrumentensammlung der
Universität, Göttingen 018753
Musikinstrumentensammlung Hans
und Hede Grumbt, Wasserburg Haus
Kemnade, Hattingen 019100
Musikinstrumentensammlung,
Friedrich-Alexander-Universität
Erlangen-Nürnberg, Erlangen . 018262
Musikmuseet, Stockholm 041023
Musikmuseum, Historisches
Basel, Basel 041281
Musikvariatet, Oslo 143407
Musin, Genevieve, Larressingle . 069192
MUSIS, Verein zur Unterstützung der
Museen und Sammlungen in der
Steiermark, Graz 058292
Muskee, Thim, Groningen 083303
Muskegon County Museum,
Muskegon 051136
Muskegon Museum of Art,
Muskegon 051137
Muskogee War Memorial Park and
Military Museum, Muskogee .. 051140
Muskoka Antique Show, Port
Carling 098038
Muskoka Heritage Place,
Huntsville 006022
Muskoka Lakes Museum, Port
Carling 006678
Muskoka's Steamship Museum,
Gravenhurst 005922
Musolff, Claudia, Münster ... 130428
Musquodoboit Railway Museum,
Musquodoboit 006441
Mussallem, Jacksonville 094015
Mussallem, Phoenix 096225
Mussard, Nathalie, Paris 071700
Musselshell Valley Historical Museum,
Roundup 052695
Musselwhite, Carol, Wilton,
Salisbury 091855
Mussenden & Son, G.B.,
Bournemouth 088538
Musta Taide, Helsinki 138623
Mustafa Kamil Museum, Cairo . 010552
Mustansiriya School Collections,
Baghdad 024805
Mustard Seed, Hong Kong 102097
The Mustard Seed Gallery,
Cleethorpes 117595
Mustard Shop Museum, Norwich 045402
Mustialan Maataloushistoriallinen
Museo, Mustiala 011172
Musumeci, Quart 137724
Muswellbrook Regional Arts Centre,
Muswellbrook 001332
Muszyńska, Magdalena, Gdynia 084520
Muszyński, Marcin, Poznań ... 113742,
                            132863
Mutare Museum, Mutare 054857
Mutascio Ronconi, Verona 081927
Mutel, René, Oullins 070828
Muthesius-Gesellschaft e.V.,
Gesellschaft zur Förderung der
bildenden Künste der Muthesius-
Hochschule, Kiel 059196
Muthesius Kunsthochschule,
Fachhochschule für Kunst und
Gestaltung, Kiel 055609
Mutinès Muziejus, Vilnius 030535
Mutlu Sanat Odasi, İstanbul ... 117068
Mutscher, Hartmut & Eva,
Neißeaue 130440
Mutsen en Poffermuseum Sint-
Paulusgasthuis, Sint Oedenrode 032877
Muttart Art Gallery, Calgary ... 005529
Mutter Museum, College of Physicians
of Philadelphia, Philadelphia . 052071
Mutter. Museum für Kunst. Sammlung
Berger, Amorbach 016578

Mutterbusch, Hamminkeln 076352
Muttner, Marcel, Le Landeron .. 087553
Muttonchop, Petworth 144786
Mutua Museum, Imphal 024311
Muurhuizen Antiek, Amersfoort . 082653
Muuri, Risto, Helsinki 066345
Muurlan Kotiseutumuseo, Muurla 011173
Muurtooi, Arnhem 082944, 112335, 132429
Muurtooi, Leeuwarden 083417
Muuruveden Kotiseutumuseo, Muuruvesi 011174
Muuseum, Tartu 010697
Művészetek Háza – Vass Gyűjtemény, Veszprém 024042
Művészetek Kincsesháza – Vaszary Emlékház, Kaposvár 023709
Művészetek Kincsesháza – Vaszary Képtár, Kaposvár 023710
Művészeti Kar, Pécsi Tudományegyetem, Pécs 055771
MUWA – Museum der Wahrnehmung, Graz 001983
Muwayire, Edward, Cape Town . 114771
Muza, Tbilisi 105916
Muzealnictwo, Warszawa 138973
MuZee Scheveningen, Scheveningen 032849
Muzei Geroičnoї Oborony i Vyzvolennja Sevastopolja, Sevastopol 043292
Muzei Permskogo Gosudarstvennogo Universiteta, Perm 037641
Muzej- Usadba F.M. Dostoevskogo Darovoe, Darovoe 036966
Muzej "Kobzarja" T.G. Ševčenka, Čerkas'kyj oblasnyj krajeznavčyj muzej, Čerkasy 043118
Muzej "V Mire Skazok", Smolenskij Muzej-Zapovednik, Smolensk . 037864
Muzej 700 let – Landskrona, Nevskoe Ustjë, Nienšanc, Sankt-Peterburg 037762
Muzej 'A Muzy ne Molčali', Sankt-Peterburg 037763
Muzej A.A. Kiseleva, Tuapse ... 037957
Muzej A.V. Suvorova, Velikij Novgorod 038014
Muzej Abaja, Abaj 029787
Muzej Afričke Umetnosti, Beograd 038154
Muzej Aleksandăr Stambolijski, Slavovica 005164
Muzej Aleksandăr Stambolijski, Sofia 005182
Muzej Al'manachu Rusalki Dnistrovoї, Lviv'ka Galereja Mystectv, Lviv 043245
Muzej Alpinizma, Turizma i Istorii Kurorta Teberda, Teberda 037933
Muzej Anatomii, Rossijskij Gosudarstvennyj Agrarnyj Universitet – Moskvskaja selskochosjajstvennaja akademija im K.A. Timirjazeva, Moskva 037353
Muzej Antropologii i Etnografii im. Petra Velikogo (Kunstkamera), Sankt-Peterburg 037764
Muzej Antropologii im. D.N. Anučina Moskovskogo Gosudarstvennogo Universiteta im M.V. Lomonosova, Naučno-issledovatelskij Institut im. D.N. Anučina, Moskva 037354
Muzej Archeologii i Etnografii, Dalnevostočnyj gosudarstvennyj universitet, Vladivostok 038037
Muzej Archeologii i Etnografii, Otdel Narodov Ufimskogo Naučnogo Centra Rossijskoj Akademii Nauk, Ufa 037981
Muzej Archeologii i Etnografii, Syktyvkarskij gosudarstvennyj universitet, Syktyvkar 037912
Muzej Archeologii Moskvy, Muzej istorii Moskvy, Moskva 037355
Muzej Archeologii Prikamja, Perm 037642

Muzej Archeologii, Akademija Nauk Respubliki Kazachstan – Institut Archeologii im A.Ch. Margulana, Almaty 029803
Muzej Archeologii, Institut jazyka, literatury i istorii Komi naučnogo centra Rossijskoj akademii nauk, Syktyvkar 037911
Muzej Archeologija, Étnografija i Ékologija Sibiri Kemerovskogo Gosudarstvennogo Universiteta, Kemerovo 037126
Muzej Baki Urmanče, Kazan ... 037106
Muzej Banjičkog Logora, Beograd 038156
Muzej Bitola, Bitola 030610
Muzej Brdovec, Prigorje Brdovče 008970
Muzej Brodskog Posavlja, Slavonski Brod 008992
Muzej I.A. Bunina, Gosudarstvennyj literaturnyj muzej im. I.S. Turgeneva, Orël 037602
Muzej Butyrskoj Tjurmy, Moskva 037356
Muzej V.I. Čapaeva, Čuvašskij Nacionalnyj Muzej, Čeboksary . 036933
Muzej Carskoselskaja Kollekcija, Puškin 037677
Muzej Cerkov Dekabristov, Oblastnoj kraevedčeskij muzej im. A.A. Kuznecova, Čita 036963
Muzej Cerkov Petra i Pavla, Tjumen 037937
Muzej Cetinjske Krajine, Sinj .. 008989
Muzej Cetinjskog Manastira, Cetinje 031752
Muzej Chlebnoe Delo v Moskve, Moskva 037357
Muzej-chram Svjatitelja Nikolaja v Tolmačach, Gosudarstvennaja Tretjakovskaja galerja, Moskva 037358
Muzej Chrama Christa Spasitelja, Muzej istorii Moskvy, Moskva . 037359
Muzej Chudožestvennych Tkanej, Moskovskij gosudarstvennyj tekstilnyj universitet im. A.N. Kosygina, Moskva 037360
Muzej Cirkovogo Iskusstva, Sankt-

Peterburg 037766
Muzej-čitalnja N.V. Fëdorova, Moskva 037361
Muzej-čitalnja na Nacionalnych Jazykach, Penza 037624
Muzej Čovek i negovoto Zdrave, Varna 005239
Muzej Crne Metalurgije Sartid A.D., Smederevo 038237
Muzej Đakovština, Đakovo 008896
Muzej V.I. Dalja, Moskva 037362
Muzej Dekabristov, Minusinsk .. 037230
Muzej Dekabristov – Dom Trubeckogo, Irkutsk 037028
Muzej Dekabristov – Dom Volkonskich, Irkutskij oblastnoj kraevedčeskij muzej, Irkutsk 037029
Muzej Dekorativno-Prikladnogo i Promyšlennogo Iskusstva, Moskva 037363
Muzej Dekorativno-prikladnogo Iskusstva Urala, Čeljabinsk ... 036937
Muzej Dekorativno-prikladnogo Iskusstva, Sankt-Peterburgskaja Gosudarstvennaja Chudožestvenno-promyšlennaja Akademija, Sankt-Peterburg 037767
Muzej Dekorativno-prykladnogo Mystectva, Sums'kyj chudožnij muzej im. N. Onac'kogo, Sumy 043304
Muzej Derevjannogo Zodčestva Malye Karely, Archangelsk 036865
Muzej Detskogo Muzykalnogo Teatra im. N. Sats, Moskva 037364
Muzej Detskogo Tvorčestva, Sankt-Peterburg 037768
Muzej Dom A.M. Gorkogo – Domik Kaširina, Gosudarstvennyj Literaturnyj Muzej A.M. Gorkogo, Nižnij Novgorod 037548
Muzej-Diorama Orlovskaja Nastupatelnaja Operacija, Orlovskij oblastnoj kraevedčeskij muzej, Orël 037603
Muzej Diplomatičeskogo Korpusa, Vologda 038056
Muzej Dom Mašarova, Tjumenskij oblastnoj kraevedčeskij muzej im. I.Ja. Slovcova, Tjumen 037938
Muzej Dominikanskog Samostana, Bol 008886
Muzej Dominikanskog Samostana, Dubrovnik 008906
Muzej Drugoe Iskusstvo, Moskva 037365
Muzej A.A. Durova, Taganrogskij gosudarstvennyj literaturnyj i istoriko-architekturnyj muzej-zapovednik, Taganrog 037918
Muzej Durylina S.N., Korolev .. 037151
Muzej Dvorec A.D. Menšikova, Gosudarstvennyj Érmitaž, otdel istorii russkoj kultury, Sankt-Peterburg 037769
Muzej Ékslibrisa, Meždunarodnyj Sojuz Knigoljubov, Moskva 037366
Muzej Emfiedžieva Kăšta, Istoričeskij Muzej, Kjustendil 005065
Muzej Étnografičeskogo Kostjuma na Kuklach, Moskva 037367
Muzej Étnografii, Istorii i Kultury Narodov Baškortostana, Ufa ... 037982
Muzej Évoljucii Mozga, NII mozga RAMN, Moskva 037368
Muzej Évolucije i Nalazište Pračovjeka Hušnjakovo, Klanjec 038524
Muzej-Filial Nadi Ruševog, Tuvinskij Respublikanskij Kraevedčeskij Muzej im. Aldan Maadyr, Kyzyl 037191
Muzej Filiki Éterija, Hellenic Foundation for Culture, Odesa . 043272
Muzej Fizičeskoj Kultury i Sporta, Moskva 037369
Muzej Fizičke Kulture, Beograd . 038157
Muzej FK Crvena Zvezda, Beograd 038158

Muzej Ford i Bălgarija, Sofia ... 005183
Muzej Franjevačkog Samostana, Dubrovnik 008907
Muzej Franjevačkog Samostana, Rovinj 008981
Muzej Franjevačkog Samostana Košljun, Punat 008975
Muzej Gabdulla Tukaja, Nacionalnyj muzej Respubliki Tatarstan, Kazan 037108
Muzej Gacka, Otočac 008963
Muzej-Galerija Kavadarci, Kavadarci 030612
Muzej Gasilstva Dravinjske doline, Slovenske Konjice 038615
Muzej Geodezii i Kartografii, Moskva 037370
Muzej Geologii, Almaty 029804
Muzej Geologii, Toshkent 054657
Muzej Geologii Centralnoj Sibirii, Krasnojarsk 037168
Muzej Geologii, Čeljabinskij oblastnoj kraevedčeskij muzej, Čeljabinsk 036938
Muzej Geologii, Nefti i Gaza, Tjumenskij oblastnoj kraevedčeskij muzej im. I.Ja. Slovcova, Tjumen 037939
Muzej Georgi Velčev, Varna 005240
Muzej Get'manstva, Kyїv 043203
Muzej Glavna Kvartira na Ruskata Armija 1877–1878, Pordim ... 005135
Muzej V.N. Gorjaeva, Moskva .. 037371
Muzej A.M. Gorkogo, Moskva .. 037372
Muzej Goroda Jurjevca, Jurjevec 037077
Muzej Goroda Penzy, Penzenskij gosudarstvennyj obedinennyj kraevedčeskij muzej, Penza .. 037625
Muzej Gorodskaja Duma, Tjumenskij oblastnoj kraevedčeskij muzej im. I.Ja. Slovcova, Tjumen 037940
Muzej Gosudarstvennogo Akademičeskogo Bolšogo Teatra Rossii, Moskva 037333
Muzej Gosudarstvennogo Akademičeskogo Malogo Teatra, Moskva 037334
Muzej Gosudarstvennogo Akademičeskogo Marijnskogo Teatra Opery i Baleta, Sankt-Peterburg 037770
Muzej Gosudarstvennogo Akademičeskogo Teatra Opery i Baleta imeni Mussorgskogo, Sankt-Peterburg 037771
Muzej Gosudarstvennogo Prirodnogo Nacionalnogo Parka Losinogo Ostrova, Moskva 037375
Muzej Grada Beograda, Beograd 038159
Muzej Grada Koprivnice, Koprivnica 008931
Muzej Grada Novog Sada, Novi Sad 038215
Muzej Grada Pazina, Pazin 008966
Muzej Grada Perasta, Perast ... 031761
Muzej Grada Rijeke, Rijeka 008976
Muzej Grada Sarajeva, Sarajevo 004276
Muzej Grada Splita, Split 008999
Muzej Grada Trogira, Trogir ... 009009
Muzej Grada Zagreba, Zagreb .. 009043
Muzej Grada Zenice, Zenica ... 004293
Muzej Gradostroitelstva i Byta Taganroga, Taganrogskij gosudarstvennyj literaturnyj i istoriko-architekturnyj muzej-zapovednik, Taganrog 037919
Muzej Grebnych Vidov Sporta, Moskva 037376
Muzej Hercegovine, Mostar ... 004269
Muzej Herzegovine-Trebinje, Trebinje 004286
Muzej Hmeljarstva, Žalec 038634
Muzej Hrvatske Književnosti i Kazalište Umjetnosti, Zagreb 009044
Muzej Hrvatskih Arheoloških Spomenika, Split 009000

Muzej Hvarske Baštine, Hvar .. 008919
Muzej i Obščestvennyj Centr Mir,
Progress, Prava Čeloveka im.
Andreja Sacharova, Moskva .. 037377
Muzej I. Karpenka-Karogo,
Kirovograd 043184
Muzej Igora Stravinskogo,
Ustyloog 043310
Muzej Igorja Talkova, Moskva .. 037378
Muzej Ikona, Korčula 008934
Muzej im. Auèzova Kzachskogo
Dramatičeskogo Teatra, Almaty 029805
Muzej im. S.M. Kirova, Sankt-
Peterburg 037772
Muzej im. B.M. Kustodieva,
Kostromskij gosudarstvennyj
obedinennyj chudožestvennyj muzej,
Ostrovskoe 037615
Muzej-institut Semji Rerichov, Sankt-
Peterburg 037773
Muzej Instituta Biologii Morja,
Dalnevostočnoe otd. Rossijskoj
Akademii Nauk, Vladivostok .. 038038
Muzej internacionalnoj Družby,
Saransk 037836
Muzej Ipavčeva Hiša, Šentjur .. 038605
Muzej Iskusstva Narodov Vostoka,
Moskva 037379
Muzej Istočne Bosnè, Tuzla 004289
Muzej Istorii Afganskoj Vojny,
Moskva 037380
Muzej Istorii Aksaja, Aksaj 036842
Muzej Istorii Architektury i
Architekturnop Obrazovanija
Sibiri, Novosibirskaja architekturno-
chudožestvennaja akademija,
Novosibirsk 037579
Muzej Istorii Architektury i
Promyšlennoj Techniki Urala,
Ekaterinburg 036985
Muzej Istorii Bolnicy im. N.A.
Alekseeva, Moskva 037381
Muzej Istorii Burjatii imeni Changalova,
Ulan-Udè 037988
Muzej Istorii Černomorskogo Flota,
Moskva 037382
Muzej Istorii Deneg i Bankovskogo
Dela, Moskva 037383
Muzej Istorii Doma XIX–XX vekov,
Tjumen 037941
Muzej Istorii Ekaterinburga,
Ekaterinburg 036986
Muzej Istorii Goroda Irkutska,
Irkutsk 037030
Muzej Istorii Goroda Joškar-Oly,
Joškar-Ola 037071
Muzej Istorii Goroda-Kurorta Soči,
Soči 037877
Muzej Istorii Goroda Naberežnye
Čelny, Nacionalnyj muzej respubliki
Tatarstan, Naberežnye Čelny .. 037531
Muzej Istorii Goroda Šlisselburga,
Šlisselburg 037862
Muzej Istorii Goroda Stavropolja,
Stavropol 037898
Muzej Istorii Goroda Tjumeni,
Tjumenskij oblastnoj kraevedčeskij
muzej im. I.Ja. Slovcova,
Tjumen 037942
Muzej Istorii i Kultury Goroda
Votkinska, Votkinsk 038072
Muzej Istorii i Kultury Moskovskogo
Rajona, Nižnij Novgorod 037549
Muzej Istorii i Kultury Srednego
Prikamja, Sarapul 037840
Muzej Istorii i Rekonstrukcii Goroda
Almaty, Almaty 029806
Muzej Istorii Izučenija i Osvoenija
Evropejskogo Severa, Apatity . 036857
Muzej Istorii Jaroslavlja, Jaroslavskij
gosudarstvennyj istoriko-
architekturnyj i chudožestvennyj
muzej-zapovednik, Jaroslavl . 037065
Muzej Istorii Kamreznogo

i Juvelirnogo Iskusstva,
Ekaterinburg 036987
Muzej Istorii Kazanskogo Universiteta,
Kazanskij gosudarstvennyj universitet,
Kazan 037109
Muzej Istorii Kolskich Saamov,
Murmanskij Oblastnoj Kraevedčeskij
Muzej, Lovozero 037206
Muzej Istorii Kraja, Belozersk .. 036894
Muzej Istorii Milicii, Sankt-
Peterburg 037774
Muzej Istorii Mista Chmel'nyc'kogo,
Chmel'nyc'kyj 043144
Muzej Istorii Mista Dniprodzeržins'ka,
Dniprodzeržins'k 043145
Muzej Istorii Mista Kolomyï,
Kolomija 043186
Muzej Istorii Mista Kyjeva, Kyïv . 043204
Muzej Istorii Morskogo Universiteta im.
G.I. Nevelskogo, Vladivostok . 038039
Muzej Istorii Moskovskogo
Gosudarstvennogo Universiteta im.
M.V. Lomonosova, Moskva ... 037384
Muzej Istorii Moskovskogo
Metropolitena, Moskva 037385
Muzej Istorii Moskovskoj
Selskochosjajstvennoj Akademii im
K.A. Timirjazeva, Moskva ... 037386
Muzej Istorii Moskvy, Moskva .. 037387
Muzej Istorii Narodnych
Chudožestvennych Promyslov, Nižnij
Novgorod 037550
Muzej Istorii OAO Gaz, Nižnij
Novgorod 037551
Muzej Istorii OAO Nižegorodskij
Televizionnyj Zavod im. V.I. Lenina,
Nižnij Novgorod 037552
Muzej Istorii Orenburga,
Orenburg 037607
Muzej Istorii Osvoenija i Razvitija
Norilskogo Promyšlennogo Rajona,
Norilsk 037568
Muzej Istorii Pečati Sankt-Peterburga,
Sankt-Peterburg 037775
Muzej Istorii Permskogo Rajona,
Perm 037643
Muzej Istorii Peterburgskoj
Konservatorii, Sankt-Peterburg 037776
Muzej Istorii Političeskoj Policii
Rossii – Gorochovaja 2,
Gosudarstvennyj muzej političeskoj
istorii Rossii, Sankt-Peterburg . 037777
Muzej Istorii Političeskoj Ssylki,
Bratskij gorodskoj objedinennyj
muzej istorii osvoenija Angary,
Bratsk 036916
Muzej Istorii Politssylki v Jakutii,
Jakutskij gosudarstvennyj muzej
istorii i kultury narodov severa im.
Jaroslavskogo, Jakutsk 037056
Muzej Istorii Poljarnych Olimpiad,
Murmanskij Oblastnoj Kraevedčeskij
Muzej, Murmansk 037529
Muzej Istorii Preobraženskoj Bolnicy
im. V.A. Giljarovskogo, Moskva 037388
Muzej Istorii Priokskogo Rajona, Nižnij
Novgorod 037553
Muzej Istorii Prosveščenija Komi Kraja,
Syktyvkarskij gosudarstvennyj
Universitet, Syktyvkar 037913
Muzej Istorii Razvitija Mediciny,
Astrachanskij muzej-zapovednik,
Astrachan 036877
Muzej Istorii Razvitija Mosenergo,
Moskva 037389
Muzej Istorii Revoljucionno-
demokratičeskogo Dviženija
1880–1890 gg, Sankt-
Peterburg 037778
Muzej Istorii Rossijskogo
Gosudarstvennogo Pedagogičeskogo
Universiteta im A.I. Gercena, Sankt-
Peterburg 037779
Muzej Istorii Rossijskogo

Predprinimatelstva, Moskva .. 037390
Muzej Istorii Russkoj Počty,
Moskva 037391
Muzej Istorii Rybnoj Promyšlennosti
Dagestana, Machačkala 037210
Muzej Istorii Sankt-Peterburgskogo
Universiteta, Sankt-Peterburg . 037780
Muzej Istorii Svjazi Respubliki
Tatarstan, Kazan 037110
Muzej Istorii ta Etnografii Grekiv
Priazov'ja, Mariupol's'kij krajeznavčyj
muzej, Mariupol' 043260
Muzej Istorii ta Kul'tury Mista
Lugans'ka, Lugans'k 043232
Muzej Istorii Teatrov Dagestana,
Dagestanskij gosudarstvennyj
objedinennyj istoriko-architekturnyj
muzej, Machačkala 037211
Muzej Istorii Tekstilnoj Fabriki
Danilovskaja Manufaktura,
Moskva 037392
Muzej Istorii Turizma, Sankt-
Peterburg 037781
Muzej Istorii UVD Kalužskoj Oblasti,
Kaluga 037092
Muzej Istorii Vojsk Moskovskogo
Voennogo Okruga, Moskva ... 037393
Muzej Istorii Vyčislitelnoj Techniki,
Moskva 037394
Muzej Istorije Jugoslavije,
Beograd 038160
Muzej Istoryčnych Koštovnostej
Ukraïny, Kyïv 043205
Muzej Ivana Gončara, Kyïv 043206
Muzej Ivanovskogo Sitca,
Ivanovo 037043
Muzej Izobraziteľnych Iskusstv,
Ašgabat 043088
Muzej Izobrazitelnych Iskusstv,
Komsomolsk-na-Amure 037147
Muzej Izobrazitelnych Iskusstv, Rostov-
na-Donu 037694
Muzej Izobrazitelnych Iskusstv im. A.S.
Puškina, Moskva 037395
Muzej Izobrazitelnych Iskusstv Karelii,
Petrozavodsk 037655
Muzej Izobrazitelnyh Iskusstv
Respubliki Marij Èl, Joškar-Ola 037072
Muzej Jantarja, Kaliningrad ... 037084
Muzej Jevreja Bosne i Hercegovine,
Muzej Grada Sarajeva, Sarajevo 004277
Muzej Josipa Jurčiča, Muljava . 038580
Muzej Jovana Cvijića, Beograd . 038161
Muzej Jovana Jovanovića Zmaja,
Sremska Kamenica 038243
Muzej Jugoslavenska Kinoteka,
Beograd 038162
Muzej Jugoslovenskih
Vazduhoplovstva, Beograd ... 038163
Muzej Junye Zaščitniki Rodiny,
Kurskij gosudarstvennyj oblastnoj
kraevedčeskij muzej, Kursk ... 037190
Muzej Južne Železnice, Šentjur . 038606
Muzej K. È. Ciolkovskogo, Aviacii i
Kosmonavtiki, Kirov 037133
Muzej Kajuma Nasyri, Nacionalnyj
muzej Respubliki Tatarstan,
Kazan 037111
Muzej Käkrinsko Chanče, Käkrina 005043
Muzej Karamzinskaja Obščestvennaja
Biblioteka, Uljanovsk 037995
Muzej Kazakstan Respublikasynyn
Prezidenttik Mèdeniet Ortalyfy,
Astana 029811
Muzej Kazanskoj Chimičeskoj Školy,
Kazanskij gosudarstvennyj universitet,
Kazan 037112
Muzej Keramiki i Usadba Kuskovo
XVIII Veka, Moskva 037396
Muzej Kino, Moskva 037397
Muzej Kinoiskusstva Uzbekistana,
Toshkent 054658
Muzej Kinostudii im. A.M. Gorkogo,
Moskva 037398

Muzej Kinostudii Mosfilm,
Moskva 037399
Muzej Klassičeskogo i Sovremennogo
Iskusstva – Dom Burganova,
Moskva 037400
Muzej V.O. Ključevskogo, Penza 037626
Muzej Knigi, Almaty 029807
Muzej Knigi i Drukarstva Ukraïny,
Kyïv 043207
Muzej Knigi, Rossijskaja
gosudarstvennaja biblioteka,
Moskva 037401
Muzej Književnosti i Pozorišne
Umjetnosti Bosne i Hercegovine i
Galerija MAK, Sarajevo 004278
Muzej A. Kolcova, Voronežskij
oblastnoj literaturnyj muzej im I.S.
Nikitina, Voronež 038066
Muzej Kolju Fičeto, Drjanovo .. 005024
Muzej Konevodstva, Moskovskaja
selskochosjajstvennaja akademija im
K.A. Timirjazeva, Moskva ... 037402
Muzej Konspirativnaja Kvartira
Simbirskoj Gruppy RSDRP
1904–1906, Uljanovskij oblastnoj
kraevedčeskij muzej im. I.A.
Gončarova, Uljanovsk 037996
Muzej Koški, Moskva 037403
Muzej Kosmonavtiki im. S. Koroljova,
Žytomyr 043327
Muzej Kraevedenija g. Elabuga,
Gosudarstvennyj istoriko-
architekturnyj i chudožestvennyj
muzej-zapovednik, Elabuga ... 036998
Muzej Krajine Negotin, Negotin . 038210
Muzej Krapinskog Pračovjeka,
Krapina 008939
Muzej Krejsera Avrora, Centralnyj
Voenno-morskoj muzej, Sankt-
Peterburg 037782
Muzej-krepost Baba Vida, Vidin . 005262
Muzej Kriminalistiki, Alktajskij
gosudarstvennyj universitet,
Barnaul 036885
Muzej-kripta Aleksandăr Nevski,
National Art Gallery, Sofia 005184
Muzej Kronštadtskaja Krepost,
Kronštadt 037179
Muzej Kronštadtskogo Morskogo
Zavoda, Kronštadt 037180
Muzej Krutickoe Podvore,
Gosudarstvennyj Istoričeskij muzej,
Moskva 037404
Muzej Krylova, Tulskij muzej
izobrazitelnych iskusstv, Tula . 037962
Muzej Kuca Bore Stankovica,
Vranje 038255
Muzej Kukol i Detskoj Knigi
"Strana Čudes", Municipalnyj
objedinennyj muzej pisatelej Urala,
Ekaterinburg 036988
Muzej Kukol 'Kukolnyj Dom',
Moskva 037405
Muzej Kul'turnoї Spadščyny, Muzej
Istorії Mista Kyjeva, Kyïv 043208
Muzej Kultury Astrachani,
Astrachanskij muzej-zapovednik,
Astrachan 036878
Muzej Kupečeskogo Byta,
Ekaterinburg 036989
Muzej Kutuzovskaja Izba, Moskva 037406
Muzej-kvartira Aktèrskoj Semi M.V. i
A.A. Mironovych i A.S. Menakera,
Gosudarstvennyj centralnyj teatralnyj
muzej im. Bachrušina, Moskva 037407
Muzej-kvartira Alliluevych,
Gosudarstvennyj istoriko-memorialnyj
muzej Smolnyj, Sankt-Peterburg 037783
Muzej-kvartira I.I. Brodskogo,
Research Museum of the
Russian Academy of Arts, Sankt-
Peterburg 037784
Muzej-kvartira A.S. Čaplygina, Sankt-
Peterburg 037785

Muzej-kvartira M.I. Cvetaevoj v
Bolševo, Korolev 037152
Muzej-kvartira Dirižëra N.S. Golovanov,
Moskva 037408
Muzej-kvartira F.M. Dostoevskogo,
Gosudarstvennyj Literaturnyj muzej,
Moskva 037409
Muzej-kvartira Elizarovych, Sankt-
Peterburg 037786
Muzej-kvartira A.B. Goldenvejzera,
Gosudarstvennyj centralnyj muzej
muzykalnoj kultury im. M.I. Glinki,
Moskva 037410
Muzej-kvartira A.M. Gorkogo,
Moskva 037411
Muzej-kvartira A.M. Gorkogo,
Gosudarstvennyj Literaturnyj Muzej
A.M. Gorkogo, Nižnij Novgorod 037554
Muzej-kvartira G.D. Krasilnikova,
Nacionalnyj muzej Udmurtskoj
Respubliki im. K. Gerda, Iževsk 037047
Muzej-kvartira A.I. Kuindži, Muzej
Rossijskoj Akademii Chudožestv,
Sankt-Peterburg 037787
Muzej-kvartira L. Lapcuja, Jamalo-
Neneckij okružnoj kraevedčeskij
muzej im. I.S. Šemanovskogo,
Salechard 037703
Muzej-kvartira V.S. Mejercholda,
Moskva 037412
Muzej-kvartira Musa Džalilja,
Nacionalnyj muzej Respubliki
Tatarstan, Kazan 037113
Muzej-kvartira Naziba Žiganova,
Nacionalnyj muzej Respubliki
Tatarstan, Kazan 037114
Muzej-kvartira V.I. Nemiroviča-
Dančenko, Muzej Moskovskogo
Chudožestvennogo Akademičeskogo
Teatra, Moskva 037413
Muzej-kvartira A.V. Neždanovoj,
Moskva 037414
Muzej-kvartira A. Puškina,
Vserossijskij Muzej Puškina, Sankt-
Peterburg 037788
Muzej-kvartira Svjatoslava Richtera,
Muzej Izobrazitelnych Iskusstv im.
A.S. Puškina, Moskva 037415
Muzej-kvartira A.D. Sacharova, Nižnij
Novgorod 037555
Muzej-kvartira F.I. Šaljapina, Sankt-
Peterburgskij Muzej Teatralnogo
i Musykalnogo Iskusstva, Sankt-
Peterburg 037789
Muzej-kvartira T.G. Ševčenko, Muzej
Rossijskoj Akademii Chudožestv,
Sankt-Peterburg 037790
Muzej-Kvartira Ju.P. Spegalskogo,
Pskov 037668
Muzej-kvartira I.D. Sytina,
Moskva 037416
Muzej-kvartira K.A. Timirjazeva,
Moskovskaja selskochozjajstvennaja
akademija im. K.A. Timirjazeva,
Moskva 037417
Muzej-kvartira A.N. Tolstogo,
Gosudarstvennyj Literaturnyj muzej,
Moskva 037418
Muzej-kvartira A.T. Tvardovskij
v Smolenske, Smolenskij Muzej-
Zapovednik, Smolensk 037865
Muzej-kvartira V. Tytova, Muzej
istoriï ta kul'tury mista Lugans'ka,
Lugans'k 043233
Muzej Lavka Čechovych, Taganrogskij
gosudarstvennyj literaturnyj i istoriko-
architekturnyj muzej-zapovednik,
Taganrog 037920
Muzej-ledokol Angara, Irkutskij
oblastnoj kraevedčeskij muzej,
Irkutsk 037031
Muzej-ledokol Bajkal, Angarsk . 036855
Muzej Lefortovo, Muzej istorii Moskvy,
Moskva 037419

Muzej V.I. Lenina, Gosudarstvennyj
istoričeskij muzej, Moskva . . . 037420
Muzej Lesi Ukraïnki, Kyïv 043209
Muzej Lesovodstva, Moskovskaja
selskochozjajstvennaja akademija im.
K.A. Timirjazeva, Moskva 037421
Muzej Ličnych Kollekcij, Moskva 037422
Muzej Literaturnaja Žizn'
Urala XIX Veka, Municipalnyj
objedinennyj muzej pisatelej Urala,
Ekaterinburg 036990
Muzej Literaturnoj Žizni Jaroslavskogo
Kraja, Gosudarstvennyj muzej-
zapovednik N.A. Nekrasova –
Karibicha, Jaroslavl 037066
Muzej Literatury Burjatii im. Choca
Namsaraeva, Ulan-Udè 037989
Muzej Literatury XVII – XIX vekov,
Gosudarstvennyj Literaturnyj muzej,
Moskva 037423
Muzej Ljudevit Gaj, Krapina . . . 008940
Muzej Ljudske Revolucije Trbovlje,
Zasavski Muzej, Trbovlje 038623
Muzej M.V. Lomonosova, Muzej
Antropologii i Etnografii im. Petra
Velikogo, Sankt-Peterburg 037791
Muzej M.V. Lisenka, Kyïv 043210
Muzej-majsternja I.P. Kavaleridze,
Kyïv 043211
Muzej Maksimilijana Vanka,
Korčula 008935
Muzej Marcialnye Vody, Karelskij
gosudarstvennyj kraevedčeskij muzej,
Petrozavodsk 037656
Muzej-masterskaja Gapara Ajtieva,
Biškek 030076
Muzej Matija Skurjeni, Zaprešič 009056
Muzej Mebeli v Moskve, Moskva 037424
Muzej Mecenatov i Blagotvoritelej
Rossii, Moskva 037425
Muzej Medjimurja, Čakovec . . . 008891
Muzej Mengeš, Mengeš 038575
Muzej Meteorologii Glavnoj
Geofizičeskoj Observatorii im. A.I.
Voejkova, Sankt-Peterburg . . . 037792
Muzej Meterologičeskaja Stancija
Simbirska, Gosudarstvennyj
zapovednik Rodina V.I. Lenina,
Uljanovsk 037997
Muzej Mimara, Zagreb 009045
Muzej Mineralov, Rud, Samocvetov –
Udivitelnoe v Kamne, Teberda . 037934
Muzej Miniaturnych Kukol Malenkij
Mir, Semipalatinsk 037854
Muzej Mir Vody Sankt-Peterburga,
Sankt-Peterburg 037793
Muzej Mirovogo Okeana,
Kaliningrad 037085
Muzej Moderne i Suvremene
Umjetnosti, Museum of Modern and
Contemporary Art, Rijeka 008977
Muzej Mologskogo Kraja, Rybinskij
gosudarstvennyj istoriko-
architekturnyj i chudožestvennyj
muzej-zapovednik, Rybinsk . . . 037699
Muzej Mordovskoj Kultury,
Saransk 037837
Muzej Morskogo Flota
Dalnevostočnogo Morskogo
Parochodstva, Vladivostok . . . 038040
Muzej Morskogo Flota, Mintrans RF,
Moskva 037426
Muzej Moskoncerta, Moskva . . . 037427
Muzej Moskovskogo Chudožestvenno-
promyšlennogo Universiteta im. S.G.
Stroganova, Moskva 037428
Muzej Moskovskogo
Chudožestvennogo Akademičeskogo
Teatra, Moskva 037429
Muzej Moskovskogo
Gosudarstvennogo Techničeskogo
Universiteta im. N.È. Baumana,
Moskva 037430
Muzej Moslavine, Kutina 008943

Muzej Muzykalnoj i Teatralnoj Kultury
na Kavkazskich Mineralnyh Vodah,
Kislovodsk 037142
Muzej Muzykalnych Instrumentov,
Sankt-Peterburg 037794
Muzej Mystectv imeni Bogdana ta
Varvari Chanenkiv, Kyïv 043212
Muzej Mystectva Davnoï Ukraïns'koï
Knigi, Lvivs'ka Galereja Mystectv,
Lviv 043247
Muzej na Aviacijata, Krumovo . . 005080
Muzej na Bivolarstvoto i
Konevãdstvoto, Nacionalen
Selskostopanski Muzej Sofia,
Šumen 005217
Muzej na Blejskem Gradu, Bled 038495
Muzej na Chudožestvenata Akademija,
Sofia 005185
Muzej na Chudožestvenite Zanajati i
Priložnite Izkustva, Trojan . . . 005232
Muzej na Grad Kratovo, Kratovo 030614
Muzej na Grad Negotino,
Negotino 030620
Muzej na Grad Skopje, Skopje . 030628
Muzej na Gradskij Bit Oslekova Kãšta,
Koprivštica 005073
Muzej na Gradskija Bit, Isoričeski
Muzej Ruse, Ruse 005153
Muzej na Kotlenskite Vazroždenci
i Panteon na Georgi S. Rakovski,
Kotel 005076
Muzej na Makedonija – Archeološki,
Etnološki i Istoriski, Skopje . . . 030629
Muzej na Ministerstvoto na
Vnatrešnite raboti, Sofia 005186
Muzej na Mozaikite, Devnja . . . 005011
Muzej na Nacionalnata Opera i Balet,
Sofia 005187
Muzej na Naroden Teatar Ivan Vazov,
Sofia 005188
Muzej na Poklonnoj, Moskva . . 037431
Muzej na Prostem, Rogatec . . . 038603
Muzej na Prostem Jurjeva Domačija,
Dobrnič 038507
Muzej na Prostem pri Kartuziji Pleterje,
Drča 038512
Muzej na Rezbarskoto i Zografsko
Izkustvo, Trjavna 005230
Muzej na Rodopskija Karst,
Čepelare 005000
Muzej na Rudodobiva i Minnoto Delo,
Madan 005087
Muzej na Sovremenata Umetnost
Skopje, Skopje 030630
Muzej na Tekstilnata Industrija,
Nacionalen Politechničeski Muzej
Sofia, Sliven 005168
Muzej na Tetovskiot Kraj, Tetovo 030642
Muzej na Umetnosta na Makedonija,
Skopje 030631
Muzej na Väzraždaneto, Varna . 005241
Muzej na Väzraždeneto i Nacionalno-
osvoboditelnite Borbi, Elena . . 005026
Muzej V.V. Nabokova, Sankt-
Peterburg 037795
Muzej Nacionalnoj Kultury, Kulturnyj
Centr Kazan, Kazan 037115
Muzej Naivne Umetnosti,
Jagodina 038194
Muzej Naivne Umetnosti Ilianum,
Šid 038235
Muzej Najdavnišich Pamjatok Lvova,
Lvivs'ka Galereja Mystectv, Lviv 043248
Muzej Narode Osvoboditve Maribor,
Maribor 038571
Muzej Narodnogo Chudožnika Rossii
A.I. Morozova, Ivanovskoj oblastnoj
chudožestvennyj muzej, Ivanovo 037044
Muzej Narodnogo Dekorativnogo
Mystectva, Oblasnyj istoryčnyj muzej
im. V. Tarnovs'kogo, Černigiv . 043121
Muzej Narodnogo Derevjannogo
Zodčestva Vitoslavlicy, Velikij
Novgorod 038015

Muzej Narodnogo Iskusstva,
Moskva 037432
Muzej Narodnogo Mystectva
Slobožanščyny, Charkivs'kij Chudožnij
Muzej, Charkiv 043136
Muzej Narodnogo Obrazovanija Oblasti,
Penza 037627
Muzej Narodnogo Pobutu,
Mariupol's'kij krajeznavčyj muzej,
Mariupol' 043261
Muzej Narodnogo Tvorčestva,
Penza 037628
Muzej Narodnogo Tvorčestva,
Uljanovsk 037998
Muzej Narodnoho Mystectva
Gucul'ščini ta Pokuttja im. J.
Kobryns'koho, Kolomija 043187
Muzej Narodnoï Architektury i Pobutu
u Lvovi, Lviv 043249
Muzej Narodnoï Architektury ta Pobutu
Ukraïny, Kyïv 043213
Muzej Narodnoï Architektury ta Pobutu,
Nacional'nyj istoriko-etnografičnyj
zapovidnyk Perejaslav, Perejaslav-
Chmel'nyc'kyj 043279
Muzej Narodnoj Grafiki, Moskva 043433
Muzej Narodnoje Obrazovanie
Simbirskoj Gubernii v 70–80
gody 19 veka, Gosudarstvennyj
zapovednik Rodina V.I. Lenina,
Uljanovsk 037999
Muzej Narodnych Muzykalnych
Instrumentov Kazachstana im.
Ychlasa, Almaty 029808
Muzej Nauke i Tehnike, Beograd 038164
Muzej Nekropol Demidovych i
Vystavočnyj Zal Tulskij Metall,
Tula 037963
Muzej-nekropol Literatorskie Mostki,
Gosudarstvennyj muzej gorodskoj
skulptury, Sankt-Peterburg . . . 037796
Muzej Nikole Tesle, Beograd . . 038165
Muzej Nižegorodskaja
Radiolaboratorija, Radiolabaratory
Museum, Nižnij Novgorod 037556
Muzej Nižegorodskij Ostrog,
Nižegorodskij gosudarstvennyj
istoriko-architekturnyj muzej-
zapovednik, Nižnij Novgorod . . 037557
Muzej Nižegorodskoj Intelligencii,
Nižnij Novgorod 037558
Muzej NOB – Pustova Hiša, Slovenske
Konjice 038616
Muzej Nonkonformistskogo Iskusstva,
Sankt-Peterburg 037797
Muzej Novejše Zgodovine Celje,
Celje 038501
Muzej Novejše Zgodovine Slovenije,
Ljubljana 038550
Muzej 'Novoj Akademii Izjaščnych
Iskusstv', Sankt-Peterburg 037798
Muzej Obretaja Svobodu,
Gosudarstvennyj centralnyj
muzej sovremennoj istorii Rossii,
Moskva 037434
Muzej Obščestvennogo Pitanija,
Moskovskaja associacija kulinarov,
Moskva 037435
Muzej Ochotovedenija, Irkutskaja
Selskochozjajstvennaja Akademija,
Irkutsk 037032
Muzej Ochoty i Rybolovstva,
Moskva 037436
Muzej Oktjabrskogo Vooruženogo
Vosstanija, Voronežskij oblastnoj
kraevedčeskij muzej, Voronež . 038067
Muzej Oleksi Dovbuša, Ivano-
Frankivs'kyj Krajeznavčyj Muzej,
Ivano-Frankivs'k 043158
Muzej Olimpijskoj Slavy,
Toshkent 054659
Muzej Orijentalnih Rukopisa,
Prizren 038229
Muzej Osvoboždenieto na Pleven

1877, Pleven 005118
Muzej Otečestvennaja Kultura,
Moskva 037437
Muzej Ozkotirne Železnice Parenzana,
Izola 038518
Muzej Paje Jovanovića, Beograd 038166
Muzej Palomničestva po Svjatym
Mestam, Moskva 037438
Muzej Pamjati Žertv Repressij,
Toshkent 054660
Muzej-Pamjatnik Chram Voskresenija
Christova (Spas na Krovi), Sankt-
Peterburg 037799
Muzej-panorama Borodinskaja Bitva,
Moskva 037439
Muzej Parapsichologii, Moskva . 037440
Muzej Parka Iskusstv na Krymskoj
Naberežnoj, Moskva 037441
Muzej-Parochod Svjatitel Nikolaj,
Krasnojarskij kraevedčeskij muzej,
Krasnojarsk 037169
Muzej Pčelarstva Porodice Živanović,
Sremski Karlovci 038247
Muzej Pčelovodstva, Moskovskaja
selskochosjajstvennaja akademija im
K.A. Timirjazeva, Moskva 037442
Muzej Pejzaža, Plesskij
gosudarstvennyj istoriko-
architekturnyj i chudožestvennyj
muzej-zapovednik, Ples 037661
Muzej Pidpil'nyj ta Partizans'kyj Ruch
na Mykolaïvščyni u Roki Velikoï
Vitčyznjanoï Vijny 1941–1945 rr,
Mykolaïv 043263
Muzej Pisanka, Kolomija 043188
Muzej Pisatelej-orlovcev,
Gosudarstvennyj literaturnyj muzej
im. I.S. Turgeneva, Orël 037604
Muzej Pisatelja A.P. Bibika, Mineralnye
Vody 037226
Muzej Podviga Ivana Susanina,
Susanino 037904
Muzej Poėta S. Orlova, Belozersk 036895
Muzej Policijske Akademije,
Ljubljana 038551
Muzej Političnih Zapornikov,
Internirancev in Izgnancev Grad
Rajhenburg, Brestanica 038498
Muzej Polović, Dobova 038505
Muzej Ponišavlja Pirot, Pirot . . . 038224
Muzej Pošte in Telekomunikacij,
Tehniški Muzej Slovenije, Škofja
Loka 038610
Muzej Pozorišne Umetnosti Srbije,
Beograd 038167
Muzej Predprinimatelej, Mecenatov i
Blagotvoritelej, Moskva 037443
Muzej Premogovništva Slovenije,
Velenje 038628
Muzej Prigorja, Sesvete 008986
Muzej Prikladnogo Iskusstva
Uzbekistana, Toshkent 054661
Muzej Primenjene Umetnosti,
Beograd 038168
Muzej Prioratskij Dvorec,
Gosudarstvennyj Muzej-zapovednik
Gatčina, Gatčina 037014
Muzej Priroda Morja i eë
Ochrana, Institut biologii morja,
Dalnevostočnoe otd. Rossijskoj
Akademii Nauk, Vladivostok . . 038041
Muzej Prirody, Čerepovec 036943
Muzej Prirody Burjatii, Ulan-Udė 037990
Muzej Prirody, Akademija Nauk
Respubliki Kazachstan, Almaty 029809
Muzej S.S. Prokofjeva, Krasne . 043189
Muzej S.S. Prokofjeva, Moskva . 037444
Muzej Pryrody, Žytomyr 043328
Muzej A.S. Puškina, Toržok . . . 037955
Muzej A.S. Puškina v Bernove,
Tverskoj gosudarstvennyj
obedinennyj muzej, Bernov . . . 036897
Muzej Radio im. A.S. Popova,
Ekaterinburg 036991

Muzej Radničkog Pokreta i Narodne
Revolucije, Novi Sad 038216
Muzej Vladimira Raevskogo,
Bogoslovka 036906
Muzej Razvitija Metallurgii,
Moskva 037445
Muzej Rečnog Brodarstva,
Beograd 038169
Muzej Rečnogo Flota, Volžskaja
Gosudarstvennaja Akademija
Vodnogo Transporta, Nižnij
Novgorod 037559
Muzej Remesla, Architektury i Byta,
Kaluga 037093
Muzej Republike Srpske, Banja
Luka 004256
Muzej Revolucionnogo Dviženija,
Altanbulag 031725
Muzej Rossijskogo Krasnogo Kresta,
Moskva 037446
Muzej Rossijskoj Akademii Chudožestv,
Sankt-Peterburg 037800
Muzej Nikolaja Rubcova, Moskva 037447
Muzej N.G. Rubinštejna – Muzej
Moskovskoj Konservatorii,
Moskva 037448
Muzej Rudarstva i Metalurgije,
Bor 038187
Muzej Rumänski Voin, Voenno-
istoričeski Muzej Pleven,
Pordim 005136
Muzej Russkaja Literatura XX Veka,
Gosudarstvennyj Literaturnyj muzej,
Moskva 037449
Muzej Russkie Valenki, Moskva . 037450
Muzej Russko-Armjanskoj Družby,
Rostovskij oblastnoj muzej
kraevedenija, Rostov-na-Donu . 037695
Muzej Russkoj Garmoniki Alfred
Mireka, Muzej istorii Moskvy,
Moskva 037451
Muzej Russkoj Pesni im. A. Averkina,
Sasovo 037849
Muzej Russkoj Usadebnoj Kultury
Kuzminki, Muzej istorii Moskvy,
Moskva 037452
Muzej Šachmata, Moskva 037453
Muzej Sakena Sejfullina, Astana 029812
Muzej Sakral'noï Skul'ptury XVIII st. –
Tvorčist' Ivana Pinzelja, Lviv'ska
Galereja Mystectv, Lviv 043250
Muzej Salicha Sajdaševa, Nacionalnyj
muzej Respubliki Tatarstan,
Kazan 037116
Muzej M.E. Saltykova-Ščedrina,
Tverskoj gosudarstvennyj
obedinennyj muzej, Tver 037972
Muzej Samocvety, Moskva 037454
Muzej Samostana Sestara od Sigurate,
Dubrovnik 008908
Muzej Sarajeva 1878–1914, Muzej
Grada Sarajeva, Sarajevo . . . 004279
Muzej Šarifa Kamala, Nacionalnyj
muzej Respubliki Tatarstan,
Kazan 037117
Muzej Savremene Likovne Umetnosti,
Novi Sad 038217
Muzej Savremene Umetnosti,
Beograd 038170
Muzej Sceničeskogo Iskusstva im. V.Ė.
Mejercholda, Penza 037629
Muzej Seljačkih Buna, Muzeji
Hrvatskog Zagorja, Gornja
Stubica 008915
Muzej Semji Benua, Gosudarstvennyj
Muzej Zapovednik Petergof, Sankt-
Peterburg 037801
Muzej Serebrjanogo Veka – Dom
V.Ja. Brjusova, Gosudarstvennyj
Literaturnyj muzej, Moskva . . . 037455
Muzej Simbirskaja Čuvašskaja
Škola – Kvartira I.Ja. Jakovleva,
Gosudarstvennyj zapovednik Rodina

V.I. Lenina, Uljanovsk 038000
Muzej Simbirskaja Klassičeskaja
Gimnazija, Gosudarstvennyj
zapovednik Rodina V.I. Lenina,
Uljanovsk 038001
Muzej Skulptury, Sankt-Peterburg 037802
Muzej Skulptury S.T. Konenkova,
Smolenskij Muzej-Zapovednik,
Smolensk 037866
Muzej Skulptury pod Otkrytym Nebom,
Biškek 030077
Muzej Slavonije Osijek, Osijek . . 008962
Muzej Smolenščina v Gody Velikoj
Otečestvennoj Vojny 1941 -1945,
Smolenskij Muzej-Zapovednik,
Smolensk 037867
Muzej Smolenskij Len, Smolenskij
Muzej-Zapovednik, Smolensk . 037868
Muzej Snovidenij Zigmunda Frejda,
Sankt-Peterburg 037803
Muzej Solane Tuzla, Tuzla 004290
Muzej Sovremennogo Iskusstva
Evrejskoj Avtonomnoj Oblasti,
Birobidžan 036901
Muzej Sovremennogo Izobrazitelnogo
Iskusstva im. A.A. Plastova,
Uljanovsk 038002
Muzej Sovremennoj Istorii i Kultury
Severnoj Osetii, Vladikavkaz . . 038025
Muzej Spasskaja Cerkov, Tjumenskij
Oblastnoj Kraevedčeskij Muzej im.
I.Ja. Slovcova, Tjumen 037943
Muzej Sporta v Lužnikach,
Moskva 037456
Muzej Sportivnogo Obščestva Dinamo,
Moskva 037457
Muzej Sportivnoj Slavy CSKA,
Moskva 037458
Muzej Srema, Sremska Mitrovica 038245
Muzej Srpske Pravoslavne Crkve,
Beograd 038171
Muzej Srpske Pravoslavne Crkve,
Dubrovnik 008909
Muzej Srpske Prvoslavne Crkve,
Beograd 038172
Muzej Štab-kvartira V.K. Bljuchera,
Tjumenskij oblastnoj kraevedčeskij
muzej im. I.Ja. Slovcova,
Tjumen 037944
Muzej Stara Livnica, Zastava
Namenski Proizvodi, Kragujevac 038201
Muzej Staražytna Belaruskaj Kultury,
Minsk 003252
Muzej Stare Srpske-Pravoslavne Crkve,
Sarajevo 004280
Muzej Staro Selo, Sirogojno . . . 038236
Muzej-studija Radioteatra,
Gosteleradiofond, Moskva . . . 037459
Muzej Sudnobuduvannja ta Flotu,
Mykolaïv 043264
Muzej Suvremene Povijesti,
Dubrovnik 008910
Muzej Suvremene Umjetnosti,
Museum of Contemporary Art,
Zagreb 009046
Muzej Svjazi Amurskoj Oblasti,
Blagoveščensk 036903
Muzej Talcev, Muzeji Radovljiške
Občine, Begunje na Gorenjskem 038493
Muzej S.I. Taneeva v Djutkovo,
Zvenigorod 038084
Muzej Andreja Tarkovskogo, Muzej
goroda Jurjeva, Jurjevec 037078
Muzej-teatr Lednikovyj period,
Moskva 037460
Muzej Teatra, Sankt-Peterburg . 037804
Muzej Teatra im. Mossoveta,
Moskva 037461
Muzej Teatra im. K.S. Stanislavskogo,
Moskva 037462
Muzej Teatra im. Evg. Vachtangova,
Moskva 037463
Muzej Teatra Junogo Zritelja,
Moskva 037464

Muzej Teatra Operetty, Moskva . 037465
Muzej Teatra Rossijskoj Armii,
Moskva 037466
Muzej Teatra Satiry, Moskva . . . 037467
Muzej Teatralnoj Kukly A.A Veselova,
Voronež 038068
Muzej Teatralnych Kukol, Moskva 037468
Muzej L.N. Tolstogo na stancii Lev
Tolstoj, Gosudarstvennyj muzej L.N.
Tolstogo, Stancija Lev Tolstoj . 037888
Muzej A.K. Tolstogo, Brjanskij
gosudarstvennyj obedinennyj
kraevedčeskij muzej, Krasnyj
Rog 037176
Muzej Tomaža Godca, Gorenjski Muzej,
Bohinjska Bistrica 038496
Muzej Tome Rosandića, Beograd 038173
Muzej Tropinina i Moskovskich
Chudožnikov ego Vremeni,
Moskva 037469
Muzej Tulskie Samovary, Tula . . 037964
Muzej Tulskij Kreml, Tula 037965
Muzej Turopolja, Velika Gorica . . 009016
Muzej Tverskogo Byta, Tverskoj
gosudarstvennyj obedinennyj muzej,
Tver 037973
Muzej Tvorčestva i Byta v GULAGe,
Obščestvo Memorial, Moskva . 037470
Muzej u Arandjelovcu,
Arandjelovac 038124
Muzej u Majdanpeku, Majdanpek 038209
Muzej u Smederevu, Smederevo 038238
Muzej Uezdnogo Goroda, Nacionalnyj
muzej Respubliki Tatarstan,
Čistopol 036960
Muzej Ugolok Deduški Durova,
Moskovskij teatralnyj kompleksnyj
centr Strana čudes deduški Durova,
Moskva 037471
Muzej Unikalnych Kukol, Moskva 037472
Muzej Unsko-Sanskog Kantona,
Bihać 004258
Muzej Uralskogo Gosudarstvennogo
Tehničeskogo Universiteta,
Ekaterinburg 036992
Muzej-usadba 'A.T. Tvardovskij
na Chutore Zagorje', Smolenskij
gosudarstvennyj muzej-zapovednik,
Zagorje 038078
Muzej-usadba "Muranovo" im. F.I.
Tjutčeva, Lugovskoe 037207
Muzej-usadba A.T. Bolotova,
Dvorjaninovo 036975
Muzej-usadba V.E. Borsova-Musatova,
Chudožestvennyj muzej im. A.N.
Radiščeva, Saratov 037844
Muzej-usadba N.G. Černyševskogo,
Saratov 037845
Muzej-usadba N.A. Durovoj, Elabužskij
muzej-zapovednik, Elabuga . . 036999
Muzej-usadba M.I. Glinki,
Novospasskoe 037584
Muzej-usadba im. N.K. Reericha,
Izvara 037051
Muzej-usadba Izmajlovo,
Gosudarstvennyj istoričeskij muzej,
Moskva 037473
Muzej-usadba Ostafjevo "Russkij
Parnas", Ostafjevo 037612
Muzej-usadba poėta D. Venevitinova,
Novoživotinnoe 037586
Muzej-usadba I.E. Repina,
Repino 037684
Muzej-usadba Roždestveno,
Roždestveno 037697
Muzej-usadba V.I. Surikova,
Krasnojarsk 037170
Muzej-usadba A.V. Suvorova,
Končanskoe-Suvorovskoe . . . 037148
Muzej-usadba L.N. Tolstogo
Nikolskoe Vjazemskoe, Nikolskoe-
Vjazemskoe 037543
Muzej-Usadba L.N. Tolstogo v
Chamovnikach, Gosudarstvennyj

muzej L.N. Tolstogo, Moskva . 037474
Muzej-usadba A.N. Tolstogo,
 Literaturno-memorialnyj muzej M.
 Gorkogoo, Samara 037712
Muzej Usadba Vučetiča, Moskva 037475
Muzej Vasil Levski, Karlovo 005049
Muzej Vasil Levski, Istoriceski Muzej,
 Loveč 005086
Muzej Konstantina Vasiljeva,
 Moskva 037476
Muzej Văzraždane i Učreditelno
 Săbranie, Veliko Tărnovo 005255
Muzej Văzraždenski Ikoni,
 Radomir 005145
Muzej Velenje na Velenjskem Gradu,
 Velenje 038629
Muzej Vereščaginych, Čerepovec 036944
Muzej Visoko, Visoko 004292
Muzej Vjatskie Narodnye
 Chudožestvennye Promysly,
 Kirov 037134
Muzej Vody, Moskva 037477
Muzej Voenno-morskogo Flota,
 Moskva 037478
Muzej Vojniško Văstanie 1918,
 Radomir 005146
Muzej Vojvodine, Novi Sad 038218
Muzej Volyns'koj Ikony, Volyns'kyj
 Krajeznavčyj Muzej, Luc'k 043227
Muzej Voskovych Figur, Sankt-
 Peterburg 037805
Muzej Vrbovec – Muzej Gozdarstva in
 Lesarstva, Nazarje 038583
Muzej Vuka i Dositeja, Beograd 038174
Muzej Vydatnych Dijačiv Ukraïns'koï
 Kul'tury, Lesi Ukraïnki, Mychajla
 Staric'koho, Mikoli Lisenka, Panasa
 Saksagans'koho, Kyïv 043214
Muzej Vyzvol'nych Zmagan'
 Prikarpats'koho Kraju, Ivano-
 Frankivs'kyj Krajeznavčyj Muzej,
 Ivano-Frankivs'k 043159
Muzej za Istorija na Medicinata,
 Varna 005242
Muzej za Istorija na Varna, Varna 005243
Muzej za Umjetnost i Obrt,
 Zagreb 009047
Muzej Zapadna Makedonija vo NOV,
 Kičevo 030613
Muzej-zapovednik Abramcevo,
 Abramcevo 036840
Muzej-zapovednik Dmitrovskij Kreml,
 Dmitrov 036969
Muzej-zapovednik Kazanskij Kreml,
 Kazan 037118
Muzej-zapovednik Pamjatniki
 Materialnoj Kultury Mangystau i
 Ustjurta, Aktau 029789
Muzej-zapovednik Pisatelja S.T.
 Aksakova, Aksakovo 036843
Muzej Zavoda Crvena Zastava,
 Kragujevac 038202
Muzej Zavoda Moskvič, Moskva 037479
Muzej Zdravoochranenija, Sankt-
 Peterburg 037806
Muzej Zdravoochranenija,
 Toshkent 054662
Muzej Železniki, Železniki 038636
Muzej Železnodorožnoj Techniki, Sankt-
 Peterburg 037807
Muzej Železnodorožnoj Techniki,
 Toshkent 054663
Muzej Zemlevedenija Moskovskogo
 Gosudarstvennogo Universiteta im.
 M.V. Lomonosova, Moskva ... 037480
Muzej Žrtava Genocida, Beograd 038175
Muzeji i Galerije Podgorica,
 Podgorica 031766
Muzejno-vystavočnyj Centr,
 Dalnegorsk 036965
Muzejno-vystavočnyj Centr,
 Ivanovo 037045
Muzejno-vystavočnyj Centr – Diorama,
 Kirovskij oblastnoj kraevedčeskij

muzej, Kirov 037135
Muzejno-vystavočnyj Centr Ego, Sankt-
 Peterburg 037808
Muzejno-vystavočnyj Centr Tulskie
 Drevnosti, Gosudarstvennyj voenno-
 istoričeskij i prirodnyj muzej-
 zapovednik Kulikovo Pole, Tula 037966
Muzejno-vystavočnyj Kompleks
 Volokolamskij Kreml,
 Volokolamsk 038060
Muzejnyj – Vystavočnyj Kompleks
 Školy Akvareli Sergeja Andrijaki,
 Moskva 037481
Muzejnyj Centr Nasledie Čukotki,
 Anadyr 036850
Muzejnyj Kompleks Tomskogo
 Politechničeskogo Universiteta,
 Tomsk 037950
Muzejnyj Kulturno-vystavočnyj
 Kompleks, Dnipropetrovs'k ... 043151
Muzejs "Ebreji Latvijā", Rīga ... 030210
Muzejska Jedinica Odžaci,
 Odžaci 038221
Muzejska Zbirka Ajdovščina .. 038491
Muzejska Zbirka, Imotski 008921
Muzejska Zbirka, Knič 038198
Muzejska Zbirka Franjevačkog
 Samostana, Imotski 008922
Muzejska Zbirka Laško, Laško . 038537
Muzejska Zbirka NOB, Zagorje ob
 Savi 038632
Muzejske Zbirke Fojničkog Samostana,
 Fojnica 004262
Muzejski Dokumentacijski Centar,
 Zagreb 009048
Muzejski Društvo na Makedonija,
 Skopje 059561
Muzejsko Društvo SR Crne Gore –
 Cetinje, Cetinje 059583
Muzejsko Društvo Srbije,
 Beograd 059788
Muzeju Vestnesis, Rīga 138897
Muzéo, Villeurbanne 074219
Muzerelle, Claude, Rochefort-en-
 Terre 072502, 072503
Muzeu Arkeologjik, Durrës 000013
Muzeu Arkeologjik, Museum of
 Archaeology, Tiranë 000031
Muzeu Etnografik, Berat 000010
Muzeu Etnografik, Elbasan 000015
Muzeu Etnografik, Gjirokastër . 000018
Muzeu Etnografik, Kavaja 000019
Muzeu Etnografik, Krujë 000025
Muzeu Etnografik, Vlorë 000034
Muzeu Gjergj Kastrioti Skënderbeu,
 Krujë 000026
Muzeu Historik Fier, Fier 000016
Muzeu Historik i Lushnjës,
 Lushnja 000027
Muzeu Historik i Vlorës, Vlorë . 000035
Muzeu Historik Kombëtar, Tiranë 000032
Muzeu Historik Shkodër, Shkodër 000028
Muzeu i Arsimit, Korçë 000020
Muzeu i Artit Mesjetar, Korçë .. 000021
Muzeu i Filmit Shkodër, Shkodër 000029
Muzeu i Kosoves / Muzej Kosova i
 Metohije, Priština 038227
Muzeu i Kulturës Popullore,
 Durrës 000014
Muzeu i Pavarësise, Vlorë 000036
Muzeu i Shkencave Natyrore,
 Tiranë 000033
Muzeu Inkonografik Onufri, Berat 000011
Muzeu Kombëtar Arkeologjik,
 Korçë 000022
Muzeu Prehistorik, Korçë 000023
Muzeu Agriculturii din România,
 Slobozia 036722
Muzeu Apelor Mihai Băcescu,
 Fălticeni 036459
Muzeu Arheologic Adamclisi, Muzeul
 de Istorie Naţional şi Arheologie,
 Adamclisi 036189
Muzeu Arheologic Istria, Istria . 036524

Muzeul Arheologic Sarmizegetusa,
 Muzeul Civilizaţiei Dacice şi Romane
 Deva, Sarmizegetusa 036681
Muzeul Arhitecturii Populare din Gorj,
 Muzeul Judeţean Gorj, Bumbeşti-
 Jiu 036329
Muzeul Aristide Caradja,
 Grumăzeşti 036484
Muzeul Arta lemnului, Wooden
 Art Museum, Câmpulung
 Moldovenesc 036348
Muzeul Aurului, Brad 036258
Muzeul Aviaţiei, Bucureşti 036288
Muzeul Badea Cârţan, Cârţişoara 036360
Muzeul Banatului, Timişoara .. 036782
Muzeul Banatului Montan, Reşiţa 036655
Muzeul Banatului. Secţia de Artă,
 Timişoara 036783
Muzeul Banatului. Secţia de Ştiinţele
 Naturii, Timişoara 036784
Muzeul Băncii Naţionale a României,
 Bucureşti 036289
Muzeul Brăilei, Brăila 036263
Muzeul Brăilei-Secţia de Etnografie şi
 Artă Populară, Brăila 036264
Muzeul Brăilei-Secţia Ştiinţele Naturii,
 Brăila 036265
Muzeul Bran, Bran 036266
Muzeul Breslelor, Muzeul National
 Secuiesc S.F. Gheorghe, Tîrgu
 Secuiesc 036787
Muzeul Căilor Ferate Române,
 Bucureşti 036290
Muzeul Câmpiei Băileştitor,
 Băileşti 036224
Muzeul Câmpiei Boianului, Drăgăneşti-
 Olt 036446
Muzeul Carpaţilor Răsăriteni, Sfântu
 Gheorghe 036693
Muzeul Casa Mureşenilor, Braşov 036270
Muzeul Casa Săsească, Livezile 036539
Muzeul Castelul Corvineştilor,
 Hunedoara 036497
Muzeul Ceasului Nicolae Simache,
 Ploieşti 036622
Muzeul Civilizaţiei Dacice şi Romane,
 Deva 036437
Muzeul Civilizatiej Gumelniţa Olteniţa,
 Olteniţa 036592
Muzeul Colecţiilor de Artă, Muzeul
 Naţional de Artă al României,
 Bucureşti 036291
Muzeul Comunei Buteni, Buteni 036332
Muzeul Conacul Bellu, Urlaţi .. 036808
Muzeul Cornel Medrea, Muzeul
 Municipiului Bucureşti,
 Bucureşti 036292
Muzeul Curtea Veche, Palaţul
 Voievodal, Bucureşti 036293
Muzeul de Arheologie, Săveni . 036686
Muzeul de Arheologie, Topalu . 036790
Muzeul de Arheologie Callatis
 Mangalia, Mangalia 036558
Muzeul de Arheologie şi Artă
 Religioasă, Băile Govora 036222
Muzeul de Arheologie şi Etnografie,
 Corabia 036404
Muzeul de Arheologie şi Istorie, Baia
 Mare 036217
Muzeul de Artă, Arad 036206
Muzeul de Artă, Botoşani 036255
Muzeul de Artă, Braşov 036271
Muzeul de Artă, Drobeta-Turnu
 Severin 036452
Muzeul de Artă, Iaşi 036510
Muzeul de Artă, Medgidia 036551
Muzeul de Artă, Râmnicu Vâlcea 036645
Muzeul de Artă, Roman 036660
Muzeul de Artă, Sfântu Gheorghe 036694
Muzeul de Artă, Târgovişte ... 036750
Muzeul de Artă, Târgu Jiu 036757
Muzeul de Artă, Tulcea 036796
Muzeul de Artă Centrul Artistic Baia
 Mare, Baia Mare 036218

Muzeul de Artă Comparată, Sângeorz-
 Băi 036677
Muzeul de Artă Constanţa,
 Constanţa 036395
Muzeul de Artă Craiova, Craiova 036417
Muzeul de Artă Decorativă,
 Dorohoi 036442
Muzeul de Artă Dinu şi Sevasta Vintilă,
 Topalu 036791
Muzeul de Artă Ion Irimescu,
 Fălticeni 036460
Muzeul de Artă Piatra-Neamţ,
 Complexul Muzeal Judetean Neamţ,
 Piatra-Neamţ 036611
Muzeul de Artă Populară,
 Constanţa 036396
Muzeul de Artă Populară, Ploieşti 036623
Muzeul de Artă Populară Prof. Dr.
 Nicolae Minovici, Bucureşti ... 036294
Muzeul de Artă Populară şi Etnografie,
 Tulcea 036797
Muzeul de Artă Religioasă al
 Mănăstirii Sfinţii Petru şi Pavel,
 Huşi 036500
Muzeul de Artă Religioasă Nicolae
 Iorga, Vălenii de Munte 036813
Muzeul de Artă şi Artă Populară,
 Calafat 036340
Muzeul de Artă Veche Apuseană
 Dumitru Minovici, Bucureşti . 036295
Muzeul de Artă Vizuală, Galaţi . 036472
Muzeul de Artă, Muzeului Judeţean
 Satu Mare, Satu Mare 036683
Muzeul de Cultură şi Artă Relgioasă,
 Borzeşti 036251
Muzeul de Etnografie, Botoşani 036256
Muzeul de Etnografie, Comarnic 036392
Muzeul de Etnografie, Topliţa . 036793
Muzeul de Etnografie al Văii
 Teleajenului, Vălenii de Munte . 036814
Muzeul de Etnografie Braşov,
 Braşov 036272
Muzeul de Etnografie Casa Băniei,
 Craiova 036418
Muzeul de Etnografie Gheorghe
 Cernea, Rupea 036669
Muzeul de Etnografie Piatra-Neamţ,
 Complexul Muzeal Judetean Neamţ,
 Piatra-Neamţ 036612
Muzeul de Etnografie Săsească Emil
 Sigerus, Sibiu 036699
Muzeul de Etnografie şi a
 Regimentului de Graniţă,
 Caransebeş 036356
Muzeul de Etnografie şi Artă,
 Drăgăşani 036447
Muzeul de Etnografie şi Artă
 Bisericească, Fântânele 036462
Muzeul de Etnografie şi Artă
 Contemporană, Comăneşti ... 036391
Muzeul de Etnografie şi Artă Populară,
 Baia Mare 036219
Muzeul de Etnografie şi Artă Populară,
 Floreşti-Stoeneşti 036464
Muzeul de Etnografie şi Artă Populară,
 Reghin 036652
Muzeul de Etnografie şi Artă Populară
 Orăştie, Orăştie 036599
Muzeul de Etnografie Universală Franz
 Binder, Complexul Naţional Muzeal
 Astra, Sibiu 036700
Muzeul de Icoane şi Obiecte
 Bisericesti, Sibiu 036701
Muzeul de Istorie, Bicaz 036243
Muzeul de Istorie, Oneşti 036593
Muzeul de Istorie, Scorniceşti . 036688
Muzeul de Istorie, Sibiu 036702
Muzeul de Istorie, Siret 036715
Muzeul de Istorie, Târgovişte .. 036751
Muzeul de Istorie, Târgu Jiu .. 036758
Muzeul de Istorie, Teremia Mare 036778
Muzeul de Istorie a Clisurii Dunării,
 Moldova Nouă 036574
Muzeul de Istorie a Evreilor din

România, Bucureşti 036296
Muzeul de Istorie a Oraşului Hârlău,
Hârlău 036490
Muzeul de Istorie a Teatrului şi a
Oraşului Oraviţa, Oraviţa 036601
Muzeul de Istorie al Moldovei,
Iaşi 036511
Muzeul de Istorie Augustin Bunea,
Blaj 036248
Muzeul de Istorie Gherla, Gherla 036476
Muzeul de Istorie Locală şi Etnografie,
Brad 036259
Muzeul de Istorie Militară,
Constanţa 036397
Muzeul de Istorie Naţional şi
Arheologie, Constanţa 036398
Muzeul de Istorie Naturală din
Sibiu, Muzeul Naţional Brukenthal,
Sibiu 036703
Muzeul de Istorie Naturală Iaşi,
Universităţii Iaşi, Iaşi 036512
Muzeul de Istorie şi Arheologie,
Tulcea 036798
Muzeul de Istorie şi Artă Roman,
Roman 036661
Muzeul de Istorie şi Etnografie, Târgu
Neamţ 036769
Muzeul de Istorie Sighişoara,
Sighişoara 036710
Muzeul de Istorie Turda, Turda . 036802
Muzeul de Istorie Valea Hârtibaciului,
Agnita 036194
Muzeul de Istorie, Etnografie şi Artă
Plastică, Lugoj 036543
Muzeul de Mineralogie Baia Mare,
Baia Mare 036220
Muzeul de Sculptură Ion Jalea,
Constanţa 036399
Muzeul de Ştiinţe Naturale, Complexul
Muzeal Judeţean Neamţ, Piatra-
Neamţ 036613
Muzeul de Ştiinţele Naturii,
Craiova 036419
Muzeul de Ştiinţele Naturii,
Dorohoi 036443
Muzeul de Ştiinţele Naturii,
Focşani 036465
Muzeul de Ştiinţele Naturii,
Suceava 036737
Muzeul de Ştiinţele Naturii Aiud,
Aiud 036195
Muzeul de Ştiinţele Naturii şi
Cinegetică, Vatra Dornei 036820
Muzeul de Ştiinţele Naturii,
Complexul Muzeal Judeţean Neamţ,
Roman 036662
Muzeul de Ştiinte Naturale,
Institutul de Cercetări Eco-Muzeale,
Tulcea 036799
Muzeul de Vânătoare August von
Spiess, Sibiu 036704
Muzeul Deltei Dunării, Tulcea .. 036800
Muzeul Doftana, Doftana 036440
Muzeul Dunării de Jos, Călăraşi 036341
Muzeul Episcopiei Romano-Catolice,
Oradea 036595
Muzeul Etnografic, Vatra Dornei 036821
Muzeul Etnografic al Bucovinei,
Suceava 036738
Muzeul Etnografic al Moldovei,
Iaşi 036513
Muzeul Etnografic al Transilvaniei, Cluj-
Napoca 036380
Muzeul Etnografic Ioan Tugui,
Câmpulung Moldovenesc 036349
Muzeul Etnografic Lupşa, Lupşa 036547
Muzeul Etnografic Tehnici Populare
Bucovinene, Rădăuţi 036640
Muzeul Etnografic, Muzeului de
Etnografie Braşov, Săcele ... 036670
Muzeul Fierului, Hunedoara 036498
Muzeul Grăniceresc Năsăudean,
Năsăud 036579
Muzeul Haáz Rezsö, Odorheiu

Secuiesc 036590
Muzeul Haszmann Pál, Cernatu de
Sus 036367
Muzeul Hrandt Avakian, Muzeul
Naţional de Artă al României,
Bucureşti 036297
Muzeul în Aer Liber, Negreşti-
Oaş 036581
Muzeul în Aer Liber din Dumbrava
Sibiului, Sibiu 036705
Muzeul Ion Luca Caragiale,
Ploieşti 036624
Muzeul Judeţean Alexandru Ştefulescu,
Târgu Jiu 036759
Muzeul Judeţean Argeş, Piteşti . 036619
Muzeul Judeţean Bistriţa-Năsăud,
Bistriţa, Năsăud 036245
Muzeul Judeţean Botoşani,
Botoşani 036257
Muzeul Judeţean Buzău, Buzău 036337
Muzeul Judeţean de Artă,
Ploieşti 036625
Muzeul Judeţean de Artă şi Etnografie,
Bacău 036214
Muzeul Judeţean de Etnografie,
Slatina 036720
Muzeul Judeţean de Istorie,
Galaţi 036473
Muzeul Judeţean de Istorie Braşov,
Braşov 036273
Muzeul Judeţean de Istorie şi
Arheologie Iulian Antonescu Bacău,
Bacău 036215
Muzeul Judeţean de Istorie şi
Arheologie Prahova, Ploieşti .. 036626
Muzeul Judeţean de Istorie şi Artă
Zalău, Zalău 036835
Muzeul Judeţean de Istorie Teleorman,
Alexandria 036200
Muzeul Judeţean de Istorie, Complexul
Muzeal Judeţean Neamţ, Piatra-
Neamţ 036614
Muzeul Judeţean de Ştiinţele Naturii
Prahova, Ploieşti 036627
Muzeul Judeţean Ialomiţa,
Slobozia 036723
Muzeul Judeţean Maramureş, Baia
Mare 036221
Muzeul Judeţean Mureş, Târgu
Mureş 036762
Muzeul Judeţean Mureş-Secţia de
Artă, Târgu Mureş 036763
Muzeul Judeţean Mureş-Secţia de
Etnografie şi Artă Populară, Târgu
Mureş 036764
Muzeul Judeţean Mureş-Secţia de
Ştiinţele Naturii, Târgu Mureş . 036765
Muzeul Judeţean Olt, Slatina . 036721
Muzeul Judeţean Satu Mare, Satu
Mare 036684
Muzeul Judeţean Ştefan cel Mare,
Vaslui 036819
Muzeul Judeţean Teohari Antonescu,
Giurgiu 036479
Muzeul Judeţean Vâlcea, Râmnicu
Vâlcea 036646
Muzeul Krikor H. Zambaccian, Muzeul
Naţional de Artă al României,
Bucureşti 036298
Muzeul Literaturii Române, Iaşi . 036514
Muzeul Locomotivelor cu Abur,
Reşiţa 036656
Muzeul Maghiar, Bogdand 036250
Muzeul Mănăstirii Bistriţa, Bistriţa,
Neamţ 036246
Muzeul Maramureşului, Sighetu
Marmaţiei 036708
Muzeul Mării, Constanţa 036400
Muzeul Marinei Române,
Constanţa 036401
Muzeul Memorial Ady Endre,
Căuaş 036363
Muzeul Memorial Ady Endre,
Oradea 036596

Muzeul Memorial Alexandru Ioan Cuza,
Ruginoasa 036668
Muzeul Memorial Alexandru Vlahuţă,
Dumbrăveni 036454
Muzeul Memorial Aurel Vlaicu,
Geoagiu 036474
Muzeul Memorial Avram Iancu, Avram
Iancu 036209
Muzeul Memorial Bojdeuca Ion
Creangă, Iaşi 036515
Muzeul Memorial C.I. & C.C. Nottara,
Bucureşti 036299
Muzeul Memorial Calistrat Hogaş,
Piatra-Neamţ 036615
Muzeul Memorial Costache Negri,
Costache Negri 036408
Muzeul Memorial Dimitrie & Aurelia
Ghiaţă, Bucureşti 036300
Muzeul Memorial Dr. Victor Babeş,
Bucureşti 036301
Muzeul Memorial Emil Isac, Cluj-
Napoca 036381
Muzeul Memorial George Călinescu,
Bucureşti 036302
Muzeul Memorial George Coşbuc,
Coşbuc 036407
Muzeul Memorial George Enescu,
Dorohoi 036444
Muzeul Memorial Gheorghe M.
Tattarascu, Bucureşti 036303
Muzeul Memorial Henri Catargi,
Scăeşti 036687
Muzeul Memorial Ioan Slavici şi Emil
Monţia, Şiria 036716
Muzeul Memorial Ion Pop Reteganul,
Reteag 036657
Muzeul Memorial Iosif Vulcan,
Oradea 036597
Muzeul Memorial János Arany,
Salonta 036673
Muzeul Memorial Liviu Rebreanu,
Năsăud 036580
Muzeul Memorial Mihail Kogălniceanu,
Iaşi 036516
Muzeul Memorial Nicolae Bălcescu,
Nicolae Bălcescu 036584
Muzeul Memorial Nicolae Grigorescu,
Câmpina 036343
Muzeul Memorial Octavian Goga,
Ciucea 036376
Muzeul Memorial Paul Constantinescu,
Ploieşti 036628
Muzeul Memorial Poni-Cernătescu,
Iaşi 036517
Muzeul Memorial Sándor Petöfi,
Coltău 036388
Muzeul Mihai Eminescu, Iaşi ... 036518
Muzeul Militar Dolj, Craiova ... 036420
Muzeul Militar Naţional,
Bucureşti 036304
Muzeul Minerilului, Petroşani .. 036608
Muzeul Mineritului Avrifer Roşia
Montană, Roşia Montană 036665
Muzeul Mitropolit Antim Ivireanul,
Bucureşti 036305
Muzeul Municipal, Huşi 036501
Muzeul Municipal, Râmnicu
Sărat 036642
Muzeul Municipal Câmpulung,
Câmpulung 036345
Muzeul Municipal Câmpulung-Secţia
de Etnografie şi, Artă Populară,
Câmpulung 036346
Muzeul Municipal Curtea de Argeş,
Curtea de Argeş 036429
Muzeul Municipal de Artă, Turnu
Măgurele 036803
Muzeul Municipal de Istorie, Roşiorii
de Vede 036666
Muzeul Municipal Dej şi Galeria de
Artă, Dej 036434
Muzeul Municipal Ioan Raica Sebeş,
Sebeş 036689
Muzeul Municipal Medias,

Mediaş 036562
Muzeul Municipiului Bucureşti,
Bucureşti 036306
Muzeul Naţional al Bucovinei, Secţia
de Artă, Suceava 036739
Muzeul Naţional al Hărţilor şi Cărţii
Vechi, Bucureşti 036307
Muzeul National al Literaturii Române,
Bucureşti 036308
Muzeul Naţional al Petrolului,
Ploieşti 036629
Muzeul Naţional al Pompierilor,
Bucureşti 036309
Muzeul Naţional al Satului Dimitrie
Gusti, Bucureşti 036310
Muzeul Naţional al Unirii, Alba
Iulia 036199
Muzeul Naţional Brukenthal,
Sibiu 036706
Muzeul Naţional Cotroceni,
Bucureşti 036311
Muzeul Naţional de Artă al României,
Bucureşti 036312
Muzeul Naţional de Artă Cluj-Napoca,
Cluj-Napoca 036382
Muzeul Naţional de Artă
Contemporană al României,
Bucureşti 036313
Muzeul National de Arte Plastice,
Chişinău 031710
Muzeul National de Etnografie si
Istorie Naturala, Chişinău ... 031711
Muzeul Naţional de Istorie a Moldovei,
Chişinău 031712
Muzeul Naţional de Istorie a României,
Bucureşti 036314
Muzeul Naţional de Istorie a
Transilvaniei, Cluj-Napoca 036383
Muzeul Naţional de Istorie Naturala
Grigore Antipa, Bucureşti ... 036315
Muzeul Naţional Geologic,
Bucureşti 036316
Muzeul Naţional George Enescu,
Bucureşti 036317
Muzeul Naţional Peleş, Sinaia . . 036713
Muzeul National Secuisec, Sfântu
Gheorghe 036695
Muzeul Nicolae Iorga, Vălenii de
Munte 036815
Muzeul Obiceiurilor Populare din
Bucovina, Gura Humorului ... 036487
Muzeul Olteniei, Craiova 036421
Muzeul Operei Române,
Bucureşti 036318
Muzeul Orăşanesc Fălticeni,
Fălticeni 036461
Muzeul Orăşenesc, Adjud 036190
Muzeul Orăşenesc, Baraolt ... 036230
Muzeul Orăşenesc, Carei 036357
Muzeul Orăşenesc, Lipova 036536
Muzeul Orăşenesc, Târnăveni . 036771
Muzeul Orăşenesc, Tăşnad ... 036772
Muzeul Orăşenesc, Tecuci ... 036773
Muzeul Orăşenesc de Istorie,
Aiud 036196
Muzeul Orăşenesc Molnar Istvan,
Cristuru Secuiesc 036425
Muzeul Oraşului Avrig, Avrig .. 036210
Muzeul Parohial de Icoane pe Sticlă,
Sibiel 036696
Muzeul Pedagogic Republican,
Chişinău 031713
Muzeul Podgoriei Drăgăşani,
Drăgăşani 036448
Muzeul Poliţiei Române,
Târgovişte 036752
Muzeul Pompierilor, Teremia
Mare 036779
Muzeul Portului Constanţa,
Constanţa 036402
Muzeul Primei Şcoli Româneşti din
Scheii Braşovului, Braşov ... 036274
Muzeul Regiunii Porţilor de Fier,
Drobeta-Turnu Severin 036453

Muzeul Rezervaţiei Bucegi,
Sinaia 036714
Muzeul Romanaţiului, Caracal . . 036352
Muzeul Romanaţiului Secţia de Artă
Plastică, Caracal 036353
Muzeul Sării, Slănic Prahova . . . 036719
Muzeul Sătesc, Arcani 036207
Muzeul Sătesc, Bănia 036229
Muzeul Sătesc, Bărbăteşti 036232
Muzeul Sătesc, Bârca 036233
Muzeul Sătesc, Bechet 036238
Muzeul Sătesc, Botoşana 036252
Muzeul Sătesc, Broşteni 036277
Muzeul Sătesc, Castranova . . . 036361
Muzeul Sătesc, Celaru 036364
Muzeul Sătesc, Cernăteşti 036366
Muzeul Sătesc, Cornu 036405
Muzeul Sătesc, Dănciuleşti 036430
Muzeul Sătesc, Drăcşenei 036445
Muzeul Sătesc, Dranic 036450
Muzeul Sătesc, Frăteşti 036468
Muzeul Sătesc, Gighera 036478
Muzeul Sătesc, Goicea 036480
Muzeul Sătesc, Gornea 036481
Muzeul Sătesc, Gostavăţu 036482
Muzeul Sătesc, Grumezoaia . . . 036485
Muzeul Sătesc, Leleşti 036530
Muzeul Sătesc, Leu 036533
Muzeul Sătesc, Mehadica 036563
Muzeul Sătesc, Sălişte 036672
Muzeul Sătesc, Tăcuta 036743
Muzeul Sătesc, Teiu 036774
Muzeul Sătesc, Valea Stanciului 036809
Muzeul Sătesc, Vultureşti 036831
Muzeul Sătesc, Vutcani 036832
Muzeul Sătesc Constantin T.
Cumpănaşu, Orlea 036602
Muzeul Sătesc Prăjeşti, Traian . 036764
Muzeul Satului Bănăţean-Pădurea
Verde, Timişoara 036785
Muzeul Satului din Bucovina,
Suceava 036740
Muzeul Satului Vâlcean, Bujoreni 036327
Muzeul Scriitorilor Târgovişteni,
Târgovişte 036753
Muzeul Secuiesc al Ciucului,
Miercurea-Ciuc 036565
Muzeul şi Rezervaţia Arheologică
de la Castrul Roman Jidava,
Câmpulung 036347
Muzeul Sportului din România,
Bucureşti 036319
Muzeul Staţiunii Băile Herculane, Băile
Herculane 036223
Muzeul Ştiinţei şi Tehnicii Ştefan
Procopiu, Iaşi 036519
Muzeul Şvăbesc, Petreşti 036607
Muzeul Tăranului Român,
Bucureşti 036320
Muzeul Ţăranului Român, Secţia de
Artă Populară, Hotarele 036496
Muzeul Tarii Beiusului, Beiuş . . 036239
Muzeul Ţării Codrului, Valea
Vinului 036811
Muzeul Ţării Crişurilor, Oradea . 036598
Muzeul Ţării Făgăraşului Valeriu
Literat, Făgăraş 036456
Muzeul Ţării Oaşului, Negreşti-
Oaş 036582
Muzeul Tarisznyás Márton,
Gheorghieni 036475
Muzeul Teatrului, Iaşi 036520
Muzeul Teatrului Naţional,
Bucureşti 036321
Muzeul Teatrului Naţional,
Craiova 036422
Muzeul Tehnic Prof. Ing. Dimitrie
Leonida, Bucureşti 036322
Muzeul Ţesutului şi Postăvăritului,
Buhuşi 036326
Muzeul Theodor Aman, Muzeul
Municipiului Bucureşti,
Bucureşti 036323
Muzeul Theodor Pallady, Muzeul

Naţional de Artă al României,
Bucureşti 036324
Muzeul Tiparului şi al Cărtii Vechi
Româneşti, Târgovişte 036754
Muzeul Unirii, Iaşi 036521
Muzeul Universitar de Istoria Medicinei
şi Farmaciei Victor Gomoiu, Târgu
Jiu 036760
Muzeul Universitatii Babeş-Bolyai, Cluj-
Napoca 036384
Muzeul Universităţii Bucureşti,
Bucureşti 036325
Muzeul Văii Superioare a Teleajenului,
Măneciu 036556
Muzeul Vasile Pârvan, Bârlad . . 036235
Muzeul Viei şi Vinului, Miniş . . . 036568
Muzeul Viticol, Huşi 036502
Muzeul Viticol, Odobeşti 036589
Muzeul Vivariul Bacău, Bacău . . 036216
Muzeul Vrancei, Focşani 036466
Muzeul Zoologic al Universităţii
Babeş-Bolyai, Cluj-Napoca . . 036385
Muzeum, Koszalin 034853
Muzeum – Františkánský klaster,
Kadaň 009613
Muzeum 24. Pułku Ułanów, Muzeum
Regionalnego w Kraśniku,
Kraśnik 034915
Muzeum 7 Pułku Ułanów Lubelskich,
Mińsk Mazowiecki 035015
Muzeum a Galerie Litomyšl,
Litomyšl 009674
Muzeum a Galerie Orlických Hor,
Rychnov nad Kněžnou 009876
Muzeum a Galerie Severního Plzeňska,
Mariánská Tynice – Kralovice . 009690
Muzeum a Pojizerská Galerie,
Semily 009880
Muzeum Adwokatury Polskiej,
Warszawa 035324
Muzeum Afrykanistyczne im. Dr.
Bogdana Szczygła, Olkusz . . . 035064
Muzeum Akademii Medycznej,
Poznań 035120
Muzeum Akademii Sztuk Pięknych,
Warszawa 035325
Muzeum Akademii Sztuk Pięknych,
Wrocław 035416
Muzeum Aloise Jiráska, Hronov 009581
Muzeum Aloise Jiráska a Mikoláše
Alše, Praha 009812
Múzeum Andreja Kmeťa, Slovenské
Národné Múzeum, Martin 038391
Múzeum Andreja Sládkoviča Krupina,
Lesnícke a Drevárske Múzeum
Zvolen, Krupina 038375
Muzeum Andrzeja Struga, Muzeum
Literatury w Warszawie,
Warszawa 035326
Muzeum Antického Sochařstvi a
Architektury, Litomyšl 009675
Muzeum Antonína Dvořáka,
Praha 009813
Muzeum Archeologiczne, Gdańsk 034728
Muzeum Archeologiczne, Poznań 035121
Muzeum Archeologiczne i
Etnograficzne, Łódź 034965
Muzeum Archeologiczne w
Biskupinie, Państwowe Muzeum
Archeologicznego w Warszawie,
Gąsawa 034721
Muzeum Archeologiczne w Krakowie,
Kraków 034876
Muzeum Archeologiczne w Krakowie,
Oddział w Nowej Hucie, Kraków 034877
Muzeum Archeologiczne, Muzeum
Miejskie Wrocławia, Wrocław . 035417
Muzeum Archeologiczno-Historyczne,
Zamek Książąt Głogowskich,
Głogów 034759
Muzeum Archidiecezjalne,

Katowice 034815
Muzeum Archidiecezjalne,
Poznań 035122
Muzeum Archidiecezjalne,
Przemyśl 035145
Muzeum Archidiecezjalne,
Szczecin 035248
Muzeum Archidiecezjalne,
Wrocław 035418
Muzeum Archidiecezjalne Lubelskie,
Lublin 034993
Muzeum Archidiecezjalne w Krakowie,
Kraków 034878
Muzeum Archidiecezjalne w Oliwie,
Gdańsk 034729
Muzeum Archidiecezji Gnieźnieńskiej,
Gniezno 034763
Muzeum Archidiecezji Łódzkiej,
Łódź 034966
Muzeum Archidiecezji Warszawskiej,
Warszawa 035327
Muzeum Architektury, Wrocław . 035419
Muzeum Armii Krajowej im. Generała
Emila Fieldorfa "Nila", Kraków 034879
Muzeum Armii Poznań, Wielkopolskie
Muzeum Walk Niepodległościowych
w Poznaniu, Poznań 035123
Muzeum Arthura Fleischmanna,
Mestské Múzeum v Bratislave,
Bratislava 038315
Muzeum Artystów, Łódź 034967
Muzeum Azji i Pacyfiku,
Warszawa 035328
Múzeum Bábkarských Kultúr a
Hraček Hrad Modry Kameň,
Slovenské Národné Múzeum, Modrý
Kameň 038402
Muzeum Barwy i Oręża Arsenał,
Zamość 035452
Múzeum Beckov, Trenčianske
Múzeum, Beckov 038293
Muzeum Bedřicha Hrozného, Lysá nad
Labem 009686
Muzeum Bedřicha Smetany,
Praha 009814
Muzeum Benátky nad Jizerou,
Benátky nad Jizerou 009441
Muzeum Beskidzkie im. Andrzeja
Podżorskiego, Muzeum w Cieszynie,
Wisła 035394
Muzeum Beskyd Frýdek-Místek,
Frýdek-Místek 009549
Muzeum Betlémů, Karlštejn . . . 009622
Muzeum Biblioteki Kórnickiej, Polish
Academy of Sciences, Kórnik . 034848
Muzeum Biograficzne Władysława
Orkana, Niedźwiedź 035035
Muzeum Bitwy Grunwaldzkiej,
Muzeum Warmii i Mazur w Olsztynie,
Stębark 035224
Muzeum Bitwy Legnickiej, Muzeum
Miedzi w Legnicy, Legnickie
Pole 034941
Muzeum Bitwy nad Bzurą, Muzeum
Regionalne, Kutno 034930
Muzeum Blansko, Blansko 009449
Muzeum Bolesława Prusa,
Nałęczów 035029
Muzeum Borów Tucholskich,
Tuchola 035291
Muzeum Boskovicka, Boskovice 009451
Muzeum Boženy Němcové, Česká
Skalice 009501
Muzeum Bratří Čapků, Malé
Svatoňovice 009688
Muzeum Broni Pancernej,
Poznań 035124
Muzeum Broumovska, Broumov 009485
Muzeum Bučovice, Muzeum
Vyškovska, Bučovice 009488
Muzeum Budownictwa Ludowego,
Sanok 035191
Muzeum Budownictwa i Techniki
Wiejskiej, Muzeum w Gorzowie

Wielkopolskim, Bogdaniec 034658
Muzeum Budownictwa Ludowego –
Park Etnograficzny w Olsztynku,
Olsztynek 035071
Muzeum Byłego Obozu Zagłady w
Chełmnie nad Narem, Chełmno nad
Nerem 034683
Múzeum Čachtice, Čachtice . . . 038336
Muzeum Cechu Rzemiosł Skórzanych
im. Jana Kilińskiego, Warszawa 035329
Muzeum Ceramiki, Bolesławiec . 034659
Muzeum Ceramiki Kaszubskiej Neclów,
Chmielno 034685
Múzeum Červený Kameň, Slovenské
Národné Múzeum, Častá 038339
Múzeum Červený Kláštor, Červený
Kláštor 038340
Muzeum Českého granátu,
Třebenice 009918
Muzeum Českého Krasu,
Hořovice 009572
Muzeum Českého Krasu v Berouně,
Beroun 009445
Muzeum Českého Krasu v Berouně –
pobačka muzeum v Žebráku,
Žebrák 009978
Muzeum Českého Lesa, Tachov 009903
Muzeum Českého Porcelánu, Klášterec
nad Ohří 009628
Muzeum Českého Ráje, Turnov . 009925
Muzeum Českého Venkova, Národní
Zemědělské Muzeum Praha, Kačina
u Kutné Hory 009610
Muzeum Chełmskie, Chełm 034678
Muzeum Chełmskie Dział Sztuki
Dawnej, Chełm 034679
Muzeum Chodska, Domažlice . . 009540
Muzeum Cystersów, Wąchock . . 035297
Muzeum Częstochowskie,
Częstochowa 034706
Muzeum Czynu Niepodległościowego,
Kraków 034880
Muzeum Czynu Partyzanckiego,
Muzeum Okręgowego w Piotrkowie
Trybunalskim, Polichno 035112
Muzeum Czynu Powstańczego,
Muzeum Śląska Opolskiego w Opolu,
Góra św. Anny 034769
Muzeum Czynu Zbrojnego
Pracowników Huty im. Tadeusza
Sendzimira, Kraków 034881
Muzeum Dawnego Kupiectwa,
Świdnica 035237
Muzeum Diecezjalne, Płock 035104
Muzeum Diecezjalne, Siedlce . . 035193
Muzeum Diecezjalne, Tarnów . . 035256
Muzeum Diecezjalne, Włocławek 035399
Muzeum Diecezjalne Sztuki Kościelnej,
Sandomierz 035187
Muzeum Diecezjalne w Opolu,
Opole 035076
Muzeum Diecezji Pelplińskiej,
Pelplin 035096
Muzeum Dom Urbańczyka, Muzeum w
Chrzanowie im. Ireny i Mieczysława
Mazarakich, Chrzanów 034696
Muzeum Dom Wincentego Witosa,
Wierzchosławice 035393
Múzeum Dopravy, Slovenské
Technické Múzeum, Bratislava . 038316
Muzeum Dr. Aleše Hrdličky,
Humpolec 009584
Muzeum Dr. Bohuslava Horáka,
Rokycany 009863
Muzeum Dr. Bohuslava Horáka,
Pobočka vodni hamr Dobřív,
Dobřív 009536
Muzeum Drukarstwa Warszawskiego,
Muzeum Historyczne Warszawy,
Warszawa 035330
Muzeum Dworek Zabytkowy, Oddział
Zamku Królewskiego na Wawelu,
Stryszów 035226
Muzeum Dworu Polskiego,

Plochocin 035103
Muzeum Dyplomacji i Uchodźstwa
Polskiego, Instytut Historii Akademii
Bydkoskiej, Bydgoszcz 034667
Muzeum Dzieje i Kultura Żydów,
Muzeum Historyczne Miasta
Krakowa, Kraków 034882
Muzeum Emeryka Hutten-Czapskiego
w Kamienicy Łozińskych, Muzeum
Narodowe w Krakowie, Kraków 034883
Muzeum Emila Zegadłowicza,
Wadowice 035299
Muzeum Energetyki Jeleniogórskiej,
Szklarska Poręba 035259
Muzeum Etnograficzne im. Franciszka
Kotuli, Muzeum Okręgowego w
Rzeszowie, Rzeszów 035183
Muzeum Etnograficzne im. M.
Znamierowskiej-Prüfferowej,
Toruń 035284
Muzeum Etnograficzne im. Seweryna
Udzieli, Kraków 034884
Muzeum Etnograficzne w Zielonej
Górze z Siedzibą w Ochli, Ochla 035059
Muzeum Etnograficzne, Muzeum
Narodowe w Poznaniu, Poznań 035125
Muzeum Etnograficzne, Muzeum
Narodowe w Wrocławiu,
Wrocław 035420
Muzeum Etnograficzne, Ziemi
Kujawskiej i Dobrzyńskiej,
Włocławek 035400
Muzeum Ewolucji Instytutu
Paleobiologii PAN, Warszawa . 035331
Muzeum Farmacji, Poznań . . . 035126
Muzeum Farmacji im. Antoniny
Leśniewskiej, Warszawa 035332
Muzeum Farmacji przy Aptece pod
Łabędziem, Bydgoszcz 034668
Muzeum Farmacji, Collegium Medicum
Uniwersytetu Jagiellońskiego,
Kraków 034885
Muzeum Feliksa Nowowiejskiego,
Barczewo 034626
Múzeum Ferdiša Kostku, Stupava 038458
Muzeum Filumenistyczne, Bystrzyca
Kłodzka 034674
Muzeum Fojtství, Kopřivnice . . 009638
Muzeum Fryderyka Chopina,
Warszawa 035333
Muzeum Gazownictwa, Paczków 035094
Muzeum Gazownictwa, Sobótka 035205
Muzeum Gazownictwa przy Gazowni
Warszawskiej, Warszawa 035334
Muzeum Geologiczne Państwowego
Instytutu Geologicznego,
Warszawa 035335
Muzeum Geologiczne Uniwersytetu
Szczecińskiego, Szczecin . . . 035249
Muzeum Geologiczne, Instytutu Nauk
Geologicznych PAN, Kraków . . 034886
Muzeum Geologiczne, Instytutu
Nauk Geologicznych Uniwersytetu
Wrocławskiego, Wrocław 035421
Muzeum Geologii Złóż im. Czesława
Poborskiego, Politechnika Śląska,
Gliwice 034755
Muzeum Górnictwa i Hutnictwa
Kopalina Złota, Złoty Stok . . . 035466
Muzeum Górnictwa Podziemnego,
Nowa Ruda 035038
Muzeum Górnictwa Węglowego,
Zabrze 035440
Muzeum Górnicze, Bogatynia . 034657
Muzeum Górnośląskie, Bytom . 034675
Muzeum Gross-Rosen, Wałbrzych 035302
Muzeum Gross-Rosen – Obóz
Koncentracyjny, Rogoźnica . . . 035175
Múzeum-Ház, Csongrád 023589
Muzeum Henryka Sienkiewicza, Wola
Okrzejska 035406
Muzeum Henryka Sienkiewicza,
Muzeum Narodowe w Kielcach,
Oblęgorek 035058

Muzeum Historické Hasičské Techniky,
Městské Muzeum, Chrastava . 009526
Muzeum Historickych Motocyklů,
Firma Brand a Balek, Kašperské
Hory 009624
Muzeum Historii Akademii Górniczo-
Hutniczej im. S. Staszica,
Kraków 034887
Muzeum Historii Fotografii,
Kraków 034888
Muzeum Historii Fotografii im. Prof.
Władysława Bogackiego, Krakowskie
Towarzystwo Fotograficzne,
Kraków 034889
Muzeum Historii i Tradycji Żołnierzy
Suwalszczyzny, Suwałki 035232
Muzeum Historii Kanału
Augustowskiego, Augustów . . . 034621
Muzeum Historii Katowic,
Katowice 034816
Muzeum Historii Katowic – Dział
Etnologii Miasta, Katowice . . . 034817
Muzeum Historii Medycyny i Farmacji,
Szczecin 035250
Muzeum Historii Miasta, Zduńska
Wola 035457
Muzeum Historii Miasta – Podziemna
Trasa Turystyczna, Muzeum
Okręgowego w Rzeszowie,
Rzeszów 035184
Muzeum Historii Miasta Łodzi,
Łódź 034968
Muzeum Historii Miasta Lublina,
Muzeum Lubelskie, Lublin . . . 034994
Muzeum Historii Miasta Poznania,
Muzeum Narodowe w Poznaniu,
Poznań 035127
Muzeum Historii Miasta, Muzeum
Narodowe w Szczecinie,
Szczecin 035251
Muzeum Historii Młynarstwa i
Wodnych Urządzeń Przemysłu
Wiejskiego w Jaraczu, Jaracz . 034785
Muzeum Historii Polskiego Rucha
Ludowego, Piaseczno koło
Gniewa 035097
Muzeum Historii Polskiego Ruchu
Ludowego, Warszawa 035336
Muzeum Historii Polskiego Ruchu
Ludowego w Warszawie, Oddział w
Sandomierzu, Sandomierz . . . 035188
Muzeum Historii Przemysłu,
Opatówek 035072
Muzeum Historii Ratusza i Trybunału
Koronnego, Muzeum Lubelskie,
Lublin 034995
Muzeum Historii Spółdzielczości w
Polsce, Nałęczów 035030
Muzeum Historii Włocławka, Muzeum
Ziemi Kujawskiej i Dobrzynskiej,
Włocławek 035401
Muzeum Historii Żydów Polskich,
Warszawa 035337
Muzeum Historyczne – Pałac w Dukli,
Dukla 034717
Muzeum Historyczne Miasta Gdańska,
Gdańsk 034730
Muzeum Historyczne Miasta Krakowa,
Kraków 034890
Muzeum Historyczne Miasta
Krakowa – Oddział Pomorska,
Kraków 034891
Muzeum Historyczne Miasta
Stołecznego Warszawy,
Warszawa 035338
Muzeum Historyczne Miasta
Tarnobrzega, Tarnobrzeg 035265
Muzeum Historyczne w Sanoku,
Sanok 035192
Muzeum Historyczne w Wrocławiu,
Muzeum Miejskie Wrocławia,
Wrocław 035422
Muzeum Historyczne, Muzeum
Okręgowe w Ostrołęce,

Przasnysz 035144
Muzeum Historyczne, Muzeum
Podlaskie w Białymstoku,
Białystok 034640
Muzeum Historyczno-Archeologiczne,
Ostrowiec Świętokrzyski 035086
Muzeum Historyczno-Etnograficzne,
Chojnice 034690
Muzeum Hlavního Města Prahy,
Praha 009815
Muzeum Hraček, Praha 009816
Muzeum Hrnčířství, Kostelec nad
Černými Lesy 009641
Muzeum Hutnictwa i Górnictwa Rud
Żelaza, Częstochowa 034707
Muzeum Hymnu Narodowego,
Muzeum Narodowe w Gdańsku,
Nowy Barkoczyn 035045
Muzeum Hymnu Narodowego,
Muzeum Narodowe w Gdańsku,
Nowy Karczma 035046
Muzeum i Instytut Zoologii PAN,
Warszawa 035339
Muzeum Ignacego Jana
Paderewskiego i Wychodźstwa
Polskiego w Ameryce, Muzeum
Narodowe w Warszawie,
Warszawa 035340
Muzeum im. Adama Mickiewicza w
Śmiełowie, Muzeum Narodowe w
Poznaniu, Żerków 035460
Muzeum im. Aleksandra Kłosińskiego,
Kęty 034830
Muzeum im. Anny i Jarosława
Iwaszkiewiczów w Stawisku,
Podkowa Leśna 035109
Muzeum im. Edmunda Bojanowskiego
w Grabonogu, Grabonóg 034773
Muzeum im. J.G. Herdera w Morągu,
Oddział Muzeum Warmii i Mazur w
Olsztynie, Morąg 035019
Muzeum im. Jacka Malczewskiego w
Radomiu, Radom 035164
Muzeum im. Jana Dzierżona,
Kluczbork 034839
Muzeum im. Jana Kasprowicza,
Inowrocław 034782
Muzeum im. Jana Nikodema Jaronia,
Olesno 035063
Muzeum im. Jerzego Dunin-
Borkowskiego, Krośniewice . . . 034920
Muzeum im. Juliana Fałata, Bystra
Śląska 034673
Muzeum im. Kazimierza Pułaskiego,
Warka-Winiary 035306
Muzeum im. Michała Kajki, Muzeum
Okręgowego w Suwałkach,
Ogródek 035060
Muzeum im. Orła Białego, Skarżysko-
Kamienna 035199
Muzeum im. Oskara Kolberga,
Muzeum Wsi Radomskiej w Radomiu,
Przysucha 035149
Muzeum im. prof. Władysława Szafera,
Ojców 035061
Muzeum im. Stanisława
Noakowskiego, Muzeum Ziemi
Kujawskiej i Dobrzyńskiej,
Nieszawa 035037
Muzeum im. Stanisława Staszica,
Muzeum Okręgowego w Zamościu,
Hrubieszów 034779
Muzeum im. Władysława Orkana,
Rabka 035161
Muzeum im. Władysława
St. Reymonta, Muzeum
Okręgowe w Żyrardowie, Lipce
Reymonotowskie 034949
Muzeum im. Wojciecha Kętrzyńskiego,
Kętrzyn 034829
Muzeum Inkluzji w Bursztynie,
Uniwersytet Gdański, Gdynia . . 034746
Muzeum Instrumentów Muzycznych,
Muzeum Narodowe w Poznaniu,

Poznań 035128
Muzeum Inżynierii Miejskiej,
Kraków 034892
Múzeum J.M. Petzvala, Slovenské
Technické Múzeum Košice, Spišská
Béla 038450
Muzeum Jana Ámoše Komenského,
Uherský Brod 009932
Muzeum Jana Kasprowicza,
Zakopane 035447
Múzeum Jána Kollára, Mošovce 038403
Múzeum Janka Kráľa, Liptovský
Mikuláš 038382
Muzeum Jihovýchodní Moravy,
Zlín 009984
Muzeum Jindřicha Jindřicha,
Domažlice 009541
Muzeum Jindřicha Šimona Baara,
Klenčí pod Čerchovem 009630
Muzeum Jindřicho Hradecka,
Jindřichův Hradec 009450
Muzeum Jižního Plzeňska v Blovicich,
Blovice 009450
Muzeum Josefa Dobrovského,
Chudenice 009530
Muzeum Josefov, Jaroměř II.
Josefov 009595
Muzeum Józefa Ignacego
Kraszewskiego, Romanów . . . 035177
Muzeum Juszkiewiczów, Mława 035017
Muzeum Kamieni Szlachetnych,
Polanica Zdrój 035110
Muzeum Kamienica Łozińskich,
Muzeum Narodowe, Kraków . . 034893
Muzeum Kamiena Orsettich w
Jarosławiu, Jarosław 034787
Muzeum Karkonoskie, Jelenia
Góra 034795
Muzeum Karola Szymanowskiego
w Willi Atma w Zakopanem,
Zakopane 035448
Muzeum Karykatury, Warszawa . 035341
Muzeum-Kaszubski Park Etnograficzny
w Wdzydzach Kiszewskich, Wdzydze
Kiszewskie 035385
Muzeum Kaszubskie, Kartuzy . 034809
Muzeum Kaszubskie,
Kościerzyna 034850
Muzeum Katedralne Jana Pawła II,
Kraków 034894
Muzeum Katyńskie, Warszawa . 035342
Múzeum Keramiky Mestské Múzeum,
Holíč 038354
Muzeum Kinematografii, Łódź . 034969
Muzeum Klasztorne na Jasnej Górze,
Częstochowa 034708
Muzeum Klasztorne OO Cystersów,
Szczyrzyc 035257
Muzeum Klimkovice, Okresní
Vlastivědné Muzeum Nový Jičín,
Klimkovice 009631
Muzeum Knihy, Národní muzeum
Praha, Žďár nad Sázavou 009976
Muzeum Kolei Wąskotorowej,
Sochaczew 035208
Muzeum Kolei Wąskotorowej w
Wenecji, Muzeum Ziemi Pałuckiej w
Żninie, Wenecja 035388
Muzeum Kolejnictwa, Warszawa 035343
Muzeum Kolekcji im. Jana Pawła
II, Fundacji Carroll Porczyńskich,
Warszawa 035344
Muzeum Komenského v Přerově,
Přerov 009850
Muzeum Komunismu, Praha . . 009817
Muzeum Koněšpřežné Cráhy v
Bujanově, Kaplice 009616
Muzeum Kornela Makuszyńskiego,
Muzeum Tatrzańskie im. Tytusa
Chałubińskiego w Zakopanem,
Zakopane 035449
Muzeum Kościuszkowskie Dworek
Zacisze, Miechów 035007
Muzeum Kouřimska, Kouřim . . 009642

Muzeum Kowalstwa w Warszawie,
Warszawa 035345
Muzeum Krajky, Prachatice 009777
Muzeum Krajky, Vamberk 009942
Muzeum Kroměřížska, Kroměříž 009653
Muzeum Krupka, Regionální muzeum
v Teplicích, Krupka 009654
Muzeum Książąt Czartoryskich,
Muzeum Narodowe w Krakowie,
Kraków 034895
Muzeum Książki Artystycznej,
Łódź 034970
Múzeum Kultúry Karpatských Nemcov,
Slovenské Národné Múzeum,
Bratislava 038317
Muzeum Kultury Kurpiowskiej,
Ostrołęka 035083
Muzeum Kultury Łemkowskiej w
Zyndranowej, Tylawa 035294
Muzeum Kultury Ludowej,
Kolbuszowa 034841
Muzeum Kultury Ludowej,
Węgorzewo 035386
Muzeum Kultury Ludowej, Muzeum
Okręgowego w Pile, Osiek nad
Notecią 035081
Múzeum Kultúry Maďarov na
Slovensku, Slovenské Národné
Múzeum, Bratislava 038318
Múzeum Kultúry Rómov na Slovensku,
Slovenské Národné Múzeum v
Martine, Martin 038392
Muzeum Kultury Szlacheckiej,
Łopuszna 034983
Múzeum Kysuckej Dediny, Nová
Bystrica 038414
Muzeum Lachów Sądeckich im. Zofii i
Stanisława Chrząstowskich, Sądecki
Park Etnograficznego w Nowym
Sączu, Podegrodzie 035108
Muzeum Lasu i Drewna przy Leśnym
Zakładzie Doświadczalnym SGGW,
Rogów 035174
Muzeum Lat Szkolnych Stefana
Żeromskiego, Muzeum Narodowe,
Kielce 034832
Muzeum Latarnia Morska,
Władysławowo 035397
Muzeum Łazienki Krolewskie,
Warszawa 035346
Múzeum Letectva Košice, Slovenské
Technické Múzeum, Košice . . 038366
Muzeum Lidových Staveb,
Kouřim 009643
Muzeum Literackie Henryka
Sienkiewicza, Poznań 035129
Muzeum Literackie im. Józefa
Czechowicza, Muzeum Lubelskie,
Lublin 034996
Muzeum Literatury im. Adama
Mickiewicza, Warszawa 035347
Muzeum Literatury im. Jarosława
Iwaszkiewicza, Muzeum Okręgowego
w Sandomierzu, Sandomierz . . 035189
Muzeum Litovel, Litovel 009677
Muzeum Lotnictwa Polskiego,
Kraków 034896
Muzeum Loutek Milana Knížáka,
Okresní muzeum, Děčín 009533
Muzeum Loutkářských Kultur,
Chrudim 009528
Muzeum Łowiectwa i Jeździectwa,
Warszawa 035348
Muzeum Lubelskie, Lublin 034997
Múzeum Ľudovíta Štúra, Modra 038401
Muzeum Ludowych Instrumentów
Muzycznych, Szydłowiec 035264
Muzeum Luhačovického Zálesí,
Muzeum Jihovýchodní Moravy ve
Zlíně, Luhačovice 009685
Múzeum M.R. Štefánika,
Košariská 038365
Múzeum Maďarskej Kultúry a
Podunajska, A Magyar Kultúra

és Duna Mente Múzeuma,
Komárno 038364
Muzeum Małego Miasta, Bieżuń 034654
Muzeum Marii Dąbrowskiej,
Warszawa 035349
Muzeum Marii Konopnickiej w
Żarnowcu, Jedlicze 034792
Muzeum Marii Konopnickiej, Muzeum
Okręgowego w Suwałkach,
Suwałki 035233
Muzeum Marii Skłodowskiej Curie,
Warszawa 035350
Muzeum Martyrologiczne w Żabikowie,
Luboń 035004
Muzeum Martyrologiczno-Historyczny,
Muzeum Południowego Podlasia,
Biała Podlaska 034633
Muzeum Martyrologii, Bielsk
Podlaski 034646
Muzeum Martyrologii (Sonnenburg),
Słońsk 035201
Muzeum Martyrologii Aliančkich
Jeńców Wojennych, Żagań . . 035443
Muzeum Martyrologii i Walki
Radogoszcz, Muzeum Tradycji
Niepodległościowych, Łódź 034971
Muzeum Martyrologii Pod Zegarem,
Muzeum Lubelskie, Lublin . . 034998
Muzeum Martyrologii Wielkopolan Fort
VII, Wielkopolskie Muzeum Walk
Niepodległościowych w Poznaniu,
Poznań 035130
Muzeum Marynarki Wojennej,
Gdynia 034747
Muzeum Mazowieckie w Płocku,
Płock 035105
Muzeum Mazurskie, Ostrzeszów 035087
Muzeum Mazurskie, Muzeum Warmii i
Mazur w Olsztynie, Szczytno . . 035258
Muzeum Města Brna, Brno . . 009477
Muzeum Města Duchcova,
Duchcov 009542
Muzeum Města Ústí nad Labem, Ústí
nad Labem 009936
Muzeum Městec Králové, Městec
Králové 009693
Muzeum Městské Ústí nad Orlicí, Ústí
nad Orlicí 009937
Muzeum MHD – vozovna Střešovice,
Praha 009818
Muzeum Miar Głównego Urzędu Miar,
Warszawa 035351
Muzeum Miasta Gdyni, Gdynia . 034748
Muzeum Miasta i Rzeki Warty,
Warta 035383
Muzeum Miasta Jaworzna,
Jaworzno 034791
Muzeum Miasta Kołobrzegu, Muzeum
Oręża Polskiego, Kołobrzeg . . 034843
Muzeum Miasta Ostrowa
Wielkopolskiego, Ostrów
Wielkopolski 035085
Muzeum Miasta Pabianice,
Pabianice 035093
Muzeum Miasta Zgierza, Zgierz . 035461
Múzeum Michala Greisigera, Spišská
Béla 038451
Muzeum Miedzi, Legnica 034940
Muzeum Miejskie, Nowa Sól . . 035041
Muzeum Miejskie,
Świętochłowice 035240
Muzeum Miejskie, Zabrze 035441
Muzeum Miejskie im. Maksymiliana
Chroboka, Ruda Śląska 035178
Muzeum Miejskie Sztygarka im
Stanisława Staszica, Dąbrowa
Górnicza 034710
Muzeum Miejskie Wrocławia,
Wrocław 035423
Muzeum Mikołaja Kopernika,
Frombork 034720
Muzeum Mikołaja Kopernika, Muzeum
Okręgowego w Toruńiu, Toruń . 035285
Múzeum Milana Dobeša,

Bratislava 038319
Muzeum Militariów Arsenal Miejski,
Muzeum Miejskie Wrocławia,
Wrocław 035424
Muzeum Mineralogiczne, Szklarska
Poręba 035260
Muzeum Mineralogiczne, Instytutu
Nauk Geologicznych Uniwersytutu
Wrocławskiego, Wrocław 035425
Muzeum Misji Karmelitańskich w
Afryce, Krceszowice 034919
Muzeum Misyjne, Krosno 034921
Muzeum Misyjne Misjonarzy Oblatów
MN, Święty Krzyż 035241
Muzeum Misyjne Ojców Sercanów,
Polanica Zdrój 035111
Muzeum Misyjno-Etnograficzne,
Seminarium Duchownego Księży
Werbistów, Pieniężno 035098
Muzeum Mladaboleslavska, Mladá
Boleslav 009699
Muzeum Młynarstwa Powietrznego,
Bęsia 034631
Múzeum moderného umenia Andy
Warhola, Medzilaborce 038399
Muzeum Mohelnice, Mohelnice . 009703
Muzeum Morskie, Muzeum Narodowe
w Szczecinie, Szczecin 035252
Múzeum Mošovských Remeselníkov,
Mošovce 038404
Muzeum Motocyklů, Lesná u
Znojma 009663
Muzeum Motoryzacji, Otrębusy . 035090
Muzeum Motoryzacji Automobilklubu
Wielkopolskiego, Poznań 035131
Muzeum Nadwiślański Park
Etnograficzny, Muzeum w
Chrzanowie im. Ireny i Mieczysława
Mazarakich – Skansen w
Wygiełzowie, Babice 034623
Muzeum Nadwiślańskie, Kazimierz
Dolny 034824
Muzeum Narodowe, Szczecin . . 035253
Muzeum Narodowe Rolnictwa i
Przemysłu Rolno-Spożywczego w
Szreniawie, Komorniki 034846
Muzeum Narodowe w Gdańsku,
Gdańsk 034731
Muzeum Narodowe w Kielcach,
Kielce 034833
Muzeum Narodowe w Krakowie,
Kraków 034897
Muzeum Narodowe w Poznaniu,
Galeria Malarstwa i Rzeźby,
Poznań 035132
Muzeum Narodowe w Poznaniu,
Oddział w Rogalinie, Rogalin . . 035172
Muzeum Narodowe w Warszawie,
Warszawa 035352
Muzeum Narodowe we Wrocławiu,
Wrocław 035426
Muzeum Narodowe Ziemi Przemyskiej,
Przemyśl 035146
Muzeum Neolitycznego Górnictwa i
Rezerwat Archeologiczny, Krzemionki
koło Ostrowca 034926
Muzeum Niepodległości,
Warszawa 035353
Muzeum Nikifora, Galeria Sztuki
Romanówka, Krynica 034925
Múzeum Obchodu, Ilava 038359
Múzeum Obchodu Bratislava,
Bratislava 038320
Muzeum Oceanograficzne Mir, Morski
Instytut Rybacki, Gdynia 034749
Muzeum Odlewnictwa Artystycznego,
Muzeum w Gliwicach, Gliwice . 034756
Muzeum Odona Bujwida, Kraków 034898
Muzeum Okręgowe, Konin . . . 034847
Muzeum Okręgowe, Krosno . . 034922
Muzeum Okręgowe, Nowy Sącz 035050
Muzeum Okręgowe, Ostrołęka . 035084
Muzeum Okręgowe, Piotrków
Trybunalski 035101

Muzeum Okręgowe, Rzeszów . . 035185
Muzeum Okręgowe, Sandomierz 035190
Muzeum Okręgowe, Suwałki . . 035234
Muzeum Okręgowe, Zamość . . 035453
Muzeum Okręgowe – Galeria Rynek,
Tarnów 035267
Muzeum Okręgowe – Muzeum
Etnograficzny, Tarnów 035268
Muzeum Okręgowe – Ratusz,
Tarnów 035269
Muzeum Okręgowe im. Leona
Wyczółkowskiego, Bydgoszcz . 034669
Muzeum Okręgowe Mazowsza
Zachodniego, Żyrardów 035472
Muzeum Okręgowe w Lesznie,
Leszno 034942
Muzeum Okręgowe w Lesznie Dział
Judaistyczny, Leszno 034943
Muzeum Okręgowe w Sieradzu,
Sieradz 035196
Muzeum Okręgowe w Tarnowie,
Tarnów 035270
Muzeum Okręgowe w Toruniu,
Toruń 035286
Muzeum Okręgowe Ziemi Kaliskiej,
Kalisz 034801
Muzeum Okresu Benešov, Historie
města Benešova, Benešov . . 009442
Muzeum Okręt Błyskawica,
Gdynia 034750
Muzeum Olomouckých Tvarůžků,
Loštice 009680
Muzeum Olomučanské Keramiky,
Olomučany 009733
Múzeum Oravskej Dediny,
Zuberec 038488
Muzeum Orderu Uśmiechu,
Rabka 035162
Muzeum Oręża Polskiego,
Kołobrzeg 034844
Muzeum Oręża Polskiego – Pawilon
Wystawowy, Kołobrzeg 034845
Muzeum Oświatowe, Szkolne
Schronisko Młodzieżowe
Muzeum Oświatowe w Puławach,
Puławy 035155
Muzeum Oświaty, Bydgoszcz . . 034670
Muzeum Oświaty Ziemi Łódzkiej,
Łódź 034972
Muzeum Otokara Březiny, Jaroměř
nad Rokytnou 009597
Muzeum Pałac w Wilanowie,
Warszawa 035354
Muzeum Pamiątek po Janie Matejce
"Koryznówka", Nowy Wiśnicz . . 035056
Muzeum Pamięci Narodowej,
Kraków 034899
Muzeum Pamięci Narodowej lata 1939
-1956, Kielce 034834
Muzeum Pamięci św. Maksymiliana
Kolbego, Zduńska Wola 035458
Muzeum Papiernictwa, Duszniki
Zdrój 034718
Muzeum Parafialne, Widawa . . 035389
Muzeum Parafialne im. Jana Piechoły,
Filia Muzeum Diecezjalnego w
Tarnowie, Iwkowa 034784
Muzeum Parafialne Sztuki Ludowej im.
Ks. Edwarda Nitki, Filia Muzeum
Diecezjalnego w Tarnowie,
Paszyn 035095
Muzeum Parafialne, Filia Muzeum
Diecezjalnego w Tarnowie,
Tropie 035289
Muzeum Parafialne, Muzeum
Diecezjalne w Tarnowie, Dobra 034714
Muzeum Parafialne, Muzeum
Diecezjalnego w Tarnowie,
Grybów 034777
Muzeum Parafialne, Muzeum
Diecezjalnego w Tarnowie,
Złota 035464
Muzeum Pawla Stellera, Muzeum
Historii Katowic, Katowice 034818

Register der Institutionen und Firmen

Muzeum Pedagogiczne im. Prof. K. Sośnickiego, Gdańsk
– Muzeum Stanisława Wyspiańskiego w Kamienicy

Muzeum Pedagogiczne im. Prof. K. Sośnickiego, Gdańsk 034732
Muzeum Piastów Śląskich, Brzeg 034662
Muzeum Pienińskie im. Józefa Szalaya, Muzeum Okręgowego w Nowym Sączu, Szczawnica ... 035245
Muzeum Pierwszych Piastów na Lednicy, Lednogóra 034938
Muzeum Piśmiennictwa i Muzyki Kaszubsko-Pomorskiej, Wejherowo 035387
Muzeum Plakatu w Wilanowie, Muzeum Narodowe w Warszawie, Warszawa 035355
Muzeum Początków Państwa Polskiego, Gniezno 034764
Muzeum Poczty i Telekomunikacji, Wrocław 035427
Muzeum Poczty i Telekomunikacji Oddział w Gdańsku, Gdańsk .. 034733
Muzeum Podhalańskie PTTK, Nowy Targ 035054
Muzeum Podkrkonoší, Trutnov .. 009924
Muzeum Podlasie w Białymstoku, Białystok 034641
Muzeum Pódrożników, Muzeum Okręgowe w Toruniu, Toruń . 035287
Muzeum Pojezierza Łęczyńsko-Włodawskiego, Włodawa 035404
Muzeum Policie České Republiky, Praha 009819
Múzeum Polície Slovenskej Republiky, Bratislava 038321
Muzeum Politechniki Warszawskiej, Warszawa 035356
Muzeum Polná, Muzeum Vysočiny Jihlava, Polná 009773
Muzeum Północno-Mazowieckie, Łomża 034982
Muzeum Polskiej Wojskowej Służby Zdrowia, Łódź 034973
Muzeum Południowego Podlasia, Biała Podlaska 034634
Muzeum Pomorza Środkowego, Słupsk 035203
Muzeum Porcelánu Pirkenhammer, Březová u Karlových Var 009459
Muzeum Powstania Chochołowskiego, Chochołów 034686
Muzeum Powstania Poznańskiego – Czerwiec 1956, Wielkopolskie Muzeum Walk Niepodległościowych w Poznaniu, Poznań 035133
Muzeum Powstania Warszawskiego, Muzeum Historycznego Miasta Starego Warszawy, Warszawa . 035357
Muzeum Powstania Wielkopolskiego 1918–1919, Wielkopolskie Muzeum Walk Niepodległościowych w Poznaniu, Poznań 035134
Muzeum Pożarnictwa, Alwernia . 034619
Muzeum Pożarnictwa, Lidzbark . 034946
Muzeum Pożarnictwa Ziemi Olkuskiej, Olkusz 035065
Muzeum Pozytywizmu im. Aleksandra Świętochowskiego, Gołotczyzna 034765
Muzeum-Prącownia Literacka Arkadego Fiedlera, Puszczykowo 035158
Muzeum Prasy Śląskiej, Pszczyna 035151
Muzeum Prof. Josepha Wittiga, Nowa Ruda 035039
Muzeum Prostějovska, Prostějov 009856
Muzeum Prowincji Ojców Bernardynów, Leżajsk 034945
Muzeum Průmyslových Železnic Brno, Brno 009478
Muzeum Prywatne Diabła Polskiego Przedpiekle, Warszawa ... 035358
Muzeum Przemysłu i Techniki, Muzeum w Wałbrzychu, Wałbrzych 035303
Muzeum Przemysłu Naftowego w

Bóbrce, Chorkówka 034692
Muzeum Przemysłu, Muzeum Techniki, Warszawa 035359
Muzeum Przyrodnicze, Jelenia Góra 034796
Muzeum Przyrodnicze Babiogórskiego Parku Narodowego, Zawoja ... 035455
Muzeum Przyrodnicze Bieszczadzkiego Parku Narodowego, Ustrzyki Dolne 035295
Muzeum Przyrodnicze im. Prof. Andrzeja Myrchy, Instytut Biologii Uniwersytetu w Białymstoku, Białystok 034642
Muzeum Przyrodnicze Instytutu Systematyki i Ewolucji Zwierząt PAN, Kraków 034900
Muzeum Przyrodnicze Karkonoskiego Parku Narodowego, Jelenia Góra 034797
Muzeum Przyrodnicze Uniwersytetu Łódzkiego, Łódź 034974
Muzeum Przyrodnicze Uniwersytetu Wrocławskiego, Wrocław 035428
Muzeum Przyrodnicze Wielkopolskiego Parku Narodowego, Puszczykowo 035159
Muzeum Przyrodnicze Wolińskiego Parku Narodowego im. Prof. Adama Wodziczki, Międzyzdroje ... 035010
Muzeum Przyrodnicze, Muzeum Nadwiślańskie, Kazimierz Dolny 034825
Muzeum Przyrodnicze, Nadmorski Park Krajobrazowy, Władysławowo 035398
Muzeum Przyrodniczo-Leśne Słowińskiego Parku Narodowego, Smołdzino 035204
Muzeum Przyrodniczo-Leśne Świętokrzyskiego Parku Narodowego, Święty Krzyż 035242
Muzeum Przyrodniczo-Leśne, Białowieski Park Narodowy im. Jana Miklaszewskiego, Białowieża 034635
Muzeum Puszczy Kampinoskiej, Kampinos 034805
Muzeum Reformacji Polskiej, Muzeum Okręgowego w Suwałkach, Mikołajki 035013
Muzeum Regionalne, Barlinek .. 034628
Muzeum Regionalne, Bełchatów 034630
Muzeum Regionalne, Brodnica . 034661
Muzeum Regionalne, Brzeziny . 034663
Muzeum Regionalne, Cedynia . 034677
Muzeum Regionalne, Chojnów . 034691
Muzeum Regionalne, Człuchów . 034709
Muzeum Regionalne, Głogówek . 034760
Muzeum Regionalne, Główno .. 034761
Muzeum Regionalne, Gostyń .. 034772
Muzeum Regionalne, Iłża 034780
Muzeum Regionalne, Jarocin .. 034786
Muzeum Regionalne, Jasło 034788
Muzeum Regionalne, Jawor ... 034790
Muzeum Regionalne, Kościan .. 034849
Muzeum Regionalne, Kozienice . 034855
Muzeum Regionalne, Krasnystaw 034918
Muzeum Regionalne, Kutno ... 034931
Muzeum Regionalne, Lubań ... 034986
Muzeum Regionalne, Łuków ... 035005
Muzeum Regionalne, Międzychód 035008
Muzeum Regionalne, Mielec .. 035012
Muzeum Regionalne, Nowe Miasto nad Pilicą 035043
Muzeum Regionalne, Nysa 035057
Muzeum Regionalne, Ojców ... 035062
Muzeum Regionalne, Opoczno . 035074
Muzeum Regionalne, Pińczów . 035100
Muzeum Regionalne, Pułtusk .. 035157
Muzeum Regionalne, Pyzdry .. 035160
Muzeum Regionalne, Radomsko 035167
Muzeum Regionalne, Siedlce .. 035194
Muzeum Regionalne, Sławków . 035200
Muzeum Regionalne, Środa

Śląska 035216
Muzeum Regionalne, Szczecinek 035256
Muzeum Regionalne, Szydłów .. 035263
Muzeum Regionalne, Wągrowiec 035301
Muzeum Regionalne, Wiślica ... 035395
Muzeum Regionalne, Wodzisław Śląski 035405
Muzeum Regionalne, Wronki ... 035436
Muzeum Regionalne Dom Grecki, Myślenice 035025
Muzeum Regionalne Dom Szewca, Gminny Ośrodek Kultury Pszczew, Pszczew 035150
Muzeum Regionalne im Wojciechy Dutkiewicz, Rogozińskie Centrum Kultury, Rogoźno 035176
Muzeum Regionalne im. Adama Fastnachta, Brzozów 034664
Muzeum Regionalne im. Albina Nowickiego, Ryn 035181
Muzeum Regionalne im. Andrzeja Kaubego, Wolin 035407
Muzeum Regionalne im. Antoniego Minkiewicza, Olkusz 035066
Muzeum Regionalne im. Dzieci Wrześińskich, Września 035437
Muzeum Regionalne im. Hieronima Ławaińczaka, Krotoszyn 034924
Muzeum Regionalne im. Janusza Petera, Tomaszów Lubelski ... 035274
Muzeum Regionalne im. Marcina Rożka, Wolsztyn 035409
Muzeum Regionalne im. Seweryna Udzieli, Stary Sącz 035222
Muzeum Regionalne im. Władysława Kowalskiego, Dobczyce 034713
Muzeum Regionalne Kromerówka – Baszta Kowalska, Biecz 034645
Muzeum Regionalne PTTK, Golub-Dobrzyń 034766
Muzeum Regionalne PTTK, Muszyna 035023
Muzeum Regionalne PTTK, Puławy 035156
Muzeum Regionalne PTTK, Starachowice 035218
Muzeum Regionalne PTTK, Zagórze Śląskie 035444
Muzeum Regionalne PTTK i Urzędu Gminy, Iwanowice 034783
Muzeum Regionalne PTTK im. Ignacego Łukasiewicza, Gorlice 034770
Muzeum Regionalne w Sosnowcu, Sosnowiec 035213
Muzeum Regionalne w Stęszewie, Stęszew 035225
Muzeum Regionalne w Świebodzinie, Świebodzin 035239
Muzeum Regionalne Ziemi Limanowskiej, Limanowa ... 034948
Muzeum Regionalne Ziemi Sadowieńskiej, Sadowne 035186
Muzeum Regionalne Ziemi Sulmierzyckiej, Sulmierzyce .. 035229
Muzeum Regionalne Ziemi Zbąszyńskiej PTTK, Zbąszyń .. 035456
Muzeum Regionalne, Filia Muzeum Okręgowego w Koninie, Słupca 035202
Muzeum Regionalne, Muzeum Lubelskiego w Lublinie, Kraśnik 034916
Muzeum Regionalne, Muzeum Lubelskiego w Lublinie, Łęczna 034936
Muzeum Regionalne, Muzeum Lubelskiego w Lublinie, Lubartów 034987
Muzeum Regionalne, Ośrodek Edukacji Plastycznej i Historycznej, Myślibórz 035026
Muzeum Regionalne, Zespół Pałacowo-Parkowy, Przeworsk 035035
Muzeum Regionu Valašsko ve Vsetíně – Zámek Vsetín, Vsetín 009962
Muzeum Rekordů a Kuriozit, Pelhřimov 009752

Muzeum Roberta Kocha, Wolsztyn 035410
Muzeum Rolnictwa im. Ks. Krzysztofa Kluka, Ciechanowiec 034699
Muzeum Romantyzmu, Opinogóra 035073
Muzeum Romské Kultury, Brno . 009479
Muzeum Rumburk, Okresní muzeum Děčín, Rumburk 009875
Muzeum Rybackie, Jastarnia ... 034789
Muzeum Rybołówstwa Morskiego, Świnoujście 035243
Muzeum Rybołówstwa, Centralne Muzeum Morskie w Gdańsku, Hel 034778
Muzeum Rzemiosł Ludowych – Zagroda Sitarska, Biłgoraj 034655
Muzeum Rzemiosła, Krosno ... 034923
Muzeum Rzemiosła Tkackiego, Turek 035292
Muzeum Rzeźby Alfonsa Karnego, Muzeum Podlaskie w Białymstoku, Białystok 034643
Muzeum Rzeźby im. Xawerego Dunikowskiego Królikarnia, Muzeum Narodowe w Warszawie, Warszawa 035360
Muzeum Sakralne Kolegiaty Zamojskiej, Zamość 035454
Múzeum Samuela Jurkoviča Sobotište-Nyáryiovský kaštieľ', Sobotište 038449
Muzeum Skansen Kolejnictwa, Kościerzyna 034851
Muzeum Skansen-Zagroda Pszczyńska, Pszczyna 035152
Muzeum Skla a Bižuterie, Jablonec nad Nisou 009588
Muzeum Skla, Sklárna Novosad & Syn, Harrachov 009552
Múzeum Školstva a Pedagogiky Bratislava, Bratislava 038322
Muzeum J.V. Sládka, Zbiroh ... 009973
Muzeum Śląska Cieszyńskiego w Cieszynie, Cieszyn 034701
Muzeum Śląska Opolskiego, Opole 035077
Muzeum Śląskie, Katowice 034819
Muzeum Ślężańskie im. Stanisława Dunajewskiego, Sobótka 035206
Múzeum Slovenského Červeného Kríža, Martin 038393
Muzeum Slovenského Národného Povstania, Banská Bystrica ... 038279
Múzeum Slovenskej Dediny, Martin 038394
Múzeum Slovenských Národných Rád, Slovenské Národné Múzeum, Myjava 038406
Muzeum Sopotu, Sopot 035210
Múzeum Špeciálneho Školstva, Levoča 038378
Múzeum Spiša, Spišská Nová Ves 038454
Muzeum Społeczne, Strzyżów . 035227
Muzeum Sportu i Turystyki, Łódź 034975
Muzeum Sportu i Turystyki, Warszawa 035361
Muzeum Sportu i Turystyki Regionu Karkonoszy, Karpacz 034806
Muzeum Sportu i Turystyki Ziemi Gdańskiej, Muzeum Historii Miasta Gdańsk, Gdańsk 034734
Muzeum Sprzętu Gospodarstwa Domowego, Ziębice 035462
Muzeum Śremskie, Śrem 035215
Muzeum Środowiska Przyrodniczego i Łowiectwa, Muzeum Narodowe Rolnictwa i Przemysłu Rolno-Spożywczego w Szreniawie, Uzarzewo 035296
Muzeum Stanisława Staszica, Piła 035099
Muzeum Stanisława Wyspiańskiego w Kamienicy Szołayskich, Muzeum

Narodowe w Krakowie, Kraków 034901
Muzeum Starego Teatru, Kraków 034902
Muzeum Starożytnego Hutnictwa
Mazowieckiego, Pruszków . . . 035143
Muzeum Starożytnego Hutnictwa
Świętokrzyskiego im. prof. M.
Radwana, Muzeum Techniki w
Warszawie, Nowa Słupia 035040
Muzeum Stefana Żeromskiego,
Nałęczów 035031
Muzeum Středniho Pootavi,
Strakonice 009892
Muzeum Strojíren Poldi, Kladno . 009626
Muzeum Stylu Zakopiańskiego im.
Stanisława Witkiewicza – Willa
Koliba, Muzeum Tatrzańskie im.
Tytusa Chałubińskiego w Zakopanem,
Zakopane 035450
Muzeum Šumavy, Kašperské
Hory 009625
Muzeum Šumavy, Sušice 009899
Muzeum Svatopluka Čecha, Liteň 009670
Muzeum Szlachty Mazowieckiej,
Ciechanów 034698
Muzeum Sztuk Użytkowych, Muzeum
Narodowe w Poznaniu, Poznań 035135
Muzeum Sztuki Cmentarnej, Muzeum
Miejskie Wrocławia, Wrocław . 035429
Muzeum Sztuki Dziecka,
Warszawa 035362
Muzeum Sztuki Książki, Muzeum
Narodowe, Wrocław 035430
Muzeum Sztuki Medalierskiej,
Muzeum Miejskie Wrocławia,
Wrocław 035431
Muzeum Sztuki Mieszczańskiej,
Muzeum Miejskie Wrocławia,
Wrocław 035432
Muzeum Sztuki Nowoczesnej w
Niepołomicach, Niepołomice . . 035036
Muzeum Sztuki Sakralnej, Bardo 034627
Muzeum Sztuki Sakralnej, Żnin . 035468
Muzeum Sztuki w Łodzi, Łódź . 034976
Muzeum Sztuki Współczesnej,
Muzeum im. Jacka Malczewskiego w
Radomiu, Radom 035165
Muzeum Sztuki Złotniczej, Muzeum
Nadwiślańskie, Kazimierz Dolny 034826
Muzeum T.G. Masaryka,
Rakovník 009860
Muzeum T.G. Masaryk's House,
Čejkovice 009495
Múzeum Tatranského Národného
Parku, Tatranská Lomnica . . . 038465
Muzeum Tatrzańskie im. Tytusa
Chałubińskiego, Zakopane . . . 035451
Muzeum Teatralne, Warszawa . 035363
Muzeum Teatralne im. Stanisława
Wyspiańskiego, Muzeum
Historycznego m. Krakowa,
Kraków 034903
Muzeum Technik Ceramicznych,
Koło 034842
Muzeum Techniki w Warzwie,
Warszawa 035364
Muzeum Techniki Włókienniczej,
Muzeum w Bielsku-Białej, Bielsko-
Biała 034652
Muzeum Telč, Telč 009905
Múzeum Telesnej Kultúry v SR,
Bratislava 038323
Muzeum Těšínska, Český Těšín . 009512
Muzeum těžby a dopravy vápence
v Českém krasu, Svatý Jan pod
Skalou 009900
Muzeum Tkactwa Dolnośląskiego
w Kamiennej Górze, Kamienna
Góra 034803
Muzeum Tradycji Niepodległościowych,
Łódź 034977
Muzeum Tradycji Szlacheckiej w
Waplewie Wielkim, Muzeum
Narodowe w Gdańsku, Stary
Targ 035223

Muzeum Třešť, Třešť 009922
Muzeum Třineckých Železáren a
Města Třince, Třinec 009923
Muzeum Týn nad Vltavou, Týn nad
Vltavou 009929
Muzeum Tynecké Keramiky, Týnec
nad Sázavou 009930
Muzeum Ubezpieczeń, Kraków . 034904
Múzeum Ukrajinsko-Rusínskej Kultúry,
Slovenské Národné Múzeum,
Svidník 038462
Múzeum Umeleckých Remesiel,
Mestské Múzeum v Bratislave,
Bratislava 038324
Muzeum Umění, Benešov 009443
Muzeum Umění Olomouc,
Olomouc 009731
Muzeum Uniwersytetu im. Ks. Jana
Władzińskiego, Katolickiego
Uniwersytetu Lubelskiego,
Lublin 034999
Muzeum Uniwersytetu
Jagiellońskiego – Collegium Maius,
Kraków 034905
Muzeum Uniwersytetu Marii Curie-
Skłodowskiej, Lublin 035000
Muzeum Uniwersytetu Warszawskiego,
Warszawa 035365
Muzeum Uniwersytetu Wrocławskiego,
Wrocław 035433
Muzeum Uzbrojenia – Cytadela
Poznańska, Wielkopolskie Muzeum
Walk Niepodległościowych w
Poznaniu, Poznań 035136
Muzeum v Bruntále, Bruntál . . . 009487
Muzeum v Ivančicích, Ivančice . 009586
Múzeum v Kežmarku a Mestský Hrad,
Kežmarok 038362
Múzeum Včelárstva na Slovensku,
Kráľová pri Senci 038371
Muzeum ve Štramberku,
Štramberk 009893
Muzeum ve Žluticích, Žlutice . . 009988
Muzeum Velké Meziříčí, Velké
Meziříčí 009946
Múzeum Vín, Prešov 038431
Múzeum vo Svätý Antone, Svätý
Anton 038459
Múzeum Vojtecha Löfflera,
Košice 038367
Muzeum Voskových Figurín České
Historie, Karlštejn 009623
Muzeum Voskových Figurín České
Historie, Praha 009820, 009821
Muzeum Východních Čech v Hradci
Králové, Oddělění Archeologické,
Historické, Přírodovědecké, Hradec
Králové 009577
Muzeum Vyškovska, Vyškov . . 009963
Muzeum Vysočiny Jihlava,
Jihlava 009604
Muzeum Vysočiny Pelhřimov,
Pelhřimov 009753
Muzeum w Bielsku-Białej, Bielsko-
Biała 034653
Muzeum w Bielsku Podlaskim,
Muzeum Podlaskie w Białymstoku,
Bielsk Podlaski 034647
Muzeum w Bochni im. Prof.
Stanisława Fischera, Bochnia . 034656
Muzeum w Chorzowie, Chorzów 034695
Muzeum w Chrzanowie im. Ireny
i Mieczysława Mazarakich,
Chrzanów 034697
Muzeum w Darłowie, Zamek Książąt
Pomorskich, Darłowo 034711
Muzeum w Dobrzycy, Dobrzyca . 034716
Muzeum w Elblągu, Elbląg . . . 034719
Muzeum w Gliwicach, Gliwice . 034757
Muzeum w Gorzowie Wielkopolskim,
Gorzów Wielkopolski 034771
Muzeum w Grudziądzu,
Grudziądz 034776
Muzeum w Kwidzynie, Muzeum

Zamkowego w Malborku,
Kwidzyń 034932
Muzeum w Lęborku, Lębork . . . 034935
Muzeum w Łęczycy, Łęczyca . 034937
Muzeum w Łowiczu, Łowicz . . 034984
Muzeum w Lubaczowie,
Lubaczów 034985
Muzeum w Międzyrzeczu,
Międzyrzecz 035009
Muzeum w Przeworsku, Zespół
Pałacowo-Parkowy, Przeworsk 035148
Muzeum w Raciborzu, Racibórz 035163
Muzeum w Rybniku, Rybnik . . 035180
Muzeum w Stargardzie, Stargard 035219
Muzeum w Szklarskiej Porębie
Hoffmanna, Szklarska Poręba . 035261
Muzeum w Tarnowskich Górach,
Tarnowskie Góry 035271
Muzeum w Tomaszowie Mazowieckim,
Tomaszów Mazowiecki 035275
Muzeum w Tykocinie, Muzeum
Okręgowego w Białymstoku,
Tykocin 035293
Muzeum w Wałbrzychu,
Wałbrzych 035304
Muzeum w Żywcu, Żywiec . . . 035473
Muzeum Waldemara Babinicza,
Nałęczów 035032
Muzeum Walewskich w Tubądzinie,
Muzeum Okręgowe w Sieradzu,
Wróblew 035412
Muzeum Walk o Wał Pomorski przy
Miejsko-Gminnym Ośrodku Kultury,
Mirosławiec 035016
Muzeum Walki i Męczeństwa
1939–1945, Muzeum Niepodległości
w Warszawie, Warszawa 035366
Muzeum Walki i Męczeństwa w
Palmirach, Muzeum Historyczne
Miasta Stołecznego Warszawy,
Warszawa 035367
Muzeum Walki i Męczeństwa w
Treblince, Muzeum Regionalne w
Siedlcach, Kosów Lacki 034852
Muzeum Warmii i Mazur, Olsztyn 035069
Muzeum Warmińskie, Muzeum Warmii
i Mazur w Olsztynie, Lidzbark
Warmiński 034947
Muzeum Wartownia nr 1 na
Westerplatte Sucharskiego, Muzeum
Historyczne Miasta Gdańska,
Gdańsk 034735
Muzeum Weterynarii, Muzeum
Rolnictwa, Ciechanowiec 034700
Muzeum Wiedzy o Środowisku,
Poznań 035137
Muzeum Więzienia Pawiak, Muzeum
Niepodległości, Warszawa 035368
Muzeum Wikliniarstwa i Chmielarstwa,
Muzeum Narodowe Rolnictwa i
Przemysłu Rolno-Spożywczego w
Szreniawie, Nowy Tomyśl . . . 035055
Muzeum Wikliny, Olkusz 035067
Muzeum Wisły, Centralny Muzeum
Morskiego w Gdańsku, Tczew . 035273
Muzeum Władysława Broniewskiego,
Muzeum Literatury im. Adama
Mieckiewicza w Warszawie,
Warszawa 035369
Muzeum Władysława Sikorskiego,
Warszawa 035370
Muzeum Wnętrz Dworskich, Muzeum
Ziemi Wieluńskiej, Ożarów . . . 035092
Muzeum Wnętrz Pałacowych w
Choroszczy, Muzeum Podlaskie w
Białymstoku, Choroszcz 034693
Muzeum Wnętrz Stylowych,
Oporów 035079
Muzeum Wnętrz Zabytkowych,
Muzeum Okręgowego w Tarnowie,
Dębno 034712
Muzeum Wojska, Białystok . . . 034644
Muzeum Wojska Polskiego,
Warszawa 035371

Muzeum Woli, Muzeum Historycznego
Miasta Starego Warszawy,
Warszawa 035372
Muzeum Wsi Kieleckiej, Kielce . 034835
Muzeum Wsi Lubelskiej, Lublin . 035001
Muzeum Wsi Mazowieckiej,
Sierpc 035198
Muzeum Wsi Opolskiej, Opole . 035078
Muzeum Wsi Radomskiej, Radom 035166
Muzeum Wsi Słowińskiej w Klukach,
Muzeum Pomorza Środkowego w
Słupsku, Kluki 034840
Muzeum X Pawilonu Cytadeli
Warszawskiej, Muzeum
Niepodległości, Warszawa . . . 035373
Muzeum Zabawek ze Zbiorow
Henryka Tomaszewskiego,
Karpacz 034807
Muzeum Zabawkarstwa, Kielce . 034836
Muzeum Zachodnio-Kaszubskie,
Bytów 034676
Muzeum Zagłębia, Będzin 034629
Muzeum Zagłębia Staropolskiego
w Sielpi, Muzeum Techniki w
Warszawie, Sielpia 035195
Muzeum Zakładu Historii Farmacji,
Lublin 035002
Muzeum-Zamek Górków, Państwowa
Instytucja Kultury, Szamotuły . . 035244
Muzeum-Zamek w Bolkowie, Muzeum
Okręgowego w Jeleniej Górze,
Bolków 034660
Muzeum Zamek w Gołuchowie,
Oddział Muzeum Narodowego w
Poznaniu, Gołuchów 034767
Muzeum Zamek w Łańcucie,
Łańcut 034934
Muzeum Zamek, Zamek w Baranowie
Sandomierskim, Baranów
Sandomierski 034625
Muzeum Zamkowe, Pszczyna . . 035153
Muzeum Zamkowe w Malborku,
Malbork 035006
Muzeum Zamoyskich w Kozłówce,
Kamionka 034804
Muzeum Zbiorów Geologicznych,
Kielce 034837
Muzeum Zbrojownia, Liw 034950
Muzeum Zdeňka Buriana,
Štramberk 009894
Múzeum Ždiarsky Dom, Ždiar . . 038483
Muzeum Zegarów Wieżowych,
Muzeum Historyczne Miasta
Gdańska, Gdańsk 034736
Muzeum Zegarów, Rzemiosł
Artystycznych i Precyzyjnych,
Warszawa 035374
Muzeum Zespoł Zamkowy w Niedzicy,
Niedzica 035034
Múzeum Židovské Komunity,
Tovačov 009915
Múzeum Židovskej Kultúry, Slovenské
Národné Múzeum, Bratislava . 038325
Muzeum Ziemi Augustowskiej,
Augustów 034622
Muzeum Ziemi Chełmińskiej,
Chełmno 034682
Muzeum Ziemi Dobrzyńskiej,
Rypin 035182
Muzeum Ziemi Kępińskiej, Kępno 034828
Muzeum Ziemi Kłodzkiej, Kłodzko 034838
Muzeum Ziemi Kociewskiej, Starogard
Gdański 035220
Muzeum Ziemi Krajeńskiej, Nakło nad
Notecią 035028
Muzeum Ziemi Krzeszowickiej,
Krzeszowice 034928
Muzeum Ziemi Kujawskiej i
Dobrzyńskiej, Włocławek 035402
Muzeum Ziemi Lubawskiej, Nowe
Miasto Lubawskie 035042
Muzeum Ziemi Lubuskiej, Zielona
Góra 035463
Muzeum Ziemi Mrągowskiej, Muzeum

Register der Institutionen und Firmen

Muzeum Ziemi Nadnoteckiej im. Wiktora Stachowiaka, Trzcianka
– Nacionalen Park-muzej Samuilova krepost, Istoričeski

Warmii i Mazur w Olsztynie,
Mrągowo 035021
Muzeum Ziemi Nadnoteckiej im.
Wiktora Stachowiaka, Trzcianka 035290
Muzeum Ziemi Otwockiej,
Otwock 035091
Muzeum Ziemi Pałuckiej, Żnin . 035469
Muzeum Ziemi Piskiej, Muzem
Okręgowego w Suwałkach, Pisz 035102
Muzeum Ziemi Polskiej Akademii
Nauk, Warszawa 035375
Muzeum Ziemi Prudnickiej,
Prudnik 035142
Muzeum Ziemi Puckiej, Puck . . 035154
Muzeum Ziemi Rawickiej, Rawicz 035170
Muzeum Ziemi Rawskiej, Rawa
Mazowiecka 035169
Muzeum Ziemi Sochaczewskiej i Pola
Bitwy nad Bzurą, Sochaczew . 035209
Muzeum Ziemi Średzkiej Dwór w
Koszutach, Koszuty 034854
Muzeum Ziemi Suskiej, Sucha
Beskidzka 035228
Muzeum Ziemi Wałeckiej, Wałcz 035305
Muzeum Ziemi Wieluńskiej,
Wieluń 035392
Muzeum Ziemi Wschowskiej,
Wschowa 035438
Muzeum Ziemi Zaborskiej i Galeria
Sztuki Ludowej, Wiele 035390
Muzeum Ziemi Zawkrzańskiej,
Mława 035018
Muzeum Ziemi Złotowskiej,
Złotów 035465
Muzeum Zlatý Klíč, Karlovy Vary 009620
Muzeum Zofii i Wacława Nałkowskich,
Wołomin 035408
Muzeum Zoologiczne Instytutu Zoologii
Uniwersytetu Jagiellońskiego,
Kraków 034906
Muzeum Żup Krakowskich Wieliczka,
Wieliczka 035391
Muzeum Żuraw, Centralne Muzeum
Morskie, Gdańsk 034737
Muzeum Żydowskiego Instytutu
Historycznego w Polsce,
Warszawa 035376
Muziejus Čepkelių Valstybiniame
Rezervate, Kauno Tado
Ivanausko Zoologijos Muziejus,
Marcinkonys 030415
Muziejus Žuvinto Valstybiniame
Rezervate, Kauno Tado Ivanausko
Zoologijos Muziejus, Aleknoniy 030311
Muziejus, Kintų Vydūno Kultūros
Centras, Kintai 030384
Muziekinformatie- en
documentatiecentrum Ton Stolk,
Vlaardingen 032989
Muziekinstrumenten-Museum,
Peer 003947
Muziekinstrumentenmakersmuseum,
Tilburg 032927
Mūzikas Muzejs, Rīga 030211
Muzikologické Pracoviště, Slezské
Zemské Muzeum, Opava 009735
Muzikų Brolių Motiejaus, Broniaus
ir Antano Budriūnų Gimtinė,
Pabiržė 030433
Muzium Alor Gajah, Alor Gajah . 030659
Muzium Angkatan Tentera, Malaysian
Armed Force Museum, Kuala
Lumpur 030687
Muzium Arkeologi, Merbok . . . 030719
Muzium Brunei, Bandar Seri
Begawan 004964
Muzium dan Galeri Seni, Universiti
Sains Malaysia, Minden 030720
Muzium Darul Ridzwan, Ipoh . . 030665
Muzium Diraja Abu Bakar, Johor
Bahru 030668
Muzium Diraja Perak, Kuala
Kangsar 030680
Muzium Jasin, Jasin 030667

Muzium Kecantikan, Melaka . . . 030708
Muzium Layang-Layang, Melaka 030709
Muzium Negara, Kuala Lumpur . 030688
Muzium Negeri Kedah, Alor Setar 030662
Muzium Negeri Kelantan, Kota
Bharu 030674
Muzium Negeri Perlis, Kangar . . 030669
Muzium Negeri Sarawak,
Kuching 030701
Muzium Negeri Selangor, Shah
Alam 030729
Muzium Negeri Sembilan,
Seremban 030726
Muzium Negeri Trengganu, Kuala
Terengganu 030699
Muzium Numismatik Maybank, Kuala
Lumpur 030689
Muzium Perak, Taiping 030732
Muzium Polis Diraja Malaysia, Kuala
Lumpur 030690
Muzium Rakyat, Melaka 030710
Muzium Samudera, Melaka 030711
Muzium Sastera, Melaka 030712
Muzium Sejarah & Etnografi,
Melaka 030713
Muzium Sejarah National, Kuala
Lumpur 030691
Muzium Seni Asia, Kuala Lumpur 030692
Muzium Teknologi Melayu, Bandar
Seri Begawan 004965
Muzium Tekstil, Kuala Lumpur . 030693
Muzium Tentera Laut Diraja Malaysia,
Melaka 030714
Muzium Tuan Yang Terutama Melaka,
Melaka 030715
Muzj Istorii Rossijskich Voennych
Učilišč, Voenno-istoričeskij Muzej
Artillerii, Inženernych Vojsk i Vojsk
Svjazi, Sankt-Peterburg 037809
Muzyčno-memorialnyj Muzej Solomiï
Krušel'nic'koï, Lviv 043251
Muzyka i Vremja – Častnyj
Muzej Džona Mostoslavskogo,
Jaroslavl 037067
Muzzleloaders Etcetra,
Minneapolis 094935
Muzzy, San Francisco 097291
MVD Galería de Arte, Montevideo 119432,
119433
MVG-Museum, München 020559
MvK Fine Art, Crauthem 082351
My Antique Trading Post, Tucson 097681
My Art Prospects, New York . . . 122905,
122906
My Art Studio, Washington 124938
My Asian Gifts, San Francisco . . 124351
My Back Yard Nursery Art, San
Diego 124147
My Collection, Verona 111192
My Daughter's Place, Oklahoma
City 095995
My House Antiques, Los Angeles 094414
My Old Kentucky Home,
Bardstown 046788
My Second Home, Nashville . . . 095017
My Secret Garden, Memphis . . . 094657
My Sisters Closet, Saint Paul . . 096841
My Sisters Stuff, Portland 096446
My Store, Moree 061884
My-Tho, Mons 063797
My Thuat, Ho Chi Minh City . . . 139540
My Turn, Saint Paul 096842
Myanan Sankyaw Shwe Nandaw,
Mandalay 031816
Mycènes, Champeix . . 067810, 128541
Myers, Charlotte 120002
Myers Museum of Egyptian and
Classical Art, Eton 044307
Myers School of Art, University of
Akron, Akron 056629
Myers, Myrna, Paris 071701
Myers & Son, R.N., Gargrave . . 089322
Myerscough, Sarah, London . . . 118468
Myfair Carpet Gallery, London . 090370

Mykolaïvs'kyj Filial Kyïvs'kogo
Nacional'nogo Universytetu Kul'tury i
Mystectv, Mykolaïv 056497
Mykolaïvs'kyj Oblasnyj Chudožnij
Muzej im. Vereščagina,
Mykolaïv 043265
Mykolaïvs'kyj Oblasnyj Krajeznavčyj
Muzej, Mykolaïv 043266
Mykolo Žilinsko Dailės Galerija,
Nacionalinis M.K. Čiurlionio Dailės
Muziejus, Kaunas 030370
Mykonos Folklore Museum,
Mykonos 023181
Myles Hall, Toronto 064707
Myles, Eve, Houston 121099
Mylin, J., List 077136
Mylin, J., Sylt 078661
Mylin, Jens, Sylt 078662
Mylin, Jens, Wyk auf Föhr 079125
Myllarheimen, Rauland 033500
Myllykolu, F.E. Sillanpään Syntymäkoti,
Hämeenkyrö 010784
Myllymuseo, Juuka 010927
Myllysaaren Museo, Valkeakoski 011482
Myndlistaskólinn Akureyri,
Akureyri 055772
Myndlistaskólinn í Reykjavík,
Reykjavík 055774
Myntkabinett, Universitetets
Kulturhistoriske Museer, Oslo . 033844
Myntsafn Seðlabanka og
Thjóðminjasafns, Reykjavík . . . 024102
Myohyang-san Museum,
Hyangsan 029874
Myong Ji University Museum,
Seoul 030027
Myopic, Chicago 145048
Myopic, Providence 145469
Myoung, Seoul 111696
Myra Powell Art Gallery and Gallery at
the Station, Ogden 051715
Myra's Antiques, Cape Town . . . 085636
Myra's Antiques, Columbus 093037
Myrbergsgården Local Museum,
Vörå 011520
Myrése, Maastricht 083472
Myreton Motor Museum,
Aberlady 043367
Myri, B., Berlin 074886
Myriad, London 090371
Myriad, Salisbury 091224, 135000
Myriade, Paris 071702
Le Myricae, Roma 081470
Myrskylän Kotiseutumuseo,
Myrskylä 011177
Myrtleville House Museum,
Brantford 005471
Mysore Silk Museum, Bangalore 024151
Mystère de Chine, Saint-Denis . 085067
Mysteria, Regina 101534
Mysteria Products, Fort Worth . . 093465
Mysterious Bookshop, New York 145359
Mysterium, Las Vegas 121369
Mystery Cat, Honolulu 093563
The Mystery of the Egyptian History,
Dallas 093182
Mystery One Bookshop,
Milwaukee 145268
Mystic Ancient Art Gallery,
Singapore 085473
Mystic Art Center, Mystic 051145
Mystic Coins, San Diego 097111
Mystic Seaport, Mystic 051146
Mystique Designs, San Francisco 097242
Mystique Salon, Indianapolis . . 121206
Mystiques Antiques and Collectable,
Mentone 061840
Mythes et Légendes, Paris 071703
Mythos Art, Merimbula 099329
Myto, México 111975
Mytton Antiques, Atcham 088318
MyVisit, Berlin 106380
Mzelaynius, San Antonio 097003
MZM – Museumszentrum

Mistelbach, Hermann Nitsch
Museum, Lebenswelt Weinviertel,
Messweinarchiv, Mistelbach an der
Zaya 002393
Mzuvele Art School, Kwamashu . 056349
Mzuzu Regional Museum, Mzuzu 030656
N 10 Antiquités, Châteaudun . . 067883
N Gallery, Tbilisi 105917
N&D, Tbilisi 105918
N. Hadjikyriakos-Ghikas Gallery,
Benaki Museum, Athinai 023018
N.K.L. Artisanat, Québec 101511
N.Z, Inglewood 083961
N7 Teen Art Gallery, Tuam 109701
Na Bolšom, Sankt-Peterburg . . . 114349
Na Francouzské, Praha . . . 065511
NA Gallery, Seoul 111697
Na Kazimierzu, Kraków 143480
Na Královce, Praha 065512
Na Kramarskiej, Poznań 113743
Na Litejnom, Sankt-Peterburg . . 085252,
143671
Na Ochozu, Karlovy Vary 102606
Na Pięterku, Kraków 113619
Na Pięterku, Warszawa 113871
Na Podhoří, Zlín 065567
Na Rynku, Hradec Králové 140473
Na Solnym, Wrocław 113946
Na Staronevskom, Sankt-
Peterburg 085253, 143672
Na.de., Roma 081471
Naaimachinemuseum Dordrecht,
Dordrecht 032249
Naaji, Milwaukee 121938
Naantalin Museo, Naantali 011178
NAB Gallery, Chicago 047682
Nabaa, Elie, Beirut 082296
Naber, Groningen 112497
Nabis Gallery, Tokyo 111477
Nábytek – Obrazy – Technické
Starožitnosti, Příbram 065548
Nábytkové Múzeum, Markušovce 038387
Naccari, Elvira, Venezia 081862
Nacetova Hiša, Škofja Loka . . . 038611
Nachbar, Richard, Buffalo 092548
Nacionalen Archeologičeski Rezervat
Kabile, Jambol 005041
Nacionalen Cărkoven Istoriko-
archeologičeski Muzej, Sofia . . 005189
Nacionalen Etnografski Muzej,
Bălgarska Akademija na Naukite,
Sofia 005190
Nacionalen Istoričeski Muzej,
Sofia 005191
Nacionalen Istoriko-archeologičeski
Rezervat i Muzej Veliki Preslav, Veliki
Preslav 005248
Nacionalen Istoriko-archeologičeski
Rezervat Pliska, Pliska 005123
Nacionalen Literaturen Muzej,
Sofia 005192
Nacionalen Muzej Bojanska Crkva,
Sofia 005193
Nacionalen Muzej Georgi Dimitrov,
Sofia 005194
Nacionalen Muzej na Bălgarskata
Architektura, Veliko Tărnovo . . 005256
Nacionalen Muzej na Fizičeskata
Kultura i Sport, Sofia 005195
Nacionalen Muzej na Obrazovanieto,
Gabrovo 005032
Nacionalen Muzej na Transporta i
Săobščenijata, Ruse 005154
Nacionalen Muzej Parachod Radecki,
Kozloduj 005079
Nacionalen Muzej Rilski Manastir,
Rilski Manastir 005150
Nacionalen Muzej Roženski Manastir,
Melnik 005091
Nacionalen Muzej Zemjata i Chorata,
Sofia 005196
Nacionalen Park-muzej Samuilova
krepost, Istoričeski Muzej Petrič,
Petrič 005114

Nacionalen Park-muzej Šipka-
Buzludža, Kazanläk 005059
Nacionalen Politechničeski Muzej,
Sofia 005197
Nacionalen Prirodonaučen Muzej,
Sofia 005198
Nacionalen Selskostopanski Muzej,
Sofia 005199
Nacionalen Voenno-morski Muzej,
Varna 005244
Nacionalinis Mikalojaus Konstantino
Čiurlionio Dailės Muziejus,
Kaunas 030371
Nacionalna Akademija na Iskustvo,
Sofia 055110
Nacional'na Akademija Obrazotvorčogo
Mystectva i Architektury, Kyïv . 056493
Nacionalna Chudožestvena Galerija,
Sofia 005200
Nacionalna Galerija za Čuždestranno
Izkustvo, Sofia 005201
Nacionalna Galerija za Dekorativni
Izkustva, Sofia 005202
Nacionalna i Sveučilišna Knjižnica,
Zagreb 136756
Nacionalna Izložba na Narodnite
Chudožestveni Zanajati i Priložni
Izkustva, Orešak 005100
Nacionalnaja Galereja Respubliki Komi,
Syktyvkar 037914
Nacionalnij Muzej u Lvovi, Lviv 043252
Nacionalnij Muzej u Lvovi – Viddil
Narodnogo Mystectva, Lviv . 043253
Nacionalnij Muzej u Lvovi – Viddil
Starodrukiv ta Rukopysiv, Lviv . 043254
Nacionalnij Muzej u Lvovi – Viddil
Sučasnogo Mystectva, Lviv ... 043255
Nacionalnyj Chudožestvennyj Muzej
Respubliki Belarus, Minsk ... 003253
Nacionalnyj Chudožestvennyj
Muzej Respubliki Sacha (Jakutija),
Jakutsk 037057
Nacional'nyj Chudožnij Muzej Ukraïny,
Kyïv 043215
Nacional'nyj Istoriko-etnografičnyj
Zapovidnyk Perejaslav, Perejaslav-
Chmel'nyc'kyj 043280
Nacional'nyj Kyïvo-Pečerskyj Istoryko-
kul'turnyj Zapovidnyk, Kyïv ... 043216
Nacional'nyj Muzej Černobyl',
Kyïv 043217
Nacional'nyj Muzej Istoriï Ukraïny,
Kyïv 043218
Nacional'nyj Muzej Istoriï Velykoï
Vitčyznjanoï Vijny 1941–1945 Rokiv,
Kyïv 043219
Nacionalnyj Muzej Kabardino-
Balkarskoj Respubliki, Nalčik . 037536
Nacional'nyj Muzej Literatury Ukraïny,
Kyïv 043220
Nacional'nyj Muzej Medycyny Ukraïny,
Kyïv 043221
Nacionalnyj Muzej Respubliki
Baškortostan, Ufa 037983
Nacionalnyj Muzej Respubliki Komi,
Syktyvkar 037915
Nacionalnyj Muzej Respubliki Marij Él
im. T. Evseeva, Joškar-Ola ... 037073
Nacionalnyj Muzej Respubliki
Tatarstan, Kazan 037119
Nacional'nyj Muzej Tarasa Ševčenka,
Kyïv 043222
Nacionalnyj Muzej Udmurtskoj
Respubliki im. K. Gerda, Iževsk 037048
Nacional'nyj Muzej Zbrojnych Syl
Ukraïny, Kyïv 043223
Nacional'nyj Zapovidnyk Chersones
Tavrijs'kyj, Sevastopol 043293
Nacional'nyj Zapovidnyk Chorticja,
Zaporižžja 043323
Nack, Saitama 111370
Nacke, Dieter, Bartenshagen-
Parkentin 106122
Nad Niprem, Bielsko-Biała ... 084491

NADA Art Fair Miami, Miami ... 098477
Al-Nada, Katr, Cairo 103062
Nadaï, Michel de, Toulouse ... 129310
Nadalini, Gianpaolo, Paris 129097
Nádasdi Ferenc Múzeum, Sárvár 023876
Nádasdy-Kastély, Nádasdladány 023801
Nadbaltyckie Centrum Kultury w
Gdansku, Gdańsk 034738
Nadeau, Windsor 127498
Nadelburgmuseum, Franz Gehrer
Heimatmuseum, Lichtenwörth . 002304
Nadelmann, Johann, Berlin 141557
Nader Mausoleum, Mashad ... 024722
Nader, Gary, Miami 121831
Nadiani, Antonio, Roma 081472, 131989
Nadia's Antiques, Toronto 064708
Nadir, Annecy 103250
Nadkarni, R.P., Mumbai 130824
Nadlerhaus, Stützengrün 022020
Nadrag, Nikolaus, Bielefeld ... 074984
NAEA News, Reston 139448
Nääs Slott & Slöjdseminarium,
Floda 040629
Naegele, Amyas, New York ... 122907,
135965
Nähmaschinen-Museum,
Steckborn 042028
Nähmaschinen-Museum Gebrüder
Mey, Sammlung Albrecht Mey,
Albstadt 016495
Näitustesaal Rotermanni Soolalaos,
Eesti Kunstimuuseum, Tallinn . 010681
Näkövammaismuseo, Jiris 010916
Närkekonservatorerna, Kumla . 133541
Närpes Hembygdsmuseum,
Närpes 011179
Næs Jernverksmuseum,
Tvedestrand 034047
Naesa, Singapore 114576
Naeshirogawa Minkan-yaki Bijutsukan,
Higashi-Ichiki 028618
Naeslund, Malmö 086857
Nässjö Auktionshall, Malmbäck . 126632
Næstved Antikvariat, Næstved . 140627
Næstved Museum, Næstved ... 010233
Naffouj, Landstuhl 107795
Nafplion Folklore Museum,
Nafplion 023194
Naftalina, Porto Alegre 063985, 128107
Naga, New York 095602
Naga, Singapore 085474
Nagai, Tokyo 111478
Nagano-ken Shinano Bijutsukan,
Higashiyama Kaiikan, Nagano . 029074
Nagano-kenritsu Rekishi
Hakubutsukan, Chikuma 028513
Nagaoka Gear Museum,
Nagaoka 029076
Nagaoka Municipal Science Museum,
Nagaoka 029077
Nagaoka-shiritsu Hakabutsukan,
Nagaoka 029078
Nagaoka-shiritsu Kogyo Hakurankai,
Nagaoka 029079
Nagasaki Atomic Bomb Museum,
Nagasaki 029095
Nagasaki-ken Bijutsukan,
Nagasaki 029096
Nagasaki Kite Museum,
Nagasaki 029097
Nagasaki Rekishi no Bunka
Hakubutsukan, Nagasaki 029098
Nagasaki Science Museum,
Nagasaki 029099
Nagasaki-shiritsu Hakabutsukan,
Nagasaki 029100
Nagasaki-shiritsu Nyokodo Nagai
Kinenkan, Nagasaki 029101
Nagashima Museum, Kagoshima 028748
Nagatoro Kyukokan, Nagatoro . 029105
Nagel, Beijing 125400
Nagel, Bruxelles 125322
Nagel, Hong Kong 125404
Nagel, Phoenix 096226

Nagel, Santiago de Chile 064979
Nagel, Tokyo 126248
Nagel, Verona 126242
Nagel Auktionen, Wien 125280
Nagel Auktionen, Nagel Auctions,
Stuttgart 126121
Nagel, B., West Hollywood 097912
Nagel, Brigitta, Hamburg 076285
Nagel, Christian, Berlin 106381
Nagel, Daniel, Neustadt an der
Weinstraße 108261
Nagelmuseum, Löchgau 020096
Nagelschmiede Sulz, Ewalt Kalt,
Sulz 042040
Nagelschmiedemuseum, Brandstätter
Hammer, Losenstein 002344
Naghi, New Orleans 095135
Nagi-cho Gendai Bijutsukan,
Nagi 029107
Nagi Museum, Cairo 010553
Nagler & Eberle, Wien 125281
Nagon, Los Angeles 094415
Nagornaja Galereja, Moskva ... 114284
Nagoya-Boston Museum of Fine Arts,
Nagoya 029120
Nagoya City Hideyoshi and Kiyomasa
Memorial Museum, Nagoya . 029121
Nagoya Gallery, Nagoya 111325
Nagoya Geijutsu Daigaku,
Nishiharu 056022
Nagoya-shi Bijutsukan, Nagoya . 029122
Nagoya-shi Daigaku Hakubutsukan,
Nagoya 029123
Nagoya-shi Hakubutsukan,
Nagoya 029124
Nagoya-shi Kagakukan, Nagoya 029125
Nagoya Zokei Geijutsu Daigaku,
Nagoya 056020
Nagswood, El Paso 120837
Nagtegaal, Hans Peter, Apeldoorn 132425
Nagtegaal, Roos, Bruntinge 132461
Nagu Hembyggdsmuseum/Nauvon
Kotiseutumuseo, Local Heritage
Museum of Nagu/Nauvo, Nauvo 011183
Nagy Ferenc Galéria, Tab 023989
Nagy István Képtár, Türr István
Múzeum, Baja 023417
Nagy Lajos Emlékkiállítás,
Apostag 023407
Nagy László Emlékház, Iszkáz .. 023699
Nagy, Michael, Woollahra 099919
Nagy, Richard, London 118469
Nagy, Tibor de, New York 122908
Nagyházi, Budapest . 109165, 126154
Nagykunsági Tájház, Györffy István
Nagykun Múzeum, Karcag ... 023718
Nagytétényi Kastélymúzeum,
Iparművészeti Múzeum,
Budapest 023529
Naharro, Madrid 133361
Naharro Reyes, Luis, Zaragoza . 086600
Naharro, José-Luis, Lausanne .. 087577
Nahas, Sharjah 088204
Nahkl Fort Museum, Nahkl 034102
Nahmad, Helly, London 118470
Nahmad, Helly, New York 122909
Nahmani, Daniel, Mont-de-
Marsan 070173
Nahnella, São Paulo 064196
Nahsholim Museum, Hof Dor .. 025063
Nahum Gutman's Museum, Tel
Aviv 025208
Naicam Museum, Naicam 006442
Naïfs du Monde Entier, Paris ... 105120
Naigai, Tokyo 142932
Nail, Kathy, Seattle 124573
Naim, Anne-Maria, Paris 071704
Naim, Pascal, Paris 105121
Naimeh, Yousef, Damascus ... 088044
Naimski, Michael, London 118471
Nain Museum, Nain 006443
Naini, Cockayne Hatley 117606
Nairn Fishertown Museum, Nairn 045309
Naismith Memorial Basketball Hall of

Fame, Springfield 053462
Naito Kinen Kusuri Hakubutsukan,
Kawashima 028812
Naiv Antikvarnaja Galereja, Sankt-
Peterburg 085254
Najana, Warszawa 113872
Nájera Núñez, Rosa, Alajuela . 102464
Nakagawa Chaini-zu Kindai Bijutsukan,
Fukuyama 028563
Nakagawa Photo Gallery, Kyoto . 028969
Nakama-shiritsu Minzoku
Hakubutsukan, Nakama 029134
Nakambale Museum, Ondangwa 031849
Nakamiya Gallery, Osaka 111364
Nakamura Kinen Bijutsukan,
Kanazawa 028776
Nakamura Kogei, Kyoto 082035
Nakamura, Kobijutsu, Hiroshima 081983
Nakano Historical Museum,
Tokyo 029558
Nakano-shi Rekishi Minzoku
Shiryokan, Nakano 029138
Nakanoshima Bijutsu Gaikun,
Osaka 056024
Nakao Shosen-do, Osaka 142908
Nakao Shoten, Osaka 142909
Nakashian-O'Neil, Saint Paul .. 096843
Nakata Bijutsukan, Onomichi .. 029234
Nake, Erich, Wehrheim 108852
Naked Art, Birmingham 119810
Nakhchivan Autonomous Republic
Djalil Mammadkulizade Literature
Museum, Naxçivan 003190
Nakhchivan Carpet Museum,
Naxçivan 003191
Nakhchivan State History Museum,
Naxçivan 003192
Nakkilan Kotiseutumuseo ja
Juustomeijerimuseo, Nakkila .. 011181
Naklada Naprijed, Zagreb 136757
Nakota Curios, Hastings 089466
Nakusp & District Museum,
Nakusp 006444
Nalbandian & Son, Loutfic, Beirut 082297
Nalbandian, Serge & Gaby, Beirut 082294
Nall, George, Louisville 121661
Nalšios Muziejus, Švenčionys . 030485
Nam June Paik Art Center,
Yongin 030064
Nam Nao, Petchabun 042708
Nam, Kyeong, Seoul 111698
Nam, Phuong, Ho Chi Minh City 125135,
125136
Nam, Son, Ha Noi 125087
Nam & Co., Thye, Singapore ... 085475
Namarook Antiques, Windsor,
Victoria 062437
Namaste, Fort Worth 093466
Namban Bunkakan, Osaka 029251
Nambara Arts & Crafts,
Nhulunbuy 061972
Nambrok Antiques, Sale 062183
Nambu Toji Denshokan,
Ishidoriya 028700
Namgyal Institute of Tibetology,
Gangtok 024258
Namık Kemal Zindanı ve Müzesi,
Gazimağusa 033453
Namikawa Banri Bukkyo Bijutsu
Shashin Gallery, Hiroshima .. 028656
Namikawa Cloisonne Museum of
Kyoto, Kyoto 028970
Namiraz, Monaco 082523
Namjin Art Museum, Imhoe 029950
Namkyung, Seoul 111699
Namoc-National Art Museum of China,
Beijing 007634
Namsdalsmuseet, Namsos 033796
Namsskogan Bygdatun,
Namsskogan 033797
Namy-Caulier, Martine, Paris .. 105122,
136968
Namyslok, Aschersleben 106019
Namyslok, Steffen, Werdau 108871

Nan Yue Wang Palace Museum,
Guangzhou 007812
Nanaimo Art Gallery, Nanaimo . 006449
Nanaimo District Museum,
Nanaimo 006450
Nan'ao Coast Defence City Museum,
Nan'ao 008094
Nanao-shiritsu Bijutsukan, Nanao 029141
Nana's Imported Arts, Milwaukee 121939
Nana's Treasures and Antiques, El
Paso 093409
NANB Nursing History Centre,
Fredericton 005842
Nanba Bijutsu Gakuin, Osaka . . 056025
Nanbando, Tokyo 082185
Nanbooks, Settle 091265
Nance Museum, Lone Jack 050367
Nancéide, Nancy 140978
Nanchang Museum, Nanchang . 008096
Nanchang Uprising Museum,
Nanchang 008097
Nancheng Museum, Nancheng . 008101
Nancy Antiquités, Montréal . . . 064446
Nancy Island Historic Site, Wasaga
Beach 007357
Nancy's Christmas Collectibles,
Denver 093317
Nandan Museum-Vichittra and Art
Gallery, Santiniketan 024466
Nandewar Historical Museum,
Barraba 000824
Nanfa, Antonio, Firenze 131221
Nanfria, Leonardo, Genova . . . 131330
Nanhai Museum, Nanhai 008102
Nanich Castañe, Maria Teresa,
Barcelona 085809
Naninck & Lengaigne, Saint-
Omer 072996
Naniwa-no Umino Jikukan,
Osaka 029252
Nanji Guan, Qingdao 008173
Nanjiang Museum, Nanjiang . . . 008103
Nanjing Calligraphy and Painting
Gallery, Nanjing 102151
Nanjing Children's Painting and
Calligraphy Association, Nanjing 058501
Nanjing College of Fine Arts,
Nanjing 055225
Nanjing Folklorish Museum,
Nanjing 008110
Nanjing International Culture and Arts
Exchange Center, Nanjing . . . 008111
Nanjing Museum, Nanjing 008112
Nanjing Museum, Nanjing,
Zhangzhou 008119
Nanjing Museum of Palaeontology,
Nanjing 008113
Nanjingdi Zhen Science Museum,
Nanjing 008114
Nanjingdi Zhi Museum, Nanjing 008115
Nankang County Museum,
Nankang 008120
The Nanking Porcelain &Co.,
London 090372
Nanni, Milano 080624
Nanni, Manuela, Milano 110388
Nanning Museum, Nanning . . . 008123
Nanny's Web, Bentleigh East . . 139642
Nanortallip Katersugaasivia,
Nanortalik 023332
Nanping Museum, Nanping . . . 008125
Nan's Hands, Indianapolis 093910
Nanshan Painting Gallery,
Shenzhen 102232
Nansong Guanstove Museum,
Hangzhou 007842
Nansong Qian Currency Museum,
Hangzhou 007843
Nantais, Darryl, Linton 118122
Nante, Eggenfelden 075576
Nantenshi, Tokyo 111479
Nantgarw China Works Museum,
Nantgarw 045310
Nanto, Nara 082072

Nanton Lancaster Air Museum,
Nanton 006452
Nantong Bowu House, Nantong . 008126
Nantong Museum, Nantong . . . 008127
Nantong Textile Museum,
Nantong 008128
Nantou Ceramic Museum,
Nantou 042405
Nantou County Natural History
Museum, Ren-Ai 042427
Nantou History Museum, Nantou 042406
Nantucket Historical Association
Museum, Nantucket 051157
Nantucket Life-Saving Museum,
Nantucket 051158
Nantucket Lightship Basket Museum,
Nantucket 051159
Nantwich Antiques, Nantwich . . 090798
Nantwich Museum, Nantwich . . 045311
Nanxiong Museum, Nanxiong . . 008129
Nanyang Academy of Fine Arts,
Singapore 056329
Nanyang Guan, Nanyang 008130
Nanzhang Museum, Nanzhang . 008133
Nanzhao Pithecanthrope Museum,
Nanzhao 008134
Nanzhou Nationality Museum,
Kaili 007985
Naomi Wood Collection at Woodford
Mansion, Philadelphia 052072
Naomi's Antiques, San Francisco 097293
Napa Valley Museum, Yountville 054601
Naper Settlement Museum,
Naperville 051161
Naphtaline, Bayeux 066970
Naphtaline, Saint-Amand-les-
Eaux 072663
Naphtaline, Valenciennes 073947
Napier Antique Centre, Napier . 084008
Napier House, Sudbury, Suffolk 091503
Napiórkowska, Katarzyna,
Poznań 113744
Napiórkowska, Katarzyna,
Warszawa 113873, 113874
Naples Art and Antiques Show,
Naples 098478
Naples Museum of Art, Naples . 051163
Napoleon- und Heimatmuseum,
Deutsch Wagram 001810
Napoléon Antiques, Montréal . . 064447
Napóleon-ház, Városi Muvészeti
Múzeum, Győr 023664
Napoleonic Society of America,
Clearwater 047794
Napoleonmuseum Thurgau,
Schloss und Park Arenenberg,
Salenstein 041916
Napoli Capitale, Napoli 080904
Napoli Nobolissima, Roma . . . 138848
Napoli, Tommaso, Milano 080625
Napotnikova Galerija, Soštanj . 038619
Naprawa Szkla Antycznego,
Warszawa 132893
Náprstek Museum of Asian,
African and American Cultures,
Liběchov 009665
Náprstkovo Muzeum Asijských,
Afrických a Amerických Kultur,
Národní muzeum, Praha 009822
Nara City Museum of Calligraphy in
Honor of Kason Sugioka, Nara 029147
Nara City Museum of Photography,
Nara 029148
Nara-kenritsu Bijutsukan, Nara . 029149
Nara-kenritsu Kashihara Kokogaku
Kenkyujo Fuzoku Hakubutsukan,
Kashihara 028792
Nara-kenritsu Man'yo Bunkakan,
Asuka 028495
Nara-kenritsu Mingeikan, Yamato-
koriyama 029704
Nara Koeki, Nara 082073
Nara-kokuritsu Hakubutsukan,
Nara 029150

Naracoorte Art Gallery,
Naracoorte 001338
Naramata Heritage Museum Society,
Naramata 006455
Naranjestan Museum, Shiraz
University, Shiraz 055852
Naranjo, Albuquerque 119496
Naranjo Fernandez, Adolfo Leon,
Medellín 102454
Narberth Museum, Narberth . . . 045313
Il Narciso, Roma 110836
Narcisos, Los Angeles 135735
Narcissa Prentiss House,
Prattsburgh 052350
Narcisse, Surry Hills 062266
Nardi, Francesco Neri, Venezia . 081863
Nardin, Damascus 088045
Nardo, Michela, Brescia 131041
Narducci, Largs 089748
Narducci, Saltcoats 091229
Naregatsi Art Institute, Jerewan 058181,
098604
Narek, Queanbeyan 099603
Narek, Tanja 099808
Narembeen Historical Museum,
Narembeen 001339
Naritasan Calligraphy Museum,
Narita 029154
Naritasan Reikokan and Reikoden
Museum, Narita 029155
Narkose Schmerztherapie Museum,
Potsdam 021192
Narlıkuyu Mozaik Müzesi, Silifke 043058
Naroden Muzej, Gevgelija 030611
Naroden Muzej, Ohrid 030622
Naroden Muzej, Sveti Nikole . . 030641
Naroden Muzej, Veles 030643
Naroden Muzej D-r Nikola Nezlobinski,
Struga 030637
Naroden Muzej Kumanovo,
Kumanovo 030617
Narodna Galerija, Ljubljana . . . 038552
Narodna Galerija Čifte Amam,
Skopje 030632
Národní Galerie v Praze – Palác
Kinských, Praha 009823
Národní Kulturní Památka Vyšehrad,
Praha 009824
Národní Kulturní Památník – Valy,
Mikulčice 009694
Narodni Muzej, Bački Petrovac . 038125
Narodni Muzej, Bela Crkva . . . 038127
Narodni Muzej, Kragujevac . . . 038203
Narodni Muzej, Leskovac 038208
Narodni Muzej, Pančevo 038222
Narodni Muzej, Požarevac 038226
Narodni Muzej, Smederevska
Palanka 038239
Narodni Muzej, Štip 030636
Narodni Muzej, Užice 038252
Narodni Muzej, Vranje 038232
Narodni Muzej, Zadar 009022
Narodni Muzej, Zaječar 038258
Narodni Muzej, Zrenjanin 038230
Narodni Muzej Čačak, Čačak . . 038188
Narodni Muzej Crne Gore, Cetinje 031753
Narodni Muzej Kikinda, Kikinda . 038196
Narodni Muzej Kraljevo, Kraljevo 038205
Narodni Muzej Labin, Labin . . . 008945
Narodni Muzej Niš, Niš 038212
Narodni Muzej Novi Vinodolski, Novi
Vinodolski 008956
Narodni Muzej Ras, Novi Pazar . 038213
Narodni Muzej Sabac, Sabac . . 038232
Narodni Muzej Slovenije,
Ljubljana 038210
Narodni Muzej Toplice, Prokuplje 038230
Narodni Muzej u Beogradu,
Beograd 038176
Narodni Muzej Valjevo, Valjevo . 038254
Národní Muzeum, Praha . . 009825
Národní Technické Muzeum,
Praha 009826
Národní Zemědělské Muzeum,

Praha 009827
Národní Zemědělské Muzeum, Úsek
Valtice, Valtice 009941
Narodnyj Geologičeskij Muzej
Volgageologija, Nižnij Novgorod 037560
Narodnyj Istoriko-kraevedčeskij Muzej
Tatarska, Novosibirskij oblastnoj
kraevedčeskij muzej, Tatarsk . 037932
Národopisná Expozícia SARM
Bardejovské Kúpele, Bardejovské
Kúpele 038291
Národopisná Pobočka, Městské
Muzeum v Železném Brodě, Železný
Brod 009980
Národopisné Muzeum, Nejdek . 009713
Národopisné Múzeum, Smižany 038448
Národopisné Múzeum Liptova,
Liptovský Hrádok 038380
Národopisné Muzeum Plzeňska,
Západočeské Muzeum, Plzeň . 009760
Národopisné Muzeum Třebíz,
Třebíz 009920
Národopisné Oddéleni, Historické
Muzeum, Národní Muzeum v Praze,
Praha 009828
Narok Museum, Narok 029869
Narooma Lighthouse Museum,
Narooma 001340
Narosny, Anne, Cleveland 092957
Narovčatskij Rajonnyj Kraevedčeskij
Muzej, Penzenskij gosudarstvennyj
obedinennyj kraevedčeskij muzej,
Narovčat 037540
Narrabri Antiques and Art,
Narrabri 061944
Narrabri Old Gaol Museum,
Narrabri 001341
Narragansett Historical Society
Museum, Templeton 053711
Narrenmuseum, Gengenbach . . 018638
Narrenmuseum, Rheinfelden . . . 021398
Das Narrenschiff, Amsterdam . . 143053
Das Narrenschiff, Chur 144061
Narrenschopf, Fastnachtsmuseum,
Bad Dürrheim 016755
Dat Narrenschyp, Vinkeveen . . 143256
Narrogin Gallery, Narrogin 001344
Narrogin Old Courthouse Museum,
Narrogin 001345
Narrow Gauge Railway Museum,
Tywyn 046013
Narryna Heritage Museum, Battery
Point 000831
Narsaq Museum, Narsaq 023333
Le Narthex, Paris 071705
Narukawa Bijutsukan, Hakone . . 028594
Narumi, San Francisco 097294
Naruyama, Tokyo 111480
Narva Museum, Narva 010642
Narvik Auksjonsforretning, Narvik 126362
Narwhal Inuit Art Gallery, London 045055,
118472
Naš Realizm, Moskva 114285
NASA Glenn Research Center's Visitor
Center, Cleveland 047819
Nasan, Seoul 082262
Nascimento, Marcia, Belo
Horizonte 100897
Do Nascimento, Toninho José,
Brasilia 063936
Nasenberg, Beate, Sprockhövel . 108658
Nasere, Reza, Los Angeles . . . 121563
Nash, E. William, Jacksonville . 094016,
135660
Nash, Garry, Auckland 112797
Nash, John, Ledbury . . 089764, 134532
Nash, Paul, Cirencester 088919
Nashan, Manama 063155
Al-Nashef, Mohammed Said,
Damascus 088046
Nasher Museum of Art at Duke
University, Durham 048410
Nash's House and New Place,
Shakespeare Birthplace Trust,

Stratford-upon-Avon 045905
Nashua Center for Arts, Nashua 051169
Nashville Gold Market, Nashville . 095018
Nasi & Co. SAS, Modena 080812
Nasionale Afrikaanse Letterkundige
Museum en Navorsingsentrum,
Bloemfontein 038660
Nasiotis Sotiris, G., Athinai 142539
Al-Nasir Sons, Abdullatif S., Al-
Khobar 114386
Nasiri, A., Los Angeles 094416, 135736
Nasiriya Museum, Nasiriya 024811
Nasjonalgalleriet, Nasjonalmuseet
for Kunst, Arkitektur og Design,
Oslo 033845
Nasjonalmuseet for Kunst, Arkitektur
og Design, Oslo 033846
Naske, S., Wolfenbüttel 108958
Nasledie, Kazan 085151
Naso, R.C., Cleveland 120445
Nass, Fulda 075930
Naß, Andreas, Waldems 078834
Naß, Dr. Markus, Berlin 074887
Nass, Frank-Dietmar, Gersfeld . . 075987
Al-Nassar, Al-Khobar 085294
Nassau County Museum of Art,
Roslyn 052685
Nassau Public Library and Museum,
Nassau 003207
Nassau, Lillian, New York 095603,
122910
Nassauer, Armin, Siegen 142362
Nasser, Gießen 075995
Nasser, Awao, Wien . . 063036, 127956
Nasser, Marhoon, Al-Khobar . . . 126481
Nassini, Kazim, Lampertheim . . . 107782
Nassler, Neuburg an der Donau . 108239
Nast à Paris, Paris 105123
Nast, Hartmund, Elmlohe 129787
Nasta, Helsinki 066346
Nastolan Kotiseutunuseo,
Nastola 011182
Nastran, O., Kempfeld 076747
Nasu Classic Car Museum, Nasu 029162
Nasu Orugoru Hakubutsukan,
Nasu 029163
Nasu Queen's Museum, Nasu . . 029164
Nasu Warehouse – a contemporary
art field, Roentgenwerke AG,
Tochigi 111394
Nasze Atelier, Kraków 113620
Natacha, Saint-Denis 085068
Natadera Fumon-kaku Homotsukan,
Komatsu 028881
Natal Museum, Pietermaritzburg 038801
Natale Labia Museum, Iziko Museum,
Muizenberg 038787
Natale, Antonietta, Napoli 080905
Natale, Sonia, Roma 142828
Natali, Roma 081473, 131990
Natalie and James Thompson Gallery,
San Jose State University, San
Jose 053040
Nat'Antic, Menton 070062
Natchez Trace Parkway, Tupelo . 053854
Nategh, Basel 087251
Natesan, Bangalore 079356
Nathan Arts d'Asie, Saint-Ouen . 073161
Nathan Denison House, Forty
Fort 048910
Nathan Manilow Sculpture Park,
University Park 053896
Nathan, Ann, Chicago 120212
Nathan, Dr. Johannes, Zürich . . 116840
Nathan, John, Exeter . 089234, 134377
Nathanaelle, Neffe, Honfleur . . . 103839
Nathaniel W. Faison Home and
Museum, La Grange 050037
Nathaniel, Byron, Dallas 093183
Nation Museum, Beijing 007635
Nationaal Ambulance- en Eerste
Hulpmuseum, Kampen 032536
Nationaal Archief, Den Haag . . . 032191
Het Nationaal Autominiaturen Museum,

Asperen 032032
Nationaal Baggermuseum,
Sliedrecht 032881
Nationaal Beiaardmuseum, National
Carillon Museum, Asten 032039
Nationaal Bevrijdingsmuseum
1944–1945, Groesbeek 032374
Nationaal Borstelmuseum,
Izegem 003734
Nationaal Brandweermuseum,
Hellevoetsluis 032458
Nationaal Brilmuseum Amsterdam,
Amsterdam 031969
Nationaal Coöperatie Museum,
Schiedam 032855
Nationaal Fietsmuseum Velorama,
Nijmegen 032689
Het Nationaal Gevangenismuseum,
Veenhuizen 032975
Nationaal Glasmuseum, Leerdam 032568
Nationaal Historisch Orgelmuseum
Elburg, Elburg 032296
Nationaal Houtambachtenmuseum De
Wimpe, Herenthout 003695
Nationaal Jenevermuseum,
Hasselt 003683
Nationaal Landschapskundig Museum,
Dordrecht 032250
Nationaal Likeur- en
Frisdrankenmuseum Isidorus Jonkers,
Hilvarenbeek 032484
Nationaal Luchtvaart-Themapark
Aviodome, Schiphol 032860
Nationaal Modelspoor Museum,
Sneek 032890
Nationaal Monument Kamp Vught,
Vught 033014
Nationaal Museum en Archief
van Douane en Accijnzen,
Antwerpen 003303
Nationaal Museum Historisch
Landbouwtechniek, Wageningen 033019
Nationaal Museum van de Speelkaart,
Turnhout 004098
Nationaal Museum van Speelklok tot
Pierement, Utrecht 032951
Nationaal Museum Verpleging en
Verzorging, Zetten 033096
Nationaal Natuurhistorisch Museum
Naturalis, Leiden 032582
Nationaal Oorlogs- en Verzetsmuseum,
Overloon 032760
Nationaal Reddingmuseum Dorus
Rijkers, Den Helder 032209
Nationaal Rijtuigmuseum, Leek . 032564
Nationaal Ruimtevaart Museum,
Lelystad 032591
Nationaal Schaakmuseum, Den
Haag 032192
Nationaal Scheepvaartmuseum,
Antwerpen 003304
Nationaal Schoeiselmuseum,
Izegem 003735
Nationaal Schoolmuseum,
Rotterdam 032854
Nationaal Scoutsmuseum,
Leuven 003785
Nationaal Sleepvaart Museum,
Maassluis 032620
Nationaal Smalspoormuseum –
Stoomtrein Valkenburgse Meer,
Valkenburg, Zuid-Holland 032967
Nationaal Tabaksmuseum,
Tabaksmuseum, Wervik 004146
Nationaal Tinnen Figuren Museum,
Ommen 032716
Nationaal Vakbondsmuseum De
Burcht, Amsterdam 031970
Nationaal Veeteelt Museum, Beers bij
Cuyk 032053
Nationaal Visserijmuseum van
Oostduinkerke, Oostduinkerke . 003923
Nationaal Vlas-, Kant- en
Linnenmuseum, Kortrijk 003758

Nationaal Vlechtmuseum,
Noordwolde 032702
Nationaal Volkssportmuseum voor
de Vinkensport, Vinkenmuseum,
Harelbeke 003675
Nationaal Wielermuseum,
Roeselare 003975
National 1798 Visitor Centre,
Enniscorthy 024921
National Aboriginal Cultural Centre,
Harbour 061595
National Academy Museum, New
York 051444
National Academy of Arts, Seoul 056103
National Afro-American Museum and
Cultural Center, Wilberforce . . . 054402
The National Agricultural Center and
Hall of Fame, Bonner Springs . 047084
National Air and Space Museum,
Smithsonian Institution,
Washington 054138
National Alliance for Media Arts &
Culture, San Francisco 060609
National Antique and Art Dealers
Association of America, New
York 060610
National Antique Centre,
Mornington 061887
National Apple Museum,
Biglerville 046984
National Archives of Zambia,
Lusaka 054840
National Army Museum, London 045056
National Army Museum Sandhurst
Departments, Sandhurst,
Berkshire 045732
National Army Museum, Sandhurst
Outstation, Camberley 043833
National Art Association of Denver,
Denver 060611
National Art Connection,
Minneapolis 122015
National Art Education Association,
Reston 060612
National Art Gallery, Bhaktapur . 031868
National Art Gallery, Colombo . . 040478
National Art Gallery, Dhaka 003225
National Art Gallery, Islamabad . 034117
National Art Gallery, Kabul 000004
National Art Gallery of Armenia –
Sisian Branch, Sisian 000712
National Art Gallery of Namibia,
Windhoek 031862
National Art League, Douglaston 060613
National Art Museum of Sport,
Indianapolis 049696
National Art School, Darlinghurst 054903
National Artists Equity Association,
Pacific Grove 060614
National Arts, Al-Ain 117170
National Arts Club, New York . . 051445
National Arts Coalition, Newtown 059820
National Arts Education Archive,
University od Leeds, Wakefield 046030
National Arts Strategies,
Washington 060615
National Assembly of State Arts
Agencies, Washington 060616
National Association for the Visual
Arts, Woolloomooloo 058209
National Association of Artists
Organizations – NAAO,
Birmingham 060617
National Association of Decorative and
Fine Arts Societies, London . . . 060192
National Association of Fine Arts,
Kathmandu 059589
National Association of Miniature
Enthusiasts, Carmel 047436
National Association of Schools of Art
and Design, Reston 060618
National Association of Women Artists,
New York 060619, 139449
National Association of Women Artists

Gallery, New York 051446
National Atomic Museum,
Albuquerque 046366
National Automobile Museum of
Tasmania, Launceston 001213
National Automobile Museum, Harrah
Collection, Reno 052508
National Automotive and Truck
Museum of the United States,
Auburn 046649
National Baseball Hall of Fame and
Museum, Cooperstown 047975
National Birendra Art Gallery,
Kathmandu 031873
National Border Patrol Museum and
Memorial Library, El Paso 048514
National Broadcasting Museum,
Minsiong 042401
National Bronze Art Museum,
Lalitpur 031879
National Building Museum,
Washington 054139
National Campaign for the Arts,
London 060193
National Canal Museum, Easton 048457
National Capital Trolley Museum,
Silver Spring 053342
National Capitol Winter Antiques Show,
Washington 098479
National Cartoonists Society, Winter
Park 060620
National Center for the American
Revolution, Valley Forge 053934
National Central University Art Center,
Jhongli 042345
National Cheng Kung University
Museum, Tainan 042488
National Chengchi University Museum
of Ethnology, Taipei 042536
National Chiang Kai-shek Memorial
Hall, Taipei 042537
National Chiao Tung University Art
Center, Hsinchu 042336
National Children's Museum,
Delhi 024227
National Children's Museum,
Washington 054140
National Chung Hsing University Art
Center, Taichung 042475
National Civil Rights Museum,
Memphis 050787
National Coal Mining Museum for
England, Wakefield 046031
National Coin and Stamp Company,
Omaha 096053
National College of Art and Design,
Dublin 055862
National College of Arts, Lahore 056216
National Commission for Museums
and Monuments, Abuja 059651
National Conference of Artists,
Detroit 060621
National Coracle Centre, Cenarth 043899
National Corvette Museum, Bowling
Green 047158
National Council of Science Museums,
Kolkata 059368
National Council on Education for the
Ceramic Arts, Erie 060622
National Cowboy and Western
Heritage Museum, Oklahoma
City 051732
National Cowgirl Museum and Hall of
Fame, Fort Worth 048907
National Cryptologic Museum, Fort
Meade 048843
National Cultural History Museum,
Northern Flagship Institute,
Tshwane 038853
National Culture and Arts Foundation,
Taipei 059994
National Cycle Collection, Llandrindod
Wells 044875
National Czech and Slovak Museum,

Cedar Rapids 047490
National D-Day Museum, New
  Orleans 051299
National Dairy Museum, Ashurst,
  Southampton 043448
National Dinosaur Museum,
  Nicholls 001359
National Doukhobour Heritage Village,
  Verigin 007312
National Dr. Sun Yat-sen Memorial
  Hall, Taipei 042538
National Dragonfly Biomuseum,
  Ashton, Oundle 043444
National Endowment for the Arts,
  Washington 060623
National English Literary Museum,
  Grahamstown 038725
National Exhibits by Blind Artists,
  Philadelphia 052073
National Fairground Archive, University
  of Sheffield, Sheffield 045775
National Farm Toy Museum,
  Dyersville 048416
National Fine Arts and Antiques Fair,
  Birmingham 098391
National Firearms Museum,
  Fairfax 048666
National Fishing Heritage Centre,
  Great Grimsby 044472
National Folk Museum, Seoul .. 030028
National Folk Museum of Lesbos,
  Mytilini 023188
National Football Museum,
  Canton 047412
National Football Museum,
  Preston 045567
National Fresh Water Fishing Hall of
  Fame, Hayward 049422
National Frontier Trails Center,
  Independence 049670
National Gallery, Bangkok 042669
National Gallery, London 045057
National Gallery Bulawayo,
  Bulawayo 054846
National Gallery Company,
  London 138231
National Gallery of Armenia,
  Yerevan 000743
National Gallery of Armenia – Alaverdi
  Branch, Alaverdi 000689
National Gallery of Art,
  Washington 054141
National Gallery of Australia,
  Canberra 000937
National Gallery of Canada,
  Ottawa 006594
National Gallery of Corfu,
  Skripero 023268
National Gallery of Ireland Publications,
  Dublin 137605
National Gallery of Ireland, Gailearaí
  Náisiúnta na hÉireann, Dublin . 024900
National Gallery of Jamaica,
  Kingston 028449
National Gallery of Modern Art,
  Delhi 024228
National Gallery of Modern Art,
  Mumbai 024406
National Gallery of Scotland,
  Edinburgh 044255
National Gallery of Scotland,
  Publishing Department,
  Edinburgh 138157
National Gallery of Victoria,
  Melbourne 001279
National Gallery of Zimbabwe,
  Harare 054851
National Gallery Shops, London . 138232,
                    144654
The National Gallery Technical Bulletin,
  London 139211
National Gandhi Museum and Library,
  Delhi 024229
The National Gas Museum,

Leicester 044798
National Glass Centre,
  Sunderland 045921
National Great Rivers Museum,
  Alton 046424
National Guild of Community Schools
  of the Arts, New York 060624
National Hall of Fame for Famous
  American Indians, Anadarko .. 046464
National Handicrafts, Kampala . 088155
National Hearing Aid Museum,
  London 045058
National Heisey Glass Museum,
  Newark 051504
National Heritage Museum, Van
  Gorden-Williams Library and
  Archives, Lexington 050256
National Heritage, The Museums
  Action Movement, Hawkhurst . 060194
National Hispanic Art Museum,
  Albuquerque 046367
National Historic Oregon Trail
  Interpretive Center, Baker City . 046716
National Historical Museum,
  Athinai 023019
National Horseracing Museum,
  Newmarket 045360
National Infantry Museum, Fort
  Benning 048780
National Inventors Hall of Fame,
  Akron 046330
National Liberty Museum,
  Philadelphia 052074
National Library, Valletta 030752
National Library Museum,
  Thimphu 004202
National McKinley Birthplace Memorial,
  Niles 051564
National Magazine Company,
  Sale 138281
National Maritime Museum,
  Alexandria 010521
National Maritime Museum, Galle 040483
National Maritime Museum,
  London 045059
National Maritime Museum,
  Mokpo 029972
National Maritime Museum Cornwell,
  Falmouth 044326
National Maritime Museum of Ireland,
  Dún Laoghaire 024916
National Maritime Museum, Affiliated
  to Haifa Museums, Haifa 025055
National Medal of Honor Museum of
  Military History, Chattanooga . 047596
National Memorabilia, Miami ... 094777
National Memorial of the Fortress of
  Breendonk, Willebroek 004152
National Mining Hall of Fame and
  Museum, Leadville 050201
National Mining Museum,
  Kwekwe 054854
National Model Aviation Museum,
  Academie of Model Aeronautics,
  Muncie 051116
National Motor Museum,
  Beaulieu 043540
National Motor Museum,
  Birdwood 000861
National Motor Racing Museum,
  Bathurst 000830
National Motorcycle Museum,
  Bickenhill 043590
National Motorcycle Museum,
  Nabiac 001335
National Museum, Aden 054824
National Museum, Bangkok ... 042670
National Museum, Luang
  Prabang 030087
National Museum, Malé 030733
National Museum, Monrovia .. 030290
National Museum, Nagara
  Pathama 042699
National Museum, Oshogbo ... 033441

National Museum, Ras-Al-
  Khaimah 043337
National Museum, Riyadh 038111
National Museum, Sanaa 054831
National Museum, Tehrän 024776
National Museum, Victoria ... 038262
National Museum and Art Gallery
  of Trinidad and Tobago, Port-of-
  Spain 042733
National Museum at Ruwi, Ruwi 034105
National Museum Bauchi, Bauchi 033417
National Museum Benin, Benin
  City 033418
National Museum Bloemfontein,
  Bloemfontein 038661
National Museum Directors'
  Conference, London 060195
National Museum Esie, Esie ... 033421
National Museum for Egyptian
  Civilization, Cairo 010554
National Museum for Trading,
  Riyadh 038112
National Museum Gidan Makama,
  Kano, Kano 033432
National Museum Ibadan, Ibadan 033423
National Museum Ile-Ife, Ife ... 033425
National Museum-Institute of
  Architecture, Yerevan 000744
National Museum Kabul, Kabul . 000005
National Museum Kaduna,
  Kaduna 033431
National Museum Makurdi,
  Makurdi 033435
National Museum Minna, Minna 033436
National Museum Nairobi,
  Nairobi 029868
National Museum of African
  Art, Smithsonian Institution,
  Washington 054142
National Museum of American Art,
  Washington 054143
National Museum of American
  History, Smithsonian Institution,
  Washington 054144
National Museum of American
  Illustration, Newport 051536
National Museum of American Jewish
  History, Philadelphia 052075
National Museum of American Jewish
  Military History, Washington .. 054145
National Museum of Archaeology,
  Valletta 030753
National Museum of Art and History,
  Yangon 031834
National Museum of Australia,
  Canberra 000938
National Museum of Australian Pottery,
  Wodonga 001651
National Museum of Bhutan,
  Paro 004200
National Museum of Calabar,
  Calabar 033419
National Museum of Catholic Art and
  History, New York 051447
National Museum of Ceramic Art and
  Glass, Baltimore 046765
National Museum of Cham Sculpture,
  Da Nang 054785
National Museum of Chinese History,
  Beijing 007636
National Museum of Civil War
  Medicine, Frederick 048947
National Museum of Colonial History,
  Aba 033410
The National Museum of
  Communications, Irving 049725
National Museum of Contemporary Art,
  Athinai 023020
National Museum of Contemporary Art,
  Gwacheon 029939
National Museum of Damascus,
  Damascus 042258
National Museum of Dance and Hall
  of Fame, Saratoga Springs ... 053156

National Museum of Enugu,
  Enugu 033420
National Museum of Ethiopia, Addis
  Ababa 010724
National Museum of Ethnology, Lok
  Virsa-National Institute of Folk and
  Traditional Heritage, Islamabad 034118
National Museum of Fine Arts,
  Valletta 030754
National Museum of Flight, National
  Museums of Scotland, North
  Berwick 045385
National Museum of Funeral History,
  Houston 049591
National Museum of Gardening,
  Helston 044562
National Museum of Ghana,
  Accra 022925
National Museum of Health and
  Medicine, Armed Forces Institute of
  Pathology, Washington 054146
National Museum of History,
  Taipei 042539
National Museum of Ilorin, Ilorin 033427
National Museum of India, Delhi 024230
National Museum of Industrial History,
  Bethlehem 046963
National Museum of Kashan,
  Kashan 024705
National Museum of Korea,
  Seoul 030029
National Museum of Man,
  Bhopal 024171
National Museum of Mandalay,
  Mandalay 031817
National Museum of Marine Biology,
  Checheng 042285
National Museum of Marine Science
  and Technology, Keelung ... 042373
National Museum of Modern Chinese
  Literature, Beijing 007637
National Museum of Mongolian History,
  Ulaanbaatar 031738
National Museum of Music,
  Beijing 007638
National Museum of Namibia,
  Windhoek 031863
National Museum of Natural History,
  Colombo 040479
National Museum of Natural History,
  Delhi 024231
National Museum of Natural History,
  Mdina 030741
National Museum of Natural History,
  Washington 054147
National Museum of Natural Science,
  Taichung 042476
National Museum of Naval Aviation,
  Pensacola 051981
National Museum of Nepal,
  Kathmandu 031874
National Museum of Pakistan,
  Karachi 034123
National Museum of Photography, Film
  and Television, Bradford 043696
National Museum of Polo and Hall of
  Fame, Lake Worth 050092
National Museum of Prehistory,
  Taitung 042567
National Museum of Racing and Hall
  of Fame, Saratoga Springs .. 053157
National Museum of Roller Skating,
  Lincoln 050297
National Museum of Science and
  Technology, Lahore 034134
National Museum of Science, Planning
  and Technology, Haifa 025056
National Museum of Singapore,
  Singapore 038522
National Museum of Taiwanese
  History, Tainan 042489
National Museum of Taiwanese
  Literature, Tainan 042490
National Museum of the American

Indian, Washington 054148
National Museum of the American
Indian, Smithsonian Institution, New
York 051448
National Museum of the Morgan
Horse, Shelburne 053299
National Museum of the Ogun State,
Abeokuta 033411
National Museum of the Ondo State,
Akure 033414
National Museum of Wildlife Art,
Jackson 049764
National Museum of Women in the
Arts, Washington 054149
National Museum of Woodcarving,
Custer 048100
National Museum Oron, Oron .. 033438
National Museum Resources
Exhibition, Birmingham 098392
National Museum Uyo, Uyo 033447
National Museum, Abuja, Abuja 033413
National Museum, Imo State,
Owerri 033442
National Museum, Lagos, Onikan 033437
National Museums and Galleries of
Northern Ireland, Belfast 060196
National Museums and Monuments of
Zimbabwe, Harare 060831
National Museums of Scotland,
Edinburgh 044256
National Music Museum,
Bloemfontein 038662
National New York Central Railroad
Museum, Elkhart 048534
National Oil and Acrylic Painters
Society, Osage Beach 060625
National Ornamental Metal Museum,
Memphis 050788
National Packard Museum,
Warren 054066
National Palace Museum, Taipei 042540
The National Palace Museum Bulletin,
Taipei 139105
The National Palace Museum Monthly
of Chinese Art, Taipei 139106
National Palace Museum Newsletter &
Gallery Guide, Taipei 139107
National Palace Museum Research
Quarterly, Taipei 139108
National Philatelic Museum,
Delhi 024232
National Pioneer Women's Hall of
Fame, Alice Springs 000790
National Police Museum and Ancient
Police Museum, Qasr al-Gawharah,
Cairo 010555
National Portrait Gallery,
Canberra 000939
National Portrait Gallery, London 045060
National Portrait Gallery,
Montacute 045289
National Portrait Gallery,
Washington 054150
National Postal Museum, Smithsonian
Institution, Washington 054151
National Presidential Wax Museum,
Keystone 049965
National Print Museum, Dublin . 024901
National Rail Museum, Delhi ... 024233
National Railroad Museum, Green
Bay 049228
National Railroad Museum and Hall of
Fame, Hamlet 049330
National Railway Museum, Port
Adelaide 001413
National Railway Museum, The
National Museum of Science and
Industry, York 046226
National Railways of Zimbabwe
Museum, Raylton 054858
National Resistance Museum,
Karpenissi 023115
National Road/Zane Grey Museum,
Norwich 051653

National Roman Legionary Museum,
Caerleon 043822
National Route 66 Museum and Old
Town Museum, Elk City 048528
National Science and Technology
Museum, Kaohsiung 042365
National Science Centre, Delhi . 024234
National Science Museum of Korea,
Daejeon 029921
National Scout Museum of New
Zealand, Kaiapoi 033231
National Scouting Museum,
Irving 049726
National Sculpture Society, New
York 051449
National Slate Museum,
Llanberis 044873
National Soaring Museum, Elmira 048567
The National Soccer Hall of Fame,
Oneonta 051777
National Society for Education in Art
and Design, Corsham 060197
National Society of Mural Painters,
New York 060626
National Society of Mural Painters.
Newsletter, New York 139450
National Society of Painters in Casein
& Acrylic, Whitehall 060627
National Society of Painters, Sculptors
and Printmakers, London ... 060198
National Society of the Children of
the American Revolution Museum,
Washington 054152
National Society of Tole and
Decorative Painters, Wichita .. 060628
National Softball Hall of Fame and
Museum Complex, Oklahoma
City 051733
National Space Centre, Leicester 044799
National Sportsmuseum, Patiala 024431
National Sprint Car Hall of Fame and
Museum, Knoxville 050003
National Steinbeck Center,
Salinas 052900
National Taiwan Museum, Taipei 042541
National Taiwan Museum of Fine Arts,
Taichung 042477
National Taiwan Science Education
Center, Taipei 042542
National Taiwan University of Arts,
Banchiao 056451
The National Time Museum,
Chicago 047683
The National Toy Train Museum,
Strasburg 053563
National Tramway Museum,
Matlock 045235
National Transport Museum,
Dublin 024902
National Trust for Historic Preservation,
Washington 054153
National Trust Museum, Streaky
Bay 001524
National Tsing Hua University Arts
Center, Hsinchu 042337
National Union of Artists of Ukraine,
Kyïv 060019
National Vietnam Veterans Art
Museum, Newhaven 001358
National Vintage, Los Angeles .. 094417
National Vintage Wireless and
Television Museum, Harwich . 044529
National Wall Art Gallery, Tampa 124666
National Wallace Monument,
Stirling 045872
National War Museum of Scotland,
National Museums of Scotland,
Edinburgh 044257
National War Museum of Umahia,
Umauhia 033446
National Warplane Museum,
Horseheads 049560
National Watch and Clock Museum,
Columbia 047904

National Watercolor Society, San
Pedro 060629
National Waterfront Museum,
Maritime and Industrial Museum,
Swansea 045936
National Waterways Museum,
Gloucester 044434
National Waterways Museum –
Ellesmere Port, Ellesmere Port 044285
National Women's Hall of Fame,
Seneca Falls 053270
National Wood Working Museum,
Bhaktapur 031869
National Wool Museum, Geelong 001095
National Wool Museum, National
Museum Wales, Llandysul 044880
National Wrestling Hall of Fame,
Stillwater 053532
National Yiddish Book Center,
Amherst 046453
Det Nationale Fotomuseum,
København 010186
La Nationale Galerie d'Art Alyssa, Le
Bardo 042735
Nationalgalerie Friedrichswerdersche
Kirche – Schinkelmuseum, Staatliche
Museen zu Berlin – Stiftung
Preußischer Kulturbesitz, Berlin 017259
Det Nationalhistoriske Museum på
Frederiksborg, Hillerød 010119
Nationalmuseet, København ... 010187
Nationalmuseum Stockholm,
Stockholm 041024
Nationalpark Haus,
Berchtesgaden 017040
Nationalpark-Haus, Museum
Butjadingen, Butjadingen 017705
Nationalparkhaus Altenau-Torfhaus,
Altenau 016528
Nationalparkhaus Sankt Andreasberg,
Sankt Andreasberg 021587
Nationalparkzentrum, Schweizerischer
Nationalpark, Zernez 042177
Nations Art Gallery, Pittsburgh . 123566
Native American Art, Tulsa .. 124809
Native American Exhibit, National
Kateri Shrine, Fonda 048773
Native American Heritage Museum at
Highland Mission, Highland ... 049468
Native American Images, Austin 119685
Native American Museum, Terre
Haute 053717
Native American Resource Center,
Pembroke 051969
Native American Trading, Denver 120751
Native Art of Australia, Alice
Springs 098636
Native Arts & Crafts, Rochedale 062157
Native Books and Beautiful Things,
Honolulu 093564, 145156
Native Cultural Arts Museum,
Grouard 005930
Native Investment, Vancouver .. 101814
Native Nations Art, Oklahoma
City 123290
Nativelle, Françoise, Caen ... 067576
Nativelle, Françoise, Paris 071706
Nativi, Berto, Firenze 080174
Natris, Wim de, Nijmegen 112610
Natsagdorj Museum, Ulaanbaatar 031739
Natsara – La Maison des Hommes,
Paris 071707
Natsume, Saitama 111371
Náttúrufræðaistofnun Íslands,
Akureyrarsetur, Akureyri 024065
Natturufræðistofa Kopavogs,
Kópavogur 024088
Náttúrufræðilistofnun Íslands,
Reykjavik 024103
Náttúrugripasafnið í Neskaupstad,
Neskaupstadur 024089
Natur- und Heimatmuseum Vellberg,
Vellberg 022313
Natur-Heimat-Museum,

Langenburg 019907
Natur-Museum Luzern, Luzern . 041710
Natur-Museum mit Bauerngarten,
Goldberg 018760
Natural Elements Craft Gallery,
Inveraray 118018
Natural History Museum, Bangor,
Gwynedd 043491
Natural History Museum, Belfast 043558
Natural History Museum,
Colchester 043991
Natural History Museum, Gorgan 024695
Natural History Museum, Ife .. 033426
Natural History Museum, Ilam .. 024698
Natural History Museum, Karachi 034124
Natural History Museum,
Kathmandu 031875
Natural History Museum,
Kingston 028450
Natural History Museum, Lahore 034135
Natural History Museum, Las
Cuevas 004178
Natural History Museum, London 045061
Natural History Museum, Medani 040499
Natural History Museum,
Nottingham 045416
Natural History Museum, Port
Louis 030794
Natural History Museum, Port
Sudan 040502
Natural History Museum,
Portsmouth 045556
Natural History Museum, Santa
Maria 053127
Natural History Museum, Tehrän 024777
Natural History Museum, Tring . 045998
Natural History Museum, Urmieh 024790
Natural History Museum, Yangon 031835
Natural History Museum of Amarousio,
Amarousio 022955
Natural History Museum of Bombay
Natural History Society, Mumbai 024407
Natural History Museum of Cephalonia
and Ithaca, Argostolion 022968
Natural History Museum of Elassona,
Elassona 023068
Natural History Museum of Los
Angeles County, Los Angeles . 050462
Natural History Museum of Oeta,
Ipati 023094
Natural History Museum of the
Aegean, Samos 023255
Natural History Museum of the
University of Basrah, Basrah .. 024807
Natural History Museum of Tilos,
Tilos 023301
Natural History Museum of Zimbabwe,
Bulawayo 054847
Natural History Museum, Bu-Ali
University, Hamedan 024697
Natural History Museum,
Chulalongkorn University,
Bangkok 042671
Natural History Museum, Department
of Antiquities, Tripoli 030298
Natural History Museum, Eastern New
Mexico University, Portales ... 052272
Natural History Museum, Iranshahr
High School, Yazd 024793
Natural History Museum, Khorram-
Abaad University, Khorram-
Abaad 024713
Natural History Museum, University of
Kansas, Lawrence 050186
Natural History Society and Folk
Museum, Bacup 043463
The Natural Science Center of
Greensboro, Greensboro ... 049256
Natural Science Collections Alliance,
Washington 060630
Natural Science Museum,
Marawi 034551
Natural Science Museum,
Victoria 030763

Natural Sciences Collections Association, Scunthorpe 060199
Natural Sciences Museum, Kibbutz Hulata 025135
Naturalia, Nembro 142769
Naturama, Svendborg 010335
Naturama Aargau, Aarau 041189
Nature by Design, Milwaukee . 121940
Nature in Art, Gloucester 044435
Nature Morte, Delhi 109335
Nature Museum at Grafton, Grafton 049171
Nature Park and Galleries, Jerusalem 025097
Nature Roots, Houston 121100
Nature-Study and Folk Art Museum of Loutra Almopias, Loutraki 023159
Naturens Hus, Stockholm 041025
Natures Image Photography, Montville 099365
Nature's Reflection, Houston . 093751
Natureum Niederelbe, Balje ... 016941
Naturhiscope, Oropesa del Mar . 039851
Naturhistorisches Museum, Admont 001686
Naturhistorisches Museum, Heiden 041589
Naturhistorisches Museum, Nürnberg 020872
Naturhistorisches Museum, Wien 003051
Naturhistorisches Museum Basel, Basel 041282
Naturhistorisches Museum der Burgergemeinde Bern, Bern . 041321
Naturhistorisches Museum Schloss Bertholdsburg, Schleusingen . 021670
Naturhistorisches Museum, Landessammlung für Naturkunde Rheinland-Pfalz, Mainz 020232
Naturhistorisches Museum, Städtische Museen Heilbronn, Heilbronn . 019162
Naturhistorisk Museum, Århus . 010008
Naturhistorisk museum, Universitetet i Oslo, Oslo 033847
Naturhistoriska Riksmuseet, Stockholm 041026
Naturhistoriske Samlinger, Bergen Museum, Bergen 033536
Naturkunde-Museum, Bielefeld . 017383
Naturkunde-Museum, Coburg .. 017753
Naturkunde-Museum Bamberg, Bamberg 016953
Naturkundemuseum, Dobersberg 001820
Naturkundemuseum, Glees 018712
Naturkundemuseum, Kassel ... 019544
Naturkundemuseum, Potsdam . 021193
Naturkundemuseum, Reutlingen 021384
Naturkundemuseum Erfurt, Erfurt 018249
Naturkundemuseum Freiberg, Freiberg, Sachsen 018472
Naturkundemuseum im Marstall, Paderborn 021066
Naturkundemuseum im Tierpark Hamm, Hamm, Westfalen 019032
Naturkundemuseum Leipzig, Leipzig 020011
Naturkundemuseum Ludwigslust, Ludwigslust 020143
Naturkundemuseum Niebüll, Niebüll 020787
Naturkundemuseum Ostbayern, Regensburg 021328
Naturkundemuseum, Kollegium Borromäus, Altdorf 041212
Naturkundliche Bildungsstätte Nordeifel, Simmerath 021867
Naturkundliche Sammlung, Altusried 016566
Naturkundliche Sammlung, Königsbrunn 019726
Naturkundliche Sammlung Oberhasli, Meiringen 041735
Naturkundliches Bildungszentrum Stadt Ulm, Ulm 022265

Naturkundliches Museum, Wiesenfelden 022644
Naturkundliches Museum in der Harmonie, Städtische Sammlungen Schweinfurt, Schweinfurt 021786
Naturkundliches Museum und Schulungsstätte "Alte Schmiede", Handeloh 019044
Naturmuseum, Sankt Gallen .. 041926
Naturmuseum des Kanton Thurgau, Frauenfeld 041492
Naturmuseum Lüneburg, Lüneburg 020183
Naturmuseum Olten, Olten 041824
Naturmuseum Solothurn, Solothurn 042006
Naturmuseum und Forschungsinstitut Senckenberg, Frankfurt am Main 018447
Naturmuseum Winterthur, Winterthur 042159
Naturmuseum, Kunstsammlungen und Museen Augsburg, Augsburg . 016664
Naturmuseum, Sammlung Schliefsteiner, Neuberg an der Mürz 002432
Naturparkhaus Stechlin, Stechlin 021970
Naturschau in der Zehntscheune, Zweigstelle Pfalzmuseum für Naturkunde, Thallichtenberg .. 022140
Naturschutzstation, Neschwitz . 020666
Naturschutzzentrum, Nettersheim 020669
Naturum Stendörren, Tystberga . 041095
Naturwissenschaftliche Sammlung, Stans 042023
Naturwissenschaftliche Sammlungen am Landesmuseum Joanneum, Abteilungen Zoologie, Botanik, Geologie und Paläontologie, Mineralogie, Graz 001984
Naturwissenschaftliche Sammlungen, Stiftung Stadtmuseum Berlin, Berlin 017260
Naturwissenschaftliche Sammlungen, Tiroler Landesmuseum Ferdinandeum, Innsbruck 002132
Naturwissenschaftliches Museum, Aschaffenburg 016622
Naturwissenschaftliches Museum der Stadt Flensburg, Flensburg ... 018373
Naturwissenschaftliches Museum Duisburg, Duisburg 018062
Naturzentrum Ternell, Forstzoologisches Museum, Eupen 003588
Natuur Historisch Museum en Heemkunde Centrum, Meerssen 032643
Natuurdiorama Holterberg, Holten 032498
Natuurhistorisch en Volkenkundig Museum, Oudenbosch 032750
Natuurhistorisch Museum Boekenbergpark, Deurne 003545
Natuurhistorisch Museum de Peel, Asten 032040
Natuurhistorisch Museum het Diorama, Nunspeet 032705
Natuurhistorisch Museum Maastricht, Maastricht 032626
Natuurhistorisch Museum Rotterdam, Rotterdam 032815
Natuurmuseum Ameland, Nes .. 032666
Natuurmuseum Brabant, Tilburg 032928
Natuurmuseum de Wielewaal, Lopik 032604
Natuurmuseum Dokkum, Dokkum 032236
Natuurmuseum E. Heimans, Zaandam 033079
Natuurmuseum Enschede, Enschede 032309
Natuurmuseum Groningen, Groningen 032382
Natuurmuseum het Drents-Friese

Woud, Wateren 033028
Natuurmuseum Mar en Klif, Oudemirdum 032749
Natuurmuseum Nijmegen, Nijmegen 032690
Natuurpunt Museum, Turnhout . 004099
Natuurwetenschappelijk Museum, Antwerpen 003305
Nau 55, Porto 085039
NAU Art Museum, Northern Arizona University, Flagstaff 048741
Naucelle-Lepeltier, Caen 067577
Nauck & Co, Albert, Köln 137314
Naučno-memorialnyj Muzej N.E. Žukovskogo, Moskva 037482
Naučnyj Morskoj Muzej, Atlantičeskij naučno-issledovatelskij institut morskogo rybnogo chozjajstva i okeanografii, Kaliningrad .. 037086
Naud, Daniel, Saint-Maixent-l'Ecole 072908
Naudet, Marguerite, Rilhac-Xaintrie 072467
Naudet, Michel, Limoges 069526
Naudin, Abbeville 066411, 103170
Naudin, Michel, Beaune 067030
Nauert, Dierk, Wien .. 063037, 125282
Naughton, Dún Laoghaire ... 142607
Naujasis Arsenalas, Lietuvos Nacionalinis Muziejus, Vilnius . 030536
Naujenes Novadpētniecības Muzejs, Naujene 030164
Naujieji Skliautai, Vilnius 111805
Nauka i Izkustvo, Sofia 136672
Naumann-Museum Köthen, Köthen 019748
Naumann, J., Könnern 076932
Naumann, Joachim, Jena 076645
Naumann, Ulrich, Cape Town .. 143703
Naumburg, Barbara, Frankfurt am Main 129852
Naumburger Domschatzgewölbe, Naumburg, Saale 020645
Naumertat, Peter, Dresden ... 129715
Naumkeag House, Stockbridge . 053536
Nauportus, Vrhnika 085626
Nausicaa, Centre National de la Mer, Boulogne-sur-Mer 012141
Nautelankosken Museo, Lieto .. 011113
Nautical Antique Centre, Weymouth 091814
Nautical Antiques, Norfolk 095869
Nautical Living, Kansas City ... 094086
Nautical Museum of Galaxidi, Galaxidi 023082
Nautical Museum, Manx National Heritage, Castletown 043891
Nauticals Limited, Dallas 093184
Nauticus, Hampton Roads Naval Museum, Norfolk 051579
Nautiko Mouseio Chiou, Chios .. 023047
Nautikon Mouseiontis Ellados, Piraeus 023224
Nautilus, Milwaukee 121941
Nautilus, Warszawa 084739
Nautilus Salon Antykwaryczny, Kraków 084589, 137895, 143481
Nauvoo Historical Society Museum, Nauvoo 051203
Nava-Hopi, Tucson 124759
Navajo Nation Museum, Window Rock 054469
Naval Academy Museum, Jinhae 029962
Naval and Maritime Museum, Ballina 000820
Naval Museum, Vancouver 007291
Naval Museum of Alberta, Calgary 005530
Naval Shipyard Museum, Portsmouth Museums, Portsmouth 052326
Naval Undersea Museum, Keyport 049961
Naval War College Museum, Newport 051537

Navan Centre, Armagh 043433
Navarra, Enrico, New York 122911
Navarra, Enrico, Paris . 105124, 105125, 105126, ... 136969
Navarra, G., Napoli 080906
Navarre-Duplantier, Hasparren . 068796
Navarrete Alvárez, Isaac, Jerez de la Frontera 133278
Navarro, Buenos Aires 060917
Navarro, Megève 070037
Navarro, Paris 129098
Navarro, Toronto 064709
Navarro County Historical Society Museum, Corsicana 048009
Navarro Miroirs Anciens, Paris . 071708
Navarro Santa Cruz, Amparo, Murcia 086315
Navarro Zavaleta, Julio, Bogotá . 128274
Navarro, Ceferino, Granada ... 115092
Navarro, Françoise, Attin 066770
Navarro, Leandro, Madrid .... 086227, ... 115264
Navas & Navas, Bogotá 102351
Navas, Enrique, Bogotá 102350
La Nave, Valencia 115511, 138005
La Nave, Vigo 086563
Nave Museum, Victoria 053982
Naveri, Marius, Beaumont-Monteux 067005
La Navicella, Milano 110389
La Navicella, Savona 110917
Il Navicello, Pisa 110667
Navier, Thierry, Azay-sur-Cher .. 066897
Navigator Contemporary Arts, Saint Just 119043
Naviglio, Milano 080626, 110390
Naviglio, Venezia 111145
Naviglio Piú, Milano .. 080627, 080628
Navikula Arts, Sankt-Peterburg . 114350
Naviliat, Oscar Luis, Buenos Aires 139590
Naville, Claude, Genève 126726
Navire, Brest 103463
Naviscope Alsace, Strasbourg .. 015888
Navona Antiquariato, Roma .... 081474, ... 131991
Navoni, Roberto, Brescia ... 079950, 079951, 131042, 131043
Navrátil, Miroslav, Praha 102710
The Navy Museum, Washington . 054154
Al-Nawras, Sharjah 117200
Naxos, Wellington 113078
Nayarit, San Salvador 066278
Naylor, Timothy, Chertsey 134248
Nayong Pilipino Museum, Philippine Museum of Ethnology, Pasay . 034566
Nayuma Museum, Limulunga .. 054835
Nazar, Fortaleza 140282
Nazarchuk, Regina 128197
Nazarri, Laura, Milano 080629
Naze, Michel, Doudeville 068397
Nazia Sultana, Brierfield 117432
Nazraeli Press, Portland 138442
NB Style, Toowoomba 062332
NBS-Múzeum Mincí a Medalí, Kremnica 038374
NC Gallery, Busan 111574
NCA Art Gallery, Lahore 113310
NCCL Galleries of Justice, Nottingham 045417
NE Galerie, Rotterdam 112653
Néa Moni Collection, Chios 023048
Neal Sons & Fletcher, Woodbridge 127099
Neal Auction Company, New Orleans 127336
Neal, David, Oxted 144775
Neale, Joseph, Erina Heights . 099003
Neales, Nottingham 127001
Neal's Gallery, Kansas City ... 121308
Nealway, Houston 093752
Neanderthal Museum, Mettmann 020395
Neary, C.F., Auckland 132694

Neary, Geoff, Huddersfield 089599, 134483
Nease, Joseph, Kansas City .. 121309
Neat Things, Toronto 064710
Neate, Gary C., Sudbury, Suffolk 091504
Neath Museum, Neath 045316
Nebbett, Andrew, London 090373
Nebehay, Christian M., Wien .. 063038, 140112
Nebehay, Dr. Stefan, Wien 063039, 140113
Nebel Verlag, Utting am Ammersee 137546
Nebelsieck, Gunther & Astrid, Berlin 074888
Nebelspalter-Verlag, Horn 138055
Nebenaigoching Heritage, Batchawana Bay 005393
Nebengaus, Berlin 129540
Nébra, Vic-en-Bigorre 105854
Nebraska Art Association, Lincoln 060631
Nebraska Conference United Methodist Historical Center, Lincoln 050298
Nebraska Museums Association, Aurora 060632
Nebraska Prairie Museum, Phelps County Historical Society, Holdrege 049503
Nebreda & Cía., Valparaíso ... 064993
Neby Eiendom, Grue Finnskog . 126350
Nečekal, František, Cheb 140467
Necessary Angel, Keswick 118038
Neckarschiffahrts-Museum und Weinbau, Stadt- und Industriegeschichte, Städtische Museen Heilbronn, Heilbronn . 019163
Nécropole de Louis Cauvin, Gareoult 013081
Ned Shed Museum, Meldrum Bay 006286
Nedelcu, Stelian, Bucureşti .. 085104, 132998
Neder, Cornelia, Hammelburg .. 076350
Nederlands Architectuurinstituut, Rotterdam 032816
Nederlands Artillerie Museum, 't Harde 032407
Nederlands Bakkerijmuseum Het Warme Land, Hattem 032420
Nederlands Baksteen en Dakpanmuseum de Panoven, Zevenaar 033099
Nederlands Biermuseum de Boom, Alkmaar 031892
Nederlands Centrum voor Handwerken, Breda 032104
Nederlands Drogisterij Museum, Dutch Drugstore Museum, Maarssen 032615
Nederlands Economisch Penningkabinet, Rotterdam ... 032817
Nederlands Fotomuseum, Rotterdam 032818
Nederlands Goud-, Zilver- en Klokkenmuseum, Schoonhoven 032864
Nederlands Graanmuseum/ Olie- en Korenmolen Woldzigt, Roderwolde 032794
Nederlands Instituut voor Beeld en Geluid-Colleties, Hilversum .. 032486
Nederlands Militair Kustverdedigingsmuseum, Fort aan den Hoek van Holland, Hoek van Holland 032492
Nederlands Museum van Knipkunst,

Schoonhoven 032865
Nederlands Openluchtmuseum, Arnhem 032028
Nederlands Parfumflessenmuseum, Winkel 033050
Nederlands Politie Museum, Apeldoorn 032012
Nederlands Scheepvaartmuseum, Amsterdam 031972
Nederlands Spoorwegmuseum, Utrecht 032952
Nederlands Steendrukmuseum, Valkenswaard 032970
Nederlands Stoommachinemuseum, Medemblik 032640
Nederlands Strijkijzer-Museum, Noordbroek 032698
Nederlands Tegelmuseum, Otterlo 032741
Nederlands Textielmuseum, Tilburg 032929
Nederlands Uitvaartmuseum, Amsterdam 031973
Het Nederlands Vestingmuseum, Naarden 032661
Nederlands Waterleidingmuseum, Utrecht 032953
Nederlands Wijnmuseum, Arnhem 032029
Nederlands Zouavenmuseum, Oudenbosch 032751
Nederlandsche Vereeniging van Antiquaren, Den Haag 059615
Nederlandse Galerie Associatie, Den Haag 059616
De Nederlandse Museumvereniging, Amsterdam 059617
Nedervetil Hembygdsgård och -museum, Nedervetil 011184
Nedev, Heidelberg 107395
Nedobeck, Milwaukee 121942
Nedomansky Brüder, Wien 063040
Nedre Manilla, Stockholm 115914
Née, Jean-Luc, Pierrecourt 072155
Née, Stéphane, Rozay-en-Brie . 072634
Neeb, Hildegard, Westerburg ... 078944
Need to Read, Omaha 145402
Needful Treasures, Oklahoma City 095996
Needles, Elva, Kansas City ... 094087
Neef, Helga, Köln 107668
Neel, Birmingham 135355
Neel, Le Golfe-Juan 103941
Neel, Saint-Paul-de-Vence 105596
Néel, Albert, Chazelles-sur-Lyon 103587
Neeltje Twiss, Den Haag 083109
Nees Morfes, Athinai 109045
Nees, Bartley, Wellington 113079
Neeses Farm Museum, Neeses . 051210
Neesh, Mumbai 109455
NEF, Kyïv 117147
La Nef des Fous, Paris 141247
Nef des Jouets, Soultz-Haut-Rhin 051855
Nefedov, D.V., Sankt-Peterburg . 085255, 143673
Nefertiti, Medellín 102455
Neff, Beat, Stäfa 087829
Neff, Charles, Mulhouse 070394, 140974
Neffe-Degandt, London 118473
Nefzger, Franz, Saaldorf-Surheim 130613
Jörg C.G. von Negelein & Alexander von Negelein, Kiel 076785
Negev Museum, Be'ersheva .. 025026
Il Negozietto, Bologna 079857
Negrel, Austin 092253, 092254
Negri, Alberto, Roma 131992
Negri, Gabriella, Roma 131993
Negri, Sergio, Bardolino 109766
Negri, Vittorio, Milano 131483
Negrini, Maria Cristina, Bologna 079858
Negrini, Maurizio, Verona 081928
Negrini, Raffaele, Milano 080630
The Negros Museum, Bacolod . 034428
Negulescu, Sabina & Jean,

Bucureşti 114149
Neher, Essen 106926
Nehru Centre Art Gallery, Mumbai 109456
Nehru Children's Museum, Kolkata 024364
Nehru Memorial Museum and Library, Delhi 024235
Nehru Science Centre, Mumbai . 024408
Nei Mongol Zhelimu Meng Guan, Tongliao 008339
Neidhardt, Fritz, Böblingen 141609
Neidhardt, S., Wiesbaden 130762
Neidhart Fresken, Wien Museum, Wien 003052
Les Neiges d'Antan, Paris 071709
Neighborhoood Gallery, New Orleans 122208
Neighbor's Antiques, Cleveland . 092958
Neil, Jacksonville 135661
Neil Armstrong Air and Space Museum, Wapakoneta 054056
Neill-Cochran Museum House, Austin 046698
Neill Museum, Fort Davis 048801
Neilmarie, Albuquerque 119497
Neils, E., Heidenheim an der Brenz 107402
Neilson, T. & J.W., Edinburgh . 089183
Neiraku Bijutsukan, Nara 029151
Neiriz, H.S., Berlin ... 074889, 129541
Neisse Galerie, Görlitz 107142
Neisser Haus – Heimathaus und Archiv, Hildesheim 019249
Neitzert, Astrid, Oberraden ... 077871
Neixiang Museum, Neixiang ... 008135
Nej, Beograd 085342
Nejjoum, Ahmed Barghach, Casablanca 082574
Nejjoum, Larbi, Casablanca ... 082575
Nejman, Laure, Paris 071710
Neka Museum and Gallery, Gianyar 024561
Nel Corso del Tempo, Firenze .. 080175
Neleh, Chicago 120213
Nelholt, Dennis, København ... 065938
Neligh Mills, Neligh 051212
Nelimarkka Museo . Etelä-Pohjanmaan Aluetaidemuseo ja Nelimarkka-resedenssi, Alajärvi 010752
Nelis, Jack, Den Haag 083110
Nelis, Jean, Bruxelles 100645
Nelis, Thom & Lenny, Haarlem . 083345
Nelis, Thom & Lenny, Amsterdam 082814
Nelis, Vincent, Amsterdam ... 082815
Nell-Buch, Fürth 129874
Nellessen, S., Simmerath ... 078473
Nell'Incanto, Napoli 110539
Nello, Quiberon 105407
Nels Berggran Museum, Imperial 006030
Nelson, Jacksonville 135662
Nelson, New York 095604
Nelson & District Museum, Art Gallery & Historical Society, Nelson .. 006459
Nelson & Nelson, New York ... 095605
Nelson & Tavares, Lisboa 084925, 132944, 143586
The Nelson-Atkins Museum of Art, Kansas City 049888
Nelson Collection at Lloyd's, London 045062
Nelson-Freeman, Paris 105127
Nelson Mandela Metropolitan Art Museum, Port Elizabeth 038807
Nelson Mandela Museum, Umtata 038871
Nelson Monument, Edinburgh .. 044258
Nelson Museum and Local History Centre, Monmouth 045288
Nelson Pioneer Farm and Museum, Oskaloosa 051831
Nelson Provincial Museum, Nelson 033253

Nelson Tower, Forres 044363
Nelson, John, Selkirk 119066
Nelson, John J., West Hollywood 097913
Nelson's Antiques, Preston ... 091056
Nemany, Dorit & Gerhard, Dachau 075287
Nemati, New York 095606
Němec, Vaclav, Ingolstadt ... 107462
Nemes-Nagel Emlékszoba, Pécsvárad 023861
Nemes, Tamás, Budapest 130812
Német Nemzetiségi Gyüjtemény, Corvin János Múzeum, Gyula . 023678
Németh László Emlékkiállítás, Németh László Városi Könyvtár, Hódmezovásárhely 023691
I Nemici del Tarlo, Roma 081475, 131994
Nemiranda, Angono 113373
NEMO, Amsterdam 031974
Nemo, San Antonio 097004
NEMO – Kunst in Nordeuropa, Eckernförde 106856, 137169
Nemorina, Helsinki 066347
Nemours Mansion and Gardens, Wilmington 054440
Nemzeti Történeti Emlékpark, Opusztaszer 023815
N'en Faites Pas un Roman, Liège 140246
Nenagh Heritage Centre, Nenagh 024973
Nencioni, Daniele, Firenze ... 131222
Nene Antigüedades, Rosario ... 060953
Nené Piatti Antichità, Milano ... 080631
Nene Valley Railway, Stibbington 045868
Neneckij Okružnoj Kraevedčeskij Muzej, Narjan-Mar 037537
Nénert, Didier-Jean, La Baule-Escoublac 068952
Nenert, Didier-Jean, Paris 071711
Nenova, S.-R., Sofia 064233
Nentai Gallery, Bucureşti 114150
Nenthead Mines Heritage Centre, Nenthead 045320
Neo Art Gallery, Roma 110837
Néo Galerie, Lagny-sur-Marne . 103912, 128650
Neo Galleri, Stavanger 113273
Neoclassica, Bologna 079859
Neoclassica, Milano 080632
Néon, Lyon 104123
Neon – Campobase, Bologna .. 109856
Neon – fdv, Milano 110391
Neon Graphics, Las Vegas 121370
Neon Lady, Saint Louis 096746
Neon Works & Nostalgia World, San Diego 097112
Neot Kedumim, The Biblical Landscape Reserve in Israel, Lod 025151
Nepal Antiques, Bhaktapur 082611
Nepal Galerie, Frankfurt am Main 107017
Nepcosons, Mumbai 079406
Nepean Historical Museum, Sorrento 001500
Nepean Museum, Nepean 006462
Nepente Art Gallery, Milano ... 110392, 137688
Népi Iparművészeti Gyüjtemény, Kecskemét 023730
Népi Műemlékház, Zalalövo ... 024054
Nepisiquit Centennial Museum, Bathurst 005396
Népmüvészetek Háza, Ferenczy Múzeum, Szentendre 023958
Népmüvészeti Kiállítás, Heves .. 023686
Nepomuk, Johann, Krefeld 076960
Néprajzi Ház, Blaskovich Múzeum, Tápiószele 023993
Néprajzi Kiállítás, Janus Pannonius Múzeum, Pécs 023848
Néprajzi Kiállítás, Soproni Múzeum, Sopron 023895
Néprajzi Múzeum, Budapest ... 023530
Neptune, Nonancourt 070743

Neptune Art Marine, Biarritz ... 067206
Neptune Gallery, Dublin 079575
Neptune Gallery, Saint Blazey Par 119011
Neptune Township Historical Museum,
Neptune 051216
Neptune, Christine, New York ... 122912
Neptunus, Anjum 143081
Népviseletes Babamúzeum –
Történelmi Panoptikum,
Keszthely 023740
Ner Hole Museum, Åndalsnes .. 033494
Néré, Henri, Saint-Malo (Ille-et-
Vilaine) 072921
Nérel, Georges, Perpignan 072089
Néret-Minet, Paris 125743
Nergal Gate Museum, Mosul ... 024810
Neri & Chas, Philadelphia 096142
Neri, Alberto, Perugia 131757
Neri, Anna Maria, Firenze 080176
Neri, G., Cagliari 079992
Neri, Giuseppe, Cagliari 079993
Nerilee, Jamberoo 061656
Nerima Bijutsukan, Tokyo 029559
Neriman Nerimanovun Xatire Muzeyi,
Bakı 003173
Neringos Istorijos Muziejus, Nida 030430
Neriton, Warszawa 137908
Nerman, Winnipeg 064930, 140435
Nermesan, Elixhausen 062557
Nero, Bianco, Stokesley 119152
Nero's Antiques, Norfolk 095870
Nes Lensemuseum, Skiptvet ... 033946
Nesbit & Co. Southsea Salerooms,
Portsmouth 127017
Nesburg, Fort Worth 093467
Nesch-Museet, Ål 033483
Nesha, Jakarta 109512
Nesher, Barcelona 114943, 114944
Nesle, New York 095607, 135966
Nesna Bygdemuseum, Nesna .. 033804
Nesplak, Karl, Walhausen 078854
Ness Historical Society Museum, Port-
of-Ness 045533
Nessim, Gabriel & Isabelle,
Genève 087487
Nesstofusafn, Seltjarnarnes 024114
The Nest, Fresno 093507
Nest, Leicester 118095
Nest, San Francisco 097295
Nest Ume Tzimbari Gabizza, Centro di
Cultura Cimbrou, Giazza-Selva di
Progno 026438
Nester, M. & B., Rottweil 078301
Nestler, Reiner, Eisenach 075592
Nestor, Houston 135612
Net Loft Gallery, Porthleven ... 118938
Netaji Museum, Kolkata 024365
Nethen, Heinz von, Oldenburg .. 077935
Netherland Inn House Museum,
Kingsport 049979
Nettesheim, Gisela, Essen 075691
Nettis, Le Touquet-Paris-Plage . 103976
Nettkunst, Kristiansand 084267
Netuschil, Claus, Darmstadt ... 106660,
137136
Netuschil, Nicole,
Wermelskirchen 078931, 108876
Netwerk Galerij, Aalst 100475
Network of European Museum
Organisations – NEMO, Berlin . 059198
Network Space, Holmfirth 117944
Network, Kunstverein Baden-Baden
e.V., Gernsbach 059197
Netzbuch, Burgdorf 141667
Neu, Annette, Fröndenberg 075919
Neubacher, Toronto 101669
Neubarth, H.-D., Berlin 129542
Neubart, Yves-Marthus,
Villeurbanne 074220
Neubarth, Michael, Hamburg ... 076286
Neuberger Museum of Art, Purchase
College, State University of New York,
Purchase 052418
Neuberin-Museum, Reichenbach im

Vogtland 021344
Neubert, Tobias, Halsbrücke ... 129943
Neubistritzer Heimatstube,
Reingers 002586
Neudorfer Suppenmuseum,
Sehmatal 021825
Neue Alte Brücke, Frankfurt am
Main 107018
Neue Galerie, Erlangen 106898
Neue Galerie am Landesmuseum
Joanneum, Graz 001985
Neue Galerie der Karl-Hofer-
Gesellschaft, Berlin 017261
Neue Galerie im Artforum,
Künstlerkreis Ortenau,
Offenburg 108331
Neue Galerie im Höhmannhaus,
Kunstsammlungen und Museen
Augsburg, Augsburg 016665
Neue Galerie New York, New
York 051450
Neue Galerie, Malerei und Skulptur ab
1750, Museumslandschaft Hessen
Kassel, Kassel 019545
Neue Galerie, Zweckverband Dachauer
Galerien und Museen, Dachau 017798
Neue Gesellschaft für Bildende Kunst
e.V., Berlin 059199
Neue Holländerei, Meldorf 107990
Neue Kammern, Stiftung Preußische
Schlösser und Gärten Berlin-
Brandenburg, Potsdam 021194
Neue Kunstschule, Basel 056415
Neue Kunstschule Zürich, Zürich 056447
Neue Münchener
Künstlergenossenschaft,
München 059200
Neue Nationalgalerie, Staatliche
Museen zu Berlin – Stiftung
Preußischer Kulturbesitz, Berlin 017262
Neue Pinakothek, Bayerische
Staatsgemäldesammlungen,
München 020560
Neue Residenz, Schloss- und
Gartenverwaltung Bamberg,
Bayerische Verwaltung der
staatlichen Schlösser, Gärten und
Seen, Bamberg 016954
Neue Sächsische Galerie,
Chemnitz 017738
Neue Sammlung, Passau 021084
Neue Wiener Werkstätte, Wien . 100356
Neuendorf, Memmingen 107992
Neuenschwander-Asuncion,
Gockhausen 116351
Neuer Aachener Kunstverein,
Aachen 059201
Neuer Berliner Kunstverein,
Berlin 017263, 059202
Neuer Kunstverein Aschaffenburg e.V.,
Aschaffenburg 016620
Neuer Kunstverein Gießen e.V.,
Gießen 059203, 107129
Neuer Leipziger Kunstverein e.V.,
Leipzig 059204
Neuer Pavillon im Schlossgarten
Charlottenburg, Stiftung Preußische
Schlösser und Gärten Berlin-
Brandenburg, Berlin 017264
Neuer Sächsischer Kunstverein,
Dresden 018003, 059205
Neues Glas/New Glass, Frechen 138742
Neues Kunsthaus Ahrenshoop,
Ahrenshoop, Ostseebad 016481
Neues Kunstquartier, Berlin ... 059206,
138743
Neues Museum Schwechat,
Schwechat 002744
Neues Museum Weimar, Klassik
Stiftung Weimar, Weimar ... 022502
Neues Museum Weserburg Bremen,
Bremen 017603
neues museum, Die Österreichische
Museumszeitschrift, Leonding . 138529

Neues Museum, Staatliches Museum
für Kunst und Design in Nürnberg,
Nürnberg 020873
Neues Palais, Stiftung Preußische
Schlösser und Gärten Berlin-
Brandenburg, Potsdam 021195
Neues Schloss Bayreuth, Bayerische
Verwaltung der staatlichen Schlösser,
Gärten und Seen, Bayreuth ... 017000
Neues Schloss Herrenchiemsee,
Schloss- und Gartenverwaltung
Herrenchiemsee,
Herrenchiemsee 019225
Neues Schloß Meersburg,
Meersburg 020331
Neues Schloss Schleißheim,
Bayerische Verwaltung der
staatlichen Schlösser, Gärten und
Seen, Oberschleißheim 020922
Neues Schloss Wallerstein,
Wallerstein 022410
Neues Schloß, Schloßmuseum,
Tettnang 022136
Neues Stadtmuseum Landsberg am
Lech, Landsberg am Lech ... 019879
Les Neuf Muses, Paris 141248
Neufeld Verlag, Schwarzenfeld . 137503
Neuffer, Simone, Lausanne ... 087578
Neugablonzer Industrie- und
Schmuckmuseum im Isergebirgs-
Museum, Kaufbeuren 019563
Neugebauer, München 130374
Neugebauer, Monika, Dresden . 129716
Neugebauer, Wolfgang, Satteins 140039
neugerriemschneider, Berlin ... 106382
Neuhäuser, Roland G., Mainz ... 077242
Neuhart, Denver 093318
Neuhaus, Albrecht, Würzburg .. 079087
Neuhaus, Beat, Fribourg 087434, 133747
Neuhaus, Lindy, Houston 093753
Neuhausen, Angelika, Krefeld .. 130194
Neuhauser Kunstmühle, Salzburg 100173
Neuheisel, Gernot, Saarbrücken . 078330
Neuhoff, New York 122913
Neujack, Bernd, Krauchenwies . 130190
Neuland-Bund und Verlag,
München 137417
Neumann, Duisburg 075554
Neumann, Krefeld 107742
Neumann, Lübeck 077181
Neumann, Peine 108370
Neumann, San Francisco 136338
Das Neumann Haus Museum,
Laidley 001203
Neumann, C., Berlin 106383
Neumann, Doris, Hofheim am
Taunus 130055
Neumann, Ernst, Günzburg 076122
Neumann, F., Saterland 078356
Neumann, Gerhard, Mühlenbecker
Land 077425
Neumann, Jan, Praha 140526
Neumann, Käthe, Hesel 107426
Neumann, Karsten, Erfurt 129804
Neumann, M., Gutach 141845
Neumann, Manfred, Neuruppin . 130451
Neumann, Marc, Remscheid ... 078199
Neumann, Nicolas, Gmund am
Tegernsee 076014
Neumarkter Brauereimuseum im
Glossner-Bräu, Neumarkt in der
Oberpfalz 020723
Neumeister, Berlin 125883
Neumeister, Köln 126010
Neumeister, New York 127355
Neumeister, Wien 063041, 125283
Neumeister, Zürich 126771
Neumeister, Münchener
Kunstauktionshaus, München . 126061
Neuner, Hefersweiler 076427
Neuner, Friedrich, Oberhofen im
Inntal 062756
Neunlist, André & Yannick,
Battenheim 066961

Neunlist, Nathalie, Horbourg Wihr 068836
Neunte Kunst, Osnabrück 137462
Neuparth, Heike-Gabriele & Reimar,
Pößneck 078062
Neupert, Herold, München 077578
Neuroth, Conny, Lindenfels 107877
Neuss Antik, Neuss 077749
Neusser Auktionshaus, Neuss .. 126080
Neusser, Franziska, Wien 100357
Neustadt Museum of Tiffany Art, New
York 051451
Neustädter Rathaus, Hanau 019042
Neut Gallery, Wellington 113080
Neutral Art, Singapore 114577
Neutral Ground, Regina 006771
Neuville Franck, Vaux-sous-
Aubigny 073992
Neuville, Jean, Boyer 067441
Neuwerk, Oldenburg 108343
Neuwerk Kunsthalle, Konstanz . 019762
Neuwerth, Helmut, Ruderting ... 078302
Neuwirth, C., Bockenem 075047
Neuwirth, Dimitrij, Brno 128321
Neuwirth, Dr. Waltraud, Wien .. 136601
Neuy, H., Steinfurt 108670
Neva-Ant, Sankt-Peterburg 085256
Nevada Art, Las Vegas 121371
Nevada Historical Society Museum,
Nevada Department Museums and
Library, Reno 052509
Nevada Museum of Art, Reno .. 052510
Nevada Pastel Society, Las
Vegas 060633
Nevada State Fire Museum and
Comstock Firemen's Museum,
Virginia City 054002
Nevada State Museum, Carson
City 047444
Nevada State Museum, Las
Vegas 050170
Nevada State Museum Newsletter,
Carson City 139451
Nevada State Museum. Carson City.
Popular Series, Carson City .. 139452
Nevada State Railroad Museum,
Carson City 047445
Nevado Cortes, Eva Maria,
Pamplona 086395
Neve, Gunnar, Leck 077040
Never Ending Books, North
Shore 143337
Never Too Old, Penola 062071
Neversink Valley Area Museum,
Cuddebackville 048082
Neves, Ana, Lisboa 084926
Neves, Guilherme Gameiro,
Lisboa 114047
Nevill, Canterbury 117519, 134208
Nevill, Fred M., Houston 093754
Neville Blakey Museum of Locks, Keys
and Safes, Brierfield 043730
Neville Budhais Gallery, Montego
Bay 111232
Neville Public Museum of Brown
County, Green Bay 049229
Nevin, Skalko, Los Angeles ... 094418
Nevograf, Sankt-Peterburg 114351
Nevra Antik Galeri, İstanbul ... 088132
Nevşehir Devlet Güzel Sanatlar
Galerisi, Nevşehir 043034
Nevşehir Müzesi, Nevşehir 043035
Nevska, Irina, San Francisco .. 124352
Nevyantik, Mladá Boleslav 065423,
140486
Nevyjel & Raggazoni, Trieste .. 132224
New & Old Shop, San Antonio . 097005
New Age Expressions, Buffalo .. 119962
New American Paintings, Boston 139453
New Archaeological Museum,
Mytilini 023189
New Art Agency i Stockholm,
Stockholm 115915
New Art Centre, East Winterslow 044213
New Art Collectors, London 118474

New Art Dealers Alliance, New
York 060634
New Art Gallery, Arezzo 079749
New Art Gallery, Milano 080633
New Art Gallery, Roma 081476
The New Art Gallery Walsall,
Walsall 046041
New Art on Newbury, Boston .. 119902
New Art Portfolio, Stroud ... 119174
New Art Publications, Brooklyn . 138310
New Art, Galerie und Kunsthandel,
Henstedt-Ulzburg .. 076495, 107416
New Artform, Singapore 137938
New Arts Gallery, Toulon (Var) .. 105716
New Arts Program, Kutztown .. 060635
New Ashgate Gallery, Farnham . 117788
New Asia Art Shop, Dallas 135513
New Avenue Fine Art, Miami .. 121832
New Bedford Art Museum, New
Bedford 051224
New Bedford Whaling Museum, New
Bedford 051225
New Beginning, Jacksonville ... 135663
The New Berlin Art Company,
Berlin 106384
The New Berlin Historical Society
Museum, New Berlin 051228
New Braunfels Conservation Society
Museum, New Braunfels 051234
New Braunfels Museum of Art &
Music, New Braunfels 051235
New Bridge Gallery, Truro .. 119263
New Brighton Art Gallery,
Christchurch 112853
New Britain Museum of American Art,
New Britain 051240
New Britain Youth Museum, New
Britain 051241
New Britain Youth Museum at
Hungerford Park, Kensington .. 049928
New British Artists, Bedford .. 117325
New Brunswick Art Bank,
Fredericton 005843
New Brunswick College of Craft and
Design, Fredericton 055128
New Brunswick College of Craft and
Design Gallery, Fredericton ... 005844
New Brunswick Healthcare Museum,
Fredericton 005845
New Brunswick Internment Camp
Heritage Museum, Minto ... 006328
New Brunswick Sports Hall of Fame,
Fredericton 005846
New Bytes and Old Bits, Tucson 097682
New Canaan Historical Society
Museum, New Canaan 051248
New Castle Historical Society Museum,
New Castle 051250
New Century Antiquarian Books,
Kew 139794
New Century Art, Cleveland ... 120446
New Century Artists, New York . 060636,
                                122914
New Century Collector, Salt Lake
City 096912
New Circle, Miami 121833
New Craftsman, Saint Ives,
Cornwall 119026
New Creations, Dallas 093185
New Curiosity Shop, Tonbridge . 091609,
                                135139
New Denmark Memorial Museum,
New Denmark 006463
New Design University,
Privatuniversität der Kreativwirtschaft,
Sankt Pölten 054963
New Detroit Science Center,
Detroit 048315
New Eltham Galleries, London .. 118475
New England Air Museum of the
Connecticut Aeronautical Historical
Association, Windsor Locks ... 054474
New England Antique Emporium,

Uralla 062382
New England Antiques Journal,
Palmer 139454
New England Appraisers Association,
Ludlow 060637
New England Brass and Iron Lace
Foundry Museum, Uralla ... 001606
New England Carousel Museum,
Bristol 047212
New England College Gallery,
Henniker 049446
New England Fire and History
Museum, Brewster 047193
New England House Antiques,
Tring 091636, 135151
New England Maple Museum,
Pittsford 052185
New England Maple Museum,
Rutland 052716
New England Museum Association,
Arlington 060638
New England Quilt Museum,
Lowell 050511
New England Regional Art Museum,
Armidale 000804
The New England School of Art
& Design, Suffolk University,
Boston 056771
New England School of Art and
Design, Boston 056772
New England Ski Museum,
Franconia 048920
New England Wireless and Steam
Museum, East Greenwich 048428
New English Art Club, London .. 060200
New Era, New York 095608
New Farm Antique Centre, New
Farm 061951
New Farm Art, New Farm 099427
New Farm Framing and Gallery,
Newstead 099444
New FMR Magazine, Villanova di
Castenaso 138849
New Forest Fine Art Gallery, Ashurst,
Southampton 117260
New Forest Museum and Visitor
Centre, Lyndhurst 045176
The New Gallery, Birmingham .. 117377
New Gallery, Ha Noi 125088
New Gallery, Houston 121101
New Gallery, Portscatho 118941
The New Gallery, Woolacombe . 119390
New Gleanings, Albany 139455
New Grafton Gallery, London .. 118476
New Grounds Print Shop,
Albuquerque 119498
New Hall, Dymchurch 044194
New Hall Art Collection,
Cambridge 043846
New Hampshire Antiquarian Society
Museum, Hopkinton 049556
New Hampshire Antiques Monthly,
Farmington 139456
New Hampshire Art Association,
Boscawen 060639
New Hampshire Arts, Concord . 139457
New Hampshire Farm Museum,
Milton 050906
New Hampshire Institute of Art,
Manchester 050629, 057399
New Harmony Gallery of
Contemporary Art, University
of Southern Indiana, New
Harmony 051260
New Harmony State Historic Site, New
Harmony 051261
New Haven Colony Historical Society
Museum, New Haven 051264
New Horizons, Ho Chi Minh City 125137
New Iceland Heritage Museum,
Gimli 005876
New Ideal Art, Singapore 114578

New Image Art, Philadelphia ... 123428,
        123429,   123430
New Image Art, West Hollywood 054309,
                                125017
New Image Gallery, Calgary ... 101249
New International Cultural Center,
Antwerpen 003306
New Jersey Association of Museums,
Newark 060640
New Jersey Center for Visual Arts,
Summit 053599
New Jersey Children's Museum,
Paramus 051913
New Jersey Historical Society
Museum, Newark 051498
New Jersey Museum of Agriculture,
North Brunswick 051597
New Jersey Naval Museum,
Hackensack 049301
New Jersey State House, Trenton 053798
New Jersey State Museum,
Trenton 053799
New Jersey Watercolor Society,
Parsippany 060641
New Juaben Palace Museum,
Koforidua 022931
The New Kelvin Gallery, East
Kilbride 117706
New Lanark Visitor Centre, New
Lanark 045323
New Land Gallery, Port Adelaide 099574
New Langton Arts, San Francisco 053017
New Life, San Francisco 097296
New Life, Tucson 136421
New Life Art, Birmingham 119811
New Lisbon Antiques,
Indianapolis 093911
New Literature and Fine Arts
Publishing House, Changchun . 136704
New London County Historical Society
Museum, New London 051278
New London Historical Society
Museum, New London 051280
New London Public Museum, New
London 051281
New Look, Minneapolis 135833
The New Los Angeles Art Fair for
Contemporary and New Art, Santa
Monica 098480
New Marine Centre, Society of
Saanich Peninsula Museums,
Sidney 007034
New Market Battlefield State Historical
Park, New Market 051283
New Mexico Art League,
Albuquerque 060642
New Mexico Association of Museums,
Albuquerque 060643
New Mexico Bureau of Geology
Mineral Museum, Socorro 053375
New Mexico Farm and Ranch Heritage
Museum, Las Cruces 050155
New Mexico Museum of
Natural History and Science,
Albuquerque 046368
New Mexico Museum of Space
History, Alamogordo 046334
New Mexico State University Art
Gallery, Las Cruces 050156
New Mexico State University Museum,
Las Cruces 050157
New Mexico Wing-Commemorative Air
Force, Hobbs 049496
New Milford Historical Society
Museum, New Milford 051284
New Millennium Gallery, Saint Ives,
Cornwall 119027
New Mills Heritage and Information
Centre, New Mills 045324
New Moment, Ljubljana 139021
New Museum Members' Bulletin, New
York 139458
New Museum of Contemporary Art,
New York 051452

New Museum of European and
Eastern Art, Athinai 023021
New Nautilus, Parma . 081120, 142783
New Nepal Art Palace, Lalitpur . 112091
New Norcia Museum and Art Gallery,
New Norcia 001351
New Orleans Academy of Fine Arts,
New Orleans 057533
New Orleans Artworks Gallery, New
Orleans 051300
New Orleans Auction Galleries, New
Orleans 127337
New Orleans Auction Saint Charles
Gallery, New Orleans 095136, 127338
New Orleans Fire Department
Museum, New Orleans 051301
New Orleans Glassworks and
Printmaking Studio, New
Orleans 122209
New Orleans Images, New
Orleans 122210
New Orleans Museum of Art, New
Orleans 051302
New Orleans Pharmacy Museum, New
Orleans 051303
New Otani Art Museum, Tokyo . 029560
New Pagoda Specialty, New York 095609
The New Place, Kitchener ... 128177
The New Press, New York 138411
New Purva Elegance at Home,
Hyderabad 079393
New Quay Gallery, Docklands .. 098970
New Realms, London 045063
New Realms at Adams Street,
London 045064
The New Rembrandt Gallery, Newport,
Isle of Wight 118801
New Satguru Art 'n' Frames, Jaipur,
Rajasthan 109385
New Selection, Knokke-Heist . 100751
New Shakya Art, Kathmandu .. 112081
New South Wales Lancers Memorial
Museum, Parramatta 001398
New Space Art Gallery, Hue ... 125162
New Space Art Gallery II, Hue .. 125163
New Street Antique Centre,
Plymouth 091022
New Street Bookshop, Penzance 144780
New Street Gallery, Plymouth .. 118922
New Studio for the Art of Painted
Finishes, New York 122915
New Sweden Farmstead Museum,
Bridgeton 047204
New Sweden Historical Museum, New
Sweden 051316
New Tingeling, Amsterdam .. 112261
New to You, Kuranda 061706
New Treasure Art Gallery, Yangon 082610,
                                112056
New Tribal Arts, Toledo 124692
New Urban Art, Phoenix 123501
New Urban Arts, Providence .. 057728
New Visions Gallery, Marshfield 050719
New Visions Gallery, Salt Lake
City 123975
New Walk Museum and Art Gallery,
Leicester 044800
New Windsor Cantonment State
Historic Site and National Purple
Heart Hall of Honor, Vails Gate 053923
New World Antique Gallery,
Detroit 093370, 135543
New World School of the Arts,
Miami 057433
New York Antiquarian Book Fair, New
York 098481
New York Art Expo, New York .. 122916
New York Art Gallery, New York 122917
New York Artists Equity Association,
New York 060644
New York Arts of Pacific Asia Show,
New York 098482
New York Ceramics Fair, New
York 098483

New York City Fire Museum, New York 051453
New York City Police Museum, New York 051454
New York Galleries Antiques, New York 095610
New York Hall of Science, Flushing 048767
The New York Historical Society Museum, New York 051455
New York Institute of Photography, New York 057558
New York International Tribal and Textile Arts Show, New York .. 098484
New York Museum of Transportation, West Henrietta 054306
New York Poster Fair, New York 098485
New York Public Library for the Performing Arts, New York .. 051456
New York Restoration Center, New York 135967
New York School of Interior Design, New York 057559
New York Society of Etchers, New York 122918, 138412
New York State College of Ceramics, Alfred University, Alfred 056641
New York State Military Museum, Saratoga Springs 053158
New York State Museum, Albany 046344
New York Studio School of Drawing, Painting and Sculpture, New York 057560
New York Studio School of Drawing, Painting and Sculpture – Gallery, New York 051457
New York Transit Museum, Brooklyn 047264
New York Vintage, New York ... 095611
New Zealand Academy of Fine Arts, Wellington 056201
New Zealand Academy of Fine Arts Gallery, Wellington 033352
New Zealand Antique Dealers' Association, Epsom 059638
New Zealand Centre for Photography, Wellington 033353
New Zealand Cricket Museum, Wellington 033354
New Zealand Fighter Pilots Museum, Wanaka 033332
New Zealand Journal of Photography, Wellington 138942
New Zealand Marine Studies Centre, Dunedin 033192
New Zealand National Maritime Museum, Auckland 033156
New Zealand Numismatic Journal, Wellington 138943
New Zealand On Art, Christchurch 083886, 112854
New Zealand Police Museum, Porirua 033285
New Zealand Portrait Gallery, Wellington 033355
New Zealand Society of Sculptors, Painters and Associates Inc., Auckland 059639
New Zealands Rugby Museum, Palmerston North 033274
New Zone Art Collective, Eugene 060645, 120847
New Zone Virtual Gallery, Eugene 048623
Newark Air Museum, Newark-on-Trent 045331
Newark Antiques Centre, Newark-on-Trent 090814
Newark Antiques Warehouse, Newark-on-Trent 090815, 134872
Newark International Antiques and Collectors Fair, Newark-on-Trent 098393
The Newark Museum, Newark . 051499
Newark Museum, Newark-on-Trent 045332

Newark Museum Association, Newark 060646
Newark School of Fine and Industrial Art, Newark 057571
Newark Town Treasures and Art Gallery, Newark-on-Trent 045333
Newark Valley Depot Museum, Newark Valley 051509
Newarke Houses Museum, Leicester 044801
Newart's Millgate Museum, Newark-on-Trent 045334
Newaygo County Museum, Newaygo 051510
Newberry Art, Dunsborough ... 098989
Newbridge House, Donabate .. 024864
Newburgh Antiques, Newburgh . 090819
Newburn Hall Motor Museum, Newburn 045339
Newbury Fine Arts, Boston 119903
Newburyport Maritime Society Museum, Amesbury 046445
Newby Hall, Ripon 045620
Newcastle Antique Centre, Hamilton, New South Wales 061579
Newcastle Antique Centre, Newcastle-upon-Tyne 090828, 134877
Newcastle Antiques & Collectables, Hamilton, New South Wales .. 061580
Newcastle Art Gallery, Newcastle, Co. Down 118780
Newcastle Art Space, Newcastle 099434
Newcastle Gaol Museum, Toodyay 001581
Newcastle Region Art Gallery, Newcastle 001355
Newcastle Regional Museum, Newcastle 001356
Newcomb Art Gallery, New Orleans 051304
Newcomb-Macklin, Chicago ... 120214
Newcomen Engine House, Dartmouth 044071
Newel, New York 095612
Newell Museum, Newell 051523
Newfoundland and Labrador Health and Community Services Archives and Museum, Saint John's .. 006905
Newfoundland Science Centre, Saint John's 006906
The Newgate-East, Seoul ... 111700
Newgate Gallery, Barnard Castle 117288
Newham, Richmond, Victoria .. 127669
Newham Grange Leisure Farm Museum, Middlesbrough 045262
Newham Heritage Centre, London 045065
Newhampton Road Antiques, Wolverhampton 091905
Newhaven Fort, Newhaven .. 045357
Newhaven Heritage Museum, Edinburgh 044259
Newhouse, New York 122919
Newhouse, Jill, New York 122920
Newington Antiques, Newington 090830, 134879
Newland Rennie Wilkins, Monmouth 126990
Newlove, B.M. & E., Surbiton .. 091511
Newlyn Antiques, Newlyn 061959
Newlyn Art Gallery, Penzance .. 045481
Newman, Toronto 101670
Newman & Co., W., Kolkata ... 142590
Newman & Popiashvili, New York 122921
Newman House, Dublin 024903
Newman, Addie, West Hollywood 097914
Newman, Nicholas J., West Ewell 135204
Newman, Walter Andrews, Philadelphia 123431
Newmarket Antiques and Collectors Fair, Newmarket 098394
Newport Art Museum, Newport . 051538

Newport Art Museum School, Newport 057573
Newport Avenue Antique Center, San Diego 097113
Newport Avenue Antiques, San Diego 097114
Newport Beach Gallery, Newport 099439
Newport Gallery, Newport, Essex 090833
Newport Mansions, Newport ... 051539
Newport Pagnell Historical Society Museum, Newport Pagnell ... 045368
Newport Roman Villa, Newport, Isle of Wight 045366
Newport School of Art, Media and Design, University of Wales, Newport, Newport, Gwent 056585
Newport Temple Antiques, Newport 061960
Newquay Galleries, Newquay .. 118805
Newry and Mourne Arts Centre and Museum, Newry 045371
Newsantandrea, Savona 110918
Newsbank Review of the Arts: Fine Arts and Architecture, Naples . 139459
Newseum, Arlington 046540
Newsletter East Asian Art & Archaeology, Ann Arbor ... 139460
Newsletter, Russian Museums Association, Tula 139016
Newsmall & Sterling, Vancouver 101815
Newsome House Museum, Newport News 051547
Newspace, Rozelle 099657
Newspark – The Japan Newspaper Museum, Yokohama 029721
Newsstand Museum, Hirano Hakabutsukan, Osaka 029253
Newstead Abbey Historic House and Gardens, Nottingham 045418
Newstudio Gallery, Sacramento . 123776
Newton, Tokyo 082186
Newton Abbot Town and Great Western Railway Museum, Newton Abbot 045372
Newton Andrews, Auckland ... 083810, 143292
Newton Andrews, North Shore . 084037, 143338
Newton History Museum at Jackson Homestead, Newton 051553
Newton Museum, Newton 051551
Newton, Barry, Goolwa 099088
Newton, Barry, Unley 099846
Newtonmore Museum, Highland Folk Museum, Newtonmore 045376
Newtons, Bury, Lancashire ... 088738, 134189
Newtown Historic Museum, Newtown 051558
Newtown Textile Museum, Newtown, Powys 045377
Newzones, Calgary 101250
Nexø Museum, Nexø 010234
Nexos, Wellington 113081
Next, Roma 138850
Next Graphics, Bucureşti 114151
Next Monet, San Francisco ... 124353
Next to New Shop, Austin 092255
Next to Nix Antiques, Fitzroy .. 061469
Nexus Antiques, Wellington ... 084176
Nexus Foundation for Today's Art, Philadelphia 052076
Nexus Gallery, Adelaide 098614
Nexus Native Art, Vancouver .. 101816
Neyer, Durdica-Julie, Bludenz . 099964
Neylon, Richard, Ultimo 139941
Neyrand, Bruno, Sarry 073420
Neyrand, Bruno, Villeurbanne . 074221
Nez Perce County Museum, Lewiston 050230
Nez Perce National Historical Park, Spalding 053426
Nezu Bijutsukan, Tokyo 029561
NFA Space Exhibits, Chicago .. 120215

Ng Viet Gallery, Ho Chi Minh City 125138
Nga, Bo Kok, Hong Kong 065086
Ngai Leng, Hong Kong 065087
Ngarak, Swifts Creek 139020
Ngatea Antiques, Ngatea 084026
Ngaumatau Fine Art, Dunedin . 112889
Nghe-Tinh Museum, Vinh 054821
NGI of New York, New York ... 095613
Ngo, An, Ho Chi Minh City ... 125139
Ng's Art School, Hong Kong ... 055219, 102098, 136724
Ngurdoto Gate Museum, Arusha 042626
Nguyen, Kim, Tulsa 124810
Nguyen, Serge, Paris 129099
Nguyên, T.P., Genève 087488
Nguyen, Than, Ho Chi Minh City 125140
Nhabe Museum, Nhabe 004299
Nhat, Le, Ho Chi Minh City ... 125141
NHK Hoso Hakubutsukan, Tokyo 029562
NHK Publishing, Tokyo 137786
Ni Tan Usados, Medellín 065300
Niagara Apothecary, Niagara-on-the-Lake 006489
Niagara Art and More, Düsseldorf 075501
Niagara County Historical Center, Lockport 050354
Niagara Falls Art Gallery – William Kurelek Art Collection, Niagara Falls 006483
Niagara Fire Museum, Niagara-on-the-Lake 006490
Niagara Galleries, Richmond, Victoria 099634, 136488
Niagara Gorge Discovery Center, Niagara Falls 051559
Niagara Historical Museum, Niagara-on-the-Lake 006491
Niagara Power Project Visitors' Center, Lewiston 050234
Niagara Upholstery, Las Vegas . 135686
Niakan, Los Angeles 094419
Niakan, West Hollywood 097915
Nianhua Museum, Mianzhu ... 008084
Niavaran Palace, Tehrän 024778
Nibelungenmuseum, Worms ... 022766
Nic Nic, Amsterdam 082816
Nicaise, Pascal, Ons-en-Bray . 070786
Niccoli, Parma 110633
Nice & Fit, Berlin 106385
Nice Old Stuff, Wallacia 062403
Niceties, Los Angeles 094420
La Niche Art Gallery, West Burleigh 099879
La Nichée, Saint-Vit 073334
Nichi-Mongoru Mingeikan, Tantou 029468
Nichido, Paris 105128
Nichido, Tokyo 111481
Nichido Contemporary Art, Tokyo 029563
Nichol, Gerry, New York 095614
Nichol, Gordon, Dublin 079576
Nichola, Simon, Damascus ... 088047
Nicholas, New York 095615
Nicholas Roerich Museum, New York 051458
Nicholas, Antonio, Las Vegas . 094163
Nicholas, Thaete, Philadelphia . 136096
Nicholls, John, Leigh, Staffordshire 089799
Nicholls, Rene, Malmesbury .. 090682
Nichols, Indianapolis 093912
Nichols, Monroe 127327
Nichols Antique Centre, Sheffield 091288, 135027
Nichols House Museum, Boston 047126
Nichols, Scott, San Francisco .. 124354
Nicholson, Cincinnati 092879
Nicholson, Launceston 139800
Nicholson Museum, Sydney ... 001548
Nicholson, Richard A., Chester . 144332
Nici Self Museum, Centennial . 047496
Nick Nacks andre, Saint Louis . 096747
Nick, H. & D., Aubais 013282

Nickel, Heinz, Zweibrücken 137569,
142533
Nickels-Sortwell House,
Wiscasset 054508
The Nickle Arts Museum,
Calgary 005531
Nickleby, Kyneton 061711
Nicko, Nicole, Blois 067258
Nick's Art Gallery, Phoenix ... 123502
Nick's Baseball Cards, Dallas .. 093186
Nick's Upholstery, Chicago ... 135417
Nicky's Antiques, Brighton,
Victoria 061204
NicNac Shop, Saint Louis 096748
Nico Gallery 619 Western Av,
Seattle 124574
Nicod, Gérard, L'Isle-sur-la-
Sorgue 069581
Nicod, Xavier & Gérard, L'Isle-sur-la-
Sorgue 069582
Nicol, Jacques-Henri,
Montmorency 070279
Nicola Restauri, Aramengo ... 130900
Nicola Restauri, Torino 132173
Nicola Valley Museum, Merritt . 006295
Nicola von Senger, Zürich ... 116841
Nicolai Verlag, Berlin 137087
Nicolaï, Geneviève, Ménerbes . 104218
Nicolai, Maria Ludovica, Firenze 131223
Nicolas & Osvaldo, Las Vegas .. 094164
Nicolas Antiquités, Aubenas .. 066781
Nicolas Denys Museum, Saint
Peters 006932
Nicolas Galeria de Artà,
Bucureşti 114152
Nicolas, Michéle, Aizac 066489
Nicolas, Rainer R., Berlin 141558
Nicolau Relotgeria, Jose Maria,
Barcelona 085810, 133138
Nicolau, Christophe, Nice 070638
Nicolaus August Otto Museum,
Holzhausen an der Haide ... 019325
Nicolaus-Schmidt-Künzel-Gedenkstätte,
Tanna 022104
Nicolay, Raymond de, Paris ... 125744
Nicolaysen Art Museum, Casper 047456
Nicole Gallery, Chicago 120216
Nicole's Art Gallery, Tea Gardens 099811
Nicolet, Chantal, Uzès 129325
Nicoletti, Roberto, Roma 081477
Nicolier, Josette, Limoges 069527
Nicolini, Mauro, Brescia 131044
Nicolle, Gerard, Nice 070639
Nicollet Book Store, Minneapolis 145281
Nicollier, Mainz 077243
Nicols, Düsseldorf 106812
Nicosia Municipal Arts Centre,
Lefkosia 009421
Nidderdale Museum, Pateley
Bridge 045467
Nidwaldner Museum, Stans ... 042024
Niece & Kimton, Tucson 124760
Niedenhuber, Ernst, Linz 062719
Nieder, B., Kerpen 076762
Nieder, Markus, Winterberg ... 079027
Niederbayerisches Feuerwehrmuseum,
Eggenfelden 018099
Niederbayerisches
Landwirtschaftsmuseum, Regen 021312
Niederbayerisches
Vorgeschichtsmuseum,
Archäologische Staatssammlung
München, Landau an der Isar . 019869
Niederbergisches Museum Wülfrath,
Wülfrath 022782
Niederdeutsche Beiträge zur
Kunstgeschichte, München ... 138744
Niederebersdorfer Heimat- und
Archivstuben, Tutzing 022237
Niederecker, Ursula, München .. 077579
Niederreichholz, Dieter, Bochum . 075039
Niederer, Boris, Genève 087489, 116334
Niederer, Götz, Stuttgart 078622
Niederhauser, Catherine,

Lausanne 116425
Niederhauser, Thomas, Auslikon 087204
Niederhoffer, Annecy 066647
Niederkorn, Philadelphia 096143
Niederlausitz-Museum Luckau, Luckau,
Niederlausitz 020122
Niederlausitzer Heidemuseum,
Regionale Forschungs- und
Bildungsstätte für Geschichte,
Volks- und Landschaftskunde,
Spremberg 021940
Niederlausitzer Sorbisches
Dorfmuseum Bloischdorf,
Felixsee 018347
Niederlintner, Frankfurt am Main 075810
Niedermaier, Böhmenkirch 016519
Niedermaier, Erika, Barcelona .. 085811
Niedermann, Urs, Sankt Gallen . 133910
Niedermayr, Josef, Punzing 127836
Niedermeier, Silvester,
Essenbach 075707
Niederöst-Brigger, Michael,
Aarburg 087175, 133610
Niederoest, Urs S., Hamburg ... 076287
Niederösterreich-Gesellschaft für Kunst
und Kultur, Sankt Pölten 058293
Niederösterreichisches
Dokumentationszentrum für Moderne
Kunst, Sankt Pölten 002696
Niederösterreichisches
Feuerwehrmuseum, Tulln 002845
Niederösterreichisches
Freilichtmuseum, Haag, ... 002034
Niederösterreichisches Landesmuseum,
Sankt Pölten 002697, 136563
Niederösterreichisches für
Volkskultur, Groß Schweinbarth 002000
Niederösterreichisches Pressehaus,
Sankt Pölten 136564
Niederrheinischer Kunstverein e.V.,
Hamminkeln 059207
Niederrheinisches Freilichtmuseum,
Grefrath 018802
Niederrheinisches Motorradmuseum,
Moers 020445
Niederrheinisches Museum für
Volkskunde und Kulturgeschichte,
Kevelaer 019598
Niedersächsisches Deichmuseum,
Dorum 017968
Niedersächsisches Kleinbahn-Museum
Bruchhausen-Vilsen, Bruchhausen-
Vilsen 017629
Niedersächsisches Kutschenmuseum,
Lilienthal 020072
Niedersächsisches Landesmuseum
Hannover, Hannover 019062
Niedersächsisches Münzkabinett der
Deutschen Bank, Hannover ... 019063
Niedersächsisches Museum für Kali-
und Salzbergbau, Ronnenberg . 021471
Niedersächsisches Spargelmuseum,
Museum Nienburg, Nienburg . 020808
Niedersätz, Olaf, Berlin 141559
Niedorf, Peter, Schwerin 078430
Niehaus, Berlin 074890
Niehaus, Hans, Claremont 085639
Nielsen, Boston 119904
Nielsen, Holbæk 128391
Nielsen, Las Vegas 094165
Nielsen, Minneapolis .. 094936, 135834
Nielsen Slovakia, Bratislava ... 114657
Nielsen & Søn, Anker, Vejle ... 066198
Nielsen, Bengt, Herlev 065748
Nielsen, Børge, Hvidovre 065804
Nielsen, Jørgen, Gage 065971
Nielsen, Karl E., Skive 066122
Nielsen, Morten, Frederiksberg . 140572
Nielsen, Sejr, Viby 066211
Nielsen, Susanne, Kjellerup ... 065816
Nielsen, Ulla, Gilleleje 065700
Nielstrup Museum, Hadsten ... 010094
Niemann, Bärbel, Jever 107483

Niemann, Detlev, Hamburg 076288
Niemeijer, H., Zwolle 083736
Niemer, Korschenbroich 076946
Niemeyer, Bingen am Rhein .. 129593
Niemeyer Tabaksmuseum,
Groningen 032383
Niemeyer, Anna Maria, Rio de
Janeiro 101051
Niemeyer, Jan Hendrik,
Padingbüttel 078011, 142292
Niemeyer, Lüder H., Padingbüttel 078012,
142293
Niendorff, Austin 119686
Niepce, Christine, Grenoble 068736
Nierat, Bertrand, Sucy-en-Brie . 073604
Nierendorf, Berlin ... 106386, 137088
Nierengarten-Smith, B.J., Saint
Louis 123847
Niermeier, Johann, Landshut ... 077016
Nies, Marcel, Antwerpen 100506
Niese, de, Den Haag 083111
Niesen, Hubert, Rodgau 142333
Niespor, Franz, Zell, Mosel 130799
Nieszawer, Nadine, Paris 105129
Niethammer, Prof. Dr. Reiner,
Mötzingen 077411
Nietsch, Sigurd, München 077580,
077581
Nietzsche-Archiv, Klassik Stiftung
Weimar, Weimar 022503
Nietzsche-Haus, Naumburg,
Saale 020646
Nietzsche-Haus, Sils Maria 041989
Nieuw Land Poldermuseum,
Lelystad 032592
Nieuw Schoten, Amsterdam ... 112262
Nieuw Schoten, Haarlem 112514
De Nieuwe Linie, Bodegraven .. 082969,
132442
Nieuwe Republiek Museums – Vryheid,
Vryheid 038876
Nieuwenhuizen Segaar, G.J., Den
Haag 083112
Nieuweschans Naaimachinemuseum,
Nieuweschans 032676
Nieuwlaat, Peter, Roosendaal . 083544,
132618
Nievergelt, Felix, Zürich 116842
Niewerth, S., Reken 078191
NIFCA – Nordic Institute for
Contemporary Art, Helsinki ... 010849
Niğde Müzesi, Niğde 043036
Nigel, West Winterslow 135208
Nigg, Franz, Schaan 082335
Niggemann, Jutta, Köln 130160
Nigro, Gérald, Cormeilles-en-
Vexin 068156
Nigro, Horacio F., Montevideo . 135273
Niguliste Museum, Eesti
Kunstimuuseum, Tallinn 010682
Nihon Bijutsu Kyokai, Tokyo .. 059504
Nihon Bijutsu Nenkan, Tokyo .. 138893
Nihon Bijutsuin, Tokyo 059505
Nihon Bijutsuka Renmei, Tokyo . 059506
Nihon Biyoin Hakubutsukan,
Kyoto 028971
Nihon Camera Hakubutsukan,
Tokyo 059564
Nihon Chokokukai, Tokyo 059507
Nihon Ga-In, Tokyo 059508
Nihon Gangu Hakubutsukan,
Kodera 028868
Nihon Geijutsu Daigaku, Tokyo . 056049
Nihon Geijutsu Gakuen, Tokyo .. 056050
Nihon Hakubutsukan Kyokai,
Tokyo 059509
Nihon Hanga Kyoukai,
Sagamihara 059510
Nihon Jidosha Hakubutsukan,
Futatsunashi 028570
Nihon Kindai Bungakukan, Tokyo 029565
Nihon Kogeikan, Tokyo National
Museum, Tokyo 029566
Nihon Kyodogangu-Kan,

Kurashiki 028904
Nihon Mangaka Kyokai, Tokyo . 059511
Nihon Mingeikan, Osaka 029254
Nihon Mingeikan, Tokyo 029567
Nihon Minka Shuraku Hakubutsukan,
Toyonaka 029647
Nihon Monki Senta Hakubutsukan,
Inuyama 028693
Nihon Oni Hakubutsukan, Oe .. 029203
Nihon Shodo Bijutsukan, Tokyo . 029568
Nihon Stained Glass Art Gallery,
Tokyo 111482
Nihon Ukiyo-e Hakubutsukan,
Matsumoto 029012
Nihotte, Pascal, Charleville-
Mézières 067849
Nihoul, Pierre, Lille 069498
Niigata-kenritsu Bandaijima Bijutsukan,
Niigata-kenritsu Kindai Bijutsukan,
Niigata 029169
Niigata-kenritsu Kindai Bijutsukan,
Nagaoka 029080
Niigata-kenritsu Rekishi Hakubutsukan,
Nagaoka 029081
Niigata Science Museum, Niigata 029170
Niigata-shi Hakubutsukan,
Niigata 029171
Niigata-shiritsu Aizu Yaichi Kinenkan,
Niigata 029172
Niigata-shiritsu Bijutsukan,
Niigata 029173
Niimi Bijutsukan, Niimi 029180
Niimi Sankodo, Tokyo 082187
Niitsu House, Niigata 029174
Nijenhuis, Hollandscheveld ... 132544
Nijhof & Lee, Amsterdam 143054
Nijinskas of Winnipeg, Winnipeg 101888
Nijmeegs Volkenkundig Museum,
Nijmegen 032691
Nika, Chimki 114178
Nika, Moskva 143648
Nika, Sankt Silvester 116626
Nikau Gallery, Motueka 112950
Nikau Gallery, Whangarei 113109
Nike, Milano 080634
Nike Restauro Opere d'Arte,
Firenze 131224
Nikh, Tilburg 112683
Niki Bijutsukan, Nasu 029165
Nikitiny, Sankt-Peterburg 133020
Nikko Futarasan-jinja Hakubutsukan,
Nikko 029182
Nikko Toshogu Treasure Museum,
Nikko 029183
Nikman, Varna 064243
Niko Pirosmanashvili State Museum,
Tbilisi 016417
Nikolaos Perantinos Museum of
Sculpture, Athinai 023022
Nikolaou, Lola, Thessaloniki ... 109093
Nikolaus Matz Bibliothek
(Kirchenbibliothek), Michelstadt 020401
Nikolov-Asja, Maksim, Sofia .. 064234
Nikolov, G., Sofia 101166
Nikolov, V., Halle, Saale 107210
Nikolaz Baratashvili Memorial
Museum, Tbilisi 016418
Nikos Nikolaides Theatrical Archives,
Lemesos 009428
Nilantik, İstanbul 088133
Nile Art Gallery, Baltimore ... 119760
Nili Antiques, Miami 094778
Nilma Collectables Emporium,
Nilma 061974
Nils Aas Kunstverksted, Inderøy 033680
Nilsen, Arendal 084211
Nilsiän Kotiseutumuseo, Nilsiä . 011188
Nilsson et Chiglien – NEC, Paris 071712
Nilsson, Bernt, Sunne 126683
Nilstad, Oslo 143408
Nimberlee, Wyndham 099924
Nimbin Museum, Nimbin 001360
Nimbus Antiques, Whaley Bridge 091817,
135218

NIMBUS Kunst und Bücher, Wädenswil 144146
Nimmergut, Florian, Nördlingen . 077792
Nimmo & Spooner, London 090374
Nimoth, Torsten, Dresden 129717
N'Importe Quoi, Gent 063652
Nina Buceti, Messina 080373
Ninaber, Paulus, Granada 086039
Ninart Centro de Cultura, México 031267
Ninas Antik, Göteborg 086713
Nine, Gwangju 111599
Nine Arts Gallery, Antwerpen .. 100507
Nine Clarendon Cross Gallery, London 118477
Nineteenth Century Antiques, Norfolk 095871
Nineteenth Century Imprints, Portland 145459
Ninety, Charenton-le-Pont 138652
Ninety-Six National Historic Site, Ninety-Six 051566
Ninfea, Venezia 081864
Ning, Anna, Hong Kong 102099
Ningbo Baoguo Si Display Center, Ningbo 008137
Ningbo Museum, Ningbo 008138
Ningbo Painting Gallery, Ningbo 102155
Ningde Museum, Ningde 008140
Ningde Zone Shezu Museum, Ningde 008141
Ningdu Museum, Ningdu 008142
Ningxia Hui Autonomous Region Museum, Yinchuan 008486
Ningyang Museum, Ningyang .. 008143
Ninni Esposito, Bari 109772
El Niño Seise, Sevilla 086460
Ninohe-shiritsu Rekishi Mingeikan, Ninohe 029185
Ninth Inning, Saint Paul 096844
Ninth Street Gallery, Saint Louis 123848
Ninth Street Market Books, Yeadon 145646
Nipawin Living Forestry Museum, Nipawin 006497
Nipigon Museum, Nipigon 006498
Nipissing Township Museum, Nipissing 006499
Nippon Design Special Training College, Tokyo 056051
Nippon Doreikan, Shirotori 029397
Nippon International Contemporary Art Festival, Tokyo 098231
Nippon Kokeshikan, Naruko ... 029156
Nippon Kokogakkai, Tokyo 059512
Nippon Shuppan Hanbai, Tokyo . 137787
Nippondo, Double Bay 098977
Nippur, Clairac 013596
Niquelor, Badalona 133089
Nirmitee, Mumbai 109457
Nirmitee Interiors, Pune 079415
Nirvana Art Gallery, Kathmandu 112082
Nisart, Launceston 099221
Niselli, Mario, Bassano del Grappa 109781
Nisha Art, Pune 109498
Nishida, Nara 111337
Nishiharu, Kyoto 082036
Nishijin Textile Center, Kyoto .. 028972
Nishimura, Niigata 137772
Nishinomiya-shiritsu Otani Bijutsu Kinenkan, Nishinomiya 029188
Nishio, Washington 136460
Nishna Heritage Museum, Oakland 051686
Nisista, Haarlem 083346
Nisple, Agathe, Appenzell 116027
Niss Art, Nice 070640, 128867
Nissan, Genève 087490
Nissen, Sylvie, Cannes 067646, 103519
Nissl, S., Crewkerne 111776
Nissou, Michel, Sarlat-la-Canéda 105653
Nisters, Josef, Speyer 108653
Nistri-Lischi, Pisa 137719
Nitobe Kinenkan, Hanamaki ... 028608

Nitrianska Galéria, Nitra 038409
Nitro Auction House, Nitro 127369
Nitsch, Carolina, New York 138413
Nitsche, Dr. Peter, Lübeck 077182, 107919
Nitsche, Tanja, Seefeld 130635
Nittel, Dietmar, Graz 127750
Nitten – Japan Fine Arts Exhibition, Tokyo 029569, 059513
Nitzschmann, Regina, Nienburg . 130461
NIU Art Museum, Northern Illinois University, DeKalb 048225
Nivågårds Malerisamling, Nivå . 010235
Nivalan Museo Katvala, Nivala . 011190
Nivarlet, Mons 063798
Niven, Janet, Woollahra 062477
Niwa Ceramic Gallery Ltd, Vancouver 101817
Nizam, Samer, Damascus 088048
Nizami ad Dövlet Azerbaycan Edebiyyati Muzeyi, Bakı ... 003174
The Nizam's Museum, Hyderabad 024300
Nižegorodskaja Starina, Nižnij Novgorod 085221
Nižegorodskij Chudožestvennyj Muzej, Nižnij Novgorod 037561
Nižegorodskij Gosudarstvennyj Architekturno-stroitelnyj Universitet, Nižnij Novgorod 056290
Nižegorodskij Gosudarstvennyj Istoriko-architekturnyj Muzej-zapovednik, Nižnij Novgorod 037562
Nižegorodskij Kreml s Archangelskim Soborom, Nižnij Novgorod ... 037563
Nizhoni, Albuquerque 119499
Nižnetagilskij Municipalnyj Muzej Izobrazitelnych Iskusstv, Nižnij Tagil 037566
Nižnetagilskij Muzej-zapovednik Gornozavodskoj Ural, Nižnij Tagil 037567
Nižyns'ke Učylyšče Kul'tury i Mystectv im. M. Zan'kovec'koï, Nižyn .. 056498
Nizza, Giulio, Torino 081736
Njærheim Antik, Nærbø 084284
Njegošev Muzej, Narodni Muzej Crne Gore, Cetinje 038192
Njepila-Hof Rohne, Rohne 021463
Njudungs Hembygdsmuseum, Vetlanda 041162
NKKM-nytt, Oslo 138947
Nk'Mip Desert & Heritage Centre, Osoyoos 006568
NKV – Nassauischer Kunstverein e.V., Wiesbaden 059208
NM Gallery, Yangon 112057
NN-fabrik, Oslip 100134, 136540
NN Gallery, Stockholm 115916
N'Namdi, G.R., Chicago 120217
N'Namdi, G.R., Detroit 120814
N'Namdi, G.R., New York 122922
No 9 Antiques, Wolverhampton . 091906
No 90, Moskva 143649
No Man's Land Historical Museum, Oklahoma Historical Society, Goodwell 049160
No Meio do Caminho, Rio de Janeiro 101052
No Name Exhibitions, Minneapolis 122016
No. 1 Castlegate Antiques, Newark-on-Trent 090816
Noa Gallery, Washington 124939
Noack, Hermann, Berlin 106387, 106388
Noack, Joachim, Dresden 075422
Noah Webster House – Museum of West Hartford History, West Hartford 054303
Noahs, Crewkerne ... 089006, 134306
Noahs Ark Fine Art, Los Angeles 121564
Noah's Art, San Antonio 124053
Noank Museum, Noank 051567
Noass, Rīga 111761

Nob Hill Gallery, Albuquerque .. 119500
Nobel, Oslo 084330
Nobel 17, Plauen 078057
De Nobele Baas, 's-Hertogenbosch 132536
Nobele, F. de, Paris ... 136970, 141249
Nobelmuseet och Bofors Industrimuseum, Karlskoga ... 040737
NobelMuseum, Stockholm 041027
Nobi, Cairo 103063
Nobile, Angelo, Roubaix 072551
Nobile, Maurizio, Bologna 079860
Nobili, Mara S.F., Rio de Janeiro 064093
Nobili, Noris, Firenze 080177
Noble Antiques, København 065939
Noble Art, Bega 098725
Noble Collectibles, Albuquerque 092071
Noble Fireplaces, Cork 079503
The Noble Maritime Collection, Staten Island 053500
Noble Upholstery, Hamilton ... 128173
NobleArt, Cambridge 117508
Nobleboro Historical Society Museum, Nobleboro 051568
Noblecourt, Marie, Paris 129100
Nobles County Historical Society, Worthington 054552
Noblewoods, Woollahra 062478, 127700
Nobre, José M. Rosário, Lisboa 132945
Nobrega, São Paulo .. 064197, 101104
Nobynäs Säteri, Aneby 040528
Nocchi, Jean-François, La Ferté-Bernard 069026
Il Noce, Napoli 080907
Noceti, Angela, Genova 080273
Noctalis-Welt der Fledermäuse, Bad Segeberg 016877
Nocturnus Antiquarios, Rio de Janeiro 064094
Noczinski, Anke, Leipzig 130226
Noda, Nagoya 111326
Noddy, Port Noarlunga 062099
Il Nodo Antico, Milano 080635
Nodrum, Charles, Richmond, Victoria 099635
Noeau, Hale O., Honolulu 120953
Der Nöck, Waiblingen 141248
Nøglen Ting og Antik, Roskilde . 066094
Noehles, Martina, Mühlheim am Main 130318
Noel, Charlotte 120003
Noël, Daniel, Bruxelles 128032
Noël, Marcel, Bourges 067431
Noël, Robert, Vitry-le-François .. 074260
Noelanders Horizon, Knokke-Heist 100752
Noelani, Honolulu 120954
Nöll, Nina, Hannover 076386
Noëlle & Gérard Antiquaires, Couhe 068178
Nørgaard, Henning, Hjørring ... 065765
Nöring, Hans, Dortmund 075391
Noero, Franco, Torino 110990
Nørregaard, Ebeltoft 102791
Nöske, Dr. Margareta, Dresden . 106743
Nøstetangen Museum, Hokksund 033660
Nöth, Dr. Michael, Ansbach ... 074417
Noetzel, Florian, Wilhelmshaven 137554
El Nogal, Bogotá 065216
Nogal, Oviedo 115327
Nogarotto, Danilo, Padova 080993
Nogata-shi Sekitan Kinenkan, Nogata 029195
Nogatsch, Zürich 087993
Nógrádi Történeti Múzeum, Nógrádi Megyei Múzeum Szervezet, Salgótarján 023870
Noguchi Hideyo Kinenkan, Inawashiro 028686
Nogueras & Blanchard, Barcelona 114945
Nogurot, Eleonora, Trieste 081798
Nogyo Hakubutsukan, Miyazaki Daigaku, Miyazaki 029045

Nohea, Honolulu ... 120955, 120956, 120957
Nohl, Peter, Bern 133686
Nohle, Wolfram, Berlin 141560
Nohrden, Jörg, Neumünster ... 077740
Nohwike Bagowa, White Mountain Apache Cultural Center and Museum, Fort Apache 048778
Noir d'Ivoire, Paris 105130
Noir Ebène, Paris ... 071713, 105131
Noir sur Blanc, Bruxelles 140192
Noir, Alain, La Verpillière 069121
Noirhomme, Alain, Bruxelles .. 100646
Noirmont Command Bunker, Saint Brelade 045682
Nokian Kotiseutumuseo Hinttala, Nokia 011192
Nokomis and District Museum, Nokomis 006500
Nolan Gallery, Tharwa 001572
Nolan-Rankin, Houston 121102
Nolan, David, New York 122923
Nolan, John, Dublin 109609
Nolan, John, Southport 091381
Nold, R., Muggensturm 142215
Nolde Stiftung Seebüll, Neukirchen, Kreis Nordfriesland 020715
Nolde Stiftung Seebüll, Dependance Berlin, Berlin 017265
Noles-Davis, Dallas 135514
Nolhin Torppa, Laitila 011085
Noll, Ingrid & Günter, Bad Ems . 074507
Nollan, Christophe, Chantilly ... 067833
Noller, Bertram, Reigate 091090, 134965
Noller, Reinhold, Heilbronn ... 076475
Nollet, Jean-Pierre, Gallardon . 068624
Nollinger, Marie-Christine, Versailles 129341
Nolte, Bielefeld 074985
Nolte-Schmidt, U. & W., Düsseldorf 129752
Nolte, Edelgard, Warburg 108846
Nolte, Klaus, Lippetal 077131
Nolte, Michael, Münster 077672
Noltekuhlmann & Mayr, Berlin .. 074891
Nolting, Horst Joachim von, Bad Soden am Taunus 141435
Nolting, Nicole, Kassel 076724
Nolton Antiques, Bridgend 088596
Nolton Art Gallery, Bridgend Mid 117421
Nomad Gallery, Cairo 103064
Nomad Gallery, Saint Paul 123921
Nomadenschätze, Zürich 087994, 116843
Nomadic Rug Traders, Pyrmont . 062128, 127666
Le Nombre d'Or, Rontalon 072534
Nomel, Sofia 064235
Nomicos, Timişoara 085137
Nominé, Delphine, Reims 072397
Nomis Arts France, Paris 071714
Nomisma, Serravalle 085290
Nomura Bijutsukan, Kyoto 028973
Non Solo Antico, Torino 081737
Non Ti Scordar Di Me, Milano .. 131484
Nonnahûs, Akureyri 024066
Nonnenberg, Kerstin, Freiburg im Breisgau 129861
Nonnenbruch, Barbara, Bielefeld 074986, 106496
Nonnenmacher, T., Freiburg im Breisgau 141807
Nonomura, Washington 097853, 136461
Nonseum, Herrnbaumgarten ... 002079
Nonsolotarli di Mulas, Cagliari .. 079994
Nonsuch Antiques, London 090375
Nonthaburi Natural History Museum, Nonthaburi 042706
Nooit Gedacht, Amersfoort 082654
Nook, Cashel 079472
Nooks & Crannies, Phoenix ... 091227
Nooks & Crannies, Ryde 091157
Noonan, Limerick 079658
Noor Arts, Monaco 112012
Noord-Hollands Motoren Museum,

Nieuwe Niedorp 032672
Noordbrabants Museum, 's-
  Hertogenbosch 032472
Noordelijk Busmuseum,
  Winschoten 033052
Noordelijk Scheepvaartmuseum,
  Groningen 032384
Het Noorderhuis, Zaandam ... 033080
Noordermeer, Birmingham ... 092425
De Noordwester, Vlieland ... 032997
Noortman Master Paintings,
  Maastricht 083473, 112580
Noosa Regional Art Gallery,
  Tewantin 001570
Noosa Shire Museum, Pomona . 001412
Noott, John, Broadway 088679, 088680,
                               134172
Nor West Bend Museum, Morgan 001311
Nor' Wester and Loyalist Museum,
  Williamstown 007419
Nora Antik, Nora 086877
Nora Auktionskammare, Nora . 126646
Nora Barnacle House Museum,
  Galway 024927
Nora Dora, Saint Louis 096749
Nora Eccles Harrison Museum of Art,
  Logan 050357
Nora Museum, Gyttorp 040675
Nora's Porch, San Francisco ... 097297
Norba: Arte, Cáceres 139064
Norberg, Charlotte, Paris 105132
Norbert, Houston 093755
Norbert G Collections, Paris ... 071715
Norbottens Antik, Boden 086643
Nord-Antique, Hamburg 098158
Nord Antique, Lille 069499
Nord Antique, Seclin 073464
Nord-Jarlsbergmuseene,
  Holmestrand 033664
Nord Ouest Antiquités, Bondues 067285
Nord-Trøndelag Fylkesgalleri,
  Namsos 033797
Nord-Troms Museum, Sørkjosen 033958
Nordamerika Native Museum-NONAM,
  Zürich 042223
Nordanstigs Konstförening,
  Bergsjö 059898
Nord'Antic, Liessies 069461
Nordantik, San Antonio 097006
Nordberg, Uddevalla 143983
Norddal Museum, Valldal 034066
Norddeutsches Antiquariat,
  Rostock 142337
Norddeutsches Auto- und Motorrad-
  Museum, Bad Oeynhausen ... 016839
Norddeutsches Spielzeugmuseum und
  Stiftung Spiel, Soltau ... 021912
Norddeutsches Vogelmuseum,
  Sammlung Dr. Baumeister, Osterholz-
  Scharmbeck 021028
Norden, D., Antwerpen 063227
Norden, Jacob, Antwerpen ... 063228,
                               125297
Nordén, Johan, Stockholm ... 126674
Norden, Peter, Tetbury 091577
Nordenfjeldske Kunstindustrimuseum,
  Trondheim 034034
Nordenhake, Stockholm 115917
Nordenhake, Claes, Berlin 106389
Nordens Ark, Hunnebostrand ... 040711
Nordera, Bruno, Beaune 067031
Norderneyer Fischerhausmuseum,
  Norderney 020822
Nordfjord Folkemuseum,
  Sandane 033919
Nordfriesisches Museum Ludwig-
  Nissen-Haus, Zweckverband
  Museumsverbund Nordfriesland,
  Husum 019367
Nordfriesland Galerie, Braderup . 106553
Nordfyns Museum, Bogense ... 010023
Nordhagen, Jan, Oslo 126376
Nordhallands Hembygdsförening,
  Kungsbacka 040764

Nordhannoversches Bauernhaus-
  Museum, Wöhler-Dusche-Hof,
  Isernhagen 019431
Nordhausen, H., Bruchhausen-
  Vilsen 106599
Nordia Art, Turku 125482
Nordic Art Promotion, Singapore 114579
Nordic Art School, Kokkola 055346
Nordic Heritage Museum, Seattle 053241
Nordica Homestead Museum,
  Farmington 048704
Nordico – Museum der Stadt Linz,
  Linz 002326
Nordico, Verlagsabteilung, Linz . 136534
Nordin, Cheptainville 067959
Nordische Botschaften
  Gemeinschaftshaus/Felleshus –
  Galerie, Berlin 017266
Nordisk Antik, Gislinge 065703
Nordisk Antik og Handelshuset,
  Holbæk 065776
Nordisk Konservatorforbund, Den
  Danske Afdeling, Kongens
  Lyngby 058560
Nordisk Kunstauktion, Vejle ... 125473
Nordiska Akvarellmuseet,
  Skärhamn 040949
Nordiska Konservatorförbundet,
  Svenska Sektionen, Stockholm 059899
Nordiska Museet, Stockholm ... 041028
Nordiska Travmuseet i Årjäng,
  Årjäng 040535
Nordjyllands Kunstmuseum,
  Aalborg 009999
Nordjysk Auktion, Hjørring 125443
Nordkappmuseet, Honningsvåg . 033668
Nordland Røde Kors
  Krigsminnemuseum, Narvik ... 033800
Nordlandsmuseet, Bodø 033549
Nordli Bygdemuseum, Nordli ... 033807
Nordlings Antikbod, Stockholm . 087035
Nordmann, Bissendorf 075009
Nordmarken, San Diego 124148
Nordmøre Museum, Kristiansund 033720
Nordnorsk Fartøyvernsenter og
  Båtmuseum, Gratangen 033622
Nordnorsk Kunstmuseum,
  Tromsø 034027
Norðoya Fornminnissavn,
  Klaksvík 010734
Nordpfälzer Heimatmuseum,
  Rockenhausen 021442
Nordqvist, Erik, Esbjerg 065641
Nordre Husan, Alvdal 033491
Nordseemuseum, Bremerhaven . 017610
Nordsjællandsk Folkemuseum,
  Hillerød 010120
Nordsjøfartmuseet, Tælavåg ... 034008
Nordsømuseet, Hirtshals 010012
Nordsten, Malmö 115766
Nordstrand Visual Arts Gallery, Wayne
  State College, Wayne 054249
Nordun, Birgit & Olav, Hvidovre . 065805,
                               128393
Nordvärmlands Jakt och
  Fiskemuseum, Sysslebäck .. 041073
Nordwestdeutsches Schulmuseum
  Friesland, Zetel 022867
Nordwolle Delmenhorst,
  Nordwestdeutsches Museum für
  Industriekultur, Delmenhorst . 017851
Nordys, Birmingham 119812
Nore og Uvdal Bygdetun, Uvdal i
  Numedal 034060
Norenska Galleri, Kristianstad . 086790
Nores, Tampa 136393
Noreview Folk and Heritage Museum,
  Bennettsbridge 024833
Norfolk and Suffolk Aviation Museum,
  East Anglia's Aviation Heritage
  Centre, Flixton 044351
Norfolk Antique Company,
  Norfolk 095872
Norfolk Arts Centre, Simcoe ... 007043

Norfolk Contemporary Art Society,
  Norwich 060201
Norfolk Historical Museum,
  Norfolk 051572
Norfolk Historical Society, Norfolk 051580
Norfolk House Galleries, Dorking 089090
Norfolk Island Museum, Kingston 033449
Norfolk Polyphon Centre,
  Bawdeswell 088409
Norfolk State University Art
  Department, Norfolk 057584
Norges Birøkterlags Museet,
  Billingstad 033541
Norges Fiskerimuseet, Bergen . 033537
Norges Hjemmefrontmuseum,
  Oslo 033848
Norges Museumsforbund, Oslo . 059663
Norges Olympiske Museum,
  Lillehammer 033743
Norgrove, P.W., Haslingden ... 089462
Noria, Binic 103389
Norick Art Center, Oklahoma City
  University, Oklahoma City 057613
Noriega, Alejandro, Buenos Aires 060918,
                               125171
Norilskaja Chudožestvennaja Galereja,
  Norilsk 037569
Norlin, Kirk, Denver 120752
Norlis Antikvariat, Oslo 143409
Normal's Book and Records,
  Baltimore 144987
Norman Cleveland County Historical
  Museum, Norman 051587
Norman No. 1 Oil Well Museum,
  Neodesha 051214
Norman R. Eppink Art Gallery, Emporia
  State University, Emporia ... 048580
Norman Rockwell Museum at
  Stockbridge, Stockbridge ... 053537
Norman Wells Historical Centre,
  Norman Wells 006501
Norman, Glenn, Woolloomooloo . 099922
Norman, Jennifer & Raymond,
  Needham 090802
Norman, Jeremy, San Francisco 145563
Norman, Michael, Brighton 088628
Norman, Peter, Burwell 088737, 134188
Normanby Hall, Scunthorpe ... 045747
Normanby Park Farming Museum,
  Scunthorpe 045748
Normandie Antiquités, Montréal 064448
Normandie Détection, Le Havre . 069287
Normandie Terre des Arts, La
  Chapelle-Urée 103881
Normandy, Honfleur 103840
Normannischer Turm auf dem
  Ruinenberg, Stiftung Preußische
  Schlösser und Gärten Berlin-
  Brandenburg, Potsdam 021196
Normanns, Rælingen 137887
Normanton Church Museum,
  Oakham 045428
Norminton, David, München ... 077582
Normisist Folk Inventions Museum,
  Northern Mindanao State Institute of
  Science and Technology (NORMISIST),
  Butuan 034446
Noronha, José Maria Pacheco,
  Lisboa 084927
Noronha, José Maria Pachero,
  Lisboa 084928
Noronha, Pedro José, Lisboa .. 084930
Noronha, Pedro Mendia, Lisboa . 084929
Norori, Luis, San Francisco 136339
Norr Antik, Sundsvall 087091
Norr-Bygdens Auktionsbyrå,
  Skelleftehamn 126662
Norræna Húsið, Reykjavík 024104
Norrahammars Industrimuseum,
  Norrahammar 040874
Norrbottens Järnvägsmuseum,
  Luleå 040803
Norrbottens Museum, Luleå ... 040804
Norrbyskärs Museum, Hörnefors 040704

Norre Lorraine, Veneux-les-
  Sablons 074014
Norris & Co., Newbury 134875
Norris Arm Heritage Museum, Norris
  Arm 006502
Norris Museum, Saint Ives,
  Cambridgeshire 045695
Norris, B.J., Harrietsham 126917
Norrköpings Konstmuseum,
  Norrköping 040879
Norrköpings Stadsmuseum,
  Norrköping 040880
Norrmalm, Stockholm 087036
Norrtälje Auktionskammare,
  Norrtälje 126647
Norrtälje Konsthall, Norrtälje ... 040882
Norseman Historical and Geological
  Collection, Norseman 001362
Norsewood Pioneer Cottage Museum,
  Norsewood 033260
Norsk Antikvarbokhandlerforening,
  Oslo 059664
Norsk Barnemuseum, Stavanger 033977
Norsk Bergverksmuseum,
  Kongsberg 033708
Norsk Bremuseum, Fjærland ... 033598
Norsk Emballasje museum, Oslo 033849
Norsk Farmasihistorisk Museum,
  Oslo 033850
Norsk Fiskeindustrimuseum,
  Melbu 033768
Norsk Fiskerværsmuseet,
  Sørvågen 033961
Norsk Fjellmuseum, Lom 033753
Norsk Folkemuseum, Norwegian
  Museum of Cultural History,
  Oslo 033851
Norsk Grafisk Museum,
  Stavanger 033978
Norsk Hagebruks Museum,
  Grimstad 033625
Norsk Hermetikkmuseum,
  Stavanger 033979
Norsk Husflidsmuseum, Moen .. 033777
Norsk ICOM, Oslo 059665
Norsk Industriarbeidermuseet,
  Rjukan 033894
Norsk Jernbanemuseum, Hamar 033635
Norsk Kartmuseum, Hønefoss . 033655
Norsk Kjøretøyhistorisk Museum,
  Lillehammer 033744
Norsk Klippfiskmuseum, Nordmøre
  Museum, Kristiansund 033721
Norsk Landbruksmuseum, Ås .. 033507
Norsk Luftfartsmuseum, Bodø .. 033550
Norsk Oljemuseum, Stavanger . 033980
Norsk Reiselivsmuseum,
  Balestrand 033519
Norsk Senmuseum, Stange ... 033971
Norsk Sjøfartsmuseum, Oslo ... 033852
Norsk Skogmuseum, Elverum .. 033586
Norsk Speidermuseum, Oslo ... 033853
Norsk Teknisk Museum, Oslo ... 033854
Norsk Telemuseum – Kristiansand,
  Kristiansand 033716
Norsk Telemuseum – Lødingen,
  Lødingen 033747
Norsk Telemuseum – Sørvågen,
  Norwegian Telecommunication
  Museum, Sørvågen 033962
Norsk Telemuseum – Stavanger,
  Stavanger 033981
Norsk Telemuseum – Trondheim,
  Trondheim 034035
Norsk Tindemuseum, Åndalsnes 033495
Norsk Tollmuseum, Oslo 033855
Norsk Trikotasjemuseum, Salhus 033916
Norsk Utvandrermuseum,
  Ottestad 033874
Norsk Vasskraft- og
  Industristadmuseum, Tyssedal 034053
Norsk Vegmuseum, Fåberg ... 033591
Norsk Veterinærmedisinsk Museum,
  Oslo 033856

Den Norske Auksjonsforretning,
Oslo 126377
Norske Billedkunstnere, Oslo . . . 059666
Norske Grafikere, Oslo 059667
Norske Grenselosers
Museum, Stiftelsen-Museum,
Bjørkelangen 033545
Norske Kunst- og Kulturhistoriske
Museer, Oslo 059668
Norske Kunstforeninger, Oslo . . 059669
Norske Museumspedagogiske
Forening, Bodø 059670
Norske Tekstilkunstnere, Oslo . . 059671
Norstrand, Lise, Svendborg 066159
Nortan, Sharjah 088205
North American Black Historical
Museum, Amherstburg 005332
North Andover Historical Museum,
North Andover 051591
North Anna Nuclear Information Center,
Mineral 050930
North Atlantic Aviation Museum,
Gander 005857
North Ayrshire Museum,
Saltcoats 045729
North Bay and Area Museum, North
Bay 006508
North Beach Gallery, San
Francisco 124355
North Bengal Science Centre,
Siliguri 024483
North Berwick Museum, North
Berwick 045386
North Carolina Central University Art
Museum, Durham 048411
North Carolina Maritime Museum,
Beaufort 046847
North Carolina Museum of Art,
Raleigh 052450
North Carolina Museum of Forestry,
Whiteville 054377
North Carolina Museum of History,
Raleigh 052451
North Carolina Museum of Life and
Science, Durham 048412
North Carolina Museum of Natural
Sciences, Raleigh 052452
North Carolina Museums Council,
Raleigh 060647
North Carolina Pottery Center,
Seagrove 053225
North Carolina State University Gallery
of Art and Design, Raleigh . . . 052453
North Carolina Transportation Museum,
Spencer 053436
North Carr Lightship, Dundee . . 044166
North Castle Historical Society
Museum, Armonk 046543
North Central Washington Museum,
Wenatchee 054281
North Coast Rare Books, Brunswick
Heads 139667
North Cornwall Museum and Gallery,
Camelford 043853
North Country Museum of Arts, Park
Rapids 051924
North Dakota Museum of Art, Grand
Forks 049177
North Devon Antiques Centre,
Barnstaple 088363
North Devon Maritime Museum,
Appledore, Devon 043422
North Down Heritage Centre, Bangor,
County Down 043489
North East Aircraft Museum,
Sunderland 045922
North Florida Community College Art
Gallery, Madison 050581
North Fremantle Antiques, North
Fremantle 061989
North Gallery, Saint Louis 123849
North Haven Historical Society
Museum, North Haven 051606
North Hertfordshire Museums,

Letchworth 044814
North Hertfordshire Museums
Resources Centre, Hitchin 044592
North Highland Community Museum,
Cape North 005553
North Hills Museum, Granville
Ferry 005918
North House Gallery,
Manningtree 118713
North House Museum, Lewisburg 050229
North Huron District Museum,
Wingham 007430
North Kerry Museum, Ballyduff . 024826
North Lakes Antiquités, Saint-Martin-
de-Ré 072940
North Lanark Regional Museum,
Almonte 005322
North Lane Antiques Market,
Brighton 088629
North Lee County Historic Center and
Santa Fe Depot Museum Complex,
Fort Madison 048841
North Leverton Windmill, North
Leverton 045387
North Light Books, Cincinnati . . 138326
North Light Gallery, Huddersfield 118004
North Lincolnshire Museum,
Scunthorpe 045749
North London Clock Shop,
London 090376, 134736
North Marysburgh Museum,
Picton 006647
North Mountain Books, Phoenix 145430
The North Museum of Natural History
and Science, Lancaster 050124
North Norfolk Fine Art and Antiques
Fair, Burnham Market 098395
North Norfolk Railway Museum,
Sheringham 045784
North Oak Gallery, Columbus . . 120505
North of England Museums Libraries
and Archives Council, Newcastle-
upon-Tyne 060202
North Otago Museum, Oamaru . 033262
North Pacific Cannery Village Museum,
Port Edward 006684
North Paddock, Laura 061728
North Perth Antique Centre, North
Perth 061995
North Plainfield Exempt Firemen's
Museum, North Plainfield 051612
North Platte Valley Museum,
Gering 049084
North Point Gallery, San
Francisco 124356
North Shore Art League,
Winnetka 058121, 060648
The North Shore Arts Association,
Gloucester 049135
North Shore Arts Association,
Gloucester 060649
North Shropshire Reclamation,
Burlton 088718
North Somerset Museum, Weston-
super-Mare 046101
North Star Galleries, New York . 095616
North Star Mining Museum, Grass
Valley 049213
North Star Museum of Boy Scouting
and Girl Scouting, North Saint
Paul 051615
North Stradbroke Island Historical
Museum, Dunwich 001041
North Sydney Museum, North
Sydney 006514
North Thompson Museum,
Barriere 005387
North Tonawanda History Museum,
North Tonawanda 051619
North Vancouver Museum, North
Vancouver 006518
North View Gallery, Portland
Community College, Portland . 052297
North Wales Antiques, Colwyn

Bay 088963
North Weald Airfield Museum and
Memorial, Epping 044299
North West Books, Liverpool . . . 144546
North West Federation of Museums
and Art Galleries, Ashton-under-
Lyne 060203
North West Film Archive,
Manchester 045214
North West Museum of Road
Transport, Saint Helens 045688
North West Sound Archive,
Clitheroe 043970
North West Style, Seattle 124575
North Western Agricultural Museum,
Warracknabeal 001621
North Wind Undersea Institute,
Bronx 047236
North Woolwich Old Station Museum,
London 045066
North Yorkshire Moors Railway,
Pickering 045503
North, Desmond & Amanda, East
Peckham 089140, 134350
Northampton and Lamport Railway,
Chapel Brampton 043903
Northampton County Historical and
Genealogical Society Museum,
Easton 048458
Northampton Museum and Art Gallery,
Northampton 045391
Northart Community Arts Centre,
Auckland 033157
Northborough Historical Museum,
Northborough 051628
Northcliffe Pioneer Museum,
Northcliffe 001369
Northcote Gallery, London 118478,
118479
Northcote Pottery Gallery,
Thornbury 001576
Northcote Road Antiques Market,
London 090377
Northcote, Robine, Saint-Quay-
Portrieux 073272
Northcutt Steele Gallery, Montana
State University Billings, Billings 046986
Northeast Classic Car Museum,
Norwich 051652
Northeast-Journal of Antiques & Art,
Palmer 139461
Northeast Louisiana Delta African
American Heritage Museum,
Monroe 050999
Northeast Mississippi Museum,
Corinth 047992
Northeast Mississippi Museums
Association, Corinth 060650
Northeast Oakland Historical Museum,
Oxford 051861
Northeast Texas Rural Heritage
Museum, Pittsburg 052152
Northeastern Illinois University Art
Gallery, Chicago 047684
Northeastern Montana Threshers and
Antique Association, Culbertson 048085
Northeastern Nevada Museum,
Elko 048541
Northern Antiques Company, Norton,
North Yorkshire 090863
Northern California Association of
Museums, Chico 060651
Northern California Coin Exchange,
Oakland 095904
Northern Clay Center Galleries,
Minneapolis 050946
Northern Clocks, Worsley 091950,
135260
Northern Culture Museum,
Niigata 029175
Northern Eye Gallery, Ceres . . . 117547
Northern Galleries, Northern State
University, Aberdeen 046300
Northern Gallery, Saint Paul . . . 123922

Northern Gallery for Contemporary
Art, City Library and Arts Centre,
Sunderland 045923
Northern Gateway Museum, Denare
Beach 005699
Northern Illinois University Art Gallery
in Chicago, Chicago 047685
Northern Image, Milwaukee . . . 121943
Northern Image, Minneapolis . . 122017
Northern Image Studio of Fine Art,
Ardersier 117247
Northern Images, Edmonton . . . 101294
Northern Images, Winnipeg . . . 101889
Northern Indiana Arts Association,
Munster 051120
Northern Indiana Center for History,
South Bend 053393
Northern Ireland Museums Council,
Belfast 060204
Northern Kentucky University Art
Galleries, Highland Heights . . . 049471
Northern Life Museum and National
Exhibition Centre, Fort Smith . . 005831
Northern Light Gallery, Brønshøj 065620
Northern Lights Art Gallery,
Mayville 050747
Northern Lights Centre, Watson
Lake 007368
Northern Lights, Keswick 118039
Northern Lights Military Museum,
Happy Valley 005985
Northern Lights Studies in Creativity,
Presque Isle 139462
Northern Pacific Railway Museum,
Toppenish 053776
Northern Paiwan Art and Culture
Museum, Sandimen 042433
Northern Taiwan Visitors Center,
Wanli 042581
Northern Traders, West
Melbourne 062421
Northern Virginia Fine Arts Association,
Alexandria 060652
Northfield Arts Guild, Northfield . 060653
Northfield Farm Museum, New
Pitsligo 045326
Northfield Historical Society Museum,
Northfield 051633
Northfleet Hill Antiques,
Northfleet 090858
Northgate Auction Rooms, Newark-on-
Trent 126994
Northgate Gallery, Hixson 127246
Northgate Gallery of Nashville,
Brentwood 127148
Northlake Coin and Stamp,
Atlanta 092162
Northland Books, Saskatoon . . 140386
Northland Exposure Artists Gallery,
Kansas City 121310
Northland Historical Society Museum,
Lake Tomahawk 050087
Northland Poster Collective,
Minneapolis 122018
Northland Upholstery, Columbus 135482
Northport Historical Society Museum,
Northport 051637
Northside Antiques, Clayfield . . . 061336
Northside Antiques, Glenorchy . . 061533
Northside Restoration, Atlanta . . 135314
Northumberland Antique Restorations,
Terrigal 127686
The Northumberland County Historical
Society Museum, Sunbury . . . 053604
Northumberland Fisheries Museum,
Pictou 006650
Northumbria Pine, Whitley Bay . 091834
Northwest Antiques, San Antonio 097007
Northwest Architectural Salvage, Saint
Paul 096845
Northwest Art Center, Minot State
University, Minot 050951
Northwest Camera, Portland . . . 096447
Northwest Collectibles and Antique

Paper, Portland 096448
Northwest Gallery and Sinclair Gallery,
Northwest College, Powell 052343
Northwest Museum of Arts and
Culture, Spokane 053442
Northwest Pastel Society, Gig
Harbor 060654
Northwest Railway Museum,
Snoqualmie 053372
Northwest Tribal Art, Seattle . . . 124576
Northwest Watercolor Society,
Bellevue 060655
Northwestern Mutual Art Gallery,
Cardinal Stritch University,
Milwaukee 050922
Northwestern Oklahoma State
University Museum, Alva 046430
Northwestern Ontario Sports Hall of
Fame, Thunder Bay 007143
Northwich Antiques Center,
Northwich 090862
Northwind Arts Center, Port
Townsend 052260
Northwood, Bournemouth . . . 117405
Northwood Art Space, Columbus 047940
Norton & Roberts, London 144655
Norton Antiques, Twyford,
Norfolk 091681
Norton Conyers, Ripon 045621
Norton Museum of Art, West Palm
Beach 054317
Norton Priory Museum, Runcorn 045651
Norton Simon Museum,
Pasadena 051935
Norton, Pauline, Bridgnorth . . . 117424
Norton, Richard, Chicago 092754,
120218
Norton & Co., W.W., London . . 138233
Norton's Fine Art Studios, Saint
Louis 123850
Norup, Næstved 102933
Norway Art, Minneapolis 122019
Norwegian Federation of Friends of
Museums, Oslo 059672
Norwegian Laft Hus, Red Deer . 006758
Norwich and District Museum,
Norwich 006527
Norwich Arts Council Gallery,
Norwich 051649
Norwich Collectors Toyshop,
Norwich 090880
The Norwich Decorative and Fine Arts
Society, Norwich 060205
Norwich Gallery, Norwich 045403
Norwich Historical Society,
Norwich 051655
Norwich School of Art and Design,
Norwich 056587
Norwich University Museum,
Northfield 051635
Norwood, München 108167
Norwood Historical Association
Museum, Norwood 051656
Norwood Museum, Holm 044594
Norwood Park Artists Guild, Des
Plaines 060656, 120795
Norwood, John, New York 095617
Nosaka O-tomatto Hakubutsukan,
Ito 028712
Nosbaum & Reding, Luxembourg 111861
Nosbüsch, Manfred, Euskirchen 141759
Nose Creek Valley Museum,
Airdrie 005307
Nosel, Doug, Jamestown 127270
Nosorog, Beograd 085343, 114431
Nossereau-Hauff, Christine, L'Isle-sur-
la-Sorgue 069543
Nossi College of Art,
Goodlettsville 057141
Nosta, Bruxelles 063515
Nostalga Kunst- und Antiquitätentage,
mit Mineralien- und Edelsteinbörse,
Oldenburg 098159
Nostalgi Haderslev, Haderslev . . 065721

Nostalgia, Aix-en-Provence 066472
Nostalgia, Annecy 066648
Nostalgia, Blackpool 088499
Nostalgia, Bologna 079861
Nostalgia, Budapest 079270
Nostalgia, Cairo 103065
Nostalgia, Christchurch 083887
Nostalgia, Clevedon 088931
Nostalgia, Dubai 088181
Nostalgia, Los Angeles 094421
Nostalgia, Nelson 118776
Nostalgia, Nijmegen 083516
Nostalgia, Solna 086947
Nostalgia, Stockport 091445
Nostalgia Antik, Offenburg 077909
Nostalgia Antiques, Moonee
Ponds 061879
Nostalgia Antiques, Porthcawl . 091039,
134950
Nostalgia Factory, Kangaroo
Valley 061670, 139786
Nostalgia Factory, Manly 061787
Nostalgia Gallery, Teddington . . 119215
Nostalgia Möbler och Ting, Falun 086673
Nostalgia on 19th Street,
Houston 093756
Nostalgia Schmuck, Pforzheim . 108377
Nostalgia Shop, Saint Louis . . . 096750
Nostalgia Shoppe, Toledo 097614
Nostalgia Traders, Mittagong . . 061867
Nostalgia World, Memphis 094658
Nostalgiaville, Miami 094779
Nostalgie, Pont-a-Marcq 072236
Nostalgie, Tilburg 083620
Nostalgie, Winterthur 087910
Nostalgie & Cottage, Buchholz in der
Nordheide 075209
Nostalgie Antique, Hamburg . . . 076289
Nostalgie auf Rädern –
Fahrzeugmuseum, Großklein . 002009
Nostalgie Museum,
Beetzseeheide 017016
Nostalgie-Museum, Wörth an der
Donau 022722
Nostalgipalatset, Stockholm . . . 087037
Nostalgirundan, Varberg 087152
Nostalgisches Musikparadies,
Oberhasli 041810
Nostalgoteket, København 065940
Nostell Priory, Wakefield 046032
Nostitz, Charles von, New York . 135968
Nostradamus Antiques,
Weybridge 091808
Nostradamus II, East Molesey . . 089138
Nostromo Books, Auckland . . . 143293
Noszlopy Kúria, Újvárfalva 024022
Nosztalgia, Budapest 124565
Not Just Antiques, Tetbury 091578
Not Just Dolls, Dallas 093187
Not Just Silver, Weybridge 091809
Nota, La Paz 100848
Nota Bene, Düsseldorf 075502
Notable Accents, Dallas 093188
Notaphilie Dresden, Dresden . . 075423
Notari, Otello, Roma 131995
Notarishuis Arnhem, Arnhem . . 126277
Notburga-Museum, Eben am
Achensee 001830
Notebaart, P.C., Bussum 143102
De Notelaer, Hingene 003700
Nothe Fort Museum of Coastal
Defence, Weymouth 046108
Nothelfer, Georg, Berlin 106390
Nothing Definite, Saint Louis . . . 096751
Nothing Trivial, Calgary 064263
Notions Antique Centre,
Grantham 089357, 134417
Notizie da Palazzo Albani, Urbino 138851
Notizie degli Scavi di Antichità,
Roma 138852
Notodden Auksjonsforretning,
Notodden 126364
Notodden Brukt og Antikk,
Notodden 084288

Notojima Glass Art Museum,
Notojima 029196
Notranjski Muzej Postojna,
Postojna 038591
Notre Dame of Jolo College Museum,
Sulu Culture Museum, Jolo . . . 034483
Nottbohm Galerie & Kunsthaus,
Göttingen 107147
Nottingham Antiques, Atlanta . . 092163
Nottingham Castle, Nottingham . 045419
Nottingham Transport Heritage Centre,
Ruddington 045644
Nottingham Trent School of Art and
Design, Nottingham Trent University,
Nottingham 056589
Nottingham University Museum of
Archaeology, Nottingham 045420
Nou Cents, Barcelona 085812
Nouchka, La Bastide-des-
Jourdans 068943
Nougues, Henri, Pontorson . . . 072274,
129177
Nouh, Damascus 088049
Noujaim, Juliette, Paris 071716
Noujaim, Paul, Paris 071717
Nouna R, Bruxelles 063516
Nourbakhsch, Giti, Berlin 106391
Nouri-Lokoschus, Ellen, Potsdam 078080
Nourisson, Philippe, Vern-sur-
Seiche 129336
Nourry-Joannès, Claudine, Verneuil-
sur-Avre 074031
Noury, Jacques, Tours 105765
Noûs-Verlag, Tübingen 137537, 142431
Nousbaum, Marie-France,
Campagne 067622
Nousiaisten Kotiseutumuseo
ja Topoisten Ulkomuseoalue,
Nousiainen 011194
Nouveaux Brocanteurs, La Chapelle-
en-Serval 068974
Nouvelle Arche de Noë Editions,
Paris 136971
Nouvelle Art, Tours 113238
Nouvelle Galerie, Grenoble 103790
Nouvelle revue d'esthétique,
Paris 138653
Nouvelle Saison, Grenoble 140852
Nouvelle Tendance, Paris 129101
Nouvelles de l'Estampe, Paris . . 138654
Nouvelles Galeries de Paris,
Bruxelles 125323
Nouvelles Hybrides, Courbevoie . 068186
Nouvelles Images, Den Haag . . 112403
Nouvelles Images, Paris 105133, 136972
Nouvelles Images, Villemandeur 136988,
137016
Nouvellet, Olivier, Paris 105134
Nova Conservação, Lisboa 132946
A Nova Eclética, Lisboa 143587
Nova Gallery, Calgary 101251
Nova Kodl, Praha 102711
Nova Morada, Porto 085040
Nova Rúa, Lugo 115128
Nova Scotia College of Art and Design,
Halifax 055131
Nova Scotia Museum, Halifax . . 005958
Nova Scotia Museum of Natural
History, Halifax 005959
Nova Scotia Printmakers Association,
Centreville 058455
Nova Terra Antiek, Apeldoorn . . 082928
Novagraphics, Tucson 124761
Novaja Galereja, Moskva 114286
Novaja Galereja, Ufa 114376
Novak, Ljubljana 085612, 114722
Novak-Spurny, Irma, Salzweg . . 108553
Novak, C. & B., Gomaringen . . . 076059
Novák, Karel, Kralupy nad
Vltavou 140477
Novalis-Literaturgedenkstätte,
Weißenfels 022545
Novalis-Museum, Forschungsstätte für
Frühromantik, Wiederstedt . . . 022624

Novanta Nove, Amsterdam 112263
Novantatrè, Bologna 109857
Novarese, Domenico, Torino . . . 081738
Novaria Restauri, Genova 131331
Novarino, Enrico, Torino 081739
Novarino, Fabien, Cassis 103538
Novato History Museum, Novato 051659
Novecento, Barcelona 085813
Il Novecento, Manfredonia 110198
Novecento, Modena 131576
Novecento, Padova 080994
Novecento, Rovereto 110881
Novecento & Dintorni, Roma . . . 081478
Il Novecento di Pandora, Firenze 080178
Novecento Editrice, Palermo . . . 137711
Novel Art, Männedorf 116487
Novellan, Hondarribia 115099
Novello, Michele, Venezia 111146
Novelty Art, Mumbai 079407
Noves, Laure de, La Ciotat 068993
Novgorod, Kouvola 140733
Novgorod-Siverskyj Istoriko-kulturnyj
Muzej-zapovidnyk Slovo o Polku
Igorevim, Novgorod-Sivers'kyj . 043268
Novgorodskij Gosudarstvennyj Muzej-
zapovednik, Velikij Novgorod . . 038016
Novi Hram, Sarajevo 100868
Novi, Marco, Bologna 130982
Novità Bibliografiche, Roma . . . 138853
Novo Arts, New York 122924
Novo Galería de Arte, Palma de
Mallorca 115362
Novo Século, Lisboa 114048
Novočerkasskij Muzej Istorii Donskich
Kazakov, Novočerkassk 037572
Novodevičij Monastyr, Gosudarstvennyj
istoričeskij muzej, Moskva . . . 037483
Novoe Prostranstvo, Odesa . . . 117160
Novohradská Galéria, Lučenec . 038384
Novohradské Múzeum, Lučenec 038385
Novohradsky, Wolfgang,
Freigericht 075893
Novokuzneckij Chudožestvennyj Muzej,
Novokuzneck 037575
Novokuzneckij Kraeveďeskij Muzej,
Novokuzneck 037576
Novorossijskij Gosudarstvennyj
Istoričeskij Muzej Zapovednik,
Novorossijsk 037577
Novosibirskaja Architekturno-
chudožestvennaja Akademija,
Novosibirsk 056292
Novosibirskaja Kartinnaja Galereja,
Novosibirsk 037580
Novosibirskij Muzej Prirody,
Novosibirsk 037581
Novosibirskij Oblastnoj Kraevedčeskij
Muzej, Novosibirsk 037582
Novotroickij Istoriko-kraevedčeskij
Muzej, Orenburgskij oblastnoj
kraevedčeskij muzej, Novotroick 037585
Novum Forum – Schule der Sinne,
Neumarkt in Steiermark 002446
Nový Antik Bazar, Praha 065513
Nový Zamok Banská Štiavnica,
Banská Štiavnica 038287
Novyj Ėrmitaž, Moskva 085194
Now & Again, Auckland 083811
Now & Again, Baltimore 092349
Now & Again at 57th Street,
Sacramento 096637
Now & Forever Images,
Christchurch 112855
Now & Then, Pittsburgh 096315
Now & Then, Vancouver 064853
Now & Then Collectibles, Miami 094780
Now & Then Shop, Indianapolis 093913
Now-Again, Los Angeles 094422
Now and Again, Atlanta 092164
Now and Again, Tampa 097580
Now and Then, Edinburgh 089184
Now and Then, Saint Louis 096752
Now Gallery, Vichy 105858
Nowa Galeria, Lublin 113701

Nowa Oficyna – Galeria Rysunku ASP,
Gdańsk 113481
Nowack & Roger, Düsseldorf ... 129753
Nowack, W., Bredstedt 125906
Nowacki, Edwin, Poznań 084646
Nowacki, Peter, Iserlohn 076620
Nowak, Berlin 129543
Nowak, Andreas, Petershagen .. 130510
Nowak, Axel, Hohen Neuendorf . 130062
Nowak, Günter, Aalen 129366
Nowak, Herbert, Augsburg 074468
Nowak, Norbert, Dortmund 075392
Nowak, Roman, Toruń 113793
Nowata County Historical Society
Museum, Nowata 051660
Nowes, Howard, New York 095618
Nowhere Land, Bruxelles 100647
Nowicki, Andrzej Marek,
Warszawa 132894
Nowicki, Piotr Tadeusz,
Warszawa 113875
Nowosądecka Mała Galeria, Nowy
Sącz 035051
Nowra Museum, Nowra 001375
Noyes and Read Gallery and Herndon
Gallery, Antioch College, Yellow
Springs 054576
Noyes Art Gallery, Lincoln 050299
The Noyes Museum of Art,
Oceanville 051706
Noyes, Michael, El Paso 135551
Noyrigat, Jacques, Millau 070122
Nozzoli, Empoli 110007
NRW-Forum Kultur und Wirtschaft,
Düsseldorf 018041
NRW-Kunstkreis, Vettweiß 059209
NS-Dokumentationszentrum der Stadt
Köln, EL-DE-Haus und Gedenkstätte
(Gestapo-Gefängnis), Köln ... 019707
NSCC Art Gallery, North Seattle
Community College, Seattle .. 053242
nt art gallery, Bologna 109858
NTT InterCommunication Center,
Tokyo 029570
Nu Black Art Gallery, Charlotte . 120004
Nuada, Chicago 120219
Nuages, Milano 110393
Nuances-Aripa, Montreuil 138655
Nuances Obernoises, Obernai .. 104420
Nuantta Sunakkutaangit Museum,
Iqaluit 006040
Nubbekullens Hembygdsmuseum,
Västra Ny Hembygdsförening,
Motala 040863
Nubia Museum, Aswan 010525
Nubien Museum, Aswan 010526
Núcleo Arqueológico da Rua dos
Correeiros, Lisboa 035819
Núcleo de Arte Sacra, Museu de
Évora, Évora 035680
Nucleo de Informatica e Cultura Itau
Galeria, Belo Horizonte 100898
Núcleo do Moinho de Maré de
Corroios, Ecomuseu Municipal do
Seixal, Corroios 035642
Núcleo Etnográfico de Cedros,
Cedros 035610
Núcleo Museológico da Bemposta,
Bemposta 035556
Núcleo Museológico da Capela do
Espírito Santo dos Mareantes,
Sesimbra 036025
Núcleo Museológico da Casa do
Infante, Porto 035954
Núcleo Museológico da Cultura do
Pastel dos Açores, Ribeira Chã 035981
Núcleo Museológico da Faculdade de
Engenharia, Universidade do Porto,
Porto 035955
Núcleo Museológico da Faculdade de
Farmácia, Universidade do Porto,
Porto 035956
Núcleo Museológico da Junta de
Freguesia da Orada, Borba ... 035565

Núcleo Museológico da Misericórdia
de Aveiro, Aveiro 035539
Núcleo Museológico da Nacional,
Lisboa 035820
Núcleo Museológico da Paróquia da
Beirã, Beirã 035552
Núcleo Museológico da Quinta do
Prado, Celorico de Basto 035612
Núcleo Museológico da Roda dos
Expostos, Almeida 035509
Núcleo Museológico das Ruínas de
Milreu, Faro 035690
Núcleo Museológico das Salinas
Naturais de Rio Maior, Rio
Maior 035983
Núcleo Museológico de Alverca,
Alverca do Ribatejo 035518
Núcleo Museológico de Amieira do
Tejo, Nisa 035872
Núcleo Museológico de Arte Sacra,
Nisa 035873
Núcleo Museológico de Cachopo,
Cachopo 035582
Núcleo Museológico de Espinho,
Espinho 035657
Núcleo Museológico de Metrologia,
Évora 035681
Núcleo Museológico de S. Cucufate,
Vila de Frades 036078
Núcleo Museológico do Castelo do
Lindoso, Braga 035575
Núcleo Museológico do Centro Cultural
de Borba, Borba 035566
Núcleo Museológico do Hospital de
Santo António dos Capuchos,
Lisboa 035821
Núcleo Museológico do IBTAM,
Instituto do Bordado, Tapeçarias e
Artesanato da Madeira, Funchal 035710
Núcleo Museológico do Lagar de
Azeite, Centro Militar de Educação
Física e Desportos, Mafra ... 035843
Núcleo Museológico do Lagido, São
João 036011
Núcleo Museológico do Santuário do
Senhor Jesus da Pedra, Óbidos 035877
Núcleo Museológico do Tempo,
Santarém 036002
Núcleo Museológico dos Escuteiros,
Torres Vedras 036065
Núcleo Museológico dos Moinhos de
Carreço, Carreço 035599
Núcleo Museológico Etnográfico José
Ramires Saraiva, Ervidel 035653
Núcleo Museológico Medieval /
Moderno, Almada 035507
Núcleo Rural de Ervidel, Museu de
Aljustrel, Ervidel 035654
Núcleo Técnico Artístico, São
Paulo 064198
Núcleos Museológicos da Lagoa,
Lagoa 035741
Nudgee Road Antiques and Arts
Centre, Hendra 061616
Nuding, Angelika, Sinsheim 078484
Nuding, Ursula, Pulheim 078102
Nüans Sanat Galerisi, İstanbul . 117069
Nüchter, Albrecht, Schlitz 078374
Nüdling, Charlotte, Fulda .. 075931,
107090
Nünlist, Octavè, Amriswil .. 087192,
133618
Nürnberger Antique & Art,
Nürnberg 098160
Nüssel, Erich F., München 130375
Nuestro Taller, Montevideo 119434
Nütt, Sybille, Dresden 106744
Nueva Era Arte y Diseño, Bogotá 102352
Nueva Roma, Buenos Aires ... 139591
Nueva Street Gallery, San
Antonio 124054
Nufer, Christoph, Aarwangen ... 087177,
133612
Nugget, Hong Kong 065088

Nugut, Tugun 062351
Nuijamaan Museo, Lappeenranta 011092
La Nuit de Varennes, Marseille . 069954
Nukarin Koulumuseo, Nukari ... 011195
Nukke- ja Lelumuseo, Helsinki . 010871
Nukke- ja Pukumuseo – Hatanpää
Mansion, Tampere 111400
Nukkemuseo Suruton,
Savonlinna 011347
Nukketeatteri- ja Lelumuseo Unelma,
Pieksämäki 011248
Nukumuuseum, Tallinna
Linnamuuseum, Tallinn 010683
Nullkommanix, Hamburg 107302
NUMA-Revue internationale de
numismatique, Luxembourg .. 137813
Numaga, Colombier (Neuchâtel) 116223,
138037
Numan, Sergio, Santiago de
Chile 101916
Numark, Cheryl, Washington ... 124940
Number 1 Royal Crescent Museum,
Bath 043523
Number 24 of Frinton, Frinton-on-
Sea 089310
Number 28 Antiques, London .. 090378
Number 8, Auckland 112798
Number Nine-The Gallery,
Birmingham 117378
Numen, Revista de Arte, Madrid 139065
Numis, Praha 065514
Numis 21, Dijon 068339
Numis Invest, Macerata 080342
Numis-Post & HMZ, Schweizer
Magazin für Münzen, Bad
Ragaz 139096
Numis'Aisne, Soissons 073527
Numisami, Nice 070641
Numisart – Ancient Art,
München 077583
Numiscarte, Saintes 073375
Numisma, Madrid 086228
Numismaatikko, Helsinki 066348
NUMISMATA Berlin, Berlin 098161
Numismata International Erich Modes,
München 077584
NUMISMATA München, München 098162
Numismatic Museum,
Kathmandu 031876
Numismatic Museum, Athens,
Athinai 023023
Numismatic News, Iola . 139463
Numismatic Review, New York . 139464
Numismatic Society of India,
Varanasi 059369
La Numismatica, Brescia 079952,
138854
Numismatica, Madrid 086229
Numismatica 2P, Verona 081929
Numismatica Ajuriaguerra, Bilbao 085882
Numismatica Antichità Filatelia, Gasparri
& Figli, Livorno . 080329, 131348
Numismatica Ars Classica,
London 090379
Numismatica Ars Classica,
Zürich 087995
Numismatica Associacão de Artes e
Antiguidades, Rio de Janeiro .. 064095
Numismatica Cuneese, Cuneo .. 080064
Numismatica Dracma, Logroño . 086076
Numismatica Filatelia Bustese, Busto
Arsizio 079969
Numismatica Filatelia Tevere,
Erba 080067
Numismatica Fiorentina, Firenze 080179
Numismatica Ghia, Seregno ... 081622
Numismatica Max Manna, Roma 081479
Numismatica Metró, Milano ... 080636
Numismatica Milanese, Milano . 080637
Numismatica Pacchiega, Torino . 081747
Numismatica Ranieri, Bologna . 079862
Numismatica Siciliana Conti,
Palermo 081061
Numismatica Terziani, Bologna . 079863

Numismatica Testa, Pulsano ... 081178
Numismatica Varesi Alberto,
Pavia 081136
Numismatica Varesina, Varese . 081814
Numismatica Villafranchese,
Villafranca di Verona 081951
Numismática, Monedas e Notas
Antigas, Belo Horizonte 063913
Numismatique du Port, Toulon
(Var) 073724
Numismatische Gesellschaft Speyer
e.V., Speyer 059210
Numismatische Gesellschaft zu Berlin
e.V., Berlin 059211
Numismatische Kommission der
Länder in der Bundesrepublik
Deutschland, Hannover 059212
Numismatische Studien,
Hamburg 138745
Numismatische Zeitschrift, Wien 138530
Numismatischer Verein, Bern .. 059952
Numismatisches Nachrichtenblatt,
Speyer 138746
Numisporto, Porto 085041
Nummus et Ars, Venezia ... 081865,
126240
Numthong, Bangkok 116955
Nun, Ghislaine, Elne 068440
Nunatta Katersugaasivia
Allagaateqarfialu, Nuuk 023334
Nundah and Districts Historical
Museum, Nundah 001376
Nundah Antiques, Nundah 062020
Nunes, Lisboa 143588
Nunes, Malakoff 069821
Nunes, Affonso, Rio de Janeiro . 064096
Nunes, Haroldo, Rio de Janeiro . 128117
Nunes, Joaquim Pereira, Lisboa 084931
Nunes, Marilene B., Rio de
Janeiro 064097
Nuñez Narbona, Jose Maria,
Toledo 086491
Nuni, Los Angeles 121565
Nunziati, Gianna, Firenze 131225
Nunzio, Salamone, Catania ... 080041,
131093
Nuova Accademia di Belle Arti,
Milano 055929
Nuova Arcadia, Padova 080995
Nuova Artesegno, Udine 111547
Nuova Bianchi d'Espinosa, Napoli 110540,
126215
Nuova Bottega, Bologna 130983
La Nuova Città, Brescia 109912
Nuova Decoritalia, Prato 131780
Nuova Era, Bari 109773
Nuova Galleria, Cortina
d'Ampezzo 109986
Nuova Galleria Campo dei Fiori,
Roma 110838
Nuova Gissi, Torino 110991
Nuova-Icona, Associazione Culturale
per le Arti, Venezia 028319
Nuova Icona, Associazione Culturale per
le Arti, Venezia ... 059470, 111147
Nuova Museologia, Brescia ... 138855
Nuova Rassegna, Cosenza ... 138856
Nuovo Fondaco, Genova 080274
Nuovo Istituto Italiano d'Arti Grafiche,
Bergamo 055875, 137621
Nuovo Museo Provinciale Sannitico,
Campobasso 025730
Nuovo Segno, Forli 110081
Nur, Sarajevo 063869
Nurihiko, Tokyo 111483
Nurkkaus, Ranskalainen, Helsinki 066349
Nurmeksen Museo, Nurmes ... 011197
Nurmon Museo, Nurmo 011198
Nurom Hat Museum, Suame ... 022690
Nurse Bennett Heritage House,
Daniel's Harbour 005677
NUS Museum, Singapore 038273
Nusraty, New York 095619
Nussbaum, Louisville 111662

Nussbaum, Zeist 112724
Nussbaum, Scott F., Louisville . . 094575,
. . . . 145247
Nusser & Baumgart Contemporary,
München . . .077585, 108168, 137418
Nusser, Ursula, München 126062
Nussknackermuseum, Neuhausen /
Erzgebirge 020710
Nussle, Sarah, Paris 071718
Nussli, Stefan, Ittigen 133778
The Nut Point Gallery,
Christchurch 112856
Nutcote Museum, Home of May Gibbs,
Neutral Bay 001349
Nuti Loredana, Roma 081480
Nutida Art, Stockholm 115918
Nutida Svenskt Silver, Stockholm 115919
Nutley Antiques, Nutley 090901
Nutmeg, West Hartford 127492
Nuttall, Pennant Hills 062070
Nutzfahrzeugmuseum,
Hartmannsdorf 019081
Nuwaubian, London 144656
NVA Museum Prora in der
Kulturkunststatt Prora, Prora . . 021230
Nvo Millenium, Scorzè 110922
Ny Ålesund By- og Gruvemuseum, Ny-
Ålesund 033811
NY Arts, New York 139465
NY Arts Beijing, Beijing 101980, 138602
Ny Carlsberg Glyptotek,
København 010188
Nya Kalmar Konst- och Antikvitetsaffär,
Kalmar 086769
Nya och Gamla Ting i Lövånger,
Lövånger 126626
Nyabinghi African Gift Shop, New
York 122925
Nyahokwe Ruins Site Museum,
Inyanga 054853
Nyborg, Oslo 084331
Nyborg Antik, Nyborg 066011
Nyborg og Omegns Museer,
Nyborg 010242
Nye Galleri 22, Trøgstad 084379
NYFA Quarterly, New York 139466
Nyfors Antik och Kuriosa,
Ludvika 086825
Nygatans Antik och Kuriosa,
Örebro 086905
Nyhuis, Emlichheim 075607
Nyírségi Erdészeti Zrt. Debreceni
Erdészet – Bánki Tájház,
Debrecen 023596
NYIT Finearts, New York Institute of
Technology, Old Westbury 057615
NYK Kaiun Hakubutsukan,
Yokohama 029722
Nykarleby Museum och
Konstsamlingen Kalevala,
Nykarleby 011201
Nykjær, Bent, Vejle 066199
Nylander Museum, Caribou 047426
Nýlistasafnið, Reykjavík 024105
Nymindegab Museum, Nørre
Nebel 010237
Nymindegab Redningsbådsmuseum,
Nørre Nebel 010238
Nynäs Konst och Hantverk,
Nynäshamn 115788
Nynäs Slott, Tystberga 041096
Nynäshamns Järnvägsmuseum,
Nynäshamn 040889
Nynorsk Antikvariat, Oslo 143410
Nyoirin-ji Homotsukan, Yoshino . 029746
Nyons Antic, Nyons 070768
Nysä, Paris 071719
Nyslotts Landskapsmuseum,
Nyslott 011202
Nyströmska Gården, Köpings Museum,
Köping 040753
Nyt Nordisk Forlag, København . 136775
Nyugat, Budapest 142566
Nyúzó Gáspár Fazekas Tájház, Kiss

Pál Múzeum, Tiszafüred 024012
NZ Art Online, Christchurch . . . 112857
NZ Arts, Auckland 112799
NZH Vervoers Museum, Noord-
Zuid-Hollandsche Stoomtramweg,
Haarlem 032396
O', Milano 059471
O, Wanju 111745
O & O, Amsterdam 082817
La O Antiquités, Sallanches . . . 073385
O Bijutsukan, Tokyo 029571
O Museu do Marajó, Cachoeira do
Arari 004376
O. Henry Museum, Austin 046699
O. Kelaptrishvili Museum, Tbilisi 016419
O.C.C.M.P, Bordeaux 067349
O.G.N, Paris 071720
O.H.M.M. Impex, Bucureşti . . . 085105
O.p.t. Studio Gallery, Hong Kong 102100
Oadby Antiques, Foxton 083937
Oak & Kauri, Kumeu 083977
Oakleigh, London 001380
Oakley Antique Mall, Cincinnati . 092880
Oakley Interiors, Chicago 135418
Oakley Pioneer Museum, Oakley 051688
Oakridge, El Paso 093410
Oakridge, Louisville 094576
Oaks House Museum, Jackson . 049755
Oaktree Antiques, Lubenham . . 090641
Oakville Galleries, Centennial,
Oakville 006535
Oakville Galleries, Gairloch,
Oakville 006536
Oakville Museum at Erchless Estate,
Oakville 006537
Oakwell Hall Country Park,
Birstall 043634
Oakwood Antiques, Hinuera . . . 083956
Oakwood Gallery, Edwinstowe . 117745
Oakwood Gallery, Leeds 089771, 134536
Oakwood House Gallery, Maldon 118699
Het Oale Meestershuus,
Slagharen 032880
Oale Smederie, Hellendoorn . . . 032455
OAN-Oceanie Afrique Noire Books,
New York 145361
Oasen, Valby 140668
Oasi Antichità, Roma 081481
Oasis, Manawatu 112939
Oasis, Seattle 097477
Oasis Antique, Sharjah 088206
Oasis Art Gallery, Portland . . . 123647,
. . . . 127411
The Oasis Company, Colombo . 086614
Oasis Gallery, Marquette 050706
Oasis Superette, Tauranga 143350
Oatlands, Leesburg 050215
Ob-Art, Barcelona 114946
Ob Luang, Chiang Mai 042687
Obadia, Nathalie, Paris 105135
Obalne Galerije – Galerija Loža,
Koper 038528

Obalne Galerije – Mestna Galerija,
Piran 038586
Obam, Toulon (Var) 105717
Oban Antiques, Oban 090907
Oban Art Gallery, Oban 118846
Oban War and Peace Museum,
Oban 045431
Obara-mura Kami no Bijutsukan,
Obara 029198
Obecné múzeum a Galéria Jana Hálu,
Galéria Petra Michala Bohúňa
Liptovský Mikuláš, Važec 038480
Obecní Dům, Praha 009829
Obedinenie Fotografov Aserbaidžana,
Baku 058342
Obelisk Antiques, Petworth . . . 091002,
. . . . 134940
Obelisk Antiques, Warminster . . 091750
Obelisk Museum, Al-Matariyya . 010510
Obelisque, Oslo 084332
Obéniche, Honfleur 103841
Oberacker, Rolf, Wiesbaden . . . 078989
Oberbaumverlag, Berlin 137089
Oberdiek, H., Melle 077338
Oberdiek, Harald, Melle 077339
Oberer Torturm-Ausstellung, Marbach
am Neckar 020261
Oberfränkisches Bauernhofmuseum,
Zell, Oberfranken 022859
Oberfränkisches Textilmuseum,
Helmbrechts 019182
Oberharzer Bergwerksmuseum,
Museum für Technik- und
Kulturgeschichte, Clausthal-
Zellerfeld 017744
Oberhausmuseum Passau,
Passau 021085
Oberkärntner Brauchtums- und
Maskenmuseum, Oberdrauburg 002459
Oberländer, Heike, Berlin 074892
Oberländer, Konrad, Augsburg . 106032
Oberlander, Eberhard,
Wurmannsquick 130798
Oberlausitzer Kunstverein e.V.,
Zittau 059213
Oberlausitzer Sechsstädtebund-
und Handwerksmuseum Löbau,
Löbau 020095
Oberle, Aline, Feldkirchen in
Kärnten 127731
Oberle, Frieder, Bad
Schussenried 074591
Oberlé, Gérard, Montigny-sur-
Canne 140960
Oberlin College Gallery,
Cleveland 047820
Oberlin Heritage Center, Oberlin 051692
Obermaier, R., Altötting 074399
Obermair, Rudolf, Wels 062876, 127896
Obermayr, Axel, Aumühle 074475
Obermayr, Günther, Piberbach . . 127833
Obermeier, Rosemarie & Johann,
Poing 078064
Obermeier, Rudolf, Regensburg . 130568
Obermeyer & Reschke, Bonn . . 075082
Obermoser, J., Kitzbühel 062665
Oberösterreichische Fotogalerie,
Linz 002957
Oberösterreichische Landesmuseen,
Linz 002328
Oberösterreichischer Künstlerbund,
Dornach 058294
Oberösterreichischer Kunstverein,
Linz 058295, 100111
Oberösterreichischer Musealverein,
Gesellschaft für Landeskunde von
Oberösterreich, Linz 058296
Oberösterreichischer Steingarten,
Vorchdorf 002878
Oberösterreichisches
Feuerwehrmuseum St. Florian, Sankt
Florian 002650
Oberösterreichisches Jagdmuseum,
Schloß Hohenbrunn, Sankt

Florian 002651
Oberösterreichisches Schulmuseum,
Bad Leonfelden 001756
Oberpfälzer Fischereimuseum,
Tirschenreuth 022158
Oberpfälzer Flußspat-
Besucherbergwerk Reichhart-
Schacht, Stulln 022022
Oberpfälzer Freilandmuseum Neusath-
Perschen, Nabburg 020633
Oberpfälzer Handwerksmuseum,
Rötz 021461
Oberpfälzer Künstlerhaus,
Schwandorf 021763
Oberpfälzer Volkskundemuseum,
Burglengenfeld 017696
Oberrheinische Narrenschau,
Kenzingen 019592
Oberrheinisches Bäder- und
Heimatmuseum, Bad Bellingen 016712
Oberrheinisches Tabakmuseum,
Mahlberg 020216
Oberschlesisches Landesmuseum,
Ratingen 021289
Oberschwäbische Barockgalerie,
Ochsenhausen 077890
Oberschwäbische Galerie Kloster
Heiligkreuztal, Altheim 016545
Oberschwäbisches Museumsdorf
Kreisfreilichtmuseum Kürnbach, Bad
Schussenried 016871
Obersimmentaler Heimatmuseum,
Heimathus am Chilchstalden,
Zweisimmen 042253
Oberto, Martino, Genova 131332
Oberweger, Renate, Schruns . . 100197
Oberweißbacher Bergbahn,
Mellenbach-Glasbach 020355
Oberwelt, Stuttgart 108724
Oberwemmer, Günter, Hamburg 129980
Obieg, Warszawa 138974
Obinitsa Põhikooli Muuseum,
Obinitsa 010643
Obir-Tropfsteinhöhlen, Bad
Eisenkappel 001738
Objectif Lune, Paris 071721
Objects of Bright Pride, Seattle . 124577
Objects of Desire, Louisville . . . 121663
Objedinenie Fotocentr Sojuza
Žurnalistov, Moskva 037484
Objektief, Enschede 112479
L'Objet Curieux, Genève 087491
Objet Trouvé, Stuttgart 078623
Objetos Diretos Multiplos Artes, Rio de
Janeiro 101053
Objets D'art, East Grinstead . . . 117705
Objets d'Art, Paris . . . 071722, 071723
Objets d'Hier, Casablanca 082576
Objets Inanimés, Perpignan . . . 072090
Objets Plus, New York 095620
Objets Trou Ve', New Orleans . . 122211
Objets Trouvés, Paris 071724
Objex Studio & Gallery,
Wellington 113082
Oblasnyj Istoryčnyj Muzej im. V.
Tarnovs'kogo, Černigiv 043122
Oblasnyj Krajeznavčyj Muzej,
Žytomyr 043329
Oblasnyj Krajeznavčyj Muzej,
Cherson 043139
Oblasten Istoričeski Muzej,
Jambol 005042
Oblastnaja Kartinnaja Galereja Obraz,
Kaluga 037094
Oblastní Galerie, Liberec 009666
Oblastní Galerie Vysočiny v Jihlavě,
Jihlava 009665
Oblastní Muzeum, Most 009709
Oblastní Muzeum v Chomutově,
Chomutov 009521
Oblastní Muzeum v Lounech,
Louny 009683
Oblastnoj Chudožestvennyj Muzej,
Petropavlovsk-Kamčatskij 037649

Oblastnoj Istoriko-Kraevedčeskij Muzej, Aktöbe 029791
Oblastnoj Istoriko-Kraevedčeskij Muzej, Semey 029845
Oblastnoj Istoriko-kraevedčeskij Muzej, Taraz 029851
Oblastnoj Kraevedčeskij Muzej, Ekaterinburg 036993
Oblastnoj Muzej im. N.M. Prževalskogo, Prževalsk 030082
Oblastnoj Muzej Izobrazitelnych Iskusstv, Semey 029846
Oblastnoj Muzej Literatury i Iskusstva im. Buchar Žyrau, Pavlodar ... 029833
Oblastnoj Muzejnyj Centr, Sankt-Peterburg 037810
Obligato, Francesco, Torino 081741
Oboro, Montréal 101436
Oboronna Bašta Malachova Kurgana, Muzej geroičnoï oborony i vyzvolennja Sevastopolja, Sevastopol 043294
Obra y Arte, Bucaramanga 102397
Obradović, Petar, Beograd 114432
Obrazárna Pražského Hradu, Praha 009830
O'Brien, Liz, New York 095621
O'Brien, Michael & Ivan, Dublin 137606
O'Brien and Son, R.J., Manchester 090695
O'Brien, Stephen B., Boston .. 119905
O'Brien's, Masterton 112949
Obrist, Torsten, Essen 106927
Obrock, Helmut, Rödinghausen . 078258
Obščinski Istoričeski Muzej, Etropole 005027
Obscura, Kiel 141994
Obscura, Zeszyty fotograficzne PFFiAST, DSAFiTA, Wrocław .. 138975
Observatoire Galerie, Bruxelles . 100648
Observatoriemuseet, Stockholm 041029
Observatory Museum, Grahamstown 038726
Obsession, Holt, Norfolk 138171
Obsession Antiques, Montréal .. 064449
Obsession Art Gallery, Kolkata .. 109403
The Obsession of Art, Bergen .. 112342
Obsidian, London 118481
Obsidian Gallery, Tucson 124762
Obsis, Gennevilliers 103767
Obsoleto, Alicante 085700
Obstbaumuseum, Werder 022572
Obstmuseum, Sörup 021890
Obu, Tartu 103123
Obu-shiritsu Mingeikan, Obu ... 029200
Óbuda, Budapest 142567
Óbudai Múzeum, Budapest ... 023531
Óbudai Pincegaléria, Budapest . 023532
Óbudai Társaskör Galéria, Budapest 023533, 109186
Obuvnické Muzeum, Zlín 009985
OCAD Gallery, Toronto 007192
Ocak Köyü Özel Müzesi, Kemaliye 042995
Ocampo, Luis Alberto, Medellín . 065301
La Ocasion, Gijón 086019
Occasion, Helsinki 066350
Occultique, Northampton 144747
Occupation Tapestry Gallery and Maritime Museum, Saint Helier 045694
Occurence, Montréal 101437
Ocean Art, Indianapolis 121207
Ocean Art Gallery, Virginia Beach 124837
Ocean Beach Antique Mall, San Diego 097115
Ocean Beach Books, San Diego 145523
Ocean City Art Center, Ocean City 051699
Ocean City Historical Museum, Ocean City 051700
Ocean Contemporary, Falmouth . 117785
Ocean County Historical Museum, Toms River 053765
Ocean Drive Space and Gallery,

Miami 121834
Ocean Point, Atlanta 092165
Ocean Star Offshore Drilling Rig & Museum, Galveston 049038
Oceania Design, Wellington 033356
Oceanic Arts, Paddington, New South Wales 062042
Oceanographic Museum, Institute of Oceanography, Nha Trang 054816
Océanopolis, Brest 012211
Oceanside Museum of Art, Oceanside 051705
Océarium du Croisic, Le Croisic 013626
Ochanomizu Bijutsu Gakuin, Tokyo 056052
Ochrona Zabytków, Krajowy Ośrodek Badań i Dokumentacji Zabytków, Warszawa 138976
Ocmulgee National Monument, Macon 050573
O'Connell, Adelaide 139617
O'Connell, J., Douglas . 079514, 130841
O'Connell, Michael, Crookstown 130838
O'Connor, Toronto 101671
O'Connor Bros., Windsor 091878, 135243
O'Connor & Co., Cynthia, Swords 109698
O'Connor, Herbie, Castlebar .. 079474
O'Connor & Assoc., Tom, San Francisco 124357
Oconto County Historical Society Museum, Oconto 051707
Ocotillo, Houston 121103
Ocre, Lisboa 084932, 132947
Les Ocres du Sud, Carpentras . 067696
Oct, Hiratsuka 111257
Octagon Books, New York 145362
The Octagon Center for the Arts, Ames 046442
Octagon Galleries, Bath 043526
The Octagon House, Hudson ... 049604
Octagon House, Watertown ... 054219
The Octagon Museum, Washington 054155
Octava del Corpus, Exposicion Museográfica Permanente, Peñalsordo 039932
Octave Chanute Aerospace Museum, Rantoul 052468
Octavia, Saint-Michel-Loubejou . 072983
Octavia Antiques, Petworth ... 091003
Octavia's Haze Gallery, San Francisco 124358
Octo Studio, Philadelphia 096144
October, Cambridge 139467
October Gallery, London 118482
October Gallery, Philadelphia ... 123432
Oda Kolleksiyon, İstanbul 088134
Odalstunet Gårdsmuseum, Skarnes 033939
Odalys, Prados del Este 097976, 127503, 138485, 145649
Odapark, Venray 033821
OD'Art, Delft 112364
Odawara, Tokyo 142933
Odawara-jo Rekishi Kenbunkan no Tenshukaku, Odawara 029202
Odd Book Shop, Kent Town ... 139793
The Odd Lot, Aberdeen 088223, 134011
Oddělení dějin tělesné výchovy a sportu, Historické Muzeum Národního Muzea, Praha 009831
Oddentunet, Os Museum, Os i Østerdalen 033821
Odder Museum, Odder 010252
Oddlie Limited, Chicago 135419
Odds & Ends, Jacksonville ... 094017
Odds & Ends, Nairobi 082250
Oddział Etnografii – Spichlerz Opacki, Muzeum Narodowego w Gdańsku, Gdańsk 034739
Oddział Literacki im. Marii Dąbrowskiej w Russowie, Muzeum Okręgowe Ziemi Kaliskiej w Kaliszu, Russów 035179

Oddział Sztuki Współczesnej – Pałac Opatów w Oliwie, Muzeum Narodowe w Gdańsku, Gdańsk 034740
Oddział Zbiorów Graficznych i Kartograficznych, Biblioteka Jagiellońska, Kraków 034907
Oddział Zielona Brama, Muzeum Narodowe w Gdańsku, Gdańsk 034741
Ode, To, Singapore 114580
Odélene Likovne i Musičke Umetnosti, Srpske Akademije Nauka i Umetnosti, Beograd 056322
Odeljenje za Preventivnu Zaštitu Diana, Narodni muzej u Beogradu, Beograd 056323
O'Dell House Museum, Annapolis Royal 005340
Odense Antik, Odense 066042
Odense Auktionerne, Odense .. 125463
Odenwälder Freilandmuseum, Walldürn 022407
Odenwald- & Spielzeugmuseum, Michelstadt 020402
Odeon, Dublin 079577
Odeon Antiques, Leek . 089785, 134541
Oderlandmuseum, Bad Freienwalde 016768
Oderljunga Mölla, Perstorp 040913
Odermatt & Rotz, Grafenort ... 087513
Odermatt, Erna, Unterägeri ... 116675
Odermatt, Franz, Hergiswil ... 087522
Odermatt, Peter & Beate, Schifferstadt 108562
Odes'ke Učylyšče Mystectv i Kul'tury im. K. Dankeviča, Odesa 056499
Odes'kyj Chudožnij Muzej, Odesa 043273
Odes'kyj Istoriko-krajeznavčyj Muzej, Odesa 043274
Odes'kyj Literaturno-memorial'nyj Muzej K. Paustovskogo, Odes'kyj literaturnyj muzej, Odesa ... 043275
Odes'kyj Literaturnyj Muzej, Odesa 043276
Odes'kyj Muzej Zachidnogo ta Schidnogo Mystectva, Odesa . 043277
Odessa Archaeological Museum, Odesa 043278
Odesser, Roman, München ... 077586
Odigia Ikonen-Museum, Den Haag 032194
Odiham Gallery, Odiham 090910
Odin Antiques, Brighton 088630
Odinani Museum, Awka 033416
Odisseus i Stockholm, Nacka .. 086876
Odlin Gallery, Wellington 113083
Odom, Fort Worth 145151
Odon Antik, Budapest 079271
O'Donnell, Chris & Lin, Wallingford 091713
O'Donoghue Books, London .. 144657
O'Donovan, Douglas .. 079515, 130842
Odontological Museum, London 045067
Odsherreds Kulturhistoriske Museum, Nykøbing Sjælland 010250
Odsherreds Kunstmuseum, Asnæs 010012
Odsherreds Museum, Asnæs ... 010013
ODTÜ Müzesi, Ankara 042825
Oduduwa, Chicago ... 120220, 120221
Oduon, Cairo 103066
O'Dwyer, P., Sehnde 078448
Ody's Antiques, Miami 094781
L'Odyssée, Lyon 140928
L'Odyssée, Paris 141250
Odyssée, Saint-Étienne 141332
Odyssée des Bâtisseurs, Alma . 005319
Odyssey Books, Montréal 140369
Odyssey Fine Arts, London ... 090381
Odyssey Maritima Discovery Museum, Seattle 053243
Odyssia Gallery, New York ... 122926
Odyssium, Edmonton 005444
Oebel, Hildegard, Detmold ... 075341
Öberg, Göteborg 115661

Øbro, København 065941
Oechsle, Karl-Heinz, Lübeck ... 077183
Ödemiş Müzesi, Ödemiş 043038
Ödenburger Heimatmuseum, Bad Wimpfen 016909
Oedipus Rex Gallery, Auckland . 112800
Ödmann, Gudrun, Stockholm .. 087038
Oegg, Margarete, Gumpoldskirchen 062620
Öhman, Styrbjörn, Göteborg .. 143850
Oehmen-Krieger, Angelika, Neuss 077750
L'Oeil de Poisson, Québec 101512
L'Oeil Fauve, Pau 105330
L'Oeil Soleil, Cliousclat 068042, 103611
Öjskogsparkens Museiapoteket, Närpes 011180
Økomuseum Grenseland, Sarpsborg 033928
Økomuseum Samsø, Tranebjerg 010344
Øksendal, Kristiansand 113172
Øksnes Bygdemuseum, Alsvåg . 033488
Oekumenische Sammlung Gertrud Weinhold 'Das Gottesjahr und seine Feste', Zweigstelle des Bayerischen Nationalmuseums, Oberschleißheim 020923
Öle-Museum, Münsingen 041777
Oelgart, Hans D., Krefeld 130195
Ølgod Museum, Ølgod 010264
Oellermann, Eike, Heroldsberg . 130037
Ölmühle Dörzbach, Dörzbach . 017922
Ölmühle Michelau, Rudersberg . 021531
Ölmühle Neckargröningen, Remseck am Neckar 021368
Ölmühlenmuseum, Ottweiler ... 021822
Der Ölschlag, Kirchberg in Tirol 002173
Önningebymuseet, Mariehamn . 011146
Öntödei Múzeum, Országós Muszaki Múzeum, Budapest 023534
von Oepen & Werner, Essen ... 075692
Örebro Konsthallen, Örebro 040894
Örebro Läns Museum, Örebro .. 040895
Øregaard-Museum, Hellerup ... 010101
Oerka Irene Verbeek Museum, Raard 032775
Örnsköldsviks Museum & Konsthall, Örnsköldsvik 040900
Örökség Galéria, Kulturális Örökségvédelmi Hivatal, Budapest 023535
Oerters, R., Gütersloh 107187
Oertijdmuseum De Groene Poort, Boxtel 032096
Ösbyholm Art, Norrtälje 115785
Oeschger von Radecke, Waldshut-Tiengen 078845
Oeß, Michael, Konstanz 107718
Östasiatiska Museet, Stockholm 041030
Østerbro Antik, København 065942
Österreichisches Brückenbaumuseum, Edelsbach bei Feldbach 001840
Östergötlands Länsmuseum, Linköping 040779
Österholm, Sundsvall 126680
Österlens Museum, Simrishamn 040947
Östermalm, Stockholm 087039
Österreichische Akademie für Facheinrahmer, Wien 058297
Österreichische Filmgalerie, Krems 002238
Österreichische Gesellschaft für Christliche Kunst, Wien 058298
Österreichische Gesellschaft für Photographie, Wien 058299
Österreichische Kunsttopographie, Wien 138531
Österreichische Nationalbibliothek, Wien 003053
Österreichische Numismatische Gesellschaft, Wien 058300
Österreichische Zeitschrift für Kunst und Denkmalpflege, Horn ... 138532
Österreichischer Ärztekunstverein, Wien 058301

Oesterreichischer Alpenverein-
Museum, Innsbruck 002133
Österreichischer Kunst- und
Kulturverlag, Wien 136602
Österreichischer Kunstsenat,
Wien 058302
Österreichischer Restauratorenverband,
Wien 058303
Österreichischer Skulpturenpark,
Landesmuseum Joanneum,
Unterpremstätten 002856
Österreichisches Donau- und
Fischereimuseum, Orth an der
Donau 002482
Österreichisches Felsbildermuseum,
Spital am Pyhrn 002764
Österreichisches Filmmuseum Wien,
Wien 003054
Österreichisches Filmmuseum,
Verlagsabteilung, Wien 136603
Österreichisches Forstmuseum
Silvanum, Großreifling 002014
Österreichisches Freilichtmuseum
Stübing, Stübing bei Graz . . 002804
Österreichisches Freimaurermuseum,
Schloß Rosenau 002723
Österreichisches Gartenbaumuseum,
Wien 003055
Österreichisches Gebäckmuseum,
Stadtmuseum Wels-Burg, Wels 002919
Österreichisches Gesellschafts- und
Wirtschaftsmuseum, Wien . . . 003056
Österreichisches Institut für
Photographie und Medienkunst,
Wien 058304
Österreichisches Jüdisches Museum,
Eisenstadt 001862
Österreichisches Kriminalmuseum,
Scharnstein 002714
Österreichisches Kulturzentrum,
Wien 058305
Österreichisches Luftfahrtmuseum,
Feldkirchen bei Graz 001884
Österreichisches Museum für
Volkskunde, Wien 003057
Österreichisches Papiermacher- und
Druckereimuseum, Steyrermühl 002793
Österreichisches Pfahlbaummuseum
und Museum Mondseeland,
Mondsee 002413
Österreichisches Spiele Museum,
Leopoldsdorf im Marchfelde . . 002301
Österreichisches Sprachinselmuseum,
Wien 003058
Österreichisches Straßenbahn-
und Lokalbahnbetriebsmuseum,
Museumstramway Mariazell-
Erlaufsee, Mariazell 002364
Österreichisches Theatermuseum,
Wien 003059
Österreichisches Wandermuseum und
Roseggers Waldschule, Alpl . . 001698
Österreichisches Zuckermuseum,
Tulln 002846
Östersunds Auktionsverk, Oviken 126654
Østfold Auksjonsforretning,
Mysen 126361
Östgöta Auktionshall, Ljungsbro 126625
Östhammars Antik och Kuriosa,
Östhammar 086910
Östra Skånes Konstnärsgille,
Simrishamn 059900, 115811
Östra Södermanlands Museijärnväg,
Mariefred 040839
Östra Tullhuset, Varberg 087153
Oestraka, Bernay 103364
Oestreicher, Marguerite, New
Orleans 122212
Østsamisk Museum, De Samiske
Samlinger, Neiden 033802
Østsjællands Museum, Stevns
Museum, Store Heddinge . . . 010329
Oeteldonks Gemintemuzejum, 's-
Hertogenbosch 032473

Ötztaler Heimat- und Freilichtmuseum,
Längenfeld 002258
L'Oeuvre Copiée, Paris 105136
Över Stens Antik och Kuriosa,
Uppsala 087123
Överjärva Antik, Solna 086948
Övermark Hembygdsmuseum,
Övermark 011204
Övertorneå Hembygdsmuseum,
Övertorneå 040906
Övralids Lantbruksmuseum, Västra Ny
Hembygdsförening, Motala . . 040864
Oexmann, Knuth H., München . . 077587,
130376
Øyfjell Bygdemuseum, Vinje . . . 034091
Özden Sanat Galerisi, İstanbul . . 117070
Of All Things, Toronto 064711
Ofakim Art Gallery, Tel Aviv . . . 109728
O'Fallon Historical Museum,
Baker 046715
Ofedo, Paris 071725, 071726
Ofen- und Keramikmuseum Velten,
Velten 022315
Ofenmuseum und Baumgalerie,
Allmersbach im Tal 016512
Ofenplattensammlung der
Schwäbische Hüttenwerke,
Aalen 016451
Off Beat Antiques, Pittsburgh . . 096316
Off Broadway Antique Dealers,
Oakland 095906
Off Broadway Antiques,
Saskatoon 064565
Off Campus Book Store,
Providence 145470
Off-Centre Gallery, Bedminster . 117327
Off Ramp Antiques, Los Angeles 094423
Off the Beaten Way, New
Orleans 122213
Off the Rails Collectables, Watsons
Bay 062415
Off the Shelf Creations,
Katherine 061671
Off the Wall, Austin 092256
Off the Wall, Ayr 117265
Off the Wall, Denver 093319
Off the Wall, Hemel Hempstead 089496
Off the Wall, Lanark 118069
Off the Wall, Los Angeles 094424
Off the Wall, Minneapolis 120020
Off the Wall, San Francisco 124359
Off the Wall Antiques, Saskatoon 064566
Off the Wall Gallery, Houston . . . 121104
Off the Wall Hangings, Salt Lake
City 123976
Offa, Knokke-Heist 100753
Offa's Dyke Antique Centre,
Knighton 089727
Offay, Anthony d', London 118483
d'Offay, Caro, Chicago 120222
Offbeat Gallery, Wellington 113084
Offeney, Burkhard, München . . . 077588
Offer, Waterman & Co., London . 090382,
118484, 134737
Offermann, Jens, Hamburg 107303
Offermanns, Natalia, Mannheim . 077267
Office Art, Boston 119906
Office for Contemporary Art Norway,
Oslo 084333
Office Wide, Tokyo 111484
Officina, Trapani 111021
Officina '96, Pomáz 137588
Officina Antiquaria, Milano 131485
L'Officina del Restauro, Firenze . 131226
Officina del Restauro Mobili Antichi,
Bologna 130984
Officina del Tempo, Roma 081482
Officina Grafica, Bolzano 109887
Officina Restauri Piccola Società
Cooperativa, Milano 131486
Officina Restauro Tappeti e Tessili,
Pisa 131767
Officinaarte, Magliaso 116488
Officinarte Conservazione e Restauro,

Bologna 130985
Offizin Zürich Verlag GmbH,
Zürich 138115
Oficina dos Anjos, Belo Horizonte 100899
Oficina Museu da Universidade Estácio
de Sá, Rio de Janeiro 004764
Oficina Paulista Arte, São Paulo 101105
Ofir, Kraków 113621
Ofoten Museum – Museum Nord,
Narvik 033801
Oftring, Reinhard, Hainburg . . . 107203
O'Gallerie, Portland . . 096449, 127412
O'Gara & Wilson, Chicago 145049
O'Gara, John J., Belleville 144994
Ogawa, Yamato 111530
Ogden, Harrogate 089437
Ogden Historical Society Museum,
Spencerport 053437
Ogden House, Fairfield 048669
Ogden Union Station Museums,
Ogden 051716
Ogden, Richard, London 090383
Oger & Semont, Commissaires
Priseurs Habilités, Paris 125745
Oggetti d'Arte, Milano 080638
Oggetti del Passato, Modena . . . 080813
L'Oggetto Ritrovato, Milano 080639,
110394, 131487
Oggiano, Paolo, Cagliari 131072
Ogham Gallery, Belfast 117342
Ogier, Jacques, Lyon 069770
Ogilvie & Perti, Chicago 120223
Oginskių Kultūros Istorijos Muziejus,
Rietavas 030454
Ogle, Portland 123648
Oglethorpe University Museum of Art,
Atlanta 046629
Ogni Moskvy – Muzej Istorii
Gorodskogo Osveščenija,
Moskva 037485
Ognibeni, Günter, Braunschweig 129625
Ogniwo Polish Museum,
Winnipeg 007455
Ogólnopolskie Targi Antykwaryczne i
Sztuki Współczesnej, Krakow . 098271
O'Gorman, Lance, North Shore . 112973
O'Gorman, P.J., Charleville 079480
Ogres Vēstures un Mākslas Muzejs,
Ogre 030167
Ogrodowicz, Piotr, Gdynia 132824
Ogrzey, C. & K., Bad Soden am
Taunus 074600
Ogunquit Museum of American Art,
Ogunquit 051720
Ogura, Tokyo 082188
Ogurek, Ferdinande, Lohmar . . . 077147
OH+T Gallery, Boston 119907
O'Hanlons, New York 122927
O'Hara, Chicago 092755
Ohara Art Reference Center,
Kobe 028853
Ohara Bijutsukan, Kurashiki . . . 028905
O'Hara, Jonathan, New York . . . 122928
Ohbayashi, Sendai 111389
Ohef Sholom Temple Archives,
Norfolk 051581
Ohi Museum & 10th Ohi Chozaemon
Ware Gallery, Kanazawa 028777
Ohio Book Store, Cincinnati . . . 145076
Ohio Ceramic Center, Roseville . 052684
Ohio Estates Coins-Jewelry,
Cleveland 092959
Ohio Historical Society Museum,
Columbus 047941
Ohio Murals, Columbus 120506
Ohio Museums Association,
Columbus 060657
Ohio Railway Museum,
Worthington 054553
Ohio River Museum, Marietta . . 050682
Ohio University Art Gallery, Saint
Clairsville 052759
Ohio University Department of Arts,
Saint Clairsville 057804

Ohk, Manfred, Kaiserslautern . . 076662
Ohl, Lutz, Schefflenz . . 078358, 130619
Ohlbrock, Christoph, Gütersloh . 076124
Ohlert, Rose Bay 062166
Ohlídal, Vladimir, Luhačovice . . . 102618
Ohlin, Anders, Göteborg 133530
Ohlinger, Jerry, New York 095622
Ohlsson, Kalmar 086770
Ohm, Bad Schwartau 074593
Ohme, Peter, München 142178
Ohno, Bryan, Seattle 124578
The Ohr-O'Keefe Museum of Art,
Biloxi 046994
Ohrnberger, Nußloch 108313
Ohsabanan Museijärnväg, Växjö 041145
Ohse, Rolf, Bremen 106591
The Ohsha'joh Museum of Art,
Hiroshima 028657
Ohya-Shobo, Tokyo . . 082189, 142934
Oi Ling, Hong Kong 065089
Oiffer-Bomsel, Laurent, Paris . . . 071727
Oijens, Göteborg 115662
Oikawa Museum of Sound
Instruments, Higashi-kurume . 028619
Oikos Antikon Gkeka, Athinai . . 079175
Oil and Gas Exhibition Centre, Mina Al-
Fahal 034095
Oil Capital Rare Coins, Tulsa . . . 097752
Oil Museum of Canada, Oil
Springs 006543
Oil On Canvas, Athelstone 061068
Oil Painters of America, Crystal
Lake 060658
Oil Painting Picture Gallery, Canning
Vale 098858
Oil Pastel Association, Stockholm 060659
Oil Pastel Society of Maine, Ocean
Park 060660
Oil Patch Museum, Russell 052709
Oil Sands Discovery Centre, Fort
McMurray 005825
Oinoussian Maritime Museum,
Oinoussai 023204
Oise, Nagoya 111327
Oise Collections, Lamorlaye . . . 069155
Oiseau d'Or, Montmirail 070274
L'Oiseau Livre, Paris 141251
Oishi, Fukuoka 111251
Oisín Gallery, Dublin . 109610, 137607
Õisu Tehnikumi Muuseum,
Halliste 010611
Õisu Valla Muuseum, Õisu 010645
Oita-kenritsu Geijutsu Kaikan,
Oita 029210
Oita-kenritsu Rekishi Hakubutsukan,
Usa 029682
OJ Art Gallery, New York 122929,
122930
OJ Art Gallery, Philadelphia 123433
Ojai Center for the Arts, Art Branch,
Ojai 051723
Ojai Valley Museum, Ojai 051724
Ojala, Los Angeles 121566
Ojāra Vācieša Memoriālais Muzejs,
Rīga 030212
Ojeblikket, København 010189
Ojeda, Manuel, Las Palmas de Gran
Canaria 115381
Ojitos, Miami 094782
Ojo al Arte Galería, Bogotá 102353
OjodePez, Madrid 138562
OK Offenes Kulturhaus Oberösterreich,
Linz 002329
Okains Bay Maori and Colonial
Museum, Okains Bay 033263
Okaiwa, Barcelona 133139
Okamoto, Kitakyushu 081993
Okanagan Falls Heritage and Museum
Society, Okanagan Falls 006545
Okanagan Military Museum,
Kelowna 006076
Okanoyama Bijutsukan,
Nishiwaki 029191
O'Karma Jones, Atlanta 119596

Okaukuejo Museum, Okaukuejo 031846
Okawa, Osaka 082088, 132309
Okawa Bijutsukan, Kiryu 028818
Okawa-suji Samurai Residence,
   Kochi 028860
Okaya Museum, Okaya 029211
Okayama-kenritsu Bijutsukan,
   Okayama 029214
Okayama-kenritsu Bizen-yaki
   Hakubutsukan, Bizen 028505
Okayama-kenritsu Hakubutsukan,
   Okayama 029215
Okayama-shiritsu Oriento Bijutsukan,
   Okayama 029216
Okaz, Rouen 072605
Okazaki-shiritsu Bijutsukan,
   Okazaki 029219
Okazja, Warszawa ... 113876, 126423
O'Keefe, Clonmel 079488
Okefenokee Heritage Center,
   Waycross 054245
O'Kelly, Dromcollogher 079516, 130843
Okeresztény Mauzóleum, Janus
   Pannonius Múzeum, Pécs 023849
Okinawa-ken Heiwa Kinen Shiryokan,
   Itoman 028718
Okinawa-kenritsu Geijutsu Daigaku,
   Naha 056021
Okinawa-kenritsu Hakubutsukan &
   Gendai Bijutsukan, Naha 029131
Okkupasjonsmuseet, Eidsvoll Museum,
   Eidsvoll Verk 033583
Oklahoma City Museum of Art,
   Oklahoma City 051734
Oklahoma Firefighters Museum,
   Oklahoma City 051735
Oklahoma Forest Heritage Center
   Forestry Museum, Broken Bow 047222
Oklahoma Indian Art Gallery,
   Oklahoma City 123291
Oklahoma Museum of African
   American Art, Oklahoma City . 051736
Oklahoma Museum of History,
   Oklahoma City 051737
Oklahoma Museums Association,
   Oklahoma City 060661
Oklahoma Route 66 Museum,
   Oklahoma Historical Society,
   Clinton 047845
Oklahoma Territorial Museum,
   Oklahoma Historical Society,
   Guthrie 049298
Okovprom, Sarajevo 063870
Okräzen Istoričeski Muzej,
   Chaskovo 005004
Okräzen Istoričeski Muzej,
   Montana 005095
Okräzen Istoričeski Muzej, Pernik 005111
Okräzen Istoričeski Muzej,
   Plovdiv 005130
Okräzen Istoričeski Muzej, Sliven 005169
Okräzen Istoričeski Muzej,
   Smoljan 005171
Okräzen Istoričeski Muzej, Veliko
   Tärnovo 005257
Okräzen Istoričeski Muzej, Vidin 005263
Okräzen Istoričeski Muzej, Vraca 005265
Okräžen Muzej na Väzraždaneto i
   Nacionalno-osvoboditelnite Borbi,
   Plovdiv 005131
Okräžna Chudožestvena Galerija,
   Kardžali 005045
Okräžna Chudožestvena Galerija, Stara
   Zagora 005210
Okräžna Chudožestvena Galerija,
   Vraca 005266
Okräžna Chudožestvena Galerija
   Dimitär Dobrovič, Sliven 005170
Okräžna Chudožestvena Galerija
   Vladimir Dimitrov-Majstora,
   Kjustendil 005066
Okresní Muzeum, Děčin 009534
Okresní Muzeum, Litoměřice .. 009672
Okresní Muzeum, Vlašim 009954

Okresní Muzeum Praha-východ,
   Brandýs nad Labem 009453
Okresní Vlastivědné Muzeum, Nový
   Jičín 009727
Okroy, Peter, Hilden 076530
Oktagon, Köln 137315
Oktibbeha County Heritage Museum,
   Starkville 053493
Oku Kyodokan, Oku 029221
Okukiyotsu Electric Power Museum,
   Yuzawa 029753
Okupatsioonimuuseum, Kistler-Ritso
   Fond, Tallinn 010684
Okura, Tokyo 082190
Okura Shukokan Museum of Fine Arts,
   Tokyo 029572
Okuuchi Bijutsukan, Toyonaka .. 029648
Okvir, Ivanč-Grad 128312
Okvir, Zagreb 128315
Okyo no Rosetsu Bijutsukan,
   Kushimoto 028913
Ola Bua Møbelverksted, Skien . 084357
Olabarri, Begoña, Bilbao 133197
Olaf-Gulbransson-Museum
   für Graphik und Karikatur,
   Graphische Sammlungen München,
   Tegernsee 022116
Olana State Historic Site, Hudson 049602
Olander, Ann, Falun 133523
Olav Bjaaland Museum,
   Morgedal 033786
Olav Holmegaards Samlinger,
   Mandal 033764
Olavinlinna Castle, Olavinlinna .. 011206
Olcese, Francesca, Genova 131333
Olczak, Dorota Ewa, Berlin ... 129544
OLD, Torino 081742
Old & Art, Warszawa . 084740, 113877
Old & Asnew Book Traders, South
   Auckland 143344
Old & Beautiful, Genova 080275
Old & Curious, South Auckland . 143345
Old & Gold, Durham ... 089122, 134344
The Old Aberdeen Book Shop,
   Aberdeen 144213
Old Aircraft Museum,
   Arnemuiden 032018
Old America, Milano 080640
Old and Curious Booksellers,
   Papakura 143340
Old and New Antiques, Highbury 061619
Old Armory, Cleveland 092960
Old Art, Roeselare 128087
Old Arts Gallery, Tshwane 038854
The Old Bakehouse, Chichester . 117581
Old Bakehouse Antiques, Broughton
   Astley 088689
Old Bakehouse Antiques,
   Parramatta 062063, 127658
The Old Bakehouse Galleries,
   Edinburgh 117734
The Old Bakery, Woolhampton . 091936
Old Bakery Antiques, Brasted . 088584
Old Bakery Antiques, Cundletown 061384
The Old Bakery Antiques,
   Wheathampstead 091821
Old Bakery Antiques,
   Wymondham 091961
Old Bakery Cottage Book Shop,
   Warradyte 139951
Old Bakery Gallery, Lane Cove . 099218
Old Ball Park, Oklahoma City .. 095997
Old Bank Antiques, Dungog 061416
Old Bank Antiques, Goomeri 061542
Old Bank Antiques Centre, Bath 088396
Old Bank Artel, McLaren Vale .. 099246
Old Bank Corner Collectables,
   Holbrook 061640
The Old Bank Craft Studio,
   Bewdley 117553
Old Bank Gallery, Pambula 099547
Old Bank of New Brunswick Museum,
   Riverside-Albert 006811
Old Barn Antiques, Sutton Bridge 091516

Old Barn Antiques Warehouse, Sutton
   Bridge 091517
Old Barracks Heritage Centre,
   Cahersiveen 024839
Old Barracks Museum, Trenton . 053800
Old Bell Museum, Montgomery . 045290
Old Bell's Antiques, Kurrajong .. 061709
Old Bethpage Village Restoration, Old
   Bethpage 051745
The Old Blacksmith's Gallery,
   Dulverton 117690
Old Blue House Antiques,
   Houston 093757
Old Blythewood, Pinjarra 001406
Old Bohemia Historical Museum,
   Warwick 054074
Old Book Shop, Jacksonville ... 145189
Old Bookshop, Langport 144514
The Old Bookshop, Lisburn 144542
The Old Bookshop,
   Wolverhampton 144930
Old Borroloola Police Station Museum,
   Borroloola 000869
Old Bridge House Museum,
   Dumfries 044155
Old Brigade, Kingsthorpe 089697
Old Brooklyn Antiques, Cleveland 092961
Old Brown's Mill School,
   Chambersburg 047516
Old Brush Gallery, Brunkerville . 098811
Old Brutus Historical Society Museum,
   Weedsport 054263
Old Butter Factory Art Gallery,
   Denmark 098966
Old Button Shop, Lytchett
   Minster 090670
The Old Byre Heritage Centre,
   Dervaig 044094
Old Cable Station Museum, Apollo
   Bay 000796
Old Cairo, Cairo 066256
Old Capitol Museum of Mississippi
   History, Jackson 049756
Old Capitol, University of Iowa, Iowa
   City 049709
Old Carleton County Court House,
   Woodstock 007489
Old Castle Museum, Baker University,
   Baldwin City 046722
Old Cathedral Museum, Saint
   Louis 052817
Old Central School Art Centre,
   Auckland 112801
Old Channel, Delhi 079387
Old Chapel Antique Centre,
   Tutbury 091669
Old Chapel Antique Mall, Salt Lake
   City 096913
Old Charm Antiques, Knokke-
   Heist 063706
Old Charms, Genk 063620
The Old Children's Book,
   Edinburgh 144397
The Old Church Bookshop,
   Carlingford 139688
Old Church Galleries, London . 090384,
                        118485,   144658
The Old Cinema Antique Department
   Store, London 090385
Old City Arts Association,
   Philadelphia 060662
Old City Park — The Historical Village
   of Dallas, Dallas 048121
Old Classics, North Perth 061996
Old Clock Museum, Pharr 052022
Old Clock Shop, West Malling .. 091789
Old Coach House Gallery,
   Gillingham 117825
Old Codgers Antiques, Ulmarra . 062368
Old Colonial, Tunbridge Wells .. 091655
Old Colonial Bank of Antiques, Mirboo
   North 061862
Old Colony Historical Museum,
   Taunton 053688

Old Constitution House, Windsor 054473
Old Conway Homestead Museum
   and Mary Meeker Cramer Museum,
   Camden 047378
The Old Cop Shop, Torquay .. 091616
The Old Corn Store, Stalbridge . 119136
Old Corner House, Wittersham . 091892
Old Cornstore Antiques, Arundel 088288
The Old Cottage, Hilversum ... 083405
Old Cottage Antiques, London .. 090386
Old Council Chambers Museum,
   Cleve 000974
Old Council Chambers Museum, East
   Torrens Historical Society, Norton
   Summit 001371
Old Country, Manawatu 083983
Old Court House Museum,
   Drysdale 001035
Old Court House Museum,
   Durban 038703
Old Court House Museum,
   Guysborough 005937
Old Court House Museum-Eva
   Whitaker Davis Memorial,
   Vicksburg 053975
Old Courthouse Gallery, Ipswich 099155
Old Courthouse Heritage Museum,
   Inverness 049705
Old Courthouse Museum, Santa
   Ana 053087
Old Courtyard Museum, Kibbutz Ein
   Shemer 025129
Old Cowtown Museum, Wichita . 054391
Old Craft, Warszawa 084741
Old Crofton School Museum,
   Crofton 005665
Old Crown Court and Cells,
   Dorchester 044119
Old Cuba Collection, Miami ... 094783
Old Curiosity Shop, Chennai ... 079376
The Old Curiosity Shop, Dalkey . 079511
Old Curiosity Shop, King's Lynn . 089692
Old Curiosity Shop, Málaga 086300
The Old Curiosity Shop,
   Sidmouth 091341
Old Curiosity Shop, Stockport . 091446,
                                   135088
Old Curiosity Shop & Art Galleries,
   Southampton 119121
Old Custom House, Penzance .. 090967
The Old Dancehall Antiques,
   Lispole 079663
The Old Day's, Dordrecht 083190
Old Days, Kauniainen 066377
Old Depot Museum, Ottawa ... 051843
Old Dominion Railway Museum,
   Richmond 052558
Old Dominion University Gallery,
   Norfolk 051582
Old Door Store, San Antonio .. 097008
Old Dorchester State Historic Site,
   Summerville 053598
Old Dubbo Gaol, Dubbo 001038
Old Dutch Parsonage, Somerville 053385
Old Economy Village Museum,
   Ambridge 046437
Old England Gallery, Torino ... 081743
Old English Oak & Pine,
   Sandgate 091242
Old Erie Street Bookstore,
   Cleveland 145088
Old Exchange and Provost Dungeon,
   Charleston 047558
Old Falls Village, Menomonee
   Falls 050204
Old Fashioned Restoration,
   Toronto 064712, 128213
Old Fashioned Things, Houston . 093758
Old Father Antiques, B'Kara ... 082409
Old Father Time Clock Centre,
   London 090867
Old Fire Engine House Gallery,
   Ely 117749
Old Fire House and Police Museum,

Superior 053613
Old Fire Station Gallery,
Caboolture 098833
Old Firehouse Museum,
Greenville 049262
Old Flames, Easingwold 089127
Old Flour Mill Gallery, Euroa . . 099010
The Old Forge Antiques, Stretton-
under-Fosse 091499
Old Forge Studio, Pembridge . . . 118884
Old Fort Bissell, Phillipsburg . . 052108
Old Fort Erie, Niagara Falls . . . 006485
Old Fort Garland, Fort Garland . 048810
Old Fort Harrod Museum,
Harrodsburg 049380
Old Fort House Museum, Fort
Edward 048808
Old Fort Jackson, Savannah . . . 053178
Old Fort Johnson, Fort Johnson . 048820
Old Fort Lauderdale Museum of
History, Fort Lauderdale 048829
Old Fort Meade Museum, Fort
Meade 048844
Old Fort Niagara, Youngstown . . 054597
Old Fort Number 4 Associates,
Charlestown 047566
Old Fort Western, Augusta 046668
Old Fort William Historical Park,
Thunder Bay 007144
Old Fox Antiques, Casino 061293
The Old French Mirror Company,
Henley-on-Thames 089511
Old Fresno Auction, Fresno . . . 127230
Old Friends, Portland . 096450, 145460
Old Friends Antiques, Seattle . . 097478
Old Fuel Store Gallery, Wilcannia 099885
Old Furniture House, Claremont 061325
Old Gala House, Galashiels . . . 044378
Old Gallery, Tbilisi 074321, 105919
Old Gaol Museum, Albany 000777
Old Gaol Museum, Buckingham 043781
Old Gaol Museum, York 054582
Old Gladstone Post Office Antiques,
Gladstone 061522
Old Glow Company, Cincinnati . 092881
Old Gold Antiques, Croydon . . . 061379
Old Government House,
Fredericton 005847
Old Government House,
Paramatta 001394
The Old Governor's Mansion,
Frankfort 048930
Old Graine Store, East Rudham . 089142
Old Grammar School, Manx National
Heritage, Castletown 043892
Old Granary Antiques, King's
Lynn 089693
Old Green Cupboard,
Jacksonville 094018
Old Greer County Museum and Hall of
Fame, Mangum 050634
The Old Guard Museum, Fort
Myer 048849
Old Guildhall Museum, East Looe 044208
Old Haa, Yell 046214
Old Hall Antiques, Burton Salmon 088732
Old Hall Bookshop, Brackley . . . 144259
Old Hall Farm, Long Clawson . . 090611,
134807
Old Harbour Masters Gallery,
Porthleven 118939
Old Harbour Museum, Hermanus 038735
Old Hastings Mill Store Museum,
Vancouver 007292
Old Hat Antiques, Thornbury . . . 062318
Old Highercombe Hotel Museum, Tea
Tree Gully 001566
Old Home Supply House, Fort
Worth 093468
Old House, Hereford 044574
The Old House, Saarburg 078337
Old House, Seaford . . 091255, 135014
Old House Antique Centre,
Adversane 088238

Old House Gallery, Cairo 103067
Old House Gallery, Oakham . . . 090902,
134901,   144762
Old House in Hog Hollow, Saint
Louis 096753
Old House Museum, Bakewell . . 043469
Old House Museum, Durban . . . 038704
Old House of Keys, Manx National
Heritage, Castletown 043893
The Old House Revival Company,
Winnipeg 064931
Old Hoxie House, Sandwich . . . 053074
Old Independence Regional Museum,
Batesville 046808
Old Inn Gallery, Port Wakefield . 099592
Old Iron County Courthouse Museum,
Wisconsin State Historical Society,
Hurley 049639
The Old Ironmongers Antiques Centre,
Lechlade 089762
Old is Beautiful, Calgary 064264
Old Jaffa Gallery, Tel Aviv 109729
The Old Jail Art Centre, Albany . 046350
Old Jail House & Old Court House,
Croydon 001005
Old Jail Museum, Albion 046354
The Old Jail Museum,
Crawfordsville 048055
The Old Jail Museum, Warrenton 054072
The Old Jameson Distillery,
Dublin 024904
Old Japan, New York 095623
Old Jefferson Town, Oskaloosa . 051832
Old Khan, Cairo 066257
Old Lancaster Antiques, Fort
Worth 093469
Old Library Gallery, Burgh-le-
Marsh 117478
Old Lifeboat Museum, Poole . . . 045527
Old Lighthouse Museum, Michigan
City 050852
Old Lighthouse Museum,
Stonington 053547
Old Lincoln County Courthouse
Museum, Lincoln 050305
Old Log Church Museum,
Whitehorse 007406
Old Log Jail and Chapel Museums,
Greenfield 049243
Old London, Milano 080641
Old Los Angeles Antiques, Los
Angeles 094425
Old Mackinac Point Lighthouse &
Colonial Michilimackinac, Mackinaw
City 050560
Old Maitland Mill Antique Centre,
Metford 061846
The Old Malthouse, Hungerford . 089612
Old Maltings Antique Centre,
Louth 090637, 134818
Old Manchester Museum,
Manchester 050623
The Old Manor, Bourtange 132450
The Old Manse, Concord 047953
Old Map Gallery, Denver 145132
Old Maps & Atlases, Murs 140975
Old Maps and Prints, London . . 144659
Old Market House, Galena 049016
Old Massett Village Council's Heritage,
Old Massett 006548
Old Master Gallery, Toronto . . . 101672
Old Masters Art Shop, Portland . 096451
Old Mearns Gallery, Glasgow . . 117853
Old Meeting House, Barrington . 005390
Old Melbourne Gaol and Penal
Museum, Melbourne 001280
Old Memories, Toronto 064713
Old Merchant's House and Row 111
Houses, Great Yarmouth 044476
Old Mill Antique Centre,
Bridgnorth 088601, 134146
Old Mill Museum, Cimarron . . . 047734
Old Mill Museum, El Molino Viejo
Museo, San Marino 053054

The Old Mint House, Pevensey . 091013
Old Mission Antiques,
Albuquerque 092072
Old Mission Santa Ines Collection,
Solvang 053378
Old Money Antiques, Sacramento 096638
Old Monterey Jail, Colton Hall
Museum, Monterey 051015
Old Mornington Post Office Museum,
Mornington 001314
Old Museum Arts Centre, Belfast 043559
Old Music Shop, Chicago 092756
Old New York Book Shop, Atlanta 144970
Old Oak Tree, Minneapolis 094937
Old Oak Tree Antique Mall,
Albuquerque 092073, 135287
Old Operating Theatre, Museum and
Herb Garret, London's Museums of
Health and Medicine, London . 045068
Old Ordinary House Museum,
Hingham 049489
Old Palace Antiques, Lostwithiel 090635
Old Parliament House, Canberra 000940
Old Pine House, Leamington Spa 089757
Old Plank on Huron, Chicago . . 092757
Old Police Station Museum, Mount
Barker 001317
Old Police Station Museum Brookton,
Brookton 000898
Old Pontville Shop, Pontville . . . 062089
Old Post House, Axbridge 088326
Old Post Office Antiques, Fitzroy
North 061475
Old Post Office Museum, Turriff 046005
Old Post Office Seymour,
Seymour 099685
The Old Post Office Traders,
Warkworth 084147
Old Print Gallery, Broadway . . . 117458
Old Print Gallery, Washington . . 097854,
136462,   145640
Old Print Shop, New York 095624,
145363
The Old Prison, Victoria 030764
Old Prison Museum, Deer Lodge 048216
Old Pueblo Gallery, Tucson 124763
The Old Queenslander Antiques Center,
Cairns 061249
Old Racing Collection, Dvůr Králové
nad Labem 009546
Old Railway Station Museum,
Northam 001366
Old Reading Room, Kelling 118034
Old Rectory Museum,
Loughborough 045154
Old Saint Edward's Anglican Church,
Clementsville 005626
Old Saint Ferdinand's Shrine,
Florissant 048761
Old Salem, Winston-Salem 054492
Old Sau's Antique Shop, Chicago 092758
Old Savannah Auction, Savannah 127450
Old School – Squawk, London . . 090388
Old School Antiques, Penryn . . . 090960
Old School Arts Workshop,
Middleham 118742
The Old School Gallery, Muker . 118763
Old School House Gallery,
Cleveland 098897
The Old School House Gallery,
Roscrea 109686
Old School House, Museums of Old
York, York 054583
Old School House, Merimbula 001286
Old Schoolsrooms Antiques,
Wincanton 091859, 135234
Old Schwamb Mill, Arlington . . 046531
Old Science Renovation,
Minneapolis 135835
Old Scotch House, Settimo
Milanese 081624
Old Scotland, Roma 081483
Old Seattle Paperworks, Seattle 097479
Old Shanghai, Saint-Étienne . . . 072770

Old Shawnee Town, Shawnee . . 053293
The Old Shop, Dún Laoghaire . . 079608
Old Smithy, Feniscowles 089271, 134389
The Old Smithy, West Burton . . 091783
The Old Smithy Studio,
Bethlehem 117349
Old South Antiques, Saint Louis 096754
Old South Books, Nashville . . . 145298
Old South Meeting House,
Boston 047127
Old South Restoration, San
Antonio 136267
Old Spanish Fort Museum,
Pascagoula 051940
Old Speech Room Gallery,
London 045069
Old Springsure Hospital Museum,
Springsure 001515
Old State Bank, Vincennes 053991
Old State Capitol, Springfield . . 053454
Old State House, Boston 047128
Old State House, Hartford 049389
Old State House Museum, Little
Rock 050337
The Old Station, London 090389
The Old Steam Bakery, Redruth 091086,
134961
Old Steamer, Turku 066405
Old Stone Fort Museum Complex,
Schoharie 053198
The Old Stone House,
Washington 054156
Old Stone House Museum,
Windsor 054470
The Old Stone House Museum,
Orleans County Historical Society,
Brownington 047277
The Old Stone Jail Museum,
Palmyra 051902
Old Stone Mill Museum, Delta . . 005698
Old Strathcona Model and Toy
Museum, Edmonton 005759
Old Stuff, San Francisco 097298
Old Stuff Two, Vancouver 064854
Old Sturbridge Village, Sturbridge 053579
Old Style, Napoli 080908
The Old Surgery, Clun 117603
The Old Tampa Book Company,
Tampa 145616
Old Tavern Antiques, Tuscaloosa 053859
Old Teazle House, Nailsworth . . 118768
Old Technology Shop, Seattle . . 097480
Old Templemore, Nannup 061938
Old Ten Chain Sawmill Company,
Benaraby 098732
The Old Time, 's-Hertogenbosch 083393
Old Time, San Marino 085287
Old Time New England, Boston . 139468
Old Time Wooden Nickel, San
Antonio 097009
Old Timers, Swinford 091534
Old Timers Antiques, Harrisville 061597,
127610
Old Timers Traeger Museum, Alice
Springs 000791
Old Times, Helsinki 066351
Old Times, Perugia 142786
Old Times Antiguidades, Brasília 063937
Old Times Antiques, Montréal . . 064450
The Old Tin Toy Shop, Auckland 083812
Old Tobacco Sheds, Bright 061195
The Old Town, Cairo 066258
Old Town Antique Market,
Portland 096452
Old Town Antiques Centre,
Eastbourne 089150
Old Town Bookshop, Edinburgh . 144398
Old Town Hall Antique Centre, Needham
Market 090804, 134866
Old Town Hall Antiques,
Falmouth 089251
Old Town Hall Arts Centre, Hemel
Hempstead 044564

Old Town Hall History House,
Portland 001430
Old Town Hall Museum, Fifield . 048729
Old Town Hall Museum,
Thamesville 007137
Old Town Museum, Old Town .. 051751
Old Town Studio Gallery, San
Diego 124149
Old Towne Antiques, Memphis . 094659
Old Toy Shop, Ringwood 091108
Old Toy Soldier, Pittsburgh ... 139469
Old Trail Museum, Choteau 047730
Old Troon Sporting Antiques,
Troon 091637
The Old Trouts Treasures, Bulls . 083845
Old Tythe Barn, Blackheath 061164
Old Valve Radios, South
Auckland 132742
Old Versailles, New York 095625, 095626
Old Victoria Post Office Gallery, Mount
Victoria 099408
Old Village Hall Museum,
Lindenhurst 050314
Old Ware House Gallery, Maryborough,
Queensland 001259
The Old Warehouse Antiques,
Chester 088867
Old Warehouse Auction House,
Evansdale 127218
The Old Warehouse Gallery,
Dorchester 117681
Old Wares, Burra 061239
Old Wares Curiosity Connection,
Amphitheatre 060997
Old Washington Museum,
Washington 054083
Old West Museum, Sunset 053609
Old West Wax Museum,
Thermopolis 053730
Old Westbury Gardens, Old
Westbury 051752
Old Westend Antiques, Toledo .. 097615,
136409
Old Wood, Scandiano . 081621, 132079
Old Wooden Pew, Tyabb 062361
Old Woodworks, Edmonton ... 128167
Old Wool and Grain Store Museum,
Beachport 000833
Old World, Columbus 135483
Old World Accents, Richmond .. 096561
Old World Antiques, Minneapolis 094938
Old World Antiques, Salt Lake
City 096914
Old World Antiques, Stepney ... 062245
Old World Auctions, Sedona ... 127456
Old World Books, Venezia 142884
Old World Charm, Moe 061870
Old World do Brasil, Belo
Horizonte 063914
Old World Galleries, New York .. 122931
Old World History, Miami 094784
Old World Market, Detroit 093371
Old World Market Place,
Birmingham 092426
Old World Prints, New Orleans . 095137
Old World Refinishing,
Farmington 135555
Old World Restoration, Saint Paul 136223
Old World Restorations,
Cincinnati 135447
Old World Trading, London 090390
Old World Upholstery, Chicago . 135420
Old World Warehouse,
Philadelphia 096145
Old World Wisconsin, Eagle ... 048417
Old Wyche Gallery, Nantwich .. 118772
Old York Historical Museum, York 054584
Old York Road Historical Society,
Jenkintown 049811
Old Zhi Display Center, Xinsijun
Military Headquarter, Nanchang 008098
Oldani, Hans-Rudolf, Mägenwil . 087656
d'Olde Bakkerieje, Deventer ... 083159
Olde Books, Gulgong 139756

Olde Bostonian, Boston 092487, 135366
Olde Chicago Antiques, Chicago 092759
Olde Colonial Courthouse,
Barnstable 046794
Olde Favorites, Portland 096453
Olde Gallery, Jacksonville 094019
Olde Good Things, Chicago ... 092760
Olde Good Things, New York .. 095627,
095628, 135969
Olde Music and Cokery Books,
Hobart 139776
'T Olde Ras, Doesburg 032232
The Olde Red Lion, Bedingfield . 088426,
134088
Olde Shanghai Antiques,
Singapore 085476
The Olde Shoppe, Windsor,
Ontario 064905
Olde Stuff, Carrara 061290
Olde Stuff Antiques, Ashmore .. 061067
Olde Tyme Antiques, Calgary .. 064265
Olde Tyme Clock Shoppe,
Toronto 064714
Olde Tyme Clocks, Cincinnati .. 092882,
135448
Olde World Charm II, Houston .. 093759
Olden Asia, Beirut 082299
Olden Days, Dobwalls . 089065, 134326
Olden Days, Waikato .. 084134, 126330
Oldenbourg, Christoph W., Bristol 134162
Oldenburg-Museum, Unna 022274
Oldenburg, Ulf, Berlin 074893
Oldenburg, Wilfried, Wiesbaden . 078990
Oldenburger Antiquariat,
Wardenburg 142463
Oldenburger Fahrradmuseum,
Oldenburg 020989
Oldenburger Kunstverein e.V.,
Oldenburg 059214
Oldenburger Münzstube,
Oldenburg 077936
Oldentime Furniture, Hawkes Bay 132722
The Oldest House Museum, Key
West 049958
Oldest House Museum Complex, Saint
Augustine 052746
Oldest Stone House Museum,
Lakewood 050103
Oldfield, Portsmouth . 091045, 134953
Oldham County History Center, La
Grange 050035
Oldham, Elinor, Albuquerque .. 119501
Oldham, Thomas A., Sacramento 123777
Oldies, Norwood 062015
Oldies & Goodies, Pittsburgh .. 096317
Oldies, Goldies & Moldies, New
York 095629
Olding, Manfred, Osnabrück ... 077976
Oldman River Cultural Centre,
Brocket 005486
Oldswinford Gallery, Oldswinford 118850
Oldtimer-Galerie, Toffen 087852, 126756
Oldtimer Museum, Sternberg .. 021998
Oldtimercentrum Select, Joure . 032530
Oldtimermuseum, Blindenmarkt 001787
Oldtimermuseum, Kleinwetzdorf 002201
Oldtimermuseum De Rijke,
Oostvoorne 032728
Oldtimermuseum De Ronkel,
Kloosterburen 032548
Oldtimermuseum Kröpfl, Hartberg 002057
Oldtimermuseum Meßkirch,
Meßkirch 020387
Oldtimermuseum Stift Ardagger,
Ardagger 001719
Oldwoods, Bristol .. 088663, 134163
Dat ole Hus, Aukrug 016676
Ole Moon and Other Tales,
Dallas 120634
Ole Rømer Museet, Taastrup .. 010339
Olefjord, Simrishamn 115812
Olejak, Bernhard, Neufahrn bei
Freising 130447
Olejárová, Mária, Poprad 114666

Olens Radiomuseum, Olen 003918
Oleos and More, Miami 121835
Oleos, Sara, Getxo 115067
Olesen, Carsten, Frederiksberg . 065671
Olevimäe Antiik, Tallinn 066283
Oliemolen de Zoeker, Zaandam . 033081
Oliemolen Het Pink, Koog aan de
Zaan 032553
Oliewenhuis Art Museum,
Bloemfontein 038663
Olim, La Varenne-Saint-Hilaire . 069120
Olin Art Gallery, Gambier 049040
Olin Fine Arts Center, Washington and
Jefferson College, Washington 054185
Olin Hall Galleries, Roanoke College,
Salem 052894
Oliodigomito, Bologna 130986
Oliva, Parma 081121
Oliva, Cinzia, Torino 132174
Olivares, Jose Manuel, Spagna . 086480
Olivastri, Renato, Firenze 131227
Olive Abella, Cristina, Barcelona 133140
Olive Avenue Antiques, Fresno . 093508
Olive DeLuce Art Gallery, Northwest
Missouri State University,
Maryville 050730
Olive Hyde Art Gallery, Fremont 048973
Olive Screiner House, National
English Literary Museum,
Cradock 038694
Oliveira, Braga 084782
Oliveira, Emídio Moreira, Porto 132972
Oliveira, João Esteves, Lisboa .. 114049
Oliveira, Joaquim Pereira, Porto 132973
Oliveira, Pedro, Porto 114100
Oliveira, Rui, Lisboa 084933
Oliveira, Rui Quintela, Lisboa .. 084934
Oliver Bros., Boston 119908
Oliver, Skopje 111888
Oliver Contemporary, London .. 118486
Oliver Cromwell's House, Ely ... 044290
Oliver H. Kelley Farm, Elk River 048531
Oliver Heritage Society Museum,
Oliver 006553
Oliver Hoare Limited, London .. 118487
Oliver House Museum, Saint
Martinville 052828
Oliver House Museum, Yates County
Genealogical and Historical Society,
Penn Yan 051978
Oliver-Maneu, Joan, Palma de
Mallorca 115363
Oliver Tucker Historic Museum,
Beverly 046968
Oliver, J., Minneapolis 094939
Oliver, Lyon, Oaksey 090906
Oliver, Patrick, Cheltenham ... 088839
Oliveria & Gelderblom,
Herrischried 076510
Oliveria, Lizabeth, Oakland ... 123258,
123259
Olivers Antique, Kerava 066379
Olivers Bookshop, Whitley Bay . 144921
Oliver's Collections, Auckland .. 083813
Olive's Oddments, Hamilton .. 064326
Olivier, Pierre, Annot 066659
Les Oliviers, Saint-Vincent-de-
Tyrosse 073331
Oliwska, Gdańsk 113482
Olkijoen Rauhanpirtti, Pattijoki . 011239
Olla Podrida, Minneapolis 094940
Olle Olsson-Huset, Solna 040977
Ollier, Jacques, Paris . 071728, 105137
Ollivary-Gauthier, Paris 071729
Ollivier-Henry, Anne-Marie,
Paimpol 104437
Olmes, Cincinnati 120362
Olmos, Gaspar, Murcia 086316
Olmstead Place State Park,
Ellensburg 048547
Olmstead, Daniel, Newfields .. 127366
Olmsted County Museum,
Rochester 052611
O'Local, Estavayer-le-Lac 116260

Olofsfors Bruksmuseum,
Nordmaling 040873
Olofsson, Claes, Uppsala 143993
Olofströms Antikaffär, Olofström 086912
Oloneckaja Kartinnaja Galereja, Muzej
izobrazitelnych iskusstv Respubliki
Karelija, Olonec 037589
Oloneckij Nacionalnyj Muzej Karelov-
Livvikov im. N.T.Prilukin, Olonec 037590
Olorgesailie Prehistoric Site Museum,
Olorgesailie 029870
O'Loughlin, Timaru 084112
Olschki, Leo S., Firenze 137653
Olsen, Bundaberg 061233
Olsen, Roskilde 102964
Olsen, Eskil, Mo i Rana 113177
Olsen, Hakon, Vanløse 066180
Olsen, Lucie, Montréal 064451
Olsen, Marianne, Hedehusene . 065726
Olsen & Co, Oddvar, Drammen . 126344
Olsen, Tim, Paddington, New South
Wales 099532
Olsgården, Mats, Borlänge ... 086652
Olson, Minneapolis 135836
Olsson, Aalborg ... 065586, 102768
Olsson & Uddenberg, Göteborg . 115663
Olsson, G.N., Kuwongbah 127625
Olsson, Gunnar, Stockholm ... 115920
Olsson, Lars, Uppsala 087124
Olsson, Stig, Stockholm 087040
Olstad Larsen, Preben, Odense . 066417
Olthuysen, Henk, Huijbergen .. 132547
Oltmanns, Teo, Unkel . 108805, 137541
Oltre il Giardino, Milano 080642
Oltremare, Genova 110125
Olustee Battlefield, Olustee ... 051754
Olustvere Museuum, Olustvere . 010646
Olympia, Kraków 113622
Olympia Fine Art and Antiques Spring
Fair, London 098395
Olympia Fine Art and Antiques
Summer Fair, London 098397
Olympia Fine Art and Antiques Winter
Fair, London 098398
Olympic Antique, Los Angeles .. 094426
Olympic Gallery, Southampton . 091373
Olympic Hall of Fame and Museum,
Calgary 005532
Olympic Museum, Wellington .. 033357
Olympic National Park Visitor Center,
Port Angeles 052243
Olympus Cove Antiques, Salt Lake
City 096915
Om Art Gallery, Ahmedabad .. 109251
Om El Hana, Cairo 066259
Om igen – 1900-tal, Hägersten 086728
Oma-Freese-Huus, Dornum 017946
Omachi Alpine Museum, Omachi 029223
Omaha Antique Company,
Omaha 096054, 136065
Omaha Auction Center, Omaha . 127380
Omaha Center for Contemporary Art,
Omaha 051760
Omaha Children's Museum,
Omaha 051768
Omaha Clay Works, Omaha ... 123334
Omaha Gallery, Omaha 112978
O'Mahony, Diana, Cork 079504
O'Mahony, R & R, Portlaw 130889
O'Malley, C.H., Baltimore 092351
Oman, Dublin 079578
Oman Museum, Muscat 034099
Oman Natural History Museum,
Muscat 034100
Oman, James & Angela, Dublin 079579
Omani-French Museum Bait Fransa,
Muscat 034101
Omani Heritage Gallery, Al-
Qurum 084399
Omar, Cairo 103068
Omar, Knaresborough . 089724, 134522
Omaruru Museum, Omaruru ... 031847
Oma's Schatzkammer, Bad Homburg
v.d.Höhe 074524

Ó'Mathúna Pádraig, Caiseal
Mumhan 109539
Ombalantu Baobab Tree,
Ombalantu 031848
A l'Ombre du Clocher, La Colle-sur-
Loup 069002
Ombre et Lumière, Orléans 104431
Ombre et Lumière, Saint-Malo (Ille-et-
Vilaine) 105551
Ombres & Lumières, Paris 071730
Ombretta, Gardella, Genova 080276
Ome Kimono Museum, Ome .. 029225
Ome-shiritsu Bijutsukan, Ome .. 029226
Ome-shiritsu Hakubutsukan,
Ome 029227
Omega, Newport, Essex 090834, 134881
Omega, Warszawa 143530
Omega Antykwariat, Poznań .. 084647
Omega Art Expo, New York 122932
Omega Gallery, Carson-Newman
College, Jefferson City 049806
Omell, Ascot 088291, 134038
Omell 5060 N.R., London 118488
Omell Galleries, Ascot 117253
O'Melveny, Don, West Hollywood 125018
Omi Natural History Museum,
Omi 029229
Omilos Dimitreli, Thessaloniki .. 055756
Het Ommelander Antiquariaat,
Appingedam 143084
Omni Art, Dallas 120635
Omni Chinatown, Singapore ... 085477
Omni Gallery, Portland 123649
Omni-Trading, Apeldoorn 082929,
137849
Omni View, Chicago 120224
Omnia, Sappemeer 126298
Omniarte, Catania 131094
Omnisphere and Science Center,
Wichita 054392
Omoide Museum, Kyoto 028974
Oms, Odile, Céret 103550
Omskij Gosudarstvennyj Istoriko-
kraevedčeskij Muzej, Omsk .. 037593
Omskij Gosudarstvennyj Literaturnyj
Muzej im. F.M. Dostoevskogo,
Omsk 037594
Omskij Oblastnoj Muzej Izobrazitelnych
Iskusstv im. M.A. Vrubelja,
Omsk 037595
Omuletz, Wilma, Eisenstadt 062556
Omuta-shi Tanko oyo Kagakukan,
Omuta 029232
On a Shoestring, Birmingham .. 092427
On Aura Tout Vu, Paris 105138
On Canvass, Sydney 099791
On Deck, Oklahoma City 095998
On-Line Gallery, Southampton .. 119122
On Paper, Providence 123708
On Porpoise, San Diego 124150
On Sight, Toronto 101673
On the Air, Hawarden . 089474, 134448
On the Avenue, Buffalo 092549
On the Hill Cultural Arts Center,
Yorktown 054591
On the Plaza Gallery, San
Antonio 124055
On the Shelf, Wagga Wagga ... 139949
On the Wild Side, Minneapolis .. 122021
On Yedinci Yüzyıl Osmalı Evi,
Bursa 042863
Onas, Bolzano 079899
Once Again, Wellington 084177
Once Read Books, Long Beach . 145207
Once Upon A Child, Canal
Winchester 145026
Once Upon a Crime, Minneapolis 145282
Once Upon a Time, Auckland .. 083814
Once Upon a Time, Clarkesville . 127181
Once upon a Time, Miami 094785
Once upon a Time, New York .. 095630
Onda, Portland 123650
Ondar-Bide, Donostia-San
Sebastián 085982

Onde, Danielle, L'Isle-sur-la-
Sorgue 069584
Onder de Sint Maarten,
Zaltbommel 143268
Ondergronds Museum, Kanne .. 003746
't Onderhuys, Nieuw Beijerland . 132592
Onderwijskundig, Hilversum ... 143175
Onderwijsmuseum Educatorium,
Ootmarsum 032730
One Bell, Lavenham 089752
The One Below Gallery, London 118489
One Club for Art & Copy, New
York 060663
One Eye Gallery, Wellington ... 084178
One-Forty-Eight Shop, Hamilton 064327
One Hannard Mile & District Historical
Society, 100 Mile House ... 005300
One Home, Denver 093320
One of a Kind Antiuqes, San
Diego 097116
One Price Store, Singapore ... 085478
One Step Back, Bungay 088700
One Stop Antiques, Clayfield ... 061337
One Stroke, San Jose 124456
One To Z, Delft 112365
One Tree Two Bears Trading Post,
Long Beach 094213
One Twenty Eight, New York ... 122933
One18, Edinburgh 117735
Oneida Community Mansion House,
Oneida 051773
Oneida County Historical Society
Museum, Utica 053917
Oneida Nation Museum, De Pere 048185
O'Neil, R.A., Toronto 064715
O'Neill, Tam, Denver 093321
Onesti, Olivia, Reims 072398
Oneten, Houston 121105
Ongakudori, Yokohama 111541
Ongerup and Needilup District
Museum, Ongerup 001383
Onggi Folk Museum and Institute,
Seoul 030030
ONI Gallery, Boston 119909
Onichimiuk, Halina, Amiens ... 066580,
128438
Onion Museum, Checheng 042286
Oniris, Rennes (Ille-et-Vilaine) . 105439
Onishi Seiwemon Museum,
Kyoto 028975
Onićko, Jan, Warszawa 113878
Online Exhibits, Earl Shilton ... 117704
Only Collectables, Saint Paul .. 096846
The Only Place for Pictures,
London 118490
Onnen, Ewald, Bamberg 129441
Ono, Nagoya 111328
Ono, Tokyo 111485
Ono Deco, Wien 063042
Onomato Kunstverein, Düsseldorf 059215,
137163
Onomichi-shiritsu Bijutsukan,
Onomichi 029235
Onondaga Historical Association
Museum, Syracuse 053637
Oñoro-Ortiz 1890, Madrid 133362
Onrust, Amsterdam 112264
Ons Museum, Giessenburg ... 032346
Ons Museum, Wintelre 033054
Ons Winkeltje, Hilversum 083406
Onsberg, Ulf, Malmö 115767
Onset Contemporary Art, Chester 117571
Onslow, Stourpaine 127057
Onslow Antiques, Shenton Park 062204
Onslow County Museum,
Richlands 052530
Onslow Goods Shed Museum,
Onslow 001384
Ontario Association of Art Galleries,
Toronto 058456
Ontario College of Art & Design,
Toronto 055161
Ontario County Historical Society
Museum, Canandaigua 047398

Ontario Electric Railway Museum,
Milton 006319
Ontario Museum Association,
Toronto 058457
Ontario Police College Museum,
Aylmer 005366
Ontario Provincial Police Museum,
Orillia 006555
Ontario Science Centre, Toronto 007193
De Ontdekhoek Kindermuseum,
Rotterdam 032819
Ontem & Hoje, Rio de Janeiro .. 064098
Ontmoetingscentrum de Veldkei,
Expositie, Havelte 032422
Ontonagon County Historical Society
Museum, Ontonagon 051784
Onyang Minsok Pakmulgwan,
Asan 029889
Onyx Images, Indianapolis ... 121208
Le Onze, Paris 071731
Ooltewah Auction, Ooltewah ... 127381
Oomes, P., Hilversum 132542
Oorlogs- en Verzetsmuseum Johannes
Post, Ridderkerk 032782
Oorlogsmuseum, Bloemfontein . 038664
Oorlogsverzetsmuseum Rotterdam,
Rotterdam 032820
Oost-Indisch Huis, Amsterdam . 031975
Oost Leeshal, Amsterdam 143055
Oostends Historisch Museum De Plate,
Oostende 003934
Oosterkade Kunstonderneming,
Groningen 112498
Oosterlaar, Johan, Luttenberg . 132571
De Oosterport, Bunde 141663
Het Oosters Antiquarium, Leiden 143185
De Oostindische Compagnie,
Schoten 063827
Oostplein Antiek, Rotterdam ... 083584
Op de Beeck, J., Mechelen ... 100791
Op de MuurEN, Deventer 059618
Op den Berghe, Deventer 083160
OPA Graphics, New Orleans ... 122214
Opal Antik och Auktion, Umeå . 087114
Opalescences, Saint-Ouen ... 073162
L'Opaline, Perros-Guirec 072105
Opaline, Tarbes 073626
Opaline Rose, Tehrän 079426
Opatska Riznica Sv. Marka,
Korčula 008936
Opbouw, Amsterdam 143056
Opbouw, Rotterdam 143223
Opcje, Katowice 138977
OPD Restauro, Rivista dell'Opificio
delle Pietre Dure e Laboratori di
Restauro di Firenze, Firenze . 138857
Opelousas Museum, Opelousas . 051787
Opelousas Museum of Art,
Opelousas 051788
Open-Air Water-Power Museum,
Dimitsana 023061
Open Art, Prato 110693
Open Art Galerie, Borken,
Westfalen 106549
Open Atelier, Leiden 083439
The Open Book, Wellington ... 143368
Open Books a Poem Emporium,
Seattle 145594
Open Care, Milano 131488
Open Door Studio, Los Angeles . 094427
The Open Eye Gallery, Edinburgh 089185,
117736,   134363
Open Eye Gallery, Liverpool .. 044862
Open Haven Museum,
Amsterdam 031976
Open Hunt, New York 095631
Open Museum, Migdal-Tefen .. 025157
Open Museum, Omer 025165
Open Press, Denver 138331
Open Space Gallery, Victoria .. 007333
Open Studio, Toronto 007194
Openart, Cormano 109978
Openbaar Vervoer Museum,
Doetinchem 032234

Openbaar Vervoer Museum,
Rotterdam 032821
Opening Night Gallery,
Minneapolis 122022
Openlucht Laagveenderij Museum
Damshûs, Nij Beets 032682
Openluchtmuseum Bachten de Kupe,
Alveringem 003267
Openluchtmuseum De Duinhuisjes,
Rockanje 032790
Openluchtmuseum Ellert en Brammert,
Schoonoord 032867
Openluchtmuseum Erve Kots,
Lievelde 032599
Openluchtmuseum Front Line Hooghe,
Ieper 003727
Openluchtmuseum Het Hoogeland,
Warffum 033021
Openluchtmuseum Nieuw Amsterdam,
Nieuw Amsterdam 040504
Openluchtmuseum Ootmarsum,
Ootmarsum 032731
Openluchtmuseum voor
Beeldhouwkunst Middelheim,
Antwerpen 003291
Opera Galeria, Warszawa .. 113879,
113880
Opera Gallery, Ha Noi 125089
Opera Gallery, Hong Kong ... 102101
Opera Gallery, London 118491
Opera Gallery, Miami 121836
Opera Gallery, New York 122934
Opera Gallery, Paris 105139
Opera Gallery, Singapore 114581
Opera Laboratori Florentini,
Firenze 131228
Operadora Claveles, Cancún .. 132360
Operkuch, Rolf, Stuttgart 078624
Ophelia's Art Gallery,
Birmingham 119813
Ophey, Norbert, Geldern 075963
Opifex et Artifex, Milano 131489
Opificio delle Arti Tessili, Milano 080643
Opitz, H, Brühl 075205
Opitz, Jörg, Greiz 076087, 076088
Opitz & Geschw., Manfred, Greiz 076089
Opium, Ajaccio 066498
Opium, Warszawa 084742
Opium Antik, København 065943
Opleiding Kunstgeschiedenis,
Leiden 056177
Oplenova Hiša, Gorenjski Muzej,
Srednja Vas v Bohinju 038621
Het Oplettende Lezertje,
Bredevoort 143099
De Oplichterij, Rotterdam 112654
O'Plus, Anglet 066623
Opolski Rocznik Muzealny, Opole 138978
La Oportunidad, Buenos Aires .. 060919
Opos, Milano 059472
Opotiki Museum, Opotiki 033264
Oppdal Bygdemuseum, Oppdal . 033819
Oppenberg, Bärbel, Dinslaken .. 106685
Oppenheimer, Joel, Chicago ... 092761,
120225, 135421, 138321, 145050
Opper, Uwe, Kronberg im Taunus 107759
Opposite, Paris 071732
Oprandi, Boltiere 109872
Opravy & Renovace, Opatovice nad
Labem 128334
Opravy Starožitnosti, Svitavy .. 065905,
128371
Oprechte Veiling Haarlem,
Haarlem 126288
Oprescu, Ion, Timişoara 133010
Opsmuk, Amsterdam 082818
De Opsteker, Amsterdam 112265
Opti Art Galerie, Boppard 106542
Optica, Montréal 101438
Optik Industrie Museum,
Rathenow 021287
Optima Gestion, Madrid 133363
Optimum Fine Arts, Hornchurch 118001
Option Art, Oostende 063818

Register der Institutionen und Firmen
Optisches Museum der Ernst-Abbe-Stiftung Jena, Jena – Origine – Kunst Antiek Design, Haarlem

Optisches Museum der Ernst-Abbe-Stiftung Jena, Jena 019452
Opulent Era Gallery IV, Miami . . 094786
Opus, Wrocław 113947
Opus 21, Paris 071733
Opus 39 Gallery, Lefkosia 102539
Opus 39 Gallery, Lemesos 102552
Opus 391, Amsterdam 143057
Opus 40 and the Quarryman's Museum, Saugerties 053162
Opus 57, Charlottenlund 065630
Opus Gallery, Ashbourne 117255
Opus II, Kitchener 101327
Opus One Gallery, Atlanta 119597
Opus Restauri, Parma 131733
Opus Zwei, Sankt Gallen 087786
Opuscula Pompeiana, Roma . . . 138858
Oquendo, Madrid 086230
Or & Bleu, Rouen 072606, 129211
D'Or Antiquitäten, Hamburg . . . 076290
L'Or du Temps, Amiens 140764
L'Or du Temps, Paris 141252
L'Or en Décor, Paris 129102
Or et Change, Bayonne 066978
Or et Change Comptoir des Minerais Précieux, Tarbes 073627
Or et Monnaies Bordeaux, Bordeaux 067350
L'Or Limousin, Limoges 069528
Oracle Junction, Buffalo 145021
Al-Oraifi, Rashid, Manama 100444
Orana Googars CDEP, Dubbo . . . 061411
Orandajin, 's-Hertogenbosch . . . 083394
Orang Asli Museum, Melaka . . . 030716
Orange, Los Angeles 094428
Orange Bleue Mogador, Le Havre 103950
Orange County Historical Museum, Hillsborough 049483
Orange County Museum of Art, Newport Beach 051544
Orange County Museum of Art – Orange Lounge, Costa Mesa . . 048024
Orange County Regional History Center, Orlando 051806
Orange Empire Railway Museum, Perris 051992
Orange Group, New York 095632
Orange Regional Gallery, Orange 001386
Orange Street Gallery, Uppingham 119288
Orange Tree Antiques, Jacksonville 094020
Orangedale Railway Museum, Orangedale 006554
De Orangerie, Delft 083035
Orangerie, 's-Hertogenbosch . . . 032474
Orangerie Belvedere, Klassik Stiftung Weimar, Weimar 022504
Orangerie-Reinz, Köln 107669
Orangerie, Kunstsammlung Gera, Gera 018649
Orangerieschloss und Turm, Stiftung Preußische Schlösser und Gärten Berlin-Brandenburg, Potsdam . 021197
Orangeriet Antik Trädgårdskonst, Lund 086837
Orangetown Historical Museum, Pearl River 051958
Oranien-Nassau-Museum, Diez . 017895
Oranien-Nassauisches Museum, im Wilhelmsturm, Dillenburg 017898
Oranmore Antiques, Dallas 093189
Oravská Galéria, Dolný Kubín . . 038345
Oravská Galéria, Námestovo . . . 038407
Oravské Múzeum Pavla Országha Hviezdoslava, Dolný Kubín . . . 038346
Orawski Park Etnograficzny, Zubrzyca Górna 035470
Orban & Streu, Frankfurt am Main 141785
Orbán Ház, Szilvásvárad 023976
Orbel, Torino 132175
Orbel Art, Sofia 101167
Orbetello, Los Angeles 121567
Orbicon, Cáceres 133220
Orbital Arts, Toronto 101674
Orblin, Ivan, Reims 072399
Orbost Gallery, Dunvegan 117702
ORCA Aart Gallery, Chicago 120226
Orcas Island Historical Museum, Eastsound 048459
Orcel, Jean-Paul, Vouvray 074280
Orchard Gallery, Londonderry . . 045143
Orchard Gallery, New York 122935
Orchard Gallery, Singapore 114582
Orchard House, Concord 047954
Orchard Park Historical Society Museum, Orchard Park 051797
Orchid Art Gallery, Dallas 120636
Orchid Pavilion Art Gallery, Singapore 114583
Orci, Bucaramanga 102398
Ordemann, D., Schwalmstadt . . 108589
Orden und Militaria, Hamburg . . 076291
Die Ordenssammlung-Historia Antiquariat, Berlin 141561
Order of Saint John Museum, Christchurch 033174
Ordronneau, Yolande, Allouis . . . 066540
Ordrupgaard, Charlottenlund . . . 010029
Ordsall Hall Museum, Salford . . 045721
Ordu Devlet Güzel Sanatlar Galerisi, Ordu 043039
Ordu Etnografya Müzesi, Ordu . 043040
Ordubad Regional History Studies, Ordubad 003195
Les Oréades, Bagnères-de-Luchon 066910, 103318, 103319, 103320, 136800
Les Oréades, Moskva 114287
Les Oréades, Paris 105140
Les Oréades, Toulouse 073787, 105750, 129311
Oredaria, Roma 110839
O'Regan, Donal, Cork 079505
O'Regan's, Carmel, Gortnamona Schull 126175
Oregon Air and Space Museum, Eugene 048624
Oregon Coast History Center, Newport 051528
Oregon College of Arts and Craft, Portland 057711
Oregon Electric Railway Historical Society Museum, Lake Oswego 050084
Oregon Electric Railway Museum, Salem 052893
Oregon Gallery, Portland 123651
Oregon Historical Society, Portland 052298
Oregon-Jerusalem Historical Museum, Oregon 051798
Oregon Military Museum, Clackamas 047754
Oregon Museum of Science and Industry, Portland 052299
Oregon Museums Association, Portland 060664
Oregon Scenics, Portland 123652
Oregon Sports Hall of Fame and Museum, Portland 052300
Oregon Trail Museum, Gering . . 049085
Oregon Trail Regional Museum, Baker City 046717
O'Reilly, Dublin 126171
O'Reilly & Co., William, New York 122936
Orellana, Luis, Santiago de Chile 064980
Orelya, Paris 071734
Orenburgskij Oblastnoj Kraevedčeskij Muzej, Orenburg 037608
Orenburgskij Oblastnoj Muzej Izobrazitelnych Iskusstv, Orenburg 037609
Orense Esculturas, Rio de Janeiro 128118
Orenstein, Reed, New York 145364
Orez, Den Haag 112404, 137852
Orfèo, Luxembourg 111862
Organ Pipe Cactus Museum, Ajo 046327
Organization of Independent Artists, New York 060665
Organization of Military Museums of Canada, Gloucester 058458
Organization of Museums, Monuments and Sites of Africa OMMSA, Accra 059323
Orgel-Art-Museum Rhein-Nahe, Windesheim 022678
Orgelbaumuseum, Ostheim vor der Rhön 021032
Orgelmuseet i Fläckebo, Salbohed 040935
Orgelmuseum Altes Schloss, Valley 022300
Orgelmuseum Borgentreich, Borgentreich 017512
Orgelmuseum Kelheim, Kelheim 019575
Orginalen Antikk, Førde 084237
Orhan Kemal Müzesi, İstanbul . . 042951
Oriamu Museum, Izumiotsu . . . 028730
Oriande, Bruxelles 063517
Oricchio, Fausto, Roma 081484
Oricha, Los Angeles 121568
Oriel, Dublin 109611
Oriel Davies Gallery, Newtown, Powys 045378
Oriel Gallery, Ascot 098672
Oriel Gallery, Mold 045284
Oriel Gallery, Picton 112990
Oriel Glanymor Gallery, Fishguard 117799
Oriel Mostyn Gallery, Llandudno 044879
Oriel Newydd, Llanerchymedd . . 118137
Oriel Plas Glyn-Y-Weddw, Llanbedrog 118133
Oriel-y-Felin Gallery, Trefin 119254
Oriel Ynys Môn, Llangefni 044885
Orient-Antiquariat, Schönried . . 144135
Orient Antiques, Bruxelles 063518
D'Orient et d'Ailleurs, Paris 105141
Orient Express, Cairo 066260
The Orient Express, Richmond, Victoria 062147
Orient Express the apartment, Cairo 103069
Orient Expressed Imports, New Orleans 095138
Orient House, Glebe 061525
Orient Interiör, Göteborg 086714
Orient Longman Private, Hyderabad 137595
Orient-Occident, Paris 071735
Orient-Teppich, Sankt-Peterburg 085257
Orient USA, New York 095633
Oriental Antique, Dallas 093190
Oriental Antique Gallery, Armadale, Victoria 061043
Oriental Antique House, Singapore 085479
Oriental Antiques, Abu Dhabi . . 088161
Oriental Antiques Center, Dubai . 088182
Oriental Art, Amsterdam 082819
Oriental Art, Cairo 103070
Oriental Art, Milano 080644
Oriental Art Gallery, San Francisco 124360
Oriental Arts, Chennai 079377
Oriental Arts, Saint Louis 123851
Oriental Bazaar, Tokyo 082191
Oriental Bronzes, Paris 071736
Oriental Ceramic Society, Cambridge 060206
The Oriental Collection, Ottawa . 064511
Oriental Danny, Dallas . 093191, 093192
Oriental Decorations, New York . 095634
Oriental Furniture and Arts, London 090391, 134738
Oriental Furniture Repair, San Diego 136297
Oriental Gallery, Vancouver 064855
Oriental Handicrafts Exhibition, Manama 063156
The Oriental Haveli, Bangalore . 079357
Oriental Heritage, Los Angeles . 094429
Oriental House, Riyadh 085320
Oriental Imports, Oklahoma City 095999
Oriental Institute Museum, University of Chicago, Chicago 047686
Oriental Interiors, Vancouver . . . 064856
Oriental International Fine Arts, New York 095635
Oriental Museum Durham, Durham 044191
Oriental Rug Gallery, Eton 089220
Oriental Rug Gallery, Guildford . 089396
Oriental Rug Gallery, Oxford . . 090935
Oriental Rug Gallery, Sacramento 096639
Oriental Rug Gallery, Saint Albans 091176
Oriental Rug House and Custom Furniture Center, Jacksonville . 094021, 135664
Oriental Rug Shop, Sheffield . . . 091289
Oriental Rug Treasures, Long Beach 094214
Oriental Shop, Nashville 095019
Oriental Treasure Box, San Diego 097117
Oriental Treasures, Dallas 093193
Oriental Treasures, Singapore . . 085480
Oriental Vista Art Collections, Shanghai 102194
The Orientalia Journal, Little Neck 139470
Orientalisches Münzkabinett Jena, Institut für Sprachen und Kulturen des Vorderen Orients, Jena . . . 019453
Orientalist, London . . . 090392, 134739
l'Orientaliste, Marrakech 082593
L'Orientaliste, Thônex 087846
Orientalna, Warszawa 113881
Orientation, London 090393
Orientations, Hong Kong 138603
Orientations, San Francisco 097299
Oriente, Pradamano 110686
Orientteppich-Museum, Hannover 019064
Origami, Paris 105142
Origen, Budapest 079272
Origin, Dublin 109612
Original- und Abgußsammlung der Universität Trier, Trier 022201
Original & Authentic Aboriginal Art, Melbourne 061826
Original & Authentic Aboriginal Art, Sydney 062296
The Original Aboriginal Art Company, Narangba 061940
Original Architectural Salvage, Dublin 079580
Original Art, Indianapolis 121209
Original Art Galleries, Subiaco . . 099729
The Original Art Shop, Colchester 117613
Original Art Shop, Stoke-on-Trent 119150
The Original Art Shop, Truro . . . 119264, 135154
The Original Artshop, Manchester 118710
Original Baltimore Antique Arms Show, Timonium 098486
Original Design, Colmar 068092
The Original Dreamtime Art Gallery, Alice Springs 098637
The Original Dreamtime Art Gallery, Cairns 061250
Original Gallery, Nice 104381
Original Garo, Sapporo 082103
Original Miami Beach Antique Show, Miami Beach 098487
Original Oz Art Gallery, Mount Martha 099402
Original Print Gallery, Dublin . . . 109613
The Original Wall Art, Langley . . 118073
Originality, Totnes 119249
Originals, Bridgnorth 117425
Original's in Glass, Tampa 136394
Origine, Beauvais 067055
Origine – Kunst Antiek Design, Haarlem 138937

Origines, Houdon 068841, 140858
Origini, Verona 081930
Origins, Oostende 063819
Origrafica Konsthandel, Malmö . 086858
Orillia Museum of Art and History,
  Orillia 006556
Orima, Stockholm 087041
Orimattilan Kotiseutumuseo,
  Orimattila 011209
Orimattilan Taidemuseo,
  Orimattila 011210
Orinasukan, Kyoto 028976
Oriol, Barcelona 085814, 114947
Orion, Miami 094787
Orion Antique, Dallas 093194
Orion Antique, Oklahoma City .. 096000
Orion Book Store, Toronto 140413
Orion Press, Tokyo ... 137788, 142935
The Orion Publishing Group,
  London 138234
Oriot, Jean-Marcel, Bagnoles-de-
  l'Orne 066914
Oripään Kotiseutumuseo, Oripää 011212
Oriskany Battlefield, Oriskany .. 051802
Orissa, Milano 080645
Orissa State Museum,
  Bhubaneshwar 024175
Orit Art Gallery, Tel Aviv 109730
Oriti Niosi, Paolo Franco, Cagliari 079995
Oriveden Paltanmäen Museo,
  Orivesi 011213
Orkdal Bygdemuseum, Svorkmo 034006
Orkney Farm and Folk Museum,
  Birsay 043633
Orkney Farm and Folk Museum,
  Harray 044514
Orkney Gallery, Kirkwall 118059
Orkney Museum, Kirkwall 044734
Orkney Wireless Museum,
  Kirkwall 044735
Orland Historical Museum,
  Orland 051803
Orlandini, Mauro, Torino 132176
Orlando & Almeida, Lisboa 084935
Orlando Brown House, Liberty Hall
  Museum, Frankfort 048931
d'Orlando Griffe, Paris 071737
Orlando Museum of Art, Orlando 051807
Orlando Science Center, Orlando 051808
Orlando, John Baptist, Tarxien . 132344
Orlaque, Paris 129103
Orleans, Toronto 064716
Orleans House Gallery,
  Twickenham 046011
Orlické Muzeum, Choceň 009519
Orlogsmuseet, København 010190
Orlovskij Gosudarstvennyj Institut
  Iskusstv i Kultury, Orël 056295
Orlovskij Oblastnoj Kraevedčeskij
  Muzej, Orël 037605
Orlovskij Oblastnoj Muzej
  Izobrazitelnych Iskusstv, Orël . 037606
Ormeau Baths Gallery, Belfast . 043560
Ormond Memorial Art Museum,
  Ormond Beach 051814
Ornabi, São Paulo 140310
Ornament, Vista 139471
Ornament and Crime, Los
  Angeles 094430
Ornamet Connection, Austin ... 092258
Ornatto, Porto Alegre . 063986, 063987,
                               063988
Ornis Jagt- & Naturbøger, Skjern 140657
Ornithology and Entomology Museum,
  Cairo 010556
Ornunga Museum, Vårgårda ... 041157
Oro de Monte Alban, San
  Antonio 097010
Oro e Colore, Firenze 131229
ORO Fine Art Gallery,
  Albuquerque 119502
Oro Fino, Düsseldorf 097999
Oro Gema, Valencia . 086527, 126560
Oro Incenso e Mirra, Milano ... 080646

Oro y Plata, Sevilla 086461
Orobs, Toledo 124693
Orologio, Antwerpen 063229
Orono Museum of Art, University of
  Maine, Bangor 046779
Orońsko, Orońsko 138979
Oroszlányi Bányász Múzeum
  Alapítvány, Központi Bányászati
  Múzeum, Oroszlány 023822
Oroville Chinese Temple, Oroville 051820
Orphan Blues Magazine, Detroit 120815
Orphée, Paris 071738
Orpiment, Paris 071739
Orplid, Icking 076593
Orr, San Diego 097118
Orr, C., Minneapolis 145283
Orr, Michael, Columbus 120507
Orrill, J., Los Angeles 127294
Orrú, Salvatore, Lippstadt 107883
Orsa Auktionsbyrå, Orsa 126652
Őrségi Népi Műemlékegyüttes –
  Szalafo-Pityerszer, Savaria Múzeum
  Szombathely, Szalafő 023914
Orsi, Carlo, Milano 080647
Orsi, Francesco, Varese 081815
Orsini Arte e Libri, Milano 080648,
                               142735
Orsini, Alessandro, Roma 081485
Orsini, Roberto, Milano 142734
Orskij Istoriko-kraevedčeskij Muzej,
  Orsk 037610
Orssich, Paul, London 144660
Országos Földtani Múzeum,
  Budapest 023536
Országos Műszaki Múzeum,
  Budapest 023537
Országos Pedagógiai Könyvtár és
  Múzeum, Budapest 023538
Országos Széchényi Könyvtár,
  Budapest 023539
Országos Színháztörténeti Múzeum és
  Intézet, Budapest 023540
Orta Dogu Teknik Universitesi Sanat
  Galleri, Ankara 042826
Orta Oyunlari Sanat Gösteri Yan,
  İstanbul 117071
Ortega, Jacksonville 094022
Ortega, Santander 086426
Ortega, Sevilla 086462
Ortega Machuca, Jose Ramon,
  Madrid 115265
Ortega Motilla, L., Denia 115042
Ortega Ortega, Enrique, Córdoba 133230
Ortega Ruiz, Manuel, Barcelona 085815
Ortega, Alfonso, Alicante 114831
Ortega, Gilbert, Phoenix 123503
Ortega, Lola, Sevilla 086463
Ortega, Marie-Cristina, Versailles 074064
Ortega, Morales, Sevilla 086464
Orteu, Jacques, Saint-Girons .. 105527
Orth, Seattle 124579
Orth & Thierfeld, Lohmar 077148
Orth, Peter J., Bonn 106535
Orthopedics Museum, Chulalongkorn
  University, Bangkok 042672
Ortiz & Bello, Miami 094788
Ortiz & Dagenbela, San Antonio 124056
Ortiz Aragon, Alicia, Granada . 115093
Ortiz Comas, Adriano Santiago,
  Madrid 115266
Ortiz un Mundo Diferente, León 115112
Ortiz, Pierre, Calvisson 067610
Ortiz, Rafael, Sevilla 115453
Ortmann, Jan-Stefan, Paris 129104
Ortodoks Antik, Moskva 085195
Ortodoksinen Kirkkomuseo,
  Kuopio 011055
Ortodox Egyházművészeti Gyűjtemény,
  Kecskemét 023731
Ortolan, Giuseppe, Padova 080996
Orton Geological Museum, Ohio State
  University, Columbus 047942
Orton, Stephen, Watlington 091761,
                               135195

Orts- und Weinbaumuseum,
  Neftenbach 041789
Orts- und Wohnmuseum,
  Marthalen 041721
Orts Vidal, Alfredo, Alicante ... 085701
Orts, Gérard, Paris 071740
Orts, Gérard, Pau 072046
Ortsgeschichtliche Sammlung,
  Bovenden 017525
Ortsgeschichtliche Sammlung,
  Oberrieden 041814
Ortsgeschichtliche Sammlung
  Rebstein, Ortsmuseum Spritzenhaus,
  Rebstein 041864
Ortsgeschichtliches Museum,
  Siegenburg 021856
Ortskundliches Museum Jois,
  Jois 002150
Ortsmusem Wittenbach im Schloss
  Dottenwil, Wittenbach 042166
Ortsmuseum, Amriswil 041218
Ortsmuseum, Andwil 041222
Ortsmuseum, Bergün 041306
Ortsmuseum, Binningen 041344
Ortsmuseum, Brachwitz-
  Saalkreis 017528
Ortsmuseum, Bülach 041378
Ortsmuseum, Bütschwil 041381
Ortsmuseum, Häggenschwil ... 041579
Ortsmuseum, Jona 041610
Ortsmuseum, Kaltbrunn 041611
Ortsmuseum, Laax 041633
Ortsmuseum, Liesberg 041672
Ortsmuseum, Meilen 041733
Ortsmuseum, Merenschwand . 041743
Ortsmuseum, Nackenheim ... 020635
Ortsmuseum, Neuburg am Rhein 020687
Ortsmuseum, Nürensdorf 041804
Ortsmuseum, Oftringen 041820
Ortsmuseum, Opfikon 041828
Ortsmuseum, Rafz 041856
Ortsmuseum, Roggwil 041887
Ortsmuseum, Rüti 041898
Ortsmuseum, Sankt Stephan ... 041934
Ortsmuseum, Schänis 041946
Ortsmuseum, Thalwil 042052
Ortsmuseum, Unterengstringen . 042079
Ortsmuseum, Untersiggenthal . 042083
Ortsmuseum, Vaz 042100
Ortsmuseum, Vnà 042115
Ortsmuseum, Wängi 042118
Ortsmuseum, Walenstadt 042121
Ortsmuseum, Weikertschlag ... 002902
Ortsmuseum Albisrieden, Zürich 042224
Ortsmuseum Altes Rathaus,
  Balgach 041258
Ortsmuseum Altstetten, Altstetten 041215
Ortsmuseum Beringen, Beringen 041307
Ortsmuseum Brittnau, Brittnau . 041367
Ortsmuseum Diessenhofen,
  Diessenhofen 041441
Ortsmuseum Dietikon, Dietikon . 041443
Ortsmuseum Eglisau, Eglisau .. 041455
Ortsmuseum Erlenbach,
  Erlenbach 041469
Ortsmuseum Eschlikon,
  Eschlikon 041475
Ortsmuseum Flawil, Flawil 041485
Ortsmuseum Frenkendorf,
  Frenkendorf 041495
Ortsmuseum für Heimatkunde,
  Schlieren 041957
Ortsmuseum Gaiserwald, Altes
  Pfarrhaus, Sankt Josefen 041928
Ortsmuseum Grächen, Grächen . 041558
Ortsmuseum Hinwil, Hinwil 041559
Ortsmuseum Höngg, Zürich ... 042225
Ortsmuseum Kefiturm, Belp ... 041304
Ortsmuseum Kilchberg, Kilchberg
  (Zürich) 041614
Ortsmuseum Küsnacht, Küsnacht 041630
Ortsmuseum mit Karl-Jauslin-
  Sammlung, Muttenz 041787
Ortsmuseum Mollis, Mollis 041747

Ortsmuseum Mühle, Museen Maur,
  Maur 041732
Ortsmuseum Neunkirch,
  Neunkirch 041799
Ortsmuseum Oberes Bad,
  Marbach 041720
Ortsmuseum Oberuzwil,
  Oberuzwil 041816
Ortsmuseum Oetwil, Oetwil am
  See 041818
Ortsmuseum Pratteln, Jacquard-Stübli,
  Pratteln 041849
Ortsmuseum Rüschlikon,
  Rüschlikon 041897
Ortsmuseum Schleitheim,
  Schleitheim 041955
Ortsmuseum Schmitten, Schmitten,
  Albula 041959
Ortsmuseum Spreitenbach,
  Spreitenbach 042016
Ortsmuseum Steinmaur, Sünikon 042038
Ortsmuseum Sust, Horgen 041599
Ortsmuseum Trotte, Arlesheim . 041231
Ortsmuseum Urdorf, Urdorf ... 042086
Ortsmuseum Ursulastift,
  Gerstetten 018670
Ortsmuseum Vaz, Lenzerheide . 041668
Ortsmuseum Vechigen, Boll ... 041352
Ortsmuseum Wallisellen,
  Wallisellen 042122
Ortsmuseum Weiach, Weiach .. 042127
Ortsmuseum Wetzikon, Wetzikon 042135
Ortsmuseum Wiesendangen,
  Wiesendangen 042138
Ortsmuseum Wila, Wila 042141
Ortsmuseum Wilchingen,
  Wilchingen 042142
Ortsmuseum Wilderswil,
  Wilderswil 042144
Ortsmuseum Wollishofen, Zürich 042226
Ortsmuseum Zollikon, Haus zum
  Felsengrund, Zollikon 042186
Ortsmuseum zur Farb, Stäfa ... 042019
Ortsmuseum zur Hohlen Eich,
  Wädenswil 042117
Ortssammlung Wettelsheim,
  Treuchtlingen 022189
Ortsstube, Bolligen 041353
Ortstein, Bremerhaven 075190
Ortus, Bonnieux 103408
Orugoru no Chisana Hakubutukan,
  Tokyo 029573
Orust, Henån 086744
Oružejnaja Palata, Gosudarstvennyj
  muzej-zapovednik Moskovskij Kreml,
  Moskva 037486
Orvieto, Paolo, Milano 080649
Orville-Fasid, C., San Antonio . 097011
Orvos- és Gyógyszerészettörténeti
  Kiállítás, Katona József Múzeum,
  Kecskemét 023732
Orvostörténeti és Patikakiállítás,
  Karcag 023719
Orwell Corner Historic Village, Prince
  Edward Island Museum and Heritage
  Foundation, Orwell 006561
Orwell Furniture, Ipswich 089650,
                               134504
Orzcy, Budapest 079273
Osage County Historical Society,
  Lyndon 050537
Osage County Historical Society
  Museum, Pawhuska 051848
Osage Gallery, Beijing 101981
Osage Gallery, Hong Kong 102102,
                       102103,   136725
Osage Historic Village, Lamar .. 050106
Osage Singapore, Singapore .. 059801,
                               114584
Osage Suojiacun Gallery, Beijing 101982
Osaka Castle Museum, Osaka .. 029255
Osaka-furistu Sayamaike
  Hakubutsukan, Osakasayama . 029274
Osaka-furitsu Gendai Bijutsu Senta,

Osaka 029256
Osaka-furitsu Senboku Koko
Shiryokan, Sakai 029312
Osaka Geijutsu Daigaku, Kanan 056004
Osaka Human Rights Museum,
Osaka 029257
Osaka International Peace Center,
Osaka 029258
Osaka-kenritsu Chikatsu-Asuka
Hakubutsukan, Kanan 028768
Osaka-kenritsu Chu Hakabutsukan,
Mino 029026
Osaka-kenritsu Kokuritsu Bijutsukan,
Osaka 029259
Osaka-kenritsu Waha Kamigata
Bungeikan, Osaka 029260
Osaka Lottery Dream Museum,
Osaka 029261
Osaka Museum of History, Osaka 029262
Osaka Museum of Housing and Living,
Osaka 029263
Osaka Museum of Natural History,
Osaka 029264
Osaka Nippon Mingeikan, Suita . 029411
Osaka Prefectural Museum of Yayoi
Culture, Izumi, Osaka 028729
Osaka-shi Kagaku Hakubutsukan,
Osaka 029265
Osaka-shiritsu Bijutsukan, Osaka 029266
Osaka-shiritsu Kindai Bijutsukan
Kensetsu Jumbishitsu, Osaka . 029267
Osaka-shiritsu Toyo Toji Bijutsukan,
Osaka 029268
Osaka Zokei Center, Osaka 056026,
111365
Osakarovskij Rajonnyj Istoriko-
Kraevedčeskij Muzej,
Osakarovka 029832
Osare, Camberwell 098852
Osbane Museet, Fana 033594
Osborn-Jackson House, East
Hampton Historical Society, East
Hampton 048435
Osborn, Richard, London 118492
Osborne, Kirkcudbright 089718
Osborne & Allen, London 118493
Osborne Gallery, Nowra 099493
Osborne House, East Cowes ... 044201
The Osborne Studio Gallery,
London 118494
Osbourne Samuel, London 118495,
126968
Oscar Anderson House Museum,
Anchorage 046473
Oscar Getz Museum of Whiskey
History, Bardstown Historical
Museum, Bardstown 046789
Oscar Howe Art Center, Mitchell 050971
Oscar Publications, Delhi 142589
Oscar, B., Miami 135803
Oscars, Haslingden 117923
Oscarsborg Festningsmuseum,
Forsvarsmuseet, Oscarsborg .. 033822
Oscarshall Slott, Oslo 033857
Osceola, Emeryville 120844
Osé, Hasselt 063686
Osebro, Porsgrunn 084345
Osen Bygdemuseum, Osen 033823
Osen Bygdetun, Steinsdalen ... 033989
Osenat, Jean-Pierre,
Fontainebleau 125587
Osenat, Jean-Pierre, Paris ... 071741,
125746
Osgoode Township Museum, Vernon,
Ontario 007316
Oshawa Community Museum and
Archives, Oshawa 006564
Oshawa Sydenham Museum,
Oshawa 006565
O'Shea, London 090394, 144661
Oshkosh Public Museum,
Oshkosh 051829
Osinskij Kraevedčeskij Muzej,
Osa 037611

Osiris, Helsinki 066352
Osiris, Tokyo 137789
Osiris Antiques, Southport 091382
Osiris Galerie, München 108169
Oskar Luts Majamuuseum, Tartu 010698
Oskar Schlemmer Archiv, Staatsgalerie
Stuttgart, Stuttgart 022053
Oskar Schlemmer Theatre Estate and
Collection, Oggebbio 027086
Oskara Kalpaka Muzejs Airišu
Pieminas Vietā, Zirņi 030260
Oskarshamns Sjöfartsmuseum,
Oskarshamn 040909
Osler Historical Museum, Osler . 006567
Oslo Auktionsforretning, Oslo .. 126378
Oslo Bymuseum, Oslo 033829
Oslo Kunstforening, Oslo 059673
Oslo Mynthandel, Oslo 084334
Oslo Nye Antikvariat, Oslo 143411
Osma, Guillermo de, Madrid ... 115267
Osman Hamdi Bey Müzesi,
Gebze 042906
Osmanbey Art Galeri, İstanbul . 117072
Osmani Museum, Sylhet 003235
Osmoz, Bulle 116180
Osmundo Antique Gallery, Makati 084476
Osona, Roberto, Cincinnati 120363
Osorkhonai Akademiiai Ulumi Jumhurii
Tojikiston, Dušanbe 042617
Osorkhonai Bostonshinosiii Khujand,
Chudžand 042615
Osorkhonai Davlatii Ta"rikhi –
Kishvarshinosi va Sa"nati Tasvirii
Jumhurii Tojikiston ba nomi
Kamoliddin Behzod, Dušanbe . 042618
Osorkhonai Jumkhuriiavii Ta"rikhi –
Kishvarshinosii ba nomi Abuabdullo
Rudaki, Pandžakent 042623
Osorkhonai Ta"rikhi – Kishvarshinosii
Isfara, Isfara 042619
Osorkhonai Ta"rikhi – Kishvarshinosii
Shahri Uro-Teppa, Uro-Teppa . 042624
Osorkhonai Ta"rikhi – Kishvarshinosii
Viloiati Khujand, Chudžand ... 042616
Osorkhonai Ta"rikhi – Kishvarshinosii
Viloiatii Külob, Külob 042621
Osorkhonai Tibb va Giёhhoi
Shifobakhshi Tojikiston, Isfara . 042620
Osorkhonai Viloiatii Ta"rikhi –
Kishvarshinosii, Chorog 042614
Osoyoos Art Gallery, Osoyoos .. 006569
Osoyoos Museum, Osoyoos ... 006570
Osper, Knut, Köln 107670
Ośrodek Biograficzny Komisji Turystyki
Górskiej, Gabinet prof. Kazimierza
Sosnowskiego, Kraków 034908
Ośrodek Muzealno-Dydaktyczny
Wielkopolskiego Parku Narodowego,
Mosina 035020
Osrodek Muzealny Bydgoskiego Wezla
Kolejowego, Bydgoszcz 034671
Ośrodek Postaw Twórczych,
Wrocław 113948
Ośrodek Propagandy Sztuki, Miejska
Galeria Sztuki, Łódź 034978
De Ossekop, Leiden 083440
Ossining Historical Society Museum,
Ossining 051835
Osškij Objedinennyj Istoriko-Kulturnyj
Muzej-Zapovednik, Oš 030081
Ossowski, London 090395, 090396,
134740,   134741
Osswald, Harald, Immenstaad am
Bodensee 141957
Ost-, Südostdeutsche Heimatstuben,
Ulm 022266
Ost-West, Timişoara 085138
Ostarrichi-Kulturhof, Neuhofen an der
Ybbs 002436
Ostasiatika-Sammlung Ehrich,
Museum Bochum, Hattingen .. 019101
Ostasiatische Zeitschrift, Neue Serie,
Berlin 138747
Ostdeutsche Heimatsammlung,

Ravensburg 021301
Ostdeutsche Heimatstube, Bad
Zwischenahn 016919
Ostdeutsche Heimatstube,
Fellbach 018350
Ostdeutsche Heimatstube,
Zweibrücken 022905
Ostdeutsche Kultur- und Heimatstuben
mit Schönbacher Stube,
Heppenheim 019194
Ostdeutsche Kunstauktionen,
Berlin 106392, 125884
Ostdeutsche Museumsräume in der
Heimatstube, Kaiserslautern .. 019481
Osted Antik, Roskilde 066095
Osted Antikhused, Roskilde 066096
Osten, Andrea von der,
Hilpoltstein 141947
Ostendorff, Münster .. 108223, 142207
Ostendorff, Wiesbaden 108926
Ostendorff, Birthe, Heidelberg .. 107396
Oster Antik, Wolfratshausen ... 098163
Oster, Ralf, Kleve 130133
Osterburger Antiquariat,
Osterburg 142283
Osterburgsammlung, Bischofsheim an
der Rhön 017398
Ostereimuseum, Sonnenbühl ... 021917
Osterley Park House, Isleworth . 044659
Ostermann, Wuppertal 108995
Osterøy Museum, Lonevåg 033755
Osterson, Vancouver 101818
Ostertag, Mühlheim am Main .. 108036
Ostertag, Freddy, Oberaach ... 087717,
133878
Osterville Historical Museum,
Osterville 051836
Osterwalder, Hamburg 107304
Osterzgebirgsmuseum, Geising . 018611
Ostfriesen Bräu, Großefehn 018845
Das Ostfriesische Teemuseum und
Norder Heimatmuseum, Norden 020819
Ostfriesischer Kunsthandel,
Remels 078195, 130580
Ostfriesisches Landesmuseum und
Emder Rüstkammer, Emden .. 018179
Ostfriesisches Landwirtschaftsmuseum,
Krummhörn 019813
Ostfriesisches Schulmuseum
Folmhusen, Westoverledingen . 022602
Ostharz-Antiquitäten,
Aschersleben 074444
Osthaus Museum Hagen, Hagen 018919
Osthoff, Daniel, Würzburg 142515
Ostholstein-Museum, Eutin ... 018329
Ostholstein-Museum Neustadt,
Neustadt in Holstein 020769
Ostholt, K., Ahaus 105962
Ostia, New York 095636
Ostindiefararen Götheborg,
Göteborg 040658
Ostindiska Kompaniet, Stockholm 087042
Ostler, München 077589
Ostler, Herbert & Andreas,
Mittenwald 108008
Ostner, Schwabmünchen 078402
Osto- ja Myyntikeskus, Kotka . 140731
Osto- ja Myyntiliike Kellari, Oulu 066391
Ostowar, Berlin 141966
Ostoya, Warszawa ... 126424, 126425
Ostoya – Galeria na Grodzkiej,
Kraków 084590, 126411
Ostpreußisches Landesmuseum,
Lüneburg 020184
Ostravská Antikva, Ostrava ... 065437
Ostravské Muzeum, Ostrava ... 009743
Ostreicher, Max, New York ... 095637
Ostrin, Joyce, New York 095638
Ostrobothnia Australis, Biological
Museum, Vaasa 011470
Ostrogožskij Rajonnyj Istoriko-
chudožestvennyj Muzej im. I.N.
Kramskogo, I.N. Kramskoy

Art Museum of Ostrogoshsk,
Ostrogožsk 037614
Ostrov Sokrovišč, Moskva 085196
Ostsee-Antik-Baabe, Baabe,
Ostseebad 074480
Ostsee-Antik-Zingst, Zingst,
Ostseebad 079143
O'Sullivan, Dublin 079581
O'Sullivan, New York 095639
O'Sullivan, A.J., Darlton 089029, 134314
Osuna, Ramon, Washington ... 124941
Osuuspankkimuseo, Helsinki .. 010851
Osvald, Oliver von, Sankt Gallen 144129
Oswald, Austin 119687
Oswald & Kalb, Wien 063043
Oswald, K., Lörrach 077144
Oswego Historical Museum,
Oswego 051837
Oswestry Transport Museum,
Oswestry 045444
Oświatowe Muzeum, Szczecin . 035254
OSZK-Soros, Budapest 130813
Oszlai-Tájház, Cserépfalu 023585
Ota, Tokyo 111486
Ōta Kinen Bijutsukan, Tokyo ... 029574
Ota-kuritsu Mingeikan, Tokyo .. 029575
Otago Art Society Gallery,
Dunedin 112890
Otago Military Museum, Dunedin 033193
Otago Museum, Dunedin 033194
Otago Settlers Museum, Dunedin 033195
Otago Studios, Glasgow 117854
Otago Vintage Machinery Club
Museum, Dunedin 033196
Otakou Marae Museum, Dunedin 033197
Otar Lordkipanidze Vani Archaeology
Miseum-Reserve, Vani 016429
Otaru Music Boxes Museum & Antique
Museum, Otaru 029275
Otaru-shi Hakubutsukan, Otaru . 029276
Otaru-shi Seishonen Kagakugijutsukan,
Otaru 029277
Otaru-shiritsu Bijutsukan, Otaru 029278
Otaru Venice Museum, Otaru .. 029279
Otautau and District Local History
Museum, Otautau 033265
Otello Kalesi, Gazimağusa 033454
Otemar, Jacques d', La Tour-
d'Aigues 069106
Otende, Jocelyne, Amsterdam . 066885
Otende, Jocelyne, Le Crès ... 069258
Oterelo, Claude, Paris 141253
Otero, Saint-Paul-de-Vence ... 105597
Otford Antiques Centre, Otford . 090925,
134913
Other Shop II, San Francisco .. 097300
Other Side, Portland 096454
The Other Side, Tampa 097581, 136395
The Other Side of the Mirror,
Cholsey 117587
Other Times Antiques, Portsoy . 091046
Other Times Books, Los Angeles 145234
Otis College of Art and Design, Los
Angeles 057370
Otley Antique Centre, Otley ... 090927
Otley Museum, Otley 045445
Otmezguine, Jane, Nice 104382
Otoe County Museum of Memories,
Syracuse 053631
Otoko, Los Angeles 094431
O'Toole, Kay, Houston 093760
O'Toole, Noel, Limerick 079569
Otori, Nara 082074
Otranto, Barcelona 085816
Otrar Memlekettik Archaeologicalik
Korik Muzej, Čimkent 029818
Otrarskij Gosudarstvennyj
Archeologičeskij Muzej-Zapovednik,
Šaulder 029842
Otraženie – Muzej Voskovych Figur,
Moskva 037600
Otsu-e Bijutsukan, Otsu 029280
Otsuka Bijutsukan, Naruto 029158
Otsuka Textile Design Institute,

Tokyo 056053
Ott, Egbert, Wassenaar 083693
Ott, Herbert, Isen 107469
Ott, J.B.A.M., Zutphen 083723
Ott, Joseph, Colmar 068093
Ott, Wilhelm, Düsseldorf 106813, 137164
Ottavi Pacitti, Paris 071743
Ottavi, Marc, Paris 071742
Ottaviani, Pier Giuseppe, Roma . 081486
Ottawa Antique Market, Ottawa 064512
Ottawa Art Gallery, Ottawa 006595
Ottawa Conservation Studio,
   Ottawa 128192
Ottawa County Historical Museum,
   Port Clinton 052247
Ottawa House By-The-Sea Museum,
   Parrsboro 006614
Ottawa Scouting Museum,
   Ottawa 051842
Otte, Achim, Hamburg 129981
Ottenby Naturum, Degerhamn . 040591
Ottenhausen, Irene, Neumagen-
   Dhron 137443
Otter, Amsterdam 143058
Otter & Bear, Albuquerque . . . 119503
Otter Antiques, Honiton 089572
Otter Tail County Museum, Fergus
   Falls 048723
Otterstetter, Claus, Bensheim . . 141462
Otterton Mill Centre and Working
   Museum, Budleigh Salterton . . 043784
Otterup Museum, Otterup 010268
Ottesen, Susanne, København . . 102909
Ottevanger & Weiman, Den Haag 083113
Ottin, Michel, Paris 071744
Ottmar-Mergenthaler-Museum, Bad
   Mergentheim 016823
Otto Dill-Museum, Neustadt an der
   Weinstraße 020761
Otto-Ditscher-Galerie, Neuhofen 020712
Otto-Dix-Haus Hemmenhofen,
   Gaienhofen 018583
Otto-Dix-Haus, Kunstsammlung Gera,
   Gera 018650
Otto Dix Stiftung, Vaduz 059549
Otto-Flath-Kunsthalle, Bad
   Segeberg 016878
Otto-Galerie, München 077590, 108171
Dat Otto Huus, Emden 018180
Otto-Klages-Sammlung, Königslutter
   am Elm 019735
Otto-König von Griechenland-Museum,
   Ottobrunn 021049
Otto-Lilienthal-Museum, Anklam 016587
Otto Modersohn Museum,
   Ottersberg 021046
Otto-Pankok-Museum, Bad
   Bentheim 016716
Otto Pankok Museum, Hünxe . . 019361
Otto-Schwabe-Museum, Hochheim am
   Main 019271
Otto-Ubbelohde-Haus, Lahntal . 019850
Otto von Bylandt-Gesellschaft,
   Mönchengladbach 059216
Otto Wagner Hofpavillion Hietzing,
   Wien Museum, Wien 003060
Otto Wagner Pavillion Karlsplatz, Wien
   Museum, Wien 003061
Otto, Conrad Claus, Bremen . . 141653
Otto, Jutta & Günter, Gevelsberg 075992
Otto, Kurt, Bad Sassendorf . . . 074587
Otto, Manfred, Berlin 074894
Otto, Manfred, Frankfurt am
   Main 075811
Otto, Rainer, Regensburg 078175
Otto, Rositta, Kraichtal 130187
Otto, Rudolf, München 108170
Otto, Ulrich, Roßdorf 130601
Dell'Ottocento, Livorno 110185
Ottocento, Messina 080374
Ottocento, Recco 110703
Ottocento, Roma 131996
Ottocento e Novecento, Milano . 080650
Ottolina Antik, Billinge 086642

Gli Ottomani, Trento 111032
Ottorino Nonfarmale, Venezia . 132252
Ottrey, Mike, Wallingford 091714
Otumpa, London 090397
Otwarta Pracownia Jerzy Hanusek,
   Kraków 084591
Otway, Anglesea 098648
Où Galerie, Marseille 104191
Ouaiss, Jacques, Beirut 082300
Ouaknine, Nantes 070473
Oud Amelisweerd Museum,
   Bunnik 032116
Oud Cortrijcke, Kortrijk 063724
Oud en Nieuw, Breda 083000
Oud Holland, Den Haag 138938
Oud Holland, Utrecht 083656
Ouddorps Raad- en Polderhuis,
   Ouddorp 032747
De Oude Aarde, Giethoorn . . . 032352
Oude Ambachten en
   Speelgoedmuseum, Terschuur . 032919
Oude Bakkerij, Medemblik 032641
De Oude Borze, Antwerpen . . . 140151
De Oude Deel, Zutphen 083724
De Oude Doos, Eindhoven 083224
De Oude Doos, Utrecht 083657
De Oude Drostdy, Tulbagh 083866
Oude Fortuyne, Antwerpen 127997
Het Oude Glashuis, Nijmegen . . 083517
Het Oude Huis, Bergen op Zoom 082965
Oude Kerk Volksmuseum,
   Tulbagh 083867
De Oude Lantaerne, Leeuwarden 083418
Oude Pijpelade, Utrecht 083658
De Oude Prenten, Rotterdam . . 083585
't Oude Raadhuis, Beek en Donk 032051
Het Oude Raadhuis, Leerdam . 032569
Het Oude Raadhuis, Oud-
   Beijerland 032742
Het Oude Raadhuis, Urk 032940
Het Oude Slot Heemstede,
   Heemstede 032425
Oude Smederij, Enschede 083256
Het Oude Stadhuis, Hasselt . . . 032418
De Oude Stal, Amsterdam 082820
Oude Vlaamse Galerij, Sint-
   Amands 100814, 128088
De Oude Wand, Zutphen 083725
Den Ouden Huls, 's-
   Hertogenbosch 083395
Oudhede & Geskenkwinkel,
   Tshwane 085682
Oudheidkamer, Bruinisse 032113
Oudheidkamer, Den Burg 032166
Oudheidkamer, Koudekerk 032555
Oudheidkamer, Steenwijk 032903
Oudheidkamer, Wijngaarden . . . 033048
Oudheidkamer Beilen, Beilen . . 032055
Oudheidkamer Bleiswijk,
   Bleiswijk 032071
Oudheidkamer Boerderij Strunk,
   Raalte 032773
Oudheidkamer Brederwiede,
   Vollenhove 033003
Oudheidkamer Buisjan Enter,
   Enter 032313
Oudheidkamer Dantumadeel,
   Zwaagwesteinde 033120
Oudheidkamer de Oude Pastory,
   Heiloo 032445
Oudheidkamer Doorn, Doorn . . 032241
Oudheidkamer Eemnes, Eemnes 032272
Oudheidkamer Geervliet,
   Geervliet 032334
Oudheidkamer Heerde, Heerde . 032426
Oudheidkamer Hellevoetsluis,
   Hellevoetsluis 032459
Oudheidkamer Hoolt'n, Holten . . 032499
Oudheidkamer Horst, Horst . . . 032513
Oudheidkamer IJsselstreek,
   Wijhe 033046
Oudheidkamer in het Stadhuis,
   Bolsward 032078
Oudheidkamer Leiderdorp,

Leiderdorp 032588
Oudheidkamer Lemster Fiifgea,
   Lemmer 032595
Oudheidkamer Leshuis,
   Dedemsvaart 032151
Oudheidkamer Lunteren,
   Lunteren 032609
Oudheidkamer Nederlandse Kaap-
   Hoornvaarders, Hoorn 032508
Oudheidkamer Nunspeet,
   Nunspeet 032706
Oudheidkamer Renswoude,
   Renswoude 032779
Oudheidkamer Ridderkerk,
   Ridderkerk 032783
Oudheidkamer Rozenburg,
   Rozenburg 032833
Die Oudheidkamer tot Medenblick,
   Medemblik 032642
Oudheidkamer Wassenaar,
   Wassenaar 033027
Oudheidkamer Weststellingwerf,
   Wolvega 033061
Oudheidkamer Wieringerwaard het
   Polderhuis, Wieringerwaard . . 033044
Oudheidkamer Willem Van Strijen,
   Zevenbergen 033100
Oudheidkamer/ Museum Mr. Andreae,
   Kollum 032551
Oudheidkundeige Verzameling,
   Schinveld 032858
Oudheidkundig Museum, Sint
   Michielsgestel 032873
Oudheidkundig Museum Arnemuiden,
   Arnemuiden 032019
Oudheidkundig Museum Sint-
   Janshospitaal, Damme 003532
Oudheidkundig Museum van het
   Begijnhof, Begijnhofmuseum,
   Turnhout 004100
Oudheidkundige Verzameling,
   Sluis 032888
Oudheidskamer Vreeswijk,
   Nieuwegein 032675
Oudin, Gilles, Paris . . . 071745, 105143
Oudot & Fils, Fougerolles 128604
Ouellette, Pierre-François,
   Montréal 101439
Ouestampille, Saint-Martin-de-
   Mailloc 072933
Oulaisten Kotiseutumuseo,
   Oulainen 011214
Oulan Bator, Orléans 104432
Ould Chibani, Casablanca 082577
Ouley, Eric, Izon 128629
d'Oultremont, Bruxelles 063519
Oulun Automuseo, Oulu 011216
Oulun Kirja-Antikvariaatti, Oulu . 140739
Oulun Taidemuseo, Oulu 011217
Oulun Yliopiston Eläinmuseo,
   Oulu 011218
Oulun Yliopiston Geologinen Museo,
   Oulu 011219
Oulun Yliopiston Kasvimuseo,
   Oulu 011220
Oulunsalon Kotiseutumuseo,
   Oulunsalo 011224
Oumow, Aude, Saint-Germain-en-
   Laye 105520
Oundle Museum, Oundle 045446
Our Dream Gallery, Portland . . 123653
Our Gallery, Sedbergh 119065
Our Heritage Museum, Adelaide 038643
Our House, Honolulu 093565
Our House State Memorial,
   Gallipolis 049028
Our India Project Museum,
   Narendrapur 024422
Our Lady of Mercy Museum, Port-au-
   Port 006676
Our Talented Friends,
   Albuquerque 092074
OurnaC, Eric, Clamart 068002
Ouro Brunido, Lisboa 132948

Ouroboros, Tucson 124764
A l'Ours d'Argent, Saint-André-de-
   Sangonis 072671
Ouse Valley Books, Sharnbrook . 144833
Oushabti, Toulouse 073788
Oushuu-shokoku Mingeikan,
   Teshikaga 029477
Ouss'Art, Clermont-Ferrand . . . 103606
Out of Africa, Philadelphia 123434
Out of Africa Gallery, Thetford . . 119226
Out of Asia, Los Angeles 094432
Out of Bounds, Seattle 124580
Out of Print, Caulfield South . . 139694
Out of Print Books, Cleveland . . 145089
Out of the Blue Gallery, Marazion 118717
Out of the Blue Interiors,
   Townsville 062340
Out of the Closet, New Orleans . 095139
Out of the Closet, Foundation Inc., New
   York 095640, 122937, 145365
Out of the East, Singapore . . . 085481
Out of the Past, Austin 092259
Outagamie Museum, Appleton . 046513
Outback Aboriginal Art, Hawthorn
   East 061609
Outback Art, Mayfield 099290
Outback Colours Galleries, Quorn 099608
Outback Galleries, Silverton . . . 099688
Outback Regional Gallery, Waltzing
   Matilda Centre, Winton 001648
Outcast Book's, Hereford 144474
The Outcrop Gallery, Alice
   Springs 098638
Outeniqua Railway Museum, Transnet
   Heritage Foundation, George . . 038719
Outer Space Museum, Garni Outer
   Space Research Centre, Garni . 000697
Outerlimits, Mount Hawthorn . . 139842
Outfitters Gallery, Margate 118718
Outhwaite & Litherland, Liverpool 126951
Outjo Museum, Outjo 031851
Outlines, Bourne 117399
Outlook and District Heritage Museum
   and Gallery, Outlook 006601
Outokummun Kaivosmuseo,
   Outokumpu 011225
Outra Era, Lisboa 084936
Outrageous Art, Edinburgh . . . 117737
Outre Antiques, Long Beach . . . 094215
Outre Gallery, Melbourne 099315
Outre Jewel, Chicago 120227
Outre-Mer, Genève 087492
Outre Monde, Caen 103486
Outremer, Saint-Renan 105613
Outremer, Morlaine, La Baule-
   Escoublac 103872
Outros Tempos, Porto 085042
Outside the Lines, Nashville . . . 122092
Outsider Art Fair, New York . . . 098488
Outters, Christophe, Reims . . . 072400
Ouvrage de la Ligne Maginot,
   Leutenheim 013772
Ouvrage Décoration, Paris 071746
Ouyen Local History Resource Centre,
   Ouyen 001388
Ovarantigo Antiguidades, Ovar . 085003
Ovens Natural Park and Museum,
   Riverport 006810
Over the Moon, Saint Agnes . . 119000
Over The Moon Antiques,
   Adelaide 060971
Over Tijd, Eindhoven 083225
Overbeck-Gesellschaft, Verein von
   Kunstfreunden e.V., Lübeck . . 059217
Overbeck-Museum, Bremen . . . 017596
Overbeck, Elisabeth, Hanstedt . 107377
Overbeck's Museum, Salcombe . 045773
Overfield Tavern Museum, Troy . 053814
Overhoff, Jürgen, Freiburg im
   Breisgau 075874, 129826
Overholser Mansion, Oklahoma
   Historical Society, Oklahoma
   City 051738
Overland Trail Museum, Sterling 053519

Overlook Press, New York 138414
Overlord, Saint Louis 096755
Overmans, R.M., Heusden 132539
Overmantels, London . 090398, 134742
Overstreet-Kerr Historical Farm,
Keota 049940
Overton, John, Atlanta 092166
Óvodamúzeum, Martonvásár .. 023771
Owatonna Arts Center, Owatonna 051849
Owela Display Centre, National
Museum of Namibia, Windhoek 031864
Owen, New York 122938
Owen & Ramsey, Bon Air 145004
Owen Sound Marine and Rail Museum,
Owen Sound 006605
Owen, Debra, San Diego 124151
Owen, Faye W., Bonny Hills ... 098760
Owen, Patrick, Philadelphia ... 123435
Owen, Peter, London 138235
Owens & Co., Spokane 127463
Owens, Wellington 084179
Owens Art Gallery, Sackville .. 006840
Owens-Thomas House,
Savannah 053179
Owensboro Area Museum of Science
and History, Owensboro 051853
Owensboro Museum of Fine Art,
Owensboro 051854
Owlets, Hythe, Kent 089629
Owls Head Transportation Museum,
Owls Head 051855
Owl's Rook Books, Calgary ... 140336
Owl's Roost Antiques, Miami ... 094789
Owo Museum, Owo 033443
O'Wyatt, Dion, New York 122939
Owyhee County Historical Museum,
Murphy 051127
Oxbow, Oxford 144771
Oxbridge & Co., Amsterdam .. 143059
Oxenfurter Antiquitätenhandel,
Ochsenfurt 077889
Oxford Antique Restorers, New
York 135970
Oxford Antiques, Unley 062377, 127696
Oxford Architectural Antiques,
Faringdon 089256
Oxford Art Journal, Oxford 139212
Oxford Bus Museum Trust, Long
Hanborough 045146
Oxford Court Antique Centre,
Waikato 084135
Oxford Gallery, Arcachon 066703
Oxford Gallery, Mount Hawthorn 099399
Oxford Library Museum, Oxford 051860
Oxford Museum, Oxford 033268, 051859
Oxford Restoration, New York .. 135971
The Oxford Story, Oxford 045456
Oxford Street Antique Centre,
Leicester 089792
Oxford University Museum of Natural
History, Oxford 045457
Oxford University Press, New
York 138415
Oxford University Press, Oxford . 138275
Oxford University Press Museum,
Oxford 045458
The Oxfordshire Museum,
Woodstock 046188
Oxley, Laurence, Alresford .. 088258,
134022,   144219
Oxley, Marc, Uppingham 091693, 119289,
135172
Oxley, P.A., Cherhill 088848
Oxmoor Circle Used Furniture,
Birmingham 092428
Oxon Cove Park Museum, Oxon
Hill 051873
Oxoxo Gallery, Baltimore 119761
Oyamazumi-jinja Kokuhokan,
Omishima 029230
The Oyster and Maritime Museum of
Chincoteague, Chincoteague . 047723
Oyu, Almaty 111552
Oyuncak Müzesi, Ankara 042827

Oz Images Gallery, Cairns 098844,
127575
OZ Sculptures, Louisville 121664
Ozanne-Marcellin, Sophie, Paris 071747
Ozark Folk Center, Mountain
View 051106
Ozark Pastel Society, Rogers ... 060666
Ozaukee Art Center, Cedarburg . 047492
Ozaukee County Historical Society
Pioneer Village, Saukville 053165
Özbekiston Respublikasi San'at
Akademiyasi, Toshkent 054664
Özbekiston San'at Davlat Muzeyi,
Toshkent 054665
Özbekiston Tarikhi Davlat Muzeyi,
Toshkent 054666
Özdi Városi Múzeum, Özd 023823
Ozgul, Paris 141254
Oziol, Denis, Marvejols 069989
Ozon, Athinai 138768
Ozone Art and Craft Gallery,
Renmark 099622
Ozu Yasujiro Seihunkan Museum,
Matsusaka 029015
P & L Furniture, Austin 135330
P & R Antiques, Fresno 093509, 135571
P & R Antiques, Halesworth ... 089410
P-Art, Gent 100700
P Dünya Sanati Dergisi, İstanbul 139115
P-P-O-W Gallery, New York ... 122940
P Street Pictures, Washington . 124942
P. and Al. Canellopoulos Museum,
Athinai 023024
P. Ramlee Memorial, Kuala
Lumpur 030694
P. Stulgos Lietuvių Tautinės Muzikos
Muziejus, Kaunas 030372
P. Terlemezyan State College of Arts,
Jerewan 054888
P.Broc, Saint-Vigor-le-Grand ... 073329
P.e.t.ô., Budapest 079274
P.H. Sullivan Museum and Genealogy
Library, Zionsville 054618
P.J.K. Gallery, Skipton 119102
P.L. Déco, Saint-Médard-en-
Jalles 105559
P.L.C, Tallard 073615
P.M. Gallery, Columbus 120508
P.O.M., Paris 105144
P.P.I., Pont-Saint-Martin 105387
P.S. 1 Contemporary Art Center,
Museum of Modern Art, Long Island
City 050382
P.S. Art Studios and Gallery,
Fremantle 099063
P.S. Arts Factory, Tokyo 082192
P.S. Galleries, Dallas 120637
P.T. Boat Museum, Germantown 049087
P.T. Color, Firenze 131230
P.W. Vorster Museum,
Middelburg 038784
Pa-Lin Arts School, Singapore .. 056330
Pa Objetos de Artes, Rio de
Janeiro 101054
Paal Kunst und Konzept,
München 108172
Paalman, F., Bruxelles 128033
Paamiune Katersugausivik, Paamiut
Museum, Paamiut 023335
Paape, Lewis, Hutt City 112920, 132723
Paarl Museum, Paarl 038795
Paas, Kurt, Lüchow ... 077196, 077197
Paaß, Marlene, Ochtendung .. 142258
Paattisten Kotiseutumuseo,
Paattinen 011226
Paavolan Kotiseutumuseo,
Paavola 011227
Pabel, Reinhold, Hamburg 141889
Pabetland Sanat Galerisi,
İstanbul 117073
Pabian, Stéphane, Paris 141255
Pablitschko, Sieglinde, Ingolstadt 107463
Pabst, Eichenau 075580
Pabst, Neustadt an der Aisch .. 077757

Pabst, Josef, Puchheim 078100
PAC – Padiglione d'Arte
Contemporanea, Milano 026825
Pacaud, C., Audierne 103286
Pacault, Philippe, Vezinnes 074086
Pacchera, Roberto, Roma 081487
Pace, New York 095641
Pace, Saint Louis 123852
Pace & MacGill, New York ... 122941
Pace Master Prints, New York .. 122942
Pace Primitive, New York 122943
Pace Prints, New York 122944
Pace Wildenstein, New York .. 122945,
122946,   122947, 138416
Pace, Charles E., Houston ... 135613
Pace, Edwin, Mqabba 132338
Pacem in Terris, Warwick 054075
Paces Bend Gallery, Atlanta .. 119598
Pacheco Art Gallery,
Mandaluyong 113400
Pachinger, M., München 077591, 130377
Pachmann, Roderich, München . 077592
Pachner, Siegfried, Steyr 127880
Pacht, Andreas, Kiel 076786
Pacific Access California, Los
Angeles 094433
Pacific American Gallery,
Honolulu 120958
Pacific Art, North Shore 084038
Pacific Arts Association, Sherman
Oaks 060667
The Pacific Arts Association
Newsletter, New York 139472
Pacific Arts Association, Europe,
Genève 059953
Pacific Arts Center, Seattle ... 053244
Pacific Asia Museum, Pasadena 051936
Pacific Book House, Antiques and
Collectibles, Honolulu 093566, 145157
Pacific Bookshop, Christchurch . 143309
Pacific Coast Air Museum, Santa
Rosa 053142
Pacific Coin, Tucson 097683
Pacific Collection, Baltimore ... 092352
Pacific County Historical Museum,
South Bend 053397
Pacific Design Center, Museum
of Contemporary Art, West
Hollywood 054310
Pacific Galleries, Billings 127140
Pacific Galleries, Centralia ... 092570,
127156
Pacific Galleries, Great Falls .. 127234
Pacific Galleries, Seattle 127454
Pacific Gallery, Honolulu 120959
Pacific Gallery, Saskatoon 101547
Pacific Glass Auction,
Sacramento 127424
Pacific Grove Art Center, Pacific
Grove 051879
Pacific Grove Museum of Natural
History, Pacific Grove 051880
Pacific Heights Antiques and Art, San
Francisco 097301
Pacific Heights Gallery, San
Francisco 124361
Pacific Interiors, Chicago 092762
Pacific Islands Museums Association,
Suva 058574
Pacific Museum, Yongin 030065
Pacific Northwest College of Art,
Portland 057712
Pacific Northwest Truck Museum,
Brooks 047272
Pacific Orient Traders, Honolulu 093567
Pacific Orient Traders, Portland . 096455
Pacific Orient Traders, Vancouver 064857
Pacific Rim Focus, Pyrmont ... 062129
Pacific Science Center, Seattle . 053245
Pacific University Museum, Forest
Grove 048774
Pacific Woodblock Prints,
Vancouver 064858, 101819
Pacini Fazzi SRL, Maria, Lucca . 137664

Pacita Abad Foundation, Manila 059690,
137890
Pacitti, Carla Adriana, Torino .. 081744
Packer Schopf, Chicago 120228
The Packhouse, Runfold 091153
Packmor, Heiko, Gronau,
Westfalen 076098
Packnut, Indianapolis 093914
Packrat, Denver 093322
Packwood House, Lapworth ... 044760
Packwood House Museum,
Lewisburg 050225
Paço dos Duques de Bragança,
Guimarães 035733
Paço Imperial, Rio de Janeiro . 004765
Paço Imperial Antiguidades, Rio de
Janeiro 064099
Padasjoen Kotiseutumuseo,
Padasjoki 011228
Padasjoen Museo Nyystölä,
Nyystölä 011203
Padasjoen Museo Torittu, Torittu 011427
Paddington Antique Centre,
Paddington, Queensland 062054
Paddle Steamer Industry Museum,
Renmark 001448
Paddock, Woodhouse Eaves ... 091924
Paddy's Emporium, Panmure ... 062059
Padgett, Paul, Cincinnati 092883
Padovani, Luri 069668
Padovani, Penny, London 144662
Padovani, Walter, Milano 080651
Padre Burgos National Museum,
Vigan 034614
The Padstow Gallery, Pimm van der
Donk, Padstow 118874
Padstow Museum, Padstow ... 045462
Padua Art Gallery, Padova 110573
Padua, Maria Lucia, Rio de
Janeiro 128119
Päämajamuseo, Mikkeli 011161
Päbst, Helma, Maxdorf 142119
Pädagogisches Museum der
Universität Bielefeld, Bielefeld . 017384
Paekakariki Arts, Paekakariki .. 084049
Paekakariki Rail and Heritage
Museum, Paekakariki 033270
Pängelanton-Eisenbahnmuseum,
Münster 020606
Päpke & Janowitz, Düsseldorf . 075503
Pärnu Muuseum, Pärnu 010648
Paeroa and District Historical Society
Museum, Paeroa 033271
Paeroa Antiques, Paeroa 084050
Pärups Antik, Hörby 086748
Paes de Barros, Lya, São Paulo 101106
Pätärin Talomuseo, Ylijärvi ... 011525
Pätzold & Einmahl, Aachen ... 074342
Paetzold, Ariane, Oberursel ... 130483
Pätzold, Kurt, Gams 133751
Páez Vilaró, Carlos, Punta del
Este 119448
PAF Museum, Karachi 034147
Paffenroter Dorfmuseum,
Marxzell 020312
Paffrath, G., Düsseldorf 106814
Paffrath, Stephanie, Düsseldorf . 139735
Pagan, Vigo 086564, 133480
Paganelli & C., G., Genova ... 110126
Paganini, Stéphane, Dunkerque . 068417
Pagazaurtundia, Paz Gimeno,
Zaragoza 086601, 133505
Page Antiques, Armadale,
Victoria 061044
Page Antiques, Leek 089786
Page Conservation, Washington 136463
Page Two, Fitzroy 139735
Page, A. & H., London 090399, 134743
Page, Colin, Brighton 144267
Page, Kevin, London 090400
Page, Roger, Yallambie 139968
Pageafterpage, Box Hill 139655
Pageant, New York 145366
Pageau, Colette, Tours 073855

Pagenwood, Portland . 096456, 136156
Pages & Prints, Cincinnati 120364
Pages Art, North Shore 084039, 132734
Pages d'Histoire, Paris 141256
Pages of Yesteryear, Sacramento 145490
Pages, Frank, Baden-Baden 106111
Pagès, Gilbert, Badaroux 066903
Pagès, Jean, Bordeaux 140800
Pagès, Robert, Lunel 069661
Paginas Antigas Livros, Belo
Horizonte 140270
La Paglia, Modena 080814
Pagliani, Massimo, Padova 131671
Pagliarini, Enrico, Parma 110634
Pagliuca, Stefano, Napoli 080909
Pagnier, Jean, Bonnefontaine . . 128477
Pagoa Restauración, Bilbao 133198
La Pagoda International, Padova 080997
Pagoda Red, Chicago 092763
Pagode de Chanteloup, Amboise 011624
Pagodenburg Rastatt, Rastatt . . 108430
Pagter, Kolding 102921
Pagter, de, Middelburg 083498
Pahiatua and District Museum,
Pahiatua 033272
Pahk, New York 122948
Pahl Museum, Mainhardt 020221
Pahner, List 077137
Paier College of Art, Hamden . . 057172
Paignton and Dartmouth Steam
Railway, Paignton 045465
Paik, Hae Young, Seoul 111701
Paikkarin Torppa, Sammatti . . . 011343
Pailet, David E., New Orleans . . 122215
Paillard, Alain, Morez 070348
Pailler, Stéphane, Tours 073856
Paillette, Gisèle, Vimoutiers . . . 074237
Pailleux, Daniel, Abbeville 066412
Paillous, Jacques, Villefranche-de-
Rouergue 074145
Paimion Kotiseutumuseo, Paimio 011229
Paine Art Center, Oshkosh 051830
Paine-Gillam-Scott Museum, Saint
Johns 052776
Paine, D., Gorseinon 089344
Paint Creek Center for the Arts
Exhibitions, Rochester 052609
Paint 'n' Palette Club, Saint
Augustine 060668
Paint Pot Gallery, Yeppoon 099944
Paint Rag, Honolulu 139473
Paint & Sons, N.S. John, Saint Peter
Port 091206, 134994
Paintbox Gallery, New York 122949
Painted Bride Art Center Gallery,
Philadelphia 052077
Painted City Gallery, Toronto . . 101675
Painted Lady's Christmas Plums, Long
Beach 094216
Painted Planet, Memphis 121705
Painted With Oil, Phoenix 123504
The Painting Center, New York . 051459
Painting Conservation of Philadelphia,
Philadelphia 136097
Painting Lady, Chicago 092764
Painting Mart, San Jose 124457
Paintings Direct, Boston 119910
Paintings Galore, Victoria Point . 062390
Paintings of the World, Tampa . 124667
Paintings Plus, Phoenix 123505, 123506
Paintingsandphotos Art Gallery,
Washington 124943
Painvin, Olivier, Savigny-le-
Temple 073451
Paio, João Algarvio, Lisboa . . . 084937
Paipoonge Historical Museum,
Thunder Bay 007145
Pairel, Daniel, Angers 066613
Pairone Mario, La Mure-d'Isère . 069054
Pairone, Mario, Nantes-en-Ratier 070486
Paisano Publications, Agoura
Hills 138297
Paisley, Toronto 064717
Paisley, Tshwane 085683

Paisley Fine Books, Paisley 144777
Paisley Museum and Art Galleries,
Paisley 045466
Paisnel, London 090401, 118496
Paiva, Maria Helena de Campos, São
Paulo 064199
Paiwan Carving Gallery, Pingtung 042416
Pajaro Valley Historical Museum,
Watsonville 054229
Pajdowska, E., Warszawa 113882
Pakan, Phoenix 136119
't Pakhuis, Den Haag 083114
Pakistan Air Force Museum Peshawar,
Peshawar 034144
Pakistan Army Museum,
Rawalpindi 034150
Pakistan Museum of Natural History,
Islamabad 034119
Pakistani American Arts Council, Los
Angeles 060669, 121569
Pákozdi Csata Emlékmüve,
Pákozd 023825
Pakruojo Dvaro Rūmai, Šiaulių Aušros
Muziejus, Pakruojis 030434
Paksi Képtár, Paks 023826
Pakzad, Freydoun, Hannover . . . 076387
Pál & Társa, Budapest 079275
Pal de Rosa, Lleida 086069
Pałac Radziwiłłów w Nieborowie,
Muzeum Narodowe w Warszawie,
Nieborów 035033
Pałac Sztuki w Krakowie,
Kraków 034909
Pałac w Rogalinie, Świątniki . . . 035236
Palace, Den Haag 126281
Palace, Toronto 101676
Palace and Museum of Antiquities,
Padmanabhapuram 024426
Palace Antiques, Bucureşti 085106
The Palace Armoury, Grandmaster's
Palace, Valletta 030755
Palace Art, Karama 088192
Palace Editions, Bad Breisig . . . 137033
Palace Galerie, Sankt Moritz . . . 087795
Palace Galleries, New York 095642
The Palace Museum, Beijing . . . 007639
Palace Museum, Ulaanbaatar . . 031740
Palace Museum, Zanzibar 042641
Palace of Holyroodhouse, Royal
Collection, Edinburgh 044260
Palace of Royal Jewelleries,
Zizinia 010588
Palace of the Governors, Museum of
New Mexico, Santa Fe 053118
Palace of the Grand Masters,
Rhodos 023250
Palace of Westminster, Houses of
Parliament, London 045070
Palacete del Embarcadero,
Santander 040116
Palacin, Jean-Pierre, Bordeaux . 128499
Palacio Arzobispal, Cusco 034267
Palácio Conde dos Arcos, Goiás 004480
Palacio Condesa de Lebrija,
Sevilla 040180
Palácio da Bolsa, Porto 035957
Palácio das Laranjeiras, Rio de
Janeiro 004766
Palácio de Estói, Estói 035663
Palacio de Exposiciones Kiosko
Alfonso, A Coruña 115036
Palacio de la Granja, Orihuela . . 039848
Palácio de Monserrate, Sintra . . 036042
Palacio de Montemuzo, Zaragoza 040456
Palacio de Viana, Córdoba 039311
Palacio del Mar, Donostia-San
Sebastián 039355
Palacio del Mate, Posadas 000502
Palácio do Correio Velho, Lisboa 126450
Palácio dos Coruchéus, Centro de
Artes Plásticas, Lisboa 035822
Palácio dos Leilões, Belo
Horizonte 100900
Palácio dos Marqueses de Fronteira,

Lisboa 035823
Palácio Guanabara, Rio de
Janeiro 004767
Palácio Gustavo Capanema, Rio de
Janeiro 004768
Palacio La Quinta, Madrid 039688
Palácio Nacional da Pena, Sintra 036043
Palacio Nacional de Queluz,
Queluz 035971
Palácio Nacional de Sintra, Sintra 036044
Palacio Real de Aranjuez,
Aranjuez 038980
Palacio Real de El Pardo, El
Pardo 039915
Palacio Real de Madrid, Madrid 039689
Palacio Real de Riofrío, Museo de la
Caza, Riofrío 040006
Palacio Real y Museos de Tapices,
San Ildefonso 040061
Palacio Taranco, Museo de Artes
Decorativas, Montevideo 046285
Palácio Valenças, Sintra 036045
Pałacyk, Warszawa 113883
Palacz, Krzysztof, Poznań 084648
Paladin Trading, New York 095643
Palaeobotanisch Museum,
Utrecht 032954
Paläontologisches Museum der
Universität Zürich, Zürich 042227
Paläontologisches Museum Dr. Herbert
Schaffer, Offenhausen 002478
Paläontologisches Museum München,
München 020561
Paläontologisches Museum Nierstein,
Nierstein 020809
Palaepaphos Museum, Kouklia . 009403
Palæsamlingerne, Roskilde 010289
Palagraziussi, Giuliano, Venezia . 111148
Palaiopoleio o Sakis,
Thessaloniki 079213
Palaios, Milano 080652
Palais Bénédictine, Fécamp . . . 012961
Palais de Glace, Palacio Nacional de
las Artes, Buenos Aires 000281
Palais de la Découverte, Paris . . 014793
Palais de la Miniature, Saint-Pourçain-
sur-Sioule 015520
Palais de l'Ile, Centre d'Interprétation
Urbain, Annecy 011672
Palais de Tokyo, Paris 014794
Le Palais des Arts, Trouville-sur-
Mer 073894
Palais des Beaux-Arts, Bruxelles 003496
Le Palais des Beaux-Arts,
Bruxelles 003495
Palais des Expositions Chinagora,
Alfortville 011598
Palais des Papes, Avignon 011842
Le Palais du Chocolat, La Côte-Saint-
André 013400
Palais du Roure, Bibliothèque et
Collections, Avignon 011843
Palais du Tau, Centre des Monuments
Nationaux, Reims 015057
Palais Ducal, Nevers 014470
Le Palais Francais, Cairo 066261
Palais für aktuelle Kunst,
Glückstadt 018719
Palais Grand Ducal, Luxembourg 030595
Palais Jacques Coeur, Bourges . 012167
Palais Liechtenstein, Forum
für zeitgenössische Kunst,
Feldkirch 001881
Palais Minéral, Composition avec des
Minéraux, Granville 013153
Palais Montcalm, Québec 006744
Palais Palffy, Wien 003062
Palais Princier, Monaco 031724
Palais Synodal Musée, Sens . . . 015772
Palamas Museum, Athinai 023025
Palamka, Lyon 104124
Palamuse O. Lutsu
Kihelkonnakoolimuuseum,
Palamuse 010650

Il Palanchino, Milano . . 080653, 131490
Palanga, Pierina, Taranto 081637
Palangka, Le Castellet 103936
Palangos Gintaro Muziejus, Lietuvos
Dailės Muziejus, Palanga . . . 030438
Palanquin, Roscoff 072543
Palas, Varna 101197
Palata na Fizikata, Nacionalen
Politehničeski Muzej Sofia,
Kazanlăk 005060
Palatina, Buenos Aires 098575
Palatino, Roma 081488
Palatino, San Francisco 124362, 138459
Palatul Scroviştea, Periş 036606
Palaty Starogo Anglijskogo Dvora,
Muzej istorii Moskvy, Moskva . 037488
Palau Antiquitats, Barcelona . . . 085817
Palau de Caramany, Girona . . . 115082
Palau Ducal dels Borja, Gandía . 039426
Palau Reial de l'Almudaina, Palma de
Mallorca 039896
Palawan Museum, Puerto
Princesa 034573
Palawan National Museum,
Quezon 034575
Palawan State University Museum,
Puerto Princesa 034574
Palazzina Marfisa, Ferrara 026219
Palazzo Apostolico ed Archivio Storico
della Santa Casa, Loreto 026627
Palazzo Armeria, Zurrieq 030773
Palazzo Attems-Petzenstein, Musei
Provinciali di Gorizia, Gorizia . 026455
Palazzo Barolo, Torino 028141
Palazzo Belgiojoso, Musei Civici,
Lecco 026586
Palazzo Belmonte Riso, Museo
Regionale d'Arte Moderna e
Contemporanea, Palermo . . . 027190
Palazzo Bianco, Musei di Strada
Nuova, Genova 026423
Palazzo Borromeo, Stresa 028025
Palazzo Bricherasio, Torino 028142
Palazzo Castelmur, Stampa 042022
Palazzo Castiglioni, Cingoli 025988
Palazzo del Principe, Genova . . 026424
Palazzo del Senato, Pinerolo . . . 027339
Palazzo della Ragione, Padova . 027160
Palazzo della Triennale, Milano . 026826
Palazzo delle Esposizioni, Roma 027656
Palazzo delle Papesse – Centro Arte
Contemporanea, Siena 027960
Palazzo Ducale, Genova 026425
Palazzo Ducale, Musei Civici Veneziani,
Venezia 028320
Palazzo Falson Historic House
Museum, Mdina 030742
Palazzo Fieschi, Genova 080277
Palazzo Foro Boario, Modena . . 026865
Palazzo Grassi, Venezia 028321
Palazzo Litta, Milano 080654
Palazzo Medici Riccardi, Firenze 026299
Palazzo Orlandi, Busseto 025664
Palazzo Parisio, Naxxar 030744
Palazzo Pfanner, Lucca 026644
Palazzo Pubblico, Opera di Francesco
Azzurri 1884–1894, San Marino 038105
Palazzo Reale, Torino 028143
Palazzo Reale-Reggia di Caserta,
Caserta 025799
Palazzo Rosso, Musei di Strada Nuova,
Genova 026426
Palazzo Sandretta Re Rebaudengo,
Guarene 026490
Palazzo Silva, Fondazione Galletti,
Domodossola 026140
Palazzo Taffini d'Acceglio,
Savigliano 027053
Palazzo Tozzoni, Collezioni Comunali
d'Arte, Imola 026506
Palazzo Tursi, Musei di Strada Nuova,
Genova 026427
Palazzo Vertemate Franchi, Piuro 027371
Palazzo Vescovile, Reggio Emilia 027506

Palazzo Zabarella, Padova 027161
Palazzo, Paolo, Palermo 131705
Palbert, Torino 081745, 110992
Pale Ink Antiquities, Phoenix . . . 096228
't Paleis, Knokke-Heist 063707
Paleis aan de Blijmarkt, Museum de
   Fundatie, Zwolle 033126
Paleis Het Loo – Nationaal Museum,
   Apeldoorn 032013
Paleologo di Lemnos, Messina . 131358
Paleontological and Geological
   Museum of the University of Athens,
   Athinai 023026
Paleontologičen Muzej,
   Asenovgrad 004971
Paleontologičeskij Muzej im. Ju.A.
   Orlova, Paleontologičeskij institut
   RAN, Moskva 037489
Paleontologisches Museum, Burg-
   Reuland 003504
Paleontologisk Museum, Oslo . . 033847
Paleontologo-stratigrafičeskij Muzej
   pri Kafedre Istorii Geologii, Sankt-
   Peterburgskogo Gosudarstvennogo
   Universiteta, Sankt-Peterburg . 037811
Paleorama – Museo Itinerante, Don
   Torcuato 000357
Palépés Galerija, Kintų Vydūno
   Kultūros Centras, Kintai 030385
Palermo, Rocco, Rio de Janeiro 064100
Palestinian Archaeological Museum,
   Birzeit 034161
Palestinian Artist House, Al-Khalil 113311
Paleta, Lisboa 114050
Paletta, Athinai 109046
La Palette, Lorient 104046
Palette, North Shore 112974
Palette, Philadelphia 123436
Palette, Saint-Étienne 072771
Palette, Toulon (Var) 105718
Palette & Chisel Academy of Fine Arts
   Gallery Exhibition, Chicago . . . 047687
Palette & Zeichenstift, Koblenz . 138748
The Palette and Chisel Academy,
   Chicago 047688
Palette Bleue, Saint-Junien 072872
La Palette d'Or, Paris . 105145, 129105
Palette du Boucher, Brest 067476
Palette Messine, Metz 104229
Paletten, Göteborg 139085
Paletten Kunst, Harstad 113158
Paletti, Parma 081122
Paley, Maureen, London 118497
La Palferine, Paris 071748
Palimpsiston, Lemesos 065330, 065331
Palio Mikro Chorio Museum, Mikro
   Chorio 023173
Pálio Objetos de Arte, Rio de
   Janeiro 101055
Palisander, New York 095644
Palissandre, Bersillies . 067116, 128467
Palissandre, Bruxelles 063520
Palissandres, Paris 129106
Palitra, Charkiv 117108
Palitra, Sankt-Peterburg 114352, 114353
Palix, Yvonamor, México 111976
Paljas, Den Haag 083115, 083116
Palk Antik, Degeberga 086656
Palka, Kazimierz, Kraków 132846
Pall, William, New York 122950
Palladio, Memphis 094660
Palladio, Padova 131672
Palladio, M., Memphis 094661
Palladion, Toulouse 105751
Pallamar, Friederike, Wien 063044,
                                       063045
Pallant House Gallery, Chichester 043945
Pallas, Breda 083001
Pallas, Praha 102712
Pallas & Janos, Den Haag 143118
Pallas Books, Leiden 143186
Pallas Gallery, Turramurra 099843
Pallas-Palacio, Bruno, Bordeaux 128500
Pallas Versandantiquariat, Neu-

Ulm 142220
Palleau, Jean-Pierre, Paris 071749
Pallen, F., Brugge 063285
Pallesi, Alfredo, Roma 081489
Pallini, Roberto, Firenze 131231
Pallocca, Franco, Roma 081490
Pallokat, Anke, Schwerin 078431
Palloni, A., Firenze 080180
Pallotta, Giuseppe, Roma 131997
Palm, Falsterbo 115610
Palm Beach | America's International
   Fine Art & Antiques Fair, West Palm
   Beach 098489
Palm Beach Institute of Contemporary
   Art, Lake Worth 050093
Palm Beach Photographic Center,
   Delray Beach 048233
Palm Springs Art Museum, Palm
   Springs 051896
Palm Street Gallery, Westport . . 113102
Palma Arte, Alseno 109749
Palma Ceia, Tampa 097582
Palma Fellenberg, Haydée, Santiago
   de Chile 101918
Palma Rosa Museum, Hamilton,
   Brisbane 001133
Palma, Cecilia, Santiago de Chile 101917
Palmaverde, Bologna 142637
palmbeach 3, Contemporary-
   Photography-Art+Design, West Palm
   Beach 098490
Palmdale Gallery, Motukarara . . 112951
Palmeblads Förlag, Göteborg . . 138008
Palmer, Birmingham 119814
Palmer House, Northfield 051634
Palmer Museum, Jewell 049821
Palmer Museum of Art, The
   Pennsylvania State University,
   University Park 053899
Palmer Wirfs & Assoc., Portland 096457
Palmer, David, Ely 117750
Palmer, Dermot & Jill, Brighton . 088631
Palmer, Herbert, West Hollywood 125019
Palmer, John, Houston 121106
Palmer/Gullickson Octagon Home,
   West Salem 054324
Palmers Green Antiques Centre,
   London 090402
Palmerston North Community Arts
   Council, Palmerston North 112982
Palmett, Strängnäs . . . 087083, 143973
Palmieri, Saint-Ouen 073163
Palmieri Sauveur, Nice 070642
Palmieri, Vezio, Napoli 080910
Palmirotta, Ottavio, Torino 110993
Palmrose, David, Portland 096458
Palmse Möis, Viitna 010717
Palmu, Juhani, Helsinki 103154
Palmyra, Ygrande 074310
Palmyre, Bruxelles . . . 063521, 128034
Palmyre-Saint-Honoré, Paris . . . 105146
Palo Alto Art Center, Palo Alto . . 051903
Palo Alto Junior Museum, Palo
   Alto 051904
Palóc Múzeum – Palóc Ház, Nógrádi
   Megyei Múzeum Szervezet,
   Balassagyarmat 023422
Palóc Népművészet, Eger 023621
Palóc Néprajzi Magángyűjtemény és
   Palóc Baba Kiállítás, Mátrafüred 023774
Paloma Art, Tucson 124765
Paloma Art at El Cortijo, Tucson 124766
Palombi Editori, Roma 137738
Palomuseo, Helsinki 010852
Palomuseo, Sairaalamäki 011339
Palookaville, Portland 096459
Palos Verdes Art Center, Rancho Palos
   Verdes 052461
Palosanto Antiguedades, Almeria 085706
Palotavárosi Skanzen,
   Székesfehérvár 023938
Palpura, Lisboa 114051
Pålsjö Kvarn, Helsingborg 040696
Paltamon Kotiseutumuseo,

Paltamo 011231
Paltenghi, Guerino, Château-
   d'Oex 116203
Palthey, Jean-Christophe, Paris . 071750
Paltrockwindmühle Brehna,
   Brehna 017570
Palu, Seattle 097481
Pałucka Izba Muzealna, Muzeum
   Okręgowego im. Leona
   Wyczółkowskiego w Bydgoszczy,
   Kcynia 034827
Paludan, København 140614
Palumbo, Buenos Aires 139592
Palumbo, New York 095645
Palumbo Scalzi, Lydia, Latina . . 110162
Palumbo, Christina, Buenos Aires 139593
Palumbus, Antwerpen 100508
Palz, M., Beckingen 074703
Památná Izba Franza Schuberta,
   Tekovské Múzeum Levice,
   Želiezovce 038466
Památná Izba Frica Kafendu,
   Mošovce 038405
Památná Izba Jozefa Murgaša,
   Tajov 038464
Památná Izba Sama Chalupku, Horná
   Lehota 038355
Památník Ludwiga van Beethovena,
   SNM-Hudobné Múzeum Bratislava,
   Dolná Krupá 038343
Pamätný Dom Kálmána Mikszátha,
   Múzeum Kultúry Maďarov na
   Slovensku, SNM, Sklabiná 038446
Pamätný Dom Pavla Jozefa Šafárika,
   Kobeliarovo 038363
Památky a Umění, Znojmo 065572
Památník Adolfa Heyduka, Písek 009756
Památník Adolfa Kašpara, Loštice 009681
Památník Antonína Dvořáka, Vysoká u
   Příbrami 009964
Památník Antonína Dvořáka,
   Zlonice 009986
Památník Antonína Dvořáka,
   NM-České Muzeum Hudby,
   Nelahozeves 009714
Památník Bedřicha Smetany,
   Jabkenice 009587
Památník Bedřicha Smetany,
   Obřistvi 009729
Památník Bible Kralické, Moravské
   Zemské Muzeum Brno, Kralice nad
   Oslavou 009645
Památník Bitvy 1866 na Chlumu,
   Muzeum Východních Čech v Hradci
   Králové, Všestary 009961
Památník Bohuslava Martinů, Městské
   muzeum a galerie, Polička 009772
Památník Bratří Mrštíků, Diváky 009535
Památník Dr. Emila Axmana,
   Chropyně 009527
Památník Dr. Emila Holuba, Holice v
   Čechách 009566
Památník Dr. Františka Křižíka,
   Plánice 009758
Památník Františka Palackého,
   Hodslavice 009564
Památník II. Světové Války,
   Hrabyně 009575
Památník Jana Ámoše Komenského,
   Bílá Třemešná 009446
Památník Jana Ámoše Komenského,
   Horní Branná 009569
Památník Jana Ámoše Komenského,
   Vlastivědné Mmuzeum Nový Jičín,
   Fulnek 009551
Památník Jiráskova Pokladu, Muzeum
   Orlických Hor, Rychnov nad Kněžnou,
   Potštejn 009775
Památník Jiřího Melantricha z
   Aventýna, Polabské muzeum
   Poděbrady, Roždalovice 009868
Památník Josefa Suka, NM-České
   Muzeum Hudby, Křečovice . . . 009650
Památník Karla Čapka, Stará Hut u

Dobříše 009889
Památník Karla Havlíčka Borovského,
   Muzeum Vysočiny Havlíčkův Brod,
   Havlíčkův Brod 009557
Památník Karla Hynka Máchy,
   Doksy 009538
Památník Karla Jaromíra Erbena,
   Miletín 009696
Památník Karla Václava Raise, Lázně
   Bělohrad 009661
Památník Krále Jiřího z Poděbrad
   a Lapidárium, Polabské Muzeum,
   Poděbrady 009766
Památník Leoše Janáčka,
   Hukvaldy 009583
Památník Leoše Janáčka, Moravské
   zemské muzeum, Brno 009480
Památník Lidice, Lidice 009669
Památník Města, Police nad
   Metují 009768
Památník Mikoláše Alše a Matěje
   Kopeckého, Mirotice 009698
Památník Mistra Jana Husa,
   Husinec 009585
Památník Mohyla Míru, Prace u
   Brna 009776
Památník Národního Písemnictví
   (Muzeum České Literatury),
   Praha 009832
Památník Otakara Štáfla, Muzeum
   Vysočiny Havlíčkův Brod, Havlíčkův
   Brod 009558
Památník Petra Bezruče, Muzeum
   Prostějovska, Kostelec na Hané 009640
Památník Petra Bezruče, Slezské
   zemské muzeum, Opava 009736
Památník Prokopa Diviše,
   Jihomoravské muzeum ve Znojmě,
   Znojmo 009991
Památník Svatopluka Čecha,
   Obřiství 009730
Památník Svatopluka Čecha,
   Ostředek 009745
Památník Terezín, Terezín 009908
Památník Terezín – Bývalé
   Krematorium na Židovském Hřbitově,
   Terezín 009909
Památník Terezín – Kolumbárium,
   Obřadní Místnosti a Ústřední Márnice,
   Terezín 009910
Památník Terezín – Magdeburská
   Kasárna, Terezín 009911
Památník Terezín – Malá Pevnost,
   Terezín 009912
Památník Terezín – Muzeum Ghetta,
   Terezín 009913
Památník Terezy Novákové, Proseč u
   Skutče 009855
Památník Václava Beneše Třebízského,
   Třebíz 009921
Památník W.A. Mozarta a Manželů
   Duškových, Bertramka, Praha . . 009833
Památník Zapadlých Vlastenců,
   Krkonošské Muzeum, Paseky nad
   Jizerou 009751
Památník Životické Tragedie, Muzeum
   Těšínska, Havířov 009554
Pambasang Museo ng Pilipinas,
   Manila 034541
Pambianco, Alberto, Milano . . . 131491
Pamětní Síň Jana Ámoše Komenského,
   Brandýs nad Orlicí 009455
Pametnik na Kulturata Zemenski
   Manastir, Zemen 005267
Pami Antichitá, Roma 081491
Pammer, Evelyn, Wien 063046
Pamminger, Peter, Hamburg . . . 076292
Pampa Centro de Artes y Diseño,
   Buenos Aires 060920
Pampaloni, Paolo, Firenze . 131232,
                                   142683
Pampaloni, Roberto, Firenze . . . 131233
Pampeana, Buenos Aires 139594
Pampille, Bernay 067110

Pamplin Historical Park and the
National Museum of the Civil War
Soldier, Petersburg 052013
Pamukbank Fotoğraf Galerisi,
İstanbul 117074
Pan African Art Gallery, Long
Beach 121407
Pan American Gallery, Dallas . . 120638
PAN Amsterdam, Amsterdam . . 098263
Pan-Art, Hong Kong 102104
Pan-Dan, Moskva 114288
PAN kunstforum niederrhein,
Emmerich am Rhein 018189
Pan-o-Rama, Bremen 141654
Panache, Los Angeles 094434
Panache, West Hollywood 097916
Panache Antiques, Vancouver . . 064859
Panachrantos Monastery, Andros 022961
Panaddaman-Cagayan State University
Museum, Tuguegarao 034610
Panalba International, Marseille . 104192
Panamericana-Escola de Arte e
Design, Campus Angelica, São
Paulo 055106
Panamericana-Escola de Arte e
Design, Campus Groenlândia, São
Paulo 055107
Panamin Museum, Pasay 034567
Panarcadic Archaeological Museum of
Tripolis, Tripoli 023308
Panasia Gallery, Davos Platz . . 087392,
116240
Panasia Gallery, Zürich 087966
Panay, Gérard, Bormes-les-
Mimosas 067370
Panayi, Nicolas, Lefkosia 102540
La Pancarte, Nort-sur-Erdre . . . 070746
Panchavati, Bangalore 079358
La Panchetta, Bari 109774
Pancho Lepina, Vigo 086565
Pancrazi, Jean-François, Toulon
(Var) 129290
Panda, Smederevo 114462
Pandanus Palace, Noosaville . . 061979
Pandiani, Mario, Torino 132177
Pandiris, André, Paris 071751
't Pandje, Utrecht 112708
Pandolfini Casa d'Aste, Firenze . 126193
Pandolfino, Daniele, Roma 131998
Pandora, Antwerpen 136623
Pandora Decor, Firenze 080181
Pandora Galéria, Budapest 109167
Pandora Old Masters, New York . 095646
Pandora's Box, Totnes 091623
Pandoras Gallery, Calgary 101252
Pánek, Bohumil, Říčany u Prahy 128370
Paneriu Memorialinis Muziejus,
Valstybinis Vilniaus Gaono Žydų
Muziejus, Vilnius 030537
Panevėžio, Panevėžys 111792
Panevėžio Kraštotyros Muziejus,
Panevėžys 030441
Pangloss Antikvariat, Visby 144006
Panhandle-Plains Historical Museum,
Canyon 047414
Panhellenic Union of Antiquities and
Art Object Restorers, Athinai . . 059331
De Panhoeve, Berlicum 132439
Panicali, Carla, Roma 110840
Panico, Rocco, Kemmelbach . . . 062661
Panier, Francine, Saint-Tropez . . 073316
Panini, Franco Cosimo, Modena 137700
Panini, Luca, Modena 142753
Panizza, Enrico Enzo, Milano . . 080655
Panje Tala, Tehrän 079427
Panke Museum, Berlin 017267
Pankhurst Centre, Manchester . 045215
Panlong Art Cooperative,
Shanghai 102195
Pannemann, K. & H.,
Westerstede 108890
Panneterie, Saint-Ouen 073164
Pannetier, Gérard, Malemort-sur-
Corrèze 069826

Pannhuis, Harald, Gummersbach 076131
Pannon Antik, Budapest 079276
Pannonhalmi Főapátság Gyűjteménye,
Pannonhalma 023829
Pannonisches Heimatmuseum,
Neusiedl am See 002450
Christiane von Pannwitz & Carolin
Zuber, Berlin 129545
Panoptikum, Gelsenkirchen 075968
Panorama, Exeter 117771
Panorama, Wien 100358, 136604
Panorama Antiquariat, Lüneburg 142094
Panorama Antiques, Saint Louis 096756
Panorama Art Gallery, Warszawa 126426
Panorama de la Bataille de Waterloo,
Braine-l'Alleud 003393
Panorama Galleries, New York . 095647
Panorama Großantiquariat,
Wiesbaden 142491
Panorama Kreuzigung Christi,
Einsiedeln 041458
Panorama Mesdag, Den Haag . . 032195
Panorama Museum, Bad
Frankenhausen 016765
Panorama Numismatique, Paris . 071752
Panorama Oborona Sevastopolja
1854–1855 rr., Muzei geroičnoï
oborony i vyzvolennja Sevastopolja,
Sevastopol 043295
Panorama Plevenska Epopeja 1877,
Pleven 005119
Panorama Racławicka, Muzeum
Narodowe, Wrocław 035434
Panorama Šturm Sapun-gori 7
Travnja 1944 r., Muzei geroičnoï
oborony i vyzvolennja Sevastopolja,
Sevastopol 043296
Panorama, Salisbury University,
Salisbury 139474
Panoramamuseum, Salzburg Museum,
Salzburg 002632
Panos Aravantinos Theatrical Museum
of Painting and Stage Design,
Piraeus 023225
Pansarmuséet, Axvall 040555
Pansi, Heinrich u Anita,
Schwanberg 100198
Panssarimuseo, Parola 011238
Państwowa Galeria Sztuki w Sopocie,
Sopot 035211
Państwowe Muzeum Archeologiczne,
Warszawa 035377
Państwowe Muzeum Etnograficzne,
Warszawa 035378
Państwowe Muzeum im.
Przypkowskich, Jędrzejów . . . 034793
Państwowe Muzeum na Majdanku,
Lublin 035003
Państwowe Muzeum Stutthof,
Sztutowo 035262
Państwowe Muzeum Stutthof w
Sztutowie, Sopot 035212
Państwowe Zbiory Sztuki,
Zamek Królewski na Wawelu,
Suloszowa 035230
Panta Rei Antiek, Leeuwarden . 083419
Panta-Res, Milano 131492
Panta Rhei, Madrid 143779
Pantaleo, Fari, Zürich 133995
Pantaleón Rodríguez, Isabel,
Salamanca 133416
Pantano, Grazia, Modena 131577
Pantazi, Agni & Georgiou,
Piraeus 079201
Panteão da Pátria Tancredo Neves,
Brasília 004370
Pantechnicon Art Gallery,
Daylesford 098957
Pantelejmonovskij, Sankt-
Peterburg 085258, 143674
Panter & Hall, London 118498
Panter, Richard, Bischheim 128470
Pantera Roz, Bucureşti 114153
Pantet, René, Genève 126727

Panthéon, Paris 014795
Pantheon Arte, Roma 110841
Pantheon Books, New York 138417
Panther Decor, Vancouver 064860
Pantiles Antiques, Tunbridge
Wells 091656
Pantiles Oriental Carpets, Tunbridge
Wells 091657, 135162
Pantiles Spa Antiques, Tunbridge
Wells 091658, 135163
Panting & Stahl, London 118499
Pantoja, Eliseo, Santiago de
Chile 101919
Pantokrátor, Budapest 079277
Panton, Milano 080656, 137689
Panton, Praha 136766
Panzano, Gian Carlo, Genova . 080278
Panzano, Mario, Genova 080279
Pao & Moltke, Toronto . 064718, 101677
Pao Galleries at Hong Kong Art Centre,
Hong Kong 007885
Paola Shop, Lima 084466
Paolella, Fabio, Napoli . 080911, 131633
Paoletti, Paolo, Firenze 080182
Paoli, Anna Loretta, Prato 081175
Paoli, de, Windsor, Ontario 101859
Paoli, Enrico, Prato 110694
Paolo, Cincinnati 120365
Paolucci Numismatica, Padova . 080998
Pap, Mária, Budapest 130814
Papa Re, Roma 131999
Papadam, Nikos, Thessaloniki . . 109094
Papadato, Roma 081492
Papadopol's Art, Constanța 085119
Papadopoulos Picture Gallery,
Evangelistria Holy Foundation,
Tinos 023305
Papadopoulos, Charis,
Thessaloniki 079214
Papadopoulos, Georgios,
Thessaloniki 109095
Papadopoulos, Kleanthis,
Thessaloniki 109096
Papadopoulos, Pavlos Georgios,
Bologna 079864
Papadopoulos, Stephan,
Kornwestheim 130185
Papagena, Le Vésinet 103978
Papageno, Oviedo 086331
Papakura and District Historical
Society Museum, Papakura . . . 033277
Papakura Art Gallery, Auckland . 112802
Papaleo, Vincenzo, Roma 132000
Papamoa Art Gallery, Tauranga . 113024
Papasideris, Antwerpen 127998
Papazian, Paris 071753
Pape, Barbara, Rinteln 078244
Pape, Hans-Werner, Berlin 129546
Papelaria Moderna, Lisboa 143589
Papeler, Heinz, Krefeld 076961
Paper Antiquities, San Diego . . 097119,
145524
The Paper Bag Princess, West
Hollywood 097917
Paper Chain, Mackay 139813
Paper Conservation Studio, New
York 135972
The Paper Conservator, Upton-upon-
Severn 139213
Paper Heart Gallery, Phoenix . . 123507
Paper Moon, Portland 145541
Paper Peddlers Out-Of-Print Book,
South Euclid 145607
Paper-Rock-Scissors, Baltimore 119762
Paper Sculpture, San Francisco . 124363
Paper-Weight, Łódź 084615
Paperback Trading Post, Buffalo 145022
Paperbark Gallery, Evans Head . 099011
Papergraphica, Christchurch . . . 083888
Paperimuseo, Espoo 010775
Paperworks Gallery, Vancouver . 101820
Paphos District Archaeological
Museum, Paphos 009433

De Papieren Korf, Gent . 063653, 140225
Papierfabrik Hohenofen, Sieversdorf-
Hohenofen 021860
Papiergeschichtliche Sammlung,
Stiftung Zanders, Bergisch
Gladbach 017053
Papiermacher- und Heimatmuseum,
Frankeneck 018394
Papiermolen de Schoolmeester,
Westzaan 033042
Papiermühle Weddersleben,
Weddersleben 022445
Papiermuseum, Düren 018022
Papiers d'Antan, Bruxelles 063522,
140193
Papilio Starozitnosti, Brno 065351
Papiliorama-Nocturama Foundation,
Kerzers 041612
Papillon, Bern 116146
Papillon, Los Angeles 094435
Papillon, Joël & Paulette, Le
Fidelaire 069267
Papiri Brignoni, Adriana, Lugano 087620,
087621
Papiri, Antonello, Ponte
Capriasca 133887
Papiro, Palermo 131706
Papiros, Denia 133238
Papita, Dubai 117184
Papon, Jean, Vevey 087882
Papp, Dr. Magdalene, Wien . . . 100359,
127957
Papp, Florian, New York 095648
Papp, Th., Niddatal 108269
Pappabello, Cleveland 092962
Pappalardo, Valentina, Catania . 109953
Pappe, Bernd, Bern 133687
Papplewick Pumping Station,
Ravenshead 045586
Pappot, Peter, Amsterdam 082821
Pappy, Richmond 096562
Paprsek, Kutná Hora 102611
Papua New Guinea Display Centre,
Crafers 000998
Papua New Guinea National Museum
and Art Gallery, Boroko 034189
Papunya Tula Artists, Alice
Springs 098639
Papworth, Barry, Lymington . . . 090661,
134823
Papworth, Chris L., Kelvedon . . 089666
Papyrud Secret, Cairo 066262
Papyrus, Brugge 063286
Papyrus, Bruxelles 140194
Papyrus, Voorburg 143259
Papyrus Antique Shop, San
Diego 097120
Papyrus Books, New York 145367
Papyrus Salus, Heukelom 132538
papyrus Zeitungsantiquariat,
Wuppertal 142527
Papyrusmuseum und
Papyrussammlung, Österreichische
Nationalbibliothek, Wien 003063
Le Paquebot, La Flotte 069036
Paquebot, Saint-Ouen 073165
Para-Börse, Basel 087252
Para/Site Art Space, Hong Kong 007886
Parabol AM, Wien 138533
Paracelso, New York 095649
Parachute, Montréal 138587
La Parada del Restaurador,
Barcelona . . . 085818, 085819, 133141
The Parade, Edmonton 064304
Parade, Toulouse 073789
Parade Antiques Market,
Plymouth 091023
Parade Galerie, Amsterdam . . . 112266
Paradigm Art, New York 122951
Paradigma, Moskva 114289
Le Paradis de l'Occasion, Nice . 070643
Paradisboden, Bandhagen 086638
Paradise Antique Arts, Honolulu 093568
Paradise Books, Surfers Paradise 139917

Paradise Found, Winnipeg 101890
Paradise Galleries, San Diego . . 124152
Paradise Gallery, Westcourt 099884
Paradise Mill Silk Industry Museum,
   Macclesfield 045186
Paradise Road, Colombo 086615, 086616
Paradise Road – The Gallery Shop,
   Colombo 086617
Paradise Road Studio, Colombo 086618
Paradise So Nice Interprizes, Los
   Angeles 121570
Paradise Valley Antiques,
   Phoenix 096229
Paradiso, Baltimore 092353
Paradox, Edam 137854
Paradox, Harderwijk 143166
Parafia Ewangelicko – Augsburska
   Wang, Krapacz 034808
Parafialne Muzeum Regionalne,
   Krasnobród 034917
Parafita, Isabel Mota, Porto . . 085043
Paragon, Auckland 132695
Paragon Antiques, Birmingham . 092429
Paragon Antiques and Collectors
   Market, Bath 088397
Paragon Book Gallery, Chicago . 145051
Paragon Press, London 138236
Paragona, Massimo, Napoli . . . 080912,
   080913
Paragone Gallery, West
   Hollywood 125020
Parainen District Museum,
   Parainen 011232
Parainen Industrial Museum,
   Parainen 011233
Paraíso, C., Lisboa 126451
Paralax, London 118500
Parallels Gallery, Philadelphia . 123437
Parametro, Faenza 138859
Paramita Museum, Komono . . . 028882
Paramount Antiques, New York . 095650
Paramount Book Exchange,
   Manchester 144727
Paramythioti, Dominique, Paris . 105147
Parant, Marie, Paris 129107
Paraphernalia, Harrogate 089438
Paraphernalia, Sheffield 091290
Paraphernalia, Stockton-on-Tees 091449,
   135091
Paraplu, Nijmegen 112611
Parasite Museum, Faculty of Medicine,
   Mahidol University, Bangkok . . 042673
Paraskevova, Dr. Jaroslava, Frankfurt
   am Main 107019
Parasol Press, Portland 138443
Parasol Unit, Foundation for
   Contemporary Art, London . . . 045071
Paratiritis, Thessaloniki 109097, 109098
Paravent, London 090403, 134744
Parazaar, Antwerpen 063230
Le Parc, München 077593
Parc & Château de Valençay,
   Valençay 016090
Parc Archéologique de la Pointe-du-
   Buisson, Melocheville 006289
Parc Archéologique de Larina, Hières-
   sur-Amby 013264
Parc Archéologique Européen de
   Bliesbruck-Reinheim, Bliesbruck 012073
Parc Ardèche Miniature, Soyons 015859
Parc Botanique de la Fosse, Montoire-
   sur-le-Loir 014284
Parc de la Préhistoire, Tarascon-sur-
   Ariège 015911
Parc de l'Art Préhistoire, Tarascon-sur-
   Ariège 015912
Parc de l'Artillerie, Québec . . . 006745
Parc de Préhistoire de Bretagne,
   Malansac 013980
Parc de Reconstitution Archéologique,
   Villeneuve-d'Ascq 016261
Parc Historique Pointe-du-Moulin,
   Notre-Dame-de-l'Île-Perrot . . . 006528
Parc Howard Museum and Art Gallery,

Llanelli 044881
Parc Miniature Alsace-Lorraine, Saint-
   Ame 015215
Parc Monceau, Atlanta 092167
Parc National de la Culture des
   Ethnies, Vientiane 030091
Parc Préhistorique des Grottes de
   Fontirou, Castella 012352
Parc Préhistorique Imaginaire,
   Chamoux 012428
Parc Zoologique Arche de Noé et
   Amazonia, Saint-Clément-des-
   Baleines 015268
La Parchemine, Lyon 140929
Parchman & Stremmel, San
   Antonio 124057
Parco, Nagoya 111329
Parco Archeologico di Castel Seprio,
   Castelseprio 025849
Parco Monumentale di Pinocchio,
   Collodi 026038
Parco Museo Minerario, Abbadia San
   Salvatore 025228
Parco Nazionale delle Incisioni
   Rupestri, Capo di Ponte 025753
Parco Scherrer, Morcote 041758
Parcul Etnografic Romulus Vuia, Cluj-
   Napoca 036386
Pardaf, Antwerpen 063231
Pardee Home Museum, Oakland 051682
Pardo, Lugo 086084
Pardo, Paris 105148
Pardo Sanchez, Jorge Antonio,
   Barcelona 085820
Pardoes, Amsterdam 082822
Parduhn, Ute, Düsseldorf 106815
La Pared Galería, Bogotá 102354
Parent, Michelle, Montréal 064452
Parent, Xavier, Marseille 128781
Parente, Madrid 086231
Parenthou, Jacques & Richard,
   Lille 069500
Parenthou, Pierre & Jacques,
   Lille 069501
Parenti, Lucia, Firenze 131234
Parenza, Gianfranco & C., Roma 081493,
   132001
Parenza, Roberto, Roma 081494
La Parete, Toronto 101678
La Parete, Wien 063047, 127958
Pareyn, Jan, Tielt, West-
   Vlaanderen 063837
Parfait, Yvon, Hudimesil 064844
Pargon, D. & H., Nancy 070427
Parham Airfield Museum and Museum
   of the Resistance Organisation
   Museum, Framlingham 044369
Parham House & Gardens,
   Storrington 045890
Pari & Dispari, Reggio Emilia . . 110711
Pari Nadimi, Toronto 101679
Pariente, Emé, Biarritz 067207
Parienté, Xavier, Paris 071754
Parienté, Xavier, Saint-Émilion . 072742
Parigi, Daniela, Torino 081746
Parikian, D., London 144663
Parikkalan Kotiseutumuseo,
   Parikkala 011235
Parin, Marie-Madeleine, Paris . . 129108
Paris & Pajares, Lleida 115118
Paris Américan Art, Paris 105149,
   129109
Paris Antique Restoration, Los
   Angeles 135737
Paris Antiques, Ampthill 088277, 134032
Paris Antiques Gallery, Los
   Angeles 094436
Fabio Paris Art Gallery, Brescia . 109913
Paris Gibson Square Museum of Art,
   Great Falls 049219
Paris Manaus, Paris 071755
Paris Map Fair & Travel Books,
   Paris 098090
Paris-Musées, Paris 136973

Paris Photo, Paris 098091
Paris to Province, New York . . . 095651
Paris, Nicole, Agon-Coutainville . 066432
Paris, Nicole, Rouen 072607
Parish, Norman, Washington . . . 124944
Parishville Museum, Parishville . 051920
Parisi, Franco, Roma 132002
Parisi, Giuseppe, Roma 132003
Parisi, Salvatore, Napoli 080914
Parisian Galerie, Singapore . . . 114585
Parisini, Wien 127959
Parisot, René, Briançon 067488
Parissi, Franco, Firenze 131235
Parizot, Alain, La Chapelle-
   Rablais 068979
Park & Tilford, Vancouver 101821
Park Antique Restoration, Sherwood
   Park 128203
Park Antiques, Detroit . 093372, 135544
Park Antiques, Menston 090749
Park Avenue Antiques,
   Minneapolis 094941
Park Avenue Gallery, Saint Louis 123853
Park Boulevard Art Works, San
   Diego 124153
Park Boulevard Gallery, San
   Diego 097121
Park City Antiques Mall, Dallas . 093195
Park City Historical Society & Museum,
   Park City 051922
Park Coin, Minneapolis 094942
Park Farm Museum, Milton
   Abbas 045272
Park Gallery, Falkirk 044323
Park Gallery, Glasgow 117855
Park Gallery, Lalitpur 112092
Park Gallery, London 118501
The Park Gallery, Mosman Park 099392
Park Gallery and Bookshop,
   Wellingborough 144912
Park Hill Cooperative Book Store,
   Denver 145133
Park House Antiques, Bladon . . 088500
Park House Museum,
   Amherstburg 005333
Park Lane Antiques Centre, Armadale,
   Victoria 061045
The Park-McCullough House, North
   Bennington 051595
Park-muzej General V.Z. Lavrov, Gorni
   Dabnik 005037
Park-muzej im. A.K. Tolstogo,
   Brjansk 036922
Park-Muzej Vladislav Varnenčik,
   Varna 005245
Park South Gallery, New York . . 122952
Park Street Antique Center, Saint
   Petersburg 127435
Park Street Antiques,
   Jacksonville 094023
Park View Antiques, Wadhurst . 091705,
   135179
Park Walk Gallery, London 118502,
   134745
Park, Dr. Ho Kil, Gangha 111595
Park, Kyu Hyeong, Seoul 111702
Park, Ryu Sook, Seoul 111703
Parkanon Kotiseutumuseo,
   Parkano 011236
Parkanon Metsämuseo, Parkano 011237
Parkdale-Maplewood Community
   Museum, Barss Corner 005392
Parke-West, San Diego 097122
Parkens Antik, Falköping 086672
Parker, Herzogenaurach 107424
Parker, Hunters Hill 061642
Parker, London 090404, 134746
Parker, Nashville 122093
Parker Middleton, London 118503
Parker-O'Malley Air Museum,
   Ghent 049094
Parker Tavern Museum, Reading 052479
Parker, Jack, Saint Louis 096757,
   123854

Parker, Kevin, Hendra 061617
Parker, William W., Washington . 124945,
   136464
Parkerson, Houston 121107
Parkett, Zürich 139097
Parkett Publishers, New York . . 138418
Parkett Verlag, Zürich 138116
Parkfields Gallery, Ross-on-Wye 118988
Parkham Place Gallery, Surry
   Hills 062267
Parkhouse Antiques, Dallas . . . 093196
Parkhouse, Mark, Barnstaple . . 088364
Parkhus Galleri, Esbjerg 102795
Parkin, Michael, London 090405
Parkinson, Anthony, Southport . . 144859
Parkland College Art Gallery,
   Champaign 047520
Parks Canada Visitor Reception Centre,
   Churchill 005619
Parksepa Keskkooli Muuseum,
   Parksepa 010651
Parkside Antiques, Melton
   Mowbray 090744
Parkside Cottage Museum,
   Narrandera 001342
Parkstone, New York 138419
Parktown, Cannes 067647
Parkville Gallery, Flemington . . 099039
Parkway Galleries, Kansas City . 121311
Parkzicht, Ermelo 032320
Parlamentsausstellung des Deutschen
   Bundestages "Wege, Irrwege,
   Umwege", Deutscher Dom,
   Berlin 017268
Parliament House Art Collection,
   Canberra 000941
Parlor, San Francisco 124364
Parlour Gallery, Wolumla 099896
Parmakkapı İş Sanat Galerisi,
   İstanbul 117075
Parmantica, Parma 081123
Parme, Biarritz 067208
Parmegiani, Elisabetta, Milano . 080657
Parmentier, Daniel, Egry 068432
Parmer, J., San Diego 145525
Parmiggiani, Marina, Modena . . 131578
Parnas, Katowice 113514
Parnas, Kraków 113623
Parnass, Wien 138534
Parnass Verlag, Wien 136605
Le Parnasse, Chartres 103574
Parnassos, Wassenaar 143264
Parnasszus, Budapest 142568
Parnell Antiques, Enniscorthy . . 079618
Parnell Gallery, Auckland 112803
Parnell Museum, Rathdrum 024980
Parnitzke, G., Bessenbach 074965
Parochialkirche, Berlin 017269
Parodi, Livia, Sanremo 110905
Paroles, Avignon 140773
Parolini, Jean-Jacques, Lausanne 087519
La Parpaïolle, Marseille 069955
Parparov, Iva-I., Sofia 064236
Parpin, Parviz, Senden, Kreis
   Coesfeld 078454, 130638
Parque Arqueológico Cueva Pintada,
   Gáldar 039422
Parque Cultural, Zerain 040462
Parque de las Ciencias, Granada 039471
Parque Etnográfico do Río Arnoia,
   Allariz 038951
Parque-Museo La Venta,
   Villahermosa 031665
Parr, Stuart, New York 095652
Parra, Carlos Gil de la, Zaragoza 086602
Parra, Suzanne & Thierry,
   Montpellier 070310
Parrado, Arévalo 133086
Parramatta Antique Centre,
   Parramatta 062054
Parrasch, Franklin, New York . . 122953
Parreira, Candida de Azevedo,
   Brasília 100922
Parren, Jan, Thorn 132639

Parreño Navarro, Jose Joaquin,
Madrid 133364
Parrin & Co., Kansas City 094088
Parris Island Museum, Parris
Island 051929
Parrish & Reinish, New York . . . 122954
The Parrish Art Museum,
Southampton 053420
Parrish, Susan, New York 095653
Parriss, Sheringham . . 091316, 135040
Parronchi Dipinti '800-'900,
Firenze 080183
Parronchi Dipinti '800-'900,
Milano 080658
Parroquial Arxiprestal, San Mateo 040070
Parruzot, William, Gap   068631, 128606
Parry Mansion Museum, New
Hope 051273
Parry, H. & R.L., Sutton Coldfield 091520
Pars, El Paso 093411
Pars, Miami 094790
Pars Museum, Shiraz 024737
Parsa, Saint-Ouen 073166
Parsa, Tucson 097684
Parschau, Günther, Frankfurt am
Main 141786
Parshall, Margaret, Louisville . . 094577
Parson, Washington 136465
Parson Fisher House, Jonathan Fisher
Memorial, Blue Hill 047057
Parson Historical Commission
Museum, Parsons 051932
Parsons, Atlanta 092168
Parsons School of Design, New
York 057561
Parsonsfield-Porter History House,
Porter 052275
Parsy, Thierry, Paris 071756
La Part des Anges, L'Isle-sur-la-
Sorgue 104040
Parteaguas, Revista del Instituto
Cultural de Aguascalientes,
Aguascalientes 138916
Parter Galeria, Łódź 113679
Parthas Verlag, Berlin 137090
The Parthenon, Nashville 051183
Le Parthénon, Paris 105150
Particular, London 118504
Particular Art Gallery, Ho Chi Minh
City 125142
Partisan Pictures, New York . . 122955,
                                    122956
Partizanska Bolnica Franja, Mestni
Muzej Idrija, Dolenji Novaki . . 038510
Partners and Crime, New York . 145368
Partners in Pine, Coggeshall . . 088947
Partners in Time, Portland 096460
Partners in Time, Seattle 097482
Partnership, Dallas 093197
Partridge, London 090406
Partridge Antique Shop, Glenelg 061532
Partridge, Loren, Oakland 123260
Partridge, Scott, Richmond . . . 145476
Partridge, Timothy, Eastbourne . 089151
Parts, Toronto 101680
Parts and Pieces, Terborg 132638
Partyka, Kraków 113624
Parudi, Pasching 100136
Paruyr Sevak House-Museum,
Zangakatun 000752
Parva Creativa, Cárdenas 138917
Parvulescu, Dan Adrian,
Bucureşti 132999
Parys Museum, Parys 038796
Pasadena City College Art Gallery,
Pasadena 051937
Pasadena Museum of California Art,
Pasadena 051938
Pasadena Museum of History,
Pasadena 051939
Pasaj, Andrzej, Kraków 113625
Pasaje del Restauro, Buenos
Aires 127528
Pascack Historical Society Museum,

Park Ridge 051926
Pascal, Fukuoka 142894
Pascal, Anne-Mieke, Marseille . 069956
Pascal, Jean-Claude, Le Soler . . 069352
Pascal, Jean-Pierre, Chambéry . 067797
Pascal, Patrick, Fontvieille 068584
Pascale, Michèle, Joigny (Yonne) 068906
Pascaline, Rueil-Malmaison . . . 072640
Pascalis, Alexandre, Sainte-Olive 073362
Paschal, Nicole, Estang 068475
Pasco & Co., Miami 094791
Pascual Ansotegui, Marisa,
Pamplona 086396
Pascucci, Mario & Figli, Ferente,
Roma 081495
Paseka, Litomyšl 140483
Paseka, Praha 102713, 136767
Paseo de Compras San Telmo, Buenos
Aires 060921
Pash & Sons, Arnold, London . . 090407
Pasha Antiques, West Hollywood 097918
Pasinger Fabrik – Ausstellungen,
Kultur und Bürgerzentrum der
Landeshauptstadt München,
München 020562
Pasino, Debora, Roma 081496
Paslack, Armin, Friedrichshafen 075913
Pasmans Huus, Ruinen 032835
Pasquale Iannetti Art Galleries, San
Francisco 097302, 124365
Pasquali, Bassano del Grappa . . 109782
Pasquarella, Sheri L., New York 122957
Pasquer, Priska, Köln 107671
Pasquesi, Umberto, Kerken . . . 076759
Pasquier, Yvette, Besançon . . . 067134
Pasquier, Yvonne, Saint-Pierre-les-
Nemours 073263
Pass Collection, New York 095654
Pass of Killiecrankie Visitor Centre,
Pitlochry 045506
Pass, Jules L., Saint Louis 096758
Passado Composto, São Paulo . 064200
La Passage, La Croix-Valmer . . 103887
Le Passage, Paris 071757
De Passage, Zaandam 083698
Passage 125 Blaes, Bruxelles . . 063523
Passage Antiques, London . . . 090408,
                                    134747
Passage de Retz, Paris 105151
Passage du XXè Siècle, L'Isle-sur-la-
Sorgue 069585
Passage Saint-Michel, Bordeaux 067351
Passage-Verlag, Leipzig 137347
Passagen Verlag, Wien 136606
Passages Antique Gallery, New
Orleans 095140
Passages Centre d'Art Contemporain,
Troyes 016056
Passageway Gallery, Omaha . . 123335
Passaic County Community College
Galleries, Paterson 051943
Passaic County Historical Society
Museum, Paterson 051944
Passarella, Giordano, Padova . . 131673
Passato Prossimo, Parma 081124
Passato Prossimo, Roma 081497
La Passe Antiques, Dallas 093198
Passé au Présent, Iwuy 068888
Le Passé au Présent, Pau 072047
Passé Composé, Biarritz 067209
Passe Compose, Biarritz 067210
Passé Composé, Billy-sur-Ourcq 067222
Passé Composé, Craponne
(Rhône) 068222
Passé-Composé, Montréal 064453
Le Passé Composé, Paris 141257
Passé Composé, Tournai 063847
Le Passé d'Aujourd'hui, Paris . . 071758
Le Passé de la Montagne, Les
Houches 069411
Passé Intérieur, Nice 070644
Passé Passion, Bruxelles 063524

Le Passé Présent, Lausanne . . . 087580
Passé-Présent, Montréal 064454
Le Passé Présent, Nîmes 070701
Passé Présent, Paris . . 071759, 071760
Au Passé Présent, Toulon (Var) . 073725
Passé Recomposé, Bordeaux . 128501,
                                    128502
Le Passé Recomposé, Lyon . . . 128725
Le Passé Retrouvé, Herblay . . . 068808
Le Passé Retrouvé, Mortagne-au-
Perche 070357
Passé Simple, Blérancourt 067245
Passé Simple, Boulogne-
Billancourt 067393
Le Passé Simple, Bruxelles . . . 063525
Passé Simple, Came 067619
Le Passé Simple, Cannes 067648
Passé Simple, Coignières 068072
Passé Simple, Eaubonne 068425, 128589
Le Passé Simple, Grenoble . . . 068737
Passé Simple, Guesnain 068765
Le Passé Simple, La Cadière-
d'Azur 068959
Le Passé Simple, Montluçon . . 070271
Passé Simple, Orléans 070819
Passebon, Pierre, Paris 071761, 105152
Passelard, François, Riom 105447
Passer, Hanne, Tranekær 102992
Passer, Marchen, Dragør 065634
La Passerelle, Auxerre 066851, 103296
Passerelle, Lyon 104125
Passeri, Leonardo, Firenze 131236
Passerin, Maurice, Nice 070645
Passerose, Châteauneuf-sur-
Charente 067896
Passers Buy, Llangollen 089872
Le Passeur de Lumière,
Valbonne 105794
Passiatore, Caterina, Bari 130917
Passiflora, Brewood 088593
Passiflora, Saint-Ouen 073167
The Passing Show Antiques, Tawa
Wellington 084108
Passion, Brugge 063287
Passion d'Antan, Chargey-les-
Gray 067841
Passion des Meubles, Dijon . . . 068340
Passion du Passé, Montréal . . . 064455
Passion et Collection, Limoges . 069529
Passion for the Past, Toronto . . 064719
Passion Maison, Toulon (Var) . . 073726
Passion Sanat Galerisi, İstanbul 117076
Passione, Zürich 087997
Passions Past & Present, Palmerston
North 084056
Passman, Bernard K., New
Orleans 122216
Past & Present, Amsterdam . . . 082823
Past & Present, Belfast 088435
Past & Present, Kortrijk 063725
Past & Present, Lakeside 127276
Past & Present, Perranporth . . . 090970
Past & Present, Saskatoon 064567
Past & Present Shop, Long
Beach 094217
Past & Presents, Polperro 091029
Past and Present, Bromley 088686
Past and Present, Clitheroe . . . 088934
Past and Present, Hingham . . . 089536
Past and Present, Honolulu . . . 093569
Past and Present, Philadelphia . 096146
Past and Present Furniture, Las
Vegas 094166
Past and Presents, Haarlem . . . 083347
Past Caring, Holt 089544
Past Centuries Pieces, Los
Angeles 094437
Past Connection Antiques,
Houston 093761
Past Connections, Mosman . . . 061896
Past Era Antique Jewelry,
Houston 093762
Past Perfect, San Francisco . . . 097303
Past Present Future, Minneapolis 094943

Past Students Association Museum,
University of Queensland,
Gatton 001090
Past Times, Albuquerque 092075
Past Times, Cincinnati 092884
Past Times, Dunedin 083929
Past Times, Long Beach 094218
Past Times, Singapore 085482
Past Times Antiques, Ormond . . 062035
Past Times Antiques, San Gwann 082429
Past Times Jewellery Workshop,
Perth 062079
Pasteger, Yvan, Mulhouse 070395
Pastel Artists Canada, Flesherton 058459
Pastel Artists of New Zealand,
Blenheim 059640
The Pastel Journal, Cincinnati . . 139475
Pastel Painters of Maine,
Westbrook 060670
Pastel Painters Society of Cape Cod,
Barnstable 060671
Pastel Society, London 060207
Pastel Society of America, New
York 060672
Pastel Society of Australia,
Moorooka 058210
Pastel Society of Canada, Ottawa 058460
Pastel Society of Colorado,
Denver 060673
Pastel Society of Eastern Canada,
Montréal 058461
Pastel Society of El Paso, El
Paso 060674
Pastel Society of New Mexico,
Albuquerque 060675
The Pastel Society of North Carolina,
Chapel Hill 060676
Pastel Society of North Florida, Fort
Walton Beach 060677
Pastel Society of Oregon,
Roseburg 060678
Pastel Society of San Diego, San
Diego 060679
Pastel Society of Southeast Texas,
Houston 060680
Pastel Society of the Northern Rockies,
Idaho Falls 060681
Pastel Society of the Southwest,
Dallas 060682
Pastel Society of the West Coast,
Sacramento 060683
Pastel Society of Victoria,
Canterbury 058212, 098861
Pastel Society of Victoria, Doncaster
Heights 058211
Il Pastello, Roma 110842
Pasti & Bencini, Firenze 080184, 110063
Pastiche Antiques, Los Angeles . 094438
Pastimes, Bristol 088664
Pastimes Antiques, South
Auckland 084082
Pastimes Vintage Toys, Glasgow 089163
Pastl, Dr. Ute, Linz . . 062720, 125262
Pastor-Lloret, André, Nice 070646
Pastor, Monique, Blois 067259
Pastor, N., Nice 104383
Pastoral Gallery, Queanbeyan . . 099604
Pastorale, La Colle-sur-Loup . . 069003
La Pastorale, Paris 071762
Pastorale, Quimper 072348
Pastorale Antiques, Lewes 089817
Pastorini, Alina, Torino 132578
Pastorius, Den Haag 083117
Pastre, Pierre, Béziers 103378
Pastrello, Francesco, Venezia . . 132253
Pasvalio Krašto Muziejus,
Pasvalys 030444
Pásztói Múzeum, Nógrádi Megyei
Múzeum Szervezet, Pásztó . . . 023834
Pásztormúzeum, Bugac 023575
Pata Gallery, Beijing 101983
Pata Gallery, Taipei 116935
Patachine, Paris 071763
Pataka, Porirua 033286

Pataky Körgaléria, Budapest ... 109168
Patan, Lalitpur 112093
Patan Museum, Lalitpur 031880
Patch Magic, Singapore 085483
Patches Gallery, Lower
  Beechmont 099239
Patchings Farm Art Centre,
  Calverton 117499
Pate Museum of Transportation,
  Cresson 048060
Patee House Museum, Saint
  Joseph 052786
Patek Philippe Museum, Genève 041540
Patel, Ketna, Singapore 114586
Pateley Bridge Antiques, Pateley
  Bridge 090947
Patella, Palermo 081062
Patelli, Nicola, Milano 131493
Pateniemen Sahamuseo, Pohjois-
  Pohjanmaan Museo, Oulu .. 011221
Patenschaftsmuseum Goldap in
  Ostpreußen, Stade 021948
Patent Model Museum, Fort
  Smith 048875
Pater Haw-Museum, Schweich an der
  römischen Weinstraße 021779
Pater Peter Singer-Museum,
  Salzburg 002633
Paterae, New York 095655
Paterlini, Paola, Bologna 079865
Paternina, Alvaro de, Madrid .. 086232
Paternuosto, Luciano, Roma ... 132004
Paterson Court House,
  Adamstown 000758
Paterson Museum, Paterson ... 051945
Paterson, A. & M., Paddington,
  Queensland 062055
Paterson, M., Ballarat 061089
Paterson, Simon, Chichester .. 134256
Path, Aalst 100476
Pathe, Kobe 082001
Pathein Museum, Pathein 031825
Pathological Museum, London .. 045072
Pathologisch-anatomisches
  Bundesmuseum, Wien 003064
Pathology and Bacteriology Museum,
  Lucknow 024384
Pathology Museum, Tshwane .. 038855
Pathology Museum, Department of
  Pathology, Mumbai 024409
Pathology Museum, University of
  Queensland, Herston 001143
Pathway Antiques, Vancouver .. 064861
Pati, Giampaolo, Roma 132005
Patika-ház, Soproni Múzeum,
  Sopron 023896
Patina, Bonn 075083
Patina, Budapest 079278
Patina, Madrid 133365
Patina, New York 095656
Patina, Santiago de Chile 064981
Patina, Szeged 079326
Patina Antik, Veszprém 079329
Patina Antikvitás, Budapest .. 079279
Patina Antiquitäten, Bern ... 087295
Patina Antiquitäten, Netphen .. 077701,
                              130441
Patina Art, Helsinki 066353
Patina Atelier, San Francisco .. 097304
Patina Consignment, Tucson .. 097685
Patina Studio, New York 135973
Patinations, Gilmore 127602
Patine, Coimbra 084791
La Patine, Hollange 082355
Patine, Maastricht 083474
Patine Antiquites, Chindrieux .. 067975
Patine du Temps, Bordeaux .. 067352
La Patine du Temps, Paris ... 071764
Patine et Vieilles Dentelles,
  Lucciana 069656
Le Patio, Antibes-Juan-les-Pins . 066677
El Patio, Sevilla 086465
Patio-Antiques, Hamburg 076293
Patio Art, Monaco 112013

Al-Patio Art Gallery, Cairo 103071
Patio Centrum Sztukii, Wyższa Szkoła
  Humanistyczno-Ekonomiczna,
  Łódź 034979
Patio-Verlag, Dreieich 137146
Patkó Imre Gyűjtemény, Xantus János
  Múzeum, Győr 023665
Patmos Verlagshaus, Düsseldorf 137165
Patna Museum, Patna 024433
Paton, Saint Albans 144811
Patonai Fazekasház, Sümeg ... 023908
Patou, Alain, Saint-Christoly-de-
  Blaye 072713
Patrascu, Manole, Galaţi 133008
Patrassi, Anna, Milano 080659
Patres, Eerbeek 132496
Patrice, Brno 065352
Patrice-Alexis, Thonon-les-Bains 105706
Patrich, Simon, Vancouver ... 101822
Patrician Design, Albuquerque .. 092076
Patricia's Antiques, Buffalo ... 092550
Patricie, Brno 065353
The Patrick and Beatrice Haggerty
  Museum of Art, Marquette University,
  Milwaukee 050923
Patrick Collection Motor Museum,
  Birmingham 043624
Patrick Taylor Cottage Folk Museum,
  Albany 000778
Patrick, Cidnee, Dallas 120639
Patrick, Jolly, Denver 120753
Patrick, Joseph, Toronto 064720, 140414
Patrick, Pierre, Cassis 133539
Patrick, Pierrejean, Châlons-en-
  Champagne 067787
Patrijaršie Palaty, Gosudarstvennyj
  muzej-zapovednik Moskovskij Kreml,
  Moskva 037490
Patrikiadis, George N., Athinai .. 079176
Patrimoine Historique et Études du
  Repassage, Sebourg 015746
Patrimonio Atelier de Conservacao e
  Restauracao, Belo Horizonte .. 128102
Patriots Point Naval and Maritime
  Museum, Mount Pleasant ... 051099
Patris, Françoise, Brest 067477
Patris, Gilles, Gouesnou 068685
Patron, Denver 120754
Patrone, Luigi, Roma 110843
Pat's Antiques, Cleveland 092963
Patscher, Stephan, Bodenheim .. 129603
Patten, Keith & Nancy, Saint
  Louis 145498
Patterson Homestead, Dayton .. 048173
Patterson Liddle, Bath 144232
Patterson, Isla Jean, Hawker,
  Australian Capital Territory ... 099111
Patterson, Liam, Dublin 130861
Patterson, Nicholas, Tübingen .. 078727
Patterson, W.H., London 090409, 118505
Patterson's Spade Mill,
  Templepatrick 045955
Pattijoen Kotiseutumuseo Ojala,
  Pattijoki 011240
Pattinson, Betty, Hawker 099941
Pattison, Tony, Chorleywood ... 088903,
                              134269
Patton, Sainte-Savine 073370
Patton Museum of Cavalry and Armor,
  Fort Knox 048822
Patton Picture Company, New
  York 122958
Patton, Frank, Saint Louis 096759
Patty's, Paddington, Queensland 139866
Patz, Robert, Magdeburg 107943,
                              142098
Patzelt, Rheine 003976
Patzelt, Lambert, Ofterschwang . 130488
Patzelt, Michael, Wölfersheim .. 108955
Patzelt, W., Baddeckenstedt ... 074626
Patzer & Trenkle, Konstanz ... 142031
Patzwall, Klaus D., Norderstedt .. 137447
Pau-Santoantiguidades, Lisboa . 084938
Pauer, A., Leipzig 077064

Pauget, André, Cormoz 068158
Paul, Krems 127796
Paul, Saint Kilda, Victoria 099669
Paul, Singapore ... 085484, 133041
Paul Anton Keller-Museum,
  Lockenhaus 002337
Paul-Art Antique, Roye 072633
Paul Bunyan Logging Camp, Eau
  Claire 048465
Paul Dresser Memorial Birthplace,
  Terre Haute 053718
Paul Ernst-Gedenkstätte, Sankt
  Georgen an der Stiefing ... 002654
Paul-Galerie, Bremerhaven ... 106597
Paul Gugelmann-Museum,
  Schönenwerd 041961
Paul H. Karshner Memorial Museum,
  Puyallup 052424
Paul Jönska Gården, Viken 041167
Paul Laurence Dunbar State Memorial,
  Dayton 048174
The Paul Mellon Centre for Studies in
  British Art, London 056566
Paul Morris Gallery, New York .. 122959
Paul Revere House, Boston .. 047129
Paul-Röder-Museum,
  Marktoberdorf 020300
Paul W. Bryant Museum, University of
  Alabama, Tuscaloosa 053860
Paul Whitney Larson Art Gallery,
  University of Minnesota, Saint
  Paul 052850
Paul, Angela, Düsseldorf 106816
Paul, Anne-Marie le, Quimper .. 072349
Paul, Christoph & Brigitte,
  Stuttgart 078625
Paul, Frédéric, Larroque-sur-
  l'Osse 069194
Paul, Fritz, Nuthetal 077852
Paul, Jon, Saint Louis 096760
Paul, Manuel Santiago, Valladolid 086551,
                              115528
Paul, Maurizio, Berlin 129547
Paul, R.K., Dunedin 132719
Paula Modersohn-Becker-Stiftung,
  Bremen 017597
Paula Stradiņa Medicīnas Vēstures
  Muzejs, Rīga 030213
Paulan Taidemyynti, Tammela .. 066394
Paulette's Antiques, Cleveland .. 092964
Pauletti, Seattle 097483
Pauli, Monique, Marseille 069957
Pauli, Stefan, Bern 087296
Pauline E. Glidden Toy Museum,
  Ashland 046567
Paulinerkirche, Niedersächsische
  Staats- und Universitätsbibliothek,
  Göttingen 018754
Pauline's Art, Las Vegas 121372
Paulise de Bush Costume Collection,
  Broadclyst 043758
Paul's Antiques, Croydon 061380
Paul's Art Gallery, Houston ... 121108
Paul's Auction Mart, Southport . 125235
Paul's Clock Repair, Pittsburgh . 136134
Pauls, K., Stemmen 078557
Pauls, Rudi, Lübeck 077184
Paulsen, Harm, Schleswig 130622
Paulus, A., Liège 063777
Paulus, Gottfried, Esslingen ... 141756
Paulus, Marcel, Toulouse 073790
Paulusgalerie, Albstadt 105972
Paupa, Donostia-San Sebastián . 115055
Pause Rétro, Montréal 064456
Pauselijk Zouavenmuseum,
  Roeselare 003976
Pautot, Alain, Clermont-Ferrand . 068540
Pauw-Müller, de, Gent 063655
Pauw, H. de, Gent 063654
Pauw, M., London 090410
Pauwels, Kortrijk 063726
De Pauweveer, Middelburg ... 112596
Pavan, Stresa 110938
Pavanati, Cinzia, Padova 110574

Pavani, Andrea, Bologna 130987
Le Pavé dans la Mare, Besançon 103370
Paved Arts, Saskatoon 006981
Pavesi Fine Art, Milano 080661
Pavesi, Arnaldo, Milano 080660
Pavia, Baden-Baden 074640
Paviglianiti, Patrizio, Firenze ... 110064,
                              110065
Pavignano, Claudia, Milano ... 080662
Pavilhão do Conhecimento,
  Lisboa 035824
The Pavilion, Abu Dhabi 117166
Pavilion, Leeds 060208
Pavilion Antiques, Chicago 092765
Pavilion des Antiquaires et des Beaux-
  Arts, Paris 098092
Pavilion Gallery, Winnipeg 007456
Paviljon Cvijeta Zuzorića,
  Beograd 038177
Paviljon Katalnoj Gorki, Muzej-
  zapovednik Oranienbaum,
  Lomonosov 037204
Pavillion Horta-Lambeaux, Musées
  Royaux d'Art et d'Histoire,
  Bruxelles 003497
Pavillon Chinois, Tour Japonaise
  et Musée d'Art Japonais,
  Bruxelles 003498
Pavillon de la Brocante, Sainte-
  Adresse 073338
Pavillon de l'Arsenal, Paris ... 014796
Pavillon de Vendôme, Aix-en-
  Provence 011565
Pavillon des Arts, Les Musées de la
  Ville de Paris, Paris 014797
Pavillon Flaubert, Canteleu ... 012317
Pavillon-Presse Druckgrafisches
  Museum Weimar, Weimar ... 022505
Pavillon Saint-Arnaud, Trois-
  Rivières 007251
Pavilon Anthropos, Moravské zemské
  muzeum, Brno 009481
Pāvilostas Novadpētniecības Muzejs,
  Pāvilosta 030168
Paviot, Alain & Françoise, Paris . 105153
Paviot, Eric, Pont-de-l'Arche .. 072249
Pavito, Lagos 084809
Pavlik, Großschönau 076114
Pavlik, Nicholas, New York ... 135974
Pavlodarskij Oblastnoj
  Chudožestvennyj Muzej,
  Pavlodar 029834
Pavlodarskij Oblastnoj Istoriko-
  Kraevedčeskij Muzej im. G.N.
  Potanina, Pavlodar 029835
Pavlos Vrellis Museum of Greek
  History, Ioannina 023093
Pavlova, M., Miami 121837
Pavlovskij Istoričskij Muzej,
  Pavlovo 037621
Pavo Real Gallery, Boston 119911
Pavo Sanat Galerisi, İstanbul .. 117077
Pavone, Antonio, Palermo 131707
Pavone, Federico, Ragusa 081181
Pawdora Galéria, Budapest 109169
Pawelzik, Andreas, Hannover .. 076388
Pawelzik, Martin, Basel 133654
Pawilon Wystawowy – Centrum
  Gemini, Muzeum Miasta Gdyni,
  Gdynia 034751
Pawlak, Caen 067578
Pawlak, Beata, Warszawa 113884
Pawley, Jody, Drummoyne ... 098980
Pawlik, Nürnberg ... 077834, 130470
Pawlowski, Lünen 077215
Pawnee Bill Ranch, Oklahoma
  Historical Society, Pawnee .. 051951
Pawnee City Historical Society
  Museum, Pawnee City 051952
Pawnee Indian Village, Republic 052517
Pawsey & Payne, London 090411
Pawson, T.W., Somersham ... 091359,
                              135058
Pax, Knokke-Heist 100574

Pax, Kirsten, Berlin 074895
Paxson Gallery, Missoula 050961
Paxton House Grannery, Berwick-upon-Tweed 043579
Paxton, Loyd, Dallas 093199
Payer, Fritz, Zürich 087998
Payer, Ursula, Leoben 100093
Paykel, Al, Sacramento 123778
Payne & Son, Oxford . 090936, 134918
Payne & Sons, Tunbridge Wells . 091659,
135164
Payne Gallery, Moravian College,
Bethlehem 046964
Payne, Aaron, New York 122960
Payne, Clive, Chipping Norton . . 134264
Paynes Creek Historic State Park,
Bowling Green 047156
Payonne, Philippe, Rieux 072463
Payot, Lausanne 138062, 144094
Payrissat, Annie, Toulouse 073791
Paysan, Moritz, Stuttgart 130682
La Paz, Taranaki 084092
Paz Feliz, Madrid 115268
Paz, Fernando, Belo Horizonte . . 100901
Pazanis, Saint-Didier-au-Mont-d'Or 105503
Pazienza & C., Gennaro, Milano 131494
Paznauner Bauernmuseum,
Mathon 002368
Pazyryk, Aix-en-Provence 103200
Pazyryk, Saint Paul 096847
PB Gallery, New York 095657
PDX Gallery, Portland 123654
AO Pe da Letra, São Paulo 140311
Pe de Boi Galeria de Arte Popular
Brasileira, Rio de Janeiro 101056
Pé Palito & Cia, Porto Alegre . . . 063989
Pea Configuracion Creativa,
Madrid 115269
Peabody Essex Museum, Salem 052880
Peabody Museum of Archaeology
and Ethnology, Harvard University,
Cambridge 047370
Peabody Museum of Natural History,
Yale University, New Haven . . . 051265
Peabody Place Museum,
Memphis 050789
Peace Memorial Museum,
Zanzibar 042642
Peace Museum, Bradford 043697
The Peace Museum, Chicago . . . 047689
Peace Museum, Heping 007859
Peace River Centennial Museum,
Peace River 006618
Peach Blossom Furniture, San
Diego 097123
Peach Gallery, Stafford 119534
Peachgrove Antiques, Hamilton . 083947
Peachtree Hill Gallery, Atlanta . . 119599
Peacock Cottage Gallery, Linton 099232
Peacock Gallery, Middle Georgia
College, Cochran 047851
Peacock Heritage Centre,
Chesterfield 043937
Peacock Visual Arts, Aberdeen . 043360,
117210
Peacock, Deborah, Austin 119688
Peacock, Jay, San Francisco . . . 097305
Peacock, Pristine, San Antonio . 097012
Peacock, W. & H., Bedford . . . 126814
Peacock, W. & H., Saint Neots . 127029
Peak District Mining Museum, Matlock
Bath 045237
Peak Gallery, Toronto 101681
Peak Rail Museum, Matlock 045236
Peake, Barbara, Torquay 099836
Pear, Portland 123655
Pear Tree Gallery, Wahroonga . . 099857
Pearce Collections Museum, Navarro
College, Corsicana 048010
Pearce Institute, Glasgow 044410
Pearce, C., Hobart 139777
Pearce, Sara, London 118506
Pearl, Hannover 130013

Pearl, Vebron 073994
Pearl Button Museum, Muscatine 051133
Pearl Gallery, Baltimore 119763
Pearl Gallery, Honolulu 120960
Pearl Publishing, Portland 123656,
138444
Pearl S. Buck Birthplace,
Hillsboro 049480
Pearl S. Buck House, Perkasie . . 051990
Pearle Dorrace, Dallas 093200
Pearlers Row Gallery, Broome . . 098809
Pearn's Steam World, Westbury . 001636
Péaron, Monique, Orgeval 070801
Pearsall, Louisville 094578
Pearse Lukies, Aylsham 088329
Pearse Museum, Dublin 024905
Pearson, Phoenix 096230
Pearson Air Museum, Vancouver 053948
The Pearson Museum,
Springfield 053455
Pearson, Duncan, Dartford 117662
Pearson, Jim, Albuquerque 119504
Pearson, John A., Horton 089591
Pearson, Michael, Canterbury . . 088779,
134209
Pearson, Sue, Brighton 088632, 134154
The Peary-MacMillan Arctic Museum,
Bowdoin College, Brunswick . . 047286
Peasantries, Vancouver 064862
Pease-Cobb, Fort Worth 093470
Peasenhall Art and Antiques Gallery,
Peasenhall 090950, 134926
Pebble Hill Plantation,
Thomasville 053737
Pebbles Yard Gallery, Saint
David's 119013
Peca, Severino, Lausanne 087581
Pecanins, México 111977
Peças Soltas, Faro 113971
Pecci, Laura, Milano 110395
Peccolo, Roberto, Livorno 110186
Péchaud, Camille, Crézancy-en-Sancerre 068242
Pechermuseum, Hernstein 002077
Pêcheur d'Images, La Trinité-sur-Mer 103908
Pecheur D'Images, Roscoff 105460
Peck & Hill, Birmingham 092430
Pecka, Praha 102714
Peckover House, Wisbech 046169
Peco, Hampton 089419, 134433
Pecorella, Napoli 080915
Pecorella, Luigi, Napoli 080916
Pecorini & C., Loredana, Milano 142736
Pecos Park, Pecos 051959
Pėda, Vilnius 111806
Pedagočeskij Muzej A.S. Makarenko,
Moskva 037491
Pedagogické Muzeum Jana Ámoše
Komenského, Praha 009834
Pedagoginis Muziejus, Kaunas . . 030373
Pedagoški Muzej, Beograd 038178
Peddler's Daughter, Milwaukee . 094876
Peddler's Daughter, Saint Louis . 096761
Peddlers Rack, Los Angeles 094439
Peddler's Village, Columbus 093038
Peddler's Wagon, San Antonio . 097013,
127438
Peden, Ludij, Yungaburra 099950
Pedersen, Porto Alegre 100994
Pedersen, Randers 066057
Pedersen, Albert, Randers 140644
Pedersen, Erling, Stege 102985
Pedersen, Franz, Horsens 102851
Pedestrian Gallery, Jacksonville . 121260
Pedlar, Maylands, South Australia 061804
Pedlar's Pack Books, Totnes . . . 144887
Pednault, Montréal 064457
Pedrazzi Cané, Angela, Bologna 079866
Pedrero, Rafael Jimenez, Sevilla 086466
Pedretti, Felice, Roma 132006
Pedretti, Jean-Jacques,
Porrentruy 087740
Pedro & Castro, Lisboa 084939

Pedrocchi, Sabrina, Losone 133815
Pedroni, Carlo, Muralto 087689, 144113
Pedrosa Garcia, Concha, Sevilla 115454
Peebles, Tulsa 136433
Peege, Frank, Freiburg im
Breisgau 125953
Peek in the Attic, Dallas 093201
Peek, Ton, Utrecht . . . 112709, 143248
Peekskill Museum, Peekskill 051962
Peel Castle, Manx National Heritage,
Peel 045472
Peel Heritage Complex,
Brampton 005457
Peel Mansion Museum,
Bentonville 046925
Peeler, Alan, Memphis 121706
Peelmuseum, Ospel 032736
Peensalu, Stockholm 087043
Peep, Laine, Tartu 140683
't Peerd, Oostende 063820
Peerlings, J., Kampen . 107503, 137330
Peerlings, Josef, Krefeld 107743, 137330
Peeter I Maja, Tallinna Linnamuuseum,
Tallinn 010685
Peeters, Leuven . . . 136660, 140236
Peetsold, L.W., Hoofddorp 132545
Peetz, Sebastian, Hannover 107373
Peg Alston Fine Arts, New York . 122961
The Peg General Store,
Vancouver 064863
Pegaso, Milano 080663
Pegaso, Roma 081498
Pegaso, West Hollywood 097919
Pegasus, Bari 079776
Pegasus, Bucureşti 133000
Pegasus Antiques, Geelong 061512
Pegasus Book Exchange, Seattle 145595
Pegasus Books, Wellington 143369
Pegasus Kunst og Antikvitetshandel,
Højbjerg 065769
Pegaz, Warszawa 115885
Pegeot, Arlette, Besançon 067135
Peggottys, Teynham 091593
Peggy Guggenheim Collection,
Venezia 028322
Peggy Notebaert Nature Museum,
Chicago Academy of Sciences,
Chicago 047690
Peggy's Antiques, Indianapolis . 093915
Peghaire, Michel, Magny-en-Vexin 069804
Pegoud, Henri, Saint-Martin-de-Vaulserre 072943
Pehböck, Erwin, Perg 100137
Pehr, Siggi Klaus-Peter, Weilrod 108859
Pehrs, Magdalena, Münster 077673
Pei County Museum, Peixian . . . 008144
Pei Kuan Crab Museum,
Toucheng 042579
Peierl, Marcela, Traisen 062860
Peiffer, Düsseldorf 075504
Peiffer, Jacques, Herserange . . . 128620
Pein, Thomas, Raubling 078141
Peinelt, Scheibenberg 078359
Peintre des Paquebots, Le Havre 103901
Peintre, Christian, Paris 071765
Les Peintres Québécois, Québec 101513
Peintres sans Frontières, Paris . 105154
Peinture Fraîche, Paris 105155
Peira Arte Contemporanea, Bra . 109894
Peironcely, Madrid 115270
Peithner-Lichtenfels, Praha 102715
Peithner-Lichtenfels, Georg &
Christina, Wien 010360
Peitzmeier, Klaus, München 130378
Pejaten, Jakarta 109513
Pejepscot Historical Society Museum,
Brunswick 047287
Peka Verlagsgesellschaft,
Gauting 137222
Pekel, Barbara, Köln 142020
Pékin Fine Arts, Beijing 101984
Peking Arts and Antiques, San
Francisco 097306

Pékmúzeum, Soproni Múzeum,
Sopron 023897
Pelastusarmeijan Museo, Helsinki 010853
Pelatan, Benoît, Suresnes 073607
Pelati, B. & C., Parma 081125
Pelati, Romano, Milano 110396
Pelavin, Cheryl, New York 122962
Pelayo, Seattle 097484
Pêle-Mêle, Bruxelles 140195
Pêle Mêle, Montmorot 070281
Pelee Island Heritage Centre, Pelee
Island 006620
Pelée, Guy, Trélivan 073878
Pele's Dream, Honolulu 120961
Pelgoron, Milano 080664
Pelgrom Museum Poorterswoning,
Antwerpen 003308
Pelgrom, J., Haarlem 083348
Pelham Art Center, Pelham 051963
Pelham Galleries, London 090412
Pelham Galleries, Paris 071766
Pelham Historical Resource Centre,
Fenwick 005801
Pelias, Marilyn, New Orleans . . . 095141
Pelican Art Gallery, Saint Michael 100465
Pelican Gallery, Forster 099047
Pelican Publishing, Gretna 138336
Peligro, New Orleans 122217
Peliti Associati, Roma 110844
Pelkosenniemen Kotiseutumuseo,
Pelkosenniemi 011241
Pella Historical Village, Pella . . . 051964
Pellas & Blaise, Paris 129110
Pellas Ateljé och Galleri, Hjo . . . 115689
Pellas, Aldo, Paris 071767
Pellat de Villedon, Jean-Yves,
Versailles 074065
Pellé, Francis, Bordeaux 128503
Pellegrin, Chamonix-Mont-Blanc 067803
Pellegrin, Jacques, Paris 071768
Pellegrini, Alberto, Roma 132007
Pellegrini, Mario, Mendrisio 087673
Pellegrini, Bartolomeo, Torino . . 081747
Pellegrino, Luciana, Perugia 131758
Pellegrino, Marianna, Taranto . . 081638,
110942
Pellenz-Museum für Archäologie,
Nickenich 020780
Pellerin, Claude & Stéphane,
Longueville-en-Barrois 069621
Pelles Lusthus, Nyköping 115787
Pellet, Patrice, Paris 129111
Pelletier, Annick, Vence 074005
Pelletier, Claude, Quiers-sur-Bezonde 072336
Pelletier, Pauline, Québec 101514
Pelligra Arredamenti, Catania . . 080042
Pellisier House Museum, Bethulie 038654
Pellitteri, Frank, New York 135975
Pellitteri, Salvatore, Palermo . . . 081063
Pellmont, Felix, Binningen 133696
Pellon Kotiseutumuseo, Pello . . . 011242
Pellonpään Talonpoikaistalo,
Inkere 010905
Pelludat, Inga, München 130379
Pelly, William, Lyme Regis 118677
Pelmama Academy, Dobsonville 056342
Pelorus Antiques, Havelock 083948
Pels-Leusden, Berlin 106394
Pels-Leusden, Zürich 116844
Pelster, Martin, Köln 076900
Peltier-Quatrefages Bérangère,
Sarzeau 073424
Peltier, Patrice, Le Mans 069310
Peltoina Mäkitupalaismuseo,
Tammijärvi 011392
Peltotyökalumuseo, Somero 011366
Peltriaux, Bernard, Reims 105425
Peltz, Milwaukee 121944
Peluffo, Lilia, Buenos Aires 127529
Pelz, Wunstorf 079098, 130791
Pelzer, Alice Hélène & Gérard,
Bougival 067379, 128511
Pelzer, Gérard, Bougival 067380

Pelzer, N. & N., Raesfeld 078121
Pema Arts Gallery, Kathmandu . 112083
Pemba Museum, Chake Chake . 042629
Pember Museum of Natural History,
Granville 049207
Pemberton & District Museum,
Pemberton 006622
Pembery, Michael, Bakewell 088336
Pembina Lobstick Historical Museum,
Evansburg 005797
Pembina State Museum,
Pembina 051967
Pembina Threshermen's Museum,
Winkler 007431
Pembroke, Nashville 095020
Pembroke Antiques Centre,
Pembroke 090951
Pembroke Historical Society Museum,
Pembroke 051968
Pembroke Hydro Museum,
Pembroke 006624
Pembroke Lodge Patterson Heritage
Museum and Family History Centre,
Birchington 043603
De Pen, Peer 140257
Pen & Pencil Grafix, Portland . 123657
Pen and Brush Museum, New
York 051460
Pena, Rio de Janeiro 064101
Pena-Peck House, Saint
Augustine 052747
Peña, J., Madrid 086233
Peñalosa Camargo, Consuelo,
Bogotá 065217
Penang Forestry Museum, Pulau
Pinang 030724
Peñas Ruiz, Enrique, Barcelona . 085821
Pencarrow House, Bodmin 043666
Pence Gallery, Davis 048162
Pence, John, San Francisco ... 124366
Penchée, Mondeville 070159
Pendaries, Claude, Malemort-sur-
Corrèze 069827
Pendariès, Virginie, Brive-la-
Gaillarde 067514
Pendarvis Historic Site, Wisconsin
Historical Society, Mineral Point 050931
Pendeen Lighthouse, Pendeen . 045475
Pender, Virginia 062393
Pendle Antiques Centre, Sabden 091166
Pendle Heritage Centre,
Barrowford 043504
Pendlebury Antiques, Dunedin . 083930
Pendleton District Agricultural
Museum, Pendleton 051973
Pendon Museum, Long
Wittenham 045148
Pendragon-Verlag, Bielefeld ... 137111
La Pendule, Dortmund   075393, 129700
La Pendule, Stolberg, Rheinland 078567
La Pendulerie de Lyon, Lyon .. 069771,
128726
Pendulo, Lugo 086085
Pendulum, Fort Worth . 093471, 135558
Pendulum, Hamburg 076294
Pendulum Gallery, Sydney 099792
Pendulum of Mayfair, London .. 090413
Peneloux, Jocelyne, Bains-sur-
Oust 066921
Penetanguishene Centennial Museum,
Penetanguishene 006626
Penfold, Brisbane 061212
Pengelly, Mike, Mitcham 139830
Penglai Art Cooperative,
Shanghai 102196
Pengsjö Nybyggarmuseum, Umeå 041104
Penguin Books, London 138237
Penha, Belo Horizonte 100902
Penhaven Gallery, Saint Ives,
Cornwell 119039
Peninsula, Eindhoven . 112465, 137855
Peninsula and Saint Edmunds
Township Museum, Tobermory 007152

Peninsula Fine Arts Center, Newport
News 051548
Peninsula Painting Restoration,
Karingal 127622
Penkill Castle, Girvan 044388
Penland Gallery, Penland School of
Crafts, Penland 051975
Penland School of Crafts,
Penland 057656
Penlee House Gallery and Museum,
Penzance 045482
Penman, Spalding 091396, 135068
Penn, Saskatoon 064568
Penn Barn, Penn 144778
Penn Dutchman Antiques,
Chicago 092766
Penn Village Antique Centre,
Penn 090954
Penna, Sandra Carvalho, Belo
Horizonte 063915
Pennard House Antiques, East
Pennard 089141
Pennarz, Magdalena, Pfaffenhofen an der
Ilm 108373, 142300
Pennello, Cleveland 120447
Pennellotto Restauro, Firenze .. 131237
Penneshaw Maritime and Folk
Museum, Penneshaw 001400
Pennestri, Santa, Messina 131359
Penney, David, Bishop's Stortford 088490
Pennica, Gino, Modena 131579
Pennies, Topsham 091614
Pennies from Heaven, Göteborg 143851
Pennings, Eindhoven 112466
Pennings, Willy, Delft 112366
Pennisi, Milano 080665
Pennsbury Manor, Morrisville . 051071
Pennsylvania Academy of the Fine
Arts, Philadelphia 057674
Pennsylvania Academy of the Fine
Arts Gallery, Philadelphia 052078
Pennsylvania Anthracite Heritage
Museum, Scranton 053218
Pennsylvania College of Art and
Design, Lancaster 057320
Pennsylvania College of Art and
Design Galleries, Lancaster .. 050125
Pennsylvania Dutch Folk Culture
Society Museum, Lenhartsville 050218
Pennsylvania Federation of Museums
and Historical Organizations,
Harrisburg 060684
Pennsylvania German Cultural Heritage
Center, Kutztown 050022
Pennsylvania Lumber Museum,
Galeton 049024
Pennsylvania Military Museum and
28th Division Shrine, Boalsburg 047064
Pennsylvania Pine, Ashburton . 088310,
134041
Pennsylvania State University Press,
University Park 138477
Pennsylvania Trolley Museum,
Washington 054186
Penny Farthing, North Berwick . 144744
Penny Farthing Antiques, North
Cave 090848
Penny Farthings Antiques Arcade,
Swindon 091531
Penny Lane, Frederiksberg 140573
Penny Lane Antiques, Los
Angeles 094440
Penny Post Antiques, Buxton ... 088749
Pennyfarthing Antiques, Boston . 088520
Pennypacker Mills,
Schwenksville 053200
Pennyroyal Area Museum,
Hopkinsville 049555
Penny's Antiques, Northampton . 090856
Penny's Collectibles, Saint Louis 096762
Penobscot Marine Museum,
Searsport 053227
Penrhyn Castle, Bangor,
Gwynedd 043492

Penrith Coin and Stamp Centre,
Penrith 090958, 134928
Penrith Farmers' & Kidd's,
Penrith 127009
Penrith Museum, Penrith 045478
Penrith Regional Gallery and The
Lewers Bequest, Emu Plains .. 001064
Pensacola Historical Museum,
Pensacola 051982
Pensacola Museum of Art,
Pensacola 051983
Il Pensatoio, Albavilla 142617
Penselmuseet i Bankeryd,
Bankeryd 040557
Pensez au Passé pour vos Présents,
Casablanca 082578
Penshurst Place and Toy Museum,
Penshurst 045479
Pensieri nel Tempo, Milano 080666
Pensler, Washington 124946
Pentagon Gallery, Cleveland .. 120448
Pentecoste, Daniel, Villeurbanne 074222
Pentes, Charlotte 120005
Pentiment, Internationale
Sommerakademie für Kunst und
Gestaltung, Hamburg 055574
Pentimenti, Philadelphia 123438
Penwith Galleries, Saint Ives,
Cornwall 045697
Penzance Art Gallery, Penzance 118898
Penzance Auction House,
Penzance 127011
Penzance Rare Books, Penzance 144781
Penzenskaja Kartinnaja Galerija im.
K.A. Savickogo, Penza 037630
Penzenskij Gosudarstvennyj
Obedinennyj Kraevedčeskij Muzej,
Penza 037631
Penzenskij Muzej Odnoj Kartiny,
Penza 037632
Penzils, Saint Heliers 113005
Pénzügytörténeti Gyűjtemény, OTP
Bank, Budapest 023541
Peola, Alberto, Torino 110994
Peony Antiques, Norwood 062016
Peoples Art Hall, Dublin 109614
People's Fine Arts Publishing House,
Beijing 136699
The Peoples Gallery, Cork 109558
People's Gallery, El Paso 048515
The People's Gallery, Stalybridge 045856
The Peoples Gallery, Stowmarket 119165
People's History Museum,
Manchester 045216
The People's Museum, Belfast . 043561
People's Palace Museum,
Glasgow 044411
People's Story Museum,
Edinburgh 044261
Peoria Art Guild, Peoria 060685, 123345
Peoria Historical Society, Peoria 051988
Pep & No Name, Basel 116093, 144028
Pepe Merino, Lugo 086086, 133282
De Peperbusse, Oostende 100803
Pepe's Special, Vantaa 140757
The Pepin Press, Agile Rabbit Editions,
Amsterdam 137838
Pepin, Bruno, Paris .. 071769, 071770
Pépin, Guy, Amboise 066556
Pepino, Timişoara 085139
Pepita, Oslo 143412
Pepper Gallery, Boston 119912
Pepper Street Gallery, Magill .. 099251
Pepperall, Noel & Eva-Louise,
Arundel 134036
Peppercorn Collectibles, Hall ... 061574
Peppergreen Trading, Berrima .. 061141
Peppers Art Gallery, University of
Redlands, Redlands 052496
Peppers Creek Antiques,
Pokolbin 062086
Peppitt, Judith, Snargate 091355
Pepys Antiques, Beckenham ... 088421
Pequeña, Asunción 113335

Pequeña Inglaterra, Málaga 086301
Le Pequod, Lyon 104126
Pera Sanat Galerisi, İstanbul ... 117078
Pera, Maurizio Giovanni, Lucca . 142702
Perard, André, Arles 066725
Perault, Sophie, Toulouse 129312
Perazzone-Brun, Paris 071771
Perbadanan Muzium Melaka,
Melaka 030717
Perc-Peretz & Ball, Saarbrücken 078331,
126110
Perc Tucker Regional Gallery,
Townsville 001590
Perch Proshyan House-Museum,
Yegishe Charents Literature and Arts
Museum, Ashtarak 000691
Perchinelli, Felice, Genova 131334
Percival, R. & S.M., Ruthin 091156
Perco, Mario, Wien 063048
Percossi, Luisa, Milano 110397
Percy House Gallery,
Cockermouth 117608
Percy Pilcher Museum,
Lutterworth 045169
Percy, Kevin, Hawkes Bay 112916
Perdijk, Tilburg 083621
Perdikakis, Gus, Cincinnati ... 092885
Père, Christophe, Semur-en-
Auxois 073487, 129257
Peregrine Galleries, Santa
Barbara 097383
Pereira Rodrigues, Oscar, Lisboa 143590
Pereira, Ângelo Augusto, Lisboa 132949
Pereira, António Ferreira, Lisboa 084940
Pereira, Atta, Rio de Janeiro ... 064102
Pereira, Christina Cantinho,
Lagos 084810
Pereira, Maria I. Costa, Lisboa . 114052
Pereira, Mónica, Porto 114101
Peremiansky, Alberto, Buenos
Aires 139595
Perera, Robert, Lymington 090662,
118682
Peres, Berlin 106395
Pères Blancs – Saint Anne,
Jerusalem 025098
Peres, Cornelia M., Roma 132008
Peres, Javier, Los Angeles 121571
Pereslavl-Zalesskij Gosudarstvennyj
Istoriko- architekturnyj i
Chudožestvennyj Muzej-zapovednik,
Pereslavl-Zalesskij 037636
Peretz, Jean-Claude, Villeneuve-sur-
Lot 074170
Perey, Wollongong 139963
Perez, London 090414, 134748
Perez, San Antonio 136268
Perez & Pires, Porto Alegre ... 100995
Perez Arrans, Jose Antonio,
Sevilla 086468
Pérez Barón, Elias, Bogotá 065219
Perez Carrasco, Tomas, Madrid . 086234
Perez Cejas, Alberto, Montevideo 135274
Perez de Albeniz, Moises,
Pamplona 115390
Perez Gonzalez, Rafael,
Valladolid 115529
Perez-Hita Cugat, Pedro,
Barcelona 085822
Perez Paya, Valencia 086528
Pérez-Paya, Valencia 143808
Perez Perez, Guillermo, Zaragoza 133506
Pérez Sanleón, Antonio, Valencia 086529
Pérez, Andrés, Sevilla 086467
Perez, Ildefonso Lorite, Almeria . 114839
Pérez, Luis, Bogotá 102355
Perez, Mickael, Les Ulmes 069437
Pérez, Roberto, Medellín 065302, 128305
Pérez, Trinidad de, Bogotá 065218
Perfect Touch, Philadelphia ... 123439
Perfecto Da Lama, Saint-Pey-
d'Armens 073254
Perfekta, Kraków 084592
Perform Arte Contemporanea, La

Spezia 110158
Performance Space, Eveleigh . . 001069
Performing and Creative Arts
  Department, College of Staten Island
  Cuny, Staten Island 057944
Performing and Visual Arts
  Department, Marian College,
  Indianapolis 057235
Performing and Visual Arts
  Department, Wagner College, Staten
  Island 057945
Performing Arts Collection of South
  Australia, Adelaide 000769
Performing Arts Collection, The Arts
  Centre, Melbourne 001281
Perfume Museum, Niagara-on-the-
  Lake 006492
Pergamentum, Granada 133268
Pergamino, Madrid 086235
Pergamon, Bredevoort 143100
Pergamon Antique, Spišská Nová
  Ves 085592
Perge Müzesi, Aksu 042803
Perhirin, Jacques, Le Relecq-
  Kerhuon 069350
Perhon Kotiseutumuseo, Perho . 011244
Peribo, Mount Kuring-Gai 136484
Pericoi, Nathalie, Saint-Ouen . . . 073168
Perides, Maria, Newstead 099445
Perier, Antoine de, Nantes 140989
Périer, Charles, Le Chesnay . . . 069250
Périès, Jean-Claude, Montauban 070198,
  070199

Perimeter Gallery, Chicago 120229
Perine, Paris 071772
Perini, Alberto & Marcus, Verona 142887
Perini, Maurizio, Milano 131495
Period Antiques & Clocks, Seoul 082263
Period Classics, Singapore 085485
Period Collections, Houston 121109
Period Features, Leek 089787
Period Fireplaces, Bristol 088665,
  134164
Period Furniture Showrooms,
  Beaconsfield 088414, 134081
Period Gallery, Omaha 123336
Period Picture Company,
  Washington 124947
Period Style Lighting, Enfield . . 089211
Perkey, Cheryl, Los Angeles . . . 121572
Perkin, Bruce, Cork 079506
Perkins Bull Collection,
  Peterborough 006640
Perkins Center for the Arts,
  Moorestown 051044
Perkins County Historical Society
  Museum, Grant 049203
Perkins History Museum,
  Watertown 054214
Perkins, Bryan, Wellingborough . 091769,
  135197
Perkins, R.G., Toronto 064721
Perkins, Steve, Whangarei 132771
Perkinson Gallery, Millikin University,
  Decatur 048203
La Perla, Buenos Aires 060922
La Perle des Puces, Cagnes-sur-
  Mer 067590
Perler & Co., Wabern . 087889, 133959
Perlerin, Alain, Saint-Valéry-sur-
  Somme 073325
Perles & Terres, Cannes 103520
Perlès, Christophe, Paris 071773
Perlin, Amy, New York 095658
Perlinger, Magarethe, München . 108173
Perlini Arte, Reggio Calabria . . . 110704
Perlman, Trelleborg 087103
Perlo, Jean-Louis, Reims 072401
Perlow Rich, Katharina, New
  York 122963
Permanenten Vestlandske
  Kunstindustrimuseum, Bergen . 033538
Permanenze Centro di Restauro,
  Milano 131496

Permskaja Gosudarstvennaja
  Chudožestvennaja Galereja,
  Perm 037644
Permskij Gosudarstvennyj Institut
  Iskusstv i Kultury, Perm 056296
Permskij Oblastnoj Kraevedčeskij
  Muzej, Perm 037645
Pernáth, Praha 102716
Perner, Udo, Hof 141950
Pernes, Civray 067993
Pernès, Michel, Beaune 103354
Pernička, Vojtěch, Brno 128322
Perniön Museo, Perniö 011245
Pernkopf, Dr. Johannes, Berlin . 106396
Pernod, Jean-Pierre, Saint-Martin-du-
  Fresne 072947
Pernot, Alain, Vic-sur-Seille . . . 074100
Pero, Beograd 114433
Péroline, L'Hospitalet 069455
Peron, Los Angeles 094441
Peron – Galerie Českého Uměni,
  Praha 102717
Perpetuum Art, Kostroma 114185
Perpignan Numismatique,
  Perpignan 072091
Perplies, Aachen 105944
Perquier, Jacques, Pont-Audemer 072241
Perra, Andrea, Catania 080043
Perramond, Alain, Tain-
  l'Hermitage 073612
Perranzabuloe Folk Museum,
  Perranporth 045484
Perras, Bernard, Paris 071774
Perrault, Gilles, Paris . 071775, 129112
Perrault's Museum, Val Marie . . 007267
Perreau-Saussine, Paris 071776
Perreau, Patrick, Châteauroux . . 128551
Perret, Gustave, Saint-Ouen . . . 073169
Perret, Paul, Saint-Just-Saint-
  Rambert 072876
La Perricholi, Buenos Aires . . . 060923
Perrier, Jean-Claude, Laval
  (Mayenne) 069205
Perrier, Michel, La Boisse 068955
Perrier, Suzanne, Régny 072373
Perrin, Montbéliard . . . 104253, 140959
Perrin, Jacqueline, Nice 070647, 104384
Perrin, Jacques & Philippe, Paris 071777
Perrin, Josiane, Aix-les-Bains . . 066483
Perrin, Marie-Jo, Lyon 069772
Perrin, Michelle, Tournus 073828
Perrin, Noël, Jasseron 068896
Perrin, Patrick, Paris 071778
Perrins Art Gallery, London . . . 118507
Perritaz, Christian, Fribourg . . . 133748
Perron, Ralph, Saint-Jeoire . . . 072857,
  129237
Perrone, Antonio, Biel . 087321, 133695
Perrot, Bernard, Châteauneuf-en-
  Thymerais 067893
Perrotin, Emmanuel, Miami 121838
Perrotin, Emmanuel, Paris 105156
Perrusson, Christian, Sennecey-le-
  Grand 073493
Perry, Charlotte 092618
Perry, Jacksonville 121261
Perry & Nicole, Memphis 121707
Perry & Phillips, Bridgnorth . . . 126835
Perry County Lutheran Historical
  Society, Altenburg 046422
Perry-Downer House and Costume
  Gallery, Monterey 051016
Perry Oldtimers Museum,
  Emsdale 005783
Perry, George, Tauranga 113025
Perryman, Ian & Belinda,
  Woollahra 062479, 127701
Perry's Cider Mills, Ilminster . . . 044629
Perry's Mart Masterton,
  Masterton 083991
Perry's Victory and International Peace
  Memorial, Put in Bay 052419
Perryville Battlefield Museum,
  Perryville 051997

Per's Afsyring og Antik, Roskilde 066097
Persano, Giorgio, Torino 110995
Persault, Michel, Albertville . . . 066502
Il Perseo, Palermo 081064
Persepolis, Isfahan 079418
Persépolis, Palma de Mallorca . 086361,
  133409
Persian and Oriental Rug Gallery,
  Pittsburgh 096318
Persian Antiques Exhibition, Abu
  Dhabi 088162
Persian Arts & Crafts, Vancouver 064864
Persian Carpet Gallery of Petworth,
  Petworth 091004
Persian Carpet Museum, Raas . 030069
The Persian Carpet Studio, Long
  Melford 090625, 134811
Persian Carpets and Antiques, Abu
  Dhabi 088163
Persian Gallery, Boston 092488
Persian Gallery, New York 095659
Persian Shop, New York 095660
La Persiana, Palermo 110612
Persico Rosalba, Milano 080667
Persico, Didier, La Valette-du-Var 069113
Persico, Didier, Pignans 072163
Persmuseum, Amsterdam 031977
Person, Catherine, Seattle 124581
Personal Museum, Paris 105157
Personal Touch, Toledo 124694
Persone, Maria Adele, Bari . . . 130918
Personé, Maria Adele, Bari . . . 079777
Perspective Art, Ottawa 101478
Perspective Gallery, Blacksburg . 047029
Perspectives, Pont-Aven 105383
Persson, Malmö 143897, 143898
Persson, Åke, Piteå 126655
Persson, Lennart, Falun 133524
Persson, Oskar H., Helsingborg . 115687
Perth Antiques, Perth 062080
Perth Bookshop, Perth 144784
Perth Galleries, Subiaco 099730
Perth Institute of Contemporary Arts,
  Northbridge, Western Australia . 001368
Perth Museum, Perth 006632
Perth Museum and Art Gallery,
  Perth 045486
Perthuis, Serge, Saint-Arnoult-des-
  Bois 072677
Perticaro, Gaetano, Genova . . . 131335
Perttelin Kotiseutumuseo, Inkere 010906
Peruggi, Andrea, Padova 110575
Pesbar, Barbate 126500
Pesca, Nashville 095021
Pescali, Daniele, Milano 110398
Pesce Maurizio, Genova 080281
Pesce, Elena, Genova 080280
Pesce, Sabrina, Bari 079778
Pescetta, Verona 081931, 132924
Pescetta, Alessandro, Verona . . 081932
Pescetto, Rosa, Genova 080282
Peschar, Hannah, Ockley 118847
Pescheteau-Badin, Paris 125747
Peschke, J., Paderborn 078008
Pesciaioli, Luciano, Roma 081499
Pesenti, Bruno, Rochefort (Charente-
  Maritime) 072498
Peshawar Museum, Peshawar . 034145
Peshtigo Fire Museum, Peshtigo 052004
Peshwa Satishchandra, Paris . . 071779
Pešková, Jitka, Český Krumlov . 070612
Pesnel, Francette, Montgeron . . 070246
Pessah, Daniel, Fuengirola 086001
Pessin, Marc, Saint-Laurent-du-
  Pont 105542
Peštera Rabiša Muzej,
  Belogradčik 004982
Pesterzsébeti Múzeum, Budapest 023542
Petalax Hembygdsmuseum,
  Petalax 011247
Petaluma Wildlife and Natural Science
  Museum, Petaluma 052005
Pete and Susan Barrett Art Gallery,
  Santa Monica College, Santa

Monica 053134
Peter, Buxtehude 075239
Peter & Martine, L'Isle-sur-la-
  Sorgue 069586
Peter Anson Gallery, Buckie . . . 043779
Peter Benoitmuseum, Harelbeke 003676
Peter Chapman Antiques & Restoration,
  The Studio, London . . 090415, 134749
Peter Conser House, Oklahoma
  Historical Society, Heavener . . 049427
Peter et Martine, Beaumettes . . 067001
Peter J. McGovern Little League
  Baseball Museum, South
  Williamsport 053419
Péter Pál Polgárház, Koszta József
  Múzeum, Szentes 023966
Peter Pan's Bazaar, Gosport . . . 089350
Peter Paul Luce Gallery, Cornell
  College, Mount Vernon 051103
Peter Raczeck Fine Art, New
  York 122964
Peter Rice Homestead, Marlboro 050699
Peter Rosegger-Geburtshaus,
  Alpl 001699
Peter Rosegger-Museum,
  Krieglach 002244
Peter Scott Gallery, Lancaster . . 044754
Peter Van den Braken Centrum,
  Sterksel 032907
Peter Wentz Farmstead,
  Worcester 054550
Peter Whitmer Farm, Waterloo . 054210
Peter-Wiepert-Museum, Fehmarn 018343
Peter Yegen Jr. Yellowstone County
  Museum, Billings 046987
Peter-Zirbes-Haus Niederkail,
  Landscheid 019881
Peter, Margrit, Basel 087253
Pētera Upīša Piemiņas Muzejs,
  Dobele 030115
Peterborough Arthouse,
  Peterborough 118908
Peterborough Centennial Museum and
  Archives, Peterborough 006641
Peterborough Historical Society
  Museum, Peterborough 052006
Peterborough Museum and Art Gallery,
  Peterborough 045489
Peterburg, Sankt-Peterburg . . . 085259
Peterburgskaja Restavracionnaja
  Kompanija, Sankt-Peterburg . . 133021
Petergof – Gosudarstvennyj Muzej
  Zapovednik, Sankt-Peterburg . 037812
Peterhof, Paris 071780
Peterman, Kenneth O., Baltimore 092354
Petermandl, Herbert, Ybbsitz . . 127986
Peterpol, Sankt-Peterburg 137932
Peters, Geesthacht 075957
Peters Antik Aurorahuset,
  Göteborg 086715
Peters Antikk og Interiør,
  Kristiansand 084268
Peter's Auction Sales, Winnipeg 125393
Peters-Barenbrock, Sabine,
  Ahrenshoop, Ostseebad 105968
Peters Bookshop, Sheringham . . 144844
Peter's Galerie, Bad Segeberg . 106092
Peter's Gallery, Lemesos 102553
Peters Gallery & Framing,
  Calgary 101253
Peter's Oldies but Goodies, Saint
  Paul 096848
Peters, Bodo, Bad Breisig 074498
Peters, Christopher, Warwick . . 091747
Peters, Flo, Hamburg 107305
Peters, Gerald, Dallas 120640
Peters, Gerald, New York 122965
Peters, Gerald, Santa Fe 124467
Peters, Nina P, Hamburg 129268
Peters, P.L., Den Haag 083118
Peters, Paul M., Harrogate 089439
Peters, T., Rees 078159
Petersburg Area Art League,
  Petersburg 052014

The Petersburg Museums,
Petersburg 052015
Petersburg National Battlefield,
Petersburg 052016
Petersen & Partner, Berlin 106397
Petersen, Drochtersen 106750
Petersen, Salt Lake City 123977
Petersen Automotive Museum, Los
Angeles 050463
Petersen, Anke, West Sneem .. 109705
Petersen, Christiane, Neuburg .. 077716
Petersen, Hans J., Aabenraa 065576
Petersen, Helge, Middelfart 140626
Petersen, J., Miami 135804
Petersen, Kirsten & John,
Svendborg 066160
Petersen, Vibeke G., Svaneke .. 066153
Petersens, Omeo 099503
Petersfield Antiques Fair,
Petersfield 098399
Petersfield Bookshop, Petersfield 134937,
144785
Petersfield Museum, Petersfield 045495
Petersohn, Jürgen, Waldbronn .. 130730
Peterson, Houdan 068839
Peterson, Giles, South Auckland 113008
Peterstormuseum, Zwiefalten .. 022912
Pete's Furniture, Bangalore 079359
Pete's Furniture Refinishing,
Fresno 135572
Petes Museigård, Havdhem 040689
Pete's Upholstery Shop, Fort
Worth 135559
Petetin, Nicole, Saint-Claude ... 072715
Petéus, J. T. F., Göteborg 133531
Peth, Carl, Lübeck 077185
Petherick, Auburn 061070
Pethő, Budapest 079280, 109170
Petit Antiquites, Paris 071782
Petit Broc, Senaud 073489
Petit Dénicheur, Carcassonne .. 067674
Le Petit Fouillis, Hayange 068801
Petit Galerie, Amsterdam 112267
La Petit Galerie, Christian Starck,
Konstanz 107719
Le Petit Jaunais, Nantes 104342
Petit Musée, Conakry 023383
Petit Musée, Montréal 064458
Petit Musée Ambulant de la Musique
Mécanique, Une Journée au Cirque,
Villeneuve-la-Garenne 016266
Petit Musée d'Art Camerounais,
Yaoundé 005299
Petit Musée d'Automat et du Vin de
Pauillac, Pauillac 014815
Petit Musée de France, Albaret-Sainte-
Marie 011575
Le Petit Musée de la Bicyclette, La
Batie-Montgascon 013362
Petit Musée de la Chartreuse Saint-
Hugon, Arvillard 011756
Petit Musée de la Table, Paris .. 014798
Petit Musée de l'Argenterie, Ateliers
du Cuivre et de l'Argent, Paris 014799
Petit-Musée de Vin au Cave de la
Dive Bouteile, Bourgueil 012172
Petit Musée des Abeilles et du Miel-
De la Fleur à la Table, Alle-sur-
Semois 003262
Petit Musée des Chailloux, Chevagny-
les-Chevrieres 012553
Petit Musée des Silos de Jouques,
Jouques 013332
Petit Musée du Bizarre, Villeneuve-de-
Berg 016264
Petit Musée Fantastique de Guignol,
Lyon 013955
Le Petit Musée Maison de Mariette,
Saint-Laurent-du-Pont 015413
Petit Musée Rural, Varaire 016126
Petit Musée, Société Historique de
Clair, Clair 005621

Petit Museu Parroquial, Calonge de
Mar 039207
Petit Palais, Musée d'Art Moderne,
Genève 041541
Petit-Palais, Musée des Beaux Arts de
la Ville de Paris, Paris 014800
Le Petit Saint-Ouen, Montréal .. 064459
Petit Schorp, Stéphanie, Saint-Martin-
de-Nigelles 072934
Le Petit Soldier Shop, New
Orleans 095142
Petit, André François, Paris 105158
Petit, Francis, Brévonnes 067485
Petit, Francis, Clermont-Ferrand 103607
Petit, Marcel, Biécourt 067217
Petit, Pierre, Saint-Malo (Ille-et-
Vilaine) 105552
Petit, Thomas, Paris 071781
Petitcollot, Bruno, Paris 071783
La Petite, Paris 105159
La Petite, Peine 108371
La Petite Anglicane, Forestville . 005813
Petite Brocante, Mer 070066
La Petite Brocante, Rochefort
(Charente-Maritime) 072499
Petite Brocante, Saint-Malo (Ille-et-
Vilaine) 072922
La Petite Curieuse, L'Isle-sur-la-
Sorgue 069587
La Petite Galerie, Bruxelles 100649
La Petite Galerie, Paris 105160
Petite Galerie, Reims 105426
Petite Galerie, Rueil-Malmaison . 105489
Petite Hirondelle, Chartres 067861
Petite Maison, Antibes-Juan-les-
Pins 066678
Petite Maison de Le Corbusier,
Corseaux 041428
La Petite Provence du Paradou, Le
Paradou 013677
Petite Souris, Den Haag 132478
Les Petites Ecuries, Maisons-
Laffitte 069816
Les Petites Folies, Tournai 063848
Les Petites Franciscaines de Marie,
Baie-Saint-Paul 005371
Petites Puces, Boulogne-Sur-Mer 067404
Petitnicolas, Robert, Nice 070648
Petiton, Jean-Jacques, Eulmont 068488
Petkov, Stan, Stockholm 133577
Petkov, Vasil, Šumen 140327
Petković, Beograd 085344, 114434
Petley-Jones, Vancouver 101823, 128228
Petley, Roy, London 118508
Petley, Roy, Monaco 112014
Petőfi Emlékmúzeum, Kiskőrös . 023742
Petőfi Irodalmi Múzeum,
Budapest 023543
Petőfi Múzeum, Aszód 023409
Petőfi Sándor Emlékház, Dömsöd 023604
Petőfi Sándor Emlékmúzeum,
Szalkszentmárton 023915
Petőfi Sándor Emlékszoba,
Kiskunfélegyháza 023746
Petone Settlers Museum, Te
Whare Whakaarao o Pito-one,
Wellington 033358
Petra Fine Art, Baltimore 119764
Petra Museum, Petra 029783
Petraglia, Antonella, Roma 081500
Petran Museum, Haltdalen 033632
Petrangeli, Valter, Roma 081501
Petrefaktensammlung, Hanns-Seidel-
Stiftung, Bad Staffelstein 016884
Petrej, Peter, Zürich 144192
Petrequin, Daniel, Paris 141258
Petri Veijalainen, Verhoomo,
Helsinki 128415
Pétrides, Gilbert & Paul, Paris .. 105161
N. Petridis & Ch. Petridou,
Thessaloniki 079215
Petrie Museum of Egyptian
Archaeology, University College
London, London 045073

Petrified Creatures Museum of Natural
History, Richfield Springs 052526
Petrified Forest of the Black Hills,
Piedmont 052128
Petrik & Rinnenburger, Frankfurt am
Main 075812
Petrikat, Peter, Lüdenscheid ... 077200
Petrini, Giovanna, Firenze 080185
Petro Cvirkos Memorialinis Muziejus-
Sodyba, Veliuona 030502
Petro, Paul, Toronto 101682
Pétroff-Rançon, Dinan 068352
Petroff-Rançon, Louis, Rennes (Ille-et-
Vilaine) 105440, 129204
The Petroleum Museum, Midland 050879
Petrolia Discovery, Petrolia 006644
Petropavlovskaja Krepost – Istoriko-
kulturnyj Zapovednik, Sankt-
Peterburg 037813
Petropol, Sankt-Peterburg 114354
Petrosemolo, Alberto, Roma ... 132009
Petrosso, Davide, Torino 132179
Petrouchka, Paris 071784
Petrović, Roman, Sarajevo 100869
Petrovskij Vernisaž, Moskva ... 114290
Petroz Fine Art, Genève 126728
Petroz, Jules & Aïcha, Genève . 087493
Petroz, Jules & Aïcha, Saint-Julien-en-
Genevois 072867
Petrů, M., Praha 140527
Petruni, Pietro, Milano 131497
Petryńska, Ewa, Wrocław 132916
Pettaquamscutt Museum,
Kingston 049987
Petteplace, Allan, Hamilton ... 128174,
140349
Petter Dass-Museet,
Sandnessjøen 033925
Petter, Eva-Maria, Nußbach ... 062750
Petter, Rainer, Berlin 074896
Petters, Nils, Båstad 115580
Pettersen, Inge, Molde 126359
Petterson, Charleston 127157
Petterson Museum of Intercultural Art,
Claremont 047757
Petterson Topdanmark, Flemming,
Horsens 065798
Pettersson, Inge, Sigtuna 115808
Petticoat Lane, Ostfildern 077992
Petticoat Lane, Subiaco 062259
Pettigrew & Mail, Glasgow 117856
Pettinelli, Alberto, Milano 131498
Pettini, Antonio, Roma 142829
Pettis County Historical Society,
Sedalia 053259
Pettit, Anthony, Flatts . 063857, 140265
Pettler & Lieberman, Los Angeles 145235
Petviashvili, Rusudan, Tbilisi ... 105920
Petworth Antique Centre,
Petworth 091005
Petworth Antiques and Fine Art Fair,
Duncton 098400
The Petworth Cottage Museum,
Petworth 045498
Petworth House and Park,
Petworth 045499
Petzel, Friedrich, New York ... 122966,
122967
Petzold, Augsburg 125852
Petzuch, Johannes, München .. 130380
Petzwinkler, Gerhard, Sittendorf bei
Wien 127874
Peucetia, Giovinazzo 142696
Peuchet, Daniel, Pleyber-Christ . 072710
Peuker, A., Rödinghausen 078259
Peuples du Monde, Paris 136974
Peute, P., Dordrecht 083191
Pevnost Josefov – Podsemní Chodby,
Jaroměř II. Josefov 009596
Pew Corner, Guildford 089397
Pewabic Pottery Museum, Detroit 048316
Pewsey Heritage Centre, Pewsey 045501
Pewter Society, Monmouth 060209
Peyer, Zürich 116845

Peyman, Los Angeles 121573
Peyralade, Marie-Thérèse, Paris 071785
Peyre-Cornu, Elyette, Béziers .. 067180
Peyre, Jean-Gabriel, Paris 071786
Peyrecave, Anglet 066624, 128444
Peyrègne, Patrick, Saverdun ... 073448
Les Peyrolliers, Genève 087494
Peyrolliers Muller, Genève 087495
Peyron, Béatrix, Souppes-sur-
Loing 073552
Peyronnet, Michel, Lanobre ... 069184
Peyronny, Jacqueline-Paule,
Courtenay 068198
Peyrot, Torino 142867
Peyrou, Eric, Argentat 066711
Peyroulet & Cie., Gilles, Paris .. 105162
Peyton-Smith, Houston 127254
Pezellier, Jérôme, Tourcoing ... 073816
Pezon, Fabrice, Saussignac ... 073441
Pezzino, Fabrizio, Palermo 131708
Pfadenhauer, A., Hof 076550
Pfälzisches Bergbaumuseum,
Imsbach 019398
Pfaff, Weinheim 078910
Pfaffenweiler Presse,
Pfaffenweiler 137471
Pfahlbaumuseum Unteruhldingen,
Uhldingen-Mühlhofen 022256
Pfahlbausammlung Dr. h.c. Carl Irlet,
Twann 042073
Pfalzmuseum, Forchheim,
Oberfranken 018381
Pfalzmuseum für Naturkunde,
POLLICHIA-Museum, Bad
Dürkheim 016751
Pfanne-Dreesen, Sabine,
Seevetal 130637
Pfannkuche, Christoph Oldendorf 078089
Pfannstiel, Axel, Baden-Baden . 129434
Pfanzelt, Linde, Villach 140050
Pfarr- und Wallfahrtsmuseum,
Mariapfarr 002362
Pfarr, Peter, Bad Münder am
Deister 129421
Pfarrmuseum, Dießen am
Ammersee 017889
Pfarrmuseum, Flintsbach am Inn 018374
Pfarrmuseum, Sankt Georgen im
Attergau 002658
Pfefferkorn, Anne, Berlin 129548
Pfefferle, Karl, München 108174, 130381,
137419
Pfefferminzmuseum, Eichenau . 018114
Pfeifenmuseum, Sankt Aegyd am
Neuwalde 002645
Pfeifenmuseum, Scheifling 002719
Pfeifenmuseum, Windigstein ... 003124
Pfeifer, Edgar, Schleiz 078368
Pfeifer, Erwin, Hamburg 141890
Pfeifer, Hildegard, Mühlacker .. 077423
Pfeifer, Robert A., Karlsruhe ... 107521
Pfeifer, Sven, Berlin 129549
Pfeifermuseum im alten Rathaus,
Werbach 022567
Pfeiffer, Leonberg 107859, 130239
Pfeiffer, Lörrach 142080
Pfeiffer, Philadelphia 123440
Pfeiffer, A., Kipfenberg 076798
Pfeiffer, C., Bruxelles 063526
Pfeiffer, Karl-Heinz, Schwäbisch
Hall 078409
Pfeiffer, Olivier, Colmar 068094
Pfeiffer, Renate, Zeuthen 130802
Pfeil, Heilbronn 076476
Pfeil, Antje Gräfin von, Potsdam 078081
Pfeilburg Museum, Fürstenfeld . 001912
Pfeiler, Anton, Simbach am Inn . 142364
Pfennig, Ingeborg, München ... 077594
Pferde-Eisenbahn Museum, Rainbach
im Mühlkreis 002563
Pfeuffer, Georg, Landshut 077017
Pfiffner, Sandra, Erlen 087411
Pfingst Antik, Radolfzell am
Bodensee 098164

Pfingsten Publishing, Seven Hills 138473
Pfingstritt-Museum Kötzting,
  Kötzting 019749
Pfinzgaumuseum, Karlsruhe ... 019515
Pfister, Amstelveen 132387
Pfister, A.-M., Basel 144029
Pfister, Michael, Weiterstadt ... 078919
Pfister, Oswald, Eibelstadt ... 129778
Pfister, Peter, Untrasried 130717
PFL, Padova 080999
Pflaster, Lola, Dallas 093202
Pfleger, Lothar, Mannheim 130280
Pflüger, Fernando, Kassel 076725
Pflügl, Johnny von, London 090416
Pfo, Busan 111575
Pfortune, Indianapolis 121210
Pforzheimer Kunst-und Auktionshaus,
  Pforzheim 126095
Pfrang, Würzburg 079088
Pfrang, L. & P., Estenfeld 075715
Pfrang, Walter, Würzburg 079089
PFT Trading, Christchurch 083889
Pfundt, Sabine, Berlin 106398
PG Antiques, Los Angeles 094442,
                             135738
PG ART, İstanbul 117079
PGM Art World, Garching bei
  München ... 107100, 137220, 137221
phæno-Die Experimentierlandschaft,
  Wolfsburg 022745
Phänomenta Lüdenscheid,
  Lüdenscheid 020171
Phaidon Press, London 138238
Phal, Christine, Paris 105163
Phaliashvili Museum, Kutaisi ... 016392
Phalippou, Marc, Rouen 072608
Phalsbourg d'Autrefois,
  Phalsbourg 072149
Phantique, Exeter 089235
Phantom of the Attic, Pittsburgh 096319
Pharaoh's Finds, Buffalo 092551
Pharaoh's Gallery, Calgary 101254
Pharaoh's Tomb & Burra Antique
  Centre, Burra 061240
The Pharaonic Village, Giza ... 010573
Pharma- und Drogistenmuseum,
  Wien 003065
La Pharmacopée Antiquités, Lyon 069773
Pharmaka Art Group, Los
  Angeles 060686
Pharmaka Gallery, Los Angeles . 050464
Pharmazie-Historisches Museum der
  Universität Basel, Apothekenmuseum,
  Basel 041283
Pharoahs Antiques Centre,
  Eastbourne 089152
Phase, Bologna 130988
Phase II Gallery, Chicago 120230
Phathai Cave, Lampang 042694
The Phatory, New York 122968
Pheasant Valley Gallery, Katikati 083972
Phelan, Gerard, Portlaoise 079042
Phelps, Twickenham .. 091677, 135167
Phelps House, Burlington 047318
Phelps Store Museum, Palmyra . 051901
Phelps, T.L., Scotton 135012
Phenix 1880–1990, Antwerpen . 063232
Phenotype Books, Penrith 144779
Phi Editions, Herborn 137808
Phidias, México 111978
Phila China, Hong Kong 065090
Philabaum, Tucson 124767
Philadelphia Antique Exchange,
  Philadelphia 096147
Philadelphia Art Alliance,
  Philadelphia 052079, 060687
Philadelphia Estate Antiques,
  Philadelphia 096148, 127394
Philadelphia Mummers Museum,
  Philadelphia 052080
Philadelphia Museum of Art,
  Philadelphia 052081
Philadelphia Museum of Judaica,
  Congregation Rodeph Shalom,

Philadelphia 052082
Philadelphia Print Shop,
  Philadelphia 136098, 145417
Philadelphia Sketch Club,
  Philadelphia 060688
Philadelphia Society for the
  Preservation of Landmarks,
  Philadelphia 052083
Philadelphia Trading Post,
  Philadelphia 096149, 127395
Philadelphia Upholstery,
  Philadelphia 136099
Philbrook Museum of Art, Tulsa 053850
Philicia Antiques, Kings Park ... 061693
Philim, Limoges 069530
Philip and Muriel Berman
  Museum of Art, Ursinus College,
  Collegeville 047875
Philip L. Wright Zoological Museum,
  University of Montana, Div. of
  Biological Sciences, Missoula . 050964
Philip & Sons, Trevor, London . 090417,
                  118509,   134750
Philipp-Matthäus-Hahn-Museum,
  Albstadt 016496
Philipp Reclam jun., Ditzingen . 137142
Philipp-Reis-Gedächtnisstätte,
  Museum für Post und
  Kommunikation Frankfurt am Main,
  Friedrichsdorf 018525
Philipp-Reis-Haus Friedrichsdorf,
  Friedrichsdorf 018526
Philipp Schäfer II Museum,
  Heimatmuseum Erfelden,
  Riedstadt 021425
Philipp, G., Quakenbrück 108421
Le Philippard, Narbonne 070494
Philippe, Croix 068247
Philippe, Auguste, Châteauneuf . 067889
Philippe, Catherine, Sancerre . 073401,
                             105650
Philippe, Eric, Paris 071787
Philippe, Monique, Paris 071788
Philippe's French Antiques,
  Auckland 083815
Philippi, Julia, Dossenheim 106712
Philippi, Markus, Scheidegg 130620
Philippine Air Force Aerospace
  Museum, Pasay 034568
Philippine Army Museum, Makati 034507
Philippine House Museum,
  Shepparton 001494
Philippine Navy Museum, Cavite
  City 034454
Philippine Presidential Museum,
  Malacañang 034510
Philippine Radiation Science Museum,
  Manila 034542
Philippine Science Centrum,
  Manila 034543
Philippine Science Centrum,
  Marikina 034552
Philips, Malvern, Victoria 125211
Philips, Manchester 118711
Het Philips Gloeilampenfabriekje anno
  1891, Eindhoven 032289
Philips Museum Eindhoven,
  Eindhoven 032290
Philipse Manor Hall, Yonkers ... 054578
Philipsen, Bent, Hjørring 065766
Philipsen, Piet, Tilburg 083622
Phillip Antiques, Phillip 062084
Phillip Island Community Art and Craft
  Gallery, Cowes 000096
Phillip, Allen, Eastwood 061432
Phillips, Fort Worth 120880
Phillips, Mosman 099389
Phillips, Mumbai 079408, 142592
Phillips, Roleystone 099649
Phillips, Saint Louis 096763
Phillips, Salt Lake City 123978
Phillips, Tulsa 097753
The Phillips Collection,
  Washington 054157

Phillips Countryside Museum,
  Brokerswood 043767
Phillips County Museum, Helena 049429
Phillips County Museum, Holyoke 049517
Phillips Historical Society Museum,
  Phillips 052107
The Phillips Museum of Art,
  Lancaster 050126
Phillips of Hitchin, Hitchin 089538,
                             144479
Phillips, Brian, Christchurch ... 143310
Phillips, de Pury & Co., Berlin . . 135808
Phillips, de Pury & Co., Genève 126729
Phillips, de Pury & Co., London . 126969
Phillips, de Pury & Co., München 126063
Phillips, de Pury & Co., New
  York 127356
Phillips, de Pury & Co., Paris ... 127356
Phillips, Peter, San Diego 124154
Phillips, Ronald, London 090418
Phillips, S.J., London . 090419, 134751
Philly Coins, Philadelphia 096150
The Philmont Museum, Seton
  Memorial Library and Kit Carson
  Museum, Cimarron 047735
Philo Fine Arts, Kunst & Theorie,
  Hamburg 137251
Philobiblon, Firenze 131238
Philobiblon, Milano 142737
Philobiblon, Roma 142830
Philosophische Fakultät I,
  Sozialwissenschaften und historische
  Kulturwissenschaften, Halle,
  Saale 055553
Philothée, Annecy 066649
Philp, Richard, London 090420
Phils Arts and Antiques, Quezon
  City 134755
Phil'z, Toronto 064722, 064723
Phimasone, Tarbes 073628
Phiphitthaphanthasathan Haeng
  Chat Ram Khamhaeng Sukhothai,
  Sukhothai 042721
Phipitapan Haeng Chart Phra Pathom
  Chedi, Nakhon Pathom 042700
Phipitapan Laeng Chart, Suphan
  Buri 042723
Phipitapantasatan Haeng Chart
  Nakhon Si Thammarat, Nakhon Si
  Thammarat 042703
Phipittapan Tasatan Haengchart Khon
  Kaen, Khon Kaen 042692
Phippen Museum of Western Art,
  Prescott 052351
Phippen, Guy, Bois-Guillaume . 067270
Phippen, Hubert, Malaunay 069823
Phô, Co, Ha Noi 125090
Phô, Hoa, Ho Chi Minh City ... 125143
Phô, Ngán, Ha Noi 125091
Phocéa, Marseille 104193
Phoebe Apperson Hearst Museum of
  Anthropology, University of California,
  Berkeley, Berkeley 046941
Phoebus, Rotterdam 112655
Phoebus, Tempe 139476
Phoénix, Fort-de-France 082455, 132350
Phoenix, Levin 083978
Phoenix, Paris 071789
Phoenix 369, Edinburgh 117738
Phoenix Ancient Art, Genève ... 087496
Phoenix Ancient Art, New York . 095661
Phoenix Antiques, Fordham 089292
Phoenix Antiques, Sherborne ... 091305
Phoenix Antiques, Tunbridge
  Wells 091660
Phoenix Art Conservation, San
  Antonio 136269
Phoenix Art Gallery, Lavenham . 118079
Phoenix Art Group, Phoenix 123508,
                             138439
Phoenix Art Museum, Phoenix . 052120
Phoenix Books and Records, San
  Francisco 145564
Phoenix Cabinet &Co., Phoenix . 136120

Phoenix Cultural Artifacts Center,
  Tainan 042491
Phoenix Design, Washington ... 124948
Phoenix Egyptian Imports,
  Phoenix 096231
Phoenix Fine Arts, Richmond, North
  Yorkshire 118967
Phoenix Gallery, New York 122969
Phoenix Museum of History,
  Phoenix 052121
Phoenix Potteries, Saint Louis . 123855
Phoenix Rising Gallery, Seattle . 124582
Phoenix Stamp and Coin,
  Phoenix 096232
Phoenix Trading, Frinton-on-Sea 089311
Phonothèque Québécoise – Musée du
  Son, Montréal 006412
Phorkys, Athinai 137576
Photage, Fort Worth 120881
Photantiques, Bowral 061183
Photantiques, Mittagong 061868
Photo Antiquities, Museum of
  Photographic History, Pittsburgh 052170
Photo Art Studio, Dubai 117185
Photo España, Madrid 059871
Photo Factory Gallery, San Diego 124155
Photo Forum, Pittsburgh 123567
Photo Gallery On Line, Miami .. 121839
Photo-London, London 098401
Photo Posters, Cincinnati 120366
Photo Suisse – Association
  Suisse pour la Photographie,
  Schweizerischer Verband für die
  Fotografie, Aedermannsdorf . 059954
Photocollect, New York 122970
PhotoFolio, Saltaire 119056
Photoglob, Zürich 138117
Photograph Scotland, Callander . 117497
Le Photographe, Paris 138656
Photographers Gallery, Denver . 120755
The Photographers' Gallery,
  London 045074
Photographers Gallery, Saskatoon 101548
Photographers' Gallery, South
  Yarra 099708
Photographia-Museu Vicentes,
  Funchal 035711
Photographic Archives, University of
  Kentucky, Lexington 050250
Photographic Art Gallery, Marino 099283
Photographic Center Northwest 900
  12th Av, Seattle 124583
Photographic Gallery, Dallas ... 120641
Photographic Image Gallery,
  Portland 123658
Photographic Investments Gallery,
  Atlanta 046630
Photographic Recollections, Broken
  Hill 098800
Photographic Resource Center,
  Boston 047130
Photographic Works of Art,
  Chicago 120231
Des Photographies, Paris 105164,
                             141259
Photographische Gesellschaft,
  Wien 058306
Photographs Do Not Bend, Dallas 126642
Photography Gallery, Indianapolis 121211
Photography Gallery and Studio,
  Indianapolis 121212
Photography Monika Mohr Galerie,
  Hamburg 107306
Photography Now, Journal für
  Fotografie und Videokunst,
  Berlin 138749
Photography Platinum Gallery, New
  York 122971
The Photography Show, New
  York 098491
Photohistorische Sammlung,
  Penk 002492
Photology, Milano 110399
Photolucida, former: Photo Americas,

Portland 098492
Photomuseum, Zarautz 040458
Photomuseum des Landes
  Oberösterreich, Oberösterreichische
  Landesmuseen, Bad Ischl 001753
Photonet, Wiesbaden 108927
Photoplatz, Berlin 106399
Phot'Optic, Marseille 069958
Photoscapes, Urangan 099850
Photospace, Wellington 113085
Photowork e.V., Wiesbaden ... 059218
Photoworld, Houston 121110
Phoung, Dong, Ho Chi Minh City 125144
Phra Borommathat National Museum,
  Nakhon Si Thammarat 042704
Phra Chetuponwimonmangkhalaram
  National Museum, Bangkok .. 042674
Phra Phuttachinnarat National
  Museum, Phitsanulok 042713
Phukradung, Loei 042697
Phulsanda Religious Arts Museum,
  Phulsanda 024435
Phuong, Ha Noi 125092
Phuthadikobo Museum, Mochudi 004298
Phyletisches Museum, Friedrich-
  Schiller-Universität, Jena 019454
Phyl's Memory Lane,
  Jacksonville 094024
Physics Museum, University of
  Queensland, Saint Lucia 001476
Pi Gallery, Kraków 113626
Pi Yandong Xue Yi Zhi Display Center,
  Guilin 007821
Pi.Ga., Roma 081502
Pia, Francesca, Zürich 116846
Piacenti, Prato 131781
Piacenti Art Gallery, Firenze ... 080186
Il Piacere e il Dovere, Vercelli .. 142885
Piage & Pieta, Atlanta 119600
Piaget, Jean-Marc et Eva, Lutry 087627,
  133827
La Pialassa, Marina Romea 110204
La Pialla, Bologna 079867, 130989
La Piana, Massimo, Palermo ... 110613
Pianet' Arte, Massagno 087666
Pianguan Museum, Pianguan .. 008146
Piani, Pieri, Bologna 142638
El Pianillo, Sevilla 086469
Piano Auctions, London 126970
Piano Collection, Hereford 044575
Piano Nobile, Genève 116335
Piano Nobile, Kraków 113627
Piano Nobile, London . 090421, 118510,
  134752
Piano Shop, Leeds ... 089772, 134537
Pianoforte-Museum, Meißen ... 020348
Pianola Museum, Amsterdam .. 031978
Pianos, Saint-Ouen 073170
Pianos Auctions, Bedford 126815
Pianos Galore, Little Horsted ... 089846,
  134565
Piano's Unlimited, Las Vegas .. 135687
Piaohao Museum, Pingyao 008157
Pia's Antique Gallery, Nashville . 095022
PIASA, Paris 125749
Piat, Bernard, Périgueux 072060, 072061
Piatt Castles, West Liberty 054312
Piatt County Museum, Monticello 051031
Piatti, Luca, Milano ... 080668, 142738
Piave Antichitá, Milano 080669
Piazza on the Boulevard, Surfers
  Paradise 099739
Piazza Vecchia Moveis e Decoracoes,
  Rio de Janeiro 064103
Piazzini, Roma 081503, 132010
Piazzola, Alain, Ajaccio 140762
Piazzotto Montevecchio, Bassano del
  Grappa 109783
Pibe Mølle, Helsinge 010105
Pic Aepli Priska, Steinach 116663
Pic du Midi, La Mongie 013441
Le Pic Verre, Bruxelles 063527
Pica Pica, Sundbyberg 143974
Pica, Alvaro, Roma 132011

Picalle, Oostende 100804
Picard, A. & J., Paris . 136975, 141260
Picard, C., Ecouis 068426
Picard, Henri & Fils, Paris 141261
Picard, Jean-Louis, Paris 125750
Picard, Patrice, Rouen 072609
Picasso Art Gallery, Lefkosia .. 102541
Picasso Art Gallery and Art Coaching
  Centre, Hyderabad 109368
Picasso Brichetto, Marcello,
  Modena 080815
Picasso Gallery, Washington .. 124949
Picasso Graphic Station,
  Hiroshima 111261
Picasso-Museum, Luzern 041713
Picasso, Carlo, Genova 131336
Piccadaci, Lilo, Montigny-lès-
  Metz 070256
Piccadilly, Narbonne 070495
Piccadilly Antiques, Batheaston . 088407,
  134080
Piccadilly Arts, Houston 121111
Piccadilly Jewellers, Birmingham 088484
Piccapane, Raymond, Nice 128868
Picchio, Arcegno 116029
Picciati Armi, Roma 081504
Fratelli Piccinini, Bologna 079868,
  130990
Piccioni, Cristiana, Roma 131012
Piccirilli, Francesco, Roma ... 081505
La Piccola, Bologna 109859
Piccola Galleria, Pieve di Teco .. 110658
La Piccola Galleria, Savona 110919
Piccolezze, Bologna 079869
Piccoli Emmer, Nicoletta, Venezia 081866
Piccolo Antiquarium, Fossato di
  Vico 026340
Piccolo Mondo Antico, Milano .. 080670
Piccolo Mondo Antico, Verona .. 081933
Piccolo Museo dei Popoli, Lodi . 026618
Piccolo Museo della Moneta,
  Vicenza 028371
Piccolo Museo di San Paolo, Reggio
  Calabria 027499
Piccolo Museo Navale, Laives .. 026545
Piccolo Museo Parrocchiale,
  Torgnon 028109
Piccolo Museo Parrocchiale,
  Valsavarenche 028265
Al Piccolo Naviglio, Milano ... 080671
Picconi, Jean-Maxime, Lausanne 087582
Piccot, François, Yvoire 074316
Pich, Peter, München 108175
Pichard, Julien, Neuville-de-
  Poitou 070527
Pichat, Marcel, Charmes-sur-
  l'Herbasse 067854
Pichlbauer, Rudolf, Mödling .. 062741,
  125263
Pichler, Anton, Saalfelden am
  Steinernen Meer 127842
Pichler, Eva M., Wien 063049
Pichler, Josef Balthasar,
  Großkirchheim 127760
Pichler, Manfred, Wien 063050
Pichot, Patrice, Trets 129323
Pici Galerie, Seoul 111704
Piciucchi, Massimiliano, Roma . 081506
Pick & Poke, Clarence Park ... 061328
Pick Szalámi és Szegedi Paprika
  Múzeum, Szeged 023929
Pickard, Oklahoma City 123292
Pickard, Nancy, Saint Just 119044
Pickens County Museum of Art History,
  Pickens 052126
Picker Art Gallery, Hamilton .. 049327
Pickering & Chatto, London .. 144664
Pickering Antique Centre,
  Pickering 091015, 134943
Pickering Museum Village,
  Pickering 006645
Picket Fence Antiques,
  Sacramento 096640
Pickett & Watts, Jacksonville . 121262

Pickett, Graham, Stamford 091406
Pickford, Paul, Hayfield 089481
Pickford's House Museum, Derby 044084
Pickles, Parramatta 062065
Pickles Auctions, Fyshwick 125195
Pickles Auctions, Salisbury Plain 125232
Pickup, David, Burford 088708
Pickwick, Katoomba 139788
Pickwick Antiques, Preston 062121,
  127663
Pickwick's Choice, Toronto 064724
Picma, Padova 110576
Picma, Sona 110935
Picma, Torri di Quartesolo 111040
Picma, Verona 111193, 111194
Picoche, La Bresse 068958
Picpus, Montreux 116513
Picta, Firenze 131239
Picton Community Museum,
  Picton 033282
Picton Street Traders, South
  Auckland 084083
Pictor, Carcassonne 103529
Pictorials, Shrewsbury 119096
Pictorica Galega, Vigo 115538
Pictura, Dordrecht 111440
Pictura, Gent 063656, 125336
Pictura, Groningen 112499
Pictura, Praha 102718, 125424
Pictural, Monaco 112015
Picture & Pottery, Cape Town .. 114772
The Picture Business, Bristol ... 117450
Picture Conservation and Restoration
  Studios, Bosham 134126
Picture-Esq, Pocklington 118928
The Picture Factory, Heidelberg
  West 099125
Picture Frames, Eschwege 106901
Picture Gallery, Bhaktapur 031870
The Picture House, Crewe 117644
Picture House, Iklin 111914
The Picture House, Loughton .. 118665
The Picture House, Penryn 118891
The Picture House, Saint Ives,
  Cornwall 119028
The Picture House, Saint Mawes 119045
Picture House Gallery, Bury Saint
  Edmunds 117492
Picture House Gallery, Charlotte 120006
Picture House Gallery, Padstow . 118875
Picture It, Albuquerque 119505
Picture Land, Austin 119689
Picture Map Company, Honolulu 120962
Picture Nook Gallery, Prebbleton 112994
Picture Options Gallery, Winnipeg 101891
Picture Parade, Sheffield 119085
Picture Paradise, Miami 121840
The Picture Parlour, Boscastle .. 117396
Picture Parts, Virginia Beach .. 124838
Picture Perfect, Indianapolis ... 121213
Picture Perfect, Milwaukee 121945
Picture Perfect, Phoenix 123509, 123510
Picture Perfect, Saint Paul 123923
Picture Perfect Art, Miami 121841
Picture Perfect Gallery, Charlotte 120007
Picture Perfect Gallery, Ilford .. 118009
Picture Perfect Gallery, London . 118511
Picture Restoration Studio, San
  Francisco 136340
Picture Restoration Studios,
  Arundel 134037
Picture Restoration Studios,
  Oxford 134919
The Picture Shop, Chorley 117588
Picture Shop, Wellington 113086
Picture Source Northwest,
  Seattle 124584
Picture This, Hong Kong 065091, 140442
Picture This, Indianapolis 121214
Picture This, Providence 123709, 145471
Picture This, Sandown 119058
Picture This, Sherwood Park .. 101554
Picture This, Wimborne 119356

Picture This, Windsor, Ontario .. 101860
Picture Us, Buddina 098814
Picturecraft Gallery, Holt, Norfolk 117986
Pictures, Bordeaux 140801
Pictures & Frames, Camberley . 117500
Pictures & Frames, Craigavon .. 117636
Pictures & Frames, Crowthorne . 117652
Pictures & Frames, Farnham ... 117789
Pictures & Frames, Fleet 117801
Pictures & Frames, Guildford .. 117896
Pictures & Frames, Maidenhead 118695
Pictures & Frames, Wokingham 119376
Pictures & Things, Auckland ... 112804
Pictures Are Beautiful, Houston . 121112
Pictures Perfect – Downtown,
  Seattle 124585
Pictures Perfect – North, Seattle 124586
Pictures Plus, Honolulu 120963
Pictures To Delight, Newport,
  Essex 118797
Pictures to Go, Minneapolis ... 122023
Picturesque, Billericay 117363
Picturesque Photography,
  Pittsburgh 123568
Pictus Verlag, Schobüll 137498
Piddig Museum, Piddig 034570
Piderit, Kurt, Zürich 133996
Pie Riera, Eugenio, Barcelona .. 085823
Piece Hall Art Gallery, Halifax .. 044503
Piece Hall Fine Art, Halifax 117904
Pièce Unique, Paris 105165, 105166
Pieces of Olde Antiques,
  Louisville 094579
Piechura, Jan W., Warszawa ... 113886
Piécourt, Marie, Rueil-Malmaison 072641
Piedmont, Charlotte 135393
Piedmont Arts Museum, Piedmont Arts
  Association, Martinsville ... 050726
Piedmont Lane Gallery, Oakland 123261
Piedra Lumbre Visitors Center, Ghost
  Ranch Living Museum, Abiquiu 046312
Piedrahita, Felipe, Bogotá 065220
Piehet, Ray J., New Orleans ... 095143
Piel, Michel, Cabourg 103478
Pielaveden Kotiseutumuseo,
  Pielavesi 011250
Pielisen Museo, Lieksa 011112
Pien, Amersfoort 082655
Pienmäen Talomuseo, Niemisjärvi 111185
Piepo & Partner, Hannover 130014
The Pier-2, Art District Kaohsiung,
  Kaohsiung 042366
Pier 21, Halifax 005960
Pier Arts Centre, Stromness ... 045914
Pier One Imports, Houston 093763
Pier Pander Museum,
  Leeuwarden 032574
Pierce Art Center, Minneapolis . 122024
Pierce College Art Gallery, Woodland
  Hills 054523
Pierce Manse, Concord 047960
Pierce's, Houston 093764, 127255
't Pierement, Leiden 083441
't Pierement, Middelburg 083499
Pierhuff, Indianapolis 093916
Pierides Museum of Ancient Cypriot
  Art, Athinai 023027
Pierides Museum, Laïki Cultural
  Centre, Larnaka 009408
Pierides Sculpture Garden,
  Larnaka 009409
Pierina Lutti, Buenos Aires 060924
Pierini, Alberto, Firenze 080187
Pierini, Filippo, Firenze 131240
Pierini, Robert, Biot 103395
Piero, Yokohama 111542
Pierogi, Leipzig 107845
Pierogi 2000, Brooklyn 119943
Pierogi 2000, Leipzig 107846
Pieroth, Heinz, Frankfurt am
  Main 075813
Pierre, 's-Hertogenbosch 112530
Pierre & Nicolas, Cannes 103521
Les Pierre Antiques, New York . 119356

Pierre aux Loups, Cheval de Bois,
Conflans-Sainte-Honorine 068133
Pierre d'Agate, Reims 072402
Pierre Gildesgame Maccabi Sports
Museum, Ramat Gan 025176
Pierre Menard Home, Ellis Grove 048556
Pierre Michel D., Paris 105167
Pierre, Alain, Paris ... 129113, 129114
Pierre, Isabelle, Monaco 082524
The Pierrepoint Gallery, Bridport 117430
Pierrepont, Oxford 118865
Pierrepont Museum, Canton ... 047407
Pierres et Vestiges, Suresnes .. 073608
Pierres Passion, Vic-le-Comte .. 074096
Pierres Précieuses, Strasbourg .. 073595
Pierron, Laurent, Saint-Georges-du-
Bois 072800
Le Pierrot, Saint-Pierre-de-
Clages 144186
Pierrot, Warszawa 113887
Piers Pisani Antiques, Sherborne 091306,
135034
Pierson, Marie-Françoise, Villers-la-
Chèvre 074178
Pierson, Pat, Sainte-Foy-la-
Grande 073349
Piert-Borgers, Barbara, Köln .. 130161
Piesk, Friedrich, Wiesbaden ... 142492
Pietarin Radan Museo, Lahti ... 011076
Pieter Vermeulen Museum,
IJmuiden 032521
Pieternel, Deventer 083161
Pieters, Herzogenrath 076517
Pieters, Guy, Knokke-Heist ... 100755
Pieters, Guy, Saint-Paul-de-
Vence 105598
Pieters, Luc, Knokke-Heist ... 100756
Pietershuis, Haarlem 112515
La Pietra d'Agata, Milano 080672
Pietro Wyżej, Górnośląskie Centrum
Kultury, Katowice 034820
Pietroforte, Manlio, Roma 132013
Piet's, Northampton 118819
Pifaçanes, Barcelona 133142
Piffl, Markus, Graz .. 062598, 127751
The Pig Yard Gallery, Wexford . 109707
Pigeon Island, Castries 038089
Pigeon Valley Steam Museum,
Wakefield 033331
Pigeon & Co, Lomas, Chelmsford 134230
Piggelmee, Zwolle 083737
Pighouse Folk Museum,
Cornafean 024861
Pigmalion, Postojna 114741
Pignatelli, Domenic R., North
Providence 136038
Pignede, Yves, Maisons-Laffitte . 069817
Pignolo, Giuliana, Milano 131499
Pigot, Jean-Michel, Sens 073500
Pig's Wings Bookshop of Dragon Hill,
Coolangatta 139702
Pihan, Jenny, Frankston 099058
Pihay, Georgette, Québec 101515
Pihlajaveden Kotiseutumuseo,
Keuruu 010998
Pihlmark, Århus 102779, 140556
Pihtiputaan Kotiseutumuseo,
Pihtipudas 011251
Pijpenkabinet & Smokiana,
Amsterdam 031979
Pikaflor, Honfleur 068830
Pikakannu Põhikooli Muuseum,
Pikakannu 010652
Pikari, Helsinki 066354
Pike County Museum, Milford .. 050894
Pikes Peak Pastel Society, Colorado
Springs 060689
Pikinasso, Roanne 105452
Piktogram, Warszawa 138980
Piktúra Galéria, Budapest 109171
Pila Museum, Pila 034571
Pilar Martí, Barcelona 085824
Pilar Parra & Romero, Madrid .. 115271
Pilardus, Cannes 103522

Pilat, Bianca, Chicago 120232
Pilat, Bianca, Milano 110400
Pilchuck Glass School, Seattle . 057898
Pile, Ronald, Ely 117751
Pilegaard, Aalborg ... 065587, 140547
Pilgrim Antiques, Ampthill 088278,
134033
Pilgrim Antiques, Honiton 089573
Pilgrim Books, Calgary 140337
Pilgrim Hall Museum, Plymouth 052211
Pilgrim Monument and Provincetown
Museum, Provincetown 052399
Pilgrims Antique Centre, Dorking 089091
Pilgrims Antiques Centre,
Gainsborough 089317
Pilgrim's Rest Museum, Pilgrim's
Rest 038805
Pili, Rosetta, Cagliari 079996
Pilies Muziejus, Mažosios Lietuvos
Istorijos Muziejus, Klaipėda .. 030393
Pilkington, Yelvertoft 138295
Pillantás, Budapest 109172
Pillars, Lyneham 090667
Pillars & Pomegranates, Fresno . 093510
Piller & Assoc., Stanley M.,
Oakland 095907
Piller, Yves, Corminboeuf 087382
Pillet, Lyons-la-Forêt 069787
Pilling, Detmold 075342, 125917
Pillon, Trieste 081799
Pilloni, Nunzia & C., Cagliari .. 079997
Pilloni, Rosaria, Cagliari 131073
Pillsbury, Michel, Houston 093765
Pillu, François, Paris 105168
Pilney, Saint Paul 096849
Pilonova Galerija, Ajdovščina .. 038492
Pilorgé, Jean, Carnac 067688
Pilotgalerie, Galerie für Junge Kunst &
Edition, Luzern ... 116479, 138072
Pilots Cottage Museum, Kiama . 001187
Pils, Erich, Krems 062692
Pils, John, Saint Louis 123856
Pilsen, Chicago 092767, 135422
Pilton Art Gallery, Barnstaple .. 117290
Piltzer, Gerald, Barbizon 103333
Piltzer, Gérald, Macherin 104140
Piltzer, Gérald, Paris 105169
Piltzer, Gérald, Roma 110845
Pilz-Insekten-Vogel-Museum, Ribnitz-
Damgarten 021410
Pilz, Evelyn, Münster 077674
Pilz, Roman J., Mainburg 130263
Pilzmuseum Taura, Schildau,
Gneisenaustadt 021649
Pim, Heerlen 083372
Pima Air and Space Museum,
Tucson 053834
Pimeria Alta Historical Society
Museum, Nogales 051570
Pimlico, Chicago 092768
Pimlico Books, London 144665
Le Pin Bleu, Beaune 067032
Le Pin Perdu, Carcassonne ... 067675
Pin, Chen Tang, Hong Kong ... 065092
Pin, Xuan Ge, Singapore 114587
Pin.Arte, Torino 110996
Pina, C., Charleroi 063605
La Pinacoteca, Barcelona 114948
La Pinacoteca, Bogotá 102356
Pinacoteca, Chiomonte 025973
Pinacoteca, Córdoba 000339
La Pinacoteca, Napoli 080917
Pinacoteca, Teramo 028070
Pinacoteca Ambrosiana, Milano . 026827
Pinacoteca Andrea Alfano,
Castrovillari 025865
Pinacoteca Barão de Santo Ângelo,
Universidade Federal do Rio Grande
do Sul, Porto Alegre 004653
Pinacoteca Benedicto Calixto,
Santos 004821
Pinacoteca Cantonale Giovanni Züst,
Rancate 041857
Pinacoteca Carmelita del Convento del

Carmen, Trujillo 034415
Pinacoteca Cascella, Museo Civico
d'Arte Contemporanea, Ortona 027114
Pinacoteca Chiesa di San Giovanni,
Pieve Torina 027330
Pinacoteca Chiesa di San Tomaso
Becket, Padova 027162
Pinacoteca Civica, Abano Terme 025226
Pinacoteca Civica, Ascoli Piceno 025390
Pinacoteca Civica, Baiardo 025443
Pinacoteca Civica, Cento 025897
Pinacoteca Civica, Cepagatti ... 025899
Pinacoteca Civica, Crotone ... 026112
Pinacoteca Civica, Fermo 026202
Pinacoteca Civica, Iesi 026498
Pinacoteca Civica, Imola 026507
Pinacoteca Civica, Imperia 026512
Pinacoteca Civica, Monza 026973
Pinacoteca Civica, Pieve di Cento 027327
Pinacoteca Civica, San
Gimignano 027755
Pinacoteca Civica, Santarcangelo di
Romagna 027841
Pinacoteca Civica, Savona 027894
Pinacoteca Civica, Spello 027999
Pinacoteca Civica A. Ricci, Monte San
Martino 026914
Pinacoteca Civica Amedeo Modigliani,
Follonica 026316
Pinacoteca Civica Bruno Molajoli,
Fabriano 026168
Pinacoteca Civica d'Arte Moderna,
Latina 026574
Pinacoteca Civica e Gipsoteca U. Gera,
Ripatransone 027533
Pinacoteca Civica F. Duranti,
Montefortino 026940
Pinacoteca Civica Francesco Podesti e
Galleria Comunale d'Arte Moderna,
Ancona 025323
Pinacoteca Civica G. Cattabriga,
Bondeno 025589
Pinacoteca Civica Melozzo degli
Ambrogi, Forlì 026330
Pinacoteca Civica Tosio Martinengo,
Brescia 025637
Pinacoteca Civica, Palazzetto del
Podestà, Montelupone 026949
Pinacoteca Civica, Villa Grappallo,
Vado Ligure 028247
Pinacoteca Comunale, Assisi .. 025403
Pinacoteca Comunale, Bevagna 025505
Pinacoteca Comunale, Bosa ... 025608
Pinacoteca Comunale, Castiglion
Fiorentino 025853
Pinacoteca Comunale, Cesena . 025930
Pinacoteca Comunale, Città di
Castello 026004
Pinacoteca Comunale, Deruta .. 026130
Pinacoteca Comunale, Manciano 026689
Pinacoteca Comunale, Massa
Fermana 026728
Pinacoteca Comunale, Massa
Marittima 026739
Pinacoteca Comunale, Matelica . 026745
Pinacoteca Comunale, Narni ... 027024
Pinacoteca Comunale, Ostra ... 027131
Pinacoteca Comunale, Quistello . 027465
Pinacoteca Comunale, Ripe San
Ginesio 027534
Pinacoteca Comunale, San Benedetto
del Tronto 027726
Pinacoteca Comunale, San Severino
Marche 027800
Pinacoteca Comunale, Sarnano . 027857
Pinacoteca Comunale, Spoleto . 028008
Pinacoteca Comunale, Veruccho 028354
Pinacoteca Comunale A. Moroni, Porto
Recanati 027416
Pinacoteca Comunale Alberto Martini,
Oderzo 027079
Pinacoteca Comunale D. Stefanucci,
Cingoli 025989
Pinacoteca Comunale d'Arte Antica e

Moderna, Faenza 026178
Pinacoteca Comunale e Museo A. De
Felice, Terni 028078
Pinacoteca Comunale E. Giannelli,
Parabita 027201
Pinacoteca Comunale F. Galante
Civera, Margherita di Savoia .. 026711
Pinacoteca Comunale Foresiana,
Portoferraio 027423
Pinacoteca Comunale Francesco
Cozza, Stilo 028019
Pinacoteca Comunale V. Bindi,
Giulianova 026449
Pinacoteca Comunale, Casa Rusca,
Locarno 041684
Pinacoteca Comunale, Museo Civico,
Mondavio 026886
Pinacoteca D. Inzaghi, Budrio .. 025655
Pinacoteca Dantesca Fortunato
Bellonzi, Torre de Passeri 028149
Pinacoteca d'Arte Antica, Gemona del
Friuli 026384
Pinacoteca d'Arte Contemporanea,
Smerillo 027975
Pinacoteca d'Arte Francescana,
Lecce 026582
Pinacoteca d'Arte Moderna,
Avezzano 025428
Pinacoteca d'Arte Moderna e
Contemporanea "AM International",
Bivongi 025523
Pinacoteca Davide Bergh, Calice al
Cornoviglio 025694
Pinacoteca de la Escuela Provincial de
Bellas Artes Dr. Figueroa Alcorta,
Córdoba 000340
Pinacoteca de la Facultad de Bellas
Artes, Manizales 008710
Pinacoteca de la Universidad de
Concepción, Concepción 007521
Pinacoteca de Nuevo León,
Monterrey 031299
Pinacoteca dei Cappuccini,
Voltaggio 028424
Pinacoteca del Ateneo, Universidad
Autónoma de Coahuila, Saltillo 031428
Pinacoteca del Banco Central de
Nicaragua, Managua 033391
Pinacoteca del Centro de la Profesa,
México 031268
Pinacoteca del Duomo, Cittadella 026007
Pinacoteca del Estado Juan Gamboa
Guzmán, Mérida 031140
Pinacoteca del Pio Monte della
Misericordia, Napoli 027019
Pinacoteca del Templo de la
Compañía, Guanajuato 031032
Pinacoteca dell' Accademia Albertina
di Belle Arti, Torino 028144
Pinacoteca della Cassa di Risparmio,
Cesena 025931
Pinacoteca della Certosa, Firenze 026300
Pinacoteca della Rocca Ubaldinesca,
Sassocorvaro 027873
Pinacoteca dell'Accademia Carrara,
Bergamo 025498
Pinacoteca dell'Accademia dei
Concordi e del Seminario,
Rovigo 027689
Pinacoteca dell'Accademia di Belle
Arti, Carrara 025785
Pinacoteca di Brera, Milano ... 026828
Pinacoteca di Corinaldo, Civica
Raccolta d'Arte Claudio Ridolfi,
Corinaldo 026074
Pinacoteca di San Silvestro,
Montecompatri 026929
Pinacoteca Diego Rivera, Xalapa 031670
Pinacoteca Diocesana di Arte Sacra,
Senigallia 027919
Pinacoteca do Estado de São Paulo,
São Paulo 004910
Pinacoteca d'Vinci, Medellín .. 102456
Pinacoteca e Biblioteca Rambaldi,

Register der Institutionen und Firmen
Pinacoteca e Musei Civici, Jesi – Pirate's Alley, Oklahoma City

Sanremo 027811
Pinacoteca e Musei Civici, Jesi . 026532
Pinacoteca e Musei Comunali, Macerata 026663
Pinacoteca e Museo Civici, Camerino 025717
Pinacoteca e Museo Civico, Palazzetto del Podestà 3, Bettona 025503
Pinacoteca e Museo Civico, Palazzo Minucci-Solaini, Volterra 028427
Pinacoteca e Museo de Napoli, Terlizzi 028071
Pinacoteca e Museo delle Arti di Locri Epizephiri, Locri 026613
Pinacoteca E. Notte, Ceglie Messapica 025891
Pinacoteca Eduardo Ramírez Castro, Aranzazu 008570
Pinacoteca G. A. Levis, Racconigi 027467
Pinacoteca G. Spano, Ploaghe .. 027376
Pinacoteca Giovanni e Marella Agnelli, Torino 028145
Pinacoteca Giovanni Morscio, Dolceacqua 026137
Pinacoteca Giuseppe Stuard, Parma 027221
Pinacoteca in Palazzo Volpi, Como 026051
Pinacoteca Internazionale dell'Età Evolutiva A. Cibaldi, Rezzato .. 027513
Pinacoteca L. Répaci, Palmi ... 027198
Pinacoteca Larraín, Santiago de Chile 101920
Pinacoteca Malaspina, Pavia ... 027237
Pinacoteca Manfrediana, Venezia 028323
Pinacoteca Municipal de Guadalupe, Guadalupe 031019
Pinacoteca Municipal de la Provincia de Corongo, Corongo 034256
Pinacoteca Municipal Ignacio Merino, Lima 034370
Pinacoteca Municipal Leoncio Lugo, Paucartambo 034381
Pinacoteca Municipal Miguel Ângelo Pucci, Franca 004468
Pinacoteca-Museo Beato Sante, Mombaroccio 026879
Pinacoteca Nazionale, Bologna . 025574
Pinacoteca Nazionale, Cagliari .. 025683
Pinacoteca Nazionale, Ferrara .. 026220
Pinacoteca Nazionale, Siena ... 027961
Pinacoteca Parrocchiale, Castroreale 025862
Pinacoteca Parrocchiale, Corridonia 026078
Pinacoteca Provinciale Corrado Giaquinto, Bari 025463
Pinacoteca Repossi, Chiari 025951
Pinacoteca Rossetti Valentini, Santa Maria Maggiore 027821
Pinacoteca Universitaria, Colima 030894
Pinacoteca Vaticana, Musei Vaticani, Città del Vaticano 054695
Pinacoteca Virreinal de San Diego, México 031269
Pinacoteca Zelantea, Acireale .. 025234
Pinacoteca, Chiesa di Santa Verdiana, Castelfiorentino 025821
Pinacoteca, Palazzo dei Musei, Varallo Sesia 028273
Pinakoteket, Hellerup 065734
Pinakothek der Moderne, München 020563
Pinakotheke, Rio de Janeiro ... 101057
Pinakothiki Kouvoutsaki, Kifissia 023129
Pinal County Historical Museum, Florence 048756
Piñanes Tena, Manuel, Sevilla .. 086470, 086471
Pinard Etellin, Colette, Clermont-Ferrand 068035
Pin'Art Antiques, Bruxelles 063528
Pincas, Nathalie, Paris 129115
Pinceladas de Málaga, Málaga . 115305
Pinchbeck Marsh Engine and Land Drainage Museum, Spalding .. 045844
Pincher Creek & District Museum, Pincher Creek 006653
Pindar Press, London 138239
Pine & Things, Shipston-on-Stour 091319
Pine and Decorative Items, Ashbourne 088301
Pine and Design Imports, Chicago 092769
Pine and Period Furniture, Grampound 089352
Pine Antiques, Olney .. 090916, 134909
Pine Antiques Workshop, Doveridge 089102
Pine-Apple Antiques, Kingston-upon-Hull 089704
The Pine Barn, Crawley 088998
Pine Cellars, Winchester 091871
The Pine Collection, Saint Peter Port 091207
Pine Collection and Ranelagh Antiques, Dublin 079582
The Pine Cottage, Jacksonville . 094025
Pine County Historical Society Museum, Askov 046578
Pine Creek Museum, Pine Creek 001405
The Pine Emporium, Hemel Hempstead 089497
The Pine Emporium, Hursley ... 089622
Pine Forest Art Centre, Dublin .. 055863
Pine Grove Historic Museum, Pontiac 052238
Pine House, Breda 083002
Pine Islet Lighthouse, Mackay .. 001238
Pine Memories, Ballina 079445
The Pine Mine, London 090422
Pine Mine, Los Angeles 094443
Pine Parlour, Ampthill 088279
The Pine Shop, Fremantle 061490
Pine Street Revolution Pantechnicon, Los Angeles 094444
Pine Stripping, Lewisham 134555
Pine Studio, Sankt-Peterburg .. 085260
Pine Woodworks, Denver 093323, 135534
Pineapple House Antiques, Alresford 088259
Pineau, Michel, Nantes 070474
Pineau, William, La Crèche 069018
Pineau, William, Saint-Maixent-l'Ecole 072909
Pineda, Bogotá 065221
Pinel & Partner, Bruxelles 063529
Pines, Sylvia, New York 095663
Pinetum, Canterbury .. 088780, 134210
Pinetz, Walter, Brunn am Gebirge 062541, 127720
Pinewood Furniture Studio, Halesowen 134432
Pinewood, Justin, Burton-upon-Trent 088735
Ping Art Space, Taipei 116936
Ping Jin Zhan Yi, Old Zhi Display Center, Yangliuqing 008462
Ping Yang Wood Sculpture Museum, Hualien 042341
Pingdu Museum, Pingdu 008148
Pingeon, Christian, Paris 071790
Pinggu Shangzhai Cultural Display Center, Beijing 007640
Pinghe Museum, Pinghe 008149
Pinghu Museum, Pinghu 008151
Pingliang Museum, Pingliang .. 008152
Pingliang Zone Museum, Pingliang 008153
Pinglin Tea Museum, Pinglin ... 042411
Pinglu Museum, Pinglu 008154
Pingshun Museum, Pingshun .. 008155
Pingtung Fo Guang Yuan Art Gallery, Pingtung 042417
Pingtung Hakka Museum, Jhutian 042349
Pingtung Museum of Natural History, Hengchun 042324
Pingvin, Maffra 139816
Pingxiang Museum, Pingxiang .. 008156
Pingyuan Museum, Pingyuan .. 008159
Pinhey's Point, Ottawa 006596
Pinies, Patrick, Pointis-Inard ... 072215
Pink & Sons, Malvern, Victoria . 061784
Pink Flamingos Antique Mall, Phoenix 096233
Pink Gallery, Helensburgh 117945
Pink Giraffe, San Antonio 097014
Pink Porch Antiques, Tucson ... 097686
Pink, Marilyn, Los Angeles 121574
Pinkerton, Heerlen 083373
Pinkhart, Gertraud, Wien 063051
Pinkney, Herbert, Memphis 121708
Pinkus, E., Haifa 079728
Pinn & Sons, W.A., Sible Hedingham 091338
Pinna, Henri, Plomodiern 072194
The Pinnacle, Cork 079507
Pinnacles Gallery, Thuringowa .. 001577
Pinnaroo Heritage Museum, Pinnaroo 001408
Pinnaroo Printing Museum, Pinnaroo 001409
Pinner, Stefan, Chemnitz 075259
Pinnsoffan, Halmstad 086734
Pino Pascual, S., Málaga 115306
Les Pins Maison des Arts, Cologny 116222
Pinson Mounds State Archaeological Area, Pinson 052144
Pinson, Françoise, Avranches .. 066892
Pinson, Joël, Saint-Quentin-sur-le-Homme 073282
Pinson, Yvon, Conques 068139
Pint, Carlo, Napoli 080918
Pintar, Maria, Salzburg 062812
Pintér, Budapest 079281
Pintér Muvek Hadtörténeti Múzeum, Hadtörténeti Park, Hadtörténeti Múzeum Budapest, Kecel 023720
Pinto e Bourdain, Lisboa 126452
Pinto Torrijos, Francisco, Madrid 086236
Pinto Torrijos, Valentin, Madrid . 086237
Pinto, António Reis, Lisboa 132950
Pintuformas, Medellín 065303
Pintura Contemporánea, Lima .. 113366
Pinturas Pueblo, Alicante 133083
Pinus, Leuven 063735, 128070
Pinx, Bochum 106514
Pinx, Helsinki 103155
Pinxit, Nashville 122094
Pinxit, Torino 110997
Pinxit, Vichy 105859
Pinzgauer Heimatmuseum, Saalfelden am Steinernen Meer 002617
Pio Cabello, Cristobal, Barcelona 085825
Pioneer Antique Auction, Cherokee Village 127160
Pioneer Antiques, Phillip 062085
Pioneer Antiques, Tanunda 062306
Pioneer Arizona Living History Village, Phoenix 052122
Pioneer Auto Museum, Murdo .. 051121
Pioneer Books, Seacombe Gardens 139907
Pioneer Corner Museum, New Holstein 051271
Pioneer Cottage Antiques, North Geelong 061992
Pioneer Farm Museum and Ohop Indian Village, Eatonville 048462
Pioneer Florida Museum, Dade City 048103
Pioneer Heritage Center, Cavalier 047469
Pioneer Heritage Center, LSU Shreveport, Shreveport 053321
Pioneer Historical Connors Museum, Connors 005644
Pioneer-Krier Museum, Ashland 046561
Pioneer Log Cabin, Manhattan . 050641
Pioneer Memorial Association of Fenton and Mounty Townships, Fenton 048722
Pioneer Museum, Fabius 048656
Pioneer Museum, Fairbanks ... 048661
Pioneer Museum, Sayward 006994
Pioneer Museum, Silverton 038828
Pioneer Museum, Watford City . 054223
Pioneer Museum, Wild Rose ... 054403
Pioneer Museum and Vereins Kirche, Gillespie County Historical Society, Fredericksburg 048950
Pioneer Museum of Alabama, Troy 053809
Pioneer Museum of Motorcycles, Tacoma 053648
Pioneer Museum Society of Grande Prairie and District, Grande Prairie 005913
Pioneer Rum Museum, Beenleigh 000843
Pioneer Settlement Museum, Swan Hill 001531
Pioneer Square Antique Mall, Seattle 097485
Pioneer Stripping and Refinish, Milwaukee 135819
Pioneer Tower Antiques, Kitchener 064339
Pioneer Town, Wimberley 054461
Pioneer Tracks of Queensland Gallery, Nebo 099423
Pioneer Village, Farmington 048708
Pioneer Village, Silverdale 033301
Pioneer Village Museum, Burnie 000914
Pioneer Woman Statue and Museum, Oklahoma Historical Society, Ponca City 052234
Pioneerimuseo, Ala-Pihlaja 010751
Pioneers of Aviation Museum, McGregor Museum, Kimberley 038767
Pioneers, Trail and Texas Rangers Memorial Museum, San Antonio 052942
Pioniermuseum, Klosterneuburg 002210
Piontkowitz, S., Haan 076142
Piopio and District Museum, Piopio 033283
Pio's Antiques, Victoria 082447
Pip Antiques, Virginia 062394
Pipaongo Etnografia Museoa, Museo Etnográfico de Pipaón, Pipaón . 039945
Pipat & Morier, Blaye . 067242, 067243
Pipat & Morier, Bordeaux 067353
La Pipe, Oostende 063821
La Pipe Eric, Bruxelles 063530
Piper Antiques, Auckland 083816
Piper Aviation Museum, Lock Haven 050351
Piper Barn, Kyneton 061712
Piper, Ulrike Margarete, Stuttgart 130683
Piperaud, Pascale, Saint-Ouen . 073171
Pipestone County Historical Museum, Pipestone 052145
Pipino, Vincenzo, Napoli 131634
Pippa, Luigi, Roma 132014
Pippa's Art Gallery, Whangarei . 113110
Pippi, Firenze 080188
Pippin, Clayfield 061338
Pippin, Kenneth, Baltimore 092355
Piqua Historical Area State Memorial, Piqua 052146
Pique-Puces, Bruxelles 140196
Piquet, Coudekerque-Branche .. 068175
Piqueux, Pierre, La Rochette ... 069096
Piraino, Pietro, Palermo 081065
Piramal Gallery, Mumbai 109458
Pirâmide, Lisboa 114053
La Piramide, Milano 080673
Piramidon, Centre d'Art Contemporani, Barcelona 114949
Piras, Maria Giovanna, Torino .. 110998
Pirate, Denver 120756
Pirate Contemporary Art Oasis, Denver 048272
Piratenamüseum, Wilhelmshaven 022662
Pirate's Alley, Oklahoma City ... 123293

Pirmoradi, Hamburg 076295
Pirngruber, Alfred, Linz 140013
Piro, Hélène, Martel 069982
La Pirogue Museum, Cheticamp 005612
Pirontidi, Ernesto, Barcelona . . . 085826
La Pirouette, Hyères 068859
Pirovano, Fabio, Magliaso 087657,
133852
Pirovano, Fabio, Pura 087745
Pirra, Stefano, Torino 110999
Pirri, Angela, Roma 081507
Pirzkall, Isabella, Worpswede . . 130780
Pisanelli, Tiziana, Roma 081508
Pischetola, Giuseppe, Torino . . . 081748
Piskarevskoe Memorialnoe Kladbišče –
Muzej, Sankt-Peterburg 037814
Pismo, Denver 120757
Pisonero Ortega, Eduard,
Barcelona 133143
Pistil, Seattle 145596
Piston, Hyman E., New York . . . 095664
Pistone, Mario Giuseppe, Catania 131095
Pistor, Lorenz, Rodenbach 142332
Pisu, Giovanni, Cagliari 079998, 131074
Pit 21, Milano 110401
Pitäjämuseo ja Nymann Talo ja
Apteekkimuseo, Tohmajärvi . . . 011424
Pitäjäntupa-Museo, Joutseno . . 010924
Pitcairn, Scott 2207 Second Av,
Seattle 124587
Pitchal, Virginie, Paris 105170
Pitcher, Nicholas S., London . . . 090423
Piteå Museum, Piteå 040915
Piteå Sparkmuseum, Piteå 040916
Pitfield, David, Sunninghill Ascot 119181
Piti Arte Ibiza, Eivissa 115060
Pitkänpellon Talomuseo,
Kangasniemi 010958
Pitlochry Festival Theatre Art Gallery,
Pitlochry 045507
Il Pitocchetto, Brescia 109914
Piton, Miriam, Saint-Raphaël . . 073289
Pitot House Museum, New
Orleans 051305
Pitow, Bruck an der Mur 099970
Pitre, Honolulu 120964
Pitrelli, Vincenzo, Milano 131500
Pitsch & Geissbühler, Zofingen . 116735
Pitstone Green Farm Museum,
Pitstone 045508
Pitt Meadows Museum, Pitt
Meadows 006654
Pitt Rivers Museum, Oxford . . . 045459
Pittà-Settelmeyer, Rosa-Maria,
Konstanz 130182
Pittarello, Gastone, Venezia . . . 132254
Pittencrieff House Museum,
Dunfermline 044173
Pittera, Jean, Marseille 128782
Pitti, Monaco 082525
Pittock Mansion, Portland 052301
Pitton, Fabrizio, Torino 081749
Pittsburg State University Natural
History Museum, Pittsburg . . . 052150
Pittsburgh Antique Mall,
Pittsburgh 096320
Pittsburgh Center for the Arts,
Pittsburgh 052171
Pittsburgh Children's Museum,
Pittsburgh 052172
Pittsburgh Pastel Artists League,
Pittsburgh 060690
Pittsburgh Plating and Lamp Service,
Pittsburgh 136135
Pittsburgh Print Group,
Pittsburgh 060691
Pittsburgh Society of Illustrators,
Monroeville 060692
Pittsford Historical Society Museum,
Pittsford 052186
Pittsworth Pioneer Historical Museum,
Pittsworth 001410
Pitz, Ludwig, Stolberg, Rheinland 078568,
108675

Pitzen, Christoph, Esslingen . . . 129825
Pitzhanger Manor-House and Gallery,
London 045075
Piumaccio d'Oro, Firenze 131241
Piva & C, Milano 080676
Piva, Francesco, Milano 080674
Piva, Giuseppe, Milano 080675
Pivano, Pietro, Torino 132180
Pivot, Bernard, Basel 133655
Pivotal Galleries, Richmond,
Victoria 099636
Pivovarské Muzeum, Plzeň 009761
Pivovarski Muzej, Ljubljana 038554
Pixel Gallery – Pixel Galerie, Frankfurt
am Main 107020
Pixel Place, Hamilton 101321
Pixelprojekt_Ruhrgebiet, Essen . 018309
Pixi & Cie., Paris 105171
Pixian Juan Calligraphy and Painting
Gallery, Chengdu 102014
Pixian Museum, Pixian, Sichuan 008161
Pixtory, Marseille 069959, 104194,
136837
Pizhou Museum, Pixian, Jiangsu 008160
Piziarte, Teramo 110944
Pizorno Azuara, Laura, Ourense 086324
Pizzi, Amilcare, Cinisello
Balsamo 137637
Pizzi, Gabrielle, Melbourne 099316
Pizzinelli, Gianfranco, Roma . . . 132015
Pizzini, Rovereto 110882
Pjatigorskij Kraevedčeskij Muzej,
Pjatigorsk 037659
Pjazza Antiques, San Gwann . . . 082430
Pjóðmenningarhúsið, Reykjavík . 024106
Pjóðminjasafn Islands, Reykjavík 024107
Pjotr, Tilburg 112684
PLA Naval Museum, Qingdao . . 008174
Pla, Alfons, Barcelona 133144
Plaar, Jonas, Zürich . . 087999, 133997
Plaaswinkel, Tshwane 085684
Plac-Art, Bremen 075178
Plac Zamkowy, Warszawa 113888
The Place Antiques, Golden
Square 061537
Place Concore Antiquario, Rio de
Janeiro 064104
Place des Arts, Paris . . 105172, 136976,
141262
Place des Vosges, New York . . . 122972
La Place du Soleil, San
Francisco 097307
Place Royale, Montréal 101440
Placemaker, Miami 121842
Placencio, Francisco, Santiago de
Chile 064982
Placentia Area Museum,
Placentia 006655
Placentia Arte, Piacenza 110652
Places and Spaces, London . . . 090424
Plachot, Christian, Ypreville-
Biville 074311, 129361
Plada, Maribor 085619
Pladeck, F., Bielefeld 074987
Plages, Boulogne-Billancourt . . 138657
Plain Farm Workshop, Alton . . . 134025
Plaine des Dinosaures, Mèze . . 014163
Plains Art Museum, Fargo 048699
Plains Indians and Pioneers Museum,
Woodward 054539
Plains Vintage Railway and Historical
Museum, Ashburton 033138
Plainsman Museum, Aurora . . . 046679
Plaisio, Lagkadikia 109074
Plaisir d'Antan, La Roche-Guyon 069065
Plaisirs d'Intérieur, Saint-Jean-de-
Luz 072846
PlakART am Dreiecksplatz, Kunst-
Drucke-Rahmen, Kiel 076787
Plakat am Markt, Seligenstadt . 108611
Plakatmuseum am Niederrhein,
Emmerich am Rhein 018189
Plakity, Michael, Dresden 075424
Plambeck, Rosamunde, Kiel . . . 130120

Plamondon and District Museum,
Plamondon 006658
Plan 2 B, Portland 139878
Plan Incliné de Ronquières,
Ronquieres 003983
Plan-Weseritzer Heimatstuben mit
Archiv, Tirschenreuth 022159
Plan.d.-Produzentengalerie,
Düsseldorf 106817
Plana, Ricard, Terrassa 115467
Planage, Binoist, Rouen 072610
La Plancha, Heerlen 083374
Plane Space, New York 122973
Plane Tree, Christchurch 083890
Planet Anime, Houston 093766
Planet Art, Cape Town 114773
Planet Art, Miami 121843
Planet Corroborce, Byron Bay . . 061244
Planet Earth Museum, Newhaven 045358
Planet Publishing, New Orleans . 122218
Planet Rugs, Houston 093767
Planetario, Trieste 111053
Planète Bordeaux, Beychac-et-
Caillau 012016
Planète de la Découverte, Musée des
Sciences pour Enfants, Beirut . 030270
Planetron – Aards Paradijs,
Dwingeloo 032259
Il Planisfero, Seriate 110926
Plank, Wien 100361
Plank, Alexander, Mauerbach . . 136535
Planque, Claudine, Lausanne . . 116426
Planšarski Muzej, Gorenjski Muzej
Kranj, Bohinjsko Jezero 038497
Plantard, Guer 068757
Plantation Agriculture Museum,
Scott 053208
Plantation Antiques, Toronto . . . 064725
Plantation Furniture & Antiques,
Calgary 064266
Plantation Historical Museum,
Plantation 052193
Plantation of Ulster Visitor Centre,
Draperstown 044139
Plantation Shop, Atlanta 092169
Planteau du Maroussem, Eric, Aix-en-
Provence 103201, 128428
Plantener Pedersen, Tine, Holte . 065787
Planter, Singapore 085486
Plantier, Le Cheylard 069251
Plantijn, Breda 143094
Plantijn, Den Haag 143119
Plantiques, San Antonio 097015
Plantureux, Serge, Paris 141263
Plas Mawr, Conwy 044004
Plas Newydd, Llanfairpwll 044884
Plas Newydd, Llangollen 044891
Plasberg, Ralf, Nürnberg 077835
Plasma, Prahran 062118
Plassard, Yves, Paris 071791
Plassart, Ch., Poitiers 125780
Plasse, Agnès, Boulogne-
Billancourt 067394
Plassenburg, Bayerische Verwaltung
der staatlichen Schlösser, Gärten und
Seen, Kulmbach 019832
Plassmann, Mignon, Attendorn . 106021
Plast-Art, Černigiv 117104
Plastic Club, Philadelphia 052084,
060693
Plastics Historical Society, Sidcup 060210
Plastique Kinetic Worms,
Singapore 114588
Plastyka, Kraków 113628
Plastyka, Warszawa 113889
Plat Art, Buenos Aires 127530
Plat, B., Hardegarijp 132525
Plat, Jean-Louis, Tours 073857
El Plata, Sevilla 086472
De Plataan, Heerlen 083375
Plâté, Colombo 086619, 086620, 115566,
115567
Plateaux Gallery, London 118512
Platenhäuschen, Erlangen 018263

Platero, Buenos Aires 139596
Platform 1 Heritage Farm Railway,
Littlehampton 001226
Platform Artists Group,
Melbourne 099317
Platform Gallery, Clitheroe 043971,
117601
Plati Galeria, Wrocław 113949
Platini, Sylvie, Portland 132659
Platini, Sylvie, Veyrier-du-Lac . . 105851
Platinum Collectibles, Las Vegas 094167
Platou, Bærums Verk 126338
Platt R. Spencer Memorial Archives
and Special Collections Area,
Geneva 049070
Platt, Alan Quentin, Chicago . . . 120233
Platt, Kevin, Padstow 118876
Platt, Susan, Seattle 124588
Platt, Trevor, Wynnum 099927
Platte, E. & H., Preußisch
Oldendorf 108413
Platten, Frank, Suderburg 130696
Platter og Kunst, Viborg 066210
Plattner Bienenhof Museum,
Soprabolzano 027988
Plattner, Nicolas & Marianne,
Liestal 087594
Plattner, Rosina, Schladming . . 100193
Plattsburgh Art Museum, State
University of New York,
Plattsburgh 052199
Platz, Danielle, Baden-Baden . . 129435
Plauener Spitzenmuseum, Plauen 021149
Play-Gallery for still and motion
pictures, Berlin 106400
Play House Gallery, Baltimore . . 119765
PLAY_platform for film and video,
Castagnola 059955
Players Press Inc., Studio City . 138475
Playhouse Gallery, Academy of
Performing Arts, Orleans 051812
Plaza, Salamanca 086412
Plaza & Janés, Barcelona 137967
Plaza Antiques, Minneapolis . . . 094944
Plaza de la Raza – Boathouse Gallery,
Los Angeles 050465
Plaza, Francisco, Madrid 086238
Plaza, Marina, Madrid 086065
Plazanet, Nicole, Malicorne-sur-
Sarthe 069828
Plazanet, Yvonne, Le Chambon-sur-
Lignon 069245
Plazzotta, London 118513
Pleasant Place Antiques,
Wellington 084180
Pleasant Point Railway and Historical
Society Museum, Pleasant Point 033284
Please Touch Museum,
Philadelphia 052085
Pleasures of Past Times, London 144666
Plečnikova Zbirka, Arhitekturni Muzej
Ljubljana, Ljubljana 038555
Pledger, Robert E., Houston . . . 121113
Pléiade, Dunkerque 103708
La Pléiade, Grenoble 103791
Pleiades Gallery, New York 122974
Le Pleiadi, Milano 110402
Le Pleiadi, Roma 081509
Plein Air Painters of America, Santa
Fe 060694
Plein Air Scene, Valley 058037, 060695
Plein Sud, Vaison-la-Romaine . . 073931
Plejada Galeria, Gdynia 113503
Plenty and District Museum,
Plenty 006660
Pier, P., Kornwestheim 076943
Pleša, Zlatko, Frankfurt am Main 076814
Plescoff, Nicolas, Paris 105173
Pleser, Cornelius, München . . . 108176
Plesse-Antiquariat, Bovenden . . 141625
Plessis, Charles-Yves, Nantes . . 104343
Plessis, du, Richmond, South
Australia 125228
Plesskij Gosudarstvennyj Istoriko-

Architekturnyj i Chudožestvennyj
  Muzej-Zapovednik, Ples 037662
Plester, Randy, London 064354, 128178
Plestor Barn Antiques, Liss 089842
Plett, Zeist 112725
Pleus, G., Oldenburg 077937
Plexihouse, Miami 094792
Plietzsch, Freiburg im Breisgau . 141808
Plimsoll Gallery, Hobart 001147
Plischke, Dr. Georg, Unterwössen 137545
Plitt, Rostock 078281
Ploch, Florentina, San Antonio . 097016
Płóciennik & Sobczak, Łódź ... 143488
Plocka Galeria Sztuki, Płock .. 035106
Plöbst, Stefan, Dorfgastein 127723
Plöger, T., Bremen 141655
De Ploegh, Amersfoort 112127
Het Ploeghuis, Groningen 112500
Ploenes, Sabine, Wrohm 108972
Plöttner, Jonas, Leipzig 137348
Plötz-Peters, Hannelore, Berlin . 074897
Plof, Denver 120758
La Ploma – Llibreria de Vell de la
  Geltrú, Vilanova i la Geltrú ... 143813
Plott, H.-G., Leipzig 142062
Plough House Interiors,
  Chepstow 088847, 134246
Ploutis, Loizou, Egkomi 065317
Plouviez, Martine, Montmorillon 070280
Plowden & Smith, London 134753
Plückbaum, Bonn 125899
Plüm, H.-Günter, Neuenkirchen, Kreis
  Soltau-Fallingbostel 077723
Plümacher, Helmut, Burgwald .. 075230
plug in-Kunst und Neue Medien,
  Basel 041284
Pluimvee Museum, Barneveld .. 032048
Pluk, London 139214
Plum Antik, Frederiksberg 065672
Plum Blossoms, Hong Kong ... 102105
Plum Blossoms, Singapore 114589
Plum Coulee and District Museum,
  Plum Coulee 006661
Plum Grove Historic Home, Iowa
  City 049710
Plumas County Museum, Quincy 052428
Plumb Farm Museum, Greeley . 049223
Plumb, Albert, Chichester 134257
Plumbline Gallery, Saint Ives,
  Cornwall 119040
Plume Antiques, Campbell Town 061263
La Plume Marinoise, Le Marin . 142982
Plumershuuske, Diepenheim ... 032223
Plumettaz, Fernand, Payerne ... 087732
Plummer, Warren, Arncliffe 098670
Plums, Fresno 120902
Plunder Inn Antiques,
  Sacramento 096641
Plunder-Kabinett, Winterthur ... 087911
Plunket Museum of Irish Education,
  Dublin 024906
Plunkett, Thomas, Saint Albans . 119005
Pluntke, Erdmut, Berlin 074898
Pluntke, Leslie, Mainz-Kostheim 130271
Pluperfect, Melton 061837
Plurabelle Books, Cambridge .. 144300
Pluraebello, Milano 110403
Pluriel, Narbonne 070496, 104349
Plus Art, Monaco 082526
Plus Fours Gallery, Hyderabad . 109369
Plus Loin, Paris 071792
Plus One Gallery, London ... 118514
Plus One Plus Two Galleries,
  London 118515
Pluskwa, David, Marseille ... 104195
Plymoth Plantation Museum,
  Plymouth 052212
Plymouth Antiquarian Society Museum,
  Plymouth 052213
Plymouth Arts Centre, Plymouth 045513
Plymouth Historical Museum,
  Plymouth 052215
Plynárenské Múzeum, OZ SPP,
  Bratislava 038326

Pm Tung Arts, New York 095665
PMP Art, Timişoara 085140
PNT Prom, Zagreb 102527
PO Gallery, Providence 123710
Po Nu Fa, Liptovský Mikuláš .. 085576
Po, Kwai, Hong Kong 136726
Po, Wing, Hong Kong 065093
Poble Espanyol, Barcelona 114950
Poble Espanyol, Palma de
  Mallorca 039897
Poblet Antiguitats, Tarragona .. 086485
Pocahontas County Iowa Historical
  Society Museum, Laurens 050181
Pocahontas County Museum,
  Marlinton 050703
Pocarjeva Domačija v dolini Radovne,
  Kranjska Gora 038535
Poch, A., Bielefeld 074988
Pochopova, Lidmila, Praha ... 065515
Pochyla, Iwona, Augsburg 129397
Pociao, Sans Soleil, Bonn 137117
Pocker & Son, J., New York ... 122975
Pocket Bookshop, Paignton 144776
O Poço Velho, Lisboa 084941
Počta, Josef, Praha 140528
Počvenno-agronomičeskij Muzej
  im. Akademika V.R. Viljamsa,
  Moskovskaja selskochozjajstvennaja
  akademija im. K.A. Timirjazeva,
  Moskva 037492
Pod Kasztanami, Konstancin-
  Jeziorna 084544
Pod Kasztanami, Warszawa ... 113890
Pod Kinskou, Praha 065516
Pod Okiwkiem, Wrocław 113717
Pod trzema Gracjami, Gdańsk . 084509,
                               113483
Pod Vintage, Toronto 064726
Pod Złotym Lewkiem, Kraków . 084593,
                               113629
Podany, Pittsburgh 136136
Podbrzezie, Kraków 113630
Podestà Umberto Antiquario,
  Milano 080677
Podgórze Dom Kultury Galeria,
  Kraków 034910
Podhofer, Horst M., Düsseldorf . 075505
Podhorácké Muzeum – Klášter Porta
  Coeli, Předklášteří 009848
Podium Galerie, Kaufbeuren ... 107542
Podium Kunst, Schramberg ... 021731
Podium Kunst e.V., Schramberg 059219
Podjavorinské Múzeum, Nové Mesto
  nad Váhom 038415
Podještědské Muzeum, Český
  Dub 009508
Podlaskie Stowarzyszenie Promocji
  Sztuki, Białystok 059700
Podlipanské Muzeum, Regionální Kolín,
  Český Brod 009507
Podpolianske Múzeum Detva,
  Detva 038342
Podpolnaja Tipografija CK RSDRP
  1905–1906, Gosudarstvennyj
  centralnyj muzej sovremennoj istorii
  Rossii, Moskva 037493
Podřipské Museum, Roudnice nad
  Labem 009866
Podska, Hawkes Bay 112917
Podszus, E.M., Düsseldorf 106818
Podtatranské Múzeum, Poprad . 038427
Podvin, Roselyne, Hazebrouck .. 068804
Podwysocka, Krystyna,
  Warszawa 132895
Poe, Steven, Emeryville 120845
Poe, Timothy, Birmingham 119815
Poech, Dieter & Erika, Trostberg 108782
Poeh Museum, Pueblo of Poloaque,
  Santa Fe 053119
Põljan Kotiseutumuseo, Põlja .. 011255
Pöllath, Bernadette, Erbendorf . 075628
Poelman, Hengelo, Overijssel .. 132533
Poeplau, Hans-Peter, Lengerich . 130237
Pörtom Hembygdsmuseum,

Pörtom 011256
Pöschl, Innsbruck 127777
Poestgens, Horst, Übach-
  Palenberg 126131
Poete-Näscht, Liestal 144098
Pöter, A., Sulzheim 078650
Poeto Jono Mačiulio-Maironio Gimtinė,
  Raseinių Krašto istorijos muziejus,
  Bernotai 030320
Poeto Jovaro Namas, Šiaulių Aušros
  Muziejus, Šiauliai 030468
Poetry Renaissance, Dallas ... 120643
Poets Gallery, Boston 119913
Pötter, Geert, Rosdorf 142334
Poettler, Wien 063052
Pöytyän Kotiseutumuseo, Pöytyä 011257
Poggendorf, Renate, München . 130382
Poggi, Eliane, Grenoble 103792
Poggi, Laura, Genova 080283
Poggi, Patrizia, Ravenna 110702
Poggi, Sandro, Roma 081510
Poggiali e Forconi, Firenze ... 110066
Pogliano, Tommaso, Torino ... 132181
Pohjanmaan Antikvariaatti, Vaasa 140754
Pohjanmaan Museo, Vaasa 011471
Pohjois-Karjalan Museo, Joensuu 010918
Pohjois-Pohjanmaan Museo, Oulu 011222
Pohjoismaisen Konservaattoriliiton
  Suomen Osasto Ry, Helsinki . 058576
Pohl, Annette, Warstein 108848
Pohl, Joachim, Berlin 106401
Pohl, U., Aldenhoven .. 074371, 141405
Pohlkötter, Münster 108224
Pohronské Múzeum v Novej Bani,
  Nová Baňa 038413
Poï, Jan Ken, Genève 116336
Poignant, Jean-Marc, Nyons ... 070769,
                               128877
Poignant, Jean-Marc, Valréas .. 073955
Poilblan, Michel, Coucy-le-Château-
  Auffrique 068173
Le Poilu, Paris 071793
Poindessault, Bernard, Paris ... 071794
Poinot, Anne, Agon-Coutainville . 066433
Poinsignon Numismatique,
  Strasbourg 073596
Point Amour Lighthouse, L'Anse
  Amour 006162
Le Point d'Accroche, Paris 129116
Point du Jour, Douarnenez 103705
Le Point du Jour, Paris 105174
Point Ellice House, Victoria 007334
Point in Time, Vancouver 101824
Point Leamington Heritage
  Interpretation Centre, Point
  Leamington 006663
Point Light, Surry Hills 099751
Point Loma Books, San Diego . 145526
Point of Art, Hanau 107339
Point of Art Gallery, Calgary ... 101255
Point of Honor, Lynchburg 050534
Point Pelee Natural Park Visitor Centre,
  Leamington 006179
Point Pleasant Galleries, Point
  Pleasant Beach 127407
Point, J., Paris 071795
Pointe-à-Callière – Musée
  d'Archéologie et d'Histoire de
  Montréal, Montréal 006413
Pointe Coupee Museum, New
  Roads 051313
Pointe-Noire – Parc Marin
  du Saguenay, Baie-Sainte-
  Catherine 005372
Poirier, Nicole, La Saulsotte ... 069098
Poirier, Sylviane, Montréal 141441
Poisa, Giuseppe, Brescia 131045
Poisat, Daniel, Paris 129117
Poissant Gallery, Houston 121114
Le Poisson à Plumes, Theoule-sur-
  Mer 105703
Poisson, Anne-Marie, Crépon .. 068234
Poisson, Jean-François, Saint-Laurent-
  des-Bois 072884

Poisson, M.H., Fontenay-sour-
  Bois 128603
Poitou Tout'Occas, Châtillon-sur-
  Thouet 067925
Pokaz, Warszawa 138981
Pokolbin Gallery, Pokolbin 099571
Pokrajinski Muzej, Murska
  Sobota 038582
Pokrajinski Muzej, Strumica ... 030638
Pokrajinski Muzej Celje, Celje .. 038502
Pokrajinski Muzej Kočevje,
  Kočevje 038527
Pokrajinski Muzej Koper, Koper . 038529
Pokrajinski Muzej Ptuj, Ptujski grad,
  Ptuj 038594
Pokrajinski Muzej, Mariborski grad,
  Maribor 038572
Pokrovskij Sobor-chram Vasilija
  Blažennogo, Gosudarstvennyj
  istoričeskij muzej, Moskva ... 037494
Pokstaavi, Helsinki 140714
Pol-Art, Warszawa 113891
Pol, H.J., Borne 132447
Pol, Simon, Reims 072403
Pola Fuzoku Hakubutsukan,
  Tokyo 029576
Pola Museum of Art, Hakone .. 028595
Polabské Muzeum, Poděbrady . 009767
Polack, Bérengère, Paris 129118
Polad-Hardouin, Paris 105175
Polajnarjeva Galerija, Bohinjska
  Bistrica 114677
Polak, Paris 141264
Polak, J., Amsterdam 082824
Polak, R., Den Haag 112405
Polanco, San Francisco 124367
Polansky & Assoc., New York . 122976
Polar Art, Stockholm 115921
Polar Arts of Asia, Singapore . 085487
Polar, Pascal, Bruxelles 100650
Polaris, Paris 105176
Polarmuseet, Andenes 033498
Polarmuseet i Tromsø, Tromsø . 034028
Polarolo & Sons, E., New York . 135976
Poldark Mine and Heritage Complex,
  Helston 044563
Polder, J.A., Borne 112348, 132448
Poldermuseum, Antwerpen ... 003309
Poldermuseum, Puttershoek ... 032771
Poldermuseum de Hooge Boezem
  Achter Haastrecht, Haastrecht . 032403
Poldermuseum Den Huijgen Dijck,
  Heerhugowaard 032432
Poldermuseum Het Grootslag –
  Nationaal Saet en Cruytmuseum,
  Andijk 032005
Poldowrian Museum of Prehistory,
  Coverack 044022
Polemiko Mouseio, Athinai ... 023028
La Polena, Genova 110127
Polenmuseum Rapperswil,
  Rapperswil 041862
Polenz-Museum, Cunewalde .. 017787
Poleschi Arte, Forte dei Marmi . 110082
Poleschi Arte, Lucca 110192
Poleschi Arte, Milano 110404
Polesden Lacey House, Dorking 044124
Poleski Ośrodek Sztuki, Łódź .. 113680
Poletti, Carlos, Milano 131501
Poley, Dennis, Alreswas 088261
Polgár, Budapest 079282, 109173,
                               126155
Poli Museo della Grappa, Bassano del
  Grappa 025474
Poli, Eugenio, Firenze 131242
Poli, Paolo, Firenze 131243
PoliArt Studio d'Arte, Milano .. 110405
Poliarts 2000, Milano 080678
Polibinskij Memoralnyj Muzej-usadba
  S.V. Kovalevskoj, Polibino ... 037664
The Police Museum, Belfast ... 043562
Police Museum, Hong Kong ... 007887
Police Museum, Imphal 024312
Police Station and Courthouse

Collections, Auburn 000808
Polidori, Alberto, Perugia 131759
Poliester, Cuauhtémoc 138918
Il Polifilo, Milano 142739
Polifroni, Felice, Torino 132182
Polígrafa, Barcelona 137968
Polimento e Restauraçoes São
Domingos, São Paulo 128129
Polish American Museum, Port
Washington 052264
Polish Association of Members of the
Museum Professions, Kraków . 059701
Polish Cultural Institute, London 045076
Polish Museum of America,
Chicago 047691
Polish Sculpture, Orońsko 138982
Polisher, Tulsa 136434
Polishistoriska Museet,
Stockholm 041031
Polistekniska Museet, Solna . . . 040978
Politeama Gallery, Roma 110846
Politechničskij Muzej, Moskva . . 037495
Political Life Museum, Amman . 029777
Political Memorbilia Marketplace, Las
Vegas 094168
Politie-Petten Museum,
Slochteren 032886
Politiemuseum, Wommelgem . . 004156
Politiemuseum Oudaan,
Antwerpen 003310
Politikens Galleri, København . 102910
Politimuseet i Oslo, Oslo 033860
Polizeihistorische Sammlung Berlin,
Berlin 017270
Polizeimuseum, Aarau 041190
Polizeimuseum, Basel 041285
Polk County Heritage Gallery, Des
Moines 048284
Polk County Historical Museum,
Bartow 046803
Polk County Historical Museum,
Cedartown 047493
Polk County Historical Museum,
Osceola 051824
Polk County Historical Society, Des
Moines 048285
Polk County Memorial Museum,
Livingston 050347
Polk County Museum, Balsam
Lake 046725
Polk County Museum, Crookston 048070
Polk Museum of Art, Lakeland . 050097
Polk Place Antiques, Nashville . 095023
Polkadot Gallery, Exeter 117772
Poll, Eva, Berlin 106402
Pollak, Tel Aviv 142615
Pollak, Patrick, South Brent . . 144850
Polland, Günther, Wien 140114
Pollard, Indianapolis 093917
Pollazzon, Alexandre, London . 118516
Poller Media GmbH, Frankfurt am
Main 107021, 137205
Pollermann, Ch., Balingen 074655
Pollet & Desrumeaux, Lille 069502
Pollet Thiollier, Guy, La Clusaz . 068998
Pollinger, Roger, Turtmann . . . 087861,
133944
Pollismolen, Bos-en Pijpenmuseum,
Bree 003401
Pollmächer, Dr. Stefan, Bad
Zwesten 106103
Pollock, Los Angeles 094445
Pollock Gallery, Southern Methodist
University, Dallas 048122
Pollock, Charles, West Hollywood 097920
Pollock's Toy Museum, London . 045077
Pollok House, Glasgow 044412
Polloni & C., Guido, Firenze . . . 131244
Polly, Venlo 112715
Polnecq, Catherine, Paris 129119
Polnische Buchhandlung – Ksiegarnia
Polska, Wien 140115
The Polo Gallery, Galway 109638
Polo-Lodge Antiques, Tulsa . . . 097754

Polokwane Art Museum,
Polokwane 038806
Polonia Art, Katowice . 084538, 126405
Polos, Brunswick Heads 061225
Polplakat, Warszawa 113892
Polska Federacja Fotograficznych
i Audiowizualnych Stowarzyszeń
Twórczych, Wrocław 059702
Polska Izba Artystów Konserwatorów
Dzieł Sztuki, Warszawa 059703
Polskie Foto, Wrocław 138983
Polskie Towarzystwo Archeologiczne
i Numizmatyczne Zarzad Główny,
Warszawa 059704
Polson Park and Museum,
Hoquiam 049557
Polstjernan Antik, Stockholm . . 087044
Poltavs'kyj Chudožnij Muzej,
Poltava 043282
Poltavs'kyj Krajeznavčyj Muzej,
Poltava 043283
Poltavs'kyj Literaturno-memorial'nyj
Muzej I.P. Kotljarevskogo,
Poltava 043284
Poltavs'kyj Literaturno-memorial'nyj
Muzej Panasa Myrnogo, Poltava 043285
Poltavs'kyj Literaturno-memorial'nyj
Muzej V.G. Korolenka, Poltava 043286
Poltavs'kyj Muzej Aviacii ta
Kosmonavtiki, Poltava 043287
Põltsamaa Muuseum, Põltsamaa 010653
Poluch, J., Livinské 133054
Poluschkin, S., Much 108034
Põlva Talurahvamuuseum,
Karilatsi 010624
Polvere di Stelle, Padova 110577
Polvo Art Studio, Chicago 047692
Polvo Magazine, Chicago 139477
Poly International Auction, Beijing 125401
Poly Produzentengalerie e.V.,
Karlsruhe 107522
Poly, M.D., Floirac 103745
Polyedre, Antwerpen 063233
Polygnotos Vagis Museum,
Thassos 023281
Polygraphicum, Eberbach 075558,
141730
Polylustre, Fortaleza 063949
Polynesian Cultural Center, Laie 050068
Polyphon, Heemstede 143167
Polytechnique, Bruxelles 140197
Polyteekkarimuseo, Espoo 010776
Polzin, Saskia Gesa, Paderborn . 130504
Pom Pom, Los Angeles 094446, 094447
Poma, Paolo, Morcote 116515
O Pomar dos Artistas, Lisboa . . 132951
Pomar, María, Palma de Mallorca 115364
Pomares Esparza, J., Pamplona 115391
Pomarkun Kotiseutumuseo,
Pomarkku 011263
Pomázer Heimatstube, Sinsheim 021881
Pomegranate Europe, Warwick . 138291
Pomeroy Living History Farm,
Yacolt 054564
Pomes Penyeach, Newcastle-under-
Lyme 144739
Pomez, Olivier, Paris 071796
Pomfret Gallery, Pontefract . . . 118929
Pomier, Lyon 069774
Pomiès, Yveline, Toulouse 129313
Pomme d'Ambre, Nancy 070428
Pomme de Pin, Bergerac 067105
La Pomme d'Or, Genève 087497
Pommerat, Gérard, Toulouse . . 073792
Pommersches Bettenmuseum,
Flugplatzmuseum Peenemünde,
Peenemünde 021089
Pommersches Landesmuseum,
Greifswald 018808
Pommier, Gérard, Ferney-Voltaire 068524,
133739
Pommois, Isabelle, Paris 071797
Pomona College Museum of Art,
Claremont 047758

Pomone, Lutry 116464
Pomorski i Povijesni Muzej Hrvatskog
Primorja, Rijeka 008978
Pomorski Muzej, Zadar 009023
Pomorski muzej Orebić, Orebić . 008959
Pomorski Muzej Sergej Mašera,
Piran 038587
Pomorski Muzej u Kotoru, Kotor 031757
Pomorski Muzej, Dubrovački Muzeji,
Dubrovnik 008911
Pomorske Muzeum Wojskowe,
Bydgoszcz 034672
Pomp, Thomas, Bottrop 137121
Pompallier House, Russell 033299
Pompei, Franco, Roma 132016
Pompejanum, Bayerische Verwaltung
der staatlichen Schlösser, Gärten und
Seen, Aschaffenburg 016624
Pompey Museum of Slavery and
Emancipation, Nassau 003208
Pompous Peasant, Tampa 097583
Pomskizillious Museum of Toys,
Xaghra 030769
Ponca City Art Association, Ponca
City 060696
Ponce DeLeon Inlet Lighthouse, Ponce
Inlet 052235
Ponce, Raquel, Madrid 115272
Ponceblanc, Jouannin, Saint-Rambert-
en-Bugey 073284
Poncelet, David, Cambrai 067616
Poncet, Michel, Essertines-sur-
Yverdon 087418
Pond, London 118517
Pond Cottage Antiques,
Stillington 091426
Pond, Lily, Geelong 061513
Ponder, Richard, Wellington . . 113087
Pondicherry Museum,
Pondicherry 024438
Pongauer Heimatmuseum im Schloß
Goldegg, Goldegg 001960
Pongratz, Alois, Zwiesel 079155
Pongratz, Wolfgang, Graschuh . 127740
Poniger Clocks and Watches,
Armadale, Victoria 061046
Ponitrianske Múzeum, Nitra . . . 038410
Pons, Chantal & Patrick, Lyon . 069775,
128727
Ponse, Charenton-le-Pont 067839
Ponsford, A.J. & R.L., Barnwood 088368,
134068
Ponshu-kan-Echigo Sake Museum,
Yuzawa 029754
Pont & Plas, Gent 100701
Pont-Dépôt Ventes, Pont-Saint-
Esprit 072261
Le Pont des Arts, Narbonne . . . 104350
Le Pont du Diable, Céret 067751
Le Pont Traversé, Paris 141265
Pont, Michel, Biarritz 067211
De Pont, Museum voor Hedendaagse
Kunst, Tilburg 032930
Pont, Philippe, Saint-Loubès . . 072896
Ponte, Beograd 085345
A Ponte – Galeria de Arte, São
Paulo 101107
Il Ponte Contemporanea, Roma . 110847,
137739
Il Ponte, Casa d'Aste, Milano . 080679,
110406,
126210
Pontecorboli, Angelo, Firenze . 137654
Ponteduro, Carlo, Bologna 079870
Pontefract Castle Museum,
Pontefract 045518
Pontefract Museum, Pontefract . 045519
Ponterio & Assoc., San Diego . . 097121
Ponti, Alessio, Roma 110848
Ponti, Francesco, Roma 142831
Ponticq, Michèle, Duras 068419
Ponticq, Michèle, Monpazier . . 070169
Pontifical Biblical Institute Museum,
Jerusalem 025099
Pontiki Thalia, Lefkosia 102542

Pontis'Antic, Valbonne 073935
Ponto de Antiguidades, Porto
Alegre 128108
Pontremoli, Milano 142740
Pontual, Mauricio, Rio de Janeiro 101058
Pontypool and Blaenavon Railway,
Blaenavon 043650
Pontypridd Museum, Pontypridd 045524
Pony Express Museum, Saint
Joseph 052787
Ponziani – Lo Studiolo, Firenze . 131245
Pool Dickynson, Madrid 133366
Poole, Douglas, Toronto 064727
Poor Man's Gallery, Virginia
Beach 124839
Poor Richard's Auctions, Tacoma 127471
Poor Richard's Books, Felixstowe 144425
Poorhosaini, Seeheim-Jugenheim 078464,
126115
Poorhosaini, S.N., Darmstadt . . . 075308
Poorter, Johanna de, Antwerpen 100509
Pop Antik, Helsinki 066355
Pop-Art, Čačak 114449
Pop Art UK, Sheffield 119086
Pop Culture Classics, Phoenix . . 096234
Pop Indiana, Indianapolis 121215
Pope, Charlotte 120008
Pope County Historical Museum,
Glenwood 049130
Pope, Philippe, Nice 070649
Pope's Tavern Museum, Florence 048753
Popescu, Victor, Düsseldorf . . . 075506
Pophal, Veit, Dresden 129718
Poplar Antiques, Memphis 094662
Poplar Grove Historic Plantation,
Wilmington 054448
Popmount, Richmond 123738
Popmuseum, Muzeum a archiv
populární hudby, Praha 009835
Popoff & Cie., Paris 105177
Popolini, Massimo, Roma 132017
Popp, Ludwig, München 108177
Popp, Margit & Matthias, Zell,
Oberfranken 079135
Popp, O., Bremen 075179
Poppa, Austin 119690
Poppelsdorfer Heimatsammlung
Stöcker, Bonn 017498
Poppen- en Speelgoedmuseum,
Tilburg 032931
Poppen en Poppenhuis Museum 't
Duvelke, Breda 032105
Poppen, A., Esens 075667
Poppenhuis Carmen, Utrecht . . 032955
Poppenhuismuseum, Heesch . . 032438
Poppenmuseum, Ter Apel 032008
Poppenmuseum, Maastricht . . . 032627
Poppenmuseum Christus Koning,
Wetteren 004149
Poppenspe(e)lmuseum, Vorchten 033007
Poppentheatermuseum, Tienen . 004068
Poppin, Merry, Tshwane 085685
Poppo, Okutama 111344
Popppenhuismuseum Alida's Kleine
Wereldje, Veendam 032971
Poptaslot Heringa-State,
Marssum 032638
Popular Art Centre, Al-Bireh . . 034156
Popular Arts Museum, Instituto
de Cultura Puertorriqueña, San
Juan 036171
Popular Culture, Denver 093324
Popular Literature and Art Publishing
House, Beijing 136700
Popular Traditions Museum
Qasrelazem, Damascus 042259
Populart, Sevilla 086473
Poraskas, Peter, Flensburg 141765
Porcelænsgården, Hjørring 065767
Porcelænskælderen, København . 065669
Porcelain Repairs, Stockport . . . 135089
Porcelain Room, New York 095666
Porcelánmúzeum, Hollóháza . . . 023693
Porceleynboom, de, Breda 083003

Porcella, Antonio, Roma 110849
Porch, Oakland 095908
Porch Cellar, Philadelphia 096151
Porch House Antiques, Tetbury . 091579
Porcha, Detroit 093373
Porcher, Marseille 069960
Porcher, Dominique, Vern-sur-
  Seiche 074025
Porchet, Carmen, Lausanne ... 087583
The Porcine Gallery, Whangarei . 113111
Porcini, Vincenzo, Napoli 080919
Porcu, Lucia, Cagliari 079999
Porcu, Orlando, Milano 131502
Porcupine Plain and District Museum,
  Porcupine Plain 006670
Porcupines, Bideford 144247
Pordes, Henry, London 144667
Porfirul Antichitati, Cluj-Napoca . 085117
Porfyrmuseet, Älvdalen 040512
Porilaismuseo, Säkylä 011335
Porin Taidemuseo, Pori 011265
Porirua Hospital Museum, Porirua 033287
Porkeris Bygdasavn, Porkeri ... 010739
Porky's Bits & Pieces, Ohingaiti . 084045
Porlammin Kotiseutumuseo,
  Porlammi 011269
Porro, Dijon 068341
Porro & C., Milano ... 080680, 126211
Porro, Amedeo, Milano 110407
Porro, Francis, Dijon .. 068342, 128583
Porrua Hnos. y Cia., México ... 142990
Porrua, Manuel, México 142991
Porsanger Museum, Lakselv ... 033728
Porsche Automuseum Helmut
  Pfeifhofer, Gmünd, Kärnten .. 001945
Porsche Museum, Stuttgart 022054
Porsche Traktorenmuseum,
  Hitzendorf 002091
Porselein Dierenpark, Wildervank 033049
De Porselijne Lampetkan, Delft . 083036
Porsgrunn Bymuseum, Porsgrunn 033879
Porslinsdoktorn Dimitar, Malmö . 133557
Porst, Angelika, München ... 130383
Port Albert Maritime Museum, Port
  Albert 001417
Port Angeles Fine Arts Center, Port
  Angeles 052244
Port Antiques, Napier 084009
Port Arthur Historic Site, Port
  Arthur 001418
Port-au-Choix, National Historic Site,
  Port-au-Choix 006675
Port Chalmers Museum, Port
  Chalmers 033288
Port Clements Museum, Port
  Clements 006679
Port Colborne Historical and Marine
  Museum, Port Colborne 006680
Port Columbus National Civil War
  Naval Center, Columbus 047924
Port Community Art Centre, Port
  Adelaide 001414
Port Coquitlam Heritage and Cultural
  Museum, Port Coquitlam 006681
Port de Québec au XIXe Siècle, Saint-
  André 006844
Port Discovery – Children's Museum
  in Baltimore, Baltimore 046766
Port Douglas Gallery of Fine Art, Port
  Douglas 099583
Port Douglas Oceanic Art Studio, Port
  Douglas 099584
Port Dover Harbour Museum, Port
  Dover 006683
Port Elizabeth Museum, Port
  Elizabeth 038808
Port Fairy Historic Lifeboat Station,
  Port Fairy 001420
Port Fairy History Centre, Port
  Fairy 001421
Port Gamble Historic Museum, Port
  Gamble 052248
Port Hardy Museum, Port Hardy 006685
Port Huron Museum, Port Huron 052251

Port Jackson Press, Fitzroy 099029
Port-la-Joye-Fort Amherst, Rocky
  Point 006822
Port MacDonnell and District Maritime
  Museum, Port MacDonnell ... 001424
Port Macquarie Antique Centre, Port
  Macquarie 062097
Port Macquarie Historic Museum, Port
  Macquarie 001425
Port McNeill Museum, Port
  McNeill 006691
Port Moody Art Association, Port
  Moody 058462
Port Moody Station Museum, Port
  Moody 006693
Le Port-Musée, Douarnenez ... 012822
Port Natal Maritime Museum,
  Durban 038705
Port of Morgan Historic Museum,
  Morgan 001312
Port Phillip City Collection, Saint Kilda,
  Victoria 001472
Port Pirie National Trust Museum, Port
  Pirie 001426
Port Pirie Regional Art Gallery, Port
  Pirie 001427
Port Royal, Annapolis Royal 005341
Port Royal Store, Los Angeles . 094448
Port-Saïd Museum, Port Saïd ... 010582
Port Stephens Shell Museum,
  Corlette 000992
Port Sunlight Heritage Centre, Port
  Sunlight 045537
Port Townsend Marine Science Center,
  Port Townsend 052261
Port Union Museum, Port Union
  South 006696
Port Victor Gallery, Victor Harbor 001608
Port Victoria Maritime Museum, Port
  Victoria 001428
La Porta Antica, Milano 080681
Porta Borsari, Firenze 080189
Porta Cristina, Cagliari 080000
La Porta Michele, Catania 080044,
                           131096
Porta Romana, Firenze 080190
Porta, Luce, Knokke-Heist 100757
Portabello Antiques, Limerick .. 079660
Portage County Historical Society
  Museum, Ravenna 052475
Portage County Museum, Stevens
  Point 053525
Portakal, Raffi, İstanbul 088135, 126791
Portal, Gdańsk 084510, 113484, 132818
Portal Artes Galeria, Brasília ... 100923
Le Portal Artour, Québec 101516
Portal Galeria de Arte, São Paulo 101108
Portal Gallery, London 118518
Portal of the Folded Wings, North
  Hollywood 051607
Portal, Catherine & Frédéric, Saint-
  Jean-de-Luz 105535
Portalibros, Ajijic 142986
Portals, Chicago 129770, 120234
Portanje's Vespa Scooter en Nostalgie
  Collectie, Bunnik 032117
La Portantina, Milano 080682
Portas do Sol, Lisboa 084942
Portas, Danielle, Nîmes 070702
Portchester Castle, Portchester . 045540
Portculli, Craven Arms 088997
La Porte Antique, Cairo 066263
La Porte County Historical Society
  Museum, La Porte 050047
Porte d'Amont, Meung-sur-Loire 070106
Porte des Gaultiers, Mirmande . 104241
Porte Heureuse, Roussillon
  (Vaucluse) 105484
Porte Plume, Le Baule-Escoublac 103924
Porte Rouge, Chicago 092771
Porte, Jacques, Paris 071798
Portéka-Antique, Budapest 079283
Portelli, Simon, Nadur 082420
Porter, Coffs Harbour 061352

Porter Almoneda, Madrid 086240
The Porter Bookshop, Sheffield . 144838
Porter County Museum,
  Valparaiso 053939
Porter County Old Jail Museum,
  Valparaiso 053939
Porter Davis, Seattle 097486
Porter-Phelps-Huntington House
  Museum, Hadley 049308
Porter Thermometer Museum,
  Onset 051780
Porter, David, Claremont 085640
Porter, James, Sandwich 091247
Porter, Les, Newcastle 099435
Porter, R.E., Bournemouth 088539
Porterville Historical Museum,
  Porterville 052276
Portfolio, Auckland 112805
Portfolio Fine Art and Antiques,
  Braemar 061186
Portfolio Fine Arts, Anna Hodson
  Pressinger, London .. 118519, 134754
Portfolio Gallery, Saint Louis ... 123857,
                           123858
Porthcawl and District Art Society,
  Porthcawl 060211
Porthcawl Museum, Porthcawl .. 045541
Porthcurno Telegraph Museum,
  Porthcurno 045542
Porthilly Crafts, Wadebridge ... 119295
Porthmadog Maritime Museum,
  Porthmadog 045543
Porthminster Gallery, Saint Ives,
  Cornwall 119029
Portholes Into the Past, Medina . 050764
Portia Gallery, Chicago 120235
Portia Paperweights, Chicago .. 120236
Il Portico, Cagliari 109925
Portico Fine Art, New York 122977
Portico Gallery, Helensburgh .. 044556
Portico Gallery, Rotorua 113001
Portico Library and Gallery,
  Manchester 045217
Portico Librerias, Zaragoza ... 143817
Portier, Thierry, Paris 071799
Portikus, Frankfurt am Main ... 018444
Portis, Will, Fresno 120903
Portland Antique Company,
  Portland 096461
Portland Antiques and Curios, West
  Bridgford 091781
Portland Art Museum, Portland . 052302
Portland Basin Museum, Ashton-under-
  Lyne 043447
Portland Castle, Portland 045545
Portland Children's Museum,
  Portland 052303
Portland Gallery, London 090425, 118520
Portland Harbor Museum, South
  Portland 053415
Portland House Antiques, Market
  Deeping 090707
Portland Maritime Discovery Centre,
  Portland 001431
Portland Museum, Louisville ... 050500
Portland Museum, Portland ... 045546
Portland Museum of Art, Portland 052283
Portland State University Galleries,
  Portland 052304
Portlandia Productions, Portland 123660
Portmann, Alfred, Ebikon 087400
Portmeuneum – Muzeum Josefa
  Váchala, Litomyšl 009676
Porto Bello, Den Haag 083119
Porto Belo, Porto 085044
Porto Coroa, Porto 085045
Porto Franco, Livorno 080330
Porto Novo, Nijmegen 112612
Porto Restauro, Porto 132974
Porto Velho Arte e Decoracao, Rio de
  Janeiro 064105
Portobello, Århus 065605
Portobello, Catania 080045
Portobello, København 065945

Portobello, Lemesos 065332
Portobello, Paris 071800
Portobello, Philadelphia 096152
Portobello Antique Dealers Association,
  London 060212
Portobello Antique Store, London 090426
Portobello Antiques, Auckland .. 083817
Portobello Antiques, Christchurch 083891
Portobello Antiques, Vancouver . 064865
Portobello Bookshop, London .. 144668
Portobello Gallery, Pittsburgh .. 123569
Portobello Gallery, Saint James . 003238
Portobello Road and Camel Antiques,
  Kansas City 094089
Portobello Road Antiques,
  Hamilton 064328
Portobello Row Antique & Collector's
  Centre, Boston 088521
Portofino, Portofino Mare 110681
Portola Railroad Museum, Portola 052308
Porton de San Pedro, Montevideo 119435
Los Portones, Fuenlabrada 086002
Portrait Brokers of America,
  Birmingham 119816
Portrait Galery, Dordrecht 112441
Portrait Gallery, Blakesley 117388
Portrait Source, Fort Worth ... 120882
Portraits, New York 122978
Portraiture, Atlanta 119601
Portret na Życzenie, Warszawa . 113893
Portretgalerie, Den Haag 112406
Portrush Antiques, Nelson 084019
Portsmouth Athenaeum,
  Portsmouth 052314
Portsmouth Museum, Portsmouth 052320
Portsmouth Museum of Military
  History, Portsmouth 052327
Portsmouth Museums,
  Portsmouth 052328
Portsteffen, Prof. Hans, Köln ... 130162
The Portstewart Galleries,
  Portstewart 118944, 118945
Portugall, Dieter, Koblenz 107576
Portuondo Wakonigg, Jorge,
  Bilbao 085883
Portuondo Wakonigg, Ramón,
  Madrid 086241
Portz, Sascha, Niederzissen ... 077781
Porvoon Museo, Porvoo 011275
Porzellan-Klinik, Hamburg 129983
Porzellan Klinik Leipzig, Leipzig . 130227
Porzellan-Museum Höchst, Kronberger
  Haus, Frankfurt am Main ... 018445
Porzellanmuseum Reichmannsdorf,
  Reichmannsdorf 021347
Porzellansammlung, Staatliche
  Kunstsammlungen Dresden,
  Dresden 018004
Porzellanwerkstatt, Freiburg im
  Breisgau 129863
Porzellanwerkstatt, Köln 130163
Pos-Art, Montréal 101442
Posada Alvarez, María Teresa,
  Medellín 102457
Posada Art Books, Bruxelles ... 140198
Posadovský, Bedřich, Praha ... 128360
Posavski Muzej Brežice, Grad Brežice,
  Brežice 038499
Posch, Gabriele, Birkfeld 062536
Poscharnig, Brigitta, Langenlois 140006
Posion Kotiseutumuseo, Posio . 011279
Posipal, Veit, Dresden 129719
Positive, Dijon 068343
Positive Images Art, Austin 119691
POSK Gallery, London 045078
Pospíšil, Josef, Kutná Hora ... 128329
Pospíšilová, Dr. Marie, Praha .. 102719,
                           128361
Poß, Ingo, Hetzerath 076521
Poßegger, Stefan, Trautenfels . 062862
Posselt, M. I., Bonn 106536
Posset, 's-Hertogenbosch 112531
Possibilities, Jacksonville 094026
Possibilities, New Orleans 095144

Possum Art Gallery, Beenleigh . 098723
Possum Brush Gallery, Taree ... 099810
Post, Enschede 083257
Post, Los Angeles 121575
Post & Tele Museum, København 010191
Post 10 Antiques, Oklahoma City 096001
Post-Archaeological Institute,
Jackson 049758
Post Art, Amsterdam 082825
Post Gallery, Houston 121115
Post Horn Books, Settle 144830
Post House Antiques,
Bletchingley 088508
Post Museum, Fort Leavenworth
Historical Society, Fort
Leavenworth 048832
Post Rock Museum, Rush Co
Historical Society, La Crosse . 050027
Post Scriptum, Bologna 130991
Post, Bert, Noorden 137866
Post, Constantin, Köln . 137316, 142021
Post, Emmanuel, Leipzig 107847
Post, J.G., Breda 132457
Post, L.A., Benningbroek 132437
Post, Wilfried, Freiburg im
Breisgau 107062, 107063
Postakürt Galéria, Miskolc ... 023792
Postal and Philatelic Museum, Tel
Aviv 025209
Postal Museum, Amman 029778
Postal Museum, Cairo 010557
Postal Museum, Cheonan 029904
Postal Museum, Taipei 042543
Postal Museum and Philatelic Library,
Manila 034544
Postal Museum Botswana,
Gaborone 004296
Postamúzeum, Balatonszemes . 023428
Postamúzeum, Budapest 023544
Postamúzeum, Hollókő 023695
Postamúzeum, Nagyvázsony .. 023808
Postamúzeum, Opusztaszer ... 023816
Postapalota, Pécs 023850
La Poste d'Antan, Paris 071801
La Poste d'Autrefois, Paris ... 141266
Poste de Traite Chauvin,
Tadoussac 007124
Poste du Village, Rennes (Ille-et-
Vilaine) 141314
Poster Art, Buffalo 119963
Poster Art, San Diego 124156
Poster Art Place, San Antonio . 124058
Poster Arts, Los Angeles 121576
Poster Dizajn, Moskva 114291
Poster-Galerie, Hamburg 107307
Poster Galerie, Innsbruck 100053
Poster-Galerie, Nürnberg 108305
Poster Galerie Wien beim
Hundertwasserhaus, Wien 100362
Poster Gallery, Albuquerque . 119506
Poster Gallery, Kansas City .. 121312
Poster Planet, Sacramento ... 123779
Poster Plus, Chicago 120237
Poster Portfolio, New York ... 122979
Poster Restoration Studio, Los
Angeles 135739
The Poster Shack, Norfolk 123234
Poster Shoppe Gallery, Columbus 120509
Poster Source, San Francisco . 124368
Poster Store, San Diego 124157
Poster World, Phoenix 123511
Postergalerie, Dortmund 106708
Postergalerie Flensburg,
Flensburg 106959
Posteritati, New York 122980
Posterity Graphics, Toronto .. 101683
Posterman, Louisville 121665
Posteroptics, Windsor, Ontario . 101861
Posterplanet, Portland 123661
Posters, Abu Dhabi 117167
Posters, San Antonio 124059
Posters and Frames, San Diego 124158
Posters Etc, Toronto 101684

Posters Posters, Atlanta 119602
Posters Publication GmbH,
Staufenberg 137512
Posterworld, Saint Louis 123859
Posthoorn, Rotterdam 083586
Postimuseo, Helsinki 010854
Postmark Gallery, Kansas City . 121313
Postmasters, New York 122981
Póstminjasafn Íslands,
Hafnarfjörður 024080
Postmuseum, Stockholm 041032
Postmuseum des Fürstentums
Liechtenstein, Vaduz 030309
Postmuseum Mettingen,
Mettingen 020391
Postmuseum Rheinhausen,
Oberhausen-Rheinhausen 020899
Il Posto delle Briciole, Milano . 080683
Postojannaja Požarno-techničeskaja
Vystavka, Sankt-Peterburg ... 037815
Postons, Neil, Leominster 134549
Poštové Múzeum Banská Bystrica,
Banská Bystrica 038280
Poštovní Muzeum, Praha 009836
Poštovní Muzeum, Poštovní Muzeum
Praha, Vyšší Brod 009967
Postrotemuseet, Storby 011375
Postville Courthouse Museum,
Lincoln 050282
Le Potager, Jacksonville 094027
Potala Art Gallery, Kathmandu . 112084
Potamianos, Johann, München . 077595
Potard, Simone, Dierre 068300
Potashnik, David, Coulsdon ... 088981,
134299
Potboard Antiques, Tenby 091558
Potburys, Sidmouth 127046
Potchefstroom Museum,
Potchefstroom 038810
Le Poteau Rose, Lambersart .. 103917
Potemska, Maria, Łódź 132851
Potennec, Jean-Claude, Verdun
(Meuse) 074021
Potereau, Châteauroux 103583
Poterie Dubois-Musée la Grange aux
Potiers, Bouffioulx 003387
Pothmann, Willi, Siegen 078464
't Pothuis, Haarlem 083349
't Pothuis, Zutphen ... 083726, 132675
Potiron, Marc, Saint-Sébastien-sur-
Loire 073306
Potland Museum, Basildon ... 043507
Potluck, Nashville 095024
Potoroo, South Melbourne 099702
Potpourri, Lörrach 077145
Potpourri, Pittsburgh 123570
Potrebny, H. & M., Horb am
Neckar 076574
Potrony, Miami 121844
Potsdam-Museum, Geschichte und
Kunst, Potsdam 021198
Potsdam Public Museum,
Potsdam 052332
Potsdamer Kunstsalon, Potsdam 078082,
126099
Potstada, Heike, Kiel 076788
Pottenbakkerij De Brinksteen,
Dwingeloo 112448
Pottenbakkerij Het Ovenhuis,
Ruinerwold 032837
Pottenbakkerij Museum,
Kattendijke 032539
Potter, Houston 093768
Potter, Lewes 089818, 134554
Potter, Merredin 127639
Potter, Vancouver 064866
The Potter Clarke Gallery, Saint Ives,
Cornwall 119030
Potter County Historical Society
Museum, Coudersport 048032
Potter, Barry, Stockton-on-Tees . 127053
Potter, Jonathan, London 144669
Potter, Nick, London 118521

Potter & Son, W.G., Axminster .. 088327,
134046
Potteries Antique Centre, Stoke-on-
Trent 091453
The Potteries Museum and Art Gallery,
Stoke-on-Trent 045884
Potteries Specialist Auctions,
Cobridge 126865
Potteries Specialist Auctions, Stoke-on-
Trent 127055
Potters, Gladstone 099078
Potter's Antiques and Coins,
Bristol 088666
Potters Bar Museum, Potters Bar 045562
Potters Cottage Gallery,
Warrandyte 099860
Potter's Guild of San Diego, San
Diego 060697
Potters Shop, Wellington 113088
Potters, Chelsea, Auckland ... 112806
Potterton, New York 145369
Potterton, Thirsk 144882
Pottery Art Cooperative, Wuxi . 058502,
102254
The Pottery Buying Centre, Stocke-on-
Trent 091437
Pottery Northwest Seattle Center, 305
Harrison St, Seattle 124589
Pottery Stall, Hong Kong 102106
Pottery Studio, Seattle 124590
Pottery Workshop, Hong Kong . 102107
Pottier, Jean-Roger, Vitteaux .. 074264
Potton, Craig, Nelson 137883
Potts Inn Museum, Pottsville .. 052335
Potts, Ludovic, Ely 134366
Pottsgrove Manor, Pottstown .. 052334
Potuguesa, Tostedt 108771
Potzmann, Franz, Litzelsdorf .. 127813
Pou d'Art, Sant Cugat del Vallès 115413
Pou, Aura & S., Barcelona ... 133145
Poubeau, Anne, Sancerre 073402
Pouce Coupe and District Museum,
Pouce Coupe 006699
Pouch Cove Museum, Pouch
Cove 006700
Pouch, M., Český Krumlov ... 140465
Pouffier, Elisabeth, Dijon 068344
Pouillier, Michel, Saint-Maixme-
Hauterive 072910
Pouillon, François, Versailles ... 074066
Poulain, Emile, Baud 066965
Poulain, Pierre, Paris 071802
Le Poulbot, Porto-Vecchio 072290
Poulin, Steve & Jeannine,
Fairfield 127220
Poulmarch, Armelle, Brest 103464
Poulsen, Kolding 065978
Poulsen, Margit, Fanø 065648
Poulsen, Svend, Silkeborg 066112
Poulter & Son, H.W., London .. 090427,
134755
Pound Gallery 1216 10th Av,
Seattle 124591
Pound Ridge Museum, Pound
Ridge 052342
Pound Sreat Gallery, Carshalton 117539
Pounder Kone Art Space, Los
Angeles 121577
Pounder, Nicholas, Double Bay . 139717
Pounds, Linda, Houston 093769
Poupard, Marguerite, Sers 073511
Poupée Merveilleuse, Marcy .. 069869
Poupée Passion, Longueil 128693
Poupées Retrouvées, Paris ... 071803,
129120
Pourhossein, Hossein, Darmstadt 075309
Pourny, Christian, Fayence 068518
Pourrier, Nathalie, Arras 066740
Pouryun Museum, Guanyun ... 007815
Pousse-Cornet, Marie-Edith, Blois 125518
Poussier, Charles, Nanterre ... 070437
La Poussière du Temps, Paris . 141267
Poussières, Bruxelles 063531
Poussières d'Empires, Paris ... 071804

Poutrain, Pascal, Douai 068393
Poutsma, J., Kampen 143177
Pouvreau, Cathy, Lille 104015
Pouyet, Gérard, Cluses 068052
Povážská Galéria Umenia, Žilina 038485
Povážské-Múzeum, Žilina 038486
Povijesni Muzej Istre, Pula ... 008974
Powder Magazine, Beechworth . 000839
Powder Magazine, Charleston .. 047559
Powderham Castle, Kenton ... 044677
Powderhouse Antiques, New
York 095667
Powell, Chicago 145052, 145053,
145054
Powell, Columbus 093039, 120510
Powell Bros. & Sons, Salt Lake
City 096916, 136241
Powell County Museum, Deer
Lodge 048217
Powell River Historical Museum,
Powell River 006701
Powell Street Gallery, San
Francisco 124369
Powell, Brian, Ackworth 083818, 126307
Powell, James, Austin 092260
Powell, M., Portland 145462
Power Art, Rio de Janeiro 101059
Power Group, Vancouver 101825
Power House Books, Brooklyn . 138311
Power Plant Contemporary Art Gallery,
Toronto 007195
Power, Robyn, Kaleen 099164
PowerHouse Arena, Brooklyn . 119944
PowerHouse Books, Brooklyn . 138312
Powerhouse House and Car Museum,
Portland 001432
PowerHouse Magazine, Brooklyn 139478
Powerhouse Museum Yanco,
Yanco 001667
Powerhouse Museum, Museum
of Applied Arts and Sciences,
Ultimo 001600
Powers Museum, Carthage 047452
Powers, Pamela, Seattle 124592
Powis Castle, Welshpool 046082
Pownalborough Court House,
Dresden 048369
Powysland Museum and Montgomery
Canal Centre, Welshpool 046083
Pożegnanie z Afryką, Łódź ... 084616
Pozza, Dario, Trieste 081800
Pozza, Neri, Vicenza 137764
Pozzi & Fontana, Milano 080684
Pozzi, Aldo, Riehen 133896
Pozzi, Luigi, Genova 131337
Il Pozzo, Modena 080816
Pozzolini, Simonetta, Roma ... 132018
PP Gallery, Stafford 119135
PR Mønter, Århus 065606
PRA Antikvariat, Stockholm ... 143959
Pra Site Museum, Bogor 024550
Prabang Antiques, Los Angeles . 094449
Prabhas Patan Museum, Prabhas
Patan 024440
Praça das Artes, Rio de Janeiro 101060
Praça Paris Obras de Arte, Rio de
Janeiro 064106
Prachatické Muzeum, Prachatice 009772
Prácheňské Muzeum, Písek ... 009757
Pracher, Peter Rudolf & Britta,
Würzburg 130785
Prachner im MQ, Wien 140116
Pracownia – Muzeum Józefa
Ignacego Kraszewskiego, Biblioteka
Raczyńskich, Poznań 035138
Pracownia Konserwacji Dzieł Sztuki,
Szczecin 132870
Pracownia Konserwacji Rzeźby,
Warszawa 132896
Practiqon Antik, Djursholm ... 086657
Pradarshak, Mumbai 109459
Le Pradel, Bruxelles 063532
Pradelle, Hugues, Toulouse ... 129314
Pradilla, Fernando, Madrid ... 115273

Prado, Brugge 100537
Del Prado, Madrid 143780
Prado Antiguedades, Madrid ... 086242
El Prado Antiquarium, Rio de
 Janeiro 064107
Prado, Vasco, Porto Alegre 100996
Praedicantenbibliothek der
 Nikolaikirche, Isny im Allgäu .. 019435
Prähistorische Privatsammlung Walz,
 Mühlen 002418
Prähistorische Siedlung Pestenacker,
 Weil, Kreis Landsberg am Lech 022475
Prähistorisches Heimatmuseum,
 Veringenstadt 022319
Prähistorisches Museum,
 Hallstatt 002052
Prämonstratenser-Chorherrenstift mit
 Stiftssammlungen, Geras 001931
Präsenz, Mössingen 108025
Präsenz Kunst & Buch, Anna-Donata
 Freifrau von Lerchenfeld,
 Hünfelden 107450, 137284
Præstø Antik, Præstø 066050
Præstø Antikvariat, Præstø 140641
Prag, Ken, San Francisco 097308
Praga, Moskva 085197
Prager Kunstsalon, Nürnberg ... 108306
Pragnell, George, Stratford-upon-
 Avon 091494
Prague Biennale, Praha 098060
Prahart, Sale 119051
Prahse, G., Erlangen 075655
Le Praillot, Bonnay 103406
Prairie Acres Museum, Picture
 Butte 006651
Prairie Art Gallery, Grande Prairie 005914
Prairie Avenue Bookshop,
 Chicago 145055
Prairie Homestead, Philip 052105
Prairie Memories Museum, Irvine 006044
Prairie Mountain Regional Museums
 Collection, Shoal Lake 007029
The Prairie Museum of Art and History,
 Colby 047864
Prairie Panorama Museum, Czar 005675
Prairie Pioneer Museum, Craik . 005661
Prairie River Museum, Prairie
 River 006702
Prairie Trails Museum of Wayne
 County, Corydon 048021
Prairie Village, Madison 050591
Prairie West Historical Centre,
 Eston 005793
Prairie Wind, Basel 116094
Prakrit Art Gallery, Chennai 079378,
 109290
Prandi, Dino & Paolo, Reggio
 Emilia 142792
Pranee, Dallas 093203
Prangenberg, Thomas, Köln ... 076901
Prano Gudynas Restauration Centre,
 Vilnius 132323
Prantl & Boch, Dornbirn 099978, 136501
Prantl, Martin, Haiming 127764
Prarin, Lavaur 069208
Prasad & Co, Giriraj, Jaipur,
 Rajasthan 109386
Praska, Warszawa 113894
Prassler, Cornelia Caroline,
 Bologna 130992
Prastiati, Doetinchem 083172
Prat, Alfred, Linz 100112, 127809
Prata, Mario da, São Paulo 128130
Sri Pratap Singh Museum,
 Srinagar 024485
Pratella & Sepe, Livorno 080331
Prater & Rouse, Norfolk 123235
Pratermuseum, Wien Museum,
 Wien 003066
Praters Mill, Varnell 053951
Prates, António, Lisboa 114854, 114055
Pratesi, Giovanni, Firenze 080191
Prati, Franca, Palermo 110614
Pratley, Bathurst 061106

Prats, Joan, Barcelona 114951
Pratt, San Diego 124159
Pratt, Sevenoaks 119070, 138282
Pratt Building, Cohasset 047861
Pratt Contemporary Art, Ightham 118008
Pratt County Historical Society
 Museum, Pratt 052349
Pratt Fine Arts Center, Seattle .. 057899
Pratt Institute, Brooklyn 056800
Pratt Institute, Munson-Williams-
 Proctor Arts Institute, Utica ... 058032
Pratt Manhattan Gallery, New
 York 051461
Pratt Museum, Homer 049521
Pratt & Son, Leo, South
 Walsham 091368
Pratter, Josef, Erfurt 129805
Pratzon, Jill, New York 135977
Praud, Dominique, Pornichet ... 072282
Prawdzic, Gdynia 084521
Praxis Art Actuel, Sainte-Thérèse 006961
Praxis Arte Internacional, Buenos
 Aires 098576
Praxis Arte Internacional, México 111979
Praxis Galería de Arte, Santiago de
 Chile 101921
Praxis Gallery, Toronto 101685
Praxis Gallery Cafe, Bideford ... 117360
Praxis International Art, Miami . 121845
Praxis International Art, New
 York 122982
Praz-Delavallade, Paris 105178
Pražské Starožitnosti, Praha ... 065517
PRD Antiques, Houston 093770
Pré aux Sources, Bruxelles ... 100651
Pre Loved Books, Torquay 139934
Pre to Post Modern, Nashville .. 095025
Pré, Jean-Marc, Limonest 069535
Préard, Yannick, Nantes 070475, 128839
Precania, Stefana, Palermo 081066
Prechtl, Bischofsgrün 141602
Prechtl, Peter, Rottach-Egern .. 078295
Precious Little Things, Las Vegas 094169
Precious Metal and Stone Unlimited,
 Philadelphia 096153
Precious Metals of Arizona,
 Tucson 097687
Precious Moments, Honolulu ... 093570
Precuious Arts, Lalitpur 112094
Predela, Granada 133269
La Predelle, Besançon 103371
Predjamski Grad, Postojna 038592
Préférences, Saint-Ouen 073172
The Preferred Perch, Winnipeg . 101892
Prefix Institute of Contemporary Art,
 Toronto 058463
Prefix Photo, Toronto 138588
Pregão e Leilão, Lisboa 126453
Pregler, Sylvia & Andreas, Neukirchen
 beim Heiligen Blut 108243
Pregliasco, Umberto, Torino ... 142868
Préhisto-Parc et Grottes, Réclère 041865
Préhistoire et Protohistoire,
 Jujurieux 013339
Le Préhistorama, Musée des Orgines
 de l'Homme, Rousson 015181
Prehistory Museum, Department of
 Antiquities, Tripoli 030299
Preilu Vēstures un Lietišķās Mākslas
 Muzejs, Preiļi 030171
Preiner, Graz 100017, 127752
Preis & Preis, Cham 129650
Preis & Preis, Parsberg 130506
Preis & Preis, Regensburg 130569
Preisegger, Saarbrücken 078332
Preiss, Los Angeles 121578
Preiss, Wolfgang, Overath 078001
Preißing, Ute, Flensburg 075749
Preißinger, Eva, Deggendorf ... 075319
Preißinger, Martin, Roding 130592
Préjean, Jean-Jérôme, Paris ... 071805
Prell, Lohne 107889
Preller, M.A. M.C., München ... 077596,
 130384

Premier Antikvitás, Budapest .. 079284
Premier Art, Nashville 122095
Premier Artworks, San Antonio . 124060
Premier Auction Gallery, Jasper 127271
Premier Gallery, Minneapolis ... 122025
Premier Gallery of Art, Salt Lake
 City 123979
Premier Prints, Milton Keynes .. 118747
Premium Quality Coin,
 Minneapolis 094945
Premoli-Lorandi, Milano 080685
Prempeh Jubilee Museum,
 Kumasi 022936
Premsela & Hamburger,
 Amsterdam 082826
Premsela, Robert, Amsterdam . 143060
Premselaar, Amsterdam 082827
Prenner, Ilse, Wien 127960
Prensa Cicuta, Cuevas del
 Almanzora 137974
't Prentarium, Leiden 083442
't Prentenkabinet, Ridderkerk .. 143214
Prentenkabinet Universiteit Leiden,
 Leiden 032583
Prentwerk, Groningen 143151
Prenzel, Dieter, Ellrich 075596
Prenzlow, A., Leverkusen 107862
Preporod, Sarajevo 063871
Le Presbytère, Aujac 066811
Presbyterian Historical Society
 Museum, Montreal 051039
Presbyterian Historical Society
 Museum, Philadelphia 052086
Prescot Museum, Prescot 045563
Prescott Fine Arts Association,
 Prescott 060698
Prescott House Museum, Port
 Williams 006697
Presekan, Reutlingen 078213
Présence du Passé, Marseille . 069961
Presenhuber, Eva, Zürich 116847
Present, České Budějovice 102578
Present Moment Books,
 Minneapolis 145284
Presentatieruimte de Overslag,
 Eindhoven 032291
Presentation House Gallery, British
 Columbia Photography and Media
 Arts Society, Vancouver 007293
Presente Passado com Antiguidades,
 São Paulo 064201
Les Présents du Passé,
 Montfermeil 070243
Prešernov Spominski Muzej,
 Kranj 038533
Preservation Hall Antiques,
 Portland 096462
Preservation Press, Bognor Regis 138141
Preservation Services, Perth ... 127659
Preservation Station, Nashville . 095026,
 135848
Preservation Tips, Columbia ... 139479
Presidency Museum, Tehrän ... 024779
President Andrew Johnson Museum,
 Greeneville 049241
President Benjamin Harrison Home,
 Indianapolis 049697
President Chester A. Arthur Historic
 Site, Fairfield 048675
President Pretorius Museum,
 Potchefstroom 038811
The Presidential Museum,
 Odessa 051710
Presidential Pet Museum,
 Williamsburg 054415
President's Cottage Museum, White
 Sulphur Springs 054375
Presidio La Bahia, Goliad 049156
Presnja Istoriko-memoralnyj Muzej,
 Gosudarstvennyj centralnyj
 muzej sovremennoj istorii Rossii,
 Moskva 037496
Presque Isle County Historical
 Museum, Rogers City 052664

Presqu'Île Provincial Park Museum,
 Brighton 005478
The Press Gallery, Chippendale . 098888
Press, Richard L., Sacramento . 145491
Presses Universitaires de France,
 Paris 136977
Preßler, Bernd, Kassel 141981
Pressler, Dr. Karl H., München . 142179
Pressler, Guido, Hürtgenwald . 137285
Pressmar, Johanna, München .. 108178
Pressoirs des Ducs de Bourgogne,
 Chenôve 012547
Presswood, Joe, Houston 127256
Prest, Nathalie, Mougon 128817
Prestel, München 137420
Prestel, New York 138420
Prestel, F.A.C., Frankfurt am
 Main 107022
Prestes, Luiz Cesar da Silva, Porto
 Alegre 100997, 100998
Prestige, İstanbul 088136
Le Prestige, Paris 105179
Prestige Art Gallery, Nashville .. 122096
Prestige Art Gallery Amsterdam,
 Amsterdam 112268
Prestige Arts, Dallas 120644
Prestige de l'Art, Grignon 068744
Prestige du Passé, Paris 129121
Prestige Framing Gallery,
 Tauranga 113026
Prestige Gallery, College Park . 098908
Prestige Gallery, Salt Lake City . 123980
Prestige Pictures, Craigavon ... 117637
Prestige Rental, Fort Worth 120883
Prestige, John, Brixham 088670
Prestinoni, Piero, Milano 131503
Preston Antique Centre, Preston 091057
Preston Art Center, Dallas 120645
Preston Auctions, Preston 125227
Preston Book, Preston 144792
Preston Hall Museum, Stockton-on-
 Tees 045879
Preston Manor, Brighton 043738
Preston Mill and Phantassie Doocot,
 East Linton 044207
Preston, Antony, Stow-on-the-
 Wold 091477
Prestongrange Museum,
 Prestongrange 045569
Prestwould Foundation,
 Clarksville 047779
Prète, Antoine, Lyon 069776
Pretérito Mais Que Perfeito,
 Lisboa 084943
Preti, Bruno, Milano 080686
Pretoria Art Museum, Tshwane . 038856
Prêtres de Saint-Sulpice de Montréal,
 Montréal 006414
Pretti, Solange, Belo Horizonte . 100903
Pretty Art Center, Taipei 116937
Pretty Bizarre, Deal 089037
Pretty Price Antikvariat, Oslo ... 143413
Pretzsch, Göttingen 076051
Pretzsch, Peter, Göttingen 141837
Preus Museum, Horten 033672
Preuschl, Monika, Kallmünz .. 076666
Preußen-Museum Nordrhein-Westfalen,
 Museum für Preußische Geschichte
 im Rheinland, Wesel 022591
Preußen-Museum Nordrhein-Westfalen,
 Museum für Preußische Geschichte
 in Westfalen, Minden 020419
Preuve par Trois, Angoulême .. 066629
Preux, Marie-José, Saint-Ouen . 073173
Prevarin, Milano 131504
Preveli Monastery, Myrthios ... 023184
Previati, Ernesto, Torino 142869
Preview Berlin, The Emerging Art Fair,
 Berlin 098165
Prévost, Paris 071806
Prévot, Buzancais 067539
Prevot, Philippe, Ruoms 072648
Preyß, Dorothea, München ... 111889
Prezent, Skopje 111889

Preziophil, Genova 080284
Prialnic, Pierrette, Fontenay-sur-
Loing 068579
Pribila, Paul, Turramurra 127692
Price Glover, New York 095668
Price Tower Arts Center,
Bartlesville 046800
Price, A.R., Coventry 144358
Price, Andrew, Carnegie 061288, 127579
Price, Graham, Polegate 091025, 134944
Prichard, Winchcombe 091865
Prichard Art Gallery, University of
Idaho, Moscow 051076
Prichard, Pat, San Antonio . . . 097017
Pricketts Fort, Fairmont 048678
Prickwillow Drainage Engine Museum,
Prickwillow 045570
Priddy, W.D., Chichester 088876
Pride and Joy Antiques, Houston 093771
Prides of London, London 090428
Priener Antikstube, Prien am
Chiemsee 078095
Priental-Museum, Aschau im
Chiemgau 016629
Prienų Krašto Muziejus, Prienai . 030449
Priešgaisrinės Apsaugos Muziejus,
Šiauliai 030469
Priest House, West Hoathly 046089
Priest Popple Books, Hexham . . 144476
Priester Poppe Museum, Temse . 004054
Priester, Jürgen, Bonn 075084
Priester, Pierre de, Rennes (Ille-et-
Vilaine) 072441
Priester, Pierre de, Saint-Quentin
(Aisne) 073279
Priestley & Ferraro, London . . . 118522
Priest's House, Easton-on-the-Hill 044224
Priest's House Museum, Wimborne
Minster 046141
Prieuré du Vieux Logis, Nice . . . 014492
Prigent, Alain, Hardricourt 068792
Prignitz-Museum Havelberg,
Havelberg 019115
Prijić, Aleksandar, Herceg-Novi . 112022
Prili, David, Roma 081511, 132019
Prili, Franco, Roma 081512
Prili, Luciano, Roma 081513
Prili, Marcello, Roma 081514
Prili, Mario, Roma 081515
Prima Art, Bratislava 143689
Prima Galleri, Örebro 086906
Prima Vista, Maastricht 083475
Primard Création, Fitilieu 103740
Primarily Portraits, Philadelphia . 123441
Primart Gallery, Claremont 114780
Primarte, Valencia 115512
Primary Source, Los Angeles . . 121579
Primavera, Cambridge 117509
Primavera Gallery, New York . . . 095669
Prime Art 2000, Perth 099562
Prime Canvas, Hayfield 117939
Prime Gallery, Toronto 101686
Primer Museo de las Abejas en
México, Colonia Condesa . . . 030896
Primerano, Francesco, Milano . 131505
Primetime Watches, München . . 077597
Primitive, Chicago 092772, 092773,
120238
Primitive Kitchen, Cincinnati . . . 092886
Primona Art Gallery, Mumbai . . 109460
Primorskaja Gosudarstvennaja
Kartinnaja Galereja, Vladivostok 038042
Primorskij Gosudarstvennyj Muzej im.
V.K. Arseneva, Vladivostok . . . 038043
Primrose Antiques, Jacksonville 094028
Primrose Hill Books, London . . . 144670
Primrose Hill Press, Newville . . . 138433
Primstaven, Nesttun . . 084285, 113186
Primula Textilkonservering,
Vällingby 133601
Primus Verlag, Darmstadt 137137
Le Prince, Cairo 103072
Prince & Pauper, San Diego . . . 145527
Prince Albert Arts Centre, Prince

Albert 006711
Prince Albert Historical Museum,
Prince Albert 006712
Prince Arthur Fine Arts, Toronto 101687
Prince d'Orange, Ukkel 063853
Prince Edward Island Heritage
Foundation, Charlottetown . . . 058464
Prince Edward Island Museum, Musée
et Patrimoine Île du Prince Édouard,
Charlottetown 005597
Prince Edward Island Potato Museum,
Musée de la Pomme de Terre de l'Île
du Prince Édouard, O'Leary . . 006551
The Prince Edward Island Regiment
Museum, Charlottetown 005598
Prince Edward Island Shellfish
Museum, Bidford 005422
Prince Georg Railway and Forestry
Museum, Prince George 006716
Le Prince Noir, Aubagne 066775
Prince of Serendip, Toronto . . . 064728
Prince of Wales Fort, Churchill . . 005620
Prince of Wales Martello Tower,
Halifax 005961
Prince of Wales Museum of Western
India, Mumbai 024410
Prince of Wales Northern Heritage
Centre, Yellowknife 007496
Prince Rupert Fire Museum, Prince
Rupert 006720
Prince Street Gallery, New York . 122983
Prince Township Museum, Sault
Sainte Marie 006991
Prince, Ghislain, Caudry 067732
Princess Antique Centre,
Shrewsbury 091332
Princess Gracie's Dowery,
Jacksonville 094029
Princess of Wales' Own Regiment
Military Museum, Kingston . . . 006110
Princess of Wales's Royal Regiment
and Queen's Regiment Museum,
Dover 044133
Princess Patricia's Canadian Light
Infantry Museum, Calgary 005533
Princesse de Kiev, Nice 104385
Princessehof Leeuwarden, Nationaal
Keramiekmuseum, Leeuwarden 032575
Princeton and District Museum,
Princeton 006722
Princeton Museum, Princeton . . 052367
Princeton University Art Museum,
Princeton 052368
Princeton University Museum of
Natural History, Princeton . . . 052369
Princeton University Press,
Princeton 138446
Principal Antiques, Mount
Gambier 061911
Principal Sombrerers, Barcelona 114952
Principia Fine Art, Hungerford . . 089613
El Principito, Burgos . 137972, 143737
Pringle Antiques, Fochabers . . . 089288
Pringle Cottage Museum,
Warwick 001626
Pringle Gallery, Philadelphia . . 123442
Pringle, Andrew, Edinburgh . . . 144399
Prins Eugens Waldemarsudde,
Stockholm 041033
Prins Hendrik de Zeevaarder Museum,
Egmond aan Zee 032278
Prinsheerlijk Antiek, Amsterdam 082828
Print, Cincinnati 139480
Print and Drawing Council of Canada,
Edmonton 058465
Print and Picture Collection,
Philadelphia 052087
Print Arts Northwest, Portland . . 060699,
123662
The Print Bin, Glen Osmond . . . 099080
The Print Bin, Glenelg 099086
The Print Center, Philadelphia . . 052088
The Print Center Gallery Store,
Philadelphia 123443

Print Club of Cleveland,
Cleveland 060700
Print Consortium, Kansas City . . 049889
Print Council of America,
Cambridge 060701
Print Gallery, Christ Church 100463
Print Quarterly, London 138240, 139215
The Print Works, Haslingden . . . 117924
Printed Image Photographic Bookshop,
Prahran 139887
Printed Project, Dublin 138782
Printed Words, Frankston 139738
Printers Row, Chicago 145056
Printfolio Gallery, Sydney 062297
Printing House Museum,
Cockermouth 043984
Printique Gallery, Woollahra . . . 099920
Printmaker, Honolulu . . 120965, 135580
Printmaker Gallery, Fitzroy 099030
The Printmakers, Dublin 109615
Printmakers Council of Prince Edward
Island, Prince Edward Island . . 058466
Printmakers Council, Clerkenwell
Workshop, London 060213
Printmakers Gallery, Saint Ives,
Cornwall 119031
Printmakers of Cape Cod, East
Falmouth 060702
Printmaking Council of New Jersey,
Somerville 053386, 060703
Printmaking Today, Witney 139216
Prints and Perspectives, Saint
Paul 123924
Prints Charming, Calgary 101256
Prints Charming, El Paso 120838
Prints Charming Gallery, Saint
Louis 123860
Prints Old and Rare, Pacifica . . 145403
Prints Plus, Albuquerque 119507
Prints Plus, Austin 119692
Prints Plus, Baltimore 119766
Prints Plus, Buffalo 119964
Prints Plus, Cincinnati . 120367, 120368
Prints Plus, Dallas 120646
Prints Plus, Fresno 120904
Prints Plus, Las Vegas 121373
Prints Plus, Memphis 121709
Prints Plus, Miami 121846
Prints Plus, Minneapolis 122026
Prints Plus, Portland . . 123663, 123664
Prints Plus, Saint Louis 123861
Prints Plus, Salt Lake City 123981
Prints Plus, San Diego 124160
Prints Plus, San Jose 124458
Prints Plus, Tampa 124668
Prints Unlimited, Chicago 120239,
120240, 120241
Prints World, Belo Horizonte . . . 063916
Printsonweb, Milano 110408
Printworks Fine Art Gallery,
Dallas 120647
Printworks Gallery, Chicago . . . 120242
Printziples Fine Art, Toronto . . . 101688
Prinz-Dunst, Karin, Berlin 074899,
125886
Prinz, Josef, Salzburg . 062813, 127855
Prinz, Kay, Felde 106948
Prior Restorations, Saint Paul . . 136224
Prior, Margret, Bochum 075040
Prior, Walter, Kassel 076726
Priors Reclamation Unit, Ditton
Priors 089064
Priory Antiques, Bridlington 088604
Priory Antiques, Godalming 089341
Priory Bookshop, Great Malvern 144445
Priory Gallery, Bishop's Cleeve . 088489
Priory Gallery, Broadway 117459
Priou, Paul, Civray 067994
Prioul, La Mézière 128643
Prirodnjački Muzej, Beograd . . . 038179
Prirodnjački Muzej Crne Gore,
Podgorica 031767
Prirodnyčyj Muzej, Oblasnyj
krajeznavčnyj muzej, Cherson . 043140

Prirodonaučen Muzej, Burgas . . 004999
Prirodonaučen Muzej, Černi
Osăm 005001
Prirodonaučen Muzej, Kotel 005077
Prirodonaučen Muzej, Plovdiv . . 005132
Prirodonaučen Muzej, Varna . . . 005246
Prirodonaučen Muzej na Makedonija,
Skopje 030633
Prirodonaučen Muzej, Istoričeski
Muzej, Belogradčik 004983
Prirodoslovni Muzej, Rijeka 008979
Prirodoslovni Muzej, Split 009001
Prirodoslovni Muzej Slovenije,
Ljubljana 038556
Prirodovedčeskij Muzej, Smolenskij
Muzej-Zapovednik, Smolensk . 037869
Přírodovědecké Muzeum, Národní
muzeum, Praha 009837
Přírodovědecké Oddělení, Slezské
zemské muzeum, Opava 009737
Prírodovedné Múzeum, Slovenské
Národné Múzeum, Bratislava . 038327
Priscilla, Paris 129122
Il Prisma, Cuneo 109998
Il Prisma, Milano 110409
Prison Arts Program, Saint Louis 123862
Prisoner's of Taste, Nashville . . 095027
Prithvi Art Gallery, Mumbai 109461
Pritschow, Garmisch-
Partenkirchen 107103
Pritt, Columbus 093040
Prittlewell Priory Museum, Southend-
on-Sea 045834
Priuli & Verlucca, Pavone 137716
Privat, Paris 071807
Privat, B. & D., Paris 141268
Private Art Dealers's Association, New
York 060704
Private Collection, Knokke-Heist 063708
Private Collection, Manama . . . 063157,
100445
Private Collection-Antiques,
Austin 092261
Private Collections, Nokomis . . . 127370
Private Collector, San Diego . . . 097125
Private Collector Gallery,
Innishannon 109645
Private Connection, New Orleans 122219,
122220
Private Sammlung Mechanischer
Musikinstrumente,
Markneukirchen 020290
Private Screening, Seattle 097487
Private View, Paris 105180, 105181
Privatizējamās Valsts Vukā Muzejs,
Kuldīga 030150
Privatmuseum Hans Klein,
Prichsenstadt 021220
Privatmuseum Im Blauen Haus,
Appenzell 041226
Privatmuseum im Reckturm, Wiener
Neustadt 003107
Privatsammlung Bergkristallmuseum,
Saalbach 002616
Privatsammlung Burg Runkel,
Runkel 021551
Privatsammlung Hermine Brandstetter,
Ostermiething 002483
Privatsammlung Leo Gesell,
Triefenstein 022194
Privatsammlung Piaty, Waidhofen an
der Ybbs 002891
Priz, Sofia 101168
Prize, San Francisco 097309
Pro-Art, Bottrop 106552
Pro Art, Gundelfingen, Breisgau 076134
Pro Arte, Bucureşti 139004
Pro Arte, Szczecin 113783
Pro-Arte-Edition, Hamburg 137252
Pro Arte Medii Aevi – Freunde des
Museum Schnütgen e.V., Köln . 059220
Pro Arts Gallery, Oakland 051683
Le Pro des Particuliers,
Montagnole 070179

Pro Domo, Graz 062599
Pro Hart Gallery, Broken Hill . . . 098801
Pro-Jex Gallery, Austin 119693
Pro Lab Imaging, Norwood 127654
Pro Litteris, Schweizerische
Urheberrechtsgesellschaft für
Literatur und bildende Kunst,
Zürich 059956
Pro Marco, Roma 132020
Pro Rodeo Hall of Fame and Museum
of the American Cowboy, Colorado
Springs 047890
Pro Sagi Samstagern,
Samstagern 041920
Pro Urba, Mosaiques Romaines,
Orbe 041830
Pró Velho, Lisboa 084944
Pró Velho, Queluz 085058
Proarta, Zug 088022
Probst, Andrea, Gotha 076064
Probst, Hedi, Nonnenhorn 108281
Probst, Kenneth, Chicago 092774
Probst, Peter, Meerbusch 077319
Probst, Walter, Chambésy 087356,
133715
Probstei-Museum, Schönberg,
Holstein 021699
Probsthain, Arthur, London . . . 144671
El Procante, Donostia-San
Sebastián 085983
Process Posters, Atlanta 119603
Procháska, Antonín, Brno 102571
Prochaska, Patrick, Spring
Branch 127464
Prochazkova, Radka, Frankfurt am
Main 129853
Otto Prochnow & Petra Georg-
Prochnow, Simmern 108626
Prock, C., Bad Buchau 074501
Procopio, Franco, Torino 132183
Procopius, Leuven 140237
Proctor House Museum, Brighton 005479
Proctor, C.D. & H.M., Puddletown 144793
Proctor, Ralph, Pittsburgh 123571
Prodejní Výstava Starožitností a Umění,
Praha 098061
Productive Living, Saint Louis . . 123863
Prøysenhuset, Rudshøgda 033910
Prof. Aleksandra Bieziņa Muzejs,
Madona 030161
Prof. Bertlein Schulmuseum,
Worms 022767
Prof. Eidenberger-Museum,
Niederwaldkirchen 002456
Prof. G.R. Sharma Memorial
Archaeology Museum,
Allahabad 024138
Prof. Hans Gerstmayr – Museum,
Mauthausen 002380
Profanierte Marienkapelle,
Ludwigsstadt 020146
Profesorius Povilo Matulionio
Memorialinis Muziejus, Kupiškis 030402
Profesorius Stasio Ušinsko
Memorialinis Kambarys,
Pakruojis 030435
Professional Arte Moderno,
Miami 121847
Professional Fine Arts,
Washington 124950
Professional Photography and
Commercial Art, Graphic Design,
Antonelli Institute, Erdenheim . 057056
Professionalno-chudožestvennyj

Fakultet, Karagandinskij
gosudarstvennyj universitet im. E.A.
Buketova, Qaraghandy 056074
Professor Li Mei-Shu Memorial
Museum, Sansia 042437
Professoriaus Adomo Hrebnickio
Memorialinis Muziejus, Dūkšto 030336
Proffitt, San Antonio 124061
Profil, Skopje 111890
Profil Auto- und Technikmuseum,
Mittelbiberach 020422
Profil Scène, Perreuil 105342
Profila, Budapest 126156
Profilen, Gävle 115620
Profiles Public Art Gallery, Saint-
Albert 006843
Profilkolor, Biarritz 067212
Profiri, Maria-Létizia, Paris 129123
Profolio, Denise, Camperdown . . 061268
Profumo d'Antico, Napoli 080920
Profumo, Maria, Genova 131338
Progetto Restauro, Saonara . . . 138860
Progetto Venezia, Venezia 111149
Program for Art on Film,
Brooklyn 060705
Programa Bellas Artes, Universidad
de Bogotá Jorge Tadeo Lozano,
Bogotá 055264
Programme for Museum Development
in Africa, Mombasa 059524
Programs in the Visual Arts, University
of Wisconsin Superior, Superior 057962
Progressive Editions, Toronto . . 111689,
136688
Progressive Painters Association,
Chennai 059370
Progresso & Sviluppo, Napoli . . 110541
Project Art, London 118523
Project Art-University of Iowa and
Clinics, Iowa City 049711
Project Arts Center, Dublin 024907
Project Contemporary Artspace,
Wollongong 099895
Project Gallery, Los Angeles . . . 121580
Projection, Bruxelles 100652
Projects in Architecture, Los
Angeles 121581
Projects United, Tägerwilen . . . 116666,
138093
Projek, Amsterdam 082829
Projektraum 4, Mannheim 107973
Projektraum Fischerinsel, Berlin 106403
Projektraum Nering & Stern,
Berlin 106404
Projektraum Viktor Bucher, Wien 100363
Prolithos Stenkonservering,
Stockholm 133578
Prom-Galerie, München 108180
Prom-Galerie, Showroom,
München 108181
Promenade Antiques, Great
Malvern 089374, 144446
Promoart, Roma 110850
Promoarte, México 111980
Promoção Arte, Lisboa 114056
Promocoupes, Caen 067579
Promont House Museum, Milford 050893
Promotionart, Bergamo 109800
Promotora Internacional de Arte
Grafico, Madrid 115274
Promyk, Gdańsk 113485
Pron, Alessandro, Torino 081750, 111000
Pronkkamer Uden-Museum
Hedendaagse Kunst, Uden . . . 032934
Proofs, London 118524
Prop Art, Chicago 120243
Prop-It, Toronto 064729
The Propeller Centre for the Visual
Arts, Toronto 101690
Properly Staged Interiors,
Albuquerque 092077
Property Antiques, Roseneath . . 062170
Proportio Divina, Arnhem 082945
The Proposition, New York 122984

Propst, Innsbruck 100054, 127778
Propst House and Marple Grove,
Hickory 049459
Propst Objektdesign, Innsbruck . 100055
Propyläen Verlag, Berlin 137091
Prora Tierreich, Prora 021231
Prošek, Dr. Václav, Praha 125425,
140529
Proserpine Historical Museum,
Proserpine 001434
Proshine, Columbus 135484
Proske-Van Heerdt, Drs. Dorine,
Amsterdam 143061
Prosotowitz, Hans,
Schwegenheim 078421
Prospect Antique Centre,
Prospect 062125
Prospect Books and Prints,
Llanrwst 144548
Prospect Gallery, Nailsworth . . . 099417
Prospect Hill Historical Museum,
Meadows 001263
Prospectus Art Gallery, Chicago 120244
Prosper Decoracoes, Rio de
Janeiro 064108
Prospero, Kansas City 145195
Prospero, Gianantonio, Verona . 132295
Prospettiva, Rivista di Storia dell'Arte
Antica e Moderna, Firenze . . . 138861
Prospettive d'Arte, Milano 110410,
137690
Prosser, Ronnie L., Oakland . . . 122528
Prost, Alain, Chalon-sur-Saône . 067782
Prost, Jean-Pierre, Bron 067516
Prosveta, Beograd 114435, 137933
Prot, Didier, Tours 073858
Protain, Frédéric, Jaux 068900
Protectart, Paris 071808
Protée, Paris 105182
Protestants Historisch Museum
Abraham Hans, Horebeke 003709
Protetch, Max, New York 122985
The Proud American, Washington 097855
Proud Camden, London 118525
Proud Central, London 118526
Proudfoot, Peter & Penny,
Berkeley 088443, 134094
Proulx & Synnett, Montréal . . . 064460
Proussos Monastery Museum,
Prousos 023236
Prouté, Paul, Paris . . . 105183, 141269
Prouty-Chew Museum, Geneva
Historical Society, Geneva . . . 049068
Prouvot, Benoit, Boulogne-sur-
Mer 125533
Prova d'Artista, Venezia 111150
Provan Hall House, Glasgow . . . 044413
Provand's Lordship, Glasgow . . 044414
Provasek, Frank, Fort Worth . . . 093472
Provenance Fine Art, San Diego 124161
Provenance Trade, Atlanta 092170
Provence & Fils, Gigean 068663
Provence Antique, Montauroux . 070207
Provence Portes Anciennes,
Roaix 072482
Provence Retrouvée, L'Isle-sur-la-
Sorgue 069588
Proviant-Eisen Museum, Gresten 001993
Providence, Atlanta 092171
Providence Art Club, Providence 052389,
060706
Providence Athenaeum,
Providence 052390
Providence Children's Museum,
Providence 052391
Providence Preservation Society,
Providence 052392
Providence University Art Center,
Shalu 042445
Province House, Charlottetown . 005599
Provinces de France Antiques, San
Antonio 097018
Provincetown Art Association and

Museum, Provincetown 052400, 060707
Provincetown Arts, Provincetown 139481
Provincetown Heritage Museum,
Provincetown 052401
Provinciaal Archeologisch Museum,
Velzeke-Ruddershove 004103
Provinciaal Archeologisch Museum-
Ename, Oudenaarde 003940
Provinciaal Archeologisch Museum
van Zuid-Oost-Vlaanderen,
Zottegem 004168
Provinciaal Domein Rivierenhof Galerij,
Deurne 003546
Provinciaal Museum Bulskampveld,
Beernem 003351
Provinciaal Museum Stijn Streuvels,
Ingooigem 003730
Provinciaal Openluchtmuseum Bokrijk,
Genk 003618
Provinciaal Suske en Wiske
Kindermuseum, Kalmthout . . . 003745
Provinciaal Utrechts Genootschap
van Kunsten en Wetenschappen,
Utrecht 059619
Provinciaal Veiligheidsinstituut,
Antwerpen 003311
Provincial Antiques, Albert Park . 060976
Provincial Antiques, Milton 083996
Provincial Museum of Newfoundland
and Labrador at The Rooms, Saint
John's 006907
Provincial Natuurhistorisch Museum
Natura Docet, Denekamp 032213
Provinsens Kunstauktioner Kolding
Auktionshus, Kolding 125458
Provinzmuseum der Franziskaner,
Wien 003067
Provost Skene's House,
Aberdeen 043361
Provost, Elizabeth, Salt Lake City 136242
Provot-Liénard, Laons 069186
Proyecta Arte, Bogotá . 102357, 102358
Prucha, Th., Berlin 125887
Prudence Crandall Museum,
Canterbury 047404
Prütz, Bad Pyrmont 074571
Prunier, Louviers 125618
Pruskin, London 090429
Pruvost, Michel, Crécy-la-
Chapelle 068229
Pruy, Silke, Schwarzenberg . . . 078418
Prylgruvan, Visby 087164
Pryors of Liverpool, Liverpool . . 089858
Prywatne Muzeum Etnograficzne, Góra
Kalwaria 034768
Prywatne Muzeum Etnograficzno-
Historyczne, Bielsk Podlaski . . 034648
Pryzmat, Galeria ZPAP, Kraków . 113631
Przibylla, Wolfgang J., Düren . . 075448
Przy Piwnej, Warszawa 113895
PS, Amsterdam 112269
PS 122 Gallery, New York 122986
Psaculia, Vassiliki, Milano 080687
Psarrou, Zoe, Chalandri 109059
Psenský, Dr. Petr, Praha 065518
Psieniečių Muziejus, Valstybės Sienos
Apsaugos Tarnybos prie LR VRM
Muziejus, Vilnius 030538
Psiman, Oklahoma City 123294
Psimar, Bruxelles 124887
Pskovskij Gosudarstvennyj Istoričesko-
architekturnyj i Chudožestvennyj
Muzej-zapovednik, Pskov 037669
Pskovskij Memorialnyj Muzej V.I.
Lenina, Pskov 037670
Psotka, Walter, Toronto 128214
Psyche, Budapest 142569
Psyche, Sapporo 082104
Psyché Antiquité, Quesnoy-sur-
Deule 072332
Psychiatrie-Museum, Bern 041322
Psychiatrie-Museum,
Emmendingen 018185
Psychiatrie Museum im

Philippshospital, Riedstadt 021426
Psychiatriemuseum Haina, Haina
(Kloster) 018930
Psykiatrihistoriska Museet,
Uppsala 041116
Psykiatrihistoriskt Muséum, Växjö 041146
Psykiatrimuseet, Medicinhistoriska
Museet, Hisings Backa 040698
Pszczulny, Ewa, Düsseldorf 129756
PTE Fine Arts, New York 122987
Ptolemais Museum, Department of
Antiquities, Tolmeitha 030292
PTT Muzej Zajednice Jugoslovenskih
Pošta, Telegrafa i Telefona,
Beograd 038180
Pu 1 Xuan Art Gallery, Shanghai 102197
Pu Le Temple, Chengde 007690
Pubblidono di Aliprandi, Cles ... 109965
Pubei Museum, Pubei 008162
Public Art Fund, New York 139482
Public Art Gallery, Imst 100038
Public Art Gallery of Chios, Chios 023049
Public Art Review, Saint Paul ... 139483
Public Corporation for the Arts, Long
Beach 060708
Public Illumination Magazine, Low-
Tech Press, Long Island City .. 139484
Public Library Gallery, Llanelli .. 044882
Public Security Museum, Gongan 007794
Publicaciones de Estética y
Pensamiento, Madrid 137994
Publications d'Histoire de l'Art et
d'Archéologie, Louvain-la-Neuve 138558
Publick, Philadelphia 096154
Publix Books, East Cleveland .. 145140
Pucci, A., Bruxelles 063534
Pucci, Bruno, Napoli 142767
Puccio, Salvatore, Palermo 081067
Puce à l'Oreille, Beirut 082301
Puce à l'Oreille, La Madeleine . 069053
La Puce a l'Oreille, Lille 069503
La Puce d'Argent, Paris 071809
O Puces, Beirut 082302
Puces, Bouville 067439
Puces, Villejuif 074153
Puces de Cherbourg, Équeurdreville-
Hainneville 068461
Puces de la Côte, Grasse 068705
Puces d'Oc, Toulouse 073793
Les Puces Lascaris, Nice 070650
Les Puces Laurentaises, Saint-Laurent-
en-Gâtines 072886
Les Puces Montluçonnaises,
Montluçon 070272
Pucheng Museum, Pucheng ... 008163
Pucher, Dr. Christina, Maria
Lankowitz 127815
Puchert, Sonja, Gröditz 107166
Puchol, Valencia 115513
Puck, Amsterdam 082830
Pucker, Bernard H., Boston 119914
Pucuk Rebung-Gallery Museum, Kuala
Lumpur 030695
Pudasjärvin Kotiseutumuseo,
Pudasjärvi 011280
Pudelko, Bonn 106537
Pueblarte Art Gallery, Cartago .. 102467
Pueblo County Historical Society
Museum, Pueblo 052409
El Pueblo de Los Angeles, Los
Angeles 050466
Pueblo Grande Museum, Phoenix 052123
Puente, José, Madrid 115275
Puerta de Alcalá, Belo Horizonte 063917
Puerta de Isabel II Gallery,
Manila 034545
Puerta Real, Granada 086040
La Puerta Verde de Españoleto,
Madrid 133367
Puerta, Jesus, Granada 115094
Puertas Herrero, Sevilla 133437
Puerto Galera National Museum,
Puerto Galera 034572
Puerto Rican Work Shop, New

York 122988
Puerto Rico Art Week, San Juan 098277
Püschner, Karsten, Hartmannsdorf-
Reichenau 130020
Püspöki Kastély, Fertőrákos ... 023639
Püspöki Palota, Sümeg 023909
Püttner, Hans-Jürgen, Hof 130054
Pütz, Berthold & Paul, Berg, Kreis
Ahrweiler 074725
Pütz, G., Freiburg im Breisgau . 075875
Pütz, Gottfried, Müllheim 077439
Pueyo, Sevilla 115455
Puffing Billy Railway Museum,
Belgrave 000845
Puga, George, Portland 123665
Puget Sound Coast Artillery Museum
at Fort Worden, Port Townsend 052262
Puget, Sabine, Fox-Amphoux .. 103753
Pugh, Jan K., Pershore 144782
Pugh, Rosemary, Salisbury 144822
Pugh's Farm Antiques, Monkton 090767
Pugin, Christine, Seewen 133919
Pugin, Max & Olivier, Genève .. 087498,
133759
Puhoi Historical Society Museum,
Puhoi 033289
Puhze, Günter, Freiburg im
Breisgau 075876
Pui Kee, Hong Kong 065094
Puig Balaguer, Maria del Carmen,
Castello de la Plana 115019
Puig Vidal, Manuel, Barcelona .. 085827
Pujol & Martin, Toulouse 073794
Pujol, Etienne, Perpignan 125778
Pujol, Joan, Castelnou 067713
PuK- Museum für PuppentheaterKultur,
Museen im Rittergut Bangert, Bad
Kreuznach 016804
Puke Ariki, New Plymouth 033258
Pukeberg, Stockholm 115922
Pukkilan Kartanomuseo, Piikkiö . 011252
Pukkilan Kotiseutumuseo, Pukkila 011281
Pukownik, Michael, Duisburg ... 075555
Pulandian Museum, Dalian 007721
La Pulce d'Acqua, Bari 079779
Pulgarín, Jorge, Medellín 128306
El Pulguero, Cali 065252
Puli Wine Museum, Puli 042421
Pulkauer Gewerbemuseum,
Pulkau 002546
Pulkautaler Weinbaumuseum,
Hadres 002039
Pulkkilan Kotiseutumuseo,
Pulkkila 011282
Pulliam Deffenbaugh, Portland . 123666
Pullman, London 090430
Pullman Museum, Historic Pullman
Foundation, Chicago 047693
Pulp Fiction, Brisbane 139660
La Pulpería de Arte, San José . 102512
La Pulperie de Chicoutimi,
Chicoutimi 005615
Puls Contemporary Ceramics,
Bruxelles 100653
Pulse Contemporary Art Fair, Miami,
Miami 098493
Pulse Contemporary Art Fair, New
York, New York 098494
Pulszky Society – Hungarian Museums
Association, Budapest 059343
Pult, Michael, Mölln 077381
Pulver Zajfman Collection,
Toronto 064730
Pulylac, Castellón de la Plana . 133226
Pulzzart, Castres 103544
Pummer, Erich, Rossatz 127840
Pump House Center for the Arts,
Chillicothe 047720
Pump House Gallery, London ... 045079
Pump House Regional Arts Center, La
Crosse 050029
Pump House Steam Museum,
Kingston 006111
Pumphouse Educational Museum,

London 045080
Punavuoren, Puu, Helsinki 128416
Punavuori, Helsinki 140715
Punch, Shaftesbury ... 091272, 135021
Punch Gallery, Balmain 098699
Punchinello, Paris 071810
Punct Muzeal, Balș 036228
Punct Muzeal, Bezdead 036242
Punct Muzeal, Bunești 036330
Punct Muzeal, Cața 036362
Punct Muzeal, Comana de Jos . 036390
Punct Muzeal, Cuciulata 036426
Punct Muzeal, Fundata 036469
Punct Muzeal, Gostavățu 036483
Punct Muzeal, Maieru 036549
Punct Muzeal, Malu cu Flori ... 036552
Punct Muzeal, Mateiaș 036559
Punct Muzeal, Moroeni 036575
Punct Muzeal, Moțăieni 036576
Punct Muzeal, Pucheni 036650
Punct Muzeal, Râu Alb 036650
Punct Muzeal, Romoș 036664
Punct Muzeal, Rotbav 036678
Punct Muzeal, Sângeorz-Băi .. 036678
Punct Muzeal, Tărlungeni 036770
Punct Muzeal, Ulmi 036807
Punct Muzeal, Văleni 036812
Punct Muzeal, Viștea 036824
Punct Muzeal, Vlădeni 036825
Punct Muzeal Dimitrie Anghel,
Miroslava 036570
Punct Muzeal Gospodăria Țărănească,
Enisala 036455
Punct Muzeal Turnu Măgurele, Turnu
Măgurele 036804
Pundole, Mumbai 109462
Pundole, Neville, Canterbury ... 117520
Puning Museum, Puning 008164
Punjab Archives Museum, Lahore 034136
Punkaharjun Kotiseutumuseo,
Punkaharju 011284
Punkt Etnograficzny w Rogierówku,
Muzeum Pierwszych Piastów
na Lednicy w Lednogórze,
Rogierówko 035173
Punkt Muzealny w Pałacu Opatów,
Supraśl 035231
Punt de Llom, Palma de Mallorca 115365
Punti, A., Terrassa 086488
Puntland, Minneapolis 122027
Punto Arte, La Spezia 110159
Punto Arte, Modena 110503
Punto Arte, Montevideo 092010
El Punto de las Artes, Madrid .. 139066
Il Punto, Associazione d'Arte e Cultura,
Genova 059473, 110128
Puolakan Talomuseo, Valkeala . 011485
Pupilles et Papilles, Valbonne . 105795
Puppen- und Spielzeugmuseum,
Lichtenstein/Sa. 020065
Puppen- und Spielzeugmuseum,
Mödling 002403
Puppen- und Spielzeugmuseum,
Rottweil 021528
Puppen- und Spielzeugmuseum,
Solothurn 042007
Puppen- und Spielzeugmuseum,
Wien 003068
Puppen- und Spielzeugmuseum mit
Zinfigurenkabinett, Dornstetten 017944
Puppen- und Spielzeugmuseum
Sammlung Katharina Engels,
Rothenburg ob der Tauber ... 021509
Puppen-Stuben-Museum,
Zimmerschied 022879
Das Puppenhaus, Nürnberg 077836
Puppenhausmuseum, Basel ... 041286
Puppenmuseum, Bad Breisig .. 016739
Puppenmuseum, Bad Schönborn 016868
Puppenmuseum, Künzelsau 019819
Puppenmuseum, Lienz 002309
Puppenmuseum, Nordrach 020834
Puppenmuseum, Schmallenberg 021683
Puppenmuseum, Villach 002863

Puppenmuseum Falkenstein,
Sammlung Elke Dröscher,
Hamburg 019016
Puppenmuseum Hanisch, Waldkirchen
an der Thaya 002894
Puppenmuseum Jeannine,
Kreuzlingen 041626
Puppenmuseum Kärntner Eisenwurzen,
Hüttenberg 002109
Puppenmuseum Laufeld, Laufeld 019933
Puppenmuseum Sasha Morgenthaler,
Museum Bärengasse, Zürich . 042228
Puppenstubenmuseum der 30er und
50er Jahre, Jakobsweiler 019442
Puppentheatermuseum,
Kaufbeuren 019564
Puppentheatermuseum, Mistelbach an
der Zaya 002394
Puppet Centre, Ontario Puppetry
Association, North York 006526
The Puppet Forge, Minneapolis . 122028
Puppetry Art Center of Taipei,
Taipei 042544
Puratatva Sangrahalaya, Rahul
Sankratayan Sansthan,
Gorakhpur 024264
Purcell, Francis J., Philadelphia . 096155
Purdon, Richard, Burford 088709,
134179
Purdue University Calumet Library
Gallery, Hammond 049332
Purdue University Galleries, West
Lafayette 054311
Purdy's Furniture, Briar Hill 061192
Pure Design, Bangkok 088083
Pure Design Classics, Brighton . 088633
Pure Galerie Art et Design, Caen 103487
Pure Nostalgia, Waihi 084119
Puri, Kassel 107537
Puri Gamelan Suar Agung,
Denpasar 024557
La Purisima Mission, Lompoc .. 050363
Purist Gallery, Yallingup 099930
Puritan Value, Southwold 091391
Purmerends Museum,
Purmerend 032769
Purnananda Memorial Museum,
Hazarikapra 024291
Purnell, Baltimore 119767
Purple, Paris 138658
The Purple Cactus, Avalon 139629
Purple Cow, Saint Louis 096764
Purple Parrot, Jacksonville 121263
Purple Picture Gallery, Reading . 118961
Purple Picture Gallery, Windsor . 119366
Purrmann-Haus, Speyer 021934
Purschwitz, Manfred,
Dossenheim 075402
Purshottam & Sons, Sharjah ... 088207
Purtee, Indianapolis 135635
Purton Museum, Purton 045573
Purutan Restorations, Delhi ... 079388,
130821
Purvattatva Sangrahalaya Madhya
Pradesh Shasan, Damoh 024207
De Pury & Luxembourg, Zürich . 116848
Pusaka, West Hollywood 097921
Pusat Latihan Kesenian &
Pertukangan Tangan Brunei, Bandar
Seri Begawan 004966
Puschner, Kerstin, Altenburg ... 105977
Puškinskaja Observatorija – Galereja
103, Sankt-Peterburg 114355
Pusset, Jean-Jacques, Beaune . 067033
Pussinen, Kristiina, Helsinki ... 128417
Pusta, Górnośląskie Centrum Kultury,
Katowice 034821
Pustet, Anton, Salzburg 136556
Pustet, Friedrich, Regensburg . 137843
Pustozerskij Kompleksnyj Istoriko-
prirodnyj Muzej, Narjan-Mar . 037538
Pusulan Museo, Pusula 011286
Pusz, Janina, Wrocław 143544
Putian Museum, Putian 008165

Putiikki, Helsinki 140716
Putkiradiomuseo, Kouvola 011036
Putman, Catherine, Paris 105184, 136978
Putman, Robert, Amsterdam . . . 143062
Putnam Cottage, Greenwich . . 049272
Putnam County Historical Society and Foundry School Museum, Cold Spring 047865
Putnam County Historical Society Museum, Kalida 049862
Putnam Museum of History and Natural Science, Davenport . . . 048155
Putney Historical Society, Putney 052420
Putney School Gallery, Putney . . 052421
Putot, Jacques, Le Perreux-sur-Marne 069327
Putte, René van de, Reims 129195
Putti, Toronto 064731
Putti, Albion, Groningen 112501
Puttkamer, Jesco von, Berlin . . . 106405
Il Putto, Amsterdam . . 082831, 082832
Putz, Gerhard & Beate, Eschwege 075663, 075664
Putz, Hubert, Untersteinach 078766
Putze, Matthias, Münster 142208
Puuhkalan Talomuseo, Ylivieska 011528
Puukohola Heiau Ruins, Kawaihae 049905
Puupaja, Helsinki 066356, 128418
Puur, Rotterdam 112656
Puy, Patrick, Bessieres 067145
Puyang Museum, Puyang 008166
Puyo, Antonia, Zaragoza 115558
Puyol, Marcelle, Auch 125499
Puyun, New York 122989
Puzin, Georges, Versailles 141382
Puzo, Christian, Bayonne 066979
Půžová, Marie, Jihlava 102601
Pybus, M., Whitby 119337
Pydar Gallery, Truro 119265
Pygmalion, Saint-Nazaire (Loire-Atlantique) 105562
Pygmalion, Stockholm 143960
Pygmalion and Galatea, Toronto 101691
Pyhäjärven Kotiseutumuseo, Pyhäjärvi 011288
Pyhältö, Heikki, Tampere 103167
Pyhämaan Kotiseutumuseo, Pyhämaa 011290
Pyhäntän Kotiseutumuseo, Pyhäntä 011291
Pyhtään Kotiseutumuseo, Pyhtää 011292
Pyka, Henryk, Łódź 084617
Pykäri Nuutajärven Lasimuseo, Designor Oy Ab, Nuutajärvi . . 011199
Pyke, Beverley J., Alderney 088244
Pyl Vekov, Moskva 085198
Pylle's Kram, Kolding 065979
Pyms Gallery, London . 090431, 118527
Pyo, Beijing 101985
Pyo, Seoul 111705
Pyongyang Institut of Fine Arts, Pyongyang 056077
Pyramid Gallery, York 119412
Pyramid Hill and District Historical Museum, Pyramid Hill 001435
Pyramid Hill Sculpture Park and Museum, Hamilton 049329
Pyramida, Praha 102720
La Pyramide, Saint-Sauveur (Drôme) 073302
Les Pyramidions, Paris 071811
Pyro Gallery, Louisville 121666
Pyromania, Tauranga 113027
Pythagoras Museum, Norrtälje . 040883
Pyynikinlinna, Tampere 011401
Q & C Militaria, Cheltenham . . 088840, 134241
Q Antiques, Dún Laoghaire . . . 079609
Q Antiques, Willingham 091847, 135228
Q Art Trading, New York 122990
Q Books, Nelson 143333
Q Street Fine Crafts, Salt Lake

City 123982
Q.'s Books, Hamilton, New South Wales 139761
Qalipakkanik Katersugaasivik, Ilulissat 023330
QANTAS Founders Outback Museum, Longreach 001231
Qaqortup Katersugaasivia, Qaqortoq Museum, Qaqortoq 023337
Qashqadaryo Viloyat Ülkashunoslik Muzeyi, Qarshi 054638
Qasigiannguit Katersugaasiviat, Christianshåb Lokalmuseum, Qasigiannguit 023338
Qasr Al-Gawharah, Cairo 010558
Qasr Al-Ibrahim, Cairo 010559
Qasr Al-Ibrahim, Cairo 010560
Qassim, Yousif, Manama 100446
Qatar National Museum, Doha . 036181
QCA Gallery, Brisbane 000888
QCC Art Gallery, Bayside 046838
Qdos, Lorne 061752
Qedem Museum, Kedumim . . . 025117
Qeqertarsuaq Museum, Qeqertasuaq 023339
Qfer Galeria, Poznań 113745
Qi County Folkways Museum, Qixian 008194
Qi Gucheng Yi Zhi Guan, Linzi . . 008035
Qian Long Antiques Art Australia, Surry Hills 062268
Qianjiang Museum, Qianjiang . . 008167
Qiankou Minzai Museum, Huangshan 007910
Qianling Museum, Qianxian 008170
Qiannan Zhousandu Nationality Museum, Sandu 008214
Qiannanzhou Museum, Douyun . 007751
Qianshan Museum, Qianshan, Anhui 008168
Qijiang Stone Engraving Museum, Gunan 007824
Qijinshang Culture Museum, Qixian 008195
Qimen Museum, Qimen 008171
Qin Dai Zhan Guan, Lintong . . . 008028
Qin Shi Huan Bingmayong Museum, Xian 008397
Qin Shi Huang Terracotta Warriors and Horses Museum, Lingtong . . 008020
Qinan Museum, Qinan 008172
Qing Mi Ge Art Gallery, Beijing . 101986
Qing Tycoon Yashu Bowu Yuanji, Hohhot 007868
Qingan Art Cooperative, Ningbo 102156
Qingdao Art Gallery, Qingdao . . 008175
Qingdao City Folklorish Museum, Qingdao 008176
Qingdao Municipal Museum, Qingdao 008177
Qingdao Oil Painting Factory, Qingdao 102160
Qingdao Oil Painting Institute, Qingdao 055228, 102161
Qingdao Painting Gallery, Qingdao 008178
Qinghai Culture and Arts School, Xining 055255
Qinghai Logroll Museum, Huzhu 007927
Qinghai Provincial Museum, Xining 008436
Qingliu Museum, Qingliu 008179
Qingpu Museum, Shanghai 008232
Qingpu Painting Gallery, Shanghai 102198
Qingsong Art Service Center, Beijing 101987
Qingyang Museum, Qingyang . . 008181
Qingyang Zone Museum, Xifeng 008423
Qingyuan Museum, Qingyuan . . 008185
Qingyunpu Museum, Nanchang . 008099
Qingzhou Museum, Qingzhou . . 008187
Qinhuai Art Cooperative, Nanjing 102152
Qinshihuang Military Forces Yong

Museum, Lintong 008029
Qinyang Museum, Qinyang 008188
Qinzhou Museum, Qinzhou 008190
Qiongshan Museum, Qiongshan 008191
Qishan Museum, Qishan 008192
Qiz Qalasy Gallery, Bakı 003175
QL, Galleria d'Arte moderna e contemporanea, Trapani 111022
Qom Museum, Qom 024728
Qo'qon Viloyat Ülkashunoslik Muzeyi, Qo'qon 054642
Qoraqalpogʻiston Respublikasi Ülkashunoslik Muzeyi, Nukus . 054635
Qoraqalpogʻiston Tarikhi Muzeyi, Nukus 054636
Qozhaya Museum, Zgharta 030285
Quaco Museum and Archives, Saint Martins 006924
Quadart, Christchurch 112858
Quaderni di Storia dell'Arte, Parma 138862
Quadraat, Amsterdam 082833
Quadrart Dornbirn, Ort zeitgenössischer Kunst, Dornbirn 099979
Quadrata, Medellín 102458
Il Quadrato, Cesenatico 109961
Il Quadrato, Milano 137691
Quadreria Cesarini, Fossombrone 026342
Quadreria Comunale, Offida . . . 027085
Quadreria Comunale, Prato 027452
Quadreria Comunale, San Constanzo 027742
Quadreria dei Girolamini, Napoli 027020
Quadreria dell'800, Milano 080688, 110411
Quadreria della Cassa Depositi e Prestiti, Roma 027657
Quadreria della Società Economica, Chiavari 025957
Quadreria d'Epoca, Acqui Terme 109735
Quadreria e Pinacoteca Civica, San Giovanni in Persiceto 027761
Quadrerie, Nice 070651
Quadreto, Pinner 118914
Quadri, Bruxelles 100654
Quadri Arte, Trento 111033
Il Quadrifoglio, Livorno 080332
Il Quadrifoglio, Milano 080689
Quadrige, Nice 104386
Quadrille, London 090432
Quadrivium, Sydney 099793
Quadro, Bülach 116174, 133708
Quadro Zenit, Enna 110008
Quadrum, Rolle 126748
Quadrum Galeria de Arte, Belo Horizonte 100904
Quagga Art, Cape Town 114774
Quagliotti, Vittorio, Torino 132184
Quai 1, Espace Dédié à l'Image Contemporaine, Vevey 042110
Quai de la Gare, L'Isle-sur-la-Sorgue 069589
Quaid-i-Azam Birthplace and Museum, Karachi 034126
Quaink, Thomas, Köln 130164
Quaint Notions, Jacksonville . . 094030, 127268, 135665
Quaker Meeting House, Mount Pleasant 051098
Quaker Museum, Ballytore 024830
Quaker William Ray House, Dartmouth 005682
Qualantoni, Salvatore, Milano . . 080690
Qualitas Antiquitas, Budapest . . 079285
Quality Antiques, Chicago 092775
Quality Art, New York 122991
Quality Coins, El Paso 093412
Quality Collectibles, Miami 094793
Quality Fine Arts, London 118528
Quality Frames and Art, Austin . 119694
Quality Furniture Refinishing, Philadelphia 136100
Quality Furniture Repair Service,

Tulsa 136435
Quan Hua Yuan, Wuxi 102255
Quandt-Verein Dittersbach zur Förderung der Künste, Dürröhrsdorf-Dittersbach 059221, 137152
Quàng Ninh Museum, Ha Long . 054786
Quantum Commercial, Wellington 084181
Quantum Contemporary Art, London 118529
Quanz, Herfried, Altenstadt, Wetteraukreis 105979
Quanzhou Maritime Museum, Quanzhou 008197
Quanzhou Museum, Quanzhou . 008198
Quanzhou Overseas-Relations History Museum, Quanzhou 008199
Quanzhou Painting Gallery, Quanzhou 008200
Quaranta, Renata, Roma 081516
Quaritch, Bernard, London 144672
Quarr Gallery, Swanage 119190
Quarry Bank Mill & Styal Estate, Styal 045917
Quart Verlag, Luzern 138073
Quartair, Den Haag 112407
Quartana, Luca, Milano 131506
Quartel de Santa Cruz, Rio de Janeiro 004769
Quarterjack Antiques, Fontmell Magna 089291
Quarterly Polish Culture, Warszawa 138984
Quartermaine, Paul, Morley 061885
Quartet Books, London 138241
Quartier Ephémère, Montréal . . 101443
Quartilho, Jorge Luiz Fernandes, Rio de Janeiro 101061
Quartilho, S., Rio de Janeiro . . . 064109
Quarzsandbergwerk-Museum, Buchs (Zürich) 041377
Quasar Antichitati, Bucureşti . . . 085107
Quaschinski, B.W., Runkel 078312
Quaschinski, Brunhild, Limburg an der Lahn 077114
Quast, Anett, München 130386
Quaternary Glacial Vestige of China Exhibition Hall, Beijing 007641
Quatrain, West Hollywood 097922
Quatre A, Nice 070652
Quatrefoil, Fordingbridge 089293
Quatro Montras, Lisboa 132952
Quattrini, Anna, Roma 081517
Quattrocento, Madrid 133368
Quattrocento, Vannes (Morbihan) 073973
Quattroemme, Ponte San Giovanni 137721
Quay Gallery, Sydney 099794
Quay Gallery Antiques Emporium, Exeter 089236
Quay School of the Arts, Universal College of Learning, Wanganui 056199
Quayside Antique Centre, Shrewsbury 091333, 135047
Quayside Antiques, Wanganui . . 084145
Quayside Art, Singapore 114590
Que Gallery, Ha Noi 125093
Que Milagro, Houston 121116
Que Van, Knokke-Heist 100758
Queanbeyan and District Historical Museum, Queanbeyan 001437
Quebec Art, Québec 101517
Quebec House, National Trust Chartwell, Westerham 046098
Queen, New York 138421
Queen Anne Antiques, Lefkosia . 065325
Queen Anne Antiques & Raven's Nest Treasure, Seattle 097488
Queen Anne House, Chesham . . 088852
Queen Antiques, Toronto 064732
Queen City Architectural Salvage, Denver 093325, 093326
Queen Elizabeth II Army Memorial Museum, Waiouru 033322
Queen Elizabeth National Park

Museum, Lake Katwe 043109
Queen Elizabeth's Hunting Lodge,
London 045081
Queen Emma Gallery, Honolulu . 049542
Queen Emma Summer Palace,
Honolulu 049543
Queen Mary Museum, Long
Beach 050373
Queen of Hearts, Miami 121848
Queen of the South, Genk 063621
Queen Street Fine Arts, Blenheim 112823
Queen Street Mill Textile Museum,
Burnley 043793
Queen Victoria Antiques Centre,
Melbourne 061827
Queen Victoria Museum, Harare 054852
Queen Victoria Museum and Art
Gallery at Inveresk, Launceston 001214
Queen Victoria Museum and
Art Gallery at Royal Park,
Launceston 001215
Queen Victoria Rifles Museum – Hill
60, Zillebeke 004161
Queen West, Toronto 064733
Queens County Farm Museum, Floral
Park 048751
Queens County Museum,
Liverpool 006203
Queens County Museum, Tilley House
and Court House, Gagetown .. 005855
Queen's Falcon Antiques,
Tewksbury 127474
Queen's Gallery, Bangkok .. 116956
Queen's Gallery, Charlotte .. 120009
Queens Gallery, Queenscliff .. 062132
The Queen's Gallery, Royal Collection,
Edinburgh 044262
The Queen's Gallery, Royal Collection,
London 045082
Queens Garden, Atlanta 092172
Queens Historical Society,
Flushing 048768
Queen's House, Teignmouth .. 091552
Queen's Key, Toronto . 064734, 101692
Queen's Lancashire Regiment
Museum, Preston 045568
Queens Library Gallery, Jamaica 049783
Queens Museum of Art, Flushing 048769
Queen's Own Cameron Highlanders
of Canada Regimental Museum,
Winnipeg 007457
The Queen's Own Hussars Museum,
Warwick 046058
Queen's Own Rifles of Canada
Regimental Museum, Toronto . 007196
Queen's Own Royal West Kent
Regimental Museum, Maidstone 045194
Queens Parade Antiques, Stow-on-the-
Wold 091478
Queens Park Conservation Studios,
Manchester 134832
Queens Road Gallery, Aberdeen 117211
Queen's Royal Irish Hussars Museum,
Eastbourne 044217
The Queen's Royal Lancers
Regimental Museum, Grantham 044463
Queen's Royal Surrey Regiment
Museum, Guildford 044491
Queen's Royal Surrey Regiment
Museum, West Clandon 046088
Queen's York Rangers Regimental
Museum, Toronto 007197
Queenscliff Antiques, Queenscliff 062133
Queenscliffe Historical Museum,
Queenscliff 001439
Queenscliffe Maritime Museum,
Queenscliff 001440
Queensferry Museum, South
Queensferry 045818
Queensland Aboriginal Creations,
Brisbane 061213
Queensland Aboriginal Creations,
South Brisbane 062222

The Queensland Aborigine,
Cairns 061251
Queensland Air Museum,
Caloundra 000926
Queensland Art Gallery, South
Brisbane 001503
Queensland Fire Museum,
Kurwongbah 001200
Queensland Maritime Museum, South
Brisbane 001504
Queensland Military Memorial
Museum, Fortitude Valley 001079
Queensland Museum, South
Bank 001501
Queensland Police Museum,
Brisbane 000889
Queensland University of Technology
Art Museum, Old Government House,
Brisbane 000890
Queenstown and Frontier Museum,
Queenstown 038819
Queenstown Gallery, Dunedin . 112891
The Queenstown Story, Cobh . 024854
Quéhen, Catherine, Equihen-
Plage 068462
Queichtalmuseum, Offenbach an der
Queich 020974
Queirós, Isabel, Porto 085046
Quel Che Manca, Venezia 081867
Quelle Belle Journée, Paris ... 071812
Quellen Museum, Bad Wildungen 016904
Quelque part, Limoges 104035
Quentin, Berlin . 074900, 125888
Quentin, Paris 071813
Quentin, Mélanie, Barbizon ... 103334
Quentin, Yannick, Saint-Florentin 072786
Querciagrossa, Modena 131580
Quercus, Debenham . 089045, 134318
Quercus, Noordlaren 132599
Quereix, Bernard, Bordeaux ... 067354
Quesnel and District Museum,
Quesnel 006751
Quesnel, Pierre, Genouillac ... 068651
Quest, New Orleans 095145
Quest – Center for Earth Science And
Discovery, Makati 034508
Questa, Enrico, Torino 081751
Questacon-The National Science and
Technology Centre, Canberra . 000942
Questarte, Dolo 110005
Questel, Bruno, Rouen . 072611, 072612
Questroyal, New York 122992
Quetico Provincial Park Heritage
Pavilion, Atikokan 005358
Quetsche, Verlag für Buchkunst,
Witzwort 137558
Quetzal, Knokke-Heist 100759
Quetzalli, Denver 120759
Quex Museum, House and Gardens,
Birchington 043604
Queyrel, Luc, Paris 105185
Quezel Mouchet, Andrée, Villeneuve-
lès-Avignon 074162
Quezonia Museum, Manila 034546
Quick Strip, Cincinnati 135449
Quicksilver, San Diego 097126
Quidam de Revel, Paris 071814
Quidi Vidi Battery, Saint John's . 006908
Quiet Street Antiques, Bath 088398
Quiet Valley Living Historical Farm,
Stroudsburg 053575
The Quiet Woman Antiques Centre,
Chipping Norton 088892
Quigley, Harry & Judith, Adelaide 060972
Quigley, Michael A., San Antonio 097019
Quijano Museum, Mandaue ... 034519
Quik, L., Den Haag 083120
Quilichini, Ludovic, La Crau ... 069017
Quilis-Semper, Brigitte, Grenoble 068738
Quill, Albuquerque 135288
Quill Antiques, Deal 089038
Quill Pen Antiques, Louisville .. 094580
Quillet, Loix-en-Ré 128692
Quillon, Tetsworth 091587

Quilpie Museum, Quilpie 001442
Quilt Patch, Long Beach 094219
Quilted Cat, Birmingham 092431
Quilted Teddy, Denver 093327
Quilter, John, Wellington 143370
Quilters Hall of Fame, Marion . 050688
Quilters' Ruild Resource, Halifax 044504
Quimera Taller de Arte, Ciudad de
Panamá 113323
La Quincailerie, Veules-les-
Roses 105850
Quincy and Adams County Museum,
Quincy 052432
Quincy Art Center, Quincy ... 052433
Quincy Historical Museum,
Quincy 052437
The Quincy Museum, Quincy .. 052434
Quincy Society of Fine Arts,
Quincy 060709
Quinéville Military Collectable,
Quineville 072354
Quiniou, Jessica, Quimper 072350
Quinlan Craft & Design,
Macroom 109673
Quinn, Camperdown 061269
Quinn, Falls Church 127221
Quinns, Taranaki 126324
El Quinque Antiguedades,
Zaragoza 086603
Quinquis, Jean-Jacques, Ile-Tudy 103852
Quinta Galería, Bogotá 102359
Quinta Municipal de Nossa Senhora
da Piedade, Póvoa de Santa Iria 035964
Quintana, Portland 123667
Quintana Diaz, Francisco Angel,
Vitoria-Gasteiz 086576
Quintana Loscos, Francisco,
Barcelona 133146
Quintana, Fernando, Bogotá .. 102360
Quintanilla, Alonso de, Gordes . 068684
Quintavalla, Giovanna, Parma .. 081126
Quintavalla, Lorenzo, Milano .. 131507
Le Quinte di Via dell'Orso,
Milano 080691
Quinte Educational Museum,
Ameliasburgh 005327
Quintero, Sofia, Toledo 124695
Quintessens, Utrecht 112710
Quintessential Antiques, Saint
Louis 096765
Quintessenz, Lich 077098
Quinto Bookshop, London 144673,
                      144674
Quintrec, Yvon le, Lesneven .. 069439
Le Quinze, Saint-Ouen 073174
Il Quirinale, Villanova di
Castenaso 138863
Quirindi and District Historical Cottage
and Museum, Quirindi 001443
Quiros Galdon, Marta Maria,
Madrid 133369
Quisqualis Art, Paddington,
Queensland 099542
Quitard, Pierre, Bazainville ... 066984
Quittenbaum, Askan, München . 126064
Quivrin, Olivier, Etaples 068481
Quixotics Gallery, Omaha 123337
Qujiang Museum, Qujiang 008202
Qun-Yi, Singapore 114591
Quorn Pine, Quorn . 091063, 134956
Quorum, Madrid 115276
Quorum Art Gallery, Bogotá .. 102361
Qurieus, Almelo 082635
Qurm Museum, Qurm 034104
Qurum Design, Seeb 084409
Qusaibati, Mohammed Adib,
Damascus 088050
Qutb Shahi Museum, Hyderabad 024301
Quwo Museum, Quwo 008203
Quynh, Ho Chi Minh City 125145
Quzhou Museum, Quzhou 008204
Qvartinge Antik, Mariefred ... 086864
R & F Antiques, Houston 093772

R & K Woodworks, Virginia
Beach 136445
R & Kim's Art Gallery, San Jose 124459
R & R Antiques, Fresno 093511
R & R Antiques, Jolimont 061661
R & R Coins, San Jose 097374
R & R Galerie, Berlin 106406
R & R Studios, Chicago 120245
R & S of Jacksonville,
Jacksonville 121264
R & T International, New York . 135978
R 20th Century Design, New
York 095670
R Art, Capinghem 103527
R-Art, Praha 065519
R M Studio Gallery, Toronto .. 101693
R. & C., Salzburg 127856
R. Galerie, Český Krumlov ... 102584
R.B. Ferguson Museum of Mineralogy,
Winnipeg 007458
R.B. Instruments, Mount Eliza . 127648
R.C., Torino . 081752, 132185
R.C. Associates, Cullompton .. 089026
R.E. Olds Transportation Museum,
Lansing 050140
R.E. Stevenson Museum,
Colenso 038609
R.G. Antiques, Dunham-on-Trent 089118
R.G.V. Arte, Padova 110578
R.J. Antiques, Bendigo 061133
R.J. Galleries, Ballarat 061090
R.M. Bohart Museum of Entomology,
University of California, Davis . 048163
R.N. Atkinson Museum, Penticton 006628
R.P. Strathearn Museum, Simi
Valley 053345
R.R., Madrid 086243
R.R.1919, Genova 110129
R.S. Barnwell Memorial Garden and
Art Center, Shreveport 053322
R.S.V., Logroño 086077
The R.W. Norton Art Gallery,
Shreveport 053323
Ra, Amsterdam 112270
Ra, Skopje 111891
Ra del Rey, Madrid 115277
RA Magazine, London 139217
RA Manufacturing, New York .. 135979
Råå Museum för Fiske och Sjöfart,
Råå 040917
Raab, Helene, Fulda . 075932, 107091
Raab, I., Neustadt an der
Weinstraße 077761
Raab, Ingrid & Julia, Berlin .. 106407
Raabe-Haus, Braunschweig ... 017564
Raadvej, København 065946
Raad, Jan de, 's-Hertogenbosch 083396
Raadts & Godecke, Wesel 078940
Raahen Museo, Raahe 011293
Raaijmakers, Marijke, Nijmegen 083518
Raasay Heritage Museum, Kyle . 044745
Raasch, D., Landsberg am Lech 107789
Raasi, Chennai 109291
Raastedbro, Randers 066058
Rabacoins, Rabat 082605
Rabah, A., Gießen 129904
Rabah, Salma, Damascus 088051
Rábaközi Muzeális Gyűjtemény,
Kapuvár 023715
Rabasova, Rakovník 102734
Rabasová, Kateřina, Praha ... 102721
Rabb, Murphy, Chicago 120246
Rabbethge-Schiller, Hella,
Rosenheim 108515
Der Rabe, Erlangen 141746
Der Rabe, Frankfurt am Main . 141787
Rabeez, Riyadh 143680
Rabekk Museum, Moss 033368
Rabel, Monaco 082527
Rabel, Serravalle 085291
Rabelaise, Nice 141004
Raben, Axel, New York 122993
Rabenou, Yris, New York 095671
Rabier, Bruxelles 063535

Rabindra Bharati Museum,
Kolkata 024366
Rabindra Bhavan Art Gallery,
Delhi 024236
Rabindra Bhavana, Santiniketan 024467
Rabisch, Helmut Otto, Dresden . 129720
Rabrooha, Ottawa 064513
Rabu, Rémi-Claude, Paris 129124
Rabun & Claiborne, New York . 095672
Raby Castle, Darlington 044067
Raccolta Comunale G.C. Corsi,
Cantiano 025742
Raccolta Agorà dell'Arte, Sersale 027934
Raccolta Archeologica, Narni ... 027025
Raccolta Archeologica, Terni .. 028079
Raccolta Archeologica, Tolentino 028104
Raccolta Archeologica Alberto Pisani
Dossi, Corbetta 026065
Raccolta Archeologica Comunale,
Celenza Valfortore 025894
Raccolta Archeologica e
Paleontologica, Corciano 026068
Raccolta Civica, Città di Castello 026005
Raccolta Civica d'Arte Contemporanea,
Molfetta 026875
Raccolta Civiche di Storia, Albino 025274
Raccolta Comunale, Acquasparta 025237
Raccolta Comunale, Sigillo 027963
Raccolta Comunale d'Arte,
Frontone 026352
Raccolta d'Arte C. Lamberti,
Codogno 026028
Raccolta d'Arte Contemporanea R.
Pastori, Calice Ligure 025695
Raccolta d'Arte della Provincia,
Modena 026866
Raccolta d'Arte e Archeologia, Piazza
Armerina 027307
Raccolta d'Arte Pagliara, Napoli 027021
Raccolta d'Arte Sacra, Figline
Valdarno 026231
Raccolta d'Arte Sacra, Scarperia 027903
Raccolta d'Arte Ubaldiana,
Piandimeleto 027303
Raccolta d'Arte, Cattedrale, Città della
Pieve 025998
Raccolta d'Arte, Oratorio Santissima
Crocifisso, Roccalbegna 027546
Raccolta dei Padri Passionisti,
Paliano 027193
Raccolta della Chiesa di Santa Maria
Maggiore, Alatri 025264
Raccolta dell'Avifauna delle Marche,
Montefortino 026941
Raccolta dell'Avifauna Lombarda,
Arosio 025374
Raccolta delle Piastrelle di Ceramica,
Sassuolo 027879
Raccolta dell'Opera del Duomo,
Oristano 027106
Raccolta di Cose Montesine,
Montese 026962
Raccolta di Fisarmoniche d'Epoca,
Camerano 025708
Raccolta di Fossili Francesco
Angellotti, Ostra 027132
Raccolta di Ingegneria Navale,
Università di Genova, Genova . 026428
Raccolta di Opere d'Arte,
Solarolo 027976
Raccolta di Sant'Urbano, Apiro . 025343
Raccolta di Scienze Naturali,
Vicenza 028372
Raccolta E. Guatelli, Collecchio . 026034
Raccolta Etnografica del Centro Studi
Pugliesi, Manfredonia 026694
Raccolta Etnografica della
Civiltà Contadina, San Giorgio
Piacentino 027757
Raccolta G. Manzù, Ardea 025353
Raccolta Guatelli, Ozzano Taro . 027138
Raccolta Kalefati, Oria 027102
Raccolta Lapidaria Capitolina,
Roma 027658

Raccolta Materiale Archeologico,
Bevagna 025506
Raccolta Memoria e Tradizioni
Religiose, Serramanna 027928
Raccolta Naturalistica, Tagliolo
Monferrato 028051
Raccolta O. Recchione, Palena . 027175
Raccolta Osservatorio di Apicoltura
Giacomo Angeleri, Pragelato . 027435
Raccolta Paleontologica e Preistorica,
Narni 027026
Raccolta Parocchiale, Valpelline 028264
Raccolta Privata Toraldo di Francia,
Tropea 028221
Raccolta Russo-Ortodossa,
Merano 026766
Raccolta Storica della Vita Materiale
dell'Antico, Altopascio 025303
Raccolta Temporanea, Convento di
San Francesco, Cosenza ... 026089
Raccolte d'Arte dell'Ospedale
Maggiore di Milano, Milano ... 026829
Raccolte del Dipartimento di Scienze
Radiologiche, Roma 027659
Raccolte dell'Istituto di Clinica
Otorinolaringoiatrica, Roma .. 027660
Raccolte Frugone Villa Grimaldi-Fassio,
Genova 026429
Race Fan Depot, Austin 092262
Racemotormuseum Lexmond,
Lexmond 032596
Racers, Ricky, Fresno 093512
Rachel Carson Homestead,
Springdale 053448
Rachel, Matthias, Heidelberg .. 076456
Rachela, Kraków 084594
Rachel's, Houston 093773
Rachinger, Paul, Dießen am
Ammersee 106677
Rachtian, Shahla, Venezia ... 081868
Racim, Mohamed, Alger 098537
Racine, Bruxelles 136646
Racine Art Museum, Racine ... 052440
Racine Heritage Museum, Racine 052441
Racine, José, Lonrai 069625
Racines Architecturales,
Casablanca 082579
Les Racines du Ciel, Rouen ... 105478
Racional Moveis e Decoracoes, Rio de
Janeiro 064110
Rack, Anne & Jürgen, Bad Orb . 074570
Rack, K. & N., Kefenrod 076738
Rackam, Paris 141270
Rackey, Joachim, Bad Honnef . 106069
Rackstadmuseet, Arvika 040544
Racó del Coleccionista,
Barcelona 143729
Rada Galerii České Republiky,
Praha 058541
Ráday-Kastély, Pécel 023835
Radbrook Culinary Museum,
Shrewsbury 045796
Radcliffe, J., Hastings 089467
Radda, Brigitte, Klagenfurt ... 062674,
                                127788
Råde Bygdetun, Råde 033883
Radeckas, Brone & Bronius,
Kuršénai 111789
Rademacher, Olpe 130493
Rademacher, Jürg, Leer 077042
Rademachersmedjorna,
Eskilstuna 040608
Radermacher, Mara M.,
Rheinbach 108495
Radermacher, Wolfgang, Kleve . 076824
Rades, Marita, Schwerin 078432
Radetzky & Co., Wien 063053
Radetzky-Gedenkstätte Heldenberg,
Kleinwetzdorf 002202
Radford University Art Museum,
Radford 052442
Radgowska, Jolanta,
Michałowice 132853
Radi-um von Roentgenwerke AG,

Tokyo 111487
Radial Access, Cincinnati 092887
Radice, Anna, Milano 080692
Radici Antiquari, Milano 080693
Radicke, Jutta, Sankt Augustin . 108554
Radico, Cairo 066264
Radijo ir Televizijos Muziejus, Šiaulių
Aušros Muziejus, Šiauliai 030470
Radin, Bairnsdale 098684
Radinger, Rud & Fritz, Scheibbs 140040
Radio- en Speelgoed Museum,
Onstwedde 032719
Radio- ja TV-Museo, Lahti 011077
Radio- og Motorcykel Museet,
Stubbekøbing 010331
Radio Amateur Museum, Reusel 032780
Rádió és Televízió Múzeum,
Diósd 023602
Radio Matterhorn Museum,
Zermatt 042176
Radio Museum Köln, Köln 019708
Radio Ranch, Austin 092263
Radio-Television Museum, Bowie 047155
Radiomuseet i Göteborg,
Göteborg 040659
Radiomuseet i Jönköping,
Jönköping 040720
Radiomuseum, Bad Bentheim . 016717
Radiomuseum, Bad
Tatzmannsdorf 001765
Radiomuseum, Borculo 032083
Radiomuseum, Grödig 001996
Radiomuseum, Linsengericht . 020093
Radiomuseum, Rottenburg an der
Laaber 021523
Radiomuseum, Waldbronn 022382
Radiomuseum, Zürich 042229
Rádiómúzeum, Balatonfüred .. 023425
Radionica Duše, Beograd 114436
Radionica Duše, Beograd 114437
Rádiós Hírközlési Múzeum,
Siófok 023884
Radius Gallery, Saint Paul 123925
Radix Gallery, Phoenix 123512
Radlett Fine Art and Framing Gallery,
Radlett 118953
Radley, Boo, Los Angeles 094450
Radloff, Horst, Wismar 079028
Radman, Barbara, Witney 144927
Radmuseum anno dazumal,
Altmünster 001709
Radnor House, Grampound ... 089353
Radnorshire Museum, Llandrindod
Wells 044876
Radstädter Museumsverein,
Radstadt 002561
Radstock Museum, Radstock .. 044576
Radua Miralles, Maria Rosa, Palma de
Mallorca 115366
Raduga, Moskva 143650
Radulescu, Paul, Paris 129125
Raduno Invernale di Antiquariato e
Modernariato, Baganzola 098215
Radvilų Rūmai, Lietuvos Dailės
Muziejus, Vilnius 030539
Radwan, Anna, Poznań 113746
Radwin, Jackie, San Antonio .. 097020
Radzuweit, H. & M., Sylt 078663,
                                108749
Rae-Smith, London ... 090433, 118530
Rae, Dennis, San Francisco .. 124370
Rae, Dorian, Vancouver 101826
Rae, Jan, Virginia Beach 136446
Rae, Peter, Dunedin . 112892, 112893
Rääkkylän Kotiseutumuseo,
Rääkkylä 011295
Raeantiik, Tallinn 066284, 140682
Raeber, Luzern 144109
Räber, Alexander E., Zürich ... 116849,
          116850,   138118
Räber, Th., Burgdorf 087342
Räch, Thomas, Gönnheim 076026
Räderscheidt, Gisèle, Köln ... 107672
Rädle, Hans-Peter, Singen 078479

Raejuveel, Tallinn 066285
Rælingen Bygdetun, Fjerdingby . 033599
Ränkimäen Ulkomuseo, Lapua . 011098
Rätisches Museum, Archäologische,
historische und volkskundliche
Sammlungen Graubündens,
Chur 041417
Rättviks Antik och Auktion,
Rättvik 086917, 126656
Rättviks Konstmuseum och
Naturmuseum, Rättvik 040918
Rätzel, Georg, Gießen . 075996, 107130
Rätzer, Meißen 077333
Rätzer, Thomas, Dresden 075425
Raevangla Fotomuuseum, Tallinna
Linnamuuseum, Tallinn 010686
Rafael, New York 122994
Rafael, Tehrän 079428
Rafael, Marc, Paris 071815
Rafanelli, Fabio, Genova 110130
Rafanelli, Luca, Firenze 131246
Rafelman, Marcia, Toronto ... 101694
Rafer, Aline, Saint-Étienne ... 072772
Raff, Erwin, Denkendorf 129676
Raffael-Verlag, Ittigen 138056
Raffaelli, Giordano, Milano ... 110412
Raffaello, Providence 123711
Raffaghello, Sabrina, Ovada .. 110553
Raffan & Kelaher, Leichardt .. 125208
Raffenel, Jean-Pierre, Domagné 068370
Raffety & Walwyn, London 090434
Raffin, Jean-François, Paris ... 071816
Raffles, Singapore ... 114592, 126486
Rafo Antigüedades, Lima 084467
Raft Artspace, Parap 099550
Rag and Bone Books,
Minneapolis 145285
Ragailong Museum, Imphal 024313
La Rage, Lyon 104127
Råger's Antik, Malmö 086859
Rageth, Jürg, Riehen 116591
Ragged School Museum, London 045083
Raggedy, Anne, Oklahoma City . 096002
Raggl-Thurner, Schönwies ... 127869
Ragici, Dr. Luigi, Milano 080694
Ragini Art Village, Lalitpur ... 112095
Ragione e Sentimento, Roma .. 081518
Raglan and District Museum,
Raglan 033292
Raglan Dealers, Raglan 084065
Raglan Gallery, Cooma 098921
Raglan Gallery, Manly 099275
Raglan's Right Up My Alley,
Raglan 084066
Ragley Hall, Alcester 043379
La Ragnatela, Milano 080695
La Ragnatela, Roma .. 081519, 132021
Ragona, Gioacchino, Palermo .. 081068
Ragosta, Alfonso, Lecce 126199
Ragot, Jones, Paris 071817
Ragtime, Bari 079780
Ragusa, Valerio, Padova 081000
Raguseo, Edward, New York .. 095673
Rah-Coco, Providence 096512
Raha- ja Mitalikokoelma, Helsinki 010855
Rahäuser, H., Breisach am Rhein 075134
Rahardt, Mike, Leipzig 130228
Rahbarnia-La Persia, San Marino 133029
Al-Rahiq, Damascus 088052
Rahm-Nebling, Monika,
Kaiserslautern 076663
Rahmanan, New York 095674
RahmenArt, Hamburg 107309
Rahmi M. Koç Müzesi, İstanbul . 042952
Rahn, C., Berlin 074901
Rahn, Dr. P., Leipzig 107848
Rahr West Art Museum,
Manitowoc 050648
Raicri, Preto, Belo Horizonte .. 100905,
                                100906
Raiffeisen-Museum,
Flammersfeld 018370
Raigné, Raymond, Fourcès ... 068600
Raike, Tucson 097688

Rail Museum, Egyptian National
  Railways Museum, Cairo 010561
Rail Transport Museum,
  Bassendean 000826
Railings Gallery, London 118531
Railroad and Heritage Museum,
  Temple 053709
Railroad House Historical Museum,
  Sanford 053083
Railroad Memories, Denver 093328
Railroad Museum of Long Island,
  Riverhead 052589
Railroad Museum of Pennsylvania,
  Strasburg 053564
Railroad Sales Agent, Milwaukee 094877
Railroader's Memorial Museum,
  Altoona 046425
Rail's End Gallery – Haliburton
  Highlands Guild of Fine Arts,
  Haliburton 005943
Railway Age, Crewe 044040
Railway Bizarre, Ringwood 062151
Railway Coastal Museum, Saint
  John's 006909
The Railway Exposition,
  Covington 048043
Railway Museum, Kalamata 023102
Railway Museum, Livingstone . . . 054837
Railway Museum, Ravenglass . . . 045583
Railway Museum, Saitama 029303
Railway Museum, Wylam 046209
Railway Preservation Society of
  Ireland, Whitehead 046122
Railway Square Antique Centre,
  Toowoomba 062333
Railway Station Gallery, Pomona 099573
Railway Station Museum,
  Castlegar 005578
Railway Station Museum,
  Uitenhage 038870
Railway Station Museum,
  Wonthaggi 001657
Railways to Yesterday, Rockhill Trolley
  Museum, Rockhill Furnace . . . 052641
Railworld, Peterborough 045490
Raimbourg, Thierry, Barentin . . . 066947
Raimo, Antonio, Atlanta 092173, 119604,
                                       144971
Raimondi, Bassano del Grappa . 109784
Raimondi di Pettinaroli, Milano . 110413
Raimondo, Guglielmo, Palermo . 110615
Raimund-Gedenkstätte,
  Gutenstein 002029
Raimundo Requixa, Maria Concepcion,
  Alicante 085702
Rain Forest Art Gallery, Barrio
  Turnón 102465
Raiņa Literatūras un Mākslas Vēstures
  Muzejs, Rīga 030214
Raiņa Memoriālais Muzejs Jasmuiža,
  Preiļi 030172
Raiņa Muzejs Tadenava, Dunavas 030118
Raiņa un Aspazijas Māja, Rīga . 030215
Raiņa un Aspazijas Memoriālā
  Vasarnīca, Raiņa Literatūras
  un Mākslas Vēstures Muzejs,
  Jūrmala 030140
Rainbird, London 118532
Rainbird Art Gallery, Montville . 099366
Rainbow Art, Tampa 124669
Rainbow Arts Gallery, Singapore 114593
Rainbow Books, Brighton 144268
Rainbow Books, Wellington 143371
Rainbow Fine Art, New York . . . 122995
Rainbow Galerie, Riehen 116592
Raindre, Guy, Paris 071818
Raine, Harry, Consett 088970
Rainermuseum, Salzburg 002634
Rainforest Arts, Atherton 098674
Rainforest Gallery, Canungra . . . 098863
Rainham Hall, Rainham 045577
Rainhof-Presse, Buchenbach . . 106607,
                                       137130
Raintree Aboriginal Art Gallery,

Darwin 098951
Raintree Aboriginal Art Gallery,
  Darwin 061388
Rainy Hills Pioneer Exhibits,
  Iddesleigh 006024
Rainy River District Women's Institute
  Museum, Emo 005782
Rainyday Gallery, Penzance 118899
Rais Ali Delvary Museum,
  Bushehr 024690
Rait Village Antiques Centre, Rait 091065
Raitioliikennemuseo, Helsingin
  Kaupunginmuseo, Helsinki . . . 010856
Raïtz von Frentz, Jürgen Freiherr,
  Wiesbaden 108928
Raiz, Donostia-San Sebastián . . 085984
Raj, Gent 063657, 063658
Raj, New York 095675
Raja Dinkar Kelkar Museum,
  Pune 024450
Rajabhat Institute, Bangkok 056467
Rajaju, South Fremantle 099696
Rajalahden Talomuseo, Virrat . . 011516
Rajamuseo, Imatra 010901
Rajcevic, Marinko, Neuwied . . . 077774
Rajesk Chitrakar, Lalitpur 112096
Rajic-Cretegny, Hélène, Genève . 087499
Rajkiya Sangrahalaya, Udaipur . 024504
Rajoloteca Salvador Miquel,
  Vallromanes 040354
Rajpol, Warszawa 084743
Rajputana Museum, Ajmer 024131
Rajskij Sad, Moskva 114292
Rakennuskulttuuritalo Toivo ja
  Korsmanin talo, Pori 011266
Rakhine State Museum, Sitture . 031826
Rakhtshuy-Khaneh Museum,
  Zanjan 024796
Rakiura Museum, Stewart Island 033302
Rakkestad Bygdetun, Rakkestad 033884
Rákóczi Múzeum, Magyar Nemzeti
  Múzeum, Sárospatak 023871
Rákospalotai Múzeum, Budapest 023545
Rakstniecības, Teātra un Mūzikas
  Muzejs, Rīga 030216
Rakstnieka Herberta Dorbes Māja-
  Muzejs Senču Puteķļi, Ventspils 030251
Raku, Singapore 114594
Raku Museum, Kyoto 028977
Rakushu Rakuchu Gallery, Kyoto 028978
Rakvere Linnakodaniku Muuseum,
  Rakvere 010654
Rakvere Linnus-Muuseum,
  Rakvere 010655
Rakvere Näitustemaja-Muuseum,
  Rakvere 010656
RAL Productions, Chicago 120247
Raleigh City Museum, Raleigh . . 052454
Raleigh Hill's Upholstery, Portland 136157
Ralf's Antiques, West Hollywood 097923
Sri Rallabandi Subbarao Government
  Museum, Rajahmundry 024457
Rallarmuseet, Moskosel 040855
Ralli Museum, Caesarea 025029
Ralls Historical Museum, Ralls . 052455
Ralls, Marsha, Washington 124951
Ralph Allen Memorial Museum,
  Oxbow 006607
Ralph Foster Museum, College of the
  Ozarks, Point Lookout 052224
Ralph M. Chait Galleries, New
  York 095676, 122996
Ralph Milliken Museum, Los
  Banos 050476
Ralph Waldo Emerson House,
  Concord 047955
Ralph's Auctions, Greer 127236
Ram, Rotterdam 112657
Ram, Valencia 086530, 133468
Ram Art, Niš 114453
Ram Mala Museum, Comilla . . . 003214
Rama, Bangalore 079360
Rama Antiques, Bangkok 088084
Rama IX Art Museum, Bangkok . 042675

Ramadier, Ludovic, Avignon . . . 066886
Ramage, Christiane, Longpré-Corps-
  Saints 069618
Ramakers, Den Haag 112408
Ramalhete, Lisboa 126454
Raman Science Centre, Nagpur . 024420
Ramani, Utrecht 083659
Ramania Art Emporium,
  Kathmandu 112085
Rambaldi, Paolo, Molinella 142755
Rambert, Michel, Lyon 125631
Rambling Rose Antiques, Fort
  Worth 093473
Rambo Antigüedades, Buenos
  Aires 060925
Rambourg, Eric, Bordeaux 128504
Le Rameau d'Or, Lyon 140930
Ramee, Washington 124952
Rameix, Christophe, La Celle-
  Dunoise 068964, 103877
Ramer, Andreas, Schnaitsee . . . 078380
Ramer, Andreas, Seefeld 078439
Ramesberger, Andrea, Steyr . . . 062851
Ramex Galerie, Kassel 076727
Ramez, Cairo 066265
Rami, Marinella, Bologna 079871
Ramirez Perez-Lucas, Tomas,
  Madrid 115278
Ramirez, Angel, Miami 121849
Ramirez, Gabriel, Bogotá 128275
Ramirez, M., Phoenix 145431
Ramirez, Mario, Los Angeles . . . 094451,
                                       135740
Ramiro, Madrid 086244, 086245, 126547,
                                       143781
Ramiro & Leitão, Lisboa 084946
Ramkema, Dik, Overveen 143212
Ramla Museum, Ramla 025179
Ramlingappa Lamture Museum,
  Osmanabad 024425
Rammal, Damascus 088053
Rammehåndværk, København . . 128397
Rammerscales, Lockerbie 044896
Ramón i Cabrera, Castello de la
  Plana 085933
Ramón Magsaysay Memorabilia
  Museum, Manila 034547
Ramona's Antiques, Atlanta 092174
Ramondon, Guéthary 103803
Ramoneda Soler, Maria Rosa,
  Barcelona 133344
Ramoni, Roberto, Roma 081520
Ramon's Antique Store, Los
  Angeles 094452
Ramon's Antique Store, West
  Hollywood 097924
Ramos Iznata, Jose Luis, Málaga 086302
Ramos, Dário, Porto 114102
Ramos, Noémio Santos, Faro . . 113972
Ramose, Maastricht 083476
Rampa, Ivano, Almens 133617
Rampal, Jean-Jacques, Paris . . 071819,
                                       129126
Rampazzo, Luciana, Roma 132022
Rampouch, Jaroslav, Olomouc . 065433
Ramsauer, Angela, Geiselhöring 129889
Ramsay, London 090435
Ramsay House, Alexandria 046395
Ramsay Museum, Honolulu 049544
Ramsay, John, Vancouver 101827
Ramsdell, Nowa Nowa 099491
Ramsden Morrison, San Diego . 124162
Ramseier, Ernst, Bern 087297
Ramsey, Saint Paul 123926
Ramsey Center for Arts,
  Shoreview 053317
Ramsey Garland, Philadelphia . . 096156
Ramsey House Museum Plantation,
  Knoxville 050015
Ramsey Rural Museum,
  Huntingdon 044618
Ramsey, Jeremy, Singapore . . . 114595
Ramsgate Maritime Museum,
  Ramsgate 045578

Ramsgate Museum, Ramsgate . 045579
Ramunni, Francesco, Bari 079781
Ramzi, Albuquerque 092078
Ramzia, Riyadh 085321
RAN Art Gallery, New York 122997
Ran In-Ting Gallery, Taipei 116938
Ran-In-Ting Museum, Taipei . . . 042545
Ran, Mary, Cincinnati 120369
Rana Museum, Cultural History
  Department, Mo i Rana 033772
Ranacher, Hansjörg, Wiesbaden 130763
Ranaldi Freschi, Viviane, Lyon . . 104128
Ranasim Art Gallery, Hyderabad 109370
Ranby Hall, Retford 091093
Ranchetti, Guido, Roma 132023
El Ranchito, Oklahoma City 096003
Rancho Los Alamitos, Long
  Beach 050374
Rancho Los Cerritos, Long Beach 050375
Ranchu, Michèle & Serge,
  Manosque 069839
Randaberg Bygdemuseum,
  Randaberg 033885
Randall, Las Vegas 121374
Randall House Museum, Wolfville 007483
Randall Museum, San Francisco 053018
Randall, Penelope, Woollahra . . 062480
Randeck-Museum, Mannweiler-
  Cölln 020257
Randee & Co., Salt Lake City . . 096917
Randel, New York 122998
Randel, Walter, New York 122999
Randell, Colin, Newcastle 139855
Randelzhofer, Kurt, Ingolstadt . . 130074
Randers Kunstmuseum, Randers 010274
Randier, Henri, Paris 071820
Randolph County Historical Museum,
  Winchester 054463
Randolph County Museum,
  Beverly 046970
Randolph Gallery, Edinburgh . . . 117739
Randolph Historical Society,
  Randolph 052465
Randolph, Suzanne, New York . . 123000
Random House Publishing,
  London 138242
Randy's Antiques, Columbus . . . 093041
Randy's Antiques, Saint Paul . . . 096850
Randy's Auction Gallery, Norfolk 127371
Ranfurly Antiques, Auckland . . . 083819
Rangau-Handwerkermuseum, Markt
  Erlbach 020294
Range Antiques, Toowoomba . . 062334
Rangiora Museum, Rangiora . . . 033293
Rangl, Peter, Bürs 127721
Rangpur Archaeological Museum,
  Rangpur 003233
Rani Laxmi Bai Palace Sculpture
  Collection, Jhansi 024330
Ranieri & Tenti, Milano 131508
Rank, Piotr, Bremen 129636
Rankin & Conn, London 090438
Rankin House Museum,
  Miramichi 006331
Rankin House State Memorial,
  Ripley 052578
Rankin Museum, Ellerbe 048550
Rankin Museum, Rankin 052466
Rankin, George, London 090436
Rankin, Piers, London 090437
Rannarootsi Muuseum, Aibolands
  Museum, Haapsalu 010609
Rannikkotykistömuuseo, Helsinki 010857
Rano, Cairo 066266
Ransberger, Fort Worth 093474, 127226
Ransdell, Alan C., Jacksonville . 094031
Ransick, Gat, New York 135980
Ransom County Historical Society
  Museum, Fort Ransom 048862
Ransom, Mark, London 090439
Rantasalmen Museo, Rantasalmi 011298
Rantsilan Kotiseutumuseo,
  Rantsila 011299
Ranuan Pappila- ja Pitäjämuseo,

Ranua 011300
Raoli, Milena, Roma 081521
Raoping Museum, Raoping 008205
Raos, Peter, North Shore 112975
Raphael, Birmingham 119817
Raphael, Napier 084010
Raphael, New York . . . 095677, 095678, 135981
Raphael et Co, Saint-Ouen 073175
Raphaël's Auctions, Sion 126755
Raphaely, Miami 121850
Rapid City Museum, Rapid City . 006753
Rapid Eye Books, Battery Point . 139636
Rapin, Philippe, Paris 071821
Rapin, Renée, Corcelles-près-
Payerne 116227
Rapinett, Alfred, Siggiewi 132342
Rapisarda, Giovanni, Livorno . . . 080333
Rapisardi, Mario, Nice 128869
Rapp, Marianne, Wil . . 116718, 126725
Rapp, Yvonne, Louisville 121667
Rappaport, Roma 142832
Rappo, Georges, Schaffhausen . 087802
Rappresentazione Scenografica
Gioacchino Murat, Pizzo 027375
Rapsodija, Sankt-Peterburg 085261
Rapuano, Antonio, Napoli 080921
Rapuano, Pasquale, Napoli 080922
Rapuano, Stefano, Napoli 131635
Raque Montagut, Barcelona . . . 085828, 114953
Rara-Avis, Portland 096463
Rare & Unusual, Long Beach . . . 094220
Rare and Racy, Sheffield 135028, 144839
Rare Art, New York 123001
Rare Books, Leeds 144520
Rare Coin Buyers of Virginia,
Richmond 096563
Rare Coin Galleries of Seattle,
Seattle 097489
Rare Collection, Hong Kong . . . 065095
Rare Find Country Antiques,
Metford 061847
Rare Finds, Memphis 094663
Rare Finds, New Orleans 095146
Rare Old Times, Brighton, South
Australia 061196
Rare Oriental Book Company,
Aptos 144957
Rare Posters, New York 095679
Rare Posters at DBA Art Wise,
Brooklyn 138313
Rare Prints, Austin 092264
Rare Rabbit, Cincinnati 092888
Rare Reads, Atlanta 144972
Rarebit Galeria de Artes, São
Paulo 101109
Rarebits, Enniscorthy 079619
Rarebooks Dotcom, Brisbane . . 139661
Rarekiek, Amsterdam 082834
Raridade, Lisboa 084947
Raritäten-Kabinett, Cottbus . . . 141676
Raritätensammlung Bruno Gebhardt,
Zwönitz 022919
Raritätenstüberl, Augsburg 074469
Rariteiten- en Ambachtenmuseum 't
Krekelhof, Koksijde 003752
Raritet, Biškek 137800, 142952
Raritet, Moskva 085199
Raritet, Murmansk 085218
Raritet Antikvarno-juvelirnyj Salon,
Samara 085226
Raritet Rosvuzdizajn, Sankt-
Peterburg 114356
Ras-al-Tin Palace Museum, Ras-al-
Tin 010583
Rasan, Toronto 064735
Rasch, Bramsche 137123
Rasch, Mario, Altenburg 074392
Rasch, Roger, Hohen Neuendorf 130063
Rasche, Stefan, Münster 108225
Raschella, Vincent, Montréal . . . 064461
Raschoffer, Jacques, Cannes . . 067649

Rase, Lothar, Saarbrücken 078333
Raseiniu Krašto Istorijos Muziejus,
Raseiniai 030451
Rasen, Kunitachi 111288
Rasha, Damascus 088054
Rashid, Manama 063158
Rashid Al-Oraifi Museum,
Muharraq 003211
Rasht Museum, Rasht 024729
Rashtriya Lalit Kala Kendra,
Kolkata 109404
Rasken, Malmköping . 086841, 126634
Rasmus, København 102911
Rasmus, Alain, Besse-et-Saint-
Anastaise 067140
Rasmussen & Gad, Næstved . . . 066006
Rasmussen Art Gallery, Pacific Union
College, Angwin 046484
Rasmussen, Kent, Frederiksberg 065673
Rasoylis, San Francisco 124371, 124372
Raspail, Richard, Saint-Ouen . . 073176
Raspante, Marcello, Palermo . . 081069
Rassan Arabzadeh Carpet Museum,
Tehrân 024780
Raßbach, Christa, Loxstedt 077152
Rassbach, H., Bremerhaven . . . 075191
Rassegna Attrezzi e Oggetti del
Passato, Belluno 025482
Rassegna d'Arte, Napoli 138864
Rassouli, M., Bern 116147
Rast, Bielefeld 074989
Rastelli Gioielli, Cremona 080061
Raster – tygodnik, Warszawa . . 138985
Rastrakavi Govind Pai Museum,
Udupi 024505
Rastrell, Valencia 086531
El Rastrillo, Bilbao 085884
El Rastro, Logroño 086078
El Rastro, Medellín 065304
Il Rastro, Padova 081001
El Rastro, Palermo 081070
Rastro Crystty's, Valencia 086532
El Rastro Sur, Málaga 086303
Rašytojo Tomo Mano Memorialinis
Muziejus, Nida 030431
Rašytojos Šatrijos Raganos
Memorialinis Muziejus, Mažeikių
Muziejus, Židikai 030561
Rašytojos Žemaitės Memorialinis
Muziejus, Žemaičių Dailės Muziejus,
Bukante 030327
Rat für Formgebung – German Design
Council, Frankfurt am Main . . 059222
Le Rat Mort, Oostende 100805
Ratcliff, G.T., Kelvedon 089667
Ratel, Michèle, Poses . 072291, 105398
Ráth György Museum, Iparművészeti
Múzeum, Budapest 023546
Rath, Sonja, Hamburg 076296
Rathaus Galerie, Euskirchen . . . 106943
Rathaus-Galerie, Kunstamt
Reinickendorf, Berlin 017271
Rathausgalerie, Aalen 105953
Rathausgalerie, Bad Harzburg . . 016782
Rathausgalerie, Castrop-Rauxel . 017718
Rathausgalerie Attendorn,
Attendorn 106022
Rathausgalerie München,
München 020564
Rathausgalerie Munster, Munster 020622
Rathausgalerie Vellmar, Vellmar 022314
Rathausmuseum, Sempach 041979
Rathausmuseum Nieder-Ohmen,
Mücke 020476
Rathauspassage Antiquariat,
Hamburg 141891
Rathaussammlung, Stein am
Rhein 042035
Rathbone Law, Chichester 088877
Rathgeber, R., Ludwigshafen am
Rhein 077164
Rathgen-Forschungslabor, Staatliche
Museen zu Berlin – Stiftung
Preußischer Kulturbesitz, Berlin 017272

Rathke, Dr. Ewald, Frankfurt am
Main 107023
Rathmines Bookshop & Art Gallery,
Dublin 109616, 142603
Ratier Frères, Olemps 070777
Rátkay Endre-Átlók Galéria,
Pesterzsébeti Múzeum,
Budapest 023547
Ratliff, Las Vegas 094170
Ratnapura National Museum,
Ratnapura 040488
Raton Museum, Raton 052474
Ratsuväkimuseo, Lappeenranta . 011093
Rattlesnake Museum,
Albuquerque 046369
Ratton & Ladrière, Paris 071822, 071823
Ratton-Hourdé, Philippe, Paris . . 071824
Ratvolden, Røros 033899
Ratz, J. & B., Grafenau, Kreis
Böblingen 141840
Ratzki, Mario, Boston 092489
Rau, Hans-Gerd, Rottenburg am
Neckar 130606
Rau, Klaus-J., Münster bei
Dieburg 130433
Rau, Natascha, Hilden 076531
Rau, Peter, Staig 130651
Raub, Brest 103465
Raubach, Jürg, Sankt Gallen . . . 087787, 116615, 133911
Rauber, Ludwig, Vechta 108815
Raubiet, Philippe, Lille 128684
Raucci, Amedeo, Roma 142833
Raucci, Oreste, Roma . 081522, 142834
Rauch, Hans Dieter, Wien 063054, 125284
Rauchfangkehrermuseum, RFK-Innung,
Wien 003069
Raud, Christian, Paris 071825
Rauguth, Johann, Nürnberg . . . 077837
Rauh, Helmut, Straubing 071826
Rauhaniemen Kotiseutumuseo,
Sulkava 011376
Raule, Rolf, Deidesheim 075323
Raum 1, Hofheim am Taunus . . 107440
Raum 45, Schwäbisch Hall 078410
raum 5 Galerie Berlin, Berlin . . 106408
Raum und Kunst, Kassel 107538
Rauman Antikvaarinen Kirjakauppa,
Lappi 140735
Rauman Museo, Rauma 011305
Rauman Taidemuseo, Rauma . . 011306
Rauminhalt – Harald Bichler,
Wien 063055
RaumTraum, Bayreuth 106125
Raunet, Jean-Louis, Moncley . . 070155
Rauoff, Bin Abdul, Riyadh 085322
Rauprich, Peter, Gönnheim 076027
Rauriser Talmuseum, Rauris . . . 002576
Rauscher, Elfriede Maria Christina,
Wien 063056, 100364
Rauscher, H., Kirchseeon 076812
Rautalammin Museo, Rautalampi 011308
Rautatieläiskotimuseo, Kouvola . 011037
Rautenberg, Lothar, Haag an der
Amper 076139
Rautenstrauch-Joest-Museum,
Kulturen der Welt, Köln 019709
Rauter, Barcelona 137969
Raution Kotiseutumuseo, Rautio 011309
Raux Brothers, Colombo 086621
Rauzada, Alain, Saint-Ouen . . . 073177
Rava & Co, Torino 132186
Ravá, Tobia, Venezia 111151
Ravaioli, Carlo, Bologna 130993
Raval Arte y Diseño, Zaragoza . 115559
Ravalli County Museum,
Hamilton 049322
Ravan, Atlanta 092175
Ravanel, Nathalie, Firenze 131247
Ravelingien, T., Gent 063659
Ravelli, Luigi, Brescia 131046
Ravelo e Hijos, Catalina de,
Bogotá 065222

Raven, Kraków 084595, 113632
Raven, Minneapolis 122029
Raven, Portland 096464
Raven, San Francisco 124373
Raven & the Bear, Vancouver . . 101828
Raven Press Gallery, Dunvegan . 117703
Ravenel, Saint Pierre 114380
Ravenel, Taipei 116939, 126782
Ravenna, Seattle 145597
Raven's Den Native Art Gallery,
Vancouver 101829
Raven's Nest Antiques,
Wellington 084182
Ravenswood, Chicago 092776
Raverty's Motor Museum,
Echuca 001049
Ravestijn, Amsterdam 082835
Ravet, Marc, Boiscommun 067272
Raveyre, Antoine, Saint-Étienne 072773
Ravier, Joëlle & Daniel, Corps-
Nuds 068161
Ravindra Art Palace, Delhi 109336
Ravintolamuseo, Lahti 011078
Ravizza, Norbert, Stuttgart 078626
Ravmuseet, Varde Museum,
Oksbøl 010267
Ravnsbjerghus Galleri, Viby . . . 103009
Ravon, Ph., Saintes . . 073376, 141341
Ravon, Xavier & Philippe, Dax . . 068280
Raw Art Gallery, Berlin 106409
Raw Art Gallery, Chicago 120248
Raw Art UK, Chester 117572
Raw Vision, Radlett 139218
Rawhide Old West Museum,
Scottsdale 053211
Rawls Museum Arts, Courtland . 048040
Al-Rawwas, Mohammed Mwafaq,
Damascus 088055
Ray & Scott, Saint Sampson . . 091210, 134995
The Ray Drew Gallery, New Mexico
Highlands University, Las Vegas 050176
Ray E. Powell Museum,
Grandview 049201
Ray & Co, Harry, Welshpool . . . 127083
Ray, Janette, York 144942
Ray, Simon, London 118533
Raya, Singapore 114596
Raym'Art, Les Sables-d'Olonne . 103990
Rayment, Derek & Tina, Barton,
Cheshire 088373, 134070
Raymond, Cincinnati 120370
Raymond Avenue Gallery, Saint
Paul 123927
Raymond M. Alf Museum of
Paleontology, Claremont 047759
Raymond, Christophe, Auterive . 066831
Raymond, Laurence, Persac . . . 072110
Raymond's Design Studio, New
York 135982
Raymore Pioneer Museum,
Raymore 006754
Raymundo's Art Gallery, Long
Beach 121408
Raynalds Mansion, Much
Wenlock 090791
Raynaud, Jean, La Roche-sur-
Yon 125604
Raynaud, Jean-Louis, Aix-en-
Provence 066473
Rayner, Port Macquarie 099590
Rayner, Michael, Cheltenham . . 144327
Raynham Hall Museum, Oyster
Bay 051876
Raynor, Tom, New Orleans 095147
Rayo Rojo, Buenos Aires 139597
Le Rayon Vert, Nantes 104344
Rayons Verts, Niort 104407
Ray's Antiques, Callington 061253
Ray's Hub, Long Beach 094221
Rayuela, Madrid 115279
Rayvec, Dallas 120648
Rayxart Invertigaciòn, Madrid . . 115280
Razstavna Galerija, Ljubljana . . 114723

Razstavna Galerija, Škofja Loka   114746
Razstavni Salon Rotovž, Maribor   038573
Razzano, Mario, Napoli   080923
Razzle Dazzle, Tucson   097689
RB Restauro Cose Antichi,
Verona   132296
RBM Fine Art Designs, Cincinnati 120371
RB's Settlers Antiques, Ulmarra   062369
RC Arts, Poole   118932
RCAF Memorial Museum, Trenton 007233
RCMP Centennial Celebration Museum,
Fairfield   005798
RCR Antique and Arts, New York   095680
RCS Libri, Milano   137692
RDS Diffusion, Paris   071826
Re, Praha   128362
RE-AL Kunstverlag, Nürnberg   137451
Re-Art, Moskva   114293
Re Artú, Padova   081002
RE Create, Chicago   092777
Re Find, Cleveland   092965
Re Nova, Warszawa   132897
Re-Soul, Seattle   097490
Re-Vu, Lille   069504
Re, Francesco, Napoli   080924
Re.Ve.Ma., Torino   132187
Rea, Marcia, Freeman's Reach .   099059
Read Heeler, Echuca   139723
Read, Alan, Horncastle   098402
Read, John, Martlesham 090731, 134845
Reader's Book Shop, Regina . . .   140384
Readers Den, Windsor, New South
Wales   139961
Reader's Digest Art Gallery,
Pleasantville   052206
Readers Realm, Castle Hill   139693
Readers Rendezvous, Kingston .   139797
Readers Rest, Lincoln   144541
Readers Teahouse, Kenmore . .   139790
Reader's Town, Mar del Plata .   139606
Readers World, Birmingham . . .   144250
Reading Abbay Gateway,
Reading   045591
Reading Historical Society,
Reading   052486
Reading Historical Society Museum,
Reading   052480
Reading Public Museum and Art
Gallery, Reading   052484
Reading Railroad Heritage Museum,
Reading   052485
Read's, Johannesburg   085647
Read's Rare Book Shop,
Brisbane   139662
Readsboro Historical Society,
Readsboro   052488
Reaktion Books, London   138243
Real-73, Almería   085707
Real Academia de Bellas Artes de la
Purísima Concepción, Valladolid 056387
Real Academia de Bellas Artes de San
Carlos, Valencia   056386
Real Academia de Bellas Artes
de San Fernando, Museo de la
Real Academia de Bellas Artes,
Madrid   056374
Real Academia de Bellas Artes de San
Telmo, Málaga   056376
Real Academia de Bellas Artes de
Santa Isabel de Hungría, Sevilla 056383
Real Academia de Córdoba de
Ciencias, Bellas Letras y Nobles
Artes, Córdoba   056361
Real Academia de Nobles y Bellas
Artes de San Luis, Zaragoza .   056389
Real Art, Chicago   120249
The Real Art Gallery, Shrewsbury 119097
Real Cartuja, Valldemosa   040350
Real Cartuja de Miraflores,
Burgos   039174
Real Casa del Labrador, Aranjuez 038981
Real Fábrica de Tapices, Madrid 039690
Real Monasterio de San Jerónimo,
Granada   039472

Real Monasterio de Santa María de El
Puig, El Puig de Santa María . 039986
Real Mother Goose, Portland . . . 123668,
123669
Real Museo Mineralogico, Napoli 027022
Real Pottery, Tring   139219
Real Reviews, Fort Worth   093475
Real TART Community Gallery, New
Plymouth   033259
Real World Computer Museum,
Boothwyn   047095
Reale & Figli, Catania   131097
Reale, Carmelo, Catania   131098
Reali Arte Contemporanea,
Brescia   109915
Realidade Galeria de Arte, Rio de
Janeiro   101062
Really Cool Stuff, Portland   096465
Really Good Stuff, Portland   096466
The Realm Imports, Chicago .   120250
Rearasa, Zamora .   086583, 133482
Reart, Bratislava   085555, 133047
Reart, Poprad   133058
Reason, C., Oxted 090939, 134920
Rebassa Guardiola, Gabriel, Palma de
Mallorca   086362
Rebbaumuseum am Bielersee-Musée
de la Vigne au Lac de Bienne,
Ligerz   041679
Rebecca Gallery, Toronto   101695
Rebecca Hossack Art Gallery,
London   118534, 118535
Rebecca Nurse Homestead,
Danvers   048141
Rebel, Marco, Luzern   133838
Rebel, Walter, Hamburg   141892
Rebelo, Antonio Silva, Lisboa .   084948
Rebelo, Nuno A. Costa, Braga .   084783
Rebensburg, Kirsten & Marcus, Bad
Kreuznach   129414
Rebillot, Jean-Louis, Nohant-Vic 070731
Reblandmuseum, Baden-Baden . 016927
Rebok, T.H., Observatory   133074
Rebolledo Cuerno, Maria Antonia,
Getxo   086012
Rebollini, Agnese, Genova   080285
Rebollo, Pepe, Zaragoza   115560
Reborn Antiques, Port Adelaide . 062091
Reborn Antiques, West
Hollywood   097925
Reboucas, Heriberto C., Fortaleza 063950
Reboul, Jean-Michel, Paris   071827
Reboul, Patrice, Paris . 071828, 071829
Rebound, Dunedin   143315
Rebound, Pittsburgh   096321
Rebourg, Claude, Neufchâteau . 070510
Rebours, Johann, Sainte-Croix-sur-
Aizier   073345
Rebuffel, Jean-Marie, Nice   070653
Rebuffel, Laurent, San Francisco 097310
Rebuttini, Sergio, Modena   131581
Le Recamier, Bruxelles   063536
Récamier, Nice   070654
Recasens Art, Barcelona   114954
Recchi, Nicoló, Roma   081523
Rech, Almine, Paris   105186
Rech, Manfred, Haan . 076143, 107194
Rechberger, Rainer, Altmünster . 139970
Rechem, Flandre van, Lille 128685
Rechnermuseum der Gesellschaft für
wissenschaftliche Datenverarbeitung,
Göttingen   018755
Rechtsmuseum, Hochschule
Furtwangen Universität, Furtwangen
im Schwarzwald   018572
Rechnitzer, Berlin   129550
Rechtshistorisches Museum,
Karlsruhe   019516
Recinto Casa Benito Juárez,
Saltillo   031429
Reciprocity, Dallas   120649

Recke, Hilmar, Magdeburg . . . 077223,
130261
Recker, H., Warendorf   078868
Recker, Michael, Mainz   130268
Reckermann, Köln   107673
Reckhaus, Rheda-Wiedenbrück . 108489
Recktenwald, L., Eppelborn . . . 075624
Reclaimed, Seattle   097491
Le Recoin, Voiron   074274
Recollect Dolls Hospital, Burgess
Hill   088716, 134184
Recollection-Antiques, Los
Angeles   094453
Recollections, Brighton 088634, 134155
Recollections, Omaha   096055
Recollections, Poynton   091052
Recollections, Saint Paul   096851
Recollections, Springston   084085,
132744
Recollections Antiques,
Indianapolis   093918
Recollections Antiques, London . 090440
Recom, Ostfildern   137464
Recon, Dallas   093204
Record of the Princeton University Art
Museum, Princeton   139485
Record Office for Leicestershire,
Leicester and Rutland,
Leicester   044802
Re'creation, Céret   103551
Recto-Verso, Strasbourg   073597
Rector, Tulsa   124811
Rectory Bungalow Workshop,
Elton   134365
Recuerdos de Bacolod Arts and
Antiques, Manila   084481
Recuerdos Gallery, Zamboanga . 113425
O' Recuncho, A Coruña   085958
Recupero e Conservazione,
Milano   138865
Les Recyclables, Genève   144080
Red Art Gallery, Nelson 112964, 132731
Red Barn Antique Market,
Indianapolis   093919
The Red Barn Antiques,
Warkworth   084148
Red Barn Auction, Phoenix . . . 127399
Red Barn Mall, Louisville   094581
Red Barn Museum, Morristown . 051068
Red Baron Antiques, Atlanta . . 092176
Red Bay National Historic Site, Red
Bay   006755
Red Biddy Gallery, Shalford . . . 119072
Red Brick Arts Centre and Museum,
Edson   005772
Red Bridge Antiques, Providence 096513
Red Bridge Books, Kansas City . 145196
Red Bridge Gallery, Shanghai . 102199
The Red Bull Gallery, Longnor,
Derbyshire   118659
Red Bull Hangar-7, Salzburg . . 002635
Red Car Gallery, Milwaukee . . 121946
Red Cedar Gallery, Montréal . . 101444
Red Centre Art, Coolbinia   061362
Red Chair Antiques, Long Beach 094222
Red Cheery Art Gallery,
Singapore   114597
Red Clay State Historical Park,
Cleveland   047829
Red Cloud's Auction Barn,
Chicago   127168
Red Corner, Vancouver   064867
Red Corridor Gallery, Fulda . . . 107092
Red Deer and District Museum, Red
Deer   006759
Red Desert Gallery, Eumundi . . 099008
Red Door Antiques, Oklahoma
City   096004
Red Door Bookshop, Warragul . 139952
Red Door Gallery, Edinburgh . 117740
Red Door on Charles Gallery,
Launceston   099222
Red Door Studio, Chicago   120251
red dot design museum Deutschland,

Essen   018310
Red Dot Gallery, Cowbridge . . . 117631
Red Dot Gallery, Deer Isle 120676
The Red Dot Gallery, Holt   117982
Red Eagle Gallery, Portland . . . 123670
The Red Gallery, Dubai   117186
Red Gallery, Fitzroy North   099037
The Red Gallery, Marlow   118724
Red Gallery, Rosebank   114813
The Red Gallery, Thame   119224
Red Gate, West Hollywood . . . 097926
Red Gate Gallery, Beijing   101988
Red Geranium, Albuquerque . . 092079
Red Geranium Shop, Louisville . 094582
Red Hat Gallery, Johannesburg . 114799
The Red Head Gallery, Toronto . 101696
Red Hill Gallery, Red Hill   099614
Red Hill Gallery 141, Red Hill . 099615
Red Hill-Patrick Henry National
Memorial, Brookneal   047271
Red Hill Studio Gallery, Telegraph
Point   099812
Red House, Singapore   133042
Red House Antiques, Yoxford . 091985
The Red House Antiques Centre,
York   091980
Red House Books, San Francisco 145565
Red House Gallery, Adelong . . 098616
Red House Glass Cone Museum,
Stourbridge   045891
Red House Museum, Gomersal . 044443
Red House Museum and Art Gallery,
Christchurch   043957
Red Indian Art Deco, Toronto . 064736
Red Kiln Connection, Salt Lake
City   123983
Red Lake Regional Heritage Centre,
Red Lake   006761
Red Lantern, San Francisco . . 097311
Red Latern Antiques,
Sacramento   096642
The Red Lion Antiques, London . 090441
Red Location Museum, New
Brighton   038789
Red Lodge, Bristol   043753
Red Lodge Antiques, Screveton . 091253
Red Mansion Foundation, London 060214
Red Mill Museum Village, Clinton 047842
Red Oak Gallery, Winnipeg . . . 101893
The Red Pale, Eton   117766
The Red Peach, Singapore . . . 085488
Red Peach Gallery, Hawkes Bay 112918
Red Quill Antiques, Virginia
Beach   097791
Red Rag Gallery, Stow-on-the-
Wold   119162
Red Ramia, Myrtleford   061935
Red River Art Gallery, Ha Noi . 125094
Red River Book Shop, Winnipeg 140436
Red River Gallery, Ha Noi   125095
Red River Historical Society Museum,
Clay City   047781
Red River Valley Museum,
Vernon   053964
Red Rock Art, Kununurra   099204
Red Rock Museum, Church Rock 047733
Red Room, Lyon   104129
Red Room Antiques, Kansas City 094090
Red Rooster Antique Mall, Las
Vegas   094171, 094172
Red Rose Antiques, Saint-Ouen . 073178
Red Rose Antiques, Vancouver . 064868
Red Rose Vintage Clothing,
Indianapolis   093920
Red Sand Art Gallery, Alice
Springs   098640
Red Sand Art Gallery, Ti Tree . 099825
Red Sea Gallery, Singapore . . 114598
Red Sea Maritime Museum, Eilat 025034
Red Shop, Wolverhampton . . . 091907
Red Sofa, Pittsburgh   096322
Red Star, Waikato   084136
Red Venus Gallery, San Diego . 124163
Red Wagon Antiques, Nashville . 095028

Reda, Alexandria 066222
Redaud, Anne, Nery 070507
Redback Art Gallery, Gordon Park 099092
Redbridge Museum, Ilford 044625
Redbud, Houston 121117
Redcliff Museum, Redcliff . . . 006762
Redcliffe Museum, Redcliffe . . . 001445
Redcoat Soldier Antiques, Sale . 062184
Redd, San Diego 124164
Reddel, C., Melle 107991
Redding, Richard, Zürich 088000
Reddingsmuseum Abraham Fock,
Hollum 032497
Redecorate, Norwood 062017
Redfern, London 118536, 138244
Redford, Robert, Altrincham . . . 088267
Redhill Museum, Redhill 001446
Redin, Uppsala 143994
Reding 41, Málaga 086304
Redington Museum, Waterville . 054221
Rediscovered Treasures, San
Antonio 097021
Redisson SAS Slavjanskaja – Galereja,
Moskva 114294
Rediviva, Sélestat 073482
Redivivus Antiquariat,
Regensburg 142324
Redjala, Mohand, Paris 071830
Redland Museum, Cleveland . . . 000975
Redlands Antique Auction,
Redlands 127417
Redlands Art Association,
Redlands 060710
Redleaf Gallery, Tunbridge Wells 091661,
119272
Redlitz, E., Konstanz 107720
Redmonds, Melbourne 061828
Redningsselskapets Museum,
Horten 033673
Redoña Residence Museum,
Tacloban 034601
Redondo House, Long Beach . . 094223
Redoubt Fortress – Military Museum
of Sussex, Eastbourne 044218
Redpath Museum, Montréal . . . 006415
Redpath Sugar Museum, Toronto 007198
Redpaths' Antiques, Newtown,
Victoria 061969
Redpoint, Port Kembla 127660
Redrock Gallery, Southbank . . . 099710
Red's Antique Galleries, Denver 093329
Redsky, Charlotte 120010
Reduta, Bratislava 085556
Redwater and District Museum,
Redwater 006763
Redwitz, K. Freiin von,
Rennertshofen 130582
Redwood County Museum, Redwood
Falls 052500
Redwood Library and Athenaeum,
Newport 051540
Redwood Museum, Crescent City 048059
Ree, H., Amsterdam 082836
Reece, New York 123002
Reece, James, Salt Lake City . . 096918
Reed, Allansford 127550
Reed & Reed, Atlanta 092177
Reed Arts, Columbus 120511
Reed Gallery, University of Main at
Presque Isle, Presque Isle . . . 052355
Reed Gold Mine, Midland 050873
Reed Gold Mine State Historic Site,
Stanfield 053487
Reed Whipple Cultural Center, City of
Las Vegas, Las Vegas 050171
Reed, Alan, Ponteland 118930
Reed, Jim, Birmingham 145002
Reed, Michael, Seattle 097492
Reedus, Dave, Detroit . 093374, 135545
Reedville Fishermen's Museum,
Reedville 052502
Reef Gallery, Cairns 098845
Reek, O. & M., Ihlow 076596
Reel, Chicago 120252

Reeman Dansie Auctions,
Colchester 126867
Rees & O'Neill, London 144675
Rees, Margot Edwina, Armidale 061053
Reese, Philadelphia 096157
Reese Bullen Gallery, Humboldt State
University, Arcata 046521
Reese, Margaret K., Houston . . 093774
Reeve, James, Warwick 091748
Reeves, Houston 093775
Reeves, New York 123003
Reeves Art Studio, Athlone 109526
The Reeves Center, Lexington . 050267
Reeves, Alexander, Richmond . . 123739
The Refectory, Dorking 089092
Refectory of the Hosios Loukas
Monastery, Boeotia 023035
Référence Espace 13, Orgeval . 104425
Reference Works, Swanage . . . 144872
Referência Galeria de Arte,
Brasilia 100924, 100925
Refettorio dell'ex-Convento di
Sant'Onofrio detto di Fuligno,
Firenze 026301
Reffert, G., Mannheim 077268
Refinishing Shop, San Antonio . 136270
Refinishing Shoppe, Austin 135331
Reflections, London 101346
Reflections Antiques, Surry Hills 062269
Reflections Art Studio, Chicago . 120253
Reflections Gallery, Paarl 085655,
126495
Reflections on Maling,
Canterbury 098862
Reflet d'Antan, Dormans-en-
Champagne 068389
Le Reflet des Temps, Portet-sur-
Garonne 077288
Reflet du Passé, Issoudun 068883
Reflet du Temps, Brassac 067451
Reflets d'Antan, Bordeaux 067355
Reflets d'Antan, Compiègne . . . 068116
Reflets d'Autrefois, Aix-les-Bains 066484
Reflets d'Epoques, Paris 071831
Reflets du Temps, Biarritz 067213
Reflets du Temps, La
Rochefoucauld 069077
Reflex, Breda 083004
Reflex, Mercin-et-Vaux 070067, 128795
Reflex Miniatuur Museum
voor Hedendaagse Kunst,
Amsterdam 031980
Reflex Modern Art Gallery,
Amsterdam 112271, 137839
The Reflex, West Australian
Photographic Federation, Swan
View 138498
Reform, Los Angeles 094454
Reform Vintage Modern,
Philadelphia 096158
Reforma, Lisboa 084949
Reformas y Obras, Madrid 133370
Református Kollégiumi és
Egyházművészeti Múzeum,
Debrecen 023597
De Refter, Tilburg . . . 083623, 143234,
143235
Le Refuge de Grasla, Les
Brouzils 013744
Refuge Fortifié Muséalisé,
Dossenheim-sur-Zinsel 012819
Refugio County Museum, Refugio 052503
Regal, San Francisco 124374
Regal Antiques, Westerham . . . 091800,
135212
Regal Art Gallery, Hamilton . . . 100829
Regal Art Gallery, Las Vegas . . 121375
Regal, Guy, New York 095681
Regalia Antiques and Militaria, South
Plympton 062225
Regaliza, Palencia 086333
Regalo Antiques, Atlanta 092178
Regalos Atlas, Málaga 086305
Regan, David M., Southport . . . 091383

Regard, Belo Horizonte 100907
Regard, Nice 070655
Regard, Paris 105187
Regard 76, Bruxelles 100655
Regard Antiquités, Carcassonne 067676
Regard Contemporain, Aix-en-
Provence 103202
Regard sur le Passé, Troyes . . . 073908
Regards, Lorient 069640
Regards, Metz 104230
Reg'Art, Lys-lez-Lannoy 104139
Reg'Art des Toiles, Nice 104387
Reg'Arts, Saint-Pol-de-Leon . . . 105604
Regatta, Richmond, Surrey 118972
Regatta Museum, Quidi Vidi Lake 006752
Regele, Josef, Stuttgart 078627, 126122
Regementsmuseet med 91:an,
Halmstad 040685
Regen Projects, Los Angeles . . 121582
Regen Projects II, Los Angeles . 121583
Regence Comércio de Artes, Belo
Horizonte 063918
Regency Antique, Stourbridge . 091457
Regency Antique Restoration,
Oakland 136044
Regency Antiques, Baltimore . . 092356
Regency Antiques, Boston 092490
Regency Art and Antiques Gallery,
West Hollywood 097927
Regency Auction Gallery, New
York 127357
Regency Gallery, Cambridge . . . 117510
Regency House, Genève 087500
Regency House Antiques, New
Orleans 095148
Regency Paperbacks, Auckland . 143294
Regency Restoration, Etobicoke 128171
Regency Wood Refinishing,
Calgary 128154
Regeneration Furniture, New
York 095682, 095683
Régent, Paris 071832
Regent Antiques, London 090442,
134756
Regent Art Studio, Christchurch 083892
Regent Auction, Winnipeg 125394
Regent Galleries, Worcester . . . 119395
Regent House Gallery, London . 090443,
118537
Régészeti Osztály és Állandó
Régészeti Kiállítás, Janus Pannonius
Múzeum, Pécs 023851
Reggiani, Andrea, Milano 080696
Reggiardo, T., San Francisco . . 097312
Reggio, Gabriele, Torino 132188
Reghion, Roma 081524
Reghizzi, Enzo, Modena 131582
Reghthuys, Giessenburg 032347
El Regillero Valiente, Madrid . . . 133371
Regiment, Bruxelles 063537
Regimental Museum of 1st
The Queen's Dragoon Guards,
Cardiff 043867
Regimental Museum of the 13th/
18th Royal Hussars and Kent
Dragoons, Cannon Hall Museum,
Cawthorne 043897
Regimental Museum of the 9th/12th
Royal Lancers, Derby 044085
Regimental Museum of the Black
Watch, Perth 045487
Regimental Museum of the
Oxfordshire and Buckinghamshire
Light Infantry, Oxford 045460
Regimental Museum of the Royal
Dragoon Guards and the Prince of
Wales's Own Regiment of Yorkshire,
York 046227
Regimental Museum of the
Royal Scots Dragoon Guards,
Edinburgh 044263
Regimental Museum of The Royal
Welsh (Brecon), Brecon 043714
Regimental Museum of the Sherwood

Foresters, Nottingham 045421
Regimental Museum The Highlanders,
The Queen's Own Highlanders
Collection, Inverness 044642
Regimentals Antiques, Victor
Harbor 062384
Régimódi bolt, Budapest 079286
Régimódi Lakberendezések boltja,
Budapest 079287
Le Regina, Liège 063778
The Regina A. Quick Center for the
Arts, Saint Bonaventure 052753
Regina Antiguidades, Rio de
Janeiro 064111
Regina Firefighters Museum,
Regina 006772
Regina Gallery, Moskva 114295
Regina General Hospital Museum,
Regina 006773
Regina Gouger Miller Gallery, Carnegie
Mellon University, Pittsburgh . . 052173
Regina Plains Museum, Regina . 006774
Reginald F. Lewis Museum of
Maryland African American History
and Culture, Baltimore 046767
Reginalde, Serrière-de-Briord . . 073510
Reginalds Tower, Waterford . . . 024998
Regina's Gallery, Norfolk 095873
Régine Décors, Benfeld 067089
Regio Galerie, Basel 116095
Regio Galerie, Grenzach-Wyhlen 107159
Regionaal Museum 1940–1945
Schagen en Omstreken,
Schagen 032845
Regionaal Natuurmuseum Westflinge,
Sint Pancras 032878
Regional- und Telegrafenmuseum,
Stegersbach 002776
Regional Arts and Culture Council,
Portland 060711
Regional Galleries Association of
Queensland, Fortitude Valley . . 058213
Regional Museum of Natural History,
Bhopal 024172
Regional Museum of Spartanburg
County, Spartanburg 053429
Regional Museum of Yeghegnadzor,
Yeghegnadzor 000718
Regional Science and Discovery
Center, Horseheads 049561
Regional Science Centre, Bhopal 024173
Regional Science Centre,
Guwahati 024279
Regional Science Centre,
Lucknow 024385
Regional Science Centre, Tirupati 024496
Regional Science Centre and
Planetarium, Kozhikode 024372
Regional Science Centre Bhubaneswar,
Bhubaneswar 024176
Regionale Heemmusea Bachten
de Kupe, Openluchtmuseum,
Izenberge 003736
Regionalen Istoričeski Muzej,
Kardžali 005046
Regionalen Istoričeski Muzej,
Kjustendil 005067
Regionalen Istoričeski Muzej,
Pazardžik 005109
Regionalen Istoričeski Muzej,
Pleven 005120
Regionalen Istoričeski Muzej,
Ruse 005155
Regionalen Istoričeski Muzej, Stara
Zagora 005211
Regionalen Istoričeski Muzej,
Šumen 005218
Regionalen Istoričeski Muzej Varna,
Varna 005247
Regionales Heimatmuseum für das
Renchtal, Oppenau 020997
Regionalgeschichtliches Museum,
Lutherstadt Eisleben 020195
Regionalmuseum, Bad

Lobenstein 016820
Regionalmuseum Alsfeld, Alsfeld 016519
Regionalmuseum Alte Schule,
Kaufungen 019566
Regionalmuseum Bad Frankenhausen,
Bad Frankenhausen 016766
Regionalmuseum Binn, Binn . . . 041342
Regionalmuseum Burg Beeskow,
Beeskow 017015
Regionalmuseum Chüechlihus,
Langnau im Emmental 041640
Regionalmuseum des Amtes Malliß,
Neu Kaliß 020675
Regionalmuseum Fritzlar, Fritzlar 018535
Regionalmuseum Harkerode,
Harkerode 019075
Regionalmuseum Leben und Arbeiten
im Blauen Ländchen, Nastätten 020640
Regionalmuseum Neubrandenburg,
Neubrandenburg 020683
Regionalmuseum Reichelsheim
Odenwald, Reichelsheim 021341
Regionalmuseum Schlangenhaus,
Werdenberg 042129
Regionalmuseum Schwarzwasser,
Schwarzenburg 041967
Regionalmuseum Stockstadt,
Stockstadt am Rhein 022002
Regionalmuseum Wolfhager Land,
Wolfhagen 022734
Regionalna Izba Muzealna,
Poddębice 035107
Regionalne Muzeum Młodej Polski –
Rydlówka, Kraków 034911
Regionalni Muzej, Djakovica . . . 038193
Regionalni Muzej Doboj, Doboj . 004261
Regionální Muzeum, Českǐ
Krumlov 009510
Regionální Muzeum, Mělník . . . 009692
Regionální Muzeum, Mikulov na
Moravě 009695
Regionální Muzeum, Náchod . . . 009711
Regionální Muzeum, Žďár nad
Sázavou 009977
Regionální Muzeum a Galerie,
Jičín 009602
Regionální Muzeum K.A. Polánka,
Žatec 009970
Regionální Muzeum v Chrudimi,
Chrudim 009529
Regionální Muzeum v Jílovém u Prahy,
Jílové u Prahy 009607
Regionální Muzeum v Kolíně,
Kolín 009635
Regionální Muzeum v Teplicích,
Teplice 009907
Regionální Muzeum ve Vysokém Mýtě,
Vysoké Mýto 009965
Regionalnyj Muzej Severnogo
Priladožja, Sortavala 037883
Regionmuseet Kristianstad,
Kristianstad 040760
Regionmuseum Västra Götaland,
Vänersborg 041131
Régipénz-Régiség Bolt, Pécs . . . 079320
Régiposta Antiquitás, Budapest . 079288
Régis & Thiollet, Argenteuil . . . 125495
Regis-Saunier, Laurence, Paris . 071834
Régis, Gilbert, Nice 070656
Régis, Jean-François, Paris . . . 071833
Régiség & Használtcikk Kereskedés,
Budapest 079289
Régiség & Mütárgy Kereskedés,
Györ 079310
Régiségbolt, Eger 079306
Régiségbolt, Mosonmagyaróvár . 079315
Regler, Thomas, München 130387
Regner, Johann, Graz 139989
Regni, Daniela, Parma 131734
Régnier, Jean-Claude, Saint-
Ouen 073179
Regoli & Giorno, Brescia 079953
El Regreso del Sibarita,
Salamanca 086413

Regresso Moveis Antigos, São
Paulo 064202
Regroupement pour la Promotion
de l'Art Imprimé, ARPRIM,
Montréal 058467
Het Regthuijs, Abbekerk 031887
Regthuis Oudkarspel, Oudkarspel 032756
Regthuis t'Schou, Schipluiden . . 032862
Reguera Libreria, Sevilla 143797
Řeháček, Ivan, Hradec Králové . 128325
Rehbein, Thomas, Köln 107674
Rehberg, Dagmar, Mainz 107953
Rehberger, R., Heidelberg 141929
Reheis, Erich, Rosenheim 130600
Rehhoff, Otterup 066049
Rehm, Georg, Augsburg 125853
Rehm, P., Ochsenhausen 108327
Rehn, Stockholm 087045
Rehnström, Mats, Stockholm . . 143961
Rehoboth Art League, Rehoboth
Beach 052505, 060712
Rehoboth Museum, Rehoboth . . 031852
Rehs, New York 095684
Rei Desentupidora, Porto Alegre 063990
Reial Acadèmia Catalana de Belles Arts
de Sant Jordi, Barcelona . . . 039107,
137970
Reiartis Gestion, Valladolid 115530
Reiat-Museum, Thayngen 042053
Reibetanz, Herbert, Siegen . . . 142363
Reich, Curt, Baden-Baden 137040
Reich, Daniel, New York 123004
Reich, Dieter Bernd, Schärding . 062835
Reichard, Hermann, Wiesbaden . 108929,
130764
Reichardt, Lothar, Hilders 107429
Reichardt, Roland, Markkleeberg 130287
Reichel, Pobershau 108390
Reichel, Robert, Nürnberg 077838,
130471
Reichelt, Braunschweig 075131
Reichelt, Eberhard & Karin,
Gundelsheim, Oberfranken . . 129931
Reichelt, Edda, Wietzen 075069
Reichenauer, Christoph, Wien . . 063057
Reichenbach, Jürgen, Freiburg im
Breisgau 075877
Reicher, Arthur, Düsseldorf . . . 075507
Reichert-Pinzl, Barbara,
Königsdorf 130177
Reichert, A., Freiburg im
Breisgau 075878
Reichert, Chr., Burghausen . . . 075222
Reichert, Dr. Ludwig, Wiesbaden 137553
Reichert, Peter, Bad Staffelstein 074605
Reichert, Siegmut, Kupferzell . . 137857
Reichl, Erich, Salzburg 127857
Reichlin, Ernst & Urs, Küssnacht am
Rigi 087551, 116403, 133788
Reichmann, Alois, Wien 140117
Reichmuth, Bernhard, Ibach . . . 087533,
133775
Reichold, Gabriele, Haimhausen 107202,
125964
Reichsdorf-Museum Gochsheim,
Gochsheim 018722
Reichskammergerichtsmuseum,
Gesellschaft für
Reichkammergerichtsforschung
Wetzlar, Wetzlar 022613
Reichspräsident-Friedrich-Ebert-
Gedenkstätte, Heidelberg . . . 019142
Reichsstadtmuseum, Rothenburg ob
der Tauber 021510
Reichsstadtmuseum im Ochsenhof,
Bad Windsheim 016914
Reichsstadtmuseum Weißenburg,
Weißenburg in Bayern 022537
Reichsstädtisches Archiv, Rathaushalle
und Ratsstube, Mühlhausen/
Thüringen 020487
Reichsstädtisches Museum, Bad
Wimpfen 016910
Reichstagsmuseum, Museen der Stadt

Regensburg, Regensburg 021329
Reichwald, Helmut F., Stuttgart . 130684
Reid & Reid, Edinburgh 144400
Reid Concert Hall Museum of
Instruments, Edinburgh University
Collection of Historic Musical
Instruments, Edinburgh 044264
Reid, Douglas, Kingston 111226
Reid, S., Nortrup 077805
Reidy, James, Houston 135614
Reif, Erkrath 075649
Reiff, Regstrup 102957
Reiff, Günther, Bad Urach 129426
Reiffenstein, Wien 100365
Reigate Priory Museum, Reigate 045603
Reigate School of Art Design
and Media, East Surrey College,
Redhill 056601
Reijinsha, Osaka 137773
Reijnboek, Apeldoorn 143082
Reik, Karl, Göppingen 076034
Reilly, Naas 079681
Reilly, Frances, Eumundi 099009
Reilly, Jenny, Rockhampton . . . 062163
Reim, Susanne, Altenburg 129382
Reimann, Mühlhausen/Thüringen 078765
Reimann-Wehr, Cornelia, Werl . 078929
Reimann, Detlef, Kirchzarten . . 130128
Réimecher, Remich 082383
Reimer, Höör 143863
Reimer, W., Wolfsburg 079061
Reimers, Dresden 141706
Reimert, Irma, Groningen 112502
Reimex, Mülheim an der Ruhr . . 077436
Reims Décor, Reims 129196
Reimus, Dagmar, Essen 106928
Rein, Jacques, Paris 071835
Rein, Michel, Paris 105188
Reinart, Oss 112617
Reinberger Galleries, Cleveland
Institute of Art, Cleveland . . . 047821
Reinbold, Friederike,
Berchtesgaden 129456
Reincarnation, Cleveland 092966
Reindeer Antiques, London . . . 090444
Reindeer Antiques, Potterspury . 091048
Reinders, J., Würselen 079262
Reindollar, Sheryl G., Charlotte . 092619
A la Reine Victoria, Ruille-La
Gravelle 072644
Reinecke, Gudrun, Hamburg . . 129984
Reinecke, S., Bad Salzdetfurth . 074578
Reinegger, Peter,
Fürstenfeldbruck 075921
Reineke-Fuchs-Museum, Friedrich von
Fuchs, Linden-Leihgestern . . . 020086
Reinelt, Elisabeth & Wolfgang,
Buxheim 106621
Reiner, Thomas, Reutlingen . . . 130583
Reinfelder & Neumann, Nürnberg 077839
Reinhard, Helsingør . . . 065744, 102828
Reinhard-Blauth-Museum,
Weilerbach 022483
Reinhard, Anita, Worms 108961
Reinhard, Susanne, Darmstadt . 075673
Reinhardt, Erich, Ravensburg . . 078148
Reinhardt, Friedrich, Basel 138025
Reinhardt, Harry, Korbach 076941
Reinhardt, J., Halle / Westfalen . 141848
Reinhardt, S., Rinteln 078245
Reinhardt, T., Extertal 075725
Reinhardt, Ute B., Schongau . . 108574
Reinhardt, Wilhelm, Bruchhausen-
Vilsen 129641
Gebr. Reinher, Fußach . 062565, 127735
Reinhold, Emden 141740
Reining, N., Borken, Westfalen . 075103
Reinisch, Sabine, Wien 127961
Reinke, Stefan, Enger 106885
Reinkober, T., Winnigstedt 079022
Reinold fils, Paris 071836, 129127
Reinshagen, W., Bad
Oeynhausen 108390
Reinsma, Gabrielle, La Garenne-

Colombes 069049
Reinstadler & Bebi, Küblis 087548
Reinwald, Joachim, Schwaigern 078412
Reipin Museo, Pirkkala 011254
Reis, Worms 108962
Reis, Norbert, Hohenems 062639
Reis, Roland, Nürnberg 077840
Reis, Vítor Mendes, Lisboa 084950
Reisch, Elmar, Friedrichshafen . 075914
Reisch, Sonja, Wien . . . 063058, 063059,
136607
Reischl, Alfred, Amstetten 062520
Reischl, Friederike, Wieselburg an der
Erlauf 140137
Reiser, Hans, Bochum 075041
Reiser, K.-H., Spaichingen 078510
Reisjärven Kotiseutumuseo,
Reisjärvi 011310
Reisler, Jo Ann, Vienna 145628
Reiss & Sohn, Königstein im
Taunus 126015, 142026
Heide Reiß & Jacqueline de Latour,
Lippstadt 077134
Reiss-Engelhorn-Museen, Museum
Zeughaus C5 mit Forum
Internationale Photographie,
Museum Weltkulturen D5, Museum
Schillerhaus B5,7, ZEPHYR – Raum
für Fotografie C4,9, Curt-Engelhorn-
Zentrum Kunst- und Kulturgeschichte
C4,9, Curt-Engelhorn-Zentrum
Archäometrie D6 An-Institut der
Universität Tübingen, Mannheim 020255
Reiß, Florian, Steinweiler 130658
Reiss, Kurt, Enns 127728
Reisser Kunstverlag, Wien 100366,
136608
Reisz, Gilles, Vitry-sur-Seine . . 141392
Reitbauer, Henrike, Freistadt . . 139982
Reither, Eugène, Colmar 068095
Reitinger, Hedwig, Grödig 062612
Reitmeier, Dr. Harald, Frankfurt am
Main 075815
Reito, San Jose 124460
Reitz, Amsterdam 082837
Reitz Home Museum, Evansville 048645
Reitz, Monika, Frankfurt am
Main 107024
Reitzel, Ann, Hørsholm 065773
Reitzel, Daniel, Saint Paul 096852,
127432
Reitzin Säätiön Kokoelmat,
Helsinki 010858
Reix, Chantal, La Rochelle 069093,
128646
Reixach Campanya Restauració,
Barcelona 133148
Reizan, Kyoto 111300
Rejans, İstanbul 088137
Rejuvenations, Port Pirie 062100
Rękawka, Kraków 113633
Rekezsus, Heiner, Wiesbaden . . 142493
Rekishi Minzoku Shiryokan,
Shimosuwa 029386
Rekishi no Bunka Hakubutsukan,
Matsusaka 029016
Rekishi Shiryokan, Ranzan,
Ranzan 029285
Rekow, André, Crest . 103645, 140830
Le Relais d'Alésia, Le Vaudioux . 069372
Relais de l'Antiquité, Saint-Jean-de-
Gonville 072837
Le Relais des Arts, Puycelsi . . . 072328,
105403
Relais des Arts, Sassenage . . . 105658
Relais Rétro, Paris 071837
Relart, Firenze 131248
Relate Art, Meilen 116498
Relf, Ian, Tunbridge Wells 091662
Relic Antiques, London 090445
Relic Antiques Trade, London . . 090446
Relicário Antiguidades, Belo
Horizonte 063919
O Relicario Antiguidades, Rio de

Register der Institutionen und Firmen
O Relicário Antiguidades e Decorações, Brasília – Research Centre for Islamic History, Art and Culture,

Janeiro 064112
O Relicário Antiguidades e Decorações, Brasília 063938
Relicario San Laureano, Bucaramanga 065237
Relics, New Orleans 095149
Relics – Pine Furniture, Bristol . 088667
Relics Antique Shop, Kingston . . 081956
Relics Antiques, North Fremantle 061990, 127653
Relics Antiques, Wellington 084183
Relics Rescued, Strathalbyn . . . 062249
Relier, Bertrand, Vemars 074000
Religieus Museum Kijk-je Kerk-Kunst, Gennep 032344
Religionskundliche Sammlung der Philipps-Universität, Marburg . 020276
Religious and Church Art, Eindhoven 083226
Religious Museum, Ano Mera . . 022963
Relin, Jean-Luc, Langres 069178
Le Reliquaire, Casablanca 082580
Reliquaire, Latrobe 061722
Relita Antiguedades, Punta del Este 119449
Reliure D'Art laTranchefile, Montréal 101445
Relojería Losada, Madrid 086246, 133372
Relojeria Renovación, Cali 065253
El Relox de Arena, México 082483
Relux, Torino 132189
Rem-Art, Kraków 132847
REM Gallery, San Antonio 124062
Remains Architectural Antiques, New York 095685
Remains Collectables, Fitzroy . . 061470
Remains To Be Seen, West Hollywood 097928
Remanguille, Madrid 133373
Remarkable Refinishing, Sacramento 136192
El Remate, Madrid 126548
Remates, Medellín 065305
Remba, West Hollywood 125021
Rembar, Armadale, Victoria 098666
Rember Iberica, Madrid 133374
Rembert & Sons, Washington . . 136466
Rembold, T., Regensburg 078176
Rembrandt, Eindhoven 083227, 143134
Rembrandt, Salamanca 115405
Rembrandt, Santiago de Chile . . 064983
Rembrandt Antiek Galerie, Dallas 093205
Rembrandt Art Corp, Ciudad de Panamá 113324
Rembrandt Galéria, Budapest . . 079290
Rembrandt Galeria de Arte, Belo Horizonte 100908
Rembrandt Van Rijn Art Gallery, Stellenbosch 038837
Rembrant, Skopje 111892
REME Museum of Technology, Royal Electrical and Mechanical Engineers, Arborfield 043424
Remedia, Sopot 113762
Los Remedios, Málaga 086306
Remember Museum 39–45, Thimister-Clermont 004061
Remember That, Hobart 061635
Remember That, Richmond 096564
Remember the Alibi, San Antonio 145512
Remember This?, Balwyn North 061098
Remember When, Gosford 061547
Remember When, Los Angeles . 094455
Remember When, Saint Louis . . 096766
Remember When, San Antonio . 097022
Remember When Antiques, Tampa 097584
Remember When Shop, Dallas . 093206, 120650
Remenber, Triel-sur-Seine 073885
Remenyik Zsigmond Emlékszoba, Dormánd 023605
Remick Country Doctor Museum and Farm, Tamworth 053674
Remigius Antiquariat, Betzdorf . 141595
Remille, Andrèe, Amponville . . . 066584
Remillet, Guy, Beaune 067034
Remington, Cincinnati 092889
Remington Carriage Museum, Cardston 005564
Remington Firearms Museum, Ilion 049658
Reminiscence, Vancouver 064869
Reminiscence Antiques, Miami . 094794
Remise, Trigance 073886
Remmert & Barth, Düsseldorf . . 106819
Remmler, Trude, Moers 142130
Remnants of the Past, Toronto . 064737
Remnisce Antique, Saint Paul . . 096853
Remolina & Assoc., Miami 121851
Remond, Jean-Louis, Le Havre . 069288
Remont, Beograd 137934, 139019
Rempex, Kraków 126412
Rempex, Warszawa 126427
Rempex Galeria, Warszawa 084744, 113896
Remshard, F., Neustadt an der Aisch 077758
Remuera, Auckland 083820
Remus, Gdańsk 132819
Rémy, Albert, Nancy 140979
Remy, J., Baillonville 082348
Remzi, İstanbul 117080
Ren, Nagoya 111330
REN Antiques, West Hollywood . 097929
Rena, Nara 111338
Rena Bransten Gallery, San Francisco 124375
Rena Centrum, Praha 128363
El Renacimiento, Madrid 086247, 143782
Renaissance, Dunedin 112894
La Renaissance, Granges-Paccot 133765
La Renaissance, Randan 072364
Renaissance, Sherborne 091307, 135035
Renaissance, Solihull . 091356, 135055
Renaissance Antique Prints and Maps, Brighton, Victoria 139658
Renaissance Antiques, Cowbridge 088988
Renaissance Art and Design Gallery, Boston 119915
Renaissance Book Shop, Milwaukee 145269
Renaissance Collection, Dallas . 093207
La Renaissance du Passé, Thaon-les-Vosges 073642
The Renaissance Furniture Store, Glasgow 089333
Renaissance Gallery, Bangalore . 109263
Renaissance ou l'Amour de l'Art, Blacé 067234
Renaissance Relics, Jacksonville 094032
Renaissance-Schloss Demerthin, Gumtow 018890
Renaissance Schloss Greillenstein, Röhrenbach 002601
Renaissance Shop, New Orleans 095150, 135864
The Renaissance Society at the University of Chicago, Chicago 047694
Renaissance Square, Himeji . . . 111256
Renaissances, Versailles 074067
Renaissanceschloss Nöthnitz, Bannewitz 016962
Renaissanceschloss Rosenburg, Rosenburg am Kamp 002609
Renaissence, Dublin 079583
Renan, Rio de Janeiro 064113, 128120
Renard, Mons 100794
Renard, Alexis, Paris 071838
Renard, Gaston, Ivanhoe 139784
Renard, Jean-Claude, Gien 125589
Renard, Jean-Claude, Paris 125751
Renard, Joël, Montmorot 128806
Renard, Olivier, Beaulieu-sur-Mer 067000
Des Renards, Bruxelles 140199
Renassia, Jacques, Saint-Flour . 072787
Renata, Kazan 085152
Renaud-Giquello, Paris 125752
Renaud, Catherine, Saint-Denis-de-l'Hôtel 129221
Renaud, Christophe, Paris 071839
Renaud, Gilbert, Guyonnière . . . 128617
Renaud, Sophie, Saint-Junien . . 072873
Renaudier, Serge, Mougins 104305
Renaudot, Les Sables-d'Olonne . 069430
Renault, Claude, Chatellerault . . 067907
Renault, Frédéric, Dinard 068358
Renault, Jack, Beaulieu-sur-Dordogne 066996
Renavo Rūmai, Mažeikių Muziejus, Renavas 030453
Rendant, Taunusstein 108760
Rendells, Ashburton . . 126800, 126801
Rendelsmann, Heijo, Wiesbaden 078991
Rendez Vous, Angers 066614
Rendez-vous, Strasbourg 105695
Rendez-Vous du Collectionneur, Québec 064540
Rendezvous Art Gallery, Vancouver 101830
Rendezvous Gallery, Aberdeen . 088224
Rendon, San Antonio 124063
Rendorség-történeti Múzeum, Budapest 023548
René, Denise, Paris . . 105189, 105190, 136979
Renel, Milano 080697
Renes, Poznań 113747
Renes International, Winnipeg . . 064932
Renesans, Skopje 111893
Renesans, Wolsztyn 132910
Renessans, Kazan 085153, 114183
Renessans, Sankt-Peterburg . . . 085262, 085263
Reneszánsz Kőtár, Janus Pannonius Múzeum, Pécs 023852
Renford Studio, Bridport 117431
Renfrew Museum, Waynesboro . 054256
Rengebu-ji Temple, Kondo, Kotsudo and Kobodo Halls, Ogi 029206
Renggli, Carla, Zug 116886
Renhua Museum, Renhua 008206
Renishaw Antiques, Sheffield . . 091291
Renishaw Hall Museum and Art Gallery, Renishaw 045604
Renk, Jürgen, Karlsruhe 076700
Renke, Christoph, Wels 125267
Renkert, Gaston, Montbertrand . 070215
Renmin Kang Ri Zhan Zhen Memorial Hall, Beijing 007642
Renn, Paris 105191
Rennebu Bygdemuseum, Rennebu 033888
Rennemeier, A., Warendorf 078869
Renner, François, Velars-sur-Ouche 073996
Renner, Gerhard, Albstadt 141404
Renner, Kurt, Langenlois 062707
Renner, W., Heidelberg 076457
Renners Mechanisches Musikinstrumenten Museum, March 020279
Rennesøy Bygdemuseum, Rennesøy 033889
Renney Antiques, Hexham 089524
Rennhofer, Udo, Zürich 088001
Rennies, London 090447
Renninger's Antique Guide, Lafayette Hill 139486
Rennotte, Paris 129128
Reno County Museum, Hutchinson 049646
Reno-Styl, Warszawa . 084745, 132898, 132899
Reno, Jean-Claude, Montpellier . 104283
Renoantik, Jona 133781
Renoir, Porto Alegre 100999
Renoir, San José 102513
Renoir, Skopje 111894
Renoir Fine Art, Vancouver 101831
Renon, Violette, Saint-Ouen . . . 073180
Renoncourt, Jean, Paris 071840
Renou & Poyet, Paris 105192
Renoud-Grappin, Jean-Paul, Besançon 125515
Renouf, Maria-Rosario, Montamy 070181
Renoux, Yves, Saint-Genis-de-Saintonge 072794, 129231
Renovace Stylového Nábytku, Praha 128364
Renov'Cuir, Fort-de-France 132351
Renowacja i Naprawa Przedmiotów Artystycznych i Zabytkowych, Warszawa 132900
Renowacja Mebli Pozłotnictwo, Poznań 132862
Renowacja Mebli Stylowych, Warszawa 132901
Renowacja Zegarów Antycznych, Warszawa 132902
Rensselaer County Historical Society Museum, Troy 053813
Rensselaer Russell House Museum, Waterloo 054205
Rent-A-Center, Cleveland 120449
Rentoileur, Andréone, Reims . . . 129197
Renton Museum, Renton 052516
Rentz, Richmond 123740
Renu-Grain, Los Angeles 135741
Renville County Historical Society Museum, Mohall 050986
Renwick Gallery, Smithsonian American Art Museum, Washington 054158
Renwick Museum, Renwick . . . 033294
Renz, Stuttgart 130685
Renz, Ingrid, Stuttgart 108725
Renz, Josef, Wilhelmsburg an der Traisen 063117
Repair Plus, Tampa 136396
Repartee Gallery, Salt Lake City 123984
Repeat Antiques, Charlotte 092620
Repeat Consignment, Houston . 093776
Repeat Performance, Los Angeles 094456
Repères, Paris 138659
Repetto & Massuco, Acqui Terme 109736
Repetto, Francesca, Milano 110414
Replicas El Antaño, Medellín . . . 065306
Replikate der Welt-Kunst im Schloss, Miltach 020406
Replot Hembygdsmuseum, Replot 011312
Reporcel, Cali 128287
Reppel, Thomas, Iserlohn 076621
Repphun, Konstanz . . . 107721, 130183
Reppuniemen Ulkomuseo, Pöytyä 011258
Reprise, Amsterdam 082838
Repro Art Workshop, Singapore 114599
Reproduction Prints, Preston . . . 118948
Repslagarmuseet, Älvängen . . . 040510
Repton Clocks, Repton 134966
Republic County Historical Society Museum, Belleville 046881
Republican College of Arts, Academy of Arts of Uzbekistan, Toshkent 058148
Republika Antico, Warszawa . . . 084746, 126428
République Numismatique, Paris 071841
Requena Lozano, Jesus, Madrid 133375
Rerat, Alexis, Montbéliard 070211
Res Allestimenti, Roma 081525
Res Artis, Bucureşti 133001
Res Centro Restauro, Perugia . . 131760
Resa Termiño, Vitoria-Gasteiz . . 086577
Resart, Napoli 131636
Resch & Resch-Schachten, Oberdolling 077858
Resch, Hans, Ruhstorf an der Rott 130612
Reschke, Torsten, Willich 079015
Rescon, Madrid 133376
Research Centre for Islamic History, Art and Culture, İstanbul 060011

Research Institute of Art Criticism,
Academy of Arts of Uzbekistan,
Toshkent 058149
Research Museum of the Institute
of Zoology, Academia Sinica,
Taipei 042546
Resena Revista de Critica Cultural,
Madrid 139067
Resenhof, Bernau im
Schwarzwald 017330
Reserva Pessoal, São Paulo . . 064203
Reservation, San Antonio . . . 124064
La Reserve, Knokke-Heist 100760
La Réserve, Paris 105193
La Reserve de Peau d'Ange,
Paris 071842
Réserve Géologique de Haute-
Provence, Digne-les-Bains . . 012774
Réserve Naturelle Géologique,
Sisteron 015818
Réserves de l'Antiquaire,
Florensac 068551
Réserves Nationales des Arts et
Traditions Populaires Picardie, Saint-
Riquier 015550
Reserviupseerikoulun Museo,
Hamina 010798
Réservoir d'Images, Paris 105194
Resetarits, Wien 140118
Resettlers Market Place,
Charlotte 092621
Reshaw Exhibit, Evansville . . . 048646
Resi, Guglielmo, Milano 080698
La Résidence, Den Haag 083121
Résidence d'Artistes, Dompierre-sur-
Besbre 012810
Residence Galeria, Kraków . . . 113634
Residency Museum, York 001675
Residenz Ansbach, Bayerische
Verwaltung der staatlichen Schlösser,
Gärten und Seen, Ansbach . . 016596
Residenz Ellingen, Bayerische
Verwaltung der staatlichen Schlösser,
Gärten und Seen, Ellingen . . . 018155
Residenz Kempten, Kempten . . 019589
Residenz und Hofgarten Würzburg,
Bayerische Verwaltung der
staatlichen Schlösser, Gärten und
Seen, Würzburg 022798
Residenzgalerie Salzburg, Salzburger
Landessammlungen, Salzburg . 002636
Residenzmuseum im Celler Schloss,
Celle 017725
Residenzmuseum, Bayerische
Verwaltung der staatlichen Schlösser,
Gärten und Seen, München . . 020565
Residenzschloß Arolsen, Stiftung des
Fürstlichen Hauses Waldeck und
Pyrmont, Bad Arolsen 016706
Residenzschloß Oettingen, Oettingen
in Bayern 020968
Résillot, Michel, Vesoul 074078
Resim ve Heykel Müzesi, Ankara 042828
Resim ve Heykel Müzesi,
İstanbul 042953
Resim ve Heykel Müzesi, İzmir . 042975
Resin, Singapore 085489
Resin Wood, Bromma . 086654, 133519
Resistance Museum, Akko 025014
Resma, Moskva 133013
Resonans, Gent 100702
Resource Centre for the Arts, Saint
John's 006910
Resource for Art, San Francisco 124376
La Respelido, Saint-Rémy-de-
Provence 073291
Rèspublikanskaja Mastackaja Galerèja,
Minsk 003254
Respublikanskij Pedagogičeskij Muzej,
Almaty 029810
Respublikanskij Vystavočnyj Zal, Muzej
izobrazitelnych iskusstv, Joškar-
Ola 037074
Respublikinis Unikalių Amenų

Muziejus, Mosėdis 030424
Resselmark, Jan, Sundbyberg . . 087087
Resserre, Le Mans 069311
Rešt-Art, Bratislava . . 085557, 133048
Restart, Bari 130919
Restart, Beograd 085346
Restau-Moder, Barcelona 133149
RestauArte, Kiel 076789, 130121
Restaudecor, Alicante 133084
Restaur Arte-1997, Zaragoza . . 133507
Restaura, Getxo 086013, 133259
Restaura Cartagena, Cartagena . 133223
Restauració de Llums, Barcelona 133150
Restauracion de Antigüedades,
Lima 132811
Restauracion de Muebles Elizasu,
Donostia-San Sebastián 133256
Restauracion de Tomo y Lomo,
Madrid 133377
Restauración y Arte, Bogotá . . 065223,
128276
Restauración y Venta de Muebles
Antigos, Medellín . . . 065307, 128307
Restauraciones, Barcelona 133151
Restauraciones Albarium,
Zaragoza 133508
Restauraciones Artísticas,
Medellín 128308
Restauraciones Bilbao, Bilbao . 133199
Restauraciones Emilio, Madrid . 133378
Restauraciones Fotograficas Vihuli,
Cali 128288
Restauraciones Larrea, Buenos
Aires 060926, 127531
Restauraciones Plumiers, Madrid 133379
Restauraciones Redó, Barcelona 133152
Restauraciones Religiosas de
Imágenes, Medellín 128309
Restauraciones Siglo XXI, Bilbao 133200
El Restaurador de Antigüedades,
Marbella 133405
La Restauradora, Barcelona . . . 085829,
133153
Restauradora Arte Antiga, São
Paulo 128131
Los Restauradores, Buenos Aires 127532
Restaur'Art, Lyon 128728
Restaurart, Milano 131509
Restaurart-Restauração de Obras de
Arte, Porto Alegre 128109
Restauarte, Roma 132024
Restaurarts, São Paulo 128132
Restauration de vos Meubles, Essey-
lès-Nancy 128596
Restaurations Werkstatt,
Emmedingen 129791
Restaurator, München 138750
Restauratore Mobili Iusco, Bari . 130920
Restauratori in Torino, Torino . . 132190
Restauretieatelier Mösenbacher, Den
Haag 132479
Restauri Artistici Monumentali,
Firenze 131249, 131250
Restauri del Sole, Bari 130921
Restauri Resta, Bari 130922
Restaurierung am Oberbaum,
Berlin 129551
Restaurierungs Werkstätten Berlin,
Berlin 129552
Restaurierungsatelier Friedrich,
Dresden 129721
Restaurierwerkstatt Irenenta, Unter-
Tullnerbach 127889
Restauro, Berlin 129553
Il Restauro, Bologna . . 130994, 130995
Il Restauro, Firenze 131251
Il Restauro, Milano 131510
Il Restauro, Napoli 131637
Restauro, Warszawa 132903
Restauro – Forum für Restauratoren,
Konservatoren und Denkmalpfleger,
München 138751
Restauro Antichità, Roma 132025
Il Restauro Antichità, Torino . . . 132191

Restauro Antigo, Porto 132975
Restauro Archeologico Ed
Architettonico, Milano 131511
Restauro Art, Bologna 130996
Restauro Conservativo di Opere Su
Carta, Firenze 131252
Restauro d'Arte, Roma 132026
Restauro di Mosaici e Materiali
Lapidei, Firenze 131253
Restauro Dipinti, Bolzano 131009
Restauro Dipinti Studio 4,
Firenze 131254
Restauro Due, Barcelona 133154
Restauro e Antiquariato, Verona 081934,
132297
Restauro e Arte, Torino 132192
Restauro Mobili, Milano 131512
Restauro Modena, Modena . . . 131583
Restauro San Giorgio, Roma . . 132027
Restauro van Dyck, Madrid . . . 133380
Restauros, Porto 132976
Restavracija EAD, Burgas 128135
Restavracija EAD, Varna 128141
Restavracija EAD, Veliko Tărnovo 128142
Restavracija i Konservacija,
Gabrovo 128136
Restavrator Anfilada, Sankt-
Peterburg 133022
Restavrator-Peterburg, Sankt-
Peterburg 133023
Reste-Langelier, Micheline, Rennes
(Ille-et-Vilaine) 072442
Restelo, Madrid 086248
Restigouche Regional Museum,
Dalhousie 005676
Resto Meubles, Saint-Denis . . . 132986
Reston and District Historical Museum,
Reston 006784
Restor Bois, Saméon . 073391, 129250
Restorart, Toronto 128215
Restoration & Conservation of Art,
Auckland 132696
Restoration and Design, Toledo . 136410
Restoration and Design Studio, New
York 135983
Restoration Antiques, New York 135984
Restoration Arts, Miami 135805
Restoration Atelier, New York . . 135985
Restoration House, Mantorville . 050660
Restoration Shop, Austin 135332
The Restoration Studio, London . 134757
Restorations, Waikato 132753
Restore It, San Francisco 136341
Restored Radios and Antiques,
Deepdene 061394
The Restorer, Santa-Venera . . . 132341
Restorers Barn, Castlemaine . . . 061298,
127581
Restoring Australia, Yarck 062499
Resurgam, Baltimore 119768
Resurrected Furniture Gallery, San
Diego 097127
Rét Galéria – Gazdagréti Kösségi Ház,
Budapest 109174
El Retablo, Bogotá 102362
Retablo Arte, Pamplona 115392
El Retaule, Terrassa 086489
Retch, Cairo 066267
Retelet, Pascal, Saint-Paul-de-
Vence 073246
Rethberg, Ronny, Dresden 075426
Reti, Thomas, Saint-Ouen 073181
El Retiro Art Gallery, Fyshwick . 099068
Rétközi Múzeum, Kisvárda 023751
Reto a la Esperanza, Bilbao . . . 085885
Reto a la Esperanza, Santander 086427
Retoc Restauracio, Barcelona . . 133155
Retoque Rustilo, Lisboa 132953
Retort, Amsterdam 112272
Retratos, Asunción 113336
Retriever Books, Dee Why 139714
Retro, Almaty 082238, 111553
Retro, Gdynia 084522
Retro, Rio de Janeiro 064114

Retro, Stourbridge 091458
Rétro Actif, Thémines 073643
Rétro Activ, Paris 071843
Retro-Active, Northcote 062004
Retro-Art, Warszawa . . 113897, 137909
Rétro Broc, La Gaude 069050
Retro Collectibles, Norfolk 095874
Rétro Désign, Rouen 072613
Retro Fun, New Orleans 095151
Retro Gallery, Hasselt 063687
Retro Gallery, Houston 121118
Retro Gallery, Los Angeles 094457
Retro Home, London 144676
Retro Inferno, Kansas City 094091
Rétro Lumières Antiquités, Durtal 068422
Rétro-Mobile Club Drouais, Musée
Automobile, Dreux 012839
Retro Modern Studio, New York 095686
Retro Nynäshamn, Nynäshamn . 086894
Retro Photo, Rognonas 072517
Retro Rose, Salt Lake City 096919
Retro-Styl, Sopot . . . 084665, 143503
Retro Verso, Bayonne 066980
Rétro Verso, Toulouse 073795
Retro Ville, Calgary 064267
Rétro Ville, Montréal 064462
Retroactive, New Orleans 095594
Retromania, Toronto 064738
Retrospect, Dublin 079584
Retrospect, San Francisco 097313,
136342
Retrospection, Portland 096467
Retrospective, Wellington 084184
Retrospective Antiques and Toys,
Singapore 085490
Retrostyl, Gdynia 084523
Retrouvailles, Saint-Étienne . . . 072774
Les Retrouvailles, Saint-Sylvestre-sur-
Lot 073311
Retrouvailles, Stadtbredimus . . 082392
Les Retrouvailles, Toulon (Var) . 073727
Retsas, Dimitris, Athinai 142540
Rettel, Andreas, Kriftel 130200
Retter, Frankfurt am Main 107025
Retterath, Pia, Boos 075097
Retucci, Bolzano 109888
Retureau, Alain, Châteauneuf . . 067890
Returned Services League War
Museum, Stawell Sub-Branch,
Stawell 001519
Réty, Michel, Flavigny-sur-
Ozerain 068536
Retzer Erlebniskeller, Retz 002590
Reubacher Heimatmuseum, Rot am
See 021498
Reuben and Edith Hecht Museum,
Haifa 025057
Reuben H. Fleet Science Center, San
Diego 052969
Reubens Gallery, Leeds 118089
Reuel B. Pritchett Museum,
Bridgewater 047206
Gebr. Reuer, Gladbeck 076005
Reul, Ronnie, Waldsolms 078848
Réunion des Musées Nationaux,
Paris 058663, 136980
Réunion des Musées Nationaux –
Nice, Nice 058664
Reunion Gallery, Felixstowe . . . 117794
Reure, Dominique, Sainte-Foy-les-
Lyon 073350
Reure, Dominique, Villeurbanne . 074223
Reus, Frankfurt am Main 107026
Reus, Hanau 107340
Reuß, Bernhard, Gießen 075997
Reußner, Reinhard, Essen 075693
Reuter-Villa mit Wagner-Sammlung,
Thüringer Museum Eisenach,
Eisenach 018134
Reuter, Marianne, Salzburg . . . 062814
Reuter, Michel, Saint-Tropez . . 073317
Reuter, R. & G., Essen 075694
Reuter, Walter, Düsseldorf 129757
Reuther, Fürth 075925

Reuther, Christel, München 077598
Reutlinger Kunstkabinett,
Reutlingen 078214, 108484
Reutovskaja Gorodskaja Kartinnaja
Galereja, Reutov 037685
Reutter, S., Kempten 076753
Rev' & Culture, Limoges 104036
Revak & Co., M., Tucson 124768,
145624
Reval Antiik, Tallinn 066286
Revalor, La Montagne 128645
Revans & Waite, Greytown ... 083946
Revantic, Paris 071844
Rêve et Plaisirs, Mazières-de-
Touraine 070022
Au Réveil du Temps, Remiremont 072414,
129199
Réveillaud, Richard, Chef-
Boutonne 067953
Revel, Sylvain, Lyon 140931
Revelation Studios, San
Francisco 124377
Revell, Sheila, Hartley Wintney . 089454
Revellin, Logroño 086079
Revelstoke Firefighters Museum,
Revelstoke 006785
Revelstoke Museum, Revelstoke 006786
Revelstoke Railway Museum,
Revelstoke 006787
Reventlow-Museet Pederstrup,
Horslunde 010145
Le Rêvèrbère, Lyon 104130
Rêverie, Cannes 067650
Revers, Berlin 141562
Rêves d'Autrefois, Richelieu ... 072458
Rêves de Poupées, Cormatin .. 068152
Rêves d'Enfants, Nice 070657
Reviejo Antigüedades, Bahía
Blanca 060843, 127504
Review, London 139220
Revillon d'Apreval, Paris 071845
Revillon Frères Museum,
Moosonee 006425
Revisita, Roma 138866
La Revista, Hermosillo 138919
Revista Artecubano, La Habana . 138611
Revista de Estética, Buenos
Aires 138489
Revista de Muséologia, Madrid . 139068
Revista muzeelor, Bucureşti ... 139005
Revival, Abridge 088234, 134016
Revival, Boston 092491
Revival, Milton 061858
Revivals Antiques, San Diego . 097128
Revive Art Gallery, Auckland ... 112807
Reviver, Porto 085047
Revol & Hermonville, Le Havre . 125611
Revol & Hermonville-Magnier,
Bolbec 125519
Revolution Books, Detroit 145139
Revolution House, Chesterfield . 043938
Revolution Museum, Harbin 007849
Revolutionary Museum Ho Chi Minh
City, Ho Chi Minh City 054806
Revolver Revue, Praha 136768, 138614
Revolving Museum, Lowell 050512
Revue Belge de Numismatique et de
Sigillographie, Bruxelles 138559
La Revue de la Céramique et du Verre,
Vendin-le-Vieil 137013, 138660
Revue du Louvre et des Musées de
France, Noisiel 138661
Revue Noire, Paris 138662
Revue Refinishing Studio,
Oakland 136045
REW Antiguidades, Rio de
Janeiro 064115
Rewa's Village, Kerikeri 033236
Rex, Bratislava 085558
Rex, Komárno 085562
Rex Antiques, Chagford 088815
Rex, Verena, Markneukirchen .. 077298
Rey, Clermont-Ferrand 068036
Rey, Espoey 068471

Rey, Firenze 080192
Rey Bello, Emilio, A Coruña ... 133236
Rey Camara, Jose Maria del,
Madrid 086249
Rey Don Pedro, Burgos 133214
Rey Valon, Denis, Saint-Cannat . 072705
Rey, Jacques, Châtillon-sur-
Chalaronne 067919
Rey & Cie., Jean, Lyon 069777
Rey & Cie., Jean, Paris 071846
Rey, Marie-Madeleine, Cahors .. 125542
Rey, Michel, Montpellier 070311
Rey, Patrice, Capdenac-le-Haut . 067661
Rey, Patrice, Figeac 068531
Reyburn House Art Gallery,
Whangarei 113112
Reyer, Harry, Grieskirchen 062610,
127756
Reyes Espinola, Cecilio, Granada 133270
Reyes Villalba, Daniel, Cali ... 128289
Reyes, Gonzalo, Granada 086041
Reyghere, de, Brugge 140161
Reymond, Charly, Censeau ... 067745
Reymond, Charly, Nozeroy 070762
Reymond, Gérard, Genève 126730
Reymond, Jacques, Carouge ... 133714
Reymond, Loic, Saint-Étienne .. 072775
Reymond, Tristan, Genève 126731
Reynaert, François, Remaugies . 072409
Reynal & Hervouet, Paris 071847
Reynaud, Marie-Claire, Cabrieres-
d'Avignon 067549
Reynaud, Robert, Boulieu-les-
Annonay 067388
Reyne, Cincinnati 092890
Reynold Rapp Museum, Spalding 007063
Reynolda House, Museum of American
Art, Winston-Salem 054493
Reynolds, Nashville 095029, 135849
Reynolds, Richmond 123741
Reynolds-Alberta Museum,
Wetaskiwin 007391
Reynolds Gallery at Jeannette Powell
Art Center, University of the Pacific,
Stockton 053542
Reynolds Homestead, Critz 048067
Reynolds Museum, Wetaskiwin . 007392
Reynolds, Anthony, London 118538
Reynolds, B.A.S., Easingwold . 089128
Reynolds, C., Oakham 090903
Reynolds, C.H., Pickering 091016
Reynolds, Douglas, Vancouver .. 101832
Reynolds, Thomas R., San
Francisco 124378
Reyre, Robert, Aix-en-Provence . 066474
Reyum Institute of Arts and Culture,
Phnom Penh . 005287, 055112, 101204,
136673
Reza Abbasi Museum, Tehrän .. 024781
Reza Shah Museum,
Johannesburg 038756
Rezag, Mohamed, Oran 060841
Rezai, A., London 090448
Reza's Antiques, Dallas 093208
Reza's Rug Gallery, Chicago ... 092778
Řezbářství a Restaurátorské Práce,
Hradec Králové 128326
Rezek, Thomas, München 142180
Rezervaţia Arheologică, Cucuteni 036427
Rezervaţia Arheologică Cotnari,
Cotnari 036412
Rezervaţia Arheologică Târgşoru Vechi,
Târgşoru Vechi 036755
Rezervaţia de Arhitectură Vrânceană,
Focşani 036467
Rezerwat Archeologiczny na Zawodziu,
Muzeum Okręgowego Ziemi Kaliskiej
w Kaliszu, Kalisz 034802
Rezvanian, Strasbourg 105696
Rezydencja Księży Młyn, Muzeum
Sztuki w Łodzi, Łódź 034980
RG Interiors, Birmingham 092432
Rhea Gallery, Leominster 118104
Rhein-Museum Koblenz, Koblenz 019665

Rheiner Auktionshaus, Rheine .. 126104
Rheingauer Auktionshaus, Eltville am
Rhein 075604, 125936
Rheingauer Weinmuseum
Brömserburg, Rüdesheim am
Rhein 021539
Rheinische Münzfreunde e.V., Bergisch
Gladbach 059223
Rheinischer Kunsthändler-Verband
(RKV) e.V., Köln 059224
Rheinisches Bildarchiv, Museen der
Stadt Köln, Köln 019710
Rheinisches Eisenkunstguss Museum,
Bendorf 017025
Rheinisches Fastnachtsmuseum
Koblenz, Koblenz 019666
Rheinisches Feuerwehrmuseum,
Erkelenz 018252
Rheinisches Freilichtmuseum und
Landesmuseum für Volkskunde
Kommern, Mechernich 020323
Rheinisches Industriemuseum,
Bergisch Gladbach 017052
Rheinisches Industriemuseum,
Außenstelle Euskirchen,
Euskirchen 018327
Rheinisches Industriemuseum,
Außenstelle Oberhausen,
Oberhausen 020897
Rheinisches Industriemuseum,
Schauplatz Solingen, Solingen 021903
Rheinisches Industriemuseum,
Zentrale, Oberhausen 020898
Rheinisches Landes Museum Bonn,
Landschaftsverband Rheinland,
Bonn 017499
Rheinisches Landesmuseum für
Industrie- und Sozialgeschichte, LVR-
Industriemuseum, Schauplatz
Ratingen, Ratingen 021290
Rheinisches Landesmuseum Trier,
Trier 022202
Rheinisches Malermuseum, Bonn 017500
Rheinisches Schützenmuseum,
Neuss 020749
Rheinland-Pfälzisches Freilichtmuseum
Bad Sobernheim, Bad
Sobernheim 016879
Rheinmuseum, Emmerich am
Rhein 018191
Rheinstädtler, M., Mannheim ... 077269
Rhinebeck Aerodrome Museum,
Rhinebeck 052522
Rhinegold Publishing, London .. 138245
Rhinelander Logging Museum,
Rhinelander 052523
Rhinestone, Dublin 079585
Rhinestone, Rosie, Seattle 097493
Rhino, New Orleans .. 122221, 122222
Rhino Books, Nashville 145299
Rhinopolis, Gannat 013076
Rho, Seung Jin, Seoul 111706
Rhode Island Black Heritage Society
Museum, Providence 052393
Rhode Island Foundation Art Gallery,
Providence 052394
Rhode Island Historical Society
Exhibition, Providence 052395
Rhode Island Museum Network,
Pawtucket 060713
Rhode Island School of Design,
Providence 057729
Rhode Island Watercolor Society,
Pawtucket 060714
Rhodes College, Memphis 050790
Rhodes Cottage Museum,
Muizenberg 038788
Rhodes Memorial Museum and
Commonwealth Centre, Bishop's
Stortford 043639
Rhodes Museum of Classical
Antiquities, Grahamstown .. 038727
Rhodes University School of Fine Art,
Grahamstown 056345

Rhodes, Charles, South Auckland 132743
Rhodes, Isobel, Woodbridge ... 091918
Rhodes, Peter, Southampton ... 144855
Rhodes, Philip, Palmerston North 084057
Rhön-Museum, Fladungen 018369
Rhöner Naturmuseum, Tann ... 022103
Rhok – Academie voor Beeldende
Kunsten Brussel, Bruxelles ... 055013
Rhomberg, Clemens, Innsbruck . 100056
Rhondda Heritage Park, Trehafod 045995
Rhondda Museum, Llwynypia .. 044895
Rhos Books, Colwyn Bay 144353
Rhuddlan Castle, Rhuddlan 045606
Rhudle Mill, Kilmichael Glassary 089689,
134513
Rhumba Antiques, Chicago 092779
Rhyl Library, Museum and Arts Centre,
Rhyl 045607
Rialto, Seattle 097494
Rialto Museum, Rialto 052524
Riarte, Venezia 111152
Riash Gallery, Cairo 103073
RIBA Drawings and Archives
Collections, Royal Institute of British
Architects, London 045084
RIBA Publishing, London 138246
Ribalta, Budapest 079291
Ribarsko-Biološki Muzej, Nov
Dojran 030621
Ribaux, Sankt Gallen 144130
Ribbentrop, Christiane von, Aschau im
Chiemgau 074441
Ribbhagen, Kurt, Stockholm ... 087046
Ribblefort, Warton 119310
Ribbons, Delhi 079389, 109337
Ribchester Roman Museum,
Ribchester 045608
Ribe Antikvariat, Ribe 140645
Ribe Kunstmuseum, Ribe 010276
Ribe Raadhussamling, Ribe 010277
Ribeiro de Sousa, Porto 085049, 126466
Ribeiro, Cecilia, Rio de Janeiro . 101063
Ribeiro, Manuel Santos, Porto . 085048
Ribeli, Stefan, Bülach 087338
Ribeli, Stefan, Gräslikon 087512, 133764
La Ribera, Murcia 115317
Ribera Corona, Barcelona 133156
Ribes Garin, Salvador, Valencia . 086533,
086534
Ribes Vikinger Museum, Ribe .. 010278
Ribes, Jean-Charles, Paris 071848
Ribière & Tuloup, Marseille ... 125639
Ribolzi, Adriano, Monaco 082528,
112016
Ribot, Michel, Anglès (Brassac) . 066618
Rica Basagoiti, Bilbao 085886
Rica Basagoiti, Madrid 086250
Rica Basagoiti, Juan, Bilbao ... 085887
Ricard, Paul, Paris 105195
Ricart, Anna, Barcelona 085830, 114955
Ricaut, Guy, Pessac 072117
Riccadag Antiquities, Calgary .. 064268
Riccadonna, Monaco 112017
Riccardi Antichità, Assisi 079755
Riccardi, Angelo, Roma 081526
Riccardi, Maresca, Roma 142835
Ricchetto, Giuliana, Torino 142870
Ricci, Livorno 110187
Ricci, Anna, Roma 081527
Ricci, Corrado, Marzabotto ... 110254
Ricci, Jacques-André,
Chapareillan 067834
Ricci, Jean, Soisy-sur-Ecole ... 073534
Ricci, Sara, Firenze 131255
Ricciardi, Lucienne, Grenoble .. 068739
Ricciotti, Alain, Nice 070658
Riccius, Göteborg 086716
Ricco/Maresca Gallery, New York 123005
Rice, Denver 120760
Rice, Hamilton, New South
Wales 139762
Rice, Houston 093777
Rice County Museum of History,
Faribault 048700

The Rice Museum, Crowley 048078
Rice Museum, Dajia 042298
The Rice Museum, Georgetown 049083
Rice University Art Gallery,
Houston 049592
Rice, Harold B., Buffalo 092552
Ricerca d'Arte, Roma 110851
Ricerche Sul '600 Napoletano,
Napoli 138867
Ricewold Museum, Los Baños . 034498
Rich, Den Haag 083122
Rich Man Poor Man, Sacramento 096643
Rich, David, Denmark 098967
Rich, Ernsting, Indianapolis 121216
Rich, Michael & Steven, London 090449,
118539
Richard, Appledore, Kent 088285
Richard, Bevaix 087312
Richard, Houston 093778
Richard, Long Jetty 139809
Richard, Rio de Janeiro 064116
Richard Billinger-Gedenkraum, Sankt
Marienkirchen bei Schärding .. 002679
Richard-Brandt-Heimatmuseum
Bissendorf, Wedemark 022449
Richard E. Peeler Art Center, DePauw
University, Greencastle 049236
Richard F. Brush Art Gallery and
Permanent Collection, Saint
Lawrence University, Canton .. 047408
Richard Gallery and Almond Tea
Gallery, Cuyahoga Falls 048101
Richard H. Schmidt Museum of
Natural History, Emporia State
University, Emporia 048581
Richard-Haizmann-Museum,
Niebüll 020788
Richard Jefferies Museum, Coate 043981
Richard L. Nelson Gallery and the
Fine Arts Collection, University of
California, Davis, Davis 048164
Richard Salter Storrs House,
Longmeadow 050389
Richard-Simoncic-Museum,
Volkskundemuseum,
Rabensburg 002556
Richard Sparrow House,
Plymouth 052214
Richard-Wagner-Museum Graupa,
Pirna 021139
Richard-Wagner-Museum, Haus
Wahnfried, Bayreuth 017001
Richard Wagner-Museum,
Städtische Sammlung historischer
Musikinstrumente Tribschen,
Luzern 041712
Richard, Anne, Bethancourt-en-
Vaux 067148
Richard, C. & M., Neuchâtel ... 087706
Richard, F., Gent 100703
Richard, François, Saint-Ouen . 073182
Richard, Jean-Luc & Takako,
Paris 105196
Richard, Jean-Philippe, Mirabel-aux-
Baronnies 104239
Richard, M., Cottenchy 068169
Richard, Marcel, Amiens 066581
Richard, Marie-Line, Saint-Pierre-
Lafeuille 073262
Richard, Michel R., Montréal ... 064463,
128184
Richard, Nicolet, Coustellet 136037
Richard, R. & F., Bruxelles 063538
Richards Antiques, Vancouver .. 064870
Richard's Oriental Rug Gallery,
Pittsburgh 096323
Richards, D., Pittsburgh 145437
Richards & Sons, David, London 090450,
134758
Richards, Glenn, Seattle 097495
Richards, J., Charlotte 120011
Richards, Lois, New York 123006
Richards, Margo, Sydney 062298

Richards, Richard, Hong Kong .. 102108
Richardson, Nantwich . 090799, 134864
Richardson & Kailas, London ... 090451
Richardson & Smith, Whitby ... 127086
Richardson-Bates House Museum,
Oswego 051839
Richardson-Clarke, Boston 119916
Richardson, Nanette, San
Antonio 124065
Richardsons, Bourne 126826
Richborough Castle – Roman Fort,
Sandwich 045737
Richco, San Diego 124165
Riche, Cairo 103074
Richel, Philippe, Pessac 072118
Richelieu, Constanţa 085120
Richelieu, Gijón 086020, 126524
Richelieu Numismatique, Paris . 071849
Richens, Andrew, Toronto 064739
Riches, Cottonend 134298
Richet, Maurice, Tigy 073681
Richey Historical Society, Richey 052525
Richfield, New York .. 095687, 135986
Richibucto River Museum,
Richibucto 006790
Richland, Nashville 122097
Richland County Historical Museum,
Wahpeton 054026
Richland County Museum,
Lexington 050261
Richmond, London 118540
Richmond, Virginia Beach 124840
Richmond Antiques, Bowdon .. 088549
Richmond Antiques, Nelson 084020
Richmond Antiques, Richmond,
Tasmania 062143
Richmond Art Center, Richmond 052532
Richmond Art Gallery, Richmond 006796
Richmond Art Museum,
Richmond 052535
Richmond Barbers, Richmond,
Surrey 118973
Richmond Book Shop, Richmond 145477
Richmond County Historical Society
Museum, Melbourne 006285
Richmond Desk Leathers,
Richmond 136175
Richmond Gaol Museum, Richmond,
Tasmania 001449
Richmond Hill Gallery, Richmond,
Surrey 118974
Richmond Main Mining Museum,
Pelaw Main 001399
Richmond Museum, Richmond . 006797,
038821
Richmond Museum of History,
Richmond 052533
Richmond National Battlefield Park,
Richmond 052559
Richmond Relics, Richmond,
Victoria 062148
Richmond River Historical Museum,
Lismore 001224
Richmond Vale Railway Museum,
Richmond Vale 001450
Richmond, David, Hyde Park ... 061646,
127617
Richmond, David, Unley 062378
Richmondshire Museum, Richmond,
North Yorkshire 045611
Richomme, New York 135987
Richon, Christian, Balma 066926
Rich's Antiques, Columbus 093042
Rich's Oldies and Goodies,
Chicago 092780
Richter, Nashville 095030
Richter, Varel 108810
E. Richter & H. Graf,
Lüdenscheid 107925
Dr. Richter & Kafitz, Bamberg . 125862
Dr. Richter & Kafitz, Coburg .. 125915
Richter & Masset, München ... 077599,
142181
Richter-Armgart, R., Bremen .. 106592

Richter-Levsen, Martha, Wyk auf
Föhr 079126
Richter Múzeum, Budapest 023549
Richter Verlag, Düsseldorf 137166
Richter, B. & J., Baden-Baden . 141442
Richter, Günter A., Rottach-Egern 108530,
137491
Richter, H., Braunshorn 075136
Richter, Helmuth, Weißenburg in
Bayern 130742
Richter, Horace, Jaffa 109713
Richter, Jürgen, Willingshausen . 079017
Richter, K., Wolgast 079062
Richter, L., Lörrach 077146
Richter, M., Leverkusen 107863
Richter, Marianne, Hamburg ... 076297
Richter, Reiner, Nordhorn 077803
Richter, Roland, Rennerod 078208
Richter, U., Neustadt am
Rübenberge 077755
Richtermeier, Ilona Marieluise,
Berlin 129554
Richters, E., Hilden 076532
Richy, Pierre-Maurice, Chalon-sur-
Saône 067783
Richy, Pierre-Maurice, Montbellet 070213
Rick, R., Frechen 075839
Rickard, Fred, Toledo . 097616, 136411
Ricke, Walter, Kottgeisering ... 142034
Ricker Pewter, Micheal, New
Orleans 122223
Rickert, Hans, Stenten 108673
Rickett & Co., Shepperton 091299,
135031
Rickety Rix, Traralgon 062343
Rick's Furniture Refinishing, San
Antonio 136271
Ricky's Galleria, Seattle 097496
Rico, Sainte-Foy-d'Aigrefeuille . 129248
Ricochet, Toronto 064740
Ricoletto, Bogotá 102363
Riconti, Bernard, Paris 071850
Ricordi Rubati, Milano 131513
Il Ricordino, Napoli 080925
Ricordo, Porto Alegre 063991
Ricotta, Sergio, Roma . 081528, 132028
Ricour-Dumas, Christian, Paris . 071851
Ricros, Jean-Luc, Aurillac 066829
Ric's Art Boat, Bruxelles 003499
Rid, Katia, München 108182
Ridard, Claudine, Jossigny 068913
Ridard, Claudine, Saint-Mandé . 072926
Riddagshausen, Braunschweig . 106568
Riddarholmskyrkan, Stockholm . 041034
Riddarhuset, Stockholm 041035
Riddell, Dallas 145118
Ridder, Thierry de, Saint-Ouen . 073183
Ridder, Willem de, Antwerpen . 127999
Ridderhof Martin Gallery,
Mary Washington College,
Fredericksburg 048959
Ridderhofstad Gunterstein,
Breukelen 032108
Riddiford Corner, Wellington .. 084185
Riddoch Art Gallery, Mount
Gambier 001318
Rideau Canal Museum, Smiths
Falls 007050
Rideau District Museum,
Westport 007388
Rideau Hall, Ottawa 006597
Ridelius, Visby 115999
Ridente, Nicola, Bari 130923
Rider University Art Gallery,
Lawrenceville 050192
Ridge House Museum,
Ridgetown 006801
Ridgeway Antiques, Westcliff-on-
Sea 091795
Ridglea, Fort Worth 093476
Ridi & Madon, Roma 132029
Riding Mountain National Park
Interpretive Centre, Riding Mountain
Historical Society, Wasagaming 007358

Ridna Chata, Kyïv 117148
Ridnik, Versailles 074068
Ridžina, Moskva 114296
Rieber, Bernd, Stuttgart 126123
Rieck, Wagersrott 078822
Rieck-Haus, Vierländer
Freilichtmuseum, Hamburg .. 019017
Riedel, Beeskow 129450
Riedel, Christa, Frankenthal,
Pfalz 106965
Riedel, Friedrich, Marktl 130290
Riedel, Gerhard, Mühlhausen/
Thüringen 077428
Riedenburger Bauernhofmuseum,
Riedenburg 021417
Riedi, Ursula, Zürich 088002
Riedl-Seletzky, Wilhelm von,
Wien 063060
Riedl, Lothar Joachim,
Niederaichbach 130458
Riedmuseum, Stadtmuseum Rastatt,
Rastatt 021281
Riefel, Joseph, Maria Enzersdorf am
Gebirge 100117
Rieffel, Paris 141291
Riegamer, Dietbert, Hamburg .. 076298
Riegel, Gerhard, Nürnberg 077841
Rieger, New York 135988
Riegner, Miami 121852
Riehl, Franck, Guidel 068767
Riehle, Brigitte, Esslingen 129826
Rieke, Barbara Maria, Kerpen . 076763
Rieker, Manfred, Heilbronn ... 107405
Riel, København 065947
Riel House, Saint-Vital 006943
De Riemkap, Heemstede 132529
Riemland Museum, Heilbron .. 038734
Rien de Neuf, Montréal 064464
Rienzo, Robert, New York 123007
Riepe, Gerhard, Herford 076504
Riera Borrell, Manel, Barcelona . 143730
Ries, H., Heerlen 083376
Ries, Judith, Zürich 133998
Ries, Sandi, Charlotte 092622
Ries, W., Waibstadt 078831
Riese, Hannover 076389
Riese, Horst, Bremen 075180
Riesengebirgsmuseum,
Marktoberdorf 020301
Riesengebirgsstube, Würzburg . 022799
Riesenrundgemälde Schlacht am
Bergisel, Innsbruck 002134
Rieser, Mayen 077311
Rieser Bauernmuseum,
Maihingen 020217
Rieser, Dr. Michael, Ravensburg 078149
Rieskrater-Museum, Nördlingen 020814
Riesner, Paris 071852
Rießbeck, A., Hof 076551
Rießelmann, Lohne 107890
Riessen, Paul, Falun 115647
Riestra, Florencia, México 111981
Rietgaverstede, Streekmuseum van
het Land van Nevele, Nevele . 003909
De Rietgors, Museum voor Moderne
Kunst, Papendrecht 032764
Rietjetoe, Den Haag 083123
Rietpengalerie, Bergen op Zoom 112346
Rietveld, Leiden ... 083443, 083444
Rietveld Schröderhuis, Centraal
Museum Utrecht, Utrecht ... 032956
Rieubon, Jean-Marc, Singapore . 085491
Rieunier, Bailly & Pommery, Paris 125753
't Riever, Nuis 032704
Rievaulx Abbay, Rievaulx 045617
Rieve, Haarlem 083350
Rifaï, Osman Shehadeh, Tripoli . 082329
Rifaldi, Richard, Nice . 070659, 128870
Rıfat Ilgaz Kültür Merkezi,
Kastamonu 042989
Rifaut, Gilbis, Cambrai 067617
Riffe Gallery, Ohio Arts Council,
Columbus 047943
Riffmuseum im Bahnhof,

Gerstetten 018671
Rifle Creek Museum, Rifle 052573
Rig-Veda, Cluj-Napoca 143622
Riga School of Art, Rīga 056112
Rigaldo & Fils, Allaman 087187
Rigaldo & Fils, Lausanne 087584
Rigaldo & Fils, Prilly 133889
Rigamonti, Franco, Roma 132030
Rīgas Galerija, Rīga 111762
Rīgas Kino Muzejs, Rīga 030217
Rīgas Motormuzejs, Rīga 030218
Rīgas Porcelana Muzejs, Rīga . . 030219
Rīgas Tehniskās Universitātes Muzejs,
   Rīga 030220
Rīgas Ūdensapgādes Muzejs,
   Ādaži 030094
Rīgas Vēstures un Kuģniecības muzejs,
   Rīga 030221
Rīgas Vīni, Rīga 111763
Rigassi, Raphael, Bern 116148
Il Rigattiere, Bolzano 079900
Rigaud, Alexis, Lembras 069384
Rigault, Patrice, Lattes 069198
Rigaut, Fabien, Reims 129198
Rigenera, Modena 131584
Riggauer, Konrad, München . . . 077600,
                                  130388
Riggenbach, Dr. H., Basel 144030
Riggiardi, Davide, Milano 131514
Righi, Roberto & Daniele, Firenze 131256
Right Angle, Brackley . 088555, 134134
Right Brain Art Gallery, Atlanta . 119605
Rightlines Art Gallery, Bangalore 109264
Rigione, Napoli 131638
Rigoleto Curioso, Buenos Aires . 139598
Rigorini, Luigi, Torino 132193
Rigo's Antiques, San Antonio . 097023
Riibe, Trondheim 084388
Riihimäen Kaupunginmuseo,
   Riihimäki 011313
Riihimäen Taidemuseo, Riihimäki 011314
Riihimuseo, Peräseinäjoki 011243
Riihipiha Museo, Vuolijoki 011521
Riise-Birger, Anita, Göteborg . . 133532
Riisfeldt, Odder 102944
Rijbroeck, Amsterdam 112273
Rijdend Electrisch Tram Museum,
   Amsterdam 031981
de Rijk, Den Haag . . . 083124, 112409
Het Rijk van Kralingen,
   Rotterdam 112658
Rijk, Walt de, Amsterdam 112274
Rijkeboer, W., Enschede 083258
Rijken, Tonnie, Geulle 132506
Rijks, Amstelveen 132388
Rijksakademie van Beeldende Kunsten,
   Amsterdam 056169
Rijksarchief, Brugge 003423
Rijksbureau voor Kunsthistorische
   Documentatie, Den Haag 056172
Rijksmuseum, Amsterdam 031982
Rijksmuseum Hendrik Willem Mesdag,
   Den Haag 032196
Rijksmuseum Twenthe, Museum
   voor Oude en Moderne Kunst,
   Enschede 032310
Rijksmuseum van Oudheden,
   Leiden 032584
Rijksmuseum voor Volkenkunde,
   Leiden 032585
Rijksprentenkabinet, Rijksmuseum,
   Amsterdam 031983
Rijnders & Zoon, B.A.H., Arnhem 082946
Rijnders, B.H., Arnhem 082947
Rijsdijk, Jan, Dordrecht 083192
Rijsdijk, Saskia, Haarlem 132521
Rijssenbeek, Suzanne, Monaco . 112018
Rijssens Museum, Rijssen 032786
Rijtuigmuseum, Bree 003402
Rijwiel- en Bromfietsmuseum,
   Zoutkamp 033109
De Rijzende Zon, Tilburg 143236
Rik, Leiden 083445
Rika Gallery, Waikato 113041

Rike, K., Dallas 120651
Rikosmuseo, Vantaa 011492
Riksföreningen Sverigekontakt,
   Göteborg 059901
Riksidrottsmuseet, Stockholm . . 041036
Rikuzentakata-shi Hakubutsukan,
   Rikuzentakata 029286
Riley County Historical Museum,
   Manhattan 050642
Riley House Museum of African
   American History and Culture,
   Tallahassee 053663
Riley Orrin, H., New York 135989
Riley-Smith, Crispian, London . 090452,
                                  118541
Riley-Smith, Hamish, Norfolk . . 144743
Riley, Peter, Cambridge 144301
Riley, Thomas R., Cleveland . . . 120450
Riley, Thomas R., Columbus . . . 120512
Rilton, Stockholm 087047
Rímal, Pavel, Praha 065520
Rimanque, Oostende 063822
Rimari India, Mumbai 109463
Rimbaud Verlag, Aachen 137023
Rimbaud, Jean-Jacques, Paris . 071853
Rimbault-Joffard, Robert, Charly-sur-
   Marne 067852
Rimbault, Jean Yves, Toucy . . . 073693
Rimmele-Marzullo, Angela, Ulm 078756
Rimrocker Historical Museum of West
   Montrose County, Naturita . . . 051201
Rinaldelli, Paolo, Fermo 142659
Rinaldi, Barbara, Napoli 080926
Rinaldi, Ciro, Brescia 131047
Rinaldi, Luisa, Genova 080286
Rinard, Guthrie, New Orleans . . 122224
Rinaudo, Nicola, Milano 080699
Rinck, Heiko, Fitzen 075737
Rincón Artístico, Buenos Aires . 098577
Rincón de Historia, Salto 000563
El Rincón de María Castaña,
   Caracas 097964
Rincón Mexicano, Medellín . . . 102459
El Rincón Vidriero, San Salvador 103102
Rindal Bygdemuseum, Rindal . . 033890
Rindlisbacher, Zürich 088003
Rinetti, Barbara, Torino 132194
Le Ring, Nantes 058665, 104345
Ring of Fire, Milwaukee 121947
Ringborg, Eva, Vreta Kloster . . 133605
Ringe Museum, Ringe 010279
Ringel, Felix, Düsseldorf 106820
Ringel, Michael, Düsseldorf . . . 075508
Ringerike Kunst, Hønefoss 084255
Ringerikes Museum, Hønefoss . 033656
Ringkamp, Bernhard, Delbrück . 075326
Ringkøbing Museum, Ringkøbing 010280
Ringles Cross Antiques, Uckfield 091686
Ringling School of Art and Design,
   Sarasota 057876
Rings, Wellington 084186
Ringsaker Vekter- og Brannhistoriske
   Museum, Moelv 033775
Ringstead Gallery, Ringstead . . 118976
Ringstead Village Antique Centre,
   Ringstead 091105
Ringsted Antik, Ringsted 066075
Ringsted Galleriet, Ringsted . . . 102960
Ringsted Museum, Ringsted . . . 010281
Ringstrøm, Bjørn, Oslo 143414
Ringve Museum, Trondheim . . . 034036
Ringwood Antique Market,
   Ringwood 062152
Ringwood Miners Cottage,
   Ringwood 001451
Rink, E., Wiesloch 079003
Rinn-Art Finland, Helsinki 066357
Rinn, Ch., Wetzlar 078960
Rinne & Schmidt, Rinteln 078003
Rinnelt, C., Wiesbaden 078992, 142494
Rinno-ji Jokodo Homotsukan,
   Nikko 029184
Rinsen Book, Kyoto . . . 137768, 142900
Rinsen Books, Tokyo 142936

Rinsing Museum, Qingshui 008180
Rio Grande, Düsseldorf 075509
Rio Grande Valley Museum,
   Harlingen 049361
Rio Hondo College Art Gallery,
   Whittier 054381
Rio Macanaz, Pedro del, Cádiz . 085924
Riocreux Michèle, Saint-Étienne 072776
Riolo, Vincenzo, Palermo 131709
Riondet, Bogdan, Lyon 069778
Riordan Mansion, Flagstaff . . . 048742
Rios, Guatemala 109122
Rios, Avelino, Genève 087501
Riot, Paris 129129
Ripa Antica, Milano 080700
Ripert Grac, Isabelle, Marseille . 069962
Ripley Castle, Ripley, North
   Yorkshire 045619
Ripley County Historical Society
   Museum, Versailles 053970
Ripley, Dan, Indianapolis 093921, 127262
Ripley's Believe It or Not Museum,
   Blackpool 043645
Ripley's Believe It or Not Museum,
   Guadalajara 031015
Ripley's Believe It or Not Museum,
   Hong Kong 007888
Ripley's Believe It or Not Museum,
   København 010192
Ripley's Believe It or Not Museum,
   México 031270
Ripley's Believe It or Not Museum,
   Niagara Falls 006486
Ripley's Believe It or Not Museum,
   Pattaya 042707
Ripley's Believe It or Not Museum,
   Shah Alam 030730
Ripley's Believe It or not!, Key
   West 049959
Ripley's Believe It or Not! Museum,
   Surfers Paradise 001529
Ripoll, Palma de Mallorca 143790
Ripon College Art Gallery, Ripon 052580
Ripon Historical Society Museum,
   Ripon 052581
Ripon Prison and Police Museum,
   Ripon 045622
Ripon Workhouse Museum,
   Ripon 045623
Ripp, Ellen, Chicago 120254
Rippert, Houston 093779
Ripping Yarns, Caloundra 139679
Ripping Yarns, London 144677
Rippl-Rónai Emlékház, Kaposvár 023711
Rippon Lea Estate, Elsternwick . 001059
Rippon, Boswell & Co,
   Wiesbaden 126141
Rippon, Boswell & Co., Zug . . . 126777
Rippstein, Pierre, Basel 087254
Riquet, Sylvie, Soissons 073528
Riquet'Broc, Toulouse 079096
Rira, Tiranë 098530
Risager, Jørgen, Odense 066044
Risborough Art Gallery, Princes
   Risborough 118950
The Risby Barn, Risby 091122
Risby Cove, Strahan 099721
Risch, Verena, Schaffhausen . . 116632
Rischkopf, Günter, Starnberg . . 078540
Riserva Naturale Speciale del Sacro
   Monte di Varalio, Varallo Sesia 028274
Rishon Le-Zion Museum, Rishon Le-
   Zion 025182
Rising Sun Tavern,
   Fredericksburg 048960
Risley, David, London 118542
Risør Museum, Festningsanlegget
   Urheia, Risør 033891
Risorgimento, Legnano 110165
Rispal, Antonin, Paris 071854
Risque Art, Solihull 119112
Riss, Samedan 116606
Rissa Bygdemuseum, Rissa . . . 033893

Risse, Ildikó, Weßling 108885
Risse, Martine, Bayonville-sur-
   Mad 066982
Risse, Martine, Nancy 070429
Rissoan, Alain, Romans-sur-Isère 072523
Rissone, Gatti, Viganello 116700
Ristea, Mihai, København 065948
Risten-Lakviks Järnväg,
   Åtvidaberg 040553
Risteski, Riste, Neustadt bei
   Coburg 077765
Ristori, Albenga 109747
Ristović, Beograd 085347
Ristovski, Fulda 075933
RiSuPia Panasonic Digital Network
   Museum, Tokyo 029577
Rita's Antique Cafe, New York . . 095688
Rita's Antiques, San Francisco . 097314
Rita's Relics, Atlanta 092179
Ritax, Helsinki 066358
Ritchie & Co, Colin, Toronto . . . 064741
Ritchie, A. & R., Saint Helier . . . 091185
Ritchies, Toronto 125384
Rite-Way, Tampa 136397
Ritorna Moveis, São Paulo 064204
Ritorno, Casablanca 082581
Ritorno Bis, Bologna 079872
Ritorto, Enzo, Roma 132031
Ritorto, Ferdinando, Roma 132032
Ritorto, Ivan, Roma 081529
Ritschard, M. & U., Bern 087298
Helga Ritscher-Sandmeier & Alexander
   Sandmeier OHG, Berlin 074902
Rittenhouse, Philadelphia 123444
Ritter, Düsseldorf 075510
Ritter, Klagenfurt 136527
Ritter, New York 095689
Ritter & Temme, München 077601
Ritter & Zamet, London 118543
Ritter, C., Bruxelles 063539
Ritter, G., Rinteln 078247
Ritter, Hans-Dieter & Waltraud,
   Jork 107485
Ritter, Luc, Bruxelles 063540
Ritter, Marion, Hamburg 076299
Ritter, P., Marienheide 077296
Ritter, Patrice, Zürich 088004
Ritterbach, Frechen 137210
Rittergallery, Klagenfurt 100074
Rittergut Haus Laer, Bochum . . 017439
Ritterhaus Bubikon, Bubikon . . 041374
Ritthaler, Albert, Hamburg 107310
Rittman Historical Society Museum,
   Rittman 052582
Ritu Creations, Mumbai 109464
Ritual Art Gallery, Kathmandu . . 112086
Ritums – Kulturas un Tautas Makslas
   Centrs, Rīga 111764
Ritzenhoff, Chicago 092781
Ritzenthaler, Gassin 068638
Ritzy Bits, Auckland 083821
Rius Garrich, Claudio, Barcelona 085831
Riuttalan Talonpoikaismuseo,
   Karttula 010967
Riva, Adriano, Padova 081003
Riva, Maria Patrizia, Firenze . . 131257
RIVAA Gallery, Roosevelt Island Visual
   Arts Association, New York . . 123008
Rivals Renaud, Villenouvelle . . 074171
Rivas Martin, Juan, Madrid . . . 086251,
                                  133381
Rivchun, Heide, Cleveland 092967,
                                  135470
Rive de Sorgues, L'Isle-sur-la-
   Sorgue 069590
Rive Gauche, Auxerre 066852
Rive Gauche, Den Haag 083125
Rive Gauche Galerie d'Art
   Contemporain, Namur 100796
Rive, H., Düsseldorf 075511
Rivenburg, Nicholson 127368
Riveneuve, Marseille 140947
River and Rowing Museum, Henley-on-
   Thames 044566

River Art Gallery, Limerick 109669
River Art Group, San Antonio . 060715,
124066
River Arts Center, Clinton 047836
River Brink, The Samuel E. Weir
Collection, Queenston 006749
River City Antiques, Saint Paul . 096854
River City Art and Antique Centre,
Bangkok ...088085, 116957, 126786,
144205
River City Auction, Vicksburg . 127485
River East Art Center, Chicago . 120255
River Emporium, Woodburn 062449
River Gallery, Saint Paul 123928
River Heritage Museum, Paducah 051885
River Hills Park Museum,
Wildwood 054404
River House Antiques, Pittsburgh 096324
River House Antiques,
Washington 097856
River Legacy Living Science Center,
Arlington 046535
River Market Antique Mall, Kansas
City 094092
River Museum, Wellsville 054276
River Read, Noosaville 139859
River Square Antiques, San
Antonio 097024
River Valley School Museum,
Virden 007345
Riverbank, Petworth 091006
Riverbank Antiques, Warrandyte 062409
Riverbank Studios, Portsmouth . 118943
Riverbend, Alderson 127117
RiverBrink – Home of the Weir
Collection, Niagara-on-the-Lake 006493
Riverchase Antique Gallery,
Birmingham 092433
Riverchase Gallery, Birmingham 119818
Riverdale Mansion, Riverdale
Park 052587
Rivère-Pagnon, Nicole, Saint-
Lizier 072894
Rivergum Art Gallery, Warrandyte 099861
Riverhouse, Steamboat Springs . 138474
Riverhouse Gallery, Saskatoon . 101549
Riverhurst F.T. Hill Museum,
Riverhurst 006809
Rivero Lake, Rodrigo, México . 082484
Rivers Gallery, Nailsworth 118769
Riverside Antiques, München .. 077602
Riverside Antiques,
Sawbridgeworth 091250
Riverside Architectural Antiques, Saint
Louis 096767
Riverside Art Museum, Riverside 052595
Riverside Auction House,
Bangkok 126787
Riverside Book Company, New
York 138422
The Riverside Gallery, Golant . 117869
The Riverside Gallery, Inverness 118022
Riverside Gallery, Looe 118662
Riverside Gallery, Toronto 101697
Riverside Gallery, Uppermill ... 119285
Riverside Municipal Museum,
Riverside 052596
Riverside Museum, La Crosse . 050030
Riverside Museum at Blake's Lock,
Reading 045592
Riverstone, New Orleans 122225
Riverton Museum, Riverton ... 052599
Riverton Transportation & Heritage
Centre, Riverton 006812
Rivertown Galleries, London .. 101347
Rivertown Gallery, Memphis .. 121710
Riverview Antique Market,
Milwaukee 094878
Riverview at Hobson Grove, Bowling
Green 047159
Riverview Ethnographic Museum, Bear
River 005403
Riverwest Artists Association,
Milwaukee 060716

Riverwest Arts Center,
Milwaukee 121948
Rivet, Gabriele, Köln 107675
Rivett, Sue, Fakenham 089248
Rivi, Riccardo, Casalgrande ... 080013
Riviera Books, Harleston 144452
Rivière & Desnoe, Saint-Leu .. 132987
Rivière, Christophe, Tournai ... 063849,
128090
Rivière, Colette, Maisons-Laffitte 069818
Rivière, Guy, Charmoy 067856
Riviere, Jean-Claude, Peujard . 072120
Rivière, Michel, Toulon (Var) .. 073728
Riviere, Roger, Peujard 072121
Rivilla Calvo, T., Segovia 115440, 115441
Rivington Arms, New York 123009
Rivington Gallery, London 118544
Rivista degli Studi Orientali,
Roma 138868
Rivista di Psicologia dell'Arte,
Roma 138869
Rivista di Storia della Miniatura,
Firenze 138870
Rivkind, Robert, Toronto 064742
Rivnens'kyj Krajeznavčyj Muzej,
Rivne 043288
Rivoli, Gladesville 061521
Rivoli, Howard, Neutral Bay ... 061949
Rivollier Alain, Vienne 074116
Rivon, Alexandre, Parent 070852
Riyahi Gallery, London 090453
Riza, Sergiev Posad 037857
Rizal Technological and Polytechnic
Institute Museum, Morong .. 034553
Rize Devlet Güzel Sanatlar Galerisi,
Rize 043044
Rizeanu, Dardu, Craiova 133007
Rizhao Museum, Rizhao 008207
Riznica Crkvene Umjetnosti – Zlato i
Srebro Nina, Nin 008954
Riznica Franjevačkog Samostana Split,
Split 009002
Riznica i Galerija Svetista Gospe
Trsatske, Rijeka 008980
Riznica Katedrale Svetog Tripuna,
Kotor 031758
Riznica Samostana Svetog Frane,
Zadar 009024
Riznica Stolne Crkve u Dubrovniku,
Dubrovnik 008912
Riznica Zagrebačke Katedrale,
Zagreb 009049
Rizo, Bernard, Nîmes 070703
Rizzato, Bruno, Venezia 132255
Rizzi, Arcangelo, Milano 131515
Rizzitiello, Pasquale, Milano ... 131516
Rizzo Antiquari, Palermo 081071, 081072,
081073
Rizzo, Gaetano, Palermo 131710
Rizzo, Giovanni, Venezia 081869
Rizzoli, Milano 137693
Rizzotti, Stefania, Milano 142741
RJ Collectibles, Chicago 092782
Rjabov, Valentin, Moskva 114297
Rjazanskij Gosudarstvennyj Istoriko-
architekturnyj Muzej-zapovednik,
Rjazan 037689
Rjazanskij Oblastnoj Chudožestvennyj
Muzej im. I.P. Požalostina,
Rjazan 037690
Rjazanskij Zaočnyj Institut
Moskovskogo Gosudarstvennogo
Universiteta Kultury i Iskusstv,
Rjazan 056297
RL Fine Arts, New York 123010
RLB Kunstbrücke, Innsbruck ... 002135
RM Antiquités, Pontoise 072273, 129176
RM Fine Art, London 118545
RMIT Gallery & RMIT University's Art
Collection, Melbourne 001282
Ro Gallery, Long Island City .. 121418,
127289
ROA Estudio, Madrid 133382
Roach Tackle Folk Art, Toronto . 101698

Road, Sendai 082119
Roadhouse Relics, Austin 119695
Roadshow Antique Jewellery, Nobby
Beach 061975
Roadshow Antiques, Southport . 062230
Roadside Antiques, Greystoke . 089389
Roald Amundsens Hjem Uranienborg,
Svartskog 034003
Roald Amundsens Minne, Torp . 034022
Roan Bygdetun, Roan 033896
Roanoke Island Festival Park,
Homeport of the Elizabeth II,
Manteo 050658
Roar Gallery, Wellington 113089
Roaring Twenties Antique Car
Museum, Hood 049549
Rob Piercy Gallery, Porthmadog 118940
Rob Roy and Trossachs Victor Centre,
Callander 043829
Robache, Gérard, Vieille-Eglise . 074115
Robarick, Dorit, Lübben 107907
Robb Street Gallery, Bairnsdale . 098685
Robbe, M., Osnabrück 142279
Robben, Cincinnati 135450
Robben Island Museum, Robben
Island 038823
Robbins Hunter Museum, Avery-
Downer House, Granville 049211
Robbins Museum of Archaeology,
Middleborough 050854
Robbins, Ivelin & Craig, Miami
Beach 121885
Robbins, Martha, New Orleans . 122226
Robbins, Tucker, New York 095690
Robbinsville, Miami 094795
Roberson Museum and Science
Center, Binghamton 046998
Robert, Swansea 144875
Robert & Li, Tainan 116898
Robert A. Bogan Fire Museum, Baton
Rouge 046820
Robert A. Peck Gallery, Central
Wyoming College, Riverton ... 052600
Robert A. Vines Environmental Science
Center, Houston 049593
Robert and Mary Montgomery Armory
Art Center, West Palm Beach . 054318
Robert Brady Museum,
Cuernavaca 030931
Robert Burns Centre, Dumfries . 044156
Robert Burns House, Dumfries . 044157
Robert C. Williams American Museum
of Papermaking, Atlanta 046631
Robert Crown Center for Health
Education, Hinsdale 049491
Robert E. Lee Memorial Association,
Stratford 053571
Robert E. Wilson Art Gallery, Merillat
Centre for the Arts, Huntington 049615
Robert Edward Hart Memorial
Museum, Souillac 030795
Robert Frost Farm, Derry 048280
Robert Fulton Birthplace,
Quarryville 052426
Robert Galerie, Pöcking 108392
Robert Gordon University Collection,
Aberdeen 043362
Robert Hamerling – Museum,
Kirchberg am Walde 002171
Robert Hull Fleming Museum,
Burlington 047328
Robert Hytha-Museum,
Sondermuseum Wildtiere der Welt,
Oberwölbling 002472
Robert-Koch-Museum, Berlin .. 017273
Robert Langen Art Gallery,
Waterloo 007366
Robert Louis Stevenson House,
Monterey 051017
Robert Louis Stevenson Memorial
Cottage, Saranac Lake 053146
Robert Louis Stevenson Museum,
Vailima 038095
Robert Louis Stevenson Silverado

Museum, Saint Helena 052771
Robert McLaughlin Gallery,
Oshawa 006566
Robert Mills Historic House and Park,
Columbia 047914
Robert Musil-Literatur-Museum,
Klagenfurt 002197
Robert Newell House, DAR Museum,
Saint Paul 052854
Robert O'Hara Burke Memorial
Museum, Beechworth 000840
Robert Owen Memorial Museum,
Newtown, Powys 045379
Robert R. McCormick Museum at
Cantigny, Wheaton 054365
Robert S. Kerr Museum, Poteau 052331
Robert S. Peabody Museum of
Archaeology, Phillips Academy,
Andover 046481
Robert-Schumann-Haus Zwickau,
Zwickau 022911
Robert-Sterl-Haus, Struppen .. 022019
Robert-Struppmann-Museum, Kunst-
und Heimatmuseum Lorch, Lorch,
Rheingau 020115
Robert Toombs House,
Washington 054179
Robert V. Fullerton Art Museum,
California State University, San
Bernardino, San Bernardino .. 052957
Robert W. Ryerss Museum,
Philadelphia 052089
Robert Wölfle KG, Buch- und
Kunstantiquariat, München .. 108183,
142182
Robert, Ariadna, Sant Cugat del
Vallès 115414
Robert, Brian, San Francisco ... 097315
Robert, Claude, Paris . 125754, 125755
Robert, Didier, Saint-Étienne .. 129228
Robert, J.C., Honfleur 103842
Robert, Jacqueline, Villeurbanne 074224
Robert, Jay, Chicago 092783
Robert, Marie-Françoise, Paris . 125756
Robert, Olivier, Capbreton 067659
Robert, Olivier, Paris 129130
Robert, Roland, Paris 105197
Robert, Y., Agde 066420
Roberta e Basta, Milano 080701, 110415
Roberta's Antique Jewelry, New
York 095691
Robertello, Thomas, Chicago .. 120256
Robertinum, Archäologisches Museum
der Martin-Luther-Universität, Halle,
Saale 018968
Roberto Mata Taller de Fotografía,
Caracas 054738
Roberto Pedoni Numismatica Arte,
Roma 081530
Roberto, Basilío Augusto, Lisboa 114058
Roberto's & Juliette's Antiques,
Merbein 061844
Roberts, Nicholls 099459
Roberts, Tampa 124670
Roberts & Tilton, Los Angeles . 121584
Robert's Antiques, Saint Helier . 091186
Robert's Antiques, Tulsa 097755
Roberts' Auction Service,
Fleischmanns 127224
Roberts Auktioner, Kungsör .. 126619
Roberts County Museum, Miami 050844
Roberts Emporium, Cardiff ... 088789,
134216
Roberts Gallery, Toronto 101699
Robert's Quality Upholstering,
Hamilton 128175
Roberts, Andrew, Gilles Plains . 061519
Roberts, Bryan H., Columbus . 120513
Roberts, Catherine, Washington 097857
Roberts, Derek, Tonbridge 091610
Roberts, Elizabeth, Washington . 124953
Roberts, John, Dallas 120652
Roberts, Mark, Welbeck 135196
Roberts, Peter, New York 095692

Roberts, R., Jacksonville 121265
Roberts, T.J., Uppingham 091694
Robertson, Glenrothes 117864
Robertson, Melbourne 061829
Robertson Museum, Millport ... 045269
Robertson Recollections,
  Robertson 062155
Robertson, Raymond, Dorchester 134332
Robeson Center, Rutgers University,
  Newark 051500
Robespierre, Charleroi 063606
Robic, Albert, Vezins 074087
Robichon, François, Saint-Tropez 073318
Robilant & Voena, London 090454
Robillard, Clamart 068003, 128558
Robin Hood Antiques, Canterbury 061281
Robin Hood's Bay and Fylingdale
  Museum, Robin Hood's Bay .. 045625
Robin, Eloi, Liège 063779
Robin, Jeanne & Roland, Cérilly 103552
Robin's Gallery, Nashville 122098
Robin's Nest, Miami 094796
Robin's Nest, Mount Tamborine . 099406
Robin's Nest, Richmond 096565
Robinson, New Orleans 095153
Robinson & Reeves, Miami ... 121853
Robinson Blaxill, Neill,
  Sevenoaks 091270
Robinson Visitors Center,
  Hartsville 049396
Robinson, Carol, New Orleans .. 122227
Robinson, Chris, Plymouth 118923
Robinson, E.N., Indianapolis ... 093922
Robinson, James, New York 095693
Robinson, John, Wigan 091845
Robinson, Mariah, Richmond ... 096566
Robinson, Peter, Heacham 089485
Robinson, Peter, Wragby 119397
Robinson, Robert, Pittsburgh ... 123572
Robinson, Thomas, Will & Pascal,
  Houston 121119
Robinsons, Knokke-Heist 100761
Robinsons, New York 123011
Robischon, Denver 120761
Robison, Holmfirth 117980
Robledo, Madrid 086252
Robles Piazzini, Lima 113367
Robotnicze Stowarzyszenie Twórców
  Kultury, Lubin 059705
Robson, Barnard Castle 088357, 134063
Robson, Northleach 090861
Robson Gallery, Halliwells House
  Museum, Selkirk 045758
Robson, Samuel, Oakham 118845
Robuck, Jennifer, Austin 092265
Robundo, Tokyo 137790
Robur House Antiques, Perth,
  Tasmania 062083
Roby, Saint Paul 136225
Robyr, Annie, Crans-Montana . 116232
Le Roc des Harmonies: Palais de
  Mineralogique, Musée-Aquarium
  Marin du Roc, Granville 013154
Roca, Bucureşti 085108, 143616
Rocamora Brocante, La Bisbal
  d'Empordá 085905
Rocanville and District Museum,
  Rocanville 006817
Rocca, Venezia 081870
Rocca Borromeo, Museo
  dell'Abbigliamento Infantile,
  Angera 025328
Rocca Sanvitale, Fontanellato .. 026319
Rocca Sforzesca, Collezioni Comunali
  d'Armi e Ceramiche, Imola .. 026508
Rocca, Antoine, L'Haye-les-Roses 069454
Rocca, Laura, Torino .. 081753, 132195
Rocchi, Sylvie, Blan 067237
Rocchini, Franca, Firenze 080193
Rocco, Vincent, New York 095694
Roccolta d'Arte Sacra, San Piero a
  Sieve 027792
Rocfort, Tony, La Baule-
  Escoublac 103873

Rocfort, Tony, Rennes (Ille-et-
  Vilaine) 105441
Rocha, Barbizon 066946
Rocha, Paris 105198
Rocha, José Manuel M., Porto . 132977
Da Rocha, Roselaine Bertoletti, Porto
  Alegre 063992
Rochan, Jeddah 085305
Rochas, Sylvain, Versailles 074069
Rochdale Art and Heritage Centre,
  Rochdale 045626
Rochdale Book Company,
  Rochdale 144802
Rochdale Pioneers Museum,
  Rochdale 045627
Roche, Bremen 075181
Roche-Laclau, Francine, Laroin . 069189
Roche, Byron, Chicago 120257
Roche, Joseph, Marseille 069963
Roche, Lysiane, L'Isle-sur-la-
  Sorgue 069591
Roche, Michel, Saint-Vincent-le-
  Paluel 073332
Roche, Stéphane, Bordeaux 067356,
                               140802
Rochebonne, Paris 105199
Rochelle & Co, Toronto 064743
Rocher Saint-Léon, Dabo 012742
Rocher, Alain, Paris 129131
Rochester Antiques, Rochester . 062158
Rochester Art Center, Rochester 052612
Rochester Hills Museum at Van
  Hoosen Farm, Rochester ... 052610
Rochester Historical and Pioneer
  Museum, Rochester 001454
Rochester Museum, Rochester . 052620
Rochet, Christophe, Bourg-en-
  Bresse 067412
Rochford, Minneapolis 135837
Rochow-Museum Reckahn, Kloster
  Lehnin 019653
Rock and Art, Halmstad 115678
The Rock Antiques, The Rock . 062159
Rock County Historical Society
  Museum, Janesville 049794
Rock Creek Gallery, Washington 124954
Rock Creek Station State Historic Park,
  Fairbury 048664
Rock House Museum, Wytheville 054561
Rock Island Arsenal Museum, Rock
  Island 052631
Rock Island County Historical Museum,
  Moline 050988
Rock 'n Roll & Blues Heritage
  Museum, Clarksdale 047772
Rock 'n Roll Collectibles, New
  Orleans 095154
Rock 'n' Roll Hall of Fame and
  Museum, Cleveland 047822
Rock 'n' Rustic, Maylands, South
  Australia 061805
Rock Pool Fine Art, Sidmouth .. 119098
Rock Springs Historical Museum, Rock
  Springs 052633
Rock, Rudolf, Zürich 088005
Rockarchive Gallery, London .. 118546
Rockart International, Montville . 099367
Rockart International, Noosa
  Heads 099462
Rockbourne Roman Villa,
  Fordingbridge 044356
Rockbridge Historical Society,
  Lexington 050268
Rockcavern, Beechworth 000841
Rockefeller Archaeological Museum,
  Israel Museum, Jerusalem,
  Jerusalem 025100
Rockefeller Museum, Jerusalem 025101
Rockefeller's Antiques,
  Jacksonville 094033
Rocket Gallery, London 118547
Rocket Projects, Miami Beach . 121886
Rocket to Mars, Providence ... 096514
Rocketart Gallery, Newcastle .. 099436

Rockford Art Museum, Rockford 052638
Rockford College Art Gallery/Clark Arts
  Center, Rockford 052639
Rockglen Rolling Hills Museum,
  Rockglen 006819
Rockhampton and District Historical
  Museum, Rockhampton 001456
Rockhampton Art Gallery,
  Rockhampton 001457
Rockhill Trolley Museum, Rockhill
  Furnace 052642
Rocking Chair Antiques,
  Warrington 091739
Rocking Horse, Oklahoma City . 136054
Rocking Horse Antiques, Goomeri 061543
Rockingham, Kingston 049981
Rockingham, Princeton 052370
Rockingham Art Gallery,
  Rockingham 099646
Rockingham Castle, Market
  Harborough 045230
Rockingham Free Public Library and
  Museum, Bellows Falls 046897
Rockland Center for the Arts, West
  Nyack 054314
Rockland County Museum, New
  City 051254
rock'n'pop museum, Gronau,
  Westfalen 018832
Rockoxhuis, Rockox House,
  Antwerpen 003312
Rockport Art Association,
  Rockport 060717
Rockport Art Association Galleries,
  Rockport 052646
Rockport Publishers, Beverly .. 138304
Rockridge Antiques, Oakland ... 095909
Rocks Centre Art Gallery, Sydney 099795
Rockston, Lusaka 125165
Rockwell Museum of Western Art,
  Corning 047995
Rockwood Museum, Wilmington 054441
Rocky Ford Historical Museum, Rocky
  Ford 052655
Rocky Lane School Museum, Fort
  Vermilion 005833
Rocky Mount Arts Center, Rocky
  Mount 052658
Rocky Mount Children's Museum,
  Rocky Mount 052659
Rocky Mount Museum, Piney
  Flats 052142
Rocky Mountain Art, Denver ... 120762
Rocky Mountain Coin, Denver .. 093330
Rocky Mountain College of Art and
  Design, Lakewood 057315
Rocky Mountain College of Art and
  Design Galleries, Denver 048273
Rocky Mountain Conservation Center,
  University of Denver, Denver .. 048274
Rocky Mountain Doll Fantasy,
  Denver 093331
Rocky Mountain House Museum,
  Reunion Historical Society, Rocky
  Mountain House 006821
Rocky Mountain National Park
  Museum, Estes Park 048614
Rocky Mountain Quilt Museum,
  Golden 049149
Rocky Mountain Rangers Museum,
  Kamloops 006058
Rocky Reach Dam Museum,
  Wenatchee 054280
Rocky Valley Gallery, Tintagel . 119237
Rocky's Antiques and Books, San
  Diego 097129, 145528
Rococo Antiques, Weedon 091764
The Rocque, Hamilton 064329
Rocznik Muzeum Narodowego w
  Kielcach, Kielce 138986
Roczniki Sztuki Śląskiej, Wrocław 138987
Rod-Pol, Kraków 084596
De Rode Cirkel, Leiden 112559
Rode, Joëlle, Cadaujac 067552

Rode, Karl, Knüllwald 142001
Rode, Volker, Gelnhausen 129892
Rodecka, A., Gdynia 113504
Rodenberg, Geschw., Borken,
  Westfalen 075104
Rodenburg en Keus, Dordrecht . 083193
Roderick, London 090455, 134759
Rodetoren4, Zutphen 112729
Rodgers & Nash, Malvern East . 061772
Rodgers Tavern, Perryville 051998
Rodgers, Paul, New York 123012
Rodhain, Bruno, Paris 141272
Rodin, Seoul 111707
Rodin Museum, Philadelphia Museum
  of Art, Philadelphia 052090
Rodinné muzeum pohledů – Dům U
  Červené židle, Praha 009838
Rodman Hall Arts Centre, Saint
  Catharines 006860
Rodney, Siegen 078465
Rodney Fox Shark Experience-
  Museum, Glenelg 001106
Rodney, Arthur, Shoreham-by-
  Sea 091322, 135043
Rodný Dom Jána Hollého, Záhorské
  Múzeum Skalica, Borský
  Mikuláš 038297
Rodný Dom Ľudovíta Štúra a
  Alexandra Dubčeka, Uhrovec .. 038479
Rodný Dom Pavla E. Dobšinského,
  Slavošovce 038447
Rodný Domek Adalberta Stiftera,
  Okresní muzeum Český Krumlov,
  Horní Planá 009570
Rodný Domek Aloise Jiráska,
  Hronov 009582
Rodný Domek Otokara Březiny,
  Městské muzeum, Počátky ... 009765
Rodný Dům Josefa Hoffmanna,
  Brtnice 009486
Rodolfo, Viola, Milano 110416
Rodon, Gent 063660
Rodrigue, New Orleans 122228
Rodrigue, Michel, Paris 071855
Rodrigues & Lopes, Lisboa ... 084952
Rodrigues, Alberto Dias, Lisboa . 114059
Rodrigues, Arnaldo, Rio de
  Janeiro 101064
Rodrigues, José Manuel, Lisboa 143591
Rodrigues, José Ribeiro, Lisboa 084951
Rodrigues, Luizandro A.G., Rio de
  Janeiro 064117
Rodrigues, Marco Antonio Silva, São
  Paulo 064205
Rodrigues, Maria Clara, Porto . 085050
Rodriguez Alvarez, Sevilla 133438
Rodriguez Carballo, Avelina,
  Valencia 086535
Rodriguez Juarez, Ruben,
  Guadalajara 082469
Rodriguez, Albert, Saint-Ouen . 073184
Rodriguez, Carlos Fernando,
  Bogotá 128277
Rodriguez, José, Zürich 088006
Rodriguez, Lilian, Montréal ... 101446
Rodriquez, Nancy, San Antonio . 097025
Roduner, Bruno, Wetzikon 087901,
                             087902
Rodziewicz, Andrzej, Wrocław .. 084772,
                             143545
Roe, Dallas 120653
Roe & Moore, London 144678
Roe, John, Islip 089652
Röbbig, München 077603
Röben, Wilhelmshaven 079013
Röber, Hamburg 076300
Roebling Hall, New York 123013
Roebourne Old Goal Museum,
  Roebourne 001458
de Roeck-Van Dijck, Antwerpen 063234
Rød Bygdetunet, Uskedalen ... 034056
Röda Bergets Antik, Stockholm . 087048
Röda Rummet, Uppsala 143995
Röda Stugans Antik, Torslanda . 087098

Roedde House Museum,
Vancouver 007294
Røde Lade, Helsingør 065745
Röder & Coumont, Rodershausen 078252
Röder, Hermann, Rodershausen 078251
Roeder, Margarete, New York . . 123014
Röder, Michael & Antje, Bergisch
Gladbach 074733
Roeder, Roland, Uhldingen-
Mühlhofen 126133
Rödner, Helmut, Haarlem 143162
Roehr, Mirosław, Poznań 132863
Röhsska Museet, Göteborg . . 040660
Roelants, H., Genk 063622
Roelantsmuseum Oude Abdij
Heemmuseum, Hemiksem . . 003691
Røldal Bygdemuseum, Røldal . . 033897
Röling, Amsterdam 082839
Röll, Dr. Josef, Dettelbach 137139
Roellin & Duerr, Sankt Gallen . . 116616
Roelofs, Jan, Amsterdam . . 082840
Roelofsz, Charles, Amsterdam . . 082841
Roels, Rudolf, Gent 128054
Roemer- und Pelizaeus-Museum,
Hildesheim 019250
Römer Galerie, Weißenburg in
Bayern 108869
Römer und Bajuwaren Museum Burg
Kipfenberg, Kipfenberg 019617
Römer, Heidi, Zürich . . 088007, 116851
Römer, K. & J., Borna 075106
Römerbad, Schwangau 021766
Römerbadmuseum, Hüfingen . . 019355
Römerhalle, Museen im Rittergut
Bangert, Bad Kreuznach 016805
Römerhaus Walheim, Walheim . . 022404
Römerkastell Saalburg, Bad Homburg
v.d.Höhe 016789
Römerkeller Ottmarsheimer Höhe,
Mundelsheim 020617
Römermuseum, Heitersheim . . . 019178
Römermuseum, Mainhardt 020222
Römermuseum, Mengen 020365
Römermuseum, Obernburg am
Main 020905
Römermuseum, Tulln 002847
Römermuseum Augst, Augst . . . 041243
Römermuseum Bedaium, Heimat-
und Geschichtsverein Bedaium in
Seebruck e.V., Seebruck 021812
Römermuseum Favianis-Sankt Severin,
Mautern 002375
Römermuseum im Haus Isenburg,
Eisenberg, Pfalz 018140
Römermuseum in der Grundschule
Rißtissen, Ehingen 018107
Römermuseum Kastell Boiotro,
Passau 021086
Römermuseum Multerer,
Grabenstätt 018781
Römermuseum Osterburken,
Zweigmuseum des Archäologischen
Landesmuseums Baden-
Württemberg, Osterburken . . . 021024
Römermuseum Stettfeld, Ubstadt-
Weiher 022238
Römermuseum Teurnia, Lendorf 002289
Römermuseum Wallsee-Sindelburg,
Wallsee 002895
Römermuseum Weißenburg,
Weißenburg in Bayern 022538
Römermuseum Xanten,
Archäologischer Park Xanten,
Xanten 022838
Römermuseum, Wien Museum,
Wien 003070
Römerpark Köngen, Köngen . . . 019716
Römersiedlung, Friesenheim . . . 018532
Römerthermen Zülpich – Museum der
Badekultur, Zülpich 022902
Römervilla e.V., Möckenlohe . . . 020430
Römerzeitliches Museum,
Ratschendorf 002572
Römisch-Germanisches-Museum

Jülich, Stadtgeschichtliches Museum
Jülich, Jülich 019473
Römisch-Germanisches Museum,
Museen der Stadt Köln, Köln . . 019711
Römisch-Germanisches
Zentralmuseum, Forschungsinstitut
für Vor- und Frühgeschichte,
Mainz 020233
Römische Badruine, Badenweiler 016931
Römische Bäder, Stiftung Preußische
Schlösser und Gärten Berlin-
Brandenburg, Potsdam 021199
Römische Baureste Am Hof, Wien
Museum, Wien 003071
Römische Kalkbrennerei Iversheim,
Bad Münstereifel 016829
Das Römische Kastell in
Waldmüssingen, Schramberg . 021732
Römische Thermen, Weißenburg in
Bayern 022539
Römische Villa Nennig, Stiftung
Saarländischer Kulturbesitz, Perl 021104
Römische Villa Urbana Longuich,
Longuich 020114
Römischer Keller in Sulz am Neckar,
Sulz am Neckar 022081
Römischer Weinkeller
Oberriexingen, Zweigmuseum des
Württembergischen Landesmuseums
Stuttgart, Oberriexingen 020917
Römisches Freilichtmuseum,
Hechingen 019124
Römisches Haus, Klassik Stiftung
Weimar, Weimar 022506
Römisches Museum Remagen,
Remagen 021359
Römisches Museum und Naturkunde-
Museum, Zumsteinhaus,
Kempten 019590
Römisches Museum,
Kunstsammlungen und Museen
Augsburg, Augsburg 016666
Rönesans Sanat Galerisi, İstanbul 117081
Roenland, Jevnaker 113165
Rønne, Palle, Næstved 066007
Rönnell, Stockholm 143962
Rönnquist & Rönnquist, Malmö . 115768
Röntgen-Gedächtnisstätte,
Würzburg 022800
Roentgen-Museum, Neuwied . . 020777
Röpke, Stefan, Köln 107676
Rösch, Franz, Otting 077997
Roesch, Jane, Pittsburgh 096325
Roesch, Sonja, Houston 121120
Rösch, T., Gütersloh 076125
Röseler, Jutta, Niendorf bei Neuhaus,
Elbe 077784
Rösener, Eva, Wuppertal 079112
Roesinger, Christa E., Köln 107677
Röske, Horst, Alfdorf 129374
Roeske, Rainer, Berlin 106410
Rösler & Oswald, Wädenswil . . 087890
Roesler, Nara, São Paulo 101110
Rösli Antiquitäten, Flaach 087423
Rösner, Duisburg 075556
Roess, Michel, Munster 070398, 128818
Rößig, Günter, Bad Blankenburg 106044
Rössler, Sankt Moritz 116625
Rößler Museum, Untermünkheim 022280
Rössler, Berthold, Ravensburg . 108444,
018445
Rössli, Balsthal 116053
Rössli, Klosters 087543
Rösti, Andreas, Gümligen 087520,
133766
Rötger, Anton, Hamburg 129985
Röthel, M., Großpösna 129924

Ria Röthinger & Michaela von Welck,
Köln 130165
Röttges, Pforzheim 078042
Rötzel & Co., Frankfurt am Main 075816
Rötzer, Franz & Elke,
Schwarzenbruck 078419
Röver, Ralf E., Donaueschingen . 075372
Röver, Sabine, Nürnberg 108307
Røyrvik Bygdatun, Limingen . . . 033746
Roffi, Tommaso, Bologna 142639
Rogai, Ennio, Roma 081531
Rogaland Bilmuseum, Vigrestad 034086
Rogaland Fiskerimuseum,
Åkrehamn 033481
Rogaland Krigshistorisk Museum,
Sola 033966
Rogaland Kunstmuseum,
Stavanger 033982
Rogaland Kunstsenter, Stavanger 113274
Rogall, Wilma, Toronto 128216
Rogalla, F. & K., Heide 076430
Rogé, Emmanuel, Pertuis 105345
Rogendorfer-Ausstellung,
Sonderausstellung des
Museumsvereins, Pöggstall . . . 002519
Rogeon, Pierre-Marie, Paris . . . 125757
Roger, Den Haag 083126
Roger, Utrecht 112711
Roger Guffey Gallery, Kansas
City 049890
Roger Raveelmuseum, Machelen 003836
Roger Tory Peterson Institute
of Natural History Museum,
Jamestown 049787
Roger Williams National Memorial,
Providence 052396
Roger, Arthur, New Orleans . . . 122229,
122230
Roger, Claude-Henri, Dreuil-les-
Amiens 068406
Roger, Hans-Peter, Neuss 130453
Rogers & Co, London 090456
Roger's Antiques Gallery, London 090458
Rogers de Rin, London 090459
Rogers Historical Museum,
Rogers 052662
Rogers House Museum and Gallery,
Ellsworth 048558
Roger's Upholstering, Pittsburgh 136137
Rogers, Alla, Washington 124955
Rogers, Anthony, Ebikon 116252
Rogers, Clive, London 090457
Rogers, Jolly, Hamilton 064330
Rogers, Robert D., Omaha 123338
Rogers, Wendy, Perth 062081
Rogg & Trachsler, Basel 087255
Rogge, Berlin 129555
Rogge, B., Berlin 129556
Roggenbuck & Schmid, Diedorf . 075351
Roggenbuck, Dr. Stefan,
Gelsenkirchen 107112
La Roggia, Pordenone 110677
Rogier Lauriance, London 090460,
134760
Rogin, Frank, New York 095695
Rogner & Bernhard, Berlin 137092
Rogue Art Space, San Francisco 124379
Rogue Buddha Gallery,
Minneapolis 122030
Rogue Gallery and Art Center,
Medford 050757
Rogue Valley Art Association,
Medford 060718
Rogues, Phoenix 096235
Rohain, Oostende 063823
Rohan, Landerneau 140864
Rohaut Delassus, Françoise, Blonville-
sur-Mer 067263
Rohaut, Françoise, Honfleur . . . 068831
Dr. Rohde & Kornatz, Kassel . . . 076728
Rohde, E., Potsdam 078083
Rohde, Eberhard, Sylt 108750
Rohe, Dr. Michael, München . . . 077604

Rohen, Karen, Hagen 129934
Rohleder, Ilona, Berlin 074903
Rohloff-Haluzán, V., Nürnberg . . 142248
Rohnstädter Heimatstube,
Weilmünster 022486
Rohr, F., Bern 133688
Rohr, Hans, Bern 144052
Rohr, Susan, Luzern 133839
Rohrbach-Hugenberg, Rottach-
Egern 078296
Rohrer, Fred, Zäziwil 087919
Rohrer, Ingrid, Konstanz 107722
Rohs, L., Düren 106754
Le Roi David, Montauban 070200
Le Roi Fou, Paris 071856
Roider, Beatrix, München 130389
Roig, Barcelona 085832
Roig-Sabater, J., Mannheim . . . 077270
Roijen, J., Banholt 132434
Roinet, Jacques, Plouguenast . . 072204
Florence Rois, Granville 125590
Roiter, Marcio, Rio de Janeiro . . 064118
Roja, Warszawa 113898
Rojahn, Karl-Heinz, Bandenitz . . 141452
Rojas, Miami 135806
Rojas Jūras un Zvejniecības Muzejs,
Roja 030225
Rojaus Kalnas, Vilnius 111807
Rojo Pinturas Española, Madrid . 086253
Rojstna Hiša Blažnega Antona
Martina Slomška, Šentjur 038607
Rojstna Hiša Matije Čopa,
Gornjesavski Muzej Jesenice,
Žirovnica 038638
Rojstna Hiša Pesnika Dr. Franceta
Prešerna, Gornjesavski Muzej
Jesenice, Žirovnica 038639
Rojstna Hiša Pisatelja Frana Saleškega
Finžgarja, Gornjesavski Muzej
Jesenice, Žirovnica 038640
Rojstna Hiša Rista Savina, Žalec 038635
Rojstna Hiša Simona Gregorčiča,
Tolminskij Muzej, Vrsno 038631
La Roka, Merano 080356
Roka, Henriette, Røyken 084347, 132795
Rokeby Museum, Rowland Evans
Robinson Memorial Association,
Ferrisburgh 048726
Rokiškio Krašto Muziejus,
Rokiškis 030455
Rokko Heart, Kobe 111285
Rokoko, Moskva 085200
Rokuzan Bijutsukan, Hotaka . . . 028664
Rol, Alain, Beaune 067035
Rolan, Riyadh 085323
Roland, Bremen 125910
Roland Gallery, Bath 088399
Roland Gibson Gallery, State University
of New York College at Potsdam,
Potsdam 052333
Roland, C., Mons 063799
Roland, Guy, Mons 063800
Roland, Rob & Bill, New York . . 095696
Roland, Victor, Bruxelles 063541
Roldán Galería, Sevilla 115456
Roldhøj Kunst, Skørping 066129
Roldo, Agostino, Modena 080817
Rolevink, Bert, Armadale, Victoria 061047
Rolf Bergendorffs Radio Museum,
Lessebo 040777
Rolfe, Susan, New York 123015
Rolies, Grimbergen 063677
Rolin, Gérard, Mazerulles 070020
Rolla Minerals Museum, University of
Missouri, Rolla 052669
Rollag Bygdetun, Rollag 033905
Rollan & Reche, Madrid 133383
Rolland, Emile, Courbevoie 068187
Rolland, Jean-François, Sologny 073537
Rolland, Pierre, La Coquille 069005
Rolland, René, Evron 068497
Rolland, Roger, Saint-Ouen . . . 073185
Rolle, Madrid 086254
Rollero, Piero, Torino 081754

Rolleston, Brian, London 090461
Rollettmuseum, Baden bei Wien 001773
Rollf's Gallery, Fresno 120905
Rolli, Bernard, Bern .. 087299, 133689
Rollin Art Centre, Port Alberni .. 006674
Rollin, Philippe, Limoges 125616
Rollin, Pierre, Rouen .. 105479, 129212
Rollo Jamison Museum,
Platteville 052196
Rolls-Royce Museum, Dornbirn . 001827
Ronald Rolly & Ronald R. Michaux,
Boston 119917
Rolly, Pierre, Coulonges-les-
Sablons 068181
Rol'nicky Dom Vlkolínec, Liptovské
Múzeum, Ružomberok 038440
Rolnik, Robert, Pfaffenhofen an der
Ilm 130513
Rolo & Majoli, Roma 132033
Rom-Art-Galerie, Braunschweig . 075132
Roma, Buenos Aires 127533
Roma Color, Roma ... 081532, 132034
Roma Numismatica, Roma 081533
Roma Roma Roma, Roma 110852
Romagne, Raymond, Fontclaireau 068571
Romagnoli, Federica, Parma ... 131735
Romahn, Annemarie, Brüggen .. 106604
Római Katolikus Egyházi Gyűjtemény,
Sárospatak 023872
Római Kötár, Ferenczy Múzeum,
Szentendre 023959
Római Kori Kötár, Soproni Múzeum,
Sopron 023898
Romain De Saeghermuseum, Sint-
Amands 004014
Romain Rea, Paris ... 071857, 129132
Romain & Sons, A.J., Wilton,
Salisbury 091856
Roman Army Museum,
Greenhead 044481
Roman Bath House, Lancaster . 044755
Roman Baths Museum, Bath ... 043525
Roman Gil, Fernando, Barcelona 133157
Román Kori Kötár, Pécs 023853
Roman Museum, Canterbury ... 043858
Roman Painted House, Dover .. 044134
Roman Theatre of Verulamium, Saint
Albans 045669
Roman Villa Museum at Bignor,
Pulborough 045572
Román, Oscar, México 111982
Romana Antichitá, Roma 081534,
081535
Romanazzi, Luca, Bari 079782, 130924
Romanet, Jeannine, Montrichard 070331
Romaní, Zamora 086584, 133483
Romani, Alberto, Roma 132035
Romanian Ethnic Art Museum,
Cleveland 047823
Romann, Henri, München 077605,
130390
Romano, Adliya 063129, 100408
Romano, Alessandro, Firenze .. 080194
Romano, Claudia, Bologna 130997
Romano, Laura, Palermo 110616
Romano, Lidia, Milano 131517
Romano, Paolo, Firenze 080195
Romano, Vittorio, Firenze 080196
Romanos, Casablanca 082582
Romanov Gallery, Vancouver .. 101833
Romanovskie Palaty v Zarjade,
Gosudarstvennyj istoričeskij muzej,
Moskva 037497
The Romantic Agony Book and Print
Auctions, Bruxelles .. 125324, 140200
Romantic Notions, Las Vegas . 094173
Romantica Aina's, Jørpeland .. 084259
Romantica Antiek, Antwerpen . 063235
Romanticismo e Dintorni, Napoli 138871
Romantika '96, Budapest 079292
Romantikerhaus, Jena 019455
Romantique, Auckland 083822
Romantique, Frederiksværk 065687
La Romantique, Saint Paul 096855

Romantique Art och Antik,
Aplared 086629
Romantiques, Brookvale 061219
Romantiques, Manawatu 083984
Romantiques, Mona Vale 061872
Romantiques, Trevor .. 091632, 135148
Romany Folklore Museum,
Selborne 045754
Romark, Letchworth 134550
Romart, Amsterdam 112275
Rombach, Freiburg im Breisgau 075879,
141809
Romberger, S., Landshut 077018
Rombourg, Franck, Isernhagen . 076625
Rombouts & Zoon, H., Casteren 083024,
132465
Romé, Auxerre 066853
Rome Antique-Antiquités, Romorantin-
Lanthenay 072528
Rome Art and Community Center,
Rome 052674
Rome Historical Society Museum,
Rome 052675
Rome University of Fine Arts,
Roma 055974
Rome, de, Bradford 126832
Romeira, José, Sion 087816
Romela, Toronto 064744
Romeleit, M. L., Köln 076902
Romeny, Regi, Amsterdam 132412
Romeo, Fresno 093513
Romere, Daniele, Malo 110196
Roméro et Fils, Bayonne 066981, 103345
Romero Zazueta, Enrique,
México 111983
Romero, Barbara, Caen 067580
Rometti, Olivier, Nice 141005
Romfeia, Plovdiv 101131
Romigioli, Giovanni, Legnano .. 080311
Romiley Antiques, Romiley 091135
Romiti, Rio de Janeiro 064119
Romito, Alfredo & Norberto,
Rosario 060954
Rommel Museum, Blaustein .. 017418
Romney Toy and Model Museum,
Romney Hythe and Dymchurch
Railway, New Romney 045327
Romsdalsmuseet, Molde 033783
Romsley Picture Gallery, Romsley 118986
Romstedt, Kornelia, Potsdam .. 108409
Ron Morel Memorial Museum,
Kapuskasing 006064
Rona Gallery, London 118548
Rona Gallery, Wellington 113090
Ronald Reagan Boyhood Home,
Dixon 048333
Ronald Reagan Presidential Library
and Museum, Simi Valley 053346
Ronan Gallery, Perth 118904
Ronathahon:ni Cultural Centre,
Cornwall 005649
Ronchi, Giorgio, Bologna 079873
Ronchini Arte Contemporanea,
Terni 110945
Ronconi, Maurizio, Parma 131736
Ronda-Brou, Tours 073860
Ronda, Jean, Tours 073859
La Ronde des Bois, Auberives-sur-
Varèze 066784
Ronde des Objets, Seilhac 073479
Rondeau Provincial Park Visitor Centre,
Morpeth 006429
Rondom 1920, Den Haag 083127
Rondula, Dölsach 062544, 099974,
127722,   136498
Ronen, Amsterdam 082842
Rong County Museum, Rongxian 008209
Rong Xin, Zhai, Singapore 085492
Rongier, Maurice, Vic-sur-Cère . 074099
Rongione, Sandro, Milano 131518
Rongshui Museum, Rongshui .. 008208
Rongyu Guan, Jiangyin 007940
Roniger Memorial Museum,
Cottonwood Falls 048029

Ronin, New York 095697, 123016
Ronny Auto Dealer, Zebbug ... 082452
Ron's Attic, Baltimore 092357
Ron's Big Book, Southport ... 139912
Ron's Book Shop, Macquarie . 139815
Ronssin, Jean-Loup, Meung-sur-
Loire 070107
Rontonton, Utrecht 083660
Ronvaux, J., Bruxelles 063542
Roodepoort Museum, Roodepoort 038825
Rooijen, H. den, Dordrecht ... 083194
Rook & Co., Willemstad 065315
Rookie, Jacksonville 094034
Rooksmoor, Bath 117311
Roola, Nagoya 111331
Roolf, Bernadette, Wismar 108947
Room, Megève 070038
Room 125, Boston 092492
Room 403, Melbourne 099318
Room Service Vintage, Austin .. 092266
Room4, Ryde 118993
Rooma, Delhi 109338
Rooms and Gardens, New York . 095698
Rooms and Gardens, San
Antonio 097026
Rooms That Bloom, Kansas City . 094432
Roopa-Lekha, Delhi 138778
Roos, Den Haag 083128
De Roos van Tudor, Leeuwarden 112551
Roos, M. & D., Frankfurt am
Main 075817
Roosens, Bruxelles 063543
Rooseum, Center for Contemporary
Art, Malmö 040832
Roosevelt Campobello, Lubec .. 050519
Roosevelt County Museum, Eastern
New Mexico University, Portales 052273
Rooster, Saint Louis 123864
Root, Baiersbronn 106113
Root House Museum, Marietta . 050677
Root Seller, Phoenix 096236
Root, Sidney, Bruxelles 063544
Rooted in Wood, Baltimore ... 092358,
135346
Rootenberg, Robert, West
Hollywood 125022
Roots & Relics, Honolulu 093571
Roots of Norfolk at Gressenhall,
Gressenhall 044483
Rop, G. De, Aachen 105945
Ropac, Thaddaeus, Paris 105200
Ropac, Thaddaeus, Salzburg .. 100174,
136557
Ropars, Philippe, Rennes (Ille-et-
Vilaine) 129205
Roparshaugsamlinga, Isdalstø .. 033683
Rope, Ted, Dargaville 112871
Ropers Hill Antiques, Staunton
Harold 091418
Ropert, Claude, Lyon 069779
Roppongi Antique Fair, Tokyo .. 098232
Rops, Namur 100797, 125351
Rops, Saint-Servais 125355
Roq la Rue Gallery, Seattle ... 124593
Roque Roque, Josefa, Barcelona 133158
Roques, Nice 070660
Roques, Fabienne, Sens 129258
Roques, Pierre-Edmond, Pau ... 072048
Roques, Teddy, Mondonville .. 070160
Roquigny, Bruno, Saint-Valery-en-
Caux 125809
Rori, Roma 081536
Rori, Parioli, Roma 081537
Rori, Priscilla, Roma 081538
Rorschach Galerie, Rorschach . 116595
De Ros, Valencia 086536
Ro's Oriental Rugs, Nashville .. 095031
Ros, Mercedes, Teia 138003
Rosa, Jacksonville 094035
Rosa Studio, Indianapolis 135636
Rosa, Andrea, Roma 132036
Rosa, José, Carcavelos 126433
Rosa, José, Lisboa 126455
Rosa, Roberto de la, Los Angeles 094458

Rosa, Umberto, Roma 132037
Rosa, Walter, Untersteinach ... 078767
Rosado, Manuel, Nantes 070476
Rosales, Madrid 086255
Rosalie House, Eureka Springs . 048632
Rosalie House Museum, Natchez 051196
Rosalie Shire Historical Museum,
Goombungee 001110
Rosalie Whyel Museum of Doll Art,
Bellevue 046890
Rosalux Gallery, Minneapolis .. 122031
Rosart, Québec 101518
Rosarta, Rotterdam 112659
Rosatelli, Francesco, Roma 081539
Rosati, Renata, Firenze 080197
Roscoe Village Foundation,
Coshocton 048023
Roscommon County Museum,
Roscommon 024982
Roscrea Heritage Centre,
Roscrea 024983
Rose, Krefeld 076962, 130196
Rose, Tampa 136398
Rose and Crown Antiques, West
Malling 091790, 135207
Rose Antiques, Ashbourne 088302
Rose Antiques, Cleveland 092968
Rose Antiques, Toronto 064745
Rose Art, Vernantois 105840
Rose Art Museum, Brandeis University,
Waltham 054051
Rose Atelier, Molde 084279
Rose Center Museum,
Morristown 051070
Rose Cottage, Denver 120763
Rose Cottage Antiques, Birdwood 061156
Rose Cottage Antiques, Fyshwick 061500
La Rose des Vents, Le Cannet . 069234
Rose Fine Art and Antiques,
Stillington 091427, 135082
Rose Galleries, Little Canada .. 127287
Rose Gallery, Northampton ... 118820
The Rose Garden, Baltimore .. 092359
Rose Hawk, Gainesville 127232
Rose Hill Mansion, Geneva Historical
Society, Geneva 049069
Rose Hill Museum, Bay Village . 046836
Rose Hill Plantation State Historic Site,
Union 053891
Rose Lawn Museum, Cartersville 047449
Rose Museum at Carnegie Hall, New
York 051462
Rose Passé, Troyes 073909
Rose Rose, São Paulo 064206
Rose Seidler House, Wahroonga 001613
Rose Street Gallery, Armadale,
Victoria 098667
Rose Valley and District Heritage
Museum, Rose Valley 006824
De Rose Winkel, Deventer 083162
Rose, Anthony, Yass 099943
Rose, Charles, Melbourne 061830
Rose, Dianne, Fort Worth 093477
Rose, Dr. Ulrich, Greifswald ... 141841
Rose, Giovanni, Bayreuth 074697
Rose, Marjorie, Pittsburgh 123573
Rose, Marjorie, San Francisco .. 097316
Rose, Michael, London 090462, 090463,
090464
Rose, Peter, New York 123017
Rose, R.L., Edinburgh 089186
Roseau County Historical Museum,
Roseau 052678
Roseberry, Calgary 064269
Rosebery's, London 126971
Rosebud, El Paso 093413
Rosebud Centennial Museum,
Rosebud 006825
Rosebud County Pioneer Museum,
Forsyth 048777
Rosebud Studios, Surf Beach . 099732
Rosebud's, Tampa 097585
Rosedale, Albany 112740
Rosedale Street Gallery, Dulwich

Hill 098985
Rosefeld, Libbesdorf 077097
Rosegg, Sikoronja, Sankt
Lamprecht 100185
Rosehr, Joachim, Kiel . 076790, 130122
Roseland Antiques, Toronto . . 064746
Rosemann, Susanna & Frank,
Vilseck 108824
Rosemary and Time, Thame . . . 091594,
135133
Rosemont Art Gallery, Regina . . 006775
Rosemont Plantation, Woodville . 054535
Rosemount Gallery, Aberdeen . . 117212
Rosemount Lodge Art Gallery,
Richmond, New South Wales . 099625
Rosemount Museum, Pueblo . . 052410
Rosen & Chadick, New York . . 135990
Rosen Kavalier, Houston 093780
Rosen, Andrea, New York 123018
Rosenbach Museum,
Philadelphia 052091
Rosenbach, Detlev, Hannover . . 107374
Rosenbach, Louis, Dallas 093209
Rosenbad, Stockholm 143963
Rosenbaum, Dallas 120654
Rosenbaum, Houston 121121
Rosenberg, Höganäs 115690
Rosenberg, Mayen 077312
Rosenberg, Milano 110417
Rosenberg & Kaufman, New York 123021
Rosenberg Gallery, Goucher College,
Baltimore 046768
Rosenberg, A., Bremerhaven . . 075192
Rosenberg, Alex J., New York . . 123019
Rosenberg, C., Heswall 089521
Rosenberg & Co., Paul, New York 123020
Rosenblatt, Minna, New York . . 095699
Rosenblom, Mölndal 115777
Rosenborg Slot, Danske
Kongers Kronologiske Samling,
København 010193
Rosenbruch Wildlife Museum, Saint
George 052768
Rosendahler Kunsthaus,
Rosendahl 078271
Den Rosendahlske Bibelsamling,
Esbjerg 010046
Rosendal, Durbanville 114783
Rosendals Slott, Stockholm . . . 041037
Roseneath Gallery, Armidale . . 061054
Rosenfeld, Bad Kissingen . . . 106073
Rosenfeld, Philadelphia 123445
Rosenfeld, Tel Aviv 109731
Rosenfeld, Michael, New York . . 123022
Rosenfeld, Samuel, New York . . 123023
Rosengren, Hauge i Dalane . . . 113159
Rosenhagen, Johann, Öhringen . 077894
Rosenhahn, Willi, Daaden 075285
Rosenhauer, Ursula, Göttingen . . 107148
Rosenheimer Verlagshaus,
Rosenheim 137489
Rosenholm Slot, Hornslet 010139
Die Roseninsel, Künstler und
Kunstfreunde am Starnberger See
e.V., Starnberg 059225
Rosenkranz, Magdeburg 077224
Rosenkranz, Ilona & Stefan,
Thiessow 108767
Rosenlew-museo, Pori 011267
Rosenlöfs Tryckerimuseum,
Kungsgården 040765
Rosenmuseum Steinfurth, Bad
Nauheim 016830
Rosenrod, Göteborg 086717
Rosenstock, Denver . . 120764, 120765,
145134
Rosenthal, Berlin 106411
Rosenthal, Minneapolis 122032
Rosenthal, New Orleans 095155
Rosenthal Gallery of Art,
Caldwell 047345
Rosenthal, Dennis, Chicago . . . 120258
Rosenthal, Ludwig,
Leidschendam 143190

Rosenthal, Marcia, Pittsburgh . . 123574
Rosenthal, Marie-Eve, Saint-
Ouen 073186
Rosenthal, Stan, Hastings 117927
Rosenvinge Kunst, Bergen . . . 084222
Rosenwald-Wolf Gallery, University of
the Arts, Philadelphia 052092
Rosenzweig, M., Bonn 106538
Rosequist, Tucson 124769
Rosersbergs Slott, Rosersberg . 040922
Rose's Bookshop, Hay-on-Wye . 144463
Rosetown & District Museum,
Rosetown 006827
Rosetzky, Marion, Red Hill . . . 099616
Rosewall Memorial Shell Museum,
Port Lincoln 001423
Rosewood, Bologna 079874
Rosewood, Oakland 095910
Rosewood Antiques, Traralgon . 062344,
127690
Rosewood Furniture, Melbourne 061831
Rosewood Scrub Historical Museum,
Marburg 001252
Roseworthy Agricultural
Museum, University of Adelaide,
Roseworthy 001460
Rosgartenmuseum, Konstanz . 019763
Rosh, West Hollywood 097930
Rosicrucian Egyptian Museum, San
Jose 053041
Rosies från Gammalt till Nytt,
Spånga 086951
Rosily Arts Museum, Taipei . . . 042547
Rosin, Douglas, Chicago 092784
Rosina's, Falmouth . . 089252, 134383
Rosing, H. & M., Bocholt 075029
Roskamp, Hyma, Hamburg . . . 129986
Roskell, Plymouth 118924
Roski School of Fine Arts, University
of Southern California, Los
Angeles 057371
Roskilde Auktionshuss, Roskilde 125467
Roskilde Domkirkemuseum,
Roskilde 010290
Roskilde Museum, Roskilde . . . 010291
Roskilde Museums Købmandsgård,
Roskilde 010292
Roskildegalleriet, Roskilde . . . 102965
Roslagens Sjöfartsmuseum,
Väddö 041129
Roslagsmuseet, Norrtälje 040884
Roslaviskij Istoriko-chudožstvennyj
Muzej, Roslavl 037691
Roslyn Oxley9 Gallery, Paddington,
New South Wales 099533
Rosmosegaard Antik, Gilleleje . 065701
Rosoff, Charles, New York 123024
Rosot, Jean, Pont-Aven 072244
Rosrestavracija, Moskva 133014
Ross, Goulburn 061552
Ross, Washington 136467
Ross Abbey, Headford 079636
Ross Abbey Gallery, Sassafras . 099680
Ross Antiques, Ross 062171
The Ross C. Purdy Museum of
Ceramics, Westerville 054335
Ross Castle, Killarney 024942
Ross County Historical Society
Museum, Chillicothe 047721
Ross Farm Museum, New Ross . 006471
The Ross Gallery, Belfast 117343
Ross House Museum, Winnipeg . 007459
Ross Memorial Museum, Saint
Andrews 006850
Ross Museum, Foresters Falls . . 005812
Ross Old Book and Print Shop, Ross-
on-Wye 144804
Ross-on-Wye Antiques Gallery, Ross-
on-Wye 091141
Ross-Thomson House Museum,
Shelburne 007016
Ross, Bernie, San Jose 136358
Roß, Betina, Ellerau 129786
Ross, Douglas, Woburn 127094

Ross, Hardie, Chipping Campden 088883
Ross, Jane, Carlton South 061284
Ross, John, Tucson 097690
Ross, Luise, New York 123025
Ross, Marcus, London 090465
Ross, Morgan, Seattle 097497
Ross, Nancy, Chicago 092785
Ross, Schmidt & Partner,
Hofolding 130061
Ross, Simon, New York 095700, 135991
Rossaert, Antwerpen 140152
Rossberg, R., Bonn 075085
Roßberg, Reinhardt,
Markkleeberg 130288
Rossberg, Werner, Davos Platz . . 087393
Rossburn Museum, Rossburn . . 006828
Rosselli, John, New York 095701,
095702
Rosselli, John, Washington 097858
Rossendale Footware Heritage
Museum, Rawtenstall 045587
Rossendale Museum, Rossendale 045636
Rosset, Hugues, Genève 126732
Rosseto, Volmir Luiz, Porto
Alegre 101000
Rossetti, Luigi, Salsomaggiore
Terme 110890
Rossetto, Roland, Chaumont (Haute-
Marne) 067936
Rossfelder Heimatstuben,
Crailsheim 017771
Rossi, Milano 131519
Rossi, San Francisco . . 097317, 136343
Rossi, Sankt-Peterburg 114357
Rossi, Toulon (Var) 073729
Rossi & Rossi, London 090466
Rossi e Onesti, Assago 079754
Rossi, Anna, Bologna 079875
Rossi, Carlo, Domodossola 080065
Rossi, D., Bad Honnef 074532
Rossi, Daniele, Firenze 131258
Rossi, Flora, Roma 081540
Rossi, Jean-Pierre, Saint-Michel-
l'Observatoire 072982
Rossi, Julcir, São Paulo 064207
Rossi, Laura, Roma 132038
Rossi, Leopoldo, Milano 080702
Rossier, Pittsburgh 123575
Rossier, David & Colette,
Rougemont 087761
Rossignol & Fils, Céret 128534
Rossignol, Daniel, Les Arcs . . . 140868
Rossignol, Gabriel, Paris 141273
Rossignol, Gérard, Paris 105201
Rossignol, Henri, Cannes 140815
Rossignoli Picotti, Giuseppina,
Verona 132298
Rossijskij Antikvarnyj Salon,
Moskva 098286
Rossijskij Étnografičeskij muzej, Sankt-
Peterburg 037816
Rossijskij Komitet Meždunarodnogo
Soveta Muzeev (IKOM Rossii),
Moskva 059775
Rossijskij Muzej Lesa, Moskva . 037498
Rossini, Paris 125758
Rossiters, Crescent Head 098933
Rossland Historical Museum,
Rossland 006829
Rosslyn, Dunedin . . 083931, 112895,
132720
Rosslyn Chapel, Roslin 045635
Rosso, Fulvio, Trieste 081801
Rosso, Mano, Oakland 095911
Rosso, Sonia, Pordenone 110678
Rosso, Sonia, Torino 111001
Rosso, Yolande, Nice 070661
Rossolacca, Milano 080703
Ross's, Belfast 126817
Rosta-Collection, Münster 142209
Rosteck, Lemgo . . 077077, 077078,
077079
Rostellan Antiques, Midleton . . 079674
Rosthal, Meerbuscher

Kunstauktionshaus, Meerbusch 126035
Rostovskij Filial Sankt-Peterburgskogo
Gosudarstvennogo Universiteta
Kultury i Iskusstv, Rostov-na-
Donu 056298
Rostovskij Kreml – Gosudarstvennyj
Muzej-zapovednik, Rostov . . . 037693
Rostovskij Oblastnoj Muzej
Kraevedenija, Rostov-na-Donu 037696
The Rostra Gallery, Bath 117312
Rostrum, South Petherton 091366,
135062
Rostworowski, Dominik, Kraków 113635
Roswell Museum and Art Center,
Roswell 052692
Rosy's Nostalgieladen, Zürich . . 088008
Roszkov, Budapest 109175
Rot-Kreuz-Museum, Geislingen an der
Steige 018614
Rot-Kreuz-Museum, Regenstauf 021332
Rota, Nurieux-Volognat 070765
Rota, Bertram, London 144679
Rotary Museum of Police and
Corrections, Prince Albert . . . 006713
Rotas de Arte, Lisboa 114060
Rotation, Los Angeles 094459
Rotch-Jones-Duff House and Garden
Museum, New Bedford 051226
Rotenberg, Judi, Boston 119918
Rotenhahn, Rumohr 078311
Roter Haubarg, Witzwort 022715
Rotermund, Thole, Hamburg . . . 076301,
107311
Rotermundt, Amsterdam 082843
Rotes Antiquariat, Berlin 141563
Roth, Leipzig 077065
Roth, Minneapolis 122033
Roth Horowitz, West Hollywood . 125023
Róth Miksa Emlékház, Budapest 023550
Roth, Andrew, New York 123026
Roth, Artur, Mainhardt 077228
Roth, Heidy, Basel 087256
Roth, Hildegard, Gelsenkirchen . 075969
Roth, Joachim, Thees 108766
Roth, Kajetan, Nastätten 077688
Roth, M., Gifhorn 076001
Roth, Michael, Kitchener 101328
Roth, Monika, Linz 127810
Rothäuser, Michael, Essen 129820
Rothas Art Gallery, Islamabad . . 113301
Rothballer, Markus, Weiden in der
Oberpfalz 078886
Rothe, Erfurt 075641
Rothe House Museum, Kilkenny 024939
Rothe, Maria, Frankfurt am Main 107027,
137206
Rothen, Königs Wusterhausen . 130176
Rothenbach, D. & R., Bad
Kreuznach 074546
Rothenbücher, P., Bayreuth . . . 125865
Rothenburger Künstlerbund e.V.,
Rothenburg ob der Tauber . . . 059226
Rothenhäusler, Ulrike,
Kirchentellinsfurt 130124
Rotherham Art Gallery,
Rotherham 045639
Rothfuss-Stein, Elwine, Karlstein am
Main 130102
Rothko Chapel, Houston 049594
Rothmann, Kippenheim 076799
Rothqvist, Linköping 143881
Rothschild, Montréal 064465
Rothschild, New Orleans 095156
Rothschild, Saint Louis 096768
Rothschild House, Port Townsend 052263
Rothschild Petersen Patent Model
Museum, Cazenovia 047472
Rothwell & Dunworth, Dulverton 144376
Rotini, Eleonora, Livorno 110188
Rotkreuz-Museum, Nürnberg . . 020874
Rotkreuz-Museum Berlin, Berlin 017274
Rotkreuzmuseum, Hofheim in
Unterfranken 019295
La Rotonda, Montecchio Emilia . 110508

Rotonda di Via Besana, Milano . 026830
Rotor – Association for Contemporary
Art, Graz 058307
Rotorua Museum of Art and History,
Te Whare Taonga o te Arawa,
Rotorua 033296
Rotstab, Liestal 116443
Rotta Farinelli, Roberto, Genova 110131
Rottar, Martina, Berlin 141564
Rottauer Museum für Fahrzeuge,
Wehrtechnik und Zeitgeschichte bis
1948, Pocking 021165
Rotten Verlag, Visp 138096
Rotterdams Radio Museum,
Rotterdam 032822
Rotterdamse Antiekhallen,
Rotterdam 083587
Rottingdean Grange Art Gallery and
Museum, Brighton 043739
Rottinghaus, Cincinnati 120372
Rottler, D., Riegel am Kaiserstuhl 078236
Rottloff, Karlsruhe .. 107523, 137291
Rottmann, Steinfurt 078551
Rottmann, Andreas, Wülfershausen an
der Saale 130781
Rottmann, Frank, Blomberg, Kreis
Lippe 075024
Rottneros Park, Rottneros 040923
Rottnest Island Museum, Rottnest
Island 001462
The Rotunda, Charlottesville ... 047584
Rotunda, Vilnius 111808
Rotunda Gallery, Brooklyn 047265
Rotunda Museum of Archaeology and
Local History, Scarborough ... 045744
Rotušės, Alytus 111782
Rotwand, Zürich 116852
Rouayroux, Jean-Pierre,
Montpellier 070312
Roubaud, Sylvia, München 108184
Roubin, Marc, Saint-Affrique ... 072659
Rouch, Yves, Carcassonne 067677
Rouchaléou, Montpellier 140969
Roucka, Wolfgang, München ... 108185
Roudaut, Landerneau . 069164, 128651
Rouden, André, La Seyne-sur-
Mer 069101
Roudillon, Jean, Paris 071858
Roudillon, Michel, Paris 071859
Roue Libre, Grignan 140854
Rouen Détection Numismatique,
Rouen 072614
Rouenson, Marie-Josèphe, Coye-la-
Forêt 068216
Rouffet-Goudier, Ascain 066756
Rouffignac, Colin de, Wigan ... 091846
Rouflay, Bernard, Perpignan ... 072092
Rouflay, Pierrette, Nîmes 070704
Rouge 91, Brunoy 103475
Rouge-Pullon, Paris 129133
Rouge, Pierre, Mâcon 069796
Rougeau, Richard, Bracieux ... 067445
Rougeaux Bayle, Martine, Pau ... 072049
Rouges, Jean-Marie, Montauban 070201
Rouget de Lisle Antiquités,
Courlans 068189
Rough and Tumble Engineers Museum,
Kinzers 049991
Rough Point, Newport Restoration
Foundation, Newport 051541
Rough Rustique, Kyneton 061713
Roughton, Dallas 120655
Rougier, Laurent, Bordeaux 128505,
140803
Rouhani, Saint Louis 096769
Rouillac, Philippe, Paris 125759
Rouillac, Philippe, Vendôme ... 125831
Rouilly, Orléans 070820
Rouje, Québec 101519
Rouleau and District Museum,
Rouleau 006831
Roullet, Dominique, Rouen 072615
Roulmann, F., Paris 141274
Roulston Museum, Carstairs ... 005572

Roumet, Paris 071860, 125760
Roumieu, Claude, Cavaillon 067739
Round Corner Antiques, Dural .. 061420
The Round House, Penzance ... 118900
Round Top Antiques Fair, Round
Top 098495
Round Top Center for the Arts – Arts
Gallery, Damariscotta 048134
Roundabout Antiques, Wondai .. 062447
Roundabout Antiques Centre,
Haywards Heath 089484
Roundell, James, London 118549
Rounds, Al, Salt Lake City 123985
Roura Comas, Juan Arturo,
Barcelona 085833
Rourke, Peter, Vancouver 128229
Rouse Hill Estate, Rouse Hill ... 001463
Roussard, André, Paris 105202
Rousseau, Marcq-en-Barœul ... 069867
Rousseau, Florent, Paris 129134
Rousseau, Nadine, Bourges 140805
Rousseaux, Dominique, Paris ... 071861
Rousseaux, Dominique, Saint-
Ouen 073187
Roussel, Toronto 064747
Roussel-Demaretz, Michèle, Nampont-
Saint-Martin 070405
Roussel, Thérèse, Perpignan ... 072093,
105339
Rousselle, Henrianne, Ris-
Orangis 072476
Rousselot, Bouxières-aux-Dames 067440
Roussil, Montréal 101447
Roussillot, Yves & Marie-Claude,
Drevant 068408
Roussos, Nikolaos, Athinai 079177
Roussou, Marilena, Athinai 079178,
130806
Route 66 Antique Connection,
Albuquerque 092080
La Route d'Alexandre, Paris ... 071862
Route de la Soie et ses Merveilles,
Paris 105203
Rouveloux, Christophe, Limoges 069531
Rouvière, Gérard, Le Cailar ... 069232
Le Roux & Morel, Paris 125761
Roux & Troostwijk, Paris 125762
Roux, Claude, Gannat 068628
Roux, Gilles, Mouriès 070375
Roux, Michel, Limay 069511
Roux, Philippe, Beaune 067036
Rouyer, Michel, Liverdun 069602
Rouzaud, Jean-Marie, Le
Cheylard 069252
Rovani, Genova 110132
Rovaniemen Kotiseutumuseo,
Rovaniemi 011323
Rovaniemen Taidemuseo,
Rovaniemi 011324
Rove, London 118550
Rovegno & C., Andrea, Genova . 080287
Rovello, Milano 142742
Rovers, Haarlem 083351
Roversi, Giorgio, Bologna 079876
Rovi, Lugo 086087
Rovira, Barcelona 085834, 114956
Rovira, Jacques, Pia 072151
Rovira, Monique, Saint-Féliu-
d'Avall 072782
Row, Charlotte 092623
Rowan, Drymen 117688
Rowan Museum, Salisbury 052905
Rowan Oak, William Faulkner's Home,
Oxford 051862
Rowan, H., Bournemouth 144256
Rowe, Martyn, Chacewater 126850
Rowe, Phil, Chicago 092786
Rowland, Mark, Chicago 120259
Rowland, Michael, Stow-on-the-
Wold 091479
Rowland, Simon H., Chelmsford 126852
Rowles & Co., Calgary 101257
Rowles, Welshpool ... 091777, 119318
Rowles & Co., Edmonton 101295

Rowletts of Lincoln, Lincoln ... 089833
Rowley, Ely 126891
Rowley, London 118551
Rowley Gallery, Holybourne ... 117987
Rowley Historical Museum,
Rowley 052697
Rowntree Antiques, Montréal .. 064466
Rowsell House Interpretation Center,
Harrington Harbour 005988
Roxane's, Caen 067581
Roxborough, Newport Beach ... 061963
Roxburgh, U., Manawatu 132728
Roxbury, Houston 093781
Roxenstein, Helsingborg 086740
Roxy, Dallas 093210
Roxy Klassik & Roxy Antik,
Frederiksberg 065674
Roy Boyd Gallery, Roy, Chicago . 047695
Le Roy Patrick, La Neuville-Roy 069056
Roy Rogers-Dale Evans Museum,
Branson 047181
Roy, Elke, Bad Mergentheim ... 074550
Roy, Joël, Montamise 070180
Roy, Joël-Denis, Royan 072632
Roy, Martine, Paris 129135
Roy, Paule, Bourg-la-Reine 067415
Royal Academy of Arts, London 045085,
045086
Royal Academy Shop, London . 138247,
144680
Royal Air Force Air Defence Radar
Museum, Norwich 045404
Royal Air Force Museum, London 045087
Royal Air Force Museum, Shifnal 045786
Royal Air Force Museum 201
Squadron, Saint Peter Port ... 045714
Royal Air Force Museum Reserve
Collection, Stafford 045851
Royal Albert Memorial Museum and
Art Gallery, Exeter 044312
Royal Alberta Museum,
Edmonton 005760
Royal Anglian Regiment Museum,
Imperial War Museum, Duxford 044193
Royal Antikvitás, Budapest 079293
Royal Antiques, Ciudad de
Panamá 084416
Royal Antiques, New Orleans .. 095157
Royal Antiques, New York 095703
The Royal Arcade Watch Shop,
London 090467
Royal Architectural Institute of Canada,
Institut Royal d'Architecture du
Canada, Ottawa 058468
Royal Armouries, London 045088
Royal Armouries at Fort Nelson,
Fareham 044328
Royal Armouries Museum, Leeds 044784
Royal Army Dental Corps Historical
Museum, Aldershot 043388
Royal Art, Aalst 063177
Royal Art, Brasília 100926
Royal Art, Frankfurt am Main .. 017028
Royal Art Galleries, Sliema 111921
Royal Art Gallery, New Orleans . 122231
Royal Art Museum, Museum für
zeitgenössische Kunst, Berlin . 017275
Royal Art Society of New South Wales,
North Sydney 058214
Royal Art Society of NSW Art School,
North Sydney 054913
Royal Arts Foundation, Newport 051542
Royal-Athena Galleries, New York 095704,
123027
Royal Australian Air Force Museum,
Point Cook 001411
Royal Australian Infantry Corps
Museum, Singleton 001497
Royal Barges National Museum,
Bangkok 042676
Royal Berkshire Yeomanry Cavalry
Museum, Windsor 046165
Royal Birmingham Society of Artists,

Birmingham 060215, 117379
Royal Borough Museum Collection,
Windsor 046166
Royal Brierley Crystal Museum,
Dudley 044148
Royal British Columbia Museum,
Victoria 007335
Royal British Society of Sculptors,
London 045089, 060216
Royal Brunei Armed Forces Museum,
Bandar Seri Begawan 004967
Royal Cambrian Academy of Art,
Academi Frenhinol Gymreig,
Conwy 044005
Royal Cameo Glass, New Orleans 122232
Royal Canadian Academy of Arts,
Toronto 058469
Royal Canadian Artillery Museum,
Shilo 007028
Royal Canadian Military Institute
Museum, Toronto 007199
Royal Canadian Mounted Police
Museum, Regina 006776
The Royal Canadian Regiment
Museum, London 006801
Royal Carriage Museum, Bulaq-el-
Dakrur 010527
Royal Ceremonial Dress Collection,
London 045090
Royal Coin and Jewelry, Houston 093782
Royal Collectibles, Honolulu ... 093572
Royal Collectibles, San Francisco 097318
Royal College of Art, London .. 045091,
056567
Royal College of Art Society,
London 060217
Royal College of Obstetricians
and Gynaecologists Collection,
London 045092
Royal College of Surgeons in Ireland
Museum, Dublin 024908
Royal Cornwall Museum, Truro . 046001
Royal Crafts and Collectibles,
Columbus 093043
Royal Crescent Antiques, New
Orleans 095158
Royal Crown Derby Museum,
Derby 044086
Royal Doulton Specialty Shop,
Buderim 061229
Royal Engineers Museum,
Chatham 043913
Royal Engineers Museum, Gillingham,
Kent 044386
Royal Estates Gallery, New York 095705
Royal Exchange Art Gallery,
London 118552
Royal Fine Art, Tunbridge Wells . 119273
Royal Flying Doctor Service Visitors
Centre, Edge Hill 001052
Royal Furniture Refinishing,
Edmonton 128168
Royal Furniture Refinishing, North
York 128166
Royal Fusiliers Museum, London 045093
Royal Gallery, Montréal 101448
Royal Gallery of San Francisco, San
Francisco 124380
Royal Glasgow Institute Kelly Gallery,
Glasgow 044415
Royal Glasgow Institute of the Fine
Arts, Glasgow 060218
Royal Gloucestershire, Berkshire
and Wiltshire Regiment Museum,
Salisbury 045726
The Royal Governor's Mansion, Perth
Amboy 052000
Royal Green Jackets Museum,
Winchester 046153
Royal Ground Art Gallery,
Oakland 123263
Royal Gunpowder Mills, Waltham
Abbey 046046
Royal Hamilton Light Infantry Heritage

Museum, Hamilton 005977
Royal Hampshire Regiment Museum,
Winchester 046154
Royal Highland Fusiliers Regimental
Museum, Glasgow 044416
Royal Holloway College Picture Gallery,
Egham 044281
Royal Hospital Chelsea Museum,
London 045094
Royal Inniskilling Fusiliers Regimental
Museum, Enniskillen 044297
Royal Institute of Oil Painters,
London 060219
Royal Institute of Painters in
Watercolours, London 060220
Royal Irish Fusiliers Museum,
Armagh 043434
Royal Irish Regiment Museum,
Ballymena 043479
Royal Jewellery Studio, Auckland 112808
Royal Lincolnshire Regiment Museum,
Lincoln 044834
Royal Logistic Corps Museum,
Deepcut 044079
Royal London Hospital Archives and
Museum, London 045095
Royal London Wax Museum,
Victoria 007336
Royal Malaysian Air Force Museum,
Kuala Lumpur 030696
Royal Malaysian Navy Museum,
Lumut 030702
Royal Marines Museum,
Portsmouth 045557
The Royal Mews, Royal Collection,
London 045096
Royal Mile Curios, Edinburgh . 089187
Royal Mile Gallery, Edinburgh .. 089188
Royal Military Academy Sandhurst
Collection, Camberley 043834
Royal Military College of Canada
Museum, Kingston 006112
Royal Military Police Museum,
Chichester 043946
Royal Mint Museum, Llantrisant 044894
Royal Museum, Seoul 030031
Royal Museum and Art Gallery,
Canterbury City Museums,
Canterbury 043859
Royal Museum Sri Menanti, Muzium
Negeri Sembilan, Seremban . 030727
Royal Museum, National Museums of
Scotland, Edinburgh 044265
Royal National Lifeboat Institution
Headquarters Museum, Poole . 045528
Royal Naval Museum Portsmouth,
Portsmouth 045558
Royal Naval Patrol Service Association
Museum, Lowestoft 045163
Royal Navy Submarine Museum,
Gosport 044453
Royal New Zealand Army Medical
Corps Museum, Christchurch . 033175
Royal New Zealand Navy Museum,
Devonport 033185
Royal Newfoundland Constabulary
Museum, Saint John's 006911
Royal Norfolk Regimental Museum,
Norwich 045405
Royal Nova Scotia Historical Society,
Halifax 005962
Royal Numismatic Society,
London 060221
Royal Numismatic Society of New
Zealand, Wellington 059641
Royal Oak Traders, Auckland .. 083823
Royal Observatory Greenwich, National
Maritime Museum, London .. 045097
Royal Observer Corps Museum,
Eastleigh 044222
Royal Ontario Museum, Toronto 007200
Royal Over-Seas League, London 060222,
118553
Royal Pavilion, Brighton 043740

Royal Pharmaceutical Society Late
Victorian Pharmacy, Edinburgh 044266
Royal Photographic Society, Bath 060223
The Royal Photographic Society,
Fenton House, Bath 043526
Royal Prince Alfred Hospital Museum,
Camperdown 000934
Royal Pump Room Museum,
Harrogate 044519
Royal Queensland Art Society,
Brisbane, Brisbane 058215
Royal Queensland Art Society, Gold
Coast, Gold Coast 058216
Royal Regiment of Fusiliers Museum,
Royal Warwickshire, Warwick . 046059
Royal Research Ship Discovery,
Dundee 044167
Royal Saint John's Regatta Museum,
Saint John's 006912
Royal Saskatchewan Museum,
Regina 006777
The Royal Scots Regimental Museum,
Edinburgh 044267
Royal Scottish Academy,
Edinburgh 044268, 056529
Royal Scottish Academy Collections,
Edinburgh 044269
Royal Shakespeare Company
Collection, Stratford-upon-Avon 045906
Royal Signals Museum, Blandford
Forum 043656
Royal Society for the Encouragement
of Arts, Manufactures and
Commerce, London 060224
Royal Society of Antiquaries of Ireland,
Dublin 059383
Royal Society of Arts and Sciences of
Mauritius, Réduit 059570
Royal Society of British Artists,
London 060225
Royal Society of Marine Artists,
London 060226
The Royal Society of Miniature
Painters, Sculptors and Gravers,
Gillingham, Dorset 060227
Royal Society of Painter-Printmakers,
London 060228
Royal Society of Portrait Painters,
London 060229
Royal South Australia Society of Arts,
Adelaide 058217
Royal Sporting Antiques,
Woodlands 062451
Royal Standard Antiques, Cowes 088990,
134302
Royal Street Gallery, Paris ... 105204
Royal Tapis, Lausanne 087585, 133804
Royal Thai Navy Museum, Samut
Prakan 042716
Royal Tyrrell Museum of Palaeontology,
Drumheller 005719
Royal Ulster Rifles Regimental
Museum, The Royal Irish Rangers,
Belfast 043563
Royal Vending, Saint Louis ... 123865
Royal Watercolour Society,
London 060230
Royal Welch Fusiliers Regimental
Museum, Caernarfon 043825
Royal West of England Academy,
Bristol 056514
Royal West of England Academy New
Gallery, Clifton 043968
Royal Western Australian Historical
Society Museum, Nedlands .. 001348
Royal Wiltshire Yeomanry Museum,
Swindon 045939
Royal Winnipeg Rifles Museum,
Winnipeg 007460
Royal York Auction Gallery,
Pittsburgh 127404
Royale Silver Company, Los
Angeles 094460
Royalis, Vaduz 126256

Royalk Galeria, Gdynia 113505
Royall House, Medford 050753
Royall, E. & C., Medbourne ... 134847
Royals Museum, Fort York Armoury,
Toronto 007201
Royalton Historical Society,
Royalton 052699
Royalty Coins and Jewelry, San
Antonio 097027
Royalty Fine Arts, Atlanta ... 119606
Le Royaume des Puces,
Bruxelles 063545
Le Royaume du Dahomey,
Venezia 081871
Royce Myers, Tulsa 124812
Roycroft, Kew East 139795
Royden Prior Gallery, Eton ... 117767
Royden Smith, L., Lichfield ... 089824,
144538
Royer, Alain, Paris 071863
Royer, Elisabeth, Paris 105205
Royer, Elizabeth, Paris 105206
Royer, Jean-Pierre, Paris 129136
Royer, Pierre-Richard, Paris ... 071864
Royet, Auguste, Saint-Just-Saint-
Rambert 072877, 072878
Royet, Georges, Panissieres ... 070847
Roykstovan, Kirkjubøur 010733
Royo, Jésus, Espoey 068472
Roy's Antiques, Clifton Hill ... 061344
Roy's Refinishing, Buffalo 135378
Roy's Upholstery, Ottawa 128193
Royston and District Museum,
Royston 045642
Roz, Maria, Quintin 072357
Roza Azora, Moskva 114298
Różańska, Ewa, Gdynia 084524
Rożdestvenskaja Jarmarka
Antikvariata, Moskva 098287
De Roze Patio, Sint-Niklaas ... 063835
Rozelle House Galleries, Ayr ... 043462
Rozenburg, Delft 083037
Rozendaal, Joop, Noordwyk ... 112614
Rozet, Frédéric, Opoul-Perillos . 070788
Rozic, B., Bad Herrenalb 074517
Rožmberský Dům, Husitske Muzeum v
Táboře, Soběslav 009886
Rózsa Naiv Art Galéria, Budapest 109176
RP-GA Lerie, Montpellier 104284
RR & H Specialty and Collectables,
Cleveland 092969
RS Fine Art, Tulsa 124813
RSA Journal, London 139221
RSM Art Conservation, Fortitude
Valley 127600
RSM Art Conservation, Mapleton 127636
RT Biz, Faversham 117792
RT Modern and Contemporary, Hong
Kong 102109
Rubaiyat, Calgary 101258
Les Rubans Bleus, Vincennes . 074245
Rubbers, Buenos Aires 098578, 098579
Rubbish, Los Angeles 094461
Rube, Eberhard, Koblenz 107577
Rubell Family Collection, Miami
Beach 050848
Rubelle and Norman Schafler Gallery,
Pratt Institute, Brooklyn ... 047266
Ruben, Los Angeles 135742
Rubenianum, Antwerpen 003313
Rubens Maskinhistoriska Samlingar,
Götene 040664
Rubens Schilderijen, Enschede . 112480
Rubenshuis, Antwerp 003278
Rubenstein, Paula, New York .. 095706
Ruberl, Richard, Wien . 063061, 100367
Rubi, Girona 086029
Rubicon Gallery, Dublin 109617
Rubin Museum, Tel Aviv 025210
Rubin, Chuck, Louisville 094583
Rubin, Lawrence, Milano 110418
Rubin, Lawrence, Zürich 116853

Rubinacci, Genova ... 080288, 126197
Rubini, Fabrizio, Roma 081541, 132039
Rubio, Mar del Plata 139607
Rubio Herrera, Jose, Palma de
Mallorca 086364
Rubio, Pepe, Palma de Mallorca 086363
Ruby Green Contemporary Arts
Foundations, Nashville 051184
Ruby, J., Donaueschingen 075373
Ruby's Book Sale, New York .. 145370
Rucellai, Roma 110853
Ruch, René, Paris 129137
Ruchill Centre for Contemporary Arts,
Glasgow 044417
Ruchi's Collection, Hyderabad .. 109371
Ruck, M., Gengenbach 075973
Rucker, Charles, San Diego ... 124166
Ruckers Hill Antiques, Northcote 062005
Ruckli, Andrea, Engelberg 116255
Ruckstuhl, Josef, Ruswil 087764
Ruční Papírna – Muzeum Papíru,
Velké Losiny 009945
Rudarska in Etnologoška Zbirka v Črni
na Koroškem, Koroški Muzej, Črna
na Koroškem 038504
Rudd House Museum, Kimberley 038768
Rudd, Winnie, Jacksonville ... 094036
Ruddell Gallery, Black Hills State
University, Spearfish 053435
Ruddick's Folly, Suffolk 053589
Ruddington Framework Knitters'
Museum, Ruddington 045645
Ruddington Village Museum,
Ruddington 045646
Rudenel, Rio de Janeiro 064120
Rudic, Acapulco 111932
Rudigier, Gertrud, München ... 077606
Rudloff, Claudia & Thomas,
Scheinfeld 108560
Rudno-petrografičeskij Muzej, IGEM
Rysskoj akedemii nauk, Moskva 037499
Rudolf, Sylt 108751
Rudolf-Scharpf-Galerie, Ludwigshafen
am Rhein 020136
Rudolf Steiner Archiv, Dornach . 041447,
138039
Rudolf Weinwurm-Museum,
Göpfritz 001957
Rudolf, Annegret, Oldenburg ... 077938
Rudolf, Hans-Jürgen, Mannheim 077271,
130281
Rūdolfa Blaumaņa Memoriālais Muzejs
Braki, Ērgļi 030122
RudolfV, Amsterdam 112276
Rudolph, Cottbus 075275
Rudolph E. Lee Gallery, Clemson 047800
Rudolph Projects Vine Street Studios,
Houston 121122
Rudolph Tegners Museum,
Dronningmølle 010036
Rudolph, Jette, Berlin 106412
Rudolph, Nicole, Zwickau 079153
Rudolph, Werner, Neuburg ... 077717
Rudomino – Bibliografičeskaja
Knižnaja Lavka, Moskva 143651
Rü-Bu Antik, Kirchhain 076804
Rue de Rosa, Oss 083528
Rue de Siam, Donville-les-Bains 068384
Rue Jadis Antiguidades, Rio de
Janeiro 064121
Rue Pouchkine, Strasbourg ... 105697
Rueb, Amsterdam 082844
Rübeländer Tropfsteinhöhlenmuseum,
Elbingerode, Harz 018151
Rübesams Da Capo Oldtimermuseum,
Leipzig 020012
Rübner, Hans-Georg, Aschau im
Chiemgau 074442
La Rueca, Buenos Aires 127534
Rückert-Christ, G., Grünwald .. 107180
Rüder, Bielefeld 074990
Rüdiger Belter Kunstvermittlung,
München 108186
La Ruée Vers l'Art, Perthes ... 072112

Rüesch, Marianne, Zürich 133999
Ruef, Hugo, München 126065
Rügen-Museum in der Kulturkunststatt
Prora, Prora 021232
Rügener Puppen- und
Spielzeugmuseum, Putbus .. 021238
Rüger, Wolfgang, Frankfurt am
Main 141788
Ruehl, Adrian, New York 095707
Rühmling, M., Kampen 076675
Ruellan, Jack-Philippe, Vannes
(Morbihan) 125830
Ruelle, Jean-Luc, Creil 068231
Rue's Antiques, Austin 092267
Rüstkammer, Staatliche
Kunstsammlungen Dresden,
Dresden 018005
Rüter, Franz-Jürgen, Oberhausen 077864
Rüter, Volker, Hannover 076390
Rüthemann, Roland, Hinwil 087525,
133769
Rüttger, Goldbach 129914
Ruetz, Barbara, Augsburg 106033
Ruetz, Dr. Karl, Basel 144031
Rueylong Museum, Cionglin ... 042294
Rufener, Caroline, Winterthur .. 144154
Rufer, San Antonio 097028
Ruff and Ready, Washington ... 097859
Ruff, Oskar, Stuttgart 078628
Ruffini, Romedio, Fällanden ... 087421
Rufin, Sylvie, Versailles 105846
Rufini, Perugia 081144
Rug and Roll, New York 095708
Rug Expo, Houston 093783
Rug Gallery, Cincinnati 092891
Rug Gallery, Leicester 089793
Rug Gallery, Tucson ... 097691, 136422
Ruga, A., Padova 131674
Rugao Museum, Rugao 008210
Rugby Art Gallery and Museum,
Rugby 045648
Rugby Decorative and Fine Arts
Society, Daventry 060231
Rugby School Museum, Rugby .. 045649
Rugeley Antique Centre, Brereton 088592
Ruggeri, Alessandro, Genova ... 080289
Ruggles House, Columbia Falls .. 047921
Rugs and Art, Los Angeles 094462
Rugs of the World, Tampa 097586
Ruh, Bernhard, Baden-Baden ... 074641
Ruhl, Helmut, Neustadt, Hessen 077768
Ruhlaer Uhrenmuseum, Ruhla ... 021542
Ruhmeshalle und Bavaria, Bayerische
Verwaltung der staatlichen Schlösser,
Gärten und Seen, München ... 020566
Ruhnu Muuseum, Ruhnu 010658
Ruhr-Rhein-Antiquariat, Mülheim an
der Ruhr 142136
Ruhrlandmuseum Essen, Essen .. 018311
Ruhrtalmuseum, Schwerte 021799
Ruichang Museum, Ruichang ... 008211
Ruicheng Museum, Ruicheng ... 008212
Ruichengzhai Biao Hua Jin Xia
Cooperative, Beijing 101989
Ruimte Morguen, Antwerpen ... 100510,
136624
Ruinas del Tazumal, Chalchuapa 010589
Ruine der Künste Berlin e.V.,
Berlin 106413
Ruïne van Brederode, Santpoort 032840
Ruis, Michel, Saint-Ouen 073188
Ruiskumestarin Talo, Helsingin
Kaupunginmuseo, Helsinki ... 010859
Ruitenberg, Jan, Zwolle 083738
Ruiz-Clavijo Fernandez, Palmira,
Barcelona 133159
Ruiz Linares, Granada 086042
Ruiz Linares, C.B., Granada 086043
Ruiz Linares, Carmen, Granada . 086044
Ruiz Linares, Hijos de, Granada 086045
Ruiz Manfredi, Lazaro, Sevilla .. 086474
Ruiz-Vernaca Restauradores,
Madrid 133384

Ruiz, Dominique, Saint-Ouen ... 073189
Ruiz, J., Madrid 086256
Rüjienas Izstāžu Zāle, Rūjiena .. 030227
Rukaj, Nikola, Toronto 101700
Rukmini Devi College of Fine
Arts, Kalakshetra Foundation,
Tiruvanmiyur 055841
Rukmini Devi Museum,
Tiruvanmiyur 024498
Ruks, Helmut, Potsdam 130537
Rule, Denver 120766
Rule & Cale, Windsor 091879
Rullier, Guillaume, Liguge 069468
Rullier, Michel, Paris 071865
Rullier, Michel, Poitiers 072225
Rullon, Jean-Michel, Tours 073861
Rulon-Miller, R., Saint Paul 145503
Rumänisches Institut, Freiburg im
Breisgau 018487
Rumale, Bangalore 109265
Rumble, Simon & Penny,
Chittering 088898
Rumbler, Helmut H., Frankfurt am
Main 141789
Rumbullion, Kamakura 111271
Rumelihisarı Müzesi, İstanbul Hisarlar
Müzeleri, İstanbul 042954
Rumer, B., Los Angeles 094463
Rumford Historical Museum, North
Woburn 051623
Rumiz, Bardwell Park 098709
Rumma, Lia, Milano 110419
Rumma, Lia, Napoli 110542
Rumners Wobble, Kitchener ... 064340
Rumolo, Raffaele, Modena 131586,
131587
Rumours, Nashville 122099
Rump, Münster 142210
Rumpff, Tanya, Haarlem 112516
Rumšiškių Jono Aisčio (Jono
Aleksandravičiaus) Muziejus,
Rumšiškės 030457
Runako, Louisville 121668
Rundāles Pils Muzejs, Pilsrundāle 030169
Runde, Tecklenburg 078670
Rundetaarn, København 010194
Rundetårn Antik, København ... 065949
Rundfunkmuseum "Manfred von
Ardenne", Backnang 016694
Rundfunkmuseum der Stadt Fürth,
Fürth 018549
Rundfunkmuseum Schloß Brunn,
Emskirchen 018202
Rundgang Urchigs Terbil, Törbel 042063
Rundlet-May House, Portsmouth 052315
Rundlingsmuseum Wendlandhof-
Lübeln, Küsten 019823
Runebergin Antikvariaatti,
Helsinki 140717
Runebergs stugan, Jakobstad .. 010912
Runes Antikvariat, Jönköping .. 143869
Runeson, Västerås 133602
Runestone Museum, Alexandria 046380
Runge, Eferding 062552
Runge, Kiel 107554
Runge, Linz 062721
Runge, Renee, Zittau 079145
Runge, Wolfgang H., Wertheim . 078937
Rungehaus, Wolgast 022753
Runkewitz, Carmen, Berlin 129557
Runnels Gallery, Eastern New Mexico
University, Portales 052274
't Runnertje, Amsterdam 082845
Runnquist & Runnquist, Toronto 064748
Runogalleria, Helsinki 140718
Runstuck, M., Neunkirchen-
Seelscheid 077742
Runte, Carl, Wuppertal 079113, 130795,
142528
Ruocco, Gabriele, Napoli 080927
Ruokolahden Kotiseutumuseo,
Ruokolahti 011325
Ruoveden Kotiseutumuseo,
Ruovesi 011327

Rup Silver, Mumbai 079409
Rupert Museum, Stellenbosch .. 038837
Rupf, C., Waldheim 078840
Rupf, Christof, Döbel 129690
Rupieć, Szczecin 084677
Rupp, E. & G., Sankt Gallen .. 116617
Rupp, Lydia, Paris 071866
Ruppert, Bochum 106515
Ruppert, Engelbert, Eltville am
Rhein 075605
Ruppert, Tilo, Landau in der
Pfalz 107786
Rupprecht, Neumünster 077741
Rupprich, Emmendingen 075610
Rural Life Centre, Old Kiln Museum,
Farnham 044336
Rural Life Museum and Windrush
Gardens, Baton Rouge 046821
Rus-Teak, Toronto 064749
Rusant, Sankt-Peterburg 085264
Rusch, Günter, Hamburg 076302
Rusche, Egon, Oelde 077896
Ruschman, Indianapolis 121217
Rush & Wicker, Cane, Virginia
Beach 136447
Rush Arts Gallery, New York ... 123028
Rush County Historical Society
Museum, Rushville 052704
Rush, John, Hamilton 140350
Rushcutters Bay Gallery, Rushcutters
Bay 099659
Rushen Abbey, Manx National
Heritage, Ballasalla 043472
Rushton, Malcolm, London 090468,
134761
Rushworth Museum, Rushworth 001465
Rusk County Historical Society
Museum, Ladysmith 050056
Ruskin, Stow-on-the-Wold 091480
Ruskin Library, Lancaster 044756
Ruskin Museum, Coniston 044002
Ruskin School of Drawing and Fine
Art, University of Oxford, Oxford 056592
Ruskon Kotiseutumuseo ja Rindellin
Mäkitupamuseo, Rusko 011328
Russ, Bissee 075006
Ruß, Dr. Wolfgang, Langen,
Hessen 142047
Russ, Paul, Waterford 127490
Ruß, Sibylle, Bamberg 129442
Russakoff, J., Philadelphia 145418
Russeck, New York 123029
Russeck, San Francisco 124381
Russell, Austin 119696
Russell, Inglewood 061650
Russell, London 118554, 144681
Russell, New Orleans 135865
Russell Cave Museum,
Bridgeport 047198
Russell-Clarke, Peter, Drysdale . 098982
Russell-Cotes Art Gallery and Museum,
Bournemouth 043682
Russell Court Studios, Salcombe 119049
Russell Gallery, Peterborough .. 006642
Russell Hill Rogers Gallery, Southwest
School of Art and Craft, San
Antonio 052943
Russell Museum, Russell 033300
Russell, Beverly, Montréal 064467
Russell, Brightwalls, Leominster 126941
Russell, Davis, Queenscliff 099605
Russell, Jack, Chipping Sodbury 117585
Russell, Jean, New York 095709
Russell, John, Ipswich 118026
Russell, M. & J., Yazor 091967
Russell, Sarah, Bath .. 088400, 144233
Russell, Sigi, New Orleans 095159
Russian Academy of Arts, Rossijskaja
Akademija Chudožestv, Moskva 056287
Russian Art Gallery, Portland ... 123671,
123672
Russian Creations, Chicago ... 120260
Russian Hill Antiques, San
Francisco 097319

Russian Hill Bookstore, San
Francisco 145566
Russian Painting Collectors Club,
Sankt-Peterburg 059776
The Russian Place, Chicago ... 092787
Russkaja Ėmal', Moskva 085201
Russkaja Galereja na Vozdviženke,
Moskva 114299
Russkaja Gallereja, Milano 080704
Russkaja Ikona, Moskva 085202
Russkaja Restavracija, Sankt-
Peterburg 133024
Russkaja Starina, Sankt-
Peterburg 085265
Russkaja Usadba, Moskva 085203
Russkie Galerei Sovremennoj Živopisi,
Moskva 114300
Russkie Kraski, Sankt-Peterburg 114358
Russkij Gosudarstvennyj Muzej Arktiki
i Antarktiki, Sankt-Peterburg .. 037817
Russkij Modern, Sankt-Peterburg 133025
Russkij Muzej Fotografii, Nižnij
Novgorod 037564
Russkoe Iskusstvo, Sankt-
Peterburg 114359
Russkoe Pole, Moskva 114301
Russo, La Coucourde 069013
Russo, Nürnberg 130472
Russo, Aldo, Milano 131520
Russo, Donatella, Torino 132196
Russo, Felice, Milano 131521
Russo, Francesco Paolo, Palermo 131711
Russo, Gianni, Milano 080705
Russo, Isabelle, Unter-
Tullnerbach 127890
Russo, John, Los Angeles 121585
Russo, Laura, Portland 123673
Russo, Roberto, Roma 142836
Russo, Rosa, Prora 108416
Russo, Vittorio, Napoli 131639
Russow, Roland, Berlin 074904
Rust en Vreugd Museum, Iziko
Museums of Cape Town, Cape
Town 038683
Rust 'n Dust, Atlanta 092180
Rust, Iris, Köln 076903
Rust, Yvonne, Whangarei 113113
Rustam Mustafayev ad. Azerbaycan
Dövlət İncəsənət Muzeyi, Bakı . 003176
Rusterholtz, Basel 116096, 126709
Rustic, Dunedin 083932
Rustic and Industrial Antiques
Warehouse, Newtown, New South
Wales 061965
Rustic Good Looks, Lockhart .. 061746
Rustic Rose Antiques, Loch 061745
Rustic Steel Creations, Tampa .. 124671
Rustica Antiques, Calgary 064270
Rustica Antiques, Los Angeles . 094464
Rusticana, Küsnacht 087550
Rusticas del Noroeste,
Salamanca 086414
Rusticas y Antigüedades El Roble,
Bogotá 065224
Rustico, Zaragoza .. 086604, 133509
Rustington Heritage Exhibition Centre,
Rustington 045652
Rustinoff, Vladimir, Saint-Paul-de-
Vence 105599
Rustler, Adolf, Düsseldorf 129758
Rusty Relics Museum, Carlyle .. 005567
Rusų Meno, Vilnius 111809
Rutberg, Jack, Los Angeles ... 131586
Rutgers Center for Innovative Print
and Paper, Mason Gross School of
the Arts, New Brunswick 138367
Ruth & Earl's Antiques, Baltimore 092360
Ruth and Carles Gilb Arcadia Historical
Museum, Arcadia 046518
Ruth Chandler Williamson Gallery,
Scripps College, Claremont ... 047760
Ruth Eckerd Hall, Clearwater .. 047795
Ruth Hall Museum of Paleontology,
Abiquiu 046313

Ruth Youth Wing, The Israel Museum,
  Jerusalem 025102
Ruth, Peter, Bad Schönborn . . . 129423
Ruthenberg von Klein, Corinna,
  Berlin 129558
Rutherford, Melbourne 061832
Rutherford, Nelson . . . 084021, 112965
Rutherford B. Hayes Presidential
  Center, Fremont 048976
Rutherford House, Edmonton . . . 005761
Rutherglen and District Artists Society,
  Rutherglen 058218
Rutherglen and District Artists
  Society – Members Gallery,
  Rutherglen 099660
Ruthie, Miami 094797
Ruthin Goal, Ruthin 045653
Ruthmere House Museum,
  Elkhart 048535
Ruth's, San Diego 097130
Ruthsatz, Bückeburg 075210
Ruthven Park, Cayuga 005584
Rutland Antiques Centre,
  Uppingham 091695
Rutland Cottage, Christchurch . . 083893
Rutland County Museum,
  Oakham 045430
Rutland Railway Museum,
  Cottesmore 044013
Rutten, Paul, Amsterdam 082846
Ruttleigh, Philip A., Crudwell . . 089019
Ruttmann, Friedrich, Frankfurt am
  Main 075818
Rutz, Maria, Düsseldorf 075512
Ruud, Jon, Mysen 143391
Ruud, Jon, Oslo 143415
Ruunnmylly, Hämeenlinna 010791
Ruurd Wiersma Hûs, Burdaard . 032120
Ruusmäe Koduloomuuseum,
  Ruusmäe 010659
Ruys, T., Almere 082646
Ruyter, Lisa, New York 123030
Ruyter, Pieter de, Delft 132468
Růžička, Jindřich, Praha 128365
Ruzicska, Nikolaus, Salzburg . . 100175
Ruzza, Aureana, Padova 081004
RV/MH Heritage Foundation,
  Elkhart 048536
RW Fine Arts Collection,
  Winterthur 116723
Ryan, Las Vegas 121376
Ryan, London 118555
Ryan Fine Arts Center, Abilene . 034513
Ryan-Wood, Liverpool . 089859, 134568
Ryan, Glenroy & Violet H.,
  Tucson 097692
Ryan, Karlene K., Fresno 120906
Ryan, Mary, New York 123031
Ryan, Peter K., Windsor, Ontario 064906
Ryan, William Henry, Tulsa 124814
Ryan's All Glass, Cincinnati . . . 092892,
  135451
Rybár, Antonín, Karlovy Vary . . 065403
Rybinskij Gosudarstvennyj Istoriko-
  architekturnyj i Chudožestvennyj
  Muzej-zapovednik, Rybinsk . . 037700
Rybski, John & Mary, Chicago . 145057
Ryckeboer, Danièle, Wimereux . 074299
Rycklösa, Linghem 126622
Rydal Mount, Ambleside 043414
Rydals Museum, Rydal 040925
Ryde Bookshop, Ryde 144806
The Ryde Gallery, Ryde 118994
Ryder Street Gallery, London . . 118556
Ryder, Georgina, Frampton . . . 089306,
  134396
Rye Art Gallery, Rye 045656
Rye Castle Museum, Rye 045657
Rye Heritage Centre Town Model,
  Rye 045658
The Rye Historical Society and Square
  House Museum, Rye 052717
Ryecliff Antiques, Clayfield 061339
Ryedale Folk Museum, Hutton-le-

Hole 044621
Ryerson, Tampa 124672
Ryerson Gallery, Toronto 101701
Ryes Antikvariat/Fredericia Antikvariat,
  Fredericia 140565
Ryesgades Möntgalleri, Århus . . 065607
Ryfylkemuseet, Sand 033918
Rygg, Kari, Bergen 084223
Rygge Museum, Larkollen 033731
Rygnestadtunet, Rysstad 033912
Ryhope Engines Museum,
  Sunderland 045924
Rymättylän Kotiseutumuseo,
  Rymättylä 011329
Rynek Sztuki, Łódź 113681
Ryös Antikvariat, Stockholm . . . 143964
Ryoma's Hometown Museum,
  Kochi 028861
Ryosan-cho Rekishi Hakubutsukan,
  Kyoto 028979
Ryotsu Folk Museum, Ryotsu . . 029288
Ryrie Street Art Gallery, Geelong 099074
Ryšių Istorijos Muziejus, Kaunas 030374
Ryton Galeria, Sopot 113763
Rytter, Horsens 065799
Ryu, Tokyo 111488
Ryuga-Do Cave Museum,
  Tosayamada 029631
Ryukodo, Yokohama 082231
Ryusaido, Kyoto 111301
Ryushi Kinenkan, Tokyo 029578
Ryuzen, Delhi 109339
Ryuzendo, Kyoto 082037
Ryvarden Galleri, Sveio 113279
rzeczy: o dizajnie, Kraków 138988
S & S Firearms, Fort Worth . . . 093478
S & S International, Amberg . . . 074403
S-D Coins, Oakland 095912
S-L-1, Milano 110420
S. & G. Antiques, London 090469
S. Merkurov House-Museum,
  Gyumri 000706
S. Ray Miller Auto Museum,
  Elkhart 048537
S.A.G.E.M., Angoulême 066630
S.A.I.E. Editrice, Torino 137750
S.A.R. Restauro, Firenze 131259
S.Art, Moskva 114302
S.C.R.A.P. Gallery, Indio 049700
S.D. Galerie, Paris 071867
S.D.A.C., Paris 136981
S.E.I. 1983, Roma 132040
S.E.M.A., Saint-Ouen 073190
S.E.P.A. – Site Expérimental Pratiques
  Artistiques, Dr., Rennes (Ille-et-
  Vilaine) 105442
S.E.R.O.D., Paris 129138
S.F.E, Saint-Lunaire 072903
S.H. Ervin Gallery, Sydney 001549
S.I.Ag., Firenze 110067
S.I.R.A., Roma 132041
S.N.A.B., Paris 136982
S.N.C. de Corbez, Saint-Germain-les-
  Corbeil 058666
S.N.E.C., Annonay 066658
S.O.G.A.C., Marseille 069964
S.P. Fine Arts, Dallas 120656
S.R. Furnishing and Antiques,
  Halesowen 089408
S.S. Meteor Maritime Museum,
  Barkers Island 046790
S.S. Moyie National Historic Site,
  Kaslo 006067
S.T.A.C., Montesson 070240
S.W. Antiques, Pershore 090974
S2 Art Center and Atelier, Las
  Vegas 121377
S2 Art Gallery, Chicago 120261
S2 Art Group, Chicago 120262
SA Breweries World of Beer,
  Johannesburg 038757
Sa Costa, Palma de Mallorca . . 086365
Sa Nostra, Ciutadella de
  Menorca 115024

Sa Nostra, Eivissa 115061
Sa, Hangang, Seoul 082264
SA3 Art Gallery, Seregno 110925
Saab Museum, Trollhättans Museum,
  Trollhättan 041091
Ibn Al-Saad, Dubai 088183
Saad Zaghlul Museum, Munira . 010581
Saad, Ahmad, Beirut 082303
SAAF Museum Swartkop,
  Tshwane 038857
SAAF Museum Ysterplaat, South
  African Air Force, Ysterplaat . . 038881
Saal, Klaus, Ludwigshafen am
  Rhein 107901, 137357
Saale-Galerie, Saalfeld/Saale . . 108539
Saalfelder Feengrotten, Saalfeld/
  Saale 021552
Saam, Ludwig, Kirchheim unter
  Teck 076809
Saamelaismuseo Siida, Inari . . 010904
Saanich Historical Artifacts Society,
  Saanichton 006837
Saar, Martin, Zusmarshausen . . 130804
Saaremaa Muuseum, Kuressaare 010631
Saarijärven Museo, Saarijärvi . . 011331
Saarländische Galerie – Europäisches
  Kunstforum, Berlin 106414
Saarländischer Museumsverband e.V.,
  Ottweiler 059227
Saarländisches Bergbaumuseum,
  Bexbach 017355
Saarländisches Künstlerhaus e.V.,
  Saarbrücken 021559
Saarländisches Schulmuseum,
  Ottweiler 021053
Saarländisches Uhrenmuseum,
  Püttlingen 021234
Saarländisches Zweiradmuseum,
  Wadgassen 022364
Saarlandmuseum, Stiftung
  Saarländischer Kulturbesitz,
  Saarbrücken 021560
Saaser Museum, Saas Fee 041901
Saatchi Gallery, London 045098
Saatkamp, M., Lengerich 077083
Saavedra, Víctor, Barcelona . . . 114957
Saba-Arifa, Bayima, Mühlheim am
  Main 077430
Saba, Joël, Lille 128686
Sabah Art Gallery, Kota Kinabalu 030676
Sabama, Eivissa 085990
Sabancı Kültür Sitesi, İmit 042915
Sabâ's House, Tehrân 024782
Sabatelli, Franco, Milano 080706,
  131522
Sabathier-Habrial, Anne-Armelie,
  Bergnicourt 067106
Sabatier, Annie, Lille 140871
Sabatier, Eduard, Verden 078790,
  108817
Sabatier, Jean-Frédéric, Apt . . . 066694
Sabatier, Nadine, Saint-Ouen . . 073191
Sabatini, Buenos Aires 060927
Al-Sabbar, Dubai 088184
Sabelli, Fabrizio, Bolzano 131010
Saber ver, Arte y Recreación para
  toda la Familia, Miguel Hidalgo 138920
Saber ver, Lo contemporáneo del arte,
  Miguel Hidalgo 138921
Sabera, London 090470
Sabet, Berlin 074905
Sabet, Los Angeles 094465
Sabiac, Pierre, Les Halles 069410
Sabiedriba, Rīga 111765
Sabin, Frank T., London 090471, 118557
Sabina Antiquitäten, Essen . . . 075695
Sabina's Gallery, Singapore . . . 085493
Sabine, C.E., Stock . . 091429, 135084
Sable, Los Angeles 094466
Sable, V., New Orleans 122233
Le Sablier, Bruxelles 063546
Sablier, Montréal 064468
Sablon, Essen 075696
Sablone, Fabrizio, Roma 132042

Sablonier, Jacqueline, Schaan . . 082336
Saborit, Eugenia, Málaga 086307
Sabot, Henri, Bordeaux 067357
Sabot, Henri, Podensac 072212
Sabots de Bois, La Bourdiniere-Saint-
  Loup 068956
Saboudjian, Lucie, Paris 071868
Sabourin, Châtellerault 125554
Saby, Lazi, Stuttgart 108726
Saçaklı Ev – Kültür ve Sanat Merkezi,
  Lefkoşa 033475
Sacao Artes Utilidades, Fortaleza 100956
Sacchet, Anna, Modena 080818
Sacchetti, Giuseppe, Roma 081542
Sacchetti, Jensen Giovanna,
  Ascona 116036
Sacchi, Mariangela, Milano . . . 131523
Sacchi, Paolo, Firenze 142684
Saccomani, Antonio, Verona . . . 081935
Saccotelli, Gaetana, Pisa 142787
Sacha, Ménerbes 070049
Sacha Gallery, Surry Hills 099752
Sacharow, Renate, Bad Neuenahr-
  Ahrweiler 074561
Sachem Historical Society Museum,
  Holbrook 049501
Sachs Harbour Museum, Sachs
  Harbour 006839
Sachs, A., Minden 108007
Sachs, Karin, München 108187
Sachse & Heinzelmann, Heinz-Walter
  Hebestreit, Hannover 141919
Sachse, Ruth, Hamburg 107312
Sachsenmaier, Henry, Kufstein . 062704
Sacilotto, Sergio, Prato 081176, 081177
Sack, Israel, New York 095710
Sackets Harbor Battlefield, Sackets
  Harbor 052719
Sackolah, Cincinnati 120373
Sacks, Beverly, New York 123032
Sacks, Kim, Johannesburg 114800
Sacksofsky, Uwe, Heidelberg . . 107397
Saco Museum, York Institute Museum,
  Saco 052720
Saco, Miguel, New York 095711, 135992
Sacramento Mountains Historical
  Museum, Cloudcroft 047848
Sacramento Museum Collection Center,
  Sacramento 052731
Sacramento Surplus Book Room,
  Sacramento 145492
Sacred Circle Gallery of American
  Indian Art, United Indians of All
  Tribes Foundation, Seattle . . . 053246
Sacred Earth Arts, Kansas City . 121314
Sacred Rose, Chicago 092788
Sacrewell Farm and Country Centre,
  Thornhaugh 045973
Sacro Convento Castillo de Calatrava
  la Nueva, Aldea del Rey 038937
Sa'dabad Museums, Tehrân . . . 024783
Al-Sadawi, Faisal Abdul Hamid,
  Riyadh 085324
Sadberk Hanım Müzesi, İstanbul 042955
Sadde, Philippe, Dijon 125570
The Saddle Room Antiques,
  Cookstown 088976
Saddlers Court Gallery, Richmond,
  Tasmania 099628
Saddleworth Museum and Art Gallery,
  Uppermill 046022
Sądecki Park Etnograficzny, Muzeum
  Okręgowego, Nowy Sącz 035052
Sadega-Wafa, Gisela, Kerpen . . 076764
Sadek-Rosshap, Monika, Paudorf 127831
Sadek-Rosshap, Monika, Wien . 127962
Sadelmakare Öbergs Hus, Solna 040979
Saden, Al-Khuwair 084397
Sadigh, New York 095712
Sadika/ Espace d'Art, Marsa
  Erriadh 116978
Sadler Street Gallery, Wells . . . 091773,
  119317
Sadler, Stanley J., Houston . . . 093784

Sadnikarjeva Muzejska Zbirka v
Kamniku, Kamnik 038523
Sado Hakabutsukan, Sawata . . . 029352
Sado Kokokan, Ogi 029207
Sado-koku Ogi Folk Museum,
Ogi 029208
Sado Nohgaku-no-sato, Ryotsu . 029289
Sadowski & Farchmin,
Ahrensfelde 074364
Sadowski & Farchmin, Bernau bei
Berlin 074958
Sadowski, Peter, Berlin 074906
Sadri, Oklahoma City . 096005, 136055
Sadriddin Aini Yodgorlik Muzeyi,
Samarqand 054646
Sadza, A., Neunkirchen-
Seelscheid 077743
Sääksmäen Kotiseutumuseo,
Sääksmäki 011332
Säätyläiskotiseutumuseo,
Kolkanlahti 011021
Sæby Museum and Sæbygaard Manor
Museum, Sæby 010299
Säbytorp Antik och Kuriosa, Kil . 086781
Sächsische Kunstwerk, Leipzig . 077066,
107849
Saechsische Kunstwerk GmbH,
Leipzig 107850
Sächsischer Künstlerbund e.V.,
Dresden 059228
Sächsischer Museumsbund e.V.,
Dresden 059229
Sächsisches Apothekenmuseum
Leipzig, Leipzig 020013
Sächsisches Auktionshaus und
Antiquariat, Leipzig . . 126026, 142063
Sächsisches Bademuseum Bad Elster,
Bad Elster 016757
Sächsisches Buch- und
Graphikkabinett, Radebeul . . . 142313
Sächsisches Eisenbahnmuseum,
Chemnitz 017739
Sächsisches Feuerwehrmuseum
Zeithain, Zeithain 022849
Sächsisches Industriemuseum
Chemnitz, Chemnitz 017740
Sächsisches Lehr- &
Besucherbergwerk-Himmelfahrt-
Fundgrube, TU Bergakademie
Freiberg, Freiberg, Sachsen . . 018473
Sächsisches Psychiatriemuseum,
Leipzig 020014
Sächsisches Schmalspurbahn-
Museum Rittersgrün,
Breitenbrunn 017577
Sächsisches Strafvollzugsmuseum,
Waldheim 022394
Säffle Antikt och Konst, Säffle . 086918
Säffle Marinmotor Museum,
Säffle 040927
Sägehammer Hofwies, Windhaag bei
Freistadt 003122
Sägemühle, Thallwitz 022141
Sägereimuseum-Alte Sagi,
Richterswil 041874
Sägmühlmuseum Marhördt,
Oberrot 020918
Saegusa, Tokyo 111489
Saeid, Aziz Mohammadi, Milano 080707
Säimenen Mylly, Rökönvaara . . . 011319
Säkylän Kotiseutumuseo, Säkylä 011336
Sämann, Wolfgang & Ursula, Frankfurt
am Main 141790
Saendig, Vaduz 137805, 142960
Sändig, Dr. Martin, Wiesbaden . 078993
Sändig, Elmar, Wiesbaden 142495
Sängermuseum, Feuchtwangen . 018359
Säntis, Gossau (Sankt Gallen) . 116355
Saenz de Maturana Aguirreurreta,
Maria Carmen, Vitoria-Gasteiz . 086578
Saenz, Arana & Bernhard,
Schaffhausen 087803
Saenz, Benjamin E., San Antonio 097029
Säräisniemen Kotiseutumuseo,

Säräisniemi 011337
Särestad Lantbruksmuseum,
Grästorp 040669
Särestöniemi-Museo, Kaukonen 010976
Särö Konst och Ram, Särö 115803
Säters Hembygdsmuseum, Säter 040930
Sætersgårds Samlinger Dølmotunet,
Tolga 034021
Saetta, Maria Letizia, Roma . . . 081543
Sättler, Harald, Erbach,
Odenwaldkreis 129797
Säyneisten Kotiseutumuseo,
Säyneinen 011338
Saez Angulo, Rosa, Santander . 086428
Saez Concha, Carlos A., Santiago de
Chile 128252
Al-Safa, Cairo 066268
Safani, New York 095713
Ramsey Safar & Osama Ismail
Abudawood, Jeddah 085306
Safar Khan Gallery, Cairo 103075
Safari Imports, Houston 121123
Safarof, Maryam, Heidelberg . . . 107398
Al-Safavi, Dubai 088185
Safe Haven Museum & Education
Center, Oswego 051840
Safety Harbor Museum of Regional
History, Safety Harbor 052736
Saffa-Janssen, Bruxelles 128035
Saffell, Michael & Jo, Bath 088401
Saffron Gallery, Saint Agnes . . 119001
Saffron Walden Antiques Centre,
Saffron Walden 091171
Saffron Walden Auctions, Saffron
Walden 127025
Saffron Walden Museum, Saffron
Walden 045661
Saffronart Gallery, Mumbai 109465
Saffronart Gallery, New York . . . 123033
Saffroy, Gaston, Paris 141275
Safi, Carcassonne 067678
Safier Heimatmuseum, Safien . . 041903
Safita, Dubai 088186
Safn Ásgríms Jónssonar,
Reykjavik 024108
Safn Rafmagnsveitu Reykjavíkur,
Reykjavik 024109
Safnahús Borgarfjardar,
Borgarnes 109223
Safnahús Skagfirðinga,
Borgarnes 109224
Safnahúsid Húsavík, Húsavík . . 024084
Safont Molinero, Maria Luisa,
Zaragoza 115561
Safor Galerie, Katzelsdorf, Leitha 100063
Safranek, Leogušen 109719
Sag Harbor Whaling and Historical
Museum, Sag Harbor 052738
Saga, Boulogne-Billancourt . . . 067395
Saga-kenritsu Bijutsukan, Saga . 029294
Saga-kenritsu Hakubutsukan,
Saga 029295
Saga-kenritsu Kyushu Toji Bunkakan,
Arita 028480
Saga-kenritsu Nagoya Jou
Hakubutsukan, Chinzei 028517
Sagalunds Museum, Kimito 011004
Sagamihara Citizen's Gallery,
Sagamihara 029298
Sagamihara-shi Hakubutsukan,
Sagamihara 029299
Sagamore Hill National Historic Site,
Oyster Bay 051877
Sagan, Irene, Essen 106929
Sagar Mahal Art Gallery, Mumbai 109466
Sagarana, Belo Horizonte 140271
Sagasco Historical Group Museum,
Brompton 000895
Sagasser, Dr. Jürgen, Amorbach 074410
Sagasta, Logroño 086080
Sagawa Bijutsukan, Moriyama . . 029057
Sage, New York 123034
Sage, Ripley, Surrey . 091114, 134969
Sagemonoya, Tokyo 082193

Sagep, Genova 137659
Sage's Quest, Buffalo 092553
Saget, Eric, Paris 071869
Saggi e Memorie di Storia dell'Arte,
Firenze 138872
Il Saggiatore, Roma 110854
Al-Saghah, Sheikh Usamah,
Damascus 088056
Saginaw Art Museum, Saginaw . 052740
Sagittaria, Pordenone 110679
Il Sagittario, Messina 110224
Sagmeister, Ulrich, Pitten 062766
Sagot-Le Garrec, S., Paris 071870,
105207
Sagrera, Miguel, Palma de
Mallorca 086366
Sagstua Skolemuseum, Sigurd Hoels
Hjem, Sagstua 033914
Saguache County Museum,
Saguache 052741
Säguddens Museum, Arvika . . . 040545
Saguer, Vigneux-sur-Seine 074122
Sahan Museo, Puumala 011287
Sahara Imports, Long Beach . . . 121409
Sahara Shoten, Yokohama 142950
Şahbuz District Regional History
Museum, Şahbuz 003197
Saheb, Alexandra, Berlin 106415
Sahitya Parishad Museum,
Midnapore 024398
Sahlender, Michael-Heinz,
Stegaurach 130655
Sahrai Tappeti Antichi, Milano . . 080708
Sahuc, Louis, New Orleans 122234
Sahuquet, Gerard, Bordeaux . . . 125531
Sahuquet, Philippe, Millau 070123
SAI, Legionowo 113654
Sai Artefacts, Singapore 085494
SAI Oriental Art Corporation, Las
Vegas 121378
Sai Swiss Auction International,
Berikon 126711
Said, Emami, Bonn 129612
Said, G., Hamrun 132336
Said, Mohamed, Bamberg 129443
Said, Nasser Bin Mohamed Bin,
Muttrah 084405
Said, Y., Berlin 074907
Saieva & Maneschi, Pisa 131768
Saigo Kinenkan, Kagoshima . . . 028749
Saigon Art, Winchester 119361
Saigon Art Gallery, Ho Chi Minh
City 125146
Saigon d'Art, Ho Chi Minh City . 125147
Saigon Fine Arts, Den Haag . . . 083129
Saigon Open City Studio Space, Ho
Chi Minh City 054807
Saijo-shi Minzokugaku Hakubutsukan,
Saijo 029301
Sailak Arkitektura, Euskal Herriko
Unibertsitatea, Leioa 056365
Sailer, Innsbruck . . . 062651, 100057
Sailor's Memorial Museum,
Islesboro 049734
Saimaan Kanavan Museo,
Lappeenranta 011094
Sainsbury & Mason, London . . . 090472
Sainsbury Centre for Visual Arts,
Norwich 045406
Sainsbury, John, Carlton, Victoria 139691
Sainsbury's, Camberwell 139681
Sainsburys of Bournemouth,
Bournemouth 088540
Sainsere, Jacqueline, Saint-Ouen 073192
Saint Albans Gallery, Saint
Albans 099661
Saint Albans Historical Museum, Saint
Albans 052742
Saint Albans Organ Museum, Saint
Albans 045670
Saint-André Antiquités, Saint-André-
lez-Lille 072672
Saint Andrews Cathedral Museum,
Saint Andrews 045675

Saint Andrews College Museum,
Christchurch 033176
Saint Andrews Fine Art, Saint
Andrews 091178, 119008
Saint Andrews Gallery, Hamilton 112905
Saint Andrews Museum, Saint
Andrews 045676
Saint Andrews Preservation Museum,
Saint Andrews 045677
Saint Andrews University Museum
Collections, University of Saint
Andrews, Saint Andrews 045678
Saint Andrews Workshop, Saint
Andrews 099662
Saint Angela's Museum, Prelate 006703
Saint Asaph Cathedral Treasury, Saint
Asaph 045680
Saint Augustin Art Association, Saint
Augustin 060719
Saint Augustine, Porto Alegre . . 063993
Saint Augustine Historical Society
Museum, Saint Augustine 052748
Saint Augustine Lighthouse and
Museum, Saint Augustine 052749
Saint Augustine's Abbey,
Canterbury 043860
Saint Barbe Museum and Art Gallery,
Lymington 045175
Saint Barnabas Icon and
Archaeological Museum,
Gazimağusa 033455
Saint Bartholomew's Hospital Museum,
London 045099
Saint Benedicts Studio,
Wallington 119299
Saint Bernadette Institute of Sacred
Art, Albuquerque 119508
Saint-Boniface Museum,
Winnipeg 007461
Saint Brannocks Gallery,
Mundesley 118767
Saint Breock Gallery, Wadebridge 091703,
135177
Saint Bride's Crypt Museum,
London 045100
Saint Catharines Museum, Saint
Catharines 006861
Saint Catherine's National School
Museum, Aughrim 024820
Saint Cecilia's Hall Museum of
Instruments, Edinburgh University
Collection of Historic Musical
Instruments, Edinburgh 044270
Saint Charles County Historical Society,
Saint Charles 052758
Saint Charles Heritage Center, Saint
Charles 052754
Saint Chrispin Antique Centre,
Beverley 088452
Saint-Christophe, Soissons . . . 073529
Saint-Clair Antiques, Toronto . . 064750
Saint Clair County Historical Museum,
Belleville 046880
Saint-Clar Galerie, Sarlat-la-
Canéda 105654
Saint Claude Museum, Claude . 005625
Saint Clements Antiques, Oxford 090937
Saint Clements Island-Potomac River
Museum, Colton Point 047895
Saint-Cornély Antiquités, Carnac 067689
Saint-Cricq Antic, Auch 066802
Saint David's Hall, Cardiff 043868
Saint-Dizier, Montréal 101449
Saint Dominic Church Museum,
Macau 008074
Saint Francis County Museum, Forrest
City 048776
Saint Francis Enterprise,
Singapore 085495
Saint Gaudens National Historic Site,
Cornish 047996
Saint George Art Museum, Saint
George 052769
Saint George Gallery, Boston . . . 119919

Saint George's, Stamford 091407
Le Saint-Georges, Toulouse 073797
Saint Georges Antiques, Battery
 Point 061108
Saint-Georges Antiquités Brocante,
 Lyon 069780
Saint George's Heritage Museum,
 Brigus 005482
Saint George's Historical Society
 Museum, Saint George's 004197
Saint Georges Regional Museum,
 Hurstville 001160
Saint George's United Methodist
 Church Museum, Philadelphia . 052093
Saint-Germain Antiquités,
 Ménerbes 070050
Saint-Germain Antiquités, Saint-
 Germain-en-Laye 072810
Saint Graal Gallery, Kraków 113636
Saint Helens, Swansea 135112
Saint Helens History Room, Saint
 Helens 001467
Le Saint-Hubert, Nancy 070430
Saint Ives Gallery, Saint Ives,
 Cornwall 119032
Saint Ives Museum, Saint Ives,
 Cornwall 045698
Saint Ives Society of Artists, Saint Ives,
 Cornwall 119033
Saint Ives Society of Artists
 Members Gallery (Norway Gallery)
 and Mariners Gallery, Saint Ives,
 Cornwall 045699
Saint-Jacques, Pierre, Montréal 064469
Saint James Antiques, Little
 Malvern 089847
Saint James Cavalier, Centre for
 Creativity, Valletta 030756
Saint James' Gallery, Uitikon
 Waldegg 116674
Saint James' House,
 Fredericksburg 048961
Saint James House Antiques,
 Boroughbridge 088517, 134125
Saint James Museum and Library,
 Jerusalem 025103
Saint James's Art Books, London 144682
Saint James's Gallery, Bath 117313
Saint James's Gallery, Saint Peter
 Port 091208
Saint-Job Auctions, Uccle 125357
Saint John and Coningsby Medieval
 Museum, Hereford 044576
Saint John Antiques, York 091981
Saint John Jewish Historical Museum,
 Saint John 006894
Saint John Sports Hall of Fame, Saint
 John 006895
Saint John Street Gallery,
 Ashbourne 117256
Saint John's Church and Parish
 Museum, Hampton 049343
Saint John's Co-Cathedral and
 Museum, Valletta 030757
Saint John's College Museum,
 University of Manitoba,
 Winnipeg 007462
Saint John's House, Warwick .. 046060
Saint John's Northwestern Military
 Academy Museum, Delafield .. 048228
Saint John's Schoolhouse Museum,
 Reid 001447
Saint John's Theatre and Arts Centre,
 Listowel 024958
Saint Johnsbury Athenaeum, Saint
 Johnsbury 052778
Saint Joseph and the Angels
 Emporium, Edmonton 140345
Saint Joseph College Art Gallery, West
 Hartford 054304
Saint Joseph Island Museum Village,
 Richard's Landing 006789
Saint Joseph Museum, Saint
 Joseph 052788

Saint Joseph's College Museum,
 Tiruchirappalli 024494
Saint Joseph's University Gallery,
 Philadelphia 052094
Saint Lambert, Paris 141276
Saint-Laurent + Hill, Ottawa ... 101479
Saint Lawrence College Art Gallery,
 Kingston 006113
Saint Lawrence Miner's Museum,
 Saint Lawrence 006918
Saint Leonards Antique and Craft,
 Newton Abbot 090839, 134884
Saint Lewis Heritage Museum, Saint
 Lewis 006920
Saint-Louis, Lorient 104047
Saint Louis Architectural Art, Saint
 Louis 096770
Saint Louis Art Museum, Saint
 Louis 052818
Saint Louis Art Museum. Bulletin,
 Saint Louis 139487
Saint Louis Art Prints, Saint
 Louis 123866
Saint Louis Artists' Guild, Saint
 Louis 060720
Saint Louis Artists' Guild Museum,
 Saint Louis 052819
Saint Louis Car Museum, Saint
 Louis 052820
Saint Louis County Historical Museum,
 Duluth 048386
Saint Louis Irish Arts, Saint Louis 123867
Saint-Louis Posters, Paris 105208
Saint Louis Science Center, Saint
 Louis 052821
Saint Louis University Museum of Arts
 and Culture, Baguio 034433
Saint-Loup, Namur 125352
Saint-Loup, Saint-Honoré-les-
 Bains 072833
Saint Lucie County Historical Museum,
 Fort Pierce 048857
Saint Luis University, Louisiana . 050484
Saint Luke's Gallery, Washington 124956
Saint Luke's Gallery, Anna Maria
 College, Paxton 051955
Saint Mamas Manastırı İkon Müzesi,
 Güzelyurt 033463
Saint Margaret's Museum, Saint
 Margaret's Bay 045703
Saint-Martin, Paris 071871
Saint Martins Antiques Centre,
 Stamford ...091408, 135074, 144860
Saint Martins Books,
 Bournemouth 144257
Saint Mary Magdalene Treasury,
 Newark-on-Trent 045335
Saint Marys and Benzinger Township
 Museum, Saint Marys 052830
Saint Mary's Chapel Antiques,
 Devizes 089057
Saint Marys Community Arts Centre,
 Saint Marys 001478
Saint Marys Museum, Saint
 Marys 006926
Saint Mary's Museum of Maxstone,
 Maxstone 006278
Saint Mary's University Art Center, San
 Antonio 052944
Saint Mary's University Art Gallery,
 Halifax 005963
Saint Mawes Castle, Truro 046002
Saint-Max Antiquités, Saint-Maximin-
 la-Sainte-Baume 072973
Saint Mel's Diocesan Museum,
 Longford 024960
Saint Michael's Loft Museum,
 Christchurch 043958
Saint Michael's Museum,
 Miramichi 006332
Saint Michel, Toulouse 073798
Saint Mullins Heritage Centre, Saint
 Mullins 024986
Saint Mungo Museum of Religious Life

 and Art, Glasgow 044418
Saint Neots Museum, Saint
 Neots 045706
Saint Nicholas Galleries, Carlisle 088794
Saint Nicholas Markets, Bristol . 088668
Saint-Nicolas, Châteaubriand .. 067878
Saint Norbert Arts Centre,
 Winnipeg 007463
Saint Norbert Provincial Heritage Park,
 Winnipeg 007464
Saint Ouen, Puckeridge 091059, 134955
Saint Pancras Antiques,
 Chichester 088878, 134258
Saint Patrick's Church Museum, Cape
 Breton Centre for Heritage and
 Science, Sydney 007116
Saint Patrick's Trian, Armagh . 043435
Saint-Paul, Saint-Félix-Lauragais 072784
Saint Paul Apostolat des Editions,
 Lyon 136831
Saint Paul Gallery, Saint Paul .. 123929
Saint Paul Museum, Randleman 052462
Saint-Paul, Eric, Toulouse 073799
Saint Pauls Gallery, Birmingham 117380
Saint Paul's Street, Stamford . 144861
Saint-Père, Paris 105209
Saint Peter Hungate Church Museum,
 Norwich 045407
Saint Peter's Church, Sunderland 045925
Saint Peter's College Art Gallery,
 Jersey City 049820
Saint Petersbourg Antiquités,
 Paris 071872, 141277
Saint Petersburg Gallery,
 Melbourne 061833
Saint Petersburg Museum of History,
 Saint Petersburg 052864
Saint Philips Books, Oxford ... 144772
Saint Photios Greek Orthodox National
 Shrine, Saint Augustine 052750
Saint-Remy, New York 095714
Saint Robert's Cave,
 Knaresborough 044739
Saint Seine, Isabelle de, Paris . 071873
Saint Sharbel Museum, Annaya 030261
Saint-Simond, Aix-les-Bains .. 066485
Saint Simons Island Lighthouse
 Museum, Saint Simons Island . 052871
Saint Sophia of Kiev, Kyïv 043224
Saint Tammany Art Association,
 Covington 060721
Saint Tammany Art Gallery,
 Covington 048044
Saint Theresa's College Folklife
 Museum, Cebu 034463
Saint Thomas-Elgin Public Art Centre,
 Saint Thomas 006938
Saint Thomas' Hospital Mini Museum,
 London 045101
Saint Thomas More Art Gallery,
 Saskatoon 006982
Saint Tropez Gallery, Surfers
 Paradise 099740
Saint-Viance Antiquités, Saint-
 Viance 073326
Saint Vigeans Museum, Arbroath 043428
Saint-Vincent, Châtillon-sur-
 Chalaronne 067920
Saint-Vincent Antiquités, Chalon-sur-
 Saône 067784
Saint Vincent Church Museum,
 Isandlwana 038738
Saint Vital Historial Society Museum,
 Winnipeg 007465
Saint Volodymyr Ukrainian Catholic
 Centre Museum, Winnipeg ... 007466
Saint Walburg and District Historical
 Museum, Saint Walburg 006945
Saint, Philippe, Bruxelles 100656
St. Martin's Press, New York ... 138423
Sainté Anciens ets Bacqueville,
 Argenteuil 066715
Sainte-Anne, Armentières 066726,
                              128447

Sainte-Anne Antiquités,
 Peymeinade 072124
La Sainte-Aubaine, Saint-Aubin-sur-
 Mer 072679
Sainte Genevieve Museum, Sainte
 Genevieve 052874
Sainte-Hélène, Six-Fours-les-
 Plages 073526
Sainte Maison de la Flocellière, La
 Flocellière 013417
Sainte-Marie Among the Hurons,
 Midland 006306
Sainte Marie among the Iroquois,
 Liverpool 050342
Sainteff, Denis, Strasbourg 073598
Saintfield Antiques & Fine Books,
 Saintfield 091216, 144821
Saints and Martyrs, Albuquerque 092081
Saints Gallery, Sans Souci 099677
Sainty Antique, Monetay-sur-
 Allier 070161
Sainz López, Carlos, Zaragoza . 133510
Sairaalamuseo, Lahti 011079
Saire, Cusco 084423
La Saison des Pluies, Liège 063780
Saison, Raymond, Antibes-Juan-les-
 Pins 066679
De Saisset Museum, Santa Clara
 University, Santa Clara 053096
Sait Faik Müzesi, Burgaz Adası . 042857
Saitama-kenritsu Hakubutsukan,
 Saitama 029304
Saitama-kenritsu Kindai Bijutsukan,
 Saitama 029305
Saitama-kenritsu Sakitama Shiryokan,
 Gyoda 028575
Saitama-kenritsu Shizenshi
 Hakubutsukan, Nagatoro 029106
Saitama-shiritsu Manga Bijutsukan,
 Saitama 029306
Saito, Yokohama 142951
Saito Ho-on Kai Shizenshi
 Hakubutsukan, Sendai 029359
Saizowa, Jörg, Frankfurt (Oder) . 075765
Saja, Jeddah 085307
Sajam Umjetnosti i Umjetničke
 Opreme, Sarajevo 098027
Sajanskaja Kartinnaja Galereja,
 Sajansk 037701
Sajot, Thérèse, Touques 073812
SAK Galerie, Amsterdam 112277
Sakae, Hayama 081969
Sakai, New York 095715
Sakai, Yokohama 111543
Sakai Antique Fair, Sakai 098233
Sakai Greenery Museum Harvest Hill,
 Sakai 029313
Sakai House, Hirano Hakubutsukan,
 Osaka 029269
Sakai Kokodo, Yokohama 082232
Sakai-shi Hakubutsukan, Sakai . 029314
Sakai-shiritsu Hamono Hakubutsukan,
 Sakai 029315
Sakaide-shi Bijutsukan, Sakaide 029317
Sakaimachi Art Gallery, Kyoto .. 111302
Sakakini, Hamburg 076303
Sakalhe's Cultural Art Museum,
 Majia 042393
Sakamoto Ryoma Kinenkan,
 Kochi 028862
Sakarya Müzesi, Sakarya 043046
Sakata-shiritsu Bijutsukan,
 Sakata 029321
Sakata, B. Garo, Sacramento .. 123780
Sake, Richmond 123742
Sake Museum, Sakata 029322
Saker och Ting, Strängnäs 087084
Sakıp Sabancı Müzesi, Sabancı
 Üniversitesi, İstanbul 042956
Sakkmúzeum, Heves 023687
Sakr, Charles, Paris 071874
Sakrale Kunst in der Sankt-Anna-
 Kapelle, Cham 017727
Sakralna Baština Muzej, Senj .. 008985

Saks, Denver 120767
Saks, Pune 109499
Sakshi, Bangalore 109266
Sakshi, Mumbai 109467
Saksische en Museumboerderij Erve
  Brooks Niehof, Gelselaar 032338
Saku-shiritsu Bijutsukan, Saku . 029323
Sakura-shiritsu Bijutsukan,
  Sakura 029326
Sakura-shiritsu Kampo Arai Kinenkan,
  Ujiie 029678
Sakuraya, Tokyo 082194
Sala 1 – Centro Culturale, Roma 110855
Sala A. De Carolis, Montefiore
  dell'Aso 026938
Sala Arte de Porto Alegre, Porto
  Alegre 101001
Sala Auktionskammare, Sala .. 126658
Sala Braulio, Castello de la Plana 115020
Sala Cadafe, Caracas 054739
Sala Cister, Málaga 115307
Sala Cristina, Milano 110421
Sala Dalmau, Barcelona 114958
Sala d'Arqueologia i Gabinet
  Numismàtic, Fundació Pública
  Institut d'Estudis Ilerdencs,
  Lleida 039584
Sala d'Art Arimany, Grup d'Art Escolà,
  Tarragona 115465
Sala de Arqueología del Municipio de
  Mújica, Múgica 031314
Sala de Arqueología e Pré-Historia,
  Faculdade de Ciências da Univsidade
  do Porto, Porto 035958
Sala de Arte Carlos F. Sáez,
  Montevideo 046286
Sala de Arte Cuchilleros, Madrid 115281
Sala de Arte Jabalcuz, Jaén ... 115101
Sala de Arte Murillo, Oviedo ... 115328
Sala de Arte Prehispánico, Santo
  Domingo 010403
Sala de Arte Zeus, Zaragoza .. 086605,
                              115562
Sala de Cultura Carlos III,
  Pamplona 115393
Sala de Espera, Bogotá 102364
Sala de Exhibitiones Temporales
  del Banco Central, Museos del
  Banco Central de Costa Rica, San
  José 008864
Sala de Exposiciones, Barakaldo 039050
Sala de Exposiciones Conde Duque,
  Madrid 115282
Sala de Exposiciones de la
  Universidad del Pacífico, Lima 034371
Sala de Exposiciones de las Alhajas,
  Fundación Caja Madrid, Madrid 039691
Sala de Exposiciones Edificio Historico,
  Universidad de Oviedo, Oviedo 039864
Sala de Exposiciones Rekalde,
  Bilbao 039149
Sala de los Bomberos, La
  Habana 009216
Sala de Memorias Culturales,
  Mogotes 008727
Sala de Memórias de Chapada
  dos Guimarães, Chapada dos
  Guimarães 004411
Sala de Recuerdos de la Escuela
  Militar de Montaña, Jaca ... 039521
Sala de Troféus do S.L. Benfica,
  Lisboa 035825
Sala de Zoologia, Faculdade de
  Ciências da Univsidade do Porto,
  Porto 035959
Sala del Centro Cultural Provincial,
  Málaga 039710
Sala del Costume e delle Tradizioni
  Popolari, Corinaldo 026075
Sala Derenzi, Castello de la
  Plana 115021
Sala d'Exposicions, Ponts 039961
Sala Diaz, San Antonio 052945
Sala Exposição Museu José Brito da

Luz, Estremoz 035672
Sala Exposición Permanente Municipal,
  Ciduadad de Cáceres, Cáceres . 039185
Sala Ferrani Martí, Sant Cugat del
  Vallès 040084
Sala Fotográfica El Pasaje, Alcázar de
  San Juan 038921
Sala Fratelli, Milano 080709
Sala Gabernia, Valencia 115514
Sala Gestalguinos, Valencia 115515
Sala Girona de la Fundació la Caixa,
  Girona 039450
Sala Goya, Alicante ... 085703, 114832
Sala Histórica de la Facultad de
  Medicina, Universidad Autónoma de
  Nuevo León, Mitras 031280
Sala Histórica del Regimiento de
  Infantería Mecanizado 24, Río
  Gallegos 000535
Sala Historica General Savio, Buenos
  Aires 000282
Sala Homenaje a Juárez, Guelatao de
  Juárez 031034
Sala Ipostel, Museo de Arte
  Contemporáneo de Caracas Sofía
  Imber, Caracas 054740
Sala Josefa Rodriguez del Fresno,
  Santa Fé 000622
Sala Llotja del Peix, Alicante 114833
Sala Maior, Porto 114103
Sala Melchor Ocampo, Universidad
  Michoacana, Morelia 031312
Sala Mendoza, Caracas 054741
Sala Montcada de la Fundació la
  Caixa, Barcelona 039108
Sala Montseny, Mario, Barcelona 143731
Sala Moyua, Bilbao ... 085888, 126519
Sala Muncunill, Terrassa 040224
Sala Museo de la Casa de la Cultura
  Martín Gamarra, Gamarra 008683
Sala Museo Literario, Archivo
  Histórico del Instituto Caro y Cuervo,
  Bogotá 008635
Sala Museo Pablo Sarasate,
  Pamplona 039912
Sala Museu de Arqueologia,
  Mogadouro 035858
Sala Museu Fialho de Almeida,
  Cuba 035646
Sala Museu Jenny Mendes de Brito,
  Entroncamento 035650
Sala Nacional de Cultura Mariano
  Fuentes Lira, Escuela Superior
  Autónoma de Bellas Artes,
  Cusco 034268
Sala Nacional de Exposiciones, San
  Salvador 010599
Sala Numismática del Banco Central
  de Nicaragua, Managua 033392
Sala Numismática del Banco de
  Crédito del Peru y Sala de Acuarelas
  Pancho Fierro, Lima 034372
Sala Padre Mariano, Santiago de
  Chile 101922
Sala Parés, Barcelona . 085835, 114959
Sala Pelaires, Palma de Mallorca 115367
Sala Piano Museum, Cebu 034464
Sala Pietro Canonica, Palazzo dei
  Congressi, Stresa 028026
Sala Raffaello Pagliaccetti,
  Giulianova 026450
Sala Retiro, Madrid 126549
Sala Rusiñol, Sant Cugat del
  Vallès 115415
Sala Sergio Larrain, Santiago de
  Chile 007550
Sala Traktormuseum, Sala 040933
Sala, Fausto, Parma 131737
Sala, Xavier, La Seyne-sur-Mer . 069102
Salacgrivas Muzejs, Salacgrīva .. 030229
Salachas, Gilbert, Paris 136983
Saladin, Erich & Gaby, Blauen . 087327,
                                133702
Saladino, Leonardo, Roma 081544

Salagoudi, S., Aachen 105946
Salah, Jacky, Appoigny 066691
Salah, Mohammad, Saint-Étienne 072777
Salama, Cairo 103076
Salamah, Talebzen M.H., Madina 085310
Salamanca Arts Centre, Hobart . 001148
The Salamanca Collection,
  Hobart 099137
Salamanca Collection, Hobart .. 061636
Salamanca Rail Museum,
  Salamanca 052876
De Salamander, Almere 143004
Salamander, Christchurch 083894,
                                112859
Salamandra, Zaragoza 086606, 133511
Salamon, Matteo, Milano 080710,
  110422,   131524
Salamone, Nunzio, Catania 080046,
                                131099
Salander & O'Reilly, New York . 123035
Salangen Bygdetun, Sjøvegan .. 033937
Salão, Rio de Janeiro 101065
Salão Nacional de Antiguidades e Arte
  Sacra, Braga 098276
Salar Gallery, Hatherleigh 117928
Salar Jung Museum, Hyderabad 024302
Salas Municipales de Arte Sanz-Enea,
  Zarautz 040459
Salatino, Eugenia Correia, São
  Paulo 064208
Salaün, Anna, Caen 067582
Salaun, André, Brest 067478
Salaz, Medellín 065308
Salazar, Belo Horizonte 063920
Salazar, San Francisco 097320
Salbitani, Gabriella, Roma 081545
Salchow, Hamburg 141893
Salcombe Maritime Museum,
  Salcombe 045718
Saldarriaga, Manolo, Medellín .. 065309
Salduba, Grup d'Art Escolà,
  Zaragoza 115563
Sale Historical Museum, Sale .. 001480
Sale Monumentali della Biblioteca
  Marciana, Venezia 028324
Saleh Kaki, Mustafa Yousuf,
  Jeddah 085308
Saleh, Terry, Kurwongbah 061710
Salek, Mehdi, Basel 087257
Salem Art Association, Salem .. 060722
Salem County Historical Society
  Museum, Salem 052885
Salem Historical Society Museum,
  Salem 052886
Salem Maritime Houses, Salem
  Maritime National Historic Site,
  Salem 052881
Salem Museum, Salem 052895
Salem Pioneer Village, Salem .. 052882
Salem Witch Museum, Salem .. 052883
Salen, Jægerspris 065807, 102853
Salerno, Michael, Los Angeles . 121587
Salerno, Paolo, Milano 080711
Sales by Sylvia, Cincinnati 092893
Sales, Alejandro, Barcelona ... 114960
Salese, Frank, Philadelphia ... 136101
Salet, Patrice, Saint-Ouen 073193
La Saletta, Milano 080712
La Saletta Gries, Bolzano 109889
Saleya, Nice 070662
Salford Museum and Art Gallery,
  Salford 045722
Salgo Trust for Education, Port
  Washington 052265
Salibello, John, New York 095716,
                                095717
Salida Museum, Salida 052896
Salies Antiquités, Salies-de-
  Béarn 073380
Salihli Kültür Merkezi, Salihli ... 043047
Salimbeni, Firenze 142685
Salina Art Center, Salina 052897
Salinas Pueblo Missions National
  Monument, Mountainair 051107

Salinas, Allan, Santiago de Chile 064984
Salinas, Joaquín & Allan, Santiago de
  Chile 064985
Saline- und Heimatmuseum, Bad
  Sulza 016889
Saline County Historical Society
  Museum, Dorchester 048340
Salinenmuseum im Gradierbau, Bad
  Dürkheim 016752
Salinenmuseum Unteres Bohrhaus,
  Rottweil 021529
Saling, Paris 071875
Salinpuoli, Helsinki 066359
Salis, Anif 099955
Salis & Vertes, Salzburg 100176, 100177
Salis, Christiane, Condom 068130
Salis, Michel, Nérac 070503
Salisbury and South Wiltshire
  Museum, Salisbury 045727
Salisbury Antiques Market,
  Salisbury 091225, 144823
Salisbury Antiques Warehouse,
  Salisbury 091226, 135001
Salisbury Folk Museum,
  Salisbury 001481
Salisbury Historical Society Museum,
  Salisbury 052903
Salisbury House, Des Moines .. 048286
Salisbury Mansion, Worcester .. 054547
Salisbury State University Galleries,
  Salisbury 052901
Salisci, Francesco, Cagliari ... 080001,
                                131075
Salit, Jean-Norbert, Paris 071876
Salix, Banff 117284
Sallan Kotiseutumuseo, Salla . 011340
Sallands Landbouwmuseum de
  Laarman, Luttenberg 032611
Sallaz, Michel, Paris 071877
Salle Alfred Pellan, Maison des Arts
  de Laval, Laval 006173
Salle Augustin-Chénier, Ville-
  Marie 007343
Salle de Traditions de la Garde
  Républicaine, Paris 014801
Salle de Vente Elisabeth, Namur 125353
Salle de Vente Regina, Liège .. 125348
Salle de Vente Saint-Jean, Bordères-
  sur-l'Echez 067368
Salle de Ventes du Béguinage,
  Bruxelles 125325
Salle de Ventes Européenne,
  Charleroi 125331
La Salle des Étiquettes, Musée de la
  Cave des Champagnes de Castellane,
  Épernay 012883
Salle des Souvenirs, Lisieux .. 013830
Salle des Ursulines, Vesoul ... 105849
Salle des Ventes, Aubervilliers . 125498
Salle des Ventes, Coutras 068213
Salle des Ventes, Pau 072050
Salle des Ventes, Péronne 125777
Salle des Ventes Alésia, Paris . 125763
Salle des Ventes des Particuliers,
  Joigny (Yonne) 068907
Salle des Ventes des Particuliers,
  Montmorot 070282
Salle des Ventes des Particuliers,
  Paris 125764
Salle des Ventes du Particulier,
  Paris 125765, 125766
Salle des Ventes du Particulier de
  l'Ariège, Mirepoix 070128
Salle des Ventes Guérandaise,
  Guérande 125594
Salle d'Exposition, Lure 104056
Salle d'Exposition Espace Lumière,
  Hénin-Beaumont 103811
Salle Exposition, Ile-d'Ouessant . 103850
Salle Mémorial Franco Américaine,
  Sommepy-Tahure 015832
Salle Panoramique de Notre-Dame-de-
  Monts, Notre-Dame-de-Monts . 014527
La Salle Raspail, Paris 071878

Salle Souvenir du 24ème BCA,
Villefranche-sur-Mer 016249
Salle Ventes Champion Gondran,
Bourgoin-Jallieu 125536
Salle Ventes Champion Gondran,
Vienne 125840
Salle Ventes Enchères Publiques,
Etampes 125584
Salles de Vente de l'Ain,
Replonges 072444
Les Salles du Palais, Genève . . 116338
Salles du Tresor, Prades 014988
Salles, Pierre, Ambleville 066552
Salling-Fjends Museum, Fur . . . 010077
Sallmann, Berlin 129559
Sallou, Hervé, Perros-Guirec . . 072106
Sallou, Hervé, Trégastel 073875
Salloum, Maroun H., Paris 071879
Sally, Cairo 066269
Sally Lunn's Refreshment House and
Kitchen Museum, Bath 043527
Sally, Marie, Saint Paul 123930
Sally's Collectables, Sacramento 096644
Salmagundi Club, New York . . . 060723
Salmagundi Museum of American Art,
New York 051463
Salmagundi West, Vancouver . . 064871
Salmen, Thomas, Hemer 141939
Salmo Museum, Salmo 006963
Salmon Arm Museum, Salmon
Arm 006964
Salmon Brook Historical Society
Museum, Granby 049172
Salmon Galleries, McMahons
Point 099249
Salmon, Richard, London 118558
Saló d'Antiquaris i Brocanters a Lleida,
Lleida 098314
Saloga Restauracion, Barcelona 133160
Saloisten Kotiseutumuseo,
Arkkukari 010756
Salome, Nils, Bruxelles 063547
Saloméjos Néries Memorialinis
Muziejus, Maironio Lietuviy
Literatūros Muziejus, Kaunas . 030375
Salomon, Amstelveen 082667
Salomon Lilian, New York 095718
Salomon Lilian, Dr., Amsterdam 082847
Salomon, Gabrielle, Paris 105210
Salomon, Richard, New York . . . 123036
Salomon, Wolfgang, Regensburg 130570
Salomons Museum, Memento Rooms,
Southborough 045831
Salon 5, Honolulu 120966
Salon 94, New York 123037
Salon 94 Freemans, New York . . 123038
Salon Antičnega Pohištva,
Maribor 085620
Salon Antykwaryczny i Sztuki
Współczesnej, Kielce 098272
Le Salon d'Art, Bruxelles 100657
Salon d'Art et d'Antiquités,
Luxemburg 098242
Salon d'Automne – Antiquité et
Brocante, Nancy 098093
Salon de Arte del Casino Nervion,
Bilbao 114996
Salón de la Fama del Beisbol
Profesional de México,
Monterrey 031300
Salón de la Fama Nicaragüense,
Galeria de la Mar, La Virgen,
Rivas 033402
Salon de la Société Nationale des
Beaux-Arts, Paris 098094
Le Salon de l'Art et de la Décoration,
Beirut 098240
Salon de l'Automobile Miniature,
Ally 098095
El Salon de los Anticuarios, Cali 065254
Salón del Cuadro y sus Componentes,
Madrid 098216
Salon des Antiquaires, Besançon 098098
Salon des Antiquaires, Colmar . 098096

Salon des Antiquaires, Grenoble 098095
Salon des Antiquaires, Lille 098097
Salon des Antiquaires, Metz . . . 098101
Salon des Antiquaires, Nancy . . 098099
Salon des Antiquaires, Toulouse 098100
Salon des Antiquaires de Biarritz,
Biarritz 098102
Salon des Antiquaires de Lausanne,
Lausanne 098332
Salon des Antiquaires et de la Belle
Brocante, Tournai 098023
Salon des Antiquaires et de la
Brocante, Dijon 098103
Salon des Antiquaires et de l'art
contemporain, Bordeaux 098104
Salon des Antiquités, Orléans . . 098105
Salon des Antiquités et Brocante,
Paris 098106
Salon des Antiquités et de la Brocante,
Mouscron 098024
Salon des Antiquités et de la Brocante,
Quimper 098107
Salon des Artisans d'Art de Toulouse,
Toulouse 098108
Salon des Créateurs et des Ateliers
d'Art, Lille 098109
Salon du Collectionneur Paris, Salon
International des Antiquaires
Spécialisés, Paris 098110
Salon du Dessin, Paris 098111
Salon du Livre et Papiers Anciens,
Paris 098112
Salon du Nord, Saint Paul 123931
Salon Dzieł Sztuki, Kraków 084597
Salon Dzieł Sztuki i Antykow Paga,
Kielce 084541
Salon Dziel Udanych-Hall of Succesful
Works, Wrocław 113951
Salón Electo Brizuela, Biblioteca
Popular El Porvenir, Concepción del
Uruguay 000314
Salon für zeitgenössische Fotokunst,
Berlin 106416
Salon-Galereja Éstetika, Saratov 114369
Salon Grange, Moskva 114303
Salón Internacional de Grabado y
Ediciones de Arte Contemporáneo,
Madrid 098316
Salon International d'Art Contemporain,
Libramont 098025
Salon International de la Céramique
Ancienne, Enghien 098026
Salon Muzeja Savremene Umetnosti,
Beograd 038181
Salon Muzyczny im. Fryderyka
Chopina, Centrum Kultury i Sztuki w
Kaliszu, Antonin 034620
Salón Nacional de Antigüedades,
Sevilla 098317
Salon Natasha, Ha Noi 125096
Salon National des Antiquaires,
Angers 098114
Salon National des Antiquaires, Bourg-
en-Bresse 098113
Salon Oven-Venera, Sankt-
Peterburg 085266
Salon Peinture et Sculpture,
Colmar 098115
Salon s'Art, Ha Noi 125097
Salon Sprzedaży Antýkow,
Warszawa 084747
Salon Sztuki Antyki i Sztuka
Współczesna, Puck 084659
Salon Utrechtse Kunstenaars,
Utrecht 098264
Salon Verlag, Köln 137317
Salon XX–XXI, Sankt-Peterburg . 114360
Salone Aste, Cornici, Accessori,
Grafica e Tecnologie, Bologna . 098216
Salone d'Arte Moderna e
Contemporanea, Forlì 098217
Salone della Numismatica,
Medaglistica e Cartamoneta,
Vicenza 098218

Salone dell'Arte del Restauro e della
Conservazione dei Beni Culturali e
Ambientali, Ferrera 098219
Salone Internazionale d'Arte
Contemporanea, Bari 098220
Salone Italiano d'Antichità, Bari . 098221
Salone Villa Romana, Firenze . . 026302
Salonik Chopinów, Muzeum Fryderyka
Chopina, Warszawa 035379
Les Salons Démodés, Québec . . 064541
Salons voor Schone Kunsten, Sint-
Niklaas 004022
Salonul de Toamnă al Antichităților și
Artelor Plastice, Cluj-Napoca . 098279
Salotto, Arnhem 112336
Il Salotto, Como 109972
Il Salotto, Milano 110423
Salpalinja-Museo, Miehikkälä . . 011155
Salsi, Pier Antonio, Parma 131738
Salt Archaeological Museum,
Salt 029784
Salt Box Collection, Portland . . 096468
Salt Folklore Museum, Salt 029785
Salt House Gallery, Saint Ives,
Cornwall 119034
Salt Lake Antiques, Salt Lake
City 096920
Salt Lake Art Center, Salt Lake
City 052916
Salt Museum, Liverpool 050343
Salt Museum, Northwich 045394
The Salt Queen Foundation, New
York 051464
Salt Spring Island Museum,
Ganges 005858
Salt Water Gallery, Burnham
Market 117483
Salt, D.W., Cincinnati 135452
Salt, N.P. & A., Sheffield 091292
Salt, Steven, Crows Nest,
Queensland 061378
Saltar i Parar, La Bisbal
d'Empordá 085906
Saltash Heritage Centre, Saltash 045728
Saltburn Artists Projects, Saltburn-by-
the-Sea 119057
Saltdal Museum, Rognan 033904
Salter Museum, Argonia 046528
Salter, F.D., Clare 088925, 134279
Salter, Nicholas, London 090473, 144683
Saltgate Antiques, Beccles 088419
Saltgrass Gallery, Lymington . . . 118683
Salti, Lorenzo, Nice 128871
Saltiel, Le Castellet . . . 069239, 133937
The Saltire Society, Edinburgh . . 060232
Saltram House, Plymouth 045514
Saltrødbukta kunstformidling,
Saltrød 084348
Saltstraumen Galleri,
Saltstraumen 113252
Saltstraumen Museum,
Saltstraumen 033917
Saltworks Gallery, Atlanta 119607
Saltyning Zone Museum,
Xianning 008412
Saltzaroen Aretoa, Bilbao 098318
Salud Segundino Antiguedades,
Sevilla 086475
Saluden, Quimper 105417
Saluhallen, Sundsvall 126681
Salusbury, Tom, Hitchin 089539, 134465
Salva, Christian, Toulouse 129315
Salvadeo, Angelo, Mezzana Bigli 080376
Salvador, Buenos Aires 139599
Salvador, Zaragoza 086607
Salvador Dalí Museum, Saint
Petersburg 052865
Salvadore, Carlotta, Firenze . . . 131260
Salvadori, Roberto, Genova . . . 080290
Salvage Fisherman's Museum,
Salvage 006965
Salvage Heaven, Milwaukee . . . 094879
Salvage One, Chicago 092789
The Salvage Shop, Toronto 064751

Salvai, Giorgio, Cannes 067651
Salvai, Giorgio, Paris 071880
Salvan, Christian, Le Dorat . . . 069263,
069264
Salvany, Clermont-Ferrand 103608
Salvat, Lisboa 137920
Salvat Montserrat, Maria,
Barcelona 114961
Salvat, J., Miami 138360, 145258
Salvation Army George Scott Railton
Heritage Centre, Toronto 007202
Salvation Army International Heritage
Centre, London 045102
The Salvation Army Museum,
Toronto 007203
Salvation Army Southern Historical
Museum, Atlanta 046632
The Salvation Army Territorial Archives
and Museum, Wellington 033359
Di Salvatore, Puyricard 072331
Salvatori, Maurizio, Roma 132043
Salvatorianerkloster, Dom zu Gurk,
Gurk 002024
Salve 97, Sofia 101169
Salverda, Kiel-Windeweer 137862
Salveymühle, Gartz 018595
Salvi Garcia, Ruben, Oviedo . . . 115329
Salvi, Giorgio, Asiago 109758
Salvò, Padova 131675
Salwender, H., Eppelheim 075625
Salzano, Giancarlo, Torino 111002
Salzbergwerk Berchtesgaden,
Berchtesgaden 017041
Salzburg Museum, Salzburg . . . 002637
Salzburger Antiquitäten, Oberalm 062754,
127825
Salzburger Barockmuseum, Sammlung
Rossacher, Salzburg 002638
Salzburger Barockmuseum,
Verlagsabteilung, Salzburg . . 136558
Salzburger Freilichtmuseum,
Großgmain 002004
Salzburger Kunstverein, Salzburg 058308,
100178,   136559
Salzburger Landes-Skimuseum,
Werfenweng 002926
Salzburger Museumsverein,
Salzburg 058309
Salzin, Namur 100798
Salzkammer, Pratteln 041850
Salzkammergut-Lokalbahn-Museum,
Mondsee 002414
Salzmagazin, Nidwaldner Museum,
Stans 042025
Salzmann KEG, Tischlerei Airluxbetten,
Dornbirn 127725
Salzmuseum, Bad Sooden-
Allendorf 016882
Salzmuseum, Bad Sülze 016887
Salzwelten Altaussee, Altaussee 001701
Salzwelten Hallstatt, Hallstatt . . 002054
Salzwelten Salzburg-Bad Dürrnberg,
Bad Dürnberg 001737
SAM – Schweizerisches
Architekturmuseum, Basel . . . 041287
Sam Bell Maxey House, Paris . . 051919
Sam Brown Log House, Browns
Valley 047278
Sam Houston Historical Schoolhouse,
Maryville 050732
Sam Houston Memorial Museum,
Huntsville 049637
Sam Noble Oklahoma Museum
of Natural History, University of
Oklahoma, Norman 051588
Sam Rayburn House Museum,
Bonham 047082
The Sam Rayburn Museum, Center for
American History of the University of
Texas at Austin, Bonham 047083
Sam Roi Yod, Phachuab Khiri
Khan 042710
Sam the Chandelier Man,
Toronto 064752

Sam Tung Uk Museum, Hong Kong ... 007889
Sam Waller Museum, The Pas . 007138
SAMAB, South African Museums Association Bulletin, Port Alfred 139022
Samagra, Paris ... 105211
Samancıoğlu S. Resim Galerisi, İstanbul ... 117082
Samaniego Caballero, Pedro Antonio, Burgos ... 085915
Samans, Dr. Rainer, Eschweiler . 075665
Samantics, Newsletter of the South African Museums Association, Port Elizabeth ... 139023
Samara, Amiens ... 066582
Samarcanda, Perugia ... 081145
Samarcanda com e loc de livros, Brasília ... 140276
Samarkand, Marseille ... 104196
Samarkand Galleries, Edinburgh 089189
Samarkand Galleries, Stow-on-the-Wold ... 091481, 135097
Samarkanda Otros Mundos, Girona ... 086030
Samarqand Ülkashunoslik Muzeyi, Samarqand ... 054647
Samarra Museum, Samarra ... 024814
Samarskaja Chudožestvennaja Galereja Marija, Samara ... 037713
Samarskaja Gosudarstvennaja Akademija Iskusstv i Kultury, Samara ... 056299
Samarskij Eparchialnyj Cerkovno-istoričeskij Muzej, Samara ... 037714
Samarskij Oblastnoj Chudožestvennyj Muzej, Samara ... 037715
Samarskij Oblastnoj Istoriko-kraevedčeskij Muzej im. P.V. Alabina, Samara ... 037716
Samawat, Christa, Berlin ... 074908
Samb, Abdoulaye, Caen ... 067583
Sambataro, Placida, Catania ... 131100
Sambeau, Saint Louis ... 096771
Sambourne House Antiques, Swindon ... 091532
Samdan, İstanbul ... 088138
Samford and District Historical Museum, Samford ... 001482
Sami, Austin ... 092268
Saminger, Erich, Linz . 062722, 127814
Saminger, Erich, Wels ... 127897
Saminger, Ferdinand, Linz ... 062723, 127812
Samir Barakat Gallery, Cairo ... 103077
De Samiske Samlinger, Karasjok 033693
Samlarkällan, Upplands Väsby . 087115
Samler Børsen, Århus ... 065608
Samler og Antikk Messen, Lillestrøm ... 098268
Samlerhuset, Hillerød ... 065756
Samler'n Antikk, Kristiansund . 084272
Samlersenteret, Oslo ... 143416
The Samlesbury Decorative and Fine Arts Society, Poulton-le-Fylde . 060233
Samlesbury Hall, Samlesbury . 091230
Sammann, Jesteburg ... 076648
Sammeck & Sammeck, München 108188
Sammel Surium, Nürnberg ... 077842
Sammelpunkt, Bünde ... 075215
Sammelpunkt-Antiquariat für Trivialliteratur, Zürich ... 144193
Sammelsurium, Witzenhausen . 079041, 142501
Sammenslutningen af Grønlandske Lokalmuseer, Sisimiut ... 059333
Sammer Sidi, London ... 118559
Sammet, Detlef, Waldsassen ... 078844
Die Sammler, Glonn ... 076008
Sammler Journal, Reichertshausen ... 138752
Sammlerstübchen, Osterode am Harz ... 077990
Sammlung + Picasso Donation Rosengart, Stiftung Rosengart,

Luzern ... 041713
Sammlung alter Musikinstrumente, Kunsthistorisches Museum Wien, Wien ... 003072
Sammlung Angewandte Kunst, Museumslandschaft Hessen Kassel, Kassel ... 019546
Sammlung Antiker Kleinkunst, Institut für Altertumswissenschaften/ Lehrstuhl für Klassische Archäologie der Friedrich-Schiller-Universität Jena, Jena ... 019456
Sammlung Bodingbauer, Steyr . 002785
Sammlung Boros, Kunst im Bunker, Berlin ... 106417
Sammlung Crous im Alten Kurhaus, Aachen ... 016444
Sammlung des Seminars für Ägyptologie, Universität Heidelberg, Heidelberg ... 019143
Sammlung Domnick, Nürtingen . 020882
Sammlung Doris Nöth, Kirchheim unter Teck ... 107561
Sammlung Dr. Berkowitz, Münzkabinett der Stadt Hannover, Hannover ... 019065
Sammlung Dr. Irmgard von Lemmers-Danforth, Europäische Wohnkultur aus Renaissance und Barock, Wetzlar ... 022614
Sammlung Ernst Schneider, Hetjens Museum, Düsseldorf ... 018042
Sammlung Falckenberg, Phoenix Kulturstiftung, Hamburg ... 107313
Sammlung Fossilien des Jura, Lichtenfels ... 020062
Sammlung Friedhof Hörnli, Riehen ... 041881
Sammlung Friedrichshof, Zurndorf ... 003142
Sammlung für Kunsthandwerk und Plastik, Kassel ... 019546
Sammlung für Papier- und Druckgeschichte, Frankenberg, Sachsen ... 018392
Sammlung Gauselmann, Deutsches Automatenmuseum, Espelkamp 018285
Sammlung Geyer-zu-Lauf, Emmendingen ... 018186
Sammlung Goetz, München ... 020567
Sammlung handgeschriebener Choralbücher, Kiedrich ... 019599
Sammlung Hauser & Wirth, Henau ... 041591
Sammlung historischer Biscuitdosen, Läckerli-Huus, Basel ... 041288
Sammlung historischer Maschinen und Geräte, Calau ... 017711
Sammlung historischer Mikroskope von Ernst Leitz, Wetzlar ... 022615
Sammlung historischer Prägestempel, Kunsthistorisches Museum, Wien ... 003073
Sammlung historischer Tasteninstrumente, Wassenach 022436
Sammlung historischer Tasteninstrumente Neumeyer-Junghanns-Tracey, Bad Krozingen ... 016808
Sammlung Hoffmann, Berlin ... 017276
Sammlung Holzinger, Privatmuseum für Laienkunst, München ... 020568
Sammlung industrielle Gestaltung, Deutsches Historisches Museum, Berlin ... 017277
Sammlung Irmgard Friedl, Bad Griesbach ... 016777
Sammlung landwirtschaftlicher Geräte, Unterpleichfeld ... 022281
Sammlung Ludwig in Bamberg, Museen der Stadt Bamberg, Bamberg ... 016955
Sammlung Meinrad Burch-Korrodi, Sarnen ... 041940

Sammlung Mittelalterliche Kunst in Thüringen, Thüringer Museum Eisenach, Eisenach ... 018135
Sammlung Moderne Kunst, Bayerische Staatsgemäldesammlungen, München ... 020569
Sammlung Nuss, Weinstadt ... 022528
Sammlung Oskar Reinhart 'Am Römerholz', Winterthur ... 042160
Sammlung Patt, Musik-Instrumente von gestern und heute, Willisau 042146
Sammlung Prinzhorn, Psychiatrische Universitätsklinik Heidelberg, Heidelberg ... 019144
Sammlung Scharf-Gerstenberg, Berlin ... 017278
Sammlung Schloß Aichberg, Rohrbach an der Lafnitz ... 002605
Sammlung und Kunstraum Grässlin, Sankt Georgen im Schwarzwald 021594
Sammlung Volkskunde, Museumslandschaft Hessen Kassel, Kassel ... 019548
Sammlung Waagen und Gewichte, Albstadt ... 016497
Sammlung zur Geschichte der Benediktinerabtei Tholey, Tholey ... 022153
Sammlung zur Geschichte der Geburtshilfe, Göttingen ... 018756
Sammlung zur Landeskunde Ost- und Westpreußens "Es war ein Land. . . ", Zweigstelle des Bayerischen Nationalmuseums, Oberschleißheim ... 020924
Sammlungen Burg Bernstein, Bernstein ... 001781
Sammlungen der Benediktiner-Abtei Braunau, Rohr in Niederbayern 021464
Sammlungen der Stadt Amstetten auf Schloß Ulmerfeld, Ulmerfeld-Hausmening ... 002850
Sammlungen der Technischen Universität Dresden, Dresden . 018006
Sammlungen der Universität für angewandte Kunst Wien mit Oskar-Kokoschka-Zentrum, Wien ... 003074
Sammlungen des Archäologischen Instituts der Universität, Göttingen ... 018757
Sammlungen des Augustiner Chorherrenstiftes, Sankt Florian 002652
Sammlungen des Hochschulkunde der Deutschen Gesellschaft für Hochschulkunde e.V., Würzburg ... 022801
Sammlungen des Kunstvereins, Ibbenbüren ... 019372
Sammlungen des Marktamtes, Wien ... 003075
Sammlungen des Stiftes Sankt Lambrecht, Volkskundemuseum, Galerie und Vogelsammlung, Sankt Lambrecht ... 002671
Sammlungen des Winckelmanns-Instituts – Lehrsammlung Klassische Archäologie, Berlin ... 017279
Sammlungen Schloß Moosham, Unternberg ... 002855
Sammlungen Zisterzienserstift Stams, Stams ... 002775
Sammons, Timothy, London ... 118560
Sammy Marks Museum, Kungwini ... 038775
Sammy Miller Motorcycle Museum, New Milton ... 045325
Samoborski Muzej, Samobor ... 008983
Samocvet, Kiriši ... 114184
Samoticha, Mario, Bochum ... 075042
Samotracia, Buenos Aires ... 127535
Samovar Antiques, Hythe, Kent . 089630
Samoy Art Museum, Bacoor ... 034429
Sampl, Ennetbaden ... 087410, 133731
Sampler Antiques, Oklahoma City 096006

Sampson, Alistair, London ... 090474
Sampson, Anthony, Dulverton . 089113
Samrhea, Genolier ... 116345
Sams, Bad Ischl ... 099957
Sam's Antique Rugs, Dallas ... 093211, 135515
Sam's Cottage, Houston ... 093785
Sam's Place Antiques, New York 095719
Sam's Souk, New York ... 095720
Sams, P., Aalborg ... 065588
Samson, Boston ... 119920
Samson, Paris ... 071881, 129139
Samson-Haus, Leer ... 019951
Samson V Maritime Museum, New Westminster ... 006475
Samson, Jean-François, Amboise 066557
Samsun Arkeoloji ve Etnografya Müzesi, Samsun ... 043049
Samsun Atatürk Müzesi, Samsun 043050
Samsun Devlet Güzel Sanatlar Galerisi, Samsun ... 043051
Samsung Children's Museum, Seoul ... 030032
Samsung Foundation of Culture, Seoul ... 030033
Samtuh, Gangsang ... 111598
Samtuh, Seoul ... 111708
Samu Géza Múzeum, Pécsvárad 023862
Samuel Cupples House, Saint Louis ... 052822
Samuel Dorsky Museum of Art, State University of New York New Paltz, New Paltz ... 051309
Samuel Johnson Birthplace Museum, Lichfield ... 044825
Samuel K. Fox Museum, Dillingham ... 048328
Samuel P. Harn Museum of Art, University of Florida, Gainesville 049004
Samuel Palmer School of Fine Art, Shoreham Sevenoaks ... 056607
Samuel S. Fleisher Art Memorial, Philadelphia Museum of Art, Philadelphia ... 052095
Samuel, Philippe, Paris ... 105212
Samuelis Baumgarte Galerie, Bielefeld ... 106497
Samuel's Gallery, Oakland ... 123264
Samuil, Skopje ... 111895
San Agustin Museum Intramuros, Manila ... 034548
San Angel Folk Art, San Antonio 124067
San Angel Art Club, San Angelo 060724
San Angelo Museum of Fine Arts, San Angelo ... 052927
San Antonio Art League, San Antonio ... 060725
San Antonio Children's Museum, San Antonio ... 052946
San Antonio Conservation Society Museum, San Antonio ... 052947
San Antonio Cultural Arts, San Antonio ... 124064
San Antonio Missions Museum, Jolon ... 049837
San Antonio Missions Visitor Center, San Antonio ... 052948
San Antonio Museum of Art, San Antonio ... 052949
San Antonio Rare Coins and Jewelry, San Antonio ... 097030
San Antonio Upholstering Company, San Antonio ... 136272
San Antonio's Center for Antiques, San Antonio ... 097031
San Barandan, Córdoba ... 085945
San Barnaba, Milano . 080713, 110424
San Bartolomeo, Venezia ... 111153
San Benito County Museum, Hollister ... 049613
San Bernardino Art Association, San Bernardino ... 060726
San Bernardino Art Association Gallery, Sturges Center for the Fine Arts, San

Bernardino 052958
San Bernardino County Museum,
Redlands 052497
San Bernardo, Genova 110133
San Buenaventura Mission Museum,
Ventura 053955
San Camillo, Rapallo 110697
San Caterina, Brescia 079954
San Cozy Ceramic Museum, Puli 042422
San Diego Aerospace Museum, San
Diego 052970
San Diego Antique Tour, San
Diego 097131
San Diego Art Institute, Museum of
the Living Artists, San Diego .. 052971
San Diego Automotive Museum, San
Diego 052972
San Diego Coin and Bullion, San
Diego 097132
San Diego Coin and Gold Exchange,
San Diego 097133
San Diego Hall of Champions Sports
Museum, San Diego 052973
San Diego Historical Society Museum,
San Diego 052974
San Diego Maritime Museum, San
Diego 052975
San Diego Model Railroad Museum,
San Diego 052976
San Diego Museum of African Art,
San Diego 052977
San Diego Museum of Art, San
Diego 052978
San Diego Museum of Man, San
Diego 052979
San Diego Natural History Museum,
San Diego 052980
San Diego Poster Distributor, San
Diego 124167
San Diego Sculptors Guild, San
Diego 060727, 124168
San Diego Shipping Antiguidades, São
Paulo 064209
San Diego W&S, San Diego ... 145529
San Diego Watercolor Society, San
Diego 060728, 124169
San Dieguito Heritage Museum,
Encinitas 048583
San Domenico Stringed Instruments,
Cardiff 088790, 134217
San-Ee, Singapore 114600
San el-Haggar Museum,
Sharqiya 010584
San Fedele Arte, Milano 110425
San Fernando Historical Museum,
Mission Hills 050954
San Fernando Valley Historical Society
Museum, Mission Hills 050955
San Francisco African American
Historical and Cultural Society
Museum, San Francisco 053019
San Francisco Airport Museums, San
Francisco 053020
San Francisco Antique and Design
Mall, San Francisco 097321
San Francisco Art Dealers, San
Francisco 124382
San Francisco Art Exchange, San
Francisco ... 124383, 124384, 138460
San Francisco Art Institute, San
Francisco 057857
San Francisco Arts of Pacific Asia
Show, San Francisco 098496
San Francisco Camerawork, San
Francisco 053021
The San Francisco Fine Print Fair, San
Francisco 098497
San Francisco Fire Department
Museum, San Francisco 053022
San Francisco Maritime National
Historical Park, San Francisco . 053023
San Francisco Museum of
Contemporary Hispanic Art, San
Francisco 053024

San Francisco Museum of Modern Art,
San Francisco 053025
San Francisco Poster Fair, San
Francisco 098498
San Francisco Rock Posters, San
Francisco 124385
San Francisco Tribal, Folk and Textile
Arts Show, San Francisco ... 098499
San Francisco Women Artists, San
Francisco 060729
San Gabriel Mission Museum, San
Gabriel 053034
San Giorgio, Ospedaletti 110551
San Giorgio, Pordenone 110680
San Giorgio, Portofino Mare .. 110682
San Giorgio Fine Art, Firenze ... 080198
San Gregorio, Venezia 111154
San Jacinto Museum, San
Jacinto 053035
San Jacinto Museum of History, La
Porte 050049
San Jaime, Palma de Mallorca . 086367
San Joaquin County Historical
Museum, Lodi 050355
San Joaquin Delta College, Fine Arts
Division, Stockton 057956
San Jose Antique Mall,
Jacksonville 094037
San Jose Art League, San Jose 060730
San Jose Coin Shop, San Jose . 097375
San Jose Collectibles, San Jose 097376
San Jose Institute of Contemporary
Art, San Jose 053042
San Jose Museum of Art, San
Jose 053043
San Jose Museum of Quilts & Textiles,
San Jose 053044
San Juan Bautista Mission Museum,
San Juan Bautista 053046
San Juan County Historical Museum,
Silverton 053344
San Juan County Museum,
Bloomfield 047035
San Juan Garcia, Federico,
Valencia 133469
San Juan Historical Society Museum,
Friday Harbor 048982
San Juan Island National Historical
Park, Friday Harbor 048983
San Juan School of Design, Hato
Rey 056253
San Luca, Bologna 109860
San Luis Obispo Art Center, San Luis
Obispo 053049
San Luis Obispo County Historical
Museum, San Luis Obispo .. 053050
San Marco, Diano Marina 110004
San Marco, Milano 080714
San Marco Book Store,
Jacksonville 145190
San Marco Gallery, Dominican
University of California, San
Rafael 053063
San Marcos de Apalache Historic
State Park, Saint Marks 052826
San Martin, Carlos, Bordeaux . 128506
San Mateo County Historical Museum,
Redwood City 052499
San Michele, Savona 110920
San Miguel Mission Church, Santa
Fe 053120
San Niccolò, Venezia 111155
San Pantalon, Venezia 111156
San Pedro Market Antiques, San
Antonio 097032
San Pedro Museo de Arte,
Puebla 031401
San Reus, Palma de Mallorca . 115368
San Román, Logroño 133281
San Salvatore, Modena 110504
San Sebastiano, Genova 110134
San Telmo, Madrid ... 086257, 086258
Sana Hastakala, Lalitpur 112097
Sanaa Exhibition, Sharjah ... 088208

Sanace, Venezia 081872
Sanaga-Njiensi, Tours 105766
Al-Sanam, Manama 100447
Sanandaj Museum, Sanandaj .. 024731
Sanat Dünyamız, İstanbul 139116
Sanati Museum, Kerman 024708
Al-Sanaya, Manama 100448
Sancarlos, Lisboa 084953
Sancerne, Pascal, Paris 071882
Sanchez, Barcelona 133161
Sanchez, Lima 084468
Sanchez, Miami 121854
Sanchez & Hijos, Juan,
Barcelona 085836
Sánchez & Priebe, Paloma,
Valencia 133470
Sanchez Adobe, Pacifica 051883
Sanchez Chacon, Ismael, Lima . 084469
Sanchez Palma, Carmen,
Granada 115095
Sanchez Pinazo, Málaga 115308
Sanchez Rigal y Otra, Murcia .. 115318
Sanchez Suarez, L., Gijón 115072
Sanchez Tapiz, Santiago, Donostia-
San Sebastián 115056
Sanchez, Dana, Austin 119697
Sanchez, Maika, Valencia 115516
Sanchez, Pedro, Madrid 086259
Sancho Pallisa, Francisco,
Barcelona 085837
Sancta, Vilnius 082342, 142968
Sancta Birgitta Klostermuseum,
Vadstena 041126
Sanctuary Cove Antique Gallery,
Sanctuary Cove 062186
Sanctuary Wood Museum – Hill 62,
Zillebeke 004162
Sand, Oslo 113239, 132789
Sand Bakkens Medisinske Museum,
Flekkefjord 033603
Sand Castle Gallery, Sorrento . 099693
Sand Point Antiques, Seattle .. 097498
Sand Springs Cultural and Historical
Museum, Sand Springs 053065
Sand Tchou Mengès, Paris ... 136984
Sand, Gérard, Bruxelles 063548
Sand, Jean-Marie, Aubevoye .. 066787
Sandaga Market, Dallas 093212
Sandal Castle, Sandal 045730
Sandauer Heimatstube, Arzberg 016612
Sandberg, Vellinge 133603
Sandberg, Patrick, London 090475
Sandblom, Victor, Seattle 124594
Sande Historielags Bygdesamlinger,
Sande 033920
Sandefjord Bymuseum,
Sandefjord 033922
Sandefjord Sjøfartsmuseum,
Sandefjord 033923
Sandellmuseet, Ramkvilla 040919
Sandels, Napoli 080928
Sandelsches Museum, Kirchberg an
der Jagst 019619
Sandelson, Robert, London ... 118561
Sanden Antikk og Brukt,
Trondheim 084389
Sander, Linköping 115729
Sander, Norrköping 115783
Sander, Gunnar, Lund 143890
Sander, H.A., Müllheim 077440
Sander, Henning, Lehrte 077050
Sander, Lothar, Goch 076022
Sander, R., Edemissen 106860
Sander, Stig, Frederiksberg ... 065675,
102806
Sanders, Capelle aan den IJssel 132463
Sanders, Nashville ... 095032, 135850
Sanders, Oxford ... 090938, 144773
Sanders, Tucson 124770
Sanders, Ken, Salt Lake City .. 145507
Sanders, R.J., Auckland 132697
Sanders & Sons, Robin,
Woodstock 091932
Sanders, Tony, Penzance 090968

Sanders, Yvonne, Auckland 083824
Sanderson, George & Peach,
Holmfirth 117981
Sanderus, Gent 140226
Sandfield, Norman L., Chicago . 092790
Sandford, Dublin 109618
Sandgate, Ayr 126805
Sandgate and District Museum,
Sandgate 001484
Sandgate Antiques, Penrith 090959
Sandhofer, Innsbruck . 062652, 100058
Sandi, Venezia 132256
Sandilands Forest Centre and
Museum, Hadashville 005938
Sandjon, J.P., Düren 075449
Sandler, Hudson, Atlanta 119608
Sandmann, Dr. Marina, Berlin .. 106418
Sandmore, Calgary 064271
Sandnes Kunstforening, Sandnes 084356
Sandnes Museum, Sandnes ... 033924
Sando, Sendai 082120
Sandon Museum, New Denver . 006464
The Sandor Teszler Gallery, Wofford
College, Spartanburg 053430
Sandorfi, Maria, Paris 071883
Sandoval, Phoenix 123513
Sandoval Sarrias, Olga de,
Barcelona 085838
Sandown Historical Museum,
Sandown 053068
Sandpiper Gallery, Mousehole .. 118760
Sandpoint Lighthouse, Escanaba 048602
Sandra & Márcio Objetos de Arte,
Belo Horizonte 063921
Sandra Ainsley Gallery, Toronto . 101702
Sandra and Mary's Doll Shop, San
Diego 097134
Sandras, Langenbernsdorf 142049
Sandra's Sensations, Miami 094798
Sandraz, Didier, Villeurbanne ... 074225
Sandrelle, Lyon 104131
Sandri, Paola, Modena 080819
Sandringham Antiques, Kingston-upon-
Hull 089705
Sandringham House Museum,
Sandringham 045735
Sandroni Rey, Los Angeles ... 121588
Sands Bygdasavn, Sandur 010740
Sands & Company Fine Art, New
York 123039
Sands-Willets House, Cow Neck
Peninsula Historical Society, Port
Washington 052266
Sandstein, Michael, Dresden .. 137149
Sandsteinmuseum, Bad
Bentheim 016718
Sandsteinmuseum Kloster Cornberg,
Cornberg 017761
Sandström, Magnus, Tingsryd . 133587
Sandtorg Bygdetun, Sørvik 033963
Sanduotang Museum, Taigu ... 008305
Sandved, Kjell B., Washington .. 124597
Sandvoss, Dieter-Christian,
Hannover 076391
Sandwich Auction House,
Sandwich 127447
Sandwich Fine Books, Sandwich 144825
Sandwich Glass Museum,
Sandwich 053075
Sandwich Historical Society Museum,
Center Sandwich 047498
Sandwich Historical Society Museum,
Sandwich 053072
Sandy Bay Museums, Rockport . 052647
Sandy K.'s Art Sales,
Minneapolis 122034
Sandy Spring Museum, Sandy
Spring 053079
Sandycove Fine Arts, Dublin ... 079586
Sandy's Antiques, Bournemouth 088541
Sandy's Gallery, Jacksonville ... 094038
Sandz Gallery, Hamilton 112906
Sanesi, Marinella, Milano 080715
Sanfeliu Gimeno, Maria Luz,

Valencia 133471
Sanfilippo, Giuseppa, Genova . . 080291
Sanford Museum, Cherokee . . . 047605
Sanford Museum, Sanford 053081
Sang, Moon Dang, Seoul 111709
Sangat, Ufa 114377
Sangenis, Albert, Barcelona . . . 085839,
. . . . 133162
Sangerberger Heimatstube,
Regenstauf 021333
Sangiorgi, Laigueglia 110161
Sangji Vocational College Museum,
Andong 029886
Sangleboeuf, Honfleur 103843
Sangli State Museum, Sangli . . 024465
Sango, Leuven 100777
Sangre de Cristo Arts Center
and Buell Children's Museum,
Pueblo 052411
Sanguin, Alberto, Genestrerio . . 087441,
. . . . 133752
Sanguin, Alberto, Mendrisio . . . 087674,
. . . . 133859
Sanguinetti, San Isidro 060961
Sani, Antonio, Firenze 080199
Sanilac County Historical Museum,
Port Sanilac 052257
Sanin Arenas, Ricardo, Cali . . . 128290
Sanitärhistorisches Museum,
Wien 003076
Sanjay Sharma Museum, Jaipur,
Rajasthan 024320
Sanjaya, Singapore 085496
Sanjo, Paris 105213
Sankara, Sylvie, Combs-la-Ville . 068110
Sankei-en Garden, Yokohama . . 029723
Sankofa, Cleveland 120451
Sankowsky, Lutz, Euskirchen . . 129828
Sankt Andrä, Salzburg 062815
Sankt Annegatans Bokbinderi och
Konservering, Lund 133550
Sankt Galler Brocante, Messe für
Antiquitäten, Kuriositäten und
Raritäten, Sankt Gallen 098333
Sankt Georg, Erfurt 075642
Sankt Georgs Antiquariat,
Amstetten 139971
Sankt Georgs Presse, Sankt Georgen
an der Gusen 136561
Verlag Sankt Gertrude, Hamburg 137253
Sankt Hans Gades Antik,
København 065950
Sankt-Jürgen Antiquariat, Lübeck 142088
Sankt Marien, Müntzergedenkstätte,
Mühlhäuser Museen, Mühlhausen/
Thüringen 020488
Sankt Matthäus-Kirche im Kulturforum,
Berlin 017280
Sankt Nikolaus-Hospital Cusanusstift,
Bernkastel-Kues 017336
Sankt Pauli-Museum, Museum
für Hamburgische Geschichte,
Hamburg 019018
Sankt-Peterburgskij Akademičeskij
Institut Živopisi, Skulptury i
Architektury im I.E. Repina,
Sankt-Peterburgskaja Akademija
Chudožestv, Sankt-Peterburg . 056301
Sankt-Peterburgskij Gosudarstvennyj
Universitet Kultury i Iskusstv, Sankt-
Peterburg 056302
Sankt-Peterburgskij Gosudarstvennyj
Universitet Technologii i Dizajna,
Sankt-Peterburg 056303
Sankt-Peterburgskij Muzej Chleba,
Sankt-Peterburg 037818
Sankt-Peterburgskij Muzej Teatralnogo
i Musykalnogo Iskusstva, Sankt-
Peterburg 037819
Sankt-Peterburgskij Pastelnyj Sojuz,
Sankt-Peterburg 059777
Sankt-Peterburgskij Sojuz Dizajnerov,
Sankt-Peterburg 059778
Sankt Pöltner Künstlerbund, Sankt

Pölten 058310
Sankt Veiter Geschichtenhaus, Sankt
Veit im Mühlkreis 002704
Sankta Karins Kyrkoruin, Visby . 041177
Şanlıurfa Devlet Güzel Sanatlar
Galerisi, Şanlıurfa 043052
Şanlıurfa Müzesi, Şanlıurfa . . . 043053
Sanmarino Selezione d'Arte Antica,
San Marino 085288
Sanmei, Nagoya 082062
Sanmen Museum, Sanmen 008215
Sanmenxia Museum, Sanmenxia 008216
Sanming City Art Cooperative,
Sanming 102165
Sanming City Calligraphy and Painting
Gallery, Sanming 102166
Sanming Museum, Sanming . . . 008217
Sanna, Gianfranco, Cagliari . . . 080002
Sannas Antik, Svedala 087094
Sannidal Bygdetun, Sannidal . . 033926
Sano Bijutsukan, Mishima 029030
Sano Gallery, Kokubunji 028877
Sano Studio, New York 135993
Sanquhar Tolbooth Museum,
Sanquhar 045739
Sans Ennui, San Francisco 124386
Sans Tache, Mumbai 109468
Sans Titre Ni Signature,
Coustellet 103638
Sanseido, Tokyo 142937
Sanselmo, Ciro, Napoli 080929, 131640
La Sanseverina, Parma 110635
Sanshui Museum, Sanshui 008218
Šanšieva, Karina, Moskva 114304
Sanskar Cendra Musiam,
Ahmedabad 109252
Sanske, Mary Anne, Zürich 116854
Sanskriti, Kolkata 109405
Sansom & Co., Bristol 138143
Sanson, Eric, La Baule-Escoublac 125602
Sanson, Jacqueline, Nice 070663
Sanson, Xavier, Le Mans 125612
Sansone, Giovanni, Verona 081936
Sansoni, Milano 137694
Sansovini, Luigi, Roma 132044
Sansu Museum, Meishan 008079
Sansum, James, New York 095721
Sant-Yago, Madrid 086260
Santa Agueda, Zaragoza 133512
Santa Ana College Art Gallery, Santa
Ana 053088
Santa Babara Maritime Museum,
Santa Barbara 053089
Santa Barbara Contemporary Arts
Forum, Klausner Gallery and Norton
Gallery, Santa Barbara 053090
Santa Barbara Historical Museum,
Santa Barbara 053091
Santa Barbara Museum of Art, Santa
Barbara 053092
Santa Barbara Museum of Natural
History, Santa Barbara 053093
Santa Cruz Art League, Santa
Cruz 053102
Santa Cruz Museum of Natural History,
Santa Cruz 053103
Santa Fe Children's Museum, Santa
Fe 053121
Santa Fe Colonial, Miami 094799
Santa Fe Community Museum, Santa
Fe 034592
Santa Fe Connections, San
Antonio 124069
Santa Fe Gallery, Santa Fe Community
College, Gainesville 049005
Santa Fe Trail Museum, Larned . 050150
Santa Fe Trail Museum, Trinidad
History Museum, Trinidad 053807
Santa Fe Trail Museums of Gray
County, Ingalls 049701
Santa Giulia – Museo della Città,
Brescia 025638
Santa Lucia Antichitá, Firenze . . 080200
Santa Maria Museum of Flight, Santa

Maria 053128
Santa Maria Valley Historical Society
Museum, Santa Maria 053129
Santa Monica Museum of Art, Santa
Monica 053135
Santa Rosa Junior College Art Gallery,
Santa Rosa 053143
Santacruz, Carlos, Bogotá 102365
Santagostini, Giorgio, Milano . . 131525
Sant'Agostino, Ventimiglia 111172
Santágueda Conservation y
Restauration, Madrid 133385
Santala, Ari, Joensuu 140725
Santamaria, Carlo, Napoli 110543
Santamaria, Claudia, Genova . . 131339
Santamaria, Nicolas Nicolaevich,
Santander 133425
Santamaria, Paola, Padova 081005
Santamarías, Bogotá 065225
Santangelo, Barbara, Palermo . . 131712
Santangelo, Massimiliano, Napoli 080930
Sant'Antonio, Rapallo 110698
Santarella Museum, Tyringham
Gingerbread House, Tyringham 053877
Santa'S, Barcelona 085840
Sante, Tania, Miami 094800
Santel, Bogotá 065226
Santi, Jakarta 109514
Santiago, Lisboa 143592
Santiago Contreras, Jerez de la
Frontera 086058
Santiago Cuadrado Rodriguez,
Madrid 133386
Santiago do Cacém Museu Municipal,
Santiago do Cacém 036004
Santiago, Antoine, Descartes . . 068288
Santiago, Ray, Louisville 135763
Santiaoshi History Museum,
Tianjin 008322
Santik Möbler, Stockholm 087049
Santillan, Buenos Aires 127536
Santo Calcagno, Catania 109954
Santo Campione, Milano 131526
Santo Niño Shrine and Heritage
Museum, Tacloban 034602
Santo Rossi, Catania 080047
Santolin, Giustiniano, Milano . . 080716
Santolini, Antonio, Roma 132045
Santolini, Gérard, Paris 071884
Santom, William, Pittsburgh . . . 136138
Santomauro, Buenos Aires 060928
Santos Júnior & Filhos, Lisboa . 132954
Santos, Miami 121855
Santos & Marcos, Lisboa 084956
Gonçalves Santos & Silveira,
Lisboa 143593
Santos Gonzalez, Carlos,
Barcelona 133163
Santos Vega, Rogelio, Madrid . . 133387
Santos, A.V., London 090476
Santos, Carlos Janeiro, Lisboa . 084954,
. . . . 084955
Santos, Fernando, Porto 114104
Santos, Fernando Ferreira, Braga 143547
Santos, J.F.J., Braga 084784
Santos, Maria Geraldo, Belo
Horizonte 063922
Santos, Maria Isabel Ribeiro,
Coimbra 084792
Santos, Silvia Lima, Fortaleza . . 100957
Santos, Virgínia, Coimbra 113966
Santosh Art Gallary, Jaipur,
Rajasthan 109387
Santostefano, Roberto, Buenos
Aires 060929
Santuario Casa Santa Caterina,
Siena 027962
Santuario de Nuestra Senora de
Guadalupe, Santa Fe 053122
Santuario della Beata Vergine della
Salute, Solarolo 027977
Santuario Eucaristico, Basilica di
Santa Cristina, Bolsena 025578
Sanuki Folk Art Museum,

Takamatsu 029434
Sanuki Museum of Soy Sauce Art,
Sakaide 029318
Sanvido, Danilo, Milano 131527
Sanxingdui Museum, Guanghan 007797
Sanya Museum, Sanya 008219
Sanying Pagoda and Museum of
Dinosaur, Shaoguan 008250
Sanyuan Bowu Guan, Sanyuan . 008220
Sanz, Madrid 143783
Sanz, Daniel, Nogent-le-Rotrou . 070727
Sao, Paris 105214
São Gabriel Antiguidades, Lisboa 084957
São Marcos Restauracao Artistica em
Prata, São Paulo 128133
São Roque, antiguidades e Galeria
de Galeria de Arte, Espaço de São
Bento, Unipessoal Lda., Lisboa 084958
Saona, Miami 121856
Saône Antiquités, Saint-Laurent-sur-
Saône 072888
SAP Galeria – Spółdzielni Artystów
Plastyków, Łódź 113682
Sapet, Mirmande 104242
Sapet, Valence (Drôme) 105802
Sapet, Christian, Saint-Ouen . . 073194
Saphan, Linda, Phnom Penh . . . 101205
Saphir, Paris 105215, 141278
Sapho, New York 095722
Sapiens, Lugano 138068
Sapino, Patrick, Briançon 067489
Sapjo, Monaco 082529
Sapone, Antonio, Nice 104388
Sappey, Huguette, Puget-Ville . . 072321
Sappey, Jacques, Puget-Ville . . 072322
Sappho, Glebe 139753
Sapporo Clock Tower Exhibition,
Sapporo 029336
Sapporo Geijutsu no Mori Bijutsukan,
Sapporo 029337
Sapporo Museum of Sculpture,
Sapporo 029338
Sapporo Science Center, Sapporo 029339
Sapporo-shiritsu Hakubutsukan,
Sapporo 029340
Sapporo Waterworks Memorial
Museum, Sapporo 029341
Sapporo Winter Sports Museum,
Sapporo 029342
Sapulpa Historical Museum,
Sapulpa 053145
Sapumal Foundation Gallery,
Colombo 040480
Sara, Sendai 082121
Sara Frances Antiques, Alderley
Edge 088242, 134018
Sara Galleries, Houston 121124
Sara Hildénin Taidemuseo,
Tampere 011402
Sara, Veneziano, Roma 081546
The Saracen's Lantern,
Canterbury 088781
Sarad, Nadine A., Köln 130166
Sarafkina Kašta Gradski Bit, Veliko
Tărnovo 005258
Sarah, Middelburg 083500
Sarah Campbell Blaffer Foundation,
Houston 049595
Sarah Campbell Blaffer Gallery, Art
Museum of the University of Houston,
Houston 049596
Sarah Joan Antiques, North
Rockhampton 061998
Sarah Moody Gallery of Art, University
of Alabama, Tuscaloosa 053861
Sarah Orne Jewett House, South
Berwick 053399
Sarah Spurgeon Gallery,
Ellensburg 048548
Sarakatsani Folklore Museum,
Serres 023258
Saramito, Marie-Dominique,
Cannes 067652
Sarang Art Gallery, Nairobi 111563

Sarasota Classic Car Museum,
Sarasota 053150
Saraswati, Freiburg im Breisgau 075880
Saraswaty, Kolkata 137596
Sarat Smriti Granthagar, Panitras 024429
Saratoga Museum, Saratoga . . 053154
Saratoga National Historical Park,
Stillwater 053530
The Saratoga Springs History Museum,
Historical Society of saratoga
Springs, Saratoga Springs . . . 053159
Saratovskij Étnografičeskij Muzej,
Saratovskij oblastnoj muzej
kraevedenija, Saratov 037846
Saratovskij Gosudarstvennyj Muzej
Boevoj Slavy, Saratov 037847
Saratovskij Kraevedčeskij Muzej,
Saratov 037848
Saravia, San Salvador . 103103, 103104
Saray, Manama 063159, 063160
Sarazin, L'Isle-sur-la-Sorgue . . 069592
Sarcander, Arthur, Gummersbach 076132
Sarda Ukil School of Arts, Delhi 055792
Sardana's Art Centre, Delhi . . . 109340
Sardar Patel University Museum,
Vallabhvidyanagar 024511
Sardar Vallabhbhai Patel Museum,
Surat 024487
Sardine, Noirmoutier-en-l'Ile . . 070737
Sareban, Verona 081937
Sarehole Mill, Birmingham 043625
Sarenhes, Québec 101520
Sargadelos, Madrid 115283
Sargadelos, Sevilla 115457
Sargeant, Alex, Seattle 124595
Sargeant, D. & A., Sevenoaks . . 091271,
135019
Sargent, Los Angeles 121589
Sargent House Museum,
Gloucester 049136
Sargent, Peter M., Portland . . . 096469
Sarginson, K.J., York 135271
Sarguinetti House Museum,
Yuma 054610
Sarington College of Arts, University
College Falmouth, Falmouth . . 056533
Šarišská Galéria, Prešov 038432
Šarišské Múzeum, Bardejov . . . 038290
Sarjeant Gallery – Te Whare o Rehua,
Wanganui 033334
Şark Sigorta Sanat Galerisi,
İstanbul 117083
Sarkel, Kerteminde 065813
Sarkis, Saint-Ouen 073195
Sarkis, Khalil Jean, Beirut 082304
Sarkisian, Denver 093332
Sarkissian, Serge, Bordeaux . . . 103439
Sarkoezy, Helene von, München 077607
Sarm-Sart, Crans-Montana 087387
Sarnelli, Antonio, Napoli 131641
Sarno, Palermo 081074, 126218
Sarno, Allen, Seattle 145598
Saroj, Jaipur, Rajasthan 079396
Sarojini Naidu School of Performing
Arts, Fine Arts and Communication,
University of Hyderabad,
Hyderabad 055802
Saron, Porto Alegre 101002
Sárospataki Képtár, Herman Ottó
Múzeum, Sárospatak 023873
Sárospataki Református Kollégium
Tudomanyos Gyüjteményei Múzeum,
Sárospatak 023874
Sarpy County Historical Museum,
Bellevue 046884
Sarrance, Bernard, Pau 072051
Sarrat Museum, Sarrat 034595
Sarraute, Dominique,
Carcassonne 067679
Sarrazin, François, Orléans 070821
Sarti, Enrico, Firenze 131261
Sarti, Giovanni, Paris . . 071885, 105216
Sarti, Marco, Bologna 130998
Sartori & Cerveau, Paris 105217

Sartori, Rossana, Padova 081006
Sartun Sociedade de Artes Unidas, Rio
de Janeiro 101066
Sarue Tatuagem, Fortaleza 063951
Şarur District Regional History
Museum, Şarur 003198
SASA Gallery, University of South
Australia, Adelaide, Adelaide . . 000770
Sasayama-shiritsu Bijutsu oyo Rekishi
Hakubutsukan, Sasayama 029346
Sasfra, Belo Horizonte 063923
SASH Book and Papercare Restauratie,
Maastricht 083477, 132577
Saskatchewan Arts Board Collection,
Regina 006778
Saskatchewan Baseball Hall of Fame
and Museum, Battleford 005400
Saskatchewan Craft Council Gallery,
Saskatoon 006983
Saskatchewan Military Museum,
Regina 006779
Saskatchewan Pharmacy Museum,
Regina 006780
Saskatchewan River Valley Museum,
Hague 005939
Saskatchewan Science Centre,
Regina 006781
Saskatchewan Sports Hall of Fame
and Museum, Regina 006782
Saskatchewan Western Development
Museum, Yorkton 007500
Saskatoon Antique, Saskatoon . 064569
Saskatoon Furniture Restorables,
Saskatoon 128202
Saslow, Judy A., Chicago . . . . . 120263.
Saß, Linde & Karl Heinz, Lübeck 130256
Sassen, Edith, Planegg 108385
Sassi-Milici, Vallauris . 105811, 137011
Sassin, Bernard, L'Isle-sur-la-
Sorgue 069593
Sassnitzer Fischerei- und
Hafenmuseum, Sassnitz 021619
Il Sasso, Don Torcuato 060943
Sassolini, Alvaro, Firenze 131262
Sasson, Buenos Aires 060930
Sasson, David, Mandelieu-La
Napoule 069834
Sassone, Francesca, Milano . . . 080717
Satakunnan Museo, Pori 011268
Satani, Kazuhiko, Tokyo 111490
Satch Kiely, Westport 079721
Satellite Space Studio, Fortitude
Valley 099053
The Satguru Art 'n' Frames, Jaipur,
Rajasthan 109388
Satguru's, Mumbai 109469
Satkatar, Pune 079416
Sato Art Museum Toyama,
Toyama 029637
Sato Institute of Typography,
Yokohama 056066
Sato-Stone, Tokyo 082195
Satoe Kindai 21-seiki Bijutsukan,
Kazo 028814
Satoru, Honolulu 120967
Satrosphere, Aberdeen 043363
Sattar Bahlulzadeh Exhibition Center,
Baki 003177
Satterlee Clark House, Horicon . 049558
Sattler, Manfred, Schwangau . . 130633
Sattmann, Helmut, Bremen . . . 075182
Satura, Genova 059474, 110135
Satwa Art Gallery, Pune 109500
Satya Art Gallery, Mumbai 109470
Saubot, Roger, Bordeaux 067358
Saubot, Roger, Saint-Jean-de-
Luz 072847
Sauc, Surry Hills 099753
Saudemont, Dominique, Arques (Pas-
de-Calais) 066735
Saudou, Jean-Paul, Eysines . . . 068499
Sauer, Susanne, Wien 063062
Sauerbrei, K.-H., Einbeck 075588

Sauerland-Museum, Arnsberg . . 016608
Sauermann & Sauermann-Kern,
Ammerbuch 074407
Sauermann, Gottfried, Radebeul 142314
Saugus Iron Works, Saugus . . . 053163
Sauhutupamuseo, Somero 011367
Sauk County Historical Museum,
Baraboo 046787
Sauk Prairie Area Historical Society
Museum, Prairie du Sac 052348
Saukkojärven Kotiseutu- ja
Koulumuseo, Ranua 011301
Saul, Julie, New York 123040
Sauli, Joseph, Châtillon (Hauts-de-
Seine) 067917
Saulneron, Patrick, Montréjeau . 070322
Sault Sainte Marie Canal, Sault Sainte
Marie 006992
Sault Sainte Marie Museum, Sault
Sainte Marie 006993
Saunders, Scullin . . . 062196, 139906
Saunders County Historical Museum,
Wahoo 054025
Saunders Memorial Museum,
Berryville 046950
Saunders, Charles, London . . . 090477
Saunders, Lorrie M., Norfolk . . . 095875
Saunier, Maryvonne, Saint-Jean-de-
Braye 129235
Saunier, Pierre, Paris 141279
Saunier, Raymond, Yvetot 074315
Saunier, Sylvain, Issoudun 068884
Saupe, Andre, Potsdam 078084
Saura Lluch, Vicenta, Barcelona 133164
Saura Miret, Gabriel, Ciutadella de
Menorca 085938
Saurer Oldtimer-Museum, Arbon 041229
Sauri Pardo De Atin, Barcelona . 133165
Saurier-Museum, Frick 041506
Sauriermuseum Aathal, Aathal-
Seegräben 041193
Saurierpark Jurassic Island,
Traismauer 002833
Sauriol, H., Montréal 064470
Sauro, Luciano, Schio 110921
Sauser, Pierre, Peseux 087734, 133886
Saussez, Patrick, Tocane-Saint-
Apre 073686
Sausverd, Patrick, Joigny (Yonne) 125598
Sautee-Nacoochee Museum,
Sautee 053172
Sauter, Adolf Val.,
Langenenslingen 130213
Sauthier, Michel, Sion 087817
Sautter & Lackmann, Hamburg . 141894
Sauvage, Yannick, Andouillé . . . 066590
Sauvageot, Daniel, Longjumeau 069654
Sauvageot, Jean-Claude,
Couches 068171
Sauvant, Segonzac 073474
Sauveboeuf, Emmanuel de, Uzès 073926
Sauvon Kotiseutumuseo, Sauvo . 011346
Sauwald-Heimathaus, Sankt
Roman 002699
SAV, Bologna 109861
SAVAC – South Asian Visual Arts
Center, Toronto 058470
Savacou, New York 123041
Savage, Agoura Hills 139488
Savage, Portland 123674
Savage Carter, Nancy, Fort Worth 120684
Savage Fine Art, Harlestone . . . 117909
Savageau, Denver 120768
Savannah College of Art and Design,
Savannah 057883
Savannah Gallery, Atlanta 119609
Savannah House, Baltimore . . . 092361
Savant, Vancouver . . 064872, 064853
Savard, Jean-Christophe, Castelnau
d'Estrétefonds 067708, 128531
Savaria Múzeum, Szombathely . 023983
Save On South, Edmonton 064305
Savelli, Roma 110856
Savenvalajamuseo, Somero . . . 011368

Savery Antiques, Brighton 088635
Savery Books, Brighton 144269
Savez Muzejskih Društva Hrvatske,
Zagreb 058526
Savez Udruženja Likovnih Umetnika,
Beograd 059789
Savill Galleries, Paddington, New
South Wales 099534
Savill, H.E., Scarborough 135011
Saving Grace, Fort Worth 093479
Savings Banks Museum,
Ruthwell 045654
Savings World Auction, Al-Wadi Al-
Kabir 126398
Savino, Esposito, Firenze 080201
Savio, Costantino, Torino 132197
Savio, Alberto, Milano 080718
Savio, Costantino, Torino 132197
Savio, Sonia, Torino 132198
Savitsky nomidagi Qoraqalpogiston
Respublikasi Davlat San"at Muzeyi,
Nukus 054637
Savoi, Bolzano 079901
Savoir de Marie, Haguenau . . . 068778
Savonlinnan Maakuntamuseo,
Savonlinna 011348
Savoy Antiques, New York 095723
Savoy, Christoph, Seon 133922
Savoy, William, Richmond 136176
Savremena Galerija Umetničke
Kolonije Ečka, Zrenjanin 038261
Savremena Galerija, Centr za kulturu
Olga Petrova, Pančevo 038223
Savreux, Marine, Biot 067227
Savupirttimuseo, Kuopio 011056
Savvy Chic, Detroit 093375
Savvy Collector, Phoenix 123514
SAW Gallery, Ottawa 006598
Sawa, Nagoya 111332
Sawanoi Museum of Traditional
Japanese Ornaments, Ome . . . 029228
Sawanworanayok National Museum,
Sukhothai 042722
Sawers, Robert G., London . . . 090478,
144684
Sawhill Gallery, Duke Hall,
James Madison University,
Harrisonburg 049378
Sawhney, Dr. Jagdev S., Bonn . 141622
Sawin Museum, Dover 048352
Saworski & Stöppler, Berlin . . . 141565
Sawtooth School for Visual Art,
Winston-Salem 058127
Sawyer's Sandhills Museum,
Valentine 053930
Saxida, Paolo, Trieste 081802
Saxon, Dallas 093213
Le Saxon, Tours 073862
Saxon Cross Antiques Emporium,
Sandbach 091231
Saxon, Daniel, West Hollywood . 125024
Saxonia, alte und neue Kunst,
München . . . 077608, 108189, 142183
Saxony Imports, Cincinnati 092894,
120374
Saxtead Green Post Mill, Saxtead 045741
Al-Sayed Al-Khatat, Hanafi, Cairo 066270
Al-Sayed, Hosam Abdul Majeed, Al-
Khobar 085295
Sayell, Graham, Vancouver . . . 101834
Sayer, M. & R., New York 123042
Sayger, Indianapolis 093923
Sayk, Frederiksberg 065676
Saylor Byrd, Louisville 094584
Sayme Arte, Las Palmas de Gran
Canaria 133410
Sayn-Wittgenstein, New York . . 123043
Sayner Gießhalle, Bendorf 017026
Sayville Historical Society Museum,
Sayville 053187
Sayward-Wheeler House, York
Harbor 054589
SB, Paris 105218
SB Antik, Göteborg 086718

SBA, Le Epouville 128656
Sbaffi, Hedy, Neunkirch 087708
SBC Pioneer Museum, San
  Francisco 053026
SBS Antiques, Denver 093333
SC Librària Corina, Arad 143608
SCA Gallery, University of Sydney,
  Rozelle 001464
Scaccabarozzi, Giorgio, Bergamo 079794
Scaccianoce, Giuseppe, Palermo 110617
SCAD Museum of Art, Savannah 053180
Scaglia, Giulio, Brescia 079955
Scagliola, Madeleine, Paris 129140
Scala, London 138248
Scala, New York 138424
Scala, Paris 136985
Scala Arte, Verona 111195
La Scala Gallery, San Diego .. 124170
Scala Group, Antella 137615
Scala, Giovanni, Napoli 080931
Scalabrino, Jean, Paris 071886
Scalabrino, Jean-Christophe,
  Paris 071887
Scalae, Documentos Periódicos de
  Arquitectura, Barcelona 139069
Scalamandre, Long Island City . 050383
La Scaletta, San Polo d'Enza . 110899,
                               137741
Scaletta di Vetro, Milano 110426
Scalia, Carminella, Messina .. 131360
Scalloway Museum, Scalloway . 045742
Lo Scalpellino, Parma 131739
Scalvini, Lino, Brescia 131048
Scalzotto, Claudio, Roma 081547
Scammell, Norwood 125220
Scanabissi, Giorgio, Bologna .. 130999
Scanabissi, Matteo, Bologna .. 079877
Scandice, Tullamarine 127691
Scandinavia House, New York . 051465
Scandinavian Antiques, Denver . 093334
Scandinavian Auction Systems,
  Oslo 126379
Scandinavian House, København 065951
Scandinavian Museum Association,
  Icelandic Section, Reykjavik . 059350
scaneg Verlag, München 137421
Scantic Academy Museum, East
  Windsor 048452
Scantik, København 065952
Scantique, Stockholm 087050
Scapa Flow Museum, Hoy 044612
Scaplen's Court Museum, Poole 065529
Scarab Oracel, Miami 094801
Scarabantia Fóruma, Régészeti
  Kiállítóhely, Sopron 023899
Le Scarabée d'Or, Le Mans ... 069312
Scarabellin, Andrea, Roma 081548
Lo Scarabeo, Livorno . 080334, 080335
Fratelli Scarano, Milano 131528
Scaravilli, Cleveland 120452
Scarborough Art Gallery,
  Scarborough 045745
Scarborough Fair Art Gallery,
  Scarborough 099682
Scarborough Historical Museum,
  Scarborough 006997, 053188
Scarborough Perry Fine Art, Hove 126928
Scarfó Assunta, Palermo 081075
Scarfó, Rosina, Torino 081755
Scarfone and Hartley Galleries,
  University of Tampa, Tampa .. 053670
Scaringi, Amalia, Napoli 080932
Scarlatella, Giacomo, Milano .. 131529
Scarlett, Paris 071888
Scarlett, Rouen 072616
Scarlett, Warszawa 113899
Scarpa, Paolo, Venezia 132257
Scarpa, Pietro, Venezia 081873
Scarpelli, Stefano, Firenze ... 131263
Scarpignato, Aldo, Roma 081549,
                               142837
Scarscelli, Dante, Nice 070664
Scarsdale Historical Society Museum,
  Scarsdale 053189

Scartazzini, Alex, Jegenstorf ... 087538,
                               133779
Scarthin Books, Cromford 144364
Scatizzi, Mauro, Trieste 081803
Scauri, Elisabeth, Parma 081127
Scavengers, New York 095724
Scavi Archeologici di Velia,
  Ascea 025381
Scavi di Ercolano, Ercolano ... 026157
Scavi di Oplontis, Torre
  Annunziata 028148
Scavi di Ostia, Ostia Antica ... 027128
Scavi di Pompei, Pompei 027393
Scavi di Stabiae, Castellammare di
  Stabia 025826
Scellier, Michel, Andainville ... 128440
Scène Antique, Paris 071889
Scerra, Paolo, Firenze 080202
Schaaf, Dr. Hildegard, Köln .. 107678
Schaár Erzsébet – Utca, Janus
  Pannonius Múzeum, Pécs .. 023854
Schaár Erzsébet Gyüjtemény,
  Szent István Király Múzeum,
  Székesfehérvár 023939
Schabel, Manfred, Berlin 074909
Schabert, Klaus, Moosburg an der
  Isar 108031
Schach, Horst, Karlsruhe 076701
Schachen-Galerie, Oensingen ... 087722,
                               116552
Schacherbauer, Erika, Aschau im
  Chiemgau 106018
Schachmuseum Ströbeck, Ströbeck
  Schachdorf 022017
Schacht, Stefan, Kastorf 130106
Schack-Galerie, Bayerische
  Staatsgemäldesammlungen,
  München 020570
Schack, E., Slagelse 066132
Schade, Iris, Altenburg 074393
Schade, Ralf, Naumburg, Saale . 077692,
                               130436
Schade, S., Naundorf 077694
Schaden, Wien 140119
Schaden, Markus, Köln 142022
Schadow, Norbert, Berlin 074910
Schäbel, Gerlinde, Fischach ... 075736
Schaede, P.A., Kiel 076791
Schäfer, Otterberg 077995
Schaefer Franck, La Frette-sur-
  Seine 069040
Schäfer, Beatrix, Stuttgart 130686
Schäfer, Bernhard, Bad
  Karlshafen 141426
Schäfer, Christian, Trendelburg . 142416
Schäfer, Claudia & Eduard, Utting am
  Ammersee 078775
Schäfer, Hermann, Roth 078284
Schaefer, Iris, Köln 130167
Schäfer, Karl Günther, Gießen . 107131,
                               137224
Schäfer, M., Sankt Augustin ... 078349
Schäfer, Michel, Wien 127963
Schäfer, Torsten, Lich 077099
Schäferling, Dillingen an der
  Donau 075358
Schäferling, A., Bissingen,
  Schwaben 075011
Schäferling, M., Bissingen,
  Schwaben 075012
Schäfertanz-Kabinett, Rothenburg ob
  der Tauber 021511
Schaeffer House Museum,
  Grafton 001119
Schäffer, Ralf, Falkensee 129832
Schaer & Wildbolz, Solothurn . 116654
Schärer, Trudi & René, Sankt
  Gallen 087788
Schärz & Frey, Aarau . 087172, 087173,
                       133608,   133609
Schätz, Sankt Pölten 062831
Schaffer, A., Halle, Saale 076170
Schaffer, Corky, Denver 093335
Schaffer, Ingrid, Wien 127964

Schafferer, Hermann, Innsbruck 100059
Schaffernicht, Christian, Schwäbisch
  Gmünd 142353
Schaffhauser Weinbaumuseum,
  Hallau 041581
Schaffrath-Larché, Susanne,
  Pforzheim 078043
Schaffrin, Ruth, Konstanz 130184
Schafschetzy, Günther, Graz .. 062600,
                               100018
Schaible, Hilde, Stuttgart 078629
Schain, Cincinnati 120375
Schake, Walter,
  Mönchengladbach 077398
Schakenraad, Jos, Schijndel .. 132628
Schaks, H., Ahrensburg 105967
Schalk, Gerhard, München 077609
Schall, Alzey 074401
Schalla, G., Karlsruhe 107524
Schaller, E. & N., Würzburg ... 079090
Schallmeiner, Karl, Ried im
  Innkreis 062776
Schamong, E. & A., Heinsberg . 076484
Schamser Talmuseum, Tgea da
  Schons, Zillis 042181
Schandog, Michael, Cottbus .. 075276
Schantl, Gerlinde, Wartberg im
  Mürztal 062871
Schanzenbächer, Günter,
  Neidenfels 077699
Schaper, Dietrich, Hamburg ... 076304,
                               141895
Schaper, Eckehardt, Alsfeld ... 129379
Schaper, Klaus, Frankfurt am
  Main 075819
Schaper, M. & H., Alfeld 137025
Schaper, Maya, New York 095725
Scharbach, V., Kempten 076754
Schardt, Mario, Michelau in
  Oberfranken 077368
Scharegg, Karl, Chur 087372
Scharf, Wolfsberg 063121
Scharfenberger, Hartmut,
  Meiningen 077330
Scharff, A., Kerpen 076765
Scharinger, Franz, Vorchdorf .. 127893
Scharmann, Marion, Köln 107679
Scharold, H. Karl, Houston ... 093786,
                               135615
Schatkamer Sint Lambertuskerk,
  Horst 032514
Schatkamer Sint Walburgis Basiliek,
  Arnhem 032030
Schatkamer van de Basiliek van Onze
  Lieve Vrouwe, Maastricht ... 032628
Schatkamer van de Kathedrale
  Basiliek Sint-Bavo, Haarlem .. 032397
De Schatkamer van de Onze-Lieve-
  Vrouwekerk, Sint-Truiden ... 004029
Schatkamer van de Sint-
  Salvatorskathedraal, Brugge . 003424
Schatkamer van de Sint-
  Servaasbasiliek, Maastricht ... 032629
Schatkamer Sint-Catharina,
  Maaseik 003835
Schatt, Bruno, Einsiedeln 087406
Schatt, Christian, Dresden 129722
Schattenburg Museum, Heimatpflege-
  und Museumsverein Feldkirch,
  Feldkirch 001882
Schattleitenmühle, Garsten .. 001927
Schatz-Truhe, Saarburg 078338
Schatz, Holger, Fürth 129875
Schatz, Peggy, Chicago 120264
Schatzhaus in der Lausitz-Museum für
  Morgenlandfahrer, Göda ... 018723
Schatzkammer, Altötting 016557
Schatzkammer, Volkach 078814
Schatzkammer – Turmmuseum der
  Pfarrkirche, Sankt Wolfgang im
  Salzkammergut 002707
Schatzkammer der Basilika St.
  Ludgerus, Essen 018312
Schatzkammer der Evangelisch-

Lutherischen Sankt Andreaskirche,
  Weißenburg in Bayern 022540
Schatzkammer der Katholischen
  Pfarrkirche Sankt-Servatius,
  Siegburg 021849
Schatzkammer der Residenz München,
  Bayerische Verwaltung der
  staatlichen Schlösser, Gärten und
  Seen, München 020571
Schatzkammer der Wallfahrtskirche,
  Hohenpeißenberg 019307
Schatzkammer der Wallfahrtskirche
  Maria Schutz, Maria Schutz am
  Semmering 002360
Schatzkammer des Münsters,
  Reichenau 021343
Schatzkammer Grafenrheinfeld,
  Grafenrheinfeld 018790
Schatzkammer Kloster Sankt
  Marienstern, Panschwitz-
  Kuckau 021070
Schatzkammer und Museum des
  Deutschen Ordens, Wien .. 003078
Schatzkammer, Kunsthistorisches
  Museum Wien, Wien 003077
Schatzkiste, Drochtersen 075435
Schatzl, Wolfgang, Salzburg .. 062816
Schatztruhe, Berlin 074911
Schatztruhe, Duisburg 075557
Schatztruhe, Frankfurt am Main 075820
Die Schatztruhe, Hamburg ... 076305
Schatztruhe, Wittlich 079036
Schau- und Erlebnissägewerk
  Mühlauersäge, Fusch an der
  Glocknerstraße 001916
Schaub, Gun, Babenhausen,
  Hessen 074482
Schaubergwerk, Leogang 002297
Schaubergwerk – Historische
  Silbergruben, Oberzeiring ... 002474
Schaubergwerk Arzberg, Arzberg 001724
Schaubergwerk Büchenberg,
  Elbingerode, Harz 018152
Schaubergwerk Grillenberg,
  Payerbach 002489
Schaubergwerk Hochfeld, Neukirchen
  am Großvenediger 002437
Schaubergwerk Kupferplatte,
  Jochberg 002149
Schaubergwerk Seegrotte,
  Hinterbrühl 002083
Schaubergwerk Teufelsgrund,
  Münstertal 020613
Schaubergwerk Zum Tiefen Molchner
  Stolln, Pobershau 021163
Schaudenkmal Gaszentrale,
  Unterwellenborn 022283
Schauer, Edeltraud, Saal an der
  Donau 108538
Schauer, Franz K., Krems 062693
Schauer, Markus, Frankfurt am
  Main 075821
Schaufuß, Hans-Joachim, Kassel 076729
Schauglashütte, Uetendorf .. 042075
Schaulager, Laurenz-Stiftung,
  Münchenstein 041775
Schauland, H., Neuhausen,
  Enzkreis 077732
Schaum, Günter, Bad
  Schwalbach 074592
Schaumanufaktur Ostsee-Schmuck,
  Leopoldshagen 020046
Schaumburg, Stade 142372
Schaumuseum Alte Schmiede, Huf-
  und Wagenschmiede, Gmünd,
  Niederösterreich 001949
Schaupp, Reinhard,
  Untergruppenbach 078763
Schauraum für angewandte Kunst,
  Wien 100368
Schauraum im ehemaligen Stift Göß,
  Leoben 002294
Schauraum K3, Simbach am Inn 108623,
                               108624

Schauwerkstatt Historische
  Handweberei, Oederan 020950
Scheck, Peter, Wien 127965
Scheding, Berry, Northwood ... 062006
Schedler, Amanda, Birmingham 119819
Scheel & Melkonian, Amsterdam 082848
Scheel-Ungerleider, Kathrin, New
  York 135994
Scheel von Rodenberg, Jan,
  München 130391
Scheel, Elisabeth, Wien 127966
Scheel, Norbert, Tübingen 078728,
                               126130
Scheele, von, Malmö 126639
Scheelen, Wolfgang, Düsseldorf 075513
Scheepens & Vink, Amsterdam . 082849
Scheepers, H.P.M., Maastricht . 083478
Scheepswerf De Delft, Rotterdam 082823
Scheepvaartmuseum, Bornem .. 003384
Scheepvaartmuseum,
  Gasselternijveen 032332
Scheepvaartmuseum Baasrode,
  Baasrode 003336
Scheer, Sandra, Roost 082386
Scheers, P., Kortrijk 063727
Scheffel, Christian K., Bad Homburg
  v.d.Höhe 106066
Scheffel, Walter, Bad Homburg
  v.d.Höhe 074525
Schefflenztal-Sammlungen,
  Schefflenz 021636
Scheherazade, Abu Dhabi 117168
Schehrezâd, Weinfelden 144147
Scheib, Volker, Biblis 129586
Scheibeler, Gundula,
  Neubrandenburg 142222
Scheible, Gunter, Ludwigsburg . 077157
Scheibler, Aurel, Berlin 106419
ScheiblerMitte, Berlin 106420
Scheibner, Sabine, Herrsching am
  Ammersee 130038
Scheich, Heinrich, Flieden 129842
Scheicher, Alois, Dirnbach 062543
Scheidegger, Richterswil 116583
Scheidemann, Stephan,
  Friedrichroda 129867
Scheider, Herbert, Bitburg 075014
Scheidwimmer, Xaver, München . 077610
Scheil, Peter, Hamburg 125974
The Schein-Joseph International
  Museum of Ceramic Art, Alfred 046399
Scheinast, Andreas, Korneuburg 100081,
                               127793
Scheiner, John, Ingolstadt 076614
Scheinowitz, H. & J., Wien 063063
Scheldemuseum de Notelaer,
  Bornem 003385
Schell, Christine, London 090479,
                               134762
Schell, Gerfried, Rees 078160
Schell, Sybille, Lengerich 077084
Schellbach, Doris, Grasbrunn .. 076080
Scheller, Oliver, Wörthsee 130775
Schellmann, New York 123044, 138425
Schellmann Art Production Munich,
  München 108190
Schelma, Bruxelles ... 063549, 128036
Schelpen Museum, Zaamslag .. 033073
Schemel, K., Schorndorf 108576
Schemm, J., Münster 108226
Schempp, Kornwestheim 130186
Schenectady County Historical Society
  Museum, Schenectady 053196
Schenectady Museum,
  Schenectady 053197
Schenk, Edmonton 128169
Schenk, Haßfurt 107380
Schenk-Röcker, Evamaria,
  Zofingen 087924
Schenk zu Schweinsberg, Ekkehard,
  Marburg 142115
Schenk, Katryn, Niederwangen bei
  Bern 116537

Schenk, Klaus, Calw 075244
Schenk, Lutz, Hamburg 141896
Schenk, M. W., Engelskirchen .. 075619
Schenker, Zürich 116855
Schepenzaal Stadhuis, Zwolle .. 033127
Schepers, Dr. K. & R., Münster . 077675
Schepke, Marek, Poznań 084649,
                               143497
Schepp, N.J.R., Rotterdam 132624
Scheppe, K., Hügelsheim 076583
Scheppler & Müller, Neubiberg . 142221
Scherb, Wolfgang, Freiburg im
  Breisgau 075881
Scherer & Fils, Saint-Maurice-sur-
  Huisne 072971
Scherer & Vandenburg,
  Hilversum 083407
Schérer, Hélène-Elisabeth, Gironde-sur-
  Dropt 068668
Scherer Sohn, J. Otto, Bern .. 087300,
                               133690
Scherer, Paulo Frederico, Rio de
  Janeiro 064122
Scherer, Pedro Jose, Rio de
  Janeiro 064123
Scherf, D. & I., Nienstädt 077788
Schering-Museum Scheringianum,
  Berlin 017281
Scheringa Museum voor Realisme,
  Spanbroek 032897
Scheringer, Georg, Konstanz ... 142032
Scherini, Lydia, Balzers 082332
Scherjon's Klompenmakerij en
  Museum, Noardburgum 032694
Scherler, Hans & Antje, Berlin .. 074912
Schernig, Ursula, Fürth 107086
Scherz, Wolfgang, München ... 130392
Schettini, Giancarla, Faenza ... 110012
Schettino, Umberto, Roma 081550
Scheuch, Andreas, München ... 130393
Scheumüller, Walter, Ried im
  Innkreis 062777
Scheuring-Rojewski, Georg,
  Würzburg 079091
Scheuringer, Linz 140014
Scheuringer, Werner, Wien 063064
Schiaffino, Michela, Genova ... 080292
Schiaparelli, Giampaolo, Roma . 081551
Schiavon, Vittorino, Padova ... 131676
Schiavone, Torino 081756
Schibli, Robert, Riehen 087749, 144123
Schicht, Renate, Düsseldorf ... 106821
Schichtwechsel, Aktionsraum für
  neue Kunst und Kommunikation,
  Triesen 059550
Schick Art Gallery, Skidmore College,
  Saratoga Springs 053160
Schick, Johann B., Koblenz 130137
Schick, Silke, Metzingen 130503
Schied, Jürgen, Kornwestheim . 076944
Schiedams Prentenkabinet,
  Schiedam 143227
Schiefer- und Ziegelmuseum Dörfles-
  Esbach, Dörfles-Esbach 017921
Schiefermuseum, Ludwigsstadt . 020147
Schiefke, Reinhard, Husum 107455
Schiel, Sankt Peter-Ording 108557
Schiele Museum of Natural History,
  Gastonia 049058
Schierding, Patrick, Delmenhorst 129674
Schierenberg, Dieter, Amsterdam 143503
Schierl, Ludwig, München 077611
Schierstaedt, Eckehard von,
  Münster 130429
Schieske, Lothar, Herten 076514
Schiess, Patricia, Basel 144032
Schiffahrts- und Schiffbaumuseum,
  Wörth am Main 022719
Schiffahrtsmuseum der
  Oldenburgischen Weserhäfen,
  Brake 017533
Schiffahrtsmuseum mit
  Nordseeaquarium Nordseeheilbad
  Langeoog, Langeoog 019917

Schiffahrtsmuseum Nordfriesland,
  Husum 019368
Schiffbau- und Schifffahrtsausstellung
  Roßlau, Roßlau 021483
Schiffbau- und Schifffahrtsmuseum,
  Rostock 021495
Schiffer, Atglen 138301
Schifferings, Neustadt am
  Rübenberge 108254
Schiffermuseum, Zehdenick 022846
Schiffers, Ewald, Aachen 129363
Schiffers, P.M., Warszawa 132904
Schifferstadt Architectural Museum,
  Frederick 048948
Schifffahrtsmuseum, Haren (Ems) 019074
Schifffahrtsmuseum, Nierstein .. 020810
Schifffahrtsmuseum der
  Marktgemeinde Spitz an der Donau,
  Spitz 002767
Schifffahrtsmuseum Haßmersheim,
  Haßmersheim 019094
Schifffahrtsmuseum Heimethus, Villa
  Liebegg, Männedorf 041716
Schifffahrtsmuseum im Schloss
  Greinburg, Oberösterreichische
  Landesmuseen, Grein an der
  Donau 001991
Schifffahrtsmuseum im Schlossturm,
  Stadtmuseum Düsseldorf,
  Düsseldorf 018043
Schiffleutmuseum, Stadl-Paura . 002768
Schiffmacher, Amsterdam 082850,
                               082851
Schiffmann, Reinhard, Forchheim,
  Oberfranken 137191
Schiffmühle Höfgen, Grimma .. 018820
Schiffmühle Minden, Minden ... 020420
Schiffshebewerk Henrichenburg,
  Waltrop 022414
Schiffsmuseum
  Seitenradschleppdampfer
  "Württemberg", Magdeburg .. 020212
Schikowski, Richard, Berlin 141566
Schilder- en Bakkerijmuseum 't
  Steenhuis, Niebert 032667
De Schildercamer, Amsterdam Fine
  Arts Centre, Amsterdam 132413
De Schilderijenschuur, Zeist ... 112726
Schilderijenzaal Prins Willem V, Den
  Haag 032197
Schildgen, R., Bonn 075086
Schildger, Peter, Frankfurt am
  Main 129854
Schildt, Burghard, Kiel 107555
Schilke, Pfungstadt 078048
Schilke, Anette, Hannover 076392
Schillay, Richard J., New York . 123045
Schiller & Bodo, New York ... 123046
Schiller Antiquariat, Marbach am
  Neckar 142110
Schiller-Museum Bauerbach, Klassik
  Stiftung Weimar, Bauerbach . 016972
Schiller-Nationalmuseum, Deutsches
  Literaturarchiv Marbach, Marbach
  am Neckar 020263
Schiller, Ludwig, Freiburg im
  Breisgau 141810
Schiller, Rosemarie, Nachrodt-
  Wiblingwerde 077685
Schiller, Soetkin, Berlin 129560
Schillerhäuschen, Stadtmuseum
  Dresden, Dresden 018007
Schillerhaus, Ludwigshafen am
  Rhein 020137
Schillerhaus, Stadtgeschichtliches
  Museum Leipzig, Leipzig 020015
Schillers Gartenhaus der Friedrich-
  Schiller-Universität, Jena 019457
Schillers Geburtshaus, Marbach am
  Neckar 020263
Schillers Wohnhaus, Klassik Stiftung
  Weimar, Weimar 022507
Schilling, Agnes, Schorndorf .. 108577
Schilling, Helmut, Stuttgart ... 078630

Schilling, Lilo, Bonn 129613
Schillinger, Arnold, Breisach am
  Rhein 075139
Schilp, Thomas, Hottenbach ... 130071
Schiltaere, Maastricht 083479
Schimanski, B., Steinhagen ... 078552
Schimmer, Reinhard, Nürnberg . 077843
Schimpfle, Jörg-Dietr.,
  Haimhausen 129937
Schimtz, Nathalie, Paris 071890
Schinart, Taranto 110943
Schindel, Wolfgang, Freiburg im
  Breisgau 107064
Schindelhauer, Manfred,
  München 130394
Schindler, Bernhard, Goldiwil .. 116352,
                               138053
Schindler, Bettina, Firenze 080203,
                               131264
Schindler, Christine, Bayreuth . 074698
Schindler, Daniel, Bremgarten
  (Bern) 133704
Schindler, Eric, Richmond 123743
Schindler, Karin, Bamberg 074670
Schindler, L., Regensburg 078177
Schindling, Eltville am Rhein ... 106877
Schingoethe Center for Native
  American Cultures, Aurora University,
  Aurora 046675
Schinkenmuseum, Apen 016598
Schinko, Gerd, Wien 063065
Schinner, Mattersburg 100118
Schino, Michele, Bari 079783
Schiølde's Sjarm og Sjel,
  Stavanger 084372
Schipper, Esther, Berlin 106421
't Schippertje, Leiden 083446
Schirmer-Mosel Verlag, München 137422
Schirmer, Friedemann, Winterbach,
  Rems-Murr-Kreis 079026
Schirmer/Mosel Showroom,
  München 108191
Schirn Kunsthalle Frankfurt, Frankfurt
  am Main 018446
Schirra, Saarlouis ... 078339, 126111
Schitke, Wolfgang, Weimar 130739
Schitter, Jacqueline, Plouharnel . 072206
Schitter, René, Dossenheim-sur-
  Zinsel 068390
Schkolnyk, Laurent, Nantes ... 104346
Schlächter, P., Münster 077676
Schlaf, A., Giesen 075993
Schlafke, Gaby, Braunfels 129621
Schlager, Melbourne 061834
Schlapka, München 077612
Schlapp, H. L., Darmstadt 141681
Schlatter Navarro, Ángel Luis,
  Sevilla 133439
Schlee, Ute Friederike, Karlsruhe 130099
Schleer, Irene, Neuburg an der
  Donau 108240
Schlegel, Holger, Dresden 129723
Schlegel, Pascal, Nancy 070431
Schlegel, Volker, Mengen 130297
Schlégl, Nicole, Zürich 116856, 138119,
                               144194
Schleich, Dieter, Wiesloch 079004
Schleicher & Lange, Paris 105219
Schleicher, T., Würzburg 142516
Schleichers Buchhandlung, Berlin 141567
Schleif, Bensheim 106131
Schleif, Schwetzingen 108599
Schleifer, Ronald, Wien 063066, 100369,
                               140120
Schleifermuseum, Solingen 021904
Gebr. Schleiffenbaum, Krefeld . 076963,
                               130197
Schleifmühle Schwerin,
  Stadtgeschichts- und Museumsverein
  Schwerin e.V., Schwerin 021795
Schleiminger, Birgit & Uwe,
  Solingen 078500
Schleimuseum, Kappeln 019498
Schleipehammer, Kierspe 019614

Schleiper, Bruxelles 128037
Schleisner-Borgsmidt,
Charlottenlund 065631
Schlemer & Schatz, Hallstadt .. 129942
Schlemme, H., Erfurt 075643
Schlemmer, Dr., Deggendorf ... 075320
Schlemmer, Rudolf, Kapfenberg 062660
Schlensog, Crimmitschau 075282
Der Schlern, Bolzano 138873
Schleser, Weimar 078899
Schlesinger, Alex, Zürich 116857
Schlesinger, Stephen L., New
York 123047
Schlesisch-Oberlausitzer Dorfmuseum
Markersdorf, Markersdorf ... 020285
Schlesisches Museum zu Görlitz,
Görlitz 018737
Schless, Antwerpen 063236
Schleswig-Holsteinischer Kunstverein
e.V., Kiel 059230
Schleswig-Holsteinisches
Freilichtmuseum, Molfsee 020451
Schleswig-Holsteinisches Künstlerhaus,
Eckernförde 059231, 106857
Schleswig-Holsteinisches
Landwirtschaftsmuseum und
Dithmarscher Bauernhaus,
Meldorf 020352
Schletterer, Raphael, Telfs 127882
Schleuse 16 – Forum für
experimentelle Arbeiten, Böblinger
Kunstverein für Stadt
und Landkreis Böblingen e.V.,
Böblingen 017454
Schleweis, W., Gelsenkirchen .. 075970
Schlicht & Geister, Berlin 141568
Schlichte Bergen, Amsterdam .. 082852
Schlichtenmaier, Stuttgart ... 078631,
108727,   137528
Schlichting, Dietfried, Göttingen 076052,
129912
Schlichting, Horst, Göttingen ... 076053
Schlickeisen, Dortmund 075394
Schliekau-Museum, Bad
Bevensen 016732
Schlienger, Valérie, Le Puy-Sainte-
Reparade 069347
Schlimme, Kay-Uwe, Hamburg . 076306,
076307,   129987
Schlipf, Thomas, Rottweil 130609
Schlissinger, Michèle, Paris 129141
Schlobach, Hansjürgen & Marga, Bad
Camberg 106047
Schlöder-Kassner, Neustadt am
Rübenberge 130455
Schlögl, Bruckneudorf 099971
Schlögl, Johannes Stephan,
Innsbruck 127779
Schlögl, Josef, Gundertshausen 127763
Schlösschen Borghees, Emmerich am
Rhein 018192
Schlößchen Vorder Bleichenberg
Biberist, Biberist 041336
Schlösser, M., Rösrath 078264
Schlötzer, Michael, Forchheim,
Oberfranken 075757, 129845
Schloß- und Goldbergbaumuseum,
Großkirchheim 002005
Schloß- und Heimatmuseum,
Sigharting 002755
Schloss- und Kirche Sankt Bartholomä,
Berchtesgaden 017042
Schloß- und Spielkartenmuseum,
Residenzschloss Altenburg,
Altenburg 016536
Schloss Achberg, Achberg 016460
Schloss Agathenburg,
Agathenburg 016473
Schloß Ambras, Kunsthistorisches
Museum Wien, Innsbruck .. 002136
Schloß Augustusburg, Brühl ... 017644
Schloss Babelsberg, Stiftung
Preußische Schlösser und Gärten
Berlin-Brandenburg, Potsdam . 021200

Schloß Baldern, Bopfingen 017509
Schloß Belvedere, Klassik Stiftung
Weimar, Weimar 022508
Schloß Berleburg, Bad Berleburg 016727
Schloß Brake – Das
Weserrenaissance-Museum,
Lemgo 020031
Schloss Britz, Gutshaus, Schlosspark,
Gutshof, Berlin 017282
Schloß Bürgeln, Schliengen 021671
Schloß Bürresheim, Sankt Johann bei
Mayen 021600
Schloss Burg, Bergisches Museum,
Solingen 021905
Schloß Burgau, Burgau 001806
Schloss Cappenberg, Selm 021839
Schloss Caputh, Stiftung Preußische
Schlösser und Gärten Berlin-
Brandenburg, Schwielowsee .. 021806
Schloss Cecilienhof, Stiftung
Preußische Schlösser und Gärten
Berlin-Brandenburg, Potsdam . 021201
Schloss Charlottenburg, Stiftung
Preußische Schlösser und Gärten
Berlin-Brandenburg, Berlin ... 017283
Schloss Charlottenhof, Stiftung
Preußische Schlösser und Gärten
Berlin-Brandenburg, Potsdam . 021202
Schloss Deutschkreutz,
Deutschkreutz 001813
Schloss Drachenburg,
Königswinter 019740
Schloß Egg, Bernried, Kreis
Deggendorf 017341
Schloss Ehrenburg, Bayerische
Verwaltung der staatlichen Schlösser,
Gärten und Seen, Coburg ... 017754
Schloß Faber-Castell, Stein bei
Nürnberg 021971
Schloß Falkenlust, Brühl 017645
Schloß Farrach, Zeltweg 003139
Schloß Favorite, Staatliche Schlösser
und Gärten Baden-Württemberg,
Rastatt 021282
Schloss Freienwalde mit Walther-
Rathenau-Gedenkstätte, Bad
Freienwalde 016769
Schloss Freudenberg, Wiesbaden 022642
Schloss Glienicke, Stiftung Preußische
Schlösser und Gärten Berlin-
Brandenburg, Berlin 017284
Schloss Glücksburg, Glücksburg 018717
Schloss Greinburg, Sammlung
Herzoglicher Kunstbesitz Sachsen-
Coburg-Gotha, Grein an der
Donau 001992
Schloss Großkühnau – Verwaltung,
Kulturstiftung Dessau-Wörlitz,
Dessau 017868
Schloss Güstrow, Staatliches Museum
Schwerin, Güstrow 018886
Schloß Hämelschenburg,
Emmerthal 018194
Schloss Halburn, Halburn 002045
Schloß Hallwyl, Seengen 041975
Schloss Harburg, Harburg 019070
Schloss Hegi, Winterthur 042161
Schloß Heidelberg, Heidelberg .. 019145
Schloß Herberstein, Sankt Johann bei
Herberstein 002668
Schloß Heubach, Heubach 019240
Schloss Höchstädt, Bayerische
Verwaltung der staatlichen Schlösser,
Gärten und Seen, Höchstädt an der
Donau 019277
Schloss Hof, Schloßhof 002724
Schloss Hohenschwangau,
Schwangau 021767
Schloss Hundisburg-Sammlung Apel,
Haldensleben 018944
Schloß Jegenstorf, Jegenstorf .. 041609
Schloss Johannisburg mit
Schlossgarten und Pompejanum,
Bayerische Verwaltung der

staatlichen Schlösser, Gärten und
Seen, Aschaffenburg 016625
Schloss Kirchheim, Kirchheim unter
Teck 019627
Schloss Kochberg mit
Liebhabertheater, Klassik Stiftung
Weimar, Großkochberg 018852
Schloss Königs Wusterhausen, Stiftung
Preußische Schlösser und Gärten
Berlin-Brandenburg, Königs
Wusterhausen 019718
Schloss Köpenick, Dependance des
Kunstgewerbemuseums, Berlin 017285
Schloß Leitzkau, Stiftung Dome
und Schlösser in Sachsen-Anhalt,
Leitzkau 020026
Schloss Lichtenstein, Lichtenstein 020064
Schloss Linderhof, Bayerische
Verwaltung der Staatlichen Schlösser,
Gärten und Seen, Ettal 018323
Schloß Loosdorf mit Prunkräumen und
Zinnfigurensammlung, Loosdorf, Bez.
Mistelbach 002342
Schloss Ludwigsburg,
Residenzschloss, Schloss Favorite,
Ludwigsburg 020129
Schloss Ludwigslust, Ludwigslust 020144
Schloss Luisium, Kulturstiftung
DessauWörlitz, Dessau 017869
Schloss Lustheim, Bayerische
Verwaltung der staatlichen
Schlösser, Gärten und Seen,
Oberschleißheim 020925
Schloß Marienburg, Pattensen . 021087
Schloß Miltach, Miltach 020407
Schloß Mörsburg, Stadel,
Winterthur 042018
Schloss Moritzburg und
Fasanenschlösschen, Staatliche
Schlösser, Burgen und Gärten
Sachsen, Moritzburg 020468
Schloss Mosigkau, Kulturstiftung
DessauWörlitz, Dessau 017870
Schloss Museum, Pram 002532
Schloß-Museum, Reidling 002584
Schloss-Museum, Wolfshagen . 022749
Schloss Neuburg, Bayerische
Verwaltung der staatlichen Schlösser,
Gärten und Seen, Neuburg an der
Donau 020690
Schloß Neunhof, Nürnberg 020875
Schloss Neuschwanstein, Staatl.
Schlossverwaltung Neuschwanstein,
Schwangau 021768
Schloss Ökonomie Gern,
Eggenfelden 018100
Schloss Oranienbaum, Kulturstiftung
DessauWörlitz, Oranienbaum .. 020999
Schloss Paretz, Stiftung Preußische
Schlösser und Gärten Berlin-
Brandenburg, Ketzin 019597
Schloß Parz, Grieskirchen 001994
Schloss Pfaueninsel, Stiftung
Preußische Schlösser und Gärten
Berlin-Brandenburg, Berlin ... 017286
Schloss Plüschow, Mecklenburgisches
Künstlerhaus, Plüschow 108389
Schloß Reinbek, Reinbek 021350
Schloß Rheda-Wiedenbrück, Rheda-
Wiedenbrück 017391
Schloss Rheinsberg, Stiftung
Preußische Schlösser und Gärten
Berlin-Brandenburg, Rheinsberg 021404
Schloss Rosenau, Schloss- und
Gartenverwaltung Coburg,
Rödental 021452
Schloss Sachsenburg, Frankenberg,
Sachsen 018393
Schloss Sacrow, Stiftung Preußische
Schlösser und Gärten Berlin-
Brandenburg, Potsdam 021203
Schloß Salem, Salem, Baden .. 021572
Schloß Sanssouci, Stiftung Preußische
Schlösser und Gärten Berlin-

Brandenburg, Potsdam 021204
Schloss Sargans – Schlossmuseum
und Schlossgalerie, Sargans .. 041938
Schloß Schallaburg, Loosdorf, Bez.
Melk 002341
Schloß Schallaburg, Schallaburg 002711
Schloss Schlettau, Schlettau .. 021669
Schloß Schönbrunn, Wien 003079
Schloss Schönhausen, Berlin .. 017287
Schloß Schwetzingen,
Schwetzingen 021802
Schloß Seehof, Bayerische Verwaltung
der staatlichen Schlösser, Gärten und
Seen, Memmelsdorf 020360
Schloß Solitude, Stuttgart 022055
Schloß Spiez, Spiez 042013
Schloss Stolzenfels, Koblenz .. 019667
Schloß Strünkede, Emschertal-
Museum, Herne 019220
Schloß Tarasp, Tarasp 042047
Schloß Thürnthal, Fels am
Wagram 001887
Schloß Thunstetten, Thunstetten 042061
Schloss Tiefurt, Klassik Stiftung
Weimar, Weimar 022509
Schloss Tratzberg, Jenbach ... 002147
Schloss Trebsen, Trebsen 022180
Schloss und Hofgarten Dachau,
Bayerische Verwaltung der
staatlichen Schlösser, Gärten und
Seen, Dachau 017799
Schloss und Park Schönbusch,
Bayerische Verwaltung der
staatlichen Schlösser, Gärten und
Seen, Aschaffenburg 016626
Schloss und Schlossgarten Weilburg,
Weilburg 022480
Schloß und Schloßpark Bad Homburg
Bad Homburg v.d.Höhe ... 016790
Schloss und Schlosspark
Charlottenburg mit Schinkelpavillon,
Belvedere und Mausoleum, Stiftung
Preussische Schlösser und Gärten
Berlin-Brandenburg, Berlin ... 017288
Schloß und Schloßpark Wilhelmsthal,
Museumslandschaft Hessen Kassel,
Calden 017713
Schloss Unteraufsess, Aufseß .. 016645
Schloß Urbach, Urbach 022285
Schloss Veitshöchheim, Bayerische
Verwaltung der staatlichen
Schlösser, Gärten und Seen,
Veitshöchheim 022307
Schloß Vianden, Burgmuseum,
Vianden 030606
Schloß vor Husum, Museumsverbund
Nordfriesland, Husum 019369
Schloß Weikersheim,
Weikersheim 022465
Schloss Weissenstein,
Pommersfelden 021170
Schloss Werdenberg,
Werdenberg 042130
Schloß Wernigerode, Zentrum für
Kunst- und Kulturgeschichte des 19.
Jahrhunderts, Wernigerode ... 022585
Schloss Wiederau, Pegau 021092
Schloss Wildegg, Wildegg 042143
Schloß Wilflingen,
Langenenslingen 019910
Schloss Wilhelmshöhe-
Weißensteinflügel,
Museumslandschaft Hessen Kassel,
Kassel 019549
Schloß Wiligrad, Lübstorf 020166
Schloss Wörlitz, Kulturstiftung
DessauWörlitz, Wörlitz 022718
Schloss Wolkenburg, Limbach-
Oberfrohna 020076
Schlossberg-Cohen, Baltimore .. 119769
Das Schlossberg-Lädchen, Neustadt
an der Weinstraße 077762
Schlossberger, Barbara,
Germering 129899

Schlossbergmuseum Chemnitz,
Chemnitz 017741
Schlosser, Augsburg 106034
Schloßgalerie Kastenscheuer,
Eberdingen 018073
Schloßgalerie Schärding,
Schärding 100192
Schloßmuseum, Arnstadt 016610
Schloßmuseum, Bad Bentheim . 016719
Schlossmuseum, Bad Iburg 016793
Schlossmuseum, Burgdorf 041387
Schloßmuseum, Ellwangen 018162
Schloßmuseum, Erbach, Alb-Donau-
Kreis 018228
Schlossmuseum, Höchstädt im
Fichtelgebirge 019278
Schloßmuseum, Kirchberg an der
Jagst 019620
Schlossmuseum,
Kirchentellinsfurt 019622
Schlossmuseum, Kronburg 019806
Schlossmuseum, Nauders 002430
Schlossmuseum, Oberhofen am
Thunersee 041813
Schloßmuseum, Ortenburg 021008
Schloßmuseum, Schrozberg . . . 021741
Schlossmuseum, Texing 002821
Schloßmuseum Amerang,
Amerang 016575
Schloßmuseum Aulendorf-
Kunst des Klassizismus-Altes
Spielzeug, Zweigmuseum des
Württembergischen Landesmuseums
Stuttgart, Aulendorf 016677
Schlossmuseum Beromünster,
Heimatmuseum und
Buchdruckerstube, Beromünster 041332
Schlossmuseum Darmstadt,
Darmstadt 017828
Schloßmuseum der Stadt
Aschaffenburg, Aschaffenburg . 016627
Schloßmuseum Heidegg,
Gelfingen 041510
Schloßmuseum Hohenlohe-
Schillingsfürst, Schillingsfürst . 021652
Schloßmuseum Hubertusburg,
Wermsdorf 022577
Schloßmuseum Ismaning,
Ismaning 019433
Schloßmuseum Jagsthausen,
Jagsthausen 019441
Schloßmuseum Jever, Jever . . . 019464
Schloßmuseum Landeck,
Landeck 002261
Schlossmuseum Langenburg,
Langenburg 019908
Schloßmuseum Mespelbrunn,
Mespelbrunn 020384
Schloßmuseum Murnau, Murnau am
Staffelsee 020625
Schloßmuseum Niederleis,
Niederleis 002453
Schlossmuseum Oranienburg,
Stiftung Preußische Schlösser
und Gärten Berlin-Brandenburg,
Oranienburg 021002
Schloßmuseum Rimpar, Rimpar . 021435
Schlossmuseum Schwerin, Staatliches
Museum Schwerin, Schwerin . 021796
Schlossmuseum Sondershausen,
Sondershausen 021914
Schloßmuseum Thun, Thun 042057
Schlossmuseum Weitra, Weitra . 002910
Schlossmuseum,
Bauernkriegsmuseum,
Oberösterreichische Landeskrippe
und Georg-von-Peuerbach-
Ausstellung, Peuerbach 002505
Schlossmuseum, Klassik Stiftung
Weimar, Weimar 022510
Schlossmuseum, Oberösterreichische
Landesmuseen, Linz 002330
Schloßmuseum, Städtische Museen
Quedlinburg, Quedlinburg 021249

Schlossmuseum, Stiftung Schloss
Friedenstein Gotha, Gotha 018777
Schloßparkmuseum, Museum
im Rittergut Bangert, Bad
Kreuznach 016806
Schlottau, Wolfgang, Hamburg . 141897
Schlotterbeck, J., Horb am
Neckar 076575
Schlürmann, Merzig 077358
Schlürmann, Mettlach 077363
Schlürmann & Sönnichsen-
Schlürmann, Sollerup 108636
Schlüter, Carl F., Hamburg 076308,
125975
Schlüter, H., Delbrück 075327
Schlüter, Horst, Delbrück 075328
Schlüter, Marita, Everswinkel . . 129829
Schlütter, J., Köln 107680
Schlumberger Wein- und Sektkellerei,
Wien 003080
Schlummer, Karin, Münster 077677
Die Schlumper, Hamburg 107314
Schmahl, Karl-Heinz, Halberstadt 076159
Schmalfuß, Eberhard, Plauen . . 142305
Schmalfuß, Michael W., Marburg 107978
Schmallery, Portland 123675
Schmalspurbahn-Museum,
Frojach 001910
Schmalstieg, Burgwedel 129647
Schmalstieg, Zetel 109005
Schmalz, Bad Gandersheim . . . 074515
Schmautz, Dieter, Ludwigsburg . 130249
Schmeja, Irmgard, Lippstadt . . . 077135
Schmela, Düsseldorf 106822
Schmelenhaus Museum,
Windhoek 031865
Schmelter, Eckhard, Münster . . 130430
Schmelter, Hubert, Münster . . . 130431
Schmelz, Bruno, Roden 078248
Schmerbach, Matthias,
Falkensee 129833
Schmetz, Bernhard, Aachen . . . 141397
SCHMI-DAU Antiquitäten,
Dortmund 075395, 129701
Schmid, Schwäbisch Gmünd . . . 078405
Schmid & Rothhaar, Arnstein . . . 129386
Schmid & Rothhaar, Würzburg . . 079092
Schmid-Wachter, Jürg, Chur . . . 126717,
133721
Schmid, Alfred, Augsburg 141416
Schmid, Beate, Neustadt an der
Weinstraße 137446
Schmid, Georg, Baden bei Wien . 062532
Schmid, H., Illingen, Enzkreis . . 076600
Schmid, Heidi, Grebin 076082
Schmid, Helmut, Obertraubling . 108319
Schmid, Karl, Bad Wörishofen . . 074618
Schmid, Konrad, Chur 087373
Schmid, Rudolf-Ernst, München . 130395
Schmider, Andreas, Einhausen . 106870
Schmidgall, Markus, Bad
Mergentheim 129417
Schmidt, Reichenbach im
Vogtland 078182
Schmidt & Dean, Philadelphia . . 123446
Schmidt & Green, Wuppertal . . . 079114,
142529
Schmidt & Günther, Kelkheim . . 141983
Schmidt & Maczollek, Köln 107682
Schmidt & Premke, Westerholt . 078948
Irmgard Schmidt & Mercedes De
Rabena, Hallein 100032
Schmidt-Felderhoff, Markus,
Bamberg 074671
Schmidt House Museum, Grants
Pass 049205
Schmidt-Kühl, Sophia-Therese,
Dresden 106745
Schmidt Kunstauktionen,
Dresden 125923
Schmidt-Rottluff, Chemnitz 106636
Schmidt, Anton J., Sent 087811, 133921
Schmidt, Carmilla, Cully 116237
Schmidt, Ch., Zittau 079146

Schmidt, Daniel, Bern 087301
Schmidt, Dr. Rainer, Tübingen . . 078729
Schmidt, Elizabeth, Porto Alegre 063994
Schmidt, Friedrich Otto, Wien . . 063067,
127967
Schmidt, George, New Orleans . 122235
Schmidt, Gérald, Rouen 072617
Schmidt, Hans-Christian,
Dresden 075427
Schmidt, Hans-Jürgen, Freiberg,
Sachsen 075843
Schmidt, Hartmut, Gießen 075998
Schmidt, Heinrich, Burgwald . . . 075231
Schmidt, Heinrich, Grenzach-
Wyhlen 107160
Schmidt, Herwart, Bisingen 141603
Schmidt, Johann, Wien 063068, 127968
Schmidt, Johann-Gottfried,
Rostock 130603
Schmidt, Josef, Meßstetten 077360
Schmidt, Joseph-Dieter,
Klipphausen 130134
Schmidt, Karin, Glaubitz 076007
Schmidt, L., Göttingen 107149
Schmidt, Lilo, Schneverdingen . . 108570
Schmidt, M., Leißling 077071
Schmidt, Martin, Satteldorf 130618
Schmidt, Monika, München 142184
Schmidt, Natalie, Dreieich 129703
Schmidt, P. & A., Bad
Oeynhausen 074566
Schmidt, Paul, Frankenberg, Eder 075761
Schmidt, R., Dessau 075337
Schmidt, Rainer, Frankfurt (Oder) 075766
Schmidt, Raymond, Metz 104231
Schmidt, Ruth, Berlin 074913
Schmidt, Sabine, Köln 107681
Schmidt, Sebastian, Panker 108358
Schmidt, Sylvia, New Orleans . . 122236
Schmidt, Thomas, Leisnig 130235
Schmidt, Udo, Lüdenscheid 107926
Schmidt, Valérie, Paris 105220
Schmidt, Veronika & Lukas, Blomberg,
Kreis Lippe 141605
Schmidtchen, Arno, Gastrose-
Kerkwitz 129886
Schmidtgen, H., Stuttgart 142401
Schmidts, Hans-Joachim, Kressbronn
am Bodensee 076966
Die Schmiede, Amsterdam 143064
Schmiede- und Schlossermuseum
Schwalbach, Schwalbach, Saar 021759
Schmiede im Hammergraben,
Helfenberg 002073
Schmiedegesellenhaus, Erlebniswelt
Mendlingtal, Göstling an der
Ybbs 001959
Schmiedemuseum, Fulpmes . . . 001914
Schmiedemuseum, Hatten 019096
Schmiedemuseum, Hohen
Demzin 019300
Schmiedemuseum, Lölling 002339
Schmiedemuseum, Sersheim . . 021845
Schmiedemuseum Arfeld, Bad
Berleburg 016728
Schmiedemuseum Bremecker
Hammer, Lüdenscheid 020172
Schmiedemuseum Kirchlauter,
Kirchlauter 019630
Schmiedmaier, Wien 127969
Schminck Memorial Museum,
Lakeview 050099
Schmit, Manuel & Robert, Paris 105221,
136986
Schmitsberger, Gottfried, Steyr . 140044
Schmitt & Mortimer, Hamburg . . 076309
Schmitt-Jüssen, Bonn 075087
Schmitt, Gerhard, Ingelheim am
Rhein 076608
Schmitt, Ingrid & Volker, Immenstadt
im Allgäu 076605
Schmitt, Jean-Pierre, Haguenau 068779
Schmitt, Klaus, Schwäbisch Hall 108587

Schmitt, Michel, Paris . 071891, 105222
Schmitt, Norbert, Lahnstein 130205
Schmitt, Sibylle, Köln 130168
Schmitter, Th., Bornhöved 075107
Schmittling, Gérard, Ceyrat 067759
Schmitz, Hilpoltstein 076535
Schmitz, Ratingen 142316
Schmitz, Willich 079016
Schmitz & Steinfeld, Neuss 077751
Schmitz-Avila, Dr. Thomas, Bad
Breisig 074499
Schmitz-Avila, Regine,
Wiesbaden 078994
Schmitz-Call, Simmerath 108625
Schmitz-Schneider, Renate,
Remscheid 078200
Schmitz, Hans-Martin, Köln 076904
Schmitz, Karl-Heinz, Ratingen . . 142317
Schmitz, Lieselotte, Moers 077409
Schmitz, Martin, Berlin 137093
Schmitz, Susanne, Berlin 129561
Schmitz, Uwe, Moers 108023
Schmock Broc, Paris 071892
Schmölz, Cornelia,
Dinkelscherben 129688
Schmölz, Helmut, Pirmasens . . . 078051
Schmoller, Dr. Bernd, Hamburg . 076310
Schmolling, W., Hatzfeld 107383
Schmolt, Axel, Krefeld 126019
Schmuck Bros., New York 095726
Schmuck & Antiquitäten,
Esslingen 075713
Schmuckli, Wädenswil 133960
Schmuckmuseum Pforzheim im
Reuchlinhaus, Pforzheim 021121
Schmücker, Anita, Stuhr 108683
Schmutzler, H., Berlin 074914
Schnabel, R., Großschwabhausen 129927
Schnabl, Bernhard, Abensberg . 074350
Schnaller, Fritz, Wolfratshausen . 079060
Schnappauf, Ralf, Ransbach-
Baumbach 078126
Schnapsglas-Museum,
Echsenbach 001837
Schnapsmuseum, Kötzting 019750
Schnarch-Museum, Alfeld 016506
Schnauferlstall, Ruhpolding 021550
Schneeberger, Christian, Neukirch,
Egnach 116534
Schneeberger, R., Straubing . . . 078580
Schneeberger, W. & J.,
Osnabrück 077977
Schneebergmuseum, Puchberg . 002544
Schneeweis, Elke, Geboltskirchen 062567
Schnegg, Emile, La Chaux-de-
Fonds 087360
Schneider, Berlin 106422
Schneider, Chicago 120265
Schneider, Nettphen 077702
Schneider-Henn, Dietrich,
München 126066, 142185
Schneider Museum of Art, Southern
Oregon University, Ashland . . . 046570
Schneider-Steidle, Spiegelberg . 142370
Schneider, Carl, Frankfurt am
Main 075822
Schneider, Dr. Gerhard, Olpe . . . 142273
Schneider, Dr. Hans, Tutzing . . . 078738,
142435
Schneider, Dr. Thomas Franz,
Basel 144033
Schneider, Dr. Wolfgang,
München 130396
Schneider, E.D., Meerbusch . . . 077320
Schneider, F., Bonn 075088
Schneider, F., Düsseldorf 106823
Schneider, Gerhard, Köln 130169
Schneider jr., J.P., Frankfurt am
Main 107029
Schneider, Joseph, Kitchener . . 101329
Schneider, Jutta,
Niederbrombach 077776
Schneider, Karl, Volkach 130728

Schneider, Lothar, Vaduz 082339, 132321
Schneider, Marc, Essen 075697
Schneider, Michael, Bonn 106539
Schneider, Thomas, Augsburg .. 141417
Schneider, Thomas, Dortmund . 075396
Schneider, Thomas, Leipzig .. 130229
Schneider, Ullrich, München ... 077613
Schneider, W., Lauchringen 077026
Schneider, Wilfried, Aachen 105947
Schneidereit-Gast, Monika,
  Solingen 130646
Schneiter, Dolores, Bern 116149
Schnell-Stöger, Irmgard,
  Oberammergau 130476
Schnell, Marion, Essen 075698
Schnell, N. & M., Ellwangen 075600
Schnellbächer, Hans-Peter, Mainz 130269
Schneph, Thomas E., Louisville . 121669
Schnez, Günther, Remagen 078194
Schnider, Ini, Graz 100019
Schniedermeier, Essen 075699
Schnitzler, Heinrich, Rheinbach . 130587
Schnitzschule Elbigenalp, Fachschule
  für Bildhauerei, Vergolder und
  Schilderhersteller, Elbigenalp .. 054926
Schnitzschule Oberammergau,
  Oberammergau 055674
Schnoebelen-Jolidon, Gérard,
  Delémont 087394
Schnödt, Weiden in der
  Oberpfalz 078887
Schnopfhagen-Museum,
  Oberneukirchen 002463
Schnupftabakmuseum, Grafenau, Kreis
  Freyung-Grafenau 018786
Schobner, Hans-Günter,
  Geisenheim 075962
Schocken Institute of Jewish
  Research, Jerusalem 025104
Schoder, Albrecht, Rheda-
  Wiedenbrück 130585
Schöchlin & Klaentschi,
  Sarmenstorf 087798, 133915
Schöchlin, Margareta, Luzern . 133840
Schöckel, Bredstedt 106572
Schödel, Ottfried, Kernen im
  Remstal 076760
Schöfegger, Rupert, Wien 140121
Schöfer, Peter, Haßloch 076415
Schöffer Museum, Kalocsa 023705
Schoefflen, Paris 071893
Schoefflen, Guy, Crécy-en-
  Ponthieu 068226
Schoega & Schoeg, Rüthen 078309
Schoeler, Clemens von, München 130397
Schoeller, Düsseldorf 106824
Schoeller, Innsbruck 062653
Schoeller, Wien 063069
Schoeller, Thomas, München .. 130398
Schoemaker, R.J., Den Haag ... 143120
Schoemansdal Museum, Louis
  Trichardt 038778
Schömer, Anita & Johann, Hinzert-
  Pölert 076540
Schön, Bonn 106540
Schoen & Nalepa, Berlin 106423
Schön, Brigitte, Bodenheim ... 129604
Schoen, Jason, Miami 121857
Schoenawa, Hartmut,
  Werlaburgdorf 078930
Schönbeck, Heinz-D., Minden .. 077374
Schönberg-Haus in Mödling,
  Mödling 002404
Schoenbrunn Village State Memorial,
  New Philadelphia 051310
Schöndorf, Manfred, Ottweiler . 130501
Schöne Alte Möbel, Berlin 074915
Schöne-Mühlen, Kristina,
  Westendorf 130750
Schöne, Joachim, Andernach .. 105992
Schöne, Peter, Halle, Saale ... 129939
Schöneborn, Klaus, Würzburg .. 142517
Schönenberger, Brigitte, Kirchberg

(Sankt Gallen) 116388
Schönenberger, Daniela, Kirchberg
  (Sankt Gallen) 116389
Schönewald, Paul, Xanten 109001
Schönfelder, Lindau 077120
Schönherr, Gabriele, Görlitz ... 107143
Schönhof, Jade 107476
Schönhuber, Hildegard, Dietfurt an der
  Atlmühl 075355
Schoeni, Hong Kong .. 102110, 102111
Schoeni, Markus, Zürich 134000
Schöni, Peter, Bern 087302
Schöning, Stuttgart 078632
Schöninger & Co., München ... 126067
Schöninger, H., München 108192
Schöningh, F., Münster 142211
Schönlau, Rolf, Dörentrup 129689
Schönleiter, Albersdorf 074368
Schönleiter, Torsten, Albersdorf . 074369
Schoenmakers, Borstel 075108
Schönpflug, Ingo, Innsbruck ... 062654,
                  062655, 125258
Schoentgen, Marie-José, L'Epine 069390
Schöpf, Daniel, Winterthur 133969
Schöpke & Lange, Hannover ... 076393
Schöppl, Gerhard, Salzburg ... 062817
Schöss, Kronach 076969
Schösser, Franz, Leuchtenberg . 077090
Schofield, Anne, Woollahra ... 062481
Schoharie Colonial Heritage
  Association Museum, Schoharie 053199
Schoharie Crossing, Fort Hunter 048816
Schokoladenmuseum Köln, Museum
  für Geschichte und Gegenwart der
  Schokolade, Köln 019712
Schokoladenmuseum-Musée du
  Chocolat, Eupen 003589
Schokoland Alprose, Caslano . 041393
SchokoMuseum in SchokoLand,
  Peine 021094
Schola San Zaccaria-Studio d'Arte,
  Venezia 111157
Scholes Gallery, Rotorua 084071
Scholl, Sonnfriede, München .. 130399
Scholl, Winfried, Wunstorf 142520
Schollaert, Aalst 063178
Scholler, Lotte von, Starnberg . 078541
Des Schollum, Wellington 143372
Scholte House Museum, Pella . 051965
Scholten, New York ... 123048, 135995
Scholten & Sohn, H., Nordhorn . 077804
Scholtes, Joël, Nice ... 070665, 104389
Scholtissek, R., Bissendorf ... 075010
Scholtz, H.-G., Berlin . 074916, 141569
Scholz, Baden-Baden 074642
Scholz Park Museum, Riverton . 001452
Scholz, A. von, Berlin 106424
Scholz, Camille von, Avignon .. 136799
Scholz, Jens, Köln 126011
Scholz, Sylvia, Bad Dürkheim .. 074506,
                      129406
Schomburg Center for Research in
  Black Culture, New York 051466
Schommer, Rolf, Beckingen ... 074704
Schon, Kurt E., New Orleans .. 095160
De Schone Kunsten Galerie,
  Haarlem 112517
Schonfeld & Co., R.D., New York 123049
School 33 Art, Baltimor 056699
School Arts, Worcester 139489
School Days' Museum,
  Fredericton 005848
School for Arts in Learning, WVSA
  Arts Connection, Washington .. 058068
School for Studies in Art and Culture:
  Art History, Carleton University,
  Ottawa 055149
School for the Contemporary
  Arts, Simon Fraser University,
  Burnaby 055118
School House Antiques, Chipping
  Campden 088884, 134260
School House Gallery, Bethany . 098744

School House Gallery, Wighton . 119351
School House Museum, Deep
  River 005693
School of Animation, Arts and Design,
  Sheridan Institute of Technology and
  Advanced Learning, Oakville . 055148
School of Architecture and Fine Arts,
  Dalian University of Technology,
  Dalian 055206
School of Architecture and Planning,
  Massachusetts Institute of
  Technology, Cambridge 056825
School of Architecture and Urban
  Planning, Shahid Beheshdi University,
  Tehrän 055854
School of Architecture, McGill
  University, Montréal 055145
School of Architecture, Montana State
  University, Bozeman 056778
School of Architecture, National
  Technical University of Athens,
  Athinai 055753
School of Architecture, Pratt Institute,
  Brooklyn 056801
School of Architecture, Princeton
  University, Princeton 057723
School of Architecture, Rensselaer
  Polytechnic Institute, Troy ... 058009
School of Architecture, Southern
  University A & M College, Baton
  Rouge 056711
School of Architecture, Syracuse
  University, Syracuse 057969
School of Architecture, Tulane
  University, New Orleans 057534
School of Architecture, University of
  British Columbia, Vancouver . 055170
School of Architecture, University of
  Detroit Mercy, Detroit 057012
School of Architecture, University
  of Southern California, Los
  Angeles 057372
School of Architecture, University of
  Texas, Austin 056694
School of Architecture, Yale University,
  New Haven 057524
School of Art, Guwahati 055797
School of Art, Tequesta 057988
School of Art, Winnipeg 055176
School of Art and Art History Galleries,
  Victoria H. Myhren Gallery, Gallery
  023, Denver 048275
School of Art and Art History,
  James Madison University,
  Harrisonburg 057184
School of Art and Art History,
  University of Denver, Denver . 057004
School of Art and Art History,
  University of Florida, Gainesville 057119
School of Art and Art History,
  University of Iowa, Iowa City . 057240
School of Art and Art History,
  University of South Florida, College
  of Visual and Performing Arts,
  Tampa 057981
School of Art and Cultures, Fine Art,
  Newcastle-upon-Tyne 056584
School of Art and Design, Auckland
  University of Technology,
  Auckland 056187
School of Art and Design,
  Buckinghamshire New University,
  High Wycombe 056540
School of Art and Design,
  Changzhou Institute of Technology,
  Changshu 055200
School of Art and Design, Coventry
  University, Coventry 056523
School of Art and Design, East
  Carolina University, Greenville . 057163
School of Art and Design, Endicott
  College, Beverly 056737
School of Art and Design, Georgia
  State University, Atlanta 056683

School of Art and Design, Northern
  Arizona University, Flagstaff .. 057087
School of Art and Design, Pratt
  Institute, Brooklyn 056802
School of Art and Design,
  Rochester Institute of Technology,
  Rochester 057780
School of Art and Design, San Jose
  State University, San Jose ... 057859
School of Art and Design, Southern
  Illinois University, MC 4301,
  Carbondale 056835
School of Art and Design, Syracuse
  University, Syracuse 057970
School of Art and Design, University of
  Hertfordshire, Hatfield 056539
School of Art and Design, University
  of Illinois at Urbana-Champaign,
  Champaign 056852
School of Art and Design, University of
  Michigan Ann Arbor, Ann Arbor 056660
School of Art and Design, University of
  Salford, Salford 056604
School of Art and Design, University of
  Ulster, Belfast 056509
School of Art and Design, University of
  Wolverhampton, Wolverhampton 056614
School of Art and Science, University
  of Richmond, Richmond 057759
School of Art Design and Media, East
  China University of Science and
  Technology, Shanghai 055232
School of Art Education, Faculty
  of Fine Arts, Xiamen University,
  Xiamen 055251
School of Art Gallery, Baton
  Rouge 046822
School of Art Gallery and Museum,
  Aberystwyth 043373
School of Art Gallery, Montana State
  University, Bozeman 047167
School of Art, Arizona State University,
  Tempe 057985
School of Art, Bowling Green State
  University, Bowling Green ... 056777
School of Art, Chautauqua Institution,
  Chautauqua 056864
School of Art, Curtin University of
  Technology, Bentley 054895
School of Art, Design & Printing, Yaba
  College of Technology, Yaba .. 056208
School of Art, Design and Architecture
  (ADA), University of Huddersfield,
  Huddersfield 056541
School of Art, Design and Art History,
  San Diego State University, San
  Diego 057851
School of Art, Illinois State University,
  Normal 057586
School of Art, Illinois Wesleyan
  University, Bloomington 056752
School of Art, Kent State University,
  Kent 057283
School of Art, Kent State University –
  Stark Campus, North Canton .. 057587
School of Art, Louisiana State
  University, Baton Rouge 056712
School of Art, Louisiana Tech
  University, Ruston 057797
School of Art, Montana State
  University, Bozeman 056779
School of Art, Munson-Williams-
  Proctor Arts Institute, Utica .. 058033
School of Art, Northern Illinois
  University, DeKalb 056995
School of Art, Ohio University,
  Athens 056675
School of Art, Rhodes University,
  Grahamstown 056346
School of Art, RMIT University,
  Melbourne 054909
School of Art, University of Cincinnati,
  Cincinnati 056898
School of Art, University of Tennessee,

Knoxville 057292
School of Art, University of Tulsa,
Tulsa 058014
School of Art, University of Wales,
Aberystwyth, Aberystwyth 056506
School of Art, University of
Washington, Seattle 057900
School of Art, Victorian College of
the Arts, University of Melbourne,
Southbank 054920
School of Art, Washington University,
Saint Louis 057815
School of Art, Yale University, New
Haven 057525
School of Arts, Gwalior 055799
School of Arts and Architecture, Azad
University, Tehrän 055855
School of Arts and Humanities,
Charles Stewart Mott Community
College, Flint 057088
School of Arts and Humanities, Oxford
Brookes University, Oxford ... 056593
School of Arts and Humanities,
University of Suffolk, Ipswich . 056542
School of Arts and Media, Nelson
Marlborough Institute of Technology,
Nelson 056194
School of Arts and Media, University
of Teesside, Middlesbrough ... 056583
School of Arts and Sciences, Clark-
Atlanta University, Atlanta ... 056684
School of Arts and Sciences, Eastern
Oregon College, La Grande ... 057297
School of Arts and Sciences, Triton
College, River Grove 057764
School of Arts and Sciences, United
States International University,
Nairobi 056076
School of Arts, Brunel University,
Romford 056603
School of Arts, Charles Darwin
University, Alice Springs 054892
School of Arts, Department of
Fine Arts, Yangzhou University,
Yangzhou 055256
School of Arts, Fashion and Design,
American InterContinental University,
London 056568
School of Arts, Histories and
Cultures, University of Manchester,
Manchester 056581
School of Arts, Lanzhou University,
Lanzhou 055223
School of Arts, Manhattan College,
Bronx 056792
School of Arts, Marymount University
of Virginia, Arlington 056668
School of Arts, McMaster University,
Hamilton 055132
School of Arts, Media and Education,
University of Bolton, Bolton .. 056511
School of Arts, Northwest University,
Xian 055252
School of Arts, Soochow University,
Suzhou 055241
School of Arts, Universiti Sains
Malaysia, Minden 056130
School of Arts, Virginia
Commonwealth University,
Richmond 057760
School of Communication and
Creative Arts, Deakin University,
Burwood 054898
School of Communication and
Design, Sun Yat-Sen University,
Zhongshan 055258
School of Contemporary Arts,
Edith Cowan University, Mount
Lawley 054911
School of Contemporary Arts,
Ramapo College of New Jersey,
Mahwah 057395
School of Creative Arts, University of
Saint Francis, Fort Wayne 057100

School of Creative Arts, University of
the West of England, Bristol . 056515
School of Design & Architecture,
University of Canberra, Bruce . 054897
School of Design and Fashion,
Stephens College, Columbia .. 056935
School of Design, George Brown
College of Applied Arts and
Technology, Toronto 055162
School of Design, Queensland
University of Technology,
Brisbane 054896
School of Design, University of
Pennsylvania, Philadelphia ... 057675
School of Earth Sciences Museum,
Townsville 001591
School of Fine and Applied Art,
Bangkok University, Bangkok . 056468
School of Fine Art and Music,
University of Guelph, Guelph .. 055129
School of Fine Art, De Montfort
University, Leicester 056546
School of Fine Art, History of Art and
Cultural Studies, University of Leeds,
Leeds 056545
School of Fine Art, Huazhong Normal
University, Wuhan 055248
School of Fine Arts, Changsha . 055199
School of Fine Arts and Design,
Kunsan National University,
Kunsan 056091
School of Fine Arts Gallery,
Indiana University Bloomington,
Bloomington 047049
School of Fine Arts, American
University in Bulgaria,
Blagoevgrad 055109
School of Fine Arts, Bob Jones
University, Greenville 057166
School of Fine Arts, Classical Studies
and Archaeology, University of
Melbourne, Melbourne 054910
School of Fine Arts, Fairmont State
University, Fairmont 057074
School of Fine Arts, Memorial
University of Newfoundland, Corner
Brook 055122
School of Fine Arts, National Academy,
New York 057562
School of Fine Arts, San Jose City
College, San Jose 057860
School of Fine Arts, Southwest
University, Chongqing 055204
School of Fine Arts, University of
Canterbury, Christchurch 056192
School of Fine Arts, University of
Kansas, Lawrence 057330
School of Humanities and Social
Sciences, Design Art Department,
Southwest Jiaotong University,
Chengdu 055202
School of Humanities, Department of
Art Education, East China Normal
University, Shanghai 055233
School of Humanities, Faculties
on Arts, Zhejiang University,
Hangzhou 055214
School of Interior Design and
School of Art, Education Direct,
Scranton 057891
School of Media and Design,
Department of Arts Design, Shanghai
Jiao Tong University, Shanghai 055234
School of Media Arts, Waikato
Institute of Technology, Waikato 056198
School of Media Studies, Humber
College of Applied Arts and
Technology, Etobicoke 055127
School of Media, Arts and Design,
University of Westminster,
London 056569
School of Nations Museum,
Elsah 048568
School of Political and Social Inquiry

Museum, Clayton 000972
School of Printing, Rochester Institute
of Technology, Rochester 057781
School of Textiles and Design, Heriot-
Watt University, Edinburgh .. 056530
School of the Art Institute of Chicago,
Chicago 056889, 056890
School of the Art Institute of Chicago
Gallery, Chicago 047696
School of the Arts, College of
Charleston, Charleston 056856
School of The Arts, University of
Northampton, Northampton ... 056586
School of the Arts, University
of Wisconsin Milwaukee,
Milwaukee 057454
School of the Museum of Fine Arts,
Boston 056773
School of Visual and Performing Arts,
Hillsborough Community College,
Tampa 057982
School of Visual and Performing
Arts, Long Island University,
Brookville 056804
School of Visual and Performing Arts,
Mississippi Gulf Coast Community
College, Perkinston 057660
School of Visual and Performing
Arts, University of Texas at Tyler,
Tyler 058021
School of Visual Art and Design,
Southern Adventist University,
Collegedale 056928
School of Visual Arts, New York 057563,
057564
School of Visual Arts, Queensland
University of Technology, Kelvin
Grove 054907
School of Visual Arts, University of
Louisiana at Monroe, Division of Art,
Monroe 057470
School of Visual Arts, University of
North Texas, Denton 057000
School oft Art+Design, Purchase
College, State University of New York,
Purchase 057735
Schoolarchief en Museum, Sint-
Katelijne-Waver 004016
Schoolhouse Gallery, Rosny Park 099655
Schoolhouse Museum,
Ridgewood 052572
Schoolmuseum Michel Thiery, De
Wereld van Kina, Gent 003637
Schools of Arts and Sciences New
South Wales, Australian Catholic
University, North Sydney 054914
Schools of Arts and Sciences
Queensland, Australian Catholic
University, Banyo 054894
Schools of Arts and Sciences Victoria,
Australian Catholic University,
Fitzroy 054905
Schooltijd Schoolmuseum,
Terneuzen 032916
Schoon, Lier 140249
Schooneman, Oss 112618
Schoonens, A., Den Haag 083130
Schoonewelle, Centrum Natuur en
Ambacht, Zwartsluis 033122
Schoonheim, Utrecht .. 083661, 132649
Schoonhovens Edelambachtshuys,
Schoonhoven 032866
Schoormans, Lucas, New York . 123050
Schoots, Willy, Eindhoven .. 112467,
137856
Schopferer, Alfred, Waldkirch . 078842
Schopmann, Hamburg 076311, 125976
Schopper- und Fischermuseum,
Aschach 001725
Schorer, R., Villingen-
Schwenningen 078803
Schorer, Walter, Neuburg an der
Donau 126078
Schorlemer, Werner von, Bergheim

(Erfft) 129458
Schorr, Michael, Wandlitz 130733
Schortgen, Esch-sur-Alzette .. 111826
Schortgen, Luxembourg 111863
Schots, Frédéric, Bruxelles ... 100658
De Schotse Huizen, Veere ... 032977
Schott, Wuppertal 079115
Schott GlasMuseum, Schott Villa,
Jena 019458
Schott, Alexander, München .. 142186
Schott, Pierre & Maria, Mainz .. 077244
Schotte, T., Southwold 091392
Schotten, Manfred, Burford 088710,
134180
Schotten, R., Oberau 077856
Schottenfeld & Partner, Wien . 140122
Schouenberg, M.T.H., Maastricht 083480
Schouten, Kees, Amsterdam .. 082853
Schouten, R., Vriescheloo ... 132664
Schouteten, C., Capelle aan den
IJssel 132464
De Schouw, Rotterdam 083588
Schouwey, Christian, Nods ... 133875
Schowalter, Albersweiler 129373
Schrade & Blashofer, Karlsruhe . 107525
Schrader, Lüneburg 077211
Schrader, Walsrode 078857
Schrader, Stefanie, Lüdenscheid 077201
Schränkla, Dachsbach 075289, 129665
Schrager, A., Milwaukee 127325
Schramek, Robert, Legau 077044
Schramkó, Pál, Solymár 130818
Schramm, Kiel 126000, 141995, 141996
Schramm, H., Greven 076090
Schramm, J. & H., Roetgen ... 108511
Schramm, Jürgen, Pöcking ... 130525
Schrank, Herbert, Steyr 062852
Schraps, R., Waldenburg, Kreis
Chemnitzer Land 078838
Schrattenecker, Heinz, Steyr .. 062853
Schraube Museum, Wohnkultur um
1900, Halberstadt 018940
Schraven, Justus, Göllheim ... 076025
Schredds, London 090480
Schreiber & Leser, München .. 137424
Schreiber Collezioni, Torino ... 081757
Schreiber, Andreas, Blaustein . 075018
Schreiber, Andreas, Ulm 078757
Schreiber, Elke & Werner,
Mannheim 077272
Schreiber, F.M., Braunschweig . 075133
Schreiber, J., Mannheim 077273
Schreiber, Jürgen, Mannheim . 077274
Schreiber, Klaus-Peter & Barbara,
Oldendorf 077946
Schreiber, Silke, München 137423
Schreiber, U., Halle, Saale 076171
Schreiber, U., Köthen 076933
Schreibmaschinen-Museum, Pfäffikon
(Zürich) 041839
Schreibmaschinenmuseum,
Wattens 002899
Schreilechner, Lorenz, Salzburg 062818,
127858
Schrem, Wilfried, Neu-Ulm .. 077712,
130443
Schrettl, Herbert, Innsbruck .. 062656
Schreuder, Robert, Amsterdam . 082854
Schreuders, J., Amsterdam ... 082855
Schreurs, M.H., Breda . 083005, 132458
Schreyer, Augsburg 141418
Schreyer, Chr., Berlin 125889
Schreyer, Hanno, Bonn 141623
Schriber, Martin, Winterthur .. 087912,
133970
Schriever, R., Köln 107683
Schrift-, Heimat- und
Druckereimuseum Bartlhaus,
Pettenbach 002504
Schrift-Art.ch, Rheineck 116581
Schriftmuseum J.A. Dortmond,
Amsterdam 031984
De Schrijnwerker, Harlingen .. 132528
Schroder, Hokitika 112919

Schröder, Den Haag 112410
Schröder, Lüchow 107921
Schröder & Dörr, Bergisch
  Gladbach 074734
Schroeder Romero, New York . 123051
Schröder, Adelheid, Lübeck 077186
Schröder, Gerhard, Hamburg ... 076312
Schröder, Helga, Düsseldorf ... 075514
Schröder, Ilse, Rottach-Egern .. 078297
Schröder, Joachim A., Berlin ... 074917
Schröder, Jörg, Lienen 077104
Schröder, Klaus, Oldenburg ... 142269
Schröder, L., Hopsten 076573
Schröder, Ove, Julita 126609
Schröder, Renate,
  Mönchengladbach 108019
Schroeder, Ulrich von & Heidi von,
  Weesen 116713
Schroeder, von, Dunalley 098986
Schröer, Rolf, Glan 075999
Schroer, P. W., Amsterdam 112278
Schröter, Sabina, Berlin 129562
Schroll, Anton, Wien 136609
Schroll, Johannes, Kirchberg in
  Tirol 127786
Schrott, Werner, Landeck 062706
Schrottmeyer, Michael, Berlin ... 141570
Schryver, Michael, Dorking 134338
Schubart Stube, Blaubeuren ... 017415
Schubert, Main Beach 099256
Schubert Club Museum of Musical
  Instruments, Saint Paul 052851
Schubert Geburtshaus und Adalbert
  Stifter-Gedenkräume, Wien Museum,
  Wien 003081
Schubert Sterbewohnung, Wien
  Museum, Wien 003082
Schubert, Andrea, Milano 110427
Schubert, B., Leipzig .. 077067, 142064
Schubert, Erhard, Bürstadt 141662
Schubert, Ludwig, Sankt Pölten . 140036
Schubert, M., Neuss 077752
Schubert, Michael, Ratingen 130559
Schubert, Micky, Berlin 106425
Schubert, Richard, Röbel/Müritz . 078255
Schubert, Rudolf, Salzburg 062819
Schubert, Siegurd, Bispingen ... 075005
Schubert, Walburga, Mannheim . 142108
Schubert, Werner, Würzburg ... 079093,
  108977, ... 130786
Schuch, Linz am Rhein 107882
Schuegraf, Josef, Stein bei
  Nürnberg 078546, 130656
Schüler, F., Kükels 107770
Schüller, Elke, Remseck am
  Neckar 078201
Schüller, Norbert, Wels 062877
Schünemann, Carl Ed., Bremen . 137128
Schüpbach, Aarwangen 087178
Schüppenhauer, Galerie + Projekte,
  Köln 107684, 137318
Schürenberg Kunsthandel,
  Aachen 074343
Schürmann, Heinz, Hagen 076152,
  076153
Schürmann, Petra, Kamp-Lintfort 107500
Schürmann, R., Düsseldorf 075515
Schüßler, Alex, Ansbach 074418
Schuessler, Michael, Denver ... 120769
Schütte, Gerd, Essen 106930
Schüttesägemuseum, Schiltach . 021655
Schüttler, Hans Ernst, Bonn ... 129614
Schütz, Wien 063070, 136610
Schütz & Süß, Nürnberg 077844
Schützenmuseum der
  Königlich privilegierten
  Feuerschützengesellschaft Weilheim,
  Weilheim in Oberbayern 022484
Schützenmuseum im Kanzlerturm,
  Eggenburg 001845
Schützenscheibenmuseum,
  Scheibbs 002717
Schützenscheibensammlung,
  Mainbernheim 020218

Schuhmacher, Amsterdam 143065
Schuhmacher, Jörg, Frankfurt am
  Main 075823
Schuhmachermuseum,
  Ladbergen 019840
Schuhmann, Norbert,
  Allershausen 105975
Schul- und Heimatmuseum, Mark
  Landin 020283
Schule für Gestaltung Basel,
  Sammlungen, Basel 041289
Schule für künstlerische Photographie,
  Wien, Wien 054983
Schule für unabhängigen Film, Wien,
  Wien 054984
Schuler, Bernhard, Flüelen 087427
Schuler, F., Chur 144062
Schuler, Horst, Düsseldorf 106825
Schuler, Philippe, Zürich 126772
Schulgeschichtliche Sammlung
  Bremen, Bremen 017598
Schulgeschichtliche Sammlung im
  Main-Taunus-Kreis, Kriftel 019796
Schulgeschichtliche Sammlung,
  Kulturhistorisches Museum,
  Magdeburg 020213
Schulhausmuseum, Hollingstedt 019322
Schulheimatmuseum, Asbach-
  Bäumenheim 016614
Schulhistorische Sammlung
  Cruismannschule, Bochum ... 017440
Schullandheim Sassen,
  Heimatkundliche Einrichtung,
  Sassen 021616
Schuller, Wolfgang, Wertheim .. 078938
Schulman, Helsinki 066360
Schulman, Laurens, Bussum ... 083022,
  126279
Schulmuseum, Haldensleben ... 018945
Schulmuseum, Immerath 019397
Schulmuseum, Königsfeld im
  Schwarzwald 019731
Schulmuseum, Lockenhaus 002338
Schulmuseum, Neumark 020722
Schulmuseum, Obersulm 020932
Schulmuseum, Offenburg 020977
Schulmuseum, Schneverdingen . 021694
Schulmuseum – Werkstatt
  für Schulgeschichte Leipzig,
  Leipzig 020016
Schulmuseum Bergisch Gladbach,
  Bergisch Gladbach 017055
Schulmuseum des Bezirks
  Unterfranken, Museen Schloss
  Aschach, Bad Bocklet 016735
Schulmuseum Friedrichshafen am
  Bodensee, Friedrichshafen ... 018527
Schulmuseum Fronau, Roding .. 021448
Schulmuseum Lilienthal,
  Lilienthal 020073
Schulmuseum Mozartschule
  Rheingönheim, Ludwigshafen am
  Rhein 020138
Schulmuseum Mühlebach,
  Amriswil 041219
Schulmuseum Nordwürttemberg in
  Kornwestheim, Kornwestheim .. 019772
Schulmuseum Nürnberg, Museen der
  Stadt Nürnberg, Nürnberg ... 020876
Schulmuseum Schulstubn Glocknhäusl,
  Sankt Peter am Wimberg 002691
Schulmuseum Steinhorst, Museen des
  Landkreises Gifhorn, Steinhorst 021990
Schulmuseum und Hallenhaus,
  Mönchguter Museen,
  Middelhagen 020403
Schulok, Iserlohn 107471
Schulte, Düsseldorf 075516
Schulte-Hobein, Stefan,
  Lambsheim 076998
Schulte, Cain, San Francisco .. 124387
Schulte, Jürgen, Kassel 076730
Schulte, R., Bad Salzuflen 074582
Schulte, Sigrid, Arnsberg 106002

Schulte, W., Bad Salzuflen 074583
Schultehuis, Diever 032227
Schulten, Ton, Ootmarsum ... 112616,
  137868
Schultheiss, Gudrun, München . 077614
Schultheiß, Wolfgang, Wien ... 063071
Schultz, Malmö 143900
Schultz, Bruno, Stühlingen 078585
Schultz, Frederick, New York .. 095727
Schultz, Günter, Wernigerode . 130748
Schultz, Julia, Moringen 130315
Schultz, Michael, Berlin 106426, 137094
Schultze, Norbert & Martina,
  Düsseldorf 075517
Schulwitz, Giesbert & Irmgard,
  Plößberg 078060
Schulz, Zutphen 083727
Schulz & Dahms, Hamburg 076314
Schulz-Aßmann, Ilona, Beeskow 074711
Schulz-Falster, Susanne, London 144685
Schulz Marty, Annette,
  Gockhausen 133761
Schulz, Annerose, Hartenstein .. 076407
Schulz, Bernd, Kamp-Lintfort .. 076670
Schulz, E., Offenburg 077910
Schulz, Günter, Bad Soden am
  Taunus 074601
Schulz, Hans-Joachim, Leipzig . 077068
Schulz, Henryk, Berlin 074918
Schulz, Ines, Dresden 106746
Schulz, Johannes, Bremen ... 075183,
  106593, ... 125911
Schulz, K.-F., Mannheim 077275
Schulz, Karl-Friedrich, Mannheim 077276
Schulz, Martina, Hamburg 076313,
  129988
Schulz, Richard, Lübeck 077187
Schulz, Stefan, Poznań 132864
Schulz, U., Hünxe 076585
Schulze, Berlin 074919
Schulze-Delitzsch-Haus, Außenstelle
  des Museums Schloß Delitzsch,
  Delitzsch 017847
Schulze-Głazik, Grażyna, Kraków 132848
Schulze, Erika, Groß Grönau ... 141844
Schulze, H., Ergolding 075646
Schulze, Martin, Berlin 074920
Schulze, Thomas, Berlin 074921
Schumacher, Badenweiler 074648
Schumacher, Berlin 074922
Schumacher & Conca, Bad
  Zurzach 087208
Schumacher Gallery, Capital University,
  Columbus 047944
Schumacher, Heike, Überlingen . 078794
Schumacher, Herbert, Müllheim 108040
Schumacher, Robert L.R., Seattle 124596
Schumacher, Rolf, Luzern 078646,
  133841
Schumacker, Liège 063781
Schumann, Louisville . 094585, 135764
Schumann-Haus, Leipzig 020017
Schumann, Hellmut, Zürich ... 144195
Schumann, R., Wiesbaden 078995
Schumann & Co., Reinhold,
  Kassel 076731
Schumann, Therese, Bad
  Salzuflen 074584
Schumann, Ursula, Freiburg im
  Breisgau 075882
Schumannhaus Bonn, Bonn ... 017501
Schumitz, Friederike, Hamm,
  Westfalen 130000
Schumm, Bernd, Hannover 076394
Schunder, R., Guxhagen 076138
Schunkelhaus, Obercunnersdorf 020890
Schupp, Kronberg im Taunus .. 076973
Schupp, Kai, Hamburg 129989
Schurian, Villach 100213
Schurig, Gunter, Bad
  Liebenwerda 106074
Schurr, B., Eislingen 106871
Schuschu, Düsseldorf 106826

Schussenrieder Bierkrugmuseum, Bad
  Schussenried 016872
Schuster, Buffalo 135379
Schuster, Dresden 075428
Schuster, Gelnhausen 107108
Schuster, Torquay 144884
Schuster & Scheuermann, Berlin 106427
Schuster, Alex, Stuttgart 078633
Schuster, Claudia & Helmut, Frankfurt
  am Main 107030
Schuster, Dirk & Hiltrud, Emmerich
  am Rhein 075613
Schuster, W., Stuttgart 130687
Schut, H.J., Laag-Keppel 132554
't Schuttershofke, Antwerpen .. 063237
Schutterskamer, Sneek 032891
Schutzverband Bildender Künstler,
  München 059232
Schuwey, Luzern 133842
Schuwirth & Van Noorden,
  Maastricht 112581
Schuyler County Historical Society
  Museum, Montour Falls 051033
Schuyler-Hamilton House,
  Morristown 051067
Schuyler Jail Museum, Rushville 052703
Schuyler Mansion, Albany 046345
Schwaab, Gabriele, Köln 130170
Schwab, Eckhart, Miesbach ... 077371,
  130306
Schwab, Gerhard, Bad Nauheim 141431
Schwabach Galerie, Meilen ... 116499
Schwabacher Künstlerbund,
  Schwabach 059233
Schwabe Verlag, Basel 138026
Schwabe, Angelika, Berlin 106428
Schwabenthal, Hans-Peter,
  Potsdam 130538
Schwabenverlag, Ostfildern 137465
Schwabinger Bilderbogen,
  Eurasburg 141757
Schwabinger Münzstube,
  München 077615
Schwäbisches Bauern- und
  Technikmuseum, Eschach 018273
Schwäbisches Bauernhofmuseum
  Illerbeuren, Kronburg 019807
Schwäbisches Handwerkermuseum,
  Augsburg 016667
Schwäbisches Krippenmuseum im
  Jesuitenkolleg, Mindelheim ... 020411
Schwäbisches Schnapsmuseum
  Bönnigheim, Bönnigheim ... 017463
Schwäbisches Schützenmuseum,
  Kronburg 019808
Schwäbisches Turmuhrenmuseum,
  Mindelheim 020412
Schwäbisches Volkskundemuseum
  Oberschönenfeld,
  Gessertshausen 018680
Schwämmi's Gallery, Campo
  Blenio 116185
Schwärter, H., Bakum 074652
Schwaighofer, Günther, Salzburg 062820
Schwalm & Menzel, Gerlingen . 075983
Schwan, Fürth 075926
Schwandl, Peter, Eferding 062553
Schwandl, Peter,
  Prambachkirchen 062769
Schwandt, Jörg, Berlin 074923
Schwaneberg, H., Haan 076144
Schwanke, Hamburg 125977
Schwartz, Rio de Janeiro 101067
Schwartz Collection of Skiing Heritage,
  Tiburon 053742
Schwartz Heritage House, Altona 053523
Schwartz Judaica, San Diego .. 145530
Schwartz, Andrea, San Francisco 124884
Schwartz, Anna, Melbourne ... 099319
Schwartz, Carol, Philadelphia .. 123447
Schwartz, Gabriele, Berlin ... 129563,
  129564
Schwartz, Nancy, New York ... 095728,
  123052

Schwartzsche Villa, Berlin 017289
Schwarz, Philadelphia . 096159, 123448
Schwarz-Afrika-Museum,
 Vilshofen 022346
Schwarz-Wrobel, Monika,
 Bochum 075043
Schwarz, Anton, Trun . 087859, 133943
Schwarz, Barbara, Schorndorf . . 108578
Schwarz, Beat, Murten 087694
Schwarz, Bernd, Detmold 129680
Schwarz, Hans, Springe 130649
Schwarz, Heinz, Kriens 133784
Schwarz, Hildegard, Sylt 108752
Schwarz, Hubert, Graz 127753
Schwarz, Karl & Barbara, Aarau 087174
Schwarz, Karl-Heinz, Tübingen . 078730
Schwarz, Leonhard,
 Georgensgmünd 129895
Schwarz, Linda, Cincinnati 120376
Schwarz, M., BarBel 074682
Schwarz, Peter, Berlin 141571
Schwarz, Rudolf, Pfronten 078045
Schwarz, W., Altenkunstadt 074395
Schwarz, Wolfgang, Weißensberg 078918
Schwarzacher, Gerhard, Sankt Jakob
 am Arlberg 062826
Schwarzachtaler Heimatmuseum,
 Neunburg vorm Wald 020727
Schwarzbach, Peter, Nürnberg . 077845
Schwarzberg, Mogens,
 København 065953
Schwarze Kunstverlag, Wuppertal 137566
Schwarzenbach, Zürich 126773
Schwarzenberger Skulpturenpark,
 Schwarzenberg am
 Böhmerwald 002736
Schwarzer, Berlin 106429
Schwarzer, Düsseldorf 106827
Schwarzer Elefant, Köln 076905
Schwarzer, Dr. Rainer, Dresden . 075429
Schwarzer, Klaus, Düsseldorf . . 075518
Schwarzfischer, Rosenheim . . 142336
Schwarzkogler, Wolfgang, Wien . 127970
Schwarzmann, Peter, Starnberg 108663
Schwarzwälder Freilichtmuseum
 Vogtsbauernhof, Gutach 018901
Schwarzwälder Mühlenmuseum,
 Grafenhausen 018789
Schwarzwälder Skimuseum,
 Hinterzarten 019261
Schwarzwälder Trachtenmuseum,
 Haslach im Kinzigtal 019087
Schwarzwälder, Manfred, Bad
 Tölz 074606, 129425
Schwarzwald-Museum, Triberg im
 Schwarzwald 022191
Schwarzweiss, Steinfurt 138753
Schwazer Silberbergwerk,
 Schwaz 002742
Schweden, Oliver, München .. 108193
Schwedenspeicher-Museum,
 Kulturgeschichtliches Museum im
 Elbe-Weser-Raum, Stade 021949
Schwedler, Holzgerlingen 076564
Schween, Bochum 129600
Schweig, Neil, Saint Louis 123868
Schweighofer, Bernd,
 Marktoberdorf 130291
Schweine-Museum, Bad
 Wimpfen 016911
Schweinfurth Memorial Art Center,
 Auburn 046652
Schweingruber, Florence 142686
Schweins, Otto, Köln 107685
Schweinsteiger, Ilse, München . 077616
Schweißhelm, H., Hannover ... 130015
Schweitzer, New York 123053
Schweitzer, Doris, Bonn 075089
Schweitzer, Lucien, Luxembourg 111864
Schweitzer, Lucien, Mondorf-les-
 Bains 111872
Schweitzer, Rolf, Rothenburg ob der
 Tauber 078288
Schweizer, Bad Homburg

v.d.Höhe 074526, 106067
Schweizer Jazzmuseum, Uster . 042088
Schweizer Kamm-Museum,
 Mümliswil 041771
Schweizer Kindermuseum, Baden 041256
Schweizer Kunst – Art Suisse – Arte
 Svizzera – Swiss Art, Zürich . . 139098
Schweizer Museum für Wild und Jagd,
 Utzenstorf 042089
Schweizer Zither-Kultur-
 Zentrum, Swiss Zither Museum,
 Trachselwald 042065
Schweizer, Daniel, Cologny 087379
Schweizer, Dr. René,
 Lampertheim 077000
Schweizer, Eglantine, Villars-le-
 Terroir 116701
Schweizer, Jürgen, Bad
 Fallingbostel 074510
Schweizer, Klaus, Konstanz ... 076937,
                                107723
Schweizerische Gesellschaft Bildender
 Künstlerinnen, Winterthur 059957
Schweizerische Gesellschaft der
 Freunde von Kunstauktionen,
 Zürich 059958
Schweizerische
 Interessengemeinschaft zur Erhaltung
 von Grafik und Schriftgut, Bern 059959
Schweizerische Nationalbibliothek,
 Bern 041323
Schweizerische Theatersammlung,
 Bern 041324
Schweizerischer Fachverband für
 Glasmalerei, Romont 059960
Schweizerischer Kunstverein,
 Zürich 059961
Schweizerischer Verband Bild und
 Rahmen, Zürich 059962
Schweizerischer Verband der Graveure,
 Stansstad 059963
Schweizerischer Verband für
 Konservierung und Restaurierung,
 Bern 059964
Schweizerischer Verband Graphisches
 Gewerbe, Bern 059965
Schweizerischer Werkbund,
 Zürich 059966
Schweizerisches Agrarmuseum
 Burgrain, Alberswil 041206
Schweizerisches Alpines Museum,
 Bern 041325
Schweizerisches
 Dampfmaschinenmuseum Vaporama,
 Thun 042058
Schweizerisches Feuerwehrmuseum
 Basel, Basel 041290
Schweizerisches Freilichtmuseum für
 ländliche Kultur, Brienz (Bern) . 041364
Schweizerisches Gastronomie-
 Museum, Thun 042059
Schweizerisches Institut für
 Kunstwissenschaft, Zürich 056448
Schweizerisches Landesmuseum,
 Schweizerische Landesmuseen,
 Zürich 042230
Schweizerisches Militärmuseum Full,
 Full 041507
Schweizerisches Schützenmuseum,
 Bern 041326
Schweizerisches Tanksäulenmuseum,
 Gänsbrunnen 041508
Schweizerisches Zentrum für
 Volkskultur, Burgdorf 041388
Schwemmgut-Museum, Finsing 018363
Schwender, Jürgen, Munster .. 077678
Schwenk-Anger, Else, Alpirsbach 137027
Schwenkfelder Library and Heritage
 Center, Pennsburg 051979
Schwenkkros, A., Osten 077986
Schwennesen, Heiko, Hamburg . 129990
Schwerkolt Cottage Museum,
 Mitcham 001299
Schwetz, Claude, Saint-Ouen .. 073196

Schwicker-Tosi, Firenze 110068
Schwickerath, Udo, Langenfeld . 077024
Schwilden, Bruxelles 140201
Schwind, Karl, Leipzig 107851
Schwind, Karlheinz, Frankfurt am
 Main 107031
Schwindel, Rudolf, Augsburg .. 129398
Schwingruber, Markus, Luzern . 087647
Schwitter, Fridolin P, Glattbrugg . 138052
Schwörer & Sohn, March 077291
Schwörer, Udo, Engelsbrand .. 141741
Schwohn, D., Hörstel 076546
Schwules Museum, Berlin 017290
Schydt Fredensborg, Kirsten,
 Holte 065788
Sci-Port Discovery Center,
 Shreveport 053324
Sci-Tech Center of Northern New York,
 Watertown 054216
Sciacca, Luigi, Messina 110225
Sciacchitano, Giulio, Torino ... 132199
Sciacchitano, Pier Paolo, Torino 132200
Scialom, Jean-Pierre, Zürich ... 144196
Scianca, Luciano, Roma 081552
Sciarra, Raniero, Roma 081553
Sciascia, Salvatore, Caltanissetta 137633
Sciberras, B'Kara 082410
Sciboz-Cordier, Danielle, Nîmes . 070705
Scicali, Renata, Palermo 110618
Science Activity Centre, Sirsa .. 024484
Science Activity Corner, Gwalior 024283
Science and Industry Collections
 Group, London 060234
Science and Technology Museum of
 Atlanta, Atlanta 046633
Science Center of Connecticut, West
 Hartford 054305
Science Center of Iowa, Des
 Moines 048287
Science Center of Pinellas County,
 Saint Petersburg 052866
Science Center of Thessaloniki
 and Technology Museum,
 Thessaloniki 023297
Science Center of West Virginia,
 Bluefield 047060
Science Center Spectrum, Berlin 017291
Science Centre for Education,
 Bangkok 042677
Science City Museum, Kolkata . 024367
Science Discovery Center of Oneonta,
 Oneonta 051778
Science et Technologie de la
 Conservation et de la Restoration
 des Œuvres d'Art du Patrimoine,
 Puteaux 138663
The Science Factory Children's
 Museum, Eugene 048625
The Science Fiction Museum and Hall
 of Fame, Seattle 053247
Science Hall of Penghu County,
 Magong 042392
Science Imaginarium, Waterloo . 054206
Science Learning Center, Upton 053908
The Science Museum, Cebu ... 034465
Science Museum, Dhaka 003226
Science Museum, Giza 010574
Science Museum of China,
 Beijing 007643
Science Museum of Long Island,
 Manhasset 050635
Science Museum of Minnesota, Saint
 Paul 052852
Science Museum of Virginia,
 Richmond 052560
Science Museum of Western Virginia,
 Roanoke 052604
Science Museum Wroughton, The
 National Museum of Science &
 Industry, Wroughton 046207
Science Museum, Bibliotheca
 Alexandria, Alexandria 010522
Science Museum, The National
 Museum of Science & Industry,

London 045103
Science North, Sudbury 007094
Science Spectrum, Lubbock ... 050518
Science Station Museum, Cedar
 Rapids 047491
Science Treasures, Oklahoma
 City 096007
Sciencenter, Ithaca 049741
Scienceworks Museum, Museum
 Victoria, Spotswood 001512
Scientific Researches Museum,
 Cairo 010562
Scipilliti, Liliana, Messina 110226
Scippa, Raffaele, Napoli 080933
Scirius, Mandelieu-la-Napoule . 104149
Scirocco & Koral Edizioni,
 Terrasini 137743
Scisco, Vincent, Longwy 069624
Scitech, Aurora 046676
Scituate Historic Sites, Scituate 053201
Scituate Maritime and Irish Mossing
 Museum, Scituate 053202
Sciworks, Winston-Salem 054494
Sclanders, Andrew, London ... 144686
Scoazec, Jean-Louis, Pont-Aven 105384
Scognamiglio, Mimmo, Napoli . 110544
Scognamiglio, Rodolfo, Milano . 080719
Scohy, Serge, Antwerpen 100511
Scola Galleria, Oakland 123265
Scoles, Saskatoon 101550
Scollay Square Gallery, Boston . 047131
Scolton Manor Museum,
 Haverfordwest 044539
Scone and Upper Hunter Museum,
 Scone 001489
Scone Palace, Perth 045488
Scope Antiques, London 090481, 134763
Scope Basel, Basel 098334
Scope Hampton, Wainscott ... 098500
Scope London, London 098402
Scope Miami, Miami Beach ... 098501
Score, Stephen, Boston 092493
Le Scorpion, Les Baux-de-
 Provence 103984
Lo Scorpione, Torino 111003
Scorrano, Carlo, Roma 081554
Scorza, Angela M., Rio de
 Janeiro 101068
Scorza, Daniela Menezes, São
 Paulo 101111
Scot, Paris 105223
Scotch Mist Gallery, Tucson ... 124771
Scotch Whisky Heritage Centre,
 Edinburgh 044271
Scotchtown, Beaverdam 046864
Scotia-Glenville Children's Museum,
 Traveling Museum, Scotia ... 053206
Scotia Museum, Scotia 053204
Scotland Art, Edinburgh 117741
Scotland Art, Glasgow 117857
Scotland Heritage Chapel and
 Museum, Scotland 053207
Scotland Street School Museum,
 Glasgow 044419
Scotland Yard Antiques, Marburg 061792
Scotland's Secret Bunker, Saint
 Andrews 045679
Scotsman, Saint Louis 096772
Scott, Atlanta 092181
Scott, Edmonton 101296
Scott & Co., Norfolk 127372
The Scott Gallery, Hawick Museum,
 Hawick 044544
Scott Monument, Edinburgh .. 044272
Scott Polar Research Institute
 Museum, University of Cambridge,
 Cambridge 043847
Scott, James, Edinburgh 089190
Scott, Lee, Atlanta 092182
Scott, Mike, Salisbury 135002
Scott, Richard, Holt, Norfolk .. 089552
Scott, Sarah, Sheffield 091293
Scotti, Emidio, Roma 132046
Scottie's Guns and Militaria,

Denver 093336
Scottish Antique, Doune 089101
Scottish Antique and Arts Centre,
    Abernyte 088230, 134013
Scottish Art Gallery, Dunoon ... 117698
Scottish Arts Council, Edinburgh 060235
Scottish Fine Art Group, Norwich 090881
Scottish Fisheries Museum,
    Anstruther 043420
Scottish Football Museum,
    Glasgow 044420
The Scottish Gallery, Edinburgh . 117742
Scottish Industrial Railway Centre,
    Dalmellington 044064
Scottish Jewish Museum,
    Glasgow 044421
Scottish Maritime Museum,
    Irvine 044652
Scottish Mining Museum,
    Newtongrange 045374
Scottish Museums Council,
    Edinburgh 060236
Scottish Museums Federation,
    Hamilton 060237
Scottish National Gallery of Modern
    Art Gallery, Edinburgh 044273
Scottish National Portrait Gallery,
    Edinburgh 044274
Scottish Pottery Historical Review,
    Edinburgh 139222
Scottish Railway Museum,
    Bo'ness 043675
Scottish Rugby Union Museum,
    Edinburgh 044275
Scottish Society for Art History,
    Glasgow 060238
Scottish Society for Art History
    Journal, Dundee 139223
Scotts' Antiques of Battery Point,
    Hobart 061637
Scotts Bluff National Monument,
    Gering 049086
Scotts of Dorking, Dorking ... 089093,
                                134339
Scottsdale Artists' League,
    Scottsdale 060731
Scottsdale Artists' School,
    Scottsdale 057887
Scottsdale Museum of Contemporary
    Art, Scottsdale 053212
Scottsdale Village Antiques,
    Albuquerque 092082
Scotty's Trading Post, Southport 062231
Scout, Chicago 092791
Scout Gallery, London 118562
Scouting Museum de Ducdalf,
    Rotterdam 032824
Scouting Museum Haagse Randstad,
    Den Haag 032198
Scranton, Oklahoma City 096008
Screen Porch Arts, Houston ... 121125
Scremini, Clara, Paris 105224
Scribes Books, Dunedin 143316
Lo Scrigno, Parma 081128
Scrimin Incontri, Bassano del
    Grappa 109785
Scrimshaw, Glen, Saskatoon ... 101551
Scriporama, Spijkenisse 143230
Scripta Manent, Monaco 142997
Scripta Manent, Reggio Calabria 131787
Scriptorium, Mantova 142904
Scriptum Art Books, Schiedam . 137872,
                                143228
Lo Scrittoio, Roma 081555
Scrocco, Roberto, Livorno 080336
Scroungers Antiques, Bridgetown 061193
Scrudato, Robert, Menton 070063,
                                140953
Scryption, Museum voor Schriftelijke
    Communicatie, Tilburg 032932
Ščučinskij Oblastnoj Muzej Literatury i
    Iskusstva, Kokšetau 029822
Scuderi, Carmelo, New York ... 135996

Scuderi, Daniel, New York 095729,
                                135997
Scuderi, Marisa, Palermo 110619
Scuderie del Quirinale, Roma .. 027661
Scugog Shores Historical Museum
    Village, Port Perry 006694
Scuiller, Samy, Audierne 103287
Sculptors Guild of America, New
    York 060732
Sculptor's Society of Canada,
    Toronto 058471
Sculpture, Hamilton 139490
Sculpture at Schoenthal,
    Langenbruck 041635
Sculpture Center, Long Island
    City 050384
Sculpture Court at City Center,
    Oakland Museum of California,
    Oakland 051684
Sculpture Galena, The Chicago
    Athenaeum, Galena 049017
Sculpture Gallery, San Francisco 124389
Sculpture Gallery, Sevenoaks .. 119071
Sculpture Highlights Maastricht,
    Maastricht 098265
Sculpture International, Coconut
    Groove 098502
SculptureCenter, Long Island City 121419
sculpturesite, San Francisco .. 124390
Scuola Dalmata dei Santi Giorgio e
    Trifone, Venezia 028325
Scuola di Specializzazione in
    Archeologia, Napoli 055937
Scuola Grande Arciconfraternità di San
    Rocco, Venezia 028326
Scuola Grande di San Giovanni
    Evangelista, Venezia 028327
Scuola Internazionale d'Arte
    'La Bottega dell'Acquaforte',
    Firenze 055899
Scuola Internazionale di Grafica,
    Venezia 055994
Scuola Superiore d'Arte Applicata del
    Castello Sforzesco, Milano ... 055930
Scuola Superiore d'Arte Cristiana
    Beato Angelico, Milano 055931
Scuoppo Musso, Marseille 069965,
                                128783
Scurr, Stuart, Mount Lawley ... 061916
Scurry County Museum, Western
    Texas College, Snyder 053374
Ščusev Museum, Chişinău 031714
SCW Reprographics, San Diego . 124171
SD Galeria Mokotówa, Warszawa 113900
SD Galeria Panorama, Warszawa 113901
SDV Saarbrücker Druckerei und
    Verlag, Saarwellingen 137496
SE Art and Design, New York ... 123054
Sea Antiques, Helsinki 066361
Sea Cliff Village Museum, Sea
    Cliff 053220
Sea Pictures, McCrae 099243
Sea Shell Museum, Scarness .. 001487
Sea View Antiques, Caernarfon . 088753
Sea-Witch Antiques, Port
    Adelaide 062092
Sea World, South African Association
    for Marine Biological Research,
    Durban 038706
Seab, Bologna 142640
Seabee Museum, US Navy Civil
    Engineer Corps, Port Hueneme 052250
Seabreeze Art Gallery, Southport 119126
Seabrook Village, Midway 050883
Seaby, London 090482
The SeaChange Trust, Great
    Yarmouth 060239
Seaford Historical Museum,
    Seaford 053223
Seaford Museum of Local History,
    Seaford 045750
Seaford's Barn Collectors Market,
    Seaford 091256, 144828
Seaforth Highlanders of Canada

Regimental Museum, Vancouver 007295
Seagate Art Gallery, Toledo 124696
Seager, Arthur, Stow-on-the-
    Wold 091482
Seagull, Singapore 114601
Seahaven Galleries, Providence . 123712
Seal Island Light Museum,
    Barrington 005391
Seale, Jeremy, London 090483
Sealed Art, Hamilton 101322
Sealey, Tuam 079716, 130895
Sealife Scheveningen, Den Haag 032199
Seanic Antiques, Newport 061961
Seans, Sankt-Peterburg 139017
Seaocean Book Berth, Seattle .. 145599
Seaport Antique, Newport 061962
Seaport Art Center, Cingshui .. 042293
SEARCH-Centre, Gosport 044454
Search Press, Tunbridge Wells . 138289
Searchlight Historic Museum and
    Mining Park, Searchlight 053226
Searle & Co., London . 090484, 134164
Sears-Peyton, New York 123055
Seashore Trolley Museum,
    Kennebunkport 049922
Seasons Art, Singapore 085497
Seasons Art Gallery, Omaha ... 123339
Seasons Galleries, Den Haag .. 112411
Seasons in Time, Birmingham .. 092434
Seaton Delaval Hall, Seaton ... 045751
Seatons, Christchurch . 083895, 132711
Seattle Antiquarian Book Fair and
    Book Arts Show, Seattle 098503
Seattle Antique Market, Seattle . 097499
Seattle Art Museum, Seattle ... 053248
Seattle Asian Art Museum,
    Seattle 053249
Seattle Auction Service, Seattle 127455
Seattle Book Center, Seattle ... 145600
Seattle Coin Shop, Seattle 097500
Seattle Mystery Bookshop,
    Seattle 145601
Seattle Print Arts, Seattle 060733
Seattle Print Fair, Seattle 098504
Seauve, Philippe, Livron-sur-
    Drôme 069604
Seaver College-Fine Arts Division,
    Pepperdine University, Malibu . 057396
Seaview Antiques, Horncastle .. 089587
Seaview Artists Gallery, Moffat
    Beach 099357
Seaview Gallery, Queenscliff .. 062134,
                                099606
Seaway Gallery, Paignton 118877
Seaweed Art, Cape Town 114775
Seawinds, Columbus 093044
Seba Beach Heritage Museum, Seba
    Beach 006999
Sebag, Félix, Paris 071894
Sebão Livros Usados, Brasília .. 140277
Sebastian-Kneipp-Museum, Bad
    Wörishofen 016915
Sebastopolis Açıkhava Müzesi, Tokat
    Müzesi, Tokat 043071
Sebbelov, Ole, Farum 065651
Sèbe, Alain, Vidauban 137014
Sebestyén, Georg, Waldenbuch . 078835,
                                108838
Sebestyen, M., Aachen 105948
Sebežskij Kraevedčeskij Muzej,
    Sebež 037851
Sebley, Oswald, Norwich 090882,
                                134895
Sebnitzer Kunstblumen- und
    Heimatmuseum Prof. Alfred Meiche,
    Sebnitz 021808
Sebo do Messias Livros Usados, São
    Paulo 140312
Sebo São João, São Paulo 140313
Sebök, Johann, Bamberg 125863
Seboga, Seoul 111710
Sebourne, Jose, Philadelphia .. 123449
Sebovero, São Paulo 140314
Sebra Galerii, Tartu 103124

SECAN-Société Européenne de Conseil
    en Archivage Neutre, Paris ... 129142
Seccamani, Romeo, Brescia ... 131049
Secção Museológica da CP de
    Estremoz, Estremoz 035673
Secção Museológica da CP de Lagos,
    Lagos 035743
Secção Museológica da CP de
    Macinhata do Vouga, Macinhata do
    Vouga 035838
Secção Museológica da CP de
    Santarém, Santarém 036003
Secção Museológica da CP de
    Valença, Valença 036071
Secchiati, Roberto, Jesolo 110152
Sección Arqueología, Museo de Ceuta,
    Ceuta 039273
Sección de Bellas Artes, Museo de
    Ceuta, Ceuta 039274
Secesja, Kraków 084598
Secesja, Skawina 132867
Secession, Vereinigung bildender
    KünstlerInnen Wiener Secession,
    Wien 003083
Secker & Warburg, Martin,
    London 138249
Secluded Alley Works, Seattle .. 124597
Seco, Virgilio Rodrigues, Lisboa 084959
Secol – Art, Torino 081758
Secol-Art, Torino 081759
Second Avenue Art Expo, New
    York 123056
Second Chance, Auckland 083825
Second Chance Antiques, Omaha 096056
Second Childhood, New York ... 095730
Second City Resale, Cleveland . 092970
Second Creek Antiques,
    Charlotte 092624
Second Edition, Edinburgh 144401
Second Edition, Regina 064556, 128198
Second Edition-Books, Pambula 139868
Second Floor Antiques,
    Sacramento 096645
Second Glance Books, Portland . 145463
Second Hand Design,
    Amsterdam 082856
Second Hand Store and Antiques,
    Long Beach 094224
Second Life, Charlestown 135387
Second Look Book Store, San
    Antonio 145513
Second Notions Antiques,
    Kirkcaldy 089716
Second Story Books, Rockville .. 145481
Second Story Books, Washington 145641
Second Street Gallery,
    Charlottesville 047579
Second Thoughts Bookstore,
    Ottawa 140379
Second Time, Los Angeles 094467
Second Time Around, Providence 096515
Second Time Around, Vancouver 064874
Second Time Around Antiques,
    Buffalo 092554
Second Time Around Antiques,
    Richmond 096567, 136177
Second Time Around Books,
    Toronto 140415
Second World War Experience Centre,
    Horsforth 044604
Seconde Modernité, La Guiche . 103895
Secondhand Rose, New York ... 095731
Secondo Mano, Gouda 083269
Secord, William, New York 123057
Secret Christians Museum,
    Nagoya 029126
Secret d'Alcove, Bruxelles 063550
Secret Garden Gallery, North
    Tamborine 099479
Secrets, Port Isaac 118935
Secrets de Soie Musée, Campel 012304
Secrets of Asia, Deventer 083163
Secrist, Carrie, Chicago 120266
Secţia de Etnografie-Bastion,

Timişoara 036786
Section Artistique Janus,
Bruxelles 058387
Section d'Histoire de l'Art et
d'Archéologie, Faculté de Philosophie
et Lettres, Université Libre de
Bruxelles, Bruxelles 055014
Section d'Histoire de l'Art, Université
de Lausanne, Lausanne ... 056433
Section Française de l'Institut
International de Conservation des
Œuvres d'Art. Core, Champs-sur-
Marne 128542
Section Sciences Humaines, Lettres et
Arts, Université Charels-de-Gaulle
Lille 3, Villeneuve-d'Ascq ... 055430
Sécula, Alain-Nicolas, Beaune . 067037
Sécula, Anthony, Beaune .. 067038
Sécula, Eric, Beaune 067039
Sécula, François, Beaune .. 067040
Sécula, Jacky, Beaune 067041
Sécula, Jean-Marc, Luzy ... 069676
Secula, Jean-Pierre, Saint-Ouen 073197
Sécula, Laurent, Beaune ... 067042
Sécula, Raymond, Beaune .. 067043
Sécula, Rodolphe, Beaune . 067044
Sécula, Rodolphe, Paris ... 071895
Seculo XX Copacabana, Rio de
Janeiro 064124
Séculos Passados, Porto ... 085051
Secundo, Liège 063782
Secundus, Terhole 143232
Secwepemc Museum and Heritage
Park, Kamloops 006059
Sed Gallery, University of Georgia,
Athens 046596
Sedan Numismatique, Sedan .. 073468
Sedarca, Oleyres 133882
Sederholmin Talo, Helsingin
Kaupunginmuseo, Helsinki .. 010860
Seders, Francine, Seattle .. 124598
Sedgeberrow, Pershore 144783
Sedgewick Archives Museum and
Gallery, Sedgewick 007601
Sedgwick-Brooklin Historical Museum,
Sedgwick 053261
Sedgwick Museum of Earth Sciences,
Cambridge 043848
Sedlák, František, Praha ... 065521
Sedlak, Karl, Offenhausen .. 062759
Sedman, Bridlington 088605
Sedona Arts Center, Sedona .. 053262
Šeduvos Kraštotyros Ekspozicija,
Daugvenės kultūros istorijos
muziejus-draustinis, Seduva .. 030460
See Restoration, Atlanta ... 135315
See Science Museum,
Manchester 050630
See-Vogel-Museum Herbert Vargyas,
Rust 002611
Seebach, Jochen, Emkendorf .. 129790
Seeck, Mason City 127310
Seedamm Kulturzentrum, Stiftung
Charles und Agnes Vögele, Pfäffikon
(Schwyz) 041837
Seedorn Verlag, Seuzach ... 138089
Seegalerie, Rottach-Egern .. 108531
Seeger, Braunschweig 106569
Seeger, Brigitte, Göppingen . 076035
Seeger, Burgel, Fellbach ... 075732
Seegers, Marie-Louise, Blies-
Ebersing 067247
Seekers Books, Toronto 140416
Seelackenmuseum, Sankt Veit im
Pongau 002705
Seeland Galerie, Jens . 116380, 133780
Seelhof, Ulrich, Dillenburg .. 129685
Seelhoff, Koblenz 076840
Seemann, E.A., Leipzig 137349
Seemann, Helfried, Wien ... 140123
Seemeyer, A., Sehnde 078449
Seemeyer, Adolf, Hannover . 076395,
                         130016
Seemuseum in der Kornschütte,

Kreuzlingen 041627
Seenot-Rettungskreuzer Georg
Breusing, Emden 018181
Seeshells Antiques, Las Vegas . 094174
Seethaphone, Bangalore 079361
Şefik Bursali Müze Evi, Ankara 042829
Segal, Thomas, Baltimore ... 119770
Ségala, Badaroux 066904
Ségalen, Michel, Rennes (Ille-et-
Vilaine) 105443
Segantini-Museum, Sankt Moritz 041932
Segarra, Maria Fernandez,
Lisboa 084960
Segas, Gilbert, Paris 071896
Segedunum Roman Fort, Baths and
Museum, Wallsend 046038
Segesser, Jeannette von, Luzern 133843
Seggern, J. von, Bremen ... 075184
Seghers, P. & L., Oostende .. 100806
Il Segno, Milano 110428
Segno, Pescara 137718, 138874
Il Segno del Tempo, Milano .. 080720,
                         080721
Sego Gallery, Salt Lake City . 123986
Segonds, Jean-Claude, Limoges 069532
Segontium Roman Museum,
Caernarfon 043826
Segovia, Marita, Madrid 086261, 115284
Segrelles del Pilar, Valencia . 115517
Seguin, Villeurbanne 129356
Séguin Poirier, Bernard, Montréal 101450
Seguin, Alain, Beaune 103355
Seguin, Patrick, Paris . 071897, 105225
Seguine-Burke Plantation, Staten
Island 053501
Seguinet, Arnaud, Duclair .. 125576
Seguins Cohorn, Michel de,
Fontainebleau 068568
Seguso Viro Srl, Venezia 111158, 111159
Sehara, Sarajevo 063872
Seharsch, Peter, Lich 107865
Sehgalerie, Bordesholm 106544
Şehir Müzesi, İstanbul 042957
Seian Zokei Daigaku, Otsu .. 056028
Seibido, Nara 082075
Seibu, Tokyo 111491
Seidel & Sohn, Berlin . 074924, 074925
Seidel, Erfurt 075644
Seidel & Sohn, Potsdam 078085
Seidel & Richter, Kerstin, Berlin 141572
Seidel, Konrad, Witten 130774
Seidel, Volker, Bochum 075044
Seidenath, Otto, Bamberg .. 129444
Seidenath, Thomas, Bamberg . 129445
Seidenberg, David, New York . 095732
Seidenfaden, Max, Frederiksberg 112807
Seidl, Josef, München 077617
Seidl, Markus, Tiefenbach, Kreis
Passau 078679
Seidler, Wilhelm, Linz . 062724, 100113
Seifert, Rüdersdorf bei Berlin . 078307
Seifert, Edmund, Coburg ... 075264
Seifert, H., Köln 076906
Seifert, Rolf, Dresden 141707
Seiffer-Schulz, Ingrid, Haigerloch 107201
Seifudo, Kubo, Hiroshima .. 081984
Seigando, Jintsu, Tokyo 082196
Seigfred Gallery, Ohio University,
Athens 046600
Seigne, Eric, Boisset Saint-Priest 128474
Seigneur, J + M, Leipzig ... 077069
Seigneur, Jacques le, Ronchin . 129210
Seiho, Tokyo 111492
Seiji, Honolulu 120968
Seiji Togo Memorial Sompo Japan
Museum of Art, Tokyo 029579
Seikado Bunko Bijutsukan, Tokyo 029580
Seikado Bunko no Bijutsukan,
Tokyo 029581
Seiko, Sendai 111390
Seikosha, Osaka 082089
Seiler, Röslau 078262
Seiler Liepe, Ute, Berlin ... 129565
Seiler, Andreas, Wöschnau ... 087914

Seiler, Thomas, Heilbronn 107406
Seiler, Werner, Basel 133656
Seillé, Carcès 067680
Seiller, Noirmoutier-en-l'Ile ... 104415
Seima, Tourriers 073831
Seinäjoen Lehti- ja Kirjapörssi,
Seinäjoki 140744
Seinäjoen Luonto-Museo,
Seinäjoki 011350
Seinan Gakuin Daigaku Hakubutsukan,
Fukuoka 028551
Seinerzeit, Bünde 075216
Seippel, Dr. R.-P., Köln 107686
Seiryu-do, Fukuoka 081968
Seiryuden Bijutsukan,
Matsushima 029019
De Seis a Ocho, Gijón 086021
Seison-kaku, Kanazawa 028778
Seissian & Fils, Hagop, Beirut . 132315
Seith, H., Kempten 076755
Seitou, Hiroshima 111262
Seitre, Simone, Saint-Martin . 109113
Seitz & Partner, Berlin 106430
Seitz, New York 095733
Seitz & Co., Saint Paul 096856
Seitz, K. & H., Rheinfelden .. 078229
Seitz, Werner, München 077618
Seiyo Inn, Sapporo 082105
Seiyun in Wadi Hadhramaut Museum,
Seiyun 054832
Seize, Quimper 105418
Seize, Louis, Cairo 103078
Sejac, Poligny 129173
Sejallet, Jean, Saint-Rambert-
d'Albon 073283
Sejarah Mesjid Banten Museum,
Banten 024538
Sejong, Seoul 111711
Sejong College Museum, Seoul . 030034
Sejong Cultural Center Art Gallery,
Seoul 111712
Sejong Gallery, Seoul 111713
Sejul, Seoul 111714
Sekai Gangu Hakubutsukan, Bandai
Collection, Karuizawa 028785
Sekai Oota Kokaikan, Yokaichi . 029708
Sekaido, Sagamihara 082095
Sekaido, Tachikawa 082124
Sekaido, Tama 082126
Sekaido, Tokyo 082197, 082198, 082199
Sekcja, Warszawa 138989
Sekhmet, Vitoria-Gasteiz ... 143814
Sekido Bijutsukan, Tokyo ... 029582
Sekiguchi Koh Museum Yuzawa,
Yuzawa 029755
Sekisho Bijutsukan, Misumi . 029031
Sekisui Bijutsukan, Tsu 029655
Sekitan Kogyo Hakubutsukan,
Ube 029666
Sekkeh Museum, Kerman ... 024709
Sekkeh Museum, Tehrän ... 024784
Sekken Museum, Sekken ... 033931
Sektor I, Górnośląskie Centrum Kultury,
Katowice 034822
Sekula, Sacramento 096646
Sekunda, Sankt-Peterburg .. 085267
Selano, Benedetto, Genève .. 116339
Selapak Neari, Phnom Penh . 058412,
                         101206
Selbu Bygdemuseum, Selbu . 033932
Selbu Strikkemuseum, Selbu
Bygdemuseum, Selbu 033933
Selby Gallery, Ringling School of Art
and Design, Sarasota 053151
Selch, Horst, Wörth am Main . 108956
Selçuk Müzesi, Taş ve Istampa
Müzesi, Konya 043008
Selden Emre, İstanbul 088139
Seldom Seen Designs,
Cheltenham 117565
Selearte 1, Padova 117468
Select, Rotterdam ... 083589, 132625
Select Art, Dallas 120657

Select Art, Utrecht ... 083662, 143249
Select Stamp and Coin Studio,
Buffalo 092555
Selected Antiques & Collectables,
Barnstaple 088365, 134064
Selected Works, Chicago ... 145058
Selectionart, Torino 111004
Selections, Haarlem 083352
Selective Eye Gallery, Saint
Helier 091187, 119015, 134991,
                         144816
Selective Prints, Whitchurch . 119338
Selective Prints, Winsford .. 119368
Self Help Graphics and Art Gallery,
Los Angeles 050467
Seligmann, Maja, London ... 090485
Selims Antiques, Winnipeg .. 064933
Seling, Dr. H.W., München .. 077619
Selinheimo, Kari, Turku 140750
Seljuk, San Francisco 136344
Selkirk, Kambah 061664, 127620
Selkirk, Winnipeg 140437
Selkirk College Mansbridge Kootenay
Collection, Castlegar 005579
Sell, Forchheim, Oberfranken . 075758
Sell Antiques, Salt Lake City . 096921
Sellares, Alain, Montjoie-en-
Couserans 070260
Sellarione, Dario, Roma 081556
Sellerio, Palermo 137712
Sellers, Basil, Moruya 099383
Selli, Daniel, Gassin 068639
Selli, Daniel, Saint-Tropez .. 073319
Sellier, Eric, Neuilly-sur-Seine . 070522
Sellier, Jean-Christian, Arras .. 066741
Selliez, Domnique, Hardifort . 068790
Sellner Glashütte, Lohberg .. 020103
Sellos, Oscar B., Rio de Janeiro 064125
Sellution Vintage Furniture,
Vancouver 064875, 064876
Sellwood Antique Collective,
Portland 096470
Sellwood Antique Mall, Portland 096471
Sellwood Coins, Portland ... 096472
Selly Manor Museum,
Birmingham 043626
Selmer, Chlodwig, Emsdetten . 106884
Selonke, Hardebek 076405
Selons, R. & B., Cuckfield .. 134310
Sels, Clara Maria, Düsseldorf . 106828
Selsey Lifeboat Museum, Selsey 045760
Selskar, Wexford 079724
Selskar Abbey Antiques, Wexford 079725
Selsor Gallery of Art, Safari Museum,
Chanute 047527
Šeltozerskij Vepsskij Étnografičeskij
Muzej, Šeltozero 037852
Selva, Alberto, Milano 131530
Selvi, Carlo, Roma 081557
Selwoods, Taunton 091545
Selwyn, Marc, Los Angeles .. 121590
Selz, Beat, Perrefitte .. 116561, 138079
Selz, Dr. G., Freiburg im
Breisgau 075883
Şemaki Evi Müzesi, Yenisehir . 043084
Semaphore Antique Centre,
Semaphore 062202
Semar, Rotterdam 112660
Sembritzki, Enno, Oldenburg . 077939,
                         130490
Semed Vurğunun Ev-Muzeyi,
Baki 003178
Semenovskij Gosudarstvennyj
Kraevedčeskij Istoriko-
chudožestvennyj Muzej,
Semenov 037853
Semenzato, Patrizia, Venezia . 081874
Semerád, Petr, Klatovy 140476
Semina Rerum, Zürich 116858
Seminar für künstlerisch-ästhetische
Praxis Menzel-Dach, Humboldt-
Universität zu Berlin, Berlin .. 055481
Seminar für Kunst Kunstpädagogik,
Universität Osnabrück,

Osnabrück 055681
Seminar für Kunst und Kirche,
Philosophisch-Theologische
Hochschule Sankt Georgen, Frankfurt
am Main 055541
Seminar für Kunst, Kunstgeschichte
und Kunstpädagogik, Carl
von Ossietzky Universität,
Oldenburg 055679
Seminar für Kunstgeschichte, Heinrich-
Heine-Universität, Düsseldorf . 055525
Seminar für Kunstgeschichte,
Universität Osnabrück,
Osnabrück 055682
Seminar für materielle und visuelle
Kunst, Carl von Ossietzky Universität,
Oldenburg 055680
Seminario, Tampa 124673
Seminario de Arte Aragones,
Zaragoza 139070
Seminario de Estudios de Arte y
Arqueologia, Universidad Valladolid,
Valladolid 056388
Seminole Canyon State Historical Park,
Comstock 047949
Seminole Nation Museum,
Wewoka 054358
Seminyak, Coffs Harbour 098906
Semler, Stefan & Rudolf, Fulda . 075934
Semley Auctioneers, Shaftesbury 127037
Semlitsch & Zirngast,
Ehrenhausen 062555
Semmelweis Orvostörténeti Múzeum,
Budapest 023551
Semmingsen, Oslo 113240
Semmler, Margitta, Wernigerode 078936
Semmler, Rita Maria, Lengdorf . 107856
Semon Gakko Kuwasawa Design
Kenkyujo, Tokyo 056054
Semprich, Christian, Mönkloh . . 077401
Sen, San Francisco 097322
Sen-Garou, Tokyo 111493
Sen-oku Hakukokan, Kyoto 028980
Sen Trop, Beograd 085348
Sen. MacArthur Memorabilia,
Tacloban 034603
Fernandes Sena, Brasilia 063939
Senac, Anna, Paris 129143
Senado Antiguidades, Rio de
Janeiro 064126
Senagoo, Ahmad, Houston 093787
Senamiesčio, Vilnius 082343
Senasis Arsenalas, Lietuvos
Nacionalinis Muziejus, Vilnius . 030540
Senasis Kuparas, Vilnius 082344, 142969
Senate House, Kingston 049984
Senda, Barcelona 114962
Sendai Music Box Museum,
Sendai 029360
Sendai-shi Hakubutsukan,
Sendai 029361
Sender- und Funktechnikmuseum,
Königs Wusterhausen 019719
Sender, Rosalia, Valencia 115518
Seneca County Museum, Tiffin . 053744
Seneca Falls Historical Society
Museum, Seneca Falls 053271

Seneca-Iroquois National Museum,
Salamanca 052877
Senge, Raymond, München . . . 130400
Senger, Bamberg 074672
Sengle, Volker, Heidelberg 076458
Senhouse Roman Museum,
Maryport 045233
Seniausias Panevėžio Pastatas,
Panevėžio kraštotyros muziejus,
Panevėžys 030442
Senior & Shopmaker, New York 123058
Senior, E.E., Taringa 062309
Seniora Antichità Restauri,
Milano 080722, 131531
Senjehesten Kystforsvarsmuseum,
Midt-Troms Museum, Moen . . 033778
Senn, Gabriele, Wien 100370
Senn, Georg F., Binningen 133697
Senn, Martin, Sankt Gallen 087789
Sennacheribbo, Ida, Barcelona . 114963
Sennerei-Museum und
Heimatmuseum, Unterwasser . 042085
Senninger, Eugène, Louverne . . 069648
Seno, Paolo, Milano 110429
Senouhi, Cairo 066271
Sensburg, München 077620
Sensen- und Heimatmuseum,
Achern 016461
Sensen-Schmiede-Museum/
Klangwelten, Micheldorf 002385
Sensenmuseum Geyerhammer,
Scharnstein 002715
Senshu Bunko Museum, Tokyo . 029583
Sensi, Francesco, Roma 081558
Sensibly Chic, Salt Lake City . . 096922
Sensler Museum, Tafers 042046
Sensua, Atlanta 119610
Senthoff, T., Oldenburg 142270
Senthoff, Thomas, Edewecht . . 141732
Sentimental Journey Antiques,
Tampa 097587
Sentimento, Maylands, Western
Australia 061808
Sentimento, New York 095734
Sentou, Paris 105226
Senza Zucchero, Milano 110430
Senzatempo, Miami 094802
Seo, Ja Hyun, Seoul 111715
Seo, Jong, Seoul 111716
Seojong Gallery, Seojong 111608
Seorim, Seoul 111717
Seoshin Gallery, Jeonju 111603
Seoul Art Museum, Seoul 030035
Seoul Arts Data Center, Seoul . 059535
Seoul Auction, Seoul 126253
Seoul Auction, Gangnam, Seoul 126254
Seoul Design Museum, Seoul . . 030036
Seoul Education Museum, Seoul 030037
Seoul Furniture and Antique Company,
Seoul 082265
Seoul International Print, Photo &
Edition Works Art Fair, Seoul . . 098237
Seoul Metropolitan Museum of Art,
Seoul 030038
Seoul National Science Museum,
Seoul 030039
Seoul National University College of
Fine Arts, Seoul 056104
Seoul National University Museum,
Seoul 030040
Seoul Olympic Museum, Seoul . 030041
Seoul Station Hall, Railroad Museum,
Seoul 030042
Sepänmäen Käsityömuseo,
Mäntsälä 011137
Sepant, Bucureşti 143617
Sepheriades E., Athinai 079179
Sépia Galerie, Paris 105227
Sepia International & Alkazi Collection,
New York 123059
Sepio, Claudio, Milano 080723
Sepon Penni, Helsinki . 066362, 140719
Septem Maria Museum, Adria . . 025245
Septentria Antikvariat, Tromsø . 143425

Sepulchre, Bruno, Paris 141280
Sequeira, Mário, Braga 113958
Sequoïa, Le Havre 128663
Sequoyah Cabin, Oklahoma Historical
Society, Sallisaw 052907
Ser, Yong, Singapore 133043
Serafino, Paola, Genova 080293
Serai, Cairo 103079
Seraphico, Oscar, Rio de Janeiro 101069
Seraphin, Philadelphia 123450
Serapi, Dallas 093214
Serbian Heritage Museum of Windsor,
Windsor, Ontario 007425
Serbin, Inna, New York 123060
Serbski muzej – Sorbisches Museum,
Bautzen 016978
Serbski muzej/ Wendisches Museum,
Stadtgeschichtliche Sammlungen
Cottbus, Cottbus 017768
Serck, Ph., Gent 100704
Serdar, Nikolas, Paris 105228
Serebrjanyj Rjad, Moskva 085204
Serebrjanyj Vek, Sankt-Peterburg 085268,
114361
Šeremetevskij Dvorec – Fontannyj
Dom, Sankt-Peterburgskij Muzej
Teatralnogo i Musykalnogo Iskusstva,
Sankt-Peterburg 037820
Seren, Bridgend 138142
The Serendib Gallery, Colombo . 086622,
115568
Serendipity, Deal 089039, 134317
Serendipity, Den Haag 083131
Serendipity, Edmonton 101297
Serendipity, Ledbury . . 089765, 134533
Serendipity, Preston 062122
Serendipity, Richmond 096568
Serendipity, Wolsingham 119378
Serendipity Antique Shows,
Fresno 093514
Serendipity Antiques, Haslemere 089457,
134444
Serendipity Antiques, Newcastle 061958
Serendipity Books, Leederville . . 139802
Serendipity Collectibles, Las
Vegas 094175
Serendipity Gallery, Katoomba . 099177
Serendipity Gallery, Las Vegas . 094176
Serendipity on Magnolia, Los
Angeles 094468
Serengeti Museum, Arusha 042627
Sereni, Renato, Brescia 079956
Serenity Designs, Congleton . . . 117623
Serero, Caroline, Marseille 104197
Sergačskij Kraevedčeskij Muzej,
Sergač 037855
Sergej Parajanov House-Museum,
Yerevan 000745
Sergej Borodin Muzeyi, Toshkent 054667
Sergej Esenin Muzeyi, Toshkent 054668
Sergievo-Posadskij Gosudarstvennyj
Istoriko-chudožestvennyj Muzej
Zapovednik, Sergiev Posad . . 037858
Seri Art Center, Bangkok 116958
Serie-Centrum, Trelleborg 143981
Serieboden, Stockholm 143965
Seriegalleriet, Stockholm 115923
Seriehörnan, Göteborg 143852
Seriekällaren, Uddevalla 143984
Serieknallen, Ullared 143985
Serieknodden, Göteborg 143853
Seriemuseet Comicland, Ullared 041100
Seriesamlarna, Malmö 143901
Serieuze Zaken, Amsterdam . . . 082857
Sérignan, Avignon 066887
Sérignan, Jean-Jacques, Avignon 103315
Sérignan, Philippe, Avignon . . . 140774
Seriko, Helsinki 066363
Serindia Publications, Chicago . 138322
Serini, Marcello, Madrid 133388
Serio, Salvatore, Napoli 110545
Serious Studios, Miami Beach . . 121887
Seriously Riches Collectables,
Christchurch 083896

La Serlas, Sankt Moritz 087796
Serlick, Howard, Philadelphia . 136102
Sermeño, San Salvador 103105
Sermon, E., Bruxelles 063551
Sermoneta, Zola Pedrosa 142891
Sermoneta, Emma, Bologna . . . 142641
Seromenho, J., Paris 071898
Seron, Sylvain, Paris 071899
Serot, Quiberon 105408
Seroussi, Natalie, Paris 105229
Serpa, Luis, Lisboa 114061
Serpent Mound Museum,
Peebles 051961
Serpente, Porto 114105
Serpentine Gallery, Aix-en-
Provence 103203
Serpentine Gallery, London 045104
Serpentine Vintage Tractors and
Machinery Museum, Serpentine 001491
Serpentinsteinmuseum Zöblitz,
Zöblitz 022891
Serpuchovskij Istoriko-
chudožestvennyj Muzej,
Serpuchov 037859
Serra d'Arquet, Barcelona 133166
Serra Museum, San Diego 052981
Serra, Christian, Figeac 068532, 140838
Serra, Valeria, Torino 132201
Serra, Yvette & Marie-Françoise,
Perpignan 105340, 129165
Serraire, Patrick, Paris 071900
Serrano, Madrid 086262
Serrano Pombo, Jenaro, Madrid 086263
Serrano, Juan, Gijón . . 086022, 133261
Serrant, Pierre, Saint-André-de-
l'Epine 072668
Serrao, Napoli 080934, 080935
Serrão, Filipe, Porto 126467
Serrão, Filipe Silva, Arcozelo . . 126432
Serrão, Filipe Silva, Porto 126468
De Serre, Arnhem 112337
Serrell, Philip, Great Malvern . . 126911
Serrenho, Pedro, Lisboa 114062
Serres, Paris 071901
Serres, Michel & Noël, Lunel-Viel 069663
Serres, Robert, Toulouse 073800
Serret – Portier, Experts Joailliers,
Paris 071902
Serret & Portier, Paris 071903
Serret, Fabiani, Neuilly-sur-Seine 104353
Serrurier, Jean-Luc, Sens 073501
Serrurier, Pierre, Sens 073502
Tavares Seruca, Lisboa 084961
Seruca, Tavares, Lisboa 084962
Servalvo, Amrita, Paradiso 087730
Servas, François, Darvault 128577
Serventi, André, Grenade 068706,
103781
Servera, Francisco, Palma de
Mallorca 086368
Service Historique de la Police,
Historische Dienst van de Politie,
Bruxelles 003500
Serviceton Historic Railway Station,
Serviceton 001492
Servicio de Publicaciones,
Murcia 137998
Servientis, Rosine, Pauillac 072052
Servinantes, Nantes 070477
Il Servo di Scena, Milano 080724,
110431
Servoir, Le Puy-en-Velay 069345, 103694,
128669
Servoninė Bitininkystės Muziejus,
Stripeikiai 030484
Sesame Art, London 118563
Sesanne Taideliike, Helsinki . . . 066364,
103156
Sese, Marie, Tilburg 083624
Sesma Garcia, Julian, Bilbao . . 133201
Sesquicentennial Museum,
Toronto 007204
Sessantasei, Napoli 080936

Session Cottage Museum, Turriff 046006
Sestieri, Andrea & Lorenzo,
Roma 081559
Šestka, Trebíč 102741
Set Alkol, Manama 100449
Il Setaccio, Sestri Levante 110931
Setagaya-kuritsi Riterashi
Hakubutsukan, Tokyo 029584
Setagaya-kuritsu Bijutsukan,
Tokyo 029585
Setareh & Söhne, Düsseldorf . 075519
Setareh & Söhne, Essen 075700
Setareh, Los Angeles 094469
Setesdalsbanen Stiftelsen,
Vennesla 034080
Setesdalsmuseet, Rysstad 033913
Seti Arte, Milano 110432
Setiawan, Singapore 085498
Seto Muuseumitaro, Obinitsa . 010644
Seto-Ohashi Memorial Bridge Museum,
Kurashiki 028906
Seto Rekishi Mingeikan,
Takamatsu 029435
Seto-shi Rekishi Minzoku Shiryokan,
Seto 029366
Seton Hall University Museum, South
Orange 053412
Setor de Malacologia, Juiz de
Fora 004523
Setsuro Takahashi Bijutsukan,
Azumino, Azumino 028502
Setsuryosha Bijitsukan, Niigata . 029176
Setti Carraro, Giovanni Maria,
Milano 080725, 110433
Setti Lino & Figlio, Ozzano Taro 080955
Settlers Hardware, Houston ... 093788
Settlers Museum, Brewarrina .. 000881
Settlers Museum, Himeville 038736
Settlers Store Antiques, Mudgee 061926
Settlers West Galleries, Tucson . 124772
Settlers West Gallery, Tucson ... 124773,
145625
Setton, Michele, Saxonwold ... 114816
Setzer, Ortrud, München 108194
Seurasaaren Ulkomuseo, Helsinki 010861
Seuren, Karlsruhe 107526
Sevan Art Gallery Limited,
Toronto 101703
Sevas, Bucureşti 085109
Ševčenkivskyj Nacional'nyj Zapovidnyk,
Kaniv 043175
Sevek, Eze 103735
Seven Art, Helsinki 103157
Seven Oaks House Museum,
Winnipeg 007467
Seven Sanat Galerisi, İstanbul . 117084
Seven Sisters Antiques, Lanceston
South 061715
Sévene, Suresnes 073609
Sevenoaks Museum and Gallery,
Sevenoaks 045763
Seventeen O Eight, Richmond . 123744
Seventh Heaven, Chirk 088895
Seventh Inning, Buffalo 092556
Severance, Detroit 120816
Severiarte, Bologna 109862
Severin, Gisela & Peter, Friedberg,
Hessen 075908
Séverine, Oriot, Honfleur 068832
Severinhaus Basilika Sankt Laurenz,
Enns 001871
Severn River Auctions, Severna
Park 127457
Severn Valley Railway, Bewdley 043585
Severo-Kavkazskij Gosudarstvennyj
Institut Iskusstv, Nalčik ... 056289
Severo-Kazachstanskij Oblastnoj
Istoriko-Kraevedčeskij Muzej,
Petropavl 029836
Severo-Kazachstanskij Oblastnoj
Muzej Izobrazitelnogo Iskusstva,
Petropavl 029837
Severo-osetinskij Gosudarstvennyj
Obedinennyj Muzej Istorii,

Architektury i Literatury,
Vladikavkaz 038026
Severo-osetinskij Literaturnyj
Muzej im. K.L. Četagurova,
Vladikavkaz 038027
Severočeská Galerie Výtvarného
Umění v Litoměřice, Litoměřice 009673
Severočeské Muzeum v Liberci,
Liberec 009667
Sevestre, Pascal, Cabourg 067547
Sévigné, Cesson-Sévigné 067755
Sévigné, Saint-Jean-sur-Vilaine . 072856
Sevilla Art Studio, New York .. 135998
Sevilla Decoracion, Sevilla ... 086476
Sevilla, Victor, Paris 105230
Seville Great House and Heritage Park,
Ocho Rios 028454
Sevimce Sanat Galerisi, İstanbul 117085
Sevniški Grad, Sevnica 038608
Sevret, René, Charenton-du-Cher 067836
Sewall-Belmont House,
Washington 054159
Seward County Historical Society
Museum, Goehner 049140
Seward House, Auburn 046653
Seward Museum, Seward 053277
Sewer, Ernest, Susten 087839
Sewerby Hall Art Gallery and
Museum, Museum of East Yorkshire,
Bridlington 043726
Sex Machines Museum, Praha . 009839
Sexmuseum Amsterdam Venustempel,
Amsterdam 031985
Sextius, Aix-en-Provence 103204
Sexton, Esther, Dublin 079587
Sexton, Robert, San Francisco . 124391
Sexton's Cottage Museum, Crows
Nest, New South Wales ... 001003
Seydoux, Paris 105231
Seyed-Ghaemi, Shamsadin,
Landshut 107793
Seyfert, Gudrun, Stuttgart 108728,
108729
Seyhoun, Nader, Tehrän 109520
Seyitgazi Müzesi, Seyitgazi ... 043056
Seymour, Rutherglen 062176
Seymour, Vancouver 101835
Seymour, Waltersboro 127487
Seymour and District Historical
Museum, Seymour 001493
Seymour Art Gallery, North
Vancouver 006519
Seymour Community Museum,
Seymour 053280
Seymour, Jane, Morrinsville ... 083998
Seyun Iron Museum, Eumseong 029922
Sezam, Beograd 114438, 143683
Sezione Arte, Dipartimento di Scienza
della Letteratura e dell'Arte
Medioevale e Moderna, Pavia . 055948
Sezione di Storia dell'Arte, Università
Genova, Genova 055909
Sezon Museum of Modern Art,
Karuizawa 028786
Sezoni, Sofia 101170
SFA Galleries, Stephen F. Austin State
University, Nacogdoches 051147
De Sfeer van Weleer, Rekken .. 032777
Sfeervol Wonen, Breda 083006
Sfeir-Semler, Beirut 111773
Sfeir-Semler, Dr. Andrée,
Hamburg 107315
La Sfinge, Parma 081129
SFMOMA Artist Gallery, San
Francisco 124392
Sforzini, Roma 142838
SG Kunst, Fredrikstad 084239
SGA Antiquités, Fontenay-sous-
Bois 068578
S'Galeriele, Göppingen 107140
Sgambati, Antonio, Modena .. 131588
Sgambato, Angelique,
Kaufbeuren 130108
Sganga, Giuseppina, Roma ... 081560

Sgarban, Giovanni, Milano 142743
Sgouros, Ioannis, Thessaloniki . 109099
Sgroi, Massimo, Palermo 081076
Sha Museum, Shaxian 008255
Shaanxi Chinese Painting Gallery,
Xian 008398
Shaanxi History Museum, Xian . 008399
Shaanxi People's Fine Arts Publishing
House, Xian 136746
Shaanxilong Books Museum,
Longxian 008048
Shab, Toronto 064753
Shaban, Yousef, Alexandria ... 066223
Shabby, Hattie, Tucson 097693
Shackerstone Railway Museum,
Shackerstone 045764
Shackleton, Anthony, David & Ann,
Snainton 091353, 135053
Shadeland Antique Mall,
Indianapolis 093924
Shades of Africa, Miami 121858
Shades of Afrika, Long Beach .. 121410
Shades of Age, Clayfield 061340
Shades of Color, Portland 123676
Shades of the Past, Carbonear . 005561
Shades of the Past, Philadelphia 096160
Shadi, Damascus 088057
Shadi Abdel Salam Museum,
Bibliotheca Alexandria,
Alexandria 010523
Shadicor Gallery, Cairo 103080
Shadowland, Richmond 145478
The Shadows-on-the-Teche, New
Iberia 051275
Shaffer, Portland 123677
Shafrazi, Tony, New York 123061
Shaftesbury Abbey Museum and
Garden, Shaftesbury 045765
Shaftesbury Town Museum,
Shaftesbury 045766
Shaftsbury Historical Society,
Shaftsbury 053283
Shah Alam Art Gallery, Shah
Alam 030731
Shah Charu, Dallas 120658
Shahd, V., Wendeburg 078921
Shaheed-e-Azam Bhagat Singh
Museum, Khatkar Kalan ... 024345
Al-Shahid, Dubai 117187
Shahinian, Zare, Damascus ... 088058
Shahnama, Mumbai 109471
Shahrastan Bros., Damascus .. 088059
Shahrood Museum, Semnan ... 024733
Shahryar Museum, Tabriz 024742
Shah's Arts Collection, Brighton,
Victoria 061205
Shaiké I am Art Snir, Caulfield
South 061307
Shaikh & Son, London 090486
Shain, Charlotte 120012
Shainman, Jack, New York ... 123062
Shaka, Saint Louis 123869
Shake Rattle & Read Book Box,
Chicago 145059
Shaker Art, Milwaukee 121949
Shaker Heritage Society Museum,
Albany 046346
Shaker Historical Society Museum,
Shaker Heights 053284
Shaker Museum, New Gloucester 051257
The Shaker Museum, Old
Chatham 051747
Shaker Museum at South Union,
South Union 053418
Shaker Square Antiques,
Cleveland 092971
Shaker Village of Pleasant Hill,
Harrodsburg 049381
Shakespeare & Co., München . 142187
Shakespeare Birthplace Trust,
Stratford-upon-Avon 045907
Shakespeare Gallery, San José . 102514
Shakespeare's Globe Exhibition and
Tour, London 045105

Shakey, Philadelphia .. 096161, 136103
Shakir Ali Museum, Lahore 034137
Shaklee, Calgary 064272, 128155
Al Shakma, Cairo 103081
Shakowi Cultural Center, Oneida 051774
Shakris, San Francisco 097323
Shakti Art Gallery, Mumbai ... 109472
Shalimar Palace, Dubai 043334
Shalom, Miami 135807
Shalom Construction &Co.,
Pittsburgh 136139
Shalovelo, Regina 128199
Shalva Amiranasvili Museum of
Art, Georgian National Museum,
Tbilisi 016420
Shamal, Villeurbanne 074226
Shambala, Hong Kong 065096
Shambellie House Museum of
Costume, National Museums of
Scotland, New Abbey 045322
Shambhala, Boston 138308
Shambhalla Librairie, Lyon ... 140932
Shambhu, Priti, Königs
Wusterhausen 076921
The Shambles, Ashburton 088311
The Shambles, Hillsborough .. 117971
Shambles Antiques, Kendal ... 089672
Shambles Museum, Newent ... 045355
Shamburger, Richmond 096569
Shambyu Museum, Shambyu .. 031853
Shamian Gallery, Guangzhou .. 102030
Shammah, Suzy, Milano 110434
Shamrock and Cotehele Quay
Museum, National Maritime
Museum Outstation, Saint Dominick,
Cornwall 045684
Shamrock Antiques, Wellington . 084187
Shamrock Museum, Shamrock . 007007
Shamwari, Oakland 123266
Shan Rong Cultural Display Center,
Beijing 007644
Shan State Museum, Taunggyi . 031828
Shan Xi, Shanghai 065147
Shan, Fenglin, Guangzhou 102031
Shand Galleries, Toronto 064754
Shand House Museum, Ferry Hill 005804
Shandong Academy of Fine Art,
Jinan 055220
Shandong Arts and Crafts College,
Tai'An 055242
Shandong Literature and Art
Publishing House, Jinan 136728
Shandong Painting Gallery, Jinan 007958
Shandong Provincial Museum,
Jinan 007959
Shandong Stone Inscription Art
Museum, Jinan 007960
Shands Antiques, Christchurch . 083897
Shandy Hall, Coxwold 044027
Shandy Hall, Geneva 049071
Shaner, Mark, Denver 120770
Shanez, Monaco 112019
Shang, Singapore 085499
Shangai, Bogotá 065227
Shanggao Museum, Shanggao . 008221
Shanghai Antiques and Curios Store,
Shanghai 065148
Shanghai Art Fair, Shanghai .. 098055
Shanghai Arts & Crafts,
Singapore 085500
Shanghai Arts and Crafts,
Shanghai 102200
Shanghai Calligraphy and Painting
Publishing House, Shanghai .. 136735
Shanghai Corner, Los Angeles . 094470
Shanghai Curio, Shanghai 065149
Shanghai Fine Arts Gallery,
Singapore 114602
Shanghai Gallery of Art,
Shanghai 102201
Shanghai History Museum,
Shanghai 008233
Shanghai Literature and Art Publishing
House, Shanghai 136736

Shanghai Museum, Shanghai .. 008234
Shanghai Natural History Museum,
Shanghai Science & Technology
Museum, Shanghai 008235
Shanghai Oil Painting and Sculpture
Academy, Shanghai 055235
Shanghai People's Fine Arts Publishing
House, Shanghai 136737
Shanghai Science & Technology
Museum, Shanghai 008236
Shanghai Spring Art Salon,
Shanghai 098056
Shanghai West, Los Angeles ... 094471
Shanghai'd, Philadelphia 096162
Shanghaied, Richmond, Victoria 062149
Shanghang Museum, Shanghang 008242
Shangjing Ruins Museum,
Ning'an 008136
Shango, Dallas 120659
Shangqiu Museum, Shangqiu ... 008243
Shangqiu Zone Museum,
Shangqiu 008244
Shangrao County Museum,
Shangrao 008245
Shangrao Zone Museum,
Shangrao 008246
Shangrila Art, Lalitpur 112098
Shangrila Hakka Museum,
Zaociao 042610
Shangzhou Museum, Shangzhou 008247
Shank, Baltimore 119771
Shankar, Jaipur, Rajasthan ... 109389
Shankara, Westerheim, Alb-Donau-
Kreis 078946
Shankar's International Dolls Museum,
Delhi 024237
Shanklin Auction Rooms,
Shanklin 127039
Shanklin Gallery, Shanklin 119074,
135024, 144832
Shannon, Apollo Bay 098650
Shannon, Milford 127319
Shans, New York 095735
Shantala, Paris 071904
Shantou Archaeology Museum,
Shantou 008248
Shantou Museum, Shantou 008249
Shantytown, West-Coast-Living
Heritage, Greymouth 033214
Shanxi, Villeurbanne 074227
Shanxi Painting Gallery, Taiyuan 008314
Shaoguan Museum, Shaoguan . 008251
Shaolin Temple Religion Museum,
Yuanli 042601
Shaowu Museum, Shaowu 008252
Shaoxing Museum, Shaoxing .. 008254
Shaper, Zammie, Parramatta .. 062066
Shapero, Bernard J., London .. 144687
Shapes Collection, West
Hollywood 097931
Shapiro, Johannesburg 085648
Shapiro, San Francisco 124393
Shapiro, Andrew, Parkside 125223
Shapiro, Andrew, Toorak 125242
Shapiro, Andrew, Woollahra ... 125249
Shapiro, Estela, México 111984
Shapiro, Richard, West
Hollywood 097932
Sharadin Art Gallery, Kutztown
University of Pennsylvania,
Kutztown 050023
Sharalan, Calgary 064273, 128156
Shardlow Antiques Warehouse,
Shardlow 091276
Sharia Museum, Cairo 010563
Sharif, Manama 100450
Sharikat Kamsis Commercial
Enterprise, Penang 082406
Sharjah Antiques, Sharjah 088209
Sharjah Archaeology Museum,
Sharjah 043340
Sharjah Art Museum, Sharjah .. 043341
Sharjah Art Production, Sharjah 117201
Sharjah Heritage Museum,

Sharjah 043342
Sharjah Islamic Museum,
Sharjah 043343
Sharjah Natural History Museum,
Sharjah 043344
Sharjah Science Museum,
Sharjah 043345
Al-Shark, Cairo 066272
Sharland & Lewis, Tetbury 091580
Sharlot Hall Museum, Prescott . 052352
Sharon Arts Center, Sharon ... 053286
Sharon Museum, Emek Hefer .. 025037
Sharon Museum Emek Hefer,
Midreshet Ruppin 025156
Sharon Temple Museum Historic Site,
Sharon 007009
Sharonville Coin, Cincinnati 092895
Sharp Artists, Glasgow 117858
Sharp Artists, Uddington 119282
Sharp Shooter Studios, West
Hollywood 125025
Sharp Street Glass, Philadelphia 123451
Sharp Studio Gallery, San Diego 124172
Sharp, Ian, Tynemouth 091684
Sharpe, Bellevue Hill 061121
Sharpe Selections, Saint Louis . 096773
Sharpsteen Museum, Calistoga . 047352
Al-Sharq, Damascus 088060
Sharrington Shells, Sharrington . 091277
Sharry, Buffalo 092557, 135380
Shasmoukine, Pierre, Sarlat-la-
Canéda 105655
Shasta College Museum,
Redding 052493
Shaston Antiques, Shaftesbury . 091273,
135022
Shaughnessy, Minneapolis 135838
Shaughnessy Antique, Vancouver 064877
Shaukat, Dubai 117188
Shaw, Goulburn 061553
Shaw & Zok, Bonn 106541
Shaw a Park III, Albuquerque .. 092083,
135289
Shaw Island Historical Society
Museum, Shaw Island 053290
Shaw, Arthur, Torrington 127477
Shaw &Co., Jack, Ilkley 089635
Shaw, Lawrence, Horncastle ... 089588
Shaw, Nicholas, Petworth 091007
Shaw, Trish, Christchurch 112860
Shaw, Walter F., Phoenix 136121
Shawano County Historical Society
Museum, Shawano 053291
Shawbrook, Benalla 061126
Shawneetown Historic Site, Old
Shawneetown 051750
Shawqi Museum, Cairo 010564
Shaw's Corner, Ayot-Saint-
Lawrence 043457
Shay, Robert, Saint Louis 096774,
136205
Shazar, Maité, Lausanne 133805
Sha'Zeez, Las Vegas 121379
Shearburn, William, Saint Louis . 123870
Shears, Diana, Penzance 118901
Shearwater Aviation Museum,
Shearwater 007013
Sheather, Willy, Wagga Wagga . 099856
Sheba, New York 095736, 135999
Sheba Gallery, Cairo 103082
Sheboygan County Historical Museum,
Sheboygan 053297
The Shed, Coatesville 083915
The Shed Antiques, Dublin 079588,
130862
Shed im Eisenwerk, Ausstellungsraum
für zeitgenössische Kunst,
Frauenfeld 041493
Sheeba, Sharjah 088210
Sheehan, Portland 136158
Sheehan Gallery at Whitman College,
Walla Walla 054039
Sheehan, Kevin, Tetbury 135128
Sheehan, Steve, Bunbury 098818

Sheelin Antiques and Irish Lace
Museum, Enniskillen 044298
Sheen, London 118564
Sheen, Max, Hong Kong 065097
Sheepchandler Gallery,
Rroundstone 109687
Sheffield Bus Museum, Sheffield 045776
Sheffield Institute of Art and Design,
Sheffield Hallam University,
Sheffield 056606
Sheffield, E., Genève 087502
Shefton Museum of Greek Art and
Archaeology, Newcastle-upon-
Tyne 045352
Shehady, Pittsburgh 096326
Shehail, Abdullah Ali, Sharjah .. 088211
Sheikan Museum, Sudan National
Museum, El-Obeid 040490
Sheikh Saeed's House, Dubai .. 043335
Sheila's Gallery, Singapore 085501
Sheker, Gus, Denver 135535
Shekhtman, Irene, New York ... 136000
Shekspear Ghost Town Museum,
Lordsburg 050397
Shelander, Kay, Orland Park ... 127382
Shelburne County Museum,
Shelburne 007017
Shelburne Museum, Shelburne . 053300
Shelby, Portland 096473
The Sheldon Art Galleries, Saint
Louis 052823
Sheldon Jackson Museum, Sitka 053361
Sheldon Jaffery, R., Beachwood 144993
Sheldon Memorial Art Gallery and
Sculpture Garden, Lincoln ... 050300
Sheldon Museum, Haines 049315
Sheldon Peck Museum, Lombard 050361
Shell Exhibit at Chi Chin Seashore
Park, Kaohsiung 042367
Shell House Gallery, Ledbury .. 118084
Shell Lake Museum, Shell Lake . 007018
Shelley, Yass 062507
Shelley, Julian, Aberystwyth ... 144216
Shelter House, Emmaus 048576
Shelter Island Historical Society
Museum, Shelter Island 053306
Shemer Art Center and Museum,
Phoenix 052124
Shemtov Shani, Barcelona 114964
Shen, Chen, Singapore 085502
Shengjing Hua Yuan, Shenyang . 102215
Shenn, Singapore 114603
Shenton Park Antiques, Shenton
Park 062205
Shenton, Rita, Twickenham 091678,
144898
Shenyang Palace Museum,
Shenyang 008259
Shenzhen Art Gallery, Shenzhen 008265
Shenzhen Museum, Shenzhen . 008266
Shenzhen Painting Gallery,
Shenzhen 008267
Shepard, Joel, Seattle 136376
Shepard, Paul S., Tucson 097694
Shephela Museum, Kibbutz Kfar
Menahem 025137
Shepherd & Derom, New York . 123063,
136001
Shepherd, Peter, Hurst 089623, 144494
Sheppard Fine Arts Gallery, University
of Nevada, Reno, Reno 052511
Sheppard Street Antiques,
Richmond 096570
Sheppard, C., Durrow 079615
Shepparton Art Gallery,
Shepparton 001495
Sheppy's Farm and Cider Museum,
Bradford-on-Tone 043699
Shepton, Denver 093337
Shepton Mallet Antiques and
Collectors Fair, Shepton Mallet 098403
Shepton Mallet Museum, Shepton
Mallet 045780
Sheqi Museum, Sheqi 008269

Sher, Miami 121859
Sheraton House, Torquay 091617
Sherborne Castle, Sherborne ... 045781
Sherborne House, Regional Centre for
Visual Arts, Sherborne 045782
Sherborne Museum, Sherborne . 045783
Sherborne World of Antiques,
Sherborne 091308, 135036
Sherbrooke Art Society, Belgrave 058219,
098728
Sherbrooke Village, Sherbrooke, Nova
Scotia 007019
Sherburne County Historical Museum,
Becker 046865
Shere Antiques Centre, Shere .. 091314
Sheridan Brown, Hawthorn 061604
Sheridan, Ian, Hampton 144451
Sheriff Andrews House, Saint
Andrews 006851
Sheringham Museum,
Sheringham 045785
Sherlock Holmes Museum,
London 045106
Sherlock Holmes-Museum,
Meiringen 041736
Sherm, Sacramento 136193
Sherman, Los Angeles 135743
Sherman Galleries, Paddington, New
South Wales 099535, 099536
The Sherman House, Lancaster 050116
Sherman, Barry, Prahran 062119
Sherman, Beca & Liza, New York 095737
Sherrier Resources Centre,
Lutterworth 045170
Sherry, Houston 093789
Sherry, Kelly, Houston 093790
Sherry's Antiques, Salt Lake City 096923
Sherwin Miller Museum of Jewish Art,
Tulsa 053851
Sherwood, Houston 121126
Sherwood, Tubbercurry 079717, 130896
Sherwood-Davidson House,
Newark 051505
Sherwood Forest Plantation, Charles
City 047535
Sherwood, Anthony, Ashgrove .. 061066
Sherwood, D.W., Rushden 091155
Shetland Croft House Museum,
Dunrossnes 044179
Shetland Museum, Lerwick 044812
Shettler, Gavin, Portland 123678
Sheung Yiu Folk Museum, Hong
Kong 007890
Sheyenne Valley Arts & Crafts
Association, Fort Ransom 060734
Shi Liao Display Center, Nanjing 008116
Shi-San-Hang Museum of Archaeology,
Bali 042268
Shibayama, Tokyo 111494
Shibayama Haniwa Hakubutsukan,
Shibayama 029370
Shibden Hall Museum and West
Yorkshire Folk Museum, Halifax 044505
Shibunkaku, Kyoto ... 082038, 142901
Shibunkaku Bijutsu, Tokyo 082200,
142938
Shibunkaku Bijutsukan, Kyoto .. 028981
Shibuya Bijutsukan, Fukuyama . 028564
Shicheng Museum, Shicheng .. 008270
Shido-ji Homotsukan, Shido 029372
Shield, Richmond 096571
Shield, Robin, Swinstead 091535
Shiell, Miriam, Toronto 101704
Shifara Kunsti-ja Antiigigalerii,
Tallinn 066287
Shifflett, Los Angeles 121591
Shifrin, Judy, Cleveland 120453
Shiga-kenritsu Kindai Bijutsukan,
Otsu 029281
Shiga-kenritsu Shiryokan, Azuchi
Shiro, Azuchi 028501
Shigaraki-cho Dentouteki Bijutsu
Senta, Shigaraki 029374
Shigaraki-yaki Bijutsukan,

Shigaraki 029375
Shige, San Francisco 097324
Shigeki Komori Atelier, Gifu .. 111255
Shihezi Junken Museum, Shihezi 008271
Shijiazhuang Museum,
  Shijiazhuang 008274
Shiki, Tokyo 111495
Shikoku-mura Minka Hakubutsukan
  no Bijutsukan, Takamatsu .. 029436
Shilla, Daegu 111589
Shillong Tribal Research Institute,
  Shillong 024475
Shiloh Museum of Ozark History,
  Springdale 053447
Shiloh National Military Park and
  Cemetery, Shiloh 053313
Shilpa Bharti Institute of Arts and
  Crafts, Delhi 055793
Shilpi Kendra, Mumbai 109473
Shima Marineland, Ago 028463
Shimamoto, Tokyo 082201
Shimane-kenritsu Hakubutsukan,
  Matsue 029005
Shimane Prefectural Yakumodatsu
  Fudoki-no-oka Museum,
  Matsue 029006
Shimazu Foundation Memorial Hall,
  Kyoto 028982
Shimen Museum, Shimen 008275
Shimeru, Bairnsdale 098686
Shimizu, Tokyo 082202
Shimizu, Yokohama 111544
Shimogamo Hanki-cho Kyoto,
  Kyoto 028983
Shimonoseki-shi Bijutsukan,
  Shimonoseki 029384
Shimonoseki-shiritsu Kaikyokan,
  Shimonoseki 029385
Shin-Yokohama Ramen Hakubutsukan,
  Yokohama 029724
Shinagawa Historical Museum,
  Tokyo 029586
Shinano Drawing Museum, Ueda 029669
Shinanobashi Gallery, Osaka .. 111366
Shinchon Museum, Shinchon .. 029880
Shindler, Hutt City 126320
Shine, Nelson 112966
Shiners, Snobs Knobs, Jesmond 089660
Shing, Singapore 085503, 114604
Shingenkou Treasure Hall Historical
  Museum, Enzan 028532
Shinn, Catherine, Cheltenham .. 088841,
                         134242
Shinobazu, Tokyo 111496
Shinrone Antique Pine, Shinrone 079698,
                         130891
Shinseidoo-Hatanaka, Tokyo .. 111497
Shinseisha, Osaka 111500
Shinshushin-machi Hakubutsukan, Art
  Museum, Arishima Museum and
  Fossil Museum, Shinshushin .. 029388
Shinuiju Historical Museum,
  Shinuiju 029881
Shinwa, Tokyo 126249
Shinyanga Mazingira Museum,
  Shinyanga 042637
Shinyoung Cinema Museum,
  Namwon 029975
Shio Aragvispireli House Museum,
  Tbilisi 016421
Shio-no-michi Hakubutsukan,
  Itoigawa 028715
Shiogama-jingu Homotsukan,
  Shiogama 029389
Shiojiri Tankakan, Shiojiri 029391
Shiota, T.Z., San Francisco .. 097325,
                         124394
Shiozawa Tsumugi Commemorative
  Museum, Shiozawa 029393
Ship Model Gallery of New York, New
  York 095738
Ship to Shore, Vancouver 064878
Shipahoy, Killarney 061689
Shipley, London 144688

Shipley Art Gallery, Gateshead . 044382
Shipley Gallery, Blackheath ... 098751
Shippensburg Historical Society
  Museum, Shippensburg 053316
Ships of the Sea Maritime Museum,
  Savannah 053181
Shipshapes Maritime Prints, Thornton
  Hough 119230
The Shipwreck, Brightlingsea .. 088610
Shipwreck Heritage Centre,
  Hastings 044534
Shipwreck Museum, Bredasdorp 038667
Shirahata Shiro World Mountains
  Photograph Collection, Yuzawa 029756
Shiraishi Contemporary Art – The
  Bathhouse, Tokyo 111498
Shirakaba Bijutsukan, Onomichi 029236
Shirakawadai Shoten, Kobe .. 142897
Shiraz, Tampa 097588
Shiraz House Museum, Gyumri . 000707
Shiraz Persian Rug Gallery,
  Goodwood 061539, 127605
Shire Hall Gallery, Stafford 045852
Shire Pottery Gallery, Alnwick .. 117228
Shire Publications, Princes
  Risborough 138278
Shire, Andrew, Los Angeles .. 121592
Shirehall Museum, Little
  Walsingham 044847
Shiretoko Museum, Shari 029369
Shirley-Eustis House, Boston ... 047132
Shirley Plantation, Charles City . 111501
Shiromatsu, Sendai 111391
Shirota, Tokyo 111499
Shiseido, Tokyo 111500
Shiseido Art House, Kakegawa . 028753
Shishi Kaikan, Takayama 029459
Shishou Museum, Shishou ... 008276
Shishu, Chengdu 102015
Shishu, Guangzhou 102032
Shitamachi Museum, Tokyo ... 029587
Shizen Kyoiku En, National Science
  Museum Tokyo, Tokyo 029588
Shizenshi Hakubutsukan, Tohoku
  Daigaku, Sendai 029362
Shizenshi Hakubutsukan, Tokai
  Daigaku, Shimizu 029377
Shizukuishi-machi Rekishi Mingeikan,
  Shizukuishi 029398
Shizuoka Art Gallery, Shizuoka . 029402
Shizuoka City Toro Museum,
  Shizuoka 029403
Shizuoka-kenritsu Bijutsukan,
  Shizuoka 029404
SHM Smith Hodgkinson, Chorley 126858
Sho, Tokyo 111502
Sho Art, Tokyo 082203, 111502
Shoah Museum, Holocaust History
  Museum, Jerusalem 025105
Shoal Creek Gallery, Austin 119698
Shoal Creek Living History Museum,
  Kansas City 049891
Shoal Lake Police and Pioneer
  Museum, Shoal Lake 007030
Shoalhaven Art Gallery, Old Erowal
  Bay 099499
Shobe, Lewistown 127283
Shobhana, Bangalore 079362
Shodo Hakubutsukan, Tokyo ... 029589
Shoe Museum, Street 045912
Shofi-Hollowell, Sylvia, Tampa . 124674
Shogetsu Bijutsukan, Wakakusa 029689
Shogun's Gallery, Portland ... 096474
Shohada Museum, Ghazvin ... 024692
Shohten, Inami, Tokyo 082204
Shoji Ueda Museum of Photography,
  Kishimoto 028820
Shojudo, Tokyo 082205
Shoko Shuseikan Historical Museum,
  Kagoshima 028750
Sholem Asch House, Bat Yam . 025024
Sholom Aleichem Museum, Tel
  Aviv 025211
Al Shomou Art Gallery, Cairo ... 103083

Shonandai Gallery, Kanagawa .. 111273
The Shooting Gallery, San
  Francisco 124395
Shooting Star Art Gallery, New
  Orleans 122237
Le Shoppe, Oklahoma City ... 096009
Shore Line Trolley Museum, East
  Haven 048437
Shore Road Pumping Station,
  Birkenhead 043606
Shore Village Museum, Rockland 052643
Shoreham Aircraft Museum,
  Shoreham 045789
Shoreham Airport Visitor Centre,
  Shoreham-by-Sea 045791
Shoreline Historical Museum,
  Seattle 053250
Shoreline Studio, Aberdour ... 117213
Shoreman, Taipei 116940
Shores Space, Amsterdam ... 112279
Shores, Ken, Portland 096475
Shorewood Coins, Milwaukee . 094880
Shorey & Co., Kevin, Cairns ... 061252
Short North Gallery, Columbus . 120514
The Short Poppy, Wellington .. 084188
Shorter Mansion Museum,
  Eufaula 048616
Shortmead Antiques,
  Biggleswade 088462
Shoshone Tribal Cultural Center, Fort
  Washakie 048886
Shosoin Treasure Repository,
  Nara 029152
The Shoto Museum of Art, Tokyo 029590
Shotton, J., Durham 144377
Shotts Heritage Centre, Shotts . 045792
Shou Museum, Shouxian 008281
Shouguang Museum, Shouguang 008282
Shouhaku Bijutsukan, Nara ... 029153
Shoun, Nara 111339
Shoundo Bijutsukan, Ube 029667
Shout!, Singapore 085504
Show Gallery, Toronto 007205
Show of Hands, Denver 120771
Show Off New Zealand, Raglan . 112996
Showa Bijutsukan, Nagoya ... 029127
Showcase, Las Vegas 094177
Showcase, Richmond 096572
Showcase Antiques, Randwick . 062138
Showcase Antiques Art, Dubai . 088187
The Showplace Antiques, New
  York 095739
Showplace Gallery, Houston .. 093791
Showroom, Chicago 120267
Showroom, Montréal 064471
SHP Special Historical Products,
  Düsseldorf 075520
Shreemanthi Bai Memorial
  Government Museum,
  Mangalore 024395
Shreve, Crump & Low, Boston . 092494
Shrewsbury Antique Centre,
  Shrewsbury 091334
Shrewsbury Castle, Shrewsbury 045797
Shrewsbury Museum and Art Gallery,
  Shrewsbury 045798
Shrewsbury Windle House,
  Madison 050586
Shri Chhatrapati Shivaji Maharaj
  Museum, Satara 024469
Shri Girdharbhai Sangrahalaya,
  Amreli 024141
Shrikrishna Science Centre,
  Patna 024434
Shrine of the Book, Israel Museum,
  Jerusalem 025106
Shrishti, Delhi 109341
Shrishti, Hyderabad 109372
Shrishti Art Gallery, Lucknow . 109408
Shropshire Regimental Museum,
  Shrewsbury 045799
Shrubsole, S.J., New York ... 095740
Shtekelis Prehistory Museum,
  Haifa 025058

Shtepia muze e piktorit Vangjush Mio,
  Korçë 000024
Shua Quan, Beijing 101990
Shuanglinsi Painted Sculpture Art
  Museum, Pingyao 008158
The Shudehill Book Center,
  Manchester 144728
Shufunotomo, Tokyo 137791
Shugborough Estate Museum, Milford,
  Staffordshire 045266
Shugoarts, Tokyo 111503
Shugu, Shanghai 065150
Shuma, Vancouver 125390
Shumpudo, Tokyo 111504
Shumway, George, York 138482
Shun Da, Qingdao 102162
Shunan-shi Bijutsukan nara
  Hakubutsukan, Shunan 029407
Shunchang Museum, Shunchang 008283
Shung Ye Museum of Formosan
  Aborigines, Taipei 042548
Shungeng, Jinan 102141
Shunyam, Amsterdam 112280
Shuswap Lake Provincial Park Nature
  House, Kamloops 006060
Shutter Door Antiques, Virginia
  Beach 097792
Shuttleworth Collection, Old
  Warden 045434
Shuvai, Svenja, Duisburg 106848
Shwan, Saint-Malo (Ille-et-
  Vilaine) 105553
Shyam Art, Kathmandu 112087
Si-Am Trading, Hong Kong ... 102112
Si-Art, Windsor 119367
Si Como No, San Antonio 124070
Si, Bao Zhai, Singapore 114605
Siag, Mona, Cairo 066273
Siagne, Saint-Cezaire-sur-Siagne 072710
SIAM- Société Internationale Art
  Moderne, Paris 105232
The Siam Gallery, Plymouth .. 118925
The Siam Society under Royal Patronage,
  Bangkok 060002, 116959
Siano, Philadelphia 123452
Siatista Botanical Museum,
  Siatista 023259
Siatista Paleontological Museum,
  Siatista 023260
Siaulio, Alfonso, Vilnius 132324
Šiauliu Aušros Muziejus, Šiauliai 030471
Šiauliu Miesto Vandentiekio ir
  Kanalizacijos Muziejus, UAB Šiauliu
  vandenys, Šiauliai 030472
Šiauliu Rajono Literatūros Muziejus,
  Naisiai 030425
Šiauliu Universiteto Istorijos Muziejus,
  Šiauliai 030473
Sibbesen, Hans Chr., Sønderborg 125469
Sibbesgården Museo, Sipoo ... 011358
Sibbo Jordbruksmuseum, Nikkila 011187
Sibeliuksen Syntymäkoti,
  Hämeenlinna 010792
Sibelius Museum, Turku 011442
Sibenaler, Jean-Claude, Bondy . 067287
Sibenaler, Jean-Claude, Noisy-le-
  Sec 070740
Sibila, Revista de Arte, Música y
  Literatura, Sevilla 139071
Sibley, Landshut 077019
Sibley, Rosenheim 078276
Sibley County Historical Museum,
  Henderson 049439
Sibley House, Mendota 050794
Sibman Gallery, Paris 105233
Siboney, Santander 115432
Siboni, Michel, Sceaux 125813
Sibony, Ariel, Paris 105234
Sibrium, Milano 142744
Sibsagar College Museum, Hiranya
  Probha Memorial Library and
  Museum, Sibsagar 024480
Sicamous and District Museum,
  Sicamous 007032

Sicard-Zuvi, Buenos Aires 127537
Sicard, Christian, Saint-Julien-de-
Peyrolas 072862
Sicardi, Houston 121127
Sicare, Auxonne 103297
Sichert, Georg, Laufen 142052
Sichuan Fine Arts Institute,
Chongqing 055205
Sichuan Fine Arts Publishing House,
Chengdu 136707
Sichuan Museum, Chengdu ... 007696
Sichuan University Museum,
Chengdu 007697
Sickingen-Museum, Landstuhl . 019887
Sicolo, Luigi, Milano 131532
Sicot, Jacky, Salon-de-Provence 073389
Siculan, Mike, Saint Louis 096775
Sicurani, Ajaccio 066499, 128431
Sid Richardson Collection of Western
Art, Fort Worth 048908
Sid Vale Heritage Centre,
Sidmouth 045801
Sidac, Leiden 112560
Side Door Antiques, Tulsa . 097756,
136436
Side Door Antiques, Winnipeg . 064934
Side Müzesi, Side 043057
Side Roads Publications, Miami 121860
Side Street Projects, Pasadena . 060735
Sidenväveri Museum, Stockholm 041038
Sidewalk Gallery, Battery Point . 098715
Sidharta, Jakarta 126164
Sidmouth Antiques Centre,
Sidmouth 091342, 144847
Sidney and Gertrude Zack Gallery,
Vancouver 007296
Sidney Historical Association Museum,
Sidney 053333
Sidney Historical Museum, Society
of Saanich Peninsula Museums,
Sidney 007035
Sidney Lanier Cottage, Middle Georgia
Historical Society, Macon ... 050574
Sidney Mishkin Gallery of Baruch
College, New York 051467
Sidoti, Salvatore, Palermo 110620
Sidra, Manama 100451
Sidvik Collections, Bangalore . 109267
Sié, Henri, Saint-Tropez 105629
Sieben-Keltern-Museum,
Metzingen 020396
Sieben-Schwaben-Museum,
Türkheim 022228
Siebenberg House, Jerusalem . 025107
Siebenberg-Verlag, Lehrte 137337
Siebenbürgisches Museum
Gundelsheim, Gundelsheim,
Württemberg 018893
Siebengebirgsmuseum,
Königswinter 019741
B. & S. Siebenhaar Verlag, Berlin 137095
Sieber, Bruno, Luzern . 116480, 133844
Siebers, Yves, Stuttgart 126124
Siebert, Fritz, Frankfurt am Main 075824,
129855
Siebert, Ulrich, Wettenberg ... 078954
Siebler, Mike, Püttlingen 108417
Sieblos-Museum Poppenhausen,
Paläontologisch-geologisches
Museum, Poppenhausen 021171
Siebold Kinenkan, Nagasaki ... 029102
Siebold-Museum, Würzburg ... 022802
Siebolts, Maria, Oberbiberg ... 077857
Siebrecht, A., Westerstede 078952
Siedentop, B., Salzgitter 078343
Siedler, Georg-Wilhelm, Wien .. 063072
Siedow, Fred, Portland 136159, 136160
Sieff, Simon, Tetbury 091581
Siege Museum, Petersburg 052017
Siegel, Allan, Pittsburgh 136140
Siegel, Carsten, Zürich 134001
Siegel, Eliot, London 118565
Siegel, Evelyn, Fort Worth 120885
Siegel, Robert A., New York ... 127358

Siegemund, Reichartshausen .. 078181
Sieger, Bonn 075090
Siegerlandmuseum mit
Ausstellungsforum Haus
Oranienstraße, Siegen 021854
Siegersma, Günter, Wesel 078941
Siegert, O.W., München 130401
Siegfried Charoux Museum,
Langenzersdorf 002269
Siegfried H. Horn Archaeological
Museum, Andrews University, Berrien
Springs 046948
Siegle, Franz, Mühlhausen, Rhein-
Neckar-Kreis 142135
Siegler, Joseph, Pézenas 072142
Siegmayer, H., Seeheim-
Jugenheim 078441
Siekerman, R.P., Amsterdam .. 132414
Sieler, Andreas, Gera 129898
SiemensForum, München 020572
Siemers, Petra, Bad Fallingbostel 074511
Siemon, K.& M., Homberg (Efze) 076570
Siempre Viva Art, Denver 120772
Sień Biała Galeria Sztuki, Gdańsk 113486
Sień Mała Galeria, Gdańsk 113487
Siena, Salamanca 086415, 133417
Siena, Zaragoza 133513
Sienna, Pamela, Boston 119921
Siepen, Harald, Bücken 075211
Siepmann, Markus, Essen 075701
Siera, B.M., Sleeuwijk 132632
Sieracki, Baltimore 092362
Sierantuur, Rotterdam 083590
Sierra Arts Foundation, Reno .. 052512
Sierra Colina, Fernando de la,
Barcelona 133167
Sierra Colours, Nashville 122100
Sierra Leone National Museum,
Freetown 038263
Sierra Pastel Society, Placerville 060736
Sierra, Amparo, Bogotá 102366
Sierra, Dolores de, Madrid 115285
Sierro Sánchez, Alvaro, Santiago de
Chile 128253
Sies & Höke, Düsseldorf 106829
Sieso, Laurent, Hindisheim ... 068820,
128623
Siete Revueltas, Sevilla 086477
Sieur, Pierre, Paris 141281
Sievert, Hans-Jörg, Gmund am
Tegernsee 076015
Sievi, Brigitte, Berlin 106431
Sievin Kotiseutumuseo, Sievi .. 011353
Siewert, Uta, Bad Bevensen ... 074495
Sifang Buyun Xuan, Qingdao .. 102163
Sigaar Name, Geraardsbergse Museum,
Geraardsbergen 003650
La Sigalière, Civray 067995
De Sigarenfabriek, Delft 112367
Sigaroudinia, Parvis, Lisburn .. 089841
Sigdal og Eggedal Museum,
Prestfoss 033882
Siggebohyttans Bergmansgård,
Nora 040871
Sigi's Antiquitäten, Staad 087826
Sigle, New Orleans 095161
Siglhaus, Sankt Georgen bei
Salzburg 002656
El Siglo, Buenos Aires 060931
El Siglo XX, Barcelona 143732
Sigma, Zutphen 143273
Sigma Antiques and Fine Art,
Ripon 091120, 134972
Sigma Syneries, Villefranche-sur-
Mer 105867
Sigmund Freud Museum, Wien . 003084
Sigmund's Samlermarked,
Bergen 084224
Sign, Groningen 112503
Sign of the Times, Wellington . 084189
Sign Today, New Orleans 122238
Signal, Malmö 115769
Signal 66, Washington 124958
Signal Art Centre, Bray 109535

Signal Gallery, Auckland 112809
Signal Hill Gallery, Weyburn ... 007395
Signal Hill National Historic Site, Saint
John's 006913
Signatary Namai, Lietuvos Nacionalinis
Muziejus, Vilnius 030541
Signature Arts, Saint Louis ... 123871
Signature Gallery, Atlanta 119611
Signature Gallery, Boston 119922
Signature Gallery, Saint Paul .. 123932
The Signature Gallery, Swansea 119195
Signed Sealed Delivered,
Houston 145173
Signet, Praha 140530
Significant Books, Cincinnati .. 145077
Signman, Indianapolis 121218
Signorello, Salvatore, Havixbeck 130024
Signorini, Alberto, Milano 080726
Signorini, Erminio, Verona 132299
Signorini, Mariarita Anna, Firenze 131246
Signs by George, El Paso 120839
Signs Now, Miami 121861
Signs Now, Omaha 123340
Signstore, Sacramento 123781, 145493
Signum, Oberfeulen 082382
Signum Antiquariat, Königstein im
Taunus 142027
Sigogne & Ko, Paris 071905
Sigoloff, San Antonio 124071
Sigrid, Paris 071906
Sigtuna-Antikvariatet, Sigtuna . 143915
Sigtuna Museer, Sigtuna 040941
Sigtuna Rådhus, Sigtuna 040942
Sigui, Angers 103237
Sigurhaedir – Hús Skáldsins Museum,
Literary Museum, Akureyri .. 024067
Sigwart, Klaus, Hüfingen 130072
Sihui Museum, Sihui 008284
Siikaisten Kotiseutumuseo,
Siikainen 011354
Siivari, T., Siilinjärvi 140745
Sika, Los Angeles 121593
Sika Galerie, Liberec 065414
Sikander, Torino 081760
Sikar Museum, Sikar 024481
Sikes, Bobby, Atlanta 119612
Sikinos Archaeological Collection,
Sikinos 023264
Sikkema, Jenkins & Co., New
York 123064
Sikkens Schildersmuseum,
Sassenheim 032842
Sikorski, Gdańsk 113488
Siksika Nation Museum, Siksika 007037
Šilalės Krašto Muziejus, Šilalė . 030476
Silas Wright House, Saint Lawrence
County Historical Association,
Canton 047409
Silax, Havlíčkův Brod 065387
Silbag, Littau 133810
Silber, Jakub, Hradec Králové . 128327
Silberberg, Dr. Helmut K.,
Tulbagh 085687, 114822
Silberbergwerk, Ramingstein .. 002565
Silbereisenbergwerk Gleißinger Fels,
Fichtelberg 018360
Silberman, Jane W., Dallas 093215
Silbernagel, Susanne, Bad Homburg
v.d.Höhe 129413
Silbernagl, Volker, Denver 110001
Silbernagl, Volker, Milano 110435
Silbernberg, Drs. Ph.M.,
Amsterdam 082858
Silbersack, Wolfgang, Würzburg 130787
Silberstein, Claude, Paris 071907
Silberstollen Geising, Geising .. 018612
Silberwaren- und Bijouteriemuseum,
Ott-Pausersche Fabrik, Schwäbisch
Gmünd 021750
Silcher-Museum des Schwäbischen
Chorverbandes, Weinstadt ... 022529
Sildarminjasafnið a Siglufirdi,
Siglufjördur 024115
Silex Ediciones, Madrid 137995

Silfverberg, Olofström 086913
Silhouette Antiques, Seattle .. 097501
Siliato, Pasquale, Roma 132047
Silicon Pulp Gallery, Stanmore . 099715
Silicon Valley Clock Repair, San
Jose 136359
Silifke Atatürk Evi, Silifke 043059
Silifke Müzesi, Silifke 043060
Siliman University Anthropology
Museum, Dumaguete 034477
Silio, Fernando, Santander ... 115433
Silipo, Assunta, Reggio Calabria 131788
Silis, Fuenlabrada 086003
Siljansfors Skogsmuseum, Mora 040852
Siljustøl Museum, Edvard Grieg
Museum, Paradis 033877
Silk Mill – Derby's Museum of
Industry and History, Derby .. 044087
Silk Museum, Gangnae 029924
Silk Museum, Soufli 023270
Silk 'n Crafts, Manama 063161
Silk Road, Nashville 095033
The Silk Road, Roma 081561
Silk Road Collection, New
Orleans 095162
Silk Road Textiles, Kathmandu . 082613
Silk Skies Restoration, New York 136002
Silk Winds, Honolulu 093573
Silkeborg Kulturhistoriske Museum,
Silkeborg 010300
Silkeborg Kunstmuseum,
Silkeborg 010301
Silken Tent, Braidwood 098771
The Silky Oak Shop,
Woolloongabba 062490
Silla University Museum, Busan 029900
Sillack, Manto, Dresden 106747
Sillam, Henry J., Wien 063073
Sillaotsa Talumuuseum, Rapla . 010657
Silly Galah Gallery, Meadows .. 099293
Silly Renaud, Josnes 068911
Silme, Roma 081562
Siloam Springs Museum, Siloam
Springs 053334
Siloë, Kerdore, Laval (Mayenne) 136824
Silom Art Space, Bangkok 116960
Silom Galleria, Bangkok 088086
Silomuseum, Waidhofen an der
Thaya 002887
Silpakorn University Art Centre,
Bangkok 042678
Silsh Way, Killiney 079646
Šilutės Melioracijos Statybos Valdybos
Istorijos Muziejus, Šilutė 030477
Šilutės Muziejus, Šilutė 030478
Silva, Lisboa 126456
Da Silva, Toronto 128217
Silva & Martinho, Lisboa 084970
Silva Bastos, Lisboa 084971
Silva Carvalho Gomes Mendes,
Lisboa 084972
Silva Rafael Lopes, Lisboa 132957
Silva Rerum, Łódź 084618, 143489
Silva, Antonio J. Caldeira, Lisboa 084963
Silva, Armando, Coimbra 132925
Silva, Claudionor B., Fortaleza . 140283
Silva, Edmundo, Lisboa 132955, 132956
Silva, Eduardo Carvalho, Lisboa 084964
Silva, Ezequiel Faria, Torres
Vedras 085063
Silva, Francisco Marques, Lisboa 084965
Silva, Geraldes da, Porto 114106
Silva, Joani P., Fortaleza 100959
Silva, Júlio Gomes, Lisboa 084966
Silva, Mamede Cerqueira, Braga 113959
Silva, Manoel Bezerra, Rio de
Janeiro 064127
Silva, Maria Dulce, Lisboa 084967
Silva, Maria Idalina, Lisboa ... 084968
Silva, Nilton, Fortaleza 100958
Silva, Pedro, Lisboa 084969
Silva, Victor, Bogotá 102367
Silvana, Zurrieq 082454
Silveira, Yara Vianna, São Paulo 101112

Silver Blue Traders, Vancouver . 101836
Silver City Art Centre, Broken Hill 098802
Silver City Museum, Silver City . 053338
Silver Cliff Museum, Silver Cliff . 053340
Silver Corner, Dubai 088188
Silver Eagle and Gold Coins,
Cleveland 092972
Silver Eye-Center for Photography,
Pittsburgh 052175
Silver Fish Books, Taupo 143348
Silver for Sale, Seattle 097502
Silver Fox Gallery, London 090487
The Silver Fund, London 090488
Silver Gallery, San Francisco ... 092972
Silver Horse Antiques, Las Vegas 094178
Silver K, Upper Mount Gravatt . . 099848
Silver K Gallery, Armadale,
Victoria 061048
Silver K Gallery, Chadstone 061310
The Silver Ledge Hotel, Ainsworth Hot
Springs 005306
Silver Line, Warszawa 113902
Silver Line and More, Atlanta ... 092183
Silver Magazine, Greensboro 139491
Silver Moon Trading Company,
Vancouver 064879
The Silver Mouse Trap, London . 090489,
134765
Silver Museum, Al-Khobar 038108
Silver Owl, Albuquerque 135290
Silver River Museum, Ocala 051697
Silver Sands Art, Jacksonville . . 121266
Silver Screen, Saint Louis 123872
Silver Sensations, Atlanta 092184
Silver Services, Wellington 132763
Silver Set, El Paso 093414
Silver Shadow Picture & Prints,
Swansea 119196
Silver Shears, Seattle 097503
Silver Shop, Dublin 079589
The Silver Shop, Vancouver 064880
Silver Society, London 060240
Silver Society of Australia, Sydney
South 058220
Silver Star, Salt Lake City 096924
Silver Stream Railway Museum, Hutt
City 033225
Silver Street Gallery, Ilminster .. 118016
Silver Thimble, Kendal 089673
Silver Thistle, Houston 093793
Silver Treasures, Dubai 088189
Silver Whale Gallery, New York . 123065
Silver, Jay, Houston 093792
Silver, Tom, Bondi Beach 061171
Silveragent, Nashville 095034
Silverbell, Tucson 124774
Silvercreek Museum, Freeport . . 048970
Silvercrest and More, Toledo ... 097617
Silverenberg, Max, Eindhoven . . 143135
The Silverhalide Gallery,
Whitstable 119344
Silverleaf Antiques, San Jose .. 097377
Silverman, B., London 090490
Silverman, Manny, West
Hollywood 125026
Silverman, Michael, London 144689
Silvermine Guild Arts Center, New
Canaan 051249
Silvermine Guild of Artists, New
Canaan 060737
Silvermine School of Art, Silvermine
Guild Arts Center, New Canaan 057519
Silvermist Studio Gallery,
Healesville 099122
Silvermuseet, Arjeplog 040536
Silversmith, New Orleans 095163
Silversmiths, Werner, Milwaukee 135820
Silverstein, Bruce, New York ... 123066,
123067
Silverstream Gallery, Timaru ... 113030
Silvertime, Brecon 088589
Silverton Gaol Museum, Silverton 001496
Silverton Outdoor Mining Exhibit,
Silverton 007040

Silvery Slocan Museum, New
Denver 006465
Silvesters, Wingham 091882
Silvestre, Jean-Pierre, Marseille 104198
Silvestri, Claudio, Milano 080727
Silvestri, Luca, Napoli 080937
Silvestri & Figli, Luigi, Napoli . 131642
Silvestri, Nino, Genova 131340
Silvi's Antiques, Houston 093794, 135616
Silvy, Jean-François, Cannes ... 067653
Silz, Ursula, Budenheim 141660
Sim, Michael, Chislehurst 088897,
134267
Sim, Roy, Blairgowrie 088503
Sima, Nürnberg 108308
Simader, Herbert, Sankt Johann am
Wimberg 127865
Símaminjasafnið, Reykjavík 024110
Simancas, Santander 115434
Simão, José Vasconcelos, Lisboa 084973
Sim'Art, Montpellier 104285
Simayspace Gallery, San Diego . 124173
Simba Art, Aabenraa 065577
Simboli, Mantova 080347
Simcoe County Museum,
Minesing 006324
Simeon, Edina 135547
Simeon Perkins House, Queens County
Museum, Liverpool 006204
Simeza, Bucureşti 114154
Simferopol's'ke Učylyšče Mystectv
i Kul'tury im. P Čajkovs'kogo,
Simferopol' 056501
Simferopolskij Chudožestvennyj Muzej,
Simferopol' 043301
Simfonia Lemnului, Bucureşti . . 133002
Simil Art Gallery, Paris 105235
Simiolus, Utrecht 138939
Simmering, Calgary 128157
Simmerling, J., Chicago 120268
Simmermacher, René, Staufen im
Breisgau 078543, 142375
Simmert & Freitag, Rieseby 130590
Simmi, Stefano, Roma 110857
Simmler, Martin Jürgen, Stuttgart 078634
Simmond's & Sons, S.F.,
Catonsville 135384
Simmonite, Peter, Sheffield ... 119087
Simmons, Kirkland 121324
Simmons, Wednesbury 144911
Simmons & Simmons,
Albuquerque 092084, 144954
Simmons Visual Center, Brenau
University, Gainesville 049010
Simmons, S.L., Brighton 088636
Simmor, Los Angeles 094472
Simms, Henry S., San Antonio . 097033
Simoens, Andre, Knokke-Heist . 100762
Simões Lda, Época Antiguidades,
Lisboa 084974
Simolan Taide ja Kehys, Lahti .. 066385,
103165
Simon, Charleroi 063607
Simon, Filderstadt 106952
Simon, Fürth 129876
Simon, Hannover 076396
Simon, Madrid 133389
Simon, New Orleans 095164
Simon, Seoul 111718
Simon, Wiesbaden 078996
Simon Fraser Gallery, Burnaby . 005506
Simon-Gfeller-Gedenkstube,
Heimisbach 041590
Simon i Patiño, La Paz 100849
Simon Janashia Museum of Georgia,
Georgian National Museum,
Tbilisi 016422
Simon Kotiseutunmuseo, Simo . 011357
Simon Lee Gallery, London 118566
Simon Michael School of Fine Arts,
Rockport 057790
Simon Paneak Memorial Museum,
Anaktuvuk Pass 046467
Simon van Gijn-Museum aan Huis,

Dordrechts Museum, Dordrecht 032251
Simon, Brigitte & Karl-Heinz,
Oldenburg 077940
Simon, Dany, Uzès 073927
Simon, Didier, Héry 068815
Simon, Dr. Ulrich, Frankfurt am
Main 075825
Simon, Gilles, Nonancourt 070744
Simon, Helene, Köln 076907
Simon, Julian, London 090491
Simon, Marie-Josée, Perpignan . 072094
Simon, Volker, Schrecksbach .. 078399
Simoncini, Luxembourg 111865
Simone Romano "Galleria Ottaviani",
Firenze 080204
Simone, Francesco, Torino 081761,
132202
Simone, Neil, Whixley 119348
Simonet, Serge, Saint-Ouen ... 073198
Simonetti, Torino 081762
Simonian, N. & Dr. S., Hamburg 076315
Simonida, Niš 114454
Simonini, Michèle, Bédarieux .. 125510
Simonis & Buunk, Ede 112449, 132495
Simonis, Henricus, Düsseldorf .. 106830
Simonnæs, Bergen 084225
Simono Stanevičiaus Memorialinė
Sodyba-muziejus, Kanopenai .. 030346
Simonoff, Maisons-Alfort 104144
Simonot-Amigues, Versailles ... 074070
Simon's Town Museum, Simon's
Town 038829
Simons, Herwig, Antwerpen ... 063238
Simonson, Bruxelles 125326
Simonson, Carol & James,
Minneapolis 122035
Simotti Rocchi, Marcello, Roma 081563
Simper, Murray & Ann,
Warrnambool 139954
Simple Treasures, Manly 061788
Simpli, Sue, Toledo 124697
Simplicissimus-Haus, Renchen . 021370
Simply Art, Athgarvan 109525
Simply Gorgeous, Wellsford 084200
Simply Oak, Biggleswade 088463,
134104
Simply Unique, Los Angeles ... 094473
Simpson, Buenos Aires 060932, 127538
Simpson, Hemel Hempstead ... 117951
Simpson, Stanmore 062240
Simpson, Wylam 119398
Simpson Galleries, Houston 093795,
127257, 145174
Simpson, Andrew, Woollahra ... 062482
Simpson, Catherine, London ... 101348
Simpson & Assoc., G., New
Orleans 095165
Simpson & Sons, H.C., Oldham . 090913,
134906
Simpson, Marianne, Fochabers . 144427
Simpson, Merton D., New York . 123068
Simpson, R.J., Houston 135617
Simpson, Russell, Los Angeles . 094474
Simpsons, Brentwood 088591
Sims Reed, London ... 090492, 144690
Sims Reed Gallery, London 118567
Sims, Michael, Lowestoft 144718
Sims, Paul, Richmond 096573
Simsbury Historical Society Museum,
Simsbury 053347
Le Simserhof, Musée National de la
Fortification Maginot, Siersthal 015807
Simyo, Seoul 111719
Sin Ken, Bagnères-de-Bigorre .. 066908
Sin Sin, Hong Kong 102113
Sin, Hua, Singapore 114606
Sin, Mi, Daegu 111590
Sinai Antiques, London 090493
Sinapi & Co., Karen, New York . 123069
Sinceux, Philippe, Paris 071908
Sincheonji Art Museum, Aeweol 029883
Sinclair, Melbourne 099320
Sinclair-Barclay, Waikato 113042
Sinclair Inn, National Historic Site,

Annapolis Royal 005342
Sinclair Lewis Museum, Sauk
Centre 053164
Sinclairs, Ardross 061005
Sinclair's, Stamford 091409
Sind Provincial Museum,
Hyderabad 034114
Sindbad Antiques, Sharjah 088212
Sindbad Books, Carnegie 139692
Sindens, R.J., Calgary . 064274, 101259
Sindler, Richard, Baltimore 092363
Sindoni, Nino, Asiago 109759
Sindringer Heimatmuseum –
Stadtmühle, Forchtenberg 018384
Sinebrychoffin Taidemuseo, Valtion
Taidemuseo, Helsinki 010862
Sineo, Marina, Roma 081564
Sinesy Aboriginal Art & Design,
Ballina 061094
Sineux, Précey 072306
Sinfield, Brian, Burford 088711, 117476
Sing Kai Hong, West Hollywood 097933
Singapore Art Museum,
Singapore 038274
Singapore Art Society, Singapore 059802
Singapore National Corporation,
Singapore 114607
Singapore Science Centre,
Singapore 038275
Singapore Tyler Print Institute,
Singapore 114608
Le Singe Blanc, Paris . 071909, 105236
Au Singe Bleu, Bordeaux 067359
Singe, Violet le, Leiden 132560
Singel 91 Antiquiteiten,
Amsterdam 082859
Singer, Seattle 097504
Singer Laren, Laren, Noord-
Holland 032562
Singer, Dorothy, Wien 140124
Singer, Mary, Washington 124959
Singhofener Heimatstuben,
Singhofen 021878
Singida Museum, Singida 042638
Singidunum, Beograd 114439
Single Tree Antiques, Richmond 096574
Singleton Antiques, Skenfrith . 091345
Singleton Historical Museum,
Singleton 001498
Singsås Museum, Singsås 033935
Les Singuliers, Paris 105237
Siniawski, Janus, Kassel 076732,
130105
Sining Makiling Gallery, University
of the Philippines Los Baños, Los
Baños 034499
Siniscalco, Maurizio, Napoli ... 080938
Sink, Mark, Denver 120773
Sinka István Emlékház – Helytörténeti
kiállítás, Vésztő 024045
Sino Arts, Singapore 114609
Sinop Devlet Güzel Sanatlar Galerisi,
Sinop 043061
Sinop Müzesi, Sinop 043062
La Sinopia, Roma 081565
Sinopis, Roma 081566, 132048
Sinquin, René, Quiberon 105409
Sint-Baafskathedraal Gent, Gent 003638
Sint-Godelievemuseum, Gistel . 003654
Sint-Janshuismolen & Koeleweimolen,
Musea Brugge, Brugge 003425
Sint Jansmuseum De Bouwloods, 's-
Hertogenbosch 032475
Sint Joseph Galerie, Leeuwarden 112552
Sint-Lucas Beeldende Kunst,
Hogeschool voor Wetenschap en
Kunst, Gent 055022
Sint-Pieters Museum op de
Lichtenberg, Maastricht 032630
Sint-Plechelmusbasiliek met
Schatkamer, Oldenzaal 032714
Sinte, Kathrien, Eindhoven 083228,
132500
Síntesis, San José 102515

Sintitulo, Mougins 104306
Sintot, Arbois 066695
Sintot, Santans 073407
Sintra Galleri & Butik, Göteborg 115664
Sintra Museu de Arte Moderna,
  Colecção Berardo, Sintra 036046
Siouni, New York 095741
Sioux City Art Center, Sioux City 053348
Sioux City Public Museum, Sioux
  City 053349
Sioux Empire Medical Museum, Sioux
  Falls 053356
Sioux Indian Museum, Rapid City 052473
Sioux Lookout Museum, Sioux
  Lookout 007044
Siouxland Heritage Museums, Sioux
  Falls 053357
Sipahi Ay. Trias Bazilikası ve Kantara
  Kalesi, Kantara 033465
Sipalaseequtt Museum Society,
  Pangnirtung 006610
Il Sipario, Parma 110636
Siping Museum, Siping 008285
Sipiora, Leonard P., El Paso ... 093415
Sipiweske Museum, Wawanesa 007369
Sipma, GW, Zwolle ... 083739, 132677
Sippel, Detlef, Pforzheim 130518
Sippolan Kotiseutumuseo,
  Sippola 011359
Siqueira, Aelcio A., Belo
  Horizonte 140272
Sir Alexander Galt Museum,
  Lethbridge 006186
Sir Alfred Munnings Art Museum,
  Dedham 044078
Sir Andrew MacPhail Foundation,
  Orwell 006562
Sir Choturam Memorial Museum,
  Sangaria 024464
Sir Edgeworth David Memorial
  Museum, Kurri Kurri 001198
Sir Francis Cook Gallery, Augrès 043450
Sir Frank Hutson Sugar Museum,
  Saint James 003239
Sir Harald Art, Venlo 112716
Sir Henry Jones Museum,
  Llangernyw 044886
Sir Henry Royce Memorial Foundation,
  Paulerspury 045468
Sir Hermann Black Gallery, University
  of Sydney, Sydney 001550
Sir Isaac and Lady Edith Wolfson
  Museum, Hechal Shlomo,
  Jerusalem 025108
Sir J.J. Institut of Applied Arts,
  Mumbai 055822
Sir John Soane's Museum,
  London 045107
Sir Joseph Banks Conservatory,
  Lincoln 044835
Sir Max Aitken Museum, Cowes 044025
Sir Seewoosagar Ramgoolam
  Memorial Centre, Plaine Verte 030791
Sir Walter Scott's Courtroom,
  Selkirk 045759
Sir Wilfred Grenfell College Art Gallery,
  Corner Brook 005647
Sir William F. Coaker Heritage
  Foundation, Port Union 006695
Sirapian, Agkop, Athinai 079180
Sirçali Medrese Müzesi, Konya . 043009
La Sirena, Miami 094803
Sirénades, Dinard 103701
La Sirene, Liège 140247
Sireyzol, Serge, Belmont-
  Bretenoux 067082
Sirgiovanni, Domenico, Torino . 132203
Siri Drawing and Oil Painting Institute,
  Hyderabad 109373
Sirius-Alfa, Moskva 114305
Sirris, Wuppertal 079116
Sirrocco-Parkstone, London ... 138250
Sisal Industry Historical Exhibition,
  Hengchun 042325

Sisavangvong University,
  Vientiane 056106
Sisian Museum of History N. Adonts,
  Sisian 000713
Sisimiut Katersugaasiviat,
  Sisimiut 023340
Siskin Museum of Religious and
  Ceremonial Art, Chattanooga .. 047597
Šiškin, Leonid, Moskva 114306
Siskiyou County Museum, Yreka 054605
Siskos, Konst., Athinai 079181
Sismann, Gabriela et Mathieu,
  Paris 071910
Sismanoglou, Thessaloniki 109100
Siso Difusiones, Lleida 086070
Sissoko, Philadelphia 123453
Sissons, William, Helmsley ... 117948
Sister Dragonfly Gallery,
  Louisville 121670
Sister Studios, Saint Louis 123873
Sisters Antiques, Saint Paul 096857
Sistu, Gianna, Paris 105238
Sit'Art, Magny-en-Vexin 104142
Sitch & Co., W., London 090494, 134766
Site Archéologique La Cave Peinte,
  Brain-sur-Allonnes 012188
Site Archéologique de Fouilles
  Archéologiques, Charavines .. 012452
Site du Bois du Cazier, Musée
  d'Industrie, Marcinelle 003849
Site du Moulin des Pionniers de la
  Doré, La Doré 006133
Site et Musée du Regourdou,
  Montignac 014266
Site Fortifié des Bois de Cattenom,
  Sentzich, Cattenom 012367
Site Gallery, Winnipeg 101894
Site Gallo-Romain de Grand,
  Grand 013143
Site Gallo-Romain du Fâ, Barzan 011901
Site Gallo-Romaine de Sanxay,
  Sanxay 015679
Site Historique de l'Île-des-Moulins,
  Terrebonne 007134
Site-Historique du Banc-de-Paspébiac,
  Paspébiac 006616
Site Historique Matamajaw,
  Causapscal 005582
Site Magazin, Düsseldorf 138754
Site Mary-Travers dite – La Bolduc,
  Newport 006477
Site Odéon 5, Paris 105239
Site of the First National Congress
  of the Communist Party of China,
  Shanghai 008237
Site Paléochrétien de Saint-Seurin,
  Bordeaux 012121
Site Santa Fe, Santa Fe 053123
Siteau, Michel, Mazières-sur-
  Béronne 070023
Sitegalerie, Düsseldorf 106831
Sitges Subastas, Sitges 126557
El Sitial, Zaragoza 133514
Sitio Arqueológico Huaca Malena,
  Capilla de Asia 034239
Sitio Arqueológico de Cumbemayo,
  Cumbemayo 034257
Sitio Arqueológico Joya de Cerén, San
  Juan 010591
Sítio Roberto Burle Marx, Rio de
  Janeiro 044770
Sitjas Antiquaris, Barcelona ... 085841
Sitka National Historical Park,
  Sitka 053362
Sitorai Mochi-Khosa, Buxoro ... 054630
Sitting Pretty, Great Glen 089366,
                               134421
Sitting Pretty Antiques, Sudbury,
  Suffolk 091505, 135104
Sitting Room, Houston 093796
Sittingbourne and Kemsley Light
  Railway, Sittingbourne 045805
Sittingbourne Heritage Museum,
  Sittingbourne 045806

Situ, Y., San Francisco 136345
Situation Kunst, Bochum 017441
Sity South Dealers, Christchurch 083898
Siuntion Kotiseutumuseo, Sjundeå
  Hembygdsmuseum, Siuntio ... 011360
Sivas Devlet Güzel Sanatlar Galerisi,
  Sivas 043064
Siverts, Bergen 113136
Sivi Antichità e Restauro, Torino 081763,
                               132204
Siviglia, Annunziato, Cesano
  Boscone 080052
Six Chapel Row, Bath 117314
Six Eight Three Haight Street, San
  Francisco 124396
Six Nations Indian Museum,
  Onchiota 051771
Six Principles, Vancouver 064881
Six, Luc, Bousbecque 128519
Sixbid, Waiblingen 126136
Sixspace, Culver City 120528
Sixt, Wolfgang, Bolzano 079902
Sixteen Fifteen Art Gallery, New
  York 123070
Sixth Avenue Antiques, San
  Francisco 097327
Sixth Chamber Used Books, Saint
  Paul 145504
Sixth Floor Museum at Dealey Plaza,
  Dallas 048123
Sixtos, San Antonio 136273
Sixty West Antique Mall,
  Richmond 096575
Sixways Antique Centre,
  Cheltenham 088842, 134243
Sizad, Abdelziz, Paris 129144
Sizdah Aban Museum, Tehrān .. 024785
Sizemore, Mary, Oklahoma City 096010
Sjöfartsmuseet, Oskarshamn .. 040910
Sjöfartsmuseet,
  Kristiinankaupunki 011042
Sjöfartsmuseet Akvariet,
  Göteborg 040661
Sjøfartsmuseet i Porsgrunn,
  Porsgrunn 033880
Sjöhistoriska Institutet vid Åbo
  Akademi, Turku 011443
Sjöhistoriska Museet, Stockholm 041039
Sjökvistens Allehanda, Laholm . 086799
Sjölin, Örebro 086907
Sjølingstad Uldvarefabrik, Mandal 033765
Sjösten, Johan, Göteborg 086719
Sjöström, Hans Göran,
  Stockholm 087051
Sjóminja- og Vélsmiðjusafn,
  Reykjavík 024111
Sjóminjasafn Austurlands,
  Eskifjörður 024074
Sjóminjasafn Islands, Reykjavík . 024112
Sjóminjasafnid á Eyrarbakka,
  Eyrarbakki 024076
Sjukehusmuseet i Molde, Molde 033784
Sjukehusmuseet i Skien, Skien . 033944
SK Upholstery, San Diego 136298
Ska-Nah-Doht Iroquoian Village and
  Museum, Mount Brydges 006434
Skaalurensamlinga –
  Skipsbyggingsmuseet i Rosendal,
  Rosendal 033909
Skácelik, Olomouc 102625
Skådan, Uppsala 087125
Skärgårdens Sågverksmuseum,
  Holmsund 040706
Skärgårdsmuseet, Lappoby-Åland 011097
Skärgårdsmuseet, Stavsnäs ... 040986
Skärhamns Sjöfartsmuseum,
  Skärhamn 040950
Skagen, New York 095742
Skagen By- og Egnsmuseum,
  Skagen 010306
Skagens Museum, Skagen 010307
Skagerrak Antikvariat,
  Tvedestrand 143428
Skaggs, Glen C., San Antonio .. 097034

Skagit County Historical Museum, La
  Conner 050025
Skagway Museum, Skagway ... 053363
Skaj, Curt & Mikael, Stockholm 087052,
                               087053
Skalmierski, Sławomir, Łódź ... 084619
Skalski, Kornelia, Berlin 129566
Skan Art, Flensburg 075750
Skanderborg Museum,
  Skanderborg 010308
Skandinavisk Museumförbund –
  Skandinaavinen Museoliitto, Finnish
  Section, Helsinki 058577
Skandinavisk Museumsforbund,
  Danske Afdeling, København .. 058561
Skandinavisk Museumsforbund,
  Norwegian Section, Tønsberg . 059674
Skandinavisk Museumsforbund,
  Swedish Section, Stockholm .. 059902
Skånes Konstförening, Malmö .. 059903,
                               115770
Skanfil, Oslo 126380
Skansen, Göteborg 086720
Skansen Archeologiczny w Mrówkach,
  Muzeum Okręgowe w Koninie,
  Mrówki 035022
Skansen Budownictwa Ludowego
  Zachodniej Wielkopolski,
  Wolsztyn 035411
Skansen Górniczy Królowa Luiza,
  Muzeum Górnictwa Węglowego w
  Zabrzu, Zabrze 035442
Skansen i Muzeum Pszczelarstwa im
  prof. Ryszarda Kosteckiego, Muzeum
  Narodowe Rolnictwa i Przemysłu
  Rolno-Spożywczego w Szreniawie,
  Swarzędz 035235
Skansen Kultury Ludowej Pogórza
  Sudeckiego, Kudowa Zdrój ... 034929
Skansen Kurpiowski im. Adama
  Chętnika, Muzeum Północno-
  Mazowieckiego, Nowogród ... 035044
Skansen Lokomotyw, Zduńska
  Wola 035459
Skansen Rybołówstwa Morskiego,
  Domek Żeromskiego, Gdynia .. 034752
Skånska Lasses Hus, Mjölby ... 040847
Skanzen – Múzeum L'udovej
  Architektúry, Bardejovské
  Kúpele 038292
Skanzen Březno u Loun, Oblastní
  Muzeum v Lounech, Louny ... 009684
Skar Editions, Chagny . 103553, 136806
Skara Järnvägsmuseum, Skara . 040951
Skarabäus, Köln 076908
Skarabaeus Verlag, Innsbruck . 136521
Skarabé Mynt & Föremål från Antiken,
  Göteborg 086721
Skarbiec, Kraków 113637
Skarics, Gerhard, Baden bei
  Wien 062533
Skate Canada Hall of Fame &
  Museum, Patinage Canada Temple
  de le Renommée et Musée,
  Ottawa 006599
Skatkis, Paarl 085656
Skattkammaren, Kristinehamn . 086793
Skattkammaren, Stockholm ... 041040
Skattkammaren – Uppsala Domkyrkas
  Museum, Uppsala 041117
Skaun Bygdamuseum, Børsa .. 033552
Skazka, Moskva 085205
Skeaping, Lydford 118674
Skeaping Gallery, Lydford 090656
Skeen, Joseph, Greenbank 061556
Skellefteå Museum, Skellefteå . 040955
Skellgate Curios, Ripon 091121
Skelsey, K.L., Beverly Hills ... 061153
Skepparn, Göteborg 086722
Skeppsvarret Sjöfartsutställning,
  Sagalund Museum, Västanfjärd 011479
Skepsi on Swanston, Carlton,
  Victoria 098871
Skerryvore Museum, Story of

Skerryvore Lighthouse, Hynish   044622
Skestos, Gabriele, Chicago 120269
Ski- und Heimatmuseum,
   Oberwiesenthal, Kurort 020939
Ski- und Heimatmuseum, Sankt Anton
   am Arlberg 002647
Ski- und Heimatmuseum Upfingen,
   Sankt Johann, Württemberg . . 021601
Skibinski, Josef, Ibbenbüren . . . 076591
Skibinski, Marc, Montmirail 070275
Skibsted, Struer 066148
Skidby Windmill and Museum of East
   Riding Rural Life, Skidby 045808
Skidt og Kanel, Frederikshavn . . 065682
Skien Antikk, Skien 143420
Skiles, David, Louisville 094586
Skillbeck, Charlotte 120013
Skillen, Peter, Cockermouth 117609
Skilton, Charles, London 138251
Skimo Arte, Madrid 115286
Skimuseet, Oslo 033861
Skimuseum, Oberreute 020915
Skimuseum, Vaduz 030310
Skinkat, Sven, Flensburg 129841
Skinner, Boston 127145
Skinner Museum, Mount Holyoke
   College, South Hadley 053406
Skinner, Helen, Stockholm 133579
Skippergata Antikviteter,
   Kristiansand 084269
Skipton Antiques Centre, Skipton 091349
Skiptvet Bygdemuseum, Skiptvet 033947
Skira, Milano 137695
Skirball Cultural Center, Hebrew Union
   College, Los Angeles 050468
Skirball Museum of Biblical
   Archaeology, Jerusalem 025109
Skirball Museum, Hebrew Union
   College-Jewish Institute of Religion,
   Cincinnati 047747
Skiri, Angela, München 077621
Skirionio Centrum Kifissa, Skironio
   Museum Polychronopoulos, Néa
   Kifissia 023200
Skironio Museum Polychronopoulos,
   Megara 023166
Skissernas Museum, Lund 040814
Skive Kunstmuseum, Skive 010309
Skive Museum, Skive 010310
Skivesset, Malmö 143902
Skjalm, H., København 065954
Skjern-Egvad Museum, Skjern . . 010314
Skjern Malerier, Skjern 066126
Skjern Vindmølle, Skjern 010315
Skjold, Søren, Roskilde 066098
Skjoldborgs Hus, Vesløs 010356
Sklářská Muzeum, Kamenický
   Šenov 009615
Sklářská Muzeum Moser, Karlovy
   Vary 009621
Sklářské Muzeum Nový Bor, Nový
   Bor 009725
Sklenka Muzeum v Železném Brodě,
   Městské Muzeum v Železném Brodě,
   Železný Brod 009981
Sklep z Antykami, Poznań 084650
Sklep z Antykami, Warszawa . . . 084748
Sklerijenn, Saint-Malo (Ille-et-
   Vilaine) 105554
Sklo, Seoul 111720
Sklorz, Martin, Ludwigsburg . . . 077158
Skobelev Park-muzej, Pleven . . 005121
Škoda Auto Museum, Mladá
   Boleslav 009700
Sköldberg, Karin, Göteborg 133533
Skövde Konsthall och Konstmuseum,
   Skövde 040957
Skövde Stadsmuseum, Skövde . . 040958
Skogsmuseet i Lycksele,
   Lycksele 040816
Skogsmuseum, Åsele 040547
Skogsmuseum, Sunne 041067
Skoindustrimuseet, Kumla 040763
Skokloster Motormuseum,

Skokloster 040959
Skoklosters Slott, Skokloster . . . 040960
Skol di Arte, Sint Nicolaas 054889
Škola Primijenjenih Umjetnosti,
   Sarajevo 055065
Skolen for Mediekunst, Kongelige
   Danske Kunstakademi,
   København 055313
Skolen for Mur og Rum, Kongelige
   Danske Kunstakademi,
   København 055314
Skolmuseet, Trolle Ljungby skola,
   Fjälkinge 040627
Skolmuseum Bunge, Fårösund . 040623
Školoto Muzej, Trjavna 005231
Školská knjiga, Zagreb 136758
Skoob, Bath 144234
Skoob Russell Square, London . . 144691
Skopia, Genève 116340
Skorianetz, Dr. Werner, Genève . 144081
Skoto, New York 123071
Skottvångs Gruva, Värdshus och
   Museum, Mariefred 040840
Skovbo Antique Export Denmark,
   Ringsted 066076
Skovbrugsmagasinet på Corselitze,
   Nykøbing Falster 010245
Skovgaard Jensen, Klippinge . . 065818
Skovgaard Museet i Viborg,
   Viborg 010360
Skovle & Skovle, København . . . 140615
Skovs Antik, Silkeborg 066113
Skowhegan History House,
   Skowhegan 053364
Skowhegan School of Painting and
   Sculpture, New York 057565
Skowron, Michał, Poznań 084651,
                                     113748
Skowronek, Heinz, Insel Poel . . 107466
Skowronska, Thomas, Berlin . . . 141573
SKP Licht + Kunst, Wolfenbüttel 108959
Skrajewski, A., Dessau 106671
Skrebowski, Justin F., London . . 090495,
                                     144692
Skrętowicz & Spółka, Kraków . . 132849
Skriaudžių Buities Muziejus, Prienų
   krašto muziejus, Skriaudžiai . . 030479
Skrifter och Tal, Stockholm 143966
Skrobanek, E., Bruchköbel 075199
Skrobek, Benedikta, Gaildorf . . . 075938
Skrobek, Gerhard, Backnang . . . 074486
Skrolsvik Fiskebruksmuseum, Midt-
   Troms Museum, Moen 033779
Skrotbua, Trondheim 084390
Skrypt, Kraków 143482
Skrypzak, Georg, Berne 129582
Skude Galleri og Kunstformidling,
   Haugesund 084250, 113162
Skude Galleri og Kunstformidling,
   Skudeneshavn 084360, 113260
Skufca, Dorothy, San Francisco . 136346
Skulptoriaus Broniaus Pundziaus
   Memorialinis Muziejus, Mažeikių
   Muziejus, Pieveniai 030446
Skulpturenhain vor Marleben,
   Trebel 022179
Skulpturenmuseum Glaskasten Marl,
   Marl, Westfalen 020304
Skulpturenmuseum im Hofberg,
   Stiftung Koenig, Landshut 019885
Skulpturenpark, Bad Nauheim . . 016831
Skulpturenpark Motzen,
   Mittenwalde 108009
Skulpturenpark Salzburger
   Freilichtskulptur, Salzburg 100179
Skulpturens Hus, Stockholm . . . 041041
Skulpturensammlung, Staatliche
   Kunstsammlungen Dresden,
   Dresden 018009
Skulpturenweg, Emmendingen . . 018187
Skulpturhalle Basel, Abguss-
   Sammlung des Antikenmuseums,
   Basel 041291
Skultura Galeria de Arte, São

Paulo 101113
Skuodo Muziejus, Skuodas 030481
Skupin, Rudolf, Bremen 141656
Skupnost Muzejev Slovenije,
   Ljubljana 059808
Skuret Kulturpub, Oslo 113241
Skyhawk Rugs, Albuquerque . . . 092085
Skylark 2, London 118568
Skylark Gallery, London 118569
Skylark Studio, Cockermouth . . 117610
Skylark Studios, Tydd Gote 119280
Skylight Galleries, Cape Town . . 114776
Skyline Books and Records, New
   York 145371
Skyscraper, New York 095743
Skyscraper Museum, New York . 051468
Skyttalan Museo, Parainen 011234
Skywalk, Cincinnati 145078
Slaait'n Hoes, Onstwedde 032720
Slabsides, John Burroughs Association,
   West Park 054320
Slaby, Dietmar, Ried im Innkreis 062778
Slaby, Gerhard, Sankt Martin im
   Innkreis 062827
The Slade Centre, Gillingham . . 117826
Slade School of Fine Art, International
   University of Science and Culture,
   Ciudad de Panamá 056218
Slade School of Fine Art, University
   College London, London 056570
Slade, John, Saint-Christophe . . 072714
Slade, Michael, London 090496, 134767
Sládečkovo Vlastivědné Muzeum v
   Kladně, Kladno 009627
Sladmore Gallery, London 090497,
                               118570, 134768
Slag van de Somme Museum,
   Schagen 032846
Slagelse Museum for Handel,
   Håndværk og Industri, Slagelse 010320
Slager, Mandy, Amsterdam 132415
Slagterbutikken anno 1920,
   Roskilde 010293
Sláma, Plzeň 065452
Slama, Franz, Wien 127971
Slama, Maria, Klagenfurt 062675,
                                      100075
Slama, Michael A., Portland . . . 096476
Slam's Boker och Skivor,
   Stockholm 143967
Śląski Dom Aukcyjny, Katowice . 126406
Śląski Salon Sztuki, Wrocław . . 084773
Slate Run Living Historical Farm,
   Ashville 046577
Slate Valley Museum, Granville . 049208
Slater, Indianapolis 093925
Slater, Sacramento 096647
Slater Memorial Museum,
   Norwich 051650
Slater Mill Historic Site,
   Pawtucket 051953
Slater, Kathy, New Orleans 095166
Slater, Phillip, Barnt Green 134067
Slatkine, Genève 138051, 144082
Slaughter House Gallery,
   Pittsburgh 123576
Slaughter Ranch Museum,
   Douglas 048343
Slave Lodge, Iziko Museums of Cape
   Town, Cape Town 038684
Slavejkov Antikvariat, Sofia 140325
Slaviero, Josilane, São Paulo . . 064210
Slavik, Renate, Wien 100371
Slavjanskij Dom, Moskva 114307
Slawenburg Raddusch, Vetschau 022321
Sledmere House, Driffield 044140
Sleeping Beauty Antique Beds,
   Brighton 088637, 134156
Sleepy Poet Stuff, Charlotte . . . 092625,
                                      092626
Slegers, P.T., Venlo 083689
Sleghem, Jacques & Sylvie, Lille 069505
De Slegte, Amersfoort 143006
De Slegte, Amsterdam 143066

De Slegte, Antwerpen . 140153, 140154
De Slegte, Arnhem 143086
De Slegte, Breda 143095
De Slegte, Bruxelles 140202
De Slegte, Den Haag 143121
De Slegte, Eindhoven 143136
De Slegte, Enschede 143139
De Slegte, Gent 140227
De Slegte, Groningen 143152
De Slegte, Haarlem 143163
De Slegte, Leiden 143187
De Slegte, Maastricht 143200
De Slegte, Rotterdam 143224
De Slegte, Tilburg 143237
De Slegte, Utrecht 143250
De Slegte, Zwolle 143275
De Slegte, Roel, Amsterdam . . . 082860
Sleigh Beds, Harriet Ann,
   Rolvenden 091133
Slein, Philip, Saint Louis 123874
Sleire Skulemuseum, Hosteland 033674
Slesvigske Vognsamling,
   Haderslev 010093
Sleuth of Baker Street, Toronto . 140417
Slevarenske Muzeum Komarov,
   Komárov 009636
Slevogthof Neukastel, Sitz des
   Malers Prof. Max Slevogt mit
   Gemäldegalerie, Leinsweiler . . 019968
Slewe, Amsterdam 112281
Slezáček, České Budějovice . . . 065371
Slezak, Marcella, Chicago 100270
Slezik, Sterling 127466
Slezské Zemské Muzeum, Opava 009738
Slidell Art Center, Slidell 053365
Slidell Museum, Slidell 053366
Sliedrechts Museum, Sliedrecht 032882
Slifer House, Lewisburg 050226
Sligo Art Gallery, Sligo 109694
Slim, R.T., Halesowen 144449
Slind Antikk og Auksjoner,
   Hommelvik 084257, 126353
Slinger, Adrian, Noosa Heads . . 099463
Slingue, Michel, Angoulême . . . 066631
Šlisselburgskaja Krepost Orešek,
   Šlisselburg 037863
Slitine, Chakib, Paris 071911
Slo Mak, Skopje 111896
Sloan, Buffalo 092558
Sloan Fine Art, New York 123072
Sloan Museum & Longway
   Planetarium, Flint 048749
Sloane, Denver 120774
Sloane, Houston 121128
Sloane-Stanley Museum and Kent
   Furnace, Kent 049930
Sloans & Kenyon, Chevy Case . 127164
Slocumb Galleries, East Tennessee
   State University, Johnson City . 049827
Slöjdmuseet, Ånge 040531
Slope, Eric D., Southampton . . . 063859
Sloss Furnaces, Birmingham . . . 047007
Slot Loevestein, Poederoijen . . 032767
Slot Zeist, Zeist 033093
Slot Zuylen, Oud-Zuilen 032745
Slot, Niels & Kirsten, Ringkøbing 066072
Slough Museum, Slough 045812
Slovácké Muzeum, Uherské
   Hradiště 009931
Slovart, Praha 136769
Slovart Publishing, Bratislava . . 137940
Slovenska Kinoteka, Ljubljana . . 038557
Slovenská Národná Galéria,
   Bratislava 038328
Slovenská Národná Galéria – Galéria
   Insitného Umenia – Schaubmar Mill,
   Pezinok 038421
Slovenská Národná Galéria – Kaštieľ
   Strážky, Spišská Béla 038452
Slovenská Národná Galéria –
   Zvolenský Zámok, Zvolen 038490
Slovenská Výtvarná Únia,
   Bratislava 059805, 114658
Slovenské Banské Múzeum, Banská

Štiavnica 038288
Slovenské Múzeum A.S. Puškina,
Matica Slovenská, Brodzany . . 038334
Slovenské Múzeum Ochrany Prírody
a Jaskyniarstva, Liptovský
Mikuláš 038383
Slovenské Národné Literárne Múzeum,
Slovenská Narodná Knižnica,
Martin 038395
Slovenské Národné Múzeum,
Bratislava 038329
Slovenské Národné Múzeum –
Múzeum Betliar, Betliar 038294
Slovenské Národné Múzeum –
Múzeum Bojnice, Bojnice . . . 038296
Slovenské Polnohospodárske Múzeum,
Nitra 038411
Slovenské Sklárske Múzeum, Lednické
Rovne 038376
Slovenské Technické Múzeum,
Košice 038368
Slovenski Etnografski Muzej,
Ljubljana 038558
Slovenski Gasilski Muzej Dr. Branka
Božiča, Metlika 038577
Slovenski Gledališki Muzej,
Ljubljana 038559
Slovenski Šolski Muzej, Ljubljana 038560
Slovenski Verski Muzej, Cistercijanska
Opatija Stična, Ivančna Gorica . 038517
Slovensko Arheološko Društvo,
Ljubljana 059809
Slovensko Konservatorsko Društvo,
Ljubljana 059810
Slovensko Muzejsko Društvo,
Ljubljana 059811
Slovo, Moskva 137926
Slugfest, Austin 119699
't Sluisje, Amersfoort 082656
Sluisman, P., Gouda 083270
Sluk, Eduard, Ubstadt-Weiher . . 078744
Slunkariki, Ísafjörður 109226
Slusariuc, Freiburg im Breisgau 075884
Sluys, Cl., Bruxelles 063552
SM City Art Center, Cebu 034466
Små Kära Ting, Göteborg 086723
Smack Mellon, Brooklyn 060738
Smågårde Auktioner Tofta, Visby 126701
Smagghe, Véronique, Paris 105240
Smålands Bil, Musik och
Leksaksmuseum, Rydaholm . . 040924
Smålands Konstarkiv, Värnamo . 041135
Smålands Militärhistoriska Museum,
Eksjö 040601
Smålands Museum, Sveriges
Glasmuseum, Växjö 041147
Small & Whitfield, Parkside 062062,
125224
Small Comforts Antiques,
Vancouver 064882, 128230
Small Furniture Store, New York 095744,
136003
Small Museums Association,
Clinton 060739
The Small Plaice Art Gallery, Upper
Stowe 119283
Small Pleasures, Boston 092495, 092496
Small Pleasures, New York 095745
Small Space, Ha Noi 125098
Small Wonder Gallery,
Murrumbeena 099416
Small World, San Antonio 097035
Small World Antiques, Saint
Louis 096776
Smallwood, Memphis 094664
Smallwood's Retreat, Marbury . 050669
Smallworks Gallery, Las Vegas . 121380
Smal'ta, Charkiv 117109
The Smart Gallery, Bath 117317
Smart Museum of Art, University of
Chicago, Chicago 047697
Smart t'Art Gallery, Cleveland . . 120454
Smart, Jonathan, Christchurch . 112861
Smashed Pipi, Mangawhai 112941

Smatolánová, Zdenka, Vrchlabí . 065566,
140543
Smeby, Klas, Visby 116000
Smedemuseet, Øster-Assels . . . 010265
Smederij Museum Wijlen dhr Verkley,
Nieuwkoop 032678
Smedjan, Vellinge 087157
Smedley, Durwyn, Indianapolis . 093926
Smeets, Weil am Rhein 078890
Smeets, Rob, Milano 080728
Smelik & Stokking, Amsterdam . 112282
Smelik & Stokking, Den Haag . . 112412
Smelik & Stokking, Rotterdam . 083591
Smellie & Son, L.S., Hamilton . . 126916
Smeraldo, Verona 081938
SMH-Art Consulting, Basel 116097
Smia Galleri og Grafisk Verksted,
Kristiansand 084270
Smibert, Tony, Deloraine 098964
Smid, Klaus J., Oldenburg 077941
Smidt Múzeum, Szombathely . . 023984
Smilgiụ Etnografinė Sodyba, Panevėžio
kraštotyros muziejus, Smilgiai . 030483
Smink, Hoofddorp 132546
Smiraglia, Rocco, Roma 132049
Smit, Emmen 083244
Smit, Louise, Amsterdam 112283
Smit, N., Amsterdam 143067
Smiterlöwsche Sammlung,
Franzburg 018457
Smith, Campbells Creek 061264
Smith, Christchurch . . 112862, 143311
Smith, Cincinnati 092896
Smith, Leechburg 127282
Smith, Norfolk 136035
Smith, Sacramento 123782
Smith, Teddington 119216
Smith, Toledo 136412
Smith, Wellington 143373
Smith & Burstert, Kansas City . 094094
Smith & Kramer, Kansas City . . 121315
Smith & Robinson, London 090498
Smith & Roelants, Auckland . . . 083826
Smith & Ward, Wellington 084190
Smith & Wong, Vancouver 064883
Smith Art Gallery, Brighouse . . 043731
Smith College Museum of Art,
Northampton 051626
Smith County Historical Society,
Tyler 053874
The Smith Gallery, Colchester . . 117614
Smith-Hill, Indianapolis 093927, 135637
Smith-McDowell House Museum,
Asheville 046557
Smith Robertson Museum,
Jackson 049757
Smith Street Galleries, Honolulu 120969
Smith-Zimmermann Museum,
Madison 050592
Smith, A.H., Miami 135808
Smith & Co., A.J., Seattle 097505
Smith, Allan, Swindon . 091533, 135114
Smith, Alun Courtney, London . . 134769
Smith, Alvin G., San Antonio . . 097036
Smith, Anne, Baltimore 092364
Smith, Betty A., Dallas 120660
Smith, C.W., Minneapolis 094946
Smith, Chad, Salt Lake City . . . 136243
Smith, Daren, Norwich 144753
Smith, David, Manton 090699
Smith, E.F., Houston 135618
Smith, Elliot, Saint Louis 123875
Smith, Elsie, San Francisco . . . 097328
Smith, Eric, Darwen 134315
Smith, Frank, Newcastle-upon-
Tyne 144740
Smith, Graham, Jesmond 089661
Smith, Granny, Saint Paul 096858
Smith, J., York 091982
Smith, James, Richmond, Surrey 144799
Smith, Jannene, Rozelle 139901
Smith, Jennie, Tampa 097589
Smith, Keith, Ledbury 144517
Smith, Ken, Enderby 089209

Smith, Lana, Louisville 094587
Smith, Larry, West Hollywood . . 125027
Smith, Lesley, Nottingham 144761
Smith, Leslie, Den Haag 112413
Smith, Louise, Kew 061686
Smith, Neal R., Denver 120775
Smith, Niall, New York 095746, 095747
Smith, Peter, Sunderland 091506,
135105
Smith, Peter Bain, Cartmel 144319
Smith, Robert Dale, Omaha . . . 096057,
136066
Smith, Roger, New York 123073
Smith, T., Chalfont Saint Giles . . 088819,
134225
Smith, Tom, Cloghe 109548
Smithen, Calgary 064275
Smithers Furniture, Austin 135333
Smithers Public Art Gallery,
Smithers 007047
Smithills Hall Museum, Bolton
Museum, Bolton 043671
Smiths, Christchurch 126312
Smith's, Harrogate 089440
Smith's Cove Historical Society
Museum, Smith's Cove 007048
Smiths Falls Railway Museum of
Eastern Ontario, Smiths Falls . 007051
Smith's Fort Plantation, Surry . . 053617
Smithson International Truck Museum,
Pas-Ka-Poo Historical Park,
Rimbey 006803
Smithsonian American Art Museum,
Washington 054160
Smithsonian Institution,
Washington 060740
Smithtown Historical Society Museum,
Smithtown 053369
The Smithy-Pioneer Gallery,
Cooperstown 047976
Smitz, J., Knokke-Heist 063709
Smocked Frock, Houston 093797
Smodej, M., Baden-Baden 106112
Smøla Museum, Smøla 033953
Smoglian, Guido, Fairfield 061460
Smoki Museum, Prescott 052353
Smoky Hill Museum, Salina . . . 052898
Smoky Mountain Trains, Bryson
City 047292
Smolenskaja Chudožestvennaja
Galereja, Smolensk 037870
Smolenskaja Regionalnaja Organizacija
Sojuza Chudožnikov Rossii,
Smolensk 059779
Smolenskij Gosudarstvennyj Institut
Iskusstv, Smolensk 056304
Smolenskij Istoričeskij Muzej,
Smolenskij Muzej-Zapovednik,
Smolensk 037871
Smolenskij Muzej-Zapovednik,
Smolensk 037872
Smolenskij Sobor, Novodevičij muzej,
Smolensk 037873
Smolka, Wilhelm, Wien 063074, 100372,
127972
Smollett Antique, Albury 125176
Smollett Antique Furniture,
Albury 060985
Smoorenburg, A.W.F.M, Raalte . 143213
Smrček, Znojmo 065573
Smrčkův Dům, Husitske Muzeum v
Táboře, Soběslav 009887
Šmrnc, Bratislava 133049
Smruti Mirani, Mumbai 109474
SM's – Stedelijk Museum's-
Hertogenbosch, 's-
Hertogenbosch 032476
Smycken och Antikt, Kalmar . . . 086771
Smyckeskrinet, Borgholm 086650
Smyrnios Gallery, Prahran 099601
Smyth County Museum, Marion 050696
Smyth, J., Dún Laoghaire 079610,
130869
Smyth, Peta, London 090499

Smyth, Spencer, San Francisco . 124397
Smythe & Co., R.M., Washington 097860
Smythe, Son & Walker, Cleveleys 126864
Smythson Yeats, Chicago 092792
Snäckan, Stockholm 087054
Snake River Heritage Center,
Weiser 054265
Snap Galleries, Canons Ashby . . 117515
Snapdragon, Ottawa 101480
Snapdragon, South Molton 091365,
135061
Snape Antiques, Snape 091354
Snappertunan Forngårdsmuseum-
Talomuseo, Snappertuna 011361
Snåsa Bygdamuseum, Snåsa . . 033954
Snchanek-Orenczuk, Elzbieta,
Berlin 074926
Snekkegårdens Antik, Gislinge . 065704
Snel, Loděnice 065420
Sneller, New York 136004
Sneller, A., New York 136005
Śniadewicz, Grażyna, Milanówek 132854
Snibston Discovery Park,
Coalville 043979
Snidaro, Mario Luis, Reus 133411
Snider Plaza Antique Shops,
Dallas 093216
Snider, Desiree, Milano 131533
Sniders, Jeremy, Glasgow 089334,
134406
Snijders, Jos, Roermond 083535
Snip It, New York 123074
The Snite Museum of Art, University
of Notre Dame, Notre Dame . . 051657
Snitzer, Fredric, Miami 121862
Snivling Sibbling, Los Angeles . 094475
Snob, Bari 079784
Snook, Wadebridge 119296
Snook, Jason, Caulfield East . . 127582
Snoop Antiques, Malvern, South
Australia 061774
Snooty Wombat Antiques, Mount
Pleasant 061920
Snooz'ease Furniture, Denver . . 093338
Snoqualmie Valley Historical Museum,
North Bend 051594
Snoriguzzi, Michel, Courgenay . 087385
Snow, Tulsa 097757
Snow Goose Associates, Seattle 124599
Snow Goose Canadian Arts,
Ottawa 101481
Snow, Catherine, Nashville 095035
Snowbird Gallery, Edmonton . . 101298
Snowden House, Waterloo 054207
Snowden, Ruby, Yarm 091965
Snowdonia Antiques, Llanrwst . 089875,
134575
Snowline Collectibles, Las Vegas 094179
Snowshill Manor, Broadway . . . 043763
Snowy Mountains Galleria de Art,
Jindabyne 099161
Snuffel, Hasselt 140230
Snug Harbor Cultural Center,
Newhouse Center for Contemporary
Art, Staten Island 053502
De Snuifdoos, Rotterdam 083592
The Snyder County Historical Society
Museum, Middleburg 050855
Snyder Museum and Creative Arts
Center, Bastrop 046805
Snyder, Alice R., Pittsburgh . . . 123577
Snyder, Gary, New York 123075
Snyder, J.R., San Diego 136299
Snyder, Susan, San Diego 124174
Snyderman, Philadelphia 123454
Snyers, Alain & Caroline, Paris . 129145
SNZ Galleries, Wiesbaden 108930
So Art, Norwich 090883, 118834
So Be It, Portland 123679
So Bell Antiques, San Francisco 097329
So-Bigraphics, Omaha 123341
So Gallery, Kamakura 111272
So Much More, Columbus 093045
So, Heon, Daegu 111591

So.Ge.Fi, Nice 070666
Soames County Auctions, Witney 127093
Soap Museum, Audi Foundation,
  Saïda 030281
Soapbox Gallery, Fortitude Valley 099054
Soardi, Nice 104390
Soares & Mendonça, Lisboa . . . 084975,
  126457
Soares, António Pinto, Porto . . . 114107
Soares, Filomena, Lisboa 114063
Soares, Vítor Fernandes, Coimbra 113967
Soares, Vítor Fernandes, Lagos . 113991
Soaring Eagle, Saint Paul 096859
Soaring Eagle Gallery,
  Albuquerque 119509
Soave, Torino 142871
Soave, Fiammetta, Roma 142839
Sobel, Gerhard & Ingeborg,
  Eberdingen 106851
Sobo, Columbus 093046
Sobocinska, Elzbieta, Bonn 129615
Sobol, Leszek, Kraków 084599, 143483
Sobottka, Aachen 074344
El Sobrao de la Abuela,
  Salamanca 086416, 133418
Sobunkan, Tokyo 142939
Socabep, Saint-Just-en-Chevalet 072875
Soccodato, Ferdinando, Roma . . 132050
Soccorsi, Raffaela, Roma 132051
Sochañ, Olomouc 128332
Sochi, Saint Paul 096860
Social History Collections, Iziko
  Museums of Cape Town, Cape
  Town 038685
Social History Curators Group,
  Waltham Abbey 060241
Socialmuseet Det Gamle Syge-
  Plejehjem, Hinnerup 010121
Sociedad Argentina de Artistas
  Plásticos, Buenos Aires 058178
La Sociedad de Arte Moderno,
  México 059574
Sociedad de Exposiciones San Vicente
  Paul, Madrid 059872
Sociedad Estatal para el Desarrollo del
  Diseño y la Innovación, Madrid 059873
Sociedad Iberoamericana de Estudios
  Numismáticos, Madrid 059874
Sociedad Mayagüezana Pro Bellas
  Artes, Mayagüez 059762
Sociedade Nacional de Belas Artes,
  Lisboa 059758, 114064
Sociedade Nacional de Leilões,
  Porto 059759
Sociedade Portuguesa de Leilões a
  Pregoeira, Lisboa 059760
Società di Belle Arti, Milano . . . 059475
Società di Belle Arti, Viareggio . 081950
Società Editrice Antiquariato Bologna,
  Bologna 079878, 142642
Società Editrice Internazionale,
  Torino 137751
Società Numismatica Italiana,
  Milano 059476
Società per le Belle Arti, Milano 059477
Società per le Belle Arti ed
  Esposizione Permanente, Milano 026831
Società Promotrice delle Belle Arti,
  Torino 059478
Societätsverlag, Frankfurt am
  Main 137207
Societas Museologia, Českȳ
  Těšín 058542
Societatea Colecționarilor de Artă din
  România, București 059766
Société Antiquités Photo Verdeau,
  Saint-Ouen 105568
Société Biblique, Les Abymes . . 142549
Société Canadienne d'Éducation par
  l'Art, Toronto 007206
Société de l'Historie de l'Art Français,
  Paris 058667
Société d'Ébénisterie et de Menuiserie

de Style et Décoration, Rabat . 082606,
  132375
Société des Amis des Arts de la
  Somme, Amiens 058668
Société des Amis du Louvre,
  Paris 058669
Société des Amis du Musée National
  d'Art Moderne Centre Pompidou,
  Paris 058670
Société des Artistes Décorateurs,
  Paris 058671
Société des Artistes Indépendants,
  Paris 058672
Société des Arts, Paris 105241
Société des Auteurs dans les Arts
  Graphiques et Plastiques (ADAGP),
  Paris 058673
Société des Musées Québécois,
  Montréal 058472, 058473
Société des Peintres-Graveurs
  Français, Paris 058674
Société d'Investissement et de
  Restauration (S.I.R.), Lacave . . 103911,
  128649
Société Européenne de Diffusion de
  Peintures Artistiques, Brainville-sur-
  Meuse 058675
Société Française d'Art Contemporain,
  Paris 058676
Société francophone des arts visuels
  de l'Alberta, Edmonton 101299
Société Géographique Royale du
  Canada, Ottawa 058474
Société Mâconnaise Cheminées
  Anciennes, Mâcon 069797
Société Mutuelle Artistique,
  Carouge 059967
Société Nationale des Beaux-Arts,
  Paris 058677
Société Neubourgeoise d'Antiquité et
  de Brocantes, Le Neubourg . . . 058678
Société Normande d'Art et d'Antiquités,
  Honfleur 058679, 068833
Société Pastel de Lyon,
  Carponne 058680
Society for Art Publications of the
  Americas, San Francisco 060741
Society for Arts, Chicago 060742,
  120271
Society for Arts and Technology,
  Montréal 006416
Society for Contemporary Craft
  Museum, Pittsburgh 052176
Society for Contemporary Photography,
  Kansas City 049892, 060743
Society for Japanese Arts,
  Aerdenhout 059620, 137819
Society of American Graphic Artists,
  New York 060744
Society of Antiquaries of London,
  London 060242
Society of Antiquaries of Scotland,
  Edinburgh 060243
Society of Architectural Historians of
  Great Britain, Edinburgh 060244
Society of Arts and Crafts,
  Boston 060745
Society of Arts and Crafts-Exhibition
  Gallery, Boston 047133
The Society of Arts, Literature and
  Welfare, Chittagong 058346
The Society of Botanical Artists,
  Gillingham, Dorset 060245
Society of Contemporary Artists,
  Kolkata 059371
Society of County Museum Officers,
  Taunton 060246
Society of Decorative Art Curators,
  Liverpool 060247
Society of Fine Art Auctioneers and
  Valuers, Woking 060248
Society of Illustrators, New York 060746
Society of Jewellery Historians,
  London 060249

Society of London Art Dealers,
  London 060250
Society of Miniaturists, Ilkley . . . 060251
Society of Museum Archaeologists,
  London 060252
The Society of Scottish Artists,
  Edinburgh 060253
The Society of the Cincinnati Museum,
  Washington 054161
The Society of the Four Arts Gallery,
  Palm Beach 051893
Society of Wildlife Artists,
  London 060254
The Society of Women Artists,
  Gillingham, Dorset 060255
Society of Wood Engravers,
  Northamptonshire 060256
Sočinskij Chudožestvennyj Muzej,
  Soči 037878
Sockel, Helmut, Oestrich-Winkel 077901
Soco, Austin 119700
Socorro, de, New York 095748
Socotom, Paris 071912
Socotti, Brugge 100538
Socrates Sculpture Park, Long Island
  City 050385
Sod House Museum, Aline 046403
Sod House Museum, Arctic Bay 005348
Sod House Museum, Gothenburg 049167
soDA, Zürich 139099
SodaArt, Wien 100373
Sodal, Nancy 104325
Sodankylän Kotiseutumuseo,
  Sodankylä 011362
Sodbury Antiques, Chipping
  Sodbury 088894
Sodbuster Archives Museum,
  Strome 007088
Sodi, Massimo, Genova 131341
Sodosha, Tokyo 111505
Sodus Bay Historical Society Museum,
  Sodus Point 053376
Soe Moe, Yangon 112058
Søagergård, Smørum 102981
Søby Brunkulslejer og Brunkulsmuseet,
  Herning 010116
Søbygård, Søby Ærø 010322
Soeder, Göteborg 133534
Söderbärke Hembygdsgård,
  Ekomuseum Bergslagen,
  Söderbärke 040966
Söderhamns Auktionshall,
  Sandarne 126659
Söderhamns Stadsmuseum,
  Söderhamn 040967
Söderklenoden, Stockholm 087055
Söderlångvikin Museo,
  Dragsfjärd 010763
Söderström, Stockholm 087056
Söderström, Leif, Stockholm . . . 087057
Söderström, Werner, Helsinki . . 136793
Södertälje Konsthall, Södertälje . 040969
Södertälje Konsthandel och Galleri,
  Södertälje 115817
Södra Galleriet, Stockholm 115924
Södra Skattkammaren,
  Stockholm 087058
Søfarts og Fiskerimuseet, Læsø 010216
Søfartssamlingerne i Troense,
  Svendborg 010336
Søgne Bygdemuseum, Søgne . . 033955
Söğüt Müzesi, Söğüt 043065
Söhn-Veigl, Barbara, Düsseldorf 129759
Söhn, Fritz-Dieter, Marburg 142116
Söhn, Gerhart, Düsseldorf 106832
Sølballegård, Knebel 102858
Soelberg, Christina E, Oslo 132790
Soelberg, Jytte, København 065955
Sölje Bygdegård, Glava 040641
Søllerød Museum, Holte 010138
Söllner, Gabriele, Braunschweig 075134
Sølvboden, Skien 084358
Sölvesborgs Museum,
  Sölvesborg 040973

Sølvkælderen, København 065956
Sømna Bygdetun, Sømna 033956
Sønderjyllands Antikvariat Bogcentret,
  Sønderborg 140660
Søndhordland Motormuseum,
  Valevåg 034065
Sør-Senja Museum, Midt-Troms
  Museum, Stonglandseidet 033993
Sør-Troms Museum, Avdeling
  Klyngetunet – Skånland Museum,
  Evenskjer 033589
Sørensen, Birthe, København . . 065957
Sørensen, Finn, Næstved 102934
Sørensen, Jan, Hillerød 065757
Sørensen, Jonna, Aalborg 065589
Sørensen, Marianne, København 065958
Sörgel, Karlheinz, Pommelsbrunn 130526
Sørlandets Kunstmuseum,
  Kristiansand 033717
Sørli Museum, Sørli 033959
Sørlistøa Fløtermuseet, Osen . . 033824
Sörmlands Museum, Nyköping . 040888
Sörmlandsgården, Eskilstuna . . 040609
Sørvágs Bygdasavn, Sørvágur . . 010741
Soester Münzcabinett, Soest . . . 078490,
  126116
Søsveen Antikk, Sola 084361
Soeten, M., Gronau, Westfalen . . 107167
Sofa, Enschede 083259
Sofa, İstanbul 088140
Sofec, Saint-Brieuc 141326
Sofer, Paris 071913
La Soffitta, Venezia 081875
La Soffitta della Nonna, Palermo 081077
Sofia, Cleveland 120455
Sofia Arte, Roma 081567
Sofia Kamburova, Sofia 101171
Sofiero Slott, Helsingborg 040697
Sofie's Galerie, Flörsheim am
  Main 106961
Sofija, Almaty . 082239, 082240, 111554,
  111555
Sofijska Gradska Chudožestvena
  Galerija, Sofia 005203
Sofijski Istoričeski Muzej, Sofia . 005204
Soft 3 di Massimiliano Tretene,
  Bologna 079879
Soft, Norske Tekstilkunstneres Galleri,
  Oslo 113242
Sogetsu Bijutsukan, Tokyo 029591
Sogo-Bunka Art Gallery,
  Amagasaki 028472
Sogo Galería, Cartago 102468
Sogo Museum of Art, Yokohama 029725
Soheili, München 077622
Soho, Seoul 111721
Soho 20 Gallery, New York 123076
El Soho Antique, Bogotá 065228
Soho Arte Millenium, Bogotá . . . 102368
Soho Fine Art Gallery, Las Vegas 121381
Soho Fort Worth Art Gallery, Fort
  Worth 120886
Soho Galería, Bogotá 102369
Soho Galleries, Sydney 099796
Soho Gallery, San Francisco . . . 124398
Soho House Museum,
  Birmingham 043627
Soho Mid-Century Gallery,
  Memphis 094665
Soho Myriad, Atlanta 119613
Soho Photo Gallery, New York . . 051469
Sohrab, New York 095749
Sohyun Gallery, New York 123077
Soieries C.J. Bonnet – Musée de la
  Soie, Jujurieux 013340
Soil Artist Cooperative, Seattle . 060747,
  124600
Soinin Museo, Soini 011364
Soissons Antiquités, Soissons . . 073530
Soizic, La Ciotat 068994
Sojeon Art Museum, Siheung . . 030055
Sojo, Amsterdam 112284
Sojus Chudožnikov Kitovras,
  Moskva 059780

Sojuz Chudožnikov Rossii,
Moskva 059781
Sojuz Chudožnikov Sankt Peterburga,
Sankt-Peterburg 059782
Soka Floris Art Center / Soka Contemporary
Space, Beijing 101991
Soka Art Center / Soka Contemporary
Space, Tainan 116899
Soka Art Center / Soka Contemporary
Space, Taipei 116941
Sokang University Museum,
Seoul 030043
Sokei Bijutsu Gaikun, Kunitachi . 056010
Sokendo, Tokyo 082206
Soker, Don, San Francisco ... 124399
Sokličev Muzej, Slovenj Gradec . 038614
Soknedal Bygdemuseum,
Soknedal 033964
Soko African Stone Art,
Newtownabbey 118808
Sokoloff, San Francisco 124400
Sokolowski, Harry, Los Angeles 135744
Sokolowski, Heinz, Weimar ... 078900
Sokrovišča Peterburga, Sankt-
Peterburg 085269
Sol, Niš 114455
Sol del Rio Art Gallery, San
Antonio 124072
Sol Floris Antiques, Seattle 097506
Sol Koffler Graduate Gallery,
Rhode Island School of Design,
Providence 052397
Sol Mur Plafond, Marseille 104199
Sola Kulturhus, Sola 113261
Sola, Guido, Roma 081568
Solajian, Kivork, Damascus ... 088061
La Solana Antigüedades, Caracas 097965
Solander, Yarralumla 099939
Solanet, Lucien, Paris 125767
Solange, Fresney 068613
Solano, C., Vitoria-Gasteiz ... 115544
Solar, Lisboa 084976
Solar Boats Museum, Giza 010575
Solar Condes de Resende,
Canelas 035595
Solar de Dom João VI, Rio de
Janeiro 004771
Solar Grandjean de Montigny, Rio de
Janeiro 004772
Solar Museu Quinta dos Cónegos,
Maia 035844
Solari Arte, Milano 110436
Solaria Arte, Piacenza 110653
Solazoque, Oviedo 115330
Solberg, Esaias, Oslo 084335
Sold, Richmond 096576
Soldan, Romina, Roma 110858
Le Soldat Artiste, Nouvron-Vingre 014528
Solder, Michael, Münster 142212
Soldi, Daniele, Firenze 080205
Soldier and Scholar, Fitzroy
North 139737
Soldiers and Sailors Memorial Building,
Huntington Historical Society,
Huntington 049623
Soldiers' Memorial Military Museum,
Saint Louis 052824
Soldiers National Museum,
Gettysburg 049092
Soldiers of Gloucestershire Museum,
Gloucester 044436
Il Sole, Perugia 110642
Sole Art, Philadelphia 123455
Il Sole Cornici, Roma . 110859, 126228
Solé, Marie-José, Paris 129146
Soleado Stichting, Utrecht 083663
Soledad Arroyo Gil, Toledo ... 086492,
115471
Soledad Lorenzo, Madrid 115287
Soleil Antiquités, Le Noirmont . 087712
Le Soleil Bleu, Versailles ... 105847
Soleil d'Art, Arcachon 103267
Soleil Ethnic, Vincennes ... 105880
Soleil Rouge, Villeurbanne 074228

Le Soleil sur la Place, Lyon 104132
Solent Sky, Southampton 045827
Soler & Llach, Barcelona 126514
Soler, Silvia, Barcelona 085842
Solers, Sofia 101172
Soletti, Bruna, Milano 110437
Soli Archaeological Site, Lefke . 033468
Solidair, Rotterdam 112661
Solidor, Cagnes-sur-Mer ... 103489
Soligo, Aldo, Roma 081569
Solihull Art Gallery, Solihull ... 119113
Solikamskij Gorodskoj Kraevedčeskij
Muzej, Solikamsk 037879
Soliman, Riad, Cairo 103842
Solin, Stockholm 087059, 115925
Solinas, Salvatore, Weißenburg in
Bayern 130743
Solinger Auktionshaus, Solingen 126118
Soliño, Donostia-San Sebastián . 133257
Solitaire Galerie, Berlin 106432
Sollefteå Antik, Sollefteå 086944
Sollefteå Museum, Sollefteå ... 040974
Sollertis, Toulouse 105752
Solmi, Giuseppe, Ozzano Emilia 142770
Solna Hembygdsmuseum, Solna 040980
Solnični, Brno 065354
Solo, Lyon 104133
Solo Arte, Waterford 109703
Solo Books, Ottawa 140380
Sololaki, Tbilisi 105921
Soloman Island National Museum and
Cultural Centre, Honiara 038641
Solomon, Dublin 109619
Solomon, Indianapolis 135638
Solomon, London 090500, 134770
Solomon & Co., New York 123078
Solomon, Seattle 124601
Solomon & Whitehead, Lichfield 138178
Solomon Dubnick, Sacramento . 123783
Solomon R. Guggenheim Museum,
New York 051470
Solomon, Jerry, Los Angeles ... 121594
Solomon, Nancy, Atlanta 119614
Solomon, Richard, New York ... 123079
Solomons, John, Armidale 061055
Solomos Museum, Corfu 023057
Solonoy, Cairo 066274
Solopark, Cambridge 088765
Soloveckij Gosudarstvennyj Istoriko-
Architekturnyj i Prirodnyj Muzej-
Zapovednik, Soloveckij ... 037882
Solovy, Stephen, Chicago 120272
Solstice, Los Angeles 121595
Solstices, Lille 140872
Solvang, Eva, Stockholm 115926
Solvay Centrum Sztuki, Kraków 113638
Solvis, Minsk 063169
Solvognen, Odense 102952
Solway Aviation Museum, Crosby-on-
Eden 044050
Solway, Carl, Cincinnati 120377
Soma Gyofu Kinenkan, Itoigawa
History and Folk Museum,
Itoigawa 028716
Somaini, Giuseppe, Milano 080729
Somali National Museum,
Mogadishu 038642
Sombra Township Museum,
Sombra 007053
Somdet Phra Narai National Museum,
Lopburi 042698
Somers Historical Society Museum
and Museum of the Early American
Circus, Somers 053380
Somers Mansion, Somers Point . 053382
Somers, Axel, Liège 063783
Somers, Elaine, Belfast 117344
Somerset and Dorset Railway Trust
Museum, Washford 046065
Somerset Brick and Tile Museum,
Bridgwater 043722
Somerset County Museum,
Taunton 045950
Somerset Cricket Museum,

Taunton 045951
Somerset East Museum, Somerset
East 038832
Somerset Historical Center,
Somerset 053383
Somerset Military Museum,
Taunton 045952
Somerset Place, Creswell 048065
Somerset Rural Life Museum,
Glastonbury 044425
Somerset, B.V.M., Tonbridge ... 091611,
135140
Somerton Antiques, Somerton . 091361,
135059
Somervell County Museum, Glen
Rose 049112
Somerville, Plymouth 118926
Somerville Museum, Somerville 053384
Somesville Museum, Mount
Desert 051085
Something Different, Cleveland . 120456
Something Different Antiques,
Brighton, Victoria 061206
Something Else, New York ... 095750
Something Fishy, Cleveland 120457
Something for Everyone,
Cleveland 092973
Something Old Something New,
Bowness-on-Windermere ... 088550
Something Old Something New,
Brampton, Cumbria 088570
Something Special Antiques, Fort
Worth 093480
Something Special Gallery,
Invercargill 112923
Something Unique, Virginia
Beach 097793, 136448
Sometime Ago Antiques,
Wellington 084191
Somewhere in Time, Denver ... 093339
Somewhere in Time, Minneapolis 094947
Somix, Bucureşti 085110
Somlo, London 090501, 134771
Somma, Salvatore, Napoli 080939
Sommazzi, Paolo, Firenze 131267
Somme Heritage Centre,
Newtownards 045383
Sommer, Graz 100020
Sommer, Saarbrücken 078334
Sommer, Sankt Blasien 078350
Sommer-Blum, Bernadette,
Willisau 133964
Sommer-Fuchs, Christine, Königsberg
in Bayern 076922
Sommer, A., München 130402, 142188
Sommer, Bruno, Hanau 076354
Sommer, Gertraud, München ... 108195
Sommerhuber-Kachelofenmuseum,
Steyr 002786
Sommerset House Books,
Blackheath 139646
Sommerville, Vancouver 064884
Somnis, Girona 086031
Somogy, Paris 136987
Somogy Kereskedelme Anno... ,
Kaposvár 023712
Somogyi Megyei Múzeumok
Igazgatósága – Rippl-Rónai Múzeum,
Kaposvár 023713
Somoza & Sims, Houston 121129
Son-Et-Lumiere, Mumbai 109475
Šon Gallery, Moskva 085206
Sona, Bayreuth 106126
Sonargaon Folk Art and Craft Museum,
Sonargaon 003234
Sonath, Paris 105242
Sonawala, Madhavi, Mumbai ... 109476
Soncin, Rinaldo, Torino 132205
Sonda, Den Haag 112414
Sondaar jr., P.Y., Utrecht 132650
Sonderegger, R. & M., Bäretswil 087210
Sondre Nordheimstova, Morgedal 033787
Song, København 065959
Song Adang, Daegu 111592

Song Qingling Former Residence,
Shanghai 008238
Song Qingling Tongzhi Former
Residence, Beijing 007645
Songam Art Museum, Incheon . 029953
Songcheng, Hangzhou 102040
Songe, Montceau-les-Mines ... 104260
Songjiang Museum, Shanghai . 008239
Songline, Amsterdam 112285
Songyang Museum, Songyang . 008286
Songzi Fuwu Guan, Songzi ... 008287
Soni, Crawford 114782
Sonia, Singapore 085505
Sonic Arts Gallery, San Diego . 124175
Sonino, Ralph, Genève 087503
Sonkajärven Kotiseutumuseo,
Sonkajärvi 011371
Sonnabend, New York 123080
Sonne, Berlin 106433
Sonne + Mond Galerie,
Heinsberg 107412
Sonnen-Galerie, Beilstein 106127
Sonnen, Cornelia, Unnau 108806
Sonnenberg Mansion,
Canandaigua 047399
Sonntag, Stefan, Stuttgart 078635
Sonntag, Udo, Lübeck 077188
Sonoma County Museum, Santa
Rosa 053144
Sonoran, Phoenix 096237
Sonrier, Jean-Pierre, Colmar ... 068096
Sonrisa, Los Angeles 121596
Sons of the American Revolution,
Roselle 052682
Sons of the American Revolution
Museum, Louisville 050501
Sonsbeek, Arnhem 112338
Sontg Hippolytus, Veulden ... 042103
The Sony Gallery for Photography,
Cairo 103085
Sony Wonder Technology Lab, New
York 051471
Sonya Art, Brooklyn 047267
Sonya's Treasures, Indianapolis . 093928
Sonzogni, Luisa, Milano 080730
Soo, Seoul 111722
Soo Line Historical Museum,
Weyburn 007396
Soo San, London 090502
Soo Tze Oriental Antiques,
Prahran 062120
Soo Visual Arts Center,
Minneapolis 050947
Soobin, Chua, Singapore 114610
Sood Sangvichien Prehistoric Museum,
Bangkok 042679
Sooke Region Museum, Art Gallery
and Historic Moss Cottage,
Sooke 007054
Sookmyung Women's University
Museum, Seoul 030044
Sooner Collectibles, Oklahoma
City 096011
Sooner Ga Museum, Dhaka ... 003227
Sopel, Lyle, Vancouver 101837
Sopena, Ramón, Barcelona 137971
Sophia Antiquités, Saint-Ouen . 073199
Sophia Wanamaker, Los Yoses . 102473
Sophie & Joe's Emporium, Saint
Paul 096861
Sophie & Van Twist, Breda ... 083007
Sophienburg Museum, New
Braunfels 051236
Sophie's Books, Moss Vale ... 139840
Sophisticated Traveller, Zeist ... 083713
Sopocki Dom Aukcyjny, Gdańsk 084511,
126401
Sopocki Dom Aukcyjny, Kraków 084600,
084601, 126413, 126414
Sopocki Dom Aukcyjny, Sopot . 084666,
126416
Sopot-Art, Sopot 113764
Soppa, U., Spangenberg 108648
Soproni Evangélikus Egyházközség

Múzeuma, Sopron 023900
Soproni Múzeum, Sopron 023901
Soproni Római Katolikus
 Egyházművészeti Gyűjtemény,
 Sopron 023902
Sorai, Kyoto 082039
Soranthe, Eindhoven 083229
Sorbische Bauernstube,
 Heinersbrück 019174
Sorcha Dallas, Glasgow 117859
Sordes, Christian, Grenoble . . 068740
Sordini, Marseille 104200
Sordoni Art Gallery, Wilkes University,
 Wilkes-Barre 054407
Sorecomodata, Porto . . 085052, 132978
Sorel, Portland 096477
Sorell Antique Centre, Sorell . . 062218
Sorell Station Antiques, Sorell . . 062219
Soren, Christensen, New Orleans 122239
Soret, Hervé & Philippe, Saint-
 Ouen 073200
Sorgato, David, Milano 080731
Sorgenfrei, Gisela, Bad
 Schwartau 074594
Sorgenicht, Margarete, Berlin . . 074927
Soria, Asti 109762
Sorma, Sharjah 088213
Soro, Los Angeles 094476
Sorø Amts Museum, Sydvestsjællands
 Museum, Sorø 010325
Sorokko, Serge, San Francisco . 124401,
 124402
Sorolla Galeria, Sevilla 115458
Soroptimist Book Shop, Port Pirie 139877
Soroptimist International, Kansas
 City 121316
Soroush, Tehrãn 137601
Sorrentino, Amedeo, Napoli . . 131643
Sós, Budapest 142570
SOS Art, Bologna 131000
Sos Moveis, São Paulo 064211
Sosa-Carillo-Fremont House Museum,
 Tucson 053835
Sosa, Lourdes, México 111985
Sósaiati, Coimbra 084793
Soshoku Design, Tokyo 138894
Sosin & Sosin, Minneapolis . . . 122036
Soskine, Michel, New York . . . 123081
Sosnitza, Michael, Göttingen . . . 129913
Sosnovoborskij Chudožestvennyj Muzei
 Sovremennogo Iskusstva, Sosnovyj
 Bor 037885
Sosnowieckie Centrum Sztuki –
 Zamek Sielecki, Sosnowiec . . 035214
Sossdorf, Hans Georg, Lübeck . 077189
Sóstói Múzeumfalu,
 Sóstógyógyfürdő 023905
Sota Zorraquin, Pedro de la,
 Bilbao 114997
Sotamuseo, Helsinki 010863
Sotarns Antik och Auktion,
 Ljungby 086819, 126624
Sotavallan Tuulimylly, Lempäälä 011105
Soteau, Christian, Souvigny . . 073557
Sotelo Salazar, Manuel, Lima . . 084470
Sotheby's, Amsterdam 126273
Sotheby's, Armadale, Victoria . 125177
Sotheby's, Bangkok 126788
Sotheby's, Beverly Hills 127139
Sotheby's, Billingshurst 126820
Sotheby's, Bofferdange 126257
Sotheby's, Boston 127146
Sotheby's, Buenos Aires 125172
Sotheby's, Caracas 127502
Sotheby's, Chicago 127169
Sotheby's, Constantia 126492
Sotheby's, Coral Gables 127189
Sotheby's, Dallas 127194
Sotheby's, Dublin 126172
Sotheby's, Edinburgh 126889
Sotheby's, Frankfurt am Main . . 125950
Sotheby's, Genève 126733
Sotheby's, Hamburg 125978
Sotheby's, Harrogate 126919

Sotheby's, Haverford 127240
Sotheby's, Hong Kong 125405
Sotheby's, Houston 127249
Sotheby's, Jakarta 126165
Sotheby's, Johannesburg 126494
Sotheby's, København 125455
Sotheby's, Köln 126012
Sotheby's, Kuala Lumpur 126258
Sotheby's, London 126972
Sotheby's, Lugano 126741
Sotheby's, Madrid 126550
Sotheby's, México 126265
Sotheby's, Milano 126212
Sotheby's, Minneapolis 127326
Sotheby's, Montecito 127328
Sotheby's, München 126068
Sotheby's, New York 127359
Sotheby's, Newtownards 126996
Sotheby's, Oslo 126381
Sotheby's, Palm Beach 127384
Sotheby's, Paris 125768
Sotheby's, Praha 125426
Sotheby's, Rio de Janeiro 125361
Sotheby's, Roma 126229
Sotheby's, Saint-Geoire-en-
 Valdaine 125802
Sotheby's, San Francisco 127445
Sotheby's, Sankt Moritz 126753
Sotheby's, Santiago de Chile . . 125395
Sotheby's, São Paulo 125362
Sotheby's, Shanghai 125411
Sotheby's, Singapore 126487
Sotheby's, Stockholm 126675
Sotheby's, Sydney 125239
Sotheby's, Taipei 126783
Sotheby's, Tel Aviv 126183
Sotheby's, Tiverton 127068
Sotheby's, Tokyo 126250
Sotheby's, Torino 126235
Sotheby's, Toronto 125385
Sotheby's, Towson 127478
Sotheby's, Ullenwood 127072
Sotheby's, Warwick 125358
Sotheby's, Wien 126285
Sotheby's, Zürich 126774
Sotheby's at Auction, Bend . . . 139492
Sotheby's Belgium SA, Bruxelles 125327
Sotheby's Monaco SA, Monaco . 126269
Sotheby's Restoration, New York 136006
Sotheran, Henry, London 090503,
 134772, 144693
Sotilaslääketieteen Museo, Lahti 011080
Sotiropoulos, Athinai 109047
Sotkamon Makasiinimuseo,
 Sotkamo 011372
Soto, Boston 119923
Soto de Collins, Lucía, Bogotá . 065229
Soto Valle, Luís Miguel, Bilbao . 085889
Soto, Blanca, Madrid 115288
Sotscheck, Helmut, Potsdam . . 130539
Sotterley Plantation Museum,
 Hollywood 049514
Sotterranei dell'Arte, Monte
 Carasso 041749
Sottocorno 9, Milano 110438
Sottopiano Beaux-Arts, Cagliari 059479,
 109926, 137632
Sottoriva, Verona 081939
Sou, Sendai 111392
Soubizantik, Saint-Ouen 073201
Soubeyre, Louis, Paris 071914
Souček, Vladislav, Plzeň 065453
Souchaud, Françoise, Lyon . . . 104134
Soucheby, Nelson 084022, 112967
Souchet, Albert, Rennes (Ille-et-
 Vilaine) 129206
Souchet, Fabien, Paris 129147
Souchko, Vladimir, Rouen . . . 072618,
 129213
Souchon, Michel, Paris 071915
Souchot, Patrice, Cepoy 067746, 067747
Soudan Corners, Toronto 064755
Souders, Fort Worth 120887
Soudry-Tharaud, Annie, Carnac . 067690

Soufer, Mahboubeh, New York . 123082
Souffle du Passé, Kfarhabab . . 082326,
 142956
Soufir, Hélène, Maulette 070004
Souflard, Jean-Pierre, Ampuis . 066585
Sougyo, Aoitori, Tokyo 082207
Souillac, Michel, Paris 071916
Soukaïna, Casablanca 082583
Soul 42, Bern 087303
Soul Mates Adventures,
 Amsterdam 112286
Soul of Africa, Vancouver 101838
Soulard, Saint Louis 123876
Soulard, Claude, Saint-Georges-de-
 Montaigu 072797
Soulard, Jacqueline & Christian,
 Flayosc 068537, 140839
Soulard, Marie-Hélène, Nantes . 128840
Soulas, Didier, La Roche-sur-Yon 069074
Soulflower, Bangkok 116961
Soulflower, Mumbai 109477
Soulié, Béatrice, Paris 105243
Soulié, Jacqueline, Finhan . . . 068533
Soulie, José, Bourganeuf 067420
Soulier, Marc, Montpellier . . . 070313
Soulières, Roger, Montréal . . . 101451
Soulis, Dirk, Lone Jack 127288
Soun Museum, Ashikaga 028489
Sound School House Museum, Mount
 Desert 051086
Sound Visitor Centre, Manx National
 Heritage, Port Saint Mary . . . 045535
Soundarte-Galería, Medellín . . 102460
Soundiata, Seattle 124602
Sour, Patrick, Niort 070715
Sourbié, Didier, Sigean 073520
Source, Buffalo 092559
Source Antiques, Dubbo 061412
Source Antiques, Widcombe . . 091843
Source aux Trouvailles, Nanterre 070438
Source Enterprises, Vancouver . 064885
Source Fine Arts, Kansas City . 121317
Source Graphics and Arts, Abu
 Dhabi 117169
Source: Notes in the History of Art,
 New York 139493
El Sourdog Hex, Berlin 017292, 059234
Sourget, Patrick, Chartres 140825
Sourgnes, Michel, Ascot 098673
Sourice, Daniel, Vitry-sur-Seine . 074262
Sourillan International, Toulouse 073801,
 105753
Sourzat, Jean-Louis, Saint-Céré 072709
Sous le Ciel Tibétain, Perpignan 072095
Sous-les-Toits-de-Paris, Paris . 071918
Sousa & Filhos, Lisboa 084977
Sousa & Roquette, Lisboa 084978
Sousa, Aldo de, Buenos Aires . . 098580
Soussana, Jacques, Jerusalem . 109720,
 137614
Soustre, François, Saint-Hilaire-
 Luc 072828
Souter Johnnie's Cottage,
 Kirkoswald 044733
Souterroscope des Ardoisières,
 Caumont-l'Eventé 012372
South African Antique Dealers'
 Association, Fontainebleau . . 059821
South African Jewish Museum,
 Gardens 038716
South African Mint Museum-Coin
 World, Tshwane 038858
South African Museum, Iziko
 Museums of Cape Town, Cape
 Town 038686
South African Museums Association
 (SAMA), Centrahil 059822
South African National Association for
 the Visual Arts, Tshwane . . . 059823
South African National Gallery, Iziko
 Museums of Cape Town, Cape
 Town 038687
South African National Museum of

Military History, Johannesburg 038758
South African National Railway &
 Steam Museum, Auckland Park 038647
South African Police Service Museum,
 Tshwane 038859
South Arkansas Arts Center, El
 Dorado 048497
South Australian Aviation Museum,
 Port Adelaide 001415
South Australian Maritime Museum,
 Port Adelaide 001416
South Australian Museum,
 Adelaide 000771
South Australian Police Museum,
 Thebarton 001574
South Australian School of Art,
 Division of Education, Arts and Social
 Sciences, Adelaide 054891
South Bank Galleries, Pittsburgh 096327,
 123578
South Bank Printmakers, London 118571
South Bannock County Historical
 Center, Lava Hot Springs . . . 050184
South Bay Auctions, Long Island
 City 127290
South Bend Regional Museum of Art,
 South Bend 053394
South Boston-Halifax County Museum
 of Fine Arts and History, South
 Boston 053400
South Bronx Art Workshop,
 Baltimore 119772
South Canterbury Museum,
 Timaru 033312
South Cariboo Historical Museum,
 Clinton 005628
South Carolina Artisans Center,
 Walterboro 054048
South Carolina Confederate Relic
 Room and Museum, Columbia 047915
South Carolina Cotton Museum,
 Bishopville 047017
South Carolina Federation of
 Museums, Columbia 060748
South Carolina Historical Society
 Museum, Charleston 047560
South Carolina Law Enforcement
 Officers Hall of Fame, Columbia 047916
South Carolina Military Museum,
 Columbia 047917
South Carolina National Guard
 Museum, Sumter 053600
South Carolina State Museum,
 Columbia 047918
South Carolina Tennis Hall of Fame,
 Belton 046907
South Central Museum Association,
 Wood Mountain 058475
South Charles Street Gallery,
 Baltimore 119773
South Charleston Museum, South
 Charleston 053401
South Coast Auction, Santa Ana 127448
South Coast Railroad Museum at
 Goleta Depot, Goleta 049153
South Coast Wood Works Gallery,
 Youngs Siding 099948
South County Antique Mall, Saint
 Louis 096777
South County Art Association,
 Kingston 060749
South County Museum,
 Narragansett 051166
South Dakota Amateur Baseball Hall
 of Fame, Lake Norden 050083
South Dakota Art Museum,
 Brookings 047243
South Dakota Discovery Center,
 Pierre 052130
South Dakota Hall of Fame,
 Chamberlain 047514
South Devon Railway,
 Buckfastleigh 043775
South-East Armed Force Museum, Ho

Chi Minh City 054808
South East Art Society, Mount
  Gambier 058221
South East European Contemporary
  Art Network, Skopje 059562
South Eastern Federation of Museums
  and Art Galleries, Bromley ... 060257
South Fitzroy Antiques and Interiors,
  Fitzroy 061471, 127598
South Florida Art Institute, Dania 056978
South Florida Museum,
  Bradenton 047171
South Florida Museum of Natural
  History, Dania Beach 048139
South Florida Science Museum, West
  Palm Beach 054319
South Front Antique Market,
  Memphis 094666
South Gate Gallery, Exeter 117773
South Gippsland Auctions,
  Korumburra 125204
South Grey Museum and Historical
  Library, Flesherton 005807
South Hero Bicentennial Museum,
  South Hero 053408
South Hill Park Arts Centre,
  Bracknell 043689
South Hills Antique Center,
  Pittsburgh 096328
South Hills Antique Gallery,
  Cleveland 092974
South Hills Art Center, Pittsburgh 123579
South Holland Historical Museum,
  South Holland 053409
South Indian Society of Painters,
  Chennai 059372
South London Antique and Bookcentre,
  London 144694
South London Gallery, London .. 045108
South Louisville Antique Mall,
  Louisville 094588
South Midlands Museums Federation,
  King's Lynn 060258
South Milwaukee Historical Society
  Museum, South Milwaukee .. 053411
South Molton and District Museum,
  South Molton 045815
South Notts Hussars Museum,
  Bulwell 043785
South of Market, Atlanta 092185
South Otago Historical Museum,
  Balclutha 033161
South Park City Museum,
  Fairplay 048679
South Pass City State Historic Site,
  South Pass City 053413
South Peace Centennial Museum,
  Beaverlodge 005407
South Peace Historical Museum,
  Dawson Creek 005690
South Perth Antiques, South
  Perth 062224
South Ribble Museum and Exhibition
  Centre, Leyland 044821
South River Meeting House,
  Lynchburg 050535
South Sask Photo Museum,
  Arcola 005347
South Shields Museum and Art Gallery,
  South Shields 045820
South Shore Art Center, Cohasset 047862
South Shore Gallery, Milwaukee 121950
South Similkameen Museum,
  Keremeos 006085
The South Street Seaport Museum,
  New York 051472
South Sutton Old Store Museum,
  South Sutton 053417
South Taranaki District Museum,
  Patea 033280
South Texas Institute for the Arts,
  Corpus Christi 056957
South Texas Museum, Alice .. 046401
South Tipperary County Museum,

Clonmel 024852
South Tynedale Railway Preservation
  Society, Alston 043404
South Wales Miner's Museum, Port
  Talbot 045539
South Wales Police Museum,
  Bridgend 043717
South West Academy of Fine and
  Applied Arts, Exeter 056532
South West Museums, Libraries and
  Archives Council, Taunton ... 060259
South Western Federation of Museums
  and Art Galleries, Bournemouth 060260
South Wood County Historical
  Museum, Wisconsin Rapids .. 054509
Southam, Salt Lake City 123987
Southampton Art Gallery,
  Southampton 007062
Southampton Art School,
  Southampton 055156
Southampton City Art Gallery,
  Southampton 045828
Southampton Historical Museum,
  Southampton 053421
Southampton Maritime Museum,
  Southampton 045829
Southchurch Hall Museum, Southend-
  on-Sea 045835
Southdown Antiques, Lewes ... 089819
Southdown House Antique Galleries,
  Brasted 088585, 134141
Southeast Arts Center, Atlanta . 046634
Southeast Asia Art, Singapore . 114611
Southeast Coin, Portland 096478
Southeast Missouri State University
  Museum, Cape Girardeau ... 047418
Southeast Museum, Brewster .. 047194
Southeast Museum of Photography,
  Daytona Beach Community College,
  Daytona Beach 048181
Southeastern Center for Contemporary
  Art, Winston-Salem 054495
Southeastern College Art Conference,
  Carrboro 060750
Southeastern Museums Conference,
  Baton Rouge 060751
Southeastern Pastel Society,
  Dawsonville 060752
Southeastern Railway Museum,
  Duluth 048379
Southend Central Museum, Southend-
  on-Sea 045836
Southend Limited, Charlotte .. 092627
Southend Pier Museum, Southend-on-
  Sea 045837
Southerbee, Oklahoma City 096012
Southern African Development
  Community Association of Museums
  and Monuments, Harare ... 060832
Southern Alberta Art Gallery,
  Lethbridge 006187
Southern Alleghenies Museum of Art,
  Altoona 046426
Southern Alleghenies Museum of Art,
  Loretto 050398
Southern Alleghenies Museum of Art
  at Johnstown, Johnstown 049834
Southern Antique Centre,
  Kogarah 061698
Southern Arizona Watercolor Guild,
  Tucson 060753
Southern Artist Union,
  Jacksonville 060754
Southern Arts Federation, Atlanta 046635,
                         060755
Southern Cross Relics, Koroit .. 061704
Southern Crossing Antique Mall,
  Jacksonville 094039
Southern Exposure Gallery, San
  Francisco 053027
Southern Expressions, New
  Orleans 122240
Southern Firearms, Clovelly Park 061346
Southern Forest World Museum,

Waycross 054246
Southern Graphics Council,
  Athens 060756
Southern Highland Craft Guild,
  Asheville 060757
Southern Highland Craft Guild at the
  Folk Art Center, Asheville ... 046558
Southern Historical Showcase,
  Nashville 095036
Southern Museum of Flight,
  Birmingham 047008
Southern Newfoundland Seamen's
  Museum, Grand Bank 005898
Southern Ohio Museum and Cultural
  Center, Portsmouth 052319
Southern Oregon Historical Society
  Museum, Medford 050758
Southern Oregon University Museum
  of Vertebrate Natural History,
  Ashland 046571
Southern Plains Indian Museum,
  Anadarko 046465
Southern Settings, Birmingham . 119820
Southern Splendor, Nashville ... 095037
Southern Ute Museum and Cultural
  Center, Ignacio 049657
Southern Vermont Art Center,
  Manchester 050633
Southern Vermont College Art Gallery,
  Bennington 046921
Southern Vermont Natural History
  Museum, Marlboro 050702
Southern Watercolor Society,
  Campbell 060758
Southgate Auction Rooms,
  London 126973
The Southland Art Collection,
  Dallas 048124
Southland Art Society,
  Invercargill 059642, 112924
Southland Fire Service Museum,
  Invercargill 033228
Southland Museum and Art Gallery,
  Invercargill 033229
Southold Historical Society Museum,
  Southold 053423
Southold Indian Museum,
  Southold 053424
Southport Antique Mall,
  Indianapolis 093929
Southport Antiques, Chicago ... 092793
Southport Antiques Centre,
  Southport 091384
Southport Fine Art, Chicago ... 120273
Southsea Castle, Portsmouth ... 045559
Southside Antiques, Saint Louis 096778
Southside Antiques Centre,
  Annerley 061002
Southside Auction, Tulsa 127481
Southside House, London 045109
Southward Museum,
  Paraparaumu 033279
Southwest Antiques, Houston .. 093798
Southwest Art, Invercargill 112925
Southwest Coin, Oklahoma City 096013
Southwest Cornerhouse,
  Albuquerque 119510
Southwest Florida Museum of History,
  Fort Myers 048853
Southwest Florida Pastel Society,
  Naples 060759
Southwest Gallery, Dallas ... 120661,
                         135516
Southwest Mercado Gallery,
  Albuquerque 119511
Southwest Museum, Los Angeles 050469
Southwest Museum of Science and
  Technology, Dallas 048125
Southwest Numismatic Corporation,
  Dallas 093217
Southwest School of Art and Craft,
  San Antonio 057844
Southwest Studios, El Paso 120840
Southwest Virginia Museum, Big

Stone Gap 046981
Southwestern Art, Denver 120776
Southwestern Artists Association, San
  Diego 060760
Southwestern Saskatchewan
  Oldtimer's Museum, Maple
  Creek 006260
Southwestern University Museum,
  Cebu 034467
Southwestern Utah Art Gallery, Saint
  George 052770
Southwick Hall, Peterborough .. 045491
Southwick, David L.H.,
  Kingswear 089709
Southwold Museum, Southwold 045842
Souvenance, Bourges . 067432, 128518
Souvenances, Montreuil-sur-Mer 070329
Souvenir, Singapore 085506
Souvenir Antiques, Carlisle ... 088795
Souvenir Press, London 138252
Souverain, Michel, Nueil-sur-
  Layon 070763
Souvré, Christian, Courmelles .. 068192
Souyris, Christian, Salles-Curan 073386
Souza, Toninho de, Brasilia 100927
Sov Art, Moskva 114308
Sovenirs de Provence, Cassis .. 103540
Sovereign Antiques, Gateshead . 089323
Sovereign Antiques, Middle Park 061850
Sovereign Antiques, Seacliff Park 062197
Sovereign Collection, Portland .. 123680
Sovereign Hill, Ballarat 000819
Soviart, Kyïv 117149
Soviet Carpet and Art Galleries,
  London 090504
Sovietiniy Skulptūry Grūto Parkas,
  The Park of Soviet Sculptures,
  Druskininkai 030334
Sovremennaja Galereja-muzej Nikor,
  Sobranie N.I. Kornilova, Moskva 037500
Sovremennoe Iskusstvo, Sankt-
  Peterburg 114362
Sowade, Reimund, Dresden ... 075430
Sowamy, Steve, Luxembourg .. 111866
Soweto Art, Johannesburg ... 114801
Sowinski, Christiane, Noailles .. 070721
Sozosha Geijutsu Gaikun, Osaka 056027
Sozzani, Carla, Milano 110439
Sozzi, Romeo, Milano . 110440, 110441
SP Gallery, Seoul 111723
Spa & Assoc., Los Angeles ... 094477
Spaans Galjoen, Amsterdam ... 082861
Spaarnestad Fotoarchief,
  Haarlem 032398
Space & Form, Montréal 064472, 064473
Space 101 Gallery, Brew House
  Association, Pittsburgh 052177
Space 1026, Philadelphia ... 123456
Space 237, Toledo 124698
Space Center, Houston 121130
Space Center Houston, Houston 049597
Space Expo, Noordwijk, Zuid-
  Holland 032701
Space Farms Zoological Park and
  Museum, Sussex 053620
Space Gallery, Denver 120777
Space Gallery, Kraków 113639
Space Gallery, New Orleans .. 122241
A Space Gallery, Toronto 007207
Space Invaders, Long Beach .. 094225
Space of Art, Baabdet El-Metn . 082268
Space of Art, Beirut 082305
Space One Eleven, Birmingham 047009,
                         119821
Space Station, Denver 093340
Spaced Gallery of Architecture, New
  York 123083
Spaces, Cleveland ... 047824, 120458
Spacial Anomaly Gallery,
  Philadelphia 123457
Spadafina, Schorndorf 078392
Spadafora, Giuseppe, Torino .. 111005
Spadari, Fréjus 068607
Spadari, Sylvie, Merano 131353

Spaddeline, München 077623
Spadina Museum, Toronto 007208
Spagna Art Gallery, Firenze 110069
Spagnoletto, Pacifico, Roma 081570
Spagnoli, Elisa, Livorno 131349
Spagnoli, Sebastiano, Napoli ... 080940
Spagnolo, Mario, Roma 132052
Spagnolo, Rosaria, Foggia 131283
Spahn, R., Flieden 075751
Spaite, Monique, Ternuay-Melay-Saint-
   Hilaire 073634
Spakenburgs Museum 't Vurhuus,
   Bunschoten Spakenburg 032119
Spalding Antiques, Spalding .. 062233,
   091397
Spalding Gentlemen's Society
   Museum, Spalding 045845
SPAM Museum, Austin 046681
Span Galleries, Melbourne 099321
Spang, Cincinnati 092897
Spanierman, New York 095751, 123084
Spanish Cross Home Furnishings,
   Tucson 097695
Spanish Governor's Palace, San
   Antonio 052950
The Spanish Institute, New York 051473
The Spanish Quarter Museum, Saint
   Augustine 052751
Spanos, Konst., Athinai 142541
Spantidaki, Athinai 079182
Spare Room, Edithburgh 061438
Sparebankmuseet, Oslo 033862
Sparekassemuseet, Korsør 010212
Sparenberg, Friedrich, Göttingen 076054
Spark Gallery, Denver 048276
Sparka, S., Tuttlingen 078734
Sparkassen-Museum, Greding .. 018801
Sparkassen-Museum der Erste Bank
   AG, Wien 003085
Sparkassenmuseum Leipzig,
   Leipzig 020018
Sparks & Sparks, Bryan 127151
Sparks, Jeff, Bryan 135368
Sparnaay, Tjalf, Amsterdam ... 112287
Sparresholm Gods og Vognsamling,
   Holme-Olstrup 010133
Spartanburg County Museum of Art,
   Spartanburg 053431
Spartanburg Science Center,
   Spartanburg 053432
Sparts, Paris 105244
Sparts, Rosny-sous-Bois 105462
Sparts – Art in Sports,
   Amsterdam .. 112288, 112289, 137840,
   137841
Spårvägsmuseet, Stockholm ... 041042
Spas, Sankt-Peterburg 114363
Spathareion Museum of the Shadow
   Theatre, Maroussi 023160
Spazi d'Essenza Contemporanea,
   Torino 111006
Spazio 10, Ivrea 110150
Spazio Arte, Milano 110443
Spazio Arte, Perugia 110643
Spazio Consolo, Milano 110444
Spazio Cultura, Belluno 109790
Spazio Ergy, Milano 110445
Spazio Idea Arredamenti, Roma 081571
Spazio Oberdan, Fondazione Cineteca
   Italiana, Milano 026832
Spazio Santabarbara, Milano ... 110446
Spazio, Erasmus, Milano 110442
SPAZIOTEMPO, Firenze 110070
Spazzio, Porto Alegre 063995
Spears, Indianapolis 121219
De Specht, Zetten 083715, 132673
Special Pieces, Paddington, New
   South Wales 062043
Special T Gallery, Edmonton ... 101300
Specialcolour, Nottingham 118843
The Specialist, Shorncliffe 127675
Specialty Finishes Unlimited,
   Columbus 135485
The Speckled Hen, Townsville .. 062341

Spectacular Folk Cultural Collection
   Museum, Longtan 042378
Spectator Bookstore, Oakland .. 145388
Spector Gallery, Philadelphia .. 123458
Spector, Warren, Houston 093799
Spectrum, Euskirchen 106944
Spectrum (Amherst), Amherst .. 139494
Spectrum Books, Warrnambool .. 139955
Spectrum Fine Arts, London ... 118572
Spectrum-Friends of Fine Art,
   Toledo 060761, 124699
Spectrum Gallery, Balwyn 098702
Spectrum Gallery, Fresno 120907
Spectrum Gallery, Machynlleth .. 118691
Spectrum Gallery, Toronto 101705,
   128218
Spectrum Gallery, Washington .. 124960
Spectrum Photos, Leicester ... 118096
Spectrum plus, Liebefeld 087592
Spectrum Sotos, Zaragoza 115564
Speeckaert, E., Bruxelles 140203
Speed Art Museum, Louisville .. 050502
Speedway Antique Mall, Tucson 097696
Speedway Bookshop,
   Christchurch 143312
Speedway Coins, Indianapolis .. 093930
Speedway, Judy, Houston 121131
Speelgoedmuseum De Brug,
   Eerbeek 032274
Speelgoedmuseum De Kijkdoos,
   Hoorn 032509
Speelgoedmuseum Deventer,
   Deventer 032218
Speelgoedmuseum Mechelen,
   Mechelen 003857
Speelgoedmuseum Op Stelten,
   Oosterhout 032726
Speelman, A. & J., London 090505
Speelman, Edward, London ... 118573
Speer, Zürich 138120
Speer, Torsten, Haseldorf 076411
Speezey, Los Angeles 121597
Spegeln, Linköping 086817
Speicherstadtmuseum, Hamburg 019019
Speier, Koblenz 107578
Speight 5060 Martin, Storrington 119155
Speight, Elisabeth, Philadelphia . 136104
Speke Hall, Liverpool 044863
Spektar, Veliko Tărnovo 101202
Spektrum, Jilemnice 102602
Spektrum, Internationale
   Vierteljahresschrift für Dichtung
   und Originalgrafik Zürich,
   Regensburg 139100
Spelbos, Utrecht 132651
Spelbos, R.H., Utrecht 083664
Speldjesmuseum, Klaaswaal ... 032544
Speller, Annie, Clermont-Ferrand 068037
Spellman Museum of Stamps and
   Postal History, Weston 054351
Spelman College Museum of Fine Art,
   Atlanta 046636
Spelman, Ken, York 144943
Spelthorne Museum, Staines ... 045854
Spence, Larry, Austin 092269
Spenceola Antiques Center, Bel
   Air 092378
Spencer, Bridlington 126837
Spencer, Houston 121132
Spencer, Shankill ... 079697, 130890
Spencer Brownstone Gallery, New
   York 123085
Spencer Entomological Museum,
   Vancouver 007297
Spencer House, London 045110
Spencer Museum of Art, University of
   Kansas, Lawrence 050187
Spencer-Peirce-Little Farm,
   Newbury 051515
Spencer Street Gallery, West
   Melbourne 099881
Spencer, Des, Main Beach 099257
Spencer, Don, Warwick 091749
Spencer, R., Portland 096479

Spencer's Treasures, Phoenix .. 096238
Spengler-Museum,
   Sangerhausen 021584
Spenner, Odense 102953
Spenner, H., Waltrop 142461
Spennick, Jacques, Lens 069386
Spenster, Ole, Frederiksberg .. 065677
Spéos – Paris Photogrphic Institute,
   Paris 055405
Sperberg, Michael J., Richmond 096577,
   136178
Sperka, Peter, Paunzhausen .. 078027
Sperl, Ursula & Rainer, Potsdam 108410,
   137473
Sperlhofer, Michael, Bad Ischl .. 062526
Sperling, Reinhard, Dresden ... 129724
Sperm Whale Museum, Tainan . 042492
Spermann, Klaus, Berlin 106434
Spero, Simon, London 090506
Sperone & Westwater, New York 123086
Sperrgebiet Museum,
   Oranjemund 031850
Sperrin Heritage Centre,
   Plumridge 045510
Sperrle, Volker, Schwäbisch
   Gmünd 078406
Spertus Museum, Chicago 047698
Spessartmuseum, Lohr am Main 020109
Speth, D., Lalling 076997
Spetses Museum, Spetses 023275
Spettel, Alain, Bruguières 067523
Speybrouck, Bruno, Kortrijk ... 063728
Speyer, Darthea, Paris 105245
SPG Gallery, Richmond, Victoria 099637
Sphero – Georgian Painters Combine,
   Tbilisi 058701
Sphinx, Groningen 143153
Sphinx, Toronto 064756
Sphinx-Antiques, Zürich 088009
Spica, Martina-Erica, Lugano .. 087622
Spice, Hastings 089468
Spichlerz, Miejski Osrodek Kultury,
   Olsztyn 035070
Spicilège, Paris 105246
Spicy's Gewürzmuseum,
   Hamburg 019020
Spider Web Painting Australia,
   Buxton 098827
spiegel, München 020573
De Spiegel, Zaltbommel 083703
Spiegel Fine Framing, New York 136007
Spiegel, Theodor, Würzburg ... 130788
Spiegelbeeld, Emmen 112473
Spiegeling Art, Amsterdam 112290
Spiegelman, K., West Hollywood 097934
't Spieghel, Leiden 083447
't Spieghelhuis, Almelo 112112
Spieker, Werner, Gelnhausen .. 075965
Spielhaus Morrison Galerie,
   Berlin 106435
Spielkartensammlung Piatnik,
   Wien 003086
Spielzeug anno dazumal – Museum
   Beilngries, Beilngries 017020
Spielzeug-Museum, Aeugstertal 041197
Spielzeugmuseum, Alsfeld ... 016520
Spielzeugmuseum, Bad
   Lauterberg 016815
Spielzeugmuseum, Davos Platz . 041435
Spielzeugmuseum, Görlitz 018739
Spielzeugmuseum, Hemau 019189
Spielzeugmuseum, Hinterzarten 019262
Spielzeugmuseum, Schkeuditz . 021658
Spielzeugmuseum,
   Schmallenberg 021684
Spielzeugmuseum, Traben-
   Trarbach 022172
Spielzeugmuseum, Wittmund .. 022706
Spielzeugmuseum, Wolfurt 003130
Spielzeugmuseum, Zwiesel 022914
Spielzeugmuseum – Museum Lydia
   Bayer, Museen der Stadt Nürnberg,
   Nürnberg 020877
Spielzeugmuseum – Sammlung

Depuoz, Zürich 042231
Spielzeugmuseum im Alten
   Rathausturm, München 020574
Spielzeugmuseum im Alten
   Schloß, Jan und Manuela Kube,
   Sugenheim 022074
Spielzeugmuseum im Stadtmuseum,
   Stadtmuseum Ingolstadt,
   Ingolstadt 019413
Spielzeugmuseum Salzburg, Toy
   Museum, Salzburg 002639
Spielzeugmuseum Solz, Bebra . 017005
Spielzeugmuseum Trier, Trier ... 022203
Spielzeugmuseum, Dorf- und
   Rebbaumuseum, Riehen 041882
Spiers Gallery, Brevard College,
   Brevard 047191
Spiess & Walther, Heidelberg .. 076459
Spieß, Karin, Hamburg 076316
Spiess, Michel, Aarburg 087176, 133611
Spijker, F., Eindhoven 083230
Spik, Leo, Berlin ... 074928, 125890
Spike, Wien 138535
Spike Hughes Rare Books,
   Innerleithen 144498
Spiliotopoulos, Andreas,
   Thessaloniki 109101
Spillemands-Jagt og Skovbrugsmuseet
   i Rebild, Skørping 010316
Spiller, Dublin 109620
Spiller, Jacksonville ... 121267, 121268
Spiller, Axel, Gronau, Westfalen . 107168
Spiller, Jerome, New York 095752
Spiller, Vincenty, Jacksonville .. 121269
Spillmann Siebdruck, Zürich ... 116859,
   138121
Lo Spillo, Firenze 080206
Spin Gallery, Toronto 101706
La Spina, Pisa 110668
Spinabroc, Épinal 068456
De Spinde, Amersfoort 082657
De Spindelaar, Amersfoort 082658
Spindler, Klaus, München 077624
Spindler, Thomas, Marseille ... 128784
Spindletop and Gladys City Boomtown
   Museum, Beaumont 046860
Spine, Phoenix 096239
Spinella, Andrea, Milano 080732
Spinelli, Andrea, Milano 080733
Spink London, London 090508, 126974
Spink Shreves Galleries, Dallas . 127195
Spink Singapore, Singapore .. 126488
Spink Smythe, New York 127360
Spink York, York 127105
Spink, John, London 090507
Spinnerei Galerie, Leipzig 107852
Spinnrocken Antik och Kuriosa,
   Norrköping 086881
Spinoza, Amsterdam 143068
Het Spinozahuis, Rijnsburg ... 032785,
   059621
Spinrath, J.L., Düsseldorf 075521
Spira, Max, Paris 071919
Spira, Thierry, Paris 105244
Spiral Art Gallery, Auckland ... 112810
Spiral Gallery, Bega 098726
The Spiral Gallery, Settle 119068
Spiral Studios, Perth 099563
La Spirale 2000, Milano 110447
Spiralearte, Milano 110448
Spiralearte, Monza 110513
Spiralearte, Pietrasanta 110656
Spiralearte, Verona 111196
Spirali/Vel, Milano 137696
Spires Art, Omagh 118853
Spires Gallery, Olney 118852
Spirig, Kurt, Lugano .. 087623, 144044
Spirit Echoes Gallery, Austin .. 119701
Spirit Gallery, Vancouver 101839
Spirit of '76 Museum, Elyria ... 048574
Spirit of '76 Museum, Wellington 054271
Spirit of Oceania, Valbonne ... 105796
Spirit River & District Museum, Spirit
   River 007065

Register der Institutionen und Firmen

The Spirit Wrestler, Vancouver – Staatlicher Schloßbetrieb Schloß Nossen / Kloster Altzella,

The Spirit Wrestler, Vancouver . 101840
Spiritual Art Gallery, Louisville . . 094589
Spiro Mounds Archaeological Center, Oklahoma Historical Society, Spiro 053439
Spironello, J., Bruxelles 128038
Spišské Múzeum, Slovenské Národné Múzeum, Levoča 038379
Spišský Hrad, Spišské Múzeum – Slovenské Národné Múzeum, Spišské Podhradie 038455
Spitäle an der Alten Mainkirche, Würzburg 022803
Spital Hengersberg, Kunstsammlung Ostbayern im Spital Hengersberg, Hengersberg 019191
Spitfire and Hurricane Memorial Museum, Ramsgate 045580
Spittka, Martin, Freiburg im Breisgau 141811
Spittler, Florianne, Kaiserslautern 107495
Spitz, Jean-Marie, Charleville-Mézières 103573
Spitzner, Harald, Bamberg 129446
Spivey, Kansas City 145197
Spivey, Oklahoma City 096014
Spjst Library Archives & Museum, Temple 053708
Splavarski Muzej, Javnik 038519
SPLIA Gallery, Cold Spring Harbor 047867
Splinter, Christine, Berlin 106436
Splinter, Juffrouw, Amsterdam . 082862
Splinters, Forster 061480
Splinters Antiques, Myaree 061933
Split Rock Lighthouse, Two Harbors 053871
Spode Museum, Stoke-on-Trent 045885
Spode, David, Darlinghurst 139710
Spode, Jan, Hanstedt 076403
Spøk og Spenning, Oslo 143417
Spokane House Interpretive Center, Nine Mile Falls 051565
Spoke Wheel Car Museum, Charlottetown 005600
Społeczne Muzeum Konstantego Laszczki, Dobre 034715
Spolidoro, Gilberto, Porto Alegre 063996
Spolnik, La Varenne-Saint-Hilaire 103909
Spomen Kuća Bitke na Sutjesci, Tjentište 004284
Spomen Kuća Laze K. Lazarevića, Beograd 038182
Spomen Muzej Biskupa Josipa Jurja Strossmayera, Đakovo 008897
Spomen Muzej Ive Andrića, Beograd 038183
Spomen Park Danica, Muzej Grada Koprivnice, Koprivnica 008932
Spomen-Park Kragujevački Oktobar, Kragujevac 038204
Spomen-Zbirka Pavla Beljanskog, Novi Sad 038219
Spominska Soba Dragotina Ketteja na Premu, Prem 038593
Spominska Soba Giuseppe Tartini, Piran 038588
Spominska Soba Ivana Cankarja, Ljubljana 038561
Spominska Soba Jurija Vege, Dol pri Ljubljani 038509
Spominska Zbirka Pisatelja Ivana Cankarja, Vrhnika 038630
Spominski Muzej Prežihovega Voranca, Koroški Muzej, Ravne na Koroškem 038599
Spongs Antiques Centre, Lindfield 089837
Spontanart Artforum, Courlevon 041429, 059968
Spooner & Son, Michael, Ottawa 064514, 125373
Spoor- en Tramweg Verzameling, Loppersum 032605
Spoorwegmuseum van Miniatuurmodellen, Treinmuseum Schaalmodellenverzameling Pieter Nombluez, Heist-op-den-Berg . 003690
Sporting Antiques, Tunbridge Wells 091663
Sporting Art Gallery, Toronto . . . 101707
Sportis, Fréderic, Paris 071920
Sportmuseum Berlin, Stiftung Stadtmuseum Berlin, Berlin . . . 017293
Sportmuseum der Stadt Frankfurt/Oder, Frankfurt (Oder) 018402
Sportmuseum Leipzig, Stadtgeschichtliches Museum Leipzig, Leipzig 020019
Sportmuseum Schweiz, Basel . . 041292
Sportmuseum Vlaanderen, Hofstade 003705
Sportmúzeum, Kaposvár 023714
Sports Car Museum of Japan, Matsuda Collection, Gotemba . 028574
Sports Images, Columbus 093047
Sports Legacy Art Gallery, Pittsburgh 123580
Sports Legends, El Paso 093416
The Sports Museum of New England, Boston 047134
Sports World Specialties, Pittsburgh 096329
Le Sportsman, Charenton-le-Pont 140821
Sportsman's Edge Gallery, Houston 121133
Sportsman's Gallery, Atlanta . . . 119615
Sportsman's Gallery, West Horndon 119324
Sporveismuseet Vognhall 5, Oslo 033863
Sporvejsmuseet Skjoldenæsholm, Jystrup Midtsj 010155
Spot Antiques, Saint Louis 096779
Spotkania z Zabytkami, Warszawa 138990
Spotsylvania Historical Association and Museum, Spotsylvania 053443
Spoutnik, Montréal 064474
Spoutz, Eric I., Detroit 120817
Sprafka & O'Neill, Saint Paul . . . 096862
Spratt, Frederick, San Jose 124461
Spratt, Jack, Newark-on-Trent . 090817
Spread Eagle Antiques, London . 090509, 134773
Spread Eagle Bookshop, London 144695
Spread Eagle Gallery, London . . 090510
Sprebitz, Peter, Torgau 078687
De Spreeuwenpot, Den Haag . . 083132
Spreewald-Galerie, Lübben 107908
Spreewald-Museum Lübbenau, Lübbenau 020153
Sprengel Museum Hannover, Hannover 019066
Sprenger, G., Velburg . 078785, 078786
Spring Creative, Chicago 120274
Spring Creek Gallery, Beechworth 098722
Spring Fair Birmingham-Gallery, Birmingham 098404
Spring Gallery, New York 123087
Spring Garden Book Supply, Philadelphia 145419
Spring Gate Galleries, Tampa . . 124675, 124676, 124677
Spring London Park Lane Arms Fair, London 098405
Spring Mill State Park Pioneer Village and Grissom Memorial, Mitchell 050967
Spring Mountain Art Gallery, Las Vegas 121382, 121383
Spring Museum, Yangchun 008460
Spring Street Historical Museum, Shreveport 053325
Spring Valley Community Museums, History Museum, Church Museum, Washburn-Zittleman House, Conley Camera Collection, Spring Valley 053445
Springbank Visual Arts Centre, Mississauga 006344
Springdale Coin and Antique Shop, Cincinnati 092898
Springer & Winckler, Berlin 106437
springerin, Hefte für Gegenwartskunst, Wien 138536
Springfield, Jacksonville 121270
Springfield Armory Museum, Springfield 053463
Springfield Art and Historical Society, Springfield 060762
Springfield Art Association, Springfield 060763
Springfield Art Gallery, Springfield 053456
Springfield Art Museum, Springfield 053468
Springfield Historical Society Museum, Springfield 053469
Springfield Museum, Springfield 053473
Springfield Museum of Art, Springfield 053471
Springfield Museum of Old Domestic Life, High Point 049467
Springfield Museums Association, Springfield 060764
Springfield Science Museum, Springfield 053464
Springfields Art, Mellieha 111915
Springhill Costume Museum, Moneymore 045285
Springhill House, Magherafelt . . 045191
Springhill Miner's Museum, Springhill 007067
Springmann, Henrik, Freiburg im Breisgau 107065
Springs Museum, Springs 053476
Springtide Gallery, Bath 117315
SpringTime, Kilmore . . 079648, 130877
Springvale and District Historical Museum, Springvale 001516
Springville Museum of Art, Springville 053478
Sprinz, Claudia, Feldafing 137188
Sprookjes, Mortsel 063804
Sprossa, Else, Lillesand 132781
Sprott, W., Memphis 094667
Sprovieri Gallery, London 118574
Spruce Lane Farm, Oakville . . . 006538
Spruce Row Museum, Waterford 007360
Spruce Woods Provincial Heritage Park Museum, Carberry 005559
Spruck, Gisela, Glauburg 129905
Sprüth & Magers London, London 118575
Sprüth Magers Berlin, Berlin . . . 106438, 106439
Sprugnoli, Marie-Christine, Nice 070667, 104391
Sprugnoli, Michel, Nice 070668, 104392
Spruill, Atlanta 119616
Spruill Gallery, Atlanta 046637
Sprung, Ralf-Peter, Berlin 074929
't Spulleke, Bergen op Zoom . . 082966
Spunk, Minneapolis 122037
Spurlock Museum, University of Illinois at Urbana-Champaign, Urbana 053909
Spurn Lightship, Kingston-upon-Hull 044718
Spurrier-Smith, I., Ashbourne . . 088303, 134040
Spydeberg Bygdetun, Spydeberg 033968
Squadro, Bologna 109863, 137623
Squam Lakes Natural Science Center, Holderness 049502
Squamish Valley Museum, Garibaldi Heights 005860
The Square Gallery, Saint Mawes 119046
Squatriti, Alessandro, Roma . . . 132053
Squatriti, Mario, Roma 132054
Squatriti, Prof. Renato, Palermo 131713
Squatters Antiques, Murray Bridge 061930, 127650
Squeri, Robert, San Francisco . . 124403
Squerryes Court, Westerham . . . 046099
Squillace, Armando, Roma 132055
Squires Antiques, Altrincham . . 088268
Squires Antiques, Faversham . . 089268
Squirrel, Seattle 097507
Squirrel Collectors Centre, Basingstoke 088374
Squirrel Hill Jewelry, Pittsburgh 096330, 136141
Squirrels, Brockenhurst 088682
Sragow Gallery, New York 123088
Sri Varalakshmi Academy of Fine Arts, Mysore 055825
Sridharani Gallerie, Delhi 109342
Srinakharinwirot Art Exhibition Hall, Bangkok 042680
Srna-Modlitby, Martha, Wien . . . 063075, 100374
Srna, Manfred, Graz . . 062601, 100021, 125256
Srokol, Bogdan, Hamburg 076317, 129991
SS Great Britain Museum, Maritime Heritage Centre, Bristol 043754
SS Sicamous Restoration Society, Penticton 006629
Ssangin, Gwangju 111600
ST Antik, Gävle 086683
Staal, J., New York 095753
Staarup Hovedgaard, Højslev . . . 065772
Staas, B., Wassenberg 078872
Staatliche Akademie der Bildenden Künste Karlsruhe, Karlsruhe . . 055604
Staatliche Akademie der Bildenden Künste Stuttgart, Stuttgart 055727
Staatliche Antikensammlungen und Glyptothek, München 020575
Staatliche Bücher- und Kupferstichsammlung Greiz, Stiftung der Älteren Linie des Hauses Reuß, Greiz 018811
Staatliche Fachschule für Keramik Landshut, Landshut 055624
Staatliche Fachschule für Porzellan und industrielle Formgestaltung, Johann-Friedrich-Böttger-Institut, Selb 055708
Staatliche Fachschulen für Keramikgestaltung und Keramiktechnik, Höhr-Grenzhausen 055591
Staatliche Graphische Sammlung München, München 020576
Staatliche Hochschule für Bildende Künste, Städelschule, Frankfurt am Main 055542
Staatliche Hochschule für Gestaltung Karlsruhe, Karlsruhe 055605
Staatliche Kunsthalle Baden-Baden, Baden-Baden 016928
Staatliche Kunsthalle Karlsruhe, Karlsruhe 019517
Staatliche Kunstsammlungen Dresden, Generaldirektion, Dresden 018010
Staatliche Münzsammlung, Museum für Münzen, Papiergeld, Medaillen und Geschnittene Steine, München 020577
Staatliche Museen Kassel, Kassel 019542, 059179
Staatliche Museen zu Berlin – Stiftung Preußischer Kulturbesitz, Generaldirektion, Berlin 059236
Staatliche Museumsberatung für Nordhessen, Museumslandschaft Hessen Kassel, Kassel 019551
Staatliche Porzellan-Manufaktur Meissen, Restaurierwerkstatt, Meißen 130295
Staatliche Zeichenakademie Hanau, Hanau 055575
Staatlicher Schloßbetrieb Schloß

Nossen / Kloster Altzella,
Nossen 020837
Staatliches Museum Ägyptischer
Kunst, München 020578
Staatliches Museum für Naturkunde
Karlsruhe, Karlsruhe 019518
Staatliches Museum für Naturkunde
Stuttgart, Museum am Löwentor
und Museum Schloss Rosenstein,
Stuttgart 022056
Staatliches Museum für Naturkunde
und Vorgeschichte, Oldenburg . 020988
Staatliches Museum für Völkerkunde
München, München 020579
Staatliches Naturhistorisches Museum,
Braunschweig 017565
Staats- und Stadtbibliothek,
Augsburg 016668
Staatsarchiv Bamberg, Bamberg 016956
Staatsbibliothek Bamberg,
Bamberg 016957
Staatsbibliothek zu Berlin, Preußischer
Kulturbesitz, Berlin 017294
Staatsburgh State Historic Site,
Staatsburg 053480
Staatsgalerie im Hohen
Schloß, Bayerische
Staatsgemäldesammlungen,
Füssen 018553
Staatsgalerie im Neuen
Schloß, Bayerische
Staatsgemäldesammlungen,
Oberschleißheim 020926
Staatsgalerie in der
Benediktiner-Abtei, Bayerische
Staatsgemäldesammlungen,
Ottobeuren 021048
Staatsgalerie in der Burg, Bayerische
Staatsgemäldesammlungen,
Burghausen 017691
Staatsgalerie in der
Katharinenkirche, Bayerische
Staatsgemäldesammlungen,
Augsburg 016669
Staatsgalerie in der Neuen
Residenz, Bayerische
Staatsgemäldesammlungen,
Bamberg 016958
Staatsgalerie in der Residenz
Würzburg, Bayerische
Staatsgemäldesammlungen,
Würzburg 022804
Staatsgalerie in der
Residenz, Bayerische
Staatsgemäldesammlungen,
Ansbach 016597
Staatsgalerie Moderne
Kunst, Bayerische
Staatsgemäldesammlungen,
Augsburg 016670
Staatsgalerie Neuburg – Flämische
Barockmalerei, Bayerische
Staatsgemäldesammlungen, Neuburg
an der Donau 020691
Staatsgalerie Stuttgart, Stuttgart 022057
Staatsopernmuseum, Wiener
Staatsoper, Wien 003087
Staatspark Fürstenlager,
Bensheim 017035
Staatspark Wilhelmsbad, Hanau 019043
Staba Arte, Wiehl 078966
Stabbursnes Naturhus og Museum,
Indre Billefjord 033681
Stabile, Ivana, Brescia 079957
The Stable, Salt Lake City ... 123988
Stable Antiques, Lindfield 089838
Stable Antiques, Sedbergh ... 091260
Stable Antiques, Storrington . 091456
Stable Antiques, Thornton Dale . 091599
Stable Antiques at Houghton,
Houghton, West Sussex 089593
Stable Art, Twickenham 119278
Stable Gallery, Inveraray 118019
Stablegate Antiques, Barham . 088349

Stabler-Leadbeater Apothecary
Museum, Alexandria 046396
Stabler, Monty, Birmingham ... 119822
Stables, Grasmere 089361
Stables Antique Centre, Hatton . 089471
Stables Art Gallery of Taos Art
Association, Taos 053678
Stables Gallery, Ballymoney ... 117275
Stacey, Leigh-on-Sea 126939
Stacey, Alan & Kathy, Yeovil . 135268
Stack-Keane, Liam, Asdessd ... 079435
Stack's Rare Coins, New York . 095754,
                               127361
Stackstreet Books, Auckland ... 143295
Stacpoole, George, Adare 079434
Stacy-Marks, Helmsley 117949
Stacy-Marks, E., Helmsley 089490
Stacy-Marks, E., Polegate ... 091026
Stacy-Marks, Nigel, Auchterarder 088322
Stacy-Marks, Nigel, Perth 090979,
                        118905, 134934
Stacy's Tavern Museum, Glen
Ellyn 049109
Stadhuis, Gent 003639
Stadhuis van Antwerpen,
Antwerpen 003314
Stadhuismuseum, Zierikzee ... 033104
Stadion, Trieste 081804, 126236
Stadium, Dieulefit 068302
Stadium, Hyères 068860
Stadium, Nantes 070478
Stadius, Christian, Stockholm ... 133580
Stadler, Bad Mitterndorf 125251
Stadler Verlaggesellschaft mbH,
Konstanz 137326
Stadler, Dr. Christoph, Bremen . 075185
Stadler, Gertraud, München ... 077625,
                               130403
Stadler, Jonny W., Wilhelmsfeld 130767
Stadler, Sieglinde, Freistadt ... 099990
Stadsboerderij Het Wevershuisje,
Almelo 031898
Stadsgalerie Engels Tuin,
Ootmarsum 032732
Stadsgalerie Gouda, Gouda ... 032367
Stadsgalerij Heerlen – Museum voor
Moderne en Hedendaagse Kunst,
Heerlen 032435
Stadsmuseet och Ahlbergshallen,
Östersund 040904
Stadsmuseum Doetinchem,
Doetinchem 032235
Stadsmuseum Groenlo, Groenlo 032372
Stadsmuseum i Gråbrödraklostret,
Ystad 041183
Stadsmuseum IJsselstein,
IJsselstein 032525
Stadsmuseum STAM, Bijlokecollectie,
Gent 003640
Stadsmuseum 't Schippershof,
Menen 003865
Stadsmuseum Woerden,
Woerden 033058
Stadsmuseum Zoetermeer,
Zoetermeer 033106
Stadt- und Bädermuseum, Bad
Salzuflen 016861
Stadt- und Bädermuseum Bad
Doberan, Bad Doberan 016745
Stadt- und Bergbaumuseum Freiberg,
Freiberg, Sachsen 018474
Stadt- und Brauereimuseum Pritzwalk,
Pritzwalk 021224
Stadt- und Burgmuseum,
Eppstein 018225
Stadt- und Dampfmaschinenmuseum,
Werdau 022570
Stadt- und Fachwerkmuseum Alte
Universität, Eppingen 018222
Stadt- und Festungsmuseum im
Ludwigstor, Germersheim ... 018658
Stadt- und Freilichtmuseum Sinsheim,
Sinsheim 021882
Stadt- und Heimatmuseum,

Heidenreichstein 002068
Stadt- und Heimatmuseum,
Malchow 020239
Stadt- und Heimatmuseum der Stadt
Marl, Marl, Westfalen 020305
Stadt- und Heimatmuseum Kusel,
Kusel 019834
Stadt- und Heimatmuseum mit Galerie,
Haus des Gastes, Burg Waischenfeld,
Waischenfeld 022377
Stadt- und Hochstiftmuseum, Dillingen
an der Donau 017900
Stadt- und Industriemuseum,
Wetzlar 022616
Stadt- und Industriemuseum
Rüsselsheim, Rüsselsheim ... 021540
Stadt- und Kreisgeschichtliches
Museum, Hünfeld 019358
Stadt- und Kulturgeschichtliches
Museum Torgau, Torgau 022169
Stadt- und Landesbibliothek Dortmund,
Artothek, Dortmund 017964
Stadt- und Manfred-Kyber-Museum,
Löwenstein 020102
Stadt- und Regionalmuseum
Lübben, Museum Schloss Lübben,
Lübben 020151
Stadt- und Schulmuseum, Ronneburg,
Thüringen 021469
Stadt- und Turmmuseum, Bad
Camberg 016743
Stadt- und Wallfahrtsmuseum,
Walldürn 022408
Stadt-Galerie, Maisach 107960
Stadt-Galerie Anna von Cleve,
Kleve 107567
Stadt-Museum, Weil der Stadt . 022474
Stadt-und Bergbaumuseum der Stadt
Staßfurt, Staßfurt 021963
Stadt-und Wagenmuseum Oschatz,
Oschatz 021010
Stadtarchiv mit städtischen
Sammlungen, Lauf an der
Pegnitz 019932
Stadtarchiv und Ehemals
Reichsstädtische Bibliothek Lindau,
Lindau 020082
Stadtbibliothek Reutlingen,
Reutlingen 021385
Stadtbibliothek Trier – Schatzkammer
und Ausstellung, Trier 022204
Stadtgalerie, Osnabrück 021019
Stadtgalerie Alte Post, Sylt ... 022092
Stadtgalerie Altena, Altena ... 016526
Stadtgalerie Altötting, Altötting . 016558
Stadtgalerie am Minoritenplatz,
Wolfsberg 003128
Stadtgalerie Bamberg – Villa Dessauer,
Museen der Stadt Bamberg,
Bamberg 016959
Stadtgalerie Bern, Bern 041327
Stadtgalerie im alten Herrenhaus,
Ternitz 002818
Stadtgalerie im Elbeforum Brunsbüttel,
Brunsbüttel 017649
Stadtgalerie Kiel, Kiel 019610
Stadtgalerie Klagenfurt,
Klagenfurt 002198
Stadtgalerie Lünen, Lünen ... 020186
Stadtgalerie Saarbrücken, Stiftung
Saarländischer Kulturbesitz,
Saarbrücken 021561
Stadtgalerie Schwaz, Schwaz . 002743
Stadtgeschichtliche Sammlung,
Gaggenau 018579
Stadtgeschichtliche Sammlung,
Velbert 022310
Stadtgeschichtliches Museum,
Hilpoltstein 019254
Stadtgeschichtliches Museum,
Kehl 019572
Stadtgeschichtliches Museum,
Ottweiler 021054
Stadtgeschichtliches Museum,

Pfullingen 021127
Stadtgeschichtliches Museum,
Tangermünde 022101
Stadtgeschichtliches Museum, Waren
(Müritz) 022430
Stadtgeschichtliches Museum im
Landrichterhaus, Karlstadt ... 019524
Stadtgeschichtliches Museum
Leipzig – Altes Rathaus, Leipzig 020020
Stadtgeschichtliches Museum
Leipzig – Neubau, Leipzig 020021
Stadtgeschichtliches Museum
Schabbellhaus, Wismar 022693
Stadtgeschichtliches Museum
Spandau, Berlin 017295
Stadtgeschichts- und
Schradenmuseum, Ortrand ... 021009
Stadthaus am Dom, Wetzlar ... 022617
Stadthaus-Galerie, Luzern ... 116481
Stadthaus-Galerie, Münster ... 020607
Stadthaus Olten, Olten 041825
Stadthaus Ulm, Ulm 022267
Stadthistorisches Sole, Salz
und Kali-Bergbaumuseum,
S+K-Bergbaumuseum, Bad
Salzdetfurth 016859
Stadtmauermuseum Nördlingen,
Nördlingen 020815
Stadtmuseum, Aichach 016483
Stadtmuseum, Bad Bergzabern . 016721
Stadtmuseum, Berndorf 001777
Stadtmuseum, Gmünd,
Niederösterreich 001950
Stadtmuseum, Hornberg 019342
Stadtmuseum, Mattersburg ... 002371
Stadtmuseum, Murau 002425
Stadtmuseum, Neuburg an der
Donau 020692
Stadtmuseum, Retz 002589
Stadtmuseum, Stadtschlaining . 002771
Stadtmuseum, Traiskirchen ... 002831
Stadtmuseum, Waidhofen an der
Thaya 002888
Stadtmuseum "Alte Burg" Wittenberge,
Wittenberge 022700
Stadtmuseum Aarau, Aarau ... 041191
Stadtmuseum Alte Hofmühle,
Hollabrunn 002102
Stadtmuseum Alte Post, Ebersbach an
der Fils 018079
Stadtmuseum Amberg, Amberg . 016571
Stadtmuseum Amtsturm, Lübz . 020167
Stadtmuseum Andernach,
Andernach 016580
Stadtmuseum Apolda, Apolda . 016602
Stadtmuseum Arelape-Bechelaren-
Pöchlarn, Pöchlarn 002514
Stadtmuseum Bad Berneck, Bad
Berneck 016729
Stadtmuseum Bad Cannstatt,
Stuttgart 022058
Stadtmuseum Bad Soden, Bad Soden
am Taunus 016881
Stadtmuseum Bad Wildungen, Bad
Wildungen 016905
Stadtmuseum Baden-Baden, Baden-
Baden 016929
Stadtmuseum Battenberg,
Battenberg 016971
Stadtmuseum Baunatal, Baunatal 016975
Stadtmuseum Beckum, Beckum 017010
Stadtmuseum Bendorf im Schloss
Sayn, Bendorf 017025
Stadtmuseum Bergen, Bergen auf
Rügen 017044
Stadtmuseum Bergkamen,
Bergkamen 017058
Stadtmuseum Bocholt, Bocholt . 017430
StadtMuseum Bonn, Bonn 017502
Stadtmuseum Borken, Westfalen,
Borken, Westfalen 017514
Stadtmuseum Brakel, Brakel ... 017534
Stadtmuseum Breuberg-Neustadt,
Breuberg 017625

Stadtmuseum Brilon, Brilon 017627
Stadtmuseum Bruck an der Leitha,
Bruck an der Leitha 001801
Stadtmuseum Bruck an der Mur,
Bruck an der Mur 001803
Stadtmuseum Burgdorf, Burgdorf 017685
Stadtmuseum Burghausen,
Burghausen 017692
Stadtmuseum Burgstädt,
Burgstädt 017700
Stadtmuseum Coesfeld, Coesfeld 017757
Stadtmuseum Cottbus, Cottbus . 017769
Stadtmuseum Cuxhaven,
Cuxhaven 017791
Stadtmuseum Damme, Damme . 017810
Stadtmuseum Deggendorf,
Deggendorf 017840
Stadtmuseum der FVG Riesa,
Riesa 021429
Stadtmuseum Ditzingen,
Ditzingen 017915
Stadtmuseum Döbeln/Kleine Galerie,
Döbeln 017919
Stadtmuseum Dornbirn, Dornbirn 001828
Stadtmuseum Dresden, Museen der
Stadt Dresden, Dresden 018011
Stadtmuseum Eilenburg,
Eilenburg 018127
Stadtmuseum Einbeck mit Radhaus
im Stadtmuseum, Einbeck . . . 018130
Stadtmuseum Eisenberg, Eisenberg,
Saale-Holzland-Kreis 018141
Stadtmuseum Erlangen, Erlangen 018264
Stadtmuseum Esslingen,
Esslingen 018321
Stadtmuseum Eupen, Eupen . . 003590
Stadtmuseum Euskirchen,
Euskirchen 018328
Stadtmuseum Fellbach, Fellbach 018351
Stadtmuseum Fembohaus mit
Noricama, Museen der Stadt
Nürnberg, Nürnberg 020878
Stadtmuseum Freilassing,
Freilassing 018495
Stadtmuseum Friesach, Friesach 001909
Stadtmuseum für Kunst und
Geschichte, Hüfingen 019356
Stadtmuseum Fürstenfeldbruck,
Fürstenfeldbruck 018543
Stadtmuseum Fürth, Fürth . . . 018551
Stadtmuseum Gardelegen,
Gardelegen 018589
Stadtmuseum Gehrden, Gehrden bei
Hannover 018603
Stadtmuseum Geithain, Geithain 018616
Stadtmuseum Gera, Gera 018651
Stadtmuseum Gerlingen – Gerlinger
Heimatmuseum – Museum
der Deutschen aus Ungarn,
Gerlingen 018656
Stadtmuseum Gerolzhofen,
Gerolzhofen 018665
Stadtmuseum Giengen, Giengen an
der Brenz 018686
Stadtmuseum Grafenau, Grafenau,
Kreis Freyung-Grafenau 018787
Stadtmuseum Graz, Graz 001986
Stadtmuseum Groß-Gerau, Groß-
Gerau 018834
Stadtmuseum Gütersloh,
Gütersloh 018889
Stadtmuseum Gunzenhausen,
Gunzenhausen 018899
Stadtmuseum Hadamar,
Hadamar 018908
Stadtmuseum Hall in Tirol, Hall in
Tirol 002048
Stadtmuseum Hartberg, Hartberg 002058
Stadtmuseum Hattingen,
Hattingen 019102
Stadtmuseum Haus Kupferhammer,
Warstein 022435
Stadtmuseum Herrenmühle,
Hammelburg 019033

Stadtmuseum Herzogenaurach,
Herzogenaurach 019233
Stadtmuseum Hildburghausen,
Hildburghausen 019244
Stadtmuseum Hofgeismar,
Hofgeismar 019292
Stadtmuseum Hofheim am Taunus,
Hofheim am Taunus 019293
Stadtmuseum Holzminden,
Holzminden 019328
Stadtmuseum Horb, Horb am
Neckar 019339
Stadtmuseum Hornmoldhaus,
Bietigheim-Bissingen 017385
Stadtmuseum im Alten Forstamt,
Stockach 022000
Stadtmuseum im Augustinerkloster,
Bad Langensalza 016812
Stadtmuseum im Historischen
Rathaus, Staufen im Breisgau . 021965
Stadtmuseum im Kammerhof,
Eisenerz 001855
Stadtmuseum im Knochenhauer-
Amtshaus, Hildesheim 019251
Stadtmuseum im Prinz-Max-Palais,
Karlsruhe 019519
Stadtmuseum im Schlössle,
Weingarten, Kreis Ravensburg . 022518
Stadtmuseum im Spital,
Crailsheim 017772
Stadtmuseum Ingolstadt im Kavalier
Hepp, Ingolstadt 019414
Stadtmuseum Innsbruck,
Innsbruck 002137
Stadtmuseum Iserlohn, Iserlohn 019427
Stadtmuseum Judenburg,
Judenburg 002151
Stadtmuseum Kassel, Kassel . . 019552
Stadtmuseum Kaufbeuren,
Kaufbeuren 019565
Stadtmuseum Kelsterbach,
Kelsterbach 019579
Stadtmuseum Killingerhaus,
Idstein 019379
Stadtmuseum Klostermühle, Bad
Urach 016894
Stadtmuseum Klosterneuburg,
Klosterneuburg 002211
Stadtmuseum Kremayr, Rust . . 002612
Stadtmuseum Kröpelin, Kröpelin 019798
Stadtmuseum Landeshauptstadt
Düsseldorf, Düsseldorf 018044
Stadtmuseum Leimen – Sankt Ilgen,
Leimen 019959
Stadtmuseum Leonberg,
Leonberg 020045
Stadtmuseum Leun, Leun 020049
Stadtmuseum Lichtenfels,
Lichtenfels 020063
Stadtmuseum Lindau, Lindau . . 020083
Stadtmuseum Ludwig Erhard,
Fürth 018551
Stadtmuseum Ludwigshafen,
Ludwigshafen am Rhein 020139
Stadtmuseum Meersburg,
Meersburg 020332
Stadtmuseum Meißen, Meißen . 020349
Stadtmuseum Melk, Melk 002382
Stadtmuseum Meppen, Meppen 020369
Stadtmuseum mit Weinmuseum, Bad
Vöslau 001766
Stadtmuseum Mosbach,
Mosbach 020474
Stadtmuseum Mühlberg, Elbe,
Mühlberg 020480
Stadtmuseum Mühlheim, Mühlheim
am Main 020489
Stadtmuseum Mülheim-Kärlich,
Mülheim-Kärlich 020498
Stadtmuseum Münster, Münster 020608
Stadtmuseum Mutzschen,
Mutzschen 020629
Stadtmuseum Naumburg, Haus zur
Hohen Lilie, Naumburg, Saale . 020647

Stadtmuseum Neumarkt, Neumarkt in
der Oberpfalz 020724
Stadtmuseum Neuötting,
Neuötting 020732
Stadtmuseum Neustadt an der
Waldnaab, Neustadt an der
Waldnaab 020756
Stadtmuseum Nittenau, Nittenau 020812
Stadtmuseum Nördlingen,
Nördlingen 020816
Stadtmuseum Norderstedt,
Norderstedt 020824
Stadtmuseum Nürtingen,
Nürtingen 020883
Stadtmuseum Obermühle,
Braunfels 017550
Stadtmuseum Offenbach am Main,
Offenbach am Main 020971
Stadtmuseum Oldenburg,
Oldenburg 020991
Stadtmuseum Penzberg,
Penzberg 021100
Stadtmuseum Pforzheim,
Pforzheim 021122
Stadtmuseum Pinneberg,
Pinneberg 021132
Stadtmuseum Pirna, Pirna 021140
Stadtmuseum Pößneck, Pößneck 021166
Stadtmuseum Povelturm,
Nordhorn 020832
Stadtmuseum Pulsnitz, Pulsnitz . 021236
Stadtmuseum Purkersdorf,
Purkersdorf 002551
Stadtmuseum Quakenbrück,
Quakenbrück 021241
Stadtmuseum Rastatt, Rastatt . . 021283
Stadtmuseum Rheinfelden,
Rheinfelden 021399
Stadtmuseum Rothenburg,
Rothenburg/Oberlausitz 021515
Stadtmuseum Rottweil, Rottweil 021530
Stadtmuseum Saalfeld im
Franziskanerkloster, Saalfeld/
Saale 021553
Stadtmuseum Sachsenheim,
Sachsenheim 021567
Stadtmuseum Sankt Pölten, Sankt
Pölten 002698
Stadtmuseum Schelklingen,
Schelklingen 021639
Stadtmuseum Schladming,
Schladming 002720
Stadtmuseum Schleswig,
Schleswig 021666
Stadtmuseum Schloß Hoyerswerda,
Hoyerswerda 019350
Stadtmuseum Schloss Wolfsburg,
Wolfsburg 022746
Stadtmuseum Schlüsselfeld,
Schlüsselfeld 021676
Stadtmuseum Schongau,
Schongau 021720
Stadtmuseum Schorndorf,
Schorndorf 021727
Stadtmuseum Schramberg,
Schramberg 021733
Stadtmuseum Schwabach,
Schwabach 021744
Stadtmuseum Schwandorf mit
Falkenauer Heimatstube,
Schwandorf 021764
Stadtmuseum Schwedt, Städtische
Museen Schwedt/Oder, Schwedt /
Oder 021777
Stadtmuseum Siegburg, Siegburg 021850
Stadtmuseum Simeonstift Trier,
Trier 022205
Stadtmuseum Sindelfingen,
Sindelfingen 021874
Stadtmuseum Stadtoldendorf,
Stadtoldendorf 021957
Stadtmuseum Steinfurt, Steinfurt 021984
Stadtmuseum Stolpen, Stolpen . 022008
Stadtmuseum Sulzbach-Rosenberg,

Sulzbach-Rosenberg 022084
Stadtmuseum Teterow, Teterow 022133
Stadtmuseum Traismauer,
Traismauer 002834
Stadtmuseum Trostberg,
Trostberg 022211
Stadtmuseum Tübingen,
Tübingen 022223
Stadtmuseum und Fürstlich
Starhembergisches Familienmuseum,
Eferding 001841
Stadtmuseum und
Tuchmachermuseum Pinkafeld,
Pinkafeld 002507
Stadtmuseum Vogtturm, Zell am
See 003136
Stadtmuseum Waldkraiburg,
Waldkraiburg 022398
Stadtmuseum Wasungen,
Wasungen 022444
Stadtmuseum Wehr, Wehr . . . 022454
Stadtmuseum Weilheim, Weilheim in
Oberbayern 022485
Stadtmuseum Weimar, Weimar . 022511
Stadtmuseum Wendlingen, Wendlingen
am Neckar 022563
Stadtmuseum Werne, Werne . . 022579
Stadtmuseum Wiener Neustadt,
Wiener Neustadt 003108
Stadtmuseum Wienertor Hainburg,
Hainburg an der Donau 002043
Stadtmuseum Wil, Wil 042140
Stadtmuseum Zehentstadel,
Nabburg 020634
Stadtmuseum Zistersdorf,
Zistersdorf 003141
Stadtmuseum Zweibrücken,
Zweibrücken 022906
Stadtmuseum Zwettl, Zwettl . . 003145
Stadtmuseum, Städtische Museen
Jena, Jena 019459
Stadtraum Vilshofen, Vilshofen . 022347
Stadtresidenz Landshut, Bayerische
Verwaltung der staatlichen Schlösser,
Gärten und Seen, Landshut . . . 019886
Stadtteilmuseum Nowaweser
Weberstube, Potsdam 021205
Stadttormuseum, Wehrheim . . 022456
Stadtturmgalerie, Gmünd,
Kärnten 001946
Stadtturmgalerie, Innsbruck . . 002138
Staeck, Prof. Klaus, Heidelberg . 107399,
137274
Stäcker, Peter, Bischofszell 116162
Städel Museum, Frankfurt am
Main 018448
Städelscher Museums-Verein e.V.,
Frankfurt am Main 059237
Städtische Ausstellungshalle Am
Hawerkamp, Münster 020592
Städtische Fossiliensammlung
Bergisch Gladbach, Bergisch
Gladbach 017056
Städtische Galerie, Bietigheim-
Bissingen 017386
Städtische Galerie, Dreieich . . . 017974
Städtische Galerie, Meersburg . . 020331
Städtische Galerie, Ravensburg . 021302
Städtische Galerie, Traunstein . . 022176
Städtische Galerie, Würzburg . . 022796
Städtische galerie ada Meiningen,
Meiningen 020343
Städtische Galerie am Markt Wurzen,
Wurzen 022829
Städtische Galerie Böblingen,
Böblingen 017455
Städtische Galerie Delmenhorst, Haus
Coburg Sammlung Stuckenberg,
Delmenhorst 017852
Städtische Galerie Die Fähre, Bad
Saulgau 016863
Städtische Galerie Dresden-
Kunstsammlung, Museen der Stadt
Dresden, Dresden 018012

Städtische Galerie Eichenmüllerhaus,
Lemgo 020032
Städtische Galerie Erlangen,
Erlangen 018265
Städtische Galerie Fauler Pelz,
Überlingen 022242
Städtische Galerie Filderhalle,
Leinfelden-Echterdingen 019966
Städtische Galerie Filderstadt,
Filderstadt 018362
Städtische Galerie Fruchthalle Rastatt,
Rastatt 021284
Städtische Galerie Grimma,
Grimma 018821
Städtische Galerie Haus Seel,
Siegen 021855
Städtische Galerie im Alten Rathaus,
Fürstenwalde 018544
Städtische Galerie im Buntentor,
Bremen 017599
Städtische Galerie im Cordonhaus,
Cham 017728
Städtische Galerie im Königin-
Christinen-Haus, Zeven ... 022874
Städtische Galerie im Kornhaus,
Kirchheim unter Teck ... 019628
Städtische Galerie im Kulturforum,
Offenburg 020978
Städtische Galerie im Lenbachhaus
und Kunstbau München,
München 020580
Städtische Galerie im Park,
Viersen 022331
Städtische Galerie im Rathaus,
Balingen 016940
Städtische Galerie im Rathauspark,
Gladbeck 018704
Städtische Galerie im Schloßpark
Strünkede, Emschertal-Museum,
Herne 019221
Städtische Galerie im Spital zum
Heiligen Geist, Riedlingen ... 021419
Städtische Galerie im Theater
Ingolstadt, Museum für konkrete
Kunst, Ingolstadt 019415
Städtische Galerie in der Alten Schule,
Sigmaringen 021863
Städtische Galerie In der Badstube,
Wangen im Allgäu 022421
Städtische Galerie in der Reithalle,
Paderborn 021068
Städtische Galerie Iserlohn,
Iserlohn 019428
Städtische Galerie Kaarst, Kaarst 019477
Städtische Galerie Karlsruhe,
Karlsruhe 019520
Städtische Galerie Leerer Beutel,
Museen der Stadt Regensburg,
Regensburg 021330
Städtische Galerie Lehrte, Lehrte 019956
Städtische Galerie Liebieghaus,
Liebieghaus Skulpturensammlung,
Frankfurt am Main 018449
Städtische Galerie Lovis-Kabinett,
Villingen-Schwenningen 022340
Städtische Galerie Lüdenscheid,
Museen der Stadt Lüdenscheid,
Lüdenscheid 020173
Städtische Galerie Mennonitenkirche,
Neuwied 020778
Städtische Galerie Neunkirchen,
Neunkirchen, Saarland 020731
Städtische Galerie Nordhorn –
Kunstwegen, Nordhorn 020833
Städtische Galerie Ostfildern,
Ostfildern 021031
Städtische Galerie Rastatt,
Rastatt 021285
Städtische Galerie Reutlingen,
Städtisches Kunstmuseum
Spendhaus Reutlingen,
Reutlingen 021386
Städtische Galerie Rosenheim,
Rosenheim 021478

Städtische Galerie Sohle 1,
Stadtmuseum Bergkamen,
Bergkamen 017059
Städtische Galerie Sonneberg,
Sonneberg 021916
Städtische Galerie Stapflehus, Weil am
Rhein 022471
Städtische Galerie Villa Streccius,
Landau in der Pfalz 019873
Städtische Galerie Villa Zanders,
Bergisch Gladbach 017057
Städtische Galerie Waldkraiburg,
Waldkraiburg 022399
Städtische Galerie Wesseling,
Wesseling 022597
Städtische Galerie Wolfsburg,
Wolfsburg 022747
Städtische Galerie, (im Museum der
Stadt Ettlingen), Ettlingen ... 018325
Städtische Galerie, Museum alter
lippischer Maler, Schieder-
Schwalenberg 021644
Städtische Galerie, Sammlung des 20.
Jahrhunderts, Paderborn ... 021067
Städtische Gemäldegalerie,
Füssen 018554
Städtische Kunstgalerie
im Deutschordenshaus,
Donauwörth 017936
Städtische Kunstgalerie Torhaus
Rombergpark, Dortmund ... 017965
Städtische Kunsthalle, Museen
der Stadt Recklinghausen,
Recklinghausen 021308
Städtische Kunstsammlung,
Murrhardt 020627
Städtische Kunstsammlung Schloß
Salder, Salzgitter 021575
Städtische Museen Heilbronn,
Heilbronn 019164
Städtische Museen Wangen im Allgäu,
Wangen im Allgäu 022422
Städtische Sammlungen Freital,
Freital 018502
Städtische Sammlungen, Bauländer
Heimatmuseum, Adelsheim ... 016467
Städtische Wessenberg-Galerie,
Konstanz 019764
Städtisches Grafik-Kabinett Backnang,
Backnang 016695
Städtisches Heimatmuseum,
Meßkirch 020388
Städtisches Heimatmuseum
Ballenstedt, Ballenstedt 016942
Städtisches Heimatmuseum Höchstadt,
Höchstadt an der Aisch ... 019275
Städtisches Heimatmuseum im
Waldemarturm, Dannenberg ... 017812
Städtisches Heimatmuseum Landau,
Landau in der Pfalz 019874
Städtisches Heimatmuseum Lippstadt,
Lippstadt 020094
Städtisches Heimatmuseum Mengen,
Mengen 020366
Städtisches Heimatmuseum Naila,
Naila 020637
Städtisches Heimatmuseum Reinfeld,
Reinfeld 021351
Städtisches Hellweg-Museum,
Geseke 018679
Städtisches Hutmuseum, Lindenberg
im Allgäu 020087
Städtisches Kramer-Museum,
Kempen 019583
Städtisches Kunstmuseum Singen,
Singen 021877
Städtisches Kunstmuseum Spendhaus
Reutlingen, Reutlingen ... 021387
Städtisches Lapidarium, Stuttgart 022059
Städtisches Museum,
Aschersleben 016631
Städtisches Museum, Bad
Reichenhall 016847
Städtisches Museum, Bruchsal . 017635

Städtisches Museum, Herborn .. 019197
Städtisches Museum, Kalkar ... 019485
Städtisches Museum, Kitzingen . 019638
Städtisches Museum, Menden . 020363
Städtisches Museum, Pfungstadt 021129
Städtisches Museum, Salzgitter . 021576
Städtisches Museum, Schärding 002709
Städtisches Museum, Überlingen 022243
Städtisches Museum, Welzheim 022557
Städtisches Museum, Werl 022576
Städtisches Museum, Wiesloch . 022647
Städtisches Museum, Zeulenroda-
Triebes 022870
Städtisches Museum Braunschweig,
Braunschweig 017566
Städtisches Museum Engen und
Galerie, Engen 018211
Städtisches Museum Fürstenwalde,
Fürstenwalde 018545
Städtisches Museum Gelsenkirchen,
Gelsenkirchen 018629
Städtisches Museum Göppingen im
Storchen, Göppingen 018733
Städtisches Museum Göttingen,
Göttingen 018758
Städtisches Museum Grevesmühlen,
Grevesmühlen 018816
Städtisches Museum Hann. Münden,
Hann. Münden 019048
Städtisches Museum Haus Letmathe,
Iserlohn 019429
Städtisches Museum im ehemaligen
Heiliggeistspital, Munderkingen 020619
Städtisches Museum im Hospital,
Nidderau 020783
Städtisches Museum im Kornhaus Bad
Waldsee, Bad Waldsee ... 016897
Städtisches Museum im Spiegelschen
Palais, Halberstadt 018941
Städtisches Museum Kamen, Haus
der Kamener Stadtgeschichte,
Kamen 019487
Städtisches Museum Ludwigsburg,
Ludwigsburg 020130
Städtisches Museum Peterskirche,
Vaihingen an der Enz 022297
Städtisches Museum Rosenheim,
Rosenheim 021479
Städtisches Museum Saarlouis,
Saarlouis 021565
Städtisches Museum Schloss Rheydt,
Mönchengladbach 020440
Städtisches Museum Seesen,
Seesen 021822
Städtisches Museum Wesel,
Festungsgeschichtliche Abteilung,
Wesel 022592
Städtisches Museum Wesel, Galerie
im Centrum, Wesel 022593
Städtisches Museum Wesel, Schill-
Kasematte, Wesel 022594
Städtisches Museum Zirndorf,
Zirndorf 022886
Städtisches Museum, Abt.
Stadtgeschichte, Städtische Galerie,
Eisenhüttenstadt 018145
Städtisches Museum/ Daniel-
Pöppelmann-Haus, Herford ... 019204
Städtisches Naturkundliches Museum,
Göppingen 018734
Städtisches Ortsmuseum
Schwamendingen, Zürich 042232
Städtisches Schulmuseum, Lohr am
Main 020110
Städtisches Zentrum für Geschichte
und Kunst, Riesa 021429
Städtligalerie, Wangen an der
Aare 116712
Stähli, M. & M., Oberdiessbach . 116545
Stähli, Marc, Auvernier 133626
Stähli, Simon, Urdorf 144143
Stählin-Tschanz, Betli, Altendorf 087190
Stämpfli Verlag, Bern 138032
Ständerhaus, Buus 041389

Stærk, København 065960
Staetshuijs, Amsterdam 082863
Staetshuijs Antiquairs Meulendijks &
Schuil, Amsterdam 082864
Staf, Antwerpen 063239
Staffanstorps Konsthall,
Staffanstorp 040984
Staffelbach, Elisabeth, Aarau ... 116006
Staffin Museum, Staffin 045849
Stafford, Barcelona 085843
Stafford Castle and Visitor Centre,
Stafford 045853
Stafford, Ulla, Binfield 088466
Staffords, London 118576
Staffordshire House Antiques,
Oakland 095913
Staffordshire Regiment Museum,
Lichfield 044826
Stafforte, Andrea, Bern 087304
Stafsjö Bruksmuseum,
Falkenberg 040615
Stage, Sapporo 082106
Stage & Screen, London 144696
Stage 21 Gallery, Tokyo 111506
Stage Nine, Sacramento 096648, 123784
Stagecoach Inn Museum Complex,
Newbury Park 051516
Stages Gallery, Machynlleth ... 118692
Stahl, Hamburg 076318, 125979
Stahl, Lund 143891
Stahl, Detlef, Speichersdorf ... 078517
Stahl, Jochen, Hirrlingen 130049
Stahl, Wilfried, Memmingen ... 077342
Stahlberger, Weil am Rhein ... 108856
Stahlmach, Adelbert, Berlin ... 074930
Stahlpark Riedersbach, Museum für
zeitgenössische Metallplastik, Sankt
Pantaleon 002688
Stahlschmidt, G., Osnabrück ... 077978
Stahmer, D. & H., Meinersen ... 077325
Stahre, Västerås 087142
Stained Glass, Raytown 139495
Stained Glass Attractions,
Metford 061848
Stained Glass Museum, Ely 044291
Stainless, Frank, Schauenstein . 078357
Stair & Co., New York 095755
Staithe Antiques, Brancaster
Staithe 088571
Staithe Lodge Gallery, Swafield 091524,
135109
Stájer-házi Erdészeti Múzeum,
Köszeg 023763
Stal, Dominique, Paris 105248
Stal, Kevin, Aiguèze 103187
Stała Ekspozycja Muzealna Ziemi
Sierakowickiej, Sierakowice ... 035197
Staley & Son, Sandy Bay 099676
Staley -Wise Gallery, New York . 123089
Staley, Carolyn, Seattle 097508, 124603
Staley, Suzanne C., Houston ... 121134
Stalham Antique Gallery, Stalham 091402,
135071
Stalheim Folkemuseum, Stalheim 033970
Stalke, Kirke Saaby 102855
Stalker, Varna 101198
Stalker, Ruth, Montréal 064475
Stall-Lädeli, Dübendorf 087398
De Stalling, Zaltbommel 112722
Stallings, Steve, Las Vegas 121384
Stallion Art Decor, Chicago ... 120275
La Stalluccia Antichità, Firenze . 080207
Stalmann, Dietrich, München . 130404
Stam, H., Haarlem 083353
Stamataris, Nikos, Chalandri ... 079191
Stamati, Paul, New York 095756
Stambak, Jean-Marc, Toulouse . 129316
Stambolov, Teodor, Amsterdam . 132416
Stamford Artists Galleries,
Stamford 119138
Stamford Auction, Stamford .. 127465
Stamford Brewery Museum,
Stamford 045858
Stamford Historical Society Museum,

Stamford 053483
Stamford Museum, Stamford . . 045859
Stamford Museum and Nature Center,
Stamford 053484
Stamkunsten, København 102912
Stamm, Ronald, Friedeburg . . . 129866
Stamm, Tirma G., Wiesbaden . . . 078997
Stammegna, Marc, Marseille . . . 069966,
104201
Stammersdorfer Kleinkunst, Wien 063076
Stammhaus Gander, Lienz . . . 140007
Stamnes, Morten, Oslo 084336
Stamp and Coin Shop, Seattle . 097509
Stamp & Sons, J., Market
Harborough 090714, 134838
Stampa, Basel 116098, 144034
Stampa, Chur 087374, 116220
Stamparte, Bologna . 109864, 137624,
142643
Stampbach, Gunten 116369
Stamperia Artistica Nazionale,
Torino 137752
Stan, Saint Louis 096780
Stan Hywet Hall and Gardens,
Akron 046331
Stanbet, Moskva 114309
Standard Bank Gallery,
Johannesburg 038759
Standing Bear Antiques, Omaha 096058
Standing Conference on Archives and
Museums, Ipswich 060261
Stanford, Nashville 122101
Stanfords, Colchester 126868
Stanger, Estelle, New York 095757
Stangl, Hermann-Josef,
Bernhardswald 074961
Staniland, Stamford 144862
Stanislas, Nancy 070432, 070433
Stanjohn, Chicago 135423
Stanka, Heinz & Lilian, Rehau . . 078179
Stanko, Bernie, Johannesburg . . 085649
Stanko, Janusz, Wrocław 113952
Stankus, M., Oakland 123267
Stanley, Hyderabad 109374
Stanley, Rockdale 099645
The Stanley and Gerald Rubin Center
for the Visual Arts, University of
Texas at El Paso, El Paso . . 048516
Stanley Gallery, Hong Kong 102114
Stanley Museum, Kingfield 044973
The Stanley Picker Gallery, Kingston-
upon-Thames 044722
Stanley Spencer Gallery,
Cookham 044606
Stanley Street Buy and Sell, East
Brisbane 061429
Stanley-Whitman House,
Farmington 048703
Stanley Williams Antiques,
Chicago 092795
Stanley, Barbara, London 118577
Stanley, Julian, Ewyas Harold . . 089227,
134376
Stanley, Ken & Rosemary,
Chicago 092794
Stanly County Historic Museum,
Albemarle 046352
Stanmer Rural Museum, Brighton 043741
Stanmore, Victor Harbor 062385
Stanmore Galleries, Stanmore . . 099716
Stannary Gallery, Tavistock 127064
Stanovsky, Dallas 093218, 135517
Stanowski, Filip, Warszawa 084749,
132905
Stan's Antiques, Los Angeles . . 094478
Stansbury Museum, Stansbury . . 001517
Stansted Mountfitchet Windmill,
Stansted Mountfitchet 045861
Stanthorpe Regional Art Gallery,
Stanthorpe 001518
Stanton County Museum, Pilger . 052131
Stanton, G., Dallas 120662
Stantondale, El Paso 135552
Stantschev, Matey, Oostende . . . 128085

Stanworth, P., Market Bosworth . 090705
La Stanzina, Firenze 110071
La Stanzina, Milano . . 080734, 110449
Stanzione Ciro, Roma 081572
Lo Stanzone, Trieste 081805
Stapelmann, Hans, Sankt Wendel 108559
Stapf, Helmut, Gernsheim 075985
Staplehurst Antiques, Staplehurst 091417,
135077
Staples, Richmond 123745
Stapleton, Unley 062379
Stapper, Venlo 143255
Star Antiques, San Francisco . . 097330
Star Books, Adelaide 139618
Star City Heritage Museum, Star
City 007071
Star Furniture, El Paso 093417
Star Gallery, Godalming 117868
Star Mound School Museum,
Snowflake 007052
Star of the Republic Museum,
Washington 054190
Star Peddler, Houston 093800
Star-Spangled Banner Flag House and
1812 Museum, Baltimore 046769
Star Troc, Épinay-sur-Seine 068458
Stara Galeria Związku Polskich
Artystów Fotografików,
Warszawa 035380
Stará Huť v Josefovském Údolí,
Technické Muzeum v Brně,
Adamov 009435
Stará Plzeň, Plzeň 065454
Stará Polenská Škola, Muzeum
Vysočiny Jihlava, Polná 009774
Stara Praga, Warszawa 084750, 132906
Stara Rzeźnia, Poznań 113749
Stará Škola, Police nad Metují . . 009769
Staraja Kniga, Sankt-Peterburg . 143675,
143676,   143677
Staraja Moneta, Sankt-Peterburg 085270
Stararte Gallery, Sesto Fiorentino 110927
Starbase 21, Tulsa 097758
Starchild, Columbus 093048
Starcik, Bärbel, Frankfurt am
Main 075826
Stardust, New York 095758
Stardust Antik, Stockholm 087060
Starfish Gallery, Whangarei 113114
Starfish Gallery, Woody Point . . 099902
Stargardt, J.A., Berlin . 125891, 141574
Stargate, Richmond 123746
Stari Antik, Silkeborg 066114
Stari Bazar, Sarajevo 063873
Stari Majstori, Sofia 064237
Starina, Moskva 085207, 133015
Starinarnica Marajna, Ljubljana . 085613
Starinárstvo Jukas, Ružomberok 085588
Starinnaja Mebel', Moskva 085208
Starinnye Časy, Sankt-Peterburg 085271
Starinnyj Interjer, Moskva 085209
Starinnyj Interjer, Sankt-
Peterburg 085272
Stark, London 118578
Stark Museum of Art, Orange . . 051791
Stark, Josef-David, Bamberg . . . 074673
Starke, Christel, Homberg (Ohm) 107445
Starke, Michael, Kirkel 130129
Starkey, Milton 061859
Starkey James H., Beverley 117353
Starkey, James H., Beverley . . . 088453,
134098
Starkie, Simon, Cartmel 088805
Starlight, Wansford 091723
Starlight Room, Long Beach . . . 094226
Starling, New Orleans 145316
Staro Selo Kumrovec, Kumrovec 008942
Staročerkasskij Istoriko-
architekturnyj Muzej-zapovednik,
Staročerkassk 037892
Starocie w Sopocie, Sopot 084667
Staroladožskij Istoriko-Architekturnyj i
Archeologičeskij Muzej-Zapovednik,
Staraja Ladoga 037889

Starooskolskij Kraevedčeskij Muzej,
Staryj Oskol 037895
Starorusskij Kraevedčeskij Muzej,
Staraja Russa 037891
Staroświecki Styl, Kraków 113640,
143484
Starovetešnictví a Kramářství Antik,
Děčín 065379
Starožitnictví, České Budějovice 065372
Starožitnictví, Kolín 065407
Starožitnictví, Praha 065522
Starožitnictví MK, Praha 065523
Starožitnictví u Zlaté uličky,
Praha 065524
Starožitnosti, Bělá pod Bezdězem 065338
Starožitnosti, Benešov 065339
Starožitnosti, Brno . . . 065355, 065356,
065357
Starožitnosti, Jablonec nad Nisou 065396
Starožitnosti, Nové Zámky 085582
Starožitnosti, Praha . . 065525, 065526,
065527,   065528,   125427
Starožitnosti – Antik, Ostrava . . 065438
Starožitnosti – Kabinet, Brno . . . 065358
Starožitnosti – Militaria, Litomyšl 065419
Starožitnosti Antik, Lučenec . . . 065578
Starožitnosti Antiqua, Jihlava . . 065399
Starožitnosti Baru, Brno 065359
Starožitnosti CS, Brno 065360
Starožitnosti Déco, Prostějov . . 065549
Starožitnosti Golem, Prostějov . . 065550
Starožitnosti Matouš, Jílemnice . 065401
Starožitnosti Ninive, Domažlice . 065383
Starožitnosti Pola, Ostrava 065439
Starožitnosti Přikryl, Prostějov . . 065551
Starožitnosti R & T, Frýdek-
Místek 065386
Starožitnosti Royal, Prešov 085585
Starožitnosti Sandra, Týn nad
Vltavou 065562
Starozitný Nábytek, Praha 065529
Starr, Denver 093341
Starr, Belle, Saint Louis 127430
Starr, Jana, New York 095759
Stars Antique Malls, Portland . . 096480
Stars Our Destination, Chicago . 145060
StART SPACE, London 118579
St'Art Up, L'attrape-Coulems,
Lyon 104135
Startgallery, Wellington 113091
Starting Line Antiques,
Indianapolis 093931
Startqing Museum, Zhaoqing . . 008533
Starved Rock State Park, Utica . 053914
Starving Artistz, København . . . 102913
Starý Královský Palác, Praha . . 009840
Starý Sklep, Kraków 113641
Starý Styl, Katowice 084539
Starý Zámek, Moravské zemské
muzeum Brno, Jevišovice 009601
Starý Zamok Banská Štiavnica,
Banská Štiavnica 038289
Starye Gody, Moskva 085210
Starye Gody, Sankt-Peterburg . . 085273
Starye Kartiny, Sankt-Peterburg 085274
Staschen, Ihlow 076597
Staschen, Wilhelmshaven 079014
Staschen, Wolfgang, Berlin 141575
Staschull, Bernd,
Niederndodeleben 130459
Stasiak, Alicja, Kalisz 113508
Stasimuseum, Berlin 017296
Stasio Gireno Gimtinė, Lietuvos
Aviacijos Muziejus, Vytogala . . 030558
Stasio Vainiūno Namai-Muziejus,
Lietuvos Nacionalinis Muziejus,
Vilnius 030542
Statarmuseet, Bara 040558
Statarmuseum Mejeriet, Örebro 040896
State Agricultural Heritage Museum,
Brookings 047244
State Archaeological Museum,
Kolkata 024368
State Art Academy – Educational

Dept., Tbilisi 055437
State Art Gallery, Ahmedabad . . 024127
State Authority of Museums,
Rīga 059545
State Ayurvedic College Museum,
Hardwar 024288
State Capital Publishing Museum,
Oklahoma Historical Society,
Guthrie 049299
State Coal Mine Museum,
Wonthaggi 001658
State Folk Art Museum of Armenia,
Yerevan 000746
State Gallery of Contemporary Cypriot
Art, Lefkosia 009422
State Health Institute Museum,
Lucknow 024386
State Historical and Architectural
Museum-reserve Shirvan Shahs'
Palace, Bakı 003179
State Historical Society of Iowa
Museum, Des Moines 048288
State Historical Society of Missouri,
Columbia 047903
State Historical Society of North
Dakota Museum, Bismarck . . . 047022
State Kala Akademi Museum,
Imphal 024314
State Library and Museum,
Kyaukpyu 031815
State Library and Museum,
Mawlamyine 031821
State Museum, Hyderabad 024303
State Museum, Kohima 024350
State Museum, Mandalay 031818
State Museum, Trichur 024500
State Museum of Abkhasia,
Suchumi 016394
State Museum of Azerbaijan Musical
Culture, Bakı 003180
State Museum of Ethnography, Arts
and Crafts, Lviv 043256
State Museum of Georgian National
Music Instruments, Tbilisi 016423
State Museum of Natural History of
NAS of Ukraine, National Academy of
Sciences of Ukraine, Lviv 043257
State Museum of Nature,
Yerevan 000747
State Museum of Pennsylvania,
Harrisburg 049372
State Museum of the History of
Religion, Bakı 003181
State Museum of Theatre, Music and
Cinema, Tbilisi 016424
State Museum, Edward Museum,
Bhopal 024174
State of the Art, Kansas City . . . 121318
State of the Art, New York 123090
State of the Art Gallery, Ithaca . 049742
State of the Arts, Kings Cross . . 136483
State of the Arts Gallery, Hong
Kong 102115
State School of Fine Arts,
Mandalay 056164
State Tribal Museum,
Chhindwada 024201
Statek-Muzeum Dar Pomorza,
Centralne Muzeum Morskie w
Gdańsku, Gdynia 034753
Statek-Muzeum Sołdek, Centralne
Muzeum Morskie, Gdańsk 034742
Statements Gallery, Napier 112953
Staten Island Children's Museum,
Staten Island 053503
Staten Island Historical Society
Museum, Staten Island 053504
Staten Island Institute of Arts and
Sciences, Staten Island 053505
Statens Historiska Museum,
Stockholm 041043
Statens Museum for Kunst,
København 010195
Statens Naturhistoriske Museum

(SNM), Geologisk Museum,
København 010196
Stathopoulos, Stelios, Pagkrati . 142543
Static Line, Sainte-Mère-Eglise . 073361
Station 15, Norton 051641
The Station Gallery, Whitby 007399
Station House Studio and Gallery
Society, Williams Lake 007418
Station Master's Cottage, Toronto 099835
Station Mill Antiques, Chipping
Norton 088893, 134265
Station Museum, North
Vancouver 006520
Station Peinture, Avignon 066888
Station Road Picture Framing and Art
Studio, Margaret River 127637
Station Square Gallery,
Pittsburgh 123581
Štátna Galéria, Banská Bystrica 038281
Státní Galerie Výtvarného Umění,
Cheb 009516
Státní Hrad a Zámek, Frýdlant v
Čechách 009550
Státní Hrad a Zámek, Horšovský
Týn 009573
Státní Hrad Bouzov, Bouzov . . . 009452
Státní Zámek Duchcov, Duchcov 009543
STATT-Museum, Herbstein 019201
Statthaus Böcklerpark, Berlin . . 017297
Statue Museum Balai Arca,
Nganjuk 024620
Statue of Liberty National Monument
and Ellis Island Immigration Museum,
New York 051474
Staudacher, Maria, Neukirchen beim
Heiligen Blut 077734
Staufengalerie, Oberstaufen . . . 108318
Staufergedächtnisstätte und Museum
Wäscherschloß, Wäschenbeuren 022366
Stauffenberg-Erinnerungsstätte,
Stuttgart 022060
Stauffenberg-Gedenkstätte,
Albstadt 016498
Stauffer, Münsingen 126074
Stauffer, Frédéric, Monaco 082530
Stauffer, Pierre, Jussy 087539
Staufferhaus – Sammlung Alt
Unterentfelden, Unterentfelden 042080
Staukewitz, Gisela & Herbert,
Hannover 076397
Stauner, Markus, Monzingen . . . 077417
Staunton Augusta Art Center,
Staunton 053513
Staurothek, Domschatz und
Diözesanmuseum Limburg, Limburg
an der Lahn 020080
Stauss, Ursula, Stuttgart 130688
Stauth Memorial Museum,
Montezuma 051022
Stavanger Antikvariat, Stavanger 143422
Stavanger Brannmuseum,
Stavanger 033983
Stavanger Kunstforening,
Stavanger 059675, 113275
Stavanger Museum, Stavanger . 033984
Stavanger Sjøfartsmuseum,
Stavanger 033985
Staveley Antiques, Staveley . . . 091419,
135079
Stavropolskij Kraevedčeskij Muzej
im. G.N. Prozriteleva i G.K. Prave,
Stavropol 037899
Stavropolskij Muzej Izobrazitelnych
Iskusstv, Stavropol 037900
Stawe, Jörn, Seestermühe 078444
Stawell Antiques, Stawell 062242
Stawell Historical Society Museum,
Stawell 001520
Stawski Gallery, Kraków 113642
Stead, Geoffrey, Todenham 091605
The Steading Gallery, Borgue . . 117395
Steadman, Saint Louis 123877
Steals & Deals, Atlanta 092186
Steam Engine Museum, Mabel . 050543

Steam Gallery, Beer 117330
Steam Museum of the Great Western
Railway, Swindon 045940
Steam Museum Strabally, Port
Laoise 024976
The Steam Museum, Kew Bridge
Steam Museum, Straffan 024990
Steam Railroading Institute,
Owosso 051858
Steamboat Arabia Museum, Kansas
City 049893
Steamboat Bertrand Museum,
Missouri Valley 050965
Steamboat Dock Museum,
Keyport 049960
Steammodels.uk.com, Alfreton . 088248,
134019
Steamship Collection, University of
Baltimore, Baltimore 046770
Steamship William G. Mather Museum,
Cleveland 047825
Steamtown National Historic Site,
Scranton 053219
Steamtown Railway Museum,
Carnforth 043879
Stearns Collection of Musical
Instruments, University of Michigan,
Ann Arbor 046494
Stearns History Museum, Saint
Cloud 052763
Stearns, David, Memphis 094668
Stechl, Carola, Bad Aibling 074488
Stechow, Barbara von, Frankfurt am
Main 107032
Steckdaub, Andreas, Stuttgart . . 130689
Steckenpferd, Lauf an der
Pegnitz 077031
Steckenpferd Antik, Mainz 077245
Stecker, Eva, Paderborn 078009
Stedefreund, Berlin 106440
Stedelijk Archeologisch Museum,
Oudenburg 003942
Stedelijk Museum, Alkmaar 031894
Stedelijk Museum, Oudenaarde . 003941
Stedelijk Museum Aalst, Aalst . . 003257
Stedelijk Museum Aarschot,
Aarschot 003258
Stedelijk Museum Alfons Blomme,
Roeselare 003977
Stedelijk Museum Amsterdam,
Amsterdam 031986
Stedelijk Museum Brusselpoort,
Mechelen 003858
Stedelijk Museum Bureau Amsterdam,
Amsterdam 031987
Stedelijk Museum De Lakenhal,
Leiden 032586
Stedelijk Museum Diest, Diest . 003549
Stedelijk Museum Diksmuide,
Diksmuide 003551
Stedelijk Museum Drenthe's Veste,
Coevorden 032136
Stedelijk Museum Het Domein,
Sittard 032879
Stedelijk Museum Het Toreke,
Tienen 004069
Stedelijk Museum-Hoevemuseum, Sint-
Truiden 004030
Stedelijk Museum Hof van Busleyden,
Mechelen 003859
Stedelijk Museum Hoogstraten,
Hoogstraten 003708
Stedelijk Museum Huize Ernest Claes,
Scherpenheuvel 004003
Stedelijk Museum Ieper, Ieper . . 003728
Stedelijk Museum Kampen,
Kampen 032537
Stedelijk Museum Lokeren,
Lokeren 003824
Stedelijk Museum Roermond,
Roermond 032795
Stedelijk Museum Schatkamer van
Sint-Pieter, Leuven 003786
Stedelijk Museum Schepenhuis,

Mechelen 003860
Stedelijk Museum Schiedam,
Schiedam 032856
Stedelijk Museum Van der Kelen-
Mertens, Leuven 003787
Stedelijk Museum Vianen, Vianen 032986
Stedelijk Museum voor Actuele Kunst,
Gent 003641
Stedelijk Museum voor Folklore en
Regionale Geschiedenis, Ronse 003985
Stedelijk Museum voor Pijp en Tabak,
Harelbeke 003677
Stedelijk Museum Zutphen,
Zutphen 033119
Stedelijk Museum Zwolle, Zwolle 033128
Stedelijk Onderwijsmuseum Ieper,
Ieper 003729
Stedelijk Textielmuseum, Ronse 003986
Stedelijke Academie voor Beeldende
Kunst, Ieper 055030
Stedelijke Academie voor Beeldende
Kunst, Menen 055040
Stedelijke Academie voor Beeldende
Kunst- Filiaal Wervik, Wervik . . 055058
Stedelijke Academie voor Beeldende
Kunsten, Bilzen 054998
Stedelijke Academie voor Beeldende
Kunsten, Herentals 055027
Stedelijke Academie voor Plastische
Kunsten, Genk 055019
Stedelijke Academie voor Schone
Kunsten, Deinze 055016
Stedelijke Academie voor Schone
Kunsten, Eeklo 055018
Stedelijke Academie voor Schone
Kunsten, Hasselt 055026
Stedelijke Academie voor Schone
Kunsten, Roeselare . . 003978, 055047
Stedelijke Academie voor Schone
Kunsten, Sint-Niklaas 055048
Stedelijke Akademie voor Beeldende
Vorming, Harelbeke 055024
Stedelijke Kunstacademie, Tielt, West-
Vlaanderen 055052
Stedelijke Musea, Torhout 004079
Stedelijke Oudheidkamer,
Genemuiden 032340
Stedelijke Oudheidkamer De Roos,
Geertruidenberg 032333
Stedman, Pinellas Park 127400
Stedman Art Gallery, Camden . . 047380
Steeb, Dr. Christian, Graz 062602
Steed, Dallas 120663
Steedehuys Antiek, Haarlem . . . 083354
Steedman, Robert D., Newcastle-upon-
Tyne 144741
Steeg, Baar 074481
Steele County Museum, Hope . . 049552
Steele, Rick, Porsgrunn 132794
Steele, Robert, New York 123091
Steelelife Gallery, Chicago 120276
Steeler Antiquariat, Essen 141752
Steen, Oslo 113243
Steen, Eric, Oslo 113244
Steen, I., Almelo 082636
Steenbakkerijmuseum 't Gelleg te
Rumst, Rumst 003989
Steendrukkerij Amsterdam,
Amsterdam 082865, 112291
Steenis, Oss 083529
Steenkolenmijn Daalhemergroeve,
Valkenburg, Limburg 032965
Steenmuseum, Gent 003642
Steensma & Van der Plas,
Amsterdam 082866
Steenuil, 's-Hertogenbosch 112532
Steep and Brew Gallery, Madison 050599
Steephill Studio, Maleny 099264
Steeple Antiques, Ceres 088811
Steeple Antiques, Falkirk 089250
Stefan Fadinger Museum, Sankt
Agatha 002646
Stefan-George-Museum im Stefan-
George-Haus, Bingen am Rhein 017390

Stefanelli Naderi, Seyed Ebrahim,
Firenze 080208
Stefanelli, Annamaria, Perugia . . 110644
Stefani, Liverpool 089860, 134569
Stefani, Paolo, Firenze 080209
Stefani, Roberta, Bologna 079880
Stefanini, Aurelio, Castiglione dei
Pepoli 109943
Stefanini, Aurelio, Firenze 110072
Stefano, Betty de, Bruxelles . . . 063553
Stefanov, V., Veliko Tărnovo . . . 140330
Stefanović, Dušan, Basel 116099
Stefanović, Dušan, Beograd . . . 114440
Stefanovic, Josip, Stuttgart 142402
Stefanovska, Maja, Skopje 111897
Stefans Antikt, Hörby 086749
Steff, Tuntange 082395
Steffani, Pärnu 103108
Steffarli, Bardon 061101
Steffek, Michal, Wien 063077
Steffen, Luxembourg 111867
Steffen Thomas Museum of Art,
Buckhead 047293
Steffen, J., Lage 076991
Steffensen, Kirsten, Sønderborg 066141
Steffny, Christa & Ernst,
Merzkirchen 130299
Steffny, Thomas, München 130405
Stefka, Michel, Kevelaer 107551
Steger, H., Auerbach, Kreis
Deggendorf 074448
Steglitz Museum, Berlin 017298
Stegmaier, Ludwig & Marianne,
München 108196
Stegner, Heinz, Mainz 142103
Stehli, Erwin, Küsnacht 144089
Stehr, Volker J., Bielefeld 074991
Steib, Cochem 106640
Steibo, Stadtlohn 078534
Steidl, Bernd, Dresden 075431
Steidle, Marcus, Rottenburg am
Neckar 130607
Steiermärkischer Kunstverein
Werkbund, Graz 058311
Steiermärkischer Kunstverein
Werkbund, Werkbundgalerie,
Graz 100022
Steiger, Thilo, Bad Schwartau . . 074595
Steiglitz Court House, Steiglitz . . 001521
Steigrad, Lawrence, New York . . 095760
Steilacoom Historical Museum,
Steilacoom 053517
Stein- und Bauernmuseum
Großdöllnerhof, Rechberg 002577
Stein- und Beinmuseum,
Weiterstadt 022552
Stein- und Fossiliensammlung Albert,
Sulzdorf an der Lederhecke . . . 022088
Stein und Wein, Aspach 016634
Stein, Albert von, München 108197
Stein, Alois & Michael, Inzell . . . 130076
Stein, Angela, Wuppertal 079117
Stein, Axel, Bonn 075091
Stein, Benjamin J., Amsterdam . 082869
Stein, Christian, Milano . 110450, 110451
Stein, Daniel, San Francisco . . . 097331
Stein & Co., David, Sydney 127684
Stein, Fabien, Seichamps 073478
Stein, Horst-Detlev von,
München 077626
Stein, Irmgard & Heinz,
Gelsenkirchen 107113
Stein, Maggy, Bettembourg 111817
Stein, Margitta, Siegen 078466
Stein, Peter, Gillenfeld 107132
Stein, Robert, Saint Louis 123878
Stein, Vernice, Buffalo 119965
Steinacker, U., Koblenz 107579
Steinbach, Krefeld 076964
Steinbach, G., Lindlar 077125
Steinbach, Michael, München . . 142189
Steinbach, Wendelin, Krefeld . . . 107744
Steinbauer, Franz, Opfenbach . . 077950,
130495

Steinbaum, Bernice, Miami ... 121863
Steinbeck, Romy, Richterswil .. 087747
Steinbeißer, W., Ingolstadt 141958
Steinberg Gård, Hauge i Dalane 084247
Steinberg, Barry, Los Angeles .. 121598
Steinberg, Peter, Linz 140015
Steinbrink, H., Osnabrück 077979
Steindl, Gerhard, Getzersdorf bei
Traismauer 127737
Steinegger, Stephanie, München 130406
Steinek, Heinrich, Wien 100375
Steinek, Heinrich & Silvia, Wien 100376
Steinemann & Co., Luzern 116482
Steiner, Rolle 087755
Steiner, San Francisco 124404
Steiner, Sydney 062299
Steiner, Alain, Homburg 076571
Steiner, Emil, Stadel, Winterthur 087827
Steiner, Franz, Stuttgart 137529
Steiner, Gerhard, Graz 062603
Steiner, Max, Zürich .. 088010, 134002
Steiner, Nelly, Bern 087305
Steiner, Nordfrid, Basel 087258, 133657
Steiner, René, Bussigny-près-
Lausanne 087344
Steiner, René, Erlach .. 116256, 138041
Steiner, Silvia, Biel 116159
Steiner, Victor, Graz 127754
Steiner, Walo, Densbüren 116243
Steinfeld, Gent 063661
Steinfeld, A., Sylt 108753
Steinhauermuseum,
Randersacker 021274
Steinhauermuseum Mühlbach,
Eppingen 018223
Steinhausen-Museum, Frankfurt am
Main 018450
Steinhauser, Freiburg im
Breisgau 075885
steinhof.restaurierung, Werkstatt für
Steinsanierung und Denkmalpflege,
Berlin 129567
Steinhoff, Almere 082647
Steinhuder Spielzeugmuseum,
Wunstorf 022813
Steininger, Ric J., Cairns 098846
Steinitz-Baruch, Bernard, Saint-
Ouen 073202
Steinitz, Bernard, Paris 071921, 105249
Steinkjelleren Antikvitetshandel,
Bergen 084226
Steinkjer Auksjonsforretning,
Steinkjer 126390
Steinkohlen-Besucherbergwerk
Rabensteiner Stollen, Ilfeld ... 019382
Steinkopf, J.F., Stuttgart 142403
Steinle Contemporary, München 108198
Steinle, S., Neustadt am
Rübenberge 077756
Steinmayer, Stefanie, Lindau ... 107875
Steinmeier, E., Bünde 106614
Steinmetz, Jon, Ahrensburg ... 074363
Steinmetz, Jon, Hamburg 076319
Steinmetzmuseum,
Burgschleinitz 001807
Steinmeyer, Ulli, Ottobrunn ... 077998
Steinmuseum, Solothurn 042008
Steinreich, Berlin 106441
Steinreich, Esslingen 106937
Steinrötter, Münster 108227
Steinsburg-Museum, Römhild .. 021457
Steinscher Hof Kirberg, Freiherr-vom-
Stein-Gedenkstätte, Hünfelden 019360
Steinsiek, Tulsa 097759, 136437
Steinsland Klyngetun, Trondarnes
Distriktsmuseum, Harstad ... 033639
Steinsträsser, Manfred, Köln ... 076909
Steintor-Verlag, Lübeck 137361
Steintor-Verlag, Möhnesee 137382
Steinzeitdorf Kussow, Moor-
Rolofshagen 020460
Steinzeitmuseum, Thunstetten . 042062
Steinzeitmuseum, Sammlung Reinhard,
Korb 019766

Steirisches Feuerwehrmuseum,
Groß Sankt Florian Kunst und
Kulturzentrum, Gross Sankt
Florian 001999
Steirisches Glaskunstzentrum
und Glasmuseum Bärnbach,
Bärnbach 001775
Steirisches Holzmuseum, Sankt
Ruprecht 002700
Stelio, Davia, Trieste 081806
Stelio, J., Fortaleza 100960
Stella, Breda 112357
Stella A., Berlin 106442, 137096, 141576
Stella Books, Chepstow 144330
Stella, Giorgia, Verona 132300
Stella, L., Paris 105250
Stella, Pierangelo, Padova 142776
Stellan Holm, New York 123092
Stellenbosch Museum,
Stellenbosch 038838
Steller, Jacksonville 121271
Stelling Projecten, Leiden 112561,
132561
Stelten, Worpswede 079069
Steltman, Amsterdam . 082868, 112292,
112293, 137842
Dr. Stelzer & Zaglmaier, Halle,
Saale 107211
Stelzer, Hubert, Bielefeld 074992
Stelzhamer-Gedenkstätte,
Oberösterreichische Landesmuseen,
Pramet 002536
Stema, Brescia 131050
Stemeseder, Salzburg 062821
Stemmer, Gilsum 138334
Stemmer, Michaela, Ismaning .. 130079
Stemp, Alfred, Eichenau 129780
Stenberg, Bjarne, Tønsberg ... 084376
Stendahl, Los Angeles 121599
Stenderhoff, Theresia, Münster . 077679,
142213
Stendhal, Maya, New York 123093
De Stenen Uil, Eindhoven 112468
Stenenexpositie de Molen – Museum
Batjuchin, Borculo 032084
Stenersenmuseet, Oslo 033864
Stengalleriet, Horsens 102852
Stengel-True Museum, Marion . 050695
Stenhusets Antik, Stockamöllan 086954
Steninge Slott Kulturcenter,
Märsta 040818
Stenlander, Stockholm 115927
Stenneset Bygdetun, Rana Museum,
Mo i Rana 033773
Steno Museet, Århus 010009
Stens Antik, Mellerud 086868
Stense, Stockholm 087061
Stenstrøm, Frederiksberg 065678,
128384
Stentebjerg, Odense 066045
Stentoft, Århus 065609
Stenton Gallery, Stenton 119142
Step Back Antiques, Hawthorn . 061605
Step Back in Time, Columbus . 093049
Stepan Shahumyan House-Museum,
Stepanavan 000714
Stepan Zoryan House-Museum,
Vanadzor 000716
Stepenik – Kcb, Beograd 114441
Stephan, Grödig 062613
Stéphan, Nantes 070479
Stephan, Christian, Beaune ... 067045
Stephan, Frank, Jena 076646
Stephan, Henri, Juvigny 068929
Stéphan, Jean-Marc, Beaune .. 067046
Stephan, Josef, Heubach 076523
Stéphan, Monique, Saint-Pol-de-
Léon 073267
Stéphan, Robert, Paris 071922
Stéphan, Roger, Varennes-
Vauzelles 073981
Stephan, Sébastien, Beaune ... 067047
Stéphan, Yannick, Coatréven .. 068054
Stéphan, Yannick, Perros-Guirec 072107

Stephanie Ann Roger Gallery,
Frostburg 048990
Stephanie's Antique, New York . 095761
Stephansson House, Edmonton . 005762
Stephen, Saint Paul 123933
Stephen A. Douglas Tomb,
Chicago 047699
Stephen C. Foster State Park,
Fargo 048695
Stephen Cohen Gallery, Los
Angeles 121600
Stephen Foster Folk Culture Center,
White Springs 054373
Stephen G. Beaumont Museum,
Wakefield 046033
The Stephen Girard Collection, Girard
College, Philadelphia 052096
Stephen Lawrence Gallery, University
of Greenwich, London 045111
Stephen Leacock Museum,
Orillia 006557
Stephen Phillips Memorial House,
Salem 052884
Stephen, Jean, Minneapolis ... 122038
Stephens African-American Museum,
Shreveport 053326
Stephens Collection, London .. 045112
Stephens County Historical Society
Museum, Duncan 048392
Stephen's Gallery, Houston ... 121135
Stephens Museum, Fayette 048714
Stephens, Chris, Charlotte 092628
Stephens, Francis, Corringham . 088978
Stephens, John, Auckland 083827
Stephenson County Historical Society
Museum, Freeport 048971
Stephenson Railway Museum, North
Shields 045388
Stephenson, Glenna, Kansas City 094095
Stephenson, Joseph, Cincinnati . 092899
Stephenson, Robert, London ... 090511,
134774
Stephensons, York 127106
Stephenville Museum,
Stephenville 053518
Stepono Dariaus Gimtinė, Lietuvos
Aviacijos Muziejus, Darsaus .. 030329
Steppan, Robert, Augsburg 074470
Steppé, Bruxelles 063554
Steppes Hill Farm Antiques,
Stockbury 091436
Stepping Stones Museum for Children,
Norwalk 051645
Steppingstone Museum, Havre de
Grace 049411
Steps, Delhi 109343
Steps Gallery, Carlton, Victoria . 098872
Sterbehaus Ferdinand Raimunds,
Gedenk- und Informationsstätte,
Pottenstein 002523
Sterenberg, Marjan, Groningen . 083304
Sterenfeld, Xavier, Perpignan .. 072096,
105341
Sterkau, Ralf, Bochum 075045
Sterkstroom Museum,
Sterkstroom 038843
Sterkwerk Galerie, Tilburg 112685
Sterl, Thomas Georg, Homburg . 076572,
130067
Sterling, Charlotte 092629
Sterling, Johnson City 127272
Sterling and Francine Clark Art
Institute, Williamstown 054419
Sterling Books, Weston-super-
Mare 144914
Sterling Coins and Medals,
Bournemouth 088542
Sterling Hill Mining Museum,
Ogdensburg 051718
Sterling Historical Society and
Little Red Schoolhouse Museum,
Sterling 053521
Sterling Hundley, Richmond .. 123747
Sterling Publishing Company, New

York 138426
Sterling-Rock Falls Historical Society
Museum, Sterling 053520
Sterling Studios, London 118580, 134775
Sterling, Eden, Cincinnati 092900
Stern, New York 095762
Stern, Tel Aviv 109732, 130899
Stern Pissaro, London . 090512, 118581,
134776
Stern-Verlag, Düsseldorf 141725
Stern, Anita, New York 095763
Stern, Barry, Paddington, New South
Wales 099537
Stern, Harry L., Chicago 145061
Stern & Co., Harvey M.,
Philadelphia 136105
Stern, Louis, West Hollywood .. 125028
Sternat, Dr. Reinhold, Wien ... 063078
Sternberg Museum of Natural History,
Hays 049416
Sternberg, Heinz, Hamburg ... 076320
Sternberg, Maurice, Chicago ... 120277
Šternberský Palác, Národní Galerie v
Praze, Praha 009841
Sterne-Hoya Museum,
Nacogdoches 051148
Sternenburg, Bonn 075092
Sternitzky, Hugues, Wasserbourg 074293
Sternmann, M., Crailsheim 075277
Šternová, Viera, Trenčín 114668
Sterr, F.X., Starnberg 130654
Sterrenburg, Drs. Marius,
Amsterdam 082869
Sterry, John Henry, Louisville .. 094590
Sterz, P. & I., Wolfsburg 108960
Sterzenbach, Mathias, Longeville-lès-
Metz 069613
Sterzik, Angelika & Heiko, Bernkastel-
Kues 106475
Stetten, Isabelle Anne, Paris ... 129148
Stettler Town and Country Museum,
Stettler 007074
Steuben House Museum, River
Edge 052584
Steuben Memorial State Historic Site,
Remsen 052507
Steuer, Helmut, Worms 079065
Stevenage Museum, Stevenage . 045864
Stevens, Houston 093801, 135619
Stevens & Son, Lubenham 090642,
134819
Steven's Art, Buffalo 119966
Stevens-Coolidge Place, North
Andover 051592
Stevens County Gas and Historical
Museum, Hugoton 049607
Stevens County Historical Society
Museum, Morris 051060
Stevens Museum, Salem 052878
Stevens, Elliot, New York 123094
Stevens, John, Nashville 135851
Stevens, Larry, Baltimore 119774
Stevenson, Hamilton 064331
Stevenson Academy of Fine Arts,
Oyster Bay 057640
Stevenson Galleries, Tunbridge
Wells 119274
Stevenson-Hamilton Memorial
Information Centre, Skukuza .. 038830
Stevenson, Mandrake, Ibstock . 089631,
134496
Stevenson, R.B., San Diego ... 124176
Steves Homestead Museum, San
Antonio 052951
Steveston Museum, Richmond . 006798
Stevington Windmill, Stevington . 045809
Stewart, Bexhill-on-Sea 117357, 134102
Stewart, Eastbourne .. 089153, 134355
Stewart, Glenorie 061534
Stewart, Jacksonville 135666
Stewart Antiques, Los Angeles . 094479
Stewart Collection, Pocklington . 045515
Stewart Hay Memorial Museum,
University of Manitoba, Department

of Zoology, Winnipeg 007468
Stewart Historical Museum,
Stewart 007075
Stewart Indian Cultural Center, Carson
City 047446
Stewart M. Lord Memorial Museum,
Burlington 047320
Stewart, Charles, Atlanta 092187
Stewart, Cynthia, Tulsa 124815
Stewart, Dennis E., San Antonio 124073
Stewart, Inessa, Dallas 093219
Stewart, Laurie, London 090513
Stewart, Tabori & Chang, New
York 138427
Stewarton and District Museum,
Stewarton 045866
The Stewartry Museum,
Kirkcudbright 044730
Stewens, Palermo 110621
Stewner, Jochen, Lübeck 107920
Stey, Villemomble 074155
Steyaert, C., Bruxelles 063555
Steyaert, Claire, Minneapolis .. 094948
Steyning Museum, Steyning ... 045867
Steyrtal-Museumsbahn, Steyr . 002787
Sticht, Uwe, Gevelsberg 141829
Stichting Atlas Van Stolk,
Rotterdam 032825
Stichting De Appel, Center for
contemporary art, Amsterdam . 031988
Stichting De Brakke Grond,
Amsterdam 031989
Stichting George Grard,
Gijverinkhove 003653
Stichting Huis Bergh, 's-
Heerenberg 032428
Stichting Joden Savanna,
Paramaribo 040505
Stichting Kunst in Openbare Ruimte,
SKOR, Amsterdam 059622
Stichting Niggendijker, Groningen 112504
Stichting Noorderlicht, Groningen 112505,
137857
Stichting Oosterkerk, Amsterdam 031990
Stichting Outline, Amsterdam .. 112294
Stichting Santjes en Kantjes,
Maastricht 032631
Stichting tot Beheer Museum
Boijmans Van Beuningen,
Rotterdam 032826
Stichting Veranneman,
Kruishoutem 003761
Stichting Zeeuwse Schaapkudde,
Heinkenszand 032448
Stickan, Marion, Köln 076910
Stickereimuseum Eibenstock,
Eibenstock 018112
Stickereimuseum Lustenau,
Lustenau 002348
Stickmaskinsmuseum,
Glemminebro 040642
Sticks & Stones, Toronto 064757
Sticks and Stones, Christchurch 083899
Stida Eiendom, Oslo 084337
Stiebel, Gérald, New York 095764
Stiebeling, Klaus, Wuppertal ... 079118
Stiebner Verlag GmbH, München 137425
Stiefelhagen, Klaus, Lindlar ... 077126
Stiefelhagen, Kunibert, Siegburg 078461
Stiefelmachermuseum, Rechnitz 002578
Stiegemann-Schöpper, I., Rheine 078228
Stiegler & Co., New York 095765
Stiegler, Christa & Helmut,
Ansbach 105997
Stieglitz, Martha, Stuttgart 108730
Stiehl, Ralf, Twistetal 078739
Stieler, Breitenbrunn 075141
Stien, Daniel, San Francisco ... 097332
Stieren, San Antonio 124074
Stierhübelteichhaus, Trift- und
Holzfällermuseum, Karlstift ... 002160
Stierle, Heinz, Salzburg 140030
Sties, Paul, Kronberg im Taunus 107760,
107761

Stievano, Paolo, Padova 131677
Stiffkey, Norwich 090884
Stiffkey Antiques, Stiffkey 091424
Stiffkey Lamp Shop, Stiffkey .. 091425
Stift Fischbeck, Hessisch
Oldendorf 019236
Stift Obernkirchen, Obernkirchen 020911
't Stift, Schatkamer en
Bezoekerscentrum, Susteren .. 032912
Stiftelsen Militärfordons Museum,
Museet Malmahed, Malmköping 040823
Stiftelsen Musikkulturens Främjande,
Stockholm 041044
Stiftlandmuseum, Waldsassen .. 022401
Stifts- und Heimatmuseum
Schloß Seeburg, Seekirchen am
Wallersee 002750
Stiftsbibliothek, Einsiedeln ... 041459
Stiftsgalerie Seitenstetten,
Seitenstetten 002752
Stiftsgården, Trondheim 034037
Stiftskirchenmuseum,
Himmelkron 019258
Stiftsmuseum, Aschaffenburg .. 016628
Stiftsmuseum, Bad Buchau 016742
Stiftsmuseum, Garsten 001928
Stiftsmuseum, Mattsee 002373
Stiftsmuseum, Millstatt 002389
Stiftsmuseum, Zwettl 003146
Stiftsmuseum Praemonstratenser-
Chorherren Stift Schlägl, Aigen im
Mühlkreis 001692
Stiftsmuseum Schatzhaus Kärntens,
Sankt Paul 002690
Stiftsmuseum Steyr-Gleink, Steyr-
Gleink 002789
Stiftsmuseum, Benediktinerabtei Sankt
Georgenberg-Fiecht, Vomp ... 002874
Stiftsmuseum, Stiftsarchiv /
Stiftsbibliothek, Xanten 022839
Stiftssammlungen Lambach,
Lambach 002259
Stiftssammlungen Wilten,
Innsbruck 002139
Stiftung B.C. Koekkoek-Haus,
Kleve 019645
Stiftung Bauhaus Dessau,
Dessau 017871
Stiftung BINZ39, Zürich 116860
Stiftung Bohlenständerhaus Schrofen,
Amriswil 041220
Stiftung Brandenburgische
Gedenkstätten, Oranienburg .. 021003
Stiftung Bundeskanzler-Adenauer-
Haus, Bad Honnef 016791
Stiftung Deutsches Historisches
Museum, Zeughaus und
Ausstellungshalle von I. M. Pei,
Berlin 017299
Stiftung DKM, Duisburg 018063
Stiftung Domäne Dahlem, Landgut
und Museum, Berlin 017300
Stiftung Dr. Edmund Müller,
Beromünster 041333
Stiftung Federkiel für zeitgenössische
Kunst und Kultur, Leipzig ... 107853
Stiftung für Eisenplastik, Sammlung
Dr. Hans Koenig, Zollikon ... 042187
Stiftung für Konkrete Kunst,
Reutlingen 021388
Stiftung für Konkrete Kunst Roland
Phleps, Freiburg im Breisgau . 018488
Stiftung Fürst-Pückler-Museum, Park
und Schloß Branitz, Cottbus . 017770
Stiftung Gerhart-Hauptmann-Haus,
Deutsch-osteuropäisches Forum,
Düsseldorf 018045
Stiftung Historische Museen Hamburg,
Museum für Bergedorf und die
Vierlande, Hamburg 019021
Stiftung Hubert Aratym,
Gutenstein 002030
Stiftung Jüdisches Kulturmuseum
Augsburg-Schwaben, Augsburg 016671

Stiftung Käthe Kollwitz Haus,
Moritzburg 020469
Stiftung Keramion, Zentrum für
moderne und historische Keramik,
Frechen 018465
Stiftung Kloster Michaelstein-Museum,
Blankenburg 017409
Stiftung Kohl'sche Einhorn-Apotheke,
Weißenburg in Bayern 022541
Stiftung Mecklenburg, Ratzeburg 021295
Stiftung Moritzburg, Kunstmuseum
des Landes Sachsen-Anhalt, Halle,
Saale 018969
Stiftung museum kunst palast
mit Sammlung Kunstakademie
und Glasmuseum Hentrich,
Düsseldorf 018046
Stiftung Museum Schloss Moyland,
Sammlung van der Grinten, Joseph
Beuys Archiv des Landes Nordrhein-
Westfalen, Bedburg-Hau 017011
Stiftung Naturschutzgeschichte,
Königswinter 019742
Stiftung Neue Synagoge Berlin –
Centrum Judaicum, Berlin ... 017301
Stiftung Opelvillen, Rüsselsheim 021541
Stiftung Preußische Schlösser
und Gärten Berlin-Brandenburg,
Potsdam 059238
Stiftung Römermuseum Homburg-
Schwarzenacker, Homburg ... 019335
Stiftung S BC – pro arte, Biberach an
der Riß 017357
Stiftung Sammlung E.G. Bührle,
Zürich 042233
Stiftung Scheibler-Museum Rotes
Haus, Monschau 020458
Stiftung Schleswig-Holsteinische
Landesmuseen Schloß Gottorf,
Schleswig 021667
Stiftung Schloss Ahrensburg,
Ahrensburg 016479
Stiftung Schloss Eutin, Eutin ... 018331
Stiftung Schloss Neuhardenberg,
Neuhardenberg 020705
Stiftung Schloss und Park Benrath,
Museum für Europäische
Gartenkunst, Museum Corps de
Logis (Hauptschloss), Museum
für Naturkunde, Orangerie, Park,
Düsseldorf 018047
Stiftung Simonshof, Gleisweiler . 018714
Stiftung Stadtmuseum Berlin,
Landesmuseum für Kultur und
Geschichte Berlins, Generaldirektion
und Verwaltung, Berlin 017302
Stiftung Starke, Berlin 017303
Stiftung Stift Neuzelle, Neuzelle 020779
Stiftung Wilhelm Lehmann,
Niederhelfenschwil 041801
Stiftung Wilhelm Lehmbruck Museum,
Zentrum Internationaler Skulptur,
Duisburg 018064
Stiftung Zollverein, Essen 018313
De Stijl, Dordrecht 083195
Stiklestad Nasjonal Kultursenter,
Verdal 034082
Stiko Karoliukai, Vilnius 111810
Stil, Beograd 085349
Stil Leben, Wermelskirchen ... 108877
Stilbruch, Oldenburg 077942
Still, Chicago 120278
Still Life, Edinburgh 089191
Still National Osteopathic Museum,
Kirksville 049994
Stilla Nova, Maastricht 083481
Stille, Maastricht 143201
Stille-Nacht- und Heimatmuseum
Bruckmannhaus, Oberndorf bei
Salzburg 002462
Stille Nacht Galerie, Oberndorf bei
Salzburg 100133
Stille Nacht-Museum, Hallein .. 002051
Stille, D., Lienen 077105

Stiller, Hans-Dirk, Düsseldorf ... 141726
Stillman, Georgie, San Diego ... 097135
Stills, Nijmegen 112613
Still's Corner, Timaru 113031
Stills Gallery, Edinburgh 044276
Stills Gallery, Paddington, New South
Wales 099538
Stillwater, Seattle 124604
STIM, Nesle-Hodeng 070508
Stima, Breda 112358
Stima, Bruxelles 100659
Stingray, Columbus 120515
Stingy Lu Lu's, Bridgetown ... 100461
Stinson, New Orleans 122242
Stinzenmuseum, Tsjerke en Uniastate
Beers, Beers 032052
Stirk Cottage, Kalamunda 001175
Stirling Antiques, Abinger
Hammer 088233
Stirling Old Town Jail, Stirling .. 045873
Stirling Smith Art Gallery and Museum,
Stirling 045874
Stirnemann, M.V., Zürich 144197
Stirt, James, Vevey 126758
Stith, Norfolk 136036
Stivanin-Bouquet, Céline, Castanet-
Tolosan 128530
Stix, W., Baden-Baden 074643
Stjepan Gruber Museum,
Županja 009058
Stjernsunds Slott, Askersund ... 040550
Stjørdal Museum KF, Stjørdal .. 033990
Stoani Haus der Musik, Gasen . 001930
Stobbe, Peter, Velbert 142449
Štobi, Skopje 111898
Stock, Toronto 128219
Stock Exchange Consignment Shop,
Saint Louis 096781
Le Stock Room, Montréal 064476
Stock Solution, Salt Lake City .. 123989
Stock, Colin, Rainford 091064
Stock, Michaela, Wien 100377, 100378
Stock, Ray, Denver 120778
Stockbridge Library Historical
Collection, Stockbridge 053538
Stocker-Schmid, Dietikon 138038
Stocker, Leopold, Graz 136511
Stocker, Mechthild, Radolfzell am
Bodensee 130550
Stockermann, W., Düsseldorf .. 075522
Stockholm Antik och Konstmässa,
Stockholm 098322
Stockholm Art Fair, Stockholm . 098323
Stockholm Historical Society Museum,
Stockholm 053539
Stockholms Auktionsverk,
Göteborg 126588
Stockholms Auktionsverk, Malmö 126640
Stockholms Auktionsverk,
Stockholm 126676
Stockholms Internationella
Antikmässan, Stockholm 098324
Stockholms Läns Museum,
Stockholm 041045
Stockholms Målerikonservering,
Stockholm 133581
Stockholms Medeltidsmuseum,
Stockholm 041046
Stockholms Serie och Skivhandel,
Stockholm 143968, 143969
Stockholms Skolmuseum, Stockholms
Stadsmuseum, Stockholm ... 041047
Stockholms Stadsmuseum,
Stockholm 041048
Stockholms Universitet Schefflerska
huset, Stockholm 041049
Stocking, Sarah, San Francisco . 124605
Stockman, Bakers Swamp 098688
Stockmans, Jos, Haarlem 083355
Stockport Art Gallery, Stockport 045876
Stockport Museum, Stockport .. 045877
Stocks and Chairs Antiques,
Poole 091035, 134948
Stockselius, Skövde 115814

Stockspring, London 090514
Stockwell, David, New York 095766
Stockwood Park Museum, Luton 045167
Stockyard Galleries, Foster 099056
Stockyard Gallery, Mataranka .. 099288
Stockyards Station Gallery, Fort
   Worth 120888
Stoddard, James A., Washington 136468
Stoddard, Richard, New York ... 145372
Stodel, Inez, Amsterdam 082870
Stodel, Jacob, London 090515
Stodel, Salomon, Amsterdam .. 082871
Stodola, Václav, Praha 065530
Stöberdiele, Büsum 075221
Stöberkiste, Grambin 076078
Stöberlein, Günter, München ... 137426
Stöckli, Stans 133933
Stöger, Martin & Monika, Piding 130520
Stöhr, Wien 140125
Stöhr, Dr. H. & W., Wuppertal .. 108996
Stöhr, Edith, Wien 063079
Stöhr, Helmut, Bayreuth 129449
Stör, Alexander, Frankfurt am
   Main 075827
Störtebeker Park, Wilhelmshaven 022663
Størupgård Antik, Hjørring 065768
Stoffels, Joh., Trier 078706
Stoffer, Nicole, Gelsenkirchen .. 107114
Stofnun Arna Magnússonar,
   Reykjavík 024113
Stoilov, Valletta 111924
Stokes Books, Dublin 142604
Stokes Clocks & Watches, Cork 130837
Stokes, Val, Waikato 113043
Stokes, William H., Cirencester . 088920
Stokke Bygdetun, Stokke 033991
Stol og Stage, Hellerup 065735
Stolarski, M., Lesznowola 084605
Stolen Buick Studio, Chicago .. 120279
Stolica, Moskva 143652, 143653
Stolk, F., Amsterdam 082872
Stolker, Maud, Breda 083008
Stolla-Holten, Gisela, Neustadt an der
   Weinstraße 077763
Stolle, Wolf-Rüdiger, Berlin 074931,
                                  129568
Stollman, Steve, New York 095767
Stollnberger, Johann, Salzburg . 062822
Stolpe, Harry, Norrköping 086882
Stoltenburg, Limburg an der
   Lahn 077115
Stoltz, Virginia C., Columbus ... 093050
Stoltze-Museum der Frankfurter
   Sparkasse, Frankfurt am Main 018451
Stolz, Gerhard, Starnberg 078542,
                                  108664
Stondon Museum, Lower
   Stondon 045160
Stone Age Antiques, Miami 094804
Stone and Press Galleries, New
   Orleans 122243
Stone Art, Bakewell 117268
Stone Arts, South Auckland ... 084084
The Stone Barn Glass Gallery,
   Newcastle-upon-Tyne 118791
Stone by Stone Gallery, Dallas . 120664
Stone Cottage Antiques, Fort
   Worth 093481
Stone Cottage Antiques, Toronto 064758
Stone Fort Museum, Stephen F. Austin
   State University, Nacogdoches 051149
The Stone Gallery, Burford 088712
Stone Gallery, Dublin 109621
Stone House Gallery, Fredonia . 048962
Stone House Museum,
   Belchertown 046874
Stone Inscription Art Museum,
   Beijing 007646
Stone Mountain Art, Atlanta ... 119617
Stone Museum, Monroe
   Township 051003
Stone Soup, Museum of Children's Art,
   Santa Cruz 139496
Stone-Tolan House, The Landmark

Society of Western New York,
   Rochester 052621
Stone Trough Books, York 144944
Stone, Allan, New York 123095
Stone, Frank, Minneapolis 122039
Stone, Jeremy, San Francisco .. 124406
Stone, June & Tony, Peacehaven 090949
Stone, M. Lee, San Jose 124462
Stonecrop Fine Art Galleries, Violet
   Town 099854
Stonefield Historic Site, Cassville 047459
Stonefront, Fredericksburg 127229
Stonehouse Gallery, Warrandyte 099862
Stonehouse Museum, Vale 053928
Stoner, Ben, Chippendale 061319,
                                  127587
Stones River National Battlefield,
   Murfreesboro 051126
Stone's Tavern Museum, Ligonier 050276
Stones, Keith, Bessacarr 088447
Stones'n Bones Museum, Sarnia 006967
Stonewall Jackson House,
   Lexington 050269
Stonewall Jackson Museum at Hupp's
   Hill, Strasburg 053568
Stoneware Gallery, Winnipeg ... 101895
Stoneway Gallery, Bridgnorth .. 117426
Stoneycroft Farm, Betchworth .. 088448
Stonington, Seattle 124605
Stonnington Local History Collection –
   Malvern, Malvern, Victoria ... 001245
Stonnington Local History Collection –
   Prahran, Prahran 001433
Stonor Park, Henley-on-Thames 044567
Stony Plain and District Pioneer
   Museum, Stony Plain 007080
Stony Point Battlefield State Historic
   Site, Stony Point 053551
Stoob, Urban, Sankt Gallen 138087
Dinah Stoodley & Celia Jennings,
   Brasted 088586
De Stoof, Deventer 112430
Stoomcentrum, Maldegem 003838
Stoomgemaal Halfweg, Halfweg 032405
Stoomgemaal Hertog Reijnout en
   Bezoekerscentrum Arkemheen,
   Nijkerk 032683
Stoomgemaal Mastenbroek,
   Genemuiden 032341
Stoomhoutzagerij, Groenlo 032373
Stoomtrein Goes-Borsele, Goes . 032356
Stoottroepen Museum, Assen ... 032037
Stop Monnaie, Pointe-à-Pitre .. 079219
Stopalo, Jean & Carsten,
   Stockholm 087062
Stopin, Camille, Lille 128687
Stoppenbach & Delestre, London 090516,
                                  118582
Stor-Elvdal Museum, Koppang . 033710
Stora Nyckelvikens Gård, Nacka 040868
Stora Nygatans Antik, Stockholm 087063
Storbecher, Daniel, Millau 070124
Storch, Neuss 142227
Storch, Maria, Zirl 127987
Storchenturm-Museum, Zell am
   Harmersbach 022856
Storck, Wolframs-Eschenbach .. 079055
Stordel, Peter, Hamburg 076321, 129992
De Store, Zaltbommel 083704
The Store Antiques, Toronto ... 064759
Store II, Portland 096481
Store Mæla Gård Auktionsforretning,
   Skien 126387
Store Mæla Gård Merdin, Skien 126388
Store Verner's Mini Antikvariat,
   København 140616
Storefront for Art and Architecture,
   New York 123096
Storehouse 22, Buffalo 119967
The Stores, Great Waltham 089380
Storey, London 090517, 144697
Storgatans Antik, Ljungby 086820
Storia della Miniatura. Studi e
   Documenti, Firenze 138875

Stories, Richmond 145479
Storkower Bogen, Berlin 106443
Storkower Versandantiquariat,
   Storkow 142379
Storm, Cambridge 117511
Storm, Great Shelford . 089379, 117882
Storm King Art Center,
   Mountainville 051109
Storm P.-Museet, Frederiksberg 010066
Stormarnsches Dorfmuseum, Altes
   Bauernhaus am Thie, Hoisdorf 019317
Stormbroek, Eindhoven 083231
Stormcloud Trading, Saint Paul . 123934
Storme, Frédéric, Lille 104017
Storme, Jacqueline, Lille 104018
Stormont, Dundas and Glengarry
   Highlanders Regimental Museum,
   Cornwall 005650
Storms, Walter, München 108199,
                                  108200
De Stormvogel, Bunde 141664
Storno-ház, Soproni Múzeum,
   Sopron 023903
Storoževoj Korabl 'Krasnyj Vympel'
   1911 g., Voenno-istoričeskij
   Muzej Tichookeanskogo Flota,
   Vladivostok 038044
Storrer, Erich, Zürich 116861
Storrowton Village Museum, West
   Springfield 054327
Storstrøms Kunstmuseum,
   Maribo 010229
Story of Castle Cornet, Saint Peter
   Port 045715
Story of People, Western Development
   Museum, Yorkton 007501
Stoßberg, R., Wermelskirchen . 078932
Stossel, Willy, Thisis 129277
Stothert Old Books, Chester ... 144333
Stott Park Bobbin Mill, Newby
   Bridge 045342
Stoughton and District Museum,
   Stoughton 007082
Stoughton Historical Society Museum,
   Stoughton 053557
Stoup, John, San Diego 124177
Stour Gallery, Shipston-on-Stour 119094
Stourhead House, Stourton 045892
Stout, William, San Francisco .. 145567
Stouvenel, Marie-Rose, Niort ... 070716
Stoves, Mantels & More, Oklahoma
   City 096015
Stow Antiques, Stow-on-the-
   Wold 091483
Stow Gallery, Penn 118887
Stow House, Museum of Goleta Valley
   History, Goleta 049154
Stow West 1825 School Museum,
   Stow 053559
Stowarzyszenia Artystycznego
   Grupa Krakowska Krzysztofory,
   Kraków 059706
Stowarzyszenie Artystów fabs,
   Warszawa 059707
Stowarzyszenie Artystyczno –
   Edukacyjne "Magazyn", Poznań 059708
Stowarzyszenie KARPATY.ART,
   Rzeszów 059709
Stowarzyszenie Klub Krytyki
   Artystycznej Pokaz, Warszawa . 059710
Stowarzyszenie Miłośników Dawnej
   Broni i Barwy, Kraków 059711
Stowarzyszenie Muzeów na Wolnym
   Powietrzu, Toruń 059712
Stowarzyszenie Niepublicznych Szkół
   Artystycznych Pro Arte, Lublin . 059713
Stowarzyszenie Plastyków i
   Fotografików Dedal, Poznań .. 059714
Stowarzyszenie Polskich Artystów
   Karykatury, Warszawa 059715
Stowarzyszenie Polskich Artystów
   Kowali, Gdańsk 059716
Stowarzyszenie Promocji Artystów Lux
   Artis, Dąbrowa Górnicza 059717

Stowarzyszenie Twórców Ludowych,
   Lublin 059718
Stowarzyszenie Związek Muzeów
   Polskich, Kraków 059719
Stowitts Museum, Pacific Grove 051881
The Stoy Museum of the Lebanon
   County Historical Society,
   Lebanon 050208
Straarup, Frederiksborg 065679, 128385
Straat Antiquaren, Amsterdam . 143069
Strachan, Glasgow 089335
Strachey, Julie, Connor Downs . 088969
Strachur Smiddy Museum,
   Strachur 045897
Stracke, Dr. Gottfried, Köln ... 107687,
                                  130171
Stradbroke Gallery, Stradbroke . 119166
Straffan Antiques, Straffan 079703
Strafvollzugsmuseum,
   Ludwigsburg 020131
Stragapede-Didra, Konstanz ... 107724
Strahan Woodworks, Strahan . 099722
Strahlendorf, J., Bremen 141657
Strahovská Obrazárna, Praha .. 009842
Straits Antiques, Singapore 085508
Straits Literature and Art Publishing
   House, Fuzhou 136709
Straits, Harvest, Singapore 085507
Strakeljahn, M., Hiddenhausen . 076524
Stramowska, Teresa, Piaseczno 084634,
                                  132856
Stranahan House, Fort
   Lauderdale 048830
Strand Antiques, London 090518
Strand Antiques, Port Elliot 062093
The Strand Antiques, Taranaki . 084093
Strand Art Conservation, Lane
   Cove 127628
Strand Book Store, New York .. 145373,
                                  145374
Strand Coins, Sydney 062300
Strand Gallery, Aldeburgh 117221
The Strand Gallery, Bideford ... 117361
Strand Gallery, Southend-on-Sea 119123
Strand Objetos de Arte, Rio de
   Janeiro 101070
Stranda Hembygdsförening,
   Mönsterås 040850
Strandbergs, Stockholm 087064, 126677
Strandgaarden Museum, Ringkøbing
   Museum, Ulfborg 010345
Strandgalleriet, Södertälje 115818
Strandgut, Hamburg 076322
Strandhalle Ahrenshoop, Ahrenshoop,
   Ostseebad 016482
Strandingsmuseum St. George,
   Ringkøbing Museum, Ulfborg . 010346
Strang Print Room, University College
   London, London 045113
Strang, Hanspeter, Tobel 087850, 133938
Strang, Jan, Helsinki 140720
Strange Frames On Whyte,
   Edmonton 101301
Le Strange Old Barns Antiques,
   Hunstanton 089618
The Strange Quark Art Gallery,
   Mallow 109674
Strangers Hall Museum of Domestic
   Life, Norwich 045408
Stranieri, Paolo, Milano 080735
Straniquario, Firenze 080210
Stranraer Museum, Stranraer .. 045899
Stransky, Michael, München .. 130407
Strasbourg and District Museum,
   Strasbourg 007083
Strasburg Museum, Strasburg .. 053569
Strašifták, Peter, Pezinok 133057
Strassacker, Ernst, Süßen 108741
Straßenbahn-Museum Thielenbruch
   der Kölner Verkehrs-Betriebe,
   Köln 019713
Strassenbaumuseum, Sayda ... 021625
Strasser, Salzburg 062823
Straßer, Axel, Wörthsee 142502

Straßer, Gebhard, Bad Wurzach . 106102
Strasser, Johann, Schlößl 127868
Strasser, Wolfgang, Ravensburg . 078150
Straßgütl, Karin, Erding 075629
Strassner, Schärding 062836
Strategic Air & Space Museum,
Ashland 046565
Stratemann, Kurt-R., Göttingen . 076055
Strater Hotel, Durango 048400
Stratfield Saye House and Wellington
Exhibition, Stratfield Saye . . 045900
Stratford, Columbus 120516
Stratford and District Museum,
Stratford 001522
Stratford Antique Centre, Stratford-
upon-Avon 091495
Stratford Antiques & Interiors Centre,
Stratford-upon-Avon 091496
Stratford Bookshop, Stratford-upon-
Avon 144865
Stratford Hills Antique Market,
Richmond 096578
The Stratford Historical Society and
Catherine B. Mitchell Museum,
Stratford 053570
Stratford-Perth Museum,
Stratford 007085
Strathalbyn Antique Market,
Strathalbyn 062250
Strathalbyn National Trust Museum,
Strathalbyn 001523
Strathclair Museum, Strathclair . 007086
Strathcona County Heritage Museum,
Sherwood Park 007027
The Strathearn Gallery, Crieff . . 117649
Strathearn Gallery & Studio, Castle
Douglas 117544
Strathmore, Montréal 064477
Strathnairn Homestead Gallery,
Holt 099145
Strathnaver Museum, Thurso . . 045978
Strathpeffer Spa Pumping Room
Exhibition, Strathpeffer 045911
Strathroy Middlesex Museum,
Strathroy 007087
Strathspey Railway, Aviemore . . 043453
Strathwork Galerie, Ladbergen . 019841
Straub, Schwäbisch Gmünd . . . 078407
Straub & Lambert, Ruderting . . 130610
Straub, Hermann, Stockdorf . . . 130660
Straubhaar-Kiss, Thun 087848
Straubinger, F., Amberg 074404
Strauch, Düsseldorf 129760
Straumen, Bergen 084227
Straumsnes Bygdemuseum,
Kanestraum 033691
Strauß Gedenkstätte, Wien 003010
Strauß, Heidemarie, Berlin 074932
Strauß, Helmut, Saarbrücken . . 078335
Strauß, Johann, Schnega 108569
Stravridis, Stavros E., Athinai . . 142542
Straw and Grass Handicraft Museum,
Seoul 030045
Strawbale Gallery, Woodend . . . 099900
Strawberry Culture Museum,
Dahu 042296
Strawberry Hill Antiques,
Parkwood 085662
Strawbery Banke Museum,
Portsmouth 052316
Mr. Straw's House, Worksop . . . 046198
Streamer, J., Leigh-on-Sea 089798
Streamline Antiques, Boston . . . 092497
Strebel, Wiesbaden 024957
Streck, H. P., Bonn 075093
Strecker Museum's Collection,
Mayborn Museum Complex,
Waco 054020
Středočeské Muzeum, Roztoky u
Prahy 009872
Stredoslovenské Múzeum, Banská
Bystrica 038282
Streekheemmuseum den Aanwas,
Ossendrecht 032738

Streekhistorisch Centrum,
Stadskanaal 032900
Streeklandbowmuseum Agrimunda,
Sint Anna ter Muiden 032870
Streekmuseum Alphen, Alphen, Noord-
Brabant 031902
Streekmuseum De Groote Sociëteit,
Tiel 032925
Streekmuseum de Meesthof, Sint
Annaland 032871
Streekmuseum de Moennik,
Helden 032450
Streekmuseum De Oude Wolden,
Bellingwolde 032058
Streekmuseum de Tolbrug, Bakel 032045
Streekmuseum De Vier Ambachten,
Hulst 032520
Streekmuseum Goeree en Overflakkee,
Sommelsdijk 032896
Streekmuseum Het Admiraliteitshuis,
Dokkum 032237
Streekmuseum Het Dorp van Bartje,
Rolde 032796
Streekmuseum Het Land van Axel,
Axel 032041
Streekmuseum Het Rondeel,
Rhenen 032781
Streekmuseum Hoeksche Waard,
Heinenoord 032446
Streekmuseum Jan Anderson,
Vlaardingen 032990
Streekmuseum Land van Valkenburg,
Valkenburg, Limburg 032966
Streekmuseum Leudal, Haelen . 032404
Streekmuseum Ommen, Ommen 032717
Streekmuseum Opsterland,
Gorredijk 032361
Streekmuseum Oudheidkamer
Reeuwijk, Reeuwijk 032776
Streekmuseum Schippersbeurs,
Eisloo 032299
Streekmuseum Stevensweert/ Ohé en
Laak, Stevensweert 032908
Streekmuseum van Klein-Brabant De
Zilverreiger, Weert 004141
Streekmuseum voor de
Krimpenerwaard Crimpenerhof,
Krimpen aan den IJssel 032557
Streekmuseum West Zeeuws-
Vlaanderen, IJzendijke 032526
Streekmuseum/Volkssterrenwacht
Burgum, Burgum 032127
Street Gallery, Street 119171
Street Level, Glasgow 117860
Street Scene Gallery, New
Orleans 122244, 122245
Street Scene Studios, New
Orleans 122246
Street Store Angeles, Long
Beach 094227
Streetlife Museum, Hull Museum of
Transport, Kingston-upon-Hull . 044719
Streffer, Walter, Stuttgart 142404
Strehl, G., Weiden in der
Oberpfalz 078888
Strehler, Brigitte, Sindelfingen . . 142365
Streicher, Hana, Berlin 129569
Streichinstrumentensammlung,
Einsiedeln 041460
Streit, Gottfried, Basel 133668
Streit, Ilja, Weimar 130740
Streitenfeld, Dirk, Oberursel . . . 108324
Streiter, Franz, Soest 078491
Strempel, C.-P., Hamburg 141898
Strenz, Jacky, Frankfurt am Main 107033
Streppel, J.A.J., Amsterdam . . . 143070
Střeštíková, Jiřina, Brno 102572
Stretham Old Engine, Stretham . 045913
Stretti, Barbara, Genova 110136
Dello Stretto, Messina 110227
Stretton Antiques Market, Church
Stretton 088910
Stretton Water Mill, Farndon . . . 044332
Streubel, Peter-Paul,

Recklinghausen 142320
Strey-Froom & Strey, Baruth . . . 074685,
074686
Strey-Froom & Strey, Kloster
Zinna 076827
Striatus, Kiel 107556
Strich-Chapell-Zimmer, Sersheim 021846
Strick, Agnes, Köln 076911
Strickhausen, Helmut, Teisendorf 130700
Strickland, Hobart 099138
Strickland & Daughters, Jane,
Honiton 089574
Stricklin, D., Washington 097861
Stricoff, New York 123097
Strictly Limited Editions, San
Francisco 124407, 138461
Stride, Calgary 101260
Stride & Son, Chichester 126857
Strides Gallery, Driffield 117687
Strieder, Fürth 129877
Strieffler-Haus, Landau in der
Pfalz 019875
Striegel, W., Dietersheim 075353
Strienz, Cornelia, Uhingen 130714
Strigazzi, Dr. Giancarlo, Haan . 129932
Strigel-Museum und Antoniter-
Museum, Memmingen 020362
Strijk, H.J., Groningen 083305
Strik, J.W., Deventer 083164
Strike One, London . . . 090519, 134777
Strike Zone, Houston 093802
Strina, Luisa, São Paulo 101114
Strindberg-Museum Saxen,
Saxen 002708
Strindbergsmuseet, Stockholm . 041050
Stringer, Brian, Houston 093803
Strip-Ease Refinishing, Louisville 135765
Strip-It, Omaha 136067
Strip Joint, Norfolk 136037
Strip Shop, Charlotte 135394
Strip Shop, San Diego 136300
Striped House Museum of Art,
Tokyo 029592
Stripp Joynt, Chicago 135424
Stripper, Salt Lake City 136244
Strippers Furniture Restoration, Saint
Paul 136226
Stripping Workshop, Albuquerque 135291
Stripping Workshop, Washington 136469
Stripwinkel, Groningen 143154
Stripwinkel, Beo, Hove 140232
STRM, Saint-Ouen 073203
Strobel, Berlin 074933
Strobel, Buffalo 092560
Strobel, Irsee 141959
Strobel, Labrador 061714
Strobel, Christian, Irsee 141960
Strobelt, Harti, Muggensturm . . 077682,
130434
Strøgets Guld & Sølv, København 065961,
128398
Ströhlein, R., Bamberg 074674
Strömberg, København 065962
Strömbom, Uppsala 115974
Strömforsin Teollisuusmuseo,
Ruotsinpyhtää 011326
Strömsholms Slott, Kolbäck . . . 040756
Strømsholt, Pia Louise,
København 065963
Strömslund, Trollhättan 087106
Strømstads Museum, Strømstad 041061
Strötzbacher Mühle, Mömbris . . 020435
Stroganovskaja Roždestvenskaja
Cerkov, Nižegorodskij Istoričesko-
architekturnyj Muzej- zapovednik,
Nižnij Novgorod 037565
Stroganovskij Dvorec, Gosudarstvennyj
Russkij Muzej, Sankt-Peterburg 037821
Stroggouou, Alexandra, Athinai . 079183
Strohhaus, Muhen 041781
Strohhaus und Dorfmuseum,
Kölliken 041619
Strohmenger, A., Weilerswist . . 078892
Strohschneider, Zürich 116862

Stromboni, Dominique, Paris . . . 071923
Strominger, Den Haag 083133
Stromness Museum, Stromness . 045915
Strona, Luca, Torino 132206
Strong National Museum of Play,
Rochester 052622
Strood Library Gallery, Rochester 045630
Stroom Den Haag, Beeldende Kunst t/
m Architectuur, Den Haag . . . 032200
Strossmayerova Galerija Starih
Majstora, Hrvatska Akademija
Znanosti i Umjetnosti, Zagreb . 009050
Strotzka, Manfred, Wien 063080
Stroud House Gallery, Stroud . . 119175
Stroud Mansion, Stroudsburg . . 053576
Strouds, Southwell 091387
Strouk, Laurent, Paris 105251
Strover, Lynne, Cambridge 117512
Strube, Henning, München 130408
Struchen, H., Tagelswangen . . . 144141
Struck, Nicolaus, Berlin 141577
Struck, Vera, Atlanta 119618
Structure 17, Perpignan 129166
The Structurist, Saskatoon 138589
Strübhaus-Haus der Malkunst,
Veringenstadt 022320
Strümann, Mosbach 108033
Strümper, Hans, Erwitte 075661
Struer Museum og Johs. Buchholtz
Hus, Struer 010330
Strüver, Oppenheim 077951
Struggletown Fine Arts Complex,
Narellan 099420
Strumpelmeyer, Friedrich,
Dortmund 075397
Struś, Jan, Poznań . . . 084652, 084653
Struts Gallery and Faucet Media Arts
Centre, Sackville 006841
Struwwelpeter-Museum, (Heinrich-
Hoffmann-Museum), Frankfurt am
Main 018452
Struwwelpeter-Museum, Sammlung
der Originale Dr. Heinrich Hoffmanns
(Nachlaß), Frankfurt am Main . 018453
Strychnin, London 118583
Strychnin, New York 123098
Strychnin Gallery, Berlin 106444
Strydom, George 114784
Strzelczyk, Warszawa 084751
Strzelczyk, Tadeusz, Warszawa . 132907
Strzelecki, Krzysztof, Wrocław . . 132917
Stualängan, Laholm 086800
Stuart, Barrington 088370
Stuart, Santiago de Chile 101923
Stuart-Buchanan, Denver 093342
Stuart Collection, University of
California, San Diego, La Jolla 050041
Stuart House Antiques, Chipping
Campden 088885, 134261
Stuart House City Museum, Mackinac
Island 050557
Stuart Livraria e Galeria, Lisboa 114065,
143594
Stuart McGregor, Kinsale 130878
Stuart Street Gallery, Dunedin . . 112896
Stuart Town Gaol Museum, Alice
Springs 000792
Stuart, D.M., Bulli 061231
Stubbs, Kristy, Dallas 120665
Stubengalerie Stoetzel-Tiedt,
Goslar 107151, 137228
Stuber, Hansruedi, Hundwil . . . 087531,
133774
Stubnitz Gallery, Adrian College,
Adrian 046322
Stubtoft, Sorø 066145, 102984
Stucco, Madrid 086264
Stuchley, A.F. & J., Port
Macquarie 062098
Stuck-Museum, Hans Rich KG,
Freiburg im Breisgau 018489
Stuckert, Düsseldorf 075523
Stuckert, Anton, Düsseldorf . . . 075524
Stuckert, Herbert, Düsseldorf . . 075525

Stuckert, M., Düsseldorf 075526
Stuckhaus Cöpenick, Berlin 129570
Studebaker National Museum, South
Bend 053395
Student Art League, Washington 060765
Student Center Gallery, Asbury College,
Wilmore 054451
Studentenkarzer, Heidelberg ... 019146
Studer, Schwanau 137502
Studer, Simon, Genève 116341
Studi di Storia dell'Arte, Todi .. 138876
Studia Muzealne, Poznań 138991
Studien zu Denkmalschutz und
Denkmalpflege, Wien 138537
Studien zur Österreichischen
Kunstgeschichte, Wien 138538
Studiengalerie Busse,
Worpswede 108971
Studiengalerie der Pädagogischen
Hochschule Ludwigsburg,
Ludwigsburg 107896, 137356
Studiengang Architektur
Holzminden, Fachhochschule
Hildesheim/Holzminden/Göttingen,
Holzminden 055592
Studiengang Architektur,
Fachhochschule Lausitz,
Cottbus 055504
Studiengang Architektur,
Fachhochschule Trier, Trier ... 055731
Studiengang Architektur, Hochschule
Zittau/Görlitz (FH), Zittau 055747
Studiengang Intermediales Design,
Fachhochschule Trier, Trier ... 055732
Studiengang Kunsttechnologie,
Konservierung und Restaurierung von
Kunst- und Kulturgut, Hochschule
für Bildende Künste Dresden,
Dresden 055519
Studiengang Modedesign,
Fachhochschule Trier, Trier ... 055733
Studiengang Museologie, Fakultät
Medien, Leipzig 055629
Studiengemeinschaft Werner
Kamprath Darmstadt GmbH, Kunst
und Gestaltung, Pfungstadt ... 055689
Studienrichtung Bildhauerei, Burg
Giebichenstein-Hochschule für Kunst
und Design Halle, Halle, Saale 055554
Studienrichtung Glas, Burg
Giebichenstein-Hochschule für Kunst
und Design Halle, Halle, Saale 055555
Studienrichtung Grafik, Burg
Giebichenstein-Hochschule für Kunst
und Design Halle, Halle, Saale 055556
Studienrichtung Keramik, Burg
Giebichenstein-Hochschule für Kunst
und Design Halle, Halle, Saale 055557
Studienrichtung Konzeptkunst Buch,
Burg Giebichenstein-Hochschule
für Kunst und Design Halle, Halle,
Saale 055558
Studienrichtung Kunsterziehung/
Diplomkunstpädagogik, Burg
Giebichenstein-Hochschule für Kunst
und Design Halle, Halle, Saale 055559
Studienrichtung Malerei, Burg
Giebichenstein-Hochschule für Kunst
und Design Halle, Halle, Saale 055560
Studienrichtung Schmuck, Burg
Giebichenstein-Hochschule für Kunst
und Design Halle, Halle, Saale 055561
Studienrichtung Textil, Burg
Giebichenstein-Hochschule für Kunst
und Design Halle, Halle, Saale 055562
Studiensammlung historischer Ziegel,
Irdning 002143
Studies in Art Education, Reston 139497
Studies in Conservation, London 139224
Studies in Iconography –
Medieval Institute Publications,
Kalamazoo 139498
Studies in Modern Art, New York 139499
Studies in the History of Art,

Landover 139500
The Studio, Gateshead 117820
The Studio, Gomshall 089343
The Studio, Hyderabad 109375
Studio, Madrid 115289
The Studio, Otford 118857
Studio, Tel Aviv 138784
Studio, Toronto 138590
Studio 101, Eton 089221
Studio 107, Austin 119702
Studio 107, Norfolk 123236
Studio-13, Moskva 114310
Studio 181, Jesolo 110153
Studió 1900, Budapest 079294, 109177
Studio 2, Appledore, Devon ... 117245
Studio 2, Hong Kong 102116
Studio 2 Gallery, Austin 119703
Studio 2 Gallery, Chermside ... 098882
Studio 2000, Brampton, Cumbria 117416
Studio 21, Maryborough,
Queensland 061798
Studio 215, Auckland 112811
Studio 220, Tucson 124775
Studio 23, Cincinnati 120378
Studio 236, Hope Island 099146
Studio 24, San Francisco 124408
Studio 2507, Portland 123681
Studio 261, Montréal 101452
Studio 28, Stockholm 115928
Studio 33, Portland 123682
Studio 34, Houston 121136
Studio 34, Ourense 115321
Studio 4, Lamlash 118066
Studio 407, San Diego 124178
Studio 48, New Lambton 099429
Studio 52, Sharjah 117202
Studio 55, Delhi 109344
Studio 67, Akaroa 132678
Studio 71, Palermo 110622
Studio 730, New York 136008
Studio 8 Graphics, Toronto ... 101708
Studio 84 West, Long Beach ... 121411
Studio 86, Wanganui 113051
Studio Aesthetic, Mumbai 109478
Studio AF, Roma 132056
Studio Agram, Budapest 079295
Studio Ah Lala, Mile End 099340
Studio Amaranto, Verona 111197
The Studio Annex, New York ... 051475
Studio Annunciata, Milano 131534
Stúdió Antikvárium, Budapest .. 142571
Studio Antiques, Oklahoma City 096016
Studio Ardiglione, Firenze 131268
Studio Arga, Venezia 111160
Studio Arl, Karachi 113307
Studio Art and Frame,
Minneapolis 122040
Studio Art Centers International,
Firenze 055900
Studio Art Department, Manhattanville
College, Purchase 057736
Studio Art Gallery, Bridgetown . 100462
Studio Art II, Washington 124961
The Studio Art School, Pathhead,
Midlothian 056594
Studio Art, Middlebury College,
Middlebury 057437
Studio Arte 3, Trieste 111054
Studio Artis, Ettelbruck 111827
Studio Arts, Norfolk 123237
Studio Arts Department, Transylvania
University, Lexington 057339
Studio Arts Gallery, Lancaster . 118070
Studio ASH, Columbus 120517
Studio Associates, Philadelphia . 123459
Studio Associato Restauro Opere
d'Arte, Brescia 131051
Studio B, Columbus 120518
Studio B, Drogheda 109567
Studio B2, Genova 110137
Studio Barbieri, Venezia 111161
Studio Bellart, Herceg-Novi ... 112023
Studio Bibliografico Aesse di Sonia
Scaglioni, Bologna 142644

Studio Bibliografico Apuleio,
Trento 142874
Studio Bibliografico Benacense, Riva
del Garda 142794
Studio Bibliografico Bononia,
Bologna 142645
Studio Bibliografico Bosio, Magliano
Alpi 142703
Studio Bibliografico La Linea d'Ombra,
Roma 142840
Studio Bibliografico L'Arca,
Milano 142745
Studio Bibliografico Orfeo,
Bologna 142646
Studio Bibliografico Tornabuoni,
Firenze 142687
Studio Bibliografico Vecchi Libri,
Firenze 142688
Studio Bildende Kunst, Berlin .. 017304
Studio Bookshop, Brighton 144270
Studio Cadrea, Cagliari 131076
Studio Capricornus, Eckernförde 106858
Studio Casoli, Milano 110452
Studio Cavalieri, Bologna 079881
Studio Celtia, London 101349
Studio Copernico, Milano 110453
Studio Cristofori, Bologna 109865
Studio D, Bolzano 109890
Studio d'Arte Marco Fioretti,
Bergamo 109801
Studio d'Arte, Rovereto 110883
Studio d'Arte Barnabò, Venezia . 111162
Studio d'Arte Canale, Monticello Conte
Otto 110511
Studio d'Arte Firenze, Firenze .. 080211,
131269
Studio d'Arte GR, Sacile 110888
Studio d'Arte Grafica, Milano .. 110454
Studio d'Arte Lanza, Verbania .. 111173
Studio d'Arte Lunardi, Asiago .. 109760
Studio d'Arte Stomaso, Bergamo 109802
Studio d'Arte Verbano, Locarno . 116447
Studio d'Arte Zanoletti, Milano . 110455
Studio de l'Image, Paris 105252
Studio Delise, Portogruaro 110684
Studio Della, Ostrava 102637
Studio della Vedova, Udine 111668
Studio dell'Arte, Montréal 064478,
128185
Studio di Restauro Ligneo,
Verona 132301
Studio di Val Cervo, Roma 110860
Studio E, Castelnuovo di Sotto . 109942
Studio Eugène Ysaye, Liège ... 003814
Studio F 22, Palazzolo sull'Oglio 110584
Studio Ferrari, Milano 131535
Studio Filatelico Euro, Vertova . 081949
Studio Fine Art, San Diego 124179
Studio Fornaresio, Torino 111007
Studio Forni, Milano 110456
Studio Franck Bordas, Paris ... 105253
Studio für Bild und Rahmen,
Essen 106931
Studio Fusion Gallery, London .. 118584
Studio Fyra Årstider, Stockholm 115929
Studio G.S., San Donà di Piave . 110891
Studio G7, Bologna 109866
Studio-Galerie, Berlin 106445
Studio-Galerie, Regensburg ... 108467
Studio Galileo Ferraris, Livorno
Ferraris 080339
The Studio Gallery, Datchet ... 089034,
134316
Studio Gallery, Denver 120779
The Studio Gallery, Fresno ... 120908
Studio Gallery, Manly 099276
Studio Gallery, Oklahoma City . 123295
The Studio Gallery, Slough ... 119105
Studio Gallery, Washington ... 054162,
124962
Studio Gallery 106, Toronto ... 101709
Studio Gariboldi, Milano 110457

Studio Ghiglione, Genova 080294,
110138,   126198
Studio Gi, Gagliano del Capo .. 137657
Studio Glass Gallery, London ... 118585
Studio Grafico, San José 136750
Studio Guastalla, Milano 110458
Studio Guenzani, Milano 110459
Studio Hercolani, Bologna 131001
Studio Hermitage, New York ... 123099,
136009
Studio III Gallery, Austin 119704
Studio im Hochhaus, Berlin ... 017305
Studio K, Singapore 114612
Studio La Città, Verona 111198
Studio La Duchesse, Montréal .. 064479
Studio M, Bamberg 106119
Studio M, Warszawa 113903
Studio Maastricht, Maastricht .. 112582
Studio Maria Cilena, Milano ... 110460
Studio Moderne, Nantes 070480
Studio Mondi, Castelfranco
Veneto 109938
Studio-Museo di Giuseppe Pellizza,
Istituto Giuseppe Pellizza,
Volpedo 028423
The Studio Museum in Harlem, New
York 051476
Studio National des Arts
Contemporains de Tourcoing,
Tourcoing 055426
Studio of Artists, Newtown, New
South Wales 099453
Studio of Contemporary Art,
Auckland 112812
Studio on J, Sacramento 123785
Studio on the Quay, Yarmouth . 119400
Studio One, Cleveland 120459
Studio One, Honolulu 120970
Studio One Gallery, Mount
Gambier 099397
Studio Paul Nicholls, Milano ... 110461
Studio Per Il Restauro, Brescia . 131052
Studio Petrecca, Arte Contemporanea,
Torino 111008
Studio Plus, Geelong 099075
Le Studio PM, Montréal 101453
Studio Potter, Manchester 139501
Studio Raffaelli, Trento 111034
Studio Reggiani, Milano 110462
Studio Restauri Formica, Milano 131536
Studio Restauro, Torino 132207
Studio Restauro Antichità,
Brescia 131053
Studio Restauro Santa Trinita,
Firenze 131270
Studio Ripetta, Roma 081573
Studio Roze Antiques, Rio de
Janeiro 128121
Studio S, Pune 109501
Studio S Arte, Roma 110861
Studio Saint-Sulpice, Paris 071924
Studio San Damiano, Milwaukee 121951
Studio San Giuseppe Art Gallery,
College of Mount Saint Joseph,
Cincinnati 047748
Studio Santo Spirito di Grossi & Lotti,
Firenze 131271
Studio Santo Stefano, Bologna . 109867
Studio Siro, Tiranë 098531
Studio Six Art, Oklahoma City . 123296
Studio Sixteen Seventeen, Los
Angeles 121601
Studio Sixty, Glasgow 117861
Studio Soligo, Roma 110862
Studio Solstone, Seattle 124606
Studio Suardo, Milano 131537
Studio Swiss, Tucson 136423
Studio Techne, Firenze 131272
Studio Ten, Saint Paul 123935
Studio Tommaseo, Istituto per la
Documentazione e la Diffusione delle
Arti, Trieste 059480
Studio Trisorio, Napoli 110546
Studio Trisorio, Roma 110863

Studio Västsvensk Konservering,
Göteborg 133535
Studio Venticinque, Milano 110463
Studio Wienblick, Wien 100379
Studio Wildflower Paintings,
Denmark 098968
Studio Windsor, Mumbai 109479
Studio Works Gallery, Nailsworth 118770
Studio X, Images de Presse,
Limours 136829
Studio za Umetničku Keramiku,
Arandjelovac 114392
Studiogalerie Kaditzsch,
Denkmalschmiede Höfgen,
Grimma 018822
Studiolo, Milano 080736
Studios of Jack Richard Creative
School of Design, Cuyahoga
Falls 056969
Studios on High, Columbus ... 120519
Studo San Domenico, Milwaukee 121952
The Study Gallery, Bournemouth and
Poole College, Poole 045530
Stüber, Olaf, Berlin 106446
Stübiger, A., Grevenbroich 076091
Stülpnagel, Karl-Heinrich von,
Leipzig 130230
Stuempfig, Anthony, Philadelphia 096163
Stümpges, V., Heidelberg 076460
Stützer, Carla, Köln 107688
Stützer, Thomas, Heiligenstadt,
Heilbad 130032
Stützle, Wolfram, Waldkirch ... 130731
Stuff & Nonsense, Chesham ... 088853
Stuff & Such, Houston 093804
Stuff 2 Gold, Phoenix 096240
Stuff Antiques, Phoenix 096241, 136122
Stuhlbau- und Heimatmuseum,
Rabenau 021254
Stuhlmann, Wedel 108850
Stuhlmuseum Burg Beverungen,
Beverungen 017353
Stuhr Museum of the Prairie Pioneer,
Grand Island 049180
Stuker, Jürg, Bern 116150, 126715
Stumbaum, Regina, Herrsching am
Ammersee 130039
Stumpf, Hong Kong 102117
Stumpf-Behrens, Karin, Kronberg im
Taunus 076974
Stumpf, Christa, Forchheim,
Oberfranken 075759
Stumpf, Hermann,
Neckarsteinach 077696
Stumpy Gully Antiques, Tyabb .. 062362
Stumvoll, Herbert, Hirschegg,
Kleinwalsertal 062635
Studner, Salzburg 127859
Sturdivant Hall, Selma 053267
Stures Galleri, Tranås 115955
Sturgeon River House, Sturgeon
Falls 007089
Sturgis Station House Museum,
Sturgis 007090
Sturies, Dr. Andreas, Düsseldorf 075527,
125933
Sturm-Larondelle, Berlin 129571
Sturm, Michael, Stuttgart 108731
Sturm, von, Hamburg . 107316, 129993
Sturman, Hawes 089476
Sturman Institute Museum of Regional
Science, Afula 025011
Sturmflut Galerie, Hooge 107447
Sturmio, Sevilla 133440
Sturni, Daniela, Roma 132057
Sturt, Mittagong 099356
Sturt Street Antiques and Art,
Ballarat 061091
Stuteville, Seattle ... 097510, 136377
Stutsman County Memorial Museum,
Jamestown 049789
Stuttfeld, Köln 107689
Stuttgarter Antiquariat, Stuttgart 142405
Stuttgarter Antique & Art,

Stuttgart 098166
Stuttgarter Feuerwehrmuseum,
Feuerwehrverein Stuttgart e.V,
Stuttgart 022061
Stuttgarter Gesellschaft für Kunst und
Denkmalpflege, Stuttgart 059239
Stuttgarter Gesellschaft für Kunst und
Denkmalpflege, Studiensammlung
Möbel, Stuttgart 022062
Stuttgarter Künstlerbund,
Stuttgart 059240, 108732
Stuttgarter Kunstverein, Stuttgart 022063,
059241
Stutzhäuser Brauereimuseum,
Luisenthal 020190
Stux, Stefan, New York 123100
Stuyck, Miguel, Madrid 133390
Styl, Pruszków 132866
Styl Antikvitás, Pécs 079321
Styl'Art, Monthey 116508
Style, Le Chambon-sur-Lignon . 069246
Style 1900, Lambertville 139502
Style 25, Lausanne 087586
Style Art Publishing, Phoenix .. 138440
Style D'Art, Rio de Janeiro 064128
Style Deco, München 077627
Style French Antiques, Los
Angeles 094480
Style French Antiques, West
Hollywood 097935
Style Gallery, London 090520
Style Junkies, Nelson 084023
Style, Denis, București 114155
Styles of Stow, Stow-on-the-
Wold 091484, 135098
Styles Silver, Hungerford 089614, 134491
Styl'Vieil, Brignoles ... 067498, 128522
Stypmann, David, New York ... 095768
Styrge, Rouen 105480
Styria, Judenburg 140001
Styrmannen Antik, Vaxholm ... 087154
Styrsjöbo Stentryck, Leksand .. 115718
Su Ho Memorial Paper Museum,
Taipei 042549
Su-no-sato Hakubutsukan,
Handa 028612
Suardi, Fabio, Mirandola 110489
Suarez Doradores, Valencia ... 133472
Suarez, Graciela, Miami 094805
Suarez, Pascal, Saint-Jean-de-la
Ruelle 072838
Suarsana, Putu & Nicole, Bad
Mergentheim 074551
Suau, Robert, Saint-Paulet-de-
Caisson 073249
Suaudeau, Marie-Pierre,
Montpellier 070314
Suay y Serra, Valencia 133473
Subak, Chicago 092796
Subak Museum, Banjar
Senggulan 024536
Subal, Tony, Wien 063081, 100380
Subalyuk-Múzeum, Cserépfalu . 023586
Subarna, Barcelona 085844, 126515
Subarunoki, Sapporo 111379
Subastas, Madrid 139072
Subastas de Barcelona,
Barcelona 126516
Subastas La Villa Vieja, Alicante 126498
Subastas la Villa Vieja, Alicante . 085704
Subastas Segre, Madrid 126551
Subastas y Galería de Arte,
Valencia 086537, 115519
Subert, Alberto, Milano 080737
Subert, Alberto e Michele, Milano 080738
Subex, Barcelona 114965
Subiaco Museum, Subiaco 001526
Subjektiv, Sankt-Peterburg 139018
Sublett, R.D., Louisville 094591, 135766
Submarine Force Museum and
Historic Ship Nautilus, Groton . 049284
Subra, Catherine, Toulouse ... 073802
Subra, Eric, Paris 071925
The Substation Gallery,

Singapore 114613
Subun-so, Tokyo 142940
Suburban 312 Gallery, Brighton,
Victoria 098777
Succès, La Chevrolière 068991
Successories, Cincinnati 120379
Succotash, Saint Paul 096863
Sucesores de Gigante, Madrid . 133391
Sucharduv Dům, Městské Muzeum,
Nová Paka 009719
Suchna Kendra Gallery, Lucknow 109409
Suchomsky, Gerd, Köln 076912
Suchow & Seigel, New York ... 095769
Suciu, Emilia, Ettlingen 106938
Suckau-Pirow, K., Sylt 078664
Sucksdorff, Jürgen, Berlin 074934
Sucret, Lionel, Rodez 072515
Sud 70, Cannes 067654
Sud Antik, Montauban 070202
Sud Loire, Pont-Saint-Martin ... 072263
Sud Textiles, Bobigny 103402
Sudan National Museum,
Khartoum 040497
Sudan Natural History Museum,
University of Khartoum,
Khartoum 040498
Sudbury Region Police Museum,
Sudbury 007095
Suddaby, Blairgowrie 139648
Sudeley Castle, Winchcombe . 046143
Sudhaus Sorgendorf, Bleiburg . 001785
Sudley House, Liverpool 044864
Sudmann, Brigitte, Stuttgart .. 108733
Sue Ellen Gallery, Main Beach . 061760
Sue Ryder Art Gallery, Herceg-
Novi 112024
Sueck, Ursula, Coesfeld 075267
Sueclaytables, Unley 062380
Süd West Galerie, Hüttlingen .. 076588
Südbahn, Museum Mürzzuschlag am
Semmering, Mürzzuschlag .. 002421
Süddeutsche Verlagsgesellschaft Ulm,
Ulm 137540
Süddeutsches Eisenbahnmuseum,
Heilbronn 019165
Südharzer Eisenhüttenmuseum, Bad
Lauterberg 016816
Südhessisches Handwerksmuseum,
Roßdorf 021482
Südmähler Heimatmuseum, Laa an
der Thaya 002256
Südmährisches Landschaftsmuseum,
Geislingen an der Steige 018615
Südostbayerisches Naturkunde- und
Mammut-Museum, Siegsdorf . 021857
Südpfälzische Kunstgilde e.V., Bad
Bergzabern 059242
Südsauerlandmuseum, Attendorn 016638
Südschwäbisches Archäologiemuseum
im Jesuitenkolleg, Mindelheim 020413
Südtiroler Gesellschaft für
Photographie, Bolzano 059481
Südtiroler Landesmuseum für Kultur-
und Landesgeschichte Schloss Tyrol,
Tirolo di Merano 028090
Süess, Robert, Dierikon 116246
Sülchgau-Museum, Rottenburg am
Neckar 021521
Sümegi Vármúzeum, Sümeg .. 023910
Suermondt-Ludwig-Museum,
Aachen 016445
Suezumi, Kobe 082002
Suffolk City Council Libraries
and Heritage, Ipswich Library,
Ipswich 044648
Suffolk County Historical Society
Museum, Riverhead 052590
Suffolk County Vanderbilt Museum,
Centerport 047499
Suffolk House Antiques, Yoxford 091986
Suffolk Museum, Suffolk 053590
Suffolk Punch Heavy Horse Museum,
Woodbridge 046182

Suffolk Rare Books, Sudbury,
Suffolk 144870
Suffolk Regiment Museum, Bury Saint
Edmunds 043811
Suffolk Resolves House, Milton . 050905
The Suffolk Table Company,
Debenham 089046, 134319
Suffren, Saint-Tropez . 073320, 105630
Suga, Tokyo 111507
Sugan, Dublin 079590
Sugar, Houston 093805
Sugar Antiques, London 090521, 134778
Sugar Barrel Antiques, Vancouver 064886
Sugar Bear Antiques Mall,
Jacksonville 094040
Sugar Industry Museum, Ciaotou 042290
The Sugar Refinery, Vancouver . 101841
Sugarhill Art Center, New York . 123101
Sugarhouse, Salt Lake City 096925,
096926, 123990, 136245
Sugarman, Budd, Toronto 064760
Sugarplums, Las Vegas 094180
Suger, Paris 071926
Sugères, Bruno, Clermont-
Ferrand 068038
Sugesa, Madrid 086265
Sugihara, Tokyo 142941
Sugimoto Bijutsukan, Mihama .. 029023
Sugino Gakuen Isho Hakubutsukan,
Tokyo 029593
Sugiura, Kyoto 111303
Suhm, Walter, Thal 087842
Suhrkamp, Frankfurt am Main . 137208
Suhube, Riyadh 114388
Sui Tangyao Zhi Display Center,
Chengdu 007698
Suichuan Museum, Suichuan .. 008288
Suide Museum, Suide 008289
Suidobata Fine Arts Academy,
Tokyo 056055
Suifu-Meitoku Shokoh-kan Tokugawa
Museum, Mito 029038
Suijkerbuijk, C., Antwerpen 063240
Suining Museum, Suining,
Jiangsu 008290
Suining Museum, Suining,
Sichuan 008291
Suipacha, Buenos Aires 098581
Suishouen Toji Shiryokan, Tokuro
Kinenkan, Nagoya 029128
Suissa, Alice, Versailles 074071
Suite Lorain Antiques, Cleveland 092975
Suixi Museum, Suixi 008292
Suk Joo-sun Memorial Museum of
Korean Folk Arts, Seoul 030046
Suka Academy, Busan 056083
Suka Art Space, Busan 029901
Suka Gogoda Bijutsukan,
Nagasaki 029103
Sukagawa-shiritsu Hakubutsukan,
Sukagawa 029413
Sukanen Ship Pioneer Village and
Museum of Saskatchewan, Moose
Jaw 006422
Sukellusvene Vesikko, Helsinki . 010864
Sukhotai Antiques, Bangkok ... 088087
Sukkulenten-Sammlung Zürich,
Zürich 042234
Suksimuseo ja Pesäpallomuseo,
Vimpeli 011513
Sukuma Museum, Mwanza 042636
Sulabh International Museum of
Toilets, Delhi 024238
Sułek, Kazimierz, Kraków 132850
Sulitjelma Gruvemuseum,
Sulitjelma 033999
Sulivan, Tehrán 109521
Sulivan Gallery, Tehrän 079429
Sullivan, Kenmare 079640
Sullivan, Sacramento 096649
Sullivan County Historical Society
Museum, Hurleyville 049640
Sullivan, Michael T., Indianapolis 121220
Sullivan, Terence Paul, Ryde ... 118995

Šulo, Jozef, Prievidza 133059
Sulphide Street Station Railway and
Historical Museum, Broken Hill 000894
Sulphur Springs Station,
Ancaster 005336
Sultan Abdul Aziz Royal Gallery,
Kelang 030670
Sultan Abu Bakar Museum Pahang,
Pekan 030721
Sultan Ali Dianr Museum, Sudan
National Museum, El-Fasher . . 040489
Sultan Mahmud Badaruddin II
Museum, Palembang 024627
Sultan, Fawzi H., Safat 082267
Sultana Center, Cairo 066275
Sultane, Zeina, Beirut 082306
Sultan's Armed Forces Museum,
Ruwi 034106
Sulu National Museum, Jolo . . . 034484
Sulzer, Schrobenhausen 108580
Sum'Art, Arcachon 103268
Sumelocenna – Römisches
Stadtmuseum, Rottenburg am
Neckar 021522
Sumer, Los Angeles 094481
Sumisho Art Gallery, Tokyo 082208
Sumiti, Luca, Venezia 132258
Sumiya Motenashi Museum of Art and
Culture, Kyoto 028984
Summa, New York 123102
Summer, Polegate 091027
Summer & Dene, San Diego . . 124180
The Summer Palace, Chinese Imperial
Garden Museum, Beijing 007647
Summer, Aaron, New York 095770
Summer, Celeste, Tucson 124776
Summerfield & Co., T.R., Albury 060986
Summerland Antiques, Ballina . . 061095
Summerland Museum,
Summerland 007096
Summerleaze Gallery, East
Knoyle 117707
Summerlee Heritage Park,
Coatbridge 043980
Summerlin, Las Vegas 094181
Summers Art Gallery, Dorking . . 117684
Summers Davis, Wallingford . . . 091715
Summers, Martin, London 090522
Summers, Richard, Richmond . . 096579
Summers, Sharon, Charlotte . . . 092630
Summersgills, York 127107
Summerson, Ed, Hong Kong . . . 136727
Summersons, Warwick 091750
Summit Auction, Greenwood . . . 127235
Summit County Historical Society
Museum, Akron 046332
Sumner Art Gallery, Christchurch 112863
Sumner, David, Toorak Gardens 099832
Sumner/Redcliffs Historical Museum,
Christchurch 033177
Sumo Hakubutsukan, Nihon Sumo
Kyokai, Tokyo 029594
Sumove Museum, Suqian 008293
Sumpter, Edward, New York . . . 095771
Sumpu Museum, Shizuoka 029405
Sums'ke Učylyšče Mystectv i Kul'tury
im. D. Bortnjans'kogo, Sumy . . 056502
Sums'kyj Chudožnij Muzej im. N.
Onac'kogo, Sumy 043305
Sums'kyj Oblasnyj Krajeznavčyj Muzej,
Sumy 043306
The Sumter County Museum,
Sumter 053601
Sumter Gallery of Art, Sumter . . 053602
Sun, Nagoya 082063
Sun Antics, Magagnosc 069801
Sun Art Gallery, Los Angeles . . 121603
Sun Contemporary, Seoul 111724
Sun Craft, Singapore 114614
Sun Gallery, Minneapolis 094949
Sun Gallery, Seoul 111725
Sun Gallery, Hayward Area Forum of
the Arts, Hayward 049421
Sun-Kong Factory Museum, Su-

Ao 042469
Sun Prairie Historical Museum, Sun
Prairie 053603
Sun Valley Center for the Arts,
Ketchum 049947, 060766
Sun Yat-Sen University Museums,
Museum of Geology, Biological
Museum, Bio-digital Museum,
Sun Yat-sen Memorial Hall,
Guangzhou 007813
Sun, Gaju, Los Angeles 121602
Sūna, Agija, Rīga 111766
Sunam, Busan 111576
Sunbury Planatation House, Saint
Phillip 003242
Sunbury Shores Arts and Nature
Centre, Saint Andrews 006852
Sunce, Beograd 114442
Sund Bygdemuseum, Skogsvåg 033948
Sund Fiskerimuseum, Sund i
Lofoten 034000
Sunda, Fabrizio, Cagliari 080003, 131077
Sundance Art, Phoenix 123515
Sundaram Tagore Gallery, Beverly
Hills 119782
Sundaram Tagore Gallery, Hong
Kong 102118
Sundaram Tagore Gallery, New
York 123103
Sundberg, Norberg 143906
Sundberg, Stockholm 115930
Sundborn Antikt, Sundborn 087085
Sundby Samling Bryggergården,
Nørresundby 010240
Sundby Samling Raschgården,
Nørresundby 010241
Sundbybergs Museum,
Sundbyberg 041063
Sunderland Museum, Sunderland 045926
Sundermann, Rüdiger, Würzburg 108978
Sundial Antiques, Amersham . . 088272
Sundomin Kotiseutumuseo,
Sundom 011380
Sundquist, Poni, San Diego . . . 136301
Sundqvist, Solna 143923
Sundridge Gallery, Sundridge . . 091507,
119179, 135106
Sundsbergs Gård Museum och
Konsthall, Sunne 041068
Sundsvalls Antikvariat, Sundsvall 143976
Sundsvalls Museum, Sundsvall . 041066
Sunesis Gallery, New York 123104
Sunflower Cottage, Louisville . . 094592
Sunflower Gallery, Calgary 101261
Sung, San Francisco 097333
Sung-Am Archives of Classical
Literature, Seoul 030047
Sung Kyun Kwan University Museum,
Seoul 030048
Sungem, Southport 062232
Sungkok Art Museum, Seoul . . . 030049
Sungshin Women's University Museum,
Seoul 030050
Sunjin, Singapore 114615, 114616
Sunken History Exhibition, Fort Victoria
Museum, Yarmouth 046213
Sunland Art Gallery, El Paso . . 120841
Sunndal Bygdemuseum,
Sunndalsøra 034001
Sunne Antik, Sunne 087092
Sunnfjord Museum, Førde 033605
Sunnhordland Museum, Stord . . 033994
Sunnmøre Museum, Ålesund . . 033486
Sunny Sky, Saint Paul 096864
Sunny Systems, Chennai 109292
Sunnybrook Antique Auction Gallery,
Swedesboro 127468
Sunnybrook Farm Museum and
Interpretive Centre, Red Deer . 006760
Sunnyhurst Wood Visitor Centre,
Darwen 044072
Sunnyside Historical Museum,
Sunnyside 053606
Sunnyside Museum, Hillside . . . 044586

Sunnyvale Historical Museum,
Sunnyvale 053608
Sunporch, Baltimore 092365
Sunrise Antiques, Chesterfield . . 127162
Sunrise Art and Sculpture Gallery, Las
Vegas 121385
Sunrise Southwest, Denver 120780
Sunrise Stained Glass, Tucson . 124777
Sunrise Trail Museum,
Tatamagouche 007127
Sunritz Hattori Bijutsukan, Suwa 029417
Sunset Antiques, Miami 094806
Sunset Auction, Edmonds 127213
Sunset Bazaar, Los Angeles . . . 094482
Sunset Gallery, Miami 121864
Sunset Gallery, Milton 099352
Sunset Orange, Los Angeles . . . 094483
Sunset Studio Gallery, Cobram . 098901
Sunshine Art Gallery, Boreen
Point 098762
Sunshine Artist, Orlando 139503
Sunshine Coast Museum,
Gibsons 005873
Sunshine, Roger, Miami 094807
Suntory Bijutsukan, Tokyo 029595
Suntory Hakubutsukan, Osaka . 029270
Sunwatch Indian Village –
Archaeological Park, Dayton . . 048175
Suodenniemi Kotiseutumuseo,
Suodenniemi 011381
Suojeluskuntamuseo, Seinäjoki . 011351
Suolahden Museo, Suolahti 011383
Suomen Antikvariaattiyhdistys,
Helsinki 058578
Suomen Finnish Honeybee Museum,
Juva 010930
Suomen Ilmailumuseo, Vantaa . 011493
Suomen Jääkiekkomuseo,
Tampere 011403
Suomen Järvikalastusmuseo
ja Kerimäen Ulkomuseoalue,
Kerimäki 010993
Suomen Jalkapallomuseo,
Valkeakoski 011483
Suomen Joutsen, Turku 011444
Suomen Käsityön Museo,
Jyväskylä 010942
Suomen Kansallismuseo, Helsinki 010865
Suomen Kansansoitinmuseo,
Kaustinen 010978
Suomen Kellomuseo, Espoo . . . 010777
Suomen Koulumuseo, temporary
closed, Tampere 011404
Suomen Lasimuseo, Riihimäki . . 011315
Suomen Matkailumuseo, Imatra 010902
Suomen Merimuseo, National Board of
Antiquities, Kotka 011033
Suomen Metsästysmuseo,
Riihimäki 011316
Suomen Muinaismuistoyhdistys,
Helsinki 058579
Suomen Museoliitto, Helsinki . . 058580
Suomen Nyrkkeilymuseo,
Tampere 011405
Suomen Partiomuseo, Turku . . . 011445
Suomen Rakennustaiteen Museo,
Helsinki 010866
Suomen Rautatiemuseo,
Hyvinkää 010891
Suomen Taidegraafikot, Helsinki 058581
Suomen Taiteilijaseura, Helsinki 058582
Suomen Tietojenkäsittelymuseo,
Jyväskylä 010943
Suomen Urheilumuseo, Helsinki 010867
Suomen Värimuseo, Kemi 010987
Suomen Valimomuseo ja Högforsin
Masuuni, Karkkila 010964
Suomen Valokuvataiteen Museo,
Helsinki 010868
Suomen Viestimuseo, Riihimäki . 011317
Suomenlinna – Sveaborg,
Helsinki 010869
Suomenlinna-Museo, Helsinki . . 010870
Suomenlinnan Lelumuseo,

Helsinki 010871
Suomi Galleri, Stockholm 115931
Suomi International College of Art and
Design, Hancock 057178
Suomusjärven Kotiseutumuseo,
Suomusjärvi 011384
Suonenjoen Kotiseutumuseo,
Suonenjoki 011385
Supa-Ngwao Museum Centre,
Francistown 004294
Super Paintings, Glasgow 117862
Super Poster, Ciudad de Panamá 133325
Superior Antique Mall, Toledo . . 097618
Superior Galleries, Woodland
Hills 127500
Superlatives, Kansas City 094096
Super's Design, Miami 094808
Superstar Collectibles, Nashville 095038
Supico & Sousa, Lisboa 143595
Suppan, Martin, Wien . 063082, 125287
Suppan, Prof. Martin, Wien . . . 100381,
125286, 136611
Supper, Pforzheim 108378
Suppes, Alexander, Ober-
Ramstadt 077854
Suppes, Bernd, Wiesbaden . . . 142496
The Supreme Court of the United
States Museum, Washington . . 054163
Suquamish Museum, Poulsbo . . 052341
Suquamish Museum, Suquamish 053614
Sur, Punta del Este 119450
Sur, Rottenburg am Neckar 078300
Sur la Route du Chineur, Sens . 073503
Sur les Ailes du Temps, Paris . . 071927
Sur Meuse, Maastricht 112583
Surahammars Bruksmuseum,
Surahammar 041070
Surapon, Bangkok 116962
Surbhi Arts and Crafts, Jaipur,
Rajasthan 109390
Surcontre, Napoli 080941
Surek-Becker, Marie-Luise,
Berlin 141578
Surface Design Association,
Sebastopol 060767
Surface Design Journal,
Sebastopol 139504
Surface Gallery, Nottingham . . . 045422
Surgeons' Hall Museum,
Edinburgh 044277
Surinaams Museum, Paramaribo 040506
Surnadal Bygdemuseum,
Surnadal 034002
Surplus Store, Innishannon 079637
Surprise, Modena 080820
Surprises, Houston 121137
Surratt House Museum, Clinton . 047837
Surrey, Cincinnati 092901
Surrey Antiques, Horley 089579
Surrey Antiques Fair, Guildford . 098406
Surrey Art Gallery, Surrey 007102
Surrey Clock Centre, Haslemere 089458,
134445
Surrey Crescent Traders,
Auckland 083828
Surrey Heath Museum,
Camberley 043835
Surrey Hills Art Gallery, Surry
Hills 099754
Surrey Museum, Surrey 007103
Surroundings, Indianapolis 093932
Survey Museum, Roorkee 024462
Survivors, Kerikeri 083976
Surxondaryo Vilojati Arxeologiya
Muzeyi, Termiz 054649
Surya Gallery, Hyderabad 109376
Susan B. Anthony House,
Rochester 052623
Susan Sheehan Gallery, New
York 123105
Susan Street Fine Art Gallery, San
Diego 124181
Susane, R., Miami 094809
Susannah, Bath 088402

Susannah Place Museum,
Sydney 001551
Susan's Antiques, Hamilton 064332
Susan's Antiques, Toronto 064761
Susan's Selections, Richmond . . 096580
Sushart, Pune 109502
Sushi-Performance and Visual Art
Museum, San Diego 052982
Susini, Aix-en-Provence 103205
Susini, Bruno, Bachant 066901
Susini, Pierre, Montpellier 128813
Suslow, Larissa,
Fürstenfeldbruck 075922
Suso, Palmira, Lisboa 114066
Susquehanna, Washington 097862
Susquehanna Art Museum,
Harrisburg 049373
Susquehanna County Historical Society,
Montrose 051042
Susquehanna Museum of Havre de
Grace, Havre de Grace 049412
Sussex Combined Services Museum,
Eastbourne 044219
Sussex County Historical Society
Museum, Newton 051554
Sussex Farm Museum,
Heathfield 044550
Sussex Fine Art and Antiques Fair,
Little Horsted 098407
Sussi, Rīga 111767
Sussman, Arthur, Albuquerque . 119512
Sutama-cho Bijutsukan, Sutama 029414
Sutama-cho Rekishi Hakubutsukan,
Sutama 029415
Sutcliffe Galleries, Harrogate . . 089441,
134438
Sutcliffe Gallery, Whitby 046116
Sutcliffe Gallery Publication,
Whitby 138292
Sutcliffe, Malcolm, Penryn 118892
Sutcliffe, Tony & Anne, Southport 091385
Suter, Owen, Richmond 096581
The Suter, Te Aratoi o Whakatu,
Nelson 033254
Sutherland, New York 123106
Sutherland Art Gallery,
Sutherland 099755
Sutherland Steam Mill Museum,
Denmark 005701
Sutherland, Martha, New York . 123107
Sutter, Cleveland 092976
Sutter-Kress, Martina, Erlangen . 106899
Sutter, Karin, Basel 116100
Sutter's Fort, Sacramento 052732
Sutton, Fitzroy 099031
Sutton, New Orleans 122247
Sutton-Ditz House Museum,
Clarion 047766
Sutton Gallery, Melbourne 099322
Sutton-Grenée, Martine, Paris . . 071928
Sutton Lane, London 118586
Sutton Park, Sutton-on-the-
Forest 045928
Sutton Place Antiques, New York 095772
Sutton Valence Antiques, Charing 088824
Sutton Windmill and Broads Museum,
Sutton, Norwich 045927
Sutton, Jack, New Orleans 095167
Sutton, Lea, Tulsa 097760
Suty, Coye-la-Forêt 103639
Suzhou Antique, Suzhou 065163
Suzhou Art Gallery, Suzhou . . . 008295
Suzhou Beike Guan, Suzhou . . . 008296
Suzhou City Art Base, Suzhou . . 102241

Suzhou Folkways Museum,
Suzhou 008297
Suzhou Gardens Museum,
Suzhou 008298
Suzhou Museum, Suzhou 008299
Suzhou National Treasure
Numismatics Museum, Suzhou 008300
Suzhou Silk Museum, Suzhou . . 008301
Suzhou Sportqu Museum, Suzhou
Sportqu Art Institute, Suzhou . 008302
Suzu-yaki Kyodo Bijutsukan,
Suzu 029421
Suzuki Bokushi Kinenkan,
Shiozawa 029394
Svalbard-Museum, Longyearbyen 033757
Svaneholms Slotts Museum,
Skurup 040962
Svaneke Antikvitetshandel,
Svaneke 066154
Svaneti History-Ethnography Museum,
Mestia 016393
Svart och Vitt, Malmö 115771
Svato Verlag, Hamburg 137254
Svatopluk Čech, Louny 065421
Svaz Antikvářů ČR, Praha 058543
Svedinos Bil- och Flygmuseum,
Slöinge 040964
Sveindal Museum, Kollungtveit . 033703
Svelvik Museum, Svelvik 034004
Svendborg Antikvariat Æseløret,
Svendborg 140663
Svendborg Auktionerne,
Svendborg 125470
Svendborg og Omegns Museum,
Svendborg 010337
Svendsen, Annalena, New York . 095773
Svendsen, Birgit & Kaj,
Bjerringbro 140558
Svendsen, Kaj, Bjerringbro 140559
Svensgaard, Britta & Kim,
Herning 065753
Svensk Konst och
Antikvitetsförmedling, Malmö . 086860
Svensk Konsthandel, Uppsala . . 087126
Svensk Möbelkonservering,
Huddinge 133539
Svenska Antikvariatföreningen,
Stockholm 059904
Svenska Galleriförbundet,
Stockholm 059905
Svenska Konstgalleriet, Malmö . 115772
Svenska Konstkritikersamfundet,
Stockholm 059906
Svenska Konstnärer Galleri och
Bokförlag, Jönköping 115696, 138009
Svenska Möbler, Chicago 092797
Svenska Möbler, Los Angeles . . 094484
Svenska Rum Antikt, Stockholm 087065
Svenska Statens Porträttsamling,
Mariefred 040841
Svensson, Kurt, Stockholm 115932
Sverdén, Bo, Växjö 115995
Sveriges Allmänna Konstförening,
Stockholm 059907, 115933
Sveriges Järnvägsmuseum,
Gävle 040638
Sveriges Jörnvögsmuseum,
Ängelholm 040514
Sveriges Konst- och
Antikhandlareförening,
Stockholm 059908
Sveriges Konstföreningar Skåne,
Lund 059909
Sveriges Konstföreningars Riksförbund,
Malmö 059910
Sveriges Konsthantverkare och
Industriformgivare, Stockholm . 059911
Sveriges Museimannaförbund,
Nacka 059912
Sveriges Rundradiomuseum,
Motala 040865
Sveriges Sjömanshusmuseum,
Uddevalla 041099
Sveriges Teatermuseum, Nacka 040869

Sveriges Tennis Museum, Båstad 040559
Sveriges VVS-Museum,
Katrineholm 040746
Sverko, Mario, Toulouse 129317
Švestka, Jiří, Praha 102722
Svet-L, Kyïv 117150
Svetska Galerija na Karikaturi, Muzej
na Grad Skopje, Skopje 030634
Svettini, Adele, Milano 080739
Svijažskij Architekturno-
chudožestvennyj Muzej,
Gosudarstvennyj Muzej
izobrazitelnych iskusstv Respubliki
Tatarstan, Svijažsk 037906
Svjato-Troickij Sobor, Sankt-
Peterburg 037822
Svjatogorskij Monastyr, Puškinskie
Gory 037682
Svolis, Giannis, Zografou 109107
Svrzina Kuća, Sarajevo 004281
Swaen, Paulus, Paris . 125769, 141282
Swaffer, Spencer, Arundel 088289
Swaffham Museum, Swaffham . . 045929
Swain, Marilyn, Grantham 126907
Swainbanks, Liverpool 089861
Swakopmund Museum,
Swakopmund 031855
Swalcliffe Barn, Swalcliffe 045930
Swaledale Museum, Reeth 045600
Swan, London 090523
Swan, Moonah 139831
The Swan, Tetsworth . 091588, 135129
Swan Antiques, Baythorne End . 088410
Swan Antiques, Cranbrook 088995
Swan Books, Billericay 144248
Swan Coach House Gallery,
Atlanta 046638
Swan Fine Art, London 118587
Swan Gallery, Burford . 088713, 134181
Swan Gallery, Sherborne 091309,
119091, 135037
Swan Hill Regional Art Gallery, Swan
Hill 001532
Swan Street Antique Mall,
Louisville 094593
Swan Valley Historical Museum, Swan
River 007109
Swan, Michael, Dockyard 100821
Swan, Michael, Hamilton 100830
Swan, Milvia, Chicago 092798
Swanage Museum, Tithe Barn
Museum and Art Centre,
Swanage 045931
Swanage Railway, Swanage . . . 045932
Swanberg, Sacramento 096650
Swangrove, New York 095774
Swann, New York 127362, 145375
Swann, Nice 141006
Swans Antique Centre, Oakham 090904,
134902
Swansea Antiques Centre,
Swansea 091530
Swansea Museum, Swansea . . . 045937
Swanson, Minneapolis 122041
Swanson Cralle, Louisville 121671,
121672
Swanson, David, Petworth 091008
Swanson, Trevor V., Phoenix . . . 123516
Swanton Mill, Ashford 043442
Swarovski, Saint Louis 096782
Swarovski Kristallwelten,
Wattens 002900
Swart, J.A., Hercules 133072
De Swarte Leeuw, Middelburg . 083501
Swarthout Memorial Museum, La
Crosse 050031
Swastika, Delhi 109345
Swatland, Sally, New York 123108
Swayambunath Museum and Art
Gallery, Kathmandu 031877
Swaziland National Museum,
Lobamba 040508
Sweden Antique and Decoration,
Helsingborg 086741

Swedenborg, Boston . . 119924, 145017
Swedish American Museum
Association of Chicago, Chicago 060768
Swedish American Museum
Association of Chicago Museum,
Chicago 047700
Swedish Federation of Friends of
Museums, Enhorna 059913
Sweeney Art Gallery, University of
California, Riverside 052597
Sweeney, Kate, London 118588
Sweerts de Landas, Ripley,
Surrey 091115, 134970
Sweet Briar College Art Gallery, Sweet
Briar 053625
Sweet Briar Museum, Sweet
Briar 053626
Sweet Home, Hamburg 076323
Sweet Memories, Tampa 097590
Sweet Repeats, Charlotte 092631
Sweet Shop Museum, Donghe . 042303
Sweet Slumber Antiques,
Kempsey 061677
Sweetbriar Gallery, Helsby 089493
Sweetings, Belper . . . 088437, 134092
Sweetow, Patricia, San Francisco 124409
Sweetwater County Historical Museum,
Green River 049231
Sweety Pie Gallery, Béziers 103379
Swensbylijda, Svensbyn 041072
Swenson Memorial Museum of
Stephens County, Breckenridge 047187
Swertz, André, Utrecht 143251
Swetec, Franz, Düsseldorf 106833
Swetec, Franz, Griesheim 107164
Swets & Zeitlinger, Lisse 143194
Świętokrzyski Salon Numizmatyczny,
Kielce 084542
Świętowe Stowarzyszenie
Artystów Fotografików i Twórców
Audiowizualnych, Wrocław . . . 059720
Swift County Historical Museum,
Benson 046922
Swift Current Museum, Swift
Current 007111
Swift, K.W., Ludlow 144719
Swigart Museum, Huntingdon . . 049614
Swindler's Ridge Museum,
Benton 046923
Swindon, Hong Kong 140443
Swindon Museum and Art Gallery,
Swindon 045941
Swinford Museum, Filkins 044345
Swinkels, Berry, Eindhoven 083232
swipe, Toronto 136689
Swiridoff Verlag, Künzelsau 137333
Swisher County Archives and Museum
Association, Tulia 053846
Swiss Art, Stein am Rhein 139101
Swiss Historical Village, New
Glarus 051256
Swiss Institute – Contemporary Art,
New York 051477
Swiss Time, Atlanta . . 092188, 135316
Swissminiatur, Melide 041737
Świtalscy, Dariusz & Irena,
Gdańsk 084512, 084513
Switzerland County Historical Society
Museum, Vevay 053972
Swoboda, Bockenheim an der
Weinstraße 075048
Swope, Albuquerque 119513
Swope Art Museum, Terre Haute 053719
Sworder & Sons, Hertford 126923
Sworder & Sons, G.E., Stansted
Mountfitchet 127052
Sworders & Olivers, Sudbury,
Suffolk 127060
Swords and Ploughshares Museum,
Kars 006065
Swords Into Plowshares Peace Center
and Gallery, Detroit 048317
Swuste, Bosch en Duin 132449
Syarikat, Abdul, Melaka 082402

Syber, Saint Kilda, Victoria 139904
Syber, Windsor, Victoria 139962
Sybil Berg Contemporary Art, New
  York 123109
La Sybille, Paris 105254
Syddansk Universitetsforlag,
  Odense 136779
Sydenham Antiques Centre,
  London 090524, 134779
Sydfyns Antikvar Boghandel,
  Svendborg 140664
Sydhimmerlands Museum, Hobro 010126
Sydney Affordable Art Show, Moore
  Park 097996
Sydney and Louisburg Railway
  Museum, Louisbourg 006225
Sydney Antique Centre, Surry
  Hills 062270
Sydney Art on Paper Fair, Sydney 097997
Sydney Children's Museum,
  Merrylands 001290
Sydney Coins and Jewellery,
  Parramatta 062067
Sydney College of the Arts, University
  of Sydney, Rozelle 054919
Sydney Galleries, Sydney 099797,
                              099798
Sydney Heritage Fleet, Pyrmont 001436
Sydney Jewish Museum,
  Darlinghurst 001017
Sydney Observatory, Museum
  of Applied Arts and Sciences,
  Sydney 001552
Sydney Road Old Things,
  Balgowlah 061079
Sydney Tramway Museum,
  Loftus 001227
Sydsjællands Museum,
  Vordingborg 010364
Sydslesvigs Danske Kunstforening,
  Flensburg 059243
Sygun Museum of Wales,
  Beddgelert 043547
Sykepleiemuseum, Oslo 033865
Sykes, Christopher, Woburn ... 091896
Sykkylven Naturhistorisk Museum,
  Sykkylven 034007
Syl'Antic, Le Peage-de-Roussillon 069323
Syllavethy Gallery, Alford 117226
Syllogy Art Gallery, Toronto ... 101710
Sylter Gemälde Cabinett, Sylt .. 108754
Sylter Heimatmuseum, Sylt 022093
Sylter Kunstfreunde e.V., Sylt .. 059244
Sylvain, Jacqueline, Gonfreville-
  l'Orcher 068680
Sylvan, Kolkata 079398
Sylvan Mall Antiques, Toledo ... 097619
Sylvania Galleries, Sylvania 099800
Sylvia Art Antic, Gracay 068692
Sylvia Plotkin Judaica Museum,
  Scottsdale 053213
Sylvia's Attic, New Orleans 095168
Symboles de l'Art, Saint-Palais . 073236
Synagóga – Centrum Súčasného
  Umenia, Galéria Jána Koniarka v
  Trnave, Trnava 038475
Synagoge, Bourtange 032091
Synagoge, Worms 022768
Synagoge Saffig, Saffig 021568
Synagoge Stommeln, Pulheim-
  Stommeln 021235
Synagoge und Jüdisches Museum
  Ermreuth, Ermreuth 018271
Synagoge Wittlich, Wittlich 022703
Synagogen-Museum, Urspringen 022287
Synagogue for the Arts, New
  York 051478
Synart Art Gallery, Frankfurt am
  Main 107034
Syndicat Belge de la Librairie
  Ancienne et Moderne, Liège .. 058388
Syndicat des Antiquaires,
  Marseille 058681
Syndicat des Antiquaires et

Brocanteurs des Vosges, Bains-les-
  Bains 058682
Syndicat Français des Experts
  Professionnels en Oeuvres d'Art et
  Objets de Collection, Paris ... 058683
Syndicat Lorrain Professionnel de
  l'Antiquité, Rosières-aux-Salines 058684
Syndicat Mixte de Brouage, Corps de
  Garde, Hiers-Brouage 013265
Syndicat Nacional des Artistes
  Plasticiens, Paris 058685
Syndicat Nacional des Artistes
  Professionnels, Paris 058686
Syndicat Nacional des Graphistes,
  Paris 058687
Syndicat Nacional des Peintres
  Sculpteurs Professionnels, Paris 058688
Syndicat National de la Librairie
  Ancienne et Moderne, Paris .. 058689
Syndicat National des Antiquaires,
  Paris 058690
Syndicat National des Artistes Auteurs,
  Paris 058691
Syndicat National du Commerce
  de l'Antiquité et de l'Occasion,
  Paris 058692
Syndicat Romand des Antiquaires,
  Paudex 059969
Synergy Art Foundation, Solana
  Beach 060769
Sýningarsalur Íslenskrar Grafíkur,
  Reykjavík 109244
Synopsis m, Lausanne 116427
Synskadades Museum, Enskede 040604
Syntax. Medienproduktion + Verlag,
  Mannheim 137373
Synthèse, Bruxelles 100660
Synthèse, Paris 105255
Synthese Gesellschaft für
  Restaurierung, Budapest 130815
Syon House, London 045114
Sýpka, Brno 102573
Syplacz, Arnold, Hagen 076154, 107198
Syracuse University Art Collection,
  Syracuse 053638
Syrlin-Kunstverein e.V., Stuttgart 059245
Sysmäen Kotiseutumuseo,
  Sysmä 011387
Syvert, Mandal 084275
Syvertsen, Kristiansand 084271
Syzranskij Gorodskoj Kraevedčeskij
  Muzej, Syzran 037917
Szaal, Horst & Wolfgang, Wien . 100382
Szabadtéri Bányászati Múzeuma,
  Tatabánya-Múzeum, Tatabánya 024002
Szabadtéri Néprajzi Gyűjtemény,
  Szenna 023951
Szabadtéri Néprajzi Múzeum,
  Nagyvázsony 023809
Szabadtéri Néprajzi Múzeum,
  Szentendre 023960
Szabadtéri Néprajzi Múzeum,
  Tihany 024008
Szabadtéri Néprajzi Múzeum – Tájház,
  Bakonybél 023421
Szabó Pál Irodalmi Emlékház,
  Munkácsy Mihály Múzeum,
  Biharugra 023447
Szaivert, Dr. Eva, Wien 063083
Szalay-Gyűjtemény – Petróház,
  Miskolci Galéria, Miskolc ... 023793
Szalinski, I., Osnabrück 077980
Szálkahalmi Orháza, Hortobágyi
  Nemzeti Park, Balmazujváros . 023432
Szalom, Kraków 113643
Szántó Kovács János Területi Múzeum,
  Oroszháza 023819
Szántó Piroska Emlékmúzeum,
  Budapest 023552
Szarka, David, Luzern 133845
Szatmárcsekei Tájház,
  Szatmárcseke 023919
Szatmári Múzeum, Mátészalka . 023772
Szczepanski, G., Meppen 077346

Szczypka-Kozłowska, Grażyna,
  Łomianki 132852
Széchenyi Irodalmi és Művészeti
  Akadémia, Magyar Tudományos
  Akadémia, Budapest 055768
Széchenyi István Emlékmúzeum,
  Nagycenk 023802
Széchenyi Patika, Győr 023666
Szedlarik, Wilhelm, Bad
  Hofgastein 139973
Szegedy Róza Ház Irodalmi
  Emlékmúzeum, Badacsony .. 023415
Székesegyházi Kincstár és
  Egyházmegyei Gyűjtemény, Vác 024026
Székesfehérvári Egyházmegyei
  Múzeum, Székesfehérvár ... 023940
Szelag, Saint Louis 136206
Szent Imre Antikvárium,
  Budapest 126157, 142572
Szent István Király Múzeum,
  Székesfehérvár 023941
Szent István Király Múzeum –
  Rendház, Székesfehérvár ... 023942
Szentendrei Képtár, Ferenczy Múzeum,
  Szentendre 023961
Szepanski, Peter, Stuttgart ... 078636
Szépművészeti Múzeum,
  Budapest 023553
Szerb Egyházy Múzeum,
  Szentendre 023962
Szerecsen Patikamúzeum, Pécs 023553
Szczląg, Warszawa 113904
Sziget, Budapest 079296
Sziget Galéria, Budapest 109178
Szín-Tér Galéria, Győr 109194
Szindlowski, Kuno, Allensbach . 074377
Színháztörténeti és Színészmúzeum,
  Miskolci Galéria, Miskolc ... 023794
Szinyei-ház, Sárospatak 023875
Szkatułka, Warszawa 113905
Szlos, Lisette, Paris 071929
Szlovák ház – Emlékház, Salašské
  múzeum, Tótkomlós 024017
Szlovak Nemzetiségi Kiállítás, Palóc
  Múzeum, Nógrádi Megyei Múzeum
  Szervezet, Bánk 023434
Szlovák Tájház, Békéscsaba .. 023445
Szlovák Tájház, Szarvas 023916
Szlovák Tájház, Petőfi Emlékmúzeum,
  Kiskőrös 023743
Szlovák Tájház, Slovenský oblastný
  dom, Tótkomlós 024018
Szoborpark, Budapest 023554
Szokoly, Veszprém 142584
Szolnay-Museum, Potzneusiedl . 002525
Szolnoki Galéria, Damjanich János
  Múzeum, Szolnok 023979
Szombathelyi Képtár,
  Szombathely 023985
Szombathelyi Római Katolikus
  Egyházmegyei Gyűjtemény és
  Kincstár, Szombathely 023986
Szőnyi, Budapest 142573
Szőnyi István Emlékmúzeum,
  Zebegény 024058
Szopinski, Neustadt an der
  Weinstraße 077764
Szórakaténusz Játékmúzeum,
  Kecskemét 023733
Sztuka – Stowarzyszenie Artystyczne,
  Warszawa 059721
Sztuka Dalekiego Wschodu, Muzeum
  Okręgowe w Toruniu, Toruń . 035288
Sztuka Dawna, Sopot 084668
Sztuka Dawna, Starogard
  Gdański 084669
Sztuka i Antyki, Bielsko-Biała . 084492,
                              113436
Sztuka i Filozofia, Warszawa . 138992
Szuhay, Peter, London 118589
Szukits, Szeged 142582
Szulevicz, B. & G., Skive 066123
Szuz Mária Kegy- és Emléktárgyak
  Gyűjteménye, Gyula 023679

Szwajcer, Micheline, Antwerpen 100512
Szy, Dénes, Düsseldorf 075528
T & K French Antiques, New
  York 095775, 136010
T E S Art, Mosta 111916
T M Ateljé, Västerås 115983
T Studio, Los Angeles 121604
T. & G. Art Gallery, Hoi An 125158
T. Tumanishvili Children Toy Museum,
  Tbilisi 016425
T.Art Spazio Espositivo Polifunzionale,
  Milano 110464
T.B. Ferguson Home, Oklahoma
  Historical Society, Watonga .. 054226
T.C. Steele State Historic Site,
  Nashville 051172
T.C. Ziraat Bankası Müzesi,
  Ankara 042830
T.E.D., Le Puy-en-Velay 069346
T.F. Chen Cultural Center, New World
  Art Center, New York 051479
T.I.A., Migennes 104236
T.J.'s Gallery on Evans, Mackay 099244
T.W. Wood Gallery and Arts Center,
  Montpelier 051035
T20, Murcia 115319
Ta' Kola Windmill, Xaghra 030770
Ta Matete, Bologna 109868
Ta Matete Living Gallery, Roma . 142841
Taalintehtaan Ruukinmuseo,
  Dalsbruk 010762
Taba, Murcia 133408
Tabak- und Zigarrenmuseum,
  Menziken 041741
Tabak Musea, Geraardsbergse Musea,
  Geraardsbergen 003651
Tabak-Museum der Stadt Hockenheim,
  Hockenheim 019273
Tabakhistorische Sammlung
  Reemtsma im Museum der Arbeit,
  Hamburg 019022
Tabakmuseum, Lorsch 020118
Tabakmuseum, Oberzeiring ... 002475
Tabakmuseum, Städtische Museen
  Schwedt/Oder, Schwedt / Oder 021778
Tabako to Shio no Hakubutsukan,
  Tokyo 029596
Tabakpfeifenmuseum und Museum für
  Stadtgeschichte, Ruhla 021543
Tabakspeicher, Museum für
  Nordhäuser Handwerks-
  und Industriegeschichte,
  Nordhausen 020829
Tabán Múzeum, Budapest 023555
Tabáni Tájház, Szolnok 023980
La Tabard, Booterstown 109534
Tabasso, Maddalena, Milano .. 080740
Tabbernee, D., Rotterdam 083593
Tabene, Saint-Jean-de-Luz ... 105536
Taber, Sacramento 096651
Taber and District Museum,
  Taber 007119
Taberna archaeologica, Mainz . 020234
Taberna Libraria, Fortaleza ... 140284
Taberna Libraria, Pistoia 142789
Tabiah, Mohammed Samir,
  Damascus 088062
Tabibnia, Milano 080741
Tabla y Lienzo, Madrid 086266
Table Rock Historical Society Museum,
  Table Rock 053639
Tableaux à Tous Prix, Dax 103656
Tabler, Joseph, San Diego 145531
Tabley House, Knutsford 044743
Taboada, Austin 119705
Tabor Opera House Museum,
  Leadville 050202
Táborvárosi Múzeum, Budapesti
  Történeti Múzeum, Aquincumi
  Múzeum, Budapest 023556
Tábua Rasa, Lisboa 132958
Tabula, Zamora 086585, 133484
Tabula Buch & Kunst, Tübingen . 142432
Tabula Rasa, Paris 129149

Táburi, Santiago, Lima 113368
Tabusintac Centennial Museum,
  Tabusintac 007120
Tabutin, Hervé, Marseille 125640
Taby, Paul, Sens 073504
Tacconi, Dina, Modena 131589
Tache d'Argent, Mons 100795
Taché-Levy, Bruxelles 100661
Tache, Carles, Barcelona 114966
Tachibana, Kyoto 082040
Tachihara Michizo Kinenkan,
  Tokyo 029597
Tachmes, Leonard, Miami 121865
Tacit, Thornbury 099822
Tacken, H.W., Velserbroek 132656
Tackenberg, Hürth 107451
Tacoma Art Museum, Tacoma .. 053649
Tacticos, Timişoara 085141
Tactual Museum, Greek Lighthouse for
  the Blind, Kallithea 023106
Tacussel, Paul, Marseille 136838
Tadema, London 090525, 118590
Tadmor, Dubai 088190
Tadre Mølle, Hvalsø 010149
Tadu, Bangkok 116963
Tadzio, Kyïv 117151
Tändsticksmuseet, Jönköping .. 040721
Täuber, Claudia, Wien 063084
Täuber, Hans-Joachim, Bad
  Salzuflen 074585
Taexpolsa, Lima 113369
Taff, James F., Sacramento 096652
The Taft Museum of Art,
  Cincinnati 047749
TAG, Den Haag 112415
TAG Art Gallery, Nashville 122102
Tag Diamond Jewelry, Buffalo .. 092561
Taganrogskaja Kartinnaja Galereja,
  Taganrog 037921
Taganrogskij Gosudarstvennyj
  Literaturnyj i Istoriko-architekturnyj
  Muzej-zapovednik, Taganrog .. 037922
Taganrogskij Kraevedčeskij Muzej,
  Taganrogskij gosudarstvennyj
  literaturnyj i istoriko-architekturnyj
  muzej-zapovednik, Taganrog . 037923
Taganrogskij Muzej im. A.P. Čechova,
  Taganrog 037924
Tagawa Bijutsukan, Tagawa ... 029424
Tage, E., Philadelphia 096164
Tagelöhner- und Brennereimuseum
  Faißt, Bad Peterstal-Griesbach 016842
Taggart, Hollis, Chicago 120280
Taggart, Hollis, New York 123110
Tagliapietra, Ivano, Torino 132208
Tagliapietra, Lino, Seattle 124607
Tagliapietra, Maurizio, Verona .. 132302
Tagliareni, Salvatore J.,
  Philadelphia 123460, 136106
Taha Hussein Museum, Cairo .. 010565
Tahir Pasa Konagi, Mudanya ... 043029
Tahiti Imports, Honolulu 093574
Tahsis Museum, Tahsis 007125
Tahydromiko Moyseio, Athinai .. 023029
Tai, Hong Kong 065098
Tai Hao Ling Museum, Huaiyang 007904
Tai Sing, Hong Kong 065099
Tai Wo & Co., Hong Kong 065100
Tai & Co., J.T., New York 095776
Tai'an Museum, Tai'an 008303
Taicang Museum, Taicang 008304
Taichung Distillery Museum,
  Taichung 042478
Taide, Helsinki 138624
Taide ja Kehys Pekkala, Oulu .. 103166
Taidehistorian Seura, Föreningen för
  Konsthistoria, Helsinki 058583
Taidekeidas, Turku 066406
Taidekeskus Mältinranta,
  Tampere 011406
Taidekoti Kirpilä, Helsinki 010872
Taidemaalariliiton Galleria,
  Helsinki 103158
Taidemuseo, Kokkola 011020

Taidesalonki, Helsinki . 066366, 103159
Taidesalonki Husa, Tampere ... 066398
Taidesalonki Piirto, Helsinki 103160
Taieri Historical Museum, Outram 033266
Taifa Llibres, Barcelona 143733
Taiga, Tobetsu 082128
Taihape and District Museum,
  Taihape 033303
Taihe Museum, Taihe 008306
Taiheiyo Bijutsukai, Tokyo 059514
Taikomosios dailės, Vilnius 111811
Taikomosios dailės Muziejus, Lietuvos
  Dailės Muziejus, Vilnius 030543
Taikoo Books, York 144945
Tailai Museum, Tailai 008307
Tain and District Museum, Tain . 045943
Tainan County History Museum,
  Syuejia 042470
Tainan Municipal Cultural Center,
  Visual Art Section, Tainan ... 042493
Tainan National University of the Arts,
  Kuantien 056455
Tainan National University of the Arts
  Museum, Kuantien 042376
Tainan Natural History Museum,
  Zuojhen 042612
Taining Museum, Taining 008308
The Taipa Houses Museum,
  Macau 008075
Taipei Astronomical Museum,
  Taipei 042550
Taipei Caves Gallery, Taipei ... 116942
Taipei Fine Arts Museum, Taipei 042551
Taipei National University of the Arts,
  Taipei 056460
Taipei Story House, Taipei 042552
Taiping Kingdom History Museum,
  Nanjing 008117
Tairawhiti Museum, Gisborne .. 033208
The Taise Gallery, Ballycastle .. 117272
Taishan Doll Museum, Taishan . 042566
Taishan Museum, Taishan 008309
Taishun Museum, Taishun 008310
Taitung Aboriginal Exhibition Room,
  Taitung 042568
Taitung Museum of Natural History,
  Chenggong 042287
Taivalkosken Kotiseutumuseo,
  Taivalkoski 011389
Taivassalon Museo, Taivassalo . 011390
Taiwan Aboriginal Culture Park,
  Majia 042394
Taiwan Antique Dealers' Association,
  Taipei 059995
Taiwan Coal Mine Museum,
  Pingsi 042413
Taiwan Craft Center, Taipei 042553
Taiwan Democratic Movement
  Museum, Taiwan Tsunah Foundation,
  Wujie 042584
Taiwan Folk Arts Museum, Taipei 042554
Taiwan Folk Culture Museum,
  Bali 042269
Taiwan Folklore Museum,
  Taichung 042479
Taiwan Forestry Exhibition Hall,
  Taipei 042555
Taiwan High Speed Rail Exhibition
  Center, Jhubei 042348
Taiwan Historica, Nantou 042407
Taiwan International Visual Arts Center,
  Taipei 042556
Taiwan Medical Museum,
  Kaohsiung 042368
Taiwan Museum of Art Sanri,
  Sanyi 042442
Taiwan Mushroom Museum,
  Wufeng 042583
Taiwan Nature Lacquer Culture
  Museum, Taichung 042480
Taiwan Silk Culture Museum,
  Gongguan 042317
Taiwan Sugar Industry Museum,
  Tainan 042494

Taiwan Television Exposition,
  Taipei 042557
Taiwan Temple and Shrine Art
  Museum, Douliou 042308
Taiwan Theater Museum, Yilan . 042592
Taiyi, Shanghai 102202
Taiyuan Painting Gallery, Taiyuan 008315
Taizhou Museum, Taizhou 008316
Taizi Harada Bijutsukan, Suwa . 029419
Taizz National Museum, Taizz . 054833
Le Taj, Cairo 066276
Tajan, Lausanne 126737
Tajan, Paris 125770
Tájház, Balatonszentgyörgy ... 023430
Tájház, Buzsák 023576
Tajmyrskij Okružnoj Kraevedčeskij
  Muzej, Dudinka 036973
Taka Ishii Gallery, Tokyo 111508
Takachiho-cho Rekishi oyo Minzoku
  Hakubutsukan, Takachiho ... 029428
Takács, Dezső, Győr 079311
Takada, San Francisco 124410
Takahama-shiritsu Kawara Bijutsukan,
  Takahama 029429
Takahashi, Kyoto 082041
Takahashi, Imono Kojo, Kyoto . 082042
Takamatsu City Museum of Art,
  Takamatsu 029437
Takamatsuzuka-Hekigakan,
  Asuka 028496
Takamiya, Osaka 111367
Takamura's Cottage Memorial
  Museum, Hanamaki 028609
Takanawa Ippodo Lobby Gallery,
  Tokyo 111509
Takaoka-shi Bijutsukan, Takaoka 029441
Takaoka Shiritsu Hakubutsukan,
  Takaoka 029442
Takara, Hiroshima 111263
Takarada, T. & M., Yokohama .. 082233
Takarazuka Daigaku Hakabutsukan,
  Takaraduka 029443
Takarazuka Zokei Geijutsu Daigaku,
  Takaraduka 056034
Takasaki City Gallery, Takasaki . 029447
Takasaki-shi Bijutsukan, Takasaki 029448
Takasawa Folk Craft Museum,
  Niigata 029177
Takashi, Wada, New York 123111
Takashimaya, Singapore 114617
Takayama Jinya, Takayama ... 029460
Takayama-shi Kyoudokan,
  Takayama 029461
Takehisa Yumeji Ikaho Kinenkan,
  Ikaho 028676
Takenaka Carpentry Tools Museum,
  Kobe 028854
Takenaka, Kenji, Kyoto 137769
Takenouchi Kaido Rekishi
  Hakubutsukan, Taishi 029425
Takeuchi, Tokyo 142942
Takeyama, Eiichi, Rothenburg ob der
  Tauber 108526
Takht-e-Jamshid, Marvdashd .. 024719
Takht-e-Jamshid, Shiraz 024738
Takhté-Djamchid, Tehrān 109522
Taki, Ayse, İstanbul 117086
Takkenberg, M., Zaandam 083699
Tako no Hakubutsukan, Tokyo .. 029598
Taksu, Singapore 114618
Takumi, Tokyo 082209
Tal, Bordeaux 067360
Tal Museum, Engelberg 041466
Talabardon & Gautier, Paris ... 105256
Talacea, Vigo 086566
Talamini, Gianni, Padova 110580
Talana Museum and Heritage Park,
  Dundee 038695
Talaos, Auray 066818
Talas, New York 136011
Talavan Moraleda, Felipe, Madrid 133392
Talaverano, Manuela, Sevilla .. 115459

Talboom, Antwerpen 140155
Talbot & Co, Edwin, Carlisle ... 117534
Talbot Arts and Historical Museum,
  Talbot 001558
Talbot Court Galleries, Stow-on-the-
  Wold 091485, 119163, 135099
The Talbot Gallery, Dublin 109622
Talbot House, Poperinge 003953
Talbot Rice Gallery, Edinburgh . 044278
Talbot Walk Antique Centre, Ripley,
  Surrey 091116
Taldykorganskij Istoriko-Kraevedčeskij
  Muzej, Taldykorgan 029849
Tale of Genji Museum, Uji, Kyoto 029676
Talebloo, New Orleans 095169
Talec, Agnès, Paris 141283
Talen, K., Staphorst 132637
Talens, Leeuwarden 083420
Talent Arts Gallery, Singapore . 114619
Les Talents, Luxembourg 111868
Talents, Manama 100452
Talents Fine Arts, Malton 090684,
                               134827
Talents Fine Arts, Malton North . 118700
Tales of the Old Gaol House, King's
  Lynn 044706
The Talish Gallery, Edinburgh .. 089192
Talisman, Gillingham, Dorset .. 089326
Talisman, Washington 124963
Talisman Antiques, Oakland ... 095914
Talisman Fine Art, Looe 118663
Talisman Gallery, Little Hartley . 099235
Talisman Gallery, Portland 123683
Talk of the Town, Dallas 093220, 135518
Talk of the Town, Houston 093806
Talkeetna Historical Society Museum,
  Talkeetna 053656
Talking Machine, London 090526
Talking Point Antiques, Sowerby
  Bridge 091394
Tall Ships Bookshop, Manawatu 143326
Tallahassee Museum of History and
  Natural Science, Tallahassee .. 053664
Tallbo, Järbo 040716
Tallec, Bordeaux 067361
El Taller, Bogotá 128278
Taller 11–12, San José 102516
El Taller Antigüedades y Restauración,
  Medellín 065310, 065311, 128310
El Taller de Angelo, Lugo 133283
Taller de Arte y Galería Universal,
  Medellín 102461
Taller de Arte y Restauración El Moral,
  Bogotá 065230, 128279
Taller de Artes, San Antonio .. 124075
Taller de Artes Ecuador, Bogotá 065231
Taller de Ceramica Nerea, Cali . 128291
Taller de Esculturas Pátinas y Bronces,
  Medellín 128311
El Taller de Marcos, Castellón de la
  Plana 133227
Taller de Restauración, Cali ... 128292
Taller de Restauracion, León .. 086065,
                               133280
Taller de Restauración del Patrimonio,
  Madrid 133393
Taller del Artista, Granada 133271
Taller el Trastam, Valencia 133474
Taller Escuela de Grabado Tres en
  Raya, Madrid 056375
Taller Galería de Robles, Madrid 133394
Taller Galeria Fort in Cadaqués-
  Barcelona, Cadaqués 115001, 137973
El Taller Gallery, Austin 119706
Taller Miró, Santiago de Chile . 064986
Taller Puertoriqueno, Philadelphia 052097
Taller Puertorriqueno,
  Philadelphia 123461
Taller Restauracion Antiguedades,
  Zaragoza 133515
Taller Tanamá y Casa Jesús T. Piñero,
  Canóvanas 036126
Taller y Almacén de Artes Aplicadas el
  Arcano, Cali 128293

Taller, Adolf, Sant Cugat del
Vallès 115416
Talleres Creativos Reojo, Madrid 133395
Talleres de Restauración Arti,
Bogotá 128280
Tallgauer, G., Stuhr 078587
Tallgauer, U., Stuhr 078588
Tallinna Linnamuuseum, Tallinn . 010687
Tallkrogens Gammalt & Nytt,
Enskede 086663
Tallon, Venezia 111163
Tally Antique Auctions,
Tallygaroopna 125240
Talma Mill Art Gallery, Christ
Church 100464
Talmuseum Hüttschlag,
Hüttschlag 002110
Talmuseum Kaunertal, Feichten . 001876
Talmuseum Lachitzhof, Klein Sankt
Paul 002200
Talmuseum Samnaun, Chasa Retica,
Samnaun Compatsch 041919
Talmuseum Ursern, Andermatt . 041221
Talomuseo, Kuortane 011058
Talomuseo, Polvijärvi 011262
Taloni, Cristiano, Roma 081574
Taloni, Piero, Roma 081575
Talonpoikaismuseo Yli-Kirra,
Punkalaidun 011285
Talonpojanmuseo, Kälviä 010946
Talsu Novada Muzejs, Talsi 030236
Talton, J.W., Long Sutton,
Lincolnshire 090629, 134814
Talvisen Työläismökki, Humppila 010889
Talvisotanäyttely, Kuhmon
Talvisotamuseo, Kuhmo 011044
Talwar, New York 123112
Tam, Hong Kong ... 102119, 102120,
102121
Tam O'Shanter Experience,
Alloway 043398
Tama Bijutsu Daigaku, Tokyo . 056056
Tama County Historical Museum,
Toledo 053756
Tamagawa Kindai Bijutsukan,
Tamagawa 029466
Tamagnini, Massimo, San Marino 085289,
133030
Tamalát, Køge 065972
Taman Warisan Tasek Merimbun,
Muzium Brunei, Mukim Rambai 004968
Tamar, New York 123113
Tamara, Tel Aviv 126184
Tamara Hanum Yodgorlik Muzeyi,
Toshkent 054669
Tamarillo, Wellington 113092
Tamarind Institute, University of New
Mexico, Albuquerque 056638
Tamastslikt Cultural Institute,
Pendleton 051970
Tamba Tachikuu Traditional Ceramic
Center, Sasayama 029347
Tambaran, New York 123114
Tambasco, Montevideo 127113
Tamblyn, Alnwick 088255
Tamborine Mountain Heritage Centre,
Eagle Heights 001042
Tambovskaja Oblastnaja Kartinnaja
Galereja, Tambov 037926
Tambovskij Gosudarstvennyj
Universitet, Tambov 056305
Tambovskij Oblastnoj Kraevedčeskij
Muzej, Tambov 037927
Tamburelli, Claudio, Roma 081576
Tamburini, Joseph, Uzer 073920
Tamburini, Roland, Paris 071930
Tamenaga, Paris 105257
Tamenaga, Tokyo 111510
Tamia, Atlanta 092189
Tamking University Maritime Museum,
Tamsui 042572
Tamlyn & Son, Bridgwater 126836
Tammaro, Luciano, Napoli 131644
Tammaro, Michael, New York .. 123115

Tammen & Seitz, Berlin 106447
Tammen, R. & U., Oldenburg .. 077943
Tammiharjun Sairaalamuseo,
Ekenäs 010767
Tampa Antiquarian Book Company,
Tampa 145617
Tampa Bay History Center,
Tampa 053671
Tampa Museum of Art, Tampa . 053672
Tampa Realistic Artists, Tampa . 060770
Tampereen Luonnontieteellinen Museo,
Tampereen Museot, Tampere . 011407
Tampereen Nykytaiteen Museo,
Tampere 011408
Tampereen Saskiat, Tampere . 103168
Tampereen Taidemuseo – Pirkanmaan
Aluetaidemuseo, Tampere 011409
Tampereen Taidemuseon Muumilaakso,
Tampere 011410
Tampico, San Francisco 097334
Tamralipta Museum and Research
Center, Tamluk 024488
Tamsen Munger, Fresno 120909
Tamura, Tokyo 142943
Tamworth Castle Museum,
Tamworth 045944
Tamworth Regional Gallery,
Tamworth 001560
Tamyang Bamboo Museum,
Tamyang 030056
Tamy's Antiques, New York 095777
Tan Oe Pang Art Studio,
Singapore 038276
Tan, Ah Han, Singapore 085509
Tan, C. K., Singapore 085510
Tan, Sim, Kuala Lumpur 111907
Tana Museum, Tana 034009
Tanabe Bijutsukan, Matsue 029007
Tanagra, Montceau-les-Mines .. 070226
Tanagra, Strasbourg 073599
Tanah Art Museum, San Antonio 004179
Tanaka, Nagasaki 082054
Tanaka, Nagoya 082064
Tanaka, Nara 082076
Tanaka, Tokyo 111511
Tanaka Bijutsukan, Suzaka 029420
Tanakamaru Collection, Fukuoka 028552
Tanakami Mineral Museum, Otsu 029282
Tanakaya – Art du Japon, Paris 071931,
105258
Tanawha Antiques, Tanawha ... 062304
Tanazawa, Tokyo 142944
Tanca, Milena, Roma 081577
Tandanya, National Aboriginal Cultural
Institute, Adelaide 000772
Tandart, Bois-Colombes 067268
Tandem Press, Madison 138354
Tandy, Robert, Bristol 134165
Tandy, Robert, Clevedon 134281
Tanegashima Teppokan,
Nishinomote 029190
Tanfer, Madrid 133396
Tang, Bangkok 116964
T'ang Horse, Singapore 085511
Tang Teaching Museum and Art
Gallery, Skidmore College, Saratoga
Springs 053161
Tang, Baoqu, Hong Kong 102122
Tang, Dapiao, Shanghai 102203, 102204
Tang, G.E., Shanghai .. 065151, 065152
Tang, Sarina, New York 123116, 123117
Tangdai Art Museum, Xian 008400
Tange, Valenciennes 073948
Tångeberg, Peter, Tystberga .. 133593
Tanger, Albert, Fort-de-France . 132352
Tangerding-Dörner, Bocholt 075030
Tangjia Glaswegiansu Guan,
Xunyi 008454
Tanglewood, Vancouver 140428
Tanglin Antique Gallery,
Singapore 085512
Tango, Montréal 064480
Tango Art Deco, Warwick 091751
Tango Chirimen Rekishi Hakubutsukan,

Nodagawa 029194
Tango Photographie, Montréal .. 101454
Tangram, Cannes 103523
Tangshan Art School, Tangshan . 055243
Tangshan Museum, Tangshan . 008317
Tanguy, Jean-Pierre, Crézancy-en-
Sancerre 068243
Tania's Gallery, Atlanta 119619
Tanimura Bijutsukan, Itoigawa . 028717
Tanis, Paris 071932
Tanis, Jean-Louis, Le Castera .. 069241
Tanja, Beograd 085350
Tanjore Collection, Chennai ... 109293
Tank Museum, Bovington 043685
Tankerton Antiques, Whitstable . 091837
Tankosha, Kyoto 137770
Tankosha, Tokyo 137792
Tanks Art Centre, Edge Hill ... 098996
Tanks Museum, Avlona 023033
The Tanks of Aden, Aden 054825
Tanneberger, Rüdiger, Zwickau . 079154
Tanner, Bern 138033
Tanner-Bürki, Cathrin,
Ostermundigen ... 116557, 133884
Tanner Chaney, Albuquerque .. 119514
Tanner, Daniel, Basel 116101
Tanner, Simon, Roma 142842
Tans Place, Blenheim 143298
Tansu Design, San Francisco . 097335
Tanta Museum, Tanta 010586
Tantalus Antiques, Troon 091638
Tantaquidgeon Indian Museum,
Uncasville 053883
Tante Blanche Museum,
Madawaska 050578
Tante Pose, Bergen 084228
Tantekin Antik, İstanbul 088141
Tanti, Paul, Sliema 082433, 132343
Tantius, Hans-Gerd, Lübbecke .. 107904,
137359
Tantus Galerie, Hamburg 107317
Tanuki la Galerie, Lyon 104136
Tanums Hällristningsmuseum –
Underslös, Tanumshede 041074
Tanya Art Restoration, New York 136012
Tanyamúzeum, Szabadkígyós .. 023912
Tanybryn Gallery, Tanybryn 099809
Tanzer, Dianne, Fitzroy 099032
Tanzilli, Roland, Meximieux 070109
Tanzillo, Gaetano, Napoli 131645
Tanzimat Müzesi, İstanbul 042958
Tanzini, Armando, Malindi 082241
Tao, Mumbai 109480
Taos Center for the Arts, Taos . 053679
Taos Historic Museums, Taos .. 053680
Taoyuan County Museum of Natural
History, Taoyuan 042576
Tap net, Roma 132058
TAP Gallery, Darlinghurst 098946
TapaFa, Wellington 113093
Tapestries Etc., New York 123118
Tapestry, Cheltenham . 088843, 134244
Tapestry Antiques, Hertford 089520
Tapetenwerk-Galerien, Leipzig .. 107984
Tápies, Toni, Barcelona 114967
Het Tapijtmuseum, Genemuiden 032342
Tapinquiri, La Paz 100850
Tapiola-Antikvariaatti, Espoo .. 140685
Tapiolan Taide ja Kehys, Espoo . 103127
Tapis, Seattle 097511
Tapisseries Jean Lurçat,
Rocamadour 015114
Tapissier, Michel, Reims 072404
Tapissier, Michel, Tinqueux 073684
Tappantown Historical Society
Museum, Tappan 053681
Tapparini, Silvano, Verona 081940
Tappenden, London 118591
Tapsell, Beechworth 139637
Tapsell, Rutherglen 139902
Tapsell, C., London 090527
Tapsell, Gillian, Rye 118997
Tapulimuseo, Lempäälä 011106
Tar Tunnel, Coalport 043978

Tara, Thomastown 079704
Tara & Kys, Ho Chi Minh City .. 125148
Tara & Kys, Siem Reap 101208
Tara and District Pioneer Memorial
Museum, Tara 001562
Tara Antiques, Atlanta 092190
Tara Antiques, London 090528
Taraj, Fatmir, Saint-Ouen 073204
Taralga Historical Museum,
Taralga 001563
Taranaki Arts Centre, Taranaki . 113018
Tarangau Caravan Park, Broome 098810
Taranis, Rožnov pod Radhoštěm 065552
Taranna, Barcelona 114968
Tarantola, Lucia, Milano 131538
Taraporevala, D.B., Mumbai ... 137599
Taras H. Shevchenko Museum
& Memorial Park Foundation,
Toronto 007209
Tara's Hall, Hadleigh 089403
Tarasiève, Suzanne, Barbizon .. 103335
Tarasiève, Suzanne, Paris 105259
Tarazi, Michel Emile, Broumana 082313
Tarble Arts Center, Eastern Illinois
University, Charleston 047538
Tarbouriech, Guy & Martine, Bagnols-
sur-Cèze 066916
Tarchiano, Umberto, Firenze .. 131273
Tardieu, Loic, Bordeaux 128507
Tardif, François, Jouars-
Pontchartrain 068916, 128632
The Tardis, Christchurch 083900
Tardy, Dany, Fontaine-le-Dun .. 068561
The Tareq Rajab Museum,
Hawelli 030066
Targa, Stefano, Torino 081764
Target, Moskva 133016
Target Gallery, London 118592
Targhetta, Andrea, Venezia ... 132259
Tari László Múzeum, Csongrád . 023590
Tarih, Sarajevo 063874
Tank Zafer Tunaya Kültür Merkezi,
İstanbul 117087
Tarka Books, Barnstaple 144225
Tarkasis Pantelis, Eptalofos ... 109063
Tarkiainen, Helsinki 066367
Tarkis, Madrid 086267, 126552
Tarlee Antiques, Tarlee 062310
Tarli e Dintorni, Civitanova
Marche 098222
Il Tarlo, Catania 080048
Il Tarlo, Padova 081007
Tarnished Treasures, Phoenix .. 096242
Taro Okamoto Bijutsukan,
Kawasaki 028810
Taro Okamoto Kinenkan, Tokyo . 029599
Tarpon Springs Cultural Center, Tarpon
Springs 053684
Tarporley Antique Centre,
Tarporley 091540
The Tarporley Gallery, Tarporley 119207
Tarquini, Nathalie, Marseille ... 104202
Tarquinio, Lille 069506
Tarrade, Arnault, Saumur 073438
Tarrade, Arnault, Villebernier .. 074136
Tarrago, Nicole, Montady 070177
Tarrant County College Southeast The
Art Corridor, Arlington 046536
Tarrant, Lorraine, Ringwood ... 091109
Tarrazi, Jean-Pierre, Marseille .. 069967
Tarrytowns Museum, The Historical
Societey Sleepy Hollow and
Tarrytown, Tarrytown 053687
Tarskij Istoriko-kraevedčeskij Muzej,
Tara 037928
Tarsus Müzesi, Tarsus 043067
Tarszys, José, Buenos Aires ... 139600
Tartaglia, Gianluca, Napoli 131646
Tartari, Roberto, Modena 080821,
131590
Le Tartarughe, Roma 081578
Tarter, Ilona, Thannhausen ... 142413
Tartous Museum, Tartous 042265
Tartt, Washington 124964

Tartu Kunstimaja, Tartu 010699
Tartu Kunstimuuseum, Tartu ... 010700
Tartu Linnamuuseum, Tartu 010701
Tartu Mänguasjamuuseum, Tartu 010702
Tartu Ülikooli Ajaloomuuseum,
  Tartu 010703
Tartu Ülikooli Geoloogiamuuseum,
  Tartu 010704
Tartu Ülikooli Kunstimuuseum,
  Tartu 010705
Tartu Ülikooli Zooloogiamuuseum,
  Tartu 010706
Tartumaa Muuseum, Elva 010606
Tarusskaja Kartinnaja Galereja,
  Kalužskij oblastnoj chudožestvennyj
  muzej, Tarusa 037929
Tarusskij Kraevedčeskij Muzej,
  Tarusa 037930
Tarusskij Muzej Semji Cvetaevych,
  Kalužskij oblastnoj kraevedčeskij
  muzej, Tarusa 037931
Tarvasjoen Kotiseutumuseo,
  Tarvasjoki 011416
Tarvastu Kihelkonnamuuseum,
  Tarvastu 010707
Taryhy we Etnografiki Muzeỳ,
  Mary 043095
Taryhy we Etnografiki Muzeỳ,
  Türkmenabat 043099
Taryhy Yer Abiwerd, Kaka 043092
Taryhy Yer Gadymy Merw, Mary 043096
Taryhy Yer Gökdepe Galasy,
  Gökdepe 043091
Taryhy Yer Könürgenç,
  Könürgenç 043093
Taryhy Yer Nusaỳ, Nusaỳ 043097
Taryhy Yer Saragt, Saragt 043098
Tarzan, Köln 076913
Taş Eserler Müzesi – Lapidari,
  Lefkoşa 033476
Tas Oda Etnografya Müzesi,
  Burdur 042856
Tasaki Bijutsukan, Karuizawa .. 028787
Tasb, Las Vegas 094182
Tasbach, Elvira, Berlin 141579
Tasca, Stella, Roma 110864
Taschen, Central Hong Kong ... 136702
Taschen, Köln 137319
Taschen, London 138253
Taschen, Los Angeles 138352
Taschen, Madrid 137996
Taschen, Paris 136988
Taschen, Tokyo 137793
TASD Gallery, Miami 121866
Tasende, West Hollywood 125029
Tasiilap Katersugaasivia, Ammassalik
  Museum, Tasiilaq 023341
Tåsinge Skipperhjem og
  Folkemindesamling, Svendborg 010338
Taşkent Sanat Galerisi, İstanbul 117088
Tasmania Distillery Museum,
  Hobart 001149
Tasmania Gallery, Hobart 099139
Tasmanian Copper and Metal Art
  Gallery, Carrick 098873
Tasmanian Medical History Museum,
  South Hobart 001505
Tasmanian Museum and Art Gallery,
  Hobart 001150
Tasmanian School of Art, University of
  Tasmania, Hobart 054906
Tasmanian Wool Centre, Ross .. 001461
Tasmania's Antique Directory, Battery
  Point 061109
Tasneem Arts Gallery, Peshawar 034146
Tassart, Mireille, Verdun (Meuse) 074022
Tassenmuseum Hendrikje,
  Amstelveen 031921
Tassiriki, Strågnäs 115941
Tassotti, Bassano del Grappa . 137619
Tassy, Jean, Alès 066531
Tassy, Philippe, Barjac 066952
Taste of San Antonio, San
  Antonio 097037

Tasviriy San'at Muzeyi – Andijon
  Qal'asi, Andijon Viloyat Adabiyot va
  San'at Muzeyi, Andijon 054627
Tat, Tong, Singapore 114620
Tatabánya Múzeum, Tatabánya . 024003
Tatafiore, Napoli 131647
Tatai Vár, Tata 024000
Tatar Gallery, Toronto 101711
Tatarna Monastery Museum,
  Granitsa 023087
Tatarskij Gosudarstvennyj Muzej
  Izobrazitelnych Iskusstv, Kazan 037120
Tatarskoe Gorodišče –
  Archeologičeskij i prirodnyj Muzej-
  zapovednik, Stavropol 037901
Tate and Antique Shop, Houston 093807
Tate Britain, London 045115
Tate Etc., Tate Gallery, London . 139225
Tate Gallery, Austin 119707
Tate Geological Museum, Casper
  College, Casper 047457
Tate House, Portland 052284
Tate Liverpool, Liverpool 044865
Tate Members, London 060262
Tate Modern, London 045116
Tate Saint Ives, Saint Ives,
  Cornwall 045700
Tate Shop, London 144698
Tateshina World Arts & Crafts
  Museum, Tateshina 029469
Tateyama Hakubutsukan anna Toyama-
  ken, Tateyama, Toyoma 029471
Tatham Art Gallery,
  Pietermaritzburg 038802
Tatiana, Singapore 114621
Tatistcheff, New York 123119
Tator, Roger, Lyon 104137
Tátrai Galéria, Piliscsaba 109210
Tatranská Galéria, Poprad 038428
Tatsumiya, Sendai 082122
Tatsuno Bijutsukan, Tatsuno .. 029472
Tatsuuma Kokokan, Nishinomiya 029189
Tatsuya, Higashi-Osaka 081970, 081971
Tattersall Page, Fort Worth ... 145152
Tattersall's, Uppingham 091696, 135173
Tattershall Station Gallery,
  Tattershall 119208
Tatti, Steven, New York 136013
Tatton Park, Knutsford 044744
Tattoo Art Museum, San
  Francisco 053028
Tattoo Museum, Amsterdam .. 031991
Tatty Old Bits, Carrum 061291
Tatura Irrigation and Wartime Camps
  Museum, Tatura 001565
Tatyana Gallery, New York 123120
Tatyana Gallery, Praha 102723
Tauber, Elisabeth & Kurt, Pegnitz 078030,
  108368
Tauber, Laurence, Surbiton ... 091512
Tauberfränkisches
  Landschaftsmuseum,
  Tauberbischofsheim 022108
Tauberländer Dorfmuseum,
  Weikersheim 022466
Tauberová, Zdeňka, České
  Budějovice 140464
Tauchmann, Alfred, Söchau ... 062843
Tauchnitz, A., Leipzig 142065
Tauer-Druck, H. Tauer,
  Markkleeberg 137377
Tauernstraßen-Museum, Eben im
  Pongau 001831
Taul, Helge, Greve 102816
Taulov Kunsthandel, Fredericia . 065654
Taunton Antiques Market,
  Taunton 091546
Taunton Press, Newton 138432
Taunton Studio Gallery, Taunton 119213
Taupe, Ingrid, Villach 136567
Taupiri Antiques, Waikato 084137
Tauragnų Krašto Muziejus, Utenio
  Kraštotyros Muziejus, Tauragnai 030486

Tauranga Museum, Tauranga .. 033306
Taureilles, Annecy 066650
Taurisano, Rio de Janeiro 101071
Tauriska Galerie Verlag, Neukirchen
  am Großvenediger 136539
Tauriska-Galerie, Leopold Kohr
  Akademie, Neukirchen am
  Großvenediger 002438
Tauro Apygardos Partizanu ir Tremties
  Muziejus, Marijampolės kraštotyros
  muziejus, Marijampolė 030418
Taurus Gallery, Oxford 118866
Taurus-Kunstkarten,
  Brackenheim 137122
Taurus Verlag Leipzig, Leipzig .. 137350
Tausch, Dieter, Stade 078528
Tausch, Dieter, Innsbruck 139999
Tausendundeinbuch, Berlin ... 141580
Tauzy, Georges, Saint-Ouen ... 073205
Tavanarasi, İstanbul 088142
Tavard, Freddy, Niort 070717
Tavares, Maria Gouveia, Lisboa . 084979
Tavastilan Kotiseutumuseo,
  Mietoinen 011156
Tavecchio, Giorgio, Milano ... 131539
Tavener Colin, Timsbury 119235
Taverna, Milano 110465
Tavistock Antiques, Eaton Ford . 089154
Tavisupali Dro, Tbilisi 105922
Tawaraya, Nagoya 082065
Taweelah Museum of Mummies,
  Mahweet 054829
Tawhiti Museum, Hawera 033220
Tawi-Tawi Ethnological Museum,
  Mindanao State University,
  Bongao 034439
Taxameter-Museum, Hamburg . 019023
Taxandriamuseum, Turnhout .. 004101
Taxi Mauve, Auxerre 066854
Taxi Photo, Lima 138948
Taxter & Spengemann, New York 123121
Taylor, Belfast 117345
Taylor, Charleston 145028
Taylor, Dublin 123122
Taylor, London 090529, 118593
Taylor, Montrose 126992
Taylor, Tarragindi 062311
Taylor & Francis, Stockholm .. 138016
Taylor Brown and Sarah Dorsey House,
  Midland 050880
Taylor-Grady House, Athens ... 046597
Taylor-Jensen, Palmerston North 084058
Taylor-Pick, Beverley 117354
Taylor-Smith, Westerham 091801,
  144913
Taylor, B. R., Kelling 089664
Taylor, Brenda, New York 123122
Taylor, Christopher, Auckland .. 112813
Taylor, Duncan, Bellevue Hill .. 061122
Taylor, Erin, Minneapolis 122042, 138364
Taylor, Frederieke, New York .. 123123
Taylor, Henry, Johannesburg .. 114802
Taylor, Jardine, London 118594
Taylor, Jenny, Busselton 098826
Taylor, John, Georgensgmünd .. 129896
Taylor, Jude, Middle Swan ... 099336
Taylor, Karen, Twickenham ... 119279
Taylor, Kerry, London 126975
Taylor, Louis, Stoke-on-Trent .. 127056
Taylor, M.C., Bournemouth ... 088543,
  134129
Taylor, Martin, Wolverhampton . 091908
Taylor, Max, Bago'walah 098691
Taylor, Owen, Warwick 119312
Taylor & Son, Peter, Watford .. 144910
Taylor, Rankine, Cirencester .. 088921
Taylor, Robin, Wakefield 091707
Taylor, Véra, Liguell 069467
Taylors, Castlebaldwin 109542
Taylor's Coins, Toronto 064762
Taymori Hall, Isfahan 024702
Tayrac, Régine de, Paris 129150
Al-Tayyeb, Dubai 117189
Tazl, Graz 100023

Tazl, Wolfgang, Weiz 062872
TBA Exhibition Space, Chicago . 047701
TBB Compani, Karlskrona 086776
Tbilisi State Museum of Anthropology
  and Ethnography, Tbilisi 016426
TD Art Collection, Dallas 120666
Te Amorangi Trust Museum,
  Rotorua 033297
Te Ara, Wellington 138944
Te Awamutu Books & Stamps,
  Waikato 143356
Te Awamutu Museum, Te
  Awamutu 033307
Te Koop Gevraagd, Alkmaar ... 082622
Te Kowhai Print Trust, Whangarei 059643,
  113115
Te Kuiti and District Historical Society
  Museum, Te Kuiti 033308
Te Manawa Art, Palmerston
  North 033275
Te Manawa Life Mind, Te Whare
  Pupuri Taonga o Manawatu,
  Palmerston North 033276
Te Mei Si Art Gallery, Guangzhou 102033
Te Papa, Museum of New Zealand Te
  Papa Tongarewa, Wellington .. 033360
Te Toi Mana Maori Gallery,
  Christchurch 112864
Te Tuhi – The Mark, Auckland . 033158
Te Wa The Space Gallery,
  Wanganui 113052
Te Whare Taonga O Akaroa,
  Akaroa 033132
Te Whare Whakaaro o Pito-one, Hutt
  City 033226
Te, Jie, Beijing 101992
Tea & Antiques, Felixstowe 089270
Tea and Tattered Pages, Paris .. 141284
Tea Ceremony School Museum,
  Kyoto 028985
Tea Museum, Bois-Cheri 030786
Teaberry Press, San Francisco . 138462
Teach Ban Artists Gallery,
  Drumcliff 109568
Teackle Mansion, Princess Anne 052359
Teak Sentiments, Singapore ... 085513
Team Auctions, Portland 127413
Team Gallery, New York 123124
Teaparty, J.J., Boston 092498
Teapot World, Conwy 089682
Teardo, Carlo, Milano . 080742, 080743
Tearston, Sylvia, New York ... 095778
Tease Art Fair, Köln 098167
Teatergalleriet, Kalmar 115705
Teatermuseet i Hofteatret,
  København 010197
Teatermuseet i Oslo, Oslo 033866
Teatermuseet og Interkulturelt
  Museum, Oslo 033867
Teatralnaja Galereja na Maloj Ordynke,
  Gosudarstvennyj centralnyj teatralnyj
  muzej im. Bachrušina, Moskva 037501
Teatralnyj Muzej, Kazanskij
  akademičeskij Bolšoj dramatičeskij
  teatr im V.I. Kačalova, Kazan .. 037121
Teatralnyj Salon na Tverskom Bulvare,
  Gosudarstvennyj centralnyj teatralnyj
  muzej im. Bachrušina, Moskva 037502
Teatre-Museu Dalí, Figueres .. 039401
Teatterimuseo, Helsinki 010873
Teb, Paris 071933
Teboul, Jacqueline, Paris 071934
La Teca, Padova 081008
Tecck an Phiarsaigh, Rosmuck . 024984
Tech Museum of Innovation, San
  Jose 053045
Techne, Centre de Recherche et de
  Restauration des Musées de France,
  Paris 138664
Techni-Macedonian Artistic Association
  Kilkis, Kilkis 023131
Technichus, Härnösand 040678
Technické Muzeum Tatra,
  Kopřivnice 009639

Technické Muzeum v Brně, Brno 009482
Technické Muzeum, Muzeum Těšínska,
Petřvald u Karviné 009754
Techniek Museum Delft, Delft . 032164
Techniekmuseum Heim, Hengelo,
Overijssel 032465
Technik- und Verkehrsmuseum,
Stade 021950
Technik anno dazumal – Museum
Kratzmühle, Kinding 019615
Technik-Museum, Soest 021894
Technik Museum Speyer, Speyer 021935
Technik u. Zweiradmuseum Dargen/
Usedom, Dargen 017815
Technikmuseum Hugo Junkers,
Dessau 017872
Technikmuseum U-Boot "Wilhelm
Bauer", Bremerhaven 017611
Technikmuseum, Kuratorium
Industriekultur in der Region
Magdeburg e.V., Magdeburg .. 020214
Technikon Natal Gallery, Durban 038707
Techniksammlung Backnang,
Backnang 016696
Technikschau, Verein Technisches
Landesmuseum Mecklenburg-
Vorpommern e.V., Wismar ... 022694
Techningsmuseet, Laholm 040770
Techniquest, Cardiff 043869
Technische Sammlungen der Stadt
Dresden, Dresden 018013
Technisches Denkmal "Gaswerk",
Neustadt, Dosse 020766
Technisches Denkmal Brikettfabrik
Louise, Domsdorf bei Bad
Liebenwerda 017927
Technisches Denkmal Kupferhammer,
Thießen 022149
Technisches Denkmal Museum
Kalkwerk Lengefeld, Zweckverband
Sächsisches Industriemuseum,
Lengefeld 020033
Technisches Denkmal Neue Hütte,
Schmalkalden, Kurort 021679
Technisches Denkmal Tobiashammer,
Ohrdruf 020981
Technisches Denkmal Ziegelei
Hundisburg, Haldensleben 018946
Technisches Eisenbahnmuseum,
Lieboch 002305
Technisches Halloren- und
Salinenmuseum, Halle, Saale . 018970
Technisches Kulturdenkmal Ölmühle,
Marbach am Neckar 020264
Technisches Landesmuseum
Mecklenburg-Vorpommern,
Schwerin 021797
Technisches Museum Alte
Wasserkunst, Bautzen 016979
Technisches Museum Alte Ziegelei
und Turmwindmühle, Konradsburg
Ermsleben, Falkenstein, Harz . 018338
Technisches Museum der
Bandweberei, Großhörsdorf .. 018860
Technisches Museum der Pforzheimer
Schmuck- und Uhrenindustrie,
Pforzheim 021123
Technisches Museum Frohnauer
Hammer, Annaberg-Buchholz . 016591
Technisches Museum Holzschleiferei
Weigel, Breitenbrunn 017578
Technisches Museum Ölmühle,
Pockau 021164
Technisches Museum Papiermühle,
Zwönitz 022920
Technisches Museum Silberwäsche
Antonsthal, Breitenbrunn 017579
Technisches Museum Wien, Wien 003089
Technisches Museum Zechau,
Rositz 021480
Technisches Schaudenkmal Gießerei
Heinrichshütte, Wurzbach 022827
Technology & Conservation,
Somerville 139505

Technopolis, Mechelen 003861
Technorama Science Center,
Winterthur 042162
Techopp, Anna, Marseille 104203
Técnica, Asunción 113337
Tecnireco, Roma 132059
Tecno, Pisa 081159
Tecno-Art-Center, Sankt-
Peterburg 114364
Tecolote Tiles and Gallery,
Albuquerque 119515
Tectónica, Madrid 139073
Ted, Genova 080295
Tedden, Peter, Düsseldorf 106834
Teddy Bear, Avranches 066893
Teddy Bear Museum, Seogwipo 029981
The Teddy Bear Museum, Stratford-
upon-Avon 045908
Teddy Bear Museum of Naples,
Naples 051164
Teddy Bear Shop and Museum,
Ironbridge 044650
Teddy Bear Trail, San Diego .. 097136
Teddy Museum Berlin in Hof/Bayern,
Hof 019288
Teddybär Museum, Luzern 041714
Teddybärmuseum, Sempach .. 041980
Teddyland, Rothenburg ob der
Tauber 078289
Teddymuseum, Klingenberg am
Main 019646
Teddy's Antiques, Jacksonville . 094041
Teddy's Galerie, Luzern 116483, 133846
Tedesco, Enrica, Firenze 080212
Ted's Refinishing, Fresno 135573
Tee-Museum, Leer 019952
Tee & Son, R.H.V., Vancouver . 064887
Teekengenootschap Pictura,
Dordrecht 059623
Tees Archaeology, Hartlepool .. 044526
Teeterville Pioneer Museum,
Teeterville 007128
Teetzmann, Peter, Hamburg ... 076324
Teezer, Cincinnati 092902
TEFAF Maastricht, Maastricht . 098266
Teger Studio, London 090530, 134780
Tegfryn Art Gallery, Menai Bridge 118738
Téglamúzeum, Magyar Építőipari
Múzeum, Veszprém 024043
Teglgårdsstrædes Kunsthandel,
København 065964, 102914
Teglia, Marco & Luca, Firenze . 080213
Tegliucci, Franca Maria, Roma . 132060
Tegnerforbundet Galleri, Oslo . 113245
Tehkummah Township Little
Schoolhause and Museum,
Tehkummah 007129
Tehni Art Gallery, Tehni Macedonian
Art Association, Thessaloniki . 023298
Tehni Macedonian Art Association,
Thessaloniki 059332
Tehnički Muzej, Zagreb 009051
Tehnični Muzej Rudarstva, Zagorje ob
Savi 038633
Tehniški Muzej Slovenije, Bistra pri
Vrhniki 038494
Tehniški Muzej Slovenije,
Ljubljana 038562
Tei, Wing, Hong Kong 065101
Teifi Valley Railway, Henllan ... 044568
Teignmouth and Shaldon Museum,
Teignmouth 045953
Teignmouth & Son, Pamela,
London 090531
Teijink, Antwerpen 063241
Teijink, Den Haag 083134
Teine Memorial Museum,
Sapporo 029343
Teipei Fo Guang Yuan Art Gallery,
Taipei 042558
Teishin Sogo Hakubutsukan,
Tokyo 029600
Teiskon Kotiseutumuseo,
Tampere 011411

Teissèdre, Paris 105260
Teissèdre, Tucson 124778
Teissèdre, F., Paris ... 141285, 141286
Teissèdre, Lionel, Murat (Cantal) 070401
Teissèdre, René François, Paris . 105261
Teixeira, Armando Gonçalves,
Porto 114108
Teixido, Carlos, Barcelona 114969
Tejería Loppacher, Gustavo, Punta del
Este 119451
Tejipari Emléktár, Székesfehérvár 023943
Tekakwitha Fine Arts Center,
Sisseton 053359
Tekel Sanat Galerisi, İstanbul . 117089
Tekfur Sarayi, İstanbul 042959
Tekiho Memorial Museum of Dolls,
Ogose 029209
Tekirdağ Devlet Güzel Sanatlar Galerisi,
Tekirdağ 043068
Tekirdağ Müzesi, Tekirdağ 043069
Tekisui Bijutsukan, Ashiya-shi . 028492
Tekniikan Museo, Helsinki 010874
Teknikens Hus, Luleå 040805
Teknikens och Sjöfartens Hus, Malmö
Museer, Malmö 040833
Teknillinen Korkeakoulu, Helsinki 055344
Tekniska Kvarnen, Örebro 040897
Tekniska Museet, Stockholm .. 041051
Tekovské Muzeum v Leviciach,
Levice 038377
Tel Aviv Museum of Art, Tel Aviv 025212
Tel Hai Museum, Tel Hai 025216
Tela, Lisboa 132959
Telar Artesanía Textil, Bilbao .. 133202
Telart, Cali 128294
Teleborgs Slott, Växjö Universitet,
Växjö 041148
Telechea, Domingo Isaac, São
Paulo 128134
Telecommunications Museum, Nortel
Community Volunteers, LaSalle 006166
Telefónia Múzeum, Postamúzeum,
Budapest 023557
Telefonmuseet, Hellerup 010102
Telefonmuseum Hittfelder Bahnhof,
Seevetal 021824
Telefonmuseum im Greuterhof,
Islikon 041607
Telefoonmuseum, Drouwen ... 032257
Telegaléria, Pécs 023856
Telekessy István Patikamúzeum,
Eger 023622
Telemark Kunstnersenter,
Porsgrunn 033881
Telemark og Grenland Museum,
Skien 033945
Telemarksgalleriet, Notodden .. 113188
Telephone Museum, Ellsworth . 048559
Telephone Pioneer Museum of New
Mexico, Albuquerque 046370
Televízió Technikatörténeti és Műsor
Múzeum, Budapest 023558
Telfair Museum of Art, Savannah 053182
Teline, Concarneau 068119
Telkamp, Karoline, München .. 077628
Telkmann, Annette, Lienen 077106
Telkwa Museum, Telkwa 007130
Tell-Museum, Bürglen 041380
Tell Qasile Archaeological Site, Eretz-
Israel Museum, Tel Aviv ... 025213
Teller, Susan, New York 123125
Telles, Claudio Franco Netto,
Brasília 100928
Telles, Richard, Los Angeles .. 121605
Telling the Mennonite Story – Visitor
Centre, Saint Jacobs 006878
Tello, Damascus 088063
De Telloor, Alkmaar 112108
Telluride Historical Museum,
Telluride 053696
Tělnieka Voldemāra Jākobsona
Memoriālāis Muzejs, Bebri .. 030104
Teloglion Foundation of Art,

Thessaloníki 023299
Telstra Museum Brisbane,
Clayfield 000970
Tem Sanat Galerisi, Istanbul ... 117090
Tema Galería de Arte, Buenos
Aires 098582
Temecula Valley Museum,
Temecula 053697
Temel Sanat Kolleksiyon, İstanbul 042960
Téméraire, Colombey-les-Deux-
Eglises 068107
Temiskaming Art Gallery,
Haileybury 005941
Temlett House, Grahamstown . 038728
Temme, Karin, Düsseldorf 129761
Temmeksen Kotiseutumuseo,
Temmes 011417
Tempantigo Antiguidades, Lisboa 084980
Tempe Center for the Arts,
Tempe 053706
Tempe Historical Museum,
Tempe 053707
Tempel Kroonluchterreparaties,
Amsterdam 132417
Tempelhof Museum, Berlin 017306
Tempelhofer Münzenhaus, Berlin 074935,
                                   125892
Tempelmuseum Frauenberg,
Seggauberg 002751
Tempera, Aalst 127990
Tempéra, Cusset 103651
Tempera, Ville-d'Avray 074133
Temperance Tavern,
Newcomerstown 051521
Temperature, Surase 139102
Temperley, David, Birmingham . 134113,
                                   144251
Tempesta, Marly, Saint-Ouen .. 073206
Tempi Antichi, Bologna 079882
Tempietto Rossiniano della Fondazione
Rossini, Pesaro 027269
Il Tempio Ritrovato, Napoli ... 131648
Tempio Voltiano, Como 026052
Temple Auctions, Lisburn 126949
Temple Bar Gallery and Studios,
Dublin 109624, 137608
Temple de Diane, Aix-les-Bains . 066486
Temple des Mille Bouddhas, La
Boulaye 013367
Temple Gallery, Birmingham ... 117381
Temple Gallery, Blackrock 109531
Temple Gallery, Dunedin 083933
Temple Gallery, London 090532, 134781
Temple Gallery, Tyler School of Art,
Temple University, Philadelphia 052098
Temple Judea Museum, American
Jewish Museums, Elkins Park . 048540
Temple Museum of Religious Art,
Beachwood 046839
The Temple Museum of Religious Art,
Cleveland 047826
Temple Newsam House, Leeds
Museums and Galleries, Leeds 044785
Temple of Aphrodite, Kouklia ... 009404
Temple Réformé Mulhouse,
Mulhouse 014381
Temple, Jeanne, Milton Keynes . 090760
Temple, Steven, Toronto 140418
Templecrest Art, Cape Town ... 114777
Templeman, Loughton 118666
Templerhaus, Amorbach 016579
Templin Historical Village,
Boonah 000866
Templon, Daniel, Paris 105262
Templon, Jean-Claude,
Chanteloup 067828
Il Tempo, Artonne 066752
Il Tempo, Verona 081941, 081942
O Tempo Antiguidades, Porto
Alegre 063997
Tempo Interiores, Porto 085053
Tempo Reale, Merano 110214
Il Tempo Ritrovato, Bari 130925
Il Tempo Ritrovato, Brescia 079958

Il Tempo Ritrovato, Cagliari 080004
Il Tempo Ritrovato, Milano 110466
Temporäre Galerie Schloß Neersen,
Willich 022666
Temporale, Lugano 139103
Temporary Contemporary,
Kraków 113644
Temporis Matériaux Anciens,
Trezioux 073883
Tempos Antigos, Porto 085054, 132979
Le Temps, Mölln 077382
Les Temps Anciensdas, Châteauneuf-
du-Rhône 067892
Temps de Choses, Bougival 067381
Le Temps de Vivre, Dijon 068345
Temps des Arts, Megève 104216
Le Temps des Cérises, Charleville-
Mézières 140823
Le Temps des Cerises, Honfleur .. 068834
Le Temps du Livre, Le Marin .. 142983
Le Temps Jadis, Soisy-sous-
Montmorency 073533
Les Temps Mêlés, Albi 066518
Temps Passé, Villefranche-sur-
Saône 074151
Le Temps Perdu, Genova 080296
Temps qui Passe, Genolier 087508,
116346
Le Temps Retrouvé, Brest 067479
Le Temps Retrouvé, Champlost .. 067817
Temps Retrouvé, Isles-sur-
Suippe 068878
Temps Retrouvé, Sceaux 073458
Temptations, Ashtead 088314
Temptations, Dorking 089094
Tempus, Bad Camberg 074502
Téms Swiya Museum, Sechelt . 007000
Temuka Courthouse Museum,
Temuka 033309
Ten Boom, Den Haag 083135
Ten Broeck Mansion, Albany .. 046347
Ten Duinen 1138, Koksijde 083753
Ten Sing Ceramic Art Village, Jiji 042355
Tenala Hembygdsmuseum,
Tenala 011418
Ténat, René, Mirande 070125
Tenbury and District Museum, Tenbury
Wells 045956
Tenby Museum and Art Gallery,
Tenby 045957
Tenchi Shobo, Osaka 142910
Tenchini, Giacomo, Brescia 079959
Il Tendaggista, Genova 080297
Tendances, Paris 105263
Tendances-Mikado, Luxembourg 111869
Tender Buttons, New York 095779
Tendille, Patrick, Bonson 128478
Tenement House, Glasgow 044422
Tenenbaum, Henri, Saint-Ouen . 073207
teNeues Verlag, Kempen 137294
Teng County Museum, Tengxian 008319
Tengyu, Suita 142917
Tengzhou Museum, Tengzhou .. 008320
Teniers, Oostende 100807
Tenino Depot Museum, Tenino . 053713
Tenisa Vēstures Muzejs, Jūrmala 030141
Tenmon Bijutsukan, Hirakata .. 028640
Tennants, Harrogate 126920
Tennants, Leyburn 126945
Tennants, Oakham 127002
Tennengauer Kunstkreis e.V,
Hallein 058312
Tennent Art Foundation Gallery,
Honolulu 049545
Erna Tenner & Hans-Hermann Lindner,
Heidelberg 141930
Tennessee Antique Mall,
Nashville 095039
Tennessee Art League, Nashville 060771
Tennessee Art League Galleries,
Nashville 122103
Tennessee Arts Report, Nashville 139506
Tennessee Association of Museums,
Piney Flats 060772

Tennessee Central Railway Museum,
Nashville 051185
Tennessee Historical Society Museum,
Nashville 051186
Tennessee Museum of Aviation,
Sevierville 053275
Tennessee State Museum,
Nashville 051187
Tennessee Valley Art Association,
Tuscumbia 060773
Tennessee Valley Art Center,
Tuscumbia 053865
Tennessee Valley Railroad Museum,
Chattanooga 047598
Tennyson's Bookshop, Hawkes
Bay 143321, 143322
Tenpyo, Nara 111340
Tenpyodo, Kyoto 082043
Tenri Daigaku Sankokan, Tenri . 029475
Tenschert, Heribert,
Rotthalmünster 142339
Tenshi Bijutsukan, Nasu 029166
Tenshodo, Aki, Hiroshima 081985
Tensta Konsthall, Spånga 040982
Tent-Centrum Beeldende Kunst,
Rotterdam 112662
Tent, Eva, Koblenz 107580
La Tentation du Mandarin, Paris 071935
Tenterden and District Museum,
Tenterden 045961
Tenterden Antiques and Silver Vaults,
Tenterden 091561
Tenterfield Centenary Cottage,
Tenterfield 001567
Tentoonstellingsruimte 't Oute Hus,
Burgh-Haamstede 032126
Tentoonstellingszaal Hessenhuis,
Antwerpen 003315
Teo & Namfah, Bangkok 116965
Teo Leo, Paris 071936
Teodoro P. Resurreccion Memorial
Museum, Ilocos Museum of Music
and Fine Arts, Luna 034500
Teodra Art Gallery, Toronto 101712
Teollisuustyöväen Asuntomuseo,
Imatra 010903
Teorética, San José 102517
Tepco Denryokukan, Tokyo 029601
Tepco Denryokukan, Yokohama . 029726
Tepes, Zoran, Beograd 114443
Tepia, Kikai Sangyo Kinenkan,
Tokyo 029602
Teplahis, Atlanta 119620
Teplitz-Schönauer Heimatmuseum,
Frankfurt am Main 018454
Tepper & Takayama, Boston ... 119925
Tepper Galleries, New York ... 095780,
127363
Tepper, Burkhardt, Hamburg ... 076325,
141899
Teppich und Heimatmuseum Oelsnitz,
Museum Schloß Voigtsberg,
Oelsnitz 020961
Tequila Bookworm, Toronto 140419
Ter Braak, Peter, Groningen .. 083306,
132509
Tér Galéria, Budapest 109179
Ter Meulen, Dordrecht 083196, 132491
Ter, Daniel, Étoile-sur-Rhône ... 068485
Teranga, Kraków 113645
Terbeck, Coesfeld 106642
Terbrüggen, Wehr 078884
Tercelin de Joigny, Benoît, Mons 063801
Tercera Gallery, San Francisco . 124411
Tere Pastoriza, Miami 121867
Terenzi, Luciana, Roma 081579
Teresa Antiques, Madrid 086268
Teresia ja Rafael Lönnströmin
kotimuseo, The Teresia and Rafael
Lönnström Foundation, Rauma 011307
Tereza Antiguidades, Belo
Horizonte 063924
Teriade Museum, Library of Modern
Art, Mytilini 023190

Teri's Authentic Restoration, Los
Angeles 135745
Terlinden, Dendermonde 063610, 125332
Term City Museum, Xiangcheng 008406
Terme Saint-Eloi, Montauroux . 070208
Terme, Jean-Louis, Montpellier . 070315
Természet Háza, Kiskunsági Nemzeti
Park Igazgatósága, Kecskemét 023734
Természettudományi Osztály, Janus
Pannonius Múzeum, Pécs ... 023857
Termuridi Tarikhi Muzeyi, Özbekiston
Respublikasi Fanlar Akademiyasi,
Toshkent 054670
Ternitzer Stahlstadtmuseum,
Ternitz 002819
Ternopil's'kyj Oblasnyj Chudožnij
Muzej, Ternopil' 043308
Ternopil's'kyj Oblasnyj Krajeznavčyj
Muzej, Ternopil' 043309
TerpenTijn, Dordrecht 112442
Terpstra, Joanna, Brushgrove . 098813
Terra, Delft 112368
Terra, Donec'k 117114
Terra, Winterthur 116724
Terra à Vista, Lagos 113976
Terra Art Gallery, Drummoyne . 098981
Terra Cotta, Las Vegas 094183
Terra Cotta, Saint-Ouen 073208
Terra d'Arte, Catania 131101
Terra Ferma, Grup d'Art Escolà,
Lleida 115119
Terra Incognita, Chicago 092799
Terra Incognita, Praha 140531
Terra Incognita Antique Maps,
Calgary 140338
Terra Museum of American Art,
Chicago 047702
Terra Mystica, Bad Bleiberg ... 001735
Terra Nova Gallery, Kuranda .. 099211
Terra Nova National Park Museum,
Glovertown 005887
Terra Promessa, Bergen op
Zoom 112347
Terra Publishing House, Moskva 137927
Terra Sancta Museum, Nazareth 025162
Terra Sigillata Museum,
Rheinzabern 021406
Terra-Vinea, Portel-des-Corbières 014973
Terrace Antiques, South
Fremantle 062223
The Terrace Gallery, Loxton ... 099240
Terrace Gallery, Orlando 051809
Terrace Glory Box, Kiama 061688
Terrace Hill Historic Site and
Governor's Mansion, Des
Moines 048289
Terrace House Factory, Ultimo . 062370
Terracota Museu de Terrissa i
Ceràmica Industrial, La Bisbal
d'Empordà 039152
Terracotta Warriors Museum,
Dorchester 044120
Terrada, Ada, Cazilhac 067741
Terrades, Gabriel, Paris 105264
Terrahe, Peter, München 142190
Terrain Gallery, Aesthetic Realism
Foundation, New York 051480
Terralha, Festival et Marché Européen
des Arts Céramiques, Saint Quentin-
la-Poterie 098116
Terralong Gallery, Kiama 099191
Terrapin, New York 123126
Terrasi, Jolanda, Palermo 131714
Terre d'Arte, Catania 080049
Terre de Chine, Paris 105265
Terre de Sienne, Beirut 082307, 132316
Terre de Sienne, Paris 129151
Terre des Arts, Paris .. 105266, 136989
Terre et Feu, Paris 071937
Terre et Toiles, Alès 103224
Terre et Toiles, La Grande-Motte 103894
Terrebonne Museum, Houma ... 049569
Terreiro, Ana Paula, Lisboa 084981
Terrel, Calgary 128158

Terres, La Ciotat 103883
Terres Anciennes Antiquités,
Montpellier 104286
Terres Contemporaines, Moret-sur-
Loing 104292
Terres d'Ailleurs, Les Rosiers-sur-
Loire 069424
Terres d'Ailleurs, Paris 071938
Terres d'Ailleurs, Toulouse 129318
Terres d'Autrefois, Beire-le-
Châtel 067065
Terres Sauvages, Châlons-en-
Champagne 067788
Terret, Stéphane, Bordeaux ... 128508
Terrier, Christian, Grenoble ... 068741
Terrill, Suzanna, Cincinnati 120380
Territoires du Jazz, Marciac ... 014007
Territorial Capital-Lane Museum,
Lecompton 050210
Territorial Statehouse State Museum,
Fillmore 048731
Territory Colours, Darwin 098952
Territory Craft, Darwin . 058222, 098953
Territory Town USA, Old West
Museum, Okemah 051725
Terror Háza Múzeum, Budapest 023559
Terry's Antiques, Manila 084482
Tersløsegaard, Dianalund 010030
Tertia, Sankt-Peterburg 085275
Teruz, Rogerio A., Rio de Janeiro 101072
Tervajärven Talomuseo, Sotkamo 011373
Terwagne, Liège 063784, 128078
Terwen, Dr. Pier A., Leiden ... 132562
Terwilliger Museum, Waterloo .. 054211
Terzo Millennio, Milano 110467
Terzo Occhio, Bologna 138877
Teschner, Volker, Kromsdorf .. 130201
Tesi, Pierre, Saint-Ouen 105569
Tesio Gallery, Milano 110468
Teske, U., Echternacherbrück .. 075564
Teson, Singapore 085514
Tesoro del Duomo, Messina ... 026776
Tesoro del Duomo, Milano 026833
Tesoro del Duomo, Monreale ... 026894
Tesoro del Duomo, Vigevano ... 028382
Tesoro della Cattedrale, Palermo 027191
Tesoro della Cattedrale, Rimini . 027525
Tesoro della Santa Maria della Scala,
Chieri 025963
Tesoro di Santa Maria La Stella,
Militello in Val di Catania 026840
Tesoros Trading, Austin 119708
Tesouro da Sé Patriarcal de Lisboa,
Lisboa 035826
Tesouro-Museu da Catedral,
Viseu 036117
Tesouro-Museu da Sé de Braga,
Braga 035576
Tesouros de Santa Maria e São Pedro
de Óbidos, Óbidos 035878
Tess, Limerick 079661
Tess, Pruszków 113754
Tessa, Kościerzyna 084545
Tessai Bijutsukan, Takaraduka . 029444
Tessaido, Kyoto 082044, 082045
Tessedik Sámuel Múzeum,
Szarvas 023917
Tessier, Marc, Saint-Ouen 073209
Tessiers, London 090533
Tessin, Quickborn 142312
Tessiner Mühlenmuseum, Tessin 022130
Tesson, Robert, Paris 071939
Tesson, Serge, Parthenay 125774
Testa, Stefano, Torino 132209
Testard-Letétu, Aigues-Mortes . 103185
Testard-Letétu, Nîmes 104403
Testnevelési és Sportmúzeum,
Budapest 023560
Testnevelés és Sportmúzeum
Kiállítóterme, Budapest 023561
Testoni, Piero, Verona 111199
Teşvikiye Sanat Galerisi, İstanbul 117091
Tetbury Antiques, Tetbury 091582
Tetbury Old Books, Tetbury 144880

Tetbury Police Museum, Tetbury 045962
Tételin, Max, Rouen . 072619, 129214
Tetley's Brewery Wharf, Leeds . 044786
Tettamanti, Massimo, Firenze . . 080214
Il Tetto 87, Roma 110865
Teubner, Dr. Manfred,
Nordhausen 142234
Teubner, Ole, Otterwisch 130499
Teulère, Michel, Annecy 125492
Teulings, Den Haag 083136
Teulon and District Museum,
Teulon 007136
Teumer, Michael, Dortmund . . . 075398
Teunissen, Eindhoven 083233
Teutloff, Lutz, Bielefeld 106498, 106499
Teutoburger Münzauktion,
Borgholzhausen 125902
Teutopolis Monastery Museum,
Teutopolis 053723
Teutschbein, G., Frankfurt am
Main 075828
Teuvan Museo, Teuva 011419
Teviot District Museum,
Roxburgh 033298
Tew, Timothy, Atlanta 119621
Tewkesbury Antique Centre,
Tewkesbury 091591, 144881
Tewkesbury Museum,
Tewkesbury 045965
Texacola Collectables, San
Antonio 097038
Texan Art Schools, Auckland . . 056188,
112814
Texan, Joe, Houston 093808
Texana Museum, Edna 048485
Texarkana Museums System,
Texarkana 053725
Texas Air Museum, Rio Hondo . 052576
Texas Art Galleries, Dallas . . . 120667
Texas Association of Museums,
Austin 060774
Texas Auction Academy, Dallas . 056976
Texas Baptist Historical Center
Museum, Brenham 047190
Texas Choice, San Antonio . . 124076
Texas City Museum, Texas City . 053726
Texas Energy Museum,
Beaumont 046861
Texas Forestry Museum, Lufkin . 050524
Texas Furniture, San Antonio . . 136274
Texas Gallery, Houston 121138
Texas Gas Collectables, Fort
Worth 093482
Texas Governor's Mansion,
Austin 046700
Texas Heritage Museum,
Hillsboro 049479
Texas Historical Museum and Tobacco
Museum, Texas 001571
Texas International Fine Art Fair,
Houston 098505
Texas Junk Company, Houston . 093809
Texas Maritime Museum,
Rockport 052649
Texas Memorial Museum, Austin 046701
Texas Military Forces Museum,
Austin 046702
Texas Music Museum, Austin . . 046703
Texas Pharmacy Museum,
Amarillo 046436
Texas Ranger Hall of Fame and
Museum, Waco 054021
Texas Seaport Museum,
Galveston 049039
Texas Sports Hall of Fame, Waco 054022
Texas Woman's University Art Galleries,
Denton 048248
Text Traders, Brisbane 139663
Text und Töne, Hamburg 141900
Texte zur Kunst, Berlin 137097, 138755
Textil- és Textilruházati Ipartörténeti
Múzeum, Budapest 023562
Textil- und Heimatmuseum, Stiftung
Paul Aebi-Gaissmaier, Sennwald 041981

Textil- und Rennsportmuseum,
Hohenstein-Ernstthal 019313
Textil-Forum-Service, Hannover . 137269
Textilarkivet Västernorrland,
Sollefteå 040975
Textildruckmuseum Mittelweiherburg,
Hard 002055
Textile Conservancy Company,
Tenterden 135126
Textile Conservation, Banwell . . 134058
Textile Conservation Centre, University of
Southampton, Winchester . . 046155,
056612, 135239
Textile Gallery, London 090534
Textile Museum, Thimphu 004203
The Textile Museum, Washington 054164
Textile Museum of Canada, Canadian
Museum of Carpets and Textiles,
Toronto 007210
Textile Restoration, Chicago . . . 135425
Textile Restoration Studio,
Altrincham 134028
Textilforum, Hannover 138756
Textilforum, Herning 010117
Textilkonservering, Stockholm . . 133582
Textilkunst, Alfeld 138757
Textilmuseet, Borås 040576
Textilmuseum, Hauptwil 041584
Textilmuseum, Sankt Gallen . . . 041927
Textilmuseum Brennet, Wehr . . . 022455
Textilmuseum Die Scheune, Spinnen/
Weben + Kunst, Nettetal 020670
Textilmuseum in Bocholt, Bocholt 017431
Textilmuseum Sandtnerstiftung,
Mindelheim 020414
Textilní Muzeum TIBA, Česká
Skalice 009502
Textilsammlung Max Berk,
Kurpfälzisches Museum,
Heidelberg 019147
Texture Antiques, Athen 092270
Textures Design, Québec 101521
Texxt, München 142191
Teylers Museum, Haarlem 032399
Teys Art Gallery, Dogwood Crossing,
Miles 001294
Teysen's Woodland Indian Museum,
Mackinaw City 050561
Teyssandier, Auxerre 066855
Teyssendier de la Serve Luc,
Fleurance 140840
TG Art, Wokingham 119377
Tgeasa Rustica, Lantsch 087555
Thabuis, Jean-Paul, La Roche-sur-
Foron 069067
Thackeray Gallery, London 118595
Thackray Museum, Leeds 044787
Thai Royal Carriages and Paraphernalia,
Bangkok 042681
Thakeham Furniture, Petworth . 091009,
134941
Thalassa – Aghia Napa Municipal
Museum, Pierides Foundation, Aghia
Napa 009400
Thaler, A., Waging am See 078824
Thaler, Friedrich, Salzburg 127860
Thaler, Werner, Basel . 116102, 133659
Thalmann & Lutz, Zürich 116863
Thames & Hudson, London 138254
Thames & Hudson, New York . . 138428
Thames Art Gallery, Chatham,
Ontario 005605
Thames Auction House, Waikato 126331
Thames Museum, Newport 051543
Thames Oriental Rug Company, Henley-
on-Thames 089512, 134456
Thames Police Museum, London 045117
Thames School of Mines and
Mineralogical Museum, Thames 033310
Thames Valley Antiques, Atlanta 092191
Thames Valley Art Dealers'
Association, Dorchester-on-
Thames 060263
Thames Valley Museum School,

Burgessville 005496
Thameside Aviation Museum, East
Tilbury 044212
Thamesware, London 118596
Thamm, Küsnacht 116402
Thamshavnbanen, Orkla
Industrimuseum, Løkken Verk . 033749
Than, Mai, Ha Noi 125099
Thanassis Frissiras Museum,
Athinai 023030
Thanet Antiques, Ramsgate . . . 091070
Thang Long, Ha Noi 125100
Thanhardt & Burger, Chicago . . 120281
Thanjavur Art Gallery, Thanjavur 024489
Thanka, Frankfurt am Main 107035
Thanks for the Memories, West
Hollywood 097936
Thar Art Gallery, Bangalore . . . 079363,
109268
That About Covers it, San
Francisco 136347
That and This, Victoria Park . . . 062388
That Antique Place, Houston . . 093810,
135620
That Gypsy Woman, Daylesford . 061392
Thatcher, Margaret, New York . . 123127
Thau Auksjoner, Dokka 126342
Thavibu Gallery, Bangkok 116966
Thaw & Co., Eugene V., New
York 123128
Thea G. Korver Visual Art Center,
Northwestern College, Orange
City 051794
Theale Fireplaces, Theale 135134
Theater-Atelier, Zürich 042235
Theater Museum, Amsterdam . . 031992
Theaterfigurenmuseum, Sammlung
Fritz Fey jun., Lübeck 020164
Theatergeschichtliche Sammlung
und Hebbel-Sammlung, Christian-
Albrechts-Universität Kiel, Kiel . 019611
Theatermuseum der Landeshauptstadt
Düsseldorf, Düsseldorf 018048
Theatermuseum Hannover,
Hannover 019067
Theatermuseum in der Reithalle,
Meiningen Museen, Meiningen 020344
Theaterwissenschaftliche Sammlung,
Universität zu Köln, Köln 019714
Theatre and Fine Arts Department,
Pace University, New York . . . 057566
Theatre Antiques, Minneapolis . 094950
The Theatre Antiques Center,
Framlingham 089305, 134395
Théâtre d'Automates-Le Village
Enchanté, Bellefontaine
(Manche) 011973
Theatre Museum Canada,
Toronto 007211
Thébaud, Concarneau 068120
Thébaud, Nantes 070481
Théberge, Claude, Montréal . . . 101455
Thee Metropolitan Gallery, Detroit 093376
De Theefabriek, Houwerzijl 032515
Theel, T., Berlin 129572
Theeuwes, J.C.A.M., Oisterwijk . 132601
Theile, Johanna Maria, Santiago de
Chile 128254
Theilmann, Klaus, Pfaffenhofen an der
Ilm 126093
Theimer, François, Toucy 137009
Theisen, Ursula, Köln 107690
Theiss, Saint Louis 136207
Theiss, Hartmut, Darmstadt . . . 075311
Theiss, Kellie Rae, Minneapolis . 122043
Theiß, N., Heeßen 076426
Theißen, Helga Maria, Aachen . 105949,
105950, 141398
Theißl, Irmgard Maria, Mitterdorf an
der Raab 100119
Theleman, Ludwig, Weimar 142473
Thelen, Bonn 075094
Thelen, Gertrud, Baden-Baden . 141443

Thellier, Paris 105267, 129152
Thelma Miles Museum,
Matheson 006275
Thélot, Fontenay-le-Comte 125588
Them Day, Happy Valley 005986
Themapark Openluchtmuseum De
Spitkeet, Harkema 032412
Themes and Variations, London 118597
Themes Antiquaire, Saint-Ouen . 073210
Then & Now, Manawatu 083985
Then Again, Gosford 061548
Then, Mirko, Riedbach 078234
Theng, Petra, Trunkelsberg . . . 078718
Theo-Steinbrenner-Turm-Museum,
Sommerach 021913
Theo Swagemakers Museum,
Haarlem 032400
Theo Thijssen Museum,
Amsterdam 031993
Theobald & Jennings, London . . 118598
Theodhosi, Arben, Tiranë 098532
Theodor-Fontane-Archiv,
Potsdam 021206
Theodor-Heuss-Haus, Stuttgart . 022064
Theodor-Heuss-Museum,
Brackenheim 017530
Theodor-Mommsen-Gedächtnisstätte,
Garding 018590
Theodor-Storm-Haus, Husum . . 019370
Theodor-Zink-Museum und
Wadgasserhof, Kaiserslautern . 019482
Théodore, Biarritz 067214
Theodore Roosevelt Birthplace, New
York 051481
Theodore Roosevelt Inaugural National
Historic Site, Buffalo 047308
Theodosiou, Theod., Thessaloniki 109102
Theofilos Museum, Mytilini 023191
Theomaria Gallery, Lemesos . . . 102554
Theomin Gallery, Olveston Dunedin's
Historic Home, Dunedin 033198
Theophilos Kairis Museum,
Andros 022962
Theophilos Museum, Varia
Lesbos 023309
Theorell, Norrköping 086883
Theotokopoulos, Madrid 086269
Thereses Kunstsalong, Oslo . . . 113246
Theresienthaler Glasmuseum,
Zwiesel 022915
Theresienthaler Heimatstube,
Gaildorf 018585
Theriault, Annapolis 127121
Therien & Co., San Francisco . . 097336
Therien & Co., West Hollywood . 097937
Therisso Museum of Eleftherios
Venizelos, Chania 023043
Thermen-Museum Juliomagus,
Schleitheim 041956
Thermenmuseum, Heerlen 032436
Thermometermuseum, Geraberg 018652
Thermostat Gallery, Manawatu . 112940
Théron, Didier, Nogent-sur-Marne 070729,
128875
Thérond, Jean-Pierre, Montpellier 070316
Théry, Christophe, Le Mee-sur-
Seine 069314
Thery, Jocelyne, Cleguer 068007
Thesauros, Messina 080375
Thesaurus, Saint Helier 144817
Thesing, M., Borken, Westfalen . 075105
Thessaloniki Museum for
the Macedonian Struggle,
Thessaloniki 023300
Theta-Museet, Bergen 033539
Thetford Collection, Thetford . . 045969
Thetford Historical Society Library and
Museum, Thetford 053733
Theuer & Scherr, Mannheim . . . 077277,
130282
Theuermann, Georg, Wien 063085
Theuermann, Valentin, Wien . . . 063086
Theunissen, L., Heerlen 083377
Theunissen, O., Bruxelles 063556

Thévenod, Léon, La Tour 069105
They, Lausanne 116428
Thézan, Nice 070669
Thiam, Metz 104232
Thian Hwa, Ang, Singapore . . . 114622
Thibault, G., Tours 073863
Thibaut, Mons 063802
Thiboumery, Jean-Yves, Rennes (Ille-
et-Vilaine) 072443
Thiebaud, Paul, New York 123129
Thiebaud, Paul, San Francisco . 124412
Thiébaut, Bertrand, Paris 071940
Thiebes, Gabriele, Koblenz 107581
Thieblemont, Gabriel, Chaumont
(Haute-Marne) 067937
Thiel, J., Knokke-Heist 063710
Thiel, Karl, Wiener Neustadt . . . 140136
Thiel, L., Schleiden 078367
Thiele & Schwarz, Kassel 137293
Thiele, Luise Christine,
Hohenwestedt 076562
Thiele, Markus, Linz 100114
Thiele, Ulrich, Bad Lippspringe . 074548
Thielens, Luc, Gent 063662
Thielmann, Siegfried, Wuppertal 130796
Thielsch, Gottfried, Markersdorf 130286
Thielsen, London 101350
Thielska Galleriet, Stockholm . . 041052
Thieme, Nicoline, Leipzig 142066
Thiénot, Jacqueline, Saint-Tropez 073321
Thierry-Lannon & Assoc., Brest . 125537
Thierry, Elisabeth, Domfront . . . 068373,
068374
Thierry, Laurent, Paris 071941
Thierry, Winsall, Paris 071942
Thiersch, Brigitte, Weimar 078901
Thiersch, David, Weimar 078902
Thierstein, Daniel, Bern 144053
Thierstein, Daniel, Biel 144055
Thiery, Elisabeth Maria, Wien . . 063087
Thies, Andreas, Nürtingen 126087
Thiessen, Elke, Kandern 130090
Thijs, André, Beveren 128010
Thil, Stéphane, Épernay 068451
Thimbleby & Shorland, Reading 127018
Thimblena's Thimbles, Dallas . . 093221
Thingbaek Kalkminer Bundgaards
Museum, Skørping 010317
Things-Antiques, Los Angeles . . 094485
Things From the Past, San
Antonio 097039
Things II, Denver 093343
Things Jamaican, Montego Bay 111233
Things Japanese, New York 095781
Things To Do, Manama 063162
Think Modern, Oakland 095915
Th'ink Tank, Denver 120751
Thinkers Bookshop, Tweed Heads
South 139940
Thinkposters, San Jose 138465
Thinktank, The Museum of Science
and Discovery, Birmingham . . 043628
Thiollière, Philippe, Saint-Étienne 129229
Thion, François, Evreux 125585
Third Avenue Antiques, Mount
Lawley 061917
Third Avenue Gallery, Vancouver 101842
Third Canyon Gallery, Denver . . 120782
Third Cavalry Museum, Fort
Carson 048793
Third Eye, San Diego 124182
Third Floor Antiques, San Diego 097137
The Third Line, Dubai 117190
Third Street Gallery, Philadelphia 123462
Third Street Gallery, Seattle . . . 124608
Third Text, Third World Perspectives
on Contemporary Art and Culture,
London 139226
Thiriassa, Los Angeles 121606
Thirion, D., Bruxelles 063557
Thiriot, Alain, Saint-Ouen 073211
Thirkill Antiques, Leyburn 089820
Thirlestane Castle, Lauder 044764
Thirroul Antique Centre, Thirroul 062316

Thirsk Museum, Thirsk 045970
Thirst Shop, Larnaka 065319
Thirteen Hundred, Cleveland . . . 120460
Thirteen Langton Street Gallery,
London 118599
Thirteenth Floor Gallery, Chicago 120282
Thirty 60, Dallas 093222
Thirty Victoria Street, Potts Point 062104,
099598
This and That, Abu Dhabi 088164
This and That, Hermanville-sur-
Mer 068813
This and That, Providence 096516
This Century Art Gallery,
Williamsburg 054416
This is the Place Heritage Park, Salt
Lake City 052917
This Old House Antiques,
Albuquerque 092086
This Old Warehouse, Jacksonville 094042
Thisted Antikvariat, Thisted . . . 140666
Thistle Antique, Houston 093811
Thistle Antiques, Aberdeen 088225
Thistle Antiques, Hamilton 063858
Thistle Books, Glasgow 144436
Thistlethwaite, E., Settle 091266
Tho, Icheon 111601
Tho, Seoul 111726
Thöle, Reinhard, Süstedt 078645
Thoemmes, Bristol 144285
Thöni, Franz, Chur 116221
Thöni, Heidi, Kloten 116391
Thoma + Thoma, Basel 116103
Thoma Galerie, Pähl 108355
Thoma Galerie, Doris Welker,
Starnberg 108665
Thoma, Peter, Köln 107691
Thoman, Elisabeth & Klaus,
Innsbruck 100060
Thomann, Ursula, Leibnitz 127804
Thomas, Bruxelles 140204
Thomas, Milwaukee 135821
Thomas & Assoc., New York . . . 123130
Thomas, Paris 071943
Thomas, Waikato 113044
Thomas & Alexander, Paddington,
Queensland 062056
Thomas Bewick's Birthplace,
Mickley 045256
Thomas-Bourgneuf, Jean-Paul,
Cesson-Sévigné 067756
Thomas Carlyle's Birthplace,
Ecclefechan 044227
Thomas Center Galleries,
Gainesville 049006
Thomas Clarke House-Princeton
Battlefield, Princeton 052371
Thomas-Dubarry, Arcachon . . . 103269
Thomas E. McMillan Museum,
Brewton 047195
Thomas Edison Birthplace Museum,
Milan 050886
Thomas Edison House, Louisville 050503
Thomas Fisher Rare Book Library,
Toronto 007212
Thomas-Foreman Home, Oklahoma
Historical Society, Muskogee . . 051141
Thomas Foster Memorial,
Uxbridge 007262
Thomas Griswold House and Museum,
Guilford 049293
Thomas Handforth Gallery,
Tacoma 053650
Thomas Hart Benton Home and Studio,
Kansas City 049894
Thomas-Hines, Richmond 096582
Thomas J. Boyd Museum,
Wytheville 054562
Thomas J. Walsh Art Gallery and
Regina A. Quick Center for the Arts,
Fairfield University, Fairfield . . 048670
Thomas K. Lang Gallery, Webster
University, Wien 003090
Thomas Kloster, Hedehusene . . 065727

Thomas L. Kane Memorial Chapel,
Kane 049867
Thomas-Legler-Haus, Diesbach . 041439
Thomas-Mann-Archiv, Zürich . . . 042236
Thomas McCulloch Museum,
Dalhousie University, Halifax . . 005964
Thomas-Müntzer-Gedenkstätte,
Heldrungen 019179
Thomas Muir Museum,
Bishopbriggs 043636
Thomas Newcomen Museum,
Exton 048655
Thomas P. Kennard House,
Lincoln 050301
Thomas Price House, Woodruff . 054524
Thomas Sappington House Museum,
Crestwood 048064
Thomas Street Framing & Gallery,
Limerick 109670
Thomas T. Taber Museum,
Williamsport 054417
Thomas Warne Historical Museum, Old
Bridge Township 051746
Thomas Warne Historical Museum,
Madison Twp. Historical Society,
Matawan 050739
Thomas Whaley House Museum, San
Diego 052983
Thomas Wolfe Memorial,
Asheville 046559
Thomas, Andrew, Boston 119926
Thomas, Andrew, Stamford 091410
Thomas, B. & B., Hammond . . . 135574
Thomas, Barry, Auckland 083829
Thomas, Barry, Minneapolis . . . 122044
Thomas, C., Gent 063663
Thomas, C.E., Stockholm 143970
Thomas, Dale, Artarmon 098671
Thomas, Dominique, Pont-
à-Mousson 072239
Thomas, Frédérique, Soustons . 129262
Thomas, Hartmut, Zittau 079147
Thomas, I.S., Paris 105268
Thomas, Jeffrey, San Francisco . 145568
Thomas, Marie-Josée & Claude,
Perpignan 072097
Thomas, Olaf, Hamburg 076326
Thomas, Patrick, Dorking 089095,
134340
Thomas, Richard, San Francisco 124413
Thomas, Richard C., New
Orleans 122248
Thomas, Simone, Paris 141287
Thomas, Tamara B., Los Angeles 121607
Thomassen, Göteborg 115665
Thomasset, Jean-Pierre, L'Isle-sur-la-
Sorgue 069594
Thomasson, Gérard, Grenoble . 128615
Thomaston Historical Museum,
Thomaston 053735
Thomaston Place Auction Galleries,
Thomaston 127475
Thomastown Antiques,
Thomastown 079705
Thomasville Cultural Center,
Thomasville 053738
Thome, New York 136014
Thome, H., Königsfeld im
Schwarzwald 076925
Thomi, Albert, Basel . . 087259, 144035
Thomine Demazures Alban, Neuilly-
sur-Seine 070523
Thompson, Aldeburgh . 117222, 134017
Thompson, Atlanta . . 119622, 119623
Thompson, London . . 118600, 118601
Thompson, Louisville 094594
Thompson, Windsor, New South
Wales 062432
Thompson, Anthony, London . . . 118602
Thompson, Bob & Pam, Charlotte 127159
Thompson, Brad, New Orleans . 122249
Thompson, Carole, Memphis . . . 121711
Thompson, Damian, Yamanto . . 099932
Thompson, Gary, Salt Lake City 096927

Thompson, James, Kirkby
Lonsdale 126934
Thompson, John, Knaresborough 089725
Thompson, John, Tunbridge
Wells 091664
Thompson, John W., Sydney . . . 127685
Thompson, Michael, Los Angeles 145236
Thompson, O.O. & Jane,
Charlotte 092632
Thomsen, Erik, Bensheim 074721
Thomsen, Inge, Siegen 108620
Thomsen, Jette & Kurt, Slagelse 066133
Thomsen, Jim & James, Oakland 136046
Thomson, Hannover 130017
Thomson & Burrows, Salt Lake
City 096928
Thomson & Roddick, Edinburgh 126890
Thomson & Roddick, Dumfries . . 126883
Thomson Bagley, Whangarei . . . 126334
Thomson, C., New York 123131
Thomson, Graeme, Auckland . . . 083830
Thomson, Percy, Taranaki 113019
Thomson, Roddick & Medcalf,
Carlisle 126848
Thomson, Roddick & Medcalf,
Dumfries 126883
Thomson's, Saint Helier 091188
Thon, Matthias, Ahlen . 125848, 141399
Thonet & Karolinsky, Wien 127973
Thonet-Museum, Friedberg 001908
Thong Nhat Palace, Ho Chi Minh
City 054809
Thor, Long Beach 094228
Thoral, Pierre, Lyon 069781
The Thoreau Institute at Walden
Woods, Lincoln 050288
Le Thorel, Jocelyne, Ploisy 072192
Thorens, Daniel Blaise, Basel . . 116104
Thorma János Múzeum,
Kiskunhalas 023749
Thormannhaus, Stiftung Rebhaus
Wingreis, Twann 042074
Thorn, David J., Budleigh
Salterton 088696
Thornbury and District Museum,
Thornbury, Gloucestershire . . . 045971
Thornbury Antiques, Thornbury,
Gloucestershire 091598
Thornby Cottage Antiques, Exton 061459
Thorndon Antiques, Wellington . 084192
Thorne-Sagendorph Art Gallery, Keene
State College, Keene 049913
Thorney Heritage Museum,
Thorney 045972
Thornhill, London . . . 090535, 134782
Thornhill Gallery, Avila College, Kansas
City 049895
Thornhill Historic Site and 19th
Century Village, Chesterfield . . 047621
Thornquest Gallery, Southport . . 099712
Thornthwaite Galleries,
Thornthwaite 119228
Thornton, Auckland 126308
Thornton, Hamilton 112907
Thornton Tomasetti, Philadelphia 123463
Thornton W. Burgess Museum,
Sandwich 053076
Thornton, Andy, Halifax 089415
Thornton, Diane, York 119413
Thornton, J.W., Bowness-on-
Windermere 088551
Thornton, John, London 144699
Thornton, Joseph, Windermere . 091875
Thorntons, Harrogate 089442
Thornwood, Houston 121139
Thorp & Bros., New York 136015
Thorp, Edward, New York 123132
Thorp, Thomas, Saint Albans . . 144812
Thorsen, Grimstad 143386
Thorsheim, M.J., Minneapolis . . 122045
Thorvaldsens Museum,
København 010198
Thorvaldsens Samlingen på Nysø,
Vordingborg 010365
Thorwirth, K., Vöhl 078808

Those were the Days, Cambridge 088766
Those Were the Days,
Uddingston 119281
Thosh, Cluj-Napoca 114163
Thost, Hans-Joachim,
Dahlenburg 129666
Thoth, Bussum 137851
Thoumyre, François, Conches-en-
Ouche 068122, 103626
Thoumyre, Perrine, Oppède ... 128878
Thousand Islands Museum of Clayton,
Clayton 047786
Thouvenot, Robert, Haroué ... 068793
Thrai Kao, Battani 042684
Thrasher Carriage Museum,
Frostburg 048991
Thrasher, Angela R., Baltimore . 092366
Three Buckets Full, Hutt City .. 083959
Three Centuries Shop, Vancouver 064888
Three Dee Art, Soldiers Point .. 099690
Three East Third Street Corporation,
New York 123133
Three Georges' Imperial Art Academy,
Beijing 055196
Three Lions Antiques,
Jacksonville 094043
Three-O-Five-Shop, New Orleans 095170
Three Pagodas, Minneapolis ... 094951
Three Rivers Auction,
Washington 127489
Three Rivers Gallery, Pittsburgh 123582
Three Rivers Museum, Muskogee 051142
Three Rivers Museum and Local
History, Rickmansworth 045616
Three Rooms, Minneapolis 122046
Three Treasures Fine Arts Centre,
Singapore 085515
Three Village Historical Society
Museum, Setauket 053274
Three Women and an Armoire,
Sacramento 096653
Threshing Museum, Mullinahone 024969
Threshold, Delhi 109346
The Thrie Estaits, Edinburgh .. 089193
Thrifty Upholstery, Albuquerque . 135292
Throckmorton Fine Art, New York 095782,
123134
Throll, Berlin 141581
Thron, Reutlingen 108485
Thronateeska Heritage Foundation,
Albany 046339
Through the Looking Glass, Dún
Laoghaire 079611
Through the Looking Glass,
London 090536
Through the Quay Hole,
Salcombe 119050
Thua Thien Hue Museum, Thua Thien
Hue 054820
Thünen-Museum-Tellow,
Warnkenhagen 022434
Thüring, Lydia, Therwil 087844
Thüringer Freilichtmuseum
Hohenfelden, Hohenfelden 019302
Thüringer Korbmachermuseum, Bad
Berka 016724
Thüringer Landesmuseum
Heidecksburg, Rudolstadt ... 021532
Thüringer Museum Eisenach,
Eisenach 018136
Thüringer Universitäts- und
Landesbibliothek Jena, Jena .. 019460
Thüringer Wald – Kreativ Museum mit
1. Deutschem Kloßpressenmuseum,
Großbreitenbach 018842
Thürmer, Lutz, Potsdam 130540
Thuillier, Agathe, Decazeville .. 103662
Thuillier, Alain, Herblay 068809
Thuillier, Serge, Noirmoutier-en-
l'Ile 070738
Thuillier, William, London 090537
Thulin & Ohlson, Göteborg ... 143854
Thulla, Howth 109644
Thum, D., Meersburg 077322

Thum, Dagobert, Markdorf 077297
Thumm, Barbara, Berlin 106448
Thun-Hohenstein, Praha 065531
Thunder Bay, New York 095783
Thunder Bay Art Gallery, Thunder
Bay 007146
Thunder Bay Military Museum,
Thunder Bay 007147
Thunder Bay Museum, Thunder
Bay 007148
Thunderbird Museum, Merrillan . 050809
Thunér, Lars, Stockholm 087066
Thunnissen, Claudia, Leiderdorp 132565
Thuns, Michael, Bonn 129616
Thur, Odder 140630
Thurgauische Kunstgesellschaft,
Kreuzlingen 059970
Thurles Famine Museum, Thurles 024993
Marion Thurner & Ulrich Sagmeister,
Seebenstein 062840
Thurnhof, Horn 100036, 136516
Thurow, Gerhard, Berlin 129573
Thurrock Museum, Grays 044468
Thursford Collection, Fakenham 044321
Thurso Heritage Museum, Thurso 045979
Thury György Múzeum,
Nagykanizsa 023804
Thwaite Mills Watermill, Leeds . 044788
Thwaites, Watford 091759, 135194
Thyade, Rotterdam 112663
Thyssen-Bornemisza Art Contemporary,
T-B A21, Wien 003091
Tiafit, Emmetten 087408
Tiagarra Aboriginal Museum,
Devonport 001028
Tian Chan Xuan Art Gallery,
Beijing 101993
Tiandu, Singapore 114623
Tianjin Academy of Fine Arts,
Tianjin 055245
Tianjin Art Museum, Tianjin 008323
Tianjin Glaswegiansu Guan,
Tianjin 008324
Tianjin History Museum, Tianjin 008325
Tianjin Museum of Natural History,
Tianjin 008326
Tianjin Painting Gallery, Tianjin . 008327
Tianjin People's Fine Arts Publishing
House, Tianjin 136743
Tianjin Science Museum, Tianjin 008328
Tianjin Theatre Museum, Tianjin 008329
Tianshui Museum, Tianshui 008332
Tiantai Museum, Tiantai 008333
Tianyi Ge Museum, Ningbo 008139
Tiara, Singapore 085516
Tibaudo, François, Lyon 128729
Tiber Bookshop, Baltimore ... 144988
Tibet, Lille 069507
Tibet House Museum, Delhi 024239
Tibet House Museum, New York 051482
Tibet Museum, Lhasa 008004
Tibet Museum, Fondation Alain Bordier,
Gruyères 041574
Tibet Songtsen House, Zürich .. 042237
The Tibetan Gallery, Hong Kong 065102
Tibetan Lama Art, Hamburg ... 076327
Tibetan Traditional Thangka Arts,
Kathmandu 112088
Fabio Tiboni Arte Contemporanea,
Bologna 109869
Tibooburra Local Aboriginal
Land Council Keeping Place,
Tibooburra 001579
Tibro Museum, Tibro 041076
Tic Toc Clock Shop, Dallas ... 135519
Tichadou, Philippe, Richebourg . 072456
Tichenor, Scott, Louisville 094595
Tichmann, Andreas, Wuppertal . 079119
Ticho House, Jerusalem 025110
Tichý, O., Praha 140532
Ticktock Media 4D, Tunbridge
Wells 138290
Tid och Rum Antik i Skellefteå,
Skellefteå 086935

Tidaholms Konstförening,
Tidaholm 059914
Tidaholms Konsthall, Tidaholm . 041077
Tidaholms Museum, Tidaholm .. 041078
Tidal Access Gallery, Nelson ... 112968
Tidal Wave Gallery, Hereford ... 117965
Tiderne Skifter, København 136776
Tidlösa Rum, Helsingborg 086742
Tiedong Qian Hua Art Cooperative,
Anshan 101928
Tiefenauer, Charly, Rüti 087763
Tiefengraber, Ursula & Dietmar,
Wildermieming 127985
Tiegs, R.-Peter, Spremberg ... 078526
Tieling Museum, Tieling 008334
Tielker, Jürgen, Bochum 075046
Tiemeyer, Ria, Berlin 129574
Tiempo's Antiques Curiosidades,
Valencia 097978
De Tiempos de Antaño, Bilbao . 085890
Tiempos Modernos, Madrid 086270
Tiemuseo, Helsinki 010875
Tien, Kee, Singapore 085517
Tienda Guadalupe, San Antonio 124077
La Tienda Latina, Bruxelles ... 140205,
140206
Tiendschuur, Tilburg 132642
Tienne, Modena 080822
Tienti, D., Genova 080298
Tientong, Los Angeles 094486
Tiepolo, Singapore 085518
Tieppo, Oscar Luigi, Roma 081580
Tier, Rafaelue, Escazú 102470
Tiercin, Christian, Dijon 103682
Tierens, Hugo, Londerzeel 128079
Tiermuseum, Lenggries 020037
Tierra Adentro, Álvaro Obregón . 138922
Tiers, Serge, Marseille 128785
Tierwelt-Panorama, Ebikon 041451
Tieste, Reinhard, Bremen 075186,
137129
Tieteelliset Kokoelmat, Forssa . 010782
Tietjen & Co., Hamburg 125980
Tifani, Beograd 114444
Tiffany, Chicago 092800, 135426
Tiffany, Enschede 083260
Tiffany, Luzern 087648
Tiffany, West Hollywood 097938
Tiffany Antiques, Shrewsbury .. 091335
Tiffany Glass Gallery, Calgary .. 101262
Tiffany Inn, Kansas City 094097
Tiffany Ursta, Wrocław 084774
Tiffany's Antiques, Chennai ... 079379
Tiffany's Decor, Edmonton 064306
Tiffany's Decor on Whyte,
Edmonton 064307
Tiffins Antiques, Emsworth ... 089208
Tifloloŝki Muzej, Zagreb 009052
Tiftickjian jr., David, Buffalo ... 092563
Tiftickjian & Sons, David, Buffalo 092562
Tiger Paws, Minneapolis 094952
Tighes Hill Gallery, Tighes Hill .. 099826
Tighnabruaich Gallery,
Tighnabruaich 119233
Tigressblue Gallery, Kirkhill ... 118058
Tigris, Felis, Troyes 105778
Tigullio, Rapallo 110699
De Tijd, Doetinchem 083173
Tijdloos, Den Haag 112416
Tijdloos, Leiden 083448
De Tijdspiegel, Den Haag 083137
De Tijdspiegel, Middelburg 083502
Tijmen, Peters, Rhenen 132613
Tijssen, Venlo 083690
Tikanojan Taidekoti, Vaasa ... 011472
Tikerak, Annecy 103251
Tikøb, Charlottenlund 065632
Tikotin Museum of Japanese Art,
Haifa Museums, Haifa 025059
Tilba Woodturning Gallery, Central
Tilba 098879
Tilbury, Draguignan ... 068404, 128587
Tilbury, Genova 080299
Tilbury Fort, Tilbury 045980

Tilché, Jean-Marc, Carcassonne 103530
Tilden-Foley, New Orleans 122250
Tilden-Thurber, Providence 096517
Tile Heritage Foundation,
Healdsburg 049425
Tilk, Helina, Tallinn 103120
Till-Eulenspiegel-Museum,
Schöppenstedt 021718
Tillamook County Pioneer Museum,
Tillamook 053746
Tillamook Naval Air Station Museum,
Tillamook 053747
De Tille, Leeuwarden 143181
Tiller, Amsterdam 112295
Tiller, Wien 100383
Tilley, Sheffield 091294, 144840
Tilley, Honey, London 090538
Tilleys, Solihull 091357, 135056
Tillier, Vincent, Paris ... 071944, 105269
Tillie's Collectibles, San Antonio 097040
Tillman, Potterspury 091049
Tillou, Dana, Buffalo .. 092564, 119968
Till's Bookshop, Edinburgh ... 144402
Tillybs, Heinz, Hanstedt 076404
Tillys Kuriosa, Stockholm 087067
Tilman, Powhatan 127414
Tilmant d'Auxy, Bruno, Château-
Chervix 103575, 128548
Tilsam, Paris 071945
Tilt, Ferntree Gully 099016
Tilton, Jack, New York 123135
tim – Bayerisches Textil- und
Industriemuseum, Augsburg .. 016672
Tim & Avi's Salvage Store,
Indianapolis 093933
Timar Italia, Parma 081130
Timárház – Kézmuvesek Háza,
Debrecen 023598
Timber Creek Police Station Museum,
Timber Creek 001580
Timber Museum of New Zealand,
Putaruru 033291
Timber Village Museum, Blind
River 005431
The Timbers Antique Centre,
Lavenham 089753
Timbert, Odette, Espaubourg .. 068469
Timbertown Museum, Wauchope 001628
Timbery, Laddie, Huskisson ... 061644
Timbres et Monnaies, Paris ... 071946
Timbres et Monnaies, Tours ... 073864
Timbuctoo II, Amsterdam 143071
Timbuktu, Amsterdam 082873
Timbuktu to Kathmandu,
Noosaville 099464
Time & Again Auction Gallery,
Linden 127285
Time & Tide, Christchurch 083901
Time & Tide – Museum of Great
Yarmouth Life, Great Yarmouth 044477
Time after Time, Hemswell Cliff 089501,
134453
Time and Again, Pittsburgh ... 096331,
136142
Time and Motion, Beverley ... 088454,
134099
Time and Tide Antiques,
Portaferry 091036, 134949
Time Bandit, Baltimore 092367
Time Booksellers, Frankston .. 139739
Time Bygdemuseum, Undheim .. 033455
Time Capsule, El Paso 093418
Time Frame, Edmonton 101302
Time Gallery, New York 095784
Time Gallery, Paris 071947
Time in Hand, Shipston-on-Stour 091320,
135042
Time Machine, Merimbula 139828
The Time Machine, Virginia
Beach 097794, 136449
Time Machine, Weston-super-
Mare 046102
The Time Museum, Newark-on-
Trent 045336

Time Museum, Tehrān 024786
Time 'n' Again, Thirroul 062317
Time Pieces & Antiques, New
York 095785
Time Shop, Milwaukee 094881, 135822
Time Tested Books, Sacramento 145494
Time Traders, Boston 092499
Time Traveller, Paeroa 084051
Time Treasures Antiques,
Chicago 092801, 135427
Time Tunnel, San Jose 097378
Time Tunnel, Zürich 088011
Time Wonderland, Kawagoe . . . 081992
Timeball Tower, Deal 044077
Timecraft Clocks, Sherborne . . . 091310
Timekeepers, Chicago 092802
Timeless, Bruxelles 063558
Timeless, Hibiscus Coast 083955
Timeless, Tauranga 132751
Timeless Antiques, Basel 087260
Timeless Antiques, Chicago . . . 092803
Timeless Art Gallery, Singapore . 114624
Timeless Beds, Malvern Link . . . 090686
Timeless Books, Angaston 139624,
139625
Timeless Collectable, San Diego 097138
Timeless Gallery, Zürich 088012
Timeless Pieces, Silverton 062212
Timeless-The Art Book Studio,
Delhi 137594
Timeless Treasures, Kansas City 094098
Timeless Treasures, San
Francisco 097337
Timelines, Saint Paul 096865
Timelounge, Wien 063088
Timely Treasures, Houston 093812
Timepeice Repairs, Lincoln 089834,
134562
Timepiece, Dublin . . . 079591, 130863
Timepiece, Teignmouth 091553
Times and Chimes, Houston . . . 093813
Times Gone By, Jacksonville . . . 135667
Time's Lost Toys, Saskatoon . . . 064570
Times Oriental Rugs, Houston . . 093814
Times Past – Bygones of Yester Year,
Blindley Heath 043659
Times Past Antiques,
Auchterarder 088323, 134044
Times Past Antiques, Eton 089222,
134375
Times Past Antiques, Potter
Heigham 091047
Times Square Gallery, Hunter College
City University of New York, New
York 051483
Timespan Heritage Centre and Art
Gallery, Helmsdale 044557
Timexpo Museum, Waterbury . . 054200
Timgems, Norwich 090885
Timken Museum of Art, San
Diego 052984
Timm, Gesa, Schleswig 078369, 130623
Timme, Hannover 141920
Timmen, Harry, Borculo 132446
Timmermans-Opsomerhuis, Lier 003817
Timmermans, Juanita, Bad
Salzuflen 074586
Timmins Museum, National Exhibition
Centre, South Porcupine 007059
Timms, S. & S., Shefford 091296
Timoteo, Burgos 133215
Timothy, Slijk Ewijk 126299
Het Timpaan, Apeldoorn 082930
Tim's Book Exchange, Winnipeg 140438
Tin Angel Antiques, Baltimore . . 092368
Tin Drum Books, Leicester 144526
Tin Horse Antiques, Louisville . . 094596
Tin Man, Miami 094810
Tin Museum – Gedung Raja Abdullah,
Kelang 030671
Tin Sheds Gallery, University of
Sydney, Faculty of Architecture,
Sydney 001553
Tin Toy Museum & Motion Display

Museum, Yokohama 029727
Tinakori Antiques, Wellington . . 084193
Tinakori Gallery, Wellington . . . 084194,
132764, 132765, 143374
Tinarelli, Evelina, Bologna 079883
Tina's Antiques, Codford 088943
Tinatini, Tbilisi 105923
Tincan Alley, Tulsa 124816
Tincu, Timişoara 085142
Tindell, Nashville 135852
Tindley & Chapman, London . . . 144700
Tinebra, Carmelo, Milano 131540
El Tinell, Barcelona 085845
Tinelli, Fulvia, Milano 080744
Tinelmuseum, Sinaai-Waas . . . 004011
Tineriba Fine Art, Adelaide 060973,
098615
Tineriba Tribal Gallery and Museum,
Hahndorf 001131
Ting, Amstelveen 082668
Ting og Antik, Helsingør 065746
Ting Qin Antiques, Shanghai . . . 065153
Ting-Stedet, Thisted 066165
Tingsryds Antikvariat, Tingsryd . 143980
Tingstedet, Rudkøbing 066102
Tingstedet Antik og Genbrug,
Hillerød 065758
Tinguely, Peter, Tafers 087840
Tingvoll Museum, Tingvoll 034015
Tingwall Agricultural Museum,
Gott 044455
Tinhorn Press Gallery, San
Francisco 124414
Tinia Numismatica, Follonica . . 080225
Tinkelbel, Maastricht 112584
Tinker, Richmond 136179
Tinker Swiss Cottage Museum,
Rockford 052640
Tinley Park Historical Society Museum,
Tinley Park 053748
Tinn Museum, Rjukan 033895
Tinnert, Falköping 115609
Tinney, Martin, Cardiff 117532
Tino, Norberto, La Paz 100851
Tino's Treasures, Omaha 096059, 136068
Tinslay, Peter, Crows Nest, New South
Wales 061376, 139706
Tinsoldaten, Aalborg 065590
TinT, Thessaloniki 109103
Tintern Abbey, Chepstow 043928
Tintern Antiques, Tintern 091601
Tinti, Olivier, Rouen 072620
Tinti, Rosanna, Bologna 079884
Tintic Mining Museum, Eureka . . 048630
Tintori, Elena & Leonetto, Prato . 131782
The Tiny Clock Shop, Windsor,
Ontario 064907
Tiny Factory-Comeprima,
Toulouse 105754
Tinzl, Christoph, Salzburg 127861
Tioda, Gijón 115073
Tioga County Historical Society
Museum, Owego 051850
Tioga Point Museum, Athens . . . 046601
Tippecanoe Battlefield, Battle
Ground 046828
Tippecanoe County Historical Museum,
Lafayette 050060
Tippl, Gerd, Bamberg 129447
Tipton County Museum,
Covington 048045
Tipton-Haynes Museum, Johnson
City 049828
Tipton, Steve, Louisville 094597
Tiptree Museum, Tiptree 045981
Tipu Sahib Museum,
Srirangapatna 024486
Tiqueibles of Irvington,
Indianapolis 093934
Tirau Antique Centre, Waikato . . 084138
Tirdad, Wädenswil 116708
Tire la Chevillette, Dinard 068359
Tire Müzesi, Tire 043070
Tirelli, Luciano, Vevey 087883

Tirenna, Palermo 081078
Tires, Terry, Jacksonville 094044
Tiroche, Micky, London 118603
Tirol Edition, Innsbruck 136522
Tiroler Bauernhausmuseum,
Kitzbühel 002183
Tiroler Bergbau- und Hüttenmuseum,
Brixlegg 001799
Tiroler Kaiserjägermuseum Obergricht,
Serfaus 002753
Tiroler Künstlerschaft, Innsbruck 058313
Tiroler Künstlerschaft – Kunstpavillon,
Innsbruck 002140
Tiroler Landesmuseum Ferdinandeum,
Innsbruck 002141
Tiroler Volkskunstmuseum,
Innsbruck 002142
Tironi, Bergamo 079795
Tirotta Arte, Genova 110140
Tirotta, Ilario, Genova 110139
Tirpitz-Stillingen, Varde Museum,
Blåvand 010022
Tisato, Massimo, Verona 132303
Tisch & Stuhl, Berlin 074936
Tisch Beere Gallery, Cardiff . . . 117533
Tischbein-Haus-Museum, Kloster
Haina, Haina (Kloster) 018931
Tischlak, Paul & Helmut, Amberg 074405,
129384
Tischlerei Museum, Bremen . . . 017600
Tish, Anya, Houston 121140
Tison, Claude, Piacé 072152
Tison, Jacky, Maresché 069873
Tisserant, Paris 071948
Tissot, Gilles, La Chaux-de-Fonds 087361
Tiszeker, Krisztina, Roma 081581
TiT Art og Design, Bergen 113137
Titan Missile Museum, Green
Valley 049232
Titbits, Milano 080745
Titchener, Mary, South Yarra . . 062229
Titcher & Sons, John, Littleton,
Cheshire 089851
Titchiner, Mark, Halesworth . . . 117900
Titgemeyer-Heck, Lucia,
München 077629
Titi, Gianna, Roma 080745
Titian Gallery, Stow-on-the-Wold 091486
Titian Studio, London 134784
Titian, Rodrigo, London 134783
Titiana, Chaskovo 101123
Title Wave, Bend 144995
Titles Book Gallery, Howick . . . 143323
Tito Arte, Torino 111010
Titon, San Francisco . . 097338, 136348
Titren, Beaune 103356
Titta In, Stockholm 087068
Tittiskåpet, Sollentuna 086945
Tittot Glass Art Museum, Taipei . 042559
Titus Brandsma Museum,
Bolsward 032079
Tiverton Museum of Mid Devon Life,
Tiverton 045982
Tivoli Gallery, Salt Lake City . . . 123991
Tiwi Art Network, Darwin 061389
Tiziano, Conegliano 080055
TJ Furniture Land, Houston 093815
TJK Antiques, San Francisco . . . 097339
Tjømes Kunstsalong, Oslo 084338
Tjolöholms Slott, Fjärås 040628
Tjumenskaja Kartinnaja Galereja,
Tjumen 037945
Tjumenskij Gosudarstvennyj Institut
Iskusstv i Kultury, Tjumen . . . 056306
Tjumenskij Oblastnoj Kraevedčeskij
Muzej im. I.Ja. Slovcova,
Tjumen 037946
Tjumenskij Oblastnoj Muzej
Izobrazitelnych Iskusstv, Tjumen 037947
Tjustgalleriet, Västervik 115986
TK Asian Antiquities, New York . 095786
Tkalec, Helmut, Oberalm 062755,
127826
TKG Editions, Tokyo 137794

TKH, Bad Königshofen 074540
Tlaqueparte Monterrey, Exposicion
Latinoamericana de Artesanias,
Antiguedades, Regalos y Decoración,
Monterrey 098248
TLC Exchange, Tucson 097697
TLC Gallery, Houston . . . 121141, 121142
Tlon Books, London 144701
TM Gallery, Crestview 048063
TN Galeria Twórców
Niepełnosprawnych, Szczecin . 113594
To, Seoul 111727
To-ji Homotsukan, Kyoto 028986
Toad Hole Cottage Museum,
Ludham 045164
Toale, Bernard, Boston 119927
Toast Gallery, Paris 105270
Tobacco Farm Life Museum,
Kenly 049919
Tobačni Muzej, City Museum of
Ljubljana and Tobachna Ljubljana,
Ljubljana 038563
Tobajas, Antoni, Barcelona 085846
Tobaks- och Tändstickmuseum,
Stockholm 041053
Tobey, Peter, New York 123136
Tobias and The Angel, London . 090539
Tobias Community Historical Society
Museum, Tobias 053754
Tobias-Mayer-Museum, Marbach am
Neckar 020265
Tobias Smollett Museum,
Alexandria 043389
Tobiassen, Stavanger . 084373, 113276
Tobiassen, Svolvær 113281
Tobien, Joachim, Husum 107456
Tobilane Designs, Penrith 134929
Tobin-Hewett, Louisville 121673
Tobinodai Shiseki Koen Hakubutsukan,
Funabashi 028568
Tobogan, Paris 071949
Tobogan, Saint-Ouen 073212
Toboggan Centre Culturel, Decines-
Charpieu 103663
Tobolskaja Kartinnaja Galereja,
Tobolsk 037948
Tobolskij Gosudarstvennyj Istoriko-
architekturnyj Muzej-zapovednik,
Tobolsk 037949
Tobwabba Art, Forster 061481
Toby Jug, Deal 089040
Toby's Antiques, Torquay 091618
Toby's Antiques, Portland 096482
Toccaceli, Monica, Roma 081583
Tocchetti, Lugano 116459, 126742
Tochigi-kenritsu Bijutsukan,
Utsunomiya 029684
Tochigi-kenritsu Hakubutsukan,
Utsunomiya 029685
Today's Memories, Singapore . . 085519
Todd & Austin, Winchester 091872
Todd Madigan Gallery, California
State University Bakersfield,
Bakersfield 046720
Todd, Don, Portland 136161
Todd, Sue, Dunedin 083934
Todd's, Launceston 089750
Tode, Horst 074937, 141582
Todeschini, Attilia, Verona 132304
Todi, Tiziana, Roma 110866
Todisco, Carl, Boston 092500
Todmorden Fine Arts, Todmorden 119240
Todmorden Mills Heritage Museum
and Arts Centre, Toronto 007213
Todorov, T., Loveč 140317
Toebosch, Amsterdam 082874
Toebs, F., Berlin 074938
Töff, Essen 075702
Toei Movie Land Toei Uzumasa
Eigamura Movie Museum,
Kyoto 028987
Tøjhusmuseet, København 010199
Tölke, Christoph, Krefeld 076965, 130198
Töllinmäen Museo, Hämeenkyrö 010785

Töllstorps Industrimuseum, Gnosjö Industrimuseum, Gnosjö 040646
Toennissen, Monika, Köln 107692
Tønsberg Antikvariat, Tønsberg . 143423
Tønsberg Auksjonsforretning, Tønsberg 126391
Toepfer, Dr. Eva, Contern 082350
Toepfer, Kelly, Northwood 062007
Töpferei und Museum im Kannenofen, Höhr-Grenzhausen 019282
Töpfereimuseum, Speicher 021929
Töpfereimuseum Langerwehe, Langerwehe 019918
Töpfereimuseum Raeren, Raeren 003959
Töpfermuseum, Duingen 018049
Töpfermuseum, Kohren-Sahlis .. 019754
Töpfermuseum, Rödermark 021453
Töpfermuseum, Stoob 002798
Töpfermuseum Im alten Kannenofen, Ransbach-Baumbach 021277
Töpfermuseum Thurnau, Thurnau 022154
Törk, Darmstadt 075312
Törngrens Krukmakeri, Falkenberg 040616
Török Ház, Szigetvár 023972
Török, Ferenc, Sankt Gallen .. 116618, 133912
Tørvikbygd Bygdemuseum, Tørvikbygd 034019
Töysän Museo, Töysä 011422
Toffanelli, Andrea, Milano 131541
Tofield Historical Museum, Tofield 007153
Toft, Rungsted Kyst 066103
Toft, Elise, Kolding 102922
Toft, Peter, Odder 066025
Toftum, Lars, Valby ... 066177, 128403
Togari-Ishi Jomon Kokokan, Chino 028516
Togei Bijutsu Kenkyujo, Kawasaki 056006
Togen, Hiroshima 081986
Toggenburger Museum, Lichtensteig 041671
Toggenburger Schmiede- und Handwerksmuseum, Bazenheid 041295
Tognino, M., Essen 075703
Tognolini, Carola, Verona 132305
Caterina Tognon Arte Contemporanea, Venezia 111164
Tognon, Caterina, Bergamo 109803
Tognon, Olivia, Bari 079785
Togonon, Julina E., San Francisco 124415
Toguna, Luxembourg 111870
Toguna, Toulouse 073803
Toguna African Art and Crafts, Amersfoort 112128
Toguri Bijutsukan, Tokyo 029603
Toh, Foong, Singapore 085520
Toho, Kyoto 142902
Tohoku Daigaku Hakabutsukan, Sendai 029363
Tohoku Geijutsu Koka Daigaku, Yamagata 056064
Tohoku Kindai Toji oyo Yaki Bijutsukan, Nakaniida 029136
Tohoku Rekishi Shiryokan, Tagajou 029423
Toiles, E., Quimper 105419
Le Toit du Monde, Paris 105271
Tokai Daigaku Kaiyo Kagaku Hakubutsukan, Shizuoka 029406
Tokaji Múzeum, Herman Ottó Múzeum, Tokaji 024015
Tokat Müzesi ve Örenyerleri, Tokat 043072
Token Hakubutsukan, Tokyo ... 029604
Toki-no-Wasuremono, Tokyo ... 111512
Toki-shiritsu Mino-yaki Rekishi Hakubutsukan, Toki 029480
Tokmachi, Nemat, Albuquerque . 092087
Toko, San Diego 097139
Toko, San Francisco 124416
Tokodo Shoten, Tokyo 142945

Tokomairiro Historical Society Museum, Milton 033246
Tokomaru Steam Engine Museum, Tokomaru 033315
Tokoname-shiritsu Mingeikan oyo Toji Rekishi, Tokoname 029482
Toksværd Antik, Holmegaard ... 065778
Toku, Tokyo 082210
Tokugawa, Phoenix 123517
Tokugawa Art Sword Exhibition Fair, Nagoya, Aichi 098234
Tokugawa Bijutsukan, Nagoya .. 029129
Tokusa, Honolulu 093575
Tokushima-kenritsu Hakubutsukan, Tokushima 029483
Tokushima-kenritsu Kindai Bijutsukan, Tokushima 029484
Tokushima-kenritsu Torii Kinen Hakubutsukan, Naruto 029159
Tokushima Prefectural Local Culture Hall & Awa Deko-puppet Museum, Tokushima 029485
Tokyo Art Auction, Tokyo 126251
Tokyo Bijutsu, Tokyo 137795
Tokyo Bijutsu Club, Tokyo 059515
Tokyo Chuo-ku Bijutsukan, Tokyo 029605
Tokyo Design Academy, Tokyo . 056057
Tokyo Designer Gakuin, Tokyo . 056058
Tokyo Fuji Art Museum, Hachioji 028582
Tokyo Gallery, Beijing 101994
Tokyo Gallery, Tokyo 111513
Tokyo Geijutsu Daigaku, Tokyo . 056059
Tokyo Honten Art Gallery, Tokyo 111514
Tokyo Kaikan Gallery, Tokyo ... 111515
Tokyo Kasei Daigaku Hakubutsukan, Tokyo 029606
Tokyo Kikaku, Musashino 082052
Tokyo Kokuritsu Hakubutsukan, Tokyo 029607
Tokyo Kokuritsu Kindai Bijutsukan, Tokyo 029608
Tokyo Kokuritsu Kindai Bijutsukan Kogeikan, Crafts Gallery, Tokyo 029610
Tokyo Kokuritsu Kindai Bijutsukan, National Film Center, Tokyo .. 029609
Tokyo Kokusai Bijutsukyoukai, Tokyo 029611
Tokyo Lease, Tokyo 082211
Tokyo Modern Art Gallery, Tokyo 111516
Tokyo Noko Daigaku Kogakubu Fuzoku Sen'i Hakubutsukan, Koganei . 028874
Tokyo Opera City Art Gallery, Tokyo 029612
Tokyo Senshoku Bijutsu Gakuin, Tokyo 056060
Tokyo Station Gallery, Tokyo ... 029613
Tokyo-to Bijutsukan, Tokyo 029614
Tokyo-to Gendai Bijutsukan, Tokyo 029615
Tokyo-to Shashin Bijutsukan, Tokyo 029616
Tokyo-to Teien Bijutsukan, Tokyo 029617
Tokyo Zokei Daigaku, Hachiouji . 056002
Tol, Jaap, Bergen 112343
Tolarno, Melbourne 099323
Tolbooth Art Centre, Kirkcudbright 044731
Tolbooth Museum, Stonehaven . 045887
Tolch, Andrew, Saint Louis ... 096783
Tolco, Galway 079625
Toldedo Antique Gallery, Miami . 094811
Tole World, Fremont 139507
Toledano, Jean-Daniel & Anne, Arcachon 125494
Toledano, Jean-Daniel & Anne, Bordeaux 125532
Toledo Artists' Club, Toledo ... 060775, 124700
Toledo Museum of Art, Toledo .. 053760
Tolerancijos Centras, Valstybinis Vilniaus Gaono Žydų Muziejus, Vilnius 030544
Tolga Eti Sanatevi, İstanbul ... 117092
Tolga Woodworks Gallery, Tolga 099828

Tolgus Tin Mill and Streamworks, Redruth 045599
Tolhouse Museum and Brass Rubbing Centre, Great Yarmouth 044478
Tolis, Gerasimos, Thessaloniki .. 109104
Tolksdorf, Wilma, Berlin 106449
Tolksdorf, Wilma, Frankfurt am Main 107036
Toll Gallery, Thurso 119231
Toll House Bookshop, Holmfirth . 144482
Tolland County Jail and Warden's Home Museum, Tolland 053762
Tollemans, Bruxelles 063559
Tollhouse Gallery, Clevedon ... 117597
Tolman, Miami 121868
Tolman, Shanghai 102205
Tolman, Singapore 114625
Tolman, Norman H., Tokyo 111517, 137796
Tolminski Muzej, Tolmin 038622
Tolotti, Luigi, Venezia 081876
Tolpuddle Martyrs Museum, Tolpuddle 045985
Tolsey Museum, Burford 043788
Tolson Memorial Museum, Huddersfield 044614
Tolson, Tom, Fort Worth 135560
Toltec Mounds Archeological State Park, Scott 053209
Toluian, Palermo 110623
Tolvmansgården, Hopsala 010886
Tolvskillingen, Kungsbacka 086795
Tom, Warszawa 084752, 143531
Tom Antique, Auckland 132698
Tom Brown's School Museum, Uffington 046017
Tom Collins House, Swanbourne 001533
Tom Mix Museum, Oklahoma Historical Society, Dewey ... 048321
Tom Thomson Memorial Art Gallery, Owen Sound 006606
Tom Tits Experiment, Södertälje . 040970
Toma, San Antonio 136275
Toma, Giuseppe, Padova 110581
Toma, Nick, Heerlen 083378
Toma, Sheila, New York 095787
Tomada, Andreina, Roma 081584
Tomakomai-shi Hakubutsukan, Tomakomai 029627
Tomakomai-shi Kagagukan, Tomakomai 029628
Tomaro, Antonio, Roma 132061
Tomarps Kungsgård, Kvidinge .. 040767
Tomaru, Tokyo 142946
Tomasetig, Andrea, Vignate ... 142890
Tomasi Gioielli, Trento 081778
Tomasi & C., Carla, Roma ... 132062, 132063
Tomasulo Gallery, Union County College, Cranford 048051
Tomatis, Francis, Lézignan-Corbières 069451
Tombak, İstanbul 088143, 144209
Tomball Community Museum Center, Tomball 053763
Tombe Reali di Casa Savoia, Torino 028146
Tombeur, Josette, Saint-Paul-de-Vence 105600
Tombland Antiques Centre, Norwich 090886, 134896
Tombland Bookshop, Norwich .. 144754
Tombland Jewellers and Silversmiths, Norwich 090887
Tome Parish Museum, Tome ... 053764
Tome Pires, Los Angeles 094447
Tomedi, Irene, Bolzano 131011
Tomelilla Auktionshall, Tomelilla 126687
Tomelilla Konsthall, Tomelilla ... 041080
Tomeo, Cleveland 120461
Tomescheit, Stefan, Ingelheim am Rhein 076609
Tomimoto Kenkichi Kinenkan, Ando 028474

Tomin, Radmilla, Bad Homburg v.d.Höhe 106068
Tomintoul Gallery, Ballindalloch . 117271
Tomintoul Museum, Tomintoul .. 045986
Tomita, Fukuyama 111253
Tomita, Hiroshima 111264
Tomita, Washington 124965
Tomlin, Dallas 093223
Tomlin, Gerald, Dallas 093224
Tomlinsons, Tockwith 091604
Tomm, A., Naumburg, Saale ... 077693
Tommasi, Paola, Milano 080746
Tommy Q, Los Angeles 094488
Tommy's Furniture, Albuquerque 135293
Tomo-no-Ura Rekishi oyo Minzoku Hakubutsukan, Fukuyama 028565
Tomorrow's Antique, Helsinki .. 066368
Tomorrow's Treasures Today, Los Angeles 094489
Tomos, Vigo 133481
Tompkins, M.D., Chicago 092804
Tom's Curios, Dublin 079592
Tom's Furniture and Collectables, Tucson 097698, 097699
Tom's Terrific Antiques Shop, Omaha 096060
Tom's Trains & Miniatures, Fresno 093515
Tomsich, Saint Louis 123879
Tomskij Muzej Derevjannogo Zodčestva, Tomsk 037951
Tomskij Oblastnoj Chudožestvennyij Muzej, Tomsk 037952
Tomskij Oblastnoj Kraevedčeskij Muzej, Tomsk 037953
Tomtom, London 090540
Ton Duc Thang Museum, Ho Chi Minh City 054810
Ton Sak Yai, Uttaradit 042727
Ton Smits Huis, Eindhoven 032292
Ton Ying & Co., New York 123137
Tonawandas Museum, Historical Society of the Tonawandas, Tonawanda 053766
Tonbandmuseum, Korneuburg .. 002226
Tonbergbaumuseum Westerwald, Siershahn 021859
Tondinelli, Floriana, Roma 110867
Tondo, Elisabetta, Bari 130926
Tondorf, Düsseldorf 106835
Tone, Saitama 111372
Tonella, Danielle, Lugano 087624
Tonelli, Chicago 120283
Tonelli, Milano 110469
Tonello, Alfio & Emilio, Milano . 131542
Tonen, Erik, Antwerpen 140156
Tong, Singapore 085521
Tong-Jin Irrigation Folk Museum, Kimje 029965
Tong, Mern Sern, Singapore ... 085522
Tongan Museum, Tongan 008336
Tongarra Museum, Albion Park . 000781
Tongass Historical Museum, Ketchikan 049945
Tongcheng Museum, Tongcheng 008337
Tongdosa Museum, Yangsan ... 030059
Tongerlohuis Heyderadey, Roosendaal 112626
Tongin, Seoul 082266
Tonglushan Ancient Metallurgy Museum, Daye 007732
Tongue, Dr., Portland 096483
Tongzhou Museum, Beijing 007648
Toni-Merz-Museum, Sasbach .. 021614
Tonic General, Calgary 064276
Tonin, Paolo, Torino 111011
Tonina Hiša v Sveti Petru, Pomorski Muzej Sergej Mašera, Portorož 038590
Toninelli, Luigi Filippo, Monaco . 082531
Tonini, Matteo, Ravenna 142791
Tonkin, Ha Noi 125101, 125102
Tonkonow, Leslie, New York ... 123138
Tonne, Terje, Oslo 113247, 132791
Tonner, Jean-Marie, Saint-Mihiel 072985

Tonner, Joseph, Pons 072235
Tonnerre, Joseph, Groix 068751
Tonnesen Hansen, Arnold,
  Herning 065754
Tonnoirs, Townsville 139935
Tono Kyodo Hakubutsukan, Tono 029630
Tonolli, Rovereto 110884
Tonon, Paolo & Josy, Genève .. 087504
Tonson Gallery, Bangkok 116967
Tony, Madrid 133397
Tony Galeria de Arte e Antiguidades, São
  Paulo 064212, 101115
Tony's Antiques, Portland 096484
Tony's Furniture Refinishing,
  Cleveland 135471
The Tool Shop, Needham Market 090805
Tooley, Adams & Co., Wallingford 144906
Tools Galerie, Paris 105272
Toomey & Tourell, San Francisco 124417
Toon Art, Cincinnati 120381
Toon-in Gallery, Melbourne 099324
De Toonkamer, Amersfoort 082659
Toons On, Philadelphia 123464
Toop, Bill, Broadmayne 117452
Tooraj, Los Angeles 094490
Tooronga Hall Antiques, Armadale,
  Victoria 061049, 127558
Tooronga Hall Antiques, Caulfield
  East 061301, 127583
Toowoomba Antiques Gallery,
  Toowoomba 062335
Toowoomba Art Society,
  Toowoomba 058223
Toowoomba Regional Art Gallery,
  Toowoomba 001585
Toowoon Bay Gallery, Toowoon
  Bay 099834
Top Art, Hiroshima 081987
Top Art, Jeddah 085309
Top Art, San Diego 124183
Top Art Galeria Sztuki Współczesnej,
  Gdańsk 113489
Top Banana Antiques Mall 1,
  Tetbury 091583
Top Banana Antiques Mall 2,
  Tetbury 091584
Top Coins & Fine Art, Mall
  Darwin 061769
Top Drawer Antiques, San Diego 097140
Top Drawer Auction, New
  Orleans 095171, 127339
Top Fresh Arts Gallery, Singapore 114626
Top Hat Antiques, Nottingham .. 090899
Top of Oklahoma Historical Museum,
  Blackwell 047030
Top of the Hill, Stoke-on-Trent . 091454
Top of the Ninth, Columbus ... 093051
Top Pictures, Kingston-upon-Hull 118050
Top Rockz Gallery, Kununurra .. 099205
Top Time, Milano 080747
Top, T.T., Kloosterburen 132553
Le Tophet de Salammbô à Carthage,
  Salammbô 042766
Topkapi, Maubeuge 069996
Topkapı Sarayı Müzesi, İstanbul 042961
Topografika Antikvariat, Oslo ... 143418
Topographie des Terrors, Berlin . 017307
Topp, Hans-Jürgen, Limburg an der
  Lahn 077116, 107871, 142076
Toppenish Museum, Toppenish . 053777
Topper, Victor & Renee, Toronto 064763
Toppler-Schlößchen, Rothenburg ob
  der Tauber 021512
Topsfield Historical Museum,
  Topsfield 053778
Topsham Museum, Topsham ... 045987
Topsham Quay Antique Centre,
  Topsham 091615
Tor 1, Georgsmarienhütte 075975
Tor Sanguigna, Roma 081585
Torahiko Terada Kinenkan, Kochi 028863
Torba Sanat Evi, Muğla 117101
Torbandena, Trieste 111055
Torbay Museum, Torbay 007154

Torbey, M., Mansourih El-Metn . 082327
Torbjörns Fackantikvariat,
  Göteborg 143855
Torbogenmuseum, Königsbronn . 019723
Torch, Amsterdam 112296
Il Torchio, Torino 132210
Il Torchio – Costantini Arte
  Contemporanea, Milano 110470
Torcular, Milano 110471, 137697
Tordenskiold, Trondheim 126396
Tordenskjold, Trondheim 084391
Tore Kunst-Auksjoner, Tønsberg 126392
Torekällbergets Friluftsmuseum,
  Södertälje 040971
Torekovs Sjöfartsmuseum,
  Torekov 041081
De Toren Antiekcentrum, Breda . 083009
Torenmuseum, Goedereede 032354
Torenmuseum, Mol 003874
Torenverplaatsingsmuseum,
  Bocholt 003374
Torerísimo de México, Benito
  Juárez 138923
Toretti, Danièle, Aix-en-Provence 066475
Torfaen Museum, Pontypool ... 045523
Torfmuseum, Peat Museum, Neustadt
  am Rübenberge 020753
Torfs, Leonardo, Antwerpen ... 100513
Torfschiffwerftmuseum,
  Worpswede 022777
Torgovyj Dom Progres, Moskva . 143654
Torhalle Frauenchiemsee und
  Vikarhaus, Frauenchiemsee .. 018461
Torhaus Dölitz, Leipzig 020022
Torhaus-Galerie, Bund Bildender
  Künstlerinnen und Künstler,
  Braunschweig 106570
Torhaus Markkleeberg, Museum
  der Völkerschlacht 1813,
  Markkleeberg 020287
Torhaus Otterndorf, Sammlung Labiau/
  Ostpreußen, Otterndorf 021044
Torhaus Warnau, Warnau 078871
Torhausgalerie, Bottrop 075110
Torindo, Tokyo 111518
Torkild, Malfait, Bruxelles 063560,
                128039,   136647
Torlo Centro Antico, Torre del
  Greco 081776, 081777
Tornabuoni Arte, Firenze 080215, 110073
Tornabuoni Arte, Milano 110472
Tornar, Elena, Pisa 081160
Tornaritis Pierides Municipal Museum
  of Paleontology, Larnaka 009410
Tornberg, Lund 115738
Tornby, Linköping 086818
Tornionlaakson Maakuntamuseo,
  Tornio 011430
Torns Möbelrestaurering, Lund . 133551
Tornyai János Múzeum,
  Hódmezővásárhely 023692
Toro Mera, Margarita Rosa, Cali 128024
Toronski, William, Evere 036613
Toronto Aerospace Museum,
  Toronto 007214
Toronto Antique Centre, Toronto 064764
Toronto Art Therapy Institute,
  Toronto 055163
Toronto Board of Education, Records
  Archives and Museum Department,
  Toronto 007215
Toronto Center for Contemporary Art,
  Toronto 007216
Toronto Dominion Gallery of Inuit Art,
  Toronto 007217
Toronto International Art Fair,
  Toronto 098039
Toronto Photographers Workshop,
  Toronto 007218
Toronto Police Museum and Discovery
  Centre, Toronto 007219
Toronto School of Art, Toronto . 055164
Toronto Sculpture Garden,
  Toronto 007220

Toronto Vintage Paper Show-Canada
  Antiquarian Book Fair, Toronto . 098040
Toronto's First Post Office,
  Toronto 007221
Torossian, Armand & Gérard,
  Grenoble 125593
Torpedo Factory Art Center,
  Alexandria 046397
Torppamuseo, Somero 011369
Torquay Museum, Torquay ... 045988
The Torrance Gallery, Edinburgh 117743
De la Torre, Madrid 086271
La Torre, Milano 110473
La Torre, Rimini 110717
La Torre, Verona 081943
Torre Abbey Historic House and
  Gallery, Torquay 045989
La Torre de Lulio, Oaxaca 142992
Torre del Conde, Herrera Castilla, San
  Sebastián de La Gomera 040076
Torre dell'Orologio, Musei Civici
  Veneziani, Venezia 028328
La Torre di Giada, Trieste 081807
Torre Enrica, Genova 080300
I Torre Guaita, San Marino ... 038106
Torre Monumental, Buenos Aires 000283
Torre, de la, Santiago de Chile . 064987
Torrebianca, Trieste 081808
Torredimare, Pasquale, Milano . 080748
Torrens, Jean, Paris ... 071950, 129153
Torrent, David, Vic-le-Comte ... 074097
Torreón Fortea, Zaragoza 040457
Torres, Madrid 086272
Los Torres, San Salvador 103106
Torres Gonzalo, Pedro, Logroño . 115122
Torres Muñiz, Jesus, Granada . 115096
Torres Strait Museum, Thursday
  Island 001578
Torres, Albert, Barcelona 133168
Torres, Miguel Alberto,
  Bucaramanga 102399
Torres, Veronica, Asunción ... 113338
Torresan, Marina, Milano 131543
Torresani, Bregenz 139976
Torresdale Station Antiques,
  Philadelphia 096165
Torretta, Locarno ... 087599, 133813
La Torretta, Roma 081586
Torri, Angelo, Brescia 079960
Torrington Historical Society Museum,
  Torrington 053781
Torrington Museum, Torrington . 045991
Torris Barrois, Marc-André, Sainte-
  Cathérine-lès-Arras 073342
Torrisi, Alfio, Catania 131102
Torrisi, Massimo, Catania 131104
Torsby Finnkulturcentrum, Torsby 041082
Torsby Fordonsmuseum, Torsby 041083
Torsegno, Federico, Genova ... 131342
Torsgården Antik och Auktion,
  Vislanda 087165, 126702
Torshälla Auktionsbyrå, Torshälla 126688
Tort, Andrée, Mirepoix 070129
Torta, Giulio, Palermo 081079
Tortoise Shell, Columbus 093052
Tortora, Bari 109775
Tortora, Farrara 110015
Tortora, Antonio, Roma 132064
Tortora, Julien, Nice 128872
Tortori, Alfredo, Roma 081587
La Tortue Eléctrique, Paris 071951
Torukomen, Yokohama 082234
Toruński Antykwariat Naukowy,
  Toruń 084685, 143510
Torus, Antwerpen 063242
Toscado, David, Guadalajara ... 082470
Toscana, Murcia 086317
Toselli, Milano 110474
Toserco Gallery, Ha Noi 125103
Toshi, Tokyo 126252
Toshiba Kagakukan, Kawasaki . 028811
Toshin, Hiroshima 081988
Tot Antic, Barcelona .. 085847, 133169
Total Antiques, Sharjah 088214

Total Art Resources, Singapore . 114627
Totally Essential, London 118604
Totara Peak Gallery, Omarama . 112979
Totem Gallery, Johannesburg ... 114803
Totem Gallery, Rosebank 114814
Totem Gallery, Venezia 111165
Totem Heritage Center, Ketchikan 049946
Toten Økomuseum, Bøverbru .. 033554
Toten Økomuseum, Kapp 033692
Tóth, Amsterdam 082875
Tóth, Róbert, Budapest 079297
Totius House Museum,
  Potchefstroom 038812
Tótkomlósi Tanyamúzeum,
  Tótkomlós 024019
Totnes Costume Museum, Totnes 045993
Totnes Elizabethan Museum,
  Totnes 045994
Totonchian, Houston 093816
Totteridge Gallery, Colchester . 088952,
                117615,   134291
Totters of Knightsbridge,
  Vancouver 064889
Tottori-kenritsu Gigei Bijutsukan,
  Nakayama 029140
Tottori-kenritsu Hakubutsukan,
  Tottori 029632
Tottori Mingeikan, Tottori 029633
Touati, Le Raincy 069349
Touba, Providence 123713
Toubas & C., Alain, Milano 110475
Toubkal, Montpellier 104287
Toucan Art, Cheadle 117557
Touch of Antique, Toronto 064765
Touch of Art, Albuquerque 119516
Touch of Art, Mumbai 079410
Touch of Design, Saint Paul ... 123936
Touch of Difference, Toronto ... 064766
Touch Wood, B'Kara 082411
Touche Contemporary Art, Stroud 119176
Le Touche of Class, Long Beach 094229
Touchons du Bois, Angoulême . 103241
Touchstone Center for Crafts,
  Farmington 048707, 057079
Touchstone Gallery, Washington 124966
Tougaloo College Art Collection,
  Tougaloo 053782
Touhfa, Casablanca 082584
Touijar, Bruxelles 100662
Toujouse, Vaux-sur-Mer . 093993, 129330
Toulemont, Martin, Sauxillanges 073445
Toulouse Antique Gallery, West
  Hollywood 097939
Toulza, Daniel, Foix 068554
Tounkara, Rotterdam 112664
Tounta, Ileana, Athinai 109048
La Toupie, Kröppen 107748
La Tour Camoufle, Metz 070101
La Tour Camoufle, Paris 071952
Tour de Constance, Monument
  D'Aigues-Mortes, Aigues-Mortes 011549
Tour de la Brocante et de l'Antiquité,
  Manom 069837
Tour de la Motte-Forte, Arnay-le-
  Duc 011737
Tour de l'Horloge, Dinan 012795
Tour de Moricq, Angles 011661
La Tour des Cardinaux, Marseille 104204
Tour des Voleurs, Riquewihr ... 015105
Tour d'Oriol, Marseille . 069968, 140948
Tour du Luxembourg, Ligny-en-
  Barrois 013798
Tour-Musée de Velaux, Le Donjon,
  Velaux 016147
Tourdot, Jean-Claude, Châtenois 067913
Touré, Mag. Edith, Groß-
  Enzersdorf 127757
La Tourelle, Chaumont-sur-
  Tharonne 067941
Tourelle, Lamotte-Beuvron ... 069157
La Tourelle Castle Gallery,
  Tanawha 099807
Les Tourelles, Arnay-le-Duc ... 066731
Les Tourelles, Brioude 103472

Les Tourelles, La Chaise-Dieu .. 103878
Tourenne, Robin, Paris 129154
Touring Exhibitions Group,
London 060264
Touristik-Museum, Unterseen .. 042081
Tourmondeur, Menton 070064
Tournebize, Guy, Sainte-Maxime 073358
Tournebize, Jean-François, Vichy 074112
Tourneboeuf, Alice, Saint-Ouen . 073213
Tournelle, Marie, Paris 071953
Tournesol, Maastricht 083482
Les Tournesols, Montbrison ... 104257
Les Tournesols, Saint-Étienne .. 105513,
                                  105514
Tournevire, Tarascon (Bouches-du-
Rhône) 073622
Tournié, Françoise, Paris 105273
Tournigand, Yvette, Saint-Ouen . 073214
Tourny's, Bordeaux 067362
Tourraton, Patrick, Rouen 072621
Toussaint Antiekhandel, Den
Haag 083138
Toussaint, Daniel, Paris 071954
Toussaint, Denis, Aydoilles ... 066895
Toussaint, Françis, Paris 071955
Toussaint, Georges, Le Havre .. 069289
Toussaint, Jean-Pierre, La Ferrière-sur-
Risle 069024
Toustrup Andersen, Niels & Anette,
Glostrup 140577
Tout pour le Plaisir, Dourdan ... 068400
Tout un Art, Saint-Cyr-au-Mont-
d'Or 105501
Tout un Peu, Saint-Martin-
Lalande 072954
Toutain, Gilles, Vire 125842
Toutantroc, Aicirits-Camou-
Suhast 066434
Toutencadre, Saint-Privat-des-
Vieux 073269
Toutoccas', Noidans-lès-Vesoul . 070732
Tovar, Dallas 135520
Tovar, Leon, New York 123139
Tovar, Liliana, Stockholm 115934
Tovar, Luis, Sheffield 135029
Toveks Auktioner, Ätran 126563
Tovey, Brian, Henley-in-Arden . 117956
Tovi, Jacques, Paris 071956
Towada Kyodo Hakubutsukan,
Towada 029636
Towarzystwo Opieki nad Zabytkami,
Warszawa 059722
Towarzystwo Przyjaciół Sztuk
Pięknych, Kielce 059725
Towarzystwo Przyjaciół Sztuk
Pięknych, Kraków 059724
Towarzystwo Przyjaciół Sztuk
Pięknych, Lublin 059726
Towarzystwo Przyjaciół Sztuk
Pięknych, Nowy Sącz 059723
Towarzystwo Przyjaciół Sztuk Pięknych
w Olsztynie, Olsztyn 059727
Towe Auto Museum, Sacramento 052733
Tower Antiques, Cranborne ... 088992
Tower Art Gallery, New York ... 123140
Tower Arts, Fresno 120910
Tower Books, Taranaki 143347
Tower Bridge Antiques, London . 090541
Tower Bridge Exhibition, London 045118
Tower Fine Arts Gallery,
Brockport 047219
Tower Grove Antiques, Saint
Louis 096784
Tower Grove East Studios, Saint
Louis 123880
Tower Museum, Derry 044091
Tower Museum of Prosphorion,
Ouranoupoli 023205
Tower of David, Museum of
the History of Jerusalem,
Jerusalem 025111
Towers, Robert, Dublin 142605
Town & Country Antiques, West
Hollywood 097940

Town & Country Antiques and
Collectables, Caulfield South .. 061308
Town and Country Antiques,
Houston 093817
Town and Country Antiques,
Saintfield 091217, 134998
Town and Country Antiques, San
Antonio 097041
Town and Country Auctioneers,
Dublin 126173
Town and Country Gallery,
Yarragon 099937
Town and Country Shop,
Milwaukee 094882
Town and Crown Exhibition,
Windsor 046167
Town Creek Indian Mound Historic
Site, Mount Gilead 051088
The Town Gallery, Ballymoney . 117276
Town Gallery, Kansas City 094099
Town Gallery, Toledo 124701
Town Gallery and Japan Room,
Brisbane 098791
Town Hall Antiques, Woburn .. 091897
Town House, Culross 044056
Town House Antiques, Marple
Bridge 090729
Town House Galleries,
Morecambe 118753
Town House Gallery, Kirkby
Lonsdale 118054
Town House Gallery, Lancaster . 118071
Town House Museum of Lynn Life,
King's Lynn 044707
Town House, East Hampton Historical
Society, Clinton Academy Museum,
East Hampton 048436
Town Museum, East Grinstead . 044202
Town n'Lake Auction, Hop
Bottom 127248
Town of Clarence Museum, Historical
Society of the Town of Clarence,
Clarence 047764
Town of Manlius Museum,
Manlius 050653
Town of North Hempstead Museum,
Westbury 054330
Town of Okotoks Museum and
Archives, Okotoks 006546
Town of Warwick Museum, The
Historical Society of the Town of
Warwick, Warwick 054076
Town of York Antiques, Toronto . 064767
Town of Yorktown Museum, Yorktown
Heights 054595
Towne, Louisville ... 094598, 135767
Towne Center Gallery, Long
Beach 121412
Towneley Hall Art Gallery and
Museums, Burnley 043794
Towner Art Gallery and Local Museum,
Eastbourne 044220
Towngate Gallery & Art Centre,
Poole 118933, 128277
Townhead Antiques, Newby
Bridge 090820
Townhouse, Dublin 137609
The Townhouse Gallery, Belfast . 117346
Townhouse Gallery of Contemporary
Art, Cairo 103086
Townsend & Bateson, Tetbury .. 091585
Townsend Booksellers, Pittsburgh 145438
Townsend Gallery, Burnley 117488
Townsend Gallery, Los Angeles . 121608
Townsend Harris Kinenkan, Zuiryuzan
Gyokusenji, Shimoda 029379
Townsend, John P., Cheltenham 088844
Townsends, London ... 090542, 090543
Townsford Mill Antiques Centre,
Halstead 089417
Townshend, James, Bath 088403
Township Museum, Migrant Labour
Museum and Arts and Crafts Centre,
Lwandle 038779

Townsville Museum, Currajong . 001009
Towteck, Singapore 114628
Toxana Basement Gallery, Richmond,
New South Wales 099626
Toy and Miniature Museum,
Stellenbosch 038839
Toy and Miniature Museum of Kansas
City, Kansas City 049896
Toy Crazy, Louisville 094599
Toy Museum, Hong Kong 007891
The Toy Museum, Stow-on-the-
Wold 045893
Toy Museum, Valletta 030758
Toy Nostalgia, Auckland 083831
Toy Shop, Vancouver 064890
Toy Shoppe, Fort Worth 093483
Toy Soldier Shoppe, Milwaukee . 094883
Toy-Toy Museum, Rotterdam .. 032827
Toy Train Depot, Oakland 095916
Toyama City Institute of Glass Art,
Toyama 056063
Toyama Kenminkaikan Bijutsukan,
Toyama 029638
Toyama-kenritsu Kindai Bijutsukan,
Toyama 029639
Toyama Kinenkan, Kawajima ... 028804
Toyama Municipal Folkcraft Village,
Toyama 029640
Toyama Museum, Toyama 029641
Toyama-shiritsu Yakimono Bijutsukan,
Toyama 029642
Toyism Art Movement, Emmen . 112474
Toynbee-Clarke, London 090544
Toyo, Tokyo 082212
Toyo Bijutsu Gakko, Tokyo 056061
Toyo Calligraphy Art Association,
Hachioji 059516
Toyobi Far Eastern Art, New York 123141
Toyohashi-shi Shizenshi Hakubutsukan,
Toyohashi 029645
Toyohashi-shiritsu Bijutsukan,
Toyohashi 029646
Toyoshina-machi Kindai Bijutsukan,
Toyoshina 029649
Toyota Automobile Museum,
Nagakute 029069
Toyota Commemorative Museum
of Industry and Technology,
Nagoya 029130
Toyota Kuragaike Commemorative Hall
and Art Salon, Toyota 029650
Toyota-shi Mingei oyo Toji Shiryokan,
Toyota 029651
Toyota-shiritsu Bijutsukan, Toyota 029652
Toyozo Shiryoukan, Kani 028779
Toys of Yesteryear, Las Vegas . 094184
Tozzetti, Brunetto, Firenze 131274
Tozzi, Franco, Firenze 131275
Trabarro, Brasília 100929
Traber, Daniel, Rieseby 078239
Trabzon Devlet Güzel Sanatlar Galerisi,
Trabzon 043074
Trabzon Müzesi, Trabzon 043075
Tracce – Cahiers d'Art, Ruvo di
Puglia 138878
Tracé, Maastricht 112585
Trace Ecart, Bulle 116181
Traces, Journal of the International
Stone Sculpture Community,
Floyd 139508
Tracey's Antiques, Drummoyne . 061409
Trachtemyriv – Deržavnyj Istoriko-
kul'turnyj Zapovidnyk, Kaniv .. 043176
Trachten- und Heimatmuseum,
Weiltingen 022487
Trachten- und Volkskunstmuseum,
Seebach 021811
Trachtenhaus Jatzwauk,
Hoyerswerda 019351
Trachtenmuseum, Ochsenfurt .. 020943
Tracker d'Art Découvreur Talen,
Toulouse 105755
Trackside Antiques, Detroit ... 093377
Tracy, Columbus 120520

Tracy Maund Historical Collection,
Carlton South 000949
Trada, Chipping Norton 144341
Tradart Genève, Genève 087505
Trade Antiques, Alford,
Lincolnshire 088246
Trade Antiques, Napoli 080942, 110547,
                      131649,   142768
Trade Art, Madrid 086273
Trade Picture Services, London . 134785
Trade Wind, Rottingdean 091147
Trader Windmill, Sibsey 045800
Trader, Sam, San Francisco ... 097340
Traders of the Lost Art, Calgary 064277
Tradestone Gallery, Baltimore .. 119775
Tradewind East, Honolulu 120971
Tradewind Fine Art, Tintagel .. 119238
Tradice, Praha 065532, 128366
Tradició, Győr 079312
Tradición Revista, Albuquerque . 139509
Trading Arts & Crafts Aboriginal
Corporation, Mount Isa 061913
Trading Post, Mountain Ash ... 090787
Trading Post, Saskatoon 101552
Trading Post Antiques, Tauranga 084106
Tradisjonsbygg, Morgedal 132782
Traditiekamer Regiment Stoottroepen,
Assen 032038
Traditiekamer Typhoon, Volkel .. 033002
Traditio Antiguidades e Decoracoes,
Rio de Janeiro 064129
Tradition Faïence, Montereau-Fault-
Yonne 128802
Traditional Chinese Culture Society of
Montréal, Montréal 006417
Traditional Heritage Museum,
Sheffield 045777
Traditions, Baltimore 092369
Traditions and Treasures, San
Antonio 097042
Traditions East Gallery, Makati . 113390
Traditions Fine Furniture, Tulsa . 097761
Tradix, Bad Neuenahr-Ahrweiler . 074562
TrädgårdsAntikvariat, Linköping . 143882
Træe, Bryne 033563
Träg, Elfriede & Klaus, Nürnberg 077846
Traen, J., Brugge 063288, 128011
Tränkle, Michael, Mahlberg 077226
Trafalgar Antiques Centre,
Woodlesford 091925
Trafalgar Art, Valletta 111925
Trafalgar Bookshop, Brighton . 144271
Trafalgar Devenport, Devenport . 083917
Trafalgar Galleries, London 090545
Trafalgars Antiques, Auckland .. 084040
Trafalgars Antiques, North Shore 084040
Traffey, Moirans 070139
Tráfico de Arte, León 115113
Trafo – Galerie 5020 in Print,
Salzburg 138539
Tragara, Chicago 092805
Tragor Ignác Múzeum, Vác 024027
Trail End State Historic Museum,
Colorado-Wyoming Association of
Museums, Sheridan 053309
Trail Museum and Sports Hall of
Memories, Trail 007231
Trailer Gallery, Saint-Martin 109114
Traill County Historical Society
Museum, Hillsboro 049476
Trailside Nature and Science Center,
Mountainside 051108
Trailside Nature Center and Museum,
Cincinnati 047750
Trailside Nature Museum, Cross
River 048075
Traineau, Versailles 141383
Trainland U.S.A., Colfax 047868
Trains & Toys, Omaha 096061
Les Trains de Saint-Eutrope, Musée
Vivant, Evry 012936
Trains Limited, Saint Louis 096785
Traits Singuliers, Dijon 068346

Trakijska Grobnica, Kazanlăk . . . 005061
Traktoren-Museum Kempen, Horn-Bad
  Meinberg 019340
Traktorenmuseum, Vordernberg . 002881
Traktorenmuseum,
  Westerkappeln 022599
Traktorveteranensammlung Dorf an
  der Pram, Wendling 002921
Traky Istorijos Muziejus, Trakai . 030491
Tram Museum Rotterdam,
  Rotterdam 032828
Tram Museum Zürich, Zürich . . . 042238
Trama, Barcelona 114970
Trama, Madrid 115290
Tramer, Irma, Bünde 075217
Trammell and Margaret Crow
  Collection of Asian Art, Dallas . 048126
Trammuseum, Schepdaal 004002
Tramway Museum, Bendigo 000851
Tramway-Museum Graz, Graz . . 001987
Tramway Treasures, Blessington 079461
The Tramyard Gallery, Dalkey . 109562
Tran, Hau Tuan, Ho Chi Minh City 125149
Tranăs Auktionskammare, Tranăs 126689
Tranberg, Ure, Frederiksberg . . 128386
Tranby House, Maylands, Western
  Australia 001262
Tranchina, Pietro, Modena 131591
Trang, An, Ha Noi 125104
Tranmeçon, Armel, Les Sables-
  d'Olonne 069431
Trannod, Concarneau 103624
Tranquilino Sandalio de Noda Museo
  de Ciencias Naturales, Pinar del
  Río 009293
Trans-Art Gallery, San Antonio . 124078
Trans, New York 123142, 138429
Transamerican Antiques, Virginia
  Beach 097795
TransArt Kunst und Kultur Projekte,
  Köln 107693
Transcona Historical Museum,
  Winnipeg 007469
Transdeco, Espoo 103128
Transenantic, Nice 070670
Transference, Flinders 061478
Transference, Hampton 061594
Transformation Treasures,
  Pittsburgh 096332
Transgariep Museum, Philippolis 038798
Transição, Lisboa 084982
TransNamib Museum, Windhoek 031866
Transnet Heritage Foundation Library,
  Johannesburg 038760
Transoceanea, Chicago 120284
Transparence, Biot 103396
Transparence, Paris 105274
Transvaal Museum, Northern Flagship
  Institution, Tshwane 038860
Transylvania Museum, Lexington 050251
Tranviaser, Zaragoza 133516
Trapenat, Châteaudun 125552
El Trapero, Bern 087306
Trapezium House, The Petersburg
  Museums, Petersburg 052018
Trapholt, Museum of Modern Danish
  Art, Applied Art and Furniture Design,
  Kolding 010207
Trappers Museum, Marten River 006268
Trappin, Raffaelle, Bolzano 131012
Traquair House, Innerleithen . . . 044633
Traralgon and District Arts Society,
  Traralgon 058224, 099841
Traralgon Antiques, Traralgon . . 062345
Trasa-Urban Arts Collective, Salt Lake
  City 060776
Trash & Treasure,
  Ballaghaderreen 079443
Trash & Treasures, Manawatu . . 143327
Trash & Treasures Too,
  Manawatu 143328
Trash From the Past, Honolulu . 093576
Trash 'n' Treasure, Alsager 088262
Trash'nTreasure, Ottawa 064515

Trash or Treasure 2000,
  Baltimore 092370
Trash Palace Epilogue, Auckland 083833
Trash to Treasures, Indianapolis 093935
Traslucido, Firenze 080216
Trastad Samlinger, Borkenes . . 033556
El Trastero, Logroño 086081
La Trastienda, Burgos 085916
La Trastienda, Cali 065255
La Trastienda, León 086066
La Trastienda de Felcar, Gijón . 086023,
                                     086024
La Trastienda del Anticuario,
  Concepción 064939
Tratteggio, Santander . 086429, 133426
Traub, Johann, Eichstätt 075582
Traube & von Hardenberg, Nörten-
  Hardenberg 077794
Traube, D., Bruxelles 063561
Traudisch, I., Wiehl 108894
Traumfährte, Dresden 141708
Traunsee Antik, Altmünster . . . 062518,
                                     127706
Trauque, Betty, Lézignan-
  Corbières 069452
Trautmann, Gregor, Bamberg . . 141451
Trauzettel, Stolberg, Rheinland . 142378
Trave Antik, Lübeck . 077190, 142089
Trave Antik, Reinfeld 078188
Travel Art Galéria, Szeged 109212
Travel Thru Time, Phoenix 096243
Travel Town Museum, Los
  Angeles 050470
Traveler's Rest State Historic Site,
  Toccoa 053755
Traveller Antiques, Edmonton . . 064308
Travellers Rest Plantation and
  Museum, Nashville 051188
Traveller's Trading Post, Enmore 061448
Traver, William, Seattle 124609
Traver, William, Tacoma 124621
Traverse County Historical Society
  Museum, Wheaton 054367
Traversée, München 108201
Traversothèque, Musée du Chemin de
  Fer, Mulhouse 014382
Travis, Atlanta 092192
Travis & Emery, London 144702
Travis, Matthew, Houston 121143
Trawler Pictures, Cleethorpes . 117596
Trax In Time, Tauranga 084107
Trayecto, Vitoria-Gasteiz 115545
Trazit, Lionel, Trizac 073887
Trazos, Bogotá 102370
Trazos, Cali 102415
Trazos Tres, Santander 115435
Tre Sekler Antik och Konsthandel,
  Nyköping 086891
Tre Sekler Antikt och Inredning,
  Norrköping 086884
Tread of Pioneers Museum, Steamboat
  Springs 053747
Treadgolds Museum, Portsmouth 045560
Treadway Gallery, Cincinnati . . 120382
Tréanton, Morlaix 070350
Treasure Aisles, Saint Louis . . 096786
The Treasure Box, Hickling . . . 117967
Treasure Chest, Berwick-upon-
  Tweed 088445, 134095
Treasure Chest, Blackrock 079460
Treasure Chest, Pleasant Valley . 139510
Treasure Chest, Weymouth . . . 091815,
                                     144918
Treasure Chest, Whitley Bay . . 091835
Treasure Chest Antiques,
  Vancouver 064891
Treasure Chest by BM, Cincinnati 092903
Treasure Chest of Indy,
  Indianapolis 093936
Treasure Earth, Brunswick 061224
Treasure Gallery, Fremantle . . . 099064
Treasure House, Charlotte 092633
Treasure House of Books, Fitzroy 139736
Treasure House of Flowers & Art,

Kingston 111227
Treasure House of the Eiheiji Temple,
  Eiheiji 028531
Treasure Hunt, Cocoa 145092
Treasure Hunt, Houston 093818
Treasure Hunt, Pittsburgh 096333
Treasure Ireland, Dublin 079593
Treasure Island, Fort Worth . . . 093484
Treasure Island, Louisville 094600
Treasure Mart, Omaha 096062
Treasure of the Sierra Madre,
  Fresno 093516
Treasure Seeker, New York . . . 095788
Treasure Shop, Tucson 097700
Treasure Trove, Beirut 082308
Treasure Trove, Chicago 092806
Treasure Trove, San Diego . . . 097141,
                                     127441
Treasure Trove Antiques,
  Columbus 093053
Treasured Restorations, San
  Antonio 136276
Treasured Scarab, Denver 093344
Treasurehouse, Washington . . . 124967
Treasurer's House, York 046228
Treasures, Athlone 079442
Treasures, Den Haag 083139
Treasures, Exmouth 089239
Treasures, Four Elms 089298
Treasures & Collectibles, Houston 093819
Treasures & Pleasures, New York 095789
Treasures & Trifles, Denver . . . 093345
Treasures & Trifles, New York . . 095790
Treasures & Trifles, San Antonio 097043
Treasures Antique Mall, Salt Lake
  City 096929
Treasures From The Past, Yarra
  Glen 062501
Treasures from the Silk Road to the
  Santa Fe Trail, Philadelphia . . 098506
Treasures of Ballater, Ballater . 088344
Treasures of Ours, Saint Louis . 096787
Treasures of the Earth, Fort
  William 044365
Treasures of the Sea Exhibit, Delaware
  Technical and Community College,
  Georgetown 049078
Treasures of Time, Toronto . . . 064768
Treasures of Yarraville, Yarraville 062506
Treasures of Yesterday,
  Coorparoo 061365
Treasures Recycled, Charlotte . 092634,
                                     135395
Treasures Unlimited, Charlotte . 092635
Treats & Treasures, Coolamon . 061359
Treaty Fireplaces, Limerick . . . 079662
Treaty House, Waitangi National Trust,
  Waitangi, Auckland 033328
Treaty Site History Center, Saint
  Peter 052856
Třebechovické Muzeum Betlémů,
  Třebechovice pod Orebem . . . 009917
Trebellar, A Coruña 115037
Trebosc & Van Lelyveld,
  Amsterdam 082876
Trébuquet, Raymond, Toulon
  (Var) 129291
Trecastle Antiques Centre,
  Trecastle 091628
Treco e Breguetes Comercio de
  Antiguidades, Rio de Janeiro . 064130
Tredegar House and Park, Newport,
  Gwent 045362
Tree Frog Gallery, Maleny 099265
Tree Studio Gallery, Chicago . . 092807
Treeck, Gustav van, München . 130409
Tredale Antiques, Little Dalby . 089844,
                                     134563
Tredale Antiques, Oakham . . . 090905,
                                     134903
Treefrogs, West Hollywood . . . 097941
Treen Antiques, Manchester . . 134833
Trees of Stone, Salt Lake City . 123992
Treff, Chemnitz 075260

Treff, H.-G., Quedlinburg 078109
Treffer, Hans, Klagenfurt 062676, 100076
Treffpunkt Architekturfotografie,
  Stuttgart 108734
Treffpunkt Kunst, Ottobrunn . . 059246,
                                     108348
Trefpunt, Deurne 140217
Tregaskis, Raymond & Victoria,
  Paddington, New South Wales 062044
Treglown, Roger J., Macclesfield 144723
Tregoning Antiques,
  Bournemouth 088544, 134130
Tregoning Gallery, Derby 117674
Tregony Antiques, Tregony . . . 091629
Tregony Gallery, Tregony 119255
Treherne Museum, Treherne . . 007232
Treibholz, Düsseldorf 075529
Treichel, Gerhard, Kippenheim . 137297
Treillage, New York 095791
Treille Marine, Saint-Martin-de-
  Ré 072941
Le Treize, Bruxelles 063562
De Trekbel, Haarlem 083356
Het Trekkermuseum, Nisse . . . 032693
Trelissick Gallery, Truro 119266
Trelleborgs Museum, Trelleborg 041087
Treloar, Michael, Adelaide 139619
Trelowarren Gallery, Helston . . 117950
Trelyon Gallery, Saint Ives,
  Cornwall 119035
Trema, Lisboa 114067
Trémaud, Catherine, Nieulle-sur-
  Seudre 070680
Trémaud, Jean-Pierre, Doizieux . 068362
Tremayne, Saint Ives, Cornwall . 091192,
                                     119036
Trembath Welch, Great Dunmow 126908
Tremblay, Pierre, Toronto 064769
Trembley, London 064355
Trembowla Cross of Freedom Museum,
  Dauphin 005685
Tréméac, Jean-Marie, Pontrieux 072276
Tremeaux, Claude, Beaune . . . 067048
Tremont Gallery, Boston 047135
Trémont, Gérard Sérgio M., Porto 143606
Trench Puzzles, Stowmarket . . 091490
Trenčianske Múzeum, Trenčín . 038471
TrenDepot, Buffalo 119969
Trender, Robert, London 134786
Trenerowski, Krzysztof, Poznań . 143498
Trengarin Gallery, Narrogin . . . 099422
Trengove, Warlingham 091730
Trent, West Hollywood 097942
Trentham Agricultural and Railway
  Museum, Trentham 001593
Trentham Treasures, Wellington . 084195
Fratelli Trenti, Brescia 131054
Trenton City Museum, Trenton . 053801
Trenton Historical Museum,
  Trenton 053793
Trenwith, Tobias, Saint Ives,
  Cornwall 119041
Trepanier Baer Gallery, Calgary . 101263
Trepassey Area Museum,
  Trepassey 007234
Trèpetie, Maastricht 083483
Treptau, Axel, München 130410
Trepte, Dresden 075432
Trerice, Newquay 045370
Tres Hombres Art, Halmstad . . 115679
Tres Petite Shoppe, Saint Louis . 096788
Tres Serveis Culturals, Palma de
  Mallorca 115369
Três Xelins, Lisboa 084983
Tresfjord Museum, Tresfjord . . . 034024
Tresillian Antiques, Tresillian . . 091631,
                                     135147
Tresnak, Lenka, Campo Tures . 109929
Tresoar, Frysk Histoarysk
  en Letterkundich Sintrum,
  Leeuwarden 032576
Le Trésor, Monthey 087679
Trésor de Grand-Mère, Rodez . 072516
Trésor de la Basilique Saint-Materne,

Walcourt 004126
Tresor de la Catedral, Girona . . 039451
Trésor de la Cathédrale de Liège,
  Liège 003815
Trésor de la Cathédrale Notre-Dame,
  Moulins (Allier) 014359
Trésor de la Cathédrale Notre-Dame,
  Tournai 004088
Trésor de la Cathédrale Notre-Dame-
  du-Glarier, Sion 041997
Trésor de la Cathédrale Saint-Nicolas,
  Chapitre Cathédral de Saint-Nicolas,
  Fribourg 041505
Trésor de la Collégiale, Huy . . 003722
Trésor de la Collégiale de Visé,
  Visé 004118
Trésor de la Collegiale Sainte-Waudru,
  Mons 003886
Trésor de l'Eglise Primaire Saint-
  Sebastien, Stavelot 004050
Trésor de Paris, Paris 071957
Trésor de Pleaux, Pleaux 014887
Trésor de Sainte Foy, Conques . 012661
Trésor d'Église, Art Sacré,
  Marseille 014063
Trésor d'Église, Cathédrale Saint-Just-
  et-Saint-Pasteur, Narbonne . . 014432
Trésor des Nomades, Marrakech 082594
Trésor du Frère Hugo d'Oignies,
  Namur 003904
Trésor du Passé, Jounieh 082323
Trésor et Fouilles Archéologiques de
  l'Abbaye, Saint-Maurice 041909
Trésor et Musée de la Collegiale
  Sainte-Begge, Andenne 003273
Trésor Public, Blois 067260
Trésor Saint-Gervais, Avranches 011849
Tresorama, La Ferté-Macé 013409
Trésorerie Générale, Moulins
  (Allier) 070373
Trésorerie Générale de la Gironde,
  Bordeaux 067363
Trésorient, Paris 071958
Trésoriers de France, Montpellier 070317
Aux Trésors d'Antan, Alès 066532
Trésors d'Antan, Dômevre-en-
  Haye 068371
Trésors d'Asie, Paris 071959
Trésors de l'Afrique et de l'Orient,
  Zürich 088013
Trésors des Siècles, La Ferté-sous-
  Jouarre 069033
Trésors du Monde, Arzon 066755
Les Trésors du Passé, Montréal 064481
Tresure Room Gallery of the
  Interchurch Center, New York . 051484
Tret'ja Galereja, Moskva 114311
Treusch, F., Bad Honnef 074533
Treussier, Ronan, Nantes 070482
Treuter, Siegi, Bodman-
  Ludwigshafen 106518
Trev Deeley Motorcycle Museum,
  Richmond 006799
Treverkstedet, Inderøy 132780
Trevi Flash Art Museum, Trevi . . 028180
Trévise Antiquités, Paris 071960
Trevithick Cottage, Camborne . 043837
Trevizanutto, Sylvio, Montréal . 064482
Tri-Art, Los Angeles 121609
Tri-Cities Museum, Grand Haven 049179
Tri-County Historical Society and
  Museum, Herington 049449
Tri-State Mineral Museum, Joplin 049843
Triada, Gdańsk 113490
Triada, Sopot 113765
Triade, Barbizon 103336
Triadis, Sofia 101173
Triangelns Prylbod, Malmö 086861
Triangle Art Gallery, San
  Francisco 124418
Le Triangle Blanc, Le Croisic . . 103940
Triangle d'Art, Libourne 103996
Triangle Gallery, Calgary 101264
Triangle Rush Exhibition Hall,

Yuanli 042602
Triangle Tattoo and Museum, Fort
  Bragg 048786
Trianglen Antik, København . . . 065965
Trianglens Antikvariat,
  København 140617
Il Triangolo, Cosenza 109991
Triangolo Arte, Bergamo 109804
Triangula Bastions, Rīga 111768
El Trianon, Málaga 086308
Trianon, Marseille 069969
Trianon, Monaco 112020
Trianon Antiques, London 090546
Trianon Museum and Art Gallery,
  Denver 048277
Trianon Múzeum, Várpalota . . . 024034
TriArt Gallery, Louisville 121674
Trias Art Gallery, Toronto 101713
Trias-Museum, Ochsenfurt 020944
Tribal, Bordeaux 103440
Tribal – The Magazine of Tribal Art,
  San Francisco 139511
Tribal & Oriental Arts & Antiques,
  Lesmurdie 061733
Tribal Art, Donaghadee 117679
Tribal Art, 's-Hertogenbosch . . 112533
Tribal Art Gallery, Melbourne . . 099325
Tribal Collections, New York . . . 095792
Tribal Design, Amsterdam 082877
Tribal Eye, Toronto 101714
Tribal Gallery, Gosford 061549
Tribal Gathering London, London 118605
Tribal Hut, Houston 121144
Tribal Museum, Ahmedabad . . . 024128
Tribal Museum, Confederated Tribes of
  Grand Ronde, Grand Ronde . . 049199
Tribal Spears Gallery, New York . 123143
Tribbia, Edgar, Champ-sur-Drac 067805
Tribbia, Edgar, Laffrey 069134
Tribble, J., Atlanta 092193
Tribe & Son, Tom, Sturminster
  Newton 091500, 135102
Tribeca, New York 123144
Tribečské Múzeum Topoľčany,
  Topoľčany 038467
Tribes Gallery, New York 123145
Tribhuvan Memorial Museum,
  Kathmandu 031878
Tribuzio, Paolo, Roma 132065
Tricard, Maurice, Arras 066742
Tricia's Treasures, Birmingham . 092435
Tricia's Treasures, Gosnells . . . 061550
Trick Bijutsukan, Nasu 029167
Tricromia, Roma 110868
Tridaya Eka Dharma Museum,
  Bukittinggi 024552
Le Trident, Nice 070671
Trident, Regency Park . 062142, 139893
Trident Booksellers, Boston . . . 145018
Trident Galleries, Leicester 118097
Il Tridente, Verona 081944
Triebel, M., Bad Sachsa 074577
Trier, Dieter, Leipzig 142067
Trierer Münzfreunde e.V., Trier . 059247
Trieste Contemporanea, Trieste . 059482
Triestingtaler Heimatmuseum,
  Weissenbach an der Triesting . 002904
Triff-Daumas, Henri, Paris 105275
Trigano, Patrice, Paris . 071961, 105276
Trigg Ison, Los Angeles 121610
Trigiani, Vincenzo, Torino 132211
Triglavska Muzejska Zbirka,
  Mojstrana 038578
Trignolles – Espace Arthur Masson,
  Treignes 004093
Trigon Gallery, Plzeň 102642
Trillarium, Waffensammlung und
  Heimatmuseum, Cleebronn . . 017745
Triller, Ina-Amélie, München . . . 130411
Trilles, Marti, Castello de la
  Plana 085934
Trilogy, New Farm 099428
Trilse, Friederike, Roth 108522

Triluz, Bilbao 085891, 133203
Trim, Nagoya 082066
Trimarchi, Bologna 109870
Trimboli, Pietro, Rastignano . . . 131784
Trimbridge Galleries, Bath 088404,
  117316
Trinca, Vittorio, Roma 081588
Trincia, Piero, Roma 081589
Trindade, Porto 114109
Trindade, António, Lisboa 084984
Trindade, João Campos Pereira,
  Lisboa 084985
Trindade, Luís Miguel, Lisboa . . 132960
Trindade, Tereza, Belo Horizonte 063925
Trinder, Neil, Sheffield 135030
Trinder, P.D. & R., Sudbury,
  Suffolk 144871
Trinders' Fine Tools, Clare 088926
Tring Market, Tring 127069
Trinidad Antigüedades, Buenos
  Aires 060933
Trinité Art'Gallery, Troyes 105779
Trinity Antiques, Perth 062082
Trinity Antiques Centre,
  Colchester 088953, 144349
Trinity Arcade Bookshop, Perth . 139872
Trinity Arts Guild, Bedford 060777,
  119780
Trinity Books, Randburg 143705
The Trinity Centre, Horncastle . . 089589
Trinity Curious, Edinburgh 089194,
  134364
Trinity Fine Art, London 090547
Trinity Fine Art, Milano 080749
Trinity Fine Arts, New York 123146
Trinity Gallery, Atlanta 119624
Trinity Heights – Saint Joseph Center-
  Museum, Sioux City 053350
Trinity House, Newcastle-upon-
  Tyne 045353
Trinity House National Lighthouse
  Centre, Penzance 045483
Trinity Interpretation Centre,
  Trinity 007238
Trinity Museum, Trinity 007239
Trinity Museum of the Parish of Trinity
  Church, New York 051485
Trinta Arte Contemporánea, Santiago
  de Compostela 115438
Trintignan, Hélène, Montpellier . 104288
Trio-Antiek, Antwerpen 128000
Trio Antik, København 065966
Trio Silversmiths, Philadelphia . 136107
Triod, Dobrič 064221, 140316
Triolo, Pietrina, Messina 131361
Triomphe Arte e Antiguidades, Rio de
  Janeiro 064131, 064132
Trionfante Antichità, Palermo . . 081080
Trionfante Antichità, Taormina . . 081628
La Trionfante di Precania Stefana,
  Palermo 081081
Tripier, Antoinette, Paris 125771
Triple Candie, New York 123147
Trippel, V., Braunschweig 075135
Trippler, Magdeburg 077225
Triptiek, Breda 083010
Triquart, B., Ober-Hilbersheim . 077853
Trish, Cleveland 092977
Trish's Dungeon, North Fremantle 061991
Triskelion, Toronto 140420
Tristanns Gallery, Dundolk 119630
Tristan's, Wadebridge 119297
Triton Galleries, Exeter 117774
Triton Gallery, Cheltenham 088845
Triton Gallery, New York 123148
Triton Gallery, Paignton 118878
Triton Gallery, Torquay 119244
Triton Museum of Art, Santa
  Clara 053025
Tritonus, Krummhörn 107765
Troadec, Pierre, Perros-Guirec . 072108
Troadkasten, Freinberg 001904
Troadkasten, Sauerlach 021622
Troadkasten, Schardenberg . . . 002712

Troadkasten Fornach, Fornach . . 001896
La Trobe, Newtown, Victoria . . 061970
Troc & Puces, Châteauroux . . . 067899
Troc 34, Le Crès 069259
Troc 40, Herm 068811, 140857
Troc 79, Niort 070718
Troc 89, Avallon 066865, 066866
Troc Ambiance, Angers 066615
Troc-Antic, La Rochelle 069094
Troc Antik, Metz 070102
Troc Broc 49, Saumur 073439
Troc Broc Antiquités, Beaune . . 067049
Le Troc de la Gloire, Toulouse . . 073804
Troc de l'Ile, Chavelot 067946
Troc-de-l'Ile, Orgeval 070802
Troc de l'Ile, Troyes 073910
Troc-de-l'Ile Floride, Saint-Maur 072959
Troc de l'Ile Occase, Bourg-en-
  Bresse 067413
Troc de l'Ile Verfer, Narbonne . . 070497
Troc en Ville, Marseille 069970
Troc et Chine, Mougins 070364
Troc International, Luxembourg . 082372
Troc Pérignon, Toulouse 073805
Troc Salonais, Salon-de-Provence 073390
Troc-Vente 89, Auxerre 066856
La Trocade du Sablon, Bruxelles 063563
Trocadero Antiguidades, Lisboa . 084986
Trocadore, Isle 068877
Trocante, Béziers 067181
Trocante, Moulin-Neuf 070367
La Trocante, Nice 070672
La Trocante, Nîmes 070706
La Trocante, Saint-Avold 072683
Le Troc'Antic, Vizille 074267
Trocaz, Michel, Beaune 067050
Trocfouille, Les Mages 069413
Trochu and District Museum,
  Trochu 007240
Trocmeubles, Cannes 067655
Trocmez, Bertrand, Clermont-
  Ferrand 103609
Troc'n'Roll, Bruxelles 063564
Trococase, Avenches 087205
Trocque, Chantal & Michel,
  Brionne 067503
Troc'Tout, Cherveix-Cubas 067964
Troc'Truc, Montauban 070203
Trøgstad Bygdemuseum,
  Trøgstad 034025
Trøndelag Folkemuseum,
  Trondheim 034038
Troense Antikvitetshandel,
  Svendborg 066161
Trösch, Armin, Zürich 144198
Tröster, Herbert, Schmallenberg 130626
Trog, Denver 093346
Troge, Birgit, Hamburg 076328
Trohay, Jean-Claude, Fréjus . . . 068608
Troickij Sobor, Serpuchovskij
  istoriko-chudožestvennyj muzej,
  Serpuchov 037860
Trois Fois Passera, Montréal . . . 064483
Trois, Vittorio, Venezia 081877
Troise, Nicola, Napoli 131650
La Troisième Veille, Paris 141288
Le Troisième Oeil, Bordeaux . . . 103441
Le Troisième Oeil, Paris 105277
Trojner, Audrey, Buffalo 092565
Troll, Halle, Saale 129940
Trollenäs Slott, Eslöv 040612
Trolley Museum of New York,
  Kingston 049985
Trolley Square Gallery, Salt Lake
  City 123993
Trolleybus Museum at Sandtoft,
  Doncaster 044112
Tromayer, Prof. Erich, Hinterbrühl 062634
Tromayer, Prof. Erich, Wien . . . 100384,
  136612
Trompat, Yvette, Orléans 070822
Trompetenmuseum, Bad
  Säckingen 016857

Trompka, Gerd, Köln 076914
Het Tromp's Huys, Vlieland 032998
Troms Forsvarsmuseum, Bardu . 033521
Tromsø Folkemuseum, Folkeparken,
  Tromsø 034029
Tromsø Kunstforening, Tromsø . 059676
Tromsø Universitets Museum,
  Tromsø 034030
Tron, Lesznowola 084606
Troncin-Denis, Nancy . 104326, 128823
Troncone & Laezza, Napoli 080944
Troncone, Giovanni, Napoli 080943
Trondarnes Distriktmuseum,
  Harstad 033640
Trondheim Kunstmuseum, Trøndelag
  Kunstgalleri, Trondheim 034039
Trondheim Politimuseum,
  Trondheim 034040
Troner Art, Düsseldorf 106836
Troops of Time, Jacksonville ... 094045
Troost, Marja, Amstelveen 082669
Tropea, Sebastiano, Catania ... 080050
Tropenmuseum Junior,
  Amsterdam 031994
Trophäensaal, Altenfelden 001703
Trophée 89, Auxerre 066857
Tropical Decorators and Upholsters,
  Jacksonville 135668
Tropical Giclees, Honolulu 120972
Tropical Trading, Paris 105278
Tropico del Cancro, Bari 109776
Tropico del Cancro 2, Bari 079786
Tropique Bouquinerie, Saint-
  Étienne 141333
Trosiner, Ralf, Hamburg 076329
Trossinger Gemäldegalerie,
  Trossingen 078716
Trost, André, Juigné-sur-Loire . 068925
Trost, Doris, Bad Camberg 106048
Trost, Heiner, München 077630
Trosterud Skolemuseum,
  Rømskog 033898
Troszkiewicz, Warszawa 143532
Trotman, Ken, Huntingdon 144493
Trotot, Emmanuel, Bougival 067382
Trotters, Masterton 083992
Troubetzkoy, New York 123149, 136016
Troubetzkoy, Paris 105279
Troussier, Rémy, Thury-Harcourt 073679
Troussillé, Guy, Toulouse 073806
Trout Farm, San Francisco 097341
Trout Gallery, Dickinson College,
  Carlisle 047428
Troutmark Books, Cardiff 144312
Trouvaille, Bern 087307
Trouvaille, Cannes 067656
La Trouvaille, Marseille 069971
La Trouvaille, Milano 080750
Trouvaille, Toulouse 073807
La Trouvaille, Trouville-sur-Mer . 073895
Trouvailles, Boulogne-sur-Mer .. 067405
Trouvailles, Cambrai 067618
Trouvailles, Épagny 068446
Trouvailles, Lézignan-la-Cèbe . 069453
Trouvailles, Lyon 128730
Trouvailles, Nantes 070483
Les Trouvailles, Paris 071962
Les Trouvailles, Saint-Ouen ... 073215
Les Trouvailles de Léonard,
  Québec 064542
Trouvailles et Envies, Missillac . 070133
Trouvé Lorelay, Pisa 081161
Les Trouveres, Bruxelles 063565, 128040
Trova Foundation, Saint Louis .. 052825
Trovato, Ben, Brugge 100539
Il Trovatore, Cagliari 080005
Trowbridge, Atlanta 119625
Trowbridge, Chicago 092808
Trowbridge, Crowborough 089016
Trowbridge, New York 123150
Trowbridge Gallery, Claremont . 061326,
                            139698
Trowbridge Museum, Trowbridge 046000

Trowbridge, M., London 090548, 118606,
                            134787
Troxell-Steckel House and Farm
  Museum, Egypt 048492
Troy-Hayner Cultural Center, Troy 053815
Troy Museum and Historic Village,
  Troy 053810
Troyer, Washington 124968
TRT Müzesi ve Ulusal Kurtulus Sergisi,
  Ankara 042831
Trub, Dieter, Gundelsheim,
  Württemberg 076135
Trubarjev, Ljubljana 143698
Truchelut, Daniel, Houilles 068843
Truck Centre, Calgary 005534
Trudgett, Irene S., Winchester . 091873
True Blue Antiques, Mangerton . 061785
True Blue Fine Art Gallery, Hobart 099140
True Devotion Design, Oakland . 095917
True Frame and Art, Cobram ... 098902
True Reflections, Hamilton 100831
True to Form, Jacksonville 121272
Trübendörfer, S. & I., Weinsberg 108867
Trümper, J., Duderstadt 075439
True's Yard Fishing Museum, King's
  Lynn 044708
Trützschler's Milch- und
  Reklamemuseum,
  Hildburghausen 019245
Truffi, Rinalda, Venezia 111166
Trulla, Joël, Fontenay-le-Comte . 068574
Trulli, Massimo, Roma 110869
Truman Corners Antique Mall, Kansas
  City 094100
Trummer und Meinrad, Hamburg 137255
Trummlehus, Langenthal 041639
Trumpington, Guildford 117897
Trumpington Fine Art, Cambridge 117513
Trumpington Gallery, Cambridge 134201
Trunke, Friedrich, Memmingen . 077343
Trunz, Hartwig, Spiez 133930
Trûong Cao My Thuât Công Nghiêp,
  Ha Noi 058155
Truppe, Matthäus, Graz 139990
Truppe, Peter, Wien 140126
Truppenmuseum, Straß in
  Steiermark 002800
Truro Center for the Arts at Castle Hill,
  Truro 058010
Truscott, Christina, London 090549
Truva Müzesi Örenyeri,
  Çanakkale 042867
Truyol, Forch 087428
Try Museum, Dronninglund 010034
Tryart Gallery, Louisville Visual Art
  Association, Louisville 050504
Tryon Gallery, London . 090550, 118607
Tryon Palace, New Bern 051231
Trysil Bygdemuseum, Trysil 034043
Trystero, Szczecin 113785
Trzcina, Gdańsk 084514
Trzebiatowski, Krzysztof,
  Hamburg 076330
Tržiška Galerija Ferdo Majer,
  Tržič 038626
Tržiški Muzej z Galerijo, Tržič . 038627
TS-Auto- & Viestintämuseo,
  Turku 011446
Tsa Mo Ga Memorial Museum,
  Plains 052189
Ts'ai-Liao Fossil Museum,
  Zuojhen 042613
Tsangarakis, Kallitheaathinas .. 109068
Tsannis, Paylos, Athinai 079184
Tsantilas, Margit, Velden 062867
Tsantilis, Athinai 079185
Tsarinas, Sydney 062301
Tschauder, B., Hameln 107328
Tschechisches Zentrum – CzechPoint,
  Berlin 017308
Tschernitz, Herbert,
  Perchtoldsdorf 062764
Tscheschlog, H.-G., Schwelm . 078425
Tschetschorke, Rolf,

Kirchhundem 107562
Tschirren, René, Basel 087261, 116105,
                            144036
Tschofen & Söhne, Ignaz, Sankt
  Gallenkirch 127864
Tschoppe, Manfred & Uschi,
  Hattstedt 076420
Tschudi, Glarus 116349
Tschudi, Zuoz 116888
Tsé Wan, Lyon 069782
Tse, C.Y., Hong Kong 065103
Tsekouras, Andreas, Athinai ... 079186
Tsepelidis, Grigorios,
  Thessaloniki 079216
Tsi, Ku Chai, Hong Kong 065104
Tsilalis, Athinai 110499
Tsimpískaki, Tzeni, Thessaloniki 109105
Tsing Hua Stone Teapot Arts Center,
  Yungho 042608
Tsiridis, Georgios, Thessaloniki . 109106
Tskhinvali Museum, Tskhinvali . 016428
Tsou-Ma-Li Country Gallery,
  Danei 042300
Tsubaki, Tokyo 111519
Tsuchida Bakusen Bijutsukan,
  Sawata 029353
Tsui Museum of Art, Hong Kong 007892
Tsukamoto, Tokyo 082213
Tsukuba Bijutsukan, Ibaraki-ken
  Kindai Bijutsukan, Tsukuba .. 029657
Tsukushi, Yokohama 111545
Tsumeb Cultural Village, Tsumeb 031857
Tsumeb Museum, Tsumeb 031858
Tsumugi, Grace, London 118608
Tsurui Bijutsukan, Niigata 029178
Tsuruki & Co., Y., Kyoto 082046
Tsurumoto, Hiroshima 081989
Tsuwano Gendai Photo Gallery,
  Tsuwano 029664
Tsuyama-furitsu Rekishi
  Hakubutsukan, Tsuyama ... 029665
Ttaka Bazar, Donostia-San
  Sebastián 085985
TTD Museum, Tirupati 024497
Tu Do, Ho Chi Minh City 125150
Tu-Endie-Wei State Park, Point
  Pleasant 052227
Tuam Mill Museum, Tuam 024995
Tuapsinskij Kraevedčeskij Muzej im.
  N.G. Poletaeva, Tuapse ... 037958
Tuatapere Bushmans Museum,
  Tuatapere 033316
Tub King, Jacksonville 094046
Tubac Center of the Arts, Tubac 053818
Tube, Düsseldorf 141727
Tube Gallery, Bedlington 117326
Tuben, Askim 084213
Tuberculosemuseum Beatrixoord,
  Appelscha 032015
Tubman African-American Museum,
  Macon 050575
Tucci, Russo, Torre Pellice ... 111018
Tuch und Technik Textilmuseum
  Neumünster, Neumünster ... 020726
Tuchel, Piet, Grafschaft 038626
Tucher, Bernhard von, Kaisheim 130087
Tucherschloss mit Hirsvogelsaal,
  Museen der Stadt Nürnberg,
  Nürnberg 020879
Tuchlauben Galerie, Wien 063089,
                            100385
Tuchler, Claudia, Düsseldorf .. 106837
Tuchmacher Museum Bramsche,
  Bramsche 017535
Tuchmuseum Lennep der Anna-Hardt-
  Stiftung, Remscheid 021364
Tuck Museum, Hampton 049337
Tucker, Jonathan & Antonia Tozer,
  London 118609
Tucker House Museum, Saint
  George's 004198
Tucker Station Antiques,
  Louisville 094601
Tucker Tower Nature Center Park

Museum, Ardmore 046527
Tucker, Alan & Joan, Stroud ... 144868
Tucker, E., Pittsburgh 145439
Tucker, Phyllis, Houston 093820
Tuckerbox Antiques, Gundagai . 061567
Tuckmill Gallery, Naas 109678
Tucson Children's Museum,
  Tucson 053836
Tucson Coin, Tucson 097701
Tucson Museum of Art and Historic
  Block, Tucson 053837
Tucson Plein-Air Painters Society,
  Green Valley 060778
Tucumcari Historical Research Institute
  Museum, Tucumcari 053843
Tuddal Bygdetun, Tuddal 034044
Tudisco, Pascal, Borgo 067369, 128510
Tudo, Houston 121145
De Tudo um Pouco, Porto Alegre 063998
Tudor, Alresford 088260, 134023
Tudor Antiques, Napier 084011
Tudor Books, Weston-super-Mare 144915
Tudor Galleries, Norwich 118835
Tudor House, Barnstaple 088366, 134065
Tudor House, Bazzano 079790
Tudor House, Margate 045227
Tudor House, Stow-on-the-Wold 091487
Tudor House Antiques,
  Christchurch 088907
Tudor House Antiques,
  Halesowen 089409
Tudor House Antiques, Henley-on-
  Thames 089513
Tudor House Antiques, Ironbridge 089651
Tudor House Fine Arts and Antiques,
  Darwin 061390
Tudor House Galleries, Charlotte 120014
Tudor House Museum,
  Southampton 045830
Tudor House, Weymouth Civic Society,
  Weymouth 046109
Tudor Manor, New York 095793, 136017
Tudor Place Museum,
  Washington 054165
Tudor Rose Antiques,
  Faulconbridge 061462
Tudor Rose Antiques, New York 095794,
                            136018
Tudor Rose Antiques, Petworth . 091010
Tudor, Richard, London 118610
Tudor, Tony, Brugge 063289
Tübingen Kunstverein, Tübingen . 059248
Tüftler-Werkstatt-Museum,
  Frickingen 018514
Tüm Sanat Galerileri Derneği,
  Ankara 060012
Tuena, Filippo Maria, Roma ... 081590
Tüötten-Museum, Mettingen ... 020392
Türbeler Müzesi, İstanbul 042962
Türi Muuseum, Türi 010710
Türi Valla ja Türi Tehnika- ja
  Maamajanduskooli Muuseum,
  Särevere 010660
Türk İslam Eserleri ve Etnografya
  Müzesi, Erzurum 042896
Türk Nümismatik Derneği,
  İstanbul 060013
Türk ve Islâm Eserleri Müzesi,
  Edirne 042886
Türk ve Islam Eserleri Müzesi,
  İstanbul 042963
Türke, Erhard, Bad Salzuflen .. 106088
Türkiye Çocuk Esirgeme Kurumu
  Müzesi, Ankara 042832
Türkiyede Sanat, İstanbul 139117
Türkmenistanyn Milli Muzeỳi,
  Aşgabat 043089
Türkmenistanyt Gahrymany
  Gurbansoltan eje Adyndaky Türkmen
  Halsynyt Muzeỳi, Aşgabat ... 043090
Türr István Múzeum, Baja 023418
Tüshaus-Mühle, Dorsten 017949
Tufariello, Giuseppe, Milano ... 080751

Tuffier, Thierry, Les Andelys . . . 103982, 128672
Tuffstein-Museum "Steinmetzbahnhof", Weibern 022457
Tufts University Art Gallery, Medford 050754
Tugu Nasional Museum, Jakarta 024596
Tuileries Images, Paris 105280
Les Tuilleries, Lys-lez-Lannoy . . 069788
Tuite, Denise A., New York . . . 095795
Tuits, Dallas 093225
Tukano, Fortaleza 100961
Tukuma Audēju Darbnīcas, Tukums 030241
Tukuma Muzejs, Tukums 030242
Tula & Niles, Nelson . . 084024, 112969
Tula, Richard, Atlanta 119626
Tulane University Art Collection, Tulane University, New Orleans 051306
Tulare County Museum, Visalia . 054004
Tularosa Basin Historical Society Museum, Alamogordo 046335
Tul'čyns'kyj Filial Vinnyc'kogo Učylyšča Kul'tury i Mystectv im. M. Leontoviča, Tul'čyn 056503
Tulip Antiques, Silvan 062211
Tulip Tree Antiques, Indianapolis 093937
La Tulipe Noire, San Francisco . 097342, 136349
Tulklana Kumbi Aboriginal Galleries, Dareton 098940
Tull, Gottfried, Hostenbach 130070
Tullgarns Slott, Vagnhärad 041152
Tullie House – City Museum and Art Gallery, Carlisle 043875
Tullimuseo, Helsinki 010876
Tullio, Emilio, Milano 080752
Tullis-Toledano Manor, Biloxi . . . 046995
Tullmuseum, Stockholm 041054
Tullner Museen im Minoritenkloster, Tulln 002848
Tulloch, Launceston 125207
Tullow Museum, Tullow 024996
Tully, Denver 093347
Tulsa, Chennai 079380
Tulsa Antique Mall, Tulsa 097762
Tulsa Artists Coalition, Tulsa . . . 060779, 124817
Tulsa Card, Tulsa 097763
Tulsa Coin and Stamp, Tulsa . . . 097764
Tulsi Sangrahalaya Ramvan, Satna 024470
Tulskij Filial Moskovskogo Gosudarstvennogo Universiteta Kultury i Iskusstv, Tula 056307
Tulskij Gosudarstvennyj Muzej Oružija, Tula 037967
Tulskij Muzej Izobrazitel'nych Iskusstv, Tula 037968
Tulskij Oblastnoj Kraevedčeskij Muzej, Tula 037969
Tulum, Wetzlar 078961
Tuma, Alexander, Wien 140127
Tumbergs Skol- och Hembygdsmuseum, Vårgårda . 041158
Tumbling Tumbleweed, Tucson . 097702
Tumen Museum, Tumen 008344
Tumlaren, Teckomatorp 087096
Tummers & Tummers, Den Haag 083140
Tumut and District Historical Museum, Tumut 001597
Tun Abdul Razak Memorial, Kuala Lumpur 030697
Tunabygdens Gammelgård, Borlänge 040582
Tunander, Vittskövle 133604
Tunbridge Wells Antiques, Tunbridge Wells 091665
Tunbridge Wells Museum and Art Gallery, Tunbridge Wells 046003
Tunc, Neset, Friedrichshafen . . . 075915
Tune Bygdemuseum, Sarpsborg 033429
Túnel do Tempo, Porto Alegre . . 063999
Tung Jing, Modena 080823

Tung Salaeng Luang, Phitsanulok 042714
Tunger, Flensburg 106960
Tunghai University Art Center, Tunghai University, Taichung 042481
Tungsten, New Orleans 122251
Tungxiang Museum, Tongxiang . 008342
Tunica-Biloxi Native American Museum, Marksville 050698
Tunicia, Madrid 086274
Tunick, David, New York 095796, 123151
Tunkl, David, Los Angeles 121611
Tunkl, David, West Hollywood . . 125030
Tunku Abdul Rahman Putra-Memorial, Kuala Lumpur 030698
Tunna, Torsten, Großlehna 076108
Le Tunnel, Véretz 105839
Tunnel dell'Arte, Sanremo 110906
Tuns Press, Dalhousie University, Halifax 136674
Tunsons, Denver 046854
Tunxi Museum, Huangshan 007911
Tuoketuo Museum, Tuoketuo . . . 008345
Tuomarinkylän Museo, Helsingin Kaupunginmuseo, Helsinki . . . 010877
Tupa, Helsinki 066369, 128419
Tupou College Museum, Nuku'alofa 042732
Tupperville School Museum, Tupperville 007254
Turaidas Muzejrezervāts, Sigulda 030232
Turak, Philadelphia 123465
Turakina Antiques, Turakina . . . 084115
Al-Turath, Dubai 088191
Turati, Milano 110476
Turbulence, New York 095797
Turchet Antiquariato, Milano . . . 080753
Turchetto, Rita, Venezia 081878
Turchetto/Plurima, Milano 110477
Turchi, G., Firenze 080217
Turchi, Valerio, Roma 081591
Turčianska Galéria, Martin 038396
Turco, Maria Lucia, Foggia 131284
Turenne-Museum Sasbach, Sasbach 021615
Turetsky, Rosa, Genève 116342
De Turfschuur, Kolhorn 032550
Turiaf, Bernard, Montfort-l'Amaury 070244
Turianelli, Massimo, Roma 132066
Turin Gallery, Torino 111012
Turion, Patrick, Saint-Ouen 105570
Turk, Francine, Chicago 120285
Turk, W.F., London . . . 090551, 134788
Turkaly, Cleveland 120462
Turkana, Los Angeles 121612
Turkana, New York 095798
Turkansaaren Ulkomuseo, Pohjois-Pohjanmaan Museo, Oulu . . . 011223
Turkeyenhof, Bredene 003398
Turki Gharli, Mazen, Damascus . 088064
Turkish Sports Museum, İstanbul 042964
Turks Head Antiques, Eton 089223
Turkuvaz Sanat Galerisi, Ankara 042833
Turlej Foundation, Kraków 137896
Turlure, Emilienne, Le Havre . . . 103952
Turm 9 – Stadtmuseum Leonding, Leonding 002299
Turm der Sinne, Nürnberg 020880
Turm-Galerie-Bonn, Wachtberg . 108829
Turmbund, Innsbruck . 058314, 136523
Turmgalerie 333, Helmstedt . . . 107414
Turmmuseum, Breitenbrunn, Neusiedlersee 001797
Turmmuseum, Öhringen 020956
Turmmuseum Oetz, Sammlung Hans Jäger, Oetz 002477
Turmmuseum Sankt Magnus, Esens 018283
Turmmuseum Schloß Mengerskirchen, Mengerskirchen 020367
Turmmuseum Stadt Blankenberg, Hennef 019192
Turmuhren- und Heimatmuseum Bockenem, Bockenem 017444

Turmuhrenmuseum, Delligsen . . 017849
Turmuhrenmuseum, Naunhof . . 020648
Turmuhrenmuseum, Neulußheim 020720
Turmuhrenmuseum Rammensee, Gräfenberg 018783
Turn of the Century, New York . 095799
Turn of the Century, Sherwood . 062208
Turn of the Century Antiques, Austin 092271
Turn of the Century Lighting, Toronto 064770
Turn On Lighting, London 090552
Turnage Place Antiques, Atlanta 092194
Turnbull Bros., Milton 061860
Turnburrys, Jesmond . 089662, 134508
Turner, Birmingham 117382
Turner, New York 095800
Turner Clayton, Fresno 120911
Turner Helton, Seattle 097512
Turner House Gallery, Penarth . . 045473
Turner House Museum, Hattiesburg 049404
The Turner Museum, Sarasota . . 053152
Turner Society, London 060265
Turner Valley Gas Plant, Turner Valley 007255
Turner, A.F.J., East Molesey . . . 089139
Turner, Annmarie, Marlborough . 090726
Turner, B.J. & M.J., Exeter 117775
Turner, Dennise, Pontefract . . . 091032
Turner, Hugh & Curtis jr., Richmond 123748
Turner, R. & M., Jedburgh 089656
Turner, Robyn, San Francisco . . 124419
Turner, Rogers, London 144703
Turner, S. & E.M., Beeston 088429
Turner, Sally, Wendover 091778
Turning Page, Milwaukee 094884
Turning Point Gallery, Washington 124969
Turnip Truck Antiques, San Diego 097142
Turner, Philip, Rotherham 091146
Turnover Shop, Baltimore 092371, 092372
Turnpike Gallery, Leigh, Lancashire 044804
Turnstone Gallery, Sandsend . . . 119059
Turpault, René, Tours 073865
Turpin, Alain, Guéret 125595
Turpins Antiques, Hungerford . . 089615
Turquat, Paul, Montpellier 070318
Turquoise, Honfleur 103844
Turquoise Museum, Albuquerque 046371
Turquoise Trading Post, Austin . 119709
Turrero Lucia, Manuel, Madrid . 086275
Turret House, Wymondham, Norfolk 091963, 144939
Tursky, Heinz, Innsbruck 062657
Turszynski, Uwe, München 142192
Turtle Creek Antiques, Chicago . 092809, 092810
Turtle Dove, Clayfield 061341
Turtle Fine Art, Rye 118998
Turtle Mountain Chippewa Heritage Center, Belcourt 046875
Turtolan Kylämuseo, Turtola . . . 011454
Turton Tower, Turton Bottoms . . 046007
Turudda, Lüneburg 077212
Turun Antiikki ja Taide, Turku . . 066407
Turun Biologinen Museo, Turun Maakuntamuseo, Turku 011447
Turun Kaupungin Historiallinen Museo, Turun Linna, Turun Maakuntamuseo, Turku 011448
Turun Maakuntamuseo, Turku . . 011449
Turun Taidemuseo, Turku 011450
Turun Taideyhdistys / Konstföreningen i ÅBO, Turku 058584
Turun Terveydenhuoltomuseo, Turku 011451
Turun Tuomiokirkkomuseo, Turku 011452
Turys Antiquitäten, Oberschan . 087720
Turzer, Michael, Schwäbisch Hall 078411
Tuš i Pero, Sofia 101174
Tuscany, Manchester 118712

Tuscany Fine Arts, Letchworth . 118106
Tuscany Galleries, Healesville . . 099123
Tusculum, Wiesbaden 108931
Tushinsky-Merlette, Ella, Paris . 129155
Tushita, Duisburg 137167
Tusmore Antiques, Tusmore . . . 062357
Tut-Uncommon Antiques, Memphis 094669
Tutankhamun Exhibition, Dorchester 044121
Tutbury Mill Antiques Centre, Tutbury 091670
Tute-Natura, Bennekom 032061
Tutino, Dominique, Paris 071963
TuttAntico, Genova 098223
Tuttle, Charles E., North Clarendon 138434
Tuttlinger Haus, Museen der Stadt Tuttlingen, Tuttlingen 022236
Tutto Casa e Chiesa, Prato 131783
Tutto Diverso, Hasselt 063688
Tutzer Pattis, Helga, Bolzano . . 079903
Tuupalan Museo, Kuhmo 011045
Tuur Galerie, Venlo 112717
Tuusniemen Kotiseutumuseo, Tuusniemi 011457
Tuvikese Shop, Tartu 103125
Tuvinskij Filial Moskovskogo Gosudarstvennogo Universiteta Kultury i Iskusstv, Kyzyl 056282
Tuvinskij Respublikanskij Kraevedčeskij Muzej im. Aldan Maadyr, Kyzyl 037192
Tűzoltástörténeti kiállítás, Dráva Múzeum, Barcs 023436
Tűzoltó- és Községi Emlékház, Vál 024030
Tűzoltó Múzeum, Bonyhád 023449
Tűzoltó Múzeum, Budapest 023563
Tűzoltó Múzeum, Vörs 024048
TV-Galereja, Moskva 114312
TV Toys Dieren, Dieren 032224
Två Sekel, Uppsala 087127
TVDART, Berlin 017309
Tvedes Antikvariat, Næstved . . 140628
Tvedestrand Museum, Tvedestrand 034048
Tveit Bygdemuseum, Tveit 034049
Tveitens Samlinger, Eggedal . . . 033578
Tverskaja Oblastnaja Kartinnaja Galereja, Tver 037974
Tverskoj Gosudarstvennyj Obedinennyj Muzej, Tver 037975
Tvøroyrar Bygda- og Sjósavn, Tvøroyri 010744
Tvorčeskij Centr i Vystavočnyj Zal Fёdor, Sankt-Peterburg 037823
Tvorčestvo Rodiny, Moskva . . . 085211, 114313
Tvrz Žumberk, Památník Venkovského Lidu, Žumberk 009994
De Twee Pauwen, Den Haag . . . 112417
De Twee Wapens, Den Haag . . . 083141
Tweed and Area Heritage Centre, Tweed 007256
Tweed Museum of Art, University of Minnesota Duluth, Duluth 048387
Tweed River Art Gallery, Murwillumbah 001331
Tweeddale Museum, Peebles . . . 045470
Het Tweede Leven, Rotterdam . . 083595
Tweede Lezer, Brasschaat 140157
Tweewielermuseum Tankstop, Workum 033065
Twentieth Century, London 090553
Twentieth Century Antiques, Coolangatta 061361
Twentieth Century Collectables & Antiques, Hutt City 083960
Twentieth Century Design, Mornington 061888
Twentieth Century Gallery, London 090554, 118611, 134789
Twentieth Century Limited Goods,

Boston 092501
Twenty First Century Gallery,
  Denver 120784
Twenty Mile Post Historical Society
  Museum, Irvine 006045
Twenty Twenty, Much Wenlock . 118761
Twenty Two Gallery, Philadelphia 123466
Twerenbold, Luzern ... 116484, 133847
Twice Around, Orange 062032
Twice as Nice, Galway 079626
Twice Loved, Pittsburgh 096334
Twice Loved Tales, Jacksonville 145191
Twice Sold Tales, Seattle 145602,
  145603,   145604
Twice Told Tales, Cranston 145105
Twice Told Tales, Houston 145175
Twice Treasured, Pittsburgh 096335
Twice Upon A Time, Albion 060980
Twickenham Antique Warehouse,
  Twickenham 091679, 135168
Twickenham Homestead, Katikati 083973
Twickenham Museum,
  Twickenham 046012
Twickler, Aachen 129364
Twierdza Wisłoujście, Muzeum
  Historyczne Miasta Gdańska,
  Gdańsk 034743
Twiga, San Francisco 124420
Twiggs & Moore, Salt Lake City 123994
Twigs, New Orleans 095172
Twillingate Museum, Twillingate 007257
Twin Cities Galleries, Minneapolis 122047
Twin Falls County Museum, Twin
  Falls 053868
Twin Falls Museum, Mullens ... 051111
Twinhouse Gallery, Atlanta ... 119627
Twinkle Art Gallery, Delhi 109347
Twinsburg Historical Society Museum,
  Twinsburg 053869
Twiss, Rodney, Eastwood 061433
Twist Art & Design, London ... 118612
Two 12 Pottery, Minneapolis ... 122048
Two Angels, Seattle 097513
Two Boys and a Girl, El Paso ... 120842
Two Cans Antiques, Seaford ... 062199
Two Coats, Tucson 124779
Two Columbia Road, London ... 118613
Two Divas, Toronto 064771
Two Dollar Rare Bookstore,
  Cleveland 145090
Two Dragons, Llanerchymedd .. 089869
Two Four Seven, Petone
  Wellington 084061
Two Hills and District Museum, Two
  Hills 007258
Two Leaves Design, Clitheroe .. 117602
Two Lions, Washington 097513
Two Ninty Two, Dallas 135521
Two One Two Gallery, New York 123152
Two Rivers Gallery of Prince George
  and Region, Prince George ... 006717
Two Sisters, Charlotte 092636
Two Swans, Wodonga 062444
Two Way Books, Brighton 144272
Two World Arts, New York 136019
Two Zero, London 118614
Tworczosc, Warszawa 138993
Twórczość Ludowa, Lublin 138994
La Txatarreria, Donostia-San
  Sebastián 085986
Txindoki, Madrid 133398
Txoko'S 96, Bilbao 085892
Ty Art Show, Perros-Guirec ... 105344
Ty Aven, Pont-Aven 105385
Ty Blurt Records, Quimper 105420
Ty Korn, Brest 140806
Ty-Llwyd Antiques, Cardiff ... 088791,
  134218
Ty Mawr Wybrnant, Dolwyddelan 044107
Tyabb Antique Centre, Tyabb .. 062363
Tyabb Grain Store Antiques,
  Tyabb 062364
Tyabb Packing House Antiques,
  Tyabb 062365

Tybee Museum and Tybee Island Light
  Station, Tybee Island 053872
Tydal Museum, Tydal 034050
Tyholmen, Arendal 084212
Tyldesleys, Vancouver . 064892, 125391
Tylecek, Jiri, Paris 105281
Tyler, Charlotte 092637
Tyler, Philadelphia 096166
Tyler, Singapore 114629, 137939
Tyler Art Gallery, Oswego 051841
Tyler Galleries, Temple University, Tyler
  School of Art, Elkins 048539
Tyler Museum of Art, Tyler 053875
Tyler Nahem, Edward, New York 123153
Tyler School of Art, Temple University,
  Elkins 057045
Tyler, Matthew, Long Melford .. 090626,
  134812
Tylldalen Bygdetun, Tylldalen .. 034051
Tylstrup Antik og Kunst, Aalborg 065591
Tymiński, Jakub, Warszawa ... 084753
Tymperleys Clock Museum,
  Colchester 043992
Tyndale Galleria, Taylor University,
  Visual Art Department, Upland 053901
Tynemouth Architectural Salvage,
  Tynemouth 091685
Tynset Bygdemuseum, Tynset .. 034052
Työväenasuntomuseo ja Kotiseututalo,
  Kuusankoski 011063
Työväenasuntomuseo, Helsingin
  Kaupunginmuseo, Helsinki ... 010878
Työväenmuseo Werstas, Tampere 011412
Typaldos-Iakobatos Museum,
  Lixouri 023157
Typhaine, Courbevoie 068188
Typhoon, Pittsburgh 096336
Typical Trading, Cairo 066277
Typorama, Museum für Bleisatz und
  Buchdruck, Bischofszell 041350
Tyresö Slott, Tyresö 041094
Tyrnävän Kotiseutumuseo,
  Tyrnävä 011461
Tyrolia, Innsbruck 140000
Tyrrell, Portland 096485
Tyrvään Seudun Museo,
  Vammala 011487
Tyrwhitt-Drake Museum of Carriages,
  Maidstone 045195
Tyseley Loco Works Visitor Centre,
  Birmingham 043629
Tyson, Morecambe 090772
Tyto Alba, Vilnius 137806
tytuł roboczy, Warszawa 138995
Tytuvėnų Miško Muziejus,
  Tytuvėnai 030494
Tytuvėnų Žmės Ūkio Mokyklos
  Kraštotyros Muziejus, Tytuvėnai 030495
Tytyrin Kalkkikaivosmuseo, Lohja 011123
Tzen, Singapore 085523
TZR Galerie, Düsseldorf 106838
Tzwern, Maurice, Bruxelles ... 100663
U-A-Galerie für orientalische Kunst,
  Dortmund 075399
U Ambrože, Hradec Králové ... 065392
U Červeného Beránka, Praha .. 102724
U Draka, Znojmo 102755
U Dvou Čápů, Karlovy Vary ... 102607
U Floriána, Ostrava 065440
U Ignáce, Praha 065533
U Jakoba, Opava 102630
U Jakuba, Frýdek-Místek 102591
U Jezuitův, Poznań 113750
U Kamenného Stolu, Praha ... 065534
U Keplera, Praha 065535
U Konvice, Hradec Králové ... 065393
U Kostela, Písek 065448
U Kováře, Brno 102574
U Král'a Mateja, Poprad 085583
U Madony, Brno 065361
U Maxima, Minsk 100468
U Michala, Telč 102737
U Mloka, Olomouc 102626
U Nádraží, Hradec Králové ... 065394

U Nádraží, Šumperk 065557
U Parnasa, Sankt-Peterburg ... 143678
U Pražkeho Jezulátka, Praha .. 102725,
  140533
U R My Sunshine, Philadelphia . 096167
U Radnice, Čelákovice 102576
U Rotundy, Týnec nad Sázavou . 102742
U Sediaceho Anjela, Kežmarok . 114660
U Slona, Písek 140497
U Sveti Jošta, Český Krumlov .. 102585
U Synagogy, Liberec 065415
U-Turn E-Zine, Oak Park 139512
U Uhelného Trhu, Praha 065536
U Vinohradského Pavilonu, Praha 065537
U Vlaškého Dvora, Kutná Hora . 102612
U Zlaté Číše, Praha 140534
U Zlaté Korony, Telč 102738
U Zlaté Koruny, Muzeum města Brna,
  Brno 009483
U Zlatého Lva, Praha 065538
U Žofky, Bratislava 085559
The U. Nahon Museum of Italian
  Jewish Art, Jerusalem 025112
U. T. E. T., Torino 137753
U.A.B. Visual Arts Gallery. Selections
  from the Permanent Selection,
  Birmingham 139513
U.A.P., Constanța 114167
U.F.R d'Art et d'Archéologie, Université
  de Paris I – Panthéon-Sorbonne,
  Paris 055406
Ubach, Madrid 086276
Ubach, Maria, Barcelona 085848
UBC Fish Museum, Vancouver . 007298
Ubeda, Yolande, Valros 073958
UBMB, Thise 073659
UBR Galerie, Salzburg 100180
Ubu, Bochum 141608
Ubu, New York 123154
Ubu Fine Art, Chicago 120286
UC Productions, New York 095801
Učebno-chudožestvennyj Muzej
  Slepkov im. I.V. Cvetaeva,
  Moskva 037503
Učebno-naučen Kompleks Iskustva,
  Varnenski Svoboden Universitet
  Černorizec Chrabar, Varna ... 055111
Ucellino, Marseille 104205
Ucher, A., Köln 107694
Uchida, Kyoto 082047
Uchiko Hakubutsukan, Uchiko . 029668
Uchu Kagakukan, Ikoma 028680
Uckermärkische Heimatstuben
  Fürstenwerder, Fürstenwerder . 018546
Uckermärkisches Volkskundemuseum
  Templin, Templin 022128
UCL Museums and Collections,
  University College London,
  London 045119
Ucross Foundation Art Gallery,
  Clearmont 047793
Udell, Douglas, Edmonton 101303
Udell, Douglas, Vancouver 101843
Udmurtskij Respublikanskij Muzej
  Izobrazitelnyh Iskusstv, Iževsk 037049
Udruženje Likovnih i Umetnika
  Primenjenih Umetnosti i Dizajnera
  Vojvodine, UPIDIV, Novi Sad .. 059790
Udruženje Likovnih Umetnika
  Primenjenih Umetnosti i Dizajnera
  Srbije, Beograd 059791
Udruženje Likovnih Umetnika Srbije,
  Beograd 059792
Udruženje Likovnih Umjetnika Crne
  Gore, Podgorica 059584
Udruženje Likovnih Umjetnika
  Primjenjenih Umjetnosti Hrvatske,
  Zagreb 058527
Udruženje Samostalnih Umetnika
  Fotografije, Beograd 059793
UDT-SEAL Museum, Fort Pierce 048858
Uebele, Sylvia, Würzburg 108979
Über NormalNull, Hamburg ... 076331
Überlinger Antiquariat,

Überlingen 142438
Übersee Museum Bremen,
  Bremen 017601
Überwälder Heimatmuseum, Wald-
  Michelbach 022378
Übleis, Christian, Neumarkt im
  Hausruckkreis 127823
Uebler, Dagmar, Amberg 105986
Uehara Kindai Bijutsukan,
  Shimoda 029380
Ueker & Ueker, Basel 116106
Uelner, S., Königswinter 076930
Uelze, Dresden 129725
Ündüz, Gülay, İstanbul 088144
Ueno Royal Museum, Tokyo ... 029618
Ürgüp Müzesi, Ürgüp 043076
Uerpmann, Ulrich & Anke,
  Oerlinghausen 077900
Ürün Sanat Galerisi, İstanbul ... 117093
Uetiker Museum, Uetikon am
  See 042076
Uferhallen, Berlin 017310
Uffenheimer Gollachgaumuseum,
  Uffenheim 022253
Ufimskaja Kartinnaja Galereja,
  Ufa 037984
Ufimskij Gosudarstvennyj Institut
  Iskusstv, Ufa 056308
UFR Histoire, Arts et Archéologie,
  Université de Toulouse le Mireail,
  Toulouse 055424
Ugalde, Pedro, Getxo 086014
Uganda Artists Association,
  Kampala 060017
Uganda Museum, Kampala 043108
Ugarte Bejega, Victor, Vitoria-
  Gasteiz 086579
Ugas, Renzo, Locarno 116448
Ugbrooke House, Chudleigh ... 043959
Ughetto, Henry, Villeurbanne .. 105878
Ugly Otter Trading Post,
  Albuquerque 092088
Ugly Stuff Antiques, Birregurra . 061158
Ugolini, Daniele, Firenze 110074
Ugolini, Paolo, Milano 131544
Uğur, İstanbul 088145
Uhde, Margrit, Dassel 075313
Uhl, S., Baden-Baden 074644
Uhlig, Gabriele, Chemnitz 075261
Uhlin, Mjölby 115775
Uhr-Alte Zeiten, Berlin 074939
Uhren- und Musikgerätemuseum-
  Museum zu Putbus, Putbus .. 021239
Uhrenindustriemuseum, Villingen-
  Schwenningen 022341
Uhrenmuseum, Bad Grund 016779
Uhrenmuseum, Bad Iburg 016794
Uhrenmuseum, Regensburg ... 021331
Uhrenmuseum Beyer Zürich,
  Zürich 042239
Uhrenmuseum Glashütte,
  Glashütte 018707
Uhrenmuseum Lorenz-Berlin,
  Berlin 017311
Uhrenmuseum, Wien Museum,
  Wien 003092
Uhrensammlung Kellenberger,
  Gewerbemuseum, Winterthur . 042163
Uhrenstube Aschau, Museum für
  Turmuhren und Bratenwender,
  Oberschützen 002467
Uhrentechnische Lehrschau Hennig,
  Tharandt 022147
Uhrig, Weingarten, Baden 078904
Uhrig, B., Weingarten, Baden .. 078905
Uhrig, Gustav, Gedern 075954
Uhrig, Iris, Alsheim 129380
Uhrig, James W., Indianapolis . 121221
Uhrmachermeister Hübner GmbH,
  Wien 063090, 127974
Uilenspiegel Museum, Damme . 003533
Uiryeong Museum, Uiryeong ... 030057
Uit Steppe en Oase, Gent 063664
Uitamonkosken Myllymuseo,

Multia 011168
Het Uitgelezen Boek, Zoelen . . . 143270
Uitgeverij 010 Publishers, Rotterdam 137871
Uitgeverij SUN, Amsterdam 137843
Uivel, Amsterdam 082878
Új Képcsarnok, Budapest 109180
Új Magyar Képtár, Szent István Király Múzeum, Székesfehérvár 023944
Uji Kanbayashi Kinenkan, Uji, Kyoto 029677
Ujlaky, Richard, Palis 070846
Újlipótvárosi Klub Galéria, Budapest 109181
Újpesti Helytörténeti Gyűjtemény, Budapest 023564
Újpesti Lepkemúzeum Alapítvány, Budapest 023565
UK Arts 'N' Crafts, Lowestoft . . 118670
Ukai Museum, Kawaguchi-ko O-tomachikku Gakki Hakubutsukan, Kawaguchi-ko 028801
Ukiyo-e, New York 095802
Ukiyo-e and Pottery Museum, Hirano Hakabutsukan, Osaka 029271
Ukiyo-e Dealers Association of Japan, Tokyo 059517
Ukmergės Kraštotyros Muziejus, Ukmergė 030497
Ukraine National Gallery, Kyïv . . 043225
Ukrainian Academy of Arts and Sciences, New York 060780
Ukrainian-American Museum, Hamtramck 049345
Ukrainian-American Museum, Warren 054062
Ukrainian Art Center, Los Angeles 050471
Ukrainian Canadian Archives and Museum of Alberta, Edmonton 005763
Ukrainian-Canadian Art Foundation, Toronto 007222
Ukrainian Catholic Women's League Museum, Edmonton 005764
Ukrainian Cultural & Education Centre Oseredok, Winnipeg 007470
Ukrainian Cultural Heritage Village, Edmonton 005765
Ukrainian Heritage Association and Museum of Canada, Toronto . . 007223
Ukrainian Homestead Museum, Winnipeg Beach 007479
Ukrainian Institute of Modern Art, Chicago 047703
Ukrainian Museum, Cleveland . . 047827
Ukrainian Museum, Gardenton . 005859
The Ukrainian Museum, New York 051486
Ukrainian Museum, Torrens Park 001587
Ukrainian Museum of Canada, Saskatoon 006984
Ukrainian Museum of Canada, Vancouver 007299
Ukrainian Museum of Canada – Alberta Branch, Edmonton 005766
Ukrainian Museum of Canada – Manitoba Branch, Winnipeg . . 007471
Ukrainian National Museum, Chicago 047704
Ukrainian Peoples Home of Ivan Franco, Angusville 005338
Ukraińska Galeria Sztuki, Kraków 113646
Ulaanbaatar City Museum, Ulaanbaatar 031741
Ulbargen, Großefehn 107174
Ulbricht, Dr. Michael, Machern . 126031, 142096
Ulenspiegel Art Gallery, Oostende 100808
Uleybury School Museum, One Tree Hill 001382
Ulferts, G., Stolberg, Rheinland . 078569
Ulibarri Junior, Mario Peres, Porto Alegre 064000
Ulises Libros, Madrid 143784
Ulivi, V., Prato 110695
Uljanovskij Oblastnoj Chudožestvennyj Muzej, Uljanovsk 038003
Uljanovskij Oblastnoj Kraevedčskij Muzej im. I.A. Gončarova, Uljanovsk 038004
Ullapool Museum, Ullapool 046018
Ullas Skattkista, Trelleborg 087104
Ullensaker Bygdemuseum, Jessheim 033685
Ullevål Sykehus Museum, Oslo . 033868
Ullman, Sylvia, Cleveland 120463
Ullmann, Markus, Bayreuth 074699
Ullmann, Robert, Schaffhausen . 087804
Ullrich, Angelika C., Düsseldorf . 075530
Ullrich, Beppo, Düsseldorf 129762
Ullrich, Markus, Reichertshausen 130576
Ullrich, W. & R., Witten 142500
Ullwer, Werner, Obergünzburg . . 077860
Ulm, Winfried, Bad Windsheim . 141439
Ulmann, Angélo, Amberieux-en-Dombes 066546
Ulmer Museum, Ulm 022268
Ulmer Verein, Verband für Kunst- und Kulturwissenschaften e.V, Berlin 059249
Ulmer, Maria, Meersburg 077323
Ulricehamns Konst- och Östasiatiska Museum, Ulricehamn 041101
Ulrich, Elizabeth, Honolulu 120973
Ulrich, Hanns, Fürth 137218
Ulrich, Herbert P., Berlin 074940
Ulrich, Hugo, Buchrain 133707
Ulrich, Peter, Augsburg 074471
Ulrich, Sarah, Saint Louis 096789
Ulrika museum, Ulrika 041102
Ulrike Herrmann, Kunsthandel/ Agentur, München 077631
Ulriksdals Slott samt Orangerimuseet, Solna 040981
Ulshöfer, Jürgen, Lauda-Königshofen 077027
Ulster-American Folk Park, Omagh 045438
Ulster County Historical Society Museum, Marbletown 050668
Ulster Folk and Transport Museum, Holywood 044599
Ulster History Park, Omagh 045439
Ulster Museum, Belfast 043564
Ulsyn Töw Muzei, Ulaanbaatar . 031742
Ultiart, Sofia 101175
Ultima Thule, Warszawa 143533
Ultime Atome, Gent 100705
Ultner Talmuseum, Ulten 028231
Ultramarine, Ventnor 091702
Ultrapiktura, Moskva 114314
Ultraviolet, Southsea 119128
Ultreya, Milano 137698
Ulucam, Lüdenscheid 077202
Ulug-Beg Yodgorlik Muzeyi, Samarqand 054648
Ulukaya, İstanbul 088146
Uluslararası Plastik Sanatlar Derneği, İstanbul 060014, 117094
Ulusoy, Ankara 116991
Ulva Konst & Kuriosa, Uppsala . 143996
Ulverstone Local History Museum, Ulverstone 001601
Ulverton Wool Mill, Ulverton . . . 007259
Ulysses, Wien 100386
Ulysses Bookshop, London 144704
Ulysses S. Grant Cottage State Historic Site, Wilton 054460
Um, Eun Sook, Seoul 111728
Umanome, Nara 111341
Umatilla County Historical Society Museum, Pendleton 051971
Umbras Kuriositätenkabinett, Berlin 141583
Umbrella, Santa Monica 139514
Umbrella Studio, Townsville . . . 099840
Umbria, Amsterdam 082879
Umeå Energicentrum med Kraftverksmuséet, Umeå 041105
Umeå Konst och Ramaffär, Umeå 115968
Umekoji Steam Locomotive Museum, Kyoto 028988
Umélec, Praha 138615
Umělecké Předměty a Galerie, Poděbrady 102643
Uměleckoprůmyslové Muzeum v Praze, Praha 009843
Umělecký Bazar, Brno 065362
Umění, Praha 138616
Umetnička Galerija, Kruševac . . 038207
Umetnička Galerija Nadežda Petrović, Čačak 038189
Umetnička Galerija Strumica, Strumica 030639
Umetnička Galerija Velimir A. Leković, Bar 031745
Umetnička Galerija, Naroden Muzej Kumanovo, Kumanovo 030618
Umetnička Zbirka Flögel, Beograd 038184
Umetnostna Galerija Maribor, Maribor 038574
Umeyama Furui Yakimono Bijutsukan, Tobe 029479
Umi-no Hakubutsukan, Toba . . . 029478
U'mista Cultural Centre, Potlatch Collection and Art Gallery, Alert Bay 005312
Umjetnička Galerija, Dubrovnik . 008913
Umjetnička Galerija, Skopje 030635
Umjetnička Galerija Bosne i Hercegovine, Sarajevo 004282
Umjetnička Galerija Josip-Bepo Benkovic, Herceg-Novi 031755
Umjetnička Galerija Nikola I i Galerija Ilija Sobajić, Nikšić 031759
Umjetnički Paviljon, Podgorica . . 031768
Umjetnički Paviljon u Zagrebu, Zagreb 009053
UMKC Gallery of Art, University of Missouri-Kansas City, Kansas City 049897
Umla, Silvia, Völklingen 142455
Umlauf Sculpture Garden and Museum, Austin 046704
Umm Qais Archaeological Museum, Umm Qais 029786
Umoja Art Village, Chicago 120287
Umoona Opal Mine and Museum, Coober Pedy 000987
Umpqua Discovery Center, Reedsport 052501
Umstädter Museum Gruberhof, Groß-Umstadt 018836
Umuzi-Afrika-Haus, Freiberg am Neckar 018469
Un Autre Regard, Le Tréport . . . 103977
Un Autre Regard, Paris 105282
Un Coin du Passé, Aix-les-Bains 066487
Un Jour a Paris, Warszawa 113906
Un Meuble Une Histoire, Le Havre 128664
Un Regard Moderne, Paris 141289
Un Souffle d'Antan, La Daguenière 069021
Un Sourire de Toi, Paris 105283
Un Tuffo nel Passato, Bari 079787, 130927
Un Village se Raconte, Ecomusée de la Vallee d'Aspe, Lourdios-Ichère 013904
Unamuno Sobrino, Consuelo, Burgos 085917, 133216
Uncle Davey's Americana, Jacksonville 094047, 135669
Uncle Edgar's Mystery Bookstore, Minneapolis 145286
Uncle Hugo's Science Fiction, Minneapolis 145287
Uncle Remus Museum, Eatonton 048460
Uncle Richard's Antiques, San Diego 097143
Uncle Tom's Cabin Historic Site, Dresden 005716
Uncle Vick's, Norfolk 095876
Uncommon Objects, Austin 092272
UND, Plattform zur Präsentation von Kunstinitiativen in Karlsruhe plus internationale Gäste, Karlsruhe 098168
Underground Gallery, Seattle . . . 124610
Underground Gold Miners Museum, Alleghany 046406
Underground Passages, Exeter . 044313
Underground Prisoners Museum, 1918–1948, Jerusalem 025113
Underhill Museum, Idaho Springs 049656
Underwood Gallery, Milwaukee . 121953
Underwoodhall Antiques, Woodhall Spa 091922
Une Autre Epoque, Brie-Comte-Robert 067493
Une Maison, Paris 071964
Une Maison, Saint-Ouen 073216
Une Page d'Histoire, Caylus . . . 067740
Une Si Belle Journée, La Colle-sur-Loup 069004
UNESCO Publishing, Paris 136990
Unexpected Treasures, Richmond 096583
Unfer, Leutenheim 069442
Ungarie Museum, Ungarie 001602
Ungarndeutsche Heimatstuben, Langenau 019902
Ungarndeutsches Heimatmuseum, Backnang 016697
Unge Kunstneres Samfund, Oslo 059677
Ungefroren, Gerhard, Halle, Saale 076172
Ungelenk, Rainer, Wandersleben 078859
Unger, Siegmar, Bad Königshofen 074541
Unger, Urs von, Basel 087262
Unger, Urs von, Saanen 087766
Ungern-Sternberg, Dr. Wolfgang von, Regensburg 142325
Ungheretti, Piero, Livorno 131350
Ungurmuiža, Raiskums 030173
Uni-Antiquariat Graz, Graz 139991
Uni Mat, Chaskovo . . . 064220, 140315
Uniacke Estate Museum Park, Mount Uniacke 006437
UniBuch, Darmstadt 141682
Unicat, Oranienburg 108344
Unicon Art, Košice 114662
Unicorn, Miami 094812
Unicorn Antiques, Edinburgh . . . 089195
Unicorn Antiques, Kingswinford . 089710, 134517
Unicorn Gallery, Chippendale . . . 098889
Unicorn Gallery, Wilmslow 119354
Unicorn Studio, Baltimore 119776
Unicorn Trust Equestrian Centre, Stow-on-the-Wold 045894
Unicorn Unique, Camberwell . . . 061259
Unicum, Lecco 080310
Unidad Museológica Municipal de La Banda, La Banda 000396
Unik Kunst, Gjøvik 084240
Unikat Art, Rastede 108433
Unikat, Salon Dzieł Sztuki, Kraków 084602
Unimex International, San Antonio 124079
Union, London 118615
Union Art Gallery, Baton Rouge . 046823
Union-Art Gallery, University of Wisconsin-Milwaukee, Milwaukee 050924
Union Avenue Antique Mall, Memphis 094670
Union Bay Antiques, Seattle . . . 097514
Union County Heritage Museum, New Albany 051222
Union County Historical Complex, Creston 048062
Union County Historical Foundation Museum, Union 053892
Union County Historical Society Museum, Blairsville 047031
Union County Historical Society

Museum, Marysville 050729
Union County Museum,
  Lewisburg 050227
Union County Museum, Union ... 053890
Union County Public Library Union
  Room, Monroe 051002
Union de Arte, Barcelona 114971
Unión de Asociaciones de Artistas
  Visuales, Barcelona 059875
Union des Photographes Créateur,
  Paris 058693
Union Française des Experts,
  Paris 058694
Union Méditerranéenne d'Art
  Contemporain, Saint-Paul-de-
  Vence 058695
Union Mills Homestead and Grist Mill,
  Westminster 054343
Unión Nacional de Escritores y Artistas
  de Cuba, La Habana 058532
Union Nacionale des Peintres
  Illustrateurs, Paris 058696
Union Nationale des Arts Culturels,
  Alger 058167
Union of American Hebrew
  Congregation, New York 051487
Union of Artists of the Kyrgyz
  Republic, Biškek 059539
Union of Belarusian Artists,
  Minsk 058351, 100469
Union of Bulgarian Artists, Sofia 058406
Union of Maine Visual Artists,
  Monroe 060781
Union Pacific Railroad Museum,
  Council Bluffs 048035
Union Square Ceramic Center, New
  York 123155
Union Station Kansas City, Kansas
  City 049898
Union Street Gallery, Chicago ... 047705
Union Syndicale des Artisans des
  Métiers d'Art, Grenoble 058697
Union Trading, Chinese Antique
  Objects, Hong Kong 065105
Uniquarium, Leipzig 142068
l'Unique, Montréal 101456
Unique Antique, Tucson 097703
Unique Antiques, Pittsburgh ... 096337
Unique Antiques, San Diego ... 097144
Unique Antiques, Singapore ... 085524
Unique Design, Ulset 084392
Unique Finds, Seattle 097515
Unique Fine Art, Shenstone ... 119088
Unique Pieces, Calgary 064278
Unique Place, Singapore 085525
Unique Resale, Houston 093821
Unique Stuff, Kansas City 094101
Unique Treasure & Antiques,
  Regina 064557
Unique Treasures, Columbus ... 093054
Uniques, Cincinnati 092904
Uniques, Lesquin 103992
Uniques, Vancouver 064893
Uniques and Antiques, Aston . 127123
Uniquities Architectural Antiques,
  Calgary 064279
UNISA Art Gallery, Tshwane ... 038861
Uniseum, Freiburg im Breisgau . 018490
Unité Musée Numismatique, Banque
  de République d'Haïti, Port-au-
  Prince 023395
United Artists, Trieste 111056
United Caribbean Art Gallery,
  Jacksonville 121273
United Counties Museum in the Wood
  House, Cornwall 005651
United Independent Art Community, San
  Francisco 060782, 124421
United Kingdom National Committee
  of the International Council of
  Museums, London 060266
United Oil Painting, Virginia
  Beach 124841
United Oil Paintings, Norfolk ... 123238

United Planets – Space Art Galerie,
  München 077632, 108202
United Society of Artists, London 060267
United States Air Force Academy
  Museum, El Paso 048501
United States Air Force Museum,
  Wright-Patterson Air Force Base 054557
United States Army Air Defense
  Artillery Museum, Fort Bliss . 048784
United States Army Aviation Museum,
  Fort Rucker 048866
United States Army Center of Military
  History, Museum Division, Fort
  McNair 048840
United States Army Chaplain Museum,
  Fort Jackson 048818
United States Army Communications-
  Electronics Museum, Fort
  Monmouth 048846
United States Army Engineer Museum,
  Fort Leonard Wood 048836
United States Army Finance Corps
  Museum, Fort Jackson 048819
United States Army Heritage and
  Education Center, Carlisle 047429
United States Army Medical
  Department Museum, Fort Sam
  Houston 048868
United States Army Military Police
  Corps Museum, Fort Leonard
  Wood 048837
United States Army Museum of
  Hawaii, Fort DeRussy 048802
United States Army Ordnance
  Museum, Aberdeen Proving
  Ground 046301
The United States Army Quartermaster
  Museum, Fort Lee 048834
United States Army Signal Corps
  and Fort Gordon Museum, Fort
  Gordon 048812
United States Army Transportation
  Museum, Fort Eustis 048809
United States Army Women's Museum,
  Fort Lee 048835
United States Brig Niagara, Homeport
  Erie Maritime Museum, Erie .. 048599
United States Capitol Visitor Center,
  Washington 054166
United States Cavalry Museum, Fort
  Riley 048865
United States Coast Guard Museum,
  New London 051279
United States Coin and Currency Shop,
  Philadelphia 096168
United States Commission of Fine Arts,
  Washington 060783
United States Department of the
  Interior Museum, Washington . 054167
United States Federation of Friends of
  Museums, Baltimore 060784
United States Grant Ancestral
  Homestead, Dungannon 044175
United States Grant's Home,
  Galena 049018
United States Hockey Hall of Fame,
  Eveleth 048647
United States Holocaust Memorial
  Museum, Washington 054168
United States-International Council
  on Monuments and Sites,
  Washington 060785
United States Marine Corps Air-
  Ground Museum, Quantico ... 052425
United States Marine Corps Museum,
  Washington 054169
United States Mint-Philadelphia,
  Philadelphia 052099
United States National Committee
  of the ICOM, ICOM-US,
  Washington 060786
United States National Ski Hall of
  Fame and Museum, Ishpeming 049729
United States Naval Academy

Museum, Annapolis 046503
United States Navy Art Gallery,
  Washington 054170
United States Navy Memorial
  Foundation and Naval Heritage,
  Washington 054171
United States Navy Supply Corps
  Museum, Athens 046598
United States Porcelain,
  Albuquerque 135294
United States Senate Commission on
  Art, Washington 060787
United States Senate Commission on
  Art Collection, Washington ... 054172
United States Society for Education
  through Art, Greenville 060788
United States Space and Rocket
  Center, Huntsville 049636
Unites States Air Force Academy
  Visitor Center, USAF Academy . 053912
Unity and District Heritage Museum,
  Unity 007261
Uniunea Artiştilor Plastici din România,
  Bucureşti 059767
L'Univers, Lausanne 144095
L'Univers du Bronze, Paris .. 105284,
                105285, 141290, 141291
Univers du Livre, Paris 129156
Universal – Sociedade de Leilões,
  Porto 059761
Universal Art, Silverton 085667
Universal Arts Galerie Studio,
  München 108203
Universal Arts Gallery, Memphis 121712
Universal Books and Old Wares,
  Taralga 062307, 139928
Universal Comics, Baltimore ... 144989
Univers'Art, Le Mans 103958
Universe, Tbilisi 105924
Universe Antiques, New York ... 095803
Universeum, Göteborg 040662
Universidad de Granada. Cuadernos
  de Arte, Granada 139074
Universidad de Sevilla. Serie: Bellas
  Artes, Sevilla 139075
Universidad del Arte Ganexa, Ciudad de
  Panamá 056219, 113326
Universidad del Diseño, San José 055279
Universita degli Studi di Parma.
  Istituto di Storia dell'Arte. Cataloghi,
  Parma 138879
Universita degli Studi di Roma.
  Seminario di Archeologia e Storia
  dell'Arte Greca e Romana. Studi
  Miscellanei, Roma 138880
Università Internazionale dell'Arte –
  UIA, Firenze 055901
Universität der Künste Berlin,
  Berlin 055482
Universität für angewandte Kunst
  Wien, Wien 054985
Universitäts- und Landesbibliothek
  Bonn, Bonn 017503
Universitäts- und Landesbibliothek
  Darmstadt, Darmstadt 017829
Universitätsbibliothek, München 020581
Universitätsbibliothek der Humboldt-
  Universität zu Berlin, Berlin ... 017312
Universitätsbibliothek Johann
  Christian Senckenberg, Frankfurt am
  Main 018455
Universitätsbibliothek Salzburg,
  Salzburg 002640
Universitätsbibliothek Wien, Wien 003093
Universitätsmuseum, Altdorf bei
  Nürnberg 016522
Universitätsmuseum Heidelberg,
  Heidelberg 019148
Universitätssammlung antiker und
  nachantiker Münzen und Medaillen,
  Tübingen 022224
Universitätssammlungen
  Kunst+Technik, Technische
  Universität Dresden, Dresden . 018014

Université de Beaux-Arts, Phnom
  Penh 055113
Universiteitsbibliotheek Gent, Gent 003643
Universiteitsmuseum, Groningen 032385
Universiteitsmuseum De
  Agnietenkapel, Amsterdam .. 031995
Universiteitsmuseum Utrecht,
  Utrecht 032957
Universitetets Oldsaksamling,
  Universitetets Kulturhistoriske Museer,
  Oslo 033869
Universities Art Association of Canada,
  Salt Spring Island 058476
University Additions, Memphis .. 121713
University Archaeology Museum,
  Birmingham 043630
University Art, Sacramento ... 123786
University Art Galleries, Akron .. 046333
University Art Galleries, Murray State
  University, Murray 051129
University Art Galleries, University of
  South Dakota, Vermillion 053961
University Art Gallery, Bridgeport 047203
University Art Gallery, Carson .. 047442
University Art Gallery, Chico ... 047713
University Art Gallery,
  Jacksonville 049777
University Art Gallery, Pittsburgh 052178
University Art Gallery,
  Potchefstroom 038813
University Art Gallery,
  Stellenbosch 038840
University Art Gallery, Stony
  Brook 053550
University Art Gallery & Cabinet
  of Natural History, University of
  Massachusetts-Dartmouth, New
  Bedford 051227
University Art Gallery of the University
  of the South, Sewanee 053276
University Art Gallery, California State
  University Stanislaus, Turlock . 053855
University Art Gallery, Central
  Michigan University, Dept. of Art,
  Mount Pleasant 051096
University Art Gallery, Indiana State
  University, Terre Haute 053720
University Art Gallery, San Diego State
  University, San Diego 052985
University Art Gallery, Sonoma State
  University, Rohnert Park 052666
University Art Gallery, Texas Christian
  University, Fort Worth 048909
University Art Gallery, Union University,
  Jackson 049760
University Art Gallery, University of
  California, San Diego, La Jolla 050042
University Art Gallery, University of
  North Alabama, Florence 048754
University Art Museum,
  Binghamton 046999
University Art Museum, Santa
  Barbara 053094
University Art Museum, California
  State University, Long Beach, Long
  Beach 050376
University Art Museum, State
  University of New York at Albany,
  Albany 046348
University Art Museum, University
  of Louisiana at Lafayette,
  Lafayette 050063
University Art Museum, University of
  New Mexico, Albuquerque ... 046372
University at Buffalo Anderson Art
  Gallery, State University of New York,
  Buffalo 047309
University at Buffalo Art Galleries,
  Buffalo 047310
University College Dublin Classical
  Museum, Dublin 024909
University College Gallery, University
  of Maryland, College Park 047871
University College of Cape Breton Art

Gallery, Sydney 007117
University College of Fine Arts,
  University of Teheran, Tehrān . 055856
University College Zoological Museum,
  Cork 024860
University Fine Art Collection,
  University of Tasmania, Hobart 001151
University for the Creative Arts,
  Canterbury 056516
University for the Creative Arts,
  Epsom 056531
University for the Creative Arts,
  Farnham 056534
University for the Creative Arts,
  Maidstone 056576
University for the Creative Arts,
  Rochester 056602
University for the Creative Arts UCA,
  Maidstone 056577
University Galleries, Art Dept.,
  University of Texas Pan American,
  Edinburg 048479
University Galleries, Clarion University
  of Pennsylvania, Clarion 047767
University Galleries, Illinois State
  University, Normal 051584
University Gallery, Alexandria . . 046379
University Gallery, Boca Raton . . 047070
University Gallery, Gainesville . . 049007
The University Gallery, Newcastle-
  upon-Tyne 045354
University Gallery, Oxford 051863
University Gallery Leeds, Leeds . 044789
University Gallery, East Central
  University, Ada 046318
University Gallery, Fort Hays State
  University, Hays 049417
University Gallery, Texas A & M
  University, Commerce 047948
University Gallery, University of
  Delaware, Newark 051494
University Gallery, University of
  Massachusetts at Amherst,
  Amherst 046454
University Health Network Artifact
  Collection, Toronto 007224
University Historical Museum,
  Musuan 034555
University History Museum, Sun Yat-
  Sen University, Guangzhou . . . 007814
University Medical Museum, Medical
  University of South Carolina,
  Charleston 047561
The University Museum,
  Edwardsville 048488
University Museum, Port
  Harcourt 033445
The University Museum, Tokyo . 029619
University Museum,
  Vallabhvidyanagar 024512
University Museum – Sasol Art
  Museum, Stellenbosch 038841
University Museum and Art Gallery,
  Hong Kong 007893
University Museum Collections,
  Dundee 044168
University Museum of Anthropology,
  University of the Philippines, Quezon
  City 034586
University Museum of Archaeology
  and Anthropology, Cambridge . 043849
University Museum of Science and
  Culture, Aligarh 024133
University Museum of Zoology,
  Cambridge 043850
University Museum, Chung Ang
  University, Anseong 029887
The University Museum, Indiana
  University of Pennsylvania,
  Indiana 049676
University Museum, Paul Meek Library,
  Martin 050722
University Museum, Southern Illinois
  University, Carbondale 047425

University Museums, Oxford . . . 051864
University Museums Group,
  Birmingham 060268
University Museums, University of
  Delaware, Newark 051495
University of Alaska Museum of the
  North, Fairbanks 048662
University of Alberta Museums,
  Edmonton 005767
University of Architecture, Ho Chi
  Minh City 058159
University of Arizona Mineral Museum,
  Tucson 053838
University of Arizona Museum of Art,
  Tucson 053839
University of Arizona Student Union
  Galleries, Tucson 053840
University of Arkansas at Little Rock
  Art Galleries, Little Rock 050338
University of Arkansas Collections,
  Fayetteville 048717
University of Art and Design,
  Miami 057434
University of Art and Design Helsinki,
  Helsinki 055345
University of Brighton Gallery,
  Brighton 043742
University of Bristol Theatre Collection,
  Bristol 043755
University of Calgary Museum of
  Zoology, Calgary 005535
University of California Press,
  Berkeley 138303
University of Chicago Press,
  Chicago 138323
University of Colorado Museum,
  Boulder 047144
University of Dundee Exhibitions,
  Copper Gallery, Matthew Gallery,
  Lower Foyer and Lamb Gallery,
  Dundee 044169
University of Essex Exhibition Gallery,
  Colchester 043993
University of Exeter Press, Exeter 138160
University of Fort Hare Museum,
  Alcie 038644
University of Hawaii Art Gallery,
  Honolulu 049546
University of Illinois at Urbana-
  Champaign. School of Art and
  Design. Newsletter, Champaign 139515
University of Iowa Museum of Art,
  Iowa City 049712
University of Iowa Museum of Natural
  History, Iowa City 049713
University of Johannesburg Art Gallery,
  Auckland Park 038648
University of Kansas Ryther Printing
  Museum, Lawrence 050188
University of Kentucky Art Museum,
  Lexington 050252
University of Lethbridge Art Gallery,
  Lethbridge 006188
University of Liverpool Art Gallery, Art
  and Heritage Collections, Victoria
  Gallery and Museum, Liverpool 044866
University of Manchester Medical
  School Museum, Manchester . 045218
University of Mandalay Collections,
  Bawdigon Archaeological Museum,
  Mandalay 031819
University of Michigan Press, Ann
  Arbor 138298
University of Nebraska Art Gallery at
  Omaha, Omaha 051769
University of Nebraska Press,
  Lincoln 138346
University of Nebraska State Museum,
  Lincoln 050302
University of New Brunswick Art
  Centre, Fredericton 005849
University of New Mexico Art Museum.
  Bulletin, Albuquerque 139516
University of North Dakota Zoology

Museum, Grand Forks 049178
University of North Texas Art Gallery
  and Cora Stafford Gallery, School of
  Visual Arts, Denton 048249
University of Northern Iowa Museum
  Museums & Collections, Cedar
  Falls 047478
University of Northern Philippines
  Museum, Vigan 034615
University of Nueva Caceres Museum,
  Naga 034558
University of Pennsylvania Museum
  of Archaeology and Anthropology,
  Philadelphia 052100
University of Pretoria Art Collection,
  Tshwane 038862
University of Queensland Art Museum,
  Brisbane 000891
University of Rhode Island Fine Arts
  Center Galleries, University of Rhode
  Island, Kingston 049988
University of San Carlos Anthropology
  Museum, Cebu 034468
University of San Carlos Biological
  Museum, Cebu 034469
University of Santo Tomas Museum of
  Arts and Sciences, Manila 034549
University of South Australia Art
  Museum, Adelaide 000761
University of South Carolina
  at Spartanburg Art Gallery,
  Spartanburg 053433
University of South Carolina Beaufort
  Art Gallery, Beaufort 046849
University of Southern Philippines
  Museum, Cebu 034470
University of Southern Queensland
  Art Gallery & Art Collection,
  Toowoomba 001586
University of Swaziland Library
  Museums Collection, Kwaluseni 040507
University of Sydney Art Collection,
  Sydney 001554
University of the Arts London,
  Kingston upon Thames 056544
University of the Arts London,
  London 056571
University of the Incarnate Word, San
  Antonio 057845
University of the Visual and
  Performing Arts, Colombo 056390
University of Toronto Art Centre,
  Toronto 007225
University of Toronto Press,
  Toronto 136690
University of Virginia Art Museum,
  Charlottesville 047585
University of Wales Press, Cardiff 138149
University of Washington Press,
  Seattle 138472
University of Waterloo Art Gallery,
  Waterloo 007367
University of West Florida Art Gallery,
  Pensacola 051984
University of Winnipeg Geography
  Museum, Winnipeg 007472
University of Wisconsin Zoological
  Museum, Madison 050600
University of Wyoming Anthropology
  Museum, Laramie 050144
University of Wyoming Art Museum,
  Laramie 050145
University of Zambia Library Museums
  Collection, Lusaka 054841
University Place Art Center,
  Lincoln 050303
University Saint Thomas Art Collection
  and Exhibitions, Saint Paul . . . 052853
University Used and Rare Books,
  Seattle 145605
Universum, Moskva 085212
Universum Antiquitäten, Wien . . 063091,
                                     063092
Universum Bremen, Bremen . . . 017602

Universum, Museo de las Ciencias de
  la UNAM, México 031271
Univerzitet Umetnosti u Beogradu,
  Beograd 056324
Univerzsitatea Naţională de Arte
  Bucureşti, Bucureşti 056260
Unlimited Contemporary Art,
  Athinai 109050
UNLV Barrick Museum, Las
  Vegas 050172
Uno A, Bogotá 102371
Uno Auksjon, Sola 126389
Unrau, Sabine, Kiel 076792
Uns lütt Museum, Dargun 017816
Unsodo, Kyoto 082048, 137771
Unsodo, Tokyo 082214
UNSW Art Collection, Sydney . . 001555
Unsworth, Robin, Olney 090917
Unsworths, London . . . 144705, 144706
Untamalan Kotiseutumuseo,
  Laitila 011086
Unterer, Heidemarie, Salzburg . . 127862
Unterfränkisches Grenzmuseum, Bad
  Königshofen 016799
Unterfränkisches Verkehrsmuseum,
  Gemünden am Main 018634
Unterlöhner, Erwin, Kuchen . . . 076983
Unterrabnitzer Malergruppe,
  Unterrabnitz 058315
Untersbergmuseum Fürstenbrunn,
  Grödig 001997
Unterthurner, Erika, Merano . . . 110215
Unterwarter Heimathaus,
  Unterwart 002857
UnterWasserReich, Naturpark Schrems,
  Schrems 002730
Untitled, London 139227
Untitled Art Publishers,
  Amsterdam 082880, 137844
Untitled Fine Arts Service,
  Albuquerque 119517
Untitled Gallery, Oklahoma City . 123297
Untitled Press, Delrey Beach . . 138330
Untucht, Bernd, Magdeburg . . . 130262
Untzi Museoa – Museo Naval,
  Donostia-San Sebastián 039356
Unverzagt, Siegfried, Köln 142023
Uovo, Torino 138881
Uozu-shi Rekishi no Mingei
  Hakubutsukan, Uozo 029679
Up & Co Publications, London . . 138255
Up & Co Publications, New York 138430
Up Against the Wall, Denver . . . 120785
Up Art, Galerie & Edition, Neustadt an
  der Weinstraße 108262
Up Country, Tunbridge Wells . . 091666
Up Six, Saint Paul 096866
Up the Creek Antiques, Buffalo . 092566
Upchurch, Kenneth, Ashbourne . 088304
John Updike, Edinburgh 144403
Upernavik Museum, Upernavik . . 023342
Upham Mansion, Marshfield . . . 050720
Uphill Galleries, New Orleans . . 122252
Upholstery by Design, Portland . 136162
Upholstery of the Stars, Miami . 135809
Upholstery Restored, Heidelberg
  Heights 127615
Upholstery Shop, Calgary 128159
Upholstery Shoppe, Kansas City 135682
Upholstery Special, Saint Paul . . 136227
Upholstery Unlimited, New
  Orleans 135866
Upholstery Workroom, San Jose 136360
UPIDIV Galerija, Novi Sad 038220
Upīte Kultūrvēstures Muzejs,
  Škilbēnu 030233
UPK Universal Prints Kunstverlag,
  Hamburg 137256
Upminster Mill, Upminster 046020
Upminster Tithe Barn Museum of
  Nostalgia, Upminster 046021
Upp, Albin, Oslo 113248
Uppark Exhibition, Petersfield . . 045496
Upper Bay Gallery, Houston . . . 121146

Upper Canada Village,
Morrisburg 006431
Upper Canadian Antique Mart,
Ottawa 064516
Upper Court Manor Antiques,
Dublin 079594
Upper Court Manor Antiques,
Freshford 079623
Upper East Antiques, Portland .. 096486
Upper East Region Museum,
Bolgatanga 022926
Upper Level Antique Mall, Seattle 097516
Upper Midwest Conservation
Association, Minneapolis 060789
Upper Musselshell Museum,
Harlowtown 049362
Upper Paxton Township Museum,
Millersburg 050898
Upper Room Chapel Museum,
Nashville 051189
Upper Rust Antiques, New York . 095804
Upper Snake River Valley Historical
Museum, Rexburg 052521
Upper Waitaki Pioneer Museum and
Art Gallery, Kurow 033237
Upper Wharfedale Folk Museum,
Grassington 044466
Upper Yarra Valley Museum, Yarra
Junction 001670
Upperbridge Antiques, Holmfirth 089543
Upperlands Eco-Museum, Derry 044092
Uppertown Firefighters Museum,
Astoria 046586
Upplands Auktioner, Uppsala .. 126693
Upplandsmuseet, Uppsala 041118
Uppsala Auktionskammare,
Uppsala 126694
Uppsala Konst- och Ramaffär,
Uppsala 115975
Uppsala Konstmuseum, Uppsala 041119
Uppsala Medicinhistoriska Museum,
Uppsala 041120
Uppsala Universitet Myntkabinett,
Uppsala 041121
Uppsala Universitets Konstsamlingar,
Uppsala 041122
Uprka, J., Praha 102726
Upside-Down Turtle, Charlotte .. 092638
Upstairs & Downstairs Antiques,
Ripley, Derbyshire 091111
Upstairs Antiques, Auckland ... 083834
Upstairs Art, Christchurch ... 112865
Upstairs Art Gallery, Los Alamos 050403
Upstairs Berlin, Berlin 106450
Upstairs Books, Engadine 139725
Upstairs Downstairs, Nairne .. 061937
Upstairs Downstairs, Oamaru .. 084043
Upstairs Downstairs, Willemstadt 102531
Upstairs Downstairs Antiques,
Boston 092502
Upstairs Gallery, Lismore 099234
Upstairs Gallery, Miami 121869
Upstairs Gallery of Fine Arts,
Waikato 113045
Upstairs, Downstairs, Devizes .. 089058,
134322
Upstairs, Downstairs, Honiton .. 089575
Upsters Gallery, Eastleigh 117710
Upstream Gallery, Amsterdam .. 112297
Upton House and Gardens,
Banbury 043485
Uptown Archeology, Cleveland . 092978
Uptown Gallery, Fresno 120912
Uptown Gallery, Las Vegas ... 121386
Uptown Gallery, New York 123156
Uptown Gallery, Richmond ... 123749
Uptown Images, Saint Paul ... 123937
Uptown Restoration, New
Orleans 135867
Uptowner Antiques, New Orleans 095173,
095174
Ur- und Frühgeschichtliche Sammlung,
Friedrich-Alexander-Universität
Erlangen-Nürnberg, Erlangen . 018266

Ur- und frühgeschichtliche
Sammlungen, Landesmuseum
Joanneum, Graz 001988
Ur-Donautal-Museum, Torbogenhaus,
Wellheim 022554
Uraeus, Paris 071965, 105286
Urahoro-cho Kyodo Hakubutsukan,
Urahoro 029680
Urajärven Kartanomuseo, Urajärvi 011462
Ural Tansykbaev Yodgorlik Muzeyi,
Toshkent 054671
Uralla Gallery, Uralla 099849
Uralla Trading Post, Uralla 062383
Uralskaja Gosudarstvennaja
Architekturno-chudožestvennaja
Akademija, Ekaterinburg 056271
Uralskij Geologičeskij Muzej V.V.
Vachruševa, Ekaterinburg ... 036994
Uran Sanat Galerisi, İstanbul ... 117095
Uranga, Cástor, Donostia-San
Sebastián 085987, 143744
Uránia, Budapest 142574
Urania-Galerie im Palais
Trauttmansdorff, Graz 100024
Urart Sanat Galerisi, İstanbul ... 117096
Urawa-kuritsu Bijutsukan,
Saitama 029307
Urban, Auckland 112815
Urban, Baltimore 119777
Urban, Washington 124970
Urban Archaeology, Eastwood .. 061434
Urban Art, Los Angeles 121613
Urban Art, Seoul 111729
Urban Art Retreat, Chicago .. 120288
Urban Artifacts, Chicago 092811
Urban Artifacts, Philadelphia ... 096169
Urban Artifacts, Richmond .. 096584
Urban Artifacts Gallery, Houston 121147
Urban Chateau, San Francisco . 097343
Urban Cottage, Houston 093822
Urban Gallery, Marseille 104206
Urban Gallery, Miami 121870
Urban Institute for Contemporary Arts,
Grand Rapids 049194
The Urban Jungle, Singapore .. 085526
The Urban Settler, Vancouver .. 064894
Urban Shaman, Winnipeg 047473
Urban Treasures, Austin 092273
Urban, Georg, München 077633
Urban, Klaus-Peter, Fellbach ... 129835
Urban, Patrick, Illkirch-
Graffenstaden 068868
Urban, Zdeněk, Olomouc ... 065434,
102627
Urbana, Barcelona ... 085849, 085850
UrbanArts Institute, Boston ... 047136
Urbaneck, Wolfgang, Porta
Westfalica 130527
Urbanek, Dieter, Paderborn .. 078010
Urbanek, Dr. phil. Regina, Brühl . 129642
Urbani, Claude, Antibes-Juan-les-
Pins 103261
Urbaniak, Grzegorz, Poznań .. 084654
Urbaniak, Piotr, Poznań 084655
Urbano, Ourense 086325
Urbano Plaza, Buenos Aires .. 127539
Urbauer, Hans Joachim,
Siegsdorf 078467
Urbauer, Hans Joachim,
Vachendorf 078777
Urben, Thomas, Urnäsch ... 116681
Urbes Verlag, Gräfelfing 137231
Urbieta 66, Donostia-San
Sebastián 085988
Urbina, Vitoria-Gasteiz 086580
Urbs et Orbis, Rheinbach ... 142331
Urdang, Bertha, Jerusalem ... 109721
Urdang, Beth, Boston 119928
Urdangarin Unzueta, Aitor, Vitoria-
Gasteiz 115546
Ure Museum of Greek Archaeology,
Reading 045593
Urgeschichtliche Sammlung,
Engerwitzdorf 001868

Urgeschichtliches Museum,
Mauer 020318
Urgeschichtliches Museum,
Peggau 002491
Urgeschichtliches Museum und
Galerie Vierzigtausend Jahre Kunst,
Blaubeuren 017416
Urgeschichtssammlung und
Heimatmuseum, Jedenspeigen 002145
Urho Kekkonen Museo Tamminiemi,
Helsinki 010879
Uri and Rami Nechushtan, Kibbutz
Ashdot Yaakov 025122
Uribe, Daniel, Valenciennes 073949
Urios College Museum, Butuan . 034447
Urioste, Marta N. de, La Paz ... 063864
Urjalan Museo, Urjala 011463
Urlass, Rüdiger, Frankfurt am
Main 075829, 107037
Urlev Antik, Horsens 065800
Urli-Vernenghi, Paris 071966
Urmensch-Museum, Steinheim an der
Murr 021989
Uroomieh Museum, Uroomieh .. 024791
Urquhart Castle, Drumnadrochit 044144
Urquijo, Guillermo, Bilbao 085893
Urrutia Cutiño, Marta, Badalona 114843
Ursini, Ennio, Trieste 081809
Ursino, Carnely, Fortaleza 100962
Urskog-Hølandsbanen,
Sørumsand 033960
Urso, Josiane, Paris 071967
Ursula Blickle Stiftung, Kraichtal 019778
Ursuline Hallway Gallery, Southwest
School of Art and Craft, San
Antonio 052952
Ursus, New York 145376
Ursus, Penzance 090969, 134932
Urumchi City Museum, Urumchi 008346
Urwelt-Museum Hauff,
Holzmaden 019327
Urwelt-Museum Oberfranken,
Bayreuth 017002
Urweltmuseum, Pirna 021141
Urweltmuseum Aalen, Aalen ... 016452
Urzeitmuseum Nußdorf ob der Traisen,
Traismauer 002835
US CoinNection, Jacksonville .. 094048
US Coins and Jewelry, Houston 093823
US Patent and Trademark Office –
Museum, Arlington 046541
US Rare Coin, Austin 092274
US1 Antiques, Chicago 092812
USA 916, Sacramento 123787
L'Usage du Monde, Saint-Valéry-sur-
Somme 105633
Uşak Arkeoloji Müzesi, Uşak .. 043078
L'Usato della Contrada, Brescia . 079941
Uschkurat, U., Bad Oldesloe .. 141433
USed, New York 095805
Used Stuff, Los Angeles 094491
Usedomer Kaleidoskop, Usedom 078772
Uselding, S., Bruxelles 063566
Usha Mirchandani, Mumbai ... 109481
Usher Gallery, Lincoln 044836
Usher & Son, James, Lincoln .. 089835
Usher, Richard, Cuckfield ... 089021,
134311
l'Usine, Charleroi 063608
Usk Rural Life Museum, Usk ... 046023
USM Art Gallery, Gorham ... 049164
Usman Jusupov Yodgorlik Muzeyi,
Yangiyo'l 054674
Uspenskij Sobor, Gosudarstvennyj
muzej-zapovednik Moskovskij Kreml,
Moskva 037504
USS Alabama Battleship Memorial
Park, Mobile 050981
USS Arizona Memorial, Honolulu 049547
USS Bowfin Submarine Museum,
Honolulu 049548
USS Constellation Museum,
Baltimore 046771
USS Constitution Museum,

Boston 047137
USS Lexington Museum on the Bay,
Corpus Christi 048006
USS Radford National Naval Museum,
Newcomerstown 051522
Uss, Zdzisław, Gdańsk 132820
Ussishkin House, Kibbutz Dan .. 025125
Ust-Ilimskaja Kartinnaja Galereja, Ust-
Ilimsk 038006
Ústav dějin umění AV ČR, Praha 055299
Ústecká Bus Historické Muzeum,
Chlumec u Ústí nad Labem ... 009518
Usual and Unusual Antiques,
Milwaukee 094885
usus, Uta Schneider & Ulrike Stoltz,
Offenbach am Main 137456
Uta, Warszawa 113907
Utah Artist Hands, Salt Lake City 123995
Utah Artists, Salt Lake City 060790
Utah Book and Magazine, Salt Lake
City 145508
Utah Field House of Natural History
State Park, Vernal 053963
Utah Museum of Fine Arts, University
of Utah, Salt Lake City 052918
Utah Museum of Natural History, Salt
Lake City 052919
Utah Museums Association, Salt Lake
City 060791
Utah Potters, Salt Lake City ... 123996
Utah State Historical Society, Salt
Lake City 052920
Ute Indian Museum, Montrose . 051041
Utenio Kraštotyros Muziejus,
Utena 030499
Uter, Peter, Rüdersdorf bei Berlin 130611
Utermann, Wilfried, Dortmund .. 106709
Uthuset Antik Kuriosa, Tranås .. 087100
Utica Museum, Hobson 049497
L'Utile e il Dilettevole, Milano .. 080754
Utilidades Raclima, Lisboa ... 084987
Útisavnið í Hoyvík, Hoyvík 010732
Utopia, Bruxelles 140207
Utopia, Haarlem 083357
Utopia Antiques, Low Newton .. 090638
Utopia Art Sydney, Waterloo ... 062414,
099868
Utopia Fine Art Gallery, Alice
Springs 098641
Utopia Retro Modern, Oslo ... 084339
Utrecht, Pittsburgh 123583
Utropet, Falköping 126580
Utställningssalongen, Stockholm 115935
Utstein Kloster, Mosterøy 033790
Utsukushigahara Kogen Bijutsukan,
Takeshi 029463
Utsunomiya Bijutsukan,
Utsunomiya 029686
Uttar Pradesh State Museum,
Lucknow 024387
Uttenrodt, Michael, Benningen am
Neckar 129451
Utterly Art Exhibition Space,
Singapore 114630
Uttoxeter Heritage Centre,
Uttoxeter 046024
UTU Artworks, Singapore ... 114631
Utvandrarnas Hus, Växjö 041149
Utz, Brigitte, Dresden 106748
Utzt, Inge, Stuttgart 142406
Uudenkaupungin Kulttuurihistoriallinen
Museo, Uusikaupunki 011466
Uummannap Katersugaasivia,
Uummannaq 023343
UVAN Historical Museum,
Winnipeg 007474
Uwajima-shiritsu Date Hakubutsukan,
Uwajima 029687
Uwi-Painting, Siegen 108621
Uxbridge – Howick's Creative Centre,
Auckland 033159
Uxbridge Historical Centre Museum,
Uxbridge 007263
Uyttenhove, L., Oostende 063824

Uzunov, Dečko, Sofia 101176
Užupio Galerija, Vilnius 111812
Uzzell, Phillip, Richmond 136180
V & A Publishing, London 138256
V & R Fine Art, Tampa 124678
V & R unipress, Göttingen 137226
V Dinastia, Madrid 086277
V Podloubí, Mladá Boleslav ... 102621
V. & V., 's-Hertogenbosch 083397
V. Gross Doll Museum, Warren . 007356
V. Samadova Exhibition Center,
  Baki 003182
V.A.M.P, Pierrefitte-sur-Seine ... 072157
V.D.A., Paris 071968
V.O.C. Antiques, Woodhall Spa . 091923
V.V.95, Sevilla 086478
V+A Museum of Childhood, Bethnal
  Green, London 045120
V2 Galerie, Rotterdam 112665
Vaajakosken Työläiskotimuseo,
  Jyväskylä 010944
Vaal Teknorama Cultural Museum,
  Vereeniging 038873
Vaas, Alois, Ellwangen 075601
Vaasan Auto- ja Moottorimuseo,
  Vaasa 011473
Vaasan Merimuseo, Vaasa 011474
Vaasen, R., Jülich 107487
Vacaro Venuto, Valros 073959
Vacaville Museum, Vacaville ... 053919
Vaccaretti, Laurent, Castellane . 067705
Vaccari, Paolo, Firenze 131276
Vaccarino, San Mauro Torinese . 081615
Vaccher, Dr. Daniela, Roma 132067
Vache, Josian, Boisseron 067275
Vachel Lindsay Home, Springfield 053457
Vachott Sándor Városi Könyvtár,
  Gyöngyös 023654
Vachtanov, Moskva 114315
Vaclovo ir Elenos Venskų Etnografinė
  Sodyba, Tytuvėnai 030496
Vacquet-Journiac, Charlette,
  Angers 066616
Váczy Péter Gyűjtemény, Városi
  Muvészeti Múzeum, Győr ... 023667
Vadala, C., Frankfurt am Main . 107038
Vadertje Tijd, Leiden 083449
Vadhera, Delhi 109348
Vadillo Vadillo, Blas Jose, Sevilla 133441
Vadrot, Jean-Louis, Pont-Authou 072242
Vadsbo Museum, Mariestad ... 040844
Vadsø Museum – Ruija Kvenmuseum,
  Vadsø 034062
Vadstena Antik och Kuriosa,
  Vadstena 087133
Vadstena Slott, Vadstena 041127
Vadstena Stadsmuseum,
  Vadstena 041128
Vadstrup, Wibeke, Slagelse ... 066134,
                              102980
Vääksyn Vesimylly- ja
  Piensähkölaitosmuseo, Vääksy 011476
Väderstad Auktionshall, Mjölby . 126643
Vähänkyrön Museo, Vähäkyrö . . 011477
Väike-Männiku Talumuuseum,
  Kõpu 010629
Väinontalo – Järviseudun Museo,
  Vasikka-Aho 011502
Välkesalmi, Kimmo, Helsinki ... 140721
Vämöparken, Karlskrona 040742
Vänermuseet, Lidköping 040781
Vänersborg Konstgille,
  Vänersborg 059915, 115978
Vänersborgs Museum,
  Vänersborg 041132
Vännäs Motormuseum, Vännäs . 041133
Världskulturmuseet, Göteborg .. 040663
Værløse Museum, Værløse 010348
Värmlands Museum, Karlstad .. 040742
Värtsilän Myllymuseo, Värtsilä .. 011478
Väsby Kungsgård, Sala 040934
Västerås Flyg Museum, Västerås 041137
Västerås Könstnärsförening,

Västerås 059916, 115984
Västerås Konstmuseum, Västerås 041138
Västerås Skolmuseum, Västerås 041139
Västerbottens Museum, Umeå .. 041106
Västergötlands Museum, Västarvet,
  Skara 040952
Västerviks Auktionsbyrå,
  Västervik 126696
Västerviks Museum, Västervik . 041143
Västkustauktioner,
  Hunnebostrand 126605
Västmanlands Läns Museum,
  Västerås 041140
Västra Auktionsverket, Jönköping 126607
Vättehult – Pälle Nävers Diktarstuga,
  Vetlanda 041163
Vätterbygdens Auktionskammare,
  Jönköping 126608
Vävboden, Mönsterås 086871
Växjö Auktionskammare, Växjö . 126697
Växjö Konsthall, Växjö 041150
Vafias, Nineta, Kifissia 109073
Våga Publishers, Vilnius 137807
Våga Se, Stockholm 115936
Vagabond Books, Los Angeles . 145237,
                              145238
Vaganay, Luc, Lyon 128731
Il Vaglio, Bari 109777
Vagn Christensen, Kolding 140623
Vagnshistoriska Museet, Fristad 040632
Vagonårskål Muzeum, Studénka 009897
Vágs Bygdasavn, Vágur 010745
Vagt, Ingeborg, Berlin 106451
Vagues à l'Ame, Sucy-en-Brie .. 073605
Vai, Denver 093348
Vaiani, Viviano, Livorno 080337
Vaijnorský L'udový Dom – Etnografická
  Expozícia, Mestské Múzeum v
  Bratislave, Bratislava 038330
Vaikų Literatūros Muziejus, Maironio
  Lietuvių Literatūros Muziejus,
  Kaunas 030376
Vail Galleries, Solihull 119114
Vaila Fine Art, Shetland 119092
Vailati, Flavio, Seregno 081623
Vaile Mansion – Dewitt Museum,
  Independence 049671
Vaillant, Gérard, Le Breuil 069228
Vaillant, Michel, Avignon 103316
Vaisy, Patricia, Messery 070081
Vaitai, Vilnius 111814
Vajda János Emlékház, Szent István
  Király Múzeum, Vál 024031
Vajda Lajos Emlékmúzeum, Ferenczy
  Múzeum, Szentendre 023963
Vaje, Oliver, Hannover 076398
Vajra, Paris 071969
Vakanas, Lucien, Sault (Vaucluse) 073429
Vakgroep Kunst-, Muziek- en
  Theaterwetenschappen, Universiteit
  Gent, Fakulteit Letteren en
  Wijsbegeerte, Gent 055023
Vakıf İnşaat ve Sanat Eserleri Müzesi,
  İstanbul 042965
Vakil & Sons, Mumbai 137600
Vakko Sanat Galerisi, İstanbul .. 117097,
                              117098
Vakoilumuseo, Tampere 011413
Val d'Art, Valbonne 105797
Val-d'Oise Documents & Collections,
  Deuil-la-Barre 068292
Val House, Bucureşti 114156
val i 30, Galeria de Arte,
  Valencia 115520
Valaamskij Naučno-issledovatelskij
  Cerkovno-archeologičeskij i Prirodnyj
  Muzej-zapovednik, Sortavala . 037884
Valabrega, Vittorio, Torino 081765
Valade, François, Gençay 068646
Valade, François, Lussac-les-
  Châteaux 069669
Valahia, Bucureşti 085111
Valansi, Claudio, Rio de Janeiro 101073

Valanti, San José 102518
Valantin, Hugues, Dordives 068387
Valaria, Detroit 120818
Valašské Muzeum v Přírodě, Rožnov
  pod Radhoštěm 009871
Valat, Eric, Erquy 068465
Valberg, Ottawa 101482
Valbruna, Cattolica 109955
Valbusa, Verona 081945
Valcarenghi, Dario, Milano 080755
Valdajskij Kraevedčeskij Muzej,
  Valdaj 038008
Valdemārļis Mežu Muzejs,
  Valdemārpils 030244
Valderrama, Mia, New York 136020
Valdés Costales, José Manuel,
  Oviedo 143788
Valdez Museum, Valdez 053925
Valdivia Museo, Valdivia 010507
Valdoria, Vallauris 105812
Valdosta State University Fine Arts
  Gallery, Valdosta 053927
Valdres Folkemuseum, Fagernes 033592
Vale, Chicago 120289
Vale and Downland Museum,
  Wantage 046050
Vale Antiques, Easingwold 089129
Vale Antiques, London 090555
Vale Coins, London 090556
Vale do Paraiba Antiguidades, São
  Paulo 064213
Valeda, Felix Hector, Cancún ... 132361
Valemount and Area Museum,
  Valemount 007269
Valence House Museum and Art
  Gallery, Dagenham 044060
Valenne, Luxembourg 082373
Valens Log Cabin, Cambridge .. 005538
Valensi, Gilbert, Belgentier ... 067072
Valensija, Sankt-Peterburg 114365
Valent, Modena 080824
Valente Arte Contemporanea, Finale
  Ligure 110024
Valentia Island Heritage Centre,
  Knightstown 024949
Valentien, Dr. Freerk C., Stuttgart 108735
Valentijn, Den Haag 083142
Valentijn, Antonius, Berlin 074941,
                              129575
Valentin, Carqueiranne 067698
Valentin, Minden 077375
Valentin Antiguedades, Madrid . 086278
Valentin-Karlstadt-Musäum,
  München 020582
Valentin, Etienne, Cluny 068049
Valentin, Françoise, Serverette . 073512
Valentin, Gilbert, Vallauris ... 105813
Valentin, Jean-Pierre, Montréal . 101457
Valentin, Martin, Geldern 075964
Valentina, Manama 100453
Valentina, Winterthur 087913
Valentinarte, Bellagio 109787
Valentine Antiquités, Toulouse .. 073808
Valentine Richmond History Center,
  Richmond 052561
Valentine's Antique Gallery,
  Bendigo 061134
Valentin's Financière, Paris ... 105287
Valentis, Bordeaux 128509
Valentown Museum, Victor Historical
  Society, Victor 053980
Valenzuela & Klenner, Bogotá .. 102372
Våler Torvdriftsmuseum, Våler i
  Solør 034064
Valéra, Tarbes 105702
Valeri, Diego, Padova . 081009, 142777
Valeria Bella Stampe, Milano .. 080756,
                   110478, 142746
Valeria Belvedere, Milano 110479
Valette, Paris 141292
Valeur Collection Télécartes, Nice 070673
Valga Muuseum, Valga 010711
Valhalla Museum, National Maritime
  Museum, Tresco 045997

The Valiant Soldier, Buckfastleigh 043776
La Valigia del Viaggiatore,
  Brescia 079962
Valík, Jozef, Košice 085571
Valiño, Alicia & Jorge,
  Montevideo 092011
Valis, Apostolos, Ludwigsburg . 130250
Valjanov, Osnabrück 077981
Valkama, Jorma-Tapio, Jämsä . 103161
Valkas Novadpētniecības Muzejs,
  Valka 030245
Valks, Jan, Haarlem 083358
Valladares Rascon, Jose Ignacio,
  Salamanca 086417
Vallade, Robert, Cachan 067551
Vallanzasca, Maria Antonietta,
  Padova 110582
Vallard House Fine Art Gallery, Mount
  Tamborine 099407
Vallardi, La Spezia 110160
Vallby Friluftsmuseum, Västerås 041141
Valldeperas Ulloa, Alfonso, San
  Salvador 128405
Del Valle Anticuarios, Toledo ... 086493
Valle de los Caídos, Valle de
  Cuelgamuros 040351
Vallé, Chrstiane, Clermont-
  Ferrand 103610
Vallebona, Silvano, Genova 110141
Valledor Aguerri, Javier, Zaragoza 086608
Vallée Verte, Grandchamp ... 103776
Vallée, Florence, Guingamp ... 068772
Vallee, Yvonne, Martel 069983
Vallejo Naval and Historical Museum,
  Vallejo 053931
Valleriaux, François, Paris 141293
Vallerini, Andrea, Pisa 142788
Valley Antiques, Paeroa 084052, 132735
Valley Antiques, Wrentham ... 091957
Valley Art Center, Clarkston ... 047775
Valley Auctions, Folkestone ... 126898
Valley Coin, Minneapolis 094953
Valley Coin, Gold and Silver Exchange,
  San Jose 097379
Valley Community Historical Society
  Museum, Valley 053932
Valley County Pioneer Museum,
  Glasgow 049103
Valley Forge National Historical Park,
  Valley Forge 053935
Valley House Gallery, Dallas .. 120668
Valley Museum, McBride 006234
Valley Oriental, Las Vegas ... 121387
Valley Originals Art, Bega 098727
Valley Trophies, Phoenix 096244
Valley Upholstering, Louisville . 135768
Valli, Tuula, Tampere 066399
Vallier, Valencia 143809
Valligiano, Roma 081592
Vallin, Örebro 115793
Vallois, Paris 105288
Vallois Sculptures XX Siécle,
  Paris 105290
Vallois, Georges-Philippe & Nathalie,
  Paris 105289
Vallotton & Partenaires,
  Lausanne 087587
Vallotton, Philippe, Lausanne ... 087588
Valls, Rafael, London . 090557, 090558,
                              118616
Valluet, Christine, Paris 105291
Valmar, Miami 121871
Valmar Antiques, Stansted
  Mountfitchet 091415
Valmar Art Gallery, Jerewan ... 098605
Valmay, Hamilton 112908
Valmieras Novadpētniecības Muzejs,
  Valmiera 030248
Valmore, Vicenza 111206
Valoir, Orléans 125670
El Valor del Arte, Madrid 139076
Valorart, Viterbo 111209
Valparadiso, Palermo 081082
Valsts Mākslas Muzejs, Rīga ... 030222

Valsts Mākslas Muzejs Izstazu Zale
  Arsenāls, Rīga 030223
Valstybinis Vilniaus Gaono Žydų
  Muziejus, Vilnius 030545
Valsyra, Saint-Ouen 073217
Valta, Bernhard, Wagersfeld .. 127894
Valtakunnallinen Työväentalomuseo,
  Riihimäki 011318
Valtion Taidemuseo, Helsinki .. 010880
Valu Oria Galeria de Artes, São
  Paulo 101116
Value Plus, Columbus 093055
Value, Christian, Christchurch .. 143313
Valued History, Ely 089202
Valverde Lacambra, Fernando,
  Zaragoza 086609, 133517
Valverde Solano, Grace, San
  José 102519
Valverde, Décio, São Paulo 064214
Vám- és Pénzügyőr Múzeum,
  Budapest 023566
Vamiali, Athinai 109051
Vampulan Museo, Vampula 011488
Ván, Ha Noi 125105
Ván, Ho Chi Minh City 125151
Van Aacken, Rudolf, Oberhausen 077865,
                                130477
Van Aalst, A., Breda .. 083011, 132459
Het Van Abbemuseum,
  Bonheiden 003377
Van Abbemuseum, Eindhoven .. 032293
Van Acker, Akko, Paris 071970
Van Allemeersch, Knokke-Heist . 063711
Van Almsick, Georg, Gronau,
  Westfalen 076099, 107169
Van Alsenoy, Antwerpen 125298
Van Andel Museum Center, Public
  Museum of Grand Rapids, Grand
  Rapids 049195
Van Art Antiguidades, Rio de
  Janeiro 064133
Van Assche, S., Gent 063665
Van Balberghe, Emile, Bruxelles 140208
Van Banning, Herman,
  Amsterdam 082881
Van Basten Batenburg, Nazir,
  Maastricht 083484
Van Beek, G.G., Aalsmeer 132376
Van Beers, Jan, London 090559
Van Bekhoven, Anton W.,
  Kampen 137861, 143178
Van Benthem-Doppegieter, A.M.E.,
  Bodegraven 132443
Van Berchem, J.-S., Bruxelles .. 140209
Van Beuzekom, Cees, Gouda .. 132508
Van Blarcom, Kenneth W., Natick 127331
Van Borkulo, Loes, Amsterdam . 082882
Van Buren County Historical Society
  Museum, Keosauqua 049939
Van Bussel, M ., Amsterdam 082883
Van Camp-Magermans, Jenny,
  Antwerpen 100514
Van Campen & Rochtus,
  Antwerpen 100515
Van Canneyt, Ignace, Kortrijk .. 063729
Van Catwijck, Leiden 083450
Van Cauter, Johan, Sint-Niklaas 128089
Van Cleemput & Schietekat, Sint-
  Niklaas 063836
Van Coevorden, Doornspijk 143130
Van Cortlandt House Museum,
  Bronx 047237
Van Daele, R., Antwerpen 063243
Van Dalen, Jan, Rotterdam 083596
Van Dam, Adriaan, Utrecht 132652
Van Dam, Drs. Michiel,
  Amsterdam 082884
Van Damme, James, Bruxelles .. 100664
Van de Laar, Maurits, Den Haag 112418
Van de Laar, Michel, 's-
  Hertogenbosch 132537
Van de Loo, Marie-José,
  München 108204, 137427
Van de Loo, P., Veldhoven 132655

Van de Loo, Peter, Eindhoven .. 083234
Van de Minkelis, T., Hardinxveld
  Giessendam 132527
Van de Moosdijk, Eindhoven .. 083235
Van de Nieuwe Dingen, Tilburg . 112686
Van de Plas Ets, Bruxelles 100665
Van de Pol, Arnhem 082948
Van de Poll-Stichting, Zeist 033094
Van de Velde, Denise, Aalst .. 100477
Van de Velde, Lucien, Antwerpen 063244
Van de Velde, Ronny, Berchem . 100529,
                                136625
Van de Ven, Alexander, Den
  Haag 083143
Van de Vijver, Bruxelles 063567
Van de Water, Eindhoven 112469
Van de Weghe, Christophe, New
  York 123157
Van de Wiele, G., Knokke-Heist . 063712
Van de Wiele, Marc, Brugge .. 140162
Van Dekken, B., Rötgesbüttel .. 078267
Van Delden & Cie., Dr.,
  Osnabrück 077982
Van den Berg, Leeuwarden .. 112553
Van den Berg, E., Breda 132460
Van den Berg, Jos, Den Haag .. 112419
Van den Besselaar, Anny,
  Maastricht 112586
Van den Bogaerde, D., Bruxelles 063568
Van den Bogart, J., Roosendaal 083545
Van den Bout, H.A.C., Langerak 132555
Van den Brulle, Kristine,
  Antwerpen 128001
Van den Bulcke, R., Brugge 063290
Van den Burg, Henk, Rotterdam 143225
Van Den Bussche, Dominique,
  Martigues 069987
Van den Elsen & Zoon, Beek en
  Donk 082952
Van den Elshout, J.A.P., Den
  Haag 083144
Van den Ende, M., Den Haag .. 132480
Van den Gaag, J.A., Den Haag . 132481
Van den Hoogen, P., Köln 076915
Van den Sommen, F., Eindhoven 083236
Van der Aa & de Boer,
  Maastricht 112587
Van der Aa, Jan W., Wassenaar 132667
Van der Beek, P.J., Stadskanaal 132636
Van der Ben, Hans, Geldrop .. 143144
Van der Bos, M.W., Haarlem .. 132522
Van der Does, Dorle, Carouge . 087352
Van der Donk, Ineke, Zaltbommel 083705
Van der Donk, Pimm, Amsterdam 112298
Van der Eems, Bolsward 132444
Van der Eerden, B.L., Rotterdam 083597
Van der Eerden, L., Rotterdam . 083598
Van der Elst, Bruxelles 140210
Van der Graaf, Tom, Ottoland .. 132608
Van der Gucht, A., Bruxelles .. 063569
Van der Gulik, A.K.J., Haarlem . 083359
Van der Heijden, Arno,
  Amsterdam 082885, 132418
Van der Heyde, Paopaoe 105888
Van der Kamp, H.J.G.E., Zwolle . 083740
Van der Klift, Hans, Den Burg . 083043
Van der Koelen, Dr. Dorothea,
  Mainz 107954, 107955
Van der Linden, Amersfoort .. 082660,
                                132385
Van der Loo, Hans & Ulrike,
  Hettenleidelheim 076520
Van der Marel, A., Heemstede . 143168
Van der Meer, Datchet 138152
Van der Meer, A., Amsterdam . 082886
Van der Mel, Andrea, Amsterdam 112299
Van der Most, Jürgen & Andrea,
  Leutershausen-Wiedersbach . 130240
Van der Pas & Van de Raadt,
  Rotterdam 112666
Van der Peet, C.P.J., Amsterdam 143072
Van der Perre, Francine,
  Bruxelles 140211
Van der Perre, Micheline,

Bruxelles ... 063570, 100666, 140212
Van der Planken, Antwerpen ... 100516
Van der Plas, New York 123158
Van Der Schueren, Jean-Luc,
  Bruxelles 063571
Van der Stappen, W.C.M., Tilburg 083625
Van der Steur, A.G., Haarlem .. 143164
Van der Straeten, Amsterdam .. 112300
Van der Teen, Dordrecht 083197
Van der Veen, Geert, Toronto .. 101715
Van der Vlist, Frederiek & Ingrid,
  Leiden 112562
Van der Voort, Eivissa 115062
Van der Werff, Auke, Amsterdam 082887
Van-der-Werf's Wedgwoodmuseum,
  Zuidwolde, Groningen 033115
Van der Wis, J.C., Groningen .. 083307
Van der Woude, J.M.,
  Amsterdam 132419
Van Deudekom, Johannes,
  Monaco 132370
Van Devlet Güzel Sanatlar Galerisi,
  Van 043079
Van Dieten-d'Eendt, Ferdinand,
  Amsterdam 112301
Van Dijk & Van Schaik, Naarden 132591
Van Dijk, Jaap, Sint
  Michielsgestel 132630
Van Dijken, H.C., Haarlem 083360
Van Dishoeck, Eduard,
  Amsterdam 143073
Van Dongen, Chloë, Troyes .. 105780
Van Doorn, A.J.M., Franeker .. 143141
Van Dorssen, Irene, Yandina .. 099935
Van Dreven, Amsterdam 082888
Van Driel, Rotterdam 083599
Van Durme, O. & F., Gent 063666
Van Duyvendijk Brouwer,
  Amsterdam 082889
Van Dyck, Gijón 115074, 115075
Van Dyke Coins, Detroit 093378
Van Eck, Frank P., Triesen 137802
Van Ee, Richmond, Victoria 099638
Van Eeghen-de Laaf, H.A.C.,
  Haarlem 132523
The Van Egmond Foundation,
  Egmondville 005774
Van Eick, Manfred, Düsseldorf . 129763
Van Eijck, Maastricht 083485
Van Eijck, Rotterdam 112667
Van Elkan, Norbert & Birgit,
  Marburg 077287
Van Elst, E., Amsterdam 082890
Van Erp-Broekhans, Tilburg .. 083626
Van Eyck, Buenos Aires 098583
Van Frankfoort & Beudel,
  Hamburg 076332
Van Galen, San Francisco 136350
Van Ganzenwinkel, G.T.J.,
  Someren 132633
Van Geenen, Delft 083038
Van Gelder, Amsterdam 112302, 137845
Van Gendt, Amsterdam 126274
Van Gennep, Amsterdam 137846
Van Gogh, Amsterdam 132420
Van Gogh, Bitola 111876
Van Gogh Documentatiecentrum,
  Nuenen 032703
Van Gogh Museum, Amsterdam 031996
Van Gold, Alicante 126499
Van Gouygem, Jean-Pierre, Saint-
  Ouen 073218
Van Gybland-Oosterhoff Collection,
  Tshwane 038863
Van Haeften, Johnny, London . 090560,
                                134790
Van Ham Kunstauktionen, Köln . 076916,
                        107695, 126013
Van Harrison, Chicago 120290
Van-Hay, Geoffrey, Merstham . 090751
Van Heekern, Peter, Essen .. 129821
Van Heemst, Vis, Sommeldijk . 132634
Van Hees, L., Saarbrücken .. 108551
Van Hefflin, Kirton-in-Lindsey . 089720

Van Heijningen, Den Haag 112420,
                                112421
Van Heijst, C., Leiderdorp 132563
Van Helden, Wellington 113094
Van Herck, Antwerpen 063245, 128002
Van Herck, Jan, Antwerpen 063246,
                                128003
Van Hoevelaak & Zoon, Zeist .. 083714
Van Hoof, Eindhoven 083237
Van Hoof, Jan, 's-Hertogenbosch 112534
Van Hoorebeke, Guy, Brugge .. 063291
Van Hoorn, Nijmegen 143210
Van Hoorn, H., Lettele 132567
Van Horn, Düsseldorf 106839
Van Horn & Hayward, Houston . 093824
Van Houtte, Charenton-le-Pont . 067840
Van Houtte, Jean-Louis, Maisons-
  Alfort 069813
Van Hove, Bruxelles 063572
Van Hove, Monique, Quimper .. 072351
Van Iperen, G.H.A., Oss 083530
Van Kaam, R.G., IJmuiden 132549
Van Kammen, A., Leeuwarden . 083421
Van Kampen, K.H.P., Spaarndam
  West 132635
Van Kio, Nara 082077
Van Kleef, Oakland 123268
Van Kranendonk, Den Haag .. 112422
Van Kuik, Paul, Voorschoten .. 143260
Van Laarhoven, A., Dordrecht . 083198
Van Laer, Frank, Antwerpen .. 128004
Van Laere, Antwerpen 100517
Van Laere, R., Hennef 076493
Van Langenhove, Gent 063667, 100706,
                                125337
Van Leest, Utrecht 083665
Van Leeuwen, Amsterdam .. 143074
Van Leeuwen, Woerden 132672
Van Leeuwen, Remco, Eindhoven 083238
Van Leeuwen, S., Den Haag .. 083145
Van Leeuwenhoek, Anthonie,
  Delft 143105
Van Liew, Marcia, Seattle 097517
Van Lijf-Van Leeuwen, N .,
  Enschede 083261
Van Lindenberg, G.J., Zutphen . 083728
Van Lines, Detroit 093379
Van Loock, Claude, Bruxelles .. 136648,
                                140213
Van Loon, J.M., Tilburg 083627, 132643
Van Meeuwen, J.W.T., Den Haag 143122
Van Megchelen, Cock, Eindhoven 083239
Van Menen, L. & J., Brugge ... 063292
Van Meter State Park, Miami .. 050843
Van Middelem, Luc, Oostende . 100809
Van Mourik, W.A., Culemborg . 132466
Van Müzesi, Van 043080
Van Mullem, Jan, Brugge 063293,
                                128012
Van Nie, Frans, Amsterdam .. 082891
Van Nieuwenhuyse, H., Gent .. 128055
Van Noppen, Providence 123714
Van Nostrand-Starkins House,
  Roslyn 052686
Van Oorschot, Nico, Amsterdam 082892
Van Opdorp, M., Hoogerheide . 112541
Van Orsouw, Oss 083531
Van Orsouw, Bob, Zürich 116864
Van Os, Ignas, Arnhem 132430
Van Os, Margot, Hilversum .. 132543
Van Os, Peter, Arnhem 082949, 143087
Van Otterloo, Doorn 143129
Van Overstraeten, J., Bruxelles . 063573
Van Paddenburgh, L., Leiden .. 143188
Van Paridon, Eloy, Sassafras .. 062194
Van Peborgh, Raymonde, Sarlat-la-
  Canéda 073412
Van Peet, C P J, Amsterdam .. 143075
Van Poelgeest, W., Amsterdam . 082893,
                        132421,   143076
Van Remmen, Solingen 078501
Van Riel, Buenos Aires 098584
Van Rijn, Utrecht 083666

Register der Institutionen und Firmen

Van Rijn, Rob, Maastricht
– Vasile, Prof. Radu, Cluj-Napoca

Van Rijn, Rob, Maastricht 083486,
112588
Van Rijswijck Galerie, Antwerpen 100518
Van Rijt, F.J.K., Apeldoorn 143083
Van Rinsum, Irmgard, München 077634
Van Riper-Hopper House, Wayne 054252
Van Rompaey, Robert, Antwerpen 100519
Van Roon, Haarlem 083361
Van Rossem, Adriaan, Nijmegen 132596
Van Rossum & Co., Amsterdam 082894
Van Rossum, Egon, Emmerich am
Rhein 075614
Van Rossum, T.N.R.,
Middenbeemster 132586
Van Roy, R., Antwerpen 128005
Van Ruiten & Zoon, Leiden ... 083451
Van Sabben, Hoorn 126291
Van Schaik, Tshwane 143706
Van Schalkwyk, Kathryn,
Tshwane 085686
Van Schouten, Ruud, Amsterdam 132422
Van Spijk, Pascal, Venlo 137875
Van Staden, Gerrit, Delmenhorst 106668
Van Stekelenburg, Hein & Marion,
Amsterdam 143077
Van Stijn & Van Stijn, Egliseneuve
d'Entraigues 068431, 103711
Van Stockum-Veilingen, Den
Haag 126282
Van 't Hof, A., Haarlem 083362
Van 't Hof, Jeroen, Heemstede . 132530
Van 't Lindenhoutmuseum,
Nijmegen 032692
Van Thienen, C., Bruxelles ... 128041
Van Tilburg Collection, Tshwane 038864
Van Turnhout, Breda 143096
Van Uffelen, Palmerston North .. 112983
Van Valkengoed, Nijkerk 132595
Van Veen, Meppel 143202
Van Vliet, F., Hilversum 083408
Van Vliet, H.C., Amsterdam ... 082895
Van Vugt, J., Eindhoven 083240
Van Vuren, Marcel, Emmerich am
Rhein 106882
Van Vyve, E., Bruxelles 128042
Van Wal, Milwaukee 121954
Van Wassenaar, J.C., Rotterdam 083600
Van Weerde, Amsterdam 082896
Van Welsenes, Bernard, Bergen 112344
Van Wersch, A.F., Heerlen 083379
Van Weydevelt, Philippe, Corné . 068159
Van Weyenbergh, Los Angeles . 121614
Van Wijland, P., Dordrecht 083199
Van Wijngaarden, Amsterdam .. 112303
Van Wingerden, Zevenbergen .. 112727
Van Wisselingh & Co., E.J.,
Haarlem 083363
Van Wüllen, Legden 077045
Van Wyck Homestead Museum,
Fishkill 048736
Van Zaane, Alma, Blaricum 143092
Van Zoetendaal, Amsterdam .. 112304
Van Zomeren, Martin,
Amsterdam 112305
Vanadzor Museum of Fine Art,
Vanadzor 000717
Vanajan Kotiseutumuseo,
Hämeenlinna 010793
Vanatehnikaklubi "Retrom"
Tuletorjemuuseum, Kärevere .. 010623
Vanavara, Tallinn 066288
Vanbeselaere, René, Oudon ... 070827
Vanbienne, Mons 063803
Vanbrugh House Antiques, Stow-on-the-
Wold 091488, 144863
Vancoppenolle, Gent 063668
Vancouver Antique Centre,
Vancouver 064895
Vancouver Art Gallery, Vancouver 007300
Vancouver Arts Centre, Albany .. 000779,
098620
Vancouver Chinese Art Gallery,
Vancouver 101844
Vancouver Holocaust Education Centre,

Jewish Historical Society of British
Columbia, Vancouver 007301
Vancouver Island Military Museum,
Nanaimo 006451
Vancouver Life Drawing Society,
Vancouver 058477
Vancouver Maritime Museum,
Vancouver 007302
Vancouver Museum, Vancouver . 007303
Vancouver Naval Museum, Qualicum
Beach 006723
Vancouver Police Centennial Museum,
Vancouver 007304
Vanczak, François, Nancy 070434
Vandalia State House, Vandalia . 053950
Vandekar, Earle D., New York . 095806
Vandelanoitte, Lille 128688
Vandeleur Antiquarian Books,
Epsom 144413
Vandelft, Mornington 127646
Vandenberghe, Philippe, Crozon 103649
Vandenhaute, Rouen 072622
Vander, Bruxelles 136649
Vander, Jenny, Dublin 079595
Vanderbilt Kennedy Center Arts Gallery,
Nashville 051190
Vanderbilt Mansion National Museum,
Hyde Park 049651
Vanderbilt University Fine Arts Gallery,
Nashville 051191
Vanderbilt University Sarratt Gallery,
Nashville 051192
Vanderbor, Barcelona . 085851, 114972
Vanderhoof Museum, Vanderhoof 007306
Vanderkindere, Bruxelles 125328
Vanderleelie Gallery, Edmonton . 101304
Vandermeersch, Michel, Paris . 071971
Vandermotte Delechaud, Jacques,
Lima 084471
Vanderslyen, Bruxelles 063574
Vanderven & Vanderven Oriental Art,
's-Hertogenbosch 083398
Vandevoorde, Paris 105292
Vandewater, Seattle 145606
Vandiver-Williams, Atlanta 092195
Vane House, London 090561
Vaned Art Gallery, Miami 121872
Vanessa Art Link, Beijing 101995
Vanessa Art Link, Jakarta 109515
Vanessa Vintage Treasures,
Baltimore 092373
Vang Olsen, Mimi, New York ... 095807
Vang Wæhrens, Bjarne,
Tommerup 066169
Vangelli dé Cresci, Francesco,
Roma 110870
Vangeon, Guy, Luisant 069657
Vanghenet, E., Bruxelles 063575
Vanghi, Lucia, Bologna 131002
Vangi – The Sculpture Garden
Museum, Nagaizumi 029067
VanGorden Stedman, R., Saint
Petersburg 127436
Vangsgaard, København 136777, 140618
Vangsgaards, Århus 140557
Vanguard, Long Beach 094230
Vanguard, San Francisco 124422
Vanguard Centennial Museum,
Vanguard 007307
Vanguardia, Bilbao 114998
Vanha-Rantalan Talomuseo,
Pertunmaa 011246
Vanhaa & Kaunista, Helsinki ... 066370
Vanhaecke, D., Bruxelles 063576
Vanhaisäntä, Igor, Helsinki ... 066371
Vanhalst, Rita, De Haan 128044
Vanhat Neidit, Helsinki 066372
Vanhoecke, Pascal, Cachan ... 103479
Vanhoegaerde, Samuel, Knokke-
Heist 100763
Vanholme, Philippe, Lille 128689
Vanhoogstraat, William R., Saint
Louis 123881
Vanhove, Leuven 140238

Vania, Jakarta 109516
Vanic, Wien 100387
Vanier, Tucson 124780
Vanini, Oakland 095918
Vanishing Point Gallery Of Arts,
Benowa 098737
Vankilamuseo, Hämeenlinnan
Kaupungin Historiallinen Museo,
Hämeenlinna 010794
Vanleeuwen, Nieuwerkerk aan den
IJssel 132593
Vann House, Chatsworth 047589
Vann, Bill, Saint Louis 123882
Vannier, Jean-Paul, Gradignan . 068693
Vannini, Brescia 109916
Vanoli, Natalino, Milano 131545
Vanonkelen, L., Bruxelles 063577
Vanot, Michel, Marseille 128786
Vanrell, Palma de Mallorca ... 115370
Vansalon, Indianapolis 121222
Vanseer, C. & Y., Liège 063785
Vanstone, Peter, Los Angeles . 094492
Van't Hul, L.W., Kampen 132551
Vantaan Kaupunginmuseo,
Vantaa 011494
Vantaan Kaupunginmuseo –
Kuvataideasiat, Vantaa 011495
Vantan Design Kenkyujo, Tokyo . 056062
Vantul de Rasarit, Bucureşti ... 143618
Vanuatu Cultural Centre and National
Museum, Port Vila 054676
Vanuxem, Paris 105293
Vanuxem, Roland, La Crèche .. 103886
Vanzo, Saint-Laurent-des-Vignes 072885
Vapentekniska Samlingarna,
Eskilstuna 040610
Vaqué, Diana, Neuhaus 116531
Vaquer Neri, Maria Carmela,
Cagliari 080006
Var Dags Antikt, Kil 086782
Varadero, Porto 143607
Varalan Liikuntamuseo, Tampere 011414
Varallo, Giandomenico, Torino .. 081766
Varando, Silvio, Firenze 080218
Varanger Museum Sør,
Grenselandmuseet, Kirkenes .. 033698
Varart, Firenze 110075
Varas, Teodoro Anton, Barcelona 085852,
085853
Varbanova, V. Ju., Sofia 064238, 064239
Varde Artillerimuseum, Varde Museum,
Varde 010352
Varden Redningsmuseum, Jahrmuseet,
Narbø 033799
Vardømuseene, Vardø 034070
Varekamp, Amsterdam 082897
Varendra Research Museum,
Rajshahi 003232
Varenne Antiquités Decoration,
Paris 071973
Varenne, Alexandre, Paris 071972
Varenne, Daniel, Genève 116343
Varese, Maria Rosa, Torino ... 081767
Várfok Gálèria, Budapest 109182
Varga Mátyás Színháztörténeti
Kiállítása, Móra Ferenc Museum
Szeged, Szeged 023930
Vargas, Lima 113370
Los Vargas, New York 136021
Vargas Museum and Art Gallery,
Mandaluyong 034518
Vargas, Luis, Arequipa 084422
Varia, Liège 140248
Varia, Paarl 085657
Variant, Glasgow 139228
Variété, Nice 070674
Variety, Chicago 145062
Variety Trading, Timaru 126327
Varine-Gincourt, Paris 105294
Varini, Arnaldo, Brescia 131055
Varissuon Antikvariaatti, Turku . 140751
Varistaipaleen Kanavamuseo,
Varistaipale 011496

Varjjat Sámi Musea, Varanger Samiske
Museum, Varangerbotn 034069
Varkauden Museot, Varkaus ... 011499
Varkauden Taidemuseo, Varkaus 011500
Varlet, Claude, Paris 071974
Varley Art Gallery of Markham,
Unionville 007260
Varma Arts Gallery, Hyderabad . 079394
Varma, Ravi, Chennai . 079381, 109294
Varming, Suzanne, Odense 066046
Vármúzeum, Dunaföldvár 023606
Vármúzeum, Pécsvárad 023863
Vármúzeum, Siklós 023880
Vármúzeum, Veszprém 024044
Vármúzeum – Kotár, Móra Ferenc
Museum Szeged, Szeged 023931
Vármúzeum, Corvin János Múzeum,
Gyula 023680
Vármúzeum, Magyar Nemzeti Múzeum,
Esztergom 023633
Vármúzeum, Wosinsky Mór Megyei
Múzeum, Simontornya 023881
Varndell, E.R., Chicago 120291
Varndell, Edward, Chicago 120292
Varner-Hogg Plantation, West
Columbia 054296
Varnier, Paris 105295
Varnum Memorial Armory and Military
Museum, East Greenwich 048429
Városi Galéria, Mezőkövesd ... 023779
Városi Helytörténeti Gyűjtemény,
Szigetszentmiklós 023971
Városi Helytörténeti Múzeum,
Komló 023760
Városi Képtár, Orosháza 023820
Városi Képtár – Deák Gyűjtemény,
Székesfehérvár 023945
Városi Képtár, Corvin János Múzeum,
Gyula 023681
Városi Könyvtár és Múzeum,
Tapolca 023995
Városi Művészeti Múzeum – Városi
Képtár, Győr 023668
Városi Múzeum, Kőszeg 023764
Városi Múzeum, Paks 023827
Városmajor Galéria, Budapest . 109183
Várostörténeti Múzeum, Janus
Pannonius Múzeum, Pécs ... 023858
Varpaisjärven Kotiseutumuseo,
Varpaisjärvi 011501
Várpalotai Bányászattörténeti
Gyűjtemény, Központi Bányászati
Múzeum, Várpalota 024035
Varpo – Viurinès Mokyklos Istorijos
Muzieus, Klaipėda 030394
Varšavka, Moskva 114316
Varteig Bygdemuseum, Varteig . 034073
Vartigian, Harry, Greenfield 145154
Vartroc a'Plus, La Valette-du-Var 069114
Varusschlacht im Osnabrücker
Land, Museum und Park Kalkriese,
Bramsche 017536
Vasa, Bad Mergentheim 074552
Vasa Konst, Uppsala 115976
Vasa Lutheran Church Museum,
Vasa 053952
Vasalopps Museet Mora, Mora . 040853
Vasalopps Museet Sälen,
Transtrand 041085
Vasamuseet, Statens Maritima Museer,
Stockholm 041055
Vasarely Múzeum, Szépművészeti
Múzeum, Budapest 023567
Vasastadens Antikvariat,
Göteborg 143856
Vasciarelli, Cortina d'Ampezzo . 109987
Vasco, João, Lisboa 084988
Vasco, Pietro, Alassio 109746
Vashon Maury Island Heritage
Museum, Vashon 053953
Vasi Múzeumfalu, Savaria Múzeum,
Szombathely 023987
Vasile, Prof. Radu, Cluj-Napoca . 114164,
126470

Vasilevskij Ostrov, Sankt-
Peterburg 114366
Vasistas, Montpellier 104289
Vaskikari, Ruben A., Turku . . . 140752
Vasneto, Lisboa 084989
Vasquez Ramon, Patricia, Nîmes 070707
Vasquez, Giuseppa, Padova . . . 081010
Vassallo Paléologo, Renato, Paris 129157
Vassallo, J., Sliema 082434, 126261
Vassallo, Thomas, Valletta . . . 082443
Vassdragsmuseet Labro,
Lågdalmuseet, Skollenborg . . 033949
Vasse, Pierre, Lille . . . 104019, 128690
Vassilev, Assen, Sofia 101177
Vassy & Jalenques, Clermont-
Ferrand 125559
Vaste Choix de Curiosités,
Draguignan 068405
Vastra Kosh, Mumbai 109482
Vastseliina Gümnaasiumi Muuseum,
Vastseliina 010713
Vastseliina Koduloomuuseum,
Vastseliina 010714
Vasúti Helytörténeti Gyűjtemény,
Mátészalka 023773
Vasúti Múzeum, Paks 023828
Vat Sisaket, Vientiane 030092
Vat Xiengkhouane, Vientiane . . . 030093
Vatalaro, Lorenzo, Milano 080757
Vathi, Tiranë 098533
Vatika, Chennai 109295
Vattani, Rita, Roma 081593
Vaubourgeix, Serge, Maroilles . 069887
Vauclair, Laurence, Paris 071975
Vaucluse House, Vaucluse 001607
Vaudoyer, Dominique, Paris . . . 105296
Vaughan, Christchurch 083902, 132712
Vaughan, London 090562
Vaughan, Stamford 091411
Vaugrante, Jean-Pierre, Granville 068700
Vault Gallery, Murchison 112952
Vaupel, H., Bad Bramstedt 106045
Vaupel, P. & R., Issum 076632
Vauth-Pawlak, Nicole, Datteln . . 075315
Vautrin & Audouy, Paris 071976
Vauvre, Jean Yves de la, Cusset 068260
Vaux, Stavanger 113277
Vavasor Godkin, Auckland . . . 112816
Vaxholms Fästnings Museum,
Vaxholm 041160
Vay Adám Múzeum, Vaja 024028
Vayhinger, Werner, Radolfzell am
Bodensee 108427
Vazio, Cristina, Roma 132068
Vazquez Pousa, Alberto, Sevilla . 086479
VB Antiquités, Paris 071977
VB Gallery, Saint Louis 123883
VBKD Stichting, Dordrecht 112443
VBKW Verband Bildender Künstler und
Künstlerinnen Württemberg e.V.,
Stuttgart 059250
VC Gallery, El Paso 120843
VDA Antiquités, Flayosc 068538
VdB Galerie, Zwolle 112736
Veaux, Thierry, Labastide-
Rouairoux 069126
Veaux, Thierry, L'Isle-sur-la-
Sorgue 069595
Vebr, Dr. Jan, Praha 065539
Vecchia Europa, Philadelphia . . 096170,
136108
Vecchiato, Dante, Forte dei
Marmi 110083
Vecchio, Cannes 103524
Il Vecchio e L'Antico, Milano . . 080758,
131546
Vecchio Lotto, Firenze 080219
Il Vecchio Melo, Torino 142872
Il Vecchio Mobile, Torino 132212
Vecciarelli, Leopoldo, Roma . . . 132069
Vecdaugavas Novadpētniecības un
Zvejniecības Muzejs, Rīga 030224
Vecht, A., Amsterdam 082898
Vectis Auctions, Stockton-on-

Tees 127054
Vector Verve, Madrid 115291
Vedbæk Antik, Vedbæk 066187
Vedder, Ratingen 078139
Vedovi, Bruxelles 100667
Vedovi, Paris 105297
Vee, Hong Kong 102123
Vee & Cranston, Melrose 090742
Veeda, Washington 097864
Veen, Haarlem 083364
Veen Park, Barger Compascuum 032047
Veen, Claudia von, Freising . . . 075898
Veenendaal de la Haye,
Amsterdam 082899
Het Veenkoloniaal Museum,
Veendam 032972
Veenmuseum het Vriezenveense Veld,
Vriezenveen 033011
Veensma, Ute, Bremen 075187
Veenstra, S., Groningen 083308
Vefsn Museum, Mosjøen 033788
De la Vega, León 086067
Vega Byggdemuseum, Vega . . . 034077
Vega, Virgen de la, Valencia . . . 086538
Végállomás, Budapest 109184
Végárshei Byggdetun, Vegårshei . 034078
Végh-kúria, Thorma János Múzeum,
Kiskunhalas 023750
Veghs, Danny, Cleveland 092979
Vegmuseet – Rogaland,
Stavanger 033986
Vegreville Regional Museum,
Vegreville 007310
Vegueta, Las Palmas de Gran
Canaria 115382
Vehbi Koç Vakfı, İstanbul 060015
Vehikelsammlung Eppelheim,
Eppelheim 018219
Vehmaan Kotiseutumuseo ja
Kivityömuseo, Vehmaa 011503
Veiga, João Carlos, Lisboa . . . 084990
Veiga, Maria Conceição, Lisboa . 143596
Veiga, Sancho, Madrid 086279
Veil of Years, Morpeth 061891
La Veille Cassine, Maignelay-
Montigny 069807
Veillerette, Francis, Châteauvillain 067900
Veillon, Cédric, Honfleur 068835
Veirhanen, København 065967
Veit, Klettgau 107566
Veit, Stuttgart 078637
Vejces, Alex, Cali 065256
Vejen Kunstmuseum, Vejen . . . 010353
Vejle Antik, Vejle 066200
Vejle Antikvariat, Vejle 140671
Vejle Kunstmuseum, Vejle 010354
Vejle Museum, Vejle 010355
Vekkilän Kotiseutumuseo,
Tammela 011391
Veksølund, Veksø Sjælland . . . 066202
Velan Centro d'Arte Contemporanea,
Torino 111013
Velasco & Guerra, Santiago de
Chile 064988
Velasco, José María, México . . . 111986
Velasquez, Deventer 083165
Veldhorst, W.A., Rotterdam . . . 083601
Veletržní Palác, Národní galerie v
Praze, Praha 009844
Velez, Jesus Santos, Lisboa . . . 084991
Velfjord Bygdemuseum,
Hommelstø 033666
Velha Lapa Antiguidades, Rio de
Janeiro 064134
Velhage, U., Forst, Baden 075760
Velharias Alfarrabista, Lisboa . . 084993
Velharias Pompeia, Lisboa 084994
Velharias, Imelda, Lisboa 084992
Velho Armazém, Porto Alegre . . 064001
Velhos Tempos, Coimbra 084794, 084795
Velhustro, Coimbra 084796
Velikolukskij Kraevedčeskij Muzej,
Velikie Luki 038012
Velikoustjugskij Gosudarstvennyj

Istoriko-Architekturnyj i
Chudožestvennyj Muzej-Zapovednik,
Velikij Ustjug 038017
Velin, Minneapolis 122049
Velin, Rumen, Varna 101199
Veliuonos Krašto Istorijos Muziejus,
Veliuona 030503
Veliz, J. & A., Barcelona 133170
Vella Canonja, Girona 086032
Vellantiques, Richmond, Surrey . 091102
Velliet, Alexis, Paris 125772
Vellinga, G.J., Rotterdam 083602
Vello, G., Aachen 105951
Velmerig, Hans-Günter,
Düsseldorf 075531
Velniu Muziejus, Nacionalinis
M.K. Čiurlionio Dailės Muziejus,
Kaunas 030377
Velona Antichità & C., Firenze . . 080220
Velryba, Praha 102727
Velsen, Engelberta von, Hamburg 076333
Velten, Elke, Trier 078707
Veltia, Rio de Janeiro 101074
Veluws Museum Nairac,
Barneveld 032049
Veluws Museum van Oudheden –
Stadsmuseum, Harderwijk . . . 032410
Veluws Streekmuseum Hagedoorns
Plaatse, Klederdrachtenmuseum en
Museum Boerderij, Epe 032315
Velvet, Torino 081768
Velvet Cabbage, Miami 094813
Velvet Da Vinci, San Francisco . 124423
Velvet Swing Antiques,
Pittsburgh 096338
Velykosoročynskyj Literaturno-
memorialnyj Muzej Gogolja, Velyki
Soročynci 043316
Vemedetudo, Porto Alegre 064002
Venango Museum of Art, Science and
Industry, Oil City 051722
Venant, Robert, Luzancy 069674
Venard, Jean-Paul, Pérouges . . 072067
Venator & Hanstein, Köln 126014
Venbroek, Deventer 083166
Venbroek, Ela, Deventer 132486
Venclauskių Namai, Šiaulių Aušros
Muziejus, Šiauliai 030474
Venclovų Namai-Muziejus;, Vilniaus
Etninės Veiklos Centras, Vilnius 030546
Vende Tutti, M.L., Madrid 133399
Vendôme, Montréal 064484
Vendôme, Paris 071978, 105298
Vendramin dei Carmini, Venezia 111167
Vendsyssel Historiske Museum,
Hjørring 010123
Vendsyssel Kunstmuseum,
Hjørring 010124
Vendu Notarishuis, Rotterdam . . 126297
Vendue Huis der Notarissen, Den
Haag 126283
Venduehuis Dickhaut, Maastricht 126293
Venegas García, Alvaro, Bogotá . 065011
Venerosi Pesciolini, Lisa, Firenze 131277
Venetia's Antiques, Crows Nest, New
South Wales 061377
Veneto Arte, Conegliano 109975
Veneto Gallery, San Francisco . . 097344
Venetoulis, Genève 087506
Venetsanos Museum of Modern Art,
Piraeus 023226
Veneziano, Giuseppe, Firenze . . 080221
Veneziano, Sara, Roma 081594
Venfield, New York . . . 095808, 095809
Vengerskij Kulturnyj Centr,
Moskva 037505
Venice Design, Venezia 111168
Venier, Cerea 109957
Venis-Art & Antik, Zürich 088014
Venkatappa Art Gallery,
Bangalore 109269
Sri Venkateswara Sculpture Centre,
Tirupati 055840

Venn, Edward, Williton 091849, 135229
Vennel Gallery, Irvine 044653
Vennesla Bygdemuseum, Museene i
Vennesla, Vennesla 034081
Vennini, Casablanca 082585
Venon-Clays, Claude, Troo 073889
Venot & Chavent, Bernard,
Bordeaux 067364
Vent de Chine, Ibos 068862
Le Vent des Cimes, Grenoble . . 103793
Vent d'Est, Bordeaux 103442
Vent d'Images, Caen 067584
Vent Haven Museum, Fort
Mitchell 048845
Ventana, Sacramento 123788
Ventana Gallery, San Antonio . . 124080
El Ventanal, Buenos Aires 139601
Ventanals, L'Hospitalet de
Llobregat 115100
Ventanas y Puertas Antiguas,
Medellín 065312
Les Ventes Jacques, Bruxelles . 125329
Del Ventisette, Renzo, Milano . . 080759
Ventnor Heritage Museum,
Ventnor 046026
Ventnor Rare Books, Ventnor . . 144903
Ventosa, Grégorio, Arlanc-en-
Livradois 066718
Ventre, Francesca, Genova 131343
Vents d'Ouest, Issy-les-
Moulineaux 136822
Ventspils Jūras Zvejniecības Brīvdabas
Muzejs, Ventspils 030252
Ventspils Muzejs, Ventspils 030253
Ventura, Rio de Janeiro 138563
Ventura College Art Galleries,
Ventura 053956
Ventura County Maritime Museum,
Oxnard 051872
Ventura County Museum of History
and Art, Ventura 053957
Ventura, Muriel, Torino 081769
Venture, Tucson 124781
Venturini, Assad Cavalieri, Rio de
Janeiro 064135
Ventzi Art Gallery, Poznań 113751
Vénuat, Olivier, Limoges 069533
Venus, Haarlem 083365
Las Venus, New York 095810
Venus, Palma de Mallorca 115371
Venz, H.-J., Dresden 075433
Venzke, Günter, Berlin 074942
Ver-Art, Milano 080760
Le Ver-Vert, Nevers 140991
Le Ver-Vert, Paris 105299, 141294
ver.di, FG Bildende Kunst, Berlin 059251
Véra, Jean-Pierre, Noirétable . . 070734
Veran, Antwerpen 100520
La Véranda, Châtenay-Malabry . 067908
Veranda, Manama 100454
Veranda, Oakland 095919
La Verandah, Monaco 082532
Verandah, Oxford 118867
Verandah Gallery, Mareeba . . . 099279
Vérand'Artistes, Bougival 103444
Verarte, Roma 110871
Verband Bildender Künstler Sachsen-
Anhalt, Halle, Saale 059252
Verband Bildender Künstler Thüringen
e.V., Erfurt 059253
Verband Bildender Künstler und
Künstlerinnen Württemberg e.V.,
Region Rems-Murr, Remshalden 059254
Verband der Antiquare Österreichs,
Wien 058316
Verband der deutschen Münzhändler
e.V, Berlin 059255
Verband der Grafik-Designer
Österreichs, Wien 058317
Verband der Händler für Historische
Wertpapiere, Wiesbaden 059256
Verband der Kunst- und
Antiquitätenhändler Niedersachsen
e.V., Moringen 059257

Verband der Museen der Schweiz, Zürich ... 059971
Verband der Restauratoren e.V., Bonn ... 059258
Verband der Restauratoren, Ländergruppe Baden-Württemberg, Stuttgart ... 059260
Verband der Restauratoren, Ländergruppe Bayern, Rain ... 059261
Verband der Restauratoren, Ländergruppe Berlin, Potsdam ... 059262
Verband der Restauratoren, Ländergruppe Brandenburg, Torgau ... 059263
Verband der Restauratoren, Ländergruppe Bremen, Bremen ... 059264
Verband der Restauratoren, Ländergruppe Hamburg, Hamburg ... 059265
Verband der Restauratoren, Ländergruppe Hessen, Neu-Anspach ... 059266
Verband der Restauratoren, Ländergruppe Mecklenburg-Vorpommern, Gingst ... 059259
Verband der Restauratoren, Ländergruppe Niedersachsen, Hannover ... 059267
Verband der Restauratoren, Ländergruppe Sachsen, Dresden ... 059268
Verband der Restauratoren, Ländergruppe Sachsen Anhalt, Halle, Saale ... 059269
Verband der Restauratoren, Ländergruppe Schleswig-Holstein, Kiel ... 059270
Verband der Restauratoren, Ländergruppe Thüringen, Eisenach ... 059271
Verband der Sächsischen Kunst- und Antiquitätenhändler e.V., Dresden ... 059272
Verband der Hessischen Antiquitäten- und Kunsthandels, Fulda ... 059273
Verband Deutscher Antiquare e.V., Elbingen ... 059274
Verband Deutscher Kunsthistoriker e.V., Bonn ... 059275
Verband Norddeutscher Kunsthändler e.V., Hamburg ... 059276
Verband Österreichischer Antiquitäten- u. Kunsthändler, Wien ... 058318
Verband Österreichischer Galerien moderner Kunst, Wien ... 058319
Verband österreichischer Kunsthistorikerinnen und Kunsthistoriker, Wien ... 058320
Verband Rheinischer Museen e.V., Solingen ... 059277
Verband Saarländischer Kunst- und Antiquitätenhändler e.V., Saarbrücken ... 059278
Verband Schweizer Berufsnumismatiker, Association of Swiss Professional Numismatists, Nussbaumen ... 059972
Verband Schweizer Galerien, Basel ... 059973
Verband Schweizer Holzbildhauer, Sankt Gallen ... 059974
Verband Schweizerischer Antiquare und Kunsthändler, Zürich ... 059975
Verband Schweizerischer Auktionatoren von Kunst und Kulturgut, Luzern ... 059976
Verband Schweizerischer Bildhauer- und Steinmetzmeister, Association suisse des maîtres sculpteurs sur pierre et maîtres marbriers, Bern ... 059977
Verbeeck, Paul, Antwerpen ... 063247
Verbeelding, Amsterdam ... 137847
De Verbeelding Kunst Landschap Natuur, Zeewolde ... 033092
Verbeke, Bart, Gent ... 128056
Verbièse, P., Cluses ... 068053, 128564
Verbini Russo, Cristina, Modena ... 131592
Verbizier, Solange de, Paris ... 129158
Verbo, Lisboa ... 137921
De Verborgen Tuin, 's-Hertogenbosch ... 112535
Das Verborgene Museum, Dokumentation der Kunst von Frauen e.V., Berlin ... 017313
Verbund Oberösterreichischer Museen, Leonding ... 058321, 058322
Vercauteren, Antwerpen 063248, 128006
Vercauteren, H., Gent ... 063669
Vercel, Félix, Paris ... 105300
Vercelli, Christophe, Antibes-Juan-les-Pins ... 066680
Verchi, Marialieta, Trieste ... 081810
Verd House, Barcelona ... 085854
A Verdadeira Antiquária, Lisboa .. 084995
Verdala, Naxxar ... 082422
Verdant Works, Dundee ... 044170
Verdenet, Alain, Vuillafans ... 074283
Verdes Soto Arten Colombia, Cali 102416
Verdier, Nérac ... 070504
The Verdier House, Beaufort ... 046850
Verdier, Jacques, Paris ... 071979
Verdière Christophe, Alençon .. 066524
Verdière, René, Ménilles ... 070054
Verdini, Costantino, Roma ... 081595
Verdmont Historic House Museum, Smith's Parish ... 004199
Verdonck & Zoon, Gent ... 128057
Verdonk, Jan, Amsterdam ... 082900
Verdreau Dufresne, Paris ... 071980
Verdugo, Manuel, Sevilla ... 133442
Verein Atelier 8–80, Chemnitz .. 059279
Verein August Macke Haus e.V., Bonn ... 059280
Verein Berliner Künstler, Berlin . 059281, 106452
Verein Berner Galerien, Bern ... 059978
Verein Bildender Künstler Schaffhausen, Forum Vebikus, Schaffhausen ... 059979
Verein der Freunde des Kunsthistorischen Museums, Wien ... 058323
Verein der Freunde des Kupferstich-Kabinetts Dresden e.V, Dresden 058282
Verein der Freunde und Förderer des Clemens-Sels-Museum e.V, Neuss ... 059283
Verein der Freunde von Schloss Prangins, Association des Amis du Château de Prangins, Prangins 059980
Verein der Hobbykünstler, Bruck an der Leitha ... 058324
Verein der Kunstfreunde e.V, Bad Zwischenahn ... 059284
Verein der Kunstfreunde für Wilhelmshaven e.V., Wilhelmshaven ... 059285
Verein der Münzfreunde für Westfalen und Nachbargebiete, Münster . 059286
Verein Freunde zeitgenössischer Kunst, Kramsach ... 058325
Verein für aktuelle Kunst / Ruhrgebiet e.V., Oberhausen ... 059287
Verein für bildende Kunst – Atelier Yin Yang, Graz ... 058326, 100025
Verein für Christliche Kunst in der Evangelisch-Lutherischen Kirche in Bayern e.V., Neuendettelsau .. 059288
Verein für Geschichte und Kunst des Mittelrheins, Koblenz ... 059289
Verein für Heimatpflege, Viersen 022332
Verein für Kunst und Altertum in Ulm und Oberschwaben, Ulm ... 059290
Verein für Kunst und Kultur, Eichgraben ... 058327
Verein für Original Radierung München, München ... 059291, 108205
Verein für Originalgraphik, Zürich 059981, 116865, 138122
Verein Haus Salmegg, Rheinfelden ... 059292
Verein KulturAXE, Wien ... 058328
Verein Kunsthalle Wien, Wien .. 058329
Verein Kunstsammlung Untersee, Untersee ... 042082, 059982
Verein Shedhalle, Zürich ... 042240
Verein zur Bert Hubl Kunst-Förderung e.V., Angelburg ... 059293
Verein zur Förderung bildender Künstler Köln-Porz e.V, Köln .. 059294
Verein zur Förderung der Künstlerischen Tätigkeiten, Trier 059295
Verein zur Förderung der Kunst – Kassel e.V, Kassel, Kassel ... 059296
Verein zur Förderung der Kunst und der 'Art Innsbruck', Innsbruck . 058330
Verein zur Förderung des Kunsthistorischen Instituts in Florenz, München ... 059297
Verein zur Förderung von Grafik und Buchkunst e.V., Leipzig ... 059298
Verein zur Förderung von Kunst und Kultur Attendorn e.V., Attendorn 059299
Verein zur Förderung zeitgenössischer Kunst, Linz ... 058331
Vereinigung Bildender Künstler Steiermark, Graz ... 058332
Vereinigung Bildender Künstlerinnen Österreichs, Wien ... 058333
Vereinigung Bildender Künstlerinnen Wiener Secession, Wien ... 058334, 100388, 136613
Vereinigung der Buchantiquare und Kupferstichhändler in der Schweiz, Syndicat de la Librairie Ancienne et du Commerce de l'Estampe en Suisse, Reinach (Basel-Land) . 059983
Vereinigung der Freunde antiker Kunst, Basel ... 059984
Vereinigung der Freunde von Kunst und Kultur im Bergbau e.V., Bochum ... 059300
Vereinigung der Kunsthistorikerinnen und Kunsthistoriker in der Schweiz, Zürich ... 059985
Vereinigung der Mund- und Fussmalenden Künstler in aller Welt e.V., Schaan ... 030302, 059551
Vereinigung Deutscher Restauratorenverbände, Köln .. 059301
Vereinigung Freunde des Klingspor-Museums e.V., Offenbach am Main ... 059302
Vereinigung Kunstschaffender Unterfrankens e.V, Würzburg .. 059303
Vereinigung Malerstübchen Willingshausen e.V., Alsfeld ... 059304, 137028
Vereinigung Westfälischer Museen e.V., Münster ... 059305
Vereniging Sint Lucas, Vorstenbosch ... 059624
Vereniging van Handelaren in Oude Kunst in Nederland, Amsterdam 059625
Vereniging van Rijksgesubsidieerde Musea, Amsterdam ... 059626
Vereniging van Vrienden der Aziatische Kunst, Soest ... 059627
Vereniging voor het Museum van Hedendaagse Kunst, Gent ... 058389
Vereniging voor Penningkunst, Sleeuwijk ... 059628
Veres Péter Emlékkiállítás, Balmazújváros ... 023433
Véret, Patrick, Arras ... 103277
Verfaille, G., Tournai ... 063850
Verfmolen de Kat, Zaandam ... 033082
Vergain, Philippe, Paris ... 129159
Vergales Pagasta Muzejs, Vērgale ... 030254
Vergani, Orio, Milano ... 110480
Vergara Pizarro, Rubén, Santiago de Chile ... 064989
Vergara, Sergio, Santiago de Chile ... 101924
Verge, Linda, Québec ... 101522
Vergitsis Tsortsis, Pagkrati ... 109078
Vergnais, Colette, Vourles ... 074279
Vergon, Vincent, Cosges ... 128571
Vergougnou, André, Nantes ... 070484
Verhaeghe, M., Gent ... 063670
Verhaeren, Vincent, Bruxelles .. 100668
Verhaest, M., Oostende ... 100810
Verhage, Middelburg ... 126294
Verheijen, Albert, Goirle ... 132507
Verhoeven, Barbara, Duisburg .. 106849
Verhulst, Middelburg ... 083503
Veria Byzantine Museum, Veria .. 023312
Vérité & Mensonge, Québec ... 101523
Verjura & Corondel, Bilbao ... 133204
Verkehrsdrehscheibe Schweiz und unser Weg zum Meer, Basel .. 041293
Verkehrshaus der Schweiz, Luzern ... 041715
Verkehrsmuseum, Frankfurt am Main ... 018456
Verkehrsmuseum Dresden, Dresden ... 018015
Verkehrsmuseum Karlsruhe, Karlsruhe ... 019521
Verkerke, Bogotá ... 102373
Verkerke, Satwa ... 117194
Verkoopzaal van de Kust Romain, Oostende ... 125354
Verkoopzalen D.V.C., Gent ... 125338
Verlag & Galerie Steyrdorf, Steyr 100202, 136566, 140045
Verlag Atelier im Bauernhaus, Ottersberg ... 137466
Verlag Der Apfel, Wien ... 136614
Verlag der Kunst Dresden, Husum ... 137286
Verlag des Germanischen Nationalmuseums, Nürnberg . 137452
Verlag des Kunstgeschichtlichen Instituts der J.-W.-Goethe-Universität, Frankfurt am Main ... 137209
Verlag des Kunstvereins Coburg, Coburg ... 137133
Verlag für das Künstlerbuch, Berlin ... 137098
Verlag für moderne Kunst Nürnberg, Nürnberg ... 137453
Verlag für photographische Literatur, Wien ... 136615
Verlag für Sammler, Graz ... 136512
Verlag Gachnang & Springer, Zürich ... 138123
Verlag Galerie B.O.A., München . 137428
Verlag Haus am Checkpoint Charlie, Berlin ... 137099
Verlag IL, Köln ... 142024
Verlag im Atelierhaus Cornelius, Mainhardt ... 137366
Verlag Klaus Wagenbach, Berlin 137100
Verlag Madlinger, Art & Books, Waldenburg ... 138097
Verlag Passauer Glasmuseum, Tittling ... 137532
Verlag Puppen & Spielzeug, Duisburg ... 138758
Verlag Sammlerfreund, Hamburg 137257
Verlag Scheidegger & Spiess, Zürich ... 138122
Verlag Schnell & Steiner, Regensburg ... 137482
Verlag Urachhaus, Stuttgart ... 137530
Verlag Valentin Koerner, Baden-Baden ... 137041, 141444
Verlagsanstalt Tyrolia, Innsbruck 136524
Verlagshaus Würzburg, Würzburg 137563
Verlagsmuseum der Unternehmensgruppe Ravensburger, Ravensburg ... 021303

Verlaine Antiquités, Saint-Ouen . 073219
Verlan Tehdasmuseo, Verla 011504
Verlat Gallery, Antwerpen 100521
Verleyen, Michel & Alexandra, Seine-
Port 073481
Verloop, J., Noordwijk, Zuid-
Holland 143211
Vermar, Caracas 097966
Vermaten, Fons, Baarn 132433
Vermeer, Buenos Aires 098585, 127540
Vermeersch, P., Antwerpen 063249
Vermesova Vila, Slovenská Národná
Galéria, Dunajská Streda . . . 038347
"Vermessen in Bayern – von
der Messlatte zur Antenne",
München 020583
Vermeulen, Gent 128058
Vermeulen & Zoon, G., Dordrecht 083200,
083201
Vermilion, Atlanta 119628
Vermilion, San Francisco 124424
Vermilion County Museum,
Danville 048144
Vermillion County Museum Society,
Danville 060792
Vermom Artes em Decor, São
Paulo 064215
Vermont Folklife Center,
Middlebury 050859
Vermont Historical Society Museum,
Montpelier 051036
Vermont Marble Museum,
Proctor 052373
Vermont Museum and Gallery Alliance,
Vergennes 060793
Vermont Pastel Society, Saint
George 060794
The Vermont State House,
Montpelier 051037
Vermorel, Bernard, Lyon 140933
Vermote, Jacky, Marignier 069875
Vern Carver & Beard,
Minneapolis 122050
Vernacular Art 'n' Craft, Nagpur . 109489
Vernay-Chantrel, Bernard, Lyon . 069783,
128732
Verne, Cleveland 120464
Verner Amell, London 118617
Verner von Heidenstams Hem,
Verner von Heidenstam's Home,
Motala 040866
Verneuil, Paris 105301
Verney-Carron, Georges,
Villeurbanne 105879
Vernhes, Guillaume, Sainte-
Christie 073343
La Vernice, Bari 109778
Vernisage, Tbilisi 105925, 105926
Vernisaž Pljus, Charkiv 117110
Vernissage, Heidelberg 137275, 138759
Vernissage, Lima 113371
Vernissage, Wien 138540
Vernon County Historical Museum,
Viroqua 054003
Vernon Dath Galleries, Newstead 099446
Vernon Diab, G., Louisville 094602,
127299
Vernon Historians, Vernon 053965
Vernon Public Art Gallery, Vernon,
British Colombia 007315
Vernon, Peter, Toronto 064772
Vernon, Tony, Topsham 135141
Vernonia Historical Museum,
Vernonia 053967
Vernoze Antiquités, Champeix . . 067811
Vero Beach Museum of Art, Vero
Beach 053969
Verolino, Raffaele, Modena 080825
Veronese, Guatemala 079225
Véronèse, Nantes 104347
Veronese, F., Bologna 142647
Veronesi, Milano 080761
Verongalli, Plinio, Roma 132070
Veronica's Treasures, Swan Hill . 062272

Verp'Hier, La Verpillière 069122
Verpleeghuis Cornelia, Zierikzee 033105
Verpleegkundig Historisch Bezit,
Amersfoort 031915
Verran Museum, Malm 033761
De Verrassing, Utrecht 112712
Verrechia, Chris, Le Pré-Saint-
Gervais 069340
Verrechia, Chris, Saint-Ouen . . . 073220
Verreglass, Paris 071981
Verrengia, Paolo, Salerno 110889
Verrerie d'Art, Chexbres 116212
Verrerie La Mailloche, Québec . . 006746
Les Verres de nos Grand Mères, Saint-
Ouen 073221
Verrier, Jean, Saint-Jean-de-Luz 072848,
141334
Verrière, Meyrin 116503
Verrimst, Concarneau 103625
Verrina, Maria Giovanna, Bologna 079885
Verroios-Stergakis, Nikos, Patras 079200
Le Vers à Bois, Montréal 064485
Versailles, B'Kara 132335
Versailles Antiques, Saint-Ouen . 073222
Versailles Artes e Antiguidades, Belo
Horizonte 063926
Versailles Elegance, Toronto . . . 064773
Versailles Enchère, Versailles . . 125837
Versailles Galeria de Artes, Rio de
Janeiro 101075
Versalles, Bilbao 085894, 085895
Versantic, Vichy 074113
De Verschijning, Tilburg 112687
Verschoor, Strijen 083610, 143231
Verschoor, Olga, Thoiré-sur-
Dinan 073660
Verschueren, J.L., Amsterdam . 082901
Verschueren, Jean, Mechelen . . 128082
Version, Cluj-Napoca 139006
Versmolder Kunstkreis e.V.,
Versmold 059306, 108818
Verso Arts et Lettres, Paris 136991,
138665
Verstegen, Rosel & Willy, Herxheim bei
Landau (Pfalz) 076515
Versteigererverband Baden-
Württemberg e.V., Stuttgart . . 059307
Vert Galant, Antibes-Juan-les-
Pins 066681
Vértes Múzeum, Csákvár 023581
Vérteszőlősi Bemutatóhely, Magyar
Nemzeti Múzeum, Vértesszőllős 024038
Verticale, Venezia 111169
Vértice, Oviedo 115331
Vertico, Rhenen 132614
Vertige, Pontevès 072269
Vertigo Galleries, London 118618
Vertigo Gallery, London 118619
Vertigo Studios, Memphis 121714
Vertiko, Brietlingen 075195
Vertu, Cincinnati 120383
Verulamium Museum, Saint
Albans 045671
Vervarcke, J., Knokke-Heist 063713
Vervat, Muriel, Firenze 131278
Verve Gallery, Los Angeles 121615
Verveer, Arnold, Nieuwpoort . . . 132594
Verveer, C.B., Waal 132666
Het Vervolg Antiquariaat, Utrecht 143252
Vervondel, H., Gent 063671
Vervoordt, Axel, s'-Gravenwezel . 063676
Verwertungsgesellschaft Bild-Kunst,
Bonn 059308
Verwertungsgesellschaft Bildender
Künstler, Wien 058335
Verwood Community Museum,
Verwood 007317
Very, New York 139517
The Very Idea, New York 123159
The Very Nice Group, Paddington, New
South Wales 062045
Very Small Shop, Den Haag 083146
Very Special Arts Gallery,
Albuquerque 046373

Very Special Arts Singapore,
Singapore 059803
Veryan Galleries, Veryan Green . 119291
De Verzamelaar, Roosendaal . . . 083546
De Verzamelaar, Tilburg 083628
Verzameld Werk, Gent 100707
Verzameling Steenbakker Ten Bokkel
Huinink, Neede 032664
Verzeletti, Serge, Saint-Étienne . 072778
Verzets Museum Zuid-Holland,
Gouda 032368
Verzetsmuseum Amsterdam,
Amsterdam 031997
Verzetsmuseum Friesland,
Leeuwarden 032577
Vescovi, Luigino, Morecambe . . . 090773
Vesko, Gertrude, Bad Aussee . . 062522
Vespa, Roland, Nice 070675
Vespucci, Cattolica 109956
Vesse, Didier, Pézenas 105353
Vest-Agder Fylkesmuseum,
Kristiansand 033718
Vest-Lindsey House, Frankfort . . 048932
Vest-Telemark Museum, Dalen i
Telemark 033568
Vestal Museum, Vestal 053971
Vesterålsmuseet, Melbu 033769
Vesterheim Norwegian-American
Museum, Decorah 048208
Vestfold Fylkesmuseum,
Tønsberg 034018
Vestfyns Hjemstavnsgård,
Glamsbjerg 010082
Vesthimmerlands Museum, Års . 010010
Vestiges, Portland 096487
Vesting Bourtange, Bourtange . . 032092
Vestingmuseum, Nieuweschans . 032677
Vestingmuseum Oudeschans,
Oudeschans 032753
Vestisches Museum,
Recklinghausen 021309
Vestlandske Skolemuseum,
Stavanger 033987
Vestnik Asociace Muzei a Galerii
České Republiky, Praha 138617
Vestry House Museum, London . 045121
Vestsjællands Kunstmuseum,
Sorø 010326
Vesttry Antiques, Cranbrook . . . 088996
Vesztő-Mágor Történelmi Emlékhely,
Csolt Monostor, Vészto 024046
Vetelin Museo, Veteli 011505
Veterans Federation of the Philippines
Museum, Taguig 034605
Veterans' Memorial Military Museum,
Kensington 006081
Veterinärhistorika Museet, Skara 040953
Veterinärmedizinhistorisches Museum
der Tierärztlichen Hochschule
Hannover, Hannover 019068
Veterinærmuseet, Oslo 033856
Veterinary Museum, London . . . 045122
Veterinary Parasitology Museum,
Chulalongkorn University,
Bangkok 042682
Vetlanda Museum, Vetlanda 041164
Vetlanda Skolmuseum, Vetlanda 041165
Vetlužskij Kraevedčeskij Muzej,
Vetluga 038021
Vetmas, Buenos Aires . 060934, 098586
La Vetrata, Roma 124611
Vetri, Seattle 124611
La Vetrina del Tempo, Pordenone 098224
Větrný Mlýn, Windmill Museum,
Rudice 009874
Vetro & Arte, Dorsoduro 110006
Vetter, Düren . 106755, 137151, 141710
Vetter, Günter, München 130412
Vetter, Jörg H., Osnabrück 077983
Vétu, Patrice, Canteleu 067657
Veturimuseo, Locomotive Museum,
Toijala 011426
Vetus, Firenze 080222
Vetusta Libros, Santiago de

Compostela 143794
Il Vetusto, Milano 080762
Veurink, J.P., Middelburg 112597
Vevelstad Bygdetun, Vevelstad . 034085
Vexer, Sankt Gallen 138088
Veyrac, Henri, Nantes 125659
Veyrat, Thérèse, Bourgoin-Jallieu 103457
Veyret, Renaud & Bruno, Thônes 073663
Veysset, Jacques, Moutier-
Rozeille 070378, 140971
Veyssière, Jean-Paul, Saint-Cyr-sur-
Loire 141327
Vezzoli Cabrioli, Maria, Brescia . 079963
Vezzosi, Massimo, Firenze 080223
VF Fine Arts, Houston 121148
VHS-Photogalerie, Stuttgart . . . 108736
VI Antiques & Fine Art, Montréal 064486
Vi Gurnis-Vieux Greniers, Bertrix 003364
Via Antica, Manila 084483
Via Antigüedades, Caracas 097967
Via Appia, Timişoara 085143
ViA Art Foundation, New York . . 123160
Via Artis, Bad Aussee 058336
Via Kunst, Rotterdam 112668
Via Sollertia Jewel Art,
Christchurch 112866
Via Venezia, Tilburg 083629
Viafarini, Milano 059483, 110481
Viafarini-in-residence, Milano . . 110482
Viafrika, Indianapolis 121223
Viaindustriae, Foligno 059484
Vial, Henri, Dourdan 136813
Vial, Marie-Thérèse, Semsales . . 128787
Vial, Philippe & Louis, Semsales . 133920
Viala, Joël, La Capelle-Balaguier 068963
Vialsoubrane, Bernard, Paris . . . 105302
ViaMaestra 114, Poggibonsi . . . 110670
Vianello, Ponzano 137722
Vianello, Bruno, Venezia 132260
Viani, Nadia, Venezia 111170
Vianna, Ivanir G., Brasília 100930
Viant, André, Épinal . . 068457, 128594
Viar, Antwerpen 063250
Viarius, Frauenfeld 144067
Viart, New York 123161
Viart, Sofia 101178
Viaux Berlin, Berlin 106453
Vibe Antik, Århus 065610
Viborg Stiftsmuseum, Viborg . . . 010361
Vibrant Art, Stow-on-the-Wold . . 119164
Vibrant Fine Art, Los Angeles . . 121616
Vibroc, Bordeaux 067365
Vic Antic, Vic-en-Bigorre 074089
Vicar Antiques, Pittsburgh 096339
Vicar Lane Antique Centre,
Beverley 088455
Vice, New York 139518
Vice Versa, Berlin 137101
Vicens Piera, Barcelona 114973, 114974
Vicent, Peter, Berlin 074943
Vicente Manansala Museum,
Binangonan 034437
Vicentini, Aristide, Verona 132306
Vichitchai Antiques Co. Ltd.,
Bangkok 088088
Vichot Corboz, Montmorot 070283
Vichot, Philippe, Paris 071982
Vicino, Paul, Baltimore 135347
Vick, New York 123162
Vickers, S., Cambridge 144302
Vickery's, Ash 126799
Vickie's Collectibles, Long Beach 094231
Vicko, Hong Kong 102124
Vicksburg National Military Park,
Vicksburg 053976
Vicky's Art and Crafts Gallery,
Mumbai 109483
Vicky's Jewellery, Toronto 064774
Vico, Jesús, Madrid . 086280, 115292,
126553,   133400
Il Vicolo, Genova 110142, 110143
Il Vicolo, Milano 131547
Vicor Restauraciones, Barcelona 133171,
133172

Vic's Coin Shop, San Jose 097380
Vic's Upholstery, Edmonton 128170
Victoire, Boulogne-Billancourt .. 067396
Victor Fissé Antiquités Morges,
  Morges 087687
Victor Harbor Antique Centre, Victor
  Harbor 062386
Victor Harbor Gallery, Victor
  Harbor 099853
Victor Heritage Museum, Victor . 053978
Victor i Fills, Barcelona 085855
Victor i Fills, Madrid 086281
Victor-Schultze-Institut für christliche
  Archäologie und Geschichte der
  kirchlichen Kunst, Ernst-Moritz-
  Arndt-Universität Greifswald,
  Greifswald 018809
Victor Trading Post, Victor Harbor 062387
Victor Valley Museum and Art Gallery,
  Apple Valley 046510
Victor Vasarely Múzeum, Janus
  Pannonius Múzeum, Pécs 023859
Victoria, Caen 067585
Victoria, Praha 065540
Victoria and Albert Museum,
  London 045123
Victoria and Albert Museum,
  Mumbai 024411
Victoria and Albert Museum Bookshop,
  London 144707
Victoria Antik, Norrköping 086885
Victoria Antik, Reykjavik 079340
Victoria Antiques, Alderney 088245
Victoria Antiques, London 090563
Victoria Antiques, Praha 065541
Victoria Antiques, Ryde 091158
Victoria Antiques, Wadebridge .. 091704,
                            135178
Victoria Antiques, Wallasey ... 091709,
                            135180
Victoria Art Gallery, Bath 043528
Victoria Art Gallery & Studio Ltd.,
  Regina 101535
Victoria Book, Wanganui 143357
The Victoria Centre, Saltaire ... 091228
Victoria County Archiv and Museum,
  Baddeck 005369
Victoria County Historical Society
  Museum, Lindsay 006197
Victoria Dreams, Headley 089487,
                            134450
Victoria Falls Information Centre,
  Livingstone 054838
Victoria Galerie, Gémenos 103766
Victoria Gallery, Widnes 119350
Victoria Imports, Stockport 091447
Victoria Jubilee Museum,
  Cawthorne 043898
Victoria Jubilee Museum,
  Vijayawada 024518
Victoria Mansion, Portland 052285
Victoria Memorial Hall, Kolkata . 024369
Victoria Police Historical Museum,
  Victoria 007337
Victoria Prints, Ludlow 118673
Victoria Road Antique Centre,
  Rozelle 062174
Victoria School Museum, Carleton
  Place 005566
Victoria School Museum and Archives,
  Edmonton 005768
Victoria Tower, Huddersfield ... 044615
Victoria Trophées, Saint-
  Apollinaire 072675
Victoria West Regional Museum,
  Victoria West 038875
Victoria, Claudette, San Antonio 136277
Victorian Antique Dealer's Guild, Ascot
  Vale 058225
Victorian Artists Society,
  Melbourne 058226
Victorian Brass Bedstead Company,
  Cocking 088942

Victorian Chairman, Bournemouth 088545,
                            134131
Victorian College of the Arts Gallery,
  Margaret Lawrence Galleries,
  Southbank 001510
Victorian Fireplace, Ashbourne . 079437
Victorian Fireplace, Canterbury . 088782,
                            134211
Victorian Gallery, Dallas 120669
Victorian Gallery, Shrewsbury .. 144846
Victorian Gilt, Auckland 083835
The Victorian Ironmonger,
  Brinklow 088641
Victorian Memories, Minneapolis 094954
Victorian Railway Workshops Art and
  Antiques, Oakleigh 062022
Victorian Railway Workshops Art and
  Antiques, Windsor, Victoria ... 062438
The Victorian Shop, Saint Annes-on-
  Sea 091179
Victorian Sitting Room, Omaha . 096063
Victorian Society, London 060269
Victorian Tapestry Workshop, South
  Melbourne 099703
Victorian Village Antiques, Ballina 079446
The Victorian Village Antiques,
  Glasgow 089336
Victoriana, Parma 081131, 081132
Victoriana Antiques, Naxxar ... 082423,
                            111917
Victoriana Dolls, London 090564
Victoria's Antique, Christchurch . 083903
Victoria's Antiques, Cork 079508
Victoria's Design, Birmingham . 092436
Victoria's Home, Toledo 097620
Victoria's Tramway Museum,
  Bylands 000920
Victoria's Treasure's, Wanganui . 084146
Victorine, Den Haag 083147
Victors Konst och Ram i Slottsstaden,
  Malmö 115773
Victory, Alkmaar 143002
Victory Lane Enterprise,
  Cleveland 092980
Victory Theatre Antique Centre,
  Blackheath 061165
Vida, České Budějovice 102579
Vida, Dordrecht 112444
VIDA Museum, Borgholm 040579
Vidal, Aix-les-Bains 103208
Vidal, New York 095811
Vidal, Nîmes 070708
Vidal Saint-Phalle, Paris 105305
Vidal, Aline, Paris 105303
Vidal, Dominique, Idron 068863
Vidal, Emmanuelle, Toulouse .. 073809
Vidal, Eric, Paris 071983
Vidal, Marie, Campagnan 067621
Vidal, Marie, Pézenas 072143
Vidal, Michel, Paris 105304
Vidal, Patricia, Marne 068782
Vidar, Jónas, Akureyri 109221
Le Vide Grenier, Chasseneuil-sur-
  Bonnieure 067865
Le Vide Grenier, Orchies 070797
Video Museum, Grass Valley .. 049214
Videomixmedia, Genève 059986
Videor Art Foundation,
  Rödermark 108509
Vido, Isabella, Padova 081011
Vidovszky Béla Helytörténeti
  Gyűjtemény, Városi Képtár,
  Gyomaendrőd 023657
Vie au Vert, Gouda 083271
La Vie d'Antan, La Roche-sur-
  Foron 069068
La Vie de Château, Brest 067480
La Vie de Château, Paris 071984
Vie-en-B, Dommartin (Rhône) .. 068377
La Vie en Rose, Paris 071985
Viebahn, Worpswede 079070
Viechtauer Heimathaus Neukirchen,
  Neukirchen bei Altmünster ... 002440
Viehstrichmuseum, Wörth am

Rhein 022721
La Vieille Epoque, Reuland 082384,
                            082385
La Vieille Epoque, Saint-Ouen .. 073223
Vieille Fontaine, Obernai 070773
La Vieille Fontaine, Rolle 087756
La Vieille Forge, Vulaines 074285
Vieille Garde, Mulhouse 070396, 070397
La Vieille Maison, Belliveau Cove 005416
Vieille Maison Traditionelle, Saint-
  Véran 015586
Vieille Poste, Saint-Ouen-de-
  Thouberville 073234
Vieille Saintonge, Tesson 073638
Vieilles Choses, Nice 070676
Vieilleurop, Paris 071986
Vieira, Lisboa 114068
Vieira & Irmão, Funchal 084808
Vieira, Graciano, Porto 132980
Vieira, João Sarmento, Lisboa .. 132961
Vieira, Marcus, Belo Horizonte . 100909
Vieira, Pedro Sousa, Lisboa ... 084996
La Vieja Esquina, Tandil 060962
La Vieja Mecedora, Madrid 133401
El Viejo Patio de San Telmo, Buenos
  Aires 060935
Viekšnių Vaistinės Muziejus, Mažeikių
  Muziejus, Viekšniai 030505
Vieleers, Amsterdam 112306
Vielfalt, Enger 075622
Vielhauer, A., Coesfeld 075268
Vielhescaze, Renaud,
  Gennevilliers 068649
Viellieber, Stuttgart 078638
Vielmann, Los Angeles 094493, 135746
Vielmetter, Susanne, Culver City 120529
Vielstedter Bauernhaus,
  Heimatmuseum der Delmenhorster
  Geest, Hude 019353
Vien Bao Tang Lich Sa Viet Nam, Ha
  Noi 054797
Viena 89 Gallery, San José 102520
Vienna Roussie, Düsseldorf 075532
Vienna Zitta, Wien 063093
Vienne Collection, Vienne 074117
Viennot, Franck, Saint-Pierre-de-
  Boeuf 129242
Viento y Agua Gallery, Long
  Beach 121413
De Vier Gemeten, Middelburg .. 112598
De Vier Winden, Katwijk aan Zee 132552
Vierbaum, Willi, Bad Kissingen . 141428
Vieremän Kotiseutumuseo,
  Vieremä 011506
Vierny, Dina, Paris 105306
Vierschaar Museum, Veere 032978
Vierseitig, Frankfurt am Main .. 107039
Viersener Auktionshaus, Viersen 126134
Viertel, Daniel, Limburg an der
  Lahn 142077
Vierth, Hans-Heinrich, Düsseldorf 075533,
                            129764
Vies Privées, Paris 071987
Viestikeskus Lokki, Mikkeli ... 011163
Viet Fine Arts Gallery, Ha Noi .. 125106
Viet Nam College of Fine Arts, Ha
  Noi 058156
Viet Nam Conservatory of Culture and
  Art, Ha Noi 058157
Viet Nam Fine Arts Association, Ha
  Noi 060827
Viet Nam History Museum Ho Chi
  Minh City, Ho Chi Minh City .. 054811
Viet Nam National Fine Arts Museum,
  Ha Noi 054798
Viet Nam Oil Painting, Ho Chi Minh
  City 125152
Viet Nam Revolution Museum, Ha
  Noi 054799
Viet Nam Women's Museum, Ha
  Noi 054800
Vietnam Era Educational Center,
  Holmdel 049106
Vietnam Veterans Museum, San

Remo 001483
Au Vieux Chéroy, Chéroy 067963
Vieux Grand Port Archaeological Site,
  Old Grand Port 030790
Le Vieux Manoir, L'Isle-sur-la-
  Sorgue 069596, 069597
Le Vieux Matos, Blois 067261
Vieux Meubles de Cluny, Cluny . 068050
Le Vieux Moulin, Sainte-Jeanne-
  d'Arc 006954
Vieux Moulin de Palette, Aix-en-
  Provence 066476
Vieux Ordinaires, Toulon (Var) .. 141359
Le Vieux Porche, Ranspach ... 072365
View Design International,
  Salammbô 088090
View Point, Bern 116151
View Street Bazaar, Bendigo .. 061135
View Two Gallery, Liverpool 118132
Vieweger, Thomas, Blaslay 128471
Viewpoint Gallery, New York ... 123163
Vigado Galeria, Budapest 023568
Vigan House Museum, Vigan .. 034616
Viganò Antichità, Milano 080763
Vigatunet, Ryfylkemuseet,
  Hjelmeland 033654
Vigeland-Museet, Oslo 033871
Vigelius, Egon, Köln 076917
Viggiani, Rosastella, Roma 081596
La Vigie, Nîmes 104404
Vigie, Bruno, Marseille 069972
Vigier, Roquefort 072538
Viglia, Maurizio, Roma 081597
Viglietti, Giuseppe, Roma 132071
Viglongo, Torino 142873
Vigna, Françoise, Nice 104393
Vigna, Paolo, São Paulo 064216
Vignal, Claude, Plailly 072168
Vignal, Hélène, Saules 129253
Vignale, Maria Cristina, Bologna 131003,
                            131004
Vignali, Tshwane 114821
Vignati & Pirruccio, Milano 131548
Vigne, Jean-Bruno, Paris 071988
Vigneras, Jacques & Régis,
  Limoges 069534
Vigneron, Etienne, Levallois-
  Perret 069447
Vigneron, Françoise & Marc,
  Cholet 067986
Vignes, Henri, Paris 141295
Vignettes, San Diego 097145
Vigneux, Le Blanc 069221
Vigo County Historical Museum,
  Sage-Robison-Nagel house, Terre
  Haute 053721
Vigo-Sternberg, New York 095812
Vigoiz, Vigo 086567, 115539
Vigoureux, Antoine, Varennes-le-
  Grand 073980
Vigoureux, Philippe, Verberie ... 074016,
                            105838
Vigreux, Jean-Jacques, Rouen . 129215,
                            141325
Vigué, Jean-Luc, Béziers 067182
Viguié, Pierre, Paris 071989
Vihannin Kotiseutumuseo, Vihanti 011507
Vihavuoden Myllymuseo, Sappee 011344
Vihdin Museo, Vihti 011508
Vihorlatské Múzeum, Humenné . 038357
Viirilän Kotiseutumuseo,
  Kuninkaankylä 011050
Viiskulman, Helsinki 066373
Viitainahon Kotiseutumuseo,
  Piippola 011253
Viitasaaren Metsätyömuseo,
  Viitasaari 011511
De Vijf Eijken, Arnhem 112339
De Vijf Verguldе Flеsсhjes, Delft 083039
Vijs'kovo-istoryčnyj Muzej, Oblasnyj
  istoryčnyj muzej im. V. Tarnovs'kogo,
  Černigiv 043123
Vika Antikk, Oslo 084340, 132792
Vikedal Bygdemuseum, Vikedal . 034089

Vikens Sjöfartsmuseum, Viken . 041168
Viking Antikhandel, Stockholm . 087069
Viking City of Jorvik, York 046224
Viking Historical Museum, Viking 007339
Viking Union Gallery, Western
  Washington University,
  Bellingham 046893
Vikingemuseet Moesgård,
  Højbjerg 010128
Vikingeskibsmuseet, Roskilde . 010294
Vikingmuseet på Borg, Bøstad . 033553
Vikingsholm Museum, Tahoma . 053655
Vikingskipshuset, Kulturhistorisk
  Museum, Universitetet i Oslo,
  Oslo 033872
Vikram Kirti Mandir Museum,
  Ujjain 024506
Viks Auksjonsforretning,
  Porsgrunn 126384
Viktor Rydbergmuseet, Jönköping 040722
Viktoria Antik, Stockholm 087070
Viktoria ir Darius, Vilnius 082345, 142970
Viktorija, Nesebär 128137
Vila A, Ljubljana 114724
Vila Art, Barcelona 114975
Vila de Arte, Fortaleza 063952
Viļa Plūdoņa Memoriālā Māja Muzejs,
  Ceraukste 030109
Vila Rabella, Marta, Barcelona . 133173
Vila Tugendhat, Muzeum města Brna,
  Brno 009484
Vila, Xavier, Barcelona 085856
Vilain dit Jorges Georges, Groix 103799
Vilain, M., Deurne 140218
Vilanova Fine Art, San Salvador . 103107
Vilardo, Pietro, Roma 110873
Vilars, Claude, Vire 074249
Vilarte, Madrid 133402
Vilatte Fabre, Agnès, Grasse . . 125591
Vilbert Diffusion, Cholet 067987
Vildy, Lydie, Gièvres 068661
Vilela, José Manuel, Lisboa . . . 143597
Vilhena, Gaspar L. Manuel, Braga 132924
Vilho, Helsinki 140722
Viljamakasiinimuseo, Vihdin Museo,
  Vihti 011509
Viljandi Museum, Viljandi 010718
Viljandimaa Gümnaasiumi Muuseum,
  Viljandi 010719
Alla Villa, Carimate 080010
The Villa, Hamilton 064333
Villa Adriana, Paris 071990
Villa Adriana, Valencia 086539
Villa Amadeus Galerias,
  Monterrey 082494
Villa Antica, Paris 071991
Villa Antique Mall, Oklahoma City 096017
Villa Antiques, Christchurch 083904
Villa Arson-École Nationale Superieure
  d'Art – Residence d'Artistes, Centre
  National d'Art Contemporarin,
  Nice 055389
Villa Arson, Centre National d'Art
  Contemporain, Nice 014493
Villa Arson, Département d'Edition,
  Nice 136842
Villa Arte, Hamm, Westfalen . . 137262
Villa Augusta, Roth 108523
Villa Bagatelle, Sillery 007039
Villa Bergzicht/Brouwerijmuseum Raaf,
  Heumen 032479
Villa Bianca Expo, Seveso 110932
Villa Bougainvillea, Miami 094814
Villa Capriglio, Torino 059485
Villa Carlotta, Tremezzo 028168
Villa Cicogna Mozzoni, Bisuschio 025520
Villa Collezione Panza, Fondo per
  l'Ambiente Italiano, Varese . . 028279
Villa Concordia Bamberg,
  Internationales Künstlerhaus,
  Bamberg 106120
Villa de Bank, Enschede 112481, 143140
Villa de Bondt, Gent 100708
Villa dell'Arte, Porto Alegre . . . 101003

Villa des Arts, Casablanca 031771
Villa d'Este, Tivoli 028093
Villa du Jardin Alpin – Société Suisse
  des Beaux-Arts-Genève, Meyrin 116504
Villa du Parc Centre d'Art
  Contemporain, Annemasse . . . 011673
Villa Eschebach, Dresden 018016
Villa et Jardins Ephrussi de Rothschild,
  Saint-Jean-Cap-Ferrat 015368
Villa Farnesina, Accademia nazionale
  dei Lincei, Roma 027662
Villa Flath, Bad Segeberg 106093
Villa Flora Winterthur, Sammlung
  Hahnloser, Winterthur 042164
Villa Foscari La Malcontenta,
  Mira 026844
Villa Gallo-Romaine de Saint-Ulrich,
  Haut-Clocher 013235
Villa Gallo-Romaine de Séviac,
  Montréal-du-Gers 014302
Villa Godi Malinverni, Lugo di
  Vicenza 026652
Villa Grècque Kérylos, Beaulieu-sur-
  Mer 011942
Villa Grisebach Auctions, New
  York 127364
Villa Grisebach Auktionen, Berlin 125893
Villa Grisebach Auktionen,
  Hamburg 125981
Villa Grisebach Auktionen,
  Leopoldshöhe 126027
Villa Grisebach Auktionen,
  London 126976
Villa Grisebach Auktionen, Mülheim an
  der Ruhr 126039
Villa Grisebach Auktionen,
  München 126069
Villa Grisebach Auktionen,
  Stuttgart 126125
Villa Grisebach Auktionen, Zürich 126775
Villa Grisebach Gallery, Berlin . . 106454
Villa Gyllenberg, Helsinki 010881
Villa Hügel, Kulturstiftung Ruhr,
  Essen 018314
Villa Imperiale, Pesaro 027270
Villa Julie College, Stevenson . . 057952
Villa Julie College Gallery,
  Stevenson 053526
Villa Kruger, Clarens 041419
Villa Lattes, Istrana 026525
Villa Louis Historic Site, Prairie du
  Chien 052346
Villa Majorelle, Nancy 014409
Villa Manin, Centro d'Arte
  Contemporanea, Codroipo . . . 026029
Villa Mansi, Capannori 025748
Villa Marie-José, Lamothe-
  Montravel 069156
Villa Médicis, Bonnieux 067296
Villa Médicis, Saint-Rémy-de-
  Provence 073292
Villa Melrose, Los Angeles 094494
Villa Merkel und Bahnwärterhaus,
  Galerien der Stadt Esslingen am
  Neckar, Esslingen 018322
Villa Mirabello, Musei Civici,
  Varese 028280
Villa Mobili, Milano 080764
Villa Montezuma Museum, San
  Diego 052986
Villa Museo Puccini, Viareggio . 028360
Villa Obermeier, München 077635
Villa Oppenheim, Galerie für
  Gegenwartskunst, Berlin 017314
Villa Pelsser, Henri-Chapelle . . 003693
Villa Pisani, Strá 028022
Villa Publica Romkert, Göcseji
  Múzeum, Zalalövo 024055
Villa Reuge, Sainte-Croix 041915
Villa Romana a Mosaici, Desenzano
  del Garda 026133
Villa Romana del Casale, Piazza
  Armerina 027308
Villa Safron, Rajagiraya 086623

Villa San Michele, Anacapri 025316
Villa Serra, Bologna . . 079886, 126186,
  137625
Villa Service og Antikvarisk Verksted,
  Godvik 132776
Villa Sinnenreich – Museum der Sinne,
  Rohrbach in Oberösterreich . . 002607
Villa Strassburger, Deauville . . 012749
Villa Terrace Decorative Arts Museum,
  Milwaukee 050925
Villa Torrigiani, Camigliano . . . 025719
Villa Trips, Museum für
  Rennsportgeschichte, Kerpen . 019594
Villa Vauban, Galerie d'Art de la Ville
  de Luxembourg, Luxembourg . 030596
Villa Vauban, Musée d'Art de la Ville
  de Luxembourg, Luxembourg . 030597
Villa Verdi, Villanova sull'Arda . 028393
Villa Wessel, Iserlohn 019430
Villa Zebra Museum voor Kinderen,
  Rotterdam 028329
Villa, Dr. Paola, Milano 131550
Villa, Giuseppina, Milano 131549
Villa, Peter L., New York 123164
Villacher Fahrzeugmuseum,
  Villach 002864
Villag, Manama 100455
Village Aldest, Genève 087507
Village Antique Mall, Denver . . 093349
Village Antique Market, Weedon 091765
Village Antiques, Albuquerque . 092089
Village Antiques, Berrima 061142
Village Antiques, Buffalo 092567
Village Antiques, Cheam 088827
Village Antiques, Cobham 088938,
  134283
Village Antiques, Geelong 061514
Village Antiques, Omaha 096064
Village Antiques, Saint Louis . . 096790
Village Antiques, Tampa 097591, 136399
Village Antiques, Weybridge . . 091810,
  135217
Village Bazaar, Milwaukee 094886
Village Book Shop, Columbus . . 145101
Village Book-Tique, Hornsby . . 139779
Village Books, Dereham 144368
Village Boutique, Memphis 121715
Village Bygones Antiques,
  Christchurch 083905
Village Clocks, Coltishall 088962, 134295
Village Clocks, Long Melford . . 090627,
  134813
Village Coin, Tampa 097592
Village Coin Shop, Austin 092275
Village Collectibles, Indianapolis 093938
Village Créole, Les Trois-Ilets . . 111930
Village Crudère, Marseille 069973
Village Custom, Edmonton 101305
Village d'Artistes, Rablay-sur-
  Layon 015029
Le Village de Passy, Paris 071992
Village des Antiquaires, Beaune 067051
Village des Antiquaires,
  Besançon 067136
Le Village des Pêcheurs, Canet-en-
  Roussillon 012312
Village du Miniature, Pailhares . 014603
Village Galeria de Arte, Rio de
  Janeiro 101076
The Village Gallery, Ballymoney . 117277
The Village Gallery, Delhi 109349
The Village Gallery, Heswall . . . 117966
The Village Gallery, Hook 117996
Village Gallery, Palm Cove 099545
Village Gallery, San Antonio . . . 124081
Village Gallery, Washington . . . 124971
Village Gallery, Wien 100389
Village Gallery of Arts, Portland . 123684
Village Historical Society Museum,
  Harrison 049376
Village Historique Acadien,
  Caraquet 005557
Village Iacustre, Gletterens 041552
Village Marketplace, Nashville . . 095040

Village Miniature, Marchesieux . 014003
Village Minier de Bourlamaque, La
  Cité de l'Or, Val-d'Or 007266
Village-Musée, Fos-sur-Mer . . . 013029
Village Museum, Kfar Tavor . . . 025120
Village Museum, Stellenbosch . . 038842
Village Museum, National Museums of
  Tanzania, Dar-es-Salaam 042631
Village Notre-Dame, Bordeaux . 067366
Village of Elsah Museum, Elsah . 048569
Village Painters, Birmingham . . 119823
Village Peddler, Houston 093825
Village Périgord Antiquité,
  Trélissac 073877
Village Pump Antiques, Exeter . 061458
Village Québécois d'Antan,
  Drummondville 005721
Village Site Museum, Eddystone 005742
Village Smithy, San Antonio . . . 136278
Village Treasures, Long Beach . 094232
Village Treasures II, Oklahoma
  City 096018
Village Upholstery, Seattle 136378
Village Vault Antiques, Long
  Beach 094233
Villageois, Gaston, Ducos 132348
Villages Antiques, Houston . . . 093826,
  135621
Villagra, Cecil, Lima 084472
Villain, Jean-Paul, Paris 105307
Villalba i Badia, Maria, Barcelona 114976
Villani & C., Marco, Firenze . . . 110076
Villani, Fabio, Forres 134393
Villaniza, Caracas 097968
Villanova University Art Gallery,
  Villanova 053986
Villanueva Riu, Pedro, Burgos . 133217
Villar Jambrina, Vicente,
  Barcelona 133174
Villaret, Claire de, Paris 105308
Villarino, Jose y Otro, Bilbao . . 085896,
  133205
Villars, Bruxelles 063578
Villás Galária és Aukciós Ház,
  Debrecen 079302, 126158
Ville delle Rose, Galleria d'Arte
  Moderna, Bologna 025575
Villegas-Inagaki, San Francisco . 124425
Villemin, Catherine, Pompey . . 072230
Villeneuve, Arnaud de, Asnières-sur-
  Seine 128448
Villeneuve, Sabrice de, Atlanta . 119629
Villermin, Daniel, Besançon . . . 067137
Villiers, Cherie de, Rosebank . . 114815
Villoresi, Simonetta, Bologna . . 079887
Vilmorin, Antoine de, Paris . . . 071993
Vilniaus Apskrities Centrinio Pašto
  Muziejus, Vilnius 030547
Vilniaus Dailès Akademijos Muziejus,
  Vilnius 030548
Vilniaus Krašto Etnografijos Muziejus,
  Nemenčinès 030427
Vilniaus Paveikslų Galerija, Lietuvos
  Dailès Muziejus, Vilnius 030549
Vilniaus Praeitis, Vilnius 082346, 142971
Vilniaus Restauratorius, Vilnius . 132325
Vilniaus Sporto Istorijos Muziejus,
  Vilnius 030550
Vilniaus Universiteto Adomo
  Mickevičiaus Muziejus, Vilnius 030551
Vilniaus Universiteto Chemijos
  Fakulteto Muziejus, Vilnius . . 030552
Vilniaus Universiteto Medicinos
  Fakulteto Medicinos Istorijos
  Muziejus, Vilnius 030553
Vilniaus Universiteto Mokslo Muziejus,
  Vilnius 030554
Vilniaus Universiteto Zoologijos
  Muziejus, Vilnius 030555
Vilnius Academy of Fine Arts,
  Vilnius 056123
Vilo Distribution, Paris 136992
Vilppulan Museo, Vilppula 011512
Vimal, Emmanuel, Vic-le-Comte 074098

Vimm Trans, Bucureşti   114157, 133003
Vimonisha, Chennai 109296
Vimont, Jacques, Bollène . . . 067281
Vin- och Sprithistoriska Museet,
Stockholm 041056
Vin Acco Time Repair, Cincinnati 135453
Vina Cooke Museum of Dolls and
Bygone Childhood, Newark-on-
Trent 045337
The Vinalhaven Historical Society
Museum, Vinalhaven 053987
Vinarium, Weinbaumuseum
und Weinbaufreilichtmuseum,
Cleebronn 017746
Vina's Gallery, Sacramento 123789
Viñas, Alfredo, Málaga 115309
Vince &Son., Pittsburgh 136143
Vince-Or, Vincennes 074246
Vincent & Sons, Jacksonville . . . 135670
Vincent, New Orleans 122253
Vincent, Saint-Pierre 114119
Vincent Art Gallery, James Cook
University, Townsville 001592
Vincent Mit L'Ane, L'Isle-sur-la-
Sorgue 069598
Vincent Price Art Gallery, Monterey
Park 051020
Vincent, Bernard, Juvisy-sur-
Orge 068931, 128635
Vincent, Garth, Allington 088250
Vincenti, Marina, Firenze 131279
Vincentz, Colmar 068097
Vinchon-Guyonnet, Annette, Paris 071994
Vinciguerra, Vincenzo, Palermo . 110624
Vinck, E., Aalst 063179
Vinco Grybo Memorialinis Muziejus,
Jurbarka 030343
Vinco Krévés-Mickevičiaus
Memorialinis Muziejus, Vilnius . 030556
Vinco Mykolaičio-Putino Memorialinis
Butas-muziejus, Vilnius 030557
Vinco Mykolaičio-Putino Memorialinis
Muziejus, Marijampolės kraštotyros
muziejus, Marijampolė 030419
Vinco Mykolaičio-Putino Memorialinis
Muziejus, Prienų krašto muziejus,
Skriaudžiai 030480
Vincy, Lara, Paris 105309
Vindelådalens Antik och Auktionshall,
Vindeln 087160, 126699
Vinderen Antikvariat, Oslo 143419
Vindolanda Fort and Museum
Chesterholm, Bardon Mill . . . 043493
Viner, Jonathan, London 118620
Vineta-Museum, Barth 016968
Vingelen Kirke- og Skolemuseum,
Vingelen 034090
Vingerhoedmuseum, Veenendaal 032974
Vingt Neuf Rue Grande, Moret-sur-
Loing 070347
Vingtième Siècle, Saint-Ouen . . 073224
Vinh, Loi, Ho Chi Minh City . . . 125153
Vinh, Xuan, Ho Chi Minh City . . 125154
Vinha d'Arte, Lisboa 114069
Vinhas, António, Porto 085055
Vinhático, Porto 085056
Vinhatti, Porto Alegre 064003
Viniani Museum, Palia Viniani . . 023208
Vinings, Atlanta 092196
Vinjestoga, Vinje 034092
Vink, Kleve 076825
Vink, R., Groningen 083309
Vinke, Thomas, Herford 141941
Vinnyc'ke Učylyšče Kul'tury i Mystectv
im. M. Leontoviča, Vinnycja . . 056504
Vinnyc'kyj Literaturno-memorial'nyj
Muzej M.M. Kocjubyns'koho,
Vinnycja 043318
Vinnyc'kyj Nacional'nyj Muzej-sadiba
im. Pirogova, Vinnycja 043319
Vinnyc'kyj Oblasnyj Chudožnij Muzej,
Vinnycja 043320

Vinnyc'kyj Oblasnyj Krajeznavčyj
Muzej, Vinnycja 043321
Vinohradnícka Expozícia Apponyiho
Palác, Mestské Múzeum v Bratislave,
Bratislava 038331
Vinorama – Musée du Vin,
Bordeaux 012122
Vinstra Auksjonsforretning,
Vinstra 126397
Vint – Trenta, Barcelona 085857, 133175
Vintage, Los Angeles 094495
Vintage 429, New Orleans 145317
Vintage Aircraft Museum,
Masterton 033243
Vintage and More, Albuquerque . 092090
Vintage Antique Centre, Warwick 091752
Vintage Antiques, Charlotte 092639
Vintage Antiques, Edmonton . . . 064309
Vintage Antiques, El Paso 093419
Vintage Art Gallery, Singapore . . 114632
Vintage Art Prints, Honolulu . . . 120974
Vintage Arts, Salt Lake City . . . 136246
Vintage Cameras, London 090565
Vintage Car and Carriage Museum,
Inishowen 024933
Vintage Clobber, Bournemouth . 088546
Vintage Clothing, Philadelphia . . 096171
Vintage Country Antiques,
Willunga 062430
Vintage Designs, New York 095813
Vintage Furniture, Indianapolis . 135639
Vintage Interiors, Phoenix 096245
Vintage Isle, Honolulu 093577
Vintage Loft, New York 095814
Vintage Mania, Memphis 094671
Vintage Modern, San Francisco . 097345
Vintage Nest, Houston 093827
Vintage on College Avenue,
Oakland 095920
Vintage Pine, Chicago 092813
Vintage Poster Art, San Francisco 124426
Vintage Poster Bank, Cincinnati . 120384
Vintage Prints and Paper, Buffalo 092568,
145023
Vintage Radio & Gramaphone,
Toronto 064775
Vintage Restoration, New York . 136022
Vintage Revival, Charlotte 092640
Vintage Rose, Ballarat 061092
The Vintage Shop, Bangalore . . 079364
Vintage Sounds, Houston 093828
Vintage Stock, Tulsa 097765
Vintage to Vogue, Bath 088405
Vintage Toy and Train Museum,
Sidmouth 091343
Vintage Universe, Houston 093829
Vintage Veruca, Winnipeg 064935
Vintage Watch Company, Tunbridge
Wells 091667, 135165
Vintage Wireless Museum,
London 045124
Vintage Wireless Shop,
Nottingham 090900
Vinterhagen, Vaxholm 087155
Vintler, Bolzano 079904
Vinton Street Antiques, Omaha . 096065
Vintschger Museum, Museo della Val
Venosta, Sluderno 027972
Vinttikamari, Helsinki 066374
Viñuela Fernandez, Antonio,
Barcelona 133176
Vinyasa, Chennai 109297
Vinzentz, Düsseldorf 075534
Vinzenz, Rolf, München 130413
Viola, Aldo, Roma 132072
Viola, André, Metz . . . 070103, 140955
Viola, Carlo, Bologna 079888
Viola, Rodolfo, Milano 110483
Violaris Eitle, Lefkosia 065326
Violeau, Jean-Claude, Pezou . . . 071247
Violet, Gent 063672
Violet Town Antiques, Violet
Town 062391

La Violetera, Paris 071995
Violin Shop, Hexham 089525
Le Violon Bleu, London 118621
Le Violon Bleu, Sidi Bou Saïd . . 116987
Violon D'Ingres, Paris 071996
Vion, Jean-Paul, Pézenas 072144
VIP Antik, İstanbul 088147
VIP Antiques, Washington 097865
Vip's International Art Galleries,
Rotterdam 112669
Virago Arte, Fitzroy North 099038
Virago Arte, Torino 111014
Viranşehir Kültür Merkezi,
Viranşehir 043081
Virden Pioneer Home Museum,
Virden 007346
Virgil I. Grissom State Memorial,
Mitchell 050968
Virgilio, Carlo, Roma 110874
Virgil's Antiques, Railway Estate 062135
Virgin Gallery, Villeurbanne . . . 074229
Virginia, London 090566
Virginia Air and Space Center,
Hampton 049344
Virginia Association of Museums,
Richmond 060795
Virginia Aviation Museum, International
Airport, Richmond 052562
Virginia Baptist Historical Society,
Richmond 052563
Virginia Beach Antique Mall, Virginia
Beach 097796
Virginia Beach Art Gallery, Virginia
Beach 124842
Virginia Beach Maritime Museum,
Virginia Beach 053999
Virginia Christopher Fine Art,
Calgary 101265
Virginia City Madison County Historical
Museum, Virginia City 054001
The Virginia Discovery Museum,
Charlottesville 047586
Virginia Fine Art Gallery, Bendigo 098736
Virginia Groot A. Foundation,
Chicago 047706
Virginia Historical Museum, Virginia
Historical Society, Richmond . . 052564
Virginia Institute of Marine Science,
Fish Collection, Gloucester Point 049137
Virginia Living Museum, Newport
News 051549
Virginia Marine Science Museum,
Virginia Beach 054000
Virginia Military Institute Museum,
Lexington 050270
Virginia Museum of Fine Arts,
Richmond 052565
Virginia Museum of Fine Arts
(Calender), Richmond 139519
Virginia Museum of Fine Arts
Publishing Department,
Richmond 138448
Virginia Museum of Natural History,
Martinsville 050727
Virginia Museum of Transportation,
Roanoke 052605
Virginia Sports Hall of Fame and
Museum, Portsmouth 052329
The Virginia War Museum, Newport
News 051550
Virginia Wright Fund, Seattle . . . 053251
Virginia's, Los Angeles 094496
Virgo Labour Museum, Oravais . 011100
Viridian Artists, New York 123165
Virk, Roskilde 102966
Virkki Käsityömuseo, Helsinki . . 010882
Viroconium Museum, Wroxeter Roman
City, Wroxeter 046208
Virolahden Kotiseutumuseo,
Virolahti 011514
Virserums Konsthall, Virserum . . 041171
Virserums Möbelindustrimuseum,
Virserum 041172
Virtain Perinnekylä, Virrat 011517

Virtanen, Armadale, Victoria . . . 061050
Virtanen, J. K., Highett 061620
Virtanen, Kent & Sharen, Mosman
Park 061900
Virtasalmen Kotiseutumuseo,
Virtasalmi 011518
Virtual Art Gallery, Las Vegas . . 121388
Virtual Art Gallery, Milwaukee . . 121955
VirtuArtNet Galéria, Budapest . . 109185
Virtudes 20, Madrid 133403
Les Virtuoses de la Réclame,
Paris 105310
Virtuti Militari, Paris 071997
Virum Antikvariat, Virum 140674
Virus Arts, Toronto 101716
Vis à Vis, Chalon-sur-Saône . . . 103556
Vis-á-Vis, Karlsruhe 076702
Visafslag Elburg, Elburg 032297
Visage Galerie, Berlin 106455
Visarte-Berufsverband Visuelle Kunst
Schweiz, Société des Artistes Visuels
Suisse, Zürich 059987
Visarte Vaud, Lausanne 059988
Visca, Umberto, Roma 081598
Visceglia Art Gallery, Caldwell College,
Caldwell 047347
Visceglie, Giuseppe, Bari 079788
Vischer, Erwin, Mönchengladbach 077399
Vischpoort en Kazematten,
Elburg 032298
Viscontea Casa d'Aste, Milano . 126213
Viseum Wetzlar, Haus der Optik +
Feinmechanik, Wetzlar 022618
De Vishal, Haarlem 112518
Vishwakarma, Hyderabad 109377
Vision Graphics, Philadelphia . . 123467
Vision Quest Trading Company,
Winnipeg 101896
Visionarium, Centro de Ciência
do Europarque, Santa Maria da
Feira 035998
Visions, Croydon 117655
Visions Gallery, Camolin 109540
Visions in Black, Dallas 120670
Visions of Gold, Houston 093830
Visiting Arts, London 060270
Visken, Lyon 104138
Viski Károly Múzeum, Kalocsa . . 023706
Vismara Arte, Milano 110484
Visol, Ourense 115322
Visone, Sergio, Napoli 080945
Visser, Dick, Antwerpen 063251, 128007
Visser, Peter, Cape Town 085637
Visserij- en Cultuurhistorisch Museum,
Woudrichem 033068
Visserijmuseum, Breskens 032107
Visserijmuseum, Bruinisse 032114
Visserijmuseum, Vlaardingen . . 032991
Visserijmuseum Zoutkamp,
Zoutkamp 033110
Visseron, Bernadette, Poursac . . 072300
Vista Fine Arts, Dallas 120671
Vista Fine Arts, West Hollywood 125031
Vista Historical Museum, Vista . 054006
Visual, Tlalpan 138924
Visual – Magazine de Diseño,
Creatividad y Comunicación,
Madrid 139077
Visual Aid, San Francisco 053029
Visual and Performing Arts
Department, Fairfield University,
Fairfield 057072
Visual and Performing
Arts Department, Finger
Lakes Community College,
Canandaigua 056828
Visual and Performing Arts Division,
Cabrillo College, Aptos 056664
Visual and Performing Arts Division,
Glendale Community College,
Glendale 057135
Visual and Performing Arts Division,
Marygrove College, Detroit . . . 057013
Visual and Performing Arts Division,

Yavapai College, Prescott 057719
Visual and Performing Arts, Aims
  Community College, Greeley . 057152
Visual and Performing Arts, Inter
  American University of Puerto Rico,
  San Germán 056255
Visual Art, Edgware 117712
Visual Art Department, Dean Junior
  College, Franklin 057104
Visual Art Department, Golden West
  College, Huntington Beach .. 057223
Visual Art Department, Mitchell
  Community College, Statesville 057947
Visual Art Galleries, Singapore . 114633
Visual Art Gallery, Santa Fe
  Community College, Santa Fe . 053124
Visual Art Program, Centro Cultural de
  la Raza, San Diego 052987
Visual Art School of Trenton,
  Trenton 058004
Visual Art Studio, Richmond . 123750
Visual Artists and Galleries Association,
  New York 060796
Visual Artists Ireland, Ealaíontóirí
  Radharcacha Éire, Dublin 059384
Visual Artists' News Sheet,
  Dublin 138783
Visual Arts, Minneapolis 122051
Visual Arts Alliance of Nashville,
  Nashville 060797
Visual Arts and Digital Media
  Department, South Nottingham
  College, Nottingham 056590
Visual Arts and Galleries Association,
  Ely 060271
Visual Arts and Media Studies,
  Pasadena City College,
  Pasadena 057649
Visual Arts Annex Gallery, Guadalupe
  Cultural Arts Center, San
  Antonio 052953
Visual Arts Center of Northwest
  Florida, Panama City 051909
Visual Arts Center, Mount Hood
  Community College, Gresham . 057168
Visual Arts Center, Tidewater
  Community College, Portsmouth 052330
Visual Arts Centre, Westmount . 007387
Visual Arts Centre of Clarington,
  Bowmanville 005450
Visual Arts Department, Antioch
  College, Yellow Springs 058139
Visual Arts Department, Bemidji State
  University, Bemidji 056723
Visual Arts Department, Bergen
  Community College, Paramus . 057644
Visual Arts Department, Bowdoin
  College, Brunswick 056808
Visual Arts Department, Brewton-
  Parker College, Mount Vernon . 057493
Visual Arts Department, Colby
  Community College, Colby ... 056923
Visual Arts Department, Flagler
  College, Saint Augustine 057802
Visual Arts Department, Hutchinson
  Community College, Hutchinson 057228
Visual Arts Department, Kilgore
  College, Kilgore 057286
Visual Arts Department, Miami-Dade
  Community College, Miami ... 057435
Visual Arts Department, Middlesex
  County College, Edison 057037
Visual Arts Department, New College
  of Florida, Sarasota 057877
Visual Arts Department, Pine Manor
  College, Newton 057578
Visual Arts Department, Rio Hondo
  College, Whittier 058101
Visual Arts Department, School of Fine
  Arts, Willoughby 058113
Visual Arts Department, Seward
  County Community College,
  Liberal 057341
Visual Arts Department, Southwestern

Oregon Community College, Coos
  Bay 056953
Visual Arts Department, State
  University of New York College at Old
  Westbury, Old Westbury 057616
Visual Arts Department, University of
  California San Diego, La Jolla . 057299
Visual Arts Department, University of
  Colorado at Denver, College of Arts
  and Media, Denver 057005
Visual Arts Department, University
  of Maryland, Baltimore County,
  Baltimore 056708
Visual Arts Department, University of
  Southern Queensland, Faculty of Arts,
  Toowoomba 054921
Visual Arts Division, Bennington
  College, Bennington 056725
Visual Arts Gallery, Birmingham . 047010
Visual Arts Gallery, New York . 051488
Visual Arts Gallery, Pensacola . 051985
Visual Arts Museum, New York . 051489
Visual Arts Program, Johnson County
  Community College, Overland
  Park 057636
Visual Arts Program, University
  of Illinois at Springfield,
  Springfield 057938
Visual Arts, University of Windsor,
  Windsor, Ontario 055174
Visual Communication Department,
  Austin Community College,
  Austin 056695
Visual Communications Class, Atlanta
  Area Technical School, Atlanta 056685
Visual Galeria Artes, Brasília . 100931
Visual History Gallery, Cincinnati 092905
Visual Instinct, Dublin 109625
Visual Jazz Art Gallery, New
  Orleans 122254
Visual Resources Association Bulletin,
  Ann Arbor 139520
Visual Resources Collection, University
  of Chicago, Chicago 120293
Visual Spaces, Montréal 101458
Visual Studies Workshop,
  Rochester 057782
Visual, Performing and Applied Arts
  Division, Harford Community College,
  Bel Air 056717
Visualize, San Francisco 124427
Visuart Galéria, Nyíregyháza .. 109203
Visvesvaraya Industrial and
  Technological Museum,
  Bangalore 024152
Vit, Seoul 111730
Vita Valvet, Stockholm 087071
Vitale Family Galleries, San
  Diego 124184
Vitale, Alessandrina, Venezia .. 081879
Vitali, Luisa, Modena 080826
Vitali, Stefano, Milano 080765
Vitenskapsmuseet, Trondheim .. 034041
Vites, Armando, Rosario 139608
Vith, D., Quedlinburg 078110
Víbar, Jónas, Akureyri 109222
Vitiello, Christophe, Marseille . 128788
Vitlycke Museum, Tanumshede . 041075
Vitold, Mary, Seattle 124612
Vitoux, Pierre-Marie, Paris ... 105311
Vitovec, Basel 144037
Vitra, Varna 101200
Vitra Design Museum, Weil am
  Rhein 022472
Vitralls Hernandez, Barcelona . 085858
Vitreous, Salt Lake City 096930
Vitreous, Truro 119267
Vitrine, Bern 087308
Vitrine, Klagenfurt 062677
Vitrine 144, Mülheim an der
  Ruhr 077437
Vitrine Antigüidades, Belo
  Horizonte 063927
La Vitrine des Antiquaires,

Nevers 070542
Vitro, Skopje 111899
Vitromusée Romont, Musée Suisse
  du Vitrail er des Arts du Verre,
  Romont 041889
Vitrotto, Clotilde, Torino 081770
Vitrus, Mostar 100858
Vitry, Antoine, Marnes-la-
  Coquette 069886
Vitry, Jean-Luc, Oberwöbling . 127829
Vitt, Walter, Köln 137320
Vittet, Claude, Paris 071998
Vittori, Claudia, Venezia 132261
Vittoria Art Gallery, Manly ... 099277
Viual Arts Unit, NIE, Nanyang
  Technological University,
  Singapore 056331
Viva, Zlín 102752
Viva-Artis, Sopot 113766
Viva Galerie, Montréal 064487
Vivace, Jounieh 082324
Vivaldi Antiquités, Paris 071999
Vivant, Seattle 097518
Vivanti, David, Roma 081599
Viv'Antique im Holländischen Viertel,
  Potsdam 078086
Vivarium du Moulin, Musée de la
  Nature, Lautenbach-Zell 013585
Vivarium, Museo de Historia Natural,
  Quito 010497
Vivar600s Auktionsbyrå,
  Hammerdal 126596
Vivement Jeudi, Paris 072000
Vivendi, Paris 105312
Vives-Atsara, Jose, San Antonio 124082
Viviani, Pisa 110669, 126219
Vivianna, Leura 099231
Vivid, Rotterdam 112670
Vivier, Luxeuil-les-Bains 069672
Vivier & Christophe, Michel, Saint-
  Sauveur 073301
Vivier d'Oie, Bruxelles 063579
Vivioli Arte Antica, Genova 080301,
                          131344
Vivliofilia-Myriovivlos, Athinai .. 126146
Vivre, Nijmegen 083519
Vivre l'Art, Bruxelles 100669
Vixseboxse, Cleveland 120465
Vizcaya Museum, Miami 050840
Vízellátás-történeti Muzeális
  Kiállítóhely, Orosháza 023821
Vizija, Beograd 114445
Vízivárosi, Budapest 109186
Vizmathy, Cecil Béla, Budapest . 079298
Vizzavona, Paris 072001
Vizzotto, Mattia, Milano 131551
VJ 22, Delft 112369
Vjatskaja Kunstkamera, Kirovskij
  oblastnoj kraevedčeskij muzej,
  Kirov 037136
Vlaams Filmmuseum en -archief,
  Kortrijk 003759
Vlaams Mijnmuseum, Beringen . 003359
Vlaams Tram- en Autobusmuseum,
  Berchem 003358
Vlaamse Museumvereniging,
  Antwerpen 058390
Vlaamse Vereniging van Experten
  voor Kunst en Antiquiteiten,
  Bruxelles 058391
Vlaamse Zaal, Antwerpen 100522
Vlada, Moskva 114317
Vladimir's Antiques, New York .. 095815
Vladimirskij, Sankt-Peterburg .. 085276
Vladivostokskaja Krepost – Voenno
  istoričeskij Fortifikacionnyj Muzej,
  Vladivostok 038045
Het Vlaemsche Erfgoed, Groede 032371
Vlam, Michiel de, Den Haag ... 083148,
                          132482
Vlasbewerkingsmuseum It Braakhok,
  Ee 032268
Vlasblom, Arnhem 112340
Vlastivědné Muzeum, Bělá pod

Bezdězem 009440
Vlastivědné Múzeum, Hanušovce nad
  Topl'ou 038352
Vlastivědné Muzeum, Kamenice nad
  Lipou 009614
Vlastivědné Múzeum, Kyjov 009658
Vlastivědné Múzeum, Nové
  Zámky 038417
Vlastivědné Muzeum, Nymburk . 009728
Vlastivědné Múzeum, Považská
  Bystrica 038429
Vlastivědné Muzeum, Slany ... 009882
Vlastivědné Muzeum, Šumperk . 009898
Vlastivědné Múzeum, Trebišov .. 038469
Vlastivědné Muzeum, Zábřeh na
  Moravě 009968
Vlastivědné Muzeum, Žirovnice . 009982
Vlastivědné Muzeum a Galerie, Česká
  Lípa 009499
Vlastivědné Muzeum Dr. Hostaše v
  Klatovech, Klatovy 009629
Vlastivědné Muzeum Jesenicka,
  Jeseník 009600
Vlastivědné Muzeum pro Vysoké
  nad Jizerou a Okolí, Vysoké nad
  Jizerou 009966
Vlastivědné Múzeum v Galante,
  Galanta 038350
Vlastivědné Múzeum v Hlohovci,
  Hlohovec 038353
Vlastivědné Muzeum v Olomouci,
  Olomouc 009732
Vlcek, Maria, Landshut 077020
De Vleeshal, Middelburg 032650
Vleeshuis, Antwerpen 003316
Vleeshuismuseum, Dendermonde 003539
Vliegend Museum Lelystad,
  Lelystad 032593
Vliegend Museum Seppe,
  Bosschenhoofd 032089
Vliegenthart, Amsterdam 132423
Vloedbeld, Almelo 082637
Vloemans, John A., Den Haag . 143123
Vlok, A., Utrecht 083667
VN Art, Ho Chi Minh City 125155
Voas Museum, Minburn 050927
Voborník, Ústí nad Labem 128372
Vock, Carl, Zürich 088015, 134003
Vodné Kasárne, Slovenská Národná
  Galéria, Bratislava 038332
Vodnikova Domačija, Ljubljana . 114725
Vöbam, Stockholm ... 087072, 143971
Vögeli-Buchwalder, Ronny & Jus,
  Luzern 087649
Vögtli, Benno, Hochwald 087526
Völgységi Múzeum, Bonyhád ... 023450
Völkel, Gisela, Bergkamen 074736
Völkerkundemuseum der Universität
  Zürich, Zürich 042241
Völkerkundemuseum der von Portheim-
  Stiftung, Heidelberg 019149
Völkerkundemuseum Herrnhut,
  Staatliche Ethnographische
  Sammlungen Sachsen,
  Herrnhut 019228
Völkerkundemuseum, Archiv- und
  Museumsstiftung Wuppertal,
  Wuppertal 022824
Völkerkundesammlung, Die Lübecker
  Museen, Lübeck 020165
Völkerkundliche Sammlung,
  Winterthur 042165
Völkerkundliche Sammlung der
  Philipps-Universität, Marburg . 020277
Völkerkundliche Sammlung der Steyler
  Missionare, Aulendorf 016678
Völkerkundliches Museum,
  Witzenhausen 022713
Völkerschlachtdenkmal / Forum 1813,
  Stadtgeschichtliches Museum Leipzig,
  Leipzig 020023
Voellmy & Co., Basel 133660
Voeltzel, Thierry, Paris 072002
Völz, Rainer, Essen 075704

Register der Institutionen und Firmen
Voena, Marco, Milano – Vos, J.H., Groningen

Voena, Marco, Milano 110485
Voenno-istoričeski Muzej, Gorna Studena 005036
Voenno-istoričeski Muzej, Pleven 005122
Voenno-istoričeski Muzej Osvoboditelna Vojna 1877–1878, Bjala 004987
Voenno-istoričeskij Muzej Artillerii, Inženernych Vojsk i Vojsk Svjazi, Sankt-Peterburg 037824
Voenno-istoričeskij Muzej Tichookeanskogo Flota, Vladivostok 038046
Voenno-medicinskij Muzej, Sankt-Peterburg 037825
Voenno-morskoj Muzej Severnogo Flota, Murmansk 037530
Voergård Slot, Dronninglund ... 010035
Voerman Museum Hattem, Hattem 032421
Vörösmarty Mihály Emlékmúzeum, Kapolnásnyék 023707
Vörsi Talpasház, Vörs 024049
Voerster, J., Stuttgart 142407
De Voetboog, Amsterdam 112307
Vötters Oldtimermuseum, Kaprun 002159
Vog Art, Toulouse 073810
Vogel, Mannheim 107974
Vogel & Vogel, Garbsen 075943
Vogel, Albert, Würzburg 079094
Vogel, Donald S., Dallas 120672
Vogel, E., Den Haag 143124
Vogel, F., Karlsruhe 076703
Vogel, Friedl, Lehrberg 077047
Vogel, Hans-Detlev, Königstein im Taunus 076926
Vogel, Jürgen, Jestetten 076650
Vogel, K., Bonn 075095
Vogel, Roland, Rain 130552
Vogelmuseum, Aigen im Mühlkreis 001693
Vogelmuseum, Waging am See . 022371
Vogels, Leon, Helmond 132531
Vogelsberger Heimatmuseum, Schotten 021728
Vogelschutzmuseum, Biberach an der Riß 017358
Vogelsperger, Basel ... 116107, 138027
Voges & Partner, Frankfurt am Main 107040
Voghera, Manosque 104151
Vogl, Franz A., München 130414
Vogl, Monika, Mönchengladbach 108020
Vogler, Philippe, Basel . 087263, 126710
Voglhofer, Stefan, Schwertberg . 127871
Voglia d'Antico, Madonna del Piano 087655, 133851
La Voglia del Passato, Verona .. 081946
Voglmair, Manfred, Gmunden .. 062572
Vogt, A., Köln 107696
Vogt, Armin, Basel 116108
Vogt, Bernardete Nadal de, Meninio Deus 128105
Vogt, Bettina, Hannover 130018
Vogt, Christa, Bad Tölz 074607
Vogt, George, Houston 093831
Vogt, Inge, Hilzingen 141948
Vogt, Johannes, München 126070
Vogt, Klaus, Lieser 077107
Vogt, M., Alling 074379
Vogt, Norbert, Herbrechtingen .. 076496
Vogt, Peter, München 077636
Vogteimuseum mit Blumenauer Heimatstube, Aurach 016680
Vogtländisches Dorfmuseum, Erlbach 018267
Vogtländisches Freilichtmuseum, Erlbach 018268
Vogtländisches Freilichtmuseum Landwüst, Markneukirchen ... 020291
Vogtlandmuseum Plauen, Plauen 021150
La Vogue Antiques and Collectables, Parkhurst 085661
Vogue Antiquités, Tournai 063851
Vogue Art and Crafts Gallery, Mumbai 109484
Vohr, Erika, Reutlingen 078215, 130584
Voigt, Nürnberg 108309
Voigt, Gerard, Stuttgart 130690
Voigt, Kathrin, Oelsnitz 077898
Voigt, Markus, Wildenberg 079008
Voigtländer, Hartmannsdorf 076409
Voilà, Dordrecht 112445
Voimalamuseo, Helsingin Kaupunginmuseo, Helsinki ... 010883
Voipaalas Taidekeskus, Sääksmäki 011333
Voit, Robert, Warmensteinach .. 130735
Voithenberghammer, Hammerschmiede von 1823, Furth im Wald 018568
La Voiture Ancienne d'Alsace, Zehnacker 074318
't Voiturke, Antwerpen 063252
Vojaški Muzej Tabor, Lokev 038568
Vojčík, Štefan, Sabinov 085589
Vojenské Múzeum Trenčín, Trenčín 038472
Vojenské Múzeum, Vojenské múzeum Svidník, Piešt'any 038423
Vojenské Starožitnosti, Praha .. 065542
Vojenské Technické Muzeum, Krhanice 009651
Vojni Muzej, Logatec 038567
Vojni Muzej Beograd, Beograd .. 038185
Vokaer, Michel, Bruxelles 100670, 136650
Vol du Papillon, Lausanne 116429
Volapük, Berlin 141584
Volcan de Lemptégy, Saint-Ours-les-Roches 015488
Volcano Art Center, Volcano ... 054007
Volda Bygdetun og Garverimuseet, Volda 034093
Voldère, Florence de, Paris 072003
Voldère, Sybille de, Paris 072004, 105313
Volek, Roman, Praha .. 065543, 128367
Volendam Windmill Museum, Milford 050892
Volendams Museum, Volendam . 033001
Volets Bleus, Angresse 066632
Le Voleur d'Images, Paris 105314
Volgogradskij Gosudarstvennyj Institut Iskusstv i Kultury, Volgograd .. 056315
Volgogradskij Memorialno-istoričeskij Muzej, Gosudarstvennyj muzej-panorama Stalingradskaja Bitva, Volgograd 038051
Volgogradskij Muzej Izobrazitelnych Iskusstv, Volgograd 038052
Volgogradskij Oblastnoj Kraevedčeskij Muzej, Volgograd 038053
Volk, Bruno, Lülsfeld 077204
Volk, Konrad, Würzburg 079095, 108980
Volkening Heritage Farm at Spring Valley, Schaumburg 053194
Volker, Monika, Hildesheim 076534
Volkers, Neu-Isenburg 077709
Volkert, Traunstein 142415
Volkmer, Peter & Renate, Aichhalden 129372
Volksbuurtmuseum, Den Haag . 032201
Volksbuurtmuseum Wijk, Utrecht 032958
Volkskultur Niederösterreich, Atzenbrugg 058337
Volkskulturmuseum Raußmühle, Eppingen 018224
Volkskunde- und Freilichtmuseum, Konz 019765
Volkskunde- und Mühlenmuseum, Großschönau 018864
Volkskunde Museum Schleswig, Stiftung Schleswig-Holsteinische Landesmuseen Schloß Gottorf, Schleswig 021668
Volkskunde-Museum, Thüringer Bauernhäuser, Rudolstadt 021533
Volkskundemuseum, Antwerpen 003317
Volkskundemuseum, Drognitz .. 018019
Volkskundemuseum, Mödling .. 002405
Volkskundemuseum, Ostrach .. 021036
Volkskundemuseum am Landesmuseum Joanneum, Graz 001989
Volkskundemuseum des Bezirks Unterfranken, Museen Schloss Aschach, Bad Bocklet 016736
Volkskundemuseum des Ratzeburger Landes, Schönberg, Mecklenburg 021701
Volkskundemuseum Deurne, Deurne 003547
Volkskundemuseum Salzburg, Salzburg 002641
Volkskundemuseum Spiralschmiede, Lasberg 002273
Volkskundemuseum Treuchtlingen, Treuchtlingen 022190
Volkskundemuseum Wyhra, Borna 017521
Volkskundliche Sammlung, Mömbris 020436
Volkskundliche Sammlung alter bäuerlicher Geräte, Ludesch .. 002345
Volkskundliche Sammlung des Fichtelgebirgsvereins, Weidenberg 022462
Volkskundliche Sammlungen, Straßburg 002801
Volkskundliche Sammlungen, Erlebnispark Ziegenhagen, Witzenhausen 022714
Volkskundliches Berufe- und Handwerker-Museum, Aspang 001727
Volkskundliches Freilichtmuseum im Stadtpark Speckenbüttel, Bremerhaven 017612
Volkskundliches Gerätemuseum, Arzberg 016613
Volkskundliches Museum, Deutschlandsberg 001816
Volkskundliches Museum Alois Alphons, Hirschwang 002087
Volksmuseum Deurne, Antwerpen 003318
Voll, Karlheinz, Frankfurt am Main 075830, 107041
Le Volle Blaes, Bruxelles 063580
Volle, Bernard, Saint-Ambroix .. 072666
Volle, S., Freiburg im Breisgau . 075886
Vollmann, Florian, Hamburg ... 129994
Vollmer, A., Freiburg im Breisgau 075887
Vollmer, Karin, Karlsruhe 130100
Vollstedt, Andreas, Rendsburg .. 078206, 078207
Volmant, Linda, Cabrieres-d'Avignon 067550
Volo Antique Auto Museum, Volo 054009
Volodymyr-Volyns'kij Istoryčnyj Muzej, Volodymyr-Volyns'kij 043322
Vologodskaja Oblastnaja Kartinnaja Galereja, Vologda 038057
Vologodskij Gosudarstvennyj Istoriko-Architekturnyj i Chudožestvennyj Muzej-Zapovednik, Vologda ... 038058
Volos, Roma 110875
Voloviec, Claude, Les Alleuds .. 069395
Volpi, Patrizia, Parma 131740
Volpin Foscarini, Maristella, Padova 131678
I Volpini, Bologna 079889
VOLTA, Basel 098335
VOLTA NY, New York 098507
Volta Regional Museum, Ho ... 022930
Le Voltaire, Lillebonne 069510
Voltaire & Rousseau, Glasgow .. 144437
Voltaire Fine Art, Paris 105315
Volubilis, Casablanca 082586
Volumina, Milano 080766, 131552, 142747
Volumnia, Perugia 137717
Volunteer Committees of Art Museum, New Orleans 060798
Volunteer Firemen's Mall and Museum of Kingston, Kingston 049986
Volyns'ke Učylyšče Kul'tury i Mystectv, Luc'k 056494
Volyns'kyj Krajeznavčyj Muzej, Luc'k 043228
Volz, Stephanie, Kandern 107506, 130091
Vom Kloster zum Dorf, Creglingen 017778
Vomáčka, František, Praha 065544
Von der Heydt-Museum, Wuppertal 022825
The von Liiebig Art Center, Naples 051165
Vonarburg, Pia, Lenzburg 144097
Vonderau Museum, Fulda 018564
Vonderbank, Bad Reichenhall .. 074574
Vonderbank, Frankfurt am Main 107042
Vondst, de, Utrecht 083668
Vonesch, Roland, Ensisheim ... 103715, 128591
Vonge Nielsen, Astrid, Højbjerg . 065770
Vonholdt, Birgit, Burrweiler 106620
Vonta, Katerina, Athinai 079187
Vonthron, William, Paris 072005
Vontobel Kunstverlag, Feldmeilen 138043
Voor de Oude Jan, Delft 083040
Voorhistorisch Museum, Zonhoven 004164
Voorhuis, Arnhem 082950
Voorhuis, Doetinchem 083174
Voormalig Stoombierbrouwerij De Keijzer N.A. Bosch, Maastricht 032632
Voortrekker Monument Heritage Site, Tshwane 038865
Voortrekker Museum, Pietermaritzburg 038803
Voorvelt, F., Amsterdam 082902
Vor- und Frühgeschichtliche Sammlung, Museumslandschaft Hessen Kassel, Kassel 019553
Vor- und Frühgeschichtliches Museum, Thalmässing 022142
Vor-Ort-Ost, Projektgalerie des Bundes Bildender Künstler Leipzig e.V., Leipzig 107855
Vorarlberger Landesmuseum, Bregenz 001794
Vorarlberger Militärmuseum, Bregenz 001795
Vorarlberger Museumswelt, Frastanz 001902
Vorauer, Karl, Großwilfersdorf .. 062618
Vorbach, Gangelt 075941, 107098
Vorbasse Museum, Vorbasse ... 010363
Vorderasiatisches Museum, Staatliche Museen zu Berlin – Stiftung Preußischer Kulturbesitz, Berlin 017315
Vorderösterreichisches Museum, Endingen am Kaiserstuhl 018207
Vorgeschichtsmuseum im Grabfeldgau, Archäologische Staatssammlung München, Bad Königshofen ... 016800
Vorkink-Heeneman, Den Haag .. 143125
Vorkutinskij Mežrajonnyj Kraevedčeskij Muzej, Vorkuta 038062
Vormayr, Christian, Riedau 127837
Voronežskij Chudožestvennyj Muzej im. I.N. Kramskogo, Voronež 038069
Voronežskij Oblastnoj Kraevedčeskij Muzej, Voronež 038070
Vorselaars, Berlin 074944
Vorspohl, K., Senden, Kreis Coesfeld 078455
Vortex, London 144708
Võrumaa Muuseum, Dr. Fr.R. Kreutzwaldi Memoriaalmuuseum, Võru 010721
Vorwerk, Jürgen, Marburg 077288
Vos, C. de, Aalst 100478
Vos, C.J.M. de, Amsterdam 143078
Vos, J.H., Groningen 083310

Vos, Jacques de, Paris 072006, 105316
Vos, M . de, Knokke-Heist 100764
Vose, Boston 092503, 119929
Vosges Antiquités, Paris 072007
Voskuil, Babs, Midden-Delfland . 112599
Voss Folkemuseum, Vestlandsmuseet,
   Skulestadmo 033951
Voß-Heitbrink, R., Stemwede . . 078558
Voss, Claudia, Hamburg 076334
Voss, D., Isernhagen 076626
Voss, Gisela, Gütersloh 076126
Voss, Kurt, San Antonio 136279
Voss, P., Merzen 077354
Voß, Petra, Niederreißen 077778
Voß, U. & B., Wesselburen 142480
Voss, Viktor, Luzern . . 087650, 133848
Vosse, M.C., Liège 063786
Voßkuhl, Emsdetten 075617
Vostočnaja Galereja, Moskva . . 114318
Vostočno-Kazachstanskij
   Etnografičeskij Muzej, Özkemen 029828
Vostočno-Kazachstanskij Istoriko-
   Kraevedčeskij Muzej, Özkemen 029829
Vostočno-Kazachstanskij Muzej
   Iskusstv, Özkemen 029830
Vostočno-Sibirskaja Gosudarstvennaja
   Akademija Kultury i Iskusstva, Ulan-
   Udě 056309
Vostok Galereja, Ufa 114378
Vostry, Erfurt 106895
Le Votaire, Beirut 082309
Votat, Pierre, Paris . . 072008, 129160
Votava, Petr, Liberec 065416
Votivsammlung der Wallfahrtskirche
   Maria Kirchental, Sankt Martin bei
   Lofer 002681
Votivschatz der Wallfahrtskirche St.
   Erhard, Sankt Jakob-Breitenau 002664
Votre Galeria de Arte, Rio de
   Janeiro 101077
Vouga, Elizabeth & Claude,
   Kirchdorf 087542
La Vouivre, Paris 141296
Vouros Eftaxias Museum, Athinai 023031
Vous Etes Ici, Amsterdam . . . 112308,
                                      137848
Voûtes du Port, Royan 105487
Vox Centre, Montréal 101459
Vox Populi Gallery, Philadelphia . 052101
Voxbrunner, Gudrun, Regensburg 078178
Voyage Intérieur Nice, Nice 104394
Voyager Toys, Houston 093832
Voyageur Heritage Centre,
   Mattawa 006277
Voyer, Michel, Saint-Germain . . 072805
Voyer, Roland, Lugo 086088
Voz Alta, San Diego 124185
Vozroždenie, Moskva 114319
Vrain, Jean Claude, Paris 141297
De Vranse Gans, Amstelveen . . 082670
De Vregille & Bizouard, Dijon . . 125571
Vrienden van de Stedelijke Musea
   Oostende, Oostende 058392
Vries, B. de, Amsterdam 082905
Vries, de, Leeuwarden 083422
Vries, Fenna de, Rotterdam . . . 112672
Vries, Rob de, Haarlem 112519, 112520
Vrindavan Research Institute Museum,
   Vrindavan 024519
Vring, Cornelia, München 130415
Vrolijk, Rotterdam 083604
Vrolijk, Maurice, Dordrecht 112446
Vrouwen in Druk, Amsterdam . . 143079
Het Vrouwenhuis, Zwolle 033129

Vrublëvskaja, Elena, Moskva . . . 114320
VSA Arts, Salt Lake City 123997
VSE Elektro-Museum, Illingen, Kreis
   Neunkirchen 019387
Vserossijskij Istoriko-étnografičeskij
   Muzej, Toržok 037956
Vserossijskij Memorialnyj Muzej-
   zapovednik V.M. Šukšina,
   Srostki 037887
Vserossijskij Muzej Dekorativno-
   Prikladnogo i Narodnogo Iskusstva,
   Moskva 037506
Vserossijskij Muzej Puškina, Sankt-
   Peterburg 037826
Vserossijskij Vystavočnyj Centr,
   Moskva 037507
Vserossijskij Chudožestvennyj Naučno-
   restavracënnyj Centr im. Akademika
   Igorja Grabarja, Moskva 133017
Vsevoložskij Gosudarstvennyj
   Istoriko-kraevedčeskij Muzej,
   Vsevoložsk 038074
VTape, Toronto 101717
VU Centre de Diffusion et de
   Production de la Photographie,
   Québec 006747
Vu Sous 7 Anglès, Lourmarin . . 104051
Vuaille, Jean-Paul, Lyon 128733
Vucci, Pasquale, Bari . 079789, 142622
Vue de l'Abbaye, Brantôme . . . 067450
Vughts Historisch Museum,
   Vught 033015
Vuillemo, Jean-Marc, Besançon 067138,
                                      103372
Vukani Zulu Cultural Museum,
   Fort Nongqayi Museum Village,
   Eshowe 038711
Vuni Sarayi, Lefke 033469
Vuollerim 6000 år – Arkeologiskt
   Museum, Vuollerim 041180
Vuyk, Wout, Amsterdam 143080
De Vuyst, Lokeren 100784, 125349,
                                      140250
VW-Käfer Museum, Gaal 001918
Vychino, Moskva 114321
Východočeská Galerie, Pardubice 009749
Východočeské Muzeum v Pardubicích,
   Pardubice 009750
Východoslovenská Galéria, Košice 038369
Východoslovenské Múzeum,
   Košice 038370
Vylym-Hütte, Userin 022289
The Vyne, Basingstoke 043510
Vyskupo Motiejaus Valančiaus
   Gimtinės Muziejus, Nasrėnų . . 030426
Vysoká Škola Uměleckoprůmyslová v
   Praze, Praha 055300
Vysoká Škola Výtvarných Umení v
   Bratislave, Bratislava 056334
Vystavka Iskusstva, Krasnojarsk 098288
Vystavka Izjaščnych Iskusstv,
   Novosibirsk 098289
Výstavní Síň – Synagoga,
   Hranice 009580
Výstavní Síň Mánes, Praha 102728
Výstavní Síň Musaion, Muzeum
   Těšínska, Havířov 009555
Výstavní Síň Muzea, Ústí nad
   Orlicí 009938
Výstavní Síň Muzea Těšínska,
   Orlová 009739
Výstavní Síň Muzea Těšínska, Muzeum
   Těšínska, Český Těšín 009513
Výstavní síň Muzeum Těšínska,
   Muzeum Těšínska, Jablunkov . 009591
Vystavočnye Zaly na Starom Arbate,
   Gosudarstvennyj muzej A.S. Puškina,
   Moskva 037508
Vystavočnye Zaly Rossijskoj Akademii
   Chudožestv, Moskva 037509
Vystavočnye Zaly v Dome Aksakovych,
   Gosudarstvennyj Literaturnyj muzej,
   Moskva 037510
Vystavočnyj Centr -Galereja, Muzejno-

vystavočnyj kompleks M.T.
   Kalašnikova, Iževsk 037050
Vystavočnyj Centr Rosstrojekspo,
   Moskva 037511
Vystavočnyj Centr Sankt-
   Peterburgskogo Otdelenija Sojuza
   Chudožnikov Rossii, Sankt-
   Peterburg 037827
Vystavočnyj Zal, Čerepovec . . . 036945
Vystavočnyj Zal, Krasnoturinsk 037175
Vystavočnyj Zal – Russkie Chudožniki
   v Samarskoj Gubernii, Žigulevsk 038083
Vystavočnyj Zal Belye Palaty na
   Prečistenke, Moskva 037512
Vystavočnyj Zal Biblioteki im. A.A.
   Bloka, Sankt-Peterburg 037828
Vystavočnyj Zal Biblioteki Iskusstv im.
   A.P. Bogoljubova, Moskva 037513
Vystavočnyj Zal Bogorodskoe,
   Moskva 037514
Vystavočnyj Zal Doma Družby s
   Narodami Zurebežnych Stran,
   Moskva 037515
Vystavočnyj Zal Doma Učenych,
   Sibirskoe otdelenie Rossijskoj
   akademii nauk, Novosibirsk . . 037583
Vystavočnyj Zal g. Smolenska,
   Smolenskij Muzej-Zapovednik,
   Smolensk 037874
Vystavočnyj Zal im. L.I. Kuindži,
   Mariupol's'kij krajeznavčyj muzej,
   Mariupol' 043262
Vystavočnyj Zal Izmajlovo,
   Moskva 037516
Vystavočnyj Zal Karelskoj
   Gosudarstvennoj Filarmonii,
   Petrozavodsk 037657
Vystavočnyj Zal-masterskaja D.I.
   Nalbandjana, Moskva 037517
Vystavočnyj Zal Moskovskogo Fonda
   Kultury, Moskva 037518
Vystavočnyj Zal Moskovskogo Sojuza
   Chudožnikov, Moskva 037519
Vystavočnyj Zal Narodnaja Galereja,
   Moskva 037520
Vystavočnyj Zal Permskogo
   Oblastnogo Kraevedčeskogo Muzeja,
   Perm 037646
Vystavočnyj Zal Raduga,
   Čeboksary 036934
Vystavočnyj Zal Rossijskoj
   Gosudarstvennoj Biblioteki po
   Iskusstvu, Moskva 037521
Vystavočnyj Zal Sojuza Chudožnikov
   Rossii, Sankt-Peterburg 037829
Vystavočnyj Zal Spasskaja Cerkov,
   Irkutsk 037033
Vystavočnyj Zal Tverskoj Oblastnoj
   Kartinnoj Galerei, Tver 037976
Vystavočnyj Zal Tvorčestvo,
   Moskva 037522
Vystavočnyj Zal, Ekaterinburgskij
   Tvorčeskij Sojus Dejatelej Kultury,
   Ekaterinburg 036995
Vystavočnyj Zal, Kirovskij oblastnoj
   chudožestvennyj muzej im. Viktora i
   Apollinarija Vasnecovych, Kirov 037137
Vystavočnyj Zal, Kirovskij oblastnoj
   kraevedčeskij muzej, Kirov . . . 037138
Vystavočnyj Zal, Tulskij muzej
   izobrazitelnych iskusstv, Tula . 037970
Vystavočnyj Zal, Volgogradskij
   muzej izobrazitelnych iskusstv,
   Volgograd 038054
Vystavočnyj Zal, Vologodskij
   Gosudarstvennyj Istoriko-
   Architekturnyj i Chudožestvennyj
   Muzej-Zapovednik, Vologda . . . 038059
Vystavočnyje Zaly Vserossijskogo
   Fonda Kultury, Moskva 037523
Vytauto Didžiojo Karo Muziejus,
   Kaunas 030378
Vytauto Kazimiero Jonyno Galerija,
   Nacionalinis M.K. Čiurlionio Dailės

Muziejus, Druskininkai 030335
Vytauto Valiušio Keramikos Muziejus,
   Utenio Kraštotyros Muziejus,
   Leliūnai 030411
Výtvarná Řemesla, Praha 128368
Výtvarná Škola Václava Hollara,
   Praha 055301
Výtvarné Centrum Chagall,
   Ostrava 009744
Vzgljad, Sankt-Peterburg 114367
W & L Antiquitäten, Eicklingen . 075585
W & W Antiqua, Basel 087264
W & W Antiques, Norfolk 095877
W 139, Amsterdam 112309
W Podwórzu, Sosnowiec 113770
W.13 Antiques, London 090567
W.Art, Longué-Jumelles 104044
W.C. Handy Home Museum and
   Library, Florence 048755
W.D. Trotter Anatomy Museum,
   Dunedin 033199
W.D.O. Gallery, Charlotte 120015
W.G.B. Antiquitäten, Witzhave . 079042
W.H. Over Museum, Vermillion . 053962
W.H. Smith Museum, Newtown,
   Powys 045380
The W.H. Stark House, Orange . 051792
W.K.P. Kennedy Gallery, North
   Bay 006509
W.M. Keck Museum, Reno 052513
W.R.S. Architectural Antiques, Low
   Newton 090639
W125 Galerie, Tilburg 112688
Wa-Ni, Göteborg 115666
Wa, Kwan Kai, Hong Kong 065106
Wa, Kwan Wing, Hong Kong . . 065107
De Waag, Alkmaar 082623
De Waag, Hilversum 083409
De Waagschaal, Delft 083041
Het Waaierkabinet, Amsterdam . 031998
Waal, Hans de, Lith 132568
Het Waalres Museum, Waalre . . 033016
Waanders, Zwolle 137879
WAB-Brocante, Metzerlen 087675
Waba Cottage Museum, White
   Lake 007402
Wabash Frisco and Pacific Association,
   Glencoe 049113
Wabasha County Museum, Reads
   Landing 052487
Wabasso Historical Society Museum,
   Wabasso 054011
Wabaunsee County Historical Museum,
   Alma 046418
Wabberi, Columbus 120521
Waber, G., München 103416
Wable, Annie, Boulogne-sur-Mer 103451
Wable, Annie, Lille 104020
Wabowden Historical Museum,
   Wabowden 007347
Wace, Rupert, London . 090568, 118622
Wachaumuseum, Weißenkirchen in der
   Wachau 002907
Wachholtz Verlag, Neumünster . 137444
Wachiraprasat Museum,
   Phetchaburi 042711
Wachs, Gottfried, Dresden 129726
Wachsstöcklkabinett – Wachszieher-
   und Lebzelter-Museum,
   Viechtach 022326
Wachtelaer, Bruxelles 037914
Wachters, Sabine, Knokke-Heist 100765
Wack Galerie, Kaiserslautern . . 107496
Wackenhut, Houston 093833
Wacker, Bernd, Bad Rodach . . . 074575
Waddesdon Manor, The Rothschild
   Collection, Waddesdon 046027
Waddington, Leslie, London . . . 118623
Waddington, Trevor, Bradford-on-
   Avon 088564, 134136
Waddington's, Toronto 125386
Wade, Birmingham 135356
Wade House Stagecoach Inn and
   Wesley Jung Carriage Museum,

Greenbush 049234
Wade, Colin, Balmain 061097
Wade, Ray, Poulton-le-Fylde ... 091050
Wade, Sara, Korora 099202
Wadena and District Museum and
Nature Centre, Wadena 007348
Wadeye Arts & Crafts, Darwin .. 061391
Wadge, Chris, Salisbury 091227
Wadham, Peter, Exeter 089237, 134378
Wadsworth Atheneum, Hartford . 049390
Wadsworth-Longfellow House,
Portland 052286
Wächter, Willi, Arzberg 074427
Wächter, Winston, New York ... 123166
Wächter, Winston, Seattle 124613
Wäger, Hans, Bielefeld 141601
Wäger, Hans, Herford 141942
Wäinö Aaltonen Museo, Turku .. 011453
Wälchli, Peter, Solothurn 087822
Wälti, Thérèse H., Bern 087309
Wängelin, Thomas, Uppsala .. 133600
Waentig, Prof. Dr. Friederike,
Köln 130172
Wäschepflegemuseum, Rainbach im
Mühlkreis 002564
Wäscher, Bollschweil 075058
Wäscher, A., Freiburg im
Breisgau 075888
Wäscherei-Museum Omas
Waschküche, Eberswalde 018085
Wafangdian Museum,
Wafangdian 008349
Waffek, Nürnberg 077847
Waffenkammer im Munot-Turm,
Schaffhausen 041951
Waffenmuseum, Oberndorf am
Neckar 020908
Waffenmuseum Suhl, Suhl 022076
Waffensammlung Schloss Ebelsberg,
Oberösterreichische Landesmuseen,
Linz 002331
Waffensammlung Willibald Folger,
Bruck an der Mur 001804
Wage, Philadelphia 096172
Wagenburg im Schloss Schönbrunn,
Kunsthistorisches Museum Wien,
Wien 003094
Wagener, Heino, Leer 077043
Wagenmakersmuseum, Leiden . 032587
Wagenpfeil, C., Schwetzingen . 078436
Anna-Maria Wager & Helga Ahrend,
München 077637
Wågermanska Konsthallen,
Skärhamn 115813
Wagga Antique Centre, Wagga
Wagga 062398
Wagga Wagga Art Gallery, Wagga
Wagga 001611
Wagin Historical Village Museum,
Wagin 001612
Wagman & Son, Stanley, Toronto 064776,
101718
Wagnalls Memorial, Lithopolis . 050327
Wagner, Gemmerich 075971
Wagner, Langenlois 127800
Wagner, Paddington, New South
Wales 099539
Wagner, Walliselben 116710
Wagner & Chavanis, Lyon 128734
Wagner & Marks, Frankfurt am
Main 107043
Wagner & Skærbæk Nielsen,
Hinnerup 065760
Wagner Bazarin, São Paulo ... 101117
Wagner Free Institute of Science,
Philadelphia 052102
Wagner-Tichy, Barbara, Pörtschach am
Wörther See 100142
Wagner, Armand, Luxembourg . 082374,
082375
Wagner, C., Heidelberg 076461
Wagner, Claus, Wettingen 087899,
133962
Wagner, Corinne, Saint-Tropez . 073322

Wagner, Dr. Wolfgang, Kassel .. 125997
Wagner, Gabriele, Dresden 129727
Wagner, H., Stegersbach 140041
Wagner, Hans-Jürgen,
Regensburg 142326
Wagner, Howard B., Louisville .. 094603
Wagner, M., Ditzingen 075366
Wagner, Magdalena, Innsbruck . 062658
Wagner, Marie-Claude, Paris ... 072009
Wagner, Matthias, Berlin 141585
Wagner, Odon, Toronto 101719, 128220
Wagner, Olaf, Dresden 129728
Wagner, Richard, Den Haag ... 083149
Wagner, Rolf, Büdingen 129644
Wagner, Shirley, Palm Beach ... 062058
Wagner, Thomas E., Kindsbach . 107557
Wágnerova, Pruhonice 102733
Wagon Wheel Antiques, Louisville 094604,
135769
Wags Curios, Semaphore 062203
Wagstaff, Oxted 090940, 134921
Wah, Chun Kwok, Hong Kong .. 065108
Wah, Leong Shin, Singapore ... 114634
Wah, Tsui Shem, San Francisco 124428
Wah, Tung, Hong Kong 065109
Wahab, Manama 100456
Wahkiakum County Historical Museum,
Cathlamet 047466
Wahl, Heilbronn 141935, 141936
Wahl, Alicja & Bożena, Warszawa 113908
Wahl, Dr. J., Bernburg 141594
Wahl & Søn, Ib, Rødovre 066080
Wahl, Markus, Erwitte 075662
Wahl, O. & R., Moonah 139832
Wahlandt, Edith, Stuttgart 108737
Wahle, Exeter 117776
Wahlrab, Nürnberg 077848
Wahlström, B. & L., Stockholm . 087073
Wahrenbrock, San Diego 145532
Wai & Teng, Singapore 114635
Wai, Poon Chin, Hong Kong ... 065110
Wai, Yin & Dave, Hong Kong .. 065111,
065112
Waibel, Wolfgang, Schramberg . 078397
Waidelich, Christina, Heiden ... 116375
Waiheke Community Art Gallery,
Auckland 033160
Waiheke Island Historical Society
Museum, Waiheke Island 033317
Waihi Arts Centre and Museum
Association, Waihi 033318
Waikato Museum of Art and History,
Hamilton 033216
Waikawa District Museum,
Tokanui 033314
Waikiki Beach Gallery, Honolulu 120975
Waikouaiti Museum, Waikouaiti . 033320
Waimate Historical Museum,
Waimate 033321
Waimer, S., Reutlingen 078216
Waimer, Walfriede, Jettingen-
Scheppach 107482
Wain, Peter, Menai Bridge 090746
Wainscot, Vancouver .. 064896, 128231
Waintraub, Henri, Paris 072010
Wainwright Museum, Battle River
Historical Society, Wainwright . 007349
Waipara County Historical Society
Museum, Hawarden 033219
Waipu Art Gallery, Whangarei .. 113116
Waipu Heritage Centre, Waipu .. 033324
Waipukurau Museum,
Waipukurau 033325
Wairarapa Arts Foundation and
Print Council of New Zealand,
Masterton 059644
Wairoa Museum Koputunga, Toanga
o Te Wairoa, Wairoa 033326
Waitoki Antiques, Auckland 083836
Waitomo Museum of Caves, Waitomo
Caves 033330
Wajawithayawat Museum,
Chulalongkorn University,
Bangkok 042683

Wajima Urushi Bijutsukan,
Wajima 029688
Waka Huia Art & Craft Gallery,
Picton 112991
Wakasa Rekishi Minzoku Shiryokan,
Obama 029197
Wakaw Heritage Museum,
Wakaw 007351
Wakayama-kenritsu Hakubutsukan,
Wakayama 029692
Wakayama-kenritsu Kindai Bijutsukan,
Wakayama 029693
Wake Forest College Bithplace Society
Museum, Wake Forest 054030
Wake Forest University Fine Arts
Gallery, Winston-Salem 054496
Wakefield Antiques, Masterton . 083993
Wakefield Art Gallery, Wakefield 046034
Wakefield Historical Museum,
Wakefield 054032
Wakefield House Museum, Blue
Earth 047056
Wakefield Museum, Wakefield .. 046035
Wakefield Museum Association,
Wakefield 054031
Wakefield, Colin, Wellington ... 132766
Wakefield & Co., Robert, Houston 093834
Wakelin & Linfield, Billingshurst 088465
Wakeman & Sons, A.E., Tedburn Saint
Mary 091550
Wakeman, Frances, Carlton ... 144314
Wakim, Brunswick 127573
Wakoh Hakubutsukan, Yasugi .. 029707
Waksberg, Daniel, Neuilly-sur-
Seine 070524
Walbrecht, Julius, Wuppertal ... 079120
Walbröl, Ursula, Düsseldorf ... 106840
Walch, Peter, Krumbach 062695
Walch, René, Ingwiller 103853
Walcot Reclamation, Bath 088406,
134079
Walczuk, Eugeniusz, Jabłonna . 084525,
132825
Walczyński, Andrzej, Warszawa . 084754,
132908
Wald- und Forstmuseum,
Herschberg 019230
Wald- und Lönsmuseum,
Heinade 019173
Wald- und Moormuseum Berumerfehn,
Großheide 018851
Waldarbeits-Museum und Historische
Wappenschmiede, Elmstein .. 018165
Waldbauernmuseum, Gutenstein 002031
Waldburg-Zeil, Franz-Clemens,
Hohenems 127770
Waldeck, Dr. Hans, Gau-
Algesheim 075952
Walden, Amsterdam 112310
Walden, London 144709
Walden Pond, Oakland 145389
Waldgeschichtliches Museum
Sankt Oswald, Sankt Oswald-
Riedlhütte 021606
Waldhorn & Adler, New Orleans 095175
Waldhufen-Heimatmuseum Salmbach,
Engelsbrand 018209
Waldinger, Andreas, Ravensburg 078151
Waldmann & Busch, Bad
Segeberg 074597
Waldmuseum, Burg,
Dithmarschen 017680
Waldmuseum, Furth im Wald . 018569
Waldmuseum, Mehlmeisel 020337
Waldmuseum, Münstertal 020614
Waldmuseum, Surwold 022089
Waldmuseum, Zwiesel 022916
Waldmuseum Göhrde, Göhrde . 018724
Waldmuseum im Wildpark Neuhaus,
Holzminden 019329
Waldmuseum Reichental,
Gernsbach 018661
Waldmuseum Stendenitz,
Neuruppin 020741

Waldmuseum Wassermühle,
Wingst 022680
Waldmuseum Watterbacher Haus,
Kirchzell 019632
Waldnaabtal-Museum in der Burg
Neuhaus, Windischeschenbach 022679
Waldner-Millesi, Sigrid, Basel .. 133661
Waldner, Erich R., Ludwigsburg . 126029
Waldo Antique Center, Kansas
City 094102
Waldock, Blackrock 109532
Waldon, Minneapolis 094955
Waldron, Patrick, Cirencester .. 088922,
134276
Waldschmidt-Ausstellung,
Eschlkam 018277
Waldviertler Bauernhaus-Museum,
Weitra 002911
Waldviertler Bauernhof-Museum,
Gföhl 001935
Waldviertler Eisenbahnmuseum,
Sigmundsherberg 002756
Waldviertler Münzecke,
Unterthumeritz 062865
Walenda, Günther, Wiesloch ... 079005
Walendowski, Michał, Łódź ... 084620
Walentowski Galerien, Galerie im
Elisenhof, Binz, Ostseebad .. 106504
Walford, Leslie, Double Bay ... 061406
Walhalla, Donaustauf 017930
Walhalla Old Wares, Walhalla .. 062402
Walicki, Leszek, Kraków 113647
Walison, Marieke, Epen 132503
Walk a Crooked Mile Books,
Philadelphia 145420
Walk Gallery, London 118624
Walkabout Gallery, Leichhardt . 099225
Walkatjara Art Centre, Yulara .. 062510
Walkaway Station Museum,
Walkaway 001614
Walker, Austin 092276
Walker, Denver 120786, 120787
Walker, Harrogate 089443, 117919,
134439
Walker, Indianapolis 093939
Walker, Ottawa 125374
Walker, Toronto 064777
Walker & Hall, Wellington 084196
The Walker African American Museum,
Las Vegas 050173
Walker Art Center, Minneapolis . 050948
Walker Art Center Publications,
Minneapolis 138365
Walker Art Collection, Garnett .. 049054
The Walker Art Gallery, Liverpool 044967
Walker Galleries, Honiton 117994
Walker Hill Art Center, Seoul ... 030051
Walker Street Gallery, Dandenong 001015
Walker, Alan, Halfway 089411
Walker, Alfred J., Boston 119930
Walker, Charles, Seattle 097519
Walker, John, New York 095816
Walker, Judith, Hermagor 100035
Walker, Judith, Klagenfurt 100077
Walker, Judith, Weizelsdorf 100218
Walker, Melissa Inez, San Diego 124186
Walker, Sarah, Castletownbere . 109544
Walkers, Beechworth 061116
Walkers Gallery, Dublin 109626
Walker's Point Artists Association,
Milwaukee 060799
Walker's Point Center for the Arts,
Milwaukee 050926
Walking Stick Shop, Arundel ... 088290
Walkner, Wien 063094
Walkner, Martin, Salzburg 140031
Walkter & Hall, Auckland 083837
The Wall, Ely 117752
Wall Antiques, Ballycullane ... 079450
Wall Impressions, Jacksonville . 121274
Wall Roman Site and Museum –
Letocetum, Lichfield 044827
Wall Street Exchange, Richmond,
Victoria 099639

Wall Street Gallery, Calgary 101266
Wall, Kandy, Columbus 120522
Wall, Thomas, Djursholm 133521
Wallace, Auckland 112817
Wallace, Brookfield 061218
Wallace, Fort Worth 093485
Wallace, Memphis 135786
Wallace, Portland 145464
Wallace Area Museum, Wallace
  Bridge 007352
The Wallace Collection, London . 045125
Wallace District Mining Museum,
  Wallace 054040
Wallace Early Settlers Museum,
  Riverton 033295
Wallace Galleries Ltd., Calgary . 101267
Wallace House, Somerville 053387
The Wallace Museum Foundation,
  Montgomery 051026
Wallace, Barbara, Philadelphia . 123468
Wallace, Kevin V., Los Angeles . 121617
Wallace, Robert, Armadale,
  Victoria 061051
Wallace, Robert, Paddington, New
  South Wales 062046
Wallaceburg and District Museum,
  Wallaceburg 007353
Wallack, Ottawa 101483
Wallaroo Heritage and Nautical
  Museum, Wallaroo 001615
Wallbrecht, Dr. R.E., Hannover . 130019
Wallén, Sala 143913
Wallenburg, Gotha 107152
Wallenfang, Heidi, Rösrath 078265
Wallenfels'sches Haus, Abteilung Vor-
  und Frühgeschichte, Archäologie
  und Völkerkunde, Oberhessisches
  Museum und Gail'sche Sammlungen,
  Gießen 018694
Waller, Amsterdam 112311
Waller Antik, Bremen 075188
Waller, Blaise, Paris 072011
Wallet Fouque, Amiens 103229, 128439
Wallett, Shirley, Griffith 061563
Wallfahrts- und Heimatmuseum,
  Altötting 016559
Wallfahrtsbasilika Mariazell,
  Schatzkammern, Galerien,
  Reliquienkammer und
  Krippenkammer, Mariazell 002365
Wallfahrtsmuseum, Maria Taferl 002361
Wallfahrtsmuseum, Neukirchen beim
  Heiligen Blut 020714
Wallfahrtsmuseum, Steingaden . 021985
Wallfahrtsmuseum Inchenhofen,
  Inchenhofen 019400
Wallfahrtsmuseum Maria Steinbach,
  Legau 019953
Wallfahrtsmuseum, Katholische
  Kirchenstiftung Weihenlinden-Högling,
  Bruckmühl 017638
Wallflower Gallery, Miami 121873
Wallin, Ingemar, Uppsala 087128
Wallin, Ralph, Seattle 136379
Walling, Julia, Saint Albans ... 119006
Wallingford Gallery, Wallingford . 119298
Wallingford Historical Society Museum,
  Wallingford 054041
Wallingford Museum, Wallingford 046036
Wallington, The National Trust,
  Cambo 043836
Wallis & Wallis, Lewes 126944
Wallis Museum at Connors State
  College, Warner 054059
Walliser Reb- und Weinmuseum,
  Salgesch 041917
Walliser, Hans, Herrliberg 087524,
                                 133768
Walliser, Hans, Meilen . 087668, 133857
Wallmuseum, Oldenburg in
  Holstein 020994
Wallner, Nicolai, København ... 102915
Wallpaper History Society,
  London 060272

Wallqu Museum, Yuanqu 008505
Wallraf-Richartz-Museum & Fondation
  Corboud, Museen der Stadt Köln,
  Köln 019715
Wallrock, Woolloongabba 062491
Walls, Amsterdam 112312
Walls Art, Nashville 122104
Walls of Art, Los Angeles 121618
Wallscapes, Las Vegas 121389
Wallspace, New York 123167
Wallwitz, Angela Gräfin von,
  München 077638
Wally Findlay Galleries, New York 051490
Walmer Castle, Walmer 046039
Walmstedtska Gårdens Museivåning,
  Uppsala 041123
Walnusshof, Leverkusen 077096
Walnut Grove Plantation,
  Roebuck 052660
Walonia, Utrecht 083669
Walpole, Monaco 082533
The Walpoles, London 090569
Walpurgishalle Thale, Thale
  (Harz) 022138
Walpurgius, Theodor, Nassau .. 142217
Walrus Gallery, Wellington 113095
Walry, Gent 100709
Walsall Leather Museum, Walsall 046042
Walsall Local History Centre,
  Walsall 046043
Walsall Museum, Walsall 046044
Walschot, Roermond 083536
Walsemann, D.G.C., Utrecht .. 083670
The Walsenburg Mining Museum and
  Fort Francisco Museum of La Veta,
  Walsenburg 054046
Walser Heimatmuseum,
  Triesenberg 030304
Walser, Philippe, Mazerulles .. 128791
Walser, Rupert, München 108206
Walserhaus Gurin, Bosco 041354
Walsermuseum, Alagna Valsesia 025261
Walsermuseum, Riezlern 002600
Walsermuseum Lech-Tannberg,
  Lech 002284
Walsh, Chris, Seattle 124614
Walsh, Frank O., Atlanta 092197, 144973
Walsh & Co, John, Ossett 127003
Walsh, Julie & Ralph, Chicago . 120294
Walt Whitman Birthplace, Huntington
  Station 049631
Walt Whitman House, Camden . 047381
Walter, Oberhausen 077866
Walter-Adams, San Francisco . 124429
Walter and McBean Galleries,
  San Francisco Art Institute, San
  Francisco 053030
Walter Anderson Museum of Art,
  Ocean Springs 051702
Walter Bischoff Galerie, Berlin . 106456
Walter de Gruyter GmbH & Co. KG,
  München 137429
Walter de Gruyter GmbH & Co.
  KG, Redaktion Allgemeines
  Künstlerlexikon, Leipzig 137351
Walter Elwood Museum,
  Amsterdam 046459
Walter F. George Law Office Museum,
  Vienna 053985
Walter P. Chrysler Boyhood Home and
  Museum, Ellis 048555
Walter P. Chrysler Museum, Auburn
  Hills 046656
Walter Phillips Gallery, Banff .. 005381
Walter Rothschild Zoological Museum,
  Tring 045999
Walter Runebergin Veistoskokoelma,
  Porvoo 011276
Walter W.S. Cook Alumni Lecture,
  Locust Valley 139521
Walter Wickiser Gallery Inc., New
  York 123168
Walter, A., Freiburg im Breisgau 075889
Walter & Son, A., Hamilton ... 128176

Walter, David, Claremont 061327
Walter, Dr. Chlaus, Meilen 138075
Walter, Fabian & Claude, Basel . 116109
Walter, François, Colombier
  (Neuchâtel) 133724
Walter, Irénée, Ballersdorf ... 066925
Walter, Karl-Heinz, Weimar ... 078903
Walter, M., Pforzheim 078044
Walter, Manfred, Freigericht .. 075894
Walter, Manfred,
  Kleinschmalkalden 130132
Walter, P.J., Lindau 077121
Walter, R., Betheln 074966
Walter, Roland, Wien 063095
Walter's Antique Clock & Watch Repair,
  New York 136023
Walters Art Museum, Baltimore . 046772
Walters Art Museum. Journal,
  Baltimore 139522
Walter's Sweet Antiques,
  Minneapolis 094956
Walters, Eileen, New York 123169
Walters, J.D. & R.M., Rolvenden 091134,
                                 134977
Walterwoods Antiques, Masterton 083994
Waltham Historical Society Museum,
  Waltham 054052
The Waltham Museum, Waltham 054053
Walther, Ratingen 108436, 131477
Walther, Caroline, Stuttgart ... 130691
Walther, Evi, Schaffhausen ... 087805
Walther, Konrad, Röderland ... 078256
Walton Galleries, Guisborough . 117898
Walton Hall Heritage Centre,
  Warrington 046056
Walton Hall Museum, Linford .. 044837
Walton House Museum,
  Centerville 047501
Walton Maritime Museum, Walton-on-
  the-Naze 046047
Walton, Michael, Dublin 079596
Walton, Stephen, Bradfield Saint
  George 117410
Waltring, Isabella, Köln 130173
Waltscher Heimatstube,
  Neckargemünd 020654
Waltzer, Stewart, New York ... 123170
Waltzing Matilda Centre and Qantilda
  Museum, Winton 001649
Walvis Bay Museum, Walvis Bay 031859
Walz, Überlingen 078748
Walz, Winden 079018
Walz, Worms 079066
Walz, Ed., Feldafing 141763
Walz, Wolfgang & Marianne,
  Großwallstadt 076115
Walzinger, Anita & Axel,
  Saarlouis 108552
Wambsgans, W., München 077639
Wamco, Oakland 095921
Wan-Bo, Shanghai 065154
Wan Fang Arts Museum, Taipei . 042603
Wan Li Brick Museum, Yuanli . 042603
Wan Zhou District Museum,
  Chongqing 007709
Wanapum Dam Heritage Center,
  Beverly 046969
Wanat, M., Kraków 084603
Wandel, Düsseldorf ... 075535, 129765
Wandle, London 090570
Wandle Industrial Museum,
  Mitcham 045280
Wandsworth Museum, London . 045126
Wanecq, Jean & Getraud, Paris . 072012
Wang, Oslo 113249
Wang, San Jose 124463
Wang Hsiu Chi Sculpture Park,
  Taipei 042561
Wang, Chen, Singapore 085527, 085528
Wang, Elizabeth, New York ... 123171
Wang, Kwong, Hong Kong 065113
Wang, Lucy, San Diego 124187
Wang, Maggy, Thezac 073647
Wang, Marcel, Wien 063096

Wangal Aboriginal Arts & Crafts,
  Kuranda 061707
Wangaratta Exhibitions Gallery,
  Wangaratta 001618
Wangaratta Museum, Wangaratta 001619
Wangfu Museum, Naiman Qi ... 008093
Wangjian Tomb Museum,
  Chengdu 007699
Wangjiang Museum, Wangjiang . 008350
Wangmiao Display Center,
  Dujiangyan 007752
Wang's Fine Art Museum,
  Kaohsiung 042369
Wangsmo, Trondheim 143427
Wanhan Puoti, Oy, Orimattila .. 140737
Wanho Shrine Museum, Taichung 042482
Wannemacher, W., Langen,
  Hessen 077022
Wannenes, Milano 080767
Wannenes & C., Genova 080302
Wannenmacher-Museum,
  Emsdetten 018200
Wanner, Anita, Zürich 134004
Wanner, Anne-Marie, La Cadière-
  d'Azur 068960
Wanner, Gerda, Kloten 087544
Wanner, Rolf, Baden 144011
Wansanxia Museum, Wanxian .. 008352
Wansink, Roermond 112623
Wantagh Historical & Preservation
  Society, Wantagh 054055
Wantika, Söderhamn 086943
Wanuskewin Heritage Park,
  Saskatoon 006985
Wanxi Museum, Lu'an 008051
Wanzke, Dr. Wolfgang, Augsburg 141419
Wao, Fukuoka 111252
Waow Gallery, Bruxelles 063582
Wapello County Historical Museum,
  Ottumwa 051845
Wapler, Vincent, Paris 125773
Waquet, Patrick, Fismes 068534
War in the Pacific National Historical
  Park, Asan 023357
War in the Pacific National Historical
  Park, Piti 023359
War Memorial, Dublin 024910
War Memorial Carillon Tower
  and Military Museum,
  Loughborough 045155
War Memorial Gallery of Fine Arts,
  University of Sydney, Sydney . 001556
War Museum, Addis Ababa 010725
War Museum, Valletta 030759
War Museum, Yangon 031836
War Remnants Museum, Ho Chi Minh
  City 054812
War Room and Motor House
  Collection, Harrogate 044520
War Shop, Tampa 097593
War Souvenirs, Auckland 083838
Wara, Lewis, Seattle 124615
Warabekan, Tottori 029634
Waragil, Blackheath 098752
Warba Mirdawaji Aboriginal Art,
  Roebourne 099647
Warbinek, Siefersheim 108613
Warburg Institute Publications,
  London 138257
Ward, London 090571
Ward, Sidcup 091339
Ward, Tampa 136400
Ward & Chowen, Tavistock ... 127065
Ward & Nasse, New York 123173
Ward & Ward, Kansas City ... 121319
Ward Antique Fireplaces, London 090572,
                                 134791
Ward County Historical Society
  Museum, Minot 050952
Ward Hall, Georgetown 049080
The Ward Museum of Wildfowl Art,
  Salisbury University, Salisbury . 052902
Ward O'Hara Agricultural Museum of
  Cayuga County, Auburn 046654

Ward-Thomas, Christian,
Balsham 088345, 134053
Ward-Thompson, Willoughby ... 099888
Ward, Keith, Tulsa 136438
Ward & Son, L.L., Brandsby ... 088572
Ward, Mark, North Geelong 061993
Ward, Meredith, New York 123172
Ward, Paul, Sheffield 091295
Wardenær, Elverum 132772
Warden's House Museum,
Stillwater 053529
Wardil, Marco A., Belo Horizonte 063928
Wardown Park Museum, Luton .. 045168
Wardrobe, Brighton 088638
Ware, Kamakura 081991
Ware Museum, Ware 046051
Wareham Town Museum,
Wareham 046052
The Warehouse, Amsterdam ... 082906
Warehouse Artworks, Winnipeg . 101897
The Warehouse Gallery, Henley-on-
Thames 117962
Warehouse Gallery, Houston ... 121149
Warehouse Gallery, Lee 050211
Wares for Art, New York 123174
Warger, Doris, Frauenfeld 133738
Waringarri Arts Aboriginal,
Kununurra 099206
Warisan, Los Angeles 094497
Warkums Erfskip, Workum 033066
Warkworth and District Museum,
Warkworth 033339
Warlenius, Agneta, Stockholm . 133583
Warminster Antiques Centre,
Warminster 091737
Warmun Art Centre, Warmun .. 099858
Warndt-Heimatmuseum,
Völklingen 022348
Warnecke-Tamaschke & Prof. Dr.
Tamaschke, Ravenstein 108446
Warner, Denver 093350
Warner Archive, Milton Keynes . 045275
Warner Museum, Concord Historical
Society, Springville 053477
Warner, W.W., Brasted 088587
Warooka and District Museum,
Warooka 001620
Waroschitz, Lutz, Berlin 074945
Warracknabeal Historical Centre,
Warracknabeal 001622
Warragul and District Historical
Museum, Warragul 001623
Warrandyte Old Post Office Museum,
Warrandyte 001624
Warren, San Francisco 136351
Warren & Wignall, Leyland 126946
Warren County Historical Society
Museum, Lebanon 050207
Warren County Museum, Warren 054068
Warren Historical Museum,
Warren 054061
Warren ICBM and Heritage Museum,
Cheyenne 047630
Warren Rifles Confederate Museum,
Front Royal 048989
Warren, F.J., Luton 138261
Warren, Fizzy, Stockbridge ... 091434
Warren, Jimmy, Littlebourne ... 089848,
134566
Warren, Joan, San Diego 124188
Warren, John F., Merchantville . 145255
Warren, Linda, Chicago 120295
Warren, Maureen, Chicago 120296
Warrender & Co, S., Sutton 091513,
135107
Warrens, Leiston 089801, 134545
Warrick County Museum,
Boonville 047092
Warringah, Brookvale 098806
Warrington Museum and Art Gallery,
Warrington 046057
Warrnambool Art Gallery,
Warrnambool 001625
Warrumbungles, Coonabarabran 098922

Warsaw Historical Museum,
Warsaw 054073
Warshaw, Karen, New York 095817
Warson Woods Antique Gallery, Saint
Louis 096791
Warszawski Antykwariat Naukowy
Kosmos, Warszawa 143534
Warszawskie Centrum Numismatyczne,
Warszawa 084755, 126429
Wartburg, Eisenach 018137
Wartel, Jean-Paul, Couterne .. 068210
Warth, Willy von der, Krefeld ... 126020
Warther Museum, Dover 048355
Warumpi Arts, Alice Springs ... 060992
Warwick & Warwick, Warwick .. 127078
Warwick Antique Centre,
Warwick 091753, 144908
Warwick Art Gallery, Warwick .. 099864
Warwick Castle, Warwick 046061
Warwick Doll Museum, Warwick 046062
Warwick Gallery, Beaconsfield . 117318
Warwick Henderson, Auckland . 112818
Warwick Leadlay, London 090574,
144710
Warwick Museum of Art,
Warwick 054078
Warwick Oakman, Newtown,
Tasmania 061966
Warwick Regional Art Gallery,
Warwick 001627
Warwickshire Museum, Warwick 046063
Warwickshire Museum of Rural Life,
Moreton Morrell 045296
Warwickshire Yeomanry Museum,
Warwick 046064
Was Designs, Tucson 124782
Wasa Antik, Stockholm 087074
Wasch-en Strijkmuseum, Boxtel 032097
Waschhaus Potsdam, Potsdam . 021207
Waschkowitz, Christiane,
Aschaffenburg 074436
De Wascht en Strekt, Gilze 032353
Wasco County Historical Museum, The
Dalles 048130
Wascoe's Siding Antiques,
Glenbrook 061529
Waseca County Museum,
Waseca 054080
Waseda Daigaku Tsubouchi Hakushi
Kinen Engeki Hakubutsukan,
Tokyo 029620
Waser, Basel 144038
Washakie Museum, Worland ... 054551
Washburn County Historical Society
Museum, Shell Lake 053305
Washburn Historical Museum,
Washburn 054082
Washburn, Joan T., New York .. 123175
Washington Antiques Show,
Washington 098508
Washington Area Lawyers for the Arts,
Washington 060800
Washington Art Association,
Washington Depot 060801
Washington Center for Photography,
Washington 054173
Washington County Historical
Association Museum, Fort
Calhoun 048791
Washington County Historical Society,
Washington 054181
Washington County Historical Society
Museum, Portland 052305
Washington County Historical Society
Museum, Washington 054187
Washington County Historical Society
Museum, West Bend 054288
Washington County Museum,
Akron 046328
Washington County Museum,
Sandersville 053067
Washington County Museum of Fine
Arts, Hagerstown 049312

Washington Crossing Historic Park,
Washington Crossing 054192
Washington Dolls House and Toy
Museum, Washington 054174
Washington Framing and Art,
Washington 124972
Washington Gallery, Penarth ... 118886
Washington Green, Newcastle-upon-
Tyne 138267
Washington Green Fine Art Publishing,
Birmingham 138140
Washington Historical Museum,
Washington 054180
Washington International Arts Letter,
Washington 139523
Washington Irving, New York ... 123176
Washington Museum,
Washington 054182
Washington Museum Association,
Bellevue 060802
Washington National Cathedral,
Cathedral Church of Saint Peter and
Saint Paul, Washington 054175
Washington Old Hall, Washington 046067
Washington Pavillion of Arts and
Science, Sioux Falls 053358
Washington Printmakers,
Washington 060803
Washington Printmakers Gallery,
Washington 124973
Washington Square Gallery,
Philadelphia 096173
Washington Square Outdoor Art
Exhibit, New York 123177
Washington State Capital Museum,
Olympia 051758
Washington State History Museum,
Tacoma 053651
Washington Studio School, Chevy
Chase 056867
Washington, Sherry, Detroit ... 120819
Washington's Headquarters,
Newburgh 051513
Washington's Lands Museum and
Sayre Log House, Ravenswood 052476
Waskada Museum, Waskada ... 007359
Wasmuth, Berlin 141586
Wasmuth, Ernst, Tübingen 137538
Wassenberg Art Center, Van Wert 053944
Wassenberg, Maria Helena, Chur 087375,
133722
Wasser- und Fischkundemuseum,
Kuhlen-Wendorf 019825
Wasser, Frank, Denkendorf ... 129677
Wasserburg Haus Kemnade, Museum
Bochum, Hattingen 019103
Wasserkraftmuseum, Vattenfall Europe
Generation (VE-G), Ziegenrück . 022875
Wasserkunst von 1535, Bad
Arolsen 016707
Wasserleitungsmuseum, Reichenau an
der Rax 002580
Wasserleitungsmuseum, Museum
Wildalpen, Wildalpen 003112
Wasserlof, Irmgard, Wien 063097
Wassermann, Potsdam 078087
Wassermann, Wolfgang, Feldkirchen-
Westerham 106949
Wassermühle Höfgen, Grimma . 018823
Wassermühle Kuchelmiß,
Kuchelmiß 019814
Wasserschloß Groß Leuthen,
Märkische Heide 020204
Wasserschloß Mitwitz, Mitwitz . 020428
Wasserturm-Galerie, Mannheim . 077278
Wassmuth, Agathe, Hann.
Münden 107341
WaSu Galleries, Tucson 124783
Wasyl Negrych Pioneer Homestead,
Gilbert Plains 005874
Wat Ko Museum, Phetchaburi .. 042712
Wat Phra Thart Lampang Luang
Museum, Lampang 042695
Watabe, Sapporo 082107

Watanabe, S., Tokyo 111520
Watari-Um, Tokyo 029621
Watashi no Shigotokan, Seika . 029354
Watatu Gallery, Nairobi 111564
The Watch Doctor, Los Angeles . 094498,
135747
Watch Out, Portland 096488
Watch This Space, Alice Springs 098642
Watchet Market House Museum,
Watchet 046068
Watelet, Olivier, Paris 072013
Water and Colors Gallery, Miami 121874
Water Gallery, Angono 113374
Water Gallery, Bourton-on-the-
Water 117406
Water Meadow Barn Gallery, Isles-of-
Scilly 118030
Water Mill Museum, Water Mill . 054198
Water Ski Hall of Fame, American
Water Ski Educational Foundation,
Polk City 052230
Water Supply Museum,
Weymouth 046110
Water Tower, Louisville 121675
Water Tower Museum, Gunnedah 001127
Waterbolk, Niek, Utrecht 143253
Waterbridge Gallery, Bacup 117266
Watercolor, New York 139524
Watercolor Art Society Houston,
Houston 060804
Watercolor Gallery, Tucson 124784
Watercolor Magic, Cincinnati ... 139525
Watercolor Society of Alabama,
Birmingham 060805
Watercolors Gallery, Pittsburgh . 123584
Watercolour Society of Queensland,
Nundah 058227, 099494
Watercolour Society of Wales,
Raglan 060273
The Watercolours & Drawings Fair,
London 098408
Waterfall Antiques, Ross-on-Wye 091142
Waterfall's, Buffalo 135381
Waterfield's, Oxford 144774
Waterford Gallery, Altrincham . 117235
Waterford Historical Museum,
Waterford 054202
Waterford Museum of Treasures,
Waterford 024999
Waterfront Antiques Market,
Falmouth 089253
Waterfront Galerie, Gent 100710
Waterfront Gallery, Honolulu ... 120976
Waterfront Gallery, Westport ... 025000
Waterfront Museum, Poole 045531
Waterfront Museum and Showboat
Barge, Brooklyn 047268
Watergate Antiques, Chester ... 088868,
134254
Watergate Gallery, Washington . 124974
The Watergate Street Gallery,
Chester 117573
Waterhouse & Dodd, London ... 118625,
134793
Waterhouse Gallery, Barton-under-
Needwood 117292
Wateridge, Barry J., Hindhead .. 134464
Waterland Neeltje Jans,
Vrouwenpolder 033012
Waterlane & Dorset, Salisbury .. 144824
Waterloo Area Farm Museum, Grass
Lake 049212
Waterloo Center for the Arts,
Waterloo 054208
Waterman, London ... 090575, 118626
Waterman, K. & V., Amsterdam . 082907
Watermark, Cockermouth 117611
The Watermark, Preston 118949
Watermark Studio, Penrith 118889
Watermen's Museum, Yorktown 054592
Watermill Brooklyn Gallery, Byrd
Hoffman Watermill Foundation,
Brooklyn 119945
Waters, Geoffrey, London 090576

Waterside Antiques Centre, Ely . 089203
Waterside Arts, Edinburgh 117744
Waterside Gallery, Bideford 117362
Waterside Gallery, Saint Mawes 119047
Watersnoodmuseum, Ouwerkerk 032758
Waterstone, Portland 123685
Waterstraat, Horst, Hamburg ... 076335
Watertown Historical Society Museum,
Watertown 054212
Watervliet Arsenal's Museum of the
Big Guns, Watervliet 054222
Waterway Gallery, Armadale, Western
Australia 098669
Waterways Visitor Centre, Dublin 024911
Waterworks Museum – Hereford,
Hereford 044577
Waterworks Visual Arts Center,
Salisbury 052906
Watford Museum, Watford 046070
Watgen, Marc, Landau in der
Pfalz 077006
Watkins Bros., Denver 120788
Watkins, London 144711
Watkins College of Art and Design,
Nashville 057510
Watkins Community Museum of
History, Lawrence 050189
Watkins Gallery, American University,
Washington 054176
Watkins Institute Art Gallery,
Nashville 051193
Watkins Museum, Taylorsville .. 053691
Watkins Woolen Mill, Lawson .. 050193
Watkins, Chuck, Fort Worth ... 135561
Watkins, Islwyn, Knighton 089728
Watling Antiques, Crayford 088999
Watling Street Galleries,
Towcester 119252
Watling, Bruce, Newstead 099447,
                          127651
Watson, Christchurch 126313
Watson, Hamburg 076336
Watson & Vincent, Nelson 132732
Watson Arts Centre, Watson ... 099870
Watson Crossley Community Museum,
Grandview 005916
Watson-Curtze Mansion, Erie Historical
Museum, Erie 048600
Watson Farm, Jamestown 049791
Watson-Guptill, New York 138431
Watson Museum, Rajkot 024458
Watson, B., Johannesburg 114804
Watson, G. & J., Turramurra ... 062355
Watson, G. & J., Williamstown . 062428
Watson, Gordon, London 090577
Watson, Peter, Armadale, Victoria 061052
Watson and Son, Thomas,
Darlington 126873
Watsons Bay Art Gallery, Watsons
Bay 099871
Watson's Creek Antiques, Kangaroo
Ground 061668
Watson's Mill, Manotick 006256
Watt Space Gallery, Newcastle . 099437
The Watteau Society, London .. 060274
The Watteau Society Bulletin,
London 139229
Wattebled, Xavier, Lille 125615
Watten Cane Furniture,
Singapore 085529
Wattenbach, Philippsthal 108380
Das Wattenmeerhaus,
Wilhelmshaven 022664
Watterott, Gilbert, Hausen 130021
Watters Smith Memorial, Lost
Creek 050480
Watters, Frank, Sydney 099799
Wattis, Hong Kong 102125
Wattlebark, Mogo 099360
Watts Gallery, Compton, Surrey . 043998
Watts Towers Arts Center, Los
Angeles 050472
Watzke, Barbara, Aschaffenburg 074437
Watzl, Ludwigsburg 107897

Watzlawik, S., Regensburg 108468
Wauchope District Museum,
Wauchope 001629
Wauconda Township Museum,
Wauconda 054231
Waude, Mönchengladbach 077400
Waukesha County Museum,
Waukesha 054233
Wauman, Knokke-Heist 100766
Wave Gallery, Philadelphia ... 123469
Waveland Museum, Lexington . 050253
Wavelength Gallery, Falmouth . 117786
Waverley Antiques, Upper Largo 091690
Waverly, Falls Church . 127222, 145146
Waves, New York 095818
Wavriensia, Wavre 138560
Wawana, Maastricht 112589
Wawota and District Museum,
Wawota 007370
Wax Museum, Dublin 024912
Wax Museum, Santa Claus ... 053098
Wax Museum Český Krumlov, Muzeum
Voskových Figurín České Historie,
Český Krumlov 009511
Waxin, Marie-Thérèse, Roubaix . 072552
Way We Wore, Houston 093835
Way, R.E. & G.B., Newmarket . 144742
Way, Richard, Henley-on-Thames 144470
Wayfarer Books, Old Hall Bookshop,
Looe 144717
Wayland, Washington 124975
Wayland Historical Museum,
Wayland 054247
Wayne Art Center, Wayne 054255
Wayne Campbell Gallery, Toronto 101720
Wayne Center for the Arts,
Wooster 054541
Wayne County Historical Museum,
Richmond 052536
Wayne County Historical Society
Museum, Lyons 050541
Wayne County Historical Society
Museum, Wooster 054542
Wayne County Museum,
Honesdale 049525
Wayne, Arthur, Winnipeg 101898
Wayne, Neil, Belper 088438
Wayne, Shoshana, Santa Monica 124472
Wayne's World, Timaru 113032
Wayser, Saint-Ouen 105571
Waysi Museum, Fufeng 007773
Wayside Antiques, Duffield ... 089110
Wayside Antiques, Tattershall . 091543
Wayside Folk Museum, Zennor . 046236
Wayside Shop, Oklahoma City . 096019
Wayside Treasures, Philadelphia 096174
Wayville Latvian Museum, Brooklyn
Park 000896
Waywood, Byron Bay 098832
Waywood Furniture, Wellington . 132767
Wazee Deco, Denver 093351
Wazzau, Iris, Davos Platz 116241
WCK Galleries, Taman Melawati 111910
Weald and Downland Open Air
Museum, Chichester 043947
Wear It Again Sam, San Diego . 097146
Weardale Museum, Weardale .. 046072
Weasel Pop Studio, Saint Paul . 123938
Weather Discovery Center, North
Conway 051602
Weatherell, Paul, Harrogate .. 089444
Weatherford, David, Seattle ... 097520
Weathersby, Virginia Beach ... 097797,
                          136450
Weatherspoon Art Museum, University
of North Carolina at Greensboro,
Greensboro 049257
Weaver, Blackheath 098753
Weaver, Tucson 136424
Weaver Gallery, Chester 117574
Weaver Gallery, Northwich .. 118822
Weaver Neave & Daughter,
London 134794
Weaver, Trude, London 090578

Weaver's Cottage, Kilbarchan .. 044692
Weavers' Triangle Visitor Centre,
Burnley 043795
Weaving Craft Museum,
Fengyuan 042314
Web Fine Art, Basingstoke 117295
Webantik, Strasbourg 073600
Webb, Buffalo 092569
Webb, Winchester 091874, 119362,
                          135240
The Webb-Deane-Stevens Museum,
Wethersfield 054356
Webb House Museum, Newark . 051506
Webb, Dan F., Oakland 145390
Webb, Michael, Llangristiolus . 089873
Webb, Roy, Wickham Market . 091841
Webb's, Auckland 083839, 126309,
                          132699
Webbs Road Fine Art, London . 118627
Weber, Baltimore 092374
Weber, Firenze 131280
Weber, Fort Worth 093486
Weber, Köln 076918
Weber, Saint Louis 096792
Weber & Bläuer, Lugano 087625
Andrea Weber & Andreas Fischer,
Dresden 075434
Weber-Schwenk, Petra, Castrop-
Rauxel 106627
Weber State University Art Gallery,
Ogden 051717
Weber, Alberto, Torino 111015
Weber, Britta, Goch 076023
Weber, Carmen, Zug 116887
Weber, Carola, Wiesbaden ... 108932
Weber & Assoc., Edith, New York 095819
Weber, Frank, Saarbrücken ... 078336
Weber, H. & L., Straubenhardt . 078574
Weber, Hans, Holzkirchen 130066
Weber, Helga Maria, Hamburg . 129995
Weber, Jakob, Bad Tölz 074608
Weber, Jamileh, Zürich 116866
Weber, Jean, Wiltz 111875
Weber, Joachim, Stockdorf ... 078563
Weber, Josef, Mainz . 107956, 137371
Weber, Karin, Hong Kong 102126
Weber, L., Badem 074628
Weber, L.& M., Erkerode 075648
Weber, M., Frei-Laubersheim . 075841
Weber, Malavé von, Caracas .. 097969
Weber, Michael, Guldental 076127
Weber, Michael, Poole 118934
Weber, Norbert, Eckernförde .. 106859
Weber, Peter, Offenberg 077911, 108332
Weber, Peter, Trier 078708
Weber, Peter, Unterpleichfeld . 130716
Weber, Thomas, Niederzissen . 077782
Weber, W., Kaufungen 076736
Weber, W. & E., Mannheim ... 077279
Weberei- und Heimatmuseum
Laichingen, Laichingen 019862
Weberei- und Heimatmuseum
Ruedertal, Schmiedrued 041958
Webereimuseum, Breitenberg . 017575
Webereimuseum, Sindelfingen . 021875
Webereimuseum, Heimatverein
Haslach, Haslach 002063
Weberhaus Marlesreuth, Naila . 020638
Weberhausmuseum,
Schauenstein 021633
Webermuseum, Biberach an der
Riß 017359
Webermuseum, Framsohn Frottier,
Heidenreichstein 002069
Weberstube Jonsdorf, Jonsdorf,
Kurort 019468
Webster, Fresno 093517
Webster, Leatherhead 118083
Webster County Historical Museum,
Red Cloud 052490
Webster House Museum, Elkhorn 048538
Webster, David, New York 123178
Webster, E.T., Blythburgh 088511
Webster & Co., G., Wellington . 113096

Webster, Janice, Bacchus Marsh 098682
Webster, Sande, Philadelphia . 123470
Webtiques, Louisville 094605
Wechsler, Ruedi, Buochs 087341
Weckmans, G. & C., Bruxelles . 063583
Weckström, Björn, Espoo 103129
Wed Galerie, Utrecht 112713
Wedberg, Göteborg 115667
Wedde, Peter, Wellington 084197
Wedding and Custom Museum,
Kuah 030679
Wedel, Martina von, Hamburg . 129996
Weder, Johannes, Bayreuth ... 074700
Wedgewood Studio, Phoenix .. 136123
Wedgwood Museum, Stoke-on-
Trent 045886
Wedhorn, Jürgen, Hamburg ... 076337
Wednesbury Museum and Art Gallery,
Wednesbury 046073
Wee Jumble Shop, Sacramento . 096654
Weecke, B., Horn-Bad Meinberg 141953
Weedon Antiques, Weedon .. 091766
Weedon, Mike, London 090579
Weegmuseum de Oude Waag,
Someren 032895
De Weegschaal, Bussum 083023
Weeks Air Museum, Miami .. 050841
The Weem, Pittenweem 091018, 118916
Weemaels, P. & M., Bruxelles . 063584
Weems, Albuquerque . 119518, 119519
Weems-Botts Museum, Dumfries 048390
Weems, Michael, Baltimore ... 119778
Weeping Water Valley Historical
Society Museum, Weeping
Water 054264
Weerd, H. de, Ruinen 132626
Weerenbeck, Maastricht 083487
Weerheijm, H., Zaandam 083700
Weerts & Van Gijsel, Roermond . 083537
Weg des Friedens, Purgstall an der
Erlauf 002549
Wegener, Gitana, Burgdorf ... 141668
Wegiel, Michel, Villeurbanne .. 074230
Wegmachermuseum, Wasserburg am
Inn 022440
Wegner, Alfons, Rosenheim .. 108516
Wegner, Carl, Berlin 141587
Wegner, Josef, Köln 107697
Wehling & Fülscher, Hamburg . 076338
Wehner, Gordon, Storkow 078571
Wehr, K., Pulheim 108418
Wehr, Rainer, Stuttgart 108738
Wehrens, Aachen 074345
Wehrens, Horst, Düsseldorf ... 075536
Wehrgeschichtliches Museum,
Salzburg 002642
Wehrgeschichtliches Museum Rastatt,
Rastatt 021286
Wehrlin-Bieli, Michaela, Liestal . 087595
Wehrtechnische Studiensammlung des
Bundesamtes für Wehrtechnik und
undBeschaffung (BWB), Koblenz 019668
Wehrturm, Niederstetten 020803
Wei, Bruxelles 063585
Wei Chin-Yuan Musical Instruments
Museum, Taipei 042562
Wei County Museum, Weixian . 008360
Weibertreu-Museum, Weinsberg 022523
Weibull, Vedbæk 102998
Weicheltmühle, Hartmannsdorf-
Reichenau 019082
Weichhart, Felicitas, Wien 063098
Weick, Wilhelm, Berlin 074946
Weidenfeld & Nicolson, London . 138258
Weider, Auray 103291
Weidle, Barbara, Bonn 137118
Weidler, Herbert, Nürnberg ... 126086
Weidman, West Hollywood ... 125032
Weidmann Cottage Heritage Centre,
Muswellbrook 001333
Weidmann, Wolfgang,
Oberkotzau 077829
Weidner, S., Rosenheim 078277
Weies, Martin, Fröndenberg ... 129870

Weifang Kite Museum, Weifang . 008353
Weifang Museum, Weifang 008354
Weigand, Heinz-Martin, Ettlingen 106939
Weigel, Georg & Christa,
Gunzenhausen 107190
Weigelt, Jörg, Hannover 125986
Weigl, Erwin, Nürnberg 108310
Weigl, Gitta, Maxhütte-Haidhof . 107983
Weigt, Paul, Lübeck 077191
Weihai Museum, Weihai 008356
Weihermüller-Curylo, C.,
Leverkusen 142074
Weihermüller, C., Leverkusen .. 142073
Weihnachts-Sammlerbörse Zürich,
Zürich 098336
Weihui Museum, Weihui 008357
Weijand, Karel, Farnham 089264, 134387
Weil Antiquités, Montréal 064488
Weil Art Gallery, Texas A&M University,
Corpus Christi 048007
Weil, Alain, Paris 072014
Weil, Christiane, Mirepoix 070130
Weil, Pascal, Fegersheim 068520
Weiland, Gustav, Lübeck 142090
Weiland, Klaus, Limburg an der
Lahn 077117
Weilinger, Dr. Arthur, Salzburg .. 100181
Weilinger, Dr. Arthur, Wien 100390
Weill, Etienne, Bogny-sur-Meuse 067265
Weiller, Paris 105317
Weimann, Alexander,
Kapellendorf 130093
Weimer, Josef, Elz 129788
Wein- und Heimatmuseum,
Durbach 018067
Wein- und Heimatmuseum,
Mehring 020338
Wein- und Heimatmuseum,
Winningen 022684
Wein-und Heimatmuseum,
Leiwen 020027
Weinan Center Museum, Weinan 008359
Weinbau- und
Fahnenschwingermuseum,
Neckenmarkt 002431
Weinbau- und Heimatmuseum
Klingenberg, Klingenberg am
Main 019647
Weinbau- und
Landwirtschaftsmuseum,
Hadersdorf 002036
Weinbau- und
Landwirtschaftsmuseum, Oetwil am
See 041819
Weinbau in Mundelsheim,
Mundelsheim 020618
Weinbau-Museum, Chur 041418
Weinbau Museum Peter, Wachenheim
an der Weinstraße 022358
Weinbaum, David, New York ... 095820
Weinbaumuseum, Asperg 016636
Weinbaumuseum, Flörsheim-
Dalsheim 018375
Weinbaumuseum, Ortenberg,
Baden 021005
Weinbaumuseum, Sitzenberg-
Reidling 002760
Weinbaumuseum Alte Kelter,
Erlenbach, Kreis Heilbronn ... 018270
Weinbaumuseum am Zürichsee,
Au 041239
Weinbaumuseum im Herrenhof,
Neustadt an der Weinstraße .. 020762
Weinbaumuseum Meersburg,
Meersburg 020334
Weinbaumuseum Stuttgart-Uhlbach,
Stuttgart 022065
Günter Weinbeer & Gabriele
Guttenberger, Ilshofen 076601
Weinbeer, K., Diebach 075350
Weinberg Nature Center,
Scarsdale 053190
Weinberg, Daniel, Los Angeles . 121619
Weinberger, København 102916

Weinbrenner, Ulrike, Stuttgart .. 130692
Weindorfmuseum, Horrweiler .. 019345
Weinek, Linz 140016
Weinek, Salzburg 140032
Weiner, München 126071
Weiner, Günter, Wangen im
Allgäu 078864
Weiner, M., Elzach 141738
Weiner, Patricia, Cincinnati ... 120385
Weingärtner, Speyer 108654
Weingärtner & Koch, Bad
Wörishofen 074619
Weingart Galleries, Los Angeles 050473
Weingartner, Dominik, Luzern .. 087651
Weingreen Museum of Biblical
Antiquities, Dublin 024913
Weingutmuseum Hoflößnitz,
Radebeul 021258
Weinheimer Auktionshaus Georg
Schmidt E.K., Weinheim 126138
Weinheimer, Corinna, Trondheim 132802
Weinlandmuseum, Asparn an der
Zaya 001730
Weinman Mineral Museum,
White 054371
Weinmann, Liz, Köln 107698
Weinmuseum, Güssing 002020
Weinmuseum, Vaihingen an der
Enz 022298
Weinmuseum Burgenland – Eisenstadt,
Eisenstadt 001863
Weinmuseum im Kloster Machern,
Bernkastel-Kues 017337
Weinmuseum Kitzeck, Kitzeck im
Sausal 002184
Weinmuseum Prellenkirchen,
Prellenkirchen 002539
Weinmuseum Schlagkamp-Desoye
Senheim, Senheim 021843
Weinmuseum Schloß Gamlitz,
Gamlitz 001924
Weinstadtmuseum Krems,
Dominikanerkirche, Krems .. 002239
Weinstadtmuseum Poysdorf,
Poysdorf 002527
Weinstein, Minneapolis 122052
Weinstein, San Francisco ... 124430,
124431
Weinstein, Harvey, New York .. 095821
Weintraub, Michael, New York .. 145377
Weinviertler Kunstschule, Mistelbach
an der Zaya 054957
Weinviertler Museumsdorf,
Niedersulz 002455
Weinviertler Oldtimermuseum Poysdorf,
Poysdorf 002528
Weinzierl, Paul, Moosthenning .. 130314
Weinzingergasse 3, Wien 127975
Weir Farm, Wilton 054457
Weir Gallery, Mundaring 099414
Weird Sisters, Minneapolis ... 122053
Weird Stuff, Carnegie 061289
Weird Stuff Antique, Kansas City 094103
Weisbrod, New York 095822
Weise, Elmshorn 075602
Weise, Gitte, Paddington, New South
Wales 099540
Weise, H.-D., Weinheim 078911
Weise, Regine & Bernd,
Chemnitz 106637
Weiser-Wehner, Hochstetten-
Dhaun 076543
Weiser, Ingrid, Stetten 062848
Weiser, Therese, Frankfurt am
Main 075831
Weisgerber, Andreas, Billerbeck 129592
Weishäupl, München 142193
Weismann, Jörg, Düsseldorf ... 075537
Weiß, Altenstadt, Kreis Neu-Ulm 105978
Weiss, London 090580, 134795
Weiss, Wien 140128
Weiß, Witten 079035
Weiß, Andreas, Rothenburg ob der

Tauber 078290, 130604
Weiß, Annerose, Rodewisch ... 078253
Weiß, Brigitte, Münchberg 077441
Weiss, Brigitte, Zürich 116867
Weiss, H., Köln 076919
Weiss, H., Mödling 062742
Weiss, Helga, Graz 062604
Weiss, Jan, Ribe 066062
Weiss, L., Dettighofen 075343
Weiss, Lothar, Jestetten 076651
Weiss, Mike, New York 123179
Weiss, Sharon, Columbus 093056
Weiß, Wolfgang & Maritta,
Römerberg 078261
Weissenborn, Stephanie, Krefeld 107745
Weißenhof-Institut, Interdisziplinäres
Forum für Architektur, Raum und
Möbel, Stuttgart 055728
Weissenhofmuseum im Haus Le
Corbusier, Stuttgart 022066
Weißenhorner Heimatmuseum,
Weißenhorn 022547
Weißenrieder, Wiesbaden ... 078999,
130765
Weissenstein-Museum, Weissenstein
bei Solothurn 042128
Weissert, Friedrich, Paris 141298
Weissert, Nikolaus, Tübingen .. 142433
weissfeld, Roentgenwerke AG,
Tokyo 111521
Weißgerbermuseum, Doberlug-
Kirchhain 017917
Weißgerberwalk, Biberach an der
Riß 017360
Weissing, Alsfeld 074385
Weisův Dům, Husitské Muzeum v
Táboře, Veselí nad Lužnicí 009948
Weisz, Richard M., Chicago 092814
Weisz, Spencer, Chicago 120297
Weitwandermuseum, Eibiswald . 001850
Weitz & Coleman, New York ... 145378
Weitz, Harold B., Pittsburgh ... 096340
Weitz, Pascal, Clamart 068004
Weitze, Heinrich, Weimar 130741
Weitze, Helmut, Hamburg ... 076339,
141901
Weitze, Wolfgang, Jena 141964
Weitzel, Regina 101536
The Al Weitzel Gallery, Saskatoon 101553
Weixler, Charles, Louisville ... 135770
Weiyuan, Quanzhou 102164
Weizmann Archives and House,
Rehovot 025180
Weka, Frankfurt am Main 075832
Weka-Nautilus, Essen 141753
Welbourne House, Mitcham ... 061863
Welbry, Lemesos 122555
The Welch Regiment Museum (41st/
69th Foot) of the Royal Regiment of
Wales, Cardiff 043870
Welch, Barney K., Dallas 093226
Welch, K. & A., Warminster ... 091738
Welch, M.G., Taunton . 091547, 135122
Welco International, Chennai ... 109298
Welcome Home Antiques, New
York 095823, 136024
Weldon, Southport . 091386, 135066
Weldon, J.W., Dublin . 079597, 079598
Welholme Galleries, Grimsby .. 044484
Welker, Karl & Karin, Homburg . 130068
Welker, Rudolf & Erika, Dannstadt-
Schauernheim 141678
Welkom Museum, Welkom ... 038877
Well Art, Warszawa 113909
Well-Dressed Wall, Milwaukee . 121956
Well Federal Artist, Philadelphia 123471
Well House Antiques, Tilston ... 091600
Well Hung, London 118628
Well Read Book Shop, Wellington 143375
Wella Museum, Darmstadt ... 017830
Wellage & Schaff, Cincinnati .. 120386
Welland Historical Museum,
Welland 007372
Wellandgalerie, Aalen 105954

Wellbrook Beetling Mill,
Cookstown 044007
Wellby, H.S., Haddenham 089400,
134428
Wellcome Library for the History
and Understanding of Medicine,
London 045127
Wellcome Museum of Anatomy and
Pathology, London 045128
Weller & Cwalinski, Houston ... 121150
Weller-Plate, Peter, Ockenheim . 130485
Weller, Sam, Salt Lake City 145509
Wellesbourne Wartime Museum,
Wellesbourne 046075
The Wellesley, New York 123180
Wellesley College Friends of Art
Newsletter, Wellesley Hills ... 139526
Wellesley Historical Society, Wellesley
Hills 054267
Wellfleet Historical Society Museum,
Wellfleet 054269
Wellford, Memphis 094672
Wellhausen, Houston 121151
Welling & Co., Kansas City ... 094104
Welling, Anthony, Ripley, Surrey 091117
Wellingborough Heritage Centre,
Wellingborough 046077
Wellings Landsbymuseum,
Lintrup 010221
Wellington Arch, London 045129
Wellington Arts Centre,
Wellington 033361
Wellington Aviation Museum, Moreton-
in-Marsh 045295
Wellington B. Gray Gallery, East
Carolina University, Greenville . 049264
Wellington Community Historical
Museum, Wellington, Ontario . 007373
Wellington County Museum,
Fergus 005802
Wellington Courthouse Museum,
Wellington 001631
Wellington Gallery, Hobart ... 099141
Wellington Gallery, Kilmarnock . 118043
Wellington Salerooms, Wellington 127081
Wellington Sculpture Trust,
Wellington 059645, 113097
Wellm, Horst, Pattensen 142297
Wellmann, E., Derschen 075334
Wellmer, Elisabeth-Christiane &
Dietrich, Himbergen 130048
Wellmer, Elisabeth-Christiane &
Dietrich, Westensee 130751
Wells, Los Angeles 094499
Wells Antiques Centre, Wells-next-the-
Sea 091775
Wells Auction Rooms, Wells ... 127082
Wells Auto Museum, Wells ... 054273
Wells County Historical Museum,
Bluffton 047061
Wells County Museum,
Fessenden 048728
Wells Fargo History Exhibits,
Portland 052306
Wells Fargo History Museum, Los
Angeles 050474
Wells Fargo History Museum,
Minneapolis 050949
Wells Fargo History Museum,
Phoenix 052125
Wells Fargo History Museum,
Sacramento 052734
Wells Fargo History Museum, San
Diego 052988
Wells Fargo History Museum, San
Francisco 053031
Wells Fargo History Museum-Old
Sacramento, Sacramento ... 052735
Wells Museum, Wells . 007375, 046078
Wells Walsingham Light Railway,
Wells-next-the-Sea 046080
Wellspring Art Gallery, Tralee .. 109700
Wellview Antiques, Carnaross
Kells 079470

Welsch, Trier 078709
Welser Puppenweltmuseum,
Wels 002920
Welsh Highland Railway Museum,
Porthmadog 045544
Welsh, Jorge, Lisboa 084997
Welsh, Jorge, London . 090581, 134796
Welshpool and Llanfair Light Railway,
Llanfair Caereinion 044883
WelTec Visual Arts Programmes,
Wellington Institute of Technology,
Petone 056195
Welterbe Kloster Lorsch, Lorsch 020119
Welters, Fons, Amsterdam 112313
Weltkulturerbe Völklinger Hütte,
Europäisches Zentrum für Kunst und
Industriekultur, Völklingen . . 022349
Weltkunst, München 138760
Welvaars, Holman, Antwerpen . . 063253
Welvaars, Odette, Amsterdam . . 082908
Welwert, Christine, Kitzbühel . . 062666
Welwyn Roman Baths Museum,
Welwyn 046084
Welz, Salzburg 100182, 136560, 140033
Welz, Friedrich, Heidelberg . . . 141931
Welzer, Prof. Dr. Harald,
Hannover 107375
Wemaere, Bertrand, Versailles . . 074072
Wemaere, Frederic, Verneuil-sur-
Avre 074032
Wemken, Bad Zwischenahn . . . 074624
Wen Bao Zhai Art Shop,
Shanghai 102206
Wen Tianxiang Temple, Beijing . 007649
Wenatchee Valley Museum,
Wenatchee 054281
Wenceslao Vinzons Historical
Landmark Museum, Vinzons . . 034617
Wende, Düsseldorf 075538
Wendel, Georg, Frankfurt am
Main 075833
Wendelboe, Bergen . . 084229, 113138
Wendell Gilley Museum, Southwest
Harbor 053425
Wendeng Museum, Wendeng . . 008361
Wending, Amsterdam 082909
De Wending, Delft 083042
Wendl, Martin, Rudolstadt 126108
Wendland, Holger R., Dresden . . 106749,
137150
Wendlandt, Wuppertal 108997
Wendler, Jörg, Merchweiler 077348
Wendling, Franz-Josef,
Kastellaun 076733
Wendouree, Ballarat 139633
Wendover Antique Centre,
Wendover 091779, 135202
Wendrich, Ursula,
Steinbergkirche 108669
Wendt, Leipzig 077070
Wendt & Friedmann, Berlin 106457
Wendt, G., Meine 077324
Wendt, J., Berlin 141588
Wendys Portraits, Hazel Grove . . 117943
Wenger, Michael, Stuttgart 078639
Wengerzink, Lothar, Paderborn . 142291
Wengyuan Museum, Wengyuan . 008362
Wenham Museum, Wenham 054282
Wenig, Caroline, Wertingen 108884
Wening, J., Gronau, Westfalen . . 129920
Wening, P.B.J., Groningen 132510
Weninger, Erich, Wien 063099
Wenk, Neckargemünd 130439
Wenk, Heinz, Dortmund 106710
Wenke, Christfried, Tübingen . . 078731,
130711
Wenlock Fine Art, Much Wenlock 118762,
134862
Wenmiao Museum, Tianjin 008330
Wennekes & Zoon, J., Den Haag 143126
Wenner, H.Th., Osnabrück . . . 126091,
126092, 142280, 142281
Wenshanzhou Museum, Wenshan 008363
Wenshui Museum, Wenshui . . . 008364

Wenske & Jehmlich, Potsdam . . 130541
Wensley, Ramsbottom 118954
Wensleydale Galleries,
Northallerton 118816
Wenten, Brigitte, Xanten 079130
Wenten, Heinz, Xanten 079131
Wentholt, Leiden 083452
Wentrup, Jan, Berlin 106458
Wentscher, Jürgen, Meckenheim,
Rheinland 130293
Wentworth, Atlanta 119630
Wentworth, Boston 119931
Wentworth, Charlotte 120016
Wentworth, Columbus 120523
Wentworth, Miami 121875, 121876
Wentworth, Saint Louis 123884
Wentworth-Coolidge Mansion,
Portsmouth 052317
Wentworth Galleries, Waterloo . . 099869
Wentworth Gallery, Laguna
Beach 121326
Wentworth Gallery, Newposrt
Beach 123223
Wentworth Gardner and Tobias Lear
Houses, Portsmouth 052318
Wentzingerhaus, Museum für
Stadtgeschichte, Freiburg im
Breisgau 018491
Wenxi Museum, Wenxi 008365
Wenz, Ginsheim-Gustavsburg . . 076003,
125958
Wenz, Renate, Mainz 077246
Wenz, Wolfgang, Berlin 074947
Wenzel, Zürich 088016, 134005
Wenzel-Hablik-Museum, Itzehoe 019438
Wenzel, Dorothea, Werther 108883
Wenzel, Matthias, Bamberg 074675
Wenzhou Museum, Wenzhou . . . 008366
Wenzl, Krystyna, Nestelbach bei
Graz 062745
Weoley Castle, Birmingham 043631
Wepf & Co., Basel 144039
Werbe Design Akademie,
Innsbruck 054945
Werberger Stuben, Motten 020475
Werdenfelser Museum, Garmisch-
Partenkirchen 018593
Wereld Bodem Museum, International
Soil Reference and Information
Centre, Wageningen 033020
Wereldmuseum Rotterdam,
Rotterdam 032830
WerkART, Sankt Gallen 116619
Werkbundarchiv – Museum der Dinge,
Berlin 017316
Werkgalerie am Läuferplatz, Bern 116152
Werkladen, Köln 130174
Werkmann, Olomouc 128333
Werksmuseum Achse, Rad und
Wagen, Wiehl 022625
Werksmuseum der Firma Linke-
Hofmann-Busch, Salzgitter . . 021577
Werksmuseum der MTU,
München 020584
Werksmuseum der Waechtersbacher
Keramik, Brachttal 017527
Werksmuseum Koenig & Bauer AG,
Würzburg 022806
Werkspoor Museum, Amsterdam 031999
Werkstatt 52, Karlsruhe 130101
Werkstatt Bleichhäuschen, Rheda-
Wiedenbrück 108490
Werkstatt für Kunst, Bonn 075096
Werkstatt-Galerie, Davos Dorf . . 116238
Werkstatt Kollerschlag,
Kollerschlag 062682, 100079
Wern Mill Gallery, Nannerch . . . 118771
Werner, Holzdorf 076563
Werner, São Paulo 064217
Werner, Wiesbaden 079000
Werner Berg Museum, Bleiburg . 001786
Werner-Egk-Begegnungsstätte,
Donauwörth 017937
Werner Weckwerth Museum, Friedrich-

Wilhelm-Lübke-Koog 018522
Werner, L., München . . 142194, 142195
Werner, M., Köln 107699
Werner, Peder, Stockholm 133584
Werner, Ralf & Margit, Radolfzell am
Bodensee 078119, 130551
Werner, U., Celle 075248
Werner, Walter, Oberursel 077888
Werners Antik, Lerum 086805
Wernersche Verlagsgesellschaft,
Worms 137559
Wernher Collection at Ranger's House,
London 045130
Wernigeröder Kunst- und Kulturverein
e.V., Wernigerode . . . 059309, 108881
Werning, Wien 100391, 127976
Werra-Kalibergbau-Museum, Heringen
(Werra) 019207
Werribee and District Historical
Museum, Werribee 001633
Werstener Kreuz, Düsseldorf . . . 075539
Werth, Dr. Wolfgang, Düsseldorf 106841
Werthern, J. von, Hofheim am
Taunus 076555
Werthmüller, Lucius, Basel 144040
Wertpapierantiquariat,
Wolfenbüttel 126142, 142504
Wertpapierwelt, Olten 041826
Wertz Bros., Los Angeles 094500,
094501
De Wervel, Dordrecht 083202
Werwein, Erich, Neunkirchen am
Brand 130450
Werz, Bernd, Ludwigshafen am
Rhein 130252
Werz, J., Bolsward 132445
Weschler & Son, Adam A.,
Washington 097866, 127488
Wescott's Station Antiques, Saint
Paul 096867
Weserburg Museum für Moderne
Kunst, Bremen 017603
Weskamp, Stephan, Höxter 076547,
107438
Weslaco Museum, Weslaco 054284
Wesley's House and Museum,
London 045131
Wesner, Konstanz 076938, 107725
Wessel & O'Connor, Miami 121877
Weßel, C., Hamburg 129997
Wesselhoefft, A., Burgdorf 075224
Wesselstuerne, København 010200
Wessex, Wareham 119305
Wessex Antiques, Sherborne . . . 091311,
135038
Weßolowski, Oberhausen 130478
West 114, New York 095824
West Alabama Auctions, Eutaw . 127217
West Allis Historical Society Museum,
West Allis 054286
West Antic, Brest 067481
West Antique, Oslo 084341
West-Art, Clarence 138327
West Baton Rouge Museum, Port
Allen 052241
West Bend Art Museum, West
Bend 054289
West Bengal Academie of the
Arts, Rabindra Bharati University,
Kolkata 055814
West Berkshire Museum,
Newbury 045341
West Blatchington Windmill,
Hove 044611
West Burton Antiques, Los
Angeles 094502
West Chicago City Museum, West
Chicago 054295
West Coast Antiques, Portland . . 096490
West Coast Gallery, Osborne Park 099506
West Coast Historical Museum,
Hokitika 033222
West Coast Museum of Flying,
Sidney 007036

West Coast Peddler, Whittier . . . 139527
West Coast Pioneers Museum,
Zeehan 001678
West Coast Style, Los Angeles
Art, Design and Antiques, Santa
Monica 098509
West Cork Model Railway Village,
Clonakilty 024850
West Cork Regional Museum,
Clonakilty 024851
West Country Pictures, South
Molton 119116
West End Antiques, Saint Louis . 096793
West End Antiques Mall,
Richmond 096585
West End Exchange, Edmonton . 064310
West End Galleries, Buxton 088750
West End Gallery, Edmonton . . . 101306
West End Gallery, Houston 121152
West End Gallery, Miami 121878
West End Gallery, Milwaukee . . 121957
West End Gallery, Richmond . . . 123751
West End Gallery, Wirksworth . . 119369
West End House Gallery,
Smarden 119106
West Essex Antiques, Matching
Green 090734
West Expansion, Casablanca . . . 112041
West Gallery, Quezon City 113421
West Gallery @ Art Center,
Mandaluyong 113401
West Gate Towers, Canterbury City
Museums, Canterbury 043861
West Hants Historical Society Museum,
Windsor, Nova Scotia 007423
West Harlem Art Fund, New York 123181
West Highland Museum, Fort
William 044366
West Kilbride Museum, West
Kilbride 046090
West Los Angeles Book Center, Los
Angeles 145239
West Louisville Antiques,
Louisville 094606
West Midlands Police Museum,
Birmingham 043632
West Neuk Gallery, Forfar 117805
West of the Pecos Museum,
Pecos 051960
West Overton Museums,
Scottdale 053210
West Palm Beach Antique and
Collectables Show, West Palm
Beach 098510
West Park Museum, Macclesfield 045187
West Parry Sound District Museum,
Parry Sound 006615
West Pasco Historical Society Museum,
New Port Richey 051311
West Point, Den Haag 083150
West Point Lighthouse Museum,
O'Leary 006552
West Point Museum, West Point 054322
West Port Books, Edinburgh . . . 144404
West Portal Books, San
Francisco 145569
West River Museum, Philip 052106
West Rock Antiques, Warwick . . 091754,
135192
West Sea Company Nautical Antiques,
San Diego 097147
West Side Art, Los Angeles 121620
West Sølv og Mynt, Oslo 084342, 126382
West Somerset Railway,
Minehead 045276
West Somerset Rural Life Museum,
Allerford 043394
West Southwest Gallery, Denver 120789
West Stow Anglo-Saxon Village, Bury
Saint Edmunds 043812
West Street Antiques, Dorking . . 089096
West Street Gallery, Crewkerne . 117647
West Surrey Decorative and Fine Arts
Society, Godalming 060275

West Texas Connection, Austin . 092277
West Torrens Railway, Signal
Telegraph and Aviation Museum,
Brooklyn Park 000897
West Valley Art Museum,
Surprise 053615
West Vancouver Museum, West
Vancouver 007382
West Virginia Independence Hall,
Wheeling 054369
West Virginia Northern Community
College Alumni Association Museum,
Wheeling 054370
West Virginia State Farm Museum,
Point Pleasant 052228
West Virginia State Museum,
Charleston 047564
West Virginia University Mesaros
Galleries, Morgantown 051057
West Wales Antiques, Murton . . 090792
West Wales Arts Centre,
Fishguard 044348
West Wycombe Motor Museum, West
Wycombe 046094
West Wycombe Park House, West
Wycombe 046095
West Yorkshire Print Workshop,
Mirfield 118750
West, Brian, Sandgate 091243
West, J.R., Vancouver 128232
West, John T., Portland 096489
West, Mark J., Redhill 091084
West, Seb, Saint Ives, Cornwall . 119037
Westallgäuer Heimatmuseum, Weiler-
Simmerberg 022482
WestArt, Auburn 139528
Westbeth, New York 123182
Westbury Antiques, Melbourne . 061835
Westbury Manor Museum,
Fareham 044329
Westchester Art Workshop Gallery,
Westchester Community College,
State University of New York, White
Plains 054372
Westchester Art Workshop,
Westchester Community College,
State University of New York, White
Plains 058097
Westcliff Picture Gallery, Westcliff-on-
Sea 119329
Westcliffe Gallery, Sheringham . 091317,
135041
Westcott, Montréal 140370
Westcott Gallery, Dorking 089097,
117685
Westdale Bookworm, Hamilton . 140351
Westdeutscher Künstlerbund e.V.,
Bochum 059310
Westelynck, Serge, Flêtre 068543
Westend Antik, Stuttgart 078640
Westende Gemtime, Christchurch 083906
Westenholz, London 090582
Westerberg, Stockholm 087075
Westerhaus, Michael, Wuppertal 130797
Westerhof, J.B., Burgwerd 083013,
126278
Western Aerospace Museum,
Oakland 051685
Western Approaches, Liverpool . 044868
Western Archaeological and
Conservation Center, Tucson . . 053841
Western Art Association,
Ellensburg 060806
Western Art Gallery and Museum, The
University of Montana, Dillon . 048330
Western Association for Art
Conservation, Santa Monica . 060807
Western Australia Scout Museum,
West Perth 001634
Western Australian Maritime Museum,
Fremantle 001086
Western Australian Medical Museum,
Subiaco 001527
Western Australian Museum Geraldton,

Geraldton 001098
Western Australian Museum Perth,
Perth 001403
Western Australian School of Art
Design and Media, TAFE International
Western Australia, Perth 054917
Western Canada Aviation Museum,
Winnipeg 007475
Western Center for Conservation of
Fine Arts, Denver 135536
Western Coin, San Diego 097148
Western Collectables, Killarney
Vale 061690
Western Colorado Center for the Arts,
Grand Junction 049183
Western Development Museum,
Saskatoon 006986
Western Front, Vancouver 007305
Western Galerie, Neustadt an der
Weinstraße 108263
Western Gallery, Western Washington
University, Bellingham 046894
Western Hennepin County Pioneers
Museum, Long Lake 050387
Western Heritage Center, Billings 046988
Western Heritage Gallery,
Albuquerque 092091
Western Hotel Museum,
Lancaster 050111
Western Illinois University Art Gallery,
Macomb 050569
Western Jewelry and Coin Exchange,
Seattle 097521
Western Kentucky University Gallery,
Bowling Green 047160
Western Light Art Galleries, Achill 109524
Western Manitoulin Island Historical
Society Museum, Gore Bay . . . 005894
Western Museum, Abilene 046306
Western Museum of Mining and
Industry, Colorado Springs . . 047891
Western Museums Association,
Berkeley 060808
Western New Mexico University
Museum, Silver City 053339
Western New York Artists Group,
Buffalo 060809, 119970
Western Power World of Energy,
Fremantle 001087
The Western Railway Museum,
Suisun 053593
Western Ranger, Bridgnorth . . . 117427
Western Reserve Historical Society
Museum, Cleveland 047828
Western Rhode Island Civic Museum,
Coventry 048041
Western Skisport Museum, Donner
Pass 048338
Western Springs Museum, Western
Springs 054332
Western States Arts Federation,
Denver 060810
Western Wyoming College Art Gallery,
Rock Springs 052634
Western, C., Weingarten, Baden . 142474
Westerwald Museum Motorrad &
Technik, Steinebach 021982
Westerwolds Crashmuseum, Oude
Pekela 032748
Westflälisch-Niederländisches
Imkereimuseum, Museen der Stadt
Gescher, Gescher 018677
Westfälische Münz- und
Sammlerbörse, Dortmund . . . 098169
Westfälischer Kunstverein,
Münster 020610, 059311
Westfälisches Feuerwehrmuseum,
Hattingen 019104
Westfälisches Glockenmuseum,
Museen der Stadt Gescher,
Gescher 018678
Westfälisches Museum für Archäologie,
Herne 019219
Westfälisches Römermuseum Haltern,

Haltern am See 018973
Westfälisches Schieferbergbau-
und Heimatmuseum Holthausen,
Schmallenberg 021685
Westfälisches Schulmuseum,
Dortmund 017966
Westfalen Verlag, Bielefeld 137112
Westfalia-Auto-Museum, Rheda-
Wiedenbrück 021392
Westfield Heritage Village,
Rockton 006820
Westfries Museum, Hoorn 032510
Westgate Antiques, Clonmel . . . 079489
Westgate Books, Saskatoon . . . 140387
Westgate Gallery, North Berwick 118812
Westgate Museum, Winchester . 046156
Westhorpe Antiques, Southwell . 091388,
135067
Westin, Trollhättan 087107
Westlake Gallery, Seattle 124616
Westland London, London 090583
Westlands Museum voor Streek- en
Tuinbouwhistorie, Honselersdijk 032500
WestLicht, Schauplatz für Fotografie,
Wien 100392
Westling, Göran, Stockholm . . . 115937
Westlock and District Museum,
Westlock 007383
Westman, Örnsköldsvik 086908, 126648
Westman & Muñoz, Santiago de
Chile 064990
Westman, Inge Lise, Gudhjem . . 065715
Westmansmors stugan,
Jakobstad 010913
Westmeyer, Angelika, Essen . . . 075705
Westminster Abbey Museum, Chapter
House, London 045132
Westminster College Art Gallery, New
Wilmington 051318
Westminster Dragoons Collection,
London 045133
Westminster Historical Society,
Westminster 054344
Westminster Mint, Minneapolis . 094957
Westmoreland Museum of American
Art, Greensburg 049259
Westmorland Clocks, Kendal . . . 089674
Weston, Coventry 127190
Weston Antiques, Weston 091802,
135213
Weston Art Gallery, Cincinnati . . 120387
Weston Park, Weston-under-
Lizard 046103
Weston Park Museum, Sheffield 045778
Weston, William, London 090584, 118629
Westonzoyland Pumping Station
Museum, Bridgwater 043723
Westover Antique Gallery, Saint
Louis 096794
Westover House, Charles City . . 047537
Westpac Museum, Sydney 001557
Westpfälzer Musikantenmuseum,
Mackenbach 020203
Westphal, Berlin 074948
Westphal, Paderborn 108354
Westphal, Herbert, Horn-Bad
Meinberg 130069
Westphal, M., Appenheim 074421
Westphal, Martin, Ingelheim am
Rhein 076610
Westphalsches Haus,
Markkleeberg 020288
Westport Gallery, Dundee 089116,
134343
Westport Historical Museum,
Westport 054354
Westport House, Westport 025001
Westport Maritime Museum,
Westport 054355
Westpreussisches Landesmuseum,
Münster 020611
Westra, Irene, Groningen 083311
Westreich, Thea, New York 123183
Westrope, I., Birdbrook 088469

Westsächsisches Textilmuseum
Crimmitschau, Sächsisches
Industriemuseum, Crimmitschau 017784
Westside Antique Mall, El Paso . 093420
Westside Books, Denver 145135
Westside Moods, Los Angeles . . 094503
Westsider Bookshop, New York . 145379
Westville Historic Handicrafts Museum,
Lumpkin 050527
Westville House Antiques, Littleton,
Somerset 089852
Westwall-Bunker, Beckingen . . . 017009
Westwallmuseum, Irrel 019421
Westwallmuseum Gerstfeldhöhe,
Pirmasens 021137
Westway Pine, Helmsley 089491
Westwerk, Hamburg 107318
Westwind Art Editions, Edmonton 101307
Westwood, New York 123184
Westwood House Antiques,
Tetbury 091586
Westwood, Mark, Hay-on-Wye . 144464
Westwords, Greenock 144448
Westzaanse Bodemvondsten Kok-
Voogt, Westzaan 033043
Weswal, Tamworth 099806
Weta Design, Waikato 113046
Wetaskiwin and District Museum,
Wetaskiwin 007393
Wetering, Amsterdam 112314
Weth, R., Würzburg 130789
Wethersfield Museum,
Wethersfield 054357
Wetpaint Gallery, Cirencester . . 117592
Wetter, F., Wachtberg 078817
Wetterau-Museum, Friedberg,
Hessen 018517
Wetterling, Björn, Stockholm . . 115938
Wetterling, Teo, Singapore 114636
Wettermuseum Lindenberg, Museum
für Meteorologie und Aerologie,
Tauche 022110
Wettstein, Dierks, Düsseldorf . . 075540
Wetz Museum, Uffikon 042077
Wetzel, Christian, Erfurt 075645
Wetzlarer Kunstverein, Sinn . . . 059312
Weuffen, P., Vechta 078782
Wever Van Wijnen, Franeker . . . 143142
Weverijmuseum, Geldrop 032337
Wewerka, M. J., Berlin 106459, 137102
Wexford Arts Centre, Wexford . . 025004
Wexford County Historical Museum,
Cadillac 047341
Wexford County Museum,
Enniscorthy 024922
Wexler, Philadelphia 123472
Wexner Center for the Arts, Ohio State
University, Columbus 047945
Wey & Co., Luzern 087652
Weyers-Sampson Art Gallery,
Greenville 049266
Weygang-Museum, Öhringen . . 020957
Weymouth Museum, Timewalk and
Brewery Museum, Weymouth . 046111
Weymouth Woods-Sandhills Nature
Preserve Museum, Southern
Pines 053422
Weyrauch, I., Frei-Laubersheim . 075842
Weyrich, Albuquerque 119520
Weyrich, A., Halle, Saale 107212
Weywoda, Manfred, Wien 063100,
125288, 140129
WG Expozaal, Amsterdam 112315
Whakatane District Museum and
Gallery, Whakatane 033363
Whale Bone Gallery, Port Fairy . 099586
Whale Museum, Friday Harbor . . 048984
Whale World, Albany 000780
Whales Tale Antiques, Napier . . 084012
Whalesong Digital Arts, Honolulu 120977
Whanau Gallery, Wellington 113098
Whanganui Arts Trust, Wanganui 059646
Whanganui Regional Museum,
Wanganui 033335

Whanganui Riverboat Centre Museum,
Wanganui 033336
Whangarei Art Museum, Te
Wharetaonga o Whangarei,
Whangarei 033365
Whangarei Museum, Whangarei 033366
Whangaroa County Museum,
Kaeo 033230
Whanki Museum, Seoul 030052
Whariti Antiques, Woodville 084206
Wharton, Charlotte 120017
Wharton County Historical Museum,
Wharton 054360
Wharton Esherick Museum,
Malvern 050616
Wharton, Tim, Redbourn 091079
What Cheer Antiques, Providence 096518
What Mama Had, San Diego .. 097149
What Now Antiques, Buxton ... 088751
Whatcom Museum of History and Art,
Bellingham 046895
Whately, Rollo, London 090585, 134797
Whatnots, Strathblane 091497
What's Art Gallery, Sendai 111393
Whats in Store, Los Angeles ... 094504
Whay, Kevin, Hoylake . 089596, 134479
Wheat Ridge Sod House and Museum,
Log Cabin, First Post Office and Tool
Shed, Wheat Ridge 054361
Wheatleys, Great Yarmouth ... 089385
Wheaton History Center, Wheaton 054366
Wheatsheaf Antiques, Chester .. 088869
Wheel a Deal, Regina 064558
Wheeler Gallery, Providence ... 052398
Wheeler Historic Farm, Salt Lake
City 052921
Wheeler, Genevieve, New York . 095825
Wheeler, Jack & Kate, Austin .. 092278
Wheeler, Simon, Midhurst 144732
Wheels, Parramatta 139870
Wheels O' Time Museum, Peoria 051989
Wheelwright Museum of the American
Indian, Santa Fe 053125
Where Art Is Art, Armadale,
Victoria 098668
Where On Earth Did You Get That,
Calgary 064280
Whetung Craft Centre and Art Gallery,
Curve Lake 005673
Whichcraft Jewellery, Writtle ... 091959,
135264
Whiddons Antiques and Tearooms,
Chagford 088816
Whillock, David, San Francisco . 136352
Whillock, F., Litton Cheney 144543
Whim Antiques, Toronto 064778
Whimsey Shoppe, Dallas 093227,
093228
Whimsical Notions, Warrimoo .. 062410
Whimsy Antiques, Toronto 064779
Whipple House Museum, Ashland 046568
Whipple Museum of the History of
Science, Cambridge 043851
Whippoorwill, Atlanta . 092198, 135317
Whisnant, Pete, New Orleans .. 095176
Whisson, Kallangur 099170
Whistle Stop Treasure Shop,
Birmingham 092437
Whistler House Museum of Art, Lowell
Art Association, Lowell 050513
Whistler Museum and Archives
Society, Whistler 007397
Whistler, Barry, Dallas 120673
Whitaker Center for Science and the
Arts, Harrisburg 049374
Whitbourne Museum, Whitbourne 007398
Whitburn Community Museum,
Whitburn 046113
Whitby Abbey Museum, Whitby . 046118
Whitby Museum, Whitby 007400, 046118
Whitchurch Silk Mill, Whitchurch,
Hampshire 046119
Whitchurch-Stouffville Museum,
Gormley 005895

White & Co SL, Madrid 115293
White & Day, Houston 093836
White Box, New York 123185
White Cat Book Company,
Saskatoon 140388
White Cockade Publishing,
Dorchester 138153
White Columns, New York 123186
White County Historical Museums,
Ratcliff Inn, L.Haas Store, Matsel
Cabin, Robinson-Stewart House,
Carmi 047437
White County Historical Society
Museum, Cleveland 047803
White County Historical Society
Museum, Monticello 051032
White Crane Gallery, Omaha .. 123342
White Cube, London .. 118630, 118631
White Dacha, Minsk 100470
White Deer Land Museum,
Pampa 051905
White Dove Art Gallery,
Albuquerque 092092
White Dragon Collectibles,
Jacksonville 094049
White Elephant Antiques, Dallas 093229
White Elephant Museum,
Makkovik 006247
White Fox Museum, White Fox . 007401
White Gallery, Portland 052307
The White Gallery, Wincanton . 119357
White Gloves Art, Clayfield ... 098895
White Hall Historic Site,
Richmond 052538
White Harvest Fine Art, Holden
Hill 099143
White Hill Gallery, Red Hill 099617
White Hills Antiques, White Hills 062425
The White House, Washington .. 054177
The White House, Waverton ... 091762
White House, Westbury 001637
White House Antiques, Houston . 093837
White House Art Gallery,
Washington 124976
White House Farm Antiques,
Easingwold 089130
White House Gallery, Castle Hill 098874
White House Gallery, Manama .. 063163
White House Gallery, Oklahoma
City 123298
White Line Gallery, Exeter 117777
White Lotus, Bangkok . 138130, 144206
White Memorial Conservation Center,
Litchfield 050325
White Mill Rural Heritage Centre,
Sandwich 045738
White Oak Gallery, Minneapolis . 122054
White Park Antiques, Malvern, South
Australia 061775
White Patch Gallery, Bribie Island 098772
White Pillars Museum, De Pere . 048186
White Pine Public Museum, Ely . 048572
White River Architectural Antiques,
Indianapolis 093940
White River Museum, Meeker .. 050766
White River Valley Museum,
Auburn 046655
White Rock Museum, White Rock 007403
White Roding Antiques, White
Roding 091832
The White Room, Galway 109639
The White Room, Meltham 118736
White Room Gallery, West
Hollywood 125033
White Rose, Mumbai 079411
White Space Beijing, Beijing ... 101996
White Space Contemporary Art,
Totnes 119250
White Space Gallery, London ... 118632
White Swan Antiques, Saint
Louis 096795
White Tiger, Nashville 122105
White Trash, New York 095826
White Trash Contemporary,

Hamburg 107319
White Wall Gallery, Chapel
Allerton 117554
White Water Gallery, North Bay . 006510
White, Albert, Toronto 101721
White, Charlie, Portland 123686
White, Christopher, Uttoxeter ... 091699,
135175
White, David N., Wellington 084198,
113099
White, E. & B., Brighton 088639
White, Frank, Cincinnati 120388
White, Gordon, Hyde Park 061647
White, Harlan, San Diego 097150
White, Joan & David, Manfield . 090696
White, Joan, David & Richard, Barnard
Castle 088358
White, M. & M., Reigate 091091
White, Scott, San Diego 124189
Whitechapel Art Gallery, London 045134
Whitecliffe College of Arts and Design,
Te whaare Takiura o Wikiriwhi,
Auckland 056189, 112819
Whitecrow, San Diego 124190
Whitefield House Museum, Moravian
Historical Society, Nazareth .. 051205
Whitehall, Cheam 043915
Whitehead House Antiques,
Austin 092279
Whitehead Memorial Museum, Del
Rio 048226
Whitehern Historic House and Garden,
Hamilton 005978
Whitelmage.com, Hillsborough . 117972
Whitelaw & Sons, John,
Auchterarder 088324
Whiteley Museum, Saint Paul's
River 006931
Whitemoor, Shenton 091298
Whitemouth Municipal Museum,
Whitemouth 007410
Whiters of Leicester, Hoby 089540,
134466
Whiteshell Natural Historic Museum,
Nutimik Lake 006533
Whitespace, Auckland 112820
Whitewall, New York 139529
Whitewall Galleries, Birmingham 117383
Whitewater Canal Historic Site,
Metamora 050819
Whitewater Historical Museum,
Whitewater 054379
Whiteway-Wilkonson, G.A.,
Maidencombe 090675
Whitewood Historical Museum,
Whitewood 007411
Whitfield, London 090586
Whitfield, Laurence, Cirencester 134277
Whitford, London 118633
Whitgift Centre Gallery, Croydon 117656
Whitgift Galleries, Croydon 089018,
117657, 134309
Whithner, Baltimore 119779
Whithorn – Cradle of Christianity,
Whithorn 046124
Whithorn Priory and Museum,
Whithorn 046125
Whitingham Historical Museum,
Whitingham 054380
Whitley, Miami 094815
Whitley County Historical Museum,
Columbia City 047920
Whitman House Museum, Canso 005550
Whitman, Sandra, San Francisco 097346
Whitmore, Great Malvern 089375
Whitmore, Vancouver 064897
Whitney, Minneapolis . 094958, 145288
Whitney Museum of American Art,
New York 051491
Whitney Museum of American Art at
Altria, New York 051492
Whitney Museum of American Art at
Champion, Stamford 053485
Whitney Pier Historical Museum,

Sydney 007118
Whitney, Adam, Omaha 123343
Whitney, Susan, Regina 101537
Whitstable Fine Art, Whitstable . 119345
Whitstable Museum and Gallery,
Whitstable 046127
Whitstable Oyster and Fishery
Exhibition, Whitstable 046128
Whitt Kraus, San Diego 124191
Whittaker & Biggs Auction Room,
Congleton 126869
Whittier Fine Arts Gallery, Wichita 054393
Whittington, Henley-on-Thames . 117963
Whittlesey Museum, Whittlesey . 046129
Whittlesford Gallery, Whittlesford 119347
Whitton & Laing, Exeter 126893
The Whitworth Art Gallery,
Manchester 045219
Whitworth Historical Society Museum,
Whitworth 046130
Who New, Pittsburgh 096341
Whodunit, Philadelphia 145421
Wholesale Antique Center, Miami 094816
Wholesale Art, Minneapolis ... 122055
Wholesale Art, San Antonio ... 124083
Whorton, Henry, Chicago 120298
Who's Who in Art, Calne 139230
Who's Who in International Art,
Lausanne 138063
Who's Who International-Art
Management AG, Lausanne .. 116430
Whowhatwherewhenwhy W5,
Belfast 043565
Whyalla Maritime Museum,
Whyalla 001640
Whyte, Dublin 126174
Whyte Museum of the Canadian
Rockies, Banff 005382
Whyte, John, Edinburgh 089196
Wibaux, Hugues, Lyon . 069784, 128735
Wiberg, Alex, Stockholm 115939
Wiblinger Schlossgalerie, Ulm . 108803
Neil Wibroe & Natasha MacIlwaine,
London 090587
Wich, G. E., Fürth 129878
Wich, Joan, Houston 121153
Wiche, Lenningen 077087
Wichers, Den Haag 083151
Wichert-Pollmann, Dr. Ursula,
Altenbeken 141407
Wichert, Georg, Gangelt 075942
Wichert, P., Wedemark 078881
Wichita Art Museum, Wichita .. 054394
Wichita Center for the Arts,
Wichita 054395, 058104
Wichita Falls Museum and Art Center,
Wichita Falls 054398
Wichita-Sedgwick County Historical
Museum Association, Wichita . 054396
Wichmann, Guido, Bernau am
Chiemsee 106472
Wichmann, Petra, Bernau am
Chiemsee 106473
Wichmann, W., Aachen 074346
Wick Antiques, Lymington 090663,
134824
Wick Focacci, Veronika, Firenze 131281
Wick Focacci, Veronika,
Innsbruck 127780
Wick Heritage Centre, Wick 046132
Wick, Eugen, Solingen 078502
Wickaninnish Gallery, Vancouver 101845
Wicker, Seattle 097522
Wicker Gazebo, New Orleans . 095177
Wicker Porch, Marion 127307
Wicker Shop, Minneapolis 094959,
135839
Wicker Shop, Saint Paul 096868, 136228
Wickersham House Museum,
Fairbanks 048663
Wickert, H., Ulm 078758
Wickerwood Lamp Shop,
Houston 135622
Wicki, Thomas, Malters 116490

Wickliff, Karen, Columbus 145102
Wickliffe Mounds, Museum of Native American Village, Wickliffe 054400
Wickman Maritime Collection, Zabbar 030771
Widauer, Johann, Innsbruck ... 100061
Widdas, Roger, Moreton-in-Marsh 090784, 118757, 134860
Widder, Wien 100393
Wide Bay and Burnett Historical Museum, Maryborough, Queensland 001260
Wide Bay Gallery, Maryborough, Queensland 099285
Wide Bay Hospital Museum, Maryborough, Queensland ... 001261
Wide Horizons Gallery, Cooks Hill 098919
Wide House Antique, Middle Dural 061849
Widell, Gøran, Helsingør 065747
Widener Gallery, Hartford 049391
Widener University Art Collection and Gallery, Chester 047616
Widenfels, Saint-Paul-de-Vence . 105601
Wider, Hermann, Fribourg 087435, 133749
Widmer, Zürich 088017, 116868, 126776
Widmer & Theodoridis, Zürich ... 116869
Widmer, Anton, Bern .. 087310, 133691
Widmer, Christoph, Sankt Gallen 116620, 144131
Widmer, Hans, Sankt Gallen ... 087790, 116621, 126750
Widmerpool House Antiques, Maidenhead 090676
Widor, Paris 072015
Widukind-Museum, Enger 018212
Widurski, Maciej, Kraków 084604
Wiebe, P., Garz 075951
Wiebold & Colby, Indianapolis .. 093941, 135640
Wiebus, Arndt, Oberhausen ... 142253
Wiechern, Gardy, Hamburg ... 107320
Wiechers Woon Oase, Dwingeloo 032260
Wiechmann, Bardowick 074679
Wiechmann, Deggenhausertal .. 137138
Wiechmann, Karl-Heinz, Bad Bramstedt 074497
Wieck, Thomas, Stuttgart 130693
Wieczorek, Peter K., Düsseldorf 075541, 129766
Wieczorek, Zbigniew, Warszawa 084756
Wiedebusch, Max, Hamburg ... 141902
Wiedemann, Frank, Düsseldorf . 075542, 129767
Wiedenbrüg, Karin, München ... 077640
Wiedenfeld, Hubert, Gerolstein . 075986
Wiedenroth, Alfred, Hannover ... 076399
Wieder, Bergheim 062535
Wiederhold, Imke, Berlin 141589
The Wiegand Gallery, Notre Dame de Namur University, Belmont .. 046899
Wiegerling, Peter, Gaißach 129884
Wiegmann, K., Lübbecke 077166
Wiegmann, K. & D., Lübbecke .. 077167
Wiehl, Blanka, Dormagen 075376
Wiehl, T., Ludwigsburg 077159
Wieland-Archiv, Biberach an der Riß 017361
Wieland-Gedenkzimmer, Achstetten 016462
Wieland-Museum (mit Wieland-Archiv), Biberach an der Riß 017362
Wieland-Schauraum, Biberach an der Riß 017363
Wieland, Martin, Trier .. 078710, 108780
Wielandgut Oßmannstedt mit Wieland-Museum, Klassik Stiftung Weimar, Oßmannstedt 021020
Wielkopolski Park Etnograficzny, Muzeum Pierwszych Piastów na Lednicy, Lednogóra 034939
Wielkopolskie Centrum Numizmatyczne, Poznań 084656

Wielkopolskie Muzeum Pożarnictwa PTTK, Rakoniewice 035168
Wielkopolskie Muzeum Walk Niepodległościowych w Poznaniu, Poznań 035139
Wielkopolskie Muzeum Wojskowe, Muzeum Narodowe w Poznaniu, Poznań 035140
Wiemann, Dr. Wolfgang, Heidelberg 141932
Wiemann, Rolf, Hamburg 076340
Wiemann, Rolf von, Eyendorf .. 075726
Wien Museum Karlsplatz, Wien . 003095
Wien, Barbara, Berlin . 137103, 141590
Wienand Verlag, Köln 137321
Wienberg, Hillerød 065759
Wienbibliothek im Rathaus, Wien 003096
Wiend Books & Collectables, Wigan 144923
Wieneman, Tom, Saint Louis ... 096796
Wiener Feuerwehrmuseum, Wien 003097
Wiener Glasmuseum, Sammlung Lobmeyr, Wien 003098
Wiener Interieur, Wien 063101
Wiener Internationale Kunst & Antiquitätenmesse, im Palais Ferstel und Palais Niederösterreich, Wien 098007
Wiener Internationale Kunst & Antiquitätenmesse, im Wiener Künstlerhaus, Wien 098008
Wiener Kriminalmuseum, Museum der Bundespolizeidirektion Wien, Wien 003099
wiener kunst schule, Wien 054986
Wiener Messing Manufaktur, Wien 063102
Wiener Neustädter Künstlervereinigung, Wiener Neustadt 058338
Wiener Tramwaymuseum, Wien . 003100
Wiener Ziegelmuseum, Wien ... 003101
Wiener, Nancy, New York 095827
Wienerroither & Kohlbacher, Wien 063103
Wienerwaldmuseum, Eichgraben 001851
't Wienkeltje van Wullempje, Hoedekenskerke 032491
Wieringer Boerderij, Den Oever . 032210
Wiese, Jochen, Lübeck 077192
Wiese, O., Neubrandenburg ... 077715, 130444
Wiese, Reiner, Wuppertal 079121
Wiesenmeisterei, Neustadt-Glewe 020768
Wiesenthäler Textilmuseum, Zell im Wiesental 022857
Wieser, Helmuth & Bärbel, München 077641
Wieser, Wolfgang & Roswitha, Stadtbergen 078530
Wiesinger, Wels 062878
Wiesinger, Hubert, Wels 062879
Wiesinger, Wilfried, Breitenfurt . 139977
Wiesmeier, Jasmin, Schlangenbad 142347
Wiesner, Eduard, Wernstein am Inn 136570, 140054
Wiesner, V., Leipzig 130231
Wiessman, Chicago .. 120299, 120300
Wießner, H., Regen 078161
Wieuw, de, Antwerpen 100523, 128008
Wieża Ratuszowa, Muzeum Historycznego m. Krakowa, Kraków 034912
Wigan Pier, Wigan 046135
Wigerdal, Stockholm 087076
Wiggertaler Museum, Schötz ... 041963
Wiggins & Sons, Arnold, London 090588
Wiggins, Peter, Chipping Norton 134266
Wight Light Gallery, Ventnor ... 119290
Wightwick Manor, Wolverhampton 046180
Wigmore, London 118634
Wigmore, D., New York 123187
Wignacourt, B'Kara 082412
Wignacourt Museum, Rabat ... 030747

Wignall Museum and Gallery, Chaffey College, Rancho Cucamonga .. 052458
Wihlborg, Magnus, Lund 133552
Wihrs, Vejle 103003
Wiingaard, Lise Lotte, København 065968
Wiingaard, Lotte Lise, Frederiksberg 065680
Wijnhoven, Jos, Deventer 143128
Wijnhoven, T., Brugge 100540
Wijnklder Soniën, Overijse 003945
Het Wijnkopersgildehuys, Amsterdam 032000
Wijnmuseum Maastricht, Cadier en Keer 032131
Wijsmuller, Lara, Amsterdam ... 082910
Wikinger Museum Haithabu, Stiftung Schleswig-Holsteinische Landesmuseen Schloß Gottorf, Busdorf 017704
Wik's Mynt & Kuriosa, Kil 086783
Wiktor, Tord, Stockholm 133585
Wilber Czech Museum, Wilber . 054401
Wilberforce House, Kingston-upon-Hull 044720
Wilbert, Stanley, Calgary 101268
Wilbrand, Dieter, Köln . 107700, 137322
Wilbur D. May Museum, Reno .. 052514
Wilbur Wright Birthplace and Interpretive Center, Hagerstown 049309
Wilbye, Jonathan, Kirkby Stephen 134520
Wilce Trading, Reims 072405
Wilce, Christine, Le Touquet-Paris-Plage 069364
Wilce, Christine, Neuville-sous-Montreuil 070528
Wilcockson, Dr. John, Marburg . 077289
Wilczyński, Tomasz, Lublin 084629
Wild About Music, Austin 119710
Wild Art, Cape Town 114778
Wild Earth Images, Salt Lake City 123998
Wild Geese Heritage Museum, Portumna 024977
Wild Goose Antiques, Modbury . 090766
Wild Honey, Guildford 099102
Wild Oates Gallery, Doonan .. 098973
Wild Orchid Antiques, Richmond 096586
Wild Pitch, Portland 096491
Wild Strawberry-Muddy Wheel Gallery, Albuquerque 119521
Wild Things, Denver 120790
Wild Walls Gallery, Bentleigh East 098738
Wild Wings Gallery, Saint Paul . 123939
Wild, Adolf, Lahr 107781
Wild, Alexander, Bern 144054
Wild y Cia, Alfred, Bogotá 102374
Wild, Eva Maria, Frankfurt am Main 107044
Wild, Eva Maria, Zürich 116870
Wild, George, New York 136025
Wild, Rose, Seattle 097523
Wilde Art, Arrington 117249
Wilde-Meyer, Tucson 124785
Wilde, Chris, Harrogate 089445
Wilde, Olaf, Braunschweig 129626
Wilde, P.S., Toronto 064780
Wilde, Yvan de, Périgueux 072062
Wildeboer, Groningen 083312
Wildeboer, Toon, Hilversum ... 112540
Wildenauer, Hans, Friedenfels . 075911
Wildenstein & Co., London 090589, 118635
Wildenstein & Co., New York .. 123188
Wildenstein, Tokyo 111522
Wildenstein Institute, Paris 136993
Wilder, Hannover 141921
Wilder, Salzhemmendorf 142345
Wilder Memorial Museum, Strawberry Point 053572
Wilderer-Museum Sankt Pankraz, Sankt Pankraz 002686
Wilderness Reflections Gallery, Kansas City 121320

Wilderness Road Regional Museum, Newbern 051511
Wildfellner, Georg, Grieskirchen . 062611, 100026
Wildfowl Art Journal, Salisbury . 139530
Wilding, R., Wisbech .. 091887, 135245
Wildlife Art, Ramona 139531
The Wildlife Art Gallery, Lavenham 118080, 138174
Wildlife Education Centre, Entebbe 043103
The Wildlife Gallery, Toronto .. 101722
Wildlife Museum, Ulaanbaatar .. 031743
Wildlife Wonderlands-Giant Earthworm Museum, Bass 000825
Wildling Art Museum, Los Olivos 050479
Wildmeister, Veitshöchheim 078783
Wildner, Wolfgang, Passau 108366
Wildnispark Zürich, Besucherzentrum, Sihlwald 041987
Wildschut, Amsterdam 082911
Wildwood, Edmonton 101308
Wildwood, Fitzroy Falls 099033
Wildwood Auction House, Wildwood 127496
Wildwood Book Rack, Norman . 145382
Wildwood Center, Nebraska City 051207
Wildwood Gallery, Bury Saint Edmunds 117493
Wildwood Press, Saint Louis ... 123885, 138450
Wile Carding Mill Museum, Bridgewater 005476
Wiles, A. Allen, Woodridge ... 062452
Wilfing, Josef, Salzburg 127863
Wilfinger Brüder, Wien 127977
Wilford & Assoc., Cleveland ... 092981
Wilford, H., Wellingborough ... 127080
Wilfrid Israel Museum of Oriental Art and Studies, Kibbutz Hazorea . 025134
Wilhelm-Busch-Geburtshaus, Wiedensahl 022623
Wilhelm-Busch-Gedenkstätte, Wilhelm-Busch-Gesellschaft, Seesen . 021823
Wilhelm-Busch-Gesellschaft e.V., Hannover 059313
Wilhelm-Busch-Mühle, Ebergötzen 018075
Wilhelm-Busch-Museum Hannover, Deutsches Museum für Karikatur und kritische Grafik, Hannover 019069
Wilhelm Busch Zimmer, Dassel . 017833
Wilhelm-Fabry-Museum, Hilden . 019246
Wilhelm-Hack-Museum Ludwigshafen am Rhein, Ludwigshafen am Rhein 020140
Wilhelm-Kienzl-Museum, Paudorf 002487
Wilhelm Kienzl-Stüberl, Waizenkirchen 002892
Wilhelm-Morgner-Haus, Soest .. 021895
Wilhelm-Ostwald-Gedenkstätte, Großbothen 018841
Wilhelm-Panetzky-Museum, Rammelsbach 021271
Wilhelm Wagenfeld Haus – Design im Zentrum, Wilhelm Wagenfeld Stiftung, Bremen 017604
Wilhelm-Zimolong-Gesellschaft e.V., Gladbeck 059314
Wilhelm, Eleonore J., Ludwigshafen am Rhein 107902, 137358
Wilhelm, M., Aldingen 074373
Wilhelm, Otto, Frankfurt am Main 075834
Wilhelmietenmuseum, Huijbergen 032516
Wilk, Józef, Warszawa 113910
Wilk, M., Brühl 075206
Wilke, Nikolaus, Sinsheim 130641
Wilken, I., Rostock 108521
Wilkening, Ludmilla, Aschaffenburg 106015
Wilkerson, New Orleans 095178
Wilkes, Minneapolis 122056
Wilkes Art Gallery, North Wilkesboro 051622

Wilkie, Saskatoon 064571
Wilkie and District Museum,
Wilkie 007416
Wilkin County Historical Museum,
Breckenridge 047186
Wilkin, Elizabeth, Paraparaumu . 084060
Wilkins & Wilkins, London 090590,
                                118636
Wilkins, Brett, Lichfield 089825
Wilkins, Joan, Witney 091890
Wilkinson, Doncaster 126877
Wilkinson, London 090591, 090592,
  118637,   134798,   134799
Wilkinson & Rhodes, San Antonio 097044
Wilkinson County Museum,
Woodville 054536
Wilkinson, Marcus, Grantham .. 089358,
                           134418
Wilkinson, Marcus, Sleaford ... 091351,
                          135052
Wilkinson, Tony, Petworth 091011
Will Rogers Home, Pacific
Palisades 051882
Will Rogers Memorial Museums,
Claremore 047763
Will, Dieter, Ettlingen 075719
Will, Joël, La Chapelle-Bertrand . 068970
Will, John, Virginia Beach 097798
Willa Cather State Historic Site, Red
Cloud 052491
Willard House and Clock Museum,
Grafton 049169
Willborg, J.P., Stockholm 087077,
  133586,   138017
Willburn Antiques, Christchurch . 083907,
                         132713
Willcox & Nightingale, Penryn .. 090961
Wille, Irmgard, Füssen 075928
Willem de Kooning Academie,
Hogeschool Rotterdam,
Rotterdam 056182
Willem Prinsloo Museum, Rayton 038820
Willemoesgårdens Mindestuer,
Assens 010015
Willems, Düsseldorf 125934
Willem's Winkel, Middelburg ... 083504
Willems, Gerard, Hobart 061638, 099142
Willems, Louis, Bruxelles 128643
Willen, Jürg, Luzern .. 116485, 133849
Willenhall Museum, Willenhall
Historic Map Room and Museum,
Willenhall 046137
Willer, London 118638
Willer Brocante, Willer-sur-Thur 074297
Willi Dickhut Museum,
Gelsenkirchen 018632
Willi Sitte Galerie, Merseburg .. 020376
Willi the Clockman, Lesmurdie .. 061734
Willi, H.-P., Tübingen 142434
William, Ascott-under-Wychwood 088292
William & Joseph, New Orleans 122255
William A. Farnsworth Art Museum
and Wyeth Center, Rockland .. 052644
William A. Quayle Bible Collection,
Baker University, Baldwin City . 046723
The William Benton Museum of Art,
University of Connecticut, Storrs 053555
William Blizard Gallery, Springfield
College, Springfield 053465
William Bonifas Fine Arts Center,
Escanaba 048603
William Booth Birthplace Museum,
Nottingham 045423
William Breman Jewish Heritage
Museum, Atlanta 046639
William Carey Museum, Leicester 044803
William Clark Market House Museum,
Paducah 051886
William Cullen Bryant Homestead,
Cummington 048093
William Fehr Collection, Iziko
Museums of Cape Town, Cape
Town 038688
William H. Harrison Museum/

Grouseland, Vincennes 053992
William H. Van Every jr. and Edward
M. Smith Galleries, Davidson College,
Davidson 048158
William Hammond Mathers Museum,
Bloomington 047050
William Heath Davis House Museum,
San Diego 052989
William Henry Steeves Museum,
Hillsborough 006006
William Holmes McGuffey Museum,
Oxford 051870
William Howard Taft National Historic
Site, Cincinnati 047751
William Humphreys Art Gallery,
Kimberley 038769
William King Regional Arts Center,
Abingdon 046309
William Lamb Memorial Studio,
Montrose Museum, Montrose . 045294
William M. Colmer Visitor Center,
Ocean Springs 051703
William McKinley Presidential Library
and Museum, Canton 047413
William Morris Gallery and Brangwyn
Gift, London 045135
The William Morris Society,
London 045136
William Pryor Letchworth Museum,
Castile 047461
William Rea Museum,
Murfreesboro 051123
William S. Hart Museum, Natural
History Museum of Los Angeles
County, Newhall 051526
William S. Webb Museum of
Anthropology, Lexington 050254
William Scarbrough House, Ships of
the Sea Museum, Savannah .. 053183
William Street Gallery, South
Yarra 099709
William Trent House, Trenton ... 053802
William Tyndale Museum,
Vilvoorde 004114
William Weston Clarke Emison
Museum of Art, DePauw University,
Greencastle 049237
William Whipple Art Gallery, Soutwest
Minnesota State University,
Marshall 050713
William Whitley House, Stanford 053489
William Woods University Art Gallery,
Fulton 048997
William, Joseph, Portland 123687
William, Robert, Vancouver 064898
William'S, Barcelona 085859
Williams, Houston 121154
Williams & Son, London 090593
Williams, Mornington 099379
Williams, Nashville ... 122106, 122107
Williams & Sons, Richmond,
Surrey 091103, 118975
Williams, Salt Lake City 123999
Williams, Tulsa 097766, 127482
Williams, Wellington .. 084199, 113100
Williams & Co., Toronto 064781
Williams College Museum of Art,
Williamstown 054420
Williams County Historical Museum,
Montpelier 051034
Williams, Adam, New York 123189
Williams, Brian, Charlotte 135396
Williams, David, London 144712
Williams, E & B, Levallois-Perret 069448
Williams, Ehoodin, Cincinnati ... 120389
Williams, George, Kells 079638
Williams & Co., H., Toronto 125387
Williams, Huw, Porthmadog 091040
Williams, Jean, Seattle 097524
Williams, Melvin D ., Pittsburgh 096342
Williams, Nigel, London 144713
Williams, Peter, Boston 119932, 135367
Williams, Philip, New York 123190
Williams, Scott, Richmond 096587

Williams, Steve, Jacksonville ... 121275
Williams, Thayne A., Baltimore . 092375
Williams, Thomas, London 090594
Williams, Tom, Calgary 140339
Williams, Trevor, Plymouth 144789
Williamsburg Decorative Arts Series,
Williamsburg 139532
Williamson, New Orleans 095179
Williamson Art Gallery and Museum,
Birkenhead 043607
Williamson County Historical Society
Museum, Marion 050686
Williamson, Connie, Dallas 093230
Williamson, P., Houston 093838
Williamson, Patrice, Nantes 128841
Williamson, William, Stoke-on-
Trent 119151
Williamstown Museum,
Williamstown 001643
Willibaldsburg, Bayerische Verwaltung
der staatlichen Schlösser, Gärten und
Seen, Eichstätt 018123
Willicher Kunstverein, Willich ... 059315
Willie, Valentine, Bangsar Baru . 111902
Willinger Milchmuhseum,
Willingen 022669
Willinger, Ernst-Peter,
Mauthausen 077309
Willingham Auctions, Willingham 127088
Willis, Memphis 121716
Willi's Antikkiste, Osthofen 077993
Willis Museum, Basingstoke 043511
Willis, Henry, Sherborne 091312, 135039
Willis, James, San Francisco ... 124432
Willistead Manor, Windsor,
Ontario 007426
Willmitzer, Helmut, Hilpoltstein . 076536
Willoughby-Baylor House, Norfolk 051583
Willoughby Historical Museum,
Niagara Falls 006487
Willow, Denver 093352
Willow, Dumfries 117691
Willow, Dunedin 083935
Willow, Phoenix 123518
Willow, Sacramento 123790
Willow, Toronto 101723
Willow, Weybridge 119332
Willow Antiques & the Nettlebed Antique
Merchants, Nettlebed 090808, 134869
Willow Art Gallery, Long Beach . 121414
Willow Bunch Museum, Willow
Bunch 007420
Willow Gallery, Weybridge 091811
Willow Place Antiques, Richmond 096588
Willow Pool, Lymm 090664
Willow Street Gallery,
Washington 124977
Willow, Glen, San Jose 097381
Willowa, Warszawa 113911
Willowbank at Taggerty, Taggerty 099867
Will's Art Warehouse, London .. 118639
Wills Lane Gallery, Saint Ives,
Cornwall 119038
Willunga Antiques, Willunga ... 062431
Willy's Coins, Omaha 096066
Wilma's Workshop, Tampa 136401
Wilmer Jenning, New York 123191
Wilmersdorf Archiv, Berlin 017317
Wilmette Historical Museum,
Wilmette 054430
Wilmington Railroad Museum,
Wilmington 054449
Wilmink, H., Hengelo 132532
Wilmsen, Münster 077680
Wilmslow Antiques, Wilmslow .. 091850
Wilner & Co., Eli, New York 136026
Wilnitsky, Boris, Wien 063104
Wilnitsky, Leonid, Wien 063105
Wilno Heritage Society Museum,
Wilno 007421
Wilson, Columbus 135486
Wilson, Oklahoma City 123299
Wilson, Robe 099643
Wilson, Windsor, Ontario 101862

Wilson Castle, Proctor 052374
Wilson County Historical Society
Museum, Fredonia 048963
Wilson Historical Museum,
Wilson 054454
Wilson MacDonald Memorial School
Museum, Selkirk 007004
Wilson, A&K, Harpenden 117911
Wilson, Amanda L., Phoenix ... 096246
Wilson, Frank, Worthing 091955
Wilson, Greg, Pokolbin 099572
Wilson, H.W., Bronx 138309
Wilson, Jerome, New York 095828
Wilson, John, Hawthorn 061606
Wilson, Maralyn, Birmingham .. 119824
Wilson, O.F., London 090595
Wilson, Peter, Nantwich 126993
Wilson, Philip, London 138259
Wilson & Sons, R.S.,
Boroughbridge 088518
Wilson, Susan M., Kilmacolm . 118042
Wilson, Virginia, Darlinghurst ... 061387
Wilsons', Haverfordwest 117932
Wilson's Creek National Battlefield,
Republic 052519
Wilton Historical Museums,
Wilton 054458
Wilton House, Wilton, Salisbury . 046140
Wilton House Museum,
Richmond 052566
Wilton Windmill, Wilton,
Marlborough 046139
Wilton, Agnes, London 090596
Wiltondale Pioneer Village, Bonne
Bay 005440
Wiltshire Fire Defence and Brigades
Museum, Potterne 045561
The Wiltshire Gallery, Salisbury . 119054
Wiltshire Heritage Museum and
Gallery, Devizes 044096
Wiltshire, Helen, Mission Beach 099353
Wiltshire, Helen, Port Douglas .. 099585
Wilzek, A., Mühlbeck 142134
Wimbledon College of Art, University
of the Arts London, London .. 056572
Wimbledon Fine Art, London ... 118640,
                          134800
Wimbledon Lawn Tennis Museum,
London 045137
Wimbledon Society Museum of Local
History, London 045138
Wimbledon Windmill Museum,
London 045139
Wimetz, Yves, Templeuve 073632
Wimmer, Otto, München 130417
Wimmera Mallee Pioneers Museum,
Jeparit 001169
Wimpole Hall and Home Farm,
Royston 045643
Wimtetz, Yves, Seclin 073465
Win Art Gallery, Roubaix 072553, 105464
Winands, Ralph, Detmold 141687
Wincanton Antiques, Wincanton 091860
Wincanton Museum, Wincanton 046142
Winchbooks of Canberra,
Fyshwick 139764
Winchcombe Folk and Police Museum,
Winchcombe 046144
Winchcombe Railway Museum,
Winchcombe 046145
Winchelsea and District Historical
Records Centre, Winchelsea . 001646
Winchelsea Museum, Winchelsea 046146
Winchendon Historical Museum,
Winchendon 054462
Winchester, Minneapolis 122057
Winchester Bookshop,
Winchester 144926
Winchester Cathedral Triforium Gallery,
Winchester 046157
Winchester City Museum,
Winchester 046158
Winchester College Treasury,
Winchester 046159

Winchester-Frederick County Historical Society, Winchester 054466
Winchester School of Art, University of Southampton, Winchester 056613
Winckelmann-Museum, Stendal 021995
Wind & Water Gallery, Houston . 093839
Wind Horse Gallery, Denver 093353
Wind Margery Studio, Saint Louis 123886
Wind, E.-W., Bremen .. 075189, 129637
Wind, Mary, La Rochelle 069095, 103904
Windermere Fine Art, London .. 118641
Windermere Steamboat Museum, Bowness-on-Windermere 043687
Windermere Steamboat Museum, Windermere 046160
Windermere Valley Museum, Invermere 006036
Winders, S., Newcastle-under-Lyme 090823, 134876
Windfall Gallery, Palmerston North 112984
Windham Textile and History Museum, Willimantic 054423
Windheuser, Höxter 141949
Windhill Antiquary, Bishop's Stortford 088491
Windhorse Gallery, Bangalow .. 098705
Windibank, P.F., Dorking 126881
The Winding Stair, Galway 079627
Windjammer Gallery, Hamilton .. 100832
Windle, John, San Francisco ... 145570
Windmann, Bad Oeynhausen ... 074567
Windmill and Milling Museum, Polegate 045517
Windmill Antiques, Bembridge .. 088439
Windmill Antiques, Stafford 091400
Windmill Fine Art, High Wycombe 089528
Windmill Gallery, Birmingham .. 088485, 134114
Windmill Gallery, Kuranda 099212
Windmill Island Municipal Park Museum, Holland 049509
Windmill Wood Works, San Antonio 136280
Windmühle, Podersdorf 002510
Windmühle, Retz 002592
Windmühle Waldfeucht, Waldfeucht 022391
Window on the Past, Seattle ... 097525
Window to the Past, Singapore . 085530
Windows Gallery, Nashville 095041
Windows on Church Galleries, Richmond, Victoria 099640
Winds of Change, Winchcombe . 119358
Windsong, Omaha 123344
Windsor, Deauville 068284
Windsor Antique Centre, Windsor, New South Wales 062433
Windsor Antiques, Dublin 079599
Windsor Antiquities, New York .. 095829
Windsor Castle, Royal Collections, Windsor 046168
Windsor Collections, New Orleans 095180
Windsor Cottage Antiques, Windsor, New South Wales 062434
Windsor Fine Art, New Orleans . 122256
Windsor Gallery, Christchurch .. 083908, 112867
Windsor Gallery, Hythe 118007
Windsor Hall, Charlotte 092641
Windsor House, San Francisco . 097347
Windsor House Antiques, London 090597
Windsor House Antiques Centre, Moreton-in-Marsh 090785
Windsor Kulturgintza, Bilbao .. 114999
Windsor Market, Tulsa 097767
Windsor Productions, Long Beach 121415
Windsor Wood Carving Museum, Windsor, Ontario 007427
Windsor, Alain, Luynes 069673
Windsor, Alain, Tours 073866
Windsor, David, Bangor,

Gwynedd 088348, 134056
Windsor, Jean-Pierre, Saint-Denis-des-Monts 072731
Windsor's Community Museum, Windsor, Ontario 007428
Windward Community College, Kaneohe 057275
Windwood, Charlotte 092642
Windy City, Chicago .. 135428, 135429
Wine Museum of Greyton H. Taylor, Hammondsport 049334
Wine Street Gallery, Devizes ... 117677
Wine, Louis, Toronto 064782
Winedale Historical Center, University of Texas, Round Top 052694
Wing, Hong Kong 102127
Wing Luke Asian Museum, Seattle 053252
Wing, Cheong, Hong Kong 065114
Wing, Hing, Hong Kong 065115
Wingaba-Vision, Yangon 112059
Wingates Fine Art, Hatherleigh . 117929
Wings Metaphysical Books, Denver 145136
Wings of Freedom Museum, Huntington 049618
Wings of History Air Museum, San Martin 053055
Wings of Love, Clinton 047838
Winiger, Barbara, Zürich 116871
Wink, A., Neunkirchen-Seelscheid 077744
Wink, Bernhard, Gengenbach . 129894
Winkel, Wolfgang, Bayrischzell . 074701
Winkel, Wolfgang, Bielefeld .. 074993, 125895
Winkelbach, Jochen, Bielefeld .. 129590
Winkeler, Eva, Frankfurt am Main 107045
't Winkelke, Roermond 083538
Winkelmann, Dr. Günter, Düsseldorf 106842
Winkelmann, Erika M., München 077642
Winkelmann, Gerda, Hallstatt .. 062627
Winkelmann, Ilse-H., Lemförde . 077072
Winkelmann, P., Kaiserslautern . 076664, 107497, 141967
Winkelriedhaus, Nidwaldner Museum, Stans 042026
't Winkeltje, Nijmegen 083520
Winkens, Geilenkirchen 075961
Winkhardt, Jörg, Cuxhaven 129664
Winkleman, Edward, New York . 123192
Winkler, Crimmitschau 129663
Winkler, Erfurt 129806
Winkler, Osnabrück 077984
Winkler, Papeete 074320, 105892
Winkler, Christoph, Gütersloh .. 129929
Winkler, Dr. Hildegard, Wien ... 140130
Winkler, Kerstin, Riesa 078237
Winkler, Uli, Überlingen 108795
Winkler, W., Würzburg 130790
Winklmeier, J., Straubing 078581
Winn, Bergisch Gladbach 074735
Winnebago Area Museum, Winnebago 054479
Winner Fraisse Demey, Nantes . 070485
Winnetka Historical Museum, Winnetka 054480
Winnie Davis Museum of History, Gaffney 049000
Winnie's Antique Resale, Milwaukee 094887
Winnikes, H., Heidelberg 076462
Winnikes, K., Heidelberg 076463
Winning, Wener, München 077643
Winnipeg Art Gallery, Winnipeg . 007476
Winnipeg Police Museum, Winnipeg 007477
Winnipeg Railway Museum, Midwestern Rail Association, Winnipeg 007478
Winnipegosis Museum, Winnipegosis 007480
Winns Historic Bakehouse Museum,

Coromandel Valley 000993
Winona County Historical Museum, Winona 054483
Winram, Aberdeen 144214
Win's Gallery, Ravenshoe 099610
Winser Museumshof, Winsen, Aller 022685
Winslow Antiques Centre, Winslow 091883
Winslow Crocker House, Yarmouth Port 054572
Winslow Saleroom – Geo. Wigley & Sons, Winslow 127090
Winston, Amsterdam 082912
Winston, San Francisco 124433
Winston Churchill Memorial and Library in the United States, Westminster College, Fulton .. 048998
Winston Mac, Bury Saint Edmunds 088743, 134192
Winter, New York 123193
Winter, Plainville 127405
Winter & Königshoff, Dornum .. 075379
Winter & Winter, München ... 108207
Winter Antiques Show, New York 098511
Winter!Sport!Museum!, Mürzzuschlag 002422
Winter, Dieter, Nürnberg 077849
Winter, Douglas, Dallas 093231
Winter, Elvira & Peter, Wiesbaden 108933
Winter, Eveline, Rugeley 091151
Winter, Fritz, Dießen am Ammersee 106678
Winter, Gerlinde, Worms 079067
Winter, Heinrich, Düsseldorf ... 075543, 135935
Winter, Hubert, Wien 100394
Winter, Karl, Wien 140131
Winter, Markus, Berlin . 074949, 106460
Winter, Ursula, Ofterdingen ... 077916
Winterberg-Kunst, München ... 108208
Winterberg-Kunst, Auktionen – Galerie – Edition, Heidelberg .. 076464, 107400, 125990, 141933
Winterbranch, San Francisco .. 124434
Winterhager, M., Freiburg im Breisgau 075890
Winterhoff, Lissy, Köln 107701
Wintering, Rita, Hannover 076400
Winterset Art Center, Winterset . 054501
Wintersport-Museum, Davos Platz 041436
Wintersteiger, Magnus, Kronstorf 127798
Winterstein, Charles, Chanos-Curson 067827
Winterstein, Peter, Egloffstein .. 129775
Winterthur Museum, Winterthur . 054502
Winterton, Lichfield 126947
Winterton Boat Building Museum, Winterton 007481
Wintertons, Uttoxeter 127074
Wintgens, Elizabeth, Marlow .. 118725
Winther, Wissen 108948
Winther, Anine, Oslo 132793
Winthrop University Galleries, Rock Hill 052626
Winzermuseum Bachem, Bad Neuenahr-Ahrweiler 016836
Winzermuseum Rauenberg, Rauenberg 021296
Wipperkotten, Solingen 021906
Wir, Livorno 080338
Wired Art Gallery, London 101351
Wiregrass Gallery, Eltham 098998
Wiregrass Museum of Art, Dothan 048342
Wireless Hill Telecommunications Museum, Applecross 000798
Wirkner, Renate, Kitzingen 076816
Wirkus, Ehlscheid 141733
Wirrabara Forestry Museum, Wirrabara 001650
Wirral Museum, Birkenhead ... 043608
Wirral Transport Museum

and Bierkenhead Tramways, Birkenhead 043609
Wirrimbah Aboriginal Art, Emerald Creek 061447
Wirth, Ranskill 118956
Wirth, Angelika, Wiesbaden 130766
Wirth, Ralf, Leipzig 130232
Wirth, Renate, Kölleda 076842
Wirth, Sonia, La Chaux-de-Fonds 116209
Wirthmore, New Orleans 095181, 095182
Wirthner, Karolin, Münster 133865
Wirthwein, Mannheim 142109
Wirtschaftsgeschichtliches Museum Villa Grün, Dillenburg 017899
Wirtz, Miami 121879
Wirtz Gallery, First National Bank of South Miami, Miami 050842
Wirtz, Ilse, Essen 106932, 129822
Wirtz, Stephen, San Francisco . 124435
Wirydarz Galeria, Lublin 084630, 113702
Wisbech and Fenland Museum, Wisbech 046170
Wischer, Lars, Klettwitz 076821
Wischer, Siegfried, Altlandsberg 074397
Wischhausen, Rolf, Kiel 076793
Wisconsin Automotive Museum, Hartford 049392
Wisconsin Federation of Museums, Wausau 060811
Wisconsin Fine Arts Association, Cedarburg 060812
Wisconsin Historical Museum, Madison 050601
Wisconsin Maritime Museum, Manitowoc 050649
Wisconsin National Guard Memorial Library and Museum, Camp Douglas 047389
Wisconsin Union Gallery, University of Wisconsin-Madison, Madison . 050602
Wisconsin Veterans Museum King, King 049972
Wisconsin Veterans Museum Madison, Madison 050603
WisdOM Books, Kolkata 142591
The Wise Art Gallery, Tyler Junior College, Tyler 053876
Wise County Heritage Museum, Decatur 048206
Wise, Mary, London ... 090598, 126977
Wiseman and Firehouse Galleries, Rogue Community College, Grants Pass 049206
Wiseman, Caroline, London 118642
Wiseman, Ruth & Carl D., Dallas 120674
Wiseman, Sarah, Oxford 118868
Wish 4 Books, Groningen 126286, 143155
Wish Barn Antiques, Rye 091164
Wishes, New Orleans 095183
Wishful Things, Milwaukee 094888
Wishing Corner, Portland 096492
Wishing Well, Fort Worth 093487
Wishnow, Houston ... 093840, 135623
Wisma, Trigg 062347
Wißdorf, Olaf, Köln 107702
Wissel-Eggimann, Simone, Basel 133662
Wisselink, L., Den Haag 112423
Wissenschaftliches Kabinett, Wien 063106
Wissing, Magitte, Bocholt 106508
Wistariahurst Museum, Holyoke 049519
Wisteria, Dallas 093232
Wisteria Gallery, Wootton Bridge 119391
De Wit – Royal Manufacturers of Tapestries, Mechelen 003862
The Witch, Hasselt 063689
The Witch Ball, London 090599, 118643
Witchcliffe Art Gallery, Witchcliffe 090642
Witchhill, Sittard 083608
With Paint, Oakland 123269
With You in Mind, Washington . 097867, 097868
Witham, Norman, Shirley 091321

Withcott Antiques, Withcott 062443
Witherells, Sacramento 096655
Withernsea Lighthouse Museum,
  Withernsea 046171
The Withindale Gallery, Long
  Melford 118657
Withington, Richard W., Hillsboro 127243
Without Pier Gallery,
  Sandringham 099675
Witkin, New York 123194
Witkowski, Heinz-Georg,
  Adendorf 074355
Witmeur, J.-P., Bruxelles 063586
Witney and District Museum and Art
  Gallery, Witney 046174
Witney Antiques, Witney 091891, 135247
Wit's End, McKinnon 061756
Wits School of Arts, University of
  Witwatersrand, Johannesburg . 056347
Witt, Dordrecht 112447
Witt, Helsinge 102826
Witt, Axel, Dortmund 129702
Witt, Claude, Riquewihr 072474
Witt, Cornell de, New York 123195
Witt, H., Oldenburg 077944
Witt, Ulrik, Århus 065611, 102781
Wittamer, M., Bruxelles 100671
Wittchen, Friedhelm, Naumburg,
  Saale 130437
Witte de With, Centrum voor
  Hedendaagse Kunst, Rotterdam 032831
De Witte Eenhoorn, Utrecht ... 083671
De Witte Lelie, Delft 112370
Witte Museum, San Antonio .. 052954
De Witte Raaf, Bruxelles 138561
De Witte Rosieren, Roosendaal . 083547
De Witte Voet, Amsterdam ... 112316
Witte, C. de, Gent 063673
Witte, Donald de, Antwerpen .. 063254,
                              128009
Witte, Georges de, Gent 128059
Wittels & Co., F., Caracas 097970
Wittelsbachermuseum Aichach,
  Archäologische Staatssammlung
  München, Aichach 016484
Wittenberger Freilichtmuseum, Tollhus
  up'n Wurnbarg, Edewecht 018094
Wittenborn Art Booker, San
  Francisco 124436, 145571
Wittener Künstlerbund, Witten .. 059316
Witter, San Francisco 097348
Witter Gallery, Storm Lake ... 053552
Witte's Museum, Axel 032042
Witteveen, Amsterdam 112317
Wittgensteiner Kunst Gesellschaft '80
  e.V., Bad Berleburg 059317
Wittgrefe, Dieter, Hamburg ... 076341
Wittke, Bodo, Rosenheim 126106
Wittke, Hans-Joachim, Adliswil . 133615
Wittkowski, Evelyn, Berlin 129576
Wittmann, Dachau 075288, 141677
Wittmann, Rohrbach, Ilm 078269
Wittmann, Marianne, Rothenburg ob
  der Tauber 078291
Wittrock, Wolfgang, Berlin 074950
Wittumspalais, Klassik Stiftung
  Weimar, Weimar 022512
Wittwe, Elisabeth, Metz 104233
Witula, Alex, Bologna 079890
Witzel, Erhard, Wiesbaden 108934
Witzel, Friedrich, Frankfurt am
  Main 107046
Witzel, H., Stiege 078561
Witzenbacher, Berlin 141591
Wixhäuser Dorfmuseum,
  Darmstadt 017831
Wizani, Manfred, München 130418
Wizard of A-Z, Seattle 097526
Wizard of Was, Nashville 095042
Wizards Wood Works, Austin .. 092280,
                              135334
Wnukowska, Maria, Wrocław .. 113953
Wobst, Bad Salzungen 106089
Woburn Abbey, Woburn 046175

The Woburn Abbey Antiques Centre,
  Woburn 091898
Woburn Book Shop, London .. 144714
Woburn Fine Arts, Woburn 119372,
                              135249
Woch, Joanna, Poznań 084657
Wocher-Panorama der Stadt Thun um
  1810, Thun 042060
Wocoal Art Center, Tokyo 029622
Woefke, Mortsel 140254
Woehl, Gerald, Marburg 130284
Wöhl, H.-G. & R., Geeste 075955
Wöhlers, Bad Wimpfen 074616
Woehlert, Michael, Bad
  Bederkesa 129403
Wöhlke, H., Jork 107486
Wöhrl, Erwin, Egweil 129776
Wölbert, Otto, Stuttgart 130694
Wölfel & Zilvar, Halle, Saale .. 076173
Woellner jr., W.C., Cincinnati .. 092906
Woerdehoff, Esther, Paris 105318
Wörner, Reinhard & Petra,
  Nidderau 108270
Woeste, Jörn-Uwe & Barbara,
  Augsburg 141420
Wogrolly, Graz 127755
Wohl Archaeology Museum Herodian
  Mansions, Jerusalem 025114
Wohl, Dietmar, Münster 130432
Wohl, Lorraine, New York 095830
Wohlers & Co., Dr. Robert,
  Hamburg 141903
Wohlfarth, Washington 124978
Wohlgemuth, Jürgen, München . 077644
Wohlgemuth, Luise, Zollikon .. 133972
Wohlgenannt, Egon, Neukirch,
  Bodenssekreis 077733
Wohlhüter, Werner, Leibertingen 107803
Wohlhuter, Paris 072016
Wohlleb, Wien 127978
Wohlleben, Reinhold, Calw 106623
Wohn- und Porzellanmuseum
  Agentenhaus, Horgen 041600
Wohn-Kunst-Galerie, Rüdiger Lasch,
  Schwerin 108597
Wohngalerie P, Wien 100395
Wohnmaschine, Friedrich Loock,
  Berlin 106461
Wohnmuseum Egga, Museum of
  Interiors, Visperterminen 042112
Wohnmuseum Schellenberg,
  Aussenstelle des Liechtensteinischen
  Landesmuseums, Schellenberg 030303
Wohnwerk, Wuppertal 079122
Wojciechowski, Jerzy, Poznań .. 132865
Wojcik, Hans Kurt, Vejle 066201
Wojtachnio, Stanisław, Warszawa 084757
Wojtas, Monschau 077415
Wojtasiak, Adam, Milanówek .. 143491
Wokingham Antiques Centre,
  Wokingham 091901, 135251
Woksens Samlinger, Kystmuseet i
  Nord Trondelag Norveg, Rørvik 033902
Wolber, Kerben 076757, 130114
Wolcott House Museum Complex,
  Maumee 050742
Wold Galleries, Bourton-on-the-
  Water 117407
Wolf, Cincinnati 120390
Wolf, Cleveland 120424
Wolf, Eichstetten am Kaiserstuhl 075583
Wolf, Kierspe 076795
Wolf House Museum, Manhattan 050643
Wolf Kahlen Museum, Bernau bei
  Berlin 017327
Wolf Point Area Historical Society, Wolf
  Point 054511
Wolf-Schirmer, Juliane, Kandern 130092
Der Wolf-Verlag, Sankt Michael,
  Lavanttal 136562
Wolf, Christin, Albuquerque ... 119522
Wolf, Daniel, New York 123196
Wolf, Doris, Wien 100396
Wolf, Ernst, Wien 063107

Wolf, Günter, Karlsruhe 076704
Wolf, Hagen, Bodenheim 075049
Wolf, Hans-Werner, Göttingen .. 076056
Wolf, Hermann, Crailsheim 075278
Wolf, Hermann, Sommerhausen 108642
Wolf, James, Cincinnati 092907
Wolf, Joachim, Pforzheim 130519
Wolf, Lone, Fort Worth 120889
Wolf, Michael J., San Diego ... 124192
Wolf, Peter, Aschaffenburg ... 074438
Wolf, Peter & Eva, Sulzbach am
  Main 078648
Wolf, Sasha, New York 123197
Wolf, Steven, San Francisco ... 124437
Wolf, U., Eichstetten am
  Kaiserstuhl 075584
Wolf, Ulrich, Gross-Zimmern .. 076106
Wolf, Wolfgang, Coswig 129659
Wolfaert-Masson, Knokke-Heist . 100767
Wolfe, Ernie, Los Angeles 121621
Wolff, Den Haag 083152
Wolff, München 142196
Wolff, B., Haguenau 103805
Wolff, Frank Freiherr von,
  Oldenburg 130491
Wolff, Gérard, Liverdun 069603
Wolff, Gisela, Fulda 129883
Wolff, Jocelyn, Paris 105319
Wolff, Jürgen, Nürnberg 137454
Wolff, Klaus, Schweinfurt 078423
Wolff, Michael, Marburg 077290
Wolff, Reinhard, Berlin 074951, 106462,
                              129577
Wolff, Tatjana, Kiel 130123
Wolffersdorff, Thomas von,
  Leipzig 130233
Wolfgang-Bonhage-Museum,
  Korbach 019767
Wolfgang Hake Verlag, Köln .. 137323
Wolfgang Lettl-Verein zur Förderung
  surrealer Kunst e.V., Augsburg 059318
Wolfhound Press, Dublin 137610
Wolfram, Lohfelden 130246
Wolfram Voelcker Fine Art, Berlin 106463
Wolfrum, Hubert, Wien 100397, 136616,
                              140132
Wolfs, Maastricht 112590
Wolfsberg Arts Forum,
  Ermatingen 116258
Wolfsberg-Verlag, Zürich 138125
Wolfschmidt, Cornelia,
  Maroldsweisach 077303
Wolfsen, Aalborg 102769
Wolfson Museum, Heichal Shlomo,
  Jerusalem 025115
The Wolfsonian, Florida International
  University, Miami Beach 050849
Wolfsteiner Heimatmuseum im
  Schramlhaus, Freyung 018511
Woljeon Art Museum, Seoul ... 030053
Die Wolkenpumpe, Osnabrück . 142282
von Wolkenstein, Oswald, Selva di Val
  Gardena 110923
Wolkoffin talomuseo,
  Lappeenranta 011095
Wolkowitz, Bryce, New York .. 123198
Wollaston Museum, Wollaston .. 046177
Wollmann, Jürgen A.,
  Schwalmstadt 078413, 142354
Wollmann, Vero, Stuttgart 108739
Wollny, Katharina,
  Wolfratshausen 130778
Wollombi Barn Antiques,
  Wollombi 062445
Wollondilly Heritage Centre, The
  Oaks 001381
Wollongong City Gallery,
  Wollongong 001654
Wollscheid, Karl-Heinz, Wittlich . 079037
Wollwert, Birgel 075001
Wolm'Art, Souffelweyersheim .. 073546
Wołosz, Bogdan, Kraków 143485
Wolpert, Heiko, Aschaffenburg . 106016
Wolseley and District Museum,

Wolseley 007484
Wolseley Fine Arts, Walterstone . 091719,
                              119302
Wolter, Freiburg im Breisgau .. 141812
Wolter, Friedrich A., Roskilde .. 066099
Wolter, K., Raesfeld 078122
Wolter, Reinhard, Welt 078920
Wolter, U., Königswinter 076931
Wolters, Maastricht ... 083488, 132578
Wolters, A., Edemissen 075566
Wolters, Thomas, Berlin 129578
Wolthaus, Düsseldorf 075544
Wolverhampton Antique Market,
  Wolverhampton ... 091909, 144931
Wolverhampton Art Gallery,
  Wolverhampton 046181
Wolverine Hobby and Historical Society
  Museum, Spy Hill 007068
Womack, Kathy, Austin 119711
Woman Made Gallery, Chicago . 047707
Womankraft, Tucson 124786
Woman's Art Journal, New
  Brunswick 139533
Women and Their Work, Austin . 046705
Women in the Arts, Indianapolis 121224
Women in the Arts, Washington 139534
Women's Art Association of Canada,
  Toronto 058478
Women's Art Library – Make,
  London 045140
Women's Art Registry of Minnesota,
  Saint Paul 060813
Women's Caucus for Art, New
  York 060814
Women's History Museum, West
  Liberty 054313
Women's Inter Art Center, New
  York 123199
The Women's Museum, Dallas . 048127
Women's Museum of South Vietnam,
  Ho Chi Minh City 054813
Won, Seoul 111731
Won Kwang College Museum, Iri 029954
Won, O, Daejeon 111594
Wondai Heritage Museum,
  Wondai 001655
Wondai Shire Art Gallery, Wondai 099897
Wonder Graphics, Washington . 124979,
                              124980
Wonderful Department Store,
  Singapore 114637
Wonderful World of Antiques,
  Detroit 093380
Wonderland, Gouda 112485
Wonderland, Tilburg 083630
Wonderwall Gallery, Cirencester 117593
Wong & Sons, New York 095831
Wong & Partners, Ruwi 084408
Wong Man Chun, Hong Kong .. 065118
Wong, Edmond, Hong Kong ... 065116
Wong, Kwong, Hong Kong 065117
Wong, T., Hong Kong 102128
Wong, Wylie, San Francisco ... 097349
Wongan Ballidu Museum, Wongan
  Hills 001656
Wonnerup House and Old School,
  Busselton 000918
Wonsan Historical Museum,
  Wonsan 029882
Wonthaggi Auctions, Wonthaggi 125247
Woo-Bong Art Museum, Daegu . 029915
Woo, Jin, Seoul 111732
Woocoo Museum, Brooweena . 000901
Wood & Hogan, New York 095832
Wood Be Goods, London 134801
Wood Carving Museum, Yerevan 000748
Wood Carving Museum and Taiwan
  Sericultural Museum, Zaociao . 042611
Wood County Historical Center,
  Bowling Green 047162
Wood Creations, Austin 119712
Wood Doctor, Albuquerque ... 135295
The Wood Doctor, Las Vegas .. 135688
Wood Doctor, Phoenix 136124

Wood End Museum, Scarborough 045746
Wood Gallery, Hazmieh 082315
Wood Intentions, London 101352
Wood Jack, Cincinnati 120391
Wood Library-Museum of
Anesthesiology, Park Ridge ... 051925
Wood Mountain Post, Wood
Mountain 007485
Wood Mountain Ranch and Rodeo
Museum, Wood Mountain 007486
Wood Pile Antiques, Portland .. 096493,
136163
Wood Shop Art Gallery, Chicago 120301
Wood-Stock, Hasselt 063690
Wood Street Antiques, Myrtleford 061936
Wood Street Galleries, Pittsburgh 052179
Wood Street Gallery, Chicago .. 120302
Wood Stripping and Refinishing,
Toledo 136413
Wood Tobe-Coburn School, New
York 057567
Wood, Christopher, London 118644
Wood, Colin, Aberdeen 088226, 144215
Wood, J. & K., Barnsley, South
Yorkshire 117289
Wood, Jeremy, Billingshurst ... 117365
Wood, Marcia, Atlanta 119631
Wood, Martin Frank, Dallas 120675
Wood, Michael, Plymouth 091024,
118927
Wood, P.N., Rugby 134983
Wood, Pat & Brian, Lynton 090669
Wood, Peter, Cambridge 144303
Wood, Roger, München 130419
Wood, Vanessa, Mosman 061897
Woodall & Emery, Balcombe .. 088337
Woodard & Greenstein, New York 095833
Woodart Picture Framers,
Mowbray 085653
Woodbine, Spalding 119131
Woodbine, Whaplode Drove 119334
Woodbine International Fire Museum,
Woodbine 054516
Woodbourne, Cincinnati 120392
Woodbridge Museum,
Woodbridge 046183
Woodburn & Westcott,
Indianapolis 121225
Woodburn Plantation, Pendleton 051974
Woodbury Antiques, Woodbury . 091919
Woodchester Villa, Bracebridge . 005453
Woodchurch Village Life Museum,
Woodchurch 046184
Woodco, Jacksonville 135671
Woodcraft, Omaha 136069
Woodcraft and Decor, Singapore 085531
Woodcraft Gallery, Christchurch . 112868
Woodd, Anthony, Edinburgh 089197
Wooden Bird, Columbus 120524
The Wooden Bird, Minneapolis . 122058
Wooden Bird, Saint Paul 123940
Wooden Box Antiques, Woodville 091933
Wooden Elephant Furniture, Saint
Louis 136208
The Wooden Goose, Drouin ... 061408
Wooden Horse, Minneapolis ... 094960
Wooden Horse Antique,
Jacksonville 094050, 135672
Wooden Nickel, Tampa 097594
Wooden Nickel Antiques,
Cincinnati .. 092908, 127179, 135454
Wooden Nickel Coins, Seattle .. 097527
Wooden Nickel Historical Museum,
San Antonio 052955
Wooden Pew, Tyabb 062366
The Wooden Rose, Kaiapoi 083970
Wooden Spoon Antiques, Coffs
Harbour 061353
Wooden Things II, Norfolk 095878
Wooden Ways and Olden Days,
Omaha 096067
Woodend History Resource Centre,
Woodend 001659
Woodfield Restorations, London . 134802

Woodford, Louisville 094607
Woodha, Boston 119933
Woodhall Spa Cottage Museum,
Woodhall Spa 046185
Woodhams, W.J., Shaldon 091275,
135023
Woodhead, Geoffrey M., Honiton 144487
Woodhorn, Northumberland Museum,
Archives and Country Park,
Ashington 043443
Woodhouse, R.C., Hunstanton . 089619,
134492
Woodland Art Gallery, Glasgow . 117863
Woodland Cultural Centre,
Brantford 005472
Woodland Heritage Museum,
Westbury 046096
Woodlands Auction Rooms, Ocean
Grove 125221
Woodlands Gallery, Falkland .. 117781
Woodlands Gallery, Winnipeg .. 101899
Woodlands Pioneer Museum,
Woodlands 007488
Woodlawn Museum, Ellsworth . 048560
Woodlawn Plantation, Alexandria 046398
Woodloes Homestead Folk Museum,
Cannington 000943
Woodman's House Antiques,
Uppingham 091697, 135174
Woodmere Art Museum,
Philadelphia 052103
Woodnuts & Co., Chicago 092815
Woodone Bijutsukan, Yoshiwa . 029747
Woodrising Gallery, Broadford . 117451
Woodrow Wilson House,
Washington 054178
Woodrow Wilson Presidential Library
at his Birthplace, Staunton ... 053514
Woodruff Museum of Indian Artifacts,
Bridgeton 047205
Woods Hole Oceanographic Institution
Ocean Science Exhibit Center, Woods
Hole 054525
Wood's Refinishing and Restoration,
Richmond 136181
Woods Street Gallery, Darwin . 098954
Woods, Arthur, Stein am Rhein . 116662
Woods, Michael, Toledo 124702
Woods, Stuart, New Town 061954
Woods, Thad, Waynesville 127491
Woodshed Antiques, Ballarat ... 061093
Woodside & Braseth, Seattle .. 124617
Woodside Auction Gallery,
Farmville 127223
Woodside House, Rochester Historical
Society, Rochester 052624
Woodside National Historic Site,
Kitchener 006126
Woodside Reclamation, Berwick-upon-
Tweed 088446
Woodson County Historical Society,
Yates Center 054573
Woodson, Charlotte, Birmingham 092438
Woodstock Art, Woodstock ... 119389
Woodstock Art Gallery,
Woodstock 007490
Woodstock Artists Association,
Woodstock 060815
Woodstock Artists Gallery,
Woodstock 054530
Woodstock Historical Society,
Woodstock 054533
Woodstock Historical Society Museum,
Bryant Pond 047290
Woodstock Historical Society Museum,
Woodstock 054528
Woodstock Museum, Woodstock 007491
Woodstock Museum of Shenandoah
County, Woodstock 054534
Woodstock School of Art,
Woodstock 058132
Woodstock School of Art Gallery,
Woodstock 054531

Woodstrip International, Gent .. 063674,
128060
Woodtech, Windsor, Ontario .. 128236
Woodville Museum, Rogue River 052665
Woodville Pioneer Museum,
Woodville 033368
Woodward Antiques, Jacksonville 094051
Woodward Gallery, New York .. 123200
Woodward-Hoffinger, Martina,
Wien 100398
Woodward, Joseph, Cork 079509
Woodware Art, Skipton 119103
Woodworking Wizard, Denver .. 135537
Woodworks, Dallas 135522
Woodworm Antiques, Fairy
Meadow 061461, 127596
Woody Keihoku, Keihoku 028816
Wookey Hole Cave Diving and
Archaeological Museum, Wookey
Hole 046189
Woolaroc Museum, Bartlesville . 046801
Woolff, London 118645
Woolfsons of James Street,
Stewarton 091421, 135080
Woolgoolga Art Gallery,
Woolgoolga 099903
Woollahra Antiques Centre,
Woollahra 062483
Woollahra Art Removals, North
Melbourne 099472
Woollahra Decorative Arts Gallery,
Bondi Junction 061172
Woollahra Times Art Gallery,
Woollahra 099921
Woolley & Wallis' Salisbury
Salesrooms, Salisbury 127031
Woolley, Mark, Portland 123688
Woolpit and District Museum,
Woolpit 046190
Wooltrak Store, Lismore 139807
Woomera Heritage Centre,
Woomera 001662
Woonbootmuseum, Amsterdam . 032001
De Woonstee, Dordrecht 132492
Woori, Seoul 111733
Wooster Gallery, New York ... 095834
Wooster Projects, New York .. 123201
Wooten, Dallas 093233
Wooth, Alexander Georg, Baden-
Baden 074645, 129436
Wootton, Baltimore .. 092376, 135348
Worcester Antiques Centre,
Worcester 091949, 135259
Worcester Art Museum,
Worcester 054548
Worcester City Museum and Art
Gallery, Worcester 046194
Worcester Historical Museum,
Worcester 054549
Worcester Museum, Worcester . 038880
Worcester of Christchurch,
Christchurch 083909
Worcestershire County Museum,
Kidderminster 044690
Worcestershire Yeomanry Cavalry
Museum, Worcester 046195
Word & Image, Abingdon 139231
The Word Bookstore, Montréal . 140371
Words and Pictures Museum,
Northampton 051627
Words Worth Books, Calgary ... 140340
Wordspring Discovery Center,
Orlando 051810
Wordsworth House, Cockermouth 043985
Worgull, Wienhausen 108895
Work Gallery, University of Michigan,
Ann Arbor 046495
Work of Art, Houston 121155
Work of Art, Phoenix 123519
The Work of Our Hands, Atlanta 119632
Work.Art in progress, Trento ... 138882
Worker Gallery, Xining 102260
Workers Arts and Heritage Centre,
Hamilton 005979

Worker's Home Museum, South
Bend 053396
Workhouse Museum, Derry 044093
Working Classroom, Albuquerque 119523
Working Wood, Cobargo 098900
Workman and Temple Family
Homestead Museum, City of
Industry 047753
Workman, Wally, Austin 119713
The Works Antiques Centre,
Llandeilo 089863, 134570
Works of Art, Paisley 118881
Works of Art Collection, University of
Cape Town, Rosebank 038827
Works of Art, Bernard Descheemaeker,
Antwerpen 063255
Works on Paper, Los Angeles .. 121622
Works on Paper, New York 098512
Works on Paper, North
Melbourne 099473
Works on Paper, Philadelphia . 123473
Works San Jose Gallery, San
Jose 124464
The Works, Ohio Center for History,
Art and Technology, Newark .. 051507
The Workshop, Roma 081600
The Workshop, Westport 079722, 130897
Workshop Arts Centre,
Willoughby 099889
Workshop Editions, Carlton North 098867
Workshop Print Gallery, Chicago 120303
Workshop Wales Gallery,
Fishguard 117800
Worksop Museum, Worksop ... 046199
Workspace, Honolulu 120978
World 4 You, Missingdorf 062737
World Accents, Portland 123689
World Antique Dealers Association,
Mansfield 060816
World Antique Fair, Osaka 098235
World Art Gallery, Eindhoven ... 112470
World Art Gallery, Ho Chi Minh
City 125156
World Arts, Beijing 138604
World Class Collectibles,
Jacksonville 094052
World Coin Company of America,
Chicago 092816
World Coin News, Iola 139535
World Coins, Canterbury 088783
World Famous Portobello Market,
London 090600, 134803
World Figure Skating Museum and
Hall of Fame, Colorado Springs 047892
World Fine Art Gallery, New York 123202,
123203
World Golf Hall of Fame, Saint
Augustine 052752
World in Wax Museum, Echuca . 001050
World Kite Museum and Hall of Fame,
Long Beach 050377
World Line Arts, Dubai 117191
World Methodist Museum, Lake
Junaluska 050081
World Money Fair, Berlin 098170
World Museum Liverpool,
Liverpool 044869
World Museum of Mining, Butte 047337
World Museum of Natural History,
Riverside 052598
World O' Tools Museum, Waverly 054240
World of Antiques & Art, Bondi
Junction 138499
World of Antiques + Art, Bondi
Junction 138500
World of Art, Cairo 103087
World of Books, Belfast 144241
World of Centuries Art, Woollahra 062484
World of Country Life, Exmouth . 044315
World of Energy at Keowee-Toxaway,
Seneca 053269
World of Glass, Saint Helens .. 045689
World of Wings Pigeon Center
Museum, Oklahoma City 051739

The World Organization of China Painters'
Museum, Oklahoma City 051740,
060817
World Police Museum, Central Police
University, Gueishan 042321
World Research Company,
Washington 097869
World Sculpture News, Hong
Kong 138605
World Trade Art Gallery, New
York 123204
World Wide Art Products, Hong
Kong 102129
World Wide Arts Resources,
Columbus 120525
World Wide Gallery, Washington 124981
Worldgallery.co.uk, London 118646
Worlds End Bookshop, London . 144715
World's End Gallery, London . . . 118647
World's Treasures, Houston 093841
Worldtisic Art, Chicago 120304
Worldwide Numismatic Galleries,
Sacramento 096656
Wormsloe State Historic Site,
Savannah 053184
Worpswede Antiquariat,
Worpswede 142507
Worpsweder fotografische Werkstätten,
Worpswede 022778
Worpsweder Kunsthalle, Kunststiftung
Friedrich Netzel, Worpswede . . 022779
Worpsweder Mühle, Worpswede 022780
Worrell, Norman, Nashville 122108
Worsbrough Mill Museum, Barnsley,
South Yorkshire 043500
Worsfold, Deborah, Vancouver . . 101846
Worshipful Company of Glaziers and
Painters of Glass, London 060276
Wortart Antiquariat, Augsburg . . 141421
Worth & Co., Jacksonville 094053
Worthing Auction Galleries,
Worthing 127103
Worthing Museum and Art Gallery,
Worthing 046200
Worthington, Chicago 120305
Worthington, Los Angeles 094505
Worthington Historical Society,
Worthington 054554
Worthington, Mac, Columbus . . . 120526
Worthington, Nancy, Houston . . . 121156
Worthy's Upholstering Company,
Washington 136470
Wortley Top Forge, Thurgoland . 045977
Wortreich, Glarus 144084
Worx of Art, Hitchin 117973
Wosinsky Mór Megyei Múseum,
Szekszárd 023950
Wosnitza, Franziska, Dresden . . 129729
Wotton Auction Rooms, Wotton-under-
Edge 127104
Wotton Heritage Centre, Wotton-under-
Edge 046201
Woudenberg, Den Haag 132483
Wouldle Museum, Jiangle 007933
Woulfe, Angela, Limerick 109671
Wouters, J.-J., Dinant 063612
Wow, Denver 093354
Wow & Zen, Chicago 092817
Wow Art Gallery, Mumbai 109485
Wowan and District Museum,
Wowan 001663
Wrackmuseum, Cuxhaven 017792
Wrange, Uppsala 087129
Wrangell Museum, Wrangell . . . 054555
Wrather West Kentucky Museum,
Murray State University, Murray 051130
Wray Museum, Wray 054556
Wray, Randy, Belle Plaine 127135
Wren, Burford 088714, 134182
Wren, Portland 096494
Wren Gallery, Burford 117477
Wren House Antiques, Wrentham 091958
Wren, Christopher, Melton
Mowbray 118737

Wren's Nest House Museum,
Atlanta 046640
Wrest Park, Silsoe 045802
Wrexham Arts Centre, Wrexham 046205
Wrexham County Borough Museum,
Wrexham 046206
Wrigglesworth, Linda, London . . 090601,
134804
Wright, Albuquerque . . 092093, 127116
Wright, Chicago 127170
Wright, Christchurch 083910
Wright Brothers National Memorial, Kill
Devil Hills 049968
Wright County Historical Museum,
Buffalo 047297
Wright Historical Museum, Swift
Current 007112
Wright Manley, Tarporley 127062
Wright Museum, Wolfeboro 054514
Wright Museum of Art, Beloit College,
Beloit 046906
Wright State University Art Galleries,
Dayton 048176
Wright, Andrew, Hobart 125201
Wright, B.J. & M.A., Burra 061241
Wright, Eli, Portarlington 062102
Wright, Gary, Moreton-in-Marsh 090786,
134861
Wright, Mick & Fanny,
Minchinhampton 090762
Wright, Peter, Great Missenden . 089376,
134423
Wright, Wayne, Christchurch . . . 083911,
132714
Wright's Ferry Mansion,
Columbia 047905
Wrightway Studios, Auckland . . 132700
Wrigley-Cross, Portland 145465
Wriston Art Center Galleries, Lawrence
University, Appleton 046514
Writers Museum, Dublin 024914
The Writers' Museum, Edinburgh 044279
Wrobel, Claude, Paris 129161
Wroblewski, Richard, Hannover . 076401
Wrocławska Fundacja Studentów
Historii Sztuki, Wrocław 059728
WSA Gallery, Hamilton 112909
WTC Art Gallery, Rotterdam . . . 112673
Wu Bruce, Grace, Hong Kong . . 065119
Wu-El, Szczecin 084678, 143508
Wucher, Francis, Storckensohn . 073560
Wuchuan Museum, Wuchuan . . 008368
Wudy, Christine, Zwiesel 109010
Wüest, Josef F., Ascona 116037
Wülfingmuseum, Radevormwald 021262
Wülker, Th., Lage 076992
Wünnenberg, Eckhard, Essen . . 141754
Würfelspielhaus –
Mensch.Macht.Leben,
Frankenburg 001898
Würmtal-Antiqauriat, Gräfelfing . 141839
Würstlein, Michelau in
Oberfranken 077369
Würth Art Room, Böheimkirchen 001790
Würth, Steffi, Haßloch 076416
Würthle & Sohn Nfg., Wien . . . 063108,
140133
Württembergische Landesbibliothek,
Stuttgart 022067
Württembergischer Kunstverein
Stuttgart, Stuttgart 022068
Württembergisches Landesmuseum
Stuttgart, Stuttgart 022047
Württembergisches
Psychiatriemuseum, Zwiefalten 022913
Württembergisches Trachtenmuseum,
Pfullingen 021128
Würzburger Kunstauktionshaus,
Würzburg 126144
Wüst, Arno, Gossau (Sankt
Gallen) 087511, 133763
Wüst, Ursula, Berlin 129579
Wüstendörfer-Heid, Pedra,
Eckental 129770

Wüstenroter Glas- und
Heimatmuseum, Wüstenrot . . . 022808
Wüster, H., Köngen 107704
Wüthrich, Tony, Basel 116110
Wuhan Museum, Wuhan 008370
Wuhole Museum, Xiaogan 008417
Wuhou Museum, Chengdu 007700
Wuhua Museum, Wuhua 008373
Wuilliot Franck, L'Isle-sur-la-
Sorgue 069599
Wujiang Museum, Wujiang 008374
Wujin Museum, Wujin 008375
Wulai Atayal Museum, Wulai . . . 042585
Wulastook Museums, Fredericton 005850
Wulf, Dele, Konstanz 107726
Wulf, Gerhilt, Hamburg 107321
Wulfert, Fritz, Kevelaer 076769
Wulff, C., Rostock 142338
Wulff, N., Krummwisch 076982
Wulff's Antikviteter, Kjellerup . . 065817
Wullschleger, John R., Zürich . . 116872
Wunderbaldinger, Sankt Pölten . 140037
Wunderkammer, Arezzo 142618
Wunderkammer, Carlton, Victoria 061287
Wunderlampe, Laufen 087556
Wunderlich, Roland, Bad
Mergentheim 129418
Wunderlich, Wolfgang, München 077645
Wunderskirchner, Manfred,
Böbing 129605
Wunram, Emmerthal 075615
Wunschbild Verlag, Hamburg . . 137258
Wunschel, Hervé, Villeurbanne . 074231
Wuping Museum, Wuping 008377
Wuppertaler Uhrenmuseum,
Wuppertal 022826
Wurlitzer Antiques, Miami 094817
Wurm Laf, Gotfried, Lassee 100090
Wurm-Schleimer, Malen, Herne . 107422
Wurm, Mark, Schmitten 078378
Wurm, Thomas, Erfurt 129807
Wurzel- und Heimatmuseum,
Amel 003270
Wurzer, Gerhard, Houston 121557
Wutzke, Koblenz 107582, 130138
Wuwei Museum, Wuwei 008378
Wuwei Zone Museum, Wuwei . . 008379
Wuxi Beike Guan, Wuxi 008380
Wuxi City Calligraphy and Painting
Gallery, Wuxi 102256
Wuxi City Museum, Wuxi 008381
Wuxi Glaswegianjian Guan, Wuxi 008382
Wuxi Revolutionary Display Center,
Wuxi 008383
Wuyi Museum, Wuyi 008385
Wuyishan Museum, Wuyishan . . 008386
Wuyishan Nature Museum,
Wuyishan 008387
Wuyuan Museum, Wuyuan 008388
Wuzhangyuan-Zhu Geliangmiao
Museum, Qishan 008193
Wuzhi Museum, Wuzhi 008389
Wuzhou Museum, Wuzhou 008390
WVSA Arts Connection,
Washington 060818
Wyalkatchem C.B.H. Agricultural
Museum, Wyalkatchem 001664
Wyalong Park Private Museum,
Maddington 001239
Wyandot County Historical Society
Museum, Upper Sandusky . . . 053907
Wyandotte County Museum, Bonner
Springs 047085
Wyandotte Museum, Wyandotte 054558
Wyannie Malone Historical Museum,
Hope Town 003199
Wyards, Faversham 117793
Wyatt, Buffalo 119971, 135382
Wyatt Earp Birthplace,
Monmouth 050992
Wychwood, Taynton 091549
Wyck Museum, Philadelphia . . . 052104
The Wyckoff Farmhouse Museum,
Brooklyn 047269

Wycombe Museum, High
Wycombe 044583
Wydawnictwa Artystyczne i Filmowe,
Warszawa 137910
Wydawnictwo Artystyczne,
Wołomin 137913
Wydawnictwo Artystyczne, Michal
Adamczyk, Warszawa 137911
Wydawnictwo Konserwatorów Dzieł
Sztuki, Łódź 137898
Wydawnictwo PGS, Sopot 137902
Wydler, Kurt, Bern 133692
Wydra, Tereza E., Gdańsk 113491
Wydział Sztuk Pjęknych, Uniwersytet
Mikołaja Kopernika, Toruń . . . 056241
Wye Gallery, Hay-on-Wye 117938
Wye Marsh Wildlife Centre,
Midland 006307
Wyeth, New York 095835
Wykeham, Stockbridge 091435, 119145,
135085
Wyland, Honolulu . . . 120979, 120980
Wyland, Key West 121322
Wyland, Las Vegas 121390
Wyland, San Diego 124193
Wyland, San Francisco 124438
Wyland, Seattle 124618
Wylde, Trigg 062348
Wylder, Jonathan, London 118648
Wylie House Museum,
Bloomington 047051
Wyllie, London 090602, 118649
Wyłupek, Marek, Szczecin 084679
Wyman, New York 123205
Wymondham Antique and Collectors
Centre, Wymondham, Norfolk . 091964
Wymondham Heritage Museum,
Wymondham 046210
The Wynd Gallery, Melrose 118735
Wyndham, Omaha 096068
Wyndham Lewis Society, London 045141
Wyndham Museum, Wyndham . . 033369
Wyndham Shopping Arcade,
Kingston 111228
Wyndhams, Stow-on-the-Wold . 091489
Wynick/Tuck Gallery, Toronto . . 101724
Wynn Cato, Miles, London 090603,
118650
Wynn Collection, Las Vegas . . . 050174
Wynn, Allen, Boston 119934
Wyoming Arts Council Gallery,
Cheyenne 047631
Wyoming Dinosaur Center,
Thermopolis 053731
Wyoming Pioneer Home,
Thermopolis 053732
Wyoming Pioneer Memorial Museum,
Douglas 048346
Wyoming State Museum,
Cheyenne 047632
Wyoming Territorial Prison,
Laramie 050146
Wyrardisbury Antiques,
Wraysbury 091956, 135263
Wyreena Gallery, Croydon 098937
Wysing Arts Centre, Bourn 043679
Wyss & Partner, Zollikon 116738, 126761
Wyss, Elisabeth & Peter,
Wegenstetten 087893
Wyss, Jaime, Dübendorf 087399, 133727
Wystan Hugh Auden-
Dokumentationsräume,
Kirchstetten 002179
Wyters, Saint-Ouen 073225
Wyters, Chantal, Paris 072017
Wythenshawe Hall, Manchester City
Art Gallery, Manchester 045220
Wyższa Szkoła Nauk Humanistycznych
i Dziennikarstwa, Poznań 056238
X-21 Modern, San Francisco . . . 124439
X 21 Modern Design,
Sacramento 096657
X Antique Complex, Castlemaine 061299
X-ART ProDivision GmbH,

Pinkafeld 054959
X F Upholstery, San Jose 136361
X-Pression, Indianapolis 121226
X.M.X. Art and Antique Centre,
 Lefkosia 065327
Xá:ytem Museum, Mission 006336
Xabier Sabi i Lopez, Barcelona . 143734
Xairel, Lisboa 084998
Xanadu, Maastricht 083489
Xanadu, Sittard 083609
Xanadu Gallery, San Francisco . 097350
Xandi, Ponta Delgada 085004
Xanthippe Antichitá, Venezia . . 081880
Xanthos, Glebe 061526
Xantus János Múzeum, Győr . . 023669
Xarrie, Barcelona 085860
Xavante, Saint Kilda, Victoria . . 099670
Xavier, Sliema 082435, 111922
Xavier-Bender, Michèle, Paris . . 072018
Xavier Square Antiques, Phoenix 096247
Xavier University Museum (Museo de
 Oro), Cagayan de Oro City . . 034449
Xavier, Tereza, Rio de Janeiro . . 101078
Xavier's, Detroit 093381
Xcape, Long Beach . . 094234, 094235
Xenos, Basel 133663
Xenos, Hamburg 137259
Xhenseval, Philippe, Marseille . 128789
Xhibit Art Contemporary & Design,
 McMahons Point 099250
Xi Bao Celadon Museum,
 Shenzhen 008268
Xi County Museum, Huangshan . 007912
Xiajiang Museum, Xiajiang 008392
Xiamen Calligraphy and Painting
 Gallery, Xiamen 102258
Xiamen Museum, Xiamen 008395
Xian Academy of Fine Art, Xian . 055253
Xian Banpo Museum, Xian 008401
Xian Beilin Museum, Xian 008402
Xian Forest of Stone Tablets Museum,
 Xian 008403
Xian Highway Display Center,
 Xian 008404
Xiangfan Museum, Xiangfan . . . 008407
Xiangfen Museum, Xiangfen . . . 008409
Xiangshen, Shanghai 065155
Xiangtan Museum, Xiangtan . . . 008410
Xiangwall Museum, Xianyuan . . 008415
Xiangxi Zone Museum, Jishou . . 007976
Xiangxiang Museum, Xiangxiang 008411
Xiangyu, Indianapolis 121227
Xianyang Museum, Xianyang . . 008413
Xianyou Museum, Xianyou 008414
Xiao Xiao, Shanghai 065156
Xiaogan Museum, Xiaogan 008418
Xiaomen Geology Hall, Siyu . . . 042466
Xiaoshan Hua Institute, Hangzhou 102041
Xiaoshan Museum, Xiaoshan . . 008419
Xiaoyanta Bag Guan Museum,
 Xian 008405
Xiaoyi Museum, Xiaoyi 008420
Xibit, Mumbai 109486
Xicali, El Paso 093421
Xifeng Museum, Xifeng 008424
Xihu Calligraphy and Painting Gallery,
 Hangzhou 102042
Xijiashan Glaswegiansu Guan,
 Jiang'an 007930
Xilin Museum, Xilin 008425
Xiling, Hangzhou 102043
Xiling Shishu Painting Gallery,
 Chengdu 102016
Ximenes, Maria Pires, Setúbal . . 132983
Xin Hai Shouyi Mausoleum of Fallen
 Heroes, Wuhan 008371
Xin Metropolis Revolution Museum,
 Weihui 008358
Xin-Shenzhou, Singapore 114638
Xin, Yun, Singapore 085532
Xin'an Qiantang Zhizhi Museum,
 Xin'an 008426
Xincheng District Museum,
 Nanning 008124

Xincity Tusi Museum, Xincheng . 008427
Xinfeng Museum, Xinfeng,
 Guangdong 008428
Xinfeng Museum, Xinfeng,
 Jiangxi 008429
Xing, West Hollywood 125034
Xingan-meng Museum,
 Wulanhoate 008376
Xingan Museum, Xingan 008430
Xinghua Museum, Xinghua 008431
Xingjing Yandao Gucheng Ruins
 Museum, Yingjing 008488
Xingning Museum, Xingning . . . 008432
Xinguang Art Gallery, Shenzhen 102233
Xinhui Museum, Xinhui 008435
Xinjian Museum, Xinjian 008437
Xinjiang Painting Gallery,
 Urumchi 008347
Xinjiang Photography and Fine Arts
 Publishing House, Urumchi . . . 136744
Xinjiang Uighur Autonomous Region
 Museum, Urumchi 008348
Xinjiang Xian Museum, Xinjiang 008438
Xinle Yi Zhi Ruins Museum,
 Shenyang 008260
Xinmi Museum, Xinmi 008439
Xinping Bowuguan, Xinping . . . 008441
Xintai Museum, Xintai 008442
Xinxiang Museum, Xinxiang . . . 008443
Xinyi Museum, Xinyi, Guangdong 008444
Xinyi Museum, Xinyi, Jiangsu . . 008445
Xinyu Museum, Xinyu 008446
Xinzhou Zone Museum, Xinzhou 008447
Xippas, Athinai 109052
Xippas, Renos, Paris 105320
Xishui Museum, Xishui 008448
Xiu Rock Museum, Xiuyan 008450
Xiuning Museum, Xiuning 008449
Xizhou Yandu Ruins Museum, Shan
 Zhou Museum, Beijing 007650
Xoio, Porto 085057
Xperiment Huset, Växjö 041151
Xu Bei-Hong Memorial Hall,
 Beijing 007651
Xuanhua, Singapore 114639
Xuanzhou Museum, Xuanzhou . . 008451
Xuchang, Shanghai 065157
Xuchang Museum, Xuchang . . . 008452
Xue Lai Er, Shanghai 102207
Xue Qin, Beijing 101997
Xunyang Museum, Xunyang . . . 008453
Xunyi Museum, Xunyi 008455
Xuxa, Amsterdam 082913
Xuyi Museum, Huai'an 007898
Xuzhou Gupeng, Xuzhou 065164
Xuzhou Museum, Xuzhou 008457
XVIII. Kerületi Pedagógiai és
 Helytörténeti Gyüjtemény,
 Budapest 023569
XX Multiple, Rotterdam 112674
XX Settembre, Bergamo 109805
XXO – Xtra Xtra Original,
 Romainville 072520
Xydis, Antonis, Athinai 079188
XYLON Deutschland, Reutlingen 059319
Xylon Museum + Werkstätten,
 Schwetzingen 021803
Xylos, Wien 127979
The Y, Hong Kong 102130
Y & B Brocanteurs, Bruxelles . . 063587
Y & Z Oriental Furniture, Los
 Angeles 094506
Y P Art Expo, Paris 105321
Y.H.F. Lardy, Toulon (Var) 073730
Y.S.F. Bookshop, Sheffield 144841
Ya Hui Art Gallery, Shenzhen . . 102234
Ya Pian Zhan Zhen Museum,
 Dongguan 007746
Ya Ya Gallery, New Orleans . . . 122257
Ya, Roman, Sapporo 082108
Yabane-Sagemonoya, Tokyo . . . 082215
Yachay Wasi, Training Center of
 Conservation and Restoration,
 Lima 056225

Yacimiento-Museo Arqueologico,
 Ambrona 038969
Yacimiento Museo Arqueológico de
 Ambrona, Miño de Medinaceli . 039754
Yacimiento-Museo Arqueológico de
 Numancia, Garray 039427
Yacoubian, G.H., Lefkosia 102543
Yad Labanim Museum and Memorial
 Center, Petah Tikva 025168
Yad Lashiryon Museum, Latrun . 025150
Yad-Lebanim Memorial Center,
 Rehovot 025181
Al-Yad Al-Mahira, Baghdad . . . 079430
Yad Vashem Art Museum, Museum of
 Holocaust Art, Jerusalem . . . 025116
Yad Yaari Museum of Jewish Youth
 Movement, Givat Haviva 025039
Yaelle, Toronto 101725
Yaeyama Museum, Ishigaki . . . 028701
Yagai Hakubutsukan, Gashozukuri
 Minkaen, Shirakawa 029396
The Yager Museum, Oneonta . . 051779
Yagi, Kyoto 082049
Yagi, Tokyo 142947
Yağmur Kayabek, İstanbul 088148
Yaguarundi Galería, Cali 102417
Yague, Barcelona 085861, 133177
Yakage Museum, Yakage 029695
Yakima Valley Museum, Yakima 054566
Yaky, Roma 081601, 081602
Yakyu Taiiku Hakubutsukan,
 Tokyo 029623
Yaldhurst Museum of Transport and
 Science, Christchurch 033178
Yale Center for British Art, New
 Haven 051266
Yale Gallery, Vancouver 101847
Yale Publications in the History of Art,
 New Haven 139536
Yale University Art Gallery, New
 Haven 051267
Yale University Collection of Musical
 Instruments, New Haven 051268
Yale University Press, London . . 138260
Yale University Press, New Haven 138368
Yallingup Galleries, Yallingup . . 099931
Yalvaç Müzesi ve Örenyerleri,
 Yalvaç 043083
Yamagata Bijutsukan, Yamagata 029696
Yamagata-kenritsu Hakubutsukan,
 Yamagata 029697
Yamaguchi, Osaka 111368
Yamaguchi Hakubutsukan,
 Yamaguchi 029698
Yamaguchi Hoshun Memorial Hall,
 Hayama 028616
Yamaguchi-kenritsu Bijutsukan,
 Yamaguchi 029699
Yamaguchi Tokenten, Tokyo . . . 082216
Yamaki Art Gallery, Osaka 111369
Yamamoto, Hiroshima 111265
Yamamoto Gendai, Tokyo 111523
Yamamoto Isoroku Kinenkan,
 Nagaoka 029082
Yamamoto Kanae Memorial Museum,
 Ueda 029670
Yamana Artsalon, Tokyo 082217
Yamanashi-kenritsu Bijutsukan,
 Kofu 028869
Yamanashi-kenritsu Hakubutsukan,
 Fuefuki 028534
Yamanashi-kenritsu Kagaku Senta,
 Kofu 028870
Yamanashi-kenritsu Koko
 Hakubutsukan, Kofu 028871
Yamashita, Ito 081990
Yamaso, Kyoto 111304
Yamatane Bijutsukan, Tokyo . . 029624
Yamato, Lille 069508
Yamato, Paris 072019
Yamato Hakubutsukan, Kure . . 028910
Yamauchi Collection of Art Museum,
 Kochi 028864
Yamauchi Library, Yokohama . . 029728

Yamba Museum, Yamba 001666
Yan, Hong Kong 102131
Yanaga Hokkaido Museums,
 Sapporo 029344
Yanaginogosho Kokokan,
 Hiraizumi 028638
Yanbian Museum, Yanji 008469
Yancey, Austin 092281
Yancey Richardson, New York . . 123206
Yancheng Museum, Yancheng . . 008458
Yanchi Museum, Yanchi 008459
Yanchyshyn, Yuri, New York . . . 136027
Yanda Aboriginal Art, Alice
 Springs 060993
Yanda Aboriginal Arts Melbourne,
 Mentone 061841
Yande Meannjin Antiques,
 Narangba 061941
Yande Meannjin Antiques, Paddington,
 Queensland 062057
Yandina Trading Post, Yandina . . 062498
Yanes, Miami 094818
Yang County Antiquity Museum,
 Yangxian 008465
Yang Ming Oceanic Culture and Art
 Museum, Keelung 042374
Yang Ming Studio, Taipei 042563
Yangjiang Museum, Yangjiang . . 008461
Yangliuqing Gallery, Tianjin . . . 102244
Yangliuqing Museum, Tianjin . . 008331
Yangon State School of Music, Drama
 and Fine Arts, Yangon 056165
Yangpyong Batangol Gallery,
 Gangha 111596
Yangshan Museum, Yangshan . . 008463
Yangtze River Antiques, Seattle . 097528
Yangxin Museum, Yangxin 008466
Yangzhou Hantomb Museum,
 Yangzhou 008467
Yangzhou Museum, Yangzhou . . 008468
Yanhuang Art Museum, Beijing . 007652
Yanjing Art Cooperative, Beijing . 058503,
 101998
Yanjing Art Cooperative-Biyun St,
 Beijing 058504, 101999
Yanjing Art Cooperative Changcheng,
 Beijing 102000
Yanjing Art Cooperative Liulichang,
 Beijing 058505, 102001
Yanjing Art Cooperative-Qi Nian Hall,
 Beijing 058506, 102002
Yanjing Art Cooperative-Shisanling,
 Beijing 058507, 102003
Yanjing Art Cooperative-Wofosi,
 Beijing 058508, 102004
Yank Azman, Toronto 064783
Yankalilla District Historical Museum,
 Yankalilla 001668
Yankee Air Museum, Belleville . . 046883
Yankee Candlemaking Museum, South
 Deerfield 053403
Yankee Peddler Antiques,
 Pittsburgh 096343
Yankee Trader, Denver 093355
Yankee Trader, Fresno 093518
Yannick, David, Bruxelles 063588
Yannis Tsarouchis Museum,
 Maroussi 023161
Al-Yanora, Sharjah 088215
Yanshi Marketplace Museum,
 Yanshi 008470
Yant, Wrocław 132918
Yantai Arts Museum, Yantai . . . 008471
Yantai Museum, Yantai 008472
Yanzhou Museum, Yanzhou . . . 008473
Yao County Museum, Yaoxian . . 008475
Yao'an Museum, Yao'an 008474
Yaoban Yaostove Museum,
 Tongchuan 008338
Yaoyao Wangshan Museum,
 Yaoxian 008476
Yaozu Self-Rule Museum, Jinxiu 007973
Yapı Kredi Vedat Nedim Tör Müzesi,
 İstanbul 042966

Yard Art, San Francisco 097351
Yard Dog Folk Art, Austin 092282
Yard Gallery, Nottingham 045424
Yarliyil Art Centre, Halls Creek .. 061575
Al-Yarmouk, Manama 100457
Yarmouth County Museum,
    Yarmouth 007495
Yarmouth Gallery, Yarmouth ... 119401
Yarnton Antiques Centre, Yarnton 091966
Yarra Sculpture Gallery,
    Abbotsford 098607
Yarra Valley Antique Centre, Yarra
    Glen 062502
Yarralumla Gallery, Yarralumla .. 099940
Yarraman Heritage Centre,
    Yarraman 001671
Yarrow, Honiton 089576
Y'Artecre Faso, Ouagadougou .. 058409
Yarton, Don, San Antonio 097045
Yashi Hua Yuan, Shenzhen 102235
Yass Railway Museum, Yass ... 001674
Yasuda, New York 123207
Yasui, Osaka 082091
Yasumi, Honolulu 093578, 135581
Yataro Noguchi Bijutsukan, Nagasaki
    City Hall, Nagasaki 029104
Yatay, Asunción 113339
Yates, David & Constance, New
    York 123208
Yates, Kenneth W., Shorncliffe . 062209
Yato, San Diego 124194
Yatsugatake Bijutsukan, Haramura
    Rekishi Minzoku Shiryokan,
    Hara 028613
Yatsuhashi-An & Shishu-Yakata Gallery,
    Kyoto 028989
Yatsushiro Kyodo Hakubutsukan,
    Yashiro 029705
Yau, Yick Hong, Hong Kong 065120
Yawa, Gambrills 145153
Yaya, Tokyo 082218
Yayo Bajo, Pamplona 086397
Yayoi Gallery, Tokyo 111524
Yayoi Museum, Tokyo 029625
Yazdan, Manama 100458
Yazdi, Frank, Calgary 101269
Yazılıkaya Müzesi, Çifteler 042871
Yazoo Historical Museum, Yazoo
    City 054574
Ybor City State Museum, Tampa 053673
Ydais, Annick, Biarritz 067215
Ydessa Hendeles Art Foundation,
    Toronto 007226
Ye Auld Curiosité Boutique,
    Montréal 064489
Ye Olde Antique Furniture House,
    Edmonton 064311
Ye Olde Cupboard, Miami 094819
Ye Olde Village Clock Shop,
    Freshwater 089308, 134397
Yea, Sung, Seoul 111734
Yeager, Ira, San Francisco 124440
Year Dot, Leeds 089773
Years Gone Buy, Chicago 092818
Yeats Country Antiques, Drumcliff 079517
Yeats Tower – Thoor Ballylee,
    Gort 024930
Yedikulehisarı Müzesi, İstanbul Hisarlar
    Müzeleri, İstanbul 042967
Yee, Mock, Seoul 111735
Yefief, Santa Fe 139537
Yeghegnadzor Art Gallery,
    Yeghegnadzor 000719
Yeghishe Charents Literary House-
    Museum, Yerevan 000749
Yeghishe Charents Literature and Arts
    Museum, Yerevan 000750
Yeh, Seoul 111736
Yeiser Art Center, Paducah 051887
Yelen, Milano 080768
Yelens, Nowa Nowa 099492
Yeleswasam Pavilion, Hyderabad 024304
Yell Crafts, Shetland 119093
Yello Gallery, Kinsale 109662

Yellow Aster, Los Angeles ... 094507
The Yellow Gallery, Dublin ... 109627
Yellow Lantern Antiques,
    Brighton 088640
Yellow Medicine County Historical
    Museum, Granite Falls 049202
Yellow Moon Gallery, Holcombe . 117974
Yellowstone Art Museum, Billings 046989
Yellowstone Gateway Museum of Park
    County, Livingston 050345
Yelmo, Cartagena ... 085927, 085928,
    126520,   126521
Yelverton Paperweight Centre,
    Yelverton 046216
Yemac, Seoul 111737, 111738
Yeongil Folk Museum, Pohang .. 029978
Yepes, Los Angeles 121623
Yeppoon Collectables, Yeppoon . 062508
Yeramian, J., Den Haag 112424
Yerba Buena Center for the Arts, San
    Francisco 053032
Yerebatan Sarnıcı, İstanbul 042968
Yeremiyew, Luc, Héry 128622
Yerevan History Museum,
    Yerevan 000751
Yes Galeria, Poznań 113752
Yeshiva University Museum, New
    York 051493
Yesim, Hurlstone Park 099148
Yesong, Daegu 111593
Yester-Days, Haddington 089402
Yester-Year, Iver 089653, 134505
Yester-Year Artifacts Museum,
    Rowley 006834
Yester-Years Community Museum,
    Dinsmore 005709
Yesterday, Torino 081771
Yesterday and Today, Miami ... 094820
Yesterday and Today, Sutton-in-
    Ashfield 091521
Yesterday Militaria, Aalborg ... 065592
Yesterday Museum, Bay Roberts 005401
Yesterday Once More, Miami ... 094821
Yesterday, Today and Tomorrow,
    Atlanta 092199
Yesterday, Today, Tomorrow,
    Dundee 089117
Yesterdays, Wareham 091726
Yesterdays Antique & Curios,
    Singapore 085533
Yesterday's Antiques, Cleethorpes 088927,
    134280
Yesterday's Antiques, Victoria
    Park 062389
Yesterdays Books, Bournemouth 144258
Yesterday's Books, Wauchope .. 139956
Yesterday's Child, Denver 093356
Yesterday's Children Antique Doll &
    Toy Museum, Vicksburg ... 053977
Yesterday's Gems, Coffs Harbour 061354
Yesterdays Memories, Los
    Angeles 121624
Yesterdays Memories, Miami .. 094822
Yesterday's Memories,
    Milwaukee 145270
Yesterday's Memories and More,
    Pittsburgh 096344
Yesterday's News, Pittsburgh .. 096345
Yesterday's News, Tulsa 097768
Yesterday's Today, Perth 090980
Yesterdays Treasures, Ariah Park 061006
Yesterday's Treasures, Cleveland 092982
Yesterday's Treasures, Houston . 093842
Yesterdaze, Seattle 097529
Yesteryear, Christchurch 083912
Yesteryear, Philadelphia 096175
Yesteryear, Princeton . 138447, 139538
Yesteryear – A Step Back in Time, Los
    Angeles 094508
Yesteryear Antiques, Columbus . 093057
Yesteryear Antiques, Dublin ... 079600
Yesteryear Book Shop, Atlanta . 144974
Yesteryear Costume Gallery,
    Ororoo 001387

Yesteryear House-Central Mine,
    Clarkston 047774
Yesteryear Mart, Las Vegas 094185
Yesteryear Museum, Captains
    Flat 000944
Yesteryear Museum, Pamplin .. 051906
Yesteryears, Cincinnati 092909
Yesteryears Doll and Toy Museum,
    Sandwich 053077
Yeungnam University Museum,
    Gyeongsan 029947
Les Yeux Fertiles, Paris 105322
Les Yeux Germourmands,
    Bruxelles 100672
Yew Tree Antiques, Four Elms .. 089299
Yew Tree Antiques, Wiveliscombe 091895
Yew Treehouse Antiques, New
    York 095836
Yew, Miaw, Singapore 085534
Yewon, Seoul 111739
Yezerski, Howard, Boston 119935
Yggdrasil Antikvariat, Tønsberg . 143424
Yglesias, New York 136028
Yi City Museum, Yicheng, Hubei 008479
Yi Hai Tang, Shanghai 065158
Yi Jun Peace Museum, Den
    Haag 032203
Yi Zhi Shi Liao Guan, Shanghai . 008240
Yi, Hong, Seattle 124619
Yi, Hyung, Seoul 111740
Yi, Xuan, Singapore 114640
Yi, Xuan Bao, Singapore 114641
Yiannos K. Gallery, New York .. 123209
Yiannoulis Halepas Museum and
    Museum of Tinian Artists, Panormos
    Tinou 023210
Yibin Museum, Yibin 008477
Yichang Museum, Yichang 008478
Yicheng Museum, Yicheng,
    Shandong 008480
Yichun Museum, Yichun, Jiangxi 008482
Yifeng Museum, Yifeng 008484
Yigal Allon Center, Ginnosar .. 025038
Yiğit, İsmail, İstanbul 088149
Yih Feng, Taipei 088070
Yilan County Historical Museum,
    Yilan 042593
Yilan County Natural History Museum,
    Luodong 042390
Yilan Fo Guang Yuan Art Gallery,
    Yilan 042594
Yilan Museums Association, Yilan 059996
Yıldız Sarayı Müzesi ve Tiyatrosu
    ve Sahne Kostümleri Müzesi,
    İstanbul 042969
Yili Zone Museum, Yili 008485
Yilpinji, Nice 104395
Yin & Co., I.T., Houston 121158
Ying City Museum, Yingcheng . 008487
Ying-Min JHS Folk Art Museum,
    Kaohsiung 042370
Ying, Kee Hong, Singapore ... 085535
Yingde Cutureal Museum,
    Qingyuan 008186
Yingge Ceramics Museum,
    Yingge 042596
Yingkou Museum, Yingkou 008489
Yingtan Museum, Yingtan 008490
Yinque Shanhan Tomb Zhujian
    Museum, Linyi, Shandong ... 008301
Yip, Tim, Hong Kong 102132
Yip's Rainbow, Hong Kong 102133
Yishu, Journal of Contemporary
    Chinese Art, Vancouver 138591
Yisulang, Singapore 114642
Yiti, Al-Qurum 113296
Yitti, Ruwi 113299
Yiwu Museum, Yiwu 008491
Yixing Ceramic Museum, Yixing . 008492
Yiyang Museum, Yiyang, Hunan . 008493
Yiyang Museum, Yiyang, Jiangxi 008494
Yiyuan, Shanghai 065159, 102208
Yizheng Museum, Yizheng 008495
Yksi, Eindhoven 112471

Ylä-Savon Kotiseutumuseo,
    Iisalmi 010893
Ylämaan Jalokivimuseo, Ylämaa 011522
Yläneen Kotiseutumuseo, Yläne . 011523
Yli-Lauroselan Talomuseo,
    Ilmajoki 010896
Ylihärmän Kotiseutumuseo,
    Ylihärmä 011524
Ylikiimingin Kotiseutumuseo,
    Ylikiiminki 011526
Ylistaron Kotiseutumuseo,
    Ylistaro 011527
Ymir Museum, Ymir 007497
Ynik, Almere 082648
Yo Sin-ci Stone Sculpture Museum,
    Siioulin 042465
Yoder, Dean, Cleveland 135472
Yogi Berra Museum, Little Falls . 050331
Yojigen, Osaka 082092
Yokogaki Suikodo, Osaka 082093
Yokohama Bijutsukan, Yokohama 029729
Yokohama Doll Museum,
    Yokohama 029730
Yokohama Kaiko Shiryokan,
    Yokohama 029731
Yokohama Maritime Museum and
    Nippon-Maru Memorial Park,
    Yokohama 029732
Yokohama Museum of EurAsian
    Cultures, Yokohama 029733
Yokohama Ogimura Shikko,
    Yokohama 082235
Yokohama Portside Gallery,
    Yokohama 029734
Yokohama Rekishi Hakabutsukan,
    Yokohama 029735
Yokohama-shi Kare-Hakubutsukan,
    Yokohama 029736
Yokohama-shi Sutori-Toka-
    Hakubutsukan, Yokohama ... 029737
Yokosuka-shi Hakubutsukan,
    Yokosuka 029740
Yokota Hakubutsukan, Tsukude . 029658
Yokoyama, Kyoto 082050
Yokoyama, Tokyo 082219
Yokoyama Memorial Manga Museum,
    Kochi 028865
Yokoyama Taikan Kinenkan,
    Tokyo 029626
Yolande's Gallery, Enniskillen .. 117758
Yoldosh Oxunboboev Yodgorlik Muzeyi,
    Marg'ilon 054634
Yolk, Los Angeles 121625
Yolo County Historical Museum,
    Woodland 054522
Yomitanson Bijutsukan,
    Yomitanson 029742
Yomitanson Rekishi oyo Minzoku
    Hakubutsukan, Yomitanson ... 029743
Yonaa Antiques, Los Angeles ... 094509,
    135748
Yonago-shi Bijutsukan, Yonago . 029744
Yonan, Serge F., Beirut 082310
Yonezawa Municipal Uesugi Museum
    & Uesugi Shrine Keisyo Palace
    Museum, Yonezawa 029745
Yong, Yuan Cheng, Singapore .. 114643
Yongan Museum, Yongan 008496
Yongchun Museum, Yongchun .. 008497
Yongding Museum, Yongding .. 008498
Yongfeng, Shenzhen 102236
Yongfeng County Museum,
    Enjiang 007756
Yonghong, Shanghai 102209
Yongji Museum, Yongji 008499
Yongtai Museum, Yongtai 008500
Yoni, Borås 086649
Yonsei University Museum, Seoul 030054
Yoo, Byung Kuk, Seoul 111741
Yook Yung, Seoul 111742
Yoon, Sebastian, Calgary 101270
Yorba and Slaughter Families Adobe
    Museum, Chino 047724
The Yorck Project, Gesellschaft für

Bildachivierung, Berlin 137104
York, Roma 081603
York, San Antonio 097046
York, Sydney 062302
York and Lancaster Regimental
  Museum, Rotherham 045640
York Antiques, Launceston 061727
York Antiques Centre, York .. 091983
York Art Gallery, York 046230
York Cottage Antiques, Helmsley 089492
York County Museum, York ... 054588
York Courthouse Complex, York . 001676
York Gallery, London 090604
York Gallery, Torquay 119245
York House, Richmond, North
  Yorkshire 091094
York Minster Undercroft Treasury and
  Crypt, York 046231
York Museum, Toronto 007227
York Quay Gallery at Harbourfront
  Centre, Toronto 007228
York Racing Museum, York 046232
York Redoubt, National Historic Site,
  Halifax 005965
York Region District School
  Board Heritage Schoolhouse,
  Markham 006266
York Story, York 046233
York Sunbury Museum,
  Fredericton 005851
York, Graham, Honiton 089577, 144488
Yorke, Moonta 061881
Yorke, John H., Saint Leonards-on-
  Sea 091200
Yorks Shona Gallery, New York . 123210
Yorkshire Air Museum, Elvington 044288
Yorkshire and Humberside Federation
  of Museums and Art Galleries,
  Doncaster 060277
Yorkshire Artscene, Halifax 139232
Yorkshire Coast College,
  Scarborough 056605
Yorkshire Museum, York 046234
Yorkshire Museum of Carriages
  and Horse Drawn Vehicles,
  Wensleydale 046085
Yorkshire Museum of Farming,
  York 046235
Yorkshire Museums Council,
  Leeds 060278
Yorkshire Pine, San Diego 097151
Yorkshire Sculpture Park, West
  Bretton 046086
The Yorkshire Watercolour Society,
  Brighouse 060279
Yorkshire Waterways Museum,
  Goole 044447
Yorkton Arts Council, Yorkton . 007502
Yorktow, Launceston .. 099223, 139801
Yorktown Battlefield Visitor Center
  Museum, Colonial National Historical
  Park, Yorktown 054593
Yorktown Colonial Park, Yorktown 054594
Yorktown Historical Museum,
  Yorktown 054590
Yorkville House, Toronto 101726
Yorozu Tetsugoro Kinen Bijutsukan,
  Towa 029635
Yoseido, Tokyo 111525, 132313
The Yosemite Museum, Yosemite
  National Park 054596
Yoshida, Portland 096495, 123690
Yoshii, New York 123211
Yoshii, Osaka 082094
Yoshii, Paris 105323
Yoshii, Hiromi, Tokyo .. 111526, 137797
Yoshimoto Bungeikan, Yoshimoto
  Comedians Museum, Osaka . 029272
Yoshimura, Emile, Mitaka 132308
Yoshinogawa Sake Museum Hisago-tei,
  Nagaoka 029633
Yoshizawa Kinenkan, Joetsu .. 028738
Yossi Milo Gallery, New York ... 123212
Yossi S. Masters Workshop,

Jerusalem 079731
Youjiang Nationality Museum,
  Baise 007580
Youli Zhouyi Museum, Tangyin . 008318
Youll, Hungerford 089616
Young, Bruxelles 100673
Young, Edinburgh 089198
Young, Ha Noi 125107
Young, Hawthorn East 125200
Young & Son, London . 090605, 134805
Young, Saint Louis 123887
Young, Salt Lake City 124000
Young at Art Children's Museum,
  Davie 048159
Young Blood Galleries, Atlanta .. 119633
Young Classics, Reutlingen ... 108486
De Young Museum, Fine Arts
  Museums of San Francisco, San
  Francisco 053033
Young Originals Gallery of Fine Art,
  Paddington, Queensland ... 099543
Young, Donald, Chicago 120306
Young, Jackie, Pittsburgh 145440
Young & Son, John, Keswick .. 089676
Young, John, Toronto 064784
Young, K.C., Seattle 097530
Young, R.M., South Molton ... 144851
Young, Robert, London 090606
Young's Brewery Museum,
  London 045142
Youngs, John & Rebecca,
  Christchurch 083913
Your Art Gallery, Knokke-Heist . 100768
Your Decor, Las Vegas 121391
Your Hearts Desire, Las Vegas . 094186
Your Treasure Is Here, Tampa . 097595
Yours Truly Antiques,
  Jacksonville 094054
Youth Museum of Southern West
  Virginia & Beckley Exhibition Coal
  Mine, Beckley 046866
Youxi Museum, Youxi 008501
Youzhou Youci Museum,
  Pingdingshan 008147
Yoxall Antiques, Solihull 091358, 135057
Yozgat Devlet Güzel Sanatlar Galerisi,
  Yozgat 043085
Ypäjän Kotiseutumuseo, Ypäjä . 011529
Ypsilanti Historical Museum,
  Ypsilanti 054603
Yr Art'Thé, Bagnoles-de-l'Orne . 103321,
                          140775
Yrjar Heimbygdslag, Opphaug . 033820
Yrjö A. Jäntin Taidekokoelma,
  Porvoo 011277
Yrjö Liipolan Taidekokoelma,
  Koski 011030
Yrjönkadun Antikvariaatti,
  Helsinki 140723
Yrwing, Stockholm 115940
Ystads Konstmuseum, Ystad .. 041184
Ystads Militärmuseum, Ystad .. 041185
Ystwyth Books, Aberystwyth .. 144217
d'Ythurbide, Olivier, Saint-Ouen . 073226
Ythurbide, Olivier d', Saint-Cloud 072717
Yttilän Museokoulu, Yttilä 011530
Yturri-Edmunds Historic Site, San
  Antonio 052956
Yu, Ogaki 111342
Yu-I Institut, Bruxelles 100674
Yu Yuen Antique Market,
  Shanghai 065160
Yu, Jing, San Francisco 124441
Yuan Meng Art, Beijing 102005
Yuan Shuo Shi Han Lin Cong
  Huei Stone Sculpture Museum,
  Sincheng 042456
Yuan, Hong, Singapore 114644
Yuanling Museum, Yuanling .. 008503
Yuanping Museum, Yuanping .. 008504
Yuanyehao Animal Specimen Museum,
  Puli 042423
Yubi, Nagoya 111333
Yucaipa Adobe, Yucaipa 054608

Yucca Art Gallery, Albuquerque . 119524
Yudu Museum, Yudu 008506
Yue Gu, Zhai, Hong Kong 065122
Yue Po Chai, Hong Kong 065123
Yue, Fung, Hong Kong 065121
Yuen, Singapore 114645
Yufuin-cho Hakubutsukan, Yufuin 029749
Yufuin-cho Kindai Bijutsukan,
  Yufuin 029750
Yugawara-machi Bijutsukan,
  Yugawara 029751
Yukara-Ori Kogeikan, Asahikawa 028486
Yuki Bijutsukan, Osaka 029273
Yukon Arts Centre Gallery,
  Whitehorse 007407
Yukon Beringia Interpretive Centre,
  Whitehorse 007408
Yukon Historical and Museums
  Association, Whitehorse 058479
Yukon Transportation Museum,
  Whitehorse 007409
Yuldosh Ahunbabaev Yodgorlik Muzeyi,
  Toshkent 054672
Yulin Museum, Yulin 008507
Yuma Fine Arts Association,
  Yuma 060819
Yuma Fine Arts Museum, Yuma . 054611
Yuma Territorial Prison Museum,
  Yuma 054612
Yumak, Reutlingen 078217
Yume-Ya, Honolulu 093579
Yumeji Bijutsukan, Okayama .. 029217
Yumen Museum, Yumen 008508
Yumi, Kobe 111286
Yumiko Igarashi Bijutsukan,
  Kurashiki 028907
Yun Brocade Museum, Nanjing . 008118
Yun County Museum, Shiyanyun 008279
Yun Yu, Zhai, Beijing 065004
Yunan Museum, Yunan 008509
Yuncheng Museum, Yuncheng . 008510
Yuncheng Zone Hedong Museum,
  Yuncheng 008511
Yunfu Museum, Yunfu 008512
Yung Lee, Hong, Hong Kong .. 065124
Yuni, Kobe 111287
Yunji Zhai Arts, Wuhan 102248
Yunlin, Wuxi 102257
Yunmeng Museum, Yunmeng . 008513
Yunnan Art Institute, Kunming . 055221
Yunnan Fine Arts Publishing House,
  Kunming 136729
Yunnan Nationality Museum,
  Kunming 007991
Yunnan Provincial Museum,
  Kunming 007992
Yunus Emre Müzesi, Eskişehir .. 042900
Yupiit Piciryarait Museum, Bethel 046954
Yurara Art Gallery, Thornlands .. 099824
Yurttaş Kitabevi, İstanbul 088150
Yushan Museum, Yushan 008514
Yushe Reliquary Museum, Yushe 008515
Yushodo, Tokyo 142948
Yusif Mammadaliyev Home Museum,
  Ordubad 003196
Yutaka, Kanazawa 111274
Yuyu Yang Museum, Taipei ... 042564
Yuzawa History and Folk Museum,
  Yuzawa 029757
Yuzhang Glaswegiansu Bowu Yuan,
  Nanchang 008100
Yuzhong Museum, Yuzhong .. 008518
YvArt, Yverdon-les-Bains 116731
Yveline, Paris 072020
Yvernault, Marie-Alexandrine, Saint-
  Ouen 073227
Yverneau, Christophe, Auch ... 066803
Yvert & Tellier, Amiens 066583, 136796
Yvert & Tellier, Dury 068423
Yvert & Tellier, Lyon .. 069785, 136832
Yvert & Tellier, Paris .. 072021, 136994
Yvinec, Pierre Adrien, Paris ... 141299
Yvonne, Haarlem 083366

Yvonne Fine Art, Christchurch .. 083914,
                          112869
YY9 Gallery, Hong Kong 102134
YYZ Artists' Outlet, Toronto 007229
Z-Bau Galerie, Nürnberg 108311
Z-Galerie, Baar 116044
Z Galerie, Long Beach 121416
Z Galerie, Denver 120791
Z Gallerie, Sacramento 123791
Z Gallerie, San Diego . 124195, 124196
Z Gallerie, San Francisco 124442
Z Gallerie, Santa Monica 124473
Z Gallerie, Tucson 124787
Z. Paliashvili House Museum,
  Tbilisi 016427
Z33, Provinciaal Centrum voor
  Beeldende Kunsten-Begijnhof,
  Hasselt 003684
Z9A, Mainz 107957
Za Bramką, Poznań .. 113753, 143499
Za Piecem, Lublin 113703
Zaaijer, Pieter, Den Haag 083153
De Zaal, Delft 112371
Zaanlandse Oudheidkamer Honig
  Breet Huis, Zaandijk 033085
Zaans Museum, Zaandam 033083
De Zaanse Schans, Zaandam .. 033084
Zabala Madina, Amaia, Irún .. 086051,
                          133277
Zabari, Avner, Miami 121880
Zabbar Sanctuary Museum,
  Zabbar 030772
Zabel, S., Köln 130175
Zaberfelder Kunstkabinett und Verlag,
  Zaberfeld . ..109002, 137567, 142531
Zabern, Philipp von, Mainz ... 137372
Zabert, Gilberto, Torino 081772, 111016
Zabol Museum, Zabol 024794
Zabriskie, New York 123213
Zabytkowa Huta Żelaza, Filia
  Muzeum Techniki w Warszawie,
  Chlewiska 034684
Zabytkowa Kopalnia, Tarnowskie
  Góry 035272
Zabytkowa Kuźnia Wodna w Starej
  Kuźnicy, Filia Muzeum Techniki w
  Warszawie, Stara Kuźnica ... 035217
Zabytkowa Kuźnia Wodna, Muzeum
  Techniki w Warszawie, Gdańsk 034744
Zaccaria, Donatella, Milano ... 131553
Zaccaria, Paolo, Bologna 079891,
                          131005
Zaccaro, Carmine, Napoli 080946
Zaccaro, Luigi, Napoli 080947
Žáček, Miroslav, Praha 065545
Zach, Gerald, Graz 133992
Zacharias, Feldkirchen bei Graz 062561
Zacharias, Lemesos 102556
Zacharias, Karl, Graz 062605
Zachary, Gertrude, Albuquerque 092094
Zachęta – Narodowa Galeria Sztuki,
  Warszawa 035381
Zacke, Irene, Kritzendorf 062694
Zac's Attic, San Diego 097152
Zaczyński, Henryk, Poznań ... 143500
Zadah, A., London 090607
Zadelhoff, Hilversum .. 083410, 126290
Zadick, Lion B., Überlingen ... 126132
Zadick, Robert, München 077646,
                          108209
Zado, Portland 123691
Zadužbina Kralja Petra I
  Karadjordjevića, Topola 038251
Zährer, Peter, Wien 127980
Zähringer, Gerhard, Zürich 116873,
                          144199
Zagar, Isaiah, Philadelphia ... 123474
Zaghloul, Kamal, Cairo 103088
Zagni, Jean, Lavardac 069206
The Zaha Collection, Seoul ... 111743
Zahedan Museum, Zahedan .. 024795
Zahn, Andrea, Mainz 107958
Zahn, Hans, Aying 074479
Zahn, Hans, Peiß 078033

Zahnärztliches Museum,
Tübingen 022225
Zahnmuseum, Linz 002332
Záhorská Galéria Senica, Senica 038444
Záhorské Múzeum, Skalica 038445
Zahraa, Cairo 103089
Al-Zaim, Louai, Damascus 088065
Zaïnal, Françoise, Quiberon 105410
Zajsanskij Gorodskoj Istoriko-
Kraevedčeskij Muzej, Zajsan . . 029855
ŽAK Gallery, Berlin 106464
Žák, Milan, Brno 065363
Zakarpats'kyj Kraeznavčyj Muzej,
Užgorod 043313
Zakarpats'kyj Oblasnyj Chudožnij
Muzej im. J. Bokšaja, Užgorod 043315
Zakian Bros., Philadelphia 096176
Het Zakkendragershuisje, Historisch
Museum Rotterdam, Rotterdam 032832
Zakład Narodowy im. Ossolińskich,
Wrocław 035435
Zakurozaka Gallery, Tokyo 111527
Zakynthos Museum of the Natural
Heritage, Zakynthos 023323
Zalacain, Jean-François, Sarlat-la-
Canéda 073413
Zaldivar Museum, Albuera 034423
Zalipianki, Kraków 113648
Žaliūkių Vėjo Malūnas, Šiaulių Aušros
Muziejus, Šiauliai 030475
Záložňa, Senica 085590
Zaltbommels Stoom- en
Energiemuseum, Zaltbommel . 033087
Zalud House, Porterville 052277
Zamaan, Beirut 082311
Zamaan, Can, Cairo 103090
Zamalek Art Gallery, Cairo 103091
Zaman, Zay, Cairo 103092
Zamanda Sanat, İstanbul 117099
Zambakejian, Jiwan & Zohrab, Saint-
Ouen 073228
Zambala Art Museum, Shanhua 042446
Zambaldo, Alessandra, Verona . 132307
Zambelli, Daniela & C., Brescia . 079964
Zambezi, Amsterdam . 082914, 112318
Zambezi Gallery, Lagos 113977
Zambezi Imports, Canning Vale . 061274
Zambouanga National Museum,
Zamboanga 034618
Zamboni, Reggio Emilia 110712
Zamboni, Pierre, Congerville-
Thionville 068137
Zámecké Muzeum, Kravaře 009649
Zámecké Muzeum, Tovačov 009916
Zámek, Hradec nad Moravicí . . 009578
Zámek – Pamětní Síň Josefa Mánesa,
Čechy pod Košířem 009494
Zámek Brandýs, Brandýs nad
Labem 009454
Zámek Budišov, Moravské Zemské
Muzeum Brno, Budišov u
Třebíče 009489
Zámek Hrubý Rohozec, Expozice
Historického Nábytku 16.–19. a
Oděvů 19. Století, Turnov . . . 009926
Zámek Kinski, Chlumec nad
Cidlinou 009517
Zámek Kinských, Muzeum Regionu
Valško ve Vsetíně, Valašské
Meziříčí 009940
Zamek Królewski na Wawelu –
Państwowe Zbiory Sztuki,
Kraków 034913
Zamek Królewski w Warszawie,
Pomnik Historii i Kultury Narodowej,
Warszawa 035382
Zamek Książąt Pomorskich,
ThePomeranian Duke's Castle,
Szczecin 035255
Zámek Libochovice, Libochovice 009668
Zámek Líčkov – Galerie O. Brázda,
Žatec 009971

Zámek Lysice, Lysice 009687
Zámek Manětín, Muzeum a galerie
severního Plzeňska, Manětín . . 009689
Zámek Moravský Krumlov – Galerie
Slovanská Epopej Alfonse Muchy,
Galerie Hlavního Města Prahy,
Moravský Krumlov 009707
Zamek Piastowski, Muzeum w
Gliwicach, Gliwice 034758
Zámek Troja, Galerie Hlavního Města
Prahy, Praha 009845
Zámek Vizovice, Vizovice 009953
Zamek w Czersku, Czersk 034703
Zamek w Gniewie – Muzeum
Regionalne, Muzeum
Archeologicznego w Gdańsku,
Gniew 034762
Zamek w Lipowcu, Muzeum w
Chrzanowie im. Ireny i Mieczysława
Mazarakich, Babice 034624
Zámek, Okresní muzeum Chomutov,
Jirkov 009609
Zammit, Mosta 082419, 132337
Zamora Chacón, Dora, San José 102521
Zamoskvorečie Galereja, Moskva 114322
Zampardi, Massimiliano, Palermo 081083
Zanabazar Museum of Fine Arts,
Ulaanbaatar 031744
Zanardelli, Roma 081604
Zanardi, Roma 132073
Zanardi, Bruno, Parma 131741
Zanasi, Giuseppe, Bologna 142648
Zanavykų Krašto Muziejus,
Sakiai 030459
Zanazanian, John, Philadelphia . 136109
Zanchiello, Cosimo, Nyon 087715
Zancopé, Paolo Emanuele,
Venezia 081881
Zandbergen, Dick, Maarssen . . 143196
Zandenbos, Nunspeet 032707
Zander, Erika, Graz . . . 062606, 062607
Zander, Ingo, Wusterhausen . . . 079124
Zander, Susanne, Köln 107703
Zandi, Mehrdad, Berlin 106465
Zandlopermuseum Glanerbrug,
Enschede 032311
Zandner, Roland, Wildpoldsried . 079012
Zandoozi, Austin 119714
Zandvoorts Museum, Zandvoort 033088
Zane Grey Museum, Lackawaxen 050052
Zane Moss, New York 095837
Zanella, Carlo, Torino 132213
Zanella, F., Passau 078025
Zanella, Léon, Vaison-la-Romaine 105791
Zanella, Roberto, Torino 132214
Zanessi, Giancarlo, Verona 081947
Zanetta, Franco, Nyon 087716
Zanettini, Chiara, Padova 081012
Zanfrognini, Sergio, Milano 131554
Zangrilli & Co., London 090608
Zangs Atlandia, São Paulo 064218
Zani Ladey, Ludlow 090653
Zanichelli, Modena 142754
Zanikos, Anne, San Antonio . . . 124084,
136281
Zanin, Stefano, Venezia 132262
Zanna, Alessandro, Roma 081605,
132074
Zannettacci-Stephanopoli, Sonia,
Genève 116344
Zanni, Alessandro, Brescia 131056
Zanni, Giorgia, Modena 080827
Zannotti, Gennaro, Roma 081606
Zanol, Alain, Montréal-la-Cluse . 070320
Zanoletti, Giulia, Brescia 079965
Zanon, Mauro, Bolzano 131013
Zanoni, Peter, Eglisau . 087402, 133728
Zanotti, Gianpietro, Brescia . . . 131057
Zanotti, William, Modena 131593
Zantout, Majida, Beirut 082312
Zantt, Vanessa, Baltimore 092377
Zanutti, Giancarla, Milano 110486
Zanutto, Vincenzo, Venezia . . . 081882

Zanzibar, Memphis 121717
Zanzibar, Napoli 080948
Zanzibar National Archives and
Museums Department, Zanzibar 042643
Zaor, Québec 064543, 101524
Zaoyang Museum, Zaoyang . . . 008519
Zaozhuang Museum, Zaozhuang 008520
Zapadno-Kazachstanskij Oblastnoj
Istoriko-Kraevedčeskij Muzej,
Oral 029831
Západočeská Galeria v Plzni,
Plzeň 009762
Západočeské Muzeum, Plzeň . . 009763
Západomoravské Muzeum v Třebíči,
Třebíč 009919
Západomoravské Muzeum v Třebíči –
Muzeum Jemnice, Jemnice . . . 009598
Západoslovenské Múzeum v Trnave,
Trnava 038476
Zapata, Eduardo, Granada 133272
Zapata, Rodrigo, Cali 102418
Zapatabuch, Bordesholm 141624
Zapf, David, San Diego 124197
Zapffe, Carl, Seattle . . 097531, 136380
Zaporiz'kyj Chudožnij Muzej,
Zaporižžja 043324
Zaporiz'kyj Krajeznavčyj Muzej,
Zaporižžja 043325
Zapotec, Houston 121159
Zapovidnyk-muzej Gogolja,
Gogoleve 043155
Zapp Magazine, International
Artmagazine on Video,
Amsterdam 138940
Zappe, J., Eimen 075586
Zappoli, Tina, Porto Alegre 101004
Žar-Ptica, Belgorod 114176
ZAR Works, Johannesburg 085650
Zara Clark Folk Museum, Charters
Towers 000955
Zaralli, Alceste, Roma 132075
Zarasų Krašto Muziejus, Zarasai 030559
Zarauzko Arte eta Historia Museoa,
Museo de Arte e Historia de Zarautz,
Zarautz 040460
Zarco Antiquários, Lisboa 084999,
114070
Zard, Ghassan, Jal-el-Dib 082317,
111774
Zardin, Luciano, Venezia 081883
Zaremski, Marcin, Warszawa . . . 113912
Zargari, Gabriela, Walsrode 108842
Zarit, Jeffrey S., Dallas 093234
Zarposh, Houston 093843
Zarre, Andre, New York 123214
Zászlómúzeum, Budapest 023570
Zattin & Restano, Padova 131679
Zatzka, Bucaramanga 102400
Zaunbauer, Jörg, Mödling 127822
Zauner-Mayerhofer, Silvia, Wien 127981
Zauner, Andreas, Baden bei Wien 062524,
127716
Zauner, Christa, Leonding 100094
Zaunmüller, Karl-Heinz, Trier . . 142422
Zausch, A., München 077647
Zavelberg-Torrekens, Bruxelles . 063589
Zavelle, Philadelphia 145422
Zavičajna Muzejska Zbirka,
Čajetina 038190
Zavičajna Zbirka Đelekovec, Muzej
Grada Koprivnice, Đelekovec . 008898
Zavičajni Muzej, Aleksinac 038123
Zavičajni Muzej, Bar 031746
Zavičajni Muzej, Biograd na
Moru 008883
Zavičajni Muzej, Buzet 008890
Zavičajni Muzej, Danilovgrad . . 031754
Zavičajni Muzej, Herceg-Novi . . 031756
Zavičajni Muzej, Nikšić 031758
Zavičajni Muzej, Ogulin 008958
Zavičajni Muzej, Pljevlja 031762
Zavičajni Muzej, Ruma 038231
Zavičajni Muzej, Travnik 004285
Zavičajni Muzej, Ulcinj 031769

Zavičajni Muzej Baška, Baška . . 008881
Zavičajni Muzej Grada Rovinja,
Rovinj 008982
Zavičajni Muzej Kozara, Prijedor 004270
Zavičajni Muzej Našice, Našice . 008951
Zavičajni Muzej Ozlja, Ozalj . . . 008964
Zavičajni Muzej Poreštine, Poreč 008968
Zavičajni Muzej u Jagodini,
Jagodina 038195
Zavicajni Muzej Zemuna, Zemun 038259
Zavizajni Muzej, Varaždinske
Toplice 009015
Zavod za Zaštita na Spomenicite na
Kulturata i Muzej – Strumica,
Strumica 030640
Zavrtálek, Jan, Brno . . 065364, 102575
Zazo, Madrid 086282
Zazou, San Diego 124198
Les Zazous, Annecy 103252
Zazzaretta, Franco, Pfronten . . . 078046
Zbiory Historyczne Ziemi
Wadowickiej im. Marcina Wadowity,
Wadowice 035300
Zbiory Historyczno-Etnograficzne,
Oświęcim 035089
Zbiory Sprzętu Ratownictwa
Kolejowego, Poznań 035141
Zbiory Sztuki, Muzeum Ziemi
Kujawskiej i Dobrzyńskiej,
Włocławek 035403
Zbirci Umjetnina u Crkvi Sv. Ivana
Krstitelja, Pinakoteka, Trogir . 009010
Zbirka Anke Gvozdanović, Museum of
Arts and Crafts, Zagreb 009054
Zbirka Baltazara Bogišića, Cavtat 008893
Zbirka Crkvene Umjetnosti,
Šibenik 008987
Zbirka Drave, Muzej Grada Koprivnice,
Gola 008914
Zbirka Franjevacki Samostan,
Orebić 008960
Zbirka Franjevačkog Samostana,
Sinj 008990
Zbirka Franjevačkog Samostana
Visovac, Visovac 009018
Zbirka Klašter Sv. Nikolaus,
Korčula 008937
Zbirka Sakralne Umjetnini Župe Sveti
Blaža, Vodnjan 009019
Zbirka Umjetnina Andra Vida i
Katarine Mihičić i Zbirka Starij
Majstora Piperata, Mali Lošinj . 008948
Zbirka Umjetnina Crkve Sv. Domenika,
Trogir 009011
Zbirka Umjetnina Franjevačkog
Samostana, Hvar 008920
Zbirka Umjetnina Kairos, Benediktinski
Samostan Sv. Nikole, Trogir . . 009012
Zbirka v Jamarskem Domu, Grad
Krumperk, Domžale 038511
Zborowska, Nina, Painswick . . . 118879,
134923
Zdeb, Erich, Wien . . . 063109, 127982
Zderzak, Kraków 113649
Zdravotnické Muzeum, Praha . . 009846
Zdrenka, Barbara, Warszawa . . . 084758,
143535
ZeaYou, Waikato 113047
Zeb, Parma 081133
Zeberg, Grethe, Antwerpen 063256
Zebone, Indianapolis 093942
Zebra Crafts, Nairobi 082251
Zebra Gallery, Hahndorf 099104
Zebra One, London 118651
Zebrak, Monaco 082534
Żebrowski, Krzysztof, Warszawa . 132909
Zebulon B. Vance Birthplace,
Weaverville 054261
Zecchi Antichità, Firenze 080224
Zecchini, Napoleone, Milano . . . 080769
Zecchino, Angelo, Torino 132215
Zecchino, Ottavia, Torino 081773
Zecconi, Auguste, Saint-Bonnet-en-

Champsaur 072693
Zech, Hans jun., Ingolstadt 076615,
107464
Zeche Hannover, Bochum 017442
Zeche Nachtigall, Witten 022698
Zeche Zollern, Dortmund 017967
Zechmeister, Michael,
Lanzenkirchen 127802
Zeck, Michael, Eberbach 075559
Zedd, fine art & design, San
Francisco 124443
Zeddies, Uwe, Kulmbach 142039
Zedell, Ragnar, Stockholm 087078
Zedes, Bruxelles 100675
Zedler, Dörte Maria, Naumburg,
Saale 130438
Zee- en Havenmuseum de
Visserijschool, IJmuiden 032522
Zee Stone Gallery, Hong Kong . . 102135
Zeemuseum, Museum Scheveningen,
Scheveningen 032849
Zeeuws Biologisch Museum,
Oostkapelle 032727
Zeeuws Museum, Middelburg . . 032651
Zeeuws Poppen- en Klederdrachten
Museum, Zoutelande 033108
Zeews Maritiem MuZEEum,
Vlissingen 032999
Zeezicht, Eindhoven 083241
De Zegel, Utrecht 083672
Zeh, Wolfgang, Berlin 074952
Zehetner, Linz 062725
Zehm, Bodo, Melle 130296
Zehnthaus, Jockgrim 019465
Zehntscheune, Holzhausen an der
Heide 076565
Zehrfeld, André, Dresden 129730
Zeidel-Museum, Feucht 018355
Zeidler, Gdańsk 084515, 113492, 126402,
132821, 143454
Zeidler, Monika, Berlin 074953
Zeige, Carsten, Hamburg 125982
Zeil-Antiquariat, Frankfurt am
Main 141791
Zeile & Judish, Denver 120792
Zeiler Foto- und Filmmuseum, Zeil am
Main 022848
Zeiler, Bernd Restaurator, Pfaffenhofen
an der Ilm 130514
Zeilopleidingsschip Mercator,
Oostende 003935
Žeimelio Žiemgalos Muziejus, Šiaulių
Aušros Muziejus, Žeimelis 030560
Zeimes, Nathalie, Liège 063787
Zein, Seoul 111744
Zeiner, Franz, Hamburg 107322, 129998
Zeiringer, Franz, Ottendorf an der
Rittschein 062761
Zeiselmairhaus, Schrobenhausen 021740
Zeisig, Kurt-Georg, Berlin 141592
Zeisner, Aach 074322
Zeiss-Großplanetarium Berlin,
Berlin 017318
Zeit-Foto Salon, Tokyo 111528
Zeit für Kunst, Hamburg 076342
Zeitbrücke Museum, Gars am
Kamp 001926
Zeitgeist, Detroit 120820
Zeitgeist Gallery, Nashville 122109
Zeitgeschichtliches Forum Leipzig,
Stiftung Haus der Geschichte
der Bundesrepublik Deutschland,
Leipzig 020024
Zeitkunstgalerie, Halle, Saale . . 107213
Zeitkunstgalerie, Kitzbühel 100067
Zeitlos, Hamburg 076343
Zeitraum, Mannheim 077280
Zeitreise, Bamberg 074676
Zeitreise – Renner Villa,
Gloggnitz 001941
Zeitschrift für Kunstgeschichte,
München 138761
Zeitschrift für Kunsttechnologie und
Konservierung, Worms 138762

Zeitspurenmuseum Ochzethaus,
Altheim 001706
Zeittunnel Wülfrath, Wülfrath . . . 022783
Zeitverlag, München 137430
Zeitwandel, Bielefeld 074994
Zeitwende, München 077648
Zeitzeichen, Kleve 142000
Zejda, Rodenbach 108504
Zekelman Family Holocaust Memorial
Center, Farmington Hills 048709
Zekes Antikvariat, Simrishamn . 143916
Zelano, Washington 124982
Zelewski, Bodo von, Alfter 129375
Železnički Muzej, Beograd 038186
Železniční Muzeum Výtopna Jaroměř,
Jaroměř 009594
Železniční Muzeum Zlonice,
Zlonice 009987
Železniški Muzej, Ljubljana 038564
Zelienople Historical Museum,
Zelienople 054615
Zelko, Emmanuel, Saint-Gervais 072813
Zelko, Emmanuel, Saint-Ouen . . 072822
Zeller, Andreas & Erika, Trimstein 087857,
133941
Zeller, Andreas & Erika, Worb . . 087915
Zeller, Dr. Klaus Robert, Lindau . 142078
Zeller, Michael, Lindau 126028
Zeller, Michèle, Bern 087311
Zellermayer, Berlin 106466
Zelli, London 118652
Zellweger, Dieter, Lübeck 077193
Zellweger, Harry, Carabietta . . . 116186
Zem Goldrähmli, Pratteln 116572,
133888
Žemaičių Dailės Muziejus,
Plungė 030447
Žemaičių Muziejus Alka, Telšiai . 030488
Žemaičių Vyskupystės Muziejus,
Varniai 030500
Žemaitijos Kaimo Muziejus, Žemaičių
Muziejus Alka, Telšiai 030489
Zemaljski Muzej Bosne i Hercegovine,
Sarajevo 004283
Zemanek-Münster, Würzburg . . 079096,
126145, 142518
Zemgales Sētas un Lauksaimniecības
Mašīnu Muzejs, Uzvara 030243
Zemljepišni Muzej Slovenije, Univerza
v Ljubljani, Ljubljana 038565
Zempléni Múzeum, Herman Ottó
Múzeum, Szerencs 023968
Zemplińské Múzeum, Michalovce 038400
Zen Gallery, Bruxelles 063590
Zenatti, Tiziano, Ancona 080357
Zendron, Manfred, Wien 063110
Zenen Art, Wooburn Common . . 119383
Zenettörténeti Múzeum, Magyar
Tudományos Akadémia,
Budapest 023571
Zenetsis Vasilis, S., Athinai 109053
Zengcheng Museum, Zengcheng 008521
Zengen, von, Bonn 125900
Zenith, Maastricht 083490
Zenith, Tiranë 098534
Zenith Antiques, Pittsburgh 096346
Zenith Gallery, Washington 124983
Zenkatao Galerie, Wuppertal . . . 108998
Zenkel, Petra, Kronach 130202
Zeno Arte, Roma 110876
Zeno X Gallery, Antwerpen 100524
Zenobi, Roma 081607
Zenoficinas, Porto 132981
Zenshi, Tokyo 111529
Zentilin, Daniela, Venezia 111171
Zentique, Los Angeles . 094510, 121626
Zentral-und Landesbibliothek Berlin,
Gotisches Haus, Berlin 017319
Zentralantiquariat Leipzig,
Leipzig 142069, 142070
Zentralarchiv, Staatliche Museen
zu Berlin – Stiftung Preußischer
Kulturbesitz, Berlin 017320
Zentrales Verzeichnis Antiquarischer

Bücher, Tutzing 142436
Zentralfilmarchiv Laxenburg,
Filmarchiv Austria, Laxenburg . 002282
Zentralinstitut für Kunstgeschichte,
München 055661
Zentralschweizer Antiquitätentage und
Antik-Börse, Luzern 098337
Zentrum für Bucherhaltung,
Leipzig 130234
Zentrum für Internationale Lichtkunst,
Unna 022275
Zentrum für Kunst und
Wissenstransfer, Universität für
angewandte Kunst Wien, Wien 054987
Zentrum Paul Klee, Bern 041328
Zepf, Manfred, Karlsruhe 076705
ZEPHYR – Raum für Fotografie!,
Mannheim 020256
Zephyr Gallery, Louisville 121676
Zephyr Raum für Fotografie,
Mannheim 107975
Zeppelin-Museum, Neu-Isenburg 020674
Zeppelin Museum Friedrichshafen,
Friedrichshafen 018528
Zeppelin-Museum Meersburg,
Meersburg 020335
Zepter, Beograd 114446
Zerbini, Paola, Parma 081134
Zerfaß, Urban, Berlin 141593
Zeriz, Hägersten 115671
Zerkalo, Sankt-Peterburg 114368
Zerline, Saint-Ouen 073230
Zero Karati, Padova 110583
Zéro l'Infini, Etuz 103723
Zero to Sixties, New York 095838
Zerovnik, Annerl, Apolda 074420, 129385
Zerrenner, Rolf, Reichenbach im
Vogtland 130575
Zerrer, Stuttgart 130695
Zerrouki, Ch, Strasbourg 141353
Zerrouki, Karim, Rueil-Malmaison 072642
Zerunian, Ingrid, Wien 063111
Zervudachi, Peter, Trélex 087855
Zervudacki, Nice 104396
Zest Gallery, London 118653
Zeta, Piran 085624
Zetel, Amsterdam 112319
Zetland Lifeboat Museum,
Redcar 045595
Zetlin Museum, Yechiel Nahari
Museum of Far Eastern Art, Ramat
Gan 025177
Zettelmeier, Erika & Wilhelm,
Wachtberg 108830
Zetterquist, New York 123215
Zettl & Seemann, Karlsruhe 076706
Zettl Langer Gyűjtemény, Sopron 023904
Zettl, Brigitte, Mallersdorf-
Pfaffenberg 107862
Zettl, Dr. Friedrich, Wien 063112, 140134
Zettlmeissl-Satzinger, Thea,
Gunzenhausen 107191
Zeugs, Peißen bei Halle, Saale . 078034
ZEUS-press, Gevelsberg 137223
De Zeven Pijlen, Gouda 083272
De Zevende Hemel, Haarlem . . . 083367
Zeze-yaki Bijutsukan, Otsu 029283
Žezkazganskij Istoriko-Kraevedčeskij
Muzej, Žezkazgan 029856
Zezschwitz, von, München 126072
Zezula, Brno 125415
Zg Gallery, Chicago 120307
Zgodovinski Arhiv i Muzej Univerze v
Ljubljani, Ljubljana 038566
Zhabei Revolutionary Shi Liao Display
Center, Shanghai 008241
Zhan Tianyou Former Residence,
Wuhan 008372
Zhan Tianyou Memorial Hall,
Beijing 007653
Zhangheng Museum, Nanyang . 008131
Zhangjiachuan Museum,
Zhangjiachuan 008522
Zhangjiagang City Gallery,

Zhangjiagang 008523
Zhangjiakou Museum,
Zhangjiakou 008524
Zhangping Museum, Zhangping 008525
Zhangpu Museum, Zhangpu . . . 008526
Zhangqiu Arts and Crafts Guan,
Jinan 007961
Zhangqiu Museum, Zhangqiu . . 008527
Zhangshu Museum, Zhangshu . 008528
Zhangye Museum, Zhangye . . . 008529
Zhangzhongjing Museum,
Nanyang 008132
Zhangzhou Museum, Zhangzhou 008530
Zhanjiang Museum, Zhanjiang . . 008531
Zhao Hua Fine Arts Publishing House,
Beijing 136701
Zhaoling Museum, Liquan 008036
Zhaotong Museum, Zhaotong . . 008534
Zhe, Chen, Shenyang 102216
Zhejiang Academy of Fine Arts Press,
Hangzhou 136713
Zhejiang Museum, Hangzhou . . 007844
Zhejiang Natural Museum,
Hangzhou 007845
Zhejiang Painting Gallery,
Hangzhou 102044
Zhejiang People's Fine Arts Publishing
House, Hangzhou 136714
Zhengding County Display Center,
Zhengding 008535
Zhenghe, Nanjing 102153
Zhengjiatun Museum, Shuangliao 008282
Zhengpinzhai, Beijing 102006
Zhengzhou Museum, Zhengzhou 008539
Zhenjiang Museum, Zhenjiang Lishi
Bowuguan, Zhenjiang 008541
Zhenjiang Revolution History Museum,
Zhenjiang 008542
Zhenjianhai, Shanghai 065161
Zheong Zhong Jia Yuan,
Shanghai 065162
Zhi City Museum, Zhicheng 008543
Zhi Li Victory Museum, Baoding 007583
Zhijiang Museum, Zhijiang 008544
Zhilko, Odense 102954
Zhiyin Ren Hai Art Gallery,
Shanghai 102210
Zhongguo Jianzhu Xuéhui,
Beijing 058509
Zhongguo Qianbi Bowuguan,
Beijing 007654
Zhonghua Art, Beijing 102007
Zhonglu, Jinan 102142
Zhongshan, Jinan 102143
Zhongshan Art Cooperative,
Liaoyang 102146
Zhongshan Museum, Zhongshan 008546
Zhou Hejing County Museum,
Hejing 007856
Zhou Man Xia, Verona 081948
Zhoukou Zone Museum, Zhoukou 008547
Zhoushan Museum, Zhoushan . . 008548
Zhouyuan Museum, Fufeng 007774
Zhouzhi County Farmer Art Association,
Xian 058510, 102259
Zhuanglang Museum,
Zhuanglang 008549
Zhucheng Museum, Zhucheng . . 008550
Zhuhai Museum, Changning . . . 007671
Zhuhai Museum, Zhuhai 008552
Zhuzhou Museum, Zhuzhou . . . 008554
Zia, Karim, Frankfurt am Main . . 141792
Ziarsolo, Paul, Bilbao . 085897, 133206
Ziba, Honarhayeh, Isfahan 079419
Žibel, Minsk 100471
Zibelius, Galerie am Museum,
Hannover 076402, 107376
Zibetti, Solduno 087818, 133925
Zibetti, Pietro, Ascona . 087202, 133624
Zibetti, Sagl, Ascona . . 087203, 133625
Zibo Museum, Zibo 008555
ZIC Fine Art, Chicago 120308
Zich, Ulrike, Ennepetal 075623
Zichy Galéria, Budapest 109187

Zichy Mihály Emlékmúzeum,
Zala 024050
Zidarics, Norbert, Frankfurt am
Main 075835
Zidella, R., Oldenburg 077945
Židovské Muzeum v Praze, Praha 009847
Židovské Múzeum, Slovenské Národné
Múzeum, Prešov 038433
Ziederer, Udo, Werdau 078927
Ziegel- und Kalkmuseum, Winzer 022689
Ziegelbauer, Dieter, Maulbronn . 077308
Ziegelei Benzin, Kritzow 019797
Ziegelei-Museum, Cham 041399
Ziegeleimuseum, Jockgrim 019466
Ziegeleimuseum in Lage, Lage . 019846
Ziegeleipark Mildenberg,
Zehdenick 022847
Ziegler, Nürnberg 077850
Ziegler, Tulsa 124818
Ziegler, Eva & Peter, Bad Endorf 074508
Ziegler, Eva-Maria, Velden 127892
Ziegler, Josef, Wien 127983
Ziegler, Renée, Zürich 116874
Ziegler, Serge, Zürich 116875
ZieherSmith, New York 123216
Ziehm, Rolf, Zeulenroda-Triebes 079140,
130800
Ziehn & Dickmeis, Düren 106756
Ziekenhuis- en Verpleegkundig
Museum, Deventer 032219
Zieliński, Jerzy, Warszawa 143536
Zielke, Cornelia & Wolfgang,
Kirtorf 076815
Ziemer, Maisach 107961
Ziems, Dipl.-Restauratorin Katrin-Andrea,
Berlin 074954, 129580
Ziern & Hanon Dorothy, Saint
Louis 096797
Zietz, Rainer, London 090609
Zig Zag Antiques, Chicago 092819
Zig Zag Collectables, Toronto . . 064785
Zigafoos, Louisville 094608
Ziggurat, Warszawa 137912
The Zigler Museum, Jennings . . 049812
Zigong Salt Industry History Museum,
Zigong 008557
Zigong Saurian Museum, Zigong 008558
Zij van Hiernaast, Breda 083012
Zijdemuseum, Meliskerke 032645
Zijin Museum, Zijin 008559
Zijper Museum, Schagerbrug . . . 032847
Zika, Beograd 085351
Zikesch, Karlsruhe 107527
Žilinsko, J., Vilnius 082347
Žilka, Janoš a Petr, Brno 140461
Zilken, Elmer, Riedering 130589
Zilla Samgrahasala, Purulia 024453
Ziller, Agathon, Göttingen 076057
Ziller, Helmut, Adnet 125372
Zillertaler Goldschaubergwerk, Zell am
Ziller 003137
Zillich-Raudies, Isabell, München 130420
Zilverberg, M., Amsterdam 082915
De Zilveren Passer, Gent 063675, 128061
De Zilveren Toren, Antwerpen . . 100525
Zilvermuseum Sterckshof Provincie
Antwerpen, Antwerpen 003319
Zim & Mer, Trier 078711
Zimart Galleria, Los Angeles . . . 121627
Zimbabwe Military Museum,
Gweru 054848
Zimmer, Bitburg 106507
Zimmer, Reinheim 078189
Zimmer Children's Museum, Los
Angeles 050475
Zimmer Stewart, Arundel 117251
Zimmer, Elke & Werner,
Düsseldorf 106843
Zimmer, Hansjörg, Oldenburg . . 142271
Zimmer, U., Traben-Trarbach . . . 108772
Zimmer, Xavier, Marseille 140949
Zimmerdahl, Lund 086838
Zimmerdahl, Stockholm 087079
Zimmerei-Museum, Feldkirch . . 001883

Zimmergalerie, Schwetzingen . . 108600
Zimmerli, Rolf, Herisau 087523
Zimmermann & Heitmann,
Dortmund 106711
Zimmermann, Franz-Josef, Trier 078712,
130709
Zimmermann, Fritz, Filderstadt . 075735
Zimmermann, Fritz, München . . 130421
Zimmermann, Horst,
Neutraubling 077771
Zimmermann, Jean, New York . . 123217
Zimmermann, Jean, Paris 072022,
105324
Zimmermann, Jürgen, Raubling . 078142
Zimmermann, Kurt & Silvia,
Flaach 087424, 133737
Zimmermann, Peter, Mannheim . 107976
Zimmermann, S., Waldshut-
Tiengen 078846
Zimmermann, U. & H., Sexau . . 078458
Zimmermann, W. & R., Breisach am
Rhein 075140
Zimmermannhaus, Brugg 116169
Zimmermannshaus Lackinger,
Windhaag bei Freistadt 003123
Zimmertoren, Lier 003818
Zimports, Dallas 093235
Zinato, Maria Luisa, Venezia . . . 081884
Zinelli, Roma 081608
Zingales, Emanuela, Roma 081609
Zinger Presents, Amsterdam . . . 112320
Zingerle, Innsbruck 127781
Zingerle, Percha 131742
Zingmagazine, New York 139539
Zingoni, Elio, Firenze 131282
Zink, Michael, München 108210
Zinkhütter Hof, Museum für Industrie-
Wirtschafts- und Sozialgeschichte,
Stolberg, Rheinland 022006
Zinkwazi, Newcastle-upon-Tyne 118792
Zinn, Heidi, Wegberg 108851
Zinnfiguren-Klause im Schwabentor,
Museum für Geschichte in Dioramen
mit Zinnfiguren, Freiburg im
Breisgau 018492
Zinnfiguren-Museum, Bad
Bergzabern 016722
Zinnfiguren-Museum Goslar,
Goslar 018773
Zinnfigurenkabinett im Haspelturm,
Eberbach 018071
Zinnfigurenmuseum, Grüningen . 041570
Zinnfigurenmuseum, Zürich . . . 042242
Zinnfigurenmuseum Aach, Aach 016430
Zinnfigurensammlung,
Rheinsberg 021405
Zinnfigurenwelt, Katzelsdorf,
Leitha 002162
Zinngießerhaus, Mattighofen . . 002372
Zinngrube Ehrenfriedersdorf,
Besucherbergwerk – Mineralogisches
Museum, Ehrenfriedersdorf . . 018109
Zinniel, Astrid, Halbturn 062625
Zinnkann, Elfriede, Landau in der
Pfalz 077007
Zinnkeller, Weidenberg 022463
Zinongo, South Fremantle 099697
Zins, Michel, Paris 072023
Zintl, Herbert, Altdorf bei
Nürnberg 074389
Zintl, Herbert, Nürnberg 077851
Zintl, Maria, Ludwigshafen am
Rhein 077165
Zinzulusa, Roma 081610
Ziolkowski, Horst, Göllersdorf . . 099994
Zion, Salt Lake City 136247
Zion Historical Museum, Zion . . 054616
Zion Human History Museum,
Springdale 053449
Zip 37, Denver 120793
Zipfel, Roland, Mannheim 077281
Zippel, Christina, Essen 075706
Zipperle, D., Bruchsal 075202
Zippo and Case Visitors Center

Company Museum, Bradford . . 047173
Zipprich, D., München 142197
Zira Guichon, Montevideo 092012
Zirkelmuseum, Wilhelmsdorf . . . 022654
Zirkus und Varieté Archivsammlung
Reinhard Tetzlaff, Hamburg . . 019024
Zirngibl, Petra, Regensburg 130571
Zisska & Schauer, München . . . 126073,
142198
Zisterzienserabtei Mehrerau,
Bregenz 001796
Zisterzienserkloster Bebenhausen,
Staatliche Schlösser und Gärten
Baden-Württemberg, Tübingen 022226
Zisterziensermuseum Riddagshausen,
Braunschweig 017567
Zisterzienserstift, Lilienfeld 002312
Zisterzienserstift Rein, Rein 002585
Zita, K. & K., Sankt Wolfgang im
Salzkammergut 062832
Žitnoostrovné Múzeum, Dunajská
Streda 038348
Zitzewitz, Jutta von, Lübeck . . . 077194
Ziva Antiques, İstanbul 088151
Ziwna, Gerald, Wien 100399
Ziya Gökalp Müzesi, Diyarbakır . 042883
ZKM I Zentrum für Kunst- und
Medientechnologie Karlsruhe,
Karlsruhe 019522
Zlámal, Michal, Praha 128369
Zlatá Reneta, České Budějovice 102580
Zlatorog, Sofia 064240
Zlatý Kříž, České Budějovice . . . 102581
Zlot & Assoc., Mary, San
Francisco 124444
Zlotowski, Paris 105325
Znojemský Hrad, Jihomoravské
muzeum ve Znojmě, Znojmo . . 009992
Zoar State Memorial, Zoar 054619
Zoboli, Modena 131594
Zoco Antiguedades, Montevideo 092013
Zodiac, Dunedin 083936
Zoe, Milwaukee 121958
Zöchling, Werner, Eggenburg . . 062554
Zöhrer-Vogt, Viktoria D., Graz . . 062608
Zoellick, Scott, Milwaukee 121959
Zoetmulder, Burt, Salt Lake City 096931,
136248
Zöttl, Hermann, Ergoldsbach . . 129808
Zoffoli, Renato, Roma 081611
Zografia, Paiania 109079
Zograph, Oostende 100811
Zograph, Toronto 128221
Zohar, Haifa 079729
Zohar, Toronto 101727
Zohiko, Tokyo 082220
Zokyudo, Kyoto 111305
Zola Interni, Sankt Gallen 087791,
133913
Zola Zolu Gallery, Bandung 109503,
109504
Zolino, Jorio, Torino 111017
Zoll- und Finanzgeschichtliche
Sammlung, Oberösterreichische
Landesmuseen, Linz 002333
Zoll- und Heimatmuseum,
Perwang 002501
Zoll, S., Gemünden, Wohra 075972
Zolla & Lieberman, Chicago 120309
Zoller, Tulsa 097769
Zollikofer, Olga, Kilchberg
(Zürich) 133782
Zollinger &Co., Saint Louis 136209
Zollmuseum, Mandelbachtal . . . 020244
Zollmuseum, Wegscheid 022453
Zollmuseum Friedrichs, Aachen . 016446
Zolotaja Kniga, Moskva 143655
Zolotnik, Moskva 085213
Zolotoe Runo, Moskva 085214
Zolotoj Larec, Moskva 085215
Zoltán, Tamás, Budapest 079299
Zomerkoestal uut 't Wold,
Ruinerwold 032838
Zompenmuseum, Enter 032314

De Zon, Gouda 083273
De Zon & De Gijselman, Diemen 126284
Zon, Beate, Wettenberg 078955
Zona arkeologijke e Butrintit,
Butrinti 000012
Zona Bovisa, Milano 059486
Zona D, São Paulo 064219
Zonal, San Francisco 097352
Zonal Anthropological Museum,
Shillong 024476
Zonal Museum, Dehradun 024215
Zonca & Zonca, Milano 110487
Zonco, Carlo, Nice 070677
ZONE – chelsea center for the arts,
New York 123218
zone B, Berlin 106467
Zone France Edition d'Art, Paris 136995
Zône Vormgeving, Leiden 112563
Zonengrenz-Museum Helmstedt,
Helmstedt 019188
Zong Lian Zhai Biao Gallery,
Tianjin 102245
Zonguldak Devlet Güzel Sanatlar
Galerisi, Zonguldak 043086
Zonguldak Ereğli Müzesi, Zonguldak
Ereğli 043087
Zonin, Roma 132076, 132077
De Zonnebloem, Amstelveen . . . 082671
De Zonnehof, Centrum voor Moderne
Kunst, Amersfoort 031916
Het Zonnetje, Haarlem 083368
Zoo, Bologna 109871
Zoo Antik, København . 065969, 128399
Zoo Art Fair, London 098409
Zoo Galerie, Nantes 104348
Zoölogisch Museum Amsterdam,
Amsterdam 032002
Zoological Museum, Addis Ababa 010726
Zoological Museum, Bukittinggi . 024553
Zoological Museum, Dublin 024915
Zoological Museum, Faizabad . . 024257
Zoological Museum, Gorakhpur . 024265
Zoological Museum, Hardwar . . 024289
Zoological Museum, Jaunpur . . . 024325
Zoological Museum, Lahore 034138
Zoological Museum, Meerut . . . 024397
Zoological Museum,
Pematangsiantar 024635
Zoological Museum, Tehrän 024787
Zoological Museum, Tel Aviv . . . 025214
Zoological Museum, Zografou . . 023327
Zoological Museum at Agricultural
College, Karaj 024704
Zoologičeskij Muzej, Biškek . . . 030078
Zoologičeskij Muzej im. N.M.Kulagina,
Moskovskaja selskochozjajstvennaja
akademija im. K.A. Timirjazeva,
Moskva 037524
Zoologičeskij Muzej Kafedry Zoologii
Pozvonočnych Sankt-Peterburgskogo
Gosudarstvennogo Universiteta,
Sankt-Peterburg 037830
Zoologičeskij Muzej Moskovskogo
Gosudarstvennogo Universiteta im.
M.V. Lomonosova, Moskva . . . 037525
Zoologičeskij Muzej Rossijskogo
Gosudarstvennogo Pedagogičeskogo
Universiteta im. A.I. Gercena, Sankt-
Peterburg 037831
Zoologičeskij Muzej Rossijskoj
Akademii Nauk, Sankt-
Peterburg 037832
Zoologičeskij Muzej Tomskogo
Gosudarstvennogo Universiteta,
Tomsk 037954
Zoologičeskij Muzej, Dalnevostočnyj
gosudarstvennyj universitet,
Vladivostok 038047
Zoologičeskij Muzej, Kazanskij
gosudarstvennyj universitet,
Kazan 037122
Zoologičeskij Muzej, SGPU,
Samara 037717
Zoologičeskij Muzej, Syktyvkarskij

Register der Institutionen und Firmen

Zoologische Sammlung der Universität Rostock, Institut für
– Žytomyrs'kyj Literaturno-memorial'nyj Muzej V.G.

Gosudarstvennyj Universitet,
Syktyvkar 037916
Zoologische Sammlung der
Universität Rostock, Institut für
Biowissenschaften, Allgemeine und
Spezielle Zoologie, Rostock . . . 021496
Zoologische Sammlungen des
Institutes für Biologie/Zoologie,
Martin-Luther-Universität Halle-
Wittenberg, Halle, Saale 018971
Zoologische Schausammlung,
Tübingen 022227
Zoologisches Forschungsinstitut
und Museum Alexander Koenig,
Bonn 017504
Zoologisches Museum der Universität
Göttingen, Göttingen 018759
Zoologisches Museum der Universität
Hamburg, Hamburg 019025
Zoologisches Museum der Universität
Heidelberg, Heidelberg 019150
Zoologisches Museum der Universität
Zürich, Zürich 042243
Zoologisches Museum zu Kiel,
Christian-Albrechts-Universität,
Kiel 019612
Zoologisches und Tiermedizinisches
Museum der Universität Hohenheim,
Stuttgart 022070
Zoologisk Museum, Grindsted . . 010088
Zoologisk Museum, København . 010201
Zoologiska Museet, Lunds Universitet,
Lund 040815
Zoologiska Utställingar,
Evolutionsmuseet Uppsala Universitet,
Uppsala 041124
Zoology and Botany Museum,
Ernakulam 024252
Zoology Museum, Annamalai
Nagar 024144
Zoology Museum, Ibadan 033424
Zoology Museum, Johannesburg 038753
Zoology Museum, Kanpur 024340
Zoology Museum, Kingston 028451
Zoology Museum, Legon 022939
Zoology Museum, Lucknow 024388
Zoology Museum, Muzaffarnagar 024414
Zoology Museum of National Taiwan
University, Taipei 042565
Zoology Museum, University of
Queensland, Saint Lucia 001477
Zoolohičnyj Muzej imeni Benedikta
Dybows'koho, Lvivskij Nacional'nyj
Universitet im Ivana Franka,
Lviv 043258
Zoom Gallery, Bruxelles 100676
Zoom Kindermuseum, Wien 003102
Zoonekynd, Ludwig, Calais 067603
Zopp, Itzstedt 076639
Zoppi, Den Haag 132484
Zoppi, Giovanni, Torino 081774
The Zora Neale Hurston Museum,
Eatonville 048461
Zordan, Pierre, Toulouse 073811
Zorn, Täby 115949
Zorn, Askolf W., München 077649
Zorn, Inge, Jesteburg 076649
Zorndt, Elke & Jost, Rantzau . . . 078127
Zornmuseet, Mora 040854
Zotos, Christos, Amsterdam 082916
Zott, Peter, Memmingen 077344
Zottmann, Judendorf-Straßengel 127783
Zouari Mardochee, Fabien, Paris 072024
Zoubok, Pavel, New York 123219
Zouein, Paris 141300
Zouied, Hemiad Farag Ahmad,
Yanbu 085325
Zoumboulakis, Athinai 079189
Zoumboulakis, Daphne, Athinai . 109054
Zoumpoulaki, Athinai 109055
Zoutmuseum Delden, Delden . . . 032153
Zrinyi Miklós Vármúzeum,
Szigetvár 023973
Zscheyge, Theresia, München . . 077650

Zschiesche, Bernhard, Dresden . 129731
Zsidai, Péter, Budapest 079300
Zsolna Galéria, Budapest 109188
Zsolnay Kerámia Kiállítás, Janus
Pannonius Múzeum, Pécs 023860
Ztichlá Klika, Praha . . 102729, 140535
Zttlau, W., Offenburg 077912
Zubal, John T., Cleveland 145091
Zubelli, Sergio, Brescia 079966
Zubia, Eduardo, Montevideo . . . 092014
Zubia, Fernando Nicolas,
Montevideo 119436
Zubiate Gonzáles, Guadalupe,
Lima 084473
Zubiller, L., Freiburg im Breisgau 141813
Zubrnická Museální Železnice,
Zubrnice 009993
Žubrowski, Mieczysław, Kraków 113650
Zubrycki, Fontaine-de-Vaucluse . 103747
Zucca Centro Artistico, Udine . . 111069
Zuccaro, Madrid 115294
Zucchetti, Serenella, Perugia . . . 081146
Zucchetto, Diego, Palermo 081084
Zucchi Collection, Milano 026834
Zucker-Museum, Berlin 017321
Zucker, Werner, Ansbach 074419
Zuckmantler Heimatstube, Bietigheim-
Bissingen 017387
Züge, Stralsund 078572
Zühlsdorf, Karl-Heinz, Köln 076920
Zühmendorf, Angelika, Waldenburg,
Hohenlohekreis 078837
Zülow-Gruppe, Linz 058339
Zürcher Abwassermuseum Kläranlage
Werdhölzli, Zürich 042244
Zürcher Hochschule der Künste,
Zürich 056449
Zürcher Kunstgesellschaft, Zürich 059989
Zürcher Spielzeugmuseum,
Zürich 042245
Zürcher, Cornelia, Kirchberg
(Bern) 144088
Zürichsee Auktionen, Erlenbach 087413,
126719
Zürichsee Zeitschriftenverlag,
Stäfa 138091
Zufferey, Annick, Carouge 116199
Zuffi, Paolo, Venezia 132263
Zuger Kunstgesellschaft, Zug . . . 059990
Zuid-Limburgse
Stoomtreinmaatschappij,
Simpelveld 032869
Zuiderkerk, Amsterdam 032003
Zuiderzeemuseum, Enkhuizen . . 032305
Zuidwestbrabants Museum, Halle 003667
Zuivelmuseum, Blankenberge . . 003370
Žukowska, Likia, Kraków 113651
Žula, Peter, Maribor 114734
Zulauf, Dr. Günther, Freinsheim . 117066
Žulebinskij, Moskva 085216
Zulehner, Martin, Igls 127771
Zulu-African Art Shop, New
Orleans 122258
Zululand Historical Museum, Fort
Nongqayi Museum Village,
Eshowe 038712
Zululand Mission Station Chapel,
Fort Nongqayi Museum Village,
Eshowe 038713
Žulys, G., Klaipéda 082340
Zum Alten Laempli, Zürich 088018
Zum Arkantus, Binningen 087324,
133698
Zum Blauen Pfauen, Birrwil 087325,
116161, 133701
Zum Bücherwald, Basel 144041
Zum Gampiross, Basel 087265, 133664
Zum goldige Nüüteli, Aarau 116007
Zum Goldspiegel, Binningen . . . 133669
Zum Holzwurm, Basel . 087266, 133665
Zum Jäger und Sammler,
Bischofszell 087326
Zum Kobold, Gossau (Sankt
Gallen) 116356

Zum Puppenhaus, Immenstaad am
Bodensee 019395
Zum Taubenschlag, Linz 062726
Zum Unteren Wildenstein, Basel 087267
Zumikon Studio und Lounge, Institut
für moderne Kunst Nürnberg,
Nürnberg 020881
Zumklei, Martina, Wesel 078942
Zumkley, Beate, Wesel 130749
Zumpicchiat, Frédéric, Origny-en-
Thierache 070803
Zunfthaus zur Meisen, Schweizerische
Landesmuseen, Zürich 042246
Zunftstube, Gernsbach 018662
Zunftstube im Torli, Endingen am
Kaiserstuhl 018208
Zunhamer, Martin, Altötting 129383
Zunino, Ada, Milano 110488
Zupan, Viktor, Radolfzell am
Bodensee 078120
Županijski Muzej Šibenik, Šibenik 008988
Zuppinger, R., Altwis 116020
Zur Gilde, Titz 078684, 130704
Zur schönen Aussicht, Passade . 108359
Zurbaran, Buenos Aires 098587, 098588
Zurcher, Paris 105326
Zurell, Manfred, Remchingen . . . 130578
Zurholt, E., Altenberge 074391
Zurolo, Maria Rosaria, Brescia . . 079967
Zuzarte, José Azevedo, Lisboa . 085000,
085001
Zväz múzeí na Slovensku, Banská
Bystrica 059806
Zvejniecíbas Brívdabas Muzejs,
Jürmala 030142
Zvejniecíbas un Kugniecíbas
Vestures Muzejs, Liepājas Muzejs,
Liepāja 030156
Žvejo Etnografinė Sodyba, Neringos
Istorijos Muziejus, Nida 030432
Zvenigorodskij Istoriko-architekturnyj
i Chudožestvennyj Muzej,
Zvenigorod 038085
Zverevskij Centr Sovremennogo
Iskusstva, Moskva 037526
Zveza Društev Slovenskih Likovnih
Umetnikov, Ljubljana 059812
Zvezda, Moskva 085217
De Zwaan, Amsterdam 126275
Zwaanenburg, Peter, Dordrecht . 083203
Zwach, Käthe, Schörfling am
Attersee 100194
Zwack Gyümölcspálinka -Fozde és
Kiállítás, Kecskemét 023735
Zwack Unicum Múzeum,
Budapest 023572
De Zwaluw, Oosthuizen 137867
Zwanenbroedershuis, Illustre
Lieve Vrouwe Broederschap, 's-
Hertogenbosch 032477
Het Zwanepand, Antwerpen . . . 063257,
100526
Zwang, Christian, Berlin 106468
Zwang, Christian, Hamburg 107323,
129999
Het Zwart Huis, Knokke-Heist . . 100769
De Zwarte Panter, Antwerpen . . 100527
Zweckverband Deutsches
Porzellanmuseum, Wunsiedel . 059320
Die Zwei, Düsseldorf 075545
Zwei-Burgen Antik, Weinheim . . 078912
Zweiffel, B., Bad Münstereifel . . 074558
Zweig, Linda, New York 123220
Zweigmuseum des Staatlichen
Museum für Völkerkunde München,
Residenzschloß Oettingen, Oettingen
in Bayern 020969
Zweiradmuseum, Berlin 017322
Zweiradmuseum, Pleidelsheim . . 021152
Zweiradmuseum Havel-Auen-Werder,
Werder 022573
Zweymüller, Baden bei Wien . . . 139975
Związek Artystów Rzeźbiarzy,
Gdańsk 059729

Związek Polskich Artystów Plastyków,
Bielsko-Biała 059733
Związek Polskich Artystów Plastyków,
Bydgoszcz 059732
Związek Polskich Artystów Plastyków,
Częstochowa 059730
Związek Polskich Artystów Plastyków,
Gdańsk 059740
Związek Polskich Artystów Plastyków,
Katowice 059735
Związek Polskich Artystów Plastyków,
Kielce 059747
Związek Polskich Artystów Plastyków,
Koszalin 059741
Związek Polskich Artystów Plastyków,
Łódź 059742
Związek Polskich Artystów Plastyków,
Olsztyn 059746
Związek Polskich Artystów Plastyków,
Opole 059738
Związek Polskich Artystów Plastyków,
Poznań 059745
Związek Polskich Artystów Plastyków,
Radom 059734
Związek Polskich Artystów Plastyków,
Rzeszów 059736
Związek Polskich Artystów Plastyków,
Szczecin 059731
Związek Polskich Artystów Plastyków,
Toruń 059744
Związek Polskich Artystów Plastyków,
Warszawa 059748, 138996
Związek Polskich Artystów Plastyków,
Wrocław 059737
Związek Polskich Artystów Plastyków,
Zakopane 059739
Związek Polskich Artystów Plastyków,
Zielona Góra 059743
Związek Polskich Artystów
Plastyków – Galeria, Bielsko-
Biała 113437
Związek Polskich Artystów
Plastyków – Okręg Warszawski,
Warszawa 059749
Związek Polskich Artystów Plastyków
Okręg Lubelski, Lublin 059750
Związek Polskich Artystów Plastyków
Zarząd Okręgu Kraków, Kraków 059751
Związek Polskich Artystów Rzeźbiarzy
zar, Wrocław 059752
Zwickauer Antiquariat, Zwickau 142534
Zwiebelzwerg Verlag,
Willebadessen 137555
Zwiener, Ansbach 105998
Zwierlein, Oswin, Bad
Königshofen 074542
Zwiers, Christina, Enschede . . . 112482
Zwierzyniecki Salon Artystyczny,
Muzeum Historycznego m. Krakowa,
Kraków 034914
Zwijvekemuseum, Dendermonde 003540
Zwinger Galerie, Berlin 106469
Zwinger, Jean, Espalion 068468
Zwingligeburtshaus, Wildhaus . . 042145
Zwirner & Wirth, New York 123222
Zwirner, David, New York 123221
Zwirner, Dr. Dorothea, Berlin . . . 106470
Zwolf, Denver 120794
Zwolinska-Gasiorowska, Anna,
Düsseldorf 129768
Zwosta, Alfons, Litzendorf 077138
Zyanya, Los Angeles 094511
Zygman Voss, Chicago 092820, 120310
Zygote, Cleveland . . . 120467, 138328
Zylka, F., Kerpen 076766
Zynger, Rio de Janeiro 064136
Zynwala, Jacek, Wrocław 132919
Zypr Gallery, Toronto 101728
Žytomyr Oblasnyj Literaturnyj Muzej,
Žytomyr 043330
Žytomyrs'kyj Literaturno-memorial'nyj
Muzej V.G. Korolenka, Žytomyr 043331

Allgemeines Künstlerlexikon Online
The Artists of the World Online

AKL Online ist die umfangreichste internationale Künstlerdatenbank und enthält gegenwärtig Einträge zu mehr als einer Million Künstlern. Ab Oktober 2009 werden die Inhalte von *AKL Online* beträchtlich erweitert:

- Die Anzahl der Volltext-Biographien von A bis Z wird verdoppelt, die Inhalte werden fortwährend ergänzt und aktualisiert.
- Die Datenbank bietet „Online First"-Artikel an und die Anzahl der veröffentlichten Artikel in ihrer Originalsprache wird deutlich zunehmen.
- Die erweiterte Online-Version wird erstmalig für eine unbegrenzte Anzahl von Nutzern zugänglich sein. Die bisher üblichen Aufpreise für mehrere gleichzeitige Nutzer entfallen.

The *AKL Online* is the most comprehensive database of artists available and currently contains entries on more than 1 million artists. Starting October 2009 the content of the *AKL Online* is being substantially expanded and upgraded:

- The basic number of full text, A–Z biographies is set to double and will thereafter be constantly increased and updated.
- The database will feature edited pre pub "online first" articles and the number of articles publishing in the language in which they were originally written will be increased.
- The upgraded online version will for the first time be available for an unlimited number of simultaneous users. Surcharges for multiple simultaneous users will no longer apply.

Jahresabonnement ab Januar 2010 / Annual subscription from January 2010:
Print*+Online: € 2.990,– / **US$ 4,186.00. ISBN 978-3-598-22819-3
Online only: € 2.490,– / **US$ 3,486.00. ISBN 978-3-598-41800-6

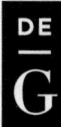 **DE GRUYTER**

www.degruyter.com/akl

*berechnet auf der Basis von 5 Bänden pro Jahr. Zzgl. Preise für Registerbände /
Calculation based on 5 print vols. p. a. Does not include index vols.

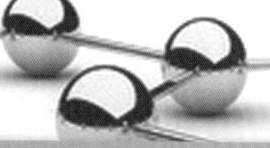
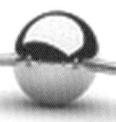
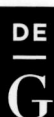